The OFFICIAL MUSEUM DIRECTORY® 2014

The 2014 edition of The Official Museum Directory® is published by National Register Publishing in partnership with the American Alliance of Museums.

Chairman	James A. Finkelstein
CEO	Fred Marks
Chief Technology Officer	Ariel Spivakovsky
Publisher	Robert Docherty
Director of Sales	Kelli MacKinnon

EDITORIAL

Managing Editor	Eileen Fanning
Content Manager	Ian Sidney O'Blenis
Content Editors	Linda Hummer
	Betty Melillo
	Mary Whitehouse

EDITORIAL SERVICES

Production Manager	David Lubanski

SALES

Advertising Account Manager	Anne Collins
Wholesale Account Representative	Gina Marie Delia

MARKETING

Business Analysis & Forecasting Manager	Kim Pappas
Creative Services Manager	Kathleen F. Ste
Marketing Analysts	Syed F. Ali
	Jeff Fitzgerald

INFORMATION TECHNOLOGY

Director of IT Development	Jeff Rooney
Director of Web Operations	Ben McClloug
Composition Programmer	Tom Haggerty
Manager of Web Development	Orlando Freda
Database Programmer	Latha Shankar

Printed and bound in the United States of America

International Standard Book Number: 978-0-87217-029-2
International Standard Serial Number: 0090-6700
Library of Congress Catalog Card Number: 79-144808

National Register Publishing has used its best efforts in collecting and preparing material for inclusion in *The Official Museum Directory*® bu does not warrant that the information herein is complete or accurate, and does not assume, and hereby disclaims, any liability to any person for any loss or damage caused by errors or omissions in *The Official Museum Directory*® whether such errors or omissions result from negligence, accident or any other cause.

Museums are essential.

At the American Alliance of Museums (AAM), we work every day to make our organization just as essential to museums, through advocacy, professional development, publishing and establishing standards and best practices. AAM's focus is embodied in our tagline: Champion Museums. Nurture Excellence.

To be the most effective museum service organization we can be, we must have all museums speaking with one strong, united voice. That's why AAM museum membership is accessible to all types and sizes of institutions–choose your level of benefits and engagement at a price you can afford.

Find out more at our website: www.aam-us.org.

THE OFFICIAL MUSEUM DIRECTORY® 2014

TABLE OF CONTENTS

INDICES

Visit www.officialmuseumdirectory.com to download additional Personnel and Collection Indices.

PRODUCTS AND SERVICES SUPPLIERS SECTION

Key to Symbols:

Accredited Museums [*]	Volunteer Chairman [Chm. (V)]
Members of the AAM [M]	Traveling Exhibit [(T)]
Volunteer President [Pres. (V)]	Telecommunications Device for the Deaf [TDD]
	Handicapped Accessible [♿]

THE OFFICIAL MUSEUM DIRECTORY®

On behalf of National Register Publishing and the American Alliance of Museums, we take great pride in presenting the 2014 edition of The Official Museum Directory®. By bringing the full resources of our two organizations and over 125 years of combined experience to bear, we are able to ensure that the OMD is the most comprehensive directory of America's museums, zoos, historic sites, and other related institutions. In this edition you will find a wealth of information on the nation's ever-expanding and influential museum community.

The 2014 Directory contains information on more than 15,000 museums operating in 87 different fields, ranging from fine arts to historic homes, from zoos to science museums. We are pleased to include information on new exhibits along with 500 new listings.

In compiling the 2014 edition, we use the same stringent quality control and collection procedures as we have in the past. Each institution receives a copy of its listing for review. Returned listings are scrupulously reviewed and updated by NRP editors. The resulting directory is both current and accurate, with most of the information coming from the museums themselves.

The Official Museum Directory® is an invaluable source for museum professionals seeking to contact directors and curators, identify unique collections and locate traveling exhibitions. Library patrons and travelers can discover important local cultural centers and identify educational travel destinations. To provide all users with alternate means of accessing this rich museum content, the directory includes institution and category indices. Additional personnel and collection indices can be found online at www.officialmuseumdirectory.com. The Products & Services section helps readers to locate suppliers who specialize in the museum field.

This directory is only one example of how the American Alliance of Museums serves the museum community and the public. We hope The Official Museum Directory® will prove to be not only an inspiration but also a valuable tool to help you take full advantage of one of America's truly great national resources – our country's diverse community of museums.

Sincerely,

Robert Docherty
Publisher
National Register Publishing
430 Mountain Avenue, 4th Floor
New Providence, NJ 07974
(800) 473-7020
www.officialmuseumdirectory.com

Ford W. Bell, DVM
American Alliance of Museums
President
1575 Eye Street, NW Suite 400
Washington, DC 20005
(202) 289-1818
www.aam-us.org

American Alliance of Museums

Champion Museums. Nurture Excellence.

Join Today!

The American Alliance of Museums is uniting the field to make the case that museums are essential in our communities, and our commitment to you remains as strong as ever. The Alliance remains your professional home, with all of the resources you need to build job skills and manage your career.

Belong to the Alliance and tap into professional development opportunities, timely updates on what's happening in the field and in Congress and a network for peer-to-peer engagement.

Add Your Voice to Our Shared Cause.

If your museum is a member, you may be eligible for a discounted or free individual Professional membership. Visit www.aam-us.org to learn more about museum membership and what it means for you.

Together We Are Stronger.

Student

As you launch your museum career, Alliance membership is an invaluable investment in future success, through job, mentoring and fellowship opportunities and practical insights and resources. Join now and start building your professional network!

Professional

Reap the rewards of a full array of benefits at steep member discounts. Participate in any or all of the 22 Professional Networks—at no additional cost.

Open to all who work (paid or unpaid) for the success of museums, including museum staff, trustees, independent professionals and staff at other nonprofit organizations. Students and retired professionals are also welcome to join at this level.

Retired Professional

Stay in the know, connected to your colleagues and keep contributing to the museum field by maintaining your Alliance membership.

	Student $50	Professional $90	Retired $50
Build Job Skills			
Online access to professional resources	✔	✔	✔
Year-round professional development programs	✔	✔	✔
Access to 22 professional networks		✔	
Membership card with benefits	✔	✔	✔
Manage Your Career			
Career development resources and programs	✔	✔	✔
Eligibility to present at Annual Meeting	✔	✔	✔
Mentoring opportunities	✔	✔	✔
Stay Informed			
Subscriptions to e-newsletters (Aviso, The Weekly and Dispatches from the Future of Museums)	✔	✔	✔
Subscription to Museum magazine (digital)	✔	✔	✔
Subscription to Museum magazine (print)		✔	✔
Save Money			
Discounts on professional development programs	✔	✔	✔
Discounts on Annual Meeting registration	✔	✔	✔
Discounts in the Alliance Bookstore	✔	✔	✔
Eligibility for fellowships	✔	✔	
Make the Case for Museums			
Free registration for Museums Advocacy Day	✔	✔	✔
Access to advocacy alerts and training	✔	✔	✔

Enhance your career, build your skills and make your voice heard.
Join the Alliance today at www.aam-us.org or call 866.226.2150.

Institutions
by State

ALABAMA

(189 listings)

Alexander City

WELLBORN MUSCLECAR MUSEUM, 124 Broad St., Alexander City, AL 35010-2689. Tel.: 256-329-8474.
Web Site: www.wellbornmusclecarmuseum.com
Institution Type/Description: Car Museum.
Collections: 1960s & 1970s musclecars; early advertising; photographs.
Activities: rental facilities.
Hours & Admission Prices: Mon.-Fri. 9-5, Sat. 10-4. Adults $10, children 7-17 $6; children 6 & under no charge.

Aliceville

ALICEVILLE MUSEUM, (M), 104 Broad St., Aliceville, AL 35442-2701. Tel.: 205-373-2363. Facebook: Aliceville Museum.
E-mail: museum@nctv.com
Web Site: www.alicevillemuseum.org
Founded: 1993.
Congressional District: 4
Key Personnel: Dir., John Gillum; Pres., Edgar Pruitt.
Personnel Profile: Full-Time Paid 1; Part-Time Volunteers 25.
Governing Authority: private; nonprofit organization. Tax-exempt: 501(c)(3).
Institution Type/Description: Museum & Cultural Arts Center.
Collections: history of German POW Camp Aliceville 1942-1945; Coca-Cola exhibit & assembly line equipment 1948-1978; Pickens Co., AL military veterans of all wars.
Activities: films; guided tours; lectures; temporary exhibitions.
Publications: quarterly newsletter, Museum News.
Hours & Admission Prices: Tues.-Sat.10-4, closed 12-1 for lunch; Sun.-Mon. & holidays by appointment. Adults $10, senior citizens, military & students $5; members no charge. &
Attendance: 3,500 (accurate)
Membership: Individual Adult $25; Family $35; Sponsor $100-$499; Director's $1,000-$4,000; President's $5,000 & up.

Anniston

* **ANNISTON MUSEUM OF NATURAL HISTORY, (M),** 800 Museum Dr., Anniston, AL 36206-2813. Mailing Address: P.O. Box 1587, Anniston, AL 36202-1587. Tel.: 256-237-6766 & 6767. Fax: 256-237-6776.
E-mail: cbragg@annistonmuseum.org
Web Site: www.annistonmuseum.org
Founded: 1930.
Congressional District: 3
Key Personnel: Exec. Dir., Cheryl H. Bragg; Chm. (V), Anna Washington; Cur. Collections, Daniel Spaulding; Program Mgr., Gina Morey; Dir. Devel., Lindie K. Brown; Mktg. Mgr., Margie Conner; Business Mgr., Regina Cooper; Facilities Mgr., Scott Williamon; Exhibits Mgr., John Parker.
Personnel Profile: Full-Time Paid 23; Part-Time Paid 4; Part-Time Volunteers 125; Interns 2.
Governing Authority: municipal. Parent Institution: City of Anniston. Tax-exempt.
Institution Type/Description: Natural History Museum & Cultural Center.
Collections: North American and African animals & their habitats; birds; William H. Werner natural habitat groups including extinct & endangered species; John B. Lagarde international mammal collection; Egyptian mummies; American Indian culture; anthropology; ethnology; entomology; paleontology; geology.
Research Fields: exhibit evaluation & visitor studies; behavioral studies on red-cockaded woodpecker.
Facilities: 2,000-vol. library of reference books & journals on all subjects related to collections available on premises; multimedia auditorium; classrooms; nature trails & outdoor lecture-demonstration areas; demonstration wildlife garden & bird of prey trail. Museum-related items for sale.
Activities: permanent & temporary exhibitions; changing exhibit gallery; guided tours; live animal programs; lectures; concerts; trips; classes in art, crafts, music & nature studies; school loan services; outreach programs. Special Events: Anniston Museum Day; Black Heritage Festival; Southeastern Indian Cultural Festival & Herpfest.
Publications: bimonthly newsletter; brochures; exhibit guides; annual report.
Hours & Admission Prices: June-Aug. Mon.-Sat. 10-5, Sun.1-5; Sept.-May Tues.-Fri. 9-5, Sat. 10-5, Sun. 1-5. Adults $6, senior citizens $5.50, children 4-17 $5; discounts to groups; members and children 3 & under no charge. Closed New Year's Day; Thanksgiving; Christmas. &
Attendance: 73,223 (accurate)

Membership: Student $10; Individual $30; Family $42; Contributor $50; Sustainer $100; Patron $250; Benefactor $500; Director's Society $1,000.

BERMAN MUSEUM OF WORLD HISTORY, 840 Museum Dr., Anniston, AL 36206-2813. Mailing Address: P.O. Box 2245, Anniston, AL 36202-2245. Tel.: 256-237-6261. Fax: 256-238-9055.
Web Site: www.bermanmuseum.org
Founded: 1996.
Congressional District: 3
Key Personnel: Dir., Cheryl Bragg; Pres. (V), Gaston O. McGinnis, M.D.; Coord. Business Devel., David Ford; Mgr. Collections, Robert Lindley; Registrar, Matt Muaubauer; Facilities Mgr., Adam Cleveland.
Personnel Profile: Full-Time Paid 4; Part-Time Volunteers 50.
Governing Authority: Tax-exempt.
Institution Type/Description: History Museum.
Collections: over 3,000 world history artifacts; bronzes; paintings; ethnographic material; weapons; historical documents; Asian art including bronze statues of Buddha, religious figures from Tibet & India, Chinese cloisonne, Southeast Asian ritual masks, Japanese bronzes & Korean textiles; WWI & II military artifacts; guns & firearms.
Hours & Admission Prices: June-Aug. Mon.-Sat. 10-5, Sun. 1-5; Sept.-May Tues.-Sat. 10-5, Sun. 1-5. Adults $5, senior citizens $4, children 4-17 $2.50; discounts to AAA members & active military; children 3 & under and members no charge. Closed New Year's Day; Thanksgiving; Christmas Eve & Day. &
Attendance: 12,250 (accurate)
Membership: Individual $15; Family $25; Contributor $50; Sustainer $100; Business $150; Patron $250; Benefactor $500 & up.

Arab

ARAB HISTORIC VILLAGE, 224 City Park Dr., Arab, AL 35016-1071. Tel.: 256-586-3138 & 4225.
Web Site: www.arab-chamber.org
Institution Type/Description: Historic Village Museum.
Collections: local history & culture; period furnishings; personal artifacts; photographs. Historic Buildings: Hunt School, 1935; Rice Church, 1910; Smith's Country Store, 1930; Boyd Homestead, 1890.
Hours & Admission Prices: March-Nov. Thurs.-Fri. 10-3, Sat. 1-4; other times by appointment. No charge.

Ashville

ASHVILLE MUSEUM AND ARCHIVES, 78 6th Ave., Ashville, AL 35953. Mailing Address: P.O. Box 1570, Ashville, AL 35953. Tel.: 205-594-2128.
Institution Type/Description: History Museum.
Collections: local history & culture; memorabilia; county records; newspapers.
Hours & Admission Prices: Mon.-Fri. 8-12 & 1-5.

Athens

ALABAMA VETERANS MUSEUM AND ARCHIVES, 100 Pryor St., Athens, AL 35611-1850. Tel.: 256-771-7578.
Key Personnel: Dir., Sandy Thompson
Institution Type/Description: Military Museum.
Collections: military artifacts; uniforms; medals; weapons; photographs; books; tapes; newspaper clippings.
Hours & Admission Prices: Call for hours. No charge; donations accepted.

ALTAR OF THE NEW TESTAMENT AND FOUNDERS HALL, Athens State Univ., 300 N. Beaty St., Athens, AL 35611. Tel.: 800-522-0272; 256-216-6671.
E-mail: sara.love@athens.edu
Web Site: www.athcns.cdu
Institution Type/Description: Religious History Museum.
Collections: wood carvings depicting the New Testament.
Hours & Admission Prices: Mon.-Fri. 8-4:30. No charge. &

DONNELL HOUSE, 601 1/2 S. Clinton St., Athens, AL 35611. Tel.: 256-232-0743 & 7370.
Institution Type/Description: Historic House Museum: housed in the former home of Rev. Robert Donnell, founder of Cumberland Presbyterian Church; built in 1851. Listed on the National Register of Historic Places.
Collections: family history; period furnishings; personal artifacts; photographs.
Activities: rental facilities; guided tours; special events.

Hours & Admission Prices: Fri. 1-3; other times by appointment.

LIMESTONE COUNTY HISTORY MUSEUM, 101 N. Houston St., Athens, AL 35611-2540. Mailing Address: P.O. Box 82, Athens, AL 35612. Tel.: 256-233-8770.
E-mail: mail@limestonecountyhistoricalsociety.org
Web Site: limestonecountyhistoricalsociety.org
Formerly: Houston Memorial Library and Museum
Founded: 1972.
Key Personnel: Dir., Jackie Leonard; Chm. (V) & Museum Shop Mgr., Rex Lewis.
Personnel Profile: Part-Time Volunteers 3.
Governing Authority: Parent Institution: Limestone County Historical Society. Tax-exempt: 501(c)(3).
Institution Type/Description: History Museum: housed in the former home of George S. Houston, attorney, member of U.S. Senate and Governor of Alabama; built in 1835.
Collections: family history; local artifacts; photographs; Native American; documents; personal artifacts; military; clothing; paintings.
Research Fields: family & local history.
Activities: tours; field trips.
Publications: quarterly, Limestone Legacy; historic book, Mary Mason's Scrapbook; Limestone County cemetery book; architectural; essay.
Hours & Admission Prices: Mon.-Fri. 10-5. No charge; donations accepted.
Attendance: 700 (estimated)
Membership: Historical Society: Senior & Student $10; Individual $15; Family $20; Life $300; Patron $500.

Attalla

TIGERS FOR TOMORROW AT UNTAMED MOUNTAIN, 710 County Rd. 345, Attalla, AL 35954. Tel.: 256-524-4150.
E-mail: untamedmountain@gmail.com
Web Site: tigerfortomorrow.org
Founded: 1999.
Personnel Profile: Full-Time Paid 2; Full-Time Volunteers 2; Part-Time Volunteers 12.
Institution Type/Description: Animal Preserve.
Collections: wildlife including cougars; leopards; tigers; lions; wolves.
Facilities: lunch pavilion.
Activities: tours.
Hours & Admission Prices: Fri.-Sun. 9-5; other times by appointment. Adults $12, children 3-11 $6. &
Attendance: 8,500 (estimated)

Auburn

BIGGIN GALLERY - AUBURN UNIVERSITY, Dept. of Art, 108 Biggin Hall, Auburn, AL 36849. Tel.: 334-844-4373. Fax: 334-844-4024.
Web Site: www.auburn.edu
Institution Type/Description: Art Gallery.
Collections: works by national & international artists; paintings; sculpture.
Hours & Admission Prices: Call for hours.

JAN DEMPSEY COMMUNITY ARTS CENTER, 222 E. Drake Ave., Auburn, AL 36830-3918. Mailing Address: 307 S. Dean Rd., Auburn, AL 36830-6105. Tel.: 334-501-2963. Fax: 334-501-2964.
E-mail: shand@auburnalabama.org
Web Site: www.auburnalabama.org/arts
Institution Type/Description: Art Museum.
Collections: works by regional artists.
Activities: concerts; theatre productions; lectures; studio art classes; theatre classes; dance classes; special events; traveling exhibitions.
Hours & Admission Prices: Mon.-Fri. 8-5. Closed major holidays. &

* **JULE COLLINS SMITH MUSEUM OF FINE ART, (M),** Auburn University, 901 S. College St., Auburn, AL 36849. Tel.: 334-844-1484.
E-mail: jcsm@auburn.edu
Web Site: jcsm.auburn.edu
Founded: 2003.
Key Personnel: Dir., Marilyn Laufer; Asst. Dir., Andy Tennant; Devel. Coord., Cindy Cox; Cur. Education, Scott Bishop Wagoner; Cur., Dennis Harper; Mktg. & Special Events, Colleen Bourdeau; Museum Shop Mgr., Carol Robicheaux; Exec. Sec., Robbin Birmingham; Office Admin., Janice Allen; IT Specialist, Mike Cortez; Cur. Education K-12, Andrew Henley; Preparator, Dan Neil; Graphic Designer, Janet Spivey Guynn; Registrar, Danielle Funderburk; Security Chief, Ethelene Jones.

Personnel Profile: Full-Time Paid 17; Part-Time Paid 10; Part-Time Volunteers 60; Interns 3.
Governing Authority: university. Parent Institution: Auburn University. Tax-exempt.
Institution Type/Description: Art Museum.
Collections: 19th-21st-century American & European art including paintings, sculpture and works on paper.
Facilities: gardens; restaurant; 120-seat auditorium. Museum-related items for sale.
Activities: special programs; lectures; walking paths; films; musical performances; receptions.
Hours & Admission Prices: Mon.-Fri. 8:30-4:45, Sat. 10-4:45. No charge; donations accepted. Closed New Year's Day; Easter; Memorial Day; Independence Day; Labor Day; Thanksgiving; Christmas. &
Attendance: 30,000 (accurate)
Membership: Individual $45; Family & Dual $80; Sustaining $125; Benefactor $250; Patron $500; Connoisseur $1,000.

Beatrice

RIKARD'S MILL HISTORICAL PARK, 4116 Hwy. 265 N., Beatrice, AL 36425. Mailing Address: 31 N. Alabama Ave., Monroeville, AL 36461. Tel.: 251-575-7433. Fax: 251-575-2513.
E-mail: mchm@frontiernet.net
Web Site: www.tokillamockingbird.com
Founded: 1994.
Congressional District: 1
Key Personnel: C.E.O. & Museum Shop Mgr., Stephanie Rogers; Chm. (V), Clark McKinley.
Personnel Profile: Full-Time Paid 1.
Governing Authority: Parent Institution: Monroe County Heritage Museums. Tax-exempt: 501(c)(3).
Institution Type/Description: History Museum: 1845 restored Grist Mill with water-powered turbine.
Collections: pioneer tools; household & agricultural implements; wagon; cane mill & furnace; blacksmith shop.
Activities: Annual Events: Ghost Storytelling & Haunted Swamp Trail in October; Cane Syrup Making Day in November; Pioneer Days in November.
Hours & Admission Prices: mid-April to mid-Nov. Sat. 9-5. Adults $4, children $2. Closed major holidays. &
Attendance: 8,500 (estimated)
Membership: Student $20; Individual $35; Family $45; Star $100; Friend $250; Patron $500; Corporate $750.

Bessemer

BESSEMER HALL OF HISTORY, 1905 Alabama Ave., Bessemer, AL 35020-5009. Tel.: 205-426-1633. Fax: 205-426-1633.
E-mail: bessemerhallofhi@bellsouth.net
Web Site: www.bessemerhallofhistory.com
Founded: 1970.
Congressional District: 7
Key Personnel: Co-Chm., Dr. Merith Byram; Co-Chm. (V), Ray Morris; Pres., Wendell Martin; Vice Pres., Bobby Cooper; Cur., Chris Eiland.
Personnel Profile: Full-Time Paid 1.
Governing Authority: bd. of directors. Tax-exempt: 170(b)(1)(A).
Institution Type/Description: History Museum: housed in the Southern R.R. Depot; built in 1916. Listed on the National Register of Historical Places.
Collections: historical items; photographs; documents of the Bessemer area; Bessemer Indian Mound Site artifacts; mining & industry artifacts & photos of the Bessemer & Birmingham area; vintage clothing; Civil War artifacts; 1,000,000 Pullman Standard Boxcar.
Research Fields: state & county history; dialect & folklore.
Facilities: 200-vol. research library of historical & arts & crafts books and newspapers on DVD from 1888-1920 available on premises.
Activities: guided tours; docent program; inter-museum, permanent & temporary exhibitions.
Hours & Admission Prices: Tues.-Sat. 9-12 & 1-4. No charge; donations accepted. &
Attendance: 6,500 (estimated)
Membership: Individual $25; Couple $35; Business/Club $70; Patron $110.

Birmingham

ALABAMA JAZZ HALL OF FAME, 1631 4th Ave. N., Birmingham, AL 35203-1903. Tel.: 205-254-2731. Fax: 205-254-2785.
E-mail: ltucker@jazzhall.com
Web Site: www.jazzhall.com
Key Personnel: Exec. Dir., Leah Tucker; Dir. Education & Community Svcs., Dr. Frank E. Adams, Sr.

Institution Type/Description: Jazz Museum.
Collections: Alabama jazz artists including Nat King Cole, Duke Ellington, Lionel Hampton & Erskine Hawkins; jazz history.
Facilities: theater.
Activities: teachers' workshop; elementary school programs; jazz classes; concerts; jazz workshops; special events.
Hours & Admission Prices: Hall of Fame: Tues.-Sat. 10-5. Guided Tours: Tues.-Wed. & Fri. 10-2, Sat. 1-5. Guided Tour: $3; Self-Guided Tour: $2.
Membership: Solo $35; Combo $50; Big Band $100.

ALABAMA MUSEUM OF THE HEALTH SCIENCES, (M), 1700 University Blvd., Birmingham, AL 35294. Mailing Address: 300 LHL - 1530 3rd Ave., S., Birmingham, AL 35294. Tel.: 205-934-4475. Fax: 205-975-8476.
Web Site: www.uab.edu/historical/museum.htm
Founded: 1981.
Congressional District: 7
Key Personnel: Dir., Michael A. Flannery; Cur., Stefanie Rookis.
Personnel Profile: Full-Time Paid 2.
Governing Authority: public university; nonprofit. Parent Institution: University of Alabama at Birmingham. Tax-exempt: 501(c)(3).
Institution Type/Description: Medical Museum.
Collections: equipment; instruments; history & development of the health sciences; southern medicine,
Facilities: library; educational facilities.
Activities: formal education programs for undergraduate or graduate college students; guided tours; study clubs; temporary & traveling exhibitions.
Publications: newsletter, Treasures.
Hours & Admission Prices: Mon.-Fri. 9-5. No charge. Closed New Year's Day; Martin Luther King Jr. Day; Memorial Day; Independence Day; Labor Day; Thanksgiving & day after; Christmas. &
Attendance: 3,000 (estimated)
Membership: Student Associate $10; Contributing Associate $25; Sustaining Associate $100; Sponsoring Associate $500; Patron $1,000 & up.

ALABAMA SPORTS HALL OF FAME, 2150 Richard Arrington Jr. Blvd. N., Birmingham, AL 35203-1102. Mailing Address: P.O. Box 10163, Birmingham, AL 35202-0163. Tel.: 205-323-6665. Fax: 205-252-2212.
E-mail: info@ashof.org
Web Site: www.ashof.org
Founded: 1967.
Key Personnel: Dir., Scott Myers
Institution Type/Description: Sports Museum.
Collections: Alabama's sports heritage; over 5,000 sports artifacts; sports legends including Jesse Owens, Hank Aaron, Joe Louis, Willie Mays, Carl Lewis.
Activities: tours; children's outreach program. Museum Sponsors: Induction Banquet; Golf Tournament; Reunion of Winners.
Hours & Admission Prices: Mon.-Fri. 9-5. Adults $5, senior citizens 60 & over $4, students $3; discounts to groups of 10 or more.

ARLINGTON, 331 Cotton Ave., S.W., Birmingham, AL 35211-1465. Tel.: 205-780-5656. Fax: 205-788-0585.
Web Site: www.informationbirmingham.com/arlington/index.htm
Founded: 1953.
Congressional District: 6
Key Personnel: Dir., Daniel F. Brooks.
Personnel Profile: Full-Time Paid 7; Part-Time Paid 1.
Governing Authority: municipal. Tax-exempt: 170(b)(1)(A).
Institution Type/Description: Historic House Museum: housed in the home of Judge William S. Mudd, one of 10 founders of Birmingham; built in 1850.
Collections: period furnishings.
Research Fields: furniture; local history; decorative arts.
Facilities: 50-seat auditorium; dining room. Museum-related gifts for sale.
Activities: guided tours; lectures; films; permanent & loan exhibitions; formally organized education programs for children; docent program.
Publications: brochures.
Hours & Admission Prices: Tues.-Sat. 10-4, Sun. 1-4. Adults $5, students 6-18 $3; discount to groups & AAM members; children under 6 no charge when accompanied by an adult. Closed holidays.
Attendance: 25,000 (estimated)
Membership: Single $7.50; Couple $10.

BIRMINGHAM BOTANICAL GARDENS, 2612 Lane Park Rd., Birmingham, AL 35223-1800. Tel.: 205-414-3950. Fax: 205-414-3966.
E-mail: akrebbs@bbgardens.org

Web Site: www.bbgardens.org
Founded: 1963.
Congressional District: 6
Key Personnel: Dir., Frederick R. Spicer, Jr.; Pres. (V), Tricia Noble.
Personnel Profile: Full-Time Paid 18; Part-Time Paid 3.
Governing Authority: municipal. Parent Institution: Branch of City of Birmingham, Birmingham Park & Recreation Board, 400 Graymont Ave., W., Birmingham, AL 35204. Tel.: 205-254-2391. Tax-exempt: 170(b)(1)(A).
Institution Type/Description: Botanical Garden & Herbarium.
Collections: Japanese garden; conservatory; rhododendron garden; fern glade; rose garden & southern living garden; bonsai house; orchid & camellia collection; wildflower garden; iris & daylily garden.
Research Fields: plant propagation.
Facilities: 5,000-vol. public library of books relating to plants; botanical garden; 300-seat auditorium; 250-seat lecture hall; 3 classrooms; herbarium, diagnostic & propagation labs. Museum-related items for sale.
Activities: guided tours; lectures; films; summer workshops for children; discovery field trips, hands-on adult classes; library activities; docent program.
Publications: newsletters & brochures.
Hours & Admission Prices: Daily sunrise-sunset. No charge. &
Attendance: 350,000 (estimated)
Membership: Student $25; Young Professional $40-$55; Trillium $45; Hydrangea $60; Magnolia $125; Oak $250; Ambassador $500; President's Circle $1,000.

✱ **BIRMINGHAM CIVIL RIGHTS INSTITUTE, (M),** 520 Sixteenth St., N., Birmingham, AL 35203-1911. Tel.: 205-328-9696, ext. 218; 866-328-9696 (toll free). Fax: 205-323-5219.
E-mail: bcri@bcri.org
Web Site: www.bcri.org
Founded: 1992.
Congressional District: 7
Key Personnel: Pres. & C.E.O., Dr. Lawrence J. Pijeaux, Jr.; Bd. Chm., Lajuana Bradford; Vice Pres. Finance & Operations, Carol Wells; Vice Pres. Institutional Programs, Priscilla Hancock Cooper; Head Archives, Wayne Coleman; Head Education, Ahmad Ward; Archivist, Laura Anderson; Head Devel., Rhonda Clark; Operations Mgr., LeRoy Simmons; Bldg. & Grounds Supt., David W. Davis; Network Admin., Michael Holland; Resource Gallery Attendant, Yvonne Williams; Museum Store Mgr., Carolyn Cunningham; Coord. Youth Programs, Michelle Craig; Coord. Group Tours, Bonnie Clark; Accounting Supvr., Dawn Garner; Head Communications, Melissa Snow-Clark; Ticketbooth Attendant, Erica Sturdivant; Ticketbooth Attendant, Verbell Brown.
Personnel Profile: Full-Time Paid 20; Part-Time Paid 6; Part-Time Volunteers 160.
Volunteer Hours: 6,649
Operating Expenses: 2,600,000
Operating Income: 2,700,000
Governing Authority: nonprofit. Tax-exempt: 501(c)(3).
Institution Type/Description: History Museum.
Collections: Civil Rights Movement; African American life, history & culture.
Major Exhibits: Numinous Neoism by John Solomon Sandridge (T), 12/13-3/23/14; Helen Keller Art Show (T), 1/23/14-2/26/14; Courage (T), 4/8/14-6/8/14; Para Todos Los Ninos (T), 4/8/14-6/8/14; Pieces of a Dream: by Carol white & India Cruse-Griffin (T), 6/17/14-8/24/14; American Boricua: Puerto Rican Life n the United States (T), 9/10/14-12/31/14.
Research Fields: Civil Rights Movement; African American life, history & culture.
Facilities: three meeting rooms; rental space available; archival research area.
Activities: public programs; conferences; workshops; storytelling for children; lectures; school tours; after school programs for students. Museum Sponsors: Annual Juneteenth Festival; King Birthday Celebration & Festival; Kwanzaa Celebration; Women's History Month Celebration.
Publications: newsletter; curriculum guide; exhibition guide; volunteer handbook; activity booklet; BCRI Teacher's Packet designed to acquaint educators with mission of institute.
Hours & Admission Prices: Tues.-Sat. 10-5, Sun. 1-5. Adults $12, senior citizens 65 & over and college students with college ID $6, Grades 4-12 (outside Jefferson County) $3; discount to adult groups of 25 or more, military & AAA members; members & Jefferson County children no charge. Closed major holidays. &
Attendance: 145,000 (accurate)
Membership: Student $15; Senior Citizen $25; Individual $35; Family $50; Friend $100; Supporter $250; Patron $500; Ambassador $1,000; Leadership Circle $2,000; Leadership Silver Circle $5,000; Gold Circle $10,000.

BIRMINGHAM HISTORY CENTER, 1731 1st Ave. N., Ste. 120, Birmingham, AL 35203-2056. Tel.: 205-202-4146. Fax: 205-202-4146.

Web Site: www.birminghamhistorycenter.org
Formerly: Birmingham History Museum
Founded: 2004.
Congressional District: 6
Key Personnel: Dir., Jerry R. Desmond; Chm. (V), Dr. Bayard Tynes; Pres. (V), A. Fox DeFuniak, III
Personnel Profile: Full-Time Paid 2.
Governing Authority: nonprofit organization. Tax-exempt: 501(c)(3).
Institution Type/Description: History Museum.
Collections: photographs; Birmingham's retail history; personal artifacts;
Hours & Admission Prices: Mon.-Fri. 9-4:30, Sat. 10-4. Adults $4, senior citizens 60 & over $3, students $2; members & children under 6 no charge.
Attendance: 5,000 (estimated)
Membership: Senior, Students & Veterans $25; Individual $35; Family & Grandparents $50.

✻　**BIRMINGHAM MUSEUM OF ART, (M),** 2000 Rev. Abraham Woods Jr. Blvd., Birmingham, AL 35203-2205. Tel.: 205-254-2566, ext. 3900. Fax: 205-254-2714.

E-mail: museum@artsbma.org
Web Site: www.artsbma.org
Founded: 1951.
Congressional District: 6
Key Personnel: Dir., Gail Andrews; Chm., Thomas L. Hamby; Sr. Cur. & Cur. Asian Art, Dr. Donald A. Wood; Deputy Dir., Amy Templeton; Cur. Decorative Arts, Dr. Anne Forschler-Tarrasch; Cur. American Art, Dr. Graham Boettcher; Cur. Modern & Contemporary Art, Ron Platt; Chief Cur. & Cur. European Art, Dr. Jeannine O'Grody; Cur. Arts of Africa & Americas, Emily Hanna; Registrar, Melissa Falkner Mercurio; Chief Security, J.R. Feagins; Graphic Designer, James Williams; Dir. Communications, Nick Patterson; Dir. Devel., Kendra Quandt; Membership Coord., Charlotte Russ; Librarian, Tatum Preston; Bldg. Supt., Wayne Blount; Designer, Terry Beckham; Events Coord., Brynne MacCann; C.F.O., Johnny McIntosh; Accountant, Ernest Hudson; Volunteer Coord., Rhonda Hethcox; Museum Shop Mgr., Kristie Allen.
Personnel Profile: Full-Time Paid 80; Part-Time Paid 5; Part-Time Volunteers 500; Interns 5.
Governing Authority: municipal. Parent Institution: City of Birmingham. Tax-exempt: 170(b)(1)(A).
Institution Type/Description: Art Museum.
Collections: European & American paintings, sculpture, decorative arts, photography, works on paper; Native American art; Asian, African & pre-Columbian art; contemporary art.
Research Fields: American & European paintings, sculpture, decorative arts, works on paper, photography; contemporary art; Asian, African, & Native American art.
Facilities: 25,000-vol. library of art history books available on premises; studios; 340-seat auditorium; restaurant; sculpture garden. Museum-related items for sale.
Activities: guided tours; lectures; films; gallery talks; concerts; arts festivals; radio programs; formally organized education programs for children; inter-museum loan, permanent, temporary & traveling exhibitions; mobile art van; art classes for adults & children.
Publications: quarterly newsletter; catalogues of special exhibitions & permanent collections; self-guided tour brochures; teacher packets.
Hours & Admission Prices: Tues.-Sat. 10-5, Sun. 12-5. No charge; donations accepted. Closed major holidays. ♿
Attendance: 193,394 (accurate)
Membership: Student $20; Nonprofit Organization $30; Individual $50; Family $70; Junior Patrons $60; Contributor $125; Fellow $300; Patron $600; Benefactor $1,000; Curator's Circle $2,500; Director's Circle $5,000.

BIRMINGHAM ZOO, 2630 Cahaba Rd., Birmingham, AL 35223-1154. Tel.: 205-879-0409. Fax: 205-879-9426.

Web Site: www.BirminghamZoo.com
Founded: 1955.
Congressional District: 6
Key Personnel: C.E.O. & Dir., Dr. William R. Foster; Mktg. & Public Rels., Beth Parmer.
Personnel Profile: Full-Time Paid 78; Full-Time Volunteers 150; Part-Time Paid 17.
Governing Authority: public/private partnership. Tax-exempt: 501(c)(3).
Institution Type/Description: Zoo.
Collections: over 700 animals representing 250 species including mammals, birds, & reptiles.
Facilities: 1,000-vol. library related to animal study; 122 acres.
Activities: guided tours; lectures; formally organized education programs for children & undergraduate college students; docent program or council.
Publications: quarterly magazine, Animal Tracks.
Hours & Admission Prices: Winter: daily 9-5; Summer: Mon.-Fri. 9-5, Sat.-Sun. 9-7. Adults $12, seniors 65 & up and children 2-12 $7; discounts on Tues., children under 2, AZA members & zoo members no charge. Closed Thanksgiving; Christmas. ♿
Attendance: 450,000 (accurate)
Membership: Individual $30; Individual Plus One $50; Family & Grandparents $75; Family Circle $125; Keeper Club $250; Curator Club $500; Director's Circle $1,000.

DON KRESGE MEMORIAL MUSEUM, 600 N. 18th St., Birmingham, AL 35203-2206. Mailing Address: Alabama Historical Radio Society, P.O. Box 131418, Birmingham, AL 35213. Tel.: 205-822-6759.

Governing Authority: Parent Institution: Alabama Historical Radio Society.
Institution Type/Description: History Museum.
Collections: period radios; radio history & memorabilia; broadcasting history.
Activities: classes; educational programs; special programs.
Hours & Admission Prices: Call for hours.

MCWANE SCIENCE CENTER, 200 19th Street, N., Birmingham, AL 35203-3117. Tel.: 205-714-8300. Fax: 205-714-8400.

E-mail: kbaasen@mcwane.org
Web Site: www.mcwane.org
Founded: 1998.
Congressional District: 6
Key Personnel: C.E.O., Amy Templeton; Vice Pres. Devel., Porynne MacCann; Vice Pres. Exhibits & Creative Svcs., Lamar Smith; Dir. Public Rels., Katie Baasen.
Personnel Profile: Full-Time Paid 70; Full-Time Volunteers 150; Part-Time Paid 45; Part-Time Volunteers 273.
Governing Authority: nonprofit organization. Parent Institution: Discovery 2000, Inc. Tax-exempt.
Institution Type/Description: Science and Technology Center.
Collections: hands-on participatory exhibits focusing on natural & physical science and paleontology.
Facilities: 154,200 sq. ft. exhibit space; large screen IMAX dome theater.
Activities: hands-on activities.
Publications: newsletter.
Hours & Admission Prices: June-Aug. Mon.-Fri. 9-6, Sat. 10-6, Sun. 12-6; Sept.-May. Mon.-Fri. 9-5, Sat. 10-6, Sun. 12-6. Parking: General Public $5, members no charge. Museum: adults $11, children $8; members no charge. IMAX DMR Movie: adults $10, children $9. IMAX Film: adults $8.50, children $7.50. Combo: adults $16, children $12. Closed New Year's Day; Easter; Thanksgiving; Christmas Eve & Day. ♿
Attendance: 500,000 (estimated)
Membership: Add-A-Guest $20; Individual Plus $70; Family $95; Family Plus $135; Family Deluxe $160.

ROBERT R. MEYER PLANETARIUM, Birmingham-Southern College, 900 Arkadelphia Rd., Birmingham, AL 35254. Tel.: 205-226-4771 & 4770.

E-mail: rbecker@bsc.edu
Web Site: www.bsc.edu/campus/planetarium
Key Personnel: Planetarium Coord., R. Becker Ingram
Institution Type/Description: Planetarium.
Collections: space science.
Hours & Admission Prices: Public Shows: one Sat. or Sun. per month 2pm; call for additional days. Adults $2, children 12 & under $1; BSC students no charge. Reserved Showings: Mon.-Fri. by appointment. Groups of 10-45 $2 per person. Closed holidays; college breaks.

RUFFNER MOUNTAIN NATURE CENTER, 1214 81st St. S., Birmingham, AL 35206-4599. Tel.: 205-833-8264. Fax: 205-836-3960.

E-mail: info@ruffnermountain.org
Web Site: www.ruffnermountain.org
Founded: 1977.
Governing Authority: nonprofit organization. Tax-exempt: 501(c)(3).
Institution Type/Description: Nature Center.
Collections: native species of plants & animals.
Facilities: nature trails.
Activities: educational programs; weekend activities; hiking; birding; workshops; scout programs.
Hours & Admission Prices: Mon.-Sat. 9-5, Sun. 1-5. No charge; donations accepted. Closed New Year's Day; Thanksgiving; Christmas Eve & Day.

SAMUEL ULLMAN MUSEUM, 2150 15th Ave. S., Birmingham, AL 35205-3920. Mailing Address: SIH, 1600 10th Ave., S, SIH22D, Birmingham, AL 35294. Tel.: 205-934-8063.
E-mail: huc@uab.edu
Web Site: www.uab.edu/ullmanmuseum
Institution Type/Description: General Museum.
Collections: personal artifacts; life & works of Samuel Ullman.
Hours & Admission Prices: By appointment only. No charge; donations accepted.

SLOSS FURNACES NATIONAL HISTORIC LANDMARK, 20 32nd St. N., Birmingham, AL 35222-1236. Tel.: 205-324-1911. Fax: 205-324-6758.
E-mail: info@slossfurnaces.com
Web Site: www.slossfurnaces.com
Founded: 1983.
Congressional District: 6
Key Personnel: C.E.O., Robert R. Rathburn, Ph.D.; Chm., Robin A. Wade, III; Cur., Karen Utz.
Personnel Profile: Full-Time Paid 7; Part-Time Paid 6; Interns 5.
Governing Authority: municipal government. Parent Institution: City of Birmingham. Tax-exempt: 170(b)(1)(A).
Institution Type/Description: Industrial Museum: c.1882-1970 ironmaking plant including blast furnaces, blowing engines, power house, boilers & related buildings.
Collections: machinery; tools; paintings; photographs; oral histories.
Research Fields: history of Sloss-Sheffield Steel & Iron Co.; Birmingham economic, technological & labor history.
Facilities: 400-vol. library on technical works on iron, steel & related industries available for research on premises; works on Alabama & Birmingham history; general works on history of technology & labor; manuscripts related to Sloss-Sheffield Steel & Iron Co.; oral histories; 1,500-seat auditorium. Museum-related items for sale.
Activities: guided tours; lectures; films; concerts; dance recitals; arts festivals; organized education programs for children; artistic metal-casting; docent program; temporary & traveling exhibitions.
Publications: brochures, Sloss Furnaces National Historic Landmark: A Self Guided Tour; The Sloss Story.
Hours & Admission Prices: Tues.-Sat. 10-4, Sun. 12-4. No charge; donations accepted. Closed New Year's Day; Thanksgiving; Christmas. &
Attendance: 140,000
Membership: Stove Tender $35; Millwright $60; Patternmaker $150; Furnace Keeper $500; Turn Foreman $1,000; James Withers Sloss Circle $5,000.

SOUTHERN ENVIRONMENTAL CENTER, 900 Arkadelphia Rd., Birmingham, AL 35254. Tel.: 205-226-4934.
Institution Type/Description: Environmental Center.
Collections: protecting & improving local environments including air pollution & water quality.
Facilities: Museum-related items for sale.
Activities: special events; EcoArt activities.
Hours & Admission Prices: By appointment.

SOUTHERN MUSEUM OF FLIGHT, 4343 73rd St. N., Birmingham, AL 35206-3642. Tel.: 205-833-8226. Fax: 205-836-2439.
E-mail: southernmuseumofflight@yahoo.com
Web Site: southernmuseumofflight.org
Founded: 1983.
Congressional District: 10
Key Personnel: Exec. Dir. & Chm., Dr. Jim Griffin; Asst. Dir., Wayne Novy; Education Coord., Deborah Watson Stone; Business Mgr., Daphne Foy; Exhibit Designer, Bruce Lucas; Business Mgr., Glenda McCarroll; Asst. Shop Mgr. & Receptionist, Sherry Greene.
Personnel Profile: Full-Time Paid 7; Part-Time Paid 2; Part-Time Volunteers 35.
Governing Authority: municipal. Parent Institution: City of Birmingham, AL. Subsidiary Institution: Southern Museum of Flight Foundation, Inc. Tax-exempt: 501(c)(3).
Institution Type/Description: Aviation Transportation Museum.
Collections: civilian, experimental & military aircraft; engines; models; aviation memorabilia.
Research Fields: aviation.
Facilities: library of aviation books & periodicals available for research on premises; reading rooms; meeting rooms; restoration workshop. Aviation-related items for sale.
Activities: guided tours; flight simulations; restoration workshops; flight instruction class; aviation-related meetings & clubs.

Publications: quarterly newsletter, monthly newsletter, 99 Neighborhood Associations in Birmingham.
Hours & Admission Prices: Tues.-Sat. 9:30-4:30. Adults $5, seniors & students $4; discounts to groups, AAA & AAM members; members, ASTC members & children 3 & under no charge. &
Attendance: 63,247 (accurate)
Membership: Family $45.

SPACE ONE ELEVEN, 2409 2nd Ave. N., Birmingham, AL 35203. Tel.: 205-328-0553. Fax: 205-254-6176.
Web Site: www.spaceoneeleven.org
Key Personnel: Co-Founder, Anne Arrasmith; C.E.O. & Co-Founder, Peter Prinz
Institution Type/Description: Art Gallery.
Collections: works by contemporary artists.
Activities: special events; workshops; classes.
Hours & Admission Prices: Tues.-Fri. 10-12 & 1-5 during exhibitions.

Blountsville

FREEMAN CABIN MUSEUM, 71406 Main St., Blountsville, AL 35031. Mailing Address: P.O. Box 232, Blountsville, AL 35031-0232. Tel.: 205-429-2338.
Formerly: Freeman Cabin Museum
Institution Type/Description: History Museum.
Collections: local history & culture; period furnishings; personal artifacts; photographs.
Activities: special events.
Hours & Admission Prices: 1st Sat. of month. No charge; donations accepted.

Brewton

THOMAS E. MCMILLAN MUSEUM, Jefferson Davis College, 220 Alco Dr., Brewton, AL 36426. Mailing Address: P.O. Box 958, Brewton, AL 36427-0958. Tel.: 251-809-1528. Fax: 251-809-1559.
E-mail: museum@jdcc.edu
Web Site: www.museum.jdcc.edu
Founded: 1978.
Congressional District: 11
Key Personnel: Museum Coord., Jerry Simmons; Maintenance Supvr., Don Odom.
Personnel Profile: Part-Time Paid 1; Part-Time Volunteers 2.
Governing Authority: state. Parent Institution: Jefferson Davis Community College. Tax-exempt: 170(b)(1)(A).
Institution Type/Description: History Museum: located near the site of 1830 Leigh plantation.
Collections: historical archaeology material & artifacts; Indian artifacts & trading wares; Civil War & 1812-1821 military excavated items; small arm weapons; camp & field equipment; insignias & hardware; early 19th century carpenter & cabinet maker tools; blacksmith tools; railroad memorabilia; 19th century lumbering & turpentining materials; print shop; mid 19th century medical, dental & pharmaceutical instruments; period cameras & photographic accessories; late 1800s & early 1900s household items.
Research Fields: genealogical; historical.
Facilities: over 1,000-vol. library of books & microfilm on local history & genealogy museum methodology and guidelines available for research on premises only; classrooms; reading room.
Activities: guided tours.
Hours & Admission Prices: Tues. & Thurs. 9-3; other times by appointment. No charge; donations accepted. Closed New Year's Eve & Day; Memorial Day; Labor Day; Thanksgiving; Christmas week. &
Attendance: 1,000 (estimated)

Bridgeport

RUSSELL CAVE NATIONAL MONUMENT, 3729 County Rd. 98, Bridgeport, AL 35740-6825. Tel.: 256-495-2672. Fax: 256-495-9220.
E-mail: shelia_reed@nps.gov
Web Site: nps.gov/ruca
Founded: 1961.
Congressional District: 5
Key Personnel: Supt., John Bundy; Park Ranger, Kenna Graham.
Governing Authority: federal. A branch of the National Park Service, U.S. Dept. of Interior, Washington, DC 20240. Tax-exempt.
Institution Type/Description: Archaeology Museum.
Collections: archaeology; geology; Southeastern archaic material; Woodland period, Mississippian period Indians.
Research Fields: archaeology; conservation.

Facilities: 450-vol. library of books & approx. 1,000 slides on archaeology; picnic grounds; nature & hiking trail. Books, slides & museum-related items for sale.
Activities: interpretive talks; demonstration of prehistoric weapons & tools, video; junior ranger program; interpretive slide program in excavation.
Publications: Life at Russell Cave.
Hours & Admission Prices: Daily 8-4:30. No charge; donations accepted. Closed New Year's Day; Thanksgiving; Christmas. &
Attendance: 30,000 (estimated)

Calera

HEART OF DIXIE RAILROAD MUSEUM, 1919 Ninth St., Calera, AL 35040. Mailing Address: P.O. Box 727, Calera, AL 35040-0727. Tel.: 205-668-3435. Fax: 205-668-9900.
Web Site: www.hodrrm.org
Founded: 1963.
Congressional District: 6
Key Personnel: Pres., Jim Garnett; Archivist, David Coombs; Treas., James Ketchersid; Museum Shop Mgr., Mark Walker.
Personnel Profile: Full-Time Paid 1; Part-Time Paid 1; Part-Time Volunteers 60.
Governing Authority: private; nonprofit organization. Subsidiary Institution: Calera & Shelby Railroad. Tax-exempt: 501(c)(3).
Institution Type/Description: Railroad Museum: housed in 1890s wooden railroad station building.
Collections: includes about 50 railroad rolling stock, steam & diesel locomotives, passenger & freight cars, cabooses; emphasis on history of railroads in Alabama; concentration on Louisville & Nashville Railroad & Southern Railway.
Facilities: 700-vol. library of railroad history & technology; 2,000 sq. ft. exhibit space. Museum-related items for sale.
Activities: passenger train rides; guided tours; outdoor exhibits. Annual Events: Rail Fest, Halloween train rides, Santa Claus train rides; Cottontail Express; North Pole Express; grandparents special; Day Out With Thomas.
Publications: monthly newsletter, Cinders from the Smokestack.
Hours & Admission Prices: Museum: Tues., Thurs. & Sat. 9-4. No charge; donations requested. Train Rides: March to mid-Dec. Sat. Adults $12, children $8; members no charge. Closed major holidays. &
Attendance: 40,000 (estimated)
Membership: Individual $51 (each additional family member $8).

Centre

CHEROKEE COUNTY HISTORICAL MUSEUM, 101 E. Main St., Centre, AL 35960. Tel.: 256-927-7835.
E-mail: museumatcentre@gmail.com
Web Site: www.museumatcentre.com
Founded: 1958.
Key Personnel: Dir., David Crum; Chm. (V), Kurt Duryea.
Personnel Profile: Full-Time Paid 1; Part-Time Paid 3; Part-Time Volunteers 5.
Governing Authority: Parent Institution: Cherokee County Historical Society, Inc. Tax-exempt.
Institution Type/Description: History Museum.
Collections: county history, art & culture; newspapers; photographs; personal artifacts; vehicles; housewares; Civil War; WWI & II; railroad artifacts; farming; education.
Publications: monthly newsletter.
Hours & Admission Prices: Tues.-Sat. 8:30-4. Adults $3, seniors over 64 & students $2, children 7-12 $1; discounts to AAM & ICOM members; members and children 6 & under no charge. Closed major holidays.
Attendance: 3,862 (accurate)
Membership: Senior & Student $11; Individual $15; Family $25; Individual Lifetime $250; Family Lifetime $425.

Chatom

WASHINGTON COUNTY MUSEUM, 443 Court St., Chatom, AL 36518. Mailing Address: P.O. Box 233, Chatom, AL 36518-0233. Tel.: 251-847-3156. Fax: 251-847-3677.
Institution Type/Description: History Museum.
Collections: county history; personal artifacts.
Hours & Admission Prices: Mon.-Fri. 8-4:30; guided tours by appointment. No charge. Closed Christmas Eve & Day.

Columbiana

KARL C. HARRISON MUSEUM OF GEORGE WASHINGTON, Mildred B. Harrison Regional Library, 50 Lester St., Columbiana, AL 35051-9477. Tel.: 205-669-8767.
E-mail: info@washingtonmuseum.com
Web Site: www.washingtonmuseum.com
Founded: 1982.
Institution Type/Description: History Museum.
Collections: history of America's first First Family; paintings; personal letters; glassware; silver; jewelry; busts; Martha Washington's prayer book; 1787 Samuel Vaughn sketch of Mt. Vernon grounds; writing instruments & tools from George Washington's survey case; 1710 handwritten will of Colonel Daniel Parke; 18th & 19th century furniture; Minton porcelain; c.1805 walnut games table.
Activities: special programs; outreach activities.
Hours & Admission Prices: Mon.-Fri. 10-3.

SHELBY COUNTY MUSEUM & ARCHIVES, 1854 Old Courthouse, Columbiana, AL 35051. Mailing Address: P.O. Box 457, Columbiana, AL 35051-0457. Tel.: 205-669-3912. Fax: 205-669-3858. Facebook; Shelby County Museum & Archives.
E-mail: schs1854@bellsouth.net
Web Site: www.schsociety.org
Formerly: Shelby County Historical Society, Inc. Museum & Archives
Founded: 1974.
Key Personnel: Dir., Bobby Joe Seales
Institution Type/Description: Historic Building Museum: built in 1854. Listed on the National Register of Historic Places.
Collections: local history, heritage & culture; photographs; period furnishings; personal artifacts.
Research Fields: genealogy.
Hours & Admission Prices: Mon.-Sat. 9-3. No charge. Closed holidays.
Membership: Individual & Couple $20; Sustaining $50; Patron $100.

Cullman

AVE MARIA GROTTO, 1600 St. Bernard Dr., S.E., Cullman, AL 35055-3057. Tel.: 256-734-4110. Fax: 256-737-8768.
E-mail: info@avemariagrotto.com
Web Site: www.avemariagrotto.com
Founded: 1934.
Congressional District: 4
Key Personnel: Dir., Rev. John O'Donnell, O.S.B.; Museum Shop Mgr., Joyce Nix
Institution Type/Description: Religious History Museum: replicas built by Benedictine Monk, Bro. Joseph Zoettl, O.S.B.
Collections: over 125 miniature reproductions of famous churches, shrines, & buildings.
Facilities: 4-acre park. Grotto gift shop.
Hours & Admission Prices: Daily 9-5. Adults $7, senior citizens $5, children 6-12 $4.50; discounts to groups; children under 6 no charge. Closed Christmas.

CULLMAN COUNTY MUSEUM, 211 Second Ave., N.E., Cullman, AL 35055-2905. Tel.: 256-739-1258; 800-533-1258. Fax: 256-737-8782.
E-mail: efuller@cullmancity.org
Web Site: www.cullmancountymuseum.com
Founded: 1975.
Congressional District: 4
Key Personnel: Cur., Elaine L. Fuller.
Personnel Profile: Full-Time Paid 1; Part-Time Paid 3; Interns 1.
Governing Authority: museum board of trustees; nonprofit organization. Tax-exempt.
Institution Type/Description: History Museum: housed in replica of 1873 home of Col. John G. Cullman, founder of Cullman.
Collections: historic items relative to Cullman & Cullman County.
Facilities: Museum-related items for sale.
Activities: guided tours; lectures; permanent exhibits.
Publications: brochures.
Hours & Admission Prices: Mon.-Fri. 9-4, Sun. 1:30-4:30. Adults $5, seniors $4, children under 12 $3; discounts to groups, AAM, AAA & ICOM members; members no charge. Closed New Year's Day; Independence Day; Thanksgiving; Christmas. &
Attendance: 15,000 (estimated)
Membership: Life $50.

CULLMAN DEPOT, 301 1st Ave., N.E., Cullman, AL 35055. Mailing Address: 211 2nd Ave., N.E., Cullman, AL 35055-2905. Tel.: 800-533-1258; 256-739-1258. Fax: 256-737-8782.
Founded: 1914.
Congressional District: 4
Institution Type/Description: Historic Building: listed on the National and State Registers of Historic Places.
Collections: railroad history & artifacts; photographs; period furnishings.
Hours & Admission Prices: Mon.-Fri. 8-4:30. No charge.

Danville

JESSE OWENS MUSEUM, 7019 County Rd. 203, Danville, AL 35619-9053. Tel.: 256-974-3636.
E-mail: jesseowensinfo@charter.net
Web Site: www.jesseowensmuseum.org
Formerly: Jesse Owens Memorial Park and Museum
Founded: 1998.
Congressional District: 5
Key Personnel: Chm. (V) & Pres. (V), Kenneth Brackins; Co Dir., James Pinion; Co Dir. & Museum Shop Mgr., Nancy Pinion.
Personnel Profile: Part-Time Paid 3; Part-Time Volunteers 4.
Governing Authority: Parent Institution: Jesse Owens Memorial Park Board, Lawrence County Commission. Tax-exempt: 501(c)(3).
Institution Type/Description: History Museum.
Collections: photographs, memorabilia & films of Jesse Owens; replica birthplace home; 1936 Olympic Torch replica; bronze statue; architecture; broad jump pit.
Facilities: visitor center; picnic area.
Hours & Admission Prices: Mon.-Sat. 10-4, Sun. 1-4. Admission for groups of 10 or more $2 per person; individuals, small family groups & children under 5 no charge; donations accepted. Closed New Year's Day; Thanksgiving; Christmas. &
Attendance: 45,000 (estimated)

Daphne

AMERICAN SPORT ART MUSEUM AND ARCHIVES, (M), US Sports Academy, One Academy Dr., Daphne, AL 36526-7055. Tel.: 251-626-3303. Fax: 251-621-2527.
E-mail: asama@ussa.edu
Web Site: www.asama.org
Founded: 1985.
Congressional District: 94
Key Personnel: C.E.O., Dr. Thomas P. Rosandich; Cur., Robert Zimlich; Coord. Communications, Leigha Bolton.
Personnel Profile: Full-Time Paid 1; Part-Time Volunteers 8; Interns 1.
Governing Authority: college; nonprofit organization. Parent Institution: U.S. Sports Academy. Tax-exempt: 501(c)(3).
Institution Type/Description: Art Museum.
Collections: bronze sculptures; original paintings; murals; lithographs; giclees, photography & film; all commemorating the artistry of sport; prints; posters; photographs.
Facilities: 100-seat auditorium; educational facilities. Books & art for sale.
Activities: guided tours; loan, temporary & traveling exhibitions. Annual Events: Sport Artist of the Year Program; Youth Sport Art Competition; Academy's Awards of Sport.
Publications: The Academy; The Sport Journal; The Sport Digest; The Sport Update; The Alumni Network.
Hours & Admission Prices: Mon.-Fri. 8-4. No charge; donations accepted. &
Attendance: 2,000 (estimated)
Membership: Bronze $100; Silver $200; Gold $500.

Dauphin Island

ESTUARIUM, Dauphin Island, AL 36528. Mailing Address: Dauphin Island Sea Lab, 101 Bienville Blvd., Dauphin Island, AL 36528-4603. Tel.: 866-403-4409.
E-mail: rdixon@disl.org
Web Site: estuarium.disl.org
Founded: 1998.
Key Personnel: Exec. Dir., Dr. John Valentine; Mgr., Robert Dixon; Registrar, Sally Brennan; Administrative Asst. to the Dir., Lori Angelo; Administrative Asst. Discovery Hall Programs, Denise Keaton; Reservations, Sara Orescan.
Governing Authority: Parent Institution: Dauphin Island Sea Lab. Tax-exempt.
Institution Type/Description: Aquarium.
Collections: plants, animals & other natural resources found in the Estuary.
Hours & Admission Prices: March-Aug. Mon.-Sat. 9-6, Sun. 12-6; Sept.-Feb. Mon.-Sat. 9-5, Sun. 1-5. Adults $7, seniors $6, children 5-18 & students

with ID $4; discounts to AAA members, groups, & military. Closed New Year's Day; Easter; Thanksgiving; Christmas. &
Attendance: 66,407 (accurate)
Membership: Individual $50; Family $100.

Daviston

HORSESHOE BEND NATIONAL MILITARY PARK, 11288 Horseshoe Bend Rd., Daviston, AL 36256-6524. Tel.: 256-234-7111. Fax: 256-329-9905.
Web Site: www.nps.gov/hobe/
Founded: 1959.
Congressional District: 3
Key Personnel: Park Ranger-Interpreter, Ove Jensen.
Personnel Profile: Full-Time Paid 10; Part-Time Volunteers 8.
Governing Authority: federal. Parent Institution: National Park Service, Washington, DC. Tax-exempt.
Institution Type/Description: Historic Site: Horseshoe Bend Battlefield, location of final battle of the Creek Indian War of 1813-1814.
Collections: Indian artifacts; military artifacts, relics & historic documents of the Creek War & War of 1812.
Research Fields: Creek War of 1813-1814 & War of 1812.
Facilities: 500-vol. library of books on national, regional & local history available for use on premises; visitor center; nature trail; tour road through battlefield. Publications related to Horseshoe Bend, postcards & other items for sale.
Activities: permanent & temporary exhibitions; living history demonstrations; Tennessee Militia demonstrations (musket and cannon firing); Creek & Cherokee cultural demonstrations.
Publications: Horseshoe Bend National Military Park Official Guide & Map.
Hours & Admission Prices: Visitor Center: daily 9-4:30. Grounds: daily 8-5. No charge; donations accepted. Closed New Year's Day; Thanksgiving; Christmas. &
Attendance: 111,864 (accurate)

Decatur

THE ART GALLERY, JOHN C. CALHOUN STATE COMMUNITY COLLEGE, 6250 U.S. Hwy. 31 N., Fine Arts Bldg., Decatur, AL 35601. Mailing Address: P.O. Box 2216, Decatur, AL 35609-2216. Tel.: 256-306-2695 & 2699. Fax: 256-306-2889.
E-mail: klv@calhoun.cc.al.us
Web Site: www.calhoun.cc.al.us/
Founded: 1965.
Congressional District: 5
Key Personnel: C.E.O., Kristine Beadle; Chm., William Godsey; Pres., Dr. Marilyn Beck; Dir., Kathryn Vaughn.
Personnel Profile: Full-Time Paid 1; Full-Time Volunteers 2; Part-Time Paid 2.
Governing Authority: state. Parent Institution: Calhoun Community College.
Institution Type/Description: College Art Gallery.
Collections: American & European graphics; student art & photographs.
Research Fields: Computer Art.
Facilities: library; student center; studios.
Activities: exhibitions for students, regional artists & traveling art exhibits.
Publications: exhibit announcements.
Hours & Admission Prices: Mon.-Fri. 9-5. No charge; donations accepted. &
Attendance: 3,600 (estimated)

BLUE & GRAY MUSEUM OF NORTH ALABAMA, 723 Bank St., N.W., Decatur, AL 35601. Tel.: 256-350-4018.
E-mail: babett@knology.net
Key Personnel: Dir. & C.E.O., Robert L. Sackheim; Museum Shop Mgr., Robert Parham.
Personnel Profile: Full-Time Volunteers 1; Part-Time Volunteers 1.
Governing Authority: board. Tax-exempt.
Institution Type/Description: Military History Museum.
Collections: Civil War military equipment; guns; swords; rifles; uniforms; personal artifacts.
Research Fields: military leaders, soldier & battles.
Facilities: 5,000 sq. ft. exhibition space.
Activities: discussions.
Hours & Admission Prices: Mon.-Sat. 10-5:30. Adults $6, students $3; discounts to seniors & groups; pre-school children no charge. &
Attendance: 1,200 (estimated)

CARNEGIE VISUAL ARTS CENTER, 207 Church St., N.E., Decatur, AL 35601-1847. Mailing Address: P.O. Box 1591, Decatur, AL 35602-1591. Tel.: 256-341-0562. Fax: 256-341-0713.
Founded: 2003.

Key Personnel: Exec. Dir., Laura Phillips
Institution Type/Description: Art Center.
Collections: works by local, regional & national artists.
Facilities: rental facilities.
Activities: special events; facilities; workshops; educational programs; lectures; temporary exhibitions; art classes.
Hours & Admission Prices: Tues.-Fri. 10-5, 3rd Sat. 10-2.

COOK'S NATURAL SCIENCE MUSEUM, 412 13th St., S.E., Decatur, AL 35601-5916. Tel.: 256-350-9347.

Web Site: www.cookspest.com/museum.html
Founded: 1980.
Institution Type/Description: Science Museum.
Collections: insects; mounted birds & animals; rocks & minerals; sea shells; coral.
Facilities: Museum-related items for sale.
Hours & Admission Prices: Mon.-Sat. 9-12 & 1-5, Sun. 2-5. No charge. Closed New Year's Day; Thanksgiving; Christmas Eve & Day. &

PRINCESS THEATRE CENTER FOR THE PERFORMING ARTS, 112 Second Ave., N.E., Decatur, AL 35601. Tel.: 256-350-1745 & 340-1778.

Governing Authority: nonprofit organization. Tax-exempt: 501(c)(3).
Institution Type/Description: Theatre Museum: housed in a former livery stable in 1887, later turned into a silent film & vaudeville playhouse. Listed on the National Register of Historic Places.
Collections: theatre history & performers; photographs; costumes.
Facilities: theatre.
Activities: guided tours.
Hours & Admission Prices: Call for hours & admission prices.

WHEELER NATIONAL WILDLIFE REFUGE, 2700 Refuge Headquarters Rd., Decatur, AL 35603. Tel.: 256-350-6639.

Institution Type/Description: Wildlife Refuge.
Collections: wildlife & their habitats; birds; ducks; fish; reptiles & amphibians.
Facilities: nature trails.
Activities: orientation video; hiking.
Hours & Admission Prices: March-Sept. Tues.-Sat. 9-4; Oct.-Feb. daily 9-5. No charge.

Demopolis

BLUFF HALL ANTEBELLUM HOUSE MUSEUM, Marengo County Historical Soc., 405 N. Commissioners Ave., Demopolis, AL 36732. Mailing Address: P.O. Box 159, Demopolis, AL 36732-0159. Tel.: 334-289-9644 & 0282. Facebook: Marengo History.

E-mail: marengohistory@bellsouth.net
Founded: 1961.
Congressional District: 7
Personnel Profile: Part-Time Paid 4.
Governing Authority: nonprofit organization. Parent Institution: Marengo County Historical Society. Subsidiary Institution: Bluff Hall. Tax-exempt: 501(c)(3).
Institution Type/Description: History Museum: house built in 1832-1850.
Collections: costumes; history; furnishings.
Research Fields: history.
Facilities: Museum-related & hand-made gifts for sale.
Activities: guided tours; permanent exhibitions.
Publications: semi-annual newsletter.
Hours & Admission Prices: Tues.-Sat. 10-5, Sun. 2-5. Adults $5, college students $4, students 6-18 $3; discounts to groups of 15 or more; members no charge. Closed New Year's Day; Easter; Independence Day; Thanksgiving; Christmas.
Attendance: 2,800 (estimated)
Membership: Individual $30; Couple $50; Contributor $100; Sustainer $250; Supporter $500; Lifetime $1,000.

GAINESWOOD, 805 S. Cedar Ave., Demopolis, AL 36732-2915. Tel.: 334-289-4846. Fax: 334-289-4846.

E-mail: gaineswd@bellsouth.net
Web Site: www.preserveala.org
Founded: 1975.
Congressional District: 7
Key Personnel: Site Dir., Eleanor W. Cunningham; Dir. Collections & Interpretation, Bruce M. Lipscombe; Maintenance, Richard Rand.
Personnel Profile: Full-Time Paid 3; Part-Time Volunteers 30.
Governing Authority: state. Affiliated with Alabama State Historical Commission. Tax-exempt.

Institution Type/Description: Historical Building & Site: housed in the former home of Nathan Bryan Whitfield; built between 1843-1861.
Collections: furnishings; 19th-century period pieces.
Research Fields: decorative arts; architecture; history; Whitfield family history; study of family letters.
Activities: guided tours; special programs.
Hours & Admission Prices: Tues.-Fri. 10-4, 1st Sat. 10-2; other times by appointment. Adults $5, youth 6-18 $3; discounts to seniors, military, college students, AAA & groups. Closed state holidays. &
Attendance: 2,732 (accurate)
Membership: Friends of Gaineswood $25.

THE LAIRD COTTAGE/GENEVA MERCER MUSEUM, 311 N. Walnut Ave., Demopolis, AL 36732. Tel.: 334-289-0282.

Governing Authority: Parent Institution: Marengo County Historical Society.
Institution Type/Description: Historic House Museum: built in 1870.
Collections: works by Geneva Mercer.
Hours & Admission Prices: By appointment.

Dora

ALABAMA MINING MUSEUM, 120 East St., Dora, AL 35062-4612. Tel.: 205-648-2442.

Founded: 1982.
Congressional District: 4
Key Personnel: Dir., Bonnie Sue Groves.
Personnel Profile: Full-Time Paid 2; Part-Time Volunteers 12.
Governing Authority: non-profit. Tax-exempt: 501(c)(3).
Institution Type/Description: Mining Museum.
Collections: social, cultural & technological aspects of coal and other types of mining in Alabama. Historic Structures: c.1930 two-story WPA project building built from hand-cut stone; c.1900 one-room schoolhouse; 1905 steam locomotive, ore car & caboose; c.1900 Oakman, Alabama railroad depot building; c.1930 U.S. post office building known as Kellerman, Alabama U.S. Post Office.
Research Fields: oral history of miners.
Facilities: library with oral history tapes; 55-capacity auditorium. Gifts & museum-related items for sale.
Activities: guided tours; lectures; films; organized education programs; hands-on room for children to learn about coal mining & living conditions; school field trips using one-room school building to simulate life in 1910 classroom.
Publications: quarterly newsletter.
Hours & Admission Prices: Tues.-Fri. 8:30-4, Sat. 10-4; school groups by appointment. No charge; donations accepted. &
Attendance: 20,000 (estimated)
Membership: Student $2; Senior Citizen $5; Individual $10; Family $15. Business & Corporation $25.

Dothan

DOTHAN AREA BOTANICAL GARDENS, 5130 Headland Ave., Dothan, AL 36303-7691. Tel.: 334-793-3224. Fax: 334-793-5275.

E-mail: info@dabg.com
Web Site: www.dabg.com
Key Personnel: Exec. Dir., Conner Vernon; Pres., Larry Dykes; Vice Pres., Russ Parrish; Groundskeeper, Charles McClendon; Office Mgr., Janie Edmondson
Institution Type/Description: Botanical Gardens.
Collections: Gardens: rose; herb; southern heirloom; demonstration; butterfly; meditation; windmill; wedding garden; camellia; azalea; daylily; hydrangea; Koi ponds; fern glade.
Hours & Admission Prices: Summer: daily 7-7; Winter: daily 7-5. No charge. &

G.W. CARVER INTERPRETIVE MUSEUM, 305 N. Foster St., Dothan, AL 36303. Tel.: 334-712-0933 & 794-3633. Fax: 334-712-0933.

Institution Type/Description: History Museum.
Collections: Dr. Carver's life & career; African American scientists, inventors, & explorers; photographs.
Hours & Admission Prices: Tues.-Fri. 10-5:30, Sat. 1-5; groups by appointment. Closed holidays & holiday weekends.

LANDMARK PARK, Hwy. 431 N., Dothan, AL 36302. Mailing Address: P.O. Box 6362, Dothan, AL 36302-6362. Tel.: 334-794-3452. Fax: 334-677-7229.

E-mail: parkinfo@landmarkpark.com
Web Site: www.landmarkpark.com

Formerly: Alabama Agricultural Museum
Founded: 1976.
Congressional District: 2
Key Personnel: Exec. Dir., William M. Holman; Dir., Dana Peters; Asst. Dir., Kathie Moore; Dir. Public Rels., Laura Stakelum; Office Mgr., Barbara Spears; Membership Coord., Janis Lepley; Teacher & Naturalist, Stephanie Vail.
Personnel Profile: Full-Time Paid 5; Part-Time Paid 2.
Governing Authority: nonprofit organization. Parent Institution: Dothan Landmarks Foundation, Inc. Tax-exempt: 501(c)(3).
Institution Type/Description: Living History Museum & Park Complex serving as the Official Museum of Agriculture for the State of Alabama.
Collections: 1890 living history farm includes buildings & artifacts; Alabama wildlife exhibits; astronomy.
Facilities: 500-vol. library on local, state, agricultural history, natural science & astronomy available on the premises; 135 acres; nature center & planetarium; nature trails; l50-seat auditorium; educational facilities; picnic area. Museum-related items for sale.
Activities: guided tours; lectures; films; concerts; arts festivals; organized education programs; docent program; participatory, loan, temporary & traveling exhibitions.
Publications: quarterly newsletter, The Lark.
Hours & Admission Prices: Summer: Mon.-Sat. 9-5, Sun. 12-6; Winter: Mon.-Sat. 9-5, Sun. 12-5. Adults $4, children 3-12 $3; members and children 2 & under no charge. Closed New Year's Day; Thanksgiving; Christmas. &
Attendance: 48,600 (accurate)
Membership: Individual $30; Senior Couple $40; Family $60; Historical $90; Heritage $150; Heirloom $250; Legacy $500; President's Circle $1,000.

WIREGRASS MUSEUM OF ART, (M), 126 Museum Ave., Dothan, AL 36303-4802. Mailing Address: P.O. Box 1624, Dothan, AL 36302-1624. Tel.: 334-794-3871. Fax: 334-792-9035.
E-mail: director@wiregrassmuseum.org
Web Site: www.wiregrassmuseum.org
Founded: 1988.
Congressional District: 2
Key Personnel: Pres., Jeff Coleman; Exec. Dir., Tara Holman; Dir. Visitor Svcs., Holly Roberts; Dir. Collections & Exhibits, Alison A. Beeson; Art Educator, Lydel Matthews; Art Educator, Jamie Richey; Grants Writer, Amber Hanson; Security & Maintenance, Mike Roberts; Dir. Devel., Tonye Frith; Membership & Administrative Asst., Dana Lemmer.
Personnel Profile: Full-Time Paid 9; Part-Time Paid 1; Part-Time Volunteers 75.
Governing Authority: municipal government. Tax-exempt: 501(c)(3).
Institution Type/Description: Art Museum: housed in 1912 electric plant.
Collections: contemporary & regional art; children's sensory exhibits.
Facilities: 300-vol. library on art instruction & history; educational facilities; conference center; catering kitchen; sculpture garden.
Activities: guided tours; lectures; arts festivals; organized education programs; docent program; participatory & traveling exhibitions.
Publications: bimonthly newsletter, Sketches.
Hours & Admission Prices: Tues.-Sat. 10-5. No charge; donations accepted. &
Attendance: 20,000 (accurate)
Membership: Student $20; Individual $35; Family $50; Supporter $100; Bronze $350; Silver $500; Gold $750; Platinum $1,000.

Elberta

BALDWIN COUNTY HERITAGE MUSEUM, (M), 25521 Hwy. 98, Elberta, AL 36530. Mailing Address: P.O. Box 356, Elberta, AL 36530-0356. Tel.: 251-986-8375 & 752-8883. Fax: 251-986-8375.
E-mail: bchm@gulftel.com
Web Site: www.baldwincountyheritagemuseum.com
Founded: 1981.
Congressional District: 4
Key Personnel: Chm. (V), Ralph Veller; Vice Chm., June Taylor; Chm. (V), Clark Cathey; Clerk, Tammy Kinney.
Personnel Profile: Part-Time Paid 1; Part-Time Volunteers 31.
Governing Authority: private; nonprofit organization. Tax-exempt: 501(c)(3).
Institution Type/Description: History Museum.
Collections: county history from Civil War to the Great Depression; farming & forestry tools; ethnic land developments c.1900; personal, recreational & communication artifacts; turpentining; early fishing & shrimping industries; period churches, schools & railroads. Historic Buildings: 1908 church; blacksmith; pole barns; potato shed; school.
Research Fields: rural history of Baldwin County c.1900; farming, forestry, turpentining, fishing industry; L&N Railroad.
Facilities: nature trails. Museum-related items for sale.
Activities: Annual Event: Winter Festival in January & December.

Hours & Admission Prices: Wed.-Sat. 10-3, 2nd & 4th Sun. 12-5. No charge. Closed all national & state holidays. &
Attendance: 2,700 (estimated)
Membership: Individual $10; Family $15; Friend $50; Sustaining $100; Lifetime $200

Eufaula

SHORTER MANSION MUSEUM, 340 N. Eufaula Ave., Eufaula, AL 36027-1518. Mailing Address: P.O. Box 486, Eufaula, AL 36072-0486. Tel.: 334-687-3793; 888-EUFAULA (383-2852). Fax: 334-687-1836.
E-mail: info@eufaulapilgrimage.com
Web Site: www.eufaulapilgrimage.com
Founded: 1965.
Congressional District: 84
Key Personnel: C.E.O., Douglas Purcell.
Personnel Profile: Full-Time Paid 1; Part-Time Paid 5; Part-Time Volunteers 5.
Governing Authority: society; board of directors. Parent Institution: Eufaula Heritage Association, Inc. Tax-exempt.
Institution Type/Description: Local History Museum: housed in the former home of Eli Sims Shorter II; built in 1884. Listed on the National Register of Historic Places.
Collections: period furnishings; period dresses; textiles; photographs; memoirs of six Alabama Governors from Barbour County; Adm. Thomas H. Moorer Room; UDC & DAR Collections.
Research Fields: genealogy; historic homes within the city of Eufaula.
Facilities: banquet facilities; meeting rooms.
Activities: group tours by appointment; weddings; receptions. Association sponsors: annual Eufaula Pilgrimage of Historic Homes; co-sponsors periodic heritage seminars.
Publications: reprints, Historic Eufaula; History of Eufaula 1875; Backtracking in Barbour County; Foundation Stone.
Hours & Admission Prices: Mon.-Sat. 10-4; group tours by appointment. Adults $5, children 5-12 $3; discounts to groups, AAM & AAA members; children under 4 no charge. &
Attendance: 10,000 (accurate)
Membership: Single $25; Couple $50; Sustaining $100-$249; Patron $250 & up.

Fairhope

EASTERN SHORE ART CENTER, 401 Oak St., Fairhope, AL 36532-2403. Tel.: 251-928-2228. Fax: 251-928-5188.
E-mail: esac@esartcenter.com
Web Site: www.esartcenter.com
Founded: 1958.
Congressional District: 94
Key Personnel: Mng. Dir., C.E.O. & Pres. (V), Kate Fisher.
Personnel Profile: Full-Time Paid 2; Part-Time Paid 3; Part-Time Volunteers 125.
Governing Authority: nonprofit organization. Parent Institution: Bd. of Dirs., Eastern Shore Art Assoc. Tax-exempt.
Institution Type/Description: Art Museum.
Collections: contemporary American paintings.
Facilities: 4 studio-classrooms.
Activities: temporary exhibits; 2 annual outdoor art shows; art classes; workshops; lectures.
Publications: The Artline Newsletter; quarterly class schedule; members directory; book, Painters Paradise.
Hours & Admission Prices: Tues.-Fri. 10-4, Sat. 10-2. No charge; donations accepted. Closed major holidays. &
Attendance: 16,000 (estimated)
Membership: Individual $25; Family $40; Business $50; Supporting $100; Patron $200; Benefactor $500; Lifetime $1,500.

Fayette

FAYETTE ART MUSEUM, 530 N. Temple Ave., Fayette, AL 35555-2211. Tel.: 205-932-8727. Fax: 205-932-8727.
E-mail: fam@watvc.com
Founded: 1969.
Congressional District: 7
Key Personnel: Dir. & Cur., Anne Perry-Uhlman.
Personnel Profile: Part-Time Paid 1; Part-Time Volunteers 15.
Governing Authority: municipal; nonprofit. Parent Institution: City of Fayette, AL. Tax-exempt.
Institution Type/Description: Art Museum: housed in c.1930 former school building.
Collections: 3,700 pieces of 20th-century art: 2,700 pieces by Lois Wilson

(1905-1980), folk art pieces by Jimmy Lee Sudduth, Benjamin F. Perkins, Sybil Gibson & Fred Webster; well-known folk artists.
Facilities: 365-seat auditorium.
Activities: guided tours; arts festivals; temporary exhibits. Annual Event: Fayette Arts Festival in August.
Hours & Admission Prices: Mon.-Fri. 9-12 & 1-4; other times by appointment. No charge; donations accepted. &

Attendance: 40,000 (estimated)

Florence

CHILDREN'S MUSEUM OF THE SHOALS, 2810 Darby Dr., Florence, AL 35630-1524. Tel.: 256-765-0500.
E-mail: cmos@shoalschildrensmuseum.org
Web Site: www.shoalschildrensmuseum.org
Founded: 2001.
Key Personnel: Dir., Peggy McCloy.
Personnel Profile: Part-Time Paid 3.
Governing Authority: Tax-exempt.
Institution Type/Description: Children's Museum.
Collections: hands-on exhibits.
Facilities: Museum-related items for sale.
Activities: educational programs; rental facilities; special events.
Hours & Admission Prices: Thurs.-Sat. 10-4:30; groups by appointment. Admission 2 & over $5; discounts to groups of 10 or more; children under 2 and ACM & museum members no charge. Closed major holidays. &
Membership: Basic $60; Four Persons $80; Unlimited Family $100.

FRANK LLOYD WRIGHT ROSENBAUM HOUSE, 601 Riverview Dr., Florence, AL 35630-6026. Mailing Address: 217 E. Tuscaloosa St., Florence, AL 35630-4724. Tel.: 256-740-8899.
Web Site: www.wrightinalabama.com
Founded: 1999.
Congressional District: 5
Key Personnel: Dir., Barbara K. Broach.
Personnel Profile: Part-Time Paid 4.
Governing Authority: municipal. Parent Institution: City of Florence. Tax-exempt: 501(c)(3).
Institution Type/Description: Historic House Museum: designed by Frank Lloyd Wright, c.1939.
Collections: Wright-designed furniture; photographs; blueprints; Rosenbaum family artifacts.
Research Fields: Frank Lloyd Wright architecture; construction, restoration & preservation of the Rosenbaum house.
Facilities: Museum-related items for sale.
Activities: guided tours.
Publications: museum brochure.
Hours & Admission Prices: Tues.-Sat. 10-4, Sun. 1-4. Adults $8, senior citizens, students & children $5; discounts to groups of 10 or more. Closed major holidays.
Attendance: 5,000 (estimated)

INDIAN MOUND & MUSEUM, 1028 S. Court St., Florence, AL 35630-6116. Mailing Address: 217 E. Tuscaloosa St., Florence, AL 35630-4724. Tel.: 256-760-6427.
Web Site: www.florenceal.org
Founded: 1968.
Congressional District: 5
Key Personnel: Dir., Barbara K. Broach; Cur., Jim Walder.
Personnel Profile: Part-Time Paid 3.
Governing Authority: municipal. Parent Institution: City of Florence. Tax-exempt.
Institution Type/Description: History Museum.
Collections: Native American artifacts including points, pottery, axes, belts, shell necklaces & blades from Paleo to historic ages.
Research Fields: Native American history; historic artifacts.
Activities: guided tours.
Publications: museum brochure.
Hours & Admission Prices: Tues.-Sat. 10-4; groups by appointment. Adults $2, students $.50. Closed major holidays. &
Attendance: 3,500 (estimated)

KENNEDY-DOUGLASS CENTER FOR THE ARTS, 217 E. Tuscaloosa St., Florence, AL 35630-4724. Tel.: 256-760-6379. Fax: 256-760-6382.
E-mail: bbroach@florenceal.org
Web Site: www.florenceal.org
Founded: 1976.
Congressional District: 5

Key Personnel: Dir., Barbara K. Broach; Program Coord., Mary Nicely; Administrative Asst., Faye Vines.
Personnel Profile: Full-Time Paid 3.
Governing Authority: municipal; nonprofit. Tax-exempt.
Institution Type/Description: Art Council: housed in 1917-18 Kennedy-Douglass House.
Collections: genealogy; historical records; art collections. Historic Houses: 1890 Southall-Moore House; 1915 Wright-Douglass House; 1917-18 Kennedy-Douglass House.
Facilities: library of books available for research on premises; reading room; classrooms. Museum-related items for sale.
Activities: guided tours; lectures; films; gallery talks; concerts; arts festivals; study clubs; formally organized education programs; docent program & council; permanent, temporary & traveling exhibitions.
Hours & Admission Prices: Mon.-Fri. 9-4; other times by appointment. No charge. &
Attendance: 12,000 (estimated)

POPE'S TAVERN MUSEUM, 203 Hermitage Dr., Florence, AL 35630-4667. Mailing Address: 217 E. Tuscaloosa St., Florence, AL 35630-4724. Tel.: 256-760-6439.
Web Site: www.florenceal.org
Founded: 1968.
Congressional District: 5
Key Personnel: Dir., Barbara K. Broach.
Personnel Profile: Part-Time Paid 3.
Governing Authority: municipal; City of Florence.
Institution Type/Description: Historic House; housed in a former stagecoach stop, tavern & inn; used as a hospital by Confederate & Union forces.
Collections: colonial period furnishings, relics, & artifacts; Civil War artifacts.
Research Fields: Revolutionary & Civil War Periods; city & county history.
Activities: tours; temporary exhibits; art & craft demonstrations. Museum Sponsors: Frontier Day in June; Open House in December.
Publications: historic brochures.
Hours & Admission Prices: Tues.-Sat. 10-4. Adults $2, children $.50. &
Attendance: 2,657 (accurate)

W.C. HANDY HOME MUSEUM AND LIBRARY, 620 W. College St., Florence, AL 35630-5360. Mailing Address: 217 Tuscaloosa St., Florence, AL 35630-4724. Tel.: 256-760-6434. Fax: 256-760-6382.
E-mail: bbroach@florenceal.org
Web Site: www.florenceal.org
Founded: 1968.
Congressional District: 5
Key Personnel: Dir., Barbara K. Broach; Cur., Sandra Ford.
Personnel Profile: Part-Time Paid 2.
Governing Authority: municipal. Parent Institution: City of Florence, Mayor Council. Tax-exempt.
Institution Type/Description: Historic House.
Collections: artifacts; memorabilia related to the life of W.C. Handy.
Research Fields: W.C. Handy papers.
Facilities: library.
Activities: guided tours. Annual Event: W.C. Handy Birthday Celebration in November.
Publications: brochures.
Hours & Admission Prices: Tues.-Sat. 10-4. Adults $2, students $.50. &
Attendance: 3,500 (estimated)

Fort Payne

ALABAMA FAN CLUB AND MUSEUM, 101 Glenn Blvd., S.W., Fort Payne, AL 35967-4963. Tel.: 800-557-8223; 256-845-1646. Fax: 256-845-5650.
Institution Type/Description: Country Music Group Museum.
Collections: country music & group history; personal artifacts; photographs.
Facilities: Museum-related items for sale.
Activities: group fan club.
Hours & Admission Prices: Memorial Day to Labor Day daily 9-6; Sept.-May Wed.-Sat. 9-6, Sun. 1-6.

COOK SOUND STUDIOS, 1419 Scenic Rd., Fort Payne, AL 35968. Tel.: 256-845-2286.
Institution Type/Description: Country Music Museum: housed in the studio of Alabama band member, Jeff Cook.
Collections: Jeff Cook's music & life; personal artifacts; photographs.
Hours & Admission Prices: By appointment.

FORT PAYNE DEPOT MUSEUM, 105 5th St., N.E., Fort Payne, AL 35967-2455. Mailing Address: P.O. Box 681420, Fort Payne, AL 35968-1615. Tel.: 256-845-5714.
Web Site: fortpaynedepotmuseum.org
Founded: 1986.
Key Personnel: Chm. (V), Eric Brisendine.
Personnel Profile: Full-Time Volunteers 1; Part-Time Paid 2; Part-Time Volunteers 5.
Governing Authority: nonprofit organization.
Institution Type/Description: History Museum: housed in historic Fort Payne railroad depot built by the Alabama Great Southern Railroad in 1891.
Collections: historical documents & artifacts; Indian artifacts; turn of the century clothing; farm equipment; household items; tools & machinery.
Hours & Admission Prices: Mon., Wed. & Fri. 10-12 & 1-3:30. Adults 19 & up $3, students 7 & up $1; groups 10 or more $2; children 6 & under no charge. Caboose $1, children 6 & under no charge. Closed major holidays.
Attendance: 4,000 (estimated)

Fort Rucker

U.S. ARMY AVIATION MUSEUM, Bldg. 6000, Fort Rucker, AL 36362. Mailing Address: P.O. Box 620610, Fort Rucker, AL 36362-0610. Tel.: 334-255-3036 & 2893. Fax: 334-255-3054.
E-mail: avnmuseum@alanet.com
Web Site: www.armyavnmuseum.org
Founded: 1962.
Congressional District: 2
Key Personnel: Dir., R.S. Maxham; Cur. Collections, Robert D. Mitchell.
Governing Authority: Army Regulation 870-20. Parent Institution: Center of Military History, Washington, DC. Tax-exempt.
Institution Type/Description: Aviation Technology and Military Museum.
Collections: aircraft, fixed wing and rotary wing; army aviation memorabilia.
Research Fields: army aviation & related activities.
Facilities: library of general & technical Army Aviation books & manuals available to persons performing Army Aviation research.
Activities: scheduled VIP, school & special group tours; permanent & temporary exhibits; classes.
Hours & Admission Prices: Mon.-Sat. 9-4, Sun. 12-4. No charge; donations accepted. Closed New Year's Eve & Day; Thanksgiving; Christmas Eve & Day &
Membership: U.S. Army Aviation Museum Foundation, Inc.: Organization, In Memoriam & Life $100; Distinguished $500; Charter $1,000.

Franklin

ALABAMA RIVER MUSEUM, Claiborne Lock & Dam, Alabama River (north of Monroeville), 31 Isaac Creek Rd., Franklin, AL 36444. Mailing Address: 31 N. Alabama Ave., Monroeville, AL 36461. Tel.: 251-575-7433. Fax: 251-575-2513.
E-mail: mchm@frontiernet.net
Web Site: www.tokillamockingbird.com
Formerly: River Heritage Museum
Founded: 2000.
Congressional District: 1
Key Personnel: C.E.O. & Museum Shop Mgr., Stephanie Rogers; Chm. (V), Clark McKinley.
Personnel Profile: Part-Time Paid 1; Part-Time Volunteers 20.
Governing Authority: Parent Institution: Monroe County Heritage Museums. Tax-exempt: 501(c)(3).
Institution Type/Description: History Museum.
Collections: tools, points & ceremonial from Native Americans of Alabama; Alabama River steamboats artifacts; eocene-age fossils from the Claiborne Bluff.
Activities: Museum Sponsors: Alabama River Festival in March.
Hours & Admission Prices: Closed until further notice. &

Gadsden

ETOWAH HISTORICAL SOCIETY - JERRY B. JONES HISTORICAL RESEARCH LIBRARY, 2829 W. Meighan Blvd., Gadsden, AL 35902. Mailing Address: P.O. Box 8131, Gadsden, AL 35902. Tel.: 256-613-6844.
E-mail: dkcrown@bellsouth.net
Institution Type/Description: Historical Society Museum & Library.
Collections: local history & culture; photographs; manuscripts; books; periodicals; maps; newspapers; audio & video materials.
Hours & Admission Prices: Mon.-Fri. 10-4. Closed holidays.

GADSDEN MUSEUM OF ART & HISTORY, (M), 515 Broad St., Gadsden, AL 35901-3719. Tel.: 256-546-7365. Fax: 256-546-7365.
E-mail: museum@cityofgadsen.com
Web Site: gadsdenmuseum.com
Founded: 1965.
Congressional District: 7
Key Personnel: Dir., Steve Temple; Graphics Dir., Dan Hampton; Cur., Rebecca Duke.
Personnel Profile: Full-Time Paid 3; Part-Time Paid 1; Interns 3.
Governing Authority: municipal; nonprofit organization. Parent Institution: Gadsden Public Library. Tax-exempt: 170(b)(1)(A).
Institution Type/Description: Art & History Museum.
Collections: paintings in oil, acrylics & watercolors; period artifacts; lithographs; Indian artifacts; china & porcelain; historical decorative arts of Gadsden area.
Major Exhibits: GMA Student Art Show, March 2014; Southeastern Plein Air Festival, April 2014.
Facilities: library of fine arts books & slides.
Activities: guided tours; permanent & temporary exhibitions.
Publications: brochure; quarterly newsletter, Museum Musings.
Hours & Admission Prices: Mon.-Sat. 10-4, 1st Fri. of month 10-8. No charge; donations accepted. &
Attendance: 8,000 (estimated)
Membership: Student $20; Individual $30; Family $60; Supporting $100; Patron $500; Corporate Sponsor $2,500.

MARY G. HARDIN CENTER FOR CULTURAL ARTS, 501 Broad St., Gadsden, AL 35901-3719. Mailing Address: P.O. Box 1507, Gadsden, AL 35902-1507. Tel.: 256-543-2787. Fax: 256-546-7435.
E-mail: bobwelch@culturalarts.org
Web Site: www.culturalarts.org
Founded: 1987.
Congressional District: 7
Key Personnel: C.E.O. & Exec. Dir., Robert M. Welch; Chm. (V), Mark Condra; Vice Chm., Daniel Sizemore; Financial Dir., Kay Moore; Deputy Dir., Tom Banks; Dir. Children's Museum & Education, Tami Brooks; School for the Arts, Maegan Burdine.
Personnel Profile: Full-Time Paid 12; Part-Time Paid 27; Part-Time Volunteers 110; Interns 1.
Operating Expenses: 1,329,197
Operating Income: 1,341,694
Governing Authority: nonprofit organization. Parent Institution: Gadsden Cultural Arts Foundation. Subsidiary Institution: Imagination Place Children's Museum. Tax-exempt: 501(c)(3).
Institution Type/Description: Art Center.
Collections: old postcards of Etowah County, AL.
Research Fields: visitor behavior.
Facilities: library; recital hall; theater; restaurant; meeting rooms; ballet studio.
Activities: guided tours; lectures; concerts; dance recitals; arts festivals; hobby workshops; organized education programs; docent program; participatory & traveling exhibitions.
Publications: monthly newsletter; Arts Sake.
Hours & Admission Prices: Mon.-Sat. 10-8, Sun. 1-5. Imagination Place: Mon.-Sat. 10-6, Sun. 1-5. Adults $6, children $5; discounts to groups & ASTC reciprocal; Tues. afternoons & members no charge. Closed New Year's Day; Independence Day; Thanksgiving; Christmas. &
Attendance: 88,285 (accurate)
Membership: Individual $50; Family $80.

Georgiana

HANK WILLIAMS' BOYHOOD HOME & MUSEUM, 127 Rose St., Georgiana, AL 36033. Mailing Address: P.O. Box 310, Georgiana, AL 36033. Tel.: 334-376-2396.
Institution Type/Description: Historic House Museum: housed in the boyhood home of Hank Williams, Sr.
Collections: Hank Williams' life & career; personal artifacts; period furnishings; photographs; 78 RPM recordings; album covers; sheet music; radio sponsor products; newspaper clippings; paintings.
Hours & Admission Prices: Mon.-Sat. 10-4.

Gilbertown

CHOCTAW COUNTY HISTORICAL MUSEUM, (M), 40 Melvin Rd., Gilbertown, AL 36908. Mailing Address: P.O. Box 162, Gilbertown, AL 36908-0162. Tel.: 251-843-2501. Facebook: Choctaw County Historical Museum.
E-mail: cchm@millry.net

Founded: 1987.
Key Personnel: Dir., Sandra Jenkins Little; Pres. (V), Tommy Campbell; Cur. & Museum Shop Mgr., Danny Roberts.
Personnel Profile: Part-Time Volunteers 3.
Volunteer Hours: 1,350
Operating Expenses: 8,500
Operating Income: 9,000
Governing Authority: Parent Institution: Choctaw County Historical Museum, Inc. Tax-exempt.
Institution Type/Description: History Museum.
Collections: local history & culture; photographs; personal artifacts; period furnishings; war memorabilia; fossils.
Hours & Admission Prices: Sat. 9-1, or by appointment. No charge; donations accepted. ♿
Attendance: 1,200 (estimated)
Membership: Individual $25; Family $35; Lifetime $500.

Grand Bay

EL CAZADOR MUSEUM, 10329 Freeland Ave., Grand Bay, AL 36541-5731. Mailing Address: P.O. Drawer 605, Grand Bay, AL 36541. Tel.: 251-865-0128. Fax: 251-865-3419.
Web Site: www.elcazador.com
Institution Type/Description: History Museum: Spanish Brigantine El Cazador lost at sea in 1784.
Collections: shipwreck artifacts including a breech loading cannon, the ship's bell, bronze sword handles, mint shipment of silver coins; gold artifacts.
Hours & Admission Prices: Thurs.-Sat. 10-4.

Graysville

MERIKS ZOO, 662 Windsor Dr. NE, Graysville, AL 35073-1231.
Key Personnel: Pres., Paul Stein; Exec. Dir., Tillian Smith; Assoc. Dir., Rey Jones; Cur. Mammals & Birds, Shawna Williams; Exhibitions, Jennifer Jackovino; Admin., Stephen Jurcsek; Zoo Keeper, Lynn Harjes.
Personnel Profile: Full-Time Paid 3; Part-Time Paid 5.
Governing Authority: state. Tax-exempt 501(c)(3).
Institution Type/Description: Zoo.
Collections: zoological collection of over 3,000 animals
Research Fields: animal research, reproduction & behavior.
Facilities: 1,000-vol. library of zoo-related books available for use by public; 250-seat theater. Gift-related items for sale.
Activities: guided tours; lectures; films; formally organized education program for children & adults.
Publications: brochures; annual report; pamphlets.
Hours & Admission Prices: Daily 9-6. Adults $15, children 3-16 $10, students with ID & seniors $5; members & children under 3 no charge. Closed New Years Day; Independence Day; Christmas.
Attendance: 10,000 (estimated)

Greensboro

MAGNOLIA GROVE-HISTORIC HOUSE MUSEUM, 1002 Hobson St., Greensboro, AL 36744-1414. Tel.: 334-624-8618. Fax: 334-624-8618.
E-mail: maggrov2@bellsouth.net
Web Site: www.preserveala.org
Founded: 1943.
Congressional District: 7
Key Personnel: Site Dir., Eleanor W. Cunningham.
Personnel Profile: Full-Time Paid 1; Part-Time Paid 2; Part-Time Volunteers 5.
Governing Authority: state. Parent Institution: Alabama Historical Commission. Tax-exempt.
Institution Type/Description: Historic House Museum: c.1840 Greek Revival style home built by Isaac Croom. Boyhood home of Rear Admiral Richmond Pearson Hobson, Spanish-American war hero.
Collections: Hobson memorabilia; family furnishings.
Research Fields: historic homes of America, Spanish American War; decorative arts.
Activities: guided tours; educational programs; community events.
Publications: pamphlets
Hours & Admission Prices: House: Tues.-Fri. 10-4, 1st Sat. each month 10-2. Grounds: daily. Adults $5, college students $4, children 6-18 $3; discount to AAA, military, seniors & groups. Closed major holidays. ♿
Attendance: 982 (accurate)

SAFE HOUSE BLACK HISTORY MUSEUM, 518 Martin Luther King Dr., Greensboro, AL 36744-2237. Tel.: 334-624-2030. Fax: 334-624-2036.
E-mail: safehouseblackhi@bellsouth.net
Formerly: Hole County Civil Rights
Founded: 2002.
Congressional District: 7
Key Personnel: Dir., David Cooper; Museum Shop Mgr., Theresa Burroughs
Governing Authority: nonprofit organization. Tax-exempt: 501(c)(3).
Institution Type/Description: History Museum: housed in the home used to conceal Dr. Martin Luther King Jr. while he organized peaceful resistance protests of segregation in Alabama during the 1960s.
Collections: local history & culture; photographs; personal artifacts.
Major Exhibits: Work and Feel the History, 12/6/13-2/14; Soldier Story by William Rembert, Artist, 2/14-3/14.
Activities: Martin Luther King Jr. Day, Black History in January, February, May & June; May Day Fest; Juneteenth; Anniversary; Back to School Gifts.
Hours & Admission Prices: By appointment. Adults $2; discounts to seniors, mentally challenged, elementary & high school students.
Attendance: 2,000 (estimated)

Grove Hill

CLARKE COUNTY HISTORICAL MUSEUM, 116 W. Cobb St., Grove Hill, AL 36451. Tel.: 251-275-8684.
E-mail: clarkemuseum@galaxycable.net
Web Site: www.clarkmuseum.com
Governing Authority: private; nonprofit organization.
Institution Type/Description: History Museum.
Collections: local history & culture; photographs; period furnishings; personal artifacts.
Hours & Admission Prices: Mon. 12-4, Tues.-Fri. 10-4, Sat. 10-2.

Gulf Shores

ALABAMA GULF COAST ZOO, 1204 Gulf Shores Pkwy., Gulf Shores, AL 36542-5908. Tel.: 251-968-5732.
E-mail: info@alabamagulfcoastzoo.org
Key Personnel: Dir., Patti Hall.
Governing Authority: Operated By: Zoo Foundation, Inc. Tax-exempt: 501(c)(3).
Institution Type/Description: Zoo.
Collections: over 290 animals including lions, tigers, bears, monkeys & macaws.
Activities: animal shows; petting zoo.
Hours & Admission Prices: Daily 9-4. Adults 12-54 $10, seniors 55 & over $8, children 3-12 $7; children 2 & under no charge. Closed New Year's Day; Thanksgiving; Christmas Eve & Day.

FORT MORGAN MUSEUM, 110 Hwy. 180 W., Gulf Shores, AL 36542-7802. Tel.: 251-540-5257. Fax: 251-540-7665.
E-mail: brianohill@fortmorgan.org
Web Site: www.fortmorgan.org
Founded: 1967.
Congressional District: 1
Key Personnel: Dir., Brian O. Hill; Educational Program Developer, Michael Bailey; Museum Shop Mgr., Jan Dillon.
Personnel Profile: Full-Time Paid 7; Part-Time Volunteers 5.
Governing Authority: society. Parent Institution: Alabama Historical Commission, 468 South Perry St., Montgomery, AL 36130-0900. Tax-exempt.
Institution Type/Description: Military History Museum.
Collections: weapons; Civil War artifacts; War of 1812 memorabilia; Spanish American War, World War I & World War II memorabilia.
Research Fields: Civil War; War of 1812; Spanish-American War; Endicott Era.
Facilities: picnic areas; bird sanctuary; concession stand.
Activities: pre-arranged tours. Museum Sponsors: Living History Program June & July.
Publications: brochures; pamphlets.
Hours & Admission Prices: Museum: daily 9-4:30. Fort & Grounds: daily 8-5. Adults $7, seniors $5, children 6-12 $4; children under 6 no charge. Closed New Year's Day; Thanksgiving; Christmas. ♿
Attendance: 89,147 (accurate)

GULF SHORES MUSEUM, 244 W. 19th Ave., Gulf Shores, AL 36542. Mailing Address: P.O. Box 299, Gulf Shores, AL 36547-0299. Tel.: 251-968-1473. Fax: 251-968-1175.
E-mail: museum@gulfshoresal.gov
Web Site: www.gulfshoresal.gov

Founded: 1999.
Congressional District: 32
Key Personnel: Chm. (V), May Alanko; C.E.O., Wendy Congiardo; Dir., Grant Brown; Museum Admin., Christie Shannon; Docent, Jan Meeks.
Personnel Profile: Part-Time Paid 1; Part-Time Volunteers 5.
Governing Authority: municipal. Parent Institution: City of Gulf Shores.
Institution Type/Description: History Museum.
Collections: local history from Native American period to present; hurricane weather; shrimping & fishing industry; boat building tools.
Research Fields: history of Gulf Shores.
Facilities: library; 1,200 sq. ft. exhibit space.
Activities: loan, participatory & traveling exhibitions; winter film series. Annual Event: Christmas Open House.
Hours & Admission Prices: Tues.-Fri. 10-12 & 1-5, Sat. 10-2. No charge; donations accepted. Closed New Year's Day; Mardi Gras Day; Memorial Day; Independence Day; Labor Day; Veterans Day; Thanksgiving; Christmas Eve & Day. &
Attendance: 4,340 (accurate)

Guntersville

GUNTERSVILLE MUSEUM, 1215 Rayburn Ave., Guntersville, AL 35976-1432. Tel.: 256-571-7597. Fax: 256-571-7584.
E-mail: info@guntersvillemuseum.org
Web Site: guntersvillemuseum.org
Founded: 1990.
Key Personnel: Dir., Julie Patton; Chm. Bd., Angela Otts.
Personnel Profile: Full-Time Paid 1; Part-Time Paid 2; Part-Time Volunteers 20; Interns 1.
Volunteer Hours: 1,000
Governing Authority: Tax-exempt.
Institution Type/Description: History Museum & Art Museum: housed in a former military armory for Company E 167th Infantry Division; built in 1936.
Collections: area history & culture; Native American artifacts; personal artifacts; photographs; paintings; over 300 pieces of art.
Hours & Admission Prices: Tues.-Fri. 10-4, Sat.-Sun. 1-4. No charge. &
Attendance: 5,000 (accurate)
Membership: Senior/Student $15; Individual $25; Family $50; Sponsor $100; Benefactor $500; Life $1,000.

Hanceville

THE EVELYN BURROW MUSEUM, (M), Burrow Center for the Fine and Performing Arts Bldg., 801 Main St., N.W., Hanceville, AL 35077. Mailing Address: P.O. Box 2000, Hanceville, AL 35077. Tel.: 256-352-8458 & 8457; 866-350-9722, ext. 8458 (Toll Free). Fax: 256-352-8098.
E-mail: donny.wilson@wallacestate.edu
Web Site: www.burrowmuseum.org
Founded: 2005.
Key Personnel: Dir., Donny Wilson
Governing Authority: Parent Institution: Wallace State Community College. Tax-exempt.
Institution Type/Description: Art Museum.
Collections: decorative arts including porcelain; pottery; fine art; period furnishings; bronzes; cut glass.
Hours & Admission Prices: Tues.-Fri. 9-5, Sat. 10-2. No charge. &
Attendance: 6,000 (estimated)

Hartselle

HARTSELLE MILITARY MUSEUM, 109 Hickory St., S.E., Hartselle, AL 35640-2547. Mailing Address: P.O. Box 992, Hartselle, AL 35640-0992. Tel.: 256-773-7271.
Founded: 2002.
Institution Type/Description: Military History Museum.
Collections: U.S. military history from the Revolutionary War to present; uniforms; equipment; medals; flags; photographs; personal artifacts.
Activities: special events.
Hours & Admission Prices: Mon.-Fri. 8-4. No charge; donations accepted.

Hillsboro

POND SPRING - THE GENERAL JOE WHEELER HOME, 12280 Alabama Hwy. 20, Hillsboro, AL 35643. Tel.: 256-637-8513 & 9291. Fax: 256-637-8513.
E-mail: wheplan@hiwaay.net
Web Site: www.wheelerplantation.org

Founded: 1994.
Congressional District: 8
Key Personnel: Dir., Melissa Beasley; Cur., Kara Long; Pres. (V), Dr. Mildred Candle; Museum Shop Mgr., Hollye Raines.
Personnel Profile: Full-Time Paid 5; Part-Time Paid 2; Part-Time Volunteers 1.
Governing Authority: state. Parent Institution: Alabama Historical Commission.
Institution Type/Description: Historic House: housed in the former home of General Joe Wheeler. National Register of Historic Places.
Collections: local history & culture; period furnishings; personal artifacts; books; military artifacts from the Civil & Spanish Wars; family portraits; photographs; decorative arts; cemeteries. Historic Buildings: 1818 dogtrot log house; c.1930 Federal-style house; 1880s Wheeler house; 8 farm-related outbuildings.
Facilities: 50 acres of land; gardens.
Activities: tours of home, gardens & grounds.
Hours & Admission Prices: Wed.-Sat. 9-4, Sun. 1-5. Adults $8, seniors $5, students $3.
Attendance: 15,000 (estimated)

Huntsville

BURRITT ON THE MOUNTAIN - A LIVING MUSEUM, (M), 3101 Burritt Dr., Huntsville, AL 35801-1142. Tel.: 256-536-2882. Fax: 256-532-1784.
E-mail: bm-recep@ci.hunstville.al.us
Web Site: www.burrittonthemountain.com/
Formerly: Burritt Museum & Park
Founded: 1955.
Congressional District: 5
Key Personnel: Exec. Dir. & C.E.O., Leslie Ecklund; Dir. Operations, Pat Robertson; Exec. Asst., Teresa O'Malley; Bd. Chm. (V), Butch Damson; Bd. Vice-Chm., LeVon Nunn; Cur. & Historic Park Coord., Stephanie Timberlake; Dir. Education, Joan Morehead; Dir. Mktg., Gina Haskins; Museum Shop Mgr., Susan Clements.
Personnel Profile: Full-Time Paid 6; Part-Time Paid 11; Part-Time Volunteers 150.
Governing Authority: municipal government. Tax-exempt: 501(c)(3).
Institution Type/Description: Historic Village & Museum: housed in a 1935 mansion. Listed on the National Register of Historic Places.
Collections: 19th-century rural life in north Alabama & south central Tennessee; local natural history; 1850 & 1900 living history farmsteads; local & nationally renowned artists; textiles.
Research Fields: north Alabama flora & fauna; folk medicines; 19th-century rural lifestyles; Tennessee Valley prehistoric Indian archeology; textiles; rural architecture; Southern decorative arts.
Facilities: research library; picnic area; nature trails; classrooms. Museum-related items for sale.
Activities: guided tours; lectures; concerts; arts festivals; hobby workshops; docent program; loan & temporary exhibitions; seasonal living history events; educational programs. Museum Sponsors: Sorghum Festival; Earth Day; Indian Heritage Festival; Earth Camp; Spring Farm Days; Fall Festival; Candlelight Christmas.
Publications: books: Faces; Shadows On The Wall; On Gossamer Threads; brochures.
Hours & Admission Prices: April-Oct. Tues.-Sat. 9-5, Sun. 12-5; Nov.-March Tues.-Sat. 10-4, Sun. 12-4. Adults $7, seniors 60 & over and military $6, students 13-17 $5, children 3-12 $4; discounts to local residents and AAM & AAA members; children 2 & under and members no charge. Closed New Year's Day; Thanksgiving; Christmas Eve & Day. &
Attendance: 100,000 (estimated)
Membership: Individual $25; Family $50; Piedmont Society $100-$249; Foothill Society $250-$999; Summit Society $1,000 & up.

EARLYWORKS FAMILY OF MUSEUMS, (M), 404 Madison St., Huntsville, AL 35801-4203. Tel.: 256-564-8100 & 8124. Fax: 256-564-8151.
E-mail: bart.williams@huntsvilleal.gov
Web Site: www.earlyworks.com
Formerly: EarlyWorks Museums
Founded: 1982.
Congressional District: 5
Key Personnel: Exec. Dir. & C.E.O., Bart Williams; Dir. Volunteers, Holly Preston; Constitution Village Museum Shop Mgr., Lora McGowan.
Personnel Profile: Full-Time Paid 16; Part-Time Paid 22; Part-Time Volunteers 250.
Governing Authority: municipal government. Parent Institution: City of Huntsville, AL. Branch Museums: Alabama Constitution Village, 109 Gates Ave., Huntsville, AL 35801; Humphreys-Rodgers House, corner Gates Ave.

& Fountain Circle, Huntsville, AL 35801; Historic Huntsville Depot, 320 Church St., Huntsville, AL 35801. Tax-exempt: 170(b)(1)(A).
Institution Type/Description: History Museum.
Collections: Earlyworks Museum: hands-on exhibits; Alabama Constitution Village: living history village & site in 1819; Historic Huntsville Depot: once used as a prison in Civil War; Humphreys-Rogers House: collection house & site for decorative arts exhibits.
Research Fields: history-related fields; social & women's history; domestic life; 19th-century crafts; architectural history; decorative arts; Federal Period; Alabama history & politics.
Facilities: research library; education classrooms; multi-purpose room; facilities available for meetings & receptions. Museum-related items for sale.
Activities: guided interpretive tours; domestic & labor skills demonstrations; special events; lectures; workshops; State History Teachers' Conference; Biscuit's Backyard preschool area; history teams outreach program; hands-on programs; organized education programs; furniture making; spinning; weaving; open hearth cooking; candle dipping; printing; soap-making; fabric dyeing; 19th-century domestic skill interpretation. Seasonal Special Programs: Holiday Open House; Folktales & Gingerbread; Dairy program; lavender & lace; boxcar children's camp; polar express.
Publications: quarterly calendar of events; brochures announcing special programs; video & activity book, The Spirit of the Times; cookbooks, Mrs. Neal's Cookbook; Society Sampler Cookbook.
Hours & Admission Prices: EarlyWorks: Tues.-Sat. 9-4. Village: March-Oct. Tues.-Sat. 10-4; Thanksgiving to Dec. 23 5 pm-9 pm. Depot: March-Dec. Tues.-Sat. 10-4. EarlyWorks: adults $10, senior citizens 55 & over and youth 4-17 $8, children 1-3 $4; discounts to AAM & ACM members; members and children 3 & under no charge. Village & Depot: adults $7, senior citizens & youth 4-17 $6; discounts to AAM & ACM members; members and children 3 & under no charge. Closed New Year's Day; Thanksgiving; Christmas Eve & Day. &
Attendance: 150,000 (accurate)
Membership: Individual & Grandparent $55 & $75; Contributing $100 & up; Corporate $100-$5,000.

HARMONY PARK SAFARI, 431 Cloud's Cove Rd., S.E., Huntsville, AL 35803-6513. Tel.: 256-723-3880; 877-726-4625 (Toll Free).
Institution Type/Description: Nature Preserve.
Collections: wildlife & their habitats from around the world.
Activities: drive-thru preserve.
Hours & Admission Prices: March-Nov. daily 10am to sunset.

HOWARD WEEDEN HOUSE MUSEUM, 300 Gates Ave., S.E., Huntsville, AL 35801-3101. Tel.: 256-536-7718.
E-mail: theweedenhouse@att.net
Web Site: https://weedenhousemuseum.com
Founded: 1981.
Key Personnel: Dir., Barbara L. Scott
Institution Type/Description: History Museum: housed in 1819 Federal style home & birthplace of artist & poet, Maria Howard Weeden.
Collections: paintings; personal artifacts; furnishings.
Hours & Admission Prices: Call for hours.

HUNTSVILLE/MADISON COUNTY BOTANICAL GARDEN, 4747 Bob Wallace Ave., Huntsville, AL 35805-3390. Tel.: 256-830-4447. Fax: 256-830-5314.
E-mail: vmhurst@hiwaay.net
Web Site: www.hsvbg.org
Founded: 1988.
Key Personnel: Dir., Paula Steigerwald
Institution Type/Description: Botanical Garden.
Collections: dogwood trail; daylily garden; herb garden; vegetable garden.
Facilities: 112 acres; cafe; nature trail. Museum-related items for sale.
Activities: group tours.
Hours & Admission Prices: May-Sept. Mon.-Wed. & Fri.-Sat. 9-6, Thurs. 9-8, Sun. 12-6; Oct.-April Mon.-Sat. 9-5, Sun. 12-5; groups by appointment. Adults $10, seniors 55 & over and military $8, children 3-18 $5; members and children 2 & under no charge. Closed New Year's Day; Thanksgiving; Christmas.
Attendance: 325,000 (accurate)
Membership: Individual $45; Family $70.

* **HUNTSVILLE MUSEUM OF ART, (M),** 300 Church St., S., Huntsville, AL 35801-4910. Tel.: 256-535-4350. Fax: 256-532-1743.
E-mail: info@hsvmuseum.org
Web Site: www.hsvmuseum.org
Founded: 1970.

Congressional District: 5
Key Personnel: Exec. Dir., Christopher Madkour; Accountant, Wendy Worley; Dir. Curatorial Affairs, Peter J. Baldaia; Registrar, Deborah S. Taylor; Cur. Exhibitions, David Reyes; Interim Dir. Education, Laura Smith; Communications, Stephanie Kelley; Facility & Event Mgr., Lil Parton; Museum Shop Mgr., Janell Zesinger; Security Supvr., Linda Berry.
Personnel Profile: Full-Time Paid 10; Part-Time Paid 17; Interns 1.
Governing Authority: municipal. Tax-exempt: 501(c)(3).
Institution Type/Description: Art Museum.
Collections: 19th- & 20th-century American Art with an emphasis on Southern art. Contemporary crafts & glass.
Research Fields: 19th- & 20th-century American art; Southern art.
Facilities: 2,000-vol. art research library; 46,000 sq. ft. exhibit space; four meeting rooms; special events area.
Activities: guided tours; films; gallery talks; organized training programs; temporary & traveling exhibitions; special exhibitions; partnership in art program; musical presentations; lectures; museum academy art classes & workshops.
Publications: quarterly newsletter; exhibition catalogues; exhibition brochures.
Hours & Admission Prices: Tues.-Wed. & Fri.-Sat. 11-4, Thurs. 11-8, Sun. 1-4. Adults $10, military, seniors & educators $8; discounts to groups; members no charge. &
Attendance: 77,652 (accurate)
Membership: Student $30; Individual $45; Dual $60; Family $70; Contributor $100; Friend $175; Sponsor $300; Patron $600; Benefactor $1,200; Artist's Circle $1,500. (Discounts to seniors, military & educators).

NORTH ALABAMA RAILROAD MUSEUM, 694 Chase Rd., Huntsville, AL 35811-1523. Mailing Address: P.O. Box 4163, Huntsville, AL 35815-4163. Tel.: 256-851-6276.
Web Site: www.northalabamarailroadmuseum.com
Formerly: North Alabama Railroad Club
Founded: 1967.
Key Personnel: Pres. (V), Jack Kluge; Dir., Hugh Dudley; Museum Shop Mgr., Amy Boren.
Personnel Profile: Full-Time Volunteers 2; Part-Time Volunteers 40.
Governing Authority: Tax-exempt.
Institution Type/Description: Railroad Museum.
Collections: railcar; restored Railway Post Office car; day coach; Pullman sleeping car; freight & passenger equipment.
Activities: train rides; guided tours; charters; school field trips.
Publications: monthly newsletter; brochures; children's coloring book school program.
Hours & Admission Prices: Daily 9-4; groups by appointment. No charge; donations accepted. Train: adults $12; children under 12 $8; discounts for guided tours & NRHS members.
Attendance: 5,000 (estimated)
Membership: Individual $49; Family $56.

SCI-QUEST HANDS-ON SCIENCE CENTER, 102-D Wynn Dr., Huntsville, AL 35805-1957. Tel.: 256-837-0606. Fax: 256-837-4536.
E-mail: information@sci-quest.org
Web Site: www.sci-quest.org
Formerly: The North Alabama Science Center
Founded: 1989.
Key Personnel: C.E.O. & Dir., Cyndy Morgan; Dir. Education, Angela Giles; Museum Shop Mgr., Kelly Brown.
Personnel Profile: Full-Time Paid 17; Part-Time Paid 11; Part-Time Volunteers 15; Interns 2.
Institution Type/Description: Science Center.
Collections: hands-on exhibits.
Facilities: 3D high definition immersive theater. Museum-related items for sale.
Activities: hands-on informal science programs; outreach; summer camps; overnight science programs in the science center; birthday party programs.
Hours & Admission Prices: Mon.-Fri. 9-5, Sat. 10-6, Sun. 1-5. Center: adults $9, seniors $8.50, children $8. Visual Vortex Show $3; discount to military. Closed New Year's Day; Thanksgiving; Christmas Eve & Day.
Attendance: 90,722 (accurate)
Membership: Family & Grandparents $75; Extended Family $95; Supporting $115; Patron $500.

THE 3 ART GALLERIES AT UAH: UNION GROVE, WILSON HALL, SALMON LIBRARY, Dept. of Art & Art History UAH, Wilson Hall 160-B, Huntsville, AL 35899. Tel.: 256-824-6114. Fax: 256-824-6438.
E-mail: art@uah.edu

Web Site: www.uah.edu/colleges/liberal/art
Key Personnel: Printmaking & Design, Brandon Gardner; Dean, College of Liberal Arts, Glenn Dasher; Painting, Kathryn Jill Johnson; Graphic Design, Keith Jones; Chair, Dept. Art & Art History, Lillian Joyce; Art History, David Stewart; Drawing, Roxie Veasey; Art History, Martha Vines; Staff Asst., Marylyn Coffey; Drawing & Painting, Monique K. Given; Photography, Jose Betancourt; Sculpture, Jimmy Kuehnle.
Personnel Profile: Part-Time Volunteers 2.
Governing Authority: university. nonprofit organization. Parent Institution: UA Huntsville. Tax-exempt: 170(b)(1)(A).
Institution Type/Description: Union Grove Gallery & Meeting House: housed in 1830s Greek Revival Chapel. Wilson Hall Gallery: Art Dept. Building. Salmon Library Gallery.
Collections: Union Grove Gallery: primarily student shows. Library Gallery: visiting artists. Wilson Hall Gallery: local & professional artists, student & faculty.
Activities: lectures; gallery talks; traveling & student exhibitions.
Hours & Admission Prices: Union Grove Gallery: Mon.-Fri. 12:30-4:30; other times by appointment. Library Gallery: Mon.-Thurs. 8am-12 midnight, Fri. 8-8, Sat. 9-6, Sun. 1-10. Wilson Hall Gallery: Mon.-Fri. 8:30-5. No charge. &
Attendance: 5,000 (estimated)

THE U.S. SPACE & ROCKET CENTER, One Tranquility Base, Huntsville, AL 35805-3371. Mailing Address: P.O. Box 070015, Huntsville, AL 35807-7015. Tel.: 256-721-7192. Fax: 256-721-7180.
E-mail: ralphb@spacecamp.com
Web Site: www.spacecamp.com
Founded: 1968.
Congressional District: 5
Key Personnel: C.E.O., Larry Capps; Chm., Jim Flynn; C.O.O., Clif Broderick; Exec. Vice Pres., Ralph Bryson; Vice Pres. Aerospace Programs, Michael Flachbart; Vice Pres. Finance, Donnie Claxton; Vice Pres. Human Resources, Vickie Henderson; Cur., Irene Willhite; Dir. Museum Operations, Samantha Peterson; Dir. Retail Sales, Scott Harbour; Vice Pres. Advancement, Brenda Carr; Corporate Donor Rels., Kelly Hatley.
Personnel Profile: Full-Time Paid 156; Part-Time Paid 400; Part-Time Volunteers 84.
Governing Authority: state. Parent Institution: State of Alabama; Alabama Space Science Exhibit Commission. Subsidiary Institution: U.S. Space & Rocket Center Foundation. Tax-exempt.
Institution Type/Description: Aeronautics & Space Museum.
Collections: missiles, rockets, & space hardware from early space history to space shuttle & other future programs; exhibits on utilizing the technology of space; manuscripts; Saturn V; SR-71 Blackbird Jet; Full Stack Space Shuttle, orbiter & solid rocket boosters; Apollo artifacts; Apollo Saturn V; 2 full-scale replicas of Saturn V; Dr. Wernher von Braun's personal artifacts.
Research Fields: history of rocketry development in United States.
Facilities: archives; 200-seat auditorium; 280-seat IMAX(R) theater; 350-seat digital theater; youth recreation area; food court; training center for U.S. Space Camp. Space-related items for sale.
Activities: museum & rocket park tours; permanent, temporary & traveling exhibits; hands-on exhibits; IMAX(R) movies; 4-G SpaceShot(TM) simulator ride; motion-based simulator; home of SPACE CAMP(R) & AVIATION CHALLENGE(R); demonstrations of life on the International Space Station; climbing wall.
Publications: Quest.
Hours & Admission Prices: March-Oct. daily 9-5; Nov.-Feb. see website for hours. Adults $24.95, children 6-12 $19.95; discount to ASTC Passport Program members & AAA members. Closed New Year's Eve & Day; Thanksgiving; Christmas Eve & Day. &
Attendance: 509,006 (accurate)
Membership: Pathfinder Individual $50; Pathfinder Family $100; Enterprise Family $145.

VETERANS MEMORIAL MUSEUM, 2060A Airport Rd., Huntsville, AL 35801-5338. Tel.: 256-883-3737. Fax: 256-883-3912.
E-mail: info@memorialmuseum.org
Web Site: www.memorialmuseum.org
Founded: 2001.
Congressional District: 5
Key Personnel: Dir. & Pres. (V), Randall Withrow.
Personnel Profile: Full-Time Volunteers 3; Part-Time Volunteers 12.
Governing Authority: Parent Institution: Alabama Center of Military History. Tax-exempt.
Institution Type/Description: Military History Museum.
Collections: ground vehicles, wheeled; ground vehicles, tracked; artillery; helicopters; patrol boat; military artifacts from American Revolutionary to present; uniforms; military reference books & documents.
Facilities: 12,000 sq. ft. exhibit space.
Activities: military history classes for school children. Museum Sponsors: Military Vehicle Preservation Assoc. Rally in May; Veterans Day Open House.
Hours & Admission Prices: Memorial Day to Labor Day Wed.-Sat. 10-5; Sept.-May Wed.-Sat. 10-4; other times by appointment. Adults $5, seniors $4, students $3; military in uniform no charge. Closed New Year's Day; Thanksgiving; Christmas. &
Attendance: 3,600 (estimated)

Jacksonville

DR. FRANCIS MEDICAL AND APOTHECARY MUSEUM, 207 Gayle Ave., S.E., Jacksonville, AL 36265-2544. Mailing Address: 650 Mountain St., N.W., Jacksonville, AL 36265. Tel.: 256-435-5091. Fax: 256-435-5410.
E-mail: calhounrsvp@jacksonville-al.org
Web Site: www.jacksonville-al.org
Founded: 1968.
Congressional District: 3
Key Personnel: Dir. Reservations, Denise Rucker.
Governing Authority: state. Affiliated with the Alabama State Historical Commission & the General John H. Forney Historical Society.
Institution Type/Description: Medical Museum: housed in 1850 Antebellum office & apothecary of Dr. Francis.
Collections: period furnishings; medical instruments; display cases; books; journals; apothecary artifacts; bottles; desks; other medical & apothecary memorabilia.
Activities: guided tours for school & other groups; loan & permanent exhibitions.
Publications: brochures.
Hours & Admission Prices: By appointment. No charge; donations accepted.
Attendance: 47 (accurate)

Lafayette

CHAMBERS COUNTY MUSEUM, 1st Ave. S.W., Lafayette, AL 36862. Mailing Address: P.O. Box 87, Lafayette, AL 36862. Tel.: 334-864-9656 & 7924.
Institution Type/Description: History Museum: housed in the former Central of Georgia railway depot.
Collections: local history & culture; regional stoneware; goat treadmill; Joe Louis' personal artifacts & memorabilia including photographs, scrapbooks, trophies & boxing gloves; spinning wheels; period dresses; river ferry Tallapoosa Queen; caboose.
Hours & Admission Prices: Wed. 9-12 & 1-4, Sat. 9-12; other times by appointment.

Leeds

BARBER VINTAGE MOTORSPORTS MUSEUM, (M), 6030 Barber Motorsports Pkwy., Leeds, AL 35094-3418. Tel.: 205-699-7275. Fax: 205-702-8700.
E-mail: bvmm@barbermuseum.org
Web Site: www.barbermuseum.org
Founded: 1994.
Key Personnel: Dir., Jeff Ray; Museum Shop Mgr., Lee L. Woehle.
Governing Authority: nonprofit. Tax-exempt: 501(c)(3).
Institution Type/Description: Motorsports Museum.
Collections: over 1,200 motorcycles & 60 cars from 20 nations with 200 manufacturers represented; history of motorcycles.
Publications: newsletter, VIN.
Hours & Admission Prices: April-Sept. Mon.-Sat. 10-6, Sun. 12-6; Oct.-March Mon.-Sat. 10-5, Sun. 12-5. Adults $15, children 4-12 $10; discounts to AAA members; children 3 & under no charge. Closed New Year's Day; Easter; Independence Day; Thanksgiving; Christmas Eve & Day. &

Lincoln

INTERNATIONAL MOTORSPORTS HALL OF FAME AND MUSEUM, 3198 Speedway Blvd., Lincoln, AL 35096-6327. Mailing Address: P.O. Box 1018, Talladega, AL 35161-1018. Tel.: 256-362-5002. Fax: 256-315-4565.
E-mail: kking@talladegasuperspeedway.com
Web Site: www.motorsportshalloffame.com/index.htm
Founded: 1983.
Key Personnel: Exec. Dir., Rick Humphrey; Mgr. Hall of Fame, Bruce Ramey; Dir. Communications & Mktg., Kristi King.

Personnel Profile: Full-Time Paid 5.
Governing Authority: nonprofit organization.
Institution Type/Description: Motorsports Museum.
Collections: racing vehicles & memorabilia from 1902 to present.
Facilities: 14,000-vol. library. Museum-related items for sale.
Activities: Annual Event: Induction Ceremony in spring.
Hours & Admission Prices: April 4 to Labor Day daily 9-5; groups of 25 or more by appointment; extended hours during race weeks at Talladega Superspeedway. Adults $10, students 7-17 $5; discounts to senior citizens, military, policemen, firemen, AAA & AARP members; children 6 & under no charge. Closed New Year's Day; Thanksgiving; Christmas.
Attendance: 900 (estimated)

Loachapoka

LEE COUNTY HISTORICAL SOCIETY, 6500 Stage Rd., Loachapoka, AL 36865. Mailing Address: P.O. Box 206, Loachapoka, AL 36865-0206. Tel.: 334-887-3007.
E-mail: angiewoad2@charter.net
Founded: 1968.
Key Personnel: Pres., Jeannette Frandsen; Vice Pres., Charles Mitchell.
Personnel Profile: Part-Time Paid 1; Part-Time Volunteers 10; Interns 2.
Governing Authority: nonprofit corporation. Tax-exempt: 501(c)(3).
Institution Type/Description: Historical Society Museum.
Collections: tools, artifacts, documents, books, records, furnishings & equipment belonging to the early settlers of the region.
Research Fields: agricultural; Creek Indians; natural resources of the area; family heirlooms & documents.
Facilities: library of material on music, genealogy, local, state & national history available for use on premises. Books & booklets for sale.
Activities: guided tours. Museum Sponsors: annual Historical Fair.
Publications: quarterly, Trails in History.
Hours & Admission Prices: By appointment only. No charge; donations accepted.
Attendance: 6,500 (estimated)
Membership: Individual $20; Family $30; Friend $50; Patron $100; Benefactor $250.

Madison

CLAY HOUSE MUSEUM, 16 Main St., Madison, AL 35758-1810. Mailing Address: 621 Eastview Dr., Madison, AL 35758-7826. Tel.: 256-325-1018.
E-mail: BrewerR@hiwaay.net
Institution Type/Description: China Museum.
Collections: Noritake dinnerware from 1850-1950.
Activities: rental facilities.
Hours & Admission Prices: By appointment.

Marion

ALABAMA WOMEN'S HALL OF FAME, Judson College, Howard Bean Hall, 302 Bibb St., Marion, AL 36756-2504. Tel.: 334-683-5100 & 5156.
E-mail: bmathews@judson.edu
Web Site: www.awhf.org
Founded: 1970.
Congressional District: 7
Key Personnel: Exec. Sec., Bill Mathews
Institution Type/Description: History Museum.
Collections: Alabama women who have made significant contributions to the state & nation including Helen Adams Keller, Julia Strudwick Tutwiler, Amelia Gayle Gorgas, Tallulah Bankhead, and Mildred Westervelt Warner; portraits; photographs; letters; bronze plaques.
Hours & Admission Prices: Call for hours. No charge; donations accepted. &

McCalla

IRON & STEEL MUSEUM OF ALABAMA, TANNEHILL IRONWORKS HISTORICAL STATE PARK, 12632 Confederate Pkwy., McCalla, AL 35111-2620. Tel.: 205-477-5711. Fax: 205-477-9400. Facebook; Iron Steel Museum.
E-mail: tannehillmuseum@bellsouth.net
Web Site: www.Tannehill.org
Founded: 1981.
Congressional District: 7
Key Personnel: C.E.O. & Museum Events Coord., Stacey Green; Museum Dir., Jennifer Watts; Chm. (V), Jim Bennett; Business Officer, Margaret Crumpton; Registrar, Carl Addison.

Personnel Profile: Full-Time Paid 6; Part-Time Paid 20; Part-Time Volunteers 50.
Governing Authority: state; nonprofit. Parent Institution: Alabama Historic Ironworks Commission. Tax-exempt: 501(c)(3).
Institution Type/Description: Industrial Museum & State Park: located on the c.1839 Tannehill Furnace Site.
Collections: c.1890s agricultural implements; iron working implements; artifacts relating to the iron industry of the 1700s-1850s; home furnishings; restored structures. Historic House & Buildings: 1820-1900 Tannehill Rural Village complex.
Research Fields: history of iron technology; architectural history; local genealogy; anthropology; agricultural history; folk life studies; historical archaeology.
Facilities: 1,000-vol. library of books & periodicals relating to the history of the iron industry, Alabama history & history of the United States available for research on premises only; nature center; reading room; classrooms; learning center. Objects related to Alabama folk life for sale.
Activities: guided tours; lectures; arts festivals; study clubs; hobby workshops; formally organized education programs for children, adults & undergraduate college students; loan, permanent, temporary & traveling exhibitions; archaeological projects.
Publications: newsletter, Tannehill Blast; book, Old Tannehill.
Hours & Admission Prices: Museum: Mon.-Fri. 8:30-4:30, Sat. 9-4:30, Sun. 12:30-4:30. Park: daily 7-sunset. Adults $2, children 6-11 and 62 & over $1; children 5 & under no charge. &
Attendance: 500,000 (accurate)

Mobile

BRAGG-MITCHELL MANSION, 1906 Springhill Ave., Mobile, AL 36607-2304. Tel.: 251-471-6364. Fax: 251-478-3800.
E-mail: info@braggmitchellmansion.com
Web Site: braggmitchellmansion.com
Founded: 1987.
Congressional District: 1
Key Personnel: Dir., Lynn Stewart.
Personnel Profile: Full-Time Paid 1; Part-Time Paid 9; Part-Time Volunteers 1.
Governing Authority: private; nonprofit organization. Branch Museum: The Explore Center, 1906 Springhill Ave., Mobile, AL 36607. Tax-exempt: 501(c)(3).
Institution Type/Description: Historic Mansion: built in 1855.
Collections: furnishings from 1725-1900.
Hours & Admission Prices: Tues.-Fri. 10-4. Adults $8, students 3-12 $5; discounts to groups & AAA members. Closed New Year's Day; Mardi Gras Day; Easter; Memorial Day; Labor Day; Thanksgiving; Christmas. &
Attendance: 3,819 (accurate)

CENTRE FOR THE LIVING ARTS, 301 Conti St., Mobile, AL 36602-2714. Mailing Address: P.O. Box 198, Mobile, AL 36601-0198. Tel.: 251-208-5671. Fax: 251-208-5655.
Web Site: www.centreforthelivingarts.com
Founded: 1999.
Congressional District: 1
Key Personnel: Exec. Dir., Robert L. Sain; Chm. (V), Mike Dow.
Personnel Profile: Full-Time Paid 8; Part-Time Paid 2.
Institution Type/Description: Contemporary visual arts organization.
Collections: contemporary works by significant regional, national & international artists of our time.
Major Exhibits: The Futures Project, 1/14-12/14.
Facilities: theater; art center.
Activities: performances; art classes & workshops; temporary exhibits.
Publications: newsletters.
Hours & Admission Prices: Wed.-Sat. 11-5, Sun 12-5. Gallery closed Mon. & Tues; adults $5, seniors $3, members & NARM no charge. &
Attendance: 122,000 (estimated)
Membership: Students $20; Individual $35; Family $50; Reciprocal $125; Conti Club $500.

GULF COAST EXPLOREUM SCIENCE CENTER, 65 Government St., Mobile, AL 36602-3107. Mailing Address: P.O. Box 1968, Mobile, AL 36633-1968. Tel.: 251-208-6870 & 6873. Fax: 251-208-6889.
E-mail: mike@exploreum.com
Web Site: www.exploreum.com
Formerly: Gulf Coast Exploreum Museum of Science
Founded: 1979.
Congressional District: 1
Key Personnel: Asst. Dir., Caroline Etherton; Business Mgr., W. Tyre Smithweck, Jr.; Mktg. Mgr., Hela Sheth.

Personnel Profile: Full-Time Paid 18; Part-Time Paid 32; Interns 2.
Governing Authority: nonprofit organization. Parent Institution: The Explore Center, Inc. Subsidiary Institution: Bragg Mitchel Mansion, Springhill Ave., Mobile, AL. Tax-exempt: 501(c)(3).
Institution Type/Description: Science Center.
Collections: interactive exhibits.
Facilities: 52,000 sq. ft. exhibit space; learning center; IMAX dome theatre. Museum-related items for sale.
Activities: films; organized education programs; overnight program; curriculum boxes for teachers; participatory & traveling exhibitions.
Publications: brochure; newsletter.
Hours & Admission Prices: Mon.-Fri. 9-5, Sat. 10-5, Sun. 12-5. Exhibits: adult $11, youth 13-18 $9.50, child 2-12 $9; discounts to ASTC members; members no charge. Closed Mardi Gras Tuesday; Easter; Thanksgiving; Christmas. &
Attendance: 200,000 (accurate)
Membership: Family $100; Explorer $250; Discoverer $500; Odyssey $1,000.

HISTORIC MOBILE PRESERVATION SOCIETY, 350 Oakleigh Place, Mobile, AL 36604-2910. Tel.: 251-432-1281. Fax: 251-432-8843.
E-mail: hmps@bellsouth.net
Web Site: www.historicmobile.org
Founded: 1935.
Congressional District: 1
Key Personnel: Exec. Dir., Rhonda P. Davis; Pres., Douglas Kearley; 2nd Vice Pres., Rob Gulledge; 3rd Vice Pres., Virginia Edington; Sec., Sally Trufant.
Personnel Profile: Full-Time Paid 3; Full-Time Volunteers 7; Part-Time Paid 5; Part-Time Volunteers 100.
Governing Authority: society. Branch Museum: Oakleigh House Museum. Tax-exempt: 501(c)(3).
Institution Type/Description: Historical Preservation Society & Historic Houses.
Collections: archives: local history; Civil War artifacts; photographs; books; maps. Museum Houses: period silver; china; period furniture; paintings; costumes; clothing; accessories; portraits. Historic Houses: 1833, Oakleigh House; 1850 Cox-Deasy House; c.1850 Cook's House.
Research Fields: archives.
Facilities: 625-vol. library of books, papers & photographs relating to the Mobile area available in archives building for research on premises. Photographs for sale.
Activities: guided tours; lectures; TV & radio programs; docent programs & council. Annual Events: Haunted Oakleigh; Candlelight Christmas at Oakleigh in December.
Publications: quarterly journal, Landmark Letter; research material & books, Church St. Graveyards; Lost Villages & Ancient Kingdoms; 19th Century Mobile Silver; Mobile: The Life & Times of a Great Southern City; triannual journal, Landmark Letter; Mobile City Directory 1867.
Hours & Admission Prices: Wed.-Sat. 10-4; other times by appointment. Adults $7, children & students $3; discounts to senior citizens, AAM & ICOM members; members no charge. Closed legal holidays; New Year's Eve & Day; Christmas Eve, Day & week. &
Attendance: 25,000 (estimated)
Membership: Junior (under 21) $12; Individual $35; Couple $45; Family $55; Contributing $100; Supporting $150; Sustaining $300; Life $1,000.

MOBILE BOTANICAL GARDENS, 5151 Museum Dr., Mobile, AL 36608-1919. Tel.: 251-342-0555. Facebook: Mobile Botanical Gardens.
E-mail: mbg2@bellsouth.net
Web Site: www.mobilebotanicalgardens.org
Formerly: Southern Alabama Botanical & Horticultural Society
Founded: 1974.
Congressional District: 1
Key Personnel: Exec. Dir., Robin Krchak; Pres. Bd., Judy Stout.
Personnel Profile: Full-Time Paid 5; Part-Time Paid 3; Part-Time Volunteers 200
Governing Authority: Tax-exempt.
Institution Type/Description: Botanical Gardens.
Collections: azaleas; camellias; Japanese maple; magnolias; roses; ferns; perennials; herbs; rhododendron; longleaf pine.
Facilities: 100-acre site; nature trails; gardens.
Activities: classes; special events.
Hours & Admission Prices: Gardens: dawn to dusk. Office: Mon.-Fri. 9-5. Adults $5; children 12 & under and APGA & museum members no charge. &
Attendance: 30,000 (estimated)
Membership: Individual $40; Family $60; Friend $100; Fellow $250; Patron $500; Benefactor $1,000.

MOBILE MEDICAL MUSEUM, (M), 1664 Springhill Ave., Mobile, AL 36604-1405. Tel.: 251-415-1109. Fax: 251-415-1110.
E-mail: med_museum@yahoo.com
Web Site: mobilemedicalmuseum.com
Formerly: Eichold Heustis Medical Museum of the South
Founded: 1962.
Congressional District: 101
Key Personnel: Dir., Sally C. Green; Pres. (V), Christopher Gill, Esq.; Vice Pres., Henry A. Callaway, III, Esq.; Sec., Shirley P. Mull; Treas., Charles S. Jones.
Personnel Profile: Part-Time Paid 3; Part-Time Volunteers 2.
Governing Authority: community; nonprofit organization. Tax-exempt: 501(c)(3).
Institution Type/Description: Medical History Museum.
Collections: early 1700s medical artifacts; history of the practice of medicine during the past 300 years; Civil War gallery of documents, artifacts, photographs; apothecary; old & rare medical books.
Research Fields: transactions of the state & city medical societies journals and city hospitals journals.
Facilities: 500-vol. medical reference library; 1,000 sq. ft. exhibit space.
Activities: guided tours; loan, temporary exhibits; lecture series. Museum Sponsors: International Museum Day.
Publications: quarterly newsletters.
Hours & Admission Prices: Tues.-Fri. 10-4. Adults $5; discounts to senior citizens, AAM members & groups; members no charge. Closed federal holidays. &
Attendance: 5,000 (estimated)
Membership: Individual $35; Family $50; Supporting $100; Sponsor $250; Patron $500; Associate $1,000; Partner $5,000.

✱   **MOBILE MUSEUM OF ART, (M),** 4850 Museum Dr., Langan Park, Mobile, AL 36608-1917. Tel.: 251-208-5200 & 5203. Fax: 251-208-5201. TDD: 251-208-5200.
E-mail: marylee.montgomery@mobilemuseumofart.com
Web Site: www.MobileMuseumOfArt.com
Founded: 1963.
Congressional District: 1
Key Personnel: Dir., Deborah Velders; Chm. Bd. (V), G. Tim Gaston; Chief Cur., Paul W. Richelson; Cur. Collections, Kurtis Thomas; Cur. Exhibitions, Donan Klooz; Registrar, Rachel Young.
Personnel Profile: Full-Time Paid 27; Part-Time Paid 11; Part-Time Volunteers 75; Interns 1.
Operating Expenses: 31,100,000
Operating Income: 3,100,000
Governing Authority: municipal; nonprofit organization. Parent Institution: City of Mobile. Tax-exempt: 501(c)(3).
Institution Type/Description: Art Museum.
Collections: fine arts; contemporary crafts; decorative arts.
Major Exhibits: Copley, Delacroix, Dali and Others: Masterworks from the Beaverbrook Art Gallery (T), 11/15/13-4/27/14; The Art & Design of Mardi Gras (T), 11/8/14-2/22/15.
Research Fields: contemporary decorative arts.
Facilities: library; cafe. Museum store.
Activities: guided tours; lectures; films; gallery talks; concerts; arts festivals; formally organized education programs; temporary exhibitions. Gallery Sponsors: urban & rural school art programs.
Publications: members' newsletter; exhibitions catalogues.
Hours & Admission Prices: Tues.-Wed. & Fri.-Sun. 10-5, Thurs. 10-9. Adults $10; discounts to City of Mobile employees, active military, students & seniors; members no charge. Closed New Year's Day; Memorial Day; Independence Day; Thanksgiving; Christmas. &
Attendance: 46,218 (accurate)
Membership: See website for current rates.

THE MUSEUM OF MOBILE, 111 S. Royal St., Mobile, AL 36602-3101. Mailing Address: P.O. Box 2068, Mobile, AL 36652-2068. Tel.: 251-208-7569. Fax: 251-208-7686.
E-mail: museum@cityofmobile.org
Web Site: www.museumofmobile.com
Founded: 1962.
Congressional District: 1
Key Personnel: Dir., David Alsobrook; Chm. Bd., Pat Edington; Cur. Exhibits & Collections, Jacob Laurence; Asst. Dir., Shelia M. Flanagan; Registrar, Shelley Berger; Education, Jennifer Fondren; Research Historian, Chuck Torrey; Special Events, Elyse Marley; Security, Donnie Curtis; Museum Tech, Lori Rockhold; Museum Shop Mgr., Sydney Betbeze.
Personnel Profile: Full-Time Paid 24; Part-Time Paid 10; Part-Time Volunteers 10.
Governing Authority: municipal; nonprofit. Parent Institution: Museum of

Mobile. Branch Museums: Phoenix Fire Museum, 203 S. Claiborne St., Mobile, AL; Historic Fort Conde, 150 S. Royal St., Mobile, AL. Tax-exempt.
Institution Type/Description: History Museum.
Collections: Museum of Mobile: French, British & Spanish Colonial history; Indian, African-American & Confederate artifacts; costumes; documents; naval; transportation; silver; glass; ceramics; manuscripts; paintings; engravings; prints; maps; Mardi Gras; steamboat & ship models; carriages; hands-on exhibits. Phoenix Museum: firefighting equipment; fire memorabilia. Ft. Conde colonial period.
Research Fields: local & regional history of the Gulf Coast area from 1702.
Facilities: 2,500-vol. library of museum reference books & manuscripts; two classrooms; discovery room.
Activities: docent program; guided tours; lectures; formal education programs for children; inter-museum loan; permanent & temporary exhibitions; special exhibits.
Publications: books: The Phoenix Volunteer Fire Company of Mobile; Military Buttons of the Gulf Coast; A Voyage to Dauphin Island in 1720; The Journal of Bertet de la Clue; Old Mobile Fort Louis de la Louisiane 1702-1711; Raphael Semmes, Rear Admiral Confederate States Navy; Brigadier General Confederate States Army; Iron Ore to Iron Lace; Confederate & Union Button of the Gulf Coast 1681-1865; Queens of Mardi Gras 1893-1986; newsletter, Timeline.
Hours & Admission Prices: Museum of Mobile: Tues.-Sat. 9-5, Sun. 1-5. Adults $5, seniors 55 & up $4, students $3; discounts to AAM members; children under 6 & members no charge. Historic Fort Conde: daily 8-5. No charge; donations accepted. Phoenix Fire Museum: Tues.-Sat. 9-5, Sun. 1-5. Closed New Year's Day; Thanksgiving; Christmas; some city holidays. &
Attendance: 90,000 (estimated)
Membership: Student $10; Individual $25; Family $35; Supporting $100; Patron $500; Benefactor $1,000.

THE NATIONAL AFRICAN-AMERICAN ARCHIVES AND MUSEUM, 564 Dr. Martin Luther King Jr. Ave., Mobile, AL 36603-5916.
Institution Type/Description: History Museum: housed in the former Davis Avenue Branch of the Mobile Public Library, the only library for African-Americans from 1932 to the mid-1960s. Listed on the National Register for Historic Places.
Collections: African-American history; cultural heritage; portraits; books; family histories; period artifacts.
Activities: group tours.
Hours & Admission Prices: Mon.-Fri. 8-4, Sat. 10-2, Sun. by appointment.

PHOENIX FIRE MUSEUM, 203 S. Claiborne St., Mobile, AL 36602-2322. Tel.: 251-208-7569.
Web Site: www.museumofmobile.com
Key Personnel: Exec. Dir., David Alsobrook
Institution Type/Description: History Museum: housed in the restored home of Phoenix Volunteer Fire Company No. 6.
Collections: Mobile volunteer fire companies history; horse-drawn steam engines; early motorized vehicles.
Hours & Admission Prices: Tues.-Sat. 9-5, Sun. 1-5. No charge.

PORTIER HOUSE, 307 Conti St., Mobile, AL 36602. Tel.: 251-441-7138.
E-mail: portierhouse@att.net
Web Site: www.portierhouse.org/
Key Personnel: Group Tours, Lynn Ellis; Group Tours, Bunky Ralph; Rentals, Karen Carlisle.
Governing Authority: Parent Institution: Archdiocese of Mobile, 400 Government St., Mobile AL 36602.
Institution Type/Description: Historic House Museum: housed in the former home of Bishop Michael Portier, the first Bishop of Mobile. Listed on the National Register of Historic Places.
Collections: Bishop Portier's life & career; period furnishings; photographs; religious artifacts.
Facilities: rental facilities. Museum-related items for sale.
Activities: rental facilities; special events.
Hours & Admission Prices: Tues.-Fri. 10-3.

RICHARDS-DAR HOUSE MUSEUM, 256 N. Joachim St., Mobile, AL 36603-6472. Tel.: 251-208-7320.
Web Site: www.richardsdarhouse.com
Founded: 1973.
Congressional District: 1
Key Personnel: Pres. & Gift Shop Mgr., Susan Tomlinson; Chm. (V), Margaret Odom.
Personnel Profile: Part-Time Paid 6; Part-Time Volunteers 126.

Governing Authority: city of Mobile; nonprofit organization. Tax-exempt: 501(c)(3).
Institution Type/Description: Historic House: built in 1860.
Collections: furnishings predating 1870.
Activities: decorative arts series. Museum Sponsors: Victorian Christmas at the Richards-DAR House Museum in December.
Publications: monthly e-newsletter.
Hours & Admission Prices: Mon.-Fri. 11-3:30, Sat. 10-4, Sun. 1-4. Adults $5, students $2; discounts to groups; children under 6 no charge. Closed New Year's Day; Mardi Gras; Easter; Independence Day; Labor Day; Thanksgiving; Christmas.
Attendance: 4,250 (estimated)

USS ALABAMA BATTLESHIP MEMORIAL PARK, 2703 Battleship Pkwy., Mobile, AL 36601. Mailing Address: P.O. Box 65, Mobile, AL 36601-0065. Tel.: 251-433-2703. Fax: 251-433-2777.
E-mail: btunnell@ussalabama.com
Web Site: www.ussalabama.com
Founded: 1963.
Congressional District: 1
Key Personnel: C.E.O., Bill Tunnell.
Personnel Profile: Full-Time Paid 34; Part-Time Paid 4; Part-Time Volunteers 2.
Governing Authority: state commission. Tax-exempt.
Institution Type/Description: Historic Ship & Military Museum.
Collections: 1942 Battleship USS Alabama; 1941 Submarine USS Drum; naval, air force, marine, army & coast guard weapons & equipment; over 20 restored military aircraft; Vietnam-era PBR river patrol boat; original documents, photographs & uniforms; graphics; silver.
Research Fields: military history.
Facilities: picnic area. Gift items for sale.
Activities: self-guided ship tours; overnight youth camping program aboard battleship; flight simulator ride; coastal birding trail stop & kiosk; wildlife observatory.
Publications: semi-annual development newsletter.
Hours & Admission Prices: March-Sept. daily 8-5; Oct.-March daily 8-4. Adults $15, children 6-11 $6; discounts to groups, AARP, AAA & Historic Naval Ships Assoc. members; children under 6 & active duty military personnel no charge. Parking: $2 per vehicle. Closed Christmas. &
Attendance: 368,003 (accurate)
Membership: Honorary $25; Individual $50; Family $100; Life $1,000.

Monroeville

MONROE COUNTY HERITAGE MUSEUM, (M), 31 N. Alabama Ave., Monroeville, AL 36460-1818. Tel.: 251-575-7433. Fax: 251-575-2513.
E-mail: mchm@frontiernet.net
Web Site: www.tokillamockingbird.com
Founded: 1990.
Congressional District: 1
Key Personnel: Exec. Dir., Stephanie Rogers; Chm. (V) & Museum Shop Mgr., Robyn Neilsen; Dir. Education, Wanda Green; Dir. Sites & Operations, Annie Hill; Mgr. Rikard's Mill Museum, Larry Tuberville; Devel. Officer, Gail Deas.
Personnel Profile: Full-Time Paid 4; Part-Time Paid 2; Part-Time Volunteers 100.
Governing Authority: county government; nonprofit. Parent Institution: Monroe Co. Museum. Tax-exempt: 501(c)(3).
Institution Type/Description: Historic Building: housed in the 1903 Old Monroe County Courthouse; setting for the film To Kill A Mockingbird.
Collections: 1900-1950s clothing; historical photographs.
Activities: Museum Sponsors: annual production of To Kill a Mockingbird in April & May.
Publications: quarterly, Mockingbird Collection.
Hours & Admission Prices: Courthouse: Mon.-Fri. 8-4, Sat. 10-2. No charge; donations accepted. Closed major holidays. &
Attendance: 25,000 (estimated)
Membership: General $50; Star Family $100; Friend $250; Patron $500; Corporate $750.

Montevallo

ALDRICH COAL MINE MUSEUM, 137 Hwy. 203, Montevallo, AL 35115-7105. Tel.: 205-665-2886.
Founded: 1989.
Key Personnel: Owner, Rose Emfinger; Owner, Henry Emfinger
Institution Type/Description: Coal Mining Museum.

Collections: area history; coal mining industry; photographs; period artifacts; simulated coal mine.
Hours & Admission Prices: Thurs.-Sat. 10-4, Sun. 1-4; other times by appointment. Adults $5, children $3. &

THE AMERICAN VILLAGE, 3727 Hwy. 119 S., Montevallo, AL 35115. Mailing Address: P.O. Box 6, Montevallo, AL 35115-0006. Tel.: 205-665-3535; 877-811-1776 (Toll Free). Fax: 205-665-7577.
Web Site: www.americanvillage.org
Key Personnel: Exec. Dir., Tom Walker
Institution Type/Description: History Museum.
Collections: period history; personal artifacts; military weapons & uniforms; photographs.
Facilities: theater. Museum-related items for sale.
Activities: educational programs; special events; rental facilities; group tours.
Hours & Admission Prices: Mon.-Fri. 10-4; other times by appointment. &

Montgomery

ALABAMA DEPARTMENT OF ARCHIVES & HISTORY, (M), 624 Washington Ave., Montgomery, AL 36130-3003. Mailing Address: P.O. Box 300100, Montgomery, AL 36130-0100. Tel.: 334-242-4435. Fax: 334-240-3433.
E-mail: debbie.pendleton@archives.alabama.gov
Web Site: www.archives.alabama.gov
Founded: 1901.
Congressional District: 2
Key Personnel: Dir., Steve Murray; Volunteer Coord. & Cur. Education, Kelly Hoomes; Asst. Dir. Government Records, Tracey Berezansky; Asst. Dir. Public Svcs., Debbie Pendleton; Education Specialist, Susan DuBose; Museum Shop Mgr., Allison Gore.
Personnel Profile: Full-Time Paid 32; Part-Time Paid 1; Part-Time Volunteers 105.
Governing Authority: board of trustees. Parent Institution: State of Alabama. Tax-exempt.
Institution Type/Description: History Museum.
Collections: state archives; state & local government records; manuscripts; portraits; artifacts of Alabama history; southeastern Indian artifacts; 19th-century Alabamiana; flags; military artifacts; civil rights photographs.
Research Fields: U.S., southern, & Alabama history, sociology, economics, culture, genealogy, anthropology & archaeology.
Facilities: reference room; auditorium.
Activities: guided tours; permanent & temporary exhibitions; monthly lecture series.
Publications: newsletter, Friends of the Alabama Archives.
Hours & Admission Prices: Museum: Mon.-Sat. 8:30-4:30. Archives: Mon.-Fri. 8:30-4:30. Reference Room: Tues.-Fri. 8:30-4:30. No charge. Closed state holidays. &
Attendance: 50,000 (estimated)
Membership: Friends of the Alabama Archives: Individual $25; Corporate $50; Sponsors $100; Sustainer $250; Benefactors $500; Patron $750; Benefactors $1,000.

DEXTER PARSONAGE MUSEUM, 309 S. Jackson St., Montgomery, AL 36104-4407. Tel.: 334-261-3270.
Key Personnel: Pres. (V), Rev. Michael F. Thurman.
Governing Authority: Subsidiary Institution: Dexter Avenue King Memorial Foundation, Inc., P.O. Box 4901., Montgomery, AL. Tax-exempt.
Institution Type/Description: Historic House Museum: housed in the former home of 12 pastors of the Dexter Avenue King Memorial Baptist Church from 1920-1992. Listed on the National Register of Historic Places.
Collections: period furnishings; photographs; religious artifacts.
Facilities: Museum-related items for sale.
Hours & Admission Prices: Tues.-Fri. 10-4, Sat. 10-2. &

FIRST WHITE HOUSE OF THE CONFEDERACY, 644 Washington Ave., Montgomery, AL 36130-3057. Mailing Address: P.O. Box 1861, Montgomery, AL 36102-1861. Tel.: 334-242-1861.
E-mail: thefirstwhitehouse@sd.alabama.gov
Web Site: www.firstwhitehousc.org
Founded: 1900.
Congressional District: 2
Key Personnel: Regent, Anne Tidmore; Supvr. Hostesses & Museum Shop Mgr., Eva Newman.
Personnel Profile: Full-Time Paid 1; Part-Time Paid 2; Part-Time Volunteers 61; Interns 1.
Volunteer Hours: 208
Governing Authority: state; association. Parent Institution: State of Alabama. Subsidiary Institution: White House Association of Alabama. Tax-exempt.

Institution Type/Description: Historic House Museum: 1832-1835 home of Jefferson Davis & his family in 1861 when Montgomery was the capital of the Confederacy.
Collections: Confederate history; furniture, books & personal property of President Davis; war relics; Civil War Discovery Trail; original furnishings & artifacts.
Research Fields: Jefferson Davis & family data.
Facilities: library; archives.
Activities: tours by appointment. Annual Events: Robert E. Lee's Birthday in January; Jefferson Davis' Birthday in June.
Publications: full color booklet, The White House of The Confederacy; no color monograph, The Struggle to Preserve the First White House of the Confederacy; no color monograph, Montgomery During the Civil War.
Hours & Admission Prices: Mon.-Fri. 8-4:30, Sat. 9-4. No charge; donations accepted. Closed state holidays. &
Attendance: 25,699 (accurate)
Membership: White House Ally $25; Stars & Bars $50; Liberty Cap $100; Robert E. Lee Ally $250; Jefferson Davis Ally $500.

FREEDOM RIDES MUSEUM, 210 S. Court St., Montgomery, AL 36104. Mailing Address: 468 S. Perry St., Montgomery, AL 36104. Tel.: 334-242-3188.
Web Site: www.freedomridesmuseum.org
Formerly: Freedom Riders Museum
Founded: 2011.
Congressional District: 2
Key Personnel: Site. Dir., Christy Carl; Exec. Dir., Frank White.
Personnel Profile: Part-Time Paid 6; Interns 1.
Governing Authority: Parent Institution: Alabama Historical Commission. Tax-exempt.
Institution Type/Description: History Museum: housed in a former Greyhound bus station; built in 1951.
Collections: local Civil Rights history; desegregation of interstate bus seating & terminal facilities; photographs; personal artifacts; paintings; bus station.
Research Fields: Civil Rights; Montgomery, AL transportation.
Activities: temporary & permanent exhibitions; gallery talks; anniversary programs; cell phone tours.
Hours & Admission Prices: Fri.-Sat. 12-4; guided tours by appointment. Adults $5; discounts to seniors, military, students & Blue Star Museums. Closed state holidays. &
Attendance: 2,155 (accurate)

THE GEORGINE CLARKE ALABAMA ARTIST'S GALLERY, Alabama State Council on the Arts, 201 Monroe St., Montgomery, AL 36104. Mailing Address: Alabama State Council on the Arts, 201 Monroe St., Ste. 110, Montgomery, AL 36130-1800. Tel.: 334-242-4076, ext. 250. Fax: 334-240-3269. Facebook: Alabama State Council on the Arts.
E-mail: elliot.knight@arts.alabama.gov
Web Site: arts.alabama.gov
Formerly: Alabama Artist's Gallery
Founded: 1966.
Key Personnel: Dir. Alabama State Council on the Arts, Albert B. Head; Dir. Gallery & Mgr. Visual Arts Program, Elliot Knight
Institution Type/Description: Art Museum.
Collections: works by Alabama artists.
Major Exhibits: Recent Fellowship Winner Showcase, 11/13-1/14; Contemporary Art in Alabama, 1/14; Celebrating Design & Architecture, 2/14-3/14; Visual Arts Achievement Program, 4/14; Selections from the Mobile Museum of Art, 5/14-6/14; PINE, 8/14-10/14; Black Belt Treasures, 11/14-12/14.
Hours & Admission Prices: Mon.-Fri. 8-5. No charge. Closed state holidays. &
Attendance: 5,000

HANK WILLIAMS MUSEUM, 118 Commerce St., Montgomery, AL 36104-2538. Tel.: 334-262-3600.
E-mail: hankwilliamsmuse@bellsouth.net
Web Site: thehankwilliamsmuseum.com
Founded: 1997.
Institution Type/Description: History Museum.
Collections: personal artifacts including Hank's 1952 Cadillac; suits designed by Nudie; boots; ties; horse saddle; piano; hats; awards; life-size portraits.
Facilities: over 6,000 sq. ft. exhibit space. Museum-related items for sale.
Hours & Admission Prices: Mon.-Fri. 9-4:30, Sat. 10-4, Sun. 1-4. Admission 15 & over $10, children 3-14 $3; discounts to groups, military & AAA members; members no charge. &
Attendance: 20,000 (estimated)
Membership: Lifetime Individual $200; Lifetime Couple $250.

LANDMARKS FOUNDATION/OLD ALABAMA TOWN, 301 Columbus St., Montgomery, AL 36104-2624. Tel.: 334-240-4500; 888-240-1850. Fax: 334-240-4519.
Web Site: www.oldalabamatown.com
Founded: 1967.
Congressional District: 2
Key Personnel: Exec. Dir., Robert McLain; Cur., Carole King; Dir. Education, Florence Giles; Dir. Mktg., Buffy Lockette.
Personnel Profile: Full-Time Paid 10; Part-Time Paid 9; Part-Time Volunteers 5.
Governing Authority: nonprofit organization. Property of the City of Montgomery. Tax-exempt: 501(c)(3).
Institution Type/Description: Historical & Preservation Society: located in Old Alabama Town Reception Center.
Collections: Historical Houses: 1820s Log Cabin; two-room Dogtrot, home of William Lowndes Yancey; 1850s Greek Revival Cottage, home of Mayor Warren S. Reese; 1856 DeWolf-Cooper Cottage; 1857 Davis-Cook House; 1880s Presbyterian Church, built by Black Presbyterians; 1897 Shotgun House, home of Grant & Vinnie Fitzpatrick; 1875 Pintlala Grange Hall; 1892 Doctor's Office; 1850s Ordeman-Shaw Townhouse & Outbuildings; 1900 Corner Grocery; 1900 one-room schoolhouse; 1840s I-House; 1850s columned mansion; 1890s cotton gin; drugstore museum; blacksmith shop; grist mill; print shop.
Research Fields: architectural history; history-related fields.
Facilities: Gifts & museum-related items for sale.
Activities: guided tours; lectures; films; formally organized education programs for children; working crafts people; docent program or council.
Publications: quarterly, Landmarks; book, The Way It Was, 1850-1930: Photographs of Montgomery & Her Central Alabama Neighbors; tape, Montgomery Landmarks & Driving Tour; A Narrative History of Cotton in Alabama' Boll Weevil Review; Essays on Central Alabama History; Old Alabama Town: An Illustrated Guide.
Hours & Admission Prices: Mon.-Sat. 9-3. Adults $10, students 6-18 $5; discount to groups, AAA & AAM members; children 5 & under and members no charge. Closed New Year's Day; Easter; Thanksgiving; Christmas. &
Attendance: 55,000 (estimated)
Membership: Individual $25; Family $40; Contributing $75; Patron $105; Grand Patron $250-$500.

MANN WILDLIFE LEARNING MUSEUM, 325 Vandiver Blvd., Montgomery, AL 36110-1815. Tel.: 334-240-4900. Fax: 334-240-4916. Facebook: Mann Wildlife Learning Museum.
Web Site: www.mannmuseum.com
Formerly: Montgomery Zoo
Founded: 2000.
Congressional District: 2
Key Personnel: Zoo Dir., Doug Goode; Deputy Zoo Dir., Marcia Woodard; Animal Care Mgr., Lisa Peek; Dir. Concessions, Debbie Harris; Cur. Education, Jennifer Murphy; Mktg. & Public Rels. Mgr., Sarah McKemey; Mgr. Program Svcs., Steven Pierce; Gift Shop Mgr., Deborah Stewart.
Personnel Profile: Full-Time Paid 65; Full-Time Volunteers 15; Part-Time Paid 1; Part-Time Volunteers 300.
Governing Authority: municipal. Parent Institution: City of Montgomery. Tax-exempt.
Institution Type/Description: Wildlife Learning Museum
Collections: wildlife.
Research Fields: wild animal medical care.
Facilities: 300-vol. library of scientific & natural history, available for research on premises only; zoological park; nature center.
Activities: guided tours; catered lunches; after hours parties offered; formally organized education programs; docent program or council.
Publications: newsletter, Jungle Drums.
Hours & Admission Prices: Daily 9-5. Zoo: adults $12, seniors 65 & over $10, children 3-12 $8; discounts to groups with two weeks advanced reservation & military; children 2 & under, AAM, AZA & society members no charge. Mann Museum & Zoo: adults $16, seniors 65 & over $14, children 3-12 $11. Closed New Year's Day; Thanksgiving; Christmas. &
Membership: Individual Plus One: Zoo or Museum $70; Family or Grandparent: Zoo or Museum $85; Family or Grandparent: Zoo & Museum Combo $115; Chimpanzee Club $250; Zebra Club $300; Giraffe Club $500; President's Club $1,000.

✱ MONTGOMERY MUSEUM OF FINE ARTS, (M), One Museum Dr., Montgomery, AL 36117-4600. Mailing Address: P.O. Box 230819, Montgomery, AL 36123-0819. Tel.: 334-240-4333. Fax: 334-240-4384.
E-mail: museuminfo@mmfa.org
Web Site: www.mmfa.org
Founded: 1930.
Congressional District: 2
Key Personnel: Dir., Mark M. Johnson; Deputy Dir. Devel., Jill Barry; Asst. Dir. Operations, Steve Shuemake; Devel. Officer, Katherine Trumble; Devel. Asst., Jennifer Pope; Dir. Mktg. & Public Rels., Haley Rennick; Cur. Art, Margaret Lynne Ausfeld; Cur. Art, Michael Panhorst; Cur. Art., Jennifer Jankauskas; Cur. Education, Tim Brown; Asst. Cur. Education, Alice Novak; Asst. Cur. Education, Donna Pickens; Registrar, Pamela Bransford; Preparator & Designer, Jeff Dutton; Special Events Coord. & Museum Shop Mgr., Tisha Rhodes.
Personnel Profile: Full-Time Paid 41; Part-Time Paid 12; Part-Time Volunteers 700; Interns 4.
Volunteer Hours: 5,082
Operating Expenses: 864,964
Operating Income: 812,023
Governing Authority: city; nonprofit organization. Parent Institution: City of Montgomery. Tax-exempt: 501(c)(3).
Institution Type/Description: Art Museum.
Collections: 19th & 20th-century American paintings & sculptures; Old Master paints; southern regional art; decorative arts; studio glass.
Major Exhibits: Amos "Ashanti" Johnson from the Paul R. Jones Collection of American Art at the University of Alabama, 11/30/13-2/2/14; David Allison Ogburn: Oggi's "Photomart" of Great African Americans from the Paul R. Jones Collection of American Art at the University of Alabama, 11/30/13-2/2/14; Nature Distilled, 1/4/14-3/9/14; Portraits of Paul R. Jones, 2/8/14-4/13/14; Creator/Created: Jerry Siegel Portraits and Artists from the MMFA Permanent Collection, 3/8/14-6/1/14; Our Town: Montgomery People and Places, 3/15/14-5/11/14; Romantic Spirits, Nineteenth Century Paintings of the South from the Johnson Collection, 4/12/14-7/6/14; Portraiture from the Paul R. Jones Collection of American Art at the University of Alabama, 4/19/14-6/15/14; Kerr Eby: Camofleur, Combat Artist, and Pacifist, 5/17/14-7/13/14; In Time We Shall Know Ourselves Photographs by Raymond Smith, 6/14/14-8/31/14; Origins: The First Twenty-Five Years of the MMFA Collection, 7/12/14-8/31/14; Forged and Fabricated: The Art and Craft of Albert Paley's Sculpture, 9/20/14-1/4/15; Reflections: African-American Life from the Myrna Colley-Lee Collection, 1/15/15-3/15/15; Masterworks on the Move American Paintings from Wesleyan College, 3/15-5/15; Living Memory: Images of Art in Transition Works by Self-Taught Artists from the MMFA Collection, 9/15-11/15; ReTooled: Highlights from the Hechinger Collection, 11/7/15-1/10/16.
Research Fields: research in support of permanent collection.
Facilities: 6,000-vol. library; 73,000 sq. ft. of space; 250-seat auditorium; orientation center; cafe. Art-related items for sale.
Activities: guided tours; lectures; films; concerts; permanent & temporary exhibitions; children's programs; outreach programs; interactive gallery.
Publications: annual report; quarterly magazine; exhibition catalogues; promotional brochures.
Hours & Admission Prices: Tues.-Wed. & Fri.-Sat. 10-5, Thurs. 10-9, Sun. 12-5. No charge; donations accepted. Closed New Year's Day; Veterans Day; Thanksgiving; Christmas. &
Attendance: 138,210 (accurate)
Membership: Student $30; Individual $45; Family $60; Contributing $100; Subscribing $150; Supporting $250; Sustaining $500; Sponsoring $800; Benefactor $1,200; Major Benefactor $2,500; Distinguished Benefactor $5,000.

THE MOOSEUM, 201 S. Bainbridge St., Montgomery, AL 36104-4332. Mailing Address: P.O. Box 2499, Montgomery, AL 36102-2499. Tel.: 334-265-1867; 888-276-3362. Fax: 334-834-5326. Facebook: The MOOseum.
E-mail: jkennedy@bamabeef.org
Web Site: www.bamabeef.org
Founded: 1996.
Key Personnel: Coord., Jessica Greene
Institution Type/Description: Children's Museum.
Collections: history of the agriculture & beef cattle industry; hands-on exhibits.
Publications: magazine, Alabama Cattleman.
Hours & Admission Prices: Mon.-Fri. 8-12 & 1-4:30 by appointment. No charge; donations accepted. Closed major holidays; special events.
Attendance: 10,000 (estimated)

ROSA PARKS LIBRARY AND MUSEUM, 252 Montgomery St., Montgomery, AL 36104-3527. Tel.: 334-241-8661 & 8615. Fax: 334-241-5435.
Founded: 2000.
Congressional District: 7
Personnel Profile: Full-Time Paid 7; Part-Time Paid 2.
Governing Authority: Parent Institution: Troy University. Tax-exempt.

Institution Type/Description: Library & History Museum.
Collections: story of Rosa Parks' life; civil rights history; bus replica; 1955 station wagon; documents.
Facilities: 103-seat auditorium.
Hours & Admission Prices: Mon.-Fri. 9-5, Sat. 9-3; groups of 10 or more by appointment. Adults over 12 $5.50, children 12 & under $3.50; discounts to AAA members & Alabama College students. Closed holidays. &
Attendance: 50,095 (accurate)

SCOTT & ZELDA FITZGERALD MUSEUM, 919 Felder Ave., Montgomery, AL 36106-1926. Mailing Address: 919 Felder Ave., Apt. B, Montgomery, AL 36106-1927. Tel.: 334-264-4222.
E-mail: info@fitzgeraldmuseum.net
Web Site: fitzgeraldmuseum.net
Founded: 1986.
Key Personnel: Exec. Dir., Thomas W. Thompson
Institution Type/Description: Literary Museum.
Collections: photographs; manuscripts; Scott's books; Zeldax family memorabilia; portraits; paintings; personal artifacts.
Hours & Admission Prices: Wed.-Fri. 10-2, Sat.-Sun. 1-5. Donations: adults $5, students & seniors $2. Closed major holidays.
Attendance: 2,500 (estimated)

UNITED STATES AIR FORCE'S ENLISTED HERITAGE HALL, Maxwell Air Force Base, Gunter Annex, 550 McDonald St., Montgomery, AL 36114. Tel.: 334-416-1110.
Institution Type/Description: Military History Museum.
Collections: military aviation history & artifacts; military memorial; Korean War; World War II; Vietnam War; Desert Shield; Desert Storm; Operation Enduring Freedom; Gulf War to present day operations; photographs; personal artifacts; audiovisual; equipment; aircraft parts; early telephones; uniforms; Glenn Miller tribute.
Hours & Admission Prices: Mon.-Fri. 8-4, Sat. 9-4. Closed holidays.

W.A. GAYLE PLANETARIUM, 1010 Forest Ave., Montgomery, AL 36106-1115. Tel.: 334-241-4799. Fax: 334-241-2301. Facebook: W. A. Gayle Planetarium.
E-mail: rlevans@troy.edu
Web Site: montgomery.troy.edu/planet
Founded: 1968.
Congressional District: 2
Key Personnel: Dir. & Museum Shop Mgr., Rick Evans.
Personnel Profile: Full-Time Paid 2; Part-Time Paid 2.
Governing Authority: municipal & university. Parent Institution: Troy University, Montgomery. Tax-exempt.
Institution Type/Description: Planetarium.
Collections: astronomy; 21 slide projectors.
Facilities: 200-seat theater. Space science & astronomy items for sale.
Activities: lectures; films; formally organized education programs for children; laser light shows.
Hours & Admission Prices: Public Programs: Sun. 2 pm, call for information. School Programs K-12: Mon.-Fri. by reservation only. Public Shows: Mon.-Thurs. 3, Sun. 2. School Programs: adults $5, students $3.50; Public Shows: admission $5; children under 5 no charge. &
Attendance: 30,000 (accurate)

Moundville

MOUNDVILLE ARCHAEOLOGICAL PARK, (M), 13075 Moundville Archaeological Park, Moundville, AL 35474-6413. Mailing Address: University of Alabama, P.O. Box 870340, Tuscaloosa, AL 35487-0340. Tel.: 205-371-2234. Fax: 205-371-4180.
E-mail: moundville@bama.ua.edu
Web Site: moundville.ua.edu
Founded: 1939.
Congressional District: 7
Key Personnel: Dir., Bill Bomar; Education Coord., Betsy Irwin; Office Asst., Angela Jones; Museum Shop Mgr., Dorothy Beckham.
Personnel Profile: Full-Time Paid 11; Part-Time Paid 2; Part-Time Volunteers 150.
Governing Authority: public university; nonprofit. Parent Institution: University of Alabama Museums, University of Alabama. Tax-exempt.
Institution Type/Description: Archaeological Site: over two dozen Mississippian mounds and 320-acre park.
Collections: archaeological collections; simulated Indian village with life-sized dioramas.

Research Fields: archaeology.
Facilities: theater; meeting space; conference center; nature trail; campground.
Activities: teaching craft huts; hiking trails. Annual Event: Moundville Native American Festival with living history encampments, artisans & performers in October.
Publications: teacher guides; fact sheets; brochures; special publications.
Hours & Admission Prices: Park: daily 8-dusk. Museum: call for hours. &
Attendance: 50,000 (estimated)
Membership: Alabama Natural History Society: Individual $20; Family $30; Associated $50-$249; Contributor $250-$499; Director's Circle $500-$999; Sustainer $1,000 and up.

Normal

STATE BLACK ARCHIVES RESEARCH CENTER AND MUSEUM, James Hembray Wilson Bldg., Alabama A&M University, Normal, AL 35762. Mailing Address: P.O. Box 595, Normal, AL 35762-0595. Tel.: 256-372-5846. Fax: 256-372-5338.
Founded: 1987.
Congressional District: 5
Key Personnel: Dir., Patricia D. Ford.
Personnel Profile: Full-Time Paid 2; Part-Time Paid 4; Part-Time Volunteers 2.
Governing Authority: Parent Institution: Alabama A&M University. Tax-exempt.
Institution Type/Description: History Museum.
Collections: African American history & culture; photographs; personal artifacts; uniforms; paintings.
Activities: permanent & traveling exhibitions.
Hours & Admission Prices: Mon.-Fri. 9-4:30. Adults 12 & over $5, senior citizens $4, children 11 & under $3; discounts to groups; children 5 & under no charge. Closed Martin Luther King Jr. Day; Memorial Day; Independence Day; Labor Day; Thanksgiving; Christmas. &
Attendance: 3,000 (accurate)

Northport

KENTUCK MUSEUM ASSOCIATION, INC., 503 Main Ave., Northport, AL 35476-4483. Tel.: 205-758-1257. Fax: 205-758-1258.
E-mail: kentuck@kentuck.org
Web Site: www.kentuck.org
Founded: 1971.
Congressional District: 6
Key Personnel: Exec. Dir., Shweta Gamble; Pres. Bd. (until 6/1/12), Edward Guy; Pres. Bd., Stephen Stabler.
Personnel Profile: Full-Time Paid 2; Part-Time Paid 1; Part-Time Volunteers 400.
Governing Authority: private; nonprofit organizations. Tax-exempt: 501(c)(3).
Institution Type/Description: Art Museum.
Collections: works by local, regional & national artists.
Facilities: artist studios; courtyard garden. Museum-related items for sale.
Activities: arts festival; children's programs. Monthly Events: Art Night; Kentucky for Kids and Art Fair; Annual Events: The Kentuck Festival of the Arts in October; Fine Crafts & Art; folk artists.
Hours & Admission Prices: Tues.-Fri. 9-5, Sat. 10-4:30. No charge; donations accepted. Closed New Year's Day; Memorial Day; Labor Day; Christmas.
Attendance: 12,000 (estimated)
Membership: Studio $50; Gallery $75; Art Center $100; Museum $200; Fine Art $500; Big Dog $1,000.

THE NORTHPORT RENAISSANCE ART GALLERY, 431 Main Ave., Northport, AL 35476-5063. Tel.: 205-752-4422. Fax: 205-752-4422.
E-mail: renaissanccartga@bellsouth.net
Web Site: www.renaissanccartgallcry.com
Formerly: Renaissance Gallery
Founded: 1994.
Key Personnel: Pres. & C.E.O., Lori Layden
Volunteer Hours: 100
Operating Expenses: 28,000
Operating Income: 26,000
Institution Type/Description: Art Gallery.
Collections: works by regional & local artists; handmade jewelry.
Hours & Admission Prices: Tues.-Sat. 11-4:30, 1st Thurs. of month 11-9; other times by appointment. No charge. &

Oneonta

BLOUNT COUNTY MEMORIAL MUSEUM, 204 2nd St. N., Oneonta, AL 35121-1740. Mailing Address: P.O. Box 45, Oneonta, AL 35121. Tel.: 205-625-6905.
E-mail: arhudy@co.blount.al.us
Founded: 1971.
Key Personnel: Dir., Cur. & Museum Shop Mgr., Amy Rhudy
Governing Authority: Parent Institution: Blount County Historical Society. Tax-exempt.
Institution Type/Description: County History Museum.
Collections: local history; county war veterans; family books & files; arrowheads; sandstone; covered bridge art; maps.
Research Fields: Blount County genealogy.
Activities: summer programs for children 6-18.
Publications: quarterly newsletter, Blount County Historical Society.
Hours & Admission Prices: Tues.-Fri. 9-3. No charge; donations accepted. &
Attendance: 2,000 (estimated)
Membership: Blount County Historical Society $10.

Opelika

THE MUSEUM OF EAST ALABAMA, 121 S. 9th St., Opelika, AL 36801-4917. Mailing Address: P.O. Box 3085, Opelika, AL 36803-3085. Tel.: 334-749-2751.
E-mail: museum@eastalabama.org
Web Site: www.eastalabama.org
Founded: 1989.
Congressional District: 3
Key Personnel: Pres. (V), Bert Harris; Treas., Mike Martin.
Personnel Profile: Full-Time Paid 1; Part-Time Paid 1; Part-Time Volunteers 100.
Governing Authority: private; nonprofit organization. Tax-exempt: 501(c)(3).
Institution Type/Description: General Museum.
Collections: Roanoke dolls; Camp Opelika (WWII) memorabilia; early 20th-century recording technology (Orr RADIO); agricultural implements; fire fighting tools & trucks; local history artifacts.
Facilities: 5,468 sq. ft. exhibit space.
Activities: docent program; guided tours; lectures; temporary exhibitions.
Hours & Admission Prices: Tues.-Fri. 10-4, Sat. 2-4. No charge; donations accepted. Closed New Year's Day; Memorial Day; Christmas Eve & Day. &
Attendance: 1,386 (estimated)
Membership: Student $10; Retired $15; Individual $20; Family $25; Associate $100; Contributor $250; Friend $500; Patron $1,000; Sponsor $2,500; Benefactor $5,000; Charter $10,000; Golden Charter $50,000.

Orange Beach

ORANGE BEACH INDIAN AND SEA MUSEUM, 25850 John Snook Dr., Orange Beach, AL 36561. Mailing Address: P.O. Box 458, Orange Beach, AL 36561-0458. Tel.: 251-981-8545. Fax: 251-981-6053.
Web Site: www.obparksandrec.com
Founded: 1995.
Key Personnel: Museum Guide, Gail Graham.
Personnel Profile: Part-Time Paid 1.
Governing Authority: city. Tax-exempt.
Institution Type/Description: Historic Building: housed in period schoolhouse.
Collections: period artifacts & furnishings; area history.
Hours & Admission Prices: Tues. & Thurs. 9-4; tours by appointment. No charge. &
Attendance: 2,000 (accurate)

Pelham

ALABAMA WILDLIFE CENTER, Oak Mountain State Park, 100 Terrace Dr., Pelham, AL 35124-4314. Tel.: 205-663-7930. Fax: 205-682-6867.
E-mail: wildlife@awrc.org
Web Site: www.awrc.org
Formerly: The Wildlife Center
Founded: 1977.
Key Personnel: Chm. (V), Richard Esposito; Exec. Dir., Doug Adair.
Personnel Profile: Full-Time Paid 3; Part-Time Volunteers 250.
Governing Authority: nonprofit organization. Tax-exempt: 501(c)(3).
Institution Type/Description: Nature & Rehabilitation Center.
Collections: over 2,500 native birds, mammals, & reptiles of over 100 species.
Facilities: nature trails; wildlife rehabilitation center.

Activities: educational programs; information stations program. Annual Events: Migration Celebration in August; Creatures of the Night in October.
Hours & Admission Prices: Daily 9-5. Center: no charge. Park: adults $3, seniors 62 & over and children 6-11 $1; children under 6 no charge. &
Attendance: 10,000 (estimated)
Membership: Annual $35-$1,000.

Prattville

AUTAUGA COUNTY HERITAGE ASSOCIATION, 102 E. Main St., Prattville, AL 36067-3114. Tel.: 334-361-0961.
E-mail: director@autaugaheritage.org
Web Site: autaugaheritge.org
Founded: 1976.
Key Personnel: Exec. Dir., Greg Duke; Pres., Louise Jennings.
Governing Authority: Tax-exempt: 501(c)(3).
Institution Type/Description: History Museum.
Collections: county history & genealogy; personal artifacts; photographs.
Activities: lectures; meetings; special events.
Hours & Admission Prices: Tues.-Fri. 10-4, Sat. 10-2. No charge; donations accepted. Closed Christmas week. &
Attendance: 1,200 (accurate)
Membership: Individual $50; Couple $75; Contributing $100.

Red Bay

RED BAY MUSEUM, 110 4th Ave., S.E., Red Bay, AL 35582. Mailing Address: 400 4th St. S.W., Red Bay, AL 35582. Tel.: 256-356-8758.
Web Site: www.redbaymuseum.org
Founded: 2004.
Governing Authority: Tax-exempt.
Institution Type/Description: History Museum.
Collections: local history, culture & heritage; photographs; personal artifacts; period furnishings.
Hours & Admission Prices: Tues. & Thurs. 1:30-4; other times by appointment. Adults $5, students $3; children under 6 no charge.
Attendance: 1,000 (estimated)

Scottsboro

SCOTTSBORO & JACKSON HERITAGE CENTER MUSEUM, 208 S. Houston St., Scottsboro, AL 35768-4318. Mailing Address: P.O. Box 53, Scottsboro, AL 35768-0053. Tel.: 256-259-2122. Fax: 256-574-6991.
Founded: 1986.
Key Personnel: Dir., Judi Weaver; Chm. (V), Kelly Goodowens
Governing Authority: Tax-exempt.
Institution Type/Description: History Museum.
Collections: Jackson County history, customs, traditions & art; Native American artifacts; family histories; public records. Historic Buildings: 1800s log cabin village & furnishings.
Research Fields: family history.
Facilities: library.
Activities: special events; research; traveling exhibits; rental facilities.
Hours & Admission Prices: Mon.-Fri. 11-4; tours by appointment. Adults $3; members no charge. &
Attendance: 1,500 (estimated)
Membership: Student & Retired $10; Individual $15; Family $20; Corporate & Business $100; Founding Family & Life $1,000.

SCOTTSBORO BOYS MUSEUM AND CULTURAL CENTER, 428 W. Willow St., Scottsboro, AL 35768. Mailing Address: P.O. Box 1557, Scottsboro, AL 35768. Tel.: 256-244-1310.
Institution Type/Description: History Museum.
Collections: history of the Scottsboro Boys trial; Civil Rights in America; photographs; press clippings.
Hours & Admission Prices: Call for hours.

Selma

NATIONAL VOTING RIGHTS MUSEUM AND INSTITUTE, 6 U.S. Hwy. 80 E., Selma, AL 36701. Mailing Address: P.O. Box 1366, Selma, AL 36702-1366. Tel.: 334-418-0800. Fax: 334-418-1991.
E-mail: nvrm1965@gmail.com
Web Site: www.nvrmi.com
Founded: 1992.

Congressional District: 7
Key Personnel: Pres. (V), Rose M. Sanders; Treas., Louretta Wimberly; Interim Dir., Pearlie L. Walker; Chm., Carolyn G. Varner; Chm. (V), Charles Mauldin; Museum Shop Mgr., Lashunda G. Brown; Tour Guide, Sam Walker.
Personnel Profile: Full-Time Paid 2; Full-Time Volunteers 2; Part-Time Paid 3; Part-Time Volunteers 15.
Governing Authority: private; nonprofit organization. Tax-exempt: 501(c)(3).
Institution Type/Description: Historical Museum: housed in the former headquarters of the White Citizen's Council of Alabama; located at the foot of the Edmund Pettus Bridge, the site of Bloody Sunday.
Collections: voting rights data; video tape library of living legends who participated in the Voting Rights Movement & the Selma-Montgomery march.
Facilities: 150-seat auditorium. Museum-related items for sale.
Activities: concerts; children's songs; films; guided tours; lectures; traveling & temporary exhibitions. Annual Event: Jubilee Festival commemorating the historic march across the Edmund Pettus Bridge.
Publications: newsletter, NVRM.
Hours & Admission Prices: Mon.-Thurs. 10-4, Fri.-Sun. by appointment only. Adults $6.50, students & seniors 65 & older $4.50; discounts to members, groups and AAM & ICOM members. Closed New Year's Day; Memorial Day; Martin Luther King Jr. Day; Christmas. &
Attendance: 150,000 (estimated)
Membership: Basic $25; Supporter & Organization $100; Dreamer $500; Mountain Top $1,000.

OLD DEPOT MUSEUM, 4 Martin Luther King St., Selma, AL 36703-3109. Tel.: 334-874-2197. Fax: 334-874-1221.
E-mail: olddepot@wwisp.com
Key Personnel: Dir. & Cur., Jean Martin
Institution Type/Description: History Museum.
Collections: period artifacts; early plantation records; 19th-century doctor's traveling kit; confederate bills.
Hours & Admission Prices: Mon.-Sat. 10-4. Adults $4, seniors $3, college students $2, children $1.

STURDIVANT HALL, 713 Mabry St., Selma, AL 36701-5521. Mailing Address: P.O. Box 1205, Selma, AL 36702-1205. Tel.: 334-872-5626. Fax: 334-872-5626.
E-mail: info@sturdivanthall.com
Web Site: sturdivanthall.com
Founded: 1957.
Congressional District: 7
Key Personnel: Dir., Manera S. Searcy; Co-Dir., Nancy Gantt; Pres., Anne F. Knight; Museum Shop Mgr., Patty DeBardeleben.
Personnel Profile: Full-Time Paid 2; Part-Time Paid 7.
Governing Authority: county; municipal. Tax-exempt: 501(c)(3).
Institution Type/Description: Historic House Museum: 1852 Sturdivant Hall.
Collections: furnishings of the antebellum period; Victorian, Chippendale & Hepplewhite furniture; Oriental rugs; china; silver; fine linens; pianos; English chests; oil paintings; dolls; toys; archival materials.
Research Fields: pertaining to the collections.
Facilities: banquet facilities.
Activities: guided tours; concerts; docent council; permanent exhibitions; rental facilities.
Hours & Admission Prices: Tues.-Sat. 10-4. Adults $5, students $2; discounts to groups of 15 or more; discounts to AAA members; members no charge. Closed major holidays. &
Attendance: 17,500 (estimated)
Membership: Single $30; Donor $50; Patron $100; Corporate $150; Lifetime $1,000.

Stevenson

STEVENSON RAILROAD DEPOT MUSEUM, 207 W. Main St., Stevenson, AL 35772-3567. Tel.: 256-437-3012.
E-mail: info@stevensondepotmuseum.com
Web Site: stevensondepotmuseum.com
Institution Type/Description: Historic Building Museum: built in 1872. Listed on the National Register of Historic Places.
Collections: railroading history; photographs; Native American culture; pioneer life; Civil War artifacts.
Hours & Admission Prices: Mon.-Fri. 8-4.

Summerdale

ALLIGATOR ALLEY, 19950 Hwy. 71, Summerdale, AL 36580. Tel.: 866-99-GATOR; 251-946-BITE.
Web Site: www.gatoralleyfarm.com
Institution Type/Description: Alligator Farm.
Collections: over 150 alligators; ospreys; owls; turtles; bull frogs; amphibians; reptiles.
Facilities: Museum-related items for sale.
Activities: group tours; field trips.
Hours & Admission Prices: Spring & Summer: daily 10-5. Adults $10, children 4-12 $8; children under 3 no charge.

Sylacauga

ISABEL ANDERSON COMER MUSEUM & ARTS CENTER, 711 N. Broadway, Sylacauga, AL 35150-2155. Mailing Address: P.O. Box 245, Sylacauga, AL 35150-0245. Tel.: 256-245-4016. Fax: 256-245-4612.
E-mail: comercenter@bellsouth.net
Web Site: comermuseum.weebly.com
Founded: 1982.
Congressional District: 3
Key Personnel: Exec. Dir., Donna Rentfrow; Bd. Pres., Don Smith; Asst. Dir., Linda Pearson.
Personnel Profile: Full-Time Paid 2; Full-Time Volunteers 18; Part-Time Volunteers 7.
Governing Authority: private; nonprofit organization. Tax-exempt: 501(c)(3).
Institution Type/Description: History & Art Museum.
Collections: fine art; mixed media; Indian artifacts; marble sculpture; period furnishings; a replica of the Hodges Meteorite & written documentation of the incident; photographs; Jim Nabors' memorabilia; costumes; gold & platinum records.
Facilities: educational facilities. Museum-related items for sale.
Activities: formal educational programs; guided tours; lectures; field trips; art classes.
Publications: quarterly newsletter, The Museum Review.
Hours & Admission Prices: Tues.-Fri. 10-5; other times by appointment. No charge; donations accepted. Closed major holidays. &
Attendance: 14,200 (estimated)
Membership: Old Master & Friend $10; Student, Artist & Teacher $20; Individual $30; Family & Couple $50; Supporter $250; Patron $500; Benefactor $1,000.

Talladega

JEMISON-CARNEGIE HERITAGE HALL, 200 South St., Talladega, AL 35161. Mailing Address: P.O. Box 1118, Talladega, AL 35161. Tel.: 256-761-1364. Facebook: Heritage Hall.
E-mail: hhmuse@bellsouth.net
Web Site: heritagehallmuseum.org
Founded: 1981.
Congressional District: 3
Key Personnel: Coord. Museum Svcs., Kelly Williams; Pres. (V), George Hartsfield.
Personnel Profile: Part-Time Paid 1; Part-Time Volunteers 50.
Operating Expenses: 50,000
Operating Income: 50,000
Governing Authority: Parent Institution: Talladega Heritage Commission. Subsidiary Institution: Jemison Carnegie Foundation. Tax-exempt.
Institution Type/Description: Historic Building: housed in the former Talladega Public Library; built in 1906.
Collections: works by local & national artists; photographs.
Major Exhibits: Kelly Fitzpatrick, Original Paintings, 1/14; Art Bacon, Original Paintings, 2/14; James Brantley, Original Paintings, 3/14; Perry Austin, Original Paintings, 4/14; Faye Perry, Art Show, 5/14; Eric Johnson, Metal Sculpture, 6/14; Jackie Stephens, Hand Carved "Hitty Dolls", 7/14.
Activities: lectures; tours; art camp for kids; art classes.
Publications: quarterly newspaper, ArtWords.
Hours & Admission Prices: Tues.-Fri. 10-4. No charge; donations accepted. &
Attendance: 4,000 (estimated)
Membership: Individual $40; Dual $65; Family $75; Bronze Sponsor $150; Silver Sponsor $300; Gold Sponsor $450 & up.

Theodore

BELLINGRATH GARDENS & HOME, 12401 Bellingrath Gardens Rd., Theodore, AL 36582-8496. Tel.: 251-973-2217, ext. 147. Fax: 251-973-0540.
E-mail: tmcgehee@bellingrath.org
Web Site: www.bellingrath.org
Founded: 1932.
Congressional District: 1
Key Personnel: Exec. Dir., William E. Barrick, Ph.D.; Museum Dir., Thomas C. McGehee.
Personnel Profile: Full-Time Paid 60; Part-Time Paid 20; Part-Time Volunteers 75; Interns 1.
Governing Authority: private association. Parent Institution: Bellingrath Morse Foundation. Tax-exempt.
Institution Type/Description: Botanical Garden & Home.
Collections: Bessie Morse Bellingrath china, silver, European porcelains, period furniture; Edward Marshall Boehm porcelain.
Facilities: 65-acre landscaped gardens; restaurant. Museum-related items for sale.
Activities: guided tours of home; films; permanent exhibitions; video show; videocassettes; scenic riverboat cruises on the Fowl River; educational programs for adults & children in summer. Museum Sponsors: Winter Wednesday programs January & February; Wonderful Wednesdays June & July. Annual Event: Christmas in Lights in December.
Publications: books, Bellingrath Gardens & Home; Mister Bell.
Hours & Admission Prices: Home: daily 9-4. Gardens, Home & River Cruise $27. Gardens & Home: $19. Gardens: $11. Closed Thanksgiving; Christmas. &
Attendance: 169,000 (accurate)
Membership: Individual $40; Couple $60; Family $100; Patron $150-$250; Belle Camp Society $500-$2,500.

Troy

JOHNSON CENTER FOR THE ARTS, (M), 300 E. Walnut St., Troy, AL 36081-3539. Mailing Address: P.O. Box 863, Troy, AL 36081-0863. Tel.: 334-670-2287. Fax: 334-808-4025.
Web Site: www.tpcac.org
Formerly: Troy-Pike Cultural Arts Center
Founded: 2000.
Key Personnel: Exec. Dir., Richard Metzger; Chm. (V), Mack Gibson.
Personnel Profile: Full-Time Paid 3; Part-Time Paid 10; Part-Time Volunteers 20.
Governing Authority: nonprofit organization. Tax-exempt: 501(c)(3).
Institution Type/Description: Art Museum.
Collections: works by regional, national & international artists.
Facilities: rental facilities.
Activities: theater; workshops; art classes; concerts; receptions.
Hours & Admission Prices: Tues.-Sat. 10-5, Sun. 1-5. No charge; donations accepted. Closed holidays. &
Attendance: 800 (estimated)
Membership: Student $25; Individual $50; Dual $90; Family $100; Associate $250.

PIONEER MUSEUM OF ALABAMA, 248 U.S. 231 N., Troy, AL 36081. Tel.: 334-566-3597.
E-mail: pioneer@troycable.net
Web Site: www.pioneer-museum.org
Formerly: Pike Pioneer Museum
Founded: 1971.
Congressional District: 2
Key Personnel: C.E.O., Jeff Kervin; Dir., Kari Barley.
Personnel Profile: Full-Time Paid 1; Part-Time Paid 4; Part-Time Volunteers 38.
Governing Authority: nonprofit organization. Tax-exempt.
Institution Type/Description: History Museum.
Collections: artifacts related to area history along with social & agricultural history of pioneers in Alabama. Historic Buildings: two pen log house; tenant house; log house; one room school; general store; log church; corn crib; smokehouse; grist mill; train depot.
Research Fields: tenant farming in Alabama; Alabama settlers.
Facilities: 35 acre complex; picnic area; nature trails; amphitheater. Museum-related items for sale.
Activities: self guided & guided tours. Museum Sponsors: Living History weekends.
Publications: brochures; calendar of events; teacher packs.
Hours & Admission Prices: Tues.-Sat. 9-5. Adults $6, seniors 60 & over $5, students $4; discounts to military, AAM & ICOM members; members no charge. Closed major holidays. &

Attendance: 10,000 (accurate)
Membership: Individual $25; Family $45; Preserver $60; Sustainer $100; Pioneer $250; Heritage $500; Underwriter $1,000.

Tuscaloosa

ALABAMA MUSEUM OF NATURAL HISTORY, (M), Smith Hall, University of Alabama Campus, 427 6th Ave., Tuscaloosa, AL 35487. Mailing Address: Box 870340, Tuscaloosa, AL 35487. Tel.: 205-348-7550 & 7551. Fax: 205-348-9292.
E-mail: natural.history@ua.edu
Web Site: museums.ua.edu
Founded: 1847.
Congressional District: 7
Key Personnel: Exec. Dir., Robert Clouse; Coord. Environmental Education & Programs, Dr. Douglas Phillips; Mgr. Collections, Mary Bade.
Personnel Profile: Full-Time Paid 15; Part-Time Paid 3; Part-Time Volunteers 50; Interns 2.
Governing Authority: university. Parent Institution: The University of Alabama. Subsidiary Institution: University of Alabama Museums. Tax-exempt: 501(c)(3).
Institution Type/Description: Natural History Museum.
Collections: anthropology; archaeology; ichthyology; mineralogy; paleontology; herpetology; botany; mammalogy; ornithology; entomology; ethnology.
Research Fields: anthropology; archaeology; ichthyology; vertebrate paleontology; natural history education; teacher education; protohistoric archaeology (DeSoto).
Activities: guided tours; lectures; permanent & traveling exhibitions; special programs for children; year-round field trips & programs; museum expedition; paleontology research site. Annual Event: Moundville Native American Festival.
Publications: bulletin; field trip guides; brochures; special publications; Bulletin of the Alabama Museum of Natural History.
Hours & Admission Prices: Smith Hall Museum: Mon.-Sat. 10-4:30. Adults $2, children $1; members no charge. Closed university holidays. &
Attendance: 32,000 (estimated)
Membership: Alabama River $40; Black Warrior River $100; Cahaba River $250; Coosa River $500; Sipsey River $1,000; Eugene Allen Smith Society $5,000.

BATTLE-FRIEDMAN HOUSE, 1010 Greensboro Ave., Tuscaloosa, AL 35401-2336. Mailing Address: P.O. Box 1665, Tuscaloosa, AL 35403-1665. Tel.: 205-758-6138. Fax: 205-758-8163. Facebook: Battle-Friedman House.
Web Site: www.historictuscaloosa.org
Founded: 1966.
Key Personnel: Dir., Katherine Richter.
Governing Authority: Operated By: the Tuscaloosa County Preservation Society. Tax-exempt.
Institution Type/Description: Historic House: built in 1835. Listed on the National Register of Historic Places.
Collections: period furnishings; personal artifacts.
Activities: rental facilities; special events; lectures; guided tours.
Publications: quarterly newsletter.
Hours & Admission Prices: Tues.-Sat. 10-12 & 1-4. Admission $5; children under 12 no charge. Closed major holidays & during scheduled events. &
Attendance: 450 (estimated)
Membership: Student $20; Regular $50; Junior Century Club (under 35) $75; Century Club $125; Grand Benefactor & Corporate $500; Life $2,500.

CHILDREN'S HANDS-ON MUSEUM (CHOM), 2213 University Blvd., Tuscaloosa, AL 35401-1541. Tel.: 205-349-4235. Fax: 205-349-4272.
E-mail: info@chomonline.org
Web Site: www.chomonline.org
Founded: 1984.
Congressional District: 7
Key Personnel: Exec. Dir. & Dir. Exhibits, Charlotte Gibson; Chm. (V), Kim S. Hudson; Visitors Svcs. Coord. & Volunteer Coord., Sherie Giles; Membership Coord., LaKesa Grey; Museum Shop Mgr., Pam Hisey.
Personnel Profile: Full-Time Paid 6; Part-Time Paid 3; Part-Time Volunteers 80; Interns 5.
Governing Authority: nonprofit organization. Tax-exempt.
Institution Type/Description: Children's Museum.
Collections: hands-on exhibits dealing with history, science, the arts, early childhood, nature, & space.
Facilities: interactive exhibit spaces; planetarium. Books & educational toys for sale.

Activities: docent program; directed tours; in-depth programs; special events; school field trips; holiday events; children's parties; rental facilities; early childhood activities.
Publications: quarterly newsletter, CHOM News.
Hours & Admission Prices: Mon.-Fri. 9-5, Sat. 10-4. Admission 3 & over $8, seniors over 60 $6, children 1-3 $5; discounts to groups & scouts; members & children under one no charge. ACM reciprocal members. Holiday or Summer Pass: $36 (3 months). Closed major holidays. &
Attendance: 59,680 (accurate)
Membership: Adventurer $75; Explorer $85; Trailblazer $140.

GORGAS HOUSE MUSEUM, (M), Capstone at McCorvy Dr., University of Alabama, Tuscaloosa, AL 35487. Mailing Address: Box 870340, University of Alabama Museums, Tuscaloosa, AL 35487. Tel.: 205-348-5906 & 7550. Fax: 205-348-9292.
E-mail: gorgashouse@ua.edu
Web Site: gorgashouse.ua.edu
Founded: 1954.
Congressional District: 7
Key Personnel: Dir., Erin Harney.
Personnel Profile: Full-Time Paid 1.
Governing Authority: state. Parent Institution: University of Alabama. Subsidiary Institution: University of Alabama Museums. Tax-exempt.
Institution Type/Description: Historic House: built in 1829 as a dining hall for students, home of Josiah & Amelia Gayle Gorgas and family from 1878-1953 & one of four buildings to survive the burning of the campus in 1865 by federal troops.
Collections: 18th & 19th century Spanish silver.
Activities: guided tours, Center for Study of University of Alabama History.
Publications: rack card.
Hours & Admission Prices: Mon.-Fri. 9-12 & 1-4:30; other times by appointment. Admission $2. University of Alabama students, faculty, staff and alumni with card no charge. Closed university holidays. &
Attendance: 1,000 (estimated)

JEMISON-VAN DE GRAAFF MANSION, 1305 Greensboro Ave., Tuscaloosa, AL 35401-2840. Mailing Address: P.O. Box 1216, Tuscaloosa, AL 35403. Tel.: 205-758-2906.
E-mail: jemison_mansion@bellsouth.net
Web Site: www.jemisonmansion.com
Governing Authority: Parent Institution: Tuscaloosa County Preservation Society and the Heritage Commission of Tuscaloosa.
Institution Type/Description: Historic House: housed in the former home of Senator Robert Jemison, Jr.; built in 1862.
Collections: period furnishings; personal artifacts.
Activities: teachers' programs; special events; rental facilities. Museum Sponsors: Christmas Open House in December.
Hours & Admission Prices: Mon.-Fri. 10-5; groups by appointment. No charge; donations accepted.

MURPHY AFRICAN AMERICAN MUSEUM, 2601 W. Paul Bryant Dr., Tuscaloosa, AL 35401-2214. Tel.: 205-758-2861. Fax: 205-758-8163.
Web Site: www.historictuscaloosa.org
Founded: 1985.
Congressional District: 7
Key Personnel: Dir. & Chm. (V), Emma Jean Melton.
Personnel Profile: Part-Time Paid 1; Part-Time Volunteers 15; Interns 3.
Governing Authority: Parent Institution: Tuscaloosa County Preservation Society. Tax-exempt.
Institution Type/Description: History Museum: housed in the home of Tuscaloosa's first licensed black mortician; built c. 1920.
Collections: lifestyle of affluent blacks in the early 1900s; personal artifacts; furnishings; African art.
Hours & Admission Prices: Tues.-Fri. 10-12 & 1-4; tours by appointment.

THE OLD TAVERN MUSEUM, 500 28th Ave.-Capitol Park, Tuscaloosa, AL 35401. Mailing Address: P.O. Box 1665, Tuscaloosa, AL 35403-1665. Tel.: 205-758-2238. Fax: 205-758-8163.
E-mail: kmautertcps@bellsouth.net
Web Site: www.historictuscaloosa.org
Founded: 1966.
Congressional District: 6
Key Personnel: Exec. Dir., Katherine Richter Mauter.
Personnel Profile: Full-Time Paid 1.
Governing Authority: nonprofit organization. Parent Institution: Tuscaloosa County Preservation Society, P.O. Box 1665, Tuscaloosa 35403. Branch Museums: Battle-Friedman House; Civic & Cultural Center. Tax-exempt: 501(c)(3).
Institution Type/Description: Historic House Museum: 1827 Old Tavern, relocated on Capitol Park, the site of the Capitol Building when Alabama's Capitol was in Tuscaloosa.
Collections: period furnishings & artifacts; folklore.
Activities: guided tours; lectures.
Publications: pamphlet, Preservationist.
Hours & Admission Prices: By appointment only. No charge; donations accepted. Closed major holidays.
Attendance: 400 (estimated)
Membership: Student $20; Regular $50; Junior Century Club (under 35) $75; Century Club $125; Grand Benefactor & Corporate $500; Life $2,500.

PAUL W. BRYANT MUSEUM, 300 Paul W. Bryant Dr., Tuscaloosa, AL 35487. Mailing Address: P.O. Box 870385, Tuscaloosa, AL 35487-0385. Tel.: 205-348-4668; 866-772-2327. Fax: 205-348-8883.
E-mail: info@bryantmuseum.com
Web Site: www.bryantmuseum.com
Founded: 1985.
Congressional District: 7
Key Personnel: Dir., Kenneth Gaddy; Visitor Svcs. Coord., Jan Scurlock; Cur., Taylor Watson; Collections Asst., Brad Green; Program Asst., Olivia Arnold; Cashier, Lynn Bobo; Museum Shop Mgr., Melinda Register; Audio Visual Tech, David Mize; Cashier, DiAnne Griffin.
Personnel Profile: Full-Time Paid 8; Part-Time Paid 3; Interns 2.
Governing Authority: public university; nonprofit. Parent Institution: University of Alabama. Tax-exempt: 501(c)(3).
Institution Type/Description: Sport Museum.
Collections: books, programs, photographs, newspapers & media guides; game films & audio tapes; sports memorabilia related to University of Alabama athletics & Southeastern Conference sports; over 100 years of Alabama football.
Research Fields: Alabama football; SEC sports; college sports.
Facilities: 7,000 sq. ft. exhibit space; 45-seat theater. Museum-related items for sale.
Activities: guided tours; films; rental facilities.
Publications: brochures; rack cards; souvenir book; newsletter, Circle of Champions; teacher's guide, Punt, Pass & Learn.
Hours & Admission Prices: Daily 9-4; call for holiday hours. Adults $2, senior citizens 60 & over, students and children 6-17 $1; children under 6, alumni & members no charge. Closed major holidays. &
Attendance: 40,000 (estimated)
Membership: Circle of Champions: Sideline: Students $15; Locker Room Club: Individual $20; Family $30; Press Box Club $100-$249; Director's Club $250-$499; William Little Club $500-$999; Bryant Gold Club $1,000 & up.

SARAH MOODY GALLERY OF ART, THE UNIVERSITY OF ALABAMA, (M), 103 Garland Hall, Tuscaloosa, AL 35487. Mailing Address: Box 870270, Tuscaloosa, AL 35487. Tel.: 205-348-1890 & 5967 (art dept.). Fax: 205-348-0287.
E-mail: wtdooley@bama.ua.edu
Web Site: art.ua.edu/site/galleries/sarah-moody-gallery-of-art/
Founded: 1967.
Congressional District: 7
Key Personnel: Dir., Bill Dooley; Exhibitions Coord., Vicki Rial.
Personnel Profile: Full-Time Paid 2; Part-Time Paid 6.
Governing Authority: university. Parent Institution: University of Alabama, Tuscaloosa, AL. Tax-exempt.
Institution Type/Description: Art Gallery.
Collections: primitive art; paintings; drawings; prints; photos; sculpture; crafts.
Activities: lectures; films; gallery talks; temporary & traveling exhibitions.
Publications: extended checklists; exhibition catalogs: In These Islands - South Carolina & Georgia; Cora Cohen: Paintings & Altered X-Rays 1983-1996; Richard Zoellner: The Continuous Quest.
Hours & Admission Prices: Sept.-June Mon.-Wed. & Fri. 9-4:30, Thurs. 9-8; Summer: Mon.-Fri. 10-12 & 2-4. No charge. Closed university holidays. &
Attendance: 10,000

STILLMAN ART GALLERY, Stillman College, Cordell Wynn Humanities and Fine Arts Center, Rm. 155B, Tuscaloosa, AL 35403. Mailing Address: P.O. Box 1430, Tuscaloosa, AL 35403-1430. Tel.: 205-248-3404.
Web Site: www.stillman.edu
Institution Type/Description: Art Gallery.
Collections: paintings.

Hours & Admission Prices: Mon.-Fri. 9-5.

UNIVERSITY OF ALABAMA ARBORETUM, (M), 4801 Arboretum Way, Tuscaloosa, AL 35404-5424. Mailing Address: Box 870344, Tuscaloosa, AL 35487. Tel.: 205-553-3278. Fax: 205-553-3728.
E-mail: arbor@bama.ua.edu
Web Site: www.arboretum.ua.edu
Founded: 1958.
Congressional District: 7
Key Personnel: Pres. (V), Julia Hartman; Asst. Dir., Mary Jo Modica; Caretaker, Kenneth Robinson.
Personnel Profile: Full-Time Paid 3; Full-Time Volunteers 1; Part-Time Paid 1; Part-Time Volunteers 25.
Governing Authority: University of Alabama. Tax-exempt.
Institution Type/Description: Arboretum & Botanical Garden.
Collections: native & exotic woody plants; native herbaceous plants.
Research Fields: applied ecological research.
Facilities: classroom; picnic area.
Activities: guided tours; lectures.
Publications: quarterly newsletter, Growings On.
Hours & Admission Prices: Daily 8am to sunset. No charge; donations accepted. Closed New Year's Day; Thanksgiving; Christmas.
Attendance: 7,000 (estimated)
Membership: Seedling (Student) $10; Wildflower $25-$49; Azalea $50-$99; Hydrangea $100-$499; Dogwood $500-$999; Magnolia $1,000-$4,999; Oak $5,000 & up.

WESTERVELT-WARNER MUSEUM OF AMERICAN ART, 8316 Mountbatten Rd., N.E., Tuscaloosa, AL 35406-1118. Mailing Address: 2700 Yacht Club Way, Tuscaloosa, AL 35406-1125. Tel.: 205-343-4540. Fax: 205-345-1493.
Institution Type/Description: Art Museum.
Collections: paintings; sculptures; period artifacts; portraits.
Hours & Admission Prices: Tues.-Fri. 12-5, Sat. 10-5. Adults $9, seniors 65 & over $8, students 10 years old to college age $7; members no charge. ♿
Attendance: 15,000 (estimated)
Membership: Student $15; Individual $30; Family $50; Pioneer $100; Patron $250; Patriot $500; Founding Father $1,000.

Tuscumbia

ALABAMA MUSIC HALL OF FAME, 617 Hwy. 72 W., Tuscumbia, AL 35674-8711. Mailing Address: P.O. Box 740405, Tuscumbia, AL 35674-7417. Tel.: 256-381-4417; 800-239-2643. Fax: 256-381-1031.
E-mail: info@alamhof.org
Web Site: www.alamhof.ocm
Founded: 1982.
Congressional District: 5
Key Personnel: Chm., Rodney Hall; Dir., Wiley Barnard; Dir. Operations, Heath Simmons.
Personnel Profile: Full-Time Paid 6; Part-Time Paid 3; Part-Time Volunteers 10; Interns 3.
Governing Authority: state; nonprofit. Parent Institution: State of Alabama. Tax-exempt.
Institution Type/Description: Music Museum.
Collections: music archives; instruments; clothing; music memorabilia including achievements & awards.
Research Fields: Southern music.
Facilities: library of Southern music heritage material. Gift items for sale.
Activities: guided tours; lectures; films; concerts; study clubs; organized education programs. Museum Sponsors: Induction Banquet; Annual Concert Series
Publications: quarterly newsletter.
Hours & Admission Prices: Wed.-Sat. 9-5. Adults $8, senior citizens & students 13-18 $7, children 6-12 $5; discounts to AAA, AAM & ICOM members and groups of 10 or more; children 5 & under no charge. Closed New Year's Day; Easter; Thanksgiving & day after; Christmas Eve, Day & week. ♿
Attendance: 35,000 (accurate)

BELLE MONT MANSION, 1569 Cook Lane, Tuscumbia, AL 35674. Mailing Address: 12280 AL Hwy. 20, Hillsboro, AL 35643-3808. Tel.: 256-381-5052 & 637-8513. Fax: 256-637-8513.
E-mail: wheplan@hiwaay.net

Web Site: www.preserveala.org
Founded: 1983.
Congressional District: 8
Key Personnel: Dir., Melissa Beasley.
Personnel Profile: Full-Time Paid 3.
Governing Authority: state. Parent Institution: Alabama Historical Commission. Tax-exempt.
Institution Type/Description: Historic House: built in 1828.
Collections: period furnishings; personal artifacts.
Activities: self-guided & group tours.
Hours & Admission Prices: Thurs.-Sat. 9-4, Sun. 1-5; groups by appointment. Adults $5, seniors & military $4, children 6-18 $3; children under 6 no charge. Closed state & federal holidays.
Attendance: 1,500 (accurate)

IVY GREEN, BIRTHPLACE OF HELEN KELLER, 300 N. Commons St., W., Tuscumbia, AL 35674-1134. Tel.: 256-383-4066. Fax: 256-383-4068.
E-mail: helenkellerbirthplace@comcast.net
Web Site: helenkellerbirthplace.org
Founded: 1952.
Congressional District: 5
Key Personnel: Mgr., Sue Pilkilton.
Personnel Profile: Full-Time Paid 1; Part-Time Paid 8.
Governing Authority: municipal. Tax-exempt: 501(c)(3).
Institution Type/Description: Historic House Museum: 1820, birthplace of Helen Keller; main house, birthplace cottage, kitchen, carriage house, memorial gardens.
Collections: books; objects connected with Helen Keller's life; period furniture.
Facilities: Museum-related items for sale.
Activities: Museum Sponsors: William Gibson's, The Miracle Worker, staged on the grounds of Ivy Green in June & July; Tennessee Valley Art Association Art Show in City Park; Helen Keller Festival & Historic Tour of Homes by Helen Keller Festival Board; musical entertainment & sports events in June.
Publications: brochures.
Hours & Admission Prices: Mon.-Sat. 8:30-4. Adults $6, AAA members & seniors $5, student 5-18 $2; discount to groups; children under 5 no charge. Closed New Year's Day; Easter; Labor Day; Thanksgiving; Christmas Eve, Day & day after. ♿
Attendance: 35,000

TENNESSEE VALLEY ART CENTER, 511 N. Water St., Tuscumbia, AL 35674-1931. Mailing Address: P.O. Box 474, Tuscumbia, AL 35674-0474. Tel.: 256-383-0533. Fax: 256-383-0535.
E-mail: tvaa@comcast.net
Web Site: www.tvaa.net/
Founded: 1962.
Congressional District: 5
Key Personnel: Exec. Dir., Mary Settle Cooney; Asst. Dir., Jim Berryman; Chm. (V), Verna Brennan; Dir. Mktg. & Devel., Kay Brackin; Program Asst., Lori Curtis; Administrative Asst., J.K. Keith McMurtrey.
Personnel Profile: Full-Time Paid 4; Part-Time Volunteers 150.
Governing Authority: nonprofit organization. Parent Institution: Tennessee Valley Art Assn., Tuscumbia, AL. Tax-exempt: 501(c)(3).
Institution Type/Description: Art Center.
Collections: Reynolds collection; Helen Keller Festival fine art & craft collection; U.S. coins; ISOM prints; 3,000 lb. prehistoric petroglyph.
Facilities: performing arts theater; classrooms.
Activities: films; gallery talks; visual & performing arts workshops; photography; monthly exhibits; community theatre presentations; children's art work; classes; meetings of community groups; juried exhibitions; formally organized educational programs; inter-museum loan & traveling exhibitions; theatrical productions for children & adults; Helen Keller Festival of art & craft national touring exhibition.
Publications: newsletter.
Hours & Admission Prices: Mon.-Fri. 9-5, Sun. 1-3. Adults $5; members no charge. Closed Easter; Memorial Day; Labor Day; Thanksgiving; Christmas. ♿
Attendance: 60,000 (accurate)
Membership: Student $10; Individual $20; Family (Dual) & Sustaining $30; Family $50; Benefactor $100; Supporter $250; Patron $500.

Tuskegee Institute

TUSKEGEE INSTITUTE NATIONAL HISTORIC SITE, TUSKEGEE AIRMEN NATIONAL HISTORIC SITE & SELMA TO MONTGOMERY NATIONAL HISTORY TRAIL, 1212 W. Montgomery Rd., Tuskegee Institute, AL 36088-1923. Tel.: 334-727-6390, 3200 & 9321. Fax: 334-727-4597, 1448 & 1178. TDD: 334-727-3201.
Web Site: www.nps.gov/tuin
Founded: 1941.
Congressional District: 3
Key Personnel: Supt., Sandy Taylor; Chief Maintenance, Charles Patrick.
Personnel Profile: Full-Time Paid 11; Part-Time Paid 2; Part-Time Volunteers 12.
Governing Authority: federal. National Park Service, Dept. of the Interior, Southeast Region, Atlanta Federal Center, 1924 Bldg. 100 Alabama St. S.W. Atlanta, GA 30303.
Institution Type/Description: History Museum: The George Washington Carver Museum; 1899 The Oaks, home of Booker T. Washington.
Collections: personal memorabilia & awards associated with Dr. George W. Carver & Booker T. Washington; items pertaining to the growth & development of Tuskegee Institute; natural history collections; paintings; pottery; needle-art; manuscripts; furnishings; products related to Dr. Carver's scientific experimentation. Historic House: 1899 The Oaks.
Facilities: 200-vol. library of African art books available for research on premise; 40-seat auditorium. Books & other museum-related items for sale.
Activities: guided tours; lectures; gallery talks; art & craft demonstrations; inter-museum loan; permanent, temporary & traveling exhibitions; films.
Publications: park brochure; site bulletins.
Hours & Admission Prices: Daily 9-4:30. No charge. Closed New Year's Day; Thanksgiving; Christmas. &
Attendance: 490,861 (estimated)

Valley

THE MUSEUM AT THE CANNERY, 61st St., Valley, AL 36854. Mailing Address: P.O. Box 186, Valley, AL 36854. Tel.: 334-756-5228. Fax: 334-756-4922.
Institution Type/Description: History Museum.
Collections: textile mill history, instruments & memorabilia; photographs; newspapers; cotton samples.
Hours & Admission Prices: Call for hours.

Vance

MERCEDES-BENZ VISITOR CENTER AND MUSEUM, 1 Mercedes Dr., Vance, AL 35490-2900. Tel.: 888-286-8762 (Toll Free); 205-507-2252.
E-mail: webteam@mbusi.com
Web Site: mbusi.com/pages/vc_home.asp
Institution Type/Description: Transportation Museum.
Collections: history of Daimler-Benz; automobile technology.
Activities: tours.
Hours & Admission Prices: Museum & Center: late June to late May Mon.-Fri. 8:30-4:30. Factory Tours: Tues. & Thurs. 9am, 9:15am, 12:30 & 12:45 by appointment. Closed holidays.

Vinemont

CROOKED CREEK CIVIL WAR MUSEUM, 516 County Rd. 1127, Vinemont, AL 35179. Tel.: 256-739-2741.
Founded: 2006.
Key Personnel: Dir., Pres. (V) & Museum Shop Mgr., Fred Wise; Chm. (V), Mike Wise.
Personnel Profile: Full-Time Volunteers 1; Part-Time Volunteers 2.
Institution Type/Description: Military Museum: housed on the site of the Crooked Creek Civil War Battle.
Collections: Civil War history, weapons & artifacts; uniforms; photographs; personal artifacts.
Hours & Admission Prices: Daily 9-6. Adults $5.
Attendance: 2,000 (estimated)

ECHOTA CHEROKEE INTERPRETIVE CENTER, 630 County Rd. 222, Vinemont, AL 35179. Tel.: 256-734-7337.
Institution Type/Description: Native American History Center.
Collections: Native American history & culture; personal artifacts.
Facilities: 1.5 mile nature trail; outdoor classrooms.
Hours & Admission Prices: Mon.-Wed. 8-4, Thurs. 8-12. No charge.

Waterloo

EDITH NEWMAN CULVER MEMORIAL MUSEUM, 501 Main St., Waterloo, AL 35677. Mailing Address: P.O. Box 251, Waterloo, AL 35677-0251. Tel.: 256-767-6081.
Founded: 1995.
Personnel Profile: Part-Time Paid 1; Part-Time Volunteers 8.
Institution Type/Description: Historic House Museum: built in 1870.
Collections: local history; period furniture; personal artifacts; Civil War; Native American; military.
Activities: Annual Event: Waterloo Heritage Days in May.
Hours & Admission Prices: Summer: Fri.-Sat. 11-5, Sun. 1-4. &

Wetumpka

FORT TOULOUSE/JACKSON PARK, 2521 W. Fort Toulouse Rd., Wetumpka, AL 36093-1112. Tel.: 334-567-3002.
E-mail: ftjack1@bellsouth.net
Web Site: www.forttoulouse.com
Key Personnel: Park Dir., Jim Parker; Living History Program Coord. & Archaeologist, Ned Jenkins.
Governing Authority: Parent Institution: Alabama Historical Commission.
Institution Type/Description: Park & Historic Site.
Collections: area history; archaeological artifacts; Native & early American artifacts. Historic Buildings: Fort Jackson; Graves House.
Facilities: 165-acre park; nature trails; picnic area; campground. Museum-related items for sale.
Activities: living history weekends; special events. Park Sponsors: French and Indian Encampment in Spring; Frontier Days in Fall.
Hours & Admission Prices: Park: April-Oct. daily 6am-9pm; Nov.-March daily 8-5. Visitor Center: daily 8-5; groups by appointment. Adults & children over 6 $1, children 6 & under and senior citizens $.50; discounts to groups. Closed New Year's Day; Thanksgiving; Christmas.

ALASKA

(89 listings)

Anaktuvuk Pass

NORTII SLOPE BOROUGII PLANNING DEPT. - TIIE SIMON PANEAK MEMORIAL MUSEUM, 341 Mekiana Rd., Anaktuvuk Pass, AK 99721. Mailing Address: P.O. Box 21085, Anaktuvuk Pass, AK 99721-0085. Tel.: 907-661-3413. Fax: 907-661-3414.
E-mail: vera.woods@north-slope.org
Web Site: spmm.nsb-ihlc.com
Founded: 1986.
Congressional District: 1
Key Personnel: C.E.O., Charlotte E. Brower; Deputy Dir., Kathy Ahgeak; Dir. Planning Dept., Rhoda Ahmaogak; Cur. & Museum Shop Mgr., Vera Woods.
Personnel Profile: Full-Time Paid 1.
Governing Authority: municipal. Parent Institution: North Slope Borough Planning Dept. Tax-exempt.
Institution Type/Description: Local History and Ethnographic Museum.
Collections: tools; skin clothing & tent; caribou; hunting; fishing; trapping; trading; Nunamiut Eskimo history & traditions.
Research Fields: ethnography; archaeology.
Facilities: library of books related to Alaska, available for public use; educational facilities. Educational materials, local native crafts & gift items for sale.
Activities: guided tours; formal education programs for children, undergraduate & graduate students; loan exhibits; special events.
Publications: Notes on Nunamiut; Nunamiut Stories, We Hunt to Live; The Hungry Summer; Anaktuvuk Pass, Land of the Nunamiut territoriality among ancient hunters, In a Hungry Country; North Alaska Chronicles. Notes from the End of Time.
Hours & Admission Prices: Mon.-Fri. 8:30-5. Adults $10, children 7-17 $5. Closed New Year's Day; Memorial Day; Seward's Day; Independence Day; Thanksgiving; Christmas. &
Attendance: 800 (estimated)

Anchorage

ALASKA AVIATION HERITAGE MUSEUM, 4721 Aircraft Dr., Anchorage, AK 99502-1080. Tel.: 907-248-5325. Fax: 907-248-6391. Facebook: Alaska Aviation Heritage Museum.
E-mail: info@alaskaairmuseum.org
Web Site: www.alaskaairmuseum.org

Founded: 1988.
Key Personnel: Exec. Dir., Shari Hart; Chm., Bill Odom.
Personnel Profile: Full-Time Paid 2; Part-Time Paid 2; Part-Time Volunteers 50.
Governing Authority: private; nonprofit organization. Tax-exempt.
Institution Type/Description: Aeronautics Museum.
Collections: 30 aircraft depicting Alaska's role in aviation history.
Research Fields: Lend-Lease aircraft & the men, American & Russian, who flew them; Lockheed Vega for replica construction to recreate Eielson polar flight in 1928; all aspects of Alaskan aviation from 1898 through Alaska's Golden Age of Aviation & World War II; the search for Sigmond Levanevsky, 1937-1994; Russian Lindbergh: disappeared in Polar flight.
Facilities: 25-seat theater; 2 9-seat mini-theaters. Museum-related items for sale.
Activities: films; formal education programs for undergraduate & graduate students affiliated with work study program at University of Alaska-Anchorage; guided tours; hobby workshops; lectures; loan, temporary & traveling exhibitions; training programs for professional museum workers.
Hours & Admission Prices: May 15-Sept. 15 daily 9-5; Sept. 16-May 14 Wed.-Sun. 9-5. Adults $10, seniors 65 & over $8, children 5-12 $6; discounts AAA members; WWII veterans, children under 5 & members no charge. Closed New Year's Day; Thanksgiving; Christmas Eve & Day. &
Attendance: 20,000 (accurate)
Membership: Individual $45; Family $60; Polaris $100; Denali $250; Juneau $500; Barrow $1,000; Alaska League $2,500.

ALASKA MASONIC LIBRARY AND MUSEUM, 518 E. 14th Ave., Anchorage, AK 99501. Mailing Address: P.O. Box 190668, Anchorage, AK 99519.
Institution Type/Description: Library & History Museum.
Collections: books; Freemasonry artifacts; Alaska history & culture; photographs.
Facilities: library of Freemasonry books.
Hours & Admission Prices: Temporarily closed for relocation. &

ALASKA MUSEUM OF SCIENCE & NATURE, (M), 201 N. Bragaw, Anchorage, AK 99508-1311. Tel.: 907-274-2400.
E-mail: webcontact@alaskamuseum.org
Web Site: www.alaskamuseum.org
Formerly: Alaska Museum of Natural History
Founded: 1992.
Key Personnel: Pres. (V), Kristine Crossen; Vice Pres., Rusty Brown; Sec., Kerri Jackson; Treas., Barbara Bach; C.E.O. & Dir., Katch Bacheller; Collections Mgr., Sam Winer; Exhibits Mgr., Judy Lima.
Personnel Profile: Part-Time Paid 1; Part-Time Volunteers 25.
Governing Authority: bd. of dirs. Tax-exempt.
Institution Type/Description: Natural History Museum.
Collections: invertebrate & dinosaur fossils of the Matanuska Formation (Cretaceous period); plant fossils of the Chickaloon Formation (Paleocene period); minerals & rocks of Alaska; large (nonliving) mammals, birds, marine & nonmarine mammal skulls; Gold Rush mining history, tools; metallic minerals exhibits; Alaskan rocks & minerals; dinosaur & invertebrate fossils; plant fossils of the Paleocene & Eocene period; seven large ecosystem dioramas; Gold Rush mining history and tools; metallic minerals; & native artifacts from coastal & interior cultures.
Research Fields: Alaska.
Facilities: Museum-related items for sale.
Activities: lectures; formal education program; natural history field tours.
Publications: quarterly newsletter.
Hours & Admission Prices: Mon.-Sat. 10-5. Adults $5, children 6-12 $4; members & children under 5 no charge. &
Attendance: 18,000 (accurate)
Membership: Student & Senior $20; Individual $25, Senior Family $30; Family $50.

ALASKA NATIVE HERITAGE CENTER, 8800 Heritage Center Dr., Anchorage, AK 99504-6100. Tel.: 907-330-8000; 800-315-6608.
Web Site: www.alaskanative.net
Founded: 1999.
Congressional District: 1
Key Personnel: C.E.O., Annette Evans Smith.
Personnel Profile: Full-Time Paid 45; Part-Time Paid 4; Part-Time Volunteers 10; Interns 2.
Governing Authority: private; nonprofit organization. Tax-exempt: 501(c)(3).
Institution Type/Description: Cultural Center.
Collections: native heritage cultures; personal artifacts; photographs.
Activities: Native dance performances; workshops; tours; demonstrations; videos.

Hours & Admission Prices: Daily 9-5. Adults $24.95, seniors & military $21.15, children 7-16 $15.95; discounts to groups; members and children 6 & under no charge. &
Attendance: 100,000 (estimated)
Membership: Jade (Senior) $45; Jade (Individual) $50; Copper (Family) $100; Baleen $250; Ivory $500; Heritage Circle $1,000.

ALASKA ZOO, 4731 O'Malley Rd., Anchorage, AK 99507-6573. Tel.: 907-346-2133. Fax: 907-346-2673.
Web Site: www.alaskazoo.org
Founded: 1969.
Key Personnel: Dir., Patrick S. Lampi; Co Pres., Christopher Kleckmen; Co Pres., Margo McCabe.
Personnel Profile: Full-Time Paid 25; Part-Time Paid 12; Part-Time Volunteers 80; Interns 2.
Governing Authority: Tax-exempt.
Institution Type/Description: Zoo.
Collections: botanical; zoological.
Research Fields: polar bear with USGS; oil spill response with Alaska Clean Seas & US Fish & Wildlife Svc.
Facilities: party room rentals; lecture hall rentals.
Activities: lectures; music; long distance education via live audio/video conferencing.
Publications: quarterly newsletter, Animal Tracks.
Hours & Admission Prices: Nov.-Feb. 10-4; March-April & Oct. 10-5; May & Sept. 10-6; June-Aug. 9-9. Adults $12, seniors & military $9, students 3-17 $6; children under 3 & passholders no charge. Closed Thanksgiving; Christmas. &
Attendance: 170,000 (accurate)
Membership: Youth $6; Grandparent $15; Adult & Unnamed Guest $35.

* **ANCHORAGE MUSEUM, (M),** 625 C St., Anchorage, AK 99501. Tel.: 907-929-9200. Fax: 907-929-9290. Facebook: Anchorage Museum.
E-mail: museum@anchoragemuseum.org
Web Site: www.anchoragemuseum.org
Formerly: Anchorage Museum at Rasmuson Center
Founded: 1968.
Key Personnel: Interim Dir., Julie Decker; Dir. Collections Dept., Monica Shah; Dir. Education & Public Programs, Monica Garcia; Enterprise & Visitor Svcs., Adam Baldwin; Museum Shop Mgr., Mark Weber.
Personnel Profile: Full-Time Paid 80; Part-Time Paid 6; Part-Time Volunteers 400; Interns 1.
Governing Authority: municipality of Anchorage. Tax-exempt: 501(c)(3).
Institution Type/Description: Art, History & Science Museum.
Collections: Alaskan art & artifacts of all periods, archaeological, ethnological, historic & contemporary; archival collection of 400,000 photographs of Alaska; manuscripts; public art.
Major Exhibits: Remarks on the Land, 5/13-2/14; Dena'Inaq' Huch'ulyeshi, 9/13-1/14; Margo Klass, 2/14-4/14; Gyre, 2/14-9/14; Nutjuitok, 4/14-9/14; Artic Desert, 5/14-9/14.
Research Fields: anthropology, history, science & art of Alaska.
Facilities: 10,000-vol. library of Alaskan anthropology, history & art, general reference, available for inter-library loan & on premises; 240-seat auditorium; classroom. Eskimo Aleut, Tlingit, & Athabascan crafts, publications & prints for sale.
Activities: guided tours; lectures; gallery talks; docent program; formally organized art & history education programs for children & adults; regional competitions; community sponsored programs & events; inter-museum loan, permanent, temporary & traveling exhibitions.
Publications: bimonthly newsletter; exhibition catalogs; A Northern Adventure: The Art of Fred Machetanz; Eskimo Drawings; John Hoover: Art and Life; Sydney Laurence, Painter of the North; Spirit of the North: The Art of Eustace Paul Ziegler; Children's Gallery exhibition catalogs; Solo exhibition catalogs; Painting in the North: Alaskan Art in the AMHA; Heaven on Earth: Orthodox Treasures from Siberia and North America; Agayuliyararput (Our Way of Making Prayer): The Living Tradition of Yup'ik Masks; Smithsonian Arctic Studies Center: Living Our Cultures, Sharing Our Heritage.
Hours & Admission Prices: May-Sept. daily 9-6; Oct.-April Tues.-Sat. 10-6, Sun. 12-6. Adults $15, residents $12, senior citizens, students & military $10, children 3-12 $7; discounts to ASTC members; Alaska Museums Assoc., museum members & children under 3 no charge. Closed New Year's Day; Thanksgiving; Christmas. &
Attendance: 200,000 (estimated)
Membership: Individual $60; Family $90.

FRATERNAL ORDER OF ALASKA STATE TROOPERS MUSEUM, 245 W. 5th Ave., Anchorage, AK 99501-2358. Mailing Address: P.O. Box 100280, Anchorage, AK 99510-0280. Tel.: 800-770-5050; 907-279-5050. Fax: 907-279-5054. Facebook: FOAST Law Enforcement Museum.
Web Site: www.alaskatroopermuseum.com
Founded: 1990.
Key Personnel: Dir., Laura L. Caperton.
Personnel Profile: Full-Time Paid 1; Part-Time Paid 1; Part-Time Volunteers 3.
Volunteer Hours: 380
Governing Authority: Parent Institution: Fraternal Order of Alaska State Troopers (F.O.A.S.T.). Tax-exempt.
Institution Type/Description: History Museum.
Collections: history of the Alaska State Troopers & law enforcement.
Facilities: Museum-related items for sale.
Activities: youth groups; safety programs; professional seminars; safety presentations.
Publications: quarterly newsletter, Banner.
Hours & Admission Prices: Mon.-Fri. 10-4, Sat. 12-4. No charge; donations accepted. &

Attendance: 4,000 (accurate)
Membership: Friends $30; Lifetime $250; Corporate $500.

HERITAGE LIBRARY MUSEUM, 301 W. Northern Lights Blvd., Ste. 103, Anchorage, AK 99503-2652. Mailing Address: Wells Fargo Historical Services, 420 Montgomery St., MAC-A0101-106, San Francisco, CA 94163. Tel.: 907-265-2834. Fax: 907-265-2860. TDD: 907-267-5678.
Founded: 1968.
Congressional District: 1
Key Personnel: Dir., Beverly Smith; Cur., Tom D. Bennett.
Personnel Profile: Full-Time Paid 1.
Governing Authority: Affiliated with Wells Fargo Bank Alaska.
Institution Type/Description: General Museum.
Collections: Alaskan native ethnology & contemporary crafts; fine arts by Alaskan artists; library of works on Alaskan subjects.
Facilities: 2,500-vol. reference library on Alaskan & Arctic topics available to the public; 3,300 sq. ft. exhibit space.
Activities: guided tours; loan exhibitions; formally organized education programs for children.
Publications: Heritage of Alaska.
Hours & Admission Prices: Mon.-Fri. 12-4. No charge. Closed bank holidays. &

INTERNATIONAL GALLERY OF CONTEMPORARY ART, 427 D St., Anchorage, AK 99501-2325. Tel.: 907-279-1116.
Institution Type/Description: Art Gallery.
Collections: works by contemporary artists.
Activities: temporary exhibitions.
Hours & Admission Prices: Tues.-Sun. 12-4.

MERIKS GALLERY, 9101 Little Creel Dr., Anchorage, AK 99507-3920.
Key Personnel: Pres. & Owner, Paul Davis; Vice Pres., Maria Stein.
Governing Authority: private. nonprofit. Tax-exempt.
Institution Type/Description: Gallery.
Collections: paintings, drawings & prints from local artists.
Facilities: 500-vol. library.
Hours & Admission Prices: Daily 9-5. No charge; donations accepted.
Attendance: 5,000 (estimated)

THE OSCAR ANDERSON HOUSE MUSEUM, 420 M St., (in Elderberry Park), Anchorage, AK 99501-1929. Tel.: 907-274-2336. Fax: 907-274-3600.
Founded: 1982.
Key Personnel: Mgr., Mary A. Flaherty.
Personnel Profile: Part-Time Paid 2; Part-Time Volunteers 10.
Governing Authority: private; nonprofit organization. Parent Institution: Anchorage Historic Properties, Inc., 645 W. Third Ave., Anchorage 99501. Tax-exempt: 501(c)(3).
Institution Type/Description: Historic House: housed in a bungalow style, one and a half story home, built in 1915.
Collections: focuses on the interpretive period from 1915-1925 depicting the time of the founding of the city of Anchorage; period furnishings; Anderson family artifacts; photos associated with early Anchorage history.
Research Fields: genealogy of the Anderson family; life in Anchorage from 1915-1925.

Facilities: 30-vol. library on local history, museum mgmt. & exhibits; 1,500 sq. ft. exhibit space.
Activities: docent program; formal education programs for children; guided tours; temporary exhibitions. Annual Events: Early Anchorage Summer Celebration including Tent City exhibit, Maypole & Swedish summer festivities; Swedish Christmas Tours in December.
Hours & Admission Prices: By appointment. Adults $3, senior citizens over 65 $2, children 5-12 $1; discounts to AAM & ICOM members and groups.
Attendance: 3,559 (accurate)
Membership: Senior & Student $15; Individual $20; Family $30; Business $50; Benefactor $100.

Anvik

ANVIK HISTORICAL SOCIETY AND MUSEUM, Main Rd., Anvik, AK 99558. Mailing Address: P.O. Box 110, Anvik, AK 99558-0110. Tel.: 907-663-6360.
Institution Type/Description: Historical Society Museum.
Collections: Native Athabascan culture & history; photographs; period artifacts; early 20th century trade goods.
Hours & Admission Prices: Summer: by appointment.

Barrow

INUPIAT HERITAGE CENTER, 5421 North Star St., Barrow, AK 99723. Mailing Address: P.O. Box 69, Barrow, AK 99723-0069. Tel.: 907-852-0422. Fax: 907-852-4224.
Web Site: www.nps.gov.inup
Founded: 1999.
Key Personnel: Dir., Patrick Glenn; Museum Shop Mgr., Debbie Suvlu.
Personnel Profile: Full-Time Paid 9; Interns 2.
Governing Authority: Parent Institution: North Slope Borough. Advisory: Inupiat History, Language, and Culture Commission.
Institution Type/Description: Heritage Center.
Collections: Inupiat heritage & history; photographs; personal artifacts; video archives.
Research Fields: Inupiat history, language & culture.
Facilities: library.
Activities: educational outreach; performances; activities.
Hours & Admission Prices: mid-May to mid-Sept. Mon.-Fri. 8:30-5. Adults $10, youth 7-17 $5, seniors 60 & over no charge. &
Attendance: 16,260 (accurate)

Bethel

YUPIIT PICIRYARAIT MUSEUM, (M), 420 Chief Eddie Hoffman Hwy., Bethel, AK 99559. Mailing Address: P.O. Box 219, Bethel, AK 99559-0219. Tel.: 907-543-1819. Fax: 907-543-1885.
Web Site: www.ypmuseum.org
Founded: 1994.
Governing Authority: Parent Institution: Association of Village Council Presidents.
Institution Type/Description: Regional Tribal Museum.
Collections: clothing; household artifacts; hunting & gathering implements; personal artifacts; Yup'ik, Cup'ik & Dene collections from the Native people who settled the Yukon-Kuskokwim Delta of Alaska.
Facilities: Museum-related items for sale.
Hours & Admission Prices: Tues.-Sat. 12-4. No charge; donations accepted. &

Central

THE CENTRAL MUSEUM, 128 Mile Steese Hwy., Central, AK 99730. Mailing Address: P.O. Box 30189, Central, AK 99730-0189. Tel.: 907-520-1893. Fax: 907-520-1893.
Web Site: www.cdhs.us
Founded: 1977.
Congressional District: 4
Key Personnel: Pres. (V), Archivist & Public Rels., Al Cook; Treas., Becky Hendrickson; Cur., Julie Cooper; Sec. & Museum Shop Mgr., Darae Murphy.
Personnel Profile: Part-Time Paid 1; Part-Time Volunteers 10.
Governing Authority: private; nonprofit organization.
Institution Type/Description: Historical Society & Mining Museum.
Collections: discovery of gold in the central area & living conditions, c.1893; fur trapping.
Facilities: library of old hard-bound novels from early 1900s, a part of Alaskana Collection, available for on-premises use only. Books, Alaskana wild plants, berries, mushrooms, cookbooks, local crafts & gold mining-related items for sale.

Activities: guided tours; study clubs. Annual Event: Christmas Bazaar.
Publications: newsletter.
Hours & Admission Prices: Memorial Day-Labor Day Fri.-Sun. 12-5; tours by appointment. Adults & students $1, children $.50; senior citizens 70 & over and members no charge. &
Attendance: 400 (estimated)
Membership: Child $1; Adult $10; Family $15; Business $25.

CIRCLE DISTRICT MUSEUM, 1275 Mile Steese Hwy., Central, AK 99730. Mailing Address: P.O. Box 1893, Central, AK 99730. Tel.: 907-520-1893.
Institution Type/Description: History Museum.
Collections: mining; local heritage & culture; personal artifacts; photographs.
Hours & Admission Prices: Memorial Day to Labor Day daily 12-5; Winter: by appointment.

Copper Center

GEORGE I. ASHBY MEMORIAL MUSEUM, Mile 101 Old Richardson Hwy., Copper Center Loop Rd., Copper Center, AK 99573. Mailing Address: P.O. Box 84, Copper Center, AK 99573-0084. Tel.: 907-822-5285.
E-mail: cvhsmus@gmail.com
Founded: 1985.
Key Personnel: Chm. (V), Geoff Bleakley; Museum Mgr., Rebecca Nelson; Treas. & Museum Shop Mgr., Barbara Sanders.
Personnel Profile: Part-Time Paid 1; Part-Time Volunteers 20.
Governing Authority: Parent Institution: Copper Valley Historical Society. Tax-exempt.
Institution Type/Description: History Museum.
Collections: concentration on Alaskan history with emphasis on Copper Valley.
Hours & Admission Prices: May 15 to Sept. 15 daily 11-5. No charge; donations accepted.
Attendance: 7,735 (accurate)
Membership: Copper Valley Historical Society $15.

Cordova

CORDOVA HISTORICAL MUSEUM, (M), 622 1st St., Cordova, AK 99574. Mailing Address: Box 391, Cordova, AK 99574-0391. Tel.: 907-424-6665. Fax: 907-424-6666.
E-mail: infoservices@cityofcordova.net
Web Site: www.cordovamuseum.org
Founded: 1966.
Key Personnel: Pres. (V), Mike Webber; Vice Pres., Ira Grindle; Dir., Cathy R. Sherman; Treas., Mimi Briggs; Collections Mgr. & Cur., Judy Fulton; Museum Shop Mgr., Sharon Ermold.
Personnel Profile: Full-Time Paid 1; Part-Time Paid 3; Part-Time Volunteers 10.
Governing Authority: municipal; society. Parent Institutions: Cordova Historical Society & City of Cordova. Tax-exempt.
Institution Type/Description: General Museum.
Collections: fishing artifacts; Copper River & Northwestern Railway artifacts; lighthouse lens; Native American (Eyak, Chugach) artifacts; photo archives.
Research Fields: oral history of pioneer Alaskans.
Facilities: library & microfilm of Cordova newspapers, 1906-2008. Books for sale.
Activities: films of old Cordova; gallery talks.
Publications: quarterly newsletter; books, Walking Tour of Cordova; From Fish & Copper: Cordova's Heritage & Buildings; Cordova to Kennecott.
Hours & Admission Prices: Museum: Memorial Day to Labor Day Mon.-Sat. 10-6, Sun. 2-4; Winter: Tues.-Fri. 10-5, Sat. 1-5. Archives: by appointment. No charge; donations accepted. Closed New Year's Day; Independence Day; Christmas. &
Attendance: 12,000 (accurate)
Membership: Student $5; Individual $10; Family $25; Life $200.

ILANKA CULTURAL CENTER, 110 Nicholoff Way, Cordova, AK 99574. Mailing Address: P.O. Box 322, 110 Nicholoff Way, Cordova, AK 99574-1388. Tel.: 907-424-7903. Fax: 907-424-3018.
E-mail: larue@eyak-nsn.gov
Web Site: www.nveyak.com
Founded: 2004.
Congressional District: 5
Key Personnel: Dir., LaRue Barnes.
Personnel Profile: Part-Time Paid 1.
Governing Authority: Parent Institution: Native Village of Eyak. Tax-exempt.
Institution Type/Description: Cultural Center.
Collections: local native culture, history & art; personal artifacts.

Research Fields: tribal artifacts; oral history.
Facilities: Native art and museum-related items for sale.
Activities: classes.
Hours & Admission Prices: Mon.-Fri. 10-5. No charge; donations accepted. Closed holidays. &
Attendance: 2,500 (estimated)
Membership: Individual $20; Family $50; Supporter $100; Patron $500; Benefactor $1,000. Corporate: $100-$499; $500-$999; $1,000-$4,999; $5,000-$9,999; $10,000 & up.

Delta Junction

SULLIVAN ROADHOUSE HISTORICAL MUSEUM, Mile 267 Richardson Hwy., Delta Junction, AK 99737. Mailing Address: P.O. Box 987, Delta Junction, AK 99737-0987. Tel.: 907-895-5068; 907-895-4415 (seasonal). Fax: 907-895-5141.
E-mail: deltacc@deltachamber.org
Web Site: www.deltachamber.org
Key Personnel: Exec. Dir., Janet Hawi; Cur. & Project Archaeologist, Jeffrey Durham
Institution Type/Description: Historic House Museum: built in 1905 by John and Florence Sullivan.
Collections: Valdez-Fairbanks Trail history; Alaska history.
Facilities: nature trails.
Activities: nature walks.
Hours & Admission Prices: May-Sept. daily 9-5. No charge; donations accepted.
Attendance: 24,000 (estimated)

Denali Park

DENALI NATIONAL PARK AND PRESERVE, Mile Post 237, George Parks Hwy., Denali Park, AK 99755. Mailing Address: P.O. Box 9, Denali Park, AK 99755-0009. Tel.: 907-683-2294. Fax: 907-683-9617. TDD: 907-683-9649.
E-mail: jane_lakeman@nps.gov
Web Site: www.nps.gov/dena
Founded: 1917.
Congressional District: 1
Key Personnel: Park Supt., Paul R. Anderson; Administrative Officer, Julie Wilkerson; Chief Ranger, Peter Armington; Museum Cur., Jane Lakeman.
Personnel Profile: Full-Time Paid 1.
Governing Authority: federal. National Park Service, Alaska Regional Office, 240 W. 5th Ave., Anchorage, AK 99503. Tel. 907-257-2543.
Institution Type/Description: Park Study Collection.
Collections: botany; zoology; geology; history; archaeology; ethnology.
Research Fields: botany; zoology; geology; archaeology.
Facilities: 2,500-vol. library of general reference books available for use on premises; reading room; outdoor museum. Publications for sale.
Activities: traditional park activities.
Publications: Mammals of Mt. McKinley; Sled Dogs of Denali; Denali Bird Finding Guide; A Back Country Companion to Denali National Park; The Denali Road Guide.
Hours & Admission Prices: Visitor Center: late May to early Sept. daily 8-6. &

Dillingham

SAMUEL K. FOX MUSEUM, 306 D St. West, Dillingham, AK 99576. Mailing Address: P.O. Box 273, Dillingham, AK 99576-0273. Tel.: 907-842-4831.
E-mail: samfoxmuseum@nushtel.net
Web Site: www.nushtel.com/~dlgchmbr
Founded: 1974.
Congressional District: 16
Key Personnel: Dir., Chm. (V) & Pres. (V), Deb Burton.
Personnel Profile: Part-Time Volunteers 9.
Governing Authority: City of Dillingham, AK. Tax-exempt: 501(c)(3).
Institution Type/Description: Alaskan Native & Indian Museum.
Collections: Southwestern Yup'ik Eskimo arts & crafts; basket weaving; skinsewing; wood, ivory & bone carving; Alaskan culture memorabilia; Aleut, Southwestern Yup'ik, Siberian Yup'ik & Inupiat Eskimo artifacts; historical photographs & prints.
Facilities: library.
Activities: lectures; formally organized education programs for children; loan, permanent, temporary & traveling exhibitions; travel slide show; community workshop.
Hours & Admission Prices: Mon.-Fri. 8-6, Sat. 10-2. No charge; donations accepted. Closed national holidays. &
Attendance: 7,500 (estimated)

Membership: Students $3; Adults $15.

Eagle

EAGLE HISTORICAL SOCIETY & MUSEUMS, 3rd & Chamberlain, Eagle, AK 99738. Mailing Address: P.O. Box 23, Eagle, AK 99738-0023. Tel.: 907-547-2325. Fax: 907-547-2325.
E-mail: ehsmdirector@gmail.com
Web Site: www.eaglehistoricalsociety.com
Founded: 1961.
Congressional District: 36
Key Personnel: C.E.O. & Pres. (V), Pat Sanders; Dir. & Cur. Education, Donna Westphal; Museum Shop Mgr., Betty Borg; Bookkeeper, Linda Nelson.
Personnel Profile: Full-Time Paid 1; Full-Time Volunteers 1; Part-Time Paid 7; Part-Time Volunteers 14.
Governing Authority: bd. of dirs.; nonprofit organization. Parent Institution: Eagle Historical Society. Tax-exempt: 170(b)(1)(A).
Institution Type/Description: Local History Museum: housed in 1900 courthouse; 1898 U.S. Army Mule Barn; 1900-1911 U.S. Army Post Fort Egbert all located in the first city incorporated in interior Alaska.
Collections: archives; photos; maps; court room; machinery; vehicles; tools; documents; living quarters; furniture; 1900-1910 restored historic buildings.
Research Fields: local history from 1890s to present.
Activities: guided tours.
Publications: Jewel On the Yukon: Eagle City; Life in the Northern Army; triannual newsletter.
Hours & Admission Prices: Tour: Memorial Day to Labor Day daily 9am; other times by appointment. Adults $7; members & children under 12 no charge.
Attendance: 6,800 (estimated)
Membership: Single $15; Family $20; Business $25; Sponsoring $100; Life $300.

Eagle River

EAGLE RIVER NATURE CENTER, 32750 Eagle River Rd., Eagle River, AK 99577. Tel.: 907-694-2108. Fax: 907-694-2119.
E-mail: info@ernc.org
Institution Type/Description: Nature Center.
Collections: local natural history & culture; wildlife & their habitats; plants; trees; flowers; ecology; geology.
Facilities: nature trails. Museum-related items for sale.
Activities: hiking; educational programs.
Hours & Admission Prices: May & Sept. Tues.-Sun. 10-5; June-Aug. daily 10-7; Oct.-April Fri.-Sun. 10-5.

Fairbanks

ALASKA HOUSE ART GALLERY, 1003 Cushman St., Fairbanks, AK 99701-4618. Mailing Address: P.O. Box 70501, Fairbanks, AK 99707-0501. Tel.: 907-456-6449.
E-mail: info@thealaskahouse.com
Key Personnel: Owner, Yolande Fejes; Owner, Ron Veliz
Institution Type/Description: Art Gallery: built in 1939.
Collections: works of Claire Fejes & other Alaskan artists; paintings; drawings; prints; sculpture; carvings; masks; fabric art.
Activities: poetry readings; special events.
Hours & Admission Prices: Summer: call for hours.

FAIRBANKS COMMUNITY/DOG MUSHING MUSEUM, 410 Cushman St., Fairbanks, AK 99701. Tel.: 907-456-6874.
Formerly: Dog Mushing Museum
Founded: 1987.
Key Personnel: Exec. Dir., Julie Fougeron.
Personnel Profile: Full-Time Paid 2; Part-Time Paid 3.
Governing Authority: Tax-exempt.
Institution Type/Description: History Museum.
Collections: vintage dog sleds; state of the art cold weather gear; Siberian skin clothing; sled dog memorabilia; photo-essay exhibit.
Research Fields: sled dog husbandry; literature research of books, periodicals & newsletters relating to dog mushing.
Facilities: library of Mushing Magazine (complete set), mushing books & miscellaneous periodicals; 1,500 sq. ft. exhibit space; 30-seat theater. Museum-related items for sale.
Activities: theater. Annual Event: autograph signing by Iditarod winners.
Publications: quarterly newsletter.
Hours & Admission Prices: May-Aug. Mon.-Sat. 10-6, Sun. 12-4; Sept.-May Tues.-Sat. 10-6. No charge; donations accepted. Closed all major holidays. &

Attendance: 8,000 (estimated)
Membership: Individual $20; Family $50; Business $100; Corporate $500.

FAIRBANKS COMMUNITY MUSEUM, 410 Cushman St., Fairbanks, AK 99701-4632. Tel.: 907-457-3669.
E-mail: info@fairbankshistorymuseum.com
Web Site: www.fairbankshistorymuseum.com
Key Personnel: Exec. Dir., Bob Eley.
Personnel Profile: Part-Time Volunteers 8.
Governing Authority: Tax-exempt.
Institution Type/Description: History Museum.
Collections: area history; photographs; mining; dog sleds; Native Alaskans; Fairbanks flood of 1967; Klondike Gold Rush.
Activities: presentations; educational programs.
Hours & Admission Prices: Tues.-Sat. 11-3 & by appointment. No charge; donations accepted. Closed major holidays. &

FAIRBANKS ICE MUSEUM, 500 2nd Ave., Fairbanks, AK 99701-4729. Tel.: 907-451-8222. Fax: 907-456-7674.
E-mail: ice@gci.net
Web Site: www.icemuseum.com
Founded: 1995.
Personnel Profile: Full-Time Paid 3.
Operating Expenses: 86,720
Operating Income: 242,000
Institution Type/Description: Historic Building: housed in the Lacey Street Theater; built in 1936.
Collections: ice sculptures; videos.
Activities: ice carvings demonstration.
Hours & Admission Prices: May-Sept. daily 10-8. Shows: 10, 2 & 8. Adult $12, military & seniors $11, children 6-12 $6, children 5 & under $2. &
Attendance: 22,526 (estimated)

PIONEER AIR MUSEUM, 2300 Airport Way, Fairbanks, AK 99701. Mailing Address: Box 70437, Fairbanks, AK 99707-0437. Tel.: 907-451-0037.
E-mail: curator@pioneerairmuseum.org
Web Site: www.pioneerairmuseum.org
Formerly: Alaskaland Pioneer Air Museum
Founded: 1983.
Congressional District: 4
Key Personnel: Dir. & Museum Shop Mgr., Peter Haggland; Chm. (V), Syd Stealy.
Personnel Profile: Full-Time Paid 1; Part-Time Paid 4; Part-Time Volunteers 4.
Governing Authority: Parent Institution: Interior & Arctic Alaska Aeronautical Foundation. Tax-exempt: 501(c)(3).
Institution Type/Description: Aeronautics Museum.
Collections: aviation artifacts pertaining to the history of interior & Arctic Alaska; aircraft; aircraft remains; engines; pictures; radios; memorabilia of bush pilots & airlines; WWII lend lease.
Publications: quarterly newsletter, IAAAF Newsletter.
Hours & Admission Prices: May 15 to Sept. 15 daily 12-8. Families of four $7, adults $3; discount to members & groups. &
Attendance: 13,000 (accurate)
Membership: Regular $20; Associate $50; Life $1,000.

PIONEER MUSEUM & THE BIG STAMPEDE, Pioneer Park, Bldg. #1, 2300 Airport Way, Fairbanks, AK 99701. Mailing Address: P.O. Box 70176, Fairbanks, AK 99707-0176.
Founded: 1964.
Key Personnel: Pres. (V) & Museum Shop Mgr., Joanne Oehring.
Personnel Profile: Part-Time Paid 9; Part-Time Volunteers 45.
Governing Authority: Parent Institution: Pioneers of Alaska Igloos #4 & #8. Tax-exempt.
Institution Type/Description: History Museum.
Collections: photographs; prospecting & mining tools and equipment; telegraphic artifacts; dog sleds; snowshoes; horse drawn sleds & stages; early aviation artifacts; period Edwardian household furnishings; handmade tools & appliances; medical, engineering, legal & other professional materials; M.C. Rusty Heurlin's paintings of Klondike & Fairbanks Gold Rushes of 1898 & 1903.
Hours & Admission Prices: March-Memorial Day Fri.-Sun. 1-5; late May to Labor Day daily 11-8; Sept. daily 1-5; Nov.-Feb. call for tours. Museum: no charge; donations accepted. Big Stampede Show: Memorial Day to Labor Day. Adults $4, youth 6-12 $2; members & children under 6 no charge. &
Attendance: 19,000 (accurate)

❋ UNIVERSITY OF ALASKA MUSEUM OF THE NORTH, (M), 907 Yukon Dr., Fairbanks, AK 99775-6960. Mailing Address: P.O. Box 756960, Fairbanks, AK 99775-6960. Tel.: 907-474-7505. Fax: 907-474-5469.

E-mail: museum@uaf.edu
Web Site: www.uaf.edu/museum
Founded: 1929.
Congressional District: 1
Key Personnel: Dir., Aldona Jonaitis; Chm. (V), Mike Cook; Exec. Asst., Emilie Nelson; Mgr. Operations, Kevin May; Cur. Archeology, Josh Reuther, Ph.D.; Cur. Mammalogy, Link E. Olson; Cur. Ornithology, Kevin S. Winker; Cur. Herbarium, Stefanie Ickert-Bond; Cur. Earth Sciences, Patrick Druckenmiller; Cur. Ichthyology, J. Andres Lopez, Ph.D.; Cur. Entomology, Derek Sykes; Dir. Alaska Center for Documentary Film, Leonard J. Kamerling; Cur. Ethnology & History, Molly Lee; Coord. Exhibits & Exhibits Designer, Steve Bouta; Cur. Fine Arts, Mareca Guthrie, M.F.A.; Mgr. Education & Public Programs, Jennifer Arseneau; Mktg. & Communications Mgr., Theresa Bakker; Mgr. Visitor Svcs. & Retail Operations, Daniel David.
Personnel Profile: Full-Time Paid 32; Part-Time Paid 53; Part-Time Volunteers 40.
Governing Authority: university. Parent Institution: University of Alaska. Subsidiary Institution: University of Alaska Fairbanks. Tax-exempt: 170(b)(2)(A).
Institution Type/Description: Natural History & Art Museum.
Collections: aquatics; archaeology; birds; botany; earth sciences; ethnology; history; film project; fine arts; herbarium; mammals; fish; insects.
Research Fields: plant ecology & taxonomy; marine invertebrate identification & distribution; Alaskan pre-Cambrian biotas; archeology; ethnohistory; tephrochronology; migratory birds & avian influenza, polar dinosaur population diversity & distribution; mammal diversity & distribution.
Facilities: Books, booklets, postcards & museum-related items for sale.
Activities: guided tours for schools; permanent, temporary & traveling exhibitions.
Publications: annual report; newsletter; exhibit catalogs; informational brochures; research monographs; visitor's guide.
Hours & Admission Prices: May 15 to Sept. 15 daily 9-9; Sept. 16 to May 14 Mon.-Fri. 9-5, Sat.-Sun. 12-5. Adults $10, seniors $9, youth 7-17 $5; children 6 & under, Univ. Alaska students w/ID & members no charge. Closed New Year's Day; Thanksgiving; Christmas. ♿
Attendance: 96,016 (accurate)
Membership: Individual $50; Couple $65; Family $75; Advocate $125; Donor $250; Patron $500.

WICKERSHAM HOUSE MUSEUM (TANANA-YUKON HISTORICAL SOCIETY), 535 2nd Ave., Ste. 201, Fairbanks, AK 99701-4770. Mailing Address: P.O. Box 71336, Fairbanks, AK 99707-1336. Tel.: 907-457-6165. Fax: 907-457-6165.

E-mail: tyhs@polarnet.com
Web Site: www.tyhsonline.org
Founded: 1959.
Congressional District: 1
Key Personnel: Pres., Ronald Inouye; Museum Mgr., Lisa Harbo; Treas., Elizabeth Cook.
Personnel Profile: Full-Time Volunteers 1; Part-Time Paid 5; Part-Time Volunteers 12.
Governing Authority: public; nonprofit. Parent Institution: Tanana-Yukon Historical Society. Tax-exempt: 501(c)(3).
Institution Type/Description: Historical Society Museum: housed in the Wickersham House Museum, home of Judge James Wickersham & the first frame home completed in the rough gold camp which became Fairbanks, AK.
Collections: 1904-1916 Alaska furnishings; photographs; history of interior Alaska following the purchase of Alaska from Russia 1867.
Facilities: Books relating to Alaska history for sale.
Activities: docent program; guided tours; temporary exhibitions. Museum Sponsors: lecture series February-April & September-November; Community Museum Day. Preservation Advocacy-Save America's Treasures Grant; Wickersham Day Celebrations.
Publications: cookbook, First Catch a Moose; monthly newsletter; Alaska: Saving Our Places; Fairbanks' First Avenue; The History & Preservation of A Streetscape; membership brochure; James Wickersham Brochure; Cold Missions -The Army Air Forces and Ladd Field in WWII.
Hours & Admission Prices: Memorial Day-Labor Day daily 12-8; other times by appointment. No charge; donations accepted. ♿
Attendance: 11,000 (estimated)
Membership: Individual $20; Family $40.

Girdwood

ALASKA WILDLIFE CONSERVATION CENTER, 79 Seward Hwy., Girdwood, AK 99587. Mailing Address: P.O. Box 949, Girdwood, AK 99587-0949. Tel.: 907-783-2025. Fax: 907-783-2370.
Governing Authority: nonprofit organization. Tax-exempt: 501(c)(3).
Institution Type/Description: Conservation Center.
Collections: bears; moose; bison; elk; deer; caribou; coyotes; foxes; bald eagles; owls; porcupines.
Hours & Admission Prices: Jan. 2-Feb. Sat.-Sun. 10-5; March to mid-May daily 10-6; mid-May to mid-Sept. daily 8-8; mid-Sept. to Jan. 1 daily 10-5. Adults $10, seniors 65 & over and children 4-12 $7.50; children under 4 no charge.

Haines

ALASKA INDIAN ARTS, INC., Historic Bldg. #13, Fort Seward, Haines, AK 99827. Mailing Address: P.O. Box 271, Haines, AK 99827-0271. Tel.: 907-766-2160. Fax: 907-766-2160.
E-mail: mail@alaskaindianarts.com
Web Site: www.alaskaindianarts.com
Founded: 1957.
Key Personnel: Exec. Dir., Lee D. Heinmiller.
Governing Authority: nonprofit organization. Tax-exempt: 501(c)(3).
Institution Type/Description: Indian Living Village Museum.
Collections: Tlinget Indian costumes; Indian art; ethnology. Historic Building: 1904 Fort William H. Seward.
Research Fields: Indian culture; Fort Seward history.
Facilities: 50-vol. library of publications on Northwest Coast Indians available for use on premises; Totem village with tribal house & totem poles. Traditional Indian art items for sale.
Activities: guided tours; dance recitals; permanent exhibitions; slides; educational kits; walking tour.
Hours & Admission Prices: Mon.-Fri. 9-5. No charge; donations accepted. Closed national holidays. ♿

AMERICAN BALD EAGLE FOUNDATION, Corner of 2nd Ave. & Haines Hwy., Haines, AK 99827. Mailing Address: P.O. Box 49, Haines, AK 99827-0049. Tel.: 907-766-3094. Fax: 907-766-3095.
E-mail: info@baldeagles.org
Web Site: www.baldeagles.org
Founded: 1982.
Key Personnel: Dir. & Museum Shop Mgr., Cheryl McRoberts.
Personnel Profile: Full-Time Paid 2; Full-Time Volunteers 1; Part-Time Volunteers 4; Interns 5.
Governing Authority: nonprofit organization. Tax-exempt.
Institution Type/Description: Natural History Museum.
Collections: over 200 wildlife specimens; 12 birds.
Research Fields: American Bald Eagle.
Facilities: Museum-related items for sale.
Activities: talks; video presentation; live bird presentations, eagles, owls, falcons & red tail hawks. Annual Event: Alaska Bald Eagle Festival in November.
Publications: quarterly newsletter.
Hours & Admission Prices: Summer: Mon.-Fri. 9-5, Sat.-Sun. 1-4; Oct.-March by appointment. Adults $10; discounts to active military & senior citizens; members no charge. ♿
Attendance: 10,000 (estimated)
Membership: Individual $35; Family $50; Silver $125; Gold $250; Lifetime $1,000.

HAMMER MUSEUM, INC., (M), 108 Main St., Haines, AK 99827. Mailing Address: P.O. Box 702, Haines, AK 99827-0702. Tel.: 907-766-2374.
E-mail: hammermuseum@aptalaska.net
Web Site: www.hammermuseum.org
Founded: 2001.
Key Personnel: Dir., David Pahl.
Personnel Profile: Full-Time Volunteers 1; Part-Time Volunteers 2; Interns 2.
Governing Authority: Tax-exempt: 501(c)(3).
Institution Type/Description: History Museum.
Collections: 2,000 types of hammers representing many different trades & uses from ancient times to present.
Publications: annual newsletter.
Hours & Admission Prices: May-Sept. Mon.-Fri. 10-5. Admission $3; children 12 & under, youth groups & local residents no charge.
Attendance: 5,000 (estimated)

* **SHELDON MUSEUM AND CULTURAL CENTER, INC.,** **(M),** 11 Main St., Haines, AK 99827. Mailing Address: P.O. Box 269, Haines, AK 99827-0269. Tel.: 907-766-2366. Fax: 907-766-2368.
E-mail: museumdirector@aptalaska.net
Web Site: sheldonmuseum.org
Founded: 1974.
Congressional District: 1
Key Personnel: Pres. Bd. Trustees (V), Frankie Perry; Dir. & Cur., Jerrie Clarke; Coord. Education, Susannah Dowds; Collections & Exhibitions, Karen Meizner; Museum Shop Mgr., Blythe Carter.
Personnel Profile: Full-Time Paid 1; Part-Time Paid 6; Part-Time Volunteers 200.
Governing Authority: bd. of trustees. Subsidiary Institution: Haines Borough. Tax-exempt: 501(c)(3).
Institution Type/Description: History & ethnographic Tlingit Indian Culture museum.
Collections: local pioneer-transportation; mining; local industries; Presbyterian mission; Tlingit & other northwest coast Indian artifacts; Eskimo & Athabascan artifacts; Eldred Rock lighthouse lens; oral histories.
Research Fields: local & Alaskan history.
Facilities: 1,500-vol. library; archives.
Activities: education programs for school classes; tours for prearranged groups; Tlingits language classes for students & adults; adult educational programs.
Publications: Journey to the Tlingits; Haines-The First Century; Song of the Chilkat People; A Personal Look at the Sheldon Museum & Cultural Center; More Than Gold: Nuggets of Haines History.
Hours & Admission Prices: mid-May to mid-Sept. Mon.-Fri. 10-5, Sat.-Sun. 1-4; call for additional hours. Adults $5; discounts for AAM, Museums Alaska & Chilkat Valley Historical Society members; children under 12 no charge. &
Attendance: 13,720 (accurate)
Membership: Sheldon Museum & Cultural Center: Individual $25; Family $50; Sustaining $100; Contributing $250; Benefactor $500; Corporate Sponsor $1,000. Chilkat Valley Historical Society: Individual $12; Family $22; Patron $50; Life $250.

Homer

ALASKA ISLANDS AND OCEAN VISITOR CENTER, 95 Sterling Hwy., Homer, AK 99603-7472. Tel.: 907-235-6961.
Key Personnel: Mgr. Visitor's Center, Marianne Aplin
Institution Type/Description: Marine Wildlife Museum.
Collections: marine wildlife; local history.
Facilities: auditorium; nature trails. Books for sale.
Activities: educational programs.
Hours & Admission Prices: April 8-May 7 Mon.-Sat. 10-5; May 8-May 11 Thurs.-Sat. 8-5, Sun. 8-4; May 12-May 23 daily 10-5; Memorial Day to Labor Day daily 9-5; Sept. 2-Sept. 21 Tues.-Sun. 10-5; Sept. 22-April 7 Tues.-Sat. 12-5. No charge.

BUNNELL STREET ARTS CENTER, 106 W. Bunnell St., Homer, AK 99603-7825. Tel.: 907-235-2662. Facebook: Bunnell Street Arts Center.
E-mail: info@bunnellarts.org
Web Site: www.bunnellarts.org
Founded: 1991.
Key Personnel: Dir., Asia Freeman.
Personnel Profile: Full-Time Paid 1; Part-Time Paid 2.
Governing Authority: nonprofit organization.
Institution Type/Description: Art Gallery.
Collections: paintings; photographs; sculpture; ceramics; jewelry; fiber arts.
Activities: workshops; lectures; concerts; artist residencies.
Hours & Admission Prices: Call for hours. No charge; donations accepted.
Attendance: 10,000 (estimated)
Membership: Individual $50; Family $100; Sustaining $150; Sponsor $250; Benefactor $500; Visionary $1,000.

FIREWEED GALLERY, 475 E. Pioneer Ave., #A, Homer, AK 99603-7622. Tel.: 907-235-3411.
E-mail: art@fireweedgallery.com
Institution Type/Description: Art Gallery.
Collections: contemporary works by Alaskan artists including paintings, sculpture, ceramics, jewelry, & photography.
Activities: special events; traveling exhibitions. Museum Sponsors: Annual Spring Show.
Hours & Admission Prices: Mon.-Sat. 10-6, Sun. 11-4.

* **PRATT MUSEUM,** 3779 Bartlett St., Homer, AK 99603-7579. Tel.: 907-235-8635. Fax: 907-235-2764.
E-mail: info@prattmuseum.org
Web Site: www.prattmuseum.org
Founded: 1955.
Congressional District: 7
Key Personnel: C.E.O., Diane Converse; Pres. (V), Mildred Martin; Dir. Education, Ryjil Christianson; Cur. Exhibits, Scott Bartlett; Bldg. Mgr., Art Koeninger; Bookkeeper, Heidi Stage; Museum Shop Mgr., Jennie Engebretsen; Office Mgr., Kim Wylde.
Personnel Profile: Full-Time Paid 5; Part-Time Paid 8; Part-Time Volunteers 50; Interns 5.
Governing Authority: nonprofit organization. Parent Institution: Homer Society of Natural History, Inc. Tax-exempt: 501(c)(3).
Institution Type/Description: General Museum.
Collections: regional Alaskan history, ethnography, archaeology, zoology, herbarium, marine aquarium, botanical garden, art; local art; art of local interest; interpretive nature trail; Homestead Cabin.
Major Exhibits: The Living Tertiary, 2/14-3/14; Key Ingredients: America By Food & Putting By: Foodland Identity on the Kenai (T), 4/14-6/14; Reliquary by Jo Going, 7/14-9/14; Communities, Disaster & Change (T), 9/14-10/14; Ritz Art Exhibit, 10/14; K-Bay Sisters in Art, 11/14-12/14.
Research Fields: local history; anthropology; ornithology; sea mammals.
Facilities: reference library. Museum-related items for sale.
Activities: permanent & special exhibitions; lecture programs; school programs; guided tours; marine aquaria demonstrations; remote controlled cameras in wilderness areas.
Publications: Pratt Museum Newsletter; A History of Kachemak Bay; Rock Paintings of South Coastal Alaska; Homer Historic Building Survey; English Bay & Port Graham Alutiiq Plantlore; Darkened Waters: A Review of the History, Science & Technology Associated with the Exxon Valdez Oil Spill & Cleanup; Kachemak Bay Bird Watch; A Birder's Guide to the Kenai Peninsula, Alaska.
Hours & Admission Prices: Feb.-April & Oct.-Dec. Tues.-Sun. 12-5; mid-May to mid-Sept. daily 10-6. Family $25, adults $8, seniors $6.50, youths 13-16 $4; discounts to military, AAM, AAA & MAM members; children under 6 & members no charge. &
Attendance: 34,300 (accurate)
Membership: Pioneer Club $25 (artist/student/senior); Kittinake Club $50; Octopus Club $100; Wild Rose Club $250; Qayak Club $500; Aurora Club $1,000; Bald Eagle Club $2,500.

Hope

HOPE-SUNRISE HISTORICAL AND MINING MUSEUM, 64851 2nd St., Hope, AK 99605. Mailing Address: P.O. Box 88, Hope, AK 99605-0088. Tel.: 907-782-3740.
E-mail: hopehistoricalsociety@gmail.com
Web Site: www.hopeandsunrisehistoricalsociety.com
Founded: 1970.
Key Personnel: Pres. (V), Scott Sherritt.
Personnel Profile: Part-Time Volunteers 25.
Volunteer Hours: 200
Operating Expenses: 18,000
Operating Income: 19,158
Governing Authority: Parent Institution: Hope and Sunrise Historical Society. Tax-exempt.
Institution Type/Description: History Museum.
Collections: 3,800 photographs; 1,100 artifacts; Turnagain Arm Gold Rush of 1896; grader; dog sled; rock crusher; blacksmith bellows; postal boxes. Historic Buildings: mine bunkhouse; blacksmith shop.
Activities: 2-4 summer programs.
Publications: newsletter.
Hours & Admission Prices: Memorial Day to Labor Day daily 12-4. No charge; donations accepted.
Attendance: 3,000 (estimated)
Membership: Individual $10; Life $100.

Juneau

* **ALASKA STATE MUSEUM, (M),** 395 Whittier, Juneau, AK 99801-1746. Tel.: 907-465-2901. Fax: 907-465-2976.
E-mail: bob.banghart@alaska.gov
Web Site: www.museums.state.ak.us
Founded: 1900.
Congressional District: 4
Key Personnel: Dir., Bob Banghart; Friends of the Alaska State Museum Pres., Ginny Palmer; Administrative Asst., Debbie McBride; Cur. Collections, Steve Henrikson; Cur. Exhibits, Jackie Manning; Cur. Museum Svcs., Scott Carrlee; Exhibitions Specialist, Aaron Elmore; Registrar, Addison Field;

Mgr. Protection & Visitor Svcs., Lisa Golisek; Asst. Protection & Visitor Svcs., Sara Lee; Supvr. Protection & Visitor Svcs., Mary Irvine; Conservator, Ellen Carrlee.
Personnel Profile: Full-Time Paid 11; Part-Time Paid 3; Part-Time Volunteers 109.
Governing Authority: state. Parent Institution: State of Alaska. Subsidiary Institution: Division of Libraries, Archives & Museums. Branch Museum: Sheldon Jackson Museum, Sitka, AK 99835. Tax-exempt.
Institution Type/Description: General Museum.
Collections: Alaskan history, art, natural history & ethnography; Russian-American; maritime; early industry.
Research Fields: Alaskan history, ethnology, fine art & natural history.
Facilities: library; conservation lab.
Activities: statewide assistance to museums in Alaska; statewide conservation services; grants-in-aid to Alaskan museums; purchase acquisitions program; statewide education services including hands-on loan program to Alaskan schools; general education programs; guided tours; temporary & permanent exhibits; statewide visual arts & traveling exhibits programs.
Publications: educational materials for schools; monographs, concepts; brochures; exhibits books; catalog; museum review.
Hours & Admission Prices: Winter: Tues.-Sat. 10-4; Summer: daily 8:30-5:30. Adults $3 (winter), $7 (summer); discounts to seniors and AAM & ICOM members; children 18 and under & members no charge. Closed holidays. Temporarily closed for construction of a new facility. &
Attendance: 60,000 (accurate)
Membership: Student & Senior $20; Individual $40; Family $50; Sustaining $125; Corporate & Life $1,000.

HOUSE OF WICKERSHAM STATE HISTORICAL SITE, 213 Seventh St., Juneau, AK 99801-1117. Mailing Address: 400 Willoughby Ave., P.O. Box 111017, Juneau, AK 99801-1783. Tel.: 907-586-9001.
Web Site: www.dnr.state.ak.us/parks/units/wickrshm.htm
Founded: 1984.
Personnel Profile: Full-Time Volunteers 1; Part-Time Volunteers 4.
Governing Authority: state government.
Institution Type/Description: Historic House: constructed in 1898 by Frank Hammond, a gold mine superintendent. The house was purchased in 1928 by James Wickersham, a judge & delegate to congress.
Collections: Alaskan memorabilia; native crafts from the Pacific Northwest.
Facilities: 1,000 sq. ft. exhibit space.
Hours & Admission Prices: May-Sept. Tues.-Sat. 10-4, call to confirm. Adults, senior citizens & students $2; children no charge. &
Attendance: 1,000 (estimated)

JUNEAU-DOUGLAS CITY MUSEUM, (M), 114 W. 4th St., Juneau, AK 99801-1758. Mailing Address: 155 S. Seward St., Juneau, AK 99801-1332. Tel.: 907-586-3572. Fax: 907-586-4512.
E-mail: jane_lindsey@ci.juneau.ak.us
Web Site: www.juneau.org/parkrec/museum
Founded: 1976.
Congressional District: 1
Key Personnel: Dir., Jane Lindsey; Cur. Collections & Exhibits, Addison E. Field.
Personnel Profile: Full-Time Paid 4; Part-Time Volunteers 35; Interns 2.
Governing Authority: municipal government; county government. Parent Institution: City and Borough of Juneau. Tax-exempt.
Institution Type/Description: Local History & Culture Museum: housed in 1951 library building.
Collections: 1880-1944 gold mining tools & equipment; fine arts; Tlingit Indian culture; pioneer decorative art; pioneer lifeways; archives; photographs; postcards; city documents; mining ruins related to the history and culture of the Juneau-Douglas area.
Research Fields: local history; Auk & Taku Tlingit Indian history; mining history.
Facilities: 525-vol. local history research library & archives; 2,500 sq. ft. exhibit space. Gift items for sale.
Activities: guided tours; lectures; meeting space; organized education programs for children; participatory & temporary exhibits; college student internship placement. Annual Event: historic walking tours May to September.
Publications: quarterly newsletter; trail guide booklet, In the Miners' Footsteps; brochure, Downtown Juneau Walking Guide to historic sites, public art & totem poles.
Hours & Admission Prices: May-Sept. Mon.-Fri. 9-5, Sat.-Sun. 10-5; Oct.-April Tues.-Sat. 10-4; other times by appointment. Summer: adults $4; discounts to AAM, AASLH & Museum Alaska members; children 18 & under no charge. Winter: no charge. Closed Thanksgiving; Christmas. &
Attendance: 23,000 (accurate)

Membership: Annual Pass $15.

LAST CHANCE MINING MUSEUM, 1001 Basin Rd., Juneau, AK 99801-1038. Tel.: 907-586-5338.
Key Personnel: Dir., Gary Gillette
Institution Type/Description: Historic Building: listed on the National Register of Historic Places.
Collections: air compressors; gold mining; mining equipment; electric locomotives & rail cars.
Hours & Admission Prices: mid-May to late Sept. daily 9:30-12:30 & 3-6:30; other times by appointment. Adults $4. Closed Independence Day.

Kenai

KENAI VISITORS & CULTURAL CENTER, 11471 Kenai Spur Hwy., Kenai, AK 99611-7757. Tel.: 907-283-1991. Fax: 907-283-2230.
E-mail: programdirector@visitkenai.com
Web Site: www.artskenai.com
Founded: 1991.
Key Personnel: Dir., Natasha Ala; Dir. Programs & Exhibitions, Zirrus Van Devere.
Personnel Profile: Full-Time Paid 1; Part-Time Paid 3; Part-Time Volunteers 2.
Governing Authority: Parent Institution: Kenai Convention and Visitors Center. Tax-exempt: 501(c)(3).
Institution Type/Description: History Museum.
Collections: Athabaskan, Aleut & Russian cultural artifacts; homesteading; mining; commercial fishing; oil industry.
Facilities: audiovisual room. Museum-related items for sale.
Activities: films.
Hours & Admission Prices: Mon.-Fri. 9-5, Sat. 11-4. No charge; donations accepted.
Attendance: 50,000 (accurate)

Ketchikan

DEER MOUNTAIN TRIBAL HATCHERY AND EAGLE CENTER, 1158 Salmon Rd., Ketchikan, AK 99901-6666. Tel.: 907-228-5278.
Institution Type/Description: Nature Center.
Collections: salmon & their life cycle; eagles; natural, cultural & environmental artifacts; hatchery.
Activities: view Salmon swimming upstream; feed Salmon; Eagle observation area; raptor program.
Hours & Admission Prices: May-Sept. daily 8-4:30. Adults $9; children 12 & under no charge.

SOUTHEAST ALASKA DISCOVERY CENTER, USDA FOREST SERVICE, 50 Main St., Ketchikan, AK 99901-6559. Tel.: 907-228-6220. Fax: 907-228-6234.
Institution Type/Description: History Museum.
Collections: totem poles; rainforest; native fish camp; Southeast Alaska's ecosystems; fishing; mining; timber; tourism.
Activities: audiovisual programs.
Hours & Admission Prices: May-Sept. daily 8-3; Oct.-April Fri. 12-8. Summer: $5; Winter: no charge. Closed Memorial Day; Independence Day; Labor Day.

TONGASS HISTORICAL MUSEUM, 629 Dock St., Ketchikan, AK 99901-6529. Tel.: 907-225-5600. Fax: 907-225-5602.
E-mail: museumdir@city.ketchikan.ak.us
Web Site: www.ketchikanmuseums.org
Founded: 1961.
Congressional District: 1
Key Personnel: Dir., Michael Naab; Administrative Sec., Karla Sunderland; Sr. Cur. Programs, Christopher Hanson; Sr. Cur. Collections, Richard H. Van Cleave; Registrar, Erika Brown; Museum Attendant, Regina Foreman.
Personnel Profile: Full-Time Paid 8; Part-Time Volunteers 14.
Governing Authority: municipal. Parent Institution: City of Ketchikan. Tax-exempt.
Institution Type/Description: General Museum.
Collections: history, art & industry of southern southeast Alaska; Indian artifacts; anthropology; ethnology; manuscripts; archaeology; objects & photos relating to the Tlingit, Haida & Tsimshian cultures; explorer, trader & pioneer history; mining, fishing, logging & marine history.
Research Fields: local area history; industry; fishing; mining; logging; marine; ethnography.
Facilities: reference library of history on Alaska; local history archives.

Activities: lectures; films; inter-museum loan; permanent, temporary & traveling exhibitions; changing exhibits on Alaskan photography, regional art, local history & Alaskan Native arts & crafts; school and special group tours; lecture series.
Publications: quarterly newsletter; descriptive brochures; historical pamphlets.
Hours & Admission Prices: mid-May to Sept. daily. 8-5; Oct. to mid-May Wed.-Fri. 1-5, Sat.-Sun. 1-4. Adults $2; Oct.-April no charge. Closed New Year's Day; Easter; Thanksgiving; Christmas. &
Attendance: 30,000 (estimated)

TOTEM HERITAGE CENTER, 601 Deermount, Ketchikan, AK 99901. Mailing Address: 629 Dock St., Ketchikan, AK 99901-6529. Tel.: 907-225-5900. Fax: 907-225-5901.
E-mail: museumdir@city.ketchikan.ak.us
Web Site: www.ketchikanmuseums.org
Founded: 1976.
Congressional District: 1
Key Personnel: C.E.O., Michael Naab.
Personnel Profile: Full-Time Paid 8; Part-Time Paid 6.
Governing Authority: municipal. Parent Institution: City of Ketchikan. Tax-exempt.
Institution Type/Description: Anthropology, Ethnology & History Museum; Indian Art Center.
Collections: Alaska Totems; contemporary N.W. Coast Indian art; photographs.
Research Fields: totem poles; Northwest Coast Indian art, culture & history.
Facilities: reference library.
Activities: Native Arts Studies program & certificate of merit in carving/design or textile arts; films, exhibits, guided tours, craft demonstrations, education collection. Lectures; workshops & demonstrations.
Publications: quarterly newsletter; catalogue of native arts classes.
Hours & Admission Prices: mid-May to Sept. daily 8-5; Oct. to mid-May Tues.-Fri. 1-5 & during classes. Adults $5. &
Attendance: 60,000 (accurate)

Kodiak

*** ALUTIIQ MUSEUM AND ARCHAEOLOGICAL REPOSITORY, (M),** 215 Mission Rd., Ste. 101, Kodiak, AK 99615-7326. Tel.: 907-486-7004. Fax: 907-486-7048.
E mail: info@alutiiqmuseum.org
Web Site: www.alutiiqmuseum.org
Founded: 1995.
Congressional District: 1
Key Personnel: Chm. Bd., William Anderson; Vice Chm. Bd., Margaret Roberts; C.E.O., Sven Haakanson, Jr.; Museum Mgr., Katie St. John; Sec., Donene Tweten; Deputy Dir., Amy Steffian; Cur., Patrick Saltonstall; Registrar, Marnie Leist; Museum Shop Mgr., Marya Halvorsen.
Personnel Profile: Full-Time Paid 6; Part-Time Paid 4; Part-Time Volunteers 20.
Governing Authority: nonprofit organization. Parent Institution: Alutiiq Heritage Foundation. Tax-exempt: 501(c)(3).
Institution Type/Description: Archaeology Museum & Ethnographic.
Collections: archaeological artifacts pertaining to coastal Alaska, with emphasis on Kodiak Archipelago.
Research Fields: archaeological fieldwork; prehistoric & historic periods of Alaska; cultural history of Alutiiq region; Russian Far East.
Facilities: 800-vol. library; 1,600 sq. ft. exhibit space; lab & classrooms. Alaska native arts & crafts, publications and other museum-related items for sale.
Activities: guided tours; lectures; temporary & traveling exhibitions; broadcast programs; docent program; training programs for professional museum workers; school loan services. Annual Events: annual archaeological excavations & surveys.
Publications: quarterly newsletter, Alutiiq Museum Bulletin; occasional papers.
Hours & Admission Prices: Winter: Tues.-Fri. 9-5, Sat. 10:30-4:30; Summer: Mon.-Fri. 9-5, Sat. 10-5. Adults $5; members no charge. &
Attendance: 6,500 (estimated)
Membership: Student & Senior Citizen $10; Individual $25; Family $40; Sponsor $100, Benefactor $250; Corporate $5,000.

BARANOV MUSEUM, Erskine House, 101 Marine Way, Kodiak, AK 99615-6307. Mailing Address: Kodiak Historical Society, 101 Marine Way, Kodiak, AK 99615-6307. Tel.: 907-486-5920. Fax: 907-486-3166. Facebook: Baranov Museum.
E-mail: baranov@ak.net
Web Site: www.baranovmuseum.org
Founded: 1954.

Congressional District: 1
Key Personnel: Dir., Tiffany Brunson; Bd. Pres., Nancy Kemp; Archivist, Alice Ryser; Cur. Collections, Anjuli Grantham; Museum Shop Mgr., Hannahgrace Lucas.
Personnel Profile: Full-Time Paid 2; Part-Time Paid 4; Part-Time Volunteers 12.
Operating Expenses: 277,734
Operating Income: 268,920
Governing Authority: nonprofit. Parent Institution: Kodiak Historical Society. Tax-exempt: 990(A).
Institution Type/Description: General Museum: housed in c.1800 fur warehouse & offices of Alexander Baranov, chief manager of Russian American Company; after 1867 used by the American Commercial Company; 1911-1948 the W.J. Erskine family home.
Collections: items from the Kodiak & Aleutian Islands; anthropology; archaeology; ethnology; folklore; geology; history; Eskimo artifacts; marine; mineralogy; natural history.
Major Exhibits: Waves of Change: 1964 Earthquake & Tsunami in Kodiak, 3/14-2/15.
Research Fields: Kodiak & Aleutian Islands archives; World War II; 1912 volcano & 1964 earthquake; grass basket weaving; photographs.
Facilities: rare book library. Russian antiques & handcrafted items for sale.
Activities: guided tours; docent program; videos; school tours.
Publications: quarterly newsletter, The Baranov Quarterly.
Hours & Admission Prices: Summer: Mon.-Sat. 10-4; Winter: Tues.-Sat. 10-3. Adults $5; discounts to AAM, ICOM & Museum Alaska members; members & children under 12 no charge. Closed national holidays. &
Attendance: 10,000 (estimated)
Membership: Senior 65 & over $20; Individual $30; Family $50-$199; Friend $200-$499; Partner $500+.

KODIAK MARITIME MUSEUM, (M), Kodiak, AK 99615. Mailing Address: P.O. Box 1876, Kodiak, AK 99615. Tel.: 907-486-0384. Fax: 907-486-0385.
E-mail: info@kodiakmaritimemuseum.org
Web Site: www.kodiakmaritimemuseum.org
Institution Type/Description: Maritime Museum.
Collections: Alaska's commercial fishing industry & maritime heritage.
Activities: fundraising events; museum without walls.
Hours & Admission Prices: Virtual Museum.

KODIAK MILITARY HISTORY MUSEUM, Fort Abercrombie State Historical Park, 1623 Mill Bay Rd., Kodiak, AK 99615. Mailing Address: 1417B Mill Bay Rd., Kodiak, AK 99615-7505. Tel.: 907-486-7015. Fax: 907-486-5541.
Web Site: www.kodiak.org/museum/museum.html
Key Personnel: Dir., Dave Ostlund; Business Mgr., Curt Law
Institution Type/Description: Military Museum: housed in the WWII Ready Ammunition bunker at Miller Point; a buried concrete structure built in 1942.
Collections: uniforms; personal artifacts; photographs; artillery; communications equipment.
Facilities: Museum-related items for sale.
Hours & Admission Prices: Fri.-Sat. & Mon. 1-4, Sun. 1:30-4; other times by appointment. Adults $3; children under 12 no charge.

Metlakatla

DUNCAN COTTAGE MUSEUM, 501 Tait St., Metlakatla, AK 99926. Mailing Address: P.O. Box 8, Metlakatla, AK 99926-0008. Tel.: 907-886-8687. Fax: 907-886-4436.
Web Site: www.metlakatours.net
Founded: 1975.
Key Personnel: Dir. Tourism, Patricia A. Beal.
Personnel Profile: Full-Time Paid 1; Part-Time Paid 1; Part-Time Volunteers 3.
Governing Authority: municipal.
Institution Type/Description: Historic House: 1891 seven room cottage that was home of Father Wm. Duncan, English missionary/teacher of the Tsimpshians of Metlakatla, AK.
Collections: period phonograph, furniture, ring-up wooden telephone; Tsimpshian Bentwood storage box with Tsimpshian designs; ceremonial drum; some 1st edition, technical, theological books; English & Canadian period bibles; magazines; 1800-1900 photographs; paintings; personal artifacts; period furnishings; turn of the century collections of music boxes; medical supplies; educational material; musical instruments.
Research Fields: Metlakatlas' historic origins; Tsimpshian pioneers, clans or crests.
Facilities: library of history books, music, autobiographies of great U.S., Great

Britain & Canadian men and women of the past; classrooms. Museum-related items for sale.
Activities: guided tours; formally organized education programs; permanent exhibitions; walking tour.
Publications: Wm. Duncan fact sheet.
Hours & Admission Prices: Mon.-Fri. 8:30-12:30; other times by appointment. Adults $2. &
Attendance: 4,000 (estimated)

Nenana

ALFRED STARR NENANA CULTURAL CENTER, 415 Riverfront, Nenana, AK 99760. Mailing Address: P.O. Box 70, Nenana, AK 99760-0070. Tel.: 907-832-5527. Fax: 907-832-5532.
E-mail: nenana1@nenana.net
Web Site: www.nenana.org
Founded: 1997.
Key Personnel: Dir., Alex Ketzler; Asst. Dir., Pamela Coghill
Institution Type/Description: Cultural Center.
Collections: history & culture of Nenana Yukon 800 riverboat racing; dog mushing; beadwork.
Facilities: Museum-related items for sale.
Hours & Admission Prices: Summer: daily 10-6; Winter: by appointment. No charge; donation accepted. &

Nome

CARRIE M. MCLAIN MEMORIAL MUSEUM, (M), 223 Front St., Nome, AK 99762. Mailing Address: P.O. Box 53, Nome, AK 99762-0053. Tel.: 907-443-6630. Fax: 907-443-7955.
Web Site: www.nomealaska.org
Founded: 1940.
Key Personnel: Dir., Laura Samuelson.
Personnel Profile: Full-Time Paid 2; Part-Time Paid 1; Part-Time Volunteers 4.
Governing Authority: Parent Institution: City of Nome, Alaska. Tax-exempt.
Institution Type/Description: History Museum.
Collections: Eskimo culture & life; Gold Rush; photographs; aviation; Nome history & culture.
Research Fields: Nome Gold Rush; Bering Sea Eskimo.
Hours & Admission Prices: June-Sept. daily 9:30-5:30; Oct.-May Tues.-Sat. 12-5:30. No charge; donations accepted. &
Attendance: 11,000 (estimated)

Palmer

COLONY HOUSE MUSEUM - PALMER HISTORICAL SOCIETY, 316 E. Elmwood Ave., Palmer, AK 99645-6621. Mailing Address: P.O. Box 1935, Palmer, AK 99645-1935. Tel.: 907-745-1935. Facebook: Palmer Historical Society Alaska.
E-mail: colony@palmerhistoricalsociety.org
Web Site: www.palmerhistoricalsociety.org
Key Personnel: Pres. & Sec., Sheri Hamming
Institution Type/Description: Historic House Museum: housed in an original Colony Farm House built for the New Deal resettlement in 1935.
Collections: Colony project history; early colonists; period furnishings.
Activities: group tours.
Hours & Admission Prices: Summer: Tues.- Sat. 10-4; Winter: by appointment only. Adults $2, children 12 & under $1.
Membership: Individual $25; Household $40; Patron $100; Lifetime $500.

MUSK OX FARM & MUSEUM, 12850 E. Archie Rd., Palmer, AK 99645. Mailing Address: P.O. Box 587, Palmer, AK 99645-0587. Tel.: 907-745-4151. Fax: 907-746-4831.
E-mail: info@muskoxfarm.org
Web Site: www.muskoxfarm.org
Institution Type/Description: Farm History Museum.
Collections: musk oxen; cows; bulls; calves; farm history.
Facilities: Museum-related items for sale.
Activities: group tours.
Hours & Admission Prices: May & Aug.-Sept. daily 10-6; June-July daily 9-7 other times by appointment. Adults 13-64 $8, seniors 65 & over $7, children 5-12 $6; children 4 & under no charge.

PALMER MUSEUM OF HISTORY AND ART, (M), 723 S. Valley Way, Palmer, AK 99645-6601. Tel.: 907-746-7668. Fax: 907-745-7882.
E-mail: director@palmermuseum.org
Web Site: www.palmermuseum.org

Founded: 2005.
Key Personnel: Dir., Selena Ortega-Chiolero; Pres. (V), Anne Lane; Museum Shop Mgr., Christy Hand.
Personnel Profile: Full-Time Paid 1; Part-Time Paid 7; Part-Time Volunteers 2.
Governing Authority: nonprofit organization. Tax-exempt: 501(c)(3).
Institution Type/Description: History & Art Museum.
Collections: Greater Palmer regions art & history; Alaska Natives; mining; farming.
Publications: monthly newsletter.
Hours & Admission Prices: May-Sept. daily 9-6; Oct.-April Wed.-Fri. 10-5, Sat. 10-2. No charge; donations accepted. &
Attendance: 35,000 (accurate)
Membership: Senior $10; Individual $15; Family $25; Organization $75; Business $100; Lifetime $200.

Petersburg

CLAUSEN MEMORIAL MUSEUM, (M), 203 Fram St., Petersburg, AK 99833. Mailing Address: P.O. Box 708, Petersburg, AK 99833-0708. Tel.: 907-772-3598. Fax: 907-772-2698.
E-mail: clausenmuseum@aptalaska.net
Web Site: www.clausenmuseum.org
Founded: 1965.
Key Personnel: C.E.O., Dir. & Museum Shop Mgr., Sue McCallum; Pres. (V), Nancy Strand.
Personnel Profile: Full-Time Paid 1; Part-Time Paid 2; Part-Time Volunteers 17.
Governing Authority: bd. of trustees. Tax-exempt.
Institution Type/Description: History Museum.
Collections: local history & residents including Tlingit, European & Asian peoples; Tlinget canoe & tools; Cape Decision lightstation lens; Fisk sculpture & fountain; cannery & fishing artifacts.
Activities: winter programs for community & school; temporary exhibits; oral history program; occasional summer programs.
Publications: newsletter; brochure; From Fish Camps to Cold Storages, a brief history of the Petersburg area to 1927; Streets to the Past.
Hours & Admission Prices: Call for hours & special events. Adults $3; Alaska members & members no charge. Closed New Year's Day; Christmas. &
Attendance: 3,902 (accurate)
Membership: Senior $10; Basic Individual $25; Basic Family $40; Supporter $100; Contributor $250; Sponsor $500; Lifetime $1,000.

Seward

ALASKA SEALIFE CENTER, 301 Railway Ave., Seward, AK 99664. Mailing Address: P.O. Box 1329, Seward, AK 99664-1329. Tel.: 907-224-6300; 800-224-2525. Fax: 907-224-6360. Facebook: Alaska SeaLife Center.
E-mail: visitaslc@alaskasealife.org
Web Site: www.alaskasealife.org
Founded: 1998.
Key Personnel: Pres. & C.E.O., Tara Riemer Jones, Ph.D
Governing Authority: Tax-exempt.
Institution Type/Description: Marine Wildlife Center.
Collections: marine wildlife.
Research Fields: Arctic & Alaska marine life.
Activities: behind the scenes tours; marine mammal, octopus & puffin animal encounters.
Hours & Admission Prices: Summer: Mon.-Thurs. 9-6:30, Fri.-Sun. 8-6:30; Winter: daily 10-5. Adults $20, student 12-17 $15, children 4-11 $10; discounts to AAM members; members and children 3 & under no charge. Closed Thanksgiving; Christmas. &
Membership: Individual $45; Couple $80; Family $125.

CHUGACH MUSEUM AND INSTITUTE OF HISTORY AND ART, Orca Bldg., 3rd & Washington St., Seward, AK 99503. Mailing Address: 3800 Centerpoint Dr., Ste. 601, Anchorage, AK 99503-5826. Tel.: 907-563-8866, ext. 4151. Fax: 907-563-8402.
E-mail: chugachmuseum@chugach-ak.com
Web Site: www.chugachmuseum.org
Key Personnel: Exec. Dir., Lora Johnson, Ph.D.
Institution Type/Description: Culturally Specific.
Collections: history & culture of the Chugach region natives; artwork.
Hours & Admission Prices: By appointment.

QUTEKCAK CULTURE CENTER, 221 Third Ave., Seward, AK 99664. Mailing Address: P.O. Box 1467, Seward, AK 99664-1467. Tel.: 907-224-3118.
Institution Type/Description: Native Heritage Museum.

Collections: Alaska Native culture, arts, & artifacts; paintings; sculpture; personal artifacts.
Facilities: Museum-related items for sale.
Hours & Admission Prices: Mon.-Fri. 12:30-4.

SEWARD MUSEUM, 336 3rd Ave., Seward, AK 99664. Mailing Address: P.O. Box 55, Seward, AK 99664-0055. Tel.: 907-224-3902.
Formerly: Resurrection Bay Historical Society Museum
Founded: 1967.
Key Personnel: Dir. & Pres. (V), Lee E. Poleske.
Personnel Profile: Full-Time Volunteers 1; Part-Time Paid 1; Part-Time Volunteers 4.
Governing Authority: society; nonprofit organization. Parent Institution: Resurrection Bay Historical Society. Tax-exempt.
Institution Type/Description: Historical Society Museum.
Collections: photographs & artifacts of Seward & Alaska.
Research Fields: local history.
Activities: guided tours; school visits; walking town tours.
Publications: newsletters.
Hours & Admission Prices: Daily 10-5. Adults $3, children $.50; members no charge. &
Attendance: 4,972 (accurate)
Membership: Annual $10; Family $15; Sustaining $25; Benefactor $50; Patron $100.

Sitka

ALASKA RAPTOR CENTER, 1000 Raptor Way, Sitka, AK 99835-9302. Tel.: 907-747-8662; 800-643-9425. Fax: 907-747-8397.
E-mail: members@alaskaraptor.org
Web Site: www.alaskaraptor.org
Key Personnel: Exec. Dir., Debbie Reeder; Avian Rehabilitation Cur., Jennifer Cedarleaf
Institution Type/Description: Environmental Conservation Center.
Collections: bald eagles; wild birds.
Facilities: Museum-related items for sale.
Activities: educational programs.
Publications: newsletter.
Hours & Admission Prices: May-Sept. Sun.-Fri. 8-4. Adults $12, children 12 & under $6.
Membership: Individual $45; Family $60.

ISABEL MILLER MUSEUM/SITKA HISTORICAL SOCIETY, (M), 330 Harbor Dr., Sitka, AK 99835-7553. Tel.: 907-747-6455. Fax: 907-747-6588.
Web Site: sitkahistory.org
Founded: 1957.
Congressional District: 3
Key Personnel: Dir., Robert Medinger; Admin., Karen Meizner.
Personnel Profile: Full-Time Volunteers 2; Part-Time Paid 5; Part-Time Volunteers 50.
Governing Authority: nonprofit organization. Parent Institution: Sitka Historical Society. Tax-exempt: 501(c)(3).
Institution Type/Description: General History Museum.
Collections: local artifacts, paintings, manuscripts, documents, photographs & scrapbooks from the pre-Russian period to the present.
Research Fields: historic buildings and locations in the area.
Facilities: 4005-vol. library of books on Alaskan history, including translations in Russian available for research on premises by appointment.
Activities: permanent exhibitions; lectures; tours; newsletter; web exhibitions.
Publications: newsletter; books, Lady Franklin Visits Sitka Alaska, 1870; Streets of Sitka; The Wreck of the Neva; A Sitka Chronology, 1867-1985; From Sitka's Past.
Hours & Admission Prices: May 7-Sept. 22 daily 8-5; Oct.-April Tues.-Sat. 10-4. Summer $1; Winter no charge. &
Attendance: 67,415 (accurate)
Membership: Student, Senior Citizen & Non-Resident Associate $20; Individual $35; Family $40; Patron $75; Business & Contributing $100; Sustaining $200; Benefactor $1,000; Life Benefactor $5,000.

RUSSIAN BISHOP'S HOUSE, 201 College Dr., Sitka, AK 99835. Mailing Address: Sitka National Historic Park, 103 Monastery St., Sitka, AK 99835-7617. Tel.: 907-747-6281.
Institution Type/Description: Historic House: the former residence of the Bishop of the Russian Orthodox Church; built in 1842. A National Historic Landmark.
Collections: period furnishings & artifacts.
Hours & Admission Prices: Summer: daily; Winter: by appointment.

* **SHELDON JACKSON MUSEUM, (M),** 104 College Dr., Sitka, AK 99835-7657. Tel.: 907-747-8981. Fax: 907-747-3004. www.museums.alaska.gov/.
E-mail: lisa.bykonen@alaska.gov
Web Site: www.museums.state.ak.us
Founded: 1887.
Key Personnel: Pres. (V), Margie Esquiro; Dir. Division, Linda Thibodeau; Museum Protection & Visitor Svcs. Supvr., Lisa Bykonen; Museum Shop Mgr., Mary Boose.
Personnel Profile: Full-Time Paid 3; Part-Time Paid 5; Part-Time Volunteers 15.
Governing Authority: state. Parent Institution: SOA. Subsidiary Institution: Div. of Libraries, Archives & Museums. Tax-exempt: 501(c)(3).
Institution Type/Description: Anthropology Museum: housed in the first concrete building built in the territory of Alaska; built in 1895.
Collections: Haida argillite carvings; Eskimo implements, ivory carvings, masks, skin clothing, baskets, kayaks, umiak; Athabaskan birchbark canoes, implements; Tlingit totem poles, shaman charms, baskets, ceremonial equipment & garments.
Research Fields: Tlingit & western Eskimo ethnography & acculturation.
Facilities: Alaskan native arts, crafts, museum-related publications, postcards & lithographs for sale.
Activities: lectures; education programs for undergraduate college students & students K-12; inter-museum loan, permanent & temporary exhibitions; native artist demonstrations.
Publications: booklet, Sheldon Jackson, The Collector; Faces, Voices, & Dreams - Centennial of SJM 1888-1988.
Hours & Admission Prices: mid-May to mid-Sept. daily 9-5; mid-Sept. to mid-May Tues.-Sat. 10-4. Summer: adults $5; seniors $4, Winter: adults $3; AAM, ASM & museum members and children 18 & under no charge. Closed holidays. &
Attendance: 16,000 (accurate)
Membership: Student $10; Individual $25; Family $35; Sustaining $60; Contributing $250; Corporate $500.

SITKA NATIONAL HISTORICAL PARK, 106 Metlakatla St., Sitka, AK 99835-7665. Mailing Address: 103 Monastery St., Sitka, AK 99835-7617. Tel.: 907-747-0110. Fax: 907-747-5938.
Web Site: www.nps.gov/sitk
Founded: 1910.
Key Personnel: Supt., Mary Miller; Cur., Sue Thorsen.
Governing Authority: federal. Parent Institution: National Park Service, Dept. of Interior. Tax-exempt: 170(b)(1)(A).
Institution Type/Description: Park Museum & Visitor Center: located near site of 1804 Battle of Sitka, fought between the Tlingit Indians & the Russians.
Collections: spruce root baskets; totem poles; Indian & Russian artifacts; 19th-century period furniture. Historic House: 1842 Russian Bishop's house.
Research Fields: Indian culture; Russian colonization; North Pacific fur trade.
Facilities: 1,000-vol. library on Alaska history available on premises; 50-seat auditorium; 1-mile totem walk.
Activities: self-guided tours; lectures; special programs; classes & demonstrations of traditional Indian arts & crafts.
Publications: continuous folder, Sitka National Historical Park & the Russian Bishop's House Project.
Hours & Admission Prices: Mon.-Sat. 8-5. Summer: $15. Winter: no charge; donations accepted. Closed New Year's Day; Thanksgiving; Christmas. &

SOUTHEAST ALASKA INDIAN CULTURAL CENTER, 106 Metlakatla St., Ste. C, Sitka, AK 99835-7665. Tel.: 907-747-8061. Fax: 907-747-8189.
E-mail: sitkaculturalcenter@gmail.com
Founded: 1969.
Congressional District: 1
Key Personnel: Exec. Dir., Gerry Bigelow; Pres. (V), Kathy Miller; Pres. Bd., Gary Lang.
Personnel Profile: Full-Time Paid 2; Part-Time Paid 2; Interns 1.
Governing Authority: Tax-exempt: 501(c)(3).
Institution Type/Description: Native American Cultural Center.
Collections: Tlingit, Haida, Tsimshean traditional cultural art forms created by contemporary artists.
Research Fields: Tlingit; Haida; Tsimshean.
Activities: demonstrations; classes.
Publications: Celebration of Weavers.
Hours & Admission Prices: May-Sept. daily 9-5; Oct.-April Mon.-Fri. 8-5. No charge; donations accepted. Closed winter holidays. &
Attendance: 130,000 (estimated)
Membership: Individual $20; Family $50.

Skagway

CORRINGTON'S MUSEUM OF ALASKAN HISTORY, Broadway & 5th, Skagway, AK 99840. Mailing Address: P.O. Box 382, Skagway, AK 99840. Tel.: 907-983-2579.
Founded: 1976.
Personnel Profile: Full-Time Paid 1; Full-Time Volunteers 1; Part-Time Volunteers 2.
Institution Type/Description: History Museum.
Collections: Alaskan history & culture; personal artifacts; scrimshaw history.
Facilities: Museum-related items for sale.
Hours & Admission Prices: Call for hours. No charge.

KLONDIKE GOLD RUSH NATIONAL HISTORICAL PARK, (M), Second & Broadway, Skagway, AK 99840. Mailing Address: P.O. Box 517, Skagway, AK 99840-0517. Tel.: 907-983-2921 & 9222. Fax: 907-983-9249. TDD: 907-983-2921.
E-mail: samantha_richert@nps.gov
Web Site: www.nps.gov/klgo
Founded: 1976.
Key Personnel: Rgnl. Dir., Sue Masica; Supt., Mike Trand.
Personnel Profile: Full-Time Paid 2; Part-Time Volunteers 2; Interns 2.
Governing Authority: federal. Parent Institution: Dept. of the Interior. National Park Service, Alaska Regional Office, 240 W. 5th Ave., Anchorage, AK 99501. Tel. 907-644-3480. Tax-exempt.
Institution Type/Description: Park Museum: housed in c.1898 White Pass & Yukon Route Railroad Depot.
Collections: over 200,000 historic archaeology artifacts associated with the Klondike gold rush; over 3,000 copies of historic photographs of the gold rush period; about 1,000 pressed plant specimens; the George & Edna Rapuzzi collection of over 10,000 Klondike Gold Rush & early Skagway artifacts; over 3,000 dried lichen specimens.
Research Fields: Klondike gold rush.
Facilities: 900-vol. library of books on Klondike gold rush & Alaskana available for research & study on premises; reading room; 100-seat auditorium.
Activities: guided tours; lectures; films; permanent exhibitions.
Publications: park brochures; archaeological reports; historic structures reports; ethnographic overview and assessment.
Hours & Admission Prices: May-Sept. daily 8-7. Moore House Tour: donations accepted. &

Attendance: 800,000 (estimated)

SKAGWAY MUSEUM & ARCHIVES, 700 Spring St., Skagway, AK 99840. Mailing Address: P.O. Box 415, Skagway, AK 99840. Tel.: 907-983-2420. Fax: 907-983-3420.
E-mail: info@skagwaymuseum.org
Web Site: www.skagwaymuseum.org
Founded: 1961.
Congressional District: 4
Key Personnel: Dir., Judith Munns.
Personnel Profile: Full-Time Paid 1; Part-Time Paid 4.
Governing Authority: nonprofit organization. Parent Institution: City of Skagway, AK. Tax-exempt.
Institution Type/Description: History Museum: housed in 1899 McCabe College, built by Methodist Church.
Collections: Alaska pioneer life; Klondike gold rush artifacts & tools; Tlingit artifacts; railroad & transportation history.
Research Fields: gold rush; local history.
Facilities: 812-vol. library.
Activities: permanent & temporary exhibitions.
Publications: brochures; newspapers.
Hours & Admission Prices: May-Sept. Mon.-Fri. 9-5, Sat.-Sun. 1-5; Oct.-April call for hours. Adults $2, students $1; children 12 & under no charge. &

Attendance: 38,000 (estimated)

Soldotna

SOLDOTNA HISTORICAL SOCIETY AND MUSEUM, Centennial Park Rd., Soldotna, AK 99669. Mailing Address: P.O. Box 1986, Soldotna, AK 99669-1986. Tel.: 907-262-3832.
Key Personnel: Pres., Margaret Mullen
Institution Type/Description: History Museum: housed in a homesteaders' village.
Collections: local history & culture; wildlife displays.
Activities: lectures.
Hours & Admission Prices: May 15-Sept. 15 Tues.-Sat. 10-4, Sun. 12-4. No charge; donations accepted. &
Attendance: 2,270 (accurate)

Talkeetna

TALKEETNA HISTORICAL SOCIETY AND MUSEUM, 22248 S. D St., Talkeetna, AK 99676. Mailing Address: P.O. Box 76, Talkeetna, AK 99676-0076. Tel.: 907-733-2487.
E-mail: museum@talkeetnahistoricalsociety.org
Web Site: www.talkeetnahistoricalsociety.org/museum.php
Founded: 1972.
Key Personnel: Pres. (V), Joellen Bye; Historic Site Mgr., Jayme Spires.
Personnel Profile: Part-Time Paid 2; Part-Time Volunteers 6.
Governing Authority: society; nonprofit organization. Tax-exempt.
Institution Type/Description: General Museum: housed in the first school house in Talkeetna. Museum Buildings: The Ole Dahl Cabin; The Three German Bachelor's Cabin & Visitors Information; The Harry Robb Cabin; The Little Red Schoolhouse, Old Railroad Section House & Old Railroad Depot.
Collections: old trading post materials; old tools; mining & trapping equipment; papers & photographs; Alaskan art; Don Sheldon-Ray Genet Memorial display; 144 sq. ft. scale model of Denali with photos by Bradford Washburn; Summit Room, view from the Summit of Denali, two mannequins (climber in modern day gear & clothing compared to Walter Harper, who climbed Denali 1913) standing on the Summit; trapper cabin display; A.R.R. section house & depot.
Research Fields: oral history projects; humanities research & projects; state and local history.
Facilities: library of local history books, documents, photos, scrapbooks & newspapers available for use by requesting xerox copies; reading room. Books & museum-related items for sale.
Activities: films; arts festivals; loan, permanent & traveling exhibitions; tours; special exhibits; in school educational program. Museum Sponsors: Moose Dropping Festival in June; walking history tour of historic sites & buildings in Talkeetna.
Publications: newsletter; brochure; map.
Hours & Admission Prices: Summer: daily 10-6; Winter: Sat.-Sun. 11-5, weekdays by appointment only. Adults $3; discount to school groups; members & children under 12 no charge. &
Attendance: 15,000 (accurate)
Membership: Senior & Student $10; Standard $25; Contributing $50; Supporting $100.

Trapper Creek

TRAPPER CREEK MUSEUM, Mile 6 W. of Parks Hwy. on Petersville Rd., Trapper Creek, AK 99683. Mailing Address: P.O. Box 13011, Trapper Creek, AK 99683-0011. Tel.: 907-733-2557.
E-mail: trappercreekmuseum@yahoo.com
Web Site: www.trappercreekmuseum.com
Key Personnel: Dir., Kenneth Marsh
Institution Type/Description: History Museum: housed in a log cabin built by the Donaldson family, members of the Michigan 59'ers.
Collections: Alaska's early pioneers including gold miners, fur trappers & homesteaders; personal artifacts; glass bottles; photographs.
Facilities: Museum-related items for sale.
Hours & Admission Prices: Summer: daily 10-4. No charge; donations accepted.
Attendance: 1,200 (accurate)

Unalaska

MUSEUM OF THE ALEUTIANS, (M), 314 Salmon Way, Unalaska, AK 99685. Mailing Address: P.O. Box 648, Unalaska, AK 99685-0648. Tel.: 907-581-5150. Fax: 907-581-6682.
E-mail: aleutians@arctic.net
Web Site: www.aleutians.org
Founded: 1997.
Key Personnel: C.E.O. & Dir., Zoya Johnson; Chm. (V), Sharon Svarny-Livingston.
Personnel Profile: Full-Time Paid 4.
Governing Authority: private; nonprofit organization. Tax-exempt.
Institution Type/Description: General Museum.
Collections: archaeological; ethnological; archives; historical items.
Research Fields: archaeology; anthropology.
Facilities: laboratory.
Activities: college credit courses; fundraisers; membership drive & auction; lectures; presentations. Museum Sponsors: Archaeological Field Season - summer dig.
Publications: newsletter, Legacy; store catalog.
Hours & Admission Prices: June-Sept. Tues.-Sat. 9-5, Sun. 11-5; Oct.-May

Tues.-Sat. 11-5. Adults $5; discounts to Museum Alaska members; members no charge. Closed New Year's Day; Memorial Day; Independence Day; Thanksgiving; Christmas. &
Attendance: 4,036 (accurate)
Membership: Student $5; Individual $25; Family $75; Patron $250; Benefactor $500; Life & Corporate $1,000; Sponsor $5,000.

Valdez

MAXINE & JESSE WHITNEY MUSEUM, (M), 303 Lowe St., Valdez, AK 99686. Mailing Address: P.O. Box 97, Valdez, AK 99686. Tel.: 907-834-1600.
Institution Type/Description: History Museum.
Collections: Native Alaskan art & artifacts; natural history; wildlife mounts; dolls; masks; books; videos; photographs.
Hours & Admission Prices: Summer: daily 9-7; Winter: by appointment. Adults $5, senior citizens & military $4, children under 12 $3.

THE VALDEZ MUSEUM & HISTORICAL ARCHIVE ASSO-CIATION, INC., (M), 217 Egan Ave., Valdez, AK 99686-0008. Mailing Address: P.O. Box 8, Valdez, AK 99686-0008. Tel.: 907-835-2764 & 5407. Fax: 907-835-5800.
E-mail: info@valdezmuseum.org
Web Site: www.valdezmuseum.org
Founded: 1976.
Key Personnel: Pres. Bd. Dir., Daniel Sparrell; Exec. Dir., Patricia Relay; Cur. Collection & Exhibits, Andrew Goldstein; Public Programs, Administrative Svcs. Mgr., & Store Mgr., Rich Dunkin; Cur. Education, Steve Richardson; Attendant, Tim Harned; Attendant, Doreen Hodges.
Personnel Profile: Full-Time Paid 3; Part-Time Paid 5; Part-Time Volunteers 35; Interns 1.
Governing Authority: Tax-exempt: 501(c)(3).
Institution Type/Description: Local History Museum.
Collections: gold & copper mining artifacts; lighthouse lens; historical artifacts of Valdez & area; aviation & transportation artifacts; firefighting apparatus; 1902 handpumper; 1921 Model T Ford Chemical fire truck; 1907 Ahrens Steam Fire Engine; Pinzon Bar; glaciation; 1964 earthquake photos & information; 1989 Exxon Valdez oil spill.
Research Fields: local history.
Facilities: local & regional history archives.
Activities: self-guided tours; permanent & temporary exhibits; local activities.
Publications: quarterly newsletter.
Hours & Admission Prices: Winter: Tues.-Sun. 12-5; Summer: daily 9-5. Adults $7, senior citizens 60 & over $6, youth 14-17 $5; discounts to groups; members & children under 13 no charge. Closed New Year's Day; Thanksgiving; Christmas. &
Attendance: 20,000 (accurate)
Membership: Fellow $20; Supporter $30; Friend $50; Partner $100; Patron & Business $250; Benefactor $500.

Wasilla

DOROTHY G. PAGE MUSEUM, 323 Main St., Wasilla, AK 99654-7021. Tel.: 907-373-9071. Fax: 907-373-9072.
E-mail: museum@ci.wasilla.ak.us
Web Site: www.cityofwasilla.com/museum
Founded: 1967.
Key Personnel: Cur., Bethany Buckingham; Museum Aide, Margaret Rogers.
Personnel Profile: Full-Time Paid 1; Part-Time Paid 2.
Governing Authority: society; nonprofit organization. Parent Institution: City of Wasilla. Subsidiary Institution: Dorothy G. Page Museum & Visitors Center. Tax-exempt: 501(c)(3).
Institution Type/Description: City of Wasilla Museum: housed in 1931 log Wasilla Community Hall.
Collections: photos, tools, household items & equipment used in the Willow Creek Mining area; historic buildings; Iditarod Trail race history.
Research Fields: early settlements of Knik & Wasilla; the Willow Creek Mining area; Iditarod Trail history; Dog Mushing history.
Facilities: historical park.
Activities: guided tours; formally organized education programs for children.
Publications: newsletter.
Hours & Admission Prices: April-Sept. Mon.-Sat. 9-5; Oct.-March Wed.-Fri. 9-5. Adults $3, senior citizens 60 & over $2.50, military $2; discounts to AAM & ICOM members; Historical Society & Museum Alaska members and children 12 & under no charge. Closed New Year's Day; Memorial Day; Independence Day; Labor Day; Thanksgiving; Christmas. &
Attendance: 4,702 (accurate)

IDITAROD TRAIL SLED DOG RACE MUSEUM, Mile 2.2 Knik Goose Bay Rd., Wasilla, AK 99654. Mailing Address: P.O. Box 870800, Wasilla, AK 99687-0800. Tel.: 907-376-5155, ext. 108. Fax: 907-373-6998.
Institution Type/Description: Dog Race Museum.
Collections: Iditarod Trail history; photographs; trophies.
Facilities: Museum-related items for sale.
Activities: videos; dog sled rides.
Hours & Admission Prices: mid-May to mid-Sept. daily 8-7; Winter: Mon.-Fri. 8-5.

INDEPENDENCE MINE STATE HISTORICAL PARK & VISI-TOR CENTER, 7278 E. Bogard Rd., Wasilla, AK 99654. Tel.: 907-745-2827.
Institution Type/Description: Park & Visitor Center.
Collections: local history; wildlife & their habitats.
Activities: educational programs.
Hours & Admission Prices: mid-June to Labor Day Wed.-Sun. 11-6 call for hours. Guided Tours: adults $6, children 6-12 $3; groups by appointment.

KNIK MUSEUM, Mile 13.9 Knik Rd., Wasilla, AK 99654. Mailing Address: 300 N. Boundary St., Wasilla, AK 99654-7128. Tel.: 907-376-2005.
E-mail: wasillaknikhistoricalsociety@yahoo.com
Key Personnel: Dir. & Pres. (V), Dale Myers.
Governing Authority: Parent Institution: Wasilla-Knik Historical Society. Tax-exempt.
Institution Type/Description: History Museum.
Collections: Knik's history & culture; clothing; furniture; period artifacts; Sled Dog Musher's Hall of Fame.
Publications: Wasilla-Knik Historical Society newsletter.
Hours & Admission Prices: June-Sept. 15 Thurs.-Sun. 1-6. Adults $5; members no charge.

MUSEUM OF ALASKA TRANSPORTATION & INDUSTRY, INC., 3800 W. Museum Dr., Wasilla, AK 99654. Mailing Address: P.O. Box 870646, Wasilla, AK 99687-0646. Tel.: 907-376-1211. Fax: 907-376-3082.
Web Site: www.museumofalaska.org/
Founded: 1976.
Key Personnel: Dir., Sherry Jackson; Pres. (V), John Strong.
Personnel Profile: Full-Time Paid 2; Part-Time Paid 2; Part-Time Volunteers 100.
Governing Authority: bd. of trustees; nonprofit organization. Tax-exempt: 501(c)(3).
Institution Type/Description: Alaskan Industrial History & Transportation Museum.
Collections: transportation used in Alaska's history: aircraft; cars; trucks; tractors; railroad rolling stock; fire trucks; boats; snow mobiles; dog sleds; tools & heavy construction equipment also photographs & related materials; railroad memorabilia.
Research Fields: Alaska transportation & Industrial history.
Facilities: 20-acres exhibit area; train yard.
Activities: guided tours; lectures; permanent exhibitions; special events.
Publications: brochures; newsletters.
Hours & Admission Prices: May-Sept. daily 10-5. Adults $8, seniors over 65 & children 3-17 $5; children under 3 & members no charge. &
Attendance: 12,000 (estimated)
Membership: Individual $25; Family $35; Sourdough $100; Pioneer $250; Grubstaker $500; Esteemed & Life $1,000; Life Family $1,500.

Wrangell

NOLAN CENTER MUSEUM, 296 Campbell Dr., Wrangell, AK 99929. Mailing Address: P.O. Box 1050, Wrangell, AK 99929-1050. Tel.: 907-874-3770.
E-mail: museum@wrangell.com
Web Site: www.wrangellalaska.org/museum/index.html
Key Personnel: Dir., Megan Clark
Institution Type/Description: Natural History Museum.
Collections: Wrangell history; natural history; logging; fishing; Native culture.
Hours & Admission Prices: May-Sept. Mon.-Sat. 10-5; Oct.-April Tues.-Sat. 1-5. Adults $5, seniors $3, children 6-12 $2.

WRANGELL MUSEUM, 296 Campbell Dr., Wrangell, AK 99929. Mailing Address: P.O. Box 1050, Wrangell, AK 99929-1050. Tel.: 907-874-3770. Fax: 907-874-3785.
E-mail: museum@wrangell.com

Web Site: www.wrangell.com
Founded: 1967.
Key Personnel: Dir. & Cur., Dennis Chapman.
Personnel Profile: Full-Time Paid 1; Part-Time Paid 3; Part-Time Volunteers 3.
Governing Authority: city. Tax-exempt.
Institution Type/Description: General Museum.
Collections: Southeast Alaska's native cultures; Wrangell history including flying under Russian, English (Hudson Bay) & American flags; wildlife; garnet ledge; Tlingit/Haida, Gold Rush exploration; early fishing (canneries); fur farming; natural history.
Research Fields: reflects Tlingit, Haida & Russian-American culture; Hudson's Bay; European, Asian & Pacific Islander's impact on area pre- and post-contact; cultural history of Wrangell; geologic, geographic & ethnographic history; Gold Rushes; mining; Sitkine River; fur trade; commercial & industrial history.
Facilities: 2,000 sq. ft. exhibit space; video & viewing area. Museum-related items for sale.
Activities: evening & weekend special events. Museum Sponsors: lecture series January to March.
Publications: Garnet Ledge Brochure; Chief Shakes Island; History of Wrangell.
Hours & Admission Prices: May-Sept. Mon.-Sat. 10-5; Oct.-April Tues.-Fri. 10-4. Family $12, adults $5, seniors 60 & up $3, children 6-12 $2; discounts to groups; Friends of Museum no charge. Closed New Year's Day; Easter; Memorial Day; Independence Day; Labor Day; Thanksgiving; Christmas. ♿
Attendance: 8,237 (accurate)
Membership: Student $5; Individual $10; Family $25.

ARIZONA

(248 listings)

Ajo

AJO HISTORICAL SOCIETY, 160 S. Mission Rd., Ajo, AZ 85321-2601. Mailing Address: P.O. Box 778, Ajo, AZ 85321. Tel.: 520-387-7105.
E-mail: walters@tabletoptelephone.com
Web Site: ajomuseum.org
Founded: 1975.
Key Personnel: Pres. (V), G. J. Walters.
Personnel Profile: Part-Time Volunteers 10.
Governing Authority: Tax-exempt.
Institution Type/Description: Historical Society Museum.
Collections: local history & culture; period furnishings; personal artifacts; photographs; blacksmith shop; dentist's office; early print shop; local newspapers & obituaries.
Hours & Admission Prices: Oct.-May daily 12-4. No charge; donations accepted.
Attendance: 2,300 (accurate)
Membership: Annual $12.50; Life $75.

ORGAN PIPE CACTUS NATIONAL MONUMENT MUSEUM, 10 Organ Pipe Dr., Ajo, AZ 85321-9626. Tel.: 520-387-6849. Fax: 520-387-7144.
E-mail: orpi_information@nps.gov
Web Site: www.nps.gov/orpi
Founded: 1937.
Congressional District: 7
Key Personnel: C.E.O. & Supt., Lee Baiza.
Governing Authority: federal. Dept. of the Interior, National Park Service. Tax-exempt.
Institution Type/Description: National Monument.
Collections: natural & cultural history specimens; photographs.
Facilities: interpretive-related items for sale.
Activities: interpretive programs including walks, talks; evening amphitheater. Museum Sponsors: programs January to March.
Publications: guide & map.
Hours & Admission Prices: Jan.-March daily 8-5; April-Dec. daily 8:30-4:30. Weekly vehicle admission $8; Golden Eagle, Golden Age & Golden Access Passport holders & children 16 under with adult no charge. Closed Thanksgiving; Christmas.
Attendance: 22,000 (estimated)

Apache Junction

SUPERSTITION MOUNTAIN LOST DUTCHMAN MUSEUM, 4087 N. Apache Trail, Apache Junction, AZ 85119-8409. Mailing Address: P.O. Box 3845, Apache Junction, AZ 85117-3845. Tel.: 480-983-4888. Fax: 480-288-6524.
E-mail: smhsgold@aol.com
Web Site: www.superstitionmountainmuseum.org
Founded: 1980.
Congressional District: 6
Key Personnel: Exec. Dir., Barbara Atkinson; Pres., Lanna Mesenbrink.
Personnel Profile: Full-Time Paid 2; Part-Time Paid 2; Part-Time Volunteers 20.
Governing Authority: Parent Institution: Superstition Mountain Historical Society. Tax-exempt: 501(c)(3).
Institution Type/Description: History Museum.
Collections: history of Superstition Mountain, Apache Junction & the surrounding region; period artifacts; minerals; mining ore cars; mining equipment; train engines.
Facilities: Elvis Memorial Chapel; The Barn.
Publications: quarterly newsletter, Museum Messenger; annual journal, Superstition Mountain Journal.
Hours & Admission Prices: Daily 9-4. Adults $5, seniors 55 & over $4; discounts to groups; children under 17 no charge. Closed New Year's Day; Thanksgiving; Christmas. ♿
Attendance: 50,000 (estimated)
Membership: Prospector $35.

Benson

KARTCHNER CAVERNS & DISCOVERY CENTER, 2980 S. Hwy. 90, Benson, AZ 85602. Mailing Address: P.O. Box 1849, Benson, AZ 85602-1849. Tel.: 520-586-4100. Fax: 520-586-4113.
Web Site: azstateparks.com
Founded: 1988.
Congressional District: 8
Key Personnel: Dir., Bryan Martyn.
Personnel Profile: Full-Time Paid 16; Part-Time Paid 15; Part-Time Volunteers 90.
Governing Authority: Parent Institution: Arizona State Parks. Tax-exempt.
Institution Type/Description: Natural History Museum.
Collections: local history & culture; natural history; photographs; geology; natural cave.
Facilities: amphitheater; nature trails; garden; cafe. Museum-related items for sale.
Activities: hiking; picnic area; cave tours.
Hours & Admission Prices: June-Oct. daily 8-5; Nov.-May daily 7:30-6. Adults $23, youth 7-14 $13; children 6 & under no charge. Park Entrance Fee: $6 per vehicle. Closed Christmas. ♿
Attendance: 125,000 (accurate)

SAN PEDRO VALLEY ARTS AND HISTORICAL SOCIETY, 180 S. San Pedro, Benson, AZ 85602. Tel.: 520-586-3070.
E-mail: bensonmuseum@theriver.com
Institution Type/Description: Historical Society Museum: housed in a former general store, c.1920.
Collections: local history & culture; early maps; period dental & mining tools; household artifacts; cattle & mining industry history; photographs; clothing.
Hours & Admission Prices: May-July & Sept. Tues.-Sat. 10-2; Oct.-April Tues.-Fri. 10-4, Sat. 10-2.

Bisbee

BISBEE MINING & HISTORICAL MUSEUM, (M), No. 5 Copper Queen Plaza, Bisbee, AZ 85603. Mailing Address: P.O. Box 14, Bisbee, AZ 85603-0014. Tel.: 520-432-7071 & 7848. Fax: 520-432-7800.
E-mail: carrie@bisbeemuseum.org
Web Site: bisbeemuseum.org
Founded: 1971.
Congressional District: 8
Key Personnel: Chm. (V) & Pres. (V), Denise Lundin; Dir., Carrie Gustavson; Cur. Collections, Annie Larkins; Office Mgr., Anna Garcia.
Personnel Profile: Full-Time Paid 2; Part-Time Paid 3; Part-Time Volunteers 70.
Governing Authority: nonprofit organization. Parent Institution: Bisbee Council on the Arts & Humanities; Friends of the Warren Ballpark. Subsidiary Institution: Muheim Heritage House. Tax-exempt: 501(c)(3).
Institution Type/Description: History Museum: housed in the Old General

Office Building of the Phelps Dodge Corp.; National Registered Landmark, a Smithsonian Affiliate.

Collections: photographs; manuscripts; costumes; mining tools & equipment; household & business artifacts; oral histories; mining artifacts. Historic House: 1905 Muheim House.

Research Fields: history of Bisbee; social history; mining; genealogy; Cochise County & Northern Sonora, Mexico.

Facilities: 1,000-vol. library of social & economic history of Bisbee, Cochise County & Arizona, available for research during regular hours or by appointment; 25,000 historic photograph archive. Gift items for sale.

Activities: school programs; lectures; workshops; slide shows; cultural events; permanent & temporary exhibitions.

Publications: Bisbee Pioneer Homes; Bisbee 1880-1920; The Photographer's View; A Century of Change; Mining Town Trolleys; newsletter; Bisbee: Urban Outpost on the Frontier; Bisbee Historic District Walking Tours; Bisbee Yesterday & Today; Meanwhile Back at the Ranch.

Hours & Admission Prices: Daily 10-4. Adults $7.50, senior citizens 60 & over $6.50, children 3-16 $3; discounts to groups of 10 or more; members & local school groups no charge. Closed Thanksgiving; Christmas. &

Attendance: 18,174 (accurate)

Membership: Individual $30; Family $40; Business, Patron & Patron Affiliate with Smithsonian Institution $75; Life $350; Heritage Keeper Annual $500 & up.

Bowie

FORT BOWIE NATIONAL HISTORIC SITE, 13 miles south of Bowie on Apache Pass Rd., Bowie, AZ 85605. Mailing Address: P.O. Box 158, Bowie, AZ 85605-0158. Tel.: 520-847-2500, ext. 2. Fax: 520-847-0113.

E-mail: larry_ludwig@nps.gov

Web Site: www.nps.gov/fobo

Founded: 1964.

Congressional District: 8

Key Personnel: Cur., Larry Ludwig.

Personnel Profile: Full-Time Paid 3; Part-Time Paid 1; Part-Time Volunteers 6.

Governing Authority: federal. Administered by National Park Service, Southern Arizona Group Office, 2120 N. Central Ave., Ste. 120, Phoenix, AZ 85004-1455. Parent Institution: Chiricahua National Monument. Tax-exempt.

Institution Type/Description: Historic Site: 1858-1894 ruins of military structures; 1858 Apache Pass Overland Mail Station.

Collections: history & natural science books pertaining to area; plants; artifacts; archeological artifacts; military records; period photographs; local history; manuscripts.

Facilities: 850-vol. library of books on history of the Indian Wars, southwest & southwestern natural histories, the Frontier Movement in American history & reference books on the park service & museums available for research on premises; Fort Bowie post records on microfilm. History books, postcards, slides & park guides for sale.

Activities: self-guided tours; lectures; permanent exhibitions.

Publications: booklet, Fort Bowie Official Map & Guide; booklet, Trail Guide to Fort Bowie.

Hours & Admission Prices: Daily 8-4:30. No charge. Closed Thanksgiving; Christmas. &

Attendance: 10,000 (accurate)

Buckeye

BUCKEYE MUSEUM, 116 E. Hwy. 85, Buckeye, AZ 85326. Mailing Address: P.O. Box 292, Buckeye, AZ 85326-0024. Tel.: 623-349-6315. Fax: 623-349-6310.

E-mail: clarson@buckeyeaz.gov

Web Site: www.buckeyeaz.gov/museum

Formerly: Buckeye Valley Museum

Founded: 1954.

Congressional District: 2

Key Personnel: Supvr., Christine Larson.

Personnel Profile: Part-Time Paid 1; Part-Time Volunteers 1.

Governing Authority: Parent Institution: Town of Buckeye. Subsidiary Institution: Buckeye Historical Society. Tax-exempt.

Institution Type/Description: History Museum.

Collections: Buckeye history.

Hours & Admission Prices: Sept.-June Mon.-Thurs. by appointment, Fri.-Sat. 11-4. No charge; donations accepted. Closed holidays; Thanksgiving weekend; Christmas weekend.

Attendance: 1,344 (accurate)

Membership: Individual $20; Family $30; Patron $75; Business $100.

LAURIDSEN AVIATION MUSEUM, Buckeye Municipal Airport, 3000 S. Palo Verde Rd., Buckeye, AZ 85326. Tel.: 480-586-7312. Fax: 480-575-6954.

E-mail: hans@lauridsenaviationmuseum.com

Web Site: www.lauridsenaviationmuseum.com

Institution Type/Description: Aviation Museum.

Collections: aircraft including CD-3, PBY-5A; C-1A; C-119G; T-28; Douglas A-26B Invader; TBM-3; T-6D; B-25J bomber.

Hours & Admission Prices: Call for hours.

Bullhead City

COLORADO RIVER MUSEUM, 2201 Hwy. 68, Bullhead City, AZ 86430. Mailing Address: P.O. Box 1599, Bullhead City, AZ 86430-1599. Tel.: 928-754-3399. Fax: 928-754-3399 (call first). Facebook: William Hardy Bullhead City AZ.

E-mail: crhsmuseum@frontier.com

Web Site: crhsmuseum.com

Founded: 1992.

Governing Authority: nonprofit organization. Tax-exempt.

Institution Type/Description: History Museum.

Collections: area history & culture; photographs; mining; ranching; Native American artifacts; mining equipment; Katherine gold mine replica.

Research Fields: Mohave County history.

Facilities: library.

Hours & Admission Prices: Call for hours. Adults $2; discount to Arizona Historical Society members; members no charge. &

Attendance: 2,839 (accurate)

Membership: Individual $15; Family $25; Business $100; Corporate $200; Club & Organization 50 or under $50, 50 & over $100.

Camp Verde

FORT VERDE STATE HISTORIC PARK, 125 E. Holloman, Camp Verde, AZ 86322. Mailing Address: P.O. Box 397, Camp Verde, AZ 86322-0397. Tel.: 928-567-3275. Fax: 928-567-4036.

E-mail: sstubler@azstateparks.gov

Web Site: www.azstateparks.gov

Founded: 1956.

Congressional District: 3

Key Personnel: Park Mgr., Shella Stubler.

Personnel Profile: Full-Time Paid 1; Part-Time Paid 1; Part-Time Volunteers 30.

Governing Authority: state. Parent Institution: Arizona State Parks. Tax-exempt.

Institution Type/Description: State Park General Museum: located on site of Fort Verde, Arizona Territory.

Collections: military artifacts; furniture; textiles; weapons; photographs; manuscripts; four historic buildings 1870-1891.

Research Fields: local military history pertaining to military activities in Arizona from 1860-1899; Indian Wars 1865-1895.

Facilities: 700-vol. library of books on history of military in Arizona available on premises.

Activities: living history presentations; permanent & temporary exhibitions. Annual Events: Reenactments in April & October.

Hours & Admission Prices: Daily 9-5. Adults 14 & over $4, children 7-13 $2; discount to groups & AZ State Park Pass holders; children 6 & under no charge. Closed Christmas. &

Attendance: 30,000 (estimated)

MONTEZUMA CASTLE NATIONAL MONUMENT, 2800 Montezuma Castle Rd., Camp Verde, AZ 86322. Mailing Address: Box 219, Camp Verde, AZ 86322-0219. Tel.: 928-567-3322 & 5276. Fax: 928-567-3597.

E-mail: moca_administration@nps.gov

Web Site: www.nps.gov/moca

Founded: 1906.

Congressional District: 3

Key Personnel: Supvr. Park Ranger, Karen Hughes.

Governing Authority: federal.

Institution Type/Description: Pre-historic Museum.

Collections: archaeology; natural history; Native American artifacts; ethnology.

Facilities: library. Publications for sale.

Activities: lectures; self-guided trail.

Hours & Admission Prices: June-Aug. daily 8-6; Sept.-May daily 8-5. Adults $5; children 15 & under no charge.

Attendance: 500,000 (estimated)

Membership: Golden Age Passport (age 62 & over) $10.

OUT OF AFRICA WILDLIFE PARK, 3505 W. Hwy. 260, Camp Verde, AZ 86322. Mailing Address: 4020 N. Cherry Rd., Camp Verde, AZ 86322. Tel.: 928-567-2840. Fax: 928-567-2839.
E-mail: ashtonooa@gmail.com
Web Site: www.outofafricapark.com
Founded: 1988.
Personnel Profile: Full-Time Paid 32; Part-Time Paid 6.
Institution Type/Description: Park Museum.
Collections: wildlife & their habitats.
Facilities: Museum-related items for sale.
Activities: guided safari tours; shows; feedings; educational activities; hands-on interaction.
Publications: book - Return to Eden; e-mail newsletter.
Hours & Admission Prices: Daily 9:30-5. Adults $29.95, seniors 65 & over $27.95, veterans & active military $22, children 3-12 $14.95; discounts to AAM members; children under 3 no charge. Closed Thanksgiving; Christmas. &
Membership: Child $50; Senior $70; Adult $80.

VERDE VALLEY ARCHAEOLOGY CENTER & MUSEUM, 385 S. Main St., Camp Verde, AZ 86322. Tel.: 928-567-0066.
E-mail: center@verdevalleyarchaeology.org
Web Site: www.verdevalleyarchaeology.org/museum
Key Personnel: Exec. Dir., Kenneth Zoll
Institution Type/Description: Archaeology Museum.
Collections: local history & culture; Native American artifacts; Sinagua culture; rock art & pottery; archaeology; Yavapai-Apache Nation.
Facilities: Gift items for sale.
Hours & Admission Prices: Aug. Thurs.-Mon. 10-4; Sept.-July Wed.-Mon. 10-4. No charge. Closed major holidays.

Carefree

DI TOMMASO GALLERIES, 30 Easy St., Carefree, AZ 85377. Tel.: 480-575-1023.
Key Personnel: Owner, John Di Tommaso
Institution Type/Description: Art Gallery.
Collections: 19th-20th century art; African art; furniture.
Hours & Admission Prices: Call for hours.

Casa Grande

CASA GRANDE VALLEY HISTORICAL SOCIETY & MUSEUM, (M), 110 W. Florence Blvd., Casa Grande, AZ 85122. Tel.: 520-836-2223. Fax: 520-836-5065.
E-mail: info@cgvhs.org
Web Site: www.cgvhs.org
Formerly: Casa Grande History Museum
Founded: 1964.
Congressional District: 5
Key Personnel: Pres., James Sommers; Archivist, Kay Benedict.
Personnel Profile: Part-Time Paid 2; Part-Time Volunteers 22.
Governing Authority: society. Affiliated with the Arizona Historical Society, 949 E. 2nd St., Tucson. Tax-exempt: 501(c)(3).
Institution Type/Description: History Museum.
Collections: local history; period artifacts; manuscripts; photographs. Historic Building: schoolhouse.
Research Fields: local history.
Facilities: library of books on Arizona history & old newspaper files available on premises; auditorium.
Activities: education program; site preservation projects; guided tours; lectures; permanent & temporary exhibitions.
Publications: newsletter to members & others interested; monographs of local history; newspaper articles on local history; annual journal.
Hours & Admission Prices: Sept. 15-May 15 Thurs.-Sun. 12-4; special tours by appointment. Adults $5, seniors 60 & over $4; children under 16 & society member no charge. Closed major holidays. &
Attendance: 5,300 (estimated)

Cave Creek

CAVE CREEK MUSEUM, (M), 6140 E. Skyline Dr., Cave Creek, AZ 85327. Mailing Address: P.O. Box 1, Cave Creek, AZ 85327. Tel.: 480-488-2764. Fax: 480-595-0838.
E-mail: info@cavecreekmuseum.com
Web Site: cavecreekmuseum.com
Founded: 1968.
Key Personnel: Dir., Evelyn Johnson; Chm. & Pres. (V), Pam DiPietro; Administrative Asst., Jean Marsh; Museum Shop Mgr., Faith Pipp; Tech Asst., Karen Friend

Institution Type/Description: History Museum.
Collections: desert foothills history; pioneer living, ranching & mining; Hohokam, Yavapai, & Apache artifacts. Historic Buildings: restored 1920s tuberculosis cabin; 1940s church.
Facilities: Museum-related items for sale.
Hours & Admission Prices: Oct.-May Wed.-Thurs. & Sat.-Sun. 1-4:30, Fri. 10-4:30. Adults $3, seniors 60 & over and students $2; discounts to groups & AAM members. Closed New Year's Day; Easter; Christmas. &
Attendance: 5,000 (estimated)

Chandler

ARIZONA RAILWAY MUSEUM, 330 E. Ryan Rd., Chandler, AZ 85224. Mailing Address: P.O. Box 842, Chandler, AZ 85244-0842. Tel.: 480-821-1108.
E-mail: azrymuseum@cox.net
Web Site: www.azrymuseum.org
Founded: 1983.
Congressional District: 5
Key Personnel: Pres. (V), Larry Benedict; Museum Shop Mgr., Karen Chimel.
Personnel Profile: Part-Time Volunteers 25.
Volunteer Hours: 10,000
Operating Expenses: 81,268
Operating Income: 20,078
Governing Authority: nonprofit organization. Tax-exempt.
Institution Type/Description: Railway Museum.
Collections: railway history of Arizona & the Southwest; railway equipment & artifacts.
Activities: special events.
Hours & Admission Prices: Labor Day to Memorial Day Sat.-Sun. 12-4; other times by appointment. Work Sessions: Sat. 8am to 12pm. No charge; donations accepted. NRHS members 10% off gift shop purchases.
Attendance: 8,000 (estimated)
Membership: Senior Citizen 62 & over $20; Regular $25; Family $30; Sustaining $50-$499; Life $300; Corporate $500 & up.

CHANDLER MUSEUM, (M), 300 S. Chandler Village Dr., Chandler, AZ 85226. Mailing Address: MS 305, P.O. Box 4008, Chandler, AZ 85244-4008. Tel.: 480-782-2717. Fax: 480-782-2875.
Web Site: www.chandlerpedia.com
Founded: 1969.
Congressional District: 6
Key Personnel: Dir., Jody A. Crago; Cur. Education, Tiffani Rigitero; Cur. Collections, Nate Meyers.
Personnel Profile: Full-Time Paid 3; Part-Time Paid 1; Part-Time Volunteers 65; Interns 3.
Governing Authority: city. Partners of Tumbleweed Ranch; Tumbleweed Ranch, 2250 S. McQueen Rd., Chandler, AZ; Chandler Sports Hall of Fame. Tax-exempt.
Institution Type/Description: History Museum.
Collections: Chandler area history, art, & cultural artifacts; period records & photographs; Indian artifacts; local artifacts; manuscripts.
Research Fields: oral history; photographs.
Facilities: Tumbleweed Ranch; McCullough-Price House.
Activities: permanent & temporary exhibitions; school & group tours; lectures; archival reference service.
Publications: Images of America: Chandler.
Hours & Admission Prices: Tues.-Wed. & Fri.-Sat. 10-4, Thurs. 10-7. No charge; donations accepted. Closed New Year's Day; Martin Luther King Jr. Day; Memorial Day; Independence Day; Labor Day; Veterans Day; Thanksgiving; Christmas. &
Attendance: 15,000 (estimated)

HUHUGAM HERITAGE CENTER, (M), 4759 N. Maricopa Rd., Chandler, AZ 85226-5203. Mailing Address: P.O. Box 5041, Chandler, AZ 85226. Tel.: 520-796-3500.
Institution Type/Description: Native American Museum.
Collections: Native American culture & history; baskets; paintings.
Hours & Admission Prices: Wed.-Fri. 10-4. Adults $5, seniors & students $3, children 6-12 $2.

VISION GALLERY, 10 E. Chicago St., Chandler, AZ 85225. Mailing Address: P.O. Box 4008, Chandler, AZ 85244-4008. Tel.: 480-782-2695. Fax: 480-782-2699.
E-mail: vision.gallery@chandleraz.gov
Web Site: www.visiongallery.org
Founded: 1996.

Personnel Profile: Full-Time Paid 2; Full-Time Volunteers 4; Part-Time Paid 1; Part-Time Volunteers 30.
Governing Authority: Parent Institution: Chandler Cultural Foundation. Tax-exempt.
Institution Type/Description: Art Gallery.
Collections: works by contemporary artists.
Activities: educational programs.
Hours & Admission Prices: Mon.-Fri. 10-5, Sat. 10-4. No charge; donations accepted. &
Attendance: 10,000 (estimated)

Chinle

CANYON DE CHELLY NATIONAL MONUMENT, 3 mi. east of Chinle, Rte. 7, Chinle, AZ 86503. Mailing Address: P.O. Box 588, Chinle, AZ 86503-0588. Tel.: 928-674-5500. Fax: 928-674-5507.
Web Site: www.nps.gov/cach
Founded: 1931.
Congressional District: 4
Key Personnel: Park Archeologist, Keith Lyons.
Personnel Profile: Full-Time Paid 1; Part-Time Paid 2.
Governing Authority: federal. Parent Institution: Dept. of the Interior; National Park Service. Tax-exempt.
Institution Type/Description: Park Museum.
Collections: archaeology; anthropology; prehistoric Navajo cultures.
Activities: ranger hikes; campfire programs; off-site interpretive programs upon request.
Publications: orientation brochure; park newspaper, Canyon Overlook.
Hours & Admission Prices: Visitor Center: daily 8-5. No charge; donations accepted. Closed Christmas. &
Attendance: 1,000,000 (accurate)

Clarkdale

TUZIGOOT NATIONAL MONUMENT, 25 W. Tuzigoot Rd., Clarkdale, AZ 86324. Mailing Address: P.O. Box 219, Camp Vere, AZ 86322-0219. Tel.: 928-634-5564 & 567-5276. Fax: 928-567-3597.
E-mail: MOCA_administration@nps.gov
Web Site: www.nps.gov/tuzi
Founded: 1939.
Congressional District: 3
Key Personnel: Supt., Kathy Davis; Supvr., Karen Hughes.
Governing Authority: federal. Parent Institution: National Park Service, Dept. of Interior, Washington, DC. Tax-exempt.
Institution Type/Description: Park Museum & Visitor Center: located on the Sinagua Indian ruins.
Collections: artifacts found during excavation of Tuzigoot & nearby ruins. Historic Building: 1125-1300 Tuzigoot National Monument.
Research Fields: archaeology; anthropology; natural sciences.
Facilities: 300-vol. library of material on archeology, anthropology, botany & zoology available for inter-library loan. Books, slides & postcards for sale.
Activities: guided tours; lectures; films; formally organized education programs for children; permanent exhibitions.
Publications: brochures, published through National Park Service.
Hours & Admission Prices: Winter: daily 8-5; Summer: daily 8-6. Adults $5; children 15 & under and senior citizens with Golden Age Passport no charge.

Clifton

GREENLEE COUNTY HISTORICAL SOCIETY, 299 Chase Creek, Clifton, AZ 85533. Mailing Address: P.O. Box 787, Clifton, AZ 85533. Tel.: 602-865-3115. Facebook: Greenlee County Historical Society.
E-mail: greenleecountyhs74@gmail.com
Web Site: www.gchistoricalsociet.com
Founded: 1974.
Congressional District: 1
Key Personnel: Pres. (V), Jo Lunt; Museum Shop Mgr., Helen Brinkley.
Governing Authority: Tax-exempt.
Institution Type/Description: Historical Society Museum.
Collections: local history & culture; period furnishings & clothing; photographs; personal artifacts; Native American artifacts; sheet music; mining; maps; rocks & minerals; early saddles; blacksmith shop; farm machinery & equipment; books; films.
Facilities: library; archives. Museum-related items for sale.
Activities: guided tours.
Publications: 3 times a year, Chronicle Paper.

Hours & Admission Prices: Tues., Thurs. & Sat. 2-4:30; tours by appointment. No charge; donations accepted. &
Attendance: 1,000 (estimated)
Membership: Single $12; Family $18.

Coolidge

CASA GRANDE RUINS NATIONAL MONUMENT, 1100 W. Ruins Dr., Coolidge, AZ 85128. Tel.: 520-723-3172. Fax: 520-723-7209.
Web Site: www.nps.gov/cagr
Founded: 1892.
Congressional District: 5
Key Personnel: Administrative Officer, Diana Mills.
Personnel Profile: Full-Time Paid 8; Part-Time Paid 6.
Governing Authority: federal. Affiliated with the National Park Service, Interior Bldg., Washington, DC. Tax-exempt.
Institution Type/Description: Archaeology & Park Museum: located on Hohokam village site dating to approx. A.D. 500-1450.
Collections: Hohokam archaeology; local ethnology; pre-Columbian Hohokam Indian artifacts. Prehistoric Sites: The Casa Grande: c.1300-1350; Hohokam village c.1200-1400.
Research Fields: Hohokam culture archaeology.
Facilities: 1,928-vol. library of archaeology books available on premises by arrangement. Archaeology, ethnology & natural history books of Arizona for sale.
Activities: permanent & temporary exhibitions; ranger talks.
Publications: Archaeology Reports; Technical Series.
Hours & Admission Prices: Daily 9-5. Adults $5; children 15 & under and educational groups no charge. Golden Eagle, Golden Age, Golden Access Passports honored. Closed Thanksgiving; Christmas. &
Attendance: 85,000 (accurate)

COOLIDGE HISTORICAL SOCIETY, 161 W. Harding Ave., Coolidge, AZ 85228. Mailing Address: P.O. Box 1186, Coolidge, AZ 85228. Tel.: 520-723-7186.
Institution Type/Description: Historical Society Museum.
Collections: local history & culture; period furnishings; photographs; personal artifacts.
Hours & Admission Prices: Sun. 1-4; other times by appointment.

Cottonwood

VERDE HISTORICAL SOCIETY, CLEMENCEAU HERITAGE MUSEUM, 1 N. Willard, Cottonwood, AZ 86326-3651. Mailing Address: P.O. Box 511, Cottonwood, AZ 86326-0511. Tel.: 928-634-2868.
E-mail: clemenceauheritagem@qwestoffice.net
Web Site: www.clemenceaumuseum.org; www.clemenceaumuseum.com
Founded: 1989.
Congressional District: 4
Key Personnel: Dir., Helen Killebrew; Pres. (V), & Museum Shop Mgr., Barbara Evans; Chm. (V), Bob Laning; Treas., Connie Phillips.
Personnel Profile: Part-Time Volunteers 62.
Volunteer Hours: 7,028
Operating Expenses: 38,995
Operating Income: 42,593
Governing Authority: private; nonprofit organization. Tax-exempt: 501(c)(3).
Institution Type/Description: Historical Society Museum: housed in 1923-1924 school building.
Collections: history of the Verde Valley; local artifacts.
Major Exhibits: Traveling Vietnam Wall Replica Exhibit, 3/13-3/14; Writing Instruments/Tools, 8/13-3/14; Men's Grooming Items, 8/13-3/14; Local Branding Irons, 10/13-7/14; Author Zane Grey Memorabilia, 9/13-3/14; Verde Valley Native American Pots/Baskets, 9/13-6/14; Die Cast Models of Farm Equipment, 8/13-6/14; Antique Weapons, 6/13-6/14; Medical Tools/Supplies, 6/13-6/14; John Wayne Memorabilia, 9/13-3/14.
Facilities: 600-vol. library; classroom. Museum-related items for sale.
Activities: guided tours. Museum Sponsors: Bar-B-Que; American Arts/Crafts Show; Early Settler Day; demonstrations & talks last Friday of month.
Publications: quarterly newsletter, The Voice; book, Cottonwood Pictorial History; Cottonwood, Clarkdale & Cornville History; A History of Verde Village; M. Clemenceau's Cookbook.
Hours & Admission Prices: Wed. 9-12, Fri.-Sun. 11-3. No charge; donations accepted. Closed New Year's Day; Easter; Independence Day; Thanksgiving; Christmas. &
Attendance: 2,410 (accurate)
Membership: Individual $10; Family $15; Business $50.

Douglas

BORDER AIR MUSEUM, 3200 E. 10th St., Douglas, AZ 85607. Mailing Address: 345 16th St., Douglas, AZ 85607. Tel.: 520-364-2478.
Institution Type/Description: Aviation History Museum.
Collections: aviation history; newspaper articles; photographs; American Airlines memorabilia; maps.
Hours & Admission Prices: Summer: Mon.-Fri. 12-4, Sat. 11-3.

THE DOUGLAS HISTORICAL SOCIETY - DOUGLAS-WILLIAMS HOUSE, 1001 D Ave., Douglas, AZ 85607.
E-mail: info.doughistsoc@qwestoffice.net
Web Site: www.douglasazhistoricalsociety.org
Founded: 1990.
Key Personnel: Dir. & C.E.O., Lavinia C. Spivey
Governing Authority: Parent Institution: The Douglas Historical Society. Tax-exempt.
Institution Type/Description: Historical Society Museum.
Collections: local history & culture; period furnishings; photographs; local genealogy.
Facilities: library of local history & genealogy materials.
Publications: quarterly newsletter.
Hours & Admission Prices: Wed. 10-4, Sat. 12-4. No charge; donations accepted.
Attendance: 600 (estimated)
Membership: Individual $10; Immediate Family $12; Business $20; Supporting $25; Sustaining $50; Patron $100; Life Sponsor $500; Benefactor $1,000.

SLAUGHTER RANCH MUSEUM, 6153 Geronimo Trail, 16 miles East of Douglas, Douglas, AZ 85608. Mailing Address: P.O. Box 438, Douglas, AZ 85608-0438. Tel.: 520-678-7935. Fax: 602-933-3777.
E-mail: sranch@vtc.net
Web Site: slaughterranch.com
Founded: 1982.
Key Personnel: C.E.O. (V), Harvey Finks; Pres. (V), Alan D. Finks; Education & Historian, Dr. Reba Grandrud.
Personnel Profile: Full-Time Paid 4; Part-Time Paid 2; Part-Time Volunteers 2.
Governing Authority: nonprofit. Parent Institution: Johnson Historical Museum of the Southwest. Tax-exempt: 501(c)(3).
Institution Type/Description: Historic Site: located on c.1893 John Slaughter Ranch.
Collections: costumes; furniture; c.1890 ranch & farm implements; 1,000 photos. Historic Buildings: ranch buildings; 1913 U.S. Army 10th Cavalry outpost from Pancho Villa era.
Facilities: 30-vol. library relating to John Slaughter & San Bernardino Valley; 20-seat theater.
Activities: guided tours; films.
Hours & Admission Prices: Wed.-Sun. 9:30-3:30. Adults $8; children no charge. Closed New Year's Day; Christmas. &
Attendance: 4,000 (estimated)

Dragoon

THE AMERIND MUSEUM, (M), 2100 N. Amerind Rd., Dragoon, AZ 85609. Mailing Address: P.O. Box 400, Dragoon, AZ 85609-0400. Tel.: 520-586-3666. Fax: 520-586-4679.
E-mail: amerind@amerind.org
Web Site: www.amerind.org
Founded: 1937.
Congressional District: 8
Key Personnel: Chm., Michael W. Hard; Pres., George J. Gumerman, Phd.; Dir., Dr. John A. Ware; Cur., Dr. Eric Kaldahl; Museum Shop Mgr., Tammy Stansberry.
Personnel Profile: Full-Time Paid 8; Part-Time Paid 11; Part-Time Volunteers 60.
Governing Authority: nonprofit organization. Tax-exempt: 501(c)(3).
Institution Type/Description: Archaeology & Ethnology Museum and Art Gallery.
Collections: archaeological collections from North America, Mesoamerica & South America; ethnological material from the Southwest, Mexico, Great Plains, Eastern Woodlands, California & the Arctic; oil paintings; sculpture; santos; ivory & scrimshaw; period furniture.
Major Exhibits: Rock Art: A to Z, 3/13-2/14.
Research Fields: anthropology; archaeology; ethnohistory of the Americas.
Facilities: 25,000-vol. library of research and technical reports, journals, periodicals, manuscripts & books on anthropology, archaeology & natural history available for use by appointment; reading room; separate laboratory operation.
Activities: guided tours; lectures; seminars; artist shows; Visiting Scholar Program; children's educational programs.
Publications: Amerind Foundation Archeology Series; Amerind New World Studies Series.
Hours & Admission Prices: Business Office: Mon.-Fri. 8-5. Museum: Tues.-Sun. 10-4. Adults $8, senior citizens $7, children 12-18 $5; members & children under 12 no charge. Closed New Year's Day; Easter; Memorial Day; Independence Day; Labor Day; Thanksgiving; Christmas. &
Attendance: 13,000 (accurate)
Membership: Individual $40; Family $50; Grandparents $60; Cochise Club $100; San Pedro Club $500; Casas Grandes Club $1,000; Dragoon Circle $2,500; Texas Canyon Circle $5,000; Fulton Society $10,000 & up.

Flagstaff

THE ARBORETUM AT FLAGSTAFF, 4001 S. Woody Mountain Rd., Flagstaff, AZ 86005. Tel.: 928-774-1442. Fax: 928-774-1441.
E-mail: tom.parker@thearb.org
Web Site: www.thearb.org
Founded: 1981.
Congressional District: 6
Key Personnel: Dir., Lynne Nemeth; Pres., Dr. Peter Jolma; Deputy Dir., Tom Parker; Dir. Horticulture, Mark Jarecki; Coord. Education, Regan Emmons; Mgr. Finance, Randi Axler; Mgr. Facilities, Dee James.
Personnel Profile: Full-Time Paid 7; Part-Time Paid 7; Part-Time Volunteers 100; Interns 4.
Governing Authority: nonprofit organization. Parent Institution: Transition Zone Horticultural Institute (TZHI). Tax-exempt: 501(c)(3).
Institution Type/Description: Arboretum.
Collections: native plants of the Colorado Plateau including rare & endangered species.
Research Fields: botanical; forestry.
Facilities: horticulture center; meeting rooms; classrooms; 200 acres of natural & cultivated landscapes; outdoor concert area; 1.6 mile nature trail; constructed wetlands; visitor center. Museum-related items for sale.
Activities: docent program; guided tours; bird walks; children's summer adventure program; science in the park; summer concert series; wildflower walks. Annual Events: Plant Sale & Penstemon Festival in July; Arbor Day Celebration; Fall Open House in October.
Publications: annual report; general brochure; membership brochure; event postcards.
Hours & Admission Prices: May-Oct. Wed.-Mon. 10-4. Guided Tours: daily 11 & 1. Adults $7, seniors $6, children 3-17 $3; children under 3 & members no charge. &
Attendance: 18,709 (accurate)
Membership: Individual $35; Family $50; Cliffrose $100; Ponderosa $250; Columbine Club $500; Penstemon Society $1,000.

ARIZONA HISTORICAL SOCIETY PIONEER MUSEUM, 2340 N. Fort Valley Rd., Flagstaff, AZ 86001-1200. Mailing Address: 949 E. 2nd St., Tucson, AZ 85719-4898. Tel.: 928-774-6272. Fax: 928-774-1596.
E-mail: ahsflagstaff@azhs.gov
Web Site: www.arizonahistoricalsociety.org
Congressional District: 1
Key Personnel: Exec. Dir., Anne I. Woosley, Ph.D.; Northern Arizona Division Dir., William Peterson.
Personnel Profile: Full-Time Paid 2; Part-Time Paid 1; Part-Time Volunteers 31; Interns 2.
Governing Authority: bd. of directors. Parent Institution: Arizona Historical Society, 949 E. 2nd St., Tucson, AZ 85719. Tel.: 520-628-5774. Tax-exempt: 501(c)(3).
Institution Type/Description: History Museum.
Collections: industrial, social, institutional & transportation history of Flagstaff & northern Arizona including logging, railroad, livestock, farming and domestic items; 1929 Baldwin logging locomotive #12; Santa Fe Railroad caboose; archives housed at Northern Arizona University. Historic Buildings: 1908 Coconino County Hospital; 1908 Ben Doney homestead cabin.
Research Fields: general history of northern Arizona; lumber industry; community development; architecture; hispanic population.
Facilities: archives of 1,300 linear feet & approximately 30,000 photographs. Museum-related items for sale.
Activities: changing exhibits; school tours. Annual Events: Independence Day Festival; Wool Festival; Heritage Festival.
Publications: quarterly, The Journal of Arizona History; books; monographs; annual report.
Hours & Admission Prices: Mon.-Sat. 9-5. Adults $6, seniors 65 & over and

students & military with ID $5, youth 7-17 $3; children 6 & under and members no charge. Closed New Year's Day; Martin Luther King Jr. Day; Presidents' Day; Columbus Day; Veterans Day; Thanksgiving; Christmas.
Attendance: 12,000 (estimated)
Membership: Student $25; Individual $50; Household $65; Sustaining $100; Patron $250; Sponsor $500; Director's Circle $1,000.

COLORADO PLATEAU BIODIVERSITY CENTER, Northern Arizona University, Flagstaff, AZ 86011. Mailing Address: Northern Arizona University, Campus Box 5640, Flagstaff, AZ 86011. Tel.: 928-523-4463. Fax: 928-523-7500.
E-mail: stefan.sommer@nau.edu
Web Site: www.mpcer.nau.edu/cpbc
Founded: 2004.
Key Personnel: Dir. & Assoc. Cur. Arthropods, Dr. Stefan Sommer; Cur. Arthropods, Dr. Neil Cobb; Cur. Botany, Dr. Randall Scott; Cur. Botany, Dr. Tina Ayers; Cur. Botany, Dr. David Hammond; Cur. Mammals, Dr. Tad Theimer; Cur. Ichthyology, Dr. Linn Montgomery; Cur. Mycology, Dr. Kitty Gehring; Cur. Genetics & Genomics, Dr. Gery Allan; Cur. Palynology, Dr. Scott Anderson; Cur. Fossil Middens, Dr. Kenneth Cole; Cur. Marine Invertebrates & Molluscs, Dr. Stephen Shuster; Mgr. Paleontology Collections, Sandra Swift; Mgr. Palynology Collections, Susan Smith; Mgr. Arthropod Collections, Jacob Higgins.
Personnel Profile: Full-Time Paid 10; Full-Time Volunteers 4; Part-Time Paid 17; Part-Time Volunteers 25; Interns 15.
Governing Authority: public university. Parent Institution: Northern Arizona University. Subsidiary Institutions: Colorado Plateau Museum of Arthropod Biodiversity; Deaver Herbarium; NAU Laboratory of Paleoecology; Marine Invertebrates & Molluscs Collection; Vertebrate Collections; Environmental Genetics & Genomics Lab. Tax-exempt: 501(c)(3).
Institution Type/Description: Natural History Museum.
Collections: entomology; botany; ichthyology; mammalogy; ornithology; mycology; DNA & tissue collections; palynology; paleontology; herpetology; marine invertebrates & molluscs; tree ring; fossil middens; fossil dung.
Research Fields: entomology; botany; paleontology; genetics; palynology; mycology; mammalogy; ornithology; ichthyology; herpetology; microbiotic crust ecology & diversity; marine invertebrate & mollusc evolution; biodiversity surveys; climate change; genes to ecosystems; environmental garden array.
Facilities: library; 3 auditoriums; educational facilities; 2,000 sq. ft. exhibit space; field research station.
Activities: films; formal education programs for adults & college students; education programs for adults & children; lectures; school loan service; temporary exhibitions; training programs for professional museum workers; broadcast programs; public programs. Annual Events: scientific conferences on the Colorado Plateau; Flagstaff Festival of Science; Mountain Campus Science Day.
Hours & Admission Prices: No charge; donations accepted. &

Attendance: 15,000 (estimated)

LOWELL OBSERVATORY, 1400 W. Mars Hill Rd., Flagstaff, AZ 86001-4499. Tel.: 928-774-3358. Fax: 928-774-6296.
E-mail: kevin@lowell.edu
Web Site: www.lowell.edu
Founded: 1894.
Congressional District: 3
Key Personnel: Dir., Jeffrey C. Hall; Trustee, William Lowell Putnam; Museum Shop Mgr., Diana Weintraub.
Personnel Profile: Full-Time Paid 90; Part-Time Paid 17; Part-Time Volunteers 40; Interns 2.
Governing Authority: private. Tax-exempt.
Institution Type/Description: Astronomy Museum & Observatory.
Collections: astronomical photographs; historic & current research instruments; astronomy exhibits; 11 astronomical telescopes.
Research Fields: astronomy; planetary science.
Facilities: 10,000-vol. library of astronomy & mathematic books available for use by special permission; visitor center.
Activities: research; guided tours; school programs; family workshops; youth science programs & nighttime telescope viewings.
Publications: Lowell Observatory Bulletin; quarterly newsletter: The Lowell Observer.
Hours & Admission Prices: Call for hours. Adults $11, children 5-17 $4; discounts to AAA members; members no charge. Closed New Year's Eve & Day; Easter; Independence Day; Thanksgiving; Christmas Eve & Day. &
Attendance: 75,130 (accurate)
Membership: Individual $35; Basic $60; Primary $100; Contributor $250; Pluto $500; Lowell Associate $1,000; Trustee Circle $2,500; Discovery $5,000.

* **MUSEUM OF NORTHERN ARIZONA, (M),** 3101 N. Fort Valley Rd., Flagstaff, AZ 86001-8348. Tel.: 928-774-5213. Fax: 928-779-1527. Facebook: Museum of Northern Arizona.
E-mail: info@mna.mus.az.us
Web Site: www.musnaz.org
Founded: 1928.
Congressional District: 1
Key Personnel: Dir., Robert G. Breunig, Ph.D.; Chm., La Velle McCoy; Colbert Cur. Vertebrate Paleontology, David D. Gillette, Ph.D.; Cur. Ecology & Conservation, Dr. Lawrence E. Stevens; Danson Chair Anthropology, Dr. Kelley Hays-Gilpin; Mgr. Education Program, Kathy Farretta; Collections Mgr., Elaine Hughes; Mktg. Mgr., Michele Mountain; Dir. Devel., David Kelner; Museum Shop Mgr., Kelly Kavanaugh.
Personnel Profile: Full-Time Paid 29; Part-Time Paid 14; Part-Time Volunteers 300; Interns 12.
Governing Authority: private; nonprofit organization. Bd. of Trustees of Museum of Northern Arizona Inc. Tax-exempt: 501(c)(3).
Institution Type/Description: History Museum: building built in 1928.
Collections: prehistoric & ethnographic Southwest Indian artifacts; ceramic repository; geological & paleontological material; herbarium; zoological specimens; Southwestern Anglo & Indian art; archives. Historic Houses: c.1886 The Homestead; Colton House complex, c.1906.
Research Fields: archaeology; ethnology; geology; paleontology; biology; American Indian art.
Facilities: 99,000-vol. reference library; auditorium; nature trail.
Activities: expeditions into backcountry of Colorado Plateau; tours; workshops; inter-museum loan, permanent, temporary & traveling exhibitions; children's programs; lectures; special arts & craft exhibits and demonstrations; lecture series; rental facilities available.
Publications: triannual members newsletter, Museum Notes; biannual magazine, Plateau; occasional research publications.
Hours & Admission Prices: Daily 9-5. Adults $10, senior citizens 65 & over $9, students w/ID $7, youth 10-17 & American Indians $6; children under 10 & members no charge. Closed New Year's Day; Thanksgiving; Christmas. &
Attendance: 60,000 (accurate)
Membership: Painted Desert Affiliate $60; Sunset Crater Colleague $100; Canyonlands Contributor $150; Chaco Sponsor $275; Black Mesa Assoc. $500; Kaibab Fellow $1,000; Mesa Verde Circle $2,500; Grand Canyon Associate $5,000 & up.

NAU ART MUSEUM, (M), Knoles & McMullen Circle, Bldg. #10, N. NAU Campus, Flagstaff, AZ 86011. Mailing Address: P.O. Box 6021, Flagstaff, AZ 86011. Tel.: 928-523-3471. Fax: 928-523-1424.
E-mail: art.museum@nau.edu
Web Site: www4.nau.edu/art_museum
Formerly: Northern Arizona University Art Museum and Galleries
Founded: 1961.
Congressional District: 3
Key Personnel: Dir., George V. Speer.
Personnel Profile: Full-Time Paid 1; Part-Time Paid 6.
Governing Authority: state. Affiliated with Northern Arizona University. Tax-exempt.
Institution Type/Description: Art Museum.
Collections: local, state, national & international historic paintings; turn of the century furniture; graphics; paintings; ceramics; sculpture.
Facilities: Artists artwork, baskets, glass & jewelry for sale.
Activities: guided tours; lectures; gallery talks; artists workshops; temporary, traveling & original exhibitions.
Hours & Admission Prices: Museum: Tues.-Sat. 12-5. Beasley Gallery: Mon.-Fri. 11-4. No charge; donations accepted. Closed legal & university holidays. &
Attendance: 40,000 (estimated)

RIORDAN MANSION STATE HISTORIC PARK, 409 W. Riordan Rd., Flagstaff, AZ 86001-6440. Mailing Address: 949 E. 2nd St., Tucson, AZ 85719-4898. Tel.: 928-779-4395. Fax: 928-556-0253.
Web Site: www.azstateparks.com
Founded: 1983.
Congressional District: 1
Key Personnel: Cur., Joe Meehan.
Personnel Profile: Part-Time Paid 4; Part-Time Volunteers 30; Interns 1.
Governing Authority: state; nonprofit. Parent Institution: State of Arizona. Subsidiary Institutions: Arizona State Parks, 1300 W. Washington, Phoenix, AZ 85007; Arizona Historical Society.

Institution Type/Description: Historic Mansion: c.1904 site containing a collection of original Craftsman furnishings.
Collections: original furnishings, c.1904; fixtures; furniture; clothing; textiles; decorations; correspondence; photographs; utensils; family mementos.
Facilities: over 13,000 sq. ft. mansion; visitor center; picnic areas.
Activities: guided tours of mansion; self-guided tours of estate surrounding mansion; slide program; children's touch table.
Publications: The Riordan Family of Flagstaff; Riordan Family Recipes.
Hours & Admission Prices: June-Aug. daily 9:30-5; Sept.-May Thurs.-Mon. 10:30-5; guided tours on the hour by appointment. Adults $10, youth 7-13 $5; children under 6 no charge. Closed Thanksgiving; Christmas. &

Attendance: 22,000 (estimated)

SUNSET CRATER VOLCANO NATIONAL MONUMENT, 6400 N. U.S. Hwy. 89, Flagstaff, AZ 86004-2759. Tel.: 928-527-0322. Fax: 928-526-4259.
Web Site: www.nps.gov
Founded: 1930.
Congressional District: 6
Key Personnel: Supt., Diane Chung; Cur., Gwenn Gallenstein.
Governing Authority: federal. Parent Institution: Dept. of the Interior, National Park System. Tax-exempt.
Institution Type/Description: Park Museum & Visitor Center.
Collections: geologic specimens of the area, the primary theme being the story of volcanic action which took place in the San Francisco Peaks Volcanic Field; pottery used by people of the period 1064-1250 A.D.
Research Fields: vulcanism; Sinagua archaeology.
Facilities: visitor center. Postcards, slides, film & publications for sale.
Activities: guided tours; films.
Publications: guide leaflet; orientation brochure.
Hours & Admission Prices: Summer: daily 8-6; Winter: daily 8-5. Admission: $5 per person; discounts to Museum Association of Arizona & AAM members; American disabled citizen and children 16 & under no charge. Golden Age passport available. Closed Christmas. &

Attendance: 300,000 (estimated)

WALNUT CANYON NATIONAL MONUMENT, 6400 N. Hwy. 89, Flagstaff, AZ 86004-2759. Tel.: 928-527-0322. Fax: 928-526-4259.
Web Site: www.nps.gov
Founded: 1915.
Congressional District: 3
Key Personnel: Supt., Diane Chung; Cur., Gwenn Gallenstein.
Governing Authority: federal. Parent Institution: Dept. of the Interior, National Park Service. Tax-exempt.
Institution Type/Description: Park Museum: located on the site of prehistoric Indian ruins of the Sinagua Indian culture.
Collections: local history & culture; photographs; period artifacts.
Facilities: library of Sinagua cultural material available under direct supervision of park personnel for on-premise use only. Traveling information, postcards, slides & film for sale.
Activities: self-guided trails.
Publications: brochure; Walnut Canyon Guide.
Hours & Admission Prices: Summer: daily 8-6; Winter: daily 8-5. Admission: $5 per person; discounts to AAM & Museum Association of Arizona members. Closed Christmas. &

Attendance: 150,000 (estimated)

WUPATKI NATIONAL MONUMENT, 6400 N. Hwy. 89, Flagstaff, AZ 86004-2759. Tel.: 928-527-0322. Fax: 928-526-4259.
Web Site: www.nps.gov
Founded: 1924.
Congressional District: 6
Key Personnel: Supt., Diane Chung; Cur., Gwenn Gallenstein.
Governing Authority: federal. Parent Institution: National Park Service, Dept. of the Interior, Washington, DC 20240. Tax-exempt.
Institution Type/Description: Historic Site: over 2,600 archaeological sites dating from approximately 1100 A.D.; Hopi ancestral homeland.
Collections: series of displays with artifacts from the sites pertaining to the Sinagua/Anasazi cultural pattern as well as historic Navajo.
Research Fields: archaeology of Sinaqua, Anasazi & historic Navajo cultures.
Facilities: 500-vol. library of material on archaeology, geology & ethnology available for use by special permission. Postcards, slides & publications on the area for sale.
Activities: self-guiding trail of Wupatki Ruin; permanent & temporary exhibits; guided tours of Wupatki Ruins.
Publications: orientation brochure; trail guide leaflets, Volcanoes.
Hours & Admission Prices: Summer: daily 8-6; Winter: daily 8-5. Admission: $5 per person; discounts to AAM & MAA members. Closed Christmas. &

Attendance: 300,000 (estimated)

Florence

MCFARLAND STATE HISTORIC PARK, 24 W. Ruggles Ave., Florence, AZ 85232. Mailing Address: P.O. Box 109, Florence, AZ 85232-0109. Tel.: 520-868-5216. Fax: 520-868-9056.
E-mail: cdemille@azstateparks.gov
Web Site: www.co.pinal.az.us/mcfarland
Formerly: McFarland Historical State Park
Founded: 1979.
Congressional District: 6
Key Personnel: Park Mgr., Christopher DeMille.
Personnel Profile: Full-Time Paid 3; Part-Time Paid 1; Part-Time Volunteers 6.
Governing Authority: state government. Subsidiary Institution: Arizona State Parks, 1300 W. Washington, Phoenix, AZ 85007.
Institution Type/Description: Park Museum: housed in 1878 Pinal County Courthouse.
Collections: local historical artifacts; archives of late U.S. Senator E.W. McFarland from private & political careers; Arizona territorial law enforcement history (1863-1912). Historic Building: 1878 county courthouse.
Hours & Admission Prices: Daily 8-5. Adults $3, students 7-13 $1; discounts to groups of 8 or more; children under 14 no charge. Closed Christmas. &

Attendance: 7,500 (estimated)

PINAL COUNTY HISTORICAL SOCIETY AND MUSEUM, 715 S. Main St., Florence, AZ 85132. Mailing Address: P.O. Box 851, Florence, AZ 85132. Tel.: 520-868-4382.
Founded: 1958.
Congressional District: 3
Key Personnel: Pres. (V), Betty Wheeler; Vice Pres., Terri Bonesteel; Treas., Larry Pfeiffer; Recording Sec., Bita Arriola; Collections, Lynn Smith; Museum Shop Mgr., H. Christine Reid.
Personnel Profile: Part-Time Paid 1; Part-Time Volunteers 15.
Volunteer Hours: 1,584
Governing Authority: nonprofit organization. Affiliated with Arizona Historical Society, Tucson, AZ. Tax-exempt: 501(c)(3).
Institution Type/Description: Historical Society Museum.
Collections: agriculture; ranching; mining; Indian, Spanish, Mexican & Anglo contributions from residents of Pinal County; Arizona State Prison archives & artifacts; rodeo.
Facilities: 1,000-vol. library; Arizona State Prison Archives. Museum-related items for sale.
Activities: guided tours; lectures; speakers.
Publications: monthly newsletter with page of local history; Florence Tour Guide; A History of Florence; Es Verdad/It Is True; Good Men Bad Men, Lawmen; Florence Images of America.
Hours & Admission Prices: Sept.-July 14 Tues.-Sat. 11-4, Sun. 12-4. No charge; donations accepted. Closed New Year's Day; Easter; Independence Day; Thanksgiving; Christmas. &

Attendance: 6,091 (accurate)
Membership: Individual $25; Family $40; Individual Life $150; Business & Family Life $200; Business Life $1,000.

Fort Apache

NOHWIKE BAGOWA, THE WHITE MOUNTAIN APACHE CULTURAL CENTER AND MUSEUM, Indian Rte. 46, Fort Apache, AZ 85926. Mailing Address: P.O. Box 507, Fort Apache, AZ 85926-0507. Tel.: 928-338-4625. Fax: 928-338-1716.
Web Site: www.wmat.nsn.us
Founded: 1969.
Key Personnel: Dir., Dr. Karl Hoerig; Tribal Chm., Ronnie Lupe; Museum Shop Mgr., Ann Skidmore.
Personnel Profile: Full-Time Paid 5; Part-Time Paid 2; Part-Time Volunteers 3.
Governing Authority: nonprofit. Parent Institution: White Mountain Apache Tribe. Tax-exempt: 501(c)(3).
Institution Type/Description: Historical Museum.
Collections: Apache culture; Fort Apache artifacts.
Research Fields: Apache culture & Fort Apache history.
Activities: Annual Event: Great Fort Apache Heritage Reunion in May.
Hours & Admission Prices: Winter: Mon.-Fri. 8-5; Summer: Mon.-Sat. 8-5. Adults $5, students $3; children under 7 & tribal members no charge. Closed major holidays. &

Attendance: 15,203 (accurate)

Fort Huachuca

FORT HUACHUCA MUSEUM, Boyd & Grierson Sts., Fort Huachuca, AZ 85613. Mailing Address: IMSW-HUA-PLT, Fort Huachuca, AZ 85613-6000. Tel.: 520-533-5736.
Web Site: www.huachuca.army.mil/site/visitor/index.asp
Founded: 1960.
Congressional District: 9
Key Personnel: Dir., Tim Phillips; Cur., Paula Ussery; Exhibits, Marty Matin; Cur. Military Intelligence, Ralph Jackson; Museum Specialist, Steve Gregory; Museum Shop Mgr., Bess Banister.
Personnel Profile: Full-Time Paid 6; Part-Time Volunteers 20; Interns 1.
Governing Authority: federal. Tax-exempt.
Institution Type/Description: History Museum: housed on 1877 Fort, National Historic Landmark.
Collections: military history of the Indian Wars; military & western artifacts.
Research Fields: military history.
Facilities: 2,000-vol. library of books on military & western history available for use on premises. Museum-related items for sale.
Activities: self-guided tour.
Publications: book, Fort Huachuca History; brochures.
Hours & Admission Prices: Mon.-Fri. 9-4, Sat.-Sun. 1-4. No charge; donations accepted. Closed Federal holidays. ♿
Attendance: 70,000 (accurate)
Membership: Individual $10; Annual $15.

Fountain Hills

RIVER OF TIME MUSEUM, 12901 N. La Montana Dr., Fountain Hills, AZ 85268-4742. Mailing Address: P.O. Box 17445, Fountain Hills, AZ 85269. Tel.: 480-837-2612. Fax: 480-836-4292.
E-mail: president@riveroftimemuseum.org
Web Site: www.riveroftimemuseum.org
Founded: 2003.
Congressional District: 6
Key Personnel: Exec. Dir., Judy Confer; Pres. (V), Debbie Skehen; Museum Shop Mgr., Pat Canning.
Personnel Profile: Full-Time Volunteers 1; Part-Time Paid 1; Part-Time Volunteers 80.
Governing Authority: Parent Institution: Fountain Hills & Lower Verde River Valley Museum & Historical Society. Tax-exempt.
Institution Type/Description: History Museum.
Collections: Verde River Valley & Fountain Hills history & culture; photographs; personal artifacts.
Activities: cultural programs & activities; 5 dinner meetings with speakers.
Publications: newsletter, Legacy.
Hours & Admission Prices: June-Aug. Fri. 1-4, Sat. 10-4; Winter: Tues.-Sat. 1-4; groups by appointment. Adults $3, seniors $2, children 5-12 $1; members no charge. ♿
Attendance: 900
Membership: Individual $25; Family $35.

Fredonia

PIPE SPRING NATIONAL MONUMENT VISITOR CENTER AND CULTURAL MUSEUM, 406 N. Pipe Spring Rd., Fredonia, AZ 86022. Mailing Address: HC 65, Box 5, 406 N. Pipe Spring Rd., Fredonia, AZ 86022. Tel.: 928-643-7105.
E-mail: jenny_leasor@nps.gov
Institution Type/Description: History Museum.
Collections: Kaibab Paiutes Indian history; local history & culture; personal artifacts. Historic Building: Winsor Castle built in 1870s.
Facilities: nature trails; picnic area.
Activities: talks; demonstrations; tours; hiking; bird watching; special events.
Hours & Admission Prices: Grounds & Museum: June-Aug. daily 7-5; Sept.-May daily 8-5. Castle: June-Aug. daily 8-4:30; Sept.-May daily 9-4. Adults $5; children 15 & under no charge. Closed New Year's Day; Thanksgiving; Christmas. ♿
Attendance: 50,000 (estimated)

Ganado

HUBBELL TRADING POST NATIONAL HISTORIC SITE, Hwy. 264, Ganado, AZ 86505. Mailing Address: P.O. Box 150, Ganado, AZ 86505-0150. Tel.: 928-755-3475. Fax: 928-755-3405.
E-mail: kathy_tabaha@nps.gov
Web Site: www.nps.gov/hutr/
Founded: 1967.
Congressional District: 1
Key Personnel: Supt., Lloyd Masayumptewa; Museum Technician, Kathleen Tabaha; Chief Visitor Svcs., Ailema Benally; Museum Shop Mgr., Edison Eskeets.
Personnel Profile: Full-Time Paid 11; Part-Time Paid 2; Part-Time Volunteers 5.
Governing Authority: federal. Parent Institution: National Park Service, Washington, DC. Subsidiary Institution: Western National Parks Association, Friends of Hubbell Trading Post NHS, Inc. Tax-exempt.
Institution Type/Description: Historic Site.
Collections: Native American arts & crafts; art; furnishings for home & trading post; ethnology; graphics; agriculture; archives; sculpture; textiles. Historic Buildings: 1880s Hubbell Trading Post; 1897 Hubbell barn; 1900-1915 Hubbell home; 1930-1943 Hubbell guest hogan.
Research Fields: ethnohistory; history; architecture & furnishings; art.
Facilities: 1,500-vol. library of general & southwestern history available for use on premises. Native American, primarily Navajo, arts & crafts for sale.
Activities: guided tours; permanent exhibitions; demonstration of Navajo weaving & silversmithing. Annual Events: Native American Art Auction in May & September; Luminaria Night in December.
Publications: orientation brochure; trail guide booklet.
Hours & Admission Prices: May-Sept. daily 8-6; Oct.-April daily 8-5. Adults $2; donations accepted. Closed New Year's Day; Thanksgiving; Christmas. ♿
Attendance: 100,000 (accurate)

Gilbert

GILBERT HISTORICAL MUSEUM, (M), 10 S. Gilbert Rd., Gilbert, AZ 85296-1047. Mailing Address: P.O. Box 1484, Gilbert, AZ 85299-1484. Tel.: 480-926-1577.
E-mail: info@gilbertmuseum.org
Web Site: www.gilbertmuseum.org
Founded: 1982.
Key Personnel: Dir., Kayla Kolar; Pres. (V), Bradley Barrett, Ph.D.
Personnel Profile: Full-Time Paid 1; Part-Time Paid 1; Part-Time Volunteers 100.
Governing Authority: Tax-exempt.
Institution Type/Description: History Museum.
Collections: area history & culture.
Hours & Admission Prices: Tues.-Sat. 9-4. Adults $5, seniors $4, children 5-12 $3; children under 5 no charge. Closed holidays. ♿
Attendance: 5,000 (accurate)
Membership: Student $10; Individual $20; Family $40; Sustaining $100; Pioneer Patron & Corporate Bronze $250; Premier & Corporate Silver $500; Corporate Gold $1,000.

Glendale

GLENDALE COMMUNITY COLLEGE ART COLLECTION, Art Department, 6000 W. Olive Ave., Glendale, AZ 85302. Tel.: 623-845-3755.
Key Personnel: Cur., Darlene Goto
Institution Type/Description: Art Museum.
Collections: paintings; sculptures; drawings; photographs; ceramic works.
Hours & Admission Prices: Fall & Spring Semesters: Mon.-Thurs. 7am-10pm, Fri.-Sat. 7-5; Summer: Mon.-Thurs. 7am-9pm, Sun. 12-5. No charge.

SAHUARO RANCH PARK HISTORIC AREA, 9802 N. 59th Ave., Glendale, AZ 85302-1203. Mailing Address: 5850 W. Glendale Ave., Glendale, AZ 85301-2563. Tel.: 623-930-4200. Fax: 623-915-7587.
Web Site: www.glendaleaz.com/srpha
Formerly: Historic Sahuaro Ranch
Founded: 1977.
Congressional District: 2
Key Personnel: Coord. Facilities & Events, Paul King; Coord. Historic Education & Outreach, John Akers.
Personnel Profile: Full-Time Paid 3; Part-Time Paid 3; Part-Time Volunteers 50.
Governing Authority: city government. Parent Institution: City of Glendale Parks & Recreation Dept.
Institution Type/Description: Historic House Museum.
Collections: agricultural history of Sahuaro Ranch & the western Maricopa County.
Research Fields: history of the William Henry Bartlett family; water use in the Salt River Valley; history of Sahuaro ranch owners, operations & laborers.
Facilities: 2,500 sq. ft. exhibit space; educational facilities.
Activities: guided tours; temporary exhibits; school programs; themed events; summer camp. Museum Sponsors: Tractor Shows.

Hours & Admission Prices: June-July Fri.-Sat. 10-2; Sept.-May Wed.-Sat. 10-2, Sun. 12-4. No charge; donations accepted. Closed Federal holidays; Easter. ♿
Attendance: 50,000 (accurate)

SOUTHWEST MUSEUM OF ENGINEERING, COMMUNICATIONS AND COMPUTATION, 5802 W. Palmaire Ave., Glendale, AZ 85301-2442. Tel.: 623-435-1522.
E-mail: info@smecc.org
Web Site: www.smecc.org
Key Personnel: Archivist, Ed Sharpe
Institution Type/Description: Technology Museum.
Collections: early engineering; communications equipment & memorabilia.
Hours & Admission Prices: Tues.-Sat. 12-3; other times by appointment. Call to confirm. No charge.

Globe

BESH-BA-GOWAH ARCHAEOLOGICAL PARK, 1324 Jesse Hayes Rd., Globe, AZ 85501. Mailing Address: 150 N. Pine St., Globe, AZ 85501-2514. Tel.: 928-425-0320. Fax: 928-402-1071.
E-mail: beshbagowah@globeaz.gov
Founded: 1987.
Personnel Profile: Full-Time Paid 3; Part-Time Paid 1.
Governing Authority: municipal government; nonprofit. Parent Institution: City of Globe. Tax-exempt.
Institution Type/Description: Archaeology Museum & Archaeological Site.
Collections: artifacts from local excavations of Hohokam & Salado prehistoric sites; 1930s & 1980s excavations at Besh-Ba-Gowah.
Activities: Annual Events: Archaeology Month Open House in March; Festival of Lights in December.
Hours & Admission Prices: Daily 9-5. Adults 12-64 $4, senior citizens $3; discounts to AAA members; children under 12 no charge. Closed New Year's Day; Thanksgiving; Christmas. ♿
Attendance: 35,000 (estimated)

GILA COUNTY HISTORICAL MUSEUM, 1330 N. Broad St., Globe, AZ 85501. Mailing Address: P.O. Box 2891, Globe, AZ 85502-2891. Tel.: 928-425-7385.
E-mail: museum@gilahistorical.com
Founded: 1955.
Congressional District: 4
Key Personnel: C.E.O., Vernon Perry; Museum Mgr., Donna Anderson.
Personnel Profile: Part-Time Paid 1; Part-Time Volunteers 16.
Governing Authority: nonprofit organization. Parent Institution: Gila County Historical Society. Tax-exempt: 501(c)(3).
Institution Type/Description: History Museum: housed in 1920 mine rescue facility.
Collections: Indian pottery; furniture; glass collections; kitchenware; ranching; mining; gems & minerals.
Research Fields: archeological; historical.
Facilities: reading room; geneology research library.
Activities: guided tours; lectures.
Publications: books, Copper Bottom Tales; Globe Arizona; Globe's Historic Buildings.
Hours & Admission Prices: Mon.-Fri. 10-4, Sat. 11-3. No charge; donations accepted. Closed major holidays. ♿
Attendance: 4,000 (estimated)
Membership: Individual $15; Family $25.

Goodyear

BIBLE MUSEUM, Hampton Inn & Suites, 2000 N. Litchfield, Goodyear, AZ 85395-1280. Tel.: 623-536-8614. Fax: 623-536-1414.
E-mail: BibleMuseum@hotmail.com
Institution Type/Description: Religious Museum.
Collections: Bible history; Bibles; theological books.
Hours & Admission Prices: Mon.-Fri. 10-4.

Grand Canyon National Park

GRAND CANYON NATIONAL PARK MUSEUM COLLECTION, 2C Albright Ave., Grand Canyon Village, South Rim, Grand Canyon National Park, AZ 86023. Mailing Address: P.O. Box 129, Grand Canyon, AZ 86023-0129. Tel.: 928-638-7769 (Museum) & 7888 (National Park). Fax: 928-638-7490.
E-mail: GRCA_Museum_Collection@nps.gov

Web Site: www.nps.gov/grca/historyculture/collections.htm
Founded: 1919.
Congressional District: 3
Key Personnel: Supt., David Uberuaga; Museum Specialist, Colleen Hyde; Cur., Kim Besom.
Personnel Profile: Full-Time Paid 2.
Governing Authority: federal. Parent Institution: Grand Canyon National Park. Exhibit Visitor Centers: Tusayan Museum; Yavapai Museum; Grand Canyon Visitor Center. Tax-exempt.
Institution Type/Description: National Park: the Grand Canyon, a deep gorge of the Colorado River measuring 277 mi. long, 1-18 mi. wide & one mile deep.
Collections: geology; history; natural history; archaeology; prehistoric Indian materials; insects; herbarium; oral history, photography; archives; manuscripts; paleontology.
Research Fields: geology; flora; fauna; history; oral history prehistory; paleontology.
Facilities: 10,000-vol. library available for research on premises only; Exhibit-Visitor Centers located in park: Tusayan Museum; Yavapai Museum; South Rim Visitor Center. Books, children's books & maps for sale.
Activities: guided tours; lectures; films; permanent & temporary exhibitions at Kolb Studio.
Publications: scientific monographs on natural & cultural history; guides to various areas & trails.
Hours & Admission Prices: Park: daily 24 hours. Library & Museum: Mon.-Fri. 8-5. Park: $12 per person; $25 per vehicle. Annual Pass: $50. ♿
Attendance: 5,000,000 (estimated)

Green Valley

TITAN MISSILE MUSEUM, 1580 W. Duval Mine Rd., Green Valley, AZ 85614-4907. Tel.: 520-625-7736 & 4598. Fax: 520-625-9845.
E-mail: ymorris@titanmissilemuseum.org
Web Site: www.titanmissilemuseum.org
Founded: 1986.
Congressional District: 5
Key Personnel: Acting Exec. Dir., Yvonne C. Morris; Chm. (V), Count Ferdinand von Galen.
Personnel Profile: Full-Time Paid 7; Part-Time Paid 4; Part-Time Volunteers 120.
Governing Authority: nonprofit organization. Parent Institution: Arizona Aerospace Foundation. Tax-exempt: 501(c)(3).
Institution Type/Description: Missile Museum: housed in a former operational Titan II ICBM complex.
Collections: 1963 Titan II intercontinental ballistic missile complex & missile related equipment.
Research Fields: Cold War & military technology.
Facilities: classroom. Gift items for sale.
Activities: guided tours.
Hours & Admission Prices: Daily 9-5. Adults $9.50, senior citizens, military & group rate of 20 or more $8.50, students 7-12 $6; discounts to AAA members; members, school groups & children under 6 no charge. Closed Thanksgiving; Christmas. ♿
Attendance: 50,000 (estimated)
Membership: See listing for Pima Air & Space Museum, Tucson, AZ.

Greer

BUTTERFLY LODGE MUSEUM, 4 County Rd. 1126, Greer, AZ 85927. Mailing Address: P.O. Box 76, Greer, AZ 85927-0076. Tel.: 928-735-7514.
E-mail: bflylodge@aol.com
Governing Authority: nonprofit organization. Tax-exempt.
Institution Type/Description: History Museum: housed in the home of Western writer, James Willard Schultz and his artist son, Hart Merriam Schultz, also known as Lone Wolf. Listed on the National Register of Historic Places.
Collections: original furnishings; personal artifacts; paintings; sculptures.
Activities: special events; education classes.
Hours & Admission Prices: Memorial Day to Labor Day Thurs.-Sun. & holidays 10-5. Adults $2, youth 12-17 $1; children no charge. ♿

Hereford

CORONADO NATIONAL MEMORIAL, 4101 E. Montezuma Canyon Rd., Hereford, AZ 85615-9376. Tel.: 520-366-5515, ext. 22. Fax: 520-366-5705.
Web Site: www.nps.gov/coro
Founded: 1952.
Congressional District: 5

Personnel Profile: Full-Time Paid 2; Part-Time Volunteers 10.
Governing Authority: federal. Parent Institution: National Park Service, Washington, DC. Tax-exempt.
Institution Type/Description: Park Museum & Visitor Center.
Collections: historical books; mid-16th century Spanish costumes documents & weapons; European exploration of the American Southwest 1540-1542.
Research Fields: early Spanish-Mexican history; natural history; Hispanic-Mexican culture.
Facilities: nature trails; picnic area. Historical & natural history books for sale.
Activities: self-guided tours; talks; permanent exhibits.
Publications: brochure, Coronado National Memorial.
Hours & Admission Prices: Daily 9-5. No charge; donations accepted. Closed Thanksgiving; Christmas. &
Attendance: 26,000 (accurate)

Holbrook

NAVAJO COUNTY HISTORICAL SOCIETY MUSEUM, 100 E. Arizona St., Holbrook, AZ 86025-2698. Tel.: 928-524-6558.
E-mail: nchs@gotouraz.com
Web Site: www.gotouraz.com/nchs
Founded: 1969.
Congressional District: 5
Key Personnel: Pres. (V) & Museum Shop Mgr., Paul DoBell; Dir., Jolynn Fox.
Personnel Profile: Full-Time Paid 1; Full-Time Volunteers 1; Part-Time Paid 1; Part-Time Volunteers 7.
Volunteer Hours: 2,250
Governing Authority: Tax-exempt: 501(c)(3).
Institution Type/Description: History Museum.
Collections: local history & culture; Native American artifacts; personal artifacts; photographs; period furnishings.
Research Fields: Holbrook & N.E. Arizona.
Facilities: 30-seat auditorium.
Hours & Admission Prices: Daily 8-5. No charge; donations accepted. Closed federal holidays. &
Attendance: 23,406 (estimated)
Membership: County Residents $10; Individual $15; Family $35; Supporting $50; Contributing $100; Benefactor $250; Lifetime $1,000.

Jerome

GOLD KING MINING MUSEUM AND GHOST TOWN, Perkinsville Rd., Jerome, AZ 86331. Mailing Address: P.O. Box 125, Jerome, AZ 86331. Tel.: 928-634-0053.
Institution Type/Description: Mining Museum.
Collections: local history; mining & mining equipment; period buildings & saw mill; mine shaft.
Facilities: Museum-related items for sale.
Hours & Admission Prices: Daily 10-5. Closed Thanksgiving; Christmas.

JEROME HISTORICAL SOCIETY MINE MUSEUM, 200 Main St., Jerome, AZ 86331. Mailing Address: P.O. Box 156, Jerome, AZ 86331-0156. Tel.: 928-634-5477.
Web Site: www.jeromehistoricalsociety.com
Founded: 1953.
Congressional District: 4
Key Personnel: Pres. (V), Allen Muma.
Personnel Profile: Full-Time Paid 3; Part-Time Paid 6; Part-Time Volunteers 3.
Institution Type/Description: Historical Society Museum.
Collections: mining equipment; ore carts; photographs; personal artifacts; maps; documents.
Facilities: Museum-related items for sale.
Activities: Annual Event: Ghost Walk.
Publications: newsletter, The Chronicle.
Hours & Admission Prices: Daily 9-5. Adults $2, seniors $1; children no charge.
Attendance: 30,112 (accurate)
Membership: Individual $15; Family $25; Business $50; Patron $100; Benefactor $500; Guardian $1,000.

JEROME STATE HISTORIC PARK, 100 Douglas Rd., Jerome, AZ 86331. Mailing Address: P.O. Box D, Jerome, AZ 86331-0097. Tel.: 928-634-5381. Fax: 928-639-3132.
E-mail: kayotte@azstateparks.gov
Web Site: www.azstateparks.com
Founded: 1962.
Congressional District: 3
Key Personnel: Supvr., Keith Ayotte.

Personnel Profile: Full-Time Paid 2; Part-Time Paid 2; Part-Time Volunteers 2.
Governing Authority: state. Parent Institution: Arizona State Parks Board, Phoenix, AZ. Tax-exempt.
Institution Type/Description: History & Mining Museum.
Collections: mining industry; mineralogy.
Facilities: picnic area. Museum-related items for sale.
Activities: self-guided tours; video program on Jerome history.
Hours & Admission Prices: Daily 8-5. Adults 14 & over $5, children 13 & under $2; children 6 & under no charge. Annual Permit: $75. Closed Christmas. &
Attendance: 70,584 (accurate)

Kayenta

KAYENTA VISITOR'S CENTER, Hwy. 160, Kayenta, AZ 86033. Mailing Address: P.O. Box 544, Kayenta, AZ 86033-0544. Tel.: 928-697-3572.
Institution Type/Description: Visitor Center.
Collections: Navajo history & culture; personal artifacts; period furnishings.
Facilities: outdoor amphitheatre. Museum-related items for sale.
Activities: demonstrations; performances; special events.
Hours & Admission Prices: Call for hours.

Kingman

KINGMAN ARMY AIRFIELD MUSEUM, 4540 Flightline Dr., Kingman, AZ 86401. Tel.: 928-757-1892.
E-mail: museum@kingmanaafhsmuseum.org
Web Site: main.kingmanaafhsmuseum.org
Founded: 1988.
Personnel Profile: Part-Time Volunteers 5.
Governing Authority: Parent Institution: Kingman Army Air Field Historical Society, Inc. Tax-exempt.
Institution Type/Description: Military History Museum.
Collections: military history & aircraft; photographs; news articles.
Research Fields: base history; personal stories.
Activities: lectures.
Hours & Admission Prices: Tues.-Sat. 10-3; other times by appointment. Donation: adults $2; children 12 & under no charge. &
Attendance: 5,000 (estimated)
Membership: Single $20; Family $30; Contributor $50; Sustaining $100; Life $1,000.

MOHAVE MUSEUM OF HISTORY AND ARTS, 400 W. Beale, Kingman, AZ 86401-5797. Tel.: 928-753-3195. Fax: 928-718-1562.
E-mail: director@mohavemuseum.org
Web Site: www.mohavemuseum.org
Founded: 1961.
Congressional District: 1 & 2
Key Personnel: Dir. & C.E.O., Shannon Rossiter; Pres. (V), William Porter.
Personnel Profile: Full-Time Paid 4; Part-Time Paid 5; Part-Time Volunteers 88.
Governing Authority: Parent Institution: Mohave County Historical Society. Tax-exempt: 501(c)(3).
Institution Type/Description: History Museum: housing Mohave County history.
Collections: Indian artifacts; archaeology; military; mining items; Andy Devine memorabilia; portraits of presidents & their wives; turquoise figurines; maps of area; c.1890 wooden caboose; ranching; wildlife replicas of western animals; 34 ft. mine tunnel replica with gem stones; mine artifacts & wildlife sanctuary murals of western wildlife; manuscripts.
Research Fields: Indian artifacts; archaeology; mining; northwestern Arizona history.
Facilities: library of books and manuscripts of local history photos available for use on premises; microfilms of newspapers & genealogy records; photograph archives relating to area; auditorium. Indian crafts, books & museum-related items for sale.
Activities: lectures; gallery talks; school tours; permanent & temporary exhibitions; monthly Mohave County Genealogical Society meetings. Museum Sponsors: Organ Concerts.
Publications: newsletter, museum exhibits guidebook; walking tour map; local history brochures.
Hours & Admission Prices: Mon.-Fri. 9-5, Sat. 1-5. Adults $4, seniors $3; discounts to AAM members; members no charge. Closed New Year's Day; Easter; Memorial Day; Labor Day; Thanksgiving; Christmas. &
Attendance: 30,000 (estimated)
Membership: General $30; Supporting $100; Business $200; Director's Circle $500.

ROUTE 66 MUSEUM, 120 W. Andy Devine Ave., Ste. 7, Kingman, AZ 86401-5807. Tel.: 928-753-9889.
Founded: 2001.
Institution Type/Description: History Museum: housed in the historical Powerhouse building.
Collections: highway travel history; murals; photographs; dioramas; U.S. Army & Native American artifacts.
Facilities: theater.
Activities: guided tours; school group tours. Museum Sponsors: Fun Hunt.
Hours & Admission Prices: Daily 9-5. Adults $4, seniors 60 & over $3; discounts to groups; children under 12 no charge. &

Lake Havasu City

LAKE HAVASU MUSEUM OF HISTORY, 320 London Bridge Rd., Lake Havasu City, AZ 86403-4645. Tel.: 928-854-4938.
E-mail: lhmuseum@npgcable.com
Web Site: www.havasumuseum.com
Founded: 1991.
Congressional District: 3
Key Personnel: Pres. (V), Ed Walker; Museum Shop Mgr., Sally McClure.
Personnel Profile: Part-Time Paid 2; Part-Time Volunteers 30.
Governing Authority: Parent Institution: Lake Havasu City Historical Society. Tax-exempt.
Institution Type/Description: History Museum.
Collections: local history; Chemehuevi Tribe history & culture; mining; steamboats; maps & documents of the lower Colorado river; Parker Dam; London Bridge.
Research Fields: lower Colorado river disputes; local tribes; city's economic development.
Activities: lectures; tours; school field trips.
Publications: quarterly newsletter.
Hours & Admission Prices: Tues.-Sat. 10-4. Adults $5; members and children 12 & under no charge. Closed New Year's Day; Independence Day; Thanksgiving; Christmas. &
Attendance: 3,500 (accurate)
Membership: Single $20; Family $30; Business $60; Patron $100.

Litchfield Park

WILDLIFE WORLD ZOO & AQUARIUM, 16501 W. Northern, Litchfield Park, AZ 85340-9466. Tel.: 623-935-9453. Fax: 623-935-9499.
Web Site: www.wildlifeworld.com
Founded: 1984.
Key Personnel: Dir., Mickey Ollson; Deputy Dir., Jack Ewert; Education, Josh Jarnagin; Museum Shop Mgr., Anna Milts.
Personnel Profile: Full-Time Paid 45; Part-Time Paid 4; Part-Time Volunteers 15; Interns 2.
Governing Authority: private; profit.
Institution Type/Description: Zoo & Aquarium.
Collections: 400 species of exotic animals.
Facilities: 580-vol. library of zoological books; aquarium; 400-seat outdoor theater; zoological park. Zoo-related items for sale.
Activities: formal education programs for children; guided tours; lectures; mobile vans; broadcast programs.
Publications: quarterly newsletter.
Hours & Admission Prices: Daily 9-6. Adults $27.95, senior citizens $26.95, children $13.95. &
Attendance: 450,000 (estimated)
Membership: Children $60; Adult $100.

Maricopa

AK-CHIN HIM-DAK ECO-MUSEUM, 47685 N. Eco-Museum Rd., Maricopa, AZ 85239-2850. Tel.: 520-568-9480. Fax: 520-568-9557.
E-mail: epeters@ak-chin.nsn.us
Key Personnel: C.E.O. & Dir., Elaine Peters
Institution Type/Description: Cultural History Museum.
Collections: cultural history; personal artifacts.
Hours & Admission Prices: Mon.-Fri. 9-5, Sat. 8-4 by appointment. No charge; donations accepted.

Mesa

ARIZONA MUSEUM OF NATURAL HISTORY, 53 N. MacDonald St., Mesa, AZ 85201-7325. Tel.: 480-644-2230. Fax: 480-644-3424. Facebook: Arizona Museum of Natural History.
E-mail: azmnh.info@mesaaz.gov
Web Site: www.azmnh.org
Formerly: Mesa Southwest Museum
Founded: 1977.
Congressional District: 6
Key Personnel: Dir., Dr. Thomas H. Wilson; Pres. (V), Lynn Johnson; Cur. Natural History, Dr. Robert McCord; Cur. Anthropology, Dr. Jerry Howard; Lead Collections Specialist, Margaret MacMinn-Barton; Coord. Exhibits, Tim Walters; Exhibits Preparator, Mike Keller; Collections Specialist, Gavin McCullough; Cur. Education, Kathy Eastman; Budget Specialist, Sandra Williamson; Administrative Asst., Barbara Dixon; Museum Graphics/Multimedia Specialist, Michael Ramos; Museum Shop Mgr., Terri Karl; Volunteer Coord., Yvonne Petersen; Membership Coord., Heather Jones.
Personnel Profile: Full-Time Paid 15; Part-Time Paid 3; Part-Time Volunteers 150.
Volunteer Hours: 14,794
Operating Expenses: 1,453,106
Operating Income: 1,453,106
Governing Authority: municipal. Parent Institution: City of Mesa. Subsidiary Institution: Mesa Grande Cultural Center. Tax-exempt: 501(c)(3); 170(B)(1)(A).
Institution Type/Description: Natural History Museum.
Collections: natural & cultural history of the American Southwest; over 47,000 objects including: 14,000 photographs; paleontology & mineralogy; archaeology of the Hohokam & other prehistoric cultures of the Southwest & Mesoamerica; ethnology; ethnographic dolls; Spanish mission period; pioneer, Frontier & 1950 domestic life; farming; ranching; transportation; Arizona movie memorabilia; historic photographs; comparative collections: osteology; Arizona geology; prehistoric ceramics. Historic House: Sirrine House.
Research Fields: Southwestern archaeology & ethnology; western paleontology; regional & local history.
Facilities: 7,000-vol. reference library; over 80,000 sq. ft. exhibit space; 150-seat auditorium/theater; Mesa Grande archaeological site. Museum-related items for sale.
Activities: formal docent program; traveling & temporary exhibits; gallery tours; school loan services; children's workshops; lectures; teen volunteer program; symposia; theater; educational outreach & inhouse programs.
Publications: gallery guides; festival programs; docent training manuals; educational school packets; archaeological & paleontological newsletters; scientific publications; archaeological site reports; monograph series; historic surveys; postcards; brochures; volunteer newsletter; symposia proceedings.
Hours & Admission Prices: Tues.-Fri. 10-5, Sat. 12-5, Sun. 1-5. Adults $10, senior citizens 65 & over $9, students $8; children 3-12 $6; discounts to AAA & AAM members; children under 3 & members no charge. &
Attendance: 144,712 (accurate)
Membership: General $70; Friend $125; Contributor $250; Patron $500; Curator's Circle $1,000; Director's Circle $1,500. Corporate: Friend $1,000; Contributor $2,500; Patron $5,000.

ARIZONA WING COMMEMORATIVE AIR FORCE MUSEUM, Falcon Field, 2017 N. Greenfield Rd., Mesa, AZ 85215. Tel.: 480-924-1940.
E-mail: info@azcaf.org
Web Site: www.azcaf.org
Institution Type/Description: Military History Museum.
Collections: military aircraft including B-17G Flying Fortress; TB-25N Mitchell; AF-2S Guardian; F4N Phantom II; MiG-15bis; MiG-21 Fishbed; A-26 Invader; P-47 Thunderbolt; RAF S.E.5a; SNJ T-6 Texan; TG-3A Glider; MiG-15 UTI Midget; PT-17 Stearman Kaydet; Twin Beech C-45; L-16 Grasshopper; Sikorsky S-55 Whirlwind.
Activities: special events; rental facilities.
Hours & Admission Prices: June-Sept. Wed.-Sun. 9-3; Oct.-May daily 10-4. Summer: adults 13 & over $7, children 5-12 $3; children under 5 no charge. Winter: adults 13 & over $10, seniors 62 & over $9, children 5-12 $3, children under 5 no charge. Closed New Year's Day; Thanksgiving; Christmas.

I.D.E.A. MUSEUM, (M), 150 W. Pepper Place, Mesa, AZ 85201-7326. Tel.: 480-644-4332. Fax: 480-644-2466.
E-mail: ideamuseum@mesaaz.gov

Web Site: ideamuseum.org
Formerly: Arizona Museum For Youth
Founded: 1977.
Congressional District: 1
Key Personnel: Chm., Carmen Guerrero; Membership Information & Museum Shop Mgr., Darlene Zajda; Exec. Dir., Sunnee D. O'Rork; Cur., Jeffory Morris; Cur. Education, Dena Milliron; Exhibitions Coord., Rex Witte; Mktg. & Public Rels., Latonya Jordan-Smith.
Personnel Profile: Full-Time Paid 8; Part-Time Paid 12; Part-Time Volunteers 25.
Governing Authority: partnership. Parent Institution: City of Mesa, AZ. Tax-exempt: 501(c)(3).
Institution Type/Description: Children's Fine Arts Center.
Collections:
Major Exhibits: Art Speaks: A Way With Words (T), 1/14-12/14; Art of the Robot, 2/7/14-5/25/14.
Research Fields: early childhood.
Facilities: classrooms; facility rentals.
Activities: guided tours; formally organized education programs for children.
Publications: annual newsletters; pre-visit & follow-up brochures for teachers.
Hours & Admission Prices: Tues.-Sat. 10-4, Sun. 12-4. Admission $7; discounts to AAM & Association of Children's Museums members; children 1 & under and members no charge. &
Attendance: 70,546 (accurate)
Membership: 10 Visit Pass $50; Family of 2 $60; Family of 4 $75; Family of 6 $125; Family of 8 $180.

MESA ARTS CENTER, One E. Main St., Mesa, AZ 85211. Mailing Address: P.O. Box 1466, Mesa, AZ 85211-1466. Tel.: 480-644-6560.
Web Site: www.mesaartscenter.org
Founded: 1976.
Key Personnel: Exec. Dir., Johann Zietsman; Cur., Patty Haberman
Institution Type/Description: Contemporary Art Museum.
Collections: paintings; sculpture; photography; decorative arts.
Hours & Admission Prices: Tues.-Wed. & Fri.-Sat. 10-5, Thurs. 10-8, Sun. 12-5. Admission $3.50; children 7 & under and Thurs. no charge. &

MESA HISTORICAL MUSEUM, 2345 N. Hornc, Mcsa, AZ 85203-1823. Mailing Address: P.O. Box 582, Mesa, AZ 85211-0582. Tel.: 480-835-7358.
E-mail: info@mesahistoricalmuseum.org
Web Site: www.mesaaz.org
Founded: 1966.
Key Personnel: Pres. & C.E.O., Lisa Anderson; Chm. Bd. (V), Victor Linoff.
Personnel Profile: Full-Time Paid 3; Part-Time Paid 4; Part-Time Volunteers 120.
Governing Authority: private; nonprofit organization. Tax-exempt: 501(c)(3).
Institution Type/Description: History Museum
Collections: history of Mesa, Arizona & the surrounding area.
Facilities: library; 400-seat auditorium; classrooms. Museum-related items for sale.
Activities: docent program; guided tours; lectures; temporary exhibitions.
Publications: monthly members newsletter; books.
Hours & Admission Prices: Tues.-Sat. 9-1. Adults $5, seniors over 65 $4, youth 3-12 $3; discounts to groups; children under 3 & members no charge. Closed major holidays. &
Attendance: 28,000 (estimated)
Membership: Individual $35; Family $50; Business $500; Friend $1,000.

WINGSPAN AIR MUSEUM, Superstition Springs Mall, 6555 E. Southern Ave., Ste. 1106, Mesa, AZ 85206. Mailing Address: Wingspan Air Heritage Foundation, P.O. Box 21268, Mesa, AZ 85277. Tel.: 480-924-5543.
E-mail: info@wingspanair.org
Web Site: www.wingspanair.com
Key Personnel: Pres. & Exec. Dir., Jeff Furnari; Dir. Aircraft Collections, Robert Kropp
Institution Type/Description: Military History Museum.
Collections: military history, artifacts & aircraft; military veterans' memorial; photographs; personal artifacts; uniforms.
Activities: special events.
Hours & Admission Prices: Tues.-Fri. 10-6, Sat. 10-5.

Miami

BULLION PLAZA CULTURAL CENTER & MUSEUM, 150 N. Plaza Circle, Miami, AZ 85539. Mailing Address: P.O. Box 786, Miami, AZ 85539-0786. Tel.: 928-473-3700. Fax: 928-473-9097.
E-mail: az.terr1912@yahoo.com
Web Site: www.bullionplazamuseum.org
Founded: 1997.
Congressional District: 1
Key Personnel: Exec. Dir., Thomas N. Foster; Pres. (V), Joe Sanchez.
Personnel Profile: Part-Time Paid 1; Part-Time Volunteers 15.
Governing Authority: bd. directors. Tax-exempt: 501(c)(3).
Institution Type/Description: Art & History Museum: housed in a former grammar school; built in 1923.
Collections: local history & culture; photographs; paintings; personal artifacts; period furnishings; mining; ranching.
Activities: monthly speakers program; monthly Miami Hardscrabble Series.
Publications: Bullion Plaza Cultural Center & Museum.
Hours & Admission Prices: Thurs.-Sat. 11-3, Sun. 12-3; other times by appointment. No charge; donations accepted. &
Attendance: 2,000 (estimated)
Membership: Individual $10; Family $25; Friend $50; Donor $100; Sponsor $250; Patron $500; Benefactor $1,000.

Mount Lemmon

UNIVERSITY OF ARIZONA MOUNT LEMMON SKYCENTER, 9800 Ski Run Rd., Mount Lemmon, AZ 85619. Tel.: 520-626-8792.
E-mail: skycenter@as.arizona.edu
Web Site: www.skycenter.arizona.edu
Key Personnel: Dir., Alan Strauss
Institution Type/Description: Science Center: located at an altitude of 9,157.
Collections: science & astronomy; hands-on exhibitions.
Activities: educational adventures; youth programs; interactive exhibitions.
Hours & Admission Prices: Call for hours & admission prices.

Nogales

PIMERIA ALTA HISTORICAL SOCIETY, 136 N. Grand Ave., Nogales, AZ 85621-3211. Mailing Address: P.O. Box 2281, Nogales, AZ 85628-2281. Tel.: 520-287-4621. Fax: 520-287-5201.
Web Site: Facebook: Pimeria Alta Museum
Founded: 1948.
Congressional District: 2
Key Personnel: Pres., Kathleen Escalada; Project Dir., Sigrid Maitrejean.
Personnel Profile: Full-Time Paid 1; Full-Time Volunteers 7; Part-Time Paid 2; Part-Time Volunteers 5.
Governing Authority: nonprofit organization. Tax-exempt: 501(c)(3).
Institution Type/Description: Historical Society Museum.
Collections: artifacts of prehistoric & historic Indians of Pimeria Alta; Mexican & American artifacts from period of settlement including items from mining, ranching, local & household industries; 19th- & 20th-century settlement of south Arizona & north Sonora, Mexico; 10,000 regional historic photographs.
Research Fields: history of northwest Sonora, Mexico & southern Arizona.
Facilities: 1,000-vol. library of history & geography books & newspapers dating from 1893, available for use only on premises; archives of Nogales.
Activities: guided tours; lectures; films; permanent exhibitions; trips into Mexico; self guided walking tour; educational & public programming. Museum Sponsors: Historic Home Tour in February.
Publications: newsletter; annual historic calendar; archaeology reports; local history books; Pimeria Post.
Hours & Admission Prices: Tues.-Sun. 11-4. No charge; donations accepted. Closed New Year's Day; Thanksgiving; Christmas. &
Attendance: 3,359 (accurate)
Membership: Student $1; Individual $20; Family $30; Patron $75; Centennial $100; Silver $250; Gold $500; Medallion $1,000.

Oracle

ORACLE HISTORICAL SOCIETY, INC. - ACADIA RANCH MUSEUM, 825 E. Mt. Lemmon Hwy., Oracle, AZ 85623. Mailing Address: P.O. Box 10, Oracle, AZ 85623. Tel.: 520-896-9609.
E-mail: oraclehistoricalsociety@gmail.com
Web Site: oraclehistoricalsociety.org
Founded: 1978.
Congressional District: 1

Key Personnel: Pres., Monica Chavez.
Personnel Profile: Part-Time Paid 2; Part-Time Volunteers 11.
Governing Authority: Tax-exempt: 501(c)(3).
Institution Type/Description: Historic Building: housed on the ranch of Edwin S. and Lillian Dodge; built c.1882.
Collections: ranch history; cattle ranching; mining; early photographs; early pioneers; signed books, letters, & photographs from authors who wrote about Oracle.
Research Fields: local history & buildings.
Facilities: library; archives.
Activities: lectures; special events; permanent & temporary exhibitions.
Publications: OHS Newsletter.
Hours & Admission Prices: Sat. 1-5; other times by appointment. No charge; donations accepted. &

Attendance: 500 (estimated)
Membership: Individual $40; Family $60; Friend $100; Benefactor $250.

Page

JOHN WESLEY POWELL MEMORIAL MUSEUM, (M), 6 N. Lake Powell Blvd., Page, AZ 86040. Mailing Address: Box 547, Page, AZ 86040-0547. Tel.: 928-645-9496. Fax: 928-645-3412.
E-mail: director@powellmuseum.org
Web Site: www.powellmuseum.org
Founded: 1969.
Congressional District: 1
Key Personnel: Dir., Roy Boughton; Pres. (V), Mark Law; Museum Shop Mgr., Joy Dennis.
Personnel Profile: Full-Time Paid 1; Part-Time Paid 4; Part-Time Volunteers 15.
Governing Authority: nonprofit organization. Tax-exempt.
Institution Type/Description: History Museum.
Collections: archives; anthropology; archaeology; ethnology; geology; Indian artifacts; mineralogy; paleontology; philatelic; John Wesley Powell memorial items; films & written records of river runners on the Colorado River & its tributaries; manuscripts.
Major Exhibits: Trading Post Treasures: The Blair Family Collection, 3/13-2/14.
Research Fields: John Wesley Powell; Colorado Plateau; Glen Canyon Dam; city of Page.
Facilities: research library of books on Colorado River, reclamation, Glen Canyon Dam, Lake Powell, city of Page, John Wesley Powell's life & works, guidebooks of the West & special collections; archives. Bookstore.
Activities: guided tours; permanent & temporary exhibitions.
Hours & Admission Prices: April-Oct. Mon.-Sat. 9-5; Nov.-March Mon.-Fri. 9-5; call for additional hours. Adults $5, senior citizens 61 & over $3, children 6-13 $1; military & their families during the summer, children 5 & under, museum & Blue Star Museums' members no charge. Closed Thanksgiving; Christmas. &

Attendance: 16,000 (estimated)
Membership: Child, Student & Visitor $10; Individual $35; Family $50; Business $150.

Parker

BILL WILLIAMS RIVER NATIONAL WILDLIFE REFUGE, 60911 Hwy. 95, Parker, AZ 85344-9528. Tel.: 928-667-4144.
Key Personnel: Mgr., Richard Gilbert
Institution Type/Description: Wildlife Refuge.
Collections: wildlife & their habitat; natural history; trees; plants; flowers.
Facilities: nature trails.
Activities: hiking.
Hours & Admission Prices: Call for hours.

COLORADO RIVER INDIAN TRIBES MUSEUM, 1007 Arizona Ave., Parker, AZ 85344. Mailing Address: 26600 Mohave Rd., Parker, AZ 85344. Tel.: 928-669-8970. Fax: 928-669-1925.
E-mail: wilene.fisher-holt@crit-nsn.gov
Web Site: coloradoriverindiantribes.com
Founded: 1970.
Congressional District: 3
Key Personnel: Dir., Michael Tsosie; Librarian & Archivist, Wilene Fisher-Holt.
Personnel Profile: Full-Time Paid 4; Part-Time Paid 2; Part-Time Volunteers 1.
Governing Authority: Colorado River Indian Tribes. Tax-exempt.
Institution Type/Description: American Indian Museum.
Collections: anthropology; archaeology; Mohave; Chemehuevi; Navajo & Hopi; Anasazi; Hohokam & Patayan tribes; Colorado River area history &

ethnology; Japanese Memorial, World War II, Poston, AZ. Historic Buildings: 1917 Old Presbyterian Indian Church; 1860 La Paz, Arizona territorial mining town.
Research Fields: archaeology; linguistics; history; oral history.
Facilities: 2,000-vol. library of books pertaining to the American Indians, Mohave & Chemehuevi archives, microfilm & photograph collection of the reservation available for inter-library loan. Indian-made arts & crafts for sale.
Activities: permanent exhibitions; oral history; crafts workshops; linguistics workshops; La Paz archaeological excavation & reconstruction.
Hours & Admission Prices: Mon.-Fri. 8-5, Sat. 10-3; call for holiday hours. No charge; donations accepted. &
Attendance: 4,000 (estimated)

PARKER AREA HISTORICAL SOCIETY, 1214 California Ave. (Hwy. 95), Parker, AZ 85344-1500. Mailing Address: P.O. Box 1500, Parker, AZ 85344. Tel.: 928-669-8077.
Institution Type/Description: Historical Society Museum.
Collections: local history & culture; period furnishings; photographs; personal artifacts.
Hours & Admission Prices: Thurs.-Fri. 10-2.

Payson

NORTHERN GILA COUNTY HISTORICAL SOCIETY, INC. - RIM COUNTRY MUSEUM, (M), 700 Green Valley Pkwy., Payson, AZ 85547. Mailing Address: P.O. Box 2532, Payson, AZ 85547-2532. Tel.: 928-474-3483.
E-mail: ngchs1@gmail.com
Web Site: www.rimcountrymuseums.com
Founded: 1986.
Congressional District: 4
Key Personnel: Museum Shop Mgr., Betty Berryman; Financial Dir., Nancy Purkey; Pres. & Archivist, Sandy Carson; Security, Peter Bernard.
Personnel Profile: Part-Time Volunteers 45.
Governing Authority: private; nonprofit organization. Parent Institution: Northern Gila County Historical Society, Inc., Payson, AZ. Tax-exempt: 501(c)(3).
Institution Type/Description: Cultural History Museum.
Collections: archaeological exhibit of Risser Ruin Site; Tonto Apache-Yavapai Indian; Indian/Anglo interaction (advent of cavalry); early pioneer artifacts; local industry.
Research Fields: local histories; landmarks; Tonto Apaches & Yavapai.
Facilities: 300-vol. library of oral histories; 3,200 sq. ft. exhibit space. Gift items for sale.
Activities: docent program; guided tours; lectures; loan & participatory exhibitions.
Publications: bimonthly newsletter, Northern Gila County Historical Society; books, Rim Country Press.
Hours & Admission Prices: Mon. & Wed.-Sat. 10-4, Sun. 1-4. Requested Donations: adults $5, seniors 55 & over $4, students 12-18 $3; discounts to groups & children under 11; members no charge. Closed New Year's Day; Easter; Thanksgiving; Christmas. &
Attendance: 4,500 (estimated)
Membership: Individual $25; Family $35; Lifetime $500.

Peoria

CHALLENGER SPACE CENTER OF ARIZONA, 21170 N. 83rd Ave., Peoria, AZ 85382-2458. Tel.: 623-322-2001. Fax: 623-322-3716.
E-mail: information@azchallenger.net
Web Site: www.azchallenger.org
Key Personnel: Exec. Dir., Kari Sliva; Museum Shop Mgr., Carole LaConte.
Governing Authority: non-profit. 501 (c)(3).
Institution Type/Description: Space Museum.
Collections: history of space, flight and science & technology; meteorites; iridium satellite; shuttle model.
Facilities: planetarium; theater. Museum-related items for sale.
Activities: education programs; simulated space missions; youth summer camps; preschool program; scout & youth group activities; stargazing & planetarium shows.
Hours & Admission Prices: Daily 10-4. Adults $8, military and seniors 55 & over $7, students 3-12 $6; discounts to AAA members; children 2 & under and members no charge. Closed New Year's Day; Thanksgiving; Christmas. &
Membership: Individual $45; Couple $60; Family $75.

PEORIA ARIZONA HISTORICAL SOCIETY, 10304 N. 83rd Ave., Peoria, AZ 85345. Mailing Address: P.O. Box 186, Peoria, AZ 85345. Tel.: 623-487-8030. Fax: 623-487-8688.
E-mail: pahsoffice@aol.com
Institution Type/Description: Historical Society Museum.
Collections: local history & culture; period furnishings; personal artifacts; photographs; agricultural tools; cotton farming; ginning; blacksmith shop; dairy farming; sheep herding; school furnishings; local industry & government. Historic Buildings: 1906 school; 1910 agricultural building.
Activities: demonstrations; hands-on exhibitions.
Hours & Admission Prices: Sept.-May Sun. & Tues. 1-3, Wed. 12-2, Thurs.-Sat. 10-2; other times by appointment. No charge.

PEORIA JAIL HOUSE MUSEUM, 8322 W. Washington St., Peoria, AZ 85345. Mailing Address: c/o Peoria Arizona Historical Society, P.O. Box 189, Peoria, AZ 85380. Tel.: 623-487-8030.
Institution Type/Description: History Museum: housed in a former jail house; built c.1930.
Collections: local history; jail cell; photographs.
Activities: special events.
Hours & Admission Prices: Sept.-May Sun. 2-4; other times by appointment.

WEST VALLEY ART MUSEUM, Peoria, AZ 85385. Mailing Address: P.O. Box 6377, Peoria, AZ 85385. Tel.: 623-972-0635.
E-mail: info@wvam.org
Web Site: www.wvam.org
Formerly: West Valley Art Museum-Sun Cities Museum of Art
Founded: 1980.
Key Personnel: Bd. Pres., John Burridge; 1st Vice Pres., michael J. Thomas; Sec. & Treas., Constance McMillin.
Governing Authority: nonprofit. Tax-exempt: 501(c)(3).
Institution Type/Description: Art Museum.
Collections: ethnic dress; international fine prints; Arizona artists.
Hours & Admission Prices: No charge. Temporarily closed. &
Attendance: 5,000 (estimated)
Membership: Family $50; Curator $150; Director $250; Benefactor $500; Nonprofits $100; Business & Corporations $200; Hoover $1,000.

Peridot

SAN CARLOS APACHE CULTURAL CENTER, Hwy. 70 at Milepost 272, Peridot, AZ 85542. Mailing Address: P.O. Box 760, Peridot, AZ 85542-0760. Tel.: 928-475-2894.
Governing Authority: Tax-exempt.
Institution Type/Description: Cultural Center.
Collections: San Carlos Apaches Indian history & culture; personal artifacts; baskets; paintings; sculptures & carvings.
Activities: educational programs; demonstrations.
Hours & Admission Prices: Mon.-Fri. 9-5. Adults $3, seniors $1.50, students $1; children under 12 no charge.
Attendance: 2,497 (accurate)

Petrified Forest National Park

PETRIFIED FOREST NATIONAL PARK, One Park Rd., Petrified Forest National Park, AZ 86028-9997. Mailing Address: P.O. Box 2217, Petrified Forest National Park, AZ 86028-2217. Tel.: 928-524-6228, ext. 267. Fax: 928-524-3567.
E-mail: patricia_thompson@nps.gov
Web Site: www.nps.gov
Founded: 1906.
Congressional District: 1
Key Personnel: C.E.O. & Supt., Cliff Spencer; Chief Resource Mgmt., Pat Thompson; Museum Shop Mgr., Paul DoBell.
Personnel Profile: Full-Time Paid 1.
Governing Authority: federal. Parent Institution: U.S. Dept. of the Interior. Subsidiary Institution: National Park Service, U.S. Dept. of the Interior. Tax-exempt.
Institution Type/Description: National Park.
Collections: vertebrate & invertebrate paleontology specimens; paleobotanical; zoological; geology & archaeology of Petrified Forest National Park. Historic Building: 1924 Painted Desert Inn.
Research Fields: geology, paleontology, archaeology of Petrified Forest National Park.
Facilities: library of material relating to the collections; 40-seat auditorium; 93,533 acres.
Activities: self-guided tours; park orientation film.
Publications: book, Petrified Forest National Park: The Story Behind The Scenery; book, This is Painted Desert; Dawn of the Dinosaur; book, The Painted Desert Land of Light & Shadow; book, The Petrified Forest Thru the Ages; Tapamveni: The Rock Art Galleries of Petrified Forest & Beyond.
Hours & Admission Prices: Summer & Winter: daily 8-5; Spring & Fall call for hours. Car: $10, Bus $5; Golden Eagle, Golden Age & Golden Access passes honored. Closed Christmas. &
Attendance: 580,000 (estimated)

Phoenix

ADOBE MOUNTAIN RAILROAD MUSEUM & DESERT RAILROAD, 23280 N. 43rd Ave., Phoenix, AZ 85310. Mailing Address: 9186 W. Grovers Ave., Peoria, AZ 85382. Tel.: 623-670-1904.
E-mail: jerryoy147@msn.com
Founded: 1998.
Congressional District: 3
Key Personnel: Dir., Jerry Oyler.
Personnel Profile: Full-Time Volunteers 1.
Governing Authority: Parent Institution: Sahuaro Central Heritage. Tax-exempt.
Institution Type/Description: Railroad Museum.
Collections: model railroad layouts; small, narrow-gauge 1884 locomotive.
Activities: train rides.
Hours & Admission Prices: Sept.-May Sun. 12-4; other times by appointment. No charge; donations accepted. &
Attendance: 4,350 (estimated)
Membership: Individual $40; Family $45.

ARIZONA CAPITOL MUSEUM, (M), 1700 W. Washington St., Phoenix, AZ 85007-2812. Tel.: 602-926-3620. Fax: 602-256-7985.
E-mail: capmus@azlibrary.gov
Web Site: www.azlibrary.gov/museum
Founded: 1974.
Congressional District: 1
Key Personnel: Interim Dir., Joan Clark; Admin. Asst., Taylor Arrazola; Museum Shop Mgr., Micki Henningsen.
Personnel Profile: Full-Time Paid 7; Part-Time Paid 6; Part-Time Volunteers 60.
Governing Authority: state. Parent Institution: Arizona State Library, Archives & Public Records. Tax exempt: 170(b).
Institution Type/Description: History Museum: housed in restored Capitol Building.
Collections: Arizona government history.
Research Fields: Arizona government history.
Facilities: 54,000 sq. ft. of exhibits. Museum-related items for sale.
Activities: guided tours; permanent & temporary exhibits; seminars; lectures.
Hours & Admission Prices: Mon.-Fri. 9-4. No charge; donations accepted. &
Attendance: 60,000 (accurate)
Membership: Active & Associate $20; Patron $40.

ARIZONA DOLL AND TOY MUSEUM, 602 E. Adams St. at Heritage Sq., Phoenix, AZ 85004-2351. Tel.: 602-253-9337.
E-mail: kathylanford@cox.net
Founded: 1988.
Key Personnel: Pres. (V), Kathleen Lanford; Museum Shop Mgr., Dale Cantrell.
Personnel Profile: Full-Time Paid 1; Part-Time Volunteers 15.
Governing Authority: nonprofit. Tax-exempt.
Institution Type/Description: Doll and Toy Museum.
Collections: period dolls & toys; modern playthings; 1912 schoolroom.
Facilities: Museum-related items for sale.
Activities: tours.
Hours & Admission Prices: Sept.-July Tues.-Sat. 10-4, Sun. 12-4. Call for admission prices; discounts to handicapped. &
Attendance: 7,500 (estimated)

ARIZONA JEWISH HISTORICAL SOCIETY, 122 E. Culver St., Phoenix, AZ 85004-1720. Tel.: 602-241-7870. Fax: 602-264-9773.
E-mail: azjhs@aol.com
Web Site: www.azjhs.org
Founded: 1981.
Congressional District: 4
Key Personnel: Exec. Dir., Lawrence Bell, Ph.D.; Pres. (V), Stu Siefer.
Personnel Profile: Full-Time Paid 1; Part-Time Paid 1; Part-Time Volunteers 10.
Governing Authority: private; nonprofit organization. Tax-exempt: 501(c)(3).
Institution Type/Description: Jewish Historical Society.
Collections: photos; documents; maps; artifacts; oral histories.

Research Fields: biography project creating a database of Jewish population of Greater Phoenix from 1860 to the present; oral histories.
Facilities: 300-vol. library of Jewish & Arizona history available for loan; conference room.
Activities: lectures; loan, temporary & traveling exhibitions. Historical Society Sponsors: Fundraising Dinner; Legacy Programs.
Publications: bimonthly newsletter, Heritage.
Hours & Admission Prices: Aug.-June Mon.-Fri. 9:30-3:30. No charge; donations accepted. Closed Memorial Day; Independence Day; Labor Day; Jewish Holidays; Thanksgiving; Christmas.
Attendance: 2,000 (estimated)
Membership: Family $50; Oral History $100; Permanent Home $250; Archival Preservation $500; Exhibits $1,000.

ARIZONA MILITARY MUSEUM, Papago Park Military Reservation, 5636 E. McDowell Rd., Phoenix, AZ 85003-2668. Mailing Address: 9014 N. Wealth Rd., Maricopa, AZ 85139. Tel.: 602-267-2676 & 253-2378. Fax: 602-253-3342.
E-mail: joeabo@qwestoffice.net
Founded: 1981.
Congressional District: 24
Key Personnel: Dir. & Pres. (V), Joseph E. Abodeely, Col. (Ret.)
Personnel Profile: Part-Time Volunteers 14.
Governing Authority: Parent Institution: Arizona National Guard Historical Society. Tax-exempt.
Institution Type/Description: Military Museum.
Collections: uniforms; vehicles; artillery; military artifacts from the Civil War to the Indian and Spanish-American Wars, WWI, WWII, Korea, Vietnam & Desert Storm; POWs & MIAs; military history; Spanish Colonial; conquistadors; pre-Civil War; Desert Storm; War on Terror; Operation Iraqi Freedom; Huey honoring Vietnam Veterans service.
Facilities: library.
Publications: quarterly, The Courier.
Hours & Admission Prices: Sept.-June Sat.-Sun. 1-4. No charge; donations accepted. &
Attendance: 1,200 (estimated)
Membership: Annual $25; Life $250.

ARIZONA SCIENCE CENTER, 600 E. Washington, Phoenix, AZ 85004-2394. Tel.: 602-716-2007. Fax: 602-716-2099.
Web Site: www.azscience.org
Founded: 1984.
Congressional District: 2
Key Personnel: C.E.O. & Pres., Chevy Humphrey; Sr. Dir. Guest Experience, Patrick Weeks; Sr. Dir. Finance, Dwight Middendorf.
Personnel Profile: Full-Time Paid 51; Part-Time Paid 25; Part-Time Volunteers 400; Interns 8.
Governing Authority: nonprofit organization. Tax-exempt: 501(c)(3).
Institution Type/Description: Science Center.
Collections: interactive exhibits: human body, psychology, networks, weather, aerospace, geology & physics.
Facilities: 40,000 sq. ft. exhibit space; 285-seat giant screen IMAX theater; 200-seat Digistar; Nano Seam Dome planetarium; classroom.
Activities: live science demonstrations; education programs; museum excursions.
Publications: quarterly newsletter, Elements; triannual newsletter, Educators Planner.
Hours & Admission Prices: Daily 10-5. Exhibits: adults $12, senior citizens 62 & over & children 3-17 $10. IMAX & Planetarium: additional fee; discounts to members. Closed Thanksgiving; Christmas. &
Attendance: 300,000 (accurate)
Membership: Basic $70; Basic Plus $85; Explorer $130; Voyager $250; Adventurer $500; Director's Circle $1,000 & up.

ARIZONA STATE PARKS BOARD, 1300 W. Washington, Phoenix, AZ 85007-2929. Tel.: 602-542-4174. Fax: 602-542-4188. Facebook: Arizona State Parks.
Web Site: azstateparks.com
Founded: 1957.
Key Personnel: Exec. Dir., Bryan Martyn; Chm., Walter D. Armer; Public Information Officer, Ellen Bilbrey; Asst. Dir., Jay Ream; Museum Cur., Michael A. Freisinger; Park Mgr. Fort Verde State Historic Park, Sheila Stubler; Park Mgr. Homolovi Ruins State Park, Chad Meunier; Park Mgr. Jerome State Historic Park, Keith Ayotte; Park Mgr. Riordan Mansion State Historic Park, Joe Meehan; Park Mgr. Tonto Natural Bridge State Park, Steve Jakubowski; Park Mgr. Yuma Crossing State Historic Park, Tina Clark; Park Mgr. Yuma Territorial Prison State Historic Park, Mike Guertin; Park Mgr. McFarland State Historic Park, Jennifer Evans; Park Mgr.

Tombstone Courthouse State Historic Park, Kelly Schreiner; Park Mgr. Tubac Presidio State Historic Park, Shaw Kinsley; Park Mgr. Kartchner Caverns State Park, Chris DeMille; Dir. & Park Mgr. Boyce Thompson Arboretum State Park, Mark Siegwarth
Governing Authority: state. Branch Museums: Fort Verde State Historic Park, Camp Verde, AZ; Jerome State Historic Park, Jerome, AZ; McFarland State Historic Park, Florence, AZ; Riordan Mansion State Historic Park, Flagstaff, AZ; Tombstone Courthouse State Historic Park, Tombstone, AZ; Tubac Presidio State Historic Park, Tubac, AZ; Yuma Territorial Prison State Historic Park, Yuma, AZ; Yuma Crossing State Historic Park, Yuma, AZ.
Institution Type/Description: Historical Parks, Archaeological Park, Museums & Arboretum.
Collections: 28 parks.
Research Fields: history; nature.
Facilities: 50-vol. library of books pertaining to Southwestern history & specific sites available on premises; visitor center; archives; research facility. Books & museum-related items for sale.
Activities: guided tours; lectures; audiovisuals; permanent, temporary & traveling exhibitions.
Publications: printed handouts on historic sites; trails; natural areas.
Hours & Admission Prices: Office: daily 8-5. Parks: adults $4-$10; children under 6 no charge. Phoenix Office closed Christmas. &
Attendance: 600,000 (estimated)
Membership: 5 Visits $15; Limited $35; Unlimited $65

ARIZONA ZOOLOGICAL SOCIETY, DBA THE PHOENIX ZOO, (M), 455 N. Galvin Pkwy., Phoenix, AZ 85008-3431. Tel.: 602-273-1341. Fax: 602-286-3886.
E-mail: zooqna@thephxzoo.com
Web Site: www.phoenixzoo.org
Founded: 1961.
Key Personnel: Chm. Bd., Mary Alexander; C.E.O. & Pres., Norberto J. (Bert) Castro; Volunteer Coord., Mike Foley; C.F.O., Bonnie Mendoza; Exec. Vice Pres. Conservation & Visitor Experiences, Ruth Allard; Dir. Strategic Partnership, Patricia Bump; Registrar, Gretchen Bickert; Security, Steve Roberson; Retail Mgr., Jennifer Hall.
Personnel Profile: Full-Time Paid 204; Part-Time Paid 149; Part-Time Volunteers 400.
Governing Authority: society; nonprofit organization. Parent Institution: Arizona Zoological Society. Tax-exempt: 501(c)(3).
Institution Type/Description: Zoo.
Collections: over 1,400 mammals; birds; reptiles; amphibians; invertebrates.
Research Fields: conservation & science; animal health; animal behavior; visitor studies.
Facilities: 1,440-vol. library pertaining to zoology, science, husbandry & nutrition; zoological park.
Activities: guided tours; lectures; docent program; formal education programs for children; mobile vans; participatory exhibits; zoomobile; extensive school & family educational programming; continuing education.
Publications: bimonthly, Wild Times.
Hours & Admission Prices: Jan. 10-May & Sept.-Nov. 2 daily 9-5; June-Aug. Mon.-Fri. 7-2, Sat.-Sun. 7-4; Nov. 3-Jan. 9 daily 9-4. Adults $18, senior citizens $13, children $9; discounts to groups; children 2 & under and members no charge. Closed Christmas. &
Attendance: 1,450,377 (accurate)
Membership: Individual $50; Couple $80; Family $15 each child; Caretaker Club $150; Keeper's Club $225; Curator's Club $450; Director's Circle $1,250; President's Club $2,500; Chairman's Club $5,000; Robert Maytag Circle $10,000.

BARBARA ANDERSON GIRL SCOUT MUSEUM, Girl Scouts - Arizona Cactus-Pine Council, Inc., 3806 N. 3rd St., Ste. 200, Phoenix, AZ 85012. Mailing Address: 119 E. Coronado Rd., Phoenix, AZ 85004-1512. Tel.: 602-452-7000; 800-352-6133. Fax: 602-452-7100.
E-mail: council@girlscoutsaz.org
Web Site: www.girlscoutsaz.org
Key Personnel: Chm. (V), Nancy E. Buell
Governing Authority: nonprofit organization. Parent Institution: Girl Scouts-Az Cactus Pine Co. Tax-exempt.
Institution Type/Description: Girl Scout Museum.
Collections: Girl Scout uniforms; handbooks; memorabilia from 1912 to present; Arizona Girl Scouting.
Research Fields: Arizona Girl Scouts.
Activities: tours; G.S. badge work; displays.
Hours & Admission Prices: By appointment. No charge; donations accepted. &
Attendance: 400 (estimated)

CHILDREN'S MUSEUM OF PHOENIX, 215 N. 7th St., Phoenix, AZ 85034. Tel.: 602-253-0501. Fax: 602-307-9833.
E-mail: information@childmusephx.org
Web Site: www.childrensmuseumofphoenix.org
Founded: 1998.
Institution Type/Description: Children's Museum: housed in the former Monroe School building.
Collections: hands-on exhibitions.
Activities: special events; rental facilities; birthday parties; educational programs; classes.
Hours & Admission Prices: Tues.-Sun. 9-4. Admission $11, seniors 62 & over $10; children under one no charge.
Membership: Climber $125; Creator $160; Learner $195; Painter $230; Player $265.

DEER VALLEY ROCK ART CENTER, (M), 3711 W. Deer Valley Rd., Phoenix, AZ 85308-2038. Tel.: 623-582-8007. Fax: 623-582-8831. Facebook: Rock Art Center.
E-mail: dvrac@asu.edu
Web Site: dvrac.asu.edu
Founded: 1994.
Congressional District: 3
Key Personnel: Dir., Richard Toon; Interpretation Coord. & Programs, Casandra Hernandez; Operations Coord., Sherri Starkey.
Personnel Profile: Full-Time Paid 2; Part-Time Paid 3; Part-Time Volunteers 65.
Governing Authority: public university; nonprofit. Parent Institution: Arizona State University, School of Human Evolution and Social Change, Tempe, AZ 85281. Tax-exempt: 501(c)(3).
Institution Type/Description: General Museum.
Collections: early rock art site of more than 1,500 boulders with petroglyphs; period archives; photographs.
Research Fields: rock art recording & interpretation; archaeology; native gardening.
Facilities: library; art activity/classroom; nature center; nature/petroglyph outdoor trail; outdoor amphitheater; heritage garden. Gifts for sale.
Activities: docent program; films; formal education programs for undergraduate or graduate college students affiliated with university; guided tours; lectures; participatory & temporary exhibitions; children's & family programs.
Publications: quarterly newsletter, Glyph Gazette.
Hours & Admission Prices: Oct.-April Tues.-Sat. 9-5. Adults $7, seniors & students $4, children 6-12 $3; discounts to volunteers, AAA & DVRAC members; children 5 & under no charge. &
Attendance: 16,000 (accurate)
Membership: Individual $25; Family $40; Friend $100; Director's Circle $500.

＊ **DESERT BOTANICAL GARDEN, (M),** 1201 N. Galvin Pkwy., Papago Park, Phoenix, AZ 85008-3437. Tel.: 480-941-1225. Fax: 480-481-8124. TDD: 480-481-8143.
E-mail: administration@dbg.org
Web Site: www.dbg.org
Founded: 1937.
Congressional District: 1
Key Personnel: Exec. Dir., Ken Schutz; Pres. Bd. of Trustees, Barbara Hoffnagle; Pres. (V), Archer V. Shelton; Cur. Living Collection, Raul Puente; Dir. Research, Dr. Joe McAuliffe; Dir. Devel., Beverly Duzik; Deputy Dir., MaryLynn Mack; Dir. Education, Tina Wilson; Dir. Human Resources, Mary Catellier; Dir. Planning & Exhibits, Elaine McGinn; Dir. Visitor Svcs., Kathleen McAlpine; Dir. Finance & Admin., Michael Olson; Dir. Horticulture, Brian Kissenger; Librarian, Beth Brand; Dir. Security & Site Operations, Kevin Cullens; Dir. Mktg., John Sallot; Museum Shop Mgr., Kristen Fruen.
Personnel Profile: Full-Time Paid 86; Part-Time Paid 78; Part-Time Volunteers 1,036.
Volunteer Hours: 72,655
Operating Expenses: 10,769,954
Operating Income: 13,036,870
Governing Authority: incorporated nonprofit educational institution. Tax-exempt: 501(c)(3).
Institution Type/Description: Botanical Garden.
Collections: arid land plants of the world; cactus & leaf succulents; 50,000-sheet herbarium covering Southwest United States & Mexico.
Major Exhibits: Chihuly in the Garden (T), 1/14-5/14; Spring Butterfly Exhibit, 3/14-5/14; Mariposa Monarca, 9/14-11/14; Fish Out of Water, 9/14-11/14.
Research Fields: basic & applied research into arid land plants; economic botany & plant conservation.
Facilities: library of botanical & arid land related books; 145 acres; 3-acre ethnobotanical area; study areas of natural desert; herbarium; auditorium; educational facility; indoor & outdoor classrooms; rental facility. Museum-related items & botanical prints for sale.
Activities: guided tours; audio tours; lectures; classes; workshops; field trips; permanent displays; special exhibitions; concerts; cultural programs.
Publications: quarterly, The Sonoran Quarterly; plant information leaflets; guide booklet; calendar; event promotional pieces.
Hours & Admission Prices: May-Sept. daily 7-8; Oct.-April daily 8-8. Adults $22, senior citizens over 60 $20, students $12, children 3-12 $10; discounts to AAA & AAM members; members no charge. Closed Independence Day; Thanksgiving; Christmas. &
Attendance: 380,259 (accurate)
Membership: Senita Club $90; Cholla Club $125; Agave Century Club $200; Boojum Tree Club $350; Golden Barrel Club $700; Saguaro $1,250; Curator's Circle $2,500; Director's Circle $5,000; President's Club $10,000; Founder's $25,000.

GEORGE WASHINGTON CARVER MUSEUM AND CULTURAL CENTER, 415 E. Grant St., Phoenix, AZ 85004-2659. Mailing Address: P.O. Box 20491, Phoenix, AZ 85036-0491. Tel.: 602-254-7516. Fax: 602-258-7050.
E-mail: gwcmccphax@qwestoffice.net
Web Site: www.gwcmuseumculturalcenter.org
Key Personnel: Exec. Dir., Princess Crump
Institution Type/Description: History Museum: housed in Arizona's first Black high school, Phoenix Union Colored High, built in 1926.
Collections: local Black history; cultural heritage; works by local & regional visual and performing artists.
Facilities: auditorium; cafe.
Activities: monthly concerts; lectures; workshops; temporary exhibitions; rental facilities.
Hours & Admission Prices: Mon.-Fri. 10-3 by appointment. Adults $3, seniors 60 & over $2.50, youth 6-12 $2, children 3-5 $1.

HALL OF FLAME MUSEUM OF FIREFIGHTING, 6101 E. Van Buren, Phoenix, AZ 85008-3421. Tel.: 602-275-3473. Fax: 602-275-0896.
E-mail: petermolloy@hallofflame.org
Web Site: www.hallofflame.org
Founded: 1961.
Congressional District: 1
Key Personnel: Exec. Dir., Dr. Peter M. Molloy; Pres. (V), George F. Getz; Vice Pres., Bert A. Getz; Vice Pres., Lynn Getz; Sec. & Treas., Michael Olsen; Cur. Public Programs & Education, Mark Moorhead; Restorator, Donald G. Hale; Docent, Grace Deutsch; Volunteer Pres., Ron Deutsch.
Personnel Profile: Full-Time Paid 3; Part-Time Paid 6; Part-Time Volunteers 24.
Governing Authority: nonprofit organization. Parent Institution: National Historical Fire Foundation, 6730 N. Scottsdale Rd., Suite 250, Scottsdale, AZ 85253. Tax-exempt: 501(c)(3).
Institution Type/Description: History & Fire Fighting Museum.
Collections: over 130 major pieces of firefighting equipment dating from 1700-1970; over 10,000 objects & graphic materials relating to the history of firefighting; over 40,000 photographs plus 6,000 books, serials & trade catalogs; National Firefighting Hall of Heroes.
Research Fields: history of fire fighting technology; social history of fire fighting in U.S.; structural & wildland firefighting.
Facilities: library; visitor center. Museum-related items for sale.
Activities: guided tours; permanent & temporary exhibitions; workshops & fire safety exhibits for children.
Publications: exhibit catalogue; quarterly newsletter; informational brochure; book, Van Der Heyden's Treatise on Firefighting in Amsterdam.
Hours & Admission Prices: Mon.-Sat. 9-5, Sun. 12-4. Adults $7, senior citizens $6, students 6-17 $5, children 3-5 $2; discount to groups & AAA members; members & children under 3 no charge. Closed New Year's Day; Thanksgiving; Christmas. &
Attendance: 35,000 (accurate)
Membership: Individual $25; Family $35; Black Helmet Brigade & Fire Professional $60; Red Helmet Brigade $100; White Helmet $250.

＊ **HEARD MUSEUM, (M),** 2301 N. Central Ave., Phoenix, AZ 85004-1323. Tel.: 602-252-8848. Fax: 602-252-9757. Facebook: Heard Museum.
E-mail: contact@heard.org
Web Site: www.heard.org
Founded: 1929.
Congressional District: 4
Key Personnel: Vice Pres. Curation & Education, Ann E. Marshall, Ph.D.; Interim C.E.O., Lee Peterson; Dir. Institutional Advancement, Jim Weaver;

Chm., Mark Bonsall; Guild Pres., Rod Passmore; Dir. Finance, Carlos Rojas; Sr. Communications Mgr., Debra Krol; Dir. Educational Svcs., Jaclyn Roessel; Cur. Collections, Diana Pardue; Registrar, Sharon Moore; Assoc. Registrar, Marcus Monenerkit; Dir. Library & Archives, Mario Nick Klimiades; Creative Dir., Caesar Chaves; Vice Pres. Retail Sales, Bruce McGee; Cur. Community Museum & Operations Mgr. Heard Museum North Scottsdale, Janet Cantley.

Personnel Profile: Full-Time Paid 80; Part-Time Paid 20; Part-Time Volunteers 700; Interns 4.

Governing Authority: nonprofit. Tax-exempt: 501(c)(3).

Institution Type/Description: Native Cultures & Art Museum.

Collections: works by American Indians; native arts from the cultures of Africa, Asia, Oceania & upper Amazon; archaeology; ethnology; paintings; sculpture; anthropology.

Research Fields: anthropology; American Indian art; cross-cultural education.

Facilities: 45,000-vol. library of reference material available to the public for reference only. Southwestern Indian crafts, fine art, books & other museum-related items for sale.

Activities: guided tours; lectures; films; gallery talks; arts festivals; regular weekend programs; workshops; permanent, temporary, changing & traveling exhibitions. Annual Events: 23rd Hoop Dance Contest in February; Heard Museum Guild Indian Fair & Market in March; A Gathering of Carvers: Katsina Doll Marketplace in April; Heard Museum Spanish Market; Holidays at the Heard in December.

Publications: newsletter; occasional catalogs & books on Indian arts and crafts, biannual magazine.

Hours & Admission Prices: Mon.-Sat. 9:30-5, Sun. 11-5. Adults $18, senior citizens 65 & over $13.50, students w/ID & children 6-12 $7.50; discounts to AAM members; children under 6, Native Americans with proof of tribal heritage & members no charge. Closed Christmas. &

Attendance: 200,000 (estimated)

Membership: Friend $60; Supporter $90; Patron $150; Sustaining Patron $250; Curator's Society $500; Director's Circle $1,250.

MUSICAL INSTRUMENT MUSEUM, (M), 4725 E. Mayo Blvd., Phoenix, AZ 85050. Tel.: 480-478-6000. Fax: 480-481-2459.

E-mail: guestservice@themim.org

Web Site: www.themim.org

Founded: 2005.

Congressional District: 3

Key Personnel: Dir., Dr. Billie R. DeWalt; Chm. (V), Robert J. Ulrich; Vice Pres. External Rels., Christopher Bell; Dir. Exhibits & Multimedia, April Salomon; C.F.O., Rhonda Boyle; Chief Cur. & Dir. Collections, Manuel Jordan; Dir. Operations & Pub. Programs, Justin Karim; Registrar, Katie Anderson; Museum Shop Mgr., Lacey Hauser.

Personnel Profile: Full-Time Paid 70; Part-Time Paid 30; Part-Time Volunteers 300; Interns 5.

Governing Authority: private; nonprofit organization. Tax-exempt.

Institution Type/Description: Musical Instrument Museum.

Collections: over 15,000 instruments & associated objects from around the world; audio & video of instruments.

Research Fields: use, conservation, preservation, performance, history, significance and cultural context of musical instruments.

Facilities: library; 75,000 sq. ft. exhibit space; restaurant; classrooms; music theater. Museum-related items for sale.

Activities: educational programs; special events; theater performances.

Publications: biweekly e-mail newsletter.

Hours & Admission Prices: Mon.-Wed. & Sat. 9-5, Thurs.-Fri. 9-9, Sun. 10-5. Adults 18-64 $15, seniors 65 & over $13, youth 6-17 $10; discounts to AAM members; children under 6 no charge. &

Attendance: 200,000 (estimated)

PHOENIX AIRPORT MUSEUM, (M), 3400 Sky Harbor Blvd., Terminal 3, Level 3 W., Phoenix, AZ 85034-4403. Tel.: 602-683-3647.

Founded: 1987.

Personnel Profile: Full-Time Paid 6.

Governing Authority: city.

Institution Type/Description: History & Art Museum.

Collections: Arizona's history & cultural heritage.

Activities: permanent & temporary exhibits.

Hours & Admission Prices: Daily 24 hours. No charge. &

∗ **PHOENIX ART MUSEUM, (M),** 1625 N. Central Ave., Phoenix, AZ 85004-1685. Tel.: 602-257-1880. Fax: 602-253-8662.

E-mail: info@phxart.org

Web Site: www.phxart.org

Founded: 1959.

Congressional District: 1

Key Personnel: Chm., James Patterson; Dir. & Chief Curator, James K. Ballinger; Asst. to Dir., Samantha Klick; Deputy Dir. Finance & Administration, Gary Egan; Chief Devel. Officer, Ron Miller; Assoc. Cur. Latin American Art, Vanessa Davidson, Ph.D.; Cur. Asian Art, Dr. Janet Baker; Cur. Fashion Design, Dennita Sewell; Research Cur. of Asian Art, Dr. Claudia Brown; Cur. American & Western Art, Jerry Smith; Registrar, Leesha Alston; Librarian, Abigail Nersesian; Museum Store Mgr., Jennifer Barnella.

Personnel Profile: Full-Time Paid 103; Part-Time Paid 41; Part-Time Volunteers 2,442; Interns 15.

Governing Authority: nonprofit. Tax-exempt: 501(c)(3).

Institution Type/Description: Art Museum.

Collections: American, European & Asian paintings and ceramics; sculpture; graphics; decorative arts; 18th, 19th & 20th-century fashion design; Latin & Western American art; Thorne miniatures; modern & contemporary art.

Research Fields: American, European & Oriental paintings & ceramics; sculpture; graphics; decorative arts; fashion design.

Facilities: 50,000-vol. library of art reference books available for use on premises; 55,000 slide collection; 300-seat auditorium; cafe; classrooms; banquet facilities. Museum-related items for sale.

Activities: guided tours; lectures; films; gallery talks; concerts; arts festivals; formally organized education programs; docent program; inter-museum loan, permanent, participatory, temporary & traveling exhibitions; films; rental facilities.

Publications: membership magazine & brochures; exhibition schedule & catalogues; brochures, exhibition highlights; education.

Hours & Admission Prices: Wed. 10-9, Thurs.-Sat. 10-5, Sun. 12-5, 1st Fri. of month 6-10. Adults $15, senior citizens $12, students $10, youth $6; members & children under 6 no charge. Discounts to AAM & ICOM members. &

Attendance: 210,000 (accurate)

Membership: General $75; Contributor $100; Supporter $200; Sponsor $400; Patron $750; Director's Circle $1,500; Connoisseurs' Circle $2,500; Trustees' Circle $5,000; President's Circle $10,000; Founders' Circle $25,000.

PHOENIX CENTER FOR THE ARTS, 1202 N. Third St., Phoenix, AZ 85004-1812. Tel.: 602-262-4627.

Institution Type/Description: Art Center.

Collections: photographs; paintings; sculpture.

Activities: educational programs; special events.

Hours & Admission Prices: Mon.-Fri. 9-5.

PHOENIX POLICE MUSEUM, Historic City Hall, 17 S. 2nd Ave., 1st Fl., Phoenix, AZ 85003. Tel.: 602-534-7278. Fax: 602-495-2491.

E-mail: mike.nikolin@phoenix.gov

Web Site: phoenixpolicemuseum.com

Founded: 1994.

Key Personnel: Dir. & Cur., Michael Nikolin.

Governing Authority: Tax-exempt.

Institution Type/Description: Police Museum.

Collections: history & heritage of the police department; hands-on exhibits; memorial room.

Facilities: Museum-related items for sale.

Publications: newsletter.

Hours & Admission Prices: Mon.-Fri. 9-3. No charge; donations accepted. Closed city & federal holidays. &

Attendance: 8,700

Membership: Annual $48; Lifetime $150.

PHOENIX TROLLEY MUSEUM/ARIZONA STREET RAILWAY MUSEUM, 1218 N. Central Ave., Phoenix, AZ 85004. Mailing Address: P.O. Box 13521, Phoenix, AZ 85002. Tel.: 602-254-0307.

Web Site: www.phoenixtrolley.com

Key Personnel: Pres., Ernest Workman; Sec., Tom Amrheim

Institution Type/Description: Railway Museum.

Collections: restored 1928 Phoenix Street Railway System trolley car; Phoenix trolley system history.

Hours & Admission Prices: Oct.-May Sat. 9-4. Suggested Donation: $1 plus donation; discounts to CAMA members; members no charge. Closed New Year's Day; Thanksgiving; Christmas; holiday weekends.

Attendance: 300 (estimated)

Membership: Students $15; Individual $20; Family $35; Sustaining $100; Corporate $500.

PIONEER ARIZONA LIVING HISTORY VILLAGE & MU-SEUM, 3901 W. Pioneer Rd., Exit 225, Interstate 17, Phoenix, AZ 85086. Mailing Address: 3901 W. Pioneer Rd., Phoenix, AZ 85086-7020. Tel.: 623-465-1052. Fax: 623-465-0683.
E-mail: pioneervillageaz@gmail
Web Site: pioneeraz.org
Founded: 1956.
Congressional District: 3
Key Personnel: Dir., Stephanie Luster; Pres. (V), C.J. Smith.
Personnel Profile: Full-Time Paid 4; Part-Time Paid 2; Part-Time Volunteers 120.
Governing Authority: nonprofit organization. Parent Institution: Pioneer Foundation, Inc. Tax-exempt: 501(c)(3).
Institution Type/Description: Living History Museum Complex.
Collections: over 20 late 19th-century homes & shops; 19th-century pioneer life artifacts; agricultural items & machinery; Arizona history memorabilia.
Research Fields: social & technological history of 19th-century Arizona; territorial education in rural Arizona.
Facilities: day theatre; restaurant. Crafts & other museum-related items for sale.
Activities: self-guided tours; living history interpretation; special events; children's activities.
Publications: members newsletter; museum guide.
Hours & Admission Prices: June-Sept. Wed.-Sun. 8-2; Oct.-May Wed.-Sun. 9-5. Adults $9, seniors & vets $6, children $7; children under 5 no charge. Guided Tour: $1 extra. Closed Easter; Thanksgiving; Christmas. &
Attendance: 65,000 (estimated)

PIONEER TELEPHONE MUSEUM, 3640 E. Indian School Rd., Phoenix, AZ 85018. Tel.: 602-630-2060.
E-mail: pioneertelephonemuseum@gmail.com
Institution Type/Description: Communications Museum.
Collections: history of telecommunications from the 1870s to present day; telephone equipment & memorabilia; early coin phones; phone booths; photographs.
Hours & Admission Prices: Mon.-Fri. 9-4; tours by appointment.

∗ PUEBLO GRANDE MUSEUM AND ARCHAEOLOGICAL PARK, (M), 4619 E. Washington, Phoenix, AZ 85034-1909. Tel.: 602-495-0901; 877-706-4408 (toll free). Fax: 602-495-5645.
E-mail: pueblo.grande.museum.pks@phoenix.gov
Web Site: www.pueblogrande.com
Founded: 1929.
Congressional District: 1
Key Personnel: C.E.O. & Dir., Roger Lidman; Chm. (V), Daniel Zuczek; Visitor Svcs., Laura Andrew; Cur. Collections, Holly Young; Museum Shop Mgr., Francine Kavanaugh.
Personnel Profile: Full-Time Paid 6; Part-Time Paid 5; Part-Time Volunteers 50; Interns 4.
Volunteer Hours: 5,800
Operating Expenses: 1,091,200
Operating Income: 1,091,200
Governing Authority: municipal. Parent Institution: City of Phoenix, Parks & Recreation Dept. Tax-exempt.
Institution Type/Description: Archaeological Site Museum.
Collections: Pubelo Grande artifacts; greater Southwest archaeological excavation artifacts; greater Southwest Native American artifacts; territorial Phoenix archaeological artifacts; prehistoric platform mound; ethnology; early Phoienix history.
Major Exhibits: Arizona Ghost Towns (T), 9/13-7/14.
Research Fields: pertaining to museum's collections, the Hohokam of the Salt River Valley & southern Arizona; native peoples of Arizona; territorial Phoenix.
Facilities: 4,000-vol. library of Southwestern archaeology, history, natural history & cultures of central Arizona, community room; outdoor trail to prehistoric ruin.
Activities: lectures; permanent & changing exhibits; workshops; guided tours for groups by appointment; outreach exhibits & programs. Annual Event: Indian market in December.
Publications: brochures, Pueblo Grande Museum; Hohokam; Anthropological Papers No. 1, 2 & 4.
Hours & Admission Prices: May-Sept. Tues.-Sat. 9-4:45; Oct.-April Mon.-Sat. 9-4:45, Sun. 1-4:45. Adults $6, senior citizens $5, children $3; discounts to Western Museums Assoc., Central Arizona Museum Assoc. & AAM members; members & children under 6 no charge. Closed New Year's Day; Martin Luther King Jr. Day; Memorial Day; Independence Day; Labor Day; Thanksgiving; Christmas. &
Attendance: 40,890 (accurate)

Membership: Student $10; Individual $20; Family $25; Contributing $60; Patron $125; Corporate $250.

ROSSON HOUSE MUSEUM, (M), 7th St. & Monroe, Phoenix, AZ 85004. Mailing Address: 113 N. 6th St., Phoenix, AZ 85004-2328. Tel.: 602-261-8063.
E-mail: rhhs@rossonhousemuseum.org
Web Site: www.rossonhousemuseum.org
Founded: 1982.
Key Personnel: Pres. (V), Donna Reiner; Dir., Michelle Reid; Museum Shop Mgr., Bobby Bonner.
Personnel Profile: Full-Time Paid 3; Part-Time Paid 2; Part-Time Volunteers 35.
Governing Authority: Tax-exempt.
Institution Type/Description: Historic House Museum: built in 1895.
Collections: personal artifacts; period furnishings.
Major Exhibits: Victorian Secrets: A Glimpse of the Unmentionable, 9/12-7/14.
Activities: school tours; history lecture series.
Publications: The Transom.
Hours & Admission Prices: Wed.-Sat. 10-4, Sun. 12-4. Adults $7.50, seniors & students $6, children 6-12 $4; active military, members and children 5 & under no charge. Closed New Year's Day; Easter; Independence Day; Thanksgiving; Christmas. &
Attendance: 8,000 (estimated)
Membership: Student & Active Volunteer $20; General $25; Family $35; Supporter $50; Sponsor $100.

SHEMER ART CENTER & MUSEUM, 5005 E. Camelback Rd., Phoenix, AZ 85018-3015. Tel.: 602-262-4727.
E-mail: info@shemerartcenterandmuseum.org
Web Site: www.shemerartcenterandmuseum.org
Founded: 1984.
Congressional District: 6
Key Personnel: Dir., Jocelyn Hanson; Graphic & Office, Lacasa Michelena; Graphic & Office, Emily Costello.
Personnel Profile: Full-Time Paid 1; Part-Time Paid 2; Part-Time Volunteers 60; Interns 1.
Governing Authority: municipal. Tax-exempt: 501(c)(3).
Institution Type/Description: Art Museum.
Collections: paintings; photographs; sculpture.
Facilities: classrooms.
Activities: Annual Events: Sunday at Shemer; Art in the Garden in March; Family Arts Festival in November.
Publications: quarterly newsletter; class schedules; annual exhibit schedule; exhibition invitations.
Hours & Admission Prices: Tues.-Wed. & Fri.-Sat. 10-3, Thurs. 10-3 & 6pm-8pm. No charge; donations accepted. Closed New Year's Day; Martin Luther King Jr. Day; Presidents' Day; Cesar Chavez's Birthday; Memorial Day; Independence Day; Labor Day; Veterans Day; Thanksgiving & day after; Christmas. &
Attendance: 25,000 (estimated)
Membership: Student $20; Individual $40; Family $75; Supporter $100; Contributor $500; Sponsor $1,000; Patron $2,500; Trustee $5,000.

SUNNYSLOPE HISTORICAL SOCIETY MUSEUM, 737 E. Hatcher Rd., Phoenix, AZ 85020-2506. Tel.: 602-331-3150. Fax: 602-331-3163.
E-mail: shsociety1@qwestoffice.net
Web Site: www.sunnyslopehistoricalsociety.org
Founded: 1989.
Key Personnel: Pres. (V), Pat Wilkerson; Museum Shop Mgr., Juanita Reeves.
Personnel Profile: Part-Time Volunteers 30.
Governing Authority: Tax-exempt.
Institution Type/Description: History Museum.
Collections: Sunnyslope history.
Major Exhibits: Palo Verde Festival, 4/14.
Research Fields: community artifacts.
Activities: quarterly educational programs.
Publications: quarterly, The Sunnyslope Heritage.
Hours & Admission Prices: Sept.-May Wed.-Fri.-Mon. 12-4. No charge; donations accepted. &
Attendance: 1,000 (estimated)
Membership: Senior $25; Individual $30; Family $50; Patron $60-$99; Palo Verde $100-$299; "S" Mountain $300-$399; 1945 House $400-$499; Viking $500 & up.

WELLS FARGO HISTORY MUSEUM, 100 W. Washington, Phoenix, AZ 85003-1805. Mailing Address: Wells Fargo Historical Services, 420 Montgomery St., MAC-A0101-106, San Francisco, CA 94163. Tel.: 602-378-1578.
Founded: 2003.
Key Personnel: Cur., Connie Whalen.
Governing Authority: profit-making organization. Affiliated with Wells Fargo Bank.
Institution Type/Description: Company History Museum.
Collections: Concord Stagecoach; Wells Fargo banking & express history; fine art, including paintings by N.C. Wyeth; mining; staging; early Phoenix history.
Activities: guided group tours; audiovisual programs; imaginary rides on replica stagecoach.
Publications: scholarly pamphlets.
Hours & Admission Prices: Mon.-Fri. 9-5. No charge. Closed bank holidays.

Pima

EASTERN ARIZONA MUSEUM AND HISTORICAL SOCIETY OF GRAHAM COUNTY INC., 2 N. Main St., Pima, AZ 85543. Mailing Address: P.O. Box 274, Pima, AZ 85543-0274. Tel.: 928-485-3032.
E-mail: edresbarney@yahoo.com
Founded: 1963.
Congressional District: 1
Key Personnel: Pres., Nick Bingham; Vice Pres., Shawn Wright; Sec., Anna Jane Jarvis; Dir. & Treas., Edres Barney.
Personnel Profile: Part-Time Volunteers 12.
Governing Authority: society; nonprofit organization. Parent Institution: Arizona Historical Society, Tucson. Tax-exempt.
Institution Type/Description: History Museum.
Collections: pioneer & Indian artifacts; photographs; personal & community histories; reproduction of a pioneer home; original notes & copies of trial transcript. of the 1889 Wham robbery Army payroll. Historic House: c.1882 Old Cluff Hall.
Facilities: reading & research room.
Activities: guided tours; permanent & temporary exhibitions.
Publications: booklet, Wham Paymaster Robbery; book, Centennial; Graham County Profiles, Vol. II; Garden of Eden & How It Grew, Mt. Graham; Museum Historical Cookbook.
Hours & Admission Prices: Thurs.-Sat. 10-3; other times by appointment. No charge; donations accepted. ♿
Attendance: 1,824 (accurate)
Membership: Student $3; Family $15; Business $40; Lifetime $100.

Pine

PINE-STRAWBERRY MUSEUM, 3886 Hwy. 87, Pine, AZ 85544. Mailing Address: P.O. Box 564, Pine, AZ 85544-0564. Tel.: 928-476-3547.
Institution Type/Description: History Museum.
Collections: local history & culture; Native American artifacts; Mormon Church; farming implements; period clothing & furniture.
Facilities: Museum-related items for sale.
Hours & Admission Prices: mid-May to mid-Oct. Mon.-Thurs. 10-2, Fri.-Sun. 10-4; mid-Oct. to mid-May Mon.-Sat. 10-2. Adults $1; children 11 & under no charge. Closed New Year's Day; Easter; Thanksgiving; Christmas. ♿
Attendance: 2,000 (estimated)

Prescott

HIGHLANDS CENTER FOR NATURAL HISTORY, 1375 S. Walker Rd., Prescott, AZ 86303-6893. Tel.: 928-776-9550. Fax: 928-776-9530.
E-mail: highlands@highlandscenter.org
Web Site: highlandscenter.org
Formerly: Community Nature Center
Founded: 1975.
Congressional District: 3
Key Personnel: Exec. Dir., Dave Irvine.
Personnel Profile: Full-Time Paid 5; Part-Time Paid 2; Part-Time Volunteers 150.
Volunteer Hours: 9,300
Operating Expenses: 407,000
Operating Income: 490,000
Governing Authority: private; nonprofit organization. Nature Center: Walker Rd. 2 Miles South of Hwy. 69, Prescott, AZ; Lynx Creek Site, 1375 S. Walker Rd., Prescott AZ 86303. Tax-exempt: 501(c)(3).

Institution Type/Description: Nature Center.
Collections: natural history of the Arizona Highlands area.
Research Fields: grasslands project; native plant arboretum; phenology.
Facilities: library; educational facilities; nature/conservation center; amphitheater; riparian (wetland) area; classroom; rented space; guided nature trail. Museum-related items for sale.
Activities: Education for children & adults; docent program; day camps; school field trips; naturalist training; guided hikes; birding; Grow Native! plant sale.
Publications: newsletter published three times annually, Highlands Vision; native plants book; book, Wildscaping; curriculum guide, Tracking the Past; trail guides.
Hours & Admission Prices: April-Sept. 7-7; Oct.-March 8-6. Charge for classes & special events only; donations accepted. Trails open to the public. ♿
Attendance: 5,000 (estimated)
Membership: Individual $35; Family $45; Donor $100.

PHIPPEN MUSEUM, (M), 4701 Hwy. 89 N., Prescott, AZ 86301-8303. Tel.: 928-778-1385. Fax: 928-778-4524. Facebook: Phippen Museum of Western Art.
E-mail: phippen@phippenartmuseum.org
Web Site: www.phippenartmuseum.org
Founded: 1974.
Congressional District: 3
Key Personnel: Dir., Kim Villalpando; Chm. (V), Dick Cornwell; Office Mgr. and Community Devel. & Communication, Edd Kellerman; Bookkeeper, Janet Hersh; Events Coord., James Ward; Cur., Mgr. Collections & Museum Shop Mgr., Lynette Tritel; Mgr. Information, Tom McCain.
Personnel Profile: Full-Time Paid 4; Part-Time Paid 1; Part-Time Volunteers 75.
Governing Authority: nonprofit organization. Parent Institution: George Phippen Memorial Foundation. Tax-exempt: 501(c)(3).
Institution Type/Description: Art of the American West Museum.
Collections: paintings; sculpture; artifacts.
Major Exhibits: National Parks of the West, 11/13-2/14; Solon Borglum Collection, 1/14-12/14; Will James, 2/14-6/14; Architecture in Art, 3/14-7/14; Hold Your Horses, 7/14-10/14; Click - The West Thru the Lens, 11/14-2/15.
Facilities: art research library; lecture hall; activity hall; multipurpose classroom. Museum-related items for sale.
Activities: guided tours; lectures; organized education programs; temporary & traveling exhibitions; docent led tours. Annual Events: Western Fine Art Show & Sale in May; Fall Gathering BBQ & Branding; Designer Home Showcase; Arizona Holiday Shopping.
Publications: quarterly newsletter; exhibit brochures.
Hours & Admission Prices: Tues.-Sat. 10-4, Sun. 1-4. Adults $7, students $5; discounts to AAM & AAA members; museum professionals, children under 12 & members no charge. Closed New Year's Day; Easter; Thanksgiving; Christmas. ♿
Attendance: 10,333 (accurate)
Membership: Individual $35; Family $60; Supporting $120; Business $150; Sustaining $250; Patron $500; Benefactor $1,000.

＊ **SHARLOT HALL MUSEUM, (M),** 415 W. Gurley St., Prescott, AZ 86301-3691. Tel.: 928-445-3122. Fax: 928-776-9053.
E-mail: gails@sharlot.org
Web Site: www.sharlot.org
Founded: 1928.
Congressional District: 1
Key Personnel: Dir., John Langellier, Ph.D.; Pres. (V), Jim Pool; Chief Cur., Mick Woodcock.
Personnel Profile: Full-Time Paid 9; Part-Time Paid 9; Part-Time Volunteers 350.
Governing Authority: state. Subsidiary Institutions: Sharlot Hall Historical Society & the Prescott Historical Society. Tax-exempt: 170(b)(1)(A).
Institution Type/Description: Regional History Museum.
Collections: regional archives; costumes; vehicles; photographs; Indian archaeological & ethnographic materials; medical; military; decorative arts; transportation. Historic Buildings: 1864 Governor's mansion; 1864 Fort Misery; 1875 John C. Fremont house; 1877 William C. Bashford house.
Research Fields: Arizona particularly the Central Mountain Region.
Facilities: library of diaries, books, manuscripts, scrapbooks, files & newspapers available for research on premises; memorial rose garden; herb garden.
Activities: guided tours by prior arrangement; permanent & temporary exhibitions; folk arts demonstrations; lecture series. Museum Sponsors: Annual Folk Arts Fair; Folk Music Festival; Arizona History Adventure; Prescott Indian Art Market; Blue Rose Theater Historical Plays.
Publications: bimonthly newsletter; annual history journal; booklets, Orejana

Bull; The Arizona Rough Rider Monument & Captain W. O. O'Neill; books, Cactus & Pine; Poems of a Ranch Woman; Arizona's First Capitol; Meeting The Four O'Clock Train; The Ernest W. McFarland Papers; Sharlot Herself; Sharlot Hall on the Arizona Strip; The Arizona Rough Riders; Rough Writings; Co-Existing with Urban Wildlife; Historic Photographs of Central Arizona Grasslands and Associated Habitats; cassette, Grand Canyon Cowboy Band; VHS video tape, Ranch Albu; The Wilderness Around Us; book, All of My People Were Killed: The Memoir of Mike Burns (Hoomothya) A Captive Indian.
Hours & Admission Prices: June-Sept. Mon.-Sat. 10-5, Sun. 12-4; Oct.-May Mon.-Sat. 10-4, Sun. 12-4. Adults $5; members and children 18 & under no charge. Closed New Year's Day; Thanksgiving; Christmas. &
Attendance: 25,285 (accurate)
Membership: Individual $25; Family $35; Family Circle $75; Heritage Circle $125; Corporate $125-$249; Curator's Circle $250; Director's Circle $500; Sharlot's Circle $1,000; Legacy Circle $5,000.

SMOKI MUSEUM - AMERICAN INDIAN ART & CULTURE, 147 N. Arizona, Prescott, AZ 86301-3184. Mailing Address: P.O. Box 10224, Prescott, AZ 86304-0224. Tel.: 928-445-1230.
E-mail: director@smokimuseum.org
Web Site: www.smokimuseum.org
Founded: 1935.
Congressional District: 3
Key Personnel: Dir., Cynthia Gresser; Pres. (V), James Christopher; Vice Pres., Ray Carlson; Treas., Kent Robinson; Museum Shop Mgr., Carol Semplice.
Personnel Profile: Full-Time Paid 1; Full-Time Volunteers 3; Part-Time Paid 1; Part-Time Volunteers 80.
Governing Authority: private; not-for-profit organization. Tax-exempt: 501(c)(3).
Institution Type/Description: Anthropology Museum.
Collections: prehistoric Southwest Native American artifacts; contemporary basketry; pottery; lithics & ceramics.
Major Exhibits: Santa Fe Indian School, 1/11-6/29/14; The Act of Peterson Yazzie, 7/5-10/17/14; An Essential Relationship: Amateurs & Professionals in Central Arizona Archaeology, 11/14-5/15.
Facilities: 5,000-vol. archaeological & ethnological library on Southwestern Native Americans; 6,000 sq. ft. exhibit space.
Activities: guided tours; lectures; docent program; research facilities. Museum Sponsors: Southwest Indian Arts Festival; Navajo Rug & Indian art auctions.
Publications: quarterly newsletter, "Talking Sun".
Hours & Admission Prices: Mon.-Sat. 10-4, Sun. 1-4. Adults $7, seniors $6, students $5; children 12 & under and Native Americans no charge. Closed Easter; Thanksgiving; Christmas. &
Attendance: 6,800 (accurate)
Membership: Viyal Student $25; Kitheeh Single $30; Hadteh Senior Plus 1 $35; Ahnaalah Family $50; A'Koh Supporting $100; Illuwii Supporter $250; Amuu Supporter $500; Qwaka Supporter $1,000; Ah'sahh Supporter $1,500; Muwadtah Supporter $2,500; Nyimidtah Supporter - Life $5,000.

THE SPOT...A CHILD'S MUSEUM, 3250 Gateway Blvd., Prescott, AZ 86302. Mailing Address: P.O. Box 3938, Prescott, AZ 86302.
Founded: 2008.
Congressional District: 1
Key Personnel: Dir., C.E.O. & Pres. (V), Judy L. Paris.
Personnel Profile: Full-Time Paid 1; Part-Time Volunteers 15.
Volunteer Hours: 7,500
Governing Authority: Parent Institution: Children's Museum Alliance, Inc. Tax-exempt: 501(c)(3).
Institution Type/Description: Children's Museum.
Collections: hands-on exhibitions.
Major Exhibits: Quantum Quizzics(TM), 9/13-9/14.
Facilities: rental facilities; cafe.
Activities: special events; birthday parties; school programs. Museum Sponsors: Science Saturdays.
Hours & Admission Prices: Fri.-Sat. 1-5. Adults $3. &
Attendance: 1,850 (estimated)

Quartzsite

TYSON'S WELL STAGE STATION MUSEUM, 161 W. Main St., Quartzsite, AZ 85346. Mailing Address: Quartzsite Historical Society, P.O. Box 331, Quartzsite, AZ 85346-0331. Tel.: 928-927-5229.
Governing Authority: nonprofit organization. Parent Institution: Quartzsite Historical Society.
Institution Type/Description: History Museum: housed in a restored adobe stage station built in 1866.

Collections: local history & culture; photographs; period mining equipment.
Facilities: Museum-related items for sale.
Activities: special events; musical concerts; fundraising events. Museum Sponsors: Quilt Show.
Publications: Quartzsite Arizona, No Ordinary Place; Quartzsite Pioneer Bill Keiser's Lost Mines and Prospector's Lore.
Hours & Admission Prices: April-Oct. Thurs. 9-12; Nov.-March Wed.-Sun. 10-4; other times by appointment. No charge; donations accepted.

Queen Creek

SAN TAN HISTORICAL SOCIETY MUSEUM, 20425 S. Old Ellsworth Rd., Queen Creek, AZ 85142. Tel.: 480-987-9380.
E-mail: info@santanhistoricalsociety.org
Web Site: www.santanhistoricalsociety.org
Institution Type/Description: Historical Society Museum.
Collections: local history & culture; period furnishings; photographs; personal artifacts.
Hours & Admission Prices: Sat. 9-1. No charge.

Roosevelt

TONTO NATIONAL MONUMENT, Hwy. 188, Roosevelt, AZ 85545. Mailing Address: HC02, Box 4602, Roosevelt, AZ 85545. Tel.: 928-467-2241. Fax: 928-467-2225.
E-mail: TONT_superintendent@nps.gov
Web Site: www.nps.gov/tont
Founded: 1907.
Congressional District: 1
Key Personnel: Supt., Terry Saunders; Park Ranger, Susan Hughes.
Personnel Profile: Full-Time Paid 12; Part-Time Volunteers 4.
Governing Authority: federal. Tax-exempt.
Institution Type/Description: Archaeology Museum.
Collections: clothing, tools, weapons & pottery of the Salado, the prehistoric Indians of the region.
Facilities: picnic area; nature trails.
Activities: guided tours; self-guided trail & tours; audiovisual program.
Publications: Tonto National Monument.
Hours & Admission Prices: Visitor Center & Museum: daily 8-5. Self-Guided Trail: daily 8-4. $3 per person; holders of National Park Pass, Golden Age or Golden Access passes & children under 16 no charge. &
Attendance: 80,021 (accurate)

Safford

DISCOVERY PARK CAMPUS, 1651 W. Discovery Park Blvd., Safford, AZ 85546-3909. Tel.: 928-428-6260. Fax: 928-428-8081.
E-mail: discoverypark@eac.edu
Web Site: www.eac.edu/discoverypark/
Founded: 1995.
Congressional District: 5
Key Personnel: Dir., Paul Anger.
Governing Authority: college. Parent Institution: Eastern Arizona College. Tax-exempt.
Institution Type/Description: Science Center.
Collections: historic, scientific, agricultural & mining exhibits; simulated mine tour; wildlife habitat; research-grade 20" telescope.
Facilities: hiking trails; visitors center; theater; rental facilities.
Activities: guided tours through the Mount Graham International Observatories; formal education programs K-20; hiking trails; films; poetry & prose readings in partnership with Univ. of Arizona Steward Observatory.
Hours & Admission Prices: Mon.-Fri. 8-5, Sat. 4pm-9:30pm. No charge; donations accepted. &
Attendance: 7,000 (accurate)

Sahuarita

ASARCO MINERAL DISCOVERY CENTER, 1421 W. Pima Mine Rd., Ste. A, Sahuarita, AZ 85629-8361. Tel.: 520-625-7513 & 8233. Fax: 520-625-4756.
E-mail: amdcinfo@asarco.com
Founded: 1997.
Institution Type/Description: Mineral Museum.
Collections: mining history, equipment & process; ball mill model; photographs; haul trucks.
Facilities: theater. Museum-related items for sale.
Activities: demonstrations; guided tours; videos.
Hours & Admission Prices: June-Sept. Tues.-Sat. 9-3.

Saint Johns

APACHE COUNTY HISTORICAL SOCIETY MUSEUM, 180
W. Cleveland, Saint Johns, AZ 85936. Mailing Address: P.O. Box
146, Saint Johns, AZ 85936-0146. Tel.: 928-337-4737 & 2000.
E-mail: achs.museum@yahoo.com
Governing Authority: nonprofit organization.
Institution Type/Description: Historical Society Museum.
Collections: local history & culture; period furnishings; personal artifacts;
clothing. Historic Building: 1881 log cabin.
Hours & Admission Prices: Mon.-Fri. 8-4.
Membership: Individual $10; Family $20; Sustaining $50; Patron $100; Donor
$500; Sponsor $1,000; Endowment $2,000.

Saint Michaels

ST. MICHAELS HISTORICAL MUSEUM, 24 Mission Rd., Saint
Michaels, AZ 86511. Mailing Address: P.O. Box 680, Saint
Michaels, AZ 86511-0680. Tel.: 928-871-4171.
Institution Type/Description: History Museum.
Collections: Navajo culture & history; personal artifacts; photographs; period
furnishings.
Activities: educational programs.
Hours & Admission Prices: Memorial Day to Labor Day Mon.-Fri. 9-5.

San Manuel

SAN MANUEL HISTORICAL SOCIETY, 137 8th Ave., San
Manuel, AZ 85631-0742. Mailing Address: P.O. Box 742, San
Manuel, AZ 85631-0742.
Web Site: www.sanmanuelhistoricalsociety.com
Founded: 1990.
Key Personnel: Dir. & Museum Shop Mgr., Janice L. Rapp; Pres. (V), Frameis
Winslow.
Governing Authority: Tax-exempt.
Institution Type/Description: Historical Society Museum.
Collections: local history & culture; photographs; maps; gems; minerals;
mining.
Hours & Admission Prices: Tues. & Fri. 10-2, Sat. 10-1.
Membership: Individual $15.

Scottsdale

AFRICAN AMERICAN MULTICULTURAL MUSEUM, 617 N.
Scottsdale Rd., Ste. A, Scottsdale, AZ 85257-4207. Tel.: 480-314-
4400.
E-mail: museum@prodigy.net
Institution Type/Description: Multicultural Museum.
Collections: art; history; culture.
Activities: lectures.
Hours & Admission Prices: Thurs.-Sat. 1-5; other times by appointment. No
charge.

CELEBRATION OF FINE ART, 7700 N. Hayden Rd., Scottsdale,
AZ 85258. Mailing Address: 7900 E. Greenway Rd., Ste. 101,
Scottsdale, AZ 85260-1714. Tel.: 480-443-7695. Fax: 480-596-
8179.
E-mail: info@celebrateart.com
Web Site: www.celebrateart.com
Institution Type/Description: Art Gallery.
Collections: paintings; sculpture.
Activities: special events. Annual Event: 10 Week Gallery & Juried Show.
Hours & Admission Prices: early Jan. to late March daily 10-6. Adults $10,
seniors & military $8; children under 12 no charge.

**CONGREGATION BETH ISRAEL'S PLOTKIN JUDAICA MU-
SEUM,** 10460 N. 56th St., Scottsdale, AZ 85253-1133. Tel.:
480-951-0323. Fax: 480-951-7150.
E-mail: library@cbiaz.org
Web Site: cbiaz.org
Founded: 1966.
Congressional District: 3
Key Personnel: Chm. (V) & Librarian, Carol Reynolds.
Personnel Profile: Part-Time Volunteers 2.
Governing Authority: religious. Parent Institution: Temple Beth Israel. Tax-
exempt: 501(c)(3).
Institution Type/Description: Religious Antiques Museum: housed in Temple
belonging to oldest Jewish Congregation in the Phoenix area.

Collections: Jewish arts & ceremonials from 1600 to the present; archaeology
of Israel; Israeli philatelic collection; Tunisian period gallery of synagogue,
original artifacts; pioneer Jews of Arizona 1850-1920; Biblical garden.
Research Fields: Judaica.
Facilities: 12,000-vol. library of books on Judaic studies, art, history, housed
in adjoining Temple library; reading room.
Activities: guided tours; lectures; films; gallery talks; concerts; study clubs; TV
programs; formally organized education programs; docent program; inter-
museum loan exhibitions; permanent & traveling exhibitions.
Publications: annual, HA-OR.
Hours & Admission Prices: By appointment. Donation: adults $3.50. Closed
national & Jewish holidays. &
Attendance: 500 (estimated)

DULEY-JONES GALLERY, 7100 E. Main St., Scottsdale, AZ
85251-4343. Mailing Address: 7019 E. Vista Dr., Paradise Valley,
AZ 85253-7058. Tel.: 480-945-8475; 800-229-8475.
E-mail: info@duleyjones.com
Web Site: www.duleyjones.com
Key Personnel: Dir., Kathy Duley; Dir., Randy Jones
Institution Type/Description: Art Gallery.
Collections: works by regional & national artists; paintings; sculpture; ceram-
ics; baskets; woodcarvings; watercolors.
Activities: temporary exhibitions; special events.
Hours & Admission Prices: Mon.-Sat. 10-5.

**ELLIE & MICHAEL ZIEGLER FIESTA BOWL CENTER &
MUSEUM,** 7135 E. Camelback Rd. Ste. 190, Scottsdale, AZ
85251. Tel.: 480-350-0900.
Web Site: www.fiestabowl.org
Key Personnel: Exec. Dir., Robert Shelton
Institution Type/Description: Sports Museum.
Collections: Fiesta Bowl history & memorabilia; Football Bowl Subdivision
team helmets; college football trophies; photographs.
Hours & Admission Prices: Mon.-Fri. 8:30-5. No charge.

FRANK LLOYD WRIGHT'S TALIESIN WEST, 12621 N. Frank
Lloyd Wright Blvd., Scottsdale, AZ 85259. Tel.: 480-860-2700 &
627-5340.
Institution Type/Description: Historic House Museum: housed in the former
winter home of Frank Lloyd Wright; built in 1937. A National Historic
Landmark.
Collections: period furnishings; personal artifacts; photographs.
Activities: guided tours.
Hours & Admission Prices: Daily by appointment. Closed Easter; Thanksgiv-
ing; Christmas.

GEBERT CONTEMPORARY ART GALLERY, 7160 Main St.,
Scottsdale, AZ 85251-4316. Tel.: 480-429-0711. Fax: 480-429-
0713.
E-mail: gallery@gebertartaz.com
Web Site: gebertartaz.com
Institution Type/Description: Art Gallery.
Collections: works by contemporary artists; paintings; sculpture.
Hours & Admission Prices: Mon.-Wed. & Fri.-Sat. 10-5, Thurs. 10-9, Sun.
12-4.

HEARD MUSEUM NORTH SCOTTSDALE, 32633 N. Scottsdale
Rd., Scottsdale, AZ 85262. Tel.: 480-488-9817.
Institution Type/Description: History Museum.
Collections: local history & culture; paintings; period furnishings; Native
American artifacts; photographs; personal artifacts.
Activities: hands-on exhibitions.
Hours & Admission Prices: Mon.-Sat. 10-5, Sun. 11-5. Adults $5, seniors 65
& over $4, students & children 6-12 $2; members, Native Americans &
children under 6 no charge.

HOO-HOOGAM KI MUSEUM, 10005 E. Osborn Rd., Scottsdale,
AZ 85256-4019. Tel.: 480-850-8190.
E-mail: huhugamki.museaum@srpmic-nsn.gov
Web Site: www.srpmic-nsn.gov/history-culture/museum.asp
Institution Type/Description: Cultural Heritage Museum.
Collections: Salt River Pima-Maricopa cultural heritage & history; Pima
basketry; Maricopa pottery; native dress; period furnishings; personal
artifacts; sculpture; artwork.
Facilities: Museum-related items for sale.

Hours & Admission Prices: Mon.-Fri. 9:30-4:30. No charge. Closed federal holidays.

HOUSE OF BROADCASTING RADIO & TELEVISION MUSEUM, 7150 E. 5th Ave., Scottsdale, AZ 85251-3238. Mailing Address: 7534 N. 7th St., Phoenix, AZ 85020-4129. Tel.: 602-944-1997. Fax: 602-997-8707. Facebook: House of Broadcasting.
E-mail: bmaack@cox.net
Web Site: www.houseofbroadcasting.com
Founded: 1996.
Key Personnel: C.E.O. & Pres. (V), Mary Morrison; Chm. (V), Mike Shaldjian.
Personnel Profile: Part-Time Volunteers 2.
Governing Authority: nonprofit. Tax-exempt: 501(c)(3).
Institution Type/Description: Radio & Television History Museum.
Collections: Arizona's radio & television broadcasting history & personalities; broadcasting artifacts; personal artifacts; photographs; costumes; autographed books; equipment.
Activities: book signings by media authors. Annual Events: Golf Tournaments; Celebrity Toast.
Publications: book, HOBI Star-Studded Celebrity Media Cookbook.
Hours & Admission Prices: Mon.-Sat. 10-6. No charge; donations accepted. Closed Thanksgiving; Christmas.
Attendance: 1,500 (estimated)

LISA SETTE GALLERY, 4142 N. Marshall Way, Scottsdale, AZ 85251-3858. Tel.: 480-990-7342. Fax: 480-970-0825.
E-mail: sette@lisasettegallery.com
Web Site: www.lisasettegallery.com
Institution Type/Description: Art Gallery.
Collections: works by contemporary artists; paintings; sculpture.
Activities: temporary exhibitions; special events.
Hours & Admission Prices: Tues.-Wed. & Fri. 10-5, Thurs. 10-5 & 7-9, Sat. 12-5.

THE MARSHALL/LEKAE GALLERY, 7106 E. Main St., Scottsdale, AZ 85251-4316. Tel.: 480-970-3111. Fax: 480-970-0092.
E-mail: email@marshall-lekaegallery.com
Web Site: www.marshall-lekaegallery.com
Institution Type/Description: Art Gallery.
Collections: works by contemporary artists; paintings; sculpture.
Activities: special events. Museum Sponsors: Annual Drawing Event in October.
Hours & Admission Prices: Mon.-Wed. & Fri.-Sat. 10-5:30, Thurs. 10-9.

RIVA YARES GALLERY, 3625 N. Bishop Ln., Scottsdale, AZ 85251-5511. Tel.: 480-947-3251. Fax: 480-947-4251.
E-mail: art@rivayaresgallery.com
Web Site: www.rivayaresgallery.com
Institution Type/Description: Art Gallery.
Collections: works by modern & contemporary artists; paintings; sculpture; drawings.
Activities: special events; temporary exhibitions.
Hours & Admission Prices: Mon.-Sat. 10-5.

SCOTTSDALE HISTORICAL MUSEUM, 7333 Scottsdale Mall, Scottsdale, AZ 85251-4414. Mailing Address: P.O. Box 143, Scottsdale, AZ 85252-0143. Tel.: 480-945-4499.
E-mail: info@scottsdalemuseum.com
Web Site: www.scottsdalemuseum.com
Founded: 1991.
Personnel Profile: Part-Time Volunteers 75.
Governing Authority: Parent Institution: Scottsdale Historical Society. Tax-exempt.
Institution Type/Description: History Museum.
Collections: local history; personal artifacts; photographs; newspapers.
Hours & Admission Prices: June & Sept. Wed.-Sun. 10-2; Oct.-May Wed.-Sun. 10-5. No charge; donations accepted. Closed holidays.
Attendance: 25,000 (estimated)
Membership: Senior $10; Individual $25; Family $40; Patron $100; Benefactor $200. Business: Bronze $100-$249; Silver $250-$499; Gold $500 & up.

＊ **SCOTTSDALE MUSEUM OF CONTEMPORARY ART, (M),** 7374 E. Second St., Scottsdale, AZ 85251-5604. Mailing Address: 7380 E. Second St., Scottsdale, AZ 85251-5604. Tel.: 480-874-4666. Fax: 480-874-4655. TDD: 480-994-2787.
E-mail: smoca@sccarts.org
Web Site: www.smoca.org
Founded: 1999.
Congressional District: 1
Key Personnel: Bd. Chm., Mike Medici; C.E.O., Bill Banchs; Museum Dir., Timothy Rodgers; Asst. Cur., Claire C. Carter; Mktg. & Public Rels. Mgr., Lesley Oliver; Cur. Education, Carolyn Robbins; Assoc. Cur. Education, Laura Hales; Exhibit Designer, Laura Best; Registrar, Pat Evans; Curatorial Coord., Dana Buhl; Museum Shop Mgr., Janice Bartczak.
Personnel Profile: Full-Time Paid 17; Part-Time Paid 3; Part-Time Volunteers 180; Interns 5.
Governing Authority: nonprofit organization. Parent Institution: Scottsdale Cultural Council. Tax-exempt: 501(c)(3).
Institution Type/Description: Modern & contemporary art, architecture & design.
Collections: modern & contemporary paintings; sculpture; installations; prints; drawings; photographs; architecture and design; James Turrell Skyspace.
Research Fields: modern & contemporary art; architecture; design.
Facilities: outdoor sculpture garden. Museum-related items for sale.
Activities: guided tours; lectures; arts festivals; organized education programs; docent program; gallery talks; workshops; artists' residencies
Publications: calendar; education program brochures; exhibition catalogues.
Hours & Admission Prices: Tues.-Wed. & Sun. 12-5, Thurs.-Sat. 12-9. Adults $7, students $5; discounts for AAM & ICOM members; Thurs. 12-9, Fri.-Sat. 5-9, members, children under 15 no charge. Reciprocal membership benefits available. Closed national holidays.
Attendance: 45,030 (accurate)
Membership: Solo $50; Duet $90; Contributing $150; Sustaining $250; Supporting $500; President's Club $1,250; Circles of Giving $2,500 & up.

Second Mesa

HOPI CULTURAL CENTER, Rte. 264, Second Mesa, AZ 86043. Mailing Address: P.O. Box 67, Second Mesa, AZ 86043-0067. Tel.: 928-734-2401.
Institution Type/Description: History Museum.
Collections: Hope history & culture; Hopi arts & crafts, pottery, weavings, woodcarvings, & silver.
Facilities: restaurant.
Hours & Admission Prices: Call for hours.

Sedona

SEDONA ARTS CENTER, INC., 15 Art Barn Rd., Sedona, AZ 86336-4249. Mailing Address: P.O. Box 569, Sedona, AZ 86339-0569. Tel.: 928-282-3809; 888-954-4442.
E-mail: sac@sedonaartscenter.com
Web Site: www.sedonaartscenter.com
Founded: 1961.
Congressional District: 3
Key Personnel: Exec. Dir., Pam Frazier; Gallery Dir., Shirley Eichten Albrecht
Governing Authority: nonprofit organization. Tax-exempt: 501(c)(3).
Institution Type/Description: Art Center.
Collections: various forms of art media.
Facilities: 150-vol. library of general art books & magazines; 2,000 sq. ft. exhibit space; 160-seat theater; drawing & painting, glass & jewelry, ceramics & sculpture classroom studios. Works by local & member artists in crafts & fine art for sale.
Activities: guided tours; lectures; films; arts festivals; organized educational programs; docent program.
Publications: monthly newsletter, Previews.
Hours & Admission Prices: Daily 10-5. No charge; donations accepted. Closed major holidays.
Attendance: 62,000 (accurate)
Membership: Individual $50; Family $75; Business $150; Silver $250; Gold $500; Other $1,000 & up.

SEDONA HERITAGE MUSEUM, (M), 735 Jordan Rd., Jordan Historical Park, Sedona, AZ 86336. Mailing Address: Sedona Historical Society, P.O. Box 10216, Sedona, AZ 86339-8216. Tel.: 928-282-7038. Fax: 928-282-7038.
E-mail: sedonamuseum@esedona.net
Founded: 1998.
Congressional District: 1
Personnel Profile: Part-Time Paid 3; Part-Time Volunteers 60.
Governing Authority: Parent Institution: Sedona Historical Society. Tax-exempt.
Institution Type/Description: History Museum: housed in the former farm home of Walter & Ruth Jordan, built in 1930. Listed on the National Register of Historic Places.
Collections: local history & culture; period furnishings; personal artifacts.
Research Fields: local history.

Facilities: Museum-related items for sale.
Activities: special events.
Publications: newsletter; book, Those Early Days; Traveling By Tin Lizzie; Filmography of Sedona.
Hours & Admission Prices: Daily 11-3; tours by appointment. Adults $5; members & children under 12 no charge. &
Attendance: 12,000 (accurate)
Membership: Basic $40; Supporting & Business Associate $100; Business Colleague & Sustaining $200; Business Partner $500.

Shonto

NAVAJO NATIONAL MONUMENT, End of U.S. Highway 564, Shonto, AZ 86054. Mailing Address: P.O. Box 7717, Shonto, AZ 86054-7717. Tel.: 928-672-2700. Fax: 928-672-2703.
Web Site: www.nps.gov/nava
Founded: 1909.
Congressional District: 1
Key Personnel: Park Archaeologist, Lloyd Masayumptewa; Admin., Matthew Klozik; WNPA Mgr., Althea James; Supt., Alden Miller.
Personnel Profile: Full-Time Paid 9; Part-Time Volunteers 6.
Governing Authority: federal. Parent Institution: U.S. Dept. of Interior. National Park Service, Washington, DC. Tax-exempt.
Institution Type/Description: Archaeological Museum.
Collections: archeological materials of the Kayenta Anasazi & Navajo cultures. Archaeological Sites: 3 prehistoric cliff villages.
Facilities: 2,000-vol. library of history & archaeological books available on premises. Navajo rugs, silver work, paintings, pottery & crafts for sale.
Activities: guided tours; lectures; back country hiking permits to Keet Seel ruin; Navajo artists demonstrations including pottery, rug weaving & basket making.
Publications: orientation brochure, Voices in the Canyon.
Hours & Admission Prices: Visitor Center: May 24-Sept. 13 daily 8-6; Sept. 14-May 22 daily 9-5. No charge. Closed New Year's Day; Christmas. &
Attendance: 89,000 (estimated)

Show Low

SHOW LOW HISTORICAL SOCIETY MUSEUM, 561 E. Deuce of Clubs, Show Low, AZ 85901-4826. Mailing Address: P.O. Box 3468, Show Low, AZ 85902-3468. Tel.: 928-532-7115. Fax: 928-532-7115. Facebook: Show Low Museum.
E-mail: showlowmuseum@cableone.net
Web Site: showlowmuseum.com
Founded: 1995.
Congressional District: 5
Key Personnel: Dir., Clair Thomas; Asst. Dir., Carol Grossheim; Museum Shop Mgr., Betty Wood.
Personnel Profile: Full-Time Paid 2.
Governing Authority: Parent Institution: Show Low Historical Society. Tax-exempt.
Institution Type/Description: Historical Society Museum.
Collections: local history & culture; photographs; period furnishings; personal artifacts.
Major Exhibits: Show Low 1912-2012, 2/12-1/14.
Activities: demonstrations; educational programs; summer youth programs. Museum Sponsors: Post Office Celebration Station in February.
Hours & Admission Prices: Call for hours. No charge; donations accepted. &
Attendance: 3,000 (estimated)
Membership: Individual $20; Family $25; Patron $100.

Sierra Vista

HENRY F. HAUSER MUSEUM, (M), Ethel Berger Center, 2950 E. Tacoma St., Sierra Vista, AZ 85635-1352. Mailing Address: 1011 N. Coronado, Sierra Vista, AZ 85635. Tel.: 520-417-6980, ext. 560.
E-mail: nancy.krieski@sierravistaaz.gov
Web Site: www.sierravistaaz.gov
Founded: 2000.
Congressional District: 8
Key Personnel: Cur. & Mgr., Nancy M. Krieski; Museum Shop Mgr., Donna Hallsten.
Personnel Profile: Full-Time Paid 1; Part-Time Volunteers 30.
Governing Authority: Parent Institution: City of Sierra Vista. Tax-exempt.
Institution Type/Description: History Museum.
Collections: local history & culture; photographs; books; memorabilia; oral histories; artifacts.
Facilities: theater. Gift items for sale.
Activities: permanent & temporary exhibitions.
Hours & Admission Prices: May-Oct. Mon.-Fri. 10-4; Nov.-April Mon.-Fri. 10-4, Sat. 10-1. No charge; donations accepted. &

Attendance: 3,000 (estimated)

Snowflake

STINSON PIONEER MUSEUM, 102 S. 1st St., Snowflake, AZ 85937. Mailing Address: 113 N. Main St., Ste. A, Snowflake, AZ 85937. Tel.: 928-536-4331 & 4881.
Institution Type/Description: History Museum.
Collections: local history & culture; photographs; personal artifacts; period furnishings; historic buildings.
Hours & Admission Prices: Tues.-Sat. 10-4.

Somerton

COCOPAH MUSEUM, County 15 and Avenue G, Somerton, AZ 85350. Tel.: 928-627-1992.
Institution Type/Description: Native American History Museum.
Collections: Cocopah Indian artifacts, culture & history; clothing; musical instruments; personal artifacts.
Facilities: Museum-related items for sale.
Hours & Admission Prices: Mon.-Fri. 9-4. No charge; donations accepted.

Springerville

CASA MALPAIS VISITOR CENTER AND MUSEUM, 418 E. Main St., Springerville, AZ 85938-5220. Tel.: 928-333-5375.
E-mail: casa@springervilleaz.gov
Web Site: www.casamalpais.org
Founded: 1991.
Institution Type/Description: History Museum.
Collections: local history & culture; 13th century Mogollon ruins; photographs; period artifacts.
Facilities: visitor center.
Activities: hiking; educational programs.
Hours & Admission Prices: Tues.-Sat. 8-4. Guided Tours: 9am, 11am, & 2pm.

LITTLE HOUSE MUSEUM, S. Fork Rd., Springerville, AZ 85938. Mailing Address: P.O. Box 791, Springerville, AZ 85938-0791. Tel.: 928-333-2286.
Institution Type/Description: History Museum.
Collections: ranching; pioneer life; local history; photographs. Historic Buildings: restored cabins.
Hours & Admission Prices: By appointment. &

Sun City

DEL WEBB SUN CITIES MUSEUM, 10801 Oakmont Dr., Sun City, AZ 85351. Tel.: 623-974-2568.
E-mail: staff@delwebbsuncitiesmuseum.org
Web Site: www.delwebbsuncitiesmuseum.org
Key Personnel: Pres., Bill Pearson
Institution Type/Description: History Museum.
Collections: local history & culture; personal artifacts & histories; photographs; print advertising; memorabilia.
Hours & Admission Prices: Sept.-May Mon., Wed. & Fri. 1-4; June-Aug. by appointment only & tours of 6 or more. No charge.

Superior

* **BOYCE THOMPSON ARBORETUM, (M),** 37615 E. US Hwy. 60, Superior, AZ 85173-5100. Tel.: 520-689-2723. Fax: 520-689-5858.
E-mail: msiegwar@cals.arizona.edu
Web Site: arboretum.ag.arizona.edu
Formerly: Boyce Thompson Southwestern Arboretum
Founded: 1924.
Congressional District: 4
Key Personnel: Dir., Mark Siegwarth; Chm. (V) & Pres. (V), Ian Thompson; Museum Shop Mgr., Lynnea Spencer.
Personnel Profile: Full-Time Paid 23; Part-Time Paid 13; Part-Time Volunteers 120.
Governing Authority: cooperative, nonprofit organization. Parent Institution: Boyce Thompson Southwestern Arboretum Inc. Subsidiary Institution: University of Arizona and Arizona State Parks. Tax-exempt: 501(c)(3).
Institution Type/Description: Arboretum.
Collections: xerophytes; botany; geology; zoology; manuscripts. Historic House: c.1915 Clevenger Homestead; c.1928 Picket Post House.
Research Fields: botany; zoology; floristics; taxonomy; systematics; horticulture.

Facilities: library of biological material; botanical garden; field research station. Color slides, books, mineral specimens & cactus plants for sale.
Activities: guided tours; lectures; permanent & temporary exhibitions; workshops; monthly special events.
Publications: semi-annual journal, Desert Plants.
Hours & Admission Prices: Daily 8-5. Adults $9, children 5-12 $4.50; discounts to AABGA members; members no charge. Closed Christmas. &
Attendance: 85,000 (estimated)
Membership: Dual $45; Family $60; Patron $100; Boojum $250; Picketpost Society $500; Director's Circle $1,000.

SUPERIOR HISTORICAL SOCIETY - THE BOB JONES MUSEUM, (M), 300 Main St., Superior, AZ 85173. Mailing Address: P.O. Box 613, Superior, AZ 85173.
Institution Type/Description: History Museum.
Collections: local history; photographs; mining equipment; geology.
Publications: monthly newsletter.
Hours & Admission Prices: Fri. 1-4, Sat.-Sun. 10-4. No charge.
Membership: Individual $8; Family $15; Business $25.

WORLD'S SMALLEST MUSEUM, 1111 W. U.S. Hwy. 60, Superior, AZ 85173-3429. Tel.: 520-689-5857 & 208-0634.
E-mail: sales@smallestmuseum.com
Web Site: www.worldssmallestmuseum.com
Institution Type/Description: General Museum.
Collections: Indian pottery; 1800s wood stove; 1850s frying pan; 1930s 16mm projector; newspaper clippings; 8mm movie camera; photographs; iron kettle.
Facilities: 134 sq. ft. exhibit space. Museum-related items for sale.
Hours & Admission Prices: Wed.-Sun. 8-1:30. No charge; donations accepted. Closed major holidays.

Taylor

TAYLOR/SHUMWAY HERITAGE FOUNDATION, 2 N. Main St., Taylor, AZ 85939. Mailing Address: P.O. Box 566, Taylor, AZ 85939. Tel.: 928-536-6649.
E-mail: taylormuseum@frontiernet.net
Formerly: Taylor Pioneer Museum
Founded: 1997.
Congressional District: 1
Key Personnel: Chm. (V), Carmen Shumway; Pres. (V), Lynn Hancock; Museum Shop Mgr., Kathryn Udall.
Personnel Profile: Part-Time Volunteers 20.
Governing Authority: Subsidiary Institution: Pioneer Museum, 14 S. 4th E., Taylor, AZ 85939; Standifird Home, 304 S. Main St., Taylor AZ 85939; Margaret Hancock McCleve Log Cabin, 7 E. Willow, Taylor, AZ 85937; Shumway School House, Old Mill Rd., Taylor, AZ 85939. Tax-exempt.
Institution Type/Description: History Museum.
Collections: local history & culture. Historic Buildings: Palmer/Hatch Bros. store includes photographs; personal artifacts; a Conestoga wagon; early tools & equipment. 1930 pioneer house includes pioneer memorabilia & artifacts. 1884 Margaret McCleve Hancock log cabin includes a spinning wheel; period furnishings. 1900 Shumway schoolhouse includes period desks & furnishings. 1890 Standifird home includes period furnishings.
Activities: outreach program for elementary school students; hayrides; video histories.
Hours & Admission Prices: Call for hours. No charge; donations accepted. &
Attendance: 1,984 (accurate)
Membership: Individual $25.

Tempe

AMERICAN HEART ASSOCIATION - HALLE HEART CHILDREN'S MUSEUM, 2929 S. 48th St., Tempe, AZ 85282-3145. Tel.: 602-414-2800. Fax: 602-414-5355.
E-mail: hhcm@heart.org
Web Site: www.hhcm.org
Formerly: Halle Heart Center
Founded: 1996.
Key Personnel: Dir. Programs & Operations, Claudine M. Wessel; Chm. (V), Dr. Moschonas.
Personnel Profile: Full-Time Paid 2; Part-Time Paid 16; Part-Time Volunteers 25.
Governing Authority: private; nonprofit organization. Tax-exempt: 501(c)(3).
Institution Type/Description: Science & Health Museum.
Collections: hands-on exhibits related to heart health.
Facilities: library; 16,000 sq. ft. exhibit space; education facilities; 50-seat large screen & interactive stage; TV theater; activity center; APS electric industrial demonstration kitchen.

Activities: guided & self-guided tours; docent program; lectures; films; formal education for adults, college students & children.
Hours & Admission Prices: Sept.-July Mon.-Fri. 1-5 by appointment. Donations suggested. Closed national holidays. &
Attendance: 28,000 (estimated)
Membership: Tree of Life $50-$100; Bronze Hearts $1,000-$4,999; Silver Hearts $5,000-$9,999; Gold Hearts $10,000-$24,999; Presidential Patrons $25,000-$49,000; Presidential Partners $50,000-$99,999; Presidential Pace-Setters $100,000-$249,999; Benefactors $250,000-$350,000.

ARIZONA HISTORICAL SOCIETY MUSEUM AT PAPAGO PARK, (M), 1300 N. College Ave., Tempe, AZ 85281-1211. Mailing Address: 949 E. 2nd St., Tucson, AZ 85719-4898. Tel.: 480-929-0292. Fax: 480-967-5450.
E-mail: ahstempe@azhs.gov
Web Site: www.arizonahistoricalsociety.org
Founded: 1864.
Congressional District: 9
Key Personnel: Exec. Dir., Anne I. Woosley, Ph.D.; Central Arizona Division Dir., John Langellier; Library & Archives Division Dir., Linda Whitaker.
Personnel Profile: Full-Time Paid 10; Part-Time Paid 3; Part-Time Volunteers 25; Interns 6.
Governing Authority: bd. of directors. Parent Institution: Arizona Historical Society, 949 E. 2nd St., Tucson, AZ 85719. Tel.: 520-628-5774. Tax-exempt: 501(c)(3).
Institution Type/Description: History Museum.
Collections: 20th & 21st century artifacts; news film & visual media; decorative arts; popular culture; textiles; clothing; tools; photographs; books; documents; archives; maps; works by Arizona artists; china; silver; coins; tokens; sports; politics; food preparation; advertising; ephemera; toys; oral histories; architectural drawings.
Research Fields: people & events shaping Arizona since 1900; history of Phoenix & Central Arizona.
Facilities: 2,500-vol. library pertaining to 20th-century Arizona & Western history available for research on premises; 23,000 sq. ft. exhibition space; 272-seat auditorium; 50-seat theater; catering kitchen; reading room; nature trails.
Activities: lectures; films; social studies, history, geology & geography education programs; loan & traveling exhibitions; school loan service; special events; tours; facility rental.
Publications: quarterly, Journal of Arizona History; books; monographs; annual report.
Hours & Admission Prices: Tues.-Sat. 10-4, Sun. 12-4. Library Archives: Mon.-Thurs. 10-6, Fri. 9-4. Adults $5, seniors 60 & over and youth 12-18 $4; children 11 & under and members no charge. Closed New Year's Day; Martin Luther King Jr. Day; Presidents' Day; Memorial Day; Independence Day; Labor Day; Columbus Day; Veterans Day; Thanksgiving; Christmas. &
Attendance: 22,500 (estimated)
Membership: Student $25; Individual $50; Household $65; Sustaining $100; Patron $250; Sponsor $500; Director's Circle $1,000.

ARIZONA STATE UNIVERSITY ART MUSEUM, (M), 51 E. 10th St., Tempe, AZ 85287. Mailing Address: Box 872911, Tempe, AZ 85287-2911. Tel.: 480-965-2787. Fax: 480-965-5254.
E-mail: asuartmuseum@asu.edu
Web Site: asuartmuseum.asu.edu
Founded: 1950.
Congressional District: 1
Key Personnel: Dir., Gordon Knox; Sr. Cur., Heather Lineberry; Sr. Cur. Social Impact, Julio Cesar Morales; Sr. Cur. Ceramics, Peter Held; Registrar, Anne Sullivan; Print Collection Mgr., Jean Makin; Business Mgr., Vanessa Cornwall; Desert Initiative Dir., Greg Esser; Sr. Pub. Rels. Mgr., Deborah Susser; Museum Shop Mgr., Natalie Carroll.
Personnel Profile: Full-Time Paid 18; Full-Time Volunteers 1; Part-Time Paid 3; Part-Time Volunteers 30; Interns 4.
Governing Authority: university. Parent Institution: Arizona State University; Subsidiary Institution: Herberger Institute. Tax-exempt: 501(c)(3).
Institution Type/Description: Art Museum & Gallery.
Collections: contemporary art, including new media; Oliver B. James collection of American Art 18th to 20th century; American & European print collection 15th-century to present; crafts; Joseph & Astrid Thomas collection of 19th-century American crockery; contemporary ceramics; Latin American art; social practice.
Research Fields: American paintings; American & European prints; American ceramics.
Facilities: 500-vol. library of assorted research materials to American collection available on premises; reading room; seminar rooms; print study room. Museum-related items for sale.
Activities: guided tours; lectures; videos; gallery talks; concerts; arts festivals;

curriculum-related education programs; school outreach programs; workshops; demonstrations; docent program; inter-museum loan, temporary & traveling exhibitions; school loan service; student docent program.
Publications: catalogs of selected exhibitions.
Hours & Admission Prices: Fall: Tues. 11-8; Summer: Tues.-Sat. 11-5, Sun. 1-5. No charge; donations accepted. Closed federal & state holidays. &
Attendance: 51,000 (accurate)
Membership: Active $50; Supporting $100; Sustaining $500; Patron $1,000.

ARIZONA STATE UNIVERSITY LIBRARIES' ARCHIVES AND SPECIAL COLLECTIONS - LUHRS GALLERY, Hayden Library, ASU Main Campus, Tempe, AZ 85287-1006. Mailing Address: P.O. Box 871006, Tempe, AZ 85287-1006. Tel.: 480-954-4925.
E-mail: karrie.porterbrace@asu.edu
Founded: 1885.
Congressional District: 5
Key Personnel: Dir., Karrie Porter Brace.
Personnel Profile: Full-Time Paid 1.
Governing Authority: Parent Institution: Arizona State University; Subsidiary institution: ASU Libraries. Tax-exempt.
Institution Type/Description: History Museum.
Collections: local history & culture; personal artifacts; Native American artifacts; photographs; documents.
Hours & Admission Prices: Call for hours. No charge.
Attendance: 75,000 (estimated)

ARIZONA STATE UNIVERSITY MUSEUM OF ANTHRO-POLOGY, (M), Arizona State University, SHESC Bldg., Tempe, AZ 85287-2402. Mailing Address: P.O. Box 872402, Tempe, AZ 85287-2402. Tel.: 480-965-6224. Fax: 480-965-7671.
E-mail: anthro.museum@asu.edu
Web Site: asuma.asu.edu
Founded: 1961.
Key Personnel: Dir. & Interim Dir. Museum Studies, Gwyneira Isaac; Asst. Cur., Catherine Nichols; Archaeology & Ethnography Collections, Arleyn Simon; Cur. Physical Anthropology, Diane Hawkey; Exhibit Developer, Judy Newland.
Personnel Profile: Full-Time Paid 2; Part-Time Paid 3; Part-Time Volunteers 20; Interns 1.
Governing Authority: public university. Parent Institution: Arizona State University. Tax-exempt.
Institution Type/Description: Anthropology Museum.
Collections: archaeology; ethnology; physical anthropology.
Research Fields: archaeology; human origins; culture & society.
Facilities: classrooms; laboratories; 2,000 sq. ft. exhibit space.
Activities: changing exhibitions; museum anthropology graduate training program.
Publications: annual, Museum Anthropology Newsletter.
Hours & Admission Prices: Fall & Spring Semesters: Mon.-Fri. 11-3; Winter & Summer: by appointment. No charge. Closed New Year's Eve & Day; Martin Luther King Jr. Day; Memorial Day; Independence Day; Labor Day; Veterans Day; Thanksgiving; Christmas Day & week. &
Attendance: 3,500 (accurate)

CENTER FOR METEORITE STUDIES - ARIZONA STATE UNIVERSITY, Interdisciplinary Science & Tech Bldg. 84, 2nd Fl., 781 E. Terrace Rd., Tempe, AZ 85287. Mailing Address: P.O. Box 872504, Tempe, AZ 85287-2504. Tel.: 480-965-6511. Fax: 480-965-4907.
E-mail: meteorites@asu.edu
Web Site: meteorites.asu.edu
Founded: 1961.
Congressional District: 1
Key Personnel: Dir., Meenakshi Wadhwa.
Personnel Profile: Part-Time Paid 1.
Governing Authority: university. Parent Institution: Arizona State University. Tax-exempt.
Institution Type/Description: Meteorite Museum.
Collections: representatives of individual meteorite falls & finds.
Research Fields: meteorite chemistry, identification, history; mineralogy.
Facilities: 14,000-vol. library of books; reprints; microfilmed references available for inter-library loan & upon request; laboratory.
Activities: guided tours; lectures; formally organized education programs for undergraduate & graduate college students; inter-museum loan & permanent exhibitions; scientific & scholarly research.
Hours & Admission Prices: Mon.-Fri. 9-5. No charge; donations accepted. Closed holidays. &
Attendance: 3,600 (estimated)

SALT RIVER PROJECT HISTORY MUSEUM, 1521 N. Project Dr., Tempe, AZ 85281-1206. Tel.: 602-236-5900.
Institution Type/Description: History Museum.
Collections: state cultural heritage; photographs; documents; ceramics; stone tools; shell jewelry; grinding wheel; period equipment; films.
Activities: research.
Hours & Admission Prices: Mon.-Fri. 9-4.

SEA LIFE ARIZONA AQUARIUM, 5000 S. Arizona Mills Circle, Ste. 145, Tempe, AZ 85282. Tel.: 877-526-3960; 480-478-7598. Fax: 480-478-7609.
E-mail: arizonasealife@sealifeus.com
Web Site: www.visitsealife.com/arizona
Institution Type/Description: Aquarium.
Collections: over 5,000 sea creatures including sharks, rays, tropical fish & turtles.
Facilities: Gift items for sale.
Activities: touch pools; interactive exhibitions.
Hours & Admission Prices: Mon.-Sat. 10-7:30, Sun. 10-6. Adults 13 & over $18, children 3-12 $13.

TEMPE HISTORY MUSEUM, (M), 809 E. Southern Ave., Tempe, AZ 85282-5205. Tel.: 480-350-5100. Fax: 480-350-5150. TDD: 480-350-5050.
E-mail: museum@tempe.gov
Web Site: www.tempe.gov/museum
Formerly: Tempe Historical Museum
Founded: 1972.
Congressional District: 1
Key Personnel: Museum Admin., Amy A. Douglass; Cur. Collections, Josh Roffler; Exhibits Coord., Dan Miller; Cur. History, Jared Smith.
Personnel Profile: Full-Time Paid 4; Part-Time Paid 5; Part-Time Volunteers 50; Interns 2.
Volunteer Hours: 3,700
Operating Expenses: 552,620
Operating Income: 552,620
Governing Authority: municipal. Parent Institution: City of Tempe, AZ.
Institution Type/Description: History Museum.
Collections: late 19th-20th century domestic artifacts; farm ranch equipment; municipal archives; 20th-century business equipment; historic photographs & archives. Historic House: 1892 Niels Petersen house.
Major Exhibits: Made in Tempe, 11/8/13-9/28/14.
Research Fields: community history; historic preservation; museum studies.
Facilities: research library; photo reproduction & other archives. Gift items for sale.
Activities: lecture series; fund-raising events; docent program; school programs; adult classes; internships for university students; multicultural events; family programs; musical performances.
Publications: quarterly newsletter, Time Lines.
Hours & Admission Prices: Museum: Tues.-Sat. 10-5, Sun. 1-5. Petersen House: open for special events only. No charge; donations accepted. Closed major holidays. &
Attendance: 25,316 (accurate)

Thatcher

GRAHAM COUNTY HISTORICAL SOCIETY, 3430 W. Hwy. 70, Thatcher, AZ 85552. Mailing Address: P.O. Box 290, Thatcher, AZ 85552-0290. Tel.: 928-348-0470.
E-mail: staff@grahammuseum.org
Web Site: www.grahammuseum.org
Founded: 1962.
Congressional District: 4
Key Personnel: Pres., Ralph Smith; Museum Dir., Mel Jones.
Personnel Profile: Part-Time Volunteers 14.
Governing Authority: society. Parent Institution: Arizona Historical Society, 949 E. 2nd St., Tucson, AZ 85719. Tax-exempt: 170(b)(1)(A).
Institution Type/Description: Historical Society Museum.
Collections: Western memorabilia; pioneer & Indian artifacts; archaeology; manuscript collections.
Research Fields: archaeology; historic sites.
Facilities: 500-vol. library of books on Western history available for research on premises & for loan to members.
Activities: guided tours; lectures.
Publications: book, Mount Graham Profiles; annual symposium papers.
Hours & Admission Prices: Jan.-July & mid-Aug. to mid-Dec. Mon.-Tues. & Sat. 10-4; other times by appointment. No charge; donations accepted. Closed holidays. &
Attendance: 3,500 (accurate)

Membership: General $15.

Tombstone

BIRD CAGE THEATRE MUSEUM, 535 E. Allen St., Tombstone, AZ 85638. Mailing Address: P.O. Box 248, Tombstone, AZ 85638-0248. Tel.: 520-457-3421; 800-457-3423 (Toll Free).
E-mail: tombstonebirdcage@gmail.com
Web Site: www.tombstonebirdcage.com
Institution Type/Description: History Museum: built in the 1880s.
Collections: area history; photographs; period furnishings & clothing; theater.
Hours & Admission Prices: Call for hours.

O.K. CORRAL, 326 E. Allen St., Tombstone, AZ 85638. Mailing Address: P.O. Box 367, Tombstone, AZ 85638-0367. Tel.: 520-457-3456. Fax: 520-457-3456.
E-mail: info@ok-corral.com
Web Site: www.okcorral.com
Formerly: Historama
Founded: 1879.
Key Personnel: Dir., Robert Love
Institution Type/Description: History Museum.
Collections: Tombstone history; early Western life & culture; films.
Facilities: theater.
Activities: films; live gunfight reenactments.
Publications: newspaper, Tombstone Epitaph.
Hours & Admission Prices: Daily 9-5. Adults $10. Closed Thanksgiving; Christmas. &
Attendance: 100,000 (accurate)

ROSE TREE MUSEUM, 118 4th St., Tombstone, AZ 85635. Mailing Address: P.O. Box 808, Tombstone, AZ 85638. Tel.: 520-457-3326.
Founded: 1964.
Personnel Profile: Part-Time Paid 3.
Institution Type/Description: History Museum: housed in a former hotel; built in 1885.
Collections: local history & culture; period furnishings; personal artifacts; photographs; a rose bush planted in 1885 covering over 8,000 sq. ft.
Hours & Admission Prices: Call for hours. Adults $5; children under 4 no charge.

TOMBSTONE COURTHOUSE STATE HISTORIC PARK, (M), 223 E. Toughnut St., Tombstone, AZ 85638. Mailing Address: P.O. Box 216, Tombstone, AZ 85638-0216. Tel.: 520-457-3311. Fax: 520-457-2565.
E-mail: tombstonecourthouse@tombstonechamber.com
Web Site: www.azstateparks.com
Founded: 1959.
Congressional District: 8
Key Personnel: Dir. & Museum Shop Mgr., Kelly Schreiner; Chm. (V), Susan Wallace.
Personnel Profile: Full-Time Paid 3.
Governing Authority: state. Parent Institution: Arizona State Parks Board, 1300 W. Washington St., Phoenix, AZ. Subsidiary Institution: Tombstone Chamber of Commerce. Tax-exempt.
Institution Type/Description: Local History Museum: housed in an 1882 Cochise County Courthouse.
Collections: artifacts associated with Tombstone history; glass; china; silver; anthropology; archaeology; mineralogy; medical; natural history; guns; furniture; dolls; wood carving; frontier, pioneer, Old West artifacts until 1929; western cattlemen ranches; domestic, industrial, business & government material.
Research Fields: Tombstone history, 1877-1929.
Facilities: Books & museum-related items for sale.
Activities: guided tour of courthouse with advanced reservation; inter-museum loan & temporary exhibitions.
Publications: Nellie Cashman.
Hours & Admission Prices: Daily 9-5 Adults 14 & over $5, youth 7-13 $2; discounts to active military; children 6 & under no charge. Closed Christmas. &
Attendance: 59,889 (accurate)

TOMBSTONE WESTERN HERITAGE MUSEUM, 6th St. & Fremont St., Tombstone, AZ 85638. Mailing Address: P.O. Box 730, Tombstone, AZ 85638-0730. Tel.: 520-457-3800.
E-mail: silvrldy@yahoo.com
Web Site: thetombstonemuseum.com

Founded: 2001.
Institution Type/Description: History Museum.
Collections: local history & culture; period artifacts; photographs; documents; 1933 Dusenberg; guns; early bicycles; Wyatt Earp family artifacts.
Hours & Admission Prices: Mon.-Tues. & Thurs.-Sat. 9-6, Sun. 12:30-6. Adults $7.50, youth 18 & under $5; discounts to active military, families & groups; children under 12 no charge. &
Membership: Individual $25; Lifetime $1,000.

Topawa

TOHONO O'ODHAM NATION CULTURAL CENTER & MUSEUM, Fresnal Canyon Rd., Topawa, AZ 85639. Mailing Address: P.O. Box 837, Sells, AZ 85634-0837. Tel.: 520-383-0211.
Institution Type/Description: Native American History Museum.
Collections: Native American culture & history; personal artifacts; photographs; tribal arts & crafts; oral histories; paintings; sculpture.
Activities: educational programs.
Hours & Admission Prices: Mon.-Sat. 10-4. No charge; donations accepted.

Tuba

EXPLORE NAVAJO INTERACTIVE MUSEUM, 10 N. Main St., Tuba, AZ 86045. Tel.: 928-640-0684; 800-644-8383.
Institution Type/Description: Native American History Museum.
Collections: Navajo Indian history, language, ceremonial life & culture; personal artifacts.
Hours & Admission Prices: Mon.-Sat. 8-6, Sun. 12-6. Adults $9, seniors 65 & over $7, children 7-12 $6; children 6 & under no charge.

Tubac

TUBAC CENTER OF THE ARTS, 9 Plaza Rd., Tubac, AZ 85646-1911. Mailing Address: P.O. Box 1911, Tubac, AZ 85646-1911. Tel.: 520-398-2371. Fax: 520-398-9511.
E-mail: contactus@tubacarts.org
Web Site: tubacarts.org
Founded: 1964.
Congressional District: 5
Key Personnel: Dir. Operations, Karin Topping; Mgr. Exhibitions & Mktg., Karon Leigh; Museum Shop Mgr., Bonnie Jaus.
Personnel Profile: Full-Time Paid 2; Part-Time Paid 4; Part-Time Volunteers 200.
Governing Authority: board of directors; nonprofit. Parent Institution: Santa Cruz Valley Art Association. Tax-exempt.
Institution Type/Description: Art Center.
Collections: works by regional & national artists.
Facilities: stage. Museum-related items for sale.
Activities: temporary & permanent exhibitions; lectures; performing arts; workshops; children's summer art program; docent program; Juried art Competitions. Museum Sponsors: Christmas Holiday Art Market.
Publications: newsletter.
Hours & Admission Prices: Sept.-May Mon.-Sat. 10-4:30, Sun. 12-4:30. Offices: daily. No charge; donations accepted. Closed national holidays. &
Attendance: 39,000 (estimated)
Membership: Individual $45; Family $60; Business $85; Sponsor $100-$249; Arts Advocate $250-$499; Patron $500-$999; Benefactor $1,000-$2,499; Curator $2,500 & up.

TUBAC PRESIDIO STATE HISTORIC PARK, (M), 1 Burruel St., Tubac, AZ 85646. Mailing Address: P.O. Box 1296, Tubac, AZ 85646-1296. Tel.: 520-398-2252. Fax: 520-398-2685.
E-mail: info@tubacpresidio.org
Web Site: www.tubacpresidio.org
Founded: 1959.
Congressional District: 9
Key Personnel: Dir., Shaw Kinsley.
Personnel Profile: Full-Time Paid 1; Full-Time Volunteers 8; Part-Time Volunteers 35.
Governing Authority: state. Parent Institution: Arizona State Parks Board. Tax-exempt: 501(c)(3).
Institution Type/Description: Park Museum and Visitor Center.
Collections: Indian, Spanish, Mexican & American material culture to demonstrate the interrelationship of these ethnic groups as related to the Presidio & history of Tubac; costumes; archaeology; mining; Washington hand press print shop exhibit; reproduction of Arizona's first newspaper. Historic Buildings: 1885 Old Tubac School House; 1752 Commandant's House (ruin); c.1914 Otero Hall.
Research Fields: Spanish Colonial, Mexican Republic, Arizona territorial periods.

Facilities: picnic area.
Activities: guided tours; lectures; film presentation; demonstration of Washington hand press. Museum Sponsors: Living History program, Spanish Colonial Life at Tubac Presidio c.1752.
Publications: reproduction, The Weekly Arizonian; brochures.
Hours & Admission Prices: Daily 9-5. Adults 14 & over $4, children 7-13 $2; children 6 & under and handicapped no charge. Closed Christmas. &
Attendance: 12,853 (accurate)

Tucson

ARIZONA HISTORICAL SOCIETY/ARIZONA HISTORY MUSEUM, (M), 949 E. 2nd St., Tucson, AZ 85719-4898. Tel.: 520-628-5774. Fax: 520-628-5695.
E-mail: ahstucson@azhs.gov
Web Site: www.arizonahistoricalsociety.org
Founded: 1864.
Congressional District: 3
Key Personnel: Exec. Dir., Anne I. Woosley, Ph.D.; Southern Arizona Division Dir., Les Roe; Publications Division Dir., Bruce Dinges, Ph.D.
Personnel Profile: Full-Time Paid 11; Part-Time Paid 6; Part-Time Volunteers 167; Interns 2.
Governing Authority: bd. of directors. Parent Institution: Arizona Historical Society, 949 E. 2nd St. Tucson AZ 85719. Tel.: 520-628-5774 Tax-exempt: 501(c)(3).
Institution Type/Description: History Museum.
Collections: Spanish colonial; Mexican culture; American military weapons; transportation; ranching equipment; mining equipment; household effects; clothing.
Research Fields: American Southwest, Spanish Colonial, Mexican Republic, northern Mexico & American military history.
Facilities: library & archives including 70,000 books, 1,000,000 photographs, periodicals, newspapers, maps, microfilm, oral history recordings & transcriptions, 2,700 linear feet of vertical files & manuscripts; meeting rooms; auditorium. Museum-related items for sale.
Activities: guided tours; school tours; behind-the-scene tours; seminars; workshops; docent program; lecture series; adult & children's programming; annual events; rental facilities.
Publications: quarterly, Journal of Arizona history; books; monographs; annual report.
Hours & Admission Prices: Museum: Mon.-Sat. 10-4. Adults $5, senior citizens 60 & over and youth 12-18 $4; children 11 & under and members no charge. Library: Mon.-Fri. 9-4. Closed New Year's Day; Martin Luther King Jr. Day; Presidents' Day; Memorial Day; Independence Day; Labor Day; Columbus Day; Veterans Day; Thanksgiving; Christmas. &
Attendance: 20,000 (estimated)
Membership: Student $25; Individual $50; Household $65; Sustaining $100; Patron $250; Sponsor $500; Director's Circle $1,000.

*** ARIZONA HISTORICAL SOCIETY DOWNTOWN HISTORY MUSEUM, (M),** 140 N. Stone Ave., Tucson, AZ 85701. Mailing Address: 949 E. 2nd St., Tucson, AZ 85719-4898. Tel.: 520-770-1473.
E-mail: ahstucson@azhs.gov
Web Site: www.arizonahistoricalsociety.org
Founded: 1864.
Congressional District: 3
Key Personnel: Exec. Dir., Anne I. Woosley, Ph.D.
Personnel Profile: Part-Time Paid 1.
Governing Authority: bd. of directors. Parent Institution: Arizona Historical Society, 949 E. 2nd St., Tucson, AZ 85719. Tel.: 520-628-5774. Tax-exempt: 501(c)(3).
Institution Type/Description: History Museum.
Collections: history of downtown Tucson; barber shop, drug store, department store, school, hotel, fire & police dept., saloon, cigar store, movie theater, aviation artifacts.
Research Fields: history of downtown Tucson.
Activities: permanent exhibitions; annual events; special programming.
Publications: quarterly, The Journal of Arizona History; books; monographs; annual report.
Hours & Admission Prices: Tues.-Fri. 10-4. Adults $3, seniors 60 & over and youth 12-18 $2; children 11 & under and members no charge. Closed New Year's Day; Independence Day; Veterans Day; Thanksgiving; Christmas. &
Attendance: 1,600 (estimated)
Membership: Student $25; Individual $50; Household $65; Sustaining $100; Patron $250; Sponsor $500; Director's Circle $1,000.

*** ARIZONA HISTORICAL SOCIETY FORT LOWELL MUSEUM, (M),** 2900 N. Craycroft Rd., Tucson, AZ 85712. Mailing Address: 949 E. 2nd St., Tucson, AZ 85719-4898. Tel.: 520-885-3832.
E-mail: ahstucson@azhs.gov
Web Site: www.arizonahistoricalsociety.org
Founded: 1864.
Congressional District: 2
Key Personnel: Exec. Dir., Anne I. Woosley, Ph.D.
Personnel Profile: Part-Time Paid 1.
Governing Authority: bd. of directors. Parent Institution: Arizona Historical Society, 949 E. 2nd St., Tucson, AZ 85719. Tel. 520-628-5774. Tax-exempt: 501(c)(3).
Institution Type/Description: Military Museum: housed in a reconstructed officer's quarters from a military post active from 1873 to 1891.
Collections: fort history; period furnishings; costumes of frontier army life; uniforms; military artifacts.
Research Fields: 1873-91 Ft. Lowell; Apache wars in Arizona; frontier army; local history.
Activities: guided tours; permanent exhibits; programs for adults & children; special events. Annual Event: La Reunion de El Fuerte.
Publications: quarterly, The Journal of Arizona History; books; monographs; annual report.
Hours & Admission Prices: Fri.-Sat. 10-4. Adults $3, senior citizens 60 & over and youth 12-18 $2; children 11 & under and members no charge. Closed New Year's Day; Independence Day; Veterans Day; Thanksgiving; Christmas. &
Attendance: 1,750 (estimated)
Membership: Student $25; Individual $50; Household $65; Sustaining $100; Patron $250; Sponsor $500; Director's Circle $1,000.

*** ARIZONA HISTORICAL SOCIETY SOSA-CARRILLO-FREMONT HOUSE MUSEUM, (M),** 151 S. Granada Ave., Tucson, AZ 85701. Mailing Address: 949 E. 2nd St., Tucson, AZ 85719-4898. Tel.: 520-628-5774.
E-mail: ahstucson@azhs.gov
Web Site: www.arizonahistoricalsociety.org
Founded: 1864.
Congressional District: 3
Key Personnel: Exec. Dir., Anne I. Woosley, Ph.D.
Personnel Profile: Part-Time Paid 2.
Governing Authority: bd. of directors. Parent Institution: Arizona Historical Society, 949 E. 2nd St., Tucson, AZ 85719. Tel.: 520-628-5774. Tax-exempt: 501(c)(3).
Institution Type/Description: Historic House: housed in the former residence of pioneer families Sosa & Carrillo; also 1881 residence of John C. Fremont, Governor of Arizona Territory.
Collections: historic house.
Publications: quarterly, The Journal of Arizona History; books; monographs; annual report.
Hours & Admission Prices: Call for hours & admission prices. &
Membership: Student $25; Individual $50; Household $65; Sustaining $100; Patron $250; Sponsor $500; Director's Circle $1,000.

ARIZONA-SONORA DESERT MUSEUM, (M), 2021 N. Kinney Rd., Tucson, AZ 85743-9719. Tel.: 520-883-1380 & 2702. Fax: 520-883-2500.
E-mail: info@desertmuseum.org
Web Site: www.desertmuseum.org
Founded: 1952.
Congressional District: 2
Key Personnel: Exec. Dir., Craig S. Ivanyi; Dir. Mktg., Rosemary Prawdzik; Chm. Bd. Trustees, Archibald M. Brown, Jr.; C.F.O., Rebecca Myers; Exec. Dir. Philanthropy, Brian Bateman; Gift Shop Mgr., Jim Hills.
Personnel Profile: Full-Time Paid 87; Part-Time Paid 14; Part-Time Volunteers 400; Interns 1.
Governing Authority: nonprofit organization. Tax-exempt: 501(c)(3).
Institution Type/Description: Nature Center & Natural History Museum.
Collections: invertebrates; herpetology; ichthyology; ornithology; mammalogy; botany; geology; paleontology; mineralogy: all related to the Sonoran Desert region.
Research Fields: regional natural history research.
Facilities: 6,000-vol. library of books on natural history of the Sonoran Desert, reprints, films & slides; botanical garden; zoological park; aquarium; classrooms; restaurant. Gift items for sale.
Activities: lectures; TV program; formally organized education programs for children; volunteer program; permanent exhibitions; environmental and outreach programs; on-site interpretation activities & demonstrations;

special off-site activities & tours for members; interpretive displays & activities conducted daily; special events; Desert Museum Art Institute.

Publications: newsletter; Secret Life of Hummingbirds; Wild Foods of the Sonoran Desert: Discovering the Desert Museum; Tucson Mountains Trail Guide; Mount Lemmon Road Guide; Desert Dogs-Coyotes, Foxes & Wolves; Strangers in Our Midst-the Startling World of Sonoran Desert Arthropods; Sonorasaurus-Dinosaur of the Desert; ASDM Book of Answers; Gardening for Pollinators; (pocket series) Desert Life/Vida Desertica; Natural History of the Sonoran Desert; All About the Arizona-Sonora Desert Museum; My Nana's Remedies; Hummingbirds of the West; Venomous Critters of the Southwest; Outdoor Gizmos; Guide to Birds of the Salton Sea; The Sonoran Desert Tortoise: Natural History, Biology and Conservation; Family Go Guide; Guide to Southern Arizona Bird Nests and Eggs; Desert Museum Scrapbook; Cactaceas de Sonora, Mexico: Su Diversidad, Uso y; Invasive Species in the Sonoran Region.

Hours & Admission Prices: See website for hours & admission prices. &

Attendance: 378,489 (accurate)

Membership: Los Coatis Kids' Club $25; Individual $50; General $65; Turquoise $150; Copper $300; Silver $600; Gold $1,200.

✻ **ARIZONA STATE MUSEUM, (M),** University of Arizona, 1013 E. University Blvd., Tucson, AZ 85721. Mailing Address: P.O. Box 210026, Tucson, AZ 85721-0026. Tel.: 520-621-6302 & 6281. Fax: 520-621-2976. Facebook: Arizona State Museum.

Web Site: www.statemuseum.arizona.edu

Founded: 1893.

Congressional District: 8

Key Personnel: Dir., Patrick D. Lyons, Ph.D., RPA; Deputy Dir., Irene Bald Romano, Ph.D.; Administrative Asst., Georgine Speranzo; Head Operations, Keith Knoblock; Mktg. Dir., Darlene Lizarraga; Head Librarian, Mary E. Graham; Head Public Program & Cur. Archaeology, Michael Brescia, Ph.D.; Conservator, Nancy N. Odegaard, Ph.D.; Museum Registrar, Lisa Zimmerman; Museum Shop Mgr., Tim Price; Dir. Education, Lisa Falk; Interim Head Collections, Diane Dittemore; Antiquity Act & Repatriation Coord., Todd Pitezel, Ph.D.

Personnel Profile: Full-Time Paid 39; Part-Time Paid 15; Part-Time Volunteers 160; Interns 3.

Volunteer Hours: 7,600

Operating Expenses: 4,730,381

Operating Income: 6,556,865

Governing Authority: state. Parent Institution: University of Arizona. Tax-exempt: 701(c)(1).

Institution Type/Description: Anthropology Museum.

Collections: focus on Indian cultures of the U.S. Southwest and northern Mexico, including more than 175,000 ethnographic/archaeological specimens and 30,000 cubic feet of bulk archaeological research specimens; 350,000 photographic items; 3,000 osteological specimens; 4,000 vertebrate zooarchaeological specimens.

Major Exhibits: Curtis Reframed: The Arizona Portfolios, 11/9/13-7/15; Shonto Begay, 12/13-5/14; From Above: Adriel Heisey (T), 2/14-9/14; Re: Curtis, 10/14-2/15.

Research Fields: southwestern archaeology; ethnohistory; ethnology; physical anthropology.

Facilities: ASM Library: 90,000 volumes of special collections; 100,000 items of ephemera; 15,000 archaeological contract reports; 1,500 linear feet of paper archives related to Southwestern archaeology and ethnology; 6,000 maps; 800 original sound recordings. AZSITE, Arizona's Cultural Resource Inventory contains records on approximately 89,000 archaeological and historical sites in Arizona available to subscribers; archaeological & material conservation labs.

Activities: guided tours; docent program; inter-museum loan, permanent & temporary exhibitions; library archives; administer Arizona antiquities act & issue archaeological permits for research on state lands.

Publications: occasional booklets; exhibit catalogs; brochures; Arizona State Museum Archaeological Series.

Hours & Admission Prices: Mon.-Sat. 10-5. Adult $5; students, children & museum members no charge. Closed state & national holidays. &

Attendance: 27,214 (estimated)

Membership: Individual $40; Family $60; Patron $125; Thompson Circle $250; Haury Circle $500; Cummings Circle $1,000; George W.P. Hunt Heritage Circle $1,500.

CENTER FOR CREATIVE PHOTOGRAPHY, (M), University of Arizona, 1030 N. Olive Rd., Tucson, AZ 85721-0103. Mailing Address: University of Arizona, P.O. Box 210103, Tucson, AZ 85721-0103. Tel.: 520-621-7968. Fax: 520-621-9444.

E-mail: info@ccp.arizona.edu

Web Site: www.creativephotography.org

Founded: 1975.

Congressional District: 5

Key Personnel: Dir., Katharine Martinez; Administrative Asst., Janet Livingstone; Archivist, Leslie Squyres; Registrar, Trinity Parker; Asst. Registrar, Betsi Meissner; Dir. Devel., Ruth McCutcheon; Norton Family Cur., Rebecca Senf; Arthur J. Bell Senior Photograph Conservator, Jae Gutierrez; Mgr. Digital Collections & Services, Denise Gose; Rights & Reproductions, Tammy Carter; Design & Prep., Tim Mosman.

Personnel Profile: Full-Time Paid 20; Part-Time Paid 25; Part-Time Volunteers 3; Interns 5.

Governing Authority: university. Parent Institution: University of Arizona. Tax-exempt.

Institution Type/Description: Art Museum.

Collections: major 20th-century photographers; 19th & 20th-century photographs including 90,000 photographs by 2,200 photographers; still photography; manuscripts; archives; personal papers; negatives; ephemera.

Research Fields: photographic history.

Facilities: 20,000-vol. library of photography & photographic history; Laura Volkerding Study Center. Center publications for sale.

Activities: lectures; gallery talks; temporary & traveling exhibitions; loans; research by appointment.

Publications: The Archive; exhibition catalogs & books.

Hours & Admission Prices: Mon.-Fri. 9-5, Sat.-Sun. 1-4. No charge; donations accepted. Closed major holidays. &

Attendance: 50,000 (estimated)

CHILDREN'S MUSEUM TUCSON, (M), 200 S. Sixth Ave., Tucson, AZ 85701-2109. Mailing Address: P.O. Box 2609, Tucson, AZ 85702-2609. Tel.: 520-792-9985. Fax: 520-792-0639.

E-mail: cmt@childrenmuseumtucson.org

Web Site: www.childrensmuseumtucson.org

Founded: 1986.

Congressional District: 2

Key Personnel: Exec. Dir., Michael Luria; Pres., Bob Janus; Museum Shop Mgr., Kendra Decker.

Personnel Profile: Full-Time Paid 10; Part-Time Paid 15; Part-Time Volunteers 125; Interns 2.

Governing Authority: nonprofit organization. Tax-exempt: 501(c)(3).

Institution Type/Description: Children's Museum.

Collections: interactive hands-on educational exhibits.

Facilities: meeting room; 11,000 sq. ft. exhibition space.

Activities: guided tours; organized education programs for children; participatory exhibits.

Publications: monthly e-newsletters; community report.

Hours & Admission Prices: School Year: Tues.-Fri. 9-5, Sat.-Sun. 10-5; Summer: call for hours. Adults $8, seniors & children $6; members no charge. &

Attendance: 161,798 (accurate)

Membership: Junior Family $60; Just For Grandparents $65; My Family $85.

CONRAD WILDE GALLERY, 101 W. 6th St., Tucson, AZ 85701. Tel.: 520-622-8997. Fax: 520-622-7988.

E-mail: info@conradwildegallery.com

Web Site: www.conradwildegallery.com

Institution Type/Description: Art Gallery.

Collections: works by contemporary artists; paintings; sculpture; works on paper.

Activities: special events; temporary exhibitions.

Hours & Admission Prices: By appointment.

DAVIS DOMINGUEZ GALLERY, 154 E. 6th St., Tucson, AZ 85705-8321. Tel.: 502-629-9759.

E-mail: info@davisdominguez.com

Web Site: www.davisdominguez.com

Founded: 1976.

Institution Type/Description: Art Gallery.

Collections: works by regional & national contemporary artists; paintings; sculpture; works on paper.

Activities: special events; temporary exhibitions.

Hours & Admission Prices: Tues.-Fri. 10-5, Sat. 10-4. No charge. &

DEGRAZIA GALLERY IN THE SUN, 6300 N. Swan Rd., Tucson, AZ 85718-3697. Tel.: 520-299-9191; 800-545-2185. Fax: 520-299-1381.

E-mail: admin@degrazia.org

Web Site: www.degrazia.org

Founded: 1977.

Congressional District: 5

Key Personnel: Exec. Dir., Lance Laber; Dir. Collections & Exhibitions, Jim Jenkins; Museum Shop Mgr., Lisa Palmer.
Personnel Profile: Full-Time Paid 10; Part-Time Paid 2.
Governing Authority: nonprofit organization. Parent Institution: DeGrazia Foundation. Tax-exempt: 501(c)(3).
Institution Type/Description: Art Gallery: building listed on the National Register of Historic Places.
Collections: Southwestern art by Ted DeGrazia includes oils, watercolors, pastels, lithographs, sculpture, textiles, adobe architecture. Historic Buildings: chapel; homes.
Major Exhibits: Our Lady of Guadalupe, 5/17/13-2/16/14.
Facilities: 500-vol. library of art books; 10-acre adobe campus. Prints, note cards & books for sale.
Activities: guided tours; films; organized education programs for children; participatory, loan, temporary & traveling exhibitions.
Publications: exhibit catalogs & books.
Hours & Admission Prices: Daily 10-4. No charge; donations accepted. Closed New Year's Day; Easter; Thanksgiving; Christmas. &
Attendance: 50,000 (estimated)

DINNERWARE CONTEMPORARY ARTS, 44 W. 6th St., Tucson, AZ 85705.
E-mail: dinnerware@dinnerwarearts.com
Web Site: www.dinnerwareArts.com
Founded: 1979.
Congressional District: 2
Key Personnel: Dir., David Aguirre; Treas., Michael Contreras; Communications Coord., Lucinda Young.
Personnel Profile: Full-Time Paid 1; Interns 3.
Governing Authority: nonprofit. Tax-exempt: 501(c)(3).
Institution Type/Description: Contemporary Art Space: located in historic downtown Tucson arts district.
Collections: works by contemporary artists.
Facilities: 2,400 sq. ft. exhibit space. Exhibit catalogues & other museum-related items for sale.
Activities: guided tours; lectures; traveling exhibitions; concerts; formally organized education programs for undergraduates & graduates. Museum Sponsors: WESTAF Invitational Exhibition; Biennial 7-state Juried Exhibit; Auction Exhibit; Solo/Group shows; travelling shows.
Publications: Retrospective catalog, The First Decade of an Alternative Art Space (1979-1989).
Hours & Admission Prices: Tues.-Sat. 12-5; other times by appointment. No charge. &
Attendance: 16,000 (accurate)
Membership: Artist $50; Lifetime $350; Patron $350-$999; Benefactor $1,000.

ETHERTON GALLERY, 135 S. Sixth Ave., Tucson, AZ 85701. Tel.: 520-624-7370. Fax: 520-792-4569.
E-mail: info@ethertongallery.com
Web Site: www.artnet.com/etherton.html
Founded: 1981.
Institution Type/Description: Art Gallery.
Collections: contemporary fine art; photography; paintings; sculpture; prints.
Activities: special events; temporary exhibitions.
Hours & Admission Prices: Tues.-Sat. 11-5.

FLANDRAU SCIENCE CENTER AND PLANETARIUM, The University of Arizona, 1601 E. University Blvd., Tucson, AZ 85719. Mailing Address: P.O. Box 210091, Tucson, AZ 85721-0091. Tel.: 520-621-STAR. Fax: 520-621-8451.
E-mail: uascreservations@gmail.com
Web Site: www.flandrau.org
Founded: 1975.
Congressional District: 5
Key Personnel: Dir., Travis Huxman; Interim Asst. Dir., William Plant; Technical Svcs. Mgr., Neil McSweeney; Planetarium Mgr., Michael Magee; Program Coord., Roseann Mankel; Education Dir., Jennifer Fields.
Personnel Profile: Full-Time Paid 3; Part-Time Paid 15; Part-Time Volunteers 12.
Governing Authority: university; nonprofit organization. Parent Institution: The University of Arizona. Tax-exempt.
Institution Type/Description: Science Center, Planetarium & Observatory.
Collections: science exhibits; period & contemporary astronomy instruments; astronomical art; memorabilia from noted astronomers; meteorites; gems & minerals; 16 in. telescope; hands-on exhibits.
Facilities: 125-seat planetarium. Science education materials for sale.
Activities: star shows; lectures; films; formally organized education programs

for children & undergraduate college students; volunteer program; permanent & temporary exhibitions; summer science workshops; science demonstrations.
Publications: newsletter.
Hours & Admission Prices: Science Center & Planetarium: Mon.-Wed. 10-3, Thurs.-Fri. 10-3 & 6-9, Sat. 10-9, Sun. 1-4. Admission 15 & over $7.50, children 4-14 $5; ASTC members no charge. &
Attendance: 50,000 (estimated)
Membership: Individual $25; Family $50; Patron $100; Sponsor $500; Benefactor $1,000.

HISTORY OF PHARMACY MUSEUM, University of Arizona College of Pharmacy, 1703 E. Mabel, Tucson, AZ 85721. Mailing Address: University of Arizona College of Pharmacy, P.O. Box 210202, Tucson, AZ 85721-0202. Tel.: 520-626-1427. Fax: 520-626-4063.
Web Site: www.pharmacy.arizona.edu
Founded: 1966.
Congressional District: 7
Key Personnel: Asst. Dean, Richard Wiedhopf.
Governing Authority: Parent Institution: University of Arizona. Subsidiary Institution: College of Pharmacy. Tax-exempt.
Institution Type/Description: Pharmacy Museum.
Collections: Arizona pharmacy artifacts c.1880-1950; history of pharmacy; over 60,000 bottles; drug containers; books; store fixtures.
Facilities: Museum-related items for sale.
Activities: tours.
Hours & Admission Prices: Mon.-Fri. 8-5. No charge.

INTERNATIONAL WILDLIFE MUSEUM, (M), 4800 W. Gates Pass Rd., Tucson, AZ 85745-9600. Tel.: 520-629-0100. Fax: 520-618-3561.
Web Site: www.thewildlifemuseum.org
Founded: 1988.
Key Personnel: Dir., Richard S. White; Cur. Education, Kristine Massey.
Personnel Profile: Full-Time Paid 7; Part-Time Paid 5; Part-Time Volunteers 10.
Governing Authority: not-for-profit organization. Parent Institution: Safari Club International Foundation. Tax-exempt: 501(c)(3).
Institution Type/Description: Natural History Museum.
Collections: 62 major natural history dioramas & exhibits; International Collection of the National Heads & Horns collection; Burnham-Eagle-Macomber collection; Wayde bird collection; insects from around the world; wildlife art.
Research Fields: vertebrate paleontology.
Facilities: 100-seat theatre. Gift items for sale.
Activities: educational & interactive computer exhibits; interpretive tours; school environmental education programs & tours; teacher training & assistance; natural history movies; school outreaches; children's & family programs; classes teaching kits; docent & intern programs.
Publications: Animal Skulls: A Guide for Teachers, Naturalists and Interpreters; Wildlife Classroom Activities; members' newsletter, Tracks; previsitation guides, Discovery Safari.
Hours & Admission Prices: Mon.-Fri. 9-5, Sat.-Sun. 9-6. Adults $7, senior citizens $5.50, children 4-12 $2.50; discounts to AAM & ICOM members & military; children 3 & under and members no charge. Closed Thanksgiving; Christmas. &
Attendance: 66,173 (accurate)
Membership: Student $20; Senior $23; Individual $25; Senior Couple $40; Family $45; Friend $125; Life $500.

JEWISH HISTORY MUSEUM, 564 S. Stone Ave., Tucson, AZ 85702. Mailing Address: P.O. Box 889, Tucson, AZ 85702. Tel.: 520-670-9073.
E-mail: JHMTucson@gmail.com
Web Site: www.jewishhistorymuseum.org
Founded: 2004.
Key Personnel: Exec. Dir., Eileen Warshaw; Chm. (V), Dr. Barry Friedman.
Personnel Profile: Full-Time Volunteers 2; Part-Time Volunteers 10.
Governing Authority: Tax-exempt.
Institution Type/Description: History Museum.
Collections: American southwest Jewish history & culture; period furnishings; personal artifacts & histories; photographs; toys; religious artifacts; wedding dresses.
Research Fields: Jewish genealogy; WWII Jewish history.
Publications: quarterly, The Chronicle.

Hours & Admission Prices: Wed.-Thurs. & Sat.-Sun. 1-5, Fri. 12-3. Adults $5; discounts to Time Travelers Network; members & children under 12 no charge. &

Membership: Student $20; Senior 65 & over $50; Single $65; Senior Couple $90; Family $100; Contributor $250; Sustaining $500; Pioneer $1,000; Benefactor $1,500; Cornerstone $2,500; Founder $5,000.

KITT PEAK NATIONAL OBSERVATORY, (M), State Rte. 86 & Rte. 386, Tucson, AZ 85726-6732. Mailing Address: 950 N. Cherry Ave., Tucson, AZ 85719-4933. Tel.: 520-318-8726. Fax: 520-318-8451.

E-mail: rfedele@noao.edu

Web Site: www.noao.edu

Founded: 1958.

Key Personnel: Mgr. Public Information, Richard Fedele; Museum Shop Mgr., Nick Petrosino.

Personnel Profile: Full-Time Paid 8; Part-Time Paid 9; Part-Time Volunteers 35.

Governing Authority: nonprofit. Parent Institution: National Optical Astronomy Observatory & Aura Inc./NSF. Tax-exempt.

Institution Type/Description: Observatory.

Collections: astronomy; 4m, 2.1m & solar telescope.

Research Fields: astronomy.

Activities: night viewing; education programs; classes.

Hours & Admission Prices: Daily 9-3:45. Guided Tours: 10, 11:30 & 1:30. Museum: no charge. Guided Tours: adults $4, children 6-12 $2.50. Closed New Year's Day; Thanksgiving; Christmas. &

Attendance: 60,000 (estimated)

Membership: Individual $35; Dual $45; Family $55.

LA PILITA MUSEUM, 420 S. Main Ave., Tucson, AZ 85701-2228. Tel.: 520-882-7454. Facebook: La Pilita Museum.

E-mail: lapilita@qwestoffice.net

Web Site: lapilita.com

Institution Type/Description: History Museum.

Collections: local history & culture; photographs; period artifacts.

Facilities: Museum-related items for sale.

Activities: educational programs; special events; lectures; workshops; concerts.

Publications: A Walk Through Time, Place and Stay; self-guided tour book of Barrio Viejo.

Hours & Admission Prices: Jan. 3-May 1 & Sept.-Dec. 16 Tues.-Sat. 11-2; other times by appointment. Suggested Donation: adults $2.

Attendance: 1,305 (estimated)

Membership: Amistad de Cobre $25; Amistad de Turquesa $50; Amistad de Oro $100; Amistad de Plata $500.

LOUIS CARLOS BERNAL GALLERY, Pima Community College - West Campus, 2202 W. Anklam Rd., Tucson, AZ 85709-0015. Tel.: 520-206-6942. Fax: 520-206-6719.

E-mail: dandres@pima.edu

Web Site: www.pima.edu/cfa

Formerly: Pima Community College Art Gallery

Founded: 1970.

Congressional District: 7

Key Personnel: Dir., David Andres; Chm. (V), Ali Traut.

Personnel Profile: Part-Time Paid 25; Interns 3.

Governing Authority: Parent Institution: Pima Community College - West Campus.

Institution Type/Description: Art Gallery.

Collections: works by regional, national & international artists.

Activities: lectures; panels.

Publications: exhibition catalogues.

Hours & Admission Prices: Mon.-Thurs. 10-5, Fri. 10-3; call for confirmation of hours. No charge; donations accepted. Closed all major holidays. &

Attendance: 8,000 (estimated)

THE MINI TIME MACHINE MUSEUM, (M), 4455 E. Camp Lowell Dr., Tucson, AZ 85712. Tel.: 520-881-0606. Fax: 520-881-9307.

Institution Type/Description: General Museum.

Collections: over 275 miniature houses and room boxes.

Hours & Admission Prices: Tues.-Sat. 9-4, Sun. 12-4. Adults $9, seniors 65 & over and military $8, children 4-17 $6; children 3 & under no charge. Closed major holidays. &

Membership: E-Friend $15; Individual $35; Senior Couple $45; Family $60.

OLD PUEBLO ARCHAEOLOGY CENTER, 2201 W. 44th St., Tucson, AZ 85713-4575. Mailing Address: P.O. Box 40577, Tucson, AZ 85717-0577. Tel.: 520-798-1201. Fax: 520-798-1966.

E-mail: info@oldpueblo.org

Web Site: oldpueblo.org

Founded: 1993.

Congressional District: 8

Key Personnel: Exec. Dir., Allen Dart.

Governing Authority: nonprofit organization. Tax-exempt: 501(c)(3).

Institution Type/Description: Archaeology Education Center.

Collections: archaeology; area history & culture.

Research Fields: Southwestern archaeology, history & culture.

Facilities: library.

Activities: archaeological & historical research programs; educational programs; simulated archaeological excavation site.

Publications: quarterly electronic bulletin, Old Pueblo Archaeology; archaeological technical reports; special publications.

Hours & Admission Prices: Call for hours. Fee charged for education programs & tours. &

Attendance: 3,000

Membership: Friend $25; Individual $40; Household $80; Sustaining $100; Contributing $200; Suppporting $500; Sponsoring & Corporation $1,000.

OLD TUCSON, 201 S. Kinney Rd., Tucson, AZ 85735. Tel.: 520-883-0100.

E-mail: guestrelations@oldtucson.com

Web Site: oldtucson.com

Institution Type/Description: History Museum: housed on the site where over 300 movies & television productions have been filmed since 1939.

Collections: Old West history & life; period furnishings; movie memorabilia & sets.

Facilities: restaurant. Gift items for sale.

Activities: stunt shows; gunfights; saloon girls in musical revues; horseback rides; early carousel; special events.

Hours & Admission Prices: Call for hours. Adults 12 & over $16.95, children 4-11 $10.95; children 3 & under no charge.

OLD WEST MOVIE POSTER MUSEUM, 1300 N. Stone Ave., Tucson, AZ 85705-7338. Tel.: 520-770-1910.

E-mail: info@flamingohoteltucson.com

Web Site: www.flamingohoteltucson.com

Institution Type/Description: Western Movie Poster Museum.

Collections: over 1,000 western movie posters; lobby cards; photographs; autographed star photos; movie memorabilia.

Hours & Admission Prices: Daily 9-5. Hotel guests no charge.

OTIS CHIDESTER SCOUT MUSEUM OF SOUTHERN ARIZONA, 1937 E. Blacklidge Dr., Tucson, AZ 85719-2847. Tel.: 520-326-7669 & 795-9484.

Web Site: www.azscoutmuseum.com

Founded: 1986.

Congressional District: 2

Key Personnel: Pres. (V), James Klein; Cur., William White.

Personnel Profile: Part-Time Volunteers 20.

Governing Authority: Tax-exempt.

Institution Type/Description: Scout Museum.

Collections: southern Arizona & Southwest Boy Scouts of America; photographs; audio; personal artifacts.

Activities: tours.

Publications: quarterly newsletter, Museum Dispatch.

Hours & Admission Prices: Wed. 9am to noon; other times by appointment. No charge; donations accepted.

Attendance: 500 (estimated)

Membership: Individual $20; Family $30.

PIMA AIR & SPACE MUSEUM, 6000 E. Valencia Rd., Tucson, AZ 85756-9403. Tel.: 520-574-0462. Fax: 520-574-9238.

Web Site: www.pimaair.org

Founded: 1976.

Congressional District: 5

Key Personnel: Exec. Dir. & Dir. Titan Missile Museum, Yvonne C. Morris; Chm. (V), Count Ferdinand von Galen; Dir. Collections & Aircraft Restoration, Scott Marchand; Asst. Cur., James Stemm; Museum Shop Mgr., Beth Barksdale.

Personnel Profile: Full-Time Paid 47; Part-Time Paid 17; Part-Time Volunteers 300.

Governing Authority: nonprofit organization. Parent Institution: Arizona Aerospace Foundation. Subsidiary Institution: Arizona Aviation Hall of Fame. Tax-exempt: 501(c)(3).
Institution Type/Description: Aeronautics & Space Museum & Arizona Aviation Hall of Fame.
Collections: aviation memorabilia & 60,000 artifacts; 300 aircraft; static displays of aircraft; photographs; space gallery. Historic Structure: World War II barracks.
Facilities: library of technical repair & operations manuals available for use on premises by appointment; theater; learning center. Reproductions, museum & aviation-related items for sale.
Activities: guided tours; lectures; temporary exhibitions.
Publications: quarterly newsletter, Skywriting; books, Wonders of the Pima Air & Space Museum; Titan Missile Museum; Contrails.
Hours & Admission Prices: Pima Air & Space Museum: daily 9-5. June-Oct.: adults $13.75, seniors 62 & over and military $9.75, children 7-12 $8; discounts to AAM members; children 6 & under & members no charge. Nov.-May: adults $15.50, seniors 62 & over and military $12.75, children 7-12 $9; discounts to AAM members; children 6 & under and members no charge. &
Attendance: 154,099 (accurate)
Membership: Senior $25; Dual $60; Crew $80.

THE POSTAL HISTORY FOUNDATION, 920 N. First Ave., Tucson, AZ 85719-4818. Tel.: 520-623-6652. Fax: 520-623-3810. Facebook: The Postal History Foundation.
E-mail: info@phftucson.org
Web Site: www.postalhistoryfoundation.org
Founded: 1960.
Personnel Profile: Full-Time Paid 2; Part-Time Paid 2; Part-Time Volunteers 50.
Governing Authority: nonprofit organization. Tax-exempt: 501(c)(3).
Institution Type/Description: Postal History Museum.
Collections: postal history; photographs; postage stamps; U.S. Civil War.
Facilities: 30,000 vol. research library.
Activities: educational programs; Youth Education Thru Stamps (YES).
Publications: Stamp Tracks; catalog available on line.
Hours & Admission Prices: Mon.-Fri. 8-3, Sat 10-2. No charge, but donations accepted. &
Attendance: 2,000 (estimated)
Membership: Individual $35; Friend $100; Donor $250; Patron $500; Benefactor $1,000.

REID PARK ZOO, 1100 S. Randolph Way, Tucson, AZ 85716-5835. Tel.: 520-791-3204. Fax: 520-791-5378.
E-mail: reidzoo@tucsonaz.gov
Web Site: www.tucsonzoo.org
Founded: 1967.
Key Personnel: Dir., Susan Basford; Education & Public Rels., Vivian W. VanPeenen.
Personnel Profile: Full-Time Paid 24; Part-Time Paid 13; Part-Time Volunteers 92; Interns 5.
Governing Authority: municipal; nonprofit. Tax-exempt: 501 (c)(3). Parent Institution: Tucson Parks & Recreation Dept.
Institution Type/Description: Zoo.
Collections: exotic animals from around the world.
Research Fields: animal nutrition.
Facilities: library material available for use on premises; 100-seat cafeteria; zoological park. Zoo-related items for sale.
Activities: docent program; formally organized education programs.
Publications: newsletter, Tucson Zoological Society Quarterly.
Hours & Admission Prices: Gate: daily 9-4. Adults $6, senior citizens $4, children 2-14 $2; discounts to AZA members; members & children under 2 no charge. Closed Christmas. &
Attendance: 525,000 (estimated)
Membership: Senior Citizen $20; Single $30; Family $50.

SAGUARO NATIONAL PARK - EASTERN DISTRICT VISITOR CENTER, 3693 S. Old Spanish Trail, Tucson, AZ 85730-5601. Tel.: 520-733-5153.
Institution Type/Description: Natural History Museum.
Collections: local history & culture; Sonoran Desert plants; Native American artifacts; photographs; period furnishings; personal artifacts.
Facilities: Books for sale.
Activities: video; educational programs.
Hours & Admission Prices: Daily 9-5. Closed Christmas.

SAGUARO NATIONAL PARK - WESTERN DISTRICT VISITOR CENTER, 2700 N. Kinney Rd., Tucson, AZ 85743-9719. Tel.: 520-733-5158.
Institution Type/Description: Natural History Museum.
Collections: local natural & cultural history; Native American artifacts; period furnishings; personal artifacts; photographs.
Facilities: Books for sale.
Activities: educational programs; video.
Hours & Admission Prices: Daily 9-5. Closed Christmas.

SAN XAVIER DEL BAC MISSION, 1950 W. San Xavier Rd., Tucson, AZ 85746-7409. Tel.: 520-294-2624. Fax: 520-294-3438.
Web Site: sanxaviermission.org
Institution Type/Description: Church & Religious Museum: built in the late 1700s.
Collections: church history; religious artifacts; wall paintings.
Facilities: Museum-related items for sale.
Activities: video; tours.
Hours & Admission Prices: Self-Guided Tour: 8:30-4:30. Services: call for hours. No charge; donations accepted. &

SOUTHERN ARIZONA TRANSPORTATION MUSEUM, 414 N. Toole Ave., Tucson, AZ 85701. Tel.: 520-623-2223.
E-mail: kkarrels@aol.com
Web Site: tucsonhistoricdepot.org
Founded: 2005.
Congressional District: 10
Key Personnel: Dir. & Museum Shop Mgr., Tom McComb; C.E.O. & Chm. (V), Dr. Ken Karrels.
Personnel Profile: Full-Time Paid 1; Part-Time Volunteers 15; Interns 1.
Operating Expenses: 39,000
Operating Income: 40,000
Governing Authority: Parent Institution: Old Pueblo Trolley. Tax-exempt.
Institution Type/Description: Transportation History Museum.
Collections: local transportation history; railroad artifacts; photographs; period artifacts; locomotive #1673.
Major Exhibits: Randy Hill, 11/13-6/14.
Facilities: Museum-related items for sale.
Activities: tours. Museum Sponsors: Coming of the RR to Tucson in March; National Train Day in May; Tucson's Birthday in Aug.; Holiday Event in December.
Hours & Admission Prices: Tues.-Thurs. & Sun. 11-3, Fri.-Sat. 10-4. No charge; donations accepted.
Attendance: 12,500 (estimated)
Membership: Honorary Conductor $25; Honorary Engineer $50; Contributor $100; Donor $250; Sponsor $500; Benefactor $1,000; Patron $2,000.

390TH MEMORIAL MUSEUM, (M), 6000 E. Valencia Rd., Tucson, AZ 85756-9403. Tel.: 520-574-0287. Fax: 520-574-3030.
E-mail: director@390th.org
Web Site: www.390th.org
Founded: 1983.
Key Personnel: Dir., Emile Terry Therrien.
Personnel Profile: Full-Time Paid 3; Part-Time Volunteers 35.
Institution Type/Description: Military History Museum.
Collections: 390th unit history & heritage; military aircraft & memorabilia; General James H. Doolittle; paintings; Honor Wall; aircraft models; photographs; B-17G.
Facilities: library; archives.
Activities: research.
Publications: quarterly, Square J Bulletin.
Hours & Admission Prices: Located on the grounds of the Pima Air and Space Museum. Daily 10-4:30. Grounds: fee charged. Memorial Museum: no charge. Closed Thanksgiving; Christmas.
Attendance: 100,000 (estimated)
Membership: 390th B.G. Veterans $20; Non-Veterans $25; Life $1,000.

TOHONO CHUL PARK INC., (M), 7366 N. Paseo del Norte, Tucson, AZ 85704-4415. Tel.: 520-742-6455, ext. 210. Fax: 520-797-1213.
E-mail: joandonnelly@tohonochulpark.org
Web Site: www.tohonochulpark.org
Founded: 1985.
Congressional District: 5
Key Personnel: C.E.O., Exec. Dir. & Financial Dir., Joan E. Donnelly; Bd. Pres., Donald F. Romano; Cur. Plants, Russ Buhrow; Cur. Exhibits, Vicki

Donkersley; Dir. Public Programs, Jo Falls; Dir. Visitor Svcs., Monica Sufaro Spigelman; Dir. Gen. Svcs., Lee Mason; General Mgr. Retail, Linda Wolfe.
Personnel Profile: Full-Time Paid 15; Part-Time Paid 15; Part-Time Volunteers 400; Interns 1.
Governing Authority: nonprofit organization. Tax-exempt: 501(c)(3).
Institution Type/Description: Nature Center.
Collections: Native and adapted plants of desert southwest; limited collection of Native American arts and crafts (contemporary and traditional).
Facilities: 800-vol. library of Southwestern ecology, culture, environment, history & botany; 100-seat auditorium; ethnobotanical garden; performance garden; demonstration garden; children's garden; classroom; nature trails. Items relating to Southwestern culture & environment for sale.
Activities: birding tours, landscape & water conservation tours; art exhibit tours; concerts; docent program; films; educational programs; guided tours; lectures; loaned & temporary exhibitions; demonstrations. Annual Event: Wildflower Festival.
Publications: quarterly newsletter, Desert Corner Journal; The Official Guide to Tohono Chul Park; annual report.
Hours & Admission Prices: Grounds & Tea Room daily 8-5. Exhibits, Museum Shops & Greenhouse: daily 9-5; guided tours by appointment. Adults $7, seniors 62 & over & active military $5, students with valid ID $3, children 5-12 $2; discounts to AAM & AABGA members; members, children under 5 & 1st Tues. of month no charge. Closed New Year's Day; Independence Day; Thanksgiving; Christmas. &
Attendance: 170,000 (estimated)
Membership: Senior Single 65 & over $20; Senior Couple 65 & over and Individual $30; Family & Dual $40; Gambel's Quail $75; Desert Tortoise $100; Red-tailed Hawk $250; Bobcat $500; Paloverde $1,000; Saguaro $5,000.

TUCSON BOTANICAL GARDENS, (M), 2150 N. Alvernon Way, Tucson, AZ 85712-3199. Tel.: 520-326-9686, ext. 10. Fax: 520-324-0166.
E-mail: info@tucsonbotanical.org
Web Site: www.tucsonbotanical.org
Founded: 1968.
Congressional District: 5
Key Personnel: Exec. Dir., Michelle Conklin; Bd. Pres. (V), Spencer Smith; Dir. Horticulture, Michael Chamberland; Dir. Mktg., Communications & Gallery, Melissa D'Auria; Dir. Visitor Services, Jennifer Hampson.
Personnel Profile: Full-Time Paid 17; Part-Time Paid 12; Part-Time Volunteers 200.
Governing Authority: nonprofit organization. Tax-exempt: 501(c)(3).
Institution Type/Description: Arboretum & Botanical Garden.
Collections: plants adaptable to southern Arizona environments; xeriscape demonstration garden; iris, herb, cactus, succulent, sensory, bird attracting, vegetable & native crops gardens; historic Tucson garden; wildflower (gardens); tropical greenhouse including epiphyte collection; seasonal tropical butterflies.
Major Exhibits: Butterfly Magic, 10/13-4/14; Alien Invasion of the Planet Kind, 11/13-4/14.
Research Fields: horticulturally adapted arid land plants.
Facilities: library pertaining to horticulture available for research on premises during open hours; meeting room; education building.
Activities: guided tours; classes & workshops for adults and children; horticultural therapy programs; school gardening programs; after school intersession programs; monthly art exhibitions; facility rentals. Annual Events: Bloom & Bites in April; Home and Garden Tour in April; Feast with the Dearly Departed in October; Luminaria Nights in December; Flights of Fancy; Garden Art Fundraiser; Twilight Bridal Fair; Weird Plant Sale. Porter Hally Gallery: rotating & group exhibits.
Publications: quarterly newsletter, Tucson Botanical Gardens News; bilingual Children's Discovery Guide; self-guided tour brochure; Attracting Birds to Your Garden; A Tucson Herb Garden - Growing Tips from the Tucson Botanical Gardens.
Hours & Admission Prices: Garden: daily 8:30-4:30. Butterfly Magic: daily 9:30 3. Garden: Oct.-April adults $13, seniors, students & military $12, children 4-12 $7.50, members & children under 4 no charge; May-Sept. adults $8, seniors, students & military $7; children 4-12 $4, members & children under 4 no charge. Closed New Year's Day; Independence Day; Thanksgiving; Christmas Eve & Day. &
Attendance: 100,000 (estimated)
Membership: Student & Senior $40; Garden Friend $50; Family $60; Cholla $100; Agave $250; Ocotillo $500; Saguaro $1,000.

✱ TUCSON MUSEUM OF ART & HISTORIC BLOCK, (M), 140 N. Main Ave., Tucson, AZ 85701-8290. Tel.: 520-624-2333. Fax: 520-624-7202.
E-mail: info@tucsonmuseumofart.org

Web Site: www.tucsonmuseumofart.org
Founded: 1924.
Congressional District: 2
Key Personnel: Dir. & C.E.O., Robert E. Knight; Dir. Finance, Katie Hiedeman; Finance Assoc., Andra Allen; Collections Mgr. & Registrar, Susan Dolan; Asst. Registrar, Rachel Shand; Mgr. Membership & Special Events, Meagan Crain; Dir. Public Rels. & Mktg., Lisa Wilkinson; Membership Assoc., Jenny Balkema; Dir. Devel., Alba Rojas-Sukkar; Chief Cur. & Cur. Modern & Contemporary Art, Julie Sasse; Grants Mgr., Michael Fenlason; Preparator, David Longwell; Librarian, Lisa Waite Bunker; Cur. Education, Morgan Wells; Supvr. Bldg. & Grounds, Al Uno; Head Security, Delmar Bamborough; Retail Mgr., John McNulty.
Personnel Profile: Full-Time Paid 20; Part-Time Paid 13; Part-Time Volunteers 483; Interns 22.
Governing Authority: nonprofit organization. Tax-exempt: 501(c)(3).
Institution Type/Description: Art Museum.
Collections: pre-Columbian, Spanish colonial, 19th- & 20th-century Latin American & Mexican folk art; contemporary art; 19th- & 20th-century art of America & the American West; World Folk Art; American arts & crafts; 20th-century art of North & South America; Asian, African & European influences on the art of the Americas.
Research Fields: American art; pre-Columbian art; art of the American West; New World art; contemporary art.
Facilities: 6,000-vol. library; 24,000 slide library on primitive & general art available for use on a research basis. Arizona crafts & other museum-related items for sale.
Activities: art classes; docent training; lectures; concerts; gallery talks; tours; competitions; extension department serving Tucson school districts; traveling & changing exhibitions; travel program.
Publications: Museum School brochure; exhibition catalogues; quarterly newsletter; annual report.
Hours & Admission Prices: Tues.-Sat. 10-5, Sun. 12-5. Adults $8, senior citizens 60 & over $6, students 13 & over $3; discounts to ICOM members; AAM members, first Sun. of month, children under 12 and members no charge. Closed Thanksgiving; Christmas &
Attendance: 237,000 (estimated)
Membership: Student $25; Individual $40; Family $50; Sustaining $100; Patron $250; President's Circle $500; Director's Circle $1,000; Fellow $5,000; Benefactor $10,000.

TUCSON RODEO PARADE MUSEUM, 4823 S. Sixth Ave., Tucson, AZ 85714-3004. Mailing Address: P.O. Box 1788, Tucson, AZ 85702-1788. Tel.: 520-294-1280.
E-mail: comatoes@cox.net
Web Site: www.tucsonrodeoparade.org
Founded: 1963.
Key Personnel: Chm. (V), Bob Taylor.
Personnel Profile: Part-Time Volunteers 12.
Governing Authority: Parent Institution: Tucson Rodeo Parade Committee, Inc. Tax-exempt.
Institution Type/Description: Western Cultural Museum.
Collections: local history; over 150 horse drawn vehicles; Native American; mining artifacts; 1890-1910 western artifacts.
Activities: tours; special events. Museum Sponsors: Tucson Rodeo Parade in February.
Hours & Admission Prices: Jan.-Marchl Mon.-Sat. 9:30-3:30. Suggested donation: adults $10, seniors $8. Closed Parade Day. &
Attendance: 5,000 (estimated)

UNIVERSITY OF ARIZONA MINERAL MUSEUM, Univ. of Arizona, Flandrau Science Center, Cherry & University, Tucson, AZ 85721. Tel.: 520-621-4227. Fax: 520-621-8451.
E-mail: bailey2@email.arizona.edu
Web Site: www.uamineralmuseum.org
Founded: 1891.
Key Personnel: Cur., Dr. Robert T. Downs; Curatorial Specialist, Sven Bailey.
Personnel Profile: Part-Time Paid 1; Part-Time Volunteers 1.
Governing Authority: university. Tax-exempt.
Institution Type/Description: Mineral Museum.
Collections: over 19,000 mineral specimens representing Arizona & worldwide localities; 7,000 micromounts which may be viewed upon request; excellent meteorite collection.
Research Fields: mineralogical & crystallographic studies in conjunction with Dept. of Geosciences, University of Arizona.
Activities: guided tours; permanent exhibitions.
Hours & Admission Prices: Fri.-Sat. 9-5. School Groups: Tues.-Fri. 9-1. Adults $4; children 4 & under no charge. Closed holidays. &

*** THE UNIVERSITY OF ARIZONA MUSEUM OF ART, (M),**
1031 N. Olive Rd., Tucson, AZ 85721. Mailing Address: University
of Arizona, P.O. Box 210002, Tucson, AZ 85721-0002. Tel.:
520-621-7567 (Main) & 7568 (Mktg./Public Rels.). Fax: 520-621-
8770.
E-mail: carolp@email.arizona.edu
Web Site: www.artmuseum.arizona.edu
Founded: 1955.
Congressional District: 7
Key Personnel: Dir., Dennis L. Jones; Mktg. Specialist, Carol Petrozzello;
Cur., Lauren Rabb; Cur. Education, Olivia Miller; Registrar, Kristen
Schmidt; Business Mgr., Kathleen Kearney; Sr. Exhibits Preparator, John
Kelly; Administrative Sec., Christine Aguilar; Ld. Security Officer, Kris
Wagman; Security, Andy Leahy.
Personnel Profile: Full-Time Paid 8; Part-Time Paid 3; Part-Time Volunteers
107; Interns 16.
Governing Authority: Bd. of Regents. Parent Institution: University of Arizona.
Tax-exempt.
Institution Type/Description: Art Museum.
Collections: Samuel H. Kress collection of 14th- to 19th-century European Art
including the 15th century altarpiece of Ciudad Rodrigo; 26 paintings by
Fernando Gallego, Maestro Bartolome and assistants; Edward J. Gallagher
Memorial collection of 20th-century & contemporary American & Euro-
pean painting, sculpture & works on paper; old master & contemporary
prints; C. Leonard Pfeiffer collection of American painting; Jacques
Lipchitz sketches & models.
Research Fields: prints; contemporary art; 15th-century Hispano-Flemish art,
education.
Facilities: 14,000 sq. ft. exhibit space. Museum-related items for sale.
Activities: intern programs; tours; research; lectures; family days; permanent,
temporary & traveling exhibitions; community outreach presentations.
Publications: catalogs, Jacques Lipchitz, Sketches and Models; permanent
collection print handbook.
Hours & Admission Prices: Tues.-Fri. 9-5, Sat.-Sun. 12-4. Adults $5; discounts
to AAM members; UA faculty & staff, students with ID, children &
members no charge. Closed major holidays. &
Attendance: 24,810 (accurate)
Membership: Basic: Student & Senior $30; Faculty & Staff $40; Senior Couple
$50. Associate: Individual $45; Family $75. Partner: Individual $125;
Family $175; Sponsor $250; Donor $500; Fellow $1,000; Benefactor
$2,500.

UNIVERSITY OF ARIZONA STUDENT UNION GALLERIES,
1303 E. Univeristy Blvd., Tucson, AZ 85719. Tel.: 520-621-6142.
Fax: 520-621-6930.
E-mail: su-gallery@email.arizona.edu
Web Site: www.union.arizona.edu/involvement/galleries/
Founded: 1971.
Personnel Profile: Part-Time Paid 5; Interns 1.
Governing Authority: public university; nonprofit.
Institution Type/Description: Art Gallery.
Collections: items purchased or donated from nationally & internationally
known artists & supporting University of Arizona student artists.
Facilities: cafeteria.
Activities: guided tours; loan & participatory exhibits; internships in curating
& graphic design. Annual Events: alumni exhibit in the fall; student art
show in the spring.
Hours & Admission Prices: Union Gallery: Mon.-Fri. 10-5; Kachina Gallery:
Mon.-Fri. 6:30 a.m. to 10 p.m., Sat.-Sun. 8 a.m. to 10 p.m. No charge.
Closed major holidays & installation day. &
Attendance: 11,500 (accurate)

**WESTERN ARCHEOLOGICAL & CONSERVATION CEN-
TER, (M),** 255 N. Commerce Park Loop, Tucson, AZ 85745-2796.
Tel.: 520-791-6400. Fax: 520-791-6465.
Founded: 1952.
Congressional District: 2
Key Personnel: Archivist, Lynn Marie Mitchell; Repository Chief, Stephanie
H. Rodeffer; Asst. Conservator, Brynn Bender; Registrar, Kim E. Beckwith;
Archivist, Khaleel Saba.
Personnel Profile: Full-Time Paid 5; Part-Time Paid 2; Part-Time Volunteers
12; Interns 1.
Governing Authority: federal. Parent Institution: U.S. Dept. of the Interior,
National Park Service. Tax-exempt.
Institution Type/Description: History Museum.
Collections: Southwestern prehistoric, ethnographic artifacts & historic house-
hold items; comparative shared library; research archives pertaining to
southwest archaeology, ethnography & history; 160,000 image photographic collection; more than 100 current serials/periodicals; natural
science.

Research Fields: cultural resources management; archaeology; collections
management.
Facilities: 17,000-vol. library on Southwest archaeology, history & ethnography for use on premises; field research station; artifact conservation
laboratories; research laboratory.
Activities: work program for graduate students in museology, artifact conservation & cultural resource management; internship program for graduate
students in library science and archival management; guided tours for high
school, university & community groups by appointment.
Publications: anthropology series.
Hours & Admission Prices: Mon.-Fri. 9-3; library & other times by appointment. No charge. Closed federal holidays. &

Tumacacori

TUMACACORI NATIONAL HISTORICAL PARK, 1891 E.
Frontage Rd., Tumacacori, AZ 85640. Mailing Address: P.O. Box
8067, Tumacacori, AZ 85640-8067. Tel.: 520-398-2341, ext. 0.
Fax: 520-398-9271.
E-mail: jeremy_moss@nps.gov
Web Site: www.nps.gov/tuma
Founded: 1908.
Congressional District: 5
Key Personnel: Cur., Jeremy Moss; Supt., Ann Razor.
Personnel Profile: Part-Time Paid 1.
Governing Authority: federal. Parent Institution: U.S. National Park Service.
Tax-exempt.
Institution Type/Description: National Park, Historic Site and Museum: site
1795 San Jose de Tumacacori Spanish Mission Church, abandoned in 1848;
Los Santos Angeles de Guevavi Mission; San Cayetano de Calabazas
Mission.
Collections: Indian & Spanish Colonial artifacts.
Research Fields: Spanish colonial history; archaeology.
Facilities: 700-vol. library of material on Spanish Colonial exploration,
Arizona & Mexico history, church history & related topics available on
premises only. Publications for sale.
Activities: guided tours; lectures; films. Museum Sponsors: craft demonstra-
tions.
Publications: brochures; booklet.
Hours & Admission Prices: Daily 9-5. Adults $3; children under 17, Golden
Age, Golden Eagle & annual park pass holders no charge. Closed
Thanksgiving; Christmas. &
Attendance: 60,000 (accurate)

Vail

**COLOSSAL CAVE MOUNTAIN PARK - LA POSTA QUE-
MADA RANCH MUSEUM,** 16721 E. Old Spanish Trail Rd.,
Vail, AZ 85641. Tel.: 520-647-7275.
E-mail: info@colossalcave.com
Web Site: www.colossalcave.com
Institution Type/Description: History Museum.
Collections: natural & cultural history; Native American artifacts; photo-
graphs; personal artifacts; period furnishings; cave history.
Hours & Admission Prices: March 16-Sept. 15 daily 8-5; Sept. 16-March 15
daily 9-5.

Valentine

KEEPERS OF THE WILD NATURE PARK, 13441 E. Hwy. 66,
Valentine, AZ 86437. Tel.: 928-769-1800. Fax: 928-769-1805.
E-mail: info@keepersofthewild.org
Web Site: www.keepersofthewild.org
Founded: 1995.
Key Personnel: Founder & Dir., Jonathan Kraft; Chm. (V), Sandy Jenkins.
Personnel Profile: Full-Time Paid 6; Part-Time Volunteers 25.
Governing Authority: Tax-exempt: 501(c)(3).
Institution Type/Description: Nature Park.
Collections: tigers; lions; cougars; reptiles; monkeys; birds; leopards; jaguars;
hoofstock.
Facilities: Park-related items for sale.
Activities: guided safari tours.
Publications: Sanctuary Scoop.
Hours & Admission Prices: Wed.-Mon. 9-5. Adults $18, seniors 65 & over
$15, children 12 & under $12; discounts to groups of 20 or more; children
under 2 no charge. Closed Thanksgiving; Christmas. &
Attendance: 10,000 (estimated)

Whiteriver

FORT APACHE HISTORIC PARK AND MUSEUM - KIN-ISHBA RUINS, Fort Apache Indian Reservation, Whiteriver, AZ 85941. Mailing Address: P.O. Box 507, Fort Apache, AZ 85926-0507. Tel.: 928-338-4625.
Institution Type/Description: Native American Museum: listed on the National Register of Historic Places.
Collections: Native American culture & history; personal artifacts; photographs; period buildings; prehistoric ruins; military cemetery.
Facilities: Museum-related items for sale.
Activities: tours; art demonstrations; special events.
Hours & Admission Prices: June-Aug. Mon.-Sat. 8-5; Winter: Mon.-Fri. 8-5. Adults $5, seniors 64 & over and students $3; children under 7 no charge. Closed major holidays.

Wickenburg

* **DESERT CABALLEROS WESTERN MUSEUM, (M),** 21 N. Frontier St., Ste. A, Wickenburg, AZ 85390-3431. Tel.: 928-684-2272, ext. 100. Fax: 928-684-5794.
E-mail: info@westernmuseum.org
Web Site: westernmuseum.org
Founded: 1960.
Congressional District: 3
Key Personnel: Exec. Dir., W. James Burns, Ph.D.; Chm. (V), W. John Daub; Museum Shop Mgr., Marilu Rix; Museum Technician, Paul Hughes.
Personnel Profile: Full-Time Paid 5; Part-Time Paid 3; Part-Time Volunteers 175.
Governing Authority: nonprofit organization. Branch Museum: Cultural Cross-roads Learning Center, Boyd Ranch, Wickenburg, AZ. Tax-exempt: 501(c)(3).
Institution Type/Description: General Museum.
Collections: western art; period rooms & dioramas; Native American collection; minerals; cowboy and ranch history; mining history.
Research Fields: history of Arizona.
Facilities: 2,000-vol. library of books on western history & art available on premises; reading room. Museum-related items & publications for sale.
Activities: permanent & temporary exhibitions; lectures; workshops; seminars; classes; trips.
Publications: books, The Town on the Hassayampa a History of Wickenburg, Arizona; The Cowboy's Dream: The Mythic Life and Art of Lon Megargee; annual exhibition catalogue, Cowgirl Up!
Hours & Admission Prices: June-Aug. Tues.-Sat. 10-5, Sun. 12-4; Sept.-May Mon.-Sat. 10-5, Sun. 12-4. Adults $7.50, senior citizens over 55 $6; discounts to AAM members; members & children 16 and under no charge. Closed major holidays.
Attendance: 74,000 (estimated)
Membership: Individual $50; Family $75; Contributor $150; Supporter $250; Sponsor $500; Patron $1,000; Founder $2,500; Board of Governors $5,000.

NATURE CONSERVANCY HASSAYAMPA RIVER PRE-SERVE VISITOR CENTER, 49614 N. U.S. Hwy. 60, Wickenburg, AZ 85390. Tel.: 928-684-2772.
E-mail: bmccollum@tnc.org
Institution Type/Description: Nature Preserve.
Collections: natural history; wildlife including over 280 species of birds.
Facilities: nature trails.
Activities: hiking; guided tours; special events; educational programs.
Hours & Admission Prices: mid-May to mid-Sept. Fri.-Sun. 7am-11am; mid-Sept. to mid-May Wed.-Sun. 8-5. Adults $5, members $3; children under 12 no charge. Closed New Year's Eve & Day; Thanksgiving; Christmas Eve & Day.

Willcox

CHIRICAHUA NATIONAL MONUMENT, 12856 E. Rhyolite Creek Rd., Willcox, AZ 85643-4722. Tel.: 520-824-3560. Fax: 520-824-3421.
Web Site: www.nps.gov/chir
Founded: 1924.
Congressional District: 5
Key Personnel: Cur., Julena Campbell; Park Ranger, Suzanne Moody.
Governing Authority: federal. National Park Service. Tax-exempt.
Institution Type/Description: Natural History Museum: located in the visitor center. Faraway Ranch & Stafford Cabin offer historical artifacts.
Collections: herbarium; entomology; geology; herpetology; historic furnishings & structures 1880s-1970s.
Activities: conducted programs; guided walks; hiking; camping; birding; picnic area.

Hours & Admission Prices: Daily 8-4:30. Adults 16 & over $5; children 15 & under no charge.
Attendance: 55,000 (accurate)

REX ALLEN ARIZONA COWBOY MUSEUM, 150 N. Railroad Ave., Willcox, AZ 85643-2132. Mailing Address: P.O. Box 142, Wilicox, AZ 85644. Tel.: 520-384-4583. Facebook: Rex Allen Arizona Cowboy Museum.
E-mail: info@rexallenmuseum
Web Site: www.rexallenmuseum.org
Founded: 1989.
Key Personnel: Chmn. (V), Terry Rowden; Pres. (V), Alfred Telles; Museum Shop Mgr., Gladys Olsen; Museum Shop Mgr., Phyllis Brooks.
Personnel Profile: Part-Time Volunteers 25.
Governing Authority: Tax-exempt.
Institution Type/Description: Western Movie & Cowboy Museum.
Collections: western movie memorabilia; cowboy artifacts; personal artifacts; local history.
Facilities: theater. Museum-related items for sale.
Activities: tours; occasional book signings.
Hours & Admission Prices: Mon.-Sat. 10-4. Admission $2; children under 10 & members no charge. Closed all major holidays.
Attendance: 2,400 (estimated)
Membership: Individual $15; Family $20; Business $50; Sponsor $100-$499; Benefactor $500 & up. International members add $5.

SULPHUR SPRINGS VALLEY HISTORICAL SOCIETY, 127 E. Maley St., Willcox, AZ 85643-2127. Tel.: 520-384-2105 & 3971.
Institution Type/Description: Historical Society Museum.
Collections: local history & culture; period furnishings; personal artifacts; photographs.
Hours & Admission Prices: Daily 1-4.

Williams

THE GRAND CANYON DEER FARM, 6769 E. Deer Farm Rd., Williams, AZ 86046-8419. Tel.: 928-635-4073. Fax: 928-635-2357.
E-mail: deerfrmr@aol.com
Web Site: www.deerfarm.com
Founded: 1969.
Key Personnel: Owner, Pat George; Owner, Randy George.
Personnel Profile: Full-Time Paid 4; Part-Time Paid 4.
Institution Type/Description: Zoo.
Collections: deer; peacocks; goats; llamas; monkeys; miniature donkeys; bison; coatimundi; wallabies; camel; zebu; marmoset.
Facilities: Museum-related items for sale.
Hours & Admission Prices: mid-March to mid-Oct. daily 9-6; mid-Oct. to mid-March daily 10-5. Adults $9.95, seniors 62 & over $8.50, children 3-13 $5.95; discounts to groups of 10 or more, military & AAA members; children 2 & under no charge. Closed Thanksgiving; Christmas.

PLANES OF FAME AIR MUSEUM - GRAND CANYON, 755 Mustang Way, Williams, AZ 86046. Tel.: 928-635-1000.
Governing Authority: Tax-exempt: 501(c)(3).
Institution Type/Description: Military Plane Museum.
Collections: over 40 military aircraft; military flight personnel history.
Activities: special events.
Hours & Admission Prices: Daily 9-5. Adults $5.95, children under 12 $1.95; discounts to AAA members; active military & children under 5 no charge. Closed Thanksgiving; Christmas.

Window Rock

NAVAJO NATION MUSEUM, Hwy. 264 & Post Office Loop Rd., Window Rock, AZ 86515. Mailing Address: P.O. Box 1840, Window Rock, AZ 86515-1840. Tel.: 928-871-7941. Fax: 928-871-7942.
E-mail: manuelito@navajonationmuseum.org
Web Site: www.navajonationmuseum.org
Founded: 1961.
Congressional District: 4
Key Personnel: Dir., Manuelito Wheeler; Cur. Collections, Clarenda Begay; ASO, Michelle Henry; Archivist, Eunice Kahn; Cultural Specialist, Robert Johnson; Museum Shop Mgr., Tracey Lynch.
Personnel Profile: Full-Time Paid 11.
Governing Authority: Navajo Nation. Parent Institution: Navajo Nation Division of Natural Resource. Tax-exempt: 170(b)(1).

Institution Type/Description: Native American Museum.
Collections: Dine (Navajo) history; culture; ethnographic; traditional & contemporary fine arts; archive collection that documents the Dine richness in cultural life, history & the development of Dine bi Keyah (Navajoland).
Research Fields: Navajo history & culture of the southwest region.
Facilities: library; rental facilities; 165-seat indoor auditorium; amphitheater. Museum-related items for sale.
Activities: permanent, temporary & changing exhibitions; guided tours; presentations; educational & outreach programs; films; concerts; art festivals. Annual Events: Rock the Rocks (early summer); Nizhoni Art Market in September; Keshmish Festival in December; Navajo Moccasin Game in December.
Publications: brochures; occasional exhibit catalogs; books, Through White Men's Eyes; A Contribution to Navajo History Vol. 1-6, 1979: Navajo Nation 1950; Traditional Life in Photographs, 2006
Hours & Admission Prices: Mon. 8-5, Tues.-Fri. 8-7, Sat. 9-5. Collections & archives by appointment only. No charge; donations accepted. Closed national & tribal holidays. &
Attendance: 106,445 (accurate)

NAVAJO NATION ZOO & BOTANICAL PARK, Hwy. 264, Bldg. 36A, Window Rock, AZ 86515. Mailing Address: P.O. Box 1480, Window Rock, AZ 86515-1480. Tel.: 928-871-6574.
Institution Type/Description: Zoo & Botanical Park.
Collections: over 30 species of wild animals; trees; plants; flowers.
Activities: special events.
Hours & Admission Prices: Mon.-Sat. 10-5. No charge. Closed New Year's Day; Christmas.

Winslow

HOMOLOVI STATE PARK, 87 North, Winslow, AZ 86047-9402. Mailing Address: HCR 63 Box #5 (SR87N), Winslow, AZ 86047-9402. Tel.: 928-289-4106. Fax: 928-289-2021.
E-mail: kke2@azstateparks.gov
Web Site: www.azstateparks.com
Formerly: Homolovi Ruins State Park
Founded: 1986.
Congressional District: 4
Key Personnel: Park Ranger, Kenn Evans, III
Personnel Profile: Full-Time Paid 3; Part-Time Paid 1; Part-Time Volunteers 20.
Governing Authority: state government. Parent Institution: Arizona State Parks. Tax-exempt: 501(c)(3).
Institution Type/Description: State Park.
Collections: prehistoric artifacts from Homolovi sites & associated areas; natural history collections documenting natural resources of Homolovi State Park; Hopi art.
Facilities: visitor center; campground; nature trails. Book store & Hopi art for sale.
Activities: self-guiding trails; programs & workshops: Native American culture & concerns.
Publications: Homolovi: A Cultural Crossroads; personal notes on Hopi gardening.
Hours & Admission Prices: Park, Visitor Center & Ruins: daily 8-5. Admission $7 4 adults per vehicle, each additional person $3. Campground: Summer & Fall $18-$25, Spring & Winter $15-$20. Closed Christmas. &
Attendance: 26,238 (accurate)

METEOR CRATER DISCOVERY CENTER, Interstate 40, Exit 233, Meteor Crater Rd. 5.5 miles S., Winslow, AZ 86047. Mailing Address: P.O. Box 30940, Flagstaff, AZ 86003-0940. Tel.: 928-289-5898. Fax: 928-289-2598.
E-mail: info@meteorcrater.com
Web Site: www.meteorcrater.com
Formerly: Museum of Astrogeology, Meteor Crater
Founded: 1955.
Key Personnel: Pres., Brad Andes.
Personnel Profile: Full-Time Paid 34; Part-Time Paid 14.
Governing Authority: profit-making organization. Affiliated with Meteor Crater Enterprises.
Institution Type/Description: Astrogeology Museum.
Collections: space capsule: Astronaut Wall of Fame, commemorating all U.S. Astronauts & all U.S. manned space flights. Interactive exhibits: Create a Crater, Shoemaker/Levy 9 Comet Impacts on Jupiter, Earth at Risk, Impact Earth, Ground Zero. Other exhibits: Meteorites, rocks, gems, impact craters throughout our solar system, planetary & astrogeology impact crater mechanics.
Facilities: library of books on meteors & astrogeology. Books for sale.

Activities: lectures; films; audio-visual presentations; formally organized education programs for children, adults & undergraduate college students; permanent exhibitions; weather permitting, daily guided rim tours; interactive computer.
Publications: Meteor Crater Story.
Hours & Admission Prices: Winter: daily 8-5; Summer: daily 7-7. Adults $16, senior citizens 60 & over $15, juniors 6-17 $8; discounts to military & groups; children 5 & under no charge. Closed Christmas. &
Attendance: 213,600 (accurate)
Membership: Juniors $9; Seniors $20; Adults $23.

OLD TRAILS MUSEUM, 212 Kinsley Ave., Winslow, AZ 86047-3618. Tel.: 928-289-5861. Facebook: Old Trails Museum.
E-mail: info@oldtrailsmuseum.org
Web Site: www.oldtrailsmuseum.org
Founded: 1989.
Congressional District: 1
Key Personnel: Dir. & Museum Shop Mgr., Ann-Mary J. Lutzick; Pres. (V), Patricia Raygor.
Personnel Profile: Part-Time Paid 1.
Volunteer Hours: 2,200
Operating Expenses: 54,000
Operating Income: 54,000
Governing Authority: Parent Institution: Winslow Historical Society. Tax-exempt: 501(c)(3).
Institution Type/Description: Historical Society Museum: housed in the 1921 First National Bank Building.
Collections: Route 66; Santa Fe Railroad; Harvey Girls; La Posada Hotel; trading posts; ranching; early settlement; Native American artifacts; local history & culture.
Major Exhibits: History of the Winslow Airport, 7/14.
Research Fields: local & regional history.
Activities: research; permanent & temporary exhibitions; public programs.
Publications: biannual newsletter; Arcadia Images of America - Winslow; annual historical calendar.
Hours & Admission Prices: Call for hours; groups by appointment. No charge; donations accepted. Closed National holidays.
Attendance: 3,700 (accurate)
Membership: Student $5; Individual $15; Family & Group $30; Patron $100; Benefactor $250; Lifetime $500.

TINA MION MUSEUM AT LA POSADA HOTEL, 303 E. 2nd St., 2nd Fl., Winslow, AZ 86047. Tel.: 928-289-4366. Fax: 928-289-3873.
E-mail: tina@tinamion.com
Web Site: www.tinamion.com
Founded: 2011.
Institution Type/Description: Art Gallery.
Collections: paintings.
Facilities: 3,000 sq. ft. exhibition space.
Hours & Admission Prices: Daily 8-8.

Yuma

ARIZONA HISTORICAL SOCIETY SANGUINETTI HOUSE MUSEUM AND GARDEN, 240 S. Madison Ave., Yuma, AZ 85364-1421. Mailing Address: 949 E. 2nd St., Tucson, AZ 85719-4898. Tel.: 928-782-1841. Fax: 928-783-0680.
E-mail: ahsyuma@azhs.gov
Web Site: www.arizonahistoricalsociety.org
Founded: 1864.
Congressional District: 3
Key Personnel: Exec. Dir., Anne I. Woosley, Ph.D.; Rio Colorado Division Dir., Susan Taylor.
Personnel Profile: Full-Time Paid 2; Part-Time Paid 1; Part-Time Volunteers 51.
Governing Authority: bd. of directors. Parent Institution: Arizona Historical Society, 949 E. 2nd St., Tucson, AZ 85719. Tel.: 520-628-5774. Tax-exempt: 501(c)(3).
Institution Type/Description: History Museum: housed in 1870s residences-Sanguinetti House Museum & Adobe Annex.
Collections: Colorado River area history including photographs, maps, documents, & artifacts; period furnishings; aviaries; historical gardens.
Research Fields: steamboats; railroad; biographical data; Southwestern Arizona business, domestic, & social history.
Facilities: photographic & manuscript archives. Museum-related gifts for sale.
Activities: extension programs include desert tours, slide programs, history talks, & docent program; facility rental.
Publications: quarterly, The Journal of Arizona History; books; monographs; annual report.

Hours & Admission Prices: Tues.-Sat. 10-4. Adults $3, senior citizens 60 & over and youth 12-18 $2; children 11 & under and members no charge. Closed New Year's Day; Independence Day; Veterans Day; Thanksgiving; Christmas.
Attendance: 5,000 (estimated)
Membership: Student $25; Individual $50; Household $65; Sustaining $100; Patron $250; Sponsor $500; Director's Circle $1,000.

CASTLE DOME MINE MUSEUM, 27550 E. County 15th St. N. Sr4, Yuma, AZ 85365. Tel.: 928-920-3062.
Web Site: www.castledomemuseum.com
Founded: 1998.
Institution Type/Description: Mine Museum: housed on the site of a former mine town built in 1878.
Collections: mining history & artifacts; period furnishings; personal artifacts. Historic Buildings: stamp mill; church; five saloons; blacksmith shop; machine shop; mercantile; doctor & dentist office; post office; land office; graveyard.
Hours & Admission Prices: Nov.-April daily 10-5. Museum or Trail: adults $6, children 6-12 $3; children under 6 no charge. Museum & Trail: adults $10, children 6-12 $5; children under 6 no charge.

KOFA NATIONAL WILDLIFE REFUGE, 9300 E. 28th St., Yuma, AZ 85365. Tel.: 928-783-7861.
Institution Type/Description: Wildlife Refuge.
Collections: wildlife including white-winged dove, desert tortoise, & desert kit fox; sheep; birds; plants.
Facilities: nature trails.
Activities: hiking.
Hours & Admission Prices: Call for hours.

SAIHATI CAMEL FARM, 15672 S. Avenue 1E, Yuma, AZ 85365. Tel.: 928-627-7511.
Institution Type/Description: Wildlife Farm.
Collections: camels; miniature donkeys; cattle; deer; sheep; pheasants; quail; horses.
Activities: special events.
Hours & Admission Prices: Oct.-May Mon.-Sat. 9-5. Adults $4, seniors $3; children 3 & under no charge.

YUMA FINE ARTS ASSOCIATION, 254 S. Main, Yuma, AZ 85364-1425. Tel.: 928-329-6607. Fax: 928-329-6616.
E-mail: director@yumafinearts.org
Web Site: www.yumafinearts.org
Founded: 1962.
Congressional District: 3
Key Personnel: Pres., Teri Ingram; Exec. Dir., Carolyn J. Bennett.
Personnel Profile: Full-Time Paid 2; Part-Time Volunteers 6; Interns 1.
Governing Authority: nonprofit organization. Subsidiary Institution: Yuma Art Center. Tax-exempt: 501(c)(3) & 170(b)(1)(A).
Institution Type/Description: Visual Arts Center & Museum.
Collections: contemporary & historic Southwest paintings; ceramics; sculpture; photographs; graphics; fiber; crafts.
Research Fields: contemporary art; contemporary & historical Southwest art.
Facilities: four galleries; performance space; local artists shop.
Activities: gallery tours; guest lecturers; art festivals; concerts; temporary & traveling exhibitions.
Publications: quarterly members publications.
Hours & Admission Prices: Please call for hours. No charge; donations accepted. ♿
Attendance: 50,000 (estimated)
Membership: Senior Citizen & Student $40; Supporting $50; Individual $60; Family & Donor $100; Sponsor & Business Sponsor $250; Patron & Business Patron $500; Director's Circle & Business Director's Circle $1,000; Silver Circle & Business Silver Circle $2,500; Golden Circle & Business Golden Circle $5,000.

YUMA QUARTERMASTER DEPOT STATE HISTORIC PARK, 201 N. 4th Ave., Yuma, AZ 85364-2336. Tel.: 928-783-0071. Fax: 928-783-1897. Facebook: Yuma QMD.
Web Site: www.azstateparks.com
Founded: 1865.
Congressional District: 3
Key Personnel: Park Mgr., Tammy Snook.
Personnel Profile: Full-Time Paid 1; Part-Time Volunteers 7.
Governing Authority: Parent Institution: Arizona State Parks Board, 1300 W. Washington, Phoenix, AZ 85007. Located within the Yuma Crossing National Heritage Area Corp., 180 W. 1st St., Ste. E Yuma, AZ 85364.
Institution Type/Description: History Museum: housed in five original buildings of the Yuma Quartermaster Depot, 1865-1883; two were later used by the Bureau of Reclamation as their Yuma Project Headquarters.
Collections: period furniture; Native American artifacts; military uniforms; maps; wagons; steamboat artifacts; Model T car; Plank Road exhibit; early western development of the southwest; early military presence in the southwest; early irrigation history of Yuma; photographs chronicling the construction of irrigation works.
Research Fields: history of Fort Yuma Quartermaster Depot.
Facilities: interpretive center.
Activities: guided tours; permanent exhibits; outreach slide programs & video.
Hours & Admission Prices: June-Sept. Tues.-Sun. 9-4:30; Oct.-May daily 9-4:30; day before Thanksgiving & Christmas Eve 9-2. Adult 14 & up $2, children 7-13 $1; children 6 & under no charge. Closed Thanksgiving Day & Christmas Day. ♿
Attendance: 23,000

YUMA TERRITORIAL PRISON STATE HISTORIC PARK, One Prison Hill Rd., Yuma, AZ 85364-8792. Tel.: 928-783-4771.
E-mail: mike.guertin@yumaaz.gov
Web Site: savetheprison.com
Founded: 1940.
Congressional District: 3
Key Personnel: Park Mgr., Mike Guertin.
Personnel Profile: Full-Time Paid 3; Part-Time Paid 3; Part-Time Volunteers 10.
Governing Authority: state. Parent Institution: Arizona State Parks Board, 1300 W. Washington, Phoenix, AZ 85007.
Institution Type/Description: General Museum.
Collections: general history; photographic; domestic artifacts; geological; numismatic; photos & artifacts related to Arizona Territorial Prison (1876-1909); cemetery; period prison cells.
Research Fields: history of Arizona Territorial Prison.
Facilities: interpretive center; picnic area; hiking trails.
Activities: guided tours; permanent exhibits; outreach slide programs & video.
Publications: The Prison Chronicle; Prisoner's in Petticoats.
Hours & Admission Prices: Daily 9-5, Adults 14 & over $6, active military $3, children 7-13 $3; children 6 & under no charge. Closed Thanksgiving; Christmas. ♿
Attendance: 80,000 (accurate)

ARKANSAS

(204 listings)

Altus

ALTUS HERITAGE HOUSE MUSEUM, 106 N. Franklin, Altus, AR 72821. Mailing Address: P.O. Box 197, Altus, AR 72821. Tel.: 479-468-1310.
Institution Type/Description: Historic Building: housed in the former German-American State Bank; c.1800. Listed on the National Register of Historic Places.
Collections: local history & culture; period furnishings; early coal mining equipment.
Hours & Admission Prices: Call for hours.

Arkadelphia

CLARK COUNTY HISTORICAL MUSEUM, 750 S. 5th St., #1, Arkadelphia, AR 71923-6237. Mailing Address: P.O. Box 516, Arkadelphia, AR 71923-0516. Tel.: 870-230-1360.
Institution Type/Description: History Museum.
Collections: local history & culture; photographs; period furnishings.
Hours & Admission Prices: Call for hours.

OUACHITA BAPTIST UNIVERSITY HAMMONS GALLERY, 410 Ouachita St., Arkadelphia, AR 71998. Tel.: 870-245-5129.
Institution Type/Description: Art Gallery.
Collections: works by university students including paintings & sculptures.
Activities: temporary exhibits.
Hours & Admission Prices: Mon.-Fri. 9-5.

REYNOLDS SCIENCE CENTER PLANETARIUM, 514 N. 12th St., Arkadelphia, AR 71998. Tel.: 870-230-5006.
Institution Type/Description: Planetarium.
Collections: astronomy-related exhibits.
Activities: classes; children's programs; tours.

Hours & Admission Prices: Call for hours. Adults $3, students $1.

Ashdown

GN & A DEPOT, 180 E. Whitaker, Ashdown, AR 71822-2724. Mailing Address: Chamber of Commerce, P.O. Box 160, Ashdown, AR 71822. Tel.: 870-898-2758.
E-mail: director@littlerivercounty.org
Web Site: www.littlerivercounty.org
Governing Authority: Parent Institution: Little River County Chamber of Commerce.
Institution Type/Description: Historic Building: listed on the National Register of Historic Places.
Collections: local history & culture; railroad artifacts; photographs; period furnishings; Drovers caboose.
Hours & Admission Prices: Call for hours.

HUNTER-COULTER MUSEUM, 310 N. Second St., Ashdown, AR 71822. Mailing Address: c/o Little River County Historical Society, 2425 Wren St., Ashdown, AR 71822. Tel.: 870-898-5200.
Governing Authority: Parent Institution: Little River County Historical Society.
Institution Type/Description: Historic House Museum: built in 1918 by Henry Westbrook. Listed on the National Register of Historic Places.
Collections: local history & culture; period furnishings; early farm tools; education memorabilia; photographs.
Activities: special events. Museum Sponsors: Candlelight Dinner in December.
Hours & Admission Prices: Mon. 12-3.

LITTLE RIVER COUNTY COURTHOUSE, 310 N. Second St., Ashdown, AR 71822. Mailing Address: P.O. Box 160, Ashdown, AR 71822-0160. Tel.: 870-898-5528.
Institution Type/Description: Historic Building: built in 1907. Listed on the National Register of Historic Places.
Collections: local history & culture; photographs; period furnishings.
Hours & Admission Prices: Call for hours.

TWO RIVERS MUSEUM, 15 E. Main St., Ashdown, AR 71822-2825. Mailing Address: 5 E. Main St., Ashdown, AR 71822-2825. Tel.: 870-898-3147.
Governing Authority: Parent Institution: Little River County Historical Society.
Institution Type/Description: History Museum.
Collections: local history & culture; photographs; period furnishings; personal artifacts.
Activities: Museum Sponsors: Two Rivers Gala.
Hours & Admission Prices: By appointment.

Batesville

LYON COLLEGE KRESGE GALLERY, Highland & 22nd Sts., Batesville, AR 72503. Tel.: 870-307-7242.
Institution Type/Description: Art Gallery.
Collections: painting; photographs; sculpture.
Hours & Admission Prices: Call for hours.

MARK MARTIN MUSEUM, 1601 Batesville Blvd., Batesville, AR 72501-8372. Mailing Address: P.O. Box 2677, Batesville, AR 72503. Tel.: 870-793-4461.
Founded: 2004.
Institution Type/Description: Racing Museum.
Collections: Mark Martin's cars & trophies; photographs; helmets; firesuits; personal artifacts.
Hours & Admission Prices: Call for hours. No charge.

OLD INDEPENDENCE REGIONAL MUSEUM, 380 S. Ninth St., Batesville, AR 72501-5703. Tel.: 870-793-2121. Fax: 870-793-2101.
E-mail: oirm@oirm.org
Web Site: www.oirm.org
Founded: 1998.
Congressional District: 1
Key Personnel: Bd. Pres., MaryAnn Marshall; Cur., Twyla Wright; Museum Shop Mgr., Claudia Nobles.
Personnel Profile: Full-Time Paid 1; Full-Time Volunteers 1; Part-Time Paid 2; Part-Time Volunteers 50; Interns 1.
Governing Authority: private; nonprofit organization. Tax-exempt: 501(c)(3).

Institution Type/Description: History Museum: housed in former National Guard Armory built as a WPA project in 1936 of local sandstone.
Collections: 200 years of North Central Arkansas artifacts; furnishings; photographs; structures.
Research Fields: regional history of North Central Arkansas.
Facilities: library; 70-seat auditorium; 10,000 sq. ft. exhibit space. Gift items for sale.
Activities: homeschool days; scouting programs; family day program; day camp program; guided tours; specialized tours & workshops; monthly educational programs & speakers; docent program.
Publications: semi-annual newsletter, Old Independence Regional Museum; annual report.
Hours & Admission Prices: Tues.-Sat. 9-4:30, Sun. 1:30-4. Adults $3, senior citizens $2, students 6-12 $1; discounts for groups of 10 or more; members & children under 6 no charge. Closed New Year's Day; Easter; Independence Day; Thanksgiving; Christmas. &
Attendance: 6,000 (estimated)
Membership: Student & Senior Citizen $20; Individual $30; Family $50; Friend $100; Patron & Day Sponsor $250; Sustainer $500; Benefactor $1,000; Society $1,750.

Bauxite

BAUXITE MUSEUM, 6706 Benton Rd., Bauxite, AR 72011-9124. Mailing Address: P.O. Box 245, Bauxite, AR 72011-0245. Tel.: 501-557-9858.
Governing Authority: Parent Institution: Bauxite Historical Association and Museum. Tax-exempt: 501(c)(3).
Institution Type/Description: History Museum.
Collections: local history & culture; period furnishings; photographs; personal artifacts.
Hours & Admission Prices: Wed. 10-2, Sun. 1:30-4; other times by appointment. No charge.

Bella Vista

BELLA VISTA HISTORICAL MUSEUM, 1885 Bella Vista Way, Bella Vista, AR 72714-3810. Tel.: 479-855-2335.
E-mail: caroleharter@gmail.com
Founded: 1976.
Congressional District: 3
Key Personnel: Pres. (V), Carole Harter.
Personnel Profile: Part-Time Volunteers 15.
Governing Authority: nonprofit. Parent Institution: Bella Vista Historical Society. Tax-exempt.
Institution Type/Description: History Museum.
Collections: materials & artifacts pertaining to the village of Bella Vista.
Facilities: library of history books & genealogy of Bella Vista; meeting room. Museum-related books for sale.
Activities: guided tours.
Publications: newsletter.
Hours & Admission Prices: March-Dec. Wed.-Sun. 12-4; other times by appointment. No charge; donations accepted. &
Attendance: 600 (estimated)
Membership: Family $10; Contributing $25; Sustaining $50; Sponsor $100; Benefactor $250; Life $500; Founding $1,000.

Benton

GANN MUSEUM OF SALINE COUNTY, 218 S. Market St., Benton, AR 72015-4304. Tel.: 501-778-5513 & 860-4113.
E-mail: thegannmuseum@live.com
Founded: 1980.
Congressional District: 2
Key Personnel: Chm. (V), Dorcas Holicer; Dir., Elton Fitzhugh
Governing Authority: Parent Institution: The Gann Museum of Saline County, Inc. Tax-exempt.
Institution Type/Description: Historic House: housed in the former medical office of Dr. Dewel Gann, Sr., built in 1896 using bauxite aluminum ore.
Collections: memorabilia; photographs; Niloak pottery; period furniture & artifacts.
Publications: Gann Legacy Newsletter, quarterly.
Hours & Admission Prices: Memorial Day to Labor Day Tues.-Thurs. 10-4; Sept.-May Tues.-Thurs. 9-3; other times by appointment. No charges, donations accepted.
Attendance: 761 (estimated)
Membership: Senior & Single $15; Family $25; Contributors $50; Jerrene Holiman $100; Patron & Business $150; Sustaining $250; Benefactor $500 & up.

Bentonville

COMPTON GARDENS INTERPRETIVE CENTER, 312 N. Main St., Bentonville, AR 72712. Tel.: 479-254-3870. Fax: 479-254-3871. Facebook: Peel Compton.
E-mail: officemanager@peelcompton.net
Key Personnel: Dir. Operations, Corrin Troutman
Institution Type/Description: History Museum.
Collections: Dr. Compton's life, family, books & personal artifacts; photographs; gardens.
Facilities: nature trails; gardens. Exhibit room of Dr. Neil Compton.
Activities: walking trails; guided tours upon request.
Hours & Admission Prices: Center: Mon.-Fri. 9:30-3:30. Gardens: dawn to dusk. No charge. &

Attendance: 4,000 (estimated)
Membership: $35-$1,000

CRYSTAL BRIDGES MUSEUM OF AMERICAN ART, (M), 600 Museum Way, Bentonville, AR 72712-4947. Tel.: 479-418-5700. Fax: 479-418-5701.
E-mail: info@crystalbridges.org
Web Site: www.crystalbridges.org
Key Personnel: Dir., Rod Bigelow
Institution Type/Description: Art Museum.
Collections: paintings; sculpture; works on paper; books; outdoor sculpture.
Facilities: library; nature trails; cafe. Museum-related items for sale.
Activities: special events; permanent & temporary exhibitions; educational programs.
Hours & Admission Prices: Wed. & Fri. 11-9, Thurs. & Sat.-Mon. 11-6. Closed Thanksgiving; Christmas.
Attendance: 600,000 (accurate)

MUSEUM OF NATIVE AMERICAN HISTORY, (M), 202 S.W. O St., Bentonville, AR 72712-3641. Tel.: 479-273-2456. Fax: 479-268-6909. Facebook: Museum of Native American History.
E-mail: info@monah.us
Web Site: www.monah.us
Institution Type/Description: Native American Museum.
Collections: Native American history & culture; personal artifacts; sculpture; paintings; tools; weapons; pottery.
Facilities: Museum-related items for sale.
Hours & Admission Prices: Mon.-Sat. 9-5. No charge; donations accepted. &

PEEL MANSION MUSEUM & HISTORIC GARDENS, 312 N. Main, Bentonville, AR 72712. Tel.: 479-254-3870. Fax: 479-254-3871. Facebook: Peel Compton.
E-mail: officemanager@peelcompton.net
Web Site: www.peelcompton.org/peel/index.htm
Founded: 1992.
Congressional District: 3
Key Personnel: Dir. Operations, Corrin Troutman; Chm. (V), Lynne Walton.
Personnel Profile: Full-Time Paid 10; Part-Time Paid 3; Part-Time Volunteers 50; Interns 1.
Governing Authority: private; nonprofit organization. Tax-exempt: 501(c)(3).
Institution Type/Description: Historic House Museum: housed in 1875 Italianate villa & 1850 log cabin.
Collections: 19th-century furnishings; photographic collection of historic buildings in the county; native Ozark habitat & artifacts; gardens.
Activities: docent program; guided tours; lectures; rental gallery; temporary exhibitions; living history events; children's educational history programs; 3 day living station for 4th graders.
Hours & Admission Prices: Gardens: Mon.-Fri. sunrise to sunset. Mansion: Tues.-Sat. 10-3. Mansion: adults $5, children $2. Gardens: no charge. Closed New Year's Eve & Day; Memorial Day; Independence Day; Labor Day; Christmas Eve, Day & week. &
Attendance: 4,000 (estimated)
Membership: $35-$1,000.

21C MUSEUM - BENTONVILLE, 200 N.E. A St., Bentonville, AR 72712. Tel.: 479-286-6500.
Institution Type/Description: Art Gallery.
Collections: 21st century art by emerging & established artists.
Activities: educational programs; poetry readings; films; lectures; performances.
Hours & Admission Prices: Daily 24 hours. Video Installations: daily 7am to 1am. No charge.

WAL-MART VISITOR'S CENTER, 105 N. Main St., Bentonville, AR 72712-5341. Tel.: 479-273-1329. Facebook: Walmart Museum.
Web Site: corporate.walmart.com/our-story/heritage/visitor-center
Key Personnel: Sr. Dir. Heritage & Assoc. Mktg., Alan Dranow
Institution Type/Description: History Museum: birthplace of Wal-Mart.
Collections: history of Wal-Mart; period store artifacts; employee manuals; Sam's general headquarters office; Sam's 1979 red Ford F150; personal artifacts.
Facilities: Museum-related items for sale.
Hours & Admission Prices: Mon.-Thurs. 8 a.m-9 p.m., Fri.-Sat. 8 a.m.-10 p.m., Sun. 12-5. No charge. &
Attendance: 50,000

Berryville

HERITAGE CENTER MUSEUM, Public Square, Berryville, AR 72616. Mailing Address: P.O. Box 249, Berryville, AR 72616-0249. Tel.: 870-423-6312.
E-mail: history1880@windstream.net
Web Site: www.rootsweb.com/~arcchs
Founded: 1955.
Congressional District: 3
Key Personnel: Pres., Gordon Hale.
Personnel Profile: Part-Time Paid 2; Part-Time Volunteers 7.
Governing Authority: society. Parent Institution: Carroll County Historical Society. Tax-exempt.
Institution Type/Description: Local History Museum: housed in 1880 Carroll County Court House.
Collections: furniture; Connor post office; funeral parlor; school room; photographs; fainting couch; Civil War artifacts; local history memorabilia; miniature railroad; clock collection; pioneer printing office; orphan train display; moonshine still; Mountain Meadows Massacre display. Historic Structures: c.1800 Newberry log cabin; c.1905 log jail cell.
Research Fields: county history; genealogy of county families; Mountain Meadows Massacre exhibit.
Facilities: 1,000-vol. library of books, records & microfilm pertaining to local history available for research on premises.
Activities: guided tours; lectures; loan & permanent exhibitions.
Publications: Carroll County Historical Society Quarterly.
Hours & Admission Prices: Mon.-Fri. 9-4. Family $5, adults $2, children $1; children under 6 no charge. Closed national holidays.
Attendance: 3,000 (estimated)
Membership: Individual $20.

SAUNDERS MEMORIAL MUSEUM, 113-15 E. Madison St., Berryville, AR 72616-3954. Mailing Address: P.O. Box 227, Berryville, AR 72616. Tel.: 870-423-2563.
Web Site: www.berryville.com
Founded: 1955.
Congressional District: 3
Key Personnel: Mgr., Rose M. Garrett; Chm. (V), Don Rustuhaltz.
Personnel Profile: Full-Time Paid 1; Part-Time Paid 2.
Governing Authority: municipal. Parent Institution: City of Berryville. Tax-exempt.
Institution Type/Description: Gun & Period Furnishings Museum.
Collections: Colonel C. Burton Saunders' gun collection; silver; china; arts & crafts; Indian artifacts; teakwood furniture; oriental rugs.
Research Fields: guns; teakwood; oriental rugs; paintings; china; silver; American Indian artifacts.
Facilities: Gift items for sale.
Publications: brochures.
Hours & Admission Prices: April 15 to 1st weekend in Nov. Mon.-Sat. 10:30-5. Adults $5, children under 13 $2.50; discount to groups of 10 & over; children under 6 no charge. Closed holidays. &
Attendance: 3,000 (estimated)

Blytheville

BLYTHEVILLE HERITAGE MUSEUM, 107C Main St., Blytheville, AR 72315-3431. Mailing Address: P.O. Box 234, Blytheville, AR 72316-0234. Tel.: 870-763-2525.
Institution Type/Description: History Museum: housed in the former S. H. Kresse Department Store building. Listed on the National Register of Historic Places.
Collections: Blytheville's cotton & logging industries; local history & culture; photographs; pioneer artifacts.
Hours & Admission Prices: Call for hours.

DELTA GATEWAY MUSEUM, (M), 210 W. Main St., Blytheville, AR 72315. Tel.: 870-824-2346. Fax: 870-824-2347. Facebook: Delta Gateway Museum.
E-mail: lhester43@yahoo.com
Founded: 2009.
Congressional District: 1
Key Personnel: Dir., Leslie Hester.
Governing Authority: Tax-exempt.
Institution Type/Description: History Museum: housed in the historic Kress Building.
Collections: local history & culture; period furnishings; photographs; personal artifacts.
Activities: special events; temporary exhibitions.
Hours & Admission Prices: Wed.-Fri. 1-4, Sat. 9-3. No charge; donations accepted. Closed federal holidays. &
Attendance: 1,000 (estimated)

Brinkley

CENTRAL DELTA DEPOT MUSEUM, 100 W. Cypress, Brinkley, AR 72021-2809. Tel.: 870-589-2124.
Founded: 2003.
Key Personnel: Dir., Bill Sayger; Chm. (V), Laura Bussell; Pres. (V), Catherine Jacques.
Personnel Profile: Full-Time Volunteers 1; Part-Time Paid 1; Part-Time Volunteers 1.
Governing Authority: Parent Institution: Central Delta Historical Society. Tax-exempt.
Institution Type/Description: History Museum.
Collections: Central Delta history & culture; Louisiana Purchase; railroad history; photographs; Louis Jordan memorabilia; Southern Pacific caboose. Historic Buildings: train depot; sharecropper's house.
Activities: Annual Event: Choo Choo Ch'Boogie Delta Music Festival.
Publications: quarterly members' newsletter & journal, The Doodlebug.
Hours & Admission Prices: Mon.-Sat. 9-5, Sun. 1-4. Adults $2, children $1; members no charge. &
Attendance: 1,150 (accurate)
Membership: Individual $15; Married Couples $25.

Calico Rock

CALICO ROCK MUSEUM & VISITOR CENTER, 104 Main St., Calico Rock, AR 72519. Tel.: 870-297-4129.
Key Personnel: Exec. Dir., Gloria Gushue
Institution Type/Description: Art & History Museum: housed in the E.N. Rand Building; built in 1903.
Collections: local history & culture; paintings; sculpture; pottery; photographs.
Activities: special events.
Hours & Admission Prices: Tues.-Sat. 9-5.

Camden

MCCOLLUM-CHIDESTER HOUSE, 926 W. Washington St., Camden, AR 71701-3382. Tel.: 870-836-9243.
E-mail: ochs2003@sbcglobal.net
Web Site: ouachitacountyhistoricalsociety.org
Founded: 1963.
Key Personnel: Pres. (V), Clara Freeland; Docent, Hubert Boddie.
Personnel Profile: Part-Time Paid 1; Part-Time Volunteers 20.
Governing Authority: Parent Institution: Ouachita County Historical Society.
Institution Type/Description: Historic House Museum: built in 1847.
Collections: family history; period furnishings; personal artifacts; Civil War; photographs.
Publications: magazine, Ouachita County Historical Quarterly.
Hours & Admission Prices: Wed.-Sat. 9-4. Adults $5. Closed major holidays. &
Attendance: 1,200 (estimated)
Membership: Annual $25.

OUACHITA COUNTY HISTORICAL SOCIETY, 926 Washington St., N.W., Camden, AR 71701-3382. Tel.: 870-836-9243, 3610 & 0245.
E-mail: ochs2003@sbcglobal.net
Web Site: ouachitacountyhistoricalsociety.org
Founded: 1963.
Congressional District: 4
Key Personnel: Pres., Clara Freeland; Museum Docent, Hubert Boddie.
Personnel Profile: Part-Time Paid 1; Part-Time Volunteers 20.
Institution Type/Description: Historical Society Museum.

Collections: Victorian furniture, 1863 rare china; glassware, silver; clothing; jewelry; lamps; pictures & frames; law office; Freedmen's Bureau. Historic Buildings: 1847 historic home; 1850 Leake-Ingham library.
Activities: guided tours. Museum Sponsors: Daffodil Festival in March; All Hallows Eve Cemetery Walk in October.
Publications: historical quarterly, Quachita County.
Hours & Admission Prices: Wed.-Sat. 9-4. Adults $5, students $2. Closed major holidays. &
Attendance: 1,200 (estimated)
Membership: Individual $25.

Charleston

BELLE MUSEUM & PRESBYTERIAN CHAPEL, 322 E. Main, Charleston, AR 72933. Mailing Address: P.O. Box 261, Charleston, AR 72933-0261.
Founded: 1996.
Key Personnel: Dir., Delbert Ervin; Dir., Mary B. Ervin
Institution Type/Description: History & Church Museum.
Collections: local, state & family histories; personal artifacts.
Activities: special events.
Hours & Admission Prices: May-June call for hours. No charge; donations accepted. &

Clinton

VAN BUREN COUNTY HISTORICAL MUSEUM, 211 3rd St., Clinton, AR 72031. Mailing Address: P.O. Box 1023, Clinton, AR 72031-1023. Tel.: 501-745-4066.
Key Personnel: Chm. (V), Carol Hutto; Chm. (V), Charlotte West; Pres. (V), Dortha Borecky.
Personnel Profile: Part-Time Volunteers 9.
Governing Authority: Tax-exempt.
Institution Type/Description: Historical Museum.
Collections: family histories; photographs; personal artifacts.
Research Fields: genealogy.
Facilities: library.
Publications: quarterly newsletter; historical calendars.
Hours & Admission Prices: Mon.-Thurs. 10-3. No charge; donations accepted. Closed holidays. &
Attendance: 500 (estimated)
Membership: Individual $15.

Conway

BAUM GALLERY OF FINE ART, UNIVERSITY OF CENTRAL ARKANSAS, (M), 201 Donaghey Ave., Conway, AR 72035-5001. Mailing Address: McAlister 101, Dept. of Art, UCA, 201 Donaghey Ave., Conway, AR 72035. Tel.: 501-450-5793 & 5000. Fax: 501-450-3670.
E-mail: barclaym@uca.edu
Web Site: www.uca.edu/art/baum
Founded: 1994.
Congressional District: 2
Key Personnel: Dir. & Cur., Barclay McConnell.
Personnel Profile: Full-Time Paid 1; Part-Time Paid 6; Part-Time Volunteers 20; Interns 2.
Governing Authority: public university; nonprofit. Parent Institution: University of Central Arkansas.
Institution Type/Description: Art Gallery.
Collections: contemporary American & European, all media, artists represented include Marshall Arisman, Warren Criswell, Don Netzer, Barry Moser, Raphael Soyer, Larry Zox, Julian Stanczak, Mel Ramos & Jiri Anderle.
Major Exhibits: Deb Schwedhelm: Whispers from The Sea (T), 1/14-2/14; Kirsten Kindler: Paper Construction, 1/14-2/14; Drawing Blood & Guts: The Best of Contemporary Medical Illustration, 1/14-2/14; Kathleen Robbins: In Cotton (T), 9/14-10/14; Patrick Dougherty Installation, 9/14.
Facilities: 3,000 sq. ft. exhibit space.
Activities: docent program; formal education programs for undergraduate students; lecture series; monthly gallery talks; guided tours; labs in standard museum practices; participatory, temporary & traveling exhibits.
Publications: exhibition brochures.
Hours & Admission Prices: July-May Mon.-Wed. & Fri. 10-5, Thurs. 10-7, Sun. for receptions only. No charge. Closed spring break; summer sessions; Christmas break. &
Attendance: 7,500 (accurate)
Membership: Individual $35; Couple Subscriber $50; Associate Members $100; Patrons $500; Sustainers $1,000.

FAULKNER COUNTY MUSEUM, (M), Courthouse Square, 805 Locust St., Conway, AR 72034. Mailing Address: P.O. Box 2442, Conway, AR 72033-2442. Tel.: 501-329-5918.
E-mail: fcm@conwaycorp.net
Web Site: www.faulknercountymuseum.org
Founded: 1992.
Congressional District: 2
Key Personnel: Dir., Lynita Langley-Ware; Chm. (V), Dr. Sondra Gordy.
Personnel Profile: Full-Time Volunteers 2; Part-Time Paid 2; Part-Time Volunteers 16.
Governing Authority: society; nonprofit organization. Parent Institution: Faulkner County. Subsidiary Institution: Faulkner County Historical Society. Tax-exempt.
Institution Type/Description: History Museum.
Collections: Indian artifacts & campsite; cadron settlement; agricultural tools; medical instruments; general store; political development; cotton culture; genealogy files; Arkansas River history; 1920s kitchen & bedroom; log cabin; 1896 jail building.
Activities: guided tours.
Publications: quarterly magazine, Faulkner Facts & Fiddlings; books, Faulkner County: Its Land & People; Faulkner County, Arkansas: Census of Cemeteries as of December 31, 1987; pamphlet, The Guiding Star: A Guide Book for Catholic Emigrants to the Arkansas River Valley; Pine Mountain Americans; bimonthly newsletter, The Key.
Hours & Admission Prices: Mon.-Thurs. 9-4; groups by appointment. No charge; donations accepted. ♿
Attendance: 1,600 (accurate)
Membership: Individual $35; Family $50.

De Queen

COLLIN RAYE MUSEUM, 607 Haes St., De Queen, AR 71832. Tel.: 870-642-6642.
E-mail: seviercountymuseum@windstream.net
Institution Type/Description: History Museum: housed in the boyhood home of country music star, Collin Raye.
Collections: Collin Raye's life & career; personal artifacts; period furnishings; photographs.
Hours & Admission Prices: By appointment.

SEVIER COUNTY MUSEUM, 717 Walter J. Leeper Dr., De Queen, AR 71832. Tel.: 870-642-6642.
E-mail: seviercountymuseum@yahoo.com
Web Site: www.seviercountymuseum.org
Formerly: The Sevier County Historical Museum
Founded: 1986.
Key Personnel: Dir., Karen Mills; Bd. Pres., Marsha Buford.
Personnel Profile: Full-Time Paid 1; Part-Time Volunteers 6.
Governing Authority: Parent Institution: Sevier County Historical Society. Tax-exempt.
Institution Type/Description: History Museum.
Collections: history of De Queen & Sevier County; United States wars; personal artifacts.
Activities: classes; festivals; seminars; musicals; genealogy research. Museum Sponsors: Hoo-Rah Days Festival in October.
Publications: books, Family History; DeQueen Centennial History; Art Cookbook; book, Sevier County & It's People; quarterly member newsletter.
Hours & Admission Prices: Tues.-Sat. 12-5. No charge; donations accepted. ♿
Attendance: 3,750 (estimated)

Des Arc

LOWER WHITE RIVER MUSEUM, 2009 Main St., Des Arc, AR 72040-3135. Tel.: 870-256-3711. Fax: 870-256-9202.
E-mail: lowerwhiterivermuseum@arkansas.com
Web Site: www.arkansasstateparks.com
Formerly: Prairie County Museum
Founded: 1971.
Congressional District: 2
Key Personnel: Dir., Neva Boatright; Museum Program Asst. & Sec., Michael Yarberry.
Personnel Profile: Full-Time Paid 2; Part-Time Paid 1; Part-Time Volunteers 1.
Governing Authority: state; nonprofit. Parent Institution: Arkansas Dept. of Parks & Tourism. Branch: Arkansas State Parks Division, Dept. of Parks & Tourism, One Capitol Mall, Little Rock, AR 72201. Tax-exempt: 170(b)(1)(A).
Institution Type/Description: History Museum.
Collections: history of Lower White River & surrounding area from settlement to present; clothes; household items; photographs; farm implements.

Research Fields: history of the Lower White River with emphasis on early settlements & commerce from 1831-1931.
Activities: guided tours.
Publications: brochure.
Hours & Admission Prices: Mon.-Sat. 8-5, Sun. 1-5. Adults $3.25, children 6-12 $2. Closed New Year's Day; Thanksgiving; Christmas Eve & Day. ♿
Attendance: 3,969 (accurate)

Dover

MERIKS HISTORICAL SOCIETY, 4504 SR 27, Dover, AR 72837-8114. Mailing Address: P.O. Box 5407, Eugene, OR 97405-0407.
Key Personnel: Pres. (V), Rey Spina; Vice Pres. (V), Dana Cullen.
Personnel Profile: Full-Time Volunteers 1.
Institution Type/Description: Historical Society.
Research Fields: Genealogical studies.
Hours & Admission Prices: No charge.

Dumas

DESHA COUNTY MUSEUM, Hwy. 161 E., Dumas, AR 71639. Mailing Address: P.O. Box 141, Dumas, AR 71639-0141. Tel.: 870-382-4222.
E-mail: deshacomuseum@yahoo.com
Founded: 1979.
Congressional District: 4
Key Personnel: Bd. Dir., Charlotte Schexnayder; Pres. (V), Martha Clark; Dir., Peggy Chapman.
Personnel Profile: Part-Time Paid 1; Part-Time Volunteers 15.
Governing Authority: nonprofit organization. Tax-exempt: 501(c)(3).
Institution Type/Description: Historical & Preservation Society.
Collections: American Indian artifacts; Period farm tools; items from the early years of Desha County; log house farmstead; log house blacksmith shop; potato house & plantation commissary.
Research Fields: history of Desha County.
Facilities: 200-vol. library; reading room.
Activities: guided tours; lectures. Museum Sponsors: concerts by the Arkansas State Symphony.
Hours & Admission Prices: Tues.-Fri. 10-3:30, Sun. 2-4. No charge; donations accepted. ♿
Attendance: 4,000 (estimated)
Membership: Single $15; Family $30.

Earle

CRITTENDEN COUNTY MUSEUM, 1112 Main St., Earle, AR 72331. Mailing Address: P.O. Box 644, Earle, AR 72331. Tel.: 870-792-7374.
Institution Type/Description: History Museum: housed in the former Missouri Pacific train depot. Listed on the National Register of Historic Places.
Collections: local history; cotton farming; broom-making factory; early churches, schools & doctors; works of art by Carroll Cloar.
Hours & Admission Prices: Call for hours.

El Dorado

SOUTH ARKANSAS ARTS CENTER, 110 E. Fifth St., El Dorado, AR 71730-3822. Tel.: 870-862-5474. Fax: 870-862-4921.
E-mail: info@saac-arts.org
Web Site: saac-arts.org
Founded: 1962.
Congressional District: 4
Key Personnel: Exec. Dir., Jack Wilson.
Personnel Profile: Full-Time Paid 4; Part-Time Paid 1.
Governing Authority: nonprofit organization. Tax-exempt.
Institution Type/Description: Art Gallery.
Collections: works by regional & national artists; paintings; Oriental wood block prints; children's Oriental collection.
Facilities: library; reading room; 250-seat auditorium; community theatre; conference rooms; classrooms.
Activities: lectures; concerts; dance performances; arts festivals; art classes; theatre classes; community theatre productions; art shows & contests; art & drama programs for schools; hobby workshops; formally organized education programs; permanent, temporary & traveling exhibitions.
Publications: bimonthly newsletter.
Hours & Admission Prices: Mon.-Fri. 9-5. No charge; donations accepted. Closed holidays. ♿
Attendance: 30,000 (estimated)

Membership: Senior $15; Student $20; Individual $25; Family $60; Sponsor $100; Donor $200; Patron $750; Benefactor $1,500; Friends of Arts & Education $3,000.

Eureka Springs

AVIATION CADET MUSEUM, 542 CR 2073, Eureka Springs, AR 72632-9630. Tel.: 479-253-5008.
E-mail: av1cadet@arkansas.net
Web Site: www.aviationcadet.com
Founded: 1994.
Key Personnel: Museum Shop Mgr., E.D. Severe.
Personnel Profile: Full-Time Paid 2; Part-Time Volunteers 2.
Governing Authority: Subsidiary Institution: Cadet World. Tax-exempt.
Institution Type/Description: Aviation Museum.
Collections: military aircraft & artifacts; personal artifacts.
Facilities: Museum-related items for sale.
Publications: The Last of a Breed; I Wanna Fly.
Hours & Admission Prices: Tours: April-Oct. Mon.-Fri. 10, 1pm, 3pm, Sat. 1 & 3, Sun. 3pm; other times by appointment. Adults $15, children 6-12 $7.
Attendance: 400 (estimated)
Membership: $25-$100,000.

BLUE SPRING HERITAGE CENTER, 1537 CR 210, Eureka Springs, AR 72632. Tel.: 479-253-9244. Fax: 479-253-9256.
E-mail: info@bluespringheritage.com
Web Site: bluespringheritage.com
Institution Type/Description: Heritage Center.
Collections: local history & culture; Native American artifacts; photographs; personal artifacts; gardens.
Activities: classes; educational programs; rental facilities.
Hours & Admission Prices: mid-March to late Nov. daily 9-6. Adults $9.75, children 6-17 $6.50; children 5 & under no charge. &

CASTLE ROGUE'S MANOR, 124 Spring St., Eureka Springs, AR 72632. Tel.: 800-250-5827.
Web Site: castleroguesmanor.com
Institution Type/Description: Historic House Museum.
Collections: local history; period furnishings; personal artifacts.
Activities: special events.
Hours & Admission Prices: By appointment. Adults $20; children $10.

CRAZY BONE GALLERY, 37 Spring St., Eureka Springs, AR 72632-3147. Tel.: 479-253-6600; 888-418-8506.
Institution Type/Description: Art Gallery.
Collections: works of local & regional artists.
Hours & Admission Prices: Call for hours.

EUREKA FINE ART GALLERY, 63 N. Main St., Eureka Springs, AR 72632-3105. Tel.: 479-363-6000.
Web Site: www.eurekafineartgallery.com
Institution Type/Description: Art Gallery.
Collections: paintings; photographs; sculptures.
Hours & Admission Prices: Sun.-Sat. 10-5, Wed. by appointment.

EUREKA SPRINGS HISTORICAL MUSEUM, (M), 95 S. Main, Eureka Springs, AR 72632-3600. Tel.: 479-253-9417. Facebook: Eureka Springs Historical Museum.
E-mail: info@eurekaspringshistoricalmuseum.org
Web Site: www.eurekaspringshistoricalmuseum.org
Founded: 1971.
Congressional District: 3
Key Personnel: Dir., Steven Sinclair; Chm. (V), Gayla Wolfinbarger.
Personnel Profile: Full-Time Paid 1; Part-Time Paid 1; Part-Time Volunteers 25.
Governing Authority: bd. of directors. Tax-exempt.
Institution Type/Description: Historic House Museum: housed in the Calif House, 1889.
Collections: area history; art; personal artifacts; period furnishings; photographs; local school artifacts; postcards; genealogy.
Research Fields: genealogy; Springs.
Activities: school tours.
Hours & Admission Prices: Mon.-Sat. 9:30-4, Sun. 11-3. Self-Guided Tours: adults $5, children 18 & under and members no charge. Closed Thanksgiving; Christmas.
Attendance: 12,000 (accurate)
Membership: Single $25; Family $30; Business $50.

FAMILY HERITAGE MUSEUM, 338 Onyx Cave Ln., Eureka Springs, AR 72632. Tel.: 479-253-5444 & 5875. Fax: 479-253-7497.
E-mail: muriels@familyheritagemuseum.com
Web Site: www.familyheritagemuseum.com
Formerly: Gay 90's Button & Doll Museum
Founded: 1971.
Congressional District: 3
Key Personnel: Owner & Cur., Muriel H. Schmidt.
Governing Authority: nonprofit organization. Parent Institution: Destiny of America Foundation, Inc. Tax-exempt: 501(c)(3).
Institution Type/Description: Family Heritage Museum.
Collections: dolls; lamps; dishes; glassware; lace; household accessories; pottery; Indian artifacts; wedding dresses; fabrics; 29 albums combining family history with U.S. history.
Research Fields: vintage clothing 1880-1950; genealogy of local families; family careers.
Facilities: Museum-related items for sale.
Activities: self-guided tours; lectures; permanent exhibitions.
Publications: newsletter; booklet.
Hours & Admission Prices: By appointment. No charge; donations accepted. &
Attendance: 500 (estimated)

THE GREAT PASSION PLAY BIBLE MUSEUM, The Great Passion Play, 935 Passion Play Rd., Eureka Springs, AR 72632-9496. Mailing Address: P.O. Box 471, Eureka Springs, AR 72632. Tel.: 479-253-8559. Fax: 479-253-2302.
E-mail: akovalcik@greatpassionplay.com
Founded: 1968.
Key Personnel: Dir., Anne Kovalcik.
Personnel Profile: Full-Time Paid 2; Part-Time Paid 1.
Governing Authority: Parent Institution: Elna M. Smith Foundation.
Institution Type/Description: Religious Museum.
Collections: Bible history; over 7,000 Bibles in 100 languages; manuscripts; period artifacts; sacred art from the 1500s to present.
Hours & Admission Prices: May-Oct. Mon.-Sat. 10-8. Adults $5, children $2.50. &
Attendance: 64,000 (accurate)

ONYX CAVE & MUSEUM, 338 Onyx Cave Lane, Eureka Springs, AR 72632-9631. Tel.: 479-253-9321.
Institution Type/Description: Geology Museum.
Collections: local history; geology; period artifacts & dolls.
Facilities: Museum-related items for sale.
Hours & Admission Prices: April & Oct.-Nov. daily 9-4; May-Sept. daily 8:30-6:30. Adults $4.25, children 13 & under $2.

QUIGLEY'S CASTLE, 274 Quigley Castle Rd., Eureka Springs, AR 72632-9144. Tel.: 479-253-8311. Facebook: Quigley's Castle.
E-mail: quigleyscastle@gmail.com
Web Site: quigleyscastle.com
Founded: 1943.
Institution Type/Description: Historic House Museum: housed in the former home of Elise & Albert Quigley; built in 1943. Listed on the National Register of Historic Places.
Collections: family history; personal artifacts; sculpture; period furnishings.
Activities: guided tours.
Hours & Admission Prices: April-Oct. Mon.-Wed. & Fri.-Sat. 8:30-5. Adults $7; children 14 & under no charge.

THE ROSALIE, 282 Spring St., Eureka Springs, AR 72632-3152. Tel.: 479-253-7377.
E-mail: info@therosalie.com
Web Site: www.therosalie.com
Founded: 1889.
Congressional District: 3
Key Personnel: Owner, Charles Ragsdell; Owner, Lori Ragsdell.
Personnel Profile: Full-Time Paid 1; Full-Time Volunteers 2; Part-Time Paid 3; Part-Time Volunteers 2.
Governing Authority: individual operation.
Institution Type/Description: Historic House: built in 1889 by J. W. Hill owner of local livery & stables, and builder of Eureka Springs phone system.
Collections: period furnishings; musical instruments; Hersey memorabilia.
Facilities: banquet facilities.
Activities: tours by appointment; rental facilities.
Hours & Admission Prices: Tours: by appointment. Adults $7.50; children under 6 no charge.

Attendance: 1,000 (estimated)

TURPENTINE CREEK WILDLIFE REFUGE, 239 Turpentine Creek Lane, Eureka Springs, AR 72632-9185. Tel.: 479-253-5841.
Institution Type/Description: Wildlife Refuge.
Collections: big cats & their habitats including lions & tigers.
Activities: educational programs; guided tours.
Hours & Admission Prices: Summer: daily 9-6; Winter: daily 9-5. Adults 13 & over $15, youth 3-12, seniors & veterans $10; children under 3 no charge. Closed Christmas.

Fairfield Bay

LOG CABIN MUSEUM, 335 Snead Dr., Fairfield Bay, AR 72088. Tel.: 501-884-4899.
E-mail: wmpaar@hypertech.net
Web Site: www.arkansas.com
Institution Type/Description: Historic Building: housed in an 1850s log cabin.
Collections: local history & culture; period furnishings; personal artifacts; photographs.
Hours & Admission Prices: Call for hours. No charge.

NORTH CENTRAL ARKANSAS ART GALLERY, 337 Snead Dr., Fairfield Bay, AR 72088. Mailing Address: P.O. Box 1643, Fairfield Bay, AR 72088. Tel.: 501-884-6100.
Institution Type/Description: Art Gallery.
Collections: works by local & national artists including paintings.
Hours & Admission Prices: Call for hours. No charge.

Fayetteville

ARKANSAS AIR & MILITARY MUSEUM, 4290 S. School Ave., Fayetteville, AR 72701-8008. Tel.: 479-521-4947. Fax: 4791-521-4947.
E-mail: arkairmus@aol.com
Web Site: www.arkansasairandmilitary.com
Formerly: Arkansas Air Museum
Founded: 1986.
Congressional District: 3
Key Personnel: Pres., Ray Boudreaux, Dir., Warren Jones; Vice Pres., Rick Bailey; Treas., James Nicholson; Sec., Rick McKinney; Museum Shop Mgr., Sally Ebbrecht.
Personnel Profile: Full-Time Paid 1; Part-Time Paid 2; Part-Time Volunteers 4.
Governing Authority: private; nonprofit organization. Tax-exempt.
Institution Type/Description: Air Museum: housed in WWII era wooden truss hangar; military section housed in 2 additional buildings.
Collections: classic aircraft & artifacts from early flight to present, with emphasis on Arkansas aviators, military & civilian; military vehicles; state historic site; aircraft engine collections piston, jet & rocket; approximately 18 aircrafts, some WWII trainers modern military A-4 Jet Huey & Cobra helicopters; artifacts from Civil War to todays conflicts.
Research Fields: photographs of general aviation; WWII naval aviation; classic aircraft data.
Facilities: 1,000-vol. of books & aviation magazines; 80-vol. of aviation videos; educational facilities; 30,000 sq. ft. exhibit space; 35-seat theatre. Museum-related items for sale.
Activities: films; guided tours; lectures; visiting aircraft.
Publications: quarterly newsletter.
Hours & Admission Prices: Sun.-Fri. 11-4:30, Sat. 10-4:30. Family $20, adults $10, children 6-12 $5; discounts to groups; members no charge. Closed New Year's Day; Thanksgiving; Christmas. &
Attendance: 15,000 (estimated)
Membership: Individual $25; Family $50; Patron $100; Life $1,000.

BOTANICAL GARDEN OF THE OZARKS, 4703 N. Crossover Rd., Fayetteville, AR 72764. Mailing Address: P.O. Box 10407, Fayetteville, AR 72703-0042. Tel.: 479-750-2620. Fax: 479-756-1920.
E mail: rcox@bgozarks.org
Web Site: www.bgozarks.org
Founded: 1994.
Congressional District: 3
Key Personnel: Dir., Ron Cox; Pres. (V), Walt Eilers; Museum Shop Mgr., Judy Smith.
Personnel Profile: Full-Time Paid 8; Part-Time Paid 3; Part-Time Volunteers 194; Interns 1.
Governing Authority: Parent Institution: Botanical Garden Society of the Ozarks. Tax-exempt.
Institution Type/Description: Botanical Garden.
Collections: plants, trees, & flowers of the Ozarks.
Facilities: 9 themed gardens.
Activities: educational programs; special events; weddings; concerts.
Publications: quarterly, Garden Elements; monthly electronic bulletins.
Hours & Admission Prices: Daily 9-5. May-Oct. adults $7, children 5-12 $4; Nov.-April adults $5, children 5-12 $2.50; children under 5, & Sat. 9-12 Fayetteville residents no charge. Reciprocal admission to AHS members. &
Attendance: 35,000 (estimated)
Membership: Family $60.

CLINTON HOUSE MUSEUM, (M), 930 W. Clinton Dr., Fayetteville, AR 72701-4912. Tel.: 479-444-0066; 877-BIL-N-HIL.
Founded: 2005.
Congressional District: 3
Key Personnel: Dir., Kate Johnson.
Governing Authority: Parent Institution: Fayetteville Advertising & Promotion Commission. Tax-exempt: 501(c)(3).
Institution Type/Description: Historic House: housed in the former home of Bill & Hillary Clinton while he was a professor at the University of Arkansas.
Collections: family life & history; personal artifacts; photographs; campaign materials; First Ladies garden.
Facilities: Museum-related items for sale.
Hours & Admission Prices: Mon.-Fri. 8:30-4:30. Adults $5, students $3, children $1.

FINE ARTS CENTER GALLERY - UNIVERSITY OF ARKANSAS, Fulbright College of Arts and Sciences, 116 Fine Arts Center, Fayetteville, AR 72701. Tel.: 479-575-7987. Fax: 479-575-2062.
E-mail: stk004@uark.edu
Web Site: art.uark.edu/fineartsgallery
Founded: 1951.
Key Personnel: Dir., Shannon Dillard Mitchell.
Personnel Profile: Full-Time Paid 1; Part-Time Paid 6.
Governing Authority: Parent Institution: University of Arkansas.
Institution Type/Description: Art Gallery.
Collections: paintings; photographs; sculpture; drawings.
Activities: special events; visiting artists & scholars lecture series.
Hours & Admission Prices: Mon.-Fri. 9-5:30, Sun. 2-5. No charge.

HEADQUARTERS HOUSE MUSEUM, 118 E. Dickson St., Fayetteville, AR 72701-4207. Tel.: 479-521-2970.
E-mail: info@washcohistoricalsociety.org
Web Site: www.washcohistoricalsociety.org/hispro
Key Personnel: Pres., Vince Chadick.
Governing Authority: Branch Museum: Ridge House, 230 West Center, Fayetteville, AR 72701.
Institution Type/Description: History Museum: built in 1853 by Judge Jonas M. Tebbetts, this home served as headquarters for the Federal and Confederate armies during the Civil War.
Collections: personal artifacts; Masonic furniture. Historic Building: Archibald Yell's law office, 1835.
Facilities: period gardens. Museum-related items for sale.
Activities: living history presentations; tours.
Hours & Admission Prices: Tours by appointment. Adults $8, children $1; discounts to members. &
Attendance: 4,000 (estimated)
Membership: Senior $15; Family $25.

OZARK MILITARY MUSEUM, 4290 S. School Ave., Fayetteville, AR 72701-8008. Tel.: 479-587-1941. Fax: 479-587-0848.
Key Personnel: Dir., Leonard McCandless
Institution Type/Description: Military Museum.
Collections: military history; WWII, Korean War, & Vietnam artifacts; photographs; personal artifacts; military aircraft & vehicles.
Activities: special events.
Hours & Admission Prices: Sun.-Fri. 11-4:30, Sat. 10-4:30.

THE UNIVERSITY MUSEUM COLLECTIONS, University of Arkansas, Biomass Bldg. Rm. 125, Fayetteville, AR 72701-1201. Tel.: 479-575-3456. Fax: 479-575-7464.
E-mail: collectn@uark.edu
Web Site: www.uark.edu/~museinfo/
Founded: 1873.
Congressional District: 3
Key Personnel: Interim Dir., Dr. Jeannine Durdik; Cur. Collections, Mary Suter; Cur. Zoology, Dr. Nancy G. McCartney.

Personnel Profile: Full-Time Paid 2; Part-Time Paid 2; Part-Time Volunteers 2.
Governing Authority: university. Parent Institution: Univ. of Arkansas. Tax-exempt: 170(b)(1)(A).
Institution Type/Description: General Museum.
Collections: prehistoric Arkansas Indian artifacts; quartz crystals; bird eggs & nests; malacology; herpetology; ichthyology; geology; herbarium; mineralogy; paleontology; early American pressed glass; early textile equipment & other Americana; ethnological collections from Oceania, Africa & Amerindians; botanical (nonliving).
Research Fields: archaeology; physical anthropology; zoology; geology; paleontology; botany.
Activities: inter-museum loan; collections research; temporary exhibitions.
Hours & Admission Prices: Mon.-Fri. by appointment only. No charge. Closed university holidays. &
Attendance: 200 (estimated)

UNIVERSITY OF ARKANSAS DISCOVERY ZONE, 1564 Martin Luther King Jr. Blvd., Fayetteville, AR 72701-6203. Mailing Address: Center for Math & Science Education, 346 West Ave., Rm. 202, Fayetteville, AR 72701. Tel.: 479-575-3875.
E-mail: lhehr@uark.edu
Key Personnel: Dir., Lynne H. Hehr
Institution Type/Description: Science Museum.
Collections: hands-on exhibits.
Activities: temporary, traveling & permanent exhibits.
Hours & Admission Prices: Sept.-May Tues.-Fri. 9-5, Sat. 10-4; groups of 10 or more by appointment.

Fordyce

DALLAS COUNTY MUSEUM, 221 S. Main St., Fordyce, AR 71742. Mailing Address: P.O. Box 703, Fordyce, AR 71742-0703. Tel.: 870-352-5262. Facebook: Dallas County Museum.
E-mail: dcmus@windstream.net
Institution Type/Description: History Museum.
Collections: county history & culture; photographs; timber industry; portraits; military artifacts; Paul "Bear" Bryant memorabilia; geology; early agriculture.
Facilities: 12,000 sq. ft. exhibit space.
Activities: group tours.
Hours & Admission Prices: Tues.-Fri. 10-4, Sat. 10-2. No charge; donations accepted. Closed most holidays. &

Foreman

NEW ROCKY COMFORT MUSEUM, 3rd & Schuman, Foreman, AR 71836. Mailing Address: P.O. Box 268, Foreman, AR 71836-0268. Tel.: 870-542-7887. Fax: 870-542-6347.
Institution Type/Description: Historic Building: housed in a restored jail; built in 1902.
Collections: local history & culture; period documents.
Hours & Admission Prices: Call for hours.

Forrest City

ST. FRANCIS COUNTY MUSEUM, 603 Front St., Forrest City, AR 72335-3808. Mailing Address: P.O. Box 1332, Forrest City, AR 72336-1332. Tel.: 870-261-1744. Fax: 870-630-1210.
Web Site: www.sfcmuseum.org
Founded: 1995.
Congressional District: 1
Key Personnel: Dir., H. Wayne Parker; Chm. (V), Rush Beavers; Cur., Shelley Gervasi.
Personnel Profile: Full-Time Paid 1; Part-Time Paid 1; Part-Time Volunteers 6.
Governing Authority: Parent Institution: St. Francis County. Tax-exempt.
Institution Type/Description: Historical & Culturally Specific Museum.
Collections: military history; Native American; Afro-American; medical history; documentations.
Hours & Admission Prices: Mon.-Fri. 10-5. No charge; donations accepted. &
Attendance: 4,205 (accurate)

Fort Smith

CHAFFEE BARBERSHOP MUSEUM, 7313 Terry St., Fort Smith, AR 72916. Tel.: 479-434-6774 & 783-8888.
Founded: 2008.
Institution Type/Description: Historic Building: housed in the former barbershop where thousands of U.S. Army enlistees including Elvis Presley received their signature army buzz cut.

Collections: Fort Chaffee history & artifacts; photographs; movie memorabilia from Biloxi Blues, The Tuskegee Airmen & A Soldier's Story which were filmed at Fort Chaffee; period furnishings & signs.
Activities: special events.
Hours & Admission Prices: Mon.-Sat. 9-4. No charge; donations accepted. Closed National holidays.

CLAYTON HOUSE, (M), 514 N. Sixth St., Fort Smith, AR 72901-2006. Tel.: 479-783-3000.
Institution Type/Description: Historic House: housed in the former home of District Attorney, William Henry Harrison Clayton; built c.1850. Listed on the National Register of Historic Places.
Collections: Clayton's life & career; period furnishings & toys; personal artifacts; photographs.
Activities: tours; school field trips; rental facilities.
Hours & Admission Prices: June to Labor Day Wed.-Sat. 10-4, Sun. 1-4; Sept.-May Wed.-Sat. 12-4, Sun. 1-4. Adults $5, students & children $2.50; children 5 & under no charge. &

FORT SMITH AIR MUSEUM, Fort Smith Regional Airport, 6700 McKennon Blvd., Fort Smith, AR 72903. Mailing Address: 3 Glen Haven Dr., Fort Smith, AR 72901-6837. Tel.: 479-785-1839.
E-mail: whaver@mynewroads.com
Founded: 1999.
Key Personnel: Pres., Wayne Haver; Vice Pres., Carl Riggens; Treas., Ralph Freeman
Governing Authority: Tax-exempt.
Institution Type/Description: Air Museum.
Collections: aviation history; military aviators; aviation pioneers; aircraft; photographs; personal artifacts.
Hours & Admission Prices: Daily 5:30am-11pm. No charge. &
Attendance: 50,000 (estimated)

FORT SMITH MUSEUM OF HISTORY, 320 Rogers Ave., Fort Smith, AR 72901-1937. Tel.: 479-783-7841. Fax: 479-783-3244.
E-mail: leisa.gramlich@fortsmithmuseum.com
Web Site: www.fortsmithmuseum.com
Founded: 1910.
Congressional District: 3
Key Personnel: C.E.O., Leisa Gramlich; Pres., John Cooley; Museum Shop Mgr., Caroline Speir.
Personnel Profile: Full-Time Paid 1; Part-Time Paid 4; Part-Time Volunteers 15; Interns 1.
Governing Authority: nonprofit. Parent Institution: Fort Smith Museum of History Association. Tax-exempt: 501(c)(3).
Institution Type/Description: History Museum: housed in 1907 Atkinson-Williams Building.
Collections: period furniture; textiles; period toys; documents; military items; c.1920 drugstore with operating soda fountain; period vehicles.
Research Fields: furniture industry in Fort Smith Area; Fort Chaffee (US Army Post 1941-present); commercial history (advertising); pharmacy.
Facilities: auditorium; rental facilities.
Activities: workshops; temporary & permanent exhibitions; interpretive programs; school group tours; holiday special events.
Publications: newsletter; Fort Smith: An Illustrated History.
Hours & Admission Prices: June-Aug. Tues.-Sat. 10-5, Sun. 1-5; Sept.-May Tues.-Sat. 10-5. Adults $5, children 6-15 $2; discounts to groups; Fort Smith public schools, children under 6 & members no charge. Closed holidays. &
Attendance: 20,000 (estimated)
Membership: Individual $25; Senior Couple $30; Family $50; Investor $100; Silver Benefactor $500-$999; Gold $1,000-$4,999; Diamond $5,000 & up.

FORT SMITH NATIONAL HISTORIC SITE, 301 Parker Ave., Fort Smith, AR 72901-1938. Mailing Address: P.O. Box 1406, Fort Smith, AR 72902-1406. Tel.: 479-783-3961. Fax: 479-783-5307. TDD: 479-783-3961.
Web Site: www.nps.gov/fosm/
Founded: 1961.
Congressional District: 3
Key Personnel: Supt., Lisa Conard Frost; Chief Interpretation & Resource Mgmt., Michael Groomer; Museum Technician, Emily Lovick; Administrative Officer, Chuck Shoemaker; Facility Mgr., Darin Huggins; Administrative Asst., Quoya Waters.
Personnel Profile: Full-Time Paid 11; Part-Time Paid 2; Part-Time Volunteers 15.
Governing Authority: federal. Parent Institution: Dept. of Interior. Subsidiary Institution: National Park Service. Tax-exempt.

Institution Type/Description: Historic Site: 1817-24 Fort Smith, 1838 regarrisoned at new location, completed larger fort 1849, became the Federal Courthouse and Jail, 1871-1896.
Collections: furnishings; art; photographs; archeological material. Outdoor exhibits include 1817 fort foundations; 1846 Commissary building & reconstructed gallows used by Federal Court.
Research Fields: frontier military history; federal Indian policy; Federal Court for the Western District of Arkansas, deputy marshalls & outlaws.
Facilities: 750-vol. library of history books for use by special request; visitor center in barracks/courthouse building. Books for sale.
Activities: operational film, deputy marshall film, interactive videos on Indian removal & Indian territory; self-guided tour of historic barracks/court house/jail; self-guided walking tour along Arkansas River & Trail of Tears overlook; school tours; children's programs; living history; interpretive talks.
Publications: orientation brochure; quarterly newsletter.
Hours & Admission Prices: Daily 9-5. Individual Seven Day Pass 16 & up $4; America the Beautiful - National Parks & Federal Recreation Lands, Golden Age, Gold Access & Fort Smith NHS passholders, educational groups and children 15 & under with adult no charge. Closed New Year's Day; Thanksgiving; Christmas. &
Attendance: 100,000 (estimated)
Membership: Fort Smith National Historic Site Annual Pass $15.

FORT SMITH REGIONAL ART MUSEUM, (M), 1601 Rogers Ave., Fort Smith, AR 72902. Mailing Address: P.O. Box 1257, Fort Smith, AR 72901-1257. Tel.: 479-784-ARTS (2787). Fax: 479-784-9071.
E-mail: info@fsram.org
Web Site: fsram.org
Formerly: Fort Smith Art Center
Founded: 1948.
Congressional District: 3
Key Personnel: Exec. Dir., Lee Ortega; Pres. (V), Marta Jones; Receptionist, Shelby Richison.
Personnel Profile: Full-Time Paid 3; Full-Time Volunteers 1; Part-Time Paid 2; Part-Time Volunteers 20; Interns 10.
Governing Authority: nonprofit. Tax-exempt: 501(c)(3).
Institution Type/Description: Art Museum.
Collections: contemporary American paintings; photography; sculpture; decorative arts; works on paper.
Major Exhibits: Re:history Recent Work of James Volkert, 11/22/13-2/16/14; Valentines: The Art of Romance (T), 1/16/14-3/30/14; Annual 65th River Valley Invitational, 5/14-8/14; George Dombek, 5/14-8/14; Carol Dickie: New York, 9/14-12/14.
Facilities: art education facilities; sculpture garden; outdoor patio.
Activities: guided tours; lectures; artist workshops; gallery talks; education programs; permanent & temporary exhibitions. Annual Events: Art Competition gala.
Publications: newsletter; monthly e-newsletter; brochures.
Hours & Admission Prices: Tues.-Sat. 10-5, Sun. 11-5. No charge. Closed New Year's Day; Memorial Day; Independence Day; Labor Day; Thanksgiving & day after; Christmas Eve & Day. &
Attendance: 14,000 (estimated)
Membership: Student $35; Individual $50; Family $75; Family Plus $125; Contributor $250; Curator's Circle $500; Director's Circle $1,000; Museum Partner $2,500; Museum Benefactor $5,000.

FORT SMITH TROLLEY MUSEUM, 100 S. 4th St., Fort Smith, AR 72901-1947. Tel.: 479-783-0205.
E-mail: info@fstm.org
Web Site: www.fstm.org
Founded: 1979.
Key Personnel: C.E.O. & Pres. (V), Art B. Martin, M.D.; Museum Shop Mgr., Bradley Martin
Institution Type/Description: History Museum.
Collections: Fort Smith's trolley transportation history; 4 original 58 Fort Smith streetcars; Frisco steam engine and tender; cabooses; a former military power car; dining car; boxcars; former Fort Smith buses; 1954 Fort Smith bus used in the filming of Biloxi Blues & Tuskegee Airmen.
Facilities: Museum-related items for sale.
Activities: trolley rides; tours; special events; birthday parties. Museum Sponsors: Open House in May.
Publications: newsletter, Trolley Report.
Hours & Admission Prices: Museum: Sat. 10-5, Sun. 1-5; other times by appointment. Trolley: May-Oct. Mon.-Sat. 10-5, Sun. 1-5; Nov.-April Sat. 10-5, Sun. 1-5. Museum: no charge; donations accepted. Trolley Rides: adult $2, children $1. &
Attendance: 10,000 (estimated)

Membership: Individual $10; Annual $15; Annual Sponsor $25; Conductor $50; Sustaining $100; 205 Club $205; 224 Club $224; Benefactor $500; Life $1,000.

JANET HUCKABEE ARKANSAS RIVER VALLEY NATURE CENTER, 8300 Wells Lake Rd., Fort Smith, AR 72916. Tel.: 479-452-3993. Fax: 479-452-1334.
Web Site: www.rivervalleynaturecenter.com
Institution Type/Description: Nature Center.
Collections: wildlife & their habitats; hands-on exhibitions.
Facilities: nature trails. Gift items for sale.
Activities: walking trails; video observation area; educational programs.
Hours & Admission Prices: Tues.-Sat. 8:30-4:30, Sun. 1-5. No charge. Closed Easter; Thanksgiving; Christmas.

Garfield

PEA RIDGE NATIONAL MILITARY PARK, 15930 Hwy. 62, Garfield, AR 72732-9532. Tel.: 479-451-8122. Fax: 479-451-0219.
Web Site: www.nps.gov/peri/
Founded: 1956.
Congressional District: 3
Key Personnel: Park Ranger, Troy Banzhaf; Supt., John Scott; Museum Shop Mgr., Serena Rothfus.
Personnel Profile: Full-Time Paid 13; Part-Time Volunteers 10.
Governing Authority: federal. National Park Service.
Institution Type/Description: Historical Building: reconstructed Elkhorn Tavern.
Collections: Civil War; military.
Research Fields: Civil War.
Facilities: 300-vol. library of Civil War books.
Activities: guided tours; lectures; films; permanent & temporary exhibits.
Publications: assorted information site bulletins.
Hours & Admission Prices: Daily 8-5. 7 Day Permit: adults 16-61 $5; discounts to NPS passholders; children under 16 no charge. Closed New Year's Day; Thanksgiving; Christmas. &
Attendance: 120,000 (estimated)

Gillett

ARKANSAS POST MUSEUM, 5530 Hwy. 165 S., Gillett, AR 72055-9730. Tel.: 870-548-2634. Fax: 870-548-3003.
E-mail: arkansaspostmuseum@arkansas.gov
Web Site: www.arkansasstateparks.com
Founded: 1960.
Congressional District: 1
Key Personnel: Dir., Christy Murphy.
Personnel Profile: Full-Time Paid 3; Part-Time Paid 2; Part-Time Volunteers 12.
Governing Authority: state. Parent Institution: Arkansas Dept. of Parks & Tourism. Subsidiary Institution: Arkansas State Parks Division. Tax-exempt.
Institution Type/Description: Local History Museum.
Collections: archives; agriculture; archaeology; costumes; history; textiles; transportation; music; glass; schools; child's playhouse; Quapaw Indians; pioneer life; Civil War. Historic House: 1877 Refeld-Hinman Home.
Research Fields: Arkansas Post area history, Grand Prairie Region.
Facilities: Gifts for sale.
Activities: guided tours; programs; lectures; special events.
Hours & Admission Prices: Tues.-Sat. & Mon. holidays 8-5, Sun. 1-5. Family $10, adults $3, children 6-12 $2; children under 6 & school groups no charge. Closed New Year's Day; Thanksgiving; Christmas Eve & Day. &
Attendance: 6,000 (accurate)

ARKANSAS POST NATIONAL MEMORIAL, 1741 Old Post Rd., Gillett, AR 72055-9733. Tel.: 870-548-2207. Fax: 870-548-2431.
E-mail: arpo_superintendent@nps.gov
Web Site: www.nps.gov/arpo
Founded: 1964.
Congressional District: 1
Key Personnel: Supt., Edward Wood.
Personnel Profile: Full-Time Paid 10; Part-Time Paid 1; Part-Time Volunteers 8.
Governing Authority: federal. Parent Institution: National Park Service. Tax-exempt: 501(c)(3).
Institution Type/Description: History Museum.
Collections: prehistoric artifacts; early American (pre-Revolutionary War); Colonial (French & Spanish); Revolutionary War; artifacts of Arkansas territorial & early statehood period; Civil War.

Research Fields: Indian, French, Spanish, early American, Civil War & westward expansion.
Facilities: historical trail; picnic area; visitor center.
Activities: films; permanent exhibitions; interpretive programs.
Publications: orientation brochure; quarterly newsletter.
Hours & Admission Prices: Park daily 8-dusk. Visitor Center daily 8-5. No charge; donations accepted. Closed New Year's Day; Thanksgiving; Christmas. &
Attendance: 38,180 (accurate)

Glenwood

BILLY'S HOUSE OF GUITARS & MUSICAL MUSEUM, 201 Broadway, Glenwood, AR 71943-9200. Tel.: 870-356-4301.
Institution Type/Description: Musical Instruments Museum.
Collections: guitars, instruments & memorabilia of celebrities including Willie Nelson, Bob Dylan, George Harrison, Pete Seeger, Elvis Presley, & Johnny Cash.
Hours & Admission Prices: Call for hours.

Gravette

GRAVETTE HISTORICAL MUSEUM, 503 Charlotte St., S.E., Gravette, AR 72736. Mailing Address: P.O. Box 1421, Gravette, AR 72736. Tel.: 479-787-7334. Fax: 479-787-9910.
Founded: 1995.
Key Personnel: Chm. (V), John Mitchael; Financial Dir., Michael Von Ree; Museum Shop Mgr., Sheila Martin.
Personnel Profile: Part-Time Paid 1.
Governing Authority: municipal. Parent Institution: City of Gravette. Tax-exempt.
Institution Type/Description: Historical Museum.
Collections: to promote, share & preserve the history of the city of Gravette & its surrounding area; artifacts that depict the life styles of area Native Americans, settlers & residents.
Major Exhibits: My Collections, 1/14; Dutch Oven Cookoff, 8/14; Local Quilt Exhibit, 9/14.
Research Fields: cabin restoration; local site excavations.
Facilities: 3,700 sq. ft. exhibit space.
Activities: temporary exhibitions. Annual Events: Annual Founders Day, State Museum Day.
Hours & Admission Prices: Tues., Thurs. & Sat. 12-4; other times by appointment. No charge; donations accepted. Closed New Year's Day; Christmas. &
Attendance: 400 (estimated)

Greenbrier

RIDDLE'S ELEPHANT AND WILDLIFE SANCTUARY, Arkansas 25 off U.S. 65 N., Greenbrier, AR 72058. Mailing Address: P.O. Box 715, Greenbrier, AR 72058. Tel.: 501-589-3291. Fax: 501-589-2248.
E-mail: info@elephantsanctuary.org
Web Site: www.elephantsanctuary.org
Key Personnel: Owner, Scott Riddle; Owner, Heidi Riddle.
Governing Authority: nonprofit organization. Tax-exempt: 501(c)(3).
Institution Type/Description: Wildlife Sanctuary.
Collections: Asian & African elephants.
Facilities: 330 acres.
Activities: educational programs. Museum Sponsors: Elephant Experience Weekends; International School for Elephant Management.
Hours & Admission Prices: 1st Sat. each month 11-3.

Greenwood

OLD JAIL MUSEUM, 307 Town Sq., Greenwood, AR 72936. Mailing Address: P.O. Box 523, Greenwood, AR 72936. Tel.: 479-996-6357.
Founded: 1963.
Key Personnel: Pres. (V), Ruth McConnell; Cur., Donna Goldstein; Archivist, Sue Edwards.
Personnel Profile: Part-Time Volunteers 40.
Governing Authority: private; nonprofit organization. Tax-exempt: 501(c)(3).
Institution Type/Description: History Museum.
Collections: local history & culture; coal mining; school memorabilia; agricultural tools & implements, 1840-1930; coal miners memorial. Historic Buildings: 1892 jail; 1848 vineyard log cabin; barn; redwine pioneer schoolhouse.
Facilities: library. Museum-related items for sale.

Activities: Annual Events: Airing of the Quilts in May; Buried Treasure in October; Christmas Tea in December.
Publications: annual magazine, The Key.
Hours & Admission Prices: May-Oct. Thurs.-Sat. 11-3; other times by appointment. No charge; donations accepted.
Attendance: 1,200 (accurate)
Membership: Individual $15; Family $25; Lifetime $250.

Gurdon

HOO-HOO INTERNATIONAL FORESTRY MUSEUM, 207 Main St., Gurdon, AR 71743-1237. Mailing Address: P.O. Box 118, Gurdon, AR 71743-0118. Tel.: 870-353-4997; 800-979-9950. Fax: 870-353-4151.
E-mail: info@hoo-hoo.org
Web Site: www.hoo-hoo.org
Founded: 1981.
Congressional District: 4
Key Personnel: Exec. Sec., Beth A. Thomas; Chm. Museum Committee, Teeny Johnston.
Personnel Profile: Part-Time Paid 1.
Governing Authority: organization. Parent Institution: The Fraternal Order of Forest Products Industry. Tax-exempt.
Institution Type/Description: Logging & Lumber Museum.
Collections: woodcarvings; tools; artifacts pertaining to forestry; items pertaining to logging & lumber; manuscript collections.
Facilities: 75-seat auditorium.
Activities: permanent exhibitions; school loan service.
Publications: Log & Tally magazine.
Hours & Admission Prices: Mon.-Fri. 9-4. No charge.

Hamburg

ASHLEY COUNTY MUSEUM, 302 N. Cherry St., Hamburg, AR 71646. Mailing Address: P.O. Box 27, Hamburg, AR 71646-0027. Tel.: 870-853-2244.
Web Site: www.ashleycountymuseum.com
Institution Type/Description: Historic House Museum: housed in the former home of David E. Watson; built in 1918. Listed on the National Register of Historic Places.
Collections: local history & culture; period furnishings; photographs; personal artifacts.
Hours & Admission Prices: Tues. 10-3; other times by appointment.

Hardy

GOOD OLD DAYS VINTAGE MOTORCAR MUSEUM, INC., 301 W. Main St., Hardy, AR 72542. Mailing Address: P.O. Box 311, Hardy, AR 72542-0311. Tel.: 870-856-4884. Fax: 870-856-4884.
Founded: 1996.
Key Personnel: Dir., Ernest E. Sutherland.
Personnel Profile: Full-Time Paid 1; Part-Time Volunteers 1.
Institution Type/Description: Automobile Museum.
Collections: period automobiles; tools; gas pumps; guns; books.
Facilities: Museum-related items for sale.
Activities: antique auto sales.
Hours & Admission Prices: Mon.-Fri. 9:30-4, Sat. 9-4:30, Sun. 12-5. Adults $10, children 12 & under $5; discounts to groups and museum, AAM & ICOM members. Closed Thanksgiving; Christmas. &
Attendance: 4,500 (estimated)

Harrison

BOONE COUNTY HERITAGE MUSEUM, 124 S. Cherry St., Harrison, AR 72601-5024. Mailing Address: P.O. Box 1094, Harrison, AR 72602-1094. Tel.: 870-741-3312.
E-mail: bchm@windstream.net
Web Site: www.bchrs.org
Founded: 1987.
Key Personnel: Pres. (V), John Berry; Dir., Roz Slavik
Institution Type/Description: History Museum.
Collections: local history; period artifacts; Civil War; the Missouri & North Arkansas Railroad Co.; Native American artifacts; period clocks; medical & domestic tools.
Research Fields: genealogy.
Facilities: library.
Publications: quarterly, Boone County Historian, Oak Leaves; Boone County History Pictoral.

Hours & Admission Prices: March-Nov. Mon.-Fri. 10-4; Dec.-Feb. Thurs. 10-4. Adults $2. Closed holidays.
Membership: Individual and Husband & Wife $20; Organization & Business $25; Lifetime $250.

BUFFALO NATIONAL RIVER, 402 N. Walnut St. Ste. #136, Harrison, AR 72601-3622. Tel.: 870-365-2700 (park headquarters); 439-2502 (visitor information). Fax: 870-365-2701.
Web Site: www.nps.gov/buff/index.htm
Founded: 1972.
Congressional District: 3
Key Personnel: Supt., Kevin G. Cheri.
Personnel Profile: Full-Time Paid 1; Part-Time Paid 1.
Governing Authority: federal. Dept. of the Interior, National Park Service. Tax-exempt.
Institution Type/Description: National Park.
Collections: herbarium & other natural history collections; archaeological & historical artifacts; Ozark Mountain settlements dating back 9,000 years. Historic Buildings: c.1860 Beaver Jim Villines Farm; c.1836 Parker-Hickman cabin.
Research Fields: local history & culture; natural history.
Facilities: 700-vol. library on Ozark folklife, park recreation, resource management, available for use on premises; reading room. Books on natural history, Ozark history & folklife for sale.
Activities: guided tours; lectures; films.
Publications: orientation brochure.
Hours & Admission Prices: Daily 8:30-4:30. No charge. Closed federal holidays except Memorial Day, Independence Day & Labor Day. &
Attendance: 1,000,000 (accurate)

MARINE CORPS LEGACY MUSEUM, 127 Rush St., Harrison, AR 72601. Mailing Address: P.O. Box 2654, Harrison, AR 72602-2654. Tel.: 870-743-1680.
Institution Type/Description: Military History Museum.
Collections: Marine Corps history from 1775 to present; photographs; uniforms; vehicles; awards.
Activities: educational programs.
Hours & Admission Prices: Tues.-Sat. 10-5.

Heber Springs

OLMSTEAD FUNERAL & HISTORICAL MUSEUM, 108 S. 4th St., Heber Springs, AR 72543-3810. Tel.: 501-362-2422.
E-mail: mail@olmstead.cc
Web Site: www.olmstead.cc/museum.htm
Institution Type/Description: History Museum.
Collections: history of undertaking & funeral directing dating back to 1896; horse drawn hearse.
Activities: tours.
Hours & Admission Prices: Call for appointment.

Helena

DELTA CULTURAL CENTER, (M), 141 Cherry St., Helena, AR 72342-3501. Tel.: 870-338-4350; 800-358-0972. Fax: 870-338-4358.
E-mail: info@deltaculturalcenter.com
Web Site: www.deltaculturalcenter.com
Founded: 1990.
Congressional District: 1
Key Personnel: Dir., Katie Harrington; Chm. Policy Advisory Bd., Emma Petty; Museum Shop Mgr., Kathleen Randall.
Personnel Profile: Full-Time Paid 10; Part-Time Paid 4.
Governing Authority: state government; nonprofit. Parent Institution: Dept. of Arkansas Heritage. Tax-exempt.
Institution Type/Description: Cultural Center: housed in the former St. Louis & Iron Mountain Railroad Depot; c.1912.
Collections: Arkansas Delta history & culture; personal artifacts; photographs.
Hours & Admission Prices: Tues.-Sat. 9-5. No charge; donations accepted. Closed New Year's Day; Thanksgiving; Christmas Eve & Day. &
Attendance: 26,988 (accurate)

HELENA MUSEUM OF PHILLIPS COUNTY, 623 Pecan St., Helena, AR 72342-3298. Mailing Address: P.O. Box 38, Helena, AR 72342. Tel.: 870-338-7790. Fax: 870-338-7732.
E-mail: helenamuseum@gmail.com
Founded: 1929.
Congressional District: 1

Key Personnel: Dir., Shane Williams; Pres. (V), Joe Ann Hargraves; Sec., Jeanie Turley; Treas., Margaret H. Kirk.
Personnel Profile: Part-Time Paid 3.
Governing Authority: nonprofit organization. Tax-exempt.
Institution Type/Description: History Museum.
Collections: memorabilia of the Civil War, Spanish-American War & World Wars I & II; glassware; china; apparel & accessories covering several generations; early settlement of Phillips County; portraits; furniture; permanent loan from Edison Foundation; early American Indian & area artifacts; contemporary & historical Black art.
Activities: permanent & temporary exhibitions.
Hours & Admission Prices: Tues.-Sat. 10-4. No charge; donations accepted. Closed New Year's Day; Memorial Day; Independence Day; Labor Day; Thanksgiving; Christmas Day & two days after. &
Attendance: 3,463 (accurate)
Membership: Membership $5; Sustaining $10.

PILLOW-THOMPSON HOUSE, 718 Perry St., Helena, AR 72342-3134. Mailing Address: P.O. Box 785, Helena, AR 72342-0785. Tel.: 870-338-8535.
E-mail: dussery@pccua.edu
Governing Authority: Parent Institution: Phillips Community College of the University of Arkansas.
Institution Type/Description: Historic House: built in 1896.
Collections: period furnishings; personal artifacts.
Activities: rental facilities.
Hours & Admission Prices: Wed.-Sat. 10-4; groups by appointment. No charge. Closed New Year's Day; Easter; Thanksgiving; Christmas.

Hope

CLINTON BIRTHPLACE HOME, 117 S. Hervey St., Hope, AR 71801-4208. Mailing Address: P.O. Box 1925, Hope, AR 71802-1925. Tel.: 870-777-4455. Fax: 870-722-6929.
E-mail: clinton@arkansas.net
Web Site: www.clintonbirthplace.com
Key Personnel: Dir., Martha Berryman
Institution Type/Description: Historic House: housed in the boyhood home of President William Jefferson Clinton. Listed on the National Register of Historic Places.
Collectons: family history; period furnishings; personal artifacts; photographs; replica of the Oval Office rug; memorial garden.
Facilities: garden. Museum-related items for sale.
Activities: special events; permanent & temporary exhibitions.
Hours & Admission Prices: Daily 9-4:30; groups by appointment. No charge.

HOPE VISITOR CENTER AND MUSEUM, (M), 100 E. Division St., Hope, AR 71801. Mailing Address: P.O. Box 596, Hope, AR 71802-0596. Tel.: 870-722-2580.
E-mail: vic-depot@hopearkansas.net
Web Site: www.hopearkansas.net
Founded: 1996.
Congressional District: 4
Personnel Profile: Full-Time Paid 1; Part-Time Paid 2.
Governing Authority: Parent Institution: City of Hope. Subsidiary Institution: Clinton Birthplace Foundation.
Institution Type/Description: History Museum: housed in the former Iron Mountain/Missouri Pacific Railroad Depot; built in 1912.
Collections: Bill Clinton memorabilia; railroad history; period furnishings; personal artifacts; photographs.
Hours & Admission Prices: Call for hours. No charge; donations accepted. &

PAUL W. KLIPSCH MUSEUM, 200 E. Division St., Hope, AR 71801. Tel.: 870-777-8200.
Institution Type/Description: History Museum: housed in the former Cairo-Fulton Railroad Depot; built c.1873.
Collections: Paul Klipsch's life & career as an audio engineer & manufacturer of loudspeakers; photographs; personal artifacts; replica of Klipsch's office.
Hours & Admission Prices: By appointment

Hot Springs

THE FINE ARTS CENTER OF HOT SPRINGS, 626 Central Ave., Hot Springs, AR 71901-5331. Mailing Address: P.O. Box 6263, Hot Springs, AR 71902-6263. Tel.: 501-624-0489.
E-mail: info@hsfac.org
Founded: 1947.
Congressional District: 19

Key Personnel: Exec. Dir., Donna Dunnahoe; Pres. (V), Bob Dion
Governing Authority: nonprofit organization. Tax-exempt: 501(c)(3).
Institution Type/Description: Art Gallery.
Collections: paintings; photographs; sculpture.
Activities: workshops; classes; outdoor concerts; special events.
Hours & Admission Prices: Mon.-Sat. 10-5. No charge, but donations accepted.
Attendance: 16,000 (estimated)
Membership: Adults $35; Seniors & Students $25; Family $50.

GANGSTER MUSEUM OF AMERICA, 510 Central Ave., Hot Springs, AR 71901. Tel.: 501-318-1717.
E-mail: director@tgmoa.com
Web Site: www.tgmoa.com
Key Personnel: Dir., Robert Raines
Institution Type/Description: History Museum.
Collections: America's most notorius criminals; photographs; videos.
Facilities: theater. Gift items for sale.
Activities: audiovisual.
Hours & Admission Prices: Sun.-Thurs. 10-5, Fri.-Sat. 10-6. Adults $12, seniors $11, children 8-12 $6; children under 8 no charge.

GARVAN WOODLAND GARDENS, 550 Arkridge Rd., Hot Springs, AR 71913-8729. Mailing Address: P.O. Box 22240, Hot Springs, AR 71903-2240. Tel.: 501-262-9300; 800-366-4664.
E-mail: gardeninfo@garvangardens.org
Web Site: www.garvangardens.org
Founded: 2002.
Key Personnel: Dir. Mktg., Sherre Freeman
Governing Authority: Parent Institution: University of Arkansas. Tax-exempt.
Institution Type/Description: Gardens.
Collections: over 200 types of azaleas; roses; shrubs; trees.
Facilities: 210 acre gardens including an Asian garden, bonsai garden, & children's garden.
Hours & Admission Prices: Feb.-March & Oct.-Nov. 21. daily 10-5; Nov. 22-Dec. 12-9, April-Sept. daily 9-8. Adults $10, children 6-12 $5, children 5 & under free; discounts to groups of 20 or more. Closed New Year's Day; Thanksgiving; Christmas. &
Attendance: 140,516 (accurate)

HOT SPRINGS NATIONAL PARK VISITOR CENTER, (M), 369 Central Ave., Bathhouse Row, Hot Springs, AR 71901-3525. Mailing Address: 101 Reserve St., Hot Springs, AR 71901. Tel.: 501-620-6701. Fax: 501-624-3458. TDD: 501-623-2308.
E-mail: hosp_interpretation@nps.gov
Web Site: www.nps.gov/hosp
Founded: 1832.
Congressional District: 4
Key Personnel: Supt., Josie Fernandez; Chief Interpreter, Mike Kusch; Interpreter & Museum Shop Coord., Nalissala Allen; Interpreter & V.I.P. Coord., Toni P. McDowell; Museum Shop Mgr., Eastern National, Amy Davis.
Personnel Profile: Full-Time Paid 7; Interns 1.
Governing Authority: federal. Parent Institution: Hot Springs National Park. A branch of National Park Service, U.S. Dept. of Interior, Washington, DC. Tax-exempt: 101(6).
Institution Type/Description: History Building.
Collections: herbarium; geology items; historical memorabilia; Native American artifacts; archives; archaeology artifacts; natural science; insects; Zander electro therapy machines; art nouveau stained glass; 1912 mission style furnishings; photographs; stereographs; blueprints; maps from 1875 to present; period rooms.
Research Fields: hydrology of thermal springs; history of bathing in Hot Springs; bathhouse architecture; spas; medical hydrology & recreation; early history of the city of Hot Springs.
Facilities: 1,500-vol. library of books on human & natural history available on premises; microfilm & Microfiche library; 52-seat auditorium; visitor center. Publications & novaculite whetstones for sale.
Activities: Fordyce Bathhouse self-guided tours; ranger-led outdoor tours for thermal feature tours; summer programs; touch screen exhibit including oral history excerpts.
Publications: orientation brochures, Fire In Folded Rocks; Valley of The Vapors; The Fordyce Bath House; Buckstaff Baths; Historical Reproductions; Hot Springs National Park In Pictures, Ye Hot Springs Picture Book; The Hot Springs of Arkansas Through the Years; American Spa; Didn't All The Indians Come Here?; Geoscenic Tour Guide; Trails of Hot Springs National Park.
Hours & Admission Prices: Daily 9-5. No charge; donations accepted. Closed New Year's Day; Thanksgiving; Christmas. &

Attendance: 206,905 (accurate)

JOSEPHINE TUSSAUD WAX MUSEUM, 250 Central Ave., Hot Springs, AR 71901. Tel.: 501-623-5836.
Institution Type/Description: Wax Museum.
Collections: over 100 wax figures; casino dice; cards; slot machines.
Hours & Admission Prices: Summer: Sun.-Thurs. 9-8, Fri.-Sat. 9-9; Winter: Mon.-Thurs. 9:30-5, Fri.-Sat. 9:30-8, Sun. 9:30-5.

MID-AMERICA SCIENCE MUSEUM, 500 Mid America Blvd., Hot Springs, AR 71913-8412. Tel.: 501-767-3461; 800-632-0583 (Arkansas). Fax: 501-767-1170.
E-mail: info@midamericamuseum.org
Web Site: www.midamericamuseum.org
Founded: 1979.
Congressional District: 4
Key Personnel: Exec. Dir., Andy Marquart; Museum Shop Mgr., Noreen Killen.
Personnel Profile: Full-Time Paid 13; Part-Time Paid 8; Part-Time Volunteers 122.
Governing Authority: Affiliated with Smithsonian Institute. Tax-exempt: 501(c)(3).
Institution Type/Description: General Museum.
Collections: visitor participation exhibits focusing on broad topics of energy, life, matter & human perception.
Facilities: theatre; snack bar; nature trail. Museum-related items for sale.
Activities: education programs, virtual reality simulator; field trips; scout badge programs; home school series & kits.
Hours & Admission Prices: Winter: Tues.-Sun. 10-5; Summer: daily 9:30-6. Adults $9, children 3-12 $7; discount to AAM & ASTC members; members & children under 3 no charge. Closed New Year's Day; Thanksgiving; Christmas Eve & Day. &
Attendance: 89,151 (accurate)
Membership: Individual $40; Family $65; Smithsonian $100; Mid-America Society $150; Friends Society $250; Dr. Martin Eisele Society $500; Cecil W. Cupp, Jr. Society $1,000.

THE MUSEUM OF CONTEMPORARY ART OF HOT SPRINGS, 425 Central Ave., Hot Springs, AR 71901. Tel.: 501-609-9966. Fax: 501-609-9955.
Web Site: www.museumofcontemporaryart.com
Institution Type/Description: Art Museum.
Collections: works by contemporary artists.
Hours & Admission Prices: Tues.-Sat. 10-5, Sun. 12-3. Adults 13 & over $5; children 12 & under no charge.

Jacksonport

JACKSONPORT STATE PARK, 205 Avenue St., Jacksonport, AR 72075. Mailing Address: 205 Avenue St., Newport, AR 72112-8771. Tel.: 870-523-2143. Fax: 870-523-4620.
E-mail: jacksonport@arkansas.com
Web Site: www.arkansasstateparks.com/jacksonport
Formerly: Jacksonport State Park Courthouse Museum
Founded: 1965.
Congressional District: 1
Key Personnel: Supt., Mark Ballard.
Personnel Profile: Full-Time Paid 7; Part-Time Paid 5.
Governing Authority: state. Subsidiary Institution: Arkansas Dept. of Parks & Tourism. Tax-exempt.
Institution Type/Description: Historic Museum: 1872 Courthouse.
Collections: period furnishings; archives; agriculture; courtroom; household items; textiles; transportation; tools; 1872 Courthouse & Clerk's Office.
Facilities: picnic area; rental pavilions; wildflower walk (trail); campground; boat launch ramp. Museum-related items for sale.
Activities: self-guided & guided tours; hiking; swimming; boating on White River; camping; interpretation programs.
Publications: annual, Stream of History.
Hours & Admission Prices: Please call for hours. No charge, donations accepted & requested. Closed New Year's Day; Thanksgiving; Christmas Eve & Day. &
Attendance: 2,100 (accurate)

Jacksonville

JACKSONVILLE MUSEUM OF MILITARY HISTORY, 100 Veterans' Circle, Jacksonville, AR 72076-4344. Tel.: 501-241-1943. Fax: 501-241-1944.
E-mail: jaxmilmuseum@gmail.com

Web Site: www.jaxmilitarymuseum.org
Founded: 1972.
Congressional District: 2
Key Personnel: Pres. (V), Joan Zumwalt; Dir., Danna Kay Duggar.
Personnel Profile: Full-Time Paid 1; Part-Time Volunteers 30.
Governing Authority: Parent Institution: Little Rock Air Force Base Historical Foundation. Tax-exempt.
Institution Type/Description: Military History Museum.
Collections: military history; personal artifacts; photographs; military equipment & uniforms.
Facilities: library.
Activities: research.
Publications: newsletter.
Hours & Admission Prices: Mon.-Sat. 9-5. Adults $3, seniors & military $2, students $1; members no charge. Closed holidays. &
Membership: Individual $25; Family $45; Supporting $100; Sustaining $250; Chairman's Circle $500; Lifetime $1,000.

Jasper

BRADLEY HOUSE MUSEUM, 403 Clark St., Jasper, AR 72641. Mailing Address: P.O. Box 360, Jasper, AR 72641-0360. Tel.: 870-446-6247.
Web Site: www.newtoncountyar.com
Founded: 1992.
Congressional District: 3
Key Personnel: Dir. & Museum Shop Mgr., Donna Dodson; Pres. (V), Thomas Niswouger.
Personnel Profile: Part-Time Paid 1; Part-Time Volunteers 1.
Governing Authority: Parent Institution: Newton County Historical Society. Tax-exempt(c)(3).
Institution Type/Description: Historic House Museum: housed in the c.1900 home of Dr. W.A. Bradley.
Collections: personal artifacts; furnishings; farm implements; woodworking tools; Native American artifacts; fossils; photographs; genealogy. Historic Building: Chaney log house.
Research Fields: genealogy; Newton County history.
Publications: annual, Newton County Homestead.
Hours & Admission Prices: April-Oct. Tues.-Thurs. 11-4; Nov.-Dec. Tues. 11-4. No charge; donations accepted. Closed Independence Day. &
Attendance: 500 (estimated)
Membership: Annual $15; Lifetime $150.

HILARY JONES WILDLIFE MUSEUM, 4208 Hwy. 7 N., Jasper, AR 72641. Mailing Address: P.O. Box 277, Jasper, AR 72641-0277. Tel.: 870-446-6180.
E-mail: newtoncoinfo@ritternet.com
Institution Type/Description: Wildlife Museum.
Collections: area wildlife history; elk mounts; fish aquariums; paintings; photographs; videos.
Facilities: Museum-related items for sale.
Activities: video presentations.
Hours & Admission Prices: Daily 9-5.

Jonesboro

AFRICAN AMERICAN CULTURAL CENTER, 1005 Logan Ave., Jonesboro, AR 72401. Tel.: 870-933-4626.
Institution Type/Description: History Museum.
Collections: African American history & culture; period furnishings; personal artifacts; photographs.
Hours & Admission Prices: Call for hours.

ARKANSAS STATE UNIVERSITY ART GALLERY, 114 S. Caraway Rd., Jonesboro, AR 72467. Mailing Address: P.O. Box 1920, State University, AR 72467-1920. Tel.: 870-972-3050. Fax: 870-972-3932.
E-mail: csteele@astate.edu
Web Site: www.finearts.astate.edu/
Founded: 1967.
Congressional District: 1
Key Personnel: Chm. Gallery Committee, Tom Chaffee; Gallery Committee, Gayle Pendergrass.
Governing Authority: university. Parent Institution: Arkansas State University Dept. of Art, Jonesboro, AR 72467. Tax-exempt.
Institution Type/Description: University Art Gallery.
Collections: contemporary & historical prints; drawings; paintings; sculpture.
Facilities: auditorium; classrooms.

Activities: lectures; gallery talks; formally organized education programs for graduate & undergraduate students; permanent, temporary, traveling & loan exhibitions.
Publications: calendar.
Hours & Admission Prices: Fine Arts Gallery: Mon.-Fri. 10-5. Bradbury Gallery: Tues.-Sat. 12-5, Sun. 2-5. No charge. Closed holidays. &
Attendance: 3,600 (estimated)

*** ARKANSAS STATE UNIVERSITY MUSEUM, (M),** Museum Bldg., 320 U Loop West Cir., Jonesboro, AR 72401. Mailing Address: P.O. Box 490, State University, AR 72467-0490. Tel.: 870-972-2074. Fax: 870-972-2793. Facebook: Arkansas State University Museum.
E-mail: mallen@astate.edu
Web Site: www.museum.astate.edu
Founded: 1933.
Congressional District: 1
Key Personnel: Exec. Dir., Dr. Marti L. Allen; Cur. Education, Jill Kary; Cur., Julie MacDonald, M.A.; Office Mgr., Valerie Ponder.
Personnel Profile: Full-Time Paid 4; Part-Time Paid 2; Interns 4.
Governing Authority: university. Parent Institution: Arkansas State University. Tax-exempt: 501(c)(3).
Institution Type/Description: History & Culture Museum.
Collections: history of Arkansas; archaeology; ethnology; costumes; military; natural history; paleontology; Commemorative China & glass; research collection: fossils, minerals, prehistoric Indian artifacts, clothing, antiques.
Research Fields: paleobotany; history of northeast Arkansas.
Facilities: reference library; reading room.
Activities: guided tours; reference service for general public; summer camps for youth; cultural diversity events; iPod tours of Old Town Arkansas; museum studies classes at M.A. & Ph.D. levels.
Publications: brochures.
Hours & Admission Prices: Tues.-Fri. 9-4, Sat.-Sun. 1-5. No charge; donations accepted. Closed national holidays. &
Attendance: 48,000 (estimated)
Membership: General $50; Lifetime Benefactor $10,000.

FORREST L. WOOD CROWLEY'S RIDGE NATURE CENTER, 600 E. Lawson Rd., Jonesboro, AR 72404. Tel.: 870-933-6787. Fax: 870-932-4582.
Web Site: www.crowleysridge.org
Institution Type/Description: Nature Center.
Collections: wildlife & their habitats; plants.
Facilities: nature trails.
Activities: hiking trails.
Hours & Admission Prices: Tues.-Sat. 8:30-4:30, Sun. 1-5. No charge. Closed major holidays.

Lake Village

MUSEUM OF CHICOT COUNTY ARKANSAS, 614 Cokley St., Lake Village, AR 71653. Mailing Address: P.O. Box 762, Lake Village, AR 71653-0762. Tel.: 870-265-2868 & 2358.
E-mail: jpburge@cei.net
Web Site: homeearthlink.net/~diod
Founded: 1994.
Key Personnel: C.E.O., Judge Mack Ball, Jr.; Pres., Edward McGehee; Treas., Vera Pesaresi.
Personnel Profile: Part-Time Volunteers 3.
Governing Authority: county; nonprofit. Tax-exempt: 501(c)(3).
Institution Type/Description: Historic Site.
Collections: medical instruments; nursery equipment; surgical table; delivery table; medical supplies; Italian immigration to Chicot county to near slavery conditions shown through photographs, letters & records. Historic Buildings: c.1910 Victorian house; 1843 log cabin.
Research Fields: data on founder, building and doctors who have practiced here, experience of former patients.
Facilities: 14,775 sq. ft. exhibit space.
Activities: guided tours.
Hours & Admission Prices: Mon., Wed. & Fri. 1-4. No charge; donations accepted.
Attendance: 2,500 (estimated)

OUR LADY OF THE LAKE CHURCH MUSEUM, 314 S. Lakeshore Dr., Lake Village, AR 71653. Tel.: 870-265-5439.
Institution Type/Description: Religious Museum.
Collections: church history; photographs; documents relating to the local Italian immigration; letters from Italy to some of Lake Villages earliest settlers.

Hours & Admission Prices: Call for hours.

Lepanto

MUSEUM LEPANTO USA, 310 S. Greenwood, Lepanto, AR 72354. Mailing Address: P.O. Box 418, Lepanto, AR 72354. Tel.: 870-475-6166. Fax: 870-475-2384.
Founded: 1978.
Key Personnel: Dir. & Pres. (V), Judy Bradford; Chm. (v), Mack Howington.
Personnel Profile: Full-Time Volunteers 8; Part-Time Volunteers 8.
Governing Authority: Tax-exempt.
Institution Type/Description: Historic Building: housed in a 1915 bank building.
Collections: local history & culture; period furniture & artifacts; Native American; replicas of a general store, blacksmith shop, doctor's office; war memorabilia; replica WWII era POW camp for German prisoners.
Publications: Books of Old.
Hours & Admission Prices: Wed. & Fri. 1-4; other times by appointment. No charge, donations accepted.

Leslie

OZARK HERITAGE ARTS CENTER & MUSEUM, 410 Oak St., Leslie, AR 72645. Mailing Address: P.O. Box 217, Leslie, AR 72645-0217. Tel.: 870-447-2500. Fax: 870-447-2528.
E-mail: ohac@windstream.net
Key Personnel: Exec. Dir., Gary Hall
Institution Type/Description: History Museum.
Collections: regional history from 1820-1960; clothing; documents; mementos; photographs; furniture.
Activities: concerts; theatrical events; permanent & temporary exhibits.
Hours & Admission Prices: April-Dec. Tues.-Sat. 10-4. No charge.

Lincoln

ARKANSAS COUNTRY DOCTOR MUSEUM, 107 & 109 N. Starr Ave., Lincoln, AR 72744. Mailing Address: P.O. Box 1004, Lincoln, AR 72744-1004. Tel.: 479-824-4307. Fax: 479-824-4307.
E-mail: countrydoc@pgtc.com
Web Site: www.drmuseum.net
Founded: 1994.
Congressional District: 3
Key Personnel: Pres., Roy Horne; Vice Pres., Mike Allen; Treas. & Security, Jerry Leach; Museum Shop Mgr., Diana Hale; Sec., Carolyn McDonald.
Personnel Profile: Part-Time Paid 2; Part-Time Volunteers 26.
Governing Authority: private; nonprofit organization. Tax-exempt: 501(c)(3).
Institution Type/Description: Country Doctor Museum.
Collections: early medical instruments & furnishings; 20th-century medical equipment; iron lung with portable respirators in polio exhibit; 1924 Model T Roadster; 1886 Studebaker Doctors Buggy; carriage house; herb garden.
Research Fields: country doctors & nurses in Arkansas up to 1950.
Facilities: 400-vol. library of medical reference books; botanical garden; 3,650 sq. ft. exhibit space. Museum-related items for sale.
Activities: docent program; films; formal education for area students; guided tours; lectures; loan exhibitions. Annual Events: ice cream social; cookout; yard sale fundraiser.
Publications: newsletter 3 times per year, Arkansas Country Doctor Museum; book, My Spirit is Free - Reflections from an Iron Lung.
Hours & Admission Prices: Feb. 20-Dec. 11 Wed.-Sat. 1-4; other times by appointment. No charge; donations accepted. Closed Independence Day.
Attendance: 625 (estimated)
Membership: Individual $10; Family $15; Patron $50; Sponsor $100; Benefactor $250; Guardian $500; Major Endowment $1,000.

Little Rock

✳ **THE ARKANSAS ARTS CENTER, (M),** MacArthur Park, 9th & Commerce, Little Rock, AR 72202. Mailing Address: P.O. Box 2137, Little Rock, AR 72203-2137. Tel.: 501-372-4000. Fax: 501-375-8053.
E-mail: showell@arkansasartscenter.org
Web Site: www.arkansasartscenter.org
Founded: 1937.
Congressional District: 2
Key Personnel: Exec. Dir., Todd A. Herman, Ph.D.; Chm. (V), Charlotte Bradbury; Pres. (V), Mary Ellen Vangilder; Chief Cur. & Cur. Contemporary Craft, Brian Lang; Cur. Drawings, Ann Prentice Wagnner, PhD; Deputy Dir. Operations, Laine Harber; Dir. Education, Louise Palermo; Registrar, Thom Hall; Dir. Devel., Kelly Ford; Dir. Children's Theatre, Bradley Anderson; State Svcs., Jessica Wright; Museum Shop Mgr., Kim White; Volunteer Coord., Sinovia Mayfield.
Personnel Profile: Full-Time Paid 47; Part-Time Paid 41; Part-Time Volunteers 359; Interns 3.
Volunteer Hours: 15,244
Operating Expenses: 5,803,052
Operating Income: 5,858,521
Governing Authority: city; nonprofit organization. Subsidiary Institution: Arkansas Arts Center Foundation. Tax-exempt: 501(c)(3).
Institution Type/Description: Art Museum.
Collections: 19th-20th century American drawings & 15th to 20th-century European drawings; American & European paintings, sculpture, prints & photographs; American, European & Asian decorative arts & contemporary American crafts.
Major Exhibits: Mark Rothko in the 1940s: The Decisive Decade (T), 10/25/13-2/9/14; Face to Face: Artists' Self-Portraits From the Collection of Jackye & Curtis Finch, Jr., 10/25/13-2/9/14; Portraiture Now: Drawing on the Edge, 10/25/13-2/9/14; The Crossroads of Memory: Carroll Cloar & the American South (T), 2/28/14-6/1/14; 53rd Young Arkansas Artists, 5/9/14-7/27/14; 56th Annual Delta, 6/27/14-9/28/14; 12th National Drawing Invitational, 7/18/14-10/5/14; A Sense of Balance: The Sculptue of Stoney Lamar (T), 10/24/14-1/18/15; William Beckman Drawings: A Retrospective 1967-2013 (T), 10/24/14-2/1/15.
Research Fields: American & European drawings.
Facilities: 5,000-vol. library on art & drama available for inter-library loan; reading room; museum school; 381-seat theater; restaurant; 140-seat lecture hall. Museum-related items for sale.
Activities: permanent, temporary & circulating exhibitions; formal organized education program; guided tours; lectures; gallery talks; studio classes for adults & children in visual arts.
Publications: quarterly newsletter, Works; annual report; exhibition & permanent collection catalogues.
Hours & Admission Prices: Mon.-Sat. 10-5, Sun. 11-5. No charge, suggested donation $5; discount to AAM members; members no charge. Closed major holidays.
Attendance: 239,680 (accurate)
Membership: Individual $55; Family $65; Participating $80; Contributing $150; Associate $300; Supporting $600. Corporate Memberships: Corporate Affiliate $300; Corporate Citizen $600.

ARKANSAS GAME AND FISH COMMISSION, 2 Natural Resources Dr., Little Rock, AR 72205-1572. Tel.: 501-223-6300; 800-364-4263. Fax: 501-223-6465.
E-mail: askagfc@agfc.state.ar.us
Web Site: www.agfc.com
Founded: 1945.
Congressional District: 2
Key Personnel: Dir., Mike Knoedl.
Governing Authority: state.
Institution Type/Description: Zoology Museum.
Collections: Arkansas wildlife exhibits; zoology.
Activities: tours.
Publications: magazine, Arkansas Wildlife.
Hours & Admission Prices: Mon.-Fri. 8-4:30. No charge.

BARTON ROCK AND ROLL MUSEUM, 2600 Howard St., Little Rock, AR 72207. Tel.: 501-372-8341.
Institution Type/Description: History Museum.
Collections: Barton Coliseum history & rock bands who have performed there; guitars; 8 track player & tapes; framed records; photographs; newspaper articles.
Activities: guided tours.
Hours & Admission Prices: By appointment.

EMOBA - MUSEUM OF BLACK ARKANSAS, 12th & Louisiana, Little Rock, AR 72214. Mailing Address: P.O. Box 46754, 12th & Louisiana, Little Rock, AR 72214-6754. Tel.: 501-661-9903.
Web Site: www.onlinelittlerock.com/emoba.htm
Key Personnel: Founder & Dir., Ernie Dodson.
Governing Authority: Tax-exempt: 501(c)(3).
Institution Type/Description: Black History Museum.
Collections: history & culture of Black Arkansans & their contributions to the state; photographs; personal artifacts.
Activities: special events.
Hours & Admission Prices: Call for hours.

FIREHOUSE HOSTEL AND MUSEUM, 1201 Commerce St., Little Rock, AR 72202. Mailing Address: P.O. Box 2753, Little Rock, AR 72203. Tel.: 501-476-0294.
E-mail: firehousehostel@sbcglobal.net
Web Site: firehousehostel.org
Institution Type/Description: Historic Building: housed in the former Old Fire Station Number Two; built in 1917.
Collections: local history; firefighter equipment, memorabilia & artifacts; 1933 & 1955 fire trucks; brass pole; photographs; early records.
Facilities: 34 bed hostel.
Activities: fire safety education; group tours.
Hours & Admission Prices: Call for hours.

HEARNE FINE ART, 1001 Wright Ave., Ste. C, Little Rock, AR 72206. Tel.: 501-372-6822. Fax: 501-372-7133.
E-mail: info@hearnefineart.com
Web Site: hearnefineart.com
Institution Type/Description: Art Gallery.
Collections: paintings; sculpture.
Activities: temporary exhibitions.
Hours & Admission Prices: Mon.-Thurs. 9-5, Fri. 9-6, Sat. 11-6; other times by appointment.

✻ **HISTORIC ARKANSAS MUSEUM,** 200 E. Third St., Little Rock, AR 72201-1608. Tel.: 501-324-9351. Fax: 501-324-9345. Facebook: Historic Arkansas Museum; TDD: 501-324-9811.
E-mail: info@historicarkansas.org
Web Site: www.historicarkansas.org
Formerly: Arkansas Territorial Restoration
Founded: 1941.
Congressional District: 2
Key Personnel: Dir. & C.E.O., William B. Worthen, Jr.; Chm. (V), Frances Ross; Pres. (V), Wally Nixon; Dir. Education, Starr Mitchell; Historic Site Specialist, David Etchieson; Dir. Communications, Ellen Korenblat; Conservator, Andrew Zawacki; Cur. Research, Swannee Bennett; Dir. Devel., Louise Terzia; Registrar, Lark Buckingham; Fiscal Mgr., Rebecca Hochradel; Dir. Volunteers & Membership, Tricia Spione; Security Supvr., Mike Croy; Museum Shop Mgr., Paige James.
Personnel Profile: Full-Time Paid 21; Part-Time Paid 20; Part-Time Volunteers 27; Interns 2.
Governing Authority: state. Parent Institution: Dept. of Arkansas Heritage. Tax-exempt.
Institution Type/Description: History Museum.
Collections: period furnishings; decorative, mechanical & fine arts produced in Arkansas from 1819 to 1890; Arkansas-made art & artifacts from 1700's to present. Historic Buildings: c.1830 Hinderliter House; c.1845 Brownlee House; c.1848 McVicar House.
Research Fields: pre-Civil War Arkansas, Arkansas made Pre-history to The Present.
Facilities: 1,000-vol. library on early Arkansas; gardens; conservation lab; educational center for school children. Museum-related items for sale.
Activities: guided tours; special tours for groups & school children. Annual Events: Christmas Open House; Independence Day open house; Candlelight Gala; Territorial Fair in May.
Publications: exhibit catalogues; brochure; newsletter, Collections; A Garden Heritage; Arkansas Made: The Decorative, Mechanical & Fine Arts Produced in Arkansas, 1819-1870, Volumes I & II; The Likeness Trade: Portrait Painting in Arkansas, 1790-1900. (1996)
Hours & Admission Prices: Mon.-Sat. 9-5, Sun. 1-5. Galleries: no charge. Tour: Adults $2.50, senior citizens $1.50, children $1; discount to AAM members; members no charge. Closed New Year's Day; Easter; Thanksgiving; Christmas Eve & Day. &
Attendance: 50,000 (estimated)
Membership: Individual $35; Family $50; Supporting $100; Sustaining $250; Founder $500; Cornerstone $1,000 & up.

HISTORICAL RESOURCES AND MUSEUM SERVICES, Arkansas State Parks, One Capitol Mall, Little Rock, AR 72201-1013. Tel.: 501-682-3603. Fax: 501-682-0081.
E-mail: patricia.murphy@arkansas.gov
Web Site: arkansasstateparks.com
Founded: 1979.
Congressional District: 2
Key Personnel: Dir., Patricia Maguire Murphy; Museums Coord., William Long.
Personnel Profile: Full-Time Paid 3.
Governing Authority: state. Parent Institution: Arkansas Dept. of Parks & Tourism. Branch Museums: Prairie County Museum, DesArc, AR; Arkansas Museum of Natural Resources, Smackover, AR; Plantation Agriculture Museum, Scott, AR; AR Post Museum, Gillett, AR. Tax-exempt.
Institution Type/Description: State Agency for Museum Services.
Collections: local history.
Activities: professional & technical advice to state museums; resource library; workshops; administrative services & master planning for affiliated museums.
Publications: occasional publications on museology & museography; workshop papers.
Hours & Admission Prices: Business Office: Mon.-Fri. 8-5. &

LITTLE ROCK CENTRAL HIGH SCHOOL NATIONAL HISTORIC SITE, 2120 Daisy L. Gaston Bates Dr., Little Rock, AR 72202. Tel.: 501-374-1957. Fax: 501-396-3001.
E-mail: CHSC_visitor_center@nps.gov
Web Site: www.nps.gov/chsc/
Formerly: Central High Museum & Visitor Center
Founded: 1995.
Congressional District: 2
Key Personnel: Supt., Robin White.
Personnel Profile: Full-Time Paid 12; Part-Time Paid 3; Part-Time Volunteers 30.
Governing Authority: Parent Institution: National Park Service.
Institution Type/Description: History Museum.
Collections: Central High photographs & artifacts from 1957 crisis.
Research Fields: School Desegregation - Civil Rights.
Facilities: visitor center.
Activities: ranger-led programs; distance learning programs; curriculum-based education programs.
Publications: lesson plans; interpretive site bulletins; newsletter, Constitutional Writes.
Hours & Admission Prices: Daily 9-4:30. No charge; donations accepted. Closed New Year's Day; Thanksgiving; Christmas. &
Attendance: 76,000 (accurate)

LITTLE ROCK ZOOLOGICAL GARDENS, One Jonesboro Dr., Little Rock, AR 72205-5401. Tel.: 501-666-2406. Fax: 501-666-7040. TDD: 501-399-3451.
E-mail: mblakely@littlerock.org
Web Site: www.littlerockzoo.com
Founded: 1926.
Congressional District: 2
Key Personnel: C.E.O., Michael E. Blakely; Chm. Bd. Governors, George Mallory; Vice Chm., Blair Allen; Chm. Docent Council (V), Lisa Buehler; Museum Shop Mgr., Barbara Brown.
Personnel Profile: Full-Time Paid 37; Part-Time Paid 11; Part-Time Volunteers 12; Interns 2.
Governing Authority: municipal. Parent Institution: City of Little Rock, AR. Subsidiary Institution: Little Rock Zoo Department.
Institution Type/Description: Zoo.
Collections: 210 species mammals, birds, reptiles, amphibians, invertebrates; 757 specimen.
Research Fields: behavior; blood chemistry; reproductive cycle monitoring through urinalysis.
Facilities: library of science & natural history books; lecture hall.
Activities: guided tours; lectures; Zoo Explorer Post; docent council; miniature train ride; camel rides in summer.
Publications: map; bimonthly newsletter; teachers' guide; education brochures biannual magazine.
Hours & Admission Prices: Summer: daily 9-5; Winter: daily 9-4:30. Adults $8, seniors 60 & over and children 1-12 $6; discounts to AZA institutions; group rates with reservation; zoo members no charge. Closed New Year's Day; Thanksgiving; Christmas. &
Attendance: 283,768 (accurate)
Membership: Individual $40; Individual Plus & Dual $50; Family & Grandparents $60; Grandparents Plus $70

MACARTHUR MUSEUM OF ARKANSAS MILITARY HISTORY, Tower Building of the Little Rock Arsenal, 503 E. Ninth St., Little Rock, AR 72202-3997. Tel.: 501-376-4602.
Web Site: www.arkmilitaryheritage.com
Founded: 2001.
Congressional District: 4
Key Personnel: Exec. Dir., Stephan McAteer
Institution Type/Description: Military History Museum: birthplace of General Douglas MacArthur.
Collections: state's military heritage; artifacts; photographs; weapons; documents; uniforms; military artifacts.
Activities: tours.

Hours & Admission Prices: Mon.-Sat. 9-4, Sun. 1-4. No charge. Closed New Year's Day; Thanksgiving; Christmas Eve & Day. &
Attendance: 28,000 (estimated)

MOSAIC TEMPLARS CULTURAL CENTER, (M), 501 W. 9th St., Little Rock, AR 72201-4111. Mailing Address: 1500 Tower Bldg., 323 Center St., Little Rock, AR 72201-2603. Tel.: 501-683-3593.
E-mail: info@mosaictemplarscenter.com
Web Site: www.mosaictemplarscenter.com
Founded: 2008.
Congressional District: 2
Key Personnel: Dir., Sericia Cole; Museum Shop Mgr., Phyllis Brown.
Personnel Profile: Full-Time Paid 7; Part-Time Paid 4.
Institution Type/Description: African American History Museum.
Collections: African American life, history & culture; photographs; personal artifacts.
Activities: performing arts; conferences; special events; educational programs.
Hours & Admission Prices: Call for hours.

*   **MUSEUM OF DISCOVERY, (M),** 500 President Clinton Ave., Ste. 150, Little Rock, AR 72201-1757. Tel.: 501-396-7050, ext. 200. Fax: 501-396-7054.
E-mail: mrobertson@amod.org
Web Site: www.amod.org
Formerly: Museum of Discovery: Arkansas Museum of Science and History
Founded: 1927.
Congressional District: 2
Key Personnel: Exec. Dir., Nan Selz; Chm. (V), Robert Childress; Educator, Animal Caretakers, Nichole Ashley; Reservationist, Beth Nelsen; Dir. Collections & Research and Grantwriter, Marci Bynum Robertson; Dir. Mktg., Katie McManners; Educator, David Westbrook; Dir. Finance, Nikki Parnell; Dir. Exhibits & Facilities, Joel Gordon; Dir. Programs, Carol Couser; Dir. Devel., Meredith Poland; Network Coord., Diane LaFollette; Museum Shop Mgr., Fawn True.
Personnel Profile: Full-Time Paid 14; Part-Time Paid 20; Part-Time Volunteers 823; Interns 2.
Governing Authority: bd. trustees; nonprofit. Tax-exempt: 501(c)(3).
Institution Type/Description: Science & Technology Museum.
Collections: birds; mammals; reptiles; invertebrates; Indian artifacts; pottery; pioneer items; interactive exhibits; furniture; appliances; postal jeep; kid's gallery; multicultural masks; Kachina dolls.
Research Fields: traveling exhibits created in-house; early childhood education; immigration into Arkansas; science & technology; biology of live animals; ethnographic studies.
Facilities: 300-vol. library of science & history books; classrooms; theater. Books, crafts, rocks, & science materials for sale.
Activities: gallery talks; TV programs; formally organized education programs for children; docent program; permanent & temporary exhibitions; demonstrations; school extension services.
Publications: seasonal class schedule; Resource Manual; Museum Newsletter; Museum Guide.
Hours & Admission Prices: Tues.-Sat. 9-5, Sun. 1-5. &
Attendance: 103,103 (accurate)
Membership: Basic $45; Family $75; Contributing: $110.

*   **OLD STATE HOUSE MUSEUM, (M),** 300 W. Markham St., Little Rock, AR 72201-1423. Tel.: 501-324-9685. Fax: 501-324-9688. Facebook: Old State House Museum.
E-mail: info@oldstatehouse.com
Web Site: www.oldstatehouse.com
Formerly: The Old State House
Founded: 1951.
Congressional District: 2
Key Personnel: Dir., Bill Gatewood; Historic Sites Mgr., Ed Garretson; Dir. Devel., Brooke Malloy; Deputy Dir., Brendetta Murrell; Dir. Public Rels., Matt Rowe; Dir. Education, Georganne Sisco; Dir. Exhibits, Gail Stephens; Cur., JoEllen Maack; Museum Shop Mgr., David Kennedy.
Personnel Profile: Full-Time Paid 22; Part-Time Paid 18; Part-Time Volunteers 15; Interns 1.
Volunteer Hours: 2,242
Governing Authority: state. Parent Institution: Dept. of Arkansas Heritage. Tax-exempt.
Institution Type/Description: History Museum: built 1833-1842, first state capitol (1836-1911).
Collections: history collection; Civil War artifacts; Confederate flags; 19th-century costumes; textiles & decorative arts; Arkansas architectural drawings; wallpaper; 19th- & early 20th-century quilts by Black Arkansans;

Arkansas art pottery; Arkansas political history; Arkansas music; first families of Arkansas.
Major Exhibits: Things You Need to Hear Growing up in Arkansas (T), 2/12-2/14; Lights, Camera Arkansas: Hollywood & Arkansas (T), 4/13-3/15; Lights! Camera! Arkansas (T), 11/13-3/14; Growing up in Arkansas (T), 12/13-2/14.
Research Fields: Arkansas political & cultural history.
Facilities: 2,000-vol. library; meeting & classrooms. Museum-related items for sale.
Activities: workshops; scholarly seminars; living history interpretation; historic tours; inter-museum loan; permanent collection & traveling exhibits; formally organized activities for school groups; summer classes for children; special events for members.
Publications: brochures; newsletter, Columns; student newspaper, The Arkansas News; exhibit catalogs; exhibit books; podcasts.
Hours & Admission Prices: Mon.-Sat. 9-5, Sun. 1-5. No charge; donations accepted. Closed New Year's Day; Thanksgiving; Christmas Eve & Day. &
Attendance: 52,866 (accurate)
Membership: Old State House Museum Associates: Basic $30; Supporting $60; Contributing $100; Sustaining $250.

PINNACLE MOUNTAIN STATE PARK, 11901 Pinnacle Valley Rd., Little Rock, AR 72223-5173. Tel.: 501-868-5806. Fax: 501-868-5018.
E-mail: pinnaclemountain@arkansas.com
Web Site: www.arkansasstateparks.com
Founded: 1977.
Congressional District: 2
Key Personnel: Supt., Ron Salley; Asst. Supt., Josh Jeffers; Ranger, Lori Froeschner; PASC, Joy Daniel; Park Interpreter, Maryanne Stansbury; Park Interpreter, Richard Spilman; Museum Shop Mgr., Vernon McClain
Governing Authority: state. Arkansas Dept. of Parks & Tourism, One Capitol Mall, Little Rock, AR 72201. Tax-exempt.
Institution Type/Description: Environmental Education Center.
Collections: plants; vertebrates; invertebrates; geology; meteorology.
Facilities: 1,200-vol. library of textbooks and Natural Science Guides available for use on premises; classrooms; visitor center; environmental education & conservation center; picnic area; nature trails. Nature-related books, gifts & postcards for sale.
Activities: guided tours; lectures; demonstrations; films; concerts; drama; formally organized education programs; interpretative exhibits; festivals.
Publications: interpretative brochures.
Hours & Admission Prices: April-Sept. Mon.-Fri. 8-5, Sat.-Sun. 8-6; Oct.-March daily 8-5. No charge. Visitor Center: closed New Year's Day; Thanksgiving; Christmas. &
Attendance: 515,666 (accurate)

UNIVERSITY OF ARKANSAS AT LITTLE ROCK ART DEPARTMENT GALLERY I & II & III, 2801 S. University Ave., Little Rock, AR 72204-1000. Tel.: 501-569-3182. Fax: 501-569-8775.
E-mail: becushman@ualr.edu
Web Site: ualr.edu/art/index.php/home/gallery/
Founded: 1972.
Congressional District: 2
Key Personnel: Dir., Brad Cushman; Chm. Dept. of Art, Win Bruhl; Gallery Asst., Nathan Larson.
Personnel Profile: Full-Time Paid 1; Part-Time Paid 1; Interns 2.
Governing Authority: state university; not for profit organization. Parent Institution: University of Arkansas. Tax-exempt.
Institution Type/Description: University Art Gallery.
Collections: 20th-Century American photography; paintings; prints; drawings; sculpture.
Research Fields: Arkansas artists.
Facilities: 4,300 sq. ft. exhibit space; educational facilities.
Activities: lectures; loan, temporary & traveling exhibitions; gallery talks; docent tours; formally organized educational programs for UALR undergraduate & graduate students.
Publications: biannual, Calendar of Events; exhibition catalogs; postcards.
Hours & Admission Prices: May-Aug. Mon.-Fri. 9-5; other times by arrangement; Sept. to mid-May Mon.-Fri. 9-5, Sat. 10-1, Sun. 2-5. No charge. Closed spring break week; Christmas week; University holidays. &
Attendance: 6,000 (estimated)

WILLIAM J. CLINTON PRESIDENTIAL LIBRARY & MUSEUM, 1200 President Clinton Ave., Little Rock, AR 72201-1749. Tel.: 501-372-4242. Fax: 501-244-2883.
E-mail: clinton.library@nara.gov
Web Site: www.clintonlibrary.gov

Founded: 2001.
Key Personnel: Dir., Terri Garner; Education, Kathleen Pate; Registrar, Audra Oliver; Cur., Christine Mouw; Deputy Dir., Kurt Senn; Museum Shop Mgr., Connie Fails; Security, Steve Samford.
Personnel Profile: Full-Time Paid 40; Part-Time Paid 4; Interns 4.
Governing Authority: federal government. Parent Institution: National Archives & Records Administration, Washington, DC.
Institution Type/Description: Presidential History Museum.
Collections: life & career of Bill Clinton; Presidential papers, documents & gifts; personal artifacts; photographs; replicas of the Oval Office & the Cabinet Room.
Facilities: research archives; restaurant; educational facilities. Museum-related items for sale.
Activities: research; special events; formal educational programs; docent program; guided tours; temporary & traveling exhibitions; acoustiguide.
Hours & Admission Prices: Mon.-Sat. 9-5, Sun. 1-5. Adults 18-61 $7, college students, retired military and senior citizens 62 & over $5, children 6-17 $3; children under 6, school groups w/reservation, active duty, military reservists, National Guard, UACS faculty & staff no charge. Closed New Year's Day; Thanksgiving; Christmas. &
Attendance: 300,000 (accurate)
Membership: Individual $35; Family $50.

WITT STEPHENS JR. CENTRAL ARKANSAS NATURE CENTER, 602 President Clinton Ave., Little Rock, AR 72201. Tel.: 501-907-0636. Fax: 501-907-0638.

Web Site: www.centralarkansasnaturecenter.com
Founded: 2008.
Institution Type/Description: Nature Center.
Collections: wildlife & their habitats; plants.
Facilities: theater; aquarium. Gift items for sale.
Activities: educational programs.
Hours & Admission Prices: Tues.-Sat. 8:30-4:30, Sun. 1-5. No charge. Closed major holidays.

Lonoke

LONOKE COUNTY MUSEUM, 215 S.E. Front St., Lonoke, AR 72086. Tel.: 501-676-6750.

Key Personnel: Dir., Sherryl Miller
Institution Type/Description: History Museum.
Collections: local history & culture; period artifacts; photographs.
Activities: special events; permanent & temporary exhibitions; research.
Hours & Admission Prices: Wed.-Thurs. 8-4; other times by appointment. No charge; donations accepted. Closed major holidays.

Lowell

LOWELL HISTORICAL MUSEUM, 304 Jackson Pl., Lowell, AR 72745. Tel.: 479-770-0191.

Founded: 1976.
Congressional District: 3
Key Personnel: Dir., Liz Estes.
Personnel Profile: Full-Time Paid 1; Part-Time Paid 5.
Governing Authority: city. Tax-exempt.
Institution Type/Description: History Museum.
Collections: period artifacts; furnishings; memorabilia; photographs; newspapers.
Major Exhibits: Steele - Allens Canning, 11/13-6/14; Grandma's Doiles, 12/13-7/14.
Activities: stage coach rides.
Hours & Admission Prices: Mon.-Thurs. 9-3, Sat. 10-4. No charge; donations accepted. Closed most holidays. &
Attendance: 1,461 (accurate)
Membership: Friends of the Museum $10.

Magazine

EVANS MUSEUM, 6335 N. Arkansas 109, Magazine, AR 72943. Tel.: 479-963-3987.

Institution Type/Description: History Museum.
Collections: local history; military artifacts; dolls; Native American artifacts; toys; porcelain.
Hours & Admission Prices: Tues.-Fri. 11-4; other times by appointment.

Malvern

HOT SPRING COUNTY MUSEUM - THE BOYLE HOUSE/1876 LOG CABIN/1868 LOG CABIN, 302 E. Third St., Malvern, AR 72104-3912. Mailing Address: 12697 Hwy. 9, Malvern, AR 72104-6323. Tel.: 501-337-4775.

E-mail: janiswest@hughes.net
Founded: 1981.
Congressional District: 4
Key Personnel: Dir. & Chm. (V), Janis West; Cur. & Business Officer, Mary Waniewski.
Personnel Profile: Part-Time Paid 3; Part-Time Volunteers 30; Interns 2.
Governing Authority: Parent Institution: HSC Museum Commission. Tax-exempt.
Institution Type/Description: History Museum.
Collections: period artifacts & local history items pertaining to Hot Spring County; first lady doll collection; 2 pump organs; child's piano; accordian; quilts; Civil War. Historic Buildings: 1892 Boyle House; 1876 log cabin; 1868 log cabin.
Research Fields: local, state & national history.
Activities: guided tours; permanent & temporary exhibitions; school tours.
Publications: brochure.
Hours & Admission Prices: Wed.-Fri. 12:30-4:30; group & school tours by appointment. No charge; donations accepted. Closed holidays. &
Attendance: 1,500 (estimated)
Membership: Individual $10; Family $20; Sponsor II $50; Sponsor I $100; Patron III $200; Patron II $300; Patron I $500.

Mammoth Spring

MAMMOTH SPRING STATE PARK, DEPOT MUSEUM, U.S. 63, Mammoth Spring, AR 72554-0036. Mailing Address: P.O. Box 36, Mammoth Spring, AR 72554-0036. Tel.: 870-625-7364. Fax: 870-625-3255.

E-mail: mammothspring@arkansas.com
Web Site: www.arkansasstateparks.com
Founded: 1971.
Congressional District: 1
Key Personnel: Park Supt., Dave Jackson.
Personnel Profile: Part-Time Paid 2.
Governing Authority: state. Parent Institution: Arkansas Dept. of Parks & Tourism, One Capitol Mall, Little Rock, AR 72201.
Institution Type/Description: History Museum: housed in 1886 Frisco Railroad depot.
Collections: railroad memorabilia; early history of Mammoth Spring.
Facilities: picnic area; playground; nature trail.
Activities: guided tours; lectures; permanent exhibitions.
Hours & Admission Prices: Tues.-Sat. 8-5, Sun. 1-5. No charge; donations accepted. &
Attendance: 7,000 (estimated)

Marianna

MARIANNA-LEE COUNTY MUSEUM ASSOC. INC., 67 W. Main St., Marianna, AR 72360-2243. Mailing Address: 60 McCulloch, Marianna, AR 72360-2030. Tel.: 870-295-2439.

Founded: 1981.
Congressional District: 30
Key Personnel: Cur., Suzy Keasler.
Governing Authority: nonprofit organization. Tax-exempt: 501(c)(3).
Institution Type/Description: Museum Association: housed in 1910 Marianna Elks Club.
Collections: 1835-present, implements & tools used by or made by the first settlers; 1800-1940, kitchen utensils, stoves & cabinets; musical instruments; music box; 1933 Philco radio; early Edisons; zithers, dulcimers, mandolins, horns; costumes, wedding gowns, bustle dresses; needlework; flapper dresses; shoes; Indian artifacts. Historic Building: 1880 one-room schoolhouse.
Activities: lectures; films; permanent & temporary exhibitions.
Publications: quarterly newsletter, Museings.
Hours & Admission Prices: Mon.-Sat. by appointment. No charge; donations accepted. Closed major holidays.
Membership: Student $2.50; Individual $5; Family $7.50; Sustaining $10; Donor $25; Patron, Institutional or Corporate $50 & up; Life $250.

Marked Tree

MARKED TREE DELTA AREA MUSEUM, 308 Frisco St., Marked Tree, AR 72365. Mailing Address: P.O. Box 106, Marked Tree, AR 72365-0106. Tel.: 870-358-4998.
Institution Type/Description: History Museum.
Collections: local history & culture; American Indian pottery; early 1900s telephones; hospital replica; general store.
Hours & Admission Prices: Wed.-Fri. 1-4:30, Sat. 9:30-12:30, Sun. 1-4.

Maynard

MAYNARD PIONEER MUSEUM & PARK, Hwy. 328 W., Maynard, AR 72444. Mailing Address: P.O. Box 486, Maynard, AR 72444-0486. Tel.: 870-647-2701. Fax: 870-647-2701. Facebook: Maynard Pioneer Museum & Park.
E-mail: maynardcityhall@centurytel.net
Founded: 1979.
Congressional District: 1
Key Personnel: Chm., Wyle Greer.
Personnel Profile: Part-Time Volunteers 7.
Governing Authority: Parent Institution: City of Maynard.
Institution Type/Description: History Museum.
Collections: Randolph County history; period furnishings; photographs; documents; newspaper clippings; personal artifacts.
Facilities: T-shirts, hats & museum-related items for sale.
Activities: festivals. Annual Event: Pioneer Days in September.
Publications: visitors center brochures.
Hours & Admission Prices: May-Oct. 1 Mon.-Fri. 8-3 by appointment. No charge; donations accepted. Closed Memorial Day.
Attendance: 7,000 (estimated)

McGehee

WWII JAPANESE AMERICAN INTERNMENT MUSEUM, 100 S. Railroad St., McGehee, AR 71654. Mailing Address: P.O. Box 1263, McGehee, AR 71654. Tel.: 870-222-9168.
Institution Type/Description: History Museum.
Collections: military history & artifacts; uniforms; photographs; personal artifacts.
Activities: special events.
Hours & Admission Prices: Tues.-Sat. 10-5. No charge.

McNeil

LOGOLY STATE PARK, County Rd. 47 (Logoly Rd.), McNeil, AR 71752. Mailing Address: P.O. Box 245, McNeil, AR 71752-0245. Tel.: 870-695-3561. Fax: 870-695-3729.
E-mail: logoly@arkansas.us
Web Site: www.arkansasstateparks.com
Founded: 1978.
Congressional District: 4
Key Personnel: Supt., Jim Gann; Interpretive Naturalist, Barley Park; Museum Shop Mgr., Pat Swearingen.
Personnel Profile: Full-Time Paid 5; Part-Time Paid 2.
Governing Authority: state; nonprofit. Affiliated with Arkansas Dept. of Parks & Tourism. Tax-exempt.
Institution Type/Description: Park Museum & Nature Center.
Collections: specimens of native wildlife.
Research Fields: environmental education; history of park area.
Facilities: nature center; theater; classrooms; picnic area; nature trails; amphitheater. Gift items for sale.
Activities: guided tours; special events; films; formally organized education programs; camping.
Publications: newsletter, Friends of Logoly; self-guided trail brochure.
Hours & Admission Prices: Park: daily 8am to one hour after sunset. Visitors Center: May-Oct. daily 8-5; Nov.-April Mon.-Fri. 8-5, Sat.-Sun. 1-5. No charge. &
Attendance: 25,000 (accurate)

Mena

MENA ART GALLERY, 607 Mena St., Mena, AR 71953. Mailing Address: P.O. Box 871, Mena, AR 71953. Tel.: 479-394-3880.
Web Site: www.menaartgallery.org
Institution Type/Description: Art Gallery.
Collections: works by artists from around the world.
Activities: workshops; seminars; classes.
Hours & Admission Prices: Gallery: Wed.-Sat. 10-3. Art Day: Tues. 11-2. No charge.

Monticello

DREW COUNTY HISTORICAL MUSEUM, 404 S. Main, Monticello, AR 71655-4818. Tel.: 870-367-7446.
Web Site: www.arkansasroots.com
Founded: 1970.
Congressional District: 4
Key Personnel: Dir., Sheilla Lampkin.
Personnel Profile: Part-Time Paid 4; Part-Time Volunteers 1; Interns 1.
Governing Authority: society. Parent Institution: Drew County Historical Society. Tax-exempt.
Institution Type/Description: History Museum.
Collections: local historical objects; textiles; furniture; Indian artifacts; clothing; war memorabilia; quilts & spinning artifacts.
Research Fields: early Drew County history; Civil War; Indian archaeology.
Facilities: archives.
Activities: classes; special visitor tours.
Publications: monthly, Society and Museum Notes; annual journal.
Hours & Admission Prices: Fri. 1-5, Sat.-Sun. 2-5. No charge; donations accepted. Closed New Year's Day; Thanksgiving; Christmas. &
Attendance: 2,500 (estimated)
Membership: Individual (with yearly Journal) $25; Couple $40; Friends $100-$299.99; Associate $300-$499.99; Patron $500-$999.99; Benefactor $1,000 & up.

TURNER NEAL MUSEUM OF NATURAL HISTORY AND POMEROY PLANETARIUM, University of Arkansas at Monticello, Science Center, 397 University Dr., Monticello, AR 71656. Mailing Address: P.O. Box 3480, Monticello, AR 71656. Tel.: 870-460-1016.
Key Personnel: Dir. Museum, Dr. Jim Edson; Dir. Planetarium, Joe Guenter
Institution Type/Description: Natural History.
Collections: invertebrates; fishes; amphibians; reptiles; birds; mammals; fossils; minerals; plants.
Facilities: planetarium.
Hours & Admission Prices: Academic Year: Mon.-Fri. call for hours; other times by appointment.

Morrilton

CONWAY COUNTY HISTORICAL PRESERVATION ASSOCIATION, INC. - MORRILTON DEPOT MUSEUM, 101 E. Railroad Ave., Morrilton, AR 72110. Mailing Address: P.O. Box 417, Morrilton, AR 72110. Tel.: 501-354-4347. Facebook: Morrilton Depot Museum.
E-mail: morriltondepotmuseum@yahoo.com
Web Site: www.morrlitondepotmuseum.com
Founded: 1977.
Congressional District: 60
Key Personnel: Pres. (V), Carl Imhauser.
Personnel Profile: Part-Time Volunteers 20.
Governing Authority: Parent Institution: CCHPA. Tax-exempt.
Institution Type/Description: History Museum.
Collections: local history; photographs; Native American artifacts; Civil War; early spinning wheel; period medical artifacts.
Hours & Admission Prices: Fri.-Sat. 10-2; other times by appointment. No charge; donations accepted. &
Attendance: 1,500 (estimated)
Membership: Individual $20; Family $25.

THE MUSEUM OF AUTOMOBILES, Petit Jean Mountain, 8 Jones Lane, Morrilton, AR 72110-9353. Tel.: 501-727-5427. Fax: 501-727-6482.
E-mail: info@museumofautos.com
Web Site: www.museumofautos.com
Founded: 1964.
Congressional District: 2
Key Personnel: Dir., Buddy Hoelzeman; Pres. (V), Raymond Harrill.
Personnel Profile: Full-Time Paid 3; Part-Time Paid 9.
Governing Authority: nonprofit organization. Tax-exempt: 501(c)(3).
Institution Type/Description: Antique Automobile Museum.
Collections: antique & classic automobiles on loan from collectors.
Facilities: Auto-related items for sale.
Activities: Museum Sponsors: Antique Auto Shows & Swap Meets in June and September.
Hours & Admission Prices: Daily 10-5. Adults $10, seniors 65 & over $9, children 6-17 $5; discounts for groups of 15 & over; members & children under 6 with parents no charge. Closed Christmas. &
Attendance: 16,000 (accurate)

Membership: Member $25; Participating $35; Friend $50; Associate $100; Sustaining $250; Contributing $500; Patron $1,000.

RIALTO COMMUNITY ARTS CENTER - THE GALLERY, 215 E. Broadway, Morrilton, AR 72110-3403. Mailing Address: P.O. Box 176, Morrilton, AR 72110. Tel.: 501-477-9955.
E-mail: director@rialtoartscenter.com
Web Site: www.rialtoartscenter.com/thegallery.html
Institution Type/Description: Art Gallery.
Collections: paintings; photographs; sculpture.
Hours & Admission Prices: Fri.-Sat. 11-2.

Mount Ida

HERITAGE HOUSE MUSEUM OF MONTGOMERY COUNTY, 819 Luzerne St., Mount Ida, AR 71957. Mailing Address: P.O. Box 1362, Mount Ida, AR 71957-1362. Tel.: 870-867-4422.
E-mail: museum@hhmmc.org
Web Site: hhmmc.org
Founded: 1999.
Congressional District: 4
Key Personnel: Dir., Emilie Kinney; Pres. (V) & Museum Shop Mgr., Betty Prince; Treas., Richard Ray.
Personnel Profile: Full-Time Paid 1; Part-Time Volunteers 35.
Governing Authority: private; nonprofit organization. Tax-exempt: 501(c)(3).
Institution Type/Description: History Museum.
Collections: minerals: crystals, mining minerals; papers; photos; Montgomery County artifacts; family history; the building & impact of Lake Quachita; Montgomery County churches.
Research Fields: genealogy of Montgomery County families; county businesses; mining & timber industry; building of Lake Ouachita; Ouachita National Forest.
Facilities: 50-vol. library; 2,500 sq. ft. exhibit space; meeting room. Museum-related items for sale.
Activities: formal education programs for children; guided tours; lectures; temporary exhibitions. Annual Events: Veterans Day Celebration; Arkansas Heritage Month Activity.
Publications: quarterly newsletter; weekly newspaper column, Museum Corner.
Hours & Admission Prices: Mon.-Wed. & Fri. 9-4, Sat.-Sun. 1-4. No charge; donations accepted. Closed New Year's Eve & Day; Thanksgiving Eve & Day; Christmas Eve & Day. &
Attendance: 1,550 (estimated)
Membership: Member $25; Patron $100; Benefactor & Corporate $200.

Mountain View

THE OZARK FOLK CENTER, 1032 Park Ave., Mountain View, AR 72560-6008. Tel.: 870-269-3851. Fax: 870-269-2909.
E-mail: ozarkfolkcenter@arkansas.com
Founded: 1973.
Congressional District: 1
Key Personnel: Lodge Mgr., Iona Barham; Accountant, Mary A. Smith; Public Information Officer, Jimmy Edwards; Museum Shop Mgr., Wanda Baird.
Personnel Profile: Full-Time Paid 23; Part-Time Paid 275; Part-Time Volunteers 200.
Governing Authority: state. Parent Institution: Arkansas Dept. of Parks and Tourism.
Institution Type/Description: Folk Arts Center.
Collections: traditional crafts of the Ozark Mountain region; artifacts; manuscripts; pre-1920 sheet music; early Ozark life; recordings; field recordings of Ozark folklore; interview transcripts.
Research Fields: Ozark crafts and music.
Facilities: 20,000-vol. library of books, recordings, vertical files related to Ozark Mountain Heritage emphasizing homecrafts & music; reading room; conference center; 1,000-seat auditorium with recording studio; restaurant. Books on crafts & music, records & craft items made in Ozarks for sale.
Activities: concerts; craft & music workshops; live performance of traditional crafts & music, 1840-1940; permanent & temporary exhibitions.
Publications: brochures; workshop newsletter; special event calendar.
Hours & Admission Prices: Craft Village: mid-April to Sept. Wed.-Sat. 10-5; Oct. Tues.-Sun. 10-5. Music Theater: mid-April to Sept. Wed.-Sat. 7:30 pm; Oct.- Tues.-Sat. 7:30 pm. Music or Craft Forum: adults $10, children $6; discount to groups. Daily Combo, 3-Day, Season Tickets, & Family Passes available. &
Attendance: 160,000 (estimated)

STONE COUNTY MUSEUM, 204 School Ave., Mountain View, AR 72560. Mailing Address: Stone County Historical Society, P.O. Box 210, Mountain View, AR 72560. Tel.: 870-269-4101.
Institution Type/Description: History Museum: built in 1895.
Collections: local history & culture; period furnishings; personal artifacts; photographs.
Hours & Admission Prices: Call for hours.

Murfreesboro

CRATER OF DIAMONDS STATE PARK MUSEUM, 209 State Park Road, Murfreesboro, AR 71958-8947. Tel.: 870-285-3113. Fax: 870-285-4169.
E-mail: crater@arkansas.com
Web Site: www.craterofdiamondsstatepark.com
Founded: 1972.
Congressional District: 4
Key Personnel: Asst. Supt., Bill Henderson; Park Ranger, Matt Briley; Gift Shop Mgr., Patricia Thomas; Park Interpreter, Waymon Cox; Park Interpreter, Margaret Jenks.
Personnel Profile: Full-Time Paid 15; Part-Time Paid 33.
Governing Authority: state. Parent Institution: Department of Parks & Tourism, One Capitol Mall, Little Rock, AR 72201. Tel.: 501-682-7777. Tax-exempt.
Institution Type/Description: Park Interpretative Museum.
Collections: diamonds; semi-precious rocks & minerals; old photos; old records of past diamond mining; modern geology displays; picture displays of past mining of diamonds.
Research Fields: geology & history of Crater of Diamonds State Park area.
Facilities: indoor theater; picnic area; campground; seasonal restaurant; visitors center; amphitheater; discovery center; water park. Gift items for sale.
Activities: guided tours available on advance notice & reservation; interpretative programs.
Hours & Admission Prices: Memorial Day-Labor Day daily 8-8; Sept.-May daily 8-5. Park: no charge. Diamond Mine: adults $7, children 6-12 $4; discounts to groups of 15 or more; children under 6 no charge. Closed New Year's Day; Thanksgiving; Christmas. &
Attendance: 154,609 (accurate)

KA-DO-HA INDIAN VILLAGE MUSEUM, 281 Kadoha Rd., Murfreesboro, AR 71958. Mailing Address: P.O. Box 669, Murfreesboro, AR 71958-0669. Tel.: 870-285-4167. Fax: 870-285-4118.
E-mail: info@kadoha.com
Web Site: www.kadoha.com
Founded: 1964.
Key Personnel: Owner, C.E.O. & Public Rels. Dir., Jack Bonds; Business Officer, Cur. & Gift Shop Mgr., JoAnn Copeland.
Personnel Profile: Full-Time Paid 2; Part-Time Paid 1.
Governing Authority: individual operation.
Institution Type/Description: Archaeology Museum: housed on 1,000 A.D. Moundbuilder Village & ceremonial center.
Collections: Indian artifacts from the site as well as other areas.
Research Fields: archaeology.
Facilities: Archaeological books relating to the Moundbuilders & museum-related items for sale.
Activities: guided tours; gallery talks; permanent exhibitions.
Hours & Admission Prices: Summer: daily 9-6; Winter: daily 9-5. Adults $8, children 6-13 $4; discounts to AAM & ICOM members. Closed Thanksgiving; Christmas.
Attendance: 10,000 (estimated)

Newport

ARNETT'S DOLL MUSEUM, 2005 Eastern Ave., Newport, AR 72112. Tel.: 870-523-2194.
Institution Type/Description: Doll Museum.
Collections: over 5,000 dolls.
Hours & Admission Prices: Call for hours.

E. BOB JACKSON MEMORIAL MUSEUM OF FUNERAL SERVICES, 1900 Block Malcolm Ave., Newport, AR 72112. Tel.: 870-523-5822. Fax: 870-523-4640.
Institution Type/Description: History Museum.
Collections: funeral history & process from 1800s to present; c.1885 horse-drawn funeral coach; 1927 solid bronze Italian Renaissance casket; late 1800s wooden baby casket.
Hours & Admission Prices: By appointment.

North Little Rock

ARKANSAS INLAND MARITIME MUSEUM, 120 Riverfront Park Dr., North Little Rock, AR 72114. Tel.: 501-371-8320. Fax: 501-244-9794.
Institution Type/Description: Maritime Museum.
Collections: local history; maritime trade; inland waterways system; Arkansas River; Arkansas aqua culture.
Facilities: library; theater.
Activities: birthday parties; groups tours; rental facilities.
Hours & Admission Prices: Spring: Friday-Sat. 10-6, Sun. 1-6. Museum: adults $2. Museum & Submarine: adults $6, children 5 & over, military and seniors 62 & over $4. Submarine tour not recommended for children under 5.

ARKANSAS NATIONAL GUARD MUSEUM, Lloyd England Hall, Camp Robinson, North Little Rock, AR 72199. Tel.: 501-212-5215. Fax: 501-212-5228.
E-mail: steve.rucker@ar.ngb.army.mil
Web Site: www.arngmuseum.com
Institution Type/Description: Military Museum.
Collections: history of Arkansas National Guard, Camp Pike & Camp Robinson.
Hours & Admission Prices: Mon.-Fri. 8-3, 1st Sat.-Sun. of month 8-3; call to confirm. ID required for day pass.

ARKANSAS SPORTS HALL OF FAME, Verizon Arena, #3 Verizon Arena Way, North Little Rock, AR 72114. Tel.: 501-663-4328.
Institution Type/Description: Sports Museum.
Collections: sports history & memorabilia; photographs; personal artifacts.
Hours & Admission Prices: Mon.-Sat. 10-4:30. Adults $6, seniors 62 & over $4, children 6-17 $3; discounts to active military & groups of 15 or more; children under 6 no charge.

Ozark

OZARK AREA DEPOT MUSEUM, 103 E. River St., Ozark, AR 72949. Tel.: 479-667-5015.
Institution Type/Description: Historic Building: housed in a former Missouri-Pacific/Union Pacific depot; built in 1911. Listed on the National Register of Historic Places.
Collections: local history & culture; railroad memorabilia; photographs; period furnishings.
Hours & Admission Prices: Call for hours.

Paragould

GREENE COUNTY MUSEUM OF PARAGOULD, 130 S. 14th St., Paragould, AR 72450. Tel.: 870-215-2407.
Institution Type/Description: History Museum.
Collections: local history & culture; period furnishings; personal artifacts; photographs.
Hours & Admission Prices: Call for hours.

Paris

ARKANSAS HISTORIC WINE MUSEUM, 101 N. Carbon City Rd., Paris, AR 72855-4630. Tel.: 479-963-3990.
E-mail: cowie@cswnet.com
Web Site: www.cowiewinecellars.com
Founded: 1967.
Congressional District: 3
Key Personnel: Dir. (V), Robert G. Cowie.
Personnel Profile: Part-Time Volunteers 3.
Governing Authority: Parent Institution: Cowie Wine Cellars & Vineyard.
Institution Type/Description: Historic Wine Museum: housed at Cowie Wine Cellars three miles west of Paris.
Collections: concentration on history of bonded wineries in Arkansas; the life & work of Professor Joseph Bachman; the history of home wine making; wine making equipment; small tools; photographs; papers; educational displays.
Research Fields: the history of wine making in Arkansas.
Facilities: 1,750 sq. ft. exhibit space.
Activities: guided tours; lectures. Annual Events: Heritage Day in May; Wine fest in September.
Hours & Admission Prices: Mon.-Sat. 10-6, Sun. 12-6. Adults $2. &
Attendance: 3,200 (estimated)
Membership: Individual $20; Family $30; Contributor $50; Patron $75; Sponsor $100; Sustainer $250; Benefactor $500.

LOGAN COUNTY MUSEUM, 202 N. Vine St., Paris, AR 72855-3222. Tel.: 479-963-3936. Fax: 479-963-3936.
E-mail: logancomuseum@centurytel.net
Web Site: logancountymuseum.com
Founded: 1972.
Congressional District: 3
Key Personnel: Admin., Jeanne S. Reynolds; Pres. (V), Dana Kuper.
Personnel Profile: Part-Time Paid 2; Part-Time Volunteers 20.
Governing Authority: Logan County Museum Association. Tax-exempt.
Institution Type/Description: History Museum: housed in 1903 former Logan County Jail.
Collections: local history.
Activities: guided tours; arts festivals; group tours & activities; living history; educational programs; 1870's hanging reenactments; bus tour groups. Museum Sponsors: Independence Day Program; Christmas Tour of Homes.
Publications: newsletter; county historical publication, Wagon Wheels.
Hours & Admission Prices: Tues.-Sat. 12-4. No charge; donations accepted. Tour Groups: adults $3 for hanging reenactment. Closed most federal holidays.
Attendance: 3,000 (accurate)
Membership: Constable: Individual $15; Couple $25. Deputy: Individual $30; Couple $45. Sheriff: Individual $45; Couple $55. Marshall: Individual $60; Couple $70.

PARIS-LOGAN COUNTY COAL MINERS MEMORIAL & MUSEUM, 804 S. Elm St., Paris, AR 72855. Tel.: 479-963-6463.
Web Site: coalmemorial-paris-ar.com
Institution Type/Description: History Museum.
Collections: coal mining history & artifacts; photographs; personal artifacts; coal car; fresh air fan; a one cylinder gas engine; blacksmith shop.
Hours & Admission Prices: Thurs.-Mon. 11-5. No charge.

Perryville

PERRY COUNTY HISTORICAL MUSEUM, 408 Main St., Perryville, AR 72126. Mailing Address: P.O. Box 1128, Perryville, AR 72126. Tel.: 501-889-2855.
E-mail: information@perrycountyhistoricalmuseum.org
Web Site: www.perrycountyhistoricalmuseum.org
Founded: 2005.
Congressional District: 2
Personnel Profile: Full-Time Volunteers 9.
Volunteer Hours: 1,000
Operating Expenses: 5,000
Operating Income: 8,000
Institution Type/Description: History Museum.
Collections: local history & culture; period furnishings; photographs; personal artifacts.
Activities: special events: Veterans in Nov. annualy, Black History in Feb. annualy, Fourche River Days in April & Christmas in Dec.
Publications: semi-annual newsletter.
Hours & Admission Prices: Call for hours. No charge. &
Attendance: 300 (estimated)
Membership: Student $5; Individual $15; Family $35; Group $100.

Piggott

HEMINGWAY-PFEIFFER MUSEUM AND EDUCATIONAL CENTER, (M), 1021 W. Cherry St., Piggott, AR 72454-1419. Tel.: 870-598-3487. Fax: 870-598-1037.
Founded: 1999.
Congressional District: 1
Key Personnel: Asst. Dir. & Facilities Mgr., Dr. Ruth Hawkins; Asst. Dir. & Facilities Mgr., Diana Sanders; Education Coord., Deanna Dismukes; Administrative Asst., Johnna Redman; Tour Guide & Housekeeper, Karen Trout
Governing Authority: Parent Institution: Arkansas State University.
Institution Type/Description: Historic House & Barn: listed on the National Historic Register.
Collections: family history; period literature; 1930s world events; agriculture; Northeast Arkansas history; furnishings; Ernest Hemingway timeline.
Facilities: Museum-related items for sale.
Activities: group tours.
Publications: Friends of the Pfeiffer's News.
Hours & Admission Prices: Mon.-Fri. 9-4, Sat. 1-3. Suggested Donation: adults $5, seniors $3. Closed New Year's Eve & Day; Memorial Day; Labor Day; Thanksgiving; Christmas Eve & Day. &
Attendance: 3,170
Membership: Student $10-$24; Individual $25-$49; Family $50-$99; Century

Club $100-$249; Heritage Club $500-$999; Presidents' Council $1,000-$4,999; Corporate Partner $5,000-$9,999; Museum Benefactor $10,000 & up.

MATILDA AND KARL PFEIFFER MUSEUM AND STUDY CENTER, 1071 Heritage Park Dr., Piggott, AR 72454. Tel.: 870-598-3228.
E-mail: pfeifferfnd@centurytel.net
Web Site: www.pfeifferfoundation.com
Institution Type/Description: History Museum: home built early 1930s.
Collections: over 1,400 mineral specimens from around the world; scenes from A Face in the Crowd, a 1956 movie filmed on site.
Facilities: 1,600-vol. library.
Hours & Admission Prices: Tues.-Fri. 9-4, Sat. 11-4; groups of 10 or more by appointment. No charge; donations accepted. Closed major holidays.

Pine Bluff

ARKANSAS ENTERTAINERS HALL OF FAME, One Convention Center Plaza, Pine Bluff, AR 71601-5067. Tel.: 870-536-7660. Fax: 870-850-2105.
E-mail: pbinfo@pinebluff.com
Web Site: www.arkansasentertainershalloffame.com
Founded: 1996.
Congressional District: 4
Institution Type/Description: Hall of Fame.
Collections: Hall of Fame inductees; personal artifacts; photographs.
Hours & Admission Prices: Mon.-Fri. 9-5, Sat.-Sun. call for hours. No charge; donations accepted. &
Attendance: 25,000 (estimated)

ARKANSAS RAILROAD MUSEUM, 1700 Port Rd., Pine Bluff, AR 71601-4663. Mailing Address: P.O. Box 2044, Pine Bluff, AR 71613-2044.
Founded: 1983.
Congressional District: 4
Personnel Profile: Part-Time Volunteers 20.
Governing Authority: Parent Institution: Cotton Belt Rail Historical Society, Inc. Tax-exempt.
Institution Type/Description: Railroad Museum.
Collections: Engine #336, Baldwin 2-6-0; Engine 819; railroad memorabilia; period artifacts.
Publications: quarterly newsletter, Cotton Belt Star.
Hours & Admission Prices: Mon.-Sat. 9-2. No charge; donations accepted. &
Attendance: 12,000 (estimated)
Membership: Individual $25; Lifetime $500.

✻ THE ARTS & SCIENCE CENTER FOR SOUTHEAST ARKANSAS, (M), 701 Main St., Pine Bluff, AR 71601-4903. Tel.: 870-536-3375. Fax: 870-536-3380.
E-mail: info@artssciencecenter.org
Web Site: www.artssciencecenter.org
Founded: 1968.
Congressional District: 4
Key Personnel: Exec. Dir., Dr. Lenore Shoults; Coord. Education, Tim Rhoades; Mgr. Operations, Rebekah Ray.
Personnel Profile: Full-Time Paid 4; Part-Time Paid 4; Part-Time Volunteers 125.
Governing Authority: municipal. Tax-exempt: 170(b)(1)(A).
Institution Type/Description: Arts Cultural Center.
Collections: 19th-century European paintings; 20th-century American paintings & prints; John Howard Memorial collection of African-American artists; Arkansas artists; Elsie Mistie Sterling collection of botanical paintings; Delta region paintings & photos.
Research Fields: 19th- & 20th-century art.
Facilities: 100-vol. library of reference books available by special request letter; theater; classrooms.
Activities: guided tours; films; gallery talks; concerts; dance recitals; arts festivals; drama; workshops; rental gallery; formally organized education programs; docent program or council; inter-museum loan & traveling exhibitions.
Publications: subscription brochures; quarterly newsletters; invitations; annual calendar of events; catalogs for exhibitions; annual report.
Hours & Admission Prices: Mon.-Fri. 10-5, Sat. 1-4. No charge. Closed New Year's Day; Easter; Independence Day; Thanksgiving; Christmas Eve & Day; for detailed information on closings visit Website. &
Attendance: 37,595 (estimated)
Membership: Individual $35; Family $50; Supporter $75; Contributing $125; Patrons $250; Fellow $500; Sponsor $750; Benefactor $1,000; Philanthropist $2,000.

GOVERNOR MIKE HUCKABEE DELTA RIVERS NATURE CENTER, 1400 Black Dog Rd., Pine Bluff, AR 71611. Mailing Address: P.O. Box 8074, Pine Bluff, AR 71611. Tel.: 870-534-0011. Fax: 870-534-4422.
E-mail: info@deltarivers.com
Web Site: www.deltarivers.com
Founded: 2001.
Personnel Profile: Full-Time Paid 3; Full-Time Volunteers 3; Interns 1.
Governing Authority: Parent Institution: Arkansas Game & Fish Commission.
Institution Type/Description: Nature Center.
Collections: natural history; ecosystems; hands-on exhibits.
Facilities: nature trails.
Activities: film; simulator; educational programs; nature trails; classes.
Hours & Admission Prices: Center: year round Tues.-Sat. 8:30-4:30, Sun. 1-5. Trails: daily dawn to dusk. No charge; donations accepted. Closed New Year's Day, Easter, Thanksgiving, Christmas. &
Attendance: 40,000 (estimated)

LEEDEL MOOREHEAD-GRAHAM FINE ARTS GALLERY, 1200 N. University, Pine Bluff, AR 71601-2799. Mailing Address: Mail Slot 4925, Pine Bluff, AR 71601. Tel.: 870-575-8236. Fax: 870-575-4636.
E-mail: gaines_c@uapb.edu
Web Site: www.uapb.edu
Founded: 1967.
Congressional District: 4
Key Personnel: Chm., Henri Linton.
Personnel Profile: Part-Time Paid 2.
Governing Authority: public university. Parent Institution: University of Arkansas at Pine Bluff, Pine Bluff, AR. Tax-exempt: 501(c)(3.
Institution Type/Description: Art Museum.
Collections: African American art; history of African Americans in Arkansas; photographs.
Facilities: 3,360 sq. ft. exhibit space.
Activities: loan, traveling & temporary exhibitions; student & faculty exhibitions.
Hours & Admission Prices: Mon.-Fri. 8:30-4:30. No charge. Closed school holidays.
Attendance: 5,000 (estimated)

PINE BLUFF/JEFFERSON COUNTY HISTORICAL MUSEUM, 201 E. 4th, Pine Bluff, AR 71601-4401. Tel.: 870-541-5402. Fax: 870-541-5405.
E-mail: historydirector@cablelynx.com
Web Site: www.pbjchistory.org
Founded: 1980.
Congressional District: 4
Key Personnel: Exec. Dir., Kristi Alexander.
Personnel Profile: Full-Time Paid 1; Part-Time Paid 1; Part-Time Volunteers 21.
Governing Authority: nonprofit organization. Tax-exempt: 501(c)(3).
Institution Type/Description: History Museum: located in the restored Union Depot, built in 1906.
Collections: history, culture & lifestyles of Pine Bluff & Jefferson County; American Indian pottery & artifacts; farm implements; Civil War artifacts; clothing from the Edwardian period; household items; room setting; historical photographs; documentary items; artifacts are from families in Jefferson county; military history Civil War to Iraq.
Research Fields: county history.
Activities: loan & temporary exhibitions; school loan service.
Publications: quarterly newsletters.
Hours & Admission Prices: Mon.-Fri. 9-4, Sat. 10-2. No charge; donations accepted. Closed legal holidays. &
Attendance: 11,000 (accurate)
Membership: Individual & Family: Bronze $35; Pewter $60; Silver $80; Gold $125; Platinum $25. Corporate: Museum Circle $500; Director's Circle $1,000; President's Circle $1,500; Chairmen's Circle $2,500.

UNIVERSITY OF ARKANSAS MUSEUM AND CULTURAL CENTER, 1200 N. University Dr., Pine Bluff, AR 71601. Tel.: 870-575-8232.
E-mail: museum@uapb.edu
Institution Type/Description: History Museum.
Collections: university history; photographs; catalogs; yearbooks; letters; personal artifacts; portraits.
Activities: educational programs; temporary exhibitions.
Hours & Admission Prices: Call for hours.

Pine Ridge/Oden

LUM & ABNER MUSEUM & JOT 'EM DOWN STORE, 4562 Hwy. 88 W., Pine Ridge/Oden, AR 71961-8056. Mailing Address: General Delivery, Pine Ridge, AR 71966. Tel.: 870-326-4442.
E-mail: nlstucker@earthlink.net
Web Site: lum-abner.com
Founded: 1971.
Congressional District: 50
Key Personnel: Co-Dir. & Owner, Noah Lon Stucker; Co-Dir. & Owner, Kathryn Moore Stucker.
Governing Authority: individual operation.
Institution Type/Description: History Museum: housed in c.1904 general merchandise store, on which the Lum & Abner radio program 1931-55 was based.
Collections: photographs & literature from the careers of Lum & Abner; radio CDs; merchandise & dry goods from that era; tools; appliances; clothing; furnishings; radio history books; letters; awards; scripts; posters & advertisements; newspaper articles.
Research Fields: local history as it relates to Lum & Abner, compares to small communities all over the country (school, customs, farming, economy, stores, etc.).
Facilities: Gift items for sale.
Activities: guided tours; radio programs; permanent exhibitions. Affiliated activity-National Lum & Abner Society Convention in June.
Publications: pamphlets; booklets, Lum & Abner & Their Friends; Hello, This is Lum & Abner-A History of Pine Ridge and Lum & Abner.
Hours & Admission Prices: March-Oct. Tues.-Sat. 9-5, Sun. 12-5; other times by appointment. No charge; donations accepted.

Pocahontas

RANDOLPH COUNTY HERITAGE MUSEUM, 106 E. Everett St., Pocahontas, AR 72455-3309. Tel.: 870-892-4056. Fax: 870-892-4056.
E-mail: kpmuseum@hotmail.com
Web Site: www.randolphcomuseum.org
Founded: 2006.
Key Personnel: Pres. (V), Museum Admin., & Museum Shop Mgr., Karen Parish; Chm. (V), Bill Carroll.
Personnel Profile: Part-Time Volunteers 9.
Governing Authority: Parent Institution: Five Rivers Historic Preservation. Tax-exempt.
Institution Type/Description: Historical Museum.
Collections: Randolph County history; personal artifacts; photographs.
Hours & Admission Prices: Mon., Wed. & Fri. 10-4. No charge; donations accepted.

Pottsville

POTTS INN MUSEUM, Town Square, Pottsville, AR 72801. Mailing Address: 6368 SR 247, Pottsville, AR 72858-8952. Tel.: 479-968-8369.
Founded: 1978.
Congressional District: 3
Key Personnel: Chm. Bd., Charles Oates; Treas., Kelly Vanes.
Personnel Profile: Full-Time Volunteers 2; Part-Time Volunteers 16.
Governing Authority: county; nonprofit. Affiliated with the Pope County Historical Foundation. Tax-exempt.
Institution Type/Description: Preservation Project: housed in 1850-1858 nine-room home & stage stop of Kirkbride Potts.
Collections: Potts family period artifacts; 100 years of ladies' hats; period furniture; gifts of first settlers. Historic Buildings: milk house; doctor's office; implement barn; c.1827 one-room log cabin.
Activities: guided tours; TV & radio programs; formally organized education programs for children & undergraduate college students affiliated with Arkansas Technical University; permanent exhibitions.
Hours & Admission Prices: Wed.-Sun. 1-5. Adults $3, children $1; discounts to AAM members.
Attendance: 1,900 (estimated)

Powhatan

POWHATAN HISTORIC STATE PARK, 4414 Arkansas 25 S., Powhatan, AR 72458. Mailing Address: P.O. Box 93, Powhatan, AR 72458. Tel.: 870-878-6765. Fax: 870-878-6319.
E-mail: pwhatan@arkansas.com
Web Site: www.arkansasstateparks.com/powhatancourthouse
Institution Type/Description: State Park & Historic Buildings: listed on the National Register of Historic Places.

Collections: local history & culture; period artifacts; photographs. Historic Buildings: 1888 Victorian courthouse; 1873 jail; c.1848 log house; 1889 2-room school; c.1888 commercial bldg.
Hours & Admission Prices: Tues.-Sat. 8-5, Sun. 1-5. Family $15, adults $5, children 6-12 $3; discounts to groups of 15 or more. Closed New Year's Day; Thanksgiving; Christmas Eve & Day.

Prairie Grove

PRAIRIE GROVE BATTLEFIELD STATE PARK, 506 E. Douglas St., Prairie Grove, AR 72753-2731. Tel.: 479-846-2990. Fax: 479-846-4035.
E-mail: prairiegrove@arkansas.com
Web Site: www.arkansasstateparks.com/prairiegrovebattlefield/
Founded: 1957.
Congressional District: 3
Key Personnel: Supt., Jesse Cox; Registrar, Alan Thompson; Museum Shop Mgr., Holly Cherry.
Personnel Profile: Full-Time Paid 7; Part-Time Paid 4; Part-Time Volunteers 4.
Governing Authority: state. Arkansas Dept. of Parks & Tourism.
Institution Type/Description: Civil War Park Museum: located on site of Dec. 7, 1862, Battle of Prairie Grove.
Collections: Civil War artifacts; period furniture; Civil War documents; Civil War & middle 19th-century Ozark Arkansas cultural exhibits; historic buildings; Pioneer Village.
Facilities: library; information center; picnic grounds. Gifts for sale.
Activities: guided tours; arts festivals; audio-visual programs; living history demonstrations; automobile driving tour; walking trail.
Hours & Admission Prices: Daily 8-5. Adults $3, children $2; discounts to groups of 20 or more with appointment; children under 6 no charge. Closed New Year's Day; Thanksgiving; Christmas Eve & Day. &
Attendance: 200,000 (accurate)

Prescott

NEVADA COUNTY DEPOT MUSEUM, 403 W. 1st St. S., Prescott, AR 71857-2067. Tel.: 870-887-5821.
E-mail: online@depotmuseum.org
Web Site: www.depotmuseum.org
Key Personnel: Cur., David Sesser
Institution Type/Description: History Museum.
Collections: Nevada County history; early settlements & settlers; Indian pottery; Civil War; agriculture; railroads; fire fighting equipment; Louisiana Purchase, April 1803.
Activities: tours.
Hours & Admission Prices: Mon.-Fri. 10-4; tours & groups by appointment. No charge; donations accepted. Closed holidays.

Rogers

HOBBS PARK VISITOR CENTER, 21392 E. Hwy. 12, Rogers, AR 72756-8183. Mailing Address: P.O. Box 709, Rogers, AR 72757-0709. Tel.: 479-789-2380.
Institution Type/Description: Park Museum & Visitor Center.
Collections: local history; wildlife; natural science; trees; plants; flowers.
Facilities: nature trails; classrooms. Museum-related items for sale.
Activities: educational programs; hiking.
Hours & Admission Prices: Call for hours.

ROGERS DAISY AIRGUN MUSEUM, 202 W. Walnut St., Rogers, AR 72756-6665. Tel.: 479-986-6873. Fax: 479-986-6875.
E-mail: info@daisymuseum.com
Web Site: www.daisymuseum.com
Formerly: Daisy International Air Gun Museum
Founded: 2000.
Congressional District: 3
Personnel Profile: Full-Time Paid 1; Full-Time Volunteers 6; Part-Time Paid 2.
Governing Authority: nonprofit organization. Tax-exempt.
Institution Type/Description: Air Gun Museum.
Collections: air guns & toy guns.
Publications: brochure, It's a Daisy.
Hours & Admission Prices: Mon.-Sat. 9-5. Adults $2; children 16 & under no charge. Closed major holidays. &
Membership: Life $50.

✱   ROGERS HISTORICAL MUSEUM, (M), 322 S. Second St., Rogers, AR 72756-4512. Tel.: 479-621-1154. Fax: 479-621-1155.
E-mail: museum@rogersarkansas.com
Web Site: www.rogersarkansas.com/museum

Founded: 1975.
Congressional District: 3
Key Personnel: Dir., John Burroughs; Asst. Dir., Terrilyn Wendling; Chm., Mike Whitmore; Cur. Collections, Jami Roskamp; Registrar, Jennifer Sweet; Office Mgr., Pat Campbell; Cur. Education, Robert Rousey; Education Asst., Ashley Sayers; Adult Programs Educator, Monte Harris; Guide, S.K. Clark-Will.
Personnel Profile: Full-Time Paid 7; Part-Time Paid 4; Part-Time Volunteers 40.
Governing Authority: municipal; nonprofit. Parent Institution: City of Rogers. Tax-exempt.
Institution Type/Description: Historical Building: c.1895 five-room Hawkins House.
Collections: local history; Frisco Railroad; Coin Harvey, Betty Blake (Mrs. Will) Rogers; quilts; Victorian period artifacts.
Major Exhibits: Art from the Earth: A Pottery Exhibit, 10/26/13-2/22/14.
Research Fields: local history; Benton County history; Victorian Period.
Activities: guided tours; lectures; organized education programs for children; loan, permanent & temporary exhibitions. Museum Sponsors: Annual Holiday Open House.
Publications: monthly newsletter, The Friendly Note.
Hours & Admission Prices: Mon.-Sat. 10-5. No charge; donations accepted. Closed major holidays. &
Attendance: 23,400 (accurate)
Membership: Individual $15; Family $25; Sponsor $50; Patron $75; Benefactor $100; Endowment $500.

WAR EAGLE CAVERN, 21494 Cavern Rd., Rogers, AR 72756-7493. Tel.: 479-789-2909.
Institution Type/Description: Geology Museum.
Collections: local history & geology.
Facilities: nature trails. Museum-related items for sale.
Activities: hiking; panning; school groups.
Hours & Admission Prices: Mon.-Sat. 9:30-5, Sun. 12-5. Adults $15, children 4-11 $9; children 3 & under no charge.

Russellville

ARKANSAS RIVER VALLEY ARTS CENTER, 1001 E. B St., Russellville, AR 72801-4252. Mailing Address: P.O. Box 2112, Russellville, AR 72811-2112. Tel.: 479-968-2452. Fax: 479-968-5015. Facebook: Arkansas River Valley Arts Center.
E-mail: artscenter@centurytel.net
Web Site: www.arvartscenter.org
Founded: 1981.
Key Personnel: Exec. Dir., Betty LaGrone; Pres. (V), Emory Molitor; Chm. (V), John Gale; Admin. Asst., Phala Harrison; Teaching Artist, Winston J. Taylor.
Personnel Profile: Full-Time Paid 1; Part-Time Paid 2.
Operating Expenses: 129,000
Operating Income: 130,000
Governing Authority: Tax-exempt.
Institution Type/Description: Art Gallery.
Collections: photographs; paintings; pottery.
Major Exhibits: Oriental Art, 1/14; Collegiate Competition & Exhibition, 2/14; High School Competition & Exhibition, 3/14; PALS Watercolor, 4/14; Gallery 307, 5/14; Struttin Our Stuff, 8/3/14; Local Artist Invitational Showcase, 11/14; YAA (T), 12/14.
Facilities: Museum-related items for sale.
Activities: summer art camp; classes; workshops.
Publications: monthly newsletter.
Hours & Admission Prices: Mon.-Thurs. 10-5, Fri. 10-4; other times by appointment. No charge. Closed national holidays. &
Attendance: 2,000 (estimated)
Membership: Students & Seniors $20; Individual $35; Family $50.

ARKANSAS TECH UNIVERSITY MUSEUM, (M), 1502 N. El Paso Ave./Techionery, Russellville, AR 72801-8816. Tel.: 479-964-0826. Fax: 479-964-0872. Facebook: Arkansas Tech University Museum.
E-mail: jstewartabernathy@atu.edu
Web Site: www.atu.edu/museum
Formerly: Arkansas Tech Museum
Founded: 1989.
Congressional District: 3
Key Personnel: Museum Dir., Judith C. Stewart-Abernathy; Cur., Theresa Jureka Johnson.
Personnel Profile: Full-Time Paid 2; Part-Time Paid 7; Part-Time Volunteers 6.
Volunteer Hours: 25

Operating Expenses: 129,591
Operating Income: 129,591
Governing Authority: public university; nonprofit. Parent Institution: Arkansas Tech University.
Institution Type/Description: History Museum.
Collections: archaeological; historical; archival.
Major Exhibits: Art and Architecture: A Student Exhibition, 3/15/14-4/15/14.
Research Fields: history of ATU; establishment, development & history of agriculture school; university students & staff history; prehistory & history of Native American groups indigenous to our region; treaty relations of Native American peoples through our region; Presbyterian Native American boarding school (Dwight Mission) in our area 1820-1828; local regional history.
Facilities: 870-vol. library; 25-seat lecture hall; educational facilities; classroom; 2,000 sq. ft. exhibit space. Museum-related & archival supply items for sale.
Activities: guided tours; teaching lectures. Museum Sponsors: Open House events; Series of Discovery evening lecture & performance programs; exhibit openings.
Publications: Discovery Handbook, all copies sold.
Hours & Admission Prices: Tues.-Thurs. 8:30-4:30; other times by appointment. No charge; donations accepted. Closed during university breaks. &
Attendance: 800 (estimated)

Scott

PLANTATION AGRICULTURE MUSEUM, 4815 Hwy. 161 S., Scott, AR 72142. Mailing Address: P.O. Box 87, Scott, AR 72142-0087. Tel.: 501-961-1409. Fax: 501-961-1579.
E-mail: plantationagrimuseum@arkansas.com
Web Site: www.arkansasstateparks.com
Founded: 1989.
Congressional District: 2
Key Personnel: Dir., Linda Goza; Cur., Randy Noah; Museum Interpreter, Lydia Leatherwood; Museum Shop Mgr., Becky Jones.
Personnel Profile: Full-Time Paid 5; Part-Time Paid 2; Interns 0.
Volunteer Hours: 35
Operating Expenses: 300,000
Operating Income: 12,098
Governing Authority: state. Parent Institution: Arkansas Dept. of Parks & Tourism, Arkansas State Parks, 1 Capitol Mall, Little Rock, AR 72201. Subsidiary Institution: Historical Resources & Museum Services; Region II.
Institution Type/Description: History Museum: housed in c.1912 general store.
Collections: tools; equipment; machinery; photographs related to cotton agriculture & plantation life in Arkansas from 1836 until World War II.
Major Exhibits: Blue Mondays, Sad Tuesdays, 11/13-3/14; Bacon, Eggs & More, 11/13-12/14.
Research Fields: cotton agriculture & plantation life in Arkansas from 1836 until World War II.
Facilities: educational facilities. Museum-related items for sale.
Activities: tours; permanent & temporary exhibitions; special events.
Publications: quarterly, Boll Weevil Newsletter.
Hours & Admission Prices: Tues.-Sat. 8-5, Sun. 1-5. No charge; donations accepted. Closed Mondays; holidays; New Year's Day; Thanksgiving; Christmas Eve & Day. &
Attendance: 7,200 (estimated)

TOLTEC MOUNDS ARCHEOLOGICAL STATE PARK, 490 Toltec Mounds Rd., Scott, AR 72142. Tel.: 501-961-9442. Fax: 501-961-9221.
E-mail: toltecmounds@arkansas.com
Web Site: www.arkansasstateparks.com
Founded: 1975.
Congressional District: 2
Key Personnel: Dir. Toltec Research Station, Dr. Julie Markin; Supt., James Wilborn; Park Interpreter, Robin Gabe; Park Interpreter, Amy Griffin; Museum Shop Mgr., Keryn Cantrell.
Personnel Profile: Full-Time Paid 7; Part-Time Paid 2.
Governing Authority: state. Parent Institution: Dept. of Parks & Tourism, No. 1 Capitol Mall, Little Rock, AR 72201. Tel.: 501-682-1191 & Arkansas Archeological Survey, P.O. Box 1249, Fayetteville, AR 72702. Tel.: 479-375-3556. Tax-exempt.
Institution Type/Description: Archaeological Site.
Collections: archeological research collection.
Research Fields: local environment 1,000 years ago; human use of environment; regional resources; lifeways; social & political systems; Mississippi Valley contacts.
Facilities: field research station; audiovisual room; visitor center; archaeological laboratory; canteen. Maps, books & publications on archaeology & history of the area & gift items for sale.

Activities: tours; films & slide presentations; formally organized education programs for children & adults by group reservation; occasional excavations; rental facility; workshops.
Publications: Emerging Patterns of Plum Bayou Culture; Arkansas Archaeological Survey of 1982; Surveyors of the Ancient Mississippi Valley: Modules & Alignments in Prehistoric Mound Sites; Arkansas Archeological Survey Research Report No. 28; Crossroads of the Past; Arkansas Archeological Survey Popular Series No. 2; Paths of our Children: Historic Indians of Arkansas; Arkansas Archeological Survey Popular Series No. 3; Arkansas Before the Americas Series No. 40.
Hours & Admission Prices: Mon. holidays & Tues.-Sat. 8-5, Sun. 12-5. Tram Tours: adults $5, children 6-12 $4. Walking Tours: adults $3, children 6-12 $2; discounts to groups of 15 or more & schools with reservations; children under 6 no charge. Closed New Year's Day; Thanksgiving; Christmas Eve & Day. &
Attendance: 30,000 (accurate)

Sheridan

GRANT COUNTY MUSEUM, 521 Shackleford Rd., Sheridan, AR 72150-7074. Tel.: 870-942-4496. Fax: 870-917-2248.
E-mail: museum4@windstream.net
Web Site: www.grantcountymuseumar.com
Founded: 1970.
Congressional District: 4
Key Personnel: Chm. (V), Mary Beth Glover-Wilson; Dir., D.J. Wallace; Financial Dir., Noka Emerson; Magazine Editor, Lindsey Stanton.
Personnel Profile: Full-Time Paid 3; Part-Time Paid 1.
Governing Authority: county; nonprofit.
Institution Type/Description: History Museum.
Collections: history of Arkansas with emphasis on Grant County, Arkansas from prehistoric to present; Civil War artifacts; World War II artifacts & vehicles; archives; photographs; Indian relics; pioneer life; restored buildings including log cabins, church, cafe, corn crib & Victorian house.
Facilities: library containing genealogical material; family research reports; microfilm records & newspaper articles available to the public.
Activities: guided tours; formally organized education programs. Annual Events: Christmas Tour of Homes & Festive Programs.
Publications: historical journal 3 times annually, Grassroots.
Hours & Admission Prices: Tues.-Sat. 9-4; other times by appointment. No charge; donations accepted. Closed legal holidays. &
Attendance: 10,000 (estimated)
Membership: Individual $12; Family $15; Supporting $25; Patron $50; Sponsor $100; Life $500; Bless You $1,000.

Siloam Springs

SAGER CREEK ARTS CENTER, 301 E. Twin Springs, Siloam Springs, AR 72761. Mailing Address: P.O. Box 1127, Siloam Springs, AR 72761. Tel.: 479-524-4000. Fax: 479-524-5713.
Institution Type/Description: Art Gallery.
Collections: paintings; sculpture; drawings.
Activities: special events.
Hours & Admission Prices: Call for hours.

SILOAM SPRINGS MUSEUM, (M), 112 N. Maxwell, Siloam Springs, AR 72761-3174. Mailing Address: P.O. Box 1164, Siloam Springs, AR 72761-1164. Tel.: 479-524-4011.
E-mail: ssmuseum@centurytel.net
Web Site: www.siloamspringsmuseum.com
Founded: 1969.
Congressional District: 3
Key Personnel: Dir. & C.E.O., Donald Warden; Pres. (V), Bill Osgood.
Personnel Profile: Full-Time Paid 2; Part-Time Volunteers 4.
Governing Authority: private. Parent Institution: The Siloam Springs Museum Society. Tax-exempt.
Institution Type/Description: History Museum: housed in 1950 church.
Collections: furniture; uniforms from World War I & II; textiles; clothing; decorative arts; pressed & cut glass; tools; archives: including books, newspapers, manuscripts, photographs & postcards.
Research Fields: local history of Benton County, Arkansas; Southern decorative arts & crafts.
Facilities: archives including 9,500 articles. Museum-related gifts for sale.
Activities: permanent & temporary exhibitions. Museum Sponsors: Springs Heritage Festival; Spring Home Tour; Fall Music Fundraiser.
Publications: quarterly newsletter.
Hours & Admission Prices: Tues.-Sat. 10-5; other times by appointment. No charge; donations accepted. Closed major holidays. &
Attendance: 2,194 (estimated)
Membership: Single Senior Citizen $8; Dual Senior Citizen & Individual $15;

Family $25; Contributing $35; Associate $60; Sponsor $125; Patron $300; Capitol $750; Millennium $1,000.

Smackover

ARKANSAS MUSEUM OF NATURAL RESOURCES, 3853 Smackover Hwy., Smackover, AR 71762-9575. Tel.: 870-725-2877. Fax: 870-725-2161.
E-mail: museum@amnr.org
Web Site: amnr.org
Founded: 1977.
Congressional District: 4
Key Personnel: Chm. (V), Phoebe Sellers; Dir., Pam Beasley; Exhibit Specialist, Rhonda Millican; Dir. Education & Research, Shelly Franques; Registrar, Sheri Neely; Cur., Van Zbinden; Museum Shop Mgr., Beth Hooks.
Personnel Profile: Full-Time Paid 9; Part-Time Paid 4; Part-Time Volunteers 48.
Governing Authority: state. Parent Institution: Arkansas Dept. of Parks & Tourism. Tax-exempt.
Institution Type/Description: History & Technology Museum.
Collections: specimens; photographs; archival materials of petroleum & by-products and their relationship to Arkansas history.
Research Fields: Arkansas oil boom; oil & brine industrial technology; timber; geology; Bromine.
Facilities: education center; woodland walk interpretive trail.
Activities: formally organized education programs; permanent & temporary exhibits; oral history program; guided tours; school loan service. Museum Sponsors: The National State Chautauqua.
Publications: newsletter, The Pipeline; publication, Boom Towns of South Arkansas; brochure, Brine, Bromine, How They Affect Our Lives, Rhene Miller's Circus Truck; visitor guide book; Teacher's Handbook.
Hours & Admission Prices: Mon.-Sat. 8-5, Sun. 1-5. No charge; donations accepted. Closed New Year's Day; Thanksgiving; Christmas. &
Attendance: 24,326 (accurate)
Membership: Single $15; Pipeline Club $30; Arkansas Club $50; Resources Council $100; Ouachita River Club $150; Cross Cut Society $300; Brine Society $500; Gusher Guild $1,000.

Springdale

ARTS CENTER OF THE OZARKS, 214 S. Main St., Springdale, AR 72764-4446. Tel.: 479-751-5441.
Founded: 1966.
Institution Type/Description: Art Gallery.
Collections: photographs; paintings.
Activities: special events.
Hours & Admission Prices: Mon.-Fri. 9-5, Sat. 9-3.
Membership: $35-$1,000.

SHILOH MUSEUM OF OZARK HISTORY, (M), 118 W. Johnson Ave., Springdale, AR 72764-4313. Tel.: 479-750-8165. Fax: 479-750-8693.
E-mail: shiloh@springdalear.gov
Web Site: www.shilohmuseum.org
Founded: 1965.
Congressional District: 3
Key Personnel: Dir., Allyn Lord; Pres. Bd. Trustees, Dianne Kellogg; Collections Mgr., Carolyn Reno; Outreach Coord., Susan Young; Exhibit Designer, Curtis Morris; Education Coord., Judy Costello; Library Asst., April Griffith; Photo Archivist, Marie Demeroukas; Collections & Education Asst., Victoria Thompson; Photographer, Don House; Maintenance, Marty Powers; Museum Shop Mgr., Kathy Plume.
Personnel Profile: Full-Time Paid 11; Part-Time Paid 3; Part-Time Volunteers 21.
Volunteer Hours: 2,936
Operating Expenses: 648,952
Operating Income: 651,794
Governing Authority: municipal. Parent Institution: City of Springdale.
Institution Type/Description: History Museum.
Collections: more than half a million indexed photographs; primitive paintings; items related to poultry, fruit, timber industries, pioneer life, and rural town life; farm equipment; domestic tools; hand tools; clothing and textiles; archival material; Native American artifacts; Vaughan-Applegate collection of cameras. Historic Buildings: c.1854 log cabin; c.1871 residence; 1870s post office/general store; 1880s doctor's office; c.1930 barn & outhouse; 1871 church & Odd Fellows lodge.
Major Exhibits: Quilts from the Shiloh Museum Collections, 8/31/13-3/1/14;

Scenes from Newton County, 12/17/13-5/10/14; Just Doing My Work: Essie Ward, 1/27/14-1/10/15; A Boy's Toys, 3/24/14-3/24/15; Healing Waters, 5/12/14-12/13/14.
Research Fields: prehistory & history of northwest Arkansas; Arkansas Ozarks.
Facilities: library & photo archives dealing with prehistory & history of the region. Museum-related items for sale.
Activities: guided tours; lectures & demonstrations; video & Power Point programs; discovery boxes. Museum Sponsors: Holiday Open House; Quilt Fair, Sheep-to-Shawl.
Publications: quarterly newsletter; website; iTunes U page; annual report; monthly e-news.
Hours & Admission Prices: Mon.-Sat. 10-5. No charge; donations accepted. Closed New Year's Day; Thanksgiving; Christmas Eve & Day. &
Attendance: 51,618 (accurate)
Membership: Senior Individual $10; School classes $12.50; Individual & Senior couple $15; Family $20; Patron $50; Sponsor $100; Sustaining $250; Benefactor $500; Founding $1,000.

Stuttgart

MUSEUM OF THE ARKANSAS GRAND PRAIRIE, 921 E. 4th St., Stuttgart, AR 72160-4558. Tel.: 870-673-7001. Fax: 870-673-3959.
E-mail: ontheprairiebayou@yahoo.com
Web Site: www.stuttgartmuseum.org
Formerly: Stuttgart Agricultural Museum
Founded: 1974.
Congressional District: 2
Key Personnel: Dir., Melanie Baden; Chm. Bd. Trustees, Bruce Martin; Chm. (V), Beth Hopson; Vice Chm., Garland Demden; Sec., Jean Pollard; Financial Chm., Richard Bell; Cur. Collections, Gena Seidenschwarz; Cur. Restoration-Furniture, Kenneth Bull; Cur. Farm Equip. & Furniture, Jim Gingerich; Educational Dir., Ann Prislovsky; Gen. Asst. & Museum Shop Mgr., Frances Camp.
Personnel Profile: Full-Time Paid 4; Part-Time Volunteers 86.
Governing Authority: city; nonprofit organization. Parent Institution: City of Stuttgart; Stuttgart Agricultural Museum Association. Tax-exempt: 501(c)(3).
Institution Type/Description: Agriculture Museum.
Collections: hay & rice industry farming equipment; agricultural history mural; period duck & turkey calls; period duck & goose decoys; swamp, prairie & wooded area wildlife; crop dusting history; furnishings; Arkansas fish farming; replicas of 1890 prairie home; fire station; printing press office; 1896 2/3 scale replica Lutheran church; mounted ducks in habitats; mounted Albino Canada Goose; guns; vintage clothing; toys; early prairie life & farming; videos; rice milling; fish farming; waterfowlers; waterfowlers of Mississippi Flyway; Hall of Transportation. Historic Building: 1914 schoolhouse.
Research Fields: Rice Branch experiment station-U.S. Dept. of Agriculture. Riceland foods & Producers rice mill, marketing rice around the world; Aqua Culture Center for fish farming.
Facilities: 200-vol. research library on agricultural equipment, furniture, antiques, rice farming & the history of agriculture; 55-gal. aquarium; theatre. Museum-related items for sale.
Activities: guided tours; lectures; films; formally organized education programs; docent program; permanent & temporary exhibit.
Publications: newsletter; brochure, History of Duck Hunting in Stuttgart; booklet, Arkansas, Its Land & Its People, Of The Grand Prairie; videotapes, Dept. of Interior Fish Farm Biologists.
Hours & Admission Prices: Tues.-Fri. 8-4, Sat. 10-4. No charge; donations accepted. Closed Easter; legal holidays. &
Attendance: 13,146 (accurate)
Membership: Annual Seniors, Singles & Students Club $25; Contributor $50; Donor Club $100 & up; Brass Plate $500.

Texarkana

FOUR STATES AUTO MUSEUM, 217 Laurel St., Texarkana, AR 71854-6051. Tel.: 870-772-2886.
E-mail: antauto@texarkaam.org
Web Site: texarkaam.org
Formerly: Tex Ark Antique Auto Museum
Founded: 2004.
Key Personnel: Dir. & Pres. (V), Paul Taylor; Treas., Charlotte Tipton.
Personnel Profile: Part-Time Volunteers 10.
Governing Authority: municipal. Tax-exempt: 501(c)(3).
Institution Type/Description: Transportation Museum.
Collections: period automobiles, petroleum signs & gasoline pump; tools; literature.

Facilities: library.
Activities: guided tours.
Publications: quarterly newsletter.
Hours & Admission Prices: Sat. 10-4, Sun. 1-4. No charge; donations accepted.
Attendance: 5,000 (estimated)
Membership: Family $30.

LINDSAY RAILROAD MUSEUM, 202 E. Broad St., Texarkana, AR 71854. Tel.: 903-748-1235.
E-mail: info@texarkanabroadstreetgalleries.com
Web Site: lindsayrailroadmuseumintexarkana.com
Institution Type/Description: Railroad Museum.
Collections: railroad history & artifacts; photographs; paintings; uniforms; badges; patches; flags; hands-on exhibitions.
Facilities: theater.
Hours & Admission Prices: Sat. 11-4. Admission $5.

Tontitown

TONTITOWN HISTORICAL MUSEUM, 257 E. Henri de Tonti Blvd., Tontitown, AR 72770. Mailing Address: P.O. Box 144, Tontitown, AR 72770-0144. Tel.: 479-361-2498 & 2700.
E-mail: bcortiana@cox.net
Web Site: www.tontitown.com
Founded: 1986.
Congressional District: 3
Key Personnel: Pres. Bd., Bev Cortiana; Cur., Charlotte Piazza.
Governing Authority: city; bd. of commissioners.
Institution Type/Description: History Museum: housed in the former home of two original settlers, sisters Mary and Zelinda Bastianelli.
Collections: Italian immigration; settler families personal artifacts; photographs; period furnishings; early grape press; wine-bottling machine; early farm tools; Catholic church artifacts.
Activities: tours; presentations; research. Museum Sponsors: Tontitown Grape Festival; Arkansas Heritage Month Event in May; Tontitown Old Fashioned Reunion and Polenta Smear in November.
Publications: biannual newsletter, Tontitown Storia.
Hours & Admission Prices: June-Oct. Sat.-Sun. 1-4; other times by appointment. Extended hours during Tontitown Grape Festival. No charge; donations accepted.

Van Buren

BOB BURNS MUSEUM & RIVER VALLEY MUSEUM OF VAN BUREN, 813 Main St., Old Frisco Depot, Van Buren, AR 72956-4315. Mailing Address: P.O. Box 1518, Van Buren, AR 72957-1518. Tel.: 479-474-6164; 800-332-5889. Fax: 501-474-5084.
E-mail: vanburen@vanburen.org
Web Site: www.vanburen.org
Founded: 1994.
Congressional District: 3
Key Personnel: Dir., Maryl Koeth.
Personnel Profile: Full-Time Paid 2; Part-Time Volunteers 21.
Governing Authority: municipal; nonprofit organization. Parent Institution: Van Buren A&P Commission.
Institution Type/Description: Historical Building: 1901 Frisco Depot.
Collections: items related to entertainment career of Bob Burns, a nationally-known 1930s & 1940s radio humorist; Van Buren history.
Activities: guided tours; permanent exhibitions.
Publications: brochures.
Hours & Admission Prices: April-Nov. Mon.-Fri. 8:30-5, Sat. 9-5; Dec.-March Mon.-Fri. 8:30-5. No Charge; donations accepted. &
Attendance: 30,000 (estimated)

CENTER FOR ART & EDUCATION, 104 N. 13th St., Van Buren, AR 72956-4512. Tel.: 479-474-7767. Facebook: Center for Art & Education.
E-mail: ccartcenter@sbcglobal.net
Web Site: www.art-ed.org
Founded: 1967.
Key Personnel: Exec. Dir., Jane Owen.
Institution Type/Description: Art Gallery.
Collections: paintings; sculpture; photographs.
Activities: educational programs; classes; workshops.
Hours & Admission Prices: Tues.-Fri. 10-4.
Attendance: 2,500

Waldron

BLYTHE'S SCOTT COUNTY MUSEUM, 1205 N. Main St., Waldron, AR 72958. Tel.: 479-637-3730. Fax: 479-637-4461.
E-mail: blythe.gary1@gmail.com
Web Site: blythemuseumar.com
Institution Type/Description: History Museum.
Collections: local history & culture; Caddo Indian & pioneer history; personal artifacts; photographs; period furnishings.
Hours & Admission Prices: Call for hours.

Walnut Ridge

WINGS OF HONOR MUSEUM AKA WALNUT RIDGE ARMY FLYING SCHOOL MUSEUM, INC., 70 S. Beacon Rd., Walnut Ridge, AR 72476. Tel.: 800-584-5575.
E-mail: harold@bscn.com
Web Site: wingsofhonor.org
Founded: 1999.
Congressional District: 1
Key Personnel: Chm. & Pres. (V), Harold Johnson; Education, Sue Whitmire; Public Rels., Brett Cooper; Treas., Carolyn Propst; Cur. & Museum Shop Mgr., Judy Wilson.
Personnel Profile: Full-Time Volunteers 1; Part-Time Volunteers 8.
Governing Authority: private; nonprofit organization. Tax-exempt: 501(c)(3).
Institution Type/Description: Military History Museum: housed on the site of the former Walnut Ridge Army Flying School during WWII then used as the Warbird Disposal Operation facility, and later the USAF 725th Radar Squadron was stationed here.
Collections: history of the WRAFS; Warbird Disposal Operation; Cadet history.
Facilities: 2,000 vol. library of WWII & military history books; 11,200 sq. ft. exhibit space; classrooms. Museum-related items for sale.
Activities: formal education programs; guided tours; lectures; temporary exhibitions. Museum Sponsors: Weekly Monday Morning Coffee for Veterans. Annual Events: monthly Pot-Luck & Evening of Music; Reunion of Cadets in April; Veterans Day Program in November; Remembrance in December.
Publications: annual newsletter, Reunion Information.
Hours & Admission Prices: Mon.-Sat. 9-5, Sun. 2-5. No charge; donations accepted. Closed New Year's Day; Thanksgiving; Christmas. &
Attendance: 3,500 (accurate)

Warren

BRADLEY COUNTY HISTORICAL MUSEUM, 200 W. Ash St., Warren, AR 71671-2602. Mailing Address: P.O. Box 311, Warren, AR 71671-0311. Tel.: 870-226-5457.
E-mail: jmldevco@sbcglobal.net
Founded: 1986.
Congressional District: 4
Governing Authority: Tax-exempt.
Institution Type/Description: History Museum: housed in the John Wilson Martin House, c.1857.
Collections: local history; personal artifacts; period furnishings.
Activities: demonstrations; hands-on activities. Museum Sponsors: Open House in June & December.
Hours & Admission Prices: Call for hours. No charge; donations accepted. &
Attendance: 800 (estimated)

Washington

HISTORIC WASHINGTON STATE PARK, 103 Franklin St., Washington, AR 71862. Mailing Address: P.O. Box 129, Washington, AR 71862-0129. Tel.: 870-983-2684. Fax: 870-983-2736.
E-mail: historicwashington@arkansas.com
Web Site: www.historicwashingtonstatepark.com
Founded: 1973.
Congressional District: 4
Key Personnel: Supt., Brandon Owen; Asst. Supt., Mike Roberts; Cur., Josh Williams; Museum Shop Mgr., Debbie Martin.
Personnel Profile: Full-Time Paid 20; Part-Time Paid 8.
Governing Authority: nonprofit organization. Parent Institution: The Arkansas Dept. of Parks & Tourism, Arkansas State Parks, 1 Capitol Mall, Little Rock, AR 72201. Tax-exempt.
Institution Type/Description: History Museum: located in 1824 town.
Collections: archaeology; manuscripts; weapons; hand tools; 19th-century furnishings; documents. Historic Houses: restored examples of southern Greek revival architecture.
Research Fields: southwest Arkansas history 1824-1889.
Facilities: 2,000-vol. library of law books & newspaper clippings on the history of the people & the town; southwest Arkansas regional archives.
Activities: guided tours; permanent exhibitions. Annual Events: Jonquil Festival in March; Frontier Days in September; Civil War weekend in October; Christmas & Candlelight in December.
Publications: information booklets.
Hours & Admission Prices: Daily 8-5. No charge. Guided walking tour of historic sites: adults $8, children 6-12 $4; discount to groups of 20 or more. Closed New Year's Day; Thanksgiving; Christmas. &
Attendance: 100,000 (estimated)

SOUTHWEST ARKANSAS REGIONAL ARCHIVES (SARA), 201 Highway 195, Washington, AR 71862. Mailing Address: P.O. Box 134, Washington, AR 71862-0134. Tel.: 870-983-2633. Fax: 870-983-2636.
E-mail: southwest.archives@arkansas.gov
Web Site: www.southwestarchives.com
Founded: 1978.
Congressional District: 4
Key Personnel: Archival Mgr., Peggy S. Lloyd; Administrative Asst., Gail Martin.
Personnel Profile: Full-Time Paid 2.
Governing Authority: Parent Institution: Arkansas History Commission.
Institution Type/Description: Archives & Research Institute: housed in former Washington Elementary School.
Collections: history of southwest Arkansas; letters; diaries; newspapers; pamphlets; maps; books including rare historical books; genealogical materials; cemetery records; family histories; periodicals; manuscripts; period photographs; cassettes of oral-history interviews; family collections; ledgers of businesses; family histories; microfilm; publications by county; historical societies.
Research Fields: history of 12 counties in southwest Arkansas; Gulf Plains region.
Facilities: 2,500-vol. library of historical books pertaining to the history of Arkansas & rare volumes of the Civil War available for research on premises only; reading room.
Activities: formally organized education programs for adults & graduate students; temporary exhibitions.
Publications: brochure; quarterly newsletter.
Hours & Admission Prices: Tues.-Sat. 8-4:30. No charge; donations accepted. Closed state & federal holidays. &
Attendance: 3,000
Membership: General $25; Contributing $75; Century $100; Patron $250; Lifetime $500; Permanent & Corporate $1,000.

Wilson

HAMPSON ARCHEOLOGICAL MUSEUM STATE PARK, #2 Lake Dr., Wilson, AR 72395. Mailing Address: P.O. Box 156, Wilson, AR 72395-0156. Tel.: 870-655-8622.
E-mail: hampsonarcheologicalmuseum@arkansas.com
Web Site: www.arkansasstateparks.com/hampsonmuseum
Founded: 1961.
Congressional District: 1
Key Personnel: Park Supt., Tess Pruett.
Personnel Profile: Full-Time Paid 2.
Governing Authority: state. Parent Institution: Arkansas Dept. of Parks and Tourism, Little Rock, AR. Tax-exempt.
Institution Type/Description: Archaeology Museum.
Collections: late Mississippian culture; archaeological collections.
Research Fields: human osteology; lithic studies; ceramics.
Facilities: picnic area. Museum-related items for sale.
Activities: guided tours; consultations for other museums; group participation-education activity programs, ages 5 and up.
Publications: brochures, museum guide.
Hours & Admission Prices: Tues.-Sat. 8-5, Sun. 1-5. Museum: no charge; donations accepted. Closed New Year's Day; Thanksgiving; Christmas Eve & Day. &
Attendance: 5,100 (estimated)

Wynne

CROSS COUNTY MUSEUM & ARCHIVES, (M), 711 E. Union, Wynne, AR 72396-3029. Mailing Address: P.O. Box 943, Wynne, AR 72396-0943. Tel.: 870-238-4100.
E-mail: crossmuseum@sbcglobal.net
Web Site: www.cchs1862.org
Formerly: Cross County Historical Society
Founded: 1972.
Key Personnel: Pres. (V), Carol Brown; Bd., Phillip Noery.

Personnel Profile: Part-Time Paid 2; Part-Time Volunteers 12; Interns 1.
Governing Authority: Parent Institution: Cross County Historical Society. Tax-exempt.
Institution Type/Description: Historical Society Museum.
Collections: county history & culture; military artifacts; genealogy; cemetery records; probate, civil & criminal court records; pre-Columbian Indian pottery.
Research Fields: local history; genealogy.
Facilities: archives.
Activities: Museum Sponsors: Spring Fundraiser; Fall Family Fun Day.
Publications: quarterly newsletter, The Cross County ERA.
Hours & Admission Prices: Mon.-Fri. 10-4. No charge; donations accepted. &
Attendance: 1,000 (estimated)
Membership: Individual $15; Family $25; Institutional $35; Sustaining $50; Patron $100; Benefactor $250; Advocate $500; Life $1,000.

CALIFORNIA

(1067 listings)

Acton

SHAMBALA PRESERVE, 6867 Soledad Canyon, Acton, CA 93510-2221. Mailing Address: P.O. Box 189, Acton, CA 93510-0189. Tel.: 661-268-0380. Fax: 661-268-8809.
Governing Authority: nonprofit organization. Tax-exempt: 501(c)(3).
Institution Type/Description: Wildlife Preserve.
Collections: over 70 lions, tigers, black & spotted leopards, mountain lions, servals, bobcats, lynxes, Asian leopard cats, a Florida Panther, & a jungle cat.
Facilities: Museum-related items for sale.
Activities: educational programs; safaris.
Publications: newsletter.
Hours & Admission Prices: Safaris: one Sat.-Sun. a month by appointment. Adults 18 & over $50; children under 18 not admitted.

Agoura Hills

REYES ADOBE HISTORICAL SITE, 30400 Rainbow Crest Dr., Agoura Hills, CA 91301. Mailing Address: 30610 Thousand Oaks Blvd., Agoura Hills, CA 91301. Tel.: 818-597-7361; 805-643-2504.
E-mail: xcastillo@ci.agoura-hills.ca.us
Web Site: www.agourahillsrec.org
Institution Type/Description: Historic House: built c.1850.
Collections: local history & culture; personal artifacts; period furnishings; photographs; Native American artifacts.
Activities: educational programs. Annual Event: Reyes Adobe Days Fest.
Hours & Admission Prices: Tues. 10-2, 2nd & 4th Sat. of month 1-4. Adults $3, seniors $2, children 5-12 $1; children under 5 no charge.

Alameda

ALAMEDA MUSEUM, 2324 Alameda Ave., Alameda, CA 94501. Tel.: 510-521-1233.
Web Site: alamedamuseum.org
Founded: 1948.
Governing Authority: Parent Institution: Alameda Historical Society. Tax-exempt.
Institution Type/Description: History Museum.
Collections: local history & culture; period furnishings; personal artifacts; photographs.
Publications: quarterly newsletter.
Hours & Admission Prices: Wed.-Fri. & Sun. 1:30-4, Sat. 11-4. No charge; donations accepted. &
Membership: Docent & Volunteer $15; Associate & Senior over 65 $20; Adult $30; Business $250; Lifetime $500.

ALAMEDA NAVAL AIR MUSEUM, 2151 Ferry Point Rd., Bldg. 77 at Alameda Point, Alameda, CA 94501. Tel.: 510-522-4262.
Institution Type/Description: Military History Museum.
Collections: Alameda Naval Air Station history; photographs; military artifacts; personal artifacts.
Facilities: Museum-related items for sale.
Activities: group tours.
Publications: book, Images of America.
Hours & Admission Prices: Sat.-Sun. 10-4; groups by appointment. Adults 12 & over $5; children under 12 no charge. &
Membership: Individual $25.

MEYERS HOUSE & GARDEN, 2021 Alameda Ave., Alameda, CA 94501. Tel.: 510-521-1247.
Institution Type/Description: Historic House Museum: built in 1897.
Collections: local history & culture; period furnishings; personal artifacts; photographs.
Hours & Admission Prices: 4th Sat. of the month 1-4. Admission $5 per person.

USS HORNET MUSEUM, 707 W. Hornet Ave., Alameda, CA 94501-5006. Mailing Address: P.O. Box 460, Alameda, CA 94501-9560. Tel.: 510-521-8448. Fax: 510-521-8327.
E-mail: info@uss-hornet.org
Web Site: www.uss-hornet.org
Founded: 1987.
Key Personnel: Dir. & C.E.O., Randall Ramian; Dir. Aircraft & Museum Operations, Rich Thom; Chm. (V), Jon Stanley; Museum Shop Mgr., Marti Morris
Governing Authority: Parent Institution: Aircraft Carrier Hornet Foundation. Tax-exempt.
Institution Type/Description: Air, Sea, & Space Museum. A National & State Historic Landmark.
Collections: military & personal artifacts; naval artifacts; aircraft; space artifacts including Apollo.
Facilities: Museum-related items for sale.
Hours & Admission Prices: Daily 10-5; groups by appointment. Adults $15, military, students and seniors 65 & over $12, youth 5-17 $6; members and children 4 & under no charge. Closed New Year's Day; Thanksgiving; Christmas.
Attendance: 60,000 (estimated)
Membership: Senior & Military $40; Individual $50; Family $75.

Aliso Viejo

SOKA UNIVERSITY ART GALLERY, 1 University Dr., Aliso Viejo, CA 92656-8081. Tel.: 949-480-4108. Fax: 949-480-4110.
Web Site: www.soka.edu
Institution Type/Description: Art Gallery.
Collections: works by national & international artists; paintings; photographs; sculpture.
Facilities: 8,000 sq. ft. exhibit space.
Hours & Admission Prices: Mon.-Fri. 9-5. No charge.

Alleghany

UNDERGROUND GOLD MINERS MUSEUM, 356 Main St., Alleghany, CA 95910-9998. Mailing Address: P.O. Box 907, Alleghany, CA 95910-0907. Tel.: 530-287-3330 & 3223. Fax: 530-287-3455.
E-mail: info@undergroundgold.com
Web Site: undergroundgold.com
Founded: 1995.
Congressional District: 1
Key Personnel: C.E.O. & Museum Shop Mgr., Rae Bell Arbogast; Pres. (V), David Scinto, C.P.A.; Education, Raymond Wittkopp.
Personnel Profile: Part-Time Paid 1; Part-Time Volunteers 4.
Governing Authority: private; nonprofit organization. Tax-exempt: 501(c)(3).
Institution Type/Description: Mining Museum.
Collections: mining technology; mines of Alleghany area; history of Alleghany; mining tools & equipment.
Facilities: 25-vol. library. Museum-related items for sale.
Activities: guided tours; theater. Annual Events: Gold Show in June; membership meeting.
Publications: annual newsletter.
Hours & Admission Prices: Memorial Day to Labor Day by appointment & special events. No charge; donations accepted. &
Attendance: 1,000 (estimated)
Membership: Student & Senior $15; Annual $30; Lifetime $300.

Allensworth

COLONEL ALLENSWORTH STATE HISTORIC PARK, 4011Grant Dr., Allensworth, CA 93219. Mailing Address: Star Rte. 1, Box 148, Earlimart, CA 93219-9710. Tel.: 661-849-3433. Fax: 661-849-4013.
Founded: 1908.
Congressional District: 18
Key Personnel: State Park Interpreter III, Steven M. Ptomey.
Personnel Profile: Full-Time Paid 4; Part-Time Paid 2; Part-Time Volunteers 50.

Governing Authority: state. Parent Institution: California Dept. of Parks & Recreation, P.O. Box 2390, Sacramento, CA 95811. Tax-exempt.
Institution Type/Description: Park Museum & Interpretive Center: located on the site of 1908 town of Allensworth, created to be a place where black people could live & work without racial prejudice.
Collections: interpretive material. Historic Buildings: Allensworth School; Singleton General Store; post office; drug store; dairy; livery stable; family homes; bakery; hotel.
Research Fields: pertaining to Allensworth & its pioneers.
Facilities: visitors' center; picnic area; campsites.
Activities: guided tours; lectures; films.
Publications: educational brochures.
Hours & Admission Prices: Park: 8-sunset. Visitor Center: daily 10-4; call for holiday hours. House Museum: by appointment only. Camping Fee: Night $20, Day $6; discounts to seniors & disabled. &

Alpine

ALPINE HISTORICAL SOCIETY, 2116 Tavern Rd., Alpine, CA 91901. Mailing Address: P.O. Box 382, Alpine, CA 91903-0382. Tel.: 619-659-8740.
E-mail: info@alpinehistory.org
Web Site: www.alpinehistory.org
Founded: 1962.
Congressional District: 52
Key Personnel: Pres. (V), Carol A. Morrison.
Governing Authority: Tax-exempt.
Institution Type/Description: Historical Society Museum.
Collections: local history & culture; period furnishings; personal artifacts; photographs. Historic Houses: Dr. Nichols house & barn, 1896; Beaty house, 1899.
Major Exhibits: Justin Gruelle, 1/14-6/14.
Research Fields: documents & photo files.
Activities: Museum Sponsors: Alpine History Day.
Publications: books, History of a Mountain Settlement; Meandering Through the Journey of Life; Echoes of the Past - The Old Timer; quarterly newsletter, Tattered Tidbits.
Hours & Admission Prices: last Sat.-Sun. each month 2-4; tours by appointment. No charge; donations accepted. &
Attendance: 500 (estimated)
Membership: Seniors $15; Adults $25; Family $35; Business $50; Life $500.

Alturas

MODOC COUNTY HISTORICAL MUSEUM, 600 S. Main St., Alturas, CA 96101-4117. Tel.: 530-233-2944.
E-mail: pitriver2@yahoo.com
Founded: 1967.
Congressional District: 14
Key Personnel: Dir. & Museum Shop Mgr., Paula Murphy.
Personnel Profile: Full-Time Paid 1.
Governing Authority: county. Tax-exempt: 501(c)(3).
Institution Type/Description: Modoc County History Museum.
Collections: settlement & development of Modoc County; period firearms; early settler, pioneer artifacts & Native American artifacts.
Research Fields: local history.
Facilities: Gift items & books for sale.
Activities: guided tours. Historical Society Sponsors: field trips to local historical sites.
Publications: book, The Journal of the Modoc County Historical Society; journals.
Hours & Admission Prices: May-Oct. Tues.-Sat. 10-4. Adults $2; discounts to members; children 16 & under no charge. &
Attendance: 6,000 (estimated)
Membership: Modoc County Historical Society: Regular $20; Joint $25; Patron $30.

Amador City

AMADOR/WHITNEY, 14170 Hwy. 49, Amador City, CA 95601. Mailing Address: P.O. Box 181, Amador City, CA 95601-0181. Tel.: 209-267-9310. Fax: 209-267-9310.
Key Personnel: Dir., Joyce Davidson
Institution Type/Description: History Museum.
Collections: local history & culture; contributions of women.
Hours & Admission Prices: Fri.-Sun. 12-4. No charge; donations accepted.

Anaheim

ANAHEIM HERITAGE CENTER, Anaheim Public Library, 241 S. Anaheim Blvd., Anaheim, CA 92805. Tel.: 714-765-6453. Fax: 714-765-6469.
E-mail: jnewell@anaheim.net
Web Site: www.anaheim.net/library
Formerly: Elizabeth J. Schultz/Anaheim History Room
Founded: 1967.
Key Personnel: Library Branch Mgr. Heritage Svcs., Jane K. Newell.
Personnel Profile: Full-Time Paid 2; Part-Time Paid 4; Part-Time Volunteers 8.
Governing Authority: city. Parent Institution: Anaheim Public Library.
Institution Type/Description: History Museum.
Collections: Anaheim history; photographs; books.
Hours & Admission Prices: Mon. 12-9, Tues.-Fri. 11-6. No charge.
Attendance: 1,529 (accurate)

MOTHER COLONY HOUSE MUSEUM, 414 N. West St., Anaheim, CA 92801-5953. Mailing Address: Anaheim Heritage Center at the Museum, 241 s. Anaheim Blvd., Anaheim, CA 92805. Tel.: 714-765-6453.
Web Site: www.anaheimcolony.com/m_colony.htm
Key Personnel: Museum Mgr., Jane Newell
Institution Type/Description: Historic House Museum: State Historical Landmark.
Collections: personal artifacts; period furnishings.
Hours & Admission Prices: Call for appointment. No charge; donations accepted. Closed major holidays

MUZEO, (M), 241 S. Anaheim Blvd., Anaheim, CA 92805-3821. Tel.: 714-956-8936.
E-mail: info@muzeo.org
Web Site: www.muzeo.org
Formerly: Anaheim Museum
Founded: 2007.
Congressional District: 38
Key Personnel: Dir., John Scola; Museum Shop Mgr., Patricia Davis.
Personnel Profile: Full-Time Paid 4; Part-Time Paid 6; Interns 1.
Governing Authority: Tax-exempt.
Institution Type/Description: Art Museum.
Collections: international traveling exhibits; local history.
Facilities: Museum-related items for sale.
Activities: special events; educational programs.
Hours & Admission Prices: Daily 10-5. Adults $13, children 3-12 $9; discounts to seniors; members no charge. Closed New Year's Day; Thanksgiving; Christmas. &
Attendance: 30,000 (estimated)
Membership: Individual $75; Family & Household $99; Group $149.

OAK CANYON NATURE CENTER, 6700 E. Walnut Canyon Rd., Anaheim, CA 92807-4948. Mailing Address: 200 S. Anaheim Blvd., #433, Anaheim, CA 92805-3820. Tel.: 714-998-8380.
E-mail: ocnc@anaheim.net
Web Site: www.anaheim.net/ocnc/
Founded: 1973.
Personnel Profile: Part-Time Paid 20.
Volunteer Hours: 2,000
Governing Authority: Parent Institution: City of Anaheim.
Institution Type/Description: Nature Center.
Collections: natural history; mounted wildlife.
Facilities: 58 acre wilderness park; 4 miles hiking trails; interpretive center; picnic area; rental facilities.
Activities: hiking; group tours; weddings; birthday parties; day camps; special events; seasonal programs.
Hours & Admission Prices: Park Grounds & Trails: daily sunrise to sunset. Interpretive Center: Sat.-Sun. 10-4. No charge; donations accepted. &
Attendance: 100,000 (estimated)

Angels Camp

ANGELS CAMP MUSEUM, (M), 753 S. Main St., Angels Camp, CA 95222. Mailing Address: P.O. Box 667, Angels Camp, CA 95222. Tel.: 209-736-2963. Fax: 209-736-0709.
E-mail: museumoffice@angelscamp.gov
Web Site: www.angelscamp.gov
Formerly: Angels Camp Museum & Carriage House
Founded: 1951.
Key Personnel: Dir., Kimberly Arth; Museum Shop Mgr., Nioma Patrick.

Personnel Profile: Full-Time Paid 1; Part-Time Paid 7; Part-Time Volunteers 5.
Governing Authority: Parent Institution: City of Angels Camp. Tax-exempt: 501(c)(3).
Institution Type/Description: History Museum.
Collections: household artifacts; Pelton water wheel; mining equipment; period carriages, carts & wagons; rocks & minerals; ranching & mining from the 1800s-1900s.
Facilities: 3 acre, 3 building museum complex; picnic tables; collections. Gift items for sale.
Activities: guided tours; lectures; workshops; festivals; programs for all ages.
Hours & Admission Prices: March-Nov. Thurs.-Sun. 10-4; Dec.-Feb. Sat.-Sun. 10-4. Adults $5, children $2.50, children under 4 & members free.
Attendance: 5,100 (accurate)
Membership: Individual $35; Family $50; Friend $100; Patron $250; Lifetime $1,000.

Antioch

ANTIOCH HISTORICAL SOCIETY MUSEUM, 1500 W. Fourth St., Antioch, CA 94509-1046. Tel.: 925-757-1326.
Founded: 1998.
Congressional District: 10
Key Personnel: Pres. (V), Bob Martin
Governing Authority: bd. of directors. Parent Institution: Antioch Historical Society. Tax-exempt.
Institution Type/Description: Historical Society Museum: housed in the former Riverview Union High School; built in 1911. Listed on the National Register of Historical Places.
Collections: local history & culture; photographs; personal artifacts; Antioch Sports Legends Hall.
Publications: quarterly, Gazette.
Hours & Admission Prices: early Jan. to late Dec. Wed. & Sat. 1-4. No charge; donations accepted. Closed holidays; New Year's Eve & Day; Christmas Eve, Day & week. &
Attendance: 4,500 (estimated)
Membership: Individual $25; Couple $40.

Apple Valley

VICTOR VALLEY MUSEUM & ART GALLERY, 11873 Apple Valley Rd., Apple Valley, CA 92308-3670. Tel.: 760-240-2111. Fax: 760-240-5290.
E-mail: vvmuseum@hotmail.com
Web Site: www.vvmuseum.com
Founded: 1976.
Congressional District: 35
Key Personnel: Exec. Dir., Carol Carr; Pres., Doug Shumway; Vice Pres., Mike Davis; School Tours, Sarah Puett.
Personnel Profile: Part-Time Paid 4; Part-Time Volunteers 30; Interns 3.
Governing Authority: nonprofit organization. Tax-exempt: 501(c)(3).
Institution Type/Description: Natural History Museum.
Collections: archaeological & geographical artifacts & tools.
Research Fields: history & natural history of Mojave Desert; Serrano Indians; transportation; desert culture.
Facilities: 200-vol. library on local & natural history of California; botanical garden; 90-seat auditorium; 6,000 sq. ft. exhibit space. Books on local history & nature, educational toys & other museum-related items for sale.
Activities: guided tours; lectures; docent program; films; participatory & temporary exhibits.
Publications: quarterly bulletin, VVM.
Hours & Admission Prices: Wed.-Sat. 10-4, Sun. 1-4. Adults $4, seniors $2; members & students no charge. Imagination Station: $1 per child with paid adult general admission. Closed New Year's Day; Easter; Mother's Day; Memorial Day; Father's Day; Labor Day; Thanksgiving; Christmas. &
Attendance: 12,000 (estimated)
Membership: Senior 62 & over $25; Senior Couple 62 & over $35; Individual $40; Family & Grandparents $50. Presidents' Circle: Avant Garde $1,000-$1,999; Presidents' Circle $2,500-$4,999; Patron $5,000-$9,999; Sponsor $10,000-$24,999.

Aptos

CABRILLO GALLERY, 6500 Soquel Dr., Bldg. 1000, Rm. 1002, Aptos, CA 95003-3119. Tel.: 831-479-6308. Fax: 831-479-5045.
E-mail: gallery@cabrillo.edu
Web Site: www.cabrillo.edu/services/artgallery
Key Personnel: Gallery Dir., Tobin Keller; Asst. Dir., Rose Sellery
Institution Type/Description: Art Museum.
Collections: paintings; photographs; drawings; prints.
Hours & Admission Prices: Spring & Fall Mon.-Tues. 9-4 & 7-9, Wed.-Fri. 9-4. Adults $2, youth 12-17 $1; children no charge. &

Attendance: 5,000

Arabia

S.C.R.A.P. GALLERY, 46-350 Arabia St., Arabia, CA 92201. Tel.: 760-863-7777. Fax: 760-863-8973. Facebook: SCRAP Gallery.
E-mail: info@scrapgallery.org
Web Site: scrapgallery.org
Founded: 1996.
Key Personnel: Exec. Dir., Karen Riley; Pres. (V), Marilyn Glassman; Treas., Donna Pease.
Personnel Profile: Full-Time Paid 1; Part-Time Paid 2; Part-Time Volunteers 20; Interns 2.
Governing Authority: private; nonprofit organization.
Institution Type/Description: Art Museum.
Collections: art objects created with reused or recycled materials.
Facilities: educational facilities; conservation center; 10,000 sq. ft. exhibit space. Museum-related items for sale.
Activities: formal education programs for children; guided tours; mobile vans; participatory exhibits; school loan service.
Publications: Landfill Lunch Box; The Eco Deck; Don't Trash My Planet.
Hours & Admission Prices: Sept.-July Mon.-Fri. 9-5, Sat. special hours monthly. No charge. Closed major holidays. &
Attendance: 270,000 (accurate)

Arcadia

THE ARBORETUM LOS ANGELES COUNTY ARBORETUM & BOTANIC GARDEN, 301 N. Baldwin Ave., Arcadia, CA 91007-2697. Tel.: 626-821-3222. Fax: 626-445-1217.
Web Site: www.arboretum.org
Formerly: The Arboretum of Los Angeles County
Founded: 1947.
Congressional District: 26
Key Personnel: C.E.O., Richard Schulhof; Pres. (V) Los Angeles Arboretum Foundation, Kenneth D. Hill, Ph.D.; Supt., Timothy Phillips; Cur. Living Collections, Jim Henrich; Librarian, Susan Eubank; Communications, Nancy Yoshihara; Museum Shop Mgr., Rosemary Bullen.
Governing Authority: county; board of supervisors. Jointly Managed by: County of Los Angeles Department of Parks & Recreation and the Los Angeles Arboretum Foundation. Tax-exempt.
Institution Type/Description: Arboretum: located on the site of Santa Anita Rancho.
Collections: plants of Australia, Americas, Asia, & Mediterranean region; specialties in orchids, cycads, eucalyptus; callistemon, melaleuca, acacia, palms, bamboo; herbarium consisting of introduced ornamental woody plants of Southern California; California history; children's gardens. Historic Buildings: 1840 Adobe; 1885 Queen Anne Cottage; 1881 Coach Barn; 1890 Santa Fe-Santa Anita Depot.
Research Fields: local history; testing of plants for introduction to California horticulture.
Facilities: 28,000-vol. & 150 periodical library of plant sciences & related topics available for public use & inter-library loan & reference; reading room; greenhouses; demonstration gardens; water conservation garden; meeting rooms; interpretive centers. Books, plants & crafts for sale.
Activities: guided & self-guided tours; tram tours; lectures; concerts; arts festivals; organized educational programs; docent program; permanent & temporary exhibitions; garden walks. Museum Sponsors: flower shows; Environmental Education Fair.
Publications: quarterly newsletter for Foundation members.
Hours & Admission Prices: Daily 9-4:30. Adults $9, seniors & students $6, children 5-12 $4; children under 5 & members no charge. Closed Christmas. &
Attendance: 300,000 (accurate)
Membership: Student/Teacher $45; Senior $50; Individual $55; Family/Grandparent $75; Garden Sustainer $150; Garden Sponsor $300; Garden Benefactor $500.

THE GILB MUSEUM OF ARCADIA HISTORY, (M), 380 W. Huntington Dr., Arcadia, CA 91007. Mailing Address. P.O. Box 60021, Arcadia, CA 91066-6021. Tel.: 626-574-5440. Fax: 626-821-9057. Facebook: Arcadia Historical Museum.
E-mail: ddunn@ci.arcadia.ca.us
Web Site: museum.ci.arcadia.ca.us
Formerly: Ruth and Charles Gilb Arcadia Historical Museum
Founded: 2001.
Congressional District: 26
Key Personnel: Dir., Mary Beth Hayes; Education, Lindsey Sun; Public Rels., Darlene Bradley; Cur., Dana Dunn.

Personnel Profile: Full-Time Paid 1; Part-Time Paid 1; Part-Time Volunteers 27; Interns 2.
Governing Authority: Parent Institution: City of Arcadia. Subsidiary Institution: Arcadia Historical Museum Foundation. Tax-exempt.
Institution Type/Description: History, Art & Culture Museum.
Collections: local history; archives; photographs.
Research Fields: Tongva history; Spanish & Mexican Rancho; The E.J. "Lucky" Baldwin family; City of Arcadia civic history; Arcadia community history; movies filmed in Arcadia history; famous Arcadians; Anita Baldwin.
Facilities: research library; 2,000 sq. ft. exhibit space; Santa Anita Japanese Assembly Center; Santa Anita Park; WWI Arcadia Balloon School.
Activities: changing exhibits; children's days & festivals; school tours; guided tours; lectures; internship, volunteer & docent programs; lunchtime talks; historic movie program; puppet theatre; workshops; scout programs.
Hours & Admission Prices: Tues.-Sat. 10-4. No charge; donations accepted. Closed national holidays. &
Attendance: 5,800 (accurate)

Arcata

HUMBOLDT STATE UNIVERSITY NATURAL HISTORY MUSEUM, 1315 G. St., Arcata, CA 95521-5820. Tel.: 707-826-4479. Fax: 707-826-3201. Facebook: Humboldt State University Natural History Museum.
E-mail: natmus@humboldt.edu
Web Site: www.humboldt.edu/natmus
Founded: 1989.
Key Personnel: Interim Mgr., Julie Van Sickle; Outreach Coord., Allison Poklemba; Coord. Educational Programs, Jennifer Ortega; Technology Specialist, Michael Kauffmann; Dir. Humboldt Science & Mathematics Center, Dr. Jeffrey White.
Personnel Profile: Part-Time Paid 11; Part-Time Volunteers 10; Interns 11.
Governing Authority: public university; nonprofit. Parent Institution: Humboldt State University. Tax-exempt.
Institution Type/Description: University Natural History Museum.
Collections: Pacific Northwest natural history including butterflies, crabs, mollusks, sponges & corals; birds; minerals; insects; fossils (1.9 billion years ago to present) from around the world; live honey bee hive; 33 ft. Quetzalcoatlus replica; hominids; North American prehistoric mammals.
Facilities: 200-vol. library of field guides & educational materials; educational facilities. Animal models, natural history books & museum-related items for sale.
Activities: formal educational programs; lectures; school loan service; participatory & temporary exhibitions.
Publications: quarterly program calendar, Nature Adventures.
Hours & Admission Prices: Tues.-Sat. 10-5. Family $10, adults $3, children over 3 $2; ASTC & museum members no charge. Closed New Year's Day; Memorial Day; Independence Day; Labor Day; Thanksgiving; Christmas. &
Attendance: 18,500 (estimated)
Membership: Student & Senior $25; Individual $35; Family $60; Associate $100; Supporting $250; Sustaining $500; Benefactor $1,000.

REESE BULLEN GALLERY, (M), Humboldt State University, 1 Harpst St., Arcata, CA 95521-8222. Tel.: 707-826-5802. Fax: 707-826-3628.
Web Site: www.humboldt.edu
Founded: 1970.
Congressional District: 1
Key Personnel: Chm. & Gallery Dir., Martin Morgan; Dir. Asst., Nancy Clark.
Personnel Profile: Full-Time Paid 1; Part-Time Paid 16; Interns 3.
Governing Authority: public college. Parent Institution: Humboldt State University.
Institution Type/Description: University Art Gallery
Collections: 20th-century American & African art.
Facilities: 3,080 sq. ft. exhibit space.
Activities: guided tours; temporary & loan exhibitions; formal education programs for children; training programs for professional museum workers.
Hours & Admission Prices: Sept.-May Mon.-Fri. 11-4. Closed university holidays & semester breaks. &
Attendance: 5,400 (accurate)

Arroyo Grande

SOUTH COUNTY HISTORICAL SOCIETY, 134 Mason St., Arroyo Grande, CA 93420. Mailing Address: P.O. Box 633, Arroyo Grande, CA 93421-0633. Tel.: 805-489-8282.
E-mail: schs76@sbcglobal.net
Web Site: southcountyhistory.org

Formerly: Paulding History House
Founded: 1976.
Congressional District: 22
Key Personnel: Chm. (V), Joe Swigert; Pres. (V), Kirk Scott.
Personnel Profile: Part-Time Paid 1; Part-Time Volunteers 150.
Governing Authority: Branch Museums: Historic 100F Hall, 128 Bridge St., Arroyo Grande, CA; Heritage House Museum & Garden, 126 S. Mason St., Arroyo Grande, CA; Patricia Loomis History Library, 134 S. Mason St., Arroyo Grande, CA; Santa Manuela Schoolhouse, 127 Short St., Arroyo Grande, CA; The Barn Museum, 127 1/2 Short St., Arroyo Grande, CA; Paulding History House, 551 Crown Hill, Arroyo Grande, CA 93420. Tax-exempt: 501(c)(3).
Institution Type/Description: Historical Society Museum.
Collections: Native American artifacts; Mexican Land Grant Era; early California statehood, schools, lifestyle & agriculture from 1800-1960; period furnishings; personal artifacts; carvings; Cumash Indian baskets; barn; period vehicles. Historic Buildings: 1902 Joof Hall; 1889 Heritage House Museum; 1901 Santa Manuela Schoolhouse; 1889 Paulding History House.
Research Fields: South County of San Luis Obispo California history.
Facilities: library; research center.
Activities: research. Museum Sponsors: Summer Historic Theatre June to August.
Publications: 9 times annually, Heritage Press.
Hours & Admission Prices: Heritage Square Museums: Sat. 12-3, Sun. 1-4. Paulding History House: 1st Sat. 12-3. Patricia Loomis History Library Mon.-Fri. 1-5. No charge; donations accepted. Closed New Year's Day; Christmas. &
Attendance: 10,000 (estimated)
Membership: Individual $15; Sustaining $100; Patron $200; Life $500; Corporate $1,000, $2,500, $5,000.

Atascadero

CHARLES PADDOCK ZOO, Morro Rd., Hwy. 41, W., Atascadero, CA 93422. Mailing Address: 9305 Pismo Ave., Atascadero, CA 93422-4939. Tel.: 805-461-5080. Fax: 805-461-7625.
E-mail: zoo@atascadero.org
Web Site: charlespaddockzoo.org
Founded: 1955.
Congressional District: 22
Key Personnel: C.E.O., Alan G. Baker; Pres. Friends of the Charles Paddock Zoo, Steve Robinson.
Personnel Profile: Full-Time Paid 7; Part-Time Paid 2; Part-Time Volunteers 50; Interns 4.
Governing Authority: Parent Institution: City of Atascadero. Subsidiary Institution: Friends of the Charles Paddock Zoo. Tax-exempt.
Institution Type/Description: Zoo.
Collections: live animals; plants.
Research Fields: captive breeding & field work.
Publications: newsletter.
Hours & Admission Prices: Summer: daily 10-5; Winter: daily 10-4. Adults 12 & over $5, senior citizens 65 & over $4.25, youth 3-11 $4; AZA members & children under 3 no charge. Closed Thanksgiving; Christmas. &
Attendance: 60,000 (estimated)
Membership: Individual $30; Individual Plus One $35; Family $50; Family Plus One $60; Sponsor $100; Benefactor & Zoo Booster (Business) $250.

Atwater

CASTLE AIR MUSEUM, 5050 Santa Fe Dr., Atwater, CA 95301-5154. Tel.: 209-723-2178 & 2182. Fax: 209-723-0323.
E-mail: castleairmuseum@clearwire.net
Web Site: www.castleairmuseum.org
Founded: 1981.
Congressional District: 18
Key Personnel: C.E.O., Joe Pruzzo; Chm. Bd., John Sundgren; Financial Dir., Nelson Howlett; Sec., Caroline Venable.
Personnel Profile: Full-Time Paid 2; Full-Time Volunteers 4; Part-Time Paid 6; Part-Time Volunteers 180; Interns 1.
Governing Authority: private; nonprofit. Tax-exempt.
Institution Type/Description: Military & Aviation History Museum.
Collections: aviation history including strategic bombardment from World War II to the present including Merced Army Airfield & Castle Air Force Base.
Facilities: 20 acres; 75-seat banquet room; 1,500 sq. ft. exhibit space; picnic grounds. Aviation-related items for sale.
Activities: docent program; guided tours; guest lectures; membership drive.
Publications: quarterly newsletter.
Hours & Admission Prices: April-Sept. daily 9-5; Oct.-March daily 10-4. Adults $10, seniors 60 & up and youth 6-17 $8; active duty, children under

5, & members no charge. Closed New Year's Day; Easter; Thanksgiving; Christmas Eve & Day. &

Attendance: 30,000 (accurate)

Membership: Supporting $30; Sustaining $60; Family $90; Life $1,000; McCarthy $2,500; Eagle $5,000; LeMay $7,500; Doolittle $10,000; Macready $15,000; General Castle $25,000.

Auburn

BERNHARD MUSEUM COMPLEX, 291 Auburn-Folsom Rd., Auburn, CA 95603-5039. Mailing Address: 101 Maple St., Auburn, CA 95603-5026. Tel.: 530-889-6500. Fax: 530-889-6510.

Web Site: www.placer.ca.gov/museum

Key Personnel: Dir., Melanie Barton.

Personnel Profile: Full-Time Paid 1.

Institution Type/Description: History Museum.

Collections: furnishings; period viticulture equipment; farm wagon; zinfandel grapes; vegetables; herbs; flowers.

Hours & Admission Prices: Tues.-Sun. 11-4. Closed holidays. No charge; donations accepted. &

PLACER COUNTY MUSEUMS, 101 Maple St., Auburn, CA 95603-5026. Tel.: 530-889-6500. Fax: 530-889-6510.

Web Site: www.placer.ca.gov/museum

Founded: 1948.

Congressional District: 1

Key Personnel: Museum Admin., Melanie Barton; Program Mgr., Ralph Gibson; Cur. Collections, Kasia Woroniecka; Exhibit Preparator, Jason Adair.

Personnel Profile: Full-Time Paid 6; Part-Time Paid 3; Part-Time Volunteers 150.

Governing Authority: county. Subsidiary Institutions: Gold Country Museum; Bernhard Museum Complex; Griffith Quarry Museum; Forest Hill Divide Museum; Golden Drift Museum; Placer County Museum; Placer County Archives. Tax-exempt.

Institution Type/Description: History Museum.

Collections: mining equipment & related items; 1880-1900 furniture & household accessories; agricultural & blacksmith artifacts; logging artifacts; items of daily life; archival records of Placer County; private archival material. Historic Buildings: c.1851 Bernhard Museum; c.1881 Bernhard Wine Processing Building.

Research Fields: history of Placer County.

Activities: guided tours; living history programs; changing exhibits; outreach programs.

Publications: department brochure; newsletter, The Placer; flyers & brochures on current temporary exhibits.

Hours & Admission Prices: Gold Country & Bernhard: Tues.-Sun. 11-4. Placer County: daily 10-4. Griffith Quarry: Sat.-Sun. 12-4. Forest Hill & Gold Drift: Memorial Day-Labor Day Wed. & Sat.-Sun. 12-4. No charge; donations accepted. &

Attendance: 55,000 (accurate)

Avalon

CATALINA ISLAND MUSEUM, INC., (M), 1 Casino Way, Casino Bldg., Avalon, CA 90704. Mailing Address: P.O. Box 366, Avalon, CA 90704-0366. Tel.: 310-510-2414. Fax: 310-510-1957.

E-mail: info@catalinamuseum.org

Web Site: www.catalinamuseum.org

Founded: 1953.

Congressional District: 36

Key Personnel: Exec. Dir., Dr. Michael De Marsche; Pres., Steven C. Schreiner; Cur., Jeannine Pedersen; Museum Shop Mgr., John Boraggina; Dir. Operations, Gail Fornasiere.

Personnel Profile: Full-Time Paid 4; Part-Time Paid 5; Part-Time Volunteers 150; Interns 1.

Governing Authority: private; nonprofit corporation. Tax-exempt: 501(c)(3).

Institution Type/Description: History Museum.

Collections: Indian artifacts; historical photographs; ship models; pottery; historical memorabilia.

Research Fields: archaeology; Catalina history.

Facilities: research library on subjects pertaining to Catalina Island by appointment.

Activities: education programs for children; lectures; field trips.

Publications: books, The Casino; The Legends of Old Ben; Avalon Walkabout; The Art of Catalina Clay Products.

Hours & Admission Prices: Fri.-Wed. 10-4. Adults $5, seniors $4, children 6-15 $2; discounts to military, AAM & ICOM members and groups of 20 or more; children under 6 & members no charge. Closed Thanksgiving; Christmas. &

Attendance: 55,000 (estimated)

Membership: Individual $40; Family & Dual $60; Sponsor $100; Associate $250; Patron $500; Fellow $1,000; Mt. Banning $2,500; Mt. Black Jack $5,000; Mt. Orizaba $10,000. Business Supporter $250; Business Associate $500; Business Patron $1,000; Business Benefactor $2,500 & up.

Bakersfield

*** BAKERSFIELD MUSEUM OF ART, (M),** 1930 R St., Bakersfield, CA 93301-4815. Tel.: 661-323-7219. Fax: 661-323-7266.

Web Site: www.bmoa.org

Founded: 1987.

Congressional District: 20

Key Personnel: Exec. Dir., Bernard J. Herman; Bd. Pres., Joe Audelo; Chief Cur., Emily Falke; Asst. Dir., David Gordon.

Personnel Profile: Full-Time Paid 4; Part-Time Paid 11; Part-Time Volunteers 70.

Governing Authority: nonprofit. Parent Institution: Bakersfield Art Foundation. Tax-exempt: 501(c)(3).

Institution Type/Description: Art Museum.

Collections: paintings; sculpture; graphics; complete works of Marion Osborn Cunningham; Phil Paradise serigraphs; California regional artists; works on paper.

Research Fields: California regionalists.

Facilities: 700-vol. library of books, catalogs & magazines available for use by the public; educational facilities; reception facilities. Gift items for sale.

Activities: guided tours; lectures; films; gallery talks; arts festivals; formally organized education programs for children; inter-museum loan; participatory, permanent, temporary & traveling exhibitions; docent programs; workshops.

Publications: quarterly newsletter; exhibit catalogs; gallery posters & prints.

Hours & Admission Prices: Tues.-Fri. 10-4, Sat.-Sun. 12-4. Adults $5, seniors $4, students $2; discounts to AAM & ICOM members; children 6 & under and members no charge. Closed New Year's Day; Martin Luther King Day; Easter; Memorial Day; Independence Day; Labor Day; Thanksgiving; Christmas. &

Attendance: 25,000 (estimated)

Membership: Students & Seniors $20; Individual $30; Family $40; Active $100; Contributing $250; Sustaining $500.

BUENA VISTA MUSEUM OF NATURAL HISTORY, 2018 Chester Ave., Bakersfield, CA 93301-4420. Tel.: 661-324-6350. Fax: 661-324-7522.

E-mail: bvmnh@sharktoothhill.com

Web Site: www.sharktoothhill.com

Key Personnel: Exec. Dir., Koral Hancharick

Institution Type/Description: Natural History Museum.

Collections: earth history; paleontology; anthropology; fossils; natural history.

Facilities: Museum-related items for sale.

Activities: workshops; educational programs; geology fieldtrips.

Publications: newsletter, Sharkbites.

Hours & Admission Prices: Thurs.-Sat. 10-4, Sun. 12-4, call to confirm; other times by appointment. Adults $7, senior citizens, children & students $4; discounts to groups of 20 or more; children 5 & under no charge. Closed major holidays.

CALIFORNIA LIVING MUSEUM CALM, 10500 Alfred Harrell Hwy., Bakersfield, CA 93306-9654. Tel.: 661-872-2256. Fax: 661-872-2205.

E-mail: toanspach@kern.org

Web Site: www.calmzoo.org

Founded: 1983.

Key Personnel: Dir., Tom Anspach; Zoo Mgr., Lana Fain; Zoo Cur., Don Richardson.

Governing Authority: Parent Institution: Kern County Superintendent of Schools.

Institution Type/Description: Zoo.

Collections: native California animals; plants; fossils.

Facilities: Museum-related items for sale.

Activities: rental facilities.

Hours & Admission Prices: Feb.-Oct. daily 9-5; Nov.-Jan. daily 9-4. Adults $9, seniors $7, children 3-12 $5; discounts to groups; children under 3 & members no charge. Closed New Year's Day; Easter; Thanksgiving; Christmas Eve & Day. &

Membership: Individual $30; Family $50; Contributing $80; Sustaining $125; Golden Bear $275; Life $1,000.

CALIFORNIA STATE UNIVERSITY, BAKERSFIELD, TODD MADIGAN GALLERY, 9001 Stockdale Hwy., 15FA, Bakersfield, CA 93311-1099. Tel.: 661-654-2238. Fax: 661-654-2539.
Web Site: www.csub.edu/art/gallery
Key Personnel: Dir., Joey Kotting
Institution Type/Description: Art Gallery.
Collections: paintings; sculpture; photographs.
Activities: special events.
Hours & Admission Prices: Academic Year: Tues.-Thurs. 1-6, Sat. 1-5. Closed university breaks & holidays.

KERN COUNTY MUSEUM, 3801 Chester Ave., Bakersfield, CA 93301-1345. Tel.: 661-852-5000. Fax: 661-322-6415.
E-mail: kcmuseum@kern.org
Web Site: kcmuseum.org
Founded: 1945.
Congressional District: 18
Key Personnel: Dir., Carola Rupert Enriquez; Chm., Beth Pandol; Asst. Dir., Jeff Nickell; Cur., Lori Wear; Bldg. & Grounds, Scott Fieber; Education Programs, Jackie Brouillette; Educational Asst., Elizabeth Herrera.
Personnel Profile: Full-Time Paid 16; Part-Time Paid 11; Part-Time Volunteers 48.
Governing Authority: county. Parent Institution: Kern County Museum Authority. Subsidiary Institution: Lori Brock Children's Discovery Center. Tax-exempt.
Institution Type/Description: History Museum.
Collections: 57 structure outdoor exhibit; household artifacts; tools & equipment; firearms; Indian artifacts; photographs; historic vehicles; two acre petroleum exhibit.
Research Fields: local history; culture.
Facilities: 2,000-vol. library of local history, museology, geology, mineralogy, Western Americana, California & fiction pertaining to area available on premises by appointment.
Activities: tours for schools; one-room school programs; docent programs; permanent & temporary exhibitions; summer camp. Museum Sponsors: Safe Halloween; Lamplight Tours; Fun Days; Living History Day; Black Gold: The Oil Experience; Early California History Day; Native American Life.
Publications: A Kern County Diary: The Forgotten Photos of Carleton E. Watkins, 1881-1888; cookbook, Nuggets, Nibbles and Nostalgia; quarterly newsletter, The Courier; Historic Kern County: An Illustrated History of Bakersfield and Kern County; Chronicles of Kern County; Hard Drivin' Country: The Honky Tonks, Musicians, and Legends of the Bakersfield Sound.
Hours & Admission Prices: Mon.-Sat. & holidays 10-5, Sun. 12-5; ticket office closes daily at 3. Adults $10, senior citizens over 60 & teens $9, children 6-12 $8, children 3-5 $7; children under 3 & members no charge. Closed New Year's Eve & Day; Easter; Independence Day; Thanksgiving; Christmas Eve & Day. &
Attendance: 75,000 (accurate)
Membership: Family $65; Supporting $125; Sustaining $250; Patron $500; Benefactor $1,000.

LORI BROCK CHILDREN'S DISCOVERY CENTER, 3801 Chester Ave., Bakersfield, CA 93301-1345. Tel.: 661-852-5000. Fax: 661-322-6415.
E-mail: kcmuseum@kern.org
Web Site: www.kcmuseum.org/loribrock
Founded: 1976.
Congressional District: 18
Key Personnel: Dir. Museum Svcs., Carola Enriquez; Asst. Dir., Jeff Nickell; Tour Bookings, Lily Soto; Education Program, Jackie Brouillette.
Governing Authority: Parent Institution: Kern County Museum, which is operated by a joint powers agreement between governmental agencies. Tax-exempt: 501(c)(3).
Institution Type/Description: Children's Museum.
Collections: hands-on exhibits.
Facilities: classrooms.
Activities: gallery talks; arts festivals; drama; day camps; formally organized education programs for children.
Publications: quarterly newsletter; summer camp flyer.
Hours & Admission Prices: Mon.-Sat. & holidays 10-5, Sun. 12-5. Adults $8, senior citizens & students 13-17 $7, youth 6-12 $6, children 3-5 $5; children under 3 & members no charge. Closed New Year's Day; Presidents' Day; Memorial Day; Independence Day; Thanksgiving; Christmas Eve & Day. &
Attendance: 95,000 (accurate)
Membership: Family $65; Supporting $125; Sustaining $250; Patron $500; Benefactor $1,000.

Banning

FIRE MEMORIES MUSEUM, 5261 W. Wilson St., Banning, CA 92220. Tel.: 951-260-9434 & 809-5457.
E-mail: museum@firemuseum.org
Web Site: www.firememories.org
Key Personnel: Dir., Doug Hammer; Dir., Monte Hammer
Institution Type/Description: Firefighting History Museum.
Collections: firefighting history & equipment; fire apparatus; hand & horse-drawn wagons and hose carts; ladders; Vajen smoke mask; leather fire hose; station gongs; call boxes; telegraph fire alarm; fire extinguishers; hydrants; helmets; patches.
Facilities: library.
Hours & Admission Prices: Sat. 10-4:30; other times by appointment. Adults $4, seniors 60 & over $2.50, students 5-17 $2; children under 5, active fire, EMS & military no charge.

GILMAN HISTORIC RANCH AND WAGON MUSEUM, 1901 W. Wilson St., Banning, CA 92220. Tel.: 951-922-9200.
Web Site: riversidecountyparks.org
Institution Type/Description: History Museum: housed in the homestead ranch of James Marshall Gilman.
Collections: California history from Cahuilla Indians to early settlement of southern California & the San Gorgonio Pass.
Hours & Admission Prices: Call for hours.

MALKI MUSEUM, 11-795 Malki Rd., Morongo Indian Reservation, Banning, CA 92220. Mailing Address: P.O. Box 578, Banning, CA 92220-0017. Tel.: 951-849-7289. Fax: 951-849-3549. Facebook: Malki Museum.
E-mail: malkimuseummail@gmail.com
Web Site: www.malkimuseum.org
Founded: 1964.
Congressional District: 37
Key Personnel: Dir., Richard Rodriguez; Pres., Dr. Amara Siva; Treas., Elaine Mathews; Historian, Collections, Exhibits & Library, Nathalie Colin, M.A.
Personnel Profile: Part-Time Paid 1; Part-Time Volunteers 3.
Governing Authority: nonprofit organization. Tax-exempt: 501(c)(3).
Institution Type/Description: Native American Museum.
Collections: Cahuilla & other Southern California Indian tribe artifacts; anthropology; archaeology; history; ethnology; basketry.
Major Exhibits: Cahuilla People & Neighbors, Fall 2014; History of the Malki Museum, Fall 2014; Song, Dance, Game Among Cahuilla People, Fall 2014; Morongo Reservation History, Fall 2014; Collecting & Processing Food Among Cahuilla People in Southern California, Fall 2014.
Research Fields: California Indians primarily Cahuilla & Serrano.
Facilities: ethno-botanical garden. Museum-related gifts for sale.
Activities: school tours; lectures; workshops; fiesta; college scholarship program for Southern California Indian students. Museum Sponsors: Agave Harvest in the Spring; Memorial Day Weekend Fiesta; Agave Roast; Fall Gathering.
Publications: books, A Dried Coyote's Tail; Aboriginal Society in Southern California; Alaawwich (Our Language); Gigyayk Vo'jka! (Walk Strong!); I' isniyatam (Designs); Lost Copper; Luiseno Language; Malki Museum's Native Food-Tasting Experiences; Mirror and Pattern: Laird's World of Chemehuevi Mythology; Stalking the Wild Agave; Studies in Cahuilla Culture: Ethnography of the Cahuilla Indians; Temalpakh (From the Earth); The Cahuilla Indians; The Cahuilla Indians of Southern California; The Chemehuevi Indians of Southern California; The Chumash Indians of Southern California; The First Angelinos; The Luiseno Indians of Southern California; The Serrano Indians of Southern California; Tovangar (World): A Gabrielino Word Book; Wappo Report; Wayta' Yawa' (Always Believe); When the Animals Were People; Willie Boy: A Desert Manhunt; biannual, The Journal of California and Great Basin Anthropology; A Story of Seven Sisters; Time of Little Choice.
Hours & Admission Prices: Tues.-Sat. 10-4. No charge; donations accepted. &
Membership: Student $15; Regular $20; Family $30; Supporting $40; Business Group $50; Contributing $150; Sustaining $250; Benefactor $500.

Barstow

BARSTOW ROUTE 66 MOTHER ROAD MUSEUM, Historic Harvey House, 681 N. First Ave., Barstow, CA 92311-2201. Tel.: 760-255-1890; 877-997-8366. Fax: 760-256-6776.
E-mail: barstowmuseum@yahoo.com
Web Site: www.route66museum.com
Founded: 2000.
Congressional District: 25

Key Personnel: Mgr., Cur. & Museum Shop Mgr., Debra Hodkin; Historian & Pres. (V), Bill Tomlinson.
Personnel Profile: Part-Time Volunteers 30.
Governing Authority: Tax-exempt.
Institution Type/Description: History Museum.
Collections: history of Rte. 66 & the Mojave Desert communities; photographs; automotive history; personal artifacts.
Facilities: Gift items for sale.
Activities: tours.
Publications: quarterly newsletter.
Hours & Admission Prices: Fri.-Sun. 10-4; other times by appointment. No charge; donations accepted. &
Attendance: 15,000 (estimated)

MOJAVE RIVER VALLEY MUSEUM, 270 E. Virginia Way, Barstow, CA 92311-3923. Tel.: 760-256-5452.
E-mail: mrvm@verizon.net
Web Site: www.mojaverivervalleymuseum.org
Key Personnel: Pres., Bob Hilburn; Vice Pres., Dave Romero.
Governing Authority: nonprofit organization.
Institution Type/Description: History Museum.
Collections: local history from 1776 to present; photographs; newspapers.
Facilities: Museum-related items for sale.
Activities: field trips; meetings.
Hours & Admission Prices: Daily 11-4. No charge. Closed Christmas.

WESTERN AMERICA RAILROAD MUSEUM - WARM, 685 N. First St., Barstow, CA 92311. Mailing Address: P.O. Box 703, Barstow, CA 92312-0703. Tel.: 760-256-WARM.
E-mail: warm95@verizon.net
Web Site: www.barstowrailmuseum.org
Governing Authority: nonprofit. Tax-exempt.
Institution Type/Description: Railroad Museum.
Collections: railroading history & development; railroad artifacts; art; uniforms; tools; equipment.
Hours & Admission Prices: Fri.-Sun. 11-4. No charge; donations accepted. &
Membership: Individual $35.

Bel Air

MARJORIE AND HERMAN PLATT ART GALLERY - AMERICAN JEWISH UNIVERSITY, Familian Campus, 15600 Mulholland Dr., Bel Air, CA 90077. Tel.: 888-853-6763; 310-476-9777.
Web Site: www.ajula.edu
Institution Type/Description: Art Gallery.
Collections: works by university art students and local, national & international artists; paintings; sculpture; photographs.
Hours & Admission Prices: Call for hours.

SONDRA & MARVIN SMALLEY SCULPTURE GARDEN, American Jewish University, 15600 Mulholland Dr., Bel Air, CA 90077-1519. Tel.: 310-476-9777.
Institution Type/Description: Art Museum.
Collections: sculptures.
Hours & Admission Prices: Daily. No charge.

Belmont

THE WIEGAND GALLERY, Notre Dame de Namur University, 1500 Ralston Ave., Belmont, CA 94002-1908. Tel.: 650-508-3595. Fax: 650-508-3488.
E-mail: ehoward@ndnu.edu
Web Site: www.wiegandgallery.org
Founded: 1970.
Key Personnel: Dir., Robert Poplack; Art Chm., Betty Friedman; Gallery Coord., Sheila Longacre.
Personnel Profile: Part-Time Paid 3.
Governing Authority: college. Parent Institution: Notre Dame de Namur University. Tax-exempt: 501(c)(3).
Institution Type/Description: University Art Gallery.
Collections: paintings; sculpture; photographs.
Research Fields: San Francisco Bay area art history.
Facilities: 40-seat theater.
Activities: traveling exhibitions; lectures; gallery talks.
Publications: exhibition catalogues.
Hours & Admission Prices: Sept.-May Tues.-Sat. 12-4. No charge; donations accepted. Closed Thanksgiving; Christmas. &

Attendance: 3,600
Membership: Individual $25; Patron $50; Contributor $100; Supporter $250; Benefactor $500; Director's Circle $1,000.

Benicia

BENICIA FIRE MUSEUM, 900 E. 2nd St., Benicia, CA 94510-3349. Mailing Address: P.O. Box 1251, Benecia, CA 94510-4251. Tel.: 707-745-1688.
Institution Type/Description: Fire-Fighting Museum.
Collections: 1820 Phoenix; 1855 Solano Engine; 1860 Griffin; over 7,000 fire hats; period water grenades & fire extinguishers.
Hours & Admission Prices: 1st three Sun. of month 1-4; other times by appointment. No charge; donations accepted.

BENICIA HISTORICAL MUSEUM, 2024 Camel Rd., Benicia, CA 94510-2339. Mailing Address: 2060 Camel Rd., Benicia, CA 94510-2339. Tel.: 707-745-5435. Fax: 707-745-5869.
E-mail: info@beniciahistoricalmuseum.org
Web Site: beniciahistoricalmuseum.org
Formerly: Benicia Historical Museum and Cultural Foundation (Camel Barn Museum)
Founded: 1985.
Congressional District: 7
Key Personnel: Exec. Dir., Elizabeth d'Huart; Chm. (V), Louise Martin; Bd. Pres. (V), Dr. James Lessenger; Cur., Beverly Phelan; Coord. Elementary Education, Susan Sullivan; Museum Shop Mgr., Toni Haughey.
Personnel Profile: Full-Time Paid 1; Full-Time Volunteers 1; Part-Time Paid 2; Part-Time Volunteers 250; Interns 4.
Governing Authority: nonprofit organization. Tax-exempt.
Institution Type/Description: Local History Museum: housed in 1853-1857 U.S. Army Arsenal; actively used until 1964. Buildings listed on the National Register of Historic Places.
Collections: photographs; documents; maps; artifacts of early Benicia; artifacts from 1847; first store in Solano County; exhibits emphasizing history of the first Arsenal & Camel Auction of 1864.
Research Fields: conversion from military to civilian life for arsenal & city.
Facilities: library of documents & other printed materials available to the public on a limited basis; 10,000 sq. ft. exhibit space; gardens. Local history-related items for sale.
Activities: education programs: traveling trunk; public outreach at civic events in Benicia; guided tours; concerts; participatory, loan & temporary exhibitions; early California hands-on activities included in school tours; reception room available; concerts; receptions; meeting areas; lectures. Museum Sponsors: Spenger Gardent concert series June-October.
Publications: bimonthly newsletter, Camel Tracks.
Hours & Admission Prices: Wed.-Sun. 1-4. Adults $5, seniors & students $3; discounts to groups of 30 or more, CAM & AASLH members; children under 7 & members no charge. Closed New Year's Day; Easter; Mother's Day; Father's Day; Thanksgiving; Christmas.
Attendance: 15,500 (estimated)
Membership: Student & Senior $20; Individual $30; Family $50; Friend $75; Supporting $100; Patron $150; Camel Corp $225; Dona Benicia $500; Benefactor $1,000; Sponsor $5,000. Corporate: Barracks $250; Quartermaster $500; Arsenal $1,000; Commandant $1,500.

FISCHER-HANLON HOUSE, 135 West G St., Benicia, CA 94510-3114. Mailing Address: P.O. Box 404, Benicia, CA 94510. Tel.: 707-745-3385.
E-mail: robin.lancaster@hotmail.com
Web Site: beniciastateparksassoc.org
Founded: 1976.
Congressional District: 10
Key Personnel: Dir. & Pres. (V), Carol Berman; Chm. (V) & Museum Shop Mgr., Robin Lancaster.
Personnel Profile: Part-Time Paid 2; Part-Time Volunteers 30; Interns 10.
Governing Authority: Parent Institution: State of California, Parks Dept. Tax-exempt.
Institution Type/Description: History Museum.
Collections: personal artifacts; furnishings; 1867 square Steinway piano; wooden toys; quilts from 1823-1873 & 1920s; outbuildings. Historic Buildings: carriage barn; 1858 Victorian home; indentured servants quarters.
Facilities: Victorian garden.
Activities: tours; holiday events; school tours; historic interest tours.
Hours & Admission Prices: House: Sat.-Sun. 12-3:30. Garden: Wed.-Sun. 10-5. Adults $3, children $2. &
Attendance: 6,000 (estimated)

Berkeley

THE BADE MUSEUM OF BIBLICAL ARCHAEOLOGY, 1798 Scenic Ave., Berkeley, CA 94709-1323. Tel.: 510-849-8286. Fax: 510-845-8948.
E-mail: bade@psr.edu
Web Site: bade.psr.edu
Formerly: The Bade Institute of Biblical Archaeology and The Howell Bible Collection
Founded: 1926.
Key Personnel: Museum Dir., Dr. Aaron Brody; Assoc. Cur., Catherine P. Foster.
Governing Authority: trustees. Parent Institution: Pacific School of Religion. Tax-exempt.
Institution Type/Description: Archaeological Museum.
Collections: biblical archaeology; major collection from Tell en-Nasbeh, Israel, believed to have been the site of Mizpah; artifacts from Egypt, Syria, Cyprus, Greece & Rome; Reformation printed Bibles; European Bibles; facsimiles of early Greek Biblical codices; fragment of Papyri from Oxyrhynchus, Egypt.
Research Fields: near Eastern, especially Syro-Palestinian archaeology.
Facilities: library; 11,000 slides on near Eastern archaeology & world art.
Activities: guided tours; lectures; formally organized educational programs; permanent & traveling exhibitions.
Publications: newsletter.
Hours & Admission Prices: Tues. & Thurs.-Fri. 10-3; other times by appointment. No charge; donations accepted. &

BERKELEY ART CENTER, (M), 1275 Walnut St., Berkeley, CA 94709-1406. Tel.: 510-644-6893. Fax: 510-540-0343.
E-mail: info@berkeleyartcenter.org
Web Site: www.berkeleyartcenter.org
Founded: 1967.
Congressional District: 8
Key Personnel: Exec. Dir., Jill Berk Jiminez; Pres. (V), Susan Klee; Program Coord., Amber Stucke; Music Coord., Marvin Sanders.
Personnel Profile: Full-Time Paid 1; Part-Time Paid 2; Part-Time Volunteers 20; Interns 1.
Governing Authority: private; nonprofit corporation, Berkeley Art Center Association. Tax-exempt.
Institution Type/Description: Art Gallery.
Collections: contemporary art.
Activities: gallery talks; concerts; slides; films; performances; workshops; poetry readings; special events; temporary exhibitions; selected exhibits of Bay Area artists.
Publications: Bodies & Souls, Science Imagined; 10X10: Ten Women, Ten Prints; Asian Roots Western Soil: Japanese Influences in American Culture; Crossings: The Installation Art of Mildres Howard; Ethnic Notions: Black Images in the White Mind; The Whole World's Watching: Peace & Social Justice Movements of the 1960s & 1970s; Sacred Spaces; From Isolation to Connection: Adults Living with Psychiatric Disabilities.
Hours & Admission Prices: Wed.-Sat. 11-5. Suggested Donations: family $5, adults $3. Closed holidays. &
Attendance: 12,400 (accurate)
Membership: Student & Senior $40; Regular & Senior Artist $45; Artist $50.

BERKELEY HISTORICAL SOCIETY AND THE BERKELEY HISTORY CENTER, Veterans Memorial Bldg., 1931 Center St., Berkeley, CA 94701. Mailing Address: BHS, P.O. Box 1190, Berkeley, CA 94701. Tel.: 510-848-0181.
E-mail: berkhist@sbcglobal.net
Founded: 1978.
Institution Type/Description: History Museum.
Collections: local history & culture; period furnishings; personal artifacts; photographs; oral histories.
Activities: workshops; lectures; concerts; educational programs.
Hours & Admission Prices: Thurs.-Sat. 1-4. No charge. &
Membership: Individual $20; Family $25; Contributor $50; Patron, Professional & Business $100.

BERKELEY ROSE GARDEN, 1200 Euclid Ave., Berkeley, CA 94708. Mailing Address: 2180 Milvia St., Berkeley, CA 94704-1122. Tel.: 510-981-6700. Fax: 510-981-6710.
E-mail: parks@ci.berkeley.ca.us
Web Site: www.ci.berkeley.ca.us/parks/parkspages/berkeleyrosegarden.html
Institution Type/Description: Rose Garden.
Collections: 3,000 rose bushes representing 250 varieties.
Facilities: trails & footbridges; amphitheater.
Activities: special events; weddings.

Hours & Admission Prices: Dawn to dusk. No charge.

HABITOT CHILDREN'S MUSEUM, 2065 Kittredge St., Berkeley, CA 94704-1404. Mailing Address: PMB 326, 1563 Solano Ave., Berkeley, CA 94707. Tel.: 510-647-1111. Fax: 510-647-1110.
E-mail: habitot@lmi.net
Web Site: www.habitot.org
Founded: 1992.
Congressional District: 9
Key Personnel: Dir., Gina Moreland.
Personnel Profile: Full-Time Paid 6; Part-Time Paid 15; Part-Time Volunteers 10; Interns 2.
Governing Authority: Tax-exempt.
Institution Type/Description: Children's Museum.
Collections: hands-on exhibits.
Activities: drop-in art studio; special seasonal & cultural events; children's, parenting & educator classes and workshops; children's summer & winter camps; group visits; birthday parties; family resources; outreach programs. Annual Event: Early Childhood Safety Campaign.
Hours & Admission Prices: Spring & Fall: Mon.-Thurs. 9:30-12:30; Summer: Fri.-Sat. 9:30-4:30; Winter: Fri.-Sun. 9:30-4:30. Admission $8.50; members & children under one no charge. ACM reciprocal admission. Closed New Year's Day; Easter; Independence Day; Labor Day; Thanksgiving; Christmas. &
Attendance: 60,000 (estimated)
Membership: Parent-Child $80; Family $110; Family Circle $140.

KALA ART INSTITUTE GALLERY, 2990 San Pablo Ave., Berkeley, CA 94702. Tel.: 510-841-7000. Fax: 510-540-6914.
E-mail: kala@kala.org
Web Site: www.kala.org/mission.html
Founded: 1974.
Key Personnel: Exec. Dir. & Co Founder, Archana Horsting; Artistic Dir. & Co Founder, Yuzo Nakano
Institution Type/Description: Art Gallery.
Collections: printmaking.
Activities: lectures; classes; workshops.
Hours & Admission Prices: Tues.-Fri. 12-5:30, Sat. 12-4:30. No charge; donations accepted. &

LACIS MUSEUM OF LACE & TEXTILES, 2982 Adeline St., Berkeley, CA 94703-2503. Mailing Address: 3163 Adeline St., Berkeley, CA 94703-2401. Tel.: 510-843-7290. Fax: 510-843-5018.
E-mail: jules@lacis.com
Web Site: lacismuseum.org
Founded: 2004.
Key Personnel: Dir., Jules Kliot; Museum Shop Mgr., Erin Algeo.
Personnel Profile: Full-Time Paid 1; Full-Time Volunteers 1.
Governing Authority: Tax-exempt.
Institution Type/Description: Lace & Textile Museum.
Collections: laces; textile related tools & materials; period clothing; costumes; books; patterns; lace-making tools; sewing machines.
Major Exhibits: Machine Lace, 11/13-2/7/14; Early Italian Needlework, 11/13-2/8/14; Smocking, 3/18/14-10/4/14.
Facilities: Gift items for sale.
Activities: classes.
Publications: exhibit catalogs.
Hours & Admission Prices: Mon.-Sat. 12-6. No charge; donations accepted. &
Attendance: 5,000 (estimated)

LAWRENCE HALL OF SCIENCE, University of California, Berkeley, 1 Centennial Dr., #5200, Berkeley, CA 94720-5200. Tel.: 510-642-5132. Fax: 510-642-1055. Facebook: Lawrence Hall of Science.
E-mail: janet.noe@berkeley.edu
Web Site: www.lawrencehallofscience.org
Founded: 1968.
Congressional District: 9
Key Personnel: Dir., Dr. Elizabeth K. Stage; Deputy Dir., Susan Gregory; Dir. Public Science Center, Gretchen Walker; Dir. Center for Leadership in Science Teaching, Craig Strang; Dir. Center for Research Evaluation & Assessment, Rena Dorph; Dir. Resource Management, Flori Ramos; Dir. Center for Curriculum Devel. & Implementation, Jacquey Barber; Dir. Exhibits, Brooke Smith; Dir. Center for Mathematics Excellence & Equity, Harold Asturias; Human Resources Mgr., Sandra Colonna; Museum Shop Mgr., Seth Harthun; Mgr. Visitor Programs, Sue Guevara.
Personnel Profile: Part-Time Volunteers 250; Interns 60.

Governing Authority: university. Parent Institution: University of California. Tax-exempt: 501(c)(3).
Institution Type/Description: Science & Technology Center.
Collections: science education & curriculum materials; Ernest O. Lawrence memorabilia; science & math exhibits and programs; outdoor earth sciences; family discovery labs; live animals & physical science activities; interactive planetarium; hands-on math challenges; real-time seismic monitor.
Research Fields: science education, intellectual development of children; science & math curricula; teacher training; informal education including exhibit evaluation & visitor behavior; mathematics education; development of educational multimedia.
Facilities: 275-seat auditorium; amphitheater; planetarium; classrooms. Gift items for sale.
Activities: workshops; formally organized education programs for children & adults; permanent & temporary exhibitions; lectures. Center Sponsors: seasonal science camp; family workshops; teen internships; school programs; special events.
Publications: Curriculum publications & newsletters, GEMS; EQUALS; Family Math; FOSS; SEPUP; Marine Activities, Resources, and Education (MARE); Seeds of Science & Roots of Reading.
Hours & Admission Prices: Daily 10-5. Adults 19-61 $12, youth 7-18, full-time students, disabled and senior citizens 62 & over $9, children 3-6 $6; discounts to groups, ACM & ASTC members; UC Berkeley students, children under 3 & members no charge. Closed Thanksgiving; Christmas. &
Attendance: 200,000 (estimated)
Membership: Senior $45; Individual $50; Family & Grandparents $85; Family Plus $125; Sponsor $250; Associate $500; Partners in Science $1,000 & up.

THE MAGNES COLLECTION OF JEWISH ART AND LIFE, The Bancroft Library, University of California, Berkeley, 2121 Allston Way, Berkeley, CA 94720-6300. Tel.: 510-642-3781. Facebook: The Magnes.
E-mail: magnes@library.berkeley.edu
Web Site: www.magnes.org
Formerly: Judah L. Magnes Museum
Founded: 1962.
Congressional District: 8
Key Personnel: Dir. & Chief Cur., Alla Efimova, Ph.D.; Pres. (V), Irving Rabin, Archivist, Western Jewish History Center, Lara Michels, Ph.D.; Exhibitions Registrar, Julie Franklin; Cur. Collections, Francesco Spagnolo.
Personnel Profile: Full-Time Paid 3; Part-Time Paid 2; Part-Time Volunteers 2.
Governing Authority: nonprofit organization. Parent Institution: University of California. Tax-exempt: 501(c)(3).
Institution Type/Description: Judaica Museum.
Collections: Jewish ceremonial art, fine arts, rare books & manuscripts; ritual objects; textiles; costumes; collections from Jews of India & North Africa; Holocaust collection; Magnes & western states Jewish archives.
Research Fields: Western Jewish History Center documents & studies the influence of the Jewish population on the development, character & culture of the Far West; Jewish ceremonial art, manuscripts & rare books; Jewish fine arts.
Facilities: library.
Activities: lectures; permanent & traveling exhibits; educational workshops & programs.
Hours & Admission Prices: Call for hours. No charge; donations accepted. &

MUSEUM OF PALEONTOLOGY, 1101 Valley Life Sciences Bldg., University of California, Berkeley, CA 94720. Mailing Address: MC: 4780, 1101 Valley Life Sciences Bldg., University of California, Berkeley, CA 94720. Tel.: 510-642-1821. Fax: 510-642-1822.
Web Site: www.ucmp.berkeley.edu
Founded: 1921.
Congressional District: 9
Key Personnel: Cur., David R. Lindberg; Cur., Jere Lipps; Cur., William B. Berry; Cur., Carole S. Hickman; Cur., Kevin Padian; Cur., Walter Alvarez; Cur., Roger Byrne; Cur., William A. Clemens; Cur., James W. Valentine; Cur., Tim White; Cur., Lynn Ingram; Cur., Roy Caldwell; Principal Museum Scientist, Mark B. Goodwin; Museum Scientist, Pat Holroyd; Museum Scientist, Diane Erwin; Museum Scientist, Kenneth L. Finger; Museum Rels., Judy Scotchmoor; Museum Rels., David K. Smith; Webmaster, Josh Frankel; Museum Preparator, Jane Mason.
Personnel Profile: Full-Time Paid 12; Full-Time Volunteers 7; Part-Time Paid 13; Part-Time Volunteers 1.
Governing Authority: state. Parent Institution: University of California. Tax-exempt.
Institution Type/Description: Paleontology Museum.

Collections: fossil vertebrates; invertebrates; plants; recent molluscan shells; foraminifera; vertebrate skeletal elements; marine sediments; protists; sedimentary rock samples, amber; scanning electronic microscope.
Research Fields: all aspects of paleobiology; cytology; anatomy & physiology of protistids; malacology; paleoceanography; morphometrics; ecology; endangered invertebrate species; evolution; systematics.
Facilities: 2.8 million vol. library; molecular sequencing laboratory; vertebrate & microfossil laboratories.
Activities: formally organized education programs for undergraduate & graduate students; public exhibits; tours available for school & other groups; teacher training. Annual Event: Open House in March.
Publications: bulletin, Paleo Bios; manuscripts, U.C. Press-University Publications in Geological Sciences.
Hours & Admission Prices: Call for hours. No charge; donations accepted. Closed national holidays. &
Attendance: 15,000 (estimated)
Membership: Donor $25; Sustaining $50; Patron $100; Sponsor $500; Benefactor $1,000.

* **PHOEBE APPERSON HEARST MUSEUM OF ANTHROPOLOGY, (M),** 103 Kroeber Hall, University of California, Berkeley, CA 94720. Tel.: 510-642-3682. Fax: 510-642-6271.
E-mail: pahma@berkeley.edu
Web Site: hearstmuseum.berkeley.edu
Founded: 1901.
Congressional District: 8
Key Personnel: Interim Dir., Judson King; Information System, Michael Black; Deputy Dir., Sandra Harris; Administrative Asst., Lisa Hart; Collections Mgr., Leslie Freund; Director's Asst., Patricia Franco; Assoc. Research Anthropologist, Ira Jacknis; NAGPRA Coord., Anthony Garcia; Coord. Collections, Victoria Bradshaw; NAGPRA Scientist, Larri Friedricks; Registrar, Joan Knudsen; Business Mgr., Gail Bergunde; Sr. Artist, Marco Centin; Conservator, Madeleine Fang; Education Specialist, Akiko Minaga; Exhibit Preparator, Ben Peters; Museum Shop Mgr., Oliver Fernandez.
Personnel Profile: Full-Time Paid 22; Full-Time Volunteers 5; Part-Time Paid 6; Part-Time Volunteers 34; Interns 3.
Governing Authority: university. Parent Institution: University of California, Berkeley. Tax-exempt.
Institution Type/Description: Anthropology Museum.
Collections: archaeological & ethnological specimens from the Americas, Oceania, Europe, Asia & Africa; human skeletal material; photographic negatives, slides & prints.
Research Fields: archaeology; ethnography; physical anthropology.
Facilities: Ethnic arts & crafts for sale.
Activities: public programs; informal training programs for professional museum workers; inter-museum loan.
Publications: newsletter, Museum News; Classics in California Anthropology; exhibit catalogues.
Hours & Admission Prices: Closed until fall 2015. &
Attendance: 45,696 (estimated)
Membership: Student, Senior & Disabled $30; Individual $40; Family $50

TILDEN NATURE AREA ENVIRONMENTAL EDUCATION CENTER, 600 Canon Dr., Berkeley, CA 94708-1162. Tel.: 510-544-2233.
E-mail: tnarea@ebparks.org
Web Site: www.ebparks.org/parks/vc/tna
Institution Type/Description: Nature Center.
Collections: history of Wildcat Creek watershed; ecology; farm animals.
Facilities: theater.
Activities: programs; puppet theater.
Hours & Admission Prices: Nature Area: daily 5am-10pm. Center: Sat.-Sun. 10-4:30. Little Farm: temporarily closed. Closed New Year's Day; Thanksgiving; Christmas. &

UNIVERSITY & JEPSON HERBARIA, University of California, 1001 Valley Life Sciences Bldg., #2465, Berkeley, CA 94720. Tel.: 510-642-2465 & 643-7008. Fax: 510-643-5390.
E-mail: bbaldwin@berkeley.edu
Web Site: ucjeps.berkeley.edu/
Founded: 1872.
Congressional District: 8
Key Personnel: Dir., Prof. Brent Mishler; Research Botanist, Algae, Dr. Paul C. Silva; Research Botanist, Ferns & Grasses, Dr. Alan R. Smith; Research Botanist Compositae, Dr. John L. Strother; Research Botanist, Dr. Barbara Ertter; Cur. Jepson Herbarium, Bruce G. Baldwin.
Personnel Profile: Full-Time Paid 7; Full-Time Volunteers 6; Part-Time Paid 2; Part-Time Volunteers 10; Interns 2.

Governing Authority: state; university. Both herbariums are part of University of California at Berkeley. Tax-exempt: 501(c)(3).
Institution Type/Description: Herbaria.
Collections: worldwide plant kingdom; 2.3 million herbarium specimens in two herbariums.
Research Fields: taxonomy; biosystematics; ecology; floristics; computer methods & floristics.
Facilities: 1,350-vol. library of books & pamphlets pertaining to taxonomy.
Activities: guided tours; formally organized education programs for graduate students affiliated with the University of California.
Publications: Jepson Globe; The Jepson Manual.
Hours & Admission Prices: Mon.-Fri. 8-12 & 1-5. No charge. Closed university holidays.
Attendance: 500 (estimated)

✱ **UNIVERSITY OF CALIFORNIA BERKELEY ART MUSEUM AND PACIFIC FILM ARCHIVE, (M), (I),** 2626 Bancroft Way, Berkeley, CA 94704. Mailing Address: 2625 Durant Ave. #2250, Berkeley, CA 94720-2250. Tel.: 510-642-0808. Fax: 510-642-4889. TDD: 510-642-8734.
E-mail: bampfa@berkeley.edu
Web Site: www.bampfa.berkeley.edu
Founded: 1965.
Congressional District: 8
Key Personnel: Pres. Bd. Trustees, Noel Nellis; Dir., Lawrence Rinder; C.A.O., Richard Tellinghuisen; Security & Operations Admin., Maria Cisneros; Sr. Film Cur., Susan Oxtoby; Film Cur., Kathy Geritz; Film Collection Mgr., Mona Nagai; Cur. Video, Steve Seid; Dir. Education & Academic Rels., Sherry Goodman; Dir. Registration, Lisa Calden; Museum Shop Mgr., Doug McCallister; Dir. Business Svcs., Rebecca Hoag; Chief Cur. and Dir. Programs & Collections, Lucinda Barnes; Mgr. Media Rels., Peter Cavagnaro; Dir. Events, Dennis Love; Dir. Engagement, Aimee Chang; Dir. Foundation & Corp. Rels., Elisa Isaacson; Sr. Cur. Asian Art, Julia White; Adjunct Sr. Cur., Philippe Pirotte; Cur. Contemporary & Modern Art and Phyllis C. Wattis MATRIX, Apsara DiQuinzio.
Personnel Profile: Full-Time Paid 59; Part-Time Paid 125; Part-Time Volunteers 80; Interns 75.
Governing Authority: Regents of the University of California. Parent Institution: University of California. Subsidiary Institution: Berkeley Campus. Tax-exempt: 501(c)(3) & 101(6).
Institution Type/Description: Art Museum.
Collections: 20th-century American & European paintings, sculpture, drawings, prints & photographs; Hans Hofmann paintings; Asian paintings; pre-20th century art; Pacific Film Archive: international contemporary films, Soviet, Japanese & American avant garde, animation & documentaries; film posters; movie stills; prints; videotapes.
Research Fields: historical art & film periods.
Facilities: film reference library of books, periodicals & clippings; research screening facilities; 234-seat theater; restaurant; sculpture garden. Artbooks, catalogs & other museum-related items for sale.
Activities: permanent & temporary exhibitions; MATRIX: ongoing exhibitions of contemporary art; Pacific Film Archive film exhibition program; video screenings; lectures; gallery talks; poetry readings; public service media program. Museum Sponsors: L@TE: Friday Nights @ BAM/PFA Events.
Publications: catalogs; brochures; bimonthly calendar of events; MATRIX artists sheets; exhibition handbills.
Hours & Admission Prices: Galleries: Wed.-Sun. 11-5. Adults $10, non-UC Berkeley students, seniors 65 & over, disabled, young adults 13-17 $7; discounts to AAM & ICOM members; BAM/PFA members, and UC Berkeley staff, students, faculty & retirees no charge. PFA Theater: adults $9.50, UC Berkeley faculty, staff, & retirees, non-UC Berkeley students, seniors 65 & over, disabled persons and youth 17 & under $6.50, BAM/PFA members & UC Berkeley students $5.50. Galleries & Theater: closed university holidays. ♿
Attendance: 100,000 (estimated)
Membership: Individual $50; Dual & Family $75; Sponsor $150; Patron $300; Donor $500; Explorers' Circle $1,000-$2,499; Collectors' Circle $2,500-$4,999.

UNIVERSITY OF CALIFORNIA BOTANICAL GARDEN, 200 Centennial Dr., Berkeley, CA 94720-5045. Tel.: 510-643-2755 & 642-0849. Fax: 510-642-5045.
E-mail: garden@berkeley.edu
Web Site: botanicalgarden.berkeley.edu
Founded: 1890.
Congressional District: 8
Key Personnel: Dir., Paul Licht; Chm. (V), Kurt Hoffman; Education Coord., Christine Manoux; Assoc. Dir. Collections & Horticulture, A. Christopher

Carmichael, Ph.D.; Volunteer Coord., Perry Hall; Devel. & Mktg. Officer, Vanessa Crews; Cur., Holly Forbes; Museum Shop Mgr., Nancy Nelson.
Personnel Profile: Full-Time Paid 24; Part-Time Paid 6; Part-Time Volunteers 260; Interns 2.
Governing Authority: University of California at Berkeley. Tax-exempt: 501(c)(3).
Institution Type/Description: Botanical Garden.
Collections: 13,000 taxa; 9,700 species of plants; 20,000 accessions from around the world including ferns; gymnosperms; flowering plants; succulents; cacti; Rhododendrons; orchids; economically important species; medicinal plants, California rare species & over 2,000 endangered species.
Research Fields: systematics; ecology; evolutionary biology; conservation biology.
Facilities: 1,000-vol. library of books & periodicals on botany & horticulture available for limited access to visitors; 34-acre botanical garden; amphitheater; classroom; conference center; picnic area.
Activities: guided tours; lectures; formally organized educational programs; permanent & temporary exhibitions.
Publications: periodically updated pamphlets; biannual newsletter; biennial seed list; self-guided tour booklet; books, Water-Wise Gardening; Math in the Garden; Botany on Your Plate, Native California Plants & People.
Hours & Admission Prices: Daily 9-5; closed 1st Tues. each month. Adults $9, seniors 65 & over $7, non-Cal students & juniors 4-17 $5, children 5-12 $2; UC faculty, students & staff, children under 4 and 1st Thurs. of month no charge. Closed New Year's Eve & Day; Martin Luther King Jr. Day; Thanksgiving; Christmas Eve & Day. ♿
Attendance: 40,000 (accurate)
Membership: Student $15; UCB Affiliate Individual $30; Individual $45; UCB Affiliate Family $55; Family $65; Supporting $100; Sponsor $250; Patron $500; Benefactor $1,000.

Beverly Hills

THE ACADEMY GALLERY - ACADEMY OF MOTION PICTURE ARTS AND SCIENCES, (M), 8949 Wilshire Blvd., Beverly Hills, CA 90211-1972. Tel.: 310-247-3000, ext. 148. Fax: 310-247-3610.
Web Site: www.oscars.org
Institution Type/Description: Art Gallery.
Collections: paintings; photographs; posters.
Activities: special events; films; special events.
Hours & Admission Prices: Tues.-Fri. 10-5, Sat.-Sun. 12-6. No charge. Closed New Year's Day; Martin Luther King Jr. Day; President's Day; Academy Awards; Memorial Day; Independence Day; Labor Day; Thanksgiving; Christmas.

CALIFORNIA MUSEUM OF ANCIENT ART, Beverly Hills, CA 90213-3515. Mailing Address: P.O. Box 10515, Beverly Hills, CA 90213-3515. Tel.: 818-762-5500.
E-mail: cmaa@att.net
Web Site: cmaa-museum.org
Founded: 1983.
Congressional District: 30
Key Personnel: Pres., John D. Hofbauer; Dir. & Cur., Jerome Berman; C.F.O., Richard Gerber; Sec., Talma Zelitzki.
Personnel Profile: Full-Time Paid 1; Part-Time Volunteers 15.
Governing Authority: nonprofit organization. Tax-exempt: 501(c)(3).
Institution Type/Description: Near Eastern Art & Archaeology Museum.
Collections: 3500 B.C.-500 A.D., art & artifacts from Sumer, Babylon, Assyria, Elam, the Hittites, Canaan, Israel, Pharaonic & Coptic Egypt; Egyptian mummies.
Research Fields: ancient Near East; publication of artworks and ancient texts in the collection.
Facilities: 150-vol. library of archaeology, art & history of the ancient Near East.
Activities: lectures; symposia; temporary exhibitions; international archaeological tours.
Publications: biannual newsletter, Ancient News.
Hours & Admission Prices: Closed until 2015. Call for more information.
Membership: Individual $40; Couple & Family $65; Sponsor $150; Patron $300; Pharaoh's Circle $1,000; Corporate & Business $1,500; Lifetime $5,000.

GAGOSIAN GALLERY, 456 N. Camden Dr., Beverly Hills, CA 90210. Tel.: 310-271-9400. Fax: 310-271-9420.
E-mail: losangeles@gagosian.com
Web Site: www.gagosian.com
Institution Type/Description: Art Gallery.
Collections: paintings; photographs sculpture.
Activities: special events; temporary exhibitions.

Hours & Admission Prices: Tues.-Sat. 10-5:30.

THE PALEY CENTER FOR MEDIA, 465 N. Beverly Dr., Beverly Hills, CA 90210-4601. Tel.: 310-786-1000. Fax: 310-786-1086.
E-mail: tebright@paleycenter.org
Web Site: www.paleycenter.org
Formerly: The Museum of Television & Radio
Founded: 1975.
Key Personnel: Pres. & C.E.O. (NY), Pat Mitchell; Vice Pres. & Exec. Dir. (LA), Craig Hitchcock; Dir. Administration & External Rels., Rebecca Faez; Public Rels. Mgr. (LA), Terry Lynn Ebright.
Governing Authority: Tax-exempt.
Institution Type/Description: Communication Museum.
Collections: 75 years of TV & radio programming & advertisements; documentaries; children's programming; comedy shows, etc.
Facilities: screening rooms.
Activities: screening series; seminars; children's workshops.
Hours & Admission Prices: Wed.-Sun. 12-5. No charge; donations accepted. &
Membership: Senior Citizen & Student $35; General $50; Dual/Family $70; Contributing $150; Supporting $250; Sustaining $500; Patron's Circle $1,000 & up.

VIRGINIA ROBINSON GARDENS, (M), 1008 Elden Way, Beverly Hills, CA 90210-2805. Tel.: 310-276-5367. Fax: 310-276-5352.
E-mail: visit@robinsongardens.org
Web Site: www.robinsongardens.org
Founded: 1911.
Key Personnel: Supt., Timothy Lindsay; Pres. (V), Kerstin Royce; Museum Shop Mgr., Kathleen Huckland.
Personnel Profile: Full-Time Paid 6; Part-Time Volunteers 200; Interns 1.
Governing Authority: county.
Institution Type/Description: Historic House & Gardens: 1911 Mediterranean Classic Revival home owned by Mr. and Mrs. Harry Winchester Robinson, heirs to the J.W. Robinson department stores empire.
Collections: Gardens: Italian terrace garden; formal mall garden; rose garden; tropical palm garden; kitchen garden. Historic House: furnishings; personal artifacts.
Activities: guided tours.
Hours & Admission Prices: Guided Tours: Tues.-Fri. 10 & 1 by appointment. Adults $11, seniors 62 & over and students $6, children 5-12 $4.
Attendance: 2,000 (estimated)
Membership: Student $25; Individual $55; Family $75; Friends Level I $475; Friends Level II $775; Fellow $1,500; Patron $2,500; Benefactor $5,000 & up.

Big Bear City

BIG BEAR SOLAR OBSERVATORY, 40386 N. Shore Lane, Big Bear City, CA 92314-9672. Tel.: 909-866-5791. Fax: 909-866-4240.
E-mail: pgoode@bbso.njit.edu
Web Site: www.bbso.njit.edu
Key Personnel: Professor & Dir., Phil Goode
Institution Type/Description: Observatory.
Collections: study of the sun.
Hours & Admission Prices: By appointment.

BIG BEAR VALLEY HISTORICAL MUSEUM, 800 N. Greenway, Big Bear City, CA 92314. Mailing Address: P.O. Box 513, Big Bear City, CA 92314-0513. Tel.: 909-585-8100.
Web Site: www.bigbearhistory.org
Institution Type/Description: History Museum.
Collections: Native American artifacts; cattle ranching & lumbering; gold mining. Historic Building: 1875 log cabin.
Hours & Admission Prices: Wed. & Sat.-Sun. 10-4. Adults $3.

Bishop

LAWS RAILROAD MUSEUM AND HISTORICAL SITE, Silver Canyon Rd., Bishop, CA 93514. Mailing Address: P.O. Box 363, Bishop, CA 93515-0363. Tel.: 760-873-5950.
E-mail: lawsmuseum@aol.com
Web Site: www.lawsmuseum.org
Founded: 1966.
Congressional District: 20 & 35
Key Personnel: C.E.O., Admin. & Museum Shop Mgr., Barbara Moss; Chm. (V), Max Cox.

Personnel Profile: Full-Time Paid 4; Part-Time Paid 6; Part-Time Volunteers 10.
Governing Authority: nonprofit organization. Parent Institution: Bishop Museum & Historical Society. Tax-exempt: 501(c)(3).
Institution Type/Description: Railroad Museum Complex: housed in 1883 Laws Railroad Depot & 28 other buildings.
Collections: railroad engines; passenger & freight cars; railroad artifacts; western items; working model railroad; Indian artifacts; old bottle collections; musical instruments; farm wagons & machinery; paintings; guns; sewing machines; barbed wire; doctor's instruments & equipment; cameras; 1870 print shop; rare books; historical village; Death Valley #5.
Research Fields: railroad; local history & pioneer families.
Facilities: 11 acres of ground; reception center. Museum-related gifts for sale.
Activities: guided tours; permanent exhibitions.
Publications: quarterly bulletin.
Hours & Admission Prices: Daily 10-4. Suggested Donation: adults $5; members no charge. Closed New Year's Day; Easter; Thanksgiving; Christmas. &
Attendance: 20,000 (accurate)
Membership: Individual $20; Family $30; Supporting $55; Sustaining $120 or $10 per mo.

Blairsden

PLUMAS-EUREKA STATE PARK, 310 Johnsville Rd., Blairsden, CA 96103-9744. Tel.: 530-836-2380. Fax: 530-836-0498.
Web Site: www.parks.ca.gov
Founded: 1959.
Congressional District: 14
Key Personnel: Ranger, Scott Elliott.
Governing Authority: state; nonprofit. State of California, Dept. of Parks & Recreation, Plumas-Eureka State Park Assoc. Tax-exempt.
Institution Type/Description: Historic Site: High Sierra Mining Town.
Collections: mining & blacksmithing tools & equipment; domestic furnishings; vehicles; tack; photographs; natural history; recreational skiing; carpentry.
Research Fields: history of recreational skiing; mining; mining camp life; domestic life in a remote high-altitude mining camp; early hydro-electric production.
Facilities: picnic area; camping ground; nature trails.
Activities: guided & self-guided tours; horse-drawn sleigh rides; blacksmithing instruction & demonstration; annual living history events; interpretive hikes; docent program; intern program; biking; cross-country skiing; intermittent downhill skiing; off highway vehicle access; swimming; fishing; boating.
Publications: park brochures.
Hours & Admission Prices: Summer: daily 8-4:30; Winter: call for hours. Campgrounds: $20 per night. Museum: no charge; donations accepted.
Attendance: 40,000
Membership: Individual $5; Family $10; Life $100.

Bolinas

BOLINAS MUSEUM, (M), 48 Wharf Rd., Bolinas, CA 94924. Mailing Address: P.O. Box 450, Bolinas, CA 94924-0450. Tel.: 415-868-0330. Fax: 415-868-0607.
E-mail: info@bolinasmuseum.org
Web Site: bolinasmuseum.org
Founded: 1982.
Congressional District: 6
Key Personnel: Dir., Jennifer A. Gately; Pres., Kirsten Walker; Treas., Terry Donohue.
Personnel Profile: Full-Time Paid 1; Part-Time Paid 2; Part-Time Volunteers 30.
Governing Authority: nonprofit organization. Tax-exempt: 501(c)(3).
Institution Type/Description: Art & History Museum.
Collections: art & history of west Marin County.
Research Fields: Coastal Marin history & art.
Facilities: Museum-related items for sale.
Activities: docent program; guided tours; lectures; participatory, loan & temporary exhibitions. Museum Sponsors: Art Auction.
Publications: biannual newsletter, Bolinas Museum News.
Hours & Admission Prices: Fri. 1-5, Sat.-Sun. 12-5. No charge; donations accepted. Closed New Year's Day; Thanksgiving; Christmas. &
Attendance: 18,000 (estimated)
Membership: Individual $25; Family & Business $50; Sponsor $100; Friend $250; Patron $500; Benefactor $1,000 & up.

Bonita

BONITA MUSEUM AND CULTURAL CENTER, 4355 Bonita Rd., Bonita, CA 91902-1351. Tel.: 619-267-5141. Fax: 619-267-2143.
E-mail: bonitamuseum@sbcglobal.net
Web Site: www.bonitamuseum.org
Founded: 1993.
Congressional District: 51
Key Personnel: Dir., Mary Oswell; Pres. (V), Tom Pocklington; Treas., Barbara Scott.
Personnel Profile: Part-Time Paid 1; Part-Time Volunteers 20.
Governing Authority: private; nonprofit organization.
Institution Type/Description: History & Art Museum.
Collections: local history; personal artifacts; period furnishings; paintings; sculpture.
Facilities: 30-vol. library of local history books.
Activities: guided tours; lectures; temporary & traveling exhibitions.
Publications: bimonthly newsletter, The Bonita Bugle.
Hours & Admission Prices: Wed.-Sat. 10-4. No charge; donations accepted. Closed New Year's Day; Independence Day; Thanksgiving; Christmas.
Attendance: 5,100 (estimated)
Membership: Student $10; Individual $30; Family $50; Business & Organization $75.

Boonville

ANDERSON VALLEY HISTORICAL SOCIETY MUSEUM, 12340 Hwy. 128, Boonville, CA 95415. Mailing Address: P.O. Box 676, Boonville, CA 95415-0676. Tel.: 707-895-3207.
E-mail: sheri@campracheria.com
Web Site: www.andersonvalleymuseum.org
Key Personnel: Pres. (V), Jim Hill.
Personnel Profile: Part-Time Volunteers 20.
Governing Authority: Tax-exempt.
Institution Type/Description: Historical Society Museum.
Collections: artifacts & memorabilia pertaining to Anderson Valley in Mendocino County.
Hours & Admission Prices: Feb.-Nov. Fri.-Sun. 1-4. No charge; donations accepted.
Attendance: 600 (estimated)
Membership: Individual $10; Family $15; Friend $75; Life $125.

Boron

COLONEL VERNON P. SAXON JR. AEROSPACE MUSEUM, 26922 Twenty Mule Team Rd., Boron, CA 93516. Tel.: 760-762-6600.
Institution Type/Description: Aerospace History Museum.
Collections: aerospace history; flight research; experimental aircraft history; F-4 fighter; X-25A gyrocopter; flight suits & helmets; model aircraft; XLR-8 rocket engine; T-45 trainer engine; Corporal missile tanks & engine.
Facilities: Museum-related items for sale.
Hours & Admission Prices: Daily 10-4; groups of 10 or more by appointment. Closed New Year's Day; Thanksgiving; Christmas.

Borrego Springs

ANZA-BORREGO DESERT STATE PARK, 200 Palm Cyn Dr., Borrego Springs, CA 92004-5005. Mailing Address: P.O. Box 2001, Borrego Springs, CA 92004-2001. Tel.: 760-767-4037. Fax: 760-767-3427.
E-mail: stheriault@parks.ca.gov
Web Site: www.parks.ca.gov/default.asp?page_ID=638
Founded: 1967.
Congressional District: 52
Key Personnel: Supt., Gail Sevrens; Exec. Dir. Foundation, Linda Carson; Mgr. VC, Sally Theriault; Museum Shop Mgr., Kelley Jorgensen.
Personnel Profile: Full-Time Paid 1; Full-Time Volunteers 4; Part-Time Paid 2; Part-Time Volunteers 125.
Governing Authority: state. Parent Institution: California Dept. of Parks & Recreation, Sacramento, CA. Cooperating Association: Anza-Barrego Foundation and Institute, 586 Palm Canyon Dr., P.O. Box 2001, Borrego Springs, CA 92004. Tel. 760-767-0446; Fax: 760-767-0465. Tax-exempt.
Institution Type/Description: Archaeology & Paleontology Museum: housed inside a subterranean structure, natural face rock without windows.
Collections: paleontological collection; plio-pleistocene mammals, birds, reptiles, archaeological collection; Cahuilla & Kumeyaay pottery & tools; San Dieguito tools; Peninsular Bighorn Sheep Skull collection.
Research Fields: paleontology; archaeology.
Facilities: 3,000-vol. library of natural history of Colorado Desert of California available for research by appointment; botanical garden; separate laboratory operation; 65-seat auditorium & theater; classrooms; outdoor amphitheater. Natural history books, maps & other museum-related items for sale.
Activities: guided tours; lectures; films; docent program.
Publications: park magazine; Anza-Borrego Desert State Park.
Hours & Admission Prices: Visitor center: June-Sept. Sat.-Sun. & holidays 9-5; Oct.-May daily 9-5 No charge; donations accepted; day use fee $8 per vehicle for campgrounds. &
Attendance: 175,774 (accurate)

Boulder Creek

BIG BASIN REDWOODS STATE PARK, 21600 Big Basin Way, Boulder Creek, CA 95006-9064. Tel.: 831-338-8861. Fax: 831-338-8863.
Web Site: bigbasin.org
Founded: 1902.
Congressional District: 15
Key Personnel: Ranger, Alex Takone.
Governing Authority: state. Parent Institution: California State Parks. Tax-exempt.
Institution Type/Description: Natural History Museum & State Park.
Collections: fauna & flora of the area; botany; entomology.
Research Fields: botany; history.
Activities: guided tours; organized education programs for children. Museum Sponsors: Celebrating the 100th Anniversary of Big Basin, the oldest state park.
Hours & Admission Prices: Daily 6 a.m. to 10 p.m. Per vehicle $10, senior citizens per vehicle $9, disabled per vehicle w/CA State Park Disabled Discount Pass $5. State annual park passes accepted.
Attendance: 750,000 (estimated)

SAN LORENZO VALLEY MUSEUM, (M), 12547 Hwy. 9, Boulder Creek, CA 95006. Mailing Address: P.O. Box 576, Boulder Creek, CA 95006-0576. Tel.: 831-338-8382. Fax: 831-338-8382.
E-mail: slvhm@cruzio.com
Web Site: www.slvmuseum.com
Founded: 1974.
Key Personnel: Exec. Dir. & Museum Shop Mgr., Lynda Phillips; Pres. (V), Lisa Robinson.
Governing Authority: Parent Institution: Boulder Creek Historical Society. Tax-exempt.
Institution Type/Description: History Museum.
Collections: logging history; Native American artifacts; clothing; railroad; local schools; flume; community life.
Research Fields: geneaology.
Facilities: Museum-related items for sale.
Hours & Admission Prices: Wed. & Fri.-Sun. 12-4; other times by appointment. No charge; donations accepted.
Membership: Individual $25; Family $40; Lifetime Individual $250; Lifetime Family $400.

Brea

BREA MUSEUM AND HERITAGE CENTER, 495 S. Brea Blvd., Brea, CA 92821-5395. Mailing Address: P.O. Box 9764, Brea, CA 92822-9764. Tel.: 714-256-2283.
Institution Type/Description: History Museum.
Collections: local history & heritage; photographs; personal artifacts; period furnishings.
Hours & Admission Prices: Sat. 10-3:30; other times by appointment. No charge; donations accepted.

CITY OF BREA ART GALLERY, Brea Civic & Cultural Center, Plaza Level, 1 Civic Center Cir., Brea, CA 92821-5732. Tel.: 714-990-7730. Fax: 714-990-7736.
E-mail: breagallery@cityofbrea.net
Web Site: www.breagallery.com
Founded: 1980.
Key Personnel: C.E.O. & City Mgr., Tim O'Donnell; Dir. & Museum Shop Mgr., Thomas Ciganko; Cultural Arts Comm. Chm., Rick Clark, Arts & Human Svcs. Mgr., Emily Keller; Art Educator, Christina Hasenberg; Coord. Events, Claudia Sandoval.
Personnel Profile: Part-Time Volunteers 2; Interns 1.
Governing Authority: municipal. Parent Institution: Brea Arts Corporation. Tax-exempt.

Institution Type/Description: Art Exhibit Area.
Collections: outdoor large-scale sculptures; paintings; photographs; prints, drawings & graphic arts.
Research Fields: sculptures; public art.
Facilities: 200-seat theater; lecture halls; classrooms; meeting rooms; TV studio.
Activities: guided tours; lectures; performing arts; juried art exhibitions; artist workshops; cultural arts commission; volunteer program; TV programs; public sculpture program; consignment art program; gallery available for rental & private receptions.
Publications: newsletter; brochures; exhibit announcements; self-guided Art in Public Places tour guide & catalog.
Hours & Admission Prices: Wed.-Sun. 12-5. Adults $2; discounts to AAM members; members, children under 12 no charge. Closed holidays. &
Attendance: 20,000 (accurate)
Membership: Annual $12.

OLINDA HISTORIC MUSEUM AND PARK, 4025 Santa Fe Rd., Brea, CA 92821. Mailing Address: 1 Civic Center Circle, Brea, CA 92821-5792. Tel.: 714-671-4447.
Institution Type/Description: Historic Site.
Collections: local history; geology; Olinda Oil Well #1 drilled in 1897; a jackline pump; records; field office.
Hours & Admission Prices: Daily 9-4; tours by appointment. No charge.

Brentwood

EAST CONTRA COSTA MUSEUM, 3890 Sellers Ave., Brentwood, CA 94513. Mailing Address: P.O. Box 202, Brentwood, CA 94513-0202. Tel.: 925-625-3553.
Institution Type/Description: History Museum.
Collections: local history; period artifacts; documents; photographs.
Hours & Admission Prices: April-Oct. Sat. & 3rd Sun. of month 2-4.

Bridgeport

BODIE STATE HISTORIC PARK, Hwy. 395, Bridgeport, CA 93517. Mailing Address: P.O. Box 515, Bridgeport, CA 93517-0515. Tel.: 760-647-6445. Fax: 760-647-6486.
Web Site: www.parks.ca.gov
Founded: 1962.
Congressional District: 18
Key Personnel: Supervising Ranger, Mark Langner.
Personnel Profile: Full-Time Paid 5; Part-Time Paid 8; Part-Time Volunteers 8; Interns 1.
Governing Authority: state; nonprofit. Parent Institution: State of California, Dept. of Parks & Recreation, Sierra State Parks Foundation. Tax-exempt.
Institution Type/Description: Historic Site: 1849-1932 Gold Rush Mining Boom Town.
Collections: furnishings; clothing; household goods; mining & milling tools & equipment; vehicles; newspapers; mines; mortuary; cemetery. Historic Structures: schoolhouse; store; hotels; saloons; jail; firehouse; fraternal order buildings; mill.
Research Fields: history of emigration in the West; history of mining in California & Nevada; mining economics; mining camp life.
Facilities: library; archives.
Activities: self-guided & guided tours; school groups; photo workshops.
Publications: teacher's guide; children's activity guide; historic newspaper re-issues; feature length docu-drama historical video: Bodie, Ghost Town Frozen in Time; Self-guide tour brochure; Books: Aurora; Bodie; Esmeralda.
Hours & Admission Prices: Summer: daily 9:30-6; Winter: daily 9-4; Museum: May 15-Oct. 31 9-5. Adults $7, children 16 & under $5.
Attendance: 190,000 (estimated)
Membership: Copper $20; Silver $35; Gold $50; W.S. Bodey $100; Life $250.

Buena Park

BUENA PARK HISTORICAL SOCIETY, 6631 Beach Blvd., Buena Park, CA 90621-2904. Tel.: 714-562-3570.
E-mail: info@historicalsociety.org
Web Site: www.historicalsociety.org
Founded: 1968.
Congressional District: 30
Key Personnel: Pres. (V), Art Brown; Cur., Dean O. Dixon.
Personnel Profile: Part-Time Paid 1; Part-Time Volunteers 12.
Governing Authority: nonprofit organization. Tax-exempt.
Institution Type/Description: Historic House Museum.
Collections: period furnishings. Historic Houses: 1887 Whitaker-Jaynes House; 1884 Bacon House.

Research Fields: Buena Park history.
Activities: guided tours; lectures; permanent & temporary exhibitions.
Publications: book, The Picture Story of Buena Park; society newsletter; online, Images of America: Buena Park.
Hours & Admission Prices: Thurs. 10:30-2:30, 2nd Sun. of month 1-4. No charge; donations accepted. &
Attendance: 684 (accurate)
Membership: Students $2.50; Active $10 (spouse $2.50); Family $15; Life $100.

Burlingame

BURLINGAME MUSEUM OF PEZ & CLASSIC TOY MUSEUM, 214 California Dr., Burlingame, CA 94010-4113. Tel.: 650-347-2301. Fax: 650-347-3840.
E-mail: gary@spectrumnet.com
Web Site: www.burlingamepezmuseum.com
Founded: 1995.
Key Personnel: C.E.O., Gary R. Doss; Museum Shop Mgr., Nancy Doss
Institution Type/Description: General Museum.
Collections: pez dispensers & memorabilia; classic toys including Tinkertoy, Mr. Potato Head, Colorforms, View-Master, Erector, Lincoln Logs, Whee-Lo, Wooly Willy, Ant Factory; original advertising artwork.
Facilities: Museum-related items for sale.
Activities: tours.
Publications: newsletter.
Hours & Admission Prices: Tues.-Sat. 10-6. Adults $3, senior citizens 65 & over and children 4-12 $1; children 3 & under and 1st Thurs. of month no charge. Closed major holidays.

PENINSULA MUSEUM OF ART, (M), 1777 California Dr., Burlingame, CA 94010. Tel.: 650-594-1577.
E-mail: peninsulamuseum@gmail.com
Web Site: www.peninsulamuseum.org
Founded: 2004.
Congressional District: 12
Key Personnel: Chm. (V), Ruth Waters; Education & Cur., James Daugherty; Public Rels., Jerry Emanuel; Treas., Arabella Decker.
Personnel Profile: Full-Time Volunteers 1; Part-Time Volunteers 36.
Governing Authority: private; nonprofit organization. Tax-exempt: 501(c)(3).
Institution Type/Description: Art Museum.
Collections: paintings; sculpture; Chinese calligraphy; digital art.
Facilities: 1,400-vol. art library; 1,000 sq. ft. exhibit space. Museum-related items for sale.
Activities: lectures; training programs for professional museum workers; art classes; workshops.
Publications: quarterly newsletter, Peninsula Museum of Art; exhibition catalogues.
Hours & Admission Prices: Wed.-Sun. 12-4. No charge; donations accepted. Closed New Year's Day; Independence Day; Christmas. &
Attendance: 2,570 (estimated)
Membership: Individual $45; Family $65; Supporter $100; Benefactor $500; Patron $1,000.

Burney

MCARTHUR-BURNEY FALLS MEMORIAL STATE PARK, 24898 Hwy. 89, Burney, CA 96013-9626. Mailing Address: MBF Interpretive Association, P.O. Box 777, Burney, CA 96013. Tel.: 530-335-2777.
Web Site: www.burney-falls.com
Key Personnel: Pres., MBFIA, Bill Cummings
Institution Type/Description: Park Museum & Visitor Center.
Collections: Park: 129-foot Burney Falls. Log Cabin Visitor Center: hands-on exhibits.
Facilities: nature trails. Museum-related items for sale.
Activities: programs; hiking; camping.
Hours & Admission Prices: Daily sunrise-sunset.

Calabasas

LEONIS ADOBE MUSEUM, 23537 Calabasas Rd., Calabasas, CA 91302-1311. Tel.: 818-222-6511. Fax: 818-222-0862.
E-mail: info@leonisadobemuseum.org
Web Site: www.leonisadobemuseum.org
Founded: 1965.
Key Personnel: Dir., Diane Ramadan; Pres. (V), Don Adams.
Governing Authority: Tax-exempt.
Institution Type/Description: Historic House & Gardens: housed in the former home of Miguel Leonis; c.1880.

Collections: late 1800s California ranch life; period furnishings; photographs; personal artifacts.
Facilities: rental facilities. Museum-related items for sale.
Activities: special events; traveling trunk.
Hours & Admission Prices: Wed.-Fri. & Sun. 1-4, Sat. 10-4. Suggested Donations: adults $4, senior citizens $3, children under 12 $1. Closed New Year's Day; Thanksgiving; Christmas Eve & Day. &
Attendance: 20,000 (estimated)
Membership: Senior & Student $15; Individual $25; Family $40; Caballero $100.

Calistoga

PETRIFIED FOREST MUSEUM, 4100 Petrified Forest Rd., Calistoga, CA 94515-9527. Tel.: 707-942-6667.
E-mail: manager@petrifiedforest.org
Web Site: www.petrifiedforest.org
Institution Type/Description: Forest Museum.
Collections: petrified redwood trees; fossils; area geology; native wildflowers; live oaks, madrone & manzanita trees; 100 ft.-high Ash Fall.
Facilities: picnic area. Museum-related items for sale.
Activities: guided tours.
Hours & Admission Prices: Summer: daily 9-7; Winter: daily 8-5; call to confirm. Adults $6, seniors over 60 & children 12-17 $5, children 6-11 $3.

SHARPSTEEN MUSEUM, 1311 Washington St., Calistoga, CA 94515-1441. Mailing Address: P.O. Box 573, Calistoga, CA 94515-0573. Tel.: 707-942-5911 & 5916 (Mon.-Fri.). Fax: 707-942-6325.
E-mail: sharpsteenmuseum@att.net
Web Site: sharpsteen-museum.org
Founded: 1978.
Congressional District: 1
Personnel Profile: Part-Time Volunteers 75.
Governing Authority: municipal government; nonprofit. Tax-exempt.
Institution Type/Description: Historical Society Museum: adjacent to the museum is one of Sam Brannan's original cottages.
Collections: history of the City of Calistoga & other areas in the upper Napa Valley.
Research Fields: geothermal exhibit.
Facilities: library; 3,500 sq. ft. exhibit space. Museum-related items for sale.
Publications: bimonthly newsletter; biography pamphlet, Ben Sharpsteen; books, Sam Brannan; Early Upper Napa Valley; Calistoga Days; Brannan Saga; They Left Their Mark; Anecdotes of Calistoga.
Hours & Admission Prices: Daily 11-4. No charge; donations accepted. Closed Thanksgiving; Christmas. &
Attendance: 14,500 (estimated)
Membership: Senior Citizen $20; Individual $30; Business & Associate $50; Sponsor $100; Benefactor $2,500. Lifetime: Individual $300; Family $500; Business $1,000.

Camarillo

COMMEMORATIVE AIR FORCE SOUTHERN CALIFORNIA WING'S WWII AVIATION MUSEUM, Camarillo Airport, 455 Aviation Dr., Camarillo, CA 93010. Tel.: 805-482-0064.
Institution Type/Description: Military History Museum.
Collections: military history & artifacts; military aircraft including World War I, World War II, Korean War, & Vietnam War.
Activities: special events; fundraising events.
Hours & Admission Prices: Tues.-Sun. 10-4. Suggested Donations: adults $7, students 11-18 $4, children 6-10 $3; children under 6 & active military no charge.

STUDIO CHANNEL ISLANDS ART CENTER, 2222 Ventura Blvd., Camarillo, CA 93010. Tel.: 805-383-1368.
E-mail: sciartcenter@verizon.net
Web Site: www.studiochannelislands.org
Key Personnel: Exec. Dir., Karin Geiger
Institution Type/Description: Art Gallery.
Collections: works by regional, national & international artists.
Major Exhibits: Thread/Bare, 1/14; Less Is More, 3/14; Ventura College Student Show, 4/14; Gary Land (Artist), 5/14; Landmarks of Camarillo, 6/14; Marion Wood & Pamela Price Klebaum (Artists), 8/14; Annual Juried Membership Exhibit, 9/14; Julia Pinkham (Artist), 10/14; Clothing Optional, 11/14; Candace Biggerstaff, Eileen Hyman, Lucia Grossberger Morales (Artists), 12/14.
Facilities: 40 working artists' studios.
Activities: lectures; special events; temporary exhibitions; monthly open artists studios; workshops; classes.

Hours & Admission Prices: Tues. 11-3, Wed.-Fri. 11-5, Sat. 10-3. No charge. &
Membership: General $50; Artist $100.

Camp Pendleton

MARINE CORPS MECHANIZED MUSEUM, (M), 2612 Vandegrift Blvd., Camp Pendleton, CA 92055-5021. Tel.: 760-725-5758. Fax: 760-725-5727.
E-mail: mcbcampen_history@usmc.mil
Web Site: www.themech.org
Formerly: Camp Pendleton Museums
Founded: 2002.
Congressional District: 49
Key Personnel: Museum Officer, Faye Jonason; Chm. (V), Lt. Col. Paul Durrance, USMC Ret.; Pres. Historical Society, Col. Richard Rothwell, USMC Ret.
Personnel Profile: Full-Time Paid 4; Part-Time Volunteers 35; Interns 1.
Governing Authority: Parent Institution: United States Marine Corps. Base Museums: World War II & Korea LVT Museum, Bldg. 21561, Boat Basin. Marine Corps Mechanized Command Museum, Bldg. 2612, Vandegrift Blvd. Ranch House Complex: Ranch House Chapel, Bunkhouse Museum, Ranch House. Tax-exempt.
Institution Type/Description: Military Museum.
Collections: Marine Corps transport & battle vehicles; over 180 World War II, Korea, Vietnam, Desert Storm & later vintage vehicles. Ranch House Complex: Ranch House Chapel; Bunkhouse Museum: military history; early ranch equipment; photographs; period furnishings.
Research Fields: military armaments; artillery; transport.
Activities: tours; permanent exhibits; local community events.
Hours & Admission Prices: Mon.-Thurs. 8-3:30, Fri. 8-1. Tours: by appointment; email or call: 760-725-0770. Entrance to base requires proof of insurance, registration & current ID. Closed for Marine special liberty. No charge; donations accepted. &
Attendance: 6,000 (estimated)

SANTA MARGARITA RANCH HOUSE NATIONAL HISTORIC SITE, 24154 Vandegrift Blvd., Camp Pendleton, CA 92055. Tel.: 760-725-5758. Fax: 760-725-5727.
E-mail: mcbcampen_history@usmc.mil
Web Site: www.pendleton.usmc.mil
Founded: 2002.
Congressional District: 49
Key Personnel: Dir., Faye Jonason; Chm. (V), Lt. Col. Paul Durrance, USMC (Ret.); Pres. Historical Society, Col. Richard Rothwell, USMC (Ret.)
Personnel Profile: Full-Time Paid 4; Part-Time Volunteers 35; Interns 1.
Governing Authority: Parent Institution: United States Marine Corps. Tax-exempt.
Institution Type/Description: National Historic House Site.
Collections: early California period furnishings; ranch tools; clothing; photographs. Historic Buildings: Santa Margarita ranch house; chapel; bunk house; adobe buildings.
Research Fields: California history; cattle ranching; California pioneers; adobe construction; early CA politics; southern CA military history.
Activities: tours; temporary & permanent exhibitions.
Publications: Rancho Santa Margarita y Las Flores Gazette.
Hours & Admission Prices: Tours: by appointment. Entry to base requires current proof of insurance, registration & ID. No charge; donations accepted. Closed for Marine Special Liberty. &
Attendance: 6,000 (estimated)

Campbell

CAMPBELL HISTORICAL MUSEUM & AINSLEY HOUSE, 51 N. Central Ave., Campbell, CA 95008-2015. Tel.: 408-866-2119 & 2757. Fax: 408-866-2795.
E-mail: juliec@cityofcampbell.com
Web Site: www.cityofcampbell.com/museum/index.htm
Founded: 1964.
Congressional District: 15
Personnel Profile: Full-Time Paid 1; Part-Time Volunteers 135.
Governing Authority: municipal. Parent Institution: City of Campbell, CA. Subsidiary Institution: Campbell Museum Foundation. Tax-exempt.
Institution Type/Description: History Museum: housed in 1951 Fire station which served as the city's first office; 1925 Tudor revival; Carriage House serves as visitor center & museum store.
Collections: local history from early inhabitants to present; artifacts; photographs; archives; farm & agricultural equipment.
Major Exhibits: The Spirit of Campbell, 3/14-6/14.
Research Fields: local history; genealogy.

Facilities: library pertaining to preservation, rehabilitation, history & museology; research room. Museum-related gift items for sale.
Activities: guided tours; lectures; docent program; oral history program; historic walking tour; historic resource survey; temporary exhibitions; hands-on history exhibits; school programs; traveling trunks.
Publications: brochure, Campbell Historical Museum Brochure; newsletter, Campbell Visitor.
Hours & Admission Prices: Historical Museum: Thurs.-Sun. 12-4. Admission $2. Ainsley House: March-Dec. 20. Docent Tour: adults $6, senior citizens $4, children 7-17 $2.50; discounts to AAM members; members no charge. Combination tickets available. Closed major holidays. &
Attendance: 6,935 (accurate)
Membership: Individual $35; Family $55; Director's Circle $100; Ainsley House Circle $500; Carriage House Circle $1,000; Firehouse Circle $2,500; Museum Circle $5,000.

Campo

PACIFIC SOUTHWEST RAILWAY MUSEUM - LIVING HISTORY & TRAIN OPERATION CENTER, 750 Depot St., Campo, CA 91906. Mailing Address: 4695 Nebo Dr., La Mesa, CA 91941-5259. Tel.: 619-465-7776.
E-mail: support@sdrm.org
Web Site: www.psrm.org
Founded: 1961.
Congressional District: 52
Key Personnel: Pres. (V), Diana Hyatt; Museum Shop Mgr., David DiGiorgio.
Personnel Profile: Part-Time Volunteers 50.
Volunteer Hours: 25,000
Operating Expenses: 250,000
Operating Income: 250,000
Governing Authority: Parent Institution: Pacific Southwest Railway Museum Association, Inc. Tax-exempt.
Institution Type/Description: Railroad Museum.
Collections: full-size vintage railroad locomotives, freight & passenger cars, maintenance equipment & related artifacts; operating demonstration train; historic buildings; African American railroad heritage.
Major Exhibits: The African-American Railroad Experience, 1/14-12/14; The PSRMA Mission (T), 3/14-5/14; The PSRMA Mission (T), 11/14.
Research Fields: railroad history & technology.
Facilities: research library; 15,000 sq. ft. exhibition space.
Activities: vintage train rides; special events; family activities; docent & train crew training; community & youth activities; exhibit interpretation.
Publications: newsletter, Hot Scoop.
Hours & Admission Prices: Jan.-May & Sept.-Nov. Sat.-Sun. 9-5; June-Aug. & Dec. Sat. 9-5; other times by appointment. Museum: adults $5, children 6-12 $2.50; discounts to seniors 65 & over and active duty military; members no charge. Closed most holidays. &
Attendance: 10,350 (estimated)
Membership: Student & Senior $20; Individual $30; Family $40; Contributing $50; Supporting $75; Sustaining $150; Life $750; Gold Spike Life $1,000; Benefactor $5,000.

Canoga Park

CANOGA-OWENSMOUTH HISTORICAL MUSEUM, Canoga Park Community Center, 7248 Owensmouth Ave., Canoga Park, CA 91303-1529. Tel.: 818-340-3696 & 346-5252.
Founded: 1988.
Governing Authority: Tax-exempt.
Institution Type/Description: History Museum.
Collections: history of San Fernando Valley; photographs; documents; paintings.
Hours & Admission Prices: 2nd & 4th Sun. of month 2-4; other times by appointment. No charge.
Attendance: 250 (estimated)

ORCUTT RANCH, 23600 Roscoe Blvd., Canoga Park, CA 91304-3057. Tel.: 818-346-7449.
Web Site: www.laparks.org/dos/horticulture/orcuttranch.htm
Institution Type/Description: Historic House: home of William Warren Orcutt & his wife Mary Logan Orcutt, c.1926. Los Angeles Historic-Cultural Monument.
Collections: period furnishings; personal artifacts; gardens; citrus orchard.
Facilities: 24-acre gardens; citrus orchard.
Hours & Admission Prices: Museum: daily sunrise to sunset. Orchards: July call for hours.

Capistrano Beach

CALIFORNIA FIRE MUSEUM, 34681 Calle Fortuna, Capistrano Beach, CA 92624. Tel.: 949-493-8718. Fax: 949-493-0444.
Institution Type/Description: Firefighting History Museum.
Collections: firefighting history & equipment; early fire trucks & apparatus; photographs.
Activities: educational programs.
Hours & Admission Prices: Call for hours.

Capitola

CAPITOLA HISTORICAL MUSEUM, 410 Capitola Ave., Capitola, CA 95010-3318. Tel.: 831-464-0322. Facebook: Capitola Historical Museum.
E-mail: capitolamuseum@gmail.com
Web Site: www.capitolamuseum.org
Founded: 1967.
Congressional District: 16
Key Personnel: Dir., Frank Perry.
Personnel Profile: Part-Time Paid 2; Part-Time Volunteers 15.
Governing Authority: municipal. Parent Institution: City of Capitola.
Institution Type/Description: General Museum.
Collections: photographs.
Activities: guided tours.
Hours & Admission Prices: Wed. & Fri.-Sun. 12-4; other time by appointment. No charge. &
Attendance: 3,600 (accurate)

Carlsbad

GIA (GEMOLOGICAL INSTITUTE OF AMERICA), (M), 5345 Armada Dr., Carlsbad, CA 92008-4602. Tel.: 760-603-4157. Fax: 760-603-4056.
E-mail: terri.ottaway@gia.edu
Web Site: www.gia.edu
Founded: 2001.
Key Personnel: Cur., Terri Ottaway; Project Mgr. & Exhibit Devel., McKenzie Santimer.
Personnel Profile: Full-Time Paid 2; Part-Time Volunteers 30.
Governing Authority: private; nonprofit organization. Parent Institution: Gemological Institute of America. Tax-exempt: 501(c)(3).
Institution Type/Description: General Museum.
Collections: gems; gem minerals; jewelry; art; photographs; paintings; books; scientific literature; gemological slides.
Facilities: 44,000-vol. library; 140-seat theater; cafeteria; educational facilities.
Activities: docent program; formal education programs for adults & undergraduate or graduate college students; guided tours; lectures; loan, temporary exhibitions; jr. gemologist program.
Publications: quarterly journal, Gems & Gemology; bi-weekly electronic bulletin, The Insider.
Hours & Admission Prices: By appointment. No charge. Closed New Year's Day; Presidents' Day; Memorial Day; Independence Day; Labor Day; Thanksgiving; Christmas.
Attendance: 10,000 (estimated)

MUSEUM OF MAKING MUSIC, A DIVISION OF THE NAMM FOUNDATION, (M), 5790 Armada Dr., Carlsbad, CA 92008-4608. Tel.: 760-438-5996. Fax: 760-438-8964.
E-mail: museum@museumofmakingmusic.org
Web Site: www.museumofmakingmusic.org
Founded: 1998.
Congressional District: 50
Key Personnel: Exec. Dir., Carolyn Grant; C.E.O. & Pres., Joe Lamond; Museum Shop Mgr., Allison Hargis.
Personnel Profile: Full-Time Paid 6; Part-Time Paid 2; Part-Time Volunteers 80; Interns 1.
Volunteer Hours: 7,355
Operating Expenses: 443,743
Operating Income: 557,862
Governing Authority: private; nonprofit organization. Parent Institution: NAMM Foundation. Tax-exempt: 501(c)(3).
Institution Type/Description: Musical Instruments Museum.
Collections: musical instruments; products.
Major Exhibits: Singing the Golden State, 11/13-1/14; The Banjo, 3/14-9/14; What Music Means to Me, 11/14-8/15.
Research Fields: history of the music products industry; music education; popular music; music retail business; instrument manufacturing & distribution; music publishing.

Facilities: library; recital & performance space; classroom.
Activities: summer kids camps; film series. Annual Events: Special Exhibit Concert Series; Local Flavor Concert Series; Global Spotlight Concert Series; Annual Series; New Horizons Adult Band; North Coast Strings Adult Orchestra.
Publications: newsletter.
Hours & Admission Prices: Tues.-Sun. 10-5. Adults $8, senior citizens, students & military $5; discounts to AAM & ICOM members; members no charge. &
Attendance: 33,000 (accurate)
Membership: Student & Educator $25; Senior $30; Individual $35; Family $65; Music Maker $150; Innovator $500; Visionary $1,000.

Carmel

CARMEL HERITAGE SOCIETY'S FIRST MURPHY HOUSE, Lincoln & Sixth, Carmel, CA 93921. Mailing Address: Carmel Heritage Society, P.O. Box 701, Carmel, CA 93921-0701. Tel.: 831-624-4447. Fax: 831-624-1970.
E-mail: info@carmelheritage.org
Web Site: www.carmelheritage.org
Key Personnel: Pres., Dawn Dull
Institution Type/Description: Historical House Museum.
Collections: Carmel history; cultural heritage; garden.
Facilities: garden.
Hours & Admission Prices: House: Wed.-Sun. 1-4. Society Office: Mon.-Thurs. 10-2. No charge.
Membership: Annual $15; Family $30; Sponsor $50; Patron $100; Lifetime $500.

CENTER FOR PHOTOGRAPHIC ART, San Carlos & 9th Sts., Carmel, CA 93921. Mailing Address: P.O. Box 1100, Carmel, CA 93921-1100. Tel.: 831-625-5181. Fax: 831-625-5199.
E-mail: info@photography.org
Web Site: www.photography.org/index.html
Formerly: Friends of Photography
Founded: 1989.
Key Personnel: Exec. Dir., Nicole Garzino.
Personnel Profile: Full-Time Paid 1; Part-Time Paid 2; Part-Time Volunteers 30; Interns 2.
Institution Type/Description: Photography Museum.
Collections: photography.
Major Exhibits: 2014 Juried Exhibition, 1/11/14-3/1/14; Ben Nixon & Esmeralda Ruiz, 3/8/14-4/26/14; The Book As Subject, 6/28/14-8/9/14; Mixed Media - A Lineage, 8/23/14-10/11/14; 8 x 10 Small Works, 10/18/14-119/14; Beautiful/Horrible, 11/15/14-1/3/15.
Activities: workshops; lectures; fine print program; member nights; auction; juried competition.
Publications: exhibition catalogues.
Hours & Admission Prices: Tues.-Sun. 1-5. No charge; donations accepted. Closed New Year's Day; Thanksgiving; Christmas. &
Attendance: 4,000 (accurate)
Membership: Student $25; Senior $35; Individual $45; Senior Couple $60; Family $70; Garrapata $100; Mono Lake $250; Point Lobos $500; Sierra Nevada $750; Half Dome $1,000.

MISSION SAN CARLOS BORROMEO DEL RIO CARMELO, 3080 Rio Rd., Carmel, CA 93923. Tel.: 831-624-1271, ext. 210. Fax: 831-624-0658.
Web Site: www.carmelmission.org
Founded: 1770.
Congressional District: 5
Key Personnel: Cur., Richard J. Menn.
Governing Authority: Roman Catholic Church. Tax-exempt.
Institution Type/Description: History Museum.
Collections: sculpture; statues; postcards; textiles; Spanish Colonial art & artifacts.
Facilities: library of books published between 1615-1833. Religious articles, books & postcards for sale.
Activities: Sunday services.
Hours & Admission Prices: Mon.-Sat. 9:30-5, Sun. 10:30-5. Adults $6.50, seniors $4, children 7 & up $2; children under 6 no charge.

WESTON GALLERY, Sixth Ave. & Dolores, Carmel, CA 93921. Mailing Address: P.O. Box 655, Carmel, CA 93921. Tel.: 831-624-4453. Fax: 831-624-7190.
E-mail: info@westongallery.com
Web Site: www.westongallery.com
Founded: 1975.

Institution Type/Description: Art Gallery.
Collections: 19th-20th century photographs.
Activities: temporary exhibitions.
Hours & Admission Prices: Tues.-Sun. 10:30-5:30. No charge.

WINFIELD GALLERY, Dolores between Ocean & 7th, Carmel, CA 93921. Mailing Address: P.O. Box 7393, Carmel, CA 93921. Tel.: 831-624-3369; 800-289-1950 (Toll Free). Fax: 831-624-5618.
E-mail: chris@winfieldgallery.com
Web Site: www.winfieldgallery.com
Institution Type/Description: Art Gallery.
Collections: paintings; sculpture; ceramics.
Hours & Admission Prices: Mon.-Sat. 11-5, Sun. 12-5.

Carmichael

EFFIE YEAW NATURE CENTER, 2850 San Lorenzo Way, Carmichael, CA 95608. Mailing Address: P.O. Box 579, Carmichael, CA 95609-0579. Tel.: 916-489-4918. Fax: 916-489-4983.
E-mail: info@sacnaturecenter.net
Web Site: www.sacnaturecenter.net
Founded: 1976.
Congressional District: 3
Key Personnel: Exec. Dir., Paul Tebbel; Pres., Diana Parker; Volunteer Coord., Jamie Washington; Exhibit Dir. and Dir. Funding & Devel., Betty Cooper; Maidu Cultural Heritage Program, Brena Seck.
Personnel Profile: Full-Time Paid 2; Part-Time Paid 11; Part-Time Volunteers 200.
Governing Authority: nonprofit. Parent Institution: American River Natural History Association. Tax-exempt.
Institution Type/Description: Nature Center.
Collections: mounts & study skins of wildlife native to central California; historic & recreated artifacts of local Sacramento area Native American culture; maps; photographs; demonstration village.
Research Fields: captive hawk behavior modification.
Facilities: 625-vol. library of natural & cultural history of Sacramento region; park management, available for in-house use; 1,800 sq. ft. exhibit space; nature/conservation center; 77-acre nature area. Books, nature-related & historical items for sale.
Activities: docent program; formal education programs for children, on & off-site; guided tours; teacher workshops; weekend family programs; participatory & temporary exhibitions; Maidu Indian cultural demonstrations.
Publications: field guides, The Outdoor World of the Sacramento Region; Birds of the American River Parkway; The Lower American: Prehistory to Parkway; Biking and Hiking the American River Parkway; children's storybook, Ooti, Child of the Nisenan; curricula; The American River Parkway, a handbook for outdoor exploration and learning; The Valley Nisenan Educator's Guide.
Hours & Admission Prices: Feb.-Oct. Tues.-Sun. 9-5; Nov.-Jan. Tues.-Sun. 9-4. No charge; donations accepted. Closed New Year's Day; Thanksgiving; Christmas. &
Attendance: 83,800 (accurate)
Membership: American River Natural History Association: Student $15; Senior Citizen $20; Individual $25; Family $40; Contributor $60; Sponsor $100; Sustainer $250; Patron $500.

Carpinteria

CARPINTERIA VALLEY HISTORICAL SOCIETY & MUSEUM OF HISTORY, 956 Maple Ave., Carpinteria, CA 93013-2021. Tel.: 805-684-3112. Fax: 805-684-4721.
E-mail: info@carpinteriahistoricalmuseum.org
Web Site: www.carpinteriahistoricalmuseum.org
Founded: 1959.
Congressional District: 23
Key Personnel: Pres., Dorothy Thielges; Dir. & Cur., David W. Griggs.
Personnel Profile: Full-Time Paid 1; Part-Time Volunteers 75.
Governing Authority: nonprofit organization. Tax-exempt: 170(b)(1)(A) & 501(c)(3).
Institution Type/Description: Local History Museum.
Collections: artifacts of Chumash Indians & early pioneers of Valley; early pioneer furnishings; 1822-1850 Mexican period artifacts; agricultural tools; oral history tapes; photographs; late 19th & early 20th century school artifacts, cameras, costumes, camping & sporting goods, toys.
Research Fields: Chumash Indians; 1769-1850 Hispanic period; pioneer & family history; agriculture; Santa Barbara Channel oil development, asphalt mines.
Facilities: research library; subject & family archives; cross-referenced, triple-indexed photograph & oral history archives.

Activities: docent program; permanent & temporary exhibits; school programs; tours; research facilities; oral history project. Museum Sponsors: Spring Lecture Series; Monthly Flea Market; Annual Potluck Picnic; Holiday Faire.
Publications: bimonthly newsletter, The Grapevine.
Hours & Admission Prices: Tues.-Sat. 1-4. No charge; donations requested.
Attendance: 10,000 (estimated)
Membership: Student $15; Individual $25; Family $35; Contributing $50; Patron $100; Corporate $150; Benefactor $250; Life $500.

Carson

THE INTERNATIONAL PRINTING MUSEUM, 315 W. Torrance Blvd., Ste. A, Carson, CA 90745-1130. Tel.: 310-515-7166. Fax: 714-538-2443.
E-mail: mail@printmuseum.org
Web Site: www.printmuseum.org
Formerly: The Printing Museum
Founded: 1988.
Key Personnel: Exec. Dir., Mark Barbour; Chm. (V), John Hedlund; Pres., Paul Doucette Ernest Lindner.
Personnel Profile: Full-Time Paid 3; Part-Time Paid 1; Part-Time Volunteers 2.
Governing Authority: private; nonprofit organization. Tax-exempt: 501(c)(3) & 170(b)(1)(A).
Institution Type/Description: Typography Museum.
Collections: period printing machinery & allied trades covering 500 years with an emphasis on the 19th century; research library with 5,000 volumes detailing the history of printing & communications; traveling exhibits.
Hours & Admission Prices: Sat. 10-4. By appointment Tues.-Fri. Adults $8, students, seniors & members $7; preschool children no charge. &
Attendance: 20,000 (estimated)

UNIVERSITY ART GALLERY, CSU DOMINGUEZ HILLS, 1000 E. Victoria St., Carson, CA 90747. Tel.: 310-243-3334 & 3310. Fax: 310-217-6967.
E-mail: kzimmerer@csudh.edu
Web Site: www.cah.csudh.edu/artgallery
Founded: 1978.
Congressional District: 31
Key Personnel: Dir., Kathy Zimmerer.
Personnel Profile: Full-Time Paid 1; Part-Time Paid 2; Interns 6.
Governing Authority: university; nonprofit. Parent Institution: CSU Dominguez Hills. Tax-exempt: 501(c)(3).
Institution Type/Description: University Art Gallery.
Collections: paintings; sculpture; photographs.
Major Exhibits: Camilla Taylor: A Conversation, Relief Prints, Monoprints and Sculpture, 1/14-3/14; Ron Pippin: Crossing the Axis, A Survey 1985-2014, 1/14-3/14; Art and Design Family Exhibit, 9/14-10/14.
Research Fields: contemporary California art with emphasis on multiculturalism; historic California art.
Facilities: 2,150 sq. ft. exhibit space.
Activities: guided tours; lectures; loan exhibitions; formally organized education programs for undergraduates affiliated with CSU Dominguez Hills & surrounding communities; teacher training conferences; K-12 education programs.
Publications: annual newsletter; biannual exhibit catalogues; New Directions in California Sculpture; Painted Light: California Impressionist Paintings from the Gardena High School/LAUSD Collection; An Architectural Stylist: W. Horace Austin and Eclecticism in California; Annual Student Art Exhibition, B.A. Graduates, 2009; Annual Student Art Exhibition: B.A. Graduates, 2010; Annual Student Art Exhibition, B.A. Graduates, 2011; Space and Substance: Abstract Paintings by Craig Antrim and Ron Pippin, 2013.
Hours & Admission Prices: Sept.-May Mon.-Thurs. 10-4. No charge; donations accepted. Closed academic holidays: 2 weeks at Christmas; spring break. &
Attendance: 10,000 (estimated)
Membership: Individual $15; Patron $25; Sustaining $50; Benefactor $100; Angel $500.

Chatsworth

HOMESTEAD ACRE AND THE HILL-PALMER HOUSE, 10385 Shadow Oak Dr., Chatsworth, CA 91311-2063. Tel.: 818-882-5614.
E-mail: chatsmimi@aol.com
Web Site: www.laparks.org
Institution Type/Description: Historic House Museum.
Collections: period furnishings; gardens; fruit trees; rose bushes.
Hours & Admission Prices: 1st Sun. each month 1-4.

Cherry Valley

EDWARD-DEAN MUSEUM & GARDENS, 9401 Oak Glen Rd., Cherry Valley, CA 92223-3799. Tel.: 951-845-2626. Fax: 951-845-2628.
E-mail: edmevents@rivcoeda.org
Web Site: www.edward-deanmuseum.org
Formerly: Edward-Dean Museum of Decorative Arts
Founded: 1958.
Congressional District: 37
Personnel Profile: Full-Time Paid 1; Part-Time Paid 2; Part-Time Volunteers 77.
Governing Authority: county. Parent Institution: County of Riverside, CA. Tax-exempt.
Institution Type/Description: Historic House Museum.
Collections: European, Oriental & American decorative arts; painting; sculptures; prints; textiles; ceramics; glass; ivory & jade carvings; Asiatic bronzes; paperweights; fans; timepieces; miniatures; cloisonne; Italian creche figures; 16th to 19th-century furniture.
Facilities: landscaped gardens; classrooms. Museum-related items for sale.
Activities: guided tours; lectures; workshops; docent program; permanent & temporary exhibitions; educational programs; seminars.
Publications: museum booklet; museum catalog; Selections from the Edward-Dean Museum of Decorative Arts; Robes of China: From the Permanent Collection; catalog, Wedgewood Masterpieces.
Hours & Admission Prices: Fri.-Sun. 10-5. Adults $5; children 12 & under and members no charge. Closed national holidays. &

Chico

BIDWELL MANSION STATE HISTORIC PARK, 525 Esplanade, Chico, CA 95926-3996. Tel.: 530-895-6144. Fax: 530-895-6699. Facebook: Bidwell Mansion State Historic Park.
Founded: 1964.
Congressional District: 3
Key Personnel: Unit Supvr., Denise Rist.
Governing Authority: state. California Dept. Parks & Recreation, P.O. Box 2390, Sacramento, CA 95811. Tel. 916-445-2358.
Institution Type/Description: Historic House: 1868 three-story Victorian Italian villa country estate, home of Gen. & Mrs. John Bidwell.
Collections: Victorian era furnishings; books; manuscripts.
Research Fields: the Bidwell family; general state history.
Facilities: Books, prints & other museum-related items for sale.
Activities: guided tours; lectures; films; drama; docent program.
Publications: brochure; quarterly newsletter, Bidwell Mansion News & Notes.
Hours & Admission Prices: Mon. 12-5, Sat.-Sun. 11-5; call to confirm. Adults 18 & over $6, children 5-17 $3; members and children 4 & under no charge. Closed New Year's Day; Thanksgiving; Christmas. &
Attendance: 35,000

CENTERVILLE SCHOOL HOUSE, 13548 Centerville Rd., Chico, CA 95928. Tel.: 530-893-9667.
Institution Type/Description: Historic Building: housed in the first Centerville School; built in 1872.
Collections: local history & culture; period artifacts; photographs.
Hours & Admission Prices: Sat.-Sun. 1-4.

CHICO AIR MUSEUM, Chico Municipal Airport, 170 Convair Court, Chico, CA 95973. Tel.: 530-345-6468.
E-mail: chicoairmuseum@digitalpath.net
Web Site: www.chicoairmuseum.org
Founded: 2005.
Personnel Profile: Part-Time Volunteers 20.
Governing Authority: bd. of directors. Tax-exempt.
Institution Type/Description: Aviation History Museum.
Collections: aviation history & aircraft; photographs; personal artifacts; guns; military uniforms & artifacts.
Facilities: 1,000-vol. library. Museum-related items for sale.
Activities: special events; group tours; educational programs.
Hours & Admission Prices: Fri.-Sun. 9-4. No charge; donations accepted.
Attendance: 7,338 (accurate)

CHICO CREEK NATURE CENTER, (M), 1968 E. 8th St., Chico, CA 95928-4110. Tel.: 530-891-4671. Fax: 530-891-0837.
E-mail: info@bidwellpark.org
Web Site: bidwellpark.org
Founded: 1990.
Congressional District: 3
Key Personnel: Exec. Dir., Caitlin Reilly; Pres. (V), Chuck Nelson.

Personnel Profile: Full-Time Paid 1; Part-Time Paid 5; Part-Time Volunteers 10; Interns 12.
Governing Authority: private; nonprofit organization. Tax-exempt.
Institution Type/Description: Nature Center.
Collections: taxidermy of animals, primarily of the northern California region; non-releasable (injured) live wild animals of the region.
Facilities: small native plant garden; 2,200 sq. ft. exhibit space; Bidwell Park Information & Interpretive Center. Museum-related items for sale.
Activities: environmental educational summer camp; weekend hikes & activities; birthday parties; K-6 environmental education field trips.
Publications: quarterly newsletter, Creekside Notes; Raptors of Bidwell Park.
Hours & Admission Prices: Living Animal Museum: Wed.-Sun. 11-4; Howard Tucker Exhibit Hall: Thurs.-Sun. 11-4. Suggested donation: adults $4, children $2. Closed New Year's Day; Easter; Independence Day; Thanksgiving; Christmas. &
Attendance: 40,000 (accurate)
Membership: Student & Senior $10; Individual $30; Family $50; Annual Supporter $75; Annual Sponsor $150; Annual Patron $250.

CHICO MUSEUM, 141 Salem St., Chico, CA 95926. Mailing Address: 270 Boeing Ave., Ste. 1, Chico, CA 95973-9510. Tel.: 530-891-4336 & 892-1525. Fax: 530-892-1524 & 891-4366.
E-mail: office@farwestheritage.org
Web Site: www.chicomuseum.org
Founded: 1986.
Congressional District: 2
Key Personnel: Interim Mgr., JoAnne Vidal-Carr; Pres. (V), Roger Steel; Treas., Linda Limberg.
Personnel Profile: Full-Time Paid 1; Part-Time Volunteers 40; Interns 2.
Governing Authority: nonprofit organization. Parent Institution: Far West Heritage Assoc. Tax-exempt: 501(c)(3).
Institution Type/Description: History Museum: housed in 1904 Queen Anne & Romanesque Revival style Carnegie Library.
Collections: Chinese Taoist Temple used in Chico 1890-1939; Chico timeline featuring portraits, artifacts & history from 1840 to present.
Research Fields: Chico & the Chico area, prehistoric to modern times.
Facilities: library of local history books & bound newspapers; 3,662 sq. ft. exhibit space; mini-theatre. Museum-related items for sale.
Activities: permanent, temporary & traveling exhibits; guided tours; bus tours; lectures; workshops; films. Museum Sponsors: Members' Country Supper; Members' Museum Night Out Dinner; Volunteer Recognition Event.
Publications: quarterly membership newsletter, Museum Notes; books related to exhibits & local history.
Hours & Admission Prices: Wed.-Sun. 12-4. Adults $3, seniors & students $2; children under 14 no charge. Closed New Year's Day; Easter; Independence Day; Thanksgiving; Christmas Eve & Day. &
Attendance: 6,500 (accurate)
Membership: Senior $50; Individual $75; Family $100; Settler Society $120; Pioneer Society $300; Explorer's Society $600; Founder's Society $1,200.

COLMAN MEMORIAL COMMUNITY MUSEUM, 13548 Centerville Rd., Chico, CA 95928. Tel.: 530-893-9667.
Institution Type/Description: History Museum.
Collections: local history & culture; Civil War memorabilia; Native American baskets; early school furnishings; 1800s clothing; mining equipment; period tools & utensils; photographs.
Activities: Annual Event: 49er Faire in June.
Hours & Admission Prices: Sat.-Sun. 1-4. No charge; donations accepted.

GATEWAY SCIENCE MUSEUM - CALIFORNIA STATE UNIVERSITY, (M), 625 Esplanade, Chico, CA 95929-0545. Mailing Address: College of Natural Sciences, Chico, CA 95929-0545. Tel.: 530-898-4121.
Institution Type/Description: Science Museum.
Collections: Northern California's natural history; Native American artifacts; photographs; period artifacts.
Hours & Admission Prices: Wed.-Fri. 12-5, Sat.-Sun. 10-5. Adults $5, children $3; members no charge.

JANET TURNER PRINT MUSEUM, (M), California State University, Chico, 400 W. 1st St., Chico, CA 95929-0820. Tel.: 530-898-4476. Fax: 530-898-5581.
E-mail: csullivan@csuchico.edu
Web Site: www.theturner.org
Founded: 1981.
Key Personnel: Chm. (V), Pat Kopp; Cur., Catherine Sullivan; Asst. Collection Mgr., Adria Crossen Davis.
Personnel Profile: Part-Time Paid 5; Part-Time Volunteers 4; Interns 3.

Governing Authority: public university; nonprofit. Parent Institution: CSU-Chico. Subsidiary Institution: CSU Foundation. Tax-exempt: 501(c)(3).
Institution Type/Description: Fine Art Print Gallery & Museum Archive.
Collections: original fine art prints, historical & contemporary, from more than 40 countries; professional & student prints.
Research Fields: printmaking artists' biographies; art historical background relating to thematic exhibitions; historical & technical printmaking books.
Facilities: 150-vol. library on the historical techniques of printmaking; 500 sq. ft. exhibit space; auditorium.
Activities: docent program; lectures; intern program; gallery talks; student research projects. Museum Sponsors: student printmaking invitational & biannual National Juried Printmaking Competition & Exhibition.
Hours & Admission Prices: Sept.-May Mon.-Fri. 11-4 during exhibitions; other times by appointment. No charge; donations accepted. Closed spring break; Thanksgiving break; Christmas break. &
Attendance: 5,000 (estimated)
Membership: Student $2; Individual $15; Family $25; Patron $100.

NATIONAL YO-YO MUSEUM, 320 Broadway, Chico, CA 95928-5322. Tel.: 530-893-0545, ext. 4.
Web Site: www.nationalyoyo.org/museum/index.htm
Institution Type/Description: Toy Museum.
Collections: yo-yo artifacts & memorabilia.
Activities: Yo-Yo lessons.
Hours & Admission Prices: Mon.-Sat. 10-6, Sun. 12-5; groups by appointment.

1078 GALLERY, 820 Broadway, Chico, CA 95928. Tel.: 530-343-1973.
E-mail: info@1078gallery.org
Web Site: 1078gallery.org
Institution Type/Description: Art Gallery.
Collections: works by contemporary artists.
Hours & Admission Prices: Thurs.-Sat. 12:30-5:30.

VALENE L. SMITH MUSEUM OF ANTHROPOLOGY, (M), California State University, Chico, Chico, CA 95929-0400. Tel.: 530-898-5397. Fax: 530-898-6143.
E-mail: anthromuseum@csuchico.edu
Web Site: www.csuchico.edu/anth/museum
Founded: 1969.
Congressional District: 2
Key Personnel: Co-Dir., Stacy Schafer; Co-Dir., Georgia Fox; Cur., Adrienne Scott.
Personnel Profile: Full-Time Paid 3; Part-Time Paid 3; Part-Time Volunteers 1.
Governing Authority: state university; nonprofit organization. Tax-exempt.
Institution Type/Description: University Anthropology Museum.
Collections: archaeological & ethnographic materials from around the world; emphasis on California prehistory & history.
Research Fields: archaeological; museological.
Facilities: 1,000 sq. ft. exhibit space; educational facilities. Museum-related items for sale.
Activities: guided tours; education programs for children; formally organized education programs for undergraduate & graduate students; cultural enrichment programs; monthly lecture series. Annual Event: photo contest.
Publications: alliance newsletter; exhibit brochure.
Hours & Admission Prices: Sept.-Oct. & Dec.-July Tues.-Sat. 11-3. Office: Summer Mon.-Thurs. 12-4. No charge; donations accepted. &
Attendance: 5,000 (accurate)
Membership: Senior Citizen & Student $10; Individual $20; Family $30; Supporting $50; Sustaining $75; Sponsor $150.

China Lake

THE CHINA LAKE MUSEUM OF NAVAL ARMAMENT & TECHNOLOGY, (M), 1 Pearl Harbor Way, China Lake, CA 93555-2803. Mailing Address: P.O. Box 217, Ridgecrest, CA 93556-0217. Tel.: 760-939-3105. Fax: 760-939-0564.
E-mail: chinalakemuseum@mediacombb.net
Web Site: www.chinalakemuseum.org
Formerly: U.S. Naval Museum of Armament & Technology
Founded: 2000.
Key Personnel: Dir. & Pres., Robert Campbell; Treas., Craig Porter; Devel., Pat Connell.
Personnel Profile: Full-Time Paid 2; Part-Time Paid 1; Part-Time Volunteers 30.
Governing Authority: federal government. Tax-exempt: 501(c)(3).
Institution Type/Description: Military Museum.
Collections: Navy weapons systems from early WWII rockets to modern day smart weapons.

Facilities: 12,000 sq. ft. exhibit space. Museum-related items for sale.
Activities: guided tours; lectures; films; education & STEM programs; book signing. Annual Events: dinner auction; members meeting.
Publications: quarterly newsletter, The China Laker.
Hours & Admission Prices: Mon.-Sat. 10-4. No charge; donations accepted. Closed federal holidays. &
Attendance: 8,500 (accurate)
Membership: Regular $35; Contributing $100; Life & Business $1,000; Platinum $5,000.

Chino

CHINO'S OLD SCHOOLHOUSE MUSEUM, 5493 "B" St., Chino, CA 91710-4241. Mailing Address: P.O. Box 972, Chino, CA 91708-0972. Tel.: 909-627-6464.
Institution Type/Description: History Museum: Chino's first schoolhouse built in 1888.
Collections: early local history; photographs.
Hours & Admission Prices: Temporarily closed.

PLANES OF FAME AIR MUSEUM, 7000 Merrill Ave. #17, Chino, CA 91710-9085. Tel.: 909-597-3722. Facebook: Planes of Fame Air Museum.
E-mail: karen.hinton@planesoffame.org
Web Site: planesoffame.org
Founded: 1957.
Key Personnel: Pres., Steve Hinton; Dir. Devel., Karen Hinton.
Governing Authority: nonprofit organization. Tax-exempt: 501(c)(3).
Institution Type/Description: Aircraft Museum.
Collections: history of aviation from 1896 Chanute Hang Glider to modern space flight; test & research flight vehicles.
Publications: TAM News.
Hours & Admission Prices: Sun.-Fri. 10-5., Sat. 9-5. Adults $11, seniors 65 & up and veterans $10, children 5-12 $4; discounts to AAA members; children under 5, active duty military, police & firefighters and members no charge. Closed Thanksgiving; Christmas. &
Attendance: 50,000 (accurate)
Membership: Individual $75; Family $150; Silver $250; Gold $500; Platinum $2,000.

YANKS AIR MUSEUM, (M), 7000 Merrill Ave., Hangar A270, Chino, CA 91710-9091. Mailing Address: 7000 Merrill Ave., Hangar A270, P.O. Box 35, Chino, CA 91710-9091. Tel.: 909-597-1735.
E-mail: christen@yanksair.com
Web Site: www.yanksair.com
Founded: 1982.
Key Personnel: Dir., Christen Wright
Institution Type/Description: Aviation Museum.
Collections: over 150 flying & static aircraft from WWI to present; drones; aircraft uniforms; flight suits; ejection seats; patches; wings; civil aviation; airliner memorabilia; instrument panels; headgear; turrets; photo archives.
Facilities: Gift items for sale.
Activities: oral history program; special events.
Publications: newsletter.
Hours & Admission Prices: Mon.-Sat. 8-4. Adults $11, seniors 65 & over $10, children 5-11 $5; children 4 & under and members no charge. Closed holidays. &
Membership: Individual $50; Family $75; Restoration $150; Preservation $500; Heritage Legacy $5,000; Life $10,000.

YORBA AND SLAUGHTER FAMILIES ADOBE, 17127 Pomona Rincon Rd., Chino, CA 91708-9285. Mailing Address: c/o San Bernardino Co. Museums, 2024 Orange Tree Lane, Redlands, CA 92374. Tel.: 909-597-8332 & 307-2669. Fax: 909-307-0539.
E-mail: rmckernan@sbcm.sbcounty.gov
Web Site: www.sbcountymuseum.org
Formerly: Yorba-Slaughter Adobe Museum
Founded: 1976.
Congressional District: 35
Key Personnel: Dir., Robert McKernan; Site Mgr., Karen Buma; Cur., Michele Nielsen.
Personnel Profile: Part-Time Paid 1.
Governing Authority: county. Parent Institution: San Bernardino County Museums. Tax-exempt.
Institution Type/Description: Local History and Historic House Museum: housed in c.1853 Adobe home.
Collections: decorative arts; photographs; farm equipment.

Research Fields: local history.
Activities: guided tours; special events.
Hours & Admission Prices: Tues.-Sat. 10-3. No charge; donations requested. Closed New Year's Day; Thanksgiving; Christmas.
Attendance: 1,000 (estimated)

Chiriaco Summit

GENERAL PATTON MEMORIAL MUSEUM, 62510 Chiriaco Rd., Chiriaco Summit, CA 92201-8203. Tel.: 760-227-3483. Fax: 760-227-3483.
E-mail: contact@generalpattonmuseum.org
Web Site: www.generalpattonmuseum.com
Founded: 1988.
Congressional District: 37
Key Personnel: Dir., Richard Ramirez.
Governing Authority: not-for-profit organization. Tax-exempt: 501(c)(3).
Institution Type/Description: Military Museum: located near former headquarters of World War II Desert Training Areas.
Collections: exhibits emphasizing Gen. Patton, Desert Training Center, World War II & other eras of military history.
Facilities: 200-vol. library on Gen. Patton, World War II & the military. Museum-related items for sale.
Activities: Annual Events: Veterans Day; Patton's Birthday Celebration.
Publications: newsletter; brochure.
Hours & Admission Prices: Daily 9:30-4:30. Adults $5, seniors $4.50, children $1; American Legion & V.F.W. members and active-duty military no charge. Closed Thanksgiving; Christmas. &
Attendance: 73,500 (accurate)
Membership: Annual Individual $25; Annual Family $75; Lifetime Individual $250; Annual Organization & Lifetimie Family $500; Lifetime Organization $1,000.

Chula Vista

* **LIVING COAST DISCOVERY CENTER,** 1000 Gunpowder Point Dr., Chula Vista, CA 91910-8222. Tel.: 619-409-5900 & 5940. Fax: 619-409-5910. Facebook: Living Coast Discovery Center.
E-mail: cvncinfo@cvnc.us
Web Site: www.thelivingcoast.org
Formerly: Chula Vista Nature Center
Founded: 1987.
Congressional District: 51
Key Personnel: Dir., Dr. Brian Joseph; Chm. (V), F.L. Wergeland, Jr., M.D.; C.O.O., Ben Vallejos; Education, Wendy Spaulding.
Personnel Profile: Full-Time Paid 7; Part-Time Paid 19; Part-Time Volunteers 200; Interns 1.
Governing Authority: nonprofit organization. Tax-exempt: 501(c)(3).
Institution Type/Description: Aquarium Zoo: located on the Sweetwater Marsh National Wildlife Refuge.
Collections: representing an experiential tour of coastal wetlands; live plant & animal specimens; artifacts.
Research Fields: wetland restoration & enhancement; ecology; ornithology.
Facilities: 200-vol. non-circulating library on wetland ecology; aquarium; 125-seat auditorium; botanical garden; classrooms; labs; 6,000 sq. ft. exhibit space; field research station; nature center; 1.5 miles of nature trails; outdoor classroom; stingray touch tank; garden; composting. Natural history books & supplies for sale.
Activities: docent program; education programs; guided tours; lectures; participatory exhibits; light-footed Clapper Rail captive breeding & release program.
Publications: quarterly newsletter; monthly e-newsletters.
Hours & Admission Prices: Daily 10-4. Adults $14, seniors 65 & over, children 4-17, & students with ID $9; discount to military; children 0-3 & members no charge. Closed Thanksgiving Day & Christmas Day. &
Attendance: 70,000 (accurate)
Membership: Single Senior (65 & up) $32; Single Adult $40; Dual Senior, $48; Dual Adult $60; Family Plus (two adults & up to five children) $75 with additional children $15; call to confirm.

City of Industry

WORKMAN & TEMPLE FAMILY HOMESTEAD MUSEUM, (M), 15415 E. Don Julian Rd., City of Industry, CA 91745-1029. Tel.: 626-968-8492. Fax: 626-968-2048. Facebook: Workman & Temple Family Homestead Museum.
E-mail: info@homesteadmuseum.org
Web Site: www.homesteadmuseum.org
Founded: 1981.

Congressional District: 33
Key Personnel: Dir., Karen Graham Wade; Asst. Dir., Paul Spitzzeri; Dir. Public Programs, Alexandra Rasic; Creative Content Mgr., Lillian Diep; Facilities Coord., Robert Barron; Volunteer Coord., Steven Dugan; Programs Coord., Gennie Truelock.
Personnel Profile: Full-Time Paid 10; Part-Time Volunteers 80; Interns 1.
Governing Authority: municipal. Parent Institution: City of Industry. Tax-exempt: 501(c)(1).
Institution Type/Description: Historic Buildings Museum & Site: located on the site of the Rancho la Puente.
Collections: 1830-1930 decorative arts, furniture, textiles, costumes, & artifacts; interior decorative elements consisting of metalwork, tile, wood carvings & stained glass windows; photographic archives. Historic Buildings & Structures: mid-19th century Workman house; 1919-1925 Spanish colonial revival Temple residence; late 19th century Water Tower; 1919 classical revival mausoleum; c.1850 El Campo Santo cemetery.
Research Fields: southern California; California & the United States from 1830-1930; architecture; decorative arts.
Facilities: 600-vol. library pertaining to architecture, art, costumes, decorative arts & history available for research by appointment on premises; auditorium. Museum-related items for sale.
Activities: guided tours; lectures; films; concerts; formally organized educational programs; volunteer programs; permanent & temporary exhibitions.
Hours & Admission Prices: Wed.-Sun. 1-4; group tours by appointment. No charge; donations accepted. Closed major holidays. &

Attendance: 15,000 (estimated)

Claremont

CLAREMONT MUSEUM OF ART, (M), Claremont, CA 91711-1136. Mailing Address: P.O. Box 1136, Claremont, CA 91711-1136.
E-mail: info@claremontmuseum.org
Web Site: www.claremontmuseum.org
Founded: 2006.
Key Personnel: Pres. (V), Sany Baldonado.
Personnel Profile: Part-Time Paid 1; Part-Time Volunteers 2.
Institution Type/Description: Art Museum.
Collections: paintings; sculptures.
Activities: educational programs.
Hours & Admission Prices: Temporarily closed.
Membership: Basic $40; Premium $100; Premium Plus $250; Patron $1,000.

CLARK HUMANITIES MUSEUM-STUDY, Humanities Bldg., Scripps College, Claremont, CA 91711-3905. Mailing Address: Scripps College, 1030 Columbia Ave., Claremont, CA 91711-3905. Tel.: 909-607-3606. Fax: 909-607-7143.
E-mail: ehaskell@scrippscollege.edu
Founded: 1970.
Congressional District: 34
Key Personnel: Chm. Museum Committee, Dr. Eric T. Haskell.
Governing Authority: college. Affiliated with Scripps College. Tax-exempt: 501(c)(3).
Institution Type/Description: General Museum.
Collections: Gen. Edward Young collection of American paintings; Johnson collection of Japanese prints; Routh collection of Cloisonne; Nagel collection of art & artifacts; Wagner collection of African sculpture; prints; drawings; photographs.
Facilities: library.
Activities: formally organized education programs for undergraduate college students; temporary exhibitions.
Publications: exhibition catalogues.
Hours & Admission Prices: Mon.-Fri. 9-12:30 & 1:30-5. No charge. &

PETTERSON MUSEUM OF INTERCULTURAL ART, (M), 730 Plymouth Rd., Claremont, CA 91711. Mailing Address: 625 Mayflower Rd., Claremont, CA 91711-4222. Tel.: 909-399-5544. Fax: 909-399-5508.
E-mail: cgil@pilgrimplace.org
Web Site: www.pilgrimplace.org
Founded: 1968.
Congressional District: 33
Key Personnel: Dir., Bill Cunitz; Pres. (V), Gail Duggan; Cur., Carol Bowdoin Gil.
Personnel Profile: Part-Time Paid 1; Part-Time Volunteers 42; Interns 3.
Governing Authority: nonprofit organization. Parent Institution: Pilgrim Place. Tax-exempt: 501(c)(3).
Institution Type/Description: International Folk & Fine Art Museum.
Collections: international folk & fine art; Chinese bronzes & imperial court robes; Latin American textiles; African masks & woodcarving; intercultural costumes & puppets; masks; international dolls.
Facilities: 2,400-vol. library of books on intercultural arts.
Activities: guided tours; lectures; films; concerts; arts festivals; docent program.
Hours & Admission Prices: Guided Tours: Fri.-Sun. 2-4 or by appointment. No charge; donations accepted. Closed Easter; Thanksgiving; Christmas. &
Attendance: 1,500 (estimated)
Membership: Basic $25; Associates $50; Sustainers $100; Patrons $250; Curator's Circle $500; Petterson Visionaries $1,000 & up.

PITZER COLLEGE ART GALLERIES, Pitzer College, 1050 N. Mills Ave., Claremont, CA 91711-6101, Tel.: 909-607-3143.
E-mail: pitzer_galleries@pitzer.edu
Web Site: www.pitzer.edu/galleries
Key Personnel: Dir., Ciara Ennis.
Personnel Profile: Full-Time Paid 2; Part-Time Paid 1; Part-Time Volunteers 1; Interns 1.
Governing Authority: Parent Institution: Pitzer College. Tax-exempt.
Institution Type/Description: Art Gallery.
Collections: Nichols Gallery: works by national & international artists. Lenzner Family Gallery: works by emerging artists.
Major Exhibits: Andrea Bowers: #sweetjane, 1/14-4/14; Martin Durazo, 1/14-5/14; Arthur Dubinsky: The Life & Times of Pitzer College, 1/14-5/14; Candy Factory, 9/14; Claudia Rankine, 9/14.
Hours & Admission Prices: Tues.-Fri. 12-5; other times by appointment. No charge.

POMONA COLLEGE MUSEUM OF ART, (M), 330 N. College Ave., Claremont, CA 91711-4401. Mailing Address: 333 N. College Way, Claremont, CA 91711-4429. Tel.: 909-621-8283 & 8000. Fax: 909-621-8989.
Web Site: www.pomona.edu/museum
Founded: 1958.
Congressional District: 34
Key Personnel: Dir., Kathleen Howe; Assoc. Dir. & Registrar, Steve Comba; Sr. Cur., Rebecca McGrew; Cur. Academic Programming, Terri Geis; Administrative Asst., Barbara Coldiron; Preparator, Gary Murphy; Security & Information Officer, Anne Merten.
Personnel Profile: Full-Time Paid 6; Part-Time Paid 3; Interns 3.
Governing Authority: college. Parent Institution: Pomona College. Tax-exempt: 501(c)(3).
Institution Type/Description: College Art Museum.
Collections: Kress Renaissance paintings; prints, drawings & photographs; American Indian basketry, ceramics & beadwork.
Research Fields: drawings; prints; American Indian; Kress panels; photographs.
Facilities: Exhibition catalogues & posters for sale.
Activities: lectures; gallery talks; temporary & traveling exhibitions; inter-museum loans; organized education programs for undergraduate students. Museum Sponsors: annual student exhibitions; biennial faculty & faculty Invitation exhibitions.
Publications: exhibition catalogues.
Hours & Admission Prices: Sept.-May Tues.-Fri. 12-5, Sat.-Sun. 1-5. No charge. Closed school & national holidays. &
Attendance: 10,770 (accurate)

* **RANCHO SANTA ANA BOTANIC GARDEN, (M),** 1500 N. College Ave., Claremont, CA 91711-3157. Tel.: 909-625-8767 ext. 200. Fax: 909-626-7670.
E-mail: info@rsabg.org
Web Site: www.rsabg.org
Founded: 1927.
Congressional District: 33
Key Personnel: Chm. Bd. Trustees, Elin R. Dowd, Jr.; Exec. Dir. & Dir. Research, Lucinda McDade; Dir. Finance, Kristine Crosby; Dir. Visitor Services, Eric Garton; Dir. Horticulture, Scott LaFleur.
Governing Authority: nonprofit corp. Tax-exempt: 509(A)(3).
Institution Type/Description: Botanical Garden.
Collections: living plants; herbarium; pollen slides; wood samples; seeds; 1,000,000 specimens.
Research Fields: plant taxonomy; cyto-taxonomy; chemo-taxonomy; mycology; ecology; molecular systematics; monographic revisions; plant anatomy; palynology.
Facilities: 76,747-vol. library of general botanical & horticultural materials available for formal inter-library loan & personal requests; herbarium; laboratories; 100-seat classroom.
Activities: guided tours; formally organized education programs for children;

docent program or council; formally organized education programs for graduate students affiliated with Claremont Graduate University. Annual Events: plant sales; biannual symposia; Wildflower Show; Musical Evenings in the Garden.
Publications: biannual scientific journal, Aliso; quarterly newsletter; RSABG Occasional Publications.
Hours & Admission Prices: Daily 8-5. Adults $8, seniors & students with ID $6, children 3-12 $4; children under 3 & members no charge. Closed New Year's Day; Independence Day; Thanksgiving; Christmas. &
Attendance: 153,000 (estimated)
Membership: Individual $45; Family $75; Supporting $100; Sustaining $250; Garden Patron $500; Director's Circle $1,000; Director's Circle Patron $2,500; Director's Circle Benefactor $5,000; Chairman's Circle $10,000.

*** THE RAYMOND M. ALF MUSEUM OF PALEONTOLOGY, (M),** 1175 W. Baseline Rd., Claremont, CA 91711-2146. Tel.: 909-624-2798. Fax: 909-624-2798.
E-mail: dlofgren@webb.org
Web Site: www.alfmuseum.org
Founded: 1937.
Congressional District: 35
Key Personnel: Dir., Donald Lofgren, Ph.D.; Dir. Outreach, Kathy Sanders; Cur., Andrew Farke.
Personnel Profile: Full-Time Paid 3; Part-Time Paid 2; Part-Time Volunteers 15.
Governing Authority: nonprofit organization. Parent Institution: The Webb Schools. Tax-exempt: 501(c)(3).
Institution Type/Description: Paleontology Museum.
Collections: paleontology; rocks; minerals; archaeology; osteology.
Research Fields: paleontology; geology.
Facilities: 60-seat auditorium; classrooms. Museum-related items for sale.
Activities: guided tours; formally organized education programs; permanent & temporary exhibitions; world-wide natural history tours.
Publications: quarterly newsletter, Quest.
Hours & Admission Prices: June-Aug. Mon.-Fri. 8-12 & 1-4; Sept.-May Mon.-Fri. 8-12 & 1-4, Sat. 1-4; groups by appointment. Admission $6; children 4 & under no charge. &
Attendance: 20,000 (accurate)

RUTH CHANDLER WILLIAMSON GALLERY, SCRIPPS COLLEGE, (M), 1030 Colombia Ave., Claremont, CA 91711-3905. Tel.: 909-607-3397 & 4690. Fax: 909-607-4691.
E-mail: mmacnaug@scrippscollege.edu
Web Site: www.scrippscollege.edu/dept/gallery
Founded: 1993.
Congressional District: 34
Key Personnel: Dir., Mary Davis MacNaughton; Asst., Colleen Saloman; Registrar & Preparator, Kirk Delman; Data Specialist, Patricia Yu.
Personnel Profile: Full-Time Paid 2; Part-Time Paid 4; Part-Time Volunteers 2; Interns 4.
Governing Authority: colleges. Parent Institution: Scripps College, Claremont, CA. Tax-exempt: 501(c)(3).
Institution Type/Description: College Art Gallery.
Collections: Gen. Edward Young collection of American paintings; Dr. & Mrs. William E. Ballard collection of Japanese prints; Fred and Estelle Marer collection of contemporary American, British, Korean, Mexican & Japanese ceramics; Mrs. James Johnson collection of Japanese prints; Dorothy Adler Routh collection of cloisonne; Wagner collection of African sculpture; prints; drawings; photographs; contemporary ceramics.
Research Fields: American paintings; drawings; prints; ceramics.
Facilities: Exhibition catalogues & posters for sale.
Activities: lectures; films; gallery talks; temporary & traveling exhibitions; inter-museum loans; organized education programs for undergraduate students. Museum Sponsors: annual student exhibitions; annual exhibition of contemporary ceramics.
Publications: exhibition catalogues, Larger Than Life: Robert Rahway Zakavitch's Big Bungalow Suite; Annals of My Glass House: Photographs by Julia Margaret Cameron; Revolution in Clay: Marer Collection of Contemporary Ceramics; Johnson, Kaufmann and Coate: Partners in the California Style; Ceramic Annual 2004, 2005, 2006, 2007, 2008, 2009, 2010; Alison and Lezley Saar; In the Mind's Sky: Intersections of Art & Science; Matter and Matrix: Kris Cox, Amy Ellingson, Elizabeth Turk, Jane Park Wells; Reading and Meaning: Word and Symbolism in the Art of Squeak Carnwath, Leslie Dill, Leslie Enders Lee, and Anne Siems; Sense and Sensibility: Laurie Paintings and Fendrich Drawings, 1990-2010.
Hours & Admission Prices: Wed.-Sun. 1-5. No charge. Closed school & national holidays. &
Attendance: 5,200 (estimated)

Clayton

CLAYTON HISTORICAL SOCIETY, 6101 Main St., Clayton, CA 94517-1201. Mailing Address: P.O. Box 94, Clayton, CA 94517-0094. Tel.: 925-672-0240.
Institution Type/Description: Historical Society Museum.
Collections: local history & culture; period furnishings; personal artifacts; photographs.
Activities: special events.
Hours & Admission Prices: Wed. 2-4 & 6-8, Sun. 2-4; groups by appointment.

Clovis

CLOVIS BIG DRY CREEK HISTORICAL MUSEUM, 401 Pollasky Ave., (at 4th St.), Clovis, CA 93612-1141. Tel.: 559-297-8033. Fax: 559-297-8882.
E-mail: pbos@clovis-museum.com
Web Site: www.clovis-museum.com
Founded: 1981.
Congressional District: 21
Key Personnel: Pres. & Cur., Peggy Bos.
Personnel Profile: Part-Time Volunteers 25.
Governing Authority: Tax-exempt.
Institution Type/Description: History Museum: site of the historic 1924 bank robbery.
Collections: history of Clovis; Indian artifacts; stories & pictures of Clovis Cole, wheat king of the US; portion of 1893 42 mile flume; pictures & articles of over 420 Clovis families; 1903-1965 graduation pictures of Clovis High School; memorabilia of Kent "Festus" Curtis of "Gunsmoke"; veterans' room honoring 58 Clovis Gold Star Heroes & more.
Activities: monthly public program, Let's Talk Clovis.
Publications: bimonthly newsletter, Clovis Roundup; historical articles.
Hours & Admission Prices: Tues.-Sat. 10-2; other times by appointment. No charge; donations accepted. &
Attendance: 10,101 (accurate)
Membership: Senior Citizen $10; Regular $15; Family $20; Sustaining $40; Sponsor $80; Business $100; Patron $250 & up.

Coalinga

R.C. BAKER MEMORIAL MUSEUM, INC., 297 W. Elm St., Coalinga, CA 93210-1923. Tel.: 559-935-1914; 877-416-5849. Fax: 559-935-2339. Facebook: R.C. Baker Memorial Museum, Inc.
E-mail: curator@rcbakermuseum.com
Web Site: www.rcbakermuseum.com
Founded: 1961.
Key Personnel: Chm. (V), JoAnn Clark; Cur., Stephanie McHaney.
Personnel Profile: Full-Time Paid 2.
Governing Authority: Tax-exempt.
Institution Type/Description: History Museum.
Collections: artifacts & memorabilia pertaining to Pleasant Valley.
Hours & Admission Prices: Mon.-Fri. 10-12 & 1-5., Sat. 11-5, Sun. 1-5. No charge; donations accepted. Closed legal holidays.
Attendance: 2,500 (estimated)
Membership: Associate (non-voting) $15; Individual (voting) $25; Commercial & Business $75; Life (voting) $100.

Coarsegold

COARSEGOLD HISTORIC MUSEUM, 31899 Hwy. 41, Coarsegold, CA 93614. Mailing Address: P.O. Box 117, Coarsegold, CA 93614-0117. Tel.: 559-642-4448. Fax: 559-642-4246.
E-mail: chs@sti.net
Web Site: coarsegoldhistoricmuseum.com
Formerly: Willow Glen Museum
Founded: 1980.
Congressional District: 19
Key Personnel: Dir., Linda Core; Office Mgr., Kay Good; Pres. (V), Karen Morris.
Personnel Profile: Part-Time Volunteers 40.
Governing Authority: nonprofit. Subsidiary Institution: Coarsegold Historical Society. Tax-exempt.
Institution Type/Description: History Museum.
Collections: local history; personal artifacts; Chukchansi Indian artifacts; blacksmith shop.
Research Fields: eastern Madera County, CA.
Activities: Indian basket weaving classes. Annual Event: Art, Dinner & Music in the Meadow in June.
Publications: book, As We Were Told, Vols. I & II; quarterly member newsletter.

Hours & Admission Prices: May-Sept. Sun. 12-4, Mon. 9-11:30, Wed.-Sat. 10-2; Oct.-April Sun. 1-4, Mon. 9-11:30; other times by appointment. No charge; donations accepted. &

Attendance: 700 (estimated)

Membership: Individual $20; Family $30; Sustaining $50; Life $300.

Colfax

COLFAX AREA HISTORICAL SOCIETY - DEPOT MUSEUM, 99 Railroad St., Colfax, CA 95713. Mailing Address: P.O. Box 185, Colfax, CA 95713-0185. Tel.: 530-346-8599.

E-mail: info@colfaxhistory.org

Web Site: www.colfaxhistory.org

Formerly: Colfax Heritage Museum

Founded: 1985.

Congressional District: 5

Key Personnel: Dir., Helen Wayland; Pres. (V), Svend Miller; Museum Shop Mgr., Donna Williams.

Personnel Profile: Part-Time Volunteers 6; Interns 1.

Governing Authority: Parent Institution: Colfax Historical Society. Tax-exempt.

Institution Type/Description: Historical Society Museum.

Collections: local history & culture; period furnishings; photographs; railroad artifacts.

Research Fields: genealogy.

Facilities: meeting room.

Activities: Museum Sponsors: Heritage Trail in August; Founder's Day in September.

Publications: Cobblestones.

Hours & Admission Prices: Daily 10-3. No charge; donations accepted. Closed New Year's Day; Easter; Independence Day; Thanksgiving; Christmas. &

Attendance: 10,887 (accurate)

Membership: Junior $5; Single $15; Family $25; Business & Nonprofit $30.

Coloma

MARSHALL GOLD DISCOVERY STATE HISTORIC PARK, 310 Back St., Coloma, CA 95613. Mailing Address: P.O. Box 265, Coloma, CA 95613-0265. Tel.: 530-622-3470. Fax: 530-622-3472.

E-mail: marshallgold@parks.ca.gov

Web Site: www.parks.ca.gov

Founded: 1890.

Congressional District: 14

Key Personnel: Park Supt., Jeremy McReynolds.

Personnel Profile: Full-Time Paid 8; Part-Time Paid 4; Part-Time Volunteers 130.

Governing Authority: state. Parent Institution: California Dept. of Parks & Recreation. Tax-exempt.

Institution Type/Description: Gold Rush History Museum: located near the site of the discovery of gold in 1848.

Collections: manuscript collections. Historic Buildings: 1867 Thomas House; 1855 Churches; 1849 pioneer cemetery; 1849 Catholic cemetery; 1866-1889 Chalmers winery ruins; 1889 Marshall Monument; 1857 Chinese stores; 1855 Robert Bell's brick store ruins; 1854 Bekearts gunsmith shop; Coloma 1857 stone jail ruins.

Research Fields: history.

Facilities: 1,000-vol. library of books pertaining to the gold rush & pioneer file available on premises; 80-seat lecture room.

Activities: self-guided tours; lectures; films; formally organized education programs for children, adults, undergraduate & graduate college students; permanent & temporary exhibitions.

Publications: brochures; trail guides; teachers guide; historic account of Gold Rush & Marshall's Role, available in translation.

Hours & Admission Prices: March-Oct. Tues.-Sun. 10-4; Nov.-Feb. Tues.-Sun. 10-3. Park: $8 per car, $7 seniors; members no charge. Closed New Year's Day; Thanksgiving; Christmas. &

Attendance: 250,000 (estimated)

Membership: Individual $15; Family $25; Lifetime $250.

Colton

AGUA MANSA PIONEER MEMORIAL CEMETERY, 2001 W. Agua Mansa Rd., Colton, CA 92324-3388. Mailing Address: c/o San Bernardino Co. Museums, 2024 Orange Tree Lane, Redlands, CA 92374. Tel.: 909-307-2669. Fax: 909-307-0539.

E-mail: rmckernan@sbcm.sbcounty.gov

Web Site: www.sbcountymuseum.org

Founded: 1977.

Congressional District: 36

Key Personnel: Dir., Robert McKernan; Cur., Michele Nielsen; Docent, Jason Bowe.

Personnel Profile: Part-Time Paid 1.

Governing Authority: county. Parent Institution: San Bernardino County Museum. Tax-exempt.

Institution Type/Description: Historic Site & Museum: 1854 cemetery.

Collections: photographs & personal memorabilia of local families. Historic Building: chapel.

Research Fields: local history.

Activities: guided tours.

Hours & Admission Prices: Fri. 12-3, Sat. 11-3, 1st Sun. of month 12-2. No charge; donations requested. Closed New Year's Day; Thanksgiving; Christmas. &

Attendance: 2,300 (accurate)

COLTON AREA MUSEUM, 380 N. La Cadena Dr., Colton, CA 92324-2928. Mailing Address: P.O. Box 1648, Colton, CA 92324-0851. Tel.: 909-824-8814 & 783-8817. Fax: 909-783-9241.

E-mail: info@coltonmuseum.net

Web Site: www.coltonmuseum.net

Founded: 1984.

Congressional District: 36

Key Personnel: Pres. (V), Michael L. Murphy; Museum Shop Mgr., Pam Gregory.

Personnel Profile: Part-Time Volunteers 25.

Governing Authority: private; nonprofit organization. Tax-exempt.

Institution Type/Description: History Museum & Historic Site: housed in 1908 Andrew Carnegie Library building.

Collections: local Colton history & history of the surrounding areas; The Earps; California Portland Cement; San Salvador & Agua Mansa areas; Southern Pacific Railroad; Colton's Federal theatre posters; citrus industry; Ken Hubbs, Chicago Cubs baseball player; local family histories.

Research Fields: Colton newspapers 1877-1972.

Activities: Museum Sponsors: Discover Colton Night; Colton Unity Day; annual quilt display; Wyatt Earp Festival; Library Week; annual sidewalk sale.

Publications: bimonthly newsletter.

Hours & Admission Prices: Wed. & Fri. 1-4, Sat. 11-2; tours by appointment. No charge; donations accepted. Closed major holidays. &

Attendance: 1,000 (estimated)

Membership: Individual $15; Patron $45; Commercial $250.

Columbia

COLUMBIA STATE HISTORIC PARK, 11255 Jackson St., Columbia, CA 95310-9425. Tel.: 209-536-2888. Fax: 209-532-5064.

E-mail: calhq@parks.ca.gov

Web Site: www.parks.ca.gov

Founded: 1945.

Congressional District: 18

Key Personnel: Sector Supt., Greg Martin.

Personnel Profile: Full-Time Paid 4; Part-Time Paid 6; Part-Time Volunteers 75.

Governing Authority: state. Parent Institution: California Department of Parks & Recreation, P.O. Box 2390, Sacramento, CA 95811. Subsidiary Institution: Chaw Se (Indian Grinding Rocks) State Park; Calaveras Big Trees State Park.

Institution Type/Description: State Park Museum.

Collections: Columbia State Historic Park: 1854 Leavitt-Walker; 1856 St. Charles Saloon-Alberding; 1858 Wells Fargo building; 1856 Duchow; 1860 schoolhouse; 1854-1855 I.O.O.F. Hall-McChesney; grocery store; candy store; general store; saloon; photographs; archives.

Research Fields: 19th century-gardens; folklore; preservation projects; California Gold Rush; Colombia related genealogy.

Facilities: 19th century-garden; picnic area. Museum-related items for sale.

Activities: walking tours; narrated talks for schools & special groups; formally organized education programs for children; permanent exhibitions; stagecoach ride; special events; living history programs.

Publications: brochure.

Hours & Admission Prices: Daily 8-5. No charge. Closed Thanksgiving; Christmas. &

Attendance: 467,000 (estimated)

Corona

FENDER MUSEUM OF MUSIC AND THE ARTS, 365 N. Main St., Corona, CA 92880-2040. Tel.: 951-735-2440. Fax: 951-735-2576.

E-mail: member@fendermuseum.com

Web Site: www.fendercenter.org

Founded: 1997.

Key Personnel: Bd. Pres. (V), Jeff Bennett; Exec Dir., Debbie Shuck.

Personnel Profile: Full-Time Paid 4; Part-Time Paid 2; Part-Time Volunteers 100; Interns 1.
Governing Authority: private; nonprofit organization. Tax-exempt.
Institution Type/Description: Music & Art Museum.
Collections: guitars; amps; factory equipment; music memorabilia.
Facilities: 48-track digital recording studio; outdoor amphitheater; classrooms.
Activities: formal education programs for children.
Publications: newsletter, Plugged In.
Hours & Admission Prices: Wed.-Sat. 11-4. Adults $10, seniors & students $8; children 12 under no charge. Closed major holidays. &
Membership: Family: Gold $35; Double Gold $50; Triple Gold $100; Gold Artist $250; Platinum $500; Double Platinum $750; Triple Platinum $1,000; Platinum Artist $2,500. Corporate: Manager $2,500; Promoter $5,000; Producer $10,000.

Corona del Mar

SHERMAN GARDENS, 2647 E. Coast Hwy., Corona del Mar, CA 92625-2103. Tel.: 949-673-2261. Fax: 949-675-5458.
E-mail: info@slgardens.org
Web Site: www.slgardens.org
Founded: 1951.
Congressional District: 40
Key Personnel: C.E.O., Donald Haskell; Pres. (V), Micky Pearlman; Library Dir., Dr. William O. Hendricks; Garden Dir., Wade Roberts; Business Officer, D.T. Daniels; Sales Shop Mgr., Peggy Schmidt.
Personnel Profile: Full-Time Paid 9; Part-Time Paid 9; Part-Time Volunteers 150.
Governing Authority: public charity. Parent Institution: Sherman Foundation. Tax-exempt: 509(a)(1) & 170(b)(A)vi.
Institution Type/Description: Historic Research Institute & Botanical Garden.
Collections: history of the Pacific Southwest; books; maps; microfilmed materials; personal papers; photographs; cacti & succulents to tropical plants.
Research Fields: history of the Pacific Southwest.
Facilities: 6,000-vol. library of books, pamphlets & other printed materials, numerous maps & photographs, microfilm, papers & documents available for research on premises; botanical garden; reading room; 150-seat auditorium; Cafe Jardin, Lunch & Afternoon Tea. Gift items for sale.
Activities: guided tours; lectures, concerts, hobby workshops, formally organized education programs for adults; docent program or council; temporary exhibitions.
Hours & Admission Prices: Gardens: daily 10:30-4. Library: Tues.-Thurs. 9-4:30. Tours: adults $3, children $1; Mon., children 11 & under and members no charge. Closed New Year's Day; Thanksgiving; Christmas. &
Attendance: 64,500 (accurate)
Membership: Student & Senior Citizen $25; General $35; Sustaining $100; Contributing $250; Sponsor $500; Benefactor $1,000; Patron $2,000; Life Patron $5,000.

Costa Mesa

COSTA MESA HISTORICAL SOCIETY & MUSEUM, 1870 Anaheim Ave., Costa Mesa, CA 92628. Mailing Address: P.O. Box 1764, Costa Mesa, CA 92628-1764. Tel.: 949-631-5918.
E-mail: cmhistory@sbcglobal.net
Web Site: costamesahistory.org
Formerly: Diego Sepulveda Adobe Estancia
Founded: 1966.
Congressional District: 2
Key Personnel: Pres. (V), B. Palazzola; Vice Pres., Terry Shaw; Treas., Susan Weeks; Sec., Gladys Refakes.
Personnel Profile: Full-Time Volunteers 18; Part-Time Volunteers 12.
Governing Authority: Alternate Facility: Diego Sepulveda Adobe in Estancia Park, 1900 Adams Ave Costa Mesa, CA 92626. Tax-exempt.
Institution Type/Description: History Museum.
Collections: area history; Native American; mission; Spanish; Victorian.
Research Fields: local history on the Costa Mesa/Harbor area & the Santa Army Air Base.
Activities: docent tours 1st & 3rd Saturday of each month. Annual Event: Early California Days at Diego Sepulveda Adobe.
Publications: monthly newsletter, Fairview Register.
Hours & Admission Prices: Adobe Tours: 1st & 3rd Sat. 12-4; other times by appointment. CMHS Headquarters & Museum: Thurs.-Fri. 10-3; other times by appointment. No charge; donations accepted.
Attendance: 350 (estimated)
Membership: Single $15; Family $20; Life $1,000.

Crescent City

DEL NORTE COUNTY HISTORICAL SOCIETY, (M), 577 H St., Crescent City, CA 95531-3743. Tel.: 707-464-3922. Fax: 707-464-7186.
E-mail: manager@delnortehistory.org
Web Site: www.delnortehistory.org
Founded: 1951.
Congressional District: 1
Key Personnel: Chm. (V) & Vice Pres., Sean Smith; Pres. (V), Loren Bommelyn.
Personnel Profile: Part-Time Paid 2; Part-Time Volunteers 50.
Governing Authority: nonprofit. Branch Museums: Main Museum, 6th & H Sts., Crescent City, CA. 95531; Battery Point Light House, Battery Point Island, Crescent City, CA. Tax-exempt.
Institution Type/Description: Historical Society & Maritime Museum: housed in 1926 County Hall of Records, Jail & Sheriff's office; Battery Point Lighthouse c.1856.
Collections: archaeology; history; lumber industry; marine; mining; naval; shipping; agriculture; genealogy; preservation projects.
Research Fields: local history.
Facilities: library of local history books, photographs, diaries, manuscripts, microfilms & originals of local newspapers, oral history tapes & research files available for use on premises under supervision; reading room.
Activities: permanent & special exhibits.
Publications: quarterly, Del Norte County Historical Society Bulletin.
Hours & Admission Prices: Main Museum: April-Sept. Mon.-Sat. 10-4; Oct.-March Mon. & Sat. 10-4. Adults $3, children 5-17 $1. Lighthouse: April-Sept. daily 10-4 (tide permitting); Oct.-March Sat.-Sun. 10-4 (tide permitting). Adults $3, children 17 & under $1; members & children 4 and under no charge. Research: $5 per hour (by volunteer); $2 per hour (own research).
Attendance: 7,119 (accurate)
Membership: Senior Citizen $10; Senior Family $15; Adult $20; Family $30; Lifetime $150 Corporate $1,000.

REDWOOD NATIONAL AND STATE PARKS, 1111 Second St., Crescent City, CA 95531-4123. Tel.: 707-464-6101. Fax: 707-464-1812.
E-mail: redw_superintendent@nps.gov
Web Site: www.nps.gov/redw
Formerly: Redwood National Park
Founded: 1968.
Congressional District: 1
Key Personnel: Supt., Steve Chaney; Cur., James B. O'Barr.
Personnel Profile: Full-Time Paid 104; Part-Time Paid 63; Part-Time Volunteers 10.
Governing Authority: federal. Dept. of the Interior, National Park Service, Washington, DC. Tax-exempt.
Institution Type/Description: National Park.
Collections: items related to natural history, cultural history & history of northcoast California.
Research Fields: coastal redwood ecosystem; Tolowa, Chilula & Yurok history & prehistory; logged lands & mining history.
Facilities: 1,500-vol. library pertaining to coastal redwood ecosystem, northern California coast, history & prehistory of American population & dairy & mining history available for research on premises only; nature trails.
Activities: guided tours; formally organized educational programs; permanent & changing exhibitions.
Publications: trail guides; books, Monarchs of the Mist: the Story of Redwood National Park; Redwood: A Guide to Redwood National & State Parks.
Hours & Admission Prices: Daily 8-5. No charge; donations accepted. Closed New Year's Day; Thanksgiving; Christmas. &
Attendance: 500,000 (estimated)

Culver City

MAYME A. CLAYTON LIBRARY & MUSEUM, 4130 Overland Ave., Culver City, CA 90230-3734. Tel.: 310-202-1647. Fax: 310-202-5464.
E-mail: info@claytonmuseum.org
Web Site: www.claytonmuseum.org
Institution Type/Description: History Museum & Library.
Collections: African American history; books; manuscripts; documents; films; photographs; music; memorabilia.
Facilities: library.
Activities: facility rental.
Publications: newsletter.
Hours & Admission Prices: Tues.-Sat. 10:30-4.

MUSEUM OF JURASSIC TECHNOLOGY, 9341 Venice Blvd., Culver City, CA 90232-2621. Tel.: 310-836-6131.
E-mail: info@mjt.org
Web Site: www.mjt.org
Founded: 1989.
Key Personnel: Bd. Pres., Terry Cannon; Museum Dir., David Wilson.
Personnel Profile: Part-Time Paid 12; Part-Time Volunteers 2; Interns 1.
Institution Type/Description: Natural History Museum.
Collections: relics & artifacts from Lower Jurassic.
Hours & Admission Prices: Thurs. 2-8, Fri.-Sun. 12-6. Suggested Admission: adults $8, students & seniors $5; discounts to members and children 12 & under no charge. Closed New Year's Day; Thanksgiving; Christmas.
Attendance: 25,000 (estimated)
Membership: Student & Senior $35; Active $50; Supporter $100; Contributing $200; Sustaining $300; Donor $500; Benefactor $1,000; Founder's Circle $5,000.

STAR ECO STATION, 10101 Jefferson Blvd., Culver City, CA 90232-3519. Tel.: 310-842-8060. Fax: 310-842-8245.
E-mail: ecostation@starinc.org
Web Site: www.joineco.org
Founded: 1997.
Congressional District: 33
Key Personnel: C.E.O. & Chm. (V), Katya Bozzi; Pres. (V), Regino Chavez; Treas., Steve Nemath; Cur., Erick Bozzi, II; Education, Katiana Bozzi; Museum Shop Mgr., Martin Covarrubias.
Personnel Profile: Full-Time Paid 15; Part-Time Paid 13; Part-Time Volunteers 45.
Governing Authority: private; nonprofit organization. Parent Institution: Star Education, Inc., 10117 Jefferson Blvd., Culver City, CA 90232. Tax-exempt 501 (c)(3).
Institution Type/Description: Environmental Science Museum.
Collections: indoor & outdoor living wildlife exhibits including mammals, reptiles, birds, amphibians & marine life.
Research Fields: endangered species management & behavior of birds, reptiles & mammals.
Facilities: zoological center. Animal & science-related items for sale in gift shop.
Activities: arts festival; concerts; dance recitals; docent programs; films; formal education programs; guided tours; lectures; mobile vans; temporary exhibitions. Annual Events: Dinofaire; African American Art Festival; Children's Earth Day; Children's Art Festival; Creepy Crawly Creature Feature; Enchanted Green Hallow's Eve; World Ocean Day.
Publications: quarterly, STAR Eco Station Newsletter.
Hours & Admission Prices: Fri. 1-5, Sat.-Sun. 10-4. Adults $8, senior citizens 65 and over $7, children $6; members no charge. Closed New Year's Eve & Day; Easter; Independence Day; Thanksgiving; Christmas Eve & Day. &
Attendance: 100,000 (estimated)

THE WENDE MUSEUM, (M), 5741 Buckingham Pkwy., Ste. E, Culver City, CA 90230-6520. Tel.: 310-216-1600. Fax: 310-216-1609. Facebook: Wende Museum.
E-mail: info@wendemuseum.org
Web Site: www.wendemuseum.org
Founded: 2002.
Congressional District: 33
Key Personnel: Dir., Justinian Jampol.
Personnel Profile: Full-Time Paid 7; Part-Time Paid 2; Part-Time Volunteers 22.
Governing Authority: nonprofit organization. Tax-exempt.
Institution Type/Description: History & Cultural Museum.
Collections: cultural artifacts & personal histories of Cold War-era Eastern Europe and the Soviet Union.
Research Fields: Cold War Eastern Europe & Soviet Union.
Activities: temporary & permanent exhibitions; special events; conferences; film festivals; publications; support research; commission new art work.
Publications: conference publications; book with Taschen Books (2013).
Hours & Admission Prices: Museum: Fri. 10-5; Mon.-Thurs. by appointment. Vault Tour: 3pm. No charge. Closed national holidays.
Attendance: 750 (estimated)

Cupertino

CALIFORNIA HISTORY CENTER & FOUNDATION, 21250 Stevens Creek Blvd., Cupertino, CA 95014-5702. Tel.: 408-864-8987. Fax: 408-864-5486.
E-mail: info@calhistory.org
Web Site: www.calhistory.org
Founded: 1969.

Congressional District: 15
Key Personnel: Dir., Tom Izu.
Personnel Profile: Full-Time Paid 1.
Governing Authority: Partnership with De Anza College. Tax-exempt.
Institution Type/Description: History Museum.
Collections: California history; photographs; documents; manuscripts.
Research Fields: agricultural history of Santa Clara Valley; transformation to Silicon Valley.
Facilities: library.
Activities: lectures; workshops.
Publications: magazine, The Californian.
Hours & Admission Prices: Sept.-June Tues.-Thurs. 9:30-12 & 1-4, Fri. by appointment. Center: no charge. Library: $5 daily; students & members no charge.
Attendance: 1,800 (estimated)
Membership: Individual $30; Family $40; Supporter $50; Sponsor $100; Patron $500; Colleague $1,000.

CUPERTINO HISTORICAL SOCIETY & MUSEUM, Quinlan Community Center, 10185 N. Stelling Rd., Cupertino, CA 95014-5732. Tel.: 408-973-1495.
E-mail: cuphistsociety@sbcglobal.net
Web Site: www.cupertinohistoricalsociety.org
Founded: 1966.
Congressional District: 13
Key Personnel: Pres., Donna Austin; Volunteer Coord., Ray Bortner; Office Mgr. & Graphic Artist, Ragini Sangameswara.
Personnel Profile: Part-Time Paid 1; Part-Time Volunteers 8.
Governing Authority: nonprofit organization; society. Parent Institution: Cupertino Historical Society. Tax-exempt: 501(c)(3).
Institution Type/Description: Historical Society Museum.
Collections: historical artifacts; photographs; 1840-1985, archives; clothing; furniture; toys; farm & blacksmith tools; school objects.
Research Fields: local history; agricultural life; preservation/buildings.
Facilities: 1,000 sq. ft. exhibit space.
Activities: guided tours; lectures; organized education programs for children; docent program; temporary exhibitions. Annual Events: fundraising BBQ; visit to an historic site.
Publications: The Cornerstone; quarterly newsletter; exhibition catalogs; The Cupertino Chronicle; Images of America: Early Cupertino.
Hours & Admission Prices: Wed.-Sat. 10-4. No charge. &
Attendance: 2,704 (accurate)
Membership: Individual $35; Family $50; Patron $125; Sponsor $250; Director's Circle $500; Benefactor's Circle $1,000 & up.

EUPHRAT MUSEUM OF ART, De Anza College, 21250 Stevens Creek Blvd., Cupertino, CA 95014-5702. Tel.: 408-864-8836. Fax: 408-864-8738.
E-mail: rindfleischjanet@fhda.edu
Web Site: www.deanza.edu/euphrat
Founded: 1971.
Congressional District: 14
Key Personnel: Exec. Dir., Marie Fox Ellison; Dir., Jan Rindfleisch; Co Pres., Margaret Kung; Co Pres., Helen Lewis; Education, Diana Argabrite.
Personnel Profile: Full-Time Paid 1; Part-Time Paid 8; Part-Time Volunteers 10.
Governing Authority: public college/community partnership; nonprofit. Tax-exempt: 501(c)(3).
Institution Type/Description: Art Museum.
Collections: contemporary paintings, prints & photographs; sculpture; installations.
Research Fields: art by recent immigrants to the United States.
Facilities: 1,470 sq. ft. exhibit space.
Activities: formal education programs for children & undergraduate/graduate students affiliated with De Anza College; guided tours; lectures; participatory, traveling & community exhibitions. Annual Event: Family Day (through Arts & Schools Program).
Publications: select exhibition books.
Hours & Admission Prices: late Sept. to mid-June Mon.-Thurs. 10-4; groups by appointment. No charge; donations accepted. &
Attendance: 10,000 (estimated)
Membership: Friends of the Euphrat (no specific levels).

FUJITSU PLANETARIUM AT DE ANZA COLLEGE, De Anza College, 21250 Stevens Creek Blvd., Cupertino, CA 95014-5797. Tel.: 408-864-8814 & 8282.
Web Site: www.planetarium.deanza.edu
Formerly: Minolta Planetarium

Founded: 1970.
Governing Authority: Tax-exempt.
Institution Type/Description: Planetarium.
Collections: space science.
Hours & Admission Prices: Field Trips: Mon.-Thurs. mornings, Fri. mornings & afternoons. Call for Family Evenings & Laser Light Show hours & admission prices. &

Dana Point

OCEAN INSTITUTE, 24200 Dana Point Harbor Dr., Dana Point, CA 92629-2723. Tel.: 949-496-2274. Fax: 949-496-4296.
E-mail: oi@ocean-institute.org
Web Site: www.ocean-institute.org
Formerly: Orange County Marine Institute
Key Personnel: Pres., Dan Stetson
Institution Type/Description: Marine Science & Maritime History Center.
Collections: hands-on marine science & environmental exhibits; period tall ships.
Activities: marine science & maritime history education programs; cruises & sails.
Hours & Admission Prices: Mon.-Fri. 8-5 (summer camps & school programs only), Sat.-Sun. 10-3 (public entry). Adults 13 & over $6.50, youth 3-12 $4.50; members and children 2 & under no charge. Closed major holidays. &
Attendance: 165,000 (accurate)
Membership: Individual $35; Family $60.

Danville

BLACKHAWK MUSEUM, 3700 Blackhawk Plaza Cir., Danville, CA 94506-4652. Tel.: 925-736-2280 & 2277. Fax: 925-736-4818. Facebook: Blackhawk Museum.
E-mail: museum@blackhawkmuseum.org
Web Site: www.blackhawkmuseum.org
Formerly: Blackhawk Automotive Museum
Founded: 1988.
Key Personnel: C.E.O. & Pres., Don Williams; Exec. Dir., Timothy P. McGrane; Education, Nora Wagner; Special Events Coord., Jon Snyder; Museum Shop Mgr., Mike Zinser.
Personnel Profile: Full-Time Paid 10; Part-Time Paid 6; Part-Time Volunteers 5.
Governing Authority: private; nonprofit organization. Parent Institution: Behring-Hofmann Educational Institute Inc. Affiliate Program Member of the Smithsonian Institution. Tax-exempt: 501(c)(3).
Institution Type/Description: Automobile & History Museum.
Collections: automotive art.
Facilities: library; classrooms; 75,000 sq. ft. exhibit space; 600-seat theater. Museum-related items for sale.
Activities: docent program; films; formal educational programs; guided tours; hobby workshops; lectures; loan, participatory, temporary & traveling exhibitions; school loan service; study clubs; training programs for professional museum workers; outreach programs for area schools.
Hours & Admission Prices: Wed.-Sun. 10-5. Adults $15, senior citizens & students $7; discounts to groups; children under 6 no charge. Closed New Year's Day; Thanksgiving; Christmas. &
Attendance: 75,000 (accurate)
Membership: Individual & Senior Couple $50; Family $65; Friend $100; Patron $150; Contributor $250; Supporter $500; Patron $1,000; Benefactor $2,500; Visionary $5,000.

EUGENE O'NEILL NATIONAL HISTORIC SITE, Danville, CA 94526-0280. Mailing Address: P.O. Box 280, Danville, CA 94526-0280. Tel.: 925-838-0249 & 943-1531. Fax: 925-838-9471.
E-mail: euon_interpretation@nps.gov
Web Site: www.nps.gov/euon
Founded: 1976.
Congressional District: 8
Key Personnel: Chm. (V), Randy Harabin; Dir., Morgan Smith; Museum Shop Mgr., Nancy Pierce.
Personnel Profile: Full-Time Paid 5; Part-Time Volunteers 11.
Governing Authority: federal. Administered by National Park Service, U.S. Department of the Interior. Partnering Organization: Eugene O'Neill Foundation. Tax-exempt.
Institution Type/Description: National Historic Site: Tao House was the home of playwright Eugene O'Neill from 1937 to 1944.
Collections: clothing; jewelry; autographed books, paintings; letters; period photographs; furnishings.
Facilities: garden; orchards; visitors center. Book store.
Activities: guided tours; occasional special events & performance programs.

Publications: booklet, Eugene O'Neill at Tao House; Eugene O'Neill - official park guide.
Hours & Admission Prices: Wed.-Fri. & Sun. 10 & 2 by reservation only, Sat. 10, 12 & 2. No charge; donations accepted. &
Attendance: 3,475 (accurate)

Davis

DAVIS ART CENTER, 1919 F St., Davis, CA 95616-1163. Mailing Address: P.O. Box 4340, Davis, CA 95617-4340. Tel.: 530-756-4100. Fax: 530-756-3041.
E-mail: davisart@dcn.org
Web Site: davisartcenter.org
Founded: 1959.
Key Personnel: Dir., Erie Vitiello.
Personnel Profile: Full-Time Paid 3; Part-Time Paid 2; Part-Time Volunteers 3; Interns 1.
Governing Authority: nonprofit organization. Tax-exempt: 501(c)(3).
Institution Type/Description: Art Center.
Collections: contemporary art.
Facilities: library; educational facilities; 10,500 sq. ft. exhibit space.
Activities: lectures; concerts; dance recitals; arts festivals; organized educational programs. Annual Events: art & craft sale.
Hours & Admission Prices: Mon.-Thurs. 9:30-7, Fri. 9:30-5, Sat. 10-4. No charge. Closed New Year's Eve & Day; Memorial Day; Independence Day; Labor Day; Christmas Day & week. &
Attendance: 40,000 (estimated)
Membership: Senior & Student $20; Individual $30; Family $50; Patron $250.

EXPLORIT SCIENCE CENTER, 3141 5th St., Davis, CA 95618-6534. Tel.: 530-756-0191. Fax: 530-756-1227.
E-mail: explorit@explorit.org
Web Site: www.explorit.org
Founded: 1982.
Congressional District: 3
Key Personnel: Exec. Dir., Lou Ziskind; Pres. (V), Dr. Rick Baker; Exhibit Coord., Anna Grace; Program Dir, Megan Chiosso; Museum Shop Mgr., Anne Hance.
Personnel Profile: Full-Time Paid 10; Part-Time Paid 11; Part-Time Volunteers 200; Interns 12.
Governing Authority: not-for-profit organization. Tax-exempt: 501(c)(3).
Institution Type/Description: Hands-On Science Center.
Collections: sand; seashells; rocks; fossil paintings.
Facilities: Museum-related items for sale.
Activities: education programs for children; lectures; mobile vans; participatory & traveling exhibits; outreach education and exhibit programs for children & adults; changing topics.
Publications: quarterly newsletter, Science Centered.
Hours & Admission Prices: Tues.-Fri. 2-4:30, Sat.-Sun. 11-4:30. Admission $4; discounts to ASTC travel passport program members/Reciprocal admission program; members, teachers, children under 4 & fourth Fri. of each month no charge. Closed New Year's Day; Memorial Day; Independence Day; Labor Day; Thanksgiving; Christmas. &
Attendance: 80,000 (accurate)
Membership: Student & Senior $20; Individual $25; Family & Grandparents $50; Darwin Circle $75; Curie Circle $125; Galileo Circle $250; Einstein Circle $500.

HATTIE WEBER MUSEUM OF DAVIS, 445 C St., Davis, CA 95616-4102. Tel.: 530-758-5637 & 753-5959.
Web Site: www.dcn.davis.ca.us/go/hattie
Founded: 1992.
Key Personnel: Dir., Dennis Dingemans.
Personnel Profile: Part-Time Volunteers 8.
Governing Authority: Parent Institution: Yolo County Historical Society. Subsidiary Institution: Historical Resources Management Commission. Tax-exempt.
Institution Type/Description: History Museum.
Collections: area history & heritage; personal artifacts; 1911 library; 1938 WPA restroom.
Hours & Admission Prices: Wed. & Sat. 10-4; other times by appointment. No charge; donations accepted. &
Attendance: 3,400 (accurate)

JOHN NATSOULAS CENTER FOR THE ARTS, 521 1st St., Davis, CA 95616. Tel.: 530-756-3938. Fax: 530-756-3961.
E-mail: art@natsoulas.com
Web Site: www.natsoulas.com
Institution Type/Description: Art Gallery.

Collections: paintings.
Activities: special events.
Hours & Admission Prices: Wed.-Thurs. 11-5, Fri. 11-10, Sat.-Sun. 12-5.

PENCE GALLERY, 212 D St., Davis, CA 95616-4513. Tel.: 530-758-3370. Fax: 530-758-4670.
E-mail: penceassistant@sbcglobal.net
Web Site: www.pencegallery.org
Founded: 1975.
Congressional District: 4
Key Personnel: Dir., Natalie Nelson; Asst. Dir., Eileen Hendren; Pres., Sue Smith; Mktg. Asst., Stacey Hilton; Preparator, Winter Jenssen; Museum Shop Mgr., Cindy Ruff.
Personnel Profile: Part-Time Paid 4; Part-Time Volunteers 30; Interns 9.
Governing Authority: private; nonprofit institution. Land owned by City of Davis, CA. Tax-exempt: 501(c)(3).
Institution Type/Description: Art Association Gallery.
Collections: changing exhibitions of contemporary art.
Major Exhibits: Leslie duPratt, 1/8/14-2/16/14; The Consilience of Art & Science, 1/11/14-2/28/14; Bloom, 3/1/14-4/20/14; Ceramics: Confrontational Ceramisc, 4/30/14-6/15/14; Slice: A Juried Cross-Section of Regional Art, 6/29/14-8/24/14; Community Hang-Up, 7/11/14-8/24/14; Art Auction 2014, 9/2/14-9/20/14; Mrk Emerson: Abstractions, 9/30/14-11/9/14; Open Studio Project: Tony Natsoulas, 10/3/14-10/31/14; Holiday Market, 11/14/14-12/23/14.
Facilities: 2,200 sq. ft. exhibit space.
Activities: guided tours; gallery talks; lectures; concerts; exhibitions of works of art; shows of ethnic & folk art, children's art & competitive shows; contemporary painting, sculpture, graphics & photography; art auctions, art related tours & bus excursions; Art Smart Education Program.
Publications: quarterly publication, Pence Events; art catalogues.
Hours & Admission Prices: Tues.-Sun. 11:30-5, 2nd Fri. 6-9. No charge; donations accepted. Closed major holidays. &
Attendance: 17,500 (estimated)
Membership: Individual $45; Household $75; Sponsor $120; Patron $250; Benefactor $500; Curator's Circle $1,000; Gallery Circle $2,500.

QUAIL RIDGE WILDERNESS CONSERVANCY, 25344 County Rd. 95, Davis, CA 95616. Tel.: 530-758-1387. Fax: 530-758-1316.
E-mail: quailrid@quailridge.org
Web Site: www.quailridge.org
Key Personnel: Exec. Dir., Frank W. Maurer
Institution Type/Description: Conservancy.
Collections: native plants & animals; oak trees; native California bunchgrasses.
Activities: educational workshops; interpretive walks.
Hours & Admission Prices: Call for hours.

R.M. BOHART MUSEUM OF ENTOMOLOGY, Dept. of Entomology, University of California, One Shields Ave., Davis, CA 95616-8584. Tel.: 530-752-0493. Fax: 530-752-9464.
E-mail: bohart@ucdavis.edu
Web Site: bohart.ucdavis.edu
Founded: 1946.
Congressional District: 3
Key Personnel: Dir., Lynn S. Kimsey; Cur., Phil Ward; Collection Mgr., Steve Heydon; Museum Shop Mgr., M.F. Keller.
Personnel Profile: Full-Time Paid 5; Part-Time Paid 3; Part-Time Volunteers 2.
Governing Authority: University of California. Tax-exempt.
Institution Type/Description: Entomology Museum.
Collections: seven million arthropod specimens.
Research Fields: systematics & insect ecology & behavior.
Facilities: library of entomology texts & periodicals.
Activities: guided tours; formally organized education programs for grades K-12, undergraduate college & graduate students; temporary exhibitions.
Publications: newsletter, Bohart Museum Society.
Hours & Admission Prices: Mon.-Thurs. 8-5. No charge; donations accepted. Closed holidays. &
Attendance: 4,000 (accurate)
Membership: Student $15; Regular $25; Donor $40; Organization & Patron $100.

RICHARD L. NELSON GALLERY & THE FINE ARTS COLLECTION, UC DAVIS, Nelson Hall, Old Davis Rd., Davis, CA 95616-5270. Mailing Address: One Shields Ave., Davis, CA 95616. Tel.: 530-752-8500. Fax: 530-754-9112. Facebook: Nelson Gallery.
E-mail: nelsongallery@ucdavis.edu
Web Site: nelsongallery.ucdavis.edu
Founded: 1976.

Congressional District: 4
Key Personnel: Dir., Rachel Teague; Preparator, Kyle Monhollen; Registrar & Collections Mgr., Robin Bernhard; Asst. to the Dir., Katrina Wong.
Personnel Profile: Full-Time Paid 3; Part-Time Paid 1; Part-Time Volunteers 15; Interns 5.
Governing Authority: public university; nonprofit. Parent Institution: Univ. of California, Davis. Tax-exempt: 501(c)(3).
Institution Type/Description: Art Gallery & Museum.
Collections: 6,000 artifacts including 16th century-present paper, prints, drawings & paintings; contemporary art, emphasizing northern California artists; Chinese & southeast Asian art; European and American art; ceramics.
Major Exhibits: Another California, 10/10/13-5/3/14.
Research Fields: contemporary art, with emphasis on northern California.
Facilities: 4,200 sq. ft. exhibit space. Exhibition catalogs for sale.
Activities: guided tours; lectures; loan, traveling & temporary exhibitions; organized education programs for undergraduate & graduate students affiliated with UC Davis Art Dept.
Publications: exhibition catalogs.
Hours & Admission Prices: Sat.-Thurs. 11-5; other times by appointment. No charge; donations accepted. Closed Martin Luther King Jr. Day; Presidents' Day; Independence Day; Thanksgiving; Christmas. &
Attendance: 7,000 (accurate)
Membership: Nelson Art Friends: Student & Senior (65 & over) $35; Individual $50; Dual & Family $70; Supporter $100; Patron $250; Business & Corporate Associate $250-$1,999; Benefactor $500; Director's Circle $1,000.

UC DAVIS DESIGN MUSEUM, Cruess Hall, One Shields Ave., Davis, CA 95616-5200. Tel.: 530-752-6150. Fax: 530-752-1392.
E-mail: designmuseum@ucdavis.edu
Web Site: www.designmuseum.ucdavis.edu
Formerly: University of California Design Museum
Founded: 1975.
Key Personnel: Dir., Timothy McNeil.
Personnel Profile: Full-Time Paid 1; Part-Time Paid 2; Interns 6.
Governing Authority: Parent Institution: University of California, Davis. Tax-exempt.
Institution Type/Description: Design Museum.
Collections: national & international design-related materials; textiles; fashion; basketry; porcelain; architecture; furniture.
Research Fields: sustainable exhibition design.
Hours & Admission Prices: Mon.-Fri. 12-4, Sun. 2-4. No charge; donations accepted. Closed holidays & holiday weekends; university breaks. &
Attendance: 2,000 (estimated)

UC DAVIS ARBORETUM, Valley Oak Cottage (TB-32), La Rue Rd., UC Davis, Davis, CA 95616. Mailing Address: One Shields Ave., Davis, CA 95616-5200. Tel.: 530-752-4880. Fax: 530-752-5796.
E-mail: arboretum@ucdavis.edu
Web Site: arboretum.ucdavis.edu
Founded: 1936.
Congressional District: 4
Key Personnel: Dir., Kathleen Socolofsky; Dir. Planning & Collections, Mary T. Burke; Cur., Mia Ingolia; Supt. Emeritus, Warren G. Roberts; Dir. Horticulture, Ellen Zagory; Asst. Dir., Carmia Feldman; Asst. Dir. Horticulture, Emily Griswold; GIS Project Mgr., Brian Morgan; Academic Coord., Elaine Fingerett; Administrative & Gifts Mgr., Judy Hayes.
Governing Authority: state. Parent Institution: University of California. Tax-exempt.
Institution Type/Description: Arboretum.
Collections: living native & exotic trees & plants: oaks; conifers; eucalyptus; acacias; redwoods; California native plants; American desert plants; drought-tolerant flowering perennial garden.
Research Fields: taxonomy of flowering plants; horticulture of native California plants; California landscape history.
Facilities: 1,000-vol. library of reference material; 100-acre botanical garden.
Activities: guided tours; lectures; elementary school field trips; university courses; field trips; workshops & courses; permanent exhibitions.
Publications: pamphlets; quarterly bulletin.
Hours & Admission Prices: Gardens: daily 24 hours. Office: Mon.-Fri. 8-5. No charge. &
Attendance: 250,000 (estimated)
Membership: Student $15; Individual $40; Family $60; Manzanita Circle $100; Valley Oak Circle $250; Sequoia Circle $500.

U.S. BICYCLING HALL OF FAME, 303 3rd St., Davis, CA 95616. Mailing Address: P.O. Box 73385, Davis, CA 95617. Tel.: 530-341-3263.
E-mail: info@usbhof.org
Web Site: usbhof.org
Founded: 1986.
Key Personnel: Pres. Bd., Dawn Wylong.
Governing Authority: bd. of directors. Tax-exempt.
Institution Type/Description: Hall of Fame & Bicycling History Museum.
Collections: cycling greats; photographs; trophies; medals; newspapers; history of cycling; Frank Kramer, 1920s; Greg Lemons, 1980s; Marshall Major Taylor, 1899s.
Activities: racing events. Annual Events: Hall of Fame Dinner; USBHOF Induction Ceremony in November.
Hours & Admission Prices: Call for information.

Death Valley

DEATH VALLEY NATIONAL PARK VISITOR CENTER AND MUSEUMS, Death Valley National Park, Hwy. 190, Death Valley, CA 92328. Mailing Address: P.O. Box 579, Death Valley, CA 92328-0579. Tel.: 760-786-3200. Fax: 760-786-3283. TDD: 760-786-2471.
E-mail: blair_davenport@nps.gov
Web Site: www.nps.gov/deva
Formerly: Death Valley Visitor Center and Museums
Founded: 1933.
Congressional District: 35
Key Personnel: Supt., Sarah Craighead.
Governing Authority: federal. Parent Institution: National Park Service. Subsidiary Institution: Death Valley National Park. Tax-exempt.
Institution Type/Description: Park Visitor Center, Museum & Historic House.
Collections: archaeology; history; anthropology; geology; herbarium; natural history. Historic Buildings: 1876 charcoal kilns; 1883 Harmony borax works; 1924-1931 Scotty's Castle.
Research Fields: geology; archaeology; botany; biology; ecology; history.
Facilities: 5,000-vol. library available for research on premises; archives. Books for sale.
Activities: guided tours; lectures; permanent exhibitions.
Publications: books; pamphlet; guides; postcards. Published by Death Valley Natural History Association, a cooperating organization.
Hours & Admission Prices: Furnace Creek Visitor Center: daily 8-5; Scotty's Castle Visitor Center: daily 8-5. Tour bus $25-$200, individual vehicles $20, walk-ins & bike-ins $10. Scotty's Castle: Adults $11, seniors (62 & over) $9, children (6-15) & adults with disabilities $6; discounts to Golden Age, Golden Eagle, & Golden Access card holders; children under 6 no charge. &
Attendance: 1,000,000 (estimated)
Membership: Death Valley Natural History Association: Individual $20; Sustaining $50; Sponsor $100; Lifetime $1,000.

Delano

DELANO HERITAGE PARK, 330 Lexington, Delano, CA 93215-3602. Tel.: 661-725-6730. Fax: 661-725-2344.
E-mail: heritagepark1@aol.com
Founded: 1961.
Congressional District: 18
Key Personnel: Pres. (V), Peter Finocchiaro.
Governing Authority: nonprofit organization. Administered by Parks & Recreation Dept. of the city. Affiliated with Delano Historical Society, 330 Lexington St. Tax-exempt: 501(c)(3).
Institution Type/Description: Park Museum.
Collections: local history; agriculture; arboretum. Historic Houses: 1890 Heritage House, Garces & Lexington; 1916 Jasmine School; 1968 Replica Jasmine School Stable; 1876 Jail; 1912 Orcier Famosa Store; 1888 Weaver House; 1800 farm equipment.
Research Fields: local history; agriculture; arboretum.
Facilities: 100-vol. library of local & California history, historical pictures, slides & old books.
Activities: guided tours; lectures; films; permanent exhibitions.
Publications: book, Delano: A Land of Promise, 1965; quarterly bulletin, The Plow.
Hours & Admission Prices: Temporarily closed. Tours by appointment; guided tours for groups. No charge; donations accepted. &
Attendance: 1,750 (estimated)

Membership: Individual $25; Business $50.

Desert Hot Springs

CABOT'S PUEBLO MUSEUM, (M), 67-616 E. Desert View Ave., Desert Hot Springs, CA 92240. Mailing Address: P.O. Box 104, Desert Hot Springs, CA 92240-0104. Tel.: 760-329-7610. Fax: 760-329-2738.
Web Site: www.cabotsmuseum.org
Congressional District: 41
Key Personnel: Dir., Ginger Ridgway; Pres. (V), Mike Chedester; Museum Shop Mgr., Dean Krumme.
Personnel Profile: Full-Time Paid 2; Part-Time Paid 7; Part-Time Volunteers 14.
Governing Authority: Parent Institution: City of Desert Hot Springs, CA; Subsidiary Institution: Cabot's Pueblo Museum Foundation. Tax-exempt 501(c)(3).
Institution Type/Description: Historic House Museum: housed in the former pueblo-style home of Cabot Yerxa.
Collections: local and state cultural heritage & history; Cabot Yerxa's life & family history; Native American pottery; photographs; personal artifacts; early 1900s tools, machinery, & housewares; paintings; sculpture.
Facilities: Books for sale.
Activities: special events.
Hours & Admission Prices: Tues.-Sun. 9-4; Summer: Thurs.-Sun. 9-1. Guided Tours: adults $11, military, seniors & children 6-12 $9. Closed major holidays.
Attendance: 9,323 (accurate)
Membership: Individual $35; Household $50.

Dinuba

ALTA DISTRICT HISTORICAL SOCIETY - DEPOT MUSEUM, 289 S. "K" St., Dinuba, CA 93618. Mailing Address: P.O. Box 254, Dinuba, CA 93618. Tel.: 559-591-2144. Fax: 559-591-2144.
Founded: 1963.
Institution Type/Description: Historical Society Museum: housed in the former Dinuba Southern Pacific Depot; built in 1888.
Collections: local history & culture; period furnishings; personal artifacts, photographs; railroad artifacts; blacksmith shop.
Facilities: cultural center; gardens.
Activities: school tours. Museum Sponsors: Annual Auction; Farmer's Day; Christmas Open House.
Hours & Admission Prices: Tues.-Wed. by appointment. No charge; donations accepted.
Membership: Student $10; Individual $25; Family $35; Life $250.

Donner Pass

WESTERN SKISPORT MUSEUM, Interstate 80 Boreal Ridge Ski Area, Donner Pass, CA 95728. Mailing Address: P.O. Box 729, Soda Springs, CA 95728-0729. Tel.: 530-426-3313, ext. 113. Fax: 530-426-3501.
E-mail: bclark@inc.auburnskiclub.org
Web Site: www.auburnskiclub.org
Founded: 1969.
Key Personnel: Dir., Bill Clark; Sec., Laura Clark.
Personnel Profile: Part-Time Paid 1.
Governing Authority: nonprofit organization. Owned & operated by the Auburn Ski Club, Inc., P.O. Box 729, Soda Springs, CA 95728. Tax-exempt.
Institution Type/Description: Ski History Museum.
Collections: steel monument of Snow-Shoe Thompson; memorabilia from Western North America Skisport Hall of Fame; ski history; manuscripts.
Research Fields: pioneer times during snow bound winter months in western states, western Canada & Alaska.
Facilities: 500-vol. library of books, magazines & manuscripts relative to skisport, ski associations, international skisport & Olympic movement available for research by arrangement with custodians or director; reading room; 30-seat theatre.
Activities: guided tours; lectures; films; permanent exhibitions.
Publications: The Lost Sierra.
Hours & Admission Prices: Winter: Fri.-Sun. 10-4; call for extended hours. No charge; donations accepted. &
Attendance: 5,000 (estimated)

Downey

COLUMBIA MEMORIAL SPACE CENTER, (M), 12400 Columbia Way, Downey, CA 90242. Tel.: 562-231-1200. Fax: 562-231-1206.
E-mail: sdelong@downeyspacecenter.org
Web Site: www.columbiaspacescience.org
Founded: 2009.
Congressional District: 34
Key Personnel: Exec. Dir., Shannon DeLong; Center Supvr., Sandra Valencia; Museum Shop Mgr., Sarah Medina.
Personnel Profile: Full-Time Paid 3; Part-Time Paid 18; Part-Time Volunteers 33.
Governing Authority: city. Tax-exempt.
Institution Type/Description: Space Museum.
Collections: space exploration & aviation; human & robot space; flight & aeronautics; NASA technology; space shuttle, Columbia; local aerospace history; earth & solar system.
Facilities: theater.
Activities: space mission simulator; educational programs; videos; robotics laboratory; flight simulator; drop tower; paper airplane making station; docking & landing shuttle simulation.
Hours & Admission Prices: Tues.-Sat. 10-5. Adults $5; discounts to groups of 15 or more; members, ASTC members and children 3 & under no charge. &
Attendance: 30,918 (accurate)
Membership: Student $30; Seniors $40; Individual $50; Family $80.

DOWNEY HISTORICAL SOCIETY, 12540 Rives Ave., Downey, CA 90241. Mailing Address: P.O. Box 554, Downey, CA 90241. Tel.: 562-862-2777.
Institution Type/Description: Historical Society Museum.
Collections: local history & culture; period furnishings; personal artifacts; photographs.
Hours & Admission Prices: Wed.-Thurs. & 3rd Sat. of the month 10-2.

Downieville

DOWNIEVILLE MUSEUM, 330 Main St., Downieville, CA 95936. Mailing Address: P.O. Box 1, Downieville, CA 95936. Tel.: 530-289-3506.
E-mail: hangman@jps.net
Founded: 1932.
Congressional District: 14
Key Personnel: Dir. & Chm. (V), Earlene Folsom; Treas., Lee Adams; Volunteer, Liz Fisher; Paid Staff, Donald McIntosh; Volunteer, L. Adams, III; Volunteer, Jane Hallman; Volunteer, David Marshall.
Personnel Profile: Part-Time Paid 2; Part-Time Volunteers 5.
Governing Authority: nonprofit organization. Parent Institution: Native Daughters & Native Sons. Tax-exempt.
Institution Type/Description: General Museum: housed in 1852 store.
Collections: artifacts of Gold Rush era, 1849-1920s; replica of gold stamp mill; horse snowshoes; pictures; paper items; Indian & Chinese artifacts; personal & household items of early-day Downieville & surrounding Sierra County; local history.
Facilities: library of histories of Lassen, Plumas & Sierra Counties, La Porte Scrapbook & other volumes of local history, available on premises only.
Hours & Admission Prices: May to mid-Oct. Mon.-Fri. 11-4, Sat.-Sun. 10-5. Suggested Donation: adults $1.
Attendance: 6,500 (accurate)

Duarte

DUARTE HISTORICAL MUSEUM, 777 Encanto Pkwy., Duarte, CA 91010. Mailing Address: P.O. Box 263, Duarte, CA 91009-0263. Tel.: 626-357-9419.
Web Site: www.duartehistory.org
Founded: 1951.
Congressional District: 32
Key Personnel: Pres. (V), Claudia Heller.
Personnel Profile: Part-Time Volunteers 25.
Governing Authority: Parent Institution: Duarte Historical Society & Friends of the Library, Inc.
Institution Type/Description: History Museum.
Collections: local history & culture; period furnishings; personal artifacts; photographs; agricultural tools.
Research Fields: local history.

Publications: The Branding Iron.
Hours & Admission Prices: 1st & 3rd Wed. of the month 1-3, Sat. 1-4. No charge; donations accepted. Closed holidays.
Attendance: 1,200 (accurate)

THE JUSTICE PRIVATE AUTOMOTIVE COLLECTION, 2734 Huntington Dr., Duarte, CA 91010-2301. Tel.: 626-359-9174. Fax: 626-357-2550.
E-mail: museum@justicebrothers.com
Web Site: www.justiceprivatecollection.com
Formerly: Justice Brothers Racing Museum and Private Collection
Founded: 1985.
Key Personnel: C.E.O. & Pres., Ed Justice, Jr.; Museum Shop Mgr., Caitlin Justice.
Governing Authority: nonprofit organization.
Institution Type/Description: Racing Museum.
Collections: racing vehicles; motorcycles; passenger cars; racing memorabilia.
Facilities: Museum-related items for sale.
Hours & Admission Prices: Mon.-Fri. 9-5. No charge. Closed business holidays. &

Durham

PATRICK RANCH MUSEUM, 10381 Midway, Durham, CA 95938. Mailing Address: 270 Boeing Ave., Chico, CA 95973. Tel.: 530-892-1525. Fax: 530-892-1524. Facebook: Patrick Ranch Museum.
E-mail: office@farwestheritage.org
Web Site: www.farwestheritage.org & www.patrickranchmuseum.org
Founded: 2004.
Congressional District: 2
Key Personnel: Pres. (V), Roger Steel; Mgr., John Chambers; Treas., Linda Limberg; Office Mgr., JoAnne Vidal-Carr.
Personnel Profile: Full-Time Paid 1; Part-Time Paid 3; Part-Time Volunteers 40; Interns 2.
Governing Authority: Parent Institution: Far West Heritage Assoc. Tax-exempt.
Institution Type/Description: History Museum.
Collections: Sacramento Valley agricultural history including social, cultural and economic artifacts.
Hours & Admission Prices: Closed for restoration. No charge; donations accepted.
Membership: Senior $50; Individual $75; Family $100; Settler Society $120; Pioneer Society $300; Explorer's Society $600; Founder's Society $1,200.

Eagle Rock

EAGLE ROCK VALLEY HISTORICAL SOCIETY, The Center for the Arts, Eagle Rock, 2225 Colorado Blvd., Eagle Rock, CA 90041. Tel.: 323-257-1357.
Institution Type/Description: Historical Society Museum.
Collections: local history & culture; period furnishings; personal artifacts; photographs.
Hours & Admission Prices: Sat. 10 to noon.

Edwards AFB

AIR FORCE FLIGHT TEST MUSEUM, 405 S. Rosamond Blvd., Edwards AFB, CA 93524. Mailing Address: 412 TW/MU, 405 S. Rosamond Blvd., Edwards AFB, CA 93524. Tel.: 661-277-8050 & 3510 (Public Rels.). Fax: 805-277-8051.
E-mail: museum@edwards.af.mil
Web Site: www.edwardsmuseum.org
Founded: 1986.
Key Personnel: Dir. & Cur., George Welsh.
Personnel Profile: Full-Time Paid 2; Full-Time Volunteers 1; Part-Time Volunteers 24.
Governing Authority: federal government; nonprofit. Tax-exempt: 501(c)(3).
Institution Type/Description: Aeronautics Museum.
Collections: concentration on history of Edwards Air Force Base, its antecedents & the history of flight testing.
Facilities: 500-vol. aeronautics library; 8,500 sq. ft. exhibit space; 30-seat theater; six acres of outside exhibit space. Aviation-related items for sale.
Activities: films.
Hours & Admission Prices: Tues.-Sat. 9-5. No charge; donations accepted. Closed federal holidays. &
Attendance: 33,000 (accurate)

El Cajon

COMMEMORATIVE AIR FORCE AIR GROUP ONE MUSEUM, Gillespie Field, 1905 N. Marshall Ave., Hangar #6, El Cajon, CA 92020. Tel.: 619-259-5541.
E-mail: airgroupone@gmail.com
Web Site: www.ag1caf.org
Founded: 1986.
Congressional District: 52
Key Personnel: Air Show Chm., Robert Simon; Wing Leader, Jim McGarvie; Museum Shop Mgr., Dave Hanson.
Governing Authority: nonprofit organization. Parent Institution: Commemorative Air Force. Subsidiary Institution: Air Group One. Tax-exempt: 501(c)(3).
Institution Type/Description: Military History Museum.
Collections: World War II aviation history & aircraft.
Activities: special events; fundraising events; warbird flights. Annual Event: World War II Airshow.
Publications: newsletter, Scuttlebutt.
Hours & Admission Prices: Call for hours. No charge; donations accepted.
Attendance: 500 (estimated)

GROSSMONT COLLEGE HYDE ART GALLERY, 8800 Grossmont College Dr., El Cajon, CA 92020-1798. Tel.: 619-644-7299. Fax: 619-644-7922. Facebook: Hyde Art Gallery.
E-mail: larry.klinet@gcccd.edu
Web Site: www.grossmont.edu/artgallery
Founded: 1961.
Congressional District: 43
Key Personnel: Cur., Ben Aubert; Chm. (V) & Dean, Steve Baker; Gallery Asst., Teresa Markey; Head Art Council & Professor, Jennifer Bennett; Art Dept. Chm., Paul Turounet.
Governing Authority: college. Parent Institution: Grossmont Community College District. Tax-exempt.
Institution Type/Description: Art Gallery.
Collections: prints; ceramics; photographs; digital.
Activities: films; gallery talks; concerts; loan, temporary & traveling exhibitions.
Publications: exhibit announcements; catalogues.
Hours & Admission Prices: Mon. & Thurs. 10-6:30, Tues.-Wed. 10-8. No charge; donations accepted. Closed all legal holidays. &
Attendance: 12,000 (estimated)
Membership: Student $10; Individual $25; Art Council $35; Patron $50; Donor $75; Benefactor $100; Sponsor $250.

HERITAGE OF THE AMERICAS MUSEUM, 12110 Cuyamaca College Dr., W., El Cajon, CA 92019-4317. Tel.: 619-670-5194. Fax: 619-670-5198.
E-mail: hofam@sbcglobal.net
Web Site: www.cuyamaca.edu/museum
Founded: 1993.
Congressional District: 52
Key Personnel: Founder, Bernard Lueck; Exec. Dir., Kathleen Oatsvall; Chm. (V), Ronald Raymond.
Personnel Profile: Full-Time Paid 1; Part-Time Paid 2; Part-Time Volunteers 30.
Governing Authority: private; nonprofit organization. Tax-exempt.
Institution Type/Description: History Museum.
Collections: natural & human history of the Americas.
Activities: lectures; educational programs.
Publications: bimonthly newsletter.
Hours & Admission Prices: Tues.-Fri. 10-4, Sat. 12-4. Adults $3, senior citizens $2; discount to AAM, ICOM & AAA members; youths under 17 & accompanied by an adult & members no charge. &
Attendance: 20,000 (accurate)
Membership: Student & Senior $10; Individual $20; Family $30; Contributor $50; Patron $100; Grantor $200.

KNOX HOUSE MUSEUM, 280 N. Magnolia Ave., El Cajon, CA 92020-3906. Mailing Address: P.O. Box 1973, El Cajon, CA 92022-1973. Tel.: 619-444-3800.
E-mail: info@elcajonhistory.org
Web Site: www.elcajonhistory.org
Founded: 1973.
Institution Type/Description: Historical House Museum: built in 1876.
Collections: El Cajon area history; period furnishings.
Activities: community heritage projects; research; guided tours; outreach; fundraising.
Publications: quarterly newsletter.

Hours & Admission Prices: 2nd & 4th Sat. 12:30-3:30, 3rd Sat. 11-1:15. No charge; donations accepted.
Membership: Individual $10; Family $15; Organization $25; Business $35; Life $200; Enhanced Life $500.

El Monte

EL MONTE HISTORICAL SOCIETY MUSEUM, 3150 Tyler Ave., El Monte, CA 91731-3354. Mailing Address: 3150 N. Tyler, El Monte, CA 91731-3354. Tel.: 626-444-3813 & 580-2232. Fax: 626-444-8142.
Founded: 1958.
Congressional District: 29
Key Personnel: Cur., Donna Crippen.
Governing Authority: municipal; nonprofit organization. Parent Institution: City of El Monte. Tax-exempt: 170(b)(1)(A).
Institution Type/Description: History Museum.
Collections: art; archives; glass; furniture; household furnishings; manuscript collections: clothing; jewelry; guns; musical instruments; pictures & history of early settlers & places; newspapers; school materials; maps; land grants & deeds; journals; diaries; church & cemetery records; farm implements; natural history; Indian & Mexican artifacts; tools; blacksmith items; saloon.
Research Fields: limited local & state history of families; genealogy; archeology.
Facilities: 1,500-vol. library of books on local & state history, original manuscripts, diaries & antique books on many subjects available by appointment only; reading room; theater.
Activities: guided tours; lectures; films; drama; formally organized educational programs; docent program or council; permanent & temporary exhibitions.
Publications: quarterly pamphlet, booklet, End of the Santa Fe Trail; book, The History of El Monte; cookbook, Our Pioneer Heritage.
Hours & Admission Prices: Jan. to mid-Dec. Tues.-Fri. 10-4, Sun. 1-3, tours by appointment. No charge; donations requested. Closed national holidays. &
Membership: Active $10; Patron $15; Fellowship $30; Life $100.

El Segundo

AUTOMOBILE DRIVING MUSEUM, 610 Lairport St., El Segundo, CA 90245-5004. Tel.: 310-909-0950. Fax: 310-231-4668.
E-mail: tomz@theadm.org
Web Site: www.theadm.org
Founded: 2002.
Congressional District: 30
Key Personnel: Pres. (V) & Cur., Earl Rubenstein; Pres. (V), Mitch Reinstein; C.E.O., Tom Zimmerman; Operations, Jodee Hulsebus.
Personnel Profile: Full-Time Paid 2; Part-Time Paid 2; Part-Time Volunteers 12.
Governing Authority: nonprofit. Tax-exempt: 501(c)(3).
Institution Type/Description: Transportation Museum.
Collections: vehicles from 1904-1989; automotive literature.
Facilities: library; theater; 25,000 sq. ft. exhibit space; 250-seat banquet facilities.
Activities: vehicle rides; concerts; movies; car club meets; car enthusiast events; educational programs. Annual Event: Auto Show.
Publications: annual newsletter.
Hours & Admission Prices: Tues.-Sun. 10-4. No charge; donations accepted. Closed holidays. &
Attendance: 6,000 (estimated)
Membership: Subscribers $25, $50, $100, $300, $750. Corporate memberships available.

Encinitas

LUX ART INSTITUTE, (M), 1550 S. El Camino Real, Encinitas, CA 92024-4908. Tel.: 760-436-6611. Fax: 760-436-1400.
E-mail: info@luxartinstitute.org
Web Site: www.luxartinstitute.org
Founded: 1998.
Key Personnel: Dir., Reesey Shaw; Chm. (V), Wally Dieckmann; Museum Shop Mgr., Grace Chen; Museum Shop Mgr., Farrah Emammi.
Personnel Profile: Full-Time Paid 10; Part-Time Paid 2; Part-Time Volunteers 35; Interns 4.
Governing Authority: Tax-exempt: 501(c)(3).
Institution Type/Description: Art Museum.
Collections: works by national & international artists.
Hours & Admission Prices: Thurs.-Fri. 1-5, Sat. 11-5. Adults $5; discounts to AAM members; members & under 21 no charge. &
Attendance: 1,800 (estimated)
Membership: Individual $50; Family $100; Deluxe $500; Luminary $1,000; Visionary $5,000.

SAN DIEGO BOTANIC GARDEN, (M), 230 Quail Gardens Dr., Encinitas, CA 92024-2707. Mailing Address: P.O. Box 230005, Encinitas, CA 92023-0005. Tel.: 760-436-3036. Fax: 760-632-0917.
E-mail: info@sdbgarden.org
Web Site: www.sdbgarden.org
Formerly: Quail Botanical Gardens
Founded: 1960.
Congressional District: 43
Key Personnel: Pres., Julian Duval; Chm., Jim Ruecker; 1st Vice Chm., Frank Mannen; Dir. Operations, Patricia Hammer; Museum Shop Mgr., Roberta Dotson.
Personnel Profile: Full-Time Paid 21; Part-Time Paid 13; Part-Time Volunteers 19; Interns 1.
Governing Authority: county. Parent Institution: Quail Botanical Gardens Foundation. Tax-exempt: 501(c)(3).
Institution Type/Description: Botanical Garden.
Collections: tropical & sub-tropical trees, shrubs, herbs; fruits; cacti & other succulents; aloe; proteaceae; bamboo; drought resistant California natives & Austral-Asian plants; hibiscus.
Major Exhibits: Sculpture in the Garden, 4/13-3/14.
Research Fields: drought resistant ornamentals; sub-tropical fruits & bamboo.
Facilities: 500-vol. library of horticultural & botanical books available for research to foundation members on premises; 96-seat auditorium; botanical garden; foundation center; orientation center; bird sanctuary; concession area; demonstration garden.
Activities: guided tours; lectures; films; arts festivals; study clubs; hobby workshops; docent program; permanent & temporary exhibitions; concerts; plant sales; art classes; children's programs.
Publications: quarterly bulletin, Quail Tracks.
Hours & Admission Prices: Daily 9-5. Adults $12, seniors, students & military $8, children 3-12 $6; children under 3 no charge. Parking: $2. &
Attendance: 140,000 (estimated)
Membership: Senior 60 & over $40; Individual $50; Senior Dual 60 & over $65; Family & Dual $75.

SAN DIEGUITO HERITAGE MUSEUM, 450 Quail Gardens Dr., Encinitas, CA 92024-2711. Tel.: 760-632-9711. Fax: 760-632-5695.
E-mail: sdheritage@sbcglobal.net
Web Site: www.sdheritage.org
Founded: 1988.
Congressional District: 51
Key Personnel: Exec. Dir., Will Neblett; Pres., Fred Bruns.
Personnel Profile: Full-Time Paid 1; Part-Time Paid 1; Part-Time Volunteers 75.
Governing Authority: private; nonprofit. Tax-exempt: 501(c)(3).
Institution Type/Description: History Museum.
Collections: local history & culture including seven communities of north coastal San Diego County; Native American artifacts; the Mexican Rancho Land Grants; the Homesteaders in late 1800; a reconstructed shanty; the first settlers to the 1950s.
Facilities: Handcrafted & other museum-related items for sale.
Activities: docent program; films; guided tours; loan, participatory, temporary & traveling exhibits; videos. Annual Event: Museum Barbecue.
Publications: quarterly newsletter; book, San Dieguito Heritage.
Hours & Admission Prices: Thurs.-Sun. 12-4. Adults $4, students & seniors $2; young children, military & members no charge. Closed holidays. &
Attendance: 9,000 (estimated)
Membership: Individual $25; Family $35; Friend $100; Patron $500; Benefactor $1,000 and up.

Escondido

CALIFORNIA CENTER FOR THE ARTS, ESCONDIDO MUSEUM, 340 N. Escondido Blvd., Escondido, CA 92025-2600. Tel.: 760-839-4120. Fax: 760-739-0205.
Web Site: www.artcenter.org
Founded: 1992.
Congressional District: 51
Key Personnel: C.E.O., Vicky Basehore; Museum Dir. & Cur., Olivia Luther; Dir. Education, Tomoko Kuta; Registrar, Mary Johnson.
Personnel Profile: Full-Time Paid 5; Part-Time Paid 5; Part-Time Volunteers 40; Interns 1.
Governing Authority: nonprofit organization. Tax-exempt: 501(c)(3).
Institution Type/Description: Art Museum.
Collections: California & Latin American contemporary artists.
Research Fields: American & regional art since 1950; Latin American art since 1950.
Facilities: library; art studios in use as educational facilities; 10,000 sq. ft.

exhibit space; 2,000 sq. ft. sculpture court; 400-seat theater; 1,532-seat concert hall. Art books & catalogues for sale.
Activities: docent program; guided tours for schools K-12; university classes & adult groups; lectures; gallery talks; demonstrations; family workshops; special events; visual & performing arts classes.
Publications: exhibition catalogues.
Hours & Admission Prices: Tues.-Sat. 10-4, Sun. 1-5. Adults $5, senior citizens & active military $4, students $3; discounts to groups, AAM & ICOM members; children under 12 & members no charge. Closed major holidays. &
Attendance: 8,200 (accurate)
Membership: Individual $50; Family $100; Patron $250; Connoisseur $500; Aficionado $1,000; Conductor $2,500; California Center Club $5,000.

DEER PARK ESCONDIDO WINERY & AUTO MUSEUM, 29013 Champagne Blvd., Escondido, CA 92026-6002. Tel.: 760-749-1666.
E-mail: mail@deerparkwine.com
Web Site: www.deerparkwine.com
Founded: 1979.
Key Personnel: Owner, Clark Knapp.
Governing Authority: private.
Institution Type/Description: Winery & Automobile Museum.
Collections: collections from the American Golden Age with emphasis on the development of the convertible automobile & the growth of the American wine industry; appliances; radios; televisions; American popular trends.
Research Fields: convertible car production.
Facilities: library on automobiles & Americana; 10,000 sq. ft. exhibit space; 100-seat cafeteria; nature/conservation center. Museum-related items for sale.
Activities: art festivals; concerts; guided tours; participatory & temporary exhibitions.
Publications: annual calendar & program booklet; quarterly newsletter, Deer Park Winery & Auto Museum.
Hours & Admission Prices: Summer: Fri.-Sun. 10-5; Winter: Fri.-Sun. 10-4; call to confirm. Adults $10, seniors 55 & over, AAA members and active military $9; discounts to groups of 10 or more; children 9 & under no charge. Closed holidays.
Attendance: 75,000 (accurate)
Membership: Individual $25; Family $50; Cadillac $100.

ESCONDIDO CHILDREN'S MUSEUM DBA SAN DIEGO CHILDREN'S DISCOVERY MUSEUM, 320 N. Broadway, Escondido, CA 92025-2716. Tel.: 760-233-7755. Fax: 760-888-1934.
E-mail: info@sdcdm.org
Web Site: www.sdcdm.org
Founded: 2001.
Key Personnel: Exec. Dir., Javier Guerrero; Exec. Dir. Asst., Rebecca Greene; Office Admin., Yadira Diaz; Museum Education & Program Mgr., Lindy Villa; Exhibit Mgr., Bill Schmidt
Institution Type/Description: Children's Museum.
Collections: hands-on exhibits.
Activities: birthday parties.
Hours & Admission Prices: Mon.-Sun. 10-4, Wed. 9:30-4:30. Admission $6; discounts to military, educators & special needs children; members & children under one no charge. Closed New Year's Day; Easter; Memorial Day; Independence Day; Labor Day; Thanksgiving; Christmas.
Membership: Grandparent, Military & Disabled Child Membership $65; Family $85; Reciprocal (ACM) $125.

ESCONDIDO HISTORY CENTER, (M), 321 N. Broadway, Escondido, CA 92025-2704. Mailing Address: P.O. Box 263, Escondido, CA 92033-0263. Tel.: 760-743-8207. Fax: 760-743-8267.
E-mail: barker@escondidohistory.org
Web Site: www.escondidohistory.org
Formerly: Heritage Walk Museum
Founded: 1956.
Congressional District: 51
Key Personnel: Pres., Sally Costello; Exec. Dir., Wendy Barker; Treas., Bob Johnson; Registrar, Marie Tuck.
Personnel Profile: Full-Time Paid 2; Part-Time Paid 1; Part-Time Volunteers 70; Interns 1.
Governing Authority: nonprofit organization. Tax-exempt.
Institution Type/Description: Local History Museum: housed in 1894 library & 1888 Santa Fe Railroad Depot; 1890 Victorian house, barn & working blacksmith shop; railroad car with model train layout.
Collections: 1860-present, Escondido history; manuscripts of local residents; newspaper files; photographs.

Research Fields: oral histories & video taping of Escondido residents.
Facilities: 1,100-vol. library of local, county & American history; preservation reference works, available to the public; 6,500 sq. ft. exhibit space. Books & gift items for sale.
Activities: docent program; guided tours; temporary exhibitions; school loan service; videos; monthly lectures series; city walking tours. Annual Events: Car Show in May; Grape Day Festival in September; living history program in October.
Publications: quarterly newsletter, The Grapevine.
Hours & Admission Prices: Office: Tues.-Sat. 10-4. Museum: Tues.-Sat. 1-4. Suggested Donations: adults $3, children $1; members no charge. Closed holidays. &
Attendance: 18,000 (estimated)
Membership: Senior $20; Individual $25; Family $50; Supporting $100; Silver Patron $250; Gold Patron $500.

LAWRENCE WELK MUSEUM, 8860 Lawrence Welk Dr., Escondido, CA 92026-6403. Tel.: 760-749-3000, ext. 22146.
E-mail: box.office@welktheatre.com
Web Site: www.welktheatresandiego.com
Institution Type/Description: History Museum.
Collections: Welk family history & career; personal artifacts; photographs; posters; instruments.
Hours & Admission Prices: Daily 10-5; call to verify. No charge.

SAN DIEGO ARCHAEOLOGICAL CENTER, (M), 16666 San Pasqual Valley Rd., Escondido, CA 92027-7001. Tel.: 760-291-0370. Fax: 760-291-0371.
E-mail: info@sandiegoarchaeology.org
Web Site: www.sandiegoarchaeology.org
Founded: 1993.
Congressional District: 49
Key Personnel: Dir., Cindy Stankowski; Pres. (V), Bruce Gallagher; Devel. & Public Rels., Marie Andersen; Education, Annemarie Cox; Registrar, Chris Mirsky; Cur., Margie Burton, Ph.D.
Personnel Profile: Full-Time Paid 4; Full-Time Volunteers 10; Part-Time Paid 4; Part-Time Volunteers 30; Interns 10.
Governing Authority: private; nonprofit organization. Tax-exempt: 501(c)(3).
Institution Type/Description: Archaeology Museum.
Collections: archaeological artifacts & documents; cultural history.
Research Fields: prehistoric groundstone technology; indigenous ceramic typologies & chronology; prehistoric demography & resource use.
Facilities: library; 2,800 sq. ft. exhibit space. Museum-related items for sale.
Activities: docent program; formal education programs; internships; guided tours; hobby workshops; lectures; loan, traveling, participatory & temporary exhibitions; training programs; school loan service. Annual Event: Cultural Resource Management BBQ.
Publications: quarterly, The San Diego Archaeological Center Newsletter.
Hours & Admission Prices: Mon.-Fri. 9-4, Sat. 10-2. Suggested Donation: family $5, individual $2. Closed New Year's Day; Thanksgiving; Christmas.
Attendance: 5,000 (accurate)
Membership: Senior & Student $25; Individual $35; Family $50; Supporter $75; Sponsor $100; Patron $150; Benefactor & Corporation $250.

SAN DIEGO CHILDREN'S DISCOVERY MUSEUM, 320 N. Broadway Blvd., Escondido, CA 92025. Tel.: 760-233-7755.
Institution Type/Description: Children's Museum.
Collections: hands-on exhibitions.
Hours & Admission Prices: Tues.-Sun. 10-4. Admission 12 months & over $5; discounts to military & disabled; children under one no charge. Closed New Year's Day; Easter; Memorial Day; Labor Day; Thanksgiving; Christmas Eve & Day.

✻ SAN DIEGO ZOO SAFARI PARK, (M), 15500 San Pasqual Valley Rd., Escondido, CA 92027-7017. Mailing Address: P.O. Box 120551, San Diego, CA 92112-0551. Tel.: 760-738-5018 & 747-8702. Fax: 760-746-7081.
E-mail: csimmons@sandiegozoo.org
Web Site: www.sandiegozoo.org
Formerly: San Diego Zoo's Wild Animal Park
Founded: 1972.
Congressional District: 14
Key Personnel: C.E.O., Douglas G. Myers; Pres. (V), Rick Gulley; Dir. Wild Animal Park, Robert McClure; Corporate Dir. Merchandising, Yvonne Miles; Dir. Collections, Bob Wiese, Ph.D.; Dir. Veterinary Svcs., Donald Janssen, DVM; Dir. Mktg., Ted Molter; Cur. Birds, Michael Mace; Cur. Mammals, Randy Reiches; Horticulturist, Cary Sharp.
Personnel Profile: Full-Time Paid 800; Part-Time Paid 300.
Governing Authority: society. Parent Institution: Zoological Society of San Diego. Subsidiary Institution: Institute for Conservation Research. Tax-exempt: 501(c)(3).
Institution Type/Description: Wildlife Preserve.
Collections: birds, mammals & reptiles of the world; botanical; African marshland.
Research Fields: veterinary medicine & pathology; animal reproduction & conservation; ecology & habitat preservation.
Facilities: 3,000-vol. library of zoology & natural history; amphitheaters; restaurants; 1 1/4 mile hiking; 30-acre walking safari; aviaries; educational campsite; conservation & research center. Gifts for sale.
Activities: guided trips through preserve aboard monorail; overnight camping experience; summer Park after Dark experience; guided walking tours through behind-the-scenes areas of park; animal shows; educational programs for pre-school through adult; photo safaris via truck into large animal enclosures; Journey into Africa; African habitat tram tour.
Publications: guidebook, The San Diego Wild Animal Park; magazine, Zoonooz; children's newsletter, Koala Club News; miscellaneous newsletters.
Hours & Admission Prices: Daily 9-4. Adults $28.50, children 3-11 $18.50; discounts to qualifying groups; members and children 2 & under no charge. &
Attendance: 1,500,000 (estimated)
Membership: Koala Club 3-11 $21, 12-17 $25; Senior Passes: Single $35; Dual $50; Regular: Single $66; Dual $84; Diamond Club: New $126; Keeper's Club $150; Curator's Club $250; Director's Club $500; President's Associates $1,000; President's Partners $2,500.

Eureka

CLARKE HISTORICAL MUSEUM, (M), 240 E St., Eureka, CA 95501-0433. Tel.: 707-443-1947. Fax: 707-443-0290. Facebook: Clarke Historical Museum.
E-mail: clarkehistorical@att.net
Web Site: www.clarkemuseum.org
Formerly: Clarke Memorial Museum, Inc.
Founded: 1960.
Congressional District: 2
Key Personnel: Dir. & Cur., Ben Brown; Pres. (V), Roy Sheppard; Treas., Wendy Wahlund; Museum Coord. & Archivist, Carly Marino; Office & Events Mgr., Amber Mitchell.
Personnel Profile: Full-Time Paid 2; Part-Time Paid 2; Part-Time Volunteers 22; Interns 2.
Governing Authority: nonprofit corp. Tax-exempt: 501(c)(3).
Institution Type/Description: Regional History Museum.
Collections: Northwestern California Native American basketry & ceremonial regalia; regional & natural history, firearms, Victoriana; decorative arts, costumes, textiles.
Research Fields: Native American culture of the Yurok, Karuk, Hupa, Wiyot & Tolowa; Humboldt County history; American costume & textile.
Facilities: 800-vol. library of regional history; 20,000 photographic archive of Humboldt County scenes.
Activities: docent guided tours for adults & school groups; lectures; temporary & permanent exhibitions; special events; outreach programs to schools.
Publications: books: The Hover Collection of Karuk Baskets, 1993; Baskets and Weavers, 1996; Eureka and Humboldt County California, 2001.
Hours & Admission Prices: Wed.-Sat. 11-4. Suggested Donation: family $5, individual $3. &
Attendance: 16,000 (accurate)
Membership: Student & Senior $15; Family $25; Sponsor $50; Patron $100; Benefactor $250; Clarke Circle $500.

COLLEGE OF THE REDWOODS ART GALLERY, 7351 Tompkins Hill Rd., Creative Arts Bldg., Eureka, CA 95501-9300. Tel.: 707-476-4558 & 4137.
E-mail: cindy-hooper@redwoods.edu
Web Site: www.redwoods.edu/departments/art/gallery/index.htm
Key Personnel: Dir., Charissa Schulze; Head Art Dept., Cindy Hooper.
Personnel Profile: Part-Time Paid 1; Part-Time Volunteers 4.
Governing Authority: Parent Institution: College of the Redwoods.
Institution Type/Description: Art Gallery.
Collections: drawings; paintings; sculpture; photography; ceramics.
Hours & Admission Prices: Call for hours. No charge. Closed during academic breaks.
Attendance: 2,500 (estimated)

THE DISCOVERY MUSEUM, 501 Third St., Eureka, CA 95501. Mailing Address: 517 3rd St., Eureka, CA 95501-5105. Tel.: 707-443-9694. Fax: 707-443-7242.
E-mail: info@discovery-museum.org
Web Site: www.discovery-museum.org
Founded: 1995.
Key Personnel: C.E.O., Trey Scott; Dir., Jennifer Taylor
Institution Type/Description: Children's Museum.
Collections: hands-on exhibits.
Facilities: classrooms.
Activities: interactive exhibits; birthday parties; rental facilities; programs for preschoolers, parents & elementary school children; field trips; planetarium shows.
Hours & Admission Prices: Tues.-Sat. 10-4, Sun. 12-4. Admission 2 & over $4; discounts to members; children under 2 no charge.
Membership: Meteorite $25; Asteroid $65; Polaris $100; Comet $250; Nova $500.

GROSS-WELLS BARNUM HOUSE - HUMBOLDT COUNTY HISTORICAL SOCIETY, 703 8th St., Eureka, CA 95502. Mailing Address: P.O. Box 8000, Eureka, CA 95502. Tel.: 707-445-4342. Fax: 707-445-4146.
Institution Type/Description: Historical Society Museum: housed in the former home of Helen Wells Barnum; built in 1902.
Collections: local history & culture; period furnishings; personal artifacts; photographs.
Publications: newsletter.
Hours & Admission Prices: Tues.-Wed. & Fri. 12-4, Thurs. 4-8.

HUMBOLDT ARTS COUNCIL/MORRIS GRAVES MUSEUM OF ART, 636 F St., Eureka, CA 95501-1012. Tel.: 707-442-0278. Fax: 707-442-2040.
Web Site: www.humboldtarts.org
Founded: 1966.
Congressional District: 1
Key Personnel: Pres. (V) & C.E.O., Sally Arnot; Cur., Jemima J. Harr.
Personnel Profile: Full-Time Paid 1; Full-Time Volunteers 1; Part-Time Paid 4; Part-Time Volunteers 45; Interns 3.
Governing Authority: private; nonprofit organization. Tax-exempt: 501(c)(3).
Institution Type/Description: Art Museum.
Collections: paintings; prints; drawings; decorative arts; photographs.
Facilities: 200-seat performance rotunda; classroom; 18,000 sq. ft. exhibit space. Museum-related items for sale.
Activities: concerts; dance recitals; docent program; films; formal education programs for children & undergraduate or graduate college students; guided tours; lectures; mobile vans; school loan service; temporary exhibitions; training programs for professional museum workers; broadcast programs.
Hours & Admission Prices: Wed.-Sun. 12-5. No charge; donations accepted. Closed New Year's Day; Easter; Independence Day; Thanksgiving; Christmas.
Attendance: 26,000 (accurate)
Membership: Artist $30; Family $50; Circle of 100 $100; Patron $500; Sponsor $1,000.

WOODEN SCULPTURE GARDEN OF ROMANO GABRIEL, 315 Second St., Eureka, CA 95502-1354. Mailing Address: Eureka Heritage Society, P.O. Box 1354, Eureka, CA 95502-1354. Tel.: 707-445-8775 & 442-8937.
E-mail: info@eurekaheritage.org
Web Site: eurekaheritage.org
Institution Type/Description: Folk Art Museum.
Collections: wooden sculptures.
Hours & Admission Prices: Call for hours.

Fair Oaks

FAIR OAKS HISTORICAL SOCIETY, (M), 10340 Fair Oaks Blvd, Fair Oaks, CA 95628. Mailing Address: P.O. Box 2044, Fair Oaks, CA 95628-2044. Tel.: 916-961-6561. fairoakshistory1@gmail.com.
E-mail: webhost@fairoakshistory.org
Web Site: fairoakshistory.org
Founded: 1975.
Key Personnel: Pres. (V), Joe Dobrowolski.
Personnel Profile: Part-Time Volunteers 2.
Volunteer Hours: 500
Operating Expenses: 13,000
Operating Income: 10,000

Institution Type/Description: Historical Society Museum.
Collections: local history, heritage & culture; photographs; period artifacts; farm equipment.
Major Exhibits: Water Development in Fair Oaks, 11/13-5/14.
Publications: quarterly newsletter; book, History of Fair Oaks.
Hours & Admission Prices: Wed.-Sun. 10-4. No charge; donations accepted. &
Attendance: 520 (accurate)

Fall River Mills

FORT CROOK HISTORICAL MUSEUM, 43030 Fort Crook Museum Ave., Fall River Mills, CA 96028. Mailing Address: Box 397, Fall River Mills, CA 96028-0397. Tel.: 530-336-5110.
E-mail: fortcrook@frontiernet.net
Web Site: www.fortcrook.com
Founded: 1934.
Congressional District: 14
Key Personnel: Pres. (V), Glorianne Weigand; Cur., Dorothy Mason.
Personnel Profile: Part-Time Paid 1; Part-Time Volunteers 18.
Governing Authority: society; nonprofit organization. Parent Institution: Fort Crook Historical Society. Tax-exempt: 170(b)(1)(A).
Institution Type/Description: General Museum.
Collections: agriculture; Indian artifacts; industry; preservation project; transportation; pioneer homemaking, photos, clippings & family histories.
Research Fields: family histories.
Facilities: collection of scrapbooks available on premises; reading room. Museum-related items for sale.
Activities: permanent exhibitions.
Publications: booklet, Reminiscence of Fort Crook Society.
Hours & Admission Prices: May-Oct. Tues.-Sun. 12-4; other times by appointment. No charge; donations accepted.
Attendance: 3,200 (accurate)
Membership: Single $15; Family $25; Business $40; Life $300.

Fallbrook

FALLBROOK HISTORICAL SOCIETY, 260 Rocky Crest Rd., Fallbrook, CA 92028. Mailing Address: P.O. Box 1375, Fallbrook, CA 92088-1375. Tel.: 760-723-4125.
Institution Type/Description: Historical Society Museum.
Collections: local history & culture; personal artifacts; photographs; Native American & pioneer artifacts; model train; local agriculture, education, industry & business; military artifacts.
Hours & Admission Prices: Sun. & Thurs. 1-4. No charge; donations accepted. &
Membership: Junior $5; Adult $15; Family $25; Business $50; Life $250.

Felicity

MUSEUM OF HISTORY IN GRANITE, (M), Two Center of the World Plaza, Felicity, CA 92283-7777. Tel.: 760-572-0100. Fax: 760-572-3000.
E-mail: museumforever@gmail.com
Web Site: historyingranite.org
Founded: 1973.
Congressional District: 51
Key Personnel: Chm. (V), Jacques Andre Istel; Treas., Felicia Lee; Museum Shop Mgr., Debra Pavey.
Personnel Profile: Part-Time Paid 4; Part-Time Volunteers 11.
Governing Authority: private; nonprofit organization. Parent Institution: Hall of Fame of Parachuting, Inc. Tax-exempt: 501(c)(3).
Institution Type/Description: History Museum.
Collections: granite monuments; History of the USA engraved in granite.
Research Fields: world research.
Facilities: library; 60 acre site; 72-seat restaurant; classrooms. Museum-related items for sale.
Activities: films; guided tours. Museum Sponsors: Dedications.
Hours & Admission Prices: Outdoor Exhibits: daily. Shop & Restaurant: Thanksgiving to Easter. $3 entrance.
Attendance: 20,000 (estimated)

Ferndale

FERN COTTAGE FOUNDATION, 2121 Centerville Rd., Ferndale, CA 95536-9719. Mailing Address: P.O. Box 1286, Ferndale, CA 95536-1286. Tel.: 707-786-4835.
E-mail: ferncottage@frontier.com
Web Site: www.ferncottage.org
Founded: 1989.

Congressional District: 1
Key Personnel: Chm., Virginia Dwight.
Governing Authority: Parent Institution: Fern Cottage Foundation. Tax-exempt.
Institution Type/Description: Historic House Museum: Russ family farm house, built in 1866.
Collections: personal artifacts; furnishings.
Hours & Admission Prices: Call for hours. Open by appointment only. Adults $10.

FERNDALE MUSEUM, 515 Shaw Ave., Ferndale, CA 95536. Mailing Address: P.O. Box 431, Ferndale, CA 95536-0431. Tel.: 707-786-4466.
Founded: 1976.
Congressional District: 1
Key Personnel: Dir., Don Andersen; Pres (V), Kirk Gothier
Governing Authority: Tax-exempt.
Institution Type/Description: History Museum.
Collections: local history & heritage; period artifacts; furniture.
Hours & Admission Prices: Feb.-May & Oct.-Dec. Wed.-Sat. 11-4, Sun. 1-4; June-Sept. Tues.-Sat. 11-4, Sun. 1-4. Adults $1; members no charge. &
Attendance: 3,000 (accurate)

Fillmore

FILLMORE HISTORICAL MUSEUM, INC., 350 Main St., Fillmore, CA 93015-2040. Mailing Address: P.O. Box 314, Fillmore, CA 93016-0314. Tel.: 805-524-0948. Fax: 805-524-0516.
E-mail: fillmore.museum@sbcglobal.net
Web Site: fillmorehistoricalmuseum.com
Founded: 1971.
Congressional District: 24
Key Personnel: Dir., Pres. (V) & Administrative Officer, Martha Gentry; Research Librarian, Ynez Haase; Chm. (V), Bev Hurst.
Personnel Profile: Part-Time Volunteers 12.
Governing Authority: nonprofit corporation. Tax-exempt: 501(c)(3).
Institution Type/Description: General Historical Museum.
Collections: photos; memorabilia. Historic Buildings: c.1887 former Southern Pacific Depot; 1905 Craftsman-Style Hinckley House; 1919 Rancho Sespe Bunk House.
Research Fields: local area history (Fillmore, Piru Sespe, Bardsdale, California).
Activities: guided tours.
Publications: newsletter published three times annually.
Hours & Admission Prices: Tues.-Fri. 10-4, Sat. 10:30-3. Adults $4; members no charge; donations accepted.
Attendance: 4,402 (estimated)
Membership: Individual $30; Family $40; Business $50; Contributing $75; Sustaining $100; Patron $500; Benefactor $1,000.

Firebaugh

HERITAGE OF EAGLES AIR MUSEUM, Eagle Field, 11163 N. Eagle Ave., Firebaugh, CA 93622. Mailing Address: 5543 Mint Rd., Dos Palos, CA 93620. Tel.: 209-364-6132.
Web Site: www.b25.net/museum/hpages/museum.html
Institution Type/Description: Military History Museum: housed on the Eagle Field Army Air Forces Training Base; built in 1942.
Collections: airport & aviation history; aircraft; vehicles; uniforms; World War II command radio station; aviation memorabilia; photographs; personal artifacts.
Hours & Admission Prices: Call for hours.

Folsom

THE FOLSOM CITY ZOO SANCTUARY, 403 Stafford St., Folsom, CA 95630-2643. Tel.: 916-351-3527.
E-mail: kbanyard@folsom.ca.us
Web Site: www.folsom.ca.us/depts/parks_n_recreation/zoo.asp
Key Personnel: Zoo Mgr., Jocelyn Smeltzer; Zoo Education, Vicki Valentine
Institution Type/Description: Zoo.
Collections: wildlife & their habitats including peacocks, black bears, wolves, parrots, red foxes, pigs, horses, raccoons, sheep, snakes, & monkeys.
Hours & Admission Prices: Tues.-Sun. 10-4. Adults 13 & over $4, senior citizens 55 & over and children 2-12 $3; children under 2 no charge.

FOLSOM, EL DORADO & SACRAMENTO HISTORICAL RAILROAD ASSOCIATION, 198 Wood St., Folsom, CA 95630. Tel.: 916-985-6001.
E-mail: feds@fedshra.org
Web Site: www.fedshra.org
Founded: 1995.
Congressional District: 7
Key Personnel: Pres. (V), Bill Anderson.
Personnel Profile: Part-Time Volunteers 20.
Governing Authority: Tax-exempt.
Institution Type/Description: History Museum.
Collections: local history; railroad artifacts; photographs.
Activities: special events.
Hours & Admission Prices: Temporarily closed for renovation. No charge; donations accepted.

FOLSOM HISTORY MUSEUM, (M), 823 Sutter St., Folsom, CA 95630-2440. Tel.: 916-985-2707. Fax: 916-985-7288.
E-mail: info@folsomhistorymuseum.org
Web Site: www.folsomhistorymuseum.org
Founded: 1960.
Congressional District: 4
Key Personnel: C.E.O. & Dir., Mary Mast; Chm. (V), Patrick Maxfield; Museum Shop Mgr., Pam Conrad.
Personnel Profile: Full-Time Paid 1; Full-Time Volunteers 5; Part-Time Paid 1; Part-Time Volunteers 50.
Governing Authority: nonprofit organization. Tax-exempt: 501(c)(3).
Institution Type/Description: Historical Society Museum: site of Pony Express terminus from June 1860 to October 1861; Wells Fargo Assay Office, 1860-1871.
Collections: history of Folsom area including Pony Express; Sacramento Valley Railroad (first railroad west of the Mississippi); Folsom Powerhouse, one of first to transmit long distance electricity in the world; Mother Lode gold mining (northern mines).
Research Fields: area history.
Facilities: archives; 2,300 sq. ft. exhibit space. Museum-related items for sale.
Activities: guided tours; lectures; temporary & loan exhibits; docent program. Annual Events: Antique Quilt Show in August and September.
Publications: quarterly newsletter, Tailings.
Hours & Admission Prices: Memorial Day to Labor Day daily 11-4; Sept.-May Tues.-Sun. 11-4. Research: by appointment. Adults $4, youth $2; members no charge. Closed New Year's Day; Easter; Mother's Day; Thanksgiving; Christmas. &
Attendance: 18,500 (estimated)
Membership: Senior $15; Individual & Senior Family $25; Family & Nonprofit $35; Supporting, Business & Professional $75; Corporate $200.

Fontana

MARY VAGLE NATURE CENTER, 11501 Cypress Ave., Fontana, CA 92337. Mailing Address: Communitry Services, 16860 Valencia Ave., Fontana, CA 92335. Tel.: 909-349-6994.
E-mail: rdean@fontana.org
Web Site: www.fontana.org/index.aspx?NID=196
Key Personnel: Community Services Coord., Rick Dean
Institution Type/Description: Nature Center.
Collections: astronomy; rocks & minerals; Native American.
Facilities: trails.
Hours & Admission Prices: Wed.-Sun. 12-5. No charge; donations accepted. Closed Federal holidays.

Foresthill

FORESTHILL DIVIDE MUSEUM, 24601 Harrison St., Foresthill, CA 95631. Mailing Address: The Forest Hill Divide Historical Society, P.O. Box 646, Foresthill, CA 95631-0646. Tel.: 530-367-3988.
Web Site: mmoffet.mystarband.net/museum.htm
Governing Authority: Parent Institution: Placer County Museums. Tax-exempt.
Institution Type/Description: History Museum.
Collections: late 19th & early 20th centuries; Gold Rush artifacts; Native Americans; recreation; transportation.
Hours & Admission Prices: mid-May to mid-Oct. Sat.-Sun. & major holidays 12-4. No charge; donations accepted.
Membership: Individual $8; Family $15; Lifetime $120.

Fort Bragg

THE GUEST HOUSE MUSEUM, 343 N. Main St., Fort Bragg, CA 95437. Mailing Address: P.O. Box 71, Fort Bragg, CA 95437-0071. Tel.: 707-964-4251 & 2404.
E-mail: ds1923@hotmail.com
Founded: 1950.
Congressional District: 2
Key Personnel: C.E.O. & Pres. (V), Mark Ruedrich; Museum Shop Mgr., Denise Stenberg.
Personnel Profile: Part-Time Volunteers 24.
Governing Authority: municipal. Parent Institution: Fort Bragg. Subsidiary Institution: Mendocino Coast Historical Society. Tax-exempt.
Institution Type/Description: Logging & Lumber Museum: housed in 1892 C.R. Johnson Home and later Union Lumber Company's guest house in 1913.
Collections: logging & lumber mill artifacts; photos; Fort Bragg historical artifacts.
Activities: permanent & temporary exhibitions.
Publications: quarterly, Voice of the Past.
Hours & Admission Prices: June-Oct. Mon.-Fri. 11-2, Sat.-Sun. 10-4; Nov.-May Thurs.-Sun. 11-2. Families $5, adults $2; members & children under 12 no charge. Closed New Year's Day; Thanksgiving; Christmas.
Attendance: 8,500 (estimated)
Membership: Individual $20; Couple $30; Nonprofit Organization $90; Patron $150; Commercial $200; Lifetime Single $400; Lifetime Couple $600.

MENDOCINO COAST BOTANICAL GARDENS, 18220 N. Hwy. 1, Fort Bragg, CA 95437-8773. Tel.: 707-964-4352. Fax: 707-964-3114.
E-mail: info@gardenbythesea.org
Web Site: www.gardenbythesea.org/about
Founded: 1961.
Congressional District: 1
Key Personnel: Exec. Dir., Mary Anne Payne; Museum Shop Mgr., Cynthia Lambie.
Personnel Profile: Full-Time Paid 10; Part-Time Paid 6; Part-Time Volunteers 150.
Governing Authority: Tax-exempt.
Institution Type/Description: Botanic Garden.
Collections: coastal pine forest; fern-covered canyons; plants; flowers; birds; heath & heathers; rhododendrons; camellias; magnolias; conifers.
Facilities: cafe. Museum-related items for sale.
Activities: education programs; weddings; docent tours.
Hours & Admission Prices: March-Oct. daily 9-5; Nov.-Feb. daily 9-4. Adults 18 & over $14, seniors 65 & over $10, juniors 5-17 $5; children 5 & under and members no charge. Closed Thanksgiving; Christmas.
Attendance: 60,000 (accurate)
Membership: Senior $35; Individual $50; Senior Household $60; Household $75; Sponsor & Business $125; Business Sponsor $200; Supporter $300.

TRIANGLE TATTOO & MUSEUM, 356 B. N. Main St., Fort Bragg, CA 95437-3406. Tel.: 707-964-8814.
E-mail: chinchilla@triangletattoo.com
Web Site: www.triangletattoo.com
Founded: 1992.
Key Personnel: C.E.O. & Museum Mgr., M. Chinchilla; C.E.O. & Museum Shop, Mr. G.
Governing Authority: private; nonprofit.
Institution Type/Description: Tattoo History Museum.
Collections: tattoo art & artifacts; photographs; tattoo implements & machines; historical items pertaining to the development of the electric tattooing era; pre-electric tattooing.
Research Fields: documenting old tattoos & tattooed people.
Facilities: working tattoo parlor.
Activities: tours; historical seminars.
Publications: books, Stewed, Screwed & Tattooed (3rd revised printing 2005); Electric Tattooing by Women 1900-2003; Electric Tattooing by Men 1900-2004; Captain Don Leslie - Sword Swallower, Circus Sideshow Attraction 1937-2007.
Hours & Admission Prices: Daily 12-6. No charge; donations accepted. Closed Christmas.
Attendance: 5,000 (estimated)

Fort Jones

FORT JONES MUSEUM, 11913 Main St., Fort Jones, CA 96032. Mailing Address: P.O. Box 428, Fort Jones, CA 96032-0428. Tel.: 530-468-5568.
E-mail: fjmuseum@sisqtel.net
Web Site: fortjonesmuseum.com
Founded: 1947.
Congressional District: 2
Key Personnel: Dir., Cecelia Reuter; Chm. (V), Brenda Mendenhall Mayor Tom McCulley; City Clerk, Linda Romaine.
Personnel Profile: Full-Time Volunteers 2; Part-Time Volunteers 25.
Governing Authority: municipal. Branch Museum: The Fort Jones Museum Carriage House, Corner of Sterling & East St., Fort Jones, CA 96032. Tel.: 530-468-2281. Tax-exempt: 170(b)(1)(A).
Institution Type/Description: Historical Museum.
Collections: Indian beads, baskets; guns; clothing; tools; military, Civil War, WWI & WWII. Carriage House: 9 horse drawn vehicles.
Hours & Admission Prices: Memorial Day to Labor Day Mon.-Fri. 10-4, Sat. 11-3; other times by appointment. No charge; donations accepted.
Attendance: 3,000 (estimated)

Fortuna

CHAPMAN'S GEM AND MINERAL MUSEUM, Hwy. 101, Fortuna, CA 95540. Mailing Address: P.O. Box 32, Carlotta, CA 95528-0032. Tel.: 707-725-2714.
Founded: 1950.
Key Personnel: Co Owner, Lyle Brown; Co Owner, Sharon Brown.
Personnel Profile: Full-Time Paid 2; Part-Time Paid 1.
Governing Authority: private. Tax-exempt.
Institution Type/Description: Gem & Mineral Museum.
Collections: petrified palms; precious stones; Indian & pre-Columbian artifacts; fossils.
Hours & Admission Prices: Daily 10-5. No charge; donations accepted. Closed New Year's Day; Easter; Thanksgiving; Christmas.
Attendance: 14,500 (estimated)

FORTUNA DEPOT MUSEUM, 3 Park St., Fortuna, CA 95540-2461. Tel.: 707-725-7645.
Founded: 1976.
Key Personnel: Cur., Caroline Weed.
Personnel Profile: Part-Time Paid 1; Part-Time Volunteers 4.
Governing Authority: Parent Institution: City of Fortuna.
Institution Type/Description: History Museum.
Collections: Eel River Valley history; railroad, farm & war memorabilia; dolls; fishing gear & lures; Fortuna High School yearbooks; Indian baskets; period household artifacts; hands-on exhibits.
Hours & Admission Prices: June-Aug. daily 12-4:30; Sept.-May Thurs.-Sun. 12-4:30. No charge; donations accepted.
Attendance: 5,000 (estimated)

Fremont

ARDENWOOD HISTORIC FARM, 34600 Ardenwood Blvd., Fremont, CA 94555-3645. Tel.: 510-791-4196.
Web Site: www.ebparks.org/parks/arden.htm
Founded: 1985.
Personnel Profile: Full-Time Paid 14; Part-Time Paid 7; Part-Time Volunteers 400.
Institution Type/Description: Living History Museum.
Collections: 1870s agricultural practices; hands-on exhibits.
Facilities: Organic vegetables for sale.
Activities: educational programs; planting, tending & harvesting; farm chore demonstrations. Museum Sponsors: Christmas program.
Hours & Admission Prices: Tues.-Sun. 10-4. April to late Nov. Tues.-Wed. & Sat.: adults $3, children 4-17 $2, Thurs.-Fri. & Sun.: adults $6, seniors 62 & over $5, children 4-17 $4; children under 4 no charge; late Nov. to March Tues.-Sun. adults $3, children $2; children under 4 no charge. Closed Thanksgiving; Christmas.

LOUIE-MEAGER ART GALLERY, Louie Meager Art Gallery, Ohlone College, 43600 Mission Blvd., Fremont, CA 94539. Tel.: 510-659-6176. Fax: 510-659-6188.
E-mail: kmencher@ohlone.edu
Key Personnel: Dir. & Cur., Kenney Mencher
Institution Type/Description: College Art Gallery.

Collections: works by local & regional artists.
Activities: art workshops; school programs.
Hours & Admission Prices: Mon.-Tues. & Thurs.-Fri. 12-3, Wed. by appointment. No charge. Closed campus holidays & breaks.

MISSION SAN JOSE CHAPEL AND MUSEUM, 43300 Mission Blvd., Fremont, CA 94539-5829. Mailing Address: P.O. Box 3159, Fremont, CA 94539-0315. Tel.: 510-657-1797.
Institution Type/Description: History Museum.
Collections: Ohlone artifacts; period furnishings.
Facilities: Museum-related items for sale.
Hours & Admission Prices: Daily 10-5. Adults $3, student $2. Closed New Year's Day; Easter; Thanksgiving; Christmas.

MUSEUM OF LOCAL HISTORY, 190 Anza St., Fremont, CA 94539-5802. Tel.: 510-623-7907.
E-mail: info@museumoflocalhistory.org
Institution Type/Description: History Museum.
Collections: Fremont's history; historical books & documents.
Facilities: library.
Activities: research.
Hours & Admission Prices: Wed., Fri. & 2nd Sat.-Sun. of month 10-4; groups by appointment.

NILES DEPOT MUSEUM, 37592 Niles Blvd., Fremont, CA 94536. Mailing Address: Niles Depot Historical Foundation, P.O. Box 2716, Fremont, CA 94536-0716. Tel.: 510-797-4449.
Web Site: nilesdepot.railfan.net/ndhfhome.html
Founded: 1982.
Key Personnel: Pres., Rick Zem; Dir., Tom Nelson.
Governing Authority: Tax-exempt.
Institution Type/Description: Railroad Museum: housed in 1901 Southern Pacific Depot.
Collections: railroad & local history museum; HO & N scale model trains; photographs; track equipment; signals; locomotive artifacts & uniforms.
Facilities: library.
Hours & Admission Prices: Sun. 10-4. No charge; donations accepted. &
Attendance: 4,000 (estimated)
Membership: Annual $25.

Fresno

ARTE AMERICAS, 1630 Van Ness Ave., Fresno, CA 93721-1129. Tel.: 559-266-2623. Fax: 559-268-6130.
E-mail: grace@arteamericas.org
Web Site: www.arteamericas.org
Key Personnel: Dir., Grace Solis; Asst. Dir., Mary Ellen G. Clay; Program Dir., Diana Hernandez; Cur., Kristen Sierra
Institution Type/Description: Art Museum.
Collections: arts in Mexico, Latin America, the Southwest & California.
Facilities: library; classrooms. Museum-related items for sale.
Activities: education programs.
Publications: newsletter, Arte Americas.
Hours & Admission Prices: Tues.-Wed. & Fri.-Sat. 11-5, Thurs. 11-8. Adults $3, senior citizens & students $2; members & children under 5 no charge.

COKE HALLOWELL CENTER FOR RIVER STUDIES, 11605 Old Friant Rd., Fresno, CA 93730-9701. Tel.: 559-433-3190. Fax: 559-433-0634.
Web Site: www.riverparkway.org/rivercenter.asp
Key Personnel: Exec. Dir., Dave Koehler
Institution Type/Description: Natural History Museum.
Collections: culture & natural history of the San Joaquin River; 1890s ranch house.
Facilities: rose garden; picnic area; orchard; vineyard. Museum-related items for sale.
Activities: education programs & activities; art workshops; readings; gardening classes; kids' crafts; rental facilities.
Hours & Admission Prices: Ranch House: Fri.-Sun. 11-3; other times by appointment. Grounds: daily 8-5.

THE DISCOVERY CENTER, 1937 N. Winery Ave., Fresno, CA 93703-2828. Mailing Address: 1944 N. Winery, Fresno, CA 93703-2829. Tel.: 559-251-5533. Fax: 559-251-5531.
E-mail: office@thediscoverycenter.net
Web Site: www.thediscoverycenter.net
Founded: 1956.

Congressional District: 17
Key Personnel: C.E.O., Roni Weil; Exec. Dir., Janet Berry; Sec., Karen Perkins.
Personnel Profile: Full-Time Paid 2; Part-Time Paid 6; Part-Time Volunteers 7.
Governing Authority: nonprofit organization. Grounds owned by Fresno City. Tax-exempt: 501(c)(3).
Institution Type/Description: Participatory, Natural History & Natural Science Museum.
Collections: participatory science exhibits; central California natural history; local Native American Indian baskets & artifacts.
Facilities: six-acre science center; freshwater pond; cactus garden; picnic area; urban wildlife center. Museum-related items for sale.
Activities: hands-on exhibits; summer stargazing parties; outdoor equipment & science classes. Center Sponsors: summer science workshops PS-6th grade; Saturday science workshop Pre-K to 6th grade; Star Lab inflatable planetarium.
Publications: newsletter.
Hours & Admission Prices: Mon.-Fri. 9-5, Sat. 10-4. No charge; donations accepted. Closed New Year's Day; Easter; Independence Day; Thanksgiving; Christmas. &
Attendance: 35,000 (estimated)
Membership: Individual 25; Family $40; Friend $250; Corporate $1,000 & up.

DOWNING PLANETARIUM, 5320 N. Maple Ave. M/S DP132, California State University, Fresno, Fresno, CA 93740-8006. Tel.: 559-278-4121. Fax: 559-278-4070.
E-mail: stevenwh@csufresno.edu
Web Site: www.downing-planetarium.org
Founded: 2000.
Institution Type/Description: Planetarium.
Collections: photographs; solar system scale; elements scale; gravity well; sundial; science toys; meteorites.
Facilities: 74-seat theater. Museum-related items for sale.
Hours & Admission Prices: Call for hours.
Attendance: 24,000

FORESTIERE UNDERGROUND GARDENS, 5021 W. Shaw Ave., Fresno, CA 93722-5026. Mailing Address: P.O. Box 1062, Wilton, CA 95693. Tel.: 559-271-0734.
E-mail: tours@undergroundgardens.com
Web Site: www.undergroundgardens.com
Key Personnel: Dir. & Museum Shop Mgr., Valery Forestiere; C.E.O., Lyn Kosewski
Institution Type/Description: General Museum: Baldassare Forestiere spent 40 years sculpting this underground complex using only hand tools.
Collections: underground rooms, niches, courts, patios & passageways including a kitchen, living room, 2 bedrooms, library, bath, & chapel; fish pond, aquarium; auto tunnel; trees.
Activities: tours.
Hours & Admission Prices: Tours: March & Nov. Sat.-Sun. 11-2; April-May & Sept.-Oct. Wed.-Sun. 11-2; Memorial Day to Labor Day Wed.-Sun. 10-4. Adults $14, seniors 60 & over $12, children 5-17 $7; children 4 & under no charge.

* **FRESNO ART MUSEUM, (M),** 2233 N. First St., Fresno, CA 93703-2364. Tel.: 559-441-4221. Fax: 559-441-4227. Facebook: Fresno Art Museum.
E-mail: info@fresnoartmuseum.org
Web Site: www.fresnoartmuseum.org
Formerly: Fresno Art Center
Founded: 1949.
Congressional District: 20
Key Personnel: Exec. Dir., Linda Cano; Assoc. Dir., Eva Torres; Dir. Devel., Annie Schmidt; Registrar, Kristina Hornback; Coord. Membership, Craig Hamilton Arnold; Coord. Education, Susan Yost-Filgate; Art Instructor, Eliana Saucedo; Art Instructor, Leslie Batty; Art Instructor, Scott Macaulay; Mgr. Office & Facilities, Debbie Horton; Receptionist, Betty Peralta; Maintenance & Security, Frank Alvarado; Security, Cesar Soto; Security, Natasha Mendoza; Security, Irene Alvarado.
Personnel Profile: Full-Time Paid 7; Part-Time Paid 8; Part-Time Volunteers 25; Interns 3.
Governing Authority: nonprofit organization. Tax-exempt: 501(c)(3).
Institution Type/Description: Modern & Contemporary Art Museum.
Collections: pre-Columbian Mexican & Andean; contemporary local & California works of art; American sculpture; graphics; photographs.
Research Fields: pre-Columbian.
Facilities: art reference library; sculpture garden; classrooms; auditorium; theater.

Activities: tours; lectures; films; formally organized education programs; docent council; inter-museum loan, temporary & traveling exhibitions; school art class service; concerts; school outreach.
Publications: catalogues of exhibitions; e-newsletter.
Hours & Admission Prices: Thurs.-Sun. 11-5. Adults, seniors & students $5; discounts to AAM, WMG & NARM members; members & children under 5 no charge. Closed national holidays. &
Attendance: 21,000 (estimated)
Membership: Student, Senior & Educator $25; Individual, Senior Couple & Student Couple $50; Family $75; Contributing $150; Sustaining $300; Benefactor $500; Director's Circle $1,000.

FRESNO CHAFFEE ZOO, 894 W. Belmont Ave., Fresno, CA 93728-2807. Tel.: 559-498-5910. Fax: 559-498-5922.
E-mail: info@fresnochaffeezoo.org
Web Site: fresnochaffeezoo.org
Formerly: Chaffee Zoological Gardens of Fresno
Founded: 1929.
Congressional District: 18
Key Personnel: C.E.O., Scott Barton; Chm., Colin Doughtery; Dir. Mktg. & Devel., Terri Mejorado; Museum Shop Mgr., Chris Schiefer.
Personnel Profile: Full-Time Paid 0; Full-Time Volunteers 6; Part-Time Paid 100; Part-Time Volunteers 200.
Governing Authority: nonprofit. Parent Institution: Fresno's Chaffee Zoo Corp. Tax-exempt: 501(c)(3).
Institution Type/Description: Zoo.
Collections: 204 species; 629 specimens of mammals, birds, reptiles, amphibians & fish; herpetology building; aviary; quarantine station; Asian elephant exhibit; tropical rainforest; Australian aviary; Sunda forest.
Research Fields: artificially induced reproduction in amphibians & reptiles.
Facilities: 150-vol. library of zoology & veterinary books. Refreshments & novelties for sale.
Activities: guided tours; lectures; films; zoomobile; wildlife workshops.
Publications: Zoo News.
Hours & Admission Prices: Daily 9-4. Adults $7, senior citizens & children 2-11 $3.50; AZA reciprocal zoos & aquariums; members & children under 2 no charge. &
Attendance: 550,000 (accurate)
Membership: Senior Citizens & Students $30; Individual $40; Plus-One $50; Family & Grandparent $55; Keeper Club $150; Safari Club $500; Toucan Club $1,000.

FRESNO CITY AND COUNTY HISTORICAL SOCIETY, (M), 7160 W. Kearney Blvd., Fresno, CA 93706-9520. Tel.: 559-441-0862. Fax: 559-441-1372.
E-mail: frhistsoc@aol.com
Web Site: www.valleyhistory.org
Key Personnel: Exec. Dir., Jill Moffat
Institution Type/Description: Historical Society Museum.
Collections: city & county history and culture; pioneer records; documents; photographs; scrapbooks; letters; diaries; early local government records.
Hours & Admission Prices: Fri.-Sun. 1, 2 & 3; call to confirm. Adults $5, seniors 60 & over and students 13-17 $4, children 3-12 $3; members & children under 3 no charge.

GALLERY 25, 660 Van Ness Ave., Fresno, CA 93721. Tel.: 559-264-4092.
Web Site: gallery25.org
Founded: 1974.
Key Personnel: Dir., Barbara Van Arnam
Institution Type/Description: Art Gallery.
Collections: works by contemporary artists.
Hours & Admission Prices: 1st Thurs. each month 5pm-8pm, Fri.-Sun. 1-4. No charge.
Attendance: 3,000 (estimated)

KEARNEY MANSION MUSEUM, 7160 W. Kearney Blvd., Fresno, CA 93706-9520. Tel.: 559-441-0862. Fax: 559-441-1372.
E-mail: frhistsoc@aol.com
Web Site: www.valleyhistory.org
Founded: 1919.
Congressional District: 15
Key Personnel: Exec. Dir., Jill Moffat; Pres. (V), John Boogaert; Dir. Public Rels., Christina Perryman; Archivist, Maria Ortiz; Cur. Collections & Education Coord., Sharon Hiigel; Tour Coord. & Museum Shop Mgr., Amy Lawrence; Bookkeeper & Membership, Barbara James Higgins; Oral History Coord., Ruth Lang.
Personnel Profile: Full-Time Paid 6; Part-Time Paid 1; Part-Time Volunteers 225.

Governing Authority: nonprofit organization. FCCHS Board of Trustees, contributions from Fresno County & City. Parent Institution: Fresno Historical Society. Tax-exempt: 501(c)(3).
Institution Type/Description: Historic Site Museum: housed in 1900 original Kearney Mansion.
Collections: costumes & accessories; furniture; textiles; Native American artifacts; household & agricultural implements; musical instruments; archives: 1860s-1960s regional history including photos, negatives, business, individual & family manuscript collections; local government records; maps; biography indexes; ephemera.
Research Fields: local history; agriculture; local families; logging; fire department; regional Indian tribes; hydroelectric development.
Activities: guided tours; bus tours; docent program; temporary exhibitions; field trips; history lectures; outreach programs in schools; historic preservation advocacy. Museum Sponsors: Annual Heritage Home Tour; Traveling History Trunks; Time Travelers Day Camp; Living History at Kearney Park.
Publications: quarterly journal, Fresno Past & Present; quarterly newsletter, Grapevine; books, Evolution of Fruit Vale Estate, reprint of 1904 original; Imperial Fresno, reprint of 1897 original; Evans & Sontag; M. Theo Kearney-Prince of Fresno; California Homes & Industries; Fresno Illustrated.
Hours & Admission Prices: By appointment. Closed New Year's Day; Easter; Independence Day; Thanksgiving; Christmas. &
Attendance: 9,023 (accurate)
Membership: Student & Educator $20; Individual $35; Contributing $45; Sustaining $100; Sponsor $250; Patron $500; Benefactor $1,000 and up; Corporate Partners $1,000-$5,000.

MEUX HOME MUSEUM, 1007 R St., Fresno, CA 93721-1312. Mailing Address: P.O. Box 70, Fresno, CA 93707-0070. Tel.: 559-233-8007. Fax: 559-233-2331.
E-mail: meauhomemuseum@gmail.com
Web Site: www.meux.mus.ca.us
Founded: 1979.
Key Personnel: Pres., Bob Flynn; Vice Pres., Colleen Sethre; Interim Treas., Jan Stafford.
Personnel Profile: Part-Time Volunteers 25.
Governing Authority: private; nonprofit organization. Parent Institution: Meux Home Corp. Tax-exempt: 501(c)(3).
Institution Type/Description: Historic House: 1888 Victorian house.
Collections: Victorian lifestyle: costumes, textiles, furniture, books, pictures; family artifacts; Dr. Meux' Civil War uniform & medical books; Dr Meux' daughter's 1907 wedding gown.
Research Fields: late 19th-century Fresno; history of Meux family's migration from England to Tennessee to California; Victorian clothing.
Facilities: library on Victorian period; 4,600 sq. ft. exhibit space; educational facilities for pre-school through college tours; rose gardens; rental facilities.
Activities: guided tours; seasonal displays; docent program; weddings; special events; rental facilities. Annual Events: Civil War Events; Victorian Teas; Christmas Holiday Event.
Publications: newsletter, Meux Home; docent newsletter.
Hours & Admission Prices: Feb.-Dec. Fri.-Sun. 12-3; private & school tours during week. Adults $5, students $4, children $3; special events rates vary. Closed holidays. &
Attendance: 30,000 (estimated)
Membership: Individual $35; Close Friend of the Family $50; Distant Relative $100; Member of the Family $500; Heir $1,000.

SIMONIAN FARMS, 2629 S. Clovis Ave., Fresno, CA 93725-9307. Tel.: 559-237-2294. Fax: 559-441-1198.
E-mail: simonian@lightspeed.net
Web Site: www.simonianfarms.com
Founded: 1901.
Institution Type/Description: Farm Museum.
Collections: period bicycles; peddle cars; porcelain signs; mannequins; vintage gas pumps; model train.
Hours & Admission Prices: No charge. &

VETERANS MEMORIAL MUSEUM, HOME OF THE LEGION OF VALOR, 2425 Fresno St., Fresno, CA 93721-1841.
Formerly: Legion of Valor Veterans Museum
Founded: 1991.
Congressional District: 20
Key Personnel: Dir. & C.E.O., Robert E. Specht; Chm. (V), Judy Jones; Deputy Dir., Mike Harris; Museum Shop Mgr., Raymond Lee.
Personnel Profile: Part-Time Volunteers 40.
Governing Authority: Tax-exempt.
Institution Type/Description: Military Museum.

Collections: citations; photographs; personal artifacts; uniforms; military equipment.
Hours & Admission Prices: Mon.-Sat. 10-3. No charge; donations accepted. &
Attendance: 15,000 (estimated)

Friant

MILLERTON COURTHOUSE, Department of Parks & Recreation, 5290 Millerton Rd., Friant, CA 93626. Tel.: 559-822-2225. Fax: 209-822-2319.
Institution Type/Description: Historic Building Museum.
Collections: archaeology; basketry; birds; ethnic & tribal art; history; mammals; mineralogy.
Activities: children's classes; guided tours; school outreach program.
Hours & Admission Prices: June-Sept. Sat. 10-6. Courthouse: no charge. Parking: fee charged.

Fullerton

ANTHROPOLOGY TEACHING MUSEUM, CALIFORNIA STATE UNIVERSITY, FULLERTON, 800 N. State College Blvd., Fullerton, CA 92834-6846. Mailing Address: Dept. of Anthropology, Cal-State Univ., Fullerton, P.O. Box 6846, Fullerton, CA 92834-6846. Tel.: 657-278-3626. Fax: 657-278-5001.
E-mail: anthropology@exchange.fullerton.edu
Web Site: www.anthro.fullerton.edu
Founded: 1970.
Congressional District: 39
Key Personnel: Coord. Administrative Support, Debra Redsteer; Dept. Chm., Dr. Mitch Avila; Evolutionary Coord., Dr. John Bock; Cultural Coord., Dr. Barbra Erickson; Archaeology Coord., Dr. Carl Wendt.
Personnel Profile: Full-Time Paid 1; Part-Time Paid 1.
Governing Authority: university; nonprofit. Parent Institution: CSU Fullerton. Tax-exempt: 501(c)(3).
Institution Type/Description: Anthropology Museum.
Collections: southern California, southwest, midwest prehistoric artifacts; ethnographic specimens from South Pacific, Near East, Mexico & South America; faunal, mineral, sherd comparative collections.
Research Fields: southern California; Honduras.
Facilities: 1,640-vol. research library & local faunal, mineral, sherd research collections for use on premises; archaeology laboratory; classrooms; 1,800 sq. ft. exhibit space.
Activities: guided tours; lectures; organized educational programs for undergraduate & graduate student affiliated with CSU Fullerton; temporary exhibitions.
Hours & Admission Prices: School Year: Mon.-Thurs. & by appointment; School Vacation: holidays & weekends by appointment. No charge; donations accepted. &

BEGOVICH GALLERY, CALIFORNIA STATE UNIVERSITY, FULLERTON, 800 N. State College Blvd., Visual Arts Center, Fullerton, CA 92831. Mailing Address: 800 N. State College Blvd., Fullerton, CA 92831. Tel.: 657-278-7750. Fax: 657-278-8191.
E-mail: mmcgee@fullerton.edu
Web Site: www.fullerton.edu/arts/art
Formerly: Main Art Gallery
Founded: 1967.
Congressional District: 39
Key Personnel: Gallery Dir., Mike McGee; Asst. Dir., Jacqueline Bunge; Preparator & Technical Asst., Marty Lorigan.
Governing Authority: university; Affiliated with California State University, Fullerton Art Dept.
Institution Type/Description: University Art Gallery.
Collections: outdoor sculpture; contemporary prints.
Facilities: 1,500-vol. library of museum catalogs & publications relating to museums, their operations & studies, available for research by special permission; 150-seat auditorium; theater; classrooms; cafeteria. Exhibition catalogs for sale.
Activities: guided tours; lectures; films; gallery talks; arts festivals; formally organized education programs for children, adults, undergraduate & graduate students; training programs for professional museum personnel; temporary & traveling exhibitions.
Publications: quarterly, exhibition catalogues.
Hours & Admission Prices: Mon.-Thurs. 12-4, Sat. 12-2, call to confirm. No charge; donations accepted. Closed national holidays. &

THE FULLERTON ARBORETUM, (M), 1900 Associated Rd., Fullerton, CA 92831-1659. Mailing Address: c/o California State University, Fullerton, P.O. Box 6850, Fullerton, CA 92834-6850. Tel.: 657-278-3407. Fax: 657-278-7066.
E-mail: farboretum@fullerton.edu
Web Site: www.fullertonarboretum.org
Founded: 1972.
Congressional District: 39
Key Personnel: Dir., Gregory T. Dyment; Pres. Fullerton Arboretum Commission, Eugene C. Jones; Pres. Friends of the Fullerton Arboretum (V), Mary Dalesci.
Personnel Profile: Full-Time Paid 10; Part-Time Paid 12; Part-Time Volunteers 600.
Governing Authority: municipal & university under joint-powers agreement. Parent Institution: California State University, Fullerton. Tax-exempt: 170(b)(1)(A).
Institution Type/Description: Arboretum & Historic House: 1894 home & office of Dr. George Clark, moved to site of the Gilman Ranch, where first Valencia oranges grown in Orange County were planted.
Collections: plant collection: 3,500 accessions, emphasis on drought tolerant plants suitable to coastal plain of southern California; historical collection: house, outbuildings & artifacts including c.1890 doctor's equipment, office & pharmacy; musical instruments, family memorabilia; Victoriana; furniture & furnishings.
Research Fields: local history; suitability of plants for coastal plain of southern California.
Facilities: library of botanical material available for use on premises; botanical gardens. Plants & plant-oriented gifts for sale.
Activities: guided & self-guided tours; formally organized education programs for children; docent program; temporary exhibitions.
Publications: brochures; newsletter.
Hours & Admission Prices: Arboretum: daily 8-4:45. Heritage House: Sat.-Sun. 2-4; other times by appointment. No charge; donations accepted. Closed New Year's Day; Thanksgiving; Christmas. &
Attendance: 120,000 (estimated)
Membership: Individual $40; Family $75; Partner $100; Steward $250; Patron $500.

FULLERTON COLLEGE ART GALLERY, (M), 321 E. Chapman Ave., Bldg. 1000, Fullerton, CA 92832-2011. Tel.: 714-992-7329. Fax: 714-992-7320.
E-mail: kjohnson@fullcoll.edu
Web Site: art.fullcoll.edu
Key Personnel: Dir., Beth Solomon Marino
Institution Type/Description: Art Gallery.
Collections: paintings; photographs; sculpture.
Facilities: 2,000 sq. ft. exhibition space.
Activities: special events; permanent & temporary exhibitions.
Hours & Admission Prices: Mon. & Wed.-Thurs. 10-2, Tues. 10-2 & 5-7.

FULLERTON MUSEUM CENTER, (M), 301 N. Pomona Ave., Fullerton, CA 92832-1927. Tel.: 714-738-6545. Fax: 714-738-3124.
E-mail: danniellem@ci.fullerton.ca.us
Web Site: www.cityoffullerton.com/depts/museum
Founded: 1971.
Congressional District: 39
Key Personnel: C.E.O., Cindy Yount; Dir., Dannielle Mauk; Museum Shop Mgr., Kelly Chidester.
Personnel Profile: Full-Time Paid 6; Part-Time Paid 7; Part-Time Volunteers 12.
Governing Authority: nonprofit organization. Parent Institution: City of Fullerton Cultural Services Div. Tax-exempt.
Institution Type/Description: General Museum.
Collections: Leo Fender related collections of musical instruments & inventions.
Facilities: 4,500 sq. ft. exhibit space; 5,000 sq. ft. outdoor plaza & stage; 75-seat auditorium. Museum-related items for sale.
Activities: changing exhibitions; lectures; workshops; educational programs for region; permanent exhibitions.
Publications: newsletter; exhibition catalogs.
Hours & Admission Prices: Tues.-Wed. & Fri.-Sun. 12-4, Thurs. 12-8. Adults $4, senior citizens & students $3, children 6-12 $1; discounts for AAM & ICOM members; children 5 & under and members no charge. &
Attendance: 30,000 (estimated)
Membership: Student & Senior $30; Individual $35; Family $50. Fender Levels: Seafoam Green $100; Fiesta Red $250; Silver Mist Metallic $500; Shoreline Gold $1,000.

MUCKENTHALER CULTURAL CENTER AND MANSION MUSEUM, 1201 W. Malvern Ave., Fullerton, CA 92833-2429. Tel.: 714-738-6595 & 6340. Fax: 714-738-6366.
E-mail: info@themuck.org
Web Site: www.themuck.org
Founded: 1965.
Congressional District: 39
Key Personnel: Exec. Dir., Zoot Velasco; Pres. (V), Frederic Ouwelem, Jr.; Museum Shop Mgr., Britt Sullivan.
Personnel Profile: Full-Time Paid 6; Part-Time Paid 10; Part-Time Volunteers 600; Interns 50.
Governing Authority: foundation. Muckenthaler Cultural Center Foundation. Tax-exempt.
Institution Type/Description: Cultural Center: housed in 1923 home of Walter Muckenthaler.
Collections: paintings; photographs; sculpture.
Facilities: 8.5 acre estate; 250-seat amphitheatre; classrooms; 2,300 sq. ft. exhibit space.
Activities: guided tours; lectures; gallery talks; arts festivals; drama; rental gallery; formally organized education programs for children, adults, undergraduate students; docent programs; traveling exhibitions; arts & craft classes; theater presentations; jazz concerts. Annual Events: Spoken Word Series, Poetry Open Mic Night; Luau, Solstice Festival; Christmas Holiday Festivals.
Publications: annual brochures of 2012 programs.
Hours & Admission Prices: Wed. & Fri.-Sun. 12-4, Thurs. 12-4 & 6-9. No charge; donations accepted. Closed holidays. &
Attendance: 28,800 (estimated)
Membership: Student & Senior Citizen $30; Individual $60; Family $100; Golden Ticket $500; Millennium Club $1,000; President's Circle $5,000; Founder's Circle Mucketymucks $10,000.

Garden Grove

GARDEN GROVE HISTORICAL SOCIETY, 12174 Euclid St., Garden Grove, CA 92840. Mailing Address: P.O. Box 4297, Garden Grove, CA 92842-4297. Tel.: 714-530-8871. Fax: 714-534-2611.
E-mail: gardengrovehistsoc@att.net
Founded: 1966.
Congressional District: 38
Key Personnel: Pres., Terry Thomas; Vice Pres., Sherry Weeks; Treas., Valetta Beauchamp.
Personnel Profile: Part-Time Volunteers 40.
Governing Authority: society. Subsidiary Institution: Stanley Ranch Museum. Tax-exempt: 501(c)(3) & 170(b)(1)(A).
Institution Type/Description: Historic House Museum: building completed 1892 by E.G. Ware, one of the first pioneers of Garden Grove.
Collections: agriculture; period artifacts; medical & dental items; restored barbershop; general store; fire station; history; 1926 American La France fire truck. Historic Buildings: c.1878 Original Garden Grove Post Office; 1880's homes; 1916 craftsman style home.
Research Fields: Life in Garden Grove, Orange County & Southern California during the late 1800s and 1900s.
Facilities: library; archive; research room. Handmade crafts & museum-related books for sale.
Activities: guided tours; monthly meetings; volunteer days Mon.-Fri.
Publications: pamphlets; brochures; monthly newsletter.
Hours & Admission Prices: 1st & 3rd Sun. of month 1:30; other times by appointment. Suggested Donation: adults $5, students under 18 $1.
Attendance: 925 (accurate)
Membership: Student $5; Individual $20; Family $30; Business & Organization $40; Sustaining $50; Patron $100; Business Patron $200; Life $500; Centennial $1,874.

Gilroy

CITY OF GILROY MUSEUM, 195 Fifth St., Gilroy, CA 95020-5703. Tel.: 408-846-0446. Fax: 408-847-5604.
Web Site: www.ci.gilroy.ca.us
Formerly: Gilroy Historical Museum
Founded: 1958.
Congressional District: 12
Key Personnel: Supvr., Cathy Mirelez.
Personnel Profile: Part-Time Paid 2; Part-Time Volunteers 35.
Governing Authority: municipal government. Tax-exempt: 501(c)(3).
Institution Type/Description: Local History Museum: housed in 1910 Carnegie Library.

Collections: local history & culture from the Ohlone Indians to present; clothing; accessories; furniture; household items; Ohlone artifacts; tools; toys; archives.
Research Fields: Henry Miller, local cattle king; early California history; family histories & genealogies; local history.
Facilities: 150-vol. library of local, regional & state history; 275-vols. local newspapers; 100-vols. local municipal archives available to the public; 2,500 sq. ft. exhibit space. Museum-related printed material for sale.
Activities: guided tours; lectures; temporary exhibitions; Traveling Trunk Program for schools and community groups. Annual Events: Young Artists Show; Art & Culture Exhibition; summer concert series; Victorian Christmas.
Publications: quarterly newsletter, The Society Page.
Hours & Admission Prices: Tues. & Thurs. 10-5, Wed. by appointment, 1st Sat. of month 10-2. Call for admission information. Closed New Year's Day; Presidents' Day; Memorial Day; Independence Day; Labor Day; Veterans Day; Thanksgiving; Christmas.
Attendance: 2,415 (accurate)

Glen Ellen

JACK LONDON STATE HISTORIC PARK, 2400 London Ranch Rd., Glen Ellen, CA 95442-9749. Tel.: 707-938-5216. Fax: 707-938-5216.
Web Site: www.jacklondonpark.com
Founded: 1959.
Congressional District: 2
Key Personnel: Museum Cur., Carol Dodge.
Personnel Profile: Full-Time Paid 1; Part-Time Paid 3; Part-Time Volunteers 100.
Governing Authority: state. Parent Institution: the State of California Department Parks & Recreation, Box 942896, Sacramento, CA 94296. Tax-exempt.
Institution Type/Description: State Park Museum.
Collections: history; paintings; sculpture. Historic Houses: 1913 ruins of Jack London's Wolf House mansion; 1919 House of Happy Walls; Jack London's ranch cottage & barns.
Facilities: Books for sale.
Activities: guided tours.
Publications: brochures.
Hours & Admission Prices: Park & Museum: Fri.-Mon. 10-5. Cottage: Sat.-Sun. 10-4. Admission per car $8, senior citizens 62 & over $7. Closed New Year's Day; Thanksgiving; Christmas. &
Attendance: 70,000 (estimated)

Glendale

BRAND LIBRARY & ART CENTER, 1601 W. Mountain St., Glendale, CA 91201-1200. Tel.: 818-548-2051. Fax: 818-548-5079.
E-mail: info@brandlibrary.org
Web Site: www.brandlibrary.org
Founded: 1956.
Key Personnel: Sr. Library Supvr., Alyssa Resnick; Librarian, Cathy Billings.
Personnel Profile: Full-Time Paid 5; Part-Time Paid 15; Part-Time Volunteers 4.
Governing Authority: municipal. Parent Institution: City of Glendale. Subsidiary Institution: Glendale Public Library. Tax-exempt.
Institution Type/Description: Library & Art Center.
Collections: art & music books; videos; DVDs; slides; records; CDs; framed prints.
Facilities: library; recital hall.
Activities: professional concerts; art exhibits; dance programs; art tours; lecture series.
Publications: catalogs; Annual Art Exhibition catalog.
Hours & Admission Prices: Tues. & Thurs. 1-9, Wed. 1-6, Fri.-Sat. 1-5. No charge; donations accepted. &
Attendance: 130,000
Membership: Regular $20; Supporting $55; Patron $100; Benefactor $500 & up.

CASA ADOBE DE SAN RAFAEL, 1330 Dorothy Dr., Glendale, CA 91202-1610. Mailing Address: City of Glendale, 613 E. Broadway, Glendale, CA 91206-4391. Tel.: 818-502-9080.
Founded: 1867.
Congressional District: 22
Governing Authority: municipal. Affiliated with Parks, Recreation & Community Services Division, City of Glendale. Parent Institution: Glendale Beautiful. Tax-exempt.

Institution Type/Description: Historic Building Museum: Early Mexican-American heritage.
Collections: early California furniture; artifacts.
Activities: guided tours.
Publications: pamphlet, History of Casa Adobe De San Rafael.
Hours & Admission Prices: Daily 8 to dusk. Guided Tours: Sept.-June first Sun. of month 1-3; July & Aug. every Sun. 1-3. Group tours by appointment. No charge. ♿
Attendance: 1,400 (estimated)
Membership: Annual $7.

DOCTORS' HOUSE MUSEUM, Brand Park, 1601 W. Mountain Ave., Glendale, CA 91201-1200. Mailing Address: c/o The Glendale Historical Society, P.O. Box 4173, Glendale, CA 91202. Tel.: 818-548-2147.
E-mail: tghs@glendalehistorical.org
Web Site: www.glendalehistorical.org/doctors.html
Key Personnel: Dir., Sonia Montejano
Institution Type/Description: History Museum.
Collections: period furnishings, medical implements & supplies.
Hours & Admission Prices: Tours: Sun. 2-4, last tour at 3:40. Adults 16 & up $2; members no charge. Closed New Year's Day; Easter; Mothers Day; Fathers Day; month of July; Christmas; very rainy days.

FOREST LAWN MUSEUM, 1712 S. Glendale Ave., Glendale, CA 91205-3320. Tel.: 800-204-3131. Fax: 323-551-5329. Facebook: Forest Lawn.
E-mail: museum@forestlawn.com
Web Site: forestlawn.com
Founded: 1951.
Key Personnel: Chm. (V) & C.E.O., Darin Drabing; Dir., Joan Adan; Museum Operations Supvr., Elizabeth Bloess.
Personnel Profile: Full-Time Paid 7; Part-Time Paid 1.
Governing Authority: nonprofit organization. Parent Institution: Forest Lawn Memorial - Park Association. Tax-exempt: 501(c)(3).
Institution Type/Description: Art Museum.
Collections: American history from 1770 to present; pre-Colombian Mexican history; American bronze & marble statuary; over 2,000 stained glass windows; paintings; pre-Christian & Christian era coins; gems; autographs; manuscripts.
Major Exhibits: L.A. Woman: Yesterday, Today, and Tomorrow, 8/13-1/14; Light & Hope, 11/13-2/14.
Research Fields: American history from 1770; Renaissance art with special reference to Leonardo da Vinci & Michelangelo; pre-Colombian Mexican history.
Facilities: theater. Mementos of museum & park for sale.
Activities: films; concerts; public school field trips; permanent exhibitions.
Hours & Admission Prices: Tues.-Sun. 10-5. No charge. ♿
Attendance: 50,000 (estimated)

GLENDALE FIRE DEPARTMENT MUSEUM, Fire Station 21, 421 Oak St., Glendale, CA 91204. Tel.: 818-548-4810.
Institution Type/Description: Firefighting History Museum.
Collections: firefighting history & equipment; early fire trucks & apparatus.
Hours & Admission Prices: By appointment.

MUSEUM OF NEON ART, P.O. Box 631, Glendale, CA 91209. Tel.: 213-489-9918. Fax: 213-489-9932.
E-mail: info@neonmona.org
Web Site: www.neonmona.org
Founded: 1981.
Congressional District: 25
Key Personnel: Exec. Dir., Kim Koga.
Personnel Profile: Full-Time Paid 1; Part-Time Paid 1; Part-Time Volunteers 24.
Governing Authority: nonprofit. Tax-exempt: 501(c)(3).
Institution Type/Description: Neon, Electric & Kinetic Art Museum.
Collections: contemporary fine art in electric media; historic neon signs.
Research Fields: history of neon, electric & kinetic art; outdoor advertising art; legal & cultural ramifications of neon art.
Facilities: 500-vol. library of books, slides & videotapes pertaining to art & neon art available for research on premises only by written request; classrooms. Books, electronic jewelry & other museum related items for sale.
Activities: guided tours; lectures; films; gallery talks; concerts; workshops; rental gallery; formally organized education programs.
Publications: quarterly newsletter.
Hours & Admission Prices: Temporarily closed. ♿

Attendance: 6,000 (accurate)
Membership: Neon $35; Argon $50; Neon Mercury $60; Helium $100; Krypton $250; Plasma $500; Xenon $1,000; Sponsor $5,000.

Goleta

RANCHO LA PATERA & STOW HOUSE (GOLETA VALLEY HISTORICAL SOCIETY), (M), 304 N. Los Carneros Rd., Goleta, CA 93117-1502. Tel.: 805-681-7217 & 7216. Fax: 805-681-7217.
E-mail: info@goletahistory.com
Web Site: goletahistory.org
Founded: 1967.
Congressional District: 19
Key Personnel: Exec. Dir., James Kyriaco; Education, Jim McNay; Coord. Events, Dacia Harwood; Caretaker, Linda Foster; Caretaker, Ron Foster.
Personnel Profile: Part-Time Paid 3; Part-Time Volunteers 100; Interns 2.
Governing Authority: nonprofit organization. Tax-exempt: 501(c)(3).
Institution Type/Description: Historical Society Museum: housed in c.1872 Stow House ranch.
Collections: wedding gowns c.1860-1960; farm machinery; five Goleta cannons from 18th & 19th centuries; furnishings.
Facilities: library material available for use on premises; small acreage & lake. Grounds available for weddings & receptions. Books of local history & other museum-related items for sale.
Activities: docent program; guided tours; temporary exhibits; concerts. Annual Events: Old-Fashioned 4th of July; Fiddler's Convention in October; Holiday at the Ranch.
Publications: quarterly newsletter, GVHS Newsletter; biannual publication of Goleta Valley History.
Hours & Admission Prices: Sat.-Sun. 2-4; hours vary for special events. Museum no charge. House Tour: $5. Closed New Year's Day; Easter; Thanksgiving; Christmas. ♿
Attendance: 10,000 (estimated)
Membership: Student $20; Individual $30; Family $50; Friend $100; Pioneer $250; Preservationist $1,000.

SOUTH COAST RAILROAD MUSEUM AT GOLETA DEPOT, (M), 300 N. Los Carneros Rd., Goleta, CA 93117-1502. Tel.: 805-964-3540. Fax: 805-964-3549.
E-mail: director@goletadepot.org
Web Site: www.goletadepot.org
Founded: 1983.
Congressional District: 19
Key Personnel: Exec. Dir., Gary B. Coombs, Ph.D.; Asst. Dir., Phyllis J. Olsen; Pres. (V), Noel Langle.
Personnel Profile: Full-Time Paid 1; Part-Time Paid 2; Part-Time Volunteers 50; Interns 1.
Governing Authority: nonprofit organization. Parent Institution: Institute for American Research. Tax-exempt: 501(c)(3).
Institution Type/Description: Railroad Museum: housed in 1901 Southern Pacific railroad depot.
Collections: railroad items; Southern Pacific railroad; local history.
Research Fields: local history; railroad history.
Facilities: 250-vol. library of railroading & area history material; 14-seat theater. Railroad items & local history books for sale.
Activities: guided tours; films; organized education programs for children; docent program; temporary exhibitions; annual events.
Publications: quarterly newsletter, Depot Dispatch; annual booklet, Publications in Local History Series.
Hours & Admission Prices: Thurs.-Fri. 1-4, Sat.-Sun. 11-4. Donation Requested: adults $1. Closed New Year's Day; Easter; Thanksgiving; Christmas Day to New Year's Day. ♿
Attendance: 22,000 (estimated)
Membership: Individual/Family $15; Sustaining $25; Contributing $50; Milepost 100 $100; Life $200; Associate $500; Builder $1,000; Provider $1,500; Patron $2,500; Benefactor $5,000.

Grass Valley

GRASS VALLEY MUSEUM, 410 S. Church St., Grass Valley, CA 95945-6722. Tel.: 530 272 4725.
E-mail: stjcc@nccn.net & saintjosephsculturalcenter@gmail.com
Web Site: www.saintjosephsculturalcenter.org
Founded: 1972.
Key Personnel: Dir., Joseph Guida.
Personnel Profile: Part-Time Volunteers 10.
Volunteer Hours: 500
Governing Authority: Parent Institution: Historic Mount Saint Mary's Preservation Committee. Subsidiary Institution: St. Joseph's Cultural Center. Tax-exempt.

Institution Type/Description: History Museum: housed in the former orphanage used during Gold Rush times; established by Father William Dalton in 1865 for children orphaned by mining accidents.
Collections: local history from Gold Rush to 1930s; period clothing & furnishings; paintings; early doctor's office; Gold Rush classroom; 130 year old rose garden.
Hours & Admission Prices: April 16-Dec. 18 Wed.-Sat. 12:30-3:30; other times by appointment. No charge; donations accepted.
Attendance: 175 (estimated)

LOLA MONTEZ HOME, 248 Mill St., Grass Valley, CA 95945-6712. Mailing Address: 128 E. Main St., Grass Valley, CA 95945. Tel.: 530-273-4667.
Institution Type/Description: History Museum.
Collections: period furnishings.
Hours & Admission Prices: Mon.-Fri. 9-5. No charge; donations accepted.

Gridley

GRIDLEY MUSEUM, 601 Kentucky St., Gridley, CA 95948. Tel.: 530-846-4482. Facebook: Gridley Museum.
E-mail: gridleymuseum@gmail.com
Web Site: gridleymuseum.com
Founded: 1996.
Key Personnel: Dir., Ruth Ann King; Chm. (V), Robert Trueax.
Personnel Profile: Part-Time Volunteers 24; Interns 1.
Volunteer Hours: 2,000
Operating Expenses: 7,400
Operating Income: 7,500
Governing Authority: Tax-exempt.
Institution Type/Description: History Museum: house in c.1909 Veatch Building.
Collections: Gridley history; photographs; military uniforms; period school artifacts.
Major Exhibits: Farm Tools, 1/14-12/14; Military Exhibit, 1/14-12/14; Photos, 1/14-12/14; Pioneer Family, 1/14-12/14; Wildlife Exhibit, 1/14-12/14; Dentist's & Doctor's Office, 6/1/14-6/30/14.
Research Fields: local business, homes & families.
Publications: newsletter, Museum Musings.
Hours & Admission Prices: Tues.-Fri. 10-2. No charge; donations accepted.
Membership: Senior $15; Individual $25; Senior Family $30; Family $50; Business $100; Patron $500; Benefactor $1,000.

Groveland

GROVELAND YOSEMITE GATEWAY MUSEUM, 18990 Main St., Groveland, CA 95321-9442. Mailing Address: P.O. Box 180, Big Oak Flat, CA 95305. Tel.: 209-962-0300.
E-mail: grovelandmuseum@mlode.com
Web Site: www.grovelandmuseum.org
Founded: 2000.
Congressional District: 19
Key Personnel: Pres., Rolene Kiesling; Cur., Gordon R. Norris; Museum Shop Mgr., Dodie Harte.
Personnel Profile: Part-Time Volunteers 25.
Governing Authority: private; nonprofit organization. Parent Institution: Southern Tuolumne County Historical Society, Groveland, CA. Tax-exempt: 501(c)(3).
Institution Type/Description: History Museum.
Collections: history of southern Tuolumne County, CA; newspapers; paintings; photographs; films; furnishings; personal artifacts.
Facilities: 40-vol. library; 12-seat theater. Museum-related items for sale.
Activities: concerts; docent program; temporary exhibitions; theater. Annual Event: fundraising picnic.
Hours & Admission Prices: Daily 1-4:30. No charge; donations accepted. &
Attendance: 2,500 (estimated)
Membership: Family & Individual $20; Business $100; Life $250.

Gualala

DOLPHIN GALLERY, 39225 Hwy. One, Gualala, CA 95445. Mailing Address: P.O. Box 244, Gualala, CA 95445. Tel.: 707-884-9611 & 785-2219.
Key Personnel: Mgr., Sharon Nickodem; Exhibitions, Nancy Kyle
Institution Type/Description: Art Gallery.
Collections: works by local artists and coastal communities of Sonoma & Mendocino Counties.
Facilities: visitor center.
Activities: temporary exhibitions.

Hours & Admission Prices: Daily 10-5. Closed Thanksgiving; Christmas.

GUALALA ARTS CENTER, 46501 Gualala Rd., Gualala, CA 95445. Mailing Address: P.O. Box 244, Gualala, CA 95445. Tel.: 707-884-1138. Fax: 707-884-3038.
E-mail: info@gualalaarts.org
Web Site: www.gualalaarts.org
Governing Authority: nonprofit organization.
Institution Type/Description: Art Gallery.
Collections: paintings; drawings; sculpture.
Activities: permanent & temporary exhibitions; community events; performing arts; classes; scholarship program. Museum Sponsors: Art in the Redwoods Festival in August.
Hours & Admission Prices: Mon.-Fri. 9-4, Sat.-Sun. 12-4. No charge; donations accepted.

Gustine

GUSTINE MUSEUM, 397 Fourth St., Hwy. 33, Gustine, CA 95322-1131. Mailing Address: 803 Laurel Ave., Gustine, CA 95322. Tel.: 209-854-2344 & 3120 (Business Office).
E-mail: gustinemuseum@gustinehistoricalsociety.org
Web Site: gustinehistoricalsociety.org
Founded: 1990.
Congressional District: 18
Key Personnel: Dir. & Museum Shop Mgr., Kim Stadter; C.E.O., Patricia S. Snoke; Pres. (V), David Perry.
Personnel Profile: Part-Time Paid 2; Part-Time Volunteers 50.
Operating Expenses: 16,529
Operating Income: 17,242
Governing Authority: Parent Institution: Gustine Historical Society. Tax-exempt.
Institution Type/Description: History Museum.
Collections: Gustine history; photographs; memorial plaques; cowboy tack & gear; early dairy industry artifacts; Yokuts Indians; WWI & WWII.
Major Exhibits: Gustine High School 1951-1975, 1/15/14-7/10/14.
Facilities: history center.
Publications: newsletter, The Magpie.
Hours & Admission Prices: Thurs. & Sun. 1-4. No charge; donations accepted. Closed all holidays. &
Attendance: 1,000 (estimated)

Hacienda Heights

YOUTH SCIENCE CENTER, Wedgeworth Elementary School, 16949 Wedgeworth Dr., Hacienda Heights, CA 91745. Mailing Address: P.O. Box 5723, Hacienda Heights, CA 91745-0723. Tel.: 626-854-9825.
E-mail: ysc@youthsciencecenter.org
Web Site: www.youthsciencecenter.org
Founded: 1962.
Congressional District: 39
Key Personnel: C.E.O., Ling-Ling Chang; Chm. (V), Ron Chong; Museum Shop Mgr., Judy Chong.
Personnel Profile: Full-Time Paid 1; Full-Time Volunteers 1; Part-Time Paid 6; Part-Time Volunteers 25.
Governing Authority: Tax-exempt.
Institution Type/Description: Science Center.
Collections: hands-on science exhibits.
Facilities: video and book library. Science-related items for sale.
Activities: family activities; educational programs.
Publications: e-newsletter.
Hours & Admission Prices: Sept.-July Tues. & Thurs. 12-3:45, Sat. 11-3. No charge; donations accepted.
Attendance: 2,000 (estimated)
Membership: Sustaining $50; Business $100; Life $250.

Half Moon Bay

COASTAL ARTS LEAGUE MUSEUM, 300 Main St., Ste. 3, Half Moon Bay, CA 94019-1742. Tel.: 650-726-6335.
E-mail: coastalartsleague@gmail.com
Web Site: www.coastalartsleague.com
Founded: 1979.
Congressional District: 14
Key Personnel: Pres. (V), Randall Reid; Museum Shop Mgr., Patricia Dailey, M.D.
Personnel Profile: Full-Time Volunteers 11; Part-Time Volunteers 3.
Volunteer Hours: 15

Operating Expenses: 78,716
Operating Income: 68,062
Governing Authority: nonprofit organization. Tax-exempt: 501(c)(3).
Institution Type/Description: Art Museum.
Collections: paintings; photographs; sculpture; ceramics.
Major Exhibits: Maverick's Everest of the Sea, 1/24/14-3/2/14; Half Moon Bay High School Exhibit, 4/25/14-5/25/14; Local Pleine Air Painters, 5/30/14-6/29/14; International Kellicett Photography Show (T), 7/4/14-8/7/14; Erie Shirpira Paintings/Sculpture, 8/8/14-9/7/14; National Women's Caucus For Arts - Peninsula Chapter, 9/12/14-10/12/14; Kendra Davis, 10/17/14-11/16/14; Cal Holiday Bazar & Fund Raiser, 11/21/14-12/7/14; 30th Annual Juried Show, 12/2/14-1/11/15.
Facilities: Gift items for sale.
Activities: facility rental; educational programs; classes; art demonstrations.
Publications: newsletter.
Hours & Admission Prices: Thurs.-Mon. 11-5. No charge; donations accepted.
Attendance: 3,500 (estimated)
Membership: Junior, Student & Senior $20; Supporting $35; Family $60; Corporate $100; Patron $1,000.

Hanford

CLARK CENTER FOR JAPANESE ART AND CULTURE, 15770 Tenth Ave., Hanford, CA 93230-9533. Tel.: 559-582-4915. Fax: 559-582-9546.
E-mail: info@ccjac.org
Web Site: www.ccjac.org
Formerly: The Ruth & Sherman Lee Institute for Japanese Art
Founded: 1995.
Congressional District: 5
Key Personnel: Dir., Dr. Andreas Marks; Chm. (V), Richard L. Schafer.
Personnel Profile: Full-Time Paid 4; Part-Time Paid 1; Part-Time Volunteers 50; Interns 2.
Governing Authority: bd. of directors. Tax-exempt.
Institution Type/Description: Art Museum.
Collections: Japanese paintings, sculpture & decorative arts; Buddhist paintings & sculpture; bonsai trees.
Facilities: library.
Publications: newsletter, Tokonoma; exhibition catalogs.
Hours & Admission Prices: Sept.-July Tues.-Sat. 12:30-5. Adults $5; discounts to children 12 & under and students; members no charge. Closed major holidays.
Attendance: 5,000 (estimated)
Membership: Individual $50; Friend $150; Donor $500; Sponsor $1,200; Patron $2,500; Benefactor $5,000.

HANFORD CARNEGIE MUSEUM, 109 E. 8th St., Hanford, CA 93230-3933. Tel.: 559-584-1367.
E-mail: hanfordcarnegie@gmail.com
Founded: 1975.
Key Personnel: Pres. (V), Melanie Hill.
Personnel Profile: Part-Time Paid 1.
Governing Authority: Tax-exempt.
Institution Type/Description: History Museum.
Collections: Hanford's history; personal artifacts; photographs; furnishings; clothing; toys; Yokuts Indians.
Hours & Admission Prices: Wed.-Sat. 10-2. Adults $3, children, students & seniors $1; members no charge.
Attendance: 2,750 (estimated)
Membership: Carnegie Individual $15; Carnegie Family $25; Historian $50; Cornerstone $100; Preservationist $250; Landmark $500.

TAOIST TEMPLE AND MUSEUM, 12 China Alley, Hanford, CA 93230. Mailing Address: P.O. Box 728, Hanford, CA 93232-0728. Tel.: 559-582-4508.
Key Personnel: Pres. (V), Arianne Wing; Museum Shop Mgr., Camille Wing.
Personnel Profile: Part-Time Volunteers 10.
Governing Authority: Tax exempt.
Institution Type/Description: History Museum.
Collections: furnishings; Hanford's Chinese residents; photographs; Chinese herb store artifacts; gambling houses; kitchen artifacts from homes & restaurants; Chinese temple including altar & furnishings.
Facilities: Museum-related items for sale.
Hours & Admission Prices: Call for hours. No charge; donations accepted.
Attendance: 1,271 (estimated)
Membership: Student $5; Individual $15; Family $25; Patron $50; Sustaining $100.

Hayward

C.E. SMITH MUSEUM OF ANTHROPOLOGY, CSU East Bay, Hayward, CA 94542-3039. Tel.: 510-885-3168 & 3104. Fax: 510-885-3353.
E-mail: george.miller@csueastbay.edu
Web Site: class.csueastbay.edu/anthropologymuseum/
Founded: 1975.
Key Personnel: Dir. & Cur., Prof. George Miller.
Personnel Profile: Part-Time Paid 5.
Governing Authority: university. Parent Institution: Cal State University East Bay. Subsidiary Institution: College of Letters, Arts & Social Sciences.
Institution Type/Description: Anthropology Museum.
Collections: Jack Lee Kachina collection; American Southwest artifacts; Native American baskets; African art; Philippine artifacts; California Bay Area & Gold Rush artifacts; Andean textiles.
Research Fields: Hopi kachinas; Bay Area archaeology.
Facilities: library, available to the public.
Activities: docent program; temporary exhibitions; museum coursework.
Hours & Admission Prices: Mon.-Fri. 11-4 during exhibits. No charge; donations accepted. Closed federal holidays; university holidays.
Attendance: 2,000 (estimated)
Membership: Student $5; Individual $20; Supporting $35; Sustaining $50; Sponsor $200; Patron $500.

HAYWARD AREA HISTORICAL SOCIETY, (M), 22380 Foothill Blvd., Hayward, CA 94541-5113. Tel.: 510-581-0223. Fax: 510-581-0217.
Web Site: www.haywardareahistory.org
Founded: 1956.
Congressional District: 13
Key Personnel: Exec. Dir., Myron Freedman; Pres., Brian Morrison; Mgr. Collections, Heather Farquhar; Dir. Devel., Alison Wenz; Mgr. Facilities, Dion Griffin; Cur., Diane Curry; Coord. Education, Johanna Fassbender.
Personnel Profile: Full-Time Paid 6; Part-Time Paid 4; Part-Time Volunteers 50; Interns 8.
Governing Authority: private; nonprofit organization. Subsidiary Institution: McConaghy House, Hayward, CA. Tax-exempt: 501(c)(3).
Institution Type/Description: Historical Society.
Collections: focus on the history of Eden Township; furnishings; personal artifacts; recreational artifacts.
Research Fields: local history.
Facilities: 500-vol. library of local history books; 1,800 sq. ft. exhibit space. Victoriana, local history books & other museum-related items for sale.
Activities: docent program; formal education programs for children; lectures; participatory & temporary exhibits; walking tours. Annual Event: Preservation Gala & Awards.
Publications: quarterly newsletter, Adobe Trails.
Hours & Admission Prices: Closed for relocation.
Attendance: 10,000 (accurate)
Membership: Senior $20; Student & Teacher $25; Individual $30; Senior Family $35; Family $40; Nonprofit $50; Silver $75; Business & Club $100; Gold $125; Diamond $200; Historian $1,000.

MCCONAGHY HOUSE, 18701 Hesperian Blvd., Hayward, CA 94541-2247. Mailing Address: 22380 Foothill Blvd., Hayward, CA 94541-5113. Tel.: 510-581-0223. Fax: 510-581-0217.
Web Site: www.haywardareahistory.org
Founded: 1976.
Congressional District: 13
Key Personnel: C.E.O., Myron Freedman; Pres. (V), Brian Morrison; Cur., Diane Curry; Collections Mgr., Heather Farguhar; Dir. Devel., Alison Wenz; Education Coord., Johanna Fassbender; Facilities Mgr., Dion Griffin.
Personnel Profile: Full-Time Paid 6; Part-Time Paid 4; Part-Time Volunteers 25.
Governing Authority: private; nonprofit. Parent Institution: Hayward Area Historical Society, Hayward CA. Tax-exempt: 501(c)(3).
Institution Type/Description: Historic Home: 1886 Victorian farmhouse depicting the lifestyle of one of the area's first pioneer families.
Collections: Victorian furnishings & decorative arts; carriage house; tank house.
Research Fields: family history.
Facilities: 6,000 sq. ft. exhibit space. Victoriana for sale.
Activities: docent program; formal education programs for children; guided tours.
Publications: quarterly newsletter, HAHS Adobe Trails.
Hours & Admission Prices: Feb.-Dec. Sat. 10-4, Sun. 1-4. Adults $5, senior

citizens & students $3; discounts AAM members; children under 10 & HAHS members no charge. Closed New Year's Day; Thanksgiving; Christmas.
Attendance: 4,583 (accurate)

SULPHUR CREEK NATURE CENTER, 1801 D St., Hayward, CA 94541-4434. Tel.: 510-881-6747. Fax: 510-888-0129. Facebook: Sulphur Creek Nature Center.
E-mail: nature@haywardrec.org
Web Site: www.haywardrec.org
Founded: 1961.
Congressional District: 14
Key Personnel: Coord., Wendy Winsted.
Personnel Profile: Full-Time Paid 1; Part-Time Paid 12; Part-Time Volunteers 150.
Volunteer Hours: 18,000
Operating Expenses: 209,735
Operating Income: 182,000
Governing Authority: municipal. Parent Institution: Hayward Area Recreation & Park District, 1099 E St., Hayward, CA 94541. Tax-exempt.
Institution Type/Description: Nature Center.
Collections: native wildlife including live mammals, birds, reptiles, amphibians & various invertebrates.
Major Exhibits: Nests: Biodiversity, 1/14-12/14.
Facilities: nature center; hiking trail; picnic area.
Activities: wildlife education programs for school groups & the general public; tours; on-site field explorations; special topic programs; outreach programs; wildlife rehabilitation; animal rental library; nature study classes; summer camps; special events; volunteer programs; docent program; live animal presentations; scout programs; convalescent outreach.
Publications: bimonthly newsletter; quarterly schedule of programs & activities.
Hours & Admission Prices: Daily 10-5. No charge; donations accepted. Closed New Year's Day; Martin Luther King Jr. Day; Presidents' Day; Veterans Day; Thanksgiving & day after; Christmas Eve & Day.
Attendance: 40,000 (accurate)

SUN GALLERY, 1015 E St., Hayward, CA 94541-5210. Tel.: 510-581-4050. Fax: 510-581-3384.
E-mail: sungallery@comcast.net
Web Site: sungallery.org
Founded: 1975.
Congressional District: 10
Key Personnel: Dir., Valerie Caveglia; Pres. (V), Orlando Somoza; Asst. to Bd. of Dir., Christine Bender; Museum Shop Mgr., Audrey LePell.
Personnel Profile: Part-Time Paid 5; Part-Time Volunteers 5; Interns 2.
Governing Authority: nonprofit. Parent Institution: Hayward Area Forum of the Arts, Inc. Tax-exempt: 501(c)(3).
Institution Type/Description: Visual Arts Center.
Collections: contemporary art by northern California artists; works by artists of 1960s & 1970s.
Facilities: art classroom. Ceramics, glass, wood items & jewelry for sale.
Activities: guided tours; lectures; gallery talks; arts festivals; docent program or council; school field trips; hands-on art activities; summer art classes for children; training programs; permanent & temporary exhibitions.
Publications: newsletter, Sun Gallery
Hours & Admission Prices: Wed.-Sat. 11-5. No charge; donations accepted. Closed major holidays. ♿
Attendance: 5,000 (estimated)
Membership: Student $10; Senior Citizen $25; Individual $35; Senior Couple $40; Family $50; Sponsor $100; Silver $250; Gold $500; Platinum $1,000.

Healdsburg

HEALDSBURG MUSEUM, (M), 221 Matheson St., Healdsburg, CA 95448. Mailing Address: P.O. Box 952, Healdsburg, CA 95448-0952. Tel.: 707-431-3325. Fax: 707-473-4471.
E-mail: healdsburg@sbcglobal.net
Web Site: www.healdsburgmuseum.org
Key Personnel: Cur., Daniel Murley
Institution Type/Description: History Museum.
Collections: local history; photographs.
Facilities: library.
Activities: educational programs; special events.
Publications: newsletter.
Hours & Admission Prices: Museum: Tues.-Sun. 11-4. Research Archives: Thurs.-Sat. by appointment. No charge.

TILE HERITAGE FOUNDATION, (M), Healdsburg, CA 95448. Mailing Address: Box 1850, Healdsburg, CA 95448-1850. Tel.: 707-431-8453. Fax: 707-431-8455. Facebook: Tile Heritage Foundation.
E-mail: foundation@tileheritage.org
Web Site: www.tileheritage.org
Founded: 1987.
Congressional District: 1
Key Personnel: Pres., Joseph A. Taylor; Financial Dir., Sheila A. Menzies.
Personnel Profile: Full-Time Paid 2; Part-Time Paid 2; Part-Time Volunteers 3.
Governing Authority: public charity; nonprofit organization. Tax-exempt: 501(c)(3).
Institution Type/Description: Art Foundation: for the research & preservation of ceramic surfaces.
Collections: archival library of historic & contemporary information on ceramic surfaces in the U.S.; designers; manufacturers; showroom sites; photographs; slides; books; catalogs.
Research Fields: history & preservation of ceramic surfaces.
Facilities: 300-vol. library; 28,000 document archives; field research station. Museum-related items for sale.
Activities: guided tours of tile sites; workshops; lectures; school loan services; loan & temporary exhibits.
Publications: journal, Tile Heritage: A Review of American Tile History; newsletter, E-News.
Hours & Admission Prices: Mon.-Sat. 10-4 by appointment only. No charge; donations accepted. Closed Easter; Independence Day; Thanksgiving; Christmas week. ♿
Membership: Individuals & Families: Regular $45; Supporting $60; Sustaining $75; Friend $150; Guardian $300; Patron $500. Businesses & Corporations: Centurian $100; Sponsor $250; Donor $500; Benefactor $1,000.

Hemet

FINGERPRINTS YOUTH MUSEUM, 123 S. Carmalita St., Hemet, CA 92543-4210. Mailing Address: 418 E. Florida Ave., Hemet, CA 92543-4210. Tel.: 951-765-1223. Fax: 951-652-0064.
E-mail: director@fingerprintsyouthmuseum.com
Web Site: www.fingerprintsmuseum.com
Formerly: The KidZone Riverside County Youth Museum
Governing Authority: Tax-exempt: 501(c)(3).
Institution Type/Description: Children's Museum.
Collections: hands-on exhibits.
Hours & Admission Prices: Winter: Tues.-Fri. 11-5, Sat. 9-5; Summer: Tues.-Sat. 9-5. Admission $5, seniors 55 & over $4; members & children under 2 no charge. ♿
Attendance: 10,000 (accurate)
Membership: Family of 4 $100; Family of 8 $130.

HEMET MUSEUM, Santa Fe Depot/State & Florida, Hemet, CA 92543. Mailing Address: 1126 Griffith Way, Hemet, CA 92543. Tel.: 951-929-4409 & 5885.
Web Site: www.hemetmuseum.org
Founded: 1973.
Congressional District: 65
Key Personnel: Cur.8, Anne B. Jennings; Museum Shop Mgr., Virginia Sisk.
Personnel Profile: Part-Time Volunteers 40.
Governing Authority: private; nonprofit organization. Parent Institution: Hemet Area Museum Assoc. Tax-exempt.
Institution Type/Description: History Museum: housed in 1898 freight house of the historic Santa Fe Depot.
Collections: concentration on Hemet area history from prehistoric to modern era; photographs; documents; artifacts; memorabilia; clothes; household wares; business equipment; fossils dating to the Pleistocene era.
Research Fields: local history.
Facilities: 100-vol. library of California & local history; garden. Museum-related items for sale.
Activities: guided tours; special tours for school children. Museum Sponsors: Open House; Old Timer Gatherings; Pioneer Picnic; outreach program in elementary schools.
Publications: book, Valley, River, Mountain - Fortune Favors the Brave Revisited: A History of the Lake Hemet Water Co; San Jacinto Valley Railway; Whittier, Fuller & Co; Vignettes of the Valley.
Hours & Admission Prices: Sept.-July Tues.-Sun. 11-3. No charge; donations accepted. Closed New Year's Day; Independence Day; Labor Day; Thanksgiving; Christmas. ♿
Attendance: 4,510 (accurate)
Membership: Individual $10; Family $15; Organization, Business & Contributing $25; Lifetime Individual $150; Lifetime Couple $250.

WESTERN SCIENCE CENTER, (M), 2345 Searl Pkwy., Hemet, CA 92543-9706. Tel.: 951-791-0033. Fax: 951-791-0032.
Web Site: westerncentermuseum.org
Formerly: Western Center for Archaeology and Paleontology
Founded: 1998.
Congressional District: 45
Key Personnel: Exec. Dir., Bill Marshall, Ed.D.
Personnel Profile: Full-Time Paid 4; Part-Time Paid 6; Part-Time Volunteers 45; Interns 2.
Governing Authority: Parent Institution: Western Center Community Foundation. Tax-exempt.
Institution Type/Description: Archaeology & Paleontology Museum.
Collections: archaeology; paleontology; fossils; Diamond Valley Lake artifacts.
Facilities: classrooms; banquet facilities.
Activities: educational programs; teaching laboratory; field trips; Science Saturdays; lectures; community events; summer camps; simulated dig site programs.
Hours & Admission Prices: Tues.-Sun. 10-5. Adults $8, seniors 62 & over and students 13-22 $6.50, youth 5-12 $6; members, military and children 4 & under no charge. &
Attendance: 45,000 (accurate)
Membership: Student $30; Senior $40; Individual $50; Family $75.

Hermosa Beach

HERMOSA BEACH HISTORICAL MUSEUM, 710 Pier Ave., Hermosa Beach, CA 90254-3940. Tel.: 310-318-9421.
E-mail: hbhs@hermosabeachhistoricalsociety.org
Web Site: www.hermosabeachhistoricalsociety.org
Founded: 1987.
Personnel Profile: Part-Time Paid 1.
Governing Authority: Parent Institution: Hermosa Beach Historical Society. Tax-exempt.
Institution Type/Description: Historical Society Museum.
Collections: local history; photographs; personal artifacts.
Hours & Admission Prices: Wed. 10-12, Sat.-Sun. 2-4. No charge; donations accepted. Closed holidays.

Hollister

SAN BENITO COUNTY HISTORICAL SOCIETY MUSEUM, (M), 498 5th St., Hollister, CA 95023-3841. Tel.: 831-635-0335.
E-mail: info@sbchistoricalsociety.org
Web Site: sbchistoricalsociety.org
Founded: 1956.
Key Personnel: Pres. & Dir., Peter Sonne.
Personnel Profile: Full-Time Volunteers 2; Part-Time Volunteers 6.
Governing Authority: nonprofit organization.
Institution Type/Description: History Museum.
Collections: photographs; artifacts; clothing; farm and farm-life related tools & equipment. Historic Buildings: c.1860 home with outhouse; c.1890 one-room schoolhouse; early 1900s public bar; large dance hall; several open-faced barns; blacksmith & carpenter shops; print shop; gas engine shop; 1890s drug store; Hose Company #2, c.1875 firehouse.
Hours & Admission Prices: Historical Society Museum: Sat.-Sun. 1-3; groups by appointments. No charge; donations accepted. Historical Village: daily dawn to dusk. Parking fee: $3.
Attendance: 475 (estimated)
Membership: Junior (under 18) $3; Individual $20; Family $30; Premier $50.

Hollywood

HOLLYWOOD GUINNESS WORLD OF RECORDS MUSEUM, 6764 Hollywood Blvd., Hollywood, CA 90028-4622. Tel.: 323-463-6433. Fax: 323-462-3953. Facebook: Guinness Museum Hollywood.
Web Site: www.guinnessmuseumhollywood.com
Founded: 1991.
Key Personnel: Gen. Mgr., Lon Casey; Corp. Communications Dir., Aileen Stein, Partner, Tej Sundher.
Governing Authority: company organized for profit. Parent Institution: World of Records Exhibition, LTD (Licensor) Beaulieu Hampshire England.
Institution Type/Description: Guinness Book of World Records Museum: housed c.1913 Hollywood's first movie theatre.
Collections: world record facts, feats & record holders; photographs; personal artifacts; movie memorabilia.
Facilities: 15,000 sq. ft. exhibit space.
Hours & Admission Prices: Daily 10 a.m.-midnight. Adults $15.99, senior citizens $13.99, children 5-12 $7.99; children under 5 no charge. &

Attendance: 125,000 (estimated)

HOLLYWOOD HERITAGE MUSEUM, 2100 Highland Ave., Hollywood, CA 90068-3241. Mailing Address: P.O. Box 2586, Hollywood, CA 90078-2586. Tel.: 323-874-2276 & 4005.
Web Site: www.hollywoodheritage.org
Formerly: Hollywood Studio Museum
Founded: 1982.
Congressional District: 13
Key Personnel: Dir., Cur. & Museum Shop Mgr., George Kiel; Pres. (V), Richard Adkins.
Personnel Profile: Part-Time Volunteers 22; Interns 1.
Volunteer Hours: 2,400
Operating Expenses: 32,000
Operating Income: 36,000
Governing Authority: nonprofit. Parent Institution: Hollywood Heritage, Inc., P.O. Box 2586, Hollywood. Subsidiary Institution: Hollywood Heritage Museum in the Lasky-DeMille Barn. Tax-exempt: 501(c)(3).
Institution Type/Description: Early Hollywood History & Heritage Museum: housed in restored 1895 barn which was adapted for use as a film studio in Hollywood in 1912; became Paramount Studios in 1916; state historic landmark in 1955.
Collections: the development & history of the film industry in Southern California, predominantly Hollywood & Los Angeles; the early years of filmmaking & silent films from 1906-1931; history of the principals of the Jesse L. Laskey Feature Play Co., Famous Players Co., and Paramount Pictures Co.
Research Fields: history of film industry in Southern California; silent film industry; musical scores of silent films; excavation of film set of The Ten Commandments; architecture of historically significant Hollywood buildings.
Facilities: 600-vol. library; 3,000 sq. ft. exhibit space. Film-related items for sale.
Activities: guided tours; lectures; docent program; films; temporary exhibits; film screenings. Museum Sponsors: quarterly Paper Collectible Show & Sale.
Publications: quarterly newsletter, Hollywood Heritage Inc.
Hours & Admission Prices: Wed.-Sun. 12-4. Adults $7, children under 12 & members no charge. Closed New Year's Day; Thanksgiving; Christmas. &
Attendance: 2,000 (estimated)
Membership: Senior $25; Individual $40; Household $50; Triangle $100; Kalem $250; Bison $500; Keystone $1,000; Majestic $2,500.

HOLLYWOOD MUSEUM, 1660 N. Highland Ave., Hollywood, CA 90028-6121. Tel.: 323-464-7776. Fax: 323-464-3777.
E-mail: info@thehollywoodmuseum.com
Web Site: www.thehollywoodmuseum.com
Founded: 1997.
Congressional District: 43rd
Key Personnel: Volunteer Chm., Donelle Dadigan.
Personnel Profile: Full-Time Paid 3; Full-Time Volunteers 1; Part-Time Paid 2; Part-Time Volunteers 1.
Governing Authority: Tax exempt.
Institution Type/Description: Movie History Museum: housed in the landmark Max Factor building.
Collections: movie memorabilia; personal artifacts; photographs; clothing; automobiles; posters; sets; props.
Hours & Admission Prices: Wed.-Sun. 10-5. Adults $15, seniors & students under 21 $12, children under 5 $5. &

HOLLYWOOD WAX MUSEUM, INC., 6767 Hollywood Blvd., Hollywood, CA 90028-4623. Tel.: 323-462-8860. Fax: 323-462-3953.
E-mail: contact@hollywoodwax.com
Web Site: www.hollywoodwax.com
Founded: 1965.
Key Personnel: Financial Dir., Chanchil Sundher; Cur., Ken Horn; Public Rels., Raubi Sundher.
Personnel Profile: Full-Time Paid 30; Part-Time Paid 3.
Governing Authority: company organized for profit.
Institution Type/Description: Wax Museum: housed in c.1929 former Embassy Club.
Collections: sculptures.
Facilities: Tourist novelties for sale.
Hours & Admission Prices: Call for hours. Adults $15.95, senior citizens $13.95, children 6-12 $6.95; children under 5 no charge.
Attendance: 130,000 (estimated)

LOS ANGELES FIRE DEPARTMENT MUSEUM AND MEMO-RIAL, 1355 N. Cahuenga Blvd., Hollywood, CA 90028. Tel.: 323-464-2727. Fax: 323-464-7401.
E-mail: lafdhsmuseum@msn.com
Web Site: lafdmuseum.org/museum_hollywood
Institution Type/Description: Firefighting History Museum.
Collections: firefighting history & equipment; fire trucks; hoses; ladders; uniforms; photographs.
Facilities: library. Museum-related items for sale.
Activities: rental facilities.
Hours & Admission Prices: Sat. 10-4.

Homewood

TAHOE MARITIME MUSEUM, (M), 5205 W. Lake Blvd., Homewood, CA 96141. Mailing Address: P.O. Box 627, Homewood, CA 96141-0627. Tel.: 530-525-9253. Fax: 530-525-9283.
E-mail: info@tahoemaritime.org
Web Site: www.tahoemaritime.org
Founded: 1988.
Key Personnel: Exec. Dir., Lora Nadolski; Pres. (V), Dave Olson.
Personnel Profile: Full-Time Paid 3; Part-Time Paid 1.
Governing Authority: nonprofit organization. Tax-exempt: 501(c)(3).
Institution Type/Description: Maritime Museum.
Collections: local maritime history; boats; photographs; period artifacts.
Research Fields: Lake Tahoe; maritime; boating.
Activities: special tours; childrens programs; lectures.
Publications: Tahoe Maritimes.
Hours & Admission Prices: June-Oct. Thurs.-Tues. 10-5; Nov.-May Fri.-Sun. 10-4:30. Adults $5; members & children under 12 no charge. ♿
Membership: $40; $100; $200; $300; $500; Friend $1,000; Sustaining Friend $2,500.

Hoopa

HOOPA VALLEY TRIBAL MUSEUM, Hwy. 96, Hoopa, CA 95546. Mailing Address: P.O. Box 1348, Hoopa, CA 95546-1348. Tel.: 530-625-4110.
E-mail: museum@hoopa-nsn.gov
Web Site: bss.sfsu.edu/calstudies/hupa/hoopa.htm
Founded: 1979.
Key Personnel: Dir., Silis-chi-tawn Jackson.
Personnel Profile: Full-Time Paid 2; Part-Time Paid 1; Part-Time Volunteers 1; Interns 1.
Governing Authority: nonprofit. Parent Institution: Hoopa Valley Tribe. Tax-exempt.
Institution Type/Description: Tribal Museum.
Collections: culture & history of native people of northern California; Hupa, Yurok, & Karuk artifacts; basketry; ceremonial regalia; redwood dugout canoes; tools.
Activities: Hupa language classes; basketry classes; village tours; storytelling events; dance demonstrations.
Hours & Admission Prices: Mon.-Fri. 8-12 & 1-5, Sat. 10-12 & 1-4. Village Tours: $10 per person, groups of 6 or more $50.
Attendance: 2,518 (accurate)

Huntington Beach

BOLSA CHICA CONSERVANCY, 3842 Warner Ave., Huntington Beach, CA 92649-4263. Tel.: 714-846-1114. Fax: 714-846-4065. Facebook: Bolsa Chica Conservancy.
E-mail: info@bolsachica.org
Web Site: bolsachica.org
Key Personnel: Exec. Dir., Grace Adams.
Governing Authority: nonprofit organization.
Institution Type/Description: Conservatory.
Collections: preservation, restoration, & enhancement of Bolsa Chica Wetlands; native plants.
Facilities: library.
Hours & Admission Prices: Daily 9-4.
Membership: Student $35; Active Volunteer $55; Basic $100; Sustaining $250; Advocate $500; Benefactor $1,000; Patron $2,500; Corporate $5,000 & up.

FINE ARTS GALLERY AT GOLDEN WEST COLLEGE, 15744 Goldenwest Street, Huntington Beach, CA 92647-2748. Tel.: 714-892-7711, ext. 51032.
E-mail: dhudson@gwc.cccd.edu
Web Site: www.goldenwestcollege.edu/gallery
Institution Type/Description: Art Museum.

Collections: paintings; photographs.
Activities: special events.
Hours & Admission Prices: Mon. & Thurs. 10-2, Tues.-Wed. 10-2 & 5-8, Fri. by appointment.

HUNTINGTON BEACH ART CENTER, 538 Main St., Huntington Beach, CA 92648. Tel.: 714-374-1650. Fax: 714-374-5304.
Web Site: www.huntingtonbeachartcenter.org
Institution Type/Description: Art Gallery.
Collections: works by contemporary artists.
Activities: lectures; special events.
Hours & Admission Prices: Tues.-Thurs. 12-8, Fri. 12-6, Sat. 12-5.

INTERNATIONAL SURFING MUSEUM, 411 Olive Ave., Huntington Beach, CA 92648. Mailing Address: P.O. Box 782, Huntington Beach, CA 92648-0782. Tel.: 714-960-3483. Fax: 714-960-1434. TDD: 714-960-3483.
E-mail: info@surfingmuseum.org
Web Site: www.surfingmuseum.org
Formerly: Huntington Beach International Surfing Museum
Founded: 1987.
Congressional District: 42
Key Personnel: Chm., Brett Barnes; Vice Chm., Lee Love; Sec., Kellie Reynolds; Financial Business Admin., Paul Taylor; Treas. & Museum Shop Mgr., Tom Gibbons; Dir. At-Large, Gary Sahagen.
Personnel Profile: Full-Time Volunteers 2; Part-Time Volunteers 1; Interns 3.
Governing Authority: nonprofit organization. Tax-exempt: 501(c)(3).
Institution Type/Description: Surfing Museum: housed in restored Art Deco building.
Collections: c.1800-present surfboards; photos; art; sculpture; film; books; magazines; posters; clothing; surf guitars; albums; cameras; documents.
Research Fields: surf lifesaving; surfing history.
Facilities: 2,000 sq. ft. exhibit space. Museum-related items for sale.
Activities: docent program; films; guided tours; loan, temporary, participatory & traveling exhibitions; broadcast programs. Museum Sponsors: monthly Art Walk Exhibits.
Publications: quarterly newsletter, Shore Break.
Hours & Admission Prices: Mon.-Fri. 12-5, Sat.-Sun. 11-6. No charge; donations accepted. Closed New Year's Day; Christmas. ♿
Attendance: 10,000 (estimated)
Membership: Individual $20; Family $35; Patron $500; Lifetime $250.

NEWLAND HOUSE MUSEUM, 19820 Beach Blvd., Huntington Beach, CA 92648. Tel.: 714-962-5777.
Institution Type/Description: Historic House Museum: housed in the former home of William & Mary Newland; built in 1898.
Collections: Newland family history; period furnishings; personal artifacts; photographs.
Activities: Museum Sponsors: Holiday Tour.
Hours & Admission Prices: Sat.-Sun. 12-4. Requested Donation: $2. Closed holidays.

Idyllwild

IDYLLWILD AREA HISTORICAL SOCIETY, 54470 N. Circle Dr., Idyllwild, CA 92549. Mailing Address: P.O. Box 3320, Idyllwild, CA 92549-3320. Tel.: 951-659-2717.
E-mail: info@idyllwildhistory.org
Web Site: www.idyllwildhistory.org
Founded: 2000.
Congressional District: 45
Key Personnel: Pres. (V), Carolyn Levitski; Shop. Mgr., Nancy Borchers.
Personnel Profile: Part-Time Volunteers 83.
Governing Authority: Tax-exempt.
Institution Type/Description: Historical Society Museum.
Collections: local history & culture; photographs; personal artifacts; rock climbing equipment; early scout camping; logging. Historic Structure: cabin.
Research Fields: San Jacinto Mountain history.
Facilities: archives.
Activities: research; house tour.
Publications: newsletter, Arti-Facts.
Hours & Admission Prices: late June to Labor Day Fri.-Sun. 11-4; Sept. to late June Sat.-Sun. 11-4. No charge; donations accepted. Closed Christmas. ♿
Attendance: 4,228 (accurate)
Membership: Individual $15; Family $25; Business, Organization & Contributor $35; Sponsor $100; Benefactor $250.

IDYLLWILD NATURE CENTER, 25225 Hwy. 243, Idyllwild, CA 92549. Mailing Address: c/o Riverside County Parks, 4600 Crestmore Rd., Jurupa Valley, CA 92509. Tel.: 951-659-3850.
Web Site: www.rivcoparks.org
Institution Type/Description: Nature Center.
Collections: mountain ecology; wildlife habitats; flora & fauna; Native American artifacts; San Jacinto Mountain history; photographs.
Activities: group tours.
Hours & Admission Prices: Tues.-Sun. 9-4:30. Adults $3, children 2-11 $2; children under 2 no charge.

Imperial

PIONEERS' MUSEUM, 373 E. Aten Rd., Imperial, CA 92251-9653. Tel.: 760-352-3211. Fax: 760-352-5411.
E-mail: pioneersmuseum@beamspeed.net
Founded: 1928.
Key Personnel: C.E.O., L. Housouer; Pres., Lee Hindman; Pres. (V), Sharon Housouer.
Personnel Profile: Full-Time Paid 1; Part-Time Paid 4; Part-Time Volunteers 100.
Governing Authority: nonprofit organization. Parent Institution: Imperial County Historical Society. Tax-exempt.
Institution Type/Description: History Museum.
Collections: local history & culture; ethnic groups; archaeology; period artifacts; baskets; Native American artifacts; furniture; farm equipment; documents; photographs.
Research Fields: photos; documente; personal histories.
Activities: lectures; special events; fundraisers.
Publications: newsletter.
Hours & Admission Prices: Tues.-Sun. 10-4. Adults $6, children 6-12 $2; members & children under 6 no charge. &

Attendance: 2,000 (estimated)
Membership: Individual $25; Lifetime $100.

Independence

EASTERN CALIFORNIA MUSEUM, 155 N. Grant St., Independence, CA 93526. Mailing Address: P.O. Box 206, Independence, CA 93526-0206. Tel.: 760-878-0364 & 0258. Fax: 760-878-0412. Facebook: Eastern California Museum.
E-mail: ecmuseum@inyocounty.us
Web Site: inyocounty.us/ecmuseum/index.html
Founded: 1928.
Congressional District: 40
Key Personnel: Museum Svcs. Admin., Jon Klusmire; Cur., Roberta Harlan; Museum Shop Mgr., Heather Todd.
Personnel Profile: Full-Time Paid 2; Part-Time Paid 1.
Operating Expenses: 225,000
Operating Income: 30,000
Governing Authority: county. Parent Institution: Inyo County. Tax-exempt: 501(c)(3); 170(b)(1)(A).
Institution Type/Description: Local & Natural History Museum.
Collections: Paiute & Shoshone Indian basketry & other cultural artifacts; Inyo County Pioneer artifacts & memorabilia relating to mining & farming; Manzanar Japanese American War Relocation Center (World War II); narrow gauge steam locomotive; photographs; Sierra Nevada Mountaineering recreation. Historic Houses: 1865 Edwards House; 1883 Commander's House.
Research Fields: anthropology, history & oral history.
Facilities: non-circulating reference library; herbarium. Museum-related items for sale.
Activities: scientific, cultural programs, including history lectures; field trips.
Publications: annual newsletter; Music Box Cassette; book, Mountains to Desert.
Hours & Admission Prices: Museum: daily 10-5. No charge; donations accepted. Edward's House & Commander's House temporarily closed. Closed New Year's Day; Easter; Thanksgiving; Christmas. &
Attendance: 10,000 (accurate)
Membership: Individual $25; Family $40; Contributing $100; Business & Organization $150; Patron $250; Sustaining $500; Benefactor $1,000 & up.

Indio

COACHELLA VALLEY HISTORICAL SOCIETY MUSEUM AND CULTURAL CENTER, 82616 Miles Ave., Indio, CA 92201-4228. Tel.: 760-342-6651. Fax: 760-863-5232.
E-mail: info@cvhm.org
Web Site: www.cvhm.org
Founded: 1984.
Congressional District: 44
Key Personnel: Pres. (V), Doug York; Cur. & Archivist, Erica Ward; Administrative Asst., Janice Woodside; Museum Shop Mgr., Carolyn Daniels.
Personnel Profile: Part-Time Paid 2; Part-Time Volunteers 35.
Governing Authority: private; nonprofit organization. Parent Institution: Coachella Valley Historical Society, Inc. Tax-exempt: 501(c)(3).
Institution Type/Description: History Museum: 1926 adobe house built by Dr. Harry Smiley, known as Smiley Place.
Collections: farm tools; horse drawn farm machinery; Cahuilla Indian artifacts; railroad; pioneer families; water development; Coachella Valley artifacts; date culture; blacksmith shop. Historic House: 1909 schoolhouse.
Major Exhibits: Journey of a People: The History of the Cahuilla and Chemehuevi Tribes in the Coachella Valley, 10/13-5/14.
Research Fields: Coachella Valley history; women pioneers; date culture history.
Facilities: library & archives; art studio; 4,000 sq. ft. exhibit space; gardens.
Activities: educational tours; docent program; summer art classes for children; permanent & temporary exhibitions; special events. Annual Events: Heritage Day; Aki Matsuri Festival.
Publications: quarterly newsletter, Scratches in the Sand; annual magazine, The Periscope; pictorial history book.
Hours & Admission Prices: Thurs.-Sat. 10-4, Sun. 1-4. Adults $3, senior citizens & students $2, children 6-12 $1; AAM members, children under 5 & members no charge. Closed holidays.
Attendance: 4,000 (accurate)
Membership: Individual $20; Family $45; Contributor $75; Sponsor & Organizations $100; Benefactor & Business $150; Life $500.

Irvine

BEALL CENTER FOR ART + TECHNOLOGY, 712 Arts Plaza, Claire Trevor School of the Arts, UC Irvine, Irvine, CA 92697-2775. Tel.: 949-824-6206. Fax: 949-824-2450.
E-mail: syoungha@uci.edu
Web Site: beallcenter.uci.edu
Formerly: University of California Art Gallery and Beall Center for Art and Technology
Founded: 2000.
Key Personnel: Dir., Joseph S. Lewis, III; Dir. Programs, Samantha Younghans-Haug; Assoc. Dir., David Familian.
Personnel Profile: Full-Time Paid 2.
Governing Authority: university. Parent Institution: University of California, Irvine. Tax-exempt.
Institution Type/Description: Art Museum.
Collections: new media art.
Activities: public & private tours; robotics camps. Museum Sponsors: Family Art Days.
Publications: exhibition catalogs.
Hours & Admission Prices: Tues.-Wed. 12-5, Thurs.-Sat. 12-8. No charge. Closed New Year's Day; Easter; Christmas Eve & Day; university breaks. &
Attendance: 10,000

IRVINE FINE ARTS CENTER, 14321 Yale Ave., Irvine, CA 92604-1901. Tel.: 949-724-6880. Fax: 949-552-2137. Facebook: Irvine Fine Arts Center.
E-mail: rmcgraw@cityofirvine.org
Web Site: www.cityofirvine.org/depts/cs/finearts
Founded: 1980.
Key Personnel: Dir., Wendy Shields.
Personnel Profile: Full-Time Paid 5; Part-Time Paid 37; Part-Time Volunteers 55.
Governing Authority: Parent Institution: City of Irvine.
Institution Type/Description: Arts & Crafts Museum.
Collections: painting; ceramics; photography; jewelry; culinary arts.
Facilities: Museum-related items for sale.
Activities: workshops.
Hours & Admission Prices: Mon.-Thurs. 10-9, Fri. 10-5, Sat. 9-5. No charge. &

IRVINE HISTORICAL MUSEUM, 5 San Joaquin, Irvine, CA 92612. Tel.: 949-786-4112. Fax: 949-854-7994. Facebook: Irvine Historical Museum.
E-mail: gadaniels@cox.net
Web Site: www.irvineranchhistory.com
Founded: 1977.
Key Personnel: Pres. (V), Gail Daniels; Museum Shop Mgr., Anne D. Johnson.

Personnel Profile: Part-Time Volunteers 6.
Governing Authority: Parent Institution: Irvine Historical Society. Tax-exempt.
Institution Type/Description: Historical Society Museum.
Collections: local history & culture; period furnishings; photographs; newspapers; personal artifacts; ranching artifacts; agricultural artifacts.
Facilities: library.
Activities: research; fundraising; education.
Publications: quarterly newsletter.
Hours & Admission Prices: Tues. & Sun. 1-4; other times by appointment. No charge; donations accepted. Closed holidays.
Attendance: 200 (estimated)
Membership: Student & Senior $10; Single $20; Family $25; Sponsor $50; Patron $100; Corporate $250.

THE IRVINE MUSEUM, 18881 Von Karman Ave., Ste. 100, Irvine, CA 92612-6541. Tel.: 949-476-0294. Fax: 949-476-2437.
Web Site: www.irvinemuseum.org
Founded: 1992.
Key Personnel: Exec. Dir., Mr. Jean Stern; Museum Shop Mgr., Don Bridges.
Personnel Profile: Full-Time Paid 5; Part-Time Paid 1; Part-Time Volunteers 15.
Governing Authority: nonprofit. Tax-exempt.
Institution Type/Description: Art Museum.
Collections: early California Impressionist paintings, c.1890-1930.
Publications: A California Women's Story, 2006; California This Golden Land of Promise, 2001; The Life and Art of Paul de Longpre, 2001; All Things Bright and Beautiful, 1998; California Impressionists, 1996; Joseph Kleitsch, A Kaleidoscope of Color, 2007; Native Grandeur, Preserving California's Vanishing Landscapes, 2003; Reflections of California The Athalie Richardson Irvine Clarke Memorial Exhibition, 1994; Romance of the Bells, The California Missions in Art, 1995; Selections from The Irvine Museum, 1992; East Coast West Coast and Beyond; Colin Campbell Cooper American Impressionist, 2006; Guy Rose American Impressionist, 1996; Impressions of California, Early Currents in Art 1850-1930, 1996; Masters of Light, Plein-Air Painting in California, 1890-1930, 2002; In Nature's Temple, The Life and Art of William Wendt, 2008; Palette of Light, 1995.
Hours & Admission Prices: Tues.-Sat. 11-5. No charge. Closed major holidays. &
Attendance: 19,442 (accurate)
Membership: Patron $60; Family Patron $125; Gallery Patron $250; Plein-Air Patron $500; California Patron $3,000; Director's Circle $5,000.

PRETEND CITY CHILDREN'S MUSEUM, 29 Hubble, Irvine, CA 92618. Tel.: 949-428-3900.
Institution Type/Description: Children's Museum.
Collections: hands-on exhibitions.
Activities: classes.
Hours & Admission Prices: Mon. 10-1, Tues.-Sun. 10-5. Admission $11, military $8; one & under no charge.

UNIVERSITY OF CALIFORNIA IRVINE ARBORETUM, University of California, North Campus Dr. & Jamboree Rd., Irvine, CA 92697. Tel.: 949-824-5833. Fax: 949-824-6146.
Web Site: arboretum.bio.uci.edu
Founded: 1965.
Congressional District: 40
Key Personnel: Dir., Dr. Peter Bowler.
Governing Authority: university. Parent Institution: University of California, Irvine. Tax-exempt: 170(b)(1)(A).
Institution Type/Description: Arboretum.
Collections: African xerophytes, Petalloid monocot plants; iridaceae, liliaceae, amaryllidaceae, geraniaceae; California native plants.
Research Fields: general botany; conservation; plant propagation & breeding.
Facilities: botanical garden.
Activities: lectures; formally organized education programs for undergraduate & graduate college students affiliated with the University of California; permanent exhibitions.
Publications: book, Plant Extinctions: A Global Crisis; quarterly newsletter, Index Seminum.
Hours & Admission Prices: Mon.-Sat. 9-3. No charge; donations accepted. Closed holidays.
Attendance: 14,100
Membership: Regular $30; Active $50; Sustaining $100; Builder $250; Patron $500; Benefactor $1,000.

Jackson

AMADOR COUNTY MUSEUM, 225 Church St., Jackson, CA 95642-2303. Mailing Address: 810 Court St., Jackson, CA 95642-2132. Tel.: 209-223-6386. Fax: 209-223-0749.
E-mail: museum@volcano.net
Founded: 1949.
Congressional District: 14
Key Personnel: Cur. & Museum Shop Mgr., Georgia Fox.
Personnel Profile: Full-Time Paid 1; Part-Time Paid 1; Part-Time Volunteers 4.
Governing Authority: county. Tax-exempt.
Institution Type/Description: General Museum: housed in c.1859 house.
Collections: Western history; gold rush history, equipment & memorabilia; gold mining artifacts; furnishings; photos; mine models; Indian & Chinese artifacts; William Lemos (1910) painting of Yosemite, CA; Bordello history; gambling; jazz pianist, Dave Brubeck.
Research Fields: pioneer families; Gold Rush era.
Activities: guided tour of the Head Frame of the Kennedy Mine Model.
Hours & Admission Prices: Museum: Feb.-Nov. Wed.-Sun. 10-4. Kennedy Gold Mine Model Tours: Sat.-Sun. 11-3 on the hour; group tours by reservation. Mine Model Tours: $2. Museum: no charge; donations accepted. &
Attendance: 9,000 (estimated)

Jenner

FORT ROSS STATE HISTORIC PARK VISITOR CENTER AND MUSEUM, 19005 Coast Hwy. One, Jenner, CA 95450. Tel.: 707-847-3437. Fax: 707-847-3601.
E-mail: friaadmin@mcn.org
Web Site: www.fortrossinterpretive.org
Congressional District: 6
Key Personnel: Admin., Sarjan Holt; Pres. (V), Sarah Sweedler; Museum Shop Mgr., Lake Perry.
Personnel Profile: Full-Time Paid 1; Part-Time Paid 3; Part-Time Volunteers 10.
Governing Authority: private; nonprofit organization. Tax-exempt.
Institution Type/Description: History Museum.
Collections: local history; archaeology; baskets; Native American artifacts.
Research Fields: Russian archives, preservation & restoration of Rotchen House.
Facilities: library; visitor center. Museum-related items for sale.
Activities: presentations; environmental living program; lectures; art shows; conferences; private tours. Museum Sponsors: Cultural Heritage Day; Harvest Festival.
Publications: Fort Ross and Salt Point newsletters, books & brochures; e-newsletter; audio-visual productions.
Hours & Admission Prices: July-Aug. daily 10-4; Sept.-June Sat.-Sun. 10-4. Park: adults $8 per car, seniors $7 per car. Museum: no charge. Closed Thanksgiving; Christmas.
Attendance: 200,000 (estimated)
Membership: Senior & Student $15; Regular $20; Family $25; Organization $30.

Jolon

MISSION SAN ANTONIO DE PADUA, Mission Creek Rd., Jolon, CA 93928. Mailing Address: P.O. Box 803, Jolon, CA 93928-0803. Tel.: 831-385-4478. Fax: 831-386-9332.
E-mail: office@missionsanantonio.net
Web Site: www.missionsanantonio.net
Formerly: San Antonio Mission
Founded: 1771.
Congressional District: 29
Key Personnel: Admin., Joan Steele; Gift Shop Mgr., Ms. Franki Grau.
Personnel Profile: Full-Time Paid 3; Part-Time Paid 3; Part-Time Volunteers 20.
Volunteer Hours: 1,200
Operating Expenses: 960
Operating Income: 25,486
Governing Authority: church. Parent Institution: Catholic Diocese of Monterey. Tax-exempt: 501(c)(3).
Institution Type/Description: Historic Museum: restored old San Antonio Mission.
Collections: wine press of Spanish colonial period; Indian relics; manuscripts; archives; Mission era furniture; musical instruments; extensive collection of Native American Southwestern baskets.
Research Fields: soldiers barracks; Indian dwellings; buildings; archaeology.
Facilities: retreat center. Religious articles & mission-related items for sale.
Activities: special guided tours; occasional inter-museum loan, permanent & temporary exhibitions.

Publications: books, Padres & People of Mission San Antonio; Herbs of Mission (used by Indians); History of Mission of San Antonio.
Hours & Admission Prices: Daily 10-4. Adult $5, active military, seniors 55 & over and children under 12 $3. Closed New Year's Day; Easter; Thanksgiving; Christmas Eve & Day.
Attendance: 20,000 (estimated)

Julian

JULIAN HISTORICAL SOCIETY, 2133 Fourth St., Julian, CA 92036. Mailing Address: P.O. Box 513, Julian, CA 92036. Tel.: 760-765-0436.
Institution Type/Description: Historical Society Museum: housed in a former one-room schoolhouse; built in 1888.
Collections: local history & culture; photographs; personal artifacts.
Hours & Admission Prices: Call for hours.

JULIAN PIONEER MUSEUM, 2811 Washington St., Julian, CA 92036. Mailing Address: P.O. Box 1866, 2129 Main St., Julian, CA 92036. Tel.: 760-765-0227.
Institution Type/Description: History Museum.
Collections: personal artifacts; period clothing from 1896-1913; photographs; mining equipment; Victorian era pianos; Native American artifacts.
Hours & Admission Prices: Call for hours.

Kentfield

COLLEGE OF MARIN FINE ART GALLERY, 835 College Ave, Kentfield, CA 94904. Tel.: 415-485-9494.
Formerly: Marin Community College Art Gallery
Key Personnel: Dir., Duane Aten
Institution Type/Description: Art Museum.
Collections: paintings; sculpture.
Hours & Admission Prices: Call for hours.

Kernville

KERN RIVER VALLEY HISTORICAL SOCIETY MUSEUM, 49 Big Blue Rd., Kernville, CA 93238. Mailing Address: P.O. Box 651, Kernville, CA 93238-0651. Tel.: 760-376-6683.
E-mail: info@krvhistoricalsociety.org
Web Site: www.krvhistoricalsociety.org/museum.htm
Founded: 1967.
Key Personnel: Pres., Ron Bolyard; Cur., Jon Partin
Institution Type/Description: Historical Society Museum.
Collections: local history & culture; Old West; Native American artifacts; gold mining; farming; ranching; Western movies; works by local artists; early 1900s electrical generator stations.
Hours & Admission Prices: Thurs.-Sun. 10-4; other times by appointment. No charge; donations accepted. &
Membership: Individual, Family & Business $25; Special Corporate $100.

King City

MONTEREY COUNTY AGRICULTURAL & RURAL LIFE MUSEUM, (M), San Lorenzo County Park, 1160 Broadway, King City, CA 93930. Mailing Address: P.O. Box 644, King City, CA 93930. Tel.: 831-385-8020. Fax: 831-386-0178. Facebook: mcarlm.
E-mail: info@mcarlm.org
Web Site: www.mcarlm.org
Key Personnel: Dir., Jessica Potts; Pres. (V), Jim Spring.
Personnel Profile: Part-Time Paid 4.
Governing Authority: Tax-exempt.
Institution Type/Description: History Museum.
Collections: local history & culture; period furnishings; personal artifacts; photographs; farming; agriculture Historic Buildings: 1887 schoolhouse; 1903 depot; 1898 Spreckels house.
Major Exhibits: The Way We Worked (T), 6/14-7/14.
Hours & Admission Prices: Barn: Tues.-Fri. 10-4. House, Schoolhouse, Depot, & Blacksmith Shop: Fri. 12-4, Sat.-Sun. 11-4. House of Irrigation: by appointment. Museum: no charge. Park: Mon.-Fri. $6 per car, Sat.-Sun. $8 per car. Closed New Year's Eve & Day; Thanksgiving; Christmas Eve & Day. &
Attendance: 6,120 (accurate)
Membership: Senior $20; Individual $30; Family $40; Service Club $100; Small Business $150; Corporate $300.

Kingsburg

KINGSBURG HISTORICAL PARK, 2321 Sierra St., Kingsburg, CA 93631. Mailing Address: P.O. Box 282, Kingsburg, CA 93631-1457.
Web Site: www.kingsburghistoricalpark.org
Founded: 1969.
Congressional District: 21
Key Personnel: Pres. (V), Gary Nelson.
Personnel Profile: Part-Time Paid 1.
Institution Type/Description: History Museum.
Collections: schoolroom; period farm tools; clothing; appliances; photographs; early printing press. Historic Buildings: medical building; grocery store; schoolhouse; service station; windmill; firehouse; heritage building; bandstand.
Research Fields: local history.
Facilities: meeting hall.
Publications: annual newsletter.
Hours & Admission Prices: Feb.-Nov. Thurs.-Sat. 1-4. Family $6, adults $3, children under 12 $.50; discounts to groups of 25 or more; members no charge. &
Attendance: 700 (estimated)
Membership: Individual $25; Family $35; Business & Professional $50; Sustaining $100; Life $500.

Klamath

END OF THE TRAIL MUSEUM, 15500 Hwy. 101 N., Klamath, CA 95548-9351. Mailing Address: P.O. Box 96, Klamath, CA 95548-0096. Tel.: 707-482-2251.
E-mail: tofm@treesofmystery.net
Web Site: treesofmystery.net
Founded: 1950.
Key Personnel: Museum Shop Mgr., Debbie Thompson
Institution Type/Description: History Museum.
Collections: Native American history & culture; personal artifacts; baby carriers.
Facilities: Museum-related items for sale.
Hours & Admission Prices: No charge. Closed Christmas. &
Attendance: 300,000

La Canada Flintridge

✳ **DESCANSO GARDENS GUILD, INC., (M),** 1418 Descanso Dr., La Canada Flintridge, CA 91011-3102. Tel.: 818-949-4290. Fax: 818-790-3291.
E-mail: dbrown@descansogardens.org
Web Site: www.descansogardens.org
Founded: 1957.
Congressional District: 22
Key Personnel: Exec. Dir., David R. Brown; C.O.O., Juliann Rooke.
Personnel Profile: Full-Time Paid 49; Part-Time Paid 25; Part-Time Volunteers 400; Interns 1.
Governing Authority: nonprofit organization. Tax-exempt: 501(c)(3).
Institution Type/Description: Botanical Garden: housed in the former home of E. Manchester Boddy; c.1930s.
Collections: Camellia collection: 30,000 plants under an oak forest canopy; 9 acres of native plant area; 5-acre rose garden; 1 1/2 Descanso Lake & bird sanctuary with over 150 species of birds.
Facilities: library; botanical garden; rental facilities; 55-seat patio cafe; educational facilities; 2,800 sq. ft. exhibit space; Bird Observation Center. Horticulturally-related items for sale.
Activities: guided tours; lectures; rental facilities; films; dance recitals; arts festivals. Museum Sponsors: Plant sales; Winter Holiday Show; Horticultural events; Spring Festival of Flowers late March thru late April; Pumpkin Festival in October; Japanese Garden Festival in November; Labor Day Picnic.
Publications: quarterly, Descanso News.
Hours & Admission Prices: Daily 9-4:30. Adults $9, senior citizens & students $6, children 5-12 $4; Guild members & children under 5 no charge. Closed Christmas Day. &
Attendance: 300,000 (accurate)
Membership: Senior Couple $65; Couple $70; Family $80; Family Plus $150; Sponsoring $250; Sustaining $500; Center Circle Associates $1,000 & up.

LANTERMAN HOUSE, 4420 Encinas Dr., La Canada Flintridge, CA 91011-3113. Tel.: 818-790-1421. Fax: 818-952-8450.
E-mail: mpatton.lanterman@gmail.com
Web Site: lanternmanfoundation.org
Founded: 1993.

Congressional District: 29
Key Personnel: Exec. Dir., Melissa Patton; Pres. (V), Robert Moses.
Personnel Profile: Part-Time Paid 3; Part-Time Volunteers 40; Interns 1.
Volunteer Hours: 2,500
Operating Expenses: 108,095
Operating Income: 103,690
Governing Authority: Parent Institution: City of La Canada Flintridge. Tax-exempt.
Institution Type/Description: Historic House: c.1915.
Collections: period furnishings; hand-painted wall & ceiling ornamentation; personal artifacts; medical equipment.
Major Exhibits: Charles Pate: Pioneer Photographer, 10/13-6/5/14.
Research Fields: local history; CA political history; 20th-century music; arts and crafts architecture & decorative arts.
Facilities: archive.
Activities: lectures; period events; member tours.
Publications: newsletter; local history books.
Hours & Admission Prices: Sept.-July Tues., Thurs. and 1st & 3rd Sun. of month 1-4. Adults $5, seniors & students $3; discounts to AAM members; children under 12 no charge. Closed holidays. &
Attendance: 2,500 (accurate)
Membership: Sponsor $35; Sustainer $60; Associate $100; Benefactor $250.

La Habra

CHILDREN'S MUSEUM AT LA HABRA, 301 S. Euclid, La Habra, CA 90631-5412. Tel.: 562-905-9693 & 9793. Fax: 562-905-9698.
E-mail: museumstaff@lahabracity.com
Web Site: www.lhcm.org
Founded: 1977.
Congressional District: 39
Key Personnel: Asst. Dir., Lovely Qureshi; Outreach Program Coord., Stephanie Bobadilla; Program Development Mgr., Maria Tinaiero-Dowdle; Cur. Exhibits & Educ., Lisa Reckon; Visitor Services Coord., Jennifer Andrade; Visitor Services Asst., Lorena Altamirano.
Personnel Profile: Full-Time Paid 2; Part-Time Paid 9; Part-Time Volunteers 80; Interns 1.
Governing Authority: municipal; nonprofit organization. Parent Institution: City of La Habra, CA. Subsidiary Institution: Friends of the Children's Museum. Tax-exempt: 170(b)(1)(A); 509(A)(1); 501(c)(3).
Institution Type/Description: Children's Museum: housed in a renovated 1923 Mission Style Union Pacific Railroad Depot.
Collections: local historical artifacts; taxidermied local animals; scientific, cultural & historical artifacts; full-size railroad cars; live bee observatory; gems; minerals; model train village; children's art; Dentzel carousel; toys; dolls; theatre costumes & props; physical science exhibits; infant & toddler exhibits; transportation; dinosaur fossils; stuffee health program.
Research Fields: local history.
Facilities: permanent & temporary exhibits; multipurpose room; classroom. Gift items for sale.
Activities: arts festivals; craft workshops; art camp; hospital visits; docent program or council; permanent & temporary exhibitions; special exhibit events; outreach programs; birthday parties.
Publications: Kid's Guide to La Habra; La Habra: The Good Old Days and the Way It Was; A Bell in the Barranca; quarterly newsletter.
Hours & Admission Prices: Tues.-Fri. 10-4, Sat. 10-5, Sun. 1-5. Adults $8, La Habra residents $7; discount to ASTC members; children under 2 no charge. Closed major holidays. &
Attendance: 100,000 (accurate)

La Jolla

ATHENAEUM MUSIC & ARTS LIBRARY, 1008 Wall St., La Jolla, CA 92037-4418. Tel.: 858-454-5872. Fax: 858-454-5835.
E-mail: athdir@pacbell.net
Web Site: ljathenaeum.org
Key Personnel: Dir., Erika Torri
Institution Type/Description: Art Gallery & Library.
Collections: works by regional, national & international artists; paintings; sculpture; drawings; photographs.
Facilities: library.
Activities: lectures; temporary exhibitions; art classes; lectures; special events.
Hours & Admission Prices: Tues. & Thurs.-Sat. 10-5:30, Wed. 10-8:30.

BIRCH AQUARIUM AT SCRIPPS, SCRIPPS INSTITUTION OF OCEANOGRAPHY, UNIVERSITY OF CALIFORNIA, SAN DIEGO, 2300 Expedition Way, La Jolla, CA 92037. Mailing Address: 9500 Gilman Dr., #0207, La Jolla, CA 92093-0207. Tel.: 858-534-5301. Fax: 858-534-7114.
E-mail: heiock@ucsd.edu
Web Site: aquarium.ucsd.edu
Founded: 1905.
Congressional District: 41
Key Personnel: Exec. Dir., Nigella Hillgarth, Ph.D.; Cur. Aquarium, Fernando Nosratpour; Mgr. Mktg., Jessica Crawford; Head Facilities, Mick Curzon; Dir. Operations, Patrick Helbling; Dir. Education, Kristin Evans; Dir. Special Events, Barbara Ramsey; Mgr. Membership, Paula Smith; Mgr. Finance, Ken Steitz; Mgr. Museum Shop, Susan Malk; Mgr. Exhibits, Charles Langsett; Program Scientist, Cheryl Peach; Program Scientist, Debbie Zmarzly.
Personnel Profile: Full-Time Paid 56; Part-Time Paid 4; Part-Time Volunteers 395.
Governing Authority: university. Parent Institutions: Scripps Institution of Oceanography (SIO), University of Calif., San Diego (UCSD). Tax-exempt.
Institution Type/Description: Aquarium-Museum.
Collections: living, pacific marine life; Scripps research.
Research Fields: aquaculture; fish diseases.
Facilities: 11,000 sq. ft. aquarium; education center; seawater system; 6,000 sq. ft. exhibit court; 4,500 sq. ft. Tidepool Plaza; service areas. Museum-related items for sale.
Activities: guided tours; lectures; films; formally organized educational programs; docent programs; permanent & traveling exhibitions; career experience program for high school students; special events.
Publications: quarterly members newsletter, OnBoard; education materials; monthly e-newsletter.
Hours & Admission Prices: Daily 9-5. Adults $14, military $12, seniors 60 & over and college students with ID $10, children 3-17 $9.50; children under 2 & members no charge. &
Attendance: 400,000 (estimated)
Membership: Scripps Oceanographic Society: Individual $55; Dual $70; Family $89; Adopt-a-Fish $25-$500; Aquarium Associate $150 & up; Ocean Sponsor $250 & up; Marine Ambassador $500 & up.

GOTTHELF ART GALLERY, Lawrence Family Jewish Community Center, 4126 Executive Dr., La Jolla, CA 92037. Tel.: 858-362-1154.
E-mail: gallery@lfjcc.com
Web Site: www.sdcjc.org/gag
Key Personnel: Exec. Dir., Michael Cohen.
Governing Authority: Parent Institution: Lawrence Family Jewish Community Center. Subsidiary Institution: San Diego Center for Jewish Culture.
Institution Type/Description: Art Gallery.
Collections: Jewish culture & heritage; paintings; drawings; sculpture.
Major Exhibits: smArt: The Art of Jewish Educators, 12/11/13-2/26/14; Transformations: The Butterfly Project and Beyond, 3/12/14-6/3/14.
Facilities: 1,000 sq. ft. exhibition space.
Activities: temporary exhibitions.
Hours & Admission Prices: Sun.-Fri. 9-5. No charge; donations accepted. &
Membership: Realist $200; Surrealist $360; Impressionist $500; Expressionist $1,000; Modern Co Sponsor $2,500; Renaissance Sponsor $5,000 & up.

JOSEPH BELLOWS GALLERY, 7661 Girard Ave., La Jolla, CA 92037. Tel.: 858-456-5620. Fax: 858-456-5621.
E-mail: info@josephbellows.com
Web Site: www.josephbellows.com
Key Personnel: Dir., Joseph Bellows; Dir., Carol Lee Brosseau
Institution Type/Description: Art Gallery.
Collections: photographs.
Hours & Admission Prices: Tues.-Sat. 10-5.

LA JOLLA HISTORICAL SOCIETY, (M), 7846 Eads Ave., La Jolla, CA 92037-4211. Mailing Address: P.O. Box 2085, La Jolla, CA 92038-2085. Tel.: 858-459-5335. Fax: 858-459-0226.
E-mail: info@lajollahistory.org
Web Site: www.lajollahistory.org
Founded: 1964.
Key Personnel: Exec. Dir., Heath Fox; Pres. (V), Nell Waltz.
Personnel Profile: Full-Time Paid 3; Part-Time Paid 4; Part-Time Volunteers 200; Interns 2.
Governing Authority: Tax-exempt.
Institution Type/Description: Historical Society Museum.

Collections: local history with concentration on architectural history & historic preservation; photographs; textural materials. Historic Buildings: Wisteria Cottage; carriage house, 1904; craftsman house, 1909.
Facilities: archives.
Activities: temporary exhibitions; public programs; educational programs; community events.
Publications: Historic La Jolla Walking Tour; newsletter, Timekeeper.
Hours & Admission Prices: Mon.-Fri. 10-4 by appointment. No charge; donations accepted.
Attendance: 3,000 (estimated)
Membership: $50; $100; $250; $500; $1,000; $5,000.

*** MUSEUM OF CONTEMPORARY ART SAN DIEGO, (M),** 700 Prospect St., La Jolla, CA 92037-4291. Tel.: 858-454-3541. Fax: 858-454-6985.
E-mail: info@mcasd.org
Web Site: www.mcasd.org
Founded: 1941.
Congressional District: 53
Key Personnel: Dir., Hugh M. Davies; Deputy Dir., Charles E. Castle; Chief Advancement Officer, Edie Nehls; Chief Cur. & Head Curatorial, Kathryn Kanjo.
Personnel Profile: Full-Time Paid 41; Part-Time Paid 45; Interns 4.
Governing Authority: nonprofit organization. Additional Location: MCASD Downtown, 1001 & 1100 Kettner Blvd., San Diego, CA 92101. Tax-exempt: 501(c)(3).
Institution Type/Description: Art Museum.
Collections: contemporary art including paintings, sculptures, drawings & prints; photography; video art; installation.
Research Fields: contemporary art.
Facilities: 14,300 sq. ft. exhibit space; 500-seat auditorium; 31,000 sq. ft. outdoor sculpture garden; cafe. Museum-related items for sale.
Activities: lectures; films; seminars; gallery talks; educational programs; concerts; inter-museum loans, temporary & traveling exhibitions.
Publications: quarterly newsletter with calendar of events; exhibition catalogs; educational brochures; membership brochure; general information brochure.
Hours & Admission Prices: Thurs.-Tues. 11-5, third Thurs. of month 11-7. Adults $10, senior citizens, students & military $5; discounts to AAM & ICOM members; 25 & under and members no charge. Admission valid for 7 days at all MCASD locations. Closed New Year's Day; Thanksgiving; Christmas. ♿
Attendance: 167,000 (estimated)
Membership: Artist & Student $35; E-member $45; Dual/Family $75; Contributor $150; Patron $300; Avant Garde $500; Supporter $600; Donor $1,500; Benefactor $2,500; Contemporary Collector $5,000; International Collector $10,0000.

STUART COLLECTION, University of California, San Diego-0010, 105 Pepper Canyon Hall, La Jolla, CA 92093-0010. Mailing Address: UCSD 0010, 9500 Gilman Dr., La Jolla, CA 92093-0010. Tel.: 858-534-2117. Fax: 858-534-9713.
E-mail: mbeebe@ucsd.edu
Web Site: stuartcollection.ucsd.edu
Founded: 1981.
Congressional District: 49
Key Personnel: Dir., Mary L. Beebe; Projects Mgr., Mathieu Gregoire; Office Mgr., Jane Peterson.
Personnel Profile: Full-Time Paid 2; Part-Time Paid 1; Part-Time Volunteers 3.
Volunteer Hours: 200
Operating Expenses: 250,000
Operating Income: 250,000
Governing Authority: public university. Parent Institution: UCSD. Tax-exempt.
Institution Type/Description: University Art Collection.
Collections: 18 contemporary outdoor commissioned sculpture throughout 1,200-acre campus.
Facilities: 2,000-vol. library of catalogues & books available to public by appointment.
Activities: guided tours; lectures; study clubs; temporary exhibitions.
Publications: brochure; map; book, Landmarks: Sculpture Commissions for the Stuart Collection at the University of California, San Diego, 2001.
Hours & Admission Prices: Daily 24 hours. No charge; donations accepted. ♿
Membership: Annual Friends of the Stuart Collection $1,500.

TASENDE GALLERY, 820 Prospect St., La Jolla, CA 92037. Tel.: 858-454-3691. Fax: 858-454-0589.
E-mail: info@tasendegallery.com
Web Site: www.tasendegallery.com
Institution Type/Description: Art Gallery.

Collections: contemporary & modern paintings; drawings; sculpture.
Hours & Admission Prices: Tues.-Fri. 10-6, Sat. 11-5.

VILLAGE GALLERY, 8100 Paseo del Ocaso, Ste. B, La Jolla, CA 92037-3115. Tel.: 858-459-1196.
Web Site: www.lajollaart.org
Key Personnel: Gallery Dir., Gwen Nobil; Dir. Exhibitions, Carrie Barton
Institution Type/Description: Art Museum.
Collections: paintings; drawings; sculptures; photography.
Activities: lectures; docent tours; demonstrations.
Hours & Admission Prices: Daily 11-5. No charge.

La Mesa

HORSELESS CARRIAGE FOUNDATION & AUTOMOTIVE RESEARCH LIBRARY, 8186 Center St., Ste. F, La Mesa, CA 91942-2959. Mailing Address: P.O. Box 369, La Mesa, CA 91944-0369. Tel.: 619-464-0301. Fax: 619-464-0301.
E-mail: research@hcfi.org
Web Site: www.hcfi.org
Founded: 1984.
Key Personnel: C.E.O. & Pres. (V), Donald Sable, II; Chm. (V), Exec. Dir. & Archivist, D.A. "Mac" MacPherson; Pres. (V) & Devel., Greg Long; Treas., Thomas E. Kettenburg; Public Rels., Reid Carroll.
Personnel Profile: Full-Time Paid 2; Part-Time Paid 1; Part-Time Volunteers 9.
Governing Authority: board of directors; nonprofit. Tax-exempt: 501(c)(3).
Institution Type/Description: Research Library.
Collections: automotive historical literature from 1895-present which includes factory sales catalogs & operation manuals; technical data; automotive technical information; automotive memorabilia; automotive sales brochures.
Research Fields: technological research related to specific makes of automobiles for private individuals, museums, historical societies, the media & academic researchers, some of which are foreign clients in 35 countries.
Facilities: 18,400-vol. library; reading room; research facility.
Activities: education programs for children; guided tours; lectures; children's car rides; elementary school show. Annual Events: general meeting; Annual Reception in February; Director's Reception; Patron's Reception.
Publications: quarterly, H.C.F. Research Library Newsletter.
Hours & Admission Prices: Jan. to mid-Dec. Tues.-Fri. 10-4. No charge; donations accepted. Closed major holidays. ♿
Attendance: 100 (estimated)
Membership: Patron $35; Century $100; Benefactor $250; Life $5,000; Platinum Life $10,000.

LA MESA DEPOT MUSEUM, 4695 Nebo Dr., La Mesa, CA 91941-5259. Tel.: 619-465-7776.
E-mail: support@sdrm.org
Web Site: www.psrm.org
Founded: 1982.
Congressional District: 52
Key Personnel: Pres. (V), Diana Hyatt; Museum Coord. (V), Richard Pennick.
Personnel Profile: Part-Time Volunteers 5.
Governing Authority: Parent Institution: Pacific Southwest Railway Museum Association, Inc. Tax-exempt.
Institution Type/Description: Restored 1915 railway depot.
Collections: local railroad artifacts; steam locomotive; freight car; caboose; early telephone & telegraph equipment; railroad lanterns.
Major Exhibits: San Diego, Cuyamaca & Eastern Railway, 1/14-12/14; La Mesa Depot Preservation, 1/14-12/14.
Research Fields: local railroad history.
Facilities: restored 1915 wooden railway depot.
Activities: tours; community events; interpretation.
Publications: bimonthly, Hot Scoop.
Hours & Admission Prices: Sat. 1-4; other times by appointment. No charge; donations accepted. ♿
Attendance: 1,000 (estimated)
Membership: Senior & Student $20; Individual $30; Family $40; Contributing $50; Supporting $75; Sustaining $150; Life $750; Gold Spike Life $1,000; Benefactor $5,000.

LA MESA HISTORICAL SOCIETY - MCKINNEY HOUSE MUSEUM AND ARCHIVES, 8369 University Ave., La Mesa, CA 91941. Mailing Address: P.O. Box 882, La Mesa, CA 91944. Tel.: 619-466-0197.
E-mail: information@lamesa
Web Site: lamesahistory.com
Founded: 1975.
Key Personnel: Pres. (V), James D. Newlan

Institution Type/Description: Historical Society Museum: house built in 1908.
Collections: local history & culture; period furnishings; personal artifacts; photographs.
Hours & Admission Prices: 1st & 3rd Sat. 1-4. No charge; donations accepted.

La Puente

LA PUENTE VALLEY HISTORICAL SOCIETY, INC., 15900 E. Main St., La Puente, CA 91744-4719. Mailing Address: P.O. Box 522, La Puente, CA 91747-0522. Tel.: 626-855-1500. Fax: 626-855-4626.
E-mail: info@lpvhistoricalsociety.org
Founded: 1960.
Congressional District: 34
Key Personnel: Pres., Patricia McIntosh; Vice Pres., C. Wictor; Sec., Heather Barron; Treas., John C. Butler.
Personnel Profile: Part-Time Volunteers 30.
Institution Type/Description: Local History Museum: housed in the home of John Rowland; built in 1855.
Collections: furniture & belongings of the Rowland family & early La Puente Valley families; Indian artifacts & early ranch items; local history.
Research Fields: La Puente Valley history.
Facilities: 200-vol. library of general & California history, available on premises by appointment.
Activities: guided tours; school tours; slide shows; docent program.
Publications: bimonthly newsletter, The Bridge; book, La Puente Kaleido-scope Part I & Part II; Footsteps to the Past; booklet, John Rowland; brochure, La Puente Historic Points of Interest.
Hours & Admission Prices: Rowland & Dibble Museum: by appointment. Heritage Room: Thurs. 1-4. No charge; donations accepted. Closed holidays. &
Attendance: 1,900 (estimated)
Membership: General $10; Life $100.

Laguna Beach

LAGUNA ART MUSEUM, (M), 307 Cliff Dr., Laguna Beach, CA 92651-1696. Tel.: 949-494-8971. Fax: 949-494-1530. Facebook: Laguna Museum.
E-mail: mfarmer@lagunaartmuseum.org
Web Site: www.lagunaartmuseum.org
Founded: 1996.
Key Personnel: Dir., Dr. Malcolm Warner; Museum Shop Mgr., Michele Manda; Pres. (V), Robert Hayden, III
Personnel Profile: Full-Time Paid 13; Part-Time Paid 9; Part-Time Volunteers 50; Interns 5.
Volunteer Hours: 200
Operating Expenses: 1,500,000
Operating Income: 1,500,000
Governing Authority: private; nonprofit organization. Tax-exempt.
Institution Type/Description: Art Museum.
Collections: sculptures; paintings.
Research Fields: California Art 1900 to present.
Facilities: library; rental facilities. Museum-related items for sale.
Activities: art auctions; lecture panels; film screenings; concerts; family days.
Publications: newsletter; exhibition books & catalogs; members' magazine.
Hours & Admission Prices: Thurs. 11-9, Fri.-Tues. 11-5. Adults $7, students, seniors & active military $5; discounts to AAM & ICOM members; children under 12 & museum members no charge. &
Attendance: 50,000 (estimated)
Membership: Individual $60; Family $75; Friend $125; Advocate $250; Patron $500.

LAGUNA BEACH HISTORICAL SOCIETY - MURPHEY-SMITH HOUSE, 278 Ocean Ave., Laguna Beach, CA 92651. Tel.: 949-939-7257 & 497-6834.
Institution Type/Description: Historical Society Museum.
Collections: local history & culture; period furnishings; personal artifacts; photographs.
Hours & Admission Prices: Fri.-Sun. 1-4. No charge.

Lake Arrowhead

MOUNTAIN SKIES ASTRONOMICAL SOCIETY & SCIENCE CENTER, 2001 Observatory Way, Lake Arrowhead, CA 92352. Mailing Address: P.O. Box 1169, Lake Arrowhead, CA 92352-1169. Tel.: 909-336-1699.
E-mail: stargazersmail@mountain-skies.org
Web Site: www.mountain-skies.org

Founded: 1989.
Key Personnel: Pres. & Chm., Dr. Lorann Parker, D.Sc., Ph.D.
Institution Type/Description: Science Center & Observatory.
Collections: hands-on exhibit; Solar System models; tools; photographs and memorabilia (historical).
Facilities: library. Museum-related items for sale.
Activities: public programs; private group programs; special events.
Hours & Admission Prices: Thurs.-Tues. 10-3. Donations accepted. Sat. night programs open to public. At door $9, advance $7. Call for information. &
Membership: Classroom $15; Student $20; Senior $25; Single $30; Family $40; Business $80; Life $500.

Lake Elsinore

LEHS MUSEUM & RESEARCH LIBRARY, 183 N. Main St., Lake Elsinore, CA 92530-4005. Mailing Address: Lake Elsinore Historical Society, P.O. Box 84, Lake Elsinore, CA 92531-0084. Tel.: 951-678-1537.
Institution Type/Description: Historical Society Museum & Library.
Collections: local history & culture; personal artifacts; books.
Activities: research.
Hours & Admission Prices: By appointment.

LAKE ELSINORE MUSEUM, 106 S. Main St., Lake Elsinore, CA 92530-4109. Mailing Address: Lake Elsinore Historical Society, P.O. Box 84, Lake Elsinore, CA 92531-0084. Tel.: 951-245-4986.
Institution Type/Description: History Museum.
Collections: local history & culture; period furnishings; personal artifacts; photographs.
Hours & Admission Prices: Sat.-Sun. 11-3.

Lakeport

LAKE COUNTY HISTORIC COURTHOUSE MUSEUM, 255 N. Main St., Lakeport, CA 95453. Mailing Address: 255 N. Forbes St., Lakeport, CA 95453-4790. Tel.: 707-263-4555. Fax: 707-263-7918.
E-mail: museum@lakecountyca.gov
Web Site: www.lakecounty.com/things/museums.html
Formerly: Lake County Museum
Founded: 1936.
Personnel Profile: Part-Time Paid 5.
Governing Authority: county. Owned & operated by the County of Lake. Tax-exempt.
Institution Type/Description: History Museum: housed in 1871 Lake County Courthouse.
Collections: Pomo baskets; Native American tools; projectile points; ground-stone; minerals & gems found in the county; clothing & artifacts from the 1800s & 1900s. Genealogical & research library with reference books; a 10,000 page manuscript by Henry Mauldin on local history; photographs & family histories.
Research Fields: local history.
Facilities: research library of local history books.
Activities: genealogy society; docent program.
Publications: quarterly newsletter, The Muse; Legend of Moon Tear; Pomo Legend Coloring Book.
Hours & Admission Prices: Sun. 12-4, Wed.-Sat. 10-4. No charge; donations accepted. Closed most holidays. &
Attendance: 6,000 (accurate)
Membership: Individual $15; Friend $25; Sponsor $50; Steward $100; Patron $250; Benefactor $500.

Lakeside

BARONA CULTURAL CENTER AND MUSEUM, (M), 1095 Barona Rd., Lakeside, CA 92040-1541. Tel.: 619-443-7003, ext. 219. Fax: 619-443-0173.
E-mail: museum@baronmuseum.org
Web Site: www.baronamuseum.org
Key Personnel: Dir. & Cur., Laurie Egan-Hedley; Collections Mgr., John George; Museum Store Coord., Robin Edmonds
Institution Type/Description: History Museum.
Collections: Native American culture & history; photographs; basketry; pottery making; stone tools.
Hours & Admission Prices: Tues.-Fri. 12-5, Sat. 10-4. No charge.

LAKESIDE HISTORICAL SOCIETY & MUSEUM, 12418 Parkside St., Lakeside, CA 92040. Mailing Address: 9906 Maine Ave., Lakeside, CA 92040. Tel.: 619-561-1886. Facebook: Lakeside Historical Society & Museum.
Web Site: www.lakesidehistory.org
Founded: 1972.
Institution Type/Description: History Museum.
Collections: local history & culture; period furnishings; photographs; personal artifacts.
Hours & Admission Prices: Sat. 11-3; other times by appointment.

Lancaster

ANTELOPE VALLEY CALIFORNIA POPPY RESERVE, 15101 W. Lancaster Rd., Lancaster, CA 93536-9733. Mailing Address: 15701 E. Ave. M, Lancaster, CA 93535. Tel.: 661-946-6092 (office) & 724-1180 (reserve). Fax: 661-946-6116.
E-mail: mdic@parks.ca.gov
Web Site: www.parks.ca.gov
Founded: 1982.
Congressional District: 20
Key Personnel: District Supt., Kathy Weatherman; Pres. & Coop Assoc., Margaret Rhyne; Museum Shop Mgr., Pat Treadwell.
Personnel Profile: Full-Time Paid 1; Part-Time Paid 4; Part-Time Volunteers 30.
Governing Authority: state. Parent Institution: California Dept. of Parks & Recreation, Mojave Sector, Tehachap District, 15701 E. Ave. M, Lancaster, CA 93535. Subsidiary Institution: Poppy Reserve/Mojave Desert Interpretive Assoc. Tax-exempt: 501(c)(3).
Institution Type/Description: Nature Center.
Collections: nature exhibits; 155 Jane S. Pinheiro wildflower watercolor paintings.
Research Fields: wildflowers; poppies.
Facilities: nature & conservation center; 8 mile hiking trails; picnic area. Park-related items for sale.
Activities: guided tours; docent program; permanent exhibitions; school programs.
Publications: pamphlet.
Hours & Admission Prices: Visitor Center: call for hours. Reserve: mid-March to mid-May, $10 per vehicle, senior citizens 62 & over $9; Small Bus: $50; Large Bus $100.
Attendance: 44,913 (accurate)
Membership: Poppy Reserve/Mojave Desert Interpretive Association: Full-time Student $5; Regular $10; Sustaining & Family $15; Organization & Patron $30; Lifetime $1,000.

ANTELOPE VALLEY INDIAN MUSEUM, 15701 E. Ave. M, Lancaster, CA 93535-7059. Tel.: 661-946-3055 & 6900. Fax: 661-946-6116. Facebook; Antelope Valley Indian Museum.
E-mail: peggy.ronning@parks.ca.gov
Web Site: www.avim.parks.ca.gov
Founded: 1928.
Congressional District: 20
Key Personnel: Cur., Peggy Ronning; Coop. Assoc. & Museum Shop Mgr., Susan Martin.
Personnel Profile: Full-Time Paid 1; Part-Time Paid 2; Part-Time Volunteers 17.
Volunteer Hours: 1,238
Operating Expenses: 124,333
Operating Income: 7,878
Governing Authority: state. Parent Institution: California State Dept. of Parks & Recreation, Tehachapi District, Mojave Sector, 15701 E. Ave. M, Lancaster, CA 93535; Friends of Antelope Valley Indian Museum Cooperative Association. Tax-exempt: 501(c)(3).
Institution Type/Description: American Indian Cultural Museum.
Collections: artifacts from California, the Great Basin & the Southwest; Sea Grass Weavings; Kachina Dolls; regional American Indian items & implements; rugs; blankets; pottery; murals, paintings & portraits by Edwards & other artists; American Indian art.
Research Fields: American Indian culture groups Californian, Western Great Basin, Southwestern region; prehistoric & ethnographic.
Facilities: library; 3,000 sq. ft. exhibit space; nature trail; picnic area. Museum-related items for sale.
Activities: guided & self-guided tours; docent program; permanent & temporary exhibitions; guest artists; touch table. Museum Sponsors: American Indian educational demonstrations; workshops; occasional ceremonial & vendor events.
Publications: brochures, Nature Trail Guide; booklet, American Indian Peoples

of the Antelope Valley, Western Mojave Desert, CA: 12,000 Years of Culture & History; Guide to the Antelope Valley Indian Museum.
Hours & Admission Prices: Sat.-Sun. 11-4. Adults $3. Closed Christmas. &
Attendance: 5,758 (accurate)
Membership: FAVIM: Individual $15; Family $20; Organization $35; Patron $50; Life $200.

LANCASTER MUSEUM OF ART & HISTORY (MOAH), 665 W. Lancaster Blvd., Lancaster, CA 93534-3226. Mailing Address: 44933 N. Fern Ave., Lancaster, CA 93534-2461. Tel.: 661-723-6250. Fax: 661-723-6260.
E-mail: moah@cityoflancasterca.org
Web Site: www.lancastermoah.org
Formerly: Lancaster Museum Art Gallery (LMAG)
Founded: 1984.
Congressional District: 20
Key Personnel: Dir. Parks Recreation & Arts, Ronda Perez; Interim Cur., Andi Campognone; Recreation Supvr., Angela Riley.
Personnel Profile: Full-Time Paid 2; Part-Time Paid 2; Part-Time Volunteers 1.
Governing Authority: municipal government. Parent Institution: City of Lancaster. Tax-exempt.
Institution Type/Description: Art & History Museum.
Collections: Lancaster & Antelope Valley history; Native American artifacts; contemporary art.
Major Exhibits: Ruth Pastine: Attraction, 1993-2013, 1/18/14-3/16/14; Johannes Girardoni: Metaspace, 1/18/14-3/16/14; Yi Kai, 3/29/14-6/8/14; John Van Hamersveld: 50 years, 6/21/14-8/14; Mana: The Artists, 6/21/14-8/14; Guillermo Bert: Coded Textiles, 9/13/14-11/9/14; Linda Vallejo: Make Em All Mexican (T), 9/13/14-11/9/14; Thomas McGovern & Juan Delgado: Vital Signs, 9/13/14-11/9/14; The California Landscape, 11/22/14-1/11/15; Hollis Cooper, 11/22/14-1/11/15.
Research Fields: local history; art history; archaeology.
Facilities: 100-vol. library of local history. Museum-related items for sale.
Activities: guided tours; lectures; organized educational programs; traveling trunk program.
Publications: Exhibition Catalogues.
Hours & Admission Prices: Tues.-Wed. & Fri.-Sun. 11-6, Thurs. 11-8. Suggested donation: Adults $5; students & seniors $3; children under 6 no charge. Closed New Year's Day; Easter; Memorial Day; Independence Day; Labor Day; Thanksgiving; Christmas. &
Attendance: 20,000 (estimated)

WESTERN HOTEL/MUSEUM, 557 W. Lancaster Blvd., Lancaster, CA 93534-2533. Mailing Address: 44933 N. Fern Ave., Lancaster, CA 93534-2461. Tel.: 661-723-6250. Fax: 661-723-6260.
E-mail: moah@cityoflancasterca.org
Web Site: www.lancastermoah.org
Founded: 1986.
Congressional District: 20
Key Personnel: Cur., Andi Campognone; Dir., Ronda Perez.
Personnel Profile: Full-Time Paid 2; Part-Time Paid 12; Part-Time Volunteers 1.
Governing Authority: municipal. Parent Institution: City of Lancaster. Tax-exempt.
Institution Type/Description: History Museum: c.1880s hotel.
Collections: local Antelope Valley Native American & pioneer history artifacts; clothing; photographs; furniture.
Major Exhibits: Legendary Locals, 9/14/14-12/31/14.
Research Fields: local history; anthropology.
Activities: educational programs; guided tours; outdoor events.
Hours & Admission Prices: 2nd & 4th Fri.-Sat. 11-4. No charge; donations accepted. Closed holidays.
Attendance: 2,000 (estimated)

Larkspur

GALLERY BERGELLI, 483 Magnolia Ave., Larkspur, CA 94939. Tel.: 415-945-9454. Facebook: Gallery Bergelli.
E-mail: gallery@bergelli.com
Web Site: www.bergelli.com
Founded: 2000.
Personnel Profile: Part-Time Paid 1; Interns 1.
Institution Type/Description: Art Gallery.
Collections: works by national & international artists; paintings; sculptures; contemporary art.
Publications: artist catalogues.
Hours & Admission Prices: Thurs.-Fri. 10-4, Sat.-Sun. 11-4. No charge.

Lebec

FORT TEJON STATE HISTORIC PARK, 4201 Fort Tejon Rd., Lebec, CA 93243. Mailing Address: P.O. Box 895, Lebec, CA 93243-0895. Tel.: 661-248-6692. Fax: 661-248-8373.
Web Site: www.parks.ca.gov
Founded: 1939.
Congressional District: 18
Key Personnel: Park Supt., Stephen Bylin; Interpreter, Sean T. Malis.
Personnel Profile: Full-Time Paid 2; Part-Time Paid 3; Part-Time Volunteers 600.
Governing Authority: state. Parent Institution: California Dept. of Parks & Recreation, P.O. Box 942896 Sacramento, CA 94296-0001.
Institution Type/Description: State Historic Park: site of 1854-64 U.S. Army fort.
Collections: interpretive collections.
Facilities: visitor center; picnic area; group camp.
Activities: permanent exhibitions; guided tours. Museum Sponsors: Civil War reenactments; Dragoon Era Living History Program; California Volunteer Living History Program.
Publications: brochure.
Hours & Admission Prices: Park: daily 9-4. Park: adults $2; children 16 & under no charge. Special Events: adults $5, children $3. Closed New Year's Day; Thanksgiving; Christmas. &
Attendance: 85,000 (estimated)
Membership: Fort Tejon Historical Association: Individual $20; Family $30.

Lemoore

SARAH A. MOONEY MEMORIAL MUSEUM, 542 W. 'D' St., Lemoore, CA 93245. Mailing Address: P.O. Box 413, Lemoore, CA 93245-0413. Tel.: 559-925-0321.
Web Site: www.lemoore.com/sammm.htm
Personnel Profile: Part-Time Volunteers 20.
Governing Authority: bd. of directors.
Institution Type/Description: Historic House Museum: built in 1893.
Collections: period furnishings; clothing; personal artifacts; office equipment.
Hours & Admission Prices: Call for hours. No charge; donations accepted.

Lodi

MICKE GROVE ZOO, 11793 N. Micke Grove Rd., Lodi, CA 95240-9426. Tel.: 209-331-3010. Fax: 209-331-7271.
E-mail: info@.mgzoo.com
Web Site: www.mgzoo.com
Founded: 1957.
Congressional District: 14
Key Personnel: Zoo Mgr., James Rexroth; Cur., Avanti Mallapur.
Personnel Profile: Full-Time Paid 10; Part-Time Paid 8; Part-Time Volunteers 65.
Governing Authority: county. Parent Institution: San Joaquin County. Tax-exempt.
Institution Type/Description: Zoo.
Collections: birds, primates & carnivores representing 80 species & 160 individuals.
Research Fields: behavioral animal research.
Facilities: 150-vol. library of zoology reference materials available for staff use & to public on special request; zoological park; auditorium.
Activities: guided tours; lectures; formally organized education programs for children; docent program or council; permanent exhibitions.
Publications: bimonthly newsletter, Paw Prints; Micke Grove Zoo News.
Hours & Admission Prices: Daily 10-5. Adults 14 & up $5, children 3-13 $3; discounts to school classes, seniors & large groups; children 2 & under no charge. Closed Christmas Day. &
Attendance: 200,000 (estimated)
Membership: Individual $35; Grandparent $45; Family $60; Zoo Friend $ 100; Small Business $250; Small Business Elite $500.

* **SAN JOAQUIN COUNTY HISTORICAL MUSEUM, (M),** 11793 N. Micke Grove Rd., Lodi, CA 95240-9426. Mailing Address: P.O. Box 30, Lodi, CA 95241-0030. Tel.: 209-331-2055. Fax: 209-331-2057.
E-mail: info@sanjoaquinhistory.org
Web Site: www.sanjoaquinhistory.org
Founded: 1966.
Congressional District: 14
Key Personnel: Exec. Dir., David R. Stuart; Pres. (V), Claude Brown; Collections & Exhibits Mgr., Julie Blood; Mgr. Education & Visitors Svcs. and Museum Shop Mgr., Robin Wood; Archivist, Leigh Johnsen; Admin-istrative Mgr., Ute Gampp; Bookkeeper, Judy Rodman; Maintenance Supvr., Mike Mason.
Personnel Profile: Full-Time Paid 4; Part-Time Paid 7; Part-Time Volunteers 130.
Volunteer Hours: 27,268
Operating Expenses: 752,287
Operating Income: 716,587
Governing Authority: society; county; nonprofit organization. Parent Institution: San Joaquin County Historical Society Affiliated with San Joaquin County Historical Museum. Tax-exempt: 501(c)(3).
Institution Type/Description: Regional History Museum.
Collections: tools; anthropology; agricultural tools & implements; transportation; textiles & costumes; Crawler & wheeled tractors; trucks; California & Southern mines collections; local history; county historical documents; Charles M. Weber collection; grape, wine, fruit & nut industries; vineyard.
Research Fields: native prehistory & history; early American trappers & settlers; agricultural development; early American hand tools; cultural & ethnic groups in county.
Facilities: over 5,000-vol. library of history & maps; botanical garden of native California flora; county archives.
Activities: guided tours; craft demonstrations; summer museum youth camp; agriculture demonstrations; student environmental living program.
Publications: The Harness Maker & His Tools; William George Micke; Man of the Soil; Chinese Clothing & Theatrical Costumes; The Heritage of Claude H. Erickson; Charles M. Weber, Founder of Stockton, San Joaquin Historian, News & Notes; Land of Bright Promise; On The Road Again: Final Journey of Julia Weber Home; Visiting Our County Museum.
Hours & Admission Prices: Summer: Wed.-Sun. 10-4; Winter: Wed.-Sun. 11-4; guided tours by appointment. Adults $5, senior citizens & teens $4, children 6-12 $2; San Joaquin County Historical Society members no charge. Closed New Year's Day; Independence Day; Veterans Day; Thanksgiving & day after; Christmas. &
Attendance: 50,000 (estimated)
Membership: Individual & Family $35; Business $50; Sponsor $100; Patron $250; Sustainer $500; Benefactor $1,000.

WORLD OF WONDERS SCIENCE MUSEUM, 2 N. Sacramento St., Lodi, CA 95240. Tel.: 209-368-0969. Fax: 209-369-1290. Facebook: World of Wonders Science Museum.
Web Site: www.wowsciencemuseum.com
Founded: 2005.
Key Personnel: Pres. (V), Sally Snyder.
Personnel Profile: Full-Time Paid 4.
Institution Type/Description: Science Museum.
Collections: hands-on science exhibitions.
Activities: educational programs; birthday parties.
Hours & Admission Prices: Wed.-Sun. 10-5. Adults $6, seniors 60 & over and students $5, children 2-17 $4; teachers no charge. &
Attendance: 20,193 (accurate)
Membership: Student or Senior 60 & over $20; Individual $30; Family & Grandparent $60; Family Plus $100; Friends $250.

Loleta

WIYOT HERITAGE CENTER, 1000 Wiyot Dr., Loleta, CA 95551. Tel.: 707-733-5055; 800-388-7633. Fax: 707-733-5601.
Key Personnel: Cultural Dir., Helene Rouvier
Institution Type/Description: Heritage Center.
Collections: Wiyot culture & history; photographs; personal artifacts.
Activities: special events; educational programs.
Hours & Admission Prices: Mon.-Fri. 8-5. Closed holidays.

Lomita

LOMITA RAILROAD MUSEUM, 2137 W. 250th St., Lomita, CA 90717-2217. Mailing Address: P.O. Box 339, Lomita, CA 90717-0339. Tel.: 310-326-6255. Fax: 310-326-0690.
E-mail: c.blount@lomitacity.com
Web Site: www.lomita-rr.org
Key Personnel: Mgr., Cindy Blount; Cur., Alice Abbott
Institution Type/Description: Railroad Museum.
Collections: railroad artifacts; Southern Pacific Railroad steam locomotive; Union Pacific caboose.
Hours & Admission Prices: Thurs.-Sun. 10-5. Adults $4, children under 12 $2. Closed Thanksgiving; Christmas.

Lompoc

LA PURISIMA MISSION STATE HISTORIC PARK, 2295 Purisima Rd., Lompoc, CA 93436-9647. Tel.: 805-733-3713. Fax: 805-733-2497.
E-mail: lpminfo@parks.ca.gov
Web Site: www.lapurisimamission.org
Founded: 1935.
Congressional District: 19
Key Personnel: Chm. (V), Robert Wilson; Museum Shop Mgr., Audrey Bowman.
Personnel Profile: Full-Time Paid 8; Part-Time Paid 7; Part-Time Volunteers 95.
Governing Authority: state. California Dept. of Parks & Recreation.
Institution Type/Description: History Museum: housed in 1813-1821 restored adobe buildings.
Collections: 1787-1834 Mission period artifacts from local sites; documentary records; translations; data on architectural studies during restoration of 1934-41. Historic Building: 1934-41 Civilian Conservation Corps.
Research Fields: mission history; contemporary restoration c.1935-1941.
Facilities: 200-vol. library of books on history of California Missions, 1787-1834 natural history & interpretive materials, available for use by reservation; mission livestock display; botanical garden.
Activities: guided tours; mission craft demonstrations; self-guided tours; living history events.
Publications: booklet, J.H. Engbeck La Purisima Mission State Historic Park.
Hours & Admission Prices: Daily 9-5. $4 per vehicle. Closed New Year's Day; Thanksgiving; Christmas. &
Attendance: 200,000 (accurate)
Membership: Active Docent $3; Associate $10; Supporting $25; Business $100.

LOMPOC MUSEUM, 200 S. H St., Lompoc, CA 93436-7297. Tel.: 805-736-3888. Fax: 805-736-2840.
E-mail: lompocmuseum@gmail.com
Web Site: www.lompocmuseum.org
Founded: 1969.
Congressional District: 19
Key Personnel: Dir. & Cur. Anthropology, Dr. Lisa A. Renken; Pres. (V), Alice Down; Administrative Asst., Angie Pasquini.
Personnel Profile: Part-Time Paid 2; Part-Time Volunteers 28.
Volunteer Hours: 1,662
Operating Expenses: 106,175
Operating Income: 103,656
Governing Authority: society; board of trustees; nonprofit organization. Lompoc Museum Associates, Inc. Tax-exempt: 501(c)(3).
Institution Type/Description: Anthropology Museum: housed in 1910 Carnegie Library Building.
Collections: Chumash Indian artifacts; natural history & wildlife exhibits; mining & uses of Diatomaceous Earth; historical articles relating to the human development & customs of the area. Historic Buildings: Stone Pine Hall; Artena schoolhouse, 1876.
Research Fields: local archaeology; local history.
Activities: guided tours; films; lectures; gallery talks; permanent & temporary exhibitions; demonstrations.
Publications: newsletter 3 times a year, Galleries.
Hours & Admission Prices: Tues.-Fri. 1-5, Sat.-Sun. 1-4. Suggested Donation: adults $1; children & members no charge. Closed New Year's Day; Easter; Independence Day; Thanksgiving; Christmas.
Attendance: 3,818 (accurate)
Membership: Single $20; Family $25; Sustaining $50; Business & Group $150; Life $500.

LOMPOC VALLEY HISTORICAL SOCIETY, INC., 207 N. L St., Lompoc, CA 93436-5901. Mailing Address: P.O. Box 88, Lompoc, CA 93438-0088. Tel.: 805-735-4626.
E-mail: myra@best1.net
Web Site: lompochistory.org
Founded: 1964.
Congressional District: 19
Key Personnel: Pres., Karen Paaske; 1st Vice Pres., Ardeane Eckert; Corresponding Sec., Julie McLaughlin; Recording Sec., Debbie Manfrina; Treas., Jeannette Miller Wynne.
Personnel Profile: Part-Time Volunteers 26.
Governing Authority: nonprofit organization. Tax-exempt 501(c)(3).
Institution Type/Description: Historical Society Museum: housed in 1875 Fabing-McKay-Spanne House.
Collections: clothing; house furnishings; farm implements; vehicles; photographs; Lompoc artifacts; blacksmith shop; carriage house; genealogy files of local facilities.
Research Fields: genealogy; Lompoc history.
Facilities: library; genealogical reference room.
Activities: permanent & temporary exhibitions; genealogical research.
Publications: quarterly newsletter, Hodge Podge; quarterly bulletin, Lompoc Legacy.
Hours & Admission Prices: Mon. & Thurs. 8:30-11, 4th Sat. each month 10-1; special tours by appointment. No charge; donations accepted. Closed major holidays. &
Attendance: 500 (estimated)
Membership: Junior (under 18) $5; Single $25; Family $35; Patron & Life $250; Family Lifetime $350; Organization $500.

Lone Pine

THE BEVERLY & JIM ROGERS MUSEUM OF LONE PINE FILM HISTORY, U.S. 395, Lone Pine, CA 93543. Mailing Address: P.O. Box 111, Lone Pine, CA 93545-0111. Tel.: 760-876-9909.
E-mail: lonepinemovies@aol.com
Web Site: www.lonepinefilmhistorymuseum.org
Institution Type/Description: History Museum.
Collections: props; costumes; still & moving images; movie memorabilia; period artifacts.
Activities: special events.
Hours & Admission Prices: Wed.-Mon. 10-4.

Long Beach

AMERICAN MUSEUM OF STRAW ART, 2324 Snowden Ave., Long Beach, CA 90815-2234. Tel.: 562-431-3540. Fax: 562-598-0457.
E-mail: curator@strawartmuseum.org
Web Site: www.strawartmuseum.org
Founded: 1984.
Congressional District: 39
Key Personnel: Dir. & Cur., Morgyn Owens-Celli; Archivist, Carol Thompson; Devel., Bob McCashey.
Personnel Profile: Part-Time Paid 7; Interns 4.
Governing Authority: private; nonprofit organization. Parent Institution: American Foundation for the Straw Arts. Tax-exempt: 501(c)(3).
Institution Type/Description: Decorative & Folk Art Museum.
Collections: concentration on decorative & folk arts in straw; 12 distinct categories including straw applique, coiled straw, straw hats & bonnets, straw plait and Swiss straw work.
Facilities: library; 1,400 sq. ft. exhibit space; educational facilities. Museum-related items for sale.
Activities: guided tours; workshops; lectures; loan & traveling exhibits; docent program; educational programs; broadcast programs. Annual Event: Festival Days.
Publications: quarterly magazine, The Straw Chronicles.
Hours & Admission Prices: Temporarily closed.
Membership: Annual $15; Friend of Museum $50; Contributing $60; Patron $250; Benefactor $1,000.

AQUARIUM OF THE PACIFIC, 100 Aquarium Way, Long Beach, CA 90802-8126. Tel.: 562-590-3100. Fax: 562-951-1629. Facebook: Aquarium of the Pacific.
E-mail: aquariumofpacific@lbaop.org
Web Site: www.aquariumofpacific.org
Formerly: Long Beach Aquarium of the Pacific
Founded: 1998.
Key Personnel: Pres. & C.E.O., Jerry Schubel; Chm. (V), John Molina; Vice Pres. Finance & C.F.O., Anthony Brown; Vice Pres. Govt. Rels. & Special Projects and Corporate Sec., Barbara Long; Vice Pres. Communications & Mktg., Cecile Fisher; Vice Pres. Devel., Christopher Clinton Conway; Dir. Education, David Bader; Vice Pres. Human Resources, Kathie Nirschl; Facilities, Tom Vantress; Dir. Retail, Jeff Spofford; Vice Pres. Husbandry, Perry Hampton; Vice Pres. Operations, John Rouse; Dir. Public Rels., Marilyn Padilla.
Personnel Profile: Full-Time Paid 150; Full-Time Volunteers 1; Part-Time Paid 125; Part-Time Volunteers 1,000; Interns 50.
Volunteer Hours: 154,590
Operating Expenses: 32,618,000
Operating Income: 38,151,000
Governing Authority: nonprofit organization. Tax-exempt.
Institution Type/Description: Aquarium.
Collections: 11,000 marine animals representing 500 species.

Major Exhibits: Wonders of the Deep, 1/14-2/15.
Facilities: teacher resource center; restaurant; 170-seat auditorium; video conferencing studio; meeting space. Gift items for sale.
Activities: shows; lectures; educational programs; cultural festivals.
Publications: membership newsletter, Pacific Currents.
Hours & Admission Prices: Daily 9-6. Adults $25.95, seniors 62 & over $22.95, children 3-11 $14.95. &
Attendance: 1,500,000 (accurate)
Membership: Student & Senior $45; Individual $55; Senior Couple $80; Dual $90; Grandparents & Family $115; Family Plus $165.

∗ CALIFORNIA STATE UNIVERSITY, LONG BEACH, UNI-VERSITY ART MUSEUM, 1250 Bellflower Blvd., Long Beach, CA 90840-0004. Tel.: 562-985-5761. Fax: 562-985-7602.
E-mail: kaplan@csulb.edu
Web Site: www.csulb.edu/uam
Founded: 1949.
Congressional District: 32
Key Personnel: Dir., Chris Scoates; Pres. (V), Michael Davis; Assoc. Dir., Ilee Kaplan; Cur. Education, Brian Trimble; Registrar & Cur. Collections, Angela Barker; Dir. Publications & Public Rels., Amanda Fruta.
Personnel Profile: Full-Time Paid 7; Part-Time Paid 30; Part-Time Volunteers 50.
Governing Authority: university; nonprofit. Parent Institution: California State University, Long Beach. Tax-exempt: 501(c)(3).
Institution Type/Description: University Art Museum.
Collections: modern & contemporary works of art on paper; site-specific sculpture; the Gordon F. Hampton collection.
Research Fields: art history; museology & museography.
Facilities: library; reading room; auditorium; theatre; classrooms; restaurants on campus.
Activities: guided tours; lectures; films; gallery talks; formally organized education programs for adults, undergraduate & graduate students; training programs for museum professionals; inter-museum loan, temporary & traveling exhibitions; Certificate in Museum Studies given by the university recognizing the four semester training program in museology at the graduate level.
Publications: catalogues; brochures; newsletter.
Hours & Admission Prices: Summer: Tues.-Sat. 12-5; Sept.-May Tues.-Wed. & Fri. 12-5, Thurs. 12-8. Adults $4; discounts to AAM members; students & staff of CSULB and members no charge. Closed university holidays. &
Attendance: 60,000 (estimated)
Membership: CSULB Student $15; Faculty & Staff $25; Friend $35; Fellow $100; Partner $300; Contemporary Council $500; Benefactor $1,000.

EARL BURNS MILLER JAPANESE GARDEN, (M), California State University Long Beach, Earl Warren Dr., Long Beach, CA 90840. Mailing Address: California State University Long Beach, 1250 Bellflower Blvd., BAC Rm. 202, Long Beach, CA 90840. Tel.: 562-985-5930 & 8889.
E-mail: jgcoordinators@csulb.edu
Web Site: www.csulb.edu/~jgarden
Founded: 1981.
Key Personnel: Dir., Jeanette Schelin; Chm. (V), Kristie Koepplin.
Personnel Profile: Full-Time Paid 4; Part-Time Paid 2; Part-Time Volunteers 500; Interns 30.
Governing Authority: Parent Institution: California State University, Long Beach. Subsidiary Institution: Long Beach FDN & LB49 FDN. Tax-exempt.
Institution Type/Description: Japanese-style Garden.
Collections: bamboo; Japanese black pine; sculptures.
Research Fields: Japanese gardens outside Japan.
Facilities: 1.3 acre garden; educational pavilion.
Activities: campus & community cultural & horticultural programs & events.
Publications: newsletter, The Lantern.
Hours & Admission Prices: Tues.-Fri. 8-3:30, Sun. 12-4; call to confirm; groups of 10 or more by appointment. Tours: no charge; donations accepted. Closed spring & winter break; Independence Day; Thanksgiving & weekend after. &
Attendance: 75,000 (estimated)
Membership: CSULB Students, Faculty & Alumni $40; Basic $50; Associate $120; Guardian $300; Garden Fellows $500; Benefactor $1,000.

EL DORADO NATURE CENTER, 7550 E. Spring St., Long Beach, CA 90815-1698. Tel.: 562-570-1745 & 1748. Fax: 562-570-8530.
Web Site: www.lbparks.org
Founded: 1969.
Congressional District: 38
Key Personnel: Supervising Park Naturalist, Meaghan O'Neill.

Personnel Profile: Full-Time Paid 1; Part-Time Paid 20; Part-Time Volunteers 80.
Governing Authority: municipal. Parent Institution: City of Long Beach, Parks Recreation & Marine Dept. Tax-exempt.
Institution Type/Description: Nature Center.
Collections: natural history of region.
Facilities: 102.5 acres with trails. Museum-related items for sale.
Activities: guided tours; self-guided tours; natural history & outdoor-related classes; school outreach - environmental & cultural; stewardship, service learning opportunity.
Publications: volunteer newsletter, On the Trail.
Hours & Admission Prices: Trail: Tues.-Sun. 8-5. Museum: Tues.-Fri. 10-4, Sat.-Sun. 8:30-4. Parking: Mon.-Fri. $5 per car; Sat.-Sun. $7 per car, holidays $8 per car, school buses $27, commercial buses $32. Closed Christmas. &
Attendance: 150,000 (estimated)
Membership: Friends: $35; $50; $100; $250; $500; $1,000.

HISTORICAL SOCIETY OF LONG BEACH, (M), 4260 Atlantic Ave., Long Beach, CA 90807. Tel.: 562-424-2220.
Key Personnel: Exec. Dir., Julie Bartolotto.
Governing Authority: Tax-exempt: 501(c)(3).
Institution Type/Description: Historical Society Museum.
Collections: local history & culture; period furnishings; personal artifacts; photographs.
Hours & Admission Prices: Thurs. 1-7, Fri. 1-5, Sat. 11-5.

LONG BEACH FIRE MUSEUM, Old Station 10, 1445 Peterson Ave., Long Beach, CA 90813-2325.
E-mail: lbfdmuseum@verizon.net
Web Site: www.lbfdm.org
Key Personnel: Cur., Herb Bramley
Institution Type/Description: Fire-Fighting Museum.
Collections: fire equipment; badges; photographs; clothing.
Hours & Admission Prices: Wed. 8am-12pm, 2nd Sat. of month 10-3; other times by appointment. No charge.

∗ LONG BEACH MUSEUM OF ART, 2300 E. Ocean Blvd., Long Beach, CA 90803-2442. Tel.: 562-439-2119. Fax: 562-439-3587. Facebook: Long Beach Museum of Art.
E-mail: ronn@lbma.org
Web Site: www.lbma.org
Founded: 1950.
Congressional District: 34
Key Personnel: Dir., Ronald C. Nelson; Dir. Collections, Sue Ann Robinson; Dir. Finance, Suzanne Rivera.
Personnel Profile: Full-Time Paid 28; Full-Time Volunteers 250; Part-Time Paid 25.
Governing Authority: private. Long Beach Museum of Art Foundation. Tax-exempt.
Institution Type/Description: Art Museum.
Collections: Milton Wichner Collection; Jawlensky, Kandinsky, Feininger; Art of Southern California & European 1850-present; contemporary sculpture; decorative arts; contemporary ceramics & turned wood; English & French ceramics.
Major Exhibits: When Collecting Becomes a Collection-Sharing the Gifts of Wilfred Davis Fletcher, 10/31/13-1/26/14; Frank E. cummings III: Jewel Harmony in Wood, 11/22/13-2/23/14; Baroque Sensibilities, Barbara Strasen & Sherrie Wolf, 2/13/14-6/15/14; The Paternal Suit, Heirlooms from the F. Scott Hess Family Foundation, 7/5/14-10/5/14; Masterworks, 10/24/14-12/31/14.
Research Fields: Blue Four; modern German art; arts of Southern California; 20th-century American art; decorative arts; French & English ceramics.
Facilities: 1,200-vol. library of art history books & catalogs of exhibitions; gallery; sculpture garden. Bookstore & art-related items for sale.
Activities: lectures; performance; docent tours; art festivals; art making workshops for children.
Publications: exhibition catalogues: Milton Wichner Collection, Abraham Walkowitz; Drawings by Painters; Video, A Retrospective Long Beach Museum of Art 1974-1984; Media Arts; Southland Video Anthology; California Video; Art of Music Video; New California Video; Video Poetics; Alexej Jawlensky; Gary Hill; Selina Trieff; Bettina Brendel; Matthew Thomas; New Visions: Video 1998; Bountiful Harvest: American Decorative Arts from the Gail-Oxford Collection; For A New Nation: American Decorative Arts from the Gail Oxford Collection; Imps On A Bridge: Wedwood Fairyland & Other Lustres; Evocations: Sharon Ellis, 1991-2001; The Modernist Jewelry of Clair Falkenstein; Enigma Variations: The Sculpture of John Frame, 1980-2005; Mineo Mizuno; Painting with Fire: Masters of Enameling in America, 1930-1980; Staffordshire

Earthenware Ceramics, gift of the Estate of Dr. & Mrs. Leslie Dornfeld; The Transforming Vision, The Wood sculpture of William Hunter, 1970-2005; Evolution, Exchange & Evolution: Worldwide Video Long Beach 1974-1999; Population, Good Man Bad Man, Ray Turner; About Face: Portraiture Now; The Paternal Suit, Heirlooms from F. Scott Hess Family Foundation; Risque (dirty little pictures); Frank E. Cummings II: Jeweled Harmony in Wood.

Hours & Admission Prices: Thurs.-Sun. 11-5. Adults $7, students & seniors over 62 $6; discounts to AAM & ICOM members; members, children under 12 & Fri. no charge. Closed New Year's Day; Independence Day; Thanksgiving; Christmas. &

Membership: Student, Senior & Educator $40; Individual $50; Family & Dual $75; Patron $150; Sponsor $500; Director's Circle $1,000; Collector's Circle $1,500; President's Circle $2,500; Milbank Philanthropic Circle $5,000.

MUSEUM OF LATIN AMERICAN ART, (M), 628 Alamitos Ave., Long Beach, CA 90802-1513. Tel.: 562-437-1689. Fax: 562-216-4190.

E-mail: info@molaa.org
Web Site: www.molaa.org
Founded: 1996.
Key Personnel: Co. Chm., Mike Deoulet; Co. Chm., Burke Gumbiner; Pres., Stuart Ashman; Vice Pres. Operations, Lee Gumbiner; Chief Cur., Cecilia Fajardo-Hill; Vice Pres. Communications, Susan Golden.
Personnel Profile: Full-Time Paid 40; Part-Time Paid 5; Interns 3.
Governing Authority: private; nonprofit organization. Tax-exempt: 501(c)(3).
Institution Type/Description: Art Museum.
Collections: modern & contemporary Latin American art.
Research Fields: contemporary art of Latin America.
Facilities: 600-vol. library; 12,000 sq. ft. exhibit space; cafe; education/entertainment center. Museum-related items for sale.
Activities: art festivals; concerts; dance recitals; docent program; films; formal education programs for children; guided tours; lectures; temporary & traveling exhibitions; theater. Annual Events: Day of the Dead; Cinco de Mayo.
Publications: quarterly newsletter; monthly calendar of events.
Hours & Admission Prices: Wed. & Fri.-Sun. 11-5, Thurs. 11-9. Adults $9, seniors & students $6; children under 12 & members no charge. Closed New Year's Day; Thanksgiving; Christmas. &
Attendance: 55,000 (estimated)
Membership: Student (with ID) & Teachers $25; Senior $55; Basic $60; Family $80; Friend $125; Colleagues $250; Associate $500; Director's Circle $1,000.

PACIFIC ISLAND ETHNIC ART MUSEUM, 695 Alamitos Ave., Long Beach, CA 90802-1514. Tel.: 562-216-4170. Fax: 562-435-3052.

E-mail: info@pieam.org
Web Site: www.pieam.org
Founded: 2010.
Institution Type/Description: Art Museum.
Collections: Pacific Islands art; paintings; murals; sculptures; textiles; wooden tools; jewelry; carvings.
Activities: rental facilities; educational programs; special events; cultural demonstrations.
Hours & Admission Prices: Wed.-Sun. 11-5. Adults $5, students & seniors over 62 $3; members & children under 12 no charge. Closed New Year's Day; Independence Day; Thanksgiving; Christmas.

THE QUEEN MARY, 1126 Queens Hwy., Long Beach, CA 90802-6390. Mailing Address: c/o Evolution Hospitality LLC, 620 Newport Center Dr., Newport Beach, CA 92660. Tel.: 877-342-0738.

E-mail: marketing@queenmary.com
Web Site: www.queenmary.com
Founded: 1971.
Congressional District: 38
Personnel Profile: Full-Time Paid 500; Part-Time Paid 100; Part-Time Volunteers 17.
Governing Authority: Operated by Evolution Hospitality LLC.
Institution Type/Description: Maritime Museum: located aboard the Queen Mary, retired British ocean liner.
Collections: Art Deco architecture; 56 varieties of wood; acid etched glass; memorabilia from the 30s & 40s; two 40,000 HP steam turbine engines & power train.

Facilities: retired British ocean liner: 307 hotel staterooms; 3 restaurants; 3 lounges & bars; 17 salons & ballrooms; shops; wedding chapel; convention accommodations; meeting facilities; marketplace.
Activities: self-guided & guided tour of Queen Mary including bridge, captain's quarters, engine room; historical displays.
Publications: Mailcall!
Hours & Admission Prices: Mon.-Thurs. 10-6, Fri.-Sun. 10-7. Adults $24.95, children $13.95.
Attendance: 1,300,000 (accurate)

RANCHO LOS ALAMITOS, 6400 Bixby Hill Rd., Long Beach, CA 90815-4706. Tel.: 562-431-3541. Fax: 562-430-9694. Facebook: Rancho Los Alamitos.

E-mail: info@rancholosalamitos.org
Web Site: www.rancholosalamitos.org
Founded: 1970.
Congressional District: 32
Key Personnel: Exec. Dir., Pamela Seager; Educational Programs, Ms. Michael Powers; Cur., Pamela Young Lee; Museum Shop Mgr., Teresa Barbee.
Personnel Profile: Full-Time Paid 8; Part-Time Paid 8; Part-Time Volunteers 167.
Governing Authority: municipal government. Owned by City of Long Beach; operated by Rancho Los Alamitos Foundation. Tax-exempt.
Institution Type/Description: Historic Site Museum: Native American site, c.1800 adobe is part of the 1790 300,000-acre Manuel Nieto land grant.
Collections: local history from 500 A.D.; furniture & decorative arts; tools; farm implements and equipment; Indian artifacts; 4-acre historic garden; livestock. Historic Buildings: ranch house; stallion barn; dairy barn; blacksmith's shop; feed shed; new Rancho Center & bookshop/classroom.
Research Fields: local & California history; California ranching; landscape & garden history.
Activities: adult, school & special tours; docent program; permanent exhibitions; public events; heritage programs; lectures; Native American cultural workshops for children. Museum Sponsors: California Ranch Day; Fall Festival; Japanese & Hispanic Heritage Celebrations; Cultural Arts Showcase; Dramatic Period Presentation at Christmas.
Hours & Admission Prices: Wed.-Sun. 1-5. No charge; donations accepted. Closed national holidays. &
Attendance: 25,000 (estimated)
Membership: Rancho Los Alamitos Foundation. Friend $30; Supporter $75; Partner $200; Patron $500; Benefactor $1,000.

RANCHO LOS CERRITOS HISTORIC SITE, 4600 Virginia Rd., Long Beach, CA 90807-1916. Tel.: 562-570-1755. Fax: 562-570-1893.

E-mail: ellen.calomiris@longbeach.gov
Web Site: www.rancholoscerritos.org
Founded: 1955.
Congressional District: 32
Key Personnel: Exec. Dir., Ellen Calomiris; Education Dir., Meighan Maguire; Horticulturist, Marie Barnidge-McIntyre; Museum Shop Mgr., Cheryl Bryan; Pres. Foundation (V), Kevin Kayse.
Personnel Profile: Full-Time Paid 1; Part-Time Paid 4; Part-Time Volunteers 150.
Governing Authority: Rancho Los Cerritos Foundation. Parent Institution: Dept. of Parks, Recreation & Marine, City of Long Beach, CA 90815. Tax-exempt: 501(c)(3)
Institution Type/Description: Historic Site Museum: located on 1790 Spanish land grant to Manuel Nieto; 2-story adobe constructed 1844.
Collections: furniture; tools; costumes; quilts; coverlets; photographs; gardens of original & native plants; archaeology. Historic Building: 1844, 2-story Monterey style adobe built by John Temple & remodeled in 1930 in the mission colonial revival style; c.1850-1940 historic gardens.
Research Fields: local & California history; California ranching; John Temple, Jotham Bixby & Sarah Bixby Smith families; adobe architecture; Ralph Cornell, landscape architect.
Facilities: 5,500-vol. research library & archives, books, maps, manuscripts.
Activities: guided tours; school tours; living history tours; garden programs; formally organized education programs for children; docent program or council; permanent & changing exhibitions; annual family events, workshops & lectures; teen docent program; summer camps; concerts.
Publications: newsletter, Voices.
Hours & Admission Prices: Guided Tours: Wed.-Sun. 1-5. Garden Tours: Sat.-Sun. 2:30. No charge; donations accepted. Closed holidays. &

Attendance: 22,000

Los Altos

LOS ALTOS HISTORY MUSEUM AKA ASSOCIATION OF THE LOS ALTOS HISTORICAL MUSEUM, (M), 51 S. San Antonio Rd., Los Altos, CA 94022-3056. Tel.: 650-948-9427. Fax: 650-559-0268.
E-mail: info@losaltoshistory.org
Web Site: www.losaltoshistory.org
Founded: 1977.
Congressional District: 14
Key Personnel: Exec. Dir., Laura Bajuk, E.D.; Pres. (V), Ginger Beman; Treas. (V), Nomi Trapnell; Collections Mgr., Lisa Robinson; Museum Shop Mgr., Jean Kenny; Museum Shop Mgr., Diane Simmons.
Personnel Profile: Full-Time Paid 1; Part-Time Paid 4; Part-Time Volunteers 300.
Governing Authority: private; nonprofit organization. Tax-exempt: 501(c)(3).
Institution Type/Description: History Museum.
Collections: newspapers, photos & documents of Los Altos & Los Altos Hills; agricultural outbuildings. Historic Building: 1905 farmhouse.
Research Fields: local history.
Facilities: classroom with AV equipment; 2,400 sq. ft. exhibit space. Museum-related items for sale.
Activities: docent programs; formal third and fourth grade education programs; lectures; temporary & traveling exhibitions; cable TV programs; rental facilities. Annual Events: Antique Shows; Historical Essay Contest for 3rd - 6th grade students; Crab Feed; Train Days.
Publications: quarterly newsletter, Under the Oaks; book, Early Los Altos & Los Altos Hills (2010).
Hours & Admission Prices: Thurs.-Sun. 12-4. No charge; donations accepted. Closed New Year's Day; Easter; Independence Day; Thanksgiving; Christmas. &
Attendance: 18,863 (accurate)
Membership: Student, Senior & Individual $35; Household $50; Sponsor $100; Supporter $250; Patron $500; Benefactor $1,000.

Los Angeles

A + D ARCHITECTURE AND DESIGN MUSEUM - LOS ANGELES, (M), 6032 Wilshire Blvd., Los Angeles, CA 90036. Tel.: 323-932-9393. Fax: 323-937-0278. Facebook: A Plus D LA.
E-mail: info@aplusd.org
Web Site: www.aplusd.org
Founded: 2001.
Key Personnel: Dir., Tibbie Dunbar; Pres. (V), John Dale; Membership & Special Programs, Sarah Lane
Institution Type/Description: Architecture & Design Museum.
Collections: architecture; interior, landscape, fashion, & product design; regional, national & international designers; photographs.
Activities: educational programs; public outreach; urban hike; walking tours.
Hours & Admission Prices: Tues.-Fri. 11-5, Sat.-Sun. 12-6. Adults $10, students & seniors $5; discounts to AAM members; members no charge. &
Membership: A+D Student $25; A+D Member $50; A+D Family $75; Supporting $250; Business $500; Sustaining $1,500.

ACE MUSEUM, 400 S. La Brea Ave., Los Angeles, CA 90036. Mailing Address: 5514 Wilshire Blvd., 2nd Fl., Los Angeles, CA 90036. Tel.: 323-965-8200. Fax: 323-965-8201.
E-mail: acemuseum@acemuseum.org
Web Site: www.acemuseum.org
Institution Type/Description: Art Gallery.
Collections: works by local, national, & international contemporary artists.
Activities: temporary exhibitions.
Hours & Admission Prices: Mon.-Fri. 1-5. No charge.

THE AFRICAN AMERICAN FIREFIGHTER MUSEUM, (M), 1401 S. Central Ave., Los Angeles, CA 90021. Tel.: 213-744-1730. Fax: 213-744-1731.
E-mail: aaffmuseum@sbcglobal.net
Web Site: aaffmuseum.org
Founded: 1997.
Governing Authority: nonprofit organization. Tax-exempt.
Institution Type/Description: Fire-Fighting History Museum.
Collections: fire fighting history & equipment; early fire engines including a 1940 Pirsch ladder truck & an 1890 hose wagon; uniforms; badges; helmets; African American women firefighters; photographs.
Activities: rental facilities.
Hours & Admission Prices: Sun. 1-4, Tues. & Thurs. 10-2. No charge; donations accepted. Closed New Year's Day; Thanksgiving; Christmas. &

Attendance: 300 (estimated)
Membership: General $60; $250; $500; $1,000; $1,500; $2,000; $2,500.

THE AMERICAN FILM INSTITUTE, 2021 N. Western Ave., Los Angeles, CA 90027-1625. Tel.: 323-856-7600. Fax: 323-467-4578.
Web Site: www.afi.com
Founded: 1967.
Key Personnel: Pres. & C.E.O., Bob Gazzale; Chm. Bd. Directors, Robert A. Daly; Chm. Bd. Trustees, Sir Howard Stringer.
Personnel Profile: Full-Time Paid 135; Part-Time Volunteers 19.
Governing Authority: nonprofit organization. Branch Institution: The AFI National Film Theater, The John F. Kennedy Center for the Performing Arts, Washington, DC 20566; 1180 Ave. of the Americas, 10th Fl., New York, NY 10032. Tax-exempt: 501(c)(3).
Institution Type/Description: National Arts Organization: a national trust dedicated to preserving the heritage of film & television & presenting the moving image as an art form.
Collections: film & TV script collection; seminar & oral history transcriptions; personal papers & manuscripts of leading film figures.
Research Fields: motion pictures; television; video; education for moving image community.
Facilities: 10,000-vol. Louis B. Mayer Library containing books; periodicals; clipping files; 3,000 motion pictures, television series & long films; oral history transcripts & tapes.
Activities: arts festivals; advanced technology courses; film festivals.
Publications: monthly theater program guide, Preview; annual report; Life Achievement Award Tribute books; board newsletter, Dialogue; AFI catalog of feature films.
Hours & Admission Prices: Call for hours and prices. Library: Mon.-Fri. 9-6 no charge to use library. &
Membership: Star $60; Two-Star $125; Three-Star $250; Four-Star $500; Five-Star $1,000; Patron Circle $2,500.

ANGLES GALLERY, 2754 S. La Cienega Blvd., Los Angeles, CA 90034. Tel.: 310-396-5019. Fax: 310-202-6330.
E-mail: info@anglesgallery.com
Web Site: www.anglesgallery.com
Institution Type/Description: Art Gallery.
Collections: works by contemporary artists; paintings; sculpture; photographs; drawings; videos.
Activities: special events; temporary exhibitions.
Hours & Admission Prices: Tues.-Sat. 10-6.

* **AUTRY NATIONAL CENTER OF THE AMERICAN WEST, (M),** 4700 Western Heritage Way, Los Angeles, CA 90027-1462. Tel.: 323-667-2000. Fax: 323-660-5721. Facebook: Autry National Center.
Web Site: www.theautry.org
Formerly: Autry Museum of Western Heritage
Founded: 1984.
Congressional District: 29
Key Personnel: Pres. & C.E.O., W. Richard West; Founding Chm., Jackie Autry; Founding Pres., Joanne D. Hale; Chm. Bd. Trustees, Marshall McKay; Vice Pres. Curatorial & Exhibitions, Dr. Shelby Tisdale; Marilyn B. and Calivn B. Gross Cur. Visual Arts, Amy Scott; Gamble Cur. Western History, Popular Culture and Firearms, Jeffrey Richardson; Chief Cur., Carolyn Brucken; Dir. Exhibition Media, Paula Kessler; Dir. Collection, LaLena Lewark; Registrar & Project Mgr., Steven Walsh; Registrar, Loans & Exhibitions, Sarah Signorovitch; Registrar, Permanent Collections, Peg Brady; Autry Chair, Western History, Stephen Aron; Dir. Autry Library & Research Svcs., Marva Felchlin; Dir. Publications & Assoc. Dir. Institute, Marlene Head; Dir. Education, Erik Greenberg; Dir. Programs & Public Events, Robyn Hetrick; Dir. Major Gifts, Cathy Crowser; Dir. Corp. Partnerships, Mercedes Tondre; Dir. The Institute for the Study of the American West, David Burton; Dir. Membership & Visitor Svcs., Sara Rodriguez; Vice Pres. Communications and Visitor Experience, Stacy Lieberman; Production & Traffic Mgr., Denise Arguijo; Dir. Finance, Angelica Castro; Vice Pres. Devel., Anna Norville; Conservator, Richard Moll; Vice Pres. Planning & Special Projects, Luke Swetland; Dir. Human Resources, Valerie Nelson; Human Resources Generalist, Amanda Trease; Dir. Facilities, Mike Garcia; Dir. Security, Everett Drayton; Sr. Mgr. Retail & Buyer, Jasmine Aslanyan; Dir. Special Events, Janet Reilly; Sr. Mgr. Event Sales & Svcs., Wendy Esensten.
Personnel Profile: Full-Time Paid 129; Part-Time Paid 42; Part-Time Volunteers 300; Interns 4.
Governing Authority: nonprofit organization. Parent Institution: Autry National Center. Tax-exempt.
Institution Type/Description: Western History & Culture Museum.

Collections: 500,000 objects; art & artifacts of Native American cultures; the multicultural American West; popular culture & fine arts.
Research Fields: history, material culture & iconography of the American West.
Facilities: 40,000-vol. library; archives.
Activities: docent tour; seminars; permanent & temporary loan exhibitions of Western & Native American art & artifacts; fund raising; outreach; special events.
Publications: biannual scholarly publication, Convergence; quarterly calendar; brochures; books; gallery guides; exhibition catalogs; quarterly membership magazine, The Autry.
Hours & Admission Prices: Autry: Tues.-Fri. 10-4, Sat.-Sun. 11-5. Adults $10, senior citizens & students $6, children 3-12 $5; discounts to groups, AAM, ICOM & Southern CA Auto Club members; members & 2nd Tues. each month no charge. Southwest Museum of the American Indian: Sat. 10-4. No charge. Closed Independence Day; Labor Day; Thanksgiving; Christmas. &
Attendance: 162,000 (accurate)
Membership: Individual $45; Dual $55; Family $65; Family Plus $125; Turquoise $165; Copper $250; Shell $500; Silver $1,250; Gold $2,500.

AVILA ADOBE AT EL PUEBLO, 10 E. Olvera St., Los Angeles, CA 90012-2921. Mailing Address: 125 El Paseo de la Plaza, History Dept. Ste. 400, Los Angeles, CA 90012. Tel.: 213-628-1274. Facebook: El Pueblo LA.
E-mail: eptours@lacity.org
Web Site: www.elpueblo.lacity.org
Formerly: Avila Adobe
Founded: 1930.
Key Personnel: Cur., Suellen Chang
Institution Type/Description: Historic House Museum: housed in c.1818 home of Don Francisco Avila; also served as headquarters for Commodore Robert Stockton during the Mexican American War.
Collections: 1840s Los Angeles lifestyle.
Research Fields: Spanish colonial America; California missions; California statehood; Los Angeles history.
Hours & Admission Prices: Daily 9-4. No charge; donations accepted. &

BEN MALTZ GALLERY AT OTIS COLLEGE OF ART AND DESIGN, (M), 9045 Lincoln Blvd., Los Angeles, CA 90045-3505. Tel.: 310-665-6905 & 6800. Fax: 310-665-6908.
E-mail. galleryinfo@otis.edu
Web Site: www.otis.edu/benmaltzgallery
Formerly: OTIS Gallery, OTIS College of Art and Design
Founded: 1918.
Key Personnel: Dir., Meg Linton; Exhibitions Coord. & Registrar, Jinger Heffner; Mgr. & Outreach Coord., Kathy MacPherson.
Personnel Profile: Full-Time Paid 4; Part-Time Paid 1; Interns 5.
Governing Authority: private; nonprofit college. Parent Institution: Otis College of Art and Design. Tax-exempt: 501(c)(3).
Institution Type/Description: Art Gallery.
Collections: Emerson Woellfer paintings and drawings.
Research Fields: contemporary art and design.
Activities: traveling exhibitions; curated shows of local, national & international emerging artists; group shows; installations; one-person retrospectives; international artist-in-residence program.
Publications: exhibition catalogs; limited edition artworks.
Hours & Admission Prices: Jan. to mid-Dec. Tues.-Wed. & Fri.-Sat. 10-5, Thurs. 10-7. No charge. Closed New Year's Day; Martin Luther King Jr. weekend; Presidents' Day weekend; Memorial Day weekend; Labor Day weekend; Thanksgiving weekend. &
Attendance: 16,000 (estimated)

CALIFORNIA AFRICAN AMERICAN MUSEUM, 600 State Dr., Exposition Park, Los Angeles, CA 90037-1267. Tel.: 213-744-7432. Fax: 213-744-2050. Facebook: CAAM in LA.
Web Site: www.caamuseum.org
Founded: 1981.
Congressional District: 48
Key Personnel: Exec. Dir., Charmaine Jefferson; Chm. Bd., Eric Lawrence Frazier; Museum Cur., Sonia Brown; Cur. History, Tiffini Bowers; Exhibit Design Supvr., Edward Garcia; Program Mgr. Visual Arts, Mar Hollingsworth; Program Mgr. Visual Arts, Viola Brown; Program Mgr. Education, Elise Woodson; Program Mgr. History, Javon Johnson; Project Mgr., Markum Stansbury, Jr.; Research Librarian, Denise L. McIver; Deputy Dir. Operations, Woodburn T. Schofield.
Personnel Profile: Full-Time Paid 14; Part-Time Paid 7.
Governing Authority: state; nonprofit. Tax-exempt.
Institution Type/Description: African American Culture Museum.
Collections: African American culture: fine art; sculpture; prints; photographs;

costumes; decorative arts; documents; books; manuscripts; audiovisual materials; Pan-African art works; implements; instruments & artifacts of the people of African descent with emphasis in California and west of the Mississippi.
Major Exhibits: Geoffrey & Carmen: A Memoir in Four Movements, 10/13-3/14; Heaven and Earth: Testimony from the South, 10/13-4/14.
Research Fields: African Americans in theater & film; contemporary art, music, sports, popular culture & history.
Facilities: 8,950-vol. library of books, periodicals, audio & video tapes, records & boxes of archival material available for inter-library loan by special request & available to the public for use in-house by appointment; theatre; 40 sq. ft. exhibit space; educational facilities. Catalogues, books, jewelry, African clothing, decorative items, posters, cards, audio visual recordings for sale.
Activities: guided tours; lectures; school loan service; loan, temporary & traveling exhibitions; films; concerts; formally organized education programs for undergraduate & graduate students.
Publications: quarterly newsletter, Museum Notes.
Hours & Admission Prices: Tues.-Sat. 10-5, Sun. 11-5. No charge; donations accepted. Parking $10. Closed New Year's Day; Thanksgiving; Christmas. &
Attendance: 60,000 (estimated)
Membership: Individual $35; Family $50; Patron $100; Sponsor $250; Supporting $500; Benefactor & Corporate $1,000; Corporate & Program Support $5,000.

*** CALIFORNIA SCIENCE CENTER, (M),** 700 Exposition Park Dr., Los Angeles, CA 90037. Tel.: 213-744-7400 & 2019. Fax: 213-744-2034.
Web Site: www.californiasciencecenter.org
Founded: 1880.
Congressional District: 33
Key Personnel: Pres., Jeffrey Rudolph; Deputy Dir. Administration, Cheryl Tateishi; Deputy Dir. Education, Dr. Ron Rohovit; Vice Pres. Retail Operations, Kent Jones; Deputy Dir. Exhibits, Dr. Diane Perlov; Sr. Vice Pres. & Chief Financial Officer, Cynthia Pygin; Deputy Dir. Operations, Tony Budrovich; Sr. Vice Pres. Devel. & Mktg., William T. Harris; Cur. Ecology Programs, Dr. Chuck Kopczak; Cur. Technology, Dr. David Bibas; Cur Aerospace, Dr. Kenneth E. Phillips.
Personnel Profile: Full-Time Paid 218; Part-Time Paid 227; Part-Time Volunteers 243.
Governing Authority: state. Parent Institution: State of California. Tax-exempt.
Institution Type/Description: Science and Technology Museum.
Collections: biology; geology; air & space.
Facilities: 3-D IMAX theater; cafe.
Activities: lectures; films; formally organized education programs for children; permanent & traveling exhibitions; workshops & demonstrations; live programs; teacher professional development.
Publications: quarterly newsletter; brochures; Discovery Guide; World of Life book; educational programming.
Hours & Admission Prices: Daily 10-5. No charge; donations accepted. Closed New Year's Day; Thanksgiving; Christmas. &
Attendance: 1,300,000 (estimated)
Membership: Explorer Family $65; Discover Family $150; Adventure Family $350; Pioneer Family $550.

CALIFORNIA STATE UNIVERSITY, LOS ANGELES, LUCKMAN GALLERY, 5151 State University Dr., Los Angeles, CA 90032-8116. Tel.: 323-343-6604. Fax: 323-343-6423.
E-mail: luckmangallery@luckmanarts.org
Web Site: www.luckmanarts.org
Founded: 1994.
Key Personnel: Exec. Dir., Wendy Baker
Institution Type/Description: Art Gallery.
Collections: works by contemporary artists from around the world.
Facilities: 300-seat theater.
Activities: special events; temporary exhibitions.
Hours & Admission Prices: Mon.-Thurs. & Sat. 12-5.

CENTER FOR THE STUDY OF POLITICAL GRAPHICS (CSPG), 8124 W. Third St., Ste. 211, Los Angeles, CA 90048-4340. Tel.: 323-653-4662. Fax: 323-653-6991.
E-mail: cspg@politicalgraphics.org
Web Site: www.politicalgraphics.org
Founded: 1988.
Key Personnel: C.E.O., Dir. & Cur., Carol A. Wells; Program Dir., Mary Sutton; Archivist, Joy Novak.
Personnel Profile: Full-Time Paid 4; Part-Time Paid 2; Part-Time Volunteers 5; Interns 2.

Governing Authority: private; nonprofit organization. Tax-exempt: 501(c)(3).
Institution Type/Description: History Museum.
Collections: over 60,000 domestic & international political and protest posters from early 20th century to present including Viet Nam War Era, Civil Rights Movement, African American, Asian, Latino & Native American Rights, Women's Rights, Human Rights, Los Angeles, Black Panther Party, Che Guevara & Latin America, racism, sexism & homophobia.
Facilities: research facility. Museum-related items for sale.
Activities: guided tours; lectures; loan, participatory & traveling exhibitions. Annual Event: Party and Art Auction.
Publications: biannual newsletter, Poligrafiks.
Hours & Admission Prices: Mon.-Fri. 9-6 by appointment. No charge; donations accepted.
Attendance: 1,500 (estimated)

CHINESE AMERICAN MUSEUM, 425 N. Los Angeles St., Los Angeles, CA 90012-2939. Mailing Address: El Pueblo de Los Angeles, 125 Paseo de la Plaza, Ste. 400, Los Angeles, CA 90012. Tel.: 213-485-8567. Fax: 213-485-8238.
Web Site: www.camla.org
Founded: 1984.
Institution Type/Description: Chinese American Museum.
Collections: local Chinese American history.
Facilities: Museum-related items for sale.
Publications: annual newsletter.
Hours & Admission Prices: Tues.-Sun. 10-3. Suggested Donation: adults $3; discounts to NARM members; members no charge. Closed major holidays. &
Membership: Charter, Student & Senior $35; Individual $50; Family & Dual $75; Advocate $125; Corporate $500; 888 Sustaining Circle $888.

CORITA ART CENTER, (M), 5515 Franklin Ave., Los Angeles, CA 90028-5901. Tel.: 323-450-4650. Fax: 323-466-2150.
E-mail: sasha@corita.org
Web Site: www.corita.org
Founded: 1995.
Congressional District: 30
Key Personnel: Dir., Alexandra Carrera; Web Designer, Jorge Lopez.
Personnel Profile: Full-Time Paid 1; Part-Time Paid 2; Part-Time Volunteers 4; Interns 1.
Governing Authority: private; nonprofit organization. Parent Institution: Immaculate Heart Community, Los Angeles, CA. Tax-exempt: 501(c)(3).
Institution Type/Description: Art Museum.
Collections: serigraphs & watercolors of Corita Kent; archives.
Research Fields: life & artistic works of Corita Kent.
Facilities: 600 sq. ft. exhibit space. Prints & reproductions for sale.
Activities: tours by appointment; art education programs; permanent & traveling exhibition.
Publications: quarterly newsletter, Corita Art Center News.
Hours & Admission Prices: Mon.-Fri. 10-4; original artwork sales by appointment only. No charge; donations accepted. Closed New Year's week; Christmas week; national holidays. &
Attendance: 200 (estimated)
Membership: Primary Colors $35; Everyday Miracles $75; Different Drummers $150; Moon Flowers $500; Dancing Stars $1,000; Visionaries $5,000; Joyous Revolutionaries $10,000.

COUNTY OF LOS ANGELES FIRE MUSEUM, 1320 N. Eastern Ave., Los Angeles, CA 90063-3244. Mailing Address: P.O. Box 3325, Alhambra, CA 91803-0325. Tel.: 323-881-2411 & 357-0311. Fax: 323-267-0668.
Web Site: www.clafma.org
Founded: 1974.
Key Personnel: Dir., Paul Schneider.
Governing Authority: nonprofit organization.
Institution Type/Description: Fire-Fighting Museum.
Collections: historic fire equipment & memorabilia; 1852 button hand pumper; 1903 steam engine; c.1970 Squad 51; hand-drawn ladder truck; photographs; books; helmets; badges.
Facilities: 520 sq. ft. exhibit space.
Activities: educational programs. Museum Sponsors: Fire Service Day; Fire Musters; Open House in April.
Publications: quarterly newsletter, The Fire Warden.
Hours & Admission Prices: First Sat. each month by appointment. No charge; donations accepted. &
Attendance: 1,400 (estimated)
Membership: Retired $12; Regular $24.

COUTURIER GALLERY, 166 N. La Brea Ave., Los Angeles, CA 90036. Tel.: 323-933-5557. Fax: 323-933-2357.
E-mail: cg@couturiergallery.com
Web Site: couturiergallery.com
Founded: 1987.
Key Personnel: Dir., Darrel Couturier
Institution Type/Description: Art Gallery.
Collections: works by contemporary American & Latin American artists; paintings; sculpture; photographs; ceramics.
Major Exhibits: Andrew More - Cuba, 1/12/14-2/15/14; Rose Cabat at 100, 5/17/14-7/5/14; Roberto Fabelo, 9/13/14-10/18/14.
Activities: special events; temporary exhibitions.
Hours & Admission Prices: Tues.-Sat. 11-5. No charge.

CRAFT AND FOLK ART MUSEUM - CAFAM, 5814 Wilshire Blvd., Los Angeles, CA 90036-4501. Tel.: 323-937-4230.
E-mail: info@cafam.org
Web Site: www.cafam.org
Founded: 1973.
Key Personnel: C.E.O. & Dir., Suzanne Isken; Museum Shop Mgr., Yuko Makuuchi.
Personnel Profile: Full-Time Paid 6; Part-Time Paid 3; Part-Time Volunteers 8; Interns 4.
Governing Authority: Tax-exempt.
Institution Type/Description: Art Museum.
Collections: folk art; contemporary craft; world cultures.
Research Fields: culture craft/folk art.
Facilities: Museum-related items for sale.
Activities: K-12 school tours; internships; family art workshops; adult workshops; curator's lectures; concerts; film screenings; studio tours; dance performances; artist demonstrations.
Publications: quarterly newsletter.
Hours & Admission Prices: Tues.-Fri. 11-5, Sat.-Sun. 12-6. Adults $7, students & senior citizens $5; discounts to AAM members; members, children under 10 & first Wed. each month no charge. Closed New Year's Day; Easter; Independence Day; Thanksgiving; Christmas. &
Attendance: 15,000 (estimated)
Membership: Individual $60; Family & Dual $75; VIP $100; Adventurer $250; Worldly Traveler $500; Diplomat $1,000; Cultural Ambassador $2,500 & up.

EL PUEBLO HISTORICAL MONUMENT, 125 Paseo de la Plaza, Ste. 400, Los Angeles, CA 90012-2959. Tel.: 213-485-8437. Fax: 213-485-0428. Facebook: El Pueblo LA.
Web Site: www.elpueblo.lacity.org
Founded: 1953.
Congressional District: 25
Key Personnel: Gen. Mgr., Christopher Espinosa; Asst. Gen. Mgr., Lisa Sarno.
Governing Authority: city. Owned by City of Los Angeles, CA. Tax-exempt.
Institution Type/Description: Historic District & Museum Complex.
Collections: manuscript collections; architectural drawings; slides; archival collections; photographs; artifact & archaeological collections. Historic Buildings: 1890 Garnier Block; c.1900 Hellman-Quon Building; 1884 Firehouse Museum; 1870 Pico House; 1870 Merced Theatre; 1858 Masonic Hall Museum; 1894 Jones-Simpson Building; c.1888 Jones Building; 1910 10 W. Olvera St.; 1887 Sepulveda House Museum; c.1855 Pelanconi House; 1908 Italian Hall; 1909 Hammel Building; 1926 Plaza United Methodist Church; 1926 Biscailuz Building; 1904 Plaza Substation; c.1818 Avila Adobe Museum; 1870-1914 Winery; 1888 Vickrey/Brunswig Building; 1883 Plaza House; 1818-1822 Plaza Catholic Church; c.1912 Brunswig Warehouse; 1897 Brunswig Annex; 1924 Old Brunswig Co. Laboratory.
Major Exhibits: Sacred Memories, 10/14-11/14.
Research Fields: Los Angeles history; African American history; Chinese American history; Italian American history; development of El Pueblo Park.
Facilities: library of books, periodicals & photographs pertaining to the history of Los Angeles available for research on premises by appointment; visitors center; restaurants. Gift items for sale.
Activities: guided tours; bus tours; slide lectures; school programs; docent program; film on early history of Los Angeles; changing exhibits. Olvera St. Merchants Sponsor: Mardi Gras; Blessing of the Animals in April; Flower & Camera Day in June; Rose Queen Tour in early December; Las Posadas week before Christmas. Park Sponsors: City Birthday Celebration in September; Cinco de Mayo; Mexican Independence Weekend event.
Publications: brochures, general & on specific buildings.
Hours & Admission Prices: Avila Adobe & Visitors Center: daily 9-4. Plaza Firehouse Museum. Chinese American Museum & America Tropical Interpretive Center: Tues.-Sun. 10-3. No charge; donations accepted. For tour reservations call 213-628-1274. &
Attendance: 260,000 (accurate)

Membership: Individual $25; Donor $50; Contributing $100; Patron $250; Sustaining $500; Supporting $1,000; Corporate or Benefactor $5,000.

FERNDELL NATURE MUSEUM, 5375 Red Oak Dr., Los Angeles, CA 90068-2531. Tel.: 323-666-5046.
Institution Type/Description: Nature Museum.
Collections: California sycamores; over 50 fern species; tropical plants & flowers.
Hours & Admission Prices: Daily 6am-10pm.

FIDM MUSEUM AND LIBRARY, INC., (M), 919 S. Grand Ave., Los Angeles, CA 90015-1421. Tel.: 213-623-5821. Fax: 213-624-7617.
E-mail: info@fidmmuseum.org
Web Site: www.fidmmuseum.org
Founded: 1978.
Congressional District: 35
Key Personnel: Dir., Barbara Bundy; Pres., Tonian Hohberg; Treas., Annie Johnson; Cur., Kevin L. Jones; Assoc. Cur., Christina Johnson; Public Rels., Shirley Wilson; Registrar, Meghan Grossman; Security, Lief Nicolaisen; Museum Shop Mgr., Judy Yaras.
Personnel Profile: Full-Time Paid 10; Part-Time Paid 4; Part-Time Volunteers 1; Interns 3.
Governing Authority: private; nonprofit organization. Parent Institution: Fashion Institute of Design & Merchandising. Tax-exempt: 501(c)(3).
Institution Type/Description: Costume Museum.
Collections: 15,000 costumes from 18th century to present; 20th century couture and ready-to-wear.
Major Exhibits: 22nd Annual Art of Motion Picture Costume Design, 2/14-4/14; 8th Annual Outstanding Art of Television Costume Design, 8/14-9/14.
Facilities: 12,000-vol. library; 12,500 sq. ft. exhibit space. Museum-related items for sale.
Activities: guided tours; loan & temporary exhibitions; receptions; lectures.
Publications: annual exhibition brochures.
Hours & Admission Prices: Feb. to mid-Dec. daily call for hours. No charge; donations accepted. Closed major holidays. &
Attendance: 80,000 (estimated)

FLIGHT PATH LEARNING CENTER & MUSEUM, LAX Imperial Terminal, 6661 W. Imperial Hwy., Los Angeles, CA 90045. Tel.: 424-646-7284.
E-mail: flightpathguides@lawa.org
Web Site: www.flightpathmuseum.com
Founded: 2003.
Key Personnel: Dir., Lee Nichols.
Governing Authority: Tax-exempt.
Institution Type/Description: Aviation History Museum.
Collections: airport & aviation history; models; photographs; uniforms; aircraft manufacturers; aerospace companies.
Activities: educational programs; flight simulator training.
Hours & Admission Prices: Tues.-Sat. 10-3. No charge. Closed national holidays.

FOWLER MUSEUM AT UCLA, W. Sunset Blvd. & Westwood Plaza, 1586 Fowler Bldg., Los Angeles, CA 90095-1549. Mailing Address: Box 951549, Los Angeles, CA 90095-1549. Tel.: 310-825-4361 & 206-7004. Fax: 310-206-7007.
E-mail: fowlerws@arts.ucla.edu
Web Site: www.fowler.ucla.edu
Formerly: UCLA Fowler Museum of Cultural History
Founded: 1963.
Congressional District: 23
Key Personnel: Dir., Marla C. Berns; Deputy Dir., David Blair; Exec. Asst., Sophia Livesy; Dir. External Affairs, Stacey Ravel Abarbanel; Exhibit Designer, Sebastian Clough; Sr. Cur. Southeast Asia & The Pacific, Roy Hamilton; Dir. Education & Curatorial Affairs, Betsy D. Quick; Cur. Archaeology, Wendy Teeter; Dir. Registration & Collections Mgmt., Rachel Raynor; Dir. Publications, Daniel Brauer; Dir. Photography, Don Cole; Museum Shop Mgr., Stella Krieger.
Personnel Profile: Full-Time Paid 42; Part-Time Paid 32; Part-Time Volunteers 31; Interns 3.
Governing Authority: university; state. Parent Institution: University of California, 405 Hilgard Ave., Los Angeles, CA 90024. Tax-exempt: 501(c)(3).
Institution Type/Description: Cultural History Museum.
Collections: African, Pacific, S.E. Asian, North, Middle & South American art, archaeology & material culture.
Research Fields: ancient & non-Western art, archaeology; ethnology.

Facilities: 10,000-vol. library of books & periodicals relating to the collections available by request; 320-seat auditorium; 20,000 sq. ft. exhibit space. Museum publications for sale.
Activities: arts festivals; guided tours; hobby workshops; lectures; gallery talks; formally organized education programs for college students, children & adults; inter-museum, participatory, traveling & temporary exhibitions. Museum Sponsors: Satellite & Peripatetic Museum programs; instruction in museology.
Publications: exhibition catalogs; occasional papers; Museum Monogram Series.
Hours & Admission Prices: Office: Mon.-Fri. 8:30-5. Gallery: Wed.-Sun. 12-5, Thurs. 12-8. No charge; donations accepted. Closed university holidays. &
Attendance: 70,000 (accurate)
Membership: Student $20; UCLA Faculty & Staff $25; Senior & UCLA Alumni $35; Individual $50; Family & Dual $75; Contributing $150; Supporting $275; Patron $500; MANUS $1,500; Curator's Circle $2,500; Director's Circle $5,000. Fowler Textile Council: Individual $50; Couple $75; Student $25.

GEORGE J. DOIZAKI GALLERY, Japanese American Cultural & Community Center, 244 S. San Pedro St., Ste. 505, Los Angeles, CA 90012-3856. Tel.: 213-628-2725. Fax: 213-617-8576.
E-mail: info@jaccc.org
Web Site: www.jaccc.org
Founded: 1980.
Key Personnel: Dir., Chris Alhara; Artistic Dir., Hirokazu Kosaka.
Personnel Profile: Full-Time Paid 1.
Institution Type/Description: Art Gallery.
Collections: paintings; ceramics; sculptures; modern art; Japanese arts.
Hours & Admission Prices: Tues.-Fri. 12-5, Sat.-Sun. 11-4. No charge; donations accepted.
Membership: Senior & Student $35; Individual $45; Senior & Student Dual $60; Dual $75; Family $90; Supporter $150; Patron $300; Connoisseur $600; Ambassador's Council $1,200.

THE GRAMMY MUSEUM AT LA LIVE, 800 W. Olympic Ave., LA Live, Downtown Los Angeles Campus, Los Angeles, CA 90015-1366. Mailing Address: 800 W. Olympic Blvd., Ste. A245, Los Angeles, CA 90015. Tel.: 213-765-6800.
E-mail: info@grammymuseum.org
Web Site: grammymuseum.org
Founded: 2008.
Key Personnel: Exec. Dir., Bob Santelli; Deputy Dir., Rita George.
Governing Authority: Parent Institution: AEG. Subsidiary Institution: The Recording Academy.
Institution Type/Description: Music History Museum.
Collections: music history; recordings & recording artists; photographs; Hall of Fame.
Facilities: Museum-related items for sale.
Activities: mix & produce music; videos; hands-on exhibits; educational programs.
Hours & Admission Prices: Sun.-Fri. 11:30-7:30, Sat. 10-7:30. Adults $14.95, students and seniors 65 & over $11.95, youth 6-17 $10.95; members and children 5 & under no charge. &
Membership: Student, Senior, Educator & Traveler $40; Emerging Artist $50; Debut $75; Opening Act $90; Headliner $125; Nominee $250; Award Winner $500; Icon $1,000.

GRIER MUSSER MUSEUM, 403 S. Bonnie Brae St., Los Angeles, CA 90057-3009. Tel.: 213-413-1814.
E-mail: griermusser@hotmail.com
Web Site: griermussermuseum.org
Founded: 1984.
Congressional District: 43
Key Personnel: Dir., Susan Tejada.
Governing Authority: individual operation; organized for profit.
Institution Type/Description: Antiques Museum: housed in c.1898 Queen Ann Victorian house.
Collections: monthly holiday exhibits; postcards of old Los Angeles.
Activities: guided tours; lectures; theatre; traveling exhibitions; weekly postcard show of old Los Angeles. Museum Sponsors: Victorian Christmas Tour in November and December; Victorian Christmas House Tour in November & December.
Publications: newsletter, Parlour Talk.
Hours & Admission Prices: Wed.-Sat. 12-4, call for reservations. Adults $10, senior citizens & students $7, children $5; discounts to groups, families and AAM & ICOM members. Closed most holidays.
Attendance: 500 (accurate)

GRIFFITH OBSERVATORY, 2800 E. Observatory Rd., Los Angeles, CA 90027-1299. Tel.: 213-473-0800. Fax: 213-473-0816.
Web Site: www.griffithobservatory.org
Founded: 1935.
Congressional District: 24
Key Personnel: Dir., Dr. Edwin C. Krupp; Pres. Friends Of The Observatory, David Primes.
Personnel Profile: Full-Time Paid 31; Part-Time Paid 200.
Governing Authority: municipal. City of Los Angeles. Tax-exempt: 170(b)(1)(A).
Institution Type/Description: Public Observatory, Planetarium & Astronomy Museum.
Collections: astronomy & related sciences; artwork & illustration; photographs; astronomical instruments & artifacts.
Research Fields: astronomy.
Facilities: 6,650-vol. library of books on astronomy available by special arrangements; telescope observatory housing a 12 in. Zeiss refracting telescope.
Activities: live planetarium performances; guided tours; special lectures; films; gallery talks; observatory telescopes open to public on clear evenings.
Publications: monthly magazine, Griffith Observer.
Hours & Admission Prices: Wed.-Fri. 12-10, Sat.-Sun. 10-10; hours are subject to change. Open Tues. in summer & school breaks. Oschin Planetarium: adults $7, seniors & students $5, children 5-12 $3; discounts to members. Closed Thanksgiving; Christmas. &
Attendance: 1,000,000 (accurate)
Membership: Friends Of The Observatory: Planet $45; Star $100; Supernova $250; Galaxy $500; Celestial Circle $1,000; Copernicus Society $2,500; Galileo Society $5,000; Newton Society $10,000.

GRUNWALD CENTER FOR THE GRAPHIC ARTS, HAMMER MUSEUM, 10899 Wilshire Blvd., Los Angeles, CA 90024-4343. Tel.: 310-443-7076. Fax: 310-443-7099. TDD: 310-443-7094.
E-mail: info@hammer.ucla.edu
Web Site: www.hammer.ucla.edu
Formerly: Grunwald Center for the Graphic Arts, UCLA Hammer Museum of Art and Cultural Center
Founded: 1956.
Congressional District: 30
Key Personnel: Dir. & Chief Cur., Cynthia Burlingham; Conservation Technician, Maureen McGee; Preparator, Lynne Blaikie; Cur., Allegra Pesenti; Assoc. Registrar, Susan Chin.
Personnel Profile: Full-Time Paid 5; Part-Time Paid 1.
Governing Authority: university. Parent Institution: University of California at Los Angeles. Tax-exempt.
Institution Type/Description: Art Museum & Study Center.
Collections: 45,000 prints, drawings, photographs & artists' books.
Research Fields: history of graphic arts.
Facilities: print study room.
Activities: guided tours; lectures; gallery talks; temporary & traveling exhibitions.
Publications: selected areas of graphic arts collection; temporary & traveling exhibitions.
Hours & Admission Prices: Grunwald Center Study Room: Mon.-Fri. 10-4, by appointment only. No charge. &

HAMMER MUSEUM, 10899 Wilshire Blvd., Los Angeles, CA 90024. Tel.: 310-443-7020 & 7000. Fax: 310-443-7099.
E-mail: info@hammer.ucla.edu
Web Site: www.hammer.ucla.edu
Founded: 1990.
Congressional District: 30
Key Personnel: Dir., Ann Philbin; Chm. & Pres., John V. Tunney; Dept. Dir. Finance & Admin., Deborah Snyder; Dir. Admin., Jenni Kim; Museum Shop Mgr., Monique Fuentes.
Personnel Profile: Full-Time Paid 90; Part-Time Paid 87; Part-Time Volunteers 56; Interns 10.
Governing Authority: university. Parent Institution: University of California at Los Angeles. Subsidiary Institution: UCLA Grunwald Center for the Graphic Arts, 10899 Wilshire Blvd., 2nd Fl., Los Angeles, CA. Tax-exempt.
Institution Type/Description: Art Museum.
Collections: French 19th-century works; European paintings; American artists from 18th to 20th centuries; sculpture garden; Hammer Daumier collection; Hammer contemporary collection.
Research Fields: history of graphic arts; modern sculpture; American & contemporary art.
Facilities: video library & viewing room; public program space; theater; cafe. Bookstore.

Activities: concerts; readings; lectures; screenings; discussions; guided tours; gallery talks; temporary & traveling exhibitions.
Publications: newsletter; temporary exhibitions catalogues.
Hours & Admission Prices: Tues.-Fri. 11-8, Sat.-Sun. 11-5. Adults $10, senior citizens 65 & over and UCLA Alumni Assoc. members $5; members, students, faculty, staff, military, veterans & children no charge. Closed New Year's Day; Independence Day; Thanksgiving; Christmas. &
Attendance: 200,000 (estimated)
Membership: Full Time Student & Artist $40; Individual $50; Friend $75; Contributor $125; Supporter $350; Hammer Fellow $1,000; Hammer Patron $2,500.

HANCOCK MEMORIAL MUSEUM, University Avenue at Childs Way, Los Angeles, CA 90089-0189. Mailing Address: USC Libraries, Special Collections, 3550 Trousdale Pkwy., Los Angeles, CA 90089-0189. Tel.: 213-740-5141. Fax: 213-740-2343.
E-mail: melindah@usc.edu
Web Site: www.usc.edu
Key Personnel: Cur., Melinda Hayes
Institution Type/Description: History Museum: housed in four rooms dismantled from the original Hancock mansion.
Collections: period furnishings; personal artifacts; paintings; Steinway piano c.1910; family heirlooms.
Hours & Admission Prices: By appointment.

HERITAGE SQUARE MUSEUM, (M), 3800 Homer St., Los Angeles, CA 90031-1530. Tel.: 323-225-2700. Fax: 323-225-2725.
E-mail: administrator@heritagesquare.org
Web Site: www.heritagesquare.org
Founded: 1969.
Congressional District: 27
Key Personnel: C.E.O., Saline Davies; Exec. Dir., Jessica Maria Alicea; Mgr. Devel., John Kearns; Cur. Grounds & Gardens, Steven Ormenyi; Education Mgr., Jessica Rivas; Museum Shop Mgr., Beatrice Lima; Administrative Asst., Kim Kirui.
Personnel Profile: Full-Time Paid 3; Part-Time Paid 4; Part-Time Volunteers 90; Interns 1.
Governing Authority: private; nonprofit organization. Parent Institution: Cultural Heritage Foundation of Southern California, Inc., Los Angeles, CA 90031. Tax-exempt: 501(c)(3).
Institution Type/Description: History Museum.
Collections: 8 Victorian structures from Los Angeles & Pasadena; related furniture & personal artifacts; textiles; medical, communication, transportation & business equipment and artifacts.
Research Fields: life of John Ford, master woodcarver & former owner of one of the homes; Southern California rural and urban development, history, agriculture, horticulture and gardening after 1865; structures & families that occupied them.
Facilities: 1,200 sq. ft. exhibit space. Museum-related items for sale.
Activities: living history programs; guided tours; docent program; lectures; temporary exhibitions. Museum Sponsors: MOTA Day, Annual Open House; Silent & Classic Movie Nights in July; Halloween Mourning Tours in October; Holiday Lamplight-Christmas tours; Victorian Fashion Show.
Publications: quarterly member newsletter, On the Square; quarterly volunteer newsletter, The Chronicle.
Hours & Admission Prices: April-Oct. Fri.-Sun. & holiday Mon. 12-5; Nov.-March Fri.-Sun. & holiday Mon. 11:30-4:30. Adults $10, senior citizens $8, children 6-12 $5; discounts to AAA & Time Travelers members; members no charge. Closed New Year's Day; Thanksgiving; Christmas.
Attendance: 12,263 (accurate)
Membership: Friends of Heritage Square: Volunteer $25; Individual $35; Family $50; Sustaining $75; Contributing $100. Heritage Square Society: Member $250; Donor $500; Sponsor $1,000; Patron $2,500.

HISTORICAL SOCIETY OF SOUTHERN CALIFORNIA - LUMMIS HOME: EL ALISAL, 200 E. Ave. 43, Los Angeles, CA 90031. Mailing Address: P.O. Box 93487, Pasadena, CA 91109-3487. Tel.: 323-222-0546 & 460-5632. Fax: 323-645-7466.
E-mail: hssc@socalhistory.org
Web Site: www.socalhistory.org
Founded: 1883.
Congressional District: 25
Key Personnel: Exec. Dir., Patricia Adler-Ingram; Pres. (V), John O. Pohlmann.
Personnel Profile: Full-Time Paid 2; Part-Time Paid 3; Part-Time Volunteers 10.
Governing Authority: municipal; nonprofit organization. Headquarters of Historical Society of Southern California Museum, Los Angeles, CA 90031. Tax-exempt: 501(c)(3).

Institution Type/Description: Historic House: 1897-1912 home of Charles F. Lummis.

Collections: archaeology; ethnology; history; Indian artifacts; photography; books by builder.

Facilities: water-conserving garden. Books on local & Southern California history & gardening for sale.

Activities: guided tours; lectures; permanent exhibitions; water conservation garden.

Publications: Southern California Quarterly; Letters from the Orange Empire; California Bibliographies; Land of Fiction; Centennial History of HSSC; Manana Land; A Guide to Historical Outing in Southern California; Land of Fact; An End and a Beginning: The South Coast and Los Angeles 1850-1887; Those Powerful Years: The South Coast and Los Angeles 1887-1917; Southern California's Spanish Heritage: An Anthology; At the Arroyo's Edge: A History of Linda Vista; A Companion and Guide to the Waterwise Garden; Southern California Local History: A Gathering of the Writings of W.W. Robinson; Month by Month in a Waterwise Garden; El Alisal: Where History Lingers; The St. Francis Dam Disaster Revisited; Women in the Life of Southern California: An Historical Anthology; Angels Flight; Griffith Park; Painting with Light; Mission San Fernando Rey de Espana; John Rowland & William Workman: 1841 So. CA Pioneers; Golden Odyssey; video, Museums Along the Arroyo; Pasadena Sketchbook, Man-made Disaster; Duncan Gleason: Artist, Athlete, and Author; Founding Documents of Los Angeles: A Bilingual Edition.

Hours & Admission Prices: Fri.-Sun. 12-4. No charge; donations accepted. Closed major holidays. craft

Attendance: 10,000 (estimated)

Membership: Student $25; Individual $75; Contributing $125; Patron $300; Benefactor $600; President's Circle $1,250.

HOLLYWOOD BOWL MUSEUM, (M), 2301 N. Highland Ave., Los Angeles, CA 90068-2742. Tel.: 323-850-2058.

E-mail: museum@laphil.org

Web Site: www.hollywoodbowl.com

Founded: 1984.

Key Personnel: Dir. & Cur., Dr. Carol Merrill-Mirsky.

Personnel Profile: Full-Time Paid 1; Part-Time Paid 4; Part-Time Volunteers 60; Interns 1.

Governing Authority: Parent Institution: Los Angeles Philharmonic Association. Tax-exempt.

Institution Type/Description: Music & Performing Arts Museum.

Collections: photographs; films; audio; documents; ephemera; 20th-century musical artists of the Hollywood Bowl.

Research Fields: music & performing arts.

Facilities: resource center; permanent & temporary exhibition galleries; performance space; education area.

Activities: changing exhibits on musical topics; educational, volunteer & docent programs; concert & lecture series.

Publications: Exiles in Paradise 1991.

Hours & Admission Prices: July-Sept. 18 Tues.-Sat. 10-8; Sept. 19-June Tues.-Fri. 10-5, Sat. by appointment. No charge. Closed New Year's Day; Thanksgiving; Christmas. craft

Attendance: 27,000 (accurate)

* THE J. PAUL GETTY MUSEUM, (M), (I), 1200 Getty Center Dr., Ste. 1000, Los Angeles, CA 90049-1687. Tel.: 310-440-7330. Fax: 310-440-7751. TDD: 310-440-7305; Facebook: Getty Museum.

E-mail: administration-museum@getty.edu

Web Site: www.getty.edu

Founded: 1953.

Congressional District: 27

Key Personnel: Dir., Timothy Potts; Pres. & C.E.O. Getty Trust, James Cuno; Assoc. Dir. Collections & Sr. Cur. Manuscripts, Thomas Kren; Assoc. Dir. Exhibitions, Quincy Houghton; Asst. Dir. Education, Toby Tannenbaum; Asst. Dir. Public Affairs, John Giurini; Head, Administration, Nik Honeysett; Acting Sr. Cur. Antiquities, Claire Lyons; Acting Dept. Head, Sculpture & Decorative Arts, Anne-Lise Desmas; Mgr. Exhibition Design, Merritt Price; Sr. Cur. Paintings, Scott Schaefer; Sr. Cur. Photographs, Judith Keller; Head Retail & Merchandising, Thomas Stewart; Sr. Cur. Drawings, Lee Hendrix; Sr. Conservator, Decorative Arts & Sculpture, Brian Considine; Sr. Conservator, Paintings Conservation, Yvonne Szafran; Sr. Conservator, Antiquities, Jerry Podany; Sr. Conservator, Paper Conservation, Marc Harnly; Head Preparations, Kevin Marshall; Chief Registrar, Sally Hibbard; Head, Collection Information & Access, Stanley Smith; Head Budget, Carolyn Simmons; Publisher, Kara Kirk; Dir. Security, Bob Combs.

Personnel Profile: Full-Time Paid 200; Full-Time Volunteers 434; Part-Time Volunteers 491.

Governing Authority: nonprofit organization. Parent Institution: The J. Paul

Getty Trust. Subsidiary Institution: Getty Villa, 17985 Pacific Coast Hwy., Pacific Palisades, CA 90272. Tax-exempt: 501(c)(3).

Institution Type/Description: Art Museum.

Collections: Greek & Roman antiquities; pre-20th century Western European paintings, drawings; manuscripts; sculpture; decorative arts; 19th to 20th-century American & European photographs.

Research Fields: fields pertaining to collections & conservation.

Facilities: 450-seat auditorium; cafe; restaurant. Museum publications for sale.

Activities: lectures; gallery talks; permanent exhibitions; concert series; special exhibits; seminars; general education programs for schools & adults; theatrical performances; family programs; interactive touchscreen access system.

Publications: exhibitions catalogues; Getty Villa & Museum guides; books, art history, antiquities, conservation, photography, European paintings, drawings, manuscripts, sculpture, & decorative arts.

Hours & Admission Prices: Museum: Tues.-Fri. & Sun. 10-5:30, Sat. 10-9. Villa: Wed.-Mon. 10-5 by appointment. Museum: no charge. Villa: no charge. Parking: $15 visit to one or both locations on same day. Closed New Year's Day; Independence Day; Thanksgiving; Christmas. craft

Attendance: 1,600,000 (estimated)

* JAPANESE AMERICAN NATIONAL MUSEUM, (M), 100 N. Central Ave., Los Angeles, CA 90012. Tel.: 213-625-0414. Fax: 213-625-1770.

Web Site: www.janm.org

Founded: 1985.

Congressional District: 34

Key Personnel: Pres. & C.E.O., G.W. (Greg) Kihura, Ph.D.; Chm. (V), Joyce Lane; Vice Pres. External Rels., Cynthia Villasenor; Dir. Community Affairs, Nancy Araki; Dir. Programs & Art Dir., Clement Hanami; Dir. Retail & Visitor Svcs., Maria Kwong; Human Resources & Volunteer Svcs., Annette Miyamoto; Dir. Media Arts, John Esaki; Dir. Education, Allyson Nakauoto; Museum Shop Mgr., Randy Imoto; Controller, Latanya Alexander; Dir. Media & Communications, Janis Wong.

Personnel Profile: Full-Time Paid 50; Part-Time Paid 8; Part-Time Volunteers 100.

Governing Authority: organization; nonprofit. Subsidiary Institution: National Center for the Preservation & Democracy. Tax-exempt: 501(c)(3).

Institution Type/Description: History Museum.

Collections: Japanese American historical artifacts; three-dimensional objects; documents; printings; photographs; moving images; oral histories; 85,000 sq. ft. Pavilion designed by Gyo Obata.

Research Fields: early immigration & settlement; World War II incarceration & military service; post WWII resettlement; Japanese American communities; Japanese American art.

Facilities: 500-vol. library of books & newspapers; classrooms; 18,000 sq. ft. exhibit space; media arts center. Crafts, books of literature & scholarship for sale.

Activities: guided tours; lectures; films; formal educational programs; docent program; temporary exhibitions of your own collections; participatory, loan & traveling exhibitions; family programs; rental facility available.

Publications: annual report; quarterly, calendar of events; museum member magazine; Donor Recognition publication.

Hours & Admission Prices: Tues.-Wed. & Fri.-Sun. 11-5, Thurs. 12-8. Adults $9, seniors (62 & over), students with ID & youth 6-17 $5; discounts to groups & AAM and ICOM members; Thurs. 5-8 & every 3rd Thurs. of month 12-8, members, children 5 & under no charge. Closed New Year's Day; Independence Day; Thanksgiving; Christmas. craft

Attendance: 82,500 (estimated)

Membership: Individual $50; Student & Senior over 62 $30; Family/Dual $75; Contributing $150; Supporting $300; Sustaining $600; Director's Circle $1,00-$4,999; President's Circle $5,000-$9,999; Chairman's Circle $10,000 & up.

KOREAN AMERICAN MUSEUM, 3727 W. Sixth St., Ste. 400, Los Angeles, CA 90020-5112. Tel.: 213-388-4229. Fax: 213-381-1288.

E-mail: info@kamuseum.org

Web Site: www.kamuseum.org

Key Personnel: Program Coord., Changmii Bae, Ph.d.

Institution Type/Description: Korean American Museum.

Collections: Korean-American history & culture.

Hours & Admission Prices: Call for hours.

LA PLAZA DE CULTURA Y ARTES, 501 N. Main St., Los Angeles, CA 90012. Tel.: 213-542-6200; 888-488-8083.

Founded: 2011.

Institution Type/Description: History Museum.

Collections: Mexican & Mexican American history & culture; photographs; personal artifacts; period furnishings.
Activities: special events.
Hours & Admission Prices: Wed.-Mon. 12-7. Adults $9, seniors 60 & over and students $7, children 5-18 $5; children under 5 no charge. Closed New Year's Day; Thanksgiving; Christmas.

LABAND ART GALLERY, LOYOLA MARYMOUNT UNIVERSITY, One LMU Dr., MS 8346, Los Angeles, CA 90045-2650. Tel.: 310-338-2880. Fax: 310-338-6024. Facebook: Laband Art Gallery.

E-mail: labandinfo@lmu.edu
Web Site: cfa.lmu.edu/laband
Founded: 1984.
Congressional District: 36
Key Personnel: Dir. & Cur., Carolyn Peter.
Personnel Profile: Full-Time Paid 1; Part-Time Paid 6; Part-Time Volunteers 1.
Governing Authority: university. Parent Institution: Loyola Marymount University. Tax-exempt: 170(b)(1)(A).
Institution Type/Description: University Art Gallery.
Collections: paintings; prints; drawings; graphic arts.
Facilities: 212-seat auditorium; 2,300 sq. ft. exhibit space; classrooms.
Activities: guided tours; lectures; films; gallery talks; concerts; dance recitals; arts festivals; formally organized education programs for undergraduate college students.
Publications: catalogs & brochures in conjunction with exhibition programs.
Hours & Admission Prices: mid-Sept. to May Wed.-Fri. 12-4. No charge; donations accepted. &
Attendance: 3,877 (accurate)

LACE (LOS ANGELES CONTEMPORARY EXHIBITIONS), 6522 Hollywood Blvd., Los Angeles, CA 90028-6210. Tel.: 323-957-1777. Fax: 323-957-9025.

E-mail: info@welcometolace.org
Web Site: www.welcometolace.org
Founded: 1977.
Key Personnel: Pres., William Moreno; Exec. Dir., Carol Stakenas; Vice Pres., Synderela Peng; Sec., Vincent Ruiz-Abogado; Producer, Ben Smith; Bd. Member, Kathie Foley-Meyer; Bd. Member, Jackie sharp; Bd. Member, Marta Sullivan; Bd. Member, Jeff Cain.
Personnel Profile: Full-Time Paid 3; Part-Time Volunteers 20; Interns 10.
Governing Authority: not-for-profit organization. Tax-exempt: 501(c)(3).
Institution Type/Description: Visual Arts Center
Collections: present works of art in all media including performance art & video.
Research Fields: panels & forums in areas related to contemporary art criticism.
Activities: exhibitions of contemporary art; performances; dialogs; film & video screenings; panels; lectures & readings.
Publications: catalog, Major Shows; brochures; CLOSER: The Blog of LACE.
Hours & Admission Prices: Wed.-Sun. 12-6. Suggested Donation: $3; members no charge. &
Attendance: 15,000 (accurate)
Membership: Student & Artist $25; Friend $50; Advocate $100; Fellow $250; Contributor's Circle $500; Presenter's Circle $1,000; Director's Council $2,500; LACE Vanguard $5,000.

THE LOS ANGELES ART ASSOCIATION/GALLERY 825, 825 N. La Cienega Blvd., Los Angeles, CA 90069-4707. Tel.: 310-652-8272. Fax: 310-652-9251.

E-mail: gallery825@laaa.org
Web Site: www.laaa.org
Founded: 1925.
Congressional District: 5
Key Personnel: Exec. Dir., Peter Mays.
Personnel Profile: Full-Time Paid 3; Part-Time Volunteers 300; Interns 2.
Governing Authority: nonprofit organization. Tax-exempt: 501(c)(3).
Institution Type/Description: Southern California Art Gallery/Association.
Collections: paintings; photographs; sculpture; drawings.
Activities: changing exhibits; lectures; round tables; annual film & video event.
Publications: biannual newsletter, Artline.
Hours & Admission Prices: Tues.-Sat. 10-5. No charge; donations accepted. Closed New Year's Day; Easter; Independence Day; Thanksgiving; Christmas. &
Attendance: 5,000 (estimated)
Membership: Friends $50; Art Enthusiast $100; Artists Advocate $250; Patron $500; Benefactor $1,000; Director's Circle $2,500.

LOS ANGELES (CENTRAL) PUBLIC LIBRARY, 630 W. Fifth St., Los Angeles, CA 90071-2002. Tel.: 213-228-7000 & 7470. Fax: 213-228-7069. TDD: 800-735-2922.

Web Site: www.lapl.org
Founded: 1872.
Congressional District: 25
Key Personnel: Central Library Dir., Giovanna Mannino; Programming & Outreach Mgr., Dan Dupill; Research & Special Collections Mgr., Ani Boyadjian; Customer Service Mgr., Selena Terrazas.
Governing Authority: municipal. Tax-exempt.
Institution Type/Description: Library with Collections.
Collections: California prints; Japanese prints; Leo Politi Bunker Hill paintings; 240,000 images, historic photographs; Gladys English original children's book illustrations; 3 million photographs & 2.6 million Herald Examiner newspaper clippings.
Facilities: 2,600,000-vol. library of books & periodicals of general & special interest available for inter-library loan & on premises; reading room.
Activities: guided tours; permanent, temporary & traveling exhibits; reference services.
Publications: book lists.
Hours & Admission Prices: Mon.-Thurs. 10-8, Fri.-Sat. 10-5:30. No charge. Closed national & state holidays. &
Attendance: 2,140,620 (accurate)

*   LOS ANGELES COUNTY MUSEUM OF ART, (M), 5905 Wilshire Blvd., Los Angeles, CA 90036-4598. Tel.: 323-857-6000. Fax: 323-857-6214. TDD: 323-857-0098.

E-mail: publicinfo@lacma.org
Web Site: www.lacma.org
Founded: 1910.
Congressional District: 24
Key Personnel: Dir., Michael Govan; Pres. & C.O.O., Melody Kanschat; Chm. Bd., Andrew Gordon; C.F.O., Ann Rowland; Mktg. Assoc., Alexandra Capriotti; Mgr. Special Events, Terri Bradshaw; Gen. Counsel, Fred Goldstein; Asst. Dir. Exhibition Program, Zoe Kahr; Dir. Conservation, Mark Gilberg; Sr. Cur. Modern Art, Stephanie Barron; Cur. South & Southeast Asian Art, Stephen A. Markel; Cur. Japanese Art, Robert Singer; Dept. Head and Sr. Cur. Costumes & Textiles, Sharon Takeda; Chief Cur. European Art, J. Patrice Marandel; Dir. Membership, Alvaro Vasquez; Deputy Dir. Art Administration & Collections, Nancy K. Thomas; Dept. Head & Cur. Decorative Arts, Wendy Kaplan; Head Music Programs, Mitch Glickman; Head Film Programs, Ian Birnie; Asst. Vice Pres. Museum Education, Jane E. Burrell; Cur. Photography, Britt Salvesen; Chief Information Officer, Peter Bodell; Asst. Dir. Collections Mgmt., Renee Montgomery; Head Registrar, Nancy Russell; Asst. Vice Pres. Merchandising, Cim B. Castellon; Assoc. Vice Pres. Communications & Mktg., Barbara P. Flaumer.
Personnel Profile: Full-Time Paid 240; Part-Time Paid 30; Part-Time Volunteers 3,300; Interns 53.
Governing Authority: county; nonprofit organization. Tax-exempt: 501(c)(3).
Institution Type/Description: Art Museum.
Collections: American furniture; Chinese, Korean & Japanese paintings, sculptures, ceramics & works on paper; Egyptian & Greco-Roman sculptures & antiquities; European & American art & decorative arts; Islamic art; South, Southeast & later West Asian paintings, sculpture, textiles, jades & metal works; European, Asian, North & South American art textiles and costumes; 20th-century painting, sculpture, prints & drawings.
Research Fields: all curatorial departments.
Facilities: 117,000-vol. library available for use by appointment; 10,000 sq. ft. children's gallery; conservation center; 600-seat theatre; restaurant & cafe. Art books, catalogs, post cards, posters & other museum-related items for sale.
Activities: guided tours; lectures; films; gallery talks; concerts; art rental gallery; formally organized educational programs; docent program; 116-seat lecture hall; 10-volunteer councils; inter-museum loan, permanent, temporary & traveling exhibitions; tours for the hearing & visually impaired.
Publications: descriptive bulletins; monthly magazine, At LACMA; Inside LACMA; exhibition & permanent collection catalogs; members magazines, Connect; Insider.
Hours & Admission Prices: Mon.-Tues. & Thurs. 12-8, Fri. 12-9, Sat.-Sun. 11-8. Adults $15, senior citizens 62 & over and students 18 & over with I.D. $8; discounts to AAM & ICOM members; members, children under 17 & 2nd Tues. of month & County of LA residents after 5 no charge. Closed Thanksgiving; Christmas. &
Attendance: 905,000 (accurate)
Membership: Student $25; Active $90; Patron $200; Supporting $600; Community Partner $1,200; President's Circle Avant-Garde $1,000; President's Circle $2,500; President's Circle Patron $5,000; President's Circle Sponsor

$10,000; Director's Circle $25,000; Director's Circle Scholar $50,000. Add MUSE to any level $50.

LOS ANGELES MUNICIPAL ART GALLERY - BARNSDALL ART PARK, 4800 Hollywood Blvd., Los Angeles, CA 90027-5302. Tel.: 323-644-6269. Fax: 323-644-6271. TDD: 213-660-4254.
E-mail: cadmag@sbcglobal.net
Web Site: www.lamag.org
Founded: 1952.
Congressional District: 24
Key Personnel: Dir. & Cur., Scott Canty; Dir. Museum Education & Tours, Sara L. Cannon; Preparator, Michael Miller.
Personnel Profile: Full-Time Paid 4; Part-Time Paid 24; Part-Time Volunteers 50; Interns 4.
Governing Authority: municipal. Parent Institution: Dept. of Cultural Affairs, Los Angeles, 201 N. Figueroa Street, Los Angeles, CA 90012. Tax-exempt: 501(c)(3).
Institution Type/Description: Contemporary Art Gallery.
Collections: contemporary southern California art; original furniture; Wright Drawings. Historic House: 1921 Frank Lloyd Wright Hollyhock House.
Research Fields: contemporary southern California art history; Frank Lloyd Wright.
Facilities: library; 300-seat theater; art studio & classrooms; 10,000 sq. ft. exhibit space; slide registry of contemporary artists. Museum-related items for sale.
Activities: guided tours; changing exhibitions of predominantly contemporary southern California artists; improvisational tours; outreach programs to hospitals, schools & senior citizen center facilities; internship program; teacher workshops; conversations with exhibiting artists; spoken work reading program in the gallery; performances emphasizing inter-relationships of the arts; formal educational programs; lectures. Museum Sponsors: Barnsdall Art Park Open House; F.L. Wright Birthday Party.
Publications: exhibition catalogs; COLA Individual Artist Fellowship.
Hours & Admission Prices: No charge; donations accepted. Closed municipal, state & federal holidays. &
Attendance: 50,000 (accurate)
Membership: Student & Senior $20; Active $35; Patron $100.

LOS ANGELES MUSEUM OF THE HOLOCAUST, 100 S. The Grove Dr., Los Angeles, CA 90036. Tel.: 323-651-3704. Fax: 323-651-3706.
E-mail: info@lamoth.org
Web Site: www.lamoth.org
Founded: 1961.
Congressional District: 23
Key Personnel: Exec. Dir., Mark Rothman; Chm., E. Randol Schoenberg; Dir. Operations, Jodi Shapiro; Chief Docent, Ilaria Benzoni-Clark.
Personnel Profile: Full-Time Paid 5; Part-Time Paid 2; Part-Time Volunteers 30; Interns 10.
Governing Authority: religious institution. Parent Institution: Jewish Federation Council. Tax-exempt: 501(c)(3).
Institution Type/Description: Holocaust History Museum.
Collections: photo-narrative exhibits; artifacts; artworks; documents; oral histories; scale model of Sobibor Death Camp; videodisc presentation of archival material & oral histories.
Facilities: 2,500-vol. library of reference materials on the Holocaust period, art, insurance, claims, art of survivors & art of the holocaust and pre-war Jewish life in Europe & North Africa.
Activities: guided tours; lectures; films; organized educational programs; docent program; participatory, loan, traveling & temporary exhibitions; school loan service; annual conference; meetings; social programs; outreach programs; curriculum writing; educators credit programs. Museum Sponsors: Annual Yom Hashoah observance.
Publications: quarterly newsletter; Denying History (University of California Press); They Shall Be Counted (the Theresienstadt Ghetto Art of Erich Lichtblau-Leskly) (Los Angeles Museum of the Holocaust, 2010); Portraits in Black and White: Holocaust Survivors of Cafe Europa (Los Angeles Museum of the Holocaust, 2011).
Hours & Admission Prices: Fri. 10-2, Sat.-Thurs. 10-5. No charge; donations accepted. Closed national & Jewish holidays. &
Attendance: 40,000 (estimated)
Membership: Student $25; Supporting $100; Sustaining $250; Friend $1,000; Society of the Righteous $5,000 & up.

LOS ANGELES POLICE MUSEUM, (M), 6045 York Blvd., Los Angeles, CA 90042-3503. Tel.: 323-344-9445. Fax: 323-344-9516.
E-mail: info@lapolicemuseum.org
Web Site: www.lapolicemuseum.org

Formerly: Los Angeles Police Historical Society & Museum
Founded: 1989.
Congressional District: 14
Key Personnel: Dir., Glynn Martin; Chm. (V), Robert Taylor; Museum Shop Mgr., Amy Condit.
Personnel Profile: Full-Time Paid 2; Part-Time Paid 1; Part-Time Volunteers 20; Interns 1.
Operating Expenses: 653,253
Operating Income: 605,236
Governing Authority: private; nonprofit organization. Tax-exempt.
Institution Type/Description: Police History Museum.
Collections: memorabilia; history of Los Angeles Police Dept.
Publications: newsletter, The Hot Sheet.
Hours & Admission Prices: Mon.-Fri. 10-4, 3rd Sat. each month 9-3. Adults $8, senior citizens over 62 $7; members & children under 12 no charge. &
Attendance: 4,888 (accurate)
Membership: Auxiliary Booster $25; Regular $48; Supporting $100; Family $250; Sustaining $500; Chief's Circle $1,000.

* **THE LOS ANGELES ZOO AND BOTANICAL GARDENS,** 5333 Zoo Dr., Los Angeles, CA 90027-1451. Tel.: 323-644-4200. Fax: 323-662-9786.
Web Site: www.lazoo.org
Founded: 1912.
Congressional District: 24
Key Personnel: Pres. Greater LA Zoo Assoc., Connie Morgan; Zoo Dir., John R. Lewis; Asst. Gen. Mgr., Denise Verret; Mktg. & Public Rels. Dir., Jason Jacobs; Visitor Svcs. & Gen. Mgr., Greg Edgar; Dir. Research, Dr. Cathleen Cox; Museum Shop Mgr., Denise Demont.
Personnel Profile: Full-Time Paid 271; Part-Time Paid 89; Part-Time Volunteers 800.
Governing Authority: city. Subsidiary Institution: Greater Los Angeles Zoo Assoc. (GLAZA). Tax-exempt.
Institution Type/Description: Zoo.
Collections: mammals; birds; reptiles; amphibians; invertebrates; botanical gardens.
Research Fields: mammals; reptiles; birds; amphibians; invertebrates.
Facilities: library of zoological books available for use on grounds. Zoo-related items for sale.
Activities: guided tours; lectures; films; formally organized education programs for children; permanent exhibitions.
Publications: quarterly magazine, Zooview, monthly newsletter, Zooscape.
Hours & Admission Prices: Daily 10-5. Adults 13 & over $17, seniors 62 & over $14, children 2-12 $12; discounts to groups; members & children under 2 no charge. Closed Christmas. &
Attendance: 1,600,000 (accurate)
Membership: Regular $49; Couple $70; Family $114; Family Deluxe $149; Contributor $250; Wildlife Assoc. $500; Conservation Circle $1,000.

MARGO LEAVIN GALLERY, 812 N. Robertson Blvd., Los Angeles, CA 90069. Tel.: 310-273-0603. Fax: 310-273-9131.
E-mail: mail@margoleavingallery.com
Web Site: www.margoleavingallery.com
Institution Type/Description: Art Gallery.
Collections: works by contemporary artists; paintings; sculpture.
Activities: special events; temporary exhibitions.
Hours & Admission Prices: Tues.-Sat. 11-5.

MICHAEL KOHN GALLERY, 8071 Beverly Blvd., Los Angeles, CA 90048. Tel.: 323-658-8088. Fax: 323-658-8068.
E-mail: info@kohngallery.com
Web Site: www.kohngallery.com
Founded: 1985.
Key Personnel: Dir., Samantha Glaser
Institution Type/Description: Art Gallery.
Collections: works by artists from California, New York & Europe; paintings; sculpture.
Activities: special events.
Hours & Admission Prices: Tues.-Fri. 10-6, Sat. 11-6.

MILDRED E. MATHIAS BOTANICAL GARDEN, University of California, 777 S. Tiverton Ave., Los Angeles, CA 90095. Mailing Address: UCLA, Box 951606, Los Angeles, CA 90095. Tel.: 310-825-3620 & 1260 (office). Fax: 310-206-3987.
E-mail: rundel@biology.ucla.edu
Web Site: www.botgard.ucla.edu.
Founded: 1929.
Congressional District: 23
Key Personnel: Dir., Dr. Phil Rundel; Staff Research Assoc., Dr. Barry Prigge.

Personnel Profile: Full-Time Paid 5; Part-Time Volunteers 20.
Governing Authority: university. Parent Institution: Univ. of California at Los Angeles. Tax-exempt: 501(c)(3).
Institution Type/Description: Arboretum; Botanical Garden-Herbarium.
Collections: subtropical ornamentals.
Research Fields: landscape potentialities of plants, species & hybrids.
Facilities: library of reference material on botany available for use by approval of director.
Activities: university-related research & instruction.
Hours & Admission Prices: Mon.-Fri. 8-5, Sat. 8-4. No charge; donations accepted. Closed university holidays. &
Attendance: 36,000
Membership: Student $15; Individual $25; Family & Couple $40; Sustaining $100; Sponsor $250; Patron $1,000.

MUSEUM EDUCATION & TOURS PROGRAM FOR HOLLY-HOCK HOUSE, Barnsdall Park, 4800 Hollywood Blvd., Los Angeles, CA 90027-5302. Tel.: 323-664-6269.
E-mail: met_scannon@sbcglobal.net
Web Site: www.hollyhockhouse.net
Founded: 1971.
Key Personnel: Dir. Education, Sara L. Cannon; Cur. Historic Site, Jeffrey Herr.
Personnel Profile: Full-Time Paid 1; Part-Time Paid 4; Part-Time Volunteers 1.
Governing Authority: Parent Institution: Dept. of Cultural Affairs. Subsidiary Institution: Dept. of Recreation & Parks. Tax-exempt.
Institution Type/Description: History Museum: home of oil heiress Aline Barnsdall designed by architect Frank Lloyd Wright, c.1919.
Collections: furnishings.
Research Fields: architecture of Frank Lloyd Wright and Rudolf Schindler.
Hours & Admission Prices: Fri.-Sun. 12:30, 1:30, 2:30 & 3:30. Adults 17 & up $2; discounts to AAM & ICOM members; conservancy & trust members and youth under 17 no charge.
Attendance: 11,000 (accurate)

THE MUSEUM OF AFRICAN AMERICAN ART, 4005 Crenshaw Blvd., 3rd Fl., Los Angeles, CA 90008-2534. Tel.: 323-294-7071. Fax: 323-294-7084.
E-mail: info@maaa-la.org
Web Site: www.maaa-la.org
Founded: 1976.
Congressional District: 27
Key Personnel: Pres. (V), Belinda Fontenot-Jamerson; Mgr., Harvey Lehman.
Governing Authority: nonprofit. Tax-exempt: 501(c)(3).
Institution Type/Description: African American Art Museum.
Collections: arts of African & African-descendant peoples; soapstone sculpture of the Shona people of S.E. Africa; Makonde sculpture of E. Africa; traditional sculpture of W. African peoples; sculpture, paintings, prints, ceramics, of Caribbean & South American peoples paintings, prints & sculptures of leading contemporary American artists; Harlem Renaissance Art.
Research Fields: African & African-American art.
Facilities: 300-vol. library of African, African-American art, literature & history available for research on premises; reading room. Periodicals on African-American art & other museum-related gifts for sale.
Activities: lectures; films; gallery talks; study clubs; docent program or council; training programs for professional museum workers; loan, permanent, temporary & traveling exhibitions.
Publications: book, Art: African American; International Review of African American Art; The Art of Elizabeth Catlett.
Hours & Admission Prices: Thurs.-Sat. 11-6, Sun. 12-5. No charge; donations accepted. &
Attendance: 1,500 (estimated)
Membership: Individual $35; Family $50; Active $100; Sponsor $250; Patron $500; Trustees Circle $1,250.

*** THE MUSEUM OF CONTEMPORARY ART, LOS ANGE-LES, (M),** 250 S. Grand Ave., Los Angeles, CA 90012-3021. Tel.: 213-626-6222. Fax: 213-620-8674. TDD: 213-626-6222.
Web Site: www.moca.org
Founded: 1979.
Congressional District: 25
Key Personnel: Co-Chair (V), David Johnson; Co-Chair (V), Maria Arena Bell; Pres. (V), Jeffrey Soros; Dir., Jeffrey Deirch; C.F.O., Richard Weil; Chief Cur., Paul Schimmel; Museum Shop Mgr., Grant Breding.
Personnel Profile: Full-Time Paid 53; Part-Time Paid 33; Part-Time Volunteers 78; Interns 10.
Governing Authority: board of trustees; nonprofit organization. Tax-exempt: 501(c)(3).

Institution Type/Description: Art Museum.
Collections: contemporary works after 1940.
Research Fields: contemporary art in all media, including painting, sculpture, prints, photography, architecture, film, video & performance.
Facilities: 75,000 sq. ft. exhibit space.
Activities: educational programs; performances; special events.
Publications: quarterly newsletter, The Contemporary; museum catalogues.
Hours & Admission Prices: Mon. & Fri. 11-5, Thurs. 11-8, Sat.-Sun. 11-6. Adults $10, students & seniors $5; discounts to AAM & ICOM members; children under 12, Thurs. 5-8 & members no charge. &
Attendance: 236,000 (accurate)
Membership: Student $30; Artist $40; Member $65; Household $85; Contributing Member $155; Moca Associate $325; Art Advocate $625; Curator's Circle $1,200; Executive Forum & Director's Forum $3,000; MOCA Partners $10,000.

MUSEUM OF TOLERANCE, 9786 W. Pico Blvd., Los Angeles, CA 90035-4720. Tel.: 310-553-8403 & 9036. Fax: 310-553-4521.
E-mail: information@wiesenthal.net
Web Site: www.museumoftolerance.com
Founded: 1993.
Congressional District: 23
Key Personnel: Founder & Dean, Rabbi Marvin Hier; Assoc. Dir., Rabbi Abraham Cooper; Exec. Dir., Rabbi Meyer H. May; Museum Dir., Liebe Geft; C.F.O. & C.A.O., Susan Burden; Dir. Membership Devel., Marlene F. Hier; Dir. Public Rels, Avra Shapiro; Dir. Communications, Michele E. Alkin; Dir. Media Projects, Richard Trank; Dir. Major Gifts, Janice Prager; Dir. Library & Archive Svcs., Adaire Klein; Museum Shop Mgr., Karen Neumann.
Governing Authority: nonprofit organization. Parent Institution: Simon Wiesenthal Center, Inc., 1300 S. Roxbury Dr., Los Angeles, CA. Tax-exempt: 501(c)(3).
Institution Type/Description: Holocaust & Human Rights Museum.
Collections: Nazi Holocaust; World War II; 20th-century genocide; racism; concentration camps; Holocaust artifacts; human rights; tolerance; Civil Rights; antisemitism; art; photographs.
Research Fields: war crimes & criminals; history of the Holocaust; historical & contemporary antisemitism; contemporary human rights issues; prejudice; aggression; tolerance; ethnic relations in the U.S.; multiculturalism.
Facilities: library; archives; 3,500 sq. ft. exhibit space.
Activities: guided tours; lectures; broadcast programs; docent programs; loan, traveling & temporary exhibitions; research; reference assistance; oral history program; Moriah Films. Annual Event: Yom Hashoah Commemoration; Museum of Tolerance Film Society.
Publications: biannual publication, Simon Wiesenthal Center.
Hours & Admission Prices: April-Oct. Mon.-Fri. 10-5; Nov.-March Mon.-Thurs. 10-5, Fri. 10-3, Sun. 11-5. Adults $15.50, senior citizens 62 & over $12.50, youth 5-18 & students w/ID $11.50; discounts to AAM & ICOM members; members no charge. Closed New Year's Day; Independence Day; Labor Day; Thanksgiving; Christmas; Jewish holidays. &
Attendance: 250,000 (estimated)
Membership: Students & Seniors $30; Individual $45; Dual & Family $65; Sustainer $100; Associate $250; Colleague $500; Director's Circle $1,000.

*** NATURAL HISTORY MUSEUM OF LOS ANGELES COUNTY, (M),** 900 Exposition Blvd., Los Angeles, CA 90007-4057. Tel.: 213-763-DINO. Fax: 213-743-4843.
E-mail: info@nhm.org
Web Site: www.nhm.org
Founded: 1910.
Congressional District: 28
Key Personnel: Pres. & Dir., Dr. Jane G. Pisano; Chief Deputy Dir., Jural J. Garrett; Sr. Vice Pres. Advancement, Tom Jacobson; Legal Counsel & Vice Pres. Government Affairs, James R. Gilson; Div. Chief Research & Collections, Margaret Hardin; Deputy Dir. Administration & Operations, Leonard Navarro; Dir. Budgeting & Financial Reporting, Linda Ross; Vice Pres. Mktg. & Communications, Cynthia Wornham; Dir. Human Resources, Micah Mullens; Vice Pres. Education, Karen Wise; Volunteer Coord., Julie McAdam; Museum Shop Mgr., Trino Marquez.
Personnel Profile: Full-Time Paid 300; Part-Time Paid 5; Part-Time Volunteers 520.
Governing Authority: county. Parent Institution: Los Angeles County Museum of Natural History Foundation. Branch Museums: Page Museum at the La Brea Tar Pits, Hancock Park, 5801 Wilshire Blvd., Los Angeles; William S. Hart Park and Museum, 24151 Newhall Ave., Newhall, CA 91321. Tax-exempt: 501(c)(3) & 170(b)(1)(A).
Institution Type/Description: Natural History Museum.
Collections: ethnology; ichthyology; archaeology; anthropology; archives; botany; entomology; herpetology; California & general American history;

mineralogy; paleontology; ornithology; mammalogy; invertebrate zoology: crustacea, echinoderms, mollusks & polychaets.

Research Fields: anthropology; archaeology; botany; entomology; ethnology; geology; herpetology; California & general American history; American Indians; mineralogy; vertebrate & invertebrate paleontology; ichthyology; ornithology; mammalogy; invertebrate zoology.

Facilities: 150,000-vol. library of books & periodicals on zoology; botany; mineralogy; paleontology; archaeology; ethnology & California & general American history available for reference use only; 350-seat auditorium; classrooms. Books, prints & ethnic art objects reproductions for sale.

Activities: guided tours; lectures; films; gallery talks; classes, workshops & field trips for children & adults; outreach programs to elementary schools, special education facilities, senior citizen centers & rest homes; Discovery Center; insect zoo; teacher workshops; docent program; inter-museum, permanent & temporary exhibitions; school loan service; Earthmobile & Seamobile, mobile science museum for schools; high school research program.

Publications: Baja California Travels Series: Contributions in Science; Science Series; Technical Reports; books & catalogues in history, natural sciences, anthropology & archaeology; member magazine, Naturalist.

Hours & Admission Prices: Daily 9:30-5. Adults $9, students & seniors $6.50, children 5-12 $2; children under 5, visitors on first Tues. of month & members no charge. Closed New Year's Day; Independence Day; Thanksgiving; Christmas. ♿

Attendance: 650,000 (estimated)

Membership: Family $70; Patron $185; Naturalist $325; Explorer $600; Fellows $1,500.

OLD PLAZA FIREHOUSE AT EL PUEBLO, 501 N. Los Angeles St., Los Angeles, CA 90013. Mailing Address: El Pueblo de Los Angeles Historical Monument, 125 Paseo de la Plaza, Ste. 400, Los Angeles, CA 90012. Tel.: 213-625-3741.

Web Site: www.elpueblo.lacity.org

Formerly: Old Plaza Firehouse

Institution Type/Description: Firefighting History Museum: housed in a former firehouse; built in 1884.

Collections: firefighting history & equipment; personal artifacts; photographs.

Hours & Admission Prices: Tues.-Sun. 10-3. No charge; donations accepted.

ORAN Z'S PAN AFRICAN BLACK FACTS & WAX MUSEUM, 3742 W. Martin Luther King Blvd., Los Angeles, CA 90008-1751. Tel.: 323-299-8829.

Institution Type/Description: Black History Museum.

Collections: black history & culture; period clothing; uniforms; personal artifacts; black wax figures; autographs; photographs; postcards; sculptures.

Hours & Admission Prices: Call for hours.

PETER MENDENHALL GALLERY, 6150 Wilshire Blvd., Space 8, Los Angeles, CA 90048. Tel.: 323-936-0061. Fax: 323-936-0288.

E-mail: info@petermendenhallgallery.com

Web Site: www.petermendenhallgallery.com

Institution Type/Description: Art Gallery.

Collections: works by contemporary artists; sculpture; painting.

Activities: special events; temporary exhibitions.

Hours & Admission Prices: Tues.-Sat. 11-6.

PETERSEN AUTOMOTIVE MUSEUM, (M), 6060 Wilshire Blvd., Los Angeles, CA 90036-3605. Tel.: 323-964-6356 & 6357. Fax: 323-930-6642.

E-mail: info@petersen.org

Web Site: www.petersen.org

Founded: 1994.

Congressional District: 4

Key Personnel: Exec. Dir., Terry Karges; Chm. (V), Steven E. Young; Cur., Leslie Kendall; Membership, Paul Moritz; Special Events, Mandy Hanlon; Museum Shop Mgr., Gregg Guenthard.

Personnel Profile: Full-Time Paid 45; Part-Time Paid 4; Part-Time Volunteers 70; Interns 2.

Governing Authority: Parent Institution: Petersen Automotive Museum Foundation. Tax-exempt: 501(c)(3).

Institution Type/Description: Automotive Museum.

Collections: works on paper; automobiles & related objects in the permanent collection reflect the history of the automobile & its impact on American life & culture using Los Angeles as the prime example.

Research Fields: automobiles built in Los Angeles & southern California.

Facilities: education facilities; banquet facilities; 100,000 sq. ft. exhibit space; children's discovery center. Museum-related items for sale.

Activities: docent program; guided tours; lectures; rental gallery; traveling exhibitions; hands-on children's exhibits.

Publications: quarterly newsletter; Petersen Quarterly; Finish Line.

Hours & Admission Prices: Tues.-Sun. 10-6. Adults $12, senior citizens $8, students $5, children 5-12 $3; discounts to AAA members; members no charge. ♿

Attendance: 180,000 (estimated)

Membership: Motorhome $50; Roadster $60; Station Wagon $70; Vintage $180; Classic $300; Concourse $600; Checkered Flag 200 $1,500.

PLAZA DE LA RAZA, INC. & BOATHOUSE GALLERY, 3540 N. Mission Rd., Los Angeles, CA 90031-3195. Tel.: 323-223-2475. Fax: 323-223-1804.

Web Site: www.plazadelaraza.org

Founded: 1969.

Congressional District: 30

Key Personnel: Exec. Dir., Rose Marie Cano; Chm. (V), Armando G. Ramirez; Dir. Education, Maria Jimenez-Torres; Museum Shop Mgr., Rosalie Portillo; Coord., Kay Rosser; Dir. Devel., Tomas J. Benitez; Dir. Mktg., Melissa Richardson Banks.

Personnel Profile: Full-Time Paid 9; Part-Time Paid 27; Part-Time Volunteers 220; Interns 2.

Governing Authority: nonprofit organization. Tax-exempt: 501(c)(3), 170(b)(1)(A).

Institution Type/Description: Art Museum: housed in historic boathouse gallery.

Collections: Mexican-American folk arts of southern California; artwork by Latino artists; photos by Jose Galvez.

Research Fields: Mexican-American folk life of southern California.

Activities: films; concerts; dance recitals; art festivals; organized educational programs; loan & traveling exhibitions; Hispanic arts.

Publications: annual catalogues, Celebration; plaza newsletter; brochures.

Hours & Admission Prices: Mon.-Fri. 10-5, Sat. 10-2. Suggested Donations: adults $3, children $1. Closed Columbus Day; Veterans Day; Thanksgiving. ♿

Membership: Amigo $35; Familia $50; Maestro $100; Padrino $250; Amigo Distinguido $500.

PSYCHIATRY: AN INDUSTRY OF DEATH MUSEUM, 6616 Sunset Blvd., Los Angeles, CA 90028-7104. Tel.: 323-467-4242; 800-869-2247.

E-mail: humanrights@cchr.org

Web Site: www.cchr.org

Institution Type/Description: Psychiatry Museum.

Collections: psychiatric history, drugs & medical practices; documentaries.

Facilities: Museum-related items for sale.

Activities: films.

Hours & Admission Prices: Mon.-Fri. 10-9, Sat.-Sun. 10-6. No charge.

SAM LEE GALLERY, 3467 Waverly Dr., Los Angeles, CA 90027-2526. Tel.: 323-788-3535.

E-mail: sam@samleegallery.com

Web Site: www.samleegallery.com

Founded: 2007.

Institution Type/Description: Art Gallery.

Collections: works by contemporary artists; photographs.

Activities: special events; temporary exhibitions.

Hours & Admission Prices: Wed.-Sat. 12-6; other times by appointment. No charge.

SEPULVEDA HOUSE AT EL PUEBLO, 125 El Paseo de la Plaza, Los Angeles, CA 90012. Tel.: 213-628-1274. Facebook: El Pueblo LA.

Web Site: elpueblo.lacity.org

Formerly: Sepulveda House

Key Personnel: Cur., Suellen Chang

Institution Type/Description: Historic House Museum: housed in former boardinghouse, built in 1887.

Collections: period artifacts & furnishings.

Research Fields: 19th-century California; Los Angeles history; Victorian architecture.

Activities: self-guided audio tour.

Hours & Admission Prices: Tues.-Sun. 10-3. No charge; donations accepted. ♿

SKIRBALL CULTURAL CENTER, (M), 2701 N. Sepulveda Blvd., Los Angeles, CA 90049-6833. Tel.: 310-440-4500. Fax: 310-440-4595.
E-mail: info@skirball.org
Web Site: www.skirball.org
Founded: 1996.
Congressional District: 29
Key Personnel: Pres. & C.E.O., Uri D. Herscher; Museum Dir., Robert Kirschner; Vice Pres. & Dir. Education, Sheri Bernstein; Cur., Erin Clancey; Dir. Mktg., Jennifer Caballero; Mgr. Membership, Sabrina Wurf.
Personnel Profile: Full-Time Paid 167; Part-Time Paid 13; Part-Time Volunteers 309; Interns 6.
Governing Authority: board of trustees. Tax-exempt: 501(c)(3).
Institution Type/Description: Art & History Museum.
Collections: Kirschstein Collection of Jewish ceremonial art; Dr. Nelson Glueck Memorial Collection of archaeological artifacts; Joseph Hamburger numismatic collection; I. Solomon & L. Grossman collections of engravings & photo prints; manuscripts, paintings, sculpture, prints & drawings by artists of Jewish origins & on Jewish themes; textiles; decorative arts; Israel archaeology; graphics; American Jewish ethnography.
Research Fields: Jewish culture & ceremonial art; contemporary Jewish artists; Jewish history artifacts.
Facilities: conference center; resource center; classrooms; restaurant; amphitheater; children's museum. Museum-related items for sale.
Activities: guided tours; lectures; films; concerts; children's workshops; docent program; classes.
Publications: catalogs; exhibition brochures; museum calendar.
Hours & Admission Prices: Tues.-Fri. 12-5, Sat.-Sun. 10-5. Adults $10, students w/ID & seniors 65 & up $7, children 2-12 $5; discounts to AAM members & museum professionals w/ID; children under 2 & members no charge. Closed national & Jewish holidays.
Attendance: 500,000 (accurate)
Membership: Student & Senior $50; Individual $55; Senior Dual $65; Dual $70; Family $95; Family Plus $195; Supporting $300; Corporate $650; Curator's Circle $1,500, $2,500, $5,000.

TRAVEL TOWN TRANSPORTATION MUSEUM, 5200 Zoo Dr., Los Angeles, CA 90027-1472. Mailing Address: P.O. Box 39846, Griffith Station, Los Angeles, CA 90039. Tel.: 323-662-5874. Fax: 818-243-0041.
Web Site: www.laparks/grifmet/tt/index.htm
Founded: 1952.
Congressional District: 29
Key Personnel: Dir., Linda J. Barth; Asst. Dir., Thomas W. Breckner; Museum Shop Mgr., Nancy Gneier.
Personnel Profile: Full-Time Paid 5; Part-Time Paid 15; Part-Time Volunteers 150.
Governing Authority: municipal government; nonprofit. Parent Institution: Recreation & Parks, City of Los Angeles. Tax-exempt.
Institution Type/Description: Transportation Museum: located on the site of the Civilian Conservation Corp. camp, which was used as a P.O.W. camp during World War II.
Collections: railroad equipment with emphasis on steam locomotives used in the American West & Southwest chiefly in the period 1880-1940; wagons used locally 1860s-1910s; local & California general & transportation history; model train display; c. 1900 freight depot.
Research Fields: development of Los Angeles & southern California as it relates to railroading; role & impact of railroads in culture; history of interurbans in Los Angeles County.
Facilities: library of railroad ephemera; 10,000 sq. ft. exhibit space; 7 acre exhibit space apart from buildings. Museum-related items for sale.
Activities: docent program; formal education & exhibit programs for children; temporary exhibitions; miniature train ride. Museum Sponsors: full-size railroad equipment operations; teacher's guide on railroad history.
Publications: quarterly newsletter, Green Eye.
Hours & Admission Prices: Mon.-Fri. 10-4, Sat.-Sun. & holidays 10-5. No charge; donations accepted. Closed Christmas.
Attendance: 350,000 (estimated)

✳ **USC FISHER MUSEUM OF ART, (M),** 823 Exposition Blvd., Los Angeles, CA 90089-0292. Mailing Address: 823 Exposition Blvd. # HAR MC 126, Los Angeles, CA 90089-0292. Tel.: 213-740-4561. Fax: 213-740-7676.
E-mail: fmoa@usc.edu
Web Site: fisher.usc.edu
Founded: 1939.
Congressional District: 28
Key Personnel: Dir., Dr. Selma Holo; Assoc. Dir., Kay Allen; Administrative Asst., Raphael Gatchalian; Cur., Ariadni Liokatis; Education & Programs Coord., Vanessa Jorion; Collections Mgr. & Registrar, Stephanie Kowalick; Chief Preparator, Juan Rojas.
Personnel Profile: Full-Time Paid 7; Part-Time Volunteers 1; Interns 6.
Governing Authority: university. Parent Institution: University of Southern California. Tax-exempt: 103(6).
Institution Type/Description: University Art Museum.
Collections: paintings; sculpture; graphics; European & American paintings, prints & drawings; Armand Hammer & Elizabeth Holmes Fisher Collection of 15th- to 20th-century paintings.
Research Fields: ancient through 20th-century art.
Facilities: fine arts library; School of Fine Arts.
Activities: guided tours; lectures; films; family festivals; gallery talks; inter-museum loan, permanent, traveling & temporary exhibitions.
Publications: exhibition catalogs including most recently: The Art of Exaggeration & Piranesi's Perspectives on Rome (spring '95); LAX: The Los Angeles Exhibition, 1992; Body to Earth: 3 Artists from Brazil, 1993; Richard Diebenkorn: Works on Paper from the Harry W. & Mary Margaret Anderson Collection, 1993; Locus, Contemporary Israeli Art Exhibition, 1993; Light in Darkness: Women in Japanese Prints of Early Showa (1926-1945), 1996; Ut Poesis Pictura: J.M.W. Turner's Illustrations to the British Poets, 1997; L.A. Obscura: The Architectural Photography of Julius Shulman, 1998; Open House West: Museum Architecture and Changing Civic Identity, 1999; Lost and Found: Rediscovering Early Photographic Processes, 2001; Willie Robert Middlebrook, 2001; Global Address, 2002; Mixed Feelings, 2002; Fashion and Transgression, 2003; Human Conditions, 2003; UnNaturally, 2003; Transfictions, 2004; Peter Plagens, An Introspective, 2004; Albert Contreras, Luminous Scapes and Environments, Fall 2005; Contemporary Soliloquies on the Natural World, Winter 2005-2006; Michael Mazur and Dante: A Different Paradise, Fall 2006; The Bone-and-Bird Art of Joyce Cutler-Shaw and Sarah Perry, Spring 2007; Material Affinities: To Clay and Back, Fall 2007; Robert Graham: Body of Work, Spring 2009; Donald Bandler: A Roving Eye on Cyprus, Spring 2009; Victor Raphael: Travels and Wanderings, 1979-2009.
Hours & Admission Prices: Tues.-Sat. 12-5, during exhibitions. No charge.
Attendance: 20,000 (estimated)

WATTS TOWERS ARTS CENTER & CHARLES MINGUS YOUTH ARTS CENTER, 1727 E. 107th St., Los Angeles, CA 90002-3621. Tel.: 213-847-4646 & 485-1795. Fax: 323-564-7030.
E-mail: watts.towers1@lacity.org
Web Site: wattstowers.org
Founded: 1961.
Congressional District: 37
Key Personnel: Dir., Rosie Lee Hooks.
Personnel Profile: Full-Time Paid 3; Part-Time Paid 15; Interns 1.
Governing Authority: municipal; nonprofit organization. Parent Institution: City of Los Angeles Dept. of Cultural Affairs, 201 N. Figueroa St., Ste. 1400, Los Angeles, CA 90012. Tax-exempt: 501(c)(3).
Institution Type/Description: Arts Center & Gallery.
Collections: The Watts Towers, Simon Rodia built using seashells, tiles, bottles & various broken pieces of ceramics; works by local & world renowned contemporary artists.
Major Exhibits: Through the Eyes of Charles Dickson, 10/13-3/2/14.
Facilities: classrooms.
Activities: lectures; films; concerts; arts festivals drama; organized education programs for children; loan & temporary exhibitions; art instruction classes. Museum Sponsors: annual Simon Rodia; Watts Towers Jazz Festival; Day of the Drum Festival in September.
Hours & Admission Prices: Watts Towers Arts Center: Tues.-Sat. 10-4, Sun. 12-4. Tours: adults $7, seniors $3; military and children under 12 no charge. Simon Rodia's Watts Towers: Sat. 10-4, Sun. 12-4. Tours: Thurs.-Fri. 11-3, Sat. 10:30-3, Sun. 12:30-3; please call for information. Closed New Year's Day; Christmas.
Attendance: 25,000 (estimated)
Membership: Associate $15; Individual $30; Family $40; Sponsor $100; Patron $500.

WELLS FARGO HISTORY MUSEUM, 333 S. Grand Ave., Los Angeles, CA 90071-1504. Mailing Address: Wells Fargo Historical Services, 420 Montgomery St., MAC A0101-106, San Francisco, CA 94163. Tel.: 213-253-7166.
Founded: 1982.
Congressional District: 25
Key Personnel: Cur., Juan Colato; Asst. Cur., Ileana Bonilla.
Governing Authority: profit-making organization. Affiliated with Wells Fargo Bank.
Institution Type/Description: Company History Museum.
Collections: Concord Stagecoach; Wells Fargo banking & express history; gold scales; mining; Southern California.

Activities: guided tours; audiovisual programs; permanent & temporary exhibitions; replica stagecoach on which visitors may take imaginary ride.
Publications: scholarly pamphlets.
Hours & Admission Prices: Mon.-Fri. 9-5. No charge. Closed bank holidays. &
Attendance: 100,000 (estimated)

ZIMMER CHILDREN'S MUSEUM, 6505 Wilshire Blvd., Ste. 100, Los Angeles, CA 90048-4908. Tel.: 323-761-8998. Fax: 323-761-8990.
E-mail: info@zimmermuseum.org
Web Site: www.zimmermuseum.org
Formerly: Zimmer Children's Museum of Jewish Community Centers of Greater Los Angeles
Founded: 2005.
Key Personnel: C.E.O., Esther Netter; Dir., Julee Brooks; Pres., Barbara Fisher.
Personnel Profile: Full-Time Paid 13; Part-Time Paid 5; Part-Time Volunteers 50; Interns 3.
Governing Authority: private; nonprofit organization. Tax-exempt: 501(c)(3).
Institution Type/Description: Children's Museum.
Collections: hands-on exhibitions.
Hours & Admission Prices: Tues., Thurs. & Sun. 12:30-5, Wed. 10-5, Fri. 10-12:30. Adults $8, children 3-12 $5; discounts to AAM & museum members & groups; grandparents & children under 3 no charge. Closed all national & Jewish holidays; holiday Sundays. &
Attendance: 46,000 (accurate)
Membership: Family $75; Family Plus $100; Family Premium $185; Silver Patron $250; Gold Patron $500; Diamond Patron $1,000 and up; President's Circle $2,500 & up; Museum Founder $5,000 & up.

Los Banos

MILLIKEN MUSEUM, 905 Pacheco Blvd., Los Banos, CA 93635. Mailing Address: P.O. Box 2294, Los Banos, CA 93635-2294. Tel.: 209-826-5505.
E-mail: millikenmuseum@att.net
Formerly: Ralph Milliken Museum
Founded: 1954.
Congressional District: 16
Key Personnel: Chm. (V) & Sec., Dan Nelson.
Volunteer Hours: 350
Operating Expenses: 9,500
Operating Income: 9,500
Governing Authority: county. Support Institution: Milliken Museum Society. Tax-exempt.
Institution Type/Description: Cultural Museum.
Collections: agriculture; archaeology; military uniforms; Native American artifacts; paleontology; photographs.
Research Fields: genealogy; history; biography; photography.
Facilities: library of old historic value newspapers; account books; maps; minutes; documents available for use on premises.
Activities: permanent & rotating displays; museum society monthly meetings.
Publications: pamphlets; brochures.
Hours & Admission Prices: Tues.-Sun. 1-4; special tours by arrangement. No charge; donations accepted.
Attendance: 1,000 (estimated)

Los Gatos

MUSEUMS OF LOS GATOS, (M), 75 Church St. & 4 Tait Ave., Los Gatos, CA 95030. Mailing Address: P.O. Box 1904, Los Gatos, CA 95031-1904. Tel.: 408-395-7375 & 354-2646. Fax: 408-395-7386.
E-mail: ed@museumsoflosgatos.org
Web Site: www.museumsoflosgatos.org
Formerly: Los Gatos Museum Association
Founded: 1965.
Congressional District: 13
Key Personnel: Exec. Dir., John C. Agg; Pres., Mary Ellen Comport.
Personnel Profile: Full-Time Paid 3; Part-Time Paid 6; Part-Time Volunteers 400.
Governing Authority: nonprofit organization. Parent Institution: Los Gatos Museum Association. Branch Museum: History Museum of Los Gatos, Forbes Mill, 75 Church St., Los Gatos, CA 95030. Tel: 408-395-7375; Art & Nature Museums of Los Gatos, 4 Tait Ave., Los Gatos, CA 95030. Tel: 408-354-2646. Tax-exempt.
Institution Type/Description: Art & History Museum.
Collections: local history; art displays; cultural displays; two historic buildings.

Research Fields: local mortuary records.
Activities: guided tours; educational programs.
Publications: book, Walking Tour of Los Gatos.
Hours & Admission Prices: Wed.-Sun. 12-4. No charge; donations accepted. Closed holidays. &
Attendance: 10,000 (accurate)
Membership: Senior $30; Individual $35; Household $50; Supporting & Business $100; Sponsor $250; Benefactor $500; Curator's Circle $1,000.

Madera

MADERA COUNTY MUSEUM, 210 W. Yosemite Ave., Madera, CA 93637. Mailing Address: Madera County Historical Society, P.O. Box 150, Madera, CA 93639-0150. Tel.: 559-673-0291. Fax: 559-673-0742.
E-mail: mchs210@gmail.com
Web Site: www.maderahistory.org
Founded: 1955.
Congressional District: 18
Key Personnel: Pres. (V), Sheryl Berry; Cur., Karen Elmore; Museum Shop Mgr., Steve Parker.
Personnel Profile: Full-Time Volunteers 6; Part-Time Volunteers 15.
Governing Authority: county.
Institution Type/Description: General Museum: housed in the 1900, first Madera County Court House.
Collections: photographs; mining; local Indian artifacts; home furnishings; lumber; agriculture; period artifacts; law enforcement; clothing; businesses; military; blacksmith; livery.
Research Fields: family histories; industry; county history.
Facilities: 400-vol. library of material of historical information, education, health, law, & literature available on premises.
Activities: guided tours.
Publications: semiannual bulletin, Madera County Historian; quarterly newsletter.
Hours & Admission Prices: Early Dec. to late Nov. Sat.-Sun. 1-4. No charge; donations accepted. Closed New Year's Day; Easter; Mother's Day; Father's Day; Christmas; some summer days.
Attendance: 1,379 (accurate)
Membership: Student $2; Individual $15; Family & Sustaining $25; Life $150.

Malibu

FREDERICK R. WEISMAN MUSEUM OF ART, Pepperdine University, 24255 Pacific Coast Hwy., Malibu, CA 90263-3999. Tel.: 310-506-7257. Fax: 310-506-4556.
E-mail: michael.zakian@pepperdine.edu
Web Site: arts.pepperdine.edu/museum/
Founded: 1992.
Congressional District: 24
Key Personnel: Dir., Michael Zakian.
Personnel Profile: Full-Time Paid 1; Part-Time Paid 1.
Governing Authority: private university; not-for-profit. Parent Institution: Pepperdine University. Tax-exempt: 501(c)(3).
Institution Type/Description: University Art Museum.
Collections: 20th-century American art; contemporary California art.
Research Fields: contemporary California art.
Facilities: 3,000 sq. ft. exhibit space.
Activities: temporary exhibits.
Hours & Admission Prices: Tues.-Sun. 11-5. No charge. Closed New Year's Eve, Day & week; Memorial Day; Independence Day; Labor Day; Thanksgiving; Christmas Day & week. &
Attendance: 20,000 (estimated)
Membership: Bronze $50; Silver $100; Sterling $150; Gold $300; Platinum $500; Diamond $1,000.

MALIBU ADAMSON HOUSE FOUNDATION, 23200 Pacific Coast Hwy., Malibu, CA 90265-4937. Mailing Address: P.O. Box 291, Malibu, CA 90265-0291. Tel.: 310-456-8432 & 1770.
Web Site: www.adamsonhouse.org
Formerly: Historic Adamson House & Museum and Malibu Lagoon Museum
Founded: 1982.
Key Personnel: Dir., Devel. & Membership, Hayden Sohm; Pres. (V), Lisa Otis-Kicor; Chm. (V), Joan Hurley; Public Rels., Beverly Gosnell; Gift Shop Mgr., Joan Page.
Personnel Profile: Part-Time Volunteers 90.
Governing Authority: state; nonprofit organization. California Dept. of Parks & Recreation; Malibu State Beach Interpretive Assoc. Tax-exempt: 501(c)(3).
Institution Type/Description: Historic House & Museum; botanical gardens.
Collections: Malibu history from Chumash Indians to present day; Chumash

Indian artifacts; ceramics; photographs; coins; early ranching items; documents. Historic House: 1929 Adamson House.
Research Fields: Chumash Indians; chain of title of Malibu Land Grant; tile & ceramic products from Malibu Potteries; Rindge family history; Malibu Railroad; 1920s & 1930s California & Mediterranean architects & architecture.
Facilities: library; meeting space; botanical garden. Museum-related items & books for sale.
Activities: guided tours; lectures; organized education programs for children; docent program.
Publications: monthly newsletter, Information for Volunteers; quarterly newsletter, Information for Membership; books, Spanish tile history; poetry.
Hours & Admission Prices: Wed.-Sat. 11-3, last house tour 2; groups of 10 & over by appointment only. Adults $7; members no charge. Reciprocal admission to docents of other organizations. Closed holidays. &
Attendance: 50,000 (estimated)
Membership: Individual $25; Family $50; Sponsor $100; Associate $250; Patron $500.

Mammoth Lakes

DEVILS POSTPILE NATIONAL MONUMENT, Mammoth Lakes, CA 93546. Mailing Address: P.O. Box 3999, Mammoth Lakes, CA 93546-3999. Tel.: 760-934-2289. Fax: 760-934-4780.
Web Site: nps.gov/depo
Founded: 1911.
Congressional District: 15
Key Personnel: Park Supt., Deanna Dulen.
Governing Authority: federal. Parent Institution: National Park Service, U.S. Dept. of Interior, Washington, DC. Tax-exempt: 501(c)(3).
Institution Type/Description: Park Museum.
Collections: rocks; pressed flowers; herbarium; volcanic rock; colored slide collection.
Research Fields: geology of volcanic rocks.
Facilities: Selected books & slides for sale.
Activities: conducted walks; campfire programs.
Hours & Admission Prices: May-Oct. daily 9-5 weather permitting. Adult $7, child 3-15 $4; not to exceed $20 per carload. Park Camping: daily.

MAMMOTH SKI MUSEUM, 100 College Pkwy., Mammoth Lakes, CA 93546. Mailing Address: P.O. Box 1815, Mammoth Lakes, CA 93546. Tel.: 760-934-6592 & 3781.
E-mail: info@mammothskimuseum.org
Web Site: www.mammothskimuseum.org
Key Personnel: Dir. & Cur., Kendra Knight
Institution Type/Description: Ski Museum.
Collections: ski history; sport & culture of skiing; fine art; literature.
Facilities: library. Museum-related items for sale.
Hours & Admission Prices: Wed.-Sat. 10-5. Adults $5; discounts to AAM members; members no charge. &
Attendance: 1,000
Membership: Rope Tow $50; Flying Skis $150; Black Diamond $500.

Manhattan Beach

OCEANOGRAPHIC TEACHING STATION INC. - ROUND-HOUSE MARINE STUDIES LAB & AQUARIUM, Manhattan Beach Pier, Manhattan Beach, CA 90266. Mailing Address: P.O. Box 1, Manhattan Beach, CA 90267. Tel.: 310-379-8117. Fax: 310-937-9366.
E-mail: roundhouse.aquarium@verizon.net
Web Site: www.roundhouseaquarium.org
Founded: 1979.
Congressional District: 36
Key Personnel: Pres. OTS Bd. (V), Matt Friedman; Treas., Chuck Milam; Dir. Aquarium, Eric Martin; Dir. Aquarium, Valerie Hill.
Personnel Profile: Full-Time Paid 3; Part-Time Paid 3; Part-Time Volunteers 180; Interns 2.
Volunteer Hours: 10,093
Operating Expenses: 243,000
Operating Income: 249,000
Governing Authority: private; nonprofit organization. Tax-exempt 501 (c)(3).
Institution Type/Description: Aquarium.
Collections: Santa Monica Bay marine life & artifacts.
Facilities: aquarium; educational facilities; field research station.
Activities: summer camp; school programs & field trips; docent program; formal education programs; guided tours. Annual Events: Halloween Event; Earth Day; Coastal Clean Up Day.
Publications: monthly newsletter, Roundhouse Newsletter.

Hours & Admission Prices: Mon.-Fri. 3pm to sunset, Sat.-Sun. 10am to sunset. No charge; donations accepted. Closed New Year's Day; Thanksgiving; Christmas. &
Attendance: 49,900 (estimated)

Manteca

MANTECA HISTORICAL SOCIETY AND MUSEUM, 600 W. Yosemite Ave., Manteca, CA 95337-5402. Mailing Address: P.O. Box 907, Manteca, CA 95336-1137. Tel.: 209-825-3021.
Web Site: www.mantecamuseum.org
Founded: 1989.
Key Personnel: Chm. (V), Phyllis Abram; Pres. (V), Leon Sucht; Museum Shop Mgr., Vivian Sarina.
Personnel Profile: Part-Time Paid 1.
Governing Authority: Tax-exempt.
Institution Type/Description: Historical Society Museum.
Collections: local history & culture; photographs; personal artifacts; documents.
Activities: special events. Museum Sponsors: Gourmet Sampler; Summer Barbecue.
Publications: monthly newsletter.
Hours & Admission Prices: Tues.-Wed. 1-3, Thurs. & Sun. 1-4; groups by appointment. No charge. &
Attendance: 70,000 (estimated)
Membership: Individual $15; Couple $25; Family $50; Business Sustaining $100; Business $250; Supporter $500; Benefactor $1,000; Life $5,000.

Mariposa

CALIFORNIA STATE MINING & MINERAL MUSEUM, 5005 Fairgrounds Dr., Mariposa, CA 95338. Mailing Address: P.O. Box 1192, Mariposa, CA 95338-1192. Tel.: 209-742-7625. Fax: 209-966-3597.
E-mail: mineralmuseum@sierratel.com
Web Site: www.parks.ca.gov
Founded: 1880.
Congressional District: 19
Key Personnel: Acting Dir., Darci Moore.
Personnel Profile: Full-Time Paid 2; Part-Time Paid 4; Part-Time Volunteers 15.
Governing Authority: state government; nonprofit. Parent Institution: California State Department of Parks and Recreation, P.O. Box 942896, Sacramento, CA 94296. Tax-exempt: 501(c)(3).
Institution Type/Description: Mineralogy Museum.
Collections: California minerals; replica mine tunnel & model stamp mill.
Facilities: 4,000 sq. ft. exhibit space. Museum-related items for sale.
Activities: docent program; formal educational programs; guided tours; hobby workshops; lectures; temporary exhibitions; junior ranger programs; special events; living history.
Hours & Admission Prices: May-Sept. Thurs.-Sun. 10-5; Oct.-April Thurs.-Sun. 10-4. Adults $4; children 12 & under no charge. Closed New Year's Day; Thanksgiving; Christmas. &
Attendance: 15,229 (accurate)
Membership: General $10; Business $25; Sustaining $100; Patron $500.

MARIPOSA MUSEUM AND HISTORY CENTER INC., (M), 5119 Jessie St., Mariposa, CA 95338. Mailing Address: P.O. Box 606, Mariposa, CA 95338-0606. Tel.: 209-966-2924. Fax: 209-742-6513.
E-mail: mmhc@sti.net
Web Site: www.mariposamuseum.com
Founded: 1957.
Congressional District: 19
Key Personnel: Pres. (V), Sylvia M. Emery; Museum Shop Mgr., Phyllis I. Faust-Stephens.
Personnel Profile: Full-Time Paid 3; Part-Time Paid 4; Part-Time Volunteers 109.
Governing Authority: society. Tax-exempt: 501(c)(3).
Institution Type/Description: Historical Society Museum.
Collections: Indian artifacts; mining; Gold Rush period material; local history; manuscripts.
Research Fields: family & local history.
Facilities: 150-vol. library of books on Mariposa, California & Yosemite available on premises by request; reading room. Books & postcards for sale.
Activities: guided school tours; lectures; docent program or council.
Publications: quarterly, The Sentinel.
Hours & Admission Prices: Daily 10-4. Adults $4; children under 16 & members no charge. Closed New Year's Eve & Day; Thanksgiving; Christmas Eve & Day. &

Attendance: 11,124 (accurate)
Membership: Senior $20; Single $25; Grubsteak (Family) $40; Patron $50; Business $100; Life $500.

Markleeville

ALPINE COUNTY MUSEUM, School St., Markleeville, CA 96120. Mailing Address: P.O. Box 517, Markleeville, CA 96120-0517. Tel.: 530-694-2317. Fax: 530-694-1087. www.alpinecounty-museum.com.
E-mail: alpinemuseum@yahoo.com
Web Site: www.co.alpine.ca.us/dept/museum/museum.htm
Founded: 1963.
Congressional District: 3
Key Personnel: Cur., Wanda Coyan; Pres. (V) & Chm. (V), John Super.
Personnel Profile: Full-Time Paid 1; Part-Time Paid 1; Part-Time Volunteers 15.
Governing Authority: Parent Institution: Historical Society of Alpine County. Tax-exempt.
Institution Type/Description: History Museum.
Collections: Alpine County history; old country store; blacksmith shop; Washoe Indian basketry; watch maker's desk; doctor's black bag with tools; toys & dolls; mining; pioneer family; clothing; photographs; paintings. Historic Buildings: ore stamp mill; 1882 Old Webster School; Old Log Jail.
Research Fields: local genealogy; mining history.
Facilities: Museum-related items for sale.
Activities: Annual Events: Pioneer Day; Silver Mountain Tour.
Publications: Alpine Heritage; Images of Alpine County; various oral histories; Summit City History.
Hours & Admission Prices: Memorial Day to Oct. Thurs.-Mon. 11-4. No charge; donations accepted. &
Attendance: 5,000 (accurate)
Membership: Individual $15; Family $20; Business & Prof. $50; Benefactor $100; Life $250.

Martinez

CONTRA COSTA COUNTY HISTORICAL SOCIETY'S HISTORY CENTER, 610 Main St., Martinez, CA 94553. Tel.: 925-229-1042. Fax: 925-229-1772.
Institution Type/Description: Historical Society Museum.
Collections: local history & culture; period furnishings; personal artifacts; photographs.
Activities: special events.
Hours & Admission Prices: Tues.-Thurs. 9-4, 3rd Sat. of the month 10-2.

JOHN MUIR NATIONAL HISTORIC SITE, (M), 4202 Alhambra Ave., Martinez, CA 94553-3826. Tel.: 925-228-8860. Fax: 925-228-8192.
E-mail: JOMU_interpretation@nps.gov
Web Site: www.nps.gov/jomu
Founded: 1964.
Congressional District: 7
Key Personnel: C.E.O., Martha Lee.
Personnel Profile: Full-Time Paid 8; Part-Time Paid 4; Part-Time Volunteers 35.
Governing Authority: federal. Affiliated with National Park Service, Dept. of Interior, Washington, DC. Tax-exempt: 501(c)(3).
Institution Type/Description: Historic Site.
Collections: period artifacts; ranching & family life of John Muir; Victorian furnishings. Historic Buildings: 1882-83 John Muir house; 1849 Martinez adobe; 1885 John Muir carriage house.
Research Fields: history.
Facilities: visitor center; auditorium. Postcards, posters, slides & books for sale.
Activities: guided & self-guided tours; lectures; films; special events.
Publications: pamphlets.
Hours & Admission Prices: Wed.-Sun. 10-5. Adults $3; Golden Age, National Park Pass, Golden Eagle Pass holders & children 15 & under no charge. Closed New Year's Day; Thanksgiving; Christmas. &
Attendance: 31,000 (accurate)

MARTINEZ MUSEUM, (M), 1005 Escobar, Martinez, CA 94553. Mailing Address: P.O. Box 14, Martinez, CA 94553. Tel.: 510-228-8160.
Web Site: martinezhistory.org
Founded: 1974.
Congressional District: 7
Key Personnel: Dir., Andrea Blachman; Pres., John Curtis.

Governing Authority: Parent Institution: Martinez Historical Society.
Institution Type/Description: History Museum.
Collections: local history; period artifacts; scrapbooks of civic groups; microfilm; photographs.
Research Fields: microfilm; assessor's books; scrapbooks; ledgers; paper files; maps; family history.
Activities: tours. Museum Sponsors: dedications; open house; parades; teas; treasure hunts.
Publications: bimonthly newsletter, Martinez Historical Society; Martinez A California Town.
Hours & Admission Prices: Tues. & Thurs. 11:30-3, Sun. 1-4. No charge; donations accepted.
Attendance: 1,700 (accurate)
Membership: Student $5; Individual $10; Family $15; Contributing $25; Supporting $50; Patron $75; Life $500.

Marysville

CHINESE-AMERICAN MUSEUM OF NORTHERN CALIFORNIA, 232 1st St., Marysville, CA 95901-6002. Tel.: 510-710-2342.
Institution Type/Description: Historic Building Museum: built in 1858.
Collections: artifacts & memorabilia pertaining to Chinese history; replicas of a Chinese general store & Sanfow Bean Sprout Plant.
Hours & Admission Prices: 1st Sat. of the month 12-4. No charge.

MARY AARON MEMORIAL MUSEUM, 704 D St., Marysville, CA 95901-5319. Mailing Address: P.O. Box 1759, Marysville, CA 95901. Tel.: 530-743-1004. Facebook: Mary Aaron Museum.
Web Site: www.maryaaronmuseum.com
Founded: 1963.
Congressional District: 2
Key Personnel: Dir., Lily J. Noonan; Pres. (V), Peppie Schrader.
Personnel Profile: Part-Time Paid 3; Part-Time Volunteers 26.
Volunteer Hours: 300
Governing Authority: Tax-exempt: 501(c)(3).
Institution Type/Description: Historic House Museum: built in 1855.
Collections: photographs; clothing; furnishings.
Hours & Admission Prices: Fri.-Sat. 1-4; other times by appointment. No charge; donations accepted.
Attendance: 800 (estimated)
Membership: Student & Senior $15; Red Hat Society $18; Individual $25; Family $50; Bronze $100; Silver $250; Gold $500; Diamond $1,000.

MUSEUM OF THE FORGOTTEN WARRIORS, 5865 A Rd., Marysville, CA 95901-8017. Tel.: 530-742-3090.
E-mail: cws21779@aol.com
Web Site: museumoftheforgottenwarriors.org
Governing Authority: Tax-exempt.
Institution Type/Description: Military Museum.
Collections: over 50,000 military artifacts.
Hours & Admission Prices: Thurs. 7pm-9pm, Sat. 10-3; other times by appointment. Call for holiday hours. No charge.

McClellan

AEROSPACE MUSEUM OF CALIFORNIA, 3200 Freedom Park Dr., McClellan, CA 95652-2432. Tel.: 916-643-3192. Fax: 916-643-0389.
Web Site: www.aerospaceca.org
Founded: 1986.
Congressional District: 5
Key Personnel: Exec. Dir., Roxanne Yonn; Pres. (V), James W. Hopp, Maj. Gen. USAF (Ret.); Cur., Barry Bauer; Museum Shop Mgr., Andrea Nelson.
Personnel Profile: Full-Time Paid 5; Part-Time Paid 7; Part-Time Volunteers 197.
Governing Authority: Tax-exempt.
Institution Type/Description: Aviation & Aerospace Museum.
Collections: aviation & aerospace industries.
Hours & Admission Prices: Tues.-Sat. 9-5, Sun. 10-5. Adults $8, seniors 65 & over and youth 13-18 $6, children 6-12 $5; active military and children 5 & under no charge. &
Attendance: 75,000 (estimated)
Membership: Senior 65 & over, Teacher, & Student $45; Individual $75; Family $100; Business Bronze $250; Business Silver $500; Business Gold $1,000; Life 65 & over $1,200; Life under 65 $2,000.

Mendocino

FORD HOUSE VISITOR CENTER AND MUSEUM, 735 Main St., Mendocino, CA 95460. Mailing Address: P.O. Box 1387, Mendocino, CA 95460-1387. Tel.: 707-937-5397. Facebook: Ford House Visitor Center and Museum.
E-mail: fordhouse@mcn.org
Web Site: www.mendoparks.org
Founded: 1984.
Congressional District: 1
Key Personnel: Mgr., Jenny Heckeroth.
Personnel Profile: Part-Time Paid 1; Part-Time Volunteers 21.
Governing Authority: Parent Institution: CA State Parks & Recreation. Subsidiary Institution: Mendocino Area Parks Association. Tax-exempt.
Institution Type/Description: History Museum: house built in 1854. Listed on the National Register of Historic Places.
Collections: scale model of the 1890 village of Mendocino; Pomo Indian artifacts; natural history.
Major Exhibits: Gray Wale Exhibit, 1/1/14-3/31/14; Wildflowers of the Mendocino Coast, 4/2014-5/2014; Mushroom House, 11/14.
Hours & Admission Prices: Daily 11-4. Suggested Donation: $2-$5. &
Attendance: 25,000 (estimated)

KELLEY HOUSE MUSEUM, 45007 Albion St., Mendocino, CA 95460. Mailing Address: P.O. Box 922, Mendocino, CA 95460-0922. Tel.: 707-937-5791. Fax: 707-937-2156.
E-mail: info@kelleyhousemuseum.org
Web Site: www.kelleyhousemuseum.org
Founded: 1973.
Congressional District: 1
Key Personnel: Exec. Dir., Nancy Freeze.
Personnel Profile: Part-Time Paid 2; Part-Time Volunteers 25.
Governing Authority: nonprofit organization. Parent Institution: Kelley House Museum, Inc. Tax-exempt: 501(c)(3).
Institution Type/Description: Historical Society Museum: 1861 Kelley House.
Collections: photographs & documents relating to Mendocino history; logging items; North Coast genealogical references.
Research Fields: California History
Facilities: library; botanic garden. Museum-related items for sale.
Activities: guided tours; study clubs; docent program; temporary exhibitions. Annual Events: House & Historic Building Tour in May; BBQ & jazz on the lawn in July; Silent Auction in October; lighting of the town Christmas tree in December.
Publications: annual, MHR Review; quarterly, members newsletter.
Hours & Admission Prices: Museum: June-Aug. Thurs.-Tues. 11-3; Sept.-May Fri.-Mon. 11-3. Walking Tours: Sat. 11am. Office: Tues.-Fri. 9-4. Research by appointment. $2 per person. Tours: $10 per person; discounts on purchases of MHR, Inc. publications for members. Closed New Year's Day; Thanksgiving; Christmas. &
Attendance: 3,500 (estimated)
Membership: Associate $25; Subscribing $40; Nonprofit & Family $50; Business $70; Sustaining $100; Lifetime $1,000.

Merced

AGRICULTURE MUSEUM, 4498 E. Hwy. 140, Merced, CA 95340-9388. Mailing Address: 4505 E. State Hwy., Merced, CA 95340. Tel.: 209-383-1912.
Web Site: www.agmuseum.us/webpage_001.htm
Founded: 1993.
Institution Type/Description: Agriculture Museum.
Collections: dairy equipment; household artifacts; period pieces; toys; blacksmith shop; dental equipment; washing machines; beauty shop.
Facilities: Museum-related items for sale.
Activities: special events.
Hours & Admission Prices: Tues.-Sun. 8-5. No charge. &

APPLEGATE PARK ZOO, 1045 W. 25th St., Merced, CA 95340-3500. Mailing Address: 678 W. 18th St., Merced, CA 95340-4721. Tel.: 209-385-6840. Fax: 209-384-5805.
E-mail: johnsonl@cityofmerced.org
Web Site: www.cityofmerced.org
Founded: 1962.
Congressional District: 18
Key Personnel: Dir. Parks & Community Svcs., Mike Conway; Pres. (V), Marlene Murphy; Recreation Supvr., Lindsey Johnson; Lead Zookeeper, Donna McDowell.
Personnel Profile: Full-Time Paid 1; Part-Time Paid 2; Part-Time Volunteers 10.

Operating Expenses: 200,000
Operating Income: 100,000
Governing Authority: municipal. Parent Institution: City of Merced. Subsidiary Institution: The Merced Zoological Society, P.O. Box 408, Merced, CA 95341. Tax-exempt: 501(c)(3).
Institution Type/Description: Zoo.
Collections: over 75 native California mammals, birds & reptiles; plants.
Facilities: 2-acre zoological park; picnic facilities. Wildlife-related items for sale.
Activities: summer educational zoo camp programs; zoo tours with a zookeeper by appointment; workshops; wildlife programs; presentations; special events; zoo-themed birthday parties.
Publications: book, Take A Walk on the Wild Side.
Hours & Admission Prices: March-Oct. daily 10-5; Nov.-Dec. daily 10-4. Admission $3, children 5-15 $2, seniors 62 & up $1.50; discounts to AZA, AAZK & reciprocal institution members; museum members, Merced Zoological Society members, & children under 5 no charge. Closed New Year's Day; Thanksgiving; Christmas. &
Attendance: 52,000 (estimated)
Membership: Student & Senior Citizen $7; Individual $10; Family & Grandparents Plus $20; Organization $50; Lifetime $150; Lifetime Family $250.

MERCED COUNTY COURTHOUSE MUSEUM, 21st and N Sts., Merced, CA 95340-3790. Tel.: 209-723-2401. Fax: 209-723-8029. Facebook: Merced County Courthouse Museum.
E-mail: mercedmuseum@sbcglobal.net
Web Site: www.mercedmuseum.org
Founded: 1975.
Congressional District: 15
Key Personnel: C.E.O. & Dir., Sarah Lim; Pres., Charlie Galatro; Financial Dir., Grey Roberts; Sec., John Hofmann; Museum Shop Mgr., Ann Carson.
Personnel Profile: Full-Time Paid 1; Part-Time Paid 3; Part-Time Volunteers 114; Interns 2.
Volunteer Hours: 4,400
Operating Expenses: 104,013
Operating Income: 171,861
Governing Authority: nonprofit. Parent Institution: Merced County Historical Society. Tax-exempt.
Institution Type/Description: Local History Museum: 1875 Italianate style Courthouse.
Collections: local period photographs; clothing; 1870-1930s county assessor records; Chinese temple; blacksmith shop; cultural artifacts from prehistory to 1920s.
Major Exhibits: Weaving A Legacy: The Art History of Yokuts and Miwoks Indian Basketry, 10/13-2/14; Weaving A Legacy: The Art History of Central California Indian Basketry, 10/17/13-2/16/14; State of Change (T), 3/6/14-4/20/14; The Way We Worked (T), 5/14; Merced's 125th Anniversary, 6/26/14-10/5/14; Mexican American Experience: From Spanish Land Grants to Civil Rights Movement, 10/16/14-3/1/15.
Research Fields: Merced County & surrounding areas.
Facilities: 8,499 sq. ft. exhibit space. Museum-related items for sale.
Activities: guided tours; lectures; organized education programs for children; history trunks used in elementary classrooms; docent program; participatory, loan, traveling & temporary exhibitions. Museum Sponsors: Christmas Open House.
Publications: quarterly newsletter, For The Record; occasional local histories.
Hours & Admission Prices: Wed.-Sun. 1-4. No charge; donations accepted. Closed New Year's Eve & Day; Easter; Independence Day; Thanksgiving; Christmas Eve & Day. &
Attendance: 8,000 (estimated)
Membership: Junior $2; Senior $15; Individual $20; Family $35; Sustaining Individual $50; Sustaining Business $100; Benefactor $250; Patron $500; Founders Circle $1,000.

Mill Valley

MUIR WOODS NATIONAL MONUMENT, Muir Woods National Monument, Mill Valley, CA 94941-2696. Mailing Address: Golden Gate National Recreation Area, Bldg. 201, Fort Mason, San Francisco, CA 94123. Tel.: 415-388-2595 & 2596. Fax: 415-389-6957.
Web Site: www.nps.gov/muwo
Founded: 1908.
Congressional District: 6
Key Personnel: Supt., Brian O'Neill.
Governing Authority: federal. Parent Institution: National Park Svc. Subsidiary Institution: Golden Gate National Recreational Area. Tax-exempt.
Institution Type/Description: National Park.
Collections: herbarium; history.
Facilities: visitors center; nature trails. Books, maps & posters for sale.

Activities: daily walks; children's program.
Publications: Muir Woods: Trail Guides; Muir Woods: Redwood Refuge.
Hours & Admission Prices: Daily 8 to sunset. Adults 18 & over $5; children 15 & under, Golden Eagle and Access Passes no charge. ♿
Attendance: 1,446,256 (estimated)
Membership: Golden Gate National Park Association: $25-$1,000.

Millbrae

THE MILLBRAE MUSEUM, 450 Poplar Ave., Millbrae, CA 94030. Mailing Address: Millbrae Historical Society, P.O. Box 511, Millbrae, CA 94030. Tel.: 650-692-5786.
Institution Type/Description: Historic House Museum: built in 1895.
Collections: local history & culture; period furnishings; photographs; personal artifacts.
Hours & Admission Prices: Call for hours.

MILLBRAE TRAIN STATION, California Dr. at Murchison Dr., Millbrae, CA 94030. Mailing Address: Millbrae Historical Society, P.O. Box 511, Millbrae, CA 94030. Tel.: 650-333-1136.
Web Site: www.millbraehs.org
Founded: 1971.
Governing Authority: Parent Organization: Millbrae Historical Society. Tax-exempt.
Institution Type/Description: Historic Building: built in 1907.
Collections: local history; photographs; period artifacts; 1941 Pullman Sleeping Car; 1929 railway express agency truck; 1930 Southern Pacific fire engine; documents; books.
Publications: quarterly newsletter.
Hours & Admission Prices: Sat. 10-2; other times by appointment. Adults $2; members no charge. ♿
Membership: Family $15; Sustaining $25; Business $35; Lifetime $250.

Milpitas

SAN FRANCISCO BAY BIRD OBSERVATORY, 524 Valley Way, Milpitas, CA 95035-4106. Tel.: 408-946-6548. Fax: 408-946-9279. Facebook: San Francisco Bay Bird Observatory.
E-mail: outreach@sfbbo.org
Web Site: www.sfbbo.org
Key Personnel: Exec. Dir., Cat Burns
Institution Type/Description: Bird Conservatory.
Collections: birds & their habitats.
Activities: guided walks.
Hours & Admission Prices: Call for hours.
Membership: Student $20; Basic $40; Friend $60; Contributor $100; Sustainer $250; Sponsor $500; Partner $1,000.

Mission Hills

HISTORICAL MUSEUM-ARCHIVAL CENTER, 15151 San Fernando Mission Blvd., Mission Hills, CA 91345-1109. Tel.: 818-365-1501. Fax: 818-361-3276.
Founded: 1962.
Key Personnel: Archivist, Kevin Feeney, A.C.D.A.
Governing Authority: church. Parent Institution: Roman Catholic Archdiocese of Los Angeles, 3424 Wilshire Blvd., Los Angeles, CA 90010. Tax-exempt.
Institution Type/Description: Museum & Archival Center: located on the grounds of the San Fernando Mission.
Collections: documents; letters; historical memorabilia associated with the Catholic Church in Southern California; manuscripts.
Research Fields: ecclesial history; western Americana.
Facilities: 8,000-vol. library of material pertaining to ecclesial & western Americana.
Activities: guided tours; hobby workshops; reading room; docent program; Las Damas Archivistas; permanent exhibitions.
Publications: quarterly newsletter, Friends of Archival Center.
Hours & Admission Prices: Mon. & Thurs. 1-3; other times by appointment. No charge. Closed national holidays. ♿
Membership: Friends of Archival Center: Student & Senior Citizen $10; Supporting $25; Sustaining $100; Benefactor $500; Life $1,000.

SAN FERNANDO MISSION, 15151 San Fernando Mission Blvd., Mission Hills, CA 91345-1109. Tel.: 818-361-0186. Fax: 818-361-3276.
Founded: 1797.
Key Personnel: Admin., Msgr. Francis J. Weber; Chm., Archbishop Jose H. Gomez; Dir. & Business Mgr., Kevin Feeney; Sales Shop Mgr., Monica Mejorado.

Governing Authority: church; nonprofit organization. Parent Institution: Roman Catholic Archdiocese of Los Angeles, 3424 Wilshire Blvd., Los Angeles, CA 90010-2241. Tel: 213-637-7000. Tax-exempt.
Institution Type/Description: Museums & Mission Complex: founded in 1797 by Fray Fermin Lasuen and later restored.
Collections: Museum: mission baskets; pictorial history; pottery; santos; trade & commerce items. Mayordomo's House: 1806 home belonging to the foreman of mission ranch. Convento: 1822 adobe building with 21 Roman arches; iron grilles; painting; pipe organ; vestments; Hispanic furniture. West Garden: rare trees; wine vats; adobe displays; 1686 & 1720 bells from the Escaray collection. Fourth Mission Church: c.1804 replica of earlier edifice. East Garden; replica of Cordova fountain; rare trees; cacti; seasonal flowers; cemetery. Workshops: recreating carpentry; pottery; saddle; blacksmiths shop; weaving room; furnishings from the provincial era; manuscripts. Piczek Tableaus. Archival Center (see separate listing).
Research Fields: mission registers.
Facilities: 1,760-vol. Biblioteca Montereyensis-Angelorum Diocese library pertaining to apologetics, scripture, theology & history available for research on premises. Religious & museum-related items for sale.
Activities: self-guided tours; lectures; concerts; reading room; permanent exhibitions.
Publications: brochures; pamphlet.
Hours & Admission Prices: Daily 9-4:30. Adults $5, children 7-15 $3; children under 7 no charge. Closed Thanksgiving; Christmas. ♿

SAN FERNANDO VALLEY HISTORICAL SOCIETY, INC., 10940 Sepulveda Blvd., Mission Hills, CA 91346. Mailing Address: P.O. Box 7039, Mission Hills, CA 91346-7039. Tel.: 818-365-7810. Fax: 818-365-7810.
E-mail: sfvhs@verizon.net
Web Site: sfvhs.com
Founded: 1943.
Congressional District: 21
Key Personnel: Corresponding Sec., Tesa Becica; Recording Sec. & Cur., Dr. Richard Doyle.
Personnel Profile: Full-Time Volunteers 2; Part-Time Volunteers 8.
Governing Authority: municipal; society. Parent Institution: Los Angeles Dept. of Recreation & Parks. Tax-exempt: 501(c)(3).
Institution Type/Description: History Museum.
Collections: archives; paintings; decorative arts; costumes; history; Indian artifacts; manuscripts. Historic House: 1834 Andres Pico Adobe.
Research Fields: San Fernando Valley & California history.
Facilities: 5,000-vol. library of books & pamphlets on San Fernando Valley & California history available for use on premises by public for research; reading room.
Activities: guided tours; lectures; films; inter-museum loan, permanent & temporary exhibitions.
Publications: monthly newsletter, The Valley; guide.
Hours & Admission Prices: Mon. 10-3; 3rd Sun. of month 1-4; tours by appointment. No charge; donations accepted. Closed New Year's Day; Easter; Thanksgiving; Christmas.
Attendance: 300 (estimated)
Membership: Student $10; Individual $18; Couple & Organization $28; Sustaining $50; Business, Professional & Corporate $100; Life $250.

Modesto

GREAT VALLEY MUSEUM OF NATURAL HISTORY, 1100 Stoddard Ave., Modesto, CA 95350-5818. Tel.: 209-575-6196. Fax: 209-575-6466.
E-mail: crawfordl@mjc.edu
Web Site: yosemite.cc.ca.us/community/great-valley
Founded: 1973.
Congressional District: 12
Key Personnel: Pres. (V), Roger Gohring; Dir., Louise J. Crawford; Gift Shop Mgr., Tana Dennen.
Personnel Profile: Full-Time Paid 1; Part-Time Paid 12; Part-Time Volunteers 16.
Governing Authority: nonprofit organization. Parent Institution: Yosemite Community College District. Subsidiary Institution: Modesto Junior College. Tax-exempt: 501(c)(3) & 170(b)(1)(A).
Institution Type/Description: Natural History Museum.
Collections: mammals; invertebrates; birds; fish; shells; geological specimens.
Facilities: educational facilities. Science-related books & materials for sale.
Activities: guided tours; lectures; participatory exhibits; docent program; field trips; school loan service; annual events.
Publications: quarterly, newsletter.
Hours & Admission Prices: July daily 10-4; Sept.-June Tues.-Fri. 9-4:30, Sat.

10-4. Families $3, adults $2.50, senior citizens 65 & over and children 5-17 $1.50; members and children 6 & under no charge. Closed major holidays. &

Attendance: 3,314 (estimated)
Membership: Student $15; Senior Citizen $20; Individual $25; Family $35; Supporting $50; Contributor $100; Institution (schools) $200; Donor $250; Sponsor $500; Sustaining Friend $1,000.

HILLIER AIR MUSEUM, Modesto Airport, Hangar 7, 700 Tioga Dr., Modesto, CA 95354. Tel.: 209-526-8297 & 985-9000.
E-mail: tomhillier@msn.com
Web Site: hillierairmuseum.com
Institution Type/Description: Aviation Museum.
Collections: aviation history; aircraft from 1937 to 1947.
Activities: special events.
Hours & Admission Prices: 2nd Sat. each month 9-3.

MCHENRY MUSEUM, 1402 I St., Modesto, CA 95354-1032. Tel.: 209-577-5366. Fax: 209-491-4407.
E-mail: museum@mchenrymuseum.org
Web Site: www.mchenrymuseum.org
Founded: 1965.
Congressional District: 15
Key Personnel: Cultural Svcs. Mgr., Wayne A. Mathes; Exhibit Designer, Laura Mesa; Museum Shop Mgr., Anne Hatheway; Museum Shop Mgr., Donnelle Dilbeck.
Personnel Profile: Full-Time Paid 2; Part-Time Paid 1; Part-Time Volunteers 80.
Governing Authority: municipal government; nonprofit organization. Parent Institution: City of Modesto. Tax-exempt: 170(b)(1).
Institution Type/Description: History Museum: housed in 1911 library building.
Collections: over 5,000 photographs & documents on local history; 1880-1910 room exhibits; blacksmith shop; doctor & dentist office; gold mining; barber shop; school-room; fire display; costumes; accessories & linens.
Research Fields: local history.
Facilities: library; 90-seat auditorium. Museum-related items for sale.
Activities: guided tours; lectures; concerts; docent program; temporary exhibitions.
Publications: quarterly newsletter; brochure.
Hours & Admission Prices: Tues.-Sun. 12-4. No charge; donations accepted. Closed New Year's Day; Easter; Thanksgiving; Christmas. &
Attendance: 25,000 (estimated)
Membership: Individual $25; Family & Couple $45; Patron $100; Benefactor $250; Corporate $500-$999; Life $1,000.

Modjeska Canyon

ARDEN - THE HELENA MODJESKA HISTORIC HOUSE & GARDENS, 29042 Modjeska Canyon Rd., Modjeska Canyon, CA 92676-9793. Mailing Address: c/o Heritage Hill Historical Park, 25151 Serrano Rd., Lake Forest, CA 92630-2534. Tel.: 949-923-2230. Fax: 949-855-6321.
E-mail: ardenmodjeska@ocparks.com
Web Site: www.ocparks.com/modjeskahouse/
Founded: 1990.
Congressional District: 47
Personnel Profile: Full-Time Paid 7; Part-Time Volunteers 30.
Governing Authority: county. Tax-exempt.
Institution Type/Description: Historic House Museum: housed in the home of Polish American Shakespearean actress, Madame Helena Modjeska.
Collections: period furniture; personal artifacts.
Facilities: gardens; ponds.
Activities: docent program; formal education programs for children; guided tours.
Publications: biannual newsletter, The Helena Modjeska Foundation.
Hours & Admission Prices: Jan.-Nov. 1st & 3rd Tues. and 2nd & 4th Sat. by appointment only. Admission $5. &
Attendance: 3,600 (accurate)
Membership: Foundation $25.

Moffett Field

MOFFETT FIELD HISTORICAL SOCIETY & MUSEUM, Bldg. 126 Severyns Ave., Moffett Field, CA 94035-0016. Mailing Address: P.O. Box 16, Moffett Field, CA 94035-0016. Tel.: 650-964-4024. Fax: 650-964-4028.
E-mail: moffettmuseum@sbcglobal.net
Web Site: www.moffettfieldmuseum.org

Personnel Profile: Full-Time Volunteers 2; Part-Time Volunteers 21.
Governing Authority: Tax-exempt.
Institution Type/Description: History Museum.
Collections: airport & aviation history; aircraft; photographs; personal artifacts; hands-on exhibits.
Hours & Admission Prices: Wed.-Sat. 10-2. Adults $8, seniors $5, active & reserve military members and children 12 & under no charge. &
Attendance: 5,000 (accurate)
Membership: Annual $25; Lifetime $300.

Montebello

JUAN MATIAS SANCHEZ ADOBE, 946 Adobe Ave., Montebello, CA 90640. Tel.: 323-887-4592.
E-mail: gbrougher@sbcglobal.net
Key Personnel: Cur., Bud Sanchez.
Governing Authority: Parent Institution: Montebello Historical Society.
Institution Type/Description: Historic House Museum: built in 1844 by Dona Maria Casilda de Lobo.
Collections: period artifacts & furnishings.
Publications: The Adobe Dust; newsletter, Montebello Historical Society and Museum.
Hours & Admission Prices: Wed. & Sat.-Sun. 1-4. No charge. &
Attendance: 200 (estimated)
Membership: Student $10; Individual $15; Family $25; Historical Society $45; Business & Civic $100.

Montecito

GANNA WALSKA LOTUSLAND, (M), 695 Ashley Rd., Montecito, CA 93108. Tel.: 805-969-9990. Fax: 805-969-4423.
E-mail: dhatch@lotusland.org
Web Site: www.lotusland.org
Founded: 1993.
Congressional District: 23
Key Personnel: Exec. Dir., Gwen L. Stauffer; Pres. (V), Larry Durham; Museum Shop Mgr., Karen Kester
Governing Authority: Tax-exempt.
Institution Type/Description: Botanic Garden: housed on a 37-acre estate.
Collections: tropical & subtropical plants; garden history.
Research Fields: applied horticulture.
Facilities: Garden-related items for sale.
Activities: concerts; lectures; workshops; themed tours.
Publications: quarterly members' newsletter.
Hours & Admission Prices: Guided Tours: Feb. 15-Nov. 15 Wed.-Sat. 10 am & 1:30 pm by appointment. Adults $35; members no charge.
Attendance: 15,000 (accurate)
Membership: Individual $75; Family $125; Friend $250; Garden Advocate $500; Garden Cultivator $1,000; Garden Conservator $2,500; Garden Guardian $5,000; Garden Steward $10,000; Garden Champion $25,000.

Monterey

COLTON HALL MUSEUM AND OLD MONTEREY JAIL, (M), City Hall, 580 Pacific St., Monterey, CA 93940-2806. Mailing Address: 570 Pacific St., Monterey, CA 93940. Tel.: 831-646-5640. Fax: 831-646-3422.
E-mail: museumpt@ci.monterey.ca.us
Web Site: www.monterey.org/museum/
Founded: 1948.
Congressional District: 16
Key Personnel: Dir., Kim Bui-Burton; Museum, Cultural Arts & Archives Mgr., Dennis Copeland; Cultural Arts Asst., Chalet Booker; Museum Admin. Asst., Claire Rygg.
Personnel Profile: Full-Time Paid 1; Part-Time Paid 4; Part-Time Volunteers 35.
Governing Authority: municipal. Parent Institution: City of Monterey Museums and Cultural Arts. Subsidiary Museum: Old Monterey Jail; Presidio Museum of Monterey; Alvarado Gallery (Monterey Conference Center); Pacific Biological Laboratories, (Cannery Row).
Institution Type/Description: Regional History Museum: housed in Colton Hall, 1849 town hall & public school, site of 1849 California Constitutional Convention.
Collections: archival materials including documents; photographs; registers & bound volumes; books; artifacts & furnishings related to 1849 Constitutional Convention; local history.
Research Fields: research files on California's Constitutional Convention & delegates; local history information files; early California history.
Facilities: 200-vol. library of historical material available for research on premises; archival materials by appointment.

Activities: guided tours; special exhibits; educational programs; acoustic music concerts. Museum Sponsors: National Museum North Talks in May; 1849 Constitutional Convention Reenactment in October; Christmas in the Adobes in December.
Publications: leaflet, Path of History Guide; booklet, Historic Monterey; brochure, Explore Monterey; 1849 California Constitution.
Hours & Admission Prices: Daily 10-4. No charge. Closed New Year's Day; Thanksgiving; Christmas.
Attendance: 18,000 (estimated)

GREEN CHALK CONTEMPORARY, 616 Lighthouse Ave., Monterey, CA 93940-1100. Tel.: 202-253-4507. Fax: 831-747-1088.
E-mail: antongallery@aol.com
Web Site: www.greenchalkcontemporary.com
Formerly: Anton Gallery
Founded: 2013.
Key Personnel: Dir., Ami-Suzanne Lawless; C.E.O., Gail Enns
Institution Type/Description: Art Gallery.
Collections: works by contemporary artists; paintings; sculpture; prints.
Activities: temporary exhibitions; special events.
Hours & Admission Prices: Tues.-Sat. 12-5. No charge; donations accepted.
Attendance: 1,000 (estimated)

MONTEREY BAY AQUARIUM, (M), 886 Cannery Row, Monterey, CA 93940-1085. Tel.: 831-648-4800. Fax: 831-648-4810.
Web Site: www.montereybayaquarium.org
Founded: 1984.
Congressional District: 17
Key Personnel: Exec. Dir., Julie Packard; Chm. (V), Dr. Peter Bing; Mng. Dir., Jim Hekkers; Vice Pres. & C.F.O., Ed Prohaska; Vice Pres. Exhibitions, Don Hughes; Vice Pres. Communications, Hank Armstrong; Vice Pres. Human Resources, Teresa Merry; Vice Pres. & Dir. The Center for the Future of The Oceans, Mike Sutton; Vice Pres. Mktg., Mimi Hahn; Dir. Conservation Research, Dr. Christopher Harrold; Chief Devel. Officer, Cristina Fekeci; Vice Pres. Conservation, Education & Research, Cynthia Vernon; Vice Pres. Husbandry, Randy Hamilton; Vice Pres. Facilities, Charles Aslanian; Museum Shop Mgr., Andrew Fischer.
Personnel Profile: Full-Time Paid 386; Part-Time Paid 49; Part-Time Volunteers 1,178; Interns 42.
Governing Authority: nonprofit organization. Parent Institution: Monterey Bay Aquarium Foundation. Tax-exempt: 501(c)(3).
Institution Type/Description: Aquarium: on former site of Hovden Cannery.
Collections: live aquatic displays representing 550 species; canning industry history; California sea otters; kelp forest exhibit; live video feed of deep-sea research; periodic displays of great white sharks; jellies; giant octopus; penguins.
Major Exhibits: The Jellies Experience, 4/12-9/14.
Research Fields: tuna research & conservation; research & conservation of sea otters; health & maintenance of marine organisms; basic biology & physiology of deep-sea & open-ocean marine organisms; vessel bay sails.
Facilities: 4,200-vol. library available to public by appointment; aquarium; 273-seat auditorium; cafeteria; restaurant; classrooms; field research station. Books & gift items for sale.
Activities: guided tours; field trips; lectures; teacher training; organized educational programs; docent program; participatory exhibits; diving experiences for children 8-13; live animal feedings & shows; educational programs & presentations.
Publications: triannual newsletter, Shorelines; monthly e-newsletter, Sea Notes; annual review; conservation research report; natural history books; blog, Sea Notes.
Hours & Admission Prices: Summer: Mon.-Fri. 9:30-6, Sat.-Sun. 9:30-8, holidays 9:30-6:30; Winter: daily 10-6. Summer: adults $34.95. Winter: adults $32.95, seniors 65 & over and students 13-17 $27.95, children 3-12 & disabled $20.95; discounts to military & groups; members & children under 3 no charge. Closed Christmas.
Attendance: 1,800,000 (accurate)
Membership: Senior & Student $50; Individual $175; Ocean Advocate $250; Supporter $300; Sustainer $500; Associate $1,000; Packard's Circle $2,500.

MONTEREY FIRE DEPARTMENT HISTORICAL MUSEUM, 582 Hawthorn St., Monterey, CA 93940. Tel.: 831-646-3900.
E-mail: ventimig@ci.monterey.ca.us
Institution Type/Description: Firefighting History Museum.
Collections: firefighting history, equipment & apparatus; photographs; badges.
Hours & Admission Prices: By appointment.

MONTEREY HISTORY AND ART ASSOCIATION, 5 Custom House Plaza, Monterey, CA 93940-2430. Tel.: 831-372-2608. Fax: 831-655-3054.
E-mail: info@montereyhistory.org
Web Site: www.montereyhistory.org
Founded: 1931.
Congressional District: 16
Key Personnel: Exec. Dir., John N. Bailey; Pres. (V), William D. Curtis; Museum Shop Mgr., Christy O'Neil.
Personnel Profile: Full-Time Paid 8; Part-Time Paid 4; Part-Time Volunteers 100; Interns 15.
Governing Authority: nonprofit organization. Subsidiary Institution: Monterey Maritime & History Museum. Tax-exempt: 501(c)(3).
Institution Type/Description: Maritime Museum & History Center.
Collections: paintings; costumes; manuscripts; maritime artifacts; artwork; books; Casa Serrano Adobe period paintings & furnishings; Jo Mora sculptures. Historic Buildings: c.1845 Serrano Adobe; c.1845 Fremont House; 1876 Mayo Hayes O'Donnell Library; 1865 Francis Doud house; Perry Downer House: costumes.
Research Fields: Monterey & California history.
Facilities: 1,500-vol. library of California history books available for use on premises. Gift items for sale.
Activities: guided tours; permanent & temporary exhibitions. Association Sponsors: LaMirenda celebration of Monterey's birthday; re-enactment of historic landing by Commodore John Drake Sloat on July 7, 1846.
Publications: quarterly bulletin, Noticias del Puerto de Monterey; quarterly newsletter.
Hours & Admission Prices: Mayo Hayes O'Donnell Library: Wed. & Fri.-Sun. 1-4. No charge. Maritime Museum of Monterey: Thurs.-Tues. 10-5. No charge. Closed New Year's Day; Thanksgiving; Christmas.
Attendance: 60,000 (estimated)
Membership: Educational & Nonprofit $15; Active (Single) $35; Active (Couple & Dual) $45; Active Family $60; Sustaining (Single) $85; Sustaining (Dual) $110; Sustaining Family $125; Life (Individual) $750; Life (Couple & Dual) $1,000.

* **MONTEREY MUSEUM OF ART, (M),** 559 Pacific St., Monterey, CA 93940-2805. Tel.: 831-372-5477. Fax: 831-372-5680.
E-mail: info@montereyart.org
Web Site: www.montereyart.org
Founded: 1959.
Congressional District: 17
Key Personnel: Pres., Melissa Burnett; Exec. Dir., E. Michael Whittington; Dir. Mktg. Communications, Mary DeGroat; Dir. Education, Terry Laurents; Cur., Karen Hendon; Museum Shop Mgr., Nanci Markey; Curatorial Asst., Helaine Glick.
Personnel Profile: Full-Time Paid 14; Part-Time Paid 6; Interns 1.
Governing Authority: nonprofit organization. Branch Museum: La Mirada, 720 Via Mirada, Monterey, CA 93940. Tel: 831-372-3689. Tax-exempt: 501(c)(3).
Institution Type/Description: Art Museum.
Collections: California art from the 19th century to the present within a national & international context.
Research Fields: relating to collections & exhibits.
Facilities: library of art books, catalogues & magazines available for use on premises. Museum-related items for sale.
Activities: changing exhibitions; lectures; art workshops; bus tours; docent programs; youth activities; exhibition openings.
Publications: e-newsletter; catalogues, Henrietta Shore: A Retrospective 1900-1963; Yesterday's Artists on the Monterey Peninsula; The Monterey Photographic Tradition: The Weston Years; The Artist & the Myth; Armin Hansen: A Centennial Salute; From Old Timer to New Timer: The Life & Work of Mark M. Walker; Colors & Impressions: The Early Work of E. Charlton Fortune; Armin Hansen: The Jane & Justin Dart Collection; Dody Weston Thompson: Photographs; Henry Gilpin: Photographs.
Hours & Admission Prices: Wed.-Sat. 11-5, Sun. 1-4. Adults $10, students & military w/ID $5; discounts to NARM members; members & children under 12 no charge. Closed New Year's Day; Thanksgiving; Christmas.
Attendance: 38,500 (accurate)
Membership: Student, Teacher, & Military $20; Individual $40; Dual Out of Town $50; Dual & Family $60; Contributor $100; Supporter $250; Enthusiast $500; Curator's $1,000; Director's Circle $2,500; President's Circle $5,000; Benefactor's Circle $10,000; Connoisseurs' Circle $15,000; Patron's Circle $20,000.

MONTEREY STATE HISTORIC PARK, 20 Custom House Plaza, Monterey, CA 93940-2430. Mailing Address: California Dept. of Parks & Recreation, 2211 Garden Rd., Monterey, CA 93940-5317. Tel.: 831-649-7118. Fax: 831-647-6236.
E-mail: info@parks.ca.gov
Web Site: www.parks.ca.gov
Founded: 1938.
Congressional District: 16
Key Personnel: Dist. Cur., Kris N. Quist; Museum Cur. I, Monterey State Historic Park, Corrine Mendoza.
Governing Authority: state. Parent Institution: California State Department of Parks & Recreation, P.O. Box 942896, Sacramento, CA 94296. Tax-exempt.
Institution Type/Description: State Historic Park Museum: consisting of 12 buildings, gardens & sites in Monterey.
Collections: history; preservation projects; anthropology; archaeology; manuscripts; Native American; costumes; historic theater. Historic Houses: 1827 Custom House; 1832 Cooper-Holera Adobe; 1835 Robert Louis Stevenson House (see separate listing); 1835 Larkin House; 1844 Casa Soberanes; 1847 Pacific House; 1847 First Brick House. Concession-Operated Bldgs.: 1841 Casa Guitierrez; 1845 Casa Del Oro (Boston Store); 1847 First Theatre; 1847 Whaling Station.
Activities: guided tours; slide presentations; drama; permanent exhibitions.
Hours & Admission Prices: May-Sept. daily 9-5; Oct.-April 10-4. No charge. Closed New Year's Day; Thanksgiving; Christmas. &
Attendance: 180,000

MONTEREY STATE HISTORIC PARK/ROBERT LOUIS STEVENSON HOUSE, 530 Houston St., Monterey, CA 93940-3226. Mailing Address: 20 Custom House Plaza, Monterey, CA 93940. Tel.: 831-649-7118. Fax: 831-647-6236.
Web Site: www.historicmonterey.org
Founded: 1941.
Congressional District: 16
Key Personnel: Cur., Kris N. Quist; Guide Supvr., Stephanie Price; District Supt., Dennis Hanson.
Personnel Profile: Full-Time Paid 12; Part-Time Paid 12.
Governing Authority: state. Parent Institution: California Department of Parks & Recreation, P.O. Box 942896, Sacramento, CA 95811.
Institution Type/Description: Historic House: hotel in which Robert Louis Stevenson stayed in 1879, located in Monterey State Historic Park.
Collections: costumes; period furnishings; largest collection of Stevenson's personal materials in U.S.; manuscripts; fine arts.
Facilities: 500-vol. library of Stevenson works.
Activities: guided tours; permanent exhibitions.
Hours & Admission Prices: Daily 8-4. No charge. Closed New Year's Day; Thanksgiving; Christmas. &
Attendance: 25,000

MUSEUM OF MONTEREY, 5 Custom House Plaza, Monterey, CA 93940-2430. Tel.: 831-372-2608. Fax: 831-655-3054.
E-mail: info@montereyhistory.org
Web Site: www.museumofmonterey.org
Formerly: Maritime Museum of Monterey
Founded: 1992.
Congressional District: 16
Key Personnel: Exec. Dir., Mark Baer; Pres. (V), Christine Sinnott; Registrar, Deborah Silguero; Business Mgr. & Museum Shop Mgr., Maya Freedman.
Personnel Profile: Full-Time Paid 5; Part-Time Paid 3.
Governing Authority: nonprofit association. Parent Institution: Monterey History and Art Association. Tax-exempt: 501(c)(3).
Institution Type/Description: Local History & Art Museum.
Collections: marine artifacts; ship models; paintings; photographs; manuscripts; costumes; works of art.
Research Fields: fishing; whaling; naval history; sailing-ship era; lighthouses.
Facilities: library of maritime-related books; theater; community room. Museum-related items for sale.
Activities: guided tours; lectures; permanent & temporary exhibitions.
Publications: flyer descriptive of museum with map of local historical places; seasonal announcements; quarterly newsletter, Noticias.
Hours & Admission Prices: Tues.-Sun. 10-5. Adults $5; members no charge. Closed Thanksgiving; Christmas. &
Attendance: 73,500 (estimated)
Membership: Nonprofit $15; Single $45; Family $60.

MY MUSEUM, 425 Washington St., Monterey, CA 93940-3023. Tel.: 831-649-6444. Fax: 831-649-1304.
E-mail: info@mymuseum.org
Web Site: www.mymuseum.org

Key Personnel: Exec. Dir., Lauren Cohen; Pres., Kandis Malfyt; Vice Pres., Debra Panelli
Institution Type/Description: Children's Museum.
Collections: hands-on exhibits.
Hours & Admission Prices: Mon.-Tues. & Thurs.-Sat. 10-5, Sun. 12-5. Admission $5.50; children under 2 no charge.

Monterey Park

EAST LOS ANGELES COLLEGE VINCENT PRICE ART MUSEUM, (M), 1301 Avenida Cesar Chavez, Monterey Park, CA 91754-6099. Tel.: 323-265-8841. Fax: 323-260-8173.
E-mail: vincentpricemuseum@elac.edu
Web Site: vincentpriceartmuseum.org
Founded: 1957.
Key Personnel: Chm. (V), Julie Silliman.
Personnel Profile: Full-Time Paid 2; Part-Time Paid 2; Part-Time Volunteers 2; Interns 4.
Governing Authority: Parent Institution: East Los Angeles College.
Institution Type/Description: Art Museum.
Collections: paintings; sculpture; photographs.
Activities: guided tours.
Hours & Admission Prices: Tues.-Wed., Fri. & 2nd Sat. each month 12-4, Thurs. 12-7. No charge. Closed campus holidays. &
Attendance: 12,000 (accurate)

MONTEREY PARK HISTORICAL MUSEUM, 781 S. Orange Ave., Monterey Park, CA 91754. Mailing Address: P.O. Box 272, Monterey Park, CA 91754. Tel.: 626-307-1267.
Institution Type/Description: History Museum.
Collections: local history & culture; photographs; period artifacts.
Hours & Admission Prices: Sat.-Sun. call for hours. No charge.

Moraga

* **ST. MARY'S COLLEGE MUSEUM OF ART, (M),** 1928 St. Mary's Rd., Moraga, CA 94556-2744. Mailing Address: P.O. Box 5110, Moraga, CA 94575-5110. Tel.: 925-631-4379. Fax: 925-376-5128.
E-mail: cbrewste@stmarys-ca.edu
Web Site: www.stmarys-ca.edu/museum
Formerly: Hearst Art Gallery, St. Mary's College
Founded: 1977.
Congressional District: 10
Key Personnel: Dir., Carrie Brewster; Registrar & Collections Mgr., Julie Armistead; Public Programming & Information Mgr., Heidi Donner; Gallery Mgr. & Museum Shop Mgr., Kyla Porter; Preparator, Jim Whiteaker.
Personnel Profile: Full-Time Paid 5; Part-Time Paid 1; Part-Time Volunteers 15; Interns 6.
Governing Authority: college bd. of trustees. Parent Institution: St. Mary's College of California. Tax-exempt: 501(c)(3).
Institution Type/Description: College Art Museum.
Collections: William Keith paintings & other California landscapes; Eastern European icons; medieval sculpture; ancient ceramics; Morris Graves works; Don Quixote collection; African & Oceanic art; contemporary Northern California paintings & prints; Andy Warhol collection; 152 polaroid & silver gelatin prints; 7 screen prints.
Research Fields: William Keith; American art.
Facilities: picnic area; meeting room; food service available. Museum publications & gift items for sale.
Activities: lectures; tours; inter-museum loan & temporary exhibitions; adult plein air painting workshops. Annual Events: Biennial Master Art Tribute.
Publications: exhibition catalogues; leaflets; Modern British Art: Vorticism & the Grosvenor School; African Alchemy; Manuel Neri: A Sculptor's Drawings; The Second Golden Age of Dutch Art: 19th Century Paintings from the Beekhuis Collection; The Comprehensive Keith, The Hundred Year History of the Saint Mary's College Collection; The Nature of Collecting: The Early 20th Century Collection of Roger Epperson.
Hours & Admission Prices: Wed.-Sun. 11-4:30. Suggested Donation $5; discount to AAM members; members, military and youth 18 & under no charge. Blue Star Museum. Closed major holidays & between exhibitions. &
Attendance: 13,500 (estimated)
Membership: Senior $40; Individual $50; Family $75; Friend & Patron $100; Connoisseur $250; Director's Circle $1,000; President's Circle $1,500.

Morgan Hill

MORGAN HILL MUSEUM, Villa Mira Monte, 17860 Monterey St., Morgan Hill, CA 95038. Mailing Address: Morgan Hill Historical Society, P.O. Box 1258, Morgan Hill, CA 95038. Tel.: 408-779-5755.
Institution Type/Description: History Museum: housed in the former home of Hiram Morgan Hill; built in 1884.
Collections: local history; period furnishings; photographs.
Hours & Admission Prices: Fri. 12-3, Sat. 10-1.

Morro Bay

MORRO BAY AQUARIUM, 595 Embarcadero, Morro Bay, CA 93442-2217. Tel.: 805-772-7647.
Web Site: www.morrobay.com/morrobayaquarium
Governing Authority: nonprofit organization.
Institution Type/Description: Aquarium.
Collections: harbor seal; sea lions; fish; sharks; octopus; eels; abalone sea anemones; horseshoe crabs.
Facilities: Museum-related items for sale.
Hours & Admission Prices: Daily 9:30-5. Admission 12 & over $2, children 5-11 $1; children 4 & under no charge.

MORRO BAY NATIONAL ESTUARY NATURE CENTER, 601 Embarcadero, Morro Bay, CA 93442. Tel.: 805-772-3834. Fax: 805-772-4162.
E-mail: staff@mbnep.org
Web Site: www.mbnep.org
Institution Type/Description: Nature Center.
Collections: wildlife & their habitats; photographs.
Activities: educational programs.
Hours & Admission Prices: Call for hours.

MORRO BAY STATE PARK MUSEUM OF NATURAL HISTORY, 20 State Park Rd., Morro Bay, CA 93442-2430. Tel.: 805-772-2694, ext. 105. Fax: 805-772-7129.
E-mail: rouvaishyana@hearstcastle.com
Web Site: ccnha.org
Founded: 1962.
Congressional District: 16
Key Personnel: Chm. (V), Celeste Royer; CCNHA Exec. Dir., Mary Golden.
Personnel Profile: Full-Time Paid 1; Part-Time Paid 2; Part-Time Volunteers 100.
Governing Authority: state. Parent Institution: California State Dept. of Parks & Recreation, P.O. Box 942896, Sacramento, CA 94296. Tax-exempt.
Institution Type/Description: Natural History Museum.
Collections: Chumash Indians; birds & mammals study skins; live mounts; reptiles; insects & other invertebrates; plants including marine algae; rocks; minerals; fossils; shells; estuary below museum.
Facilities: 400-vol. library of reference books on natural & local history available on premises by appointment; 76-seat auditorium.
Activities: guided tours & walks; lectures; films; docent program; school groups.
Publications: monthly newsletter; activities schedule; booklets, Wanderer; The Monarch Butterfly; Estuary; newsletter, Nature Notes Public; monthly e-newsletter, The Bear's Den.
Hours & Admission Prices: Daily 10-5. Adults $3; discounts for State Park Pass holders; members, children 16 & under and school groups with reservation no charge. Closed New Year's Day; Thanksgiving; Christmas. &
Attendance: 60,000 (accurate)
Membership: Individual $25; Family $50; Advocate $100; Conservator $500; Guardian $1,000.

Mount Shasta

MT. SHASTA SISSON MUSEUM, #1 N. Old Stage Rd., Mount Shasta, CA 96067-9701. Tel.: 530-926-5508. Facebook: Mt. Shasta Sisson Museum.
E-mail: museum@mtshastamuseum.com
Web Site: www.mtshastamuseum.com
Formerly: Sisson Hatchery Museum
Founded: 1983.
Congressional District: 2
Key Personnel: Dir., Jean Nels; Pres., Jim McChesney; Treas., Griff Bloodhart; Museum Shop Mgr., Linda Siegel.
Personnel Profile: Part-Time Volunteers 30.
Governing Authority: private; nonprofit organization. Tax-exempt: 501(c)(3).
Institution Type/Description: History Museum.
Collections: Mt. Shasta historical artifacts; furnishings; recreational artifacts; paintings; photographs; geological; model railroad set; recreated 1850s cabin.
Major Exhibits: Lenticulars - The Spectacular Clouds of Mt. Shasta, 4/6/12-12/15/15; Joaquin Miller - Messages From His Shasta Years, 1/13-12/16.
Facilities: 80-seat auditorium; 4,000 sq. ft. exhibit space. Museum-related items for sale.
Activities: concerts; temporary exhibitions; monthly talks; lectures; films; children's activities and experiments. Annual Events: History Night; Quilt Show; Art Show; Yard Sale; Photo Show of Lenticular Clouds.
Publications: quarterly newsletter, Sisson Mirror.
Hours & Admission Prices: April-May & Oct. to mid-Dec. Fri.-Sun. 1-4; call to confirm; Memorial Day to Sept. daily 10-4. Suggested Donations: adults 16 & over $1. Closed New Year's Day; Easter; Thanksgiving; Christmas Eve & Day. &
Attendance: 8,743 (accurate)
Membership: Senior 62 & over $10; Individual $15; Family $35; Commercial $40; Senior Life $100; Life $175.

Mountain View

COMPUTER HISTORY MUSEUM, (M), 1401 N. Shoreline Blvd., Mountain View, CA 94043-1311. Tel.: 650-810-1010. Fax: 650-810-1055.
Web Site: www.computerhistory.org
Founded: 1999.
Congressional District: 14
Key Personnel: Chm. Bd., Len Shustek; Pres. & C.E.O., John Hollar
Institution Type/Description: History Museum.
Collections: history of computers; hardware; software; documentation; film; video; audio; ephemera.
Activities: tours.
Hours & Admission Prices: Wed.-Sun. 10-5. Adults $15; members no charge. &
Attendance: 75,000 (accurate)
Membership: Senior & Student $60; Individual $75; Family $150.

Murphys

OLD TIMERS MUSEUM, 470 Main St., Murphys, CA 95247. Mailing Address: P.O. Box 94, Murphys, CA 95247-0094. Tel.: 209-728-1160.
Founded: 1950.
Key Personnel: Dir. & Pres. (V), Robert Frew; Chm. (V), Tres. & Museum Shop Mgr., Michael F. Davis.
Personnel Profile: Part-Time Volunteers 10.
Governing Authority: private; nonprofit organization. Tax-exempt: 501(c)(3).
Institution Type/Description: Historic Building Museum.
Collections: local history & culture; photographs; period artifacts; documents; family artifacts.
Facilities: library; 1,200 sq. ft. exhibit space. Museum-related items for sale.
Activities: guided tours.
Hours & Admission Prices: Winter: Sat.-Mon. 12-4; Summer: Fri.-Mon. 12-4. No charge; donations accepted. Closed holidays. &
Attendance: 11,008 (accurate)

Napa

COPIA: THE AMERICAN CENTER FOR WINE, FOOD & THE ARTS, 500 First St., Napa, CA 94559-2629. Tel.: 707-259-1600; 888-512-6742. Fax: 707-257-8601.
E-mail: info@copia.org
Web Site: www.copia.org
Founded: 1990.
Congressional District: 1
Key Personnel: Pres. & C.E.O., Garry McGuire; Exec. Asst., Christi Skibbins; Chm. (V), Lauren Ackerman; C.O.O., Kurt Nystrom; Dir. Exhibitions, Neil Harvey; Chief Mktg. Officer, Larry Tsai; Dir. Human Resources, Marina Kreager, Museum Retail Mgr., Christina Taylor-Godwin, Dir. Visitor Svcs., Betty Teller; Dir. Visitor Rels., Lynn Norris; Mgr. Public Rels., Kathleen Iudice.
Personnel Profile: Full-Time Paid 77; Part-Time Paid 20; Part-Time Volunteers 204; Interns 2.
Governing Authority: private; nonprofit organization. Tax-exempt: 501(c)(3).
Institution Type/Description: Wine, Food and Art Museum.
Collections: contemporary art, photographs, design & artifacts related to food & wine; Julia Child's copper cookware.
Facilities: 13,000 sq. ft. exhibition galleries; 3.5 acre garden; 500-seat outdoor

river concert terrace; 260-seat theater; 75-seat tiered lecture & demonstration room; 3 40-seat classrooms; 75-seat dining room; children's garden; wedding pavilion; cafe. Museum-related items for sale.
Activities: formal & informal education programs for adults, children, professionals & the general public; concerts; lectures; films; garden classes; wine & food tastings; participatory & temporary exhibitions; festivals; special events; rental facilities.
Publications: quarterly calendar & catalogues.
Hours & Admission Prices: Summer: daily 10-6; Winter: Fri.-Sun. 10-6. Adults $12.50, students & seniors $10, youth $7.50; discounts to AAM & ICOM members; members no charge. &
Attendance: 180,000 (estimated)
Membership: Individual $60; Dual $75; Family $100; Reciprocal $125; Community Partner $250; Trustee Circle $500-$25,000.

DI ROSA, (M), 5200 Sonoma Hwy., Napa, CA 94559-9761. Tel.: 707-226-5991. Fax: 707-255-8934.
Web Site: www.dirosaart.org
Key Personnel: Dir., Kathryn Reasoner
Institution Type/Description: Art Gallery.
Collections: paintings; sculpture; drawings; northern California artists.
Activities: groups tours; special events; educational programs.
Hours & Admission Prices: April-Oct. Wed.-Sun. 10-6; Nov.-March Wed.-Sun. 10-4. Adults $5-$15; discounts to NARM members; members no charge. New Year's Day; Martin Luther King Jr. Day; Presidents' Day; Memorial Day; Labor Day; Veterans Day; Thanksgiving & day after; Christmas.
Membership: Minimalist $50; Conceptualist (Dual) $100; Constructivists $250; Surrealist $500; Expressionist $1,000; Regionalist $2,500; Futurist $5,000 & up.

NAPA COUNTY HISTORICAL SOCIETY, 1219 First St., Napa, CA 94559-2929. Tel.: 707-224-1739.
E-mail: director@napahistory.org
Web Site: www.napahistory.org
Founded: 1948.
Key Personnel: Dir., Kristie Sheppard.
Personnel Profile: Full-Time Paid 1; Part-Time Paid 1; Part-Time Volunteers 40; Interns 7.
Governing Authority: Tax-exempt.
Institution Type/Description: Historical Society Museum.
Collections: Napa County history & culture; photographs; books; period directories & maps; manuscripts; videos.
Facilities: library.
Activities: research.
Publications: Gleanings.
Hours & Admission Prices: Tues.-Sat. 12-4. No charge; donations accepted. &
Attendance: 3,000
Membership: Senior & Student $25; Individual $30; Dual $40; Patron $75; Business $100; Life $500; Business Life $1,000.

NAPA FIREFIGHTERS MUSEUM, INC., 1201 Main St., Napa, CA 94559-2636. Tel.: 707-259-0609.
E-mail: info@napafirefightersmuseum.org
Web Site: www.napafirefightersmuseum.org
Founded: 1997.
Congressional District: 1
Institution Type/Description: Fire-Fighting Museum.
Collections: firefighting equipment; personal artifacts; photographs.
Hours & Admission Prices: Wed.-Sat. 11-4; other times by appointment. No charge.

Needles

HAVASU NATIONAL WILDLIFE REFUGE, 317 Mesquite Ave., Needles, CA 92363. Tel.: 760-326-3853.
Institution Type/Description: Wildlife Refuge.
Collections: natural history; wildlife & their habitats; plants; trees; flowers.
Facilities: nature trails.
Activities: hiking.
Hours & Admission Prices: Call for hours.

Nevada City

MINERS FOUNDRY CULTURAL CENTER, 325 Spring St., Nevada City, CA 95959-2420. Mailing Address: P.O. Box 1991, Nevada City, CA 95959-1940. Tel.: 530-265-5040. Fax: 530-265-5462.
E-mail: info@minersfoundry.org
Web Site: www.minersfoundry.org

Founded: 1989.
Congressional District: 2
Key Personnel: Dir., Gretchen Bond; Pres. (V), Dave Iorns.
Personnel Profile: Full-Time Paid 2; Part-Time Paid 3; Part-Time Volunteers 50; Interns 2.
Governing Authority: Parent Institution: Nevada County Cultural Preservation Trust. Tax-exempt.
Institution Type/Description: Mining Museum: site of the Pelton wheel.
Collections: use of original building as a foundry; Victorian artifacts.
Activities: lectures; theatre; concerts; weddings.
Hours & Admission Prices: Mon.-Fri. 10-4; docent tours available by appointment. No charge; donations accepted. &
Attendance: 50,000 (estimated)
Membership: Hearthstone $50; Cornerstone $150; Patron $300; Founders Circle $500; Lester Pelton Society $1,000.

NEVADA COUNTY HISTORICAL SOCIETY, INC., 214 Church St., Nevada City, CA 95959-1300. Mailing Address: P.O. Box 1300, Nevada City, CA 95959-1300. Tel.: 530-265-5910.
E-mail: info@nevadacountyhistory.org
Web Site: www.nevadacountyhistory.org
Founded: 1945.
Congressional District: 14
Key Personnel: Pres. (V), Alan de Negris; Cur. Mining Museum, Glenn Jones; Librarian, Searls Historic Library, Ed Tyson; Cur. Transportation Museum, Brian Blair; Cur. Firehouse Museum, Wally Hagaman.
Personnel Profile: Full-Time Volunteers 3; Part-Time Paid 1; Part-Time Volunteers 35.
Governing Authority: society; nonprofit corporation. Branch Museums: Firehouse No. 1 Museums, 214 Main St., Nevada City 95959. Tel. 530-265-5468; Mining Museum & Pelton Wheel Exhibit, Mill St., Grass Valley, CA 95945. Tel. 530-273-4255; Searls Historical Library, 214 Church St., Nevada City 95959. Tel. 530-265-5910; Transportation Museum, 5 Kidder Court, Nevada City 95959. Tax-exempt.
Institution Type/Description: Historical Society Museums.
Collections: cultural; Native American; Chinese; railroad; communications; transportation; mining; agriculture; period furnishings; genealogies; county records; photography; videos; gold rush era history; Cornish miners history; emigrant trail history.
Research Fields: genealogy; Gold Rush history.
Facilities: library of genealogical records, archives, county records & photographs.
Activities: lectures; guided tours of the area; demonstrations.
Publications: quarterly newsletter; quarterly bulletin; pamphlets; brochures; maps; books.
Hours & Admission Prices: Firehouse No. 1 Museums, Mining Museum & Pelton Wheel Exhibit: May to Oct. 15 10-5. Searls Memorial Library: Mon.-Sat. 1-4. Research: $20 hour. Transportation Museum: May-Oct. Fri.-Tues. 10-4; Nov.-April Sat.-Sun. 10-4. No charge; donations accepted. Closed New Year's Day; Thanksgiving; Christmas. &
Attendance: 2,000 (estimated)
Membership: Individual $25; Business $50.

Newbury Park

SATWIWA NATIVE AMERICAN INDIAN CULTURE CENTER, 4126 Potrero Rd., Newbury Park, CA 91320-5239. Mailing Address: 401 W. Hillcrest Dr., Thousand Oaks, CA 91360-4223. Tel.: 805-370-2301.
Web Site: www.nps.gov/samo
Key Personnel: Museum Shop Mgr., Razsa Cruz
Governing Authority: Subsidiary Institution: National Park Service, Santa Monica Mountains National Recreation Area. Tax-exempt.
Institution Type/Description: Native American Museum.
Collections: Chumash & Gabrielino/Tongva cultures.
Activities: Native American cultural workshops & programs.
Publications: catalog, Santa Monica Mountains National Recreation Area; catalog, Outdoors.
Hours & Admission Prices: Sat.-Sun. 9-5. No charge. &
Attendance: 12,000 (estimated)

STAGECOACH INN MUSEUM COMPLEX, (M), 51 S. Ventu Park Rd., Newbury Park, CA 91320-3943. Tel.: 805-498-9441. Fax: 805-498-6375.
E-mail: stagecoach@stagecoachmuseum.org
Web Site: www.stagecoachmuseum.org
Founded: 1967.
Congressional District: 24

Key Personnel: Pres., Elaine Williams; Dir., Sandra Hildebrandt; Cur. Education, Jackie Pizitz; Cur. History, Miriam Sprankling; Cur. Anthropology & Archaeology, Dr. Thomas Maxwell; Museum Shop Mgr., Constance Clarke; Museum Shop Mgr., Carol Salyer.
Personnel Profile: Part-Time Paid 2; Part-Time Volunteers 186.
Governing Authority: society. Parent Institution: Conejo Valley Historical Society. Subsidiary Institution: Conejo Recreation & Park. Tax-exempt: 501(c)(3).
Institution Type/Description: History Museum: housed in 1876 Grand Union Hotel.
Collections: local history & culture; Chumash native people, Spanish-Mexican & Anglo-American, from 1846-1930; shells; fossils; minerals. Tri-Village Complex: Chumash hut, Spanish adobe, pioneer house, Conejo Valley history & culture; Carriage House; replica of 1889 Timber School.
Research Fields: anthropology; local history; botany; prehistoric Indian culture; costumes; decorative arts & textiles.
Facilities: approx. 1,500-vol. library of pioneer & natural history including flora, fauna, geology & paleontology; Indian & decorative arts; nature trails. Handcrafted items for sale.
Activities: docent guided tours for visitors, elementary school classes & special groups; living history program; weddings.
Publications: Chumash Indian pamphlet; Newbury Park; Mad Agnes & Pierre: A Spirited History of the Stagecoach Inn Museum.
Hours & Admission Prices: Wed.-Sun. 1-4. Adults $4, senior citizens & youth 13-21 $3, children 5-12 $1; members & children under 5 no charge. Closed New Year's Day; Easter; Thanksgiving; Christmas. &
Attendance: 10,902 (accurate)
Membership: Children under 18 & Senior Citizens $15; Individual $20; Family $30; Organization (nonprofit) $40; Business & Supporter $50; Backer $75; Friend $100; Patron $150; Sponsor $250; Sustaining $500; Benefactor $1,000.

Newhall

PLACERITA CANYON NATURE CENTER, 19152 Placerita Canyon Rd., Newhall, CA 91321. Tel.: 661-259-7721.
E-mail: info@placerita.org
Institution Type/Description: Nature Center.
Collections: local history; plants; ecology; natural history.
Facilities: nature trails.
Activities: nature trails; live animal presentations; educational programs.
Hours & Admission Prices: Tues.-Sun. 9-5.

＊ **WILLIAM S. HART COUNTY PARK & MUSEUM, (M),** 24151 Newhall Ave., Newhall, CA 91321-2908. Tel.: 661-254-4584. Fax: 661-254-6499. Facebook: William S. Hart Park and Museum.
E-mail: information@hartmuseum.org
Web Site: www.hartmuseum.org
Founded: 1958.
Congressional District: 22
Key Personnel: Dir. Natural History Museum, Dr. Jane Pisano; Park Supt., Norman Phillips; Volunteer Coord., Rachel Barnes; Gift Store Mgr., Becki Basham; Admin., Margi Bertram.
Personnel Profile: Full-Time Paid 2; Part-Time Paid 2; Part-Time Volunteers 50.
Governing Authority: county. Parent Institution: L.A. County Dept. of Parks & Recreation. Subsidiary Institution: L.A. County Natural History Museum. Tax-exempt.
Institution Type/Description: Historic Building & Site: 1927 home of silent film star William S. Hart.
Collections: paintings including Russell, De Yong, J.M. Flagg, Remington; sculpture; woodcarvings; Navajo blankets & rugs; Native American artifacts; 1920s-1930s furniture, clothing & housewares; barnyard animals; bison herd.
Research Fields: William S. Hart; early filmmaking; Western film; period architecture & furnishings.
Activities: guided tours; permanent exhibitions.
Publications: brochures.
Hours & Admission Prices: Museum: mid-June to Labor Day Wed.-Sun. 11-3:30; Sept. to mid-June 10-12:30. Park: mid-June to Labor Day daily 8-6; Sept. to mid-June daily 8-5. No charge; donations accepted. Closed New Year's Day; Thanksgiving; Christmas. &
Attendance: 35,000 (accurate)
Membership: Senior Citizens $15; Volunteer $25; Individual $35; Family $40; Sponsor $75; Patron $125; Silver Benefactor $250; Gold Benefactor $500; Diamond Benefactor $1,000; Corporate Benefactor $2,500.

Newport Beach

EXPLOROCEAN/NEWPORT HARBOR NAUTICAL MUSEUM, 600 E. Bay Ave., Newport Beach, CA 92661. Tel.: 949-675-8915.
Web Site: www.explorocean.org
Formerly: Newport Harbor Nautical Museum
Founded: 1984.
Governing Authority: Tax-exempt.
Institution Type/Description: Nautical Museum.
Collections: maritime history & artifacts; hands-on exhibitions; ship models.
Activities: touch tank; educational programs; temporary exhibitions; rental facilities; special events; educational programs.
Hours & Admission Prices: Sun.-Thurs. 11-3, Fri.-Sat. 11-6, Sun. 11-5. Adults 13 & over $4, youth 4-12 $2; members, active military and children 3 & under no charge. &
Attendance: 42,000 (accurate)
Membership: Couple $50; Family $100.

ORANGE COUNTY MUSEUM OF ART, 850 San Clemente Dr., Newport Beach, CA 92660-6399. Tel.: 949-759-1122. Fax: 949-759-5623.
Web Site: www.ocma.net
Founded: 1918.
Congressional District: 34
Key Personnel: Dir., Dennis Szakacs; Chm. & Pres., Craig W. Wells; Dir. Museum Shop, Hayley Miller; Deputy Dir. Programs & Chief Cur., Dan Cameron; Dir. Public Programs & Education, Lisa Silagyi; Dir. Devel., Beth Bradley.
Personnel Profile: Full-Time Paid 40; Part-Time Paid 10; Part-Time Volunteers 150; Interns 10.
Institution Type/Description: Visual Art Museum.
Collections: 20th & 21st century modern & contemporary art.
Research Fields: modern & contemporary art.
Facilities: library of art books & catalogues available for use by request; auditorium; classrooms. Cards, magazines, jewelry, art books & posters for sale.
Activities: guided tours; lectures; educational programs.
Publications: quarterly newsletter; catalogs of current exhibits.
Hours & Admission Prices: Wed. & Fri.-Sun. 11-5, Thurs. 11-8. Adults $12, seniors & students $10; discounts to AAA, AAM & ICOM members; members, children under 12 & 2nd Sun. each month no charge. Closed Easter; Independence Day; Thanksgiving; Christmas. &
Attendance: 40,000 (accurate)
Membership: Member $60; Family Donor $100; Friend $250; Supporter $500; Advocate $1,500; Patron $2,500; Contributing Patron $5,000; Distinguished Patron $10,000.

Nipomo

DANA ADOBE, 671 S. Oakglen Ave., Nipomo, CA 93444-9009. Tel.: 805-929-5679.
E-mail: dana@danaadobe.org
Web Site: danaadobe.org
Founded: 1997.
Key Personnel: Exec. Dir., Kathy Kubiak; Pres. (V), Herb Kandel; Museum Shop Mgr., Helen Daurio.
Personnel Profile: Full-Time Paid 1; Part-Time Paid 1; Part-Time Volunteers 30.
Governing Authority: tax-exempt.
Institution Type/Description: Historic Home & Rancho; History Museum.
Collections: period artifacts.
Publications: newsletter.
Hours & Admission Prices: Sat.-Sun. 1-4. Adults $5; members & students no charge. &
Attendance: 2,000 (estimated)
Membership: Single $25; Family $40.

North Fork

SIERRA MONO INDIAN MUSEUM, 33103 Rd. 228, North Fork, CA 93643-9442. Tel.: 559-877-2115. Fax: 559-877-6515. Facebook: Sierra Mono Indian Museum.
E-mail: monomuseum@gmail.com
Web Site: www.sierramonomuseum.org
Founded: 1969.
Congressional District: 6
Key Personnel: Dir. & Museum Shop Mgr., Stephanie Clark; Pres. (V), Kelly Marshall.
Personnel Profile: Part-Time Paid 1; Part-Time Volunteers 15.

Governing Authority: Tax-exempt.
Institution Type/Description: Native American Museum.
Collections: North Fork Mono Tribe; fishing; hunting; cooking; healing; basketmaking; games; ceremonies; Tettleton wildlife collection; California Native American baskets; Crane Valley artifacts; animal dioramas; California Native Mono Tribe beadwork; Indian village & trail.
Research Fields: anthropology; California Native Mono Tribe.
Facilities: nature trail. Museum-related items for sale.
Activities: guided tours by appointment. Museum Sponsors: Pow Wow in August.
Publications: newsletter, Nuck-a-Hee.
Hours & Admission Prices: Tues.-Sat. 10-3:30. Suggested Donation: adults $7, seniors & children $5; discounts to AAM members. &
Attendance: 5,000 (estimated)
Membership: Youth $15; Individual $25; Business $150.

North Hollywood

CAMPO DE CAHUENGA, 3919 Lankershim Blvd., North Hollywood, CA 91604-3419. Mailing Address: P.O. Box 956, North Hollywood, CA 91603-0956. Tel.: 818-762-3998, ext. 2. Fax: 818-762-2734.
E-mail: campodecahuenga1847@hotmail.com
Web Site: www.campodecahuenga.com
Key Personnel: Dir., Deuk Perrin.
Governing Authority: Tax-exempt: 501(c)(3).
Institution Type/Description: History Museum: site of the signing of the Treaty of Cahuenga in 1847.
Collections: plaques; monuments; documents.
Activities: special programs; meetings.
Hours & Admission Prices: Sat. 11-3; call to confirm. Closed holidays.

THE PORTAL OF THE FOLDED WINGS SHRINE TO AVIATION AND MUSEUM, 10621 Victory Blvd., North Hollywood, CA 91606-3918. Tel.: 818-763-9121. Fax: 818-763-3801.
Web Site: www.portalofthefoldedwings.com
Founded: 1996.
Congressional District: 24
Key Personnel: Dir., John Torres; Gen. Mgr. Cemetery, Oliver Yeo.
Personnel Profile: Part-Time Paid 1; Part-Time Volunteers 6.
Governing Authority: nonprofit.
Institution Type/Description: Memorial Park.
Collections: burial place for 13 of America's most noted aviation pioneers from the Wright brothers to Lockheed Aircraft; photographs.
Research Fields: biographies of aviators buried at Portal.
Facilities: 36 sq. ft. exhibit space.
Activities: guided tours; lectures. Annual Event: Memorial Day Celebrations.
Hours & Admission Prices: Tours 1st Sun. of each month 1-3; other times by appointment. No charge. &
Attendance: 2,000 (estimated)

Northridge

ART GALLERIES, CALIFORNIA STATE UNIVERSITY, NORTHRIDGE, 18111 Nordhoff St., Northridge, CA 91330-8299. Tel.: 818-677-2226 (gallery) & 2156 (office). Fax: 818-677-5910.
Web Site: www.csun.edu/artgalleries/
Founded: 1972.
Congressional District: 21, 23 & 26
Key Personnel: Dir., Jim Sweeters; Exhibitions Coord., Michelle Giacopuzzi; Museum Shop Mgr., Dorothy Goggin.
Personnel Profile: Full-Time Paid 2; Part-Time Paid 3; Part-Time Volunteers 80; Interns 10.
Governing Authority: university; nonprofit organization. Parent Institution: California State University, Northridge. Volunteer Institution: Arts Council for California State University, Northridge. Tax-exempt.
Institution Type/Description: University Art Gallery.
Collections: works by student & established artists.
Research Fields: contemporary art; international art.
Facilities: classrooms; 3,200 sq. ft. exhibit space. Gallery-related items for sale.
Activities: guided tours; lectures; gallery talks; formally organized education programs for children & graduate students; docent program or council; training programs for professional museum workers.
Publications: exhibition catalogues; brochures.
Hours & Admission Prices: June-Aug. Mon.-Fri. 12-4; Sept.-May Mon.-Wed. & Fri.-Sat. 12-4, Thurs. 12-8. No charge; donations accepted. Closed holidays. &

Attendance: 32,000 (estimated)
Membership: Annual $30.

CAL STATE NORTHRIDGE BOTANIC GARDEN, Biology Dept., 18111 Nordhoff St. - MC 8303, Northridge, CA 91330. Tel.: 818-677-3496. Fax: 818-677-2034.
E-mail: botanicgarden@csun.edu
Web Site: www.csun.edu/botanicgarden
Founded: 1958.
Key Personnel: Chm. (V), Brenda Kanno.
Personnel Profile: Part-Time Volunteers 12.
Governing Authority: Parent Institution: California State University, Northridge Biology Department.
Institution Type/Description: Botanical Garden.
Collections: 3,000 plant species, CA Natives; cacti; tropical plants; native grasses; New Zealand plants; palms; herbs; redwood trees; orchids; bromeliads; ferns; pond; waterfall.
Facilities: 1.5-acre garden.
Activities: tours.
Hours & Admission Prices: Mon.-Fri. 8-4:45. No charge. Closed holidays. &
Attendance: 10,000 (estimated)
Membership: Individual $30; Family $40

Norwalk

HARGITT HOUSE, 12426 Mapledale, Norwalk, CA 90650-6026. Mailing Address: 12700 Norwalk Blvd., Rm. 10, Norwalk, CA 90650. Tel.: 562-864-9663 & 929-5566.
Institution Type/Description: Historic House Museum: housed in the home of Charles & Ida Hargitt, built in 1891 by the D.D. Johnston family.
Collections: period furnishings.
Hours & Admission Prices: 1st & 3rd Sun. of month 1-4.

Novato

HAMILTON FIELD HISTORY MUSEUM, 555 Hangar Ave., Novato, CA 94947. Tel.: 415-382-8614.
Founded: 2010.
Key Personnel: Contact Mgr., Ray Dwelly
Institution Type/Description: History Museum: housed on a former air base; operational from 1935-1974; first as an Army Air Corps field and later for the U.S. Air Force.
Collections: military artifacts; personal artifacts; photographs; uniforms; pins; medals; cockpit simulator; books; posters; videos.
Facilities: library.
Activities: research.
Hours & Admission Prices: Wed.-Thurs. & Sat. 12-4; other times by appointment. No charge. Closed holidays.

MUSEUM OF THE AMERICAN INDIAN, 2200 Novato Blvd., (in Miwok Park), Novato, CA 94947-2079. Mailing Address: P.O. Box 864, Novato, CA 94948. Tel.: 415-897-4064. Fax: 415-892-7804.
E-mail: office@marinindian.com
Web Site: marinindian.com
Formerly: Marin Museum Society
Founded: 1967.
Congressional District: 6
Key Personnel: C.E.O., Colleen Hicks; Pres. (V), Arthur Scott; Dir. Education, Alicia Retes.
Personnel Profile: Full-Time Paid 1; Part-Time Paid 1; Part-Time Volunteers 20; Interns 2.
Governing Authority: private; board of directors; nonprofit organization. Parent Institution: Marin Museum Society, Inc. Tax-exempt.
Institution Type/Description: Archaeological & Anthropological Museum: located on prehistoric site once occupied by Coast Miwok Indians.
Collections: archaeological, ethnographic & archival materials ranging from Alaska to Peru, pertaining to Native Americans of Marin County, California & the Western Americas; photographic materials pertaining to California Indians; original Edward S. Curtis photogravures.
Major Exhibits: Creation Stories, 11/13-5/14; Kay Dey YA, 6/14-12/14.
Research Fields: California Indian culture.
Facilities: non-circulating reference library; classroom; outdoor classroom; native plant garden; 35-acre park with nature trail. Books, & other museum-related gifts for sale.
Activities: tours; lectures; hands on education programs for pre-kindergarten thru adult ages; formally organized classes for children; family field trips; docent training program; intermuseum loans; permanent, temporary & traveling exhibitions; interpretative kit loans for schools & organizations.

Publications: bibliographies; curriculum materials; native plant garden guide; teachers Resource Pack; Precious Cargo.
Hours & Admission Prices: Tues.-Fri. 12-5, Sat.-Sun. 12-4. Adults $5; members no charge.
Attendance: 25,000 (estimated)
Membership: Individual $25; Family $35; Patron $100; Sponsor $250; Benefactor $1,000.

NOVATO HISTORY MUSEUM, 815 De Long Ave., Novato, CA 94945-7005. Mailing Address: Novato Historical Guild, P.O. Box 1296, Novato, CA 94948. Tel.: 415-897-4320.
Web Site: www.novato.org/museum
Founded: 1976.
Congressional District: 6
Key Personnel: Pres. (V), Thomas Keena; Museum Shop Mgr., Pat Johnstone.
Personnel Profile: Part-Time Volunteers 100.
Governing Authority: municipal; nonprofit. Parent Institution: City of Novato. Subsidiary Institution: Novato Historical Guild. Tax-exempt.
Institution Type/Description: History Museum: located in 1850 Postmaster's House, a Greek Revival two-story farmhouse typical of the era.
Collections: concentration on northern Marin County, California, particularly the Novato area.
Research Fields: oral history; video projects; historic architecture of downtown Novato; agriculture & ranching life in Novato; history & impact of Hamilton airfield.
Facilities: 240-vol. library of local newspapers & regional history; 1,200 sq. ft. exhibit space. Gift items for sale.
Activities: docent program; adult programming; formal education programs for children; guided tours; lectures; four general meetings with lecturers.
Publications: quarterly newsletter & historic journal, The Novato Historian; book, Novato Township.
Hours & Admission Prices: Wed.-Thurs. & Sat. 12-4; other times by appointment. No charge; donations accepted. Closed major holidays. &
Attendance: 2,500 (estimated)
Membership: Historical Guild: Student $10; Individual $20; Family $30; Business $50; Patron $100; Life $500.

Oak Glen

HISTORIC OAK GLEN SCHOOLHOUSE, 11911 S. Oak Glen Rd., Oak Glen, CA 92399-9488. Tel.: 909-797-1691. Facebook: Oak Glen School House Museum.
E-mail: oakglenschoolmuseum@gmail.com
Formerly: Oak Glen School House Museum
Founded: 1981.
Personnel Profile: Part-Time Paid 3.
Governing Authority: San Bernardino City Special District # 63. Tax-exempt.
Institution Type/Description: History Museum.
Collections: local history; vintage desks, books & teaching materials.
Facilities: gazebo pavillion; community room; park & playground.
Activities: concerts; educational events; research; facility rental.
Publications: quarterly newsletter.
Hours & Admission Prices: Weekends 12-4; Sept. to Thanksgiving week Wed.-Sun. 12-4; other times by appointment. No charge; donations accepted. Blue Star Museum.
Attendance: 5,000 (estimated)
Membership: Seniors & Students $10; Individual $15; Family $30; Supporting $50.

Oakdale

OAKDALE COWBOY MUSEUM, 355 E. F St., Oakdale, CA 95361-4084. Mailing Address: P.O. Box 1155, Oakdale, CA 95361-1155. Tel.: 209-847-5163. Fax: 209-847-4183.
E-mail: christie@oakdalecowboymuseum.org
Web Site: www.oakdalecowboymuseum.org
Founded: 1995.
Key Personnel: Exec. Dir., Christie Camarillo
Institution Type/Description: Cowboy History Museum.
Collections: history of local rodeo cowboys & ranchers; cowboy memorabilia; World Champion Rodeo.
Hours & Admission Prices: Mon.-Sat. 10-4. Adults $1; members no charge. Closed holidays. &
Attendance: 5,000 (accurate)
Membership: Individual $25; Family $50; Business $100.

OAKDALE MUSEUM & HISTORY CENTER, 212 W. F St., Oakdale, CA 95361. Mailing Address: P.O. Box 2212, Oakdale, CA 95361. Tel.: 209-847-9229.
Web Site: friendsofoakdaleheritage.org
Founded: 1985.
Congressional District: 18
Key Personnel: Pres. (V), Donald E. Riise; Vice Pres. & Museum Shop Mgr., Barbara Torres.
Personnel Profile: Part-Time Volunteers 7.
Governing Authority: municipal government. Parent Institution: City of Oakdale. Managed by Friends of Oakdale Heritage. Tax-exempt: 501(c)(3).
Institution Type/Description: Local History Museum: housed in 1869 residence.
Collections: Haidlen historical photographs, interviews & mementos; Sandl art; microfilm collection of Oakdale newspapers, 1881-present; obituaries 1884-present; tax records 1907-1951.
Research Fields: history of Oakdale & area; people; businesses; structures; events.
Facilities: 25-vol. library of local history books, available to the public; microfilm library: Oakdale newspapers, Stanislaus Co. census.
Activities: arts festivals; docent program; formal education programs for children; guided tours; temporary & traveling exhibitions. Annual Events: fundraisers in spring and fall.
Publications: quarterly newsletter, Museum Matters.
Hours & Admission Prices: Tues. & Thurs. 1-5. No charge; donations accepted. Closed major holidays. &
Attendance: 2,500 (estimated)
Membership: Adult $25; Couple $40; Corporate & Business $100.

Oakhurst

CHILDREN'S MUSEUM OF THE SIERRA, 49269 Golden Oak Dr., Oakhurst, CA 93644-8536. Mailing Address: P.O. Box 1200, Oakhurst, CA 93644-1200. Tel.: 559-658-5656. Fax: 559-658-5656.
E-mail: cmos@sti.net
Web Site: www.childrensmuseumofthesierra.org
Founded: 1997.
Key Personnel: Dir. & Chm. (V), Jean Hand; Education, Donna Marks; Public Rels., Ronda Clarke; Treas. & Cur., Jim Elliott; Registrar, Angelo Pizelo; Museum Shop Mgr., Lian Rausch.
Personnel Profile: Full-Time Volunteers 1; Part-Time Paid 2; Part-Time Volunteers 5.
Governing Authority: private; nonprofit organization. Parent Institution: Educational Enhancement Foundation. Tax-exempt: 501(c)(3).
Institution Type/Description: Children's Museum.
Collections: hands-on exhibits.
Facilities: 150-vol. library; 3,500 sq. ft. exhibit space; party room; theater. Museum-related items for sale.
Activities: arts festivals; hobby workshops; school loan service; rodeos; local events; travel trunk. Annual Events: Breakfast with Santa; Parking Lot Sale.
Publications: quarterly newsletter, The Museum Messenger.
Hours & Admission Prices: July-Aug. Tues.-Sat. 10-5; Sept.-June Tues.-Sat. 10-4, Sun. 1-4. Admission $4, seniors 60 & over $3; discounts to groups; children under 2 no charge. Closed Mother's Day; major holidays. &
Attendance: 9,813 (accurate)
Membership: Family $50; Contributor $50-$99; Patron $100-$199; Sponsor $200-$499.

FRESNO FLATS HISTORIC VILLAGE AND PARK, 49777 Rd. 427, Oakhurst, CA 93644. Mailing Address: P.O. Box 451, Oakhurst, CA 93644-0451. Tel.: 559-683-6570. Fax: 559-658-2161.
E-mail: fresnoflatsmuseum@sti.net
Web Site: fresnoflatsmuseum.org
Founded: 1968.
Congressional District: 19
Key Personnel: Pres., Sandy Edmondson; 1st Vice Pres., Jackie Byers; 2nd Vice Pres., Fran Metzler; Treas., Suzanne Harvey; Sec., Shirley Halderman; Ways & Means, Dick Killingsworth.
Personnel Profile: Full-Time Volunteers 40; Part-Time Paid 2; Part-Time Volunteers 15.
Governing Authority: private; nonprofit organization. Parent Institution: Sierra Historic Site Association, P.O. Box 451, Oakhurst, CA 93644. Tax-exempt: 501(c)(3).
Institution Type/Description: History Museum.
Collections: the period from the California Gold Rush to 1930 with emphasis on late 19th & early 20th-century in the foothills & mountains of central California; log house; two schoolhouses; blacksmith shop; two jailhouses.

Research Fields: life in the foothills & mountains of central California 1850-1930.
Facilities: 1,500-vol. library of local history, California history, national history & museum manuals; 134,000 sq. ft. exhibit space. Museum-related items for sale.
Activities: docent programs; guided tours; rental facilities. Museum Sponsors: Film Festival in April; Heritage Days in September; Christmas at the Flats in December.
Publications: quarterly newsletter, Fresno Flats Gazette.
Hours & Admission Prices: Museum: Mon.-Fri. 10-3, Sat.-Sun. 12-4. Library: Mon.-Fri. 10-3. Grounds: daily dawn to dusk. Museum & Tours: adults $4, children $2; members no charge. ♿
Attendance: 7,000 (estimated)
Membership: General $25; Bronze $50; Silver $300; Gold $500; Platinum $1,000.

KING VINTAGE CLOTHING MUSEUM, 40680 Hwy. 41, Oakhurst, CA 93644. Mailing Address: P.O. Box 303, Oakhurst, CA 93644-0303. Tel.: 559-658-6999. Fax: 559-683-3094.
E-mail: kingvintagemuseum@sti.net
Web Site: www.kingvintagemuseum.org
Founded: 2002.
Key Personnel: Chm. (V), Toni Lagunoff; Museum Shop Mgr., Mary Ann Hutcherson.
Personnel Profile: Part-Time Paid 1; Part-Time Volunteers 10.
Governing Authority: Tax-exempt.
Institution Type/Description: History Museum.
Collections: period clothing from the 1790 to 1990s including war memorabilia.
Hours & Admission Prices: Wed.-Sat. 11-5, Sun. 1-4. Adults $3. ♿
Attendance: 925 (estimated)
Membership: Individual $25; Family $50; Business $100; Life $500.

Oakland

AFRICAN-AMERICAN MUSEUM AND LIBRARY AT OAKLAND, 659 14th St., Oakland, CA 94612-1242. Tel.: 510-637-0200. Fax: 510-637-0204. TDD: 510-652-8634.
Web Site: www.oaklandlibrary.org
Founded: 1965.
Congressional District: 7
Key Personnel: C.E.O. & Pres. Trustees (V), Melvin Terry; Cur. & Museum Mgr., Rick Moss; Museum Cur. Asst., Erica L. Watkins; Librarian, Veronica L. Lee; Librarian, Linda Jolivet.
Personnel Profile: Full-Time Paid 7; Part-Time Paid 2.
Governing Authority: nonprofit organization. Parent Institution: Oakland Public Library. Tax-exempt: 501(c)(3).
Institution Type/Description: African-American History Museum.
Collections: California's African-American history; books; photographs; documents; artifacts; videotapes & audiotapes.
Research Fields: California history of African Americans; Western, California & Northern California Black History.
Facilities: 1,000-vol. library & archives of general Black history, available for research on premises during office hours; reading room.
Activities: guided tours; lectures; films; permanent & temporary exhibitions.
Hours & Admission Prices: Tues.-Sat. 12-5:30. ♿

THE CAMRON-STANFORD HOUSE PRESERVATION ASSOCIATION, 1418 Lakeside Dr., Oakland, CA 94612-4307. Tel.: 510-444-1876 & 874-7802 (office). Fax: 510-874-7803.
E-mail: pelican@cshouse.org
Web Site: www.cshouse.org
Founded: 1971.
Congressional District: 8
Key Personnel: Vice Pres. Collections & House Committee, Frankie Rhodes; Vice Pres. (V) House Administrative, Elaine O. Oldham; Consulting Cur., Wayne Mathes.
Personnel Profile: Part-Time Volunteers 20.
Governing Authority: nonprofit organization. Parent Institution: Camron-Stanford House Preservation Assoc. Tax-exempt: 501(c)(3).
Institution Type/Description: Historic House & Preservation Project: 1876 four-story Victorian House, former Oakland Museum.
Collections: period rooms; paintings; sculpture; 1875-85 decorative arts & furnishings; 19th-century women's needle & craft work; 19th-century artifacts, history & household items; photographs.
Research Fields: 1875-1885 decorative arts; 19th-century gardens, construction, craft techniques, customs, living styles, parties & Christmas practices.
Facilities: 200-vol. library of books, magazines, slides & slide tapes pertaining to all aspects of the Victorian era available for research by special request; reading room; meeting room; reception facilities. Museum-related items for sale.
Activities: guided tours; lectures; films; docent program & council; formally organized education programs for children; permanent & temporary exhibitions; rental facilities.
Publications: Camron-Stanford House Newsletter; brochure.
Hours & Admission Prices: 3rd Wed. of month 1-5. Adults $5, senior citizens $4, juniors 12-18 $3; children under 12, school tours and 1st Sun. no charge.
Attendance: 500 (estimated)
Membership: Student $10; Individual $35; Family $50; Preservationist $100; Victorian $150; Italianate $250.

CHABOT SPACE & SCIENCE CENTER, 10000 Skyline Blvd., Oakland, CA 94619-2450. Tel.: 510-336-7373. Fax: 510-336-7491.
E-mail: visit@chabotspace.org
Web Site: www.chabotspace.org
Formerly: Chabot Observatory & Science Center
Founded: 1883.
Congressional District: 9
Key Personnel: C.E.O. & Dir., Alexander Zwissler; Chm. (V), Michael Levi; Museum Shop Mgr., Lara Miranda.
Personnel Profile: Full-Time Paid 54; Part-Time Volunteers 180; Interns 6.
Governing Authority: Tax-exempt.
Institution Type/Description: Science Center.
Collections: astronomy; space sciences; climate & energy sciences; technology; environment.
Major Exhibits: Tales of the Maya Skies, 11/10-11/14; Beyond Blastoff: Living and Walking in Space, 11/10-11/14; Destination Universe, 11/10-11/14; Bill Nye's Climate Lab, 11/10-11/15; Touch the Sun, 1/13-1/15.
Research Fields: astronomy; climate change.
Facilities: classrooms; theater; digital planetarium; three observatories. Museum-related items for sale.
Activities: simulated space mission; hands-on exhibits & activities; telescope viewing; classes.
Publications: e-Voyager.
Hours & Admission Prices: Summer: Tues.-Thurs. & Sun. 10-5, Fri.-Sat. 10-10; Winter: Wed.-Thurs. & Sun. 10-5, Fri.-Sat. 10-10. Telescope Viewing: dusk to 10:30. Adults $15.95, youth 3-12 $11.95; discounts to military; children under 3 & members no charge. ASTC reciprocal member's program. Closed Thanksgiving; Christmas. ♿
Attendance: 165,000 (accurate)
Membership: Educator $59; Earth Active & Earth Parent $75; Earth Family $99; Earth Family Plus & Smithsonian Partnership $125; Earth Orbiter $150; Neptune $250; Saturn $500; Jupiter $750.

COHEN-BRAY HOUSE, 1440 29th Ave., Oakland, CA 94601-2309. Tel.: 510-536-1703.
Institution Type/Description: Historic House Museum.
Collections: personal artifacts; period furnishings.
Activities: special events.
Hours & Admission Prices: 4th Sun. of month 2pm by appointment.

DUNSMUIR HELLMAN HISTORIC ESTATE, 2960 Peralta Oaks Ct., Oakland, CA 94605-5320. Tel.: 510-619-5994. Fax: 510-562-8294.
E-mail: jimd@dunsmuir.org
Web Site: www.dunsmuir.org
Founded: 1971.
Congressional District: 9
Key Personnel: Chm. (V), Dinah Benson; Dir., Jim DeMersman; Museum Shop Mgr., Debra Bresso.
Personnel Profile: Full-Time Paid 5; Part-Time Paid 10; Part-Time Volunteers 65; Interns 2.
Governing Authority: Tax-exempt.
Institution Type/Description: Historic Estate: a 37-room Neoclassical Revival mansion built in 1899.
Collections: furnishings; personal artifacts. Historic Buildings: milk barn; carriage house; 1928 residence; chauffeur cottage.
Research Fields: Edwardian life in America; Jewish American experience in the West.
Facilities: 50 acres. Museum-related items for sale.
Activities: Museum Sponsors: Family First Sundays; Scottish Highland Games; Oktoberfest & Pumpkin Patch; Summer Movie Under the Stars; Holidays at Dunsmuir in December.
Publications: quarterly newsletter.
Hours & Admission Prices: Grounds: Mon.-Fri. 10-4. Mansion Tours: April 5-Sept. 27 Wed. 11am. Adults $5, seniors $4; discounts to AAM members; children 13 & under no charge.

Attendance: 65,125 (accurate)
Membership: Individual $40; Family $60; Century Club $100; Peralta Club $250; Dunsmuir Club $500; Hellman Club $1,000.

JUNIOR CENTER OF ART AND SCIENCE, Lakeside Park, 558 Bellevue Ave., Oakland, CA 94610-5026. Tel.: 510-839-5777. Fax: 510-839-8102.
E-mail: jrcenter@sbcglobal.net
Web Site: www.juniorcenter.org
Founded: 1954.
Congressional District: 9
Key Personnel: Exec. Dir., Tammara Katsikas; C.E.O. & Pres. (V), Stacy Chretien; Cur., Daniel McClain.
Personnel Profile: Full-Time Paid 3; Part-Time Paid 9; Part-Time Volunteers 250; Interns 2.
Governing Authority: Tax-exempt.
Institution Type/Description: Children's Museum.
Collections: hands-on exhibits; California Native American tools, games, baskets, & artifacts.
Activities: workshops; demonstrations.
Hours & Admission Prices: Summer: Mon.-Fri. 8:30-4:30; Academic Year: Tues.-Fri. 10-6, Sat. 10-3; see website for holiday hours. No charge; donations accepted. &

KENNEDY ART CENTER GALLERY AT HOLY NAMES UNIVERSITY, 3500 Mountain Blvd., Oakland, CA 94619-1627. Tel.: 510-436-1457.
Institution Type/Description: Art Gallery.
Collections: paintings; music recordings; art history slides.
Facilities: classrooms; theater.
Activities: performances.
Hours & Admission Prices: Sat.-Sun. 12-5; other times by appointment.

MERRITT MUSEUM OF ANTHROPOLOGY, 12500 Campus Dr., Library Bldg., Oakland, CA 94619-3107. Tel.: 510-436-2607. Fax: 415-922-0905.
Founded: 1973.
Congressional District: 7
Key Personnel: Dir., Dr. Barbara Joans; Cur. Conservation, Leslie Fleming; Cur. & Museum Assoc., Lisa Valcenier.
Governing Authority: college. Parent Institution: Peralta Community College District. Tax-exempt.
Institution Type/Description: Anthropology Museum.
Collections: ethnological material from North & South America, Africa, the Pacific & Asia; prehistoric collection; European collection.
Research Fields: contemporary United States, South Africa & Mexico.
Facilities: classrooms.
Activities: guided tours; lectures; films; formally organized education programs for undergraduate college students; training programs for paraprofessional museum workers; loan, temporary & permanent exhibits. Museum Sponsors: satellite exhibits at various locations in surrounding communities.
Hours & Admission Prices: Mon.-Thurs. 7:45-7, Fri. 7:45-3. No charge. Closed academic holidays. &
Attendance: 300 (estimated)
Membership: Individual $5.

MILLS COLLEGE ART MUSEUM, (M), 5000 MacArthur Blvd., Oakland, CA 94613-1302. Tel.: 510-430-2164. Fax: 510-430-3168.
E-mail: museum@mills.edu
Web Site: mcam.mills.edu
Founded: 1925.
Congressional District: 8
Key Personnel: Dir., Dr. Stephanie Hanor; Mgr. Exhibitions & Collections, Stacie Daniels; Program Mgr., Lori Chinn.
Personnel Profile: Full-Time Paid 3; Full-Time Volunteers 0; Part-Time Volunteers 1; Interns 10.
Governing Authority: college. Parent Institution: Mills College. Tax-exempt: 501(c)(3).
Institution Type/Description: College Art Gallery.
Collections: international prints & drawings; Asian & Latin American textiles; 20th-century American ceramics; Japanese ceramics; California regionalist paintings; contemporary photography & new media.
Research Fields: art of the San Francisco Bay Area.
Activities: lectures; workshops; exhibition & collection tours.
Publications: exhibition brochures; catalogues.
Hours & Admission Prices: Tues. & Thurs.-Sun. 11-4, Wed. 11-7:30. No charge; donations accepted. &
Attendance: 10,100 (accurate)

MUSEUM OF CHILDREN'S ART - MOCHA, 538 Ninth St., Oakland, CA 94607. Tel.: 510-465-8770. Fax: 510-465-0772.
E-mail: hello@mocha.org
Web Site: mocha.org
Founded: 1990.
Institution Type/Description: Children's Museum.
Collections: hands-on exhibitions.
Activities: special events; educational programs; rental facilities.
Hours & Admission Prices: Tues.-Fri. 10-3, Sat.-Sun. 12-4.

OAKLAND AVIATION MUSEUM, 8252 Earhart Rd., Oakland, CA 94621-4548. Tel.: 510-638-7100.
E-mail: oamdirector@att.net
Web Site: www.oaklandaviationmuseum.org
Formerly: Western Aerospace Museum
Founded: 1981.
Congressional District: 9
Key Personnel: Dir. Operations, Ian Wright; Pres., John Horton.
Personnel Profile: Full-Time Paid 1; Part-Time Volunteers 54; Interns 2.
Governing Authority: private; not-for-profit. Tax-exempt: 501(c)(3).
Institution Type/Description: Aviation History Museum; housed in 1939 Boeing School of Aeronautics hangar on historic Oakland Airport, North Field.
Collections: aircraft; furnishings & personal artifacts of aviation; photographs; prints & films; history of Oakland Airport 1922-1965.
Research Fields: history of aviation in San Francisco Bay area & northern California.
Facilities: 14,000-vol. library on aviation; 30,500 sq. ft. exhibit space; classroom. Books, t-shirts, postcards, model aircraft, toys & other museum-related items for sale.
Activities: guided tours; lectures; loan exhibitions; films; docent programs; speaker series; members' meeting. Annual Events: Aviation Authors Book Signing; Open Cockpit Days.
Publications: quarterly newsletter, Chatter from the Flight Line.
Hours & Admission Prices: Wed.-Sun. 10-4. Adults $10, senior citizens 55 & over $9, teens 13-18 $7, children 6-12 $5; discounts to groups & Time Travelers; children 5 & under and members no charge. Closed New Year's Eve & Day; Thanksgiving; Christmas Eve & Day. &
Attendance: 7,000 (estimated)
Membership: Individual $35; Family $60; Golden Gater $100; Life $600.

* **OAKLAND MUSEUM OF CALIFORNIA, (M),** 1000 Oak St., Oakland, CA 94607-4892. Tel.: 510-318-8400; 888-625-6873 (Toll Free). Facebook: Oakland Museum of California; TDD: 510-451-3322.
E-mail: equist@museumca.org
Web Site: www.museumca.org
Founded: 1969.
Congressional District: 8 & 9
Key Personnel: Chm., Mike Moye; Exec. Dir., Lori Fogarty; Sr. Cur. Art, Rene De Guzman; Sr. Cur. History, Louise Pubols; Sr. Cur. Natural Sciences, Douglas Long.
Personnel Profile: Full-Time Paid 105.
Governing Authority: partnership. Parent Institution: Oakland Museum of California Foundation. Tax-exempt: 170(b)(1)(A); 501(c)(3).
Institution Type/Description: Regional Multidisciplinary Museum.
Collections: California environment, history & art; archives, fine art & historical photography; paintings, sculpture; fine crafts; decorative arts; aquatic biology; regional ecology; California Library of Natural Sounds; California Indian artifacts; California history & cultures from prehistory to present.
Major Exhibits: Peter Stackpole: Bridging the Bay, 7/13-1/14; Above & Below: Stories from our Changing Bay, 8/13-2/14; The Smallest of Worlds, 10/13-7/14; Giant Robot, 4/14-7/14.
Research Fields: California landscape & contemporary art, history, historic costume & textiles, nature sounds, biotic zones & ecology; the arts & crafts movement; Dorothea Lange photo collection.
Facilities: gardens; theater; lecture hall; classroom; restaurant. Museum-related items for sale.
Activities: guided tours; lectures; gallery talks; concerts; plays, films; ethnic festivals; formally organized education programs for children; docent program; museum-school partnerships; youth interpreter & junior docent programs; inter-museum loan; permanent, temporary & traveling exhibitions; school loan service; teacher training; travel program. Museum Sponsors: Earth Day Fest; Annual Wildflower Show.
Publications: 4-5 annual catalogs; quarterly magazine, Inside/Out; brochures.
Hours & Admission Prices: Wed.-Thurs. & Sat.-Sun. 11-5, Fri. 11-9. Adults $15, seniors & students $10, youth 9-17 $6; discounts to non-Oakland

school groups, AAM members & Fri. 5-9; children 8 & under & members no charge. Closed New Year's Day; Independence Day; Thanksgiving; Christmas. &
Attendance: 135,000 (accurate)
Membership: Golden State $50; Senior Citizen, Student & Educator $55; Dual Senior $65; Individual $75; Family $95; Supporter $175; Sponsor $300; Patron $600; Donor Forum $1,250 & up.

OAKLAND ZOO, 9777 Golf Links Rd., Oakland, CA 94605-4925. Mailing Address: P.O. Box 5238, Oakland, CA 94605-0238. Tel.: 510-632-9525, ext. 132. Fax: 510-635-5719.
E-mail: nancy@oaklandzoo.org
Web Site: www.oaklandzoo.org/
Founded: 1922.
Congressional District: 9
Key Personnel: Pres. & C.E.O., Dr. Joel J. Parrott, D.V.M.; C.F.O., Carl Nichols; Dir. Animal Care, Conservation & Research, Colleen Kinzley; Mng. Dir., Nancy Filippi; Dir. Strategic Initiatives, Nik Dehejia; Dir. Operations, Deb Menduno; Dir. Devel., Emma Lee Twitchell; Dir. Park Svcs., Bob Westfall; Supervising Mgr. Retail, F&B, Moe Perez.
Personnel Profile: Full-Time Paid 109; Part-Time Paid 17; Part-Time Volunteers 375.
Governing Authority: municipal; nonprofit organization. Operated by East Bay Zoological Society; development through the East Bay Zoological Society. Tax-exempt: 501(c)(3).
Institution Type/Description: Adult & Children's Zoo.
Collections: over 660 native & exotic animals.
Research Fields: California Condor Recovery Program; Western Pond Turtle Project; Amboseli Trust for Elephants; Budongo Snare Removal Project; Bay Area Puma Project; ARCAS Wild Animal & Rescue Rehabilitation Center; Animals Asia.
Facilities: 150-vol. library; educational facilities; amphitheater; botanical garden; 60 acres exhibit space; snack shop. Zoo-related items for sale.
Activities: guided tours; lectures; formal educational programs; docent program; Zoomobile programs; members nights. Annual Events: Walk in the Wild; Fundraiser; Earth Day; Celebrating Elephants; Discovering Primates; Boo at the Zoo; Feasts for the Beasts; Zoolights.
Publications: quarterly newsletter, Roar; annual report.
Hours & Admission Prices: Daily 10-4. Adults 15-55 $13.75, children 2-14 $9.75; discounts to groups; members no charge. Parking: bus $10, car $7. Closed Thanksgiving; Christmas. &
Attendance: 650,000 (estimated)
Membership: Student & Senior $60; Individual $70; Dual Senior $85; Individual Plus & Single Parent $90; Grandparents $95; Family $99; Family Plus $125; Sustaining $150; Contributing $250; Sponsor $500.

THE PARDEE HOME MUSEUM, 672 11th St., Oakland, CA 94607-3651. Tel.: 510-444-2187.
E-mail: info@pardeehome.org
Web Site: www.pardeehome.org
Founded: 1982.
Key Personnel: Chm. (V), Kay Cheatham; Chm. (V), Dr. Ron Bachman; Pres. (V), Ann Brown; Museum Shop Mgr., Maria Vermiglio.
Personnel Profile: Part-Time Volunteers 43.
Governing Authority: nonprofit organization. Tax-exempt: 501(c)(3).
Institution Type/Description: Historic House: c.1868 former residence of Gov. George Pardee.
Collections: 1850-1980 Pardee family furniture; ethnography; archives; photographs.
Facilities: 165-vol. library of collections management; historic preservation; collections research & identification available for use by the public; archival research room; 5,000 sq. ft. exhibit space.
Activities: docent program; guided tours. Museum Sponsors: Dessert and High Teas. Museum-related items for sale.
Publications: newsletter, Pardee Home Museum.
Hours & Admission Prices: Tours: Wed. & 2nd Sat. of month at 10:30, 2nd Sun. of month at 2; groups by appointment. Adults $5; children under 12 & members no charge. Tour & High Tea: $25. &
Attendance: 905 (accurate)
Membership: Student $15; Individual $35; Family $50; Benefactor $75; Patron $100-$249; Sustaining $250-$499; Sponsor $500-$999; Visionary $1,000 & up.

PEERLESS COFFEE & TEA COMPANY MUSEUM, 260 Oak St., Oakland, CA 94607-4587. Tel.: 800-310-KONA; 410-763-1763. Fax: 510-763-5026.
Founded: 2001.
Key Personnel: Pres. & Cur., Sonja Vukasin
Institution Type/Description: Company Museum.

Collections: coffee memorabilia; period coffee grinders; tins & tea bins; jewelry; Vukasin family history; 1922 Model T Ford Depot Hack; Peerless Coffee Company & Tea heritage.
Facilities: Museum-related items for sale.
Hours & Admission Prices: Mon.-Fri. 10-4; groups by appointment. No charge. &

PRO ARTS, 150 Frank H. Ogawa Plaza, Oakland, CA 94612. Tel.: 510-763-4361. Fax: 510-763-9470. Facebook: Pro Arts.
E-mail: info@proartsgallery.org
Web Site: www.proartsgallery.org
Founded: 1974.
Congressional District: 9
Key Personnel: Exec. Dir., Margo Dunlap
Institution Type/Description: Art Gallery.
Collections: works by contemporary artists.
Activities: special events; temporary exhibitions; artist series; arts education; arts advocacy; East Bay open studios.
Publications: annual, Directory of East Bay Arts.
Hours & Admission Prices: Tues.-Fri. 10-5, Sat. 11-4. No charge; donations accepted. &
Membership: Student & Senior $20; Artist $45.

ROTARY NATURE CENTER, 600 Bellevue Ave., Oakland, CA 94610-5000. Tel.: 510-238-3739. Fax: 510-238-7962.
E-mail: sbenavidez@oaklandnet.com
Web Site: www.lakemerritt.org
Founded: 1953.
Congressional District: 8
Key Personnel: Supervising Naturalist, Stephanie Benavidez.
Personnel Profile: Full-Time Paid 2; Part-Time Paid 13; Part-Time Volunteers 10.
Governing Authority: municipal. Parent Institution: Oakland Office of Parks & Recreation. Tax-exempt.
Institution Type/Description: Natural History Museum.
Collections: mounted birds; live specimens; mammals; reptiles; study skins.
Research Fields: water birds; aquatic marine life; botanical.
Facilities: approx. 600-vol. library of books on Western U.S. natural sciences available for use on premises; reading room; nature center; wildlife refuge; 75-seat auditorium.
Activities: guided tours; classes in nature & ecology; films; walks; nature art & day camps; school loan service. Center Sponsors: wildlife & animal demonstrations.
Publications: bimonthly program of activities; brochures.
Hours & Admission Prices: Daily & holidays 10-5. Feedings 3:30pm. No charge; donations accepted. &
Attendance: 35,000 (estimated)

Oceanside

CALIFORNIA SURF MUSEUM, (M), 312 Pier View Way, Oceanside, CA 92054. Tel.: 760-721-6876.
E-mail: csm@surfmuseum.org
Web Site: www.surfmuseum.org
Founded: 1986.
Key Personnel: Pres. (V), Jim Kempton; Operations & Museum Shop Mgr., Sam Zuegner.
Personnel Profile: Full-Time Paid 1; Part-Time Paid 3; Part-Time Volunteers 30; Interns 4.
Governing Authority: nonprofit organization. Tax-exempt: 501(c)(3).
Institution Type/Description: Surf Museum.
Collections: historically significant surfboards; surf equipment; photographs; clothing; art; posters; vintage film; surf-related board games; magazines; trophies; skateboards.
Research Fields: surfing; California beach culture.
Facilities: rental facilities. Museum-related items for sale.
Activities: surf film festival; exhibit opening parties; VIP events; concerts; book signings; lectures. Annual Events: an exhibit highlighting one pioneering legend of surfing; Gala Fundraiser; Legends Day.
Publications: quarterly newsletter, Outside.
Hours & Admission Prices: Thurs. 10-8, Fri.-Wed. 10-4. Adults $5, seniors 62 & over, military with ID and students with ID $3; members & Tues. no charge. &
Attendance: 30,000 (accurate)
Membership: Gremmie & Legend $25; Surfer $50; Ohana $75; Malibu Chip $100, $250, $500.

MISSION SAN LUIS REY MUSEUM, 4050 Mission Ave., Oceanside, CA 92057-6402. Tel.: 760-757-3651. Fax: 760-757-4613.
E-mail: museumdesk@sanluisrey.org
Web Site: www.sanluisrey.org
Founded: 1798.
Congressional District: 40
Key Personnel: Admin., Edward Gabarra; Exec. Dir., Rev. David Gaa, O.F.M.; Historic Church & Museum Coord., Raffaella Avolio; Museum Assoc. & Tour Coord., Beverly Perna.
Personnel Profile: Full-Time Paid 2; Part-Time Paid 1; Part-Time Volunteers 21.
Governing Authority: Parent Institution: Franciscan Friars of the Santa Barbara Province. Subsidiary Institution: Old Mission San Luis Rey, Inc. Tax-exempt: 501(c)(3).
Institution Type/Description: Historic Building & Site: 1798 San Luis Rey Mission.
Collections: Native American baskets; decorative & fine arts from Spanish occupation; Spanish, Mexican American & Native American ecclesiastical furnishings; church vestments; polychrome statues; books from mission period; religious vessels.
Research Fields: Spanish colonial mission period in Alta California.
Facilities: picnic grounds; meeting rooms available. Native American, mission history & religious articles for sale.
Activities: guided tours to groups with reservations.
Publications: Mission News.
Hours & Admission Prices: Mon.-Fri. 9:30-5; Sat-Sun. 10-5. Adult $4, students 6-18 & retreatants $3; children under 6 & active military no charge. Behind the Scenes Guided Tours: adults $12. Closed New Year's Day; Easter; Thanksgiving; Christmas. &
Attendance: 55,000 (accurate)
Membership: Individual $50; Dual/Family $85; Franciscan Circle $250; Patron $500; Serra Circle $1,500; Sponsor $2,500 & up.

OCEANSIDE MUSEUM OF ART, (M), 704 Pier View Way, Oceanside, CA 92054-2802. Tel.: 760-435-3720. Fax: 760-966-5819.
E-mail: danielle@oma-online.org
Web Site: oma-online.org
Founded: 1995.
Congressional District: 48
Key Personnel: Exec. Dir., Daniel Foster; Dir. Exhibitions & Collections, Danielle Susalla Deery; Mgr. Facilities, Lissette Guzman; Mgr. Programs, Mitzi Summers; Membership Mgr., Teresa Ellis; Deputy Dir., Tara Smith; Education Coord., Julia Fister; Graphic Designer & Weekend Supvr., Erika Koga; Museum Shop Mgr., Nancy Bergmann.
Personnel Profile: Full-Time Paid 5; Part-Time Paid 4; Part-Time Volunteers 50; Interns 4.
Governing Authority: private; nonprofit organization. Tax-exempt: 501(c)(3).
Institution Type/Description: Art Museum.
Collections: regional art from 1900 to present.
Major Exhibits: Charles Arnoldi, 8/13-1/14.
Research Fields: regional art.
Facilities: 9,000 sq. ft. exhibit space. Gift items for sale.
Activities: docent program; tours to other museums & cultural destinations; lectures; temporary & permanent exhibitions. Annual Events: Dinner Dance/Formal; receptions; Jazz at the Museum; Mardi Gras; Museum Ball; Art After Dark; Free Family Art Days; Artists@Work.
Publications: newsletter; catalog, James Hubbell Retrospective; Chouinard: A Living Legacy; Masterpieces of San Diego Painting: Fifty Works from Fifty Years; Damngorgeous: A Daughters Memoir of Millard Owen Sheets; Art of the WPA Era; Worn With Pride: Celebrating Samoan Artistic Heritage; Deloss McGraw: The Circus Loosely; Ethel Green: Surrealist Painter; Convergence: The Studio Furniture of Tasmania and America; James Hubbell: Meditations on Nature and Life.
Hours & Admission Prices: Tues.-Sat. 10-4, Sun. 1-4. Adults $8; discounts to AAM & ICOM members; museum staff members, members, students, and active military & their dependents, 1st Sun. each month no charge. Closed major holidays. &
Attendance: 53,000 (accurate)
Membership: Student, Military & Educator $30; Individual $45; Dual & Family $60; Patron $150; President's Circle $500; Founders Circle $1,000; Millennium Club $2,500.

Ocotillo

IMPERIAL VALLEY DESERT MUSEUM, (M), 11 Frontage Rd., Ocotillo, CA 92244. Mailing Address: P.O. Box 430, Ocotillo, CA 92259.
Institution Type/Description: History Museum.

Collections: local history; Native American artifacts; archaeology; photographs; period artifacts.
Activities: educational programs.
Hours & Admission Prices: Thurs.-Sat. 10-3.

Ojai

BEATRICE WOOD CENTER FOR THE ARTS, 8560 Ojai-Santa Paula Rd., Ojai, CA 93023-9351. Tel.: 805-646-3381. Fax: 805-646-0560.
Web Site: www.beatricewood.com
Key Personnel: Dir., Kevin Wallace
Institution Type/Description: Art Gallery.
Collections: works by Beatrice Wood; sculpture; paintings.
Activities: workshops; classes; special events; public programs.
Hours & Admission Prices: Fri.-Sun. 11-5.

OJAI VALLEY MUSEUM OF HISTORY AND ART, 130 W. Ojai Ave., Ojai, CA 93023-3212. Mailing Address: P.O. Box 204, Ojai, CA 93024-0204. Tel.: 805-640-1390. Fax: 805-640-1342.
E-mail: ojaivalleymuseum@ojai.net
Web Site: www.ojaivalleymuseum.org
Formerly: Ojai Valley Historical Society and Museum
Founded: 1966.
Key Personnel: Museum Dir., Michelle Ellis Pracy; Pres. (V) Bd. Trustees, Ann Scanlin; Museum Shop. Mgr., Bobbie Boschan; Administrative Asst., Jill Helson; Design Svcs., Fred Kidder.
Personnel Profile: Full-Time Paid 1; Part-Time Paid 2; Part-Time Volunteers 110; Interns 2.
Governing Authority: Tax-exempt.
Institution Type/Description: Ojai History, Art & Culture Museum.
Collections: concentration on Ojai Valley; pioneer settlers; natural history; historical & contemporary art & culture.
Research Fields: history of Ojai California & Ojai Valley.
Facilities: research library. Museum-related items for sale.
Activities: lectures & family programs; Chumash day camp; permanent & temporary exhibitions; town talks; field trips; school traveling exhibit program.
Publications: rack card
Hours & Admission Prices: Tues.-Wed. & Sat. 10-4, Thurs.-Fri. 1-4, Sun. 12-4. Adults $4, children 6-18 $1; discounts to NARM members; members & children under 6 no charge. Closed New Year's Day; Independence Day; Thanksgiving; Christmas. &
Attendance: 10,000 (estimated)
Membership: Individual $35; Family $50; Business & Associate $100; Suporter $250; Donor $500.

Old Sacramento

OLD SACRAMENTO STATE HISTORIC PARK, 111 I St., Old Sacramento, CA 95814-2204. Tel.: 916-445-7387. Fax: 916-327-5655. TDD: 916-324-2667.
E-mail: csrmlibrary@csrmf.org
Web Site: www.csrmf.org
Founded: 1976.
Congressional District: 3
Key Personnel: Dir., Paul J. Hammond; Chief Cur., Stephen E. Drew; Dir. Collections, Ellen L. Halteman; Maintenance Chief, Dennis Pruitt.
Personnel Profile: Full-Time Paid 40; Part-Time Paid 15; Part-Time Volunteers 900.
Governing Authority: state. Affiliated with California State Parks, P.O. Box 942896 Sacramento, CA 94296. Parent Institution: California State Railroad Museum. Tax-exempt.
Institution Type/Description: State Park Museum: located in Old Sacramento State Historic Park.
Collections: original State Supreme Court Chambers & Wells Fargo Bank; artifacts of 1st transcontinental telegraph & western terminus of the Pony Express; personal artifacts; period furnishings; Pony Express monument.
Facilities: theatre.
Activities: guided tours; talks to groups; permanent exhibits; docent program.
Hours & Admission Prices: Daily 10-5. Call for schedule of activities. No charge. Closed New Year's Day; Thanksgiving; Christmas. &
Attendance: 60,000 (estimated)

Ontario

CHAFFEY COMMUNITY MUSEUM OF ART, (M), 217 S. Lemon Ave., Ontario, CA 91761. Tel.: 909-463-3733. Facebook: Chaffey Community Museum of Art.
E-mail: info@chaffeymuseum.org
Web Site: www.chaffeymuseum.org
Founded: 1941.
Congressional District: 26
Key Personnel: Pres. (V), Nancy DeDiemar.
Personnel Profile: Part-Time Paid 1.
Volunteer Hours: 1,500
Operating Expenses: 29,800
Operating Income: 33,000
Institution Type/Description: Art Museum.
Collections: paintings.
Major Exhibits: Paperless Society, 1/14-2/14; China: Inspired Interpretations, 1/14-2/14; California Watercolorist, 1/14-5/14; Milford Zornes, 2/14-3/14; Kirk Kain, 2/14-4/14; Order Out of Chaos, 3/14-4/14; Premium Blend, 5/14-6/14; On Target, 5/14-6/14; Robert George, 5/14-9/14; Burnt Offerings, 10/14-12/14.
Publications: newsletter.
Hours & Admission Prices: Thurs.-Sun. 12-4. No charge; donations accepted.
Attendance: 2,800 (accurate)
Membership: Senior & Student $35; Individual $40; Artist $45; Family $60; Patron $125; Sponsor $250; Benefactor $500; Corporate $1,000; Life $2,500.

GRABER OLIVE HOUSE MUSEUM, 315 E. 4th St., Ontario, CA 91764-2709. Mailing Address: P.O. Box 511, Ontario, CA 91762-8511. Tel.: 909-983-1761; 800-996-5483. Fax: 909-984-2180.
Web Site: www.graberolives.com
Founded: 1894.
Key Personnel: C.E.O. & Dir., Clifford C. Graber, II
Institution Type/Description: Historic House Museum: housed in the home of the Graber Family, growers & producers of Graber Olives since 1894.
Collections: grading, curing & canning Graber Olives.
Activities: tours.
Hours & Admission Prices: Daily 9:30-5:30; group tours by appointment. No charge. Closed major holidays.

MUSEUM OF HISTORY & ART, ONTARIO, (M), 225 S. Euclid Ave., Ontario, CA 91762-3812. Tel.: 909-395-2510. Fax: 909-983-8978.
Web Site: www.ci.ontario.ca.us/index.cfm/1605
Founded: 1979.
Congressional District: 34
Key Personnel: Pres. Bd., Chris Kueng; Dir., Theresa Hanley; Pres. (V) Museum Assoc., Virginia Eaton; Cur., Steve Thomas; Cur. & Education Coord., Marcilyn Callejo; Office Asst., Loretha Nwosu; Museum Asst., Leslie Matamoros; Gallery Attendant, Irene Gapido.
Personnel Profile: Full-Time Paid 3; Part-Time Paid 5; Part-Time Volunteers 50.
Governing Authority: municipal; nonprofit. Parent Institution: City of Ontario. Tax-exempt: 501(c)(3).
Institution Type/Description: Local History & Art Museum.
Collections: history of Ontario, inland Southern California, citrus, social and home life, local businesses, viticulture, mining & assaying history; 20th-century California paintings.
Research Fields: history of agricultural & industrial development of Ontario & region; major collections include Hotpoint/General Electric appliance collection; Biane Family Winery collection; Barnes family mining, agricultural & home life collection; citrus industry; Latimer family; local citrus & pioneer family; regional & California artists collection.
Facilities: 75-seat auditorium & meeting space. Museum-related items for sale.
Activities: guided tours; lectures; temporary exhibitions school & community workshops & activities; film presentations; reading & discussion groups; lectures; docent programs.
Publications: museum newsletter; local history & California history publications.
Hours & Admission Prices: Wed.-Sun. 12-4. No charge; donations accepted. Closed New Year's Eve & Day; Easter; Memorial Day; Christmas.
Attendance: 20,000 (estimated)
Membership: Student & Senior Citizen $15; Individual $25; Family $35; Contributing $100; Sponsor $250; Benefactors $1,000; $2,500; $5,000.

Orange

CHAPMAN UNIVERSITY GUGGENHEIM GALLERY, One University Dr., Moulton Center, Orange, CA 92866-1005. Tel.: 714-997-6815.
Key Personnel: Coord., Marcus Herse
Institution Type/Description: Art Gallery.
Collections: paintings; sculpture; photographs.
Activities: special events; receptions.
Hours & Admission Prices: Call for hours.

Oroville

BOLT'S ANTIQUE TOOL MUSEUM, 1650 Broderick St., Oroville, CA 95965-4809. Mailing Address: 1735 Montgomery St., Oroville, CA 95965-4820. Tel.: 530-538-2415. Fax: 530-538-2417.
E-mail: boltmuseum@cityoforoville.org
Founded: 1999.
Congressional District: 2
Key Personnel: Chm. (V), Bud Bolt; Museum Shop Mgr., Laila Bolt.
Personnel Profile: Part-Time Paid 2; Part-Time Volunteers 20.
Governing Authority: municipal.
Institution Type/Description: Tool Museum.
Collections: over 11,000 hand tools; power tools; manufacturing documents.
Research Fields: tool history.
Facilities: library; 3,000 sq. ft. exhibit space. Museum-related items for sale.
Activities: docent program; guided tours; lectures.
Hours & Admission Prices: Tues.-Sun. 11:45-3:45. Adults $3, AAA members & groups of 15 or more $2.50; children under 12 no charge. Closed national holidays.
Attendance: 1,902 (accurate)

BUTTE COUNTY HISTORICAL SOCIETY MUSEUM, 1749 Spencer Ave., Oroville, CA 95965. Mailing Address: P.O. Box 2195, Oroville, CA 95965. Tel.: 530-533-9418.
E-mail: buttehistory@sbcglobal.net
Web Site: www.buttecountyhistoricalsociety.org
Founded: 1956.
Congressional District: 2
Key Personnel: Dir., Lucy Sperlin; Pres. (V), Nancy Brower; Museum Shop Mgr., RuthAnn King.
Personnel Profile: Full-Time Volunteers 1; Part-Time Volunteers 35.
Governing Authority: Tax-exempt.
Institution Type/Description: Historical Society Museum.
Collections: local history & culture; period furnishings; personal artifacts; photographs; documentary artifacts.
Research Fields: Butte County history.
Facilities: Museum-related items for sale.
Activities: Museum Sponsors: ISHI Gathering & Seminar in May.
Publications: quarterly, Diggin's; newsletter, Slickens.
Hours & Admission Prices: Sat. 11-3; other times by appointment. No charge; donations accepted.
Attendance: 2,000 (estimated)
Membership: Senior $20; Individual & Organization $25; Senior Family $30; Family $35; Supporting $50; Sustaining $100.

BUTTE COUNTY PIONEER MEMORIAL MUSEUM, 2332 Montgomery St., Oroville, CA 95965-4924. Mailing Address: 1735 Montgomery St., Oroville, CA 95965-4820. Tel.: 530-538-2415. Fax: 530-538-2417.
E-mail: info@cityoforoville.org
Web Site: www.cityoforoville.org/pioneermuseum.html
Founded: 1932.
Congressional District: 2
Personnel Profile: Part-Time Paid 2; Part-Time Volunteers 30.
Governing Authority: municipal government. Tax-exempt.
Institution Type/Description: History Museum.
Collections: early pioneering; Gold Rush Era of California; early area artifacts.
Facilities: 3,600 sq. ft. exhibit space. Museum-related items & books for sale.
Activities: docent program; formal education programs for children; guided tours.
Hours & Admission Prices: Feb. 2-Dec. 14 Fri.-Sun. 12-4. Adults $3, AAA members & groups of 15 or more $2.50; children under 12 no charge.
Attendance: 1,000 (estimated)

C.F. LOTT HISTORIC HOME, 1067 Montgomery St., Oroville, CA 95965. Mailing Address: 1735 Montgomery St., Oroville, CA 95965-4820. Tel.: 530-538-2415. Fax: 530-538-2417.
E-mail: info@cityoforoville.org
Web Site: www.cityoforoville.org/cflotthome.html
Founded: 1963.
Congressional District: 2
Personnel Profile: Part-Time Paid 2; Part-Time Volunteers 30.
Governing Authority: municipal government. Tax-exempt.
Institution Type/Description: History Museum.
Collections: furniture; paintings; rugs; textiles; clothes; 1849-1915 silver & glassware.
Facilities: Museum-related items for sale.
Activities: docent program; formal education programs for children; guided tours; temporary exhibitions. Annual Event: Mistletoe Party - Open House.
Hours & Admission Prices: Feb.-Dec. 14 Sun.-Mon. & Fri. 11:30-3:30. Adults $3; discounts to AAA members & groups of 15 or more.
Attendance: 891 (accurate)

MILITARY MUSEUM OF BUTTE COUNTY, 4514 Pacific Heights Rd., Oroville, CA 95965-9266. Tel.: 530-534-9956. Fax: 530-534-1170.
Institution Type/Description: Military History Museum.
Collections: military history; personal artifacts; photographs; memorabilia; military uniforms, vehicles & weapons.
Hours & Admission Prices: Mon.-Sat. 8-6. No charge.

OROVILLE CHINESE TEMPLE COMPLEX & MUSEUM, 1500 Broderick St., Oroville, CA 95965-4871. Mailing Address: 1735 Montgomery St., Oroville, CA 95965-4820. Tel.: 530-538-2415. Fax: 530-538-2417.
E-mail: info@cityoforoville.org
Web Site: www.cityoforoville.org/chinesetemple.html
Founded: 1949.
Congressional District: 2
Personnel Profile: Part-Time Paid 2; Part-Time Volunteers 30.
Governing Authority: municipal. City of Oroville, 1735 Montgomery St., Oroville, CA. Tax-exempt.
Institution Type/Description: Religious Museum: 1863 Chinese Temple.
Collections: restored Chinese Temple; religious figures; tapestries; writings; screens; altars; Chinese immigrants during California gold rush.
Facilities: library. Postcards & gifts for sale.
Activities: guided tours; permanent exhibitions; school group tours.
Hours & Admission Prices: Feb.-Dec. 15 daily 12-4; groups by appointment. Adults $3, AAA members & groups of 15 or more $2.50; children under 12 no charge.
Attendance: 4,886 (accurate)

Oxnard

CARNEGIE ART MUSEUM, 424 S. C St., Oxnard, CA 93030-5944. Tel.: 805-385-8157 & 8179. Fax: 805-483-3654.
Web Site: www.carnegieam.org
Founded: 1980.
Congressional District: 19
Key Personnel: Dir., Suzanne Bellah; Cur. Education, Jeanette LaVere; Pres. (V), Stacy Roscoe.
Personnel Profile: Full-Time Paid 3; Part-Time Paid 4; Part-Time Volunteers 20; Interns 1.
Governing Authority: municipal; nonprofit. Parent Institution: City of Oxnard. Subsidiary Institution: Carnegie Art Museum Cornerstones. Tax-exempt: 501(c)(3).
Institution Type/Description: Art Museum: housed in 1906 former library & city hall.
Collections: over 1,500 works of California art; Eastwood Collection: archaeological & anthropological artifacts; Hollywood portrait photography.
Research Fields: art history.
Facilities: classroom.
Activities: guided tours; lectures; performances; concerts; permanent, temporary & traveling exhibitions; school tour program; adult & children's art classes; workshops.
Publications: brochure, Municipal Art Collection of Oxnard; book, Sacha Moldovan, Expressionist 1901-1982; catalog, The Pastel Landscapes of Theodore Lukits; museum newsletter; exhibit gallery guides.
Hours & Admission Prices: Thurs.-Sat. 10-5, Sun. 1-5. Adults $3, senior citizens & students $2, children 6-16 $1; discounts to AAM & WMG members; Fri. 3-5, members & children under 6 no charge. Closed Easter; Memorial Day; Independence Day; Labor Day; Thanksgiving; Christmas.

Attendance: 20,000 (estimated)
Membership: Student & Senior $25; Individual $35; Family $45; Sustaining $60; Patron $100; Angel $250; Golden Circle $1,000.

GULL WINGS CHILDREN'S MUSEUM, 418 W. Fourth St., Oxnard, CA 93030-5995. Tel.: 805-483-3005. Fax: 805-483-3226.
E-mail: info@gullwings.org
Web Site: www.gullwings.org
Founded: 1989.
Key Personnel: Dir., Melissa Baffa.
Personnel Profile: Part-Time Paid 8; Part-Time Volunteers 20; Interns 2.
Governing Authority: Tax-exempt.
Institution Type/Description: Children's Museum.
Collections: hands-on exhibits.
Facilities: Museum-related items for sale.
Activities: school programs; birthday parties; special events.
Publications: newsletter.
Hours & Admission Prices: Tues.-Sun. 10-4. Adults $6, children $5; children under 2 no charge.
Attendance: 30,000 (estimated)
Membership: Seasonal $45; Individual $59; Family & Grandparent $69; Gull $160; Snowy Plover $290; Sandpiper $600; Kestrel $1,200. Add One Guest $15; Add 2 Guests $30.

MURPHY AUTO MUSEUM, 2230 Statham Blvd., Oxnard, CA 93033. Tel.: 805-487-4333. Fax: 805-487-4441.
E-mail: murphyautomuseum@gmail.com
Web Site: murphyautomuseum.org
Founded: 2004.
Congressional District: 23 & 24
Key Personnel: Dir., Dan Murphy.
Personnel Profile: Part-Time Volunteers 10.
Governing Authority: nonprofit organization. Tax-exempt.
Institution Type/Description: Auto Museum.
Collections: over 50 period cars; early clothing.
Major Exhibits: Travel On 2 Wheels, Summer 2013.
Activities: visiting auto clubs.
Hours & Admission Prices: Sat.-Sun. 10-4; other times by appointment. No charge; donations accepted. Closed New Year's Day; Thanksgiving; Christmas. &
Attendance: 5,230 (estimated)
Membership: Individual $25; Family $50; Business & Auto Club $150; Sustaining Collector $3,000.

VENTURA COUNTY MARITIME MUSEUM, INC., (M), 2731 S. Victoria Ave., Oxnard, CA 93035-2947. Tel.: 805-984-6260. Fax: 805-984-5970.
E-mail: vcmm@aol.com
Web Site: www.vcmm.org
Founded: 1991.
Congressional District: 23
Key Personnel: Exec. Dir., Bill Conroy; Trustee Chm., Joyce Nelson; Trustee Pres., William Buenger; Chm. (V), Karen Harter; Cur. Art, Jackie Cavish.
Personnel Profile: Full-Time Paid 1; Part-Time Paid 4; Part-Time Volunteers 100.
Governing Authority: Parent Institution: Ventura County Maritime Museum, Inc. Not-for-profit organization. Tax-exempt: 501(c)(3).
Institution Type/Description: Maritime Museum.
Collections: sailing ship models; 17th- to 20th-century maritime art; shipboard items; Ventura County maritime history; 18th century POW bone models.
Research Fields: California & European maritime history; American maritime history, 17th to 19th-century.
Facilities: 2,500-vol. library of maritime history & art
Activities: docent program; formal education programs for children; guided tours; school outreach programs; tall ship visits; educational field trips; ship model guild.
Publications: quarterly newsletter; books, Port Hueneme: A History; Ventura County's Maritime Legacy.
Hours & Admission Prices: Daily 11-5. No charge; donations accepted. Closed New Year's Eve & Day; Thanksgiving; Christmas. &
Attendance: 20,000 (accurate)
Membership: Individual $35; Family $50; Lieutenant $100; Business $150; Commander $250; Captain $500; Commodore $1,000.

Pacific Grove

HERITAGE SOCIETY OF PACIFIC GROVE - KETCHAM'S 1891 BARN, 605 Laurel Ave., Pacific Grove, CA 93950. Mailing Address: P.O. Box 1007, Pacific Grove, CA 93950. Tel.: 831-372-2898.
E-mail: info@pacificgroveheritage.org
Web Site: www.pacificgroveheritage.org
Founded: 1981.
Key Personnel: Pres. (V), Steve Honegger.
Personnel Profile: Part-Time Volunteers 20.
Governing Authority: nonprofit organization. Parent Institution: Heritage Society of Pacific Grove. Tax-exempt.
Institution Type/Description: History Museum.
Collections: local history & culture; period furnishings; personal artifacts; photographs.
Publications: newsletter.
Hours & Admission Prices: Sat. 1-4. No charge; donations accepted.
Attendance: 1,000 (estimated)
Membership: Senior $12; Individual $15; Family $20; Sponsor $50; Friend $100; Partner $250; Patron $500; Lifetime $1,000.

PACIFIC GROVE ART CENTER ASSOCIATES, INC., 568 Lighthouse Ave., Pacific Grove, CA 93950-2624. Mailing Address: P.O. Box 633, Pacific Grove, CA 93950-0633. Tel.: 831-375-2208. Fax: 831-375-2208.
E-mail: pgart@mbay.net
Web Site: www.pgartcenter.org
Founded: 1969.
Congressional District: 16
Key Personnel: Pres., Johnny Aliotti; Dir., Alana Puryear; Preparator, Mark Davy; Preparator, Kait Kent.
Personnel Profile: Part-Time Paid 2; Part-Time Volunteers 20.
Governing Authority: bd. of directors; nonprofit organization. Tax-exempt: 501(c)(3).
Institution Type/Description: Art Center.
Collections: paintings; graphics; photographs; sculpture.
Facilities: artists' studios; classroom.
Activities: rotating exhibits; art & rental of studios; art & dance classes; concerts; artist talks & demonstrations; artist lectures.
Publications: PGAC newsletter.
Hours & Admission Prices: Wed.-Sat. 12-5, Sun. 1-4. No charge; donations accepted. Closed legal holidays.
Attendance: 16,000 (accurate)
Membership: Student $20; Single $30; Family $40; Business $100; Sponsor $150; Patron $500; Lifetime $1,000; Corporate Sponsor $1,200.

✻ PACIFIC GROVE MUSEUM OF NATURAL HISTORY, (M), 165 Forest Ave., Pacific Grove, CA 93950-2698. Tel.: 831-648-5716. Fax: 831-648-5755.
E-mail: admin@pgmuseum.org
Web Site: www.pgmuseum.org
Founded: 1881.
Congressional District: 16
Key Personnel: Dir. & C.E.O., Lori Mannel; Pres. (V), Jason Burnett; Cur., Paul VandeCarr; Dir. Education, Annie Holdren; Museum Store Mgr., Beverly Bruno.
Personnel Profile: Full-Time Paid 5; Part-Time Paid 3; Part-Time Volunteers 100.
Governing Authority: municipal. Parent Institution: Museum Foundation of Pacific Grove. Tax-exempt.
Institution Type/Description: Natural History Museum.
Collections: birds; mammals; reptiles; amphibians; marine life; plants; insects of Monterey County; zoology; botany; geology; herbarium; ethnology. Historic House: 1853 Point Pinos Lighthouse.
Research Fields: natural history of Monterey County.
Facilities: 2,000-vol. library of reference material available on premises; 10,000 ft. gallery space; botanical garden. Books, educational toys & games, minerals and fossils for sale.
Activities: guided tours; school trips & educational programs; lectures; inter-museum loan; permanent, temporary & traveling exhibitions; monarch sanctuary. Museum Sponsors: annual wildflower show; monthly science Saturdays for children.
Publications: Supplement to the Vascular Plants of Monterey County, California; quarterly newsletter, Horizons; Wildflowers of Monterey County: A Field Companion.
Hours & Admission Prices: Tues.-Sun. 10-5. No charge; donations accepted. Closed major holidays. ♿
Attendance: 115,400 (accurate)

Membership: Individual $45; Family $65; Friend $100; Sponsor $250; Patron $500.

Pacific Palisades

WILL ROGERS STATE HISTORIC PARK, 1501 Will Rogers State Park Rd., Pacific Palisades, CA 90272-3941. Tel.: 310-454-8212, ext. 104. Fax: 310-459-2031.
E-mail: rnicholas@parks.ca.gov
Web Site: www.ugf.edu/aboutugf/galerietrinitas
Founded: 1944.
Key Personnel: District Supt., Ron Schafer; Topanga Supt., Lynette Hernandez; Museum Cur., Rochelle Nicholas-Booth.
Personnel Profile: Full-Time Paid 11; Part-Time Paid 10; Part-Time Volunteers 45; Interns 2.
Governing Authority: state. Parent Institution: California Dept. of Parks & Recreation, Sacramento, CA.
Institution Type/Description: Historic Building & Site Museum: housed in 1924-1935 original home of Will Rogers.
Collections: personal belongings of Will Rogers; cowboy souvenirs; American Indian artifacts; books; Western art, principally Charles M. Russell & Edward Borein.
Facilities: visitor center; picnic grounds. Books & museum-related items for sale.
Activities: house tours; paths & hiking trails; polo games; orientation film.
Publications: brochure, Will Rogers.
Hours & Admission Prices: Park: 8-sunset. Visitors Center & House: call for hours. Ranch Tour: Tues.-Sat. 11, 1 & 2. $7 per vehicle. Polo Games: April-Sept. Sat. 2-5, Sun. 10-1. Ranch Tours: closed New Year's Day; Thanksgiving; Christmas. ♿
Attendance: 230,000 (estimated)

Pacifica

✻ SANCHEZ ADOBE HISTORIC SITE, (M), 1000 Linda Mar Blvd., Pacifica, CA 94044-3545. Tel.: 650-359-1462. Fax: 650-359-1462.
E-mail: sanchezadobe@historysmc.org
Web Site: www.historysmc.org/sanchez.html
Founded: 1947.
Congressional District: 1
Key Personnel: Exec. Dir., Mitch Postel; Site Mgr., Linda Corwin; Chm. (V), Charline Smith; Education Coord., Carmen Blair.
Personnel Profile: Part-Time Paid 3; Part-Time Volunteers 20.
Governing Authority: county. Parent Institution: San Mateo County Historical Museum Assoc. Tax-exempt.
Institution Type/Description: Historic House Site: historic rancho built 1842-46 site of Mission Dolores farming outpost (1783-1828); Pruristac village site for Ohlone Indians.
Collections: period furnishings; interpretive exhibits; artifacts from archaeological digs.
Facilities: library.
Activities: guided tours; lectures; living history programs.
Publications: descriptive brochures, The Sanchez Adobe; historic booklet, Journal of the San Mateo County Historical Association.
Hours & Admission Prices: Tues.-Thurs. 10-4, Sat.-Sun. 1-5. No charge; donations accepted. Closed legal holidays.
Attendance: 6,103

Paicines

PINNACLES NATIONAL MONUMENT, 5000 Hwy. 146, Paicines, CA 95043-9770. Tel.: 831-389-4486, ext. 233. Fax: 831-389-4489.
Web Site: www.nps.gov
Founded: 1908.
Congressional District: 17
Key Personnel: Supt., Karen Beppler-Dorn; Volunteer Coord., Veronica Johnson; Museum Shop Mgr., Linda Regan.
Governing Authority: federal. Parent Institution: National Park Service. Subsidiary Institution: Pinnacles Partnership. Tax-exempt: 501(c)(3).
Institution Type/Description: Natural History Museum.
Collections: herbarium; herpetology; entomology; geology; zoology.
Facilities: 300-vol. library of natural history books available for use by special request; visitor center.
Activities: on-site natural history study collections.
Publications: booklets, Natural History of Pinnacles National Monument.
Hours & Admission Prices: Winter: daily 7:30-6; Summer: daily 7:30-8. Vehicle $5 for 7 days; Walk-in $3 for 7 days; children under 16 & educational groups with advanced waiver no charge. Visitor Center: daily 9:30-5. ♿

Attendance: 171,000 (estimated)

Pala

SAN ANTONIO DE PALA ASISTENCIA, 3015 Mission Rd., Pala, CA 92059. Mailing Address: P.O. Box 70, Pala, CA 92059-0070. Tel.: 760-742-1600. Fax: 760-742-3040.
Web Site: missionsanantonio.org
Founded: 1816.
Key Personnel: Museum Shop Mgr., Sr. Barbara Jackson.
Personnel Profile: Full-Time Paid 1.
Institution Type/Description: History Museum: housed in a sub-mission of Mission San Luis Rey de Francia.
Collections: Indian art; period artifacts.
Activities: religious services.
Hours & Admission Prices: Office & Store: Tues.-Fri. 9:30-4:30, Sat. 9:30-3, Sun. 9:30-1. Mass: Mon.-Fri. 8am, Sat. 5pm, Sun. 8:30am & 11:30 (English), 10:30am (Spanish). Closed Thanksgiving; Christmas.

Palm Desert

IMAGO GALLERIES, 45-450 Hwy. 74, Palm Desert, CA 92260. Tel.: 760-776-9890.
E-mail: info@imagogalleries.com
Web Site: www.imagogalleries.com
Founded: 1991.
Key Personnel: Dir., Leisa Austin; Dir., David Austin
Institution Type/Description: Art Gallery.
Collections: works by contemporary artists; paintings; sculpture.
Hours & Admission Prices: Tues.-Sat. 10-5; other times by appointment. No charge. &

THE LIVING DESERT, 47900 Portola Ave., Palm Desert, CA 92260-6156. Tel.: 760-346-5694, ext. 2102. Fax: 760-568-9685.
E-mail: sjohnson@livingdesert.org
Web Site: livingdesert.org
Founded: 1970.
Congressional District: 45
Key Personnel: C.E.O. & Pres., Mr. Stacey Johnson; Chm. Bd., Roger Snoble; Chm. (V), Bill Simpkins; Dir. Conservation & Education, Peter Siminski; Dir. Accounting, Dwight Middendorf; Dir. Park Svcs., Kerry Graves; Acting Dir. Devel., Ernest Phinney; Dir. Facilities Mgmt., Bert Buxbaum; Cur. Garden Dept., Kirk Anderson; Cur. Education Dept., Mike Chedester; Cur. Animals, Liz Hile; Museum Shop Mgr., Sharon Mays.
Personnel Profile: Full-Time Paid 91; Part-Time Paid 23; Part-Time Volunteers 500.
Volunteer Hours: 61,112
Operating Expenses: 7,363,165
Governing Authority: nonprofit organization. Tax-exempt.
Institution Type/Description: Zoo & Gardens.
Collections: 6,000 slide photographic library; desert fauna & flora; Colorado desert herbarium; inanimate zoological & botanical materials; Cahuilla Indian art & artifacts; North American & African aviaries; live animals including birds of prey; coyotes, mountain lions, Mexican wolves, bighorn sheep, giraffe, cheetahs, leopards, zebra, & nocturnal species; art.
Major Exhibits: Camel ride Desert Experience (T), 11/13-4/14; Australian Parakeet Feeding Desert Experience, 1/14-4/14.
Research Fields: zoology; botany; geology.
Facilities: 9,000-vol. reference library of natural history books available for use by written request; botanical garden; walk-through aviaries; 300-seat auditorium; 600-seat outdoor amphitheater; visitors center; picnic & party facilities. Museum-related items for sale.
Activities: guided tours; lectures; docent program; permanent & temporary exhibitions; nature & hiking trails; art craft & natural history classes; field trips; concert series; wildflower program; volunteer program; volunteer training; summer nature school. Museum Sponsors: annual desert plant sale; Earth Day Celebration.
Publications: newsletter, L.D. Newsletter; annual report; Living Desert coloring book; book, Desert Wildflower.
Hours & Admission Prices: June-Sept. daily 8-1:30 (last admission 1pm); Oct.-May daily 9-5 (last admission 4pm). Adults $17, seniors 62 & over $15.75, children over 2 $8.75; discounts to AZA & AAA members, groups & schools; children under 2 & members no charge. Closed Christmas Day. &
Attendance: 377,887 (accurate)
Membership: Individual $55; Dual $75; Family $100; Supporting $145; Donor $250; Sustaining $500; Curator $1,000; Director $2,500; President's Circle $5,000. Corporate: $350; $500; $1,000; $2,500; $5,000.

Palm Springs

AGUA CALIENTE CULTURAL MUSEUM, 219 S. Palm Canyon Dr., Palm Springs, CA 92262-6310. Mailing Address: 901 E. Tahquitz Canyon Way, Ste. C-204, Palm Springs, CA 92262-6757. Tel.: 760-778-1079. Fax: 760-322-7724.
Web Site: www.accmuseum.org
Founded: 1993.
Congressional District: 37
Key Personnel: Pres. & Chm. (V), Mildred Browne; Exec. Dir., Dr. Michael Hammond; Dir. Devel., Steve Sharp; Devel. Assoc., Cherie Courtade; Cur., Dawn Wellman; Museum Interpreter, Tina Richey; Museum Interpreter, Ursula Cripps; Museum Interpreter, Tom Cole; Museum Interpreter, Don Karvelis; Administrative Asst., Claire Victor; Office Mgr., Jackie Bagnall.
Personnel Profile: Full-Time Paid 6; Part-Time Paid 5; Part-Time Volunteers 7.
Governing Authority: private; nonprofit organization. Tax-exempt: 501(c)(3).
Institution Type/Description: History Museum.
Collections: cultural & heritage artifacts from the Agua Caliente Band of Cahuilla Indians; baskets; pottery.
Research Fields: rock art; native & non-native interactions from 1780 to present.
Facilities: 2,000-vol. library; educational facilities; 800 sq. ft. exhibit space. Museum-related items for sale.
Activities: formal education programs; lectures. Annual Events: Dinner in the Canyons (fundraiser); Native Film Fest; Bird Song & Dance Festival.
Publications: quarterly, The Spirit.
Hours & Admission Prices: Memorial Day to Labor Day Fri.-Sat. 10-5, Sun. 12-5; Sept.-May Wed.-Sat. 10-5, Sun. 12-5. No charge; donations accepted. Closed New Year's Day; Thanksgiving; Christmas.
Attendance: 14,000 (estimated)
Membership: Senior $30; Individual $40; Family $50; Paloverde $100; Mesquite $250; Palm $500; Eagle $1,000; Eagle Bronze $2,500; Eagle Silver $5,000; Eagle Gold $10,000.

MICHAEL H. LORD GALLERY, 1090 N. Palm Canyon Dr., Palm Springs, CA 92262-4435. Tel.: 760-699-8957. Fax: 760-699-8962.
E-mail: art@michaelhlordgallery.com
Web Site: www.michaelhlordgallery.com
Founded: 1978.
Institution Type/Description: Art Gallery.
Collections: works by contemporary American & European artists; paintings; sculpture; prints; photographs.
Facilities: 7,000 sq. ft. exhibition space.
Activities: temporary exhibitions.
Hours & Admission Prices: Tues.-Sat. 10-6, Sun. 12-5, Mon. by appointment. No charge.

MOORTEN BOTANICAL GARDEN AND CACTARIUM, 1701 S. Palm Canyon Dr., Palm Springs, CA 92264-8936. Tel.: 760-327-6555.
E-mail: clarkmoorten@yahoo.com
Founded: 1938.
Congressional District: 37
Key Personnel: Exec. Dir., C.E.O. & Cur., Clark Moorten.
Governing Authority: profit-making organization; individual operation.
Institution Type/Description: Botanical Garden.
Collections: botany; geology; natural history; over 3,000 varieties of worldwide desert plants, including rare & endangered species. Historic House: 1915 pioneer home, Cactus Castle.
Research Fields: botany; geology; natural history.
Facilities: 300-vol. library of botanical books of desert plants available for use on premises & by appointment; desert plant nursery.
Activities: guided tours; lectures; films; formally organized education programs for children, adults, undergraduate & graduate college students; inter museum loan, permanent, temporary & traveling exhibitions; Desert Lore classes for children; conservation education.
Publications: book, Desert Plants for Desert Gardens.
Hours & Admission Prices: Mon.-Sat. 9-4:30, Sun. 10-4. Adults $3, children age 5-15 $1.50; children under 5 no charge.

PALM SPRINGS AIR MUSEUM, 745 N. Gene Autry Trail, Palm Springs, CA 92262-5464. Tel.: 760-778-6262, ext. 223.
E-mail: fred@palmspringsairmuseum.org
Web Site: www.palmspringsairmuseum.org
Founded: 1996.
Congressional District: 45
Key Personnel: Dir., Fred Bell.
Governing Authority: nonprofit organization. Tax-exempt: 501(c)(3).

Institution Type/Description: Air Museum.
Collections: WWII military aircraft; aviation art; photographs; aircraft & ship models.
Facilities: 8,500-vol. library; 3 climate-controlled hangars; rental facilities; education center.
Activities: educational programs; seminars; flight demonstrations; temporary exhibitions; rental facilities; children's exhibitions.
Hours & Admission Prices: Daily 10-5. Adults $15, seniors 65 & over, military and youth 13-17 $13, children 6-12 $8; members & children under 6 no charge. Closed Thanksgiving; Christmas.
Attendance: 90,000 (accurate)
Membership: Educator & Active Volunteer $45; Individual $50; Crew Chief $75; Family Crew $100; Squadron Commander $250; Wing Commander $500; Air Division Commander $1,000.

✱ **PALM SPRINGS ART MUSEUM, (M),** 101 Museum Dr., Palm Springs, CA 92262-5659. Mailing Address: P.O. Box 2310, Palm Springs, CA 92263-2310. Tel.: 760-322-4800. Fax: 760-327-5069.
E-mail: info@psmuseum.org
Web Site: www.psmuseum.org
Formerly: Palm Springs Desert Museum, Inc.
Founded: 1938.
Congressional District: 37
Key Personnel: Chm., Harold J. Meyerman; Exec. Dir., Steven A. Nash, Ph.D.; Chief Cur., Katherine Hough; The Donna & Cargill Macmillan Jr. Dir. Art, Daniell Cornell; Interim Dir. Education, Irene N. Rodriguez; The Dorothy & Harold J. Meyerman Dir. Devel., Greg Polzin; C.F.O., Robert Wayne; Registration & Collections Mgmt., Alicia Thomas; Dir. Operations, Debra Preston; Dir. Mktg. Communications, Bob Bogard; C.O.O., Greg Johnson; Assoc. Dir. Merchandising, Betty Rinnig.
Personnel Profile: Full-Time Paid 85; Part-Time Paid 5; Part-Time Volunteers 76; Interns 2.
Volunteer Hours: 13,000
Operating Expenses: 10,521,530
Operating Income: 10,555,994
Governing Authority: nonprofit organization. Subsidiary Institution: 9 support councils. Tax-exempt: 501(c)(3).
Institution Type/Description: Art Museum.
Collections: 19th- & 20th-century art with emphasis on contemporary international art, classic Western American & Native American art, Mesoamerican art, Mexican art, European modern sculpture, American mid-20th century architecture, and American photography; glass studio art, architecture-design.
Major Exhibits: Secrets of the Sun: Stephen H. Willard Photos of the West, 12/13-4/14; Into the Future: New Gifts to Commemorate the Museum's 75th Anniversary, 1/14-5/14; Recline/Design: Art & the Aesthetics of Repose, 2/14-8/14; California Dreamin': Thirty Years of Collecting, 3/14-1/15; Links: Australian Glass & The Pacific N.W. (T), 10/14-1/15; An Eloquent Modernist, E. Stewart Williams, Architect, 11/14-2/15.
Research Fields: photography by Bill Anderson and Stephen Willard; Albert Frey, E. Stewart Williams architects; George Montgomery Collection.
Facilities: 12,000-vol. library of art books available for use on premises; 53,000 sq. ft. exhibition space; 90-seat lecture hall; three sculpture gardens; nature trail; 435-seat theater; 2 classrooms; cafe; educational center. Museum-related items for sale.
Activities: guided tours; lectures; films; gallery talks; concerts; dramatic productions; dance recitals; formally organized educational programs; inter-museum loan, permanent, temporary & traveling exhibitions; docent program or council.
Publications: Duane Hanson: Virtual Reality; NW x SW: Painted Fictions; Michael Todd: 25-Year Survey; California Grandeur and Genre; Transforming the Western Image; Namingha: Timeless Land and Enduring Images; American Arts & Crafts from the Collection of Alexandra and Sidney Sheldon; Agnes Pelton: Poet of Nature; Collaborations: William Allan, Robert Hudson, William Wiley; Icons and Legends: The Photography of Michael Childers; Stephen H. Willard: Photography and Archive; Contemporary Desert Photography: The Other Side of Paradise; Lights! Camera! Glamour! George Hurrell at 100; quarterly member magazine; D.J. Hall Thirty-Five Year Retrospective; Picasso to Moore: Modern Sculpture from the Weiner Collection; Portrait: Robert Mapplethorpe; 70 Years of Painting: Wayne Thiebaud; Backyard Oasis: The Swimming Pool in Southern California Photography 1945-1982; Palms Springs Art Museum 75 Years, 75 Artworks; Palm Springs Art Museum Celebrating 75 Years.
Hours & Admission Prices: Tues.-Wed. & Fri.-Sun. 10-5, Thurs. 12-8. Adults $12.50, seniors 62 & over $10.50, students & military personnel with I.D. $5; discounts to CAM, WMA, AAM, ICOM & AAMD members; NARM, members & Thurs. 4-8 no charge. Closed major holidays. ♿
Attendance: 175,000 (accurate)
Membership: Individual $50; Family & Dual $85; Supporting $125; Contributing $350; Patron $1,000; President's Circle $2,000.

RUDDY'S 1930S GENERAL STORE MUSEUM, 221 S. Palm Canyon Dr., Palm Springs, CA 92262-6310. Tel.: 760-327-2156.
Institution Type/Description: History Museum.
Collections: re-created 1930s general store; period artifacts.
Hours & Admission Prices: July-Sept. Sat.-Sun. 10-4; Oct.-June Thurs.-Sun. 10-4. Adults $.95; children under 12 no charge.

VILLAGE GREEN HERITAGE CENTER, 221 S. Palm Canyon Dr., Palm Springs, CA 92262-6310. Mailing Address: P.O. Box 1498, Palm Springs, CA 92263-1498. Tel.: 760-323-8297. Fax: 760-320-2561.
Governing Authority: nonprofit organization. Parent Institution: Palm Springs Historical Society.
Institution Type/Description: Historical Society Museum.
Collections: local history & culture; personal artifacts; photographs; Native American artifacts. Historic Buildings: two 19th century pioneer homes.
Hours & Admission Prices: mid-Oct. to May Wed. & Sun. 12-3, Thurs.-Sat. 10-4. Adults $2; children no charge.

Palmdale

BLACKBIRD AIRPARK MUSEUM, 2503 E. Avenue P, Palmdale, CA 93550. Mailing Address: 405 S. Rosamond Blvd., Edwards Air Force Base, CA 93524. Tel.: 661-277-8050.
Institution Type/Description: Air Force Flight Test Center Museum.
Collections: military aircraft including 1960s Lockheed reconnaissance plane, the A12 test plane, and the ultra secret D21 drone; Blackbird Heritage Wall.
Facilities: Museum-related items for sale.
Hours & Admission Prices: Fri.-Sun. 10-5. No charge; donations accepted.
Attendance: 12,000 (estimated)

JOE DAVIES HERITAGE AIRPARK AT PALMDALE PLANT 42, 2001 E. Ave. P, Palmdale, CA 93550. Mailing Address: 38250 Sierra Hwy., Palmdale, CA 93550. Tel.: 661-267-5300.
Institution Type/Description: Military History Museum.
Collections: military aircraft & history.
Hours & Admission Prices: Fri.-Sun. 11-4. Closed Independence Day.

Palo Alto

MUSEUM OF AMERICAN HERITAGE, (M), 351 Homer Ave., Palo Alto, CA 94301-2727. Mailing Address: P.O. Box 1731, Palo Alto, CA 94302-1731. Tel.: 650-321-1004. Fax: 650-473-6950. Facebook: MOAHPA.
E-mail: mail@moah.org
Web Site: www.moah.org
Founded: 1992.
Key Personnel: Exec. Dir., Laurie Hassett
Institution Type/Description: Technology Museum.
Collections: early 20th century life & technology; electrical & mechanical devices; inventions.
Activities: hands-on experiments; educational programs.
Hours & Admission Prices: Fri.-Sun. 11-4. No charge.

PALO ALTO ART CENTER, 1313 Newell Rd., Palo Alto, CA 94303-2909. Tel.: 650-329-2366. Fax: 650-326-6165.
E-mail: artcenter@cityofpaloalto.org
Web Site: www.cityofpaloalto.org/artcenter
Founded: 1971.
Congressional District: 12
Key Personnel: Dir., Karen Kienzle; Studio Supvr., Lynn Stewart; Operations & Sales Mgr., Rebecca Barbee; Volunteer Coord., Emily Lacroix; Children's Education, Ariel Feinberg Berson; Museum Shop Mgr., Elizabeth Evans; Publications, Sharon Fox.
Personnel Profile: Full-Time Paid 9; Part-Time Paid 21; Part-Time Volunteers 300.
Governing Authority: municipal. Parent Institution: City of Palo Alto. Affiliated with the Division of Arts & Sciences, City of Palo Alto. Tax-exempt: 501(c)(1).
Institution Type/Description: Visual Art Center.
Collections: works by national & international artists.
Research Fields: contemporary art & design; fine crafts.
Facilities: sculpture gardens; auditorium; classrooms; meeting rooms. Museum-related items for sale.
Activities: temporary exhibitions; school tours; classes; drop-in programs; family programs; lectures.
Publications: exhibition catalogues & brochures; quarterly calendar & newsletter.

Hours & Admission Prices: Tues.-Thurs. 10-5 & 7pm-10pm, Fri.-Sat. 10-5, Sun. 1-5. No charge; donations accepted. &
Attendance: 79,000 (accurate)
Membership: Individual $50; Family & Dual $75; Enthusiast $150; Advocate $250; Visionary $1,000; Director $2,500.

PALO ALTO JUNIOR MUSEUM AND ZOO, 1451 Middlefield Rd., Palo Alto, CA 94301-3351. Tel.: 650-329-2111. Fax: 650-473-1965.
Web Site: www.cityofpaloalto.org
Founded: 1934.
Congressional District: 12
Key Personnel: Dir., John Aikin; Pres. (V), Aletha Coleman; Zoo Cur., Robert Steele; Dir. Education, Alex Hamilton; Exhibits Cur., Tina Keagan; Receptionist, Ines Thiessen.
Personnel Profile: Full-Time Paid 5; Part-Time Paid 13; Part-Time Volunteers 30.
Governing Authority: municipal. Parent Institution: City of Palo Alto. Affiliated with Division of Arts & Sciences of Palo Alto, CA. Friends Group: Tax-exempt: 501(c)(3).
Institution Type/Description: Children's Museum & Zoo.
Collections: natural history; anthropological; historical; living; ethnographic.
Research Fields: informal education.
Facilities: classrooms.
Activities: lectures; art & science classes; field trips; school loan service; changing participatory science exhibition; hands-on science outreach program; animal demonstrations.
Publications: monthly, activities bulletin; quarterly, activities bulletin; current exhibition guide; brochure; quarterly membership newsletter.
Hours & Admission Prices: Tues.-Sat. 10-5, Sun. 1-4. No charge; donations accepted. Closed state & federal holidays; day after Thanksgiving. &
Attendance: 140,000 (estimated)
Membership: $100; $250; $500; $1,000.

Palomar Mountain

PALOMAR OBSERVATORY, 35899 Canfield Rd., Palomar Mountain, CA 92060-0200. Tel.: 760-742-2119.
E-mail: palomar-info@astro.caltech.edu
Web Site: www.astro.caltech.edu/palomarnew
Governing Authority: nonprofit organization. Tax-exempt: 501(c)(3).
Institution Type/Description: Observatory.
Collections: astronomical photographs; discoveries at Palomar Observatory; Hale Telescope.
Facilities: Museum-related items for sale.
Hours & Admission Prices: early March to early Nov. daily 9-4; early Nov. to early March daily 9-3. Closed Christmas Eve & Day.

Palos Verdes Peninsula

SOUTH COAST BOTANIC GARDEN, 26300 Crenshaw Blvd., Palos Verdes Peninsula, CA 90274-2515. Tel.: 310-544-1948.
E-mail: info@southcoastbotanicgarden.org
Web Site: www.southcoastbotanicgarden.org
Founded: 1960.
Key Personnel: Bd. Pres., Robin Hill; C.E.O., Adrienne L. Nakashima; Garden Supt., Tanya E. Finney
Institution Type/Description: Botanical Garden.
Collections: over 25,000 species of plants from around the world; over 200 species of birds.
Facilities: 87-acre garden. Garden-related items for sale.
Activities: classes; special events.
Hours & Admission Prices: Daily 9-5. Adults $8, seniors 62 & over and students $6, children 5-12 $3; children 4 & under no charge. Closed Christmas.
Attendance: 80,000

Paradise

GOLD NUGGET MUSEUM, 502 Pearson Rd., Paradise, CA 95969-5114. Mailing Address: P.O. Box 949, Paradise, CA 95967-0949. Tel.: 530-872-8722. Fax: 530-872-1050. Facebook: Gold Nugget Museum.
E-mail: goldnuggetmuseum@aol.com
Web Site: goldnuggetmuseum.com
Founded: 1981.
Governing Authority: Parent Institution: Gold Nugget Days, Inc. Tax-exempt.
Institution Type/Description: History Museum.
Collections: old country store; one-room schoolhouse; walk-through mine; farm & mining equipment; old West mining town replica; working blacksmith shop; gold panning sluices; in-ground Maidu grinding rock; 1800s print shop.
Major Exhibits: Holiday, Nov.-Dec.
Facilities: picnic area.
Publications: Tales of the Ridge: articles of the area history; Dogtown Nugget: annual program & stories in newspaper format about the annual celebration of 54 pound gold nugget.
Hours & Admission Prices: Wed.-Sun. 12-4. No charge; donations accepted.
Attendance: 14,500
Membership: Student & Senior $25; Individual $30; Family $40; Supporter $50; Business $100; Life $500; Sponsor $1,000.

Pasadena

ALYCE DE ROULET WILLIAMSON GALLERY, Art Center College of Design, 1700 Lida St., Pasadena, CA 91103. Tel.: 626-396-2446 & 2397. Fax: 626-405-9104.
E-mail: stephen.nowlin@artcenter.edu
Web Site: www.williamsongallery.net
Key Personnel: Dir., Stephen Nowlin.
Governing Authority: Tax-exempt.
Institution Type/Description: Art Gallery.
Collections: works of contemporary fine art & design.
Activities: educational programs.
Hours & Admission Prices: Tues.-Thurs. & Sat.-Sun. 12-5, Fri. 12-9. Closed holidays. No charge.
Attendance: 15,000 (estimated)

ARMORY CENTER FOR THE ARTS, 145 N. Raymond Ave., Pasadena, CA 91103. Tel.: 626-792-5101. Fax: 626-449-0139.
E-mail: information@armoryarts.org
Web Site: www.armoryarts.org
Key Personnel: Dir., Irene Tsatsos
Institution Type/Description: Art Gallery.
Collections: paintings; sculpture.
Activities: classes; educational programs.
Hours & Admission Prices: Tues.-Sun. 12-5. Suggested Donation: adults $5; seniors, students & members no charge.

EATON CANYON NATURE CENTER, 1750 N. Altadena Dr., Pasadena, CA 91107. Tel.: 626-398-5420.
Institution Type/Description: Nature Center.
Collections: canyon history & ecology; plants; geology; live animals.
Facilities: auditorium; classrooms. Museum-related items for sale.
Activities: hiking trails; educational programs.
Hours & Admission Prices: Sunrise to sunset Tues.-Sun. 9-5. Closed New Year's Day; Thanksgiving; Christmas.

THE GAMBLE HOUSE, (M), 4 Westmoreland Pl., Pasadena, CA 91103-3593. Tel.: 626-793-3334. Fax: 626-577-7547.
E-mail: gamblehs@usc.edu
Web Site: www.gamblehouse.org
Founded: 1966.
Congressional District: 27
Key Personnel: Dir., Edward R. Bosley; Museum Shop Mgr., Bryan Gonzales.
Personnel Profile: Full-Time Paid 5; Part-Time Paid 4; Part-Time Volunteers 180.
Governing Authority: Parent Institution: University of Southern California. Tax-exempt.
Institution Type/Description: Historic House Museum: house & decorative arts designed by architects Greene & Greene.
Collections: furnishings; personal artifacts; period fixtures.
Research Fields: Architects Greene & Greene; American arts and crafts movement.
Facilities: archives. Books for sale.
Activities: craftsman tours; lectures
Publications: quarterly, Friends of The Gamble House Update.
Hours & Admission Prices: Guided Tours: Thurs.-Sun. 12-4; last tour begins at 3. Adults $12.50, seniors 65 & over and students $10; discounts to National Historic Trust members; members & children under 12 no charge. Brown Bag Tours: Tues. 12:15 & 12:45. Adults $5. Closed national holidays. &
Attendance: 25,000 (estimated)
Membership: Individual $40; Active $125; Supporting $250; Patron $500; Director's Circle $1,000; Life $2,500.

KIDSPACE CHILDREN'S MUSEUM, 480 N. Arroyo Blvd., Pasadena, CA 91103-3269. Tel.: 626-449-9144, ext. 5215. Fax: 626-449-9985.
E-mail: maviviano@kidspacemuseum.org
Web Site: www.kidspacemuseum.org
Founded: 1979.
Congressional District: 23
Key Personnel: C.E.O., Michael Shanklin; Chm., Kris Popovich; Dir. Education, Yvonne Chavez-Lombardi; Assoc. Dir. Devel., Amy Hasquet; Dir. Business Operations, Mary Ann Viviano; Museum Shop Mgr., Susan Cardosi-Albert; Dir. Devel., Eric Blodgett.
Personnel Profile: Full-Time Paid 21; Part-Time Paid 30.
Operating Expenses: 2,823,130
Operating Income: 2,748,356
Governing Authority: nonprofit. Tax-exempt: 501(c)(3).
Institution Type/Description: Children's Participatory Museum.
Collections: interactive exhibits; raindrop climber; wisteria climber; wildlife pond.
Facilities: cafe. Museum-related items for sale.
Activities: interactive & exploratory programs in the arts, sciences, & humanities.
Publications: calendar of events; guest guide; early childhood programs brochure; education program guide; membership brochure.
Hours & Admission Prices: June-Aug. Mon.-Fri. 9:30-5, Sat.-Sun. 10-5; Sept.-May Tues.-Fri. 9:30-5, Sat.-Sun. 10-5. Admission $10; members & children under one no charge. ACM & ASTC reciprocal membership. Closed New Year's Day; Independence Day; Thanksgiving; Christmas. &
Attendance: 235,522 (accurate)
Membership: 6 Month $99; 12 Month $179.

NORTON SIMON MUSEUM, 411 W. Colorado Blvd., Pasadena, CA 91105-1825. Tel.: 626-449-6840. Fax: 626-796-4978.
E-mail: info@nortonsimon.org
Web Site: www.nortonsimon.org
Formerly: Pasadena Art Museum
Founded: 1924.
Key Personnel: Pres., Walter W. Timoshuk; Chief Cur., Carol Togneri; Cur., Gloria Williams; Cur., Leah Lehmbeck; Asst. Cur., Melody Rod-ari; Registrar, Lisa Escovedo; Rights & Reproductions, Jacqui Chambers; Public Rels., Leslie Denk; Exec. Admin., Sally Swaney; Security, Eric Mills; Museum Shop Mgr., Andrew Uchin.
Personnel Profile: Full-Time Paid 50; Part-Time Paid 114; Interns 1.
Governing Authority: nonprofit organization. Tax-exempt: 501(c)(3).
Institution Type/Description: Art Museum.
Collections: European paintings, drawings & sculpture from the early Renaissance through the 20th century; stone & bronze sculpture from India, Nepal & southeast Asia; tapestries; Galka Scheyer Blue Four Collection; 20th century American Art; Picasso, Rembrandt & Goya graphics; Degas sculpture.
Major Exhibits: In the Land of snow: Buddhist Arts of the Himalatas, 3/28/14-8/25/14; Face It: The Photographic Portrait, 4/4/14-8/11/14; Country Pursuits: Indian Miniature Paintings, 9/19/14-1/19/15; Lock, Stock & Barrel: Norton Simon's Acquisition of Dween Brothers, 10/24/14-2/16/15.
Facilities: theater; Garden Cafe. Museum-related items for sale.
Activities: guided tours; family programs; lectures; musical performances; dance performances; adult education; teen workshops.
Publications: Selected Paintings at the Norton Simon Museum; Masterpieces from the Norton Simon Museum; book, The Blue Four Collection at the Norton Simon Museum; Vincent van Gogh: Painter, Printmaker, Collector; Rembrandt Etchings; Picasso: Graphic Magician; Radical P.A.S.T.: Contemporary Art & Music in Pasadena, 1960-1974; Painted Poems: Rajput Paintings from the Ramesh and Urmil Kapoor Collection; Asian Art at the Norton Simon Museum Vol. I-III; The Collectible Moment: Photographs in the Norton Simon Museum; Handbook of the Norton Simon Museum; Nineteenth-Century Art in the Norton Simon Museum, Vol. 1; Degas in the Norton Simon Museum: Nineteenth Century Art, Vol. II; Collection Without Walls: Norton Simon and his Hunt for The Best; Proof: The Rise of Printmaking in Southern California.
Hours & Admission Prices: Wed.-Thurs. & Sat.-Mon. 12-6, Fri. 12-9; tours by appointment. Adults $10, seniors $7; discounts to AAM & ICOM members; members & students no charge. Closed Rose Parade Day; Thanksgiving; Christmas. &
Attendance: 200,000 (estimated)
Membership: Active $75; Contributing $125; Sponsoring $275; Patron $550; Benefactor $1,000.

* **PACIFIC ASIA MUSEUM, (M),** 46 N. Los Robles Ave., Pasadena, CA 91101-2071. Tel.: 626-449-2742, ext. 0. Fax: 626-449-2754. Facebook: Pacific Asia Museum.
E-mail: info@pacificasiamuseum.org
Web Site: www.pacificasiamuseum.org
Founded: 1971.
Congressional District: 22
Key Personnel: Interim Exec. Dir., Margaret Leong Checca; Chm. (V), Katherine Murray-Morse; Cur. & Mgr. Collections, Bridget Bray; Dir. Mktg., Chelsea Mason; Grants & Special Events Mgr., Daniel Johnson; Cur. Education, Amelia Chapman; Visitor Svcs. & Facility Rental, Gabriella Karsch; Mgr. Library, Sally McKay; Volunteer Coord., Kabir Singh; Museum Shop Mgr., Tai-Ling Wong.
Personnel Profile: Full-Time Paid 13; Part-Time Paid 15; Part-Time Volunteers 400; Interns 4.
Governing Authority: nonprofit organization. Tax-exempt: 501(c)(3).
Institution Type/Description: Museum of Asian and Pacific Island Art: housed in a Chinese Imperial Palace style building; built in 1924.
Collections: Asian artifacts in all media including Chinese ceramics & textiles; Japanese paintings & prints; Himalayan sculpture; Pacific Island cultural artifacts.
Major Exhibits: Tapa Cloth from the Collection, 9/28/12-1/27/14.
Research Fields: arts relating to Chinese ceramics, Pacific Island tapas, Chinese textiles, Japanese Art.
Facilities: 7,500-vol. library of material relating to Asian & Pacific Island Arts, available for research to members & scholars by appointment only; 100-seat auditorium; classrooms. Books, gift items & art objects for sale.
Activities: permanent, temporary & traveling exhibitions; guided tours; lectures; films; concerts; dance programs; gallery talks; formally organized educational programs; docent training program; travel programs; 8 Asian Arts Councils; Docent Council; Service Council; Family Festival Days.
Publications: quarterly museum newsletter; catalogues pertaining to the exhibitions; posters.
Hours & Admission Prices: Wed.-Sun. 10-6. Adults $10, seniors & students $7; members no charge. Closed New Year's Eve & Day; Independence Day; Thanksgiving; Christmas. &
Attendance: 53,275 (accurate)
Membership: Active $55; Donor $100; Friend $250; Patron $500; Benefactor $1,000; Associate $1,500-$5,000; Collector's Circle $3,000.

PASADENA MUSEUM OF CALIFORNIA ART, 490 E. Union St., Pasadena, CA 91101-1790. Mailing Address: 495 E. Colorado Blvd., Pasadena, CA 91101-2024. Tel.: 626-568-3665. Fax: 626-568-3674.
E-mail: info@pmcaonline.org
Web Site: www.pmcaonline.org
Founded: 2002.
Key Personnel: Exec. Dir., Jenkins Shannon; Chm., David Partridge; Gallery Mgr., Emmett Clements; Dir. Public Rels., Emma Jacobson-Sive; Vice Chm., Jim Crawford; Exhibition Designer, Sergio Gomez; Exhibition Mgr. & Registrar, Erin Aitali.
Personnel Profile: Full-Time Paid 5; Part-Time Paid 4; Part-Time Volunteers 5; Interns 3.
Governing Authority: private; nonprofit organization. Tax-exempt: 501(c)(3).
Institution Type/Description: Art Museum.
Collections: California art, design & architecture from 1850 to present.
Research Fields: contemporary & historic California art.
Facilities: 500-vol. library; 8,000 sq. ft. exhibit space. Museum-related items for sale.
Activities: concerts; docent program; guided tours; lectures; loan & traveling exhibitions.
Publications: exhibition catalogues; newsletters.
Hours & Admission Prices: Wed.-Sun. 12-5. Adults $7, senior citizens & students $5; 1st Fri. of month, children under 12 & members no charge. Closed New Year's Day; Independence Day; Thanksgiving; Christmas. &
Attendance: 35,000 (estimated)
Membership: Senior & Student $25; Studio $50; Guild $200; Director $500.

PASADENA MUSEUM OF HISTORY, (M), 470 W. Walnut St., Pasadena, CA 91103-3562. Tel.: 626-577-1660. Fax: 626-577-1662.
E-mail: info@pasadenahistory.org
Web Site: www.pasadenahistory.org
Formerly: Pasadena Historical Museum
Founded: 1924.
Congressional District: 44
Key Personnel: Exec. Dir., Jeannette O'Malley; Pres., Laura Thompson; Dir. Exhibitions & Public Programming, Ardis Willwerth; Archivist, Laura

Verlaque; Coord. Membership, Michelle Turner; Coord. Visitor Svcs. & Volunteers, Emily Leiserson; Museum Shop Mgr., Katie Brandon.

Personnel Profile: Full-Time Paid 6; Part-Time Paid 10; Part-Time Volunteers 200.

Governing Authority: private; nonprofit organization. Tax-exempt: 501(c)(3).

Institution Type/Description: Historic House Museum: 1906 Pasadena home belonging to the Fenyes Family & descendents; mansion served as Finnish consulate.

Collections: manuscripts; photographs; architectural records; books; pamphlets; 19-century Finnish folk art; costumes; 17th- & 18th-century period furnishings & furniture; early California paintings.

Research Fields: history.

Facilities: 10,000-vol. library for Pasadena & surrounding area; 2 acre garden; 10,000 sq. ft. exhibit space. Museum-related items for sale.

Activities: guided tours; lectures; historical exhibits; concerts.

Publications: quarterly newsletter; pamphlets.

Hours & Admission Prices: History Center & Store: Wed.-Sun. 12-5. Library & Archives: Wed.-Sun. 1-4. Fenyes Mansion Tours: Fri.-Sun. 1:30 & 3. General $5, Mansion Tours $4; discounts to seniors, students & AAM members; children under 12 no charge. &

Attendance: 10,712 (accurate)

Membership: Active $50; Friend $75; Sponsor $125; Donor $250; Business Sponsors $300; Patron $500; Founders Circle $1,000 & up.

TOURNAMENT HOUSE AND WRIGLEY GARDENS, 391 S. Orange Grove Blvd., Pasadena, CA 91184-0002. Tel.: 626-449-4100. Fax: 626-449-9066.

E-mail: rosepr@rosemail.org

Web Site: www.tournamentofroses.com/aboutus/house.asp

Institution Type/Description: Historic House Museum: housed in former home of chewing gum manufacturer William Wrigley, Jr.

Collections: personal artifacts; portraits.

Hours & Admission Prices: Feb.-Aug. House: Thurs. 2-4. Gardens: daily.

Paso Robles

ESTRELLA WARBIRDS MUSEUM, 4251 Dry Creek Rd., Paso Robles, CA 93446. Tel.: 805-238-3897 & 227-0440.

E-mail: webmaster@ewarbirds.org

Web Site: www.ewarbirds.org

Founded: 1989.

Congressional District: 23

Personnel Profile: Part-Time Volunteers 60.

Governing Authority: nonprofit organization. Tax-exempt: 501(c)(3).

Institution Type/Description: Military History Museum.

Collections: military history, aircraft & vehicles; personal artifacts; photographs.

Research Fields: military aviation.

Facilities: meeting hall; rental facilities.

Activities: monthly dinner meeting.

Publications: newsletter, Eagle.

Hours & Admission Prices: Thurs.-Sun. 10-4. Adults $10, seniors 60 & over $8, children 6-12 $5; members & children under 6 no charge. &

Attendance: 7,886 (accurate)

Membership: Single $50; Family $75; Sustaining $100; Patron $200; Life $500; Star $1,000 & up.

PASO ROBLES ART ASSOCIATION, 2208 Golden Hill Rd., Paso Robles, CA 93446-6379. Mailing Address: P.O. Box 2219, Paso Robles, CA 93447-2219. Tel.: 805-238-5473.

Web Site: www.pasoroblesartassociation.org

Formerly: Call-Booth House Gallery

Key Personnel: Dir., Jerry Boxen

Institution Type/Description: History & Art Museum.

Collections: paintings.

Hours & Admission Prices: Wed.-Sun. 11-3.

PASO ROBLES CHILDREN'S MUSEUM, 623 13th St., Paso Robles, CA 93446. Tel.: 805-238-7432.

E-mail: sarah@pasokids.org

Web Site: www.pasokids.org

Founded: 2002.

Governing Authority: nonprofit organization.

Institution Type/Description: Children's Museum.

Collections: hands-on exhibitions.

Activities: educational programs.

Hours & Admission Prices: Sun. & Thurs.-Fri. 11-4, Wed. & Sat. 10-4. Adults $7, children 1-13 $6, seniors 65 & over $5.

Penn Valley

MUSEUM OF ANCIENT & MODERN ART, 11392 Pleasant Valley Rd., Penn Valley, CA 95946-0975. Mailing Address: P.O. Box 975, Penn Valley, CA 95946-0975. Tel.: 530-432-3080. Fax: 530-272-0184.

Web Site: www.mama.org

Founded: 1981.

Congressional District: 14

Key Personnel: Chm. Bd. (V), Dr. Ann-Victoria Hopcroft; Pres. (V), Cur. Science & Technology, Dr. Claude Needham; Vice Pres. & Dir. Children's Art Academy, Zoe Alowan; C.F.O., David Franco; Dir. Public Rels., Morgan Fox; Public Rels. Coord. & Museum Shop Mgr., Jewel McInroy; Program Devel. & Cur. 19th- & 20th-Century Works on Paper, Linda Corriveau; Cur. Antiquities, Jeff Spencer; Assoc. Cur. Stellar Astronomy, Dr. James D. Wrey; Cur. Mathematics & Science, David Christie; Cur. Interactive Multimedia, Tim Elston; Conservator (Ancient), Nancy Christie; Conservator (Modern), Robbert Trice; Librarian, Nancy Burns; Computer Archivist, Wayne Hoyle.

Personnel Profile: Part-Time Volunteers 30.

Governing Authority: nonprofit organization. Tax-exempt: 501(c)(3).

Institution Type/Description: Art Museum.

Collections: dioramas of ancient history from prehistoric times; 18th-century Dynasty Amarna artifacts from 1926-27 Pendlebury digs; Theodora Van Runkel collection of ancient gold; objects of daily use; jewelry; African masks; dinosaur fossils; meteorites; early books; works on paper from the 16th century; modern & contemporary art.

Research Fields: ancient technologies; comparative lifestyles; ancient rituals; cultural influences; ancient dance; slime molds.

Facilities: 1,500-vol. library for research & reference upon special request; Children's Art Academy. Museum-related items for sale.

Activities: guided tours; Children's Art Academy; field trips; workshops; seminars; lectures; docent program; SHARE & TAP tour, Traveling Art Program offering exhibits on ancient Egypt & Dinosaurs (Mummies, Myth & More & Dinosaurs in my Garden) which travel to school; Art Appreciation program; temporary & traveling exhibitions; concerts; gallery talks. Museum Sponsors: From a Kid's Point of View Annual Children's Art Festival; special preview events.

Publications: exhibition catalogs; quarterly literary journal for children, SeeSaw; quarterly newsletter of Children's Art Academy, Art is Fun!; bimonthly newsletter, DIMENSIONS.

Hours & Admission Prices: Call for hours. No charge; donations accepted. &

Attendance: 18,000 (estimated)

Membership: Individual $25; Family $35; Supporting $100; Sustaining $1,000.

Perris

LAKE PERRIS REGIONAL INDIAN MUSEUM (HOME OF THE WIND), 17801 Lake Perris Dr., Perris, CA 92571-8400. Tel.: 951-940-5657. Fax: 951-657-0077.

Founded: 1985.

Congressional District: 44

Key Personnel: Park Supt., John Rowe.

Personnel Profile: Full-Time Paid 1; Part-Time Paid 1; Part-Time Volunteers 3.

Governing Authority: Parent Institution: California State Parks. Tax-exempt.

Institution Type/Description: Park & Museum.

Collections: Southern California Native Americans.

Facilities: nature trails.

Activities: school group tours; animal tracking; basket weaving; story tellers; cultural & nature walks; special events; camping; water sports; equestrian trails. CA Dept. Parks & Recreation: boating, fishing, camping, hiking, bicycling.

Publications: maps; bird guides; brochure, Wild Flower.

Hours & Admission Prices: Museum: Fri. 10-2, Sat.-Sun. 10-4; groups by appointment. Museum: no charge; donations accepted. Park: entrance fee charged. &

Attendance: 12,000 (accurate)

ORANGE EMPIRE RAILWAY MUSEUM, 2201 S. A St., Perris, CA 92570-9318. Mailing Address: P.O. Box 548, Perris, CA 92572-0548. Tel.: 951-943-3020. Fax: 951-943-2676. Facebook: Orange Empire Railway Museum.

E-mail: info@oerm.org

Web Site: www.oerm.org

Founded: 1956.

Congressional District: 42

Key Personnel: C.E.O. & Pres. (V), George L. Huckaby, Jr.; Chm. (V), Thomas Jacobson; Museum Shop Mgr., Donna Zanin.

Personnel Profile: Full-Time Paid 1; Part-Time Paid 2.

Governing Authority: nonprofit organization. Parent Institution: Southern California Railway Museum, Inc. Tax-exempt: 501(c)(3).
Institution Type/Description: Railway Museum: on the historical Santa Fe line from Riverside to San Diego.
Collections: steam, diesel & electric locomotives; electric streetcars & interurbans; passenger cars; freight cars, including maintenance-of-way cars and cabooses. Historical Structures: 1882 dugout-type pioneer house; 1892 Santa Fe depot.
Research Fields: railway technology; railway history.
Facilities: picnic area; rental facilities. Gift items for sale.
Activities: special events. Museum Sponsors: Rail Festival in April.; Swap Meets in spring & fall; A Day Out With Thomas in November.
Publications: monthly newsletter; web-based publications.
Hours & Admission Prices: Daily 9-5. Museum: no charge except for gated events. Weekend Rides: adults $12, children 5-11 $8; discounts to Assoc. of Railway Museums members; children under 5 no charge. Closed Thanksgiving; Christmas. &
Attendance: 40,000 (estimated)
Membership: Junior Affiliate $15; Associate $30; Individual $40; Family $60; Sustaining $100; Benefactor $250; Corporate $500; Life $800.

PERRIS VALLEY HISTORICAL MUSEUM, 120 4th St., Perris, CA 92570. Mailing Address: P.O. Box 343, Perris, CA 92572. Tel.: 951-657-0274.
E-mail: info@perrismuseum.com
Web Site: perrismuseum.com
Founded: 1964.
Congressional District: 49
Key Personnel: Pres. (V), Quinn Hawley.
Personnel Profile: Part-Time Paid 1; Part-Time Volunteers 25.
Governing Authority: Parent Institution: Perris Valley Historical & Museum Association. Tax-exempt.
Institution Type/Description: Historic Building Museum: housed in the Santa Fe Depot; built in 1892.
Collections: local history & culture; photographs; period furnishings; personal artifacts; Native American.
Major Exhibits: City of Perris Centennial Display (T), 11/13-6/14.
Facilities: Museum-related items for sale.
Publications: quarterly newsletter.
Hours & Admission Prices: Thurs.-Sun. 12-4. No charge. &
Attendance: 1,000 (estimated)
Membership: Individual $15; Family $25; Nonprofit $30; Business $50; Lifetime $100; Nonprofit Lifetime $250; Business Lifetime $300.

Pescadero

ANO NUEVO STATE RESERVE, 1 New Years Creek Rd., Pescadero, CA 94060. Mailing Address: 303 Big Trees Park Rd., Felton, CA 95018-9660. Tel.: 650-879-2025. Fax: 650-879-2031.
E-mail: info@parks.ca.gov
Web Site: www.parks.ca.gov
Founded: 1980.
Key Personnel: Supervising Ranger, Gary Strachan.
Governing Authority: state. Parent Institution: California State Parks. Tax-exempt.
Institution Type/Description: Historic Building & Park Museum: housed in c.1880 Dickerman Dairy Barn.
Collections: shipwreck artifacts; early farm equipment & tools; elephant seal pictures; photographs; archaeology; Indian boats & tools; flora & fauna exhibits.
Research Fields: early Indian; whaling; farming; natural history; elephant seals; marine mammals.
Facilities: nature/conservation center. Books, photos & other museum-related items for sale.
Activities: guided tours; lectures; formally organized education programs for children, undergraduate & graduate college students; guided disabled tours; docent program; permanent & temporary exhibitions.
Publications: brochure, Ano Nuevo State Reserve.
Hours & Admission Prices: Daily 8:30-4:30. Auto regular $10, auto senior $9. Guided Walks: Dec. 15 to March 31 daily 8:45-2:45. Per ticket $7; children 3 & under no charge. &
Attendance: 105,000
Membership: Individual $15; Family $25; Supporting $50; Sponsor $100.

Petaluma

PETALUMA ADOBE STATE HISTORIC PARK, 3325 Adobe Rd., Petaluma, CA 94954. Tel.: 707-762-4871. Fax: 707-762-4871.
E-mail: petaluma.adobe@parks.ca.gov
Web Site: www.petalumaadobe.com
Founded: 1951.
Personnel Profile: Full-Time Paid 1; Part-Time Paid 3; Part-Time Volunteers 15.
Governing Authority: state. Parent Institution: State of California, Dept. of Parks & Recreation.
Institution Type/Description: Historic Building: 1836 Vallejo's adobe ranch headquarters.
Collections: furnishings; tools pertaining to 1840 ranch life.
Facilities: picnic area.
Activities: self-guided tours; special events.
Publications: brochures.
Hours & Admission Prices: Tues.-Wed. 10-5. Adults 17 & over $3; children 6-16 $2; members no charge. Closed New Year's Day; Thanksgiving; Christmas. &
Attendance: 8,700 (accurate)
Membership: Sonoma State Historic Park Association: Individual $20.

PETALUMA HISTORICAL LIBRARY AND MUSEUM, (M), 20 Fourth St., Petaluma, CA 94952-3004. Tel.: 707-778-4398. Fax: 707-762-3923.
E-mail: pmuseum.info@petalumamuseum.com
Web Site: petalumamuseum.com
Founded: 1978.
Congressional District: 6
Personnel Profile: Part-Time Paid 1; Part-Time Volunteers 60; Interns 1.
Governing Authority: Parent Institution: City of Petaluma. Tax-exempt.
Institution Type/Description: History Museum.
Collections: local history; Miwok Indian artifacts; poultry & agricultural industries; period furnishings; vintage textiles.
Activities: guided tours; outreach programs; research; internships; special events. Museum Sponsors: Annual Tea.
Publications: quarterly newsletter, Petaluma Museum Association Newsletter.
Hours & Admission Prices: Thurs.-Sat. 10-4, Sun. 12-3; other times by appointment. General admission $5, seniors & members $3; military no charge. Closed holidays. &
Attendance: 15,000 (accurate)
Membership: Seniors, Educators & Students $20; Individual $35; Family/Dual $70; Supporter $125; Business Sponsor $150; Patron $400; Company Sponsor $500; Benefactor $750; Corporate Sponsor $1,000 & up.

PETALUMA WILDLIFE AND NATURAL SCIENCE MUSEUM, 201 Fair St., Petaluma, CA 94952-2594. Tel.: 707-778-4787. Fax: 707-778-4603.
E-mail: info@petalumawildlifemuseum.com
Web Site: www.petalumawildlifemuseum.com
Formerly: Wildlife Museum at Petaluma High School
Founded: 1990.
Key Personnel: Exec. Dir., Neal Ramus; Pres., George Grossi.
Personnel Profile: Part-Time Paid 2; Part-Time Volunteers 45.
Governing Authority: private; nonprofit. Tax-exempt: 501(c)(3).
Institution Type/Description: Wildlife & Natural Science Museum.
Collections: North American, African & Asian taxidermy specimens; minerals; fossils; live & stuffed marine animals; forestry & logging; live mammals & reptiles; African artifacts; dinosaurs.
Facilities: research & reference library; aquarium; classroom; 8,000 sq. ft. exhibit space; nature/conservation center; zoological park.
Activities: docent program; formal education program for children; guided tours; broadcast programs; blind & handicapped program. Annual Events: Pasta Feed-local dinner & tours for public; dinner, raffle & auction.
Publications: annual newsletter, Museum News & Update; monthly e-newsletter.
Hours & Admission Prices: Sat. 11-3, call to confirm. Admission $5; children 5 & under no charge. Closed major holidays. &
Attendance: 3,171 (accurate)
Membership: Silver Ram Foundation $250 & up.

Piercy

CONFUSION HILL GRAVITY HOUSE, 75001 N. Hwy. 101, Piercy, CA 95587-8805. Tel.: 707-925-6456. Fax: 707-925-6477.
E-mail: confusion@asis.com
Web Site: www.confusionhill.com
Founded: 1949.
Key Personnel: Owner & Operator, Doug Campbell; Owner & Operator, Carol Campbell
Institution Type/Description: Logging Museum.
Collections: hands-on exhibits; period logging equipment.
Facilities: Gift items for sale.
Activities: train ride; self-guided tour.
Hours & Admission Prices: House: May-Sept. daily 9-6; Oct.-April daily 9-5.

Adults $5, children 4-12 $4; children 3 & under no charge. Train Ride: June Sept. daily 10-5. Rides: adults $8.50, children 4-12 $6.50; children 3 & under no charge.
Attendance: 20,000 (estimated)

Piru

RANCHO CAMULOS MUSEUM, Rte. 126, Piru, CA 93040. Mailing Address: P.O. Box 308, Piru, CA 93040-0308. Tel.: 805-521-1501.
Key Personnel: Dir., Susan Falck.
Personnel Profile: Part-Time Paid 1; Part-Time Volunteers 50.
Governing Authority: Tax-exempt.
Institution Type/Description: Historic Site: listed on the National Register of Historic Places; a National Historic Landmark.
Collections: California history & culture; photographs; 1853 adobe.
Facilities: 1,800 acre ranch.
Activities: living history presentations; special events; docent tours.
Publications: quarterly newsletter.
Hours & Admission Prices: Sat. 1-4; other times by appointment. Adults $5; members no charge.
Attendance: 4,500 (estimated)
Membership: Individual $40; Family $50; $125; $250; $500; $1,000.

Pittsburg

LOS MEDANOS COLLEGE ART GALLERY, 2700 E. Leland Rd., Pittsburg, CA 94565-5197. Tel.: 925-439-2181, ext. 3493.
Web Site: www.losmedanos.net/groups/art/default.htm
Key Personnel: Dir., Judi Pettite; Cur., Dawn Black
Institution Type/Description: Art Gallery.
Collections: paintings; sculpture.
Hours & Admission Prices: Tues.-Thurs. 12:30-2:30 & 6:30-8:30. No charge.

Placentia

GEORGE KEY RANCH, 625 W. Bastanchury Rd., Placentia, CA 92870-2230. Tel.: 714-973-3190 & 3191.
E-mail: keyranch@ocparks.com
Web Site: www.ocparks.com/keyranch
Institution Type/Description: Historic House Museum.
Collections: farm equipment; hand tools; orange grove; garden.
Facilities: 2.2 acres; garden; orange grove; botanical preserve; picnic area.
Activities: interpretive programs.
Hours & Admission Prices: June-April Tues.-Fri. 12:30-4:30, 1st Sat. each month 2-4.

Placerville

EL DORADO COUNTY HISTORICAL MUSEUM, 104 Placerville Dr. Fairgrounds, Placerville, CA 95667-3910. Tel.: 530-621-5865. Fax: 530-621-6644.
E-mail: museum@co.el-dorado.ca.us
Web Site: www.co.el-dorado.ca.us/museum
Founded: 1974.
Congressional District: 14
Key Personnel: Museum Admin., Mary Cory.
Personnel Profile: Full-Time Paid 1; Part-Time Volunteers 70.
Governing Authority: county; nonprofit. Parent Institution: El Dorado County. Subsidiary Institution: Museums Foundation. Tax-exempt: 501(c)(3).
Institution Type/Description: History Museum.
Collections: archives; history; Indian artifacts; mine equipment; historical artifacts from El Dorado County; country store; Victorian room display; early transportation display.
Research Fields: El Dorado County history.
Facilities: collection of material pertaining to historical places and people of El Dorado County available for use on premises.
Activities: permanent & temporary exhibitions.
Publications: pamphlets; books.
Hours & Admission Prices: Wed.-Sat. 10-4, Sun. 12-4; call for extended hours. No charge; donations accepted. Closed holidays.
Attendance: 20,000 (accurate)

Pleasanton

MUSEUM ON MAIN, (M), 603 Main St., Pleasanton, CA 94566-6603. Tel.: 925-462-2766. Fax: 925-462-2779.
E-mail: info@museumonmain.org
Web Site: www.museumonmain.org
Formerly: Amador-Livermore Valley Historical Society
Founded: 1963.
Congressional District: 15
Key Personnel: Exec. Dir., Jim DeMersman; Pres. (V), Chuck Deckert; Dir. Education, Jennifer Amiel; Cur., Ken MacLennan; Administrative Asst., Bonnie Fitzpatrick.
Personnel Profile: Full-Time Paid 1; Part-Time Paid 3; Part-Time Volunteers 65; Interns 2.
Volunteer Hours: 11,000
Operating Expenses: 213,297
Operating Income: 202,108
Governing Authority: society. Parent Institution: Amador-Livermore Valley Historical Society. Tax-exempt: 501(c)(3).
Institution Type/Description: General Museum: housed in c.1914 Town Hall.
Collections: tools; household items; local artifacts; photographs; textiles; manuscripts; newspapers; film.
Major Exhibits: Multiply by Six Million (T), 1/14-2/14; Imagination Expressed, 3/14-4/14; Hacienda Business Park, 5/14-7/14; Art of China Painting, 7/14-10/14; Phobias: What Are You Afraid of?, 10/14; 100 Years @ 603 Main, 11/14-12/14.
Research Fields: local history.
Facilities: library & archives of local history available for research. Books & museum-related items for sale.
Activities: guided tours; lectures; volunteer program; permanent & traveling exhibits; rotating cultural exhibits.
Publications: books; pamphlets; prints
Hours & Admission Prices: Tues.-Sat. 10-4, Sun. 1-4. No charge; donations accepted.
Attendance: 20,071 (accurate)
Membership: Individual $30; Family $50; Friend $100; Director's Circle $250; Patron & Business $500; Life $1,000.

Point Arena

POINT ARENA LIGHTHOUSE AND MUSEUM, 45500 Lighthouse Rd., Point Arena, CA 95468. Mailing Address: P.O. Box 11, Point Arena, CA 95468-0011. Tel.: 877-725-4448; 707-882-2777. Fax: 707-882-2111. Facebook: Point Arena Lighthouse.
E-mail: palight@mcn.org
Web Site: www.pointarenalighthouse.com
Founded: 1870.
Congressional District: 2
Key Personnel: Exec. Dir., Ty Moore; Chm. (V), Nicolas Epanchin.
Governing Authority: Parent Institution: Point Arena Lighthouse Keepers, Inc.
Institution Type/Description: Lighthouse & History Museum.
Collections: structure; period artifacts.
Research Fields: lighthouse & its keepers.
Facilities: Museum-related items for sale.
Activities: walking tours; exploring 23 acres of land; 1st order Fresnel lens on display.
Publications: quarterly, Lighthouse Journal.
Hours & Admission Prices: Winter: daily 10-3:30. Adults $7.50, children $1; members no charge.
Attendance: 40,000 (estimated)
Membership: Individual $40; Family $55; Life Individual $400; Life Family $550.

Point Reyes

POINT REYES NATIONAL SEASHORE, 1 Bear Valley Rd., Point Reyes, CA 94956-9703. Tel.: 415-464-5100 & 5125 (Curator). Fax: 415-464-5229.
E-mail: carola_derooy@nps.gov
Web Site: www.nps.gov/pore
Founded: 1962.
Congressional District: 6
Key Personnel: Park Supt., Cicely Muldoon; Chief Interpretation, John Dell'Osso; Archivist, Carola DeRooy.
Governing Authority: federal. Parent Institution: National Park Service, U.S. Dept. of the Interior. Tax-exempt.
Institution Type/Description: Natural & Cultural Museum.
Collections: natural history; archaeology; history; 16th Century porcelain shards; Maritime radio. Historic Building: 1870 Point Reyes Lighthouse.
Research Fields: Applied Natural Resources Studies.
Facilities: library of natural history, local history, collections, archives & research available for use on premises. Interpretive museums & visitors centers. Books on park-related subjects for sale.
Activities: lectures; demonstration.
Publications: pamphlets; books; posters.

Hours & Admission Prices: Bear Valley Visitor Center: Mon.-Fri. 9-5, Sat.-Sun. & holidays 8-5. Ken Patrick Visitor Center: Sat.-Sun. & holidays 10-5. Lighthouse & Lighthouse Visitor Center: Thurs.-Mon. 10-4:30. Library & Study collection: Mon.-Fri. 8-4:30, by appointment only. No charge; donations accepted. &
Attendance: 500,000 (accurate)
Membership: Point Reyes National Seashore Association.

Point Richmond

GOLDEN STATE MODEL RAILROAD MUSEUM, 900-A Dornan Dr., Point Richmond, CA 94801-4126. Mailing Address: P.O. Box 71244, Point Richmond, CA 94807-1244. Tel.: 510-232-2472 & 234-4884.
E-mail: info@gsmrm.org
Web Site: www.gsmrm.org
Founded: 1985.
Congressional District: 7
Key Personnel: Gen. Mgr., Randy Smith; C.F.O., Phil Figel; Museum Shop Mgr., Floyd McCarty.
Personnel Profile: Part-Time Volunteers 50.
Governing Authority: private; nonprofit organization. Tax-exempt: 501(c)(3).
Institution Type/Description: Model Railroad Museum.
Collections: historic scale model railroad equipment & steam locomotives; modern diesel & intermodal trains; PFM - Bill Ryan brass locomotives.
Facilities: library; 10,000 sq. ft. exhibit space.
Activities: hobby workshop.
Publications: quarterly newsletter, Zephyr.
Hours & Admission Prices: April-Dec. Sat.-Sun. 12-5. Train operations only on Sun. Family $9, adults $4; discounts to Contra Costa Library members; members, Wed. 11-3 & Sat. no charge. &
Attendance: 7,161 (accurate)
Membership: Individual $24; Family $45; Assessed $120.

Pomona

ADOBE DE PALOMARES, 491 E. Arrow Hwy., Pomona, CA 91767-2264. Mailing Address: Historical Society of Pomona Valley, 585 E. Holt Ave., Pomona, CA 91767. Tel.: 909-620-0264; 909-623-2198.
E-mail: pomonahistorical@verizon.net
Web Site: www.pomonahistorical.org/palomares
Institution Type/Description: Historic House Museum.
Collections: period furniture; personal artifacts; herb garden; blacksmith shop.
Facilities: garden.
Hours & Admission Prices: Sun. 2-5. No charge. Closed Easter; Memorial Day; Labor Day; Thanksgiving.
Membership: Students & Seniors $10; Individual $15; Family $20; Life $150.

AMERICAN MUSEUM OF CERAMIC ART, (M), 399 N. Garey Ave., Pomona, CA 91767-5431. Tel.: 909-865-3146. Fax: 909-629-1067. Facebook: American Museum of Ceramic Art.
E-mail: frontdesk@amoca.org
Web Site: www.amoca.org
Founded: 2004.
Key Personnel: Pres., David Armstrong; Devel. Consultant, Carolyn Wagner; Assoc. Cur., Rody Lopez; Collection Mgr., Nicole Frazer; Education Mgr., Angie Reyes; Mktg. & IT, Toby Lam; Membership Mgr., Victor Crosett; Studio Dir., Heidi Kreitchet.
Personnel Profile: Full-Time Paid 8; Part-Time Paid 3; Part-Time Volunteers 3; Interns 3.
Governing Authority: private; nonprofit organization. Tax-exempt: 501(c)(3).
Institution Type/Description: Art Museum.
Collections: works of ceramics artists from around the world.
Major Exhibits: Best Kept Secret: The Scripps College Ceramic Collection, 1/11/14-3/30/14; kilnopening.edu / Little T Pot, 4/12/14-6/1/14; Larger As Life: Betty Davenport Ford, Elaine Katzer, Lisa Reinertson, 6/14/14-8/31/14.
Facilities: 2,200 sq. ft. exhibit space. Museum-related items for sale.
Activities: arts festivals; concerts; films; guided tours; hobby workshops; lectures; loan exhibition; theater; temporary & traveling exhibitions. Museum Sponsors: Pottery Market in summer & winter; member only events.
Publications: quarterly newsletter; exhibition catalogues.
Hours & Admission Prices: Wed.-Sat. 12-5, 2nd Sat. of month 12-9. Adults $5, senior citizens & students $4; members no charge. Closed New Year's Day; Thanksgiving; Christmas Eve & Day. &
Attendance: 10,448 (estimated)
Membership: Student $25; Active $40; Sustaining $75; Cobalt $100; Celadon $250; Silver Luster $500; Gold Luster $1,000; Armstrong Society $5,000 & up.

THE DA CENTER FOR THE ARTS, 252-D S. Main St., Pomona, CA 91766. Tel.: 909-397-9716. Fax: 909-629-8697.
E-mail: daartcenter@gmail.com
Web Site: www.dacenter.org
Institution Type/Description: Art Gallery.
Collections: paintings; sculpture; photographs.
Activities: classes; workshops.
Hours & Admission Prices: Call for hours.

LA CASA PRIMERA DE RANCHO SAN JOSE, 1569 N. Park Ave., Pomona, CA 91768-1835. Mailing Address: Historical Society of Pomona Valley, 585 E. Holt Ave., Pomona, CA 91767. Tel.: 909-623-2198.
Web Site: www.laokay.com/lacasaprimera.htm
Institution Type/Description: History Museum.
Collections: 19-century furnishings; period artifacts.
Hours & Admission Prices: Sun. 2-5. Closed Easter weekend; Memorial Day weekend; Labor Day weekend; Thanksgiving and weekend after.

LATINO ART MUSEUM, 281 S. Thomas St., Ste. 105, Pomona, CA 91766-1750. Tel.: 909-620-6009.
E-mail: latinoartmuseum0@gmail.com
Web Site: www.lamoa.net
Founded: 2001.
Key Personnel: Dir. & C.E.O., Graciela H. Nardi
Governing Authority: Tax-exempt.
Institution Type/Description: Art Museum.
Collections: paintings; photographs; Latin American art; international art.
Facilities: rental facilities. Museum-related items for sale.
Activities: special events; lectures; poetry. Museum Sponsors: Art Walk & Opening Reception; Pot-Luck.
Publications: Yearbook, 2008; Yearbook 2009; Biennale 2006; Biennale 2004; Yearbook, 2010 Yearbook, 2012.
Hours & Admission Prices: Wed.-Sat. 3:30-6:30, 2nd & last Sat. each month 3:30-9:30. No charge; donations accepted. &
Attendance: 500
Membership: Senior & Student $15; Individual $25; Family $40; Sponsor $100; Benefactor $1,000; Corporate Benefactor $2,500.

RAIL GIANTS TRAIN MUSEUM, Los Angeles County Fairgrounds, 1101 W. McKinley, Pomona, CA 91769. Mailing Address: P.O. Box 2250, Pomona, CA 91769-2250. Tel.: 909-623-0190.
E-mail: railgiants@gmail.com
Web Site: www.railgiants.org
Formerly: Southern California Chapter Railway & Locomotive Historical Society
Founded: 1954.
Key Personnel: Chm. (V), Robert Shatsnider; Pres. (V), Paul Guercio; Membership, Steve McFerson; Museum Shop Mgr., Shelley Hunter.
Personnel Profile: Full-Time Volunteers 200.
Governing Authority: private; not-for-profit organization. Parent Institution: Railway & Locomotive Historical Society. Subsidiary Institution: Southern California Chapter. Tax-exempt: 501(c)(3).
Institution Type/Description: Railroad Historical Society Museum.
Collections: railroad steam & diesel motive power including largest remaining locomotives in the world; rail signal systems; tools; telegraph systems; related hardware; reference books; photographs; original documents; employee timetables; historical steam engine display: Santa Fe Engine 3450, Southern Pacific Engine 5021, Union Pacific Engine 9000, Outer Harbor Engine 2, Fruit Growers Engine 3, Union Pacific Engine 6915, U.S. Potash Engine 3; Santa Fe Horse Express 1993; Pullman 8 Dining-Lounge Car; Santa Fe Caboose 1314; General American Reefer 67806; Arcadia Santa Fe Depot.
Activities: Museum Sponsors: Open Houses; LA County Fair in September.
Publications: monthly newsletter; biannual books, railroad history.
Hours & Admission Prices: 2nd Sat.-Sun. each month 10-4; other times by appointment. Open everyday during the Los Angeles County Fair until 9 pm. No charge; donations accepted.
Attendance: 30,000 (estimated)
Membership: Annual $52.

SCA PROJECT GALLERY, 281 S. Thomas St., #104, Pomona, CA 91766-1750. Tel.: 909-865-0252.
E-mail: scaprojectgallery@gmail.com
Web Site: www.scaprojectgallery.com
Governing Authority: nonprofit organization.
Institution Type/Description: Art Gallery.
Collections: works by contemporary artists; paintings; sculpture.

Activities: temporary exhibitions.
Hours & Admission Prices: Thurs.-Sat. 12-4, 2nd Sat. each month 12-9.

WALLY PARKS NHRA MOTORSPORTS MUSEUM, Fairplex Gate 1, 1101 W. McKinley Ave., Bldg. 3A, Pomona, CA 91768-1639. Tel.: 909-622-2133. Fax: 909-622-1206.
E-mail: themuseum@nhra.com
Web Site: www.museum.nhra.com
Founded: 1998.
Key Personnel: Exec. Dir., Tony Thacker; Cur., Greg Sharp; Mgr. Mktg. & Advertising, Rose Dickinson; Coord. Museum Svcs., Sheri Watson.
Governing Authority: Presented by Automobile Club of Southern California. Tax-exempt.
Institution Type/Description: Motorsports and Transportation Museum.
Collections: vintage & historical racing vehicles; photographs; trophies; helmets; driving uniforms; paintings; motorsports memorabilia.
Facilities: Museum-related items for sale.
Activities: Annual Events: California Hot Rod Reunion; National Hot Rod Reunion.
Hours & Admission Prices: Wed.-Sun. 10-5. Adults $8, seniors 60 & over and juniors 6-15 $6; discount to AAA members; children under 5 no charge. Closed New Year's Day; Memorial Day; Independence Day; Thanksgiving; Christmas. &
Attendance: 100,000

Port Hueneme

U.S. NAVY SEABEE MUSEUM, 99 23rd Ave., Port Hueneme, CA 93043. Mailing Address: 99 23rd Ave., Bldg. 99, Port Hueneme, CA 93043. Tel.: 805-982-5165. Fax: 805-982-5595.
E-mail: lara.godbille@navy.mil
Web Site: www.history.navy.mil
Formerly: U.S. Navy Civil Engineer Corps/Seabee Museum
Founded: 1947.
Congressional District: 21
Key Personnel: Dir., Lara Glodbille; Cur., Kimberlyn Crowell; Archivist, Gina Nichols.
Personnel Profile: Full-Time Paid 5; Part-Time Volunteers 12; Interns 2.
Governing Authority: federal. Parent Institution: Naval Historical Center, Washington, DC. Tax-exempt.
Institution Type/Description: Military History Museum.
Collections: military weapons; uniforms; models of military equipment; cultural & art displays; wearing apparel & native artifacts from many countries of the world; personal papers & books of high ranking officers; unit plaques & flags; dioramas; nuclear power plant model & displays from Antarctic & Alaska.
Research Fields: history of the Civil Engineer Corps/Seabees; projects; equipment.
Facilities: archives including records for CEC & Seabees; 20,000 sq. ft. exhibit space. Museum-related items for sale.
Activities: films; tours.
Publications: brochure.
Hours & Admission Prices: Tues.-Fri. 8-4, Sat. 10-3. No charge. Closed federal holidays; Christmas week. &
Attendance: 15,000 (estimated)

Porterville

PORTERVILLE HISTORICAL MUSEUM, 257 N. D St., Porterville, CA 93257-3622. Tel.: 559-784-2053. Fax: 559-784-4009.
E-mail: phm@pvilleca.com
Web Site: portervillemuseum.org
Founded: 1965.
Congressional District: 21
Key Personnel: Dir. & Pres., Rick Struble; Assoc. Dir., Bill Scruggs; Treas., Wayne Foltz; Sec., Judy Kover; Cur., Shiela Pickrell.
Personnel Profile: Full-Time Paid 1; Full-Time Volunteers 4; Part-Time Volunteers 16.
Volunteer Hours: 9,360
Governing Authority: board of directors. Tax-exempt.
Institution Type/Description: History Museum: housed in c.1913 Southern Pacific Railroad passenger depot.
Collections: Indian artifacts; farm implements; china; glass; vignettes.
Major Exhibits: Boy Scouts, 2/8/14-3/23/14; Kids Friendly Exhibit, 6/21/14-8/9/14; Art Exhibit, 8/16/14-9/7/14; Train Art Exhibit, 10/11/14-10/25/14; 31st Annual Toy & Train Show, 11/28/14-1/15.
Research Fields: heritage & anthropology; history of the Yokuts Indians.
Activities: guided tours; lectures by local historian; blacksmith shop; permanent & temporary exhibitions. Museum Sponsors: local model train society exhibit & Santa visit in December.

Publications: quarterly newsletter.
Hours & Admission Prices: Summer: Thurs.-Sat. 9-3; Winter: Thurs.-Sat. 10-4. Adults $4, students 6-12 $1; discounts to Friends of the Museum; members & children under 6 no charge. Closed New Year's Day; Fair Week; Thanksgiving; Christmas. &
Attendance: 8,000 (estimated)
Membership: Individual $20; Family $35; Sponsor $250, $500 & $1,000.

ZALUD HOUSE, 393 N. Hockett St., Porterville, CA 93257-3639. Mailing Address: 291 N. Main, Porterville, CA 93257-3737. Tel.: 559-782-7548. Fax: 559-791-7854.
E-mail: dmoore@ci.porterville.ca.us
Founded: 1976.
Congressional District: 21
Key Personnel: C.E.O., Donnie Moore; Cur., Heather Raymond.
Personnel Profile: Full-Time Paid 1; Part-Time Paid 2; Part-Time Volunteers 1.
Governing Authority: municipal; nonprofit organization.
Institution Type/Description: Historic House: built in 1892.
Collections: late 1800-1950 Zalud house; personal items including furniture, hats & clothing; Victorian garden; history of the City of Porterville & Tulare County.
Research Fields: preservation & conservation of items; paranormal.
Facilities: 600-vol. library of books & magazines; botanical garden.
Activities: docent program; guided tours; temporary exhibitions; weddings & events in garden; paranormal nighttime investigations. Annual Events: June Weddings; Old Fashioned Christmas.
Hours & Admission Prices: Feb.-Dec. Wed.-Sat. 10-4, Sun. 2-4. Adults $2, children $.50. Closed Easter; Independence Day; Thanksgiving; Christmas. &
Attendance: 2,888 (accurate)

Portola

WESTERN PACIFIC RAILROAD MUSEUM, (M), 700 Western Pacific Way, Portola, CA 96122-8636. Mailing Address: P.O. Box 608, Portola, CA 96122-0608. Tel.: 530-832-4131. Fax: 530-832-1854.
E-mail: info@wplives.org
Web Site: www.wplives.org
Formerly: Portola Railroad Museum
Founded: 1984.
Congressional District: 2
Key Personnel: Pres., Rod McClure; Membership, Eugene Vicknair.
Personnel Profile: Part-Time Volunteers 40.
Governing Authority: private; nonprofit organization. Parent Institution: Feather River Rail Society. Tax-exempt.
Institution Type/Description: Railroad Museum: former Western Pacific shop & service facility.
Collections: Western railroad history & equipment; including the Western Pacific Railroad & diesel-electric locomotive development; 38 diesel locomotives from 1929-1971; 99 freight, passenger cars, & cabooses; maintenance of way equipment.
Facilities: archives; 65,250 sq. ft. exhibit space; 60-seat cafeteria. Museum-related items for sale.
Activities: Annual Events: Railroad Days; Rail Fan Day; Run-A-Locomotive Program March to November.
Publications: bimonthly newsletter, Train Sheet; semiannual bulletin, The Headlight.
Hours & Admission Prices: March & Nov. daily 11-4; April-Oct. daily 10-5. Train Rides: Memorial Day to Labor Day. Adults $8, children $4; members no charge.
Attendance: 10,000 (estimated)
Membership: Associate $25; Active $50; Institutional $60; Family $80; Sustaining $150; Life $1,800; Family Life $3,000.

Presidio of San Francisco

PRESIDIO HISTORICAL ASSOCIATION, 1806 Belles St., #2A, Presidio of San Francisco, CA 94129. Mailing Address: P.O. Box 29163, San Francisco, CA 94129-0163. Tel.: 415-752-2270. Fax: 415-752-2260.
E-mail: presidio-assoc@att.net
Web Site: www.presidioassociation.org
Founded: 1959.
Congressional District: 5
Key Personnel: Pres. (V), Gary Widman; Vice Pres., Robert Cherny; Treas., Charlotte Hennessy.
Personnel Profile: Part-Time Paid 1; Part-Time Volunteers 15.
Governing Authority: nonprofit. Tax-exempt: 501(c)(3).

Institution Type/Description: Historical Society.
Research Fields: history of the west; military history; Presidio of San Francisco history.
Activities: guided tours; lectures; temporary exhibitions.
Publications: newsletter, Communique.
Hours & Admission Prices: By appointment only.
Membership: Participating $30; Associate $50; Contributing $100; Sustaining $250; Donor $500; Corporate $1,000.

Quincy

PLUMAS COUNTY MUSEUM, 500 Jackson St., Quincy, CA 95971-9412. Tel.: 530-283-6320. Fax: 530-283-6081.
E-mail: pcmuseum@digitalpath.net
Web Site: wwwplumasmuseum.org
Founded: 1964.
Congressional District: 4
Key Personnel: Dir. & C.E.O., Scott J. Lawson; Chm. (V) Bd. Dir., Norman Lamb; Pres. (V) Bd. Trustees & Chm. (V), Don Clark.
Personnel Profile: Full-Time Paid 1; Part-Time Volunteers 10.
Governing Authority: county. Subsidiary Institution: Plumas County Museum Association, Inc. Tax-exempt.
Institution Type/Description: Historical Museum.
Collections: Indian artifacts; geology; costumes; Maidu basketry; wildlife specimens; local historic artifacts; bottles; tools; mining & logging; ranching; railroad artifacts; photographs; negatives; documents; oral history tapes; historical books & maps; blacksmith shop; goldminer's cabin. Historic House: restored 1878 home.
Research Fields: local history; mining; lumbering; railroad; water rights; transportation; birth, death, marriage & naturalization records; Plumas County historic people & families.
Facilities: library; archives; garden rental facilities. Local history books & museum-related items for sale.
Activities: guided tours; lectures; permanent & temporary exhibitions; rental facilities; county history tours; annual pioneer living history school days; annual first peoples' school days.
Publications: Plumas County Historical Cookbook; History of Plumas County 1882; Historical Diaries of Plumas County Pioneers; History of Rich Bar; quarterly newsletter; Growing Up In Plumas County 1850-1920; The Diary of Sarah M. Dean, 1864-1865; books, Plumas County-History of the Feather River Region; Recollections of a 49er.
Hours & Admission Prices: Tues.-Sat. 10-4. Adults $2, children 12-17 $1; children under 12 & members no charge. Closed major holidays Oct.-April.
Attendance: 10,000 (estimated)
Membership: Individual $25; Family $35; Patron $100; Business $150; Sustaining $1,000.

Ramona

CLASSIC ROTORS, (M), Ramona Airport, 2690 Montecito Rd., Ramona, CA 92065-1638. Tel.: 760-650-9257.
E-mail: communications@rotors.org
Web Site: www.classicrotors.org
Founded: 1992.
Institution Type/Description: Helicopter Museum.
Collections: period helicopters.
Activities: Museum Sponsors: Air Shows.
Hours & Admission Prices: Fri.-Mon. 10-4, Tues.-Thurs. by appointment. No charge, donations accepted.

GUY B. WOODWARD MUSEUM, 645 Main St., Ramona, CA 92065-2043. Mailing Address: Ramona Pioneer Historical Society, P.O. Box 625, Ramona, CA 92065. Tel.: 760-789-7644.
E-mail: info@woodwardmuseum.org
Web Site: www.woodwardmuseum.org
Key Personnel: Pres., Judy Nachazel.
Governing Authority: nonprofit organization. Tax-exempt: 501(c)(3).
Institution Type/Description: History Museum.
Collections: photographs; Ramona Sentinel newspapers from 1894; period tools & equipment; period clothing; furniture; toys; historic records.
Hours & Admission Prices: Thurs.-Sun. 1-4; other times by appointment.

Rancho Cordova

SACRAMENTO CHILDREN'S MUSEUM, 2701 Prospect Park, Rancho Cordova, CA 95670. Tel.: 916-638-7228. Fax: 916-638-7245.
E-mail: info@sackid.org

Web Site: www.sackids.org
Founded: 2011.
Key Personnel: Dir., Sharon Stone Smith.
Personnel Profile: Full-Time Paid 2; Part-Time Paid 10.
Governing Authority: Tax-exempt.
Institution Type/Description: Children's Museum.
Collections: hands-on exhibitions.
Activities: birthday parties.
Hours & Admission Prices: Tues.-Sat. 9-5, Sun. 12-5. Admission $7. &
Attendance: 100,000 (accurate)

Rancho Cucamonga

CASA DE RANCHO CUCAMONGA, 8810 Hemlock St., Rancho Cucamonga, CA 91730-2319. Tel.: 909-989-4970.
Web Site: www.co.san-bernardino.ca.us/museum/branches/rains.htm
Institution Type/Description: Historic House Museum.
Collections: personal artifacts; furnishings; photographs.
Hours & Admission Prices: Tues.-Sat. 10-3. No charge; donations requested.

JOHN RAINS HOUSE, 8810 Hemlock Ave., Rancho Cucamonga, CA 91730-2319. Mailing Address: c/o San Bernardino Co. Museums, 2024 Orange Tree Lane, Redlands, CA 92374. Tel.: 909-989-4970 & 307-2669. Fax: 909-307-0539.
E-mail: rmckernan@sbcm.sbcounty.gov
Web Site: www.sbcountymuseum.org
Formerly: Casa de Rancho Cucamonga
Founded: 1972.
Congressional District: 35
Key Personnel: Dir., Robert McKernan; Site Mgr., Pam Strunk; Cur., Michele Nielsen.
Personnel Profile: Part-Time Paid 1; Part-Time Volunteers 25.
Governing Authority: county. Parent Institution: San Bernardino County Museums, 2024 Orange Tree Lane, Redlands, CA. Tax-exempt.
Institution Type/Description: Historic House Museum: housed in c.1860 John & Merced Rains home.
Collections: period furniture; local historical items; decorative arts.
Research Fields: local history.
Activities: guided tours. Special Events: Holidays, Rancho Days.
Hours & Admission Prices: Tues.-Sat. 10-3. No charge; donations accepted. Closed New Year's Day; Thanksgiving; Christmas. &
Attendance: 3,500 (accurate)

WIGNALL MUSEUM OF CONTEMPORARY ART, (M), Chaffey College, 5885 Haven Ave., Rancho Cucamonga, CA 91737-3002. Tel.: 909-652-6492. Fax: 909-652-6491.
E-mail: wignall.staff@chaffey.edu
Web Site: www.chaffey.edu/wignall
Founded: 1972.
Congressional District: 35
Key Personnel: Dir. & Cur., Rebecca Trawick; Asst. Cur., Roman Stollenwerk.
Personnel Profile: Full-Time Paid 2; Part-Time Paid 8.
Governing Authority: Parent Institution: Chaffey Community College. Tax-exempt: 501(c)(3).
Institution Type/Description: Contemporary art.
Collections: works by contemporary artists.
Facilities: 2,100 sq. ft. exhibit space.
Activities: lectures; arts festivals; concerts; organized education programs for children & undergraduate college students.
Publications: exhibition catalogs.
Hours & Admission Prices: Aug.-May Mon.-Thurs. 10-4, Sat. 12-4. No charge; donations accepted. Closed college holidays. &
Attendance: 5,000 (accurate)

Rancho Mirage

CHILDREN'S DISCOVERY MUSEUM OF THE DESERT, 71-701 Gerald Ford Dr., Rancho Mirage, CA 92270-1934. Tel.: 760-321-0602. Fax: 760-321-1605.
E-mail: jmiller@cdmod.org
Web Site: www.cdmod.org
Founded: 1987.
Congressional District: 37
Key Personnel: Pres., Scott Wilson; Chm. (V), Betty Barker; Interim Exec. Dir., Judi Miller; Sec., Stephen Christian; Mgr. Gallery Experience, Kim Hatten; Administrative Mgr., Carey Alvarez; Museum Shop Mgr. & Visitors Svcs., Debra Aiello.
Personnel Profile: Full-Time Paid 5; Full-Time Volunteers 50; Part-Time Paid 9; Part-Time Volunteers 250; Interns 6.

Governing Authority: nonprofit organization. Tax-exempt: 501(c)(3).
Institution Type/Description: Children's Museum.
Collections: hands-on exhibits: strobe light & fan; rope maze; climbing wall; archaeological dig; paint the car; supermarket; pizza parlor; attic; animation station.
Facilities: 14,500 sq. ft. exhibit space; children's garden & grass maze. Educational toys & gifts for sale.
Activities: arts festivals; adult & youth docent programs; school tours; participatory exhibits; school outreach, community service & special holiday, vacation & summer programs; children's gardens & grass maze; outdoor playscapes. Exploration gallery: race track, Power Pump It, lie detector, safe crackers & more.
Publications: newsletter, Discovery-Gram.
Hours & Admission Prices: Jan.-April daily 10-5; May-Dec. Tues.-Sun. 10-5. Admission 2 & over $8; members, ACM reciprocal members & children under 2 no charge. Closed New Year's Day; Easter; Memorial Day; Independence Day; Labor Day; Thanksgiving; Christmas. &
Attendance: 70,000 (accurate)
Membership: Individual $25.

SUNNYLANDS CENTER & GARDENS, 37977 Bob Hope Dr., Rancho Mirage, CA 92270. Mailing Address: The Annenberg Retreat at Sunnylands, 70-177 Hwy. 111, Ste. 202, Rancho Mirage, CA 92270. Tel.: 760-202-2222.
E-mail: contact@sunnylands.org
Web Site: sunnylands.org
Founded: 2012.
Key Personnel: Dir., Janice Lyle, Ph.D.; Pres., The Annenberg Foundation Trust at Sunnylands, Geoffrey Cowan; Retail Mgr., Rick Aronson.
Personnel Profile: Full-Time Paid 77; Part-Time Paid 21.
Governing Authority: Parent Institution: The Annenberg Foundation Trust at Sunnylands. Tax-exempt.
Institution Type/Description: Historic House Museum: housed in the former winter residence of Walter and Leonore Annenberg.
Collections: Annenberg family history; personal artifacts; period furnishings; paintings; sculpture; decorative arts.
Facilities: Museum-related items for sale.
Activities: center, estate & garden tours; bird watching tours; garden walks; lecture series. Museum Sponsors: Third Sunday Family Days; Music in the Gardens, Nightlife Series for Young Professionals.
Hours & Admission Prices: House Tours: by appointment. Adults $35; children under 10 not admitted. Center & Gardens: Sept.-June Thurs.-Sun. 9-4. No charge. &
Attendance: 60,000 (estimated)

Rancho Palos Verdes

PALOS VERDES ART CENTER, 5504 W. Crestridge, Rancho Palos Verdes, CA 90275. Tel.: 310-541-2479. Fax: 310-541-9520.
E-mail: info@pvartcenter.org
Web Site: www.pvartcenter.org
Founded: 1931.
Congressional District: 30
Key Personnel: C.E.O., Robert A. Yassin; Chm. (V), Loren DeRoy; Pres. (V), Nancy Cumming; Dir. Exhibits, Scott Canty; Administrative Dir., Ann Willens; Dir. Publicity, Julia Parton; Dir. Education, Gail Phinney; Coord. Education, Angela Hoffman.
Personnel Profile: Full-Time Paid 11; Part-Time Paid 60; Part-Time Volunteers 600.
Governing Authority: nonprofit organization. Tax-exempt: 501(c)(3).
Institution Type/Description: Art Gallery.
Collections: historical & contemporary California cultural exhibits.
Research Fields: art of southern California.
Facilities: 50,000 PCS, book & slide library; banquet facility; ceramic, photography, print making & fine art studios; studio classrooms; professional kitchen (not a restaurant).
Activities: changing exhibitions; programs related to exhibitions; workshops; classes.
Publications: quarterly members newsletter & class brochure; gallery exhibit announcements; special programs; exhibition catalogues.
Hours & Admission Prices: Galleries: Mon.-Sat. 10-4, Sun. 1-4. No charge; donations accepted. Closed New Year's Day; President's Day; Memorial Day; Independence Day; Labor Day; Thanksgiving; Christmas. &
Attendance: 80,000 (estimated)
Membership: Individual $50; Family $60; Supporting $100; Contributing $150; Sustaining $250; Art Patron $500; Art Patron Silver $1,000; Gold Art Patron $2,500; Platinum Art Patron $5,000.

POINT VICENTE LIGHTHOUSE, 31550 Palos Verdes Dr., W., Rancho Palos Verdes, CA 90275. Tel.: 310-541-0334.
Institution Type/Description: Historic Lighthouse: built in 1926. Listed on the National Registry of Historic Sites.
Collections: local history; photographs; period artifacts.
Hours & Admission Prices: Tower & Museum: 2nd Sat. each month 10-3; children under 7 not allowed in tower. No charge.

Rancho Santa Fe

RANCHO SANTA FE ART GUILD, 6004 Paseo Delicias, Rancho Santa Fe, CA 92067. Mailing Address: P.O. Box 773, Rancho Santa Fe, CA 92067-0773. Tel.: 858-759-3545.
Institution Type/Description: Art Gallery.
Collections: paintings; photographs; sculpture.
Activities: special events.
Hours & Admission Prices: Tues.-Sat. 11-4:30.

Randsburg

RANDSBURG DESERT MUSEUM, 161 Butte Ave., Randsburg, CA 93554. Mailing Address: P.O. Box 307, Randsburg, CA 93554-0307. Tel.: 760-371-0965.
E-mail: hafdog@aol.com
Web Site: randdesertmuseum.com
Founded: 1950.
Congressional District: 22
Key Personnel: Dir. & Pres. (V), J. Bart Parker
Governing Authority: Tax-exempt.
Institution Type/Description: General Museum.
Collections: history of Rand District; gold, silver & tungsten rush history; gems; minerals; photographs; maps.
Facilities: library.
Hours & Admission Prices: Sat.-Sun. 10-5; other times by appointment. No charge; donations accepted.
Attendance: 2,500 (estimated)
Membership: Individual $25; Family $35.

Red Bluff

KELLY-GRIGGS HOUSE MUSEUM, 311 Washington St., Red Bluff, CA 96080-3430. Mailing Address: P.O. Box 9082, Red Bluff, CA 96080-6068. Tel.: 530-527-1129.
Founded: 1965.
Congressional District: 2
Key Personnel: C.E.O. & Pres. (V), Sharon Wilson.
Personnel Profile: Part-Time Volunteers 40.
Governing Authority: nonprofit. Tax-exempt: 501(c)(3).
Institution Type/Description: Local History Museum: housed in 1880 Victorian home.
Collections: period furniture & furnishings; art collection spanning over a century; Indian artifacts including original arrowheads & possessions of Ishi; pioneer artifacts; period photographs; Victorian costumes.
Research Fields: local history.
Facilities: 1880s Victorian home.
Activities: guided tours; permanent, temporary & traveling exhibitions; bi-monthly activity: coffee hours with museum-related programs. Museum Sponsors: Members' Victorian Christmas Champagne Party; Old-Fashioned Ice Cream Social; band concert; outdoor art show.
Publications: pamphlet, Map Tour of Victorian Red Bluff, California; bi-monthly newsletter & guide schedule, Kelly-Gram; publication, A Sketch-book From Indian Ways To Victorian Days.
Hours & Admission Prices: Thurs.-Sun. 1-4; groups by appointment. No charge; donations accepted. Closed New Year's Day; Easter; Independence Day; Thanksgiving; Christmas.
Attendance: 3,500 (estimated)
Membership: Associate (Individual) $10; Sustaining $50; Charter $100 plus $10 annual; Memoriam $100; Life $200; Patron $500; Benefactor $1,000.

WILLIAM B. IDE ADOBE STATE HISTORIC PARK, 21659 Adobe Rd., Red Bluff, CA 96080-9392. Tel.: 530-529-8599. Fax: 530-529-8598. Facebook: Friends of William B. Ide State Historical Park.
E-mail: dchakarun@parks.ca.gov
Web Site: www.parks.ca.gov/?page_id=458
Founded: 1951.
Congressional District: 2
Key Personnel: Interpreter, Debbie Chakarun Judy Fessenden.
Personnel Profile: Full-Time Paid 1; Part-Time Paid 1; Part-Time Volunteers 100.

Governing Authority: state. Parent Institution: CA Dept. of Parks & Recreation. Tax-exempt.
Institution Type/Description: Park & Historic House: 1850 adobe cabin, memorial to William B. Ide, president of the California Republic.
Collections: artifacts & reproductions relating to the site & its interpretive period, 1845-1854; adobe smokehouse; carriage shed; small cattle corral; woodworker's shop; blacksmith shop.
Research Fields: local history.
Facilities: 50-vol. library relating to the history of the area. Gift items for sale.
Activities: guided tours; lectures; formally organized education programs for children; docent program; living history events; weddings; reunions; birthdays.
Publications: brochure; quarterly newsletter, The Adobe Ferry Ledger.
Hours & Admission Prices: Park: Fri.-Sun. sunrise to sunset; Visitor Center: Fri.-Sun. 10-4. $6 per vehicle; discount to seniors. POWs & disabled veterans, California State Parks Assn. members and Golden Poppy Vehicle Day Pass & California Day Pass users no charge. Closed New Year's Day; Thanksgiving; Christmas. &
Attendance: 30,000 (estimated)
Membership: Ide Adobe Interpretive Association: Individual $10; Family $25; Institutional $50; Donor $250.

Redding

BEHRENS-EATON DISPLAY MUSEUM, 1939 Butte St., Redding, CA 96001-1613. Tel.: 530-241-3454.
E-mail: eaton@c-zone.net
Founded: 1994.
Governing Authority: Parent Institution: The Eaton Gift. Tax-exempt.
Institution Type/Description: History Museum.
Collections: local & family history; period furnishings; personal artifacts; photographs. Historic House: 1895 home.
Research Fields: history from 1850 to present.
Activities: community activities.
Publications: quarterly newsletter.
Hours & Admission Prices: Tues.-Wed. 10-4, Sat. 1-4. No charge; donations accepted. Closed holidays.
Attendance: 300
Membership: Charter $20; Life $200.

BEHRENS-EATON HOUSE MUSEUM, 1520 West St., Redding, CA 96001-1624. Mailing Address: 1939 Butte St., Redding, CA 96001-1613. Tel.: 530-241-3454.
Governing Authority: Branch Museum: Behrens-Eaton Display Museum, 1939 Butte St., Redding, CA.
Institution Type/Description: Historic House Museum: housed in the home of Charles Behrens, Sheriff of Shasta County and grandfather of Judge Richard Behrens Eaton; built in 1895.
Collections: family history; Judge Eaton's life & career; personal artifacts; photographs; period furnishings; books; letters; illustrations; papers.
Hours & Admission Prices: Tues.-Wed. 10-4, Sat. 1-4. No charge; donations accepted. Closed holidays.

SHASTA COLLEGE MUSEUM & RESEARCH CENTER, 11555 Old Oregon Trail, Redding, CA 96003-7692. Mailing Address: P.O. Box 496006, Redding, CA 96049-6006. Tel.: 530-242-7520. Fax: 530-225-3946.
E-mail: dsmith@shastacollege.edu
Founded: 1970.
Congressional District: 1
Key Personnel: Vice Pres. Academic Affairs, Bill Cochran; Dean, Dr. Ralph Perrin.
Personnel Profile: Part-Time Paid 1.
Governing Authority: public school district. Affiliated with Shasta-Tehama-Trinity Joint Community College District. Tax-exempt: 501(c)(3).
Institution Type/Description: History Museum.
Collections: artifacts, papers, & photographs of local history; Shasta County coroners' reports, 1851-1939; farming equipment; 1872 Buffalo Pitts Separator-Thresher; logging tools; early logging drag saws; mining tools; commercial records of a grocery store; clothing & accessories; 6-ton petroglyph; Northern Sacramento Valley prehistory; 1920s John Deer Tractor; narrow garage railroad car; 1920s Harvester; c. 1900 buggy & wagon.
Research Fields: local history & prehistory.
Activities: temporary exhibitions; loan services; research facilities; weekend classes.
Publications: annual, Museum Report.
Hours & Admission Prices: Temporarily closed. &
Attendance: 1,000

TURTLE BAY EXPLORATION PARK, 844 Sundial Bridge Dr., Redding, CA 96001. Mailing Address: 1335 Arboretum Dr., Ste. A, Redding, CA 96003-3628. Tel.: 530-243-8850; 800-887-8532. Fax: 530-243-8898. Facebook: Turtle Bay Exploration Park.
E-mail: info@turtlebay.org
Web Site: www.turtlebay.org
Formerly: Turtle Bay Museums & Arboretum on the River
Founded: 1990.
Congressional District: 2
Key Personnel: C.E.O., Mike Warren; Chm. (V), Janice Cunningham; Exhibits Mgr., Julia Cronin; Public Rels. Mgr., Mktg. & Sales Officer, Toby Osborn; Visitor Svcs. Mgr., Carrian Harwig.
Personnel Profile: Full-Time Paid 32; Part-Time Paid 13; Part-Time Volunteers 350.
Governing Authority: private; nonprofit. Tax-exempt: 501(c)(3).
Institution Type/Description: History, Art & Nature Museum.
Collections: regional history; California art; regional Native American artifacts; archeological; photography; natural science; forestry; ecology & water resources; lorikeet aviary.
Major Exhibits: Sin in the Sagebrush (T), 10/13-1/14; Tough by Nature (T), 10/13-1/14; The Big Adventure (T), 2/14-4/14; Native Baskets of North California, 2/14-4/14; Identity: An Exhibition of You (T), 5/14-9/14; Liberty on the Border (T), 10/14-2/15.
Research Fields: Northern California history, art & natural science.
Facilities: 300-acre arboretum; gardens; amphitheatre; classrooms; nature trails. Museum-related items for sale.
Activities: temporary exhibitions; docent program; formal educational programs; guided tours; hobby workshops; lectures; school loan service; teacher training.
Publications: bi-monthly member's calendar; exhibition catalogs.
Hours & Admission Prices: May-Labor Day Mon.-Sat. 9-5, Sun. 10-5. Labor Day-April Wed.-Sat. 9-4, Sun. 10-4. Adults $14, seniors 65 & over and children 4-12 $10; discounts to ASTC members; members and children 3 & under no charge. &
Attendance: 145,000 (estimated)
Membership: College Student $30; Individual $55; Family $80; Contributor $160; Patron $300; Benefactor $500; Leader $1,000.

Redlands

ASISTENCIA: SAN GABRIEL MISSION OUTPOST, 26930 Barton Rd., Redlands, CA 92373-4312. Mailing Address: c/o San Bernardino Co. Museums, 2024 Orange Tree Lane, Redlands, CA 92374. Tel.: 909-793-5402 & 307-2669. Fax: 909-307-0539.
E-mail: rmckernan@sbcm.sbcounty.gov
Web Site: sbcountymuseum.org
Founded: 1937.
Congressional District: 35
Key Personnel: Dir., Robert McKernan; Site Mgr., Mark Turpin; Cur., Michele Nielsen.
Personnel Profile: Part-Time Paid 1; Part-Time Volunteers 3.
Governing Authority: county. Parent Institution: San Bernardino County Museums, 2024 Orange Tree Lane, Redlands, CA 92374. Tax-exempt.
Institution Type/Description: History Museum: 1930s early California mission style ranch buildings.
Collections: history of local California Mission & Rancho era.
Research Fields: California mission era.
Facilities: 100-seat auditorium.
Activities: concerts; guided tours; lectures; temporary exhibitions; rental space for private weddings & receptions. Annual Events: History Day; Astronomy Evening.
Hours & Admission Prices: Tues.-Sat. 10-3. No charge; donations accepted. Closed New Year's Day; Thanksgiving; Christmas.
Attendance: 8,500 (accurate)

HISTORICAL GLASS MUSEUM FOUNDATION, 1157 N. Orange St., Redlands, CA 92374-3218. Mailing Address: P.O. Box 921, Redlands, CA 92373-0281. Tel.: 909-798-0868.
E-mail: historicalglassmuseum@gmail.com
Web Site: www.historicalglassmuseum.com
Founded: 1976.
Key Personnel: Pres. (V), William Brakemeyer.
Personnel Profile: Part-Time Paid 2; Part-Time Volunteers 12.
Governing Authority: bd. dirs. Tax-exempt.
Institution Type/Description: Historical American Glass Museum.
Collections: 4,300 artifacts of American glassware including glass beaded purses; art glass; milk glass; cruets; perfume bottles; carnival glass; cut glass; Jadeito; slag glass; depression glass; flint glass; pressed glass; Art Nouveau glass; Burmese glass; milk-cased glass.

Facilities: Museum-related items for sale.
Publications: quarterly newsletter.
Hours & Admission Prices: Sat.-Sun. 12-4. Suggested Donation: adults $3; members no charge. Closed major holidays.
Membership: $25.

KIMBERLY CREST HOUSE & GARDENS, 1325 Prospect Dr., Redlands, CA 92373-7049. Tel.: 909-792-2111. Fax: 909-798-1716.
E-mail: info@kimberlycrest.org
Web Site: www.kimberlycrest.org
Founded: 1981.
Congressional District: 35
Key Personnel: Exec. Dir., Terri deVries.
Personnel Profile: Full-Time Paid 2; Full-Time Volunteers 1; Part-Time Paid 3; Part-Time Volunteers 110; Interns 1.
Governing Authority: private; nonprofit organization. Parent Institution: Kimberly-Shirk Association. Tax-exempt: 501(c)(3).
Institution Type/Description: Historic Site: housed in an 1897 French chateau-style house & carriage house, formal 1909 Italian gardens and citrus grove.
Collections: decorative arts (original furnishings of Kimberly family); fine arts; archives; architecture; gardens; photography; family history.
Research Fields: architecture; decorative arts; landscape design.
Facilities: 6-acre site; garden structure & pathways.
Publications: newsletter, View from Kimberly Crest.
Hours & Admission Prices: Thurs.-Sun. 1-4. Suggested Donation: adults $10, seniors & students $8, children 6-12 $5; discounts AAM & ICOM members; children 5 & under no charge. Closed Easter; Thanksgiving; Christmas. &
Attendance: 13,000 (estimated)
Membership: Friend $35; Partner $50; Contributing $100; Supporter $250; Patron $500; Benefactor $1,000; Mary Kimberly Shirk Circle $5,000.

LINCOLN MEMORIAL SHRINE, (M), 125 W. Vine St., Redlands, CA 92373-4761. Tel.: 909-798-7632 (administrative) & 7636. Fax: 909-798-7566.
E-mail: archives@akspl.org
Web Site: www.lincolnshrine.org
Founded: 1932.
Congressional District: 37
Key Personnel: Pres. (V), Boyd Nies; Cur. & Archivist, Donald McCue; Assoc. Archivist, Nathan Gonzales.
Personnel Profile: Full-Time Paid 2; Part-Time Paid 2; Part-Time Volunteers 30; Interns 2.
Governing Authority: municipal. Parent Institution: A. K. Smiley Public Library, 125 W. Vine St. Tax-exempt: 501(c)(3).
Institution Type/Description: History Museum.
Collections: rare manuscripts & documents of Lincoln & leading Civil War generals, soldiers & civilians; artifacts of Lincoln & the Civil War period; sculpture; murals; paintings; manuscript collections.
Research Fields: Lincoln and Civil War.
Facilities: 5,000-vol. library of manuscripts, pamphlets newsletters, rare & new books on the Civil War period available for use on premises. Museum-related items for sale.
Activities: guided tours; lectures; docent program or council; permanent & temporary exhibitions.
Publications: quarterly newsletter; annual keepsakes dealing with the collections & Lincoln/Civil War history.
Hours & Admission Prices: Feb. 12 & Tues.-Sun. 1-5, special hours by appointment. No charge; donations accepted. Closed holidays. &
Attendance: 14,000 (estimated)
Membership: Individual $15; Family $25; Corporate $35.

*** SAN BERNARDINO COUNTY MUSEUM, (M),** 2024 Orange Tree Lane, Redlands, CA 92374-4560. Tel.: 909-307-2669 & 798-5719. Fax: 909-307-0539.
E-mail: rmckernan@sbcm.sbcounty.gov
Web Site: www.sbcountymuseum.org
Founded: 1957.
Congressional District: 37
Key Personnel: Dir., Robert L. McKernan; Cur. History, Michele Nielsen; Senior Cur. Geological Sciences, Kathleen Springer; Research Biologist, Gerald Braden; Cur. Exhibitions, Carey Smith; Cur. Anthropology, Dr. Adella Schroth; Cur. Education, Jolene Redvale; Registrar, Andrea Morics.
Personnel Profile: Full-Time Paid 57; Full-Time Volunteers 5; Part-Time Paid 55; Part-Time Volunteers 75; Interns 1.
Governing Authority: county. Subsidiary Institution: San Bernardino County Museum Association. Branch Museums: John Rains' House, 8810 Hem-

lock, Rancho Cucamonga, CA; Mousley Museum of Natural History, 35350 Panorama Dr., Yucaipa, CA; Asistencia: San Gabriel Mission Outpost, 26930 Barton Rd., Redlands, CA; Agua Mansa Pioneer Memorial & Cemetery, 270 E. Agua Mansa Rd., Colton, CA; Yorba and Slaughter Families Adobe, 17127 Pomona Rincon Rd., Chino, CA; Yucaipa Adobe, 32183 Kentucky St., Yucaipa, CA 92399. Tax-exempt.
Institution Type/Description: General Museum.
Collections: archaeology; anthropology; art; history; natural history; geology; ornithology; paleontology; photographs; hands-on exhibits.
Research Fields: archaeology; history; natural history; paleontology; mineralogy.
Facilities: 6,000-vol. library on ornithology, anthropology & history available on premises only; botanical garden; field research station; classrooms. Museum-related items for sale.
Activities: guided tours; field trips; lectures; hobby workshops; docent program; permanent exhibitions; school loan service.
Publications: monthly newsletter, San Bernardino County Museum Assn; quarterly, & occasional papers.
Hours & Admission Prices: Tues.-Sun. 9-5. Adults $6, seniors & students $5, children $4; discounts to AAM & AAA members; members no charge. Closed New Year's Day; Thanksgiving; Christmas. &
Attendance: 100,000 (accurate)
Membership: Individual $30; Family $40; Family Plus $50; Subscriber $60; Sustaining $75; Contributing $100; Corporate-Patron $500-$999; Benefactor $1,000-$5,000.

Redwood City

LATHROP HOUSE, 627 Hamilton St., Redwood City, CA 94063. Mailing Address: P.O. Box 1273, Redwood City, CA 94064-1273. Tel.: 650-365-5564.
Institution Type/Description: Historic House: built in 1863. Listed on the National Register of Historic Places.
Collections: period furnishings.
Facilities: Museum-related items for sale.
Activities: Annual Events: Open House in July; Victorian Days in August.
Hours & Admission Prices: Sept.-July Wed. & 3rd Sat. each month 11-3. No charge; donations accepted.

MARINE SCIENCE INSTITUTE, 500 Discovery Pkwy., Redwood City, CA 94063-4746. Tel.: 650-364-2760. Fax: 650-364-0416.
E-mail: gail@sfbaymsi.org
Web Site: www.sfbaymsi.org
Key Personnel: Exec. Dir., Marilou S. Seiff
Institution Type/Description: Marine Science Museum.
Collections: hands-on science & environmental exhibits.
Activities: education programs.
Hours & Admission Prices: Call for hours.

*** SAN MATEO COUNTY HISTORICAL ASSOCIATION AND MUSEUM, (M),** 2200 Broadway, Redwood City, CA 94063-1639. Tel.: 650-299-0104. Fax: 650-299-0141.
E-mail: info@historysmc.org
Web Site: www.historysmc.org
Founded: 1935.
Congressional District: 11
Key Personnel: Pres., Mitchell Postel; Chm. Bd. Dirs., Patrick Ryan; Archivist, Carol Peterson; Cur., Dana Neitzel; Deputy Dir., Carmen Blair; Assoc. Dir. Education, Dawn Distasio; Site Mgr., Becky Christ; Site Mgr., Marilyn Murphy.
Personnel Profile: Full-Time Paid 12; Part-Time Paid 7; Part-Time Volunteers 250; Interns 1.
Governing Authority: bd. directors. Branch Museums: 1846, Sanchez Adobe, Pacifica; 1854; Woodside Store, Woodside. Tax-exempt: 501(c)(3).
Institution Type/Description: San Mateo County History Museum.
Collections: archives of local history; horse drawn vehicles; 19th-century tools; photographs; assessment records; costumes; local material culture.
Research Fields: local government, society, industry; Native American, Spanish & Mexican heritage.
Facilities: library of pamphlets, photographs, manuscripts, monographs & documents of the history of San Mateo County available on premises.
Activities: guided tours; lectures; films; gallery talks; docent programs; educational activities for children; outreach programs for schools; inter-museum loan, permanent & temporary exhibitions designed for self-touring of the visually impaired.
Publications: semiannual journal, La Peninsula; monthly newsletter, Historical Happenings; books, Sawmills in the Redwoods (republished); Carolands Hillsboro; San Mateo County: A Sesquicentennial History; exhibit catalogue, Land of Opportunity: The Immigrant Experience in San Mateo County.

Hours & Admission Prices: Tues.-Sun. 10-4. Adults $5, seniors 62 and over & students with ID $3; discounts to Time Travelers, AAM & NARM members; members & children under 5 no charge. &

Attendance: 52,000 (accurate)

Membership: Student & Senior $35; Educator $40; Individual $50; Family Time Traveler $100; Archivist $250; Historian $500; Curator $1,000.

Reedley

MENNONITE QUILTING CENTER, 1012 G St., Reedley, CA 93654-2936. Tel.: 559-638-3560.

E-mail: quiltcenter@mcc.org

Web Site: mennonitequiltcenter.org

Governing Authority: Tax-exempt: 501(c)(3).

Institution Type/Description: Quilt Museum.

Collections: quilts; wall hangings.

Facilities: Museum-related items for sale.

Activities: quilting & handmade rug making; workshops; quilting classes.

Hours & Admission Prices: Mon.-Fri. 10-5, Sat. 10-4; groups of 10 or more by appointment. No charge; donations accepted. Closed holidays.

REEDLEY MUSEUM, 1752 10th St., Reedley, CA 93654-2933. Mailing Address: P.O. Box 877, Reedley, CA 93654. Tel.: 559-638-1913.

Founded: 1979.

Governing Authority: Parent Institution: Reedley Historical Society. Tax-exempt.

Institution Type/Description: History Museum.

Collections: Reedley history & culture.

Hours & Admission Prices: Tues. 10-12, Sat. 9:30-12. Adults $1; students & children under 18 no charge

Membership: Student $10; Individual $25; Family $35; Organization $50; Life $200; Organization Life $300.

Represa

RETIRED CORRECTIONAL PEACE OFFICERS MUSEUM AT FOLSOM PRISON, 312 3rd St., Represa, CA 95671. Tel.: 916-985-2561, ext. 4589.

Founded: 1994.

Key Personnel: Operations Mgr., Jim Brown.

Personnel Profile: Part-Time Volunteers 7.

Institution Type/Description: History Museum: housed in an old prison house, c.1898.

Collections: documents; personal artifacts; handcuffs; license plates; newspaper clippings; hanging rope; weapons; artwork; crafts.

Research Fields: historical research of family trees.

Hours & Admission Prices: Daily 10-4. Adults $2; school groups, law enforcement & military no charge. Closed New Year's Day; Thanksgiving; Christmas.

Attendance: 10,000 (accurate)

Rialto

RIALTO HISTORICAL SOCIETY, 201-205 N. Riverside Ave., Rialto, CA 92376. Mailing Address: P.O. Box 413, Rialto, CA 92377-0413. Tel.: 909-875-1750 & 1175.

Founded: 1971.

Congressional District: 36

Key Personnel: Pres., Jean Randall; Vice Pres., Jo Elliott; Treas., Helen McCain; Sec., Judy Roberts; Historian, John Adams; Corresponding Sec., Shirley Knowles; Computer Technician, Richard McInnis.

Personnel Profile: Part-Time Volunteers 18; Interns 2.

Governing Authority: society; nonprofit. Tax-exempt: 501(c)(3).

Institution Type/Description: Historical Society Museum: adjacent to historic church building.

Collections: carpenters tools; medical equipment; clothing; orange crate labels; orange industry equipment; Rialto Record newspapers; photographs; local citrus labels; Victorian memorabilia & furniture; library; maps; water history; military artifacts; cameras; local artists' works; Native American. Historic Building: c.1848 early adobe in Bender Park.

Research Fields: Rialto area history.

Facilities: cultural center: reading room, meeting room; rental facilities.

Activities: school tours; arts festivals; lectures; music recitals; travelogues; reading room; permanent & temporary exhibitions; weddings. Museum Sponsors: monthly luncheon & speakers October to June.

Publications: newsletter, Rialto Landmarks & Homes Prior to 1900; books, History of Rialto; Then and Now; Rialto-Images of America; The Little Girl in the Window; Scammers, Schemers and Dreamers.

Hours & Admission Prices: Wed. 2-4, Sat. 10-2; other times by appointment. No charge; donations accepted. Closed holidays.

Attendance: 400 (estimated)

Membership: Annual $5; Contributing $10; Sustaining $25; Organizational & Benefactor $100; Life $500; Perpetual $1,000.

Richmond

NIAD GALLERY, 551 23rd St., Richmond, CA 94804. Tel.: 510-620-0290. Fax: 510-620-0326.

E-mail: admin@niadart.org

Web Site: www.niadart.org

Key Personnel: Dir., Brian Stechschulte

Institution Type/Description: Art Gallery.

Collections: paintings; drawings; prints; ceramics; textiles; sculpture.

Hours & Admission Prices: Mon.-Fri. 9-4; other times by appointment.

RICHMOND ART CENTER, 2540 Barrett Ave., Richmond, CA 94804-1600. Tel.: 510-620-6772. Fax: 510-620-6771.

E-mail: admin@therac.org

Web Site: www.therac.org

Founded: 1936.

Congressional District: 11

Key Personnel: Pres. (V), Andi Biren; Exec. Dir., Richard Ambrose; Dir. On-site Education, Kato Jaworski.

Personnel Profile: Full-Time Paid 9; Part-Time Paid 6; Part-Time Volunteers 240; Interns 1.

Volunteer Hours: 13,000

Operating Expenses: 760,000

Operating Income: 905,000

Governing Authority: nonprofit organization, board of directors. Tax-exempt: 501(c)(3) & 170(b)(1)(A).

Institution Type/Description: Art Center.

Collections: rotating exhibits.

Major Exhibits: The Art of Living Black, Winter 2014; The Breakfast Group: Jive & Java, Spring 2014; Lisa Kokin - Recent Work, Spring 2014; Pacific Rim Sculpture - Juried Exhibition, Summer 2014; Closely Considered - Dieben Korn in Berkeley, Fall 2014.

Facilities: sculpture court; classrooms; resource room.

Activities: extension exhibition program; class program; tours; lectures; temporary exhibitions. Center Sponsors: scholarship programs; outreach instruction programs to schools, community centers & the public library.

Publications: catalogues; calendar of exhibition & events; newsletter; class schedule.

Hours & Admission Prices: Tues.-Sat. 11-5. No charge; donations accepted. &

Attendance: 20,000 (estimated)

Membership: Senior Citizen $40; Individual $50; Artist $60; Family $75; Patron $100; Patron Family $150.

RICHMOND MUSEUM OF HISTORY, 400 Nevin Ave., Richmond, CA 94801-3017. Mailing Address: P.O. Box 1267, Richmond, CA 94802-0267. Tel.: 510-235-7387.

E-mail: info@richmondmuseumofhistory.org

Web Site: richmondmuseumofhistory.org

Founded: 1954.

Congressional District: 11

Key Personnel: Pres., C.E.O. (V) & Programs, Lois H. Boyle; Cur., Melinda McCrary.

Personnel Profile: Part-Time Paid 1; Part-Time Volunteers 5.

Governing Authority: nonprofit organization. Parent Institution: Richmond Museum Association. Tax-exempt: 501(c)(3)

Institution Type/Description: History Museum.

Collections: local history; Bay area prehistory; World War II Richmond shipyard years; photographs; archives; decorative arts; costumes; 1931 Model A automobile; archaeology visual arts.

Research Fields: Richmond & the surrounding areas.

Facilities: 640-vol. library.

Activities: permanent & temporary exhibitions; lectures; films; organized educational programs for youth; docent program; guided tours.

Publications: quarterly, The Mirror.

Hours & Admission Prices: Wed.-Sun. 1-4. Adults $2, seniors & students $1; children & members no charge. Closed legal holidays. &

Attendance: 2,900 (estimated)

Membership: Student & Senior $25; Individual $35; Family & Organization $50; Contributing $65; Sustaining $100; Patron $500; Benefactor $1,000.

Ridgecrest

MATURANGO MUSEUM OF THE INDIAN WELLS VALLEY, 100 E. Las Flores, Ridgecrest, CA 93555-3654. Tel.: 760-375-6900. Fax: 760-375-0479.

E-mail: matmus6@maturango.org

Web Site: www.maturango.org

Founded: 1962.

Congressional District: 22

Key Personnel: Dir., Harris Brokke; Bd. Pres., Carolyn Shepherd; Cur. Archaeology, Alexander K. Rogers; Cur. History, Elizabeth Babcock; Gallery Coord., Sylvia Winslow Art Gallery, Rosemary Lackaye; Bookkeeper, Julie Stephens; Petroglyph Tour Coord., Fran Van Valkenburgh; Education Coord., Nora Nuckles.

Personnel Profile: Full-Time Paid 1; Part-Time Paid 11; Part-Time Volunteers 150.

Governing Authority: nonprofit organization. Tax-exempt: 501(c)(3).

Institution Type/Description: Cultural and Natural History.

Collections: minerals; mining tools & equipment; history; flora & fauna of the upper Mojave desert; archaeology, paleontology; geology; entomology; mammals; reptiles; prehistoric rock art; Emma Lou Davis library & archives; artwork of local & regional artists.

Research Fields: Native American archaeology prehistoric rock art of the Coso Range & adjacent Indian Wells Valley; regional botanical & zoological species.

Facilities: 2,000-vol. library of archaeology & local history & desert natural history; video library of lectures & programs; Xeriscape garden; observatory; Death Valley Tourist Center; Northern Mojave Visitor Center. Museum-related items for sale.

Activities: guided tours; lectures; formally organized education programs for children; docent program; inter-museum, permanent & temporary exhibitions; foreign study tours; observatory star parties. Museum Sponsors: guided tours of Coso petroglyphs in Little Petroglyph Canyon; wildflower show in spring.

Publications: monthly newsletter; annual report; books, Rock Drawings of the Coso Range; Art & Poetry of Gladys Merrick; Before the Navy; Epsom Salts Monorail; Ayers Rock; Millennium Conference Proceedings; Following the Shaman's Path: A Walking Guide to Little Petroglyph Canyon, Coso Range, CA; A Festschrift Honoring the Contributions of California Archaeologist Jay Von Werlhof; DVD, Somewhere on the Edge of Nowhere; booklet, Common Sense in Desert Travel; Fossil Mammals of the Indian Wells Valley, history of the Indian Wells Valley; Adventures with a Desert Bush Pilot; Coso Rock Art: A New Prospective; Maturango Country-Discover It; recipe book, Maturango Museum Luncheons; videos on various lectures & programs available for loan to members; tourist brochures.

Hours & Admission Prices: Daily 10-5. Adults $5, Seniors & children $3; children under 6 & members no charge. Closed New Year's Day, Memorial Day; Independence Day, Labor Day, Thanksgiving, Christmas. &

Attendance: 24,119 (accurate)

Membership: Senior Citizen 55 & over $35; Individual $40; Senior Family $45; Family $50.

Rio Vista

RIO VISTA MUSEUM, 16 N. Front St., Rio Vista, CA 94571-1837. Tel.: 707-374-5169.

Institution Type/Description: History Museum.

Collections: history of Rio Vista; photographs; newspapers; clothing; farm equipment; wedding gowns from 1876-1920.

Hours & Admission Prices: Sat.-Sun. 1:30-4:30.

Riverside

BOTANIC GARDENS - UNIVERSITY OF CALIFORNIA RIVERSIDE, 900 University Ave., Riverside, CA 92521-0124. Tel.: 951-784-6962. Fax: 951-827-4437.

E-mail: ucrbg@uct.edu

Web Site: www.gardens.uct.edu

Founded: 1963.

Key Personnel: Dir., J.G. Waines.

Personnel Profile: Full-Time Paid 3; Part-Time Paid 5; Part-Time Volunteers 100; Interns 1.

Governing Authority: Parent Institution: Univ. California Riverside. Subsidiary Institution: CNAS. Tax-exempt.

Institution Type/Description: Botanic Garden.

Collections: over 3,500 plant species from around the world; 200 bird species.

Activities: special events. Annual Event: Primavera in the Gardens in May.

Publications: Deserts of the Southwest.

Hours & Admission Prices: Daily 8-5. Suggested Donation: $4; discounts to American Horticulture Society & RASP. Closed New Year's Day; Independence Day; Thanksgiving; Christmas. &

Attendance: 40,000 (estimated)

Membership: Single $35; Family $55.

JENSEN-ALVARADO HISTORIC RANCH & MUSEUM, 4307 Briggs St., Riverside, CA 92519. Tel.: 951-369-6055.

Web Site: www.riversidecountyparks.org

Institution Type/Description: Historic Building: housed on the former ranch of retired sea captain Cornelius Jensen and his wife Mercedes Alvarado; built c.1880.

Collections: family history; personal artifacts; period furnishings; farm tools & equipment. Historic Buildings: tank house; milk house; windmill.

Facilities: 30 acre site.

Hours & Admission Prices: Mon.-Fri. by appointment. Adults $3, children 3-12 $2.

JURUPA MOUNTAINS DISCOVERY CENTER, 7621 Granite Hill Dr., Riverside, CA 92509-1299. Tel.: 951-685-5818. Fax: 951-685-1240.

E-mail: info@jmdc.org

Formerly: Jurupa Mountains Cultural Center

Founded: 1964.

Congressional District: 41

Key Personnel: Exec. Dir., Mark Yeager; Pres., Mike Denholtz.

Personnel Profile: Full-Time Paid 6; Full-Time Volunteers 1; Part-Time Paid 3; Part-Time Volunteers 30; Interns 1.

Governing Authority: private; nonprofit organization. Parent Institution: Jurupa Mountains Cultural Center. Tax-exempt: 501(c)(3).

Institution Type/Description: Earth Science Museum.

Collections: rocks; minerals; fossils; fluorescent minerals; Native American artifacts; gems; dinosaur eggs from China.

Facilities: library; 86 acre site; educational facilities. Museum-related items for sale.

Activities: guided tours; formal education programs; lectures; participatory, loan & temporary exhibitions; scouting merit badges. Annual Events: Greenfaire; Olive Curing.

Publications: monthly newsletter.

Hours & Admission Prices: Center & Earth Science Museum: Tues.-Sat. 8-4. Adults $3, teens $2, children 6-12 $1. Public Tours: Sat. 9 & 12. Pre-arranged Group Tours: Tues.-Sat. 8-5, Sun. 12-5. Groups: $8-$12 per person; discounts to members. Closed New Year's Day; Easter; Independence Day; Thanksgiving; Christmas. &

Attendance: 20,000 (estimated)

Membership: Individual $20; Family $35.

* **MARCH FIELD AIR MUSEUM, (M),** 22550 Van Buren Blvd., Riverside, CA 92518-2400. Mailing Address: P.O. Box 6463, March ARB, CA 92518-0394. Tel.: 909-697-6600 & 6602. Fax: 909-697-6605.

E-mail: info@marchfield.org

Web Site: www.marchfield.org

Founded: 1979.

Congressional District: 43

Key Personnel: C.E.O., Patricia Korzec; Pres. (V), Jamil DaDa; Collections Mgr., Michelle Sifuentes; Events, Adriana Bradley; Museum Shop Mgr., Bill Berg.

Personnel Profile: Full-Time Paid 8; Part-Time Paid 2; Part-Time Volunteers 163; Interns 1.

Volunteer Hours: 36,708

Operating Expenses: 705,758

Operating Income: 733,254

Governing Authority: nonprofit. Parent Institution: March Field Museum Foundation. Tax-exempt: 501(c)(3).

Institution Type/Description: Aviation History.

Collections: 72 aircraft; over 20,000 military & aviation artifacts; uniforms; aircraft engine displays; Korea, Vietnam & Desert Storm artifacts; photographs; weapons; World War I, World War II & 15th Air Force artifacts; memorabilia of famous flyers; aviation films.

Research Fields: oral history program; intern program for local college students.

Facilities: library of aviation-related books; theater; reading area. Museum-related items for sale.

Activities: walking tours to aircraft display area daily; tour groups welcome; reunions & conferences welcome; special commemorative weeks for groups such as the Tuskegee Airmen.

Publications: foundation newsletter, Flightline.

Hours & Admission Prices: Tues.-Sun. 9-4. Adults $10; discounts to AAM members; members no charge. Closed major holidays. &

Attendance: 102,000 (accurate)

Membership: Individual $41; Family $65; Sustaining $125; Corporate $250.

MISSION INN FOUNDATION/MUSEUM, (M), 3696 Main St., Riverside, CA 92501-2839. Tel.: 951-781-8241. Fax: 951-341-6574.

E-mail: info@missioninnmuseum.com

Web Site: www.missioninnmuseum.com

Founded: 1976.

Key Personnel: Exec. Dir., John Worden; Museum Shop Mgr., Sharla Wright; Collection Mgr., Steve Spiller; Office Administration, Nanci Larsen; Dir. Mktg., Brinna Wrightsman.

Personnel Profile: Full-Time Paid 5; Part-Time Paid 5; Part-Time Volunteers 125.

Governing Authority: private; nonprofit. Tax-exempt: 501(c)(3).

Institution Type/Description: Historic House & Site: restored turn-of-the-century resort hotel now a national historic landmark.

Collections: arts & crafts furniture; paintings; sculpture; Oriental art; dolls; bells; crosses; Mission memorabilia. Inn features arcades, gardens, turrets, domes, flying buttresses, circular staircases, stained glass, wrought iron & tile.

Facilities: 2,400 sq. ft. exhibit space. Museum publications, posters, T-shirts, postcards & other museum-related items for sale.

Activities: guided tours; lectures; docent program; temporary exhibits. Annual Event: 5/10K Run in November.

Hours & Admission Prices: Daily 9:30-4. Museum: $2. Tours $12. Closed Easter; Mother's Day; Thanksgiving; Christmas.

Attendance: 45,000 (accurate)

Membership: Student $10; Individual $25; Family $35; Contributor $100; Corporate $250; Curator's Circle $1,000; Director's Circle $2,500; President's Circle $5,000.

RIVERSIDE ART MUSEUM, 3425 Mission Inn Ave., Riverside, CA 92501-3368. Tel.: 951-684-7111. Fax: 951-684-7332. Facebook: Riverside Art Museum.

Web Site: www.riversideartmuseum.org

Founded: 1931.

Congressional District: 36

Key Personnel: Exec. Dir., Drew Oberjuerge; Fin. Mgr., Shannon Kane; Art Education Dir., Nicole Tartoni; Permanent Collections & Exhibit Liaison, Kathryn Poindexter; Communications, Ai M. Kelley; Visitor Services, Katie Hernandez; Consultant Cur., Peter Frank.

Governing Authority: nonprofit; board of trustees. Subsidiary Institution: Art Alliance of Riverside Art Museum. Tax-exempt: 501(c)(3).

Institution Type/Description: Art Museum: housed in 1929 building, designed by Julia Morgan, architect for Hearst Castle.

Collections: paintings; sculpture; decorative arts; graphics.

Research Fields: contemporary, American art.

Facilities: rental gallery; courtyard restaurant. Gift items for sale.

Activities: guided tours; lectures; films; gallery talks; arts festivals; formally organized education programs for children & adults; permanent, temporary & traveling exhibitions.

Publications: quarterly newsletter; exhibition catalogues.

Hours & Admission Prices: Tues.-Sat. 10-4, Sun. 12-4. Adults $5, students, educators and seniors 65 & over $3; military families w/ID, members & children 12 and under no charge. Closed holidays. &

Attendance: 70,000 (estimated)

Membership: Student/Senior $40; Individual $55; Family $100; Friend of RAM $250; JMS Associate $500; Artist Circle Advocate $1,000.

RIVERSIDE COMMUNITY COLLEGE ART GALLERY, 4800 Magnolia Ave., Quad Room 140, Riverside, CA 92506-1201. Tel.: 951-222-8358. Fax: 909-222-8740.

E-mail: julia.buckley@rcc.edu

Web Site: academic.rcc.edu/art/exhibitions.jsp

Key Personnel: Coord. Art Gallery, Leslie A. Brown

Institution Type/Description: Art Gallery.

Collections: paintings.

Activities: film series; lectures; research; outreach to schools.

Hours & Admission Prices: Mon.-Wed. & Fri. 10-3, Thurs. 10-3 & 5:30-8.

RIVERSIDE HERITAGE HOUSE, 8193 Magnolia Ave., Riverside, CA 92504-3409. Mailing Address: 3580 Mission Inn Ave., Riverside, CA 92501-3321. Tel.: 951-826-5273.

Web Site: www.riversideca.gov/museum/heritage.asp

Institution Type/Description: Historic House Museum.

Collections: period furnishings; personal artifacts.

Activities: docent tours; special events.

Hours & Admission Prices: Sept.-June Fri.-Sun. 12-3:30. Suggested Donations: adults $5. Closed federal holidays.

* **RIVERSIDE METROPOLITAN MUSEUM, (M),** 3580 Mission Inn Ave., Riverside, CA 92501-3321. Tel.: 951-826-5273. Fax: 951-369-4970.

E-mail: smundy@riversideca.gov

Web Site: www.riversideca.gov/museum

Formerly: Riverside Municipal Museum

Founded: 1925.

Congressional District: 41

Key Personnel: Dir., Sarah Suverkrup Mundy; Chm. Bd. (V), David Barnhart; RMA Pres. (V), Jim Ferguson; Cur. Natural History, James Bryant; Cur. Collections, Brenda Buller-Focht, Ph.D.; Cur. Historic Structures & Collections, Lynn Voorheis; Cur. Educator, Teresa Belding-Woodard; Archivist, Kevin Hallaran; Maintenance, German Ponce; Sr. Office Specialist, Toni Kinsman.

Personnel Profile: Full-Time Paid 13; Part-Time Paid 1; Part-Time Volunteers 86; Interns 8.

Governing Authority: municipal. Parent Institution: City of Riverside, 3900 Main St., Riverside, 92522. Subsidiary Institution: Riverside Museum Associates. Tax-exempt: 170(b)(1)(A).

Institution Type/Description: Natural History Museum.

Collections: ethnology; archaeology; history; photo & document archives; decorative arts; rocks; minerals; fossils; mammals; birds; reptiles; insects; herbarium; Citrus label art; quilts; costumes. Historic House: 1891 Queen Anne Victorian House; Harada House, a National Historic Landmark.

Major Exhibits: John Muir & The Personal Experience with Nature, 12/2/12-10/26/14; Telling Riverside's Story in 50 Objects, 10/13-1/4/15; Force of Arms, 10/13-7/27/14.

Research Fields: history; anthropology; archaeology; zoology; geology; paleontology; botany; art & decorative arts.

Facilities: 2,000-vol. library pertaining to museum collections available for use on premises by appointment; gardens. Books, pamphlets of local importance & other museum-related items for sale.

Activities: hands on programs; lectures; education programs for children; docent program or council; inter-museum loan, permanent & temporary exhibitions; school loan service; family programs. Annex: local history research available by appointment. Historic House Sponsors: community outreach ice cream social; Christmas Open House.

Publications: books; exhibit & collection pamphlets; brochure; newsletters, Native American Basketry of Central California, American Indian Basketry of Northern California, Native American Basketry of Southern California, A History of Citrus in the Riverside Area, Fading Images: Indian Pictographs of Western Riverside County.

Hours & Admission Prices: Museum: Tues.-Wed. & Fri. 9-5, Thurs. 9-9, Sat. 10-5, Sun. 11-5. Heritage House: Sept.-June Thurs.-Fri. 12-3, Sun. 12-3:30; groups by appointment. No charge; donations requested. Closed major holidays. &

Attendance: 75,000 (estimated)

Membership: Junior $1; Active $15; Individual $20; Family $30; Associate $50; Sustaining $75; Contributing $100; Life $1,000; Patron $2,500; Benefactor $5,000 & up.

SHERMAN INDIAN MUSEUM, 9010 Magnolia Ave., Riverside, CA 92503-4431. Tel.: 951-276-6719.

E-mail: lsisquoc@charter.net

Web Site: www.shermanindianmuseum.org

Key Personnel: Cur., Lorene Sisquoc

Institution Type/Description: Native American Museum.

Collections: Native American history & culture; photographs; basketry; Navajo rugs; paintings.

Facilities: library.

Hours & Admission Prices: Tues.-Thurs. 1-4:30 by appointment.

SWEENEY ART GALLERY, UNIVERSITY OF CALIFORNIA, 3800 Main St, Riverside, CA 92501-3624. Tel.: 951-827-3755. Fax: 951-827-3798.

E-mail: krapp@pop.ucr.edu

Web Site: sweeney.ucr.edu

Founded: 1963.

Congressional District: 36

Key Personnel: Dir., Karen Rapp; Gallery Mgr., Jennifer Frins.

Personnel Profile: Full-Time Paid 2; Part-Time Volunteers 2; Interns 4.

Governing Authority: state university; nonprofit organization. Tax-exempt: 501(c)(3).

Institution Type/Description: University Art Gallery.

Collections: various portfolios on paper; Jules Cheret vintage prints; Vigango sculpture of East Africa; basic study collection; contemporary California.
Facilities: 500-vol. non-circulating library of exhibition catalogues; 2,000 sq. ft. exhibit space.
Activities: lectures; loan & traveling exhibitions.
Publications: quarterly newsletter for members; catalogues for some exhibitions.
Hours & Admission Prices: Call for hours. Closed major holidays. &
Attendance: 7,500 (estimated)
Membership: Subscriber $50; Friend $120; Fellow $500; Patron $1,000; Benefactor $2,500; Corporate $5,000.

UCR/CALIFORNIA MUSEUM OF PHOTOGRAPHY, 3824 Main St., Riverside, CA 92501-3624. Mailing Address: UCR/California Museum of Photography, Riverside, CA 92521. Tel.: 951-827-4787. Fax: 951-827-4797.
E-mail: colin.westerbeck@ucr.edu
Web Site: www.cmp.ucr.edu/
Founded: 1973.
Congressional District: 43
Key Personnel: Dir. & Prof. Art & Art History, Colin Westerbeck; Asst. Cur., Kristine Thompson; Store Contact, Emily Papavero; Preparator, Jason Chakravarby; Cur. Collections, Leigh Gleason; Cur. Digital Media, Georg Burwick; Administrative Mgr., Zelda Glenn.
Personnel Profile: Full-Time Paid 9; Full-Time Volunteers 2; Part-Time Paid 10; Part-Time Volunteers 25; Interns 10.
Governing Authority: university; nonprofit organization. Parent Institution: University of California at Riverside, Riverside, CA. Tax-exempt.
Institution Type/Description: Photography Museum.
Collections: 19th- & 20th-century photographs; Keystone-Mast Stereoview collection; stereoview negatives & prints; 19th- & 20th-century Bingham cameras & photographic apparatus; 20th century master prints & contemporary works.
Research Fields: history of photography; media; culture & optically generated images.
Facilities: library of photographically related monographs & serials available for research on premises only; digital studio; 12,000 sq. ft. exhibit space; separate laboratory operation. Museum-related items for sale.
Activities: temporary exhibitions exploring photography & video's relationship to art, society & politics; new music & performance art events; lectures & symposia; cooperative photography & video exhibitions with K-12 schools; hands-on optics & photographic technology for children; guided tours; formally organized education programs for undergraduate & graduate college students affiliated with UCR & other area universities; loan, permanent, temporary & traveling exhibitions; internet gallery.
Publications: occasional catalogues & artists' books; quarterly newsletter, FOTOTEXT.
Hours & Admission Prices: Tues.-Sat. 12-5. Adults $3; members, students & seniors no charge. Closed New Year's Day; Thanksgiving; Christmas. &
Attendance: 35,000 (estimated)
Membership: Student & Senior $25; Participating $35; Supporting $65; Contributing $125; Sustaining $500; Museum Fellow $1,000.

WORLD MUSEUM OF NATURAL HISTORY, (M), La Sierra Univ., 4500 Riverwalk Pkwy., Riverside, CA 92505-3344. Tel.: 951-785-2500 (Mon.-Thurs.); 2209 (Sat.). Fax: 951-785-2478.
E-mail: cmall@lasierra.edu
Web Site: www.lasierra.edu/wmnh
Founded: 1971.
Congressional District: 43
Key Personnel: C.E.O. & Pres. La Sierra Univ., Randal Wisbey; Cur., Dr. Billy Hankins; Cur., Dr. Virchel Wood.
Personnel Profile: Part-Time Volunteers 5.
Governing Authority: La Sierra University. Tax-exempt: 501(c)(3).
Institution Type/Description: Natural History Museum.
Collections: mounted animals including amphibians, reptiles; birds & mammals; fossils; gems; minerals fluorescent minerals; meteorites; tektites; petrified woods; American ethnology; wildlife paintings & sculpture.
Facilities: 200-seat auditorium; arboretum; 2,000 sq. ft. exhibit space; educational facilities.
Activities: videos; formal education programs for children & undergraduate or graduate college students affiliated with La Sierra University; guided tours.
Publications: occasional newsletter.
Hours & Admission Prices: Sat. 2-5; other times by appointment. No charge; donations accepted. &
Attendance: 8,000 (estimated)

Rohnert Park

UNIVERSITY ART GALLERY, SONOMA STATE UNIVERSITY, 1801 E. Cotati Ave., Rohnert Park, CA 94928-3609. Tel.: 707-664-2295. Fax: 707-664-4333.
E-mail: carla.stone@sonoma.edu
Web Site: www.sonoma.edu/artgallery/
Founded: 1977.
Key Personnel: Dir., Michael Schwager; Exhibition Coord., Carla Stone.
Personnel Profile: Full-Time Paid 1; Part-Time Paid 1; Interns 10.
Governing Authority: university; nonprofit organization. Parent Institution: Sonoma State University. Tax-exempt: 501(c)(3).
Institution Type/Description: Art Gallery.
Collections: 20th-century American art works on paper; Asian art works; Garfield collection.
Research Fields: 20th-century American art.
Activities: lectures; organized educational programs; training programs for professional museum workers; loan exhibitions; organized education programs for undergraduate or graduate students affiliated with Sonoma State University.
Publications: exhibition catalogues; posters.
Hours & Admission Prices: Sept.-May Tues.-Fri. 11-4, Sat.-Sun. 12-4. No charge. Closed holidays. &
Attendance: 2,000 (estimated)
Membership: Special $15; Active $35; Contributing $50; Patron $100; Benefactor $1,000.

Roseville

MAIDU MUSEUM AND HISTORIC SITE, 1970 Johnson Ranch Dr., Roseville, CA 95661-3749. Tel.: 916-774-5934. Fax: 916-772-6161. TDD: 916-774-5220.
E-mail: mmurphy@roseville.ca.us
Web Site: www.roseville.ca.us/indianmuseum
Formerly: Maidu Interpretive Center
Founded: 2001.
Congressional District: 4
Key Personnel: Dir., Mark Murphy; Education, Heidi Frantz; Museum Shop Mgr., Isabella Zaia; Historic Site Restoration Coord., Linda Maurer; Security, Rich Douglas.
Personnel Profile: Full-Time Paid 2; Part-Time Paid 7; Part-Time Volunteers 52; Interns 1.
Governing Authority: municipal; nonprofit. Parent Institution: City of Roseville, CA.
Institution Type/Description: Historic Site: listed on the National Register of Historic Places.
Collections: Nisenan Maidu & other Maiduan history & culture from 5000 BC to present.
Facilities: 10,000 sq. ft. exhibit space. Museum-related items for sale.
Activities: docent program; formal education programs; guided tours; lectures; participatory, temporary & traveling exhibitions. Annual Events: Leafing Out of Spring Cultural Celebration.
Publications: newsletter, Ah Kapa (Bear Rock) News.
Hours & Admission Prices: Mon.-Fri. 9-4, Sat. 9-1. Adults $4.50, children & seniors $4; discounts to groups. Closed major holidays. &
Attendance: 34,500 (accurate)
Membership: Student & Teacher $25; Individual $45; Companion $55; Family $70; Executive $150; Corporate $500.

ROSEVILLE UTILITY EXPLORATION CENTER, 1501 Pleasant Gove Blvd., Roseville, CA 95747. Tel.: 916-746-1550.
E-mail: ruec@roseville.ca.us
Web Site: www.roseville.ca.us/explore
Founded: 2008.
Key Personnel: Interim Supvr., Matthew Davis.
Personnel Profile: Full-Time Paid 3; Part-Time Paid 1; Interns 2.
Volunteer Hours: 2,585
Governing Authority: city. Subsidiary Institution: Roseville Electric & Environmental Utilities Dept. Tax-exempt.
Institution Type/Description: Environmental Preservation Museum.
Collections: environmental preservation & conservation; energy technology; recycling.
Major Exhibits: NASA - Earth from Space, 1/14-4/14.
Activities: educational programs.
Publications: annual publication, Green Living Guide.
Hours & Admission Prices: Tues.-Sat. 10-5. No charge. &
Attendance: 33,000

Sacramento

CALIFORNIA AUTOMOBILE MUSEUM, 2200 Front St., Sacramento, CA 95818-1107. Tel.: 916-442-6802. Fax: 916-442-2646. Facebook: California Automobile Museum.

E-mail: pr@calautomuseum.org
Web Site: www.calautomuseum.org
Formerly: Towe Auto Museum
Founded: 1982.
Congressional District: 5
Key Personnel: Exec. Dir., Karen McClaflin; Pres., Joe Hensler; Vice Pres., Carl Stein; Museum Shop Mgr., Angel Nunez.
Personnel Profile: Full-Time Paid 5; Part-Time Paid 5; Part-Time Volunteers 300; Interns 1.
Governing Authority: nonprofit membership organization. Parent Institution: California Vehicle Foundation. Tax-exempt: 501(c)(3).
Institution Type/Description: Transportation Museum.
Collections: period American automobiles 1900 to present day; car history; California culture.
Research Fields: Henry Ford; Ford Motor Company; historic automobiles; general automotive history; evolution of the automobile; role of the automobile in changing American society; Chrysler; Dodge; Chevrolet; import cars; exotics; EV1 Volkswagen.
Facilities: automotive research library; banquet facilities. Museum-related items for sale.
Activities: guided tours; lectures; organized education programs; docent program; participatory & temporary exhibitions; annual events; educational classes; educational tours; rental facilities; classic & antique cars on consignment.
Publications: newsletter; annual report.
Hours & Admission Prices: Daily 10-6, 3rd Thurs. each month 10-9. Adults $8, senior citizens $7, students $4; discounts to Time Travelers, AAA & NARM members and groups; children under 5, members, military on Veterans Day & Memorial Day no charge. Closed New Year's Day; Thanksgiving; Christmas. &
Attendance: 56,105 (accurate)
Membership: Teacher no charge; Student $20; Individual $40; Dual & Companion $55; Family $70; Car Clubs $125; Executive $150; Corporate $500.

THE CALIFORNIA MUSEUM, (M), 1020 O St., Sacramento, CA 95814-5704. Tel.: 916-653-7524. Fax: 916-653-0314.

E-mail: museuminfo@californiamuseum.org
Web Site: www.californiamuseum.org
Formerly: California State History Museum
Founded: 1998.
Congressional District: 5
Key Personnel: Exec. Dir., Dori Moorehead; Chm., Richard Costigan, III; Dir. Exhibitions & Programs, Amanda Meeker.
Personnel Profile: Full-Time Paid 6; Part-Time Paid 1; Part-Time Volunteers 30; Interns 4.
Governing Authority: private; nonprofit organization. Tax-exempt: 501(c)(3).
Institution Type/Description: General Museum.
Collections: archives; artifacts; manuscripts; ephemera related to California; multimedia exhibit technology; state-of-the-art technology; documents & artifacts.
Facilities: 260-seat auditorium; classrooms; 30,000 sq. ft. exhibit space.
Activities: Annual Events: family oriented events with California themes; California Hall of Fame Induction Ceremony in December.
Hours & Admission Prices: Mon.-Sat. 10-5, Sun. 12-5. Adults $8.50, senior citizens & students $7, children 6 -13 $6; discounts to AAM members & groups; members and children 5 & under no charge. Closed New Year's Day; Thanksgiving; Christmas. &
Attendance: 65,000 (accurate)
Membership: Student, Senior & Teacher $40; Individual $50; Family & Dual $85; Friend $250.

CALIFORNIA STATE CAPITOL MUSEUM, 10th & L Streets, Rm. B-27, Sacramento, CA 95814. Tel.: 916-324-0333. Fax: 916-445-3628. TDD: 916-324-2092.

E-mail: capitol@parks.ca.gov
Web Site: www.capitolmuseum.gov, www.parks.ca.gov
Founded: 1981.
Congressional District: 3
Key Personnel: Museum Dir., Casey Hayden.
Personnel Profile: Full-Time Paid 23; Part-Time Paid 19; Part-Time Volunteers 121.
Governing Authority: state. Parent Institution: California Dept. of Parks & Recreation. Subsidiary Institution: Gold Rush District. Tax-exempt.
Institution Type/Description: Historic Building Museum: c.1860-1874 restored California State Capitol.
Collections: architectural features from the State Capitol; late 19th-century office furnishings; tools & materials from the restoration of the State Capitol; documents & other items related to the state; 1900-1906 executive officers; interior & exterior images of the Capitol.
Research Fields: history of California government; State Capitol building; legislature, executive & judicial offices; general office practices of 1900-1910.
Facilities: theater; 200-seat cafeteria. Items made from original capitol architectural fragments, books & other museum-related items for sale.
Activities: guided tours; films; formally organized education programs; docent program; loan, permanent & temporary exhibitions.
Publications: California State Capitol Restoration: A Pictorial History, 1983; California's Historic Capitol, 1983; The Governors of California and Their Portraits.
Hours & Admission Prices: Tours: daily 9-4. No charge. Closed New Year's Day; Thanksgiving; Christmas. &
Attendance: 513,000 (estimated)
Membership: Volunteer Association: Active $10; Sustaining $18; Supporting $25.

THE CALIFORNIA STATE MILITARY MUSEUM, 1119 Second St., Sacramento, CA 95814-3203. Tel.: 916-442-2883. Fax: 916-442-7532.

E-mail: daniel.sebby@us.army.mil
Web Site: www.militarymuseum.org
Founded: 1991.
Congressional District: 1
Key Personnel: Dir., Dan Sebby; Deputy Dir. Library, Bill Davies; Deputy Dir. Special Projects, Ernie McPherson; Deputy Dir. Grants, Marilyn Starbuck; Administration, Butch Dixon; Museum Shop Mgr., Lynden Larsen; Security, Neil Strasbauch; Exhibit Specialist, Ricardo Davis.
Personnel Profile: Full-Time Paid 4; Part-Time Paid 4; Part-Time Volunteers 60.
Governing Authority: nonprofit. Parent Institution: California State Military Dept. Branch Museums: Camp Roberts, southern Monterey County; Camp San Luis Obispo; Fresno Air National Guard Base; Los Alamitos Joint Forces Training Base, Orange County. Tax-exempt: 501(c)(3).
Institution Type/Description: Military Museum.
Collections: military artifacts; photographs; personal artifacts.
Research Fields: California military history.
Facilities: library.
Activities: tours.
Hours & Admission Prices: Tues.-Sun. 10-4. Adults $5, children 6-17 $3; members and children 5 & under no charge. Closed New Year's Day; Easter; Thanksgiving; Christmas. &
Attendance: 70,000 (estimated)

CALIFORNIA STATE RAILROAD MUSEUM, 125 I St., Sacramento, CA 95814-2265. Mailing Address: 111 I St., Sacramento, CA 95814-2265. Tel.: 916-445-6645 & 7387. Fax: 916-327-5655.

E-mail: rrmuseuminfo@parks.ca.gov
Web Site: www.csrmf.org
Founded: 1976.
Congressional District: 3
Key Personnel: Dir., Paul J. Hammond; Cur. Exhibits, Kendra Dillard; Cur. History & Technology, Kyle K. Wyatt; Dir. Collections, Ellen L. Halteman; Archivist, Kathryn Santos; Dir. Foundation, Pamela Horan; Exhibits, Paul Brown; Librarian, Cara Randall; Maintenance Chief, Dennis Pruitt; Museum Shop Mgr., Tom Grenache.
Personnel Profile: Full-Time Paid 25; Full-Time Volunteers 10; Part-Time Paid 15; Part-Time Volunteers 1,000; Interns 3.
Governing Authority: state. Parent Institution: California State Parks. Related Organization: California State Railroad Museum Foundation. Tax-exempt.
Institution Type/Description: Railroad Museum: in Old Sacramento State Historic Park.
Collections: railroads and railroading in California, Nevada & the West; 200 historic steam locomotives, diesels & cars; artifacts; archival materials; manuscripts; Railtown 1897 State Historic Park. Historic Buildings: Big Four Building; Dingley Spice Mill; Central Pacific Railroad Passenger Station; Central Pacific Railroad Freight Depot.
Research Fields: railroads & railroading in the United States, emphasizing California, Nevada & the West.
Facilities: library of books, photographs, manuscripts, archival materials regarding railroading in California & the United States, available for research use; reading room; two 135-seat theaters; classrooms; restaurant. Books & other memorabilia for sale.
Activities: guided tours; lectures; films; gallery talks; live steam train rides;

formally organized educational programs; docent programs; permanent & temporary exhibitions; biannual living history. Museum Sponsors: National Handcar Races.
Publications: museum brochure; guidebook; On Track!
Hours & Admission Prices: Daily 10-5. Adults $10, children 6-17 $5; children under 6 & members no charge. Closed New Year's Day; Thanksgiving; Christmas. &
Attendance: 525,000 (estimated)
Membership: Children's Caboose Club $25; Fireman $35; Conductor $60; Engineer $100; Trainmaster $250; Copper Spike $250; Bronze Spike $500; Silver Spike $750; Gold Spike $1,000; Platinum Spike $5,000.

CENTER FOR SACRAMENTO HISTORY, 551 Sequoia Pacific Blvd., Sacramento, CA 95811-0229. Tel.: 916-264-7072. Fax: 916-264-7582.

E-mail: meymann@cityofsacramento.org
Web Site: www.sacramenities.com/history/
Formerly: Sacramento Archives and Museum Collection Center
Founded: 1953.
Congressional District: 3
Key Personnel: Mgr., Marcia Eymann; Sr. Archivist, Patricia Johnson; Archivist, Dylan McDonald; Cur. History, Veronica Kandl.
Personnel Profile: Full-Time Paid 5; Part-Time Paid 3; Part-Time Volunteers 25; Interns 3.
Governing Authority: municipal. Parent Institution: City of Sacramento. Subsidiary Institutions: Sacramento History Museum; Sacramento Science Center; Sacramento City Cemetery. Tax-exempt: 170(b)(1)(A).
Institution Type/Description: Historic Agency.
Collections: transportation; history; city and county records; archives; photographs; printing; California newspapers; manuscripts; business records.
Research Fields: local history; ethnic studies; textiles; theater; transportation; printing; California journalism; television news.
Facilities: 15,000-vol. library, including news collection, documentary photographs, architectural drawings, American west maps; local government archive; local television film library available for use on premises.
Activities: lectures; graduate & undergraduate student intern program; community outreach.
Publications: reports, Old Sacramento Historic Illustrations Guide; Telling the Sacramento Story; booklets, Old Sacramento, A Reference Point in Time; Sidewalk History: Pioneer Sites in Old Sacramento; book, The City of the Plain, Sacramento in the 19th Century; The Sacramento Museum; Recommendations for Planning and Development in the Old Sacramento State Historic Park; catalogue, For the Record; quarterly journal, Interchange; History of the Lower American River.
Hours & Admission Prices: Mon.-Fri. 8-12 & 1-5; other times by appointment. No charge; donations accepted. Closed city holidays. &
Attendance: 5,000 (estimated)

∗　CROCKER ART MUSEUM, (M), 216 O St., Sacramento, CA 95814-5399. Tel.: 916-808-7000. Fax: 916-808-7372. TTY CA Relay: 1-888-877-5379.

E-mail: cam@crockerartmuseum.org
Web Site: www.crockerartmuseum.org
Founded: 1885.
Congressional District: 5
Key Personnel: Dir., Lial A. Jones; Dir. Devel., Linda Farley; Cur., Scott Shields; Finance, David Separovich; Deputy Dir., Randy Roberts; Museum Store Mgr., Donna Natsoulas.
Personnel Profile: Full-Time Paid 50; Part-Time Paid 36; Part-Time Volunteers 311; Interns 12.
Governing Authority: Partnership: Crocker Art Museum Association; City of Sacramento. Tax-exempt.
Institution Type/Description: Art Museum: housed in 1872 E.B. Crocker Art Gallery; Crocker Mansion Wing; Teel Family Pavilion.
Collections: master drawings; American, Asian, African, Oceanic & California art; European art, prints, paintings & drawings; photography; international ceramics.
Major Exhibits: Sam Francis - 5 Decades, 12/13-3/14.
Research Fields: master drawings, California painting & sculpture, international ceramics; Asian art.
Facilities: 11,500-vol. library of books on art available for use on premises by appointment only.
Activities: guided tours; lectures; films; concerts; arts festivals; formally organized educational programs; docent program; formal inter-museum loan, temporary & traveling exhibitions.
Publications: exhibition catalogs; ArtLetter; annual reports; program guides.
Hours & Admission Prices: Tues.-Wed. & Fri.-Sun. 10-5, Thurs. 10-9; Mon. Holiday's call for hours. Adults $10, senior citizens $8, college & youth 7-17 $5; discounts to AAM members; members & children 6 and under no charge. Closed New Year's Day; Thanksgiving; Christmas. &
Attendance: 250,493 (accurate)
Membership: Individual $65; Family & Supporter $85; Associate $175; Contributor $300; Benefactor $600; Director's Circle Patron $1,500; Director's Circle Trustee $2,500; Director's Circle President $5,000; Crocker Council Founder $10,000. (Senior discounted rates available).

GOVERNOR'S MANSION STATE HISTORIC PARK, 1526 H St., Sacramento, CA 95814-2005. Tel.: 916-323-3047. Fax: 916-322-4775.

Web Site: www.parks.ca.gov/governorsmansion
Founded: 1967.
Congressional District: 3
Key Personnel: Supervising Lead Guide, Erin Renfre; Supervising Lead Guide, Terry Cook; Chm. (V) & Pres. (V), Al Howenstein.
Personnel Profile: Full-Time Paid 3; Part-Time Paid 6; Part-Time Volunteers 75.
Governing Authority: state. California State Dept. of Parks and Recreation, P.O. Box 2390, Sacramento, CA 95811. Subsidiary Institution: California Historic Governor's Mansion Foundation. Tax-exempt.
Institution Type/Description: Historic House Museum: housed in Victorian mansion, built in 1877.
Collections: furnishings collected by 13 California governors from 1903-1967; Albert & Clemenza Gallatin's hardware c.1877.
Research Fields: printed & oral history of residents.
Activities: guided tours; special events; rental facilities. Museum Sponsors: Living History Days; Halloween after dark tours; Gala Christmas Celebration.
Publications: Golden Notes #1; Golden Notes #2; The Historic Governor's Mansion; Governor's Mansion Cookbook; Governor's Mansion Coloring Book; Earthquake Days (1906); California Gold; Sacramento and The California Delta; Governor's Mansion Traditional Family Recipes; History books; Day Hikers Guide to California State Parks.
Hours & Admission Prices: Tours: 10-4; groups by appointment. Park: 10-5. Adults $4, children 6-16 $2; children under 6 no charge. Closed New Year's Day; Thanksgiving; Christmas. &
Attendance: 39,078 (estimated)

GREGORY KONDOS GALLERY, Sacramento City College - FA9, 3835 Freeport Blvd., Sacramento, CA 95822-1318. Tel.: 916-558-2559.

E-mail: stevens@scc.losrios.edu
Web Site: web.scc.losrios.edu/art/gallery
Key Personnel: Dir., Michael Stevens
Institution Type/Description: Art Gallery.
Collections: works by regional artists; Gregory Kondos paintings & prints.
Hours & Admission Prices: Mon.-Thurs. 12-4, Fri. 12-3.

LA RAZA/GALERIA POSADA, 2700 Front St. Ste. A, Sacramento, CA 95818-1121. Tel.: 916-446-5133. Fax: 916-446-1324.

E-mail: larazagaleria@sbcglobal.net
Web Site: www.larazagaleriaposada.org
Founded: 1972.
Congressional District: 5
Key Personnel: Dir., Marie Acosta; Chm. (V), Mario Gutierrez; Gallery Coord., Roberto Lopez.
Personnel Profile: Full-Time Paid 1; Part-Time Paid 1; Part-Time Volunteers 100.
Governing Authority: nonprofit organization. Tax-exempt: 501(c)(3).
Institution Type/Description: Art Center & Museum: housed in the Heilbron Mansion, a Victorian landmark.
Collections: poster art by national Chicano artists; traditional folk & fine art by Native American, Latino & Chicano.
Research Fields: Mexican, Chicano & Latino art movement.
Activities: guided tours; temporary & traveling exhibitions of museum collections; story-telling for children; mask-making workshops. Annual Event: Dia de los Muertos Posada Navidena.
Publications: exhibition notes.
Hours & Admission Prices: Wed.-Sat. 12-6, 2nd Sat. of month 12-9. Suggested Donation: $5. Closed Easter; Thanksgiving; Christmas. &
Attendance: 25,000 (estimated)
Membership: Student & Senior Citizen $15; Individual $30; Family $40; Contributor $50; Sponsor $100; Benefactor $500; Patron, Business & Group Sponsor $1,000.

OLD SACRAMENTO SCHOOLHOUSE MUSEUM, 1200 Front St., Sacramento, CA 95814-3247. Mailing Address: 5325 Ridgefield Ave., Carmichael, CA 95608-2109. Tel.: 916-483-8818.
E-mail: info@oldsacschoolhouse.org
Web Site: www.scoe.net/oldsacschoolhouse
Founded: 1977.
Key Personnel: Pres. (V), Suzanne Hicklin.
Personnel Profile: Part-Time Volunteers 30.
Governing Authority: Tax-exempt.
Institution Type/Description: History Museum.
Collections: 19th century school life in California; pot-bellied stove; period furnishings; student desks; antique pump organ; birds eggs c 1870.
Activities: one hour lesson from the 1800s for school children. Museum Sponsors: Gingerbread Holiday House Contest in December.
Hours & Admission Prices: Mon.-Sat. 10-4, Sun. 12-4. No charge; donations accepted. &
Attendance: 50,000 (estimated)

THE POWERHOUSE SCIENCE CENTER, Discovery Museum, 3615 Auburn Blvd., Sacramento, CA 95821-2097. Tel.: 916-808-3942. Fax: 916-808-3925.
E-mail: marlenec@thediscovery.org
Web Site: www.powerhousesciencecenter.org
Formerly: Discovery Museum, Sacramento Museum of History, Science and Technology
Founded: 1994.
Congressional District: 4
Key Personnel: Interim Exec. Dir., Michele Wong; Dir. Science Public Programs, Susan Douglas; Ri. Challenger Lead Flight, Brenta Bechler; Museum Shop Mgr., Anna Fantasia.
Personnel Profile: Full-Time Paid 14; Full-Time Volunteers 4; Part-Time Paid 2; Part-Time Volunteers 250; Interns 2.
Governing Authority: nonprofit organization. Tax-exempt: 501(c)(3).
Institution Type/Description: Science & Space Museum.
Collections: local historical objects & artifacts; insects; fossils; rocks & minerals; shells; live animals; native & drought tolerant plants.
Facilities: planetarium; classroom; amphitheatre; 14 acres of nature trail. Museum-related items for sale
Activities: guided tours; docent program; planetarium shows; hands-on temporary exhibits; traveling science programs to schools & other locations; youth classes; teacher training; rental facility.
Publications: newsletter; brochures; educational program booklet.
Hours & Admission Prices: Daily 10-5. Adults $6 (Sat.-Sun.), $5 (Mon.-Fri.), children 13-17 & seniors $4, children 4-12 $3; discounts to AAM & ASTC members; members & children under 3 no charge. Closed New Year's Day; Thanksgiving; Christmas. &
Attendance: 120,000 (accurate)
Membership: Family & Grandparents $50; Donor $75; Sustainer $150; Patron $250; Benefactor $500.

ROBERT ELSE GALLERY, California State University Sacramento, 6000 'J' St., Kadema Hall, First Fl., Sacramento, CA 95819-2605. Tel.: 916-278-6166.
Web Site: www.csus.edu/galleries/else.html
Institution Type/Description: Art Gallery.
Collections: works by student artists.
Activities: special events.
Hours & Admission Prices: Academic Year: Mon.-Fri. 12-4:30. No charge.

SACRAMENTO ZOO, 3930 W. Land Park, Sacramento, CA 95822-1123. Tel.: 916-808-5166. Fax: 916-264-5887.
E-mail: saczooinfo@cityofsacramento.org
Web Site: www.saczoo.com
Founded: 1927.
Congressional District: 5
Key Personnel: Exec. Dir. & C.E.O., Mary Healy; Pres. (V), Terry Kastanis; Cur., Jim Schnormeier; Museum Shop Mgr., Fred Klinker.
Personnel Profile: Full-Time Paid 58; Part-Time Paid 26; Part-Time Volunteers 1,400.
Governing Authority: society. Parent Institution: Sacramento Zoological Society. Tax-exempt.
Institution Type/Description: Zoo.
Collections: 413 specimens of exotic animals; 131 species.
Research Fields: behavioral research through University of California at Davis; Chimpanzoo (Jane Goodall Institute).
Facilities: 300-vol. library of zoology material not available for inter-library loan; botanical garden; zoological park; banquet facilities. Gift items for sale.

Activities: guided tours; lectures; fund raising events; outreach programs; formally organized education programs for children, adults, undergraduate & graduate students; docent program or council; permanent exhibitions.
Publications: quarterly newsletter, Maagizo.
Hours & Admission Prices: Feb.-Oct. daily 9-4; Nov.-Jan. daily 10-4. Mon.-Fri. adults $9, seniors $8.25, children 3-12 $6.50; discounts to groups of 25 or more; AZA, children 2 & under and members no charge. Sat.-Sun. adults $9.50, seniors $8.75, children 3-12 $7; discounts to groups of 25 or more; AZA, children 2 & under and members no charge. Closed Thanksgiving; Christmas. &
Attendance: 500,000 (estimated)
Membership: Individual $35; Member plus Guest $50; Family $75; Family Plus $115; Zoo Builder $250; Zoo Patron $500.

SIERRA SACRAMENTO VALLEY MEDICAL SOCIETY MUSEUM OF MEDICAL HISTORY, 5380 Elvas Ave., Ste. 100, Sacramento, CA 95819-2300. Tel.: 916-456-3152. Fax: 916-452-2690.
E-mail: ssvmsmus@winfirst.com
Web Site: www.ssvms.org/museum.aspx
Founded: 2001.
Key Personnel: Chm. (V) & Cur., Dr. Bob LaPerriere.
Personnel Profile: Part-Time Volunteers 8.
Governing Authority: Parent Institution: Community Service Education Research Fund (CSERF). Tax-exempt.
Institution Type/Description: Medical History Museum.
Collections: history of medicine including surgery, clinical diagnosis, infectious disease, pharmacy, radiology, Chinese medicine, obstetrics & gynecology, and medical quackery; books; journals; medical artifacts.
Facilities: 1,200 sq. ft. exhibit space.
Activities: guided tours.
Hours & Admission Prices: Mon.-Fri. 9-4. No charge; donations accepted. Closed holidays. &
Attendance: 1,000 (estimated)

STATE INDIAN MUSEUM, 2618 K St., Sacramento, CA 95816-5104. Tel.: 916-324-0971. Fax: 916-322-1561.
E-mail: cmcgough@parks.ca.gov
Web Site: www.parks.ca.gov/indianmuseum
Founded: 1940.
Congressional District: 3
Key Personnel: Interpretive Svcs., Connie McGough; Cur., Ileana Maestas; Museum Shop Mgr., Helen Kawello.
Personnel Profile: Full-Time Paid 3; Part-Time Paid 3; Part-Time Volunteers 45.
Governing Authority: state. Affiliated with California State Dept. of Parks & Recreation, P.O. Box 2390, Sacramento, CA 95811.
Institution Type/Description: California Native American Cultural & Historical Museum.
Collections: California Indian cultural items; contemporary California Native American art work.
Facilities: hands-on area. Native-made items for sale.
Activities: gallery talks upon request; groups by reservation. Museum Sponsors: Annual Honored Elders Day; Acorn Day; Indian Arts & Crafts Marketplace in spring; Holiday Arts & Crafts Fair in fall.
Publications: brochures.
Hours & Admission Prices: Daily 10-5. Adults 17 & up $2, youth 6-17 $1; children 5 & under no charge. Closed New Year's Day; Thanksgiving; Christmas. &
Attendance: 40,000 (estimated)

SUTTER'S FORT STATE HISTORIC PARK, 2701 L St., Sacramento, CA 95816-5613. Mailing Address: Capital District State Museums & Historic Parks, 101 'J" St., Sacramento, CA 95814. Tel.: 916-445-4422. Fax: 916-442-8613.
E-mail: info@parks.ca.gov
Web Site: www.parks.ca.gov/suttersfort
Founded: 1839.
Congressional District: 3
Key Personnel: Interpreter II, Bob Russo; Museum Tech, Judy Russo; Unit Ranger, Karen Meltzer; Museum Shop Mgr., Sparky Anderson.
Personnel Profile: Full-Time Paid 2; Part-Time Paid 2; Part-Time Volunteers 125.
Governing Authority: state. Parent Institution: California State Dept. of Parks & Recreation, P.O. Box 2390, Sacramento 95811. Tax-exempt.
Institution Type/Description: State Park Museum.
Collections: 1839-1850 period artifacts; pre-California gold rush story.
Facilities: Museum-related items for sale.
Activities: talks; electronic self-guided tour in English, German, Spanish &

Japanese; environmental living program with school groups; docent training program. Museum Sponsors: living history days & demonstrations; authentic 1840s Trade Faire in April; costumed interpreters & demonstrations in summer.

Publications: costume manual; brochures, Teacher Guide; California Historical Landmarks; Gold Rush Era.

Hours & Admission Prices: Daily 10-5. Adults $5, youth 6-17 $3; children 5 & under no charge. Closed New Year's Day; Thanksgiving; Christmas. ♿

Attendance: 200,000 (estimated)

WELLS FARGO HISTORY MUSEUM, 400 Capitol Mall, Sacramento, CA 95814-4407. Mailing Address: Wells Fargo Historical Services, 420 Montgomery St., MAC-A0101-106, San Francisco, CA 94163. Tel.: 916-440-4161.

Founded: 1992.

Key Personnel: Cur., Indulis Kalnins.

Governing Authority: profit-making organization. Affiliated with Wells Fargo Bank.

Institution Type/Description: Company History Museum.

Collections: Concord stagecoach; Wells Fargo banking & express history; gold mining; gold specimens; original gold scales; Livingston collection of Sacramento Postal; Wells Fargo express covers.

Activities: guided tours.

Publications: scholarly pamphlets.

Hours & Admission Prices: Mon.-Fri. 9-5. No charge. Closed bank holidays. ♿

WELLS FARGO HISTORY MUSEUM, 1000 Second St., Sacramento, CA 95814-3202. Mailing Address: Wells Fargo Historical Services, 420 Montgomery St., MAC-A0101-106, San Francisco, CA 94163. Tel.: 916-440-4263.

Founded: 1984.

Key Personnel: Cur., Denise M. Fellini.

Governing Authority: profit-making organization. Affiliated with Wells Fargo Bank.

Institution Type/Description: Company History Museum.

Collections: original gold scales; model Concord Stagecoach; Wells Fargo banking & express history; staging; Pony Express; mining; local items.

Activities: guided group tours; permanent & temporary exhibitions.

Publications: scholarly pamphlets.

Hours & Admission Prices: Daily 10-5. No charge. Closed New Year's Day; Easter; Thanksgiving; Christmas. ♿

Saint Helena

BALE GRIST MILL STATE HISTORIC PARK, 3369 St. Helena Hwy. N., Saint Helena, CA 94515. Mailing Address: 3801 St. Helena Hwy. N., Calistoga, CA 94515-9617. Tel.: 707-942-4575. Fax: 707-942-9560.

Web Site: www.napavalleystateparks.org

Founded: 1975.

Congressional District: 7

Key Personnel: Supvr., Sandy Jones.

Personnel Profile: Part-Time Paid 2; Part-Time Volunteers 12.

Governing Authority: nonprofit co-op assoc. state. Tax-exempt: 501(c)(3).

Institution Type/Description: History Museum: housed in a grist mill built in 1844-47.

Collections: tools specific to milling grain; photographs.

Research Fields: milling history; mill maintenance; sociological basis of mills.

Facilities: 11,000 sq. ft. exhibit space. Gift items for sale.

Activities: guided tours; informal education programs for children; school groups; milling demonstrations. Annual Events: Old Mill Days in October; Pioneer Christmas in December.

Publications: newsletter, Napa Valley State Parks Association.

Hours & Admission Prices: Park: daily Buildings: Sat.-Sun. 10-5. School Tours: Tues. by appointment. Adults $3, children $2; children under 6 no charge. Closed major holidays. ♿

Attendance: 16,000 (accurate)

Membership: Napa Valley State Parks Association: Senior & Student $15; Individual $20; Family $30; Business & Sponsor $50 & up.

ROBERT LOUIS STEVENSON MUSEUM, (M), 1490 Library Lane, Saint Helena, CA 94574-1143. Tel.: 707-963-3757. Fax: 707-963-0917.

E-mail: director@stevensonmuseum.org

Web Site: www.stevensonmuseum.org

Formerly: Robert Louis Stevenson Silverado Museum

Founded: 1969.

Congressional District: 2

Key Personnel: Chm. (V), Donald Fraser; Exec. Dir., Marissa Schleicher.

Personnel Profile: Part-Time Paid 4.

Governing Authority: nonprofit. Parent Institution: The Vailima Foundation, P.O. Box 23, Saint Helena, CA. Tax-exempt.

Institution Type/Description: History & Literary Museum.

Collections: first editions, manuscripts, letters, memorabilia of Robert Louis Stevenson; books, periodicals, paintings, photographs & sculptures relating to Stevenson.

Research Fields: life and works of Robert Louis Stevenson, his immediate family & associates.

Facilities: library of books related to Robert Louis Stevenson & his family; original letters & manuscripts; photographs; periodicals; paintings; prints; sculptures; memorabilia. Publications written & related to Stevenson for sale.

Activities: guided tours; lectures; docent program; permanent & temporary exhibitions.

Publications: The Silverado Squatters; Prayers Written at Vailima; The New Lighthouse on the Dhu Heartache Rock; Play-If I'm Spared; Treasured Recipes; books, Robert Louis Stevenson biographies.

Hours & Admission Prices: Tues.-Sat. 12-4; other times by appointment. No charge; donations accepted. Closed New Year's Day; Easter; Independence Day; Memorial Day; Thanksgiving; Christmas. ♿

Attendance: 4,000 (accurate)

ST. HELENA HISTORICAL SOCIETY, St. Helena Public Library, Saint Helena, CA 94574. Mailing Address: P.O. Box 87, Saint Helena, CA 94574-0087. Tel.: 707-967-5502.

Key Personnel: Pres., Skip Lane

Institution Type/Description: Historical Society Museum.

Collections: local history & culture; books; photographs.

Activities: research.

Hours & Admission Prices: Mon.-Fri. 9-5.

Salinas

HARVEY-BAKER HOUSE, Salinas Transportation Center, Salinas, CA 93901. Mailing Address: Monterey County Historical Society, P.O. Box 3576, Salinas, CA 93912-3576. Tel.: 831-424-7155.

Institution Type/Description: Historic House Museum: housed in the home of Isaac Harvey, the first mayor of Salinas, built in 1868.

Collections: period furnishings.

Hours & Admission Prices: 1st Sun. of month 1-4; other times by appointment. No charge.

JOSE EUSEBIO BORONDA ADOBE, 333 Boronda Rd., Salinas, CA 93907-1808. Mailing Address: P.O. Box 3576, Salinas, CA 93912-3576. Tel.: 831-757-8085.

Institution Type/Description: Historic House Museum: built in the 1840s.

Collections: local & family history; period furnishings; personal artifacts; photographs.

Hours & Admission Prices: Mon.-Fri. 10-3.

NATIONAL STEINBECK CENTER, 1 Main St., Salinas, CA 93901-3436. Tel.: 831-796-3833. Fax: 831-796-3828.

E-mail: info@steinbeck.org

Web Site: www.steinbeck.org

Founded: 1983.

Congressional District: 17

Key Personnel: Dir., Colleen Bailey; Chm. (V), Chris Steinbruner; Museum Shop Mgr., Phillip Saldana.

Personnel Profile: Full-Time Paid 12; Part-Time Paid 6; Part-Time Volunteers 100; Interns 3.

Volunteer Hours: 8,000

Operating Expenses: 2,134,000

Operating Income: 2,073,000

Governing Authority: private; nonprofit. Tax-exempt: 501(c)(3).

Institution Type/Description: Literary, Agriculture & Art Museum.

Collections: 30,000 objects, manuscripts, letters, photographs, oral history tape, truck, screen plays & reviews related to the life & art of John Steinbeck.

Research Fields: the life, legacy & art of John Steinbeck.

Facilities: archives; 215-seat meeting area; 12,000 sq. ft. exhibit space. Museum-related items for sale.

Activities: volunteer program; lectures; book group; temporary exhibitions; educational programs; festivals; literary programs; art activities.

Publications: quarterly newsletter.

Hours & Admission Prices: Daily 10-5. Adults $14.95, seniors $8.95, youth

13-17 $7.95, children $5.95; discounts to AAA members; members and children 5 & under no charge. Closed New Year's Day; Thanksgiving; Christmas. &
Attendance: 50,000 (estimated)
Membership: Individual $45; Dual & Family $65; Literary Guild $125; Pulitzer Prize Circle $250; Author's Circle $500.

THE STEINBECK HOUSE, 132 Central Ave., Salinas, CA 93901-2651. Tel.: 831-424-2735. Fax: 831-757-5806.
E-mail: steinbeckhouse@sbcglobal.net
Web Site: steinbeckhouse.com
Governing Authority: Parent Institution: Valley Guild.
Institution Type/Description: Historic House Museum: housed in the birthplace & boyhood home of author John Steinbeck; built in 1897. A National Historical Monument.
Collections: Steinbeck life & career; personal artifacts; period furnishings; photographs.
Facilities: restaurant. Museum-related items for sale.
Hours & Admission Prices: By appointment. Adults $5, students, retired military & seniors $3; active military & children under 6 no charge.

Samoa

HUMBOLDT BAY MARITIME MUSEUM, Hwy. 255, Samoa, CA 95564. Mailing Address: P.O. Box 282, Samoa, CA 95564-0282. Tel.: 707-444-9440.
E-mail: intrepid52315@gmail.com
Web Site: www.humboldtbaymaritimemuseum.com
Founded: 1977.
Congressional District: 1
Key Personnel: Chm., Joshua Smith; Operations, Dalene Zerlang; Museum Shop Mgr., Jeffry Hood.
Personnel Profile: Part-Time Paid 1; Part-Time Volunteers 3.
Governing Authority: nonprofit. Tax-exempt: 501(c)(3).
Institution Type/Description: Maritime Museum.
Collections: regional maritime history; photographs; local culture.
Research Fields: local maritime history.
Activities: educational programs; group tours.
Hours & Admission Prices: Tues.-Wed. & Fri.-Sat. 11-4. No charge; donations accepted.
Attendance: 2,500 (estimated)

SAMOA COOKHOUSE & LOGGING MUSEUM, 908 Vance Ave., Samoa, CA 95564. Mailing Address: 710 E. St., Ste. 136, Eureka, CA 95501-1853. Tel.: 707-442-1659. Fax: 707-442-1699.
Web Site: www.samoacookhouse.net/samoa-cookhouse-museum.html
Key Personnel: Mgr., Jeff Brustman; Asst. Mgr., Sharon Nichols
Institution Type/Description: Lumber & Logging Industry Museum.
Collections: cookhouse & logging history; period artifacts; photographs. Historic Building: 1893 cookhouse.
Facilities: cookhouse.
Hours & Admission Prices: Daily 7am-8pm. No charge. &

San Andreas

CALAVERAS COUNTY ARCHIVES, 891 Mountain Ranch Rd., San Andreas, CA 95249-9713.
E-mail: archives@goldrush.com
Key Personnel: Archivist, Shannon Van Zant.
Personnel Profile: Part-Time Paid 1.
Governing Authority: Parent Institution: Calaveras County.
Institution Type/Description: County History Museum.
Collections: Indian artifacts & baskets; Chinese artifacts from the California Gold Rush era; county documents, 1851-1960.
Research Fields: genealogy; Calaveras County history; Mother Lode history; mining; farming; state parks.
Activities: special historic & cultural programs; research.
Hours & Admission Prices: Mon. & Wed. 8-4. No charge. Research $10 per hour. &

CALAVERAS COUNTY HISTORICAL SOCIETY MUSEUM, 30 N. Main St., San Andreas, CA 95249. Mailing Address: P.O. Box 721, San Andreas, CA 95249-0721. Tel.: 209-754-4658 & 1058. Fax: 209-754-1086.
E-mail: cchs@goldrush.com
Web Site: www.calaverascohistorical.com
Founded: 1936.
Congressional District: 18

Key Personnel: Pres. Calaveras Historical Society, Donna Shannon; Cur., Karen Nicholson.
Personnel Profile: Full-Time Paid 1; Part-Time Paid 2; Part-Time Volunteers 5.
Governing Authority: county. Subsidiary Institution: Red Barn Museum, 891 Mountain Ranch Rd., San Andreas, CA 95249. Tel.: 209-754-0800. Tax-exempt.
Institution Type/Description: County History Museum.
Collections: Indian artifacts & baskets; 1849 gold rush mining & pioneer artifacts. Historic Buildings: 1867 courthouse & jail; 1893 Hall of Records.
Research Fields: Calaveras County & Mother Lode history; Calaveras County mining; genealogical research.
Facilities: library.
Activities: special historic & cultural programs.
Publications: quarterly published by C.C. Historical Society, Las Calaveras.
Hours & Admission Prices: Daily 10-4. Adults $3, senior citizens $2, children under 12 $1; discounts to student groups; members no charge. Closed New Year's Day; Easter; Christmas. &
Attendance: 3,000 (estimated)
Membership: County Historical Society: Individual $22; Life $440.

RED BARN MUSEUM, Government Center, 891 Mountain Ranch Rd., San Andreas, CA 95249-9713. Tel.: 209-754-0800. Fax: 209-754-1086.
E-mail: cchs@goldrush.com
Web Site: www.calaverascohistorical.com
Key Personnel: Cur., Rosemary Faulkner
Institution Type/Description: Historic Building: housed in the former dairy barn of the old County Hospital.
Collections: county history; agriculture; mining; logging; ranching; mining carts; farm wagons; paintings; tools; farming; pioneer artifacts.
Activities: hands-on exhibits.
Hours & Admission Prices: Thurs.-Sun. 10-4. Adults $3, senior citizens 60 & over $2, children under 12 $1. Closed New Year's Day; Easter; Christmas.

San Bernardino

CALIFORNIA DEPARTMENT OF FORESTRY AND FIRE PROTECTION MUSEUM, 3800 Sierra Way, San Bernardino, CA 92405. Tel.: 909-881-6984.
E-mail: cdfmuseum@yahoo.com
Key Personnel: Mgr., Jarrel B. Glover
Institution Type/Description: History Museum.
Collections: firefighting history & equipment; fire prevention; photographs.
Activities: educational programs.
Hours & Admission Prices: Sat. 10-2. Closed holidays.

✳ **ROBERT AND FRANCES FULLERTON MUSEUM OF ART (RAFFMA), CALIFORNIA STATE UNIVERSITY, SAN BERNARDINO, (M),** 5500 University Pkwy., San Bernardino, CA 92407-2318. Tel.: 909-537-7373. Fax: 909-537-7068.
E-mail: raffma@csusb.edu
Web Site: raffma.csusb.edu
Founded: 1965.
Congressional District: 36
Key Personnel: Pres. University, Dr. Tomas Morales; Dir. & Cur., Eva Kirsch; Pres. Advisory Bd., Janice Lemann.
Personnel Profile: Full-Time Paid 6; Part-Time Paid 2; Part-Time Volunteers 8; Interns 4.
Governing Authority: public university. Parent Institution: California State University System. Subsidiary Institution: San Bernardino Campus. Tax-exempt: 501(c)(3).
Institution Type/Description: University Art Museum.
Collections: Asian & Etruscan ceramics; African sculpture; Egyptian antiquities; contemporary prints; drawings; paintings; sculptures.
Research Fields: Egyptian art; African sculpture.
Facilities: educational facilities; 6,500 sq. ft. exhibit space.
Activities: formal education programs for undergraduate students; art history lectures; gallery talks. Annual Events: Summer Egyptian Art Workshop for kids; Friends Holiday Event.
Publications: annual exhibition catalogs.
Hours & Admission Prices: Sept.-July Mon.-Wed. & Sat. 10-5, Thurs. 11-7. Suggested Donation: $3. Parking: $5. Closed major holidays. &
Attendance: 10,000 (estimated)
Membership: Student, Faculty, Staff & Senior $35; Individual $50; Family $75; Associate $100; Contributor $500; Benefactor $1,000.

SAN BERNARDINO HISTORICAL AND PIONEER SOCIETY - HERITAGE HOUSE, 796 N. ”D“ St., San Bernardino, CA 92402. Mailing Address: P.O. Box 875, San Bernardino, CA 92402. Tel.: 909-885-2204.
E-mail: sbhistoricalsociety@mac.com
Web Site: www.sbhistoricalsociety.com
Institution Type/Description: Historical Society Museum: house built in 1891.
Collections: local history & culture; period furnishings; personal artifacts; photographs.
Hours & Admission Prices: Sat. 10-2. No charge; donations accepted.
Membership: Individual $30; Life $500.

San Carlos

HILLER AVIATION INSTITUTE, 601 Skyway Rd., San Carlos, CA 94070-2702. Tel.: 650-654-0200. Fax: 650-654-0220.
E-mail: museum@hiller.org
Web Site: www.hiller.org
Founded: 1995.
Congressional District: 12
Key Personnel: C.E.O. & Pres., Jeffery Bass; Chm. (V), Steve Hiller; Vice Pres. Devel., Bernadette Mellott, MPA; Vice Pres. Mktg., Willie Turner; Dir. Volunteers and Exhibits, James Lichtenstein; Education Programs Mgr., Jon Welte; Private Events Mgr., Lanie Agulay; Registrar, Katie McGee; Dir. Merchandising & Retail, Duncan Chadwick.
Personnel Profile: Full-Time Paid 6; Part-Time Paid 12; Part-Time Volunteers 140.
Governing Authority: private; nonprofit organization. Tax-exempt: 501(c)(3).
Institution Type/Description: Aviation Museum & Research Center: housed at the historic San Carlos Airport.
Collections: aviation artifacts; past, present & future aviation technology.
Research Fields: progress of air transportation.
Facilities: classrooms; 35,000 sq. ft. exhibit space; 50-seat theater; aircraft restoration shop. Museum-related items for sale.
Activities: docent program; films; formal education programs for adults, children & college students; guided tours; lectures; rental gallery; temporary & traveling exhibitions; theater; conferences & seminars for scholars and corporate leaders to exchange ideas; special events; guest speakers. Museum Sponsors: Teacher Curriculum Guides; Flight Night (paper airplane contest for ages 8-17); Explorer Scout Post; Members Previews & Receptions.
Publications: quarterly, Briefings.
Hours & Admission Prices: Daily 10-5. Adults $11, senior citizens 65 & over and youth 5-17 $7; discounts to groups of 15 or more; members & children under 5 no charge. Closed New Year's Day; Easter; Thanksgiving; Christmas; occasional special event. &
Attendance: 84,000 (accurate)
Membership: Senior $35; Individual $50; Family $75; Sustaining $100; Supporting $250; Director's Circle $500; Patron $1,000; Benefactor $2,500; Trustee's Circle $5,000; Founder's Circle $10,000 & up.

San Diego

CABRILLO NATIONAL MONUMENT, 1800 Cabrillo Memorial Dr., San Diego, CA 92106-3601. Tel.: 619-557-5450. Fax: 619-226-6311. TTY: 619-222-8211.
Web Site: www.nps.gov/cabr/
Founded: 1913.
Congressional District: 49
Key Personnel: Acting Supt., Wendy Janssen; Chief of Interpretation, Karl Pierce; Historian, Robert Munson; Museum Technician, Tracie Cobb; Museum Shop Mgr., Karen Eccles.
Personnel Profile: Full-Time Paid 15; Part-Time Paid 10; Part-Time Volunteers 300; Interns 1.
Governing Authority: federal. Parent Institution: National Park Service. Tax-exempt.
Institution Type/Description: Historic Site: commemorating the exploration of California coast & San Diego Bay by Juan Rodriguez Cabrillo in 1542.
Collections: commemorates the role that the historic Old Point Loma Lighthouse played in the development of maritime economy in San Diego & California, the WWI & WWII coastal defenses protecting San Diego Harbor; the ecologically sensitive coastal sage scrub and rocky intertidal communities & the ecology & annual migration of the Pacific gray whale. Historic Buildings: 1854 Old Point Loma Lighthouse furnished with artifacts of the period; coastal defense structures from WWI and WWII; exhibit in former army radio building.
Facilities: 1,500-vol. library of history & natural history books available for use on premises; 100-seat auditorium; reading room. Books & other museum-related items for sale.
Activities: informal introduction of film; formally organized educational programs; permanent & temporary exhibitions; environmental education films & discussions.
Publications: books, Cabrillo National Monument; Whale Primer; The Old Point Loma Lighthouse; Cabrillo's Log; The Cabrillo Era; Account of the Voyage of Juan Rodriguez Cabrillo; quarterly newsletter, Explorer; biannual park visitors guide, Cabrillo Journal; Junior Ranger Program; booklet, Just for Kids, Understanding the Life of Point Loma.
Hours & Admission Prices: Daily 9-5. $5 per vehicle, $3 per person; children under 16, seniors over 61, disabled U.S. citizens, members with annual Cabrillo, Golden Age, Golden Access, America the Beautiful Interagency, Senior or Access Pass no charge. &
Attendance: 886,620 (accurate)
Membership: Individual $35; Duo (2 people in same household) $60; Family (2 adults & 2 children) $75; Supporting $100; Sustaining $500; Benefactor $1,000.

CALIFORNIA FLIGHT MUSEUM, 1424 Continental St., Hangar #8 Brownfield Airport, San Diego, CA 92154-5717. Mailing Address: P.O. Box 181140, Coronado, CA 92178. Tel.: 619-661-2516. Fax: 619-435-3928.
E-mail: rfinch@san.rr.com
Web Site: californiaflightmuseum.org
Formerly: San Diego Flight Museum
Founded: 1990.
Key Personnel: Pres. (V), Reg Finch.
Personnel Profile: Part-Time Volunteers 9; Interns 1.
Governing Authority: Tax-exempt.
Institution Type/Description: Military Aircraft Museum.
Collections: military aviation history & aircraft.
Research Fields: aerodynamics.
Activities: lectures. Museum Sponsors: Air Shows; Open Houses.
Hours & Admission Prices: Call for hours. No charge.
Attendance: 50,000 (estimated)
Membership: Associate $20; Active Engineering, Pilot, Science or Promoter $50; Hot Shot $150; 50 Mission Veteran $1,000; Ace $2,500; Top Gun $4,950.

FLYING LEATHERNECK AVIATION MUSEUM, (M), T-4203 Anderson Ave., San Diego, CA 92145. Mailing Address: P.O. Box 452008, San Diego, CA 92145-2008. Tel.: 877-359-8762. Fax: 858-693-0037.
Web Site: www.flyingleathernecks.org
Founded: 1999.
Congressional District: 53
Key Personnel: Dir., Ed Downum; Chm. (V), Dick Miller; Cur., Steve Smith; Museum Shop Mgr., Kathy McLaughlin.
Personnel Profile: Full-Time Paid 6; Full-Time Volunteers 2; Part-Time Volunteers 60.
Governing Authority: Parent Institution: MCAS Miramar. Tax-exempt.
Institution Type/Description: U.S. Marine Corps Aviation Museum.
Collections: Marine Corps aviation & aviators; period aircraft; portraits; military clothing & weapons; Marine women; trophies; Marine Corps Aviation Assoc. awards & trophies.
Publications: quarterly, Flying Leatherneck's Log Book.
Hours & Admission Prices: Tues.-Sun. 9-3:30. No charge; donations accepted. &
Attendance: 30,000 (accurate)
Membership: Individual $35.

GASLAMP MUSEUM AT THE WILLIAM HEATH DAVIS HOUSE HOME OF THE GASLAMP QUARTER HISTORICAL FOUNDATION, 410 Island Ave., San Diego, CA 92101-6925. Tel.: 619-233-4692. Fax: 619-233-4148.
E-mail: info@gaslampquarter.org
Web Site: www.gaslampquarter.org
Founded: 1982.
Key Personnel: Exec. Dir., Bob Marinaccio; Cur., Lindy Harshberger; Mgr. Operations & Mktg., Jake Romero.
Personnel Profile: Full-Time Paid 2; Part-Time Paid 2; Part-Time Volunteers 100, Interns 2.
Governing Authority: private; nonprofit organization. Tax-exempt: 501(c)(3).
Institution Type/Description: Historic House & Museum.
Collections: William Heath Davis House, the oldest wooden structure in San Diego; each room has been decorated to represent a different period of time in the life of the house from 1850-1940.
Research Fields: genealogical study of William Heath Davis.
Facilities: urban park available for rental. Museum-related items for sale.
Activities: guided tours of Gaslamp Quarter; children's tours. Annual Event: Fall Back Festival Children's Historic Street Faire.

Hours & Admission Prices: Tues.-Sat. 10-5, Sun. 12-4. Adults 12 & over $5, seniors 55 & over $4. Gaslamp District Guided Tours: Sat. 11; group tours by appointment. Adults $15, seniors, military & students $12 (includes admission to museum); children under 12 no charge. Closed New Year's Day; Easter; Memorial Day; Independence Day; Thanksgiving; Christmas. &

Attendance: 30,000 (estimated)
Membership: Individual $25; Family $50, $100; Corporate $250, $500; Benefactor $1,000.

JAPANESE FRIENDSHIP GARDEN SOCIETY OF SAN DIEGO, (M), 2215 Pan American Rd. E., San Diego, CA 92101-1656. Mailing Address: Balboa Park Administration Bldg., 2125 Park Blvd., San Diego, CA 92101-4792. Tel.: 619-232-2721. Fax: 619-232-0917.

E-mail: jfgsd@niwa.org
Web Site: www.niwa.org
Founded: 1979.
Congressional District: 53
Key Personnel: Exec. Dir., Luanne Lao-Kanzawa; Pres. (V), Dennis Otsuji; Museum Shop Mgr., Marisa Espinosa.
Personnel Profile: Full-Time Paid 7; Part-Time Paid 7; Part-Time Volunteers 5; Interns 2.
Governing Authority: Tax-exempt.
Institution Type/Description: Japanese Gardens.
Collections: Japanese heritage & culture; plants; trees; sculptures; personal artifacts.
Activities: workshops; educational programs; classes; festivals; seminars; demonstrations.
Publications: newsletter.
Hours & Admission Prices: Memorial Day to Labor Day Mon.-Fri. 10-5, Sat.-Sun. 10-4; Sept.-May daily 10-4. Adults $4, seniors, students & military $3; children 6 & under and members no charge. &
Attendance: 100,000 (estimated)
Membership: Student $15; Individual $30; Family $45; Contributing $60; Fusion $120; Supporter $125; Associate $250; Friend $500; Advocate $1,000.

MARINE CORPS RECRUIT DEPOT COMMAND MUSEUM, SAN DIEGO, (M), Building 26 Day Hall, San Diego, CA 92140. Mailing Address: AC/S G-3 - Attn Command Museum, 1600 Henderson Ave., Suite 212, San Diego, CA 92140. Tel.: 619-524-6719. Fax: 619-524-0076.

E-mail: info@corpshistory.org
Web Site: www.corpshistory.org
Founded: 1987.
Congressional District: 41
Key Personnel: Dir., Barbara S. McCurtis.
Personnel Profile: Full-Time Paid 8; Full-Time Volunteers 25; Part-Time Paid 4; Part-Time Volunteers 30; Interns 1.
Governing Authority: federal; nonprofit. Parent Institution: Headquarters Marine Corps, Historical Center, Bldg. 58, Washington Navy Yard, Washington, DC 20374-0580 & G-3 Command Museum Sponsor. Subsidiary Institution: Commanding General, MCRD; Museum Historical Society. Tax-exempt: 501(c)(3).
Institution Type/Description: Military Museum.
Collections: USMC military equipment, battlefield souvenirs; reference material; personal papers; accoutrements & historic costumes, 1846-present; photos; artwork; film; archives; U.S. Marine Corps & the Recruit Depot and the city of San Diego.
Research Fields: role of U.S. Marine Corps in southern California & the Pacific, 1846-present.
Facilities: 5,000-vol. library of military history, tactics, & technology material; 50-seat theater; 25,000 sq. ft. exhibit space; educational facilities. Gift items for sale.
Activities: guided tours; films; undergraduate & graduate internship programs with San Diego State University & the University of San Diego; docent program; temporary exhibitions.
Publications: quarterly newsletter.
Hours & Admission Prices: Mon.-Fri. & most Saturdays 8-4. Photo ID required for admittance. No charge; donations accepted. Closed federal holidays. &
Attendance: 175,000 (accurate)

MARITIME MUSEUM OF SAN DIEGO, 1492 N. Harbor Dr., San Diego, CA 92101-3309. Tel.: 619-234-9153. Fax: 619-234-8345.

E-mail: info@sdmaritime.org
Web Site: www.sdmaritime.org
Founded: 1948.

Congressional District: 44
Key Personnel: Pres. & C.E.O., Raymond E. Ashley; Dir. Operations, Mark Gallant; Dir. Devel., Chris Berggren; Event Coord., James Davis; Dir. Mktg., Robyn Gallant; Controller, Grace Rodriquez; Cur. Models & Collections Mgr., Kevin Sheehan; (V) Coord., Jeff Loman; Curriculum Coord., Susan Sirota; Editor, Mains'l Haul, Neva Sullaway; Vice Pres., Mark Montijo; Office Mgr., Erminia Taranto; Dir. Corporate Rels., Gregg Doherty; Museum Store Mgr., Carolyn Goben.
Personnel Profile: Full-Time Paid 37; Full-Time Volunteers 15; Part-Time Paid 50; Part-Time Volunteers 400.
Governing Authority: nonprofit organization. Tax-exempt: 501(c)(3).
Institution Type/Description: Maritime Museum: housed aboard four ships.
Collections: small crafts; ship models; navigation instruments; luxury liner memorabilia; tools; photographs; U.S. Navy artifacts; marine engineering; pleasure boating; maritime research library; submarines: USS Dolphin & Soviet b-39; replica ships Californian & Surprise. Historic Ships: 1863 Star of India; 1898 ferryboat Berkeley; 1904 steam yacht Medea, 1914 Pilot.
Research Fields: San Diego, California & Pacific Ocean maritime history.
Facilities: library pertaining to maritime history and research available to accredited students & members on premises. Maritime & museum-related items for sale.
Activities: guided tours; speakers bureau; lectures; films; loan, permanent & temporary exhibits; demonstrations; membership program; historic boat rides; day sailing adventures. Educational programs available: docent guided tours (K-12 and beyond), Science of Seafaring workshops (grades K-8), Living history on the Star of India (grades 4-8), Science environmental studies and technology, field trips (Day on the Bay, Transportation Adventures) and Sea Chest Loan Program.
Publications: quarterly, Mains'l Haul: A Journal of Pacific Maritime History; newsletter, Full'n By.
Hours & Admission Prices: Daily 9-8. Adults $17, senior citizens & military $14, children 5-17 $8; discounts to groups and AAM & ICOM members; children under 5 & members no charge. &
Attendance: 165,502 (accurate)
Membership: Navigator (Senior 62 & over) & Apprentice (Student) $35; Mariner (Individual) $40; Crew (Family) $65; Boatswain $150; First Mate $500; Captain's Table $1,000; Captain's Cabin $5,000.

* MINGEI INTERNATIONAL MUSEUM, (M), 1439 El Prado, San Diego, CA 92101-1617. Tel.: 619-239-0003. Fax: 619-239-0605.

E-mail: mingei@mingei.org
Web Site: www.mingei.org
Founded: 1974.
Congressional District: 43
Key Personnel: C.E.O. & Dir., Rob Sidner; C.F.O., Katharine Wardle; Chair (V) Bd. Trustees, Carolyn Sheets Owen-Towle; Dir. Exhibitions & Chief Cur., Christine Knoke; Dir. Devel., Cathy Sang; Dir. Mktg. & Visitor Experience, Jessica Hanson York; Mgr. Membership, Lauren Tye; Mgr. Devel., James Miller; Devel. Asst., Ashley Fong; Exhibition Designer, Jeremiah Maloney; Exhibition Preparator, Jenny Merullo; Exhibition Preparator, Jim Woodard; Collectors' Gallery Asst., Tien Sada; Education Mgr., Johanna Pope; Photographer & Graphic Designer, Alexis O'Banion; Library, Archives & Digitization Mgr., Kristi Ehrig-Burges; Reception & Coord. Visitor Rels., Jill DeDominicis; Registration Asst., Sarah Winston; Mgr. Security & Facilities, Simon Leader; Security, Nhan Ha.
Personnel Profile: Full-Time Paid 26; Part-Time Paid 7; Part-Time Volunteers 150; Interns 7.
Governing Authority: nonprofit organization. Tax-exempt: 501(c)(3).
Institution Type/Description: International Folk Art, Craft & Design Museum.
Collections: arts of daily life from all cultures of the world.
Major Exhibits: Please Be Seated, 10/5/13-3/30/14; Function & Fantasy - Steven and William Ladd, 2/14-6/1/14; Huyler Chocolates - An Art and Design Story, 3/22/14-8/17/14; Surf Craft, 6/21/14-1/11/15.
Research Fields: international exhibition materials; folk art; craft; design.
Facilities: 9,000-vol. library of audiovisual art reference library; theater; multimedia education center. Exhibition-related items for sale.
Activities: docent guided tours; illustrated lectures; artist craftsmen demonstrations; volunteer council; permanent collection loans; traveling exhibitions; outreach programs for San Diego Community.
Publications: 46 documentary publications; 22 DVDs & videotapes.
Hours & Admission Prices: Tues.-Sun. 10-5. Adults $8, seniors $5, youth 6-17, students, & military with ID $4; members no charge. Closed national holidays. &
Attendance: 108,968 (accurate)
Membership: Individual $50; Friends & Family $70; Contributor $115; Sustainer $250; Patron $500; Director's Circle $1,250.

MISSION SAN DIEGO DE ALCALA, 10818 San Diego Mission Rd., San Diego, CA 92108-2498. Tel.: 619-283-7319. Fax: 619-283-7762.
E-mail: info@missionsandiego.com
Web Site: www.missionsandiego.com
Founded: 1769.
Key Personnel: Pastor, Rev. Msgr. Richard F. Duncanson; Assoc. Pastor, Rev. Bill Springer; Historian, Anthony Tarantino; Business Mgr., Pat Schmitzer; Gift Shop Mgr., Rhonda Sosa.
Personnel Profile: Part-Time Paid 1; Part-Time Volunteers 25.
Governing Authority: church. Affiliated with Roman Catholic Diocese of San Diego, 3888 Paducah Dr., San Diego 92117. Tax-exempt: 501(c)(3).
Institution Type/Description: Historic Site Museum.
Collections: Native American baskets; American & Spanish colonial period artifacts; ecclesiastical artifacts; archaeological findings; archives.
Research Fields: archaeology.
Facilities: Gift items, reproductions & postcards for sale.
Activities: tote-a-tape tour; formally organized education programs for under-graduate & graduate students; permanent exhibitions.
Publications: booklet, Mission San Diego de Alcala; annual brochure, Mission San Diego de Alcala; book, San Diego de Alcala.
Hours & Admission Prices: Daily 9-4:45. Suggested Donation: adults $3, seniors $2, children $1. Closed Thanksgiving; Christmas. &
Attendance: 75,000 (estimated)

MUSEUM OF CONTEMPORARY ART SAN DIEGO - DOWN-TOWN, 1001 & 1100 Kettner Blvd., San Diego, CA 92101. Mailing Address: 700 Prospect St., La Jolla, CA 92037-4291. Tel.: 858-454-3541. Fax: 619-814-4670.
Web Site: www.mcasd.org
Founded: 1941.
Key Personnel: Dir., Hugh M Davies; Pres. Bd. Trustees, Dr. Peter C. Farrell; Vice Pres., David C. Copley; Vice Pres., Maryanne Pfister; Vice Pres., Matthew Strauss; Deputy Dir., Charles E. Castle; Dir. External Affairs, Anne Farrell; Chief Advancement Officer, Jeanna Yoo; Sr. Cur., Stephanie Hanor; Dir. Communications, Denise Montgomery; Dir. Devel., Cynthia Tuomi; Registrar, Anne Marie Purkey Levine.
Personnel Profile: Full-Time Paid 43; Part-Time Paid 40; Part-Time Volunteers 105; Interns 6.
Governing Authority: nonprofit organization. Additional Location: MCASD, 700 Prospect St., La Jolla, CA 92037. Tax-exempt: 501(c)(3).
Institution Type/Description: Art Museum.
Collections: contemporary art including paintings, sculpture, drawings & prints; photography; video art; installation.
Research Fields: contemporary art.
Facilities: 17,000 sq. ft. exhibit space; 120-seat auditorium. Museum-related items for sale.
Activities: lectures; films; seminars; gallery talks; educational programs; concerts; inter-museum loans, temporary & traveling exhibitions.
Publications: quarterly newsletter with calendar of events; exhibition catalogs; educational brochures; membership brochure; general information brochure.
Hours & Admission Prices: Thurs.-Tues. 11-5, third Thurs. 11-7. Adults $10, senior citizens & military $5; discounts to AAM & ICOM members; 25 & under and members no charge. Closed New Year's Day; Thanksgiving; Christmas. Admission valid for 7 days at all MCASD locations. &
Attendance: 168,000 (estimated)
Membership: Artist & Student $35; Active $75; Contributor $150; Patron $300; Donor $600-$1,499; Benefactor $1,500-$2,499; Start Up $2,500; Contemporary Collectors $5,000; International Collectors $10,000. Corporate Members: $1,500 & up.

✱　**MUSEUM OF PHOTOGRAPHIC ARTS,** 1649 El Prado, Ste. 14, Balboa Park, San Diego, CA 92101-1664. Tel.: 619-238-7559. Fax: 619-238-8777.
E-mail: info@mopa.org
Web Site: www.mopa.org
Founded: 1983.
Congressional District: 49
Key Personnel: Dir., Deborah Klochko; Pres. Bd., Gary Payne; Deputy Dir., Vivienne Esrig; Operations Mgr., John Hogan; Accounting Mgr., Debra Rosacker, CPA; Registrar, Megan Clancy; Dir. Exhibitions & Design, Scott Davis; Curatorial Asst., Chantel Paul; Special Events Mgr., Melissa Pfeiffer; Membership Coord., Angela Venuti; Mgr. Mktg. & Communications, Kristine Page; Dir. Education & Public Programs and Mgr. Digital Interpretation, Joaquin Ortiz; Coord. School & Teacher's Programs, Lori Sokolonski; Film Program & Digital Interpretation, Paolo Zuniga; Mgr. Library, Holland Kessinger; Docent & Visitor Rels. Mgr., Kristie Taylor; Grant Writer, Jana Holsenback; Devel. Asst., Christina Shih; Asst. Store

Mgr., Heather Thomas; Store Buyer, Holly Martin; Asst. Visitor Rels. Mgr., Savannah Jarman; Volunteer Coord., Jazmyne Lemar; Administrative Asst., Courtney Browne.
Personnel Profile: Full-Time Paid 19; Part-Time Paid 28; Part-Time Volunteers 100; Interns 2.
Volunteer Hours: 20,000
Operating Expenses: 1,700,000
Operating Income: 2,200,000
Governing Authority: nonprofit organization. Tax-exempt: 501(c)(3).
Institution Type/Description: Photographic Arts.
Collections: photography.
Major Exhibits: CA Invitational, 10/13-2/14; Prix Pictet - Power, 2/8/14-5/18/14; Ansel Adams, 5/14-10/14; Aperture Remix, 5/24/14-10/5/14; Cy Kuckenbaker, Summer 2014; Youth Exhibition, 10/14-2/15.
Facilities: 27,000-vol. library pertaining to photography; 5,500 sq. ft. exhibit space; 228 seat theater; classrooms. Museum-related items for sale.
Activities: lectures; docent program; film programs; traveling exhibitions; workshops; school tours; educator packets for selected exhibitions; children camps; special events; photographic workshops for adults.
Publications: periodic exhibition catalogues; monthly e-newsletter.
Hours & Admission Prices: Tues.-Sun. 10-5; docent tours by appointment. Gallery: adults $8, students, seniors & retired military $6; discounts to AAM & ICOM members; children 12 and under, active military, members, & San Diego residents on 2nd Tues. each month no charge. Closed New Year's Day; Thanksgiving; Christmas. &
Attendance: 95,830 (estimated)
Membership: Student $25; Senior & Educator $40; Individual $50; Duo Senior $70; Duo & Family $75; Photographer's Circle $150; Silver Circle $250; Platinum Circle $500; Photo Forum $1,000-$4,999; Collector's Salon $5,000 & up.

MUSEUM OF SAN DIEGO HISTORY AND RESEARCH LI-BRARY, 1649 El Prado, Ste. 3, San Diego, CA 92101-1664. Tel.: 619-232-6203, ext. 109. Fax: 619-232-6297.
E-mail: angela.sieckman@sandiegohistory.org
Web Site: www.sandiegohistory.org
Formerly: San Diego Historical Society Museum and Research Archives
Founded: 1928.
Congressional District: 4
Key Personnel: Dir., David Kahn; Bd. Pres. (V), Robert Adelizzi; Museum Shop Mgr., Trina Brewer.
Personnel Profile: Full-Time Paid 22; Full-Time Volunteers 14; Part-Time Paid 14; Part-Time Volunteers 107.
Governing Authority: private; nonprofit organization. Tax-exempt: 501(c)(3).
Institution Type/Description: History Museum & Research Archives.
Collections: concentration on San Diego's history & growth since the 1840s; photographic negatives; historical & art-related materials.
Facilities: library; 100-seat theater; research archives with reading room. Books & gift items for sale.
Activities: school programs; public programs; tours; lectures; temporary & permanent exhibitions.
Publications: quarterly newsletter, The San Diego Historical Society Times; quarterly, The Journal of San Diego History; newsletter, The Times.
Hours & Admission Prices: Museum: daily 10-5. Adults $5, seniors, students & military with ID $4, children 6-17 $2; discounts to groups & AAM members; members & children under 6 no charge. Research Library: Mon.-Fri. 9:30-1. Adults $8, seniors $6, children $4; children under 6 no charge. Closed New Year's Day; Thanksgiving; Christmas Eve & Day. &
Attendance: 25,200 (accurate)
Membership: Family & Household Circle $60; Century Circle $100; Scholar's Circle $250; Curator's Circle $500; Director's Circle $1,000.

NEW AMERICANS MUSEUM, 3232 Dove St., San Diego, CA 92103-5548. Tel.: 619-255-8908. Fax: 619-269-3499.
E-mail: contactus@namuseum.org
Web Site: www.newamericansmuseum.org
Formerly: New Americans Immigration Museum & Learning Center
Founded: 2001.
Key Personnel: Exec. Dir., Gayle Hom; Founder, Deborah Szekely; Chm. (V), Stacey Goto; Program Dir., Miguel-Angel Soria.
Personnel Profile: Full-Time Paid 4; Part-Time Volunteers 28; Interns 2.
Governing Authority: private; nonprofit organization. Tax-exempt: 501(c)(3).
Institution Type/Description: Cultural Center.
Collections: American immigrants.
Hours & Admission Prices: Wed.-Sun. 11-5. No charge; donations accepted. Closed major holidays. &
Membership: Student, Senior & Military $25; Individual $50.

THE NEW CHILDREN'S MUSEUM, 200 W. Island Ave., San Diego, CA 92101-6850. Tel.: 619-233-8792. Fax: 619-233-8796.
Web Site: www.thinkplaycreate.org
Formerly: Children's Museum/Museo de Los Ninos San Diego
Founded: 1981.
Congressional District: 41
Key Personnel: Bd. Pres., Patsy Marino; Exec. Dir., Rachel Teagle; C.O.O., Donna Gough; Dir. External Affairs & Mktg., Jessica Hanson York.
Personnel Profile: Full-Time Paid 22; Part-Time Paid 30; Part-Time Volunteers 40; Interns 8.
Governing Authority: nonprofit organization. Tax-exempt: 501(c)(3).
Institution Type/Description: Children's Museum.
Collections: learning through the arts, providing hands-on, minds-on experiences for all ages.
Facilities: performance studio; cafe; classrooms. Museum-related items for sale.
Activities: organized educational programs for children; volunteer program; student internships offered to college students; rental facilities; birthday parties; participatory exhibitions; Teen Advisory Council; school visits; workshops; summer camps.
Hours & Admission Prices: Call for hours. Adults $10; discounts to AAM members. &
Attendance: 190,000 (accurate)
Membership: Family $85; Friend $150; Neighbor $500.

OLD TOWN SAN DIEGO STATE HISTORIC PARK, 4002 Wallace St., San Diego, CA 92110-2743. Mailing Address: 4477 Pacific Hwy., San Diego, CA 92110-3136. Tel.: 619-220-5422. Fax: 619-220-7387.
Web Site: www.parks.ca.gov/oldtownsandiego
Founded: 1968.
Congressional District: 41
Key Personnel: District Supt., Clay Phillips.
Personnel Profile: Full-Time Paid 10; Part-Time Paid 12; Part-Time Volunteers 75.
Governing Authority: state. Parent Institution: California Department of Parks & Recreation, P.O. Box 2390, Sacramento, CA 95814. Tax-exempt.
Institution Type/Description: State Park Museum.
Collections: preservation project. Historic Houses: 1827 Casa de Estudillo; 1835 Casa de Machado y Stewart; 1865 Mason Street school; 1868 San Diego Union building; c.1830 Casa de Machado y Silvas; 1829 Casa de Bandini; c.1869 Seeley Stables; Casa de Rodriguez; Pedrorena Adobe; c.1850 U.S. House; c.1847 Light & Freeman Adobe; c.1851 Wrightington Adobe; c.1862 McCoy House; c.1853 Robinson Rose.
Research Fields: preservation project; archaeology; history of California.
Activities: guided tours; lectures; formally organized education programs for children; permanent exhibitions; craft demonstrations; living history programs.
Publications: tour guide, Old Town San Diego State Historic Park; map, California State Park.
Hours & Admission Prices: Daily 10-5. No charge; donations accepted. Closed New Year's Day; Thanksgiving; Christmas. &
Attendance: 6,000,000 (estimated)
Membership: Individual $7-$10.

REUBEN H. FLEET SCIENCE CENTER, 1875 El Prado, Balboa Park, San Diego, CA 92101-1625. Mailing Address: P.O. Box 33303, San Diego, CA 92163-3303. Tel.: 619-238-1233. Fax: 619-685-5771.
E-mail: ssnyder@rhfleet.org
Web Site: www.rhfleet.org
Founded: 1973.
Congressional District: 42
Key Personnel: Exec. Dir., Dr. Steven L. Snyder; Bd. Pres., Gary Philips; C.O.O., Craig Blower; Consulting Astronomer, Lisa Wills; Dir. Devel., Julie Schardin; Dir. Education, Kris Mooney; Dir. Exhibitions, Paul Siboroski; Dir. Mktg. & Communications, Wendy Grant; Sr. Dir. Engineering & Facilities, Dave McGrew; Museum Shop Mgr., David Miller.
Personnel Profile: Full-Time Paid 44; Part-Time Paid 64; Part-Time Volunteers 320; Interns 2.
Governing Authority: bd. trustees. Tax-exempt: 501(c)(3).
Institution Type/Description: Science Center & IMAX Dome Theater.
Collections: interactive science exhibits; IMAX films & planetarium presentations.
Research Fields: cognitive issues relating to scientific concepts.
Facilities: hemisphere screen; 340-seat planetarium theater with Omnimax Film Projection System; 4 wheelchair spaces & 122-seat lecture hall; 125-seat community forum; learning center. Scientific books, kits & other educational items for sale.

Activities: lectures; films; science workshops; interactive exhibitions; learning center missions; traveling & permanent exhibitions; corporate special events; teacher's inquiry institute; senior programs.
Publications: monthly e-newsletter.
Hours & Admission Prices: Daily 10 a.m. (closing times varies). Galleries: adults $13, seniors 65 & over $12, juniors 3-12 $11; discounts to ASTC members; members no charge. Gallery & IMAX: adults $17, seniors 65 & over $15, juniors 3-12 $14; discounts to members & ASTC members. &
Attendance: 426,576 (estimated)
Membership: Individual & Senior $59; Exhibit Experience $64; Family $89. Discovery Society: Adventurer $150; Voyager $300; Explorer $500; Pioneer $1,000; Innovator $2,500. Corporate Partnerships: Friend $250; Associate $500; Sponsor $1,000; Patron $2,500; Benefactor $5,000; Leader $10,000 & up.

* **SAN DIEGO AIR & SPACE MUSEUM, (M),** 2001 Pan American Plaza, San Diego, CA 92101-1636. Tel.: 619-234-8291, ext. 10. Fax: 619-233-4526.
Web Site: www.sandiegoairandspace.org
Formerly: San Diego Aerospace Museum
Founded: 1961.
Congressional District: 41
Key Personnel: Pres., James Kidrick; Chm., Lynda Moerschvalcher; Exec. Asst., Jessica Packard.
Personnel Profile: Full-Time Paid 31; Full-Time Volunteers 5; Part-Time Paid 14; Part-Time Volunteers 241.
Governing Authority: nonprofit organization. Subsidiary Institutions: San Diego Air & Space Technology Center; San Diego Aerospace Foundation. Tax-exempt: 501(c)(3) & 170(b)(1)(A).
Institution Type/Description: History Museum.
Collections: aviation; space; technology; historical aircraft & engines; historical aviation library; aeronautical books; magazines; manuals; documents; photographs.
Research Fields: aviation; science; history.
Facilities: 18,000-vol. library of historical, books pertaining to aviation available on premises by special arrangements.
Activities: films; inter-museum loan & permanent exhibitions; temporary aviation art shows.
Publications: magazine; bimonthly newsletter; Air Capital of the West; quarterly newsletter, Flight Lines.
Hours & Admission Prices: Winter: daily 10-4:30; Summer: extended hours. Adults 12 & over $15, seniors, students & retired military $12, youth 3-11 $6; discounts to groups, AAM & Smithsonian Institution Museum of Flight members; children under 3, active military, & members no charge. Closed New Year's Day; Thanksgiving; Christmas. &
Attendance: 175,000 (accurate)
Membership: Senior $30; Individual $40; Family $55; Wings Club $150; Mercury $500; Apollo $1,000.

SAN DIEGO AUTOMOTIVE MUSEUM, 2080 Pan American Plaza, Balboa Park, San Diego, CA 92101-1636. Tel.: 619-231-2886. Fax: 619-231-9869.
E-mail: info@sdautomuseum.org
Web Site: www.sdautomuseum.org
Founded: 1988.
Congressional District: 49
Key Personnel: Exec. Dir., Paula Brandes; Mgr. Exhibits, Marisela Duron; Dir. Education, Kenn Colclasure; Accounting, Faye Levy; Librarian, Guy Preuss.
Personnel Profile: Full-Time Paid 5; Part-Time Paid 6; Part-Time Volunteers 48.
Governing Authority: private; nonprofit organization. Tax-exempt: 501(c)(3).
Institution Type/Description: Transportation Museum: housed in a structure built in 1935 for the Pan American Exposition as the California State Building.
Collections: international automobiles & automotive memorabilia, motorcycles, engines and related archival & attendant accoutrements.
Major Exhibits: Orphan Cars, 10/13-1/14.
Research Fields: automotive history 1890s-present.
Facilities: library; 40,000 sq. ft. exhibit space. Gift items for sale.
Activities: docent program; guided tours; hobby workshops; lectures; loan, participatory & temporary exhibitions; clubs; show roll-outs.
Publications: quarterly newsletter, Auto Museum News.
Hours & Admission Prices: Daily 10-5. Adults $8, senior citizens 65 & over and active military w/ID $6, students $5, children 6-15 $4; children under 6 & members no charge. Closed New Year's Day; Thanksgiving; Christmas. &
Attendance: 100,000 (estimated)

Membership: Individual $45; Car Club $55; Family $65; Friend of the Museum $200; Patron $1,000; Wheels 100 $10,000.

SAN DIEGO CHINESE HISTORICAL MUSEUM, 404 Third Ave., San Diego, CA 92101-6803. Tel.: 619-338-9888. Fax: 619-338-9889.
E-mail: info@sdchm.org
Web Site: www.sdchm.org
Key Personnel: Pres., Michael Lee; Cur., Murray K. Lee.
Governing Authority: nonprofit organization. Admission $2; members & children under 12 no charge.
Institution Type/Description: Culturally Specific.
Collections: Chinese culture; maps; books; period art; personal artifacts; embroidered shoes; teapots.
Facilities: library; garden.
Activities: research.
Hours & Admission Prices: Tues.-Sat. 10:30-4, Sun. 12-4. Adults $2; children under 12 no charge. Closed New Year's Day; Independence Day; Thanksgiving; Christmas.

SAN DIEGO FIREHOUSE MUSEUM, 1572 Columbia St., San Diego, CA 92101-2913. Tel.: 619-232-3473. Facebook: San Diego Firehouse Museum.
E-mail: sdfirehousemuseum@gmail.com
Web Site: www.sandiegofirehousemuseum.org
Institution Type/Description: Fire-Fighting Museum.
Collections: fire-fighting equipment & memorabilia; photographs.
Hours & Admission Prices: Thurs.-Fri. 10-2, Sat.-Sun. 10-4. Adults $3, seniors & children $2.
Membership: Active & Retired Emergency Services Personnel or Support $35; Civilians $75.

SAN DIEGO HALL OF CHAMPIONS SPORTS MUSEUM, 2131 Pan American Plaza, Balboa Park, San Diego, CA 92101-1683. Tel.: 619-234-2544. Fax: 619-234-4543.
E-mail: info@sdhoc.com
Web Site: www.sdhoc.com
Founded: 1961.
Congressional District: 44
Key Personnel: C.E.O., Pres. & Exec. Dir., Alan Kidd; Chm. (V), Ron Fowler; Cur., Anthony Lachica; Museum Shop Mgr., Gil Johnson.
Personnel Profile: Full-Time Paid 14; Part-Time Paid 10; Part-Time Volunteers 30; Interns 20.
Governing Authority: nonprofit organization. Tax-exempt: 501(c)(3).
Institution Type/Description: Sports Museum & Breitbard Hall of Fame: housed in the Federal Building in Balboa Park.
Collections: sports memorabilia & artifacts.
Research Fields: sport.
Facilities: resource center including library containing volumes, periodicals, films & tapes pertaining to sports.
Activities: guided tours; films; permanent exhibitions; seminars; clinics & counseling.
Publications: quarterly newsletter.
Hours & Admission Prices: Daily 10-4:30. Adults $8, senior citizens, military & students with ID $6, children 7-17 $4; discounts to AAM & ICOM members; members & children under 7 no charge. Closed New Year's Day; Thanksgiving; Christmas.
Attendance: 25,000 (accurate)
Membership: Friends of Bob $100; Patron $500; Bronze $1,000; Silver $2,500; Gold $5,000; Platinum $10,000; All-American $25,000; Hall of Fame $50,000.

＊　**SAN DIEGO MODEL RAILROAD MUSEUM, INC.,** 1649 El Prado, Ste. 4, San Diego, CA 92101-1664. Tel.: 619-696-0199. Fax: 619-696-0239. Facebook: Model Railroad Museum.
E-mail: info@sdmrm.org
Web Site: www.sdmrm.org
Founded: 1980.
Congressional District: 53
Key Personnel: Pres., Jon Everett; Exec. Dir., Anthony Ridenhour; Treas., Corrine Neuffer; Museum Shop Mgr., Veronique Rotsart.
Personnel Profile: Full-Time Paid 5; Part-Time Paid 28; Part-Time Volunteers 300; Interns 1.
Governing Authority: nonprofit organization. Tax-exempt: 501(c)(3).
Institution Type/Description: Model Railroad & Railroad History Museum: housed on site of 1915 Panama-California Exposition in the Casa de Balboa building.
Collections: 27,000 sq. ft. model railroad layouts depict prototypical & historic

railroads as well as toy train & fantasy displays; interactive children's exhibit; railroads throughout the Pacific Southwest.
Major Exhibits: History of Model Railroading, 1/15/14-1/15/15; Circus Rails, 2/14-8/1/14; Harvey Girls, 8/16/14-11/22/14.
Research Fields: southern California railroads, including San Diego & Arizona Eastern, Santa Fe & Southern Pacific, Union Pacific; history of model railroading.
Facilities: 5,000-vol. library pertaining to railroads. Railroad-related items for sale.
Activities: operating model trains; guided tours; films & videos on actual railroads; arts festivals; hobby workshops; traveling & participatory exhibits; field trips; study clubs; traveling slide show & lecture on the museum exhibits.
Publications: quarterly newsletter, San Diego Telegraph; self-guided tour brochure.
Hours & Admission Prices: Museum: Tues.-Fri. 11-4, Sat.-Sun. 11-5; guided tours by appointment. Library: see website for hours. Adults $8, senior citizens 65 & over $6, military $4, students $3, children 6-14 $1; discounts to CAM, Assoc. of RR Museums & SD Inter-Museum Promotion Council members; members, San Diego county residents 1st Tues. of month & children under 5 no charge. Closed Thanksgiving; Christmas.
Attendance: 137,760 (accurate)
Membership: Individual $25; Individual Plus $35; Family $45; Contributing $60; Silver Spike $150; Gold Spike $300; Life $700; Benefactor $1,000; Patron $2,500; Founder $7,500.

＊　**SAN DIEGO MUSEUM OF ART, (M), (I),** Balboa Park, 1450 El Prado, San Diego, CA 92101. Mailing Address: P.O. Box 122107, San Diego, CA 92112-2107. Tel.: 619-232-7931. Fax: 619-232-9367.
E-mail: info@sdmart.org
Web Site: www.sdmart.org
Founded: 1925.
Congressional District: 42
Key Personnel: Dir., Roxana Velasquez; Pres. (V), Frank E. Rogozienski; Deputy Dir. Curatorial Affairs, Julia Marciari-Alexander; Cur. European Art, John Marciari; C.O.O., C.F.O. & Deputy Dir. Administration, Reed Vickerman; Deputy Dir. External Affairs, Katy McDonald; Assoc. Dir., Mary Salerno; Mgr. Library, James Grebl; Museum Store Mgr., Warren Herman.
Personnel Profile: Full-Time Paid 75; Part-Time Paid 25; Part-Time Volunteers 501; Interns 5.
Governing Authority: nonprofit organization. Tax-exempt: 501(c)(3).
Institution Type/Description: Art Museum: housed in a Spanish plateresque-style building located on the site of the Panama/California International Exposition.
Collections: 14th- to 20th-century European paintings, sculpture & decorative arts including Italian Renaissance & Spanish Baroque painting; Asian arts, including Chinese, Japanese, Korean, Thai, & Indian painting, sculpture, ceramics, stone & metal work; contemporary painting & sculpture; Edward Binney 3rd Collection of Indian paintings.
Research Fields: pertaining to works in the museum collections.
Facilities: non-circulating library with 30,000 art books & catalogs; slides; subject & artists files available on premises; 400-seat auditorium; cafe. Books & museum-related items for sale.
Activities: guided tours; lectures; gallery talks; concerts; formally organized education programs for children, adults & undergraduate college students; inter-museum loan; permanent, temporary & traveling exhibitions; art school for children, teens, & adults; K-12 art exhibition.
Publications: quarterly magazine; electronic newsletter, Muse News; exhibition catalogs.
Hours & Admission Prices: Summer: Tues.-Wed. & Fri.-Sun. 10-5, Thurs. 10-9; Winter: Tues.-Sun. 10-5. Adults $12, military $10, seniors $9, students $8, children $4.50; discounts to ICOM members; members and children 5 & under no charge. Closed New Year's Day; Thanksgiving; Christmas.
Attendance: 350,000 (accurate)
Membership: Student & Educator $35; Individual $45; Dual & Household $75; Friend $175; Sponsor $500; President's Circle $1,500; Patron's Circle $3,000; Director's Circle $5,000; Benefactor's Circle $10,000. Seniors & Active Military discounts available.

＊　**SAN DIEGO MUSEUM OF MAN, (M),** 1350 El Prado, Balboa Park, San Diego, CA 92101-1681. Tel.: 619-239-2001. Fax: 619-239-2749.
E-mail: rgarniewicz@museumofman.org
Web Site: www.museumofman.org
Founded: 1915.
Congressional District: 49

Key Personnel: C.E.O., Micah Parzen; C.O.O., Rex Garniewicz; Registrar, Megan Clancy; Cur. Physical Anthropology, Tori Randall; C.O.O. & Dir. Devel., Hope Carlson; Museum Shop Mgr., Patti Pepper.

Personnel Profile: Full-Time Paid 27; Part-Time Paid 7; Part-Time Volunteers 69; Interns 16.

Governing Authority: society. Tax-exempt: 501(c)(3).

Institution Type/Description: Anthropology Museum.

Collections: over 140,000 artifacts, photographs & cultural objects relating to ethnological & archaeological collections pertaining to peoples of the Western Americas; physical anthropology primarily from California, the Southwest & Peru; Egyptian antiquities from Tell el-Amarna, Abydos & other sites. Historical Building: 1915 California Building.

Research Fields: California archaeology; physical anthropology; ethnography & western Americas.

Facilities: 13,510-vol. library of books and 72 archival manuscripts & periodicals on anthropology, archaeology, & ethnology; 20,000 sq. ft. exhibit space. Arts & craft items from the Southwest & around the world for sale.

Activities: guided tours; formally organized education programs for children; docent program or council; permanent & temporary exhibitions; contemporary discussion panels. Museum Sponsors: Tower After Hours - Celebrating World Cultures.

Publications: San Diego Museum Papers; Ethnic Technology Notes; leaflets; Journey to the Copper Age: Archeology in the Holy Land, 2007.

Hours & Admission Prices: Daily 10-4:30. Adults $12.50, seniors, students, military & youth 13-17 $8, children 3-12 $5; discounts to AAM & ICOM members; members & children under 3 no charge. Closed Thanksgiving; Christmas. &

Attendance: 156,106 (accurate)

Membership: Student $20; Individual $45; Family & Grandparent $60; Friend $125; Sponsor $250; President's Associate $500; Patron $1,000; Los Compadres $5,000.

∗ SAN DIEGO NATURAL HISTORY MUSEUM, (M), 1788 El Prado, Balboa Pk., San Diego, CA 92101. Mailing Address: P.O. Box 121390, San Diego, CA 92112-1390. Tel.: 619-232-3821. Fax: 619-232-0248.

E-mail: mhager@sdnhm.org

Web Site: www.sdnhm.org

Founded: 1874.

Congressional District: 49

Key Personnel: Pres. & C.E.O., Dr. Michael W. Hager; Bd. Chm., Virginia Crockett; Vice Pres. Institutional Advancement, Ann Laddon; C.O.O., Susan Loveall; Dir. Devel. & Membership, Eowyn Bates; Vice Pres. Research & Programs, Michael Wall; Dir. Binational Education, Doretta Winkelman; Dir. Exhibits, Tim Murray; Dir. Research Library, Margaret Dykens; Dir. Volunteer Svcs., Janet Morris; Dir. Visual Media, Cary Canning; Dir. Foundation Rels., Wendy Endsley; Cur. Botany, Dr. Jon Rebman; Cur. Paleontology, Dr. Thomas Demere; Cur. Herpetology, Dr. Brad Hollingsworth; Mgr. Ornithology & Mammalogy Collections, Phil Unitt; Mgr. AV Svcs., Alexis McKee.

Personnel Profile: Full-Time Paid 108; Part-Time Paid 167; Part-Time Volunteers 742; Interns 4.

Governing Authority: society. Parent Institution: San Diego Society of Natural History. Tax-exempt: 501(c)(3).

Institution Type/Description: Natural History Museum

Collections: botany; entomology; geology; herpetology; mammalogy; marine invertebrates; mineralogy; ornithology; regional ecology & paleontology

Research Fields: botany; entomology; herpetology; mammalogy; mineralogy; ornithology; paleontology; marine invertebrates.

Facilities: 90,000-vol. research library; research center; science center; 300-seat large-format theatre; 4 classrooms; 8,500 sq. ft. exhibit space; catering facilities. Museum-related items for sale.

Activities: traveling exhibitions; large-format films; natural history classes, field trips, & lecture series for all ages; 3D films; summer camps; school tours & lab programs; specimen loan library; school outreach programs; binational teacher education; free public nature hikes; guided hikes for the visually impaired; long-range expedition travel; activities for members & volunteers.

Publications: member newsletter, Field Notes; education programs catalog; proceedings; occasional papers.

Hours & Admission Prices: Daily 10-5. Adults $17, seniors 62 & over $15, military, youth 13-17 and students $12, children 3-12 $11; discounts to AAM members; children 2 & under and members no charge. Closed Thanksgiving; Christmas. &

Attendance: 312,907 (accurate)

Membership: Student $25; Senior $35; Dual Senior & Individual $55; Grandparents & Family $70; Kate Sessions Circle $175-$274; Daniel Cleveland Circle $275-$524; A.R. Valentien Circle $525-$1,199; John Audubon $1,200-$2,499; John Muir $2,500-$4,999; Carl Linnaeus $5,000-$9,999; Charles Darwin $10,000 & up.

∗ SAN DIEGO ZOO, (M), 2920 Zoo Dr., San Diego, CA 92101-1693. Mailing Address: P.O. Box 120551, San Diego, CA 92112-0551. Tel.: 619-231-1515 & 685-3291. Fax: 619-557-3970.

E-mail: publicrelations@sandiegozoo.org

Web Site: www.sandiegozoo.org

Founded: 1916.

Congressional District: 41

Key Personnel: C.E.O., Douglas G. Myers; Zoo Dir., John Dunlap; Collections Dir., Bob Wiese, Ph.D.; Pres., Rick Gulley; Deputy Dir., Matthew Musella; Dir. Conservation & Science, Allison Alberts, Ph.D.; Deputy Dir. Conservation & Science, Alan Lieberman; Cur. Mammals, Carmi Penny; Cur. Birds, David Rimlinger; Devel. Dir., Mark Stuart; Mktg. Dir., Ted Molter; Registrar, Toni Giezendanner; Veterinary Svcs., Patrick Morris, D.V.M.; Nutritionist, Michael Schlegel, Ph.D.; Pathologist, Bruce Rideout, D.V.M., Ph.D.; Geneticist, Oliver Ryder, Ph.D.; Assoc. Education Dir., Victoria Garrison; Architecture & Planning Dir., David E. Rice, FAIA; Reproductive Physiologist, Barbara Durrant, Ph.D.; Dir. Corporate Merchandising & Museum Shop Mgr., Yvonne Miles.

Personnel Profile: Full-Time Paid 2,000.

Governing Authority: society. Parent Institution: Zoological Society of San Diego. Subsidiary Institution: San Diego Wild Animal Park, CRES. Tax-exempt: 501(c)(3).

Institution Type/Description: Zoo & Botanical Garden.

Collections: birds, mammals, reptiles & amphibians of the world; botany; scientific & photo library collections; cytogenetic collections (cell strings, semen & eggs).

Research Fields: veterinary medicine & pathology; zoology; animal reproduction & conservation; in situ conservation.

Facilities: 400,000 image photographic library supporting museum & collections; botanical gardens; nature center; 200-seat auditorium; 1,500-seat & 3,000-seat amphitheaters; restaurant; classrooms. Books, artifacts, jewelry & other museum-related items for sale.

Activities: guided tours; TV & radio programs; formally organized educational programs; animal shows; lecture & special events for members; ecotourism; nighttime zoo experience; worldwide tours; interactive website.

Publications: monthly magazine, Zoonooz; guidebook, The San Diego Zoo; quarterly children's newsletter, Koala Club News; miscellaneous publications; books, A World of Animals, The San Diego Zoo; Mr. Zoo-The Life & Legacy of Dr. Charles Schroeder.

Hours & Admission Prices: June 21-Labor Day daily 9-10; Sept.-June 20 daily 9-4. Adults 12 & over $34, children 3-11 $24; discounts to groups; Koala Club members, children under 3 & members no charge. &

Attendance: 3,200,000 (accurate)

Membership: Koala Club 3-11 $21, 12-17 $25; Senior Passes: Single $35; Dual $50; Regular: Single $66; Dual $84; Diamond Club New: $126; Keeper's Club $150; Curator's Club $250; Director's Club $500; President's Associates $1,000; President's Partners $2,500.

SEAWORLD SAN DIEGO, 500 Sea World Dr., San Diego, CA 92109-7904. Tel.: 800-257-4268. Fax: 619-226-3996.

Web Site: www.seaworldsandiego.com

Founded: 1964.

Congressional District: 41

Key Personnel: Park Pres., John T. Reilly; Vice Pres. Zoological Operations, Michael Scarpuzzi; Vice Pres. Mktg., Marilyn Hannes; Vice Pres. Entertainment, Adrian Fischer; Vice Pres. Operations, Steve Melanese.

Personnel Profile: Full-Time Paid 500; Part-Time Paid 2,500.

Governing Authority: corporation. Owned by SeaWorld Parks and Entertainment.

Institution Type/Description: Aquarium, Marine Museum and Oceanarium.

Collections: marine life.

Research Fields: marine science.

Facilities: 500-vol. library of books on marine sciences available for research on premises by appointment; botanical garden; zoological park; aquarium; restaurants; snack bars. Gift items for sale.

Activities: guided tours; TV & radio programs; formally organized education programs for children, adults, undergraduate & graduate students; field projects; shows; rides; play area; outreach education programs.

Hours & Admission Prices: Visit website for information. &

Attendance: 4,000,000

Membership: Visit website for information.

SERRA MUSEUM/SAN DIEGO HISTORY CENTER, 2727 Presidio Dr., Presidio Park, San Diego, CA 92103. Mailing Address: 1649 El Prado, Ste. 3, San Diego, CA 92101-1664. Tel.: 619-297-3258. Fax: 619-232-6297.
E-mail: gselak@sandiegohistory.org
Web Site: www.sandiegohistory.org
Formerly: Junipero Serra Museum/San Diego Historical Society
Founded: 1928.
Congressional District: 49
Key Personnel: Interim Dir., Charlotte Cagan; Pres. (V), Hal Sadler; Archivist, Jane Kenealy; Museum Shop Mgr., Andrea Chung.
Personnel Profile: Full-Time Paid 11; Part-Time Paid 17; Part-Time Volunteers 100; Interns 5.
Governing Authority: private; nonprofit organization. Parent Institution: San Diego Historical Society. Branch Museum: San Diego History Center, 1649 El Prado, San Diego, CA 92101, Tel.: 619-232-6203. Tax-exempt: 501(c)(3).
Institution Type/Description: History Museum: housed in a mission-style 1929 structure constructed to commemorate the site where Father Junipero Serra & Captain Gaspar de Portola established the first mission and military outpost on the west coast of the U.S. & Canada.
Collections: period artifacts including housewares, period clothing, furniture & tools from the Native American, Spanish and Mexican periods through 1835.
Facilities: Museum-related items for sale.
Activities: lectures; guided tours; films; docent program; formal education programs for children; temporary exhibitions.
Publications: quarterly newsletter, The Times; quarterly magazine, The Journal of San Diego History.
Hours & Admission Prices: Sat.-Sun. 10-5. Adults $6, senior citizens, students & military with ID $4, youth 6-17 $3; discounts to AAM, AAA, KPBS, & Historic Trust/Time Travelers members; members & children under 5 no charge. Closed New Year's Day; Thanksgiving & day after; Christmas.
Attendance: 14,978 (accurate)
Membership: Basic $45; Family $60; Associate $100; Scholar Circle $250; Curator's Circle $500; Director's Circle $1,000; Chairman's Circle $2,500.

THOMAS WHALEY HOUSE MUSEUM, 2476 San Diego Ave., San Diego, CA 92110-2730. Tel.: 619-297-7511. Fax: 619-291-3576.
E-mail: soho-1@sohosandiego.org
Web Site: www.whaleyhouse.org
Founded: 1969.
Key Personnel: Exec. Dir., Bruce Coons; Pres., Curtis Drake; Museum Shop Mgr., Alana Coons.
Personnel Profile: Full-Time Paid 4; Full-Time Volunteers 1; Part-Time Paid 18; Part-Time Volunteers 11.
Governing Authority: private; nonprofit organization. Parent Institution: SOHO. Tax-exempt.
Institution Type/Description: House Museum.
Collections: 1850-1880 furnishings.
Activities: guided tours; docent program.
Publications: annual magazine, Our Heritage.
Hours & Admission Prices: Summer: daily 10-9:30; adults $6, senior citizens $5, students & children $4; members no charge. Non-Summer: Sun.-Tues. 10-5, Thurs.-Sat. 10-9:30; adults $10, children $5; children 2 & under no charge. Closed Thanksgiving; Christmas. &
Attendance: 120,000 (estimated)
Membership: Adult $35.

TIMKEN MUSEUM OF ART, (M), 1500 El Prado, Balboa Park, San Diego, CA 92101-1620. Tel.: 619-239-5548. Fax: 619-531-9640.
E-mail: info@timkenmuseum.org
Web Site: www.timkenmuseum.org
Founded: 1965.
Congressional District: 4
Key Personnel: Exec. Dir., John Wilson, Ph.D.; Pres. (V), Tim Zinn; Deputy Dir. Operations, Carrie Cottriall; Deputy Dir. Devel. & Endowment, Laurie Hawkins; Museum Admin., James Petersen; Dir. Education, Kristina Rosenberg; Controller, Jen Austgen.
Personnel Profile: Full-Time Paid 6; Part-Time Paid 12; Part-Time Volunteers 20.
Governing Authority: nonprofit organization. Parent Institution: The Putnam Foundation. Tax-exempt: 501(c)(3).
Institution Type/Description: Art Museum.
Collections: 14th to 19th-century Old Masters European paintings; 19th-century American paintings; Gobelin Tapestries; 15th to 19th-century Russian Icons.

Facilities: 200-vol. director's library of art books.
Activities: docent tours.
Publications: Timken Museum of Art, European Works of Art, American Paintings and Russian Icons in the Putnam Foundation Collection, 1996; Timken Museum Acquisitions 1995-2005, 2006; Guercino: Stylistic Evolution in Focus, 2006; Rembrandt's Apostles, 2005; Benjamin West: Allegory and Allegiance, 2004; Portraiture in Paris Around 1800: Cooper Penrose by Jacques-Louis David, 2003; The Portraits of Bartolomeo Veneto, 2002; Luca Carlevarijs: Views of Venice, 2001; John Singleton Copley and Margaret Kemble Gage: Turkish Fashion in 18th Century America, 1998; Art & Devotion in Siena after 1350: Luca de Tomme & Niccolo di Buonaccorso, 1997; Eastman Johnson: The Cranberry Harvest, Island of Nantucket, 1990; Robert Wilson Video Portraits, 2011; Rembrandt's Recession: Passion and Prints in the Dutch Golden Age, 2010.
Hours & Admission Prices: Tues.-Sat. 10-4:30, Sun. 12-4:30. No charge; donations accepted. Guided tours. Closed legal holidays. &
Attendance: 163,638 (accurate)
Membership: Friends: $100, $250 & $500; Ames Circle $1,000-$2,499; Putnam Circle $2,500-$4,999; Timken Circle $5,000-$9,999; Director's Circle $10,000-$24,999; Founder's Society $25,000 & up.

UNIVERSITY ART GALLERY SAN DIEGO STATE UNIVERSITY, 5500 Campanile Dr., San Diego, CA 92182-0003. Mailing Address: School of Art, Design & Art History, San Diego, CA 92182. Tel.: 619-594-5171. Fax: 619-594-1217.
E-mail: artgallery@sdsu.edu
Web Site: artgallery.sdsu.edu
Key Personnel: Dir., Tina Yapelli.
Governing Authority: university. Tax-exempt: 501(c)(3).
Institution Type/Description: University Art Gallery.
Collections: contemporary art; study collection of Asian sculpture; African & Mexican art.
Research Fields: contemporary art in all media.
Facilities: 2,500 sq. ft. exhibition space.
Activities: lectures; organized educational programs for adults & undergraduate or graduate college students affiliated with SDSU; traveling exhibitions; artists' lectures.
Publications: exhibition catalogues & brochures.
Hours & Admission Prices: Sept.-May Mon.-Thurs. & Sat. 12-4. No charge; donations accepted. Closed holidays. &
Attendance: 5,000 (estimated)
Membership: Annual $50.

UNIVERSITY GALLERIES: THE HOEHN FAMILY GALLERIES & PRINT STUDY ROOM/FINE ART GALLERIES AT THE INSTITUTE FOR PEACE & JUSTICE/THE DAVID W. MAY GALLERY, University of San Diego, 5998 Alcala Park, San Diego, CA 92110. Mailing Address: Dept. of Art, Architecture & Art History, Univ. of San Diego, 5998 Alcala Park, San Diego, CA 92110. Tel.: 619-260-7516 (Hoehn) & 4238 (May). Fax: 619-849-8237.
E-mail: joycea@sandiego.edu
Web Site: www.sandiego.edu/anthropology/maycollection.php
Formerly: Anthropology Museum, Founder's Gallery University of San Diego
Key Personnel: Dir., Dr. Alana Cordy-Collins; Mgr. Collections, Joyce Antorietto.
Personnel Profile: Part-Time Paid 1; Interns 1.
Governing Authority: private university.
Institution Type/Description: Anthropology Museum.
Collections: Native American art & artifacts of southwestern U.S. and California from prehistoric to modern times; Alaskan & Northwest coast cultures; Baja California; Mexico Indians; Indians of Middle & South America.
Activities: lectures; loan, temporary & traveling exhibitions; artist-in-residence.
Hours & Admission Prices: Tues. & Thurs.-Fri. 1-3; other times by appointment. No charge. Closed university holidays. &
Attendance: 200 (estimated)

USS MIDWAY MUSEUM, 910 N. Harbor Dr., San Diego, CA 92101-5811. Tel.: 619-544-9600. Fax: 619-544-9188.
E-mail: mmclaughlin@midway.org
Web Site: www.midway.org
Formerly: San Diego Aircraft Carrier Museum
Key Personnel: Pres., Mac McLaughlin; C.F.O., Rich Benard
Institution Type/Description: Military Museum.
Collections: over 19 restored aircraft; flight simulators; military history; personal artifacts.
Facilities: cafe. Museum-related items for sale.

Activities: self-guided audio tour; flight simulators.
Hours & Admission Prices: Daily 10-5; groups of 15 or more by appointment. Adults $19, seniors 62 & over $16, students 13-17 $15, retired military & youth 6-12 $10; children 5 & under no charge. Closed Thanksgiving; Christmas.

VILLA MONTEZUMA MUSEUM AKA JESSE SHEPARD/FRANCIS GRIERSON HOUSE, 1925 K St., San Diego, CA 92102-3828. Mailing Address: 657 20th St., San Diego, CA 92102-2810. Tel.: 619-255-9367.
E-mail: fovm@villamontezumamuseum.org
Web Site: www.villamontezumamuseum.org
Founded: 1972.
Congressional District: 53
Key Personnel: Chm., Louise Torio.
Governing Authority: Tax-exempt: 501(c)(3).
Institution Type/Description: Historic House Museum: housed in the former home of musician & spiritualist Jesse Shepard also known as metaphysical author Francis Grierson; built in 1887. Listed on the National Register of Historic Places.
Collections: late 19th-century furnishings & decorative arts; stained glass windows; personal artifacts; photographs.
Facilities: 4,000 sq. ft. exhibit space. Museum-related items for sale.
Activities: special events; walking tours of the Sherman Heights Historic District. Museum Sponsors: Victorian Valentine Tea; Dia de los Muertos Celebration.
Hours & Admission Prices: Closed for restoration.
Membership: Individual $25; Professional $50; Supporting $100; Patron $250; Benefactor $500.

WELLS FARGO HISTORY MUSEUM, 2733 San Diego Ave., San Diego, CA 92110-2731. Mailing Address: Wells Fargo Historical Services, 420 Montgomery St., MAC-A0101-106, San Francisco, CA 94163. Tel.: 619-238-3929.
Founded: 1990.
Key Personnel: Cur., Allan E. Peterson; Asst. Cur., Casey (William) Gill.
Governing Authority: profit-making organization. Affiliated with Wells Fargo Bank.
Institution Type/Description: Company History Museum.
Collections: Concord Stagecoach; Wells Fargo banking & express history; mining; staging; local items.
Activities: guided group tours; permanent & temporary exhibitions; audiovisual programs.
Publications: scholarly pamphlets.
Hours & Admission Prices: Daily 10-5. No charge. Closed bank holidays. &

San Dimas

PACIFIC RAILROAD MUSEUM & LIBRARY, 210 W. Bonita Ave., San Dimas, CA 91773. Tel.: 909-394-0616.
E-mail: info@pacificrailroadsociety.org
Web Site: www.pacificrailroadsociety.org/index.html
Founded: 1995.
Key Personnel: Dir. & Museum Shop Mgr., David B. Housh; Pres. (V), Neil Bjornsen.
Governing Authority: Parent Institution: Pacific Railroad Society. Tax-exempt.
Institution Type/Description: Railroad Museum & Library.
Collections: local history & railroading; railroad artifacts; railroad models/displays; books; documents; maps; photographs.
Facilities: library.
Activities: train excursions.
Publications: monthly journal, Wheel Clicks.
Hours & Admission Prices: Mon. & Wed. 12-5, Sat. 10-4; groups by appointment. No charge; donations accepted. &
Attendance: 3,000 (estimated)
Membership: Annual $45.

San Fernando

LOPEZ ADOBE, 1100 Pico St., San Fernando, CA 91340-3514. Mailing Address: 117 Macheil St., San Fernando, CA 91340-2911. Tel.: 818-898-1227. Fax: 818-898-7329.
Web Site: www.sfcity.org
Congressional District: 28
Institution Type/Description: History Museum: structure built in 1882. Listed on the National Register of Historic Places.
Collections: local history; period furnishings.
Hours & Admission Prices: 4th Sun. of month 1-4. &

San Francisco

ACADEMY OF ART COLLEGE GALLERIES, 625 Sutter St., San Francisco, CA 94105. Mailing Address: 79 New Montgomery St., San Francisco, CA 94105. Tel.: 415-274-2229. Fax: 415-263-8819.
E-mail: info@academyart.edu
Web Site: www.academyart.edu/aboutus/gallery.asp
Governing Authority: Branch Galleries: 688 Sutter St., San Francisco, CA; 410 Bush St., San Francisco, CA 94105.
Institution Type/Description: Art Gallery.
Collections: works by college students, faculty & alumni.
Activities: special events.
Publications: newsletter.
Hours & Admission Prices: Sutter St. Galleries: Mon.-Fri. 10-6, Sat. 10-5. Bush St. Gallery: Mon.-Fri. 9-6, Sat. 10-5.

ALCATRAZ ISLAND, Pier 33, San Francisco, CA 94123. Mailing Address: Natl. Park Svc., Golden Gate Natl. Rec. Area, Bldg. 201, Fort Mason, San Francisco, CA 94123-1307. Tel.: 415-981-7625.
Web Site: www.nps.gov/alcatraz
Key Personnel: Gen. Supt., Brian O'Neill
Institution Type/Description: Historic Site: federal prison 1934-1963.
Collections: island history; prison remnants; American Indian history; military history 1850-1933.
Facilities: theater; gardens.
Activities: video presentation; Alcatraz inmates & correctional officers audio tour.
Hours & Admission Prices: Ferry: departs every 30 minutes beginning at 8:45am. Island Tours: Summer: 10-6:30; Winter: 10-4:30. Closed New Year's Day; Christmas; extreme weather.

ALEXANDER F. MORRISON PLANETARIUM, Calif. Academy of Sciences, 55 Music Concourse Dr., San Francisco, CA 94118-4503. Tel.: 415-750-8000.
E-mail: planet@calacademy.org
Web Site: www.calacademy.org/planetarium
Founded: 1952.
Key Personnel: Dir., Ryan Watt.
Governing Authority: nonprofit organization. Parent Institution: California Academy of Sciences. Tax-exempt: 501(c)(3).
Institution Type/Description: Planetarium: housed in Natural History Museum.
Collections: astronomy; space science.
Facilities: planetarium.
Activities: lectures; formally organized educational programs; permanent exhibitions.
Hours & Admission Prices: Mon.-Sat. 9:30-5, Sun. 11-5. Adults $29.95, seniors 65 & over, student and youth 12-17 $19.95, children 4-11 $14.95; children 3 & under no charge. Closed Thanksgiving; Christmas. &
Attendance: 140,000 (estimated)

AQUARIUM OF THE BAY, The Embarcadero at Beach St., Pier 39, San Francisco, CA 94133. Tel.: 415-623-5300. Fax: 415-623-5324.
E-mail: info@bay.org
Web Site: www.aquariumofthebay.org
Key Personnel: Pres. & C.E.O., John Frawley; Dir. Education & Conservation, Carrie Chen; Dir. Communications, Kelly Cash; Dir. Operations, Chris Connors; Dir. Devel., Jennie Leichtling
Institution Type/Description: Aquarium.
Collections: over 20,000 local marine animals; hands-on exhibitions.
Activities: hands-on exhibitions; educational programs; special events; birthday parties.
Hours & Admission Prices: March-May & Sept.-Oct. Mon.-Thurs. 10-7, Fri.-Sun. 10-8; Memorial Day to Labor Day daily 9-8; Nov.-Feb. Mon.-Thurs. 10-6, Fri.-Sun. 10-7. Adults $16.95, seniors 65 & over and children 3-11 $10; discounts to families.

*** ASIAN ART MUSEUM OF SAN FRANCISCO, CHONG-MOON LEE CENTER FOR ASIAN ART AND CULTURE, (M),** 200 Larkin St., San Francisco, CA 94102-4734. Tel.: 415-581-3500. Fax: 415-581-4700. TDD: 415-861-2035.
Web Site: www.asianart.org
Formerly: Asian Art Museum of San Francisco, The Avery Brundage Collection
Founded: 1966.
Congressional District: 8
Key Personnel: C.O.O., Mark McLoughlin; Dir., Jay Xu; Chm. (V) Asian Art

Commission, Anthony Sun; Pres. (V), Akiko Yamazaki; Dir. Human Resources, Valerie Pechenik; Chief Cur. & Wattis Cur. of South & Southeast Asian Art, Forrest McGill; Cur. Japanese Art, Yoko Woodson; Cur. Korean Art, Cheeyun Kwon; Senior Cur. Chinese Art, Michael Knight; Dir. Education, Deborah Clearwaters; Dir. Devel., Amory Sharpe; Dir. Mktg. & Communications, Tim Hallman; Librarian, John Stucky; Museum Shop Mgr., Peri Danton.

Personnel Profile: Full-Time Paid 134; Part-Time Paid 11; Part-Time Volunteers 655; Interns 1.

Governing Authority: Asian Art Commission of San Francisco, an agency of the City & County of San Francisco. Tax-exempt: 501(c)(3).

Institution Type/Description: Art Museum.

Collections: over 15,000 art objects spanning 6,000 years of history & representing countries and cultures throughout Asia which includes ceramics; bronze; jade; lacquer; sculpture; architectural elements; metalwork; glass; ivory; textiles; applied art; painting & screens.

Research Fields: Asian art, with emphasis on China, Korea, Japan, India, Nepal, Bhutan, Tibet, Southeast Asia & Iran.

Facilities: 34,000-vol. library on Asian art & culture available for use on premises; conservation & photography laboratories; resource room; 3 multi-purpose classrooms; cafe. Museum slides for sale.

Activities: guided tours; courses & lectures; films; performances; family festivals; intern program for graduate students; docent council; speakers bureau programs available to public; inter-museum loan; permanent, temporary & traveling exhibitions; docent tours for the deaf; live demonstrations; hands-on activities.

Publications: handbooks; catalogues on museum collections & traveling shows; slides; photographs; Members' Magazine.

Hours & Admission Prices: Tues.-Sun. 10-5. Adults $12, seniors, youth & students $8; discounts to AAM & ICOM members; children 12 & under and members no charge. Closed New Year's Day; Thanksgiving; Christmas. &

Attendance: 262,482 (accurate)

Membership: Student & Teacher $40; Out of Region $50; Senior Active $55; Active and Senior Friends & Family $75; Friends & Family $100; Senior Contributor $115; Contributor $150; Donor $350; Sponsor $650; Patron $1,000; Jade Circle $3,000 & up.

CCA WATTIS INSTITUTE FOR CONTEMPORARY ARTS,

360 Kansas St., San Francisco, CA 94107. Mailing Address: 1111 Eight St., San Francisco, CA 94107. Tel.: 415-551-9305. Fax: 415-551-9209. Facebook: Wattis.

E-mail: wattis@cca.edu
Web Site: www.wattis.org
Founded: 1998.
Congressional District: 9
Personnel Profile: Full-Time Paid 4; Part-Time Paid 1; Interns 2.

Governing Authority: college. Parent Institution: California College of The Arts. Branch Museums: 1111 8th St., San Francisco, CA 94107. Tax-exempt.

Institution Type/Description: Art Gallery.

Collections: contemporary culture.

Activities: lectures; gallery talks; contemporary exhibitions of fine art, fine craft, architecture & design; education programs.

Publications: books, Searchlight: Consciousness at the Millennium; Rock My World - Recent Art and the Memory of Rock 'n Roll; After the Gold Rush; The Artist's World; Utopia Now; How Extraordinary That the World Exists!; Sudden Glory; Extra Art: A Survey of Artist's Ephemera; Reality Check: Painting in the Exploded Field; Baja to Vancouver: The West Coast and Contemporary Art; Likeness: Portraits of Artists by Other Artists; Monuments for the USA; What We Want is Free; A Brief History of Invisible Art; Spaced Out: Late 1990's Works from the Vicki and Kent Logan Collection; Prophets of Deceit; Pioneers, Apocalypse Now: The Theater of War, Amateurs; Capp Street Project: Mario Ybarra, Jr.; Capp Street Project: Tim Lee: Capp Street Project: Abraham Cruzvillegas; Capp Street: Beatriz Santiago Munoz; The Wizard of Oz; Paul McCarthy's Low Life Slow Life; Moby Dick; Huckleberry Finn; Painting Between the Lines; More American Photographs; Give Them the Picture; The Way Beyond Art: White Wide Space; Americana; When Attitudes Became Form Become Attitudes.

Hours & Admission Prices: Logan Center Galleries: Tues.-Fri. 12-7, Sat. 10-6. No charge; donations accepted. Closed major holidays; during installation. &

Attendance: 28,000 (estimated)

✳ CALIFORNIA ACADEMY OF SCIENCES, (M), 55 Music

Concourse Dr., San Francisco, CA 94118-4503. Tel.: 415-379-8000.

E-mail: info@calacademy.org
Web Site: www.calacademy.org
Founded: 1853.

Congressional District: 9

Key Personnel: Exec. Dir., Gregory Farrington, Ph.D.; Chm. Bd. (V), John C. Atwater; Pres. (V), Dr. John Hefernik; Dir. Steinhart Aquarium, Bart Shepherd; Chm. Anthropology, Dr. Zeray Alemseged; Chm. Aquatic Biology, Dr. John McCosker; Chm. Botany, Dr. Frank Almeda; Chm. Entomology, Dr. Wojciech Pulawski; Chm. Ichthyology, Dr. Tomio Iwamoto; Chm. Herpetology, Dr. Alan Leviton; Chm. Ornithology & Mammalogy, Dr. John Dumbacher; Chm. Invertebrate Zoology & Geology, Dr. Gary Williams; Public Information Officer, Stephanie Stone.

Personnel Profile: Full-Time Paid 320; Part-Time Paid 250; Part-Time Volunteers 1,100; Interns 35.

Governing Authority: nonprofit organization. Tax-exempt: 501(c)(3) & 170(b)(1)(A).

Institution Type/Description: Natural History Museum, Aquarium & Planetarium.

Collections: ichthyology; invertebrate zoology; ornithology; mammalogy; natural history; botany; entomology; herpetology; anthropology; paleontology; geology; minerals; artifacts; diatoms.

Research Fields: anthropology; botany; entomology; geology; herpetology; ichthyology; invertebrate zoology; ornithology; mammalogy; aquatic biology.

Facilities: 210,000-vol. library of science books on botany, zoology, geology & related topics available for inter-library loan & on premises; aquarium. Books, museum reproductions & other related items for sale.

Activities: guided tours; lectures; formally organized education programs for children & adults; docent program or council; inter-museum loan, permanent & temporary exhibitions.

Publications: newsletters; scientific papers; memoirs; proceedings; occasional papers.

Hours & Admission Prices: Mon.-Sat. 9:30-5, Sun. 11-5. Adults $24.95, youth 12-17, senior citizens 65 & over and students $19.95, children 7-11 $14.95; children 6 & under no charge. Closed Thanksgiving; Christmas. &

Attendance: 1,500,000 (accurate)

Membership: Senior $79; Individual $99; Family $199; Family Plus $250; Associate $500; Guild $750.

CALIFORNIA HISTORICAL SOCIETY, (M), 678 Mission St.,

San Francisco, CA 94105-4014. Tel.: 415-357-1848. Fax: 415-357-1850.

E-mail: info@calhist.org
Web Site: www.calhist.org
Founded: 1871.
Congressional District: 8

Key Personnel: Exec. Dir., David Crosson; Dir. Education & Public Programs, Lisa Eriksen; Pres. (V), Jan Berckefeldt; Dir. Library & Archives, Mary Morganti; Dir. Devel., Derlene Plumtree; Dir. Operations, Liliana Vasquez; Editor California History, Janet Fireman; Dir. Finance, Pamela Garcia; Museum Shop Mgr., Sy Russell.

Personnel Profile: Full-Time Paid 10; Part-Time Paid 7; Part-Time Volunteers 12; Interns 4.

Governing Authority: private; nonprofit organization. Partnerships with Bancroft Library & Autry National Center, USC. Tax-exempt: 501(c)(3).

Institution Type/Description: Historical Society Museum.

Collections: California history from the pre-Gold Rush days to the present; photography; fine art; books; manuscripts; artifacts; costumes.

Research Fields: exhibition-specific research on art, photography, and library collections.

Facilities: 506,500-vol. library of books, pamphlets, photographs, periodicals, archival collections & manuscripts; 3,300 sq. ft. exhibit space. Museum-related items for sale.

Activities: lectures; loan, participatory, temporary & traveling exhibitions; training programs for professional museum workers; family days; walking tours. Museum Sponsors: Holiday Party; Hats-Off Benefit.

Publications: quarterly peer-reviewed scholarly journal, California History; quarterly newsletter, California Chronicle.

Hours & Admission Prices: Wed.-Sat. 12-4:30. Suggested Donations: adults $3, senior citizens & students $1; discounts to AAM & ICOM members; members & children no charge. Closed New Year's Day; Martin Luther King Jr. Day; Presidents' Day; Memorial Day; Independence Day; Labor Day; Thanksgiving & day after; Christmas. &

Attendance: 12,000 (estimated)

Membership: Student, Teachers $40; Senior $45; Libraries & Nonprofit $50; Regular $55; Plus $75; Friend $125; Contributor $250; Benefactor $500; Silver Circle $1,000.

CARTOON ART MUSEUM, 655 Mission St., San Francisco, CA

94105-4126. Tel.: 415-227-8666, ext. 300. Fax: 415-243-8666. Facebook: Cartoon Art Museum.

E-mail: office@cartoonart.org

Web Site: www.cartoonart.org
Founded: 1984.
Key Personnel: Exec. Dir., Summerlea Kashar; Chm., Ron Evans; Cur., Andrew Farago; Bookstore Mgr., Heather Plankett.
Personnel Profile: Full-Time Paid 3; Part-Time Paid 3; Part-Time Volunteers 25; Interns 6.
Governing Authority: nonprofit organization. Tax-exempt: 501(c)(3).
Institution Type/Description: Cartoon Art Museum.
Collections: original cartoon art, flat graphics-animation cells; videos & film; toys; figurines; ephemera.
Research Fields: cartoon art; cartooning.
Facilities: library available to the public for research only; 3,400 sq. ft. exhibit space. Museum-related items for sale.
Activities: guided tours; lectures; artist appearances; loan, traveling & temporary exhibitions. Museum Sponsors: Saturday cartooning classes, one Saturday a month during school year, two Saturdays a month during summer.
Publications: e-newsletter; exhibition catalogs.
Hours & Admission Prices: Museum: Tues.-Sun. 11-5. Bookstore: Tues.-Sun. 11-5:30. Adults $8, students & seniors $6, children 6-12 $3; discounts to AAM & ICOM members; CAM members & children 5 and under no charge. Pay what you wish first Tues. each month. Closed New Year's Day; Easter; Independence Day; Thanksgiving; Christmas Eve & Day.
Attendance: 35,000 (estimated)
Membership: Student & Senior $35; Individual $45; Family $65; Friend $125; Patron $750; Sponsor $500; Benefactor $1,000.

CHILDREN'S CREATIVITY MUSEUM, (M), 221 Fourth St., San Francisco, CA 94103-3116. Tel.: 415-820-3320. Fax: 415-820-3330. Facebook: Creativity Museum.
E-mail: info@creativity.org
Web Site: www.creativity.org
Formerly: Zeum
Founded: 1998.
Key Personnel: Exec. Dir., Michael Nobleza; Mktg. Mgr., Cathy Barragan; Dir. Finance & Operations, Mindy Galoob.
Personnel Profile: Full-Time Paid 19; Full-Time Volunteers 3; Part-Time Paid 46; Part-Time Volunteers 18; Interns 16.
Governing Authority: bd. of directors. Tax-exempt.
Institution Type/Description: Children's Museum.
Collections: interactive visual arts; performing & media arts technology; community & youth artwork; 1906 Charles Looff carousel.
Facilities: technology studios; 200-seat theater. Museum-related items for sale.
Activities: digital art, animation, sound & video production; live performance; traveling international & community art exhibits; art workshops; preschool workshops; youth docent internships; professional (teacher) development programs; school field trips; carousel rides.
Hours & Admission Prices: Summer: Tues.-Sun. 10-4; Sept.-June Wed.-Sun. 10-4. Admission $11; discounts to ASTC members; members & children under 2 no charge. &
Attendance: 60,000 (estimated)
Membership: Caregiver $20; Classic Family $99; Creative Family $149.

CHINESE CULTURE CENTER OF SAN FRANCISCO, 750 Kearny St., 3rd. Fl., San Francisco, CA 94108-1887. Tel.: 415-986-1822. Fax: 415-986-2825.
E-mail: info@c-c-c.org
Web Site: www.c-c-c.org
Founded: 1965.
Congressional District: 5
Key Personnel: Co Chm., Richard Lee, M.D.; Co Chm., Colin C. Wong, D.D.S.; Pres., Russell E. Leong, M.D.; Exec. Dir., Albert Cheng; Program Dir., Abby Chen.
Personnel Profile: Full-Time Paid 2; Part-Time Paid 5; Part-Time Volunteers 20.
Governing Authority: nonprofit organization. Parent Institution: Chinese Culture Foundation. Tax-exempt: 501(c)(3).
Institution Type/Description: Art Gallery.
Collections: historical photographs on Chinese-Americans; Chinese art; Chinese folk arts; crafts.
Research Fields: pertaining to exhibits; special research.
Facilities: 400-seat auditorium; classrooms. Museum-related items for sale.
Activities: youth genealogy research programs; guided tours; lectures; gallery talks; TV & radio programs; docent program; classes & workshops; visual arts exhibitions; performing arts & festivals.
Publications: exhibition catalogues; educational brochures; newsletters.
Hours & Admission Prices: Tues.-Sat. 10-4. No charge; donations accepted. Closed major holidays. &
Attendance: 65,000 (estimated)

Membership: Senior Citizens 65 & over and Full-Time Students $25; Individual $35; Family $50; Contributing $200; Benefactor $500; Business & Corporate $1,000; Life $10,000.

CHINESE HISTORICAL SOCIETY OF AMERICA, 965 Clay St., San Francisco, CA 94108-1527. Tel.: 415-391-1188. Fax: 415-391-1150. Facebook: Chinese Historical Society of America.
E-mail: info@chsa.org
Web Site: www.chsa.org
Founded: 1962.
Key Personnel: Dir., Sue Lee; Pres., Barre Fong
Institution Type/Description: Chinese Historical Society.
Collections: Chinese history & culture; photographs; chalkware; trade cards; sheet music; toys.
Publications: Chinese America: History & Perspectives.
Hours & Admission Prices: Tues.-Fri. 12-5, Sat. 11-4. Adults $5, college students & seniors $3, children 6-17 $2; members & children 5 & under no charge. Closed New Year's Day; Veterans Day; Independence Day; Christmas.
Membership: Student, Senior & Educator $30; Individual $50; Family $60; Contributing $100; Sponsor $250; Patron $500; President's Circle & Legacy Wall $1,000.

THE CONTEMPORARY JEWISH MUSEUM, (M), 736 Mission St., San Francisco, CA 94103-3113. Tel.: 415-655-7800. Fax: 415-655-7815. Facebook: The Contemporary Jewish Museum.
E-mail: info@thecjm.org
Web Site: www.thecjm.org
Founded: 1984.
Congressional District: 8
Key Personnel: Exec. Dir., Lori Starr; Chm. (V), Mark Schlesinger; Museum Shop Mgr., Kevin Grenon.
Personnel Profile: Full-Time Paid 40; Part-Time Paid 10; Part-Time Volunteers 146; Interns 5.
Governing Authority: nonprofit organization. Tax-exempt: 501(c)(3).
Institution Type/Description: Art Museum & Center.
Collections: Jewish art & culture.
Major Exhibits: Frog and Toad and the World of Arnold Lobel, 11/13-3/14; Jason Lazarus: Live Archive, 11/13-3/14; To Build & Be Built: Kibbutz History, 10/13-7/14; The Art of the Haggadah: Arthur Szych, 2/14-6/14; Designing Home: Mid-century Modern Jewish Artists, 1930-1970, 4/14-10/14; Project Mah Jongg (T), 7/14-10/14; Arnold Newman: Master Class (T), 10/14-2/15.
Research Fields: Jewish art, culture & traditions; Jewish education; museum education.
Facilities: 11,700 sq. ft. exhibit space.
Activities: traveling exhibition; artists' invitational exhibits; community-based exhibitions; guided tours; lecture series; educational programs for children, adults & elders; intergenerational program; college guides; films; contemporaries. Museum Sponsors: family holiday celebrations; fund-raising auction.
Publications: exhibition catalogues.
Hours & Admission Prices: Thurs. 1-8, Fri.-Tues. 11-5. Adults $12; youth 18 & under and members no charge. Closed New Year's Day; Passover; Independence Day; Rosh Hashanah; Yom Kippur; Thanksgiving. &
Attendance: 125,000 (estimated)
Membership: Senior $50; Dual Senior $70; Individual $75; Household $95; Supporter $180; Sponsor $300; Patron $600; Circle of Friends $1,000 & up.

THE EXPLORATORIUM, (M), Pier 15, San Francisco, CA 94111. Mailing Address: Piers 15/17, San Francisco, CA 94110-1456. Tel.: 415-528-4367. Fax: 415-885-6001. TDD: 415-885-6033.
E-mail: pubinfo@exploratorium.edu
Web Site: www.exploratorium.edu
Founded: 1969.
Congressional District: 5
Key Personnel: Exec. Dir., Dennis M. Bartels; Chm. Bd. Trustee (V), George W. Cogan; Exec. Assoc. Dir., Robert J. Semper; C.O.O., Laura Zander; Dir. Exhibitions & Public Programs, Thomas Rockwell; Dir. Mktg., Sabrina Smith; Museum Shop Mgr., Julie Nunn.
Personnel Profile: Full-Time Paid 350; Part-Time Paid 61; Part-Time Volunteers 250; Interns 34.
Governing Authority: nonprofit organization. Tax-exempt.
Institution Type/Description: Science Museum & Center.
Collections: participatory exhibits & art works explore & illustrate the physical nature of world & sensory mechanisms through which we perceive; commissioned art works dealing with natural phenomena.
Major Exhibits: Changing Face of What Is Normal, 4/13-4/16/14.

Research Fields: exhibit-based education; exhibit & curriculum development; research & development of multi-media learning tools; visitor learning research.
Facilities: classrooms; 2 cafes; theater. Museum-related items for sale.
Activities: lectures; films; concerts; field trips; festivals; special theme exhibitions; demonstrations; performances; teachers workshops; adults only Thurs. 6pm-10pm; evening events.
Publications: monthly newsletter; Exploratorium Cookbooks (exhibit construction manuals); supplemental curriculum materials & exhibit plans for teachers, including Science Snackbook series, Scale & Structure, Human Body Explorations & Math/Science Across Cultures; reference charts, Electromagnetic Spectrum, Sound Spectrum, History of the Alphabet Cycles & Language Families of the World; trade books for children, families & curious readers, Explorabook, The Science Explorer series, Math Explorer and Exploratopia.
Hours & Admission Prices: Tues. & Fri.-Sun. 10-5, Wed.-Thurs. 10-10. Adults $25, students, seniors, teachers, youth 6-17 $19; 1st Wed. of month & members no charge. Closed Thanksgiving; Christmas. &
Attendance: 1,100,000 (estimated)
Membership: Individual Explorer $60; Dual Explorer $95; Family Explorer $150; Insiders $225; Supporters $300; Sustainers $600; Innovators $1,000 & up. (Seniors & Disabled $10 discount).

***   THE FINE ARTS MUSEUMS OF SAN FRANCISCO, DE YOUNG MUSEUM, (M),** 50 Hagiwara Tea Garden Dr., San Francisco, CA 94118-4502. Tel.: 415-750-3600. Fax: 415-750-7692.
E-mail: guestbook@famsf.org
Web Site: www.deyoungmuseum.org
Founded: 1894.
Congressional District: 8
Key Personnel: Dir., John E. Buchanan, Jr.; Pres. Bd. Trustees (V), Dede Wilsey; Assoc. Dir., Robert Futernick; Administrative Asst. to Dir., Lauren Ito; Deputy Dir. Devel., Martha Brigham; Dir. Mktg. & Communications, Susannah Stringam; Chief Cur. & Founding Cur. Photography, Julian Cox; Cur. Prints & Drawings, Achenbach Foundation for Graphic Arts, Karin Breuer; Ednah Root Cur. American Arts, Timothy Anglin Burgard; Cur. Africa, Oceania & Americas, Kathleen Berrin; Cur. European Paintings, Dr. Lynn Federle Orr; Cur. Ancient Art & Interpretation, Renee Dreyfus; Advertising & Promotion, Gina Yrrizarry; Dir. Education, Sheila Pressley; Museum Shop Mgr., Stuart Hata; Media Rels., Jill Lynch; Publications, Karen Levine.
Personnel Profile: Full-Time Paid 420; Part-Time Volunteers 600.
Governing Authority: municipal; county. Tax-exempt.
Institution Type/Description: Fine Arts Museum.
Collections: paintings, prints & drawings; sculpture & decorative arts from America; traditional arts of Africa, Oceania & the Americas; rugs; textiles; photographs.
Research Fields: European & American painting; decorative arts; conservation; African art; art of Oceania.
Facilities: auditorium; theatre; cafe; conservation laboratories. Museum-related items for sale.
Activities: temporary exhibitions; lectures; docent tours; films; concerts; community outreach programs; musical programs; school programs.
Publications: quarterly members magazine, Fine Arts; collection & exhibition catalogues.
Hours & Admission Prices: Tues.-Thurs. & Sat.-Sun. 9:30-5:15, Fri. 9:30-8:45. Adults $10; 1st Tues. each month & members no charge. Closed New Year's Day; Independence Day; Thanksgiving; Christmas. &
Attendance: 1,900,000 (accurate)
Membership: Student $55; Senior $60; Teacher $65; Out of State $70; Individual $95; Family/Dual $125; Contributing $225; Supporting $350; Sustaining $600; Friend $1,000-$2,499.

***   THE FINE ARTS MUSEUMS OF SAN FRANCISCO, LEGION OF HONOR, (M),** 100 34th Ave., San Francisco, CA 94121-1677. Mailing Address: c/o deYoung, 50 Hagiwara Tea Garden Dr., San Francisco, CA 94118. Tel.: 415-750-3600. Fax: 415-750-3656.
E-mail: guestbook@famsf.org
Web Site: www.legionofhonor.org
Formerly: California Palace of the Legion of Honor
Founded: 1924.
Congressional District: 8
Key Personnel: Dir., John E. Buchanan, Jr.; Pres. Bd. Trustees (V), Dede Wilsey; Assoc. Dir., Robert Futernick; Administrative Asst. to Dir., Lauren Ito; Dir. Mktg. & Communications, Susannah Stringam; Deputy Dir. Devel., Martha Brigham; Cur. Prints & Drawings, Achenbach Foundation for Graphic Arts, Karin Breuer; Cur. European Art, Dr. Lynn Federle Orr; Cur. Ancient Art & Interpretation, Renee Dreyfus; Advertising & Promotion, Gina Yrrizarry; Dir. Education, Shelia Pressley; Museum Shop Mgr., Stuart Hata; Media Rels., Jill Lynch; Webmaster, Andrew Fox; Publications, Karen Levine; European Decorative Arts & Sculpture, Martin Chapman.
Personnel Profile: Part-Time Volunteers 600.
Governing Authority: county. Tax-exempt.
Institution Type/Description: Fine Arts Museum.
Collections: paintings, sculpture & decorative arts from Europe and England; tapestries; prints; drawings; arts of ancient Egypt, Greece & Rome; Achenbach Graphic Arts Foundation.
Research Fields: European & American painting; decorative arts; conservation.
Facilities: auditorium; theatre; cafe; conservation laboratories. Museum-related items for sale.
Activities: temporary exhibitions; lectures; docent tours; films; concerts; community outreach programs; musical programs; school programs.
Publications: quarterly members magazine, Fine Arts; collection & exhibition catalogues.
Hours & Admission Prices: Tues.-Sun. 9:30-5:15. Adults $10, seniors over 65 $7, children 12-17 $6; discounts to AAM members; members, children 12 & under, San Francisco public and private school students K-12 with ID & 1st Tues. of month no charge. Closed New Year's Day; Independence Day; Thanksgiving; Christmas. &
Attendance: 500,000
Membership: Student $55; Senior $60; Teacher $65; Out of State $70; Individual $95; Family/Dual $125; Contributing $225; Supporting $350; Sustaining $600; Friends $1,000-$2,499.

FORT POINT NATIONAL HISTORIC SITE, Presidio of San Francisco, Bldg. 989, San Francisco, CA 94129. Mailing Address: Fort Mason, Bldg. 201, San Francisco, CA 94123-1307. Tel.: 415-556-1693.
E-mail: michele_gee@nps.gov
Web Site: www.nps.gov/fopo
Founded: 1970.
Congressional District: 5
Key Personnel: Supt., Frank Dean.
Personnel Profile: Full-Time Paid 2; Full-Time Volunteers 8.
Governing Authority: federal government. Parent Institution: Dept. of the Interior, National Park Service.
Institution Type/Description: Historic Site: c.1853-1861 Fort Point which is the only third system fort on the west coast.
Collections: items related to the history of Fort Point & the garrisons who served there.
Research Fields: military history of Fort Point.
Activities: guided tours; formally organized education programs for children; permanent exhibitions; video.
Publications: book, Fort Point; color photo book; video, Fort Point: Guardian of the Golden Gate; audio tour, A Day in the Life of a Fort Point Soldier.
Hours & Admission Prices: May-Sept. Thurs.-Mon. 10-5; Oct.-April Fri.-Sun. 10-5. No charge; donations accepted. Closed New Year's Day; Thanksgiving; Christmas. &
Attendance: 300,000
Membership: Golden Gate National Park Conservancy $35-$100.

GOLDEN GATE NATIONAL RECREATION AREA, Fort Mason, Bldg. 201, San Francisco, CA 94123-0022. Tel.: 415-561-2831. Fax: 415-441-8272. TDD: 415-556-2766.
E-mail: jonathan_bayless@nps.gov
Web Site: www.nps.gov/goga/
Founded: 1972.
Congressional District: 6, 8 & 12
Key Personnel: Supt., Brian O'Neill; Cur., Jonathan Bayless; Museum Specialist, Lulu Chye; Archivist, Susan Ewing Haley; Chief Interpretation, Howard Levitt; Registrar, John Weingardt; Security, Yvette Ruan.
Personnel Profile: Full-Time Paid 370; Part-Time Volunteers 5,000.
Governing Authority: federal government; nonprofit. Parent Institution: Dept. of the Interior. Subsidiary Institution: National Park Service. Tax-exempt.
Institution Type/Description: Park Museum: area includes Muir Woods National Monument & Fort Point National Historic Site.
Collections: artifacts represent over 10,000 years of human history & prehistory; natural resources of the north & south Pacific border regions of the California coast; 180,000 historic documents; 6,000 plants; 20,000 photographs; 700 historic structures; historical & military architecture; coastal defense systems.
Research Fields: cultural landscape of history army airfields; small mammal predators; historic furnishings & structures of World War II & Cold War era; historic furnishings of Civil War fortification; historic resources of Presidio of San Francisco & coastal fortifications.

Facilities: library related to park resources for use on premises; 4,000 sq. ft. exhibit space. Gift items for sale.

Activities: concerts; docent program; formal education programs for children; guided tours.

Publications: quarterly newsletter, The Park; quarterly event calendar, Park Events.

Hours & Admission Prices: Daily 10-5. No charge; donations accepted. Closed New Year's Day; Thanksgiving; Christmas. &

Attendance: 13,803,382 (estimated)

Membership: Golden Gate National Park Association: Student & Senior $15; Senior Dual $20; Individual $25; Associate $50; Participant $100.

HAAS-LILIENTHAL HOUSE, 2007 Franklin St., San Francisco, CA 94109-2909. Tel.: 415-441-3000. Fax: 415-441-3015.

E-mail: info@sfheritage.org

Web Site: www.sfheritage.org

Founded: 1973.

Congressional District: 5

Key Personnel: Pres., David Wessel; Exec. Dir., Mike Buhler; Operations Mgr., Barbara Roldan.

Personnel Profile: Full-Time Paid 4; Part-Time Paid 2; Part-Time Volunteers 100; Interns 1.

Governing Authority: nonprofit. Parent Institution: San Francisco Architectural Heritage, 2007 Franklin St., San Francisco, CA. Tax-exempt: 501(c)(3).

Institution Type/Description: Historic House: c.1886 Queen Anne style Victorian.

Collections: late 1800s & turn-of-the-century furniture, accessories & paintings.

Research Fields: architectural history of San Francisco; feasibility studies for renovation; adaptive reuse of significant buildings in San Francisco; preservation & conservation of city structures.

Facilities: room & house rentals.

Activities: docent-guided tours; special programs; lectures; architectural walking tours; preservation & conservation of old city buildings; heritage architectural walks; slide-lectures; school program.

Publications: bimonthly newsletter; Splendid Survivors - San Francisco's Downtown Architectural Heritage; Jessie Street Substation: Adaptive Reuse Feasibility Study & Proposal; Directory 77: Rehabilitation Advice & Useful Sources For Owners of Vintage Buildings; Victorian Sampler: A Walk in Pacific Heights & the Haas Lilienthal House; Port City, The History and Transformation of the Port of San Francisco: 1848-2010.

Hours & Admission Prices: Wed. & Sat. 12-3, Sun. 11-4. Adults $8, children 6-12 & senior citizens $5; discount to AAA & Travel with Visa members; members & children under 6 no charge.

Attendance: 6,000 (estimated)

Membership: Young Preservationist 40 & under $30; Individual $60; Fog City Family $75; Splendid Survivor $125; The Golden City Ally $500; Paris of the West Pillar $1,000.

HENRY WILSON COIL MASONIC LIBRARY & MUSEUM, 1111 California St., San Francisco, CA 94108-2252. Tel.: 415-776-7000, ext. 143.

Institution Type/Description: History Museum.

Collections: Freemasonry history; photographs.

Facilities: library.

Activities: lectures.

Hours & Admission Prices: Mon.-Tues. & Thurs. 10-3; Wed. 10-5; other times by appointment. No charge.

INTERNATIONAL ART MUSEUM OF AMERICA, 1023 Market St., San Francisco, CA 94103. Tel.: 415-376-6344. Fax: 415-255-9415.

E-mail: lhuang@internationalartmuseum.org

Web Site: www.internationalartmuseum.org; www.iamasf.org

Founded: 2011.

Key Personnel: Dir., Loretta Huang, PhD; Pres., Yu-Hua Shou Zhi Wang; Store Assoc., Phaedra Restad; Store Assoc., Ayanna Madison.

Personnel Profile: Full-Time Paid 2; Part-Time Paid 3; Part-Time Volunteers 6; Interns 4.

Volunteer Hours: 1,240

Operating Expenses: 550,000

Operating Income: 450,000

Governing Authority: Tax-exempt.

Institution Type/Description: Art Museum.

Collections: works by artists from around the world.

Publications: newsletters.

Hours & Admission Prices: Tues.-Sun. 11-5. No charge.

Attendance: 14,493 (accurate)

Membership: Student $35; Senior $45; Individual $80; Family $100; Donor

$500; Patron $1,000-$9,999; Bronze 1 Star Insignia Supporter $10,000-$29,999; Silver 2 Star Insignia Contributor $30,000-$69,999; Gold 3 Star Insignia Benefactor $70,000-$149,999; Diamond 4 Star Insignia Status $150,000-$299,999; First Class 5 Star Insignia Leader $300,000-$499,999; Freeminent 6 Star Insignia Leader $500,000 & up.

INTERNATIONAL MUSEUM OF WOMEN, 235 Montgomery St., 28th Fl., San Francisco, CA 94104. Mailing Address: P.O. Box 190038, San Francisco, CA 94119-0038. Tel.: 415-543-4669. Fax: 415-543-4668. Facebook: International Museum of Women.

E-mail: info@imow.org

Web Site: www.imow.org

Founded: 1985.

Congressional District: 12

Key Personnel: Exec. Dir., Clare Winterton; Vice Pres. Exhibitions & Programs, Catherine King.

Personnel Profile: Full-Time Paid 4; Part-Time Paid 1; Part-Time Volunteers 3.

Governing Authority: nonprofit organization. Tax-exempt: 501(c)(3).

Institution Type/Description: Virtual Women's Museum.

Collections: women's voices globally including film, paintings, photographs, prints, & drawings.

Research Fields: women's human rights.

Facilities: development office.

Activities: Speaker series; art & education program for K-12.

Hours & Admission Prices: Online museum.

Membership: Student & Senior $35; Individual $50; Family $75; Auxiliary $100-$499; Initiator $500-$4,999; Ambassador $1,000-$4,900; Emissary $5,000-$9,999.

THE MEXICAN MUSEUM, (M), Fort Mason Center, Marina Blvd., Bldg. D., San Francisco, CA 94123. Tel.: 415-202-9700.

E-mail: info@mexicanmuseum.org

Web Site: www.mexicanmuseum.org

Founded: 1975.

Congressional District: 5

Key Personnel: Dir., David J. De la Torre; Bd. Chair, Andrew Uluger; Vice Chair, Nora E. Wagner; Asst. to Dir., Marlena Cannon; Adjunct Cur. Culinary Arts, Antelmo Faria.

Personnel Profile: Full-Time Paid 2; Part-Time Paid 3; Interns 5.

Governing Authority: nonprofit organization. Tax-exempt: 501(c)3.

Institution Type/Description: Fine Arts Museum.

Collections: pre-conquest art; Colonial & Popular art; modern & contemporary Mexican and Latino art; Chicano art.

Major Exhibits: An Inspired Gift: The Rex May Collection of Mexican Popular Art, 4/26/13-5/16/14.

Research Fields: Mexican & Chicano-Mexican American fine arts; pre-conquest art; colonial art; popular art; Mexican & Latino art; Mexican American & Chicano contemporary art.

Facilities: 300-vol. library of Mexican art & culture. Books, Mexican folk art & other museum-related items for sale.

Activities: guided tours; lectures; temporary exhibitions; family day workshops & presentations.

Publications: catalogues, The Nelson A. Rockefeller Collection of Mexican Folk Art, Witness to the Self; Rodolfo Morales; Colonial Mexican & Popular Religious Art; Leonora Carrington; Patssi Valdez: A Precarious Comfort.

Hours & Admission Prices: Wed.-Sun. 12-4. No charge; donations accepted. Closed holidays. &

Attendance: 6,432 (estimated)

Membership: Student & Senior $25; Artist & Educator $30; Individual $60; Dual & Family $75; Supporting $100; Contributing $250; Builder's Society $365; Patron $500; Donor's Circle $1,000; Director's Circle $2,500; Chairperson's Circle $5,000 & up.

MISSION CULTURAL CENTER FOR LATINO ARTS, 2868 Mission St., San Francisco, CA 94110-3908. Tel.: 415-821-1155. Fax: 415-648-0933.

E-mail: info@missionculturalcenter.org

Web Site: www.missionculturalcenter.org

Founded: 1977.

Congressional District: 6

Key Personnel: Exec. Dir., Jennie E. Rodriguez; Multimedia Coord., Adrian Arias; Events Coord., Jason Wallach; Mission Grafica Coord., Marsha Shaw; Facilities Coord., J. Angel Varela; Youth Program Coord, Leticia Guzman; Gallery Assoc., Nicole Crescenzi; Gallery Coord., Maurizzio Hector Pineda.

Personnel Profile: Full-Time Paid 2; Part-Time Paid 13; Part-Time Volunteers 7.

Governing Authority: municipal; nonprofit. Parent Institution: Neighborhood Arts Program, 45 Hyde St., #319, San Francisco, CA 94102. Tel.: 415-558-3463. Tax-exempt: 170(b)(1)(A).
Institution Type/Description: Civic Art, Cultural Center.
Collections: private collection of artists work; posters of events housed at the Mission Cultural Center & from gallery shows.
Research Fields: local, Mexican, Central & South American Culture.
Facilities: 200-seat theatre; stage; dance & music studios.
Activities: performing events; lectures, films; on-going classes & workshops; receptions; site-visits, tours; permanent & temporary exhibitions.
Publications: quarterly cultural arts magazine, Humanizarte.
Hours & Admission Prices: Tues.-Sat. 10-5. Adults $2. Closed holidays. &
Attendance: 20,000 (estimated)

MISSION SAN FRANCISCO DE ASIS (MISSION DOLORES), 3321 Sixteenth St., San Francisco, CA 94114-1712. Tel.: 415-621-8203.
Institution Type/Description: Historic Building: c.1791.
Collections: archives; ethnic & tribal art; furniture; paintings.
Activities: children's classes; guided tours.
Hours & Admission Prices: Daily 9-4. Adults $5, children $3; children under 6 no charge. Closed Thanksgiving; Christmas.

MODERNISM, 685 Market St., Ste. 290, San Francisco, CA 94105. Tel.: 415-541-0461. Fax: 415-541-0425.
E-mail: info@modernisminc.com
Key Personnel: Owner, Martin Muller
Institution Type/Description: Art Gallery.
Collections: paintings; photographs; sculpture.
Activities: temporary exhibitions.
Hours & Admission Prices: Tues.-Sat. 10-5:30.

THE MUSEE MECANIQUE, Pier 45, Shed A, San Francisco, CA 94133. Tel.: 415-346-2000.
E-mail: coad01@yahoo.com
Web Site: www.museemecanique.com
Key Personnel: Owner, Daniel Galand Zelinsky
Institution Type/Description: General Museum.
Collections: over 300 coin-operated mechanical art machines.
Hours & Admission Prices: Mon.-Fri. 10-7, Sat.-Sun. & holidays 10-8. No charge. &

MUSEO ITALOAMERICANO, Ft. Mason Center, Bldg. C, San Francisco, CA 94123-1301. Tel.: 415-673-2200. Fax: 415-673-2292.
E-mail: sfmuseo@sbcglobal.net
Web Site: www.museoitaloamericano.org
Founded: 1978.
Congressional District: 5
Key Personnel: Dir., Paola Bagnatori; Pres. (V), Michael Feno; Asst. Mng. Dir., Susan Filippo.
Personnel Profile: Full-Time Paid 1; Part-Time Paid 4; Part-Time Volunteers 45; Interns 1.
Governing Authority: nonprofit. Tax-exempt.
Institution Type/Description: Contemporary Italian, Italian-American Art Museum & Italian Cultural Center.
Collections: 20th-century paintings, sculpture & photographs by Italian & Italian-American artists; changing exhibits.
Major Exhibits: Group Show - 3 Artists, 12/13-3/14.
Facilities: library. Museum-related items for sale.
Activities: Italian language & conversation classes; art & culture lectures; Culinary Arts Series; concerts; CIAO (Children's Italian Art Outreach Program); Italian film program.
Publications: newsletter; Italian American Studies Curriculum Unit; exhibition catalogs.
Hours & Admission Prices: Tues.-Sun. 12-4. No charge; donations accepted. Closed holidays. &
Attendance: 8,450 (accurate)
Membership: Senior Citizens $35; Next Generation $50; Dual Senior $60; Dual Family $75; Special Friend $100; Sponsor $150; Supporter $250; Benefactor $500; Silver Circle $1,000 & up; Gold Circle $5,000 & up; Platinum Circle $10,000 & up.

MUSEUM OF CRAFT AND DESIGN, 2569 Third St., San Francisco, CA 94107. Tel.: 415-773-0303; 877-487-3623. Fax: 415-7730303.
E-mail: info@sfmcd.org
Web Site: www.sfmcd.org

Institution Type/Description: Art Museum.
Collections: contemporary works of art.
Facilities: Museum-related items for sale.
Activities: educational programs; workshops; groups tours.
Hours & Admission Prices: Tues.-Wed. & Fri.-Sat. 11-6, Thurs. 11-7, Sun. 12-5. Closed New Year's Day; Thanksgiving; Christmas.

MUSEUM OF RUSSIAN CULTURE, 2450 Sutter St., San Francisco, CA 94115-3016. Tel.: 415-921-4082. Fax: 415-921-4082.
Founded: 1948.
Key Personnel: Pres., Nicholas Koretsky; Vice Pres., Yuri Tarala
Governing Authority: nonprofit. Tax-exempt: 501(c)(3).
Institution Type/Description: Ethnic History Museum.
Collections: archives; history; military; ethnology; numismatic; manuscripts.
Research Fields: Russian history & immigration.
Activities: lectures; inter-museum loan & permanent exhibitions.
Publications: biennial, Sbornik Museia Russkoy Kultury.
Hours & Admission Prices: Wed. & Sat. 10:30-2:30. No charge. Closed holidays.
Attendance: 300 (estimated)
Membership: Annual $6.

MUSEUM OF THE AFRICAN DIASPORA, (M), 685 Mission St., San Francisco, CA 94105-4126. Tel.: 415-358-7200. Fax: 415-358-7252.
E-mail: gstanislaus@moadsf.org
Web Site: www.moadsf.org
Founded: 2005.
Key Personnel: Chm. Bd. Directors, L. Wade Rose; Vice Chm. Bd., Deborah Santana; Dir. Devel., Leslie Lombre; Education, Lovisa Brown; Cur. Public Programs, Shiree Dyson; Museum Shop Mgr., Keirra Hall.
Personnel Profile: Full-Time Paid 13; Part-Time Volunteers 60; Interns 5.
Governing Authority: nonprofit organization. Tax-exempt: 501(c)(3).
Institution Type/Description: Art Museum.
Collections: art, history & culture of African Diaspora.
Facilities: 10,000 sq. ft. exhibit space; 85-seat auditorium; educational facilities. Museum-related items for sale.
Activities: docent program; films; formal education programs for adults; guided tours; lectures.
Hours & Admission Prices: Wed.-Sat. 11-6, Sun. 12-5. Adults $10, senior citizens 65 & over and students $5; children 12 & under and members no charge. Closed New Year's Day; Labor Day; Thanksgiving; Christmas.
Attendance: 85,000 (accurate)
Membership: Senior, Teacher & Student $45; Individual $65; Dual & Family $85; Supporter $150; Contributor $250; Donor $500; Sustaining $1,000; Patron $1,500; Sponsor $2,500; Benefactor $5,000.

MUSEUM OF VISION, 655 Beach St., San Francisco, CA 94109-1342. Tel.: 415-561-8500. Fax: 415-561-8533.
E-mail: museum@aao.org
Web Site: www.museumofvision.org
Formerly: Museum of Vision Foundation of the American Academy of Ophthalmology
Founded: 1980.
Key Personnel: Dir. Museum of Vision, Jenny Benjamin.
Personnel Profile: Full-Time Paid 1.
Governing Authority: nonprofit organization. Parent Institution: The Foundation of the American Academy of Ophthalmology. Tax-exempt.
Institution Type/Description: Medical History Museum.
Collections: history of ophthalmology from the 14th century-present; spectacles; diagnostic & surgical instruments; pharmaceuticals; stamps & coins; photography; memorabilia; literature & oral histories; archives.
Research Fields: 1600-present history of ophthalmology.
Facilities: 500-vol. library & archive pertaining to ophthalmology; 600 sq. ft. exhibit space.
Activities: historical research; exhibits; educational publications for children; preservation.
Publications: brochure; oral histories; Eye Openers; Art and Vision: Seeing in 3-D; Discover Your Eye Q!; Eye Qs: Activities with Vision; Healthy Eyes, Healthy Body.
Hours & Admission Prices: Mon.-Fri. 10-5. Tours by appointment only. No charge; donations accepted. Closed national holidays. &
Attendance: 300

NATIONAL JAPANESE AMERICAN HISTORICAL SOCIETY, (M), 1684 Post St., San Francisco, CA 94115-3604. Tel.: 415-921-5007. Fax: 415-921-5087.
E-mail: njahs@njahs.org
Web Site: www.njahs.org

Founded: 1980.
Key Personnel: Exec. Dir., Rosalyn Tonai
Institution Type/Description: Historical Society Museum & Peace Gallery.
Collections: Japanese American culture & history; WWII textiles & artifacts pertaining to the Japanese American experience; artifacts & oral histories of the Japanese American internment camp experience.
Research Fields: Japanese American WWII; 100th Infantry Battalion; Military Intelligence Service (MIS); 442nd Regiment; internment camp experience.
Facilities: library; archives. Gift items for sale.
Activities: educational programs; special events; workshops.
Publications: quarterly newsletter, Nikkei Heritage.
Hours & Admission Prices: Mon.-Fri. & 1st Sat. of month 12-5.
Membership: Student $25; Basic $40; Family & Nonprofit $50; Supporting $60; Contributing $100; Corporation $250; Patron $500; Life $1,000.

NATIONAL LIBERTY SHIP MEMORIAL/S.S. JEREMIAH O'BRIEN,

1275 Columbus Ave., Ste. 300, San Francisco, CA 94133. Tel.: 415-544-0100. Fax: 415-544-9890.
E-mail: liberty@ssjeremiahobrien.org
Web Site: www.ssjeremiahobrien.org
Founded: 1978.
Key Personnel: Chm., Carl Nolte
Governing Authority: Parent Institution: National Liberty Ship Memorial.
Institution Type/Description: Merchant Marine Ship Museum and U.S. Naval Armed Guard: built in 1943.
Collections: WWII artifacts.
Facilities: Museum-related items for sale.
Activities: tours; cruises.
Publications: quarterly, Steady As She Goes.
Hours & Admission Prices: Daily 9-4. Adults $10, seniors $5, juniors 6-14 $4; discounts to AAM & ICOM members; members, active military & children under 6 no charge. Closed New Year's Day; Thanksgiving; Christmas.
Attendance: 52,500 (accurate)
Membership: Student $20; Individual $50; Family $100; Sponsor $250; Life $800; Corporate $1,500.

OCTAGON HOUSE,

2645 Gough St., San Francisco, CA 94123-4402. Tel.: 415-441-7512.
Institution Type/Description: Historic House: octagonal in shape, used as a family residence until the late 1920s. Listed on the National Register of Historic Places.
Collections: period furnishings; portraits; silver; pewter; ceramics; documents bearing signatures of 54 of the 56 Signers of the Declaration of Independence.
Hours & Admission Prices: Feb.-Dec. 2nd Sun. & 2nd and 4th Thurs. 12-3. No charge; donations accepted. Closed legal holidays.
Attendance: 1,249

RANDALL MUSEUM,

199 Museum Way, San Francisco, CA 94114-1429. Tel.: 415-554-9600. Fax: 415-554-9609.
E-mail: info@randallmuseum.org
Web Site: www.randallmuseum.org
Founded: 1937.
Congressional District: 6
Key Personnel: Exec. Dir., Chris Boettcher; Dir. Devel., Traci McCollister; Cur. Science, Nancy Ellis; Arts Coord., Julie Dodd Tetzlaff.
Personnel Profile: Full-Time Paid 11; Part-Time Paid 1; Part-Time Volunteers 25; Interns 3.
Governing Authority: municipal; county. Parent Institution: San Francisco Recreation & Park Dept. Subsidiary Institution: Randall Museum Friends. Tax-exempt.
Institution Type/Description: Community Museum.
Collections: rocks & minerals; fossils; shells; butterflies; live animals.
Facilities: 180-seat Children's Theater; woodworking shop; art studio; ceramics studio; lapidary shop; photography darkroom; live animal room; model railroad; seismograph; classroom; greenhouse; gardens; park.
Activities: guided tours; lectures; films; workshops for children & adults; permanent & temporary exhibits.
Publications: quarterly newsletter; class flyer published five times annually.
Hours & Admission Prices: Tues.-Sat. 10-5. No charge; donations accepted. Closed New Year's Day; Independence Day; Veterans Day; Thanksgiving & day after; Christmas; state & federal holidays.
Attendance: 85,000 (estimated)
Membership: Individual $30; Family $45;Sponsor $100; Patron $250; Benefactor $500; Steward $1,000.

* SFO MUSEUM - SAN FRANCISCO INTERNATIONAL AIRPORT, (M),

San Francisco International Airport, San Francisco, CA 94128. Mailing Address: P.O. Box 8097, San Francisco, CA 94128-8097. Tel.: 650-821-6700. Fax: 650-821-6777.
E-mail: curator@flysfo.com
Web Site: sfomuseum.org
Formerly: San Francisco Airport Museums
Founded: 1980.
Congressional District: 12
Key Personnel: Pres. Airport Commission, Larry Mazzola; Dir. Airport, John L. Martin; Dir. & Chief Cur., Blake Summers; Asst. Dir., Abe Garfield; Asst. Dir. Aviation, John Hill; Cur. Exhibitions, Timothy O'Brien; Cur. Administration & Special Projects, Sonya Knudsen; Cur. Registration, Barbara Geib.
Personnel Profile: Full-Time Paid 28; Part-Time Paid 10; Part-Time Volunteers 60; Interns 2.
Governing Authority: municipal. Parent Institution: City & County of San Francisco. Subsidiary Institution: Airport Commission. Tax-exempt.
Institution Type/Description: History Museum: housed in 20 exhibition sites throughout the airport.
Collections: commercial aviation in the Pacific region; 20,000 artifacts; 10,000 books & periodicals; 62,000 archival objects.
Major Exhibits: The History of Radio, 6/13-1/14; Model Yachts, 8/13-3/14; United We Stand: United Airlines Uniforms, 9/13-4/14; Japanese Toys! From Kokeshi to Kaiju, 11/13-5/14; Lighter Than Air, 11/13-7/14; Mapping San Francisco, 12/13-7/14; Turn, Weave, Fire, & Fold: Vessels from the Forrest L. Merrill Collection, 1/14-6/14; A History of Lace, 1/14-6/14; Portals: Doors from the Fowler Museum at UCLA, 2/14-9/14; Luggage, Bag Tags, Labels, and Tickets, 3/14-10/14; Airlines of Oceania, 4/14-10/14; Korean Ceramics from the Asian Art Museum of San Francisco, 5/14-12/14; Yoruba Culture: Objects from the Phoebe Hearst Museum of Anthropology, 6/14-1/15.
Research Fields: commercial aviation.
Facilities: library; archives.
Activities: temporary exhibitions; lectures; films; school programs.
Publications: monthly newsletters; exhibition brochures; exhibition catalogues.
Hours & Admission Prices: Airport Terminals: daily twenty-four hours a day. No charge. Aviation Library & Museum: Sun.-Fri. 10-4:30. No charge.
Attendance: 4,000,000 (estimated)

SAN FRANCISCO AFRICAN AMERICAN HISTORICAL AND CULTURAL SOCIETY, INC.,

762 Fulton St., 2nd Fl., San Francisco, CA 94102-4119. Tel.: 415-292-6172.
E-mail: info@sfaahcs.org
Web Site: www.sfaahcs.org
Founded: 1955.
Congressional District: 5
Key Personnel: Exec. Dir., W. E. Hoskins; Pres. & Chm. Bd. Dirs., Alfred W. Williams; Treas. & Systems Administrator, Ellis C. Joseph; Bd. Member, Jesse J. Byrd.
Governing Authority: nonprofit organization. Affiliated with the Association for Study of Afro-American Life & History, 1528 9th St., N.W., Washington DC 20001. Subsidiary Institution: Center for African & African American Art. Tax-exempt: 501(c)(3).
Institution Type/Description: African American Cultural Center.
Collections: Sargent Johnson, Mary Ellen Pleasant, R. Alan Williams & Dorothy Haywood collections; Haitian art.
Research Fields: African American Western history; African American national history of Africa and South America; Caribbean Islands; African American curriculum development.
Facilities: 3,000-vol. library of books, pamphlets, photographs, periodicals, magazines & encyclopedia available for use on premises & for inter-library loan; reading room; 100-seat auditorium; classroom.
Activities: guided tours; lectures; films; gallery talks; concerts; arts festivals; drama; study clubs; permanent & temporary exhibitions. Museum Sponsors: youth programs; African American history classes; senior history & crafts.
Publications: quarterly newsletter.
Hours & Admission Prices: Tues.-Sat. 1-5. No charge; donations accepted. Closed New Year's Eve & Day; Memorial Day; Independence Day; Labor Day; Thanksgiving; Christmas.
Attendance: 55,000
Membership: Senior & Youth $10; Individual $25; Family $40; Sustaining $100 or more; Life $500; Organization-small $1,000; Organization-large $2,500.

SAN FRANCISCO ART INSTITUTE - WALTER & MCBEAN GALLERIES, 800 Chestnut St., San Francisco, CA 94133-2299. Tel.: 415-749-4563. Fax: 415-351-3516.
E-mail: exhibitions@artists.sfai.edu
Web Site: www.sfai.edu
Founded: 1871.
Congressional District: 6
Personnel Profile: Full-Time Paid 2; Part-Time Paid 8.
Governing Authority: nonprofit organization. Parent Institution: San Francisco Art Institute. Tax-exempt: 501(c)(3).
Institution Type/Description: Art Museum.
Collections: works by contemporary artists.
Research Fields: contemporary art.
Facilities: 27,000-vol. library; slide library of works by contemporary artists; 250-seat auditorium; 3,000 sq. ft. galleries.
Activities: contemporary exhibitions program; performance; video & film series; active visiting artists program.
Publications: exhibition catalogs/artists' books, Sue Williams; Joyce J. Scott; Lee Bul Live Forever; Sarkis, Wherever We Go; Allora & Calzadilla; Jen Haaning, 5/09; Teddy Cruz; Pedro Reyes; Yan Pei Ming; Aleksander Komarov; Atelier Bow-Wow; Cyprien Gaillard; Jimmie Durham; Knut Asdam; Mark Lewis; Sam Samore; Hamra Abbas; Jewyo Rhii; Kan Xuan; Ringo Bunoan; Cao Fei; Ursula Biemann, Dan Perjovschi, Carlos Motta, Claire Fontaine, Societe Realiste, Hector Zamora, Teresa Margolles.
Hours & Admission Prices: Tues.-Sat. 11-6. No charge. Closed holidays. &
Attendance: 18,000

SAN FRANCISCO BOTANICAL GARDEN AT STRYBING ARBORETUM, 1199 9th Ave., (at Lincoln Way), San Francisco, CA 94122-2370. Tel.: 415-661-1316. Fax: 415-661-3539.
E-mail: info@sfbotanicalgarden.org
Web Site: www.sfbotanicalgarden.org
Founded: 1937.
Congressional District: 5
Key Personnel: Dir., Brent Dennis; Pres., Ann Cameron; Supt., Susan Nervo; Exec. Dir., Michael McKechnie; Plant Collections Mgr., Mona Bourell; Volunteer Coord., Tom Laursen; Head Librarian, Barbara Pitschel; Asst. Librarian, Brandy Kuhl; Children's Education, Annette Huddle; Shop Mgr., Dennis Guttman; Membership, Stephanie Perez; Adult Education, Fred Bove.
Personnel Profile: Full-Time Paid 27; Part-Time Paid 7; Part-Time Volunteers 300; Interns 16.
Governing Authority: municipal. A branch of the Recreation & Park Dept. of the City & County of San Francisco, McLaren Lodge, Golden Gate Park, CA 94117. Subsidiary Institution: San Francisco Botanical Garden Society. Tax-exempt.
Institution Type/Description: Arboretum & Botanical Gardens.
Collections: plants from Mediterranean climates; Asiatic magnolias, Vireya rhododendrons; conifers; California natives; primitive plants; fragrance garden for blind; demonstration garden; Mexican cloud forest; periodicals.
Facilities: 25,000-vol. library; 2 lecture rooms; auditorium; 55-acres of gardens.
Activities: guided tours; lecture series; outdoor education school program, elementary through college; work study program; docent training program; training programs for professional museum workers; horticultural therapy workshops to train activity directors & volunteers.
Publications: garden guide; plant guide list; newsletter; self-guiding pamphlets.
Hours & Admission Prices: Mon.-Fri. 8-4:30, Sat.-Sun. & holidays 10-5. No charge; donations requested. &
Attendance: 500,000 (estimated)
Membership: Avid Gardener $60; Garden Friend $75; Garden Lover $125; Garden Steward $250; Garden Patron $500; Strybing Circle $1,000.

SAN FRANCISCO CAMERAWORK, 1011 Market St., 2nd Fl., San Francisco, CA 94103. Tel.: 415-487-1011.
E-mail: info@sfcamerawork.org
Web Site: www.sfcamerawork.org
Founded: 1974.
Congressional District: 6
Key Personnel: Dir., Chuck Mobley; Pres., Stephen Vance; Gallery Mgr., Jennifer Jordan.
Personnel Profile: Full-Time Paid 3; Part-Time Paid 2; Part-Time Volunteers 20; Interns 10.
Governing Authority: nonprofit organization. Tax-exempt: 501(c)(3).
Institution Type/Description: Photography Museum.
Collections: photographic reference library.
Facilities: library of artists' books & catalogs available for research on premises. Museum-related items for sale.

Activities: ongoing exhibitions program; temporary & traveling exhibitions; lectures& workshops; critique sessions; video & film screenings; book signing events.
Publications: biannual; exhibition catalogs.
Hours & Admission Prices: Tues.-Sat. 12-5. No charge; donations accepted. Closed national holidays.
Attendance: 30,000 (estimated)
Membership: Senior & Student $30; Subscribing Member $50; Household $85; Sponsor $200; Collector $350 & up.

SAN FRANCISCO FIRE DEPARTMENT MUSEUM, 655 Presidio Ave., San Francisco, CA 94115-2424. Tel.: 415-558-3546 & 563-4630.
E-mail: sffdhs@sffiremuseum.org
Web Site: www.sffiremuseum.org
Founded: 1964.
Congressional District: 5
Key Personnel: Dir., Paul L. Barry; Chief Dept., Joanne Hayes-White.
Governing Authority: municipal. Sponsored by: San Francisco Fire Dept. Historical Society. Tax-exempt: 501(c)(3).
Institution Type/Description: Fire Museum:
Collections: artifacts, memorabilia & apparatus relating to the firefighting history of San Francisco.
Research Fields: San Francisco Fire Dept. history.
Activities: firemen's musters throughout the west; special demonstrations.
Publications: newsletter, St. Francis Hook & Ladder Society.
Hours & Admission Prices: Thurs.-Sun. 1-4 by appointment. No charge; donations accepted. &
Membership: Student & Senior Citizen $15; General $25; Business & Professional $50; Life $1,000 & up.

* **SAN FRANCISCO MARITIME NATIONAL HISTORICAL PARK,** Fort Mason Ctr., Bldg. E, San Francisco, CA 94123. Tel.: 415-561-7000. Fax: 415-556-1624.
E-mail: lynn_cullivan@nps.gov
Web Site: www.nps.gov/safr
Formerly: San Francisco Maritime Museum
Founded: 1951.
Congressional District: 8
Key Personnel: Supt. & Dir., Craig Kenkel; Chief Cultural Resources, Robbyn Jackson; Acting Chief Facilities & Maintenance, Phil Erwin; Acting Chief Interpretation, John Cunnane; Chief Administration, Shelley Neidernhofer; Cur. Maritime History, Steve Canright; Cur. Small Craft, Bill Doll; Cur. Exhibits, Richard Everett; Mgr. Collections, Keri Koehler; Public Affairs Officer, Lynn Cullivan; Coord. (V), Terry Dorman.
Personnel Profile: Full-Time Paid 103; Part-Time Volunteers 500; Interns 2.
Governing Authority: federal. Parent Institution: National Park Service, Dept. of the Interior. Tax-exempt.
Institution Type/Description: Maritime Museum Complex.
Collections: over 5 million archive materials; 35,000 artifacts; 250,000 photographs & negatives of ships & shipping ports, primarily West Coast; 120,000 sheets of ship plans; 4,000 log books & charts; manuscripts; scrimshaw; paintings & other fine arts & decorative arts; ship models; 36,000 books & periodicals, tape recordings of interviews with seafaring men & ship owners; artifacts from historic vessels; steam machinery; small craft. Historic Ships: Balclutha, square-rigged ship; C.A. Thayer, schooner; Eureka, walking-beam ferry; Hercules, steam tug; Alma, scow schooner; Eppleton Hall, paddle-wheel tug, located at Hyde Street Pier; Clipper-Bowed Monterey, fishing boat.
Research Fields: maritime history with emphasis on West Coast & Pacific Area; maritime culture, its lifestyles & folkways which are unique and set its men & women apart from the majority of society.
Facilities: bayfront recreation area.
Activities: permanent exhibits; special exhibits; lectures; living history program; special tours; environmental living program for school children; volunteer program; small boat restoration program; films.
Publications: brochure, quarterly park newsletter; monthly e-newsletter.
Hours & Admission Prices: Museum: daily 10-4. No charge. Visitor Center: daily 9:30-5. Hyde Street Pier: daily 9:30-4:30. No charge. Historic Ships: daily 9:30-4:30. Adults 17 & over $5, children & federal pass holders no charge. &
Attendance: 4,224,898 (accurate)

SAN FRANCISCO MUSEUM AT THE MINT, 5th St. at Mission St., San Francisco, CA 94103. Mailing Address: P.O. Box 420470, San Francisco, CA 94142-0470. Tel.: 415-537-1105. Fax: 415-537-1108. Facebook: San Francisco Museum at the Mint.
E-mail: info@sfhistory.org
Web Site: www.theoldmint.org

Institution Type/Description: Historic Building: housed in the former United States Mint; built in 1874. A National Historic Landmark.
Collections: local history; hands-on exhibitions; photographs.
Activities: hands-on exhibitions; rental facilities.
Hours & Admission Prices: Call for hours.
Membership: Senior & Community Access $30; History Enthusiast $45; Dual Senior $50; History Enthusiast Dual $80; Sustaining $120; Contributing $275; Gold Circle $1,000; Gold Circle Dual $1,500.

*** SAN FRANCISCO MUSEUM OF MODERN ART, (M),** 151 Third St., San Francisco, CA 94103-3159. Tel.: 415-357-4000. Fax: 415-357-4037. TDD: 415-357-4154.
Web Site: www.sfmoma.org
Founded: 1935.
Congressional District: 8
Key Personnel: Chm., Charles Schwab; Dir., Neal Benezra; Deputy Dir. Collection & Exhibitions, Ruth Berson; Deputy Dir. & Admin. Finance, Ikuko Satoda; Dir. ISS, Leo Ballate; Head Registrar, Tina Garfinkel; Sr. Cur. Painting & Sculpture, Gary Garrels; Cur. Painting & Sculpture, Janet C. Bishop; Cur. Media Arts, Rudolf Frieling; Sr. Cur. Photography, Sandra S. Phillips; Cur. Architecture & Design, Henry Urbach; Dir. Collections & Conservation, Jill Sterrett; Dir. Museum Store, Jana Machin; Dir. Human Resources, Sanchie Fernandez.
Personnel Profile: Full-Time Paid 189; Part-Time Paid 49; Part-Time Volunteers 143; Interns 17.
Governing Authority: nonprofit organization. Tax-exempt: 501(c)(3).
Institution Type/Description: Art Museum.
Collections: early 20th-century to present international paintings, sculpture, graphics, photography, architecture, design, digital projects & media arts.
Research Fields: modern & contemporary paintings, sculpture, graphics, photography, architecture, design & media arts.
Facilities: 80,000-vol. library of modern art books, periodicals & exhibit catalogues available for use on premises by appointment; painting & works on paper conservation laboratory; classrooms; 299-seat theater; rental facilities; cafe. Art books, gifts & other museum-related items for sale.
Activities: guided docent tours; lectures; concerts; formally organized educational programs; inter-museum loan, permanent, temporary & traveling exhibitions; random-access (CD-ROM based) audio tours of permanent collection.
Publications: bimonthly exhibition & program guide; annual report; exhibition catalogs; Web based member newsletters; web site.
Hours & Admission Prices: Memorial Day-Labor Day Thurs. 10-9, Fri.-Tues. 10-6; Sept.-May Thurs. 11-9, Fri.-Tues. 11-6. Adults $15, senior citizens 62 & over and students with ID $9; discounts to AAM & ICOM members and on Thurs. 6-9; first Tues. of every month, AFMOMA & museum members, and children 12 & under accompanied by an adult no charge. Closed New Year's Day; Independence Day; Thanksgiving; Christmas. ᷣ
Attendance: 550,000 (estimated)
Membership: Senior $45; Dual Senior $65; Individual $80; Dual Family $105; Supporting $250; Contributing $500; Benefactor $1,000.

SAN FRANCISCO ZOO, 1 Zoo Rd., San Francisco, CA 94132. Tel.: 415-753-7080. Fax: 415-681-2039.
E-mail: webmaster@sfzoo.org
Web Site: www.sfzoo.org
Formerly: San Francisco Zoological Gardens
Founded: 1929.
Congressional District: 6
Key Personnel: Dir. Finance, Wayne Reading; Chm. (V), Dave Stanton; Pres. (V), Tanya Peterson; Veterinarian, Graham Cranford; Dir. Public Rels. & Mktg., Jill Lynch; Dir. Membership, Abbie Tuller; Dir. Animal Care & Conservation, David Bocian; Human Resources, Jean Brennan; Assoc. Cur., Corinne MacDonald.
Personnel Profile: Full-Time Paid 220; Part-Time Volunteers 350; Interns 15.
Governing Authority: Parent Institution: San Francisco Zoological Society. Tax-exempt.
Institution Type/Description: Zoo.
Collections: 210 wild species & 900 wild specimens of mammals, birds, amphibians & reptiles; 64 species & 6,500 specimens of invertebrates.
Research Fields: zoology; animal sciences.
Facilities: 600-vol. library of materials on zoology available on request; children's zoo; botanical garden; insect zoo. Gift items for sale.
Activities: guided tours; lectures; films; formally organized education programs for children, adults & undergraduate students; permanent & temporary exhibitions; zoomobile program; avian conservation program.
Publications: e-newsletter.
Hours & Admission Prices: Daily & holidays 10-5. Adults 15-64 $15, senior

citizens 65 & over $12, youth 4-14 $9; discounts to residents; AZA reciprocating zoos, children 3 & under, members & children's zoo no charge. ᷣ
Attendance: 1,000,000 (estimated)
Membership: Teacher & Student $45; Senior Citizen $60; Individual $75; Family $105; Patron $175; Supporting $250; Sustaining $500; Guardian $1,000.

THE SOCIETY OF CALIFORNIA PIONEERS, 300 Fourth St., San Francisco, CA 94107-1207. Tel.: 415-957-1849. Fax: 415-957-9858. Facebook: Society of California Pioneers.
E-mail: info@californiapioneers.org
Web Site: californiapioneers.org
Founded: 1850.
Key Personnel: Mgn. Dir., Mercedes M. Devine; Pres. (V), David Cebalo; Education & Exhibitions Coord., John Hogan; Membership & Registrar, Natasha Shannon; Bookkeeper, Lorna Buehler; Dir. Library & Archives, Patricia Keats; Collections Asst., Lacey Lieberthal; Admin. & Collectioins Asst., Michael M. Lange.
Personnel Profile: Part-Time Paid 7; Part-Time Volunteers 2; Interns 4.
Governing Authority: private; nonprofit organization.
Institution Type/Description: Art & History Museum.
Collections: California paintings, prints, drawings, artifacts, books, manuscripts, journals, newspapers, photographs & decorative arts.
Research Fields: California art, history & culture.
Facilities: 10,000-vol. library; educational facilities; 3,000 sq. ft. exhibit space; 25-seat theater. Museum-related items for sale.
Activities: concerts; films; formal education programs; guided tours; lectures; loan, temporary & traveling exhibitions. Annual Events: Discovery of Gold Party; President's Picnic; Shumate lecture; Holiday Party; Library Discovery Programs.
Publications: annual journal, The Pioneer; exhibition catalogues.
Hours & Admission Prices: Wed.-Fri. & 1st Sat. of month 10-4. Adults $3, senior citizens & students $2.50; discounts to Friends; members & children no charge. Closed New Year's Eve & Day; Martin Luther King Jr. Day; Presidents' Day; Memorial Day; Independence Day; Labor Day; Veterans Day; Thanksgiving & day after; Christmas Eve, Day & week. ᷣ
Attendance: 4,500 (estimated)
Membership: Student & Senior $25; Annual $65; Family $90; Contributor $150; Argonaut Circle $250; Barbary Coast $500; Comstock $1,000 & up.

SOUTHERN EXPOSURE GALLERY - SOEX, 3030 20th St., San Francisco, CA 94110. Tel.: 415-863-2141. Fax: 415-863-1841.
Web Site: www.soex.org
Key Personnel: Exec. Dir., Courtney Fink
Institution Type/Description: Art Gallery.
Collections: works by local, national & international artists.
Activities: educational programs; special events.
Hours & Admission Prices: Gallery: Tues.-Sat. 12-6. Office: Mon.-Fri. 10-6.
Membership: Artist & Student $30; Individual $50; Dual & Family $100.

STEINHART AQUARIUM, 55 Music Concourse Dr., San Francisco, CA 94118. Tel.: 415-379-5445. Fax: 415-379-5704.
E-mail: aquarium@calacademy.org
Web Site: www.calacademy.org/aquarium
Founded: 1923.
Personnel Profile: Full-Time Paid 31; Part-Time Paid 6; Part-Time Volunteers 125.
Governing Authority: Parent Institution: California Academy of Sciences. Tax-exempt.
Institution Type/Description: Aquarium.
Collections: over 38,000 live animals representing 900 species; penguins; sharks; stingrays; piranha; sea stars; eels; rockfish; herring; sea bass; Pacific octopus; alligators; spiders; snakes; chameleons; snapping turtles; reptiles; amphibians; insects; birds; butterflies.
Facilities: Gift items for sale.
Activities: hands-on exhibitions; educational programs; coral reef dive program; penguin feeding program.
Hours & Admission Prices: Mon.-Sat. 9:30-5, Sun. 11-5. Adults $29.95, seniors, students & youth 12-17 $24.95, children 4-11 $19.95; children 3 & under no charge. Closed Thanksgiving; Christmas. ᷣ
Attendance: 1,400,000 (accurate)

THE UNIVERSITY OF SAN FRANCISCO THACHER GALLERY, (M), Gleeson Library Geschke Center, 2130 Fulton St., San Francisco, CA 94117-1080. Tel.: 415-422-5178 & 2044.
E-mail: thachergallery@usfca.edu
Key Personnel: Dir. & Cur., Thomas Lucas, S.J.; Assoc. Dir., Glori Simmons
Institution Type/Description: Art Gallery.

Collections: sculpture; paintings; photographs.
Activities: lectures; educational programs; craft seminars; guided tours; special
  events.
Hours & Admission Prices: Call for hours.

**THE VIRTUAL MUSEUM OF THE CITY OF SAN FRAN-
  CISCO,** San Francisco, CA 94116. Mailing Address: PMB 423,
  945 Taraval St., San Francisco, CA 94116.
E-mail: curator@sfmuseum.org
Web Site: www.sfmuseum.org
Key Personnel: Cur., Gladys Hansen
Institution Type/Description: Virtual History Museum.
Collections: San Francisco history; photographs.
Facilities: Museum-related items for sale.
Hours & Admission Prices: Daily.

THE WALT DISNEY FAMILY MUSEUM, LLC, (M), 104
  Montgomery St., The Presidio of San Francisco, San Francisco, CA
  94129-1718. Tel.: 415-345-6800. Fax: 415-345-6896. Facebook:
  The Walt Disney Family Museum, LLC.
E-mail: info@wdfmuseum.org
Web Site: www.waltdisney.org
Founded: 2009.
Congressional District: 8
Key Personnel: Exec. Dir., Kirsten Komoroske; Dir. Collections, Michael
  Labrie; Facilities & Operations, Nancy Wolf; Museum Shop Mgr., Katy
  Dashiell.
Personnel Profile: Full-Time Paid 50; Part-Time Paid 10; Part-Time Volunteers
  130; Interns 4.
Governing Authority: private; nonprofit organization. Parent Institution: Walt
  Disney Family Foundation. Tax-exempt: 501(c)(3).
Institution Type/Description: Walt Disney Family History Museum.
Collections: Walt Disney's life, work & family history; animation development
  & art; 20th century theme parks; personal artifacts; period furnishings;
  family photographs & home movies; films; Oscar statuettes; drawings;
  animation cells; personal papers & belongings of Walt Disney; early movie
  posters.
Major Exhibits: Bruno Bozzetto: Animation, Maestro! (T), 11/21/13-4/7/14;
  Magic, Color, Flair: The World of Mary Blair, 3/13/14-7/14/14; The Art of
  Marc Davis, 4/3/14-10/15/14; Water to Paper, Paint to Sky: The Art of Tyrus
  Wong (T), 8/13/14-2/3/14.
Research Fields: Snow White and the Seven Dwarfs art; all animation artist
  (1901-1960); 1901-1960 art; Walt Disney life history; original plans &
  opening of Disneyland; oral history from Walt Disney's colleagues and
  employees.
Facilities: 114-seat large screen theater; 114-seat auditorium; two studios;
  restaurant. Museum-related items for sale.
Activities: animation classes; listening stations; videos; concerts; films; formal
  education programs for adults & children; guided tours; lectures; participatory, temporary, & traveling exhibits; theater.
Publications: quarterly member newsletter.
Hours & Admission Prices: Wed.-Mon. 10-6. Adults $20, seniors & students
  $15, children 6-17 $12.50; children under 6 & members no charge. &
Attendance: 120,000 (accurate)
Membership: Senior & Student $55; Individual $75; Dual $125; Family $160;
  Associate $250; Friend $500; Supporter $1,250.

WAX MUSEUM AT FISHERMAN'S WHARF, 145 Jefferson St.,
  Ste. 500, San Francisco, CA 94133-1295. Tel.: 800-439-4305.
E-mail: sales@waxmuseum.com
Web Site: www.waxmuseum.com
Founded: 1963.
Institution Type/Description: Wax Museum
Collections: over 250 wax figures.
Facilities: Museum-related items for sale.
Hours & Admission Prices: Daily 10-9. Adults $14, senior 55 & over and
  junior 12-17 $10, child 6-11 $7. &
Attendance: 436,000 (accurate)

WELLS FARGO HISTORY MUSEUM, Wells Fargo Bank, His-
  torical Services, 420 Montgomery St., MAC-A010-106, San Fran-
  cisco, CA 94163. Tel.: 415-396-2619 & 4157. Fax: 415-975-7430.
Founded: 1929.
Key Personnel: Mgr., Glen T. Myers; Cur., Joycee Wong.
Governing Authority: profit-making organization. Affiliated with Wells Fargo
  Bank.
Institution Type/Description: Company History Museum: located in Wells
  Fargo Bank's corporate headquarters.

Collections: Concord Stagecoach; Wells Fargo banking & transportation
  history; San Francisco & Gold Rush California exhibits; mining tools; gold
  specimens & coins; Wiltsee Collection of western stamps & postal history.
Activities: guided group tours; permanent & temporary exhibitions; imaginary
  rides on replica stagecoach; audiovisual programs.
Publications: scholarly pamphlets. History of Wells Fargo.
Hours & Admission Prices: Mon.-Fri. 9-5. No charge. Closed bank holidays.
  &

YERBA BUENA CENTER FOR THE ARTS, 701 Mission St., San
  Francisco, CA 94103-3138. Tel.: 415-978-2700 & ARTS (2787).
  Fax: 415-978-9635.
E-mail: kbudas@ybca.org
Web Site: www.ybca.org
Founded: 1986.
Key Personnel: Exec. Dir., Debra Cullinan; Pres. Bd., Diana Cohn; Dir.
  Performing Arts, Marc Bamuthi Joseph; Dir. Visual Arts, Betti-Sue Hertz;
  Dir. Mktg., Kathy Budas; Film Video Cur., Joel Shepard.
Personnel Profile: Full-Time Paid 50; Part-Time Paid 15; Part-Time Volunteers
  50; Interns 5.
Governing Authority: private; nonprofit organization. Tax-exempt: 501(c)(3).
Institution Type/Description: Arts Center.
Collections: visual art; performing arts; film.
Facilities: 10,000 sq. ft. exhibit space; 750, 600 & 100-seat theaters.
Activities: concerts; dance recitals; docent program; films; lectures; theater;
  exhibitions; music; dance; film/video; new media (hi-tech installations);
  educational programs.
Publications: exhibition catalogues.
Hours & Admission Prices: 1st Tues. each month & Thurs.-Sat. 12-8, Sun.
  12-6. Adults $7, senior citizens, students, teachers & children $5; members
  no charge. &
Attendance: 200,000 (accurate)
Membership: Basic $65; Household $85; Discoverer $165; Groundbreaker
  $300; Curator's Circle $500; Director's Forum $1,000.

San Gabriel

**HAYES HOUSE & MUSEUM - SAN GABRIEL HISTORICAL
  ASSOCIATION,** 546 W. Broadway, San Gabriel, CA 91776. Tel.:
  626-308-3223.
Institution Type/Description: History Museum.
Collections: local history & culture; period furnishings; personal artifacts;
  photographs.
Hours & Admission Prices: 1st Sat. of the month 1-4; groups by appointment.

SAN GABRIEL MISSION MUSEUM, 428 S. Mission Dr., San
  Gabriel, CA 91776-1252. Tel.: 626-457-3035. Fax: 626-282-5308.
E-mail: alsgm1@aol.com
Web Site: www.sangabrielmission.church.org
Founded: 1771.
Congressional District: 29
Key Personnel: Dir. & C.E.O., Rev. Bruce Wellems, C.M.F.; Business Mgr.,
  Alfred Sanchez; Cur., Helen Nelson.
Personnel Profile: Full-Time Paid 1; Part-Time Paid 3; Part-Time Volunteers
  14.
Governing Authority: church. Parent Institution: San Gabriel Mission Catholic
  Church. Tax-exempt.
Institution Type/Description: Religious Museum: fourth in a chain of 21
  California Missions founded Sept. 8, 1771, site of San Gabriel Mission.
Collections: Old Mission church; books; artifacts; furnishings of old mission;
  manuscripts; paintings; statues; photographs; plants.
Facilities: Museum-related items for sale.
Activities: self-guided tours; formally organized education programs for
  groups by appointment; permanent exhibitions.
Hours & Admission Prices: Daily 9-4:30. Adults $5, seniors 62 & over $4,
  youth 6-17 $2; discounts to inner schools; no charge to clergy & religious.
  Closed Good Friday noon, Easter, Independence Day, Thanksgiving,
  Christmas.
Attendance: 50,000 (estimated)

San Jacinto

MT. SAN JACINTO COLLEGE FINE ART GALLERY, 1499 N.
  State St., San Jacinto, CA 92583. Tel.: 951-487-3585 & 3586.
E-mail: bdillaway@msjc.edu
Web Site: www.msjc.edu/artgallery
Key Personnel: Dir., Brandelyn Dillaway
Institution Type/Description: Art Gallery.
Collections: paintings; sculpture; photographs.

Activities: special events. Museum Sponsors: Annual Holiday Sale in December.
Hours & Admission Prices: Mon.-Thurs. 8-4; other times by appointment.

SAN JACINTO MUSEUM, 695 Ash St., San Jacinto, CA 92583. Mailing Address: P.O. Box 922, San Jacinto, CA 92581-0922. Tel.: 951-654-4952. Fax: 909-654-9270.
Web Site: www.ci.san-jacinto.ca.us
Founded: 1939.
Congressional District: 37
Key Personnel: Asst., Betty Jo Dunham.
Personnel Profile: Part-Time Paid 1; Part-Time Volunteers 6.
Governing Authority: municipal. Parent Institution: City of San Jacinto. Tax-exempt.
Institution Type/Description: History Museum.
Collections: Indian archaeology; geology; mineralogy; paleontology; period artifacts; San Jacinto & Hemet area history.
Research Fields: local history.
Activities: permanent & temporary historical exhibitions.
Hours & Admission Prices: Fri.-Sun. 11-4. No charge; donations accepted. Closed New Year's Day; Thanksgiving; Christmas. &
Attendance: 3,000 (accurate)
Membership: Single $10; Couples & Family $15; Organization $25.

San Jose

BRANDENBURG HISTORICAL GOLF MUSEUM, 23600 McKean Rd., San Jose, CA 95141. Mailing Address: 1122 Willow St. #200, San Jose, CA 95125. Tel.: 408-323-7814.
Founded: 1998.
Institution Type/Description: Golf Museum.
Collections: golf equipment & artifacts; photographs; trophies; golf clubs.
Hours & Admission Prices: Call for hours. No charge.
Attendance: 100,000 (estimated)

CHILDREN'S DISCOVERY MUSEUM OF SAN JOSE, (M), 180 Woz Way, San Jose, CA 95110-2780. Tel.: 408-298-5437. Fax: 408-298-6826.
E-mail: contactus@cdm.org
Web Site: www.cdm.org
Founded: 1982.
Congressional District: 16
Key Personnel: Exec. Dir., Marilee Jennings; Chm. Bd. of Dir. (V), Mark Garrett; Vice Chm. Bd. of Dir., Mark McCaffrey; Dir. Administration & Finance, Susan Clark; Dir. Education & Programs, Jenni Martin; Dir. Devel. & Mktg., Patrica Narciso; Dir. Information & Compliance, Cheryl Blumenthal.
Personnel Profile: Full-Time Paid 36; Part-Time Paid 57; Part-Time Volunteers 100.
Governing Authority: nonprofit organization. Legal Name: San Jose Children's Discovery Museum. Tax-exempt.
Institution Type/Description: Children's Museum: housed in Ricardo Legorreta-designed structure.
Collections: San Jose regional history; health & physiognomy; energy; environmental sciences; cultural arts; early childhood education; communications; banking; postal service; media studio; digital communications; city & historical vehicles.
Research Fields: related to collections.
Facilities: 33,500 sq. ft. exhibit space; 120-seat black box theatre; 48-seat cafe; outdoor garden & amphitheatre.
Activities: films; concerts; theatre; school visitation program; outreach programs to schools; participatory exhibits; parent & child workshops; creative arts; environmental & science programs.
Publications: bimonthly calendar, Discovery Dates; annual report.
Hours & Admission Prices: Tues.-Sat. 10-5, Sun. 12-5; call for holiday hours. Admission $12, senior citizens 60 & up $11; members and infants under one no charge. Closed Thanksgiving and Christmas. &
Attendance: 295,062 (accurate)
Membership: Grandparent & Family $120; Explorer's Circle $175; Developer's Circle $250; Partner's Circle $500; Leader's Circle $1,000.

CHINESE CULTURAL GARDEN/OVERFELT GARDENS, 368 Educational Park Dr., San Jose, CA 95133-1711. Mailing Address: Overfelt Gardens c/o Prusch Farms Park, 647 S. King Rd., San Jose, CA 95116-3557. Tel.: 408-251-3323. Fax: 408-251-2865.
Web Site: chineseculturalgarden.org
Key Personnel: Program Dir., Sylvia Lowe; Park Ranger, Will Bick; Gardener, Sheila Strand; Docent, Pauline Lowe

Institution Type/Description: General Museum.
Collections: sculptures; Chinese culture & history; gardens.
Facilities: 6 acres.
Hours & Admission Prices: Tues.-Sun. 10 to sunset. No charge.

HAPPY HOLLOW PARK & ZOO, 1300 Senter Rd., San Jose, CA 95112-2520. Tel.: 408-277-4193 & 3999. Fax: 408-975-9369.
Web Site: www.hhpz.org
Founded: 1967.
Congressional District: 16
Key Personnel: C.E.O. & Zoo Dir., Gregg Owens.
Personnel Profile: Full-Time Paid 12; Part-Time Paid 5; Part-Time Volunteers 45.
Governing Authority: Parent Institution: City of San Jose. Subsidiary Institution: Happy Hollow Corporation.
Institution Type/Description: Zoo.
Collections: live animals; birds.
Facilities: snack bar; education classroom.
Activities: rides. Museum Sponsors: Feasts With the Beasts; Exhibit Grand Openings; Annual Membership Party; Earth Day; Halloween event.
Publications: quarterly newsletter.
Hours & Admission Prices: Call for hours. Admission $12, seniors 70 & over $8; discounts to groups; children one & under no charge. &
Attendance: 337,123 (accurate)
Membership: Sr. Student $20; Individual $30; Family $60; Donor $100; Patron $250; Sustaining $500.

*** HISTORY SAN JOSE, (M),** 1650 Senter Rd., San Jose, CA 95112-2599. Tel.: 408-287-2290. Fax: 408-287-2291.
E-mail: abray@historysanjose.org
Web Site: www.historysanjose.org
Founded: 1971.
Congressional District: 13
Key Personnel: Pres. & C.E.O., Alida Bray; Chm. Bd., Jack Frazer; Dir. Operations, Barbara Johnson; Dir. Devel., Michelle Powers; Archivist, James Reed; Events/Retail Coord., Juanita Lara; Finance Mgr., Maggie Williams; Facilities Mgr., Mike Brag; Sr. Accountant, Rene Foronda; Dir. Education, Barbara Johnston; Exec. Asst., Dayna Grabeklis.
Personnel Profile: Full-Time Paid 9; Part-Time Paid 19; Part-Time Volunteers 200.
Governing Authority: nonprofit organization. Tax-exempt: 501(c)(3).
Institution Type/Description: History Museum.
Collections: local historic furniture; household accessories; textiles; archaeology; vehicles; business equipment; archives of city records, maps, local history documents & photographs; medical instruments; California art; 16 historic structures; 10 recreated structures; Chinese temple; Portuguese Imperio; operating trolley restoration barn. Historic Houses: 1797 Peralta Adobe; 1855 Fallon House.
Major Exhibits: Shirlie Montgomery Photography Exhibit, 3/14-12/14; Bicycling of San Jose, 4/14-4/15.
Research Fields: local & regional history.
Facilities: cafe. Gift items for sale.
Activities: guided tours; lectures; formally organized educational programs; docent program; research activities; history makers; summer camp; adult literacy program; rental facilities. Annual Events: Spirit of 4S; Carshows; Bicycle Show; Public Archaeology Days; Holiday Tea; Holiday Children's Programming.; Heritage Days; Portuguese Festival.
Hours & Admission Prices: History Park in Kelley Park: Tues.-Sun. 11-5; Retail Store: Tues.-Sun. 11-4. Special Programming May-Oct.: adults $8, senior & students $5; Nov.-April no charge; members, children under 6 & Tues.-Fri. no charge. Closed New Year's Day; Martin Luther King Jr. Day; Presidents' Day; Memorial Day; Independence Day; Labor Day; Thanksgiving & day after; Christmas Eve & Day. &
Attendance: 100,000 (accurate)
Membership: Family $50; Associate $100; Partner $250; Advocate $500; Patron $1,000. Corporate: Concierge $500; Associate $1,000; Partner $2,500; Advocate $5,000; Patron $10,000 & up.

NATALIE & JAMES THOMPSON ART GALLERY, Department of Art and Art History, San Jose, CA 95192-0089. Tel.: 408-924-4320 & 4327. Fax: 408-924-4326.
E-mail: thompsongallery@sjsu.edu
Web Site: www.sjsu.edu
Formerly: San Jose State University Art Galleries
Founded: 1959.
Key Personnel: Gallery Dir., Jo Farb Hernandez; Asst. Dir., Theta Belcher.
Personnel Profile: Full-Time Paid 2; Part-Time Paid 6; Part-Time Volunteers 2; Interns 2.

Governing Authority: university. Affiliated with the State of California. Parent Institution: San Jose State University. Tax-exempt.
Institution Type/Description: University Art Gallery.
Collections: paintings; photographs; sculpture; prints.
Research Fields: exhibitions & projects; fine arts; traditional arts; design.
Facilities: lecture hall.
Activities: gallery talks; formally organized education programs for under-graduate & graduate college students; temporary exhibitions; visiting artists program; weekly lecture series; class presentations; workshops.
Publications: catalogues; brochures; monthly announcements & newsletters.
Hours & Admission Prices: Sept.-May Mon. & Wed.-Fri. 11-4, Tues. 11-4 & 6-7:30. No charge; donations accepted. Closed semester breaks; national holidays. &
Attendance: 15,000 (accurate)

NEW ALMADEN QUICKSILVER MINING MUSEUM, 21350 Almaden Rd., San Jose, CA 95120-4306. Tel.: 408-323-1107. Fax: 408-323-0943.
Founded: 1998.
Key Personnel: Museum Mgr. & Museum Shop Mgr., Julie Lee.
Personnel Profile: Full-Time Paid 3; Part-Time Volunteers 15.
Governing Authority: Parent Institution: Santa Clara County Parks.
Institution Type/Description: Mining Museum.
Collections: history of mercury mining; area mining communities; mine shaft replica; Cornish, Mexican & Chinese mining families.
Hours & Admission Prices: July-Aug. Fri.-Sun. 10-4; Sept.-June Fri. 12-4, Sat.-Sun. 10-4; other times by appointment. No charge; donations accepted. Closed New Year's Day; Thanksgiving; Christmas. &
Attendance: 7,500 (estimated)

PERALTA ADOBE FALLON HOUSE HISTORIC SITE, 184 W. Saint John St., San Jose, CA 95110-2434. Mailing Address: c/o History San Jose, 1650 Senter Rd., San Jose, CA 95112. Tel.: 408-287-2290. Fax: 408-287-2291.
Web Site: www.historysanjose.org/wp/plan-your-visit/peralta-fallon-historic-site
Institution Type/Description: Historic Building: built in 1797.
Collections: structure; furnishings; outside working oven; daily life during the Spanish & Mexican periods.
Hours & Admission Prices: Thurs.-Sun. 11-4. Adults $6, seniors $5, youth $3.

PORTUGUESE HISTORICAL MUSEUM, 1650 Senter Rd., San Jose, CA 95112-2599. Mailing Address: Portuguese Heritage Society of California, P.O. Box 18277, San Jose, CA 95158. Tel.: 650-964-0406.
E-mail: aldutra@aol.com
Web Site: www.portuguesemuseum.org
Founded: 1997.
Key Personnel: Bd. Pres., Al Dutra; Pres., Joseph Machado.
Personnel Profile: Part-Time Volunteers 25.
Governing Authority: Tax-exempt.
Institution Type/Description: Cultural History Museum.
Collections: Portuguese heritage & culture.
Hours & Admission Prices: Sat.-Sun. 1-5; other times by appointment. &
Attendance: 7,000 (estimated)

ROSICRUCIAN EGYPTIAN MUSEUM, (M), 1660 Park Ave., San Jose, CA 95191. Mailing Address: 1342 Naglee Ave., San Jose, CA 95191. Tel.: 408-947-3600 & 3636. Fax: 408-947-3638.
E-mail: curator@egyptianmuseum.org
Web Site: www.egyptianmuseum.org
Founded: 1929.
Congressional District: 13
Key Personnel: Dir., Julie Scott; Mgr., Nancy Leonard; Cur., Steven Armstrong.
Personnel Profile: Full-Time Paid 11; Part-Time Volunteers 16.
Governing Authority: society. Parent Institution: Rosicrucian Order AMORC. Tax-exempt: 501(c)(3).
Institution Type/Description: Egyptian History Museum.
Collections: Assyrian, Babylonian & Egyptian antiquities; astronomy; paintings; sculpture; graphics; archaeology; Coptic textiles; utensils; scarabs; jewelry; statuary; amulets; human & animal mummies; replica of an Egyptian rock tomb; 18th Dynasty Egyptian garden.
Research Fields: Egyptology.
Facilities: art gallery; temporary & permanent exhibits. Egyptian arts, crafts, statuary & books for sale.
Activities: guided tours; lectures; films; gallery talks; special tours for the blind & school groups; workshops.

Publications: catalogue, Treasures of the Rosicrucian Egyptian Museum; Hieroglyphic Coloring Book; catalogue, Women of the Nile; teacher's guide; postcard book of collection; blog, The Scribe.
Hours & Admission Prices: Mon.-Fri. 9-5, Sat.-Sun. 11-6. Adults $9, students with ID & senior citizens $7, children 5-10 $5; discounts to military, KQED, AAM & AAA members; children 4 & under & members no charge. Planetarium Showtimes: Mon.-Fri. 2, Sat.-Sun. 2 & 3:30. Closed major holidays.
Attendance: 100,000 (accurate)
Membership: Senior, Student & Teacher $20; Individual $35; Family $50; Friends of Anubis $75; Friends of NUT $125; Friends of Sekhmet $250; Friends of Isis $500; Friends of Osiris $1,000; Friends of RA $2,500.

SAN JOSE FIRE MUSEUM, 1661 Senter Rd., Bldg. D1, San Jose, CA 95112. Tel.: 408-998-6184. Fax: 408-287-0401.
E-mail: macmhalain@yahoo.com
Web Site: www.sjfivemuseum.org
Founded: 2001.
Congressional District: 19
Key Personnel: Pres. (V), John A. McMillan.
Personnel Profile: Part-Time Volunteers 20.
Volunteer Hours: 2,500
Governing Authority: Tax-exempt: 501(c)(3).
Institution Type/Description: Firefighting History Museum.
Collections: San Jose firefighting history & equipment; early fire trucks, tools & apparatus; fire memorabilia; photographs; journals & constitutions.
Major Exhibits: Antique Apparatus Display, 1/14-12/14; 1800s Lug Books, 1/14-12/14; Fire Helmets, 1/14-12/14; Historical Photographs, 1/14-12/14; Antique Equipment, 1/14-12/14.
Activities: educational programs.
Hours & Admission Prices: By appointment. No charge; donations accepted.
Attendance: 800 (estimated)

SAN JOSE INSTITUTE OF CONTEMPORARY ART, 560 S. 1st St., San Jose, CA 95113-2806. Tel.: 408-283-8155. Fax: 408-283-8157.
E-mail: info@sjica.org
Web Site: www.sjica.org
Founded: 1980.
Congressional District: 16
Key Personnel: Dir., Cathy Kimball.
Personnel Profile: Full-Time Paid 5; Part-Time Paid 3; Interns 7.
Governing Authority: nonprofit organization. Tax-exempt.
Institution Type/Description: Art Gallery.
Collections: works by contemporary artists.
Facilities: rental facilities.
Activities: educational programs; workshops.
Hours & Admission Prices: Tues.-Fri. 10-5, Sat. 12-5. No charge; donations accepted.

* **SAN JOSE MUSEUM OF ART, (M),** 110 S. Market St., San Jose, CA 95113-2383. Tel.: 408-271-6840 & 6880. Fax: 408-294-2977.
E-mail: info@sjmusart.org
Web Site: www.sanjosemuseumofart.org
Founded: 1969.
Congressional District: 16
Key Personnel: Exec. Dir., Susan Krane; Pres. (V), Hildy Shandell; Deputy Dir. Operations, Deborah Norberg; Dir. Devel., Lisa James; Treas., William Faulkner; Dir. Mktg. & Communications, Sherrill Ingalls; Cur. Education, Lucy Larson; Sr. Cur., Susan Leask; Asst. Cur., Rory Padeken; Chief Design & Installation, Richard Karson; Facilities Mgr., John Renzel; Museum Store Mgr., Pat Downward; Events Mgr., LT Beeton.
Personnel Profile: Full-Time Paid 24; Part-Time Paid 27; Part-Time Volunteers 169; Interns 3.
Operating Expenses: 5,324,769
Operating Income: 6,778,203
Governing Authority: nonprofit organization. Parent Institution: San Jose Museum of Art Association. Tax-exempt: 501(c)(3).
Institution Type/Description: Contemporary Art Museum.
Collections: 20th and 21st-century art.
Major Exhibits: Food, Culture, Creativity, 9/26/13-3/14; Hidden Heroes: The Genius of Everyday Things (T), 10/13-2/14; Jitish Kallat: Epilogue, 10/13-4/14; Judy Pfaff, 10/17/13-2/2/14; Around the Table: Food, Creativity, Community, 11/9/13-4/20/14; Initial Public Offering: New Works from SJMA's Collection, 2/20/14-8/30/14; Legacy: The Emily Fisher Landam Collection (T), 5/15/14-9/15/14; David Levinthal: Photographs, 6/5/14-11/14.

Research Fields: 20th & 21st-century art with special emphasis on contemporary art.
Facilities: 78,000 sq. ft. exhibit space; two sculpture courts; cafe. Museum-related items for sale.
Activities: docent tours; lectures; gallery talks; local school art program; art classes for children & adults; poetry readings; concerts; art demonstrations; rental facilities. Annual Events: quarterly signed exhibition tours; free hands-on workshops; community days.
Publications: triannual newsletter.
Hours & Admission Prices: Tues.-Sun.11-5. Adults $8; discounts to AAM & ICOM members; members no charge. Closed New Year's Day; Thanksgiving; Christmas. &
Attendance: 91,744 (accurate)
Membership: Individual $50; Young Professional $65; Dual & Family $75; Art Advocate $150; Contributor $300; Benefactor $600; Collector $1,000; Curator $2,500.

SAN JOSE MUSEUM OF QUILTS & TEXTILES, 520 S. First St., San Jose, CA 95113-2806. Tel.: 408-971-0323, ext. 16. Fax: 408-971-7226.
E-mail: jane@sjquiltmuseum.org
Web Site: www.sjquiltmuseum.org
Founded: 1977.
Congressional District: 10
Key Personnel: Bd. Pres., Marie Strait; Dir., Christine Jeffers; Museum Shop Mgr., Sofia Motamedi; Educational Outreach, Sylvia Carroll.
Personnel Profile: Full-Time Paid 2; Full-Time Volunteers 1; Part-Time Paid 5; Part-Time Volunteers 200; Interns 1.
Governing Authority: board of directors. Tax-exempt: 501(c)(3).
Institution Type/Description: Quilt & Textile Museum
Collections: over 500 quilts & textiles from 1850 to present.
Research Fields: contemporary fiber art & quilt history.
Facilities: Museum-related items for sale.
Activities: guided tours; lectures; organized education programs for children; docent program; participatory, loan, temporary & traveling exhibitions; education outreach to local 2nd & 5th grade students, call museum for schedule.
Publications: quarterly newsletter, Connections.
Hours & Admission Prices: Tues.-Wed. & Fri.-Sun. 10-5, Thurs. 10-8. Adults $8; discounts to AAM members; children 12 & under and members no charge. Closed major holidays. &
Attendance: 20,000 (estimated)
Membership: Individual $40; Family $55; Patron $100; Sponsor $250; Benefactor $500.

THE TECH MUSEUM, 201 S. Market St., San Jose, CA 95113-2008. Tel.: 408-294-TECH. Fax: 408-279-7167.
E-mail: info@thetech.org
Web Site: www.thetech.org
Formerly: The Tech Museum of Innovation
Founded: 1990.
Key Personnel: C.E.O. & Pres., Peter Friess; Chm., Ann Bowers; Mktg. & Public Rels., Elizabeth Williams; Mktg. & Public Rels., Roqua Montez; Operations, Bill Bailor.
Personnel Profile: Full-Time Paid 90; Part-Time Paid 44; Part-Time Volunteers 440.
Governing Authority: Tax-exempt.
Institution Type/Description: Science & Technology Museum.
Collections: technology & how it affects our lives; innovation; the internet; the human body; exploration; hands-on exhibits.
Facilities: 295-seat IMAX dome theater; center for learning; educational program area; science & technology labs; 80-seat cafe. Museum-related items for sale.
Activities: hands-on learning workshops.
Publications: Innovation calendar for members only.
Hours & Admission Prices: Daily 10-5. Admission $10; members no charge. Closed Christmas. &
Attendance: 500,000 (accurate)
Membership: Individual $25; Family $80; Sustaining $250; Explorer $500; Leadership Circle $1,000 & up.

YOUTH SCIENCE INSTITUTE, Penitencio Creek Rd., San Jose, CA 95127. Mailing Address: 296 Garden Hill Dr., Los Gatos, CA 95032-7669. Tel.: 408-258-4322. Fax: 408-358-3683.
E-mail: info@ysi-ca.org
Web Site: www.ysi-ca.org
Founded: 1953.
Congressional District: 10, 12, 13
Key Personnel: Exec. Dir., Susane Mulcahy; Pres. (V), Mark Lohbeck;

Administrative Asst., Marion Blair; YSI Center Mgr. Sanborn & Animal Cur. Sanborn, Laura Weiss; Animal Cur. Alum Rock, Dorothy Johnson; Dir. Education, Bonnie Lemat.
Personnel Profile: Full-Time Paid 7; Part-Time Paid 30; Part-Time Volunteers 25.
Governing Authority: nonprofit organization. Branch Museums: Alum Rock Nature Center, 16260 Alum Rock Ave., San Jose, CA 95127. Tel.: 408-258-4322; Vasona Park Nature Center, 296 Garden Hill Dr., Los Gatos, CA 95032. Tel.: 408-356-4945; Sanborn Park Nature Center, 16055 Sanborn Rd., Saratoga, CA 95070. Tel.: 408-867-6940. Tax-exempt: 501(c)(3).
Institution Type/Description: Children's Natural History Museum & Nature Center.
Collections: Alum Rock Center: houses a live animal collection of vertebrate species. Holmes bird collection: 300 species & hands-on exhibits. Vasona Discovery Center: physics, watershed, live animals; hands-on exhibits. Sanborn Discovery Center: insect zoo, live amphibian & reptile collection, rock & mineral collection; hands-on natural history displays.
Facilities: 200-vol. library of science material available for inter-library loan & to the public; aquarium; nature & conservation center; educational facilities; 1,000 sq. ft. exhibit space at each center. Science games & toys for sale.
Activities: lectures; organized educational programs; training programs for animal caretakers; docent program; participatory exhibits. Museum Sponsors: Wildlife Festival & Insect Fair.
Publications: quarterly newsletter; summer programs & general brochures; curriculum guide to the Viola Anderson Native Plant Trail; checklist of birds of Alum Rock Park.
Hours & Admission Prices: Summer: Tues.-Sun. 12-4:30; School Groups: Mon.-Fri. 9-5. Alum Rock: adults $1, children $.50. Sanborn & Vasona locations donations accepted. &
Attendance: 70,000 (estimated)

San Juan Bautista

OLD MISSION SAN JUAN BAUTISTA VISITOR CENTER, 406 Second St., San Juan Bautista, CA 95045. Mailing Address: P.O. Box 400, San Juan Bautista, CA 95045-0400. Tel.: 831-623-2127 & 4528. Fax: 831-623-2433.
E-mail: ann@oldmissionsjb.org
Web Site: www.oldmissionsjb.org
Founded: 1797.
Congressional District: 25
Key Personnel: Pastor, Rev. Jim Henry; Business Mgr., Ann McMahon; Visitor Center & Museum Shop Mgr., Ana Silva.
Governing Authority: church.
Institution Type/Description: Historic Site: one of the original California missions founded by Franciscan Fathers.
Collections: adobe walls with original Indian decorations; local Indian, Spanish, Mexican & religious artifacts; archives; convent wing containing Spanish colonial period artifacts; chairs; candelabra; figures of saints; baptismal fonts.
Research Fields: mission & local history.
Facilities: Museum-related items for sale.
Activities: self-guided tour.
Publications: Guide to Old Mission SJB.
Hours & Admission Prices: Office: Mon.-Fri. 9-12 & 1-4. Gift Shop/Visitor Center: daily 9:30-4:30. Suggested Donations: adults $2, students & seniors $1. Closed New Year's Day; Good Friday; Thanksgiving; Christmas. &

SAN JUAN BAUTISTA HISTORICAL SOCIETY LUCK MUSEUM, Monterey St., San Juan Bautista, CA 95045. Mailing Address: P.O. Box 1, San Juan Bautista, CA 95045. Tel.: 831-623-2001.
Institution Type/Description: Historical Society Museum: housed in the former filling station owned by Carl Luck.
Collections: local history & culture; photographs period artifacts.
Hours & Admission Prices: Sat. 10-4; other times by appointment.

SAN JUAN BAUTISTA STATE HISTORIC PARK, 2nd St., Washington & Mariposa Sts., San Juan Bautista, CA 95045-0787. Mailing Address: P.O. Box 787, San Juan Bautista, CA 95045-0787. Tel.: 831-623-4881 & 4526. Fax: 831-623-4612.
Web Site: www.parks.ca.gov/sjbshp
Founded: 1933.
Congressional District: 16
Key Personnel: Supt., Stuart Organo; Pres. (V), Bob Cable; Chm. (V), Nikki Combs; Cur., Kris N. Quist; Chief Interpreter, Pat Clark-Gray; Museum Shop Mgr., Joanna McMahon.

Personnel Profile: Full-Time Paid 7; Part-Time Paid 6; Part-Time Volunteers 42.
Volunteer Hours: 6,666
Governing Authority: state. Parent Institution: California Dept. of Parks & Recreation, P.O. Box 942896, Sacramento, CA 95811. Tax-exempt.
Institution Type/Description: State Historic Park: located next to a Franciscan mission.
Collections: furniture; house furnishings; wagons & buggies; blacksmith tools & equipment; period clothing; 1880 fire wagon. Historic Buildings: Castro/Breen Adobe; Plaza Hotel; Plaza Stable; Plaza Hall (Zanetta home); jail; settler's cabin.
Research Fields: Spanish, Mexican & early American California.
Facilities: rental facilities.
Activities: self-guided & guided tours by reservation; hornito & tortilla demonstrations; living history demonstrations; blacksmith demonstrations; video; cell phone tours. Museum Sponsors: Living History Day monthly; entertainment in Plaza Hotel Barroom in cooperation with the Plaza History Assoc.
Publications: general information brochure; video on the town of San Juan Bautista's History.
Hours & Admission Prices: Tues.-Sun. 10-4:30. Adults $3; children 16 & under no charge. Closed New Year's Day; Thanksgiving; Christmas. &
Attendance: 114,838 (accurate)
Membership: Individual $10; Family $15; Historical Association $25; Sustaining $50; Corporate $100.

San Juan Capistrano

MISSION SAN JUAN CAPISTRANO HISTORIC LANDMARK AND MUSEUM, 26801 Ortega Hwy., San Juan Capistrano, CA 92675-2601. Tel.: 949-234-1300. Fax: 949-493-8747.
Web Site: www.missionsjc.com
Congressional District: 48
Key Personnel: Exec. Dir., Mechelle Lawrence-Adams; Devel. & Mktg., Barb Beier; Museum Registrar, Jennifer Ring.
Personnel Profile: Full-Time Paid 40; Part-Time Paid 10; Part-Time Volunteers 100.
Governing Authority: Diocese of Orange County California; nonprofit. Tax-exempt.
Institution Type/Description: Historic & Archaeological Site and Museum.
Collections: historic buildings & artifacts including a 230 year old chapel; local history & culture; religious & liturgical artifacts.
Facilities: lecture & media room; educational gardens; 10.5 acres of grounds.
Activities: multi-language audio tours; special events; community activities; hands-on exhibits; children's education programs; docent programs; living history programs; summer camp programs; summer concert series.
Publications: quarterly newsletter, The Jewel.
Hours & Admission Prices: Daily 8:30-5. Adults $9, seniors 60 & over and children 4-11 $6; children 3 & under no charge. Closed Thanksgiving; Christmas.
Attendance: 330,000 (estimated)
Membership: Individual $35; Family $50 & up; Supporting $100; Supporting Plus $250; Patron $500.

San Leandro

SAN LEANDRO HISTORIC RAILWAY SOCIETY, 1302 Orchard Ave., San Leandro, CA 94577. Tel.: 510-569-2490.
Web Site: www.slhrs.org
Personnel Profile: Part-Time Volunteers 30.
Governing Authority: Tax-exempt.
Institution Type/Description: History Museum: housed in a Southern Pacific Railroad Station; c.1898.
Collections: depot history; HO scale model railroad; G scale model railroad; O scale model railroad; prototype railroad artifacts.
Activities: special events.
Hours & Admission Prices: Tues. 7:30pm-9pm, Sat. 9-1. No charge.
Attendance: 1,200 (estimated)
Membership: Individual $120.

San Luis Obispo

DALLIDET ADOBE AND GARDENS, 1185 Pacific St., San Luis Obispo, CA 93401-3301. Tel.: 805-543-6762. Fax: 805-783-2919.
Web Site: www.slochs.org/dalliet.asp
Key Personnel: Exec. Dir., Kimberly Alfaro
Institution Type/Description: History Museum.
Collections: furniture; paintings; personal artifacts.
Activities: rental facilities.
Hours & Admission Prices: Fri. 10-4, 2nd Sun. of month 1-4; call to confirm.

HISTORY CENTER OF SAN LUIS OBISPO COUNTY, 696 Monterey St., San Luis Obispo, CA 93401-3515. Tel.: 805-543-0638. Fax: 805-783-2919.
E-mail: ewighton@historycenterslo.org
Web Site: www.historycenterslo.org
Formerly: San Luis Obispo County Historical Society & Museum
Founded: 1956.
Congressional District: 20
Key Personnel: Pres., Charles Crotser; Museum Mgr., Jill Fletcher; C.A.O., Erin Wighton; Museum Shop Mgr., Sandy Carpenter.
Personnel Profile: Full-Time Paid 1; Part-Time Paid 4; Part-Time Volunteers 100; Interns 8.
Governing Authority: Tax-exempt: 501(c)(3).
Institution Type/Description: History Museum.
Collections: 40,000 miscellaneous historic artifacts; archives; textiles; decorative arts; folklore; history items; photographs; newspapers; ledgers; maps; Historic House: 1853 Dallidet Adobe.
Major Exhibits: Native Baskets: The Spirit That Binds, 1/14-12/14.
Research Fields: local & California history.
Facilities: 1,000-vol. library of history books relating to city, county & state of California available for use on the premises; reading room.
Activities: temporary exhibitions; special lectures; music events; book signings.
Publications: monthly newsletter; local history books.
Hours & Admission Prices: Thurs.-Sun. 10-4; other times by appointment. No charge; donations accepted. Closed New Year's Day; Easter; Memorial Day; Labor Day; Thanksgiving; Christmas. &
Attendance: 25,000 (estimated)
Membership: Historian: Student & Educator $20; Individual $30; Dual $50; Collector $100; Preservationist $500; Time Traveler $1,250. Carnegie Club: Curator $5,000; Visionary $10,000. Business: Good Neighbor $200; Corporate $500.

MISSION SAN LUIS OBISPO DE TOLOSA, Old Mission Parish, 751 Palm St., San Luis Obispo, CA 93401-3521. Tel.: 805-781-8220 & 543-6850 (gift shop). Fax: 805-781-8214.
E-mail: office@oldmissionslo.org
Web Site: www.missionsanluisobispo.org
Founded: 1772.
Key Personnel: Pastor, Rev. Russell Brown; Mgr. Gift Shop & Museum, Minerva Soto.
Governing Authority: church; nonprofit. Parent Institution: Diocese of Monterey. Tax-exempt: 170(b)(1)(a).
Institution Type/Description: Religious & History Museum: 1772 Mission San Luis Obispo de Tolosa.
Collections: area artifacts; Chumash Indian collection; original stations of the cross; robes; original altar pieces.
Facilities: Religious goods, books on the history of the Mission & religious themes & other museum-related items for sale.
Activities: docent guided tour by appointment.
Publications: tapes & books by Fr. Nisbet.
Hours & Admission Prices: Summer: daily 9-5. Winter: daily 9-4. Suggested Donation: $2. Closed New Year's Day; Easter; Thanksgiving; Christmas. &
Attendance: 2,000,000 (estimated)

SAN LUIS OBISPO CHILDREN'S MUSEUM, 1010 Nipomo St., San Luis Obispo, CA 93401. Tel.: 805-545-5874.
E-mail: info@slocm.org
Web Site: www.slocm.org
Institution Type/Description: Children's Museum.
Collections: hands-on exhibitions.
Activities: educational programs; birthday parties; holiday events.
Hours & Admission Prices: Winter: Tues.-Wed. 10-3, Thurs.-Sat. 10-5, Sun. 1-5; Summer: Mon.-Wed. 10-3, Thurs.-Sat. 10-5, Sun. 1-5. Admission $8, seniors 60 & over $5; discounts to groups & military families; children under 2 no charge.
Attendance: 45,000 (estimated)
Membership: Playdate $65; You Can Play $95; Play Every Day $120; Endless Play $165.

SAN LUIS OBISPO MUSEUM OF ART, (M), 1010 Broad St., San Luis Obispo, CA 93401-3505. Mailing Address: P.O. Box 813, San Luis Obispo, CA 93406. Tel.: 805-543-8562. Fax: 805-543-4518.
E-mail: office@sloma.org
Web Site: sloma.org
Formerly: San Luis Obispo Art Center
Founded: 1967.
Congressional District: 22

Key Personnel: Dir., Karen M. Kile.
Personnel Profile: Full-Time Paid 3; Part-Time Paid 6; Part-Time Volunteers 25; Interns 2.
Governing Authority: Tax-exempt.
Institution Type/Description: Art Museum.
Collections: paintings; sculpture.
Hours & Admission Prices: July 4 to Labor Day daily 11-5; Sept.-July 3 Wed.-Mon. 11-5. No charge; donations accepted. Closed New Year's Day; Easter; Independence Day; Christmas. &
Attendance: 42,000 (estimated)
Membership: Student $20; Senior $25; Individual $30; Family $45; Business $225.

San Marcos

BOEHM GALLERY, 1140 W. Mission Rd., San Marcos, CA 92069-1415. Tel.: 760-744-1150, ext. 2304. Fax: 760-744-8123.
E-mail: jbigfeather@palomar.edu
Web Site: www.palomar.edu/art/BoehmGallery.html
Founded: 1964.
Congressional District: 43
Key Personnel: Dir., Joanna Bigfeather; Pres. Palomar College, Robert Deegan; Librarian, Daniel Arnsan; Dir. Public Rels., Mike Norton.
Personnel Profile: Full-Time Paid 1; Part-Time Paid 3; Part-Time Volunteers 2; Interns 2.
Governing Authority: college. Parent Institution: Palomar College. Tax-exempt.
Institution Type/Description: Art Gallery.
Collections: over 200 pieces by major & local artists dating from the 16th century to present.
Research Fields: contemporary San Diego artists.
Activities: lectures; gallery talks; temporary exhibitions.
Publications: catalogues.
Hours & Admission Prices: Mon.-Tues. 10-4, Thurs. 10-7, Fri. 10-2. No charge; donations accepted. Closed school holidays. &
Attendance: 6,500 (accurate)

San Marino

HUNTINGTON LIBRARY, ART COLLECTIONS, AND BO-TANICAL GARDENS, 1151 Oxford Rd., San Marino, CA 91108-1218. Tel.: 626-405-2140 & 2100. Fax: 626-405-0225.
E-mail: publicinformation@huntington.org
Web Site: www.huntington.org
Founded: 1919.
Congressional District: 22
Key Personnel: Chm. Bd. Trustees, Stewart R. Smith; Chm. Bd. Overseers, Russel Kully; Pres., Dr. Steven S. Koblik; Dir. Art Collections, John Murdoch; Cur. Manuscripts, Mary Robertson; Cur. Rare Books, Alan Jutzi; Dir. Botanical Gardens, James P. Folsom; Dir. Library, David Zeidberg; Dir. Research, Steven Hindle; Vice Pres. Financial Affairs, Alison Sowden; Readers' Svcs. Librarian, Laura Stalker; Head Technical Svcs., Lorraine Perrotta; Vice Pres. Operations, Laurie Sowd; Vice Pres. Communications, Susan Turner-Lowe; Vice Pres. Advancement, Randy Shulman; Museum Shop Mgr., Janet Crockett.
Personnel Profile: Full-Time Paid 225; Part-Time Paid 70; Part-Time Volunteers 1,000; Interns 20.
Governing Authority: nonprofit organization. Tax-exempt: 501(c)(3).
Institution Type/Description: Library, Art Gallery & Botanical Gardens.
Collections: 18th-century British & European art; paintings; drawings; watercolors; manuscripts; books; sculpture; silver; miniatures; furniture; porcelain; French 18th-century paintings, decorative arts; Renaissance paintings & bronzes; c.1730-1930 American paintings; botanical collections of camellias, roses, trees and rare & endangered succulents.
Research Fields: British & American history & literature; art history, history of science.
Facilities: library; 120-acre botanical garden; 400-seat multi-purpose room; classrooms; 175-seat restaurant. Books & gift items for sale.
Activities: guided tours; gallery talks; formally organized education programs for children; permanent & temporary exhibitions; public programs; invitational programs; a teaching greenhouse.
Publications: reprints; pamphlets; scholarly research books; calendar of events; magazine, Frontiers; annual report; exhibition catalogues; gallery guides.
Hours & Admission Prices: Summer: Wed.-Mon. 10:30-4:30; Winter: Mon. & Wed.-Fri. 10:30-4:30. Call for admission prices. Closed New Year's Day; Independence Day; Thanksgiving; Christmas Eve & Day. &
Attendance: 550,000 (accurate)

OLD MILL MUSEUM, EL MOLINO VIEJO - OLD MILL FOUNDATION, 1120 Old Mill Rd., San Marino, CA 91108-1840. Tel.: 626-449-5458. Fax: 626-449-1057.
E-mail: oldmill@sbcglobal.net
Web Site: old-mill.org
Founded: 1816.
Congressional District: 22
Key Personnel: Pres. (V), Warren Weber; Vice Pres., John Quinn.
Governing Authority: nonprofit institution. Parent Institution: Old Mill Foundation.
Institution Type/Description: History Museum.
Collections: history of California; art; period paintings & furnishings; photographs; operating model of the mill.
Facilities: library; garden.
Activities: historic tours; lectures; films.
Publications: quarterly journal, California History; books; pamphlets; bulletins; catalogs.
Hours & Admission Prices: Tues.-Sun. 1-4. No charge; donations accepted. Closed holidays.
Membership: Senior Citizen & Student $35; Contributor $50; Benefactor & Guardian $100; Sponsor $250; Ambassador $500.

San Martin

WINGS OF HISTORY AIR MUSEUM, 12777 Murphy Ave., San Martin, CA 95046-9527. Mailing Address: P.O. Box 495, San Martin, CA 95046-0495. Tel.: 408-683-2290. Fax: 408-683-2291.
E-mail: wohoffice@sbcglobal.net
Web Site: wingsofhistory.org
Formerly: California Antique Aircraft Museum
Founded: 1983.
Congressional District: 12
Key Personnel: C.E.O., Pres. & Chm. (V), Dan Petroff; Education, Peter Talbot; Prop Maker, Howard Pomerantz; Restoration, Jerry Impellezzeri; Office Mgr., Susan Talbot; Museum Shop Mgr., Sita Kern; Librarian, Norm Zimmerman; Cur., Frank Nichols.
Personnel Profile: Part-Time Volunteers 40.
Volunteer Hours: 8,000
Operating Expenses: 47,000
Operating Income: 56,000
Governing Authority: private; nonprofit organization. Tax-exempt: 501(c)(3).
Institution Type/Description: Aeronautics Museum.
Collections: homebuilt & pre-war aircraft; propellers; engines; models; aviation artifacts.
Facilities: 2,000-vol. library on aviation. Gift items for sale.
Activities: docent program; films; formal educational programs; guided tours; temporary exhibitions. Annual Event: Open House and Fly-In.
Publications: quarterly newsletter, The Vintage Flyer.
Hours & Admission Prices: Tues. & Thurs. 10-3, Sat.-Sun. 11-4; other times by appointment. Donation Encouraged: adults & teens $7; children 6-12 $5; children under 6, active military & members no charge. Closed New Year's Day; Easter; July 4; Thanksgiving; Christmas. &
Attendance: 2,000 (estimated)
Membership: Student $10; Member $40; Sustaining $100; Life $1,000.

San Mateo

CURIODYSSEY, 1651 Coyote Point Dr., San Mateo, CA 94401-1097. Tel.: 650-342-7755. Fax: 650-342-7853.
E-mail: info@curiodyssey.org
Web Site: www.curiodyssey.org
Formerly: Coyote Point Museum
Founded: 1953.
Congressional District: 12
Key Personnel: Exec. Dir., Rachel Meyer; Dir. Wildlife, Nikii Finch-Morales; Exec. Asst., Myra Sinkamo; Museum Shop Mgr., Laura Ellison.
Personnel Profile: Full-Time Paid 25; Part-Time Paid 19; Part-Time Volunteers 171; Interns 6.
Volunteer Hours: 13,500
Operating Expenses: 2,798,103
Operating Income: 2,161,683
Governing Authority: Tax-exempt: 501(c)(3).
Institution Type/Description: Science Center and zoo.
Collections: living & nonliving collections: birds; mammals; insects; plants; shells; reptiles; amphibians, native to California.
Facilities: zoological park; nature & conservation center; classrooms; wildlife center. Science books, field supplies & natural science notions for sale.
Activities: in house exhibits; formally organized educational programs; volunteer programs and council; summer camp; youth training programs; zoo;

winter camp; homeschool program; weekend workshops; after school programs; outreach & PK science programs.
Publications: quarterly newsletter, visitors guides; volunteer newsletter, Spark.
Hours & Admission Prices: Tues.-Sat. 10-5, Sun. 12-5. Adults $9, seniors over 61 & students 13-18 $7, children 2-12 $6; discounts to AAM, AZA, KQED, AAA, ASTC & museum members. Closed New Year's Day; Thanksgiving; Christmas. &
Attendance: 126,516 (accurate)
Membership: Individual $55; Family $90; Supporter $275; Contributing $500; Advocate $525; Innovators Circle $1,000. (Senior $5 discount on any level).

SAN MATEO ARBORETUM SOCIETY, INC., 101 9th Ave., San Mateo, CA 94401-4202. Tel.: 650-579-0536. Fax: 650-343-8416.
Web Site: www.sanmateoarboretum.org
Founded: 1975.
Congressional District: 11
Key Personnel: Pres. (V), Brian Silk; Treas., Jack Bennett.
Personnel Profile: Part-Time Volunteers 12.
Governing Authority: municipal; nonprofit organization. Affiliated with the San Mateo Arboretum Society. Tax-exempt: 170(b)(1)(A).
Institution Type/Description: Arboretum: housed in 1874 Kohl Pump House.
Collections: approx. 300 specimen trees; flowering plants; Native American plants.
Facilities: turn-of-century lawns & flower beds; Japanese Garden; display garden; botanical garden.
Activities: guided tours; lectures, seasonal workshops & classes; demonstrations; gardening information.
Publications: quarterly newsletter, San Mateo Arboretum Society Newsletter.
Hours & Admission Prices: Green House: Tues. & Thurs. 10-2, Sun. 10 to noon. Japanese Garden: daily 9-4. No charge; donations accepted. &
Attendance: 200 (estimated)
Membership: Senior Citizen $15; Individual or Senior Citizen Couple $20; Couple $25; Garden Clubs $30; Commercial or Sponsor $50; Patron $100; Life $1,000.

San Miguel

MISSION SAN MIGUEL, 775 Mission St., San Miguel, CA 93451. Mailing Address: P.O. Box 69, San Miguel, CA 93451-0069. Tel.: 805-467-3256. Fax: 805-467-2448.
E-mail: giftshop@missionsanmiguel.org
Web Site: www.missionsanmiguel.org
Founded: 1797.
Congressional District: 16
Key Personnel: Dir., Max Hottle, O.F.M.
Personnel Profile: Full-Time Paid 1; Part-Time Paid 4.
Governing Authority: church. Parent Institution: Franciscan Friars, Santa Barbara Province. Tax-exempt.
Institution Type/Description: Religious Museum: housed in 1797 Mission.
Collections: artifacts of old mission days; original rooms; murals; paintings.
Facilities: Religious articles for sale.
Activities: self-guided tours; permanent exhibitions.
Hours & Admission Prices: Mission: daily 10-4:30; Mass: Sun. 7 & 11. No charge; donations accepted. Closed New Year's Day; Easter; Thanksgiving; Christmas. &
Attendance: 30,000 (estimated)

RIOS-CALEDONIA ADOBE, 700 S. Mission St., San Miguel, CA 93451. Mailing Address: P.O. Box 326, San Miguel, CA 93451-0326. Tel.: 805-467-3357. Facebook: Rios-Caledonia Adobe.
E-mail: hermanjah@tcsn.net
Web Site: rios-caledoniaadobe.org
Founded: 1968.
Institution Type/Description: Historic House Museum: built in 1835.
Collections: local history; personal artifacts.
Facilities: Museum-related items for sale.
Hours & Admission Prices: Fri.-Sun. 11-4. No charge.

San Pablo

ALVARADO ADOBE & THE BLUME HOUSE, 13831 San Pablo Ave., San Pablo, CA 94806-3703. Tel.: 510-215-3092.
Web Site: www.ci.san-pablo.ca.us/main/museums.htm
Governing Authority: Parent Institution: San Pablo Historical Society.
Institution Type/Description: Alvarado Adobe: reconstructed home of California Governor Juan Bautiste Alvarado, 1848-1882. Blume House: 1905 farm house.
Collections: Adobe Museum: furnishings; personal artifacts. Blume House: period artifacts & furnishing.

Hours & Admission Prices: 2nd & 4th Sun. each month 12-4; other times by appointment. No charge.

San Pedro

ANGELS GATE CULTURAL CENTER, 3601 S. Gaffey St., San Pedro, CA 90731-6969. Tel.: 310-519-0936. Fax: 310-519-8698.
E-mail: info@angelsgateart.org
Web Site: www.angelsgateart.org
Key Personnel: Exec. Dir., Nathan Birnbaum
Institution Type/Description: Art Museum.
Collections: works by local & international artists.
Facilities: Museum-related items for sale.
Activities: classes; workshops; rental facilities.
Hours & Admission Prices: Gallery: Tues.-Sat. 11-5. Office: Mon.-Fri. 10-5:30. No charge; donations accepted.

CABRILLO MARINE AQUARIUM, 3720 Stephen M. White Dr., San Pedro, CA 90731-7012. Tel.: 310-548-7562. Fax: 310-548-2649.
E-mail: mike.schaadt@lacity.org
Web Site: www.cabrillomarineaquarium.org
Formerly: Cabrillo Marine Museum
Founded: 1935.
Congressional District: 32
Key Personnel: Dir. CMA, Mike Schaadt; Exec. Dir., Friends of CMA, Paula Moore.
Personnel Profile: Full-Time Paid 34; Part-Time Paid 60; Part-Time Volunteers 600; Interns 6.
Governing Authority: city. Affiliated with the city of Los Angeles, Dept. of Recreation & Parks. Tax-exempt.
Institution Type/Description: Aquarium & Marine Museum.
Collections: southern California marine plants & animals; marine invertebrates; shore birds; marine fossils; sea shells; fish; whales.
Research Fields: population studies of intertidal invertebrates & gray whales; water quality monitoring; invertebrate & fish propagation; grunion monitoring.
Facilities: 16,500 sq. ft. exhibit space; classroom; research laboratories; 300-seat auditorium.
Activities: guided tours; lectures; gallery talks; formally organized education programs for children, adults, undergraduate & graduate college students; teacher workshops; outreach program for schools; docent program; research internship for students. Special Events: Whale Fiesta in January; Earth Day in April; Annual Autumn Sea Fair in October.
Publications: quarterly, Tidelines; education brochures; visitors guides (English & Spanish); Whales of Baja boat trip brochure; Meet the Grunion Flyers & CMA Sealife Rack cards; membership applications.
Hours & Admission Prices: Tues.-Fri. 12-5, Sat.-Sun. 10-5. Suggested Donations: adults $5, students, senior citizens & children $1. Beach Parking: $1 per hour for each car. Closed Thanksgiving; Christmas. &
Attendance: 280,000 (estimated)
Membership: Senior $40; Individual $45; Family $60; Sponsor $150; Benefactor $300; Patron $500; Partner $1,000.

FORT MACARTHUR MUSEUM, 3601 S. Gaffey St. at Battery Osgood-Farley, San Pedro, CA 90731-6969. Mailing Address: P.O. Box 268, San Pedro, CA 90733-0268. Tel.: 310-548-2631. Fax: 310-241-0847.
E-mail: director@ftmac.org
Web Site: www.ftmac.org
Founded: 1985.
Key Personnel: Dir. & Cur., Stephen R. Nelson.
Personnel Profile: Full-Time Paid 1; Part-Time Paid 4; Part-Time Volunteers 10.
Governing Authority: municipal; nonprofit. Parent Institution: City of Los Angeles Dept. of Recreation and Parks. Subsidiary Institution: Fort MacArthur Museum Association. Tax-exempt.
Institution Type/Description: Military History Museum.
Collections: 1916-1945 Coast Artillery; The Cold War; 1945-1975 NIKE Missile; operational military vehicles, radios & equipment.
Research Fields: coast artillery; local military history; local photo collection; World Wars I & II; Nike missile sites; coast artillery weapons; 1920s & 1930s Cold War.
Facilities: 12-seat theater.
Activities: docent program; films; guided tours; lectures. Museum Sponsors: Old Ft. MacArthur Days; Artillery Show; monthly living history program.
Publications: quarterly newsletter, Alert; brochures; pamphlets.
Hours & Admission Prices: Tues., Thurs. & Sat.-Sun. 12-5. Requested Donation: $3 per person. &

Attendance: 35,000 (estimated)
Membership: Individual $25; Sustaining $35; Supporting $100.

HARBOR MUSEUM, 638 Beacon St., San Pedro, CA 90731. Tel.: 323-464-2727.
Institution Type/Description: Firefighting History Museum.
Collections: firefighting history & equipment; Los Angeles Fire Department apparatus; photographs; personal artifacts.
Facilities: Museum-related items for sale.
Hours & Admission Prices: Sat. 10-3.

HARBOR MUSEUM OLD FIRE STATION 36, 638 Beacon St., San Pedro, CA 90731. Tel.: 323-464-2727.
Web Site: lafdmuseum.org/museum_sanpedro
Institution Type/Description: Firefighting History Museum.
Collections: firefighting history & equipment; fire trucks; photographs; personal artifacts; uniforms.
Facilities: Museum-related items for sale.
Hours & Admission Prices: Sat. 10-3.

LOS ANGELES MARITIME MUSEUM, Berth 84, (foot of 6th St.), San Pedro, CA 90731. Tel.: 310-548-7618. Fax: 310-832-6537. Facebook: LA Maritime Museum.
E-mail: info@lamaritimemuseum.org
Web Site: www.lamaritimemuseum.org
Founded: 1980.
Congressional District: 32
Key Personnel: Dir., Marifrances Trivelli; Admin. Sec., Franceen McClung; Registrar, Lucy Ruggirello.
Personnel Profile: Full-Time Paid 7; Part-Time Paid 12; Part-Time Volunteers 100.
Governing Authority: municipal; nonprofit. Parent Institution: City of Los Angeles. Tax-exempt: 501(c)(3).
Institution Type/Description: Maritime Museum: housed in 1941 former Los Angeles Municipal Ferry Building, built in Art Deco style.
Collections: California maritime history; shipbuilding; transportation technology; longshoring; naval history; fishing industry history; history of maritime trades, port of Los Angeles; historic harbor tug.
Research Fields: California maritime history; shipbuilding; Pacific fleet.
Facilities: research library, maritime records and archives available to the public by appointment; 45-seat auditorium. Maritime history books, nautical objects & gifts for sale.
Activities: guided tours; lectures; films; docent program; participatory & permanent exhibits.
Publications: quarterly members' newsletter, Channel Crossings.
Hours & Admission Prices: Tues.-Sun. 10-5; last admission at 4:30. Adults $3; discounts AAM & ICOM members; members, children & school groups no charge. Closed holidays. ♿
Attendance: 75,000 (estimated)
Membership: Individual $30; Family $45; Sponsor $100; Steward $250; Benefactor $500; Patron $1,000.

MULLER HOUSE MUSEUM, 1542 S. Beacon St., San Pedro, CA 90731-4849. Mailing Address: San Pedro Bay Historical Society, 638 S. Beacon St., San Pedro, CA 90731. Tel.: 310-831-1788.
E-mail: sanpedrohistory@gmail.com
Web Site: www.sanpedrochamber.com/champint/mulrhsmu.htm
Founded: 1974.
Institution Type/Description: Historic House Museum: built in 1899.
Collections: period furnishings.
Facilities: library.
Activities: Museum Sponsors: First Sunday Programs.
Hours & Admission Prices: Sun. 1-4; other times by appointment. Suggested Donation: $3. Closed holidays.
Membership: Senior & Student $25; Individual $30; Family $40; Civic Organization & Business $50; Sustaining $100; Sponsoring $300; Life $1,000.

POINT FERMIN LIGHTHOUSE, 807 W. Paseo del Mar, San Pedro, CA 90731-7131. Tel.: 310-241-0684. Fax: 310-241-0732.
E-mail: kristen.heather@lacity.org
Web Site: www.pointferminlighthouse.org
Founded: 2003.
Key Personnel: Cur., Kristen Heather.
Personnel Profile: Full-Time Paid 1; Part-Time Paid 2; Part-Time Volunteers 30.

Governing Authority: municipal. Parent Institution: Dept. of Recreation and Parks, City of Los Angeles. Tax-exempt.
Institution Type/Description: Lighthouse & Museum.
Collections: lighthouse history.
Research Fields: lighthouse history; architecture; family history.
Facilities: botanical garden; 1,000 sq. ft. exhibit space. Museum-related items for sale.
Activities: docent program; guided tours; temporary exhibitions. Annual Event: Lighthouse Birthday.
Hours & Admission Prices: Tues.-Sun. 1-4. No charge; donations accepted. Closed holidays; special events.
Attendance: 15,000 (accurate)

San Rafael

FALKIRK CULTURAL CENTER, 1408 Mission Ave., San Rafael, CA 94901-1971. Mailing Address: P.O. Box 151560, San Rafael, CA 94915-1560. Tel.: 415-485-3328. Fax: 415-485-3404.
E-mail: jane.lange@ci.san-rafael.ca.us
Web Site: www.falkirkculturalcenter.org
Founded: 1974.
Congressional District: 5
Key Personnel: Dir., Jane Lange; Cur., Beth Goldberg; Pres. (V), Margaret Farley; Public Programs & Public Rels., Cory Bytof.
Personnel Profile: Full-Time Paid 1; Part-Time Paid 4; Part-Time Volunteers 30.
Governing Authority: municipal. Parent Institution: City of San Rafael. Located on premises: Marin Poetry Center. Tax-exempt.
Institution Type/Description: Contemporary Arts Center: housed in the former home of Robert Dollar; built in 1888. Listed on the National Registry of Historic Places.
Collections: furnishings of mansion; Dollar family archivesl; 1927 vintage greenhouse; gardener's cottage.
Research Fields: contemporary arts; history preservation; Bay Area poetry.
Facilities: classroom; poetry reading room; gardens; lecture rooms; private rental facilities; outdoor kitchen; 11 acres of wooded land.
Activities: artist lectures; concerts; education programs; writers' workshops; poetry readings; symposia; art classes, docent lead tours. Annual Events: juried exhibition; summer camp; Alice in Wonderland Egg Hunt; Showcase in December.
Publications: quarterly newsletter; exhibition catalogues; special event announcements.
Hours & Admission Prices: Mon.-Fri. 1-5, Sat. 10-1. No charge; donations requested. Closed legal holidays. ♿
Attendance: 35,000 (estimated)
Membership: Student & Artist $20; Individual $30; Family $50; Copper $100; Silver $250; Gold $5,000; Platinum $5,000 and up.

MARIN HISTORY MUSEUM, (M), 1125 B St., San Rafael, CA 94901-2907. Tel.: 415-454-8538. Fax: 415-454-6137. Facebook; Marin History Museum.
E-mail: info@marinhistory.org
Web Site: www.marinhistory.org
Formerly: Marin County Historical Society
Founded: 1935.
Congressional District: 6
Key Personnel: Pres. (V), Carleton Prince; Collections Mgr., Doug DeFors; Librarian, Jocelyn Moss; Museum Mgr., Marcie Miller.
Personnel Profile: Part-Time Paid 6; Part-Time Volunteers 30; Interns 6.
Governing Authority: board of directors. Tax-exempt.
Institution Type/Description: Historical Museum; county.
Collections: memorabilia; artifacts; newspapers; photographs; books; newspaper clippings; pamphlets; brochures; maps; clothing; San Quentin Prison photos, books, papers of Clinton Duffy (warden); explorer Louise A. Boyd's photos, papers, artifacts from her 1920 to 1940 expeditions to the Arctic & Greenland.
Major Exhibits: By-Gone Art, 6/13/2/14.
Research Fields: Marin County history; genealogy; architecture.
Facilities: library of books & newspapers on history of Marin; historic photo archives.
Activities: guided tours; lectures; trips to historic sites; dinners with guest speakers on local history; family programs; films, author series; oral histories; internships.
Publications: newsletter; historic photo calendars; special publications.
Hours & Admission Prices: Tues.-Wed. & Fri.-Sat. 11-4, Thurs. 11-8. No charge; donations accepted. Closed legal holidays. ♿
Attendance: 5,500 (accurate)
Membership: Seniors, Active Military, Students & Educators $20 flat fee; Individual $30; 2 for 1 Individuals $50; Families $65; Contributor $100; Supporter $250; Patron $500; Benefactor $1,000.

MISSION SAN RAFAEL ARCANGEL, 1104 Fifth Ave., San Rafael, CA 94901-2916. Tel.: 415-454-8141, ext. 12. Fax: 415-454-8193.
E-mail: tbrunner@saintraphael.com
Web Site: www.saintraphael.com
Founded: 1817.
Key Personnel: Cur., Theresa Brunner; Museum Shop Mgr., Helen Bernardoni.
Personnel Profile: Full-Time Paid 1; Part-Time Paid 1; Part-Time Volunteers 5.
Governing Authority: Parent Institution: San Francisco Archdiocese. Tax-exempt.
Institution Type/Description: History Museum.
Collections: religious artifacts; Native American artifacts.
Activities: ongoing lecture series.
Publications: newsletter, The Mission News.
Hours & Admission Prices: Sun.-Mon. & Wed.-Fri. 11-4. No charge. &

Membership: Annual $50.

WILDCARE, 76 Albert Park Lane, San Rafael, CA 94901-3929. Tel.: 415-453-1000.
E-mail: info@wildcarebayarea.org
Web Site: www.wildcarebayarea.org
Key Personnel: Exec. Dir., Karen Wilson
Institution Type/Description: Natural History Museum.
Collections: hands-on exhibits.
Hours & Admission Prices: Daily 9-5. No charge; donations accepted.

San Simeon

* **HEARST CASTLE, (M),** 750 Hearst Castle Rd., San Simeon, CA 93452-9740. Tel.: 805-927-2020. Fax: 805-927-2031. TDD: 800-274-7275.
E-mail: curator@hearstcastle.com
Web Site: www.hearstcastle.com
Formerly: Hearst Castle-Hearst San Simeon State Historical Monument
Founded: 1957.
Congressional District: 22
Key Personnel: District Supt., Nicholas Franco; Museum Dir., Hoyt Fields.
Personnel Profile: Full-Time Paid 250; Part-Time Paid 150; Part-Time Volunteers 275.
Governing Authority: state. Parent Institution: California State Parks, P.O. Box 942896, Sacramento, CA 94296. Tel. 916-653-8380. Tax-exempt.
Institution Type/Description: Historic House.
Collections: Hearst family life & history; decorative & fine arts of Gothic & Renaissance Spain & Italy; architectural elements; silver; paintings; sculpture; antiquities; tapestries; textiles; Oriental rugs; 127 acres including gardens, pools & guest houses.
Research Fields: William Randolph Hearst; Julia Morgan; construction of Hearst castle; the art collection.
Facilities: visitor center. Museum-related items for sale.
Activities: guided tours; films; permanent exhibitions.
Hours & Admission Prices: Tours: daily 8:20-3:20; times may vary; reservations recommended, call 800-444-4445. Discounts to groups of 12 or more; children under 6 no charge. Closed New Year's Day; Thanksgiving; Christmas. &

Attendance: 860,000 (accurate)
Membership: Friends of Hearst Castle $45-$5,000.

Sanger

SANGER DEPOT MUSEUM, 1700 7th St., Sanger, CA 93657-2804. Mailing Address: P.O. Box 44, Sanger, CA 93657-0044. Tel.: 559-875-2848.
Web Site: www.sangerdepotmuseum.com
Founded: 1982.
Congressional District: 20
Key Personnel: Pres. (V), James Walton.
Personnel Profile: Part-Time Volunteers 15.
Governing Authority: Tax-exempt.
Institution Type/Description: History Museum.
Collections: artifacts & memorabilia pertaining to the city of Sanger.
Hours & Admission Prices: Fri. 9:30-12:30, Sun. 1-4. Adults $1, children $.25.
Membership: Individual $10; Life $200.

Santa Ana

* **BOWERS MUSEUM, (M),** 2002 N. Main St., Santa Ana, CA 92706-2776. Tel.: 714-567-3601. Fax: 714-567-3603.
E-mail: pkeller@bowers.org
Web Site: www.bowers.org
Founded: 1936.
Congressional District: 40
Key Personnel: Pres., Peter C. Keller, Ph.D.; Chm. Bd., Frank O'Bryan; C.F.O., Thuy Nguyen; Vice Pres. Education, Nancy Warzer Brady; Exhibit Design & Fabrication, Paul Johnson; Dir. Retail Sales, Pauline Rusterholtz.
Personnel Profile: Full-Time Paid 36; Part-Time Paid 39; Part-Time Volunteers 300.
Governing Authority: nonprofit organization. Tax-exempt: 501(c)(3).
Institution Type/Description: Art & Cultural Arts Museum.
Collections: South Pacific; Native North and South America; Pre-Columbian; African; Asian; 19th & 20th-century American paintings; Orange County historic photography; Southern California history.
Research Fields: cultures of the Americas, Pacific Rim, Oceania & Africa.
Facilities: education center; 800-seat auditorium; restaurant. Museum-related items for sale.
Activities: guided tours; lectures; films; studio workshops; community festivals; African Cultural Arts Council; California Art Council; Docent Guild; Indian Cultural Arts Council; Persian Arts Council; Hispanic Arts Council; Chinese Cultural Art Council; Collector's Council; Orange County American Italian Renaissance Foundation; Bead Society of Orange County; Plein Air Art Council; Bells; education programs for children & adults; permanent, temporary & traveling exhibitions; mobile museum exhibits; inter-museum loan; speakers bureau; mini-museum classroom kits.
Publications: monthly calendar of events; exhibit catalogs; gallery guides.
Hours & Admission Prices: Tues.-Sun. 10-4. Adults $17-$19, seniors & students $12-$14; discounts to AAM, ICOM & WMA members; children under 5 & members no charge. &

Attendance: 132,300 (accurate)
Membership: Senior & Student $45; Active $65; Contributor $175; Patron $500; Fellow $1,000; Corporate Fellow $2,000.

CALIFORNIA STATE UNIVERSITY, FULLERTON, GRAND CENTRAL ART CENTER, 125 N. Broadway, Santa Ana, CA 92701-8237. Tel.: 714-567-7233. Fax: 714-558-4145.
Web Site: www.grandcentralartcenter.com
Key Personnel: Dir., John D. Spiak.
Personnel Profile: Full-Time Paid 2.
Governing Authority: Parent Institution: CSUF Auxiliary Services Corp. Tax-exempt.
Institution Type/Description: Art Gallery.
Collections: works by contemporary artists.
Activities: temporary exhibitions.
Hours & Admission Prices: Tues.-Thurs. & Sun. 11-4, Fri.-Sat. 11-7. No charge; donations accepted. &

DISCOVERY SCIENCE CENTER, 2500 N. Main St., Santa Ana, CA 92705-6600. Tel.: 714-542-CUBE. Fax: 714-542-2828.
Institution Type/Description: Science Center.
Collections: hands-on science exhibits; environment; physics; earth; dinosaurs; air & space; hockey; perception.
Facilities: restaurant; rental facilities.
Activities: teacher programs; summer camp; camp-ins; rental facilities.
Hours & Admission Prices: Daily 10-5. Adults $12.95, children 3-17 $9.95; members and children 2 & under no charge. Closed Thanksgiving; Christmas. &

Attendance: 440,000 (accurate)
Membership: Individual $60; Family $99; Family Plus $149; Family & Friends $179.

DR. HOWE-WAFFLE HOUSE AND MEDICAL MUSEUM, 120 Civic Center Dr., W., Santa Ana, CA 92701-7505. Tel.: 714-547-9645.
E-mail: sahps@sahps.org
Web Site: www.santaanahistory.com/house.html
Founded: 1974.
Key Personnel: Chm. & Pres. (V), Alison Young.
Personnel Profile: Part-Time Volunteers 20.
Governing Authority: Parent Institution: Santa Ana Historical Preservation Society. Tax-exempt.
Institution Type/Description: Historic House & Medical Museum: home built in 1889 by Orange County physicians, Alvin & Willella Howe.
Collections: period furnishings; personal artifacts; medical equipment & supplies. Historic Buildings: Victorian home; carriage barn.
Facilities: digital photographic archives.
Activities: living history tours & events. Annual Event: Historical Cemetery Tour.
Publications: quarterly newsletter; books, Logan Barrio; O.C. Fruit Crate Labels.

Hours & Admission Prices: Feb., April, June, Aug.-Sept. & Dec. 1st Sat. of month 12-4. Adults $5; discounts to members.
Attendance: 3,000 (estimated)
Membership: $20 to $1,000.

LYON AIR MUSEUM, 19300 Ike Jones Rd., Santa Ana, CA 92707. Tel.: 714-210-4585. Fax: 714-210-4588.
E-mail: info@lyonairmuseum.org
Web Site: www.lyonairmuseum.org
Institution Type/Description: Military History Museum.
Collections: military history & aviation; aircraft World War II; vehicles; personal artifacts.
Hours & Admission Prices: Daily 10-4. Adults $12, seniors & veterans $9, children 5-17 $6; discounts to groups; children under 5 no charge. Closed Christmas & Thanksgiving.
Attendance: 30,000 (estimated)
Membership: Membership $55; Family (4) $85.

OLD COURTHOUSE MUSEUM, 211 W. Santa Ana Blvd., Santa Ana, CA 92701-7554. Tel.: 714-834-6605 & 973-6607. Fax: 714-834-2280.
E-mail: marshall.duell@ocparks.com
Web Site: www.ocparks.com
Founded: 1987.
Congressional District: 47
Key Personnel: Cur., Marshall Duell; Education Coord., Donna Brietfeller.
Personnel Profile: Full-Time Paid 2; Part-Time Volunteers 15; Interns 1.
Governing Authority: county government. Tax-exempt.
Institution Type/Description: County History Museum: housed in 1901 Richardsonian Romanesque style county courthouse.
Collections: law & government artifacts; court documents; photographs.
Research Fields: local & regional history.
Facilities: 70-seat auditorium. Gift items for sale.
Activities: docent program; guided tours; lectures; loan, temporary & traveling exhibitions.
Publications: The Court Reporter.
Hours & Admission Prices: Mon.-Fri. 9-5. No charge. Closed state holidays. &
Attendance: 18,000 (estimated)
Membership: Individual $10; Family $15; Sustaining & Organization $25: Business $35.

SANTA ANA COLLEGE ART GALLERY, 1530 W. 17th St., Santa Ana, CA 92706-3398. Tel.: 714-564-5615. Fax: 714-564-5629.
E-mail: mccabe_caroline@sac.edu
Web Site: www.sac.edu/art
Founded: 1970.
Congressional District: 47
Key Personnel: Dir., Phillip Marquez; Coord. Gallery, Caroline McCabe.
Personnel Profile: Full-Time Paid 2; Part-Time Paid 1; Part-Time Volunteers 25.
Volunteer Hours: 4,000
Governing Authority: public university. Rancho Santiago Community College District. Branch Location: SAC Arts at the Santora, Santora Arts Complex, 207 N. Broadway, Suite Q, Santa Ana, CA 92701. Tax-exempt.
Institution Type/Description: College Art Gallery.
Collections: works by student artists.
Major Exhibits: From A Familiar Place, 12/9/13-5/2/14; Home Grown, 1/4/14-3/1/14; Around the Corner, 2/20/14-4/3/14; Annual Student Art Show, 5/2/14-5/23/14; High School Invitational, 6/7/14-7/5/14.
Facilities: library of exhibit catalogs; 1,120 sq. ft. exhibit space.
Activities: guided tours; lectures; organized education programs for adults. Museum Sponsors: monthly Artwalk in Artists Village; Holiday Art Sale in December.
Publications: exhibition brochures.
Hours & Admission Prices: Santa Ana College Main Art Gallery: School Year: Mon.-Wed. 10-2, Thurs. 10-2 & 6:30pm-8:30pm. No charge; donations accepted. Closed holidays; spring, summer & winter breaks. SAC Arts at the Santora: Fri.-Sat. 12-4. No charge; donations accepted. Closed holidays; spring & winter break. &
Attendance: 15,000 (estimated)

SANTA ANA ZOO, (M), 1801 E. Chestnut Ave., Santa Ana, CA 92701-5001. Tel.: 714-647-6575. Fax: 714-953-7401.
E-mail: kyamaguchi@santa-ana.org
Web Site: santaanazoo.org
Founded: 1952.

Key Personnel: Dir., Kent Yamaguchi; Devel., Cathi Decker; Zoo Cur., Suzanne Werner.
Personnel Profile: Full-Time Paid 20; Part-Time Paid 14; Part-Time Volunteers 65.
Governing Authority: municipal; nonprofit. Parent Institution: City of Santa Ana. Subsidiary Institution: Recreation & Community Services.
Institution Type/Description: Zoo.
Collections: mammals; birds; reptiles; amphibians; fish.
Facilities: library; educational facility; zoological park. Educational supplies, animal & conservation-related items for sale.
Activities: guided tours; formal educational programs; docent program; mobile vans.
Publications: bimonthly newsletter, Animal Tales.
Hours & Admission Prices: Memorial Day-Labor Day Mon.-Fri. 10-5, Sat.-Sun.10-6; Sept.-May daily 10-5; last ticket sold 1 hour before closing. Adults $8, senior citizens 60 & over and children 3-12 $5; children 2 & under, handicapped, reciprocal zoos, AZA & members no charge. &
Attendance: 268,878 (accurate)
Membership: Individual $35; Couple $45; Family $55.

Santa Barbara

✳ **ART, DESIGN & ARCHITECTURE MUSEUM UC, SANTA BARBARA, (M),** University of California, Santa Barbara, Rm. 1626 Art Bldg., Santa Barbara, CA 93106-7130. Tel.: 805-893-2951. Fax: 805-893-3013. Facebook: ADA Museum.
E-mail: mmoore@museum.ucsb.edu
Web Site: www.museum.ucsb.edu
Formerly: University Art Museum, Santa Barbara
Founded: 1959.
Congressional District: 22
Key Personnel: Dir., Bruce Robertson; Asst. to Dir., Michaela Moore; Cur. Exhibitions, Elyse A. Gonzales; Designer, Menmet Dogu; Asst. Designer, Todd Anderson; ADC Cur., Jocelyn Gibbs; Project Archivist (ADC), Christina Marino.
Personnel Profile: Full-Time Paid 6; Part-Time Paid 6; Interns 12.
Governing Authority: university. Parent Institution: University of California. Tax-exempt: 501(c)(3).
Institution Type/Description: Art Museum.
Collections: Morgenroth Collection of Renaissance medals & plaquettes; Sedgwick Collection of Old Master paintings; Feitelson Collection of Old Master Drawings; Dreyfus Collection of Luristan Bronzes, Near Eastern ceramics & pre-Columbian art; Ala Story Graphic Arts Collection (15th-18th centuries); Ruth S. Shaffner Collection of Contemporary Art; Ken Trevey Collection of American Realist Prints; Fernand Lungren paintings & drawings; various paintings, sculpture, drawings, prints, photography & ceramics with emphasis on contemporary; separately housed collection of over 750,000 architectural drawings & related materials, largely the work of California architects.
Research Fields: contemporary art; modern architecture; Renaissance medals & plaquettes; California artists; Old Master drawings; ethnographic art; works on paper.
Facilities: Books for sale.
Activities: docent guided tours; lectures; gallery talks; permanent, temporary & traveling exhibitions; education programs for children, adults & university students.
Publications: exhibition catalogs; Lasting Impressions: MFA08.
Hours & Admission Prices: Wed.-Sun. 12-5. No charge; donations accepted. Closed major holidays. &
Attendance: 50,000 (estimated)
Membership: Gaucho $20; Household $50-$99; Contributor $100-$499; Patron $500-$999; Director's Circle $1,000-$2,500.

CARRIAGE AND WESTERN ART MUSEUM, 129 Castillo St., Santa Barbara, CA 93101-5725. Mailing Address: P.O. Box 1587, Santa Barbara, CA 93102-1587. Tel.: 805-962-2353.
Web Site: www.carriagemuseum.org/
Founded: 1972.
Key Personnel: Pres. (V), Peter Georgi; Museum Shop Mgr., Tom Peterson.
Personnel Profile: Full-Time Paid 1; Part-Time Volunteers 16.
Institution Type/Description: History Museum.
Collections: history of the west; 50 saddles once belonging to famous people including Cisco Kid, Will Rogers, & Clark Gable; mud wagons; army wagons; circus wagons; photographs; carriages; wagons.
Activities: special events.
Publications: quarterly newsletter.
Hours & Admission Prices: Mon.-Fri. 9-3; groups by appointment. No charge; donations accepted. &
Attendance: 26,000 (estimated)
Membership: Family $25.

CASA DEL HERRERO, 1387 E. Valley Rd., Santa Barbara, CA 93108-1202. Tel.: 805-565-5653. Fax: 805-969-2371.
E-mail: tours@casadelherrero.com
Web Site: www.casadelherrero.com
Formerly: The Casa del Herrero Foundation
Founded: 1995.
Key Personnel: Dir., Molly Barker; Devel., Olga Rogers; Volunteer Dir., Susannah Gordon.
Personnel Profile: Full-Time Paid 4; Part-Time Paid 1; Part-Time Volunteers 40.
Governing Authority: private; nonprofit organization. Tax-exempt: 501(c)(3).
Institution Type/Description: Historic House Museum: former home of George Fox Steedman designed by architect George Washington Smith, built in 1925. National Historic Landmark.
Collections: 14th-18th century Spanish furnishings; period artifacts; handmade silver; paintings.
Facilities: library; botanical garden; 4,000 sq. ft. exhibit space. Museum-related items for sale.
Activities: docent program; members luncheons & dinner parties; guided tours; lectures. Annual Events: Volunteer Recognition Events; Art in the Garden Fundraiser; Holiday Tours & Auction.
Publications: quarterly membership newsletter, Newsletter of the Casa del Herrero.
Hours & Admission Prices: Public Tours: year round, Wed. & Sat. 10 & 2. Group Tours: Tues.-Sat. 10 & 2 by appointment. Adults $20; discounts to National Trust for Historic Preservation; members no charge. Children under 10 not admitted. Closed New Year's Day; national holidays.
Attendance: 3,225 (estimated)
Membership: Individual (adult) $50; Household $100; Friend $250; Patron $500; Casa Society $1,000; Director's Circle of the Casa Society $2,500-$5,000.

HISTORIC SANTA BARBARA COURTHOUSE, 1100 Anacapa St., Santa Barbara, CA 93101-2099. Mailing Address: 105 E. Anapamu St., Santa Barbara, CA 93101. Tel.: 805-962-6464.
Web Site: www.santabarbaracourthouse.org/sbch/
Key Personnel: Pres., Pam Stoney
Institution Type/Description: History Museum.
Collections: functional courthouse; clock tower; gardens.
Facilities: gardens.
Activities: special events.
Hours & Admission Prices: Museum: Mon.-Fri. 8-5, Sat.-Sun. 10-4. Tours: Mon.-Tues. & Fri. 10:30 & 2, Wed.-Thurs. & Sat. 2. No charge; donations accepted. Closed Christmas.

OLD MISSION SANTA BARBARA, 2201 Laguna St., Santa Barbara, CA 93105-3611. Tel.: 805-682-4713, ext. 166. Fax: 805-687-7841 & 6067.
E-mail: museumtours@sboldmission.org
Web Site: www.santabarbaramission.org
Formerly: Santa Barbara Mission Museum
Key Personnel: Tour Administrator, Laura Foss
Institution Type/Description: Mission Museum.
Collections: religious artifacts; period furnishings.
Facilities: Museum-related items for sale.
Hours & Admission Prices: Summer: daily 9-4:30; Fall & Winter: daily 9-4:15. Self-Guided Tours: adults 16-64 $6, seniors 65 & up $5, youth 5-15 $1; Docent-Guided Tours: adults 16-64 $8, seniors 65 & up $7, youth 5-15 $4; children under 4 no charge. Closed Easter; Thanksgiving; Christmas.

＊　SANTA BARBARA BOTANIC GARDEN, 1212 Mission Canyon Rd., Santa Barbara, CA 93105-2199. Tel.: 805-682-4726, ext. 101. Fax: 805-563-0352.
E-mail: info@sbbg.org
Web Site: www.santabarbarabotanicgarden.org
Formerly: The Botanic Garden at Santa Barbara and Ojai
Founded: 1926.
Congressional District: 19
Key Personnel: Chm. Bd. Trustees, The Honorable Fife Symington, III; Pres. & C.E.O., Edward L. Schneider, Ph.D.; Vice Pres. Finance & Administration, Robert Sherwood; Vice Pres. Programs & Collections, Dieter Wilken, Ph.D.; Vice Pres. Mktg. & Government Rels., Nancy Johnson; Vice Pres. Devel., Gina Benesh; Dir. Horticulture, Andrew Wyatt; Dir. Nursery, Bruce Reed; Dir. Education, Sally Isaacson; Cur. Herbarium, Steve Junak; Librarian, Joan Ariel; Museum Shop Mgr., Gail Meadows Milliken; Human Resources & Volunteer Mgr., Cherie Welsh.
Personnel Profile: Full-Time Paid 25; Part-Time Paid 18; Part-Time Volunteers 308; Interns 1.

Governing Authority: nonprofit organization. Tax-exempt: 501(c)(3).
Institution Type/Description: Arboretum/Botanical Garden: located at site of Old Mission Dam & Aqueduct.
Collections: native California plants; plants from California offshore islands including rare & endangered California plants as part of Garden's program in rare plant conservation; over 1,000 living California plants; 15,000 books & journals; 140,000 preserved plant specimens.
Research Fields: plant systematics, evolution & ecology; structural botany; conservation biology.
Facilities: 15,000-vol. library; regional maps; series of oral history tapes; herbarium of over 140,000 specimens; 78 acre gardens. California native & Mediterranean plants, prints, stationery, books & handcrafted gifts for sale.
Activities: guided tours; lectures; symposia; workshops; formally organized programs for children, including pre-school; Master Gardening Certificate & Certificate in So. Calif. Gardening, Garden Grower Certificate (retail nursery management) programs; cooperative graduate program with the Univ. of California, Santa Barbara for M.A., M.S. & Ph.D. degrees in botanical science; docent programs; field trips; camping trips; environmental education programs.
Publications: Flora of the Santa Barbara Region; Flora of Santa Cruz Island; Trees of Santa Barbara; quarterly newsletter, Ironwood; occasional books; Alice's Garden.
Hours & Admission Prices: March-Oct. daily 9-6; Nov.-Feb. daily 9-5. Adults $8, seniors 60 & over, teens 13-17, full-time students & active military $6, children 2-12 $4; discounts to AAM members; members & children under 2 no charge. Closed New Year's Day; Thanksgiving; Christmas Eve & Day; occasional special events. &
Attendance: 96,000 (accurate)
Membership: Individual $50; Family & Dual $75; Wildflower Guild $150; Seasons Guild $300; Ironwood Circle $600; Director's Circle $1,000; President's Circle $2,500; Corporate President's Circle $5,000.

SANTA BARBARA CONTEMPORARY ARTS FORUM, (M), 653 Paseo Nuevo, Santa Barbara, CA 93101-3392. Tel.: 805-966-5373. Fax: 805-962-1421.
E-mail: lmermel@sbcaf.org
Web Site: www.sbcaf.org
Key Personnel: Exec. Dir., Miki Garcia; Communications Coord., Lauren Mermel; Dir. Operations, Margie Yahyavi
Institution Type/Description: Contemporary Arts Center.
Collections: works by local, regional, national & international artists; contemporary art.
Activities: performances.
Hours & Admission Prices: Tues.-Sat. 11-5, Sun. 12-5. No charge; donations accepted. &
Membership: Annual Individual $75; Couple $125; Collector's Council $125; Director's Council $1,000; Contemporary Council $2,500.

THE SANTA BARBARA HISTORICAL MUSEUM, (M), 136 E. De la Guerra St., Santa Barbara, CA 93101-2205. Tel.: 805-966-1601. Fax: 805-966-1603.
E-mail: director@sbhistorical.org
Web Site: www.santabarbaramuseum.com
Founded: 1932.
Congressional District: 19
Key Personnel: Chm. (V), William Burtness; Chief Cur., Daniel Calderon; Dir. Research, Michael Redmon; Mktg. Coord., Dacia Harwood; Membership Coord., Jeanne M. Buchanan; Dir. Education, Rebekah P. Beveridge; Museum Shop Mgr., Rebecca Harwood.
Personnel Profile: Full-Time Paid 8; Part-Time Paid 7; Part-Time Volunteers 50.
Governing Authority: society. Tax-exempt: 501(c)(3).
Institution Type/Description: Local History Museum.
Collections: history; clipping books; genealogical material; artifacts covering Spanish, Mexican & American Periods; local newspapers; old photographs; manuscript collections. Historic Houses: 1825 Historic Adobe; 1817 Covarrubias Adobe; 1854 Trusell-Winchester Adobe; 1861 Fernald Mansion.
Research Fields: Santa Barbara & California history & genealogy.
Facilities: 6,000-vol. library of California history books available for use on premises. Local and California history books, local gifts & jewelry for sale in museum store.
Activities: pre-arranged tours; lectures; reading room; formally organized education programs for children; docent program or council; inter-museum loan, permanent & temporary exhibitions.
Publications: quarterly, Noticias; Historic Santa Barbara; My Santa Barbara Scrapbook.
Hours & Admission Prices: Tues.-Sat. 10-5, Sun. 12-5. Suggested Donation: general $5, children 12 & under $2. Closed New Year's Day; Easter; Memorial Day; Independence Day; Thanksgiving; Christmas. &
Attendance: 20,000 (estimated)

Membership: Individual $50; Supporter $75; Family $100; Sustainer $150; Sponsor $250; Benefactor $500; Patron $1,000; Director's Circle $2,500; President's Circle $5,000.

SANTA BARBARA MARITIME MUSEUM, 113 Harbor Way, Ste. 190, Santa Barbara, CA 93109-2344. Tel.: 805-962-8404. Fax: 805-962-7634.
E-mail: museum@sbmm.org
Web Site: www.sbmm.org
Founded: 1994.
Congressional District: 22
Key Personnel: Pres., Steve Epstein; Exec. Dir., Greg Gorga; Vice Pres., Willard Thompson; Treas., Gail Anikouchine; Deputy Dir., Emily Falke.
Personnel Profile: Full-Time Paid 5; Part-Time Paid 4; Part-Time Volunteers 150; Interns 2.
Governing Authority: private; nonprofit organization. Tax-exempt: 501(c)(3).
Institution Type/Description: Maritime Museum.
Collections: nautical instruments & objects related to maritime endeavors; maritime history & archaeology; photographs; documents.
Major Exhibits: Winfield Scott, 4/11-12/14; Spills Broad Reach, 4/11-12/14; Purisima Diving Bell, 11/11-12/15; Dan Wilson 400' Dive, 11/12-12/14; Point Conception Lighthouse Lens, 9/13-12/15.
Research Fields: Santa Barbara & central coast maritime history; archaeological remains of local shipwrecks.
Facilities: library; 88-seat theater & auditorium; 8,500 sq. ft exhibit space. Museum-related items for sale.
Activities: docent program; films; formal education programs for adults; guided tours; lectures; floating & participatory exhibits; rental facilities.
Publications: Seafarers; Santa Barbara Fisheries.
Hours & Admission Prices: Memorial Day-Labor Day Thurs.-Tues. 10-6; Sept.-May Thurs.-Tues. 10-5. Adults $7, seniors, students, youth 6-17 $4, children 1-5 $2; members, CAMM & AASLH members no charge. Closed New Year's Day; Thanksgiving; Christmas. &
Attendance: 25,909 (accurate)
Membership: Students (with ID) & Senior Citizens 65 & up $25; Individual $40; Crew (family) $55; Mariner $125; Navigator $500; Explorer $1,000; Astronavigator $2,500; Circumnavigator $5,000.

* **SANTA BARBARA MUSEUM OF ART, (M),** 1130 State St., Santa Barbara, CA 93101-2746. Tel.: 805-963-4364. Fax: 805-966-6840.
E-mail: info@sbma.net
Web Site: www.sbma.net
Founded: 1941.
Congressional District: 5
Key Personnel: Chm. Bd. Trustees (V), Ken Anderson; Dir., Larry J. Feinberg; C.F.O., James Hutchinson; Dir. Education, Patsy Hicks; Cur. Asian Art, Susan Shin-tsu Tai; Cur. Contemporary Art, Julie Joyce; Cur. Photography, Karen Sinsheimer; Registrar, Cherie Summers; Chief Cur., Eik Kahng; Dir. Devel., Barbara Ben-Horin.
Personnel Profile: Full-Time Paid 59; Part-Time Paid 42; Part-Time Volunteers 215; Interns 15.
Governing Authority: nonprofit organization. Tax-exempt: 501(c)(3).
Institution Type/Description: Art Museum.
Collections: Greek, Roman, Egyptian antiquities; Asian sculpture, woodblock prints; painting, ceramic & decorative arts; American paintings from colonial period to present; works on paper from Renaissance to present; European paintings & sculpture; photographs; modern & contemporary art.
Research Fields: 19th-century French lithographs, American Modernism, Japanese Meiji woodblock prints.
Facilities: art library & education center for art classes; 153-seat auditorium. Books & cards for sale.
Activities: docent guided tours; lectures; films; gallery talks; formally organized education programs for children; docent program; permanent & temporary exhibitions; national & international museum sponsored trips.
Publications: catalogs; calendars; announcements; reports.
Hours & Admission Prices: Tues.-Sun. 11-5. Adults $10, seniors $6, students with ID & children 6-17 $6; discounts to AAM & ICOM members; members & children under 6 no charge. Closed some holidays. &
Attendance: 130,000 (estimated)
Membership: General $60; Associate Patron $125; Gallery Patron $250; Collector's Patron $500; Curator's Patron $750; Director's Patron $1,500; Benefactor's Circle $3,000.

* **SANTA BARBARA MUSEUM OF NATURAL HISTORY, (M),** 2559 Puesta del Sol Rd., Santa Barbara, CA 93105-2998. Tel.: 805-682-4711. Fax: 805-569-3170.
E-mail: info@sbnature2.org
Web Site: www.sbnature.org

Founded: 1916.
Congressional District: 19
Key Personnel: Pres. & C.E.O., Luke Swetland; C.O.O., Diane E. Wondolowski; Cur. Vertebrate Zoology, Paul W. Collins; Librarian, Terri Sheridan; Cur. Anthropology, Dr. John R. Johnson; Dir. Collections & Research, Dr. Henry Chaney; Dir. Education & Exhibits, Heather Moffat; Dir. Facilities, Gary Robinson; Dir. Visitor Svcs., Amy Carpenter; Mktg. & Public Rels Mgr., Easter Moorman; Dir. Devel., Caroline Grange.
Governing Authority: nonprofit organization. Subsidiary Institution: Sea Center. Tax-exempt: 501(c)(3).
Institution Type/Description: Natural History Museum.
Collections: natural history & prehistoric life of Pacific Coast; anthropology; invertebrate zoology; vertebrate zoology; paleontology; mineralogy; historical archive; antique & natural history art.
Research Fields: anthropology; invertebrate zoology; vertebrate zoology.
Facilities: 40,000-vol. library of science & natural history available for inter-library loan & for use on premises by special permission from librarian; planetarium; 350-seat & 100-seat auditoriums; classrooms; historical archives & nature art; marine aquariums. Books, laboratory supplies, specimens & limited specialty items for sale.
Activities: guided tours; field trips; lectures; films; study clubs; formally organized educational programs; formally organized education programs for undergraduate & graduate students affiliated with University of California, Santa Barbara campus; inter-museum loan, permanent, temporary & traveling exhibitions.
Publications: monthly public program calendars; books; teacher guides & papers.
Hours & Admission Prices: Daily 10-5. Adults $12, seniors 65 & up and teens 13-17 $8, children 2-12 $7; discounts to groups, ASTC & AAM members; children under 2 & members no charge. Closed New Year's Day; last Saturday in June; Thanksgiving Day; Christmas Day. &
Attendance: 110,000 (accurate)
Membership: Member $60; Family $80; Naturalist $150; Explorer $350; Patron's Circle $1,000.

SANTA BARBARA ORCHID ESTATE, 1250 Orchid Dr., Santa Barbara, CA 93111-2914. Tel.: 805-967-1284; 800-553-3387. Fax: 805-683-3405.
E-mail: sboe@sborchid.com
Web Site: www.sborchid.com
Institution Type/Description: Orchid Estate.
Collections: thousands of orchid species & hybrids.
Facilities: 5 acres.
Activities: special events.
Hours & Admission Prices: Mon.-Sat. 8-4:30, Sun. 11-4. Closed New Year's Day; Presidents' Day; Easter; Memorial Day; Independence Day; Labor Day; Thanksgiving; Christmas.

SANTA BARBARA TRUST FOR HISTORIC PRESERVATION, (M), 123 E. Canon Perdido, Santa Barbara, CA 93101-2215. Tel.: 805-965-0093. Fax: 805-568-1999.
Web Site: www.sbthp.org
Founded: 1961.
Key Personnel: Dir., Dr. Jarrell Jackman; Chm. (V), Craig Makela; Treas., Sally Fouhse; Devel., Kendra Rhodes; Education, Karen Anderson; Public Rels., Christa Clark Jones; Cur., Anne Peterson; Archivist, Torie Quinonez; Museum Shop Mgr., Joan Stewart.
Personnel Profile: Full-Time Paid 8; Part-Time Paid 14; Part-Time Volunteers 100.
Governing Authority: private; nonprofit organization. Casa de la Guerra, 15 E. de la Guerra St., Santa Barbara, CA. Tax-exempt.
Institution Type/Description: Historic Buildings: El Presidio de Santa Barbara State Historic Park - adobe fort built by the Spanish as they colonized California in the late 18th century. Casa de la Guerra - adobe house, home of early 19th century commandant of El Presidio.
Collections: archives pertaining to Spanish colonization of California; archaeological.
Research Fields: Spanish borderlands & colonization.
Facilities: library. Museum-related items for sale.
Activities: arts festivals; concerts; dance recitals; docent programs; films; education programs; guided tours; lectures; loan & temporary exhibitions; rental gallery; training programs for professional museum workers; fundraising events; archaeological dig. Annual Event: Living History.
Publications: quarterly newsletter, La Campana.
Hours & Admission Prices: El Presidio: daily 10:30-4:30. Casa de la Guerra: Thurs.-Sun. 12-4. Adults $5; discounts to seniors & AAM members; members no charge. Closed Thanksgiving; Christmas. &
Attendance: 20,000 (accurate)
Membership: Student $15; Individual $40; Dual $50; Preservationist $100; Steward $250.

SANTA BARBARA ZOOLOGICAL GARDENS, 500 Ninos Dr., Santa Barbara, CA 93103-3798. Tel.: 805-962-5339. Fax: 805-962-1673. Facebook: Santa Barbara Zoo.
E-mail: zooinfo@sbzoo.org
Web Site: www.sbzoo.org
Founded: 1963.
Congressional District: 19
Key Personnel: C.E.O., Richard Block; Zoo Dir., Nancy McToldridge; Dir. Animal Programs, Sheri Horiszny; Dir. Conservation & Research, Estelle Sandhaus; Cur., Mammals, Michele Green; Dir. Mktg., Dean Noble; Dir. Public Rels., Julia McHugh.
Governing Authority: nonprofit organization. Parent Institution: Santa Barbara Zoological Foundation. Tax-exempt: 501(c)(3).
Institution Type/Description: Zoo.
Collections: live zoological collection; 600 specimens including 175 species of invertebrates, amphibians, reptiles, birds & mammals.
Research Fields: reproductive physiology; behavioral.
Facilities: classrooms; picnic facilities. Animal-oriented gift items, books, art, games & film for sale; restaurant; catering.
Activities: guided tours; formally organized educational programs; docent program; permanent exhibitions; workshops; Zoo Camp.
Publications: quarterly newsletter, Zoo News; annual report.
Hours & Admission Prices: Daily 10-5, Thanksgiving and Christmas Eve & Day 10-3. Adults 13-64 $14, children 2-12 & senior citizens 65 & up $10; discounts to AZA members; children under 2 & members no charge. &
Attendance: 431,753 (accurate)
Membership: Individual $55; Individual Plus $70; Family $90; Family Plus $125; Supporting $165.

UNIVERSITY OF CALIFORNIA, SANTA BARBARA - WOMEN'S CENTER ART GALLERY, Student Resource Bldg., 1st Fl., Santa Barbara, CA 93106-7190. Tel.: 805-893-3778. Fax: 805-893-3289.
Web Site: www.sa.ucsb.edu/women/VisitourSpace/ArtGallery.aspx
Institution Type/Description: Art Gallery.
Collections: paintings; sculpture.
Activities: educational programs; classes.
Hours & Admission Prices: Mon.-Thurs. 9-9, Fri. 9-5.

WESTMONT RIDLEY - TREE MUSEUM OF ART, (M), 955 La Paz Rd., Santa Barbara, CA 93108-1099. Tel.: 805-565-6162. Fax: 805-565-7161.
E-mail: art@westmont.edu
Web Site: www.westmontmuseum.org
Formerly: Reynolds Gallery at Westmont College
Congressional District: 23
Key Personnel: Dir., Judy L. Larson, Ph.D.
Personnel Profile: Full-Time Paid 3; Part-Time Paid 5; Part-Time Volunteers 20.
Governing Authority: Parent Institution: Westmont College. Tax-exempt.
Institution Type/Description: Art Museum.
Collections: paintings; photographs; prints; drawings; graphic arts; sculpture; ceramics.
Activities: special events; receptions.
Hours & Admission Prices: Mon.-Fri. 10-4, Sat. 11-5. No charge. Closed college holidays.
Attendance: 5,000 (accurate)
Membership: Individual $50; Art Enthusiasts Circle $100; Art Collector's Circle $250; Art Patron's Circle $500; Major Benefactor $1,000; Friends of the Westmont Ridley-Tree Museum of Art $2,500; Director's Circle $10,000.

Santa Clara

✳ DE SAISSET MUSEUM, SANTA CLARA UNIVERSITY, (M), 500 El Camino Real, Santa Clara, CA 95053. Tel.: 408-554-4528. Fax: 408-554-7840.
E-mail: aefstratiou@scu.edu
Web Site: www.scu.edu/desaisset
Founded: 1955.
Congressional District: 13
Key Personnel: Dir., Rebecca M. Schapp; Cur. Exhibits & Collections, Lindsey Kouvaris; Collections Mgr., Jean MacDougall; Administrative Asst. to Dir., Anastasia Efstratiou; Preparator, Chris Sicat.
Personnel Profile: Full-Time Paid 3; Part-Time Paid 2; Part-Time Volunteers 45.
Governing Authority: university. Parent Institution: Santa Clara University. Tax-exempt: 501(c)(3).
Institution Type/Description: Art & History Museum.
Collections: 19th-century to present American art including graphics, painting, sculpture, photography & video; 15th-century to contemporary European painting, sculpture & graphic arts; American, European & Oriental decorative art; Native American (Costanoan); California Mission & University of Santa Clara History collection.
Research Fields: California art & history; modern American & European art; ecclesiastical textiles of the California missions.
Facilities: 1,200-vol. library of books, catalogs & periodicals; video archive, video art & video tapes; interviews with New Deal artists; 200-seat auditorium. Museum catalogs for sale.
Activities: guided tours; inter-museum loan program; permanent, temporary & traveling exhibitions.
Publications: exhibition catalogs; member newsletter.
Hours & Admission Prices: Tues.-Sun. 11-4. No charge; donations accepted. Closed national holidays; Martin Luther King Jr. Day; Good Friday; Thanksgiving weekend; between exhibitions. &
Attendance: 10,500 (estimated)
Membership: Faculty, Staff, Senior & Senior Dual $45; Sponsor $100; Patron $250; Benefactor $500; de Saisset Fellow $1,000; Locatelli Fellow $2,500; Presidential Fellow $5,000.

EDWARD PETERMAN MUSEUM OF RAILROAD HISTORY, 1005 Railroad Ave., Santa Clara, CA 95050-4319. Tel.: 408-243-3969.
E-mail: info@sbhrs.org
Web Site: www.sbhrs.org
Founded: 1985.
Congressional District: 17
Key Personnel: C.E.O., Robert Marshall; Pres. (V), Michael Steckwell.
Personnel Profile: Part-Time Volunteers 100.
Governing Authority: nonprofit organization. Tax-exempt: 501(c)(3).
Institution Type/Description: History Museum: housed in the Santa Clara Depot; built in 1863.
Collections: railroad heritage & history; railroad artifacts; photographs; Oregon-Washington Railroad & Navigation business car #184; HO & N scale model railroads. Historic Buildings: 1926 Southern Pacific Santa Clara Tower; 1894 Section Tool House.
Activities: guided tours; educational programs; classes; scout & school programs.
Hours & Admission Prices: Tues. 6pm-9pm, Sat. 10-3. No charge; donations accepted. &
Attendance: 5,000 (estimated)
Membership: Senior 60 & over $240; Individual $300.

INTEL MUSEUM, Robert Noyce Bldg., 2200 Mission College Blvd., Santa Clara, CA 95054. Tel.: 408-765-5050.
E-mail: museum@intel.com
Web Site: www.intel.com/museum
Founded: 1992.
Key Personnel: Cur., Tracey Mazur; Archivist, Rachel Stewart; Education, Elizabeth Jones; Technical Network Systems Engineer, Elyse Polhaupessy; Museum Shop Mgr., Richard Smith.
Personnel Profile: Full-Time Paid 3; Part-Time Volunteers 10.
Governing Authority: corporate; nonprofit. Parent Institution: Intel Corp.
Institution Type/Description: Corporate History, Science & Technology Museum.
Collections: hands-on exhibits; corporate history; industry records, artifacts, products, documents & images; 80,000 records in database.
Research Fields: early & current intel product & corporate development.
Facilities: learning lab; 10,000 sq. ft. exhibit space. Museum-related items for sale.
Activities: formal education programs; hands-on exhibits; guided tours.
Publications: monthly newsletter; brochure; anniversary publications; product timeline charts.
Hours & Admission Prices: Mon.-Fri. 9-6, Sat. 10-5. No charge. Closed New Year's Day; Washington's Birthday; Memorial Day; Independence Day; Labor Day; Thanksgiving; Christmas. &
Attendance: 116,847 (accurate)

TRITON MUSEUM OF ART, 1505 Warburton Ave., Santa Clara, CA 95050-3791. Tel.: 408-247-3754. Fax: 408-247-3796.
E-mail: staff@tritonmuseum.org
Web Site: www.tritonmuseum.org
Founded: 1965.
Congressional District: 12
Key Personnel: Exec. Dir. & Senior Cur., George Rivera; Chm. (V), Rosalie Wilson; Pres. (V), Robert Baden; Registrar & Assoc. Cur., Stephanie Learmonth; Chief Cur., Preston Metcalf; Dir. Operations, Jill Meyers; Museum Shop Mgr., Amanda McDonough.

Personnel Profile: Full-Time Paid 4; Part-Time Paid 12; Part-Time Volunteers 40; Interns 4.
Governing Authority: private; nonprofit corporation. Tax-exempt: 170(b)(1)(A) & 501(c)(3).
Institution Type/Description: Art Museum & Center.
Collections: c.1859-1939 paintings by California artist Theodore Wores; American artists; Vivan Woodward Elmer Majolica collection; contemporary American drawings & prints; 19th to 21st-century artists; Native American art & artifacts; 20th to 21st-century Bay Area art. Historic House: 1866 Jamison-Brown house.
Research Fields: pertaining to permanent collection & temporary exhibitions.
Facilities: 1,000-vol. library of books & journals available for staff, docents & members; sculpture garden. Gift items for sale.
Activities: guided tours; slide lectures; gallery talks; workshops; auctions; studio art classes for children; outreach program, ARTREACH; docent out-reach program to schools & senior centers; rental space.
Publications: catalogues; A Catalog of the Paintings by Theodore Wores in the Permanent Collection of the Triton Museum of Art; Austen D. Warburton Collection of American Indians Art & Artifacts; Expressions from the Soul: Bay Area Korean - American Women; Iku K. Nagai; Los Tres: Jose Clemente Orozco, Diego Rivera & David Alfaro Siqueiros; A Visual Heritage: Bay Area African-American Artists; A Bay Area Connection: Works from the Anderson Collection 1954-1984; brochures related to exhibitions; quarterly calendar & newsletter.
Hours & Admission Prices: Tues.-Sat. 11-5, Sun. 12-4. No charge; donations accepted. Closed national holidays. &
Attendance: 40,000 (estimated)
Membership: Student & Senior $20; Individual $40; Family $55; Curator's Circle $100; Director's Circle $250; Patron $500; Benefactor $1,000.

Santa Cruz

ARBORETUM AT UC SANTA CRUZ, University of California, 1156 High St., Santa Cruz, CA 95064-1077. Tel.: 831-427-2998. Fax: 831-427-1524. Facebook: UCSC Arboretum.
E-mail: arboretum@ucsc.edu
Web Site: ucsc.edu/arboretum/
Founded: 1964.
Congressional District: 16
Key Personnel: Dir., Brett Hall; Pres. (V), Peggy Williams; Dir. Research, Stephen McCabe; Harry O. Warren Cur. of New Zealand Collection, Thomas Sauceda; Admin. & Museum Shop Mgr., Susie Bower; Cur. Australian Collection, Melinda Johnson; Cur. Native Collection, Rick Flores.
Personnel Profile: Full-Time Paid 4; Full-Time Volunteers 25; Part-Time Paid 4; Part-Time Volunteers 200; Interns 10.
Governing Authority: Parent Institution: University of California Santa Cruz. Tax-exempt: 501(c)(3).
Institution Type/Description: Arboretum & Botanical Gardens.
Collections: living plants of scientific, conservation, educational or ornamental value.
Research Fields: horticulture; evolution of plants; conifers; bulbs; natives; land management; rare fruit collections; primitive angiosperms; Australian; South African; New Zealand; California Natives; succulent plants.
Facilities: botanical garden. Plants & gift items for sale.
Activities: guided tours; lectures; docent program or council; formally organized education programs for undergraduate college students affiliated with UCSC; permanent exhibitions; teaching; public outreach; conservation.
Publications: quarterly bulletin, Friends of the Arboretum.
Hours & Admission Prices: Daily 9-5. Adults $5; members no charge. &
Attendance: 55,000 (estimated)
Membership: Individual $50; Dual & Family $65; Kauri $100-$249; Erica $250-$499; Banksia $500-$999; Protea $1000 & up; Life $2500.

ELOISE PICKARD SMITH GALLERY, Cowell College, University of California, Santa Cruz, CA 95064. Tel.: 831-459-2953.
E-mail: lapope@ucsc.edu
Web Site: cowell.ucsc.edu/smith.gallery/main.php
Key Personnel: Dir. & Gallery Cur., Linda Pope
Institution Type/Description: Art Gallery.
Collections: Monterey Bay region contemporary art & photographs.
Hours & Admission Prices: Call for hours. &
Attendance: 3,200

MARY PORTER SESNON ART GALLERY, (M), UC Santa Cruz, Porter College, Santa Cruz, CA 95064. Mailing Address: Porter Faculty Svcs., Porter College, UC Santa Cruz, Santa Cruz, CA 95064. Tel.: 831-459-3606. Fax: 831-459-3535.
E-mail: sgraham@ucsc.edu

Web Site: arts.ucsc.edu/sesnon
Founded: 1971.
Key Personnel: Dir. & Cur., Shelby Graham; Gallery Mgr. & Asst. Cur., Mark Shunney.
Personnel Profile: Full-Time Paid 1; Part-Time Paid 4; Part-Time Volunteers 1; Interns 1.
Governing Authority: tax-exempt. Parent Institution: University of California.
Institution Type/Description: Art Gallery.
Collections: archives; drawings; video; paintings.
Publications: exhibition catalogues.
Hours & Admission Prices: Sept.-June 15 Tues.-Sat. 12-5. No charge; donations accepted. &
Attendance: 3,500 (estimated)

MISSION SANTA CRUZ, 130 Emmet St., Santa Cruz, CA 95060. Mailing Address: 126 High St., Santa Cruz, CA 95060-3711. Tel.: 831-426-5686. Fax: 831-423-1043.
Web Site: missiontour.org/santacruz/index.htm
Founded: 1791.
Institution Type/Description: Religious Museum.
Collections: replica of the original mission church.
Hours & Admission Prices: Tues.-Sat. 10-4, Sun. 10-2. No charge; donations accepted. Closed holidays.

MUSEUM OF ART AND HISTORY AT THE MCPHERSON CENTER, 705 Front St., Santa Cruz, CA 95060-4508. Tel.: 831-429-1964. Fax: 831-429-1954.
E-mail: director@santacruzmah.org
Web Site: www.santacruzmah.org
Founded: 1996.
Congressional District: 16
Key Personnel: Exec. Dir., Nina Simon; Dir. Exhibitions, Susan Leask; Cur. Education, Ashley Carniglia; Administrative Mgr., Nicole Campbell; Dir. Membership, Karen Bush; Mgr. Visitors & Volunteers, Diana Kapsner; Cur. History & Registrar, Marla Novo; Program Assoc., Emily Hope Dobkin; Dir. Community Programs, Stacey Marie Garcia.
Personnel Profile: Full-Time Paid 9; Part-Time Paid 4; Part-Time Volunteers 120; Interns 20.
Governing Authority: nonprofit organization. Tax-exempt.
Institution Type/Description: Museum of Art & History, Archives Library on Site.
Collections: art & historical artifacts.
Research Fields: Santa Cruz County, CA history.
Facilities: auditorium. Museum-related items for sale.
Activities: guided tours; lectures; films; organized education programs for adults; docent program; children's workshops; family fun days.
Publications: quarterly newsletter; yearly history journal; exhibition catalogs; Lime Kiln Legacies; Sidewalk Companion to Santa Cruz Architecture; Legal History Santa Cruz County; Surf, Sand and Streetcars; A Gathering of Voices - The Native Peoples of the Central California Coast.
Hours & Admission Prices: Tues.-Sun. 11-5. Adults $5, students 18 and over & senior citizens $4; members, students under 18 & children under 12 no charge. Closed legal holidays. &
Attendance: 36,000 (estimated)
Membership: Educator & Student $20; Individual $40; Dual & Family $60; Director's Circle $125; Artist's Circle $250; Collector's Circle $500; Trustee's Circle $1,000; President's Circle $2,500; Patron's Circle $5,000.

SANTA CRUZ ART LEAGUE, INC., 526 Broadway, Santa Cruz, CA 95060-4622. Tel.: 831-426-5787. Fax: 831-426-5789.
E-mail: cindy@scal.org
Web Site: www.scal.org
Founded: 1919.
Congressional District: 16
Key Personnel: Pres. (V), T. Mike Walker; Administrative Dir., Cindy Liebenthal.
Personnel Profile: Part-Time Paid 4; Part-Time Volunteers 50; Interns 5.
Governing Authority: nonprofit organization. Tax-exempt: 501(c)(3).
Institution Type/Description: Art Gallery.
Collections: pioneer paintings of local area artists; sculpture from the 1915 Panama Pacific Expo (San Francisco).
Research Fields: local art and artists.
Facilities: 1,800 sq. ft. main gallery; 62-seat performance space; 700 sq. ft. classroom; kitchen.
Activities: gallery talks; art classes; art demonstrations & workshops; theater performances; outdoor art fairs.
Publications: monthly members' bulletin.
Hours & Admission Prices: Wed.-Sat. 12-5, Sun. 12-4. No charge; donations accepted. &

Attendance: 20,000
Membership: Students & Seniors $50; Individual $75; Family $125; Sponsor $150; Patron & Business Sponsor $250-$500; Benefactor $500 & up.

Santa Fe Springs

HATHAWAY RANCH MUSEUM, (M), 11901 E. Florence Ave., Santa Fe Springs, CA 90670-4494. Tel.: 562-777-3444. Fax: 562-945-1892.
E-mail: hathawayranch@gmail.com
Web Site: hathaworld.com
Founded: 1986.
Key Personnel: C.E.O. & Exec. Dir., Francine Rippy; Pres. Bd., Virginia Boles.
Personnel Profile: Full-Time Paid 1; Part-Time Paid 2; Part-Time Volunteers 15.
Governing Authority: bd. directors. Tax-exempt.
Institution Type/Description: History Museum.
Collections: 19th- & 20th-century American furnishings; farm buildings & equipment; 3,000 local history photographs; oil field & feed lot artifacts; household articles; clothing.
Research Fields: Southern California oil industry; Santa Fe Springs history; early ranching; farming.
Facilities: archives; visitors center; picnic area; 30-seat meeting room; outdoor banquet facilities.
Activities: historical lectures; farm equipment demonstrations; hay wagon rides; tours; farm animals; belt driven machine shop.
Publications: pamphlet, Lovingly Yours - the Letters of Lola McCarric to Jesse Hathaway; booklet, Settlers of Southern California, Vols. 1-9.
Hours & Admission Prices: Mon.-Tues. & Thurs. 11-4, Fri.-Sun. & group tours by appointment. No charge; donations accepted. Closed major holidays.
Attendance: 1,200 (estimated)
Membership: Individual $5; Family $10.

Santa Maria

THE NATURAL HISTORY MUSEUM OF SANTA MARIA, 412 S. McClelland, Santa Maria, CA 93454-5117. Mailing Address: P.O. Box 5254, Santa Maria, CA 93456-5254. Tel.: 805-614-0806. Fax: 805-614-0806.
Web Site: www.naturalhistorysantamaria.com
Formerly: Samuel J. Perry Natural History Museum
Founded: 1996.
Congressional District: 33
Key Personnel: Pres., Lora Carter; Chm. (V), Bill Decker; Sec., Laura Dias; Treas., Tahir Masood.
Personnel Profile: Part-Time Volunteers 20.
Governing Authority: private; nonprofit organization. Tax-exempt: 501(c)(3).
Institution Type/Description: Natural History Museum: housed in the historic Hart Home.
Collections: natural history; regional wildlife dioramas & their habitats.
Activities: bird kit & curriculum; bug kit & curriculum; rock and fossil kits; videos. Special Events: Live! At the Museum each month; Earth Day Celebration in April; Conservation Awards on Earth Day.
Publications: semiannual newsletter, The Meadowlark quarterly members newsletter.
Hours & Admission Prices: Wed.-Sat. 11-4, Sun. 1-4. No charge; donations accepted. Closed holidays. &
Attendance: 4,850 (accurate)
Membership: Hummingbird $25; California Quail $50; Meadowlark $100; Great Horned Owl $250; Golden Eagle $500; Red Tailed Hawk $1,000; California Condor $2,000.

SANTA MARIA MUSEUM OF FLIGHT, INC., (M), 3015 Airpark Dr., Santa Maria, CA 93455-1821. Tel.: 805-922-8758. Fax: 805-922-8958.
E-mail: info@smmof.org
Web Site: www.smmof.org
Founded: 1984.
Congressional District: 19
Key Personnel: Pres., Michael Geddry, Sr.; Registrar, Jerry Nicholson; Museum Shop Mgr., Tom Reedy.
Personnel Profile: Full-Time Volunteers 12; Part-Time Volunteers 20.
Governing Authority: public benefit nonprofit corporation. Tax-exempt: 501(c)(3).
Institution Type/Description: Aeronautics Museum.
Collections: aircraft & aviation artifacts from early flight to current space activities with special emphasis on the aviation & space history of the central coast of California; Stinson Reliant, Fleet Model 2, Rand KR2, Sonerii Fly Baby, 1931 Great Lakes, rocket engines, satellite simulator; rotating display of aircraft; the Topping collection of miniature aircrafts; dioramas; L-5; F-4 Phantom; full scale Wright 1902 glider & F-86 sabre; H1 (Howard Hughes) full scale replica; A-4; Flight of the Phoenix scale movie prop; Folland Gnat.
Research Fields: national flying schools of the late 20s & early 30s; Flight of the Southern Cross; early aviation of central coast of California; pilot training & the original 9 primary flight training schools (1939-1945); history of North American Aviation Company.
Facilities: 2,100-vol. library & 1,600 VCR tapes on aviation; 11,000 sq. ft. exhibit space. Museum-related items for sale.
Activities: guided tours; docent training. Annual Event: Warbirds Aircraft Fly In; Ford Car Show.
Publications: quarterly newsletter, The Flight Plan.
Hours & Admission Prices: Fri.-Sun. 10-4; groups & other times by appointment. Suggested Donation: adults $5, seniors $4, students $3, children 6 & under $1; members, military & their dependents no charge. Closed New Year's Day; Christmas. &
Attendance: 10,000 (estimated)
Membership: Student & Senior $20; Individual $35; Family $50; Business $100; Life 55 & over $250; Life under 55 $350.

SANTA MARIA VALLEY DISCOVERY MUSEUM, 705 S. McClelland St., Santa Maria, CA 93454-5122. Tel.: 805-928-8414.
E-mail: programs@smdiscoverymuseum.org
Web Site: www.smvdiscoverymuseum.org
Founded: 1992.
Personnel Profile: Full-Time Paid 1; Part-Time Paid 6; Part-Time Volunteers 10.
Institution Type/Description: Children's Museum.
Collections: hands-on exhibits.
Facilities: Museum-related items for sale.
Activities: educational programs; summer camps.
Hours & Admission Prices: Mon.-Wed. & Fri.-Sat. 10-5, Thurs. 10-7. Admission $8 per person; members & children under 2 no charge. &
Attendance: 30,000 (accurate)
Membership: Basic $85; Family $100; Reciprocal $125.

SANTA MARIA VALLEY HISTORICAL MUSEUM, 616 S. Broadway, Santa Maria, CA 93454-5111. Tel.: 805-922-3130.
E-mail: richard@santamariahistory.org
Web Site: www.smvrhm.org
Formerly: Santa Maria Valley Historical Society, Inc.
Founded: 1955.
Key Personnel: Pres., Dave Carey; Museum Dir., Richard Chenoweth; Sec., Jim O'Neill.
Personnel Profile: Full-Time Paid 1; Part-Time Paid 2; Part-Time Volunteers 30.
Governing Authority: nonprofit organization. Tax-exempt: 501(c)(3).
Institution Type/Description: General Museum: housed in building constructed on site of c.1900 1st municipal water works.
Collections: Indian artifacts; Rancho period memorabilia; pioneer profile room displays; models; Santa Maria Valley history.
Research Fields: local & valley history.
Facilities: library; reading room. Books & museum-related items for sale.
Activities: guided tours; lectures; films; education programs; temporary exhibits.
Publications: quarterly newsletter; book, This Is Our Valley - Santa Maria Historical Photo Album; Santa Maria Style Barbecue.
Hours & Admission Prices: Tues.-Sat. 12-5. &
Attendance: 4,320 (estimated)
Membership: Active $20; Family $30; Patron $50; Vaquero & Townbuilder $100; Pioneer Patriot $250; Founder's Circle $500; Lifetime $1,000.

Santa Monica

ANGELS ATTIC MUSEUM, 516 Colorado Ave., Santa Monica, CA 90401. Tel.: 310-394-8331. Fax: 310-656-6865.
E-mail: info@angelsattic.com
Web Site: www.angelsattic.com
Founded: 1984.
Key Personnel: Museum Dir., Nicole Dickerson; Dir., Eleanor La Vove; Museum Shop Mgr., Susan Baker.
Personnel Profile: Full-Time Paid 1; Full-Time Volunteers 2; Part-Time Paid 3; Part-Time Volunteers 20.
Governing Authority: private; nonprofit. Tax-exempt: 501(c)(3).
Institution Type/Description: Toy & Doll Museum: housed in a c.1895 Queen Anne Victorian House.
Collections: dollhouses; miniatures; toys; dolls.
Facilities: Gift items for sale.

Hours & Admission Prices: Thurs.-Sun. 12:30-4:30. Adults $6.50; discounts to seniors & children under 12. &
Attendance: 5,000 (estimated)
Membership: Individual $35; Angel $100.

CALIFORNIA HERITAGE MUSEUM, 2612 Main St., Santa Monica, CA 90405-4002. Tel.: 310-392-8537. Fax: 310-396-0547.
E-mail: calmuseum@earthlink.net
Web Site: californiaheritagemuseum.org
Founded: 1977.
Key Personnel: Dir., Tobi Smith; Chm. (V), Mark Burton.
Personnel Profile: Full-Time Paid 3; Part-Time Paid 4; Part-Time Volunteers 35; Interns 2.
Governing Authority: Tax-exempt.
Institution Type/Description: Fine Art & Decorative Art Museum.
Collections: California decorative arts, fine art, folk art, pottery & tile 1900-1930; photo archive; photography; culture & heritage of the state.
Research Fields: California history & decorative arts; photography archive.
Publications: Hawaiian Style; Arts & Crafts Movement; International Masks, Masks of the World; Monterey Furniture - California Spanish Revival; California Tile, The Golden Era 1910-1940; Saints & Sinners - Mexican Devotional Art; Everyday Life in California - Regional Watercolors; Shortboard Revolution.
Hours & Admission Prices: Wed.-Sun. 11-4. Adults $8, senior citizens & students $5; discount to AAA members; children under 12 & members no charge. Closed major holidays &
Attendance: 17,500 (estimated)
Membership: Student & Senior $25; General $35; Patron $100; Collector's Club $250; Roy Jones Society $1,000 & up.

FIRST INDEPENDENT GALLERY, Bergamot Station G6, 2525 Michigan Ave., Santa Monica, CA 90404. Tel.: 310-829-0345. Fax: 310-829-7612.
E-mail: fig@figgallery.com
Web Site: www.figgallery.com
Institution Type/Description: Art Gallery.
Collections: contemporary works of art by Southern California artists.
Activities: temporary exhibitions.
Hours & Admission Prices: Wed.-Sat. 11-5; other times by appointment.

ROSAMUND FELSEN GALLERY, Bergamot Station B-4, 2525 Michigan Ave., Santa Monica, CA 90404-4031. Tel.: 310-828-8488. Fax: 310-828-1075.
E-mail: rosamund@rosamundfelsen.com
Web Site: www.rosamundfelsen.com
Key Personnel: Dir., Rosamund Felsen
Institution Type/Description: Art Gallery.
Collections: works by contemporary artists.
Activities: temporary exhibitions; special events.
Hours & Admission Prices: Tues.-Sat. 10-5:30.

SANTA MONICA HISTORY MUSEUM, (M), 1350 7th St., Santa Monica, CA 90401. Mailing Address: P.O. Box 3059, Santa Monica, CA 90408-3059. Tel.: 310-395-2290. Fax: 310-395-2290 (call first).
Web Site: www.santamonicahistory.org
Founded: 1975.
Key Personnel: Pres. & C.E.O., Louise Gabriel; Chm. (V), Eddie Guerboian; Museum Shop Mgr., Elaine Blaugrund.
Personnel Profile: Full-Time Paid 1; Part-Time Paid 1.
Governing Authority: Tax-exempt.
Institution Type/Description: History Museum.
Collections: local history & culture; photographs; artwork; architectural drawings; newspapers; books; period artifacts.
Publications: quarterly newsletter; monthly e-zine.
Hours & Admission Prices: Tues. & Thurs. 12-8, Wed. 10-4, Fri-Sat. 11-5. Adults $5, seniors & students $3; children 12 & under and members no charge. &
Attendance: 2,000 (accurate)
Membership: Individual $40; Family/Organization $50; Sustaining $60; Patron $100; Business $200; Life $500.

SANTA MONICA MUSEUM OF ART, (M), Bergamot Station, 2525 Michigan Ave., Bldg. G1, Santa Monica, CA 90404-4042. Tel.: 310-586-6488. Fax: 310-586-6487.
E-mail: info@smmoa.org
Web Site: www.smmoa.org

Founded: 1984.
Congressional District: 27
Key Personnel: Exec. Dir., Elsa Longhauser; Pres., Price Latimer Agah; Deputy Dir., Doug Rimerman; Dir. Education, Asuka Hisa; Deputy Dir., Claire Ruud; Dir. Mktg., Elizabeth Pezza; Registrar, Brian Briggs; Assoc. Dir. Advancement, Kim Newstadt; Gracie Cur., Amy Coane; Bookkeeper, Carole Faxon; Asst. Cur., Laura Copelin; Museum Shop Mgr., Adrienne Parker.
Personnel Profile: Full-Time Paid 12; Part-Time Paid 2; Part-Time Volunteers 50; Interns 5.
Governing Authority: nonprofit organization; board of trustees. Tax-exempt: 501(c)(3).
Institution Type/Description: Contemporary Art Museum.
Collections: works by contemporary artists.
Facilities: 6,000 sq. ft. exhibit space; education and orientation rooms.
Activities: educational & outreach programs; lectures; temporary exhibitions; intern program.
Publications: exhibition catalogs.
Hours & Admission Prices: Tues.-Sat. 11-6. Suggested Donations: adults $5, artists, senior citizens & students $3; members no charge. Closed legal holidays. &
Attendance: 40,000 (estimated)
Membership: Artist, Student & Senior $40; Individual $55; Partner $100; Family $300; SMMOA Fellows $500; Curatorial Corps $1,000; Director's Salon $3,000 & up.

SANTA MONICA PIER AQUARIUM, 1600 Ocean Front Walk, Santa Monica, CA 90401. Tel.: 310-393-6149.
Web Site: www.healthebay.org/smpa
Governing Authority: Parent Institution: Heal The Bay.
Institution Type/Description: Aquarium.
Collections: local history; marine science; marine animals & plants.
Activities: educational programs; interactive exhibits; touch tanks; birthday parties; summer camps.
Hours & Admission Prices: April-Aug. Tues.-Fri. 2-6, Sat.-Sun. 12:30-6; Sept.-March Tues.-Fri. 2-5, Sat.-Sun. 12:30-5. Adults 12 & over $5; discounts to groups; children under 12 no charge.
Attendance: 85,000 (accurate)

Santa Paula

AVIATION MUSEUM OF SANTA PAULA, Santa Paula Airport, Hwy. 126, Santa Paula, CA 93060. Mailing Address: 800 E. Santa Maria St., #E, Santa Paula, CA 93060. Tel.: 805-525-1109.
E-mail: amszp@verizon.net
Web Site: www.amszp.org
Institution Type/Description: History Museum.
Collections: aviation & airport history; early aircraft; period radios; racecars.
Facilities: Museum-related items for sale.
Hours & Admission Prices: Call for hours. No charge; donations accepted. &

CALIFORNIA OIL MUSEUM, 1001 E. Main St., Santa Paula, CA 93060-2809. Mailing Address: P.O. Box 48, Santa Paula, CA 93061-0048. Tel.: 805-933-0076. Fax: 805-933-0096. Facebook: California Oil Museum.
E-mail: info@oilmuseum.net
Web Site: www.oilmuseum.net
Formerly: Santa Paula Union Oil Museum
Founded: 1950.
Congressional District: 23
Key Personnel: Dir., Jeanne Orcutt; Chm., Mary Alice Henderson.
Personnel Profile: Full-Time Paid 1; Part-Time Paid 1; Part-Time Volunteers 15; Interns 2.
Governing Authority: city. Tax-exempt.
Institution Type/Description: Museum of the Oil Industry.
Collections: oil industry artifacts; photographs; paintings; publications.
Major Exhibits: Prehistoric California, 10/13-2/2/14.
Research Fields: oil history in California.
Facilities: facility available for receptions & special events.
Publications: Pipelines museum newsletter.
Hours & Admission Prices: Museum: Wed.-Sun. 10-4. Guided Tours: Sat.-Sun. 10-4, other days & times by appointment. Adults $4, senior citizens $3, children $1. Closed major holidays. &
Attendance: 12,000 (accurate)
Membership: Regular $35; Plus $50; Premium $100; Supreme $500; Corporate $1,000-$10,000.

SANTA PAULA ART MUSEUM, (M), 117 N. 10th St., Santa Paula, CA 93060-2877. Tel.: 805-525-5554.
Founded: 2004.
Congressional District: 24
Key Personnel: Exec. Dir., Jennifer Heighton.
Personnel Profile: Full-Time Paid 1; Part-Time Paid 1; Part-Time Volunteers 50.
Governing Authority: nonprofit organization. Tax-exempt: 501(c)(3).
Institution Type/Description: Art Museum.
Collections: over 300 works of art.
Activities: educational programs.
Hours & Admission Prices: Wed.-Sat. 10-4, Sun. 12-4. Adults $4, seniors $3; members & students no charge. ♿
Attendance: 5,000 (estimated)
Membership: Foundation $40; Portico $65; Cornice $100; Gable $150; column $250; Parapet $500; Atrium $1,000.

Santa Rosa

CALIFORNIA INDIAN MUSEUM & CULTURAL CENTER, 5250 Aero Dr., Santa Rosa, CA 95403-8069. Tel.: 707-579-3004. Fax: 707-579-9019.
E-mail: CIMandCC@aol.com
Web Site: www.cimcc.org
Key Personnel: Pres. (Pomo), Andrew Maisel; Vice Pres., Jerry Burroni; Exec. Dir., Nicole Myers-Lim; Project Mgr., David Lim; Devel. Specialist, Carol Oliva; Administrative Asst., Ramona Cruz
Institution Type/Description: Indian Museum.
Collections: art; traveling exhibits; photo displays.
Activities: lectures; walking tours; community events.
Publications: book, The California Indian.
Hours & Admission Prices: Mon.-Fri. 9-5. No charge; donations accepted.
Membership: Members $25; Sponsor $50; Patron $100; Benefactor $1,000; President's Circle $1,000 & up.

CHARLES M. SCHULZ MUSEUM AND RESEARCH CENTER, (M), 2301 Hardies Lane, Santa Rosa, CA 95403-2668. Tel.: 707-579-4452, ext. 260. Fax: 707-579-4436.
E-mail: inquiries@schulzmuseum.org
Web Site: www.schulzmuseum.org
Founded: 2002.
Congressional District: 1
Key Personnel: Pres. (V), Jean F. Schulz; Dir., Karen Johnson.
Personnel Profile: Full-Time Paid 11; Part-Time Paid 12; Part-Time Volunteers 150.
Governing Authority: Tax-exempt.
Institution Type/Description: Art Museum.
Collections: morphing Snoopy wood sculpture; tile mural featuring the images of Charlie Brown & Lucy composed of 3,588 Peanuts comic strips; 96 sq. ft. wall from early Schulz home containing characters painted by Charles Schulz; original Peanuts comic strip drawings; photographs & memorabilia related to life of Charles Schulz.
Major Exhibits: Merry Christmas Charlie Brown, 11/13-1/14.
Research Fields: cartoon art history, life & art of Charles Schulz.
Facilities: 27,384 sq. ft. exhibit space; 100 seat auditorium; classroom; outdoor gardens.
Activities: permanent & temporary exhibits.
Publications: exhibitions catalogs; quarterly newsletter, Musings.
Hours & Admission Prices: Memorial Day to Labor Day Mon.-Fri. 11-5, Sat.-Sun. 10-5; Sept.-May Mon. & Wed.-Fri. 11-5, Sat.-Sun. 10-5. Adults $10, seniors & youth $5; discounts to AAM, CAM, AAA & WMA members; members no charge. Closed New Year's Day; Easter; Independence Day; Thanksgiving; Christmas Eve & Day. ♿
Attendance: 70,000 (accurate)
Membership: Individual $40; Family $70; Fan $100; Super Fan $250; Legacy $500; Golden Legacy $1,000; Patron $2,500; Sponsor $5,000; Corporate Levels: Contributing $500; Supporting $1,000; Associate $2,500; Benefactor $5,000.

CHILDREN'S MUSEUM OF SONOMA COUNTY, 1835 W Steele Ln., Santa Rosa, CA 95403-2628. Mailing Address: P.O. Box 12323, Santa Rosa, CA 95406. Tel.: 707-546-4069. Fax: 707-658-1981.
E-mail: info@cmosc.org
Web Site: cmosc.org
Key Personnel: Found & C.E.O., Collette Michaud; C.F.O., Douglas Kay; Operations & Event Coord., Eleanor Gorman; Dir. Programs & Education, Theresa Giacomino; Assoc. Dir. Programs & Education, Anya Kennerly
Institution Type/Description: Children's Museum.

Collections: hands-on exhibitions.
Activities: educational programs; special events; workshops.
Hours & Admission Prices: Call for hours.

LUTHER BURBANK HOME & GARDENS, Santa Rosa Ave. at Sonoma Ave., Santa Rosa, CA 95402. Mailing Address: 100 Santa Rosa Ave., Rm. 10, Santa Rosa, CA 95404-4957. Tel.: 707-524-5445. Fax: 707-524-5827.
E-mail: burbankhome@lutherburbank.org
Web Site: www.lutherburbank.org
Founded: 1979.
Congressional District: 2
Key Personnel: Vice Chm., Daniel Flock; Chm. (V), Claire Borges; Treas., Toni Hower; Archivist, Rebecca Baker; Museum Shop Mgr., Sharlene McCaw.
Personnel Profile: Part-Time Paid 2; Part-Time Volunteers 150.
Governing Authority: Tax-exempt: 501(c)(3).
Institution Type/Description: Historic House & Garden: 1884 Luther Burbank Home; 1889 greenhouse.
Collections: furnishings; tools; photographs; plants; documents.
Major Exhibits: Pursuing the Perfect Plum, 4/1/13-10/31/14.
Research Fields: Burbank life & work.
Facilities: 450-vol. library of catalogs, photographs & publications available for research by application on premises only; demonstration gardens. Museum-related items for sale.
Activities: guided tours; plant sales. Annual Events: fundraiser in July; holiday open house in December.
Publications: Luther Burbank: Gardener to the World; Fifty Famous Favorites, Plant Introductions by Luther Burbank; A Gardener Touched with Genius: The Life of Luther Burbank; Here a Plant, There a Plant, Children's Biography on Luther Burbank.
Hours & Admission Prices: April-Oct. Tues.-Sun. 10-3:30; groups & children's tours by appointment. Guided tours $7; Friends of Luther Burbank & cell phone audio tours no charge. Gardens: daily 8 to dusk. No charge. ♿
Attendance: 75,000 (accurate)
Membership: Senior Citizens & Students $25; Supporter $50; other levels $100 & up

PACIFIC COAST AIR MUSEUM, N. Laughlin at Becker, Santa Rosa, CA 95403. Mailing Address: 2230 Becker Blvd., Santa Rosa, CA 95403-8275. Tel.: 707-575-7900. Fax: 707-545-2813. www-.wingsoverwinecountry.org.
E-mail: director@pacificcoastairmuseum.org
Web Site: www.pacificcoastairmuseum.org
Founded: 1990.
Key Personnel: Pres. (V), Allan Morgan; Exec. Dir., Dave Pinsky; Public Rels., Doug Clay; Museum Shop Mgr., Robert Couz.
Personnel Profile: Full-Time Paid 1; Part-Time Volunteers 150.
Governing Authority: private; nonprofit organization. Tax-exempt.
Institution Type/Description: Aircraft Museum.
Collections: aircraft including DC-6 cockpit section, F-15, D-21 Drone, T-28, F-5, S-2, A-7, IL-14, F-8, A-4, F-14, F-4, A-6, C-118, F-16, A-26, T-38, T-37, T-33, F-84, F-86, F-105, F-106, H-34 helicopter, RF86, HU-16E & a UH-1 on display; indoor exhibits of WWII material; civil artifacts.
Facilities: Museum-related items for sale.
Activities: docent program; guided tours. Annual Event: Air Show.
Publications: monthly newsletter, Straight Scoop.
Hours & Admission Prices: Tues., Thurs. & Sat.-Sun. 10-4. Adults $9, seniors 65 & over $7, child 6-17 $5; discount to California Assoc. of Museums; active & retired military and children 5 & under no charge. Closed New Year's Day; Thanksgiving; Christmas. ♿
Attendance: 23,569 (accurate)
Membership: Individual $30; Family $45.

ROBERT F. AGRELLA ART GALLERY, 1501 Mendocino Ave., Santa Rosa, CA 95401-4395. Tel.: 707-527-4298.
Web Site: www.santarosa.edu/art-gallery/index.shtml
Formerly: Santa Rosa Jr. College Art Gallery
Key Personnel: Dir., Renata Breth
Institution Type/Description: College Art Gallery.
Collections: works by student & faculty artists.
Activities: programs; temporary exhibitions. Museum Sponsors: Annual Student Art Show; Art Faculty Exhibits.
Hours & Admission Prices: Mon.-Thurs. 10-4, Sat. 12-4. No charge; donations accepted. Closed summers & all school holidays.

SAFARI WEST WILDLIFE PRESERVE, 3115 Porter Creek Rd., Santa Rosa, CA 95404-9655. Tel.: 707-579-2551; 800-626-2695. Fax: 707-579-8777.
E-mail: safariwest@safariwest.com
Web Site: www.safariwest.com
Founded: 1993.
Key Personnel: Founder, C.E.O. & Pres, Peter Lang; Gen. Dir., Nancy Lang; Dir. Mktg. & Communications, Aphrodite Caserta; Dir., Mark DeWitt
Institution Type/Description: Wildlife Preserve.
Collections: over 500 animals representing more than 80 species including cheetah, giraffe, lemurs, cape buffalo, & zebra.
Facilities: Gift items for sale.
Activities: special events; educational programs; school fieldtrips; rental facilities.
Hours & Admission Prices: Tours: 9am, 10am, 1pm, 2pm & 4pm by appointment. Adults $68, youth 3-12 $30, children 1-2 $10.

SANTA ROSA JUNIOR COLLEGE MUSEUM, 1501 Mendocino Ave., Santa Rosa, CA 95401-4332. Tel.: 707-527-4479. Fax: 707-524-1861.
Web Site: www.santarosa.edu/museum/index.html
Formerly: Jesse Peter Native American Art Museum
Founded: 1932.
Congressional District: 2
Key Personnel: Dir., Theresa Molino; Exhibit Coord., Christine Vasquez.
Personnel Profile: Full-Time Paid 1; Part-Time Paid 4; Interns 3.
Governing Authority: college. Tax-exempt.
Institution Type/Description: Multicultural Museum.
Collections: Native American cultural art; Latino, African & Asian material culture.
Research Fields: material culture of North American Indians with emphasis on California & the Southwest.
Facilities: research center.
Activities: rotating exhibits; guided tours; lectures; demonstrations. Annual Event: A Day Under the Oaks, California Native American Cultural Festival in May.
Publications: occasional monographs & teacher's guides.
Hours & Admission Prices: Mon.-Fri. 9-12 & 1-4:30. No charge; donations accepted. Closed summers & all school holidays & breaks. &
Attendance: 17,000 (accurate)

SONOMA COUNTY MUSEUM, 425 Seventh St., Santa Rosa, CA 95401-5233. Tel.: 707-579-1500. Fax: 707-579-4849.
E-mail: info@sonomacountymuseum.org
Web Site: www.sonomacountymuseum.org
Founded: 1976.
Congressional District: 2
Key Personnel: Exec. Dir., Diane Evans; Cur. Exhibitions, Eric Stanley; Cur. Education, Cynthia Conway; Business Mgr., Kirsten Olney.
Personnel Profile: Full-Time Paid 5; Part-Time Paid 3; Part-Time Volunteers 150.
Governing Authority: nonprofit organization. Parent Institution: Sonoma County Museum Foundation. Tax-exempt: 501(c)(3).
Institution Type/Description: Art & History Museum.
Collections: fine art from 19th century to present with emphasis on San Francisco Bay area & northern California; Christo & Jeanne Claude; Sonoma County history.
Research Fields: art history; Christo & Jeanne Claude.
Facilities: Books, crafts & museum-related items for sale.
Activities: guided tours; lectures; films; arts festivals; docent program; temporary & traveling exhibitions; public programs; family days; school loan service; education materials; educational releases; history & exhibitions outreach; teacher open house; school tours; fundraisers; social activities; bus tours.
Publications: newsletter, The Muse; exhibit catalogues.
Hours & Admission Prices: Tues.-Sun. 11-5. Adults $7, seniors & students $5; discounts to NARM members; children under 12, school groups, and museum, AAM & ICOM members no charge. Closed New Year's Day & day after; Martin Luther King Jr. Day; Easter; Memorial Day; Independence Day; Labor Day; Thanksgiving & day after; Christmas Eve, Day & day after. &
Attendance: 25,000 (estimated)
Membership: Discount & Individual $40; Dual $70; Supporting $125; Advocate $250; Patron $500; Curator's Circle $1,000; Director's Circle $2,500; Director's Circle $5,000.

Santa Ynez

SANTA YNEZ VALLEY HISTORICAL SOCIETY, (M), 3596 Sagunto St., Santa Ynez, CA 93460-9110. Mailing Address: P.O. Box 181, Santa Ynez, CA 93460-0181. Tel.: 805-688-7889. Fax: 805-688-1109.
E-mail: syvm@verizon.net
Web Site: www.santaynezmuseum.org
Founded: 1961.
Key Personnel: Exec. Dir., Chris Bashforth; Pres., Art Lacerte; Cur. Carriage House, John Crockett.
Personnel Profile: Full-Time Paid 2; Part-Time Paid 3; Part-Time Volunteers 40.
Governing Authority: private; nonprofit organization. Tax-exempt: 501(c)(3).
Institution Type/Description: History Museum.
Collections: local historical artifacts. Carriage House: carriages & stage coaches
Major Exhibits: Kimono...Tradition, Pattern and Symbolism, 12/13-4/1/14; History of Wine Making in Santa Ynez Valley, 4/14-10/14; Historic Kitchen Utensils, 11/14-3/15.
Facilities: 500-vol. library of local & state history books. Museum-related items for sale.
Activities: docent program; guided tours; lectures; temporary exhibitions. Annual Events: Vaquero Show; Fiesta Preludio de Santa Ynez; Old Santa Ynez Days Spaghetti Western Auction; dinner dance.
Publications: quarterly newsletter, Valley Heritage.
Hours & Admission Prices: Wed.-Sun. 12-4. Adults $4; members and children 16 & under no charge. Closed major holidays. &
Attendance: 9,000 (estimated)
Membership: Individual $35; Family $50; Sponsor $150; Supporter $300; Patron $500; Benefactor $1,000; Lifetime $5,000.

Santa Ysabel

MISSION SANTA YSABEL, 23013 Hwy. 79, Santa Ysabel, CA 92070. Mailing Address: P.O. Box 129, Santa Ysabel, CA 92070-0129. Tel.: 760-765-0810.
Founded: 1818.
Personnel Profile: Full-Time Paid 1; Part-Time Paid 1.
Institution Type/Description: Indian Museum.
Collections: Indian artifacts; period furnishings.
Facilities: Museum-related items for sale.
Hours & Admission Prices: Daily 8-3. No charge; donations accepted.

Santee

CREATION AND EARTH HISTORY MUSEUM, 10946 Woodside Ave., N., Santee, CA 92071-3272. Mailing Address: Life and Light Foundation, 9336 Abraham Way, Santee, CA 92071. Tel.: 619-596-1104.
E-mail: jayson.payne@creationsd.org
Web Site: www.creationsd.org
Formerly: ICR Museum of Creation & Earth History
Founded: 1976.
Key Personnel: Dir., Jayson Payne; C.E.O., John Von Duzer; Pres., Tom Cantor.
Governing Authority: Parent Institution: Life and Light Foundation. Tax-exempt.
Institution Type/Description: Earth History & Science Museum.
Collections: earth history & science; butterfly collection; live animals; fossils; archaeology; Middle East.
Activities: field tours; classes; tours.
Hours & Admission Prices: Mon.-Sat. 9-6. No charge; donations accepted. Guided Tours: $39. Closed New Year's Day; Thanksgiving & day after; Christmas Eve & Day. &
Attendance: 16,000 (estimated)
Membership: Child $25; Senior $30; Adult $50; Family $100.

Saratoga

HAKONE ESTATE AND GARDENS, 21000 Big Basin Way, Saratoga, CA 95070-5755. Mailing Address: Hakone Foundation, P.O. Box 2324, Saratoga, CA 95070. Tel.: 408-741-4994. Fax: 408-741-4993. Facebook: Saratoga Hakone.
E-mail: giftshop@hakone.com
Web Site: www.hakone.us
Key Personnel: Exec. Dir., Lon Saavedra; Pres. of Bd. (V), Connie Young Yu; Japanese Garden Specialist, Jabob Kellner; Event Mgr., Kyongmi Ader
Institution Type/Description: Japanese-style Gardens.

Collections: bamboo garden; tea garden; zen garden; statues.
Facilities: Museum-related items for sale.
Hours & Admission Prices: Mon.-Fri. 10-5, Sat.-Sun. 11-5. Adults $8, seniors 65 & up and students 5-17 w/ID $6; children 4 & under and members no charge. Closed New Year's Day; Christmas Day.

MONTALVO ARTS CENTER, 15400 Montalvo Rd., Saratoga, CA 95070-6327. Mailing Address: P.O. Box 158, Saratoga, CA 95071-0158. Tel.: 408-961-5800. Fax: 408-961-5850.
Web Site: www.montalvoarts.org
Founded: 1930.
Congressional District: 12
Key Personnel: Exec. Dir., Angela McConnell; Pres. (V), Cathie Thermond; Dir. Mktg. & Communications, Diane Maxwell.
Personnel Profile: Full-Time Paid 24; Part-Time Paid 18; Part-Time Volunteers 500; Interns 3.
Governing Authority: nonprofit. Parent Institution: Montalvo Association. Tax-exempt: 501(c)(3).
Institution Type/Description: Art Museum, Arboretum & Historic House/Site.
Collections: decorative arts; sculpture park. Historic Houses: 1912 Villa Montalvo; 1912 carriage house.
Facilities: 175-acre park & nature preserve; 300-seat Carriage House theatre; 1,250-seat garden theatre.
Activities: guided tours; lectures; gallery talks; concerts; artist in residence program; arts festivals; drama; rotating exhibitions; education programs; sculpture grounds.
Publications: monthly calendar; monthly e-newsletter.
Hours & Admission Prices: Park & Grounds: daily 9-5. Gallery: Thurs.-Sun. 11-3. No charge, donations accepted. Closed New Year's Day; Thanksgiving; Christmas. ♿
Attendance: 250,000 (estimated)
Membership: Advocate $100; Senator's Circle $250; Griffin Circle $500; Phelan Circle $1,000; Villa Circle $2,500; Artist Circle $5,000; Creative Circle $10,000; Director's Circle $25,000.

YOUTH SCIENCE INSTITUTE, SANBORN NATURE CENTER, 16055 Sanborn Rd., Sanborn-Skyline Park, Saratoga, CA 95070-9746. Mailing Address: Youth Science Institute, 296 Garden Hill Dr., Los Gatos, CA 95032-7669. Tel.: 408-867-6940 ext. 27.
E-mail: sanborn@ysi-ca.org
Web Site: www.ysi-ca.org
Founded: 1953.
Key Personnel: Exec. Dir., Diane Riccio.
Governing Authority: nonprofit. Parent Institution: Youth Science Institute, Vasona Park, CA. Tax-exempt: 501(c)(3).
Institution Type/Description: Youth & Science Museum.
Collections: geology items; taxidermied items & invertebrate zoo; living & nonliving insects; California Indian artifacts; redwood ecology.
Research Fields: community garden program.
Facilities: botanical garden; educational facilities; 1,000 sq. ft. exhibit space; nature center.
Activities: lectures; organized educational programs; docent program; temporary exhibitions. Museum Sponsors: Insect Fair.
Publications: quarterly newsletter Explore.
Hours & Admission Prices: Open all year, by appointment only. No charge; donations accepted. Closed New Year's Day; Thanksgiving; Christmas. ♿
Attendance: 16,500 (estimated)
Membership: Individual $20; Family $35; Supporting $50; Patron $100; Donor $250; Gavilan $500; Sponsor $750; Benefactor & Corporate $1,000.

Sausalito

BAY AREA DISCOVERY MUSEUM, (M), Fort Baker, 557 McReynolds Rd., Sausalito, CA 94965-2601. Tel.: 415-339-3900. Fax: 415-339-3901. Facebook: Bay Area Discovery Museum.
E-mail: contact@badm.org
Web Site: www.baykidsmuseum.org
Founded: 1984.
Key Personnel: C.E.O. & Exec. Dir., Karyn Y. Flynn; C.F.O., Rolf Neuweiler; Pres., Sanjay Jain; Treas., Andrew Hoybach; Museum Shop Mgr., Mark Giberson.
Personnel Profile: Full-Time Paid 45; Part-Time Paid 10; Part-Time Volunteers 100; Interns 2.
Governing Authority: nonprofit organization. Tax-exempt: 501(c)(3).
Institution Type/Description: Children's Museum.
Collections: art; nature; science; culture of the Bay Area.
Major Exhibits: Superkids Save the World (T), 1/14-5/14; Broken Fix-It (T), 6/14-8/14.
Research Fields: early childhood education; outdoor education; children's creativity.

Activities: weekend programs & workshops.
Publications: Discoveries.
Hours & Admission Prices: Tues.-Sun. 9-5. Admission 6 months & over $11; children under 6 months no charge; discounts to AAM members; family level membership no charge. ♿
Attendance: 300,000 (accurate)
Membership: Family $150; Supporting $250.

SAN FRANCISCO BAY MODEL VISITOR CENTER, 2100 Bridgeway, Sausalito, CA 94965-1764. Tel.: 415-289-3007. Fax: 415-289-3004.
Web Site: www.spn.usace.army.mil/bmvc/
Founded: 1956.
Congressional District: 6
Key Personnel: Chm. (V), Steve Machtinger; Park Mgr., Chris Gallagher.
Personnel Profile: Full-Time Paid 5; Part-Time Volunteers 30; Interns 2.
Volunteer Hours: 2,492
Operating Expenses: 904,000
Operating Income: 904,000
Governing Authority: federal; U.S. Army Corps. of Engineers. Tax-exempt.
Institution Type/Description: Park Museum: located within the Bay Model Regional Visitor Center, an hydraulic testing facility.
Collections: biota of San Francisco Bay & how hydraulic modeling is used to study the bay; interactive video program; Corps of Engineers.
Research Fields: study hydraulic problems that affect the San Francisco Bay & the Sacramento/San Joaquin Delta.
Facilities: archival library pertaining to tests that have been conducted on the Bay Model; 125-seat auditorium; 6,000 sq. ft. gallery area. Books on local environment & maritime history for sale.
Activities: guided tours; lectures; films; volunteer program; audio tours available in 6 languages: German, French, Spanish, Japanese, Russian, English; orientation program also available in 5 languages.
Publications: brochure, Bay Model User's Guide; self-guiding brochure.
Hours & Admission Prices: Winter: Tues.-Sat. 9-4; Summer: Tues.-Fri. 9-4, Sat.-Sun. & holidays 10-5. No charge, donations accepted. ♿
Attendance: 144,973 (accurate)

Scotia

PALCO, 125 Main. St., Scotia, CA 95565. Mailing Address: P.O. Box 37, Scotia, CA 95565-0037. Tel.: 707-764-4472. Fax: 707-764-4171.
Web Site: www.palco.com
Formerly: The Pacific Lumber Company
Founded: 1869.
Congressional District: 1
Key Personnel: C.E.O., George O'Brien; Chm. (V), Mary Bullwinkel.
Personnel Profile: Part-Time Paid 3; Part-Time Volunteers 1.
Institution Type/Description: Logging & Lumber Museum: housed in 1920 First National Bank of Scotia, all redwood building.
Collections: period logging equipment; pictorial history of The Pacific Lumber Co. Historic Building: 1920 Winema Theatre located nearby.
Hours & Admission Prices: June-Sept. Mon.-Fri. 8-4:30; other times by request. No charge. Closed holidays.
Attendance: 15,000 (estimated)

Seal Beach

SEAL BEACH HISTORICAL SOCIETY/RED CAR MUSEUM, Electric Ave. & Main, Seal Beach, CA 90470. Mailing Address: P.O. Box 152, Seal Beach, CA 90740-0152. Tel.: 562-430-1450.
Founded: 1971.
Congressional District: 46
Key Personnel: Pres. (V), Marie Antos.
Personnel Profile: Part-Time Volunteers 5.
Governing Authority: Tax-exempt.
Institution Type/Description: Historic Train Car: a roving machine shop used to troubleshoot problems along the Pacific Electric LA-Newport Line; built in 1925.
Collections: local history; railroad artifacts; signs; tickets; pins; photographs; video.
Activities: groups tours for schools & senior citizens; community meetings with speakers.
Publications: quarterly newsletter.
Hours & Admission Prices: 2nd & 4th Sat. 12-3. No charge; donations accepted.
Attendance: 1,000 (estimated)
Membership: Student $7; Single $15; Family $25; Ten Years $250.

Sebastopol

WEST COUNTY MUSEUM, 261 S. Main St., Sebastopol, CA 95472. Tel.: 707-829-6711.
Web Site: www.sebastopol-farm-museum.org
Founded: 1993.
Key Personnel: Dir., Rae Swanson.
Governing Authority: Parent Institution: Western Sonoma County Historical Society.
Institution Type/Description: History Museum: housed in the former Petaluma and Santa Rosa Electric Railway Depot; built in 1917.
Collections: local history & culture; period furnishings; personal artifacts; photographs.
Major Exhibits: An Eclectic Food History, 11/13-4/30/14; Clothing History, 5/14-12/14.
Hours & Admission Prices: Thurs.-Sun. 1-4.

Shafter

MINTER FIELD AIR MUSEUM, 401 Vultee St., Shafter, CA 93263. Mailing Address: P.O. Box 445, Shafter, CA 93263-0445. Tel.: 661-393-0291.
E-mail: mfam@minterfieldairmuseum.com
Web Site: www.minterfieldairmuseum.com
Founded: 1981.
Congressional District: 20
Key Personnel: Chm. Bd., Maj. Gen. James T. Whitehead, Jr. USAF Ret.; Vice Chm., David Patterson; Cur., Dr. David Day.
Personnel Profile: Part-Time Volunteers 30.
Governing Authority: private; nonprofit organization. Tax-exempt: 501(c)(3).
Institution Type/Description: Military Aeronautics Museum.
Collections: model aircraft & vehicles; historical artifacts; videos; military & aviation history.
Facilities: 500-vol. library; 5,000 sq. ft. exhibit space. Museum-related items for sale.
Activities: guided tours; BT-13L Link trainer & military vehicle restorations; parades; fly-ins; Aermachi jet trainer. Museum Sponsors: Open Houses; Annual Auction and Awards Dinner; Airshow.
Publications: quarterly newsletter, Snap Roll II.
Hours & Admission Prices: Fri.-Sat. 10-2; other times by appointment. No charge; donations accepted. Closed most holidays. &
Attendance: 12,000 (estimated)
Membership: Individual $40; Family $50; Senior Pilot $100; Command Pilot $250; Squadron Commander (Life) $500; Wing Commander (Life) $1,000; Division Commander (Life) $2,000; A.F. Commander (Life) $5,000; Supreme Commander (Life) $10,000.

SHAFTER DEPOT MUSEUM, 150 Central Valley Hwy., Shafter, CA 93263-2002. Mailing Address: Shafter Historical Society, P.O. Box 1088, Shafter, CA 93263-1088. Tel.: 661-746-4423. Fax: 661-746-1620.
E-mail: hwfarms@sbcglobal.net
Founded: 1979.
Congressional District: 20
Key Personnel: Cur., Stan Wilson; Pres. (V), Helen Goede.
Personnel Profile: Part-Time Paid 1.
Governing Authority: Parent Institution: Shafter Historical Society. Tax-exempt.
Institution Type/Description: History Museum
Collections: railroad artifacts; agriculture; local history.
Publications: quarterly newsletter.
Hours & Admission Prices: Sat. 10-2; other times by appointment. No charge; donations accepted. &
Attendance: 854 (accurate)
Membership: Student $10; Individual $15; Family $25.

Shasta

SHASTA STATE HISTORIC PARK, 15312 Hwy. 299 W., Shasta, CA 96087. Mailing Address: P.O. Box 2430, Shasta, CA 96087-2430. Tel.: 530-225-2065. Fax: 530-225-2038.
E-mail: lmartin@parks.ca.gov
Founded: 1937.
Key Personnel: Cur., Lori Martin.
Personnel Profile: Full-Time Paid 1; Part-Time Paid 2; Part-Time Volunteers 20.
Governing Authority: state.
Institution Type/Description: State Park Museum.

Collections: early California paintings; gold rush & mining materials; extensive archives pertaining to Shasta County; Native American basketry; artifacts.
Facilities: visitor center; picnic area.
Activities: permanent exhibitions.
Hours & Admission Prices: Thurs.-Sun. 10-5. Adults $3, children 17 & under $2. Closed New Year's Day; Thanksgiving; Christmas.
Attendance: 36,000 (accurate)

Shoshone

SHOSHONE MUSEUM, 118 Hwy. 127, Shoshone, CA 92384. Mailing Address: P.O. Box 38, Shoshone, CA 92384-0038. Tel.: 760-852-4524.
E-mail: shoshonemus@veawb.coop
Web Site: deathvalleychamber.org
Founded: 1986.
Key Personnel: Pres. (V), Susan Sorrells.
Personnel Profile: Part-Time Paid 3; Part-Time Volunteers 3.
Governing Authority: Tax-exempt.
Institution Type/Description: Historic Building: built in Greenwater in 1906, moved to Zabriskie & later to its present location where it served as a store & gas station.
Collections: area history; mining & farming equipment; mammoth skeleton; natural history; Native American artifacts; minerals; mining; pioneers; T&T Railroad.
Publications: triannual newsletter, The Shoshone Museum Reader.
Hours & Admission Prices: Wed.-Mon. 9-3. No charge; donations accepted. Closed Federal holidays. &
Attendance: 55,000 (accurate)
Membership: Individual $15; Family $25; Business & Nonprofit $50.

Sierra City

SIERRA COUNTY HISTORICAL SOCIETY (KENTUCKY MINE MUSEUM), 100 Kentucky Mine Rd., Sierra City, CA 96125. Mailing Address: P.O. Box 260, Sierra City, CA 96125-0260. Tel.: 530-862-1310.
E-mail: museum@sierracountyhistory.org
Web Site: sierracountyhistory.org
Key Personnel: Dir., Cur. & Museum Shop Mgr., Virginia Lutes; Pres., Mary Nourse; Vice Pres., Joleen Torris; Chm. Membership, Suzi Schoensee; Treas., William Copren; Caretaker, Toni Strine.
Personnel Profile: Part-Time Paid 2; Part-Time Volunteers 2.
Governing Authority: private; nonprofit organization. Tax-exempt: 501(c)(3).
Institution Type/Description: History Museum.
Collections: mine history; mining tools & equipment.
Research Fields: archaeology; Native American history; bat biology.
Facilities: 100-vol. library. Museum-related items for sale.
Activities: concerts; docent program; guided mine & stamp mill tours; temporary exhibitions.
Publications: quarterly membership newsletter, The Sierran.
Hours & Admission Prices: Memorial Day to Labor Day Wed.-Sun. 10-4. Adults $1. Guided Tours: adults $7, children 17 & under $3.50; children under 6 no charge.
Attendance: 1,000 (estimated)
Membership: Adults $20; Family & Institutions $25; Business & Supporting $35; Sustaining $50; Life $300.

Silverado

TUCKER WILDLIFE SANCTUARY, 29322 Modjeska Canyon Rd., Silverado, CA 92676-9784. Tel.: 714-649-2760. Fax: 714-649-2760.
E-mail: kcornell@exchange.fullerton.edu
Web Site: www.tuckerwildlife.org
Founded: 1938.
Key Personnel: Dir., Karon Cornell; Site Mgr., Marcella Gilchrist.
Governing Authority: university. Parent Institution: California State University, Fullerton, CA. Tax-exempt.
Institution Type/Description: Wildlife Sanctuary.
Collections: botany; entomology; geology; herpetology; ornithology.
Research Fields: flora & fauna of the Santa Ana Mountains; entomology; ornithology; native wildlife.
Facilities: visitor center; picnic area; amphitheater; gardens. Museum-related items for sale.
Activities: K-12 school tours; public groups tours; bird observation area; workshops; special events.
Publications: quarterly newsletter; bird list; plant guide.

Hours & Admission Prices: Tues.-Sun. 9-4. Suggested Donation: $3 per person. Naturalist Guided Group Tours: $6 per person. Closed major holidays. &

Attendance: 30,000

Membership: Senior Quail $20; Hummingbird $25; Red Tailed Hawk $50; Eagle $100.

Simi Valley

R.P. STRATHEARN HISTORICAL PARK AND MUSEUM, 137 Strathearn Place, Simi Valley, CA 93065-1605. Mailing Address: P.O. Box 940461, Simi Valley, CA 93094-0461. Tel.: 805-526-6453. Fax: 805-526-6462. Facebook: Strathearn Park.

E-mail: simimuseum@sbcglobal.net

Web Site: www.simihistory.com

Founded: 1970.

Congressional District: 20

Key Personnel: Dir., Patricia Havens; Pres. (V), Georgia Trumble.

Personnel Profile: Part-Time Paid 3; Part-Time Volunteers 100; Interns 1.

Governing Authority: society; nonprofit organization. Parent Institution: Rancho Simi Recreation & Parks District & Simi Valley Historical Society, 1692 Sycamore Dr. Tax-exempt: 170(b)(1)(A).

Institution Type/Description: Historic Buildings.

Collections: period artifacts; letters; documents; books; farm implements; furniture; papers; household items; ranch tools; school & business records; financial records; clothing. Historic Buildings: c.1800 Simi Adobe; 1890s Victorian house; 1890s Strathearn home; 1889 The Colony House; St. Rosa Lima Church.

Research Fields: history of Simi Valley, Southern California & Ventura County.

Facilities: library of books & letters; reading room.

Activities: guided tours; lectures; demonstrations; field trips for school children & other groups.

Publications: books, Straw Roads-A Story of Simi Valley from 1908-1960; Simi Grows Up; Moorpark, Star of the Valley; Simi Valley - A Journey through Time (Earliest days through 1997); Strathearn Letters - Windows on the Past, 2009.

Hours & Admission Prices: Tours: Wed.-Fri. 1 pm, Sat.-Sun. 1-4. Park: Mon.-Fri. 9-3. Requested Donation: adults $3; members and children 17 & under no charge. Closed Easter; Christmas.

Attendance: 6,000 (estimated)

Membership: Single $15; Family $25; Sustaining $100; Patron $500; Benefactor $1,000.

RONALD REAGAN PRESIDENTIAL LIBRARY AND MUSEUM, (M), 40 Presidential Dr., Simi Valley, CA 93065-0699. Tel.: 800-410-8354. Fax: 805-577-4074.

Web Site: www.reagan.utexas.edu

Founded: 1991.

Congressional District: 23

Key Personnel: Dir., R. Duke Blackwood; Deputy Dir., Tony Chauveaux; Registrar, Jennifer Torres; Gift Shop Mgr., Carolyn Mente.

Personnel Profile: Full-Time Paid 30; Part-Time Paid 10; Part-Time Volunteers 182; Interns 2.

Governing Authority: federal. A unit of the National Archives and Records Administration, Washington, D.C. 20408. Tax-exempt: 170(b)(1)(A).

Institution Type/Description: Presidential Library.

Collections: personal papers; government records; still photographs; motion pictures; audio & video tapes; head of state gifts; gifts from private citizens; political campaign items; personal & family memorabilia.

Research Fields: the Reagan presidency; the 1980s; life, times, and career of Ronald Reagan and his family.

Facilities: archival reading room; 400-seat auditorium; classroom; conference spaces.

Activities: guided tours; lectures; conferences; changing exhibits; organized educational programs for children & adults; docent program.

Publications: exhibit-related brochures & catalogs; curriculum guides; volunteer newsletter; membership newsletter.

Hours & Admission Prices: Daily 10-5. Adults $21, seniors 62 & over $18, children 11-17 $15, children 3-10 $6; children 2 & under no charge. Closed New Year's Day; Thanksgiving; Christmas. &

Attendance: 300,000 (accurate)

Membership: Friends of the Reagan Library & Museum: Individual $35; Family $50; Diplomat $100; Cabinet $250; Chairman's Club $1,000.

SANTA SUSANA DEPOT, 6503 Katherine Rd., Simi Valley, CA 93063-4786. Tel.: 805-581-3462.

E-mail: checkmate0072002@yahoo.com

Institution Type/Description: Historic Building: built by the Southern Pacific Railroad in 1903.

Collections: railroad history; period furnishings; photographs; personal artifacts.

Hours & Admission Prices: Sat.-Sun. 1-4. Closed major holidays.

Attendance: 3,200

Smith River

SHIP ASHORE RESORT MUSEUM, 12370 Hwy. 101 N., Smith River, CA 95567-9448. Tel.: 800-487-3141; 707-487-3141. Fax: 707-487-7070.

E-mail: shipashore@charterinternet.com

Web Site: www.ship-ashore.com

Institution Type/Description: Maritime Museum.

Collections: local historic artifacts; seashells; rocks & minerals; natural history.

Facilities: Museum-related items for sale.

Hours & Admission Prices: Call for hours. No charge.

Soledad

MISSION NUESTRA SENORA DE LA SOLEDAD, 36641 Fort Romie Rd., Soledad, CA 93960. Mailing Address: P.O. Box 515, Soledad, CA 93960-0515. Tel.: 831-678-2586.

Web Site: missiontour.org/soledad/index.htm

Founded: 1791.

Congressional District: 17

Key Personnel: Chm. (V) & Museum Shop Mgr., Kristen Dow; Pres. (V), Carlene Bell.

Governing Authority: Tax-exempt.

Institution Type/Description: Historical Religious Museum: an historic site.

Collections: historical & religious artifacts.

Activities: special services; weddings.

Hours & Admission Prices: Daily 10-4. Mass: 1st Sun. of month. No charge; donations requested. Closed New Year's Day; Easter; Independence Day; Thanksgiving; Christmas.

Attendance: 16,000 (estimated)

Solvang

ELVERHOJ MUSEUM OF HISTORY AND ART, 1624 Elverhoy Way, Solvang, CA 93463-2704. Mailing Address: P.O. Box 769, Solvang, CA 93464-0769. Tel.: 805-686-1211. Fax: 805-686-1822.

E-mail: info@elverhoj.org

Web Site: www.elverhoj.org

Founded: 1988.

Key Personnel: C.E.O., Esther Jacobsen Bates.

Personnel Profile: Full-Time Paid 1; Part-Time Paid 1; Part-Time Volunteers 50.

Governing Authority: nonprofit. Tax-exempt.

Institution Type/Description: History and Art Museum.

Collections: local history; Danish culture; photographs; porcelain; silverware.

Facilities: garden; picnic gazebo.

Activities: special events; tours.

Publications: newsletter.

Hours & Admission Prices: Wed.-Thurs. 1-4, Fri.-Sun. 12-4. No charge; donations accepted. Closed New Year's Day; Easter; Thanksgiving; Christmas Eve & Day. &

Attendance: 14,570 (accurate)

Membership: Individual $35; Couple & Family $50; Business $100; Benefactor $250; Patron $500.

HANS CHRISTIAN ANDERSEN MUSEUM, 1680 Mission Dr., Solvang, CA 93463-3602. Tel.: 805-688-2052.

E-mail: mail@bookloftsolvang.com

Founded: 1989.

Congressional District: 24

Key Personnel: Dir., Katheryn Mullina.

Personnel Profile: Part-Time Volunteers 2.

Governing Authority: nonprofit organization. Parent Institution: Ugly Duckling Foundation. Tax-exempt.

Institution Type/Description: General Museum.

Collections: Andersen's life & work; period tools; Andersen's first & early edition tales; travel journals; letters; photographs; Andersen's artwork.

Activities: tours; lectures. Annual Event: Andersen's Birthday Celebration in April.

Hours & Admission Prices: Daily 10-5; groups by appointment.

Attendance: 12,000 (estimated)

OLD MISSION SANTA INES, 1760 Mission Dr., Solvang, CA 93463-2625. Mailing Address: P.O. Box 408, Solvang, CA 93464-0408. Tel.: 805-688-4815. Fax: 805-686-4468.
E-mail: sheila.benedict@missionsantaines.org
Web Site: www.missionsantaines.org
Founded: 1804.
Key Personnel: Archives, Sheila Benedict; Gift Shop Mgr., Maruska Westermann.
Governing Authority: church. Roman Catholic Archdiocese of Los Angeles.
Institution Type/Description: Historic Site & Historic Building: 1804 Old Mission Santa Ines.
Collections: religious vestments; Mexican & Spanish Colonial art; tools.
Activities: recorded tours.
Publications: brochures; book, Mission Santa Ines, The Hidden Gem.
Hours & Admission Prices: Winter & Summer: daily 9-4:30. Adults $5; discounts for groups; children under 12 no charge. Closed New Year's Day; Easter; Thanksgiving; Christmas. &
Attendance: 50,000 (estimated)

WILDLING MUSEUM, (M), 1511 B Mission Dr., Solvang, CA 93463. Tel.: 805-688-1082. Fax: 805-686-8339.
E-mail: info@wildlingmuseum.org
Web Site: www.wildlingmuseum.org
Formerly: Wildling Art Museum
Founded: 1997.
Congressional District: 24
Key Personnel: Exec. Dir., Stacey A. Otte; Pres. (V), Patti Jacquemain; Treas., Marilyn Parke; Asst. Dir., Jessica McLoughlin.
Personnel Profile: Full-Time Paid 3; Part-Time Paid 1; Part-Time Volunteers 30.
Governing Authority: private; nonprofit organization. Tax-exempt.
Institution Type/Description: Art Museum.
Collections: paintings, prints, drawings, photographs & sculpture depicting America's wilderness and wildlife including flora & fauna.
Facilities: library; 900 sq. ft. exhibit space; education center. Museum-related items for sale.
Activities: formal education programs for adults; guided tours; lectures; artists workshops; art classes for children; field trips. Annual Events: Vivan los Californios BBQ; Santa Ynez Valley Artists' Studio Tour; Wildling Horse Ride Benefit.
Publications: quarterly newsletter, The Fox Tale; brochures; exhibition monographs; catalogue, Endangered Species: Flora & Fauna in Peril.
Hours & Admission Prices: Wed.-Sun. 11-5. Donation Requested: adults $5, children 6-17 $3; AAM, ICOM & museum members no charge. Closed New Year's Day; Thanksgiving; Christmas. &
Attendance: 8,000 (accurate)
Membership: Individual $45; Family $60; Wilderness Circle $150; Meadow Guild $250; Foothills Council $500; Pinnacle Group $1,000; Summit Stars $5,000.

Sonoma

SONOMA STATE HISTORIC PARK, 363 3rd St., W., Sonoma, CA 95476-5632. Tel.: 707-938-1519. Fax: 707-938-1406.
Web Site: www.parks.ca.gov
Founded: 1906.
Congressional District: 6
Key Personnel: Mission Guide, Jackie Barros; Chm. (V), Jennifer Hanson; Pres. (V), Rick Arent; State Park Ranger, Vince Anibale; Museum Cur., Carol A. Dodge; Museum Shop Mgr., Sue Vargas.
Personnel Profile: Full-Time Paid 1; Part-Time Paid 3; Part-Time Volunteers 10.
Governing Authority: state. Dept. of Parks & Recreation, P.O. Box 2390, Sacramento, CA 95811. Tax-exempt.
Institution Type/Description: State Historic Park Museum Complex.
Collections: furnishings; Jorgonsen paintings of California Missions. Historic Houses: c.1870 Toscano Hotel; 1852 Vallejo Home; 1852 Swiss Chalet; 1823 San Francisco Solano Mission; 1836 Sonoma Barracks.
Research Fields: Mission San Francisco Solano de Sonoma; California history 1823-1846; General Vallejo 1807-1890.
Facilities: visitor centers at Mission San Francisco, Swiss Chalet & Sonoma Barracks.
Activities: formally organized environmental education programs for children; permanent exhibitions; guided tours for school groups with advanced reservations.
Publications: descriptive brochure.
Hours & Admission Prices: Daily 10-5. Adults $3, children 6-17 $2. Closed New Year's Eve. & Day; Thanksgiving; Christmas.
Attendance: 602,398 (accurate)

SONOMA VALLEY HISTORICAL SOCIETY/DEPOT PARK MUSEUM, (M), 270 First St., W., Sonoma, CA 95476. Mailing Address: P.O. Box 861, Sonoma, CA 95476-0861. Tel.: 707-938-1762.
E-mail: info@depotparkmuseum.com
Web Site: www.depotparkmuseum.com
Founded: 1979.
Congressional District: 6
Key Personnel: Pres., Newton Dal Pogetto; Mgr., Dir. & Cur., Diane Smith.
Personnel Profile: Part-Time Paid 1; Part-Time Volunteers 27.
Governing Authority: council. Parent Institution: Sonoma Valley Historical Society. Tax-exempt.
Institution Type/Description: Art & History Museum.
Collections: exhibits & artifacts relating to pioneers & settlers of Sonoma Valley; native American collection; Victorian & early Sonoma period furniture & artifacts; railroad cars & equipment; station agent's office.
Research Fields: genealogy of local early Sonoma Valley families; buildings, history.
Facilities: research library.
Activities: lectures; films; permanent & temporary exhibitions; school tours; self-guided & docent guided tours.
Publications: books, Saga of Sonoma; Sonoma Mission; Sonoma Valley Legacy; monthly, historical society newsletter; Sonoma Valley Notes; Pioneer Sonoma; Schools & Scows in Early Sonoma; The Men of The California Bear Flag Revolt and Their Heritage; Images of America Sonoma Valley.
Hours & Admission Prices: Thurs.-Sun. 1-4:30. No charge; donations accepted. &
Attendance: 3,000 (estimated)
Membership: Individual $25; Family $35; Business & Supporter $50; Friend $100; Patron $200; Life $300.

SONOMA VALLEY MUSEUM OF ART, (M), 551 Broadway, Sonoma, CA 95476-6601. Mailing Address: P.O. Box 322, Sonoma, CA 95476-0322. Tel.: 707-939-7862. Fax: 707-939-1080.
E-mail: admin@svma.org
Web Site: www.svma.org
Founded: 1998.
Congressional District: 1
Key Personnel: Exec. Dir., Kate Eilertsen; Bd. Pres., Gerret Snedaker.
Personnel Profile: Full-Time Paid 4; Part-Time Paid 2; Part-Time Volunteers 100; Interns 4.
Governing Authority: private; nonprofit organization. Tax-exempt: 501(c)(3).
Institution Type/Description: Art Museum.
Collections: fine arts from around the world.
Facilities: 5,000 sq. ft. exhibit space. Museum-related items for sale.
Activities: concerts; docent program; films; formal education programs for adults & children; guided tours; lectures; loan & traveling exhibitions. Annual Event: Wet Paint Gala Auction Fundraiser.
Publications: quarterly museletter.
Hours & Admission Prices: Wed.-Sun. 11-5 during exhibitions. Adults $5; discounts to groups of 15 or more and NARM, AAM & ICOM members; members, students K-12 & Sun. no charge. Closed New Year's Day; Presidents' Day; Memorial Day; Independence Day; Labor Day; Thanksgiving; Christmas. &
Attendance: 15,000 (accurate)
Membership: Senior $30; Individual $35; Contributor $100; Sustainer $250; Benefactor $500; Director's Circle $1,000; President's Circle $3,000; Angels $5,000.

Sonora

TUOLUMNE COUNTY MUSEUM & HISTORY CENTER, 158 W. Bradford Ave., Sonora, CA 95370-4920. Tel.: 209-532-1317.
E-mail: info@tchistory.org
Web Site: tchistory.org
Founded: 1960.
Key Personnel: Chm. (V), John Brunskill.
Personnel Profile: Part-Time Paid 1; Part-Time Volunteers 26.
Governing Authority: Parent Institution: Tuolumne County Historical Society.
Institution Type/Description: Historic Building: housed in a former jail built in 1857 and rebuilt in 1865 after being destroyed by a fire that killed the inmate who allegedly started it. Listed on the National Register of Historic Places.
Collections: local history & culture; genealogy; photographs; personal artifacts; firearms; maps; railroad; lumber; mining.
Facilities: gardens; rental facilities.
Activities: rental facilities.
Publications: quarterly, CHISPA.

Hours & Admission Prices: Call for hours. No charge; donations accepted.
Attendance: 1,500 (estimated)
Membership: Annual $25-$100; Life $500.

South Lake Tahoe

LAKE TAHOE HISTORICAL SOCIETY & MUSEUM, 3058 Lake Tahoe Blvd., South Lake Tahoe, CA 96150-7810. Mailing Address: P.O. Box 18501, South Lake Tahoe, CA 96151-8501. Tel.: 530-541-5458.
E-mail: laketahoemuseum@att.net
Web Site: laketahoemuseum.org
Founded: 1968.
Congressional District: 14
Key Personnel: Chm. (V), Diane L. Johnson; Pres. (V), Catherine Whelan; Museum Shop Mgr., Kim Copel.
Personnel Profile: Part-Time Volunteers 45.
Governing Authority: society; nonprofit. Parent Institution: Lake Tahoe Historical Society. Tax-exempt: 501(c)(3).
Institution Type/Description: Historical Society Museum.
Collections: Washo Indian: photographs; artifacts related to South Lake Tahoe historical significance, industry, commerce & gaming; R/R display in the Bijou Community Park. Historic Buildings: 1930s log cabin; 1859 Toll House.
Research Fields: historic slides; photographs; oral histories.
Facilities: library & archives available by special arrangement. Books, photos & notepaper for sale.
Activities: guided tours; lectures; films; slide presentation; library; permanent & temporary exhibitions; school loan service.
Publications: books, Lake Valley's Past; Legends of Lake Tahoe; Railroads & Steamers of Lake Tahoe; The Steamer Tahoe; A Short History of the Lake Tahoe Basin; Sierra Trailblazers; Hank Monk, He'll Get You There on Time; member newsletter.
Hours & Admission Prices: Memorial Day to Labor Day Wed.-Sun. 11-3. No charge; donations accepted.
Attendance: 2,640 (accurate)
Membership: Senior $20; Individual $25; Family $40; Business $65; Individual Lifetime $250; Couple Lifetime $350; Benefactor $500 & over.

Stanford

CANTOR ARTS CENTER, STANFORD UNIVERSITY, (M), 328 Lomita Drive at Museum Way, Stanford, CA 94305-5060. Tel.: 650-723-4177. Fax: 650-725-0464. Facebook: Cantor Arts Center, Stanford University.
Web Site: museum.stanford.edu
Formerly: Iris & B. Gerald Cantor Center for Visual Arts at Stanford University
Founded: 1891.
Congressional District: 12
Key Personnel: Dir., Connie Wolf; Cur. European Art, Bernard Barryte; Cur. Modern & Contemporary Art, Hilarie Faberman; Exhibition Registrar, Katie Clifford; Cur. Education, Patience Young; Cur. Asian Art, Xiaoneng Yang; Cur. Prints, Drawings & Photographs, Elizabeth Mitchell.
Personnel Profile: Full-Time Paid 47; Part-Time Paid 20; Part-Time Volunteers 305; Interns 15.
Volunteer Hours: 23,325
Operating Expenses: 6,845,323
Operating Income: 7,655,116
Governing Authority: university. Affiliated with Stanford University. Tax-exempt: 501(c)(3).
Institution Type/Description: Art Museum.
Collections: photographs; paintings; sculpture; Rodin collection; decorative arts; archaeology; prints & drawings; Oriental; ancient; pre-Columbian; African; coins & medals; textiles; Stanford family memorabilia.
Major Exhibits: Flesh and Metal: Modern Art and the Machine, 11/13-3/14; Flesh and Metal: Body and Machine in Early 20th-Century Art, 11/13/13-3/16/14; The Honest Landscape: Photographs by Peter Henry Emerson, 11/27/13-5/4/14; Jim Dine and Claes Oldenburg: Transformations of the Ordinary, 12/11/13-4/27/14; Her Story: Prints by Elizabeth Murray, 1986-2006, 1/22/14-3/30/14; Inside Rodin's Hands: Teaching Surgery through Art and Anatomy, 4/9/14-8/3/14; Carleton Watkins: The Stanford Albums, 4/23/14-8/17/14; Night, Smoke and Shadows: The Presence of Atmosphere in the 19th Century, 5/14/14-10/5/14; Sympathy for the Devil: Satan, Sin and the Underworld, 8/20/14-11/14; Robert Frank in America, 9/10/14-1/4/15; Daumier on Art and the Theatre (T), 10/14-3/15.
Research Fields: American, Ancient, Oriental & European art from 1500.
Facilities: auditorium; cafe.
Activities: guided tours; lectures; gallery talks; concerts; inter-museum loan, permanent, temporary & traveling exhibitions.

Publications: exhibition catalogs; The Cantor Arts Center Journal.
Hours & Admission Prices: Wed. & Fri.-Sun. 11-5, Thurs. 11-8. No charge; donations accepted. Closed Thanksgiving; Christmas. &
Attendance: 200,000 (estimated)
Membership: Individual $60; Family & Dual $85; Sponsor $175; Patron $275; Benefactor $500; Artists' Circle $1,000; Connoisseurs' Circle $2,500; New Founders' Circle $5,000; Director's Circle $10,000.

Stinson Beach

AUDUBON CANYON RANCH, 4900 Hwy. 1, Stinson Beach, CA 94970. Tel.: 415-868-9244. Fax: 415-868-1699.
E-mail: acr@egret.org
Web Site: www.egret.org
Founded: 1962.
Key Personnel: Exec. Dir., J. Scott Feierabend; Pres. Bd., Bryant Hichwa; Treas., Barbara Kosnar; Education, Diane Jacobson; Museum Shop Mgr., Yvonne Pierce.
Personnel Profile: Full-Time Paid 25; Part-Time Volunteers 600.
Governing Authority: nonprofit organization. Subsidiary Institutions: Cypress Grove Research Center, Marshall, CA; Bouverie Preserve, Glen Ellen, CA. Tax-exempt: 501(c)(3).
Institution Type/Description: Nature Center.
Collections: local natural & environmental history; Coast Miwok Indian history.
Research Fields: monitoring grasses; shore birds, egrets, great blue herons; native plants & animals.
Facilities: 800-vol. library of natural history material; educational facilities; field research station. Natural history items for sale.
Activities: guided tours; lectures; organized educational programs; workshops; docent program.
Publications: semiannual seminar schedule, ACR Bulletin; monthly newsletters, The Heron & The Nutshell; e-newsletter.
Hours & Admission Prices: mid-March to mid-July Sat.-Sun. & holidays 10-4, Tues.-Fri. by appointment. No charge; donations accepted. &
Attendance: 20,000 (estimated)

Stockton

ALAN SHORT GALLERY, 928 E. Rose St., Stockton, CA 95202-1849. Mailing Address: c/o DDSO, 5051 47th Ave., Sacramento, CA 95824-4036. Tel.: 209-462-8208 & 948-5759. Fax: 209-948-9042.
E-mail: asc@ddso.org
Web Site: www.ddso.org
Institution Type/Description: Art Museum.
Collections: photographs.
Hours & Admission Prices: Daily 8-4. No charge; donations accepted. Closed national holidays

CHILDREN'S MUSEUM OF STOCKTON, 402 W. Weber Ave., Stockton, CA 95203-3108. Tel.: 209-465-4386. Fax: 209-465-4394.
E-mail: yremlinger@gmail.com
Web Site: www.childrensmuseumstockton.org
Founded: 1991.
Congressional District: 11
Key Personnel: Bd. Pres., Diane Batres; Dir., Yvette Remlinger.
Personnel Profile: Full-Time Paid 4; Part-Time Paid 10; Part-Time Volunteers 20.
Governing Authority: private; nonprofit organization. Tax-exempt: 501(c)(3).
Institution Type/Description: Children's Museum.
Collections: hands-on exhibits.
Publications: monthly calendar; bimonthly newsletter; summer camp brochure.
Hours & Admission Prices: Wed.-Fri. 9-4, Sat. 9-5, Sun. 12-5. Admission $6; discounts to school groups; children under one & members no charge. Closed New Year's Day; Thanksgiving; Christmas Eve & Day. &
Attendance: 60,000 (estimated)
Membership: Individual $50; General $85; Reciprocal $150.

* **THE HAGGIN MUSEUM, (M),** 1201 N. Pershing Ave., Stockton, CA 95203-1604. Tel.: 209-940-6311. Fax: 209-462-1404.
E-mail: info@hagginmuseum.org
Web Site: www.hagginmuseum.org
Founded: 1928.
Congressional District: 14
Key Personnel: Pres. (V), Thomas Bowe III; C.E.O. & Cur. History, Tod Ruhstaller; Dir. Devel., Susan Obert; C.F.O., Karen Richards; Cur. Archival

Collections, Kimberly Bray; Cur. Collections, Kylee Denning; Cur. Education, Lisa Cooperman; Registrar, Erin Hicks; Webmaster, Eddie Hargreaves; Museum Store Mgr., Lisa Falls; Facilities Supt., Ray Shermantine; Membership & Mktg., Kristen Anema; Education Asst., Janet Men; Admin. Asst., Merylene Merengo.
Personnel Profile: Full-Time Paid 13; Part-Time Paid 5; Part-Time Volunteers 300; Interns 6.
Governing Authority: nonprofit organization. Affiliated with the San Joaquin Pioneer & Historical Society. Tax-exempt: 509(a)(1).
Institution Type/Description: Art & History Museum.
Collections: 19th century French & American paintings; American, European & Oriental decorative arts; glass; folk art; California Central Valley history artifacts including agricultural technology, furnishing, personal effects, replicas of local historic interiors & Indian artifacts, primarily California basketry.
Research Fields: late 19th & early 20th-century agricultural technology; Stockton manufacturing history; general Stockton history.
Facilities: library of books, photographs & ephemera on California & local history & art available on the premises by appointment; Holt Archives of Industrial & Agricultural Technology; Stephens Brothers Shipbuilding Archives.
Activities: guided tours; lectures; gallery talks; concerts; art workshops for children; docent council; inter-museum loan, permanent, temporary & traveling exhibitions.
Publications: quarterly activities calendar; history publications; catalogue, Haggin Art Collection.
Hours & Admission Prices: Wed.-Sun. 1:30-5, 1st & 3rd Thurs. of month 1:30-9. Adults $5; discounts to AAM & ICOM members; members no charge. Closed New Year's Day & day after; Easter; Independence Day; Thanksgiving & day after; Christmas Eve & Day. &
Attendance: 45,000 (estimated)
Membership: Student $15; Individual $25; Couple $35; Family $45; Sustaining $60; Supporting $100; Cornerstone $300-$2,500; Business $500-$1,500; Life $1,000; Life Founder $5,000.

THE HERBARIUM OF THE UNIVERSITY OF THE PACIFIC, Classroom Bldg., Stockton, CA 95211. Mailing Address: Dept. Biological Sciences, Univ. of the Pacific, Stockton, CA 95211. Tel.: 209-946-2181. Fax: 209-946-3022.
E-mail: mbrunell@pacific.edu
Founded: 1851.
Congressional District: 14
Key Personnel: Cur., Mark Brunell.
Personnel Profile: Full-Time Paid 1; Part-Time Paid 1.
Governing Authority: university. Parent Institution: University of the Pacific.
Institution Type/Description: Herbarium.
Collections: botany.
Research Fields: herbarium; botany.
Activities: loans for research to accredited herbaria.
Hours & Admission Prices: Mon.-Fri. 8-5. No charge; donations accepted. Closed university holidays. &

HOLT-ATHERTON SPECIAL COLLECTIONS, University of the Pacific Library, 3601 Pacific Ave., Stockton, CA 95211. Tel.: 209-946-2431 & 2404. Fax: 209-946-2942.
E-mail: trichards@pacific.edu
Web Site: library.pacific.edu/ha
Key Personnel: Archivist, Michael Wurtz; Special Collections Asst., Trish Richards
Institution Type/Description: History Museum.
Collections: research materials including manuscripts, photographs & specialized books; Dave Brubeck; John Muir papers; Locke-Hammond Family papers; Jedediah Strong Smith papers; pamphlets; maps; history of Stockton.
Hours & Admission Prices: Mon.-Fri. 10-5. Closed New Year's Eve & Day; Christmas Eve & Day.

STOCKTON FIELD AVIATION MUSEUM, 7430 C.E. Dixon St., Stockton, CA 95206. Tel.: 209-982-0273. Fax: 209-982-4832.
E-mail: museum@twinbeech.com
Web Site: www.twinbeech.com/STOCKTON_FIELD_PAGE.htm
Governing Authority: nonprofit organization.
Institution Type/Description: Aviation History Museum.
Collections: aviation history; WWII aircraft & equipment; photographs; personal artifacts.
Activities: special events.
Hours & Admission Prices: Mon.-Fri. 8:30-5 by appointment. No charge, donations accepted. Closed holidays.

Suisun City

THE WESTERN RAILWAY MUSEUM, (M), 5848 State Hwy. 12, Suisun City, CA 94585-9641. Tel.: 707-374-2978. Fax: 707-374-6742.
Web Site: www.wrm.org
Founded: 1946.
Congressional District: 4
Key Personnel: C.E.O. & Chm. (V), John Coleman; Treas., Bob Towar; Gen. Mgr., Paul Zaborsky; Exec. Dir., Phil Kohlmetz; Sec., John Krauskopf.
Personnel Profile: Full-Time Paid 1; Part-Time Paid 3; Part-Time Volunteers 100.
Governing Authority: company. Parent Institution: Bay Area Electric Railroad Association, Inc. Tax-exempt: 501(c)(3).
Institution Type/Description: Transportation Museum.
Collections: trolley cars; electric interurbans; freight and work cars; Pullman cars; electric locomotives.
Major Exhibits: Traction Labs, 1/14-1/19.
Research Fields: vintage construction techniques for restoration.
Facilities: 4,000-vol. library on railways available by written request to curator; display and restoration buildings for large artifacts; photographic and corporate records files. Railway books for sale.
Activities: guided tours; films; permanent & temporary exhibitions; demonstration railway for living history experience.
Publications: quarterly newsletter, The Review.
Hours & Admission Prices: Memorial Day-Labor Day Wed.-Sun. 10:30-5; Sept.-May Sat.-Sun. 10:30-5. Adults $10, seniors $9, children 2-14 $7; discounts to CAM & AAM members; members no charge. &
Attendance: 25,000 (accurate)
Membership: Individual $50; Family $70; Sustaining $100; Patron $500; Benefactor $1,000.

Sunnyvale

THE LACE MUSEUM, (M), 552 S. Murphy Ave., Sunnyvale, CA 94086-6116. Tel.: 408-730-4695.
E-mail: lacemuseum@gmail.com
Web Site: www.thelacemuseum.org
Founded: 1981.
Congressional District: 12
Key Personnel: C.E.O. & Chm., Eleanore Schwartz; Dir. & Museum Shop Mgr., Suzanne Meyer.
Personnel Profile: Part-Time Volunteers 30.
Governing Authority: nonprofit organization. Tax-exempt: 501(c)(3).
Institution Type/Description: Lace Museum.
Collections: lace-trimmed clothing; books; old & recent laces; decorative items; lace-making tools; family heirlooms; original patterns dated from the late 1800s; costumes.
Research Fields: identification & verification of laces & lace artifacts.
Facilities: 300-vol. library pertaining to the research of lace. Threads, needles, books & patterns for sale.
Activities: guided tours; lectures; classes; films; study clubs; video programs; docent programs; participatory & loan exhibits; demonstrations & classes of several lace-making techniques.
Publications: quarterly newsletter.
Hours & Admission Prices: Tues.-Sat. 11-4; tours by appointment. No charge; donations accepted. Private Tours: $3 per person. Closed New Year's Day; Thanksgiving & two days after; Christmas. &
Attendance: 3,500 (estimated)
Membership: Lace Museum Guild: Senior Citizen & Student $8; General $15; Non US Resident $20; Patron $25 & up.

SUNNYVALE HISTORICAL SOCIETY AND MUSEUM ASSOCIATION, 570 E. Remington Dr., Sunnyvale, CA 94087-2652. Mailing Address: P.O. Box 2187, Sunnyvale, CA 94088-0187. Tel.: 408-749-0220. Fax: 408-732-4726. Facebook: Sunnyvale Museum.
E-mail: info@heritageparkmuseum.org
Web Site: www.heritageparkmuseum.org
Founded: 1973.
Congressional District: 17
Key Personnel: Pres., Jim Reynolds; Vice Pres., Leslie Lawton; Dir., Laura Babcock; Museum Shop Mgr., Margaret Lawson.
Personnel Profile: Part-Time Volunteers 120.
Governing Authority: nonprofit organization. Parent Institution: Sunnyvale Historical Society & Museum Association. Branch Museum: Orchard Heritage Park, 550 Remington, Sunnyvale, CA 94087. Tax-exempt.
Institution Type/Description: History Museum.
Collections: personal items of Ida Trubschenck, first city clerk; photographs; old newspapers, oral history tapes; furnishings from Martin Murphy, Jr. home; artifacts from Hendy Iron Works, Del Monte Seed Dept.; some early

electronics games from Atari; early farm tools & equipment; early office equipment; model of U.S.S. Macon (dirigible) stationed at Sunnyvale Naval Air Station; model of Sparrowhawk, fighter plane of U.S.S. Macon; Silicon Valley artifacts.
Major Exhibits: Depression Era, 1930's, 1/14-4/14; Low Tech, High Tech, 5/14-9/14.
Research Fields: local history; Martin Murphy, Jr. family.
Facilities: 600-vol. library of books available for use on site by prior arrangement.
Activities: lectures; gallery talks; docent program; permanent & temporary exhibitions; school docent program; general meetings with speakers. Community Events: Author's Day; Halloween Celebration; Victorian Teas.
Publications: monthly newsletter.
Hours & Admission Prices: Tues., Thurs. & Sun. 12-4; other times by appointment. No charge; donations accepted. Closed national holidays. &
Attendance: 6,000 (estimated)
Membership: Senior $25; Individual $30; Family $45; Contributing $50; Business & Organization $125; Lifetime $600; Business Lifetime $1,000.

Sunol

GOLDEN GATE RAILROAD MUSEUM, Niles Canyon Railway, Brightside Yard, 5550 Niles Canyon Rd., Sunol, CA 94586. Mailing Address: 1755 E. Bayshore Rd., Ste. 19A, Redwood City, CA 94063-4153. Tel.: 650-365-2472. Fax: 650-385-2473.
E-mail: 2472info@ggrm.org
Web Site: www.ggrm.org
Founded: 1985.
Key Personnel: Pres. (V), Dave Hensarling; Mgr. Operations, Dave Roth; Treas., Ronald Vane.
Personnel Profile: Full-Time Paid 1; Part-Time Volunteers 100.
Governing Authority: private; nonprofit organization. Tax-exempt: 501(c)(3).
Institution Type/Description: Railroad Museum: housed within the former U.S. Navy Hunters Point Shipyard.
Collections: railroad equipment & artifacts pertaining to the San Francisco Bay area's regional rail transportation history; 60 locomotives & railcars; technical drawings & blueprints from Southern Pacific Company.
Publications: monthly newsletter, Stack Talk.
Hours & Admission Prices: Call for details on hours and admissions.
Membership: Friend of GGRM $25; Regular Individual $40; Regular Family $40; Sustaining Member $75; Patron $150.

Susanville

LASSEN HISTORICAL MUSEUM, 75 N. Weatherlow St., Susanville, CA 96130. Mailing Address: P.O. Box 321, Susanville, CA 96130-0321. Tel.: 530-257-3292.
Founded: 1958.
Congressional District: 14
Key Personnel: Pres., Tony Jonas.
Personnel Profile: Part-Time Volunteers 14.
Governing Authority: municipal. Lassen County Historical Society. Tax-exempt: 501(c)(3).
Institution Type/Description: Regional History Museum.
Collections: historical items related to Lassen County history.
Research Fields: genealogy & County history.
Facilities: Historic information for sale.
Activities: permanent & special exhibitions; monthly meetings with presentations of local history; field trips.
Publications: pamphlets; books; monthly newsletter.
Hours & Admission Prices: Summer: Mon.-Fri. 10-4, Sat. 11-3; Winter: Tues.-Thurs. & Sat. 10-2; other times by appointment. No charge; donations accepted. &
Membership: Individual $10; Family $15.

Sylmar

NETHERCUTT COLLECTION, 15151 Bledsoe St., Sylmar, CA 91342. Tel.: 818-364-6464. Fax: 818-364-6466.
E-mail: info@nethercuttcollection.org
Web Site: www.nethercuttcollection.org
Founded: 1971.
Institution Type/Description: Automobile & Musical Instrument Museum.
Collections: period automobiles; mechanical musical instruments; period furniture.
Hours & Admission Prices: Thurs.-Sat. 10-1:30 by appointment. No charge. Closed New Year's Eve & Day; Memorial Day weekend; Independence Day; Labor Day weekend; Thanksgiving weekend; Christmas Eve, Day & week. &

Taft

WEST KERN OIL MUSEUM, 1168 Wood St., Taft, CA 93268-4336. Mailing Address: P.O. Box 491, Taft, CA 93268-0491. Tel.: 661-765-6664. Fax: 661-765-9175.
E-mail: curator@lightspeed.net
Web Site: www.westkern-oilmuseum.org
Founded: 1972.
Key Personnel: Dir., Don Maxwell; Chm. (V), Rick Woodson; Pres. (V), Michael McCormick; Museum Shop Mgr., Mary Wenzel.
Personnel Profile: Interns 30.
Governing Authority: Tax-exempt.
Institution Type/Description: Oil History Museum.
Collections: California's oil industry; local history; photographs; equipment; books.
Publications: quarterly newsletter, The Pumper.
Hours & Admission Prices: Thurs.-Sat. 10-4, Sun. 1-4. No charge; donations accepted. Closed New Year's Day; Independence Day; Thanksgiving; Christmas. &
Attendance: 1,057 (estimated)
Membership: Single $12.50; Family $15; Supporting $35; Patron $75; Memorial & Life $250; Benefactor $500.

Tahoe City

GATEKEEPER'S MUSEUM & WATSON CABIN MUSEUM, (M), 130 W. Lake Blvd., Tahoe City, CA 96145. Mailing Address: P.O. Box 6141, Tahoe City, CA 96145-6141. Tel.: 530-583-1762. Fax: 530-583-8992.
E-mail: info@northtahoemuseums.org
Web Site: www.northtahoemuseums.org
Formerly: Gatekeeper's Museum & Marion Steinbach Indian Basket Museum
Founded: 1969.
Congressional District: 4
Key Personnel: Museum Coord., Javier Rodriguez; Pres. (V), Nileta Morton; Vice Pres. Finance, Jim Phelan; Vice Pres. Governance, Carol Shaw; Vice Pres. Devel., Trudy Lesem.
Personnel Profile: Full-Time Paid 2; Part-Time Paid 1; Part-Time Volunteers 12.
Governing Authority: private; nonprofit organization. Parent Institution: North Lake Tahoe Historical Society. Branch Museums: Watson Cabin Museum; Marion Steinbach Indian Basket Museum; Gatekeeper's Museum. Tax-exempt: 501(c)(3).
Institution Type/Description: Historical Society Museum: site of prehistoric summer camping site & Washoe camp site.
Collections: archaeological artifacts; research library. Marion Steinbach Indian Basket Museum: over 1,000 American Indian baskets, research library, dolls & artifacts.
Major Exhibits: Ursus Among Us: The American Black Bear in the Tahoe Basin, 7/12-10/14; Washoe Baskets from Jack Bogart Collection, 7/12-10/14.
Research Fields: Lake Tahoe history; Native American basketry; American West & Emigrant Trail.
Facilities: 600-vol. library; 5,000 sq. ft. exhibit space. Museum-related items for sale. Gatekeeper's Museum/Marion Steinbach Indian Basket Museum/Watson Cabin Museum/William B. Layton Park.
Activities: guided tours; hobby workshops; temporary exhibitions; special events. Museum Sponsors: basket weaving classes; weavers market; outings.
Publications: Gatekeeper's Gazette.
Hours & Admission Prices: June-Sept. Wed.-Mon. 10-5; Oct.-May Fri.-Sat. 10-4; other times by appointment. Adults $5, seniors citizens 65 & over $4; discounts to active military; children under 12 & members no charge.
Attendance: 30,000 (estimated)
Membership: Individual $30; Family $40; Business $50; Sponsor $100; Patron $150; Benefactor $250.

Tahoma

PINE LODGE (EHRMAN MANSION), Ed Z'berg Sugar Pine Point State Park, Tahoma, CA 96142. Mailing Address: Sugarpine Point State Park, Hwy. 89, P.O. Box 266, Tahoma, CA 96142-0266. Tel.: 530-525-5055. Fax: 530-525-3380.
E-mail: ndavenport@parks.ca.gov
Web Site: www.parks.ca.gov
Founded: 1965.
Congressional District: 14
Key Personnel: Lake Sector Supt., Brian Barton; Unit Supervising Ranger, Steven Walloupe; Museum Cur. II, Natalie E. Davenport.
Personnel Profile: Full-Time Paid 2; Part-Time Paid 4; Part-Time Volunteers 4.

Governing Authority: state; nonprofit. Parent Institution: State of California, Dept. of Parks & Recreation, Sierra State Parks Foundation. Tax-exempt.
Institution Type/Description: Nature & Visitor Center.
Collections: period non-primary domestic furnishings; natural history specimens. Historic Building: c.1901 home.
Research Fields: Lake Tahoe resort living 1900-1940s; Lake Tahoe summer homes; Isaias Hellman.
Facilities: trails; picnic areas; grounds available for weddings; campground.
Activities: guided tours; interpretive hikes & campfire programs; annual living history event; cross country skiing; swimming; snowshoeing.
Publications: The Ehrman Mansion.
Hours & Admission Prices: Mansion: mid-June to Labor Day daily 11-4, tours on the hour. Adults $5, children 6-12 $3; children under 6 no charge. Parking $5.
Attendance: 20,000 (estimated)

VIKINGSHOLM, Emerald Bay State Park, Hwy. 89, Tahoma, CA 96142. Mailing Address: Emerald Bay State Park, P.O. Box 266, Hwy. 89, Tahoma, CA 96142-0266. Tel.: 530-525-5055. Fax: 530-525-3380.
E-mail: ndavenport@parks.ca.gov
Web Site: www.parks.ca.gov
Founded: 1953.
Congressional District: 14
Key Personnel: Supervising Ranger, Steven Walloupe; Museum Cur. II, Natalie Davenport; Lake Sector Supt., Brian Barton.
Personnel Profile: Full-Time Paid 2; Part-Time Paid 3; Part-Time Volunteers 3.
Governing Authority: state; nonprofit. Parent Institution: California Dept. of Parks & Recreation. Subsidiary Institution: Friends of Vikingsholm. Tax-exempt.
Institution Type/Description: Visitor Center & House Museum: house museum is a c.1929 historic Scandinavian styled resort home.
Collections: primary historic domestic furnishings; 16th-century European gate locks; Scandinavian furniture; architectural details. Historic Building: c.1929 Scandinavian styled resort home.
Research Fields: Lake Tahoe historic summer homes; arts & crafts movement.
Facilities: visitor center; trails; picnic area; campground.
Activities: guided tours.
Publications: Vikingsholm; Tahoe's Hidden Castle.
Hours & Admission Prices: mid-June to Labor Day daily 10-4, tours on the half hour. Adults $5, children 6-12 $3.
Attendance: 40,000 (estimated)

Taylorsville

INDIAN VALLEY MUSEUM, 4288 Cemetery St., Taylorsville, CA 95983. Mailing Address: P.O. Box 194, Taylorsville, CA 95983-0194. Tel.: 530-284-1046. Facebook: Indian Valley Museum.
E-mail: IVM@frontiernet.net
Web Site: www.indianvalley.net/iv-museum
Founded: 1963.
Key Personnel: Pres. (V), Francis F. Musser; Museum Shop Mgr., Katherine Iglesias.
Personnel Profile: Part-Time Volunteers 13.
Governing Authority: Parent Institution: Mt. Jura Gem & Museum Society. Tax-exempt.
Institution Type/Description: History Museum.
Collections: local history & culture; Native American artifacts from 1860-1940; gems & minerals.
Hours & Admission Prices: Memorial Day to Oct. Sat.-Sun. 1-4; other times by appointment. No charge; donations accepted. &
Attendance: 1,000 (estimated)
Membership: Individual $10.

Tehachapi

MOURNING CLOAK RANCH AND BOTANICAL GARDENS, 22101 Old Town Rd., Tehachapi, CA 93561-8886. Tel.: 661-822-1661. Fax: 661-822-5062.
Institution Type/Description: Botanical Garden.
Collections: 2,200 plants.
Hours & Admission Prices: May-Oct. Mon.-Sat. 10-3.

Tehama

TEHAMA COUNTY MUSEUM, 275 C St., Tehama, CA 96090-0275. Mailing Address: P.O. Box 275, Tehama, CA 96090-0275. Tel.: 530-384-2595.
E-mail: tcmuse@tehama.net

Web Site: tehamacountymuseum.org
Founded: 1969.
Volunteer Hours: 5,000
Institution Type/Description: History Museum.
Collections: local history & culture; photographs; personal artifacts.
Facilities: 1859 Masonic lodge / school house; Marty Graffell Annex.
Publications: quarterly newletter.
Hours & Admission Prices: Fri.-Sun. 1-4. No charge; donations accepted. &
Membership: Students, 10-18 $10; Individual $20; Family $30; Sustaining $50; Patron $100; Lifetime $300.

Temecula

PENNYPICKLE'S WORKSHOP, 42081 Main St., Temecula, CA 92590-2769. Tel.: 951-308-6370. Fax: 951-695-0636.
E-mail: phineas@pennypickles.org
Web Site: www.pennypickles.org
Formerly: Imagination Workshop - Temecula Children's Museum
Founded: 2004.
Key Personnel: Museum Mgr., Robin Gilliland; Museum Shop Mgr., Debbi Nelson.
Personnel Profile: Full-Time Paid 1; Part-Time Paid 8; Part-Time Volunteers 2.
Governing Authority: city. Tax-exempt.
Institution Type/Description: Children's Museum.
Collections: hands-on exhibits.
Facilities: Museum-related items for sale.
Hours & Admission Prices: Tues.-Thurs. & Sat. Session 1: 10-12, Session 2: 12:30-2:30, Session 3: 3-5, Fri. Session 1: 10-12, Session 2: 12:30-2:30, Session 3: 3-5, Session 4: 5:30-7:30, Sun. Session 1: 12:30-2:30, Session 2: 3-5. Admission 3 & over $4.50; discounts to groups & ACM Reciprocal members; children 2 & under no charge. &
Attendance: 50,000 (estimated)
Membership: $100.

TEMECULA VALLEY MUSEUM, 28314 Mercedes St., Temecula, CA 92590-1837. Tel.: 951-694-6455. Fax: 951-506-6871.
E-mail: tracy.frick@cityoftemecula.org
Web Site: www.temeculamuseum.org
Founded: 1985.
Congressional District: 48
Key Personnel: Museum Svcs. Mgr., Tracy Frick; Pres., Phil McElhinney; Sec., Dale Garcia; Treas., Phil Baily; Museum Shop Mgr., Jo Ann Lamb.
Personnel Profile: Full-Time Paid 1; Part-Time Paid 5; Part-Time Volunteers 57.
Governing Authority: Parent Institution: City of Temecula. Tax-exempt.
Institution Type/Description: History Museum.
Collections: focus on photographs, documents, tools & household goods from local ranches & frontier towns; artifacts & photographs relating to local Native American cultures. Special collections relate to the Roripaugh & Vail ranching families, & to Erle Stanley Gardner, author of the Perry Mason stories.
Major Exhibits: The Many Faces of George Washington (T), 1/1-2/14; Women & Spirit: Catholic Sisters in California (T), 3/1-4/14; Third Grade Perspective: Temecula History, 4/1-5/14; Bear in Mind (T), 5/1-6/14; The Grand Tour (T), 7/1-8/14; Apron Strings (T), 9/1-10/14; Miniature Trains, 11/1-12/14.
Research Fields: Luiseno history & culture; Vail Ranch history.
Facilities: 7,200 sq. ft. exhibit. Museum-related items for sale.
Activities: guided tours; lectures; docent program.
Publications: Temecula Valley Museum Docent Handbook; Temecula Valley Museum Newsletter; High Country Magazine.
Hours & Admission Prices: Tues.-Sat. 10-4, Sun. 1-4. Suggested Donation: $2; discount to AAM & ICOM members. &
Attendance: 19,008 (accurate)

Thousand Oaks

CHUMASH INDIAN MUSEUM, (M), 3290 Lang Ranch Pkwy., Thousand Oaks, CA 91362. Tel.: 805-492-8076. Fax: 805-492-8096.
E-mail: chumashindianmuseum@verizon.net
Web Site: www.chumashindianmuseum.com
Formerly: Ashbrook Chumas Interpretive Center
Founded: 1994.
Congressional District: 24
Key Personnel: Dir., Alfred Mazza; C.E.O., Miles Lang; Museum Shop Mgr., Gray Wolf.
Personnel Profile: Full-Time Paid 1; Full-Time Volunteers 1; Part-Time Paid 5; Part-Time Volunteers 12; Interns 2.

Governing Authority: Parent Institution: City of Thousand Oaks. Subsidiary Institution: Conejo Park & Recreation. Tax-exempt.
Institution Type/Description: Native American Museum.
Collections: Native American history & culture; personal artifacts; paintings; photographs.
Research Fields: archaeology; anthropology; Chumash culture; ethnography; maritime lifeways.
Facilities: lecture hall. Museum-related items for sale.
Activities: school tours; temporary & permanent exhibitions; lectures; inter-Tribal ritual meetings; seminars.
Publications: newsletters; brochures; consulting archaeologist publications.
Hours & Admission Prices: Thurs.-Sun. 12-5. Adults $5, seniors & children $3; discounts to AAM & ICOM members. Closed major holidays. &
Attendance: 18,000 (estimated)
Membership: Lizard $5; Senior & Student $15; Coyote $25; Deer $40; Bear $100; Swordfish $250; Dolphin $500; Whale $1,000.

CONEJO VALLEY ART MUSEUM, 3143 Potter Ave., Thousand Oaks, CA 91358-0616. Mailing Address: P.O. Box 1616, Thousand Oaks, CA 91358-0616. Tel.: 805-373-0054 & 492-2147. Fax: 805-492-7677. TDD: 805-492-7677.
E-mail: dessornes@earthlink.net
Web Site: www.cvam.us
Founded: 1977.
Congressional District: 21
Key Personnel: C.E.O. & Pres. (V), Maria E. Dessornes.
Personnel Profile: Full-Time Volunteers 1; Part-Time Volunteers 50; Interns 2.
Governing Authority: not-for-profit organization. Tax-exempt: 501(c)(3).
Institution Type/Description: Art Museum.
Collections: various medias of art.
Facilities: 2,500 sq. ft. exhibit space. Books, tapes, folk art, fiber arts & pottery cards for sale.
Activities: concerts; lectures; loan exhibitions; rental gallery; traveling exhibitions; formal education programs for adults. Museum Sponsors: ArtWalk (a fine art & designer craft fair) in June.
Publications: monthly newsletter; calendar of events.
Hours & Admission Prices: Temporarily closed for relocation. &
Attendance: 12,000 (estimated)
Membership: Students & Seniors $10; General & Family $25; Sponsor $50; Patron $100-$249; Silver Patron $250-$499; Gold Patron $500-$999; Benefactor $1,000-$1,499; Corporate Donor $2,500 & up.

CONEJO VALLEY BOTANIC GARDEN, Hendrix Ave., Thousand Oaks, CA 91359. Mailing Address: P.O. Box 6614, Thousand Oaks, CA 91359-6614. Tel.: 805-494-7630.
E-mail: conejogarden@hotmail.com
Web Site: conejogarden.org
Governing Authority: Tax-exempt: 501(c)(3).
Institution Type/Description: Botanic Garden.
Collections: native plants; oak trees; water-conserving plants; wildlife & their habitats.
Facilities: nature trails.
Activities: educational programs.
Hours & Admission Prices: Daily 7-5. No charge. Closed New Year's Eve & Day; Easter; Independence Day; Thanksgiving; Christmas Eve & Day.

SANTA MONICA MOUNTAINS NATIONAL RECREATION AREA, 401 W. Hillcrest Dr., Thousand Oaks, CA 91360-4223. Tel.: 805-370-2300.
Web Site: www.nps.gov/samo/
Founded: 1978.
Congressional District: 21, 23, 24, 26, 27
Key Personnel: Supt., David Szymanski; Archeologist, Gary Brown.
Personnel Profile: Full-Time Paid 1.
Governing Authority: federal. Parent institution: National Park Service, U.S. Dept. of the Interior. Tax-exempt.
Institution Type/Description: National Park: Collections consist of natural history specimens, local archaeological & historical artifacts and park archives. Satwiwa Native American Indian Culture Center.
Collections: natural & cultural history items.
Research Fields: Santa Monica Mountains history & natural history.
Facilities: cultural center; visitor center. Books & museum-related items for sale.
Activities: seminars; concerts; theme events; naturalist programs; self-guided tours; education programs for children.
Publications: quarterly calendar, Outdoors.
Hours & Admission Prices: Visitor Center: Mon.-Sat. 8-5. Park Site: daily. No charge. Closed major holidays. &
Attendance: 76 (accurate)

Membership: Friends of Satwiwa: Student & Elder $10; Individual $25; Family $35; Associate $100; Life $500; Patron $1,000; Corporate $5,000.

Three Rivers

SEQUOIA AND KINGS CANYON NATIONAL PARKS, 47050 Generals Hwy., Three Rivers, CA 93271-9599. Tel.: 559-565-3136. Fax: 559-565-3744.
Web Site: www.nps.gov/seki
Founded: 1933.
Congressional District: 17
Key Personnel: Chief Park Interpreter, Colleen Bathe; Museum Technician, Ward Eldredge.
Personnel Profile: Full-Time Paid 2; Part-Time Paid 1.
Governing Authority: federal. Parent Institution: National Park Service, U.S. Dept. of the Interior, Washington, DC. Tax-exempt: 501(c)(3).
Institution Type/Description: Park Museum & Visitor Centers: located at Ash Mountain, Lodgepole & Grant Grove.
Collections: archives; biological & geological specimens, prehistoric & historic artifacts & an extensive historic photographic file of black & white images.
Facilities: 1,500-vol. reference library of natural history books available at park headquarters. Publications for sale.
Activities: visitor centers.
Hours & Admission Prices: See website for hours. Visitor Centers: no charge; donations accepted. Parks: fee charged. &

Tiburon

ANGEL ISLAND STATE PARK, Tiburon, CA 94920. Mailing Address: P.O. Box 318, Tiburon, CA 94920-0318. Tel.: 415-435-5390. Fax: 415-435-0850. Facebook: Angel Island State Park.
E-mail: tours.angelisland@parks.ca.gov
Web Site: www.parks.ca.gov/angelisland
Formerly: U.S. Immigration Station
Founded: 1954.
Congressional District: 6
Governing Authority: Parent Institution: State of California; Department of Parks and Recreation.
Institution Type/Description: Historic Buildings & Sites.
Collections: immigration & quarantine history; Civil War through Cold War; military site.
Major Exhibits: American Civil War: 150th Anniversary, 11/13-12/14.
Activities: Annual Events: Civil War Days in June; Children's Outdoor Bill of Rights Day in July; Holidays on the Home Front in November.
Hours & Admission Prices: 8am to sunset, call to confirm. Park entrance fee included in price of ferry ticket; discounts to school groups K-12. &
Attendance: 175,000 (estimated)

BELVEDERE-TIBURON LANDMARKS SOCIETY, 1550 Tiburon Blvd., Ste. M, Tiburon, CA 94920. Tel.: 415-435-1853.
E-mail: lmsoffice@sbcglobal.net
Web Site: landmarks-society.org
Founded: 1959.
Congressional District: 6
Key Personnel: Exec. Dir., Alan Brune.
Personnel Profile: Part-Time Paid 4; Part-Time Volunteers 60.
Governing Authority: society; nonprofit organization. Branch Museums: Old St. Hilary's Landmark, Esperanza St., Tiburon CA, 94920; China Cabin, Beach Rd. Belvedere, CA 94920. Tiburon Railroad-Ferry Museum, 1920 Paradise Dr. Tiburon, CA 94920; Landmarks Art & Garden Center, site of 1870 brick kiln & bunkhouse, 841 Tiburon Blvd. Tiburon, CA 94920. Tax-exempt: 501(c)(3).
Institution Type/Description: Old St. Hilary's: Preservation Project & Museum: housed in 1888 Carpenter Gothic style church with wildflower preserve. China Cabin: restored social hall of the SS China, 1866-1886.
Collections: society archives: photographs; art collection; social and architectural history; artifacts of early California domestic life; railroad-ferry system. Historic Buildings: Tiburon Railroad-ferry Museum Depot 1884; Brick Kiln Bunkhouse, 1872.
Research Fields: local history; architecture; native flora.
Activities: botanic & historic tours; art exhibits; concerts; historic lectures; railroad-ferry museum exhibits.
Publications: books, A Garland for John Thomas Howell; Both Sides of the Track; Excerpts of Marin History; Glimpses II; Old St. Hilary's; Pictorial History of Tiburon; Shark Point, High Point Index; A Pictorial History of Belvedere; 1890-1990.
Hours & Admission Prices: April-Oct. Wed. & Sat.-Sun. 1-4; other times by appointment. No charge; donations accepted. Charge for group tours off hours. &

Attendance: 3,500 (estimated)
Membership: Friends $50; Preservationists $100; Patron $250; Historian $500; Steward $1,000; Guardian $2,500.

OLD ST. HILARY'S LANDMARK & WILDFLOWER PRE-SERVE, 201 Esperanza, Tiburon, CA 94920. Mailing Address: 1550 Tiburon Blvd., Ste. M, Tiburon, CA 94920-2529. Tel.: 415-435-1853.
E-mail: lmsoffice@sbcglobal.net
Web Site: www.landmarks-society.org
Formerly: Old St. Hilary's Church Museum & St. Hilary's Preserve
Founded: 1959.
Congressional District: 6
Key Personnel: Exec. Dir., Alan Brune.
Personnel Profile: Part-Time Volunteers 5.
Governing Authority: Parent Institution: Belvedere Tiburon Landmarks Society.
Institution Type/Description: Historical Museum: housed in a 19th-century Carpenter Gothic-style church, built in 1888.
Collections: religious artifacts. Preserve: wildflower garden containing 217 species of plants.
Publications: newsletter, The Landmark.
Hours & Admission Prices: April-Oct. Wed. & Sun. 1-4. No charge; donations accepted.
Attendance: 500 (estimated)
Membership: Friends $50; Preservationist $100; Patron $250; Historian $500; Steward $1,000; Guardian $2,500.

Tomales

TOMALES REGIONAL HISTORY CENTER, 26701 State Hwy. #1, Tomales, CA 94971. Mailing Address: P.O. Box 262, Tomales, CA 94971-0262. Tel.: 707-878-9443.
E-mail: info@tomaleshistory.com
Web Site: tomaleshistory.com
Founded: 1978.
Key Personnel: Bd. Pres., Liz Mitchell; Cur., Ginny Mackenzie Magan; Museum Shop Mgr., Nancy Conzett.
Governing Authority: nonprofit organization. Tax-exempt.
Institution Type/Description: History Center.
Collections: period artifacts; photographs; costumes; painting; prints & drawings.
Activities: children's classes; lectures; performances; living history program; research; fundraising events; yearly week-long study unit of interactive California & local history for two fourth grade classes; periodic lecture series.
Publications: quarterly history journal, TRHC Bulletin; book, Tomales Township: A History.
Hours & Admission Prices: Fri.-Sun. 1-4; other times by appointment. No charge; donations accepted. Closed major holidays. &
Membership: Individual $15; Family $20; Life $250.

Torrance

TORRANCE ART MUSEUM, (M), 3320 Civic Center Dr. N., Torrance, CA 90503-5016. Tel.: 310-618-6340. Fax: 310-618-2399.
E-mail: jramos@torranceca.gov
Web Site: www.torranceartmuseum.com
Founded: 1985.
Congressional District: 36
Key Personnel: Dir. & Head Cur., Max Presneill; Asst. Cur., Jason Ramos; Preparator, Jon Flack; Volunteer Coord., Regina Taylor.
Personnel Profile: Part-Time Paid 4; Part-Time Volunteers 40.
Governing Authority: municipal.
Institution Type/Description: Art Museum.
Collections: contemporary art.
Research Fields: contemporary art.
Facilities: 500-vol. library; educational facilities; 6,000 sq. ft. exhibit space.
Activities: docent program; lectures; temporary & participatory exhibits; film screenings.
Publications: exhibition catalogs.
Hours & Admission Prices: Tues.-Sat. 11-5. No charge; donations accepted. Closed New Year's Day; Memorial Day; Armed Forces Day; Independence Day; Labor Day; Thanksgiving & day after; Christmas week. &
Attendance: 6,821 (accurate)

TORRANCE HISTORICAL SOCIETY & MUSEUM, 1345 Post Ave., Torrance, CA 90501-2621. Tel.: 310-328-5392.
E-mail: museum@torrancehistoricalsociety.org
Web Site: www.torrancehistoricalsociety.org
Founded: 1973.
Key Personnel: Dir., Janet Payne; Pres. (V), Kurt Weideman.
Personnel Profile: Part-Time Volunteers 20.
Governing Authority: Parent Institution: city of Torrance. Tax-exempt.
Institution Type/Description: Historical Society Museum: housed in the city's first main library; built in 1936.
Collections: local history & culture; period artifacts; books; documents.
Publications: Torrance Historian newsletter.
Hours & Admission Prices: Wed.-Thurs. & Sun. 1-4 & by appointment for special tours & research. No charge; donations accepted. Closed legal holidays. &
Attendance: 5,000 (accurate)
Membership: Youth $10; Senior (65 & over) $15; Adult $20; Family $30; Business & Organization $50; Individual Life $350.

WESTERN MUSEUM OF FLIGHT, (M), 3315 Airport Dr., Red Baron #3, Torrance, CA 90505-6152. Tel.: 310-326-9544. Fax: 310-326-9556.
E-mail: info@wmof.com
Web Site: www.wmof.com
Key Personnel: Pres., Cindy Macha Skjonsby.
Governing Authority: nonprofit organization. Parent Institution: Southern California Historical Aviation Foundation.
Institution Type/Description: Aviation History Museum.
Collections: Southern California's aviation history; airplanes; target drones; piston & jet aircraft engines; World War II instruments; aircrew accessories; model aircrafts.
Facilities: Museum-related items for sale.
Activities: educational programs for children.
Hours & Admission Prices: Tues.-Sun. 10-3. Adults $3; members & children under 12 no charge.
Membership: Contributor $50; Friend $100; Family $250; Patron $500; Bronze Patron $1,200; Silver Patron $2,500; Gold Patron $5,000; Associate $10,000; Benefactor $15,000; Founder $25,000.

Tracy

TRACY HISTORICAL MUSEUM, 1141 Adam St., Tracy, CA 95376-3506. Mailing Address: P.O. Box 117, Tracy, CA 95378. Tel.: 209-832-7278.
E-mail: tracymuseum@sbcglobal.net
Web Site: www.tracymuseum.org
Founded: 1993.
Congressional District: 11
Key Personnel: Dir., Onalee Koster; Pres. (V), Larry Gamino.
Personnel Profile: Part-Time Volunteers 30.
Governing Authority: city. Operated by the West Side Pioneer Association. Tax-exempt.
Institution Type/Description: History Museum: housed in the former Tracy Post Office building; built in 1937.
Collections: local history & culture; photographs; period artifacts.
Research Fields: newsletter, The Pioneer Press; book, Images of America: Tracy; local history.
Activities: history day; art celebration; museum tours.
Hours & Admission Prices: June-Sept. Sun. & Fri. 1-4, Mon. 9-2. No charge; donations accepted. Closed most major holidays. &
Attendance: 1,000 (estimated)
Membership: Students $5; Adults & Organization $10; Couple $15; Commercial $20; Lifetime Membership $150.

Travis AFB

TRAVIS AIR MUSEUM AKA JIMMY DOOLITTLE AIR & SPACE MUSEUM, 461 Burgan Blvd., Bldg. 80, Travis AFB, CA 94535. Tel.: 707-424-5605.
E-mail: curator@travisairmuseum.org
Web Site: www.travisairmuseum.org
Institution Type/Description: Military Museum.
Collections: aircraft artifacts & memorabilia.
Facilities: Museum-related items for sale.
Hours & Admission Prices: Guided Tours: DoD ID: Tues.-Sat. 9-4. General Public: Thurs. & Sat. 9-3. No Charge. Closed federal holidays.

Trinidad

STONE LAGOON RED SCHOOLHOUSE MUSEUM AT RED-WOOD TRAILS, 265 Idlewood Lane, Trinidad, CA 95570-9641. Mailing Address: 265 Redwood Trails Cir., P.O. Box 1240, Trinidad, CA 95570. Tel.: 707-488-2061.
E-mail: info@rv4fun.com
Institution Type/Description: Historic Building: housed in former one-room schoolhouse, c.1894.
Collections: period furnishings; structure.
Hours & Admission Prices: Memorial Day to Labor Day. No charge.

Truckee

DONNER MEMORIAL STATE PARK AND EMIGRANT TRAIL MUSEUM, 12593 Donner Pass Rd., Ste. 9, Truckee, CA 96161-3856. Tel.: 530-582-7892. Fax: 530-550-2347.
Web Site: www.parks.ca.gov
Founded: 1962.
Congressional District: 14
Key Personnel: Unit Supervising Ranger, Don Schmidt; Ranger, Mike Rominger; Cur., Judith K. Polanich.
Personnel Profile: Full-Time Paid 6; Part-Time Paid 3; Part-Time Volunteers 5; Interns 1.
Governing Authority: state; nonprofit. Parent Institution: Dept. of Parks & Recreation. Subsidiary Institution: Sierra State Parks Foundation. Tax-exempt.
Institution Type/Description: Historic Site & Museum.
Collections: California Emigrant Trail-Truckee routes; Donner Emigrant Party; Central Pacific Railroad; weapons; photographs.
Research Fields: Westward Emigration; Native American culture; Chinese American culture; winter survival.
Facilities: operational; outdoor text & photo panel interpretive exhibit; self-guided nature trails; camping grounds; biking; swimming; boating (no launch); cross country skiing; fishing.
Activities: interpretive hikes & campfire programs.
Publications: trail guides; park brochure; coloring book.
Hours & Admission Prices: Daily 9-4. Museum: no charge; donations accepted. Closed New Year's Day; Thanksgiving; Christmas. No winter camping. &

Attendance: 75,000 (accurate)

KIDZONE MUSEUM, 11711 Donner Pass Rd., Truckee, CA 96161-4954. Tel.: 530-587-KIDS. Fax: 530-587-0200.
E-mail: info@kidzonemuseum.org
Web Site: www.kidzonemuseum.org
Founded: 1992.
Key Personnel: Exec. Dir., Carol Meagher; Pres. (V), Helen Pelster; Exhibits Mgr., Liz Bordner; Mgr. Operations, Romina Branje.
Personnel Profile: Full-Time Paid 4; Part-Time Paid 4; Part-Time Volunteers 120.
Governing Authority: nonprofit. Tax-exempt: 501(c)(3).
Institution Type/Description: Children's Museum.
Collections: hands-on exhibits.
Hours & Admission Prices: June-July Tues.-Sat. 9-1; Winter: Tues.-Sat. 10-5, Sun. 10-1. Children $7, adults $5, seniors $3, 1st Tues. each month $1; discounts to ACM Assoc. members; children under one no charge. &
Attendance: 22,000 (accurate)
Membership: Co-Op Parent & Child $85; Co-Op Family $100; ACM Explorer Family $125; ACM Explorer & Grandparents $135.

TRUCKEE RAILROAD MUSEUM, 10075 Donner Pass Rd., Truckee, CA 96161. Mailing Address: P.O. Box 3838, Truckee, CA 96160.
E-mail: info@truckeedonnerrailroadsociety.com
Web Site: www.truckeedonnerrailroadsociety.com
Founded: 2010.
Congressional District: 4
Key Personnel: Chm. (V), Don Davis; Pres. (V), Jim Hood; Museum Shop Mgr., Bob Bell.
Personnel Profile: Part-Time Volunteers 22.
Governing Authority: Parent Institution: Truckee Donner Railroad Society. Tax-exempt.
Institution Type/Description: History Museum: housed in a Southern Pacific Railroad caboose.
Collections: railroad history & artifacts; photographs.
Facilities: library.
Publications: quarterly newsletter, Snowshed.
Hours & Admission Prices: Sat.-Sun. 10-4. No charge; donations accepted. &

Attendance: 5,160 (accurate)

TRUCKEE'S OLD JAIL MUSEUM, Jibbom & Spring Sts., Truckee, CA 96160. Mailing Address: P.O. Box 893, Truckee, CA 96160-0893. Tel.: 530-582-0893.
E-mail: info@truckeehistory.org
Web Site: www.truckeehistory.org
Formerly: Truckee-Donner Historical Society Jail Museum
Key Personnel: Cur. & Coord. Special Group Tours, Chelsea Walterscheid.
Governing Authority: Operated by the Truckee-Donner Historical Society.
Institution Type/Description: Historic Building: served as jailhouse from 1875 to 1964.
Collections: local historical artifacts; photographs; lumbering; box manufacturing; ice harvesting; film industry; Truckee's early winter sports.
Facilities: garden.
Hours & Admission Prices: Memorial Day-Sept. Sat.-Sun. 11-4; group tours by appointment. Requested Donation: $2.

Tujunga

BOLTON HALL HISTORICAL MUSEUM, 10110 Commerce Ave., Tujunga, CA 91042-2313. Mailing Address: P.O. Box 203, Tujunga, CA 91043-0203. Tel.: 818-352-3420.
Founded: 1957.
Key Personnel: Pres. (V) & Museum Shop Mgr., Lloyd Hitt.
Personnel Profile: Part-Time Volunteers 25; Interns 2.
Governing Authority: Tax-exempt.
Institution Type/Description: Community Historic Museum.
Collections: photographs; newspapers; cemetery records; directories; Rancho Tujunga artifacts.
Facilities: library; photoarchives.
Activities: tours. Museum Sponsors: programs September to June.
Publications: newsletter.
Hours & Admission Prices: Sun. & Tues. 1-4. No charge; donations accepted. &

Attendance: 1,500 (estimated)
Membership: Individual $15; Family $25; Patron & Business $50; Lifetime $350.

Tulare

INTERNATIONAL AGRI-CENTER ANTIQUE FARM EQUIP-MENT MUSEUM AND AGVENTURES LEARNING CENTER, 4500 S. Laspina St., Tulare, CA 93274-9165. Tel.: 559-688-1030. Fax: 559-686-5527. Facebook: International Agri-Center.
E-mail: kerissa@farmshow.org
Web Site: www.internationalagricenter.org
Founded: 2000.
Congressional District: 31
Key Personnel: C.E.O., Jerry Sinift.
Personnel Profile: Full-Time Paid 24; Part-Time Volunteers 1,300.
Governing Authority: Parent Institution: International Agri-Center. Tax-exempt.
Institution Type/Description: Agriculture Museum.
Collections: period tractors; farming equipment; steam engine equipment; hands-on exhibits.
Hours & Admission Prices: Daily. No charge; donations accepted. &
Attendance: 10,000 (estimated)

TULARE HISTORICAL MUSEUM, 444 W. Tulare Ave., Tulare, CA 93274-3831. Mailing Address: P.O. Box 248, Tulare, CA 93275-0248. Tel.: 559-686-2074. Fax: 559-686-9295.
Web Site: www.tularehistoricalmuseum.org
Founded: 1985.
Congressional District: 21
Key Personnel: Exec. Dir., Terry Brazil; Pres. (V), Michael Wasnick; Vice Pres., Tina Macedo; Financial Dir., Robert L. Bandy; Devel., Cathy Mederos; Cur., Chris Harrell; Historian & Public Rels., Linda Ruminer; Docent Coord., Ron Vaughan; Museum Shop Mgr., Helen McCourt.
Personnel Profile: Part-Time Paid 3, Part-Time Volunteers 45, Interns 1.
Governing Authority: private; nonprofit organization. Parent Institution: Tulare City Historical Society. Tax-exempt: 501(c)(3).
Institution Type/Description: History Museum.
Collections: Tulare history & heroes memorabilia; period furniture; art glass; photographs; books; military uniforms & artifacts; historical piano, telephones, cameras, & medical/dental tools.
Facilities: library; 117-seat auditorium; 6,356 sq. ft. exhibit space. Museum-related items for sale.
Activities: concerts; docent program; lectures; loan & traveling exhibitions;

training programs for professional museum workers & museum volunteers. Annual Events: Lovely Way to Spend an Evening in February; Taste Treats in Tulare; Reverse Drawing.

Publications: quarterly newsletter, Then & Now.

Hours & Admission Prices: June-Aug. Thurs.-Sat. 10-4; Sept.-May Thurs.-Sat. 10-4, 3rd Sun. of month 12:30-4. Adults $5, senior citizens $3, students $2; discounts to AAA members; members, children under 5 & 3rd Sun. Sept.-May no charge. Closed New Year's Day; Easter; Mother's Day; Father's Day; Independence Day; Thanksgiving & day after; Christmas & day after.

Attendance: 4,000 (estimated)

Membership: Individual $40; Family $60; Patron $100; Business $200; Lifetime $1,000.

Tulelake

LAVA BEDS NATIONAL MONUMENT, 1 Indian Wells Headquarters, Tulelake, CA 96134. Mailing Address: P.O. Box 1240, Tulelake, CA 96134-1240. Tel.: 530-667-8101. Fax: 530-667-2737.

E-mail: labe_interpretation@nps.gov

Web Site: www.nps.gov/labe/

Founded: 1925.

Congressional District: 1

Key Personnel: Supt., Mike Reynolds.

Governing Authority: federal.

Institution Type/Description: History/Natural History Museum.

Collections: geology of shield volcanoes & lava tubes; desert ecology; history; archaeology; Modoc Indian artifacts.

Publications: publications pertaining to natural history; human history; Modoc Indian War 1872-73.

Hours & Admission Prices: Memorial Day weekend-Labor Day weekend daily 8-6; Sept.-May daily 8:30-5. Monument: $10 per car; annual pass $20; federal lands pass holders no charge.

Attendance: 100,000 (estimated)

Tustin

MARCONI AUTOMOTIVE MUSEUM, 1302 Industrial Dr., Tustin, CA 92780-6416. Tel.: 714-258-3001. Fax: 714-258-9117.

E-mail: mhanover@marconimuseum.org

Web Site: www.marconimuseum.org

Founded: 1994.

Key Personnel: C.E.O., Priscilla "Bo" Marconi; Exec. Dir., Missy Hanover; Dir. Operations, Miguel De la Cerda.

Governing Authority: nonprofit.

Institution Type/Description: Automobile Museum.

Collections: historical, exotic, & classic cars.

Facilities: rental facilities.

Activities: facility rental.

Hours & Admission Prices: Mon.-Fri. 9-4:30 by appointment. Donation: adults $5; children no charge. Closed major holidays.

Attendance: 3,000 (estimated)

Twentynine Palms

JOSHUA TREE NATIONAL PARK, 74485 National Park Dr., Twentynine Palms, CA 92277-3597. Tel.: 760-367-5500 & 5502. Fax: 619-367-6392.

Web Site: www.nps.gov/jotr

Formerly: Twentynine Palms Oasis Visitor Center, Joshua Tree National Park

Founded: 1936.

Congressional District: 40 & 44

Key Personnel: Superintendent, Mark Butler.

Governing Authority: federal. Parent Institution: National Park Service. Tax-exempt: 501(c)(3).

Institution Type/Description: Park Museum.

Collections: archaeology; geology; zoology; botany; desert flora & fauna; palm trees; rabbits; birds; reptiles.

Research Fields: natural history, cultural history.

Facilities: 750-vol. library of books on history & natural history of the area; 9 camp grounds. Educational literature of the area for sale.

Activities: orientation talks; programs on the park and its history; guided walks on nearby nature trail & historic site.

Publications: Guide to Joshua Tree National Park.

Hours & Admission Prices: Park: Open daily. Joshua Tree Visitor Center: daily 8-5. Oasis Visitor Center: daily 9-5. Cottonwood Visitor Center: daily 9-4. Black Rock Nature Center: Oct.-May Mon.-Thurs. & Sat.-Sun. 8-4, Fri. 12-8. Park entrance fees: 7-day vehicle permit $15, 7-day single entry permit $5.

Attendance: 1,256,928 (accurate)

Membership: Joshua Tree National Park Pass $30.

Ukiah

GRACE HUDSON MUSEUM & SUN HOUSE, 431 S. Main St., Ukiah, CA 95482-4923. Tel.: 707-467-2836. Fax: 707-467-2835.

E-mail: info@gracehudsonmuseum.org

Web Site: www.gracehudsonmuseum.org

Founded: 1975.

Congressional District: 2

Key Personnel: Chm. (V), Paige Poulos; Dir., Sherrie Smith-Ferri; Cur., Marvin Schenck; Museum Shop Mgr., Marian Scalmanini; Registrar, Karen Holmes.

Personnel Profile: Full-Time Paid 2; Part-Time Paid 4; Part-Time Volunteers 60.

Governing Authority: municipal. Parent Institution: City of Ukiah. Subsidiary Institution: Sun House Guild, Grace Hudson Museum Endowment Fund, Inc. Tax-exempt.

Institution Type/Description: Historic House: 1911 home of Dr. John W. Hudson & Artist Grace Carpenter Hudson.

Collections: Artwork of Grace Carpenter Hudson; Hudson-Carpenter manuscript & photographic collections; ethnographic materials of Pomo culture; 19th & 20th-century decorative art & furnishings.

Research Fields: Mendocino County history; California regional art; Pomo culture; Grace Hudson's art & life; Hudson-Carpenter family; John Hudson's research & ethnographic collection.

Facilities: public meeting room. Museum-related items for sale.

Activities: guided tours; docent program or council; permanent & temporary exhibitions; lectures; symposia related to purposes.

Hours & Admission Prices: Wed.-Sat. 10-4:30, Sun. 12-4:30. Family $10, adults $4, students & seniors $3; discount to NARM, AAM & ICOM members; members no charge. Closed holidays.

Attendance: 11,000 (accurate)

Membership: Annual $15-$1,000; Life $1,500; Endowment Society $5,000.

HELD-POAGE MEMORIAL HOME AND RESEARCH LIBRARY, 603 W. Perkins St., Ukiah, CA 95482-4726. Tel.: 707-462-6969.

E-mail: mchs@pacific.net

Founded: 1970.

Congressional District: 1

Personnel Profile: Full-Time Volunteers 1; Part-Time Volunteers 10.

Governing Authority: society. Parent Institution: Mendocino County Historical Society Inc. Tax-exempt: 501(c)(3).

Institution Type/Description: Historical Society Museum: housed in 1903 Queen Ann Victorian home of William D. L. Held.

Collections: Native American artifacts & research material; local historical artifacts; kitchen furniture; books; children's toys; manuscripts.

Research Fields: county, state & U.S. history.

Facilities: 5,000-vol. library of books; 13,000 historical photographic negatives, microfilms, documents, maps, scrapbooks, Great Registers, genealogies relating to history of northern California; county records; newspapers & magazines available for use on premises; reading room. Publications for sale.

Activities: guided tours.

Publications: quarterly newsletter, from Mendocino County Historical Society, Inc.

Hours & Admission Prices: Wed.-Fri. 1-4; other times by appointment. No charge; donations accepted.

Attendance: 600 (estimated)

Membership: Regular $20; Couple $25; Contributing $50; Life $300.

Upland

COOPER REGIONAL HISTORY MUSEUM, 217 A St., Upland, CA 91786-6024. Mailing Address: P.O. Box 772, Upland, CA 91785-0772. Tel.: 909-982-8010.

E-mail: lola@coopermuseum.org

Web Site: www.coopermuseum.org

Founded: 1965.

Congressional District: 35

Key Personnel: Exec. Dir., Marilyn Anderson; Pres., David W. Stevens.

Personnel Profile: Full-Time Volunteers 13; Part-Time Volunteers 5.

Governing Authority: nonprofit organization. Parent Institution: Chaffey Communities Cultural Center. Tax-exempt: 501(c)(3).

Institution Type/Description: History Museum.

Collections: domestic life; citrus industry; local history; hunting; Native American; seasonal exhibits.

Research Fields: history & culture of Upland, Ontario, Montclair, Mt. Baldy, and Rancho Cucamonga, California from 1800-present.

Facilities: 100-seat meeting room.
Activities: lectures; films; oral history program with local residents.
Publications: bulletin published three times annually; book, Oranges for Health-California for Wealth.
Hours & Admission Prices: Fri. 1-4, Sat.-Sun. 11-5. No charge; donations accepted. Closed national holidays. &
Attendance: 5,000 (accurate)
Membership: Individual $15; Family $25.

UPLAND FIRE COMPANY MUSEUM, 151 East ”D“ St., Upland, CA 91786. Mailing Address: Fire Administration, 475 N. 2nd Ave., Upland, CA 91786. Tel.: 909-931-4180. Fax: 909-931-4196.
E-mail: dcorbin@ci.upland.ca.us
Institution Type/Description: Fire Company Museum.
Collections: fire station history; firefighting equipment & vehicles; personal artifacts; photographs.
Hours & Admission Prices: Call for hours.

Vacaville

VACAVILLE MUSEUM, (M), 213 Buck Ave., Vacaville, CA 95688-3835. Tel.: 707-447-4513. Fax: 707-447-2661.
E-mail: vacamuseum@sbcglobal.net
Web Site: vacavillemuseum.org
Founded: 1981.
Congressional District: 1
Key Personnel: Dir., Shawn Lum; Cur. Collections, Annie Farley; Cur. Exhibits, Philip Nollar; Registrar, Heidi Casebolt; Office Mgr., Sheri Ware.
Personnel Profile: Full-Time Paid 3; Part-Time Paid 6; Part-Time Volunteers 100.
Governing Authority: nonprofit organization. Tax-exempt: 501(c)(3).
Institution Type/Description: History Museum.
Collections: pre-history through present local history items.
Research Fields: agricultural history; Japanese-American history; Spanish-American history; women's history; water history in Solano County.
Facilities: conference room; native plant garden area. Museum-related items for sale.
Activities: guided tours; lectures; workshops; TV & radio programs; organized educational programs; docent program; traveling, permanent, temporary & changing exhibitions.
Publications: quarterly newsletter, Vacaville Museum News & Notes; exhibit catalogue, Berryessa Valley: The Last Year; book, Solano's Gold - The People and Their Orchards; memories & reminiscences of Vacaville's Japanese community, Omo i de; cookbook.
Hours & Admission Prices: Wed.-Sun. 1-4:30. Adults $3, children & students $2; discounts to AAM & ICOM members; CAM & WMA members no charge. Closed holidays. &
Attendance: 10,000 (estimated)
Membership: Participating $50; Supporting & Family $75; Sponsor $145; Principal $250; Patron $500; Benefactor $1,200; Life $5,000.

Vallejo

SIX FLAGS DISCOVERY KINGDOM, 2001 Marine World Pkwy., Vallejo, CA 94589-4001. Tel.: 707-644-4000. Fax: 707-644-0241. TDD: 707-643-6769.
Web Site: www.sixflags.com
Formerly: Six Flags Marine World
Founded: 1968.
Key Personnel: Gen. Mgr., Rick Mcurley; Dir. Mktg., Dwayne McNeil; Mgr. Merchandise & Museum Shop Mgr., Mike Southern; Supvr. Education, Terran Rosenberg; Mgr. Oceanarium, Kathy France; Dir. Opers., Tim Ready; Show Productions, David Miller; Mgr. Public Rels., Jeff Jouett; Dir. Animal Operations, David Blasko.
Personnel Profile: Full-Time Paid 175; Part-Time Paid 1,500.
Governing Authority: City of Vallejo.
Institution Type/Description: Wildlife Theme Park.
Collections: live marine & land mammals, birds, reptiles, fish & invertebrates on exhibit & performing in educational shows.
Research Fields: dolphin communication; sea lions; otter conservation; killer whale vocalizations and behavior; communication & cognition; artificial insemination; reproductive cycles in killer whales & elephants.
Facilities: 1,200-vol. library of materials for staff use only; restaurants; classrooms; discovery center; show stadiums. Gift items for sale.
Activities: guided tours; lectures; TV & radio programs; formally organized education programs; rides; mobile vans; animal shows.
Publications: A Closer Look; Teacher Guides on Marine Mammals, Tropical Rain Forests, Vanishing Animals & Habitats available on request; teacher's guide on Sharks.
Hours & Admission Prices: Memorial Day to Labor Day Mon.-Thurs. 10-8,

Fri.-Sun. 10-9. Adults $44.99, senior citizens $26.99, children 4-12 $29.99; discounts to groups; children 2 & under no charge. &
Attendance: 1,950,000 (accurate)

VALLEJO NAVAL & HISTORICAL MUSEUM, (M), 734 Marin St., Vallejo, CA 94590-5992. Tel.: 707-643-0077. Fax: 707-643-2443.
E-mail: valmuse@pacbell.net
Web Site: www.vallejomuseum.org
Founded: 1974.
Congressional District: 7
Key Personnel: Dir., James E. Kern; Pres., Lloyd Chan; Museum Shop Mgr., Mimi Farone.
Personnel Profile: Full-Time Paid 1; Part-Time Paid 2; Part-Time Volunteers 50.
Governing Authority: nonprofit organization. Tax-exempt: 501(c)(3).
Institution Type/Description: Historical Museum: located in Vallejo's Old City Hall.
Collections: history of Vallejo & the nearby Mare Island Naval Shipyard from 1850-present.
Research Fields: local history.
Facilities: 6,000-vol. library pertaining to local & maritime history; 125-seat auditorium; classroom. Books & museum-related items for sale.
Activities: guided tours; lectures; concerts; organized education programs for children; docent program; temporary exhibitions.
Publications: bimonthly newsletter, Valmuse.
Hours & Admission Prices: Tues.-Sat. 12-4. Adults $5, seniors & students 12-17 $3; discounts to AAM members; members & children under 12 no charge. Closed holidays. &
Attendance: 9,000 (estimated)
Membership: Individual $30; Family $45; Sustaining $100; Supporting $150; Patron $275; Sponsor $650; Benefactor $1,000; Director's Circle $5,000.

Valley Center

RINCON BAND OF LUISENO INDIANS' MUSEUM, 1 W. Tribal Rd., Valley Center, CA 92082. Tel.: 760-297-2621.
Institution Type/Description: Native American Museum.
Collections: Native American culture & history; personal artifacts; basketry.
Activities: special events.
Hours & Admission Prices: Mon.-Fri. 2-5.

Valley Glen

LOS ANGELES VALLEY COLLEGE ART GALLERY, 5800 Fulton Ave., Valley Glen, CA 91401-4062. Tel.: 818-778-5536.
Web Site: www.lavc.edu/arts/artgallery.html
Founded: 1960.
Institution Type/Description: Art Gallery.
Collections: works by student & faculty artists; contemporary & ethnic art.
Hours & Admission Prices: Mon.-Thurs. 11-5.

Van Nuys

THE JAPANESE GARDEN, 6100 Woodley Ave., Van Nuys, CA 91406-6450. Tel.: 818-756-8166. Fax: 818-756-9648.
E-mail: betty.ethridge@lacity.org
Web Site: www.thejapanesegarden.com
Founded: 1984.
Congressional District: 27
Key Personnel: Dir., Gene Greene; Chm. (V), Hazel Graynor; Landscape Architect, Patrick Rigney; Technician, Julius Luna; Office Mgr., Betty Ethridge; Tour Desk Staff, Lori Stewart; Museum Shop Mgr., Jan E. Abrams.
Personnel Profile: Full-Time Paid 4; Part-Time Paid 2; Part-Time Volunteers 70.
Governing Authority: Parent Institution: city of Los Angeles. Subsidiary Institution: Bureau of Sanitation. Tax-exempt.
Institution Type/Description: Cultural Garden.
Collections: plants; flowers; trees; local wildlife; culturally significant structures & decorative elements.
Research Fields: Japanese culture; water reclamation process.
Facilities: rental facilities. Museum-related items for sale.
Activities: Museum Sponsors: Taiko Drumming; Origami; Ikebana; Kimono exhibition & demonstration.
Hours & Admission Prices: Mon.-Thurs. 11-4, Sun. 10-4. Tours: Mon.-Thurs. 9:30, 10 & 10:30. Adults $3, children under 10 and seniors 62 & over $2; special events admission $5; members no charge. Participates in American Horticultural Society reciprocal admission program (except for special events). Closed holidays. &

Attendance: 13,316 (accurate)
Membership: Senior $35; Individual $40; Family $60; Patron $100.

Ventura

ALBINGER ARCHEOLOGICAL MUSEUM, 113 E. Main St., Ventura, CA 93001-2606. Tel.: 805-648-5823.
E-mail: jscott@ci.ventura.ca.us
Web Site: www.albingermuseum.org
Key Personnel: Mgr., Jeanne Scott.
Personnel Profile: Part-Time Paid 1.
Institution Type/Description: Archaeology Museum.
Collections: Chumash, Spanish, Chinese, American & Mexican cultures.
Hours & Admission Prices: Call for hours. &
Attendance: 12,000

CHANNEL ISLANDS NATIONAL PARK, ROBERT J. LAGO-MARSINO VISITOR CENTER, 1901 Spinnaker Dr., Ventura, CA 93001-4354. Tel.: 805-658-5730. Fax: 805-658-5799.
E-mail: chris_interpretation@nps.gov
Web Site: www.nps.gov/chis
Founded: 1980.
Congressional District: 23
Key Personnel: Supt., Russell E. Galipeau, Jr.; Chief Interpretation, Yvonne Menard; Supervisory Park Ranger, April Rabuck.
Governing Authority: federal. U.S. Dept. of the Interior, National Park Service, Washington, DC. Subsidiary Institution: S.W. Parks & Monuments. Tax-exempt.
Institution Type/Description: National Park & Museum; Natural History Museum.
Collections: cultural, historical & archaeological objects; photographs; documents; natural preserved animals; plant collection; native plant garden; radio carbon dating 11,500 years.
Research Fields: natural & cultural resources of the California Channel Islands.
Facilities: 2,000-vol. library pertaining to diverse Channel Islands & National Park Service subjects; archival material; visitor center; tide pool; native plant garden. Publications & museum-related items for sale.
Activities: guided tours; organized educational programs for children; permanent & temporary exhibits; slide program; off-site program; evening programs in the summer.
Publications: trail guides, Anacapa: An Island, a State of Mind; Canyon View Nature Trail, Santa Barbara Island; monthly schedule of events; all island site bulletins; Park newspaper.
Hours & Admission Prices: Daily 8:30-5. Park: no charge; donations accepted. Closed Thanksgiving; Christmas. &
Attendance: 321,492 (accurate)

MUSEUM OF VENTURA COUNTY, (M), 100 E. Main St., Ventura, CA 93001-2828. Tel.: 805-653-0323. Fax: 805-653-5900. Facebook: Museum of Ventura County.
E-mail: mfreedman@venturamuseum.org
Web Site: www.venturamuseum.org
Formerly: Ventura County Museum of History & Art
Founded: 1913.
Congressional District: 26
Key Personnel: Pres. Bd. Dir. (V), Richard Pidduck; Co-Pres. Docent Council (V), Linda Kimbroug; Co-Pres. docent Council (V), Kathy Sisson; Exec. Dir., Myron Freedman; Librarian, Charles Johnson; Dir. Education, Wendy VanHorn; Cur. Collections, Anna Rios Bermudez; Dir. Devel., Robin C. Woodworth; Mktg. Coord., Holly Raftery; Museum Shop Mgr., Linden Royce.
Personnel Profile: Full-Time Paid 16; Part-Time Paid 12; Part-Time Volunteers 165; Interns 1.
Governing Authority: nonprofit organization. Subsidiary Institution: Ventura County Historical Society. Tax-exempt: 501(c)(3). Satellite Museum: Museum of Ventura County Agriculture Museum in Santa Paula, CA.
Institution Type/Description: History & Art Museum.
Collections: art & artifacts from the region before European contact to the present; George Stuart historical figures; 5,000 maps, newspapers, books & periodicals relating to regional history; 25,000 photographs; 40,000 negatives; 1,000 early & contemporary California art; 900 farm machinery; implements & tools; oil industry artifacts.
Major Exhibits: Strawberry Fields Forever? (Agriculture Museum), 11/13-2/14; Jeff Sanders: Tina's Toys and Other Things, 12/7/13-2/2/14; New Sculptures by Arlene Tepper, 12/13-2/14; Omar D'Leon, 2/14-5/14; TRAC Fresh Harvest: Paintings by Michael Lynn Adams, 2/14-4/14; Who's Minding the Store?, 6/14-8/14; The Rise & Fall of the Lima Bean Empire (Agriculture Museum), 3/14-6/14; Hey, Hey, Hay! (Agriculture Museum), 1/14-12/14; Zest! The Art of Citrus (Agriculture Museum), 4/14-6/14.

Research Fields: Ventura County, California.
Facilities: Museum-related items for sale.
Activities: special lectures & programs; monthly family arts & crafts; gallery openings; speakers bureau; school outreach; docent networking program; member travel program.
Publications: quarterly journal, Journal of Ventura County; newsletter; exhibit catalogs; books on county history; collection catalog.
Hours & Admission Prices: Tues.-Sun. 11-5. Museum: adults $4, seniors & students w/ID $3, children 6-17 $1, MVC members & children 5 & younger no charge. Closed New Year's Day; Thanksgiving; Christmas. &
Attendance: 65,000 (accurate)
Membership: Individual $50; Dual $75; Associate $100; Business & Professional $150; Supporter $250; Donor & Business Member Plus $500; Friend $1,000; Patron $2,500; Benefactor $5,000.

ORTEGA ADOBE, 215 W. Main St., Ventura, CA 93001. Mailing Address: P.O. Box 99, Ventura, CA 93002-0099. Tel.: 805-658-4728.
E-mail: greyes@cityofventura.net
Web Site: www.ventura.com/points_of_interest/ortegaadobe
Institution Type/Description: History Museum: built in 1857 by Miguel Emigdio Ortega.
Collections: period artifacts; structure.
Hours & Admission Prices: Daily 9-4 by appointment. No charge.

SAN BUENAVENTURA MISSION MUSEUM, 225 E. Main St., Ventura, CA 93001-2622. Mailing Address: 211 E. Main St., Ventura, CA 93001-2691. Tel.: 805-643-4318. Fax: 805-643-7831.
Web Site: www.sanbuenaventuramission.org
Founded: 1782.
Congressional District: 38
Key Personnel: Pastor, Rev. Thomas Elewaut; Gift Shop Mgr., Veronica Basoco.
Personnel Profile: Full-Time Paid 3; Part-Time Paid 1; Part-Time Volunteers 3.
Governing Authority: church. Roman Catholic Archbishop of Los Angeles. Tax-exempt.
Institution Type/Description: Historical Museum: housed in 1782 San Buenaventura Mission, founded by Fray Junipero Serra.
Collections: paintings; statues; Indian artifacts; military; wooden bells; Chumash baskets; Juan Camarillo II artifacts; 23 sets of vestments from the mission; books comprising the Bibliotheca Sancti Bonaventurae; historical mementos; religious artifacts; musical instruments. Historic Structures: 1829 Adobe Settling Tank; 1809 San Buenaventura Church.
Facilities: Literature on the history of the mission for sale.
Activities: guided tours by parish docents; permanent exhibitions; Latin Mass on Sundays.
Hours & Admission Prices: Mon.-Sat. 10-5, Sun. 10-4. Museum: adults $4, seniors $3, children $1; discounts to groups. Mass schedule: daily 7:30 am, Sat. 5:30 pm, 7:30 pm (Spanish), Sun. 7:30, 9, 10:30, 12:15 (Spanish). No charge; donations accepted. Closed New Year's Day; Easter; Thanksgiving; Christmas. &
Attendance: 18,500 (estimated)

VENTURA COUNTY MUSEUM OF HISTORY AND ART, 100 E. Main St., Ventura, CA 93001-2607. Tel.: 805-653-0323.
Key Personnel: Exec. Dir., Tim Schiffer
Institution Type/Description: History & Art Museum.
Collections: local history & culture; paintings; farm implements; early clothing; personal artifacts; period furnishings; Native American artifacts; musical instruments; medical equipment.
Facilities: library. Museum-related items for sale.
Activities: educational programs; special events.
Hours & Admission Prices: Tues.-Thurs. & Sat.-Sun. 11-6, Fri. 11-8. Adults $4, seniors $3, children 6-17 $1; members & children under 6 no charge.

Victorville

CALIFORNIA ROUTE 66 MUSEUM, 16825 S. D St., Victorville, CA 92395. Mailing Address: P.O. Box 2151, Victorville, CA 92393-2151. Tel.: 760-951-0436. Fax: 760-951-0509.
E-mail: cart66musm@aol.com
Web Site: califrt66museum.org
Formerly: Old Town Victorville Heritage Preservation, Inc. dba California Route 66 Museum
Founded: 1995.
Congressional District: 40
Key Personnel: Chm. (V) & Pres. (V), Gene St. John; Public Rels., Registrar & Museum Shop Mgr., Betty A. Halbe.
Personnel Profile: Full-Time Volunteers 2; Part-Time Volunteers 15.

Governing Authority: nonprofit corporation. Tax-exempt: 501(c)(3).
Institution Type/Description: History Museum.
Collections: Hula Ville folk art & panorama from early 20th century to present; Victorville community from 1800s to present.
Research Fields: Route 66 & the eight states that the highway goes through.
Facilities: library available for research on the premises; educational facilities. Museum-related items for sale.
Activities: films, formal education programs; guided tours; lectures; loan & temporary exhibitions.
Publications: Museum Musings.
Hours & Admission Prices: Thurs.-Mon. 10-4, Sun. 11-3; other times by appointment. No charge; donations accepted. Closed Memorial Day; Independence Day; Labor Day; Thanksgiving; Christmas. &
Attendance: 11,000 (accurate)
Membership: Individual $30.

VICTORVILLE FIRE DEPARTMENT MUSEUM, 15620 8th St., Victorville, CA 92392. Mailing Address: 14343 Civic Dr., P.O. Box 5001, Victorville, CA 92392-5001. Tel.: 760-955-5229.
Key Personnel: Cur., Greg Coon
Institution Type/Description: Firefighting History Museum.
Collections: firefighting history & equipment; fire trucks & apparatus; uniforms; photographs; documents.
Hours & Admission Prices: Sat. 9-12 by appointment. No charge; donations accepted.

Visalia

IMAGINEU CHILDREN'S MUSEUM, 700 E. Main St., Visalia, CA 93292-6447. Mailing Address: P.O. Box 688, Visalia, CA 93279-0688. Tel.: 559-733-5975 & 0735. Fax: 559-733-0871. Facebook: ImagineU Children's Museum.
E-mail: imagineumuseum@sbcglobal.net
Web Site: www.imagineumuseum.org
Founded: 2002.
Key Personnel: Exec. Dir., Angela Huerta-Reyna.
Governing Authority: nonprofit organization. Tax-exempt: 501(c)(3).
Institution Type/Description: Children's Museum.
Collections: hands-on exhibits.
Activities: birthday parties; classes.
Hours & Admission Prices: Tues.-Sat. 10-4. Admission $5, children under 2 $3; members no charge. &

TULARE COUNTY MUSEUM AT MOONEY GROVE PARK, 27000 S. Mooney Blvd., Visalia, CA 93277-9341. Mailing Address: 5953 S. Mooney Blvd., Visalia, CA 93277. Tel.: 559-733-6616. Fax: 559-635-4896. Facebook: Tulare County Museum.
E-mail: aking1@co.tulare.ca.us
Founded: 1948.
Congressional District: 17
Key Personnel: Dir & Cur., Amy King.
Personnel Profile: Full-Time Paid 1; Part-Time Paid 2; Part-Time Volunteers 25.
Governing Authority: county. Parent Institution: General Services, County of Tulare. Tax-exempt.
Institution Type/Description: History Museum.
Collections: sculpture of James Earle Frasher, End of the Trail; Indian relics; agricultural implements; wagons; buggies; clothing. Historical Buildings: 1854 log cabin; 1872 jail; c.1890 home; blacksmith shop; 1863 home; 1888 & 1890 schools.
Activities: guided tours by appointment.
Publications: quarterly bulletin, Los Tulares.
Hours & Admission Prices: Summer: Mon. & Wed.-Fri. 10-4, Sat.-Sun. 10-4; Spring & Fall: Thurs.-Mon. 10-4; Winter: Thurs.-Mon. 10-4, Sat.-Sun. 1-4. Car admission fee for Mooney Grove Park $6, admission to museum included in park entrance fee. Closed New Year's Eve & Day; Thanksgiving; Christmas Eve & Day. &
Attendance: 125,000 (accurate)

Vista

ANTIQUE GAS & STEAM ENGINE MUSEUM, INC., 2040 N. Santa Fe Ave., Vista, CA 92083-1534. Tel.: 760-941-1791. Fax: 760-941-0690.
E-mail: rod_agsem@yahoo.com
Web Site: www.agsem.com
Founded: 1976.
Congressional District: 43
Key Personnel: C.E.O., Jeanette Stevens; Dir., Rod Groenewold; Museum Shop Mgr., Glenda Garrison.

Personnel Profile: Full-Time Paid 2; Full-Time Volunteers 3; Part-Time Volunteers 500; Interns 2.
Governing Authority: nonprofit organization. Tax-exempt: 501(c)(3).
Institution Type/Description: Agriculture & Industrial Museum Complex: located on 40 acres of farm land.
Collections: archival materials; steam & gas powered engines; horse-drawn equipment used in farming, industrial & construction industries; blacksmith & wheelwright shops; sawmill; farmhouse with parlor; 1/3 scale train.
Activities: guided tours; organized education programs for children; participatory exhibits; antique farm equipment demonstrations; 1/3 scale train operational during special shows. Museum Sponsors: Threshing Bees & Antique Engine Shows in June & October.
Publications: newsletter, Ignitor.
Hours & Admission Prices: Daily 10-4. Adults $3. Closed Christmas. &
Attendance: 55,000 (estimated)
Membership: Student & Associate $15; Member $30; Family $50; Lifetime $600.

VISTA HISTORICAL MUSEUM, 2317 Old Foothill Dr., Vista, CA 92084. Mailing Address: P.O. Box 1032, Vista, CA 92085-1032. Tel.: 760-630-0444. Fax: 760-295-9993.
E-mail: vhm67@1882.sdcoxmail.com
Web Site: www.vistahistoricalsociety.com
Founded: 2009.
Congressional District: 5
Key Personnel: Pres. (V), Janice Klafehn; Dir. & Museum Shop Mgr., Jack Larimer; Financial Dir., Michele Moxley; Devel., Sharon Larimer.
Personnel Profile: Full-Time Paid 1; Part-Time Volunteers 20.
Governing Authority: private; nonprofit organization. Parent Institution: Vista Historical Society, Vista, CA. Tax-exempt: 501(c)(3).
Institution Type/Description: Historical Museum.
Collections: artifacts from indigenous people of North County; relating to the discovery, diversity & destiny of Vista, San Diego & California.
Facilities: 3,200 sq. ft. exhibit space.
Activities: docent program; films; guided tours; lectures; temporary exhibitions; tours for school groups & other societies; bimonthly programs.
Publications: bimonthly newsletter, Vista Historical Society.
Hours & Admission Prices: Wed.-Fri. 1st & 2nd Sat. each month 10-2:30. No charge; donations accepted. &
Attendance: 1,000 (estimated)
Membership: Individual $25; Family $45; Business $100.

Walnut

THE MT. SAN ANTONIO COLLEGE WILDLIFE SANCTUARY, 1100 N. Grand Ave., Walnut, CA 91789. Tel.: 909-594-5611, ext. 4794.
Institution Type/Description: Wildlife Sanctuary.
Collections: birds & other wildlife; trees; shrubs; flowers.
Facilities: 25-seat amphitheater.
Activities: lectures; classes; educational programs.
Hours & Admission Prices: Tours: Tues. & Thurs. 9am, 10am, 11am, 2pm, 3pm, & 4pm.

Walnut Creek

BEDFORD GALLERY AT THE DEAN LESHER REGIONAL CENTER FOR THE ARTS, 1601 Civic Dr., Walnut Creek, CA 94596-4299. Tel.: 925-295-1417. Fax: 925-295-1486.
E-mail: lederer@bedfordgallery.org
Web Site: www.bedfordgallery.org
Founded: 1963.
Congressional District: 7
Key Personnel: Dir. Cultural & Community Svcs., Gary Pokorny; Cur., Carrie Lederer; Preparator, Erik Mortensen.
Personnel Profile: Full-Time Paid 2; Part-Time Paid 4; Part-Time Volunteers 100.
Governing Authority: municipal. Tax-exempt.
Institution Type/Description: Municipal Art Gallery.
Collections: paintings; photographs; sculpture.
Facilities: slide & art library; learning center; 800-seat theatre, 300-seat theatre.
Activities: guided tours; docent program or council; gallery talks; 5-6 annual exhibitions in all media; formally organized education programs for children, adults, undergraduate & graduate college students; in-school program; lecture series; changing exhibitions of contemporary art; school programs.
Publications: calendar of exhibitions; exhibit announcements; magazine, Diablo Arts.
Hours & Admission Prices: Tues.-Wed. & Sun. 12-5, Thurs.-Sat. 12-5 & 6-8.

Adults $5, students 17 & under $3; children under 12 no charge. Closed national holidays. &
Attendance: 45,000

* **LINDSAY WILDLIFE MUSEUM,** 1931 First Ave., Walnut Creek, CA 94597-2540. Tel.: 925-935-1978. Fax: 925-935-8015.
Web Site: www.wildlife-museum.org
Founded: 1955.
Congressional District: 7
Key Personnel: Pres., Kramer Klabau; Exec. Dir., Loren Behr; Cur. Live Collections, Michele Setter; Financial Dir., Suzie Mahaffay; Dir. Wildlife Rehabilitation, Susan Heckly; Dir. Veterinary Svcs. & Research, Shannon Riggs, D.V.M.; Dir. Operations, Chris Bernard; Dir. Education, Patti Harris; Cur. Natural History & Dir. Operations, Marty Buxton; Museum Shop Mgr., Christine Garcia.
Personnel Profile: Full-Time Paid 25; Part-Time Paid 19; Part-Time Volunteers 550; Interns 5.
Governing Authority: bd. of directors; nonprofit. Tax-exempt: 501(c)(3).
Institution Type/Description: Natural History Museum.
Collections: live wild animals; taxidermied specimens; Indian artifacts; entomology; botany; marine.
Research Fields: wildlife release & rehabilitation; natural history of California.
Facilities: classrooms; discovery room. Books & other museum-related items for sale.
Activities: guided tours; trips; lectures; formally organized educational programs; docent program & council; permanent, temporary & traveling exhibitions; mobile vans; school loan service; volunteer training to work with animals; petting area.
Publications: monthly newsletter, Volunteer News; Public Program brochure.
Hours & Admission Prices: Jan.-June 15 Thurs.-Fri. 12-5, Sat.-Sun. 10-5; June 16-Aug. Wed.-Sun. 10-5; Sept.-Dec. Wed.-Fri. 12-5, Sat.-Sun. 10-5. Adults $7, senior citizens 65 & over $6, children 2-17 $5; discount to AAM, ASTC reciprocal members & groups of 15-35; members & children under 2 no charge. Closed Independence Day; Thanksgiving; Christmas. &
Attendance: 102,542 (accurate)
Membership: Individual $35; Family & Grandparents $65; Member Plus $100; Friends $250 & up.

SHADELANDS RANCH HISTORICAL MUSEUM, 2660 Ygnacio Valley Rd., Walnut Creek, CA 94598. Tel.: 925-935-7871.
E-mail: wcshadelands@sbcglobal.net
Web Site: www.walnutcreekhistory.info
Institution Type/Description: History Museum: housed in the former home of pioneer Hiram Penniman; built in 1903. National Register of Historic Places.
Collections: local history & culture; period furnishings; personal artifacts; photographs; newspapers; government records.
Activities: special events; research.
Hours & Admission Prices: early Feb. to Oct. Wed. & Sun. 1-4; groups of 10 or more by appointment. Adults $3, students 6-17 $1; members & children under 6 no charge.
Membership: Individual $25; Family $40; Sustaining $100; Life $500.

Walnut Grove

SACRAMENTO RIVER DELTA HISTORICAL SOCIETY RESOURCE CENTER, The Jean Harvie Community Center, 14273 River Rd., Walnut Grove, CA 95690. Mailing Address: P.O. Box 293, Walnut Grove, CA 95690.
Institution Type/Description: Historical Society Museum: center built in 1924.
Collections: local history & culture; period furnishings; personal artifacts; photographs; books; transcripts.
Hours & Admission Prices: Tues. by appointment.

Wasco

WASCO HISTORICAL SOCIETY MUSEUM, 918 6th St., Wasco, CA 93280-1902. Mailing Address: P.O. Box 186, Wasco, CA 93280-0186. Tel.: 661-758-8948.
Institution Type/Description: Historical Society Museum.
Collections: local history & culture; photographs; personal artifacts.
Hours & Admission Prices: Call for hours.

Watsonville

ELKHORN SLOUGH NATIONAL ESTUARINE RESEARCH, 1700 Elkhorn Rd., Watsonville, CA 95076-9218. Tel.: 831-728-2822. Fax: 831-728-1056.
E-mail: info@elkhornslough.org

Web Site: www.elkhornslough.org
Key Personnel: Reserve Mgr., Dave Feliz
Institution Type/Description: Natural History Museum.
Collections: tidal salt march; live plants; animals; birds; fish nursery.
Facilities: Museum-related items for sale.
Hours & Admission Prices: Wed.-Sun. 9-5. No charge.

PAJARO VALLEY HISTORICAL ASSOCIATION, 332 E. Beach St., Watsonville, CA 95076. Mailing Address: P.O. Box 623, Watsonville, CA 95077-0623. Tel.: 831-722-0305. Fax: 831-722-5501. Facebook: Pajaro Valley Historical Association.
E-mail: info@pajarovalleyhistory.org
Web Site: www.pajarovalleyhistory.org
Founded: 1940.
Congressional District: 16
Key Personnel: Pres., Rob Allen; Sec., Cathy Moresco-Schimpeler; Treas., Eileen Sambrailo; Online Store Mgr., Alice Leyland; Office Mgr., GeriAnne Solano-Simmons.
Personnel Profile: Part-Time Paid 3; Part-Time Volunteers 50.
Governing Authority: nonprofit organization. Tax-exempt: 501(c)(3).
Institution Type/Description: History Museum: housed in the Bockius-Orr House.
Collections: Pajaro Valley; photographs; vignettes; clothing & costumes; textiles; post cards of Pajaro Valley; slides of buildings & houses in Watsonville; archives. Historic Buildings: Volck Home & Carriage House.
Research Fields: oral histories; old homes, schools, businesses, genealogy, maps, local history & anything on Pajaro Valley history.
Facilities: over 500-vol. library of books & manuscripts available for use on premises by appointment only.
Activities: monthly board meetings; special events & fundraising activities; group tours; school group programs. Museum Sponsors: annual meeting in February; biennial Historical Haunting; Gatsby Under the Oak Gala Fundraiser; Old Timers BBQ; Christmas Tea.
Publications: quarterly newsletter.
Hours & Admission Prices: Tues.-Thurs. 11-3. No charge; donations accepted. &
Attendance: 750 (estimated)
Membership: Donor $40; Sponsor $50; Supporter $100; Benefactor $250; Patron $500 & up.

Weaverville

J.J. JACKSON MEMORIAL MUSEUM, 780 Main St., Weaverville, CA 96093. Mailing Address: P.O. Box 333, Weaverville, CA 96093-0333. Tel.: 530-623-5211. Fax: 530-623-5053.
E-mail: jake@trinitymuseum.org
Web Site: www.trinitymuseum.org
Founded: 1968.
Congressional District: 1
Key Personnel: Dir., Dero Forslund; Pres., George Chapman; Museum Shop Mgr., Pat Williams.
Personnel Profile: Part-Time Paid 6; Part-Time Volunteers 25.
Governing Authority: county; society. Parent Institution: Trinity County Historical Society. Tax-exempt.
Institution Type/Description: History Museum & Research Center.
Collections: Gold Rush artifacts; Chinese & Native American culture & history; pioneer artifacts; textiles; photographs & documents of Trinity history; blacksmith shop; tin shop; paymaster mine stamp mill; ditch tender's cabin; stagecoach & buggy barn.
Research Fields: county records; genealogical research; Trinity County history; textiles.
Facilities: history center; park & picnic area. Gift items, booklets & stationery for sale.
Activities: permanent exhibitions; blacksmith shop demonstrations; research. Museum Sponsors: stamp mill demonstrations on 3-day weekends & Independence Day; Antique Quilt Bed Turning in October; Dutch Oven Cook-Off in October.
Publications: books, Trinity County Historic Sites, Tales of the Trinity; yearbook, Trinity, From the Known to the Unknown, Flowers and Trees of the Trinity Alps.
Hours & Admission Prices: Jan.-March Tues.-Sat. 12-4; April & Nov.-Dec. daily 12-4; May-Oct. daily 10-5. Adults $2. &
Attendance: 11,750 (accurate)
Membership: Individual $15; Family $25; Supporter $50; Friend $100; Patron $200; Benefactor $500.

WEAVERVILLE JOSS HOUSE, State Historic Park, 630 Main St., Weaverville, CA 96093. Mailing Address: P.O. Box 1217, Weaverville, CA 96093-1217. Tel.: 530-623-5284.
Web Site: www.parks.ca.gov
Founded: 1956.
Key Personnel: Historic Monument Guide, Jack Frost.
Personnel Profile: Full-Time Paid 1; Part-Time Paid 2.
Governing Authority: state. Affiliated with State of California Department of Parks & Recreation, P.O. Box 942896, Sacramento, CA 94296-0001.
Institution Type/Description: Historic Building: housed in a Chinese temple; built in 1874.
Collections: history of the Taoist temple of worship; temple equipment; Chinese art objects; photographs; mining tools; wrought iron weapons used in the 1854 Tong War.
Activities: guided tours; permanent exhibitions; guided tours for educational groups of high school level or below upon written request.
Hours & Admission Prices: Wed.-Sun. 10-5. Adults $3, children 6-17 $2.
Attendance: 20,000

Weed

LIVING MEMORIAL SCULPTURE GARDEN, Hwy. 97, Weed, CA 96094. Mailing Address: P.O. Box 301, Weed, CA 96094-0301. Tel.: 530-842-2477.
Web Site: livingmemorialsculpturegarden.org
Founded: 1987.
Key Personnel: Pres. (V), Suzanne Breceda.
Personnel Profile: Part-Time Volunteers 9.
Governing Authority: Parent Institution: Kiwanis Club of Weed/Lake Shastina. Tax-exempt.
Institution Type/Description: War Memorial Sculpture Garden.
Collections: war memorial sculptures.
Activities: Annual Events: Memorial Services in May & November.
Publications: brochures; biannual newsletter.
Hours & Admission Prices: Daily sunrise-sunset. No charge.
Attendance: 6,000 (estimated)
Membership: Annual $25; Contributor $50; Lifetime Individual $400; Sponsor $1,000.

Weott

HUMBOLDT REDWOODS STATE PARK VISITOR CENTER, 17119 State Rte. 254, Weott, CA 95571. Mailing Address: P.O. Box 276, Weott, CA 95571-0276. Tel.: 707-946-2263. Fax: 707-946-2618.
E-mail: vc@humboldtredwoods.org
Web Site: www.humboldtredwoods.org
Founded: 1979.
Congressional District: 1
Key Personnel: State Park Liaison, Tyson Young; Chm. (V), Susan O'Hara; Museum Exec. Dir., David Pritchard; Museum Shop Mgr., Deborah Gardner; Dist. Supt., Dana Jones.
Personnel Profile: Full-Time Paid 1; Full-Time Volunteers 30; Part-Time Paid 1; Part-Time Volunteers 25.
Volunteer Hours: 6,441
Operating Income: 173,573
Governing Authority: nonprofit organization. Parent Institution: Humboldt Redwoods State Park. Tax-exempt.
Institution Type/Description: Park Museum & Visitor Center.
Collections: plant & animal specimens native to north coast Redwood habitat; early logging; local native artifacts.
Facilities: nature & conservation center. Museum-related items for sale.
Activities: lectures; films; docent program.
Publications: Humboldt Redwoods Trail Guide; Avenue of the Giants Auto Tour Guide; The Killer Eel; Humboldt Redwoods trail map; brochure, Founders Grove.
Hours & Admission Prices: March-Oct. daily 9-5; Nov.-Feb. daily 10-4. No charge; donations accepted. &
Attendance: 87,254 (accurate)
Membership: Student & Senior $10; Individual $25; Supporting $50; Life $250; Patron $500; Endowment $1,000; Redwood Crown $2,500.

West Covina

HURST RANCH HISTORICAL FOUNDATION, 1227 S. Orange Ave., West Covina, CA 91790-3320. Tel.: 626-813-0116. Fax: 626-919-1133.
E-mail: info@hurstranch.com
Web Site: www.hurstranch.com

Founded: 1996.
Congressional District: 28
Key Personnel: C.E.O. & Pres. (V), Bill Berger; Treas., Stan Daubenbis.
Personnel Profile: Part-Time Volunteers 28.
Governing Authority: Tax-exempt.
Institution Type/Description: History Museum.
Collections: 19th-century furniture; tools; farm implements; kitchen utensils; Coca-Cola memorabilia; books; early automobiles; fully maintained agricultural garden.
Hours & Admission Prices: Call for hours & admission prices. &
Attendance: 1,600 (estimated)
Membership: Friend $25; Family $50; Patron $100; Bronze $250; Silver $500; Gold $1,000; Platinum $2,500.

West Hollywood

MAK CENTER FOR ART AND ARCHITECTURE AT THE SCHINDLER HOUSE, 835 N. Kings Rd., West Hollywood, CA 90069-5409. Tel.: 323-651-1510, ext. 5 & ext. 10. Fax: 323-651-2340.
E-mail: office@makcenter.org
Web Site: www.makcenter.org
Founded: 1994.
Key Personnel: C.E.O., Peter Noever; Pres., Robert Sweeney, FOSH; Dir., Kimberli Meyer; Asst. Programs Coord., Janet Owen; Museum Shop Mgr., Angelica Fuentes; Security, Omar Velazquez.
Personnel Profile: Full-Time Paid 4; Part-Time Paid 1; Part-Time Volunteers 25; Interns 2.
Governing Authority: private; nonprofit organization. Parent Institution: Museum for Angewandte Kunst, Stubenring 5, Vienna, Austria A1010. Subsidiary Institution: MAK, Vienna, Austria. Tax-exempt: 501(c)(3).
Institution Type/Description: Center for Art & Architecture: R.M. Schindler's landmark Kings Road House/Studio (1921-22)
Collections: paintings; photographs; historic building.
Research Fields: history of R.M. Schindler's work for exhibition.
Facilities: library; 1,500 sq. ft. exhibit space; outdoor courtyard. Books for sale.
Activities: docent program; films; guided tours; lectures; loan & traveling exhibitions. Museum Sponsor: annual architectural tour of Schindler homes.
Publications: catalogues for exhibitions.
Hours & Admission Prices: Wed.-Sun. 11-6. Adults $7, seniors & students $6; members no charge. Closed New Year's Day; Independence Day; Labor Day; Christmas. &
Attendance: 9,400 (accurate)
Membership: Student $25; General $45; Active $100; Charitable $250; Sustaining $500; Contributor $1,500; Supporter $5,000; Donor $10,000.

Whiskeytown

WHISKEYTOWN UNIT, WHISKEYTOWN-SHASTA-TRINITY NATIONAL RECREATION AREA, Hwy. 299 W., Whiskeytown, CA 96095-0188. Mailing Address: P.O. Box 188, Whiskeytown, CA 96095-0188. Tel.: 530-242-3451 & 246-1225. Fax: 530-246-5154. Facebook: Whiskeytown National Recreation Area.
Web Site: www.nps.gov/whis
Founded: 1965.
Congressional District: 2
Key Personnel: Supt., Jim F. Milestone.
Personnel Profile: Part-Time Paid 1.
Governing Authority: federal. Administered by National Park Service, U.S. Department of the Interior. Tax-exempt.
Institution Type/Description: Anthropology, History Museum: located in former gold rush area. Tower House Historic District is on the National Register of Historic Places.
Collections: mining tools & memorabilia; Indian artifacts; natural history; archaeological sites relating to the occupation of the California Northern Wintu; historic structures dating back to the California gold rush era.
Facilities: visitor center. Postcards, slides & books for sale.
Activities: tours; gold panning; educational programs.
Hours & Admission Prices: Recreation Area: daily 10-4. Visitor Center: daily 9-6. Park Headquarter Mon.-Fri. 8-4:30. Vehicle Pass: annual $25, weekly $10, daily $5; Golden Age, Golden Access & Golden Eagle Passport members no charge. Closed New Year's Day; Thanksgiving; Christmas. &
Attendance: 715,000 (estimated)

Whittier

JONATHAN BAILEY HOME, 13421 E. Camilla St., Whittier, CA 90601-4608. Mailing Address: Whittier Historical Society, 6755 Newlin Ave., Whittier, CA 90601. Tel.: 562-945-3871. Fax: 562-945-9106.
E-mail: info@whittiermuseum.org
Web Site: www.whittiermuseum.org
Founded: 1975.
Congressional District: 38
Key Personnel: Chm., Ellen Sandoval; Office Mgr., Gabby Alan; Archival Asst., Hannah Colvin.
Personnel Profile: Part-Time Volunteers 20.
Institution Type/Description: History Museum.
Collections: period furnishings; personal artifacts; tools.
Activities: Founders Day.
Hours & Admission Prices: Guided Tours: Sun. 1-4. School tours/group tours by appointment only Tues.-Fri. 9:30 - 3. No charge; donations accepted.
Membership: ndividual $25; Family $40; Associate: Society Patron $100; Society Benefactor $200; Corporate: Business Associate $500; Corporate Affiliate $1,000.

WHITTIER MUSEUM, 6755 Newlin Ave., Whittier, CA 90601. Tel.: 562-945-3871. Fax: 562-945-9106. Facebook: Whittier Museum.
E-mail: info@whittiermuseum.org
Web Site: www.whittiermuseum.org
Founded: 1981.
Congressional District: 38
Key Personnel: Office Mgr., Gabby Alan; Archival Asst., Hannah Colvin; Pres. (V), Rosalie Dannenbaum.
Personnel Profile: Part-Time Paid 2; Part-Time Volunteers 50; Interns 20.
Operating Expenses: 147,000
Operating Income: 135,000
Governing Authority: Tax-exempt.
Institution Type/Description: History Museum.
Collections: archives; textiles; period artifacts.
Major Exhibits: Magic Exhibit (T), 9/13-6/14.
Research Fields: Whittier history.
Facilities: archives; rental space.
Activities: Museum Sponsors: Magic Shows; Flea Market; Wine Tasting; Tea & Fashion Show; Booth at Walnut Tree Festival; Founders Day.
Publications: members' monthly newsletter, Gazette.
Hours & Admission Prices: Call for hours. No charge donations accepted. Public Tours: Sat.-Sun. 1-4. School/Group Tours: Tues.-Fri. 9:30-3 by appointment only. Archives: by appointment only. Students with ID $25 deposit, returned with copy of completed assignment; non-members $25 per hour; members $18 per hour, no charge for the first hour. Site use: non-profit organizations $75; 50 or fewer people $150; more than 50 people $300; events with alcohol $500.
Attendance: 4,000 (estimated)
Membership: Regular: Individual $25; Family $40. Associate: Society Patron $100; Society Benefactor $200. Corporate: Business Associate $500; Corporate Affiliate $1,000.

Williams

SACRAMENTO VALLEY MUSEUM, 1491 E St., Williams, CA 95987. Mailing Address: P.O. Box 1437, Williams, CA 95987-1437. Tel.: 530-473-2978.
E-mail: sacvalleymuseum@frontiernet.net
Web Site: www.sacvalleymuseum.org
Founded: 1963.
Key Personnel: Dir., Kathy Manor; Archive Mgr., Emily Conrado
Institution Type/Description: History Museum: former building of Williams Union High School, built in 1911.
Collections: party line phone switchboard; sheriff's desk; landscape paintings & portraits; horse buggies; early fire equipment; California Native Tribal artifacts; pioneer quilts; toys; dolls; apothecary shop; Asian heritage; period clothing; military memorabilia; saddler & blacksmith shops; bedroom & parlor furnishings.
Publications: biannual newsletter.
Hours & Admission Prices: mid-March to mid-Nov. Thurs.-Sat. 10-4; other times by appointment. Adults $5, children 5-12 $3; members no charge.
Attendance: 1,500 (estimated)
Membership: Individual $20; Family $30; Business $50; Sponsor $100; Patron $250; Benefactor $500; Friends of the Museum $1,000.

Willits

MENDOCINO COUNTY MUSEUM, (M), 400 E. Commercial St., Willits, CA 95490-3204. Tel.: 707-459-2736. Fax: 707-459-7836.
E-mail: museum@co.mendocino.ca.us
Web Site: www.mendocinomuseum.org
Founded: 1972.
Congressional District: 1
Key Personnel: C.E.O. & Dir., Alison Glassey.
Personnel Profile: Full-Time Paid 2; Part-Time Volunteers 30; Interns 1.
Governing Authority: county. Parent Institution: County of Mendocino (CA) board of supervisors. Subsidiary Institution: Grassroots History Publications. Tax-exempt.
Institution Type/Description: History & Railroad Museum.
Collections: Mendocino County life including furniture, clothing, tools, vehicles, art, machinery & household utensils; archives; steam powered logging & railroad equipment; 1850 Frolic shipwreck; Seabiscuit exhibition; Carlos Hittel 1902 oil paintings; L.P. Latimer 1915 oil paintings; steam equipment & trains.
Research Fields: Native American & American ethnic cultural history; Pomo basketry; historic textiles; archaeology; historic preservation; rural environmental adaptation; family & community history; railroad logging history.
Facilities: research library. Books & museum-related items for sale.
Activities: volunteer projects; rotating exhibits; grassroots history publication series; workshops; artifact conservation; photograph darkroom. Annual Events: Steam-Ups with free train rides.
Publications: Grassroots History Publications.
Hours & Admission Prices: Wed.-Sun. 10-4:30. Adults $4. Closed most holidays.

Willow Creek

WILLOW CREEK - CHINA FLAT MUSEUM, 38949 Hwy. 299, Willow Creek, CA 95573. Mailing Address: P.O. Box 102, Willow Creek, CA 95573-0102. Tel.: 530-629-2653.
Web Site: bigfootcountry.net
Personnel Profile: Part-Time Volunteers 20.
Institution Type/Description: History Museum.
Collections: area historical artifacts; bigfoot/sasquatch research material.
Hours & Admission Prices: May-Sept. Wed.-Sun. 10-4; Oct. Fri.-Sun. 10-4. No charge; donations accepted.
Membership: Individual $10; Family $25.

Willows

WILLOWS MUSEUM, 336 W. Walnut St., Willows, CA 95988-2819. Tel.: 530-934-5644.
Founded: 1972.
Key Personnel: Chm. (V), Ray Crabtree.
Personnel Profile: Part-Time Volunteers 16.
Governing Authority: Tax-exempt.
Institution Type/Description: History Museum.
Collections: historical artifacts.
Hours & Admission Prices: Thurs. & Sat.-Sun. 1-4; other times by appointment. No charge; donations accepted.
Attendance: 800 (estimated)

Wilmington

BANNING MUSEUM, 401 E. M St., Wilmington, CA 90744-2610. Mailing Address: P.O. Box 397, Wilmington, CA 90748-0397. Tel.: 310-548-7777. Fax: 310-548-2644.
Web Site: www.thebanningmuseum.org
Founded: 1974.
Congressional District: 32
Key Personnel: C.E.O., Ed Beall; Pres. (V), Bonnie Winters; Dir., Michael Sanborn; Devel. & Membership, Dina Dini; Cur., Tara Fansler; Public Rels., Marty Washington.
Personnel Profile: Full-Time Paid 5; Part-Time Paid 5; Part-Time Volunteers 107; Interns 2.
Governing Authority: municipal. Parent Institution: Getty Grant Program. Subsidiary Institution: Banning Residence Museum & Friends of Banning Museum. Tax-exempt: 501(c)(3).
Institution Type/Description: History Museum: 1864 Greek Revival home of Phineas Banning.
Collections: 19th-century decorative arts; textiles & costumes; furniture; glass & ceramics; silver; books; photographs; Los Angeles harbor development & transportation 1851-1975.
Research Fields: 19th & early 20th-century history of Southern California.
Facilities: library available to public for research; 10,000 sq. ft. exhibit space. Books, paper goods & gift items for sale.

Activities: guided tours; lectures; organized educational programs; docent programs; participatory & temporary exhibitions. Museum Sponsors: Victorian Christmas.
Publications: biannual newsletter, Banning and Company.
Hours & Admission Prices: Tours: Tues.-Thurs. 12:30, 1:30 & 2:30, Sat.-Sun. 12:30, 1:30, 2:30 & 3:30; groups of 10 or more by appointment. Requested Donation: adults $5, children under 12 $1. Closed legal holidays.
Attendance: 22,000 (accurate)
Membership: Friends of Banning Museum: Individual $55; Family $100; Sponsoring $250; Patron $500; President's Circle $1,000; Preservation Circle $2,500; Vanguard Circle $5,500.

DRUM BARRACKS CIVIL WAR MUSEUM, 1052 Banning Blvd., Wilmington, CA 90744-4604. Tel.: 310-548-7509. Fax: 310-548-2946.
E-mail: susan.ogle@lacity.org
Web Site: www.drumbarracks.org
Founded: 1987.
Congressional District: 32
Key Personnel: Dir., Susan Ogle.
Personnel Profile: Full-Time Paid 1; Part-Time Paid 3; Part-Time Volunteers 18.
Governing Authority: state; nonprofit. Parent Institution: Recreation & Parks Dept. City of Los Angeles. Support Group: Drum Barracks Garrison & Society. Tax-exempt.
Institution Type/Description: History Museum.
Collections: Civil War library; Civil War weaponry, firearms, medicine, currency & uniforms; officers' sitting room & family bedroom; small scale 1863 site model; 34 star battle flag from Vicksburg; autograph book of 50 Union general's signatures.
Major Exhibits: Turning Points 1863: Gettysburg and Vicksburg, 9/13-8/14; Hispanics in the American Civil War, 9/14-8/15.
Research Fields: California participation in Civil War period & 1861-1871, American Indian Wars; officers stationed at Wilmington/San Pedro camp; American Civil War.
Facilities: library of national & local military Civil War material, including historical magazines available to the public by appointment.
Activities: guided tours; organized education programs for children, adults, undergraduate & graduate college students; docent program; temporary exhibitions; annual events; internships.
Publications: monthly newsletter, Reveille.
Hours & Admission Prices: Tours: Tues.-Thurs. 10 & 11:30, Sat.-Sun. 11:30 & 1. Requested Donation: adults $5; discounts to AAA members. Closed New Year's Day; Good Friday; Easter; Memorial Day; Independence Day; Labor Day; Thanksgiving; Christmas.
Attendance: 8,600 (accurate)
Membership: Individual $20; Family $30; Sustaining $50; Sponsor $100; Patron $250; Founder $500; Corporate $1,000.

Woodland

HAYS ANTIQUE TRUCK MUSEUM, 1962 Hays Lane, Woodland, CA 95776-6216. Mailing Address: P.O. Box 2347, Woodland, CA 95776-2347. Tel.: 530-666-1044. Fax: 530-666-5777.
Web Site: truckmuseum.org
Founded: 1982.
Key Personnel: Exec. Dir., Ed Rocha
Institution Type/Description: Transportation Museum.
Collections: period trucks; history of trucking.
Hours & Admission Prices: Mon.-Fri. 10-5, Sat. 10-6, Sun. 10-4. Adults $7, seniors 62 & over $6, children 6-14 $4; children under 6 no charge.
Membership: General $50; Company $300; Life $5,000.

YOLO COUNTY HISTORICAL MUSEUM - GIBSON HOUSE, 512 Gibson Rd., Woodland, CA 95695-4843. Tel.: 530-666-1045.
E-mail: ychmoffice@sbcglobal.net
Web Site: www.gibsonhouse.org
Founded: 1979.
Congressional District: 4
Key Personnel: Dir., David M. Flory; Pres. (V), Karen Lafferty; Event Coord., Donna Armstrong.
Personnel Profile: Part-Time Paid 2; Part-Time Volunteers 20.
Governing Authority: nonprofit organization. Parent Institution: County of Yolo. Tax-exempt: 501(c)(3).
Institution Type/Description: Historic House Museum: c.1857 Greek Revival mansion.
Collections: regional history; paintings; photographs; Native American artifacts; clothing; furniture; furnishings; agriculture.
Research Fields: Yolo County history.
Facilities: library; 2.5-acre parkland; rental facilities. Gift items for sale.

Activities: concerts; docent program; guided tours; lectures; temporary exhibitions; weddings; receptions; special events; rental facilities; festivals; family programs; school tours & activities.
Publications: exhibit catalogs; quarterly newsletter; annual report.
Hours & Admission Prices: Tours: 1st Sat. & last Sun. each month 1-4. Adults $5, children 5-17 $3; discounts to Blue Star Family, AAM & ICOM members; members & children under 4 no charge. Closed major holidays. &
Attendance: 4,000 (accurate)
Membership: Youth $15; Senior Citizen $35; Individual $45; Family $75; Sustaining $150; Donor $250.

Woodside

FILOLI CENTER, 86 Canada Rd., Woodside, CA 94062-4144. Tel.: 650-364-8300. Fax: 650-367-0724.
E-mail: friends@filoli.org
Web Site: www.filoli.org
Founded: 1976.
Congressional District: 12
Key Personnel: Pres. (V), Antoinette Barrack; Dir., Cynthia D'Agosta; Museum Shop Mgr., Cat Bishop.
Personnel Profile: Full-Time Paid 36; Part-Time Paid 28; Part-Time Volunteers 1,300; Interns 3.
Volunteer Hours: 120,884
Governing Authority: nonprofit. Property of the National Trust for Historic Preservation; operated by the Filoli Center Inc. Affiliated with The Friends of Filoli. Tax-exempt: 501(c)(3).
Institution Type/Description: Formal Garden, Historic House & Nature Preserve: 1915-1917 residence designed by Willis J. Polk for William B. Bourn, II; gardens designed by Bruce Porter.
Collections: botanical; library; furnishings; fine art; decorative arts.
Research Fields: landscape architecture; S.F. Bay Historic Estates & Gardens.
Facilities: 16 acres of gardens; nature education center.
Activities: guided & self-guided tours; docent-led nature hikes.
Publications: self-guided brochure.
Hours & Admission Prices: mid-Feb. to late Oct. Tues.-Sat. 10-3:30, Sun. 11-3:30; call for tour schedule & reservations Tel.: 650-364-8300, ext. 507. Adults $15, students 5-17 $5; discounts to National Trust for Historic Preservation; children 4 & under and members no charge. Additional fee for special events. Closed Federal holidays. &
Attendance: 102,235 (accurate)
Membership: Individual Senior (1 year) $45; Individual (1 year) $60; Individual Senior (2 year) $80; Dual Family (1 year) $85; Individual (2 year) $110; Dual Family (2 year) $160; Friend Circle $175; Filoli Circle $250; President's Circle $500; Bourn Circle $1,000; Roth Circle $2,500.

*** WOODSIDE STORE HISTORIC SITE, (M),** 3300 Tripp Rd., Woodside, CA 94062-3632. Tel.: 650-851-7615.
E-mail: woodsidestore@historysmc.org
Founded: 1954.
Congressional District: 12
Key Personnel: Pres., Mitchell P. Postel; Chm., Umang Gupta; Dir. Education, Carmen Blair; Coord. Educational Programs, Marilyn Murphy.
Personnel Profile: Part-Time Paid 2; Part-Time Volunteers 20; Interns 1.
Governing Authority: county. Parent Institution: San Mateo County Historical Association. Tax-exempt.
Institution Type/Description: Historic Building: c.1854 general store.
Collections: hands-on museum of 1880s general store.
Research Fields: Local history.
Activities: guided tours for groups; educational programs.
Hours & Admission Prices: Tues. & Thurs. 10-4, Sat.-Sun. 12-4. No charge; donations accepted. Closed national holidays. &

Yermo

CALICO GHOST TOWN AND THE LANE HOUSE & MUSEUM, 36600 Ghost Town Rd., Yermo, CA 92398. Mailing Address: P.O. Box 638, Yermo, CA 92398-0638. Tel.: 760-254-3679. Fax: 760-254-2047.
Web Site: www.calicotown.com
Governing Authority: Parent Institution: County of San Bernardino.
Institution Type/Description: History Museum.
Collections: Town: life of miners & townspeople. House & Museum: local history; natural history; furnishings; mining equipment & artifacts; photographs.
Hours & Admission Prices: Town: daily 9-5. House & Museum: daily 10-4; other times by appointment. Adults $6, youth 6-15 $3; children 5 & under no charge.

Yorba Linda

RICHARD NIXON LIBRARY & BIRTHPLACE, 18001 Yorba Linda Blvd., Yorba Linda, CA 92886-3949. Tel.: 714-993-5075. Fax: 714-528-0544.
E-mail: rexjht@msn.com
Web Site: www.nixonlibraryfoundation.org
Institution Type/Description: History Museum.
Collections: personal artifacts; presidential papers; furnishings; photographs.
Hours & Admission Prices: Mon.-Sat. 10-5, Sun. 11-5. Adult 12 & over $11.95, seniors 62 & over $8.50, students $6.95, children 7-11 $4.75; children 6 & under no charge. Closed Thanksgiving; Christmas.

Yosemite National Park

THE ANSEL ADAMS GALLERY, Village Mall, Yosemite National Park, CA 95389. Mailing Address: P.O. Box 455, Yosemite Village, CA 95389. Tel.: 209-372-4413. Fax: 209-372-4714. Facebook: The Ansel Adams Gallery.
E-mail: yosemite@anseladams.com
Web Site: www.anseladams.com
Key Personnel: Pres., Matthew Adams
Institution Type/Description: Art Gallery.
Collections: photographs.
Facilities: Museum-related items for sale.
Hours & Admission Prices: Mon.-Sun. 9-5.

THE YOSEMITE MUSEUM, NATIONAL PARK SERVICE, Museum Bldg., Yosemite National Park, CA 95389. Mailing Address: P.O. Box 577, Yosemite, CA 95389-0577. Tel.: 209-372-0297 & 0281. Fax: 209-372-0255.
E-mail: yose_museum@nps.gov
Founded: 1915.
Congressional District: 19
Key Personnel: Cur., Jonathan Bayless; Cur. Collections, Barbara L. Beroza; Archivist, Brenna Lissoway; Librarian, Linda Eade; Registrar, Miriam Luchans.
Personnel Profile: Full-Time Paid 6; Part-Time Paid 2; Part-Time Volunteers 15.
Governing Authority: federal. Parent Institution: National Park Service. Subsidiary Institution: Yosemite National Park. Tax-exempt: 101(6).
Institution Type/Description: History, Ethnography & Natural Science Museum.
Collections: anthropology; archaeology; archives; entomology; ethnology; geology; herbarium; history; Indian culture; natural history; photography; zoology. Historic Houses: 1880-1915 Pioneer Yosemite History Center; 1880-1900 park trail cabins.
Research Fields: history; ethnography; photography; art; natural history.
Facilities: 20,000-vol. library; visitor centers; history center. Museum-related items for sale.
Activities: guided tours; talks; films; self-guiding displays; formally organized education programs for children; inter-museum loan, permanent & temporary exhibitions; field seminars; living history demonstrations.
Publications: books; pamphlets; periodicals.
Hours & Admission Prices: Yosemite Valley Visitor Center: Memorial Day-Labor Day daily 8-6; Sept.-May daily 9-5. Call for hours for Pioneer Yosemite History Center; Hill's Studio; Indian Village. No charge. &
Attendance: 485,000 (estimated)
Membership: Yosemite Conservancy: Friend of Yosemite $25 & up; Sequoia Society $60 & up; John Muir Heritage Society $1,000 & up.

Yountville

NAPA VALLEY MUSEUM, (M), 55 Presidents Circle, Yountville, CA 94599. Mailing Address: P.O. Box 3567, Yountville, CA 94599-3567. Tel.: 707-944-0500. Fax: 707-945-0500.
E-mail: kristie@napavalleymuseum.org
Web Site: www.napavalleymuseum.org
Founded: 1973.
Congressional District: 1
Key Personnel: Exec. Dir., Kristie Sheppard.
Personnel Profile: Full-Time Paid 3; Part-Time Paid 3; Part-Time Volunteers 2; Interns 4.
Governing Authority: private; nonprofit organization. Tax-exempt: 501(c)(3).
Institution Type/Description: Local History, Art & Natural Science Museum.
Collections: area history, art & environment heritage including the Wappo Indians, Chinese, Mexicans, pioneers, ranchos, viticulture, Napa River; 19th century painter Sophie Alstrom Mitchell art; Henry Evans botanical prints.
Research Fields: Wappo Indians; Chinese history; history & culture in Napa Valley; cultural heritage of the Napa Valley.
Facilities: education center; collections & resource center. Museum-related items for sale.
Activities: arts festivals; concerts; dance recitals; lectures; rental gallery; organized educational programs; guided tours; workshops; school loan service; docent program; loan & temporary exhibitions; internships. Annual Events: Angel Gala Fundraiser.
Publications: quarterly newsletter.
Hours & Admission Prices: Tues.-Sun. 10-4. Adults $5, senior citizens $3.50, youth $2.50; members no charge. Closed major holidays. &
Attendance: 20,000 (estimated)
Membership: Individual $45; Family $75; Supporter $125; Benefactor $250; Sponsor $500; Patron $1,000.

Yreka

SISKIYOU COUNTY MUSEUM, (M), 910 S. Main St., Yreka, CA 96097-3373. Tel.: 530-842-3836. Fax: 530-842-3166.
E-mail: schs.main@gmail.com
Web Site: siskiyoucountyhistoricalsociety.org
Founded: 1950.
Congressional District: 1
Key Personnel: Dir., Michael Hendryx; Pres. (V), Claudia East.
Personnel Profile: Full-Time Paid 1; Part-Time Volunteers 30.
Governing Authority: county. Subsidiary Institution: Siskiyou County Historical Society. Tax-exempt: 501(c)(3).
Institution Type/Description: Local History Museum.
Collections: ethnographic Indian & Chinese artifacts; trapping paraphernalia; gold mining, agricultural & transportation artifacts; lumbering, settlement, turn-of-the-century, Depression & other 20th-century objects. Outdoor Museum: 1856 log cabin; 1870 miner's cabin; church; schoolhouse; blacksmith shop; 1912 logging skid shack; general merchandise store.
Facilities: 17,000-vol. research library including 32,000 catalogued photographs, 1864-1983 county newspapers & pamphlet files on local history, complete set Bancroft History volumes & pioneer biographies; voter registers; scrapbooks from early 1900s to present.
Activities: special interpretive exhibits; lectures; films; quarterly meetings county historical society; field trips; musical fundraising programs. Annual Events: Living History Celebration in summer.
Publications: annual, The Siskiyou Pioneer; Occasional Paper, No. 1, Siskiyou County: A Time of Change; Occasional Paper, No. 2, Gold Mining in Siskiyou County 1850-1900, technical leaflet, Siskiyou Horizontal Log Construction; Plants & the People, the Ethnobotany of the Karuk Tribe; State of Jefferson; Walking the Medicine Path.
Hours & Admission Prices: Summer: Tues.-Fri. 9-3, Sat. 10-4; Winter: Tues.-Thurs. 9-3, Sat. 10-4. Adults $3, children $.75; children under 7 no charge. Closed national holidays. &
Attendance: 18,000 (estimated)
Membership: Individual $30; Sponsor $75.

Yuba City

COMMUNITY MEMORIAL MUSEUM OF SUTTER COUNTY, (M), 1333 Butte House Rd., Yuba City, CA 95993-2301. Tel.: 530-822-7141. Fax: 530-822-7291.
E-mail: museum@syix.com
Web Site: www.co.sutter.ca.us/
Founded: 1975.
Congressional District: 4
Key Personnel: Dir. & Museum Shop Mgr., Julie Stark; Chm. (V), Suzanne West; Chm. (V), Marsha Amaro; Pres. (V), Brad Willoughby; Asst. Cur., Sharyl Simmons.
Personnel Profile: Full-Time Paid 2; Part-Time Paid 3; Part-Time Volunteers 30.
Governing Authority: county. Parent Institution: County of Sutter. Tax-exempt.
Institution Type/Description: History Museum.
Collections: Maidu Indian artifacts; Sutter County related artifacts including furniture, clothing & agricultural items; documents; photographs; cultural history; multi-cultural history.
Major Exhibits: Marysville Chinese School Books, 2/21/14-3/8/14; Marysville Snapshots 1900-1940, Summer 2014; Sutter Art & Agriculture, Fall 2014.
Research Fields: Sutter County.
Facilities: 200-vol. library of local history books, tax lists, California history, documents, & photos available for research on premises only. Books & other museum-related items for sale.
Activities: guided tours; lectures; workshops; permanent, temporary & historical exhibition; volunteer auxiliary.
Publications: membership newsletter.

Hours & Admission Prices: Wed.-Fri. 9-5, Sat. 12-4. No charge; donations accepted. Closed New Year's Day; Memorial Day; Labor Day; Thanksgiving; Christmas. ♿

Attendance: 9,000 (estimated)

Membership: Student & Senior Citizen $20; Individual $25; Organizations & Clubs $35; Family $40; Business & Sponsor $100; Corporate & Benefactor $1,000.

Yucaipa

MOUSLEY MUSEUM OF YUCAIPA HISTORY, 35308 Panorama Dr., Yucaipa, CA 92399-3532. Mailing Address: P.O. Box 297, Yucaipa, CA 92399-0297. Tel.: 909-790-4685. Fax: 909-790-4685.

E-mail: yucaipahistory@verizon.net

Web Site: www.yucaipahistory.org

Founded: 1970.

Congressional District: 37

Key Personnel: Pres., Claire Teeters; Pres., Harry Birkbeck; Vice Pres., Bob Rippy; Vice Pres., Jack Curtright.

Personnel Profile: Part-Time Volunteers 26.

Governing Authority: county. Parent Institution: Yucaipa Valley Historical Society. Subsidiary Institution: The Old Fire Station Museum, 35106 Avenue A, Yucaipa, CA. Tax-exempt.

Institution Type/Description: Local History Museum.

Collections: local history & artifacts.

Major Exhibits: Square Dancing, 1/14-2/14; First Buildings, 3/14-4/14.

Facilities: 70-seat presentation hall.

Activities: temporary & permanent exhibitions; meetings; research. Museum Sponsors: Touch & Feel Exhibit.

Publications: brochures; books on local history.

Hours & Admission Prices: Wed. 5-9, Sat. 10-3. No charge; donations accepted. ♿

Attendance: 1,000 (accurate)

Membership: Individual $20; Corporate Sponsor $30; Cornerstone $100; Landmark & Scholarship Endowment $250; Heritage $500; Founder $1,000.

YUCAIPA ADOBE, 32183 Kentucky St., Yucaipa, CA 92399-1768. Mailing Address: c/o San Bernardino Co. Museums, 2024 Orange Tree Lane, Redlands, CA 92374. Tel.: 909-795-3485 & 307-2669. Fax: 909-307-0539.

E-mail: rmckernan@sbcm.sbcounty.gov

Web Site: www.sbcountymuseum.org

Founded: 1958.

Congressional District: 35

Key Personnel: Dir., Robert McKernan; Site Mgr., Tony Webb; Cur., Michele Nielsen.

Personnel Profile: Part-Time Paid 1.

Governing Authority: county. Parent Institution: San Bernardino County Museums. Tax-exempt.

Institution Type/Description: Local History and Historic House Museum: housed in late 1850s adobe house.

Collections: local history; farm equipment; decorative arts; blacksmith shop; historic houses.

Research Fields: Local History.

Activities: guided tours; holiday open house; grounds available for weddings.

Hours & Admission Prices: Tues.-Sat. 10-3. No charge; donations suggested. Closed New Year's Day; Thanksgiving; Christmas.

Attendance: 500 (accurate)

Yucca Valley

HI-DESERT NATURE MUSEUM, (M), 57116 29 Palms Hwy., Yucca Valley, CA 92284-2930. Tel.: 760-369-7212. Fax: 760-369-1605.

E-mail: museum@yucca-valley.org

Web Site: www.hidesertnaturemuseum.org

Founded: 1964.

Congressional District: 67

Key Personnel: Cur., Stefanie Ritter; Educator, Crystal Mason.

Personnel Profile: Full-Time Paid 1; Part-Time Paid 1; Part-Time Volunteers 3.

Governing Authority: Parent Institution: Town of Yucca Valley.

Institution Type/Description: Natural History Museum.

Collections: fossils; rocks; gems; minerals; pressed desert wild flowers; Indian artifacts; points; pottery; manos; pestles; petrified wood; mini-zoo of live small desert animals; photographs; mounted golden eagle, deer, badger, bobcat & reptiles, various birds; regional history; local artwork.

Research Fields: plants; insects; geology; reptiles; birds; local history; earthquakes.

Facilities: mini zoo.

Activities: tours; education program for children; lecture series; temporary exhibits. Annual Events: Earth Day Fair; Art & Culture Wednesdays; Science Saturdays.

Publications: e-newsletter, Tortoise Tales.

Hours & Admission Prices: Thurs.-Sat. 10-5. No charge; donations accepted. Closed holidays. ♿

Attendance: 34,180 (accurate)

Membership: Individual $20; Family $35; Business $100; Contributor $250; Sponsor $500; Benefactor $1,000.

COLORADO

(287 listings)

Alamosa

ADAMS STATE UNIVERSITY LUTHER BEAN MUSEUM, Richardson Hall, #256, Alamosa, CO 81101. Mailing Address: 208 Edgemont Blvd., Alamosa, CO 81101. Tel.: 719-587-7151; 800-824-6494. Fax: 719-587-7522.

E-mail: lutherbean@adams.edu

Web Site: adams.edu/lutherbean

Founded: 1921.

Congressional District: 3

Key Personnel: Pres., Dr. David Svaldi; Treas., Bill Mansheim; Coord. Museum, Linda Relyea; Security, Joel Schultz.

Personnel Profile: Part-Time Paid 1; Interns 2.

Governing Authority: college. Parent Institution: Adams State University. Tax-exempt: 501(c)(3).

Institution Type/Description: Anthropology, Ethnology Indian & History Museum.

Collections: Paleo-Indian Folsom points; Pueblo Indian cultural artifacts, primarily pottery; Rio Grande weaving; Spanish culture; Santos; Spanish history primarily San Luis Valley & Southwest; regional art including paintings by Stephen Quiller; Bill Moyers bronzes; Native American pottery, drawings & artifacts; European & Asian decorative arts.

Research Fields: archaeology & local history; pioneer history of the San Luis Valley.

Facilities: planetarium; 500-seat auditorium; classrooms; 40 sq. ft. exhibit space.

Activities: permanent & temporary exhibitions; guided tours.

Publications: quarterly alumni magazine, A-Stater.

Hours & Admission Prices: Mon.-Fri. 2-5; other times by appointment only. No charge; donations accepted. Closed New Year's Eve & Day; Memorial Day; Labor Day; Thanksgiving; Christmas Eve, Day & week; national holidays. ♿

Attendance: 500 (estimated)

SAN LUIS VALLEY MUSEUM, 401 Hunt Ave., Alamosa, CO 81101. Mailing Address: P.O. Box 1593, Alamosa, CO 81101-1593. Tel.: 719-587-0667.

E-mail: sanluisvalleymuseum@yahoo.com

Formerly: San Luis History Center

Founded: 1992.

Key Personnel: Chm. (V), Dorothy M. Brandt; Office Mgr. & Museum Shop Mgr., Joyce Gunn.

Personnel Profile: Full-Time Paid 1; Part-Time Volunteers 15; Interns 2.

Governing Authority: Tax-exempt.

Institution Type/Description: History Museum.

Collections: local history; multicultural heritage; photographs; period artifacts; ranch & farm life; Native American; military; early railroading.

Activities: lectures; art shows.

Publications: annual report; semiannual newsletter; operational plan; museum brochure.

Hours & Admission Prices: Tues.-Sat. 10-4; other times by appointment. Adults $2; students & members no charge. ♿

Attendance: 1,800 (estimated)

Membership: Single $15; Couples $25.

Arvada

ARVADA CENTER FOR THE ARTS AND HUMANITIES, (M), 6901 Wadsworth Blvd., Arvada, CO 80003-3499. Tel.: 720-898-7200. Fax: 720-898-7217. TDD: 720-898-7203.

Web Site: www.arvadacenter.org

Founded: 1976.

Congressional District: 2

Key Personnel: Exec. Dir., Gene Sobczak; Chm. (V), Debra Havins; Administrative Clerk, Bernita Christenson; Dir. Performing Arts, Rod Lansberry; Dir. Mktg., Cynthia DeLarber; Exhibit Designer, Collin Parson.

Governing Authority: municipal. Parent Institution: City of Arvada. Tax-exempt.
Institution Type/Description: Art and History Museum.
Collections: 19th century photography; contemporary art & crafts; outdoor sculpture.
Research Fields: humanities programming in museums; Arvada area history.
Facilities: 500-seat auditorium; classrooms.
Activities: guided tours; lectures; films; concerts; dance recitals; arts festivals; theater; study clubs; organized education programs; docent program; participatory, loan & temporary exhibitions.
Publications: quarterly magazine, Artscentric.
Hours & Admission Prices: Mon.-Fri. 9-6, Sat. 9-5, Sun. 1-5. No charge; donations accepted. Closed New Year's Day; Memorial Day; Independence Day; Labor Day; Thanksgiving; Christmas. &
Attendance: 360,000 (estimated)
Membership: General $40.

ARVADA FLOUR MILL, 5580 Old Wadsworth Blvd., Arvada, CO 80002-3104. Mailing Address: Arvada Historical Society, 7307 Grandview Ave., Arvada, CO 80002. Tel.: 303-431-1261.
Governing Authority: Parent Institution: Arvada Historical Society.
Institution Type/Description: Historic Building: restored flour mill; built in 1926. Listed on the National Register of Historic Places.
Collections: mill history; late 1800s equipment.
Activities: guided tours.
Hours & Admission Prices: By appointment. Adults $1.50, children 12 & under $.50.

ARVADA HISTORICAL SOCIETY, 7307 Grandview Ave., Arvada, CO 80002-2507. Tel.: 303-431-1261.
Institution Type/Description: Historical Society Museum.
Collections: area history & culture; photographs; personal artifacts; books; documents.
Hours & Admission Prices: Tues.-Sat. 11-3.

Aspen

* **ASPEN ART MUSEUM, (M),** 590 N. Mill St., Aspen, CO 81611-1510. Tel.: 970-925-8050. Fax: 970-925-8054. Facebook: Aspen Art Museum.
E-mail: info@aspenartmuseum.org
Web Site: www.aspenartmuseum.org
Founded: 1979.
Congressional District: 3
Key Personnel: C.E.O., Dir. & Chief Cur., Heidi Zuckerman Jacobson; Deputy Dir., John-Paul Schaefer; Dir. Finance & Administrative, Karen Johnsen; Dir. Communications, Jeffrey Murcko; Dir. Special Events, Jason Hurley; Editor & Publications Mgr., Ryan Shafer; Registrar & Mgr. Exhibitions, Luis Yllanes; Chief Preparator & Facilities Mgr., Jonathan Hagman; Campaign Mgr., Grace Nims; Cur. Education, Danielle Stephens; Special Events & Devel. Coord., John Barker; Coord. Education, Annie Henninger; Community Liaison & Museum Shop Mgr., Nicole Kinsler; Visitor Svcs. Asst., Jennifer Schneider; Visitor Svcs. Asst., Cecilia Anthony; Visitor Svcs. Asst., Leslie Bixel; Cur., Jacob Proctor; Adjunct Cur., Matthew Thompson; Curatorial Asst., Kelly May; Accounting Clerk, Sallie Klein; Project Mgr., Mike O'Connor; Exec. Asst. to the Dir., Sherry Black; Devel. Asst., Amelia Russo; Graphic Designer, Michael Aberman.
Personnel Profile: Full-Time Paid 18; Part-Time Paid 6; Part-Time Volunteers 45.
Governing Authority: nonprofit. Tax-exempt: 501(c)(3).
Institution Type/Description: Contemporary Art Museum.
Collections: outdoor sculpture.
Major Exhibits: Trapping Lions in the Scottish Highlands, 11/15/13-2/2/14.
Research Fields: contemporary art.
Facilities: library of slides & periodicals available for research.
Activities: guided tours; lectures; educational programs; loan & traveling exhibitions; art workshops for children & teens.
Publications: books, One Hour Ahead: The Avant Garde in Aspen, 1945-2004; Simon Evans: How to Get About, 2006; Doug Aitken 99 cent Dreams; Yutaka Sone (co-published with The Renaissance Society at the University of Chicago, and Kunsthalle Bern); exhibition catalogues, Amy Adler; Like Color in Pictures; Sculptors' Drawings; Belief and Doubt; Jeremy Deller; Aida Ruilova: The Singles 1999-Now; Friedrich Kunath; Phil Collins; Now You See It; Fred Tomaselli; Restless Empathy; Mamma Andersson; Mark Manders; Sergej Jensen; Art in Unexpected Places; Slater Bradley & Ed Lachman: Look Up and Stay in Touch; The Anxiety of Photography; Parallel Occurrences/Documented Assignments; Simon Denny: Full Participation; Haegne Young: Wild Against Gravity; Mark Manders: Parallel Occurrences/Documented Assignments; Avner Ben-Gal; The Rainbow Hour; Mark Bradford: Merchant Posters; Tonico Lemos Anad; Disembodied.
Hours & Admission Prices: Tues.-Wed. & Fri.-Sat. 10-6, Thurs. 10-7, Sun. 12-6. Art After Hours: Thurs: 5-7. No charge. Closed New Year's Day; Thanksgiving; Christmas. &
Attendance: 35,089 (accurate)
Membership: Art Steward $35; Art Family $50; Art Addict $150-$249; Art Fanatic $250-$499; Art Scholar $500-$999; Patron's Circle $1,000-$2,999; Benefactor's Circle $3,000-$4,999; Director's Circle $5,000-$9,999; President's Circle $10,000-$24,999; Founder's Circle $25,000-$49,999; Visionaries Circle $50,000 & up.

ASPEN HISTORICAL SOCIETY, (M), 620 W. Bleeker St., Aspen, CO 81611-1230. Tel.: 970-925-3721. Fax: 970-925-5347.
E-mail: info@aspenhistory.org
Web Site: www.aspenhistory.org
Formerly: Heritage Aspen
Founded: 1963.
Congressional District: 3
Key Personnel: Pres. (V), Tony Vagneur; Dir., Georgia Hanson; Cur., Lisa Hancock; Museum Shop Mgr., Nina Gabianelli.
Personnel Profile: Full-Time Paid 7; Part-Time Paid 3; Part-Time Volunteers 40; Interns 7.
Governing Authority: society; nonprofit. Branch Museum: Holden-Marolt Ranching & Mining Museum, Aspen; Wheeler-Stallard Museum; Independence & Ashcroft Ghost Towns. Tax-exempt: 501(c)(3).
Institution Type/Description: Housed in 1888 Wheeler-Stallard House.
Collections: Aspen and Roaring Fork Valley history & lifestyles; domestic life; mining; ranching; skiing & culture; textiles; manuscripts; newspapers; photographs.
Major Exhibits: Ute Indians, 6/12-5/14.
Research Fields: Aspen & Roaring Fork Valley history.
Facilities: 1,000-vol. library of books on Aspen and Roaring Fork Valley history; 200 Aspen Times Newspapers; written materials including brochures, stocks, bonds & other forms of original documentation; 6,000 sq. ft. exhibit space.
Activities: docent program; adult & children's educational programs; films; guided tours; walking tours of historic West End & downtown Aspen; temporary exhibitions.
Publications: quarterly newsletter.
Hours & Admission Prices: Summer: Tues.-Sat. 1-5; Winter: Tues.-Fri. 1-5. Adults $6, seniors $5, children $3; children under 5 & members no charge. Closed Christmas. &
Attendance: 25,000 (estimated)
Membership: Student & Limited Income $18.79; Individual $35; Family $60; Business $125; Contributing $300; Collaborator $700; Sustaining $1,000.

BALDWIN GALLERY, 209 S. Galena St., Aspen, CO 81611. Tel.: 970-920-9797. Fax: 970-920-1821.
E-mail: baldwingallery@baldwingallery.com
Web Site: www.artnet.com/baldwin.html
Key Personnel: Dir., Richard Edwards
Institution Type/Description: Art Gallery.
Collections: works by established & emerging artists.
Hours & Admission Prices: Mon.-Sat. 10-6, Sun. 12-5.

RED BRICK CENTER FOR THE ARTS, 110 E. Hallam St., Aspen, CO 81611-1458. Tel.: 970-429-2777. Fax: 970-920-5700.
E-mail: info@aspenart.org
Web Site: www.aspenart.org
Founded: 1973.
Key Personnel: Exec. Dir., Angie Callen
Institution Type/Description: Art Gallery.
Collections: works by regional artists.
Activities: educational programs; classes; workshops; presentations; demonstrations; special events.
Hours & Admission Prices: Mon.-Fri. 10-6.

Aurora

AURORA HISTORY MUSEUM, (M), 15051 E. Alameda Pkwy., Aurora, CO 80012-1554. Tel.: 303-739-6660. Fax: 303-739-6657.
Web Site: www.auroramuseum.org
Founded: 1979.
Congressional District: 2, 6
Key Personnel: Exec. Dir., Jennifer Kuehner; Cur. Exhibits, MaryJane Valade; Cur. Collections, Jennifer Cronk; Facility Mgr., Ken Clinton; Historic Preservation Asst., Jeanne Ramsay.

Personnel Profile: Full-Time Paid 7; Full-Time Volunteers 3; Part-Time Paid 10; Part-Time Volunteers 125.
Governing Authority: municipal. Parent Institution: City of Aurora. Tax-exempt.
Institution Type/Description: History Museum.
Collections: local history collections reflecting rural Great Plains agricultural and 19th- & 20th-century suburban development; Indian culture of the Aurora area encompassing Adams & Arapahoe Counties; manuscripts. Historic Buildings: Gully Homestead; DeLaney Farm and Round Barn; Coal Creek School; Centennial House.
Research Fields: Great Plains agricultural, late Victorian suburban & 20th-century urban history; historic preservation.
Facilities: archive & library of periodicals, manuscripts & photographs on Aurora, Colorado, Western history, museology & historic preservation available for research on premises; meeting rooms.
Activities: guided tours; lectures; films; gallery talks; formally organized educational programs; permanent & temporary exhibitions.
Publications: Aurora: Gateway to the Rockies; First Sun. of Colorado, Aurora
Hours & Admission Prices: Tues.-Fri. 9-4, Sat.-Sun. 11-4; call for Gully Homestead and Centennial House hours. No Charge. Closed holidays. &
Attendance: 30,000 (accurate)
Membership: Student 18 & under $5; Senior Citizens 60 & over $10; Senior Couple & Individual $15; Family & General $25; Sustaining $50; Century Club $100; Benefactor $250.

PLAINS CONSERVATION CENTER, 21901 E. Hampden Ave., Aurora, CO 80013-5000. Tel.: 303-693-3621. Fax: 303-693-3379.
E-mail: info@plainscenter.org
Web Site: www.plainscenter.org
Founded: 1949.
Key Personnel: C.E.O. & Dir., Tudi Arneill; Pres. (V), Gordon C. Tucker.
Personnel Profile: Full-Time Paid 6; Part-Time Paid 8; Part-Time Volunteers 100; Interns 1.
Governing Authority: Parent Institution: West Arapahoe Conservation District. Tax-exempt.
Institution Type/Description: General Museum.
Collections: natural & cultural heritage; replica sod village, schoolhouse, blacksmith shop, & barn; live animals; period furnishings.
Research Fields: paleontology; zoology; prairie ecology.
Facilities: library; 6,800 acres of short grass prairie; visitor center; hiking trails.
Activities: nature walks & programs; historical interpretation programs of Plains Indians & 1880s.
Publications: quarterly newsletter.
Hours & Admission Prices: Sat. 9-5. No charge. &
Attendance: 20,000 (accurate)
Membership: Individual $20; Family $30; Business & Supporting $60.

Basalt

ANN KOROLOGOS GALLERY, 211 Midland Ave., Basalt, CO 81621. Tel.: 970-927-9668.
E-mail: art@korologosgallery.com
Web Site: www.korologosgallery.com
Key Personnel: Dir., Jay Magidson; Owner, Ann McLaughlin Korologos
Institution Type/Description: Art Gallery.
Collections: Western art; paintings; sculpture.
Hours & Admission Prices: Call for hours. No charge. &

Beaver Creek

THE MAY GALLERY, PATRON'S LOUNGE, Vilar Center for the Arts, 68 Avondale Lane, Beaver Creek Resort, Beaver Creek, CO 81620. Mailing Address: P.O. Box 3822, Avon, CO 81620-3822. Tel.: 970-845-8497; 888-920-2787. Fax: 970-748-1396.
E-mail: ksabel@vvf.org
Web Site: vilarcenter.org
Formerly: The May Gallery, A Program of the Vail Valley Arts Council
Founded: 1972.
Congressional District: 56
Key Personnel: Vice Pres. Cultural Programming, Kris Sabel; Pres. (V), Bob Jamar; Dir. Mktg., Shelley Woodworth.
Personnel Profile: Full-Time Paid 3; Part-Time Paid 3; Part-Time Volunteers 150.
Governing Authority: Vail Valley Foundation; private; nonprofit organization. Tax-exempt: 501(c)(3).
Institution Type/Description: Art Museum & Center.
Collections: paintings.
Facilities: 100-vol. library of periodicals; 2,247 sq. ft. exhibit space. Museum-related items for sale.

Activities: arts festivals; docent program; formal education programs; guided tours; hobby workshops; lectures; loan & traveling exhibitions.
Publications: quarterly newsletter; exhibition brochures & catalogues.
Hours & Admission Prices: By appointment only. No charge. &
Attendance: 20,000 (estimated)
Membership: Artist, Student & Educator $25; Friend $50; Supporting $100; Sustaining $250; Benefactor $500; Inner Circle: $1,000-$2,500 (per Year for 5 years); Collector's Group: $2,500, $3,000 & $5,000; Donor $5,000; Major Donor $10,000.

Berthoud

LITTLE THOMPSON VALLEY PIONEER MUSEUM, 224 Mountain Ave., Berthoud, CO 80513. Tel.: 970-532-2147.
E-mail: ltvpm@berthoudhistoricalsociety.org
Web Site: www.berthoudhistoricalsociety.org
Key Personnel: Pres. (V), Mark French.
Governing Authority: Parent Institution: Berthoud Historical Society.
Institution Type/Description: History Museum.
Collections: local history & culture; photographs; personal artifacts.
Facilities: Museum-related items for sale.
Hours & Admission Prices: Wed.-Sun. 1-5; other times by appointment.
Attendance: 1,500 (accurate)

Boulder

BOULDER HISTORY MUSEUM, (M), 1206 Euclid Ave., Boulder, CO 80302-7224. Tel.: 303-449-3464. Fax: 303-938-8322.
E-mail: info@boulderhistory.org
Web Site: www.boulderhistorymuseum.org
Formerly: Boulder Museum of History
Founded: 1944.
Congressional District: 2
Key Personnel: C.E.O., Nancy Geyer; Cur. Exhibits & Facilities, Julie Schumaker; Cur. Education, Laura Stroud; Mktg. & Public Rels., Susan Linde.
Personnel Profile: Full-Time Paid 5; Part-Time Paid 3; Part-Time Volunteers 30; Interns 2.
Governing Authority: society. Affiliated with Boulder Historical Society. Tax-exempt: 501(c)(3).
Institution Type/Description: History Museum; housed in the Harbeck-Bergheim house built in 1899.
Collections: 38,000 material objects; 2,500 early clothing items; 200,000 period photographs.
Research Fields: local history.
Activities: guided tours; in-house programs; lectures; permanent & rotating interpretive exhibits; outreach programs; festivals; workshops.
Publications: informational brochures; newsletter.
Hours & Admission Prices: Tues.-Fri. 10-5, Sat.-Sun. 12-4. Adults $6, seniors $4, children & students $2; discounts to AAM & ICOM members; members & children under 5 no charge. Closed major holidays. &
Attendance: 8,000 (accurate)
Membership: Senior $25; Individual $35; Senior Couple $40; Family $55; Supporting $100; Sustaining $250; Patron $500; Leadership Circle $1,000.

BOULDER MUSEUM OF CONTEMPORARY ART, 1750 13th St., Boulder, CO 80302-6226. Tel.: 303-443-2122. Fax: 303-447-1633. Facebook: Boulder Museum of Contemporary Art.
E-mail: info@bmoca.org
Web Site: www.bmoca.org
Founded: 1972.
Congressional District: 2
Key Personnel: Exec. Dir., David Dadone; Assoc. Cur., Petra Sertic; Assoc. Cur., Kirsten Gerdes Stoltz.
Personnel Profile: Full-Time Paid 6; Part-Time Paid 6; Part-Time Volunteers 100; Interns 15.
Institution Type/Description: Contemporary Art Museum.
Collections: works by contemporary artists.
Facilities: 10,000 sq. ft. exhibit space.
Activities: creative experiences; exploration; contemporary exhibitions; youth & adult education; outreach.
Publications: newsletter, BMOCA; catalogs.
Hours & Admission Prices: Tues.-Sun. 11-5. Adults $5, senior citizens, teachers & students $4; discounts to AAM & ICOM members; MOD/CO, NARM members & children under 12 no charge. &
Attendance: 30,000 (estimated)
Membership: Senior Citizen, Teacher & Student $25; Individual $35; Friends With Benefits $45; Dual & Family $55; Contributor $100; Benefactor $250; Sustainer $500.

CU ART GALLERIES, (M), 318 UCB, Sibell Wolle Fine Arts Bldg., University of Colorado/Boulder, Boulder, CO 80309. Tel.: 303-492-8300. Fax: 303-735-4197.
Web Site: www.colorado.edu/cuartmuseum
Founded: 1939.
Key Personnel: Dir., Lisa Tamiris; Collections Mgr., Bridget Carlin; Exhibitions Mgr., Stephen Martonis; Outreach Coord., Carrie Olson; Galleries & Administrative Coord., Evan Cantor.
Personnel Profile: Full-Time Paid 4; Part-Time Paid 11; Part-Time Volunteers 8; Interns 7.
Governing Authority: university. Parent Institution: University of Colorado at Boulder. Tax-exempt.
Institution Type/Description: Art Gallery.
Collections: 19th & 20th-century paintings & prints; 15th to 20th-century drawings, watercolors, sculptures & ceramics; 20th-century photographs.
Research Fields: contemporary art.
Facilities: 5,000 sq. ft. exhibit space.
Activities: lectures; participatory, loan, temporary & traveling exhibitions.
Publications: catalogs.
Hours & Admission Prices: Call for information. &
Attendance: 20,000 (accurate)

CU HERITAGE CENTER, (M), University of Colorado, Boulder Campus, 3rd Fl. Old Main, Boulder, CO 80302. Mailing Address: Campus Box 459, Boulder, CO 80309. Tel.: 303-492-6329. Fax: 303-492-1244.
E-mail: allyson.smith@colorado.edu
Web Site: cuheritage.org
Founded: 1985.
Congressional District: 2
Key Personnel: Dir., Allyson Smith; Asst. Dir., Mona Lambrecht.
Personnel Profile: Full-Time Paid 2; Part-Time Paid 3; Part-Time Volunteers 10; Interns 1.
Governing Authority: university; nonprofit. Parent Institution: University of Colorado Foundation. Tax-exempt: 501(c)(3).
Institution Type/Description: History Museum.
Collections: University of Colorado at Boulder history & alumni; photographs, textiles, furniture, objects, books, documents, architectural drawings & art.
Research Fields: university & alumni history.
Facilities: rental facilities.
Activities: permanent & temporary exhibits; tours; workshops; receptions; events.
Hours & Admission Prices: Mon.-Fri. 10-5. No charge; donations accepted. Closed all university holidays. &
Attendance: 26,300 (accurate)

THE DAIRY CENTER FOR THE ARTS, 2590 Walnut St., Boulder, CO 80302-5700. Tel.: 303-440-7826. Fax: 303-440-7104.
E-mail: bill@thedairy.org
Web Site: www.thedairy.org
Founded: 1992.
Key Personnel: Exec. Dir., Bill Obermeier; Pres., Richard Polk; Dir. Programming, Deven Shaff; Dir. Devel., Beth Smith.
Personnel Profile: Full-Time Paid 12; Part-Time Paid 14; Interns 2.
Operating Expenses: 1,400,000
Operating Income: 1,400,000
Governing Authority: Tax-exempt.
Institution Type/Description: Art Gallery.
Collections: paintings; sculpture; photographs.
Hours & Admission Prices: Daily 8:30-11. Closed holidays. &
Attendance: 200,000 (estimated)

LEANIN' TREE MUSEUM OF WESTERN ART, 6055 Longbow Dr., Boulder, CO 80301-3296. Mailing Address: Box 9500, Boulder, CO 80301-9500. Tel.: 303-530-1442; 800-777-8716.
E-mail: artmuseum@leanintree.com
Web Site: www.leanintreemuseum.com
Founded: 1974.
Congressional District: 2
Key Personnel: C.E.O. & Pres., Thomas E. Trumble; Dir. & Chm., Edward P. Trumble.
Personnel Profile: Full-Time Paid 3; Part-Time Paid 6.
Governing Authority: company organized for profit. Parent Institution: Leanin' Tree, Inc., 6055 Longbow Dr., Boulder, CO 80301.
Institution Type/Description: Art Museum of the American West.
Collections: contemporary & deceased masters, Western cowboy & Indian art; 85 Western bronze sculptures; 300 paintings.
Facilities: Gift items & paper products for sale.

Activities: guided & self-guided tours; permanent & temporary exhibits.
Publications: souvenir book, Leanin' Tree Museum of Western Art.
Hours & Admission Prices: Mon.-Fri. 8-5, Sat.-Sun. 10-5. No charge. Closed national holidays. &
Attendance: 50,000 (accurate)

SHELBY AMERICAN COLLECTION, 5020 Chaparral Ct., Boulder, CO 80301-3351. Mailing Address: P.O. Box 19228, Boulder, CO 80308-2228. Tel.: 303-516-9565. Fax: 303-447-1380.
E-mail: info@shelbyamericancollection.org
Web Site: www.shelbyamericancollection.org
Founded: 1996.
Key Personnel: C.E.O., Steven B. Volk.
Governing Authority: private; nonprofit organization.
Institution Type/Description: Automotive Museum.
Collections: race cars manufactured by Carroll Shelby from 1962-1967; Shelby Cobras; Ford GT 40s; Shelby Mustang GT 350s.
Facilities: 2,000-vol. library of automotive books; 10,000 sq. ft. exhibit space; 25-seat large screen theater. Museum-related items for sale.
Activities: guided tours. Museum Sponsors: annual fundraiser with Carroll Shelby in December.
Hours & Admission Prices: Sat. 10-4; groups & clubs by appointment. Admission $5. &
Attendance: 3,000 (estimated)

*** UNIVERSITY OF COLORADO MUSEUM OF NATURAL HISTORY, (M),** Henderson Bldg., 218 UCB, 15th & Broadway, Boulder, CO 80309-0218. Tel.: 303-492-6892 & 6297. Fax: 303-492-4195.
E-mail: cumuseum@colorado.edu
Web Site: cumuseum.colorado.edu
Founded: 1902.
Congressional District: 2
Key Personnel: Dir. & Prof. Biology, Pat Kociolek; Prof. Emeritus Anthropology, Linda S. Cordell; Cur. Museography & Prof. Emeritus, John R. Rohner; Collections Mgr. Botany, Tim Hogan; Collections Mgr. Botany, Dina Clark; Cur. & Prof. Anthropology, Stephen Lekson; Cur. Botany Emeritus, William A. Weber; Cur. Entomology & Prof. Biology, M. Deane Bowers; Collections Mgr. Entomology, Virginia Scott; Cur. Invertebrate Paleontology & Assoc. Prof. Geological Sciences, Dena Smith; Assoc. Prof. Emeritus, Natural History, Judith A. Harris; Cur. Paleontology & Assoc. Prof. Geological Sciences, Karen Chin; Cur. Paleontology & Asst. Prof., Jaelyn Eberle; Cur. Zoology Emeritus, Shi-Kuei Wu; Collections Mgr. Zoology, Mariko Kageyama; Cur. Invertebrate Zoology & Assoc. Prof Biology, Robert Guralnick; Cur. Vertebrate Zoology & Asst. Prof. Biology, Christy McCain; Coord. Exhibits, Charles Counter; Exhibits Technician, William Moats; Sr. Educator, Jim Hakala; Coord. Public Programs, Dulce Aldama; Museum & Field Studies Graduate Program Coord., Janet Bensko; Visitor Svcs., Kory Katsimpalis; Admin., Susan S. Reinke; Asst. to Dir., Susanna Drogsvold; Information Coord., Steven Reinhold; Collection Mgr. Anthropology, Christina Cain; Collection Mgr. Invertebrate Paleontology, Talia Karim; Collection Mgr. Vertebrate Paleontology, Toni Culver.
Personnel Profile: Full-Time Paid 20; Part-Time Paid 13; Part-Time Volunteers 80.
Governing Authority: state; university. Parent Institution: University of Colorado. Tax-exempt: 170(b)(1)(A).
Institution Type/Description: Natural History & Anthropology Museum.
Collections: Native American textiles; U.S. Southwestern archeology; Colorado insects, especially butterflies & bees; Colorado mammals, birds; vertebrate and invertebrate fossils. Arctic, Alpine, Colorado & Western U.S. plants.
Research Fields: anthropology; archaeology; ethnology; botany; lichenology; malacology; bryology; floristics; geology; vertebrate & invertebrate paleoecology; zoology; mammalogy; herpetology; paleontology; entomology.
Facilities: research labs.
Activities: lectures; formally organized education programs for undergraduate & graduate students affiliated with University of Colorado; Master's program in museum & field studies; education outreach kits; pamphlets; field trips; inter-museum loan; continuing, temporary & traveling exhibitions; K-12 guided tours; teacher & adult workshops; family programs.
Publications: episodic, Natural History Guides; Natural History Inventories of Colorado; leaflets; newsletter.
Hours & Admission Prices: Mon.-Fri. 9-5, Sat. 9-4, Sun. 10-4. No charge; donations accepted. Closed university holidays; New Year's Day; Easter; Independence Day; Thanksgiving; Christmas. &
Attendance: 25,600 (estimated)
Membership: Student $10; Educator $20; Individual $25; Family $45; Sponsoring $100; Sustaining $250; Curator's Circle $500; Director's Circle $1,000.

Breckenridge

BARNEY FORD HOUSE MUSEUM, 111 E. Washington Ave.,
Breckenridge, CO 80424. Mailing Address: P.O. Box 2460, Breck-
enridge, CO 80424-2460. Tel.: 970-453-9767.
E-mail: info@breckheritage.com
Web Site: www.breckheritage.com
Founded: 2006.
Congressional District: 2
Key Personnel: Dir., Larissa O'Neil.
Governing Authority: Parent Institution: Breckenridge Heritage Alliance.
Tax-exempt.
Institution Type/Description: Historic House: former home of Barney L. Ford,
an escaped slave who became a prominent entrepreneur and Black civil
rights pioneer in Colorado; built in 1882.
Collections: family history; period furnishings.
Hours & Admission Prices: See website for hours. No charge; donations
accepted.
Attendance: 5,315 (accurate)
Membership: Student $18.59; Individual $25; Family $75; Silver $100; Gold
Nugget $200.

MOUNTAIN TOP CHILDREN'S MUSEUM, 605 S. Park Ave.,
Breckenridge, CO 80424. Mailing Address: P.O. Box 4359, Breck-
enridge, CO 80424-4359. Tel.: 970-453-7878.
E-mail: mtntopmuseum@gmail.com
Web Site: www.mtntopmuseum.org
Founded: 2002.
Key Personnel: Exec. Dir., Laura Horvath; Asst. Dir., Bethany Boland; Chm.
(V), Jennifer Ruff; Chm. (V), Cheryl Rothey; Chm. (V), Lynn Ryckman;
Pres. (V), Sue O'Brien.
Volunteer Hours: 2,200
Operating Expenses: 161,511
Operating Income: 166,083
Governing Authority: nonprofit organization. Tax-exempt: 501(c)(3).
Institution Type/Description: Children's Museum.
Collections: hands-on exhibits.
Facilities: Museum-related items for sale.
Activities: rental facilities; day camp; outreach programs; birthday parties.
Publications: newsletter.
Hours & Admission Prices: April 20-May 18 & Sept. 4-Dec. 12 Fri.-Mon.
10-4:30; Dec. 12-April 20 daily 10-4:30. Children $7, adults $5; children
under one & seniors no charge. Closed New Year's Day; Christmas. &
Attendance: 10,000 (accurate)

RED, WHITE & BLUE FIRE MUSEUM, 308 N. Main St.,
Breckenridge, CO 80424. Mailing Address: P.O. Box 710, Breck-
enridge, CO 80424. Tel.: 970-453-2474. Fax: 970-453-1350.
Institution Type/Description: Firefighting History Museum.
Collections: firefighting history & equipment; early fire hoses & extinguishers;
ladder cart; pumper cart; leather bucket; uniforms & boots.
Hours & Admission Prices: Call for hours.

SUMMIT HISTORICAL SOCIETY, 111 N. Ridge St., Brecken-
ridge, CO 80424. Mailing Address: P.O. Box 745, Breckenridge,
CO 80424-0745. Tel.: 970-453-9022.
E-mail: mail@summithistorical.org
Web Site: www.summithistorical.org
Founded: 1966.
Congressional District: 13
Key Personnel: Pres. (V), John Ebright; Vice Pres., Jerry Dziedzic; Treas., Bill
Musolf.
Personnel Profile: Part-Time Paid 1; Part-Time Volunteers 40.
Governing Authority: society; nonprofit. Subsidiary Institutions: Edwin Carter
Museum, 111 N. Ridge St., Breckenridge, CO 80424; Dillon Schoolhouse
Museum, 403 LaBonte, Dillon, CO 80435. Tel.: 303-453-9022; Main St.
Museum, 111 S. Main St., Breckenridge, CO 80424. Tax-exempt.
Institution Type/Description: Historical Society Museums: 1870 Cabin -
Edwin Carter Museum, 1885 Lula Myers Cabin & 1936 honeymoon cabin.
Collections: furnishings from old Dillon, Breckenridge, Frisco, Montezuma &
Kokomo schools; old domestic furnishings; Victorian parlor; 19th-century
general store; blacksmith shop with original tools; taxidermy; photographs;
W.W. Boyd artistic & corrective horseshoes. Historic Structures: 1883
Dillon Schoolhouse, Dillon; 1936 Slate Creek Hall, Slate Creek; 1896
William Harrison Briggle House Museum, Breckenridge; 1880 Alice G.
Milne House, Breckenridge; 1880 Washington mine, Breckenridge; 1900
Gold Dredge Boat, Breckenridge; 1884 Montezuma Schoolhouse, Monte-
zuma; 1885 Lula Myers Cabin, Dillon; 1930s Honeymoon Cabin, Dillon;

Rice Born, Dillon Co., Barney Ford House Museum; Breckenridge,
Silverthorne Museum, Silverhorne.
Research Fields: local history.
Facilities: 200-vol. library of books & manuscripts relating to local history
available for use on the premises only. Books & other museum-related
items for sale.
Activities: guided tours; library; temporary exhibitions.
Publications: annual, Newsletter; books, A History of Montezuma, Sts. John &
Argentine; The Blue River Valley Wonderland; Southern Summit; Roadside
Summit; Summit County's High Altitude Cookbook; Memoirs of Monte-
zuma; Friends in Feathers and Fur; Roadside Summit II; A Strew of
Wonder; Dillon, Denver & The Dam; Small Spectaculars; Men, Mining &
Machines; Stalking the Wild Flower.
Hours & Admission Prices: Guided Tours: June-Sept. daily 10-7; Oct.-May
call for hours. Tours: $3 per person; members no charge. &
Attendance: 33,000 (accurate)
Membership: Individual $25; Family $50, Sponsor $100, Patron $250,
Historian $500.

Brighton

ADAMS COUNTY HISTORICAL SOCIETY MUSEUM, (M),
9601 Henderson Rd., Brighton, CO 80601-8127. Tel.: 303-659-
7103. Fax: 303-659-7988.
E-mail: adamscomuseum@aol.com
Founded: 1974.
Congressional District: 4
Key Personnel: Pres., Cliff Lushbough; Admin., Dixie Pierce; Sec., Casey
Hayes; Treas., Richard Hoffman.
Personnel Profile: Part-Time Paid 2; Part-Time Volunteers 20.
Governing Authority: society; nonprofit. Parent Institution: Adams County
Historical Society. Subsidiary Institution: Adams County Museum. Tax-
exempt: 501(c)(3).
Institution Type/Description: Historical Society Museum.
Collections: prehistoric life & geology; Native American periods up to 1850;
mountain men & explorers, 1820-1840; early western settlement, 1840-
1902; Adams County; World War I; earth science; blacksmith shop; new fire
station replica; newspapers 1892-1936; Japanese inspired garden; period
equipment; fire station replica. Historic Structures: 1930s era Conoco gas
station; 1920s one room schoolhouse; 1887 Victorian House; 1930 red
caboose.
Facilities: over 500-vol. library of art & art related books; 8,000 sq. ft. exhibit
space. Museum-related items for sale.
Activities: guided tours by appointment; temporary & traveling exhibitions;
school tour; field trips; craft shows. Museum Sponsors: book sale in
February & March; Mother's Day Tea in May; Family Heritage Day in
September; Christmas Past & Present in November; Christmas Tea in
December.
Publications: quarterly newsletter, Hi-Story News; craft show flyers; bro-
chures.
Hours & Admission Prices: Jan. 4-Dec. 20 Tues.-Sat. 10-4. Guided Tours: $3
each person. Closed Thanksgiving. &
Attendance: 13,860 (accurate)
Membership: Individual $15; Family $25; Supporting $35; Supporting Busi-
ness $100; Corporation $150; Life & Life Plus $500.

Broomfield

BROOMFIELD DEPOT MUSEUM, 2201 W. 10th Ave., Broom-
field, CO 80020-6713. Mailing Address: 212 Agate Way, Broom-
field, CO 80020-2312. Tel.: 303-466-3663.
Founded: 1981.
Personnel Profile: Part-Time Volunteers 4.
Governing Authority: Tax-exempt.
Institution Type/Description: History Museum.
Collections: Broomfield's history; photographs; personal artifacts; railroad
artifacts & history.
Research Fields: railroad books.
Activities: school tours.
Hours & Admission Prices: Sun. 2-4. No charge; donations accepted. Closed
holidays.

BROOMFIELD VETERANS MEMORIAL MUSEUM, 12 Garden
Center, Ste. 230, Broomfield, CO 80020-7036. Tel.: 303-460-6801.
Founded: 2003.
Key Personnel: Pres. (V), Michael Fellows.
Personnel Profile: Part-Time Volunteers 27.
Institution Type/Description: Military Museum.
Collections: military history; war memorabilia; photographs; uniforms; per-
sonal artifacts.

Hours & Admission Prices: Tues. 7pm-9pm, Sat. 1-4; other times by appointment. No charge; donations accepted.
Attendance: 600
Membership: Annual $20.

Brush

BRUSH AREA MUSEUM AND CULTURAL CENTER, 314 S. Clayton, Brush, CO 80723. Mailing Address: P.O. Box 341, Brush, CO 80723-0341. Tel.: 970-842-9879.
E-mail: engineer@kci.net
Institution Type/Description: History Museum: housed in the former Knearl School building; built in 1910.
Collections: local history & culture; photographs; period furnishings; personal artifacts.
Activities: hands-on exhibits; educational programs; permanent & temporary exhibits.
Hours & Admission Prices: Fri.-Sat. 10-4, Sun. 1-4; other times by appointment.
Membership: Single $10; Family $25; Small Business $50; Corporate $100.

Buena Vista

BUENA VISTA HERITAGE MUSEUM, 506 E. Main St., Buena Vista, CO 81211. Mailing Address: P.O. Box 1414, Buena Vista, CO 81211-1414. Tel.: 719-395-8458.
E-mail: buenavistaheritage@msn.com
Web Site: www.buenavistaheritage.org
Founded: 1974.
Personnel Profile: Full-Time Paid 1; Part-Time Paid 2.
Institution Type/Description: History Museum.
Collections: personal artifacts; clothing; furniture; rocks & minerals; schoolroom; period artifacts.
Hours & Admission Prices: Memorial Day to Sept. 10-5. Adults $5; members no charge. &
Attendance: 3,500 (accurate)
Membership: Individual $20; Family $30; Business $100; Corporate $250.

Burlington

OLD TOWN MUSEUM, 420 S. 14th St., Burlington, CO 80807-2300. Tel.: 719-346-7382; 800-288-1334. Fax: 719-346-7169.
E-mail: cheryl.jacobson@burlingtoncolo.com
Web Site: burlingtoncolo.com
Founded: 1986.
Institution Type/Description: History Museum.
Collections: period furnishings & equipment; books; photographs; dolls; saddles; tools; printing press; Native American artifacts. Old Town Buildings: barn; depot, c.1889; law office; bank; barber shop; cream station; general store; blacksmith shop, built late 1800s; caretakers house; sod house; school house, c.1911; church, c.1921; doll house; manor house, built early 1900s; drug store; leather shop; wood shop; jail house; The Longhorn Saloon; The Burlington Blade; Heritage Hall.
Activities: special events.
Hours & Admission Prices: Mon.-Sat. 9-5, Sun. 12-5. &
Attendance: 15,000 (estimated)

Canon City

DINOSAUR DEPOT MUSEUM, (M), 330 Royal Gorge Blvd., Canon City, CO 81212-6739. Mailing Address: 330 Royal Gorge Blvd. #A, Canon City, CO 81212-6740. Tel.: 719-269-7150. Fax: 719-269-7227.
E-mail: office@dinosaurdepot.com
Web Site: www.dinosaurdepot.com
Founded: 1995.
Congressional District: 5
Key Personnel: Exec. Dir., Jon P. Stone.
Personnel Profile: Full-Time Paid 3; Full-Time Volunteers 1; Part-Time Volunteers 55.
Governing Authority: private; nonprofit organization. Parent Institution: Garden Park Paleontology Society. Tax-exempt: 501(c)(3).
Institution Type/Description: Paleontology Museum.
Collections: Garden Park Fossil Area discoveries; fossils; rocks; minerals; tools & equipment.
Research Fields: Garden Park Fossil area; historical quarry mapping.
Facilities: 700-vol. library of geology & paleontology books; 4,000 sq. ft. exhibit space; classrooms. Museum-related items for sale.
Activities: docent program; formal education programs; guided tours; school loan service; temporary & traveling exhibitions; paleontology field training program.

Publications: bimonthly newsletter, Tracks in Time.
Hours & Admission Prices: Memorial Day to Labor Day daily 9-5; Winter: Wed.-Sun. 10-4. Adults $4, children $2; members no charge. Closed New Year's Day; Thanksgiving; Christmas. &
Attendance: 20,000 (accurate)
Membership: Student $5; Senior Citizens $7.50; Individual $15; Family $25; Supporting $100; Patron $250; Benefactor $500.

MUSEUM OF COLORADO PRISONS, 201 N. 1st St., Canon City, CO 81212-3219. Tel.: 719-269-3015. Fax: 719-269-9148.
E-mail: kellison@prisonmuseum.org
Web Site: www.prisonmuseum.org
Founded: 1987.
Congressional District: 5
Key Personnel: Dir., M. Kay Ellison.
Personnel Profile: Full-Time Paid 2; Part-Time Paid 2.
Institution Type/Description: Prison Museum.
Collections: Colorado Corrections history; hangman's noose used for last execution; confiscated inmate weapons & contraband; gas chamber; photographs.
Facilities: Museum-related items for sale.
Activities: Annual Event: Ghost Walk in May to September.
Publications: quarterly newsletter.
Hours & Admission Prices: May & Sept. to mid-Oct. daily 10-5; Memorial Day to Labor Day daily 8:30-6; mid-Oct. to April Wed.-Sun. 10-5; other times by appointment. Adults $7, seniors 65 & over $6, youth 6-12 $5; discounts to AAA members, Dept. of Corrections & active military. Closed New Year's Day; Easter; Thanksgiving; Christmas. &
Attendance: 15,000 (accurate)
Membership: Individual $30; Family $50; Sustaining $100; Benefactor $250; Corporate $500.

ROYAL GORGE REGIONAL MUSEUM & HISTORY CENTER, (M), 612 Royal Gorge Blvd., Canon City, CO 81212-3751. Mailing Address: P.O. Box 1460, Canon City, CO 81215-1460. Tel.: 719-269-9036.
E-mail: historycenter@canoncity.org
Web Site: www.royalgorgehistory.org
Formerly: Canon City Municipal Museum
Founded: 1928.
Congressional District: 3
Key Personnel: Dir., Susan Ooton; Archivist & Cur., Lisa Studts.
Personnel Profile: Full-Time Paid 4; Part-Time Paid 1; Part-Time Volunteers 15.
Governing Authority: municipal. Parent Institution: City of Canon City. Tax-exempt.
Institution Type/Description: History Museum.
Collections: material & natural history of the Fremont County region; textiles; decorative arts, guns, fossils; minerals. Historic Buildings: 1927 building; 1880s house; 1860s cabin.
Research Fields: local history; genealogy; labor history; environmental history.
Activities: permanent & loan exhibitions.
Publications: brochure; newsletter.
Hours & Admission Prices: Wed.-Sat. 10-4. No charge; donations accepted. Closed major holidays. &

Carbondale

MT. SOPRIS HISTORICAL MUSEUM, 499 Weant Blvd., Carbondale, CO 81623. Mailing Address: P.O. Box 2, Carbondale, CO 81623-0002. Tel.: 970-963-7041.
E-mail: mtsoprishistoricalsociety@yahoo.com
Web Site: mtsoprishistoricalsociety.org
Key Personnel: Dir., Beth White; Pres. (V), Jeannie Perry.
Personnel Profile: Part-Time Paid 1; Part-Time Volunteers 10.
Governing Authority: Tax-exempt.
Institution Type/Description: History Museum: housed in a cabin built by homesteaders in the late 1800s.
Collections: local history & culture; Native American artifacts; photographs; personal artifacts.
Facilities: library.
Activities: research; special events.
Hours & Admission Prices: Call for hours & admission prices.
Attendance: 500 (estimated)
Membership: $35; $50; $75 & up.

Castle Rock

CASTLE ROCK MUSEUM AND HISTORICAL SOCIETY, (M),
420 Elbert St., Castle Rock, CO 80104. Tel.: 303-814-3164.
Institution Type/Description: Historical Society Museum.
Collections: local history, heritage & culture; period artifacts; photographs.
Hours & Admission Prices: Wed.-Fri. 12-5, Sat. 11-4. No charge.

Centennial

GAMMA PHI BETA INTERNATIONAL HEADQUARTERS &
MUSEUM, 12737 E. Euclid Dr., Centennial, CO 80111-6445. Tel.:
303-799-1874. Fax: 303-799-1876.
E-mail: gammaphibeta@gammaphibeta.org
Web Site: www.gammaphibeta.org
Institution Type/Description: History Museum.
Collections: sorority history; personal artifacts; photographs.
Hours & Admission Prices: Mon.-Fri. 8-4:30. No charge.

Central City

CENTRAL CITY OPERA HOUSE ASSOCIATION, 124 Eureka
St., Central City, CO 80427. Mailing Address: 400 S. Colorado
Blvd., Ste. 530, Denver, CO 80246-1253. Tel.: 303-292-6500. Fax:
303-292-4958.
E-mail: vhamlin@centralcityopera.org
Web Site: www.centralcityopera.org
Founded: 1932.
Congressional District: 4
Key Personnel: Chm. Bd., J. Landis Martin; Pres. Bd., Maureen Barker; Gen.
Dir., Pelham G. Pearce; Artistic Dir. Emeritus, John Moriarty.
Personnel Profile: Full-Time Paid 18.
Governing Authority: nonprofit organization. Tax-exempt.
Institution Type/Description: Historic Opera House & Teller House Hotel
Museum; built in 1878. A National Historic Landmark.
Collections: over 30 Victorian era properties; restored parlours & rooms,
Victorian era Hotel. Historical Buildings: 1878 Opera House; 1872 Teller
House; 1876 Williams Stable; 1863 Medical Building; 1869 Weckbaugh
House; 1884 McFarlane House.
Research Fields: early theater & western Americana.
Facilities: theatre. Museum-related gifts for sale.
Activities: summer opera festival, guided tours, lectures, arts festivals, perma-
nent exhibitions. Annual Event: Summer Festival June to August.
Publications: Central City Opera High Notes; Central City Opera Season
Program.
Hours & Admission Prices: Tues.-Sun. 10-5. Tours: $6. Performance Tickets:
$20-$110. Closed major holidays. &
Attendance: 20,000 (estimated)
Membership: Associates $100; Patron's Circle $500; Artist's Circle $1,000;
Conductor's Circle $2,500; Director's Circle $5,000; President's Circle
$10,000; Chairman's Circle $25,000; Eureka Circle $50,000; Guarantor
$75,000; Grand Guarantor $100,000 & up.

GILPIN HISTORY MUSEUM, 228 E. High St., Central City, CO
80427. Mailing Address: P.O. Box 247, Central City, CO 80427-
0247. Tel.: 303-582-5283.
Web Site: www.gilpinhistory.org
Founded: 1971.
Congressional District: 4
Key Personnel: Exec. Dir., James Prochaska.
Personnel Profile: Part-Time Paid 1; Part-Time Volunteers 5.
Governing Authority: society; nonprofit. Affiliated with Gilpin County Histori-
cal Society. Branch Museum: The Thomas House. Tax-exempt: 501(c)(3).
Institution Type/Description: Historical Society Museum: housed in c.1870
Gilpin County Public School.
Collections: personal items, furniture and tools of pioneers and early residents
of Gilpin county; living room; parlor; kitchen; school room; clothing
fashions; general store; carriage collection; dolls; doctor's office; pharmacy;
law office; bank display; assay office; barber shop; sheriff's office; saloon;
tools; mineral exhibits; fire fighting equipment; maps and photos of the
railroad and tram; fraternal organizations badges and paraphernalia; opera
house. Historic Buildings: 1870 2-story stone school; 1874 Greek Revival
house; mine shaft building.
Research Fields: Gilpin county cemeteries; genealogy; mining.
Activities: guided tours; permanent exhibitions.
Publications: brochure, History of Building and Society; Mining Gold to
Mining Wallets: Central City 1854-1999; Little Kingdom of Gilpin.
Hours & Admission Prices: Memorial Day to Labor Day daily 11-4. Adults $5;
discounts to AAM members; senior citizens, children under 12 & members
no charge.

Attendance: 1,200 (accurate)
Membership: Individual $20; Family $35; Patron $50; Business Associate
$100.

Clark

HAHNS PEAK AREA HISTORICAL SOCIETY AND SCHOOL
HOUSE, Main St., Hahns Peak Village, Clark, CO 80428-9999.
Mailing Address: P.O. Box 803, Clark, CO 80428-0803. Tel.:
970-879-7291.
Web Site: www.hahnspeakhistoric.com
Founded: 1972.
Congressional District: 4
Key Personnel: Pres. (V) & Cur., Marge Eardley; Treas., Shelley Stanford.
Personnel Profile: Part-Time Volunteers 5.
Governing Authority: nonprofit organization. Parent Institution: Hahns Peak
Area Historical Society. Tax-exempt: 501(c)(3).
Institution Type/Description: History Museum: located on the site of the first
permanent settlement in Routt County & N.W. Colorado.
Collections: school furnishing including desks; piano; books; mining artifacts;
furniture. Historic Buildings: schoolhouse; Wither cabin.
Research Fields: local history; mining methods; wildlife & native plants.
Facilities: 100-vol. library of original school texts & local history books &
papers to be used on premises.
Activities: lectures; films.
Publications: book, Historic Hahns Peak.
Hours & Admission Prices: June-Sept. daily 12-4. No charge; donations
accepted.
Attendance: 1,500 (estimated)
Membership: Annual $5; Life $100.

Colorado Springs

AMERICAN NUMISMATIC ASSOCIATION MONEY MU-
SEUM, 818 N. Cascade Ave., Colorado Springs, CO 80903-3279.
Tel.: 719-632-2646. Fax: 719-634-4085.
E-mail: museum@money.org
Web Site: www.money.org
Formerly: Museum of the American Numismatic Association
Founded: 1967.
Congressional District: 5
Key Personnel: Pres., Tom Hallenbeck; Dir. Operations, Kim Klick; Museum
Dir., Tiffanie Bueschel; Controller, Carol Shuman; Treas., Larry Baber;
Cur., Douglas Mudd; Mktg. & Education Advisor, Jay Beeton; Library &
Communications Dir., RyAnn Scott; Collections Mgr., Andrew Dickes.
Personnel Profile: Full-Time Paid 33; Part-Time Paid 1; Part-Time Volunteers
9.
Governing Authority: private; nonprofit organization. Parent Institution:
American Numismatic Association. Tax-exempt: 501(c)(3).
Institution Type/Description: Numismatic Museum.
Collections: coins, medals, tokens, paper money and related materials from
c.700 B.C.
Research Fields: numismatics.
Facilities: 50,000-vol. library of numismatic books & tapes; 5,000 sq. ft.
exhibit space. Museum-related items for sale.
Activities: hobby workshops; lectures; participatory, loan, temporary & trav-
eling exhibitions; training programs for professional museum workers.
Annual Events: summer & early spring conventions; summer seminars.
Publications: monthly magazine, The Numismatist; ANA journal.
Hours & Admission Prices: Tues.-Sat. 10:30-5; groups by appointment. Adults
$5, seniors, active military & students $4; discounts to groups; members &
3rd Sat. each month no charge. Closed federal holidays. &
Attendance: 23,000 (accurate)
Membership: Junior $26; Senior $41; Regular $46.

CHEYENNE MOUNTAIN ZOOLOGICAL PARK, 4250 Chey-
enne Mountain Zoo Rd., Colorado Springs, CO 80906-5755. Tel.:
719-633-9925. Fax: 719-633-2254.
E-mail: info@cmzoo.org
Web Site: www.cmzoo.org
Founded: 1926.
Congressional District: 5
Key Personnel: Chm., Vic Andrews; Pres. & C.E.O., Bob Chastain; General
Cur., Tracy Thessing; Registrar, Randy Barker; Education Cur., Melanie
Sorensen; Horticulture Cur., Frank Haas; Dir. Conservation, Dr. Della
Garelle; Vice Pres., Tracey Gazibara; Public Rels. & Special Events Mgr.,
Erica Meyer; Dir. Operations, Mike Sukel; Dir. Mktg., Jenny Hillard; Guest
Svcs. Mgr., Jane Majeske; Museum Shop Mgr., Todd Langfield; Head
Veterinarian, Dr. Liza Dadone; Dir. Finance & Human Resources, Kathy
French.

Personnel Profile: Full-Time Paid 82; Part-Time Paid 11; Part-Time Volunteers 200; Interns 5.
Volunteer Hours: 24,000
Operating Expenses: 6,600,000
Operating Income: 8,200,000
Governing Authority: society. Parent Institution: Cheyenne Mountain Zoological Society. Tax-exempt: 501(c)(3).
Institution Type/Description: Zoo.
Collections: mammals; birds; reptiles; amphibians; breeding herd of reticulated giraffe; breeding colony of orangutan; felines & hoofed mammals; tram; period carousel.
Research Fields: animal reproduction.
Facilities: 150-seat education center. Gift items for sale.
Activities: guided tours; films; formally organized education programs for children; docent program or council; permanent exhibitions; sky ride.
Publications: Cheyenne Mountain Zoological Park Annual Report; quarterly newsletter.
Hours & Admission Prices: May to Labor Day: daily 9-6; Sept.-May daily 9-5. May to Labor Day: adults 12-64 $17.25, seniors 65 & over $15.25, children 3-11 $12.25; Sept.-May adults 12-64 $14.25, seniors 65 & over $12.25, children 3-11 $10.25; discounts to AZA & AAA members & military; members & children 2 & under no charge. &
Attendance: 575,000 (accurate)
Membership: Individual Plus $89.25; Family & Grandparent $147.50; Family Plus $190.50; Conservator $387.50; Patron $757.50; Patron Gold $1,507.50.

*** COLORADO SPRINGS FINE ARTS CENTER (TAYLOR MUSEUM), (M),** 30 W. Dale St., Colorado Springs, CO 80903-3210. Tel.: 719-634-5581. Fax: 719-634-0570.
E-mail: info@csfineartscenter.org
Web Site: www.csfineartscenter.org
Founded: 1936.
Congressional District: 5
Key Personnel: Pres. & C.E.O., Sam Gappmayer; Museum Dir., Blake Milteer; C.F.O., Debbie Linster; C.O.O., Kari Torgerson; Dir. Devel., Tom Jackson; Dir. Bemis School of Art, Tara Thomas; Dir. Performing Arts, Scott R.C. Levy; Dir. Mktg. & Public Rels., Charlie Snyder.
Personnel Profile: Full-Time Paid 46; Part-Time Paid 85; Part-Time Volunteers 225.
Governing Authority: nonprofit organization. Parent Institution: Colorado Springs Fine Arts Center. Subsidiary Institution: Taylor Museum; FAC Modern, 121 S. Tejon St., Ste. 100, Colorado Springs, CO. Tax-exempt: 501(c)(3).
Institution Type/Description: Art Museum.
Collections: historic & contemporary art & material culture of the Southwest United States including major Hispanic & Native American collections; 20th-century American & contemporary art.
Research Fields: art & culture of the Southwest United States; 20th century American & contemporary art.
Facilities: 450-seat theater; 48,000 sq. ft. exhibit space; classrooms; sculpture courtyard; art school; 140-seat recital room; conference facilities; restaurant. Museum-related items for sale.
Activities: docent tours; lectures; films; gallery talks; art classes; performing arts programming; inter-museum loan; permanent, temporary & traveling exhibitions; repertory theatre company.
Publications: Artsfocus; exhibition catalogs; scholarly publications on Southwest studies; annual report.
Hours & Admission Prices: Tues.-Sun. 10-5. Adults $10, senior citizens, students & military with ID $8.50; children 4 & under & museum employees with ID no charge. Closed holidays. &
Attendance: 200,000 (estimated)
Membership: Individual $45; Dual $55; Family $60; Contributing $110.

COLORADO SPRINGS MUSEUM, 215 S. Tejon St., Colorado Springs, CO 80903-2206. Tel.: 719-385-5990. Fax: 719-385-5645.
E-mail: cosmuseum@springsgov.com
Web Site: www.cspm.org
Founded: 1937.
Congressional District: 5
Key Personnel: Dir., Matt Mayberry; Chm. (V), Kathie Walker; Archivist, Leah Davis Witherow; Cur. & Registrar, David Ryan; Museum Shop Mgr., Carol Denning.
Personnel Profile: Full-Time Paid 6; Part-Time Paid 2; Part-Time Volunteers 190; Interns 3.
Governing Authority: city. Tax-exempt.
Institution Type/Description: History Museum: housed in the former El Paso County Courthouse, 1903.
Collections: history & culture of the Pikes Peak region; period artifacts; Van

Briggle art pottery; regional art; Native American artifacts; mining & agricultural history.
Research Fields: regional history & art.
Facilities: 200-seat auditorium; meeting rooms.
Activities: guided & self-guided tours; permanent & temporary exhibits; lecture series; special events.
Publications: quarterly newsletter; gallery guides; educational materials; pamphlets; monographs.
Hours & Admission Prices: Tues.-Sat. 10-4. No charge; donations accepted. Closed holidays; Thanksgiving week. &
Attendance: 55,818 (accurate)
Membership: Senior $15; Individual $25; Family $30; Sustaining $50; Business $250; Life $1,000.

DR. LESTER L. WILLIAMS FIRE MUSEUM, 375 Printers Pkwy., Colorado Springs, CO 80901. Mailing Address: P.O. Box 119, Colorado Springs, CO 80901-0119. Tel.: 719-385-5950.
E-mail: rayjturner@q.com
Web Site: williamsfiremuseum.com
Founded: 2001.
Institution Type/Description: History Museum.
Collections: fire memorabilia; firefighting history; personal artifacts; photographs.
Publications: newsletter.
Hours & Admission Prices: Mon.-Fri. 8-5. No charge; donations accepted.
Attendance: 2,000 (estimated)

EL POMAR FOUNDATION CARRIAGE MUSEUM, 10 Lake Cir., Colorado Springs, CO 80906-4201. Tel.: 719-577-7065.
Web Site: elpomar.org
Founded: 1941.
Key Personnel: C.E.O., Robert J. Hilbert; Cur., Jason J. Campbell.
Personnel Profile: Full-Time Paid 1; Part-Time Paid 2.
Governing Authority: nonprofit. Parent Institution: El Pomar Foundation. Tax-exempt.
Institution Type/Description: Transportation Museum.
Collections: period carriages; covered wagons; stage coaches; early race cars & autos.
Hours & Admission Prices: Mon.-Sat. 9-5, Sun. 1-5. No charge. Closed New Year's Day; Easter; Thanksgiving; Christmas. &
Attendance: 9,000 (estimated)

GALLERY OF CONTEMPORARY ART, UNIVERSITY OF COLORADO, COLORADO SPRINGS, 1420 Austin Bluffs Pkwy., Colorado Springs, CO 80918-3733. Tel.: 719-255-3567. Fax: 719-262-3183. TDD: 262-3621.
E-mail: gallery@uccs.edu
Web Site: www.galleryuccs.org
Founded: 1981.
Congressional District: 5
Key Personnel: Interim Dir., Caitlin Green.
Personnel Profile: Full-Time Paid 1; Part-Time Paid 3; Part-Time Volunteers 2; Interns 2.
Governing Authority: university. Parent Institution: University of Colorado. Tax-exempt: 501(c)(3); 170(b)(A).
Institution Type/Description: Art Museum & Center.
Collections: works by contemporary artists.
Research Fields: contemporary art; art history.
Facilities: classrooms.
Activities: guided tours; lectures; concerts; educational programs; docent program; education programs for undergraduate college students affiliated with University of Colorado; training programs for professional museum workers; museum studies; group exhibitions with a specific theme or medium; loan & traveling exhibitions.
Hours & Admission Prices: Tues.-Fri. 12-6; other times by appointment. No charge; donations accepted. Closed major holidays. &
Attendance: 28,000 (estimated)
Membership: Student & Senior $10; Individual $20; Friend $35-$99; Patron $100-$499; Donor $500-$999; Founder $1,000-$2,499; Benefactor $2,500.

GARDEN OF THE GODS VISITOR & NATURE CENTER, 1805 N. 30th St., Colorado Springs, CO 80904-1247. Tel.: 719-634-6666. Fax: 719-634-0094.
Institution Type/Description: Nature Center.
Collections: hands-on exhibits; local history; geology; plants; wildlife.
Facilities: visitor center; cafe. Museum-related items for sale.
Activities: lectures; special events; scout & school programs; guided tours; picnic area; hiking.

Hours & Admission Prices: Park: May-Oct. daily 5am-11pm; Nov.-April daily 5am-9pm. Visitor Center: Memorial Day to Labor Day daily 8-8; Winter: daily 9-5. No charge.

GHOST TOWN MUSEUM, 400 S. 21st St., Colorado Springs, CO 80904-3755. Tel.: 719-634-0696.
E-mail: history@ghosttownmuseum.com
Institution Type/Description: History Museum.
Collections: old west town history; Colorado heritage; period furnishings; hands-on exhibits; structures.
Facilities: Museum-related items for sale.
Activities: hands-on activities; film.
Hours & Admission Prices: June-Aug. Mon.-Sat. 9-6, Sun. 10-6; Sept.-May daily 10-5. Adults $7.50, children 6-16 $5; children under 6 no charge. &

MAY NATURAL HISTORY MUSEUM AND MUSEUM OF SPACE EXPLORATION, John May Museum Center, 710 Rock Creek Canyon Rd., Colorado Springs, CO 80926-9799. Tel.: 719-576-0450; 800-666-3841. Fax: 719-576-3644.
E-mail: maymuseum2001@yahoo.com
Web Site: www.maymuseum-camp-rvpark.com
Founded: 1941.
Congressional District: 5
Key Personnel: C.E.O. & Pres. (V), John M. May; Museum Shop Mgr., Louise N. Steer.
Personnel Profile: Full-Time Paid 1; Full-Time Volunteers 1; Part-Time Paid 5; Part-Time Volunteers 2.
Governing Authority: individual operation. Parent Institution: John May Museum Center. Tax-exempt: 501(c)(3) & 509(a)(2).
Institution Type/Description: Natural History, Entomology & Space Exploration Museum.
Collections: entomological specimens; primitive artifacts; geological; space photography; models; NASA films; New Guinea artifacts.
Research Fields: entomology.
Facilities: 250-vol. library of entomology and astronomy; nature center; nature trails. Curios, Indian artifacts & clothing for sale.
Activities: guided tours; permanent exhibitions; hiking; public & private school programs.
Publications: annual brochures.
Hours & Admission Prices: May-Sept. daily 9-6; Oct.-April by appointment. Adults $6, seniors 60 & over $5, children 6-12 $3; discounts to schools, senior citizens & AAA members; children under 6 with family no charge. &
Attendance: 30,000 (estimated)

MCALLISTER HOUSE MUSEUM, 423 N. Cascade Ave., Colorado Springs, CO 80903-3391. Tel.: 719-635-7925.
E-mail: curator@mcallisterhouse.org
Web Site: www.mcallisterhouse.org
Founded: 1961.
Congressional District: 5
Key Personnel: Chm., Mary Anne Sehorn; Treas., Barbara Rasmussen; Cur., Owanah Wick.
Personnel Profile: Part-Time Paid 1; Part-Time Volunteers 30.
Governing Authority: society. Parent Institution: The National Society of The Colonial Dames in the State of Colorado. Tax-exempt.
Institution Type/Description: Historic House: c.1873 Major Henry McAllister Home.
Collections: late Victorian era furnishings; Downing Gothic Style Cottage.
Facilities: banquet facilities; limited meeting space.
Activities: guided tours; lecture series. Museum Sponsors: Victorian Ice Cream Social; Victorian Christmas.
Publications: book, Henry McAllister: Colorado Pioneer; book, McAllister House Museum; newsletter.
Hours & Admission Prices: Feb.-April & Sept.-Dec. Thurs.-Sat. 10-4; May-Aug. Tues.-Sat. 10-4. Adults $5, seniors, students & members $4, children 6-12 $3; children under 6 no charge. Closed major holidays.
Attendance: 3,000 (estimated)
Membership: Individual $15; Family $25; Sponsor $50; Patron $100; Life $500.

OLD COLORADO CITY HISTORICAL SOCIETY, One South 24th St., Colorado Springs, CO 80904-3319. Tel.: 719-636-1225. Facebook: Old Colorado City Historical Society.
E-mail: info@occhs.org
Web Site: www.occhs.org
Founded: 1995.
Congressional District: 5
Key Personnel: Pres., Sharon Swint.

Governing Authority: bd. of directors. Tax-exempt.
Institution Type/Description: Historical Society Museum.
Collections: area history & culture; photographs; personal artifacts.
Research Fields: early El Paso county history & culture.
Facilities: research library. Museum-related items for sale.
Activities: research library; rental facilities; monthly educational programs; tours.
Publications: newsletter, West Word.
Hours & Admission Prices: Tues.-Sat. 11-4. Programs: adults $5; members no charge. Closed holidays. &
Membership: Single & Student $25; Family $35; Business $75; Life $250; Life Plus One $300.

PRO RODEO HALL OF FAME & MUSEUM OF THE AMERICAN COWBOY, 101 Pro Rodeo Dr., Colorado Springs, CO 80919-2396. Tel.: 719-528-4761. Fax: 719-548-4874. Facebook: Pro Rodeo Hall of Fame & Museum of the American Cowboy.
E-mail: prorodeo@prorodeo.com
Web Site: www.prorodeo.com
Founded: 1979.
Congressional District: 5
Key Personnel: C.E.O., Karl Stressman; Dir., Kent L. Sturman; Mktg. & Events Coord., Sara Tadken.
Personnel Profile: Full-Time Paid 2; Part-Time Paid 3.
Governing Authority: Parent institution: Professional Rodeo Cowboys Association, Inc.; nonprofit organization. Tax-exempt: 501(c)(3).
Institution Type/Description: Rodeo & Cowboy History Museum.
Collections: sculptures; paintings; saddles; chaps; ropes; boots; clothing & artifacts tracing history of rodeo 1865-present; memorabilia, trophies & artifacts of champions & notables; buckles; books; periodicals; Wild West show memorabilia; rodeo programs; photographs.
Major Exhibits: Tribute to Chris LeDoux, 1/14-12/14.
Research Fields: rodeo history; ranching & livestock history.
Facilities: 1,500-vol. library of rodeo records & history available for research on premises; two electronic multi-media theaters; 40,000 sq. ft. exhibit space; garden. Prints & other museum-related items for sale.
Activities: guided tours; lectures; films; loan & permanent exhibitions; demonstrations. Museum Sponsors: Pro Rodeo Week.
Publications: brochures; pamphlets; newsletter.
Hours & Admission Prices: Summer: daily 9-5; Winter: Wed.-Sun. 9-5. Adults $6, seniors $5, children 6-12 $3; discount to groups & military; members & children 5 & under no charge. Closed New Year's Day; Easter; Thanksgiving; Christmas. &
Attendance: 25,000 (accurate)
Membership: Pro Rodeo Fan Zone: Bronze $75; Silver $250; Gold $600; Platinum $1,600.

ROCK LEDGE RANCH HISTORIC SITE, 30th St. & Gateway Rd., Colorado Springs, CO 80904. Mailing Address: 1401 Recreation Way, Colorado Springs, CO 80905-1024. Tel.: 719-578-6777. Fax: 719-578-6965.
Key Personnel: Mgr., Andy Morris.
Personnel Profile: Full-Time Paid 2; Part-Time Paid 2; Part-Time Volunteers 50.
Governing Authority: city. Tax-exempt.
Institution Type/Description: Historic Site: listed on the National Register of Historic Places.
Collections: American Indian history; late 1860s homesteading. Historic Houses: 1880 farm house; 1907 estate house.
Facilities: Museum-related items for sale.
Activities: special events; horse drawn equipment demonstrations; musical performances; lectures. Annual Events: Fall Harvest Festival; Holiday Evenings Celebration.
Hours & Admission Prices: June to Labor Day Wed.-Sat. 10-5; Winter: call for special events. Adults $6, seniors 55 & over and students 13-18 $4, youth 6-12 $2; discounts to groups; children under 6 no charge. &
Attendance: 74,000 (accurate)
Membership: Senior $10; Senior Couple $15; Individual $20; Family $30; Contributing $50; Supporting $100.

SEVEN FALLS, 2850 S. Cheyenne Canyon Rd., Colorado Springs, CO 80906-2919. Tel.: 719-632-0765. Fax: 719-632-0781.
Institution Type/Description: Natural History Museum.
Collections: local history; wildlife; ecology.
Facilities: nature trails.
Activities: hiking.
Hours & Admission Prices: Temporarily closed.

U.S. OLYMPIC VISITOR CENTER, 1 Olympic Plaza, Bldg. 6, Colorado Springs, CO 80909-5781. Tel.: 719-632-5551.
Web Site: www.olympic-usa.org
Institution Type/Description: Sports Museum.
Collections: Olympic history; hands-on exhibits; Hall of Fame; photographs.
Facilities: Museum-related items for sale.
Activities: guided tours.
Hours & Admission Prices: Mon.-Sat. 9-6, Sun. 10-5. No change.

✳ **WESTERN MUSEUM OF MINING & INDUSTRY, (M),** 225 N. Gate Blvd., Colorado Springs, CO 80921-3002. Tel.: 719-488-0880; 800-752-6558. Fax: 719-488-9261.
E-mail: info@wmmi.org
Web Site: www.wmmi.org
Founded: 1970.
Congressional District: 5
Key Personnel: Dir., Richard Sauers; Chm. (V), Jeff Campbell; Museum Shop Mgr., Christine Nestlerode.
Personnel Profile: Full-Time Paid 3; Part-Time Paid 3; Part-Time Volunteers 20; Interns 1.
Governing Authority: nonprofit. Tax-exempt: 501(c)(3).
Institution Type/Description: Mining & Technology History Museum.
Collections: metal mining & milling machinery; steam, water & electrically powered prime movers; electrical equipment; machine tools; exhibit dealing with western hard rock mining; placer mining & milling; social history of miners, mining districts & art; mine reclamation & environmental science; operating reconstructed stamp mill; locomotives; steeple-compound steam stamp & walking beam engine on outdoor display. Historic House: 19th-century farmhouse & barn; blacksmith shop & hoist house; stamp mill.
Research Fields: metallurgy; industrial archaeology of the American West; mining engineering.
Facilities: 7,000-vol. library of texts, original histories, trade catalogs & serials relating to electrical, mechanical, hydraulic, civil & mine engineering; 12,000 sq. ft. exhibit space; 27 acres; picnic area.
Activities: guided tours; classes; lectures; permanent & temporary exhibitions.
Publications: quarterly newsletter; annual report; catalogue; Mining History & Technology Series Vol. I: A Concise History of Mine Hoisting.
Hours & Admission Prices: Mon.-Sat. 9-4. Adults $8, military $7, seniors 60 & over and students 13-18 $6, children 4-12 $4; discounts to AAM members; children under 3 & members no charge. ♿
Attendance: 40,000 (estimated)
Membership: Individual $30; Dual $40; Family $50; Centennial $100; Silver Medallion $250; Gold Medallion $500; Heritage $1,000.

WORLD FIGURE SKATING MUSEUM & HALL OF FAME, (M), 20 First St., Colorado Springs, CO 80906-3624. Tel.: 719-635-5200. Fax: 719-635-9548.
Web Site: www.worldskatingmuseum.org
Founded: 1965.
Key Personnel: Dir. Communications, Barb Reichert; Museum Admin. Coord. & Museum Shop Mgr., Karen Cover.
Personnel Profile: Full-Time Paid 1; Part-Time Paid 2; Part-Time Volunteers 1.
Governing Authority: nonprofit organization. Parent Institution: U.S. Figure Skating. Tax-exempt: 501(c)(3).
Institution Type/Description: Sports Museum.
Collections: figure skating history from 800 AD to present; skating in art, 16th- to 20th century; prints; films; photographs; video archive; official repository for figure skating records & memorabilia; costumes; trophies; US & World Hall of Fame.
Research Fields: skating in art as part of European & American cultures; history of the development of the skate blade over 1,200 years; study of U.S., World & Olympic champions & medalists since 1908.
Facilities: library.
Activities: films; guided tours; loan, temporary & traveling exhibitions; special event fundraising. Annual Event: U.S. & World Halls of Fame induction.
Publications: Figure Skating: A History.
Hours & Admission Prices: Tues.-Fri. 10-4. Adults $5, seniors 60 & over $3; discounts to military & AAA members; children 5 & under and U.S. Figure Skating members no charge. Closed New Year's Eve & Day; Memorial Day; Independence Day; Labor Day; Thanksgiving; Christmas; postal & bank holidays. ♿
Attendance: 7,500 (estimated)

Cortez

CORTEZ CULTURAL CENTER, 25 N. Market St., Cortez, CO 81321-3212. Tel.: 970-565-1151. Fax: 970-565-4075. Facebook: Cortez Cultural Center.
E-mail: info@cortezculturalcenter.org

Web Site: www.cortezculturalcenter.org
Founded: 1987.
Congressional District: 3
Key Personnel: Exec. Dir., Anne Beach; Pres. (V), Shelby Smith.
Personnel Profile: Full-Time Paid 1; Part-Time Volunteers 50; Interns 1.
Institution Type/Description: Native American Museum.
Collections: Native American history, culture & life; personal artifacts; photographs; paintings; mural.
Facilities: cultural center; preserve.
Activities: storytellers; lectures; special events; music & cultural programs. Museum Sponsors: Native American Dances in summer.
Hours & Admission Prices: May & Sept.-Oct. daily 10-6; June-Aug. daily 10-10; Nov.-April daily 10-5. No charge; donations accepted.
Attendance: 8,000 (estimated)
Membership: Individual $25; Family $30.

CROW CANYON ARCHAEOLOGICAL CENTER, 23390 County Rd. K, Cortez, CO 81321-9408. Tel.: 970-565-8975. Fax: 970-565-4859.
E-mail: gprior@crowcanyon.org
Web Site: www.crowcanyon.org
Founded: 1983.
Key Personnel: C.E.O., Ricky Lightfoot, Ph.D.; Chm. (V), Sue Anschutz-Rodgers.
Personnel Profile: Full-Time Paid 43; Part-Time Paid 5; Part-Time Volunteers 3; Interns 10.
Governing Authority: private; nonprofit organization. Tax-exempt: 501(c)(3).
Institution Type/Description: Archaeological Center.
Collections: archaeological artifacts.
Hours & Admission Prices: March-Oct. Mon.-Fri. 8:30-5. No charge. Day program: June-Aug. adults $50, children $25. ♿
Attendance: 3,000 (estimated)
Membership: Youth under 18 $30; Adult $50; Family $75; Research Assistant $100; Research Fellow $250; Research Benefactor $500. Chairman's Council $1,000; Director's Circle $2,500; President's Circle $5,000.

NOTAH DINEH MUSEUM, 345 W. Main, Cortez, CO 81321-3132. Tel.: 800-444-2024.
E-mail: notah@fone.net
Key Personnel: Owner, Greg Leighton
Institution Type/Description: Native American Museum.
Collections: Navajo rugs; Native American weavings; Native American art including hand carved wooden Kachinas; sculpture; sand paintings; beaded baskets; cradle boards; moccasins; beadwork; 14k & sterling silver jewelry.
Facilities: Museum-related items for sale.
Hours & Admission Prices: Mon.-Sat. 9-6:30.

Craig

MUSEUM OF NORTHWEST COLORADO, 590 Yampa Ave., Craig, CO 81625-2612. Tel.: 970-824-6360. Fax: 970-824-1098.
Web Site: www.museumnwco.org
Founded: 1964.
Congressional District: 3
Key Personnel: Dir., Dan Davidson; Pres. (V), Delaine Voloshin; Asst. Dir., Jan Gerber.
Personnel Profile: Full-Time Paid 2; Part-Time Paid 4; Part-Time Volunteers 24.
Governing Authority: county. Parent Institution: Moffat County. Tax-exempt.
Institution Type/Description: Local History Museum.
Collections: Indian artifacts; local history; home of the Cowboy & Gunfighter Museum; wildlife photography; largest stretched canvas oil painting on the western slope of Colorado, depicting Craig 1895.
Research Fields: local buildings; local genealogy.
Publications: museum brochure.
Hours & Admission Prices: Mon.-Sat. 9-5. No charge; donations accepted. ♿
Attendance: 16,500 (accurate)

WYMAN MUSEUM, 94350 E. Hwy. 40, Craig, CO 81625. Mailing Address: P.O. Box 339, Craig, CO 81626-0339. Tel.: 970-824-6346. Fax: 970-824-5890. Facebook: Lou Wyman.
E-mail: wymanmuseum@earthlink.net
Web Site: wymanmuseum.com
Founded: 2006.
Key Personnel: Co Dir., Lou Wyman; Co Dir., Paula Wyman; Chm. (V), Al Shepherd; Museum Shop Mgr., Nicky Boulger.
Personnel Profile: Full-Time Paid 1; Part-Time Paid 2.
Operating Expenses: 130,000
Institution Type/Description: Living History Museum.

Collections: chain saws; personal artifacts; trophies; farm equipment; live animals; WWI memorabilia; M47 Patton tank; WWII memorabilia; blacksmith shop; dental equipment; general store.
Facilities: picnic area; rental facilities. Museum-related items for sale.
Activities: hay rides; fishing area. Museum Sponsors: Wymans Wacky Winter Festival in February; Sheep Wagon Days in September; Pumpkin Patch.
Hours & Admission Prices: Mon.-Fri. 9-5, Sat.-Sun. 11-4. No charge; donations accepted. Closed major holidays. &
Attendance: 10,395 (accurate)

Creede

CREEDE UNDERGROUND MINING MUSEUM, 503 Forest Service Rd. #9, Creede, CO 81130. Mailing Address: P.O. Box 422, Creede, CO 81130-0422. Tel.: 719-658-0811.
E-mail: creedeminingmuseum@hotmail.com
Key Personnel: Museum Shop Mgr., Ricky Brown; Museum Shop Mgr., Dianna Brown.
Personnel Profile: Full-Time Paid 1; Part-Time Paid 3.
Institution Type/Description: Mining Museum.
Collections: mining history; mining equipment.
Facilities: Museum-related items for sale.
Hours & Admission Prices: Memorial Day to Labor Day daily 10-4; Fall & Spring Mon.-Fri. 10-3. Tour: $15 per person.

Crested Butte

CENTER FOR THE ARTS PIPER GALLERY, 606 6th St., Crested Butte, CO 81224-1819. Mailing Address: P.O. Box 1819, Crested Butte, CO 81224-1819. Tel.: 970-349-7487. Fax: 970-349-5626.
E-mail: jenny@crestedbuttearts.org
Web Site: www.crestedbuttearts.org
Founded: 1986.
Key Personnel: Exec. Dir., Jenny Birnie; Pres. (V), Gail Digale.
Personnel Profile: Full-Time Paid 3; Part-Time Paid 11; Part-Time Volunteers 80.
Governing Authority: Tax-exempt.
Institution Type/Description: Art Gallery.
Collections: works by local & regional artists; photographs; pottery; oil pastels.
Facilities: auditorium.
Activities: special events.
Hours & Admission Prices: Call for hours.
Attendance: 20,500 (estimated)

CRESTED BUTTE MOUNTAIN HERITAGE MUSEUM, (M), 331 Elk Ave., Crested Butte, CO 81224. Mailing Address: P.O. Box 2480, Crested Butte, CO 81224-2480. Tel.: 970-349-1880. Fax: 970-349-1885.
E-mail: museum@crestedbutte.cc
Web Site: www.crestedbuttemuseum.com
Founded: 1991.
Key Personnel: C.E.O., Gio Cunningham.
Personnel Profile: Part-Time Paid 3; Part-Time Volunteers 25.
Institution Type/Description: History Museum.
Collections: history of Crested Butte; Ute Indians; immigrant miners; ranchers; skiers; mountain bikers; personal artifacts; coal stove; period furniture & artifacts; photographs; model railroad; mining diorama; textiles; history of skiing; Mountain Bike Hall of Fame.
Activities: programs; special events. Museum Sponsors: Annual Black & White Ball in July.
Publications: walking tour brochure.
Hours & Admission Prices: Summer: daily 10-8; Winter: daily 12-6; Spring & Fall: call for hours. Adults $4, inactive duty military $3; discounts to AAM & ICOM members; AASLH members & active duty military no charge. &
Attendance: 27,000 (accurate)
Membership: Individual $15; Family $30; Business $50; Supporter $100; Guarantor $200; Sponsor $500; Benefactor $1,000.

MOUNTAIN BIKE HALL OF FAME, 122 Teocalli Rd., Crested Butte, CO 81224. Tel.: 970-349-6817.
Web Site: www.mtnbikehalloffame.com
Institution Type/Description: Sports Museum.
Collections: mountain bike history; Hall of Fame inductees; mountain bikes; photographs.
Activities: Annual Event: Induction Ceremony.
Hours & Admission Prices: June-Sept. daily 10-8; Dec.-March daily 12-6. Admission $4.

Cripple Creek

CRIPPLE CREEK DISTRICT MUSEUM, INC., 500 E. Bennett Ave., Cripple Creek, CO 80813. Mailing Address: P.O. Box 1210, Cripple Creek, CO 80813-1210. Tel.: 719-689-9540 & 2634.
E-mail: ccdmuseum@aol.com
Web Site: www.cripple-creek.org
Founded: 1953.
Congressional District: 6
Key Personnel: Pres., Steve Mackin; Dir., Jan Collins; Vice Pres., John Bowman; Bd. Member, Milford Ashworth; Bd. Member, Ike Hern; Bd. Member, Bonnie Mackin; Bd. Member, John Sharpe; Bd. Member, Georganna Peiffer; Bd. Member, Art Tremayne; Museum Shop Mgr., Johnna Luck.
Personnel Profile: Full-Time Paid 2; Part-Time Paid 4; Part-Time Volunteers 1.
Governing Authority: nonprofit organization. Tax-exempt: 990-A.
Institution Type/Description: Historic Buildings.
Collections: pioneer artifacts; mining & geological displays; Victorian furniture; assay office; paintings & art objects; photographic history gallery; photographs; local newspapers 1896-present; gold ore. Historic Buildings: 1895 Midland Jerminal Railroad Depot; 1893 Colorado Trading & Transfer Co.; 1890s Assay Office; 2 cabins.
Research Fields: mining & local history.
Facilities: 50-vol. collection of old books available for use by appointment. Historical books & other museum-related items for sale.
Activities: guided tours; demonstrations; permanent exhibitions.
Publications: quarterly newsletter, Friends of the Museum.
Hours & Admission Prices: May-Oct. 15 daily 10-5; Oct. 16-April Fri.-Sun. 10-4. Adults $5, locals $4, military, senior citizens & children 7-12 $3; discounts to groups; members and children 6 & under no charge. Closed Thanksgiving; Christmas. &
Attendance: 15,000 (accurate)
Membership: Friends of the Museum: Seniors 10% discount on all levels; Single Jack $50; Double Jack $75; Nugget $150; Sylvanite $250; High Grade $500.

MOLLIE KATHLEEN GOLD MINE, 9388 U.S. Hwy. 67, Cripple Creek, CO 80813. Mailing Address: P.O. Box 339, Cripple Creek, CO 80813-0339. Tel.: 719-689-2466.
Institution Type/Description: Mining Museum.
Collections: area history; coal mining industry; photographs.
Facilities: Museum-related items for sale.
Activities: mine tours.
Hours & Admission Prices: April 2-May 14 & Oct. daily 10-4; May 15-Sept. 15 daily 9-5; Sept. 16-Sept. 30 daily 10-5. Adults $18, children 3-12 $10; children 2 & under no charge.

Del Norte

RIO GRANDE COUNTY MUSEUM AND CULTURAL CENTER, 580 Oak St., Del Norte, CO 81132-2210. Tel.: 719-657-2847; 800-233-4403. Fax: 719-657-2627.
E-mail: rgmuseum@riograndecounty.org
Web Site: museumtrail.org/RioGrandeCountyMuseum.asp
Key Personnel: Dir., Louise Colville
Institution Type/Description: History Museum.
Collections: cultural & natural history of the area; early settlers; Indian rock art; Western art; period furnishings; personal artifacts; Old Spanish Trail.
Facilities: library. Museum-related items for sale.
Activities: seminars; workshops; lectures; special events.
Hours & Admission Prices: Tues.-Fri. 10-4, Sat. 10-2.

Delta

DELTA COUNTY MUSEUM, 251 Meeker St., Delta, CO 81416-1914. Tel.: 970-874-8721.
E-mail: deltamuseum@aol.com
Founded: 1964.
Congressional District: 3
Key Personnel: Chm. (V), Bernice Musser; Pres. (V), Tom Gore; Cur., James K. Wetzel.
Personnel Profile: Part-Time Paid 1; Part-Time Volunteers 40.
Governing Authority: nonprofit organization. Parent Institution: Delta County Historical Society. Tax-exempt: 170(b)(1)(A).
Institution Type/Description: History Museum.
Collections: Vivian Jones dinosaur collection; butterflies; county newspapers; manuscripts; barbed wire; cameras; photographs; agriculture; Indian artifacts; transportation; industrial; c.1883 to present.
Research Fields: local history.

Facilities: 200-vol. library of Delta County books & newspapers on microfilm, directories, town records & magazines, available for research on premises.
Activities: permanent exhibitions.
Publications: quarterly newsletter.
Hours & Admission Prices: May-Sept. Tues.-Sat. 10-4; Oct.-April Wed. & Sat. 10-4. Adults $2, seniors $1; children under 12 with adult & members no charge. Closed legal holidays. &
Attendance: 3,500 (estimated)
Membership: Senior $7.50; Individual $10; Family $20; Business $30; Sponsor $50; Lifetime $150.

FORT UNCOMPAHGRE HISTORY MUSEUM, 205 Gunnison River Dr., Delta, CO 81416-1847. Mailing Address: 360 Main St., Delta, CO 81416-1837. Tel.: 970-874-1718. Fax: 970-874-1353.
E-mail: wilma@cityofdelta.net
Web Site: www.cityofdelta.net
Founded: 1990.
Congressional District: 3
Key Personnel: C.E.O., Wilma Erven; Cur., Ken Reyher.
Personnel Profile: Full-Time Paid 1; Part-Time Paid 2; Part-Time Volunteers 15.
Governing Authority: municipal. Parent Institution: City of Delta.
Institution Type/Description: General Museum.
Collections: fur trade period items ranging from firearms & traps to tools & livestock.
Research Fields: Antoine Robidoux & the southwest fur trade.
Facilities: Museum-related items for sale.
Activities: self guided tours; hobby workshops; lectures; training programs for professional museum workers. Museum Sponsors: Seasonal Special Occasions.
Hours & Admission Prices: April-Sept. Mon.-Fri. 9-3. Admission $3.50. Closed major holidays. &
Attendance: 3,000 (estimated)

Denver

BLACK AMERICAN WEST MUSEUM & HERITAGE CENTER, 3091 California St., Denver, CO 80205-3044. Tel.: 303-482-2242. Fax: 303-382-1981.
E-mail: executivedirector@blackamericanwestmuseum.com
Web Site: www.blackamericanwestmuseum.com
Founded: 1971.
Congressional District: 1
Key Personnel: C.E.O., LaWanna Larson; Co Chm., Terry Gentry; Co Chm., Thelma Craig.
Personnel Profile: Full-Time Paid 2; Part-Time Paid 1; Part-Time Volunteers 15.
Governing Authority: state & federal; nonprofit. Tax-exempt: 501(c)(3).
Institution Type/Description: History Museum: located in the home of Dr. Justina L. Ford, the first Black female physician in Colorado.
Collections: concentration on the history, achievements & contributions of Black Americans in the building of the American West.
Facilities: Books, T-shirts, Western-oriented items & jewelry of African heritage for sale.
Activities: concerts; formal educations programs for adults; guided tours; lectures; loan, temporary & traveling exhibitions. Annual Events: Western Dance; Jamboree; Awards Ceremony.
Publications: book; Crossing over Jordon: A History of the Black Church in Colorado.
Hours & Admission Prices: June-Aug. Tues.-Sat. 10-5; Sept.-May Tues.-Sat. 10-2. Adults $8, senior citizens over 65 $7, children 12 & under $6; members no charge. Closed New Year's Day; Easter; Memorial Day; Independence Day; Labor Day; Thanksgiving; Christmas. &
Attendance: 12,500 (estimated)
Membership: Student & Senior $20; Individual $25; Family $35; Pioneer $50; Heritage $100; Organization $250; Life $1,000; Corporate $2,500.

NATIONAL BALLPARK MUSEUM / B'S BALLPARK MUSEUM, 1940 Blake St. #101, Denver, CO 80202. Tel.: 303-974-5835. Fax: 303-997-8468.
Web Site: www.ballparkmuseum.com
Formerly: B's Ballpark Museum
Founded: 1999.
Key Personnel: Dir. & Pres., Bruce S. Hellerstein.
Governing Authority: private; nonprofit organization. Tax-exempt: 501(c)(3).
Institution Type/Description: Baseball Ballpark Museum.
Collections: stadium seats; signs; postcards; stadium artifacts; bricks; lithographs.
Facilities: 100-vol. library; 600 sq. ft. exhibit space.
Activities: films; guided tours; lectures.

Hours & Admission Prices: Wed.-Fri. 10-6, Sat. 12-6. Adults $5, senior citizens $4, children under 10 free.
Attendance: 400 (estimated)

＊ **BYERS-EVANS HOUSE MUSEUM, (M),** 1310 Bannock St., Denver, CO 80204-2719. Tel.: 303-620-4933. Fax: 303-620-4795.
E-mail: kevin.gramer@state.co.us
Web Site: www.byersevanshousemuseum.org
Founded: 1990.
Congressional District: 1
Key Personnel: Admin., Kevin Gramer; Asst. Dir., Ashley Rogers; Museum Program Asst., Lindsey McCutchan.
Personnel Profile: Full-Time Paid 1; Part-Time Paid 2; Part-Time Volunteers 60; Interns 2.
Governing Authority: state government. Parent Institution: History Colorado, 1200 Broadway, Denver, CO 80203. Tax-exempt.
Institution Type/Description: Historic House: 1883 Italianate style home.
Collections: 1912-1924 Restoration period, Evans family furnishings.
Facilities: tea room & classroom; rental space for parties; theatre capabilities. Museum-related items for sale including Denver and Colorado made objects.
Activities: docent program; guided tours; educational programs for children; art gallery exhibit program; school field trip program; lecture series; Byers-Evans House theatre company; free days.
Hours & Admission Prices: Mon.-Sat. 10-4. Adults $6, senior citizens & students $5, children 6-12 $4; discounts to AAM members; members & children under 6 no charge. Closed state holidays. &
Attendance: 19,121 (estimated)
Membership: Through Colorado Historical Society.

CHILDREN'S MUSEUM OF DENVER, (M), 2121 Children's Museum Dr., Denver, CO 80211-5200. Tel.: 303-433-7444. Fax: 303-433-9520. Facebook: Children's Museum of Denver.
E-mail: information@cmdenver.org
Web Site: www.mychildsmuseum.org
Founded: 1973.
Congressional District: 1
Key Personnel: Pres. & C.E.O., Mike Yankovich; Chm. Bd., Lauren Arnold; Chief Museum Officer, Gretchen Kerr; Vice Pres. Devel. & Communications, Amy Burt; C.F.O., Cyndi Kerins.
Personnel Profile: Full-Time Paid 31; Part-Time Paid 23; Part-Time Volunteers 740.
Volunteer Hours: 10,833
Operating Expenses: 3,973,960
Operating Income: 4,109,061
Governing Authority: bd. of directors; nonprofit organization. Tax-exempt: 501(c)(3).
Institution Type/Description: Children's Museum.
Collections: hands-on exhibits.
Research Fields: early childhood education.
Facilities: resource center.
Activities: interactive exhibits for children, newborn through 8 yrs. old & their caregivers; theater performances; educational programs; special events; community outreach; school groups.
Publications: newsletter, Children's Museum of Denver Happenings; Teachable Moment articles; Grand Play articles; Recipes for Play; Playtime Pointers.
Hours & Admission Prices: Mon.-Tues. & Thurs.-Fri. 9-4, Wed. 9-7:30, Sat.-Sun. 10-5. Admission 2-59 $9, senior citizens 60 & over & children one-year old $7; discounts to AAM members; children under one & members no charge. Closed New Year's Day; Easter; Thanksgiving; Christmas Eve & Day. &
Attendance: 348,459 (accurate)
Membership: Just for Two $80; Grandparent $85; Family & Prof. Childcare Provider $100; Adventure Passport $140.

CLYFFORD STILL MUSEUM, (M), 1250 Bannock St., Denver, CO 80204-3631. Tel.: 720-354-4880. Fax: 303-534-1766.
E-mail: info@clyffordstillmuseum.org
Web Site: clyffordstillmuseum.org
Founded: 2005.
Congressional District: 1
Key Personnel: Dir., Dean Sobel; Chm. (V), Christopher Hunt; Museum Shop Mgr., Lydia Garmaier.
Personnel Profile: Full-Time Paid 7; Part-Time Paid 10; Interns 1.
Governing Authority: private; nonprofit organization. Tax-exempt: 501(c)(3).
Institution Type/Description: Art Gallery.
Collections: more than 2,400 works of art by Clyfford Still; Clyfford Still's personal archives.

Publications: members magazine, Words on Paper.
Hours & Admission Prices: Adults $10; discounts to AAM members; members no charge. &

Membership: Senior, Student, Teacher $35; Individual $45; Dual $65; Family $75; 1944 Society $250; Curator's Circle $750; Director's Circle $5,000.

COLORADO SPORTS HALL OF FAME, Field at Mile High, 1701 Bryant St., Ste. 500, Denver, CO 80204. Tel.: 720-258-3888. Fax: 303-244-1003.
E-mail: sportsauthorityfield@milehigh
Web Site: www.coloradosports.org
Founded: 1964.
Key Personnel: C.E.O., Tom Lawrence; Mktg. & Tour Dir., Laura Gee.
Personnel Profile: Full-Time Paid 2; Part-Time Paid 12.
Governing Authority: Tax-exempt.
Institution Type/Description: Sports Museum.
Collections: Colorado's athletes, coaches & sports industry leaders; photographs; personal artifacts; sports memorabilia, equipment & uniforms.
Hours & Admission Prices: June-Aug. Tues.-Sat. 10-3; Sept.-May Thurs.-Sat. 10-3. No charge. &
Attendance: 20,000 (estimated)

COLORADO STATE CAPITOL, 200 E. Colfax Ave., Denver, CO 80203-1776. Tel.: 303-866-2604.
E-mail: simon.maghakyan@state.co.us
Web Site: www.colorado.gov
Key Personnel: Asst. Visitor Svcs. Mgr., Simon Maghakyan
Institution Type/Description: Historic Building: built in the 1890s.
Collections: Colorado history; photographs; period furnishings; war artifacts; portraits; murals.
Hours & Admission Prices: June-Aug. Mon.-Fri. 9-3:30; Sept.-May Mon.-Fri. 9:15-2:30; reservations for groups & those with special needs. No charge. Closed most legal holidays. &

CORE NEW ART SPACE, 900 Santa Fe Dr., Denver, CO 80204-3937. Tel.: 303-297-8428.
E-mail: art@corenewartspace.com
Web Site: www.corenewartspace.com
Key Personnel: Chm. (V), Gene George
Institution Type/Description: Art Gallery.
Collections: works by contemporary artists.
Activities: special events.
Hours & Admission Prices: Thurs. & Sat. 12-6, Fri. 12-9, Sun. 1-4. No charge.
Attendance: 15,000 (estimated)

* **DENVER ART MUSEUM, (M),** 100 W. 14th Ave. Pkwy., Denver, CO 80204-2788. Tel.: 720-865-5000. Fax: 720-913-0001. Facebook: Denver Art Museum.
E-mail: info@denverartmuseum.org
Web Site: www.denverartmuseum.org
Founded: 1893.
Congressional District: 1
Key Personnel: Chm., J. Landis Martin; Pres. (V), Eileen Pappas; Dir. Frederick & Jan Mayer, Christoph Henrich; Deputy Dir. & Chief Mktg. Officer, Andrea Fulton; Deputy Dir. & C.F.O., Curtis Woitte; Dir. Conservation, Sarah Melching; Frederick & Jan Mayer Cur. Pre-Columbian Art, Margaret Young-Sanchez; Assoc. Dir. Communications, Kristy Bassuener; Assoc. Dir. Mktg. & Communications, Katie Ross; Dir. Petrie Institute of Western American Art, Thomas Smith; Registrar, Lori Iliff; Dir. Education, Melora McDermott-Lewis; Gates Cur. Paintings & Sculpture, Timothy J. Standring; Dr. Joseph de Heer Cur. Asian Art, Ronald Otsuka; Chief Cur. & Cur Native Art, Nancy Blomberg; Frederick & Jan Mayer Cur. Spanish Colonial Art, Donna Pierce; Cur. Photography, Eric Paddock; Avenir Foundation Cur. Textile Art, Alice Zrebiec; Assoc. Cur. Architecture, Design & Graphics, Darrin Alfred; Assoc. Cur. Paintings & Sculpture, Angelica Daneo; Cur. Berger Collection, Kathleen Stuart; Cur. Modern Art and Herbert Boyer Collection & Archive, Gwen Chanzit; Polly and Mark Addison Cur. Contemporary Art, William Morrow; Assoc. Dir. Retail Operations, Greg McKay; Assoc. Cur. American Indian Art, John Lukavic.
Personnel Profile: Full-Time Paid 200; Part-Time Volunteers 500; Interns 10.
Volunteer Hours: 39,800
Operating Expenses: 23,000,000
Operating Income: 24,400,000
Governing Authority: municipal. Subsidiary Institution: Denver Art Museum Foundation. Tax-exempt: 501(c)(3).
Institution Type/Description: Art Museum.
Collections: Native arts, including works of North America, Africa & Oceania; Asian art; architecture, design & graphics; modern & contemporary art;

photography; paintings & sculpture, featuring works by Monet, Renoir, Matisse & Degas; pre-Columbian & Spanish colonial art; textile art; Western American art.
Major Exhibits: All That Glistens: A Century of Japanese Lacquer, 11/18/12-10/5/14; Depth & Detail: Carved Bamboo from China, Japan & Korea, 8/25/13-9/15; Sovereign: Independent Voices, 9/15/13-8/14; Passport to Paris, 10/27/13-2/9/14; Cover Story, 11/13-4/27/14; Herbert Bayer: Berlin Graphics 1928-1938, 11/10/13-11/23/14; Seen in Passing: The Photographs of Chuck Forsman, 11/17/13-5/25/14; Drawn to Action: Posters from the AIGA Design Archives, 12/15/13-1/18/15; Rebranded: Polish Film Posters for the American Western, 2/15/14-6/1/14; Modern Masters: 20th Century Icons from the Albright-Knox Art Gallery (T), 3/2/14-6/8/14; The American West in Bronze, 1850-1925 (T), 5/11/14-8/14.
Research Fields: art history & education.
Facilities: over 40,000-vol. library in administrative annex, call 720-913-0100. Museum-related items for sale.
Activities: guided tours; docent tours by appointment; lectures; slide talks; seminars; symposia; youth & adult art appreciation classes; performing arts events; permanent, temporary & traveling exhibitions; special access programs for handicap.
Publications: annual report; exhibition catalogs; permanent collection catalogs; books, Herbert Bayer Collection & Archive at the Denver Art Museum; The Denver Art Museum: The First 100 Years; annual publication, Western Passages; Nature as Muse; Drawing Room; Threads of Heaven.
Hours & Admission Prices: Museum: Tues.-Thurs. & Sat.-Sun. 10-5, Fri. 10-8. Library: by appointment; located in the administrative annex. Adults in-state $10, out-of-state $13, senior citizens & students in-state $8, out-of-state $10, youth 6-18 in-state $3, out-of-state $5; discounts to AAM, AAA & ICOM members; children 5 & under and members no charge. Closed Thanksgiving; Christmas. &
Attendance: 808,000 (accurate)
Membership: Senior, Student, Teacher $45; Individual $50; Dual & Family $70; Sustaining $125; Supporting $500; Benefactor $1,000; Associate $1,000-$5,000.

* **DENVER BOTANIC GARDENS, INC., (M),** 1005 York St., Denver, CO 80206-3014. Mailing Address: 909 York St., Denver, CO 80206-3751. Tel.: 720-865-3502 & 3585. Fax: 720-865-3713. TDD: 720-865-3745.
Web Site: botanicgardens.org
Founded: 1951.
Congressional District: 1
Key Personnel: C.E.O., Brian Vogt; Chm. (V), Jerry D. Ladd; Dir. Education, Matthew Cole; Dir. Horticulture, Sarada Krishnan; C.F.O., Florence Welch; Dir. Garden Operations, Tom Aljinovich; Dir. Mktg., Robin Doerr.
Personnel Profile: Full-Time Paid 93; Part-Time Paid 10; Part-Time Volunteers 1,147; Interns 30.
Governing Authority: city & county of Denver; private, nonprofit corporation. Parent Institution: Dept. of Parks & Recreation, 1805 Bryant St., Denver 80204. Branch Facility: Denver Botanic Gardens at Chatfield, 8500 Deer Creek Canyon., Littleton, CO 80123. Tax-exempt: 501(c)(3).
Institution Type/Description: Botanical Gardens & Arboretum.
Collections: herb garden; Japanese garden, rock-alpine garden; high plains garden; scripture garden; western perennial border fragrance garden; xeriscape garden; aquatic garden; rose garden; lilac garden; Montane garden; low-maintenance garden; vegetable garden; Water-Smart garden; xeriscape collection; tropical conservatory; herbarium with 29,000 sheets; water lily gardens; endangered plants garden; research collections of endangered species; original watercolors of Colorado wildflowers; Mt. Goliath; Chatfield, 84 woody windbreak species & Western dryland perennial collections; Sacred Earth Garden (tribes of Four Corners Region); birds & bees garden; children's secret path; PlantAsia garden; Sensory garden; Cloud Forest Tree with orchids & bromeliads; cutting garden.
Research Fields: 59,000 specimen vascular plant & mycological herbarium with accompanying extensive mycological laboratory & library; 29 endangered species as a Center for Plant Conservation participant; vascular plants; alpine plant identification/conservation; grassland restoration; floristic inventories.
Facilities: 28,000-vol. library with 1,000 periodicals, 7,200 current pamphlets, 8,500 current seed & nursery catalogs, rare book room including a bromeliad-literature collection; herbarium; conservatory & greenhouses; Orchid-Bromeliad House; Cactus & Succulent House; 350-seat auditorium & show facility; accessible classrooms; mycological research lab; Mt. Goliath alpine nature trail & Dos Chapelle Nature Center at Mt. Evans; 700-acre Denver Botanic Gardens at Chatfield in Littleton; Cloud Forest Tree. Books, gardening accessories & other museum-related items for sale.
Activities: 2 certification programs, Botanical Art, Illustration & Rocky

Mountain Gardening; certificate training: applied plant conservation; community garden program; Ask The Expert plant information service; educational programs for adults & children; flower shows; guided tours of Chatfield, conservatory, outside gardens & sensory garden; classes, lectures & field trips. Botanic Gardens Sponsors: annual plant & book sale, 2-day holiday sale; summer concerts in outdoor amphitheater; pumpkin festival; corn maze; Earth Day event; winter evening light show; summer college intern program.

Publications: book, Rocky Mountain Alpines; bimonthly membership magazine; education programs; summer brochure, Summertime Kids Program; Denver Botanic Gardens at Chatfield brochure; Visitors Guide; interpretive brochures.

Hours & Admission Prices: Sept. 14-May 9 daily 9-5; May 10-Sept. 12 Wed.-Fri. 9-5, Sat.-Tues. 9-8. Call for admission prices. Closed New Year's Day; Thanksgiving; Christmas. &

Attendance: 503,000 (accurate)

Membership: Individual $40; Individual Plus One $50; General $70; General Plus $160; Supporter $325; Patron $625.

THE DENVER CENTER FOR THE PERFORMING ARTS, 1101 13th St., Denver, CO 80204-5319. Tel.: 303-893-4100. Fax: 303-595-9634.

Institution Type/Description: History Museum.

Collections: theater history; paintings; cast autographs; props; costumes; stages.

Facilities: theater.

Activities: guided tours; performances.

Hours & Admission Prices: Tours: Mon. & Sat. 10 am by appointment. Adults $8.

DENVER FIREFIGHTERS MUSEUM, 1326 Tremont Pl., Denver, CO 80204-2120. Tel.: 303-892-1436. Fax: 303-893-4835.

E-mail: info@denverfirefightersmuseum.org

Web Site: www.denverfirefightersmuseum.org

Founded: 1978.

Congressional District: 1

Key Personnel: Exec. Dir. & Museum Shop Mgr., Winifred Ferrill; Pres. (V), Robert Vallero; Treas., Mark Gregory; Sec., Diane Ensminger.

Personnel Profile: Full-Time Paid 3; Part-Time Paid 4; Part-Time Volunteers 120.

Governing Authority: private; nonprofit organization. Tax-exempt: 501(c)(3).

Institution Type/Description: History Museum.

Collections: historic photographs; firefighting equipment & artifacts; hand-drawn to motorized fire fighting vehicles; books; archives.

Research Fields: history of Denver Fire Department.

Facilities: 400-vol. library; 12,000 sq. ft. exhibit space. Museum-related items for sale.

Activities: hands-on children's museum; fire safety programs; guided tours; available to rent for parties & meetings.

Publications: newsletter, Four Rings.

Hours & Admission Prices: Mon.-Sat. 10-4. Adults $6, senior citizens & students $5, children 2-12 $4; discounts to groups & AAM members; children under 2 & members no charge.

Attendance: 35,000 (accurate)

Membership: Hook & Ladder $36-$49; Hose Reel $50-$149; Steamer $150-$499; Engine $500-$999; Truck $1,000-$4,999; Tower $5,000 & up.

DENVER MUSEUM OF MINIATURES, DOLLS AND TOYS, (M), 1880 Gaylord St., Denver, CO 80206-1211. Tel.: 303-322-1053. Fax: 303-322-3704.

E-mail: director@dmmdt.org

Web Site: www.dmmdt.org

Founded: 1981.

Congressional District: 1

Key Personnel: Dir., Wendy Littlepage; Pres. Bd., Margery Smith; 1st Vice Pres., Pat Vick; Treas., Carol Kluver; Volunteer Coord., Stasia Steele; Museum Shop Mgr., Deanna Thomas.

Personnel Profile: Full-Time Paid 1; Part-Time Paid 1; Part-Time Volunteers 125; Interns 2.

Governing Authority: nonprofit organization. Tax-exempt: 501(c)(3).

Institution Type/Description: Miniatures, Dolls & Toys Museum.

Collections: miniatures; dolls & toys from prehistoric to modern times; doll houses; board games; paper dolls; tin toys; trains; international dolls; crafts; archives; mechanical toys; doll furniture.

Research Fields: miniatures; dolls; toys.

Facilities: 100-vol. library; classroom; 6,000 sq. ft. exhibit space. Miniatures, toys & dolls for sale.

Activities: lectures; hobby workshops; temporary exhibitions; children's art programs. Annual Events: spring & fall.

Publications: quarterly newsletter.

Hours & Admission Prices: Wed.-Sat. 10-4, Sun. 1-4. Adults $6, senior citizens 62 & over $5, children 5-16 $4; discounts to groups, AAA, CAA, & CHS members; members & children under 5 no charge. Closed major holidays. &

Attendance: 7,809 (accurate)

Membership: Senior Individual $25; Individual $30; Family $45; Sponsor $100; Benefactor $125; Patron $150; Contributing $250; Fellow $500; President's Circle $1,000.

* DENVER MUSEUM OF NATURE & SCIENCE, (M), 2001 Colorado Blvd., Denver, CO 80205-5798. Tel.: 303-370-6000. TDD: 303-370-8257.

E-mail: feedback@dmns.org

Web Site: www.dmns.org

Formerly: Denver Museum of Natural History

Founded: 1900.

Congressional District: 1

Key Personnel: Pres. & C.E.O., George Sparks; Vice Pres. Finance & Business Operations, Ed Scholz; Vice Pres. Visitor Experience, Mary Hacking; Vice Pres. Strategic Partnerships & Programs, Dr. Bridget Conghlin; Vice Pres. Research & Collections, Scott Sampson; Vice Pres. Capital Construction & Facilities, Dave Noel; Archivist & Dept. Chair, Kris Haglund; Cur. Invertebrate Zoology & Dept. Chair, Dr. Paula Cushing; Cur. Anthropology & Dept. Chair Archaeology, Dr. Stephen Nash; Cur. Vertebrate Zoology, Dr. John Demboski; Cur. Archaeology, Dr. Steve Holen; Cur. of Planetary Sciences & Dept. Chair, Dr. Steve Lee; Cur. Space Science, Dr. Ka Chun Yu; Cur. Astrobiology, Dr. David Grinspoon; Librarian, Katherine B. Gully; Image Archivist, Rene Payne; Conservator & Dept. Conservation Chair, Jude Southward.

Personnel Profile: Full-Time Paid 203; Part-Time Paid 212; Part-Time Volunteers 1,467.

Governing Authority: private; nonprofit organization. Tax-exempt: 501(c)(3).

Institution Type/Description: Natural History Museum.

Collections: archaeology; ethnography; ornithology; mammalogy; herpetology; entomology; conchology; geology; mineralogy; vertebrate paleontology; paleobotany; space science; natural history art.

Major Exhibits: Maya: Hidden Worlds Revealed (T), 2/14-8/24/14; Traveling the Silk Road (T), 11/21/14-5/15; Whales Tohora (T), 10/10/14-2/16/15.

Research Fields: North American archaeology and ethnography with emphasis on Southwest & Plains; Colorado mineralogy & Western interior paleontology; ornithology, mammalogy & entomology; space sciences; planetary sciences; astrobiology; genetics.

Facilities: 25,000-vol. library of natural history & anthropology books and journals; archives & 300,000 photographic images; IMAX theater; planetarium; health education center; cafe. Museum-related items for sale.

Activities: guided tours; adults lectures & courses; children's workshops; permanent & temporary exhibitions; gallery demonstrations; teen program; IMAX & Planetarium shows; outreach programs; internships; classroom programs; learning lab.

Publications: newsletter, Museum magazine; exhibit catalogs; symposia proceedings; annual report; scientific books & papers; teacher's guides.

Hours & Admission Prices: Museum: daily 9-5. Adults $13, seniors 65 & over $10, juniors 3-18 $8; members no charge. IMAX: adults $10, juniors 3-18 and seniors 65 & over $8; discounts to AAM & museum members. Museum & IMAX Combination: adults $20, seniors 65 & over $16; juniors 3-18 $14; discounts to AAM & museum members. Closed Christmas. &

Attendance: 1,252,300 (accurate)

Membership: Senior Individual $45; Individual $50; Senior Dual $60; Dual $65; Family & Grandparent $85; Family Plus $125; Young Professional Pioneer $250; Curator $300-$749; Young Professional Innovator $500; Explorer $750-$1,499; Naturalist $1,500-$4,999; Collector $5,000-$9,999; Campion Circle $10,000 & up.

DENVER ZOOLOGICAL GARDENS, 2300 Steele St., Denver, CO 80205-4899. Tel.: 303-376-4800. Fax: 303-376-4801.

E-mail: zooinfo@denverzoo.org

Web Site: www.denverzoo.org

Founded: 1896.

Congressional District: 1

Key Personnel: C.E.O. & Pres., Craig Piper; Chm. Bd., Katie Schoelzel; C.O.O., Kyle D. Burks; C.F.O., Donna Mei Lin Driscoll; Vice Pres. Strategic Initiatives & External Affairs, Ana Bowie; Vice Pres., Guest Svcs. & New Business Devel., Amber Christopher; Vice Pres. Human Resources, Jacque Taylor; Vice Pres. Planning, George Pond; Veterinarian, Scott Larsen; Gen. Cur., Brian Aucone.

Personnel Profile: Full-Time Paid 269; Part-Time Paid 57; Part-Time Volunteers 660; Interns 26.

Governing Authority: municipal; nonprofit organization. Parent Institution:

City & County of Denver. Subsidiary Institution: Denver Zoological Foundation, Inc. Tax-exempt: 501(c)(3).
Institution Type/Description: Zoo.
Collections: 3,667 specimens; 682 species; 1,850 specimen biofact collection.
Research Fields: ethnology; nutrition; conservation policy; field conservation.
Facilities: 500-vol. research library of zoology books; 80-acre campus with 90 buildings and exhibits; cafe. Museum-related items for sale.
Activities: field trips; tours; lectures; summertime daily wildlife theatre shows.
Publications: annual report; quarterly, Zoo Review; monthly, On the Wildside.
Hours & Admission Prices: March-Oct. daily 9-6. Adults $13, seniors 65 & over $10, children $8. Nov.-Feb. daily 10-5. Adults $10, senior citizens 65 & over $8, children $6; discount to groups; AZA Professional members, members & children 2 and under no charge. &
Attendance: 1,883,167 (accurate)
Membership: Individual $55; Individual & Guest and Zoo for Two $70; Family $90; Family Plus $110; Curator's Club $275-$499; Director's Circle $500-$999; President's Circle $1,000-$2,499; Chairman's Circle $2,500-$4,999; Guardian Society $5,000-$9,999; Conservation Society $10,000 & up.

DOWNTOWN AQUARIUM, 700 Water St., Denver, CO 80211-5210. Tel.: 303-561-4450. Fax: 303-561-4465.
Institution Type/Description: Aquarium.
Collections: freshwater fish; Sumatran tigers.
Facilities: restaurant.
Hours & Admission Prices: Sun.-Thurs. 10-9, Fri.-Sat. 10-9:30. Adults 12-64 $17.99, seniors 65 & over $16.99, children 3-11 $11.99; children 2 & under no charge.

EMMANUEL GALLERY, Auraria Campus, Denver, CO 80217. Mailing Address: Auraria Campus, Campus Box 177, P.O. Box 173364, Denver, CO 80217-3364. Tel.: 303-556-8337. Fax: 303-556-2335. Facebook: Emmanuel Gallery.
Founded: 1976.
Key Personnel: Dir. & Cur., Shannon K. Corrigan.
Personnel Profile: Full-Time Paid 1.
Institution Type/Description: Art Gallery.
Collections: works by regional, national & international artists.
Activities: lectures; temporary exhibitions.
Hours & Admission Prices: Call for hours. No charge.
Attendance: 10,000 (accurate)

FORNEY MUSEUM OF TRANSPORTATION, (M), 4303 Brighton Blvd., Denver, CO 80216-3702. Tel.: 303-297-1113. Fax: 303-297-3113.
E-mail: director@forneymuseum.org
Web Site: www.forneymuseum.org
Formerly: Colorado Museum of Transportation
Founded: 1955.
Congressional District: 1
Key Personnel: Pres. Bd., Jack D. Forney; Dir. Museum, Christof Kheim; Asst. Dir., Josh Barr; Event & Education Coord., Monica Petit; Grant Coord., Krysta Hand; Financial Svcs., Ellen O'Donnell; Mgr. Visitor Svcs. & Museum Shop Mgr., Amanda Cope.
Personnel Profile: Full-Time Paid 3; Part-Time Paid 4; Part-Time Volunteers 35.
Volunteer Hours: 8,000
Operating Expenses: 353,402
Operating Income: 352,744
Governing Authority: nonprofit organization. Tax-exempt: 501(c)(3).
Institution Type/Description: Transportation Museum.
Collections: period automobiles, trains, carts, carriages, wagons, aircraft, fire engines, Denver cable car, bicycles, motorcycles, trolleys, sleighs, steam tractor, & clothing.
Major Exhibits: Chevrolet Camaro, 11/13-1/14; Air-Cooled Volkswagens, 1/14-3/14; Ford's Iconic Mustang, 2/14-4/14; Plymouth Barracuda - Fast & Furious 1964-1974, 5/14-7/14.
Research Fields: transportation.
Facilities: library; meeting room; event space. Gift items for sale.
Activities: tours; lectures; rental facilities; birthday parties. Museum Sponsors: Senior Day; Big Boy Weekend in July; Forney Fall Fest in October; Night at the Museums; Doors Open Denver.
Publications: Forney Museum of Transportation Coloring Book.
Hours & Admission Prices: Mon.-Sat. 10-4, Sun. 12-4. Adults 13-64 $9, seniors 65 & over $7, child 3-12 $5; discounts to disabled, military, AAA, AAM & ICOM members; children under 3 & members no charge. Closed New Year's Eve & Day; Thanksgiving; Christmas Eve & Day. &
Attendance: 20,000 (accurate)

Membership: Individual $30; Dual $40; Family of 4 $50; Small Group of 8 $75; Large Group of 12 $100. Sustaining: Goldbug $500; Hispano-Suiza $1,000; Big Boy $5,000 & up.

FOUR MILE HISTORIC PARK, 715 S. Forest St., Denver, CO 80246-2324. Tel.: 720-865-0800. Fax: 720-865-0801. Facebook: Four Mile Historic Park.
E-mail: info@fourmilepark.org
Web Site: www.fourmilepark.org
Founded: 1977.
Congressional District: 1
Key Personnel: C.E.O., Mackenzie Pacifico; Chm. (V), Larry Harte; Artifact Mgr., Bonnie Bowman; Dir. Educational Programs, Paul Reimer; Dir. Mktg. & Devel., MacKenzie Pacifico; Rental & Special Events Dir., Cory VanZytreldt; Bookkeeper, Connie Wyatt; Site Mgr., Bill Suhr; Coord. (V), Laura Hiniker; Museum Shop Mgr., Scotty Wilkins.
Personnel Profile: Full-Time Paid 7; Part-Time Paid 7; Part-Time Volunteers 163.
Volunteer Hours: 14,000
Governing Authority: nonprofit. Parent Institution: Four Mile Historic Park, Inc. Tax-exempt: 501(c)(3).
Institution Type/Description: Historic Site: 1859 & 1883 Four Mile House, log house, which served as stage stop, wayside inn, tavern & farmhouse.
Collections: reconstructed middle 19th-century farm/ranch, log barn, stallion barn, carriage barn; furniture & furnishings of the period 1859-1883; collection of early photographic material; objects recovered from archaeological studies & excavations.
Research Fields: archaeology; history & rural life; 1859-1883 agricultural history transportation history.
Facilities: Museum-related items for sale.
Activities: guided tours; workshops; living history demonstrations; formally organized educational programs; docent program; permanent exhibits. Museum Sponsors: special events.
Publications: Four Mile Post.
Hours & Admission Prices: April-Sept. Wed.-Fri. 12-4, Sat.-Sun. 10-4; Oct.-March Wed.-Sun. 12-4. Adults $5, senior citizens 65 & over & military with ID $4, youth 7-17 $3; discount to AAA & National Trust members; members and children 6 & under no charge. Closed New Year's Day; Thanksgiving; Christmas. &
Attendance: 47,000 (accurate)
Membership: Pioneer $35; Settler $50; Homesteader $100; 59'er $250; Wagon Master $500; Trail Blazer $1,000; Corporate memberships available.

GALLERY M, 180 Cook St., Ste. 101, Denver, CO 80206. Tel.: 303-331-8400. Fax: 303-331-8522.
E-mail: aa@gallerym.com
Web Site: gallerym.com
Key Personnel: Dir., Myrna Hayutin; Dir., Mason Hayutin
Institution Type/Description: Art Gallery.
Collections: works of contemporary fine art; paintings; sculpture; photography.
Hours & Admission Prices: Mon.-Sat. 10-6.

GRANT-HUMPHREYS MANSION, 770 Pennsylvania St., Denver, CO 80203-3619. Tel.: 303-894-2505. Fax: 303-894-2508.
E-mail: rentalsghm@chs.state.co.us
Web Site: www.granthumphreysmansion.org
Founded: 1976.
Congressional District: 1
Key Personnel: Dir., Kevin Gramer; Events Coord., Debbie Golden; Asst. Dir., Ashley Rogers; Museum Program Asst., Lindsey McCutchan.
Personnel Profile: Full-Time Paid 2; Part-Time Paid 2; Part-Time Volunteers 5.
Governing Authority: state. Parent Institution: History Colorado, 1200 Broadway, Denver, CO 80203. Tax-exempt.
Institution Type/Description: Historic House Museum: 1902 Beaux-Arts style home.
Collections: early 20th-century decorative & fine arts belonging to the Humphreys family.
Facilities: banquet facilities. Books on Denver history for sale.
Activities: special programs.
Hours & Admission Prices: Mon.-Fri. 9-5. No charge, donations accepted. &
Attendance: 20,000 (estimated)
Membership: See State Historical Society of Colorado for membership information.

***　HISTORY COLORADO, THE COLORADO HISTORICAL SOCIETY, (M),** 1200 Broadway, Denver, CO 80203-2109. Tel.: 303-477-8679. Fax: 303-866-4447.
E-mail: information@chs.state.co.us
Web Site: www.historycolorado.org

Founded: 1879.
Congressional District: 1
Key Personnel: Pres. & C.E.O., Edward C. Nichols; Vice Pres., Joseph Bell; Chm. (V), Jim McCotter; C.O.O., Kathryn Hill; Dir. Devel., Megan Mahneke; Dir. Devel., Jill Cowperthwaite; State Archaeologist & Deputy SHPO, Richard Wilshusen; Historic Preservation & Deputy SHPO, Steve Turner; Mng. Editor, Steve Grinstead; State Historian, William J. Convery; Dir. Library, Laura Ruttum Senturia; Dir. Education, J.J. Rutherford; Dir. Facilities Svcs., Andrew Stire; Cur. Decorative & Fine Arts, Alisa Zahller; Cur. Books & Manuscripts, Keith Schrum; Dir. Collections, Elisa Phelps.
Personnel Profile: Full-Time Paid 122; Part-Time Paid 14; Part-Time Volunteers 547; Interns 7.
Governing Authority: state. Parent Institution: State of Colorado. Colorado Commission on Higher Education. History Colorado. Regional Museums: Grant-Humphreys Mansion (303-894-2505), Byers-Evans House (303-620-4933), Pearce-McAllister Cottage (303-322-1053), Denver; Fort Garland & Pikes Stockade (719-379-3512), Fort Garland; Georgetown Loop Historic Mining & Railroad Park (303-569-2788), Georgetown; Healy House & Dexter Cabin (719-486-0487), Leadville; Ute Indian Museum (970-249-3098), Montrose; Fort Vasquez (970-785-2832), Platteville; El Pueblo History Museum (719-583-0453), Pueblo; Trinidad History Museum comprising Santa Fe Trail Museum, Baca House, Bloom Mansion & Historic Gardens (719-846-7217), Trinidad. Tax-exempt: 501(c)(3).
Institution Type/Description: Historical Society Museum.
Collections: history & prehistory of Colorado and the American West to present; over 140,000 artifacts including American Indian, archaeological, Hispanic, domestic life, industrial, military, transportation, crafts & industries, graphic arts & advertising, fine arts, political & community life, religious, costume & textiles, sports, firearms, furniture; over 750,000 historic photographs, moving images and negatives; Approximately 14 million documents, microfilm, books, serials and ephemera collections.
Research Fields: history of Colorado & the American West; archaeology; American Indian studies; ethnic communities.
Facilities: 14,000,000-item library including books, maps, newspapers, photographs, microfilm, manuscripts, tapes & film; reading room; classrooms; 400-seat auditorium. Museum-related items for sale.
Activities: special, permanent exhibitions; guided tours; lectures; films; gallery talks; formally organized education programs; training programs for professional museum workers & classroom teachers; school loan service; inter-museum loan service; inventory, nomination & restoration of historic & archaeological sites under National Register Program; State Historical Fund preservation grant program; review of environmental impact statements; historic markers program; publications program; friends groups & volunteer organizations.
Publications: bimonthly magazine, Colorado Heritage; monthly e-newspaper, Colorado History Now; books.
Hours & Admission Prices: Colorado History Museum: Mon.-Sat. 10-5, Sun. 12-5. Library Wed.-Sat. 10-5. Adults $10, students with I.D. & senior citizens 65 & over $8, children 6-12 $6; discounts to AAM & AAA members; members and children 5 & under no charge. Closed major holidays; some state holidays. &

Attendance: 229,509 (accurate)
Membership: Individual $65; Family $80; Explorer $150; Centennial $300; Historian $500; Preservationist $750; BanCroft $1,000.

KIRKLAND MUSEUM OF FINE & DECORATIVE ART, (M),
1311 Pearl St., Denver, CO 80203-2518. Tel.: 303-832-8576. Fax: 303-832-8404.
E-mail: info@kirklandmuseum.org
Web Site: www.kirklandmuseum.org
Formerly: Vance Kirkland Museum
Founded: 1996.
Congressional District: 1
Key Personnel: Founding Dir. & Cur., Hugh Grant; Deputy Dir., Gerald Horner; Mgr. Collections & Deputy Cur., Christopher Herron; Visitor Svcs. & Education Mgr., Shelly Bleckley; Mktg. & Communications Mgr., Maya Wright; Operations Mgr., Jacob Stauber; Registrar, Alisha Stovall; Museum Asst., Megan Sullivan; Community Rels. Liaison, Renee Albiston.
Personnel Profile: Full-Time Paid 6; Part-Time Paid 5; Part-Time Volunteers 45.
Governing Authority: private; nonprofit. Tax-exempt.
Institution Type/Description: Art Museum: housed in a 1910-1911 commercial art building, which was inaugurated as The Student's School of Art by artist Henry Read; the building was later acquired by Colorado painter Vance Kirkland, who used the structure as the Kirkland School of Art from 1932-1946 & continued to paint at this location until his death in 1981. National Trust Associate Site.
Collections: decorative art from 1880-1980 including Arts & Crafts, Art Nouveau, Glasgow Style, Wiener Werkstatte, De Stijl, Bauhaus, Art Deco, Modern and Pop Art; regional modernist art; painter, Vance Kirkland.

Facilities: 1,580-vol. library on American & European artists and American dinnerware & ceramics available to public by special arrangement; 2,100 sq. ft. exhibit space. Posters, postcards & Vance Kirkland exhibition catalogs for sale.
Activities: lectures; loan, temporary & traveling exhibitions; broadcast programs.
Hours & Admission Prices: Tues.-Sun. 11-5. Adults $7, seniors, teachers & students $6; discounts to groups, museum employees, AAA, AAM & ICOM members; members no charge. Children under 13 not admitted. Closed major holidays. &

Attendance: 14,426 (accurate)
Membership: Student/Senior/Teacher $30; Individual $35; Dual $45; Individual + Guest $50; Arts & Crafts $100; Art Nouveau $250; Bauhaus $500; Art Deco $1,000; Modern $2,500.

METROPOLITAN STATE COLLEGE OF DENVER/CENTER FOR VISUAL ART, (M),
965 Santa Fe Dr., Denver, CO 80204. Tel.: 303-294-5207. Fax: 303-294-5210.
E-mail: cva@msudenver.edu
Web Site: www.metrostatecva.org
Founded: 1991.
Congressional District: 1
Key Personnel: Exec. Dir., Greg Watts; Creative Dir., Cecily Cullen; Business & Devel. Dir., Amy Tancig.
Personnel Profile: Full-Time Paid 4; Part-Time Paid 6; Part-Time Volunteers 2; Interns 1.
Governing Authority: university; nonprofit. Parent Institution: Metropolitan State College of Denver. Tax-exempt: 501(c)(3).
Institution Type/Description: Art Gallery.
Collections: works of contemporary art.
Research Fields: contemporary art.
Facilities: 5,000 sq. ft. exhibit space.
Activities: adult & children's programs; temporary exhibits.
Publications: quarterly newsletter.
Hours & Admission Prices: Tues.-Fri. 11-6, Sat. 12-5. No charge; donations accepted. Closed all major holidays. &

Attendance: 15,000 (accurate)
Membership: Metro State Students, Faculty, & Staff $40; Individual $50; Sustaining & Family $100; Other $500 & up.

MIZEL MUSEUM, (M),
400 S. Kearney St., Denver, CO 80224-1238. Tel.: 303-394-9993. Fax: 303-394-1119. Facebook: Mizel Museum.
E-mail: mizelmuseum@mizelmuseum.org
Web Site: www.mizelmuseum.org
Founded: 1982.
Congressional District: 1
Key Personnel: Exec. Dir., Ellen Premack; Pres. (V), Larry A. Mizel; Dir. Mktg., Lisa Rimmert; Dir. Education, Jan Cooper Nadav; Special Projects, Deanne Kapnik; Database & Office Mgr., Amy Klingenberg; Cur., Georgina Kolber; Office Mgr. & Bookkeeper, Leslie English; Jewish Education Coord., Penny Nisson.
Personnel Profile: Full-Time Paid 6; Part-Time Paid 2; Part-Time Volunteers 45; Interns 1.
Volunteer Hours: 1,900
Operating Expenses: 1,700,000
Operating Income: 1,700,000
Governing Authority: nonprofit organization. Parent Institution: Mizel Institute. Subsidiary Institution: The C.E.L.L. Tax-exempt: 501(c)(3).
Institution Type/Description: Cultural Museum.
Collections: ritual & religious objects of the synagogue; historical artifacts; contemporary Judaica.
Major Exhibits: 4,000 Year Road Trip: Gathering Sparks, 1/14-12/14.
Research Fields: Jewish history & culture; multicultural education.
Facilities: 1,075-vol. library pertaining to Jewish art & ceremony.
Activities: guided tours; lectures; films; workshops; art classes; organized educational programs & exhibitions; docent program; participatory, traveling & temporary exhibitions.
Publications: quarterly newspaper, Mizel Tov; biannual education brochure.
Hours & Admission Prices: Mon., Tues, Wed. & Fri. 9-4:30, Thurs. 9-8, first Sun. each month 11-4. Adults $7; discounts to AAM & ICOM members. Closed Jewish holidays. &

Attendance: 135,000 (accurate)

MOLLY BROWN HOUSE MUSEUM, (M),
1340 Pennsylvania St., Denver, CO 80203-2417. Tel.: 303-832-4092, ext. 16. Fax: 303-832-2340.
E-mail: admin@mollybrown.org
Web Site: www.mollybrown.org

Founded: 1970.
Congressional District: 1
Key Personnel: Exec. Dir. Historic Denver, Annie Levinsky; Cur. Collections, Nicole Roush; Museum Dir., Andrea Malcomb.
Personnel Profile: Full-Time Paid 5; Part-Time Paid 5; Part-Time Volunteers 90; Interns 4.
Governing Authority: nonprofit organization. Historic House Museum & Preservation Society. Parent Institution: Historic Denver, Inc. Tax-exempt: 501(c)(3).
Institution Type/Description: Historical & Preservation Society.
Collections: furnishings; fashions. Historic House: 1889 Molly Brown House.
Research Fields: architectural; Victoriana; community revitalization; urban preservation; women's history.
Facilities: 200-vol. library of materials dealing with turn-of-the-century life in Denver.
Activities: guided tours; lectures; Victorian eating experiences; educational outreach programs.
Publications: monthly newsletters, Historic Denver News; Life of Margaret (Molly) Brown.
Hours & Admission Prices: Tues.-Sat. 10-3:30, Sun. 12-3:30. June-Aug.: Mon. 10-1:30, Tues.-Sat. 10-3:30, Sun. 12-3:30. Adults $8, senior citizens 65 & over and military $6, children 6-12 $4; Historic Denver members & members no charge. Closed major holidays. &
Attendance: 47,980 (accurate)
Membership: Annual $40; Friend $100.

MUSEO DE LAS AMERICAS, (M), 861 Santa Fe Dr., Denver, CO 80204-4344. Tel.: 303-571-4401. Fax: 303-607-9761.

E-mail: maruca@museo.org
Web Site: www.museo.org
Founded: 1991.
Congressional District: 1
Key Personnel: Exec. Dir., Maruca Salazar; Pres. Bd., George Martinez; Operations Mgr., Saniego Sanchez; Dir. Education, Claudia Meran; Public Rels., Tessa Harvey; Devel. Mgr., Christy Costello; Visitor Svcs., Yolanda Chichester.
Personnel Profile: Full-Time Paid 4; Part-Time Paid 4; Part-Time Volunteers 50; Interns 10.
Governing Authority: private; nonprofit organization. Tax-exempt: 501(c)(3).
Institution Type/Description: Art & History Museum.
Collections: focus on the art & culture of Latin America from ancient times to contemporary.
Research Fields: Latin American arts ancient, folk art, and contemporary.
Activities: tours; youth education programs; lectures; film; symposia; Spanish Happy Hour; bilingual programs; culturally responsive teacher training.
Publications: newsletter, Notitas.
Hours & Admission Prices: Tues.-Fri. 10-5, Sat.-Sun. 12-5. Adults $4, students and seniors 65 & over $3; discounts to AAM members; members & children under 13 no charge. Closed New Year's Day; Independence Day; Thanksgiving; Christmas. &
Attendance: 30,000 (accurate)
Membership: Senior & Student $30; Teacher $35; Jicama $45; Sandia Dual $60; Maiz Family $70; Pina $125; Nopal $250; Papaya $500; Cacao $800.

MUSEUM OF ANTHROPOLOGY, UNIVERSITY OF DENVER, (M), 2000 Asbury Ave. #146, Denver, CO 80208. Tel.: 303-871-2688. Fax: 303-871-2437.

E-mail: ckreps@du.edu
Web Site: www.du.edu/duma/duma.html
Founded: 1932.
Congressional District: 1
Key Personnel: Dir., Dr. Christina Kreps; Cur. Collections, Ms. Brook Rohde; Dept. of Anthropology Faculty Member, Dr. Dean Saitta; Dept. of Anthropology Faculty Member, Dr. Bonnie Clark.
Personnel Profile: Full-Time Paid 1; Part-Time Paid 1; Interns 15.
Governing Authority: university; nonprofit. Parent Institution: University of Denver. Tax-exempt: 501(c)(3).
Institution Type/Description: Anthropology Museum.
Collections: textiles; period artifacts.
Research Fields: ethnology, archaeology of the Southwest US.
Facilities: archaeology lab.
Activities: educational outreach; lecture series.
Hours & Admission Prices: Sept.-June Mon.-Fri. 9-4. No charge. Closed university holidays. &
Attendance: 1,000 (estimated)
Membership: Student $5; Non-Student $10; Patron $25; Donor $50.

MUSEUM OF CONTEMPORARY ART DENVER, 1485 Delgany, Denver, CO 80202-1100. Tel.: 303-298-7554.

Web Site: www.mcadenver.org
Founded: 1996.
Key Personnel: Dir. & Chief Animator, Adam Lerner.
Personnel Profile: Full-Time Paid 15; Part-Time Paid 11; Part-Time Volunteers 25; Interns 3.
Institution Type/Description: Art Museum.
Collections: paintings; sculpture; photography.
Activities: special events; education programs.
Hours & Admission Prices: Tues.-Thurs. 12-7, Fri. 12-8, Sat.-Sun. 10-7. Adults $8, students, military, and seniors 65 & over $5, teens 12-18 $3, youth 7-11 $1; children under 6 & members no charge. &
Membership: Individual $45; Dual Family $75; Donor $150. Heart Club: Patron $250; Innovator $500; Activator $1,000.

PIRATE: CONTEMPORARY ART, 3655 Navajo St., Denver, CO 80211. Tel.: 303-458-6058.

Web Site: www.pirateartonline.org
Founded: 1980.
Institution Type/Description: Art Gallery.
Collections: works by contemporary artists.
Activities: Annual Event: Day of the Dead in November.
Hours & Admission Prices: Fri. 6pm-10pm, Sat.-Sun. 12-5. No charge. &

RED ROCKS AMPHITHEATRE & VISITOR CENTER, 18300 W. Alameda Pkwy., Denver, CO 80465. Tel.: 720-865-2475.

Institution Type/Description: History Museum.
Collections: Red Rocks geologic & music history; Hall of Fame; wildlife.
Facilities: amphitheatre; visitor center; nature trail. Museum-related items for sale.
Activities: guided tours; hiking; hands-on exhibits; videos.
Hours & Admission Prices: May-Sept. daily 8-7; Oct.-April daily 9-4. Adults $6, seniors & children $3.

ROBISCHON GALLERY, 1740 Wazee St., Denver, CO 80202. Tel.: 303-298-7788. Fax: 303-298-7799.

Key Personnel: Dir., Jim Robischon; Dir., Jennifer Doran; Gallery Mgr. & Registrar, Debra Malik Demosthenes
Institution Type/Description: Art Gallery.
Collections: paintings; sculpture.
Hours & Admission Prices: Tues.-Fri. 11-6, Sat. 12-5.

RULE GALLERY, 3036 Osceola St., Denver, CO 80212-1441. Tel.: 303-777-9473.

E-mail: info@rulegallery.com
Web Site: www.rulegallery.com
Institution Type/Description: Art Gallery.
Collections: works by contemporary artists; paintings; sculpture; photographs; works on paper.
Hours & Admission Prices: Tues.-Sat. 12-5.

SPARK GALLERY, 900 Santa Fe Dr., Ste. #1, Denver, CO 80204-3937. Tel.: 720-889-2200.

E-mail: jwmatlack@yahoo.com
Web Site: www.sparkgallery.com
Key Personnel: Dir., John Matlack
Institution Type/Description: Art Gallery.
Collections: works by regional, national & international artists; paintings; sculpture; drawings; photographs.
Activities: Annual Event: Members Show.
Hours & Admission Prices: Thurs. & Sat. 12-5, Fri. 12-9, Sun. 1-4.

STILES AFRICAN AMERICAN HERITAGE CENTER, INC., 2607 Glenarm Place, Denver, CO 80205-3151. Tel.: 303-294-0597. Fax: 303-294-0597.

Web Site: www.stilesheritagecenter.org
Founded: 1998.
Key Personnel: Dir., Grace L. Stiles; Chm. (V), Karolette K. Greene
Institution Type/Description: Heritage Center.
Collections: African American history & culture; photographs; memorabilia; personal artifacts.
Facilities: classrooms.
Activities: guided tours; research; workshops.
Hours & Admission Prices: Mon., Wed. & Fri. 11-3, Sat. 2-4; other times by appointment. Adults $7, senior citizens $6, students $3, children under 6 $1. &

Attendance: 250 (estimated)

UNIVERSITY OF DENVER, SCHOOL OF ART & ART HISTORY, VICTORIA MYHREN GALLERY, 2121 E. Asbury Ave., Denver, CO 80208. Tel.: 303-871-3716. Fax: 303-871-4112.
E-mail: galleryinfo@du.edu
Web Site: www.myhrengallery.com
Founded: 1978.
Congressional District: 1
Key Personnel: Dir., Dan Jacobs.
Personnel Profile: Full-Time Paid 1; Part-Time Paid 5; Interns 4.
Governing Authority: private university; not-for-profit. Parent Institution: University of Denver School of Art & Art History. Tax-exempt: 501(c)(3).
Institution Type/Description: University Art Gallery.
Collections: late 19th- & early 20th-century European & regional masters; digital art; performance art; photographs.
Major Exhibits: A Decade of Gifts & Discoveries from the University Art Collections, 1/14-2/14; The Female Gaze: The Linda Lee Alter Collection at the Pennsylvania Academy of the Fine Arts (T), 3/14-4/14; BFA Senior Exhibition 2014, 5/14-6/14; Psychedelic Art of the Bay Area, 9/14-11/14.
Facilities: 2,564 sq. ft. exhibit space; educational facilities; 4,000 sq. ft. art storage & study center.
Activities: lectures; loan & participatory exhibits; educational programs for undergraduate & graduate students affiliated with the Univ. of Denver.
Publications: newsletter; exhibition catalogs.
Hours & Admission Prices: Exhibitions: daily 12-4. Office: Mon.-Fri. 9-4. No charge. Closed campus holidays. &

WALKER FINE ART, 300 W. 11th Ave. #A, Denver, CO 80204. Tel.: 303-355-8955. Fax: 303-623-0553.
E-mail: info@walkerfineart.com
Founded: 1999.
Key Personnel: Dir., Bobbi Walker
Institution Type/Description: Art Gallery.
Collections: works by regional & national artists.
Hours & Admission Prices: Tues.-Sat. 12-6, 1st Fri. each month 12-9; other times by appointment. No charge. &

WINGS OVER THE ROCKIES AIR & SPACE MUSEUM, 7711 E. Academy Blvd., Denver, CO 80230-6929. Tel.: 303-360-5360, ext. 110. Fax: 303-360-5328.
E-mail: dkerr@wingsmuseum.org
Web Site: wingsmuseum.org
Founded: 1994.
Key Personnel: C.E.O. & Pres., Greg Anderson; Cur., Matthew Burchette; Museum Shop Mgr., Nancy McCurdy.
Personnel Profile: Full-Time Paid 11; Part-Time Volunteers 160.
Governing Authority: Tax-exempt.
Institution Type/Description: Air & Space Museum: housed in World War II vintage hangar.
Collections: over 36 air & spacecraft spanning 8 decades including a full-size Star Wars X-Wing Fighter; aircraft armament; full-size mockup of a manned space station crew module; Colorado astronaut tribute.
Research Fields: history of Lowry, Air Force & WWII.
Facilities: library; photo archive; photo reproduction laboratory; welcome theater. Museum-related items for sale.
Publications: newsletter, Wingspan.
Hours & Admission Prices: Mon.-Sat. 10-5, Sun. 12-5. Adults $9, seniors 65 & over $8, children 4-12 $6; discounts to groups; members & children under 4 no charge. Closed major holidays. &
Attendance: 150,000 (accurate)
Membership: Individual $35; Family $50; Patron $300; Lifetime $1,000.

Dillon

DILLON SCHOOLHOUSE MUSEUM, 403 La Bonte St., Dillon, CO 80435. Mailing Address: P.O. Box 745, c/o Summit Historical Society, Breckenridge, CO 80424. Tel.: 970-468-2207.
Governing Authority: Tax-exempt.
Institution Type/Description: Historic Building: originally used as a schoolhouse until 1910 when it became a church; built in 1883.
Collections: local history; period artifacts; desks; slates; individual learning stations; Centennial flag; chemistry set; kerosene slide projector; phonograph; piano; organ; clothing; jewelry.
Hours & Admission Prices: Call for hours.

Dolores

BUREAU OF LAND MANAGEMENT - ANASAZI HERITAGE CENTER - CANYONS OF THE ANCIENTS NATIONAL MONUMENT, 27501 Hwy. 184, Dolores, CO 81323-9217. Tel.: 970-882-5600. Fax: 970-882-7035.
E-mail: rene_farias@blm.gov
Web Site: www.co.blm.gov/ahc
Founded: 1988.
Congressional District: 3
Key Personnel: Mgr., LouAnn Jacobson; Museum Shop Mgr., Robert DeNyke; Museum Shop Mgr., Diana Donohue.
Personnel Profile: Full-Time Paid 12; Part-Time Paid 2; Part-Time Volunteers 72; Interns 3.
Governing Authority: federal government; not-for-profit. Parent Institution: Bureau of Land Management. Tax-exempt.
Institution Type/Description: Archaeology Museum.
Collections: focus on the northern San Juan Ancestral Puebloan people from Basketmaker to Pueblo III periods; over 3 million artifacts from federal lands in SW Colorado; Native American cultures of the Four Corners region.
Research Fields: Southwest archaeology.
Facilities: research lab.
Activities: interpretation & visitor services.
Hours & Admission Prices: March-Oct. daily 9-5; Nov.-Feb. daily 10-4. March-Oct. adults $3; Nov.-Feb. no charge. America the Beautiful passes accepted. Closed New Year's Day; Thanksgiving; Christmas. &
Attendance: 22,066 (accurate)
Membership: Annual Pass $6.

Durango

ANIMAS MUSEUM, 3065 W. 2nd Ave., Durango, CO 81301-4209. Mailing Address: P.O. Box 3384, Durango, CO 81302-3384. Tel.: 970-259-2402.
E-mail: animasmuseum@frontier.net
Web Site: www.animasmuseum.org
Founded: 1978.
Congressional District: 3
Key Personnel: Dir., Carolyn Bowra; Pres. (V), Kathy McKenzie; Cur. Collections, Jan Postler; Museum Asst., Kellie G. Cheever.
Personnel Profile: Part-Time Paid 3; Part-Time Volunteers 50.
Governing Authority: private; nonprofit. Parent Institution: LaPlata Co. Historical Society. Tax-exempt: 501(c)(3).
Institution Type/Description: Historical Society Museum: housed in the former 1904 Animas City School building, a 3-story sandstone structure; a c.1870s hand hewn log cabin is also located on the grounds.
Collections: concentration on San Juan Basin history with a focus on La Plata County.
Major Exhibits: Journey Stories (T), 1/14-3/14.
Facilities: 225-vol. library of books relating to local & regional history/culture, southwest archaeology, genealogy and historical preservation; restored classroom; 3,000 sq. ft. exhibit space. Southwest Indian arts & crafts, books, cards and museum-related items for sale.
Activities: formal education programs for children; guided tours; lectures; school loan service; temporary exhibitions. Annual Events: Community Heritage Award; Christmas Bazaar; Durango Heritage Celebration; historically themed dinners; artists contests; internships.
Publications: newsletter published four times annually, Artifacts; History La Plata; cookbook.
Hours & Admission Prices: May-Oct. Mon.-Sat. 10-5; Nov.-April Tues.-Sat. 10-4. Adults $4, senior citizens $3, children 7-12 $2; discounts to groups & tours; children 6 and under & members no charge. &
Attendance: 8,230 (estimated)
Membership: Pioneer (over 65) & Student $25; Single $40; Family $50; Centennial $125; Small Commercial $150; Corporate $300; Chief Ouray (Single Life) $1,200; Otto Mears (Family Life) $1,800.

BARBARA CONRAD GALLERY - DURANGO ARTS CENTER, 802 E. Second Ave., Durango, CO 81301-5426. Tel.: 970-259-2606. Fax: 970-259-6571.
Founded: 1960.
Key Personnel: Exec. Dir., Sheri Rochford; Pres. (V), Terry Swan; Museum Shop Mgr., Mary Puller
Institution Type/Description: Art Gallery.
Collections: works by local & regional artists.
Facilities: Museum-related items for sale.
Activities: special events; educational programs.
Hours & Admission Prices: Tues.-Sat. 10-5. No charge; donations accepted. &
Attendance: 30,000 (estimated)

Membership: $50; $100; $500-$1,000; $1,001-$4,999; $5,000 & up.

CENTER OF SOUTHWEST STUDIES/FORT LEWIS COLLEGE, 1000 Rim Dr., Durango, CO 81301-3911. Tel.: 970-247-7456. Fax: 970-247-7422.
E-mail: brako_j@fortlewis.edu
Web Site: swcenter.fortlewis.edu
Founded: 1964.
Congressional District: 3
Key Personnel: Dir., Jay Harrison; FLC Pres., Dene Thomas; Cur., Jeanne Brako; Archivist, Nik Kendziorski; Coord. Special Events, Julie Tapley Booth; Asst. Librarian & Archivist, Jen Pack.
Personnel Profile: Full-Time Paid 5; Full-Time Volunteers 10; Part-Time Paid 3; Part-Time Volunteers 15; Interns 5.
Governing Authority: college; nonprofit organization. Fort Lewis College. Tax-exempt: 501(c)(3).
Institution Type/Description: Art Museum.
Collections: American Southwest with emphasis on Four Corners area; ranching heritage; ancestral Puebloan history; books; archives; art; anthropology; ethnology; archaeology; college memorabilia.
Research Fields: training Native Americans in cultural heritage management & museum studies.
Facilities: 15,000-vol. library; 105-seat auditorium; restaurant; classrooms; labs; 4,400 sq. ft. exhibit space; field research station.
Activities: arts festival; concerts; dance recitals; docent program; films; formal education programs; guided tours; hobby workshops; lectures; loan, participatory, temporary & traveling exhibitions; rental gallery; study clubs; training programs. Annual Events: Hozhoni Days in March; Hispanic Heritage Days in May.
Publications: occasional papers series; magazine, Timelines.
Hours & Admission Prices: Gallery: Mon.-Fri. 1-4, Sat. 12-4. No charge; donations accepted. Closed state & federal holidays. &
Attendance: 20,000 (estimated)
Membership: Basic $55; Sustaining $150.

DURANGO & SILVERTON NARROW GAUGE RAILROAD & MUSEUM, 479 Main Ave., Durango, CO 81301. Tel.: 970-247-2733; 877-872-4607.
Web Site: www.durangotrain.com
Institution Type/Description: History Museum: housed in a former railroad depot; built in 1882.
Collections: railroad history & artifacts; period furnishings.
Activities: train rides.
Hours & Admission Prices: Call for hours.

DURANGO DISCOVERY MUSEUM, 1333 Camino Del Rio, Durango, CO 81301. Tel.: 970-259-9234.
E-mail: info@durangodiscovery.org
Web Site: www.durangodiscovery.org
Formerly: Children's Museum of Durango
Congressional District: 3
Key Personnel: Exec. Dir., Chris Cable.
Governing Authority: Tax-exempt.
Institution Type/Description: Science Museum.
Collections: hands-on exhibits.
Facilities: 5,000 sq. ft. exhibit space.
Activities: special events; educational programs; rental facilities.
Hours & Admission Prices: Wed.-Sat. 10-5, Sun. 1-5. Adults $9.50; children 2 and under & members no charge.
Membership: Elemental Educators $75; Atom Family $80; Bunsen Buddies $160.

THE STRATER HOTEL, 699 Main Ave., Durango, CO 81301-5423. Tel.: 970-247-4431. Fax: 970-259-2208.
E-mail: mthom@strater.com
Web Site: www.strater.com
Founded: 1887.
Congressional District: 3
Key Personnel: C.E.O., Rod Barker.
Governing Authority: corp.
Institution Type/Description: Historic Building: c.1887 operative Victorian hotel & saloon.
Collections: 93 rooms of early American Victorian walnut furnishings 1860-1910 period; carved walnut beds, wash stands & armoires.
Research Fields: Victorian era.
Facilities: restaurant; theatre.
Activities: guided tours. Museum Sponsors: plays June to Sept.
Publications: book, Strater Hotel Story.

Hours & Admission Prices: Daily 24 hours. Regularly scheduled tours. No charge; donations accepted.
Attendance: 36,000 (estimated)

Eads

KIOWA COUNTY HISTORICAL MUSEUM, 1313 Maine St., Eads, CO 81036. Mailing Address: P.O. Box 100, Eads, CO 81036. Tel.: 719-438-5810 & 5847.
E-mail: plroper@plainsonline.net
Web Site: www.kiowacountycolo.com/museum.htm
Governing Authority: Parent Institution: Kiowa County Historical Society.
Institution Type/Description: Historical Society Museum.
Collections: local history & culture; period furnishings; personal artifacts; photographs.
Hours & Admission Prices: Memorial Day to mid-Sept. Mon.-Sat. 1-4:30; other times by appointment.

Eaton

A.J. EATON HOUSE MUSEUM, 207 Elm Ave., Eaton, CO 80615-3428. Tel.: 970-454-2456.
Key Personnel: Pres. (V), Roger Jordan.
Governing Authority: Tax-exempt.
Institution Type/Description: Historic House Museum: housed in the former home of A.J. Eaton.
Collections: US military, wars & flag history; uniforms; photographs; personal memorabilia; newspapers; magazines; scrapbooks.
Hours & Admission Prices: Tues., Thurs. & Sat. 2-4. No charge; donations accepted.
Attendance: 250 (accurate)
Membership: Single $10; Couple $15; Family $20; Corporate $50.

ANTIQUE WASHING MACHINE MUSEUM, 35901 WCR 31, Eaton, CO 80615. Tel.: 970-454-1856.
E-mail: lee@oldewash.com
Web Site: www.oldewash.com
Institution Type/Description: Washing Machine Museum.
Collections: over 1,400 period washing machines.
Publications: book, Save Womens Lives - History of Washing Machines.
Hours & Admission Prices: By appointment only.
Attendance: 1,500 (estimated)

Englewood

THE MUSEUM OF OUTDOOR ARTS, (M), 1000 Englewood Pkwy., Ste. #2-230, Englewood, CO 80110-2373. Tel.: 303-806-0444, ext. 301. Fax: 303-806-0504.
E-mail: thayes@moaonline.org
Web Site: www.moaonline.org
Founded: 1982.
Congressional District: 6
Key Personnel: Pres. & Exec. Dir., Cynthia Madden Leitner; C.O.O., Rodney Lontine; Administrative Dir., Tatum Hayes; Project Mgr., Timothy Vacca; Technical Dir., Kelley Bergmann; Research Coord. & Archivist, Paul Leitner; Creative Dir., Lonnie Hanzon; Construction Supvr., Schuyler Madden; Administrative Asst., Jessica Brack.
Personnel Profile: Full-Time Paid 8; Part-Time Paid 1; Part-Time Volunteers 6; Interns 20.
Governing Authority: nonprofit organization. Tax-exempt: 501(c)(3).
Institution Type/Description: Art Museum.
Collections: fine arts.
Research Fields: public art; outdoor performing arts.
Facilities: 18,000-seat amphitheatre.
Activities: concerts; arts festivals; arts education classes; outdoor public art; exhibitions; college student internships; collaborative performances.
Publications: collections & information brochure; walking tour guides.
Hours & Admission Prices: Gallery: Tues.-Thurs. 9-5, Fri. 9-4, Sat. 11-4; guided tours by arrangement. No charge; donations accepted. Closed federal holidays. &
Attendance: 250,000 (estimated)

Erie

SPIRIT OF FLIGHT CENTER & MUSEUM, 2650 S. Main St., Bldg. A, Erie, CO 80516. Tel.: 303-460-1156.
Governing Authority: nonprofit organization. Tax-exempt: 501(c)(3).
Institution Type/Description: Aviation History Museum.
Collections: aviation history & artifacts; aircraft & aircraft parts; art; uniforms; photographs; personal artifacts.

Activities: rental facilities.
Hours & Admission Prices: Mon.-Fri. 10-2; other times by appointment. Suggested Donation: $5.

Estes Park

ENOS MILLS CABIN MUSEUM & GALLERY, 6760 Hwy. 7, Estes Park, CO 80517-6404. Tel.: 970-586-4706.
E-mail: enosmillscbn@earthlink.net
Governing Authority: private.
Institution Type/Description: Historic House Museum: built in 1885 by 15 year old Kansan Enos A. Mills. Listed on the National Register of Historic Places.
Collections: personal artifacts; photographs; letters; books.
Hours & Admission Prices: By appointment. No charge; donations accepted.

ESTES PARK MUSEUM, 200 4th St., Estes Park, CO 80517-6339. Tel.: 970-586-6256. Fax: 970-577-3768.
E-mail: dfortini@estes.org
Web Site: www.estes.org/museum
Formerly: Estes Park Area Historical Museum
Founded: 1962.
Congressional District: 4
Key Personnel: Pres. Friends Bd., Patricia Washburn; Cur. Education, Alicia Mittelman; Cur. Collection, Bryon Hoerner; Mgr. & Cur. Exhibits, Derek Fortini; Museum Shop Mgr., Elaine Hunt-Downey.
Personnel Profile: Full-Time Paid 3; Part-Time Paid 1; Part-Time Volunteers 50.
Volunteer Hours: 7,234
Operating Expenses: 250,797
Operating Income: 12,400
Governing Authority: municipal. Parent Institution: Town of Estes Park. Tax-exempt.
Institution Type/Description: History Museum.
Collections: local history; photographs; textiles; archives; paintings; furnishings; Stanley Steamer automobile. Historic Buildings: early 20th century cabin; Rocky Mountain National Park administration building; 1909 Hydroplant Interpretive Center.
Research Fields: Estes Park history.
Activities: guided tours; permanent & temporary exhibits; lectures.
Publications: books: Weaving Mountain Memories; This Was Estes Park; The Ways of the Mountains; In the Vale of Elkanah; Early Narratives of Estes Park Vols. I-IV; Rocky Mountain Celts: The Scottish & Irish in Early Estes Park; Lost Links: The Search for Estes Park's Oldest Golf Course; If I Ever Grew Up & Became a Man: William Allen White's Moraine Park Years.
Hours & Admission Prices: May-Oct. Mon.-Sat. 10-5, Sun. 1-5; Nov.-April Fri.-Sat. 10-5, Sun. 1-5. No charge; donations accepted. Closed major holidays. &
Attendance: 22,687 (accurate)
Membership: Individual $30; Family $50; Business & Organization $100; Contributor $150; Sustaining $300; Patron $500; Donor $1,000.

MACGREGOR RANCH, 180 MacGregor Lane, Estes Park, CO 80517. Mailing Address: P.O. Box 4675, Estes Park, CO 80517-4675. Tel.: 970-586-3749. Fax: 970-586-1092.
E-mail: office@macgregorranch.org
Web Site: www.macgregorranch.org
Founded: 1973.
Personnel Profile: Full-Time Paid 2; Part-Time Volunteers 60.
Governing Authority: Tax-exempt.
Institution Type/Description: Historic House Museum: housed in the MacGregor family home, c.1896. Listed on the National Register of Historic Places.
Collections: period furnishings & agricultural equipment; personal artifacts; oil paintings; photographs. Historic Buildings: milkhouse; blacksmith shop; root cellar; smokehouse.
Facilities: nature trail; children's nature center.
Activities: museum & outbuilding tours; children's heritage camp; camping & hiking for children's groups. Annual Event: Children's Heritage Camp in summer.
Hours & Admission Prices: June-Aug. Tues.-Fri. 10-4; other times by appointment. Adults $5; children no charge.
Attendance: 2,700 (estimated)

RELIANCE FIRE MUSEUM, 460 Elm Rd., Estes Park, CO 80517. Tel.: 970-577-1953.
E-mail: douglink@airbits.com
Web Site: www.reliancefiremuseum.org
Institution Type/Description: Firefighting History Museum.
Collections: firefighting history & equipment; early fire trucks & apparatus;

photographs; personal artifacts; 1901 Waterous horse drawn steam pumper; 1958 Pirsch 100' tillered aerial.
Hours & Admission Prices: Call for hours.

ROCKY MOUNTAIN NATIONAL PARK, 1000 Highway 36, Estes Park, CO 80517-8311. Tel.: 970-586-1222. TTY: 970-586-1319.
Web Site: www.nps.gov/romo
Founded: 1915.
Congressional District: 2, 3 & 4
Key Personnel: Supt., Vaughn Baker.
Personnel Profile: Full-Time Paid 1; Part-Time Volunteers 3.
Governing Authority: federal. Parent Institution: Department of the Interior. Subsidiary Institution: National Park Service.
Institution Type/Description: Cultural & Natural History Museum.
Collections: herbarium; zoological; geology; botany; archaeological; historical; art.
Research Fields: Alpine ecology; archaeology history; glaciology; air quality; wildlife; natural history.
Facilities: Literature pertaining to surrounding national park for sale.
Activities: guided tours; talks; films; permanent exhibitions.
Hours & Admission Prices: Open 24 hours a day. 7-day vehicle entrance fee $20, 7-day individual entrance fee $10; Martin Luther King, Jr. Day, National Park Week, National Park Service Birthday, Public Lands Day & Veterans Day Weekend no charge. &
Attendance: 3,000,000 (estimated)

STANLEY STEAMCAR MUSEUM, 333 Wonderview Ave., Estes Park, CO 80517. Tel.: 970-577-1903. Fax: 970-577-1924.
E-mail: estespark@stanleymuseum.org
Founded: 1985.
Key Personnel: Dir., Donald Hoke
Institution Type/Description: Transportation Museum.
Collections: automotive & transportation history; 1909 Model R Stanley Roadster; personal artifacts; steam cars.
Publications: quarterly.
Hours & Admission Prices: June-Oct. daily 10-5; Nov.-May Wed.-Mon. 10-5; by appointment. No charge; donations accepted.

Evergreen

HIWAN HOMESTEAD MUSEUM, (M), 4208 S. Timbervale Dr., Evergreen, CO 80439-8456. Mailing Address: 700 Jefferson County Pkwy., Ste. 100, Golden, CO 80401-6025. Tel.: 720-497-7650. Fax: 303-670-7746.
E-mail: jsteinle@jeffco.us
Founded: 1974.
Congressional District: 25
Key Personnel: Dir., Tom Hoby; Admin., John Steinle.
Personnel Profile: Full-Time Paid 4; Part-Time Paid 14; Part-Time Volunteers 150.
Governing Authority: county. Parent Institution: Jefferson County Open Space. Tax-exempt: 501(c)(3).
Institution Type/Description: History Museum & Heritage Center: housed in 1880s 25-room log mansion, Camp Neosho, later renamed Hiwan Ranch.
Collections: Native American artifacts from the Douglas collection; Julia Douglas dolls; 6,000 catalogued historical photos; catalogued oral histories; manuscripts; Jefferson County historical items; religious objects; period furnishings.
Research Fields: Jefferson County history.
Facilities: 250-vol. library of local history & period books available for use on premises; nature trail; classrooms; Victorian gardens.
Activities: guided tours; lectures; docent program; formally organized education programs for children; special tours for blind, physically handicapped & nursing home seniors; after school specials; pioneer shop for printing & carpentry.
Publications: Indian Hills: The Place, The Times, The People; Mountain Memories; From Camp Neosho to the Hiwan Homestead; Evergreen; Upper Side of the Pie Crust.
Hours & Admission Prices: Jan. 8-May & Sept.-Dec. Tues.-Sun. 12-5; June-Aug. Tues.-Sun. 11-5. No charge; donations accepted.
Attendance: 11,986 (accurate)

HUMPHREY MEMORIAL PARK & MUSEUM, 620 Soda Creek Rd., Evergreen, CO 80439-9263. Tel.: 303-674-5429.
E-mail: angela.hmpm@yahoo.com
Founded: 1995.
Key Personnel: Dir., Angela Rayne.

Personnel Profile: Full-Time Paid 1; Part-Time Paid 1; Part-Time Volunteers 25.
Governing Authority: nonprofit organization. Tax-exempt: 501(c)(3).
Institution Type/Description: History Museum.
Collections: Humphrey family's personal artifacts; period furnishings; wall hangings; photographs; letters; scrapbooks.
Activities: living history interpretations of life 1930s & 1940s; classes; concerts; special events.
Hours & Admission Prices: Tues.-Sat. 10-4. Adults $5, seniors $4, children 2-12 $3; discounts to AAM & ICOM members. &

Fairplay

SOUTH PARK CITY MUSEUM, (M), 100 4th, Fairplay, CO 80440-0634. Mailing Address: P.O. Box 634, Fairplay, CO 80440-0634. Tel.: 719-836-2387. Fax: 719-836-9855.
E-mail: southparkhistorical@gmail.com
Web Site: www.southparkcity.org
Founded: 1957.
Congressional District: 3
Key Personnel: Finance Officer, Nancy Kreiling; Admin., Cindy Huelsman; Pres. (V), Harley Hamilton; Cur., Carol Davis.
Personnel Profile: Part-Time Paid 13; Part-Time Volunteers 9.
Volunteer Hours: 300
Operating Expenses: 233,970
Operating Income: 198,292
Governing Authority: nonprofit organization. Parent Institution: South Park Historical Foundation. Tax-exempt: 501(c)(3).
Institution Type/Description: Historic Village Museum: located on the site of a Colorado mining town.
Collections: furnishings; 34 representative pieces of architecture (1860-1900). Historic Buildings: c.1876 South Park Lager Beer Brewery; 1879 Summer Saloon. (Listed on the National Register of Historic Places).
Research Fields: culture & history of the South Park area 1860-1920.
Facilities: visitors center. Fine art reproductions & museum-related items for sale.
Activities: self-guided tours; permanent exhibitions; video.
Publications: video tour, A Town is Born; books, A Town is Born; South Park City.
Hours & Admission Prices: mid-May to late May daily 9-5; Memorial Day-Labor Day daily 9-7; Sept. to mid-Oct. daily 10-6. Adults $10, senior citizens 65 & over $8, children 6-12 $4; discounts to members of Time Travelers affiliated institutions AAM, ICOM, AAA members & groups; Park City students, members, handicapped & children under 6 no charge. &
Attendance: 14,738 (accurate)
Membership: Individual $20; Family $60.

Fleming

FLEMING HISTORICAL SOCIETY, Heritage Museum Park, Fleming, CO 80728. Mailing Address: P.O. Box 444, Fleming, CO 80728-0351. Tel.: 970-265-2591 & 3611.
E-mail: ntatencio@kcl.net
Founded: 1965.
Congressional District: 4
Key Personnel: Dir., C.E.O. & Pres. (V), Tom Atencio.
Personnel Profile: Full-Time Volunteers 6.
Governing Authority: society; nonprofit. Tax-exempt: 501(c)(3).
Institution Type/Description: Historical Society Museum.
Collections: local history; Indian artifacts. Historic Buildings: 1905 Philarado, a one-room schoolhouse; Burlington Depot.
Research Fields: history of the early settlers in this area.
Activities: permanent exhibitions. Museum Sponsors: Pioneer Day Activities in September.
Publications: books, Memories of Our Pioneers, Our Pioneer Heritage.
Hours & Admission Prices: Summer: by appointment. No charge; donations accepted. &
Attendance: 300 (estimated)

Florence

FLORENCE PIONEER MUSEUM, 100 E. Front St., Florence, CO 81226. Mailing Address: P.O. Box 131, Florence, CO 81226. Tel.: 719-784-1904.
Formerly: Price Pioneer Museum
Founded: 1964.
Congressional District: 3
Key Personnel: Pres., Richard Upton; Vice Pres., Sally Combs; Treas., Pauline Upton; Sec., Ben Combs; Museum Cur., Roberta Miller.
Personnel Profile: Part-Time Paid 1; Part-Time Volunteers 14.

Governing Authority: municipal. Parent Institution: Pioneer Day Board Assoc. Tax-exempt.
Institution Type/Description: Pioneer Museum: housed in 1894 building.
Collections: industrial; general; folklore; Indian artifacts; mineralogy; manuscripts. Historic House: 1875 first city jail.
Facilities: 300-vol. library of newspaper files, historic documents; scrapbooks; general reference files available for research on premises; reading room.
Activities: guided tours; permanent exhibitions.
Hours & Admission Prices: mid-May to Oct. Tues.-Sat. 1-4; other times by appointment. Adults $2. &
Attendance: 1,500 (estimated)
Membership: Silver Club $25; Platinum Club $50; Century Club $100; Lifetime $250.

Florissant

FLORISSANT FOSSIL BEDS NATIONAL MONUMENT, 15807 Teller County Rd. #1, Florissant, CO 80816. Mailing Address: P.O. Box 185, Florissant, CO 80816-0185. Tel.: 719-748-3253. Fax: 719-748-3164.
E-mail: keith.payne@nps.gov
Web Site: www.nps.gov/flfo
Founded: 1969.
Congressional District: 5
Key Personnel: Chief Interpretation & Resources Mgmt. & Visitor Protection, Rick Wilson; Supt., Keith Payne; Paleontologist, Dr. Herbert W. Meyer; Lead Interpreter & Volunteer Coord., Jeff Wolin; Museum Technician, Conni O'Connor.
Personnel Profile: Full-Time Paid 8; Part-Time Paid 1; Part-Time Volunteers 52; Interns 6.
Governing Authority: federal. Parent Institution: Department of the Interior, National Park Service, Washington, DC 20240. Tax-exempt.
Institution Type/Description: Historic House Museum.
Collections: fossil insects, leaves, fish, & several other categories of fossils of the Eocene period, approximately 34 to 35 million years ago; archaeological specimens 5,000 to 8,000 years old.
Research Fields: paleontology especially paleoentomology and paleobotany; geology; modern flora & fauna; history.
Facilities: library of florissant fossil & scientific books; hiking trails. Interpretive books & postcards for sale.
Activities: self-guided hiking trails year-round; ranger-guided tours during summer; off-site lectures; school programs on & off-site year-round; junior ranger.
Hours & Admission Prices: Daily 9-5. Adults $3; senior citizens, children under 16 & Federal Lands Pass holders no charge. Closed New Year's Day; Thanksgiving; Christmas. &
Attendance: 62,000 (accurate)

Fort Collins

AVENIR MUSEUM OF DESIGN AND MERCHANDISING, (M), Colorado State University, 1574 Campus Delivery, Fort Collins, CO 80523-1574. Tel.: 970-491-1983. Fax: 970-491-4376.
E-mail: linda.carlson@colostate.edu
Web Site: www.colostate.edu/depts/dm
Formerly: Gustafson Gallery/Design & Merchandising
Founded: 1986.
Congressional District: 4
Key Personnel: Cur., Linda Carlson.
Personnel Profile: Full-Time Paid 1; Part-Time Paid 1; Part-Time Volunteers 15; Interns 2.
Governing Authority: Colorado State University. Tax-exempt: 501(c)(3).
Institution Type/Description: Costume, Textiles & Interior Furnishings Museum.
Collections: 19th-20th century western dress; Asian costume & textiles; period chairs.
Research Fields: influence of films on fashion styles & trends; middle 19th-century men's dress in the American west; quilting traditions.
Facilities: 200-vol. library of costume history books & magazines; classrooms; labs.
Activities: exhibit openings with curator talks; formal education programs for undergraduate or graduate students affiliated with Colorado State University; guided tours; lectures; temporary exhibitions; traveling trunk program.
Publications: exhibit catalogs, Kimono; Women's Suits: Transformations in Form & Fabric, 1890-1990; Window to the World, 2009; Mr. Blackwell: A Retrospective, 2010.
Hours & Admission Prices: Mon.-Wed. & Fri. 11-6, Thurs. 11-8. No charge. Closed Thanksgiving; Christmas. &
Attendance: 500 (estimated)
Membership: Associate $20; General $40; Groups & Organizations $50; Sponsor $100; Benefactor $250.

AVERY HOUSE AND POUDRE LANDMARKS FOUNDA-TION, 328 W. Mountain Ave., Fort Collins, CO 80521-2702. Mailing Address: 108 N. Meldrum St., Fort Collins, CO 80521. Tel.: 970-221-0533.
E-mail: poudrelandmarks@gmail.com
Web Site: poudrelandmarks.org
Founded: 1972.
Congressional District: 4
Key Personnel: Chm. (V) Friends of Water Works, Bill Miller; Chm. (V) Avery House Historic District Guild, Lynda Lloyd; Pres. (V), Doug Ernest.
Personnel Profile: Part-Time Paid 2; Part-Time Volunteers 50.
Governing Authority: Parent Institution: Poudre Landmarks Foundation, Inc. Subsidiary Institution: Fort Collins Water Works, 2005 N. Overland Tr., Fort Collins, CO 80521. Tax-exempt.
Institution Type/Description: Historic House: former home of Franklin Avery, founder of the First National Bank; built in 1879. Listed on the National Register of Historic Places.
Collections: family history; period furnishings.
Activities: special events. Museum Sponsors: Fourth of July at Avery House; Fort Collins Historic Homes Tour; Big Splash at the Water Works; Christmas Open House at Avery House.
Publications: quarterly newsletter.
Hours & Admission Prices: Sat.-Sun. 1-4. No charge; donations requested. Closed New Year's Day; Christmas.
Attendance: 3,123 (accurate)
Membership: Brick and Mortar $35; Cornerstone $50; Keystone $100; Foundation $250; Structure $500; Architect $1,000.

BEE FAMILY CENTENNIAL FARM MUSEUM, 4320 E. County Rd., 58, Fort Collins, CO 80524-9326. Tel.: 970-482-9168. Facebook: Bee Family Farm.
E-mail: info@beefamilyfarm.com
Web Site: www.beefamilyfarm.org
Founded: 2004.
Key Personnel: Chm. (V) & Museum Shop Mgr., Liz Harrison.
Personnel Profile: Full-Time Volunteers 1; Part-Time Volunteers 15.
Governing Authority: nonprofit organization. Tax-exempt.
Institution Type/Description: Farm Museum.
Collections: family history; period farm equipment; early irrigation methods; personal artifacts; pioneer farming.
Activities: demonstrations; hands-on activities for children.
Publications: brochure; interpretive guide; introduction DVD.
Hours & Admission Prices: May-Oct. Fri.-Sat. 9-4; other times by appointment. Adults $7, seniors 60 & over $5, children 3-12 $3; members no charge. &
Attendance: 875 (accurate)
Membership: Individual & Senior Family $10; Family $15; Business $20.

BUDWEISER TOUR CENTER, 2351 Busch Dr., Fort Collins, CO 80524. Tel.: 970-490-4691.
E-mail: fortcollinsbrewerytour@budweisertours.com
Web Site: www.budweisertours.com
Institution Type/Description: Company Museum.
Collections: Budweiser company history & brewing process; photographs.
Facilities: Center-related items for sale.
Activities: guided tours; special events; beer tasting.
Hours & Admission Prices: June-Sept. daily 10-4; Oct.-May Thurs.-Mon. 10-4. No charge.

THE CENTER FOR FINE ART PHOTOGRAPHY, 400 N. College Ave., Fort Collins, CO 80524-2409. Tel.: 970-224-1010.
Institution Type/Description: Art Gallery.
Collections: photographs.
Activities: workshops; educational programs.
Hours & Admission Prices: Mon.-Fri. 9-5, Sat. 10-3.

COLORADO STATE UNIVERSITY ART MUSEUM, (M), 1400 Remington St., Fort Collins, CO 80523-1778. Mailing Address: Colorado State Univ., Campus Box 1778, Fort Collins, CO 80523-1778. Tel.: 970-491-1989.
E-mail: linda.frickman@colostate.edu
Web Site: www.artmuseum.colostate.edu
Formerly: Hatton Gallery, Colorado State University
Founded: 2008.
Congressional District: 4
Key Personnel: Dir., Linda Frickman; Program Coord., Keith Jentzsch; Collections Mgr., Suzanne Hale.

Personnel Profile: Full-Time Paid 3; Part-Time Volunteers 10; Interns 8.
Governing Authority: public university. Parent Institution: Colorado State University. Tax-exempt: 501(c)(3).
Institution Type/Description: University Art Museum.
Collections: Japanese prints; African art; South Seas art; modern & contemporary works on paper.
Research Fields: contemporary poster design; African art.; contemporary & Native American Art.
Activities: contemporary artists & art historical exhibitions; formal education programs for adults & undergraduate or graduate students affiliated with Colorado State University; K-12 partnerships; lectures; loan, temporary & traveling exhibitions.
Publications: critic & artist residency series catalogs; annual calendar.
Hours & Admission Prices: Tues.-Sat. 11-7. No charge. Closed university holidays; fall, spring & winter breaks. &
Attendance: 15,000 (accurate)

CURFMAN GALLERY, Colorado State University, Fort Collins, CO 80523. Mailing Address: 8033 Campus Delivery, Lory Student Center, Colorado State University, Fort Collins, CO 80523. Tel.: 970-491-2810. Fax: 970-491-3746.
E-mail: lsc_artsmanager@mail.colostate.edu
Web Site: www.curfman.colostate.edu
Founded: 1969.
Congressional District: 4
Key Personnel: Graduate Asst. & Dir., Nick Croghan; Designer, Jack Curfman.
Personnel Profile: Full-Time Paid 1; Part-Time Paid 7.
Governing Authority: public university. Parent Institution: Colorado State University. Tax-exempt.
Institution Type/Description: Exhibit Area & Gallery.
Collections: posters; Native American artifacts; African tribal artifacts.
Facilities: 1,700 sq. ft. exhibit area.
Activities: Annual Events: juried student exhibit; Biennial Event: Colorado International Invitational Poster Exhibition.
Publications: schedule of exhibits.
Hours & Admission Prices: mid-Jan. to May & Sept. to mid-Dec. Mon.-Thurs. 9-9, Fri. 9-9:30, Sat. 12-4. No charge; donations accepted. Closed university holidays. &
Attendance: 35,000 (estimated)

FORT COLLINS MUSEUM & DISCOVERY SCIENCE CENTER, (M), 408 Mason Ct., Fort Collins, CO 80524-4421. Tel.: 970-221-6738. Fax: 970-416-2236.
E-mail: ascott@fcgov.com
Web Site: fcmdsc.org
Founded: 1940.
Congressional District: 4
Key Personnel: City Mgr., Darin Atteberry; Mayor, Doug Hutchinson; Dir., Cheryl Donaldson; Asst. Dir., Brent Carmack; Public Rels. & Devel. Coord., Beth Higgins; Education Cur., Treloar Bower; NAGPRA Coord., Brenda Martin; Collections Mgr., Linda Moore; Exhibitions Technician, Cory Gundlach; Education Coord., Toby Swaford; Museum Shop Mgr., Amy Scott.
Personnel Profile: Full-Time Paid 5; Part-Time Paid 11; Part-Time Volunteers 60; Interns 3.
Governing Authority: municipal. Parent Institution: City of Fort Collins. Subsidiary Institution: Cultural, Library & Recreational Services. Tax-exempt.
Institution Type/Description: History Museum: housed in historic Carnegie library.
Collections: human interaction with nature through the history of the Cache La Poudre River Valley; Paleolithic material; American Indian artifacts; early military & pioneer materials of Larimer Co.: furniture, clothing, personal artifacts, tools & equipment. Historic Structures: Boxelder Schoolhouse (1905); Franz-Smith Cabin (1882); the Elizabeth Auntie Stone Cabin (1864); the Antoine Janis Cabin (1854).
Research Fields: history; Folsom archaeology; material culture; Western history.
Facilities: Museum-related items for sale.
Activities: special exhibits; guided group tours with reservation; outreach trunk program; workshops & classes; hands-on programs for families; special events. Museum Sponsors: Rendezvous in July; Carol Fest in December.
Publications: postcard notebook, Fort Collins Memories; book, Diary of Ann Sloan Sargisson.
Hours & Admission Prices: Tues.-Sat. 10-5, Sun. 12-5. Adults $4, seniors & children $3; members no charge. Closed national holidays. &
Attendance: 40,000 (accurate)
Membership: Individual $35; Grandparent $60; Family $70; Deluxe $85.

FORT COLLINS MUSEUM OF ART, INC., (M), 201 S. College Ave., Fort Collins, CO 80524-3182. Tel.: 970-482-2787. Fax: 970-482-0804.
E-mail: info@ftcma.org
Web Site: www.ftcma.org
Founded: 1990.
Congressional District: 4
Key Personnel: Chm. & Pres. (V), David Prosser; Exec. Dir., Marianne Lorenz.
Personnel Profile: Full-Time Paid 1; Part-Time Paid 2; Part-Time Volunteers 75; Interns 2.
Governing Authority: private; nonprofit. Tax-exempt: 501(c)(3).
Institution Type/Description: Art Museum.
Collections: works by regional & national artists.
Major Exhibits: Works by Warhol (T), 12/13/13-3/16/14; Masks, 4/4/14-5/2/14; Downstream: Encounters with the Colorado River (T), 5/16/14-7/3/14; The New West: Colorado Photography Now (T), 5/16/14-7/3/14; Marilyn: Celebrating an American Icon (T), 11/14-12/27/14.
Facilities: art library; art school.
Activities: gallery education programs; tours (including studio); lectures.
Publications: exhibit catalogs.
Hours & Admission Prices: Wed.-Fri. 10-5, Sat.-Sun. 12-5. Adults $4; discounts to AAM members, students & seniors; NARM & museum members no charge. Closed holidays. &
Attendance: 15,000 (estimated)
Membership: Student & Senior $25; Individual $40; Family $75; Contributor $150; Patron $500; Benefactor $1,000; Founder's Society $1,000-$5,000; Masterpiece $2,500.

Fort Garland

* **FORT GARLAND MUSEUM, (M),** 29477 Hwy. 159, Fort Garland, CO 81133. Mailing Address: P.O. Box 368, Fort Garland, CO 81133-0368. Tel.: 719-379-3512. Fax: 719-379-3479. www-.coloradohistroy.org.
E-mail: anita.mcdaniel@state.co.us
Web Site: www.fortgarlandmuseumfriends.org
Formerly: Old Fort Garland
Founded: 1945.
Congressional District: 3
Key Personnel: Chm. (V) & Pres. (V), Theresa Rudder; Museum Shop Mgr., Bella Maldonado.
Personnel Profile: Full-Time Paid 2; Part-Time Paid 2; Part-Time Volunteers 30.
Governing Authority: state. State Historical Society of Colorado, 1300 Broadway, Denver, CO 80203. Tax-exempt.
Institution Type/Description: Military & Pioneer Museum: housed in c.1858-1883 Old Fort Garland.
Collections: military; pioneer and Indian; dioramas pertaining to period of Fort operation, from Spanish exploration-1880.
Facilities: Historical publications & post cards for sale.
Activities: self-guided tours. Special Events: military reenactments.
Publications: monthly magazine, Colorado Heritage.
Hours & Admission Prices: April-Oct. daily 9-5; Nov.-March Wed.-Sat. 10-4. Adults $5, senior citizens $4.50, children $3.50; children under 6, active military & C.H.S. members no charge. Closed New Year's Day; Thanksgiving; Christmas. &
Attendance: 15,000 (accurate)
Membership: Senior $40; Seniors Family, Groups & Individual $50; Family & Group $60; Associate $65; Explorer $100; Centennial $250; Historian $500; Heritage Club $1,000; Millennium $2,100.

Fort Lupton

VINTAGE AERO FLYING MUSEUM, 7125 Parks Ln., Fort Lupton, CO 80621. Tel.: 303-502-5347.
Institution Type/Description: Military History Museum.
Collections: military aviation history; aircraft; personal artifacts; World War I & II memorabilia; photographs.
Hours & Admission Prices: Tues.-Sat. 10-4. Adults 12 & over $5.

Fort Morgan

FORT MORGAN MUSEUM, 414 Main St., City Park, Fort Morgan, CO 80701. Mailing Address: P.O. Box 184, Fort Morgan, CO 80701-0184. Tel.: 970-542-4009 & 4010. Fax: 970-542-4012.
E-mail: research@cityoffortmorgan.com
Web Site: www.cityoffortmorgan.com/museum
Founded: 1969.
Congressional District: 4
Key Personnel: Dir., Andrew Dunehoo; Cur., Nikkie Cooper.
Personnel Profile: Full-Time Paid 2; Part-Time Paid 1; Part-Time Volunteers 36; Interns 1.
Governing Authority: city.
Institution Type/Description: General Museum.
Collections: prehistoric & historic Indian relics; agricultural implements; ranching equipment; quilts; china; photographs; textiles; clothing; Fort Morgan Times on microfilm from 1884 to present; small Glenn Miller archive; small Great Western (beet) sugar archive; documents.
Research Fields: history of Morgan County & northeastern Colorado.
Facilities: research center; meeting room. Museum-related items for sale.
Activities: guided tours on request; lectures; docent program; inter-museum loan, permanent, temporary & traveling exhibitions; school loan service; school outreach program; research.
Publications: books, 111 Trees; Best of Friends Cookbook; Instructor's Delights: Cooking Schools I-IV; Early Fort Morgan; Fort Morgan; A Step Back in Time; Patchwork Chatelaine; Windows To The Past; From The Steppes To The Prairies; Memories of Morgan County; Platte Reflections; Baker...More Than A School.
Hours & Admission Prices: Mon. 9-6, Tues.-Thurs. 9-8, Fri.-Sat. 9-5. Research Room: Wed. & Fri. mornings, Tues.-Thurs. afternoons. No charge; donations accepted. Closed national holidays. &
Attendance: 13,500 (estimated)
Membership: Individual $20; Family $30; Sustaining $50; Patron $100; Benefactor $300; Philanthropist $500.

Fraser

COZENS RANCH MUSEUM, 77849 U.S. Hwy. 40, Fraser, CO 80442. Mailing Address: P.O. Box 165, Hot Sulphur Springs, CO 80451-0165. Tel.: 970-726-5488. Fax: 970-725-0129.
E-mail: crm@rkymtnhi.com
Web Site: www.grandcountymuseum.com
Founded: 1990.
Congressional District: 4
Key Personnel: Pres., Yvonne Knox; Treas. & Sec., Barbara Mitchell; Dir., Don Woster.
Personnel Profile: Full-Time Paid 3; Part-Time Paid 1; Part-Time Volunteers 25.
Governing Authority: nonprofit organization. Parent Institution: Grand County Historical Association. Tax-exempt: 501(c)(3).
Institution Type/Description: Historic House: 1874 ranch built by William & Mary Cozens, among the first homesteaders to ranch in the Fraser Valley.
Collections: c.1900 furnishings.
Research Fields: county settlement; transportation; county ranches; post offices.
Facilities: County history books & Cozens special interest items for sale.
Activities: lectures; docent program.
Publications: quarterly newsletter, The Spoke; annual journal.
Hours & Admission Prices: Memorial Day to Labor Day Tues.-Sat. 10-5; Sept.-May Wed.-Sat. 10-4. Families $10, adults $4, senior citizens $3, students $2; discounts to tour groups, seniors & locals; children under 6 & members no charge. &
Attendance: 1,836 (accurate)
Membership: Individual $25; Couple $40; Explorer $80; Historian $125; Pioneer $250; Heritage $500.

Frisco

FRISCO HISTORIC PARK & MUSEUM, (M), 120 Main St., Frisco, CO 80443. Mailing Address: P.O. Box 4100, Frisco, CO 80443-4100. Tel.: 970-668-3428. Fax: 970-668-0694.
E-mail: simoneb@townoffrisco.com
Web Site: www.townoffrisco.com
Formerly: Frisco Historical Society
Founded: 1983.
Congressional District: 2
Key Personnel: Dir. & Museum Shop Mgr., Simone Belz; Museum Coord., Kris Ann Knish; Museum Coord., Nancy Anderson.
Personnel Profile: Full-Time Paid 1; Part-Time Paid 3; Part-Time Volunteers 30.
Operating Expenses: 198,423
Operating Income: 18,290
Governing Authority: Parent Institution: Town of Frisco. Tax-exempt.
Institution Type/Description: History Museum: located in c.1900 Frisco Schoolhouse Museum.
Collections: local history, 1880s to 1960s; Town of Frisco & Summit County, CO.
Research Fields: local history.

Facilities: 50-seat Gazebo. Museum-related items for sale.

Activities: arts festivals; concerts; formal educational programs; guided tours; lectures; rental facilities. Annual Events: children's Halloween program; Founder's Day Celebration; Night at The Museum.

Hours & Admission Prices: May-Sept. Tues.-Sat. 9-5, Sun. 9-3; Oct.-April Tues.-Sat. 10-4, Sun. 10-2. No charge; donations accepted. Closed Thanksgiving; Christmas. ♿

Attendance: 35,000 (estimated)

Fruita

COLORADO NATIONAL MONUMENT, 1750 Rim Rock Rd., Fruita, CO 81521. Tel.: 970-858-3617 ext. 360. Fax: 970-858-0372.

Web Site: www.nps.gov

Founded: 1911.

Congressional District: 4

Key Personnel: Supt., Lisa Eckert; Chief Ranger, Mark Davison.

Governing Authority: federal. Parent Institution: National Park Service. Tax-exempt.

Institution Type/Description: Park Museum.

Collections: geology; cultural & natural history.

Facilities: 500-vol. library of natural history and geology books available for on-site use; study collection of plant & animal specimens.

Activities: research; interpretive programs.

Publications: NPS-CRM publication.

Hours & Admission Prices: Monument: daily 24 hours. Visitor Center: daily 9-5. 7-day vehicle $10, 7-day motorcycle & individual $5; Visitor Center no charge. Closed Christmas. ♿

Attendance: 385,000 (estimated)

✳ **DINOSAUR JOURNEY,** 550 Jurassic Court, Fruita, CO 81521-7707. Mailing Address: P.O. Box 2000, Grand Junction, CO 81502-5020. Tel.: 970-858-7282.

Web Site: www.dinosaurjourney.org

Institution Type/Description: Dinosaur Museum.

Collections: dinosaurs; dinosaur skeletons; hands-on exhibits; dinosaur history; fossils.

Hours & Admission Prices: Mon.-Sat. 10-4, Sun. 12-4. Adults $7, seniors $6, children $4; members no charge.

Gateway

GATEWAY COLORADO AUTOMOBILE MUSEUM, 43224 Hwy. 141, Gateway, CO 81522. Mailing Address: P.O. Box 339, Gateway, CO 81522-0339. Tel.: 970-931-2895.

Founded: 2006.

Key Personnel: Exec. Dir., Preston Patterson

Institution Type/Description: Automobile Museum.

Collections: over 40 cars depicting America's automobile history, science, design and social impact; hands-on exhibits.

Facilities: 30,000 sq. ft. exhibit space. Museum-related items for sale.

Activities: hands-on exhibits.

Publications: book, The Performing Art of The American Automobile.

Hours & Admission Prices: Sun.-Mon. 10-5, Tues.-Sat. 10-7. Adults $9, seniors 65 & over $7, youth 6-12 $5; discounts to groups; children 5 & under and members no charge. Closed Thanksgiving; Christmas.

Georgetown

GEORGETOWN ENERGY MUSEUM, 600 Griffith St., Georgetown, CO 80444. Mailing Address: 600 Griffith St., P.O. Box 398, Georgetown, CO 80444-0398. Tel.: 303-569-3557.

E-mail: gtnem@juno.com

Web Site: georgetownenergymuseum.org

Founded: 1900.

Governing Authority: nonprofit organization. Tax-exempt: 501(c)(3).

Institution Type/Description: History Museum: housed in fully functioning and operational hydroelectric generating plant.

Collections: hydroelectric power history; photographs; early electrical household appliances.

Activities: tours.

Hours & Admission Prices: Memorial Day to Oct. 1 Mon.-Sat. 11-4, Sun. 12-4; other times by appointment; groups by appointment. No charge; donations accepted.

HAMILL HOUSE MUSEUM, 305 Argentine, Georgetown, CO 80444. Mailing Address: P.O. Box 667, Georgetown, CO 80444-0667. Tel.: 303-569-2840. Fax: 303-569-2111.

E-mail: preservation@historicgeorgetown.org

Web Site: www.historicgeorgetown.org

Formerly: Historic Georgetown Inc.

Founded: 1970.

Congressional District: 4

Key Personnel: Exec. Dir., Sharon Rossino; Chm. (V), Dana Abrahamson; Dir. Membership, Tristen Greenleaf.

Personnel Profile: Full-Time Paid 1; Part-Time Paid 3; Part-Time Volunteers 50.

Governing Authority: society. Tax-exempt: 501(c)(3).

Institution Type/Description: Historic Houses Museum: located in the Georgetown-Silver Plume Historic District.

Collections: original furnishings & woodwork. Historic Buildings: c.1867-1881 William A. Hamill House & 1880-1881 Hamill office building, stable & carriage house; 1892 Bowman-White House; Tucker-Rutherford House; 1875-1880 Miner's cottage; 1870 log cabin.

Research Fields: general Georgetown history & architecture; biography of William A. Hamill; biographies of John H. Bowman & J. J. White; Christmas at Hamill House.

Facilities: Books & memorabilia of Colorado history & the Victorian period for sale.

Activities: tours; lectures; auctions; Biennial house tour; facility rental. Museum Sponsors: Christmas market; biennial Gala Auction Benefit.

Publications: quarterly newsletter; journal; Guide to the Georgetown-Silver Plume Historic District; William A. Hamill, The Gentleman from Clear Creek; Rise of the Silver Queen, History of Georgetown; Walking & Driving Tour of Georgetown; Star Crossings and Stone Monuments-the Wheeler Survey; annual report.

Hours & Admission Prices: Hamill House: June-Sept. daily 10-4; Winter: call for hours. Adults $6, seniors & students $4; discounts for AAA, Colorado Historical Society members & groups of 10 or more; member no charge. Closed New Year's Day; Thanksgiving; Christmas. Bowman-White House: closed for restoration. ♿

Attendance: 18,000 (estimated)

Membership: Annual Memberships, Domestic/Foreign: Senior/Student $20-$25; Regular $30-$35; Associate $50-$55; Sustaining $100-$105; Contributing $200-$205; Affiliate $350-$355; Supporting $500-$505. Lifetime Memberships: Johnson Log Cabin $1,500; Tucker/Rutherford $2,500; Kneisel House & Bowman-White House $10,000; Hamill House $25, 000.

HOTEL DE PARIS MUSEUM, 409 6th Ave., Georgetown, CO 80444. Mailing Address: P.O. Box 746, Georgetown, CO 80444. Tel.: 303-569-2311.

E-mail: kevin.kuharic@hoteldeparismuseum.org

Web Site: www.hoteldeparismuseum.org

Founded: 1954.

Congressional District: 2

Key Personnel: Chm. (V), Mary Riddle Clark; Pres. (V), Mary Anne Sehorn; Dir., Kevin Kuharic.

Personnel Profile: Full-Time Paid 1; Part-Time Paid 3; Part-Time Volunteers 35.

Governing Authority: nonprofit organization. Parent Institution: The National Society of the Colonial Dames of America in the State of Colorado (NASCDA in CO). Subsidiary Institution: National Trust for Historic Preservation. Tax-exempt.

Institution Type/Description: Historic Building Museum: built in 1875.

Collections: original furniture, dishes & kitchen utensils pertaining to a 19th-century hotel with a strong French influence; academic library.

Activities: guided tours; permanent exhibitions.

Publications: newsletter, The Hotel Register; pamphlet, The Hotel de Paris Museum; DVD, This Little Souvenir; booklets; tour booklets.

Hours & Admission Prices: Memorial Day-Labor Day daily 10-4:30; call for details on hours. Adults $5, senior citizens & children 6-12 $3, group tours over 10 people $5-$10 per person, National Trust for Historic Preservation members $2.50; discounts to AAA members; Children under 6, Les Amis Museum members & Georgetown-Silver Plume National Historic Landmark District residents no charge. ♿

Attendance: 7,000 (accurate)

Glenwood Springs

FRONTIER HISTORICAL SOCIETY AND MUSEUM, 1001 Colorado Avenue, Glenwood Springs, CO 81601-3319. Tel.: 970-945-4448. Facebook: Frontier Historical Society and Museum.

E-mail: history@rof.net

Web Site: www.glenwoodhistory.com

Founded: 1963.

Key Personnel: Exec. Dir., Cindy Hines; Pres., Bob Zanella; Vice Pres., Ann Maggard; Sec. & Treas., Sharon Haller; Archivist & Registrar, Patsy Stark; Education, Sue Plush.

Personnel Profile: Full-Time Paid 1; Part-Time Paid 2; Part-Time Volunteers 75.
Governing Authority: private; nonprofit organization. Tax-exempt: 501(c)(3).
Institution Type/Description: General Museum.
Collections: Glenwood Springs history particularly around 1900.
Facilities: 200-vol. library; 1,000 sq. ft. exhibit space. Museum-related items for sale.
Activities: formal education programs for children; guided tours; lectures; temporary exhibitions. Annual Event: Cemetery Ghost Walk.
Publications: quarterly newsletter, Frontier Times; monthly newspaper article, Frontier Diary.
Hours & Admission Prices: May-Oct. Mon.-Sat. 10-4; Nov.-April Mon. & Thurs.-Sat. 1-4. Adults $4, senior citizens $3, children 12 & under $2; discounts to groups; members no charge. Closed New Year's Day; Memorial Day; Independence Day; Labor Day; Thanksgiving; Christmas.
Attendance: 3,513 (accurate)
Membership: Old Timer $15; Old Timer Couple $20; Frontier $30; Vintage Family $50; Pioneer $100; Antique $250; Historian $500; Nostalgia Sponsor $1,000.

GLENWOOD RAILROAD MUSEUM, 413 7th St., Glenwood Springs, CO 81601-3442. Tel.: 970-945-7044.
E-mail: janandpat@sopris.net
Web Site: www.glenwoodrailroadmuseum.org
Founded: 2003.
Key Personnel: Gen. Mgr. & Museum Shop Mgr., Jan Girardot; Cur., Dick Helmke.
Personnel Profile: Part-Time Paid 4; Part-Time Volunteers 4.
Governing Authority: Parent Institution: Western Colorado Chapter, NRHS. Tax-exempt: 501(c)(3).
Institution Type/Description: Railroad Museum: housed in the historic Glenwood Springs Railroad Station; built in 1904.
Collections: railroad history, memorabilia & artifacts; photographs; large scale model railroad; telegraphs; railroad semaphore signals.
Facilities: Museum-related items for sale.
Activities: school tours.
Publications: monthly, Roaring Fork Foamer.
Hours & Admission Prices: Fri.-Mon. 11-3. Adults $1; members no charge.
Membership: National Chapter $46.

GLENWOOD SPRINGS CENTER FOR THE ARTS, 601 E. Sixth St., Glenwood Springs, CO 81601. Tel.: 970-945-2414.
Institution Type/Description: Art Gallery.
Collections: paintings; photographs; sculpture.
Activities: educational programs; summer camps; workshops.
Hours & Admission Prices: Call for hours.

Golden

BRADFORD WASHBURN AMERICAN MOUNTAINEERING MUSEUM, (M), 710 10th St., Golden, CO 80401. Tel.: 303-996-2755.
E-mail: info@mountaineeringmuseum.org
Web Site: mountaineeringmuseum.org
Institution Type/Description: Mountaineering Museum.
Collections: local history; mountaineering; photographs; personal artifacts; 135 sq. ft. scale model of Mount Everest & the Khumbu region.
Facilities: library.
Activities: educational programs; lectures; traveling exhibitions.
Hours & Admission Prices: Tues.-Fri. 10-5, Sat. 10-6, Sun. 11-4; other times by appointment. Adults $6.50, seniors & students $5, children $4.50; discounts to groups of 8 or more & members.

BUFFALO BILL MUSEUM & GRAVE, (M), 987 1/2 Lookout Mountain Rd., Golden, CO 80401-9646. Tel.: 303-526-0747 & 0744. Fax: 303-526-0197.
E-mail: buffalobill.museum@ci.denver.co.us
Web Site: www.buffalobill.org
Formerly: Buffalo Bill Memorial Museum
Founded: 1921.
Key Personnel: Dir., Steve Friesen; Educational Programs Coord., Betsy Martinson.
Personnel Profile: Full-Time Paid 5; Part-Time Volunteers 60.
Governing Authority: municipal. Affiliated with City & County of Denver, Dept. of Parks & Recreation, 1805 Bryant St., Denver, CO 80202. Tax-exempt: 170(b)(1)(A).
Institution Type/Description: History Museum: Founded by Johnny Baker, a close friend of Buffalo Bill & an important member of Buffalo Bill's Wild West Show.

Collections: items connected with the life & associations of William F. Cody, Buffalo Bill; Indian artifacts; paintings; photographs; manuscripts.
Major Exhibits: Formal, Folk and Funky: Buffalo Bill in Art, 2/26/14-1/20/15.
Research Fields: William F. Cody; Buffalo Bill's Wild West.
Facilities: 400-vol. library of Wild West Show programs, couriers, rosters & books available for use under supervision of museum personnel.
Activities: permanent & temporary exhibitions; special events; orientation video. Annual Event: Buffalo Bill's Birthday in February.
Publications: quarterly newsletter, Scout's Dispatch.
Hours & Admission Prices: May-Oct. daily 9-5; Nov.-April Tues.-Sun. 9-4. Adults $5, seniors 65 & over $4, children 6-15 $1; AAM members with membership card & children under 6 no charge. &
Attendance: 62,776 (accurate)
Membership: Individual $10; Family $15.

CLEAR CREEK HISTORY PARK, (M), 1020 11th St., Golden, CO 80401. Mailing Address: 923 10th St., Golden, CO 80401. Tel.: 303-278-3557. Fax: 303-278-8916.
Web Site: www.clearcreekhistorypark.org
Founded: 1999.
Key Personnel: Exec. Dir., Shannon Voirol
Institution Type/Description: Living History Park.
Collections: Homestead Structures: 2 cabins, barn, chicken coop, one-room schoolhouse.
Activities: guided tours; concerts; formally organized educational programs; docent programs; training program; children's living history day program.
Publications: quarterly newsletter, Dear Friends; monthly volunteer newsletter, The Friendly Reminder.
Hours & Admission Prices: May & Sept. Sat. 10-4:30; June-Aug. Tues.-Sat. 10-4:30, Sun. 11-3; Oct.-April groups by appointment. Adults $3, seniors $2.50, youth $2; discounts to AASLH & AAM members; children 5 & under and members no charge. Closed major holidays. &
Attendance: 9,610

COLORADO RAILROAD MUSEUM, (M), 17155 W. 44th Ave., Golden, CO 80403-1621. Mailing Address: P.O. Box 10, Golden, CO 80402-0010. Tel.: 303-279-4591; 800-365-6263. Fax: 303-279-4229.
Web Site: www.coloradorailroadmuseum.org
Founded: 1958.
Congressional District: 2
Key Personnel: Exec. Dir., Donald Tallman; Business Mgr., Bonnie Prater.
Personnel Profile: Full-Time Paid 6; Part-Time Paid 9; Part-Time Volunteers 360.
Governing Authority: nonprofit organization. Parent Institution: The Colorado Railroad Historical Foundation, Inc. Tax-exempt: 501(c)(3).
Institution Type/Description: Railroad Museum.
Collections: rolling stock of Colorado railroads; interior exhibits of Colorado railroad items, documents; photos; memorabilia; manuscripts.
Research Fields: railroads of Colorado & adjoining states.
Facilities: 10,000-vol. library of railroad related materials available by request; research facilities. Railroad books & museum-related items for sale.
Activities: lectures; films; slide & history shows; school talks; permanent & temporary exhibitions.
Publications: Iron Horse News & Colorado Rail Annual.
Hours & Admission Prices: Daily 9-5. Family (2 adults & children) $20, adults $10, senior citizens $8, children under 16 $5; discount to AAM members; members no charge. Closed New Year's Day; Thanksgiving; Christmas. &
Attendance: 97,506 (accurate)
Membership: Individual $30; Family $50; Family Premium $80; Patron $250; Life $1,000.

COLORADO SCHOOL OF MINES GEOLOGY MUSEUM, 1310 Maple St., Golden, CO 80401-1887. Tel.: 303-273-3815.
E-mail: geomuseum@mines.edu
Web Site: www.mines.edu/academic/geology/museum
Founded: 1874.
Congressional District: 2
Key Personnel: Dir., Bruce Geller.
Personnel Profile: Full-Time Paid 2, Part-Time Paid 12, Part-Time Volunteers 34.
Governing Authority: state. Parent Institution: Colorado School of Mines. Tax-exempt.
Institution Type/Description: Geology & Mineralogy Museum.
Collections: mineralogy; paleontology; mining; meteorites; gemstones.
Research Fields: mineralogy; paleontology.
Activities: guided tours; permanent & temporary exhibitions; teaching kits for loan; geology trail tours outdoors.
Hours & Admission Prices: Mon.-Sat. 9-4, Sun. 1-4. No charge; donations

accepted. Closed New Year's Day; Easter; Memorial Day; Independence Day; Thanksgiving; Christmas; CSM school holidays. &

Attendance: 16,000 (estimated)
Membership: Annual $30 & up.

FOOTHILLS ART CENTER, 809 15th St., Golden, CO 80401-1813. Tel.: 303-279-3922.

Web Site: www.foothillsartcenter.org
Key Personnel: Exec. Dir., Reilly Sanborn; Cur., Michael Chavez
Institution Type/Description: Art Museum.
Collections: works by local, regional & national artists.
Facilities: Museum-related items for sale.
Hours & Admission Prices: Mon.-Fri. 10-5, Sat. 12-5; call to confirm. Adults $5, seniors $3; members, children & students no charge. &
Membership: Senior $25; Individual $30; Senior Dual $35; Dual Family $40; Patron $100; Sustaining $250; Benefactor $500 & up.

GOLDEN BREWERY TOUR, 1221 Ford St., Golden, CO 80401. Tel.: 800-642-6116; 303-277-2337.

Institution Type/Description: Company Museum.
Collections: Coors company history & brewing process; photographs; neons; early beer cans, bottles & memorabilia.
Activities: guided tours; beer tasting; special events.
Hours & Admission Prices: Memorial Day to Labor Day Mon.-Sat. 10-4, Sun. 12-4; Sept.-May Thurs.-Mon. 10-4, Sun. 12-4. No charge.

GOLDEN HISTORY MUSEUMS, (M), 923 10th St., Golden, CO 80401-1112. Tel.: 303-278-3557. Fax: 303-278-8916.

E-mail: nrichie@goldenhistorymuseums.org
Web Site: www.goldenhistorymuseums.org
Formerly: Astor House Museum
Founded: 1938.
Key Personnel: Dir., Nathan Richie.
Personnel Profile: Full-Time Paid 5; Part-Time Paid 3; Part-Time Volunteers 50; Interns 1.
Governing Authority: city. Branch Museums: Golden History Center; Astor House Hotel; Clear Creek History Park. Tax-exempt.
Institution Type/Description: Historic House Museum: housed in c.1867 Astor House hotel.
Collections: decorative art of the frontier period; furniture; clothing; historical objects from collections predate 1908. Historic House: c.1867 Astor House hotel.
Research Fields: local history; architectural survey of city; late 19th-century decorative arts & architecture.
Activities: guided tours; lectures; formally organized education programs for children; docent program or council; training programs; loan, permanent & temporary exhibitions; call for information on living history demonstrations & living history day camp for children. Museum Sponsors: Tea Time at the Astor House.
Publications: semi-annual newsletter; monthly e-news.
Hours & Admission Prices: Tues.-Sat. 10-4:30. Adults $4, senior citizens 65 & over $3, children 6-16 $2; discount to AAM & ICOM members; members no charge. Closed major holidays. &
Attendance: 13,300 (accurate)
Membership: Individual $20; Family $40; Supporter (1 year) $60; Supporter (2 years) $90; Patron $250; Contributor (1 year) $150; Contributor (2 years) $270; Benefactor $500; Founder $1,000.

GOLDEN OLDY CYCLERY & SUSTAINABILITY, (M), 17224 W. 17th Place, Golden, CO 80401-2509. Tel.: 720-497-1100.

E-mail: oldbike2@comcast.net
Web Site: www.goldenoldy.org
Founded: 2000.
Congressional District: 7
Key Personnel: Dir., C.E.O. & Chm. (V), Steve Stevens.
Personnel Profile: Full-Time Volunteers 1; Part-Time Volunteers 15.
Institution Type/Description: Bicycle & Living Museum of Sustainability.
Collections: late 1800s bicycles; 1889 bicycle repair shop; early bicycle headlamps; period clothing & accessories; over 500 poems; history & people of cycling; photographs; cycling publications; living tropical plants including banana, orange, lemon, & lime.
Research Fields: Victorian bicycle poetry; built environmental sustainability; permaculture.
Facilities: library.
Activities: research; education.
Hours & Admission Prices: Call for hours. No charge, donations accepted.
Attendance: 1,100 (estimated)

ROCKY MOUNTAIN QUILT MUSEUM, 1213 Washington Ave., Golden, CO 80401-1144. Mailing Address: 651 Corporate Cir., Ste. 102, Golden, CO 80401-5652. Tel.: 303-277-0377. Fax: 303-215-1636.

E-mail: rmqm@rmqm.org
Web Site: www.rmqm.org
Founded: 1982.
Congressional District: 6
Key Personnel: Exec. Dir. & Mgr. Collections & Exhibitions, Karen Roxburgh; Pres. (V), Cindy Harp; Finance Mgr., Rhonda Schneider; Volunteer Mgr., Kathy Williams.
Personnel Profile: Full-Time Paid 1; Part-Time Paid 7; Part-Time Volunteers 111.
Governing Authority: nonprofit organization. Tax-exempt: 501(c)(3).
Institution Type/Description: Quilt Museum.
Collections: quilts from 1800s to contemporary times.
Research Fields: quilts; textiles.
Facilities: 4,000-vol. research library; 2,000 sq. ft. exhibit space. Museum related items for sale.
Activities: guided tours; temporary exhibitions. Annual Events: Fabric & Friends-live & silent auction; Hundred Club Tea; garage sale; trunk shows.
Publications: weekly E-newsletter.
Hours & Admission Prices: Mon.-Sat. 10-4. Adults $6, seniors $5, children 6-12 $3; discounts to groups over 15; children 5 & under, 1st Fri. of quarter and members no charge; active military half price. Closed New Year's Day; Thanksgiving; Christmas. &
Attendance: 12,000 (accurate)
Membership: Individual $30; Family $45; Patron $70; Sponsor $125; Associate $250; Benefactor $500; Opal Frey Sustaining $1,000; Eugenia Mitchell Lifetime $5,000.

Grand Junction

✳ **CROSS ORCHARDS HISTORIC FARM,** 3073 F Rd., Grand Junction, CO 81504. Mailing Address: P.O. Box 20000, Grand Junction, CO 81502-5020. Tel.: 970-434-9814. Fax: 970-242-3960.

Institution Type/Description: Historic Site: housed on an early 1900s apple orchard. National Register of Historic Sites & Places.
Collections: early 1900s furnishings & farm equipment; train cars; Grand Valley pioneers; bunkhouse; barn; Uintah narrow gauge railway cars.
Hours & Admission Prices: April 10 to mid-Oct. Tues.-Sat. 9-4. Adults $4, seniors $3, children $2.50.

THE MUSEUM AT ALLEN UNIQUE AUTOS, 2285 River Rd., Grand Junction, CO 81505. Tel.: 970-263-7410.

Institution Type/Description: Automobile Museum.
Collections: over 80 automobiles.
Hours & Admission Prices: Mon.-Sat. 11-4.

✳ **MUSEUM OF WESTERN COLORADO,** 462 Ute Ave., Grand Junction, CO 81501-2516. Mailing Address: P.O. Box 20000, Grand Junction, CO 81502-5020. Tel.: 970-242-0971. Fax: 970-242-3960.

Web Site: www.museumofwesternco.org
Founded: 1965.
Congressional District: 4
Key Personnel: Exec. Dir., Peter M. Booth; Chm. Bd. & Pres., Laurena Mayne Davis; Cur. Paleontology, Dr. John R. Foster; Cur. History & Dir. Western Investigations Team, David Bailey; Cur. Collections & Registrar, Zebulon Miracle; Asst. Dir. Operations, Kay Fiegel; Dir. Lloyd Files Research Library & Cur. Archives, Michael Menard; Business Mgr., Erik Vliek; Facilities Mgr., Don Kerven; Membership, Ronna Lee Sharpe; Museum Shop Mgr., Sally D'Augustino.
Personnel Profile: Full-Time Paid 12; Part-Time Paid 8; Part-Time Volunteers 265; Interns 4.
Governing Authority: private; nonprofit. Branch Sites: Cross Orchards Living History Site, 3073 F Rd., Grand Junction. Tel.: 970-434-9814; Museum of the West, 462 Ute Ave. Grand Junction, CO 81501; Whitman Education Bldg., 248 S. Fourth St.; Dinosaur Journey, 550 Jurassic Ct., Fruita, CO 81521, Tel: 970-858-7282. Tax-exempt: 501(c)(3).
Institution Type/Description: Cultural & Natural History Museum.
Collections: small arms; costumes; archaeology; paleontology; geology; mineralogy; archives; ethnology; botany; zoology; Western Colorado history.
Research Fields: western Colorado history; paleontology; anthropology; genealogy.
Facilities: auditorium; classrooms. Outdoor Sites: Rabbit Valley research natural area. Gifts & museum-related items for sale.
Activities: guided tours; lectures; inter-museum loan; permanent, temporary & traveling exhibitions; school loan service; educational programs; archival

research; children's & adult programs; living history farm; animated dinosaur exhibit; dinosaur digs; interpretive walks; Fruita Paleo area; Dinosaur Hill; Riggs Hill.

Publications: books, Familiar Insects of Mesa County; Biography of Cinosauria 1667-1986; Mesa County Colorado: A 100 Year History; Grand Heritage: A Photographic History of Grand Junction; Cross Times Cookin 1896-1996; 125 Years 125 People-A History of Grand Junction; tri-annual museum journal, Pathways; quarterly newsletter, Museum Times; monthly activity calendars; Distant Treasures in the Mist, book & movie.

Hours & Admission Prices: History Museum: Winter: Tues.-Sat. 10-3; Summer: Tues.-Sat. 9-5. Family $16, adults $5.50, seniors $4.50, children $3. Cross Orchards Living History Site: May-Oct. open for special events, school tours, & rentals. Call for hours. Dinosaur Journey: daily 9-5. Family $20, adults $7, seniors $6, children $4; discounts to AAA, CCA & Time Traveler members. ♿

Attendance: 105,000 (estimated)

Membership: Senior $20; student $25; Senior Couple $30; Individual $35; Grandparent $40; Family $50; Supporting $100; Sustaining $250; Patron $500; Benefactor $1,000.

WESTERN COLORADO CENTER FOR THE ARTS DBA THE ART CENTER, 1803 N. 7th St., Grand Junction, CO 81501-3009. Tel.: 970-243-7337. Fax: 970-243-2482.

E-mail: info@gjartcenter.org
Web Site: www.gjartcenter.org
Founded: 1953.
Congressional District: 3
Key Personnel: Dir., Cheryl McNab; Pres. (V), Robbie Breax; Museum Shop Mgr., Carolyn Gillette.
Personnel Profile: Full-Time Paid 4; Part-Time Paid 5; Interns 1.
Institution Type/Description: Art Gallery.
Collections: works by Western artists; Navajo blankets & rugs; Anasazi pottery; Paul Pletka, Harold Bryant & Alfred Nestler works; contemporary Colorado paintings & ceramics.
Facilities: Museum-related items for sale.
Activities: art classes; workshops.
Hours & Admission Prices: Tues.-Sat. 9-4. Adults $3; members, children under 12 & Tues. no charge.
Attendance: 30,000 (estimated)
Membership: Senior Citizen & Student $25; Senior Couple & Individual $35; Family $50.

Grand Lake

GRAND LAKE AREA HISTORICAL SOCIETY, (M), 407 Pitkin St., Grand Lake, CO 80447-0656. Mailing Address: Box 656, Grand Lake, CO 80447-0656. Tel.: 970-627-9644. Facebook: Smith Eslick Cottage Court.

E-mail: glhistory@rkymtnhi.com
Web Site: www.grandlakehistory.org
Founded: 1973.
Key Personnel: Pres., Jim Cervenka; Treas., Patti Stahl; Sec., Elin Capps.
Governing Authority: nonprofit organization; society. A member of the State Historical Society of Colorado, Colorado State Museum, 200 Fourteenth Ave., Denver, CO 80203. Tax-exempt: 170(b)(1)(A).
Institution Type/Description: Historical Society Museum.
Collections: furniture, furnishings & artifacts of the period; manuscripts. Historic Buildings: 1892 Ezra Kauffman Tourist Hotel/Residence; Smith Eslick Cottage Camp.
Research Fields: Grand Lake from the time of the Ute Indians-present.
Activities: guided tours; self-guided walking tours; lectures; films; gallery talks; formally organized education programs; docent program; permanent exhibitions.
Publications: Stagecoach Stops; Where the Colorado River Began; The Kauffman House; Grand Lakes 100th Year 1881-1981; A Walking Tour of Historic Grand Lake.
Hours & Admission Prices: Memorial Day to Labor Day daily 11-5; Sept. Sat.-Sun. 11-5; tours & other times by appointment. Adults $5; children 12 & under and members no charge. ♿
Attendance: 5,500 (accurate)
Membership: Individual $25; Family $40; Business $50; Sustaining $100-$1,000.

Greeley

CENTENNIAL VILLAGE MUSEUM, 1475 A St., Greeley, CO 80631-2185. Mailing Address: 714 8th St, Greeley, CO 80631-3910. Tel.: 970-350-9220. Fax: 970-350-9700.
E-mail: museums@greeleygov.com
Web Site: www.centennialvillagemuseum.com

Founded: 1976.
Congressional District: 4
Key Personnel: Museum Mgr., Daniel Perry; Cur. Education & Living History, Bill Armstrong; Archives & Digitization Specialist, Caroline Blackburn; Devel. Cur., Peggy Ford Waldo; Asst. Cur. Exhibits, Angela Alton; Registrar, JoAnna Luth Stull; Coord. Historic Sites, Randy Smith; Exhibits Cur., Nancy Lynch; Mktg. Specialist, Emily Lance; Mgr. Collections, Diane Karlson; Asst. Cur. Education, Scott Chartier.
Personnel Profile: Full-Time Paid 11; Part-Time Paid 10; Part-Time Volunteers 350; Interns 9.
Volunteer Hours: 12,476
Governing Authority: municipal. Parent Institution: City of Greeley Museums, culture, parks & recreation department, Division of Culture. Tax-exempt.
Institution Type/Description: Living History Museum.
Collections: 30 historic structures; personal artifacts; 1860-1940 furnishings.
Major Exhibits: Cattle Trails and Tales, 11/13-10/1/16.
Research Fields: local, county & regional history.
Facilities: meeting rooms; special events area; historical homes, vehicles, structures. Books, note cards & other museum-related items for sale.
Activities: festivals; guided tours; organized educational programs for children; permanent, temporary exhibitions; lectures; hands-on living history programs; demonstrations; seasonal activities; themed living history days. Museum Sponsors: History Fest; Potato Day; Weld County Fair, Centennial Celebration, Greeley Stampede Historic Village, Howl-o-ween Trick or Treat.
Publications: newsletter; book, Weld County Towns: The 1st 150 Years (2011).
Hours & Admission Prices: May-Sept. Wed.-Sun. 10-4. Adult $6, senior citizens 60 & over $4, children 3-17 $3; discounts to Time Travelers, active military, AAM, ICOM & AAA members; children 2 & under and members no charge. ♿
Attendance: 16,755 (accurate)
Membership: Student $15; Senior $30; Individual $35; Dual $40; Family $45; Contributor $100; Special Donor $250 & up.

GREELEY FREIGHT STATION MUSEUM, 680 10th St., Greeley, CO 80631. Tel.: 970-392-2934.
E-mail: info@gfsm.org
Web Site: www.gfsm.org
Key Personnel: Directing Mgr., Dave Trussell; Exec. Dir., Michelle Kempema
Institution Type/Description: Hobby Museum.
Collections: over 600 railroad-related artifacts; over 2,500 model railroad cars.
Hours & Admission Prices: Memorial Day to Labor Day Wed.-Sat. 10-4, Sun. 1-4; Sept.-May Fri.-Sat. 10-4, Sun. 1-4. Adults $8, seniors 65 & over $6, children 12 & under $4; discounts to military; children 3 & under no charge.

GREELEY HISTORY MUSEUM, 714 8th Street, Greeley, CO 80631-3910. Tel.: 970-350-9220. Fax: 970-350-9570.
E-mail: museums@greeleygov.com
Web Site: www.greeleymuseums.com
Formerly: Municipal Archives
Founded: 2005.
Congressional District: 4
Key Personnel: Mgr., Daniel Perry; Cur. Education & Living History, Bill Armstrong; Cur. Exhibits, Nancy Lynch; Registrar, JoAnna Luth Stull; Collections Specialist, Diane Karlson; Archives & Digitization Specialist, Caroline Blackburn; Asst. Cur. Exhibits, Angela Alton; Asst. Cur. Education, Scott Chartier; Devel. Cur., Peggy Ford Waldo.
Personnel Profile: Full-Time Paid 11; Part-Time Paid 10; Part-Time Volunteers 350; Interns 9.
Volunteer Hours: 12,476
Governing Authority: municipal. Parent Institution: City of Greeley, culture, parks & Recreation Department, Division of Culture. Tax-exempt.
Institution Type/Description: History Museum.
Collections: city records; photographs; archival documents; scrapbooks; maps; books.
Major Exhibits: Fight! Work! Save! 1940s Homefront, 3/16/13-12/31/14; Bicycles of Greeley, 12/1/13-2/28/15; Weld Towns: 1861-Today (T), 12/15/13-4/15/14.
Research Fields: local, county & regional history.
Facilities: 8,000-vol. library of images, genealogy records, records for local business, schools & events.
Activities: guided tours; organized education programs; first Friday presentations; seasonal activities; special events; hands-on interactives.
Publications: e-newsletter; book, Weld County Town: The First 150 Years.
Hours & Admission Prices: Wed.-Fri. 8:30-4:30, Sat. 10-4. Adults $2, seniors (60+) & Children 3-17 $1, members & children 0-2 no charge. Closed major holidays. ♿
Attendance: 5,592 (estimated)

Membership: Student (with valid ID) $15; Senior $30; Individual $35; Dual $40; Family $45; Contributor $100; Special Donor $250 & up.

MARIANI GALLERY, UNIVERSITY OF NORTHERN COLORADO, 8th Ave. & 18th St., Greeley, CO 80639. Mailing Address: School of Visual Arts Galleries, Campus Box 30, Guggenheim Hall University of Northern CO, Greeley, CO 80639. Tel.: 970-351-2184. Fax: 970-351-2299.
Web Site: www.arts.unco.edu/visarts/visarts_galleries.html
Founded: 1972.
Key Personnel: Dir., Joan Shannon-Miller.
Personnel Profile: Part-Time Paid 7.
Governing Authority: public university; not-for-profit organization. Parent Institution: University of Northern Colorado. Subsidiary Institution: Oak Room Gallery. Tax-exempt.
Institution Type/Description: University Art Gallery.
Collections: works by contemporary artists.
Hours & Admission Prices: Sept.-May Mon.-Tues. & Thurs.-Fri. 10-3, Wed. 1-6. No charge. &
Attendance: 2,500 (estimated)

MEEKER HOME MUSEUM, 1324 9th Ave., Greeley, CO 80631-4608. Mailing Address: 714 8th St., Greeley, CO 80631-3910. Tel.: 970-350-9220. Fax: 970-350-9570.
E-mail: museums@greeleygov.com
Web Site: www.greeleymuseums.com
Founded: 1929.
Congressional District: 4
Key Personnel: Mgr., Daniel Perry; Coord. Historic Sites, Randy Smith; Cur. Exhibits, Nancy Lynch; Cur. Education & Living History, Bill Armstrong; Registrar, JoAnna Luth Stull; Mgr. Collections, Diane Karlson.
Personnel Profile: Full-Time Paid 11; Part-Time Paid 10; Part-Time Volunteers 350; Interns 9.
Volunteer Hours: 12,476
Governing Authority: municipal; nonprofit organization. Parent Institution: City of Greeley, Culture, Parks, and Recreation Dept. Tax-exempt.
Institution Type/Description: Historic House: 1870 home of Nathan C. Meeker, founder of Greeley and his family.
Collections: personal artifacts; material culture of local area, 1870-1882.
Major Exhibits: Underwear to Outerwear, 1/1/14-9/30/14.
Research Fields: local, county & regional history.
Facilities: rental facilities.
Activities: guided tours; organized educational programs for children; seasonal activities; lectures; programs. Museum Sponsors: Founder's Day; First Friday Event: Robert Silbernagel, author of Troubled Trails (2012).
Publications: newsletter.
Hours & Admission Prices: May-Oct. 1st Sat. each month 10-4; other times by appointment. &
Attendance: 639 (accurate)
Membership: Student $15; Senior $30; Individual $35; Dual $40; Family $45; Contributor $100; Special Donor $250 & up.

PLUMB FARM LEARNING CENTER, 955 39th Ave., Greeley, CO 80634-1549. Mailing Address: 714 8th St., Greeley, CO 80631-3910. Tel.: 970-350-9220. Fax: 970-350-9570.
E-mail: museums@greeleygov.com
Web Site: www.greeleymuseums.com
Formerly: Plumb Farm Museum
Founded: 1997.
Congressional District: 4
Key Personnel: Mgr., Daniel Perry; Cur. Education & Living History, Bill Armstrong; Cur. Exhibits, Nancy Lynch; Coord. Historic Sites, Randy Smith.
Personnel Profile: Full-Time Paid 11; Part-Time Paid 10; Part-Time Volunteers 350; Interns 9.
Volunteer Hours: 12,476
Governing Authority: municipal. Parent Institution: City of Greeley, Department of Leisure Services, Division of Museums. Tax-exempt.
Institution Type/Description: Historic Farm Museum.
Collections: farm house; potato cellar; several out buildings; gardens; 1870 flood irrigation system; farm tools & equipment.
Research Fields: local, county & regional agricultural history.
Facilities: picnic area.
Activities: seasonal activities. Museum Sponsors: Baby Animal Days in April & May; Pets 'n Popsicles in August.
Publications: newsletter.
Hours & Admission Prices: Special Event Days: call for information. &
Attendance: 6,769 (accurate)

Membership: Student $15; Senior $30; Individual $35; Dual $40; Family $45; Contributor $100; Special Donor $250 & up.

Gunnison

GUNNISON ARTS CENTER, 102 S. Main St., Gunnison, CO 81230. Mailing Address: P.O. Box 1772, Gunnison, CO 81230-1772. Tel.: 970-641-4029.
Institution Type/Description: Art Center.
Collections: works by local artists.
Facilities: Museum-related items for sale.
Activities: visual art classes; workshops.
Hours & Admission Prices: Tues.-Fri. 10-6, Sat. 10-4. No charge.

GUNNISON COUNTY PIONEER AND HISTORICAL SOCIETY, S. Adams St. & Hwy. 50, Gunnison, CO 81230. Mailing Address: P.O. Box 824, Gunnison, CO 81230-0824. Tel.: 970-641-4530.
Founded: 1930.
Key Personnel: C.E.O. & Pres., C.J. Miller.
Personnel Profile: Full-Time Volunteers 12; Part-Time Volunteers 40.
Governing Authority: nonprofit. Tax-exempt: 501(c)(3).
Institution Type/Description: Historical Society Museum.
Collections: geology; costumes; mineralogy; transportation; railroad artifacts; agriculture; period cars; 50 vintage cars; telephones & telephone equipment. Historic Buildings: 1890 Log Barn; 1904 Paragon School House; 1890 Train Depot; c.1880 Log Post Office; 1881 narrow-gauge railroad train.
Research Fields: historic buildings.
Facilities: 500-vol. library of old textbooks from Paragon School & bound volumes of Gunnison newspapers, 1880-1950.
Activities: guided tours; permanent exhibitions.
Publications: annual news sheet, Gunnison Museum.
Hours & Admission Prices: Memorial Day to Sept. Mon.-Sat. 9-5, Sun. 11-5. Adults $7, children 6-12 $3; discounts to groups of 10 or more.
Attendance: 5,000 (accurate)
Membership: Individual $20.

Holyoke

PHILLIPS COUNTY MUSEUM, 109 S. Campbell Ave., Holyoke, CO 80734-1501. Tel.: 970-854-2129.
E-mail: pcmuseum@pctelcom.coop
Founded: 1967.
Congressional District: 4
Key Personnel: Pres., Mike Coyne; Sec., Carrie Anderson; Treas., Hilda Hassler.
Governing Authority: nonprofit society. Parent Institution: Phillips County Historical Society. Tax-exempt: 501(c)(3).
Institution Type/Description: Historical Society Museum.
Collections: agricultural tools; pioneer household items; dolls; photographs; Native American artifacts; weapons; furnishings from the 1920s; vintage filling station; one-room schoolhouse.
Research Fields: local history.
Facilities: 400-vol. library of books & microfilm of old newspapers, available for use on the premises; reading room.
Activities: lectures; special guest exhibitions; informational & demonstrations programs.
Publications: book, Those Were The Days, Vols. I, II & III.
Hours & Admission Prices: Memorial Day to Labor Day Sun. 2-4:30, Wed. 2-4; other times by appointment. No charge; donations accepted. &
Attendance: 850 (estimated)
Membership: Individual $5; Family $15; Business $25.

Hot Sulphur Springs

GRAND COUNTY MUSEUM, 110 E. Byers, Hot Sulphur Springs, CO 80451. Mailing Address: P.O. Box 165, Hot Sulphur Springs, CO 80451-0165. Tel.: 970-725-3939. Fax: 970-725-0129.
E-mail: gcha@grandcountymuseum.com
Web Site: www.grandcountymuseum.com
Founded: 1974.
Congressional District: 4
Key Personnel: C.E.O. & Dir., Don Woster; Pres. (V), Yvonne Knox; Treas. & Sec., Barbara Mitchell.
Personnel Profile: Full-Time Paid 2; Part-Time Paid 1; Part-Time Volunteers 50.
Governing Authority: nonprofit organization. Parent Institution: Grand County Historical Association. Tax-exempt.

Institution Type/Description: History Museum: housed in c.1924 school house.
Collections: pertaining to the settlement of Grand County. Historical Buildings: 1870 Grand County Pioneer Museum; 1924 Grand County Museum; 1915 ranch house; blacksmith shop; Grand County court house; Grand County jail; Eight Mile school.
Research Fields: Windy Gap Indian digs; county history.
Facilities: 7,000 photographs on file & manuscripts; library of records available for use under supervision only.
Activities: guided tours; oral history; site designations; school programs; permanent exhibitions; special displays; historic treks; lectures; receptions; programs. Museum Sponsors: two annual meetings.
Publications: pamphlets; journals; newsletter.
Hours & Admission Prices: Memorial Day-Labor Day Tues.-Sat. 10-5, Sun. 1-5; Sept.-May Wed.-Sat. 10-4. Adults $4, senior citizens 62 & over $3, children 6-18 $2; discounts for tours, locals & AAM members; members & children under 6 no charge. ら
Attendance: 2,350 (accurate)
Membership: Individual $25; Couple $40; Explorer $80; Historian $125; Pioneer $250; Heritage $500; Centennial $1,000.

Hugo

LINCOLN COUNTY MUSEUM, 617 Third Ave., Hugo, CO 80821. Mailing Address: P.O. Box 124, Hugo, CO 80821-0124. Tel.: 719-740-0106. Fax: 719-743-2447.
E-mail: twbndee@yahoo.com
Formerly: Hedlund House Museum
Founded: 1972.
Congressional District: 4
Key Personnel: Pres., Dee Ann Blevins; Treas., Garald Ensign; Representative, Gary Ensign; Contact, Terry Blevins.
Personnel Profile: Part-Time Volunteers 7.
Governing Authority: municipal. Parent Institution: Lincoln County Historical Society. Tax-exempt.
Institution Type/Description: History Museum: located on c.1880 site of the homestead of Hugo's founder.
Collections: period furniture & furnishings; utensils; glass; photos; postcards; period clothing; guns.
Research Fields: Lincoln County History.
Activities: guided tours.
Hours & Admission Prices: Memorial Day to Labor Day Sat. 10-6, Sun. 1-5; other times by appointment. No charge; donations accepted.
Attendance: 100 (estimated)
Membership: Lincoln County Historical Society: Annual $3; Lifetime $50.

Idaho Springs

ARGO GOLD MINE & MILL MUSEUM, 2350 Riverside Dr., Idaho Springs, CO 80452. Mailing Address: P.O. Box 1990, Idaho Springs, CO 80452-1990. Tel.: 303-567-2421. Fax: 303-567-9304.
E-mail: argo2350@aol.com
Web Site: www.historicargotours.com
Founded: 1977.
Key Personnel: C.E.O., James N. Maxwell; Dir., Robert N. Maxwell.
Personnel Profile: Full-Time Paid 2; Part-Time Paid 3.
Governing Authority: nonprofit organization. Tax-exempt.
Institution Type/Description: Mining & Milling Museum: housed in 1913 six-story Argo Gold Mill.
Collections: original mill equipment; historic milling & mining items; photographic displays; journals, ledgers & other written material of relevance to the history of Clear Creek County.
Facilities: Museum-related items for sale.
Activities: self-guided tours; educational & guided tours by appointment; permanent exhibitions; gold panning; demonstrations.
Publications: Argo Tailing Guide.
Hours & Admission Prices: mid-April to mid-Oct. daily 9-6; Winter: by appointment. Adults $15, children 7-12 $7.50; discounts to AAA, AAM, ICOM members & groups of 25 or more with reservation; children 6 & under no charge.
Attendance: 15,000 (estimated)

THE HERITAGE MUSEUM, 2060 Miner St., Idaho Springs, CO 80452-1318. Mailing Address: P.O. Box 1318, Idaho Springs, CO 80452-1318. Tel.: 303-567-4382.
E-mail: njohnson.historicidahosprings@gmail.com
Web Site: www.historicidahosprings.com
Founded: 1964.
Congressional District: 4
Key Personnel: Pres. (V), Robert Bowland; Museum Shop Mgr., Bonnie Hammett; Exec. Asst., Nancy Johnson.

Personnel Profile: Part-Time Paid 5; Part-Time Volunteers 25.
Governing Authority: nonprofit organization. Operated by the Historical Society of Idaho Springs. Tax-exempt: 170(b)(1)(A).
Institution Type/Description: History Museum.
Collections: local history & culture from 1860-1940s; mining & mining tools; Indian artifacts; fire department equipment; geological specimens; hand crafts; photographs; paintings; costumes; furnishings from 1800s to early 1900s; George Jackson's discovery of gold.
Research Fields: genealogy; photos.
Facilities: Gift items for sale.
Activities: permanent exhibitions; walking tour of homes & cemetery. Museum Sponsors: 150th Anniversary of the Colorado Gold Rush; Tommyknocker Mining Competition.
Publications: quarterly, The Idaho; books, Tailing, Track & Tommyknockers: History of Clear Creek County; Historical Highlights of Idaho Springs; Historical Walking Tour of Idaho Springs; The Waterwheel; Motherlode of Metal Miners' Mealtime Munchies.
Hours & Admission Prices: Daily 9-5. No charge; donations accepted. Closed New Year's Day; Easter; Thanksgiving; Christmas. ら
Attendance: 50,000 (estimated)
Membership: Senior Citizen $25; Single $35; Senior Family $40; Family $50; Business $100; Sustaining $500; Patron $1,000; Life Patron $5,000.

UNDERHILL MUSEUM, 1416 Miner St., Idaho Springs, CO 80452-1318. Mailing Address: P.O. Box 1318, Idaho Springs, CO 80452-1318. Tel.: 303-567-4382. Fax: 303-567-9188.
E-mail: njohnson.historicidahosprings@gmail.com
Web Site: www.historicidahosprings.com
Founded: 1964.
Congressional District: 4
Key Personnel: Pres. (V), Robert Bowland; Exec. Asst., Nancy Johnson; Museum Shop Mgr., Bonnie Hammett.
Personnel Profile: Part-Time Paid 4; Part-Time Volunteers 3.
Governing Authority: nonprofit organization. Operated by the Historical Society of Idaho Springs. Tax-exempt: 170(b)(1)(A).
Institution Type/Description: Historic House & Mining Engineer Office: housed in the former home & office of college professor & U.S. Mineral Surveyor, James Underhill; built in 1912.
Collections: mining equipment; early photographs; furnishings; garden.
Facilities: garden.
Hours & Admission Prices: Memorial Day to Sept. 10:30-4. No charge; donations accepted. ら
Attendance: 3,700 (accurate)
Membership: Senior Citizen $25; Single $35; Senior Family $40; Family $50; Business $100; Sustaining $500; Patron $1,000; Life Patron $5,000.

Ignacio

SOUTHERN UTE CULTURAL CENTER AND MUSEUM, (M), 503 Ouray Dr., Ignacio, CO 81137. Mailing Address: Box 737, Ignacio, CO 81137-0737. Tel.: 970-563-9583. Fax: 970-563-4641. Facebook: Southern Ute Cultural Center and Museum.
E-mail: info@succm.org
Web Site: www.sucm.org
Founded: 1972.
Congressional District: 3
Key Personnel: Acting Exec. Dir., Nathan Strong Elk; Chm., Robert Burch.
Personnel Profile: Full-Time Paid 4; Part-Time Paid 2.
Governing Authority: Parent Institution: Southern Ute Indian tribe. Subsidiary Institution: Southern Ute Cultural Center. Tax-exempt: 501(c)(3).
Institution Type/Description: Native American History & Ethnology Museum: located on Southern Ute Indian reservation.
Collections: photographs & artifacts pertaining to Ute Indians & other Native American tribes.
Research Fields: Southern Ute oral history.
Facilities: 50-vol. library of Native American history, arts & crafts available for use by the public & inter-library loan; archives; 8 acre campus.
Activities: permanent exhibitions; educational programs; demonstrations; rental facilities.
Publications: Southern Ute Tribal History; Ute Dictionary, Grammar & Narrative; educational packet, The Ute Legacy: history books, study guides & poster.
Hours & Admission Prices: Memorial Day to Labor Day Tues.-Fri. 9-4:30, Sat. 10-4, Sun. 12-4; Sept.-May Tues.-Fri. 9-4:30. Adults $7, seniors over 65 $4, children 3-14 $3; Ute tribal members no charge. ら
Attendance: 15,000 (estimated)
Membership: Senior 65 & over $35; Individual $50; Family $120; Spring $121-$499; Summer $500-$999; Fall $1,000-$2,499; Winter $2,500-$4,999.

Julesburg

DEPOT MUSEUM, (M), 201 W. 1st St., Julesburg, CO 80737. Mailing Address: Fort Sedgwick Historical Society, 114 E. 1st St., Julesburg, CO 80737. Tel.: 970-474-2264.
E-mail: history@kci.net
Web Site: www.kci.net/~history
Formerly: Pioneer Museum
Founded: 1976.
Congressional District: 4
Key Personnel: Pres. (V), Doris Heath; Vice Pres., Dallas Williams.
Personnel Profile: Full-Time Volunteers 1; Part-Time Paid 1; Part-Time Volunteers 5.
Volunteer Hours: 342
Governing Authority: Parent Institution: Fort Sedgwick Historical Society, Julesburg, CO 80737. Branch Museum: Fort Sedgwick Museum, 114 E. 1st St., Julesburg, CO 80737. Tax-exempt.
Institution Type/Description: History Museum: housed in the Union Pacific Railroad Depot; built in 1930. Listed on National Register of Historical Places.
Collections: agriculture; archaeology; archives; folklore; glass; history; Indian artifacts; mineralogy; music; natural history; transportation; railroad; Pony Express home station diorama.
Publications: History of Sedgwick County Vol. I & II; booklets; brochures; biannual newsletter.
Hours & Admission Prices: Memorial Day to Labor Day Tues.-Sat. 10-4, Sun. 1-4 & by appointment. Adults $1, children under 12 $.50; school groups & members no charge. Closed national holidays. &
Attendance: 350 (accurate)
Membership: Individual $10; Family $25; Business $50; Lifetime $200.

FORT SEDGWICK MUSEUM, 114 E. 1st St., Julesburg, CO 80737. Mailing Address: Fort Sedgwick Historical Society, 114 E. 1st St., Julesburg, CO 80737. Tel.: 970-474-2061.
E-mail: history@kci.net
Web Site: www.kci.net/~history
Formerly: Fort Sedgwick Museum & Archives
Founded: 1998.
Congressional District: 4
Key Personnel: Pres. (V), Doris Heath; Vice Pres., Dallas Williams.
Personnel Profile: Full-Time Paid 1; Full-Time Volunteers 1; Part-Time Volunteers 5.
Volunteer Hours: 2,445
Governing Authority: society. Affiliated with Fort Sedgwick Historical Society, Julesburg, CO 80737. Branch Museum: Depot Museum, 201 W. 1st, Julesburg, CO 80737. Tax-exempt.
Institution Type/Description: History Museum.
Collections: Sedgwick County history; cannons; military artifacts.
Research Fields: Sedgwick County structures; genealogy.
Facilities: research room.
Activities: research; special events; accessioning relics & documents; permanent & temporary exhibits; guest programs; tours.
Publications: biannual newsletter.
Hours & Admission Prices: Memorial Day to Labor Day Tues.-Sat. 10-4, Sun. 1-4; Sept.-May Tues.-Fri. 9-1. Adults $1, children $.50; members no charge. Closed national holidays. &
Attendance: 523 (accurate)
Membership: Individual $10; Family $25; Business $50; Lifetime $200.

HIPPODROME THEATRE, 215 Cedar St., Julesburg, CO 80737-1521. Tel.: 970-474-9977.
E-mail: hippo@pctelcom.coop
Web Site: www.rivertrailonline.org/users/hippodrome
Founded: 1996.
Governing Authority: nonprofit organization.
Institution Type/Description: Historic Building: housed in a functioning movie theater; built in 1919.
Collections: historic building; movies.
Hours & Admission Prices: Shows: Fri.-Sat. 7:30pm, Sun. 2pm & 6pm. Admission $4; children 3 & under no charge. &

Keenesburg

THE WILD ANIMAL SANCTUARY, 1946 WCR 53, Keenesburg, CO 80643. Tel.: 303-536-0118.
E-mail: information@wildlife-sanctuary.org
Governing Authority: nonprofit organization.
Institution Type/Description: Wildlife Sanctuary.
Collections: lions; tigers; bears; jaguars; leopards; mountain lions; lynx; bobcats; wolves.
Activities: observation decks.
Hours & Admission Prices: Daily 9 to sunset. Adults $15, children 3-12 $7.50.

Kiowa

ELBERT COUNTY HISTORICAL SOCIETY AND MUSEUM, 515 Comanche St., Hwy. 86, Kiowa, CO 80117. Mailing Address: P.O. Box 43, Kiowa, CO 80117-0043.
Web Site: www.elbertcountymuseum.org
Founded: 1956.
Governing Authority: Tax-exempt.
Institution Type/Description: History Museum.
Collections: county history & culture; photographs; personal artifacts; manuscripts; documents.
Hours & Admission Prices: Memorial Day to Labor Day Thurs.-Sun. 1-4. No charge; donations accepted.
Attendance: 750 (estimated)

Kit Carson

KIT CARSON HISTORICAL SOCIETY, 202 W. Hwy. 287, Kit Carson, CO 80825. Mailing Address: P.O. Box 67, Kit Carson, CO 80825-0067. Tel.: 719-962-3306.
Web Site: www.kcdr1.org
Founded: 1968.
Congressional District: 5
Key Personnel: C.E.O., Carl Randel; Pres. (V), Penny McPherson; Vice Pres. & Dir., Victor Gibbs; Dir. & Treas., Ronald White; Dir., Polly Johnson; Dir., Deb Dwyer; Dir. & Sec., Charles Oswald; Dir., Marilyn Bullock.
Personnel Profile: Full-Time Volunteers 7; Part-Time Volunteers 45.
Governing Authority: nonprofit organization. Tax-exempt: 501(c)(3).
Institution Type/Description: History Museum.
Collections: Indian artifacts; natural history; manuscripts; railroad caboose. Historic Building: 1904 Union Pacific Depot; 1928 Signal Maintainers House; steam era stone pump house.
Activities: guided tours; permanent & temporary exhibitions. Annual Events: pancake supper; Memorial Day Picnic.
Publications: books: Homesteaders & Other Early Settlers (1900-1930); Kit Carson Colorado, Home of the Famous Scout; Friendly People & Prime Beef Cattle.
Hours & Admission Prices: May & Sept. Sat.-Sun. 9-5; Memorial Day-Labor Day daily 9-5. No charge; donations accepted. &
Attendance: 877 (accurate)
Membership: Active $5; Supporting $25; Life $100.

La Junta

BENT'S OLD FORT NATIONAL HISTORIC SITE, 35110 Hwy. 194 E., La Junta, CO 81050-9523. Tel.: 719-383-5010. Fax: 719-383-2129.
E-mail: beol_cultural_resources@nps.gov
Web Site: www.nps.gov/beol
Founded: 1963.
Congressional District: 4
Key Personnel: Supt., Alexa Roberts; Cur., Rhonda Brewer; Museum Shop Mgr., Karla Linenberger.
Personnel Profile: Full-Time Paid 12; Part-Time Paid 4; Part-Time Volunteers 120.
Governing Authority: federal. Parent Institution: National Park Service. Tax-exempt.
Institution Type/Description: Living History Museum.
Collections: archaeological artifacts from excavation of original 1833-1849 trading post; reconstructed adobe fort with reproduction furnishings c.1833-1849; archival documents & photographs; herbarium of native plants.
Research Fields: American Southwest fur trade; Santa Fe Trail; Bent St. Vrain Company; Bent family; St. Vrain family; the War with Mexico.
Facilities: 1,000-vol. reference library relating to the American Southwest & early 19th-century history; picnic area. Books, reproduction trade goods & other museum-related items for sale.
Activities: guided tours; living history training program; craft demonstrations; special events.
Publications: brochures; worldwide web home page.
Hours & Admission Prices: June-Aug. daily 8-5:30; Sept.-May daily 9-4. Adults $3, children 6-12 $2; children 5 & under no charge; America the Beautiful passes honored. Closed New Year's Day; Thanksgiving; Christmas.
Attendance: 25,815 (accurate)

KOSHARE INDIAN MUSEUM, INC., 115 W. 18th, La Junta, CO 81050-3302. Mailing Address: P.O. Box 580, La Junta, CO 81050-0580. Tel.: 719-384-4411. Fax: 719-384-8836.
E-mail: jeremy.manyik@ojc.edu
Web Site: koshare.org
Founded: 1949.
Key Personnel: Program Dir. & Cur., Jeremy Manyik.
Personnel Profile: Full-Time Paid 4; Part-Time Paid 5; Part-Time Volunteers 65.
Governing Authority: college. Affiliated with Otero Junior College. Tax-exempt.
Institution Type/Description: Indian Art & Artifacts Museum.
Collections: paintings; archaeology; anthropology; plains & pueblo Indian art & artifacts; two-dimensional artwork with emphasis on southwest.
Research Fields: anthropology.
Facilities: 2,000-vol. library of Indian lore, archaeology & anthropology books available by request. Indian arts, crafts & other museum-related items for sale in Kiva Trading Post.
Activities: guided tours; gallery talks; hobby workshops; permanent & temporary exhibitions. Museum Sponsors: interpretive Indian dances. Annual Events: Kiva presents artist series featuring a different artist 1st. Sun. each month, no charge.
Publications: book, Koshare; brochure; biannual newsletter, Koshare News; artist calendar, Kiva Presents.
Hours & Admission Prices: June-Aug. daily 10-5; Sept.-May call for hours. Adults $5, students $3; members no charge. Interpretive Indian Dances: mid-June to Aug. Sat. 7:30pm; Dec. 27-Jan. 4 call for hours. Show: adults $8, children 3-18 $5. Closed major holidays. &
Attendance: 24,000 (estimated)
Membership: Jr. Feather Clan: Student (18 & under) $10; Adult (19-24) $25. Adult Feather Clan: Individual (25 & up) $50; Family $75. Kokopelli $100-$249; Chippewa $200 & up; Warriors $250-$499; Dreamcatchers $500-$999; Kira Society $1,000-$2,999; Burshears Circle $3,000 & up.

OTERO MUSEUM ASSOCIATION, 218 Anderson St., La Junta, CO 81050-0223. Mailing Address: P.O. Box 223, 218 Anderson St., La Junta, CO 81050-0223. Tel.: 719-384-7500. Fax: 719-384-7500.
E-mail: oteromuseum@centurytel.net
Web Site: www.coloradoplains.com/otero/museum
Founded: 1984.
Congressional District: 4
Key Personnel: Pres. (V), Cur. & Public Rels., Don Lowman; Treas. & Devel., Donna Aldea; Recording Sec., Marilyn Mast.
Personnel Profile: Full-Time Volunteers 2; Part-Time Volunteers 21.
Governing Authority: private; nonprofit organization. Tax-exempt: 501(c)(3).
Institution Type/Description: General Museum.
Collections: 1876-1950 local history; displays of life in early La Junta & Otero County; ranching & farming displays; transportation; railroad memorabilia; Santa Fe Railroad memorabilia; doctor's office; post office; volunteer fire dept. display building; clocks; agricultural implements; war artifacts from civil war to Vietnam era; chinaware; 137 mustache cups; guns. Historic Houses: Daniel Sciombato Home/Grocery Store; Nancy Wickham Home; 1873 railroad workers boarding house; 1876 log cabin school; Coach House: 1867 Concord stage coach, chuck wagon, 1905 Reo.
Research Fields: irrigation in the Arkansas River Valley; early fire fighting.
Facilities: 160-vol. library; 29,000 sq. ft. exhibit space. Museum-related items for sale.
Activities: guided tours; arts festivals. Annual Events: video programs April to October; Art Shows in May & November; Bean Supper (Fundraiser) in June; antique appraisal program June to October; school tours in spring.
Hours & Admission Prices: June-Sept. Mon.-Sat. 1-5; Oct.-May & groups by appointment only. No charge; donations accepted.
Attendance: 2,750 (estimated)
Membership: Family $25; Contributing $50; Sustaining $100.

La Veta

FRANCISCO FORT MUSEUM, 306 S. Main St., La Veta, CO 81055. Mailing Address: P.O. Box 263, LaVeta, CO 81055-0263. Tel.: 719-742-5501.
E-mail: director@franciscofort.org
Web Site: franciscofort.org
Founded: 1965.
Congressional District: 3
Key Personnel: Dir., Kim McKee; Contact, Peggy Arnold-Hoobler.
Personnel Profile: Part-Time Paid 2; Part-Time Volunteers 18; Interns 3.

Governing Authority: nonprofit organization. Parent Institution: Huerfano County Historical Society. Tax-exempt.
Institution Type/Description: Historical Society Museum: housed on 1862 adobe fort and other historic buildings.
Collections: American Indian artifacts of area; 1800s furniture of area settlers; old post office; general store; switchboard; doctor's office; blacksmith shop & tools; farm & ranch implements; 1860-1940 clothes; Civil War to WWII weaponry; steam engines; spinning wheels; Lincoln letter of 1850; chuck-wagon; taxidermy collection; musical instruments; one-room school; coal mining artifacts; Edison sound & light inventions; mammoth tooth, fossilized coral, dinosaur fossilized footprint; Hispanic artifacts; Hispanic exhibits & penitente; religious articles.
Research Fields: local history; projectile points; genealogy.
Facilities: library of taped oral histories, maps, photos, general & local history, old local newspapers available for research by permission from board of directors; reading room. Museum-related items for sale.
Activities: guided tours; gallery talks; permanent & temporary exhibitions; oral history project with La Veta Public Library; children's treasure hunt; special events. Museum Sponsors: Francisco Fort Days in July; 3-day programs: Native American & Hispanic Cultural Heritage.
Publications: annual newsletter; brochures, HCHS Museums; quarterly newsletter.
Hours & Admission Prices: May 1-Oct. 31 seasonal hours check website. $5 (12 years & over). &
Attendance: 6,044 (estimated)

Lafayette

LAFAYETTE MINERS MUSEUM, 108 E. Simpson St., Lafayette, CO 80026-2322. Tel.: 303-665-7030.
E-mail: minersmuseum@cityoflafayette.com
Founded: 1976.
Congressional District: 2
Key Personnel: Cur., Claudia Lund.
Personnel Profile: Part-Time Paid 1; Part-Time Volunteers 15.
Governing Authority: nonprofit society. Parent Institution: Lafayette Historical Society. Tax-exempt.
Institution Type/Description: Coal Mining Museum: housed in c.1892 Lewis House. Listed on the National Register of Historic Places.
Collections: coal mining artifacts; period furniture & furnishings; photographs; newspapers; rocks & minerals; local obituary file; mine union records; mine company time books; war memorabilia from the Spanish-American War to the Gulf War; local high school memorabilia; personal artifacts; paintings; furnishings.
Research Fields: coal mining; local history.
Facilities: archives; picnic area.
Activities: guided & self-guided tours; lectures; formally organized education programs for children; slide presentations; demonstrations; permanent & loan exhibits; bimonthly community program at Lafayette Public Library. Museum Sponsors: Peach Festival; Oatmeal Festival.
Publications: book: Lafayette Centennial History, Lafayette, Colorado Volunteer Fire Department 1893 to 2001 Squirt to Telesquirt; Slaughter in Serene: The Columbine Coal Strike Reader; History of Waneka Lake; The Price of Colorado Coal; The Cherokee Trail; area coal mining maps; Cherry Blossom Comrades; Lincoln Highway in Colorado; Remember Ludlow; The Lincoln Highway Through Lafayette; Into the Jaws of Hell; quarterly newsletter.
Hours & Admission Prices: Thurs. & Sat. 2-4; tours by appointment. No charge; donations accepted. Closed holidays. &
Attendance: 1,800 (accurate)
Membership: Individual $10; Family $15; Business $30; Life $100.

WOW! CHILDREN'S MUSEUM (WORLD OF WONDER), 110 N. Harrison Ave., Lafayette, CO 80026-2336. Tel.: 303-604-2424.
Web Site: www.wowmuseum.com
Key Personnel: Co-Founder & Exec. Dir., Lisa Atallah; Mgr., Susan Rasmussen; Mgr. Mktg., Katie MacDonald
Institution Type/Description: Children's Museum.
Collections: hands-on exhibits.
Facilities: Museum-related items for sale.
Activities: special events; parties; workshops; theater & dance performances.
Hours & Admission Prices: June-Aug. 18 Mon.-Wed. 9-5, Thurs.-Sat. 10-6; mid-Aug. to May Tues.-Wed. 9-5, Thurs.-Sat. 10-6, Sun. 12-4; other times by appointment. Children $7; discounts to groups; adults and children 14 months & under no charge. Closed New Year's Day; Memorial Day; Father's Day; Labor Day; Thanksgiving; Christmas.

Lake City

HINSDALE COUNTY HISTORICAL SOCIETY MUSEUM, 130 N. Silver St., Lake City, CO 81235. Mailing Address: P.O. Box 353, Lake City, CO 81235. Tel.: 970-944-2050.
E-mail: info@lakecitymuseum.com
Web Site: lakecitymuseum.com
Founded: 1974.
Congressional District: 3
Personnel Profile: Part-Time Paid 1; Interns 1.
Governing Authority: Tax-exempt.
Institution Type/Description: History Museum.
Collections: local history & culture; period furnishings; personal artifacts; photographs.
Hours & Admission Prices: Summer: Mon.-Sat. 10-5; other times by appointment. Adults $4, children 8-15 $2; children 7 & under no charge.
Attendance: 2,000 (accurate)
Membership: Individual $40; Family $50; Business $60.

Lakewood

LAKEWOOD'S HERITAGE CENTER, (M), 801 S. Yarrow St., Lakewood, CO 80226-4372. Tel.: 303-987-7850. Fax: 303-987-7851. TDD: 303-987-4862.
E-mail: jefmur@lakewood.org
Web Site: www.lakewood.org
Founded: 1976.
Congressional District: 6
Key Personnel: HCA Mgr., Michelle Nierling; Admin., Jeffrey Murray; Office Mgr., Julie Elan; Volunteer Coord., Karla Grahn; Mktg. Coord., Allison Scheck.
Personnel Profile: Full-Time Paid 6; Part-Time Paid 4; Part-Time Volunteers 175.
Governing Authority: municipal. Parent Institution: City of Lakewood. Tax-exempt: 170(b)(1)(A).
Institution Type/Description: History Museum.
Collections: culture, pertaining to the Foothills area of the Eastern Slope of Colorado; restored vintage farm machinery; local history; archival, local art; 18 historic structures.
Research Fields: agricultural history; 20th-Century Lakewood area history; farming; ranching; homesteading; transportation; local history; natural history.
Facilities: library of Lakewood Colorado history, Jefferson County history & early farm methodology, available for use on premises; classrooms; meeting rooms; kitchen areas; picnic areas; nature trail; outdoor amphitheater; festival area; cultural center with 300-seat indoor theater.
Activities: lectures; films; gallery talks; discovery trunks; summer camps; art classes; exhibits; tours; performances; community events.
Publications: brochures; quarterly, Bravo Lakewood; membership quarterly.
Hours & Admission Prices: Tues.-Sat. 10-4. Adults $5, seniors citizens $4, youth $3; discounts to AAM members; children 3 & under members no charge. Closed holidays. &
Attendance: 45,238 (accurate)

ROBERT H. JOHNSON PLANETARIUM, 200 Kipling St., Lakewood, CO 80226-1046. Mailing Address: Mandalay Middle School, 9651 N. Pierce St., Westminster, CO 80021. Tel.: 303-982-7278. Fax: 303-982-7277.
Formerly: Jefferson County Schools Planetarium
Founded: 1963.
Key Personnel: Dir., Kathy Miller; Coord. Scheduling, Lisa Delameter.
Governing Authority: public school district. Parent Institution: Jefferson County Public Schools, 809 Quail. Branch Observatories: Jefferson County Lab School, Upper Bear Creek; Windy Peak, Bailey, CO. Tax-exempt.
Institution Type/Description: Planetarium & Observatories.
Collections: astronomy; space science.
Facilities: approx. 500-vol. library of astronomy & related subjects; 120-seat planetarium.
Activities: lectures; films; education programs for students K-6.
Publications: curriculum guides.
Hours & Admission Prices: Temporarily closed. &
Attendance: 32,000

ROCKY MOUNTAIN COLLEGE OF ART + DESIGN GALLERIES, 1600 Pierce Ave., Lakewood, CO 80214-1897. Tel.: 800-888-ARTS. Fax: 303-759-4970. Facebook: Philip J. Steele Gallery.
E-mail: cstell@rmead.edu
Web Site: www.rmcad.edu/exhibitions

Formerly: The Philip J. Steele Gallery
Founded: 1963.
Congressional District: 4
Key Personnel: C.E.O., Maria Puzziferro; Cur., Cortney Stell; Registrar, Chuck King.
Personnel Profile: Full-Time Paid 40; Part-Time Paid 50.
Governing Authority: private college; profit. The Philip J. Steele Gallery.
Institution Type/Description: College Art Gallery.
Collections: paintings; sculptors; graffiti.
Facilities: library; educational facilities; 2,500 sq. ft. exhibit space. Art supplies for sale.
Activities: formal education programs offering Bachelor of Fine Art degrees; affiliated with Rocky Mountain College of Art & Design; lectures; gallery forums for the public.
Hours & Admission Prices: Mon.-Sat. 11-4. No charge. Closed New Year's Day; Independence Day; Labor Day; Thanksgiving; Christmas.

Lamar

BIG TIMBERS MUSEUM, 7515 US Hwy. 50, Lamar, CO 81052. Mailing Address: P.O. Box 362, Lamar, CO 81052-0362. Tel.: 719-336-2472. Fax: 719-336-2472.
E-mail: bigtimbers@prowerscounty.net
Web Site: www.bigtimbersmuseum.org
Founded: 1966.
Congressional District: 3
Key Personnel: Cur., Kathleen Scranton.
Personnel Profile: Full-Time Paid 1; Part-Time Paid 3; Part-Time Volunteers 5.
Governing Authority: nonprofit organization. Sponsored by Prowers County Historical Society. Tax-exempt: 501(c)(3).
Institution Type/Description: History Museum: housed in the original AT&T Co. repeater station.
Collections: history; WWI poster collection; Charles Frederick Worth gown exhibit; Indian arrowhead collection; furniture.
Facilities: 100-vol. library of newspaper files.
Activities: permanent & temporary exhibitions.
Publications: A Prowers Country History.
Hours & Admission Prices: June-Aug. Tues.-Sat. 10-5; Sept.-May Tues.-Sat. 1-4. Family $5, adults $3. Closed New Year's Day; Good Friday; Thanksgiving; Christmas.
Attendance: 1,300 (accurate)
Membership: Individual $7; Family $10; Patron $25; Life $100.

Las Animas

JOHN W. RAWLINGS HERITAGE CENTER & MUSEUM, 560 Bent Ave., Las Animas, CO 81054. Mailing Address: P.O. Box 68, Las Animas, CO 81054-0068. Tel.: 719-456-6066.
E-mail: bentctyheritage@centurytel.net
Web Site: bentcounty.org
Formerly: Kit Carson Museum
Founded: 1959.
Personnel Profile: Part-Time Volunteers 24.
Governing Authority: society. Affiliated with Pioneer Historical Society. Tax-exempt: 501(c)(3).
Institution Type/Description: General Museum.
Collections: Native American artifacts; cattle industry artifacts; carriages; railroad; agricultural; Fort Lyon; Santa Fe; Llewellyn Thompson memorial; local history; Bell park; telephones; bank artifacts; barber shop; candy shop; jewelry store furnishings. Historic Buildings: 1876 county jail; 1891 schoolhouse; 1860 stage coach station; blacksmith shop.
Research Fields: local history.
Facilities: Books & cards for sale.
Activities: guided tours; lectures; permanent & temporary exhibitions.
Publications: quarterly, Bent County Chronicle.
Hours & Admission Prices: Mon.-Fri. 12-4:30. Adults $3; discounts to groups of 10 or more. &
Attendance: 1,000 (accurate)
Membership: Pioneer Historical Society of Bent County: Individual $25; Sponsor $50-$100; Patron $200; Lifetime $400; Benefactor $1,000 & up.

Leadville

＊ HEALY HOUSE-DEXTER CABIN, (M), 912 Harrison Ave., Leadville, CO 80461-3321. Tel.: 719-486-0487. Fax: 719-486-2557.
E-mail: maureen.scanlon@state.co.us
Web Site: www.historycolorado.org
Founded: 1936.
Congressional District: 5

Key Personnel: C.E.O., Ed Nichols.
Personnel Profile: Part-Time Paid 5.
Governing Authority: state. Parent Institution: History Colorado, 1200 Broadway, Denver, CO 80203. Tax-exempt.
Institution Type/Description: Historic Buildings.
Collections: Victorian household items. Historic Buildings: 1878 Healy House; 1879 Dexter Cabin.
Research Fields: Leadville history.
Activities: tours of house & cabin; Leadville Mining District 4 wheel drive tours.
Hours & Admission Prices: mid-May to Oct. daily 10-4:30; other times tours by appointment. Adults $6, senior citizens $5.50, children 6-16 $4.50; discounts to AAM members, school & group tours; children under 6 & members no charge.
Attendance: 8,000

HERITAGE MUSEUM & GALLERY, 102 E. Ninth St., Leadville, CO 80461-3302. Mailing Address: P.O. Box 962, Leadville, CO 80461-0962. Tel.: 719-486-1878.
Founded: 1971.
Congressional District: 5
Key Personnel: Pres. (V), Carl Schaefer; C.E.O., Ray Stamps; Museum Shop Mgr., Nancy Manly; Museum Shop Mgr., Barbara Beck.
Personnel Profile: Full-Time Paid 1; Part-Time Paid 1; Part-Time Volunteers 20.
Governing Authority: nonprofit organization. Parent Institution: Lake County Civic Center Association. Tax-exempt: 501(c)(3).
Institution Type/Description: History Museums: located in Historic Leadville District.
Collections: mining artifacts; ore samples; diorama display-Leadville mining history; Oak Room furnishings; clothing; personal articles & household artifacts; 1/4-inch scale replica of Leadville's 1896 Ice Palace. Historic Buildings: 1880s The Old Church; 1902 Heritage Museum; 1870 Prospector's cabin.
Research Fields: Leadville/Lake county history.
Facilities: 250-vol. books; manuscripts, newspaper & census microfilm; 2,500 photographs on Lake County history available for research within the Lake County Public Library; Prospector Park. Leadville history booklets, crafts, newspaper reproductions & other museum-related items for sale.
Activities: self-guided tours; art exhibits; school & educational tours.
Publications: quarterly newsletter, The Tallyboard.
Hours & Admission Prices: May-Oct. daily 10-6. Adults $6, seniors $5, children 6-16 $3; children under 6 & members no charge. &
Attendance: 9,000 (accurate)
Membership: Annual $20; Patron $30; Contributing $50; Donor $100.

THE HISTORICAL TABOR OPERA HOUSE, 308 Harrison Ave., Leadville, CO 80461-3612. Tel.: 719-486-8409.
E-mail: info@taboroperahouse.net
Web Site: www.taboroperahouse.net
Founded: 1955.
Key Personnel: C.E.O., Dir. & Cur., Sharon Bland; Chm. (V), Marylee Mosqovoy.
Personnel Profile: Full-Time Volunteers 1; Part-Time Volunteers 12.
Governing Authority: individual operation.
Institution Type/Description: Historic Theater Museum: listed on the National Register of Historic Places.
Collections: paintings; theater; history; original scenery; Leadville & Tabor.
Research Fields: paintings; theater; history; original scenery.
Facilities: Books, pictures & museum-related items for sale.
Activities: guided tours; lectures; films; drama; permanent exhibitions.
Publications: books, The Tabor Opera House; Silver Dollar Tabor; Leadville Crystal Carnival 1896; My Search for Augusta Pierce Tabor; Leadville's First Lady.
Hours & Admission Prices: House: May-Oct. Mon.-Sat. 10-5. Adults $5; discounts to seniors, AAA & military. Tabor's Second Story: by appointment. Adults $5. House & Second Story: $7. &

HOUSE WITH THE EYE MUSEUM, 127 W. Fourth St., Leadville, CO 80461-3629. Mailing Address: P.O. Box 911, Leadville, CO 80461-0911. Tel.: 719-486-0860.
Founded: 1964.
Congressional District: 3
Key Personnel: Cur., Barbara M. Bost.
Personnel Profile: Full-Time Paid 1; Full-Time Volunteers 2.
Governing Authority: individual operation.
Institution Type/Description: Historic House Museum: 1879 House with the Eye, built by noted French architect Robitaille for his bride; features a gingerbread front & hand-crafted wainscoting.

Collections: carriages; mining tools; pianos; jewelry; guns; old bottles; picture albums; clocks; boomtown memorabilia.
Research Fields: Leadville history.
Activities: guided tours.
Publications: descriptive brochure.
Hours & Admission Prices: June-Labor Day daily 10-4. Adults $5, senior citizens $4; discounts to AAM & ICOM members.

LEADVILLE NATIONAL MINING HALL OF FAME & MUSEUM, 120 W. 9th St., Leadville, CO 80461-3403. Mailing Address: P.O. Box 981, Leadville, CO 80461-0981. Tel.: 719-486-1229. Fax: 719-486-3927.
E-mail: info@mininghalloffame.org
Web Site: www.mininghalloffame.org
Founded: 1987.
Key Personnel: Dir. & Pres., Robert Hartzell; Chm. (V), Dick Moolick.
Personnel Profile: Full-Time Paid 2; Part-Time Paid 10; Part-Time Volunteers 10; Interns 1.
Governing Authority: Tax-exempt.
Institution Type/Description: Mining Museum.
Collections: mining history; mining pioneers.
Facilities: Museum-related items for sale.
Publications: High Grad Newsletter.
Hours & Admission Prices: Memorial Day to Oct. daily 9-5; Winter: daily 11-6. Adults $7, seniors 65 & over $6, teens 13-19 $5, children 6-12 $3; discounts to groups & schools; members & children under 6 no charge. Closed New Year's Day; Thanksgiving; Christmas. &
Attendance: 25,000 (estimated)
Membership: Individual $35; Family $50; Contributing $150; Benefactor $1,000; Life $2,500.

MATCHLESS MINE MUSEUM, 120 W. 9th St., Leadville, CO 80461-3403. Mailing Address: P.O. Box 981, Leadville, CO 80461-0981. Tel.: 719-486-1229. Fax: 719-486-3927.
E-mail: director@mininghalloffame.org
Web Site: www.mininghalloffame.org
Founded: 1953.
Congressional District: 6
Key Personnel: Dir. & Chm. (V), Robert Hartzell; Opers. Mgr., Kat Neilson.
Personnel Profile: Full-Time Paid 2; Part-Time Paid 3.
Governing Authority: nonprofit organization. Parent Institution: National Mining Hall of Fame & Museum. Tax-exempt.
Institution Type/Description: Mining Museum.
Collections: period mine artifacts.
Activities: guided tours.
Hours & Admission Prices: Memorial Day to Oct. 1 daily 10-4. Adults 20-64 $7, seniors 65 & up $6, teens 13-19 $5, children 6-12 $3; discounts to groups & AAM members; members & children 5 & under no charge.
Attendance: 10,000 (estimated)

TABOR HOME, 116 E. 5th St., Leadville, CO 80461. Tel.: 719-486-3900; 855-488-1222.
Institution Type/Description: Historic House Museum: housed in the former home of Horace Tabor and his wife Augusta until 1881.
Collections: family history; photographs; period furnishings.
Hours & Admission Prices: Call for hours.

Limon

LIMON HERITAGE MUSEUM, (M), 899 First St., Limon, CO 80828. Mailing Address: P.O. Box 341, Limon, CO 80828-0341. Tel.: 719-775-9418. Fax: 719-775-8808. Facebook: Limon Heritage Museum.
E-mail: limonmuseum@hotmail.com
Web Site: limonmuseum.com
Founded: 1990.
Congressional District: 4
Key Personnel: C.E.O. & Dir., Tony Wernsman; Dir., Kevin Pickerill; Mgr. Operations, Kathryn Chittenden; Education & Devel., Lucille Reimer; Treas. & Registrar, Barbara Berry.
Personnel Profile: Full-Time Paid 1; Part-Time Paid 2; Part-Time Volunteers 35.
Operating Expenses: 67,076
Operating Income: 73,052
Governing Authority: private; nonprofit organization. Parent Institution: Limon Heritage Society. Tax-exempt: 501(c)(3).
Institution Type/Description: History Museum: housed in a restored Union Depot.
Collections: local history of the Colorado high plains, the Union Pacific &

Rock Island Railroads, ranching & farming; sheepherder's wagons; period farm machinery; one-room schoolhouse; Houtz Native American artifacts.
Major Exhibits: I-70-A Turning Point, 5/13-9/14; Dust Bowl, 5/14-9/14; Changes in Technology, 5/14-9/15.
Research Fields: homesteaders prior to 1910; ancient plains, geology of area.
Facilities: 7,000 sq. ft. exhibit space.
Activities: concerts; docent program; guided tours; temporary exhibitions. Annual Events: Heritage Celebration; Lost Arts Demonstrations & annual dinner & auction.
Publications: annual newsletter, Windbreak.
Hours & Admission Prices: Memorial Day-Labor Day Mon.-Sat. 1-8. No charge; donations accepted. &

Attendance: 5,000 (estimated)
Membership: Student $5; Adult $15; Family $25; Donor $100; Business $150.

Littleton

COLORADO GALLERY OF THE ARTS/ARAPAHOE COMMUNITY COLLEGE, 5900 S. Santa Fe Dr., 1st Fl. Annex Bldg., Littleton, CO 80120-1801. Tel.: 303-797-5649.
Founded: 1979.
Congressional District: 6
Key Personnel: Gallery Coord., Trish Sangelo.
Governing Authority: university. Parent Institution: Arapahoe Community College. Tax-exempt.
Institution Type/Description: Art Gallery.
Collections: exhibitions of loaned & leased art objects.
Facilities: 2,128 sq. ft. exhibition area.
Activities: guided tours; lectures; films; concerts; rental gallery; organized educational programs; docent program; loan & traveling exhibitions.
Publications: quarterly newsletter.
Hours & Admission Prices: Mon. & Wed.-Fri. 12-5, Tues. 12-7; other times by appointment. No charge. Closed holidays. &
Attendance: 14,000
Membership: Student & Senior Citizen $10; Individual $15; Family & Dual $20; Associate $35; Patron $100 & up; Donor $500 & up.

✱ **LITTLETON MUSEUM, (M),** 6028 S. Gallup St., Littleton, CO 80120-2703. Tel.: 303-795-3950. Fax: 303-730-9818.
E-mail: tnimz@littletongov.org
Web Site: www.littletongov.org/museum/default.asp
Formerly: Littleton Historical Museum
Founded: 1969.
Congressional District: 6
Key Personnel: Dir., Tim Nimz; Cur. Exhibits, Bill Hastings; Deputy Dir. & Cur. Collections, Lorena Donohue; Administrative Coord., Becky Kosma; Cur. Education & Interpretation, Suellen Winstead.
Personnel Profile: Full-Time Paid 12; Part-Time Paid 7; Part-Time Volunteers 165.
Governing Authority: municipal. Parent Institution: City of Littleton. Tax-exempt.
Institution Type/Description: General Museum: history museum & living history farms.
Collections: agriculture; archives; manuscripts; costumes; decorative arts, fine arts; industry; medicine; transportation. Historic Buildings: 1863 log schoolhouse; 1890 blacksmith shop; ice house; two living history farms: 1860s & 1895-1905.
Research Fields: general; agriculture; archives; costumes; decorative arts; industry; medicine; transportation.
Facilities: 600-vol. library of public & private records, papers on local history, newspapers & photographs available for research under supervision by staff.
Activities: guided tours; lectures; arts festivals; formally organized education programs for children; permanent & temporary exhibitions.
Publications: pamphlets; brochures; museum news.
Hours & Admission Prices: Tues.-Fri. 8-5, Sat. 10-5, Sun. 1-5. No charge; donations accepted. Closed holidays. &
Attendance: 132,577 (accurate)

Longmont

AGRICULTURAL HERITAGE CENTER AT THE LOHR-MCINTOSH FARM, (M), 8348 Ute Hwy. 66, Longmont, CO 80503-9232. Tel.: 303-776-8848.
Institution Type/Description: Heritage Center.
Collections: local history & heritage; photographs; period furnishings; farm equipment.
Hours & Admission Prices: April-Oct. Fri.-Sun. 10-5; Nov.-March 1st Sat. each month 10-5. No charge.

LONGMONT MUSEUM, (M), 400 Quail Rd., Longmont, CO 80501-8989. Tel.: 303-651-8374. Fax: 303-774-4780. Facebook: Longmont Museum.
Web Site: www.longmontmuseum.org
Founded: 1940.
Congressional District: 4
Key Personnel: Dir., Wesley Jessup; Bd. Chm., Bryan Bowles; Bd. Vice Chm., Richard Luke; Pres. Friends, Eric Kittelberger; Cur. Exhibits, Jared Thompson; Cur. Education, Ann Holley; Cur. Collections, Cathey Dunn; Cur. Research & Information, Erik Mason; Discovery Days Coord., Stephanie Ohlsen; Art in Public Places Admin., Lauren Greenfield; Museum Asst., Joann McCoy; Museum Shop Mgr., Christina Mehler.
Personnel Profile: Full-Time Paid 6; Part-Time Paid 8; Part-Time Volunteers 65; Interns 2.
Volunteer Hours: 2,348
Operating Expenses: 1,057,998
Operating Income: 1,120,269
Governing Authority: municipal. Parent Institution: City of Longmont, Dept. of Community Services. Tax-exempt: 170(b)(1)(A).
Institution Type/Description: Regional History Museum.
Collections: history & culture of the region; development of the Longmont area.
Research Fields: Longmont & the St. Vrain Valley cultural history prehistoric to present.
Facilities: photograph & document archives available for research on premises only. Museum-related items for sale.
Activities: guided & self-guided tours; inter-museum loan, permanent & traveling exhibitions; programs for children, adults & families; special events; volunteer program; internship program; panoramic view of the mountains & plains.
Publications: quarterly newsletter of events & exhibits.
Hours & Admission Prices: Mon.-Sat. 9-5, Sun. 1-5. Adults $5; members no charge. Closed holidays. &
Attendance: 62,024 (accurate)
Membership: Senior & Student $25; Individual $35; Senior Couple $45; Family $50; Supporting $150; Corporate $300.

Louisville

LOUISVILLE HISTORICAL MUSEUM, 1001 Main St., Louisville, CO 80027-1725. Mailing Address: City Hall, 749 Main St., Louisville, CO 80027. Tel.: 303-665-9048.
Key Personnel: Museum Coord., Bridget Bacon
Institution Type/Description: History Museum.
Collections: Louisville history; coal mining. Historic Buildings: Jacoe Store: period artifacts; photographs; coal mining; industrial equipment. Tomeo House: period furnishings.
Hours & Admission Prices: Tues.-Wed. & Sat. 10-3.

Loveland

BENSON SCULPTURE GARDEN, 2908 Aspen Dr., Loveland, CO 80538. Mailing Address: P.O. Box 7006, Loveland, CO 80537-0006. Tel.: 970-663-2940. Fax: 970-669-7390.
E-mail: lhpac@sculptureinthepark.org
Web Site: www.sculptureinthepark.org
Institution Type/Description: Sculpture Garden.
Collections: 139 sculptures.
Activities: Annual Event: Sculpture in the Park in summer.
Hours & Admission Prices: Daily dawn to dusk.

✱ **LOVELAND MUSEUM GALLERY, (M),** 503 N. Lincoln Ave., Loveland, CO 80537-5619. Tel.: 970-962-2410. Fax: 970-962-2910.
Web Site: ci.loveland.co.us
Founded: 1956.
Congressional District: 51
Key Personnel: C.E.O. & Dir. Curatorial Svcs., Susan P. Ison; Pres. Historical Society (V), Mike Perry; Cur. Education, Jenni Dobson; Cur. History, Jennifer Cousino; Cur. Art, Maureen Corey; Public Art Mgr., Suzanne Janssen; Exhibits Preparator, Quinn Johnson; Administrative Specialist, Mary Shada; Coord. Mktg., Kim Akeley-Charron; Registrar, Robert Hoot; Graphic Designer, Michelle Standiford.
Personnel Profile: Full-Time Paid 7; Part-Time Paid 3; Part-Time Volunteers 120; Interns 6.
Governing Authority: municipal; nonprofit. Tax-exempt: 170(b)(1)(A).
Institution Type/Description: Art and History Museum.
Collections: social, natural, cultural & art history of Loveland.
Research Fields: history of the Larimer County Fairground; Great Western

Sugar Co.; Germans from Russia; history of community business, architecture & social development.

Facilities: library; auditorium.

Activities: concerts; docent program; films; teen coffeehouses; Lone Tree School schoolhouse program; after school art programs; guided tours; hobby workshops; lectures; loan, participatory & traveling exhibitions; school loan service; art exhibits. Annual Events: Cherry Pie Festival; Halloween Fun Festival.

Publications: Fifth & Lincoln News; books, Loveland's Historic Downtown; Stone Quarrying in Loveland's Foothills; Historic Loveland Churches; A Guide to Historic Loveland; Exploring Loveland's Hidden Past: People and Places of Early Loveland, Colorado; Germans from Russia in the Loveland Area.

Hours & Admission Prices: Tues.-Wed. & Fri. 10-5, Thurs. 10-9, Sat. 10-4, Sun. 12-4. No charge; donations accepted. Closed New Year's Day; Independence Day; Thanksgiving & day after; Christmas. &

Attendance: 52,000 (estimated)

Membership: Senior $20; Individual $30; Individual Plus One $45; Family $65; Contributing $250; Patron $500; Corporate $1,000.

TIMBERLANE FARM MUSEUM, 2306 E. First St., Loveland, CO 80537-5906. Tel.: 970-663-7348. Fax: 970-663-7364.

Web Site: www.timberlanefarmmuseum.org

Institution Type/Description: Farm Museum.

Collections: local history & farm life from 1860 to 1940; farm equipment; restored home; farm buildings.

Activities: working farm.

Hours & Admission Prices: Call for hours. &

Lyons

THE LYONS REDSTONE MUSEUM, 340 High St., Lyons, CO 80540. Mailing Address: P.O. Box 9, Lyons, CO 80540-0009. Tel.: 303-823-5925 & 5271. Fax: 303-823-8257.

E-mail: lavern921@aol.com

Web Site: lyonsredstonemuseum.com

Founded: 1973.

Congressional District: 2

Key Personnel: Dir. & Chm. (V), LaVern M. Johnson; C.E.O., Terri Weir; Vice Pres., Jerry L. Johnson; Deputy Dir., Maxine Harkalis; Historic Asst., Calvin Schilling; Museum Shop Mgr., Kathleen Spring.

Personnel Profile: Part-Time Paid 3; Part-Time Volunteers 5; Interns 1.

Volunteer Hours: 1,500

Operating Expenses: 24,100

Operating Income: 24,500

Governing Authority: nonprofit organization. Parent Institution: Lyons Historical Society. Subsidiary Institution: Town of Lyons. Tax-exempt: 501(c)(3).

Institution Type/Description: History Museum: housed in c.1881 school building.

Collections: local artifacts; E.S. Lyon & family artifacts; sandstone tools; old books; household items; photographs; period artifacts; Lyon Post office boxes c.1890; early Arapaho Indian artifacts; square grand piano; player organ of the Loukonen family; power pole; 577 telephone insulators; horse drawn mail cart; Lyons State Bank; teller cage (1907-1983); c.1880 Friden Family, pump organ; Lyons Drug Store artifacts (1924-1970); Lyons Historic District, 15 sandstone buildings in the Lyons area from 1880-1917; Lyons area people and business files.

Research Fields: local history.

Facilities: 50-vol. library of books printed in the 1880s & old school books available for research on premises or by special permission. Books, crafts & other gift items for sale.

Activities: guided tours; historic walking tours; history programs; formally organized education programs; loan, permanent & temporary exhibitions.

Publications: special reports & events; small oral history books of area pioneers and their families on ES Lyon family founder of town of Lyons; video, history of sandstone quarries & geology; book, The Sandstone - 1800s; 2006 Lyons Timeline; Billie Welch Resort; History of Sandstone Quarries, History of Old Stone Church; History of Pella, (Refuge in the Valley) Altoona District; Big Elk Meadows; Homestead Acres; Lyons - The Gateway to the Rockies - 1900s; Lyons - The Quarry Town - 1800s; tour maps; Piecing A Town Together - The Families of Lyons, CO; Lyons County High School graduates 1925-2011.

Hours & Admission Prices: June-Sept. Sat. 9:30-4:30, Sun. 12:30-4:30. No charge; donations accepted.

Attendance: 1,000 (estimated)

Membership: Senior & Student & Family $5; Friend $10; Business Contributor $25; Sustaining $50; Supporting $100; Life $500.

Manassa

JACK DEMPSEY MUSEUM AND PARK, 412 Main St., Manassa, CO 81141. Mailing Address: P.O. Box 130, Manassa, CO 81141-0130. Tel.: 719-843-5207.

Founded: 1966.

Governing Authority: Parent Institution: town of Manassa.

Institution Type/Description: History Museum: housed in the cabin where heavyweight boxing champion, Jack Dempsey was born.

Collections: Dempsey's career; boxing gloves from his NY fight; personal artifacts; photographs.

Hours & Admission Prices: Memorial Day to Labor Day Tues.-Sat. 10-5. No charge; donations accepted. &

Attendance: 2,000 (estimated)

Manitou Springs

CAVE OF THE WINDS, 100 Cave of the Winds Rd., Manitou Springs, CO 80829. Mailing Address: P.O. Box 826, Manitou Springs, CO 80829-0826. Tel.: 719-685-5444. Fax: 719-685-1712.

Institution Type/Description: Natural History Museum.

Collections: natural history.

Activities: guided tours; rental facilities.

Hours & Admission Prices: Summer daily 9-9; Winter: daily 10-5. Adults $18, children 6-11 $9; children 5 & under no charge.

CLIFF DWELLINGS MUSEUM, U.S. Hwy. 24, 10 Cliff Dwellings Rd., Manitou Springs, CO 80829. Mailing Address: P.O. Box 272, Manitou Springs, CO 80829-0272. Tel.: 719-685-5242; 800-354-9971.

E-mail: info@cliffdwellingsmuseum.com

Web Site: www.cliffdwellingsmuseum.com

Founded: 1904.

Key Personnel: Dir., Michele Hefner

Institution Type/Description: Natural History Museum.

Collections: Anasazi cliff dwellings; American Indian culture; archaeology; natural history preserve; stone mesa-top building; Anasazi baking oven; native flowers, herbs, trees & plants.

Facilities: picnic area; nature trail. Museum-related items for sale.

Activities: tours.

Hours & Admission Prices: March-April & Oct.-Nov. daily 9-5; May-Sept. daily 9-6; Dec.-Feb. daily 10-4. Adults 12 & over $9.50, seniors 60 & over $8.50, children 7-11 $7.50; discounts to groups & military; handicapped and children 6 & under no charge. Closed Thanksgiving; Christmas.

Attendance: 100,000 (estimated)

GARDEN OF THE GODS TRADING POST, 324 Beckers Lane, Manitou Springs, CO 80829. Tel.: 800-874-4515; 719-685-9045.

Institution Type/Description: Art Gallery.

Collections: works by Colorado fine artists; traditional & contemporary jewelry; Native American jewelry; Navajo rugs; sand paintings; Pueblo pottery.

Facilities: cafe. Museum-related items for sale.

Hours & Admission Prices: Summer: daily 8-8; Winter: daily 9-5. Closed Thanksgiving; Christmas.

MIRAMONT CASTLE MUSEUM, 9 Capitol Hill Ave., Manitou Springs, CO 80829-1618. Tel.: 719-685-1011; 888-685-1011. Fax: 719-685-1985.

E-mail: miramontcastle@yahoo.com

Web Site: www.miramontcastle.org

Founded: 1976.

Congressional District: 5

Key Personnel: Pres. Bd., Lance Michels; Treas. & Museum Shop Mgr., Peggie Yager.

Personnel Profile: Full-Time Paid 3; Full-Time Volunteers 12; Part-Time Paid 3; Part-Time Volunteers 12.

Governing Authority: nonprofit organization. Parent Institution: Manitou Springs Historical Society. Tax-exempt: 501(c)(3).

Institution Type/Description: Historic Building Museum: built in 1895.

Collections: Victorian period furniture & furnishings; period artifacts, books, maps, papers & records of the Manitou Springs area; paintings; worldwide dolls; miniatures from around the world; fire department.

Research Fields: local history.

Facilities: library; banquet facilities; tea room. Museum-related items for sale.

Activities: self-guided tours; lectures; meetings; concerts; arts festivals; seminars; guided group tours; community events.

Hours & Admission Prices: Memorial Day to Labor Day daily 9-5; Sept.-May

Tues.-Sun. 10-4; groups by appointment. Adults $8, seniors over 59 $7, children 6-12 $5; active military, children under 6 & members no charge. &
Attendance: 34,943 (accurate)
Membership: Golden Age $5; Individual $10; Family $15; Life $100; Sustaining $250; Benefactor $500; Patron $1,000.

Meeker

THE WHITE RIVER MUSEUM, 565 Park St., Meeker, CO 81641. Mailing Address: Rio Blanco County Historical Society, P.O. Box 413, Meeker, CO 81641-0413. Tel.: 970-878-9982.
E-mail: wrmuseum@nctelecom.net
Web Site: www.rioblancocounty.org
Founded: 1956.
Congressional District: 3
Key Personnel: Pres. (V), Ellene Meece; Cur., Ardith Douglass; Museum Shop Mgr., Sandra Shimko.
Personnel Profile: Part-Time Paid 2.
Governing Authority: society; nonprofit organization. Parent Institution: Rio Blanco County Historical Society. Tax-exempt.
Institution Type/Description: History Museum: housed in 1880 Officer's Quarters for military cantonment.
Collections: rocks; furniture; Indian artifacts; clothing; musical instruments; stoves; wooden tools; school furnishings; early photographs including Theodore Roosevelt hunting in 1901; period money; period furnishings; historic buildings.
Research Fields: local history.
Activities: tours; quarterly meetings; special events.
Publications: quarterly newsletter.
Hours & Admission Prices: Mon.-Fri. 9-5, Sat.-Sun. 10-5. No charge; donations accepted. Closed New Year's Day; Easter; Mother's Day; Father's Day; Thanksgiving; Christmas. &
Attendance: 3,500 (estimated)
Membership: Seniors $10; Individual $20; Family $30; Corporate $50-$100.

Mesa Verde National Park

MESA VERDE NATIONAL PARK MUSEUM, Mesa Verde National Park, CO 81330. Mailing Address: P.O. Box 8, Mesa Verde National Park, CO 81330-0008. Tel.: 970-529-4465. Fax: 970-529-4637.
E-mail: meve_general_information@nps.gov
Web Site: www.nps.gov/meve
Founded: 1917.
Congressional District: 3
Key Personnel: Cur., Tara Travis, Ph.D.
Personnel Profile: Full-Time Paid 1; Part-Time Paid 1; Interns 1.
Governing Authority: federal. Parent Institution: National Park Service. Tax-exempt.
Institution Type/Description: Park Museum: archaeological & ethnographic exhibits; ancestral Puebloan archaeological sites dating from AD 550-1300.
Collections: ancestral Puebloan archaeology; Southwest ethnology; archives, archaeology & park history; photographs, excavation, stabilization & park history; historical, park, Civilian Conservation Corps.
Research Fields: archaeology; geology; biology.
Facilities: 7,200-vol. library of archaeological & ethnographic books available for research on premises & through inter-library loan; campground; cafeteria; picnic area. Slides, postcards, books, pamphlets, trail guides & museum-related items for sale.
Activities: permanent exhibitions; guided & self-guided tours; inter-museum loan.
Publications: The Story of Mesa Verde National Park; The Towers of Hovenweep; The Coming of Gray Owl; Cliff Dwellings of the Mesa Verde; Indians of the Mesa Verde; Archeological Techniques used at Mesa Verde National Park.
Hours & Admission Prices: early April to mid-Oct. daily 8-6:30; mid-Oct. to early April daily 8-5. Park: Summer $15 per car; Winter $10 per car. Tickets available at Far View Visitor Center for guided tours to 3 major sites during summer season, $1.75 per person. &
Attendance: 623,000 (estimated)
Membership: Individual $30; Family $45; Heritage $1,000.

Monte Vista

MONTE VISTA HISTORICAL SOCIETY HISTORY CENTER, 110 Jefferson St., Monte Vista, CO 81144-1700. Mailing Address: P.O. Box 323, Monte Vista, CO 81144-0323. Tel.: 719-849-9320.
E-mail: mvhs123@msn.com
Founded: 1987.
Congressional District: 3

Key Personnel: Pres. (V), Peg Schall.
Personnel Profile: Part-Time Volunteers 5.
Governing Authority: Tax-exempt.
Institution Type/Description: History Museum.
Collections: period documents; photographs; area history.
Hours & Admission Prices: May-Sept. Mon.-Fri. 2-5; Winter: by appointment. No charge; donations accepted. &
Attendance: 10 (estimated)
Membership: Individual $10; Contributing $25; Professional $50; Sustaining $100; Life $1,000.

TRANSPORTATION OF THE WEST MUSEUM, 916 First Ave., Monte Vista, CO 81144-1445. Mailing Address: P.O. Box 323, Monte Vista, CO 81144-0323. Tel.: 719-849-0974.
E-mail: mvhs123@msn.com
Founded: 1988.
Congressional District: 3
Key Personnel: Pres. (V), Peg Schall.
Personnel Profile: Part-Time Volunteers 20.
Governing Authority: Parent Institution: Monte Vista Historical Society. Tax-exempt.
Institution Type/Description: Transportation Museum.
Collections: early modes of transportation; 1880s-1900s railroad; farming; sports; photographs.
Hours & Admission Prices: By appointment. No charge; donations accepted. &
Attendance: 150 (accurate)

Montrose

MONTROSE COUNTY HISTORICAL MUSEUM, 21 N. Rio Grande, Montrose, CO 81401-3467. Mailing Address: P.O. Box 1882, Montrose, CO 81402-1882. Tel.: 970-249-2085.
E-mail: info@montrosehistory.org
Web Site: www.montrosehistory.org
Founded: 1974.
Key Personnel: Pres. (V), Zilla May Brown; Sec., Ruth Heath; Treas., Steve Gray; Cur., Sally Johnson.
Personnel Profile: Part-Time Paid 1; Part-Time Volunteers 35.
Governing Authority: society. Parent Institution: Montrose County Historical Society. Tax-exempt.
Institution Type/Description: Historical Society Museum: housed in c.1912 Denver & Rio Grande Western Depot.
Collections: farm machinery; mining & railroad memorabilia; Indian artifacts; tools; country store; children's room; homesteader's cabin; scrapbooks; tapes; old newspapers; dolls; Union Pacific railroad caboose; cowboy cabin.
Research Fields: pioneers; old time buildings; pictures; newspapers, 1896-1940.
Activities: children's workshops; monthly programs. Museum Sponsors: Open House; Historic Treks; History & Culture Fair.
Publications: book, Experience in Lieu of Education; Montrose-100 Years; quarterly newsletter.
Hours & Admission Prices: May-Sept. Mon.-Fri. 9-5, Sat. 10-2. Adults $6, students $2; members no charge. Closed holidays. &
Attendance: 3,500 (accurate)
Membership: Individual $25; Family $40; Business $50; Life $175; Family Lifetime $300; Business Lifetime $500.

* **UTE INDIAN MUSEUM/MONTROSE VISITOR CENTER, (M),** 17253 Chipeta Dr., Montrose, CO 81403-4748. Tel.: 970-249-3098. Fax: 970-252-8741.
E-mail: cj.brafford@state.co.us
Founded: 1956.
Congressional District: 3
Key Personnel: C.E.O. & Museum Shop Mgr., C.J. Brafford.
Personnel Profile: Full-Time Paid 1; Part-Time Paid 5; Part-Time Volunteers 20; Interns 1.
Governing Authority: state. Parent Institution: Colorado Historical Society, 1300 Broadway, Denver, CO 80203. Tax-exempt.
Institution Type/Description: Indian History Museum: located on the site of Chief Ouray, leader of the Uncompahgve Utes.
Collections: ethnology of the Ute Indians; Indian artifacts; history; archaeology; paintings; anthropology; clothing; Ute Indian baskets, dioramas & wickiup.
Facilities: picnic area; visitor center; 8 1/2-acre farm. Books & other museum-related items for sale.
Activities: self-guided tours.
Publications: Colorado History News.
Hours & Admission Prices: May 31-Oct. Mon.-Sat. 9-4:30, Sun. 11-4:30; Nov.-May Mon.-Sat. 9-4:30. Adults $3.50, seniors $3, students 6-16 $1.50,

school groups $1.25 per child & $1 per adult; co-historical society members & children under 6 no charge. &

Attendance: 17,000 (accurate)

Membership: Students $15; Senior Citizens $20; Individual $25; Senior Couple $30; Family $35; Centennial $100; Patron $300; Annual Corp. $1,000.

Morrison

MORRISON NATURAL HISTORY MUSEUM, (M), 501 Colorado Hwy. 8, Morrison, CO 80465. Mailing Address: P.O. Box 564, Morrison, CO 80465-0564. Tel.: 303-697-1873. Fax: 303-697-8752.

E-mail: info@mnhm.org

Key Personnel: Dir., Matthew T. Mossbrucker

Institution Type/Description: Natural History Museum.

Collections: hands-on exhibits; local history; rocks & fossils; life-size sculpture of a baby stegosaurus.

Facilities: Museum-related items for sale.

Activities: tours; special events; birthday parties; volunteer programs.

Hours & Admission Prices: March-Oct. daily 10-6; Nov.-Feb. daily 10-4. Admission 12 & over $8, children 3-11 $6; children 2 & under no charge. Closed New Year's Day; Easter; Thanksgiving; Christmas Eve & Day.

Mosca

GREAT SAND DUNES NATIONAL MONUMENT AND PRE-SERVE, 11500 Hwy. 150, Mosca, CO 81146-9502. Tel.: 719-378-6363. Fax: 719-378-6360.

E-mail: phyllis_bovin@nps.gov

Founded: 1932.

Congressional District: 3

Key Personnel: Supt., Steve Chaney; Resource Mgmt., Fred Bruch; Cur., Phyllis Pineda Bovin.

Personnel Profile: Full-Time Paid 1; Part-Time Paid 1.

Governing Authority: federal. Parent Institution: National Park Service. Tax-exempt.

Institution Type/Description: Park Museum.

Collections: geology; general; natural history; anthropology; archaeology; Indian artifacts; herbarium; insects.

Research Fields: geology; zoology; botany.

Facilities: 1,500-vol. library of natural science & resource books available for inter-library loan & for use to public upon written request to superintendent; reading room. Slides, postcards, books, pamphlets & trail guides for sale.

Activities: guided tours; lectures; gallery talks; permanent exhibits.

Publications: information handouts.

Hours & Admission Prices: Visitor Center: Summer: daily 9-6; Spring & Fall daily 9-5; Winter: daily 9-4:30. Adults 17 & over $3.

Naturita

RIMROCKER HISTORICAL SOCIETY OF WEST MON-TROSE COUNTY AND MUSEUM, 411 W. 2nd Avenue, Naturita, CO 81422. Mailing Address: P.O. Box 913, Nucla, CO 81424-0913. Tel.: 970-865-2100. Facebook: Rimrocker Historical Society.

E-mail: rimrocker@nntcwireless.net

Founded: 1977.

Congressional District: 3

Key Personnel: Chm. (V), Marie Templeton; Pres. (V), Jane Thompson.

Personnel Profile: Part-Time Paid 1; Part-Time Volunteers 8.

Volunteer Hours: 600

Operating Expenses: 2,000

Operating Income: 2,000

Governing Authority: society. Tax-exempt.

Institution Type/Description: Historical Society Museum: housed in Old Naturita Elementary School House.

Collections: Uranium mining; Indian artifacts; outside exhibit-mining & farming; blacksmith shop displaying tools of the trade; ranch artifacts; household items; dolls; costumes; tools; collection depicting history of the Pinon colony, the Colorado Cooperative Co., which built irrigation ditch to irrigate Tabequacke Park c.1894-1904; Umetco's video collection of mining & milling operations; pioneer kitchen, bedroom & business safe; hanging flume model; obituaries; file of articles of local people, schools, businesses, sports events, organizations & churches.

Research Fields: local history.

Facilities: Museum-related items for sale.

Activities: guided tours; temporary & permanent exhibitions. Museum Sponsors: Annual Mine Tour; Labor Day Picnic.

Publications: books, Spell of the Tabeguache; Visionaries; Standard Chemical Company; Naturita, Where the Past Meets the Present; Uravan: 100 Years; annual newsletter; pamphlets; CDs & DVDs.

Hours & Admission Prices: June-Labor Day Tues.-Sat. 2-4; Winter: Wed. 2-4. No charge; donations accepted. &

Attendance: 450 (estimated)

Membership: Individual $10; Family $20; Business $45.

Nunn

DRYLANDERS MUSEUM, 755 3rd St., Nunn, CO 80648. Tel.: 970-897-3125 (summer) 2671 (winter).

Institution Type/Description: History Museum.

Collections: area history & culture; period artifacts; pioneer life.

Activities: tours; lectures.

Hours & Admission Prices: Summer: Sat.-Sun. 1-4; other times by appointment.

Ouray

OURAY COUNTY HISTORICAL SOCIETY MUSEUM, 420 6th Ave., Ouray, CO 81427. Mailing Address: P.O. Box 151, Ouray, CO 81427-0151. Tel.: 970-325-4576.

E-mail: ochs@ouraynet.com

Institution Type/Description: Historical Society Museum: housed in the former St. Joseph's Miners' Hospital; built in 1886.

Collections: local history & culture; mining; ranching; railroading; minerals; Native American artifacts; photographs; personal artifacts.

Facilities: library.

Hours & Admission Prices: April 15-May 14 Thurs.-Sat. 1-4:30; May 15-Sept. Mon.-Sat. 10-4:30, Sun. 12-4:30; Oct.-Nov. Thurs.-Sat. 10-4:30. Adults $5, seniors 60 & over $3, children 12 & under $1; discounts to groups; members no charge.

Pagosa Springs

FRED HARMAN ART MUSEUM, 85 Harman Park Dr., Pagosa Springs, CO 81147. Mailing Address: P.O. Box 192, Pagosa Springs, CO 81147-0192. Tel.: 970-731-5785. Fax: 970-731-4832.

E-mail: info@harmanartmuseum.com

Web Site: www.harmanartmuseum.com

Founded: 1979.

Congressional District: 3

Key Personnel: C.E.O., Cur. & Chm. (V), Fred C. Harman, III; Financial Dir. & Treas., Marilyn Harris; Devel., Membership & Museum Shop Mgr., Norma Harman; Security, Lee Ligon.

Personnel Profile: Full-Time Volunteers 2; Part-Time Volunteers 12.

Governing Authority: nonprofit organization. Tax-exempt: 501(c)(3).

Institution Type/Description: Art Museum: home of Fred Harman, artist & cartoonist, creator of the Red Ryder & Little Beaver comics; one of the founders of the Cowboy Artists of America.

Collections: Harman oil paintings; pen & inks; books; original studio intact; saddles; firearms; photographs of rodeo & movie stars; awards; western memorabilia.

Research Fields: history of Fred Harman.

Facilities: 2,000 sq. ft. exhibit space. Museum-related items for sale.

Activities: guided tours; lectures.

Hours & Admission Prices: Feb.-March by appointment only; April-May & Sept.-Jan. Mon.-Fri. 10:30-5; Memorial Day-Labor Day Mon.-Sat. 10:30-5, Sun. 12:30-4. Adults $4, students & children $.50; discounts to CWYN members; members no charge. &

Attendance: 4,950 (estimated)

Membership: Individual $25; Family $50; Friend $100; Supporting $250; Patron $500; Sustaining $1,000; Life $10,000.

Palmer Lake

LUCRETIA VAILE MUSEUM, 66 Lower Glenway St., Palmer Lake, CO 80133. Mailing Address: Palmer Lake Historical Society, P.O. Box 662, Palmer Lake, CO 80133-0662. Tel.: 719-559-0837.

Web Site: www.palmerdividehistory.org

Founded: 1956.

Congressional District: 20

Key Personnel: Dir., Roger Davis.

Personnel Profile: Part-Time Volunteers 16.

Governing Authority: society. Parent Institution: Palmer Lake Historical Society. Tax-exempt: 501(c)(3).

Institution Type/Description: Historical Society Museum.

Collections: local relics & photographs; manuscripts; Timeline 1873 to present.

Research Fields: local history.
Facilities: 100-vol. collection of old books available for use under supervision on premises.
Activities: lectures; films.
Publications: Palmer Lake Narrative; Palmer Lake Historic Landmarks; Estemere Estate of Palmer Lake; Monument-Faded Neighbor Towns; Through the Years at Monument; Communities of the Palmer Divide.
Hours & Admission Prices: June-Aug. Wed. 1-4, Sat. 10-2; Wed. 1-3 balance of year; 1st Thurs of month 1-12. Sept.-May Sat. 10-2. No charge; donations accepted. Closed national holidays. &

Attendance: 750 (estimated)

Parker

WILDLIFE EXPERIENCE MUSEUM, 10035 S. Peoria St., Parker, CO 80134-9600. Tel.: 720-488-3300.
E-mail: info@twexp.org
Web Site: www.thewildlifeexperience.org
Institution Type/Description: Natural History Museum.
Collections: natural history; fine art; film; paintings; sculpture; photography.
Facilities: theater; conference rooms; restaurant. Museum-related items for sale.
Activities: educational programs; conservation classes; films; school tours; Meet the Artist programs; special events.
Hours & Admission Prices: Tues.-Sun. 9-5. Adults 13-64 $7.95, seniors 65 & over $6.95, children 2-12 $4.95; discounts to groups; children under 2 no charge. Additional fee for theater & shows. Closed Thanksgiving; Christmas.
Membership: Individual $35; Dual $50; Family & Grandparent $75; Patron $125; Supporting $200; Benefactor $500; Life $900-$1,800.

Peterson Air Force Base

PETERSON AIR & SPACE MUSEUM, 21st Space Wing/MU, 150 E. Ent Ave., Peterson Air Force Base, CO 80914-1303. Tel.: 719-556-4915. Fax: 719-556-8509. Facebook: Peterson Air and Space Museum.
E-mail: 21sw.mu@us.af.mil
Web Site: www.petemuseum.org
Founded: 1981.
Personnel Profile: Full-Time Paid 2; Part-Time Volunteers 40.
Governing Authority: Parent Institution: National Museum of the U.S. Air Force. Tax-exempt.
Institution Type/Description: Military Museum.
Collections: aviation & space history of Colorado Springs and Peterson Air Force Base; 16 aircraft; 4 missiles; models; uniforms; WWII.
Facilities: theater. Museum-related items for sale.
Hours & Admission Prices: Tues.-Sat. 9-4. No charge; donations accepted. Closed federal holidays. &
Attendance: 14,000 (estimated)
Membership: Annual $25; Family $50; Life $325; Sponsor $500. Eagle Wings: Kitty Hawk $1,000-$4,999; Lightning $5,000-$9,999; Spirit of St. Louis $10,000-$24,999; Thunderbolt $25,000-$49,999; Super Connie $50,000-$74,999; Eagle Wings $75,000 & up.

Platteville

FORT VASQUEZ MUSEUM, (M), 13412 U.S. Hwy. 85, Platteville, CO 80651-8017. Tel.: 970-785-2832.
Web Site: www.coloradohistory.org
Founded: 1958.
Congressional District: 4
Key Personnel: Dir., Susan Hoskinson.
Personnel Profile: Full-Time Paid 1; Part-Time Paid 1; Interns 1.
Governing Authority: state; society. Parent Institution: Colorado Historical Society, 1300 Broadway, Denver, CO 80203-2137. Tax-exempt: 501(c)(3).
Institution Type/Description: Fur Trade and Native American Museum: reconstructed adobe fort on site of original 1835 trading post. Fort Vasquez trading post archaeological site and the Works Progress Administration adobe fort are listed in the National Register of Historic Places.
Collections: items related to Rocky Mountain fur trade era, 1830-1840s; items created by Arapahoe, Cheyenne, Sioux, & Ute during late 19th century, interpretation of archaeological investigation at Fort Vasquez trading post sites in the late 1960s. Historic Structure: 1935-1936 Works Progress Administration reconstructed Fort Vasquez fur trading post.
Research Fields: Plains Indians; historic archaeology; history of Rocky Mountain fur trade.
Facilities: Museum-related items for sale.
Activities: guided tours; lectures; living history demonstrations of Plains Indians and fur trade life.

Publications: Colorado History Now.
Hours & Admission Prices: Memorial Day-Labor Day daily 9-4:30, call to confirm. Adults $1; children 4 & under no charge. Closed New Year's Day; Thanksgiving; Christmas. &
Attendance: 9,565 (accurate)
Membership: Contact Colorado Historical Society membership office.

Pueblo

COLORADO MENTAL HEALTH INSTITUTE AT PUEBLO MUSEUM, 13th & Francisco Sts., Pueblo, CO 81003. Mailing Address: 1600 W. 24th St., Pueblo, CO 81003-1411. Tel.: 719-543-2012.
E-mail: info@cmhipmuseum.org
Web Site: www.cmhipmuseum.org
Key Personnel: Dir., Bob Mitchell; Asst. Dir., Nell Mitchell; Dir. Volunteer Svcs., Eunice Wolther.
Governing Authority: nonprofit organization.
Institution Type/Description: Psychiatry Museum: listed on the National Register of Historic Places.
Collections: institute history; treatment of mentally ill; personal artifacts; historical documents; photographs; medical equipment; awards.
Activities: research.
Hours & Admission Prices: Tues. 10-4; other times by appointment. No charge.

✱ **EL PUEBLO HISTORY MUSEUM, (M),** 301 N. Union, Pueblo, CO 81003-4266. Tel.: 719-583-0453. Fax: 719-583-8214.
E-mail: deborah.espinosa@chs.state.co.us
Web Site: www.coloradohistory.org
Founded: 1959.
Congressional District: 3
Key Personnel: Dir., Deborah Espinosa; Administrative Asst., Kathleen Byers; Education Coord., Kathleen Eriksen; Trades, Truman Pooler.
Personnel Profile: Full-Time Paid 1; Part-Time Paid 3; Part-Time Volunteers 65.
Governing Authority: state; Parent Institution: Colorado Historical Society. Tax-exempt: 501(c)(3).
Institution Type/Description: Historical Society Museum.
Collections: Spanish period of occupation; prehistoric & historic Indian artifacts; artifacts related to cattle industry; early man; Plains Indians; Spanish/Mexican Period; Pueblo's industrialization & European immigration to the city until 1900; historic site, El Pueblo Trading Post; archaeological dig; 1921 flood; 20th-century Pueblo.
Research Fields: Spanish/Mexican settlement in southern Colorado; archaeology of El Pueblo Trading Post; history of Pueblo, Colorado, 19th & 20th centuries.
Activities: formally organized education programs for children. Annual Event: Mercado (Frontier Market, Living History) and Chile Festival. Museum Sponsors: Children's Day; Women's History Week; Family Saturdays.
Publications: quarterly, Colorado Heritage; monthly, Colorado History NOW.
Hours & Admission Prices: Tues.-Sat. 10-4. Adults $4; discounts to AAM members & military; children 12 & under no charge. Closed major holidays. &
Attendance: 23,000 (accurate)
Membership: Student $18; Seniors over 65 $25; Individual $30; Institution $35; Family $40; Centennial $100; Patron $300; Heritage Club $1,000.

HOSE CO. NO. 3 FIRE MUSEUM, 116 Broadway Ave., Pueblo, CO 81003. Tel.: 719-821-1273. Fax: 719-553-2831.
E-mail: gmmpf@hotmail.com
Web Site: hosecono3.com
Founded: 1979.
Key Personnel: Chm. (V), Gary Micheli; Pres. (V), Mark Pickerel.
Personnel Profile: Part-Time Volunteers 2.
Volunteer Hours: 300
Governing Authority: Parent Institution: Pueblo Five Fighters Historical Society. Tax-exempt.
Institution Type/Description: Firefighting History Museum.
Collections: firefighting history & equipment; early fire trucks & apparatus; 1882 Orman hose cart; photographs; log book from 1881-1889; c.1883 Gamewell telegraph fire alarm system base station; championship belt & ribbons; uniforms; personal artifacts; fire patches.
Research Fields: fire department history.
Facilities: historic fire stations.
Hours & Admission Prices: Call for hours. No charge; donations accepted.
Attendance: 738 (accurate)

INFOZONE NEWS MUSEUM, (M), 100 E. Abriendo Ave., Pueblo, CO 81004-4232. Tel.: 719-553-0205.
E-mail: maria.tucker@pueblolibrary.org
Web Site: www.pueblolibrary.org
Key Personnel: Dir., Maria Sanchez-Tucker.
Personnel Profile: Full-Time Paid 2; Part-Time Volunteers 10.
Governing Authority: city. Parent Institution: County Library District. Tax-exempt.
Institution Type/Description: History Museum.
Collections: Pueblo history & culture; hands-on exhibits.
Facilities: 100-seat theater.
Activities: hands-on exhibits; educational programs; films.
Hours & Admission Prices: Mon.-Thurs. 9-9, Fri.-Sat. 9-6, Sun. 1-5. No charge. &
Attendance: 30,000 (accurate)

PUEBLO ART GUILD AND GALLERY, 1500 N. Santa Fe, Pueblo, CO 81003-3700. Tel.: 719-543-2455.
E-mail: pag_enews@q.com
Web Site: www.puebloartguild.com
Founded: 1959.
Congressional District: 3
Key Personnel: Pres., Freda Moore
Institution Type/Description: Art Gallery: housed in the historic boathouse in Mineral Palace Park.
Collections: works by local artists.
Hours & Admission Prices: Spring, Summer & Fall Wed.-Sun. 12-4. Winter: call for hours. No charge; donations accepted. Closed most holidays.
Membership: Student $7; Individual $20; Family $30; Sponsor $80 & up.

PUEBLO COUNTY HISTORICAL SOCIETY MUSEUM AND EDWARD H. BROADHEAD LIBRARY, 201 W. ”B“ St., Pueblo, CO 81003-3403. Mailing Address: 203 W. ”B“ St., Pueblo, CO 81005. Tel.: 719-543-6772.
E-mail: info@pueblohistory.org
Web Site: www.pueblohistory.org
Founded: 1976.
Congressional District: 3
Key Personnel: Pres., John Ercul.
Personnel Profile: Part-Time Volunteers 16.
Volunteer Hours: 900
Governing Authority: Parent Institution: Pueblo County Historical Society. Tax-exempt.
Institution Type/Description: Historical Society Museum & Library.
Collections: railroad furniture; memorabilia; Native American; Pueblo saddlery; baseball; prehistoric artifacts; saddle collection; local baseball collection
Research Fields: meta physical.
Facilities: conference room.
Activities: monthly membership meetings; annual historical interest tours.
Publications: monthly, The Pueblo Lore; books related to the history & heritage of Pueblo & southern Colorado.
Hours & Admission Prices: Tues.-Sat. 10-4. Adult $5, seniors & military $4, members & members of assoc. member organizations $2, southern Colorado Heritage Center members no charge. Closed holidays. &
Attendance: 4,000 (estimated)
Membership: Student $12; Single $35; Family $45; Supporting $80; Small Business $150; Sponsor $200; Large Business $275; Patron $300; Corporate Sponsor $500; Benefactor $1,000.

PUEBLO WEISBROD AIRCRAFT MUSEUM, (M), 31001 Magnuson Ave., Pueblo, CO 81001-4822. Tel.: 719-948-9219. Fax: 719-948-2437.
E-mail: phas@pwam.org
Web Site: www.pwam.org
Formerly: International B-24 Museum
Founded: 1985.
Key Personnel: Pres., Don Blehm; Vice Pres., Ralph Decker.
Personnel Profile: Part-Time Volunteers 50.
Volunteer Hours: 9,600
Operating Expenses: 86,000
Operating Income: 120,000
Governing Authority: bd. of directors. Tax-exempt.
Institution Type/Description: Aircraft & Military History Museum.
Collections: US military aviation; military uniforms, aircraft & vehicles; space artifacts.
Activities: special events; fly-ins; scout; youth; STEM education.
Publications: newsletter.

Hours & Admission Prices: Mon.-Sat. 10-4, Sun. 1-4. Adults $7; military with proper ID no charge. &
Attendance: 18,000 (accurate)
Membership: Annual $25; Life Time 60 & over $125; 40-59 $150; Under 40 $175.

PUEBLO ZOO, 3455 Nuckolls Ave., Pueblo, CO 81005-1234. Tel.: 719-561-1452, ext. 100. Fax: 719-561-8686.
E-mail: zoodirector@pueblozoo.org
Web Site: www.pueblozoo.org
Founded: 1903.
Congressional District: 4
Key Personnel: Exec. Dir., Stephanie Stowell; Pres., Karen Ross; Cur., Marilyn McBirney; Mktg., Gloria Madrill; Education, Sue Hardesty; Museum Shop Mgr., Betty Wilkinson.
Personnel Profile: Full-Time Paid 19; Part-Time Paid 10; Part-Time Volunteers 75; Interns 2.
Governing Authority: private; nonprofit. Tax-exempt: 501(c)(3).
Institution Type/Description: Zoo.
Collections: over 425 exotic & domestic animals of 135 species.
Facilities: 400-vol. library covering zoological-related topics; classroom. Zoo-related items for sale.
Activities: docent program; formal education programs for children; guided tours; lectures; participatory exhibits. Annual Events: Zoofari; Senior Safari; ElectriCritters; Party for the Planet; Zoo Boo.
Publications: newsletter, Pueblo Zoo News; annual report.
Hours & Admission Prices: May Mon.-Fri. 9-4, Sat.-Sun. 9-5; June-Aug. daily 9-5; Sept.-April Mon.-Sat. 9-4, Sun. 12-4. Adults $8, children 3-16 $6; discounts to groups & AZA members; members no charge. Closed New Year's Day; Thanksgiving; Christmas. &
Attendance: 86,457 (accurate)
Membership: Student $25; Senior $35; Individual $40; Senior Couple $45; Family & Grandparents $60.

ROSEMOUNT MUSEUM, 419 W. 14th St., Pueblo, CO 81003-2707. Tel.: 719-545-5290. Fax: 719-545-5291.
E-mail: ddarrow@rosemount.org
Web Site: www.rosemount.org
Founded: 1967.
Congressional District: 3
Key Personnel: C.E.O. & Exec. Dir., Deb Darrow; Pres., Kathlyn Thatcher Vail; Collections Mgr. & Housekeeping, Susan Kittinger; Maintenance & Groundskeeper, Roger Cain; Museum Shop Mgr., Patricia Bedard.
Personnel Profile: Full-Time Paid 4; Part-Time Paid 1; Part-Time Volunteers 80.
Governing Authority: nonprofit organization. Tax-exempt: 501(c)(3).
Institution Type/Description: Historic House Museum: housed in an 1893 Victorian mansion, the John A. Thatcher residence; Henry Hudson Holly, architect.
Collections: late Victorian decorative arts; personal artifacts; period furnishings; photographs; 1904 McClelland Collection of World Curiosities.
Research Fields: late Victorian period.
Facilities: research library; restaurant. General gift items for sale.
Activities: guided tours; lectures; concerts; educational programs; community events.
Publications: brochure; quarterly newsletter; school workbooks; Victorian cookbook.
Hours & Admission Prices: Feb.-Dec. Tues.-Sat. 10-3:30. Adults $6, senior citizens $5, children 6-18 $4; discounts to active military; members & children under 6 no charge. Closed holidays.
Attendance: 7,000 (accurate)
Membership: Seniors $15; Individual $25; Family $40; Contributor $75; Patron & Corporate $150; Benefactor $250; Collectors Council $500 & up.

＊ **SANGRE DE CRISTO ARTS CENTER & CONFERENCE CENTER, (M),** 210 N. Santa Fe Ave., Pueblo, CO 81003-4133. Tel.: 719-295-7200. Fax: 719-295-7230.
E-mail: mail@sdc-arts.org
Web Site: www.sdc-arts.org
Formerly: Sangre de Cristo Arts Center & Buell Children's Museum
Founded: 1972.
Congressional District: 3
Key Personnel: Exec. Dir. & C.E.O., Dr. Dan Lere; Chm. (V), Steven Miller; Cur. Visual Arts, Christel Dussart; Cur. Children's Museum, Donna Stinchcomb; Asst. Cur. Children's Museum, Joleen Ryan; Cur. Education, Jackie Henderson; Asst. Education Cur. & Cur. Asst., Diane Pirraglia; Dir. School of Dance, Stephen Wynne; Accountant & Controller, Rochelle Spoone; Membership & Box Office Mgr., Cheryl Califano; Facilities & Beverage

Mgr., Lorrie Marquez; Accounting Asst. & Museum Shop Mgr., Julie Salas; Asst. Membership & Box Office Mgr., Dan Masterson.
Personnel Profile: Full-Time Paid 17; Part-Time Paid 41; Part-Time Volunteers 97; Interns 1.
Governing Authority: nonprofit organization. Tax-exempt: 501(c)(3).
Institution Type/Description: Arts Center & Children's Museum.
Collections: historic & contemporary art of the American West.
Research Fields: Western American paintings; Gene Kloss intaglio prints.
Facilities: library; theater; conference & banquet facilities; children's museum. Gift items for sale.
Activities: guided tours; lectures; films; concerts; theater; workshops; dance school; permanent, temporary, traveling & loan exhibitions; docent program; conferences; outdoor summer concerts; performing arts; after school art classes for students K-12.
Publications: quarterly newsletter, Town & Center Mosaic; education brochure; exhibit catalogs; books, Gene Kloss: Impressions of the Land & People; Joseph Hitchins: The Passionate Landscape; The Art of C.M. Russell & Northern Plains Indians; Colorado on Canvas: A Pictorial Survey of Early Colorado; Francis King Collection; South by Southwest: 25th Anniversary Catalog; Under Western Skies 30 Years of Collecting.
Hours & Admission Prices: Arts Center & Children's Museum: Tues.-Thurs. & Sat. 11-4, Fri. 9-4. Adults $4, children $3; discounts to military & AAM members; members no charge. Office: Mon.-Fri. 9-5. Closed legal holidays. &
Attendance: 222,000 (accurate)
Membership: Student $20-$74; Senior $25-$74; One-Person $30-$74; Two-Person $35-$74; Grandparents $45-$74; Family $45-$74; Contributor $100-$149; Patron $150-$274; Sponsor $275-$549; Benefactor $550-$1,449; Premiere Club $1,500 & up.

SOUTHEASTERN COLORADO HERITAGE CENTER AND MUSEUM, 201 W. B St., Pueblo, CO 81003-3403. Tel.: 719-295-1517. Fax: 719-295-0040.
Founded: 1997.
Key Personnel: Museum Coord., Fran Reed.
Personnel Profile: Full-Time Paid 1; Part-Time Volunteers 28.
Governing Authority: bd. of directors. Tax-exempt: 501(c)(3).
Institution Type/Description: History Museum: listed on the National Register of Historic Buildings.
Collections: southern Colorado history & culture; saddles; period telephones; railroad artifacts.
Facilities: library.
Activities: school tours; Legacy trunk education outreach; cultural organizations workshops.
Hours & Admission Prices: Tues.-Sat. 10-4. seniors 55 & over and active military w/ID $4, children 6-12 $3; children 5 & under and SCHC members no charge. Closed New Year's Day; Independence Day; Thanksgiving & day after; Christmas. &
Attendance: 9,000 (estimated)
Membership: Student 15-22 $20; Senior over 55 $25; Individual $30; Senior Couple over 55 $35; Family $40; Associate $50-$199; Patron $200-$499; Business $500 & up.

Rangely

RANGELY OUTDOOR MUSEUM, 150 Kennedy Dr., Rangely, CO 81648-3503. Mailing Address: Rangely Museum Society, P.O. Box 131, Rangely, CO 81648. Tel.: 970-675-2612.
E-mail: ramuseum@quikus.com
Formerly: Rangely Museum
Key Personnel: Dir., Brenda Hopson; Pres. (V), Tom Collins.
Governing Authority: Tax-exempt.
Institution Type/Description: History Museum.
Collections: local history; Native American; pioneers; ranching from 1883 to 1946; energy development from 1946 to present.
Activities: Annual Events: Ice Cream Social in September.
Publications: quarterly newsletter, The White River Crier.
Hours & Admission Prices: April-May & Sept.-Oct. Fri.-Sat. 10-4, Sun. 12-4; June-Aug. Mon.-Sat. 10-4, Sun. 12-4; other times by appointment. No charge; donations accepted.
Membership: Individual $10; Couple & Family $15; Business & Patron $25; Benefactor $50.

Ridgway

RIDGWAY RAILROAD MUSEUM, US Hwy. 550 & Colorado State Hwy. 62, Ridgway, CO 81432. Mailing Address: P.O. Box 588, Ridgway, CO 81432-0588. Tel.: 970-626-4239.
Web Site: www.ridgwayrailroadmuseum.org
Founded: 1998.

Key Personnel: Pres. (V), Karl Schaeffer.
Personnel Profile: Part-Time Volunteers 20.
Governing Authority: nonprofit organization. Tax-exempt: 501(c)(3).
Institution Type/Description: Railroad Museum.
Collections: railroading history & equipment; photographs; personal artifacts; rail cars.
Research Fields: history of narrow gauge railraoding in Southwestern Colorado.
Facilities: library.
Activities: educational programs; special events.
Publications: book, Narrow Gauge Railroading in the San Juan Triangle; two books in press.
Hours & Admission Prices: May daily 10-3; June-Sept. daily 9-5; Oct.-Nov. daily 10-3; Dec.-April by appointment. No charge; donations accepted.
Attendance: 7,500 (accurate)
Membership: Individual $20; Family $40; Business & Silver $100; Gold $200; Platinum $500; Life $1,000.

Rifle

RIFLE CREEK MUSEUM, 337 East Ave., Rifle, CO 81650-2333. Tel.: 970-625-4862.
Founded: 1967.
Congressional District: 3
Key Personnel: Pres. (V), Cecil Waldron; Vice Pres., Ardis Green; Dir., Kim Fazzi; Sec., Betty Waldron; Treas., Jean Mullenax.
Personnel Profile: Part-Time Paid 1; Part-Time Volunteers 10.
Governing Authority: nonprofit. Tax-exempt.
Institution Type/Description: General Museum.
Collections: natural history; history; Indian artifacts; Garrison photos; textiles.
Research Fields: local history.
Activities: guided tours; slide shows.
Hours & Admission Prices: May-Oct. Mon.-Fri. 10-4; other times by appointment. Adults $5, senior citizens $4, children 6-12 $3; children 5 & under and members no charge. &
Attendance: 2,500 (estimated)
Membership: Individual $20; Family $35; Business $50; Lifetime $1,000.

Rocky Ford

ROCKY FORD HISTORICAL MUSEUM, 1005 Sycamore Ave., Rocky Ford, CO 81067-1760. Mailing Address: P.O. Box 835, Rocky Ford, CO 81067-0835. Tel.: 719-254-6737.
Founded: 1940.
Congressional District: 3
Key Personnel: Bd. Pres., Brian Liekam; Bd. Sec., Nancy Martin; Cur., William B. Hodges, Jr.
Personnel Profile: Part-Time Paid 1; Part-Time Volunteers 4.
Governing Authority: board of trustees. Parent Institution: City of Rocky Ford.
Institution Type/Description: General Museum: housed in the form Carnegie Library; built in 1909.
Collections: Swink family; American Crystal Sugar Co.; produce farming; furnishings; first fire engine; military; photography; oral history tapes; archives; ethnography; rocks & minerals; fossils; Indian artifacts.
Research Fields: local history.
Facilities: library of school yearbooks, photographs & newspapers available for research.
Activities: guided tours.
Hours & Admission Prices: Memorial Day to Labor Day Tues.-Fri. 1-5, Sat. 10-2; Sept.-May Wed. 1-5; other times by appointment. No charge; donations accepted.
Attendance: 700 (estimated)

Saguache

HAZARD HOUSE MUSEUM, 735 Pitkin Ave., Saguache, CO 81149. Mailing Address: P.O. Box 569, Saguache, CO 81149. Tel.: 719-655-6550.
Founded: 1998.
Key Personnel: Dir., Trish Gilbert.
Personnel Profile: Part-Time Volunteers 10.
Governing Authority: Parent Institution: Saguache County Museum. Tax-exempt.
Institution Type/Description: History Museum: housed in a 1908 home.
Collections: Hazard family artifacts; period furnishings; Steinway piano; oriental rugs; silver flatware; tea sets; personal artifacts.
Hours & Admission Prices: Memorial Day to 3rd week of Sept. daily 9-4. Adults $2, children under 12 $1; discounts to groups of 40 or more.

SAGUACHE COUNTY MUSEUM, 405 8th St., Saguache, CO 81149. Mailing Address: P.O. Box 569, Saguache, CO 81149-0569. Tel.: 719-655-6550 & 2557.
E-mail: bearbasinllc@gojade.org
Web Site: www.museumtrail.org
Founded: 1958.
Congressional District: 3
Key Personnel: Chm. (V), Virginia Sutherland; C.E.O., Trish Gilbert; Co-Chm. (V), Thad Englert.
Personnel Profile: Part-Time Paid 6; Part-Time Volunteers 6.
Governing Authority: county. Parent Institution: Saguache County Museum Board. Tax-exempt: 501(c)(3).
Institution Type/Description: Pioneer Museum & National Historic Site.
Collections: history artifacts; mineralogy; geology; glass; early pioneer memorial; archaeology; Indian artifacts; zoology; family histories. Historic Houses: 1870 school & jailer's residence; 1908 Saguache County jail; c.1880 blacksmith shop.
Facilities: library of early school, law & medical books, documents, letters, pictures & papers, historical books & works by local authors, available for research by people or organizations upon permission of the board. Books on local history for sale.
Activities: inter-museum loan, permanent & temporary exhibitions; special tours for groups of children. Museum Sponsors: Memorial weekend parade; Annual Fall Festival in Saguache Co. in September; crafts sales by local artisans.
Publications: pamphlets on local history; Frontier Eyewitness - Diary of John Lawrence; Images of The Past, Vol. I, II & III - Saguache County History People, Places & Events Vol. IV; Images of Past Vol. V; Sunny San Luis Valley. 1875 Episcopal Minister; Bonanza G. Kerber Creek; Images VI, 2007.
Hours & Admission Prices: Memorial Day-Labor Day daily 9-4; 3rd week in Sept. to Oct. by appointment. Adults $5, children under 12 $1; discounts to groups of 50 or more; members no charge. &
Attendance: 1,424 (accurate)
Membership: Sustaining $15-$100.

Salida

SALIDA MUSEUM, 406 1/2 W. Rainbow Blvd., Salida, CO 81201-2236. Tel.: 719-539-7483. Facebook: Salida Museum.
E-mail: salidamuseum@gmail.com
Web Site: www.salidamuseum.org
Founded: 1954.
Congressional District: 3
Key Personnel: Museum Shop Mgr., Bob Campbell.
Personnel Profile: Part-Time Volunteers 12.
Governing Authority: municipal. Tax-exempt.
Institution Type/Description: General Museum.
Collections: local historic artifacts; Indian artifacts; mineralogy; mining; homesteading; textiles.
Research Fields: newspapers; recorded stories; pictures.
Hours & Admission Prices: Memorial Day to Labor Day daily 11-5; tours by appointment; Sept.-May Sat. 10-5, Sun. 1-5. Adults $3, children 12-17 $1.50, youth 7-11 $1; children 6 & under and members no charge. &
Attendance: 500 (estimated)
Membership: Annual $10; Lifetime $100 & up.

Silt

SILT HISTORICAL SOCIETY, 707 Orchid Ave., Silt, CO 81652. Mailing Address: P.O. Box 401, Silt, CO 81652-0401. Tel.: 970-876-5801.
E-mail: silthistorical@yahoo.com
Web Site: www.silthistoricalpark.com
Founded: 1982.
Congressional District: 3
Key Personnel: Pres. Bd., Bill Smith; Public Rels., Joan Nestor; Museum Shop Mgr., Alice Jones.
Personnel Profile: Part-Time Volunteers 12.
Volunteer Hours: 1,500
Governing Authority: private; nonprofit organization. Tax-exempt: 501(c)(3).
Institution Type/Description: Living History Museum.
Collections: Native American artifacts; horse drawn machinery. 10 Historic Buildings 1880s-1930s: school, house, store, office, saloon, shop, blacksmith shop, cow camp, & machine sheds.
Research Fields: family histories & history of towns in the area.
Facilities: 662-vol. library of National Geographic books, school books, booklets, pamphlets & scrapbooks. Museum-related items for sale.
Activities: guided tours; lectures; temporary exhibitions; artisan demonstrations. Museum Sponsors: early 1900 Chautauqua in May; Family Hey Day in August.
Publications: Garfield County, First 100 Years; Reflections, Rifle Early and Late; Reflections, Dry Hollow Ranch Memories; Reflections, Yes I Remember; Reflections, The Family History of Mary Wright; Reflections, The Century of My Life; Silt, Colorado Homesteads, 1880-1940.
Hours & Admission Prices: May-Oct. Tues.-Sat. 12-5. No charge; donations accepted. &
Attendance: 3,000 (estimated)
Membership: Family $10; Business $50; Patron $100.

Silver Cliff

SILVER CLIFF MUSEUM, 606 Main St., Silver Cliff, CO 81252. Mailing Address: P.O. Box 154, Silver Cliff, CO 81252-0154. Tel.: 719-783-2615 & 2837. Fax: 719-783-4480.
E-mail: silverclifftown@centurytel.net
Founded: 1959.
Congressional District: 3
Key Personnel: Cur., Dorothy L. Urban; Chm. (V), Carol Franta.
Personnel Profile: Part-Time Paid 2; Part-Time Volunteers 6.
Governing Authority: municipal. Tax-exempt.
Institution Type/Description: General Museum.
Collections: clothes; silver pieces; photographs; c.1800 fire cart; period furnishings; paintings; native plants.
Research Fields: local history.
Facilities: 100-vol. library.
Activities: guided tours.
Publications: brochure.
Hours & Admission Prices: Memorial Day to Labor Day. Sat.-Sun. 1-5; other times by appointment. No charge; donations accepted.
Attendance: 310 (accurate)

Silver Plume

GEORGE ROWE MUSEUM, 315 Main St., Silver Plume, CO 80476. Mailing Address: P.O. Box 935, Silver Plume, CO 80476-0935. Tel.: 303-569-2562.
Founded: 1960.
Key Personnel: Chm. (V), Judith Caldwell.
Personnel Profile: Part-Time Paid 3.
Governing Authority: Parent Institution: People for Silver Plume, Inc. Tax-exempt.
Institution Type/Description: History Museum.
Collections: area history; 19th century school room; period furnishings; mining; maps & educational materials.
Publications: Recipes-Remedies from the Kitchens of Silver Plume.
Hours & Admission Prices: Memorial Day to Labor Day daily 10-4; Sept. Sat.-Sun. 10-4. Adults $2.50; members & Silver Plume residents no charge.
Attendance: 1,210 (accurate)
Membership: Senior $10; Individual & Senior Family $15; Family $25; Sponsor $25; Sponsor $50; Sustaining $100; Life $500.

Silverton

SAN JUAN COUNTY HISTORICAL SOCIETY MUSEUM, 1557 Greene, Silverton, CO 81433. Mailing Address: P.O. Box 154, Silverton, CO 81433-0154. Tel.: 970-387-5838 & 5609. Fax: 970-387-5144.
E-mail: silvertonarchive@aol.com
Web Site: silvertonhistoricalsociety.org
Founded: 1964.
Congressional District: 3
Key Personnel: Chm. (V), Beverly Rich; Treas. & Sec., Scott Fetchenhier; Museum Shop Mgr., Duane Murphy.
Personnel Profile: Part-Time Paid 2; Part-Time Volunteers 7; Interns 1.
Governing Authority: society. Tax-exempt: 170(b)(1)(A).
Institution Type/Description: Local History Museum.
Collections: mining equipment; office & home furniture; china, glass & paper work from businesses; railroad cars; hardware; archives of photographs, mine records & oral history; records; home furnishings; clothing. Historic Buildings: 1904 San Juan County Jail; 1882 Denver & Rio Grande Western Railroad Depot; 1900 Silverton Northern Railroad Depot; 1912 Silverton Northern Railroad Engine House, Power House & Mule barn; 1929 Mayflower Gold Processing Mill.
Research Fields: history of San Juan County.
Facilities: 200-vol. library of microfilm copies of Silverton newspapers, business records & bound books.
Activities: permanent exhibitions; programs of local, regional & oral history.

Publications: yearly, San Juan Courier.
Hours & Admission Prices: Memorial Day-Sept. daily 9-5; Oct. daily 10-3. Adults $3.50. Mayflower Gold Mill: Memorial Day-Sept. 30 daily 10-4:30. Tours on half hour: adults $8.50, children 5-12 $4.50; discounts to groups; members & children under 5 no charge. &
Attendance: 14,000 (estimated)
Membership: Member $15; Family $30; Supporting $60; Lifetime $500.

Snowmass Village

ANDERSON RANCH ARTS CENTER & MUSEUM, 5263 Owl Creek Rd., Snowmass Village, CO 81615. Mailing Address: P.O. Box 5598, Snowmass Village, CO 81615-5598. Tel.: 970-923-3181. Fax: 970-923-3871.
E-mail: info@andersonranch.org
Web Site: www.andersonranch.org
Founded: 1966.
Key Personnel: Exec. Dir., Nancy Wilhelms; Chm. (V), Ann Korologos; Museum Shop Mgr., Jessica Cerise.
Personnel Profile: Full-Time Paid 25; Part-Time Paid 2; Part-Time Volunteers 20; Interns 17.
Institution Type/Description: Art School & Gallery.
Collections: works by Ranch artists.
Facilities: cafe; store featuring art work, jewelry; art supplies & unique gifts.
Activities: workshop program; residency program.
Publications: Anderson Ranch Art Center Summer Workshop Catalog.
Hours & Admission Prices: Mon.-Fri. 9-5. No charge. &
Attendance: 5,000

Steamboat Springs

K. SAARI GALLERY, 700 Yampa St., (on 7th St.), Steamboat Springs, CO 80487. Mailing Address: P.O. Box 770536, Steamboat Springs, CO 80477. Tel.: 970-819-7134. Facebook: K. Saari Gallery.
E-mail: kimberly@ksaari.com
Web Site: www.ksaari.com
Founded: 2007.
Key Personnel: Owner, Kimberly Saari
Institution Type/Description: Art Gallery.
Collections: works by contemporary artists.
Hours & Admission Prices: Call for hours.

STEAMBOAT ART MUSEUM, 801 Lincoln Ave., Steamboat Springs, CO 80488. Mailing Address: P.O. Box 883434, Steamboat Springs, CO 80488-3434. Tel.: 970-870-1755.
E-mail: sam@steamboatartmuseum.org
Web Site: www.steamboatartmuseum.org
Founded: 2005.
Key Personnel: Pres. (V), Shirley Stocks; Museum Shop Mgr., Sheila Weekly
Institution Type/Description: Art Museum: housed in an historic bank building.
Collections: western art; paintings; sculpture.
Activities: workshops.
Publications: exhibit catalogs.
Hours & Admission Prices: Tues.-Sat. 11-7. No charge; donations accepted. &
Membership: $25-$25,000.

TREAD OF PIONEERS MUSEUM, (M), 800 Oak St., Steamboat Springs, CO 80477. Mailing Address: P.O. Box 772372, Steamboat Springs, CO 80477-2372. Tel.: 970-879-2214. Fax: 970-879-6109.
E-mail: topmuseum@springips.com
Web Site: www.treadofpioneers.org
Founded: 1959.
Key Personnel: Dir., Candice Bannister; Cur., Katie Adams.
Personnel Profile: Full-Time Paid 2; Part-Time Paid 1; Part-Time Volunteers 56.
Governing Authority: bd. of directors.
Institution Type/Description: Local History Museum.
Collections: history of skiing; ranching; Routt County historical documents; southwestern Native American artifacts; pioneer & turn-of-the-century home furnishings; mining & agricultural history.
Activities: guided tours; temporary & permanent exhibitions & programs; kids scavenger hunt; computer interactive exhibit; educational programs & events.
Publications: biannual newsletter; annual report; walking tour brochure.
Hours & Admission Prices: Tues.-Sat. 11-5; call for seasonal extended hours. Adults $5, senior citizens $4, children 6-12 $1; discounts to AAA members; members & Routt County residents no charge. &
Attendance: 10,000 (estimated)

Membership: Chairlift $25; Chuck Wagon $50; Stagecoach $100; Haystack $250; Homestead $500; Pioneer Circle $1000.

Sterling

OVERLAND TRAIL MUSEUM, (M), 110 Overland Trail, Sterling, CO 80751. Mailing Address: Box 4000, Sterling, CO 80751-0400. Tel.: 970-522-3895.
E-mail: krich@sterlingcolo.com
Web Site: www.sterlingcolo.com
Founded: 1936.
Congressional District: 4
Key Personnel: Cur., Kay L. Brigham Rich; Asst., Janet Bigler; Asst., Snow Staples; Asst., Perry Johnson.
Personnel Profile: Full-Time Paid 2; Part-Time Paid 3; Part-Time Volunteers 8.
Governing Authority: municipal. Parent Institution: City of Sterling. Subsidiary Institution: Logan County Historical Society. Tax-exempt.
Institution Type/Description: General Museum: located on the site of the Overland Trail, near the Valley Station of Ben Holladay Stage line.
Collections: Indian artifacts; rock & lapidary; local natural history; mammal & marine fossils; frontier rifles; shot guns; relics of Civil, Spanish-American & later wars; cattle branding irons; frontier saddles; ranch & farm equipment; hand tools; hand-weaving, spinning wheel & loom; furniture; household items; one room schoolhouse; county church; blacksmith shop; 19th-century baby clothes; men & women's garments & accessories; musical instruments; dolls; country store; barn.
Research Fields: local history; Western history.
Facilities: education center. Postcards & books for sale.
Activities: guided tours; lectures; permanent exhibits.
Hours & Admission Prices: April-Oct. Mon.-Sat. 9-5, Sun. 1-5, Holidays 10-5; Nov.-March Tues.-Sat. 10-4. Adults $1.50-$3. &
Attendance: 14,000 (estimated)

Strasburg

COMANCHE CROSSING HISTORICAL SOCIETY & MUSEUM, 56060 E. Colfax Ave., Strasburg, CO 80136. Mailing Address: P.O. Box 647, Strasburg, CO 80136-0647. Tel.: 303-622-4322.
E-mail: csmith80136@tds.net
Web Site: cchscolorado.com
Founded: 1969.
Congressional District: 6
Key Personnel: C.E.O., Vencil Welp; Treas., Kathleen Burnet; Cur., Clifford Smith; Museum Shop Mgr., Beth Smialek.
Personnel Profile: Part-Time Volunteers 40.
Governing Authority: nonprofit organization. Parent Institution: Comanche Crossing Historical Society. Tax-exempt: 501(c)(3).
Institution Type/Description: Historical Society Museum: Comanche Crossing is the location where the first continuous chain of rails was completed by the Kansas Pacific Railroad, Aug. 15, 1870.
Collections: reconstructed first bank, post office, drug store; Union Pacific caboose; printing press; Hickok paper ruler; linotype; tools; fossils; Indian artifacts; fencing tools; impact wrenches; cameras; watches; books; maps; quilts; minerals; polished rocks; buggies; clothing; antique cars; military memorabilia; dolls; Halladay's Mill windmill; early 20th century room areas. Historical Buildings: 1891 Living Springs schoolhouse; 1910 Leslie Dyer Homestead; 1904 frame schoolhouse; 1917 Railroad Depot.
Research Fields: stage stations; trails; family history.
Facilities: library of old books, encyclopedias & maps available for loan by signing a legal request & posting an amount of funds to cover their value; classrooms. Centennial celebration & other museum-related items for sale.
Activities: special tours; lectures; guided tours; wheat thrashing; horse pull; tractor pull. Museum Sponsors: Pioneer School in July.
Publications: Comanche Crossing Centennial; History of Depot & Caboose.
Hours & Admission Prices: June-Aug. daily 1-4; other times by appointment. No charge; donations accepted. &
Attendance: 800 (estimated)
Membership: Annual $10; Lifetime $100.

Telluride

TELLURIDE GALLERY OF FINE ART, 130 E. Colorado Ave., P.O. Box 1900, Telluride, CO 81435. Tel.: 970-728-3300. Fax: 775-320-3646.
E-mail: info@telluridegallery.com
Web Site: www.telluridegallery.com
Founded: 1985.
Institution Type/Description: Art Gallery.
Collections: works by contemporary artists; paintings; sculpture; photographs.

Hours & Admission Prices: Tues.-Sat. 10-7, Sun.-Mon. 12-6.

TELLURIDE HISTORICAL MUSEUM, 201 W. Gregory Ave., Telluride, CO 81435. Mailing Address: P.O. Box 1597, Telluride, CO 81435-1597. Tel.: 970-728-3344. Fax: 970-728-6757.
E-mail: museum@telluridecolorado.net
Web Site: www.telluridemuseum.org
Founded: 1964.
Congressional District: 3
Key Personnel: Exec. Dir., Lauren Bloemsma; Asst. Dir., Kathy Rohrer; Pres., Carol Kammer.
Personnel Profile: Full-Time Paid 2; Part-Time Paid 1; Part-Time Volunteers 40.
Governing Authority: bd. of directors. Tax-exempt.
Institution Type/Description: General Museum: housed in 1896 community hospital building.
Collections: anthropology; costumes; glass; history; Indian artifacts; mineralogy; mining; photographs of San Miguel County dating from 1875 to the present; early local fraternal lodge memorabilia; the Rio Grande Southern Railroad; manuscripts; toys; 1890s bedroom & kitchen; medical equipment; L.L. Nunn & Tesla alternating current artifacts.
Research Fields: genealogy; mining; photographs; AC current electricity; Native American-early Pueblo history.
Facilities: 1,300-vol. library of books on the San Miguel area, available for research. Books of local interest for sale.
Activities: lectures; films; permanent exhibitions; school programs.
Publications: brochures; newsletter; Telluride Tales of Two Early Pioneers; Telluride's Victorian Vernacular.
Hours & Admission Prices: Summer: Tues.-Sat. 11-5, Sun. 1-5; Winter: Tues.-Sat. 11-5. Adults $5, seniors & students $3; discounts to AAM members; children 5 & under & members no charge. Closed major holidays. &
Attendance: 6,000 (estimated)
Membership: Quartz $35; Galena $60; Nickel $100; Business $125; Zinc $150; Copper $300; Silver $500; Gold $1,000; Prospector $1,500; Historian $2,500.

Trinidad

ARTHUR ROY MITCHELL MEMORIAL MUSEUM, 150 E. Main St., Trinidad, CO 81082-2709. Mailing Address. P.O. Box 95, Trinidad, CO 81082-0095. Tel.: 719-846-4224. Fax: 719-846-2004.
E-mail: mitchellmuseum@qwestoffice.net
Web Site: armitchell.org
Founded: 1979.
Congressional District: 4
Key Personnel: Dir. & Museum Shop Mgr., Paula Little; Pres., Cosette Henritz; Treas., John N. Beardan.
Personnel Profile: Full-Time Paid 1; Part-Time Paid 2; Part-Time Volunteers 60.
Governing Authority: nonprofit organization. Parent Institution: A. R. Mitchell Memorial, Inc. Tax-exempt: 501(c)(3) & 170(b)(1)(A).
Institution Type/Description: Art Museum: housed in c.1906 building.
Collections: paintings by Arthur Roy Mitchell, Harvey Dunn, Harold von Schmidt, Nick Eggenhofer, Frank Street, Ned Jacob, Frank Hoffman, Grant Reynard, Paul Milosevich, Dave Powell & Otto Kuhler; drawings; sketches; Indian artifacts; Western memorabilia; Hispanic religious folk art; Arthur Roy Mitchell's personal papers; Aultman photography, representing 106 years of continuous photography work within a family-owned studio, begun by Oliver E. Aultman in 1889.
Facilities: Books, magazines, Indian jewelry & weavings for sale.
Activities: guided tours; docent program; loan & temporary exhibitions.
Hours & Admission Prices: May-Sept. Tues.-Sat. 10-5; other times by appointment. Adult $3; members & children under 12 no charge. &
Attendance: 3,500 (accurate)
Membership: Individual $30; Family $50; Sponsor $100; Sustaining $500; Benefactor $1,000.

LOUDEN-HENRITZE ARCHAEOLOGY MUSEUM, Trinidad State Junior College, 600 Prospect St., Trinidad, CO 81082 2356. Tel.: 719-846-5508. Fax: 719-846-5050.
E-mail: loretta.martin@trinidadstate.edu
Web Site: www.trinidadstate.edu/museum
Founded: 1955.
Congressional District: 4
Key Personnel: Dir., Loretta Martin.
Personnel Profile: Part-Time Paid 1; Part-Time Volunteers 2.
Governing Authority: Trinidad State Junior College. Tax-exempt.
Institution Type/Description: Natural History Museum.

Collections: archaeology; geology; paleontology; paleogeology; ethnobotany; history.
Activities: guided tours by appointment; formally organized education programs for undergraduate college students; permanent & temporary exhibitions.
Hours & Admission Prices: Jan.-Nov. Mon.-Thurs. 10-3; groups by appointment. No charge. Closed national & state holidays. &
Attendance: 2,322 (accurate)

USAFA

U.S. AIR FORCE ACADEMY VISITOR CENTER, 2346 Academy Dr., USAFA, CO 80840-9401. Tel.: 719-333-2025. Fax: 719-333-4402.
E-mail: pa.comrel@usafa.af.mil
Web Site: usafa.af.mil
Founded: 1955.
Congressional District: 5
Key Personnel: Chief Visitor Svcs. Branch, Melissa Porter; Bldg. Mgr., Larry Wells.
Governing Authority: Parent Institution: United States Air Force Academy. Tax-exempt: 170(b)(1)(A).
Institution Type/Description: Military Museum.
Collections: pictorial & audiovisual exhibits explaining the history & current programs of the U.S. Air Force Academy.
Research Fields: aviation & aerospace history.
Facilities: theater; snack bar. Air Force Academy-related items for sale.
Activities: guided & self-guided tours; film; permanent exhibits.
Publications: self-guided tour map.
Hours & Admission Prices: Daily 9-5. No charge. closed Thanksgiving, Christmas Eve & Day, New Year's Eve & Day. &
Attendance: 430,506 (accurate)

Vail

COLORADO SKI & SNOWBOARD MUSEUM AND HALL OF FAME, 231 S. Frontage Rd. E., Level 3-Vail Village Transportation Center, Vail, CO 81657. Tel.: 970-476-1876. Fax: 970-476-1879.
E-mail: info@skimuseum.net
Web Site: www.skimuseum.net
Founded: 1976.
Congressional District: 3
Key Personnel: Exec. Dir., Susie Tjossem; Chm., David Scott; Cur., Erica Garcia Lloyd.
Personnel Profile: Full-Time Paid 1; Part-Time Paid 3; Part-Time Volunteers 40.
Governing Authority: nonprofit. Tax-exempt: 501(c)(3).
Institution Type/Description: Ski and Snowboard History Museum.
Collections: history of skiing & snowboarding in Colorado; artifacts; photographs; books; ski, boot & pole exhibit; ski clothing; cross-country skiing; military skiing & safety equipment; military skiing, 10th mountain division; Hall of Fame; chronology of skiing exhibit; World Ski Championship exhibit; snowboarding; pamphlets; videos; films; magazines.
Research Fields: Colorado history.
Facilities: 100-vol. library of books on the history & techniques of skiing; theater; banquet facilities. Museum-related items for sale.
Activities: guided tours; lectures; films; docent program; permanent, temporary & loan exhibitions; videos. Museum Sponsors: 10th Mountain Veteran Fireside Chat.
Publications: quarterly newsletter.
Hours & Admission Prices: June-Sept. & Nov.-April Tues.-Sun. 10-5; guided tours by appointment. Guided Tours: Museum: Tues. 1pm. Historic Rail Village: Tues. 2pm. No charge; donations accepted. Closed Thanksgiving; Christmas. &
Attendance: 60,000 (estimated)
Membership: Friend $50; Bronze Medal $100; Silver Medal $250; Gold Medal $500; Olympic $1,000; Patron $2,500; Pioneer $5,000; Legend $10,000.

Victor

VICTOR LOWELL THOMAS MUSEUM, 3rd St. & Victor Ave., Victor, CO 80860. Mailing Address: P.O. Box 238, Victor, CO 80860-0238. Tel.: 719-689-5509.
E-mail: museum@victorcolorado.com
Web Site: victorcolorado.com
Founded: 1954.
Personnel Profile: Part-Time Paid 3; Part-Time Volunteers 6.
Governing Authority: nonprofit organization. Parent Institution: Victor Improvement Association. Tax-exempt: 501(c)(3).

Institution Type/Description: History Museum: built in 1899.
Collections: local history & culture; photographs; personal artifacts.
Facilities: Museum-related items for sale.
Activities: gold panning; special events; mine tours.
Hours & Admission Prices: Memorial Day to Labor Day daily 9:30-5:30; Sept. to Columbus Day Sat.-Sun. 9:30-5:30. Adults $6, seniors 60 & over $5, children 12 & under $4; discounts to groups of 20 or over.
Attendance: 9,000 (accurate)
Membership: Individual $40; Senior Couple $45; Family $50; Turquoise Patron $100 & up; Silver Patron $250 & up; Sylvanite Patron $500 & up; Gold Patron $1,000 & up; Cresson Gold Patron $5,000 & up.

Walsenburg

HUERFANO COUNTY HISTORICAL SOCIETY, INCLUDING THE WALSENBURG MINING MUSEUM & FT. FRANCISCO MUSEUM OF LA VETA, 112 W. 5th St., Walsenburg, CO 81089-1941. Mailing Address: P.O. Box 134, Walsenburg, CO 81089-0134. Tel.: 719-738-1992. Fax: 719-738-6218.
Founded: 1987.
Congressional District: 3
Key Personnel: Dir. Walsenberg, Margaret Gleisberg; Museum Shop Mgr., Marge Figal.
Personnel Profile: Part-Time Volunteers 10.
Governing Authority: society; nonprofit. Parent Institution: Huerfano County Historical Society. Subsidiary Institutions: Francisco Ft. Museum of La Veta, Francisco Plaza, La Veta, CO. Tax-exempt: 501(c)(3).
Institution Type/Description: Mining Museum: housed in c.1896 jail building.
Collections: Southern Colorado coal mining industry artifacts; mining equipment; miner's personal equipment; mine records & maps; photographs; letters; UMWA, IWW items; coalfield wars & strikes items.
Research Fields: mining industry.
Facilities: 56-vol. library of mining material available for viewing on premises. Museum-related books & other gift items for sale.
Publications: books, Hiram Vasquez, All Our Yesterdays, Romance of the Spanish Peaks, In the Shadow of the Peaks.
Hours & Admission Prices: Walsenburg Mining Museum: May-Sept. Mon.-Fri. 10-4, Sat. 10-1. Adults $2, teens $1; children under 12 no charge. Ft. Francisco Museum: Mon.-Sat. 10-4, Sun. 1-4. Adults $6, children & seniors $3.
Attendance: 1,500 (estimated)
Membership: Huerfano County Historical Society: Senior & Student $10; Sustaining Individual $20; Sustaining Family $30; Sponsor Patron $50; Associate Patron $100; Corporate $500; Executive Patron $1,000.

Westminster

BOWLES HOUSE MUSEUM - WESTMINSTER HISTORICAL SOCIETY, 3924 W. 72nd Ave., Westminster, CO 80031. Tel.: 303-430-7929.
Institution Type/Description: Historical Society Museum: housed in the former home of Edward Bruce Bowles; built in 1871. Listed on the National Register of Historic Places.
Collections: local history & culture; period furnishings; personal artifacts; photographs.
Hours & Admission Prices: Tours: April 21-Dec. 30 Sat. 10-3; other times by appointment.

BUTTERFLY PAVILION, 6252 W. 104th Ave., Westminster, CO 80020-4107. Tel.: 303-469-5441. Fax: 303-657-5944. Facebook: Butterfly Pavilion.
E-mail: marketing@butterflies.org
Web Site: www.butterflies.org
Founded: 1995.
Key Personnel: Dir., Robert Bonacci; Museum Shop Mgr., Cynthia Killingbeck.
Personnel Profile: Full-Time Paid 25; Part-Time Paid 25; Part-Time Volunteers 125; Interns 4.
Governing Authority: Tax-exempt.
Institution Type/Description: Tropical Conservatory.
Collections: over 1,200 butterflies; invertebrate; aquatic animals.
Facilities: nature trails; conservatory. Museum-related items for sale.
Activities: special events; school groups; workshops; scholarships; classes; camps; story time.
Publications: newsletter, Flutterings.
Hours & Admission Prices: Daily 9-5. Adults $9.50, seniors 65 & over $7.50, children 2-12 $6.50; children under 2 no charge. Closed Thanksgiving; Christmas. &
Attendance: 250,000 (accurate)

Membership: Individual $50; Just for Two $60; Family $75; Dual Membership with Children's Museum of Denver $142; Family & Friends $150; Science & Research $200.

Wheat Ridge

WHEAT RIDGE HISTORIC PARK, 4610 Robb St., Wheat Ridge, CO 80033-2537. Mailing Address: P.O. Box 1833, Wheat Ridge, CO 80034-1833. Tel.: 303-421-9111. Fax: 303-467-0023.
Formerly: Wheat Ridge Soddy
Founded: 1970.
Congressional District: 6
Key Personnel: Pres. (V), Charlotte Whetsel; Vice Pres. & Museum Shop Mgr., Claudia Worth.
Personnel Profile: Part-Time Volunteers 3.
Governing Authority: municipal; nonprofit organization. Parks Dept., City of Wheat Ridge; Wheat Ridge Historical Society. Tax-exempt.
Institution Type/Description: History Museum.
Collections: local history & culture; period furnishings; local industry. Historic Buildings: 1859, homestead cabin; 1913 post office; 1910 farmhouse; c.1860s sod house.
Research Fields: local history; homestead families & identification of original homesteads; schools.
Facilities: library of local history available for research by request.
Activities: guided tours; lectures; craft workshops; permanent & temporary exhibitions; special events. Museum Sponsors: May Festival-Antique Road Show; Cider Day in October.
Publications: brochure; books, History of Pioneer Wheat Ridge; Activities 1976-Centennial-Bicentennial Year; Biographical Sketches-Early Settlers of Wheat Ridge; Guide to the Collection of the Wheat Ridge Historical Society Museum & Library.
Hours & Admission Prices: Fri. 10-3; other times by appointment. Adults $2. Closed national holidays. &
Attendance: 3,100 (estimated)
Membership: Individual $10; Family $15; Sustaining $25; Individual Life $50. Family Life $100.

Windsor

WINDSOR MUSEUM, (M), 6th St. & Ash St., Windsor, CO 80550. Mailing Address: 115 5th St., Windsor, CO 80550. Tel.: 970-674-2443. Fax: 970-674-2456.
E-mail: cknight@windsorgov.com
Web Site: www.windsorgov.com
Founded: 2003.
Congressional District: 4
Key Personnel: Mgr. Arts & Heritage, Carrie Knight; Cur., Elizabeth Handwerk Kurt.
Personnel Profile: Full-Time Paid 3.
Governing Authority: municipal.
Institution Type/Description: History Museum.
Collections: Windsor's residents, cultural history, businesses, & events. Historic Buildings: railroad depot; 1890s rural schoolhouse; pioneer church.
Research Fields: Western rural schoolhouses & education; immigration of Germans to Russia.
Activities: formal education programs; guided tours; temporary exhibitions. Annual Events: Octoberfest; Creepy Crawl.
Hours & Admission Prices: Memorial Day to Labor Day Tues.-Sat. 1-4. No charge; donations accepted.
Attendance: 5,000 (estimated)

Woodland Park

ROCKY MOUNTAIN DINOSAUR RESOURCE CENTER, 201 S. Fairview St., Woodland Park, CO 80863-1154. Tel.: 719-686-1820. Fax: 719-686-1399. Facebook: Rocky Mountain Dinosaur Resource Center.
E-mail: info@rmdrc.com
Web Site: www.rmdrc.com
Founded: 2004.
Key Personnel: Dir. & Pres., J.J. Triebold.
Personnel Profile: Full-Time Paid 4; Part-Time Paid 9.
Institution Type/Description: Paleontology Museum.
Collections: prehistoric life including dinosaurs; marine reptiles; flying reptiles; fish.
Facilities: Museum-related items for sale.
Activities: educational workshops.
Hours & Admission Prices: Mon.-Sat. 9-6, Sun. 10-5. Adults $11.50; members no charge. Closed New Year's Day; Easter; Thanksgiving; Christmas. &
Membership: Call for pricing.

Wray

WRAY MUSEUM, 205 E. Third St., Wray, CO 80758-1106. Mailing Address: P.O. Box 161, Wray, CO 80758-0161. Tel.: 970-332-5063.
Web Site: www.wrayco.net
Founded: 1969.
Congressional District: 4
Key Personnel: Dir., Ardith Hendrix; EYCHS Pres. (V), Lou Ann Deterding.
Personnel Profile: Full-Time Paid 1; Part-Time Paid 2.
Governing Authority: society. Parent Institution: City of Wray and the East Yuma County Historical Society. Tax-exempt.
Institution Type/Description: History Museum.
Collections: permanent Smithsonian archaeological exhibit on Paleo Indians; bison; arrowheads; diorama of Beecher Island Battle of 1868; paintings; textiles; photographs; big game animal trophies; general store display.
Research Fields: archaeology; local family history; history of trails; post offices; towns; old ranches.
Activities: tours; educational programs; permanent, traveling & temporary exhibits. Museum Sponsors: The Greater Prairie Chickens - Sandhills dancers in March & April.
Hours & Admission Prices: Tues.-Sat. 12-4. Adults $1, children $.50; members no charge. &
Attendance: 2,000 (accurate)
Membership: Individual $10; Civic Clubs & Organization $15; Commercial Business $20; Family $35; Patron & Business $50; Life Individual $100; Lifetime $125; Corporate $1,000.

CONNECTICUT

(309 listings)

Ansonia

ANSONIA NATURE & RECREATION CENTER, 10 Deerfield Ln., Ansonia, CT 06401. Tel.: 203-736-1053. Fax: 203-734-1672.
E-mail: ansnaturectr@ansonicct.org
Founded: 1977.
Congressional District: 5
Key Personnel: Dir., Donna Lindgren; Pres. (V), FANCI Rich Wade.
Personnel Profile: Full-Time Paid 2; Part-Time Paid 4; Part-Time Volunteers 30.
Governing Authority: Parent Institution: City of Ansonia. Tax-exempt.
Institution Type/Description: Nature Center.
Collections: natural history; plants; trees; butterfly & hummingbird garden; wildflowers.
Facilities: picnic pavilions; visitor center.
Activities: educational programs; special events; classes.
Publications: seasonal calendar of events.
Hours & Admission Prices: Park: sunup to sundown. Center: daily 9-5. No charge. Closed New Year's Day; Thanksgiving; Christmas.
Attendance: 10,000 (estimated)
Membership: Friends: Individual $15; Family $25; Sponsor $35; Class & Troop $100; Business & Organization $100; Individual Life $150; Family Life $250.

DERBY HISTORICAL SOCIETY & GENERAL DAVID HUMPHREYS' HOUSE, 37 Elm St., Ansonia, CT 06401-3312. Mailing Address: P.O. Box 331, Derby, CT 06418-0331. Tel.: 203-735-1908.
E-mail: info@derbyhistorical.org
Web Site: www.derbyhistorical.org/humphrey.htm
Founded: 1946.
Congressional District: 3
Key Personnel: Dir., Julia Baldini.
Personnel Profile: Part-Time Paid 9; Part-Time Volunteers 60; Interns 1.
Governing Authority: Parent Institution: Derby Historical Society. Tax-exempt.
Institution Type/Description: Historic House: built in 1698.
Collections: family history; period furnishings. Historic Houses: David Humphreys' house; Dr. John I. Howe house.
Activities: school group programs; special events.
Publications: monthly e-news.
Hours & Admission Prices: Mon.-Fri. 1-4. Adults $5.
Attendance: 2,000 (estimated)
Membership: Individual $30; Life $150.

Avon

THE AVON HISTORICAL SOCIETY, INC., 8 E. Main St., Avon, CT 06001. Mailing Address: P.O. Box 448, Avon, CT 06001-0448. Tel.: 860-678-7621.
E-mail: ahs.mail.1830@sbcglobal.net
Web Site: www.avonhistoricalsociety.org
Founded: 1974.
Congressional District: 5
Key Personnel: Pres., Terri Wilson; Vice Pres., Pam Fahey.
Personnel Profile: Part-Time Volunteers 12.
Governing Authority: nonprofit organization. Branch Museums: Pine Grove School House & The Derrin House, W. Avon Rd., Avon, CT. Tax-exempt: 501(c)(3).
Institution Type/Description: Historical Society Museum; housed in two 19th century schoolhouses & farmhouse.
Collections: concentration on local Avon history; archives.
Publications: Avon, Connecticut...An Historical Story; Avon, Connecticut - Photographs 1880-1940.
Hours & Admission Prices: Living Museum & Pine Grove Schoolhouse: June-Oct. Sun. 2-4. Derrin House: June-Oct. Sun. 2-4. Library: Tues. & Thurs. 2-4; other times by appointment. No charge; donations accepted.
Attendance: 600 (estimated)
Membership: Senior Citizen & Student $20; Individual $25; Family $35; Business $100; Life $500.

FARMINGTON VALLEY ARTS CENTER (FVAC), 25 Arts Center Lane, Avon, CT 06001-3746. Tel.: 860-678-1867. Fax: 860-674-1877. Facebook: FV Arts Center.
E-mail: fvartscenter@gmail.com
Web Site: www.artsfvac.org
Founded: 1974.
Personnel Profile: Full-Time Paid 2; Part-Time Paid 3; Part-Time Volunteers 60.
Governing Authority: Tax-exempt.
Institution Type/Description: Arts Center.
Collections: paintings; sculpture.
Facilities: Museum-related items for sale.
Activities: educational programs; classes; workshops; summer camp; birthday parties.
Publications: catalogue 3 times per year, Connections.
Hours & Admission Prices: Call for hours. No charge.
Attendance: 6,000 (estimated)
Membership: Individual $30; Household $60.

THE LIVING MUSEUM OF AVON, 8 E. Main St., Avon, CT 06001. Mailing Address: P.O. Box 448, Avon, CT 06001. Tel.: 860-678-7621.
Institution Type/Description: History Museum: housed in a 19th-century schoolhouse.
Collections: photographs; farm implements; household artifacts; period clothing.
Hours & Admission Prices: June-Sept. by appointment. No charge; donations accepted.

Bethel

BETHEL HISTORICAL FIRE FIGHTING MUSEUM, 36 South St., Bethel, CT 06801. Tel.: 203-794-8521.
Web Site: www.bvfdinc.com/bvfd/gallery
Institution Type/Description: Fire Fighting Museum.
Collections: fire fighting history & equipment; photographs; uniforms; helmets; badges; hand-drawn fire equipment.
Hours & Admission Prices: By appointment.

Bethlehem

BELLAMY-FERRIDAY HOUSE & GARDEN, 9 Main St. N., Bethlehem, CT 06751. Mailing Address: P.O. Box 181, Bethlehem, CT 06751. Tel.: 203-266-7596.
Web Site: www.ctlandmarks.org
Personnel Profile: Part-Time Paid 8.
Governing Authority: Parent Institution: Connecticut Landmarks.
Institution Type/Description: Historic House: built in 1754.
Collections: family history & culture; personal artifacts; period furnishings; photographs; botanical; archival.
Facilities: historic garden.
Hours & Admission Prices: May-Aug. Wed. & Fri.-Sun. 11-4; Sept. to

Columbus Day Sat.-Sun. 11-4; groups by appointment. Adults $7, students and seniors 65 & over $6, children 6-18 $4; members & children under 6 no charge.
Attendance: 1,500 (accurate)

OLD BETHLEHEM HISTORICAL SOCIETY, 4 Main St. N., Bethlehem, CT 06751. Mailing Address: P.O. Box 132, Bethlehem, CT 06751-0132. Tel.: 203-266-5188.
Founded: 1968.
Key Personnel: Pres. (V), Carol Ann Brown
Institution Type/Description: Historical Society Museum.
Collections: Bethlehem history & culture; photographs; personal artifacts.
Hours & Admission Prices: June-Aug. Sun. 1-4; other times by appointment. No charge; donations accepted. Closed holidays. &

Bloomfield

WINTONBURY HISTORICAL SOCIETY, 151-153 School St., Bloomfield, CT 06002-2718. Mailing Address: 151-153 School St., P.O. Box 7454, Bloomfield, CT 06002. Tel.: 860-243-1531.
E-mail: wintonbursociety@att.net
Web Site: www.bloomfieldcthistory.org
Founded: 1949.
Congressional District: 1
Key Personnel: Pres., Richard N. Pierce
Governing Authority: Parent Institution: Wintonbury Historical Society, Inc. Subsidiary Institutions: Old Farm School; Southwest District School, Simsbury Rd., Bloomfield, CT. Tax-exempt.
Institution Type/Description: Historic Building: built c.1796. Listed on the National Register of Historic Places.
Collections: local history; period furnishings; paintings; genealogy.
Facilities: library.
Publications: newsletter 5 times a year, The Wintonbury Drummer; books, From Wintonbury to Bloomfield; Over Tunxis Trails; Images of America: Bloomfield; Bloomfield and the Civil War; History of the Bloomfield Catholic Church.
Hours & Admission Prices: May 15-Oct. 15 Sun. 1-4. No charge; donations accepted. &
Attendance: 300 (estimated)
Membership: Individual $15; Family $25; Supporting $40.

Branford

BRANFORD HISTORICAL SOCIETY - HARRISON HOUSE, 124 Main St., Branford, CT 06405-3523. Mailing Address: P.O. Box 504, Branford, CT 06405-0504. Tel.: 203-488-4828.
E-mail: branfordhistory@gmail.com
Web Site: branfordhistory.org
Founded: 1960.
Congressional District: 3
Key Personnel: Pres. (V), Peter Black.
Personnel Profile: Part-Time Volunteers 40.
Governing Authority: society; nonprofit organization. Parent Institution: New England Antiquities, 141 Cambridge St. Boston, MA 02114. Tel. 617-227-3956. Operated by the Branford Historical Society. Tax-exempt: 501(c)(3).
Institution Type/Description: Historic House: c.1724 restored by architect J. Frederick Kelly.
Collections: furnishings; town memorabilia; farm tools.
Activities: guided tours; programs.
Publications: newsletter; local histories.
Hours & Admission Prices: June-Sept. Sat. 2-5; other times by appointment. No charge; donations accepted.
Attendance: 125 (estimated)
Membership: Student $5; Individual $15; Family $20; Sustaining $25; Contributing $50.

Bridgeport

THE BARNUM MUSEUM, 820 Main St., Bridgeport, CT 06604-4912. Tel.: 203-331-1104. Fax: 203-331-0079.
Web Site: www.barnum-museum.org
Founded: 1893.
Congressional District: 4
Key Personnel: Exec. Dir. & Cur., Kathleen Maher; Dir. Education, Jaime Knoedler; Administrative Asst., Debbie Saviello.
Personnel Profile: Full-Time Paid 3; Part-Time Paid 3; Part-Time Volunteers 29; Interns 3.
Governing Authority: Parent Institution: The Barnum Museum Foundation, Inc. Subsidiary Institution: City of Bridgeport, CT. Tax-exempt: 501(c)(3).

Institution Type/Description: History Museum: housed in original 1893 structure.
Collections: 19th-century historical & scientific society; biological specimens; period relics; colonial artifacts; 19th century industrial & decorative arts including paintings, furniture, and costumes; P.T. Barnum documents & personal items; 1920 miniature circus model.
Research Fields: P.T. Barnum & his genre in 19th century American history; social & material culture; American decorative arts; industrial & entertainment histories.
Facilities: archive library available by appointment; 15,000 sq. ft. exhibit space. Museum-related items for sale.
Activities: school & family programs; lectures; films; gallery talks; TV & radio programs; weekend guided tours by reservation.
Publications: newsletter, The Barnum Herald; calendar of events.
Hours & Admission Prices: Tues.-Sat. 10-4:30, Sun. 12-4:30. Adults $7, students & senior citizens $5, children $4; discounts to AAA & AAM members; children under 4 & members no charge. Closed most national holidays. &
Attendance: 25,000 (estimated)
Membership: Single $25; Family $45; Family Plus $60; Sustaining $150; Patron Circle $500; Director's Circle $1,000.

CITY LIGHTS GALLERY, 37 Markle Ct., Bridgeport, CT 06604-4816. Tel.: 203-334-7748.
Institution Type/Description: Art Gallery.
Collections: paintings; photographs; sculpture.
Activities: art programs; receptions; workshops; demonstrations; rental facilities; classes.
Hours & Admission Prices: Mon.-Fri. 9:30-5:30, Sat. 11-4. No charge.

CONNECTICUT'S BEARDSLEY ZOO, 1875 Noble Ave., Bridgeport, CT 06610-1646. Tel.: 203-394-6569. Fax: 203-394-6566.
E-mail: gdancho@beardsleyzoo.org
Web Site: www.beardsleyzoo.org
Founded: 1922.
Congressional District: 7
Key Personnel: Chm., Richard Perusi; Zoo Dir., Gregg Dancho; Cur. Education, Tedor Whitman; Dir. Animal Care & Operations, Don Goff; Chm. (V), Tracy Benham; Museum Shop Mgr., Rose Ryan.
Personnel Profile: Full-Time Paid 42; Part-Time Paid 25; Part-Time Volunteers 125; Interns 30.
Governing Authority: state; nonprofit organization. Parent Institution: Connecticut Zoological Society. Tax-exempt: 501(c)(3).
Institution Type/Description: Zoo.
Collections: endangered animals of North & South America; New England farmyard contains minor breeds of domestic animals.
Facilities: 300-vol. library of zoology references; classrooms; botanical garden; 400-seat Gazebo; zoological park. Gifts, jewelry & other items for sale.
Activities: guided tours; lectures; school loan service; mobile vans; formally organized education programs; docent program.
Publications: quarterly newsletter, The ZooTimes.
Hours & Admission Prices: Daily 9-4. Adults $11, senior citizens & children $9; discount to groups; AZA members & members no charge. Closed New Year's Day; Thanksgiving; Christmas. &
Attendance: 262,000 (estimated)
Membership: Individual $50; Family $80; Director's Circle $150; President's Circle $250; Benefactor $500.

THE DISCOVERY MUSEUM, INC., 4450 Park Ave., Bridgeport, CT 06604-1098. Tel.: 203-372-3521. Fax: 203-374-1929.
E-mail: info@discoverymuseum.org
Web Site: www.discoverymuseum.org
Formerly: Museum of Art, Science & Industry
Founded: 1958.
Congressional District: 4
Key Personnel: Chm. (V) & Exec. Dir., Joseph D'Avanzo; Dir. Corp. Rels, Dimitris Raptopoulos; Dir. Finance, Laurel Anderson; Dir. Exhibits & Public Programming, Adam Zuckerman; Dir. Facilities, Jerry McKoy; Dir. Education, Alan Winick.
Personnel Profile: Full-Time Paid 11; Part-Time Paid 12; Part-Time Volunteers 35; Interns 3.
Governing Authority: nonprofit corporation. Tax-exempt: 501(c)(3).
Institution Type/Description: Science Museum & Planetarium.
Collections: over 100 interactive physical science exhibits; planetarium.
Research Fields: scientific exploration of earth & space; sustainable energy technology; public understanding of science.
Facilities: planetarium with digital & optical projection; auditorium; toddler area; learning center; classrooms.

Activities: lectures; formally organized education programs for children; permanent & temporary exhibitions; special programs for the handicapped; volunteer training programs; summer programs for children; planetarium shows; science shows; educational programs; learning center; live science demonstrations; birthday parties; team building programs; simulated space missions; corporate rentals.

Publications: quarterly newsletter.

Hours & Admission Prices: July-Aug. daily 10-5; Sept.-June Tues.-Sun. 10-5. Museum & Planetarium: adults $9.50, children 3-17, senior citizens & students with ID $8; discounts to ASTC members; members & children under 2 no charge. Closed New Year's Day; Easter; Memorial Day; Independence Day; Labor Day; Thanksgiving; Christmas. &

Attendance: 59,344 (accurate)

Membership: Individual $35; Couple $50; Family $65; Family Plus $95; Friend $125. (Add-A-Guest $15).

HOUSATONIC MUSEUM OF ART, 900 Lafayette Blvd., Bridge-port, CT 06604-4704. Tel.: 203-332-5052. Fax: 203-332-5123.

E-mail: rzella@hcc.commnet.edu

Web Site: www.hctc.commnet.edu/artmuseum/index.html

Founded: 1967.

Congressional District: 4

Key Personnel: Dir. & Cur., Robbin Zella.

Personnel Profile: Full-Time Paid 1; Part-Time Paid 1; Part-Time Volunteers 3; Interns 4.

Governing Authority: state. Parent Institution: Housatonic Community College. Tax-exempt.

Institution Type/Description: College Museum.

Collections: 19th- & 20th-century European & American art; contemporary Latin American & Connecticut artists; ethnographic collections in African, Asian & South Seas artifacts.

Facilities: 2,000-vol. art library; classrooms; slides; cafeteria.

Activities: guided tours; lectures, films; gallery talks; inter-museum loan, permanent & temporary exhibitions.

Publications: exhibition catalogs, Housatonic Museum of Art.

Hours & Admission Prices: June-Aug. Mon.-Wed. & Fri. 8:30-5:30, Thurs. 8:30-7; Sept.-May Mon.-Wed. & Fri. 8:30-5:30, Thurs. 8:30-7, Sat. 9-3, Sun. 12-4. No charge; donations accepted. &

Attendance: 10,000 (estimated)

Bristol

AMERICAN CLOCK AND WATCH MUSEUM, INC., 100 Maple St., Bristol, CT 06010-5092. Tel.: 860-583-6070. Fax: 860-583-1862.

E-mail: info@clockmuseum.org

Web Site: www.clockandwatchmuseum.org

Founded: 1952.

Congressional District: 1

Key Personnel: Exec. Dir., Donald Muller; Pres. (V), Gary Plonski; Office Coord., Jill Godbout; Dir. Interpretation, Colleen Nicastro; Museum Shop Mgr., Jean Haines.

Personnel Profile: Full-Time Paid 2; Part-Time Paid 7; Part-Time Volunteers 40.

Governing Authority: nonprofit organization. Tax-exempt: 501(c)(3).

Institution Type/Description: Horological Museum: housed in 1801 Miles Lewis House & 1956 Ebenezer Barnes Wing & 1987 Edward Ingraham Wing.

Collections: manuscripts; clocks, watches & other horological items.

Research Fields: horology; historical; technical.

Facilities: 2,000-vol. library of horological books & journals; local & state history books available by special appointment with curator; 1801 garden.

Activities: lectures; permanent exhibitions; video presentations; special exhibits.

Publications: quarterly newsletter; semi-annual journal, The Timepiece Journal; booklets, reproductions of scarce horological trade catalogs.

Hours & Admission Prices: April-Nov. daily 10-5; Dec.-March by appointment only. Families $9, adults $5, senior citizens $4, children 8-15 $2; discounts to AAM & AAA members; members no charge. Closed Thanksgiving. &

Attendance: 5,532 (accurate)

Membership: Annual $30; Family $45; Contributing $100; Life $1,500.

BRISTOL HISTORICAL SOCIETY, 98 Summer St., Bristol, CT 06010. Mailing Address: P.O. Box 1231, Bristol, CT 06010. Tel.: 860-583-6309.

Institution Type/Description: Historical Society Museum.

Collections: local history & culture; period furnishings; personal artifacts; photographs; military uniforms, equipment & artifacts; weapons; letters.

Facilities: Museum-related items for sale.

Hours & Admission Prices: Wed. 10-2, Sat. 12-4; other times by appointment.

IMAGINE NATION MUSEUM, One Pleasant St., Bristol, CT 06010-6254. Tel.: 860-314-1400. Fax: 860-584-3608. Facebook: Imagine Nation Museum.

E-mail: info@imaginenation.org

Web Site: www.imaginemuseum.org

Founded: 2004.

Congressional District: 1

Key Personnel: Dir., Doreen Stickney; C.E.O., Michael Suchopar; Dir. Education, Kymrie Zaslow.

Personnel Profile: Full-Time Paid 6; Part-Time Paid 13; Part-Time Volunteers 150; Interns 3.

Governing Authority: nonprofit organization. Parent Institution: The Boys & Girls Club and Family Center of Bristol. Tax-exempt.

Institution Type/Description: Children's Museum.

Collections: hands-on exhibits.

Activities: educational field trips; scout badge programs; state licensed camp programs; birthday parties; workshops; vacation day programs; outreach programs. Museum Sponsors: Family Festivals.

Publications: newsletter.

Hours & Admission Prices: Winter: Wed.-Fri. 9:30-5, 1st Fri. of month 9:30-8, Sat. 11-5, Sun. 12-5; Summer: call for extended hours. Open Tues. in July & Aug. Admission $7; members & children under one no charge. Special Events: admission prices vary. Closed New Year's Day; Thanksgiving; Christmas. &

Attendance: 50,000 (accurate)

Membership: Family $125.

THE NEW ENGLAND CAROUSEL MUSEUM, 95 Riverside Ave., Bristol, CT 06010-6390. Tel.: 860-585-5411. Fax: 860-314-0483.

E-mail: info@thecarouselmuseum.org

Web Site: thecarouselmuseum.org

Founded: 1989.

Congressional District: 6

Key Personnel: Exec. Dir., Louise L. DeMars; Pres. (V), Laureen Rubino; Education, Allison Feigen; Museum Mgr., Leanne Grelish; Head of Restoration, Judith Baker; Visitor Svcs., Elaine Lipton.

Personnel Profile: Full-Time Paid 4; Part-Time Paid 18; Part-Time Volunteers 71.

Governing Authority: private; nonprofit organization. Tax-exempt: 501(c)(3).

Institution Type/Description: Carousel Museum: housed in a restored hosiery factory.

Collections: turn-of-the-century wooden carousel pieces including horses, menagerie animals, chariots, rounding boards, shields & band organs; fine arts; firefighting history.

Research Fields: history, art & science of the carousel; history of Lake Compunce & its carousel.

Facilities: reference library available to public; art room; 7,000 sq. ft. exhibit space; 33,000 sq. ft. building. Museum-related items for sale.

Activities: docent program; formal education programs for children; guided tours; lectures; loan, participatory & temporary exhibitions. Museum Sponsors: art classes for children.

Publications: biannual newsletter, The Carousel, Flying Horses.

Hours & Admission Prices: Tues.-Sat. 10-5, Sun. 12-5. Adults $6, senior citizens $5.50, children 4-14 $2.50; discounts to groups, AAM, AAA, ICOM, Lets Go Arts & Chamber of Commerce members; members no charge. Closed Jan. & Feb., New Year's Day; Easter; Independence Day; Labor Day; Thanksgiving; Christmas. &

Attendance: 10,745 (accurate)

Membership: Individual $35; Family $50.

WITCH'S DUNGEON CLASSIC MOVIE MUSEUM, 90 Battle St., Bristol, CT 06010. Tel.: 860-583-8306.

Web Site: www.preservehollywood.org/DungeonWebNew

Founded: 1966.

Key Personnel: Owner, Cortlandt Hull.

Institution Type/Description: Movie Museum.

Collections: life-size replicas of horror movie characters; movie props.

Hours & Admission Prices: Call for hours.

Brookfield

BROOKFIELD CRAFT CENTER, 286 Whisconier Rd., Brook-field, CT 06804. Mailing Address: P.O. Box 122, Brookfield, CT 06804-0122. Tel.: 203-775-4526. Fax: 203-740-7815.

E-mail: director@brookfieldcraft.org

Web Site: www.brookfieldcraft.org

Founded: 1954.
Congressional District: 8
Key Personnel: Interim Exec. Dir., Barbara Prete; Chm. (V), William Markus; Dir. Mktg., Betsy Halliday; Museum Shop Mgr., Gillian Doherty.
Personnel Profile: Full-Time Paid 2; Part-Time Paid 4; Part-Time Volunteers 45.
Governing Authority: nonprofit. Tax-exempt: 501(c)(3).
Institution Type/Description: Craft Gallery: housed in 1780 old grist mill.
Collections: contemporary national profile artists.
Research Fields: fine craft.
Facilities: Museum-related items for sale.
Activities: classes; workshops; temporary exhibitions; lectures; seminars.
Publications: online catalog of classes & workshops.
Hours & Admission Prices: Tues.-Fri. 10-5, Sat. 11-5, Sun. 12-5; call to confirm. No charge; donations accepted. Closed major holidays.
Attendance: 5,000 (estimated)
Membership: Youth 16-21 and Senior 65 & over $35; Individual $50; Family $100; Friend $250; Supporter $500; Benefactor $1,000.

BROOKFIELD MUSEUM AND HISTORICAL SOCIETY, 165 Whisconier Rd., Brookfield, CT 06804. Mailing Address: P.O. Box 5231, Brookfield, CT 06804-5231. Tel.: 203-740-8140.
E-mail: brookfieldhistsoc@snet.net
Web Site: www.brookfieldcthistory.org
Founded: 1968.
Congressional District: 5
Key Personnel: Pres. (V), John D. Furlong; Vice Pres., Robert Brown; Museum Shop Mgr., Jan Furlong.
Personnel Profile: Part-Time Volunteers 15.
Governing Authority: Tax-exempt.
Institution Type/Description: Historical Society Museum.
Collections: Brookfield history & culture; photographs.
Publications: newsletter.
Hours & Admission Prices: Sat. 12-4; other times by appointment. No charge.
Attendance: 1,500 (estimated)
Membership: Family $20; Life $1,000.

MOTHER EARTH GALLERY & MINING COMPANY, 806 Federal Rd., Brookfield, CT 06804. Tel.: 203-775-6272. Fax: 203-775-5620.
Web Site: www.motherearthcrystals.com
Founded: 1987.
Personnel Profile: Full-Time Paid 1; Part-Time Paid 2.
Institution Type/Description: Mining Museum.
Collections: mining history; natural & cultural artifacts; geology.
Facilities: Museum-related items for sale.
Activities: hands-on activities; children dig for gemstones in re-created mine.
Hours & Admission Prices: Mon. & Wed.-Sat. 10-6, Sun. 12-5. Closed New Year's Day; Thanksgiving; Christmas.
Attendance: 6,000

Brooklyn

BROOKLYN HISTORICAL SOCIETY & MUSEUM AND DANIEL PUTNAM TYLER LAW OFFICE, 25 Canterbury Rd., Brooklyn, CT 06234. Tel.: 860-774-7728.
Web Site: www.brooklynct.org
Founded: 1970.
Congressional District: 2
Key Personnel: Pres. (V), Mary Beth Leonard; Cur., Elaine R. Knowlton
Governing Authority: Parent Institution: Brooklyn Historical Society. Tax-exempt.
Institution Type/Description: History Museum.
Collections: local history & culture; 19th century law office of Daniel Putnam Tyler; period furnishings; photographs; personal artifacts.
Activities: Museum Sponsors: 2nd Annual Art Detour/Fall Festival Days in Brooklyn in October.
Publications: quarterly newsletter.
Hours & Admission Prices: May 23-Oct. 7 Wed. & Sun. 1-5; other times by appointment. No charge; donations accepted.
Attendance: 384 (accurate)
Membership: Brooklyn Historical Society: Individual $10; Family $20; Sustaining $50; Patron $100; Sponsor $500; Benefactor $1,000.

Burlington

SESSIONS WOODS WILDLIFE MANAGEMENT AREA & CONSERVATION EDUCATION CENTER, 341 Milford St., Burlington, CT 06013-1550. Mailing Address: P.O. Box 1550, Burlingston, CT 06013-1550. Tel.: 860-675-8130.
Institution Type/Description: Nature Center.
Collections: wildlife & their habitats; natural history.
Activities: demonstrations; hiking; educational programs.
Hours & Admission Prices: Call for hours.

Canterbury

PRUDENCE CRANDALL MUSEUM, 1 S. Canterbury Rd., Canterbury, CT 06331-1536. Mailing Address: P.O. Box 58, Canterbury, CT 06331-0058. Tel.: 860-546-7800. Fax: 860-546-7803. Facebook: Prudence Crandall Museum.
E-mail: crandall.museum@ct.gov
Web Site: www.cultureandtourism.org
Founded: 1984.
Congressional District: 2
Key Personnel: Exec. Dir., Christopher Bergstrom; Dir. Museum, Karin Peterson; Cur. II, Kazimiera Kozlowski.
Personnel Profile: Full-Time Paid 1; Part-Time Volunteers 35.
Governing Authority: state. Parent Institution: State Historic Preservation Office (Dept. of Economic & Community Development). Tax-exempt.
Institution Type/Description: Historic House & Site: housed in 1805 Prudence Crandall House.
Collections: life & work of Prudence Crandall; black history; books; manuscripts, newspapers.
Major Exhibits: Friends & Neighbors: Canterbury's 18th & 19th Century African American Residents, 10/13-11/14.
Research Fields: Afro-American history; Canterbury history; education; women's history; local history materials.
Facilities: 1,000-vol. library of books & pamphlets. Gift items for sale.
Activities: lectures; changing exhibitions.
Publications: book & DVD, To All On Equal Terms: The Life & Legacy of Prudence Crandall.
Hours & Admission Prices: See website for current hours. Adults $6, senior citizens, college students & youth $4; children 5 & under no charge. Closed major holidays.
Attendance: 1,800 (accurate)

Canton

GALLERY ON THE GREEN, 5 Canton Green Rd., Canton, CT 06019. Mailing Address: P.O. Box 281, Canton, CT 06019. Tel.: 860-693-4102.
Institution Type/Description: Art Gallery: housed in a c.1790 schoolhouse.
Collections: paintings; photographs; sculpture.
Facilities: Museum-related items for sale.
Hours & Admission Prices: Fri.-Sun. 1-5.

ROARING BROOK NATURE CENTER, 70 Gracey Rd., Canton, CT 06019-2113. Tel.: 860-693-0263.
E-mail: jkaplan@thechildrensmuseumct.org
Web Site: ww.roaringbrook.org
Founded: 1966.
Governing Authority: Parent Institution: Children's Museum of CT. Tax-exempt.
Institution Type/Description: Nature Center.
Collections: wildlife & their habitat; Native American artifacts; period artifacts; native plant & butterfly gardens; raptors.
Facilities: nature trails. Museum-related items for sale.
Activities: educational programs; hiking.
Hours & Admission Prices: July-Aug. Mon.-Sat. 10-5, Sun. 1-5; Sept.-June Tues.-Sat. 10-5, Sun. 1-5. Adults $6, seniors $5, children $4; members no charge.

Cheshire

THE BARKER CHARACTER, COMIC AND CARTOON MUSEUM, 1188 Highland Ave. (Rte. 10), Cheshire, CT 06410-1624. Tel.: 203-699-3822.
E-mail: museum@barkeranimation.com
Web Site: www.barkermuseum.com
Founded: 1997.
Key Personnel: Owners, Herbert Barker; Owners, Gloria Barker; Dir. & Cur., Judy Fuerst.

Personnel Profile: Full-Time Paid 1; Part-Time Paid 2.
Institution Type/Description: General Museum.
Collections: over 80,000 items including toys & character collectibles from 1873 to present; TV & advertising collection; cartoon memorabilia; Celebriducks; the California Raisins & Gumby.
Activities: tours.
Hours & Admission Prices: Wed.-Sat. 12-4; group tours for ages 8 & over by appointment. Adults $5, children 12 & under $3; discounts to Blue Star Museum members; children under 2 no charge. Closed holidays. &
Attendance: 20,000 (estimated)

Clinton

STANTON HOUSE, 63 E. Main St., Clinton, CT 06413-2036. Mailing Address: 109 E. Main St., Clinton, CT 06413. Tel.: 860-669-2132.
E-mail: curator@stantonhousect.com
Web Site: www.stantonhousect.com/
Founded: 1916.
Congressional District: 33
Key Personnel: Cur., David Perrelli.
Personnel Profile: Full-Time Volunteers 1.
Governing Authority: Parent Institution: Fleet Bank, Trustee. Tax-exempt: 501(c)(3).
Institution Type/Description: Historic House: 1789 Stanton House.
Collections: period furniture; china; glass; 18th Century store.
Activities: guided tours.
Hours & Admission Prices: By appointment. No charge; donations requested.
Attendance: 1,200 (estimated)

Collinsville

CANTON HISTORICAL MUSEUM, 11 Front St., Collinsville, CT 06019-3118. Tel.: 860-693-2793. Fax: 860-693-2793.
E-mail: cantonmuseum@gmail.com
Web Site: www.cantonmuseum.org
Founded: 1966.
Congressional District: 1
Key Personnel: Pres. (V), Paul J. Therrien; Museum Shop Mgr., Curt Edgar.
Personnel Profile: Part-Time Paid 2; Part-Time Volunteers 27.
Governing Authority: Parent Institution: Canton Historical Society. Tax-exempt.
Institution Type/Description: History Museum: housed in c.1865 building used for finishing agricultural plows.
Collections: 19th century artifacts; period fire equipment & medical instruments; Almond D. Fisk burial case; reconstructed 19th century general store, post office, barber shop & blacksmith shop.
Research Fields: genealogy; historic sites.
Activities: tours; winter lectures. Museum Sponsors: Open House; Christmas Boutique; fundraisers.
Publications: newsletter.
Hours & Admission Prices: Wed. & Fri.-Sun. 1-4, Thurs. 1-8. Adults $3. Closed Easter; Christmas.
Attendance: 2,000 (accurate)
Membership: Single $15; Family $25; Sustaining $50; Life $200.

Cos Cob

*** GREENWICH HISTORICAL SOCIETY AKA BUSH-HOLLEY HISTORIC SITE, (M),** 39 Strickland Rd., Cos Cob, CT 06807-2727. Tel.: 203-869-6899, ext. 10. Fax: 203-861-9720.
E-mail: mcouture@hstg.org
Web Site: www.hstg.org
Founded: 1931.
Congressional District: 4
Key Personnel: C.E.O., Debra Mecky; Chm., Davidde Strackbein; Museum Shop Mgr., Michele Couture.
Personnel Profile: Full-Time Paid 8; Part-Time Paid 12; Part-Time Volunteers 200; Interns 2.
Governing Authority: nonprofit corporation. Parent Institution: Greenwich Historical Society, Inc. Tax-exempt: 501(c)(3).
Institution Type/Description: History Museum: includes Bush-Holley House; site of Connecticut's first art colony, a National Historic Landmark.
Collections: 18th, 19th & 20th-century Connecticut & American decorative arts; paintings by American Impressionists; period furnishings; regional history artifacts; manuscripts. Historic Buildings: c.1730 Bush-Holley House; c.1805 Justus Luke Bush storehouse; c.1850 Brush House.
Major Exhibits: Greenwich Faces the Great War, 10/14-2/15.
Research Fields: local history & genealogy; Cos Cob art colony & architecture.

Facilities: library; archives; visitors & education center.
Activities: guided tours; lecture series; family programs; school programs; summer camp; historic markers program.
Publications: Childe Hassam in Connecticut; Chains Unbound; Slave Emancipations in the Town of Greenwich, Connecticut; Greenwich Grows Up; On Home Ground: Elmer Livingston MacRae at The Holley House. Greenwich Before 2000; Building Greenwich: Architecture and Design, 1640 to the Present (2006); Bush-Holley Historic Site Guide; Carved with Rasps and Chisels: The Sculpture of Margaret Brassler Kane.
Hours & Admission Prices: Library & Archives: Wed. 10-4; other times by appointment. Bush Holley House & Museum: Jan.-Feb. Fri.-Sun. 12-4; March-Dec. Wed.-Sun. 12-4. House Tours: 1, 2 & 3. Adults $10, senior citizens & students $8; discounts to AAM & ICOM members; children under 6 & museum members no charge. Closed New Year's Day; Easter; Independence Day; Thanksgiving; Christmas. &
Attendance: 10,000 (accurate)
Membership: Seniors & Teachers $ 30; Individual $40; Family $65; Sponsor $100; Donor $250; Patron $500; Benefactor $1,000.

Coventry

NATHAN HALE HOMESTEAD MUSEUM, 2299 South St., Coventry, CT 06238. Mailing Address: P.O. Box 760, Coventry, CT 06238-0760. Tel.: 860-742-6917.
Founded: 1948.
Key Personnel: Dir. & C.E.O., Sheryl Hack.
Personnel Profile: Part-Time Paid 18; Part-Time Volunteers 10; Interns 1.
Governing Authority: Parent Institution: Connecticut Landmarks. Tax-exempt.
Institution Type/Description: History Museum: site of Capt. Nathan Hale's birth in 1755; Hale family built present structure in 1776 & moved in a month after Nathan's death.
Collections: Hale family artifacts; period furniture.
Facilities: nature trails. Museum-related items for sale.
Activities: tours; living history & hearth cooking programs; birthday parties; summer camp; educational programs for school & scout groups; 18th-century demonstration garden; hands-on activities; horseback riding; nature walks; Sunday Farmer's Market.
Publications: newsletter.
Hours & Admission Prices: May & Sept.-Oct. Sat. 12-4, Sun. 11-4; June-Aug. Wed.-Sat. 12-4, Sun. 11-4. Family $15, adults $7, seniors $6, students $4; discounts to AAM members; members & children under 6 no charge. &
Attendance: 8,000 (estimated)
Membership: Individual $40; Family $55; Preservation Circle $100-$249; Collector's Circle $250-$499.

STRONG PORTER MUSEUM - COVENTRY HISTORICAL SOCIETY, 2382 South St., Coventry, CT 06238. Mailing Address: P.O. Box 534, Coventry, CT 06238. Tel.: 860-742-3054.
E-mail: info@ctcoventryhistoricalsociety.org
Web Site: www.ctcoventryhistoricalsociety.org
Founded: 1962.
Key Personnel: Pres. (V), James Murphy.
Governing Authority: Tax-exempt.
Institution Type/Description: Historical Society Museum: housed in the former farmhouse of Aaron Strong; built in 1730. Listed on the National Register of Historic Places.
Collections: local history & culture; period furnishings; photographs; personal artifacts. Historic Buildings: 1825 schoolhouse; carpenter shop; carriage shed; barn.
Activities: educational programs.
Hours & Admission Prices: June to early Oct. Sun. 12-3. No charge; donations accepted.
Membership: Individual $10; Family $15.

Danbury

THE DANBURY MUSEUM & HISTORICAL SOCIETY, 43 Main St., Danbury, CT 06810-8011. Tel.: 203-743-5200. Fax: 203-743-1131.
E-mail: info@danburymuseum.org
Web Site: www.danburymuseum.org
Founded: 1942.
Congressional District: 5
Key Personnel: Exec. Dir. & Museum Shop Mgr., Brigid Guertin; Pres., Robert Young.
Personnel Profile: Full-Time Paid 1; Part-Time Paid 2; Part-Time Volunteers 10; Interns 3.
Governing Authority: society. Tax-exempt: 501(c)(3).
Institution Type/Description: History Museum.
Collections: 18th-, 19th- & early 20th-century furnishings; textiles & costumes; hatting & woodworking tools; Charles Ives memorabilia; DAR &

the Revolutionary War. Historic Buildings: 1785 John & Mary Rider House; c.1790 John Dodd Shop; c.1829 Charles Ives Birthplace; Marian Anderson Studio.
Research Fields: local history; hatting; Charles Ives; genealogy.
Facilities: library & archives. Museum-related items for sale.
Activities: guided tours; lectures; films; workshops; permanent & temporary exhibits.
Publications: newsletter.
Hours & Admission Prices: Wed.-Fri. 9-4, Sat. 10-4. Research: by appointment. No charge; donations accepted. &
Attendance: 4,000 (estimated)
Membership: Student 21 & under and Senior Citizens 65 & over $20; Individual $25; Family $50; Fellowship $100; Special Friend $200; Patron $500; Corporate $1,000.

DANBURY RAILWAY MUSEUM, 120 White St., Danbury, CT 06810-6642. Mailing Address: P.O. Box 90, Danbury, CT 06813-0090. Tel.: 203-778-8337. Fax: 203-778-1836.
E-mail: info@danburyrail.org
Web Site: www.danburyrail.org
Founded: 1994.
Congressional District: 5
Key Personnel: Pres., Wade Roese; Museum Shop Mgr., Patty Osmer.
Governing Authority: nonprofit organization.
Institution Type/Description: Railway Museum.
Collections: railroad history; rolling stock; railroad artifacts.
Facilities: Museum-related items for sale.
Activities: rail excursions; hands-on railroad work at 12 inches to the foot scale; birthday parties; special events.
Publications: monthly newsletter, Railyard Local.
Hours & Admission Prices: Memorial Day to Labor Day Mon.-Sat. 10-5, Sun. 12-5; Sept.-May Wed.-Sat. 10-4, Sun. 12-4. Adults $6, seniors 60 & over $5, children 3-12 $4; children under 3 & members no charge. Closed New Year's Day; Thanksgiving; Christmas.
Attendance: 20,000 (estimated)
Membership: Retired Railroader $20; Individual $40; Family $50; Patron $200; Lifetime $400; Corporate $500.

MILITARY MUSEUM OF SOUTHERN NEW ENGLAND, 125 Park Ave., Danbury, CT 06810-7504. Tel.: 203-790-9277. Fax: 203-790-0420.
E-mail: usmilitarymuseum@yahoo.com
Web Site: www.usmilitarymuseum.org
Founded: 1985.
Congressional District: 5
Key Personnel: Pres., John V. Valluzzo; Vice Pres., Joseph Di Candido; Exec. Dir. & Sec., Samuel Johnson; Treas., John Purtill.
Personnel Profile: Full-Time Paid 1; Part-Time Volunteers 65.
Governing Authority: nonprofit organization. Tax-exempt: 501(c)(3).
Institution Type/Description: Military Museum.
Collections: World War I-present armored vehicles & weapons; special collection of World War II tank destroyer vehicles & weapons; heavy weapons, books, technical manuals, uniforms, insignia.
Research Fields: restoration of military vehicles & weapons; tank destroyer history; weapons research.
Activities: guided tours; lectures; temporary & traveling exhibitions.
Publications: quarterly newsletter.
Hours & Admission Prices: April-Nov. Tues.-Sat. 10-5, Sun. 12-5; Dec.-March Fri.-Sat. 10-5, Sun. 1-5. Adults $6, senior citizens, children 6-18 & active duty military $4; children under 5 & members no charge. &
Attendance: 7,000 (estimated)
Membership: Junior $10; Individual $20; Family $40; Sustaining $100; TD Vets Honor Roll $500; Regular Life $1,000.

Danielson

KILLINGLY HISTORICAL AND GENEALOGICAL SOCIETY, INC., 196 Main St., Danielson, CT 06239-2823. Mailing Address: P.O. Box 6000, Danielson, CT 06239. Tel.: 860-779-7250.
E-mail: keboo4924@sbcglobal.net
Web Site: www.killinglyhistory.org
Formerly: Killingly Historical Society
Founded: 1972.
Congressional District: 2
Key Personnel: Dir., Marilyn Labbe; Pres. (V), Lynn LaBerge
Governing Authority: Tax-exempt.
Institution Type/Description: History Museum.
Collections: local history & culture; genealogy; photographs.

Research Fields: New England history & genealogy.
Activities: lectures; programs.
Publications: newsletter; journal.
Hours & Admission Prices: Wed.-Sat. 10-4. No charge; donations accepted. Closed New Year's Eve & Day; Christmas Eve, Day & week.
Attendance: 200 (estimated)
Membership: Individual $20; Family $20 (1st person) $10 (each additional person); Nonprofit & Libraries $25; Business $100; Life $200.

Darien

BATES-SCOFIELD HOMESTEAD, THE DARIEN HISTORICAL SOCIETY, 45 Old King's Hwy., N., Darien, CT 06820-4607. Tel.: 203-655-9233. Fax: 203-656-3892.
E-mail: info@ukremer@darienhistorical.org
Web Site: www.darienhistorical.org
Founded: 1954.
Congressional District: 5
Key Personnel: Exec. Dir., John E. Gault, Jr.; Pres. (V), Jon Zagrodzky; Cur. Costume, Babs White; Cur. History, Ken Reiss.
Personnel Profile: Part-Time Paid 3; Part-Time Volunteers 25.
Governing Authority: society. Parent Institution: The Darien Historical Society. Tax-exempt: 501(c)(3).
Institution Type/Description: Historic House: housed in c.1736 structure.
Collections: costumes; archives; quilts; manuscripts. Historic Building: 1827 barn.
Research Fields: Darien history; Fairfield county history; decorative arts.
Facilities: 3,000-vol. library of New England historical & genealogical books available for use on premises; reading room.
Activities: guided tours; lectures; slide shows; antiques show; trips; eight formally organized education programs; temporary exhibitions.
Publications: book, Darien-Historical Sketches; quarterly newsletter.
Hours & Admission Prices: Tues.-Fri. 9-4. No charge; donations accepted. Closed major holidays.
Attendance: 4,000 (estimated)
Membership: Family $65; Friend $125; Patron $250; Sustaining $500; Benefactor $1,000.

Deep River

STONE HOUSE, 245 Main St., Deep River, CT 06417-2055. Tel.: 860-526-5811.
Founded: 1937.
Key Personnel: Pres. (V), Jeffrey D. Hostetler.
Governing Authority: Parent Institution: Deep River Historical Society.
Institution Type/Description: Historic House: built in 1840.
Collections: period furnishings; personal artifacts; portraits; musical instruments.
Hours & Admission Prices: Call for hours.

Derby

KELLOGG ENVIRONMENTAL CENTER, 500 Hawthorne Ave., Derby, CT 06418. Tel.: 203-734-2513.
Institution Type/Description: Science Center.
Collections: natural history; environment; geology.
Facilities: nature trails.
Activities: workshops; educational programs; lectures; hiking.
Hours & Admission Prices: Tues.-Sat. 9-4:30. No charge; donations accepted.

OSBORNE HOMESTEAD MUSEUM, 500 Hawthorne Ave., Derby, CT 06418-1020. Tel.: 203-734-2513. Fax: 203-922-7833.
E-mail: susan.d.robinson@ct.gov
Web Site: www.ct.gov/deep
Key Personnel: Cur., Susan Robinson; Dir., Diane Joy
Institution Type/Description: History Museum: housed in the former estate of Frances Osborne Kellogg, an accomplished businesswomen & conservationist who was dedicated to preserving land for future generations.
Collections: history of the Osborne family; period furniture; fine art.
Facilities: gardens.
Activities: guided tours; school programs; special programs; hands-on activities.
Hours & Admission Prices: May-Oct. Thurs.-Fri. 10-3, Sat. 10-4, Sun. 12-4; Nov. 29-Dec. 21 Thurs.-Sun. 10-4.

East Granby

OLD NEW-GATE PRISON AND COPPER MINE, 115 Newgate Rd., East Granby, CT 06026-9545. Mailing Address: P.O. Box 254, East Granby, CT 06026-0254. Tel.: 860-653-3563. Fax: 860-844-2142.
E-mail: newgate.museum@ct.gov
Web Site: www.cultureandtourism.org
Founded: 1969.
Congressional District: 6
Key Personnel: Exec. Dir. CCT, Karen Senich; Museum Dir., Karin Peterson; Museum Asst., Lance Kozikowski.
Personnel Profile: Part-Time Paid 6; Part-Time Volunteers 5.
Governing Authority: state. Parent Institution: Connecticut Commission on Culture & Tourism, 1 Constitution Plaza., Hartford. Tax-exempt.
Institution Type/Description: History Museum: abandoned underground copper mine used as first state prison from 1773 to 1827.
Collections: photos; mining & prison memorabilia.
Research Fields: American Revolution; mining & copper mining; prisons.
Facilities: hiking trails; picnic area. Museum-related items for sale.
Activities: guided tours; lectures; permanent exhibitions; living history programs.
Hours & Admission Prices: See website for hours. Adults $10, senior citizens 60 & over and college students $8, children 6-17 $6; discounts to groups; children under 6 no charge. Closed Memorial Day; Independence Day; Labor Day; Columbus Day. &
Attendance: 18,114 (accurate)

East Haddam

ALLEGRA FARM & HORSE-DRAWN CARRIAGE AND SLEIGH MUSEUM, 69 Town Rd., East Haddam, CT 06415. Mailing Address: P.O. Box 455, East Haddam, CT 06423-0455. Tel.: 860-537-8861.
Institution Type/Description: Transportation Museum.
Collections: horse-drawn carriages, wagons & equipment.
Activities: hay & sleigh rides; educational programs.
Hours & Admission Prices: By appointment.

GILLETTE CASTLE STATE PARK, 67 River Rd., East Haddam, CT 06423-1462. Tel.: 860-526-2336. Fax: 860-4244070.
E-mail: dep.stateparks@ct.gov
Web Site: www.ct.gov/dep
Key Personnel: Park Supvr., Scott Dawley.
Governing Authority: state. Subsidiary Institution: Department of Environmental Protection.
Institution Type/Description: Historic Building: housed in a twenty four room mansion built for actor, director, & playwright, William Hooker Gillette, c.1914.
Collections: period furnishings; personal artifacts.
Facilities: nature trails.
Activities: hiking; picnicking; camping.
Hours & Admission Prices: Park: daily 8am to sunset. No charge. Castle: Memorial Day to Columbus Day daily 10-4:30. Adults $6, children 6-12 $2; children under 6 no charge.
Attendance: 225,000 (estimated)

GOODSPEED OPERA HOUSE, 6 Main St., East Haddam, CT 06423-1302. Tel.: 860-873-8668 & 8664.
Institution Type/Description: Historic Building: housed in a Victorian theater built in 1876 by shipping magnate William Goodspeed. Listed on the National Register of Historic Places.
Collections: opera house history; period furnishings; photographs.
Activities: guided tours; performances.
Hours & Admission Prices: June-Oct. Sat. 10:30-1. Adults $5, children under 12 $1.

NATHAN HALE SCHOOLHOUSE IN EAST HADDAM, 29 Main St., Rte. 149, East Haddam, CT 06423. Mailing Address: Connecticut SAR, P.O. Box 411, East Haddam, CT 06423. Tel.: 860-873-3399.
Web Site: www.connecticutsar.org
Founded: 1889.
Governing Authority: Tax-exempt.
Institution Type/Description: Historic Building: housed in the former schoolhouse where Nathan Hale was schoolmaster; built in 1750.
Collections: local history; Nathan Hale's life & military career history; period furnishings.

Hours & Admission Prices: May-Oct. Wed.-Sun. 12-4; call to confirm. No charge; donations accepted.
Attendance: 1,000 (accurate)

East Hampton

CHATHAM HISTORICAL SOCIETY OF EAST HAMPTON, 6 Bevin Blvd., East Hampton, CT 06424. Tel.: 860-267-2442.
E-mail: podskoch@comcast.net
Web Site: www.chathamhistoricalsocietyct.org
Institution Type/Description: Historical Society Museum.
Collections: local history & culture; period furnishings; photographs; bell making history. Historic Building: 1840 schoolhouse.
Activities: educational programs; special events.
Hours & Admission Prices: 1st Sun. each month 2-4; other times by appointment.

East Hartford

EDWARD E. KING MUSEUM, Raymond Library, 840 Main St., East Hartford, CT 06108-3128. Tel.: 860-289-6429.
Key Personnel: Dir. Cultural Assets/Reference Librarian, Jason Pannone.
Personnel Profile: Full-Time Paid 1.
Institution Type/Description: History Museum.
Collections: tobacco & aviation history.
Hours & Admission Prices: Jan.-Sept. Mon.-Thurs. 9-9, Fri.-Sat. 9-5; Oct.-April Mon.-Thurs. 9-9, Fri.-Sat. 9-5, Sun. 1-4. The King Museum is currently closed as the East Hartford Public Library undergoes renovation & expansion. The collection will be available again some time in 2015.

HISTORICAL SOCIETY OF EAST HARTFORD, Martin Park, 307 Burnside Ave., East Hartford, CT 06108. Mailing Address: P.O. Box 380166, East Hartford, CT 06138-0166. Tel.: 860-568-2884; 860-528-0716.
E-mail: hseh@hseh.org
Web Site: www.hseh.org
Founded: 1964.
Congressional District: 1
Key Personnel: Dir., Pres. (V), Craig R. Johnson; Vice Pres. (V), Bette Daraskevich.
Governing Authority: Tax-exempt.
Institution Type/Description: Historical Society Museum.
Collections: local history & culture; period furnishings; personal artifacts; photographs; early tools & equipment. Historic Buildings: 1761 Makens Bemont House; 1820s Goodwin Schoolhouse; 1850s Burnham Blacksmith Shop.
Activities: educational programs; special events.
Publications: Jan., March, May, Sept., Nov. newsletters.
Hours & Admission Prices: Guided Tours: June-Aug. Sun. 1-4. No charge; donations accepted.
Membership: Student $10; Individual $15; Family $20; Patron $50.

East Haven

SHORE LINE TROLLEY MUSEUM, 17 River St., East Haven, CT 06512-2519. Tel.: 203-467-6927. Fax: 203-467-7635.
E-mail: info@shorelinetrolley.org
Web Site: www.shorelinetrolley.org
Formerly: Branford Trolley Museum
Founded: 1945.
Congressional District: 3
Key Personnel: Dir., Nathan Niekering; Chm. (V), Louis Levinson; Pres., William J. Wall; Treas., Ronald Kupin; Dir. Vehicle Collection Management, Denis Pacelli; Dir. Library, Michael Schreiber; Dir. Exhibits, Frederick Sherwood; Dir. Mktg., Katherine Slinsky; Museum Shop Mgr., David A. Cohen.
Personnel Profile: Full-Time Paid 2; Part-Time Paid 1; Part-Time Volunteers 50.
Volunteer Hours: 26,000
Operating Expenses: 280,000
Operating Income: 312,000
Governing Authority: nonprofit corporation. Operated by Branford Electric Railway Association, Inc. Tax-exempt: 501(c)(3).
Institution Type/Description: Transportation: Technology & Operating Railway Museum.
Collections: urban, suburban & interurban electric railway cars & equipment; buses; trolley bus.
Research Fields: history of electric railway operations; evolution of mass transit; urban rail transit & its role in urban life.

Facilities: reception center; picnic area.

Activities: guided tours; permanent & temporary exhibitions; rides on operating electric cars. Members: Car operation; car restoration; maintenance & operation of the museum railway & associated facilities as volunteers using authentic tools, vehicles, & procedures.

Publications: monthly newsletter; guidebook; semi-annual journal.

Hours & Admission Prices: April & Nov. Sun. 10:30-4:30; May & Sept.-Oct. Sat.-Sun. 10:30-4:30; Memorial Day to Labor Day, daily 10:30-4:30. Adults $10, senior citizens $8, children 2-15 $6; discounts to groups, AAM, ICOM, AAA & ARM members; children under 2 & members no charge. &

Attendance: 21,530 (accurate)

Membership: Associate $20; Regular $40; Family $65; Supporting $75; Sustaining $150; Benefactor $500.

East Windsor

CONNECTICUT FIRE MUSEUM, 58 North Rd. (Rte. 140), East Windsor, CT 06088. Mailing Address: P.O. Box 297, East Windsor, CT 06088-0297. Tel.: 860-627-6540. Fax: 860-627-6510.

E-mail: ctfiremuseum@hotmail.com

Web Site: ct-trolley.org/firemuseum

Founded: 1968.

Congressional District: 1

Key Personnel: Pres. (V), Bert Johanson; Treas., Devel. & Museum Shop Mgr., Alan Walker.

Personnel Profile: Part-Time Volunteers 11.

Governing Authority: private; nonprofit organization. Tax-exempt: 501(c)(3).

Institution Type/Description: Fire-Fighting Museum.

Collections: early fire apparatus & equipment; fire sleigh c.1894; 1967 Walter airport crash truck.

Facilities: library; 6,900 sq. ft. exhibit space.

Activities: guided tours. Annual Events: Parades.

Hours & Admission Prices: Memorial Day-Labor Day Mon. & Wed.-Sat. 10-4, Sun. 12-4; Sept. to Thanksgiving Sat. 10-4, Sun. 12-4. Adults $8, senior citizens $7, children $5.

Attendance: 16,250 (estimated)

Membership: Associate $10; Active $15; Family $25.

CONNECTICUT TROLLEY MUSEUM, (M), 58 North Rd., East Windsor, CT 06088. Mailing Address: P.O. Box 360, East Windsor, CT 06088-0360. Tel.: 860-627-6540. Fax: 860-627-6510.

E-mail: office@ceraweb.org

Web Site: www.ct-trolley.org

Founded: 1940.

Congressional District: 6

Key Personnel: Pres. (V), Galen Semprebon; Chm. (V), Fred Stroiney; Business Mgr., Carol Zenczak; Museum Shop Mgr., John Pelletier.

Personnel Profile: Full-Time Paid 1; Full-Time Volunteers 1; Part-Time Paid 2; Part-Time Volunteers 65.

Governing Authority: nonprofit organization. Parent Institution: Connecticut Electric Railway Association, Inc. Tax-exempt: 501(c)(3).

Institution Type/Description: Trolley Museum.

Collections: electric transportation equipment from 1880-1947; 1894 Steeple-Cab electric locomotive; streetcars; 4-wheeled ex-horse cars; interurban & suburban passenger cars; railway cars.

Research Fields: histories of street railways.

Facilities: collection of documents & historic papers of Southern New England railways & street railway companies available for use by written application. Museum-related items for sale.

Activities: guided tours (3-mile round trip); lectures; films; formally organized education programs for children; permanent exhibitions. Annual Events: Halloween & December programs.

Publications: quarterly newsletter, Connecticut Electric News; books, The Cleveland Railway; New York, New Haven & Hartford Railroad Electrification; Heisler; Metropolitan Subway & Elevated Systems.

Hours & Admission Prices: April to mid-June Sat. 10-4:30, Sun. 12-4:30; mid-June to Labor Day Mon. & Wed.-Fri. 10-3:30, Sat. 10-4:30, Sun. 12-4:30; groups of 10 or more by appointment. Unlimited rides: adults $8.50, senior citizens 62 & up $7.50, children 2-12 $5.50; discounts to AAA, MTA, ALA & ARM members & museum members with ID; children under 2 no charge. Closed Thanksgiving; Christmas.

Attendance: 20,000 (estimated)

Membership: Senior $20; Individual $30; Family $50; Contributing $100; Supporting $250; Sustaining $500; Patron $1,000.

SCANTIC ACADEMY MUSEUM, EAST WINDSOR HISTORICAL SOCIETY, INC., 115 Scantic Rd., Rte. 191, East Windsor, CT 06088-9737. Tel.: 860-623-5327.

E-mail: eastwindsorhistory@gmail.com

Founded: 1965.

Congressional District: 1

Key Personnel: Pres. (V), Michael Hunt; Treas., Larry Tribble; Sec., Jessica Bottomley.

Personnel Profile: Part-Time Volunteers 10.

Governing Authority: private. Tax-exempt.

Institution Type/Description: General Museum: housed in 1817 Scantic Academy Building.

Collections: industry; agriculture; paintings; transportation.

Research Fields: local history; genealogy.

Facilities: 500-vol. library of books on local history.

Activities: annual meetings & dinners; town community day program.

Publications: book: History of East Windsor Through the Years; newsletter, New HS.

Hours & Admission Prices: Sat. 9-12; other times by appointment. No charge; donations accepted. &

Attendance: 150 (estimated)

Membership: Individual $10; Family $15; Corporate $50.

Enfield

MARTHA A. PARSONS HOUSE MUSEUM, 1387 Enfield St., Enfield, CT 06082-5524. Mailing Address: Enfield Historical Society, P.O. Box 586, Enfield, CT 06083-0586. Tel.: 860-745-6064.

E-mail: questions@enfieldhistoricalsociety.org

Web Site: www.enfieldhistoricalsociety.org

Founded: 1960.

Institution Type/Description: Historic House Museum: built in 1782 by John Meacham.

Collections: period furnishings; personal artifacts; Shaker furniture & artifacts.

Hours & Admission Prices: May-Oct. Sun. 2-4:30; other times by appointment. No charge.

OLD TOWN HALL MUSEUM, 1294 Enfield St., Enfield, CT 06082-4928. Mailing Address: Enfield Historical Society, P.O. Box 586, Enfield, CT 06083. Tel.: 860-745-1729.

E-mail: questions@enfieldhistoricalsociety.org

Web Site: www.enfieldhistoricalsociety.org

Founded: 1960.

Personnel Profile: Part-Time Volunteers 40.

Governing Authority: Parent Institution: Enfield Historical Society. Tax-exempt.

Institution Type/Description: Historic Building: housed in the former meeting house of the First Ecclesiastical Society; c.1774. Listed on the National Register of Historic Places.

Collections: Enfield's history & culture; photographs; period furnishings; personal artifacts; farming implements; industrial machinery; Shaker artifacts.

Hours & Admission Prices: May-Oct. Sun. 2-4:30; other times by appointment. No charge. Closed major holidays.

Attendance: 700 (estimated)

Membership: Individual $10; Family $15; Contributing $20; Sustaining $40; Supporting $100; Corporate $200; Life $250; Benefactor $500.

Essex

∗ CONNECTICUT RIVER MUSEUM, (M), 67 Main St., Essex, CT 06426-1150. Tel.: 860-767-8269. Fax: 860-767-7028.

Web Site: www.ctrivermuseum.org

Founded: 1974.

Congressional District: 2

Key Personnel: Exec. Dir., Jerry Roberts; Cur., Amy Trout; Chm. Bd. Trustees, Tim Boyd; Front Desk & Shop Mgr., Helen Davis; Business Mgr., Joan Meek; Dir. Education, Jennifer White-Dobbs.

Personnel Profile: Full-Time Paid 5; Part-Time Paid 5; Part-Time Volunteers 125; Interns 4.

Governing Authority: nonprofit organization. Parent Institution: The Connecticut River Foundation at Steamboat Dock, Inc. Tax-exempt: 501(c)(3).

Institution Type/Description: Maritime & River Museum: housed in 1878 wooden warehouse.

Collections: objects related to Connecticut River Valley agriculture; industry; small craft collection; archaeology; model of American Turtle, first submarine. Historical Building: 1813 Chandlery of Capt. Richard Hayden.

Research Fields: Connecticut River maritime & River Valley history; cultural landscape of the Connecticut River; river ecology.

Facilities: library of unpublished documents on Connecticut & U.S. maritime subjects available for research by appointment.

Activities: guided tours; lectures; formally organized education programs for children; permanent & temporary exhibitions.

Publications: biannual newsletter, Steamboat Log; walking map, A Walking

Tour of Essex; books, Connecticut River Master Mariners; Potapaug Quarter, The First Settlers of Essex, Connecticut; Life in the Connecticut River Valley; booklet, Life In the Connecticut River Valley 1800-1840; prints of river scenes.

Hours & Admission Prices: Tues.-Sun. 10-5. Adults $8, senior citizens, students, & AAA members $7, children $5; discounts to AAM members; children under 6 & members no charge. &

Attendance: 16,000 (accurate)

Membership: Individual $35; Two People $50; Family $60; Supporting $100; Sustaining $250; Benefactor $500; Life $1,500.

ESSEX HISTORICAL SOCIETY, INC., 22 Prospect St., Essex, CT 06426-1021. Mailing Address: P.O. Box 123, Essex, CT 06426-0123. Tel.: 860-767-0681 & 0375.

E-mail: ehs@essexhistory.net
Web Site: essexhistory.org
Founded: 1955.
Congressional District: 2
Key Personnel: Pres. (V), Sharon Clark; Cur., Celia Francis.
Personnel Profile: Part-Time Volunteers 100.
Governing Authority: society. Tax-exempt.
Institution Type/Description: History Museum.
Collections: late 1700s furnishings displayed in a home setting; 18th-19th century records & items relating to industrial & domestic history of Essex & vicinity. Historic Houses: c.1732 Pratt House; 1832 Hills Academy.
Research Fields: local history; early New England style of living.
Facilities: 50-vol. library of local history available by arrangement.
Activities: lectures; films; individual & group tours.
Publications: The British Raid on Essex: April 8, 1814; The Oliver Cromwell; The Golden Rule Days of Essex; From Bicycles to Buicks, The Story of Behrems and Bushnell.
Hours & Admission Prices: Pratt House: June-Aug. Sat.-Sun. 1-4; Sept.-May by appointment. No charge; donations accepted.
Attendance: 400 (estimated)
Membership: Individual $25; Family $40; Business & Professional $50; Sponsor $75; Patron $150; Benefactor $500.

Fairfield

BELLARMINE MUSEUM OF ART, FAIRFIELD UNIVER-SITY, (M), (I), 1073 N. Benson Rd., Fairfield, CT 06824. Tel.: 203-254-4000, ext. 4046. Fax: 203-249-5529. Facebook: Bellarmine Museum at Fairfield University.

E-mail: museum@fairfield.edu
Web Site: www.fairfield.edu/museum
Founded: 2010.
Congressional District: 4
Key Personnel: Dir., Jill Deupi, J.D., Ph.D.; Mgr. Collections, Carey Mack Weber.
Personnel Profile: Full-Time Paid 2; Part-Time Paid 1; Part-Time Volunteers 1; Interns 4.
Governing Authority: private university. Parent Institution: Fairfield University, Fairfield, CT. Tax-exempt: 501(c)(3).
Institution Type/Description: Art Museum.
Collections: Kress study collection of Italian paintings, 1200-1800; cast collection of Greco Roman antiquities; non-western art.
Major Exhibits: In the Wake of the Butterfly: James Abbott McNeill Whistler and His Circle in Venice, 1/23/14-4/4/14; La Ragnatela/The Spiderweb: Works by Giampaolo Seguso from the Corning Museum of Glass, 4/10/14-6/13/14; Gabor Peterdi: The War Years, 10/2/14-12/19/14.
Research Fields: Kress collection of paintings; 18th century Italian art & culture; the Acropolis & Parthenon; late antique, Celtic & medieval art.
Facilities: 2,700 sq. ft. exhibit space; educational facilities.
Activities: guided tours; lectures; formal education programs for undergraduate or graduate college students; loan, permanent & temporary exhibitions; K-12 enrichment activities; family programs; lifelong learning programs.
Hours & Admission Prices: Academic Year: Mon. Fri. 9:30-4:30. No charge; donations accepted. Closed university & national holidays. &
Attendance: 7,000 (accurate)

CONNECTICUT AUDUBON BIRDCRAFT MUSEUM, 314 Unquowa Rd., Fairfield, CT 06824-5018. Tel.: 203-259-0416. Fax: 203-259-1344.

E-mail: birdcraft@ctaudubon.org
Web Site: www.ctaudubon.org
Founded: 1914.
Congressional District: 4
Key Personnel: Pres., Robert Martinez; Chm. (V), Ralph Wood; Dir., Nelson North; Museum Shop Mgr., Jane Guenther.

Personnel Profile: Full-Time Paid 2; Part-Time Paid 2; Part-Time Volunteers 10.
Governing Authority: private; nonprofit. Parent Institution: Connecticut Audubon Society, Fairfield, CT. Tax-exempt: 501(c)(3).
Institution Type/Description: Natural Science & Ornithology Museum: founded in 1914 by Mabel Osgood Wright, who played a major role in establishing the American conservation movement.
Collections: biological collections focusing on the zoology of Connecticut, particularly the coastal area; bird mounts & artifacts from around the world; African animal mounts; fossils; minerals; insects; botanical; nature & avian decorative arts & artifacts; specimens & artifacts collected by Wright & naturalist, G.B. Grinnell.
Research Fields: migratory birds of Connecticut; Mabel Osgood Wright, conservationist & author; monk parakeet distribution in Greater Bridgeport, Connecticut; bird banding station.
Facilities: library; 1,550 sq. ft. exhibit space; field research station; nature & conservation center; 6-acre sanctuary. Museum-related items for sale.
Activities: docent program; films; formal education programs; guided tours; hobby workshops; lectures; loan, participatory & temporary exhibitions. Annual Event: International Migratory Bird Day in May.
Hours & Admission Prices: Mon.-Fri. 9-1; other times by appointment. Adults $2, children $1; discounts to Connecticut Ornithological Assoc. members; members no charge. Closed major holidays. &
Attendance: 9,000 (accurate)
Membership: Individual $45; Family $55; Sustaining, Associated Organization & Supporting $75; Contributing $100; Donor $200; Patron $500; Benefactor $1,000.

CONNECTICUT AUDUBON SOCIETY OF FAIRFIELD, 2325 Burr St., Fairfield, CT 06824-1806. Tel.: 203-259-0416, ext. 109. Fax: 203-254-7673.

E-mail: bmucci@ctaudubon.org
Web Site: www.ctaudubon.org
Founded: 1898.
Congressional District: 4
Key Personnel: Pres., Robert Martinez; Chm. (V), Stephen B. Oresman; Sr. Dir. Southwest CT Operations & Dir. Connecticut Audubon Fairfield, Rgnl. Programs, Nelson North; Dir. Finance & Human Resources, Frank Thiel; Dir. Education, Carol Kratzman; Dir. Connecticut Audubon Center at Glastonbury, Sally Carbone; Dir. Devel., Ann O'Leary; Dir. Mktg. & Communications, Mara Neville; Museum Shop Mgr., Deborah Callan.
Personnel Profile: Full-Time Paid 4; Part-Time Paid 3; Part-Time Volunteers 150; Interns 2.
Governing Authority: private; nonprofit environmental education organization. Branch Museums: Connecticut Audubon Center at Fairfield, 2325 Burr St., Fairfield, CT 06430; Connecticut Audubon Center at Glastonbury, 1361 Main St., Glastonbury, CT 06033; Connecticut Audubon Coastal Center at Milford Point, 1Milford Point Rd., Milford; Connecticut Audubon Birdcraft Museum, 314 Unquowa Rd., Fairfield; Connecticut Audubon Environmental Center, 118 Oak St., Hartford; Connecticut Audubon Center at Pomfret, 189 Pomfret St. (Rte. 169), Pomfret Center, CT 06259; 19 statewide sanctuaries. Tax-exempt: 501(c)(3).
Institution Type/Description: Nature Center.
Collections: living & mounted animals.
Research Fields: ornithology; wildlife ecology & environmental issues; bird banding program; colonial waterbird survey.
Facilities: 3,000-vol. library; nature center; aquarium; 200-seat auditorium; classrooms; 160-acre wildlife sanctuary. Nature-related items for sale.
Activities: guided tours; lectures; films; summer classes; educational programs; permanent & temporary exhibitions; school loan service; local, regional, international field trips. Annual Events: Friends Event; Halloween Event; Volunteers Dinner.
Publications: quarterly, membership newsletter; bulletin, Connecticut Audubon News.
Hours & Admission Prices: Tues.-Sat. 10-3. Adults $2, children $1; members no charge. Closed major holidays. &
Attendance: 12,000 (accurate)
Membership: Individual $40; Supporting $75; Contributing $100; Donor $200; Patron $500; Benefactor $1,000.

FAIRFIELD MUSEUM AND HISTORY CENTER, (M), 370 Beach Rd., Fairfield, CT 06824-6639. Tel.: 203-259-1598. Fax: 203-255-2716.

E-mail: info@fairfieldhs.org
Web Site: www.fairfieldhistory.org
Formerly: Fairfield Historical Society
Founded: 1902.
Congressional District: 4

Key Personnel: C.E.O., Michael A. Jehle; Pres., Thomas Walsh; Dir. Education, Christine Jewell; Prog. & Volunteer Coord., Walter Matis.
Personnel Profile: Full-Time Paid 7; Part-Time Paid 4; Part-Time Volunteers 90; Interns 5.
Governing Authority: society. Parent Institution: Fairfield Historical Society. Branch Museum: Ogden House, 1520 Bronson Rd., Fairfield. Tax-exempt: 101(6).
Institution Type/Description: History Museum & Library.
Collections: personal artifacts; furnishings; paintings; photographs; maps; manuscripts.
Research Fields: regional genealogy & history; decorative arts.
Facilities: 12,000-vol. library of genealogical & historical interest for use on premises.
Activities: guided tours; seminars, workshop & lecture series; changing exhibits. Annual Event: Fall Festival.
Publications: quarterly newsletter; books, Fairfield, The Biography of a Community, 1639-2000; History of Fairfield County Men In The Revolution; Walking Through History-The Seaports of Black Rock & Southport; Cooking with Fire: Open Hearth Cooking & Recipes; Clocks & Clockmakers of Fairfield 1736-1813; Fairfield Bicycle Tour: Travel Back in Time; Fairfield Printmaker: John Taylor Arms; Visions of Home and Abroad; Images of America: Fairfield, Connecticut.
Hours & Admission Prices: Mon.-Fri. 10-4, Sat.-Sun. 12-4. Adults $5; discounts to AAM members; members no charge. &
Attendance: 20,600 (estimated)
Membership: Individual $30 Family $50; Contributing & Business $100; Sustaining $250; Patron $500; Benefactor $1,000.

THOMAS J. WALSH ART GALLERY, QUICK CENTER FOR THE ARTS AT FAIRFIELD UNIVERSITY, 1073 N. Benson Rd., Fairfield, CT 06824-5171. Tel.: 203-254-4000, ext. 2969. Fax: 203-254-4113.

E-mail: info@quickcenter.com
Web Site: www.quickcenter.com
Founded: 1990.
Congressional District: 4
Key Personnel: Exec. Dir., Thomas V. Zingarelli; Gallery Dir., Dr. Diana Dimodica Mille.
Personnel Profile: Full-Time Paid 1; Part-Time Volunteers 7; Interns 4.
Governing Authority: nonprofit. Parent Institution: Fairfield University. Tax-exempt: 501(c)(3).
Institution Type/Description: University Art Gallery.
Collections: pre-Columbian, African, Oceanic, Eastern and Indian sculpture; contemporary art-20th century; Renaissance paintings.
Research Fields: art history.
Facilities: 2,200 sq. ft. exhibit space; theater.
Activities: lectures; films; concerts; arts festivals; theater; organized education programs for children, adults, undergraduate & graduate college students affiliated with Fairfield Univ.; training programs for professional museum workers; participatory, loan & traveling exhibitions.
Publications: catalogue, Images of the Divine.
Hours & Admission Prices: Tues.-Sat. 11-5, Sun. 12-4, No charge; donations accepted. Closed Memorial Day; Labor Day; Thanksgiving; Christmas. &
Attendance: 10,000 (estimated)
Membership: Friend $50; Member $100; Sustainer $250; Patron $500; Benefactor $1,000.

Falls Village

FALLS VILLAGE-CANAAN HISTORICAL SOCIETY, 44 Railroad St., Falls Village, CT 06031. Mailing Address: P.O. Box 206, Falls Village, CT 06031-0206. Tel.: 860-824-8226. Fax: 860-824-4506.

E-mail: eec06031@comcast.net
Web Site: www.betweenthelakes.com/canaan/hist_society.htm
Founded: 1952.
Congressional District: 6
Key Personnel: Pres. (V), Judy Jacobs; Cur. South Canaan Meeting House, Mrs. H. DeVries; Cur. Falls Village Depot, Cheryl Aeschliman; Cur. Beebe Hill Schoolhouse, Mary Lu Sinclair.
Governing Authority: nonprofit organization. Tax-exempt.
Institution Type/Description: Historical Society Museum.
Collections: costumes; local history items; railroad artifacts.
Facilities: 200-vol. library of genealogy, ledgers & day books, local history, & children's books available for use on premises.
Activities: school groups; temporary exhibitions; concerts. Special Events: semiannual Peddler's Markets; Antique Tool Show; Living History Demonstrations; Lecture Series June to August.
Publications: quarterly newsletter; pamphlet, South Canaan Congregational Church; The Two Canaans-Division of the Town of Canaan; 1853 map of the Town of Canaan; Lemuel Haynes; Chronicles of Old Canaan; book, Canaan: A Small New England Town During the Revolutionary War; book, Country Depots in the Connecticut Hills; book, Exploring the Berkshire Hills; book, Iron Country Revisited; Mumbet.
Hours & Admission Prices: Falls Village: Sat. 10-1; other times by appointment. No charge; donations accepted. &
Attendance: 175 (estimated)
Membership: Single $15; Family $25; Business $50; Friends of the Society $100.

Farmington

* HILL-STEAD MUSEUM, (M), 35 Mountain Rd., Farmington, CT 06032-2304. Tel.: 860-677-4787. Fax: 860-677-0174.
E-mail: cagenelloc@hillstead.org
Web Site: www.hillstead.org
Founded: 1946.
Congressional District: 6
Key Personnel: Dir. & C.E.O., Sue Sturtevant; Cur., Melanie A. Bourbeau; Pres. (V), M. Timothy Corbett; Museum Shop Mgr., Priscilla Ramage; Dir. Communications, Cynthia Cagenello; Dir. Operations, David Perbeck; Office Mgr., Kate Ebner.
Personnel Profile: Full-Time Paid 10; Part-Time Paid 28; Part-Time Volunteers 230; Interns 10.
Governing Authority: nonprofit organization. Tax-exempt: 501(3)(c).
Institution Type/Description: Art Museum: housed in c.1901 building designed by Theodate Pope Riddle in collaboration with McKim, Mead & White. Period garden based on plan by Beatrix Farrand.
Collections: French Impressionist paintings; prints; sculpture; furniture; porcelain.
Research Fields: art; architecture.
Facilities: research library available to scholars by appointment; gardens. Museum-related books for sale.
Activities: guided tours; permanent exhibitions; nature & educational programs; poetry; music.
Publications: quarterly newsletters; booklets, House Guide; Theodate Pope Riddle: Her Life & Her Work; brochures, Garden & Grounds; Flip card; Sunken Garden guide; Theodate Pope Riddle: A Pioneer Woman Architect; Oriental Carpets in the Collections of H-S Museum; Hill-Stead, An Illustrated Museum Guide; Dearest of Geniuses, A Life of Theodate Pope Riddle; annual reports; Hill-Stead Plant Book; poetry chapbooks; Hill-Stead: The Country Place of Theodate Pope Riddle; exhibition catalogue, Wonders Revealed: Rarely Seen Original Prints from Hill-Stead's Collection.
Hours & Admission Prices: Tues.-Sun. 10-4. Adults $12; discounts for AAA members; members no charge. Closed New Year's Day; Independence Day; Thanksgiving; Christmas. &
Attendance: 46,365 (accurate)
Membership: Student $25; Individual $50; Family & Household $75; Contributor $150; Sponsor $300; Patron $500; Theodate Pope Riddle Society $1,000 & up.

STANLEY-WHITMAN HOUSE, FARMINGTON, 37 High St., Farmington, CT 06032-2314. Tel.: 860-677-9222. Fax: 860-677-7758.

E-mail: lisa@stanleywhitman.org
Web Site: www.stanleywhitman.org
Founded: 1935.
Congressional District: 6
Key Personnel: C.E.O., Lisa Johnson; Chm. (V), Jane Dalal.
Personnel Profile: Full-Time Paid 2; Part-Time Paid 5; Part-Time Volunteers 70; Interns 4.
Governing Authority: nonprofit organization. Parent Institution: Farmington Village Green & Library Association, Main St., Farmington, CT. Tax-exempt: 501(c)(3).
Institution Type/Description: Historic House Museum: c.1719-1772 Stanley-Whitman House; interpretation of 18th-century Farmington.
Collections: 1642-1880 Farmington history; 18th-century furniture & decorative arts; local artifacts; herb garden.
Research Fields: Colonial Farmington life; furniture; decorative arts; 17th- & 18th-century furniture & decorative arts; historic house investigation; 18th-century dooryards & gardens; agriculture; Judah Woodruff, constructor of houses, barns, mid-late 18th-century; 17th century New England culture & witchcraft trials.
Facilities: education building annex. Museum-related gifts for sale.
Activities: house tours; school programs; member events; exhibition previews.
Publications: newsletter; A Short History of Farmington, Connecticut; A Guide to Historic Farmington, Connecticut; The Preservation of the Stanley-Whitman House; The Stanley-Whitman House: A Dwelling House and Homestead.

Hours & Admission Prices: Wed.-Fri. 9-4, Sat.-Sun. 12-4. House Tours: Wed.-Sun. 12-4. Adults $5, senior citizens & group rate per person $4; discounts to New England Museum Assoc. members; Greater Hartford Assoc. of Historic Houses & museum members no charge. Closed national holidays. &

Attendance: 12,000 (estimated)

Membership: Full Time Student $18; Individual $20; Family $30; Friend (2 guest passes) $45; Donor (4 guest passes) $65; Sponsor (6 guest passes) $100; Patron (8 guest passes) $150; Benefactor (10 guest passes) $500.

Franklin

BLUE SLOPE COUNTRY MUSEUM, 138 Blue Hill Rd., Franklin, CT 06254-1601. Tel.: 860-642-6413.

E-mail: museum@blueslope.com

Web Site: www.blueslope.com/museum.html

Key Personnel: Dir., Sandy Staebner

Institution Type/Description: History Museum.

Collections: early farming; period farm equipment & furnishings; personal artifacts.

Facilities: Museum-related items for sale.

Activities: special events; educational programs.

Hours & Admission Prices: By appointment.

Gales Ferry

NATHAN LESTER HOUSE, Long Cove & Vinegar Hill Rds., Gales Ferry, CT 06335. Mailing Address: 153 Vinegar Hill Rd., Ledyard, CT 06335. Tel.: 860-464-8540.

Institution Type/Description: Historic House: housed in an 18th century farmhouse.

Collections: period furnishings; personal artifacts; photographs; outbuildings.

Facilities: nature trails.

Activities: hiking.

Hours & Admission Prices: House: Memorial Day to Labor Day Tues. & Thurs. 2-4, Sat.-Sun. 1-4:30; other times by appointment. Grounds: daily. No charge; donations accepted.

Gaylordsville

GAYLORDSVILLE HISTORICAL SOCIETY - THE LITTLE RED SCHOOLHOUSE, 56 Gaylord Rd., Gaylordsville, CT 06755. Mailing Address: P.O. Box 25, Gaylordsville, CT 06755. Tel.: 860-350-0300.

Web Site: www.gaylordsville.org

Founded: 1974.

Key Personnel: Pres. (V), Richard Kosier.

Governing Authority: Tax-exempt.

Institution Type/Description: Historical Society Museum: housed in a one room schoolhouse; built in 1740.

Collections: local history; period furnishings; personal artifacts; blacksmithing tools; photographs. Historic Building: 1870 Brown's Forge; 1740 Red Schoolhouse.

Hours & Admission Prices: Red Schoolhouse: July-Sept. Sun. 2-5. Brown's Forge: Aug.-Sept. Sun. 2-4. No charge; donations accepted.

Attendance: 350 (estimated)

Glastonbury

CONNECTICUT AUDUBON SOCIETY CENTER AT GLASTONBURY, 1361 Main St., Glastonbury, CT 06033-3105. Tel.: 860-633-8402. Fax: 860-657-4228.

E-mail: cbartholomew@ctaudubon.org

Web Site: www.ctaudubon.org

Formerly: Holland Brook Nature Center

Founded: 1982.

Key Personnel: Dir., Cindy Bartholomew; Pres., Robert Martinez.

Personnel Profile: Full-Time Paid 1; Part-Time Paid 2; Part-Time Volunteers 150.

Governing Authority: Parent Institution: Connecticut Audubon Society. Tax-exempt.

Institution Type/Description: Nature Center.

Collections: natural history; hands-on exhibits; wildlife mounts; live animals; Connecticut River ecosystem; birds & their habitats.

Activities: educational programs; classes; workshops; nature hikes; hands-on exhibits; birthday parties; special events.

Publications: State of the Birds.

Hours & Admission Prices: Tues.-Sat. 10-5, Sun. 1-4. No charge; donations accepted. &

Attendance: 6,000 (estimated)

Membership: Student/Teacher $25; Individual $40; Family $55; Supporting $75; Contributing $100; Donor $200; Patron $500; Benefactor $1,000.

HISTORICAL SOCIETY OF GLASTONBURY, 1944 Main St., Glastonbury, CT 06033-2901. Mailing Address: P.O. Box 46, Glastonbury, CT 06033-0046. Tel.: 860-633-6890. Facebook: Historical Society of Glastonbury.

E-mail: hsglastonbury@sbcglobal.net

Web Site: www.hsgct.org

Founded: 1936.

Congressional District: 1

Key Personnel: Exec. Dir., James Bennett; Pres., Jane Fox; Treas., David Motycka; Cur., Linda Scarduzio; Librarian, Phylis Reed.

Personnel Profile: Full-Time Paid 1; Part-Time Paid 2; Part-Time Volunteers 40.

Volunteer Hours: 1,500

Operating Expenses: 441,541

Operating Income: 271,770

Governing Authority: nonprofit organization. Branch Museum: Welles-Shipman-Ward House, 972 Main St., S., Glastonbury, CT 06073. Tax-exempt: 501(c)(3).

Institution Type/Description: Historical Society Museum: housed in c.1840 Old Town Hall.

Collections: Glastonbury history from Native American to present; historical documents; textiles; folk art; Colonial artifacts; Civil War; WW I & WW II artifacts; early industries; maps; ship building.

Research Fields: Indians; historic site surveys; local businesses.

Facilities: 200-vol. library pertaining to local history; genealogy, Native Americans & New England folkways; herb garden. Books, note cards & other related items for sale.

Activities: guided tours; lectures; organized educational programs; docent program; loan & temporary exhibitions.

Publications: quarterly newsletter; annual journal, The Publick Post; books, The Glastonbury Express; Glastonbury: From Settlement to Suburb; The Letter Kills But the Spirit Gives Life; Moments in History; Glastonbury (2012).

Hours & Admission Prices: Welles-Shipman-Ward: April-Nov. 3rd Sun. each month and 2nd & 4th Tues. each month 1-4. Adults $3. Museum on the Green: Mon.-Tues. & Thurs. 9-4, 3rd Sun. each month 1-4; other times by appointment. No charge; donations accepted. Closed national holidays.

Attendance: 3,800 (estimated)

Membership: Senior 62 & over $25; Individual $30; Family $35; Contributing $75; Sustaining $100; Patron $200; Life $500; Life Couple $750.

Goshen

ACTION WILDLIFE FOUNDATION, 337 Torrington Rd., Rte. 4, Goshen, CT 06756-2031. Tel.: 860-482-4465; 491-9191. Fax: 860-482-8337.

E-mail: info@actionwildlife.org

Web Site: www.actionwildlife.org

Key Personnel: Museum Dir., Jim Mazzarelli; Animal Mgr., Sue Tracy; Sec., Julie Mazzarelli

Institution Type/Description: Foundation.

Collections: mounted wildlife scenes; live exotic animals.

Hours & Admission Prices: Spring: Sat.-Sun. 10:30-4; Summer: Tues.-Sun. 10:30-4; Sept.-Oct. Thurs.-Fri. 11-3, Sat.-Sun. 10:30-4. Adults $10, children under 12 $8. &

GOSHEN HISTORICAL SOCIETY, 21 Old Middle St., Goshen, CT 06756. Mailing Address: P.O. Box 457, Goshen, CT 06756-0457. Tel.: 860-491-9610 & 3129.

E-mail: jvnkuq@juno.com

Web Site: goshenhistoricalct.org

Founded: 1955.

Congressional District: 5

Key Personnel: Pres. (V), Henrietta C. Horvay.

Personnel Profile: Full-Time Volunteers 5; Part-Time Volunteers 4.

Governing Authority: private. Tax-exempt.

Institution Type/Description: History Museum.

Collections: pewter; rocks & shells; history; Indian artifacts; medical; natural history; textiles; dolls; farm tools; historic flags.

Facilities: 300-vol. library of historical subjects available for research use by appointment.

Activities: lectures; formally organized education programs for children; permanent exhibitions.

Publications: annual letter; event programs.

Hours & Admission Prices: April-Oct. Tues. 10-12; other times by appointment only. No charge; donations accepted. &

Attendance: 300 (estimated)

Membership: Sustaining $15; Life $500.

Granby

SALMON BROOK HISTORICAL SOCIETY, INC., 208 Salmon Brook St., Granby, CT 06035. Mailing Address: P.O. Box 840, Granby, CT 06035-0840. Tel.: 860-653-9713.

Web Site: www.salmonbrookhistorical.org

Founded: 1959.

Congressional District: 6

Key Personnel: Pres. (V), Ken Kuhl; Cur., Archivist & Genealogist, Carol Laun.

Personnel Profile: Part-Time Volunteers 20.

Governing Authority: society. Tax-exempt: 501(c)(3).

Institution Type/Description: History Museum: housed in c.1732 Abijah Rowe House.

Collections: household & farm furniture; history; local genealogy; manuscript collections; personal & farming artifacts; documents. Historic Buildings: 1790 Weed-Enders House; 1870 Cooley Rd. School; 1900 Tobacco Barn.

Research Fields: local, area & state history.

Facilities: 1,000-vol. library of books & documents pertaining to history & genealogy of local interest available for research on premises; collection of newspapers dating back to pre-Civil War period; Reference & Educational Center open to public.

Activities: permanent & special exhibits; tours of house; lectures & classes.

Publications: books, The Heritage of Granby, 1786-1967; Collections of the Salmon Brook Historical Society Vol. I, II, III, IV & V; Granby, Connecticut: A Brief History 1786-1986, 1986; Selectmen's Records 1786-1850, 1986; Bicentennial Quilt Booklet, 1986; Tempest in a Small Town, 1996; The Holcomb Collection, 1998; Burials in the Granby Center Cemetery, 1998. Beneath These Stones, 2003; Burials in the Granby Center Cemetery, Revised 2004; Wintonbury Church Records, 2004.

Hours & Admission Prices: June-Sept. Sun. 2-4; other times by appointment. Adults $5; members no charge. Closed Independence Day; Labor Day weekend.

Attendance: 600 (estimated)

Membership: Student $3; Single $15; Family $20; Sustaining $30; Life $300.

Greenwich

AUDUBON GREENWICH, 613 Riversville Rd., Greenwich, CT 06831-2624. Tel.: 203-869-5272. Fax: 203-869-4437.

E-mail: greenwich_center@audubon.org

Web Site: greenwich.audubon.org

Founded: 1942.

Key Personnel: Exec. Dir., Tom Baptist; Center Dir., Karen A. Dixon; Environmental Education Specialist, Edward S. Gilman; Environmental Education Specialist, Lindsay DeVito; Environmental Education Specialist, James Flynn; Education Mgr., Jeff Cordulack; Nature Shop Mgr., Brian O'Toole; Office Mgr., Gigi Lombardi.

Personnel Profile: Full-Time Paid 9; Part-Time Volunteers 30; Interns 1.

Governing Authority: society; nonprofit. Affiliated with National Audubon Society, 225 Varick St., New York, NY 10014. Tel.: 212-979-3000. Tax-exempt: 501(c)(3).

Institution Type/Description: Nature Center.

Collections: natural history; wildlife & their habitats; hands-on exhibits.

Research Fields: bird banding.

Facilities: 686 acres; nature trails. Museum-related items for sale.

Activities: guided tours; lectures; films; formally organized education programs for children & adults; permanent exhibitions; children's learning center; bird observation; summer & winter camp programs for K-10.

Publications: Audubon Greenwich in Flight.

Hours & Admission Prices: Daily 10-5. Adults $3, children & senior citizens over 62 $1.50; discount to AAM members; members no charge. &

Attendance: 30,000 (estimated)

Membership: National Audubon Society $35.

BENDHEIM GALLERY, 299 Greenwich Ave., Greenwich, CT 06830-6504. Tel.: 203-862-6750. Fax: 203-862-6753.

E-mail: info@greenwicharts.org

Web Site: www.greenwicharts.org/index.asp

Founded: 1973.

Congressional District: 4

Key Personnel: Dir., Paul Master-Karnik; Pres. (V), Barbara Collier.

Personnel Profile: Full-Time Paid 3; Part-Time Paid 5; Part-Time Volunteers 200; Interns 1.

Governing Authority: Tax-exempt.

Institution Type/Description: Art Gallery.

Collections: works by accomplished & emerging artists.

Research Fields: contemporary art.

Facilities: 1,000 sq. ft. exhibit space.

Activities: temporary exhibitions; programs; classes; special events.

Publications: newsletters; brochures.

Hours & Admission Prices: Mon.-Fri. 10-5, Sat.-Sun. 12-5. No charge. &

Attendance: 50,000 (estimated)

Membership: Basic $35; Dual $100; Donor $250; Sponsor $500.

✳ **BRUCE MUSEUM, (M),** 1 Museum Dr., Greenwich, CT 06830-7157. Tel.: 203-869-0376. Fax: 203-869-0963.

E-mail: info@brucemuseum.org

Web Site: www.brucemuseum.org

Formerly: Bruce Museum of Arts and Science

Founded: 1912.

Congressional District: 4

Key Personnel: C.E.O. & Exec. Dir., Peter C. Sutton, Ph.D.; Co Chm. (V), Leora Levy; Co Chm. (V), Patricia Chadwick; Dir. Mktg. & Communications, Troy Ellen Dixon; Dir. Education, Robin Garr; Cur. Science, Gina C. Gould, Ph.D.; Assoc. Dir. Devel., Whitney Rosenberg; Dir. Exhibits, Anne von Stuelpnagel; Dir. Finance & C.F.O., Gregory Hollop; Registrar, Jack Coyle; Dir. Devel., Liz Wooster; Museum Shop Mgr., Justine Matteis; Mgr. Volunteers, Mary Ann Lendenmann.

Personnel Profile: Full-Time Paid 33; Part-Time Paid 12; Part-Time Volunteers 475; Interns 6.

Governing Authority: private. Parent Institution: Bruce Museum, Inc. Subsidiary Institution: Seaside Center. Tax-exempt: 501(c)(3).

Institution Type/Description: Arts, Science & History.

Collections: American painting; sculpture & decorative art; Native American art & artifacts; photography; costumes & textiles; North American mounted mammals & birds; fossils; minerals; shells.

Research Fields: art history; geology.

Facilities: 150-seat lecture gallery; educational workshop; Seaside Museum annex. Museum-related gifts for sale.

Activities: lectures; films; docent program; permanent, temporary & traveling exhibitions; college student interns; school & adult tours; family events & activities; afterschool programs.

Publications: bimonthly calendar of events; exhibition catalogs; e-newsletter; online teacher resources.

Hours & Admission Prices: Tues.-Sat. 10-5, Sun. 1-5. Adults $7, students & senior citizens $6; discounts to AAM & ASTC members; children under 5, members & Tues. no charge. Closed New Year's Day; Easter; Memorial Day; Independence Day; Thanksgiving; Christmas. &

Attendance: 71,282 (accurate)

Membership: Student & Educator $35; Senior $40; Individual $60; Senior Couple $60; Young Friend $150; Patron $250; Benefactor $750. Robert Bruce Circle: Bronze $2,000; Silver $3,000; Gold $6,000; Platinum $10,000; Director's Diamond $25,000.

PUTNAM COTTAGE, 243 E. Putnam Ave., Greenwich, CT 06830-4808. Tel.: 203-869-9697.

E-mail: info@putnamcottage.org

Web Site: www.putnamcottage.org

Founded: 1900.

Key Personnel: Pres. (V), Mrs. Wedgbury Jayes; Museum Shop Mgr., Mrs. Sally Bretschger.

Personnel Profile: Full-Time Volunteers 20; Part-Time Volunteers 30.

Governing Authority: organization. Parent Institution: Israel Putnam House Association. Tax-exempt.

Institution Type/Description: History Museum: housed in c.1700 Knapps Tavern.

Collections: early Americana; period Greenwich artifacts; archives; 18th century furnishings; General Putnam memorabilia; c.1690 gunstock posts; fieldstone fireplaces.

Facilities: genealogical library; herb garden.

Activities: guided tours. Museum Sponsors: American History essay contests; Historical Battle Reenactment in February.

Hours & Admission Prices: April-Dec. Sun. 1-4; other times by appointment. Adults $6; children under 12 no charge. Closed holidays.

Attendance: 300 (estimated)

Groton

ALEXEY VON SCHLIPPE GALLERY OF ART, 1084 Shennecossett Rd., Groton, CT 06340. Tel.: 860-405-9052.

Web Site: www.averypointarts.uconn.edu

Key Personnel: Dir. & Cur., Julia Pavone

Institution Type/Description: Art Gallery.

Collections: paintings; sculpture.

Activities: special events.
Hours & Admission Prices: Wed.-Sun. 12-4.

AVERY-COPP MUSEUM, 154 Thames St., Groton, CT 06340-3631. Mailing Address: P.O. Box 7011, Groton, CT 06340-7011. Tel.: 860-445-1637 & 449-1596. Fax: 860-405-1154.
E-mail: averycopphouse@sbcglobal.net
Web Site: averycopphouse.com/index.html
Institution Type/Description: Historic House Museum: built c.1800.
Collections: local history & culture; period furnishings; personal artifacts; photographs.
Facilities: Museum-related items for sale.
Activities: special events.
Publications: newsletter.
Hours & Admission Prices: By appointment. Adults $4, youth 12-18 & students $2; children under 12 no charge.

EBENEZER AVERY HOUSE, Fort Griswold Battlefield State Park, Monument St. & Park Ave., Groton, CT 06340. Mailing Address: Avery Memorial Association, P.O. Box 7245, Groton, CT 06340-7245. Tel.: 860-445-1729 & 446-9257. Facebook: Ebenezer Avery House.
E-mail: sglantiere@aol.com
Web Site: averymemorialassociation.com
Congressional District: 2
Key Personnel: Museum Shop Mgr., Stephanie Lantiere; Cur., Elizabeth (Liz) Williams.
Personnel Profile: Part-Time Paid 1; Part-Time Volunteers 20.
Governing Authority: Tax-exempt.
Institution Type/Description: Historic House Museum: built in 1750.
Collections: local history; period furnishings; paintings.
Activities: Museum Sponsors: Annual Reunion & Meeting.
Publications: triannual newsletter.
Hours & Admission Prices: Memorial Day to Labor Day Fri.-Sun. 12-4. No charge; donations accepted.
Attendance: 700 (estimated)
Membership: Individual $25; Household $35; Lifetime $500.

SUBMARINE FORCE MUSEUM AND HISTORIC SHIP NAU-TILUS, 1 Crystal Lake Rd., Groton, CT 06340-2464. Mailing Address: P.O. Box 571 NAVSUBASE, New London, Groton, CT 06349-5571. Tel.: 860-694-3174; 800-343-0079. Fax: 860-694-4150. Facebook: Submarine Force Museum and USS Nautilus.
Web Site: www.ussnautilus.org
Founded: 1954.
Key Personnel: Dir., Lt. Commander Benjamin Amdur; Mng. Dir., Stephen Finnigan; Supervisory Cur., Stephen Finnigan; Educational Specialist, Liz Murphy; Archives Technician, Wendy Gulley; Museum Aide, Martha Barber; Museum Specialist, Thaddeus Wakefield; Tour Coord. & Sec., Linda Williams.
Personnel Profile: Full-Time Paid 6; Part-Time Volunteers 1.
Governing Authority: federal. Parent Institution: Dept. of the Navy. Tax-exempt.
Institution Type/Description: Naval Museum: housed on USS Nautilus submarine & adjacent museum.
Collections: historic submarine: USS Nautilus, first nuclear-powered submarine; objects, documents & photographs related to the history of the U.S. Submarine Force; working periscopes; mini-theaters; ship models; torpedoes.
Research Fields: U.S. submarines.
Facilities: 5,000-vol. submarine research library.
Activities: self-guided tours of museum and submarine Nautilus.
Hours & Admission Prices: Museum: May-Oct. Wed.-Mon. 9-5; Nov.-April Wed.-Mon. 9-4. Library: Mon.-Fri. by appointment only. No charge. Closed New Year's Day; Thanksgiving; Christmas. ♿
Attendance: 152,192 (accurate)

Guilford

HENRY WHITFIELD STATE HISTORICAL MUSEUM, 248 Old Whitfield St., Guilford, CT 06437-3459. Tel.: 203-453-2457. Fax: 203-453-7544. Facebook: Henry Whitfield State Museum.
E-mail: whitfieldmuseum@ct.gov
Web Site: www.cultureandtourism.org
Founded: 1899.
Congressional District: 3
Key Personnel: Dir. Museum, Karin Peterson; Cur., Michael A. McBride; State Historic Preservation Officer, Daniel Forrest; Guide, Chris Collins.

Personnel Profile: Full-Time Paid 1; Part-Time Paid 1; Part-Time Volunteers 10.
Governing Authority: state. Parent Institution: State Historic Preservation Office, Dept. of Economic & Community Development, One Constitution Plaza, 2nd Fl., Hartford, CT 06103. Tax-exempt: 170(b)(1)(a).
Institution Type/Description: National Historic Landmark: 1639 oldest building in Connecticut, oldest stone house in New England; first state-owned museum in Connecticut (1899), colonial revival restorations by Norman Isham in 1902-1904 & J. Frederick Kelly in the 1930s.
Collections: 17th-19th century European & American furniture; textiles & household furnishings; firearms; manuscripts.
Major Exhibits: Guilford Goes Global, 5/14-12/14; The Old Stone House, 5/14-12/14.
Research Fields: all aspects of the 17th century & the Whitfield family; Guilford & Connecticut history.
Facilities: 600-vol. library of local historical materials available on application; local & state historical materials; visitor & tourist info center. Museum-related items for sale.
Activities: self-guided tours; programs; group tours by appointment; changing exhibits.
Publications: book, Henry Whitfield House.
Hours & Admission Prices: See website for current hours. Adults $8, seniors $6, youth $5; children 5 & under no charge. Closed major holidays. ♿
Attendance: 5,000 (accurate)

THE HYLAND HOUSE, 84 Boston St., Guilford, CT 06437-2874. Mailing Address: P.O. Box 229, Guilford, CT 06437-0229. Tel.: 203-453-9477 & 3850.
E-mail: jack@carles.us
Web Site: www.hylandhouse.com
Founded: 1918.
Key Personnel: Pres. (V) Bd. Dir. Dorothy Whitfield Historic Society, John Carles; Treas., Paul Shuell; Cur., Pamela Besse; Membership, Sally Leighton; Education, Teresa Buchannon.
Personnel Profile: Part-Time Paid 4; Part-Time Volunteers 18.
Governing Authority: private; nonprofit organization. Tax-exempt.
Institution Type/Description: General Museum.
Collections: 17th to early 18th century furniture; decorative arts; early colonial life & architecture; hearth tools; cooking implements.
Research Fields: historic structures; Dorothy Whitfield Historic Society & the colonial revival; Ebenezer Parmelee & Connecticut clockmaking.
Activities: concerts; formal education programs; guided tours; lectures; temporary exhibitions; hands-on workshops for Guilford 4th graders. Museum Sponsors: Annual Antiques Show in March.
Publications: newsletter, Hyland House Times.
Hours & Admission Prices: April 15 to May Tues.-Sat. 12-4:30; June-Columbus Day Tues.-Sun. 12-4:30. No charge; donations accepted.
Attendance: 1,835 (accurate)
Membership: Dorothy Whitfield Historic Society: Single $15; Family $20; Contributing $30; Sustaining $100.

THE THOMAS GRISWOLD HOUSE AND MUSEUM, 171 Boston St., Guilford, CT 06437. Mailing Address: P.O. Box 363, Guilford, CT 06437-0363. Tel.: 203-453-3176.
E-mail: info@guilfordkeepingsociety.com
Web Site: guilfordkeepingsociety.com
Founded: 1947.
Congressional District: 3
Key Personnel: Pres., Thomas Black; Dir., Patricia Lovelace; Vice Pres., Winnie Seibert; Treas., Thomas Williams; Programs, Robert Donahue.
Personnel Profile: Part-Time Paid 6; Part-Time Volunteers 50.
Governing Authority: private; nonprofit organization. Parent Institution: Historical & Preservations Societies. Branch Museum: The Medad Stone Tavern and Museum, 197 Three Mile Course, Guilford, CT 06437. Tel.: 203-453-2263. Tax-exempt: 501(c)(3).
Institution Type/Description: Historic House: built in 1774.
Collections: 19th & 20th century glass plates & photographs; documents; deeds; genealogical data from 1639; furniture & household items from the 18th & 19th century.
Research Fields: genealogical research on Guilford ancestors.
Facilities: library of books on Guilford historical data on families, deeds; early photos & town documents available to the public. Museum-related items for sale.
Activities: concerts; formal education for children; guided tours; temporary exhibitions. Annual Events: Antiques Festival; Memorial Day Picnic; Wine Tasting & Silent Auction.
Publications: quarterly newsletter, Society News.
Hours & Admission Prices: June-Sept. Tues.-Sun. 11-4; other times for

pre-arranged groups. Adults & students $2; discounts to NEMA members & groups; children & members no charge.
Attendance: 1,500 (estimated)
Membership: Individual $20; Family $40; Contributing $75; Sustaining $150; Life $500.

Haddam

HADDAM HISTORICAL SOCIETY - THANKFUL ARNOLD HOUSE, 14 Hayden Hill Rd., Haddam, CT 06438. Mailing Address: P.O. Box 97, Haddam, CT 06438-0097. Tel.: 860-345-2400.
E-mail: contact@haddamhistory.org
Web Site: www.haddamhistory.org
Founded: 1955.
Congressional District: 2
Key Personnel: Dir., Elizabeth Malloy; Pres. (V), Sue Costa; Pres. (V), Terry Smith.
Personnel Profile: Part-Time Paid 2; Part-Time Volunteers 50.
Institution Type/Description: Historic House Museum: built between 1794 and 1810.
Collections: local history & culture; Arnold family history; period furnishings; personal artifacts; photographs.
Hours & Admission Prices: June-Oct. Wed. 9-3, Thurs. 2-8, Fri. 12-3, Sun. 1-4; Sept.-May Wed. 9-3, Thurs. 2-8, Fri. 12-3. No charge; donations accepted.
Attendance: 1,500 (estimated)

Hamden

ELI WHITNEY MUSEUM, INC., 915 Whitney Ave., Hamden, CT 06517-4036. Tel.: 203-777-1833. Fax: 203-777-1229.
E-mail: dc@eliwhitney.org
Web Site: www.eliwhitney.org
Founded: 1976.
Congressional District: 3
Key Personnel: Pres., Stephen Latham; Dir., William Brown; Treas., Ray Fair; Dir. Design, Sally Hill; Mgr., Dana Clough.
Personnel Profile: Full-Time Paid 8; Part-Time Paid 45; Interns 2.
Governing Authority: nonprofit organization. Tax-exempt: 501(c)(3).
Institution Type/Description: Experimental Learning Workshop & Museum.
Collections: period arms; photographs; maps; history of manufacturing & the effect of technology on American society; Whitney armory firearms; A.C. Gilbert toys & artifacts. Historic Building: 1816 barn.
Major Exhibits: Erector Set at 100, 11/13-1/14.
Research Fields: history of education in non scholastic subjects.
Activities: tours; lectures; workshops; organized education programs for children; experimental learning; wood working; temporary & permanent exhibitions.
Publications: brochure; catalog, Windows on the Works: Industry on the Whitney site: 1798-1976.
Hours & Admission Prices: Memorial Day to Labor Day daily 11-4; Sept.-May Wed.-Fri. & Sun. 12-5, Sat. 10-3. No charge; donations accepted. Closed Easter; Christmas. &
Attendance: 48,000 (estimated)
Membership: Family $45.

HAMDEN HISTORICAL SOCIETY, INC., 105 Mt. Carmel Ave., Hamden, CT 06518. Mailing Address: P.O. Box 5512, Hamden, CT 06518-0512. Tel.: 203-288-0017.
E-mail: hamdenhistoricalsociety@wordpress.com
Founded: 1928.
Congressional District: 3
Key Personnel: Museum Shop Mgr., Lois Casey.
Governing Authority: society. Tax-exempt: 501(c)(3).
Institution Type/Description: History Museum: housed in 1792 Jonathan Dickerman House.
Collections: herbarium; late 18th century period furnishings & artifacts; 1800 Dickerman Talmade Cider Mill/Barn.
Research Fields: local history, structure & homes in Hamden.
Facilities: collection of history books; land deeds; photographs available for use by the public.
Activities: guided tours; lectures; films; permanent & temporary exhibitions.
Publications: Hamden: Our Architectural Heritage; Images of America: Hamden.
Hours & Admission Prices: July-Aug. Sat.-Sun. 1-4. No charge; donations accepted.
Attendance: 400 (estimated)
Membership: Individual $15; Family $25; Life $250.

Hampton

GOODWIN FOREST CONSERVATION EDUCATION CENTER, 23 Potter Rd., Hampton, CT 06247-3616. Tel.: 860-455-9534. Fax: 860-455-9857.
E-mail: sbroderick@ctwoodlands.org
Web Site: www.ct.gov/dep/goodwin
Founded: 1914.
Congressional District: 2
Key Personnel: Dir. Forest & Programs, Steve Broderick
Institution Type/Description: Conservation Center.
Collections: wildlife & forest conservation.
Activities: educational programs; self-guided trails; workshops.
Publications: Friends of Goodwin Forest quarterly e-newsletter.
Hours & Admission Prices: Call for hours. No charge; donations accepted.
Attendance: 9,200 (estimated)

TRAIL WOOD: THE EDWIN WAY TEALE MEMORIAL SANCTUARY, 93 Kenyon Rd., Hampton, CT 06274. Tel.: 860-928-4948.
Institution Type/Description: Historic House: housed in the former home of Pulitzer Prize-winning author Edwin Way Teale & his wife.
Collections: Edwin W. Teale life & career; personal artifacts; memorabilia; photographs.
Facilities: 168-acre sanctuary.
Hours & Admission Prices: House: by appointment. Grounds: daily dawn to dusk. No charge.

Hartford

BUTLER-MCCOOK HOUSE AND GARDEN, 396 Main St., Hartford, CT 06103-3001. Tel.: 860-522-1806. Fax: 860-249-4907.
E-mail: butler.mccook@ctlandmarks.org
Web Site: www.ctlandmarks.org
Founded: 1782.
Key Personnel: Dir., Sheryl Hack.
Governing Authority: private; nonprofit organization. Parent Institutions: CT Landmarks, Hartford, CT. Tax-exempt: 501(c)(3).
Institution Type/Description: Historic House.
Collections: local history & culture; family heritage; personal artifacts; period furnishings; photographs; paintings; early toys; Samurai armor; Victorian ornamental garden.
Research Fields: paranormal investigations.
Facilities: Victorian garden.
Activities: special events; educational programs; guided tours; workshops; lectures; concerts.
Hours & Admission Prices: May-Oct. call for hours; groups by appointment. Adults $7, senior citizens, students & teachers $6, children 6-18 $4; discounts to groups of 15 or more; members & children under 6 no charge.

CRRA TRASH MUSEUM, 211 Murphy Rd., Hartford, CT 06114-2100. Tel.: 860-757-7765. Fax: 860-278-8471.
Web Site: www.crra.org/pages/Trash_Museum.htm
Institution Type/Description: Trash Museum.
Collections: waste management solutions; recycling; early disposal methods; mural.
Facilities: 6,500 sq. ft. exhibit space. Museum-related items for sale.
Activities: view the Container Processing Facility.
Hours & Admission Prices: July-Aug. Tues. 10-2, Wed.-Fri. 10-4; Sept.-June Wed.-Fri. 12-4. No charge. Closed New Year's Day; Good Friday; Independence Day; Thanksgiving & day after; Christmas Day & day after. &

CHARTER OAK CULTURAL CENTER, 21 Charter Oak Ave., Hartford, CT 06106-1801. Tel.: 860-249-1207. Fax: 860-524-8014.
E-mail: laura.rozza@charteroakcenter.org
Web Site: www.charteroakcenter.org
Founded: 1979.
Institution Type/Description: Cultural Art Center.
Collections: paintings; sculpture; photographs.
Activities: special events; educational programs.
Hours & Admission Prices: July-Aug. Tues.-Fri. 10-4; Sept.-June Mon.-Fri. 10-4; other times be appointment. No charge; donations accepted. &

CONNECTICUT HISTORICAL SOCIETY, 1 Elizabeth St., Hartford, CT 06105-2292. Tel.: 860-236-5621. Fax: 860-236-2664.
E-mail: ask_us@chs.org
Web Site: www.chs.org
Founded: 1825.

Congressional District: 1
Key Personnel: Exec. Dir., Kate Steinway; Pres., James C. Williams; Florence S. Marcy Crofut Dir. Collections, Dr. Susan P. Schoelwer; Dir. Public Programs, Andrea Rapacz; Dir. Collections Access, Richard Malley; Dir. Admin., Kevin Hughes; Dir. Education, Rebecca Furer.
Personnel Profile: Full-Time Paid 28; Part-Time Paid 8; Part-Time Volunteers 300; Interns 10.
Governing Authority: Tax-exempt: 501(c)(3).
Institution Type/Description: Historical Society Museum & Library.
Collections: 33,000 artifacts including: 17th-18th century Connecticut furniture & decorative arts; painting & miniatures; 18th to early 19th-century American tavern signs; sculpture; clocks & clock tools; silver, gold & jewelry; glass & ceramics; metalware including 20th-century chromium household artifacts; industrial artifacts; costumes; textiles; 230,000 graphics including drawings, watercolors & pastels; lithographs by the Kelloggs of Hartford; 20th-century prints; 1,200 maps; 4,000 broadsides; 17th-20th century manuscripts.
Research Fields: Connecticut & New England history; genealogy; material culture; decorative arts; graphic arts; industry & technology; women's history; domestic life; Civil War.
Facilities: research library focusing on Connecticut & New England history & genealogy; print & photograph study room; hands-on activity room; meeting rooms; auditorium.
Activities: interactive exhibitions; educational outreach programs & curriculum materials; family programs; lectures & workshops; guided tours; gallery talks.
Publications: quarterly newsletter; exhibition catalogs.
Hours & Admission Prices: Exhibition Galleries: Tues.-Sat. 12-5; groups by appointment. Library Reading Room: Tues.-Sat. 10-5. Photograph Study Room: Mon.-Fri. 1-5 by appointment. Adults $6, senior citizens & students $3; discount to AAM members; children 6 & under, members & first Sat. of each month 9-1 no charge. Closed New Year's Day; Easter; Independence Day; Thanksgiving Day; Christmas. ♿
Attendance: 60,000 (estimated)
Membership: Individual $30; Family & Household $45; Sustaining $150; Benefactor $300; Sponsor $500; Presidents' Circle $1,000.

CONNECTICUT LANDMARKS, 255 Main St., 4th Fl., Hartford, CT 06106-1821. Tel.: 860-247-8996. Fax: 860-249-4907.
E-mail: info@ctlandmarks.org
Web Site: www.ctlandmarks.org
Formerly: The Antiquarian and Landmarks Society, Inc.
Founded: 1936.
Congressional District: 1
Key Personnel: Exec. Dir., Sheryl N. Hack; Exec. Sec. & Office Mgr., Tina Daly; Mktg. & Devel. Assoc., Jamie-Lynn Fontaine; Cur., Beverly Lucas; Property Mgr., Joseph Pukas; Dir. Donor Devel., April Paterno.
Personnel Profile: Full-Time Paid 5; Part-Time Paid 1; Part-Time Volunteers 10.
Governing Authority: society. Parent Institution: The Antiquarian & Landmarks Society, Inc. Branch Museums: 1678 Joshua Hempsted House, New London, CT; 1854 Isham-Terry House, Hartford, CT; 1720 Buttolph-Williams House, Wethersfield, CT; 1761 & 1794 Phelps-Hatheway House, Suffield, CT; 1776 Nathan Hale Homestead, Coventry, CT; 1816 Amasa Day House, Moodus, CT; 1759 Nathaniel Hempsted House, New London, CT; 1782 Butler-McCook House & Garden, Hartford, CT; Bellamy-Ferriday House & Garden, Bethlehem, CT. Tax-exempt: 501(c)(3).
Institution Type/Description: Historic Houses & Historic Site concerned with Connecticut's Social History.
Collections: decorative arts.
Research Fields: architecture & furnishings; state history.
Activities: guided tours; lectures; workshops; living history programs; bus tours; formally organized education programs for students; docent program; permanent exhibitions.
Publications: quarterly newsletter.
Hours & Admission Prices: Call or see website for hours. Adults $7, students, teachers & seniors $6, children 6-18 $4; discounts to groups of 10 or more, New England Museum Assoc. & AAM members; members & children under 6 no charge.
Attendance: 20,000 (estimated)
Membership: Individual $40; Family $55; Contributing $125; Preservation $250; Landmark $500; George Dudley Seymour Circle $1,000.

CONNECTICUT SCIENCE CENTER, (M), 250 Columbus Blvd., Hartford, CT 06103-2802. Tel.: 860-SCIENCE (724-3623).
Web Site: www.ctsciencecenter.org
Founded: 2008.
Key Personnel: C.E.O., Matt Fleury; Dir., Tracy Shirer.
Governing Authority: Tax-exempt.

Institution Type/Description: Science Center.
Collections: hands-on science exhibits.
Major Exhibits: Lost Egypt (T), 2/14-5/4/14; Mindbender Mansion (T), 5/24/14-8/14; Mr. Potato Head (T), 5/24/14-8/14; Grossology (T), 9/27/14-10/14.
Research Fields: robotics.
Facilities: Museum-related items for sale.
Activities: educational programs; Saturday science sessions; camps; professional development; Science in Motion; Teen Robotics Program.
Publications: digital newsletter, Science Forward.
Hours & Admission Prices: July-Aug. daily 10-5; Sept.-June Tues.-Sun. 10-5. Adults $19, senior citizens 65 & over $16.50, youth 3-17 $14; members and children 3 & under no charge. Closed Thanksgiving; Christmas. ♿
Attendance: 301,050 (accurate)
Membership: Individual $75; Two Person $105; Three Person $135; Four Person $165; Five Person $195; Six Person $225; Up to Ten People $285.

CONNECTICUT SUPREME COURT, External Affairs Div., Connecticut Judicial Branch, 231 Capitol Ave., Hartford, CT 06160. Tel.: 860-757-2270. Fax: 860-757-2215.
E-mail: James.Senich@jud.ct.gov
Web Site: www.jud.state.ct.us/external/news/SupCtTour.html
Key Personnel: Mgr. Communications, James Senich
Institution Type/Description: Historic Building: built from 1908-1910.
Collections: courthouse history; portraits; murals; books including judicial history, state registers, & manuals.
Activities: group tours; oral arguments; presentations.
Hours & Admission Prices: Mon.-Fri. by appointment.

CONNECTICUT'S OLD STATE HOUSE, 800 Main St., Hartford, CT 06103-2301. Tel.: 860-522-6766. Fax: 860-522-2812.
E-mail: ctoldstatehouse@cga.ct.gov
Web Site: www.ctoldstatehouse.com
Founded: 1975.
Congressional District: 1
Key Personnel: Dir. Old State House, Sally Whipple; Coord. Public Programs & History Day, Rebecca Taber-Conover.
Personnel Profile: Full-Time Paid 10; Part-Time Paid 10.
Governing Authority: state legislature. Parent Institution: Connecticut General Assembly, Joint Committee on Legislative Management. Tax-exempt.
Institution Type/Description: Historic Federal Building: built in 1796, designed by Charles Bulfinch; site of first meeting house, George Washington's meeting with Comte de Rochambeau and the French troops; first Amistad Trial.
Collections: Connecticut history from 1636 to present; portrait of George Washington by Gilbert Stuart; 18th-20th century furnishings & paintings; Joseph Steward's Museum of Curiosities.
Research Fields: Old State House history; Hartford history; Connecticut state government history; civics.
Facilities: 6,000 sq. ft. exhibit space; educational center; rental facilities; theater. Museum-related items for sale.
Activities: guided tours; formally organized educational programs; historical reenactments; permanent & temporary exhibitions; hands-on activities; media experiences.
Hours & Admission Prices: July 4 to Columbus Day Tues.-Sat. 10-5; Oct.-July 3 Mon.-Fri. 10-5. Guided Tours: adults $6, senior citizens, students & children 6-17 $3; children 5 & under no charge. Closed major holidays. ♿
Attendance: 32,368 (accurate)

GOVERNOR'S RESIDENCE, 990 Prospect Ave., Hartford, CT 06105-1102. Mailing Address: Office of the Governor, State of Connecticut, 990 Prospect Ave., Hartford, CT 06105-1102. Tel.: 860-566-4840.
Institution Type/Description: Historic Building: built in 1909.
Collections: local history & government; period furnishings; personal artifacts; portraits; photographs.
Activities: guided tours. Annual Event: Holiday Open House.
Hours & Admission Prices: Guided Tours: by appointment.

✱ **HARRIET BEECHER STOWE CENTER, (M),** 77 Forest St., Hartford, CT 06105-3296. Tel.: 860-522-9258. Fax: 860-522-9259.
E-mail: info@stowecenter.org
Web Site: www.harrietbeecherstowecenter.org
Founded: 1941.
Congressional District: 1
Key Personnel: Exec. Dir. & C.E.O., Katherine Kane; Chm., Christiana N. Gianopulos; Collections Mgr., Elizabeth Burgess; Mgr. Education & Visitor

Svcs., Shannon Burke; Visitor Center Coord., Kathleen Rounds; Dir. Devel., Andrea Spak; Dir. Mktg., Mary Ellen White.
Personnel Profile: Full-Time Paid 8; Part-Time Paid 20.
Governing Authority: nonprofit foundation. Tax-exempt: 501(c)(3).
Institution Type/Description: Historic Buildings: housed in the home of activist author Harriet Beecher Stowe; located on Nook Farm, the 19th-century neighborhood where Mark Twain, Isabella Beecher Hooker & William H. Gillette also resided.
Collections: 180,000 manuscripts; 5,000 photographs; 7,000 broadsides, posters, prints & drawings; 12,000 books; 4,000 pamphlets; 100 scrapbooks related to decorative art, memorabilia and art work of & by Stowe & extended Beecher family; artifacts relating to abolition, suffrage, architecture, African-Americans & women; gardens. Historic Houses: 1871 Stowe House; 1884 Katharine Seymour Day House; 1873 Carriage House.
Research Fields: abolition; slavery; African American history; suffrage movement; 19th-century domestic life; architecture; decorative arts; Nook Farm residents; 19th & early 20th-century theater; genealogy.
Facilities: 12,000-vol. library; 4,000 pamphlets & 180,000 manuscript items with emphasis on 19th-century architecture, decorative arts, history & literature; Victorian gardens; visitor center. Museum-related items for sale.
Activities: Programs using Harriet Beecher Stowe's life and impact to inspire people to work for positive change. Guided tours, lectures, community activities, programs to preserve & interpret Harriet Beecher Stowe's home & the Stowe Center's collections.
Publications: members monthly e-newsletter; Harriet Beecher Stowe's Hartford Home; Victorian Blossoms; Harriet Beecher Stowe Hartford Home; Unfolding History; Harriet Beecher Stowe in Europe; Nook Farm; Portraits of a Nineteenth Century Family; Harriet Beecher Stowe Reader; The Minister's Wooing; Poganuc People; American Woman's Home.
Hours & Admission Prices: June-Sept. Tues.-Sat. 9:30-4:30, Sun. 12-4:30; Oct.-May Wed.-Sat. 9:30-4:30, Sun. 12-4:30. Adult 17-64 $9, seniors 65 & over $8, children 5-16 & students $6; discounts to groups & AAM members; members no charge. Closed New Year's Day; Easter; Independence Day; Thanksgiving; Christmas Eve & Day.
Attendance: 22,785 (accurate)
Membership: Student $15; Individual $25; Household $45; International $50; Sustaining $100; Library $175; Associate $250; Benefactor $500 & up; HBS Society $1,000 & up.

✳ THE MARK TWAIN HOUSE & MUSEUM, 351 Farmington Ave., Hartford, CT 06105-4401. Tel.: 860-247-0998. Fax: 860-278-8148.
E-mail: info@marktwainhouse.org
Web Site: www.marktwainhouse.org
Founded: 1929.
Congressional District: 1
Key Personnel: Exec. Dir., Jeffrey Nichols; Pres. (V), Dede D. DeRosa; Supvr. Buildings & Grounds, William Perez; Beatrice Fox Averbach Chief Cur., Patti Philippon.
Personnel Profile: Full-Time Paid 17; Part-Time Paid 39; Part-Time Volunteers 8; Interns 1.
Governing Authority: nonprofit. Tax-exempt: 501(c)(3).
Institution Type/Description: Historic Building: Mark Twain House, author's home 1874-1891 designed by Edward Tuckerman Potter; Alfred M. Thorp & decorated by Louis C. Tiffany and Associated Artists in 1881.
Collections: memorabilia, photographs, furniture of Mark Twain; period house furnishings; 19th-century American paintings, prints & drawings; decorative arts; Louis C. Tiffany collection of furniture & art glass; Candace Wheeler collection of fabrics & memorabilia; study samples of 19th-century wallpaper & textiles; rare books & manuscripts.
Research Fields: Mark Twain; Louis C. Tiffany; architecture; decorative arts, history and literature from 1850-1900.
Facilities: library of 4,000 books & 3,000 manuscript items; autographed copies of first editions; inscribed copies; foreign translations; documents & photographic file on Mark Twain, his family & 19th century. Museum-related items for sale.
Activities: guided tours; lectures; symposia; family programs; musical performances; dramatic readings.
Publications: quarterly newsletter; exhibition catalogues.
Hours & Admission Prices: Jan.-March Wed.-Mon. 9:30-5:30; April-Dec. Mon.-Sat. 9:30-5:30, Sun. 12-5:30; group tours by appointment. Adults $14, seniors $12, children 6-16 $8; discounts to AAM members & groups of 10 or more; children under 6 & members no charge. Closed New Year's Day; Easter; Independence Day; Thanksgiving; Christmas Eve & Day. &
Attendance: 67,855 (accurate)
Membership: National $40; Individual $50; Household $75; Sustaining $150.

MUSEUM OF CONNECTICUT HISTORY, (M), Connecticut State Library, 231 Capitol Ave., Hartford, CT 06106-1569. Tel.: 860-757-6535. Fax: 860-757-6533.
Web Site: www.museumofcthistory.org/
Founded: 1910.
Congressional District: 1
Key Personnel: Admin., Dean E. Nelson; Cur. Education, Patrick Smith; Cur. Collections, David J. Corrigan.
Personnel Profile: Full-Time Paid 3.
Governing Authority: state. Parent Institution: Connecticut State Library. Tax-exempt: 170(c).
Institution Type/Description: History Museum.
Collections: Connecticut history; governors' portraits; industrial; military; political; numismatic; general history.
Research Fields: Connecticut & American history.
Facilities: 500,000-vol. library containing 2,000,000 manuscripts on history, law & genealogy; archives dating from 1636; books available for interlibrary loan; manuscripts available on premises; reading room.
Activities: permanent & temporary exhibitions.
Hours & Admission Prices: Mon.-Fri. 9-4, Sat. 9-2. No charge. Closed state holidays. &
Attendance: 25,000 (accurate)

PUMP HOUSE GALLERY, Bushnell Park, 60 Elm St., Hartford, CT 06123. Mailing Address: P.O. Box 230778, Hartford, CT 06123-0778. Tel.: 860-543-8874.
Institution Type/Description: Historic Building: housed in a working Tudor-style pump house, part of the Connecticut River Flood Control Project; built in 1947.
Collections: local history; works by local artists; paintings; sculpture.
Activities: temporary exhibits.
Hours & Admission Prices: Tues.-Sat. 11-2. No charge.

REAL ART WAYS, 56 Arbor St., Hartford, CT 06106-1228. Tel.: 860-232-1006. Fax: 860-233-6691.
E-mail: info@realartways.org
Web Site: www.realartways.org
Founded: 1975.
Key Personnel: Exec. Dir., Will K. Wilkins
Institution Type/Description: Art Gallery.
Collections: works by contemporary artists; videos.
Facilities: cinema/theater; cafe; rental facilities.
Activities: temporary exhibitions; films; concerts; readings; educational programs; openings; lectures; workshops; spoken work; performances.
Hours & Admission Prices: Tues.-Thurs. & Sun. 2-10, Fri.-Sat. 2-11. Suggested Donation: $3; members no charge. &
Membership: Individual $50; Household $75.

STATE CAPITOL, HARTFORD, CONNECTICUT, 210 Capitol Ave., Hartford, CT 06106-1535. Tel.: 860-240-0222. Fax: 860-240-8627.
E-mail: capitol.tours@cga.ct.gov
Web Site: www.cga.ct.gov/capitoltours
Founded: 1878.
Congressional District: 1
Key Personnel: Dir., Kimberly Fabrizio.
Personnel Profile: Full-Time Paid 2; Part-Time Paid 2; Part-Time Volunteers 21; Interns 1.
Governing Authority: Parent Institution: League of Women Voters of Connecticut Education Fund. Tax-exempt.
Institution Type/Description: Historic Building.
Collections: Hall of Flags; Charter Oak Chair; figureheads; models; Israel Putnam's tombstone; furniture.
Activities: guided tours; educational tours.
Publications: leaflet, This Is Your General Assembly; How a Bill Becomes A Law; Directory of Connecticut's Federal & State Elected Officials; Connecticut: A Guide to State Government; self-guided tour pamphlets.
Hours & Admission Prices: One-hour Tour: July-Aug. Mon.-Fri. 9:15, 10:15, 11:15, 12:15, 1:15 & 2:15; Sept.-June Mon.-Fri. 9:15, 10:15, 11:15, 12:15 & 1:15. No charge; donations accepted. Closed state & national holidays; Christmas week. &
Attendance: 30,228 (accurate)

✳ WADSWORTH ATHENEUM MUSEUM OF ART, (M), 600 Main St., Hartford, CT 06103-2990. Tel.: 860-278-2670. Fax: 860-527-0803. Facebook: Wadsworth Atheneum.
E-mail: info@wadsworthatheneum.org
Web Site: www.wadsworthatheneum.org

Formerly: Wadsworth Atheneum
Founded: 1842.
Congressional District: 1
Key Personnel: Dir. & C.E.O., Susan L. Talbott; Bd. Pres. (V), David W. Dangremond; Dir. Institutional Advancement & Communications, Kimberly Reynolds; Dir. Finance, Judy Mihalko; Georgette Averbach Koopman Dir. Education, Johanna Plummer; Dir. Facility, Alan Barton; Exec. Asst., Sherri Malinoski; Senior Cur. and Charles C. & Eleanor Lamont Cunningham Cur. European Decorative Arts, Linda H. Roth; Coord. Exhibitions, Adria Patterson; Head Museum Design, Cecil Adams; Collection Imaging Mgr., Allen Phillips; Visitors Svcs. Mgr., Susan Carey; Head Registrar, Edd Russo; Registrar, Loans & Exhibitions, Mary Busick; Curatorial Admin., Kelly Agnew; William G. DeLana Archivist & Cur., Austin House, Eugene R. Gaddis; Assoc. Frame Conservator, Zenon Gansziniec; Librarian & Cur. Special Book Collection, John Teahan; Chief Conservator, Ulrich Birkmaier; Dept. of American Paintings & Sculpture, Erin Monroe; Cur. Film & Theater, Deborah Gaudet; Mgr. Devel., Dina Silva; Coord. Devel., Brian Benibe; Mgr. Special Events, Kristen Alfonso; Museum Shop Mgr., Stacey Stachow; Museum Shop Asst. Mgr., Jane Gallagher; Coord. Membership, Kristyn Ribera; Mgr. Membership, Hilary Burrall; Exec. Dir., The Amistad Center for Art & Culture, Olivia White; Dir. Human Resources, Shawn Lewinson; Coord. Community Programs, Lauren Cross; Group Visit Assoc., Courtney Hebert; Chief Cur. & Krieble Cur. American Painting & Sculpture, Robin Jaffee Frank.
Personnel Profile: Full-Time Paid 46; Part-Time Paid 36; Part-Time Volunteers 414; Interns 34.
Governing Authority: nonprofit organization. Branch Museum: Austin House, 130 Scarborough St., Hartford, CT. Tax-exempt: 170(b)(1)(A).
Institution Type/Description: Art Museum.
Collections: classical bronzes; Renaissance & Baroque art; Spanish, Italian & 17th-century Dutch & northern European art; Wallace Nutting collection of American furniture; two fully restored American period rooms removed from their original homes; 18th-century French & German porcelains; 19th century American & European art; 20th-century art; American paintings, sculpture, drawings & decorative arts; French & American Impressionist paintings; African-American art; European & American costumes & textiles; English & American silver; contemporary paintings, sculpture; Sol LeWitt wall drawings; Matrix series.
Research Fields: paintings; sculpture; decorative arts; costumes; textiles; African-American & contemporary art.
Facilities: Auerbach Library: 32,000-vols. on art history, museology & art education; archives; 287-seat theater; cafe. Museum publications for sale.
Activities: lectures; films; gallery talks; guided group tours for students & adults; teachers workshops; children's studio programs & summer art camp; docent program; college internship program; costume & textile society; decorative arts council; permanent, temporary & traveling exhibitions. Women's Committee Organizes: Festival of Trees & Traditions; Fine Art & Flowers.
Publications: artist sheets, MATRIX; collection catalogs; exhibition catalogs; annual report; gallery & special exhibition guides.
Hours & Admission Prices: Museum: Wed.-Fri. 11-5, Sat.-Sun. 10-5, first Thurs. of month 11-8. Adults $10, senior citizens $8, students 13 to college $5; discounts first Thurs. of month & AAM members; children 12 & under and members no charge. Library: Wed.-Thurs. 11-5. Closed New Year's Day; Independence Day; Thanksgiving; Christmas. &
Attendance: 100,000 (estimated)
Membership: Select Individual $35; Individual $45; Select Household $50; Household $60; Dual Access $100; Insider Access $125; Total Access $170; Premier Access $370.

WIDENER GALLERY, AUSTIN ARTS CENTER, TRINITY COLLEGE, 300 Summit St., Hartford, CT 06106-3100. Mailing Address: 300 Summit St., Hallden Hall - Fine Arts, Hartford, CT 06106. Tel.: 860-297-5232 & 2199. Fax: 860-297-5349.
E-mail: felice.caivano@trincoll.edu
Web Site: www.trincoll.edu/artsattrinity
Founded: 1964.
Congressional District: 1
Key Personnel: Chair Fine Arts Dept. & Dir. Studio Arts, Pablo Delano; Cur., Felice Caivano.
Governing Authority: college. Tax-exempt: 170(b)(1)(A).
Institution Type/Description: College Gallery.
Collections: teaching collections; The Samuel H. Kress Collection.
Facilities: theatre.
Activities: education programs for college students affiliated with Trinity college & the public.

Hours & Admission Prices: Academic Year: Sun.-Fri. 1-6. No charge. Closed academic holidays & recesses.

Ivoryton

MUSEUM OF FIFE AND DRUM, 63 N. Main St., Ivoryton, CT 06442-0277. Mailing Address: P.O. Box 277, Ivoryton, CT 06442-0277. Tel.: 860-767-2237. Fax: 860-767-9765.
E-mail: companyhq@companyoffifeanddrum.org
Web Site: www.companyoffifeanddrum.org
Founded: 1965.
Congressional District: 2
Key Personnel: Pres. (V), John Hanewich; Vice Pres., Mark Logsdon; 2nd Vice Pres., Bill Bouregy; Treas., Sara Brown; Cur. & Archivist, Marty Sampson; Public Rels., Kevin Brown; Education, Mark Riley; Security, Daniel Riley.
Personnel Profile: Full-Time Volunteers 10; Part-Time Volunteers 15.
Governing Authority: private; nonprofit organization. Parent Institution: The Company of Fifers & Drummers. Tax-exempt: 501(c)(3).
Institution Type/Description: Musical Instruments Museum.
Collections: instruments; photographs; uniforms; documents; drums dating back to American Revolution; archives.
Research Fields: history & development of fifing & drumming.
Facilities: library available only to scholars & historians; 250-seat auditorium. Museum-related items for sale.
Activities: guided tours; formal education programs for undergraduate or graduate college students as requested; company & committee meetings; lectures as requested. Annual Events: summer concerts; Jaybird (Old-Timers) Day.
Publications: quarterly magazine, The Ancient Times.
Hours & Admission Prices: June 30-Labor Day Sat.-Sun. 1-5 by appointment only. Adults $3, youth 13-17 & seniors 60 and over $2; members & children 12 and under no charge. &
Attendance: 1,150 (estimated)
Membership: Individual $25; Drum Corps. $100; Life $1,000.

Kensington

BERLIN HISTORICAL SOCIETY, 305 Main St., Kensington, CT 06037. Mailing Address: P.O. Box 8192, Kensington, CT 06037. Tel.: 860-828-5114.
Web Site: berlincthistorical.org
Institution Type/Description: Historical Society Museum.
Collections: local history & culture; photographs; personal artifacts; period furnishings.
Hours & Admission Prices: April-Dec. Sat. 1-4; other times by appointment.

NEW BRITAIN YOUTH MUSEUM AT HUNGERFORD PARK, 191 Farmington Ave., Kensington, CT 06037-1220. Tel.: 860-827-9064. Fax: 860-827-1266.
E-mail: marketing@newbritainyouthmuseum.org
Web Site: newbritainyouthmuseum.org
Founded: 1984.
Congressional District: 6
Key Personnel: Dir., Ann F. Peabody; Chm. (V), Robert Ramsey.
Personnel Profile: Full-Time Paid 6; Part-Time Paid 5; Part-Time Volunteers 20; Interns 4.
Governing Authority: nonprofit organization. Parent Organization: New Britain Institute. Branch Museum: New Britain Youth Museum. Tax-exempt: 501(c)(3).
Institution Type/Description: Children's Museum & Nature Center.
Collections: mounted specimens; geological & agricultural tools; farm & exotic animals; pets; local history; natural history & environmental issues.
Facilities: nature & conservation center; classroom; kitchen workshop; nature trails. Items related to natural science for sale.
Activities: guided tours; lectures; films; study clubs; organized education programs for children; docent program; participatory & temporary exhibitions.
Publications: brochures, Visitors Guide; Trail Guide; quarterly newsletter, Ramblings.
Hours & Admission Prices: Tues.-Sat. 10-4.30. Adults $4, senior $3, children $2; discounts to AAM members; members & children under 2 no charge. Closed major holidays. &
Attendance: 23,732 (accurate)
Membership: Individual $25; Family $40; Supporting $50; Sustaining $100; Patron $250.

Kent

CONNECTICUT ANTIQUE MACHINERY ASSOCIATION MUSEUM, 31 Kent-Cornwall Rd., Kent, CT 06757. Mailing Address: P.O. Box 425, Kent, CT 06757. Tel.: 860-927-0050. Fax: 860-927-0050.
Web Site: www.ctamachinery.com
Founded: 1984.
Key Personnel: Dir. & Pres. (V), John Pawloski; Museum Shop Mgr., John Stauffer.
Personnel Profile: Full-Time Volunteers 6; Part-Time Volunteers 12.
Governing Authority: Tax-exempt.
Institution Type/Description: History Museum.
Collections: industrial & agricultural heritage & history; machinery; mining; minerals.
Activities: special events. Museum Sponsors: Spring Power-Up; Annual Fall Festival; Gem and Mineral Show.
Hours & Admission Prices: May-Oct. Wed.-Sun. 10-4; other times by appointment. Adults $3; children under 12 no charge. &
Attendance: 15,000 (estimated)
Membership: Individual $20; Life $300.

ERIC SLOANE MUSEUM & KENT IRON FURNACE, 31 Kent Cornwall Rd., Rte. 7, Kent, CT 06757. Mailing Address: P.O. Box 917, Kent, CT 06757-0917. Tel.: 860-927-3849. Fax: 860-927-2152. Facebook: Eric Sloane Museum.
E-mail: ericsloane.museum@ct.gov
Web Site: www.ericsloanemuseum.com
Formerly: Sloane-Stanley Museum and Kent Furnace
Founded: 1969.
Congressional District: 6
Key Personnel: Dir. Culture, State Historic Preservation Officer, Daniel Forrest; Museum Dir., Karin Peterson; Museum Asst., Barbara Russ.
Personnel Profile: Part-Time Paid 2; Part-Time Volunteers 5.
Governing Authority: state. Parent Institution: State Historic Preservation Office, Dept. of Economic & Community Devel., One Constitution Plaza, 2nd Fl., Hartford, CT 06103. Tax-exempt: 170(c)(1).
Institution Type/Description: Early American Tools & Implements Museum: located on the site of the ruins of Kent Iron Furnace.
Collections: early American tools; diorama of early iron industry at Kent Furnace; paintings by Eric Sloane; recreation of Sloane's art studio from his Connecticut home.
Research Fields: early American tools & implements; early iron industry.
Activities: lectures; permanent exhibitions.
Publications: The America of Eric Sloane: A Collector's Bibliography.
Hours & Admission Prices: See website for hours. Adults $8, senior citizens 60 & over $6, students 6-17 $5; discounts to school groups; children 5 & under no charge.
Attendance: 3,000 (accurate)

THE KENT ART ASSOCIATION, 21 S. Main St., Kent, CT 06757. Mailing Address: P.O. Box 202, Kent, CT 06757-0202. Tel.: 860-927-3989. Fax: 860-927-4218.
E-mail: info@kentart.org
Web Site: www.kentart.org
Founded: 1923.
Key Personnel: Pres. (V), Carolyn Fisher; Exec. Dir., Davia Kennedy Fink.
Governing Authority: private; nonprofit organization. Tax-exempt.
Institution Type/Description: Art Museum.
Collections: paintings; drawings; sculpture.
Activities: rental gallery; charity events. Annual Events: Spring Show; Member's Show; President's Show; Fall Show; Student Art Show; Elected Artist Show.
Publications: monthly newsletter; quarterly catalogs; show catalogs.
Hours & Admission Prices: During Shows & Holidays: Thurs.-Sun. 1-5. No charge; donations accepted.
Attendance: 5,000 (estimated)
Membership: Associate $30; Sustaining $40; Patron $65; Family $80; Life $300.

Lebanon

BEAUMONT HOMESTEAD, 844 Trumbull Hwy., Lebanon, CT 06249. Mailing Address: Lebanon Historical Society, P.O. Box 151, Lebanon, CT 06249-0151. Tel.: 860-642-6579. Fax: 860-642-6583.
E-mail: museum@historyoflebanon.org
Web Site: www.historyoflebanon.org
Formerly: William Beaumont Birthplace
Key Personnel: Dir., Donna K. Baron.
Personnel Profile: Part-Time Paid 3.
Governing Authority: Parent Institution: Lebanon Historical Society. Tax-exempt.
Institution Type/Description: Historic House Museum: housed in the childhood home of Dr. William Beaumont.
Collections: William Beaumont's life, career & family; period furnishings; early medical equipment; photographs.
Hours & Admission Prices: mid-May to Columbus Day Sat. 12-4; other times by appointment. No charge; donations accepted.
Attendance: 325 (estimated)

GOVERNOR JONATHAN TRUMBULL HOUSE & WADSWORTH STABLE MUSEUMS, 169 W. Town St., Lebanon, CT 06249-1550. Mailing Address: Connecticut Daughters of the American Revolution, P.O. Box 54, Lebanon, CT 06249-0054. Tel.: 860-642-7558.
E-mail: info@govtrumbullhousedar.org
Web Site: www.govtrumbullhouse.dar.org
Key Personnel: Chm. (V), Cecelia Messier.
Personnel Profile: Part-Time Paid 3; Part-Time Volunteers 12.
Governing Authority: Parent Institution: CT Daughters of the American Revolution. Tax-exempt.
Institution Type/Description: Historic House Museum: built c.1740. Listed on the National Registry of Historic Places.
Collections: local history & culture; period furnishings; personal artifacts; photographs. Historic Building: 1730 stable.
Activities: Museum Sponsors: Gov. Trumbull 303rd Birthday - Living History Special Events.
Hours & Admission Prices: See website for hours. charge. No charge; donations accepted.
Attendance: 900 (estimated)

JONATHAN TRUMBULL JR. HOUSE MUSEUM, 780 Trumbull Hwy., Lebanon, CT 06249-1523. Mailing Address: 579 Exeter Rd., Lebanon, CT 06249. Tel.: 860-642-6100. Fax: 860-642-7716.
Web Site: www.lebanontownhall.org/trumbulljuniormuseum.htm
Governing Authority: Parent Institution: Town of Lebanon. Tax-exempt.
Institution Type/Description: Historic House Museum: housed in the home of Jonathan Trumball, Jr., son of Connecticut's Revolutionary War Governor who served as Gen. George Washington's secretary and later Connecticut's governor from 1797-1809; house built c.1769. Listed on the National Register of Historic Places.
Collections: Trumbull family history; period artifacts & furnishings.
Hours & Admission Prices: mid-May to Columbus Day Sat.-Sun. 12-4. No charge. &

LEBANON HISTORICAL SOCIETY, 856 Trumbull Hwy., Lebanon, CT 06249-1546. Mailing Address: P.O. Box 151, Lebanon, CT 06249. Tel.: 860-642-6579. Fax: 860-642-6583.
E-mail: museum@historyoflebanon.org
Web Site: www.historyoflebanon.org
Founded: 1965.
Congressional District: 2
Key Personnel: Dir., Donna K. Baron.
Personnel Profile: Part-Time Paid 3.
Governing Authority: Tax-exempt.
Institution Type/Description: Historical Society Museum.
Collections: local history & culture; photographs; period furnishings.
Research Fields: history of Lebanon, CT; eastern CT; CT & American Revolution; genealogy.
Facilities: Museum-related items for sale.
Activities: educational programs.
Hours & Admission Prices: Wed.-Sat. 12-4. No charge; donations accepted. &

REVOLUTIONARY WAR OFFICE, 149 W. Town St., Lebanon, CT 06249. Mailing Address: CT Sons of American Revolution, P.O. Box 411, East Haddon, CT 06423. Tel.: 860-916-1804.
Governing Authority: Parent Institution: Sons of the American Revolution. Tax-exempt.
Institution Type/Description: History Museum: housed in the building where Gov. Jonathan Trumbull held meetings during the Revolutionary War; built in 1727. Listed on the National Register of Historic Places.
Collections: Revolutionary War history; period furnishings; personal artifacts; photographs.
Hours & Admission Prices: Memorial Day to Labor Day Sat.-Sun. 12-4. No charge; donations accepted.
Attendance: 1,000 (accurate)

TRUMBULL WAR OFFICE, 149 W. Town St., Lebanon, CT 06426. Mailing Address: Connecticut SAR, P.O. Box 411, East Haddam, CT 06423. Tel.: 860-873-3399.
Institution Type/Description: Historic Building: housed in the former store & office of Jonathan Trumbull. During the American Revolution the store became the headquarters of the colony of Connecticut where meeting were held from 1775-1784. Listed on the National Register of Historic Places.
Collections: local history; period furnishings & artifacts.
Hours & Admission Prices: May-Oct. Thurs.-Sun. 12-4.

Litchfield

✱ **LITCHFIELD HISTORICAL SOCIETY AND MUSEUM, (M),** 7 South St., On-the-Green, Litchfield, CT 06759. Mailing Address: P.O. Box 385, Litchfield, CT 06759-0385. Tel.: 860-567-4501. Fax: 860-567-3565.
E-mail: cfields@litchfieldhistoricalsociety.org
Web Site: www.litchfieldhistoricalsociety.org
Founded: 1856.
Congressional District: 6
Key Personnel: Dir., Catherine Keene Fields; Cur., Julie Frey; Librarian & Archivist, Linda Hocking; Museum Shop Mgr., Kate Zullo; Cur. Education, Linda Loveday.
Personnel Profile: Full-Time Paid 5; Part-Time Paid 6; Part-Time Volunteers 60; Interns 2.
Governing Authority: society. Tax-exempt: 501(c)(3).
Institution Type/Description: Historical Society Museum & Historic House: Tapping Reeve House & Law School, America's first law school.
Collections: documents & manuscripts; furniture; textiles; decorative arts; paintings; artifacts of daily life. Historic Buildings: 1774 Tapping Reeve House; 1784 Litchfield Law School.
Research Fields: local history; genealogy.
Facilities: 10,000-vol. library of books, 40,000 manuscripts available for researchers & students; four galleries of changing exhibits. Museum-related items for sale.
Activities: educational program; guided tours; walking tours; lecture series; curriculum projects; outreach programs; workshops; changing exhibitions.
Publications: quarterly newsletter, Discover Litchfield; children's workbook; catalogs, Though Inanimate They Speak: Ralph Earl Paintings in the Collection of the LHS; To Ornament Their Minds: Sarah Pierce's Litchfield Female Academy, 1792-1833.
Hours & Admission Prices: Museum & Tapping Reeve House: mid-April to Nov. Tues.-Sat. 11-5, Sun. 1-5. Adults $5, senior citizens & students $3 (combined admission to Reeve House, Law School & Museum); AAM & museum members no charge. ♿
Attendance: 13,000 (estimated)
Membership: Individual $25; Family $40; Contributing $100; Donor $250; Benefactor $500; Tapping Reeve Society $1,000.

WHITE MEMORIAL CONSERVATION CENTER, INC., 80 Whitehall Rd., Litchfield, CT 06759-3914. Mailing Address: P.O. Box 368, Litchfield, CT 06759-0368. Tel.: 860-567-0857. Fax: 860-567-2611.
E-mail: info@whitememorialcc.org
Web Site: whitememorialcc.org
Founded: 1964.
Congressional District: 6
Key Personnel: Exec. Dir., Keith R. Cudworth; Dir. Administration & Devel., Gerri Griswold; Pres., Arthur Hill Diedrick; Dir. Education, Jeffrey Greenwood; Dir. Research, James Fischer; Museum Shop Mgr., Lois Melaragno.
Personnel Profile: Full-Time Paid 5; Part-Time Paid 2; Part-Time Volunteers 75.
Operating Expenses: 428,967
Operating Income: 449,755
Governing Authority: nonprofit organization. Affiliated with White Memorial Foundation. Tax-exempt: 501(c)(3).
Institution Type/Description: Natural History Museum: located on 4,000-acre preserve of the White Memorial Foundation.
Collections: live animals; native fauna & flora; rocks & minerals; natural history interpretive exhibits; dioramas & life-like exhibits.
Research Fields: wildlife censusing & banding of birds; forestry.
Facilities: nature & conservation center; classroom; 200-seat auditorium; nature trails; food service facility. Books & educational aids related to natural history for sale.
Activities: guided tours; lectures; films; formally organized educational programs; permanent & changing exhibitions; self-guiding trails.
Publications: quarterly newsletter.
Hours & Admission Prices: Museum: Tues.-Sat. 9-5, Sun. 12-5. Suggested Donations: adults $6, children $3; members no charge. Guided Tours up to 25 people $75. Closed major holidays. ♿

Attendance: 17,800 (accurate)
Membership: Individual $35; Family $50.

Madison

MADISON HISTORICAL SOCIETY, Lee Academy, 14 Meeting House Lane, Madison, CT 06443. Mailing Address: P.O. Box 17, Madison, CT 06443-0017. Tel.: 203-245-4567.
E-mail: contact@madisoncthistorical.org
Web Site: www.madisonhistorical.org
Founded: 1917.
Congressional District: 3
Key Personnel: Dir., Pamela Allen; Pres. (V), Rick Camp.
Personnel Profile: Part-Time Paid 3; Part-Time Volunteers 25.
Governing Authority: society. Branch Museum: The Lee Academy, 14 Meeting House Ln., Madison, CT 06443. Tax-exempt: 501(c)(3).
Institution Type/Description: History Museum: housed in c.1785 Allis-Bushnell House.
Collections: costumes; dolls; furniture; toys; farming, carpentry, housekeeping, spinning & weaving tools & equipment; manuscripts; ship construction half models; shipbuilding; shoemaking; textiles.
Major Exhibits: The Business of Leisure: Madison Welcomes its Summer Colony, 6/14-8/14.
Research Fields: Madison history.
Facilities: 650-vol. library of books on history, genealogy, theology; pamphlets; maps available during regular hours; reading room.
Activities: guided tours; lectures; films; education programs for children; permanent & temporary exhibits.
Publications: books, Madison's Heritage; Daniel Hand of Madison, Connecticut, 1801-1891; Madison: Three Hundred Years by the Sea; pamphlet, Madison's Unique Contribution to Summer Theatre in America.
Hours & Admission Prices: June-Aug. Sat. 11-4; Winter by appointment. No charge; donations accepted.
Attendance: 500 (estimated)
Membership: Junior (under 18) $3; Individual $25; Couple $35; Contributing $40-$99; Business $75; Sustaining $100 & up; Life $300.

MEIGS POINT NATURE CENTER, Hammonasset Beach State Park, 1288 Boston Post Rd., Madison, CT 06443. Mailing Address: P.O. Box 271, Madison, CT 06443. Tel.: 203-245-8743; 860-462-9643.
Web Site: www.hammonasset.org
Institution Type/Description: Nature Center.
Collections: local wildlife including snakes, turtles, crabs, fish, & birds.
Activities: educational programs & classes.
Hours & Admission Prices: April-Oct. Tues.-Sat. 10-5; Nov.-March Tues.-Sun. 10-4. No charge; donations accepted. ♿
Attendance: 20,000 (accurate)

Manchester

THE FIRE MUSEUM, 230 Pine St., Manchester, CT 06040-5829. Tel.: 860-649-9436.
Web Site: www.thefiremuseum.org
Founded: 1983.
Congressional District: 1
Key Personnel: Pres. (V), Wayne Crossman; Museum Shop Mgr., Lucy Crossman.
Personnel Profile: Part-Time Paid 1.
Governing Authority: Tax-exempt.
Institution Type/Description: Fire Museum: housed in a former fire station; built in 1901.
Collections: firefighting history, vehicles & artifacts; photographs; personal artifacts; paintings.
Activities: special events.
Hours & Admission Prices: mid-April to mid-Nov. Fri.-Sat. 12-4; other times by appointment. Suggested Donation: adults $4, seniors, youth 12-16 & firefighters $2, children 6-12 $1; children under 6 no charge.
Attendance: 1,000 (accurate)
Membership: Individual $15; Family $30; Fire Department $45; Sustaining $60; Corporate $200.

LUTZ CHILDREN'S MUSEUM, 247 S. Main St., Manchester, CT 06040-6561. Tel.: 860-643-0949.
E-mail: reckert@lutzmuseum.org
Web Site: www.lutzmuseum.org
Founded: 1953.
Congressional District: 1
Key Personnel: Dir. & C.E.O., Bob Eckert; Pres. (V), Kristen Addabbo;

Operations Dir., Stephanie Radowitz; Arts Cur., Laura Diller; Volunteer Coord., Amanda Andersen; Visitor Svcs., Karyn Dillon.
Personnel Profile: Full-Time Paid 8; Full-Time Volunteers 2; Part-Time Paid 4; Part-Time Volunteers 65; Interns 4.
Governing Authority: nonprofit organization. Tax-exempt: 501(c)(3).
Institution Type/Description: Children's Museum.
Collections: live animals; history; natural history; ethnology; art.
Research Fields: child participation; educational teaching aids & kits.
Facilities: 53-acre nature center; classrooms; outdoor playscape. Museum-related gifts for sale.
Activities: guided tours; lectures; formally organized education programs for children; field trips; participatory exhibits; craft demonstrations; school loan service; family programs; parent-child workshops, classes, resource lessons; live animals.
Publications: membership newsletter; quarterly teacher newsletter; educational services catalogue.
Hours & Admission Prices: Tues.-Fri. 9-5, Sat.-Sun. 12-5; Nature Trails dawn-dusk. Admission $6; members no charge. Closed some holidays. &
Attendance: 30,000 (estimated)
Membership: Child & Senior $10; Adult $15; Family $40.

MANCHESTER HISTORICAL SOCIETY, 175 Pine St., Manchester History Center, Manchester, CT 06040-5921. Tel.: 860-647-9983.
E-mail: info@manchesterhistory.org
Web Site: www.manchesterhistory.org
Founded: 1965.
Congressional District: 1
Key Personnel: Dir., Mary M. Donohur.
Personnel Profile: Part-Time Paid 2; Part-Time Volunteers 40.
Governing Authority: nonprofit. Tax-exempt.
Institution Type/Description: Historical Society
Collections: Manchester history. Historic House: 1785 Cheney homestead; replica 1751 Keeney Schoolhouse.
Research Fields: Manchester history.
Activities: guided tours; permanent & temporary exhibitions.
Publications: quarterly newsletter, The Courier.
Hours & Admission Prices: Mon.-Fri. 10-2. No charge; donation requested. Closed legal holidays. &
Attendance: 1,200 (accurate)
Membership: Individual $20.

OLD MANCHESTER MUSEUM, (M), 126 Cedar St., Manchester, CT 06040-5839. Mailing Address: Manchester Historical Society, 175 Pine St., Manchester, CT 06040. Tel.: 860-647-9983.
E-mail: info@manchesterhistory.org
Web Site: www.manchesterhistory.org
Founded: 1965.
Key Personnel: Pres., John A. Dormer.
Personnel Profile: Full-Time Volunteers 2; Part-Time Paid 1; Part-Time Volunteers 20; Interns 1.
Governing Authority: Parent Institution: Manchester Historical Society. Tax-exempt.
Institution Type/Description: History Museum: housed in c.1859 school house.
Collections: Manchester history.
Research Fields: Manchester history; genealogy; local history.
Facilities: Museum-related items for sale.
Activities: guided tours; permanent & temporary exhibitions.
Publications: quarterly newsletter, The Courier.
Hours & Admission Prices: Sat. 10-4, Sun. 1-4; other times by appointment. Donation: adults $2; discounts to AAM members; members no charge. Closed holidays. &
Attendance: 1,500 (accurate)
Membership: Student $5; Individual $15; Family $25; Contributing Individual $45; Contributing Family $75; Corporate $200; Life Individual $250; Life Couple $400.

WICKHAM PARK AVIARY AND NATURE CENTER, 1329 W. Middle Tpke., Manchester, CT 06040. Tel.: 860-528-0856. Fax: 860-528-5156.
Web Site: www.wickhampark.org
Founded: 1961.
Key Personnel: Dir., Jeffrey Maron.
Personnel Profile: Full-Time Paid 4; Part-Time Paid 14; Part-Time Volunteers 12.
Governing Authority: Parent Institution: Wickham Park Foundation. Tax-exempt.
Institution Type/Description: Nature Center.

Collections: birds including turkeys, pheasants, peafowl, owls, waterfowl, hawks, geese, owls, & chickens; hands-on exhibits; local plants & wildlife; photographs.
Facilities: nature trails; classroom.
Activities: educational programs.
Hours & Admission Prices: April-Oct. daily 9:30 to dusk. Vehicle Fee: Mon.-Fri. $4, Sat.-Sun. $5. &
Attendance: 200,000 (estimated)

Mashantucket

MASHANTUCKET PEQUOT MUSEUM AND RESEARCH CENTER, (M), 110 Pequot Trail, Mashantucket, CT 06338-3180. Mailing Address: P.O. Box 3180, Mashantucket, CT 06338-3180. Tel.: 800-411-9671; 860-396-6945. Fax: 860-396-7013. Facebook: Pequot Museum.
E-mail: bkingsland@mptn-nsn.gov
Web Site: www.pequotmuseum.org
Founded: 1998.
Congressional District: 2
Key Personnel: Interim Dir., Travis Williams; Interim Dir., Dale Merrill; Mgr. Finance, Randy Banker; Education, Kimberly Shockley; Registrar, Meredith Vasta; Security, George Eleazor, Sr.; Museum Shop Mgr., Heather Montey.
Personnel Profile: Full-Time Paid 38; Part-Time Paid 11; Part-Time Volunteers 3.
Volunteer Hours: 1,300
Operating Expenses: 6,513
Operating Income: 2,681
Governing Authority: nonprofit. Parent Institution: Mashantucket Pequot Tribal Nation (Tribal Council). Tax-exempt: 170(b)(1)(A).
Institution Type/Description: Native American Museum.
Collections: personal artifacts; tribal history; area land history; archaeology; ethnographic. Historic House: c.1780 farmstead.
Major Exhibits: Backyard Monsters: The World of Insects (T), 2/15/14-5/17/14; Bison: American Icon (T), 6/28/14-10/18/14; Wild Music: Sounds and Songs for Life (T), 11/21/14-2/14/15.
Research Fields: ethnohistory; material culture studies; archaeology; ethnobotany.
Facilities: 45,000-vol. library; 328-seat auditorium; 268-seat restaurant; archaeology & conservation labs; classrooms; 80,000 sq. ft. exhibit space; 7 theaters; observation tower. Museum-related items for sale.
Activities: interactive computer programs; concerts; dance recitals; films; formal education programs for children, University of Connecticut & Connecticut College students; guided tours; craft workshops; lectures; theater; teacher seminars. Annual Events: Strawberry Festival; Gifts of the Land and Waters; PowWows.
Publications: quarterly calendar of events, Native Visions.
Hours & Admission Prices: Wed.-Sat. 9-5; last admission 4pm. Adults $20, senior citizens 55 & over and college students $15, children 6-18 $12; discounts to AAM members & groups; members & children 5 and under no charge. Closed New Year's Eve & Day; Thanksgiving Eve & Day; Christmas Eve & Day. &
Attendance: 60,000 (estimated)
Membership: Individual $55; Educator, Military & Homeschooling $60; Family $75; Associate Library $200; Contributing $250; Contributing Library $325; Supporting $500; Patron $1,000.

Meriden

GALLERY 53, 53 Colony St., Meriden, CT 06451-3210. Tel.: 203-235-5347.
Institution Type/Description: Art Gallery.
Collections: paintings; drawings; photographs.
Facilities: Gallery-related items for sale.
Activities: educational programs; workshops; lectures; temporary exhibitions.
Hours & Admission Prices: Tues.-Fri. 12-4, Sat. 10-2, Sun. 12-3. No charge.

MERIDEN HISTORICAL SOCIETY, INC., 1090 Hanover St., Morehouse Research Center, Meriden, CT 06451-6207. Mailing Address: P.O. Box 3005, Meriden, CT 06450-9305. Tel.: 203-639-1913.
E-mail: meridenhistoricalsociety@gmail.com
Web Site: www.meridenhistoricalsociety.org
Founded: 1893.
Congressional District: 5
Key Personnel: C.E.O. & Pres. (V), Lesley Solkoske; Treas., Kathy McMahon; Cur., Allen Weathers; Sec., Dorothy Heffernan.
Governing Authority: society. Tax-exempt: 990A(SF).

Institution Type/Description: Local History Museum & Research Library: museum housed in 1760 Andrews Homestead; Research Library: 1090 Hanover Rd., South Meriden, CT 06451.
Collections: glass & china; history; industry; dolls; furniture; manuscripts; maps; books; items manufactured locally; schools; churches; wars; art; design; photographs; period silver; International Silver Co. historic silver collection & library.
Research Fields: Victorian silverplate, early local silver manufacturers.
Facilities: history library, with catalogs, maps, patents & original photographs.
Activities: special exhibits.
Publications: annual reports; quarterly newsletter.
Hours & Admission Prices: Research Center: Wed. 2-4; tours by appointment. No charge; donations accepted. Closed holidays.
Membership: Senior $10; Regular $15; Corporate $50; Life Senior $125; Life $250.

SOLOMON GOFFE HOUSE, 677 N. Colony St., Meriden, CT 06450. Mailing Address: 1176 N. Colony St., Meriden, CT 06450. Tel.: 203-235-2192.
Founded: 1986.
Personnel Profile: Part-Time Volunteers 5.
Governing Authority: Tax-exempt.
Institution Type/Description: Historic House Museum: built in 1711. A National Historic Landmark.
Collections: local history & culture; period furnishings; photographs; herb garden.
Activities: programs on daily living in the 1700s & 1800s; hearth cooking; needlework; candlemaking.
Hours & Admission Prices: April-Nov. first Sun. of month 1:30-4:30. Admission $2.

Middletown

DAVISON ART CENTER, WESLEYAN UNIVERSITY, (M), 301 High St., Middletown, CT 06459-0487. Tel.: 860-685-2500. Fax: 860-685-2501.
Web Site: www.wesleyan.edu/dac
Founded: 1952.
Congressional District: 2
Key Personnel: Cur., Clare Rogan; Mgr. Museum Information Svcs. & Registrar Collections, Robert Lancefield; Gallery Supvr., Lee Berman.
Personnel Profile: Full-Time Paid 2; Part-Time Paid 1.
Governing Authority: university. Parent Institution: Wesleyan University. Tax-exempt.
Institution Type/Description: Art Museum: housed in c.1838-40 Alsop House, a pre-Civil War mansion.
Collections: prints from the mid-15th century to the present; photographs from the 1840s to the present; period wall decorations in Alsop House.
Research Fields: prints & photographs.
Facilities: 3,000-vol. library of art history & print reference available for use under university regulations; reading room; classrooms.
Activities: lectures; gallery talks.
Publications: exhibition catalogs; exhibition announcements.
Hours & Admission Prices: Sept.-May Tues.-Sun. 12-4. No charge; donations accepted. Closed holidays; academic vacations. &
Attendance: 3,000 (accurate)
Membership: Individual $30; Family $50; Sustaining $100; Donor $250; Patron $500; Life $1,000.

EZRA AND CECILE ZILKHA GALLERY, Center for the Arts, Wesleyan University, Middletown, CT 06459. Mailing Address: Wesleyan University, 283 Washington Terr., Middletown, CT 06459. Tel.: 860-685-2695. Fax: 860-685-2061.
E-mail: cparente@wesleyan.edu
Web Site: wesleyan.edu/cfa
Founded: 1973.
Congressional District: 2
Key Personnel: Dir., Pamela Tatge.
Personnel Profile: Part-Time Paid 2.
Governing Authority: affiliated with Wesleyan University. Tax-exempt.
Institution Type/Description: Art Gallery.
Collections: contemporary art.
Activities: lectures; gallery talks; video.
Publications: exhibition brochures & catalogs.
Hours & Admission Prices: Tues.-Sun. 12-4. No charge; donations accepted. Closed holidays; during academic recess. &
Attendance: 9,500 (accurate)

KIDCITY CHILDREN'S MUSEUM, 119 Washington St., Middletown, CT 06457-2817. Tel.: 860-347-0495.
E-mail: info@kidcitymuseum.com
Web Site: www.kidcitymuseum.com
Key Personnel: Exec. Dir., Jennifer Alexander
Institution Type/Description: Children's Museum.
Collections: hands-on exhibits.
Activities: birthday parties.
Hours & Admission Prices: Sun.-Tues. 11-5, Wed.-Sat. 9-5. Admission $8; children under one no charge. Closed New Year's Day; Easter; Thanksgiving; Christmas. &

MIDDLESEX COUNTY HISTORICAL SOCIETY, 151 Main St., Middletown, CT 06457-3423. Tel.: 860-346-0746. Fax: 860-346-0746.
E-mail: middlesexhistory@wesleyan.edu
Web Site: www.middlesexhistory.org
Founded: 1901.
Congressional District: 3
Key Personnel: Dir., Deborah D. Shapiro; Pres., Patricia Tully.
Personnel Profile: Full-Time Paid 1; Part-Time Volunteers 54; Interns 2.
Governing Authority: nonprofit organization. Tax-exempt: 501(c)(3).
Institution Type/Description: Historical Society Museum: housed in General Mansfield house; c.1817.
Collections: decorative arts; archives; manuscripts; Civil War uniforms & artifacts.
Research Fields: Civil War; genealogy; 18th-century shipping; history of Middlesex County.
Facilities: library of Frank Farnsworth Starr & Milo Wilcox African American genealogy materials; 18th-century manuscripts & account books available for use by appointment.
Activities: guided tours; lectures; permanent & temporary exhibitions, Middletown Heritage Trail (self guided walking tour with 23 historical markers throughout downtown area).
Publications: Historical Observer.
Hours & Admission Prices: Museum: Tues.-Thurs. 10-3, Fri. 10-12, 1st Sat. each month 12-3. Research: Tues.-Thurs. 10-3, Fri. 10-12 by appointment. Adults $5; discount to AAM members; members no charge. Closed holidays & holiday weekends. &
Attendance: 800 (estimated)
Membership: Student $15; Individual $25; Family $35; Business & Contributing $50; Patron $100; Life $500.

Milford

MILFORD HISTORICAL SOCIETY, 34 High St., Milford, CT 06460-4732. Mailing Address: P.O. Box 337, Milford, CT 06460-0337. Tel.: 203-874-2664.
E-mail: milfordhistorical@gmail.com
Web Site: www.milfordhistoricalsociety.org
Formerly: Eells-Stow House, Milford Historical Society
Founded: 1930.
Congressional District: 3
Key Personnel: Dir., Sandra Elgee; Pres. (V), Arthur Stowe; Museum Shop Mgr., Barbara Arndt.
Personnel Profile: Full-Time Volunteers 4; Part-Time Volunteers 50.
Governing Authority: society. Parent Institution: Milford Historical Society, Inc., P.O. Box 337, Milford, CT. Tax-exempt: 501(c)(3).
Institution Type/Description: General Museum.
Collections: period artifacts; Claude C. Coffin Indian collection; furniture. Historic Buildings: Eells-Stow house, c.1700; Bryan-Downs house, c.1785; Clark Stockade house, c.1780.
Facilities: 200-vol. library of historical books available for use on premises for reference use only; gardens. Museum-related items for sale.
Activities: guided tours; temporary & permanent exhibitions; walking tours; bimonthly lectures; summer programs; school group tours.
Publications: 6 newsletters annually, Wharf Lane Newsletter.
Hours & Admission Prices: June to mid-Oct. Sat.-Sun. 1-4; other times by appointment. Adults $3, children $1; members no charge. &
Attendance: 2,000 (estimated)
Membership: Student $5; Individual $15; Family $30; Life $400.

Monroe

MONROE HISTORICAL SOCIETY, 31 Great Ring Rd., Monroe, CT 06468-1328. Mailing Address: P.O. Box 212, Monroe, CT 06468-0212. Tel.: 203-261-8554.
Web Site: www.monroehistoricsociety.org
Founded: 1959.

Congressional District: 5
Key Personnel: Pres. (V), Nancy Zorena; Chm. (V), Karen Cardi; Historian, Edward Coffey; Vice Pres., Sue Selk.
Personnel Profile: Part-Time Volunteers 12.
Governing Authority: society. Tax-exempt.
Institution Type/Description: General Museum.
Collections: costumes; folklore; agriculture; archives; history; mineralogy. Historic Buildings: 1790 East Village Schoolhouse; 1770 Beardsley House; 1811 East Village Meeting House.
Research Fields: genealogy; title research for old homes in Monroe.
Facilities: 100-vol. library of history & genealogy books, photographs and local school registers from 1913-1965 housed in the Monroe Public Library.
Activities: guided tours; slide lectures of Old Monroe; formally organized education programs for children; traveling exhibitions; 19th-century school day programs upon request; Hands on History summer program.
Publications: quarterly newsletter.
Hours & Admission Prices: By appointment. No charge; donations accepted.
Attendance: 1,500 (estimated)
Membership: Senior $10; Individual $20; Family $25; Friend $50; Business Associate $200; Life $350.

Moodus

AMASA DAY HOUSE MUSEUM, 33 Plains Rd., Moodus, CT 06469. Mailing Address: 255 Main St., 4th Fl., Hartford, CT 06106-1821. Tel.: 860-247-8996, ext. 12.
Institution Type/Description: Historic House: built in 1816.
Collections: Day family furnishings; photographs; period artifacts.
Hours & Admission Prices: Temporarily closed.

Mystic

DENISON HOMESTEAD MUSEUM - PEQUOTSEPOS MANOR, 120 Pequotsepos Rd., Mystic, CT 06355. Tel.: 860-536-9248. Fax: 860-536-9248.
E-mail: membership@denisonsociety.org
Web Site: www.denisonhomestead.org
Institution Type/Description: Historic House Museum: housed in the former home of Captain George Denison; built in 1717. Listed on the National Register of Historic Places.
Collections: Denison family history; period furnishings; personal artifacts.
Hours & Admission Prices: mid-May to mid-Oct. Fri.-Mon. 1-5. Adults $5, seniors & students $4, children $2; discounts to active military; children under 5 no charge.

DENISON PEQUOTSEPOS NATURE CENTER, 109 Pequotsepos Rd., Mystic, CT 06355-3045. Mailing Address: P.O. Box 122, Mystic, CT 06355-0122. Tel.: 860-536-1216. Fax: 860-536-2983.
E-mail: info@dpnc.org
Web Site: www.dpnc.org
Founded: 1946.
Congressional District: 18
Key Personnel: C.E.O. & Exec. Dir., Margaret L. Jones; Pres. (V), Dorrit Castle; Treas., Theodore Liston; Treas., Daniel Brannegan; Dir. Finance, Mayada Wadsworth; Education, Kim Macklin; Public Rels., Jennifer Johnson; Museum Shop Mgr., Christine Anderson.
Personnel Profile: Full-Time Paid 5; Full-Time Volunteers 3; Part-Time Paid 6; Part-Time Volunteers 60; Interns 2.
Governing Authority: private; nonprofit. Tax-exempt: 501(c)(3).
Institution Type/Description: Natural Science & Environmental Education.
Collections: natural history of southeastern Connecticut; forest, wetland & meadow habitats, including live birds of prey, frogs, snakes & turtles; mounted birds, mammals & insects.
Research Fields: resident breeding bird populations; distribution of Pink Lady slippers in a small nature preserve; horseshoe crabs; amphibian monitoring.
Facilities: wildflower garden; 80-seat classroom; 2,000 sq. ft. exhibit space; small theater; 300-acre sanctuary with over eight miles of trails. Educational toys, games, jewelry & locally produced items for sale.
Activities: formal & early childhood education programs; lectures; participatory exhibits; ecological area field trips; workshops; seasonal farmers market featuring organically grown produce from June-October; field guides, bird feeders & seed; state licensed nature-based preschool for ages 3-5. Annual Events: Lady's Slipper Festival in May; Earth Day Celebration; Summer Nature Camp; Summer Eco-luncheon or Eco-evening; Wild Mushroom Festival in September; Halloween Spooky Nature Trail.
Publications: quarterly newsletter, The Chickadee.
Hours & Admission Prices: Mon.-Sat. 9-5, Sun. 10-4. Adults $8, seniors & children under 12 $5; discounts to military, ANCA & AAA members. Closed New Year's Day; Easter; Thanksgiving; Christmas.
Attendance: 50,000 (estimated)

Membership: Senior 65 & over $20; Individual $25; Senior Couple $35; Single Parent Family & Couple $40; Family $50; Library & Nonprofit $100.

MYSTIC AQUARIUM, 55 Coogan Blvd., Mystic, CT 06355-1997. Tel.: 860-572-5955. Fax: 860-572-5969.
E-mail: info@mysticaquarium.org
Web Site: www.mysticaquarium.org
Founded: 1973.
Congressional District: 2
Key Personnel: Pres. & C.E.O., Dr. Stephen Coan; Chm. Bd. (V), George M. Milne, Jr., Ph.D.; C.F.O., Denise H. Armstrong; Sr. Vice Pres. Mktg. & Public Affairs, Peter P. Glankoff; Vice Pres. Facilities, Keith P. Sorensen; Sr. Dir. Mktg. & Public Affairs, Andrew Wood; Dir. Devel., Megan Brown; Dir. Education & Public Conservation Programs, Kelly E. Matis; Dir. Guest & Member Svcs., Celinda A. Beaudreau; Sr. Vice Pres. Research & Zoological Operations, Dr. Tracy A. Romano.
Personnel Profile: Full-Time Paid 115; Part-Time Paid 130; Part-Time Volunteers 550; Interns 6.
Governing Authority: nonprofit organization. Parent Institution: Sea Research Foundation, Inc., 55 Coogan Blvd., Mystic, CT 06355. Tax-exempt: 501(c)(3).
Institution Type/Description: Aquarium & Marine Museum.
Collections: marine mammals; invertebrates; fishes; plants; penguins; water fowl; marine archaeological artifacts.
Research Fields: marine mammal behavior, biology & medicine; fish reproductive biology & medicine; invertebrate biology; water chemistry; deep sea archaeology & exploration.
Facilities: 1,000-vol. library of books, journals & magazines pertaining to natural history & husbandry of marine animals available for research on premises by appointment; theater; classrooms. Books, sculpture, shells & other museum-related items for sale.
Activities: public programs; formally organized education programs for children, adults, undergraduate & graduate students; training programs for museum workers; permanent exhibitions; marine mammal demonstrations.
Publications: quarterly newsletters.
Hours & Admission Prices: Daily 9-6. Adults $26; members no charge. Closed Thanksgiving; Christmas.
Attendance: 700,000 (accurate)
Membership: Individual $60; Couple $119; Single Parent Family & Single Grandparent $139; Family $169; Grandparent $169; Member Plus $249.

MYSTIC ARTS CENTER, 9 Water St., Mystic, CT 06355-2592. Tel.: 860-536-7601. Fax: 860-536-0610. Facebook: Mystic Arts Center.
E-mail: info@mysticarts.org
Web Site: www.mysticarts.org
Founded: 1913.
Congressional District: 2
Key Personnel: Exec. Dir., Karen Barthelson; Pres., Mary Anne Stets; Treas., Bill Middleton.
Personnel Profile: Full-Time Paid 3; Part-Time Paid 7; Part-Time Volunteers 150.
Governing Authority: nonprofit organization. Parent Institution: Mystic Art Association, Inc. Tax-exempt: 501(c)(3).
Institution Type/Description: Art Museum Center.
Collections: works by founding & early artists.
Activities: concerts; films; formal educational programs; lectures; temporary exhibitions; studio classes; outreach program; cultural programs. Annual Events: Mystic Photo Show; Regional Show; Members Show.
Publications: quarterly newsletter.
Hours & Admission Prices: Daily 11-5. Suggested Donation: adults $3. Closed New Year's Day; Easter; Thanksgiving; Christmas Eve & Day.
Attendance: 20,000 (estimated)
Membership: Junior $20; Individual $40; Elected Artist $45; Family $50; Elected Artist Family $60; Donor $100; Sustaining $250; Patron $500; Benefactor $1,000; Director's Circle $5,000.

* **MYSTIC SEAPORT - THE MUSEUM OF AMERICA AND THE SEA, (M),** 75 Greenmanville Ave., Mystic, CT 06355-0990. Mailing Address: 75 Greenmanville Ave., P.O. Box 6000, Mystic, CT 06355-0990. Tel.: 860-572-0711. Fax: 860-572-5326. TDD: 860-572-5319.
E-mail: info@mysticseaport.org
Web Site: www.mysticseaport.org
Founded: 1929.
Congressional District: 2
Key Personnel: Pres., Stephen C. White; Chm. Bd., J. Barclay Collins; Vice Pres. Finance & Administration, Marcy Withington; Exec. Vice Pres., Susan Funk; Vice Pres. Watercraft Preservation & Programs, Dana Hewson; Dir.

Communications, Daniel McFadden; Dir. Facilities Management, Ken Wilson; Dir. Information Technology, Michael Lehnertz; Dir. Exhibitions, Jonathan Shay; Dir. Visitor Svcs., Jenny Doak; Dir. Shipyard, Quentin Snediker; Dir. Maritime Studies, James Carlton; Cur. Photography & Dir. Intellectual Property, Mary Anne Stets; Dir. Administrative Svcs., Mark Dulin; Dir. Membership & Volunteer Svcs., Alexandra Alpert; Librarian, Paul J. O'Pecko.

Personnel Profile: Full-Time Paid 150; Part-Time Paid 200; Part-Time Volunteers 800.

Governing Authority: bd. of trustees. Tax-exempt: 501(c)(3) & 170(b)(1)(A).

Institution Type/Description: Maritime History.

Collections: America's historic & contemporary relationship with the sea; maritime art & artifacts; historic film footage & photographs; manuscripts; ships' plans; maps & charts; books & periodicals; small craft collection; historic buildings in a New England maritime village setting. Historic Ships: Charles W. Morgan; Joseph Conrad; L. A. Dunton; Emma C. Berry; Sabino; Brilliant; Roann.

Research Fields: American social & cultural history with emphasis on the relationship of America & the sea; 19th century maritime communities; vessel documentation; shoreside industries; ethnomusicology; literature; women's studies; marine biology & oceanography; coastal life; ethnic studies in maritime America.

Facilities: 70,000-vol. research library; 2,000 roles of microfilm; 1,000 ship registers; 1,000,000 manuscript pieces; 1,200 log books; 600 audiotaped interviews; 200 videotaped interviews; 9,000 charts & maps; planetarium; film & video archives; classrooms; children's museum; food service available. Museum-related items for sale.

Activities: internet home page; guided tours; films; gallery talks; concerts; education programs for special needs visitors, children, adults, undergraduate & graduate college students; inter-museum loan, permanent & temporary exhibitions; sail education programs; live interpretations; crafts demonstrations; video production; river cruises; scholarly research & symposia; planetarium classes & programs; workshops; lectures; boat livery; weekend events; music festival; daytime & overnight camps; historic holiday re-enactments; family activities & programs; music; Elderhostel programs; curriculum units for school children; teacher institutes; first & third person interpretation; maritime skills demonstrations; sign language interpreter.

Publications: biannual magazine.

Hours & Admission Prices: Ships & Exhibits: Feb. 15-23 Thurs.-Sun 10-4; March 29-Oct. 26 daily 9-5; Oct. 27-Nov. 30 daily 10-4; Dec. 1-Jan. 1, 2015 Thurs.-Sun. 10-4. Adults $24, seniors, military & students $22, youth 6-17 $15; members and children 5 & under no charge. Closed Christmas Eve & Day. &

Attendance: 250,000 (accurate)

Membership: Individual $40; Dual $70; Family & Grandparents $95; Mariner $125; Sustaining $225; Associate $300; Benefactor $500; Champion $750.

Naugatuck

NAUGATUCK HISTORICAL SOCIETY, 195 Water St., Naugatuck, CT 06770-2826. Mailing Address: P.O. Box 317, Naugatuck, CT 06770-0317. Tel.: 203-729-9039.

E-mail: naugatuckhistory@sbcglobal.net
Web Site: www.naugatuckhistory.com
Founded: 1959.
Congressional District: 5
Key Personnel: Pres. (V), Wendy Murphy; Museum Shop Mgr., Mary Doback.
Personnel Profile: Part-Time Paid 1; Part-Time Volunteers 20.
Volunteer Hours: 4,000
Operating Expenses: 12,000
Operating Income: 13,000
Governing Authority: bd. of directors. Subsidiary Institution: Naugatuck Historical Society Museum. Tax-exempt.
Institution Type/Description: Historical Society Museum: housed an historic railroad station.
Collections: Naugatuck town history; rare books; photographs; manuscripts; city directories (census) dating back to 1877; rubber industry items; local manufacturing.
Major Exhibits: Women's Club, 3/14; Postcard, 5/14-8/14.
Research Fields: rubber industry.
Activities: tours; outreach; research; educational programming.
Publications: quarterly newsletter; monthly e-newsletter, The Bridge.
Hours & Admission Prices: Wed.-Fri. 12-4, 1st Thurs. each month 12-4 & 6-8, Sat. 10-2; other times by appointment. Adults $2, children & student $1.50; active military & their families and members no charge. &
Attendance: 1,500 (estimated)
Membership: Individual $15; Senior & Student $12; Family $35; Corporate $100 & up; Lifetime Individual $150; Lifetime Couple $250.

New Britain

CENTRAL CONNECTICUT STATE UNIVERSITY ART GALLERIES, University Galleries, Maloney Hall, S.T. Chen Fine Arts Center, 1615 Stanley St., New Britain, CT 06050-2439. Tel.: 860-832-2620 & 2633. Fax: 860-832-2634.

E-mail: rodiamond2@yahoo.com
Web Site: www.art.ccsu.edu/gallery.html
Formerly: Museum of Central Connecticut State University
Founded: 1965.
Key Personnel: Co Dir., Sean Gallagher; Co Dir., Mark Strathy.
Personnel Profile: Part-Time Paid 1; Interns 1.
Governing Authority: state. Parent Institution: Connecticut State University, 1615 Stanley St., New Britain, CT. Tax-exempt.
Institution Type/Description: Art Museum.
Collections: anthropology; paintings; sculpture; graphics; decorative arts; archaeology; folklore; industry; textiles; architectural drawings & models.
Facilities: library of books and other materials related to the history of art & art education available on premises.
Activities: formally organized education programs for undergraduate and graduate college students.
Hours & Admission Prices: Mon.-Fri. 1-4. No charge; donations accepted. &
Attendance: 3,000 (estimated)

COPERNICAN OBSERVATORY & PLANETARIUM, 1615 Stanley St., Central CT State University, New Britain, CT 06053. Tel.: 860-832-3399.

Web Site: www.ccsu.edu/astronomy
Institution Type/Description: Planetarium.
Collections: hands-on activities; astronomy.
Activities: educational programs; planetarium shows; workshops; lectures.
Hours & Admission Prices: Shows: Fri.-Sat. 8:30 pm. Children's Shows: Fri. 7 pm, Sat. 1:30. No charge. Closed state holidays.

NEW BRITAIN INDUSTRIAL MUSEUM, (M), 185 Main St., ITBD Bldg., Fl. 2, New Britain, CT 06051-2296. Tel.: 860-832-8654. Facebook: NB Industrial.

E-mail: newbritainim@gmail.com
Web Site: www.nbim.org
Founded: 1995.
Key Personnel: Pres. (V), Randall Judd; Dir., Karen Hudkins; Museum Shop Mgr., Newspaper Editor & Membership Chm., Lois L. Blomstrann.
Personnel Profile: Full-Time Paid 1; Part-Time Volunteers 10.
Governing Authority: Parent Institution: New Britain Institute. Tax-exempt.
Institution Type/Description: Industrial Museum.
Collections: manufactured products from invention to production developed in New Britain, CT including The Stanley Works; Fafnir Bearing; American Hardware; and North & Judd; Landers, Frank & Clark.
Research Fields: industrial history.
Publications: quarterly newsletter featuring articles on history of New Britain Industry.
Hours & Admission Prices: Tues. & Thurs.-Fri. 2-5, Wed. 12-5, Sat. 10-4. No charge; donations accepted. &
Attendance: 900 (accurate)
Membership: Senior & Student $15; Individual $25; Family & Contributing $40; Sponsor $100; Patron $500; Benefactor $1,000.

❋　NEW BRITAIN MUSEUM OF AMERICAN ART, INC., (M), 56 Lexington St., New Britain, CT 06052-1412. Tel.: 860-229-0257. Fax: 860-229-3445.

E-mail: nbmaa@nbmaa.org
Web Site: www.nbmaa.org
Founded: 1903.
Congressional District: 5
Key Personnel: Chm. (V), Kathryn Cox; Dir., Douglas Hyland; Dir. Devel., Claudia Thesing; Dir. Finance, Patricia Levandoski; Deputy Dir. & Cur. Education, Maura O'Shea; Collections Mgr., John Urgo; Museum Shop Mgr., Donna Downes; Museum Shop Mgr., Judith Gaffney.
Personnel Profile: Full-Time Paid 18; Part-Time Paid 9; Part-Time Volunteers 420; Interns 52.
Governing Authority: nonprofit corporation. Tax-exempt: 170(b)(1)(A) & 501(c)(3).
Institution Type/Description: Art Museum.
Collections: American art from colonial times to contemporary; paintings; sculpture; graphics; Sanford Low Memorial collection of illustration.
Research Fields: American art.
Facilities: 200-vol. library; auditorium; cafe; study room; art studio; outdoor sculpture garden. Museum-related items for sale.
Activities: docent-led tours by appointment; free 1-hour Sunday Masterpiece

tours; Take 20 at 12 Friday gallery talks; art activities & teacher services; workshops & packets for pre- and post-visit classroom activities; age-appropriate classes offered weekly; lectures; gallery talks; monthly art programs; monthly films & concerts; quarterly art & literature series. Museum Sponsors: monthly First Friday jazz evenings; quarterly Museum After Dark social events for young professionals; Annual Juneteenth Celebration; Annual Gala/Art Auction; The Art of Wine & Food Benefit.
Publications: quarterly newsletter, Art & Insight; quarterly calendar of events; catalogues, New Britain Museum of American Art: Highlights of the Collection: Volumes I & II; Charles Ethan Porter: African-American Master of Still-Life; Double Lives: American Painters as Illustrators 1850-1950; John Haberle: Master of Illusion; 15-page illustrated brochure: The Christopher Hyland Collection of Photography, By Way of These Eyes: The Sublime, Exotic and Familiar; Electronic: www.nbmaa.org.
Hours & Admission Prices: Tues.-Wed. & Fri. 11-5, Thurs. 11-8, Sat. 10-5, Sun. 12-5. Adults $9, seniors $8, students & educators $7; discounts to AAM & ICOM members; members, children under 12, & Sat. 10-12 no charge. Closed New Year's Day; Easter; Memorial Day; Independence Day; Thanksgiving; Christmas. &
Attendance: 75,459 (accurate)
Membership: Basic: Educator & Student $35; Senior $40; Individual $45; Senior Household $70; Household $75. Contributing: Friend's Circle $100; Artist's Circle $250; Collector's Circle $500. Premier: American Art Circle $1,000; John Butler Talcott Society $2,500-$10,000

NEW BRITAIN YOUTH MUSEUM, (M), 30 High St., New Britain, CT 06051-4227. Tel.: 860-225-3020. Fax: 860-229-4982.
E-mail: nbymdwtn@yahoo.com
Web Site: www.newbritainyouthmuseum.org
Founded: 1956.
Congressional District: 6
Key Personnel: Acting Dir., Ann F. Peabody; Chm. Bd., Robert Ramsey; Programs & Education, Jennifer Setten; Mktg. & Special Events, Donna M. Veach.
Personnel Profile: Part-Time Paid 3; Part-Time Volunteers 1.
Governing Authority: nonprofit organization Parent Institution: New Britain Institute. Subsidiary Institution: New Britain Youth Museum at Hungerford Park. Tax-exempt: 501(c)(3).
Institution Type/Description: Children's Museum & Nature Center.
Collections: circus miniatures; children's clothing & toys; period dolls; Native American & international artifacts; live animals; farming tools; natural history specimens.
Major Exhibits: Healthy, Wealthy & Wise, 1/14-6/14.
Research Fields: natural history; period dolls & doll houses; ethnology; natural sciences; arts & cultures; Indian artifacts; illustration; cartoons; American circus.
Facilities: 300-vol. library on circus history & fiction; available for use on premises and for inter-library loan; classrooms; Hungerford park site.
Activities: guided tours by appointment; education programs for children; temporary & participatory exhibits; school loan service; docent program.
Publications: brochures; quarterly newsletter, Ramblings.
Hours & Admission Prices: Museum: Tues.10-5, Wed.-Fri. 12-5, Sat. 10-4. No charge; donations accepted. New Britain Youth Museum, Hungerford Park: Tues.-Sat. 10-5. Adults $4, seniors $3, children $2; discounts to AAM members; members no charge. Closed federal & state holidays. &
Attendance: 42,850 (accurate)
Membership: Individual $25; Family $40; Supporting $50; Sustaining $100; Patron $250.

New Canaan

NEW CANAAN HISTORICAL SOCIETY, (M), 13 Oenoke Ridge, New Canaan, CT 06840-4195. Tel.: 203-966-1776. Fax: 203-972-5917.
E-mail: newcanaan.historical@snet.net
Web Site: www.nchistory.org
Founded: 1889.
Congressional District: 4
Key Personnel: Exec. Dir., Mrs. Janet Lindstrom; Pres. (V), Prudence Parris; Dir. Education, Kim Mellin; Archivist, Victoria Ambrose; Cur. Cody Drug Store, George Cody.
Personnel Profile: Full-Time Paid 1; Part-Time Paid 5; Part-Time Volunteers 142.
Governing Authority: nonprofit organization. Tax-exempt.
Institution Type/Description: History Museum.
Collections: history; costumes; sculpture; medical; archives; manuscripts; antique tool museum; old printing press; dolls; pewter; furniture. Historic Buildings: 1764 Hanford-Silliman House; 1825 Town House; 1878 John Rogers Studio; 1845 Cody Drug Store; 1799 Rock Schoolhouse; 1960 Irwin Pavilion; 1865 Little Red Schoolhouse.

Research Fields: local history; genealogical.
Facilities: 3,000-vol. library of history & genealogical books and bound newspapers. Books, pamphlets, postcards for sale.
Activities: lectures; permanent & temporary exhibitions; school tours.
Publications: annual magazine; books, Readings in New Canaan History; Landmarks of New Canaan; Portrait of New Canaan, the History of a Connecticut Town; A New Canaan Private in the Civil War: Letters of Justus M. Silliman; Records of World War II, Vols. I, II & III; Philip Johnson in New Canaan; John Rogers & the Rogers Groups; The Merritt Parkway; Wampum to Wall Street; various pamphlets; books, Impressions of the Hour, The Diary of an Early New Canaan Teacher; New Canaan, Texture of a Community; quarterly newsletter, Tydings.
Hours & Admission Prices: Library: Tues.-Fri. 9:30-4:30, Sat. 9:30-12:30. Museums: call for hours. No charge; donations accepted. Closed Memorial Day; Easter; Independence Day; Thanksgiving; Christmas. &
Attendance: 5,000 (estimated)
Membership: Individual $35; Family $50; Supporting $100; Sustaining $250; Patron $500; Benefactor $1,000.

NEW CANAAN NATURE CENTER, 144 Oenoke Ridge, New Canaan, CT 06840-4198. Tel.: 203-966-9577. Fax: 203-966-6536.
Web Site: www.newcanaannature.org
Founded: 1960.
Congressional District: 5
Key Personnel: Pres. (V), Mary Otocka; Exec. Dir., Laura Heckman.
Personnel Profile: Full-Time Paid 15; Part-Time Paid 15; Part-Time Volunteers 500; Interns 2.
Governing Authority: nonprofit organization. Tax-exempt: 501(c)(3).
Institution Type/Description: Nature Center.
Collections: live birds of prey, native vertebrates & invertebrates; mounts of birds, mammals & insects; rock & mineral specimens; shells; 40 acres of diverse habitats; arboretum; sugar house; cider house; bird & butterfly garden; wildflower & herb gardens; naturalist's garden.
Research Fields: bird migration.
Facilities: 8 classrooms; gardens; arboretum; sanctuary; nature trails. Gift items for sale.
Activities: lectures; field trips; seminars; workshops; courses & programs; 725 programs for school classes each year; licensed nursery school; apple cider & maple syrup making demonstrations; licensed summer camp for ages 3-17. Center Sponsors: Spring Horticulture Symposium; Spring Garden Market; Fall Fair; Secret Gardens Tour; Holiday Market.
Publications: New Canaan Nature Center newsletter; annual report; programs brochures.
Hours & Admission Prices: Mon.-Sat. 10-4. No charge; donation accepted. Closed major holidays. &
Attendance: 70,000 (estimated)
Membership: Individual $35; Family $55; Supporter $100; Sustainer $250; Benefactor $500.

PHILIP JOHNSON GLASS HOUSE, Visitor Center, 199 Elm St., New Canaan, CT 06840-5328. Tel.: 203-594-9884.
Key Personnel: Cur. & Collections Mgr., Irene Shum Allen
Institution Type/Description: Architecture Museum.
Collections: house history; contemporary art; sculptures.
Facilities: 47-acre site; visitor center. Museum-related items for sale.
Activities: guided tours; educational programs.
Hours & Admission Prices: Tours: April-Oct. 10:30 & 2:30 by appointment. Admission 10 & over $30-$45; not recommended for children under 10.

SILVERMINE GUILD ARTS CENTER, 1037 Silvermine Rd., New Canaan, CT 06840-4398. Tel.: 203-966-9700. Fax: 203-966-2763.
E-mail: silvermine@silvermineart.org
Web Site: www.silvermineart.org
Formerly: Silvermine Guild Arts Center
Founded: 1922.
Congressional District: 4
Key Personnel: Exec. Dir., Leslee Asch; Gallery Dir., Jeffrey Mueller; Dir. Silvermine School of Art, Anne Connell; Chm. (V), Roger Mudre; Museum Shop Mgr., Nancy Woodward.
Personnel Profile: Full-Time Paid 10; Part-Time Paid 3; Part-Time Volunteers 300; Interns 2.
Governing Authority: Parent Institution: Silvermine Guild of Artists, Inc.; nonprofit organization. Tax-exempt: 501(c)(3).
Institution Type/Description: Contemporary Art Gallery.
Collections: works by contemporary artists; graphics; award winners of the National Print Biennial Exhibition; 160 works in all print media from 1960-present.

Facilities: sculpture gardens & walk; studios; auditorium. Art & fine crafts shop.

Activities: guided tours; lectures; gallery talks; concerts; arts festivals; formally organized education programs; temporary exhibitions; film screenings; multi-disciplinary performances; outreach programs.

Publications: monthly enewsletter; quarterly events brochure; quarterly school catalogue.

Hours & Admission Prices: Wed.-Sat. 12-5, Sun. 1-5. No charge; donations accepted. Closed New Year's Day; Easter; Independence Day; Thanksgiving; Christmas. &

Attendance: 12,000 (estimated)

Membership: Individual $55; Dual $75; Family $125; Patron $250; Benefactor $500; Silver Benefactor $1,000; Gold Benefactor $2,500; Platinum Benefactor $5,000; Borglum Circle $10,000.

New Hartford

NEW HARTFORD HISTORICAL SOCIETY, 537 Main St., New Hartford, CT 06057. Mailing Address: P.O. Box 41, New Hartford, CT 06057-0041. Tel.: 860-379-6894.

E-mail: newhartfordhistory@att.net

Web Site: www.newhartfordhistory.org

Founded: 1943.

Congressional District: 1

Key Personnel: Chm. (V) & Pres. (V), Patrick Casey.

Governing Authority: Tax-exempt.

Institution Type/Description: Historical Society Museum.

Collections: New Hartford history; photographs; personal artifacts.

Activities: lectures; meetings; school programs; research.

Publications: quarterly newsletter.

Hours & Admission Prices: Wed. 7pm-9pm, 2nd Sat. each month 12-2 by appointment. No charge; donations accepted.

Attendance: 100 (estimated)

Membership: Individual $25-$49; Patron $50-$99; NHHS Circle $100.

New Haven

CONNECTICUT CHILDREN'S MUSEUM, 22 Wall St., New Haven, CT 06511-6528. Tel.: 203-562-5437. Fax: 203-787-9414.

E-mail: info@childrensbuilding.org

Web Site: childrensbuilding.org

Key Personnel: Dir., Sandra Malmquist

Institution Type/Description: Children's Museum.

Collections: hands-on exhibits.

Activities: educational field trips; programs.

Hours & Admission Prices: Educational Field Trips: by appointment. General Public: Fri.-Sat. 12-5. Admission $7.50. &

HARVEY CUSHING/JOHN HAY WHITNEY MEDICAL LIBRARY, HISTORICAL LIBRARY, 333 Cedar St., New Haven, CT 06510-3206. Mailing Address: P.O. Box 208014, New Haven, CT 06520-8014. Tel.: 203-785-4354. Fax: 203-785-5636.

E-mail: medical.historical@yale.edu

Web Site: www.med.yale.edu/library/historical/

Founded: 1941.

Congressional District: 3

Key Personnel: Dir. Medical Library, Regina Kenny Marone; Interim Librarian Medical History, Janene Batten; Preservation Librarian, Sarah Burge; Library Asst., Historical, Florence Gillich; Research Specialist, Historical, Thomas Falco; Cur. Prints & Drawings, Susan Wheeler.

Personnel Profile: Full-Time Paid 3; Part-Time Paid 2; Part-Time Volunteers 1.

Governing Authority: university. Parent Institution: Yale University. Tax-exempt: 501(a).

Institution Type/Description: Medical Library.

Collections: graphics; paintings; weights & measures; medical instruments.

Facilities: 400,000-vol. library of books & periodicals available to health practitioners & historians; reading room.

Activities: permanent & temporary exhibitions.

Publications: catalog of print collection.

Hours & Admission Prices: Library: Mon.-Fri. 9-4:30. No charge. Closed New Year's Day; Independence Day; Thanksgiving; Christmas. &

Membership: Associates of Cushing/Whitney Medical Library: Contributing $35; Sustaining $100; Life $1,000.

JEWISH HISTORICAL SOCIETY OF GREATER NEW HAVEN, INC., Southern Connecticut State Univ., Ethnic Heritage Center, 270 Fitch St., New Haven, CT 06515. Mailing Address: P.O. Box 3251, New Haven, CT 06515. Tel.: 203-392-6125. Fax: 203-392-5140.

E-mail: jhsgnh@yahoo.com

Web Site: jhsgnh.org

Founded: 1976.

Congressional District: 3

Key Personnel: Pres. (V), Leonard Honeyman.

Personnel Profile: Part-Time Paid 2.

Institution Type/Description: Jewish History Museum.

Collections: Jewish history, heritage & culture; photographs; personal artifacts; period furnishings.

Facilities: archives.

Activities: educational programs.

Publications: Jews in New Haven (9 Vols. 1977-2010).

Hours & Admission Prices: Mon.-Fri. 8:30 to noon; other times by appointment. No charge; donations accepted. &

Membership: Single $25; Family $36; Lifetime $200.

JOHN SLADE ELY HOUSE, 51 Trumbull St., New Haven, CT 06510-1004. Tel.: 203-624-8055. Fax: 203-624-2306.

E-mail: info@elyhouse.org

Web Site: www.elyhouse.org

Founded: 1960.

Congressional District: 3

Key Personnel: Cur., Paul Clabby.

Personnel Profile: Part-Time Paid 1; Interns 2.

Governing Authority: nonprofit organization. Parent Institution: John Slade Ely Trust. Tax-exempt.

Institution Type/Description: Art Gallery.

Collections: permanent collection of paintings & sculpture of the New Haven Paint & Clay Club.

Facilities: meeting rooms.

Activities: temporary exhibitions by groups or individuals; changing exhibits featuring contemporary work by Connecticut & Greater New Haven artists; internships; concerts; lectures. Gallery Sponsors: annual juried exhibitions; Connecticut Women Artists; New Haven Paint & Clay Club; Connecticut Water Color Society; Calligraphers Guild of New Haven.

Publications: annual calendar; brochures; catalog, 15 painters & Sculptors: An Invitational Exhibition of Connecticut Commission on the Arts Grant Recipients, 1972-1985; poster, Chinese Protest Calligraphy, 1989; exhibition catalog, Artists from People's Republic of China: Min Wang, Tong Wang, Yue-Mei Zhang & Liang Wei; also poster for the 1991 exhibition; exhibition catalog: Architecture & Process 1992; John McCrillis: 50 Years in Design 1994.

Hours & Admission Prices: Wed.-Fri. 11-4, Sat.-Sun. 2-5. No charge; donations accepted. &

Attendance: 5,000 (estimated)

KNIGHTS OF COLUMBUS MUSEUM, (M), One State St., New Haven, CT 06511-6702. Tel.: 203-865-0400 & 0320. Fax: 203-865-0351.

Web Site: www.kofcmuseum.org

Founded: 1982.

Congressional District: 3

Key Personnel: Supreme Knight & C.E.O., Carl A. Anderson; Dir., Lawrence D. Sowinski; Cur. Museum, Mary Lou Cummings; Archivist, Susan Brosnan; Asst. to Dir., Kathy Cogan; Museum Shop Mgr., Olga Lapaeva; Administrative Asst., Erica Ruzbarsky.

Personnel Profile: Full-Time Paid 9; Full-Time Volunteers 1; Part-Time Paid 3; Part-Time Volunteers 12.

Governing Authority: nonprofit organization. Parent Institution: Knights of Columbus Museum, Inc. Tax-exempt: 501(c)(3).

Institution Type/Description: History Museum; Catholic Fraternal Society.

Collections: fine & decorative art; paintings; prints; sculpture; costumes; rare book collection on Christopher Columbus; memorabilia & manuscripts relating to the history, formation & activities of the K of C from 1882 to present day.

Major Exhibits: Russian Icons and Religious Treasures, 3/15/13-2/9/14.

Research Fields: history of K of C, Catholic Church & Christopher Columbus.

Facilities: archives & library with manuscripts & other archival material pertaining to the history of K of C, Catholic Church & secular history in America available for research by appointment; reading room.

Activities: guided tours; permanent & temporary exhibitions. Annual Event: Christmas Tree Festival.

Publications: brochure; post cards; posters; fact sheet; exhibit catalogues; note cards.
Hours & Admission Prices: Daily 10-5; groups by appointment. No charge, donations accepted. Closed Good Friday; Thanksgiving; Christmas Eve & Day. &
Attendance: 23,314 (accurate)

NEW HAVEN MUSEUM, 114 Whitney Ave., New Haven, CT 06510-1238. Tel.: 203-562-4183. Fax: 203-562-2002. Facebook: New Haven Museum.
E-mail: info@newhavenmuseum.org
Web Site: www.newhavenmuseum.org
Formerly: New Haven Museum & Historical Society
Founded: 1862.
Congressional District: 3
Key Personnel: Exec. Dir., Margaret Anne Tockarshewsky.
Personnel Profile: Full-Time Paid 6; Part-Time Paid 5; Part-Time Volunteers 14; Interns 2.
Governing Authority: bd. of directors. Tax-exempt: 501(c)(3).
Institution Type/Description: History Museum.
Collections: fine, decorative & inventive arts of New Haven; photographs of New Haven, The New Haven Colony & Connecticut.
Research Fields: social, cultural, economic, technological, political & ethnic-racial history of New Haven; fine, decorative & inventive arts.
Facilities: 30,000-vol. library of archival, photographic & oral history material on New Haven & Connecticut for use on premises; 100-seat auditorium; 35-seat education center. Gift items for sale.
Activities: lectures; films; gallery talks; history workshops; education programs; docent program; inter-museum loan, permanent, temporary & traveling exhibitions.
Publications: newsletter; exhibition catalogs.
Hours & Admission Prices: Tues.-Fri. 10-5, Sat. 12-5. Adults $4, senior citizens $3, students 6-17 $2; discounts to AAA, AASLH & NEMA; children under 6 & members no charge. Closed New Year's Day; Martin Luther King Day; President's Day; Memorial Day; Independence Day; Labor Day; Columbus Day; Thanksgiving; Christmas. &
Attendance: 10,000 (estimated)
Membership: Student $20; Senior $35; Individual $40; Household $65; Contributing $100; Supporting $250; Director's Circle $500; President's Cabinet $1,000.

PARDEE-MORRIS HOUSE, 325 Lighthouse Rd., New Haven, CT 06512. Mailing Address: 114 Whitney Ave., New Haven, CT 06510. Tel.: 203-562-4183. Facebook: New Haven Museum.
E-mail: info@newhavenmuseum.org
Web Site: newhavenmuseum.org
Institution Type/Description: Historic House: housed in the former farmhouse of Amos Morris.
Collections: local history; period furnishings; personal artifacts.
Activities: special events; classes; guided tours.
Hours & Admission Prices: Summer: Sun. 12-5. No charge; donations accepted.

* **PEABODY MUSEUM OF NATURAL HISTORY, (M),** Yale University, 170 Whitney Ave., New Haven, CT 06511-8118. Mailing Address: P.O. Box 208118, New Haven, CT 06520-8118. Tel.: 203-432-5050. Fax: 203-432-9816.
E-mail: peabody.director@yale.edu
Web Site: www.peabody.yale.edu
Founded: 1876.
Congressional District: 3
Key Personnel: Dir., Cur. Invertebrate Paleontology, Derek E. G. Briggs; Asst. Dir. Public Programs & Deputy Dir., Jane Pickering; Asst. Dir. Collections & Operations, Tim White; Cur. Anthropology, Michael Dove; Cur. Anthropology, Roderick McIntosh; Cur. Anthropology, Eric Sargis; Cur. Paleobotany, Leo J. Hickey; Cur. Vertebrate Paleontology & Vertebrate Zoology, Elisabeth Vrba; Cur. Vertebrate Zoology, Thomas Near; Cur. Vertebrate Zoology, David Skelly; Cur. Vertebrate Paleontology & Vertebrate Zoology, Jacques A. Gauthier; Cur. Invertebrate Paleontology, Adolf Seilacher; Cur. Entomology, Antonia Monteiro; Assoc. Cur. Entomology, Leonard E. Munstermann; Cur. Invertebrates, Leo W. Buss; Cur. Botany, Michael J. Donoghue; Cur. Mineralogy, Jay Ague; Cur. Vertebrate Zoology, Richard O. Prum; Public Rels. & Mktg. Mgr., Melanie Brigockas; Business Mgr., Susan Castaldi; Sr. Conservator, Catherine Sease; Dir. Devel., Eliza Cleveland; Head, Peabody Fellows Program, Laura Fawcett; Head, Education & Outreach, David Heiser; Museum Shop Mgr., Kathleen Sullivan; Cur. Anthropology, Anne Underhill; Cur. Anthropology, John Darnell; Cur. Paleobotany, Peter Crane.

Personnel Profile: Full-Time Paid 63; Part-Time Paid 15; Part-Time Volunteers 217; Interns 3.
Governing Authority: university. Parent Institution: Yale University. Tax-exempt: 501(c)(3).
Institution Type/Description: University Natural History Museum.
Collections: anthropology; botany; paleobotany; invertebrate paleontology; vertebrate paleontology; vertebrate & invertebrate zoology; entomology; ichthyology; herpetology; ornithology; mammalogy; mineralogy; meteorites; historic scientific instruments; The Age of Reptiles & The Age of Mammals, murals by Rudolph F. Zallinger.
Research Fields: anthropology; botany; paleobotany; invertebrate paleontology; vertebrate paleontology; invertebrate & vertebrate zoology; entomology; ichthyology; herpetology; ornithology; mammalogy; mineralogy; meteorites; historic scientific instruments.
Facilities: 12,000-vol. library; biological field station on Long Island Sound; 175-seat auditorium. Museum-related items for sale.
Activities: formally organized education programs for schools, events for children & adults in the museum; guided tours by docents; lectures; field trips; diversified volunteer program; permanent & temporary exhibitions; audio tours.
Publications: monograph series, Bulletin of the Peabody Museum of Natural History, Yale University Publications in Anthropology; shorter papers; irregular serial; quarterly newsletter, Explorer; proceedings from scientific meetings.
Hours & Admission Prices: Mon.-Sat. 10-5, Sun. 12-5. Adults $9, senior citizens 65 & over $8, children 3-18 $5; discounts to AAM, ASTC & ICOM w/ID & Friends of Museum Associates; members, children under 3, Yale Univ. faculty & students w/ID no charge. Closed New Year's Day; Easter; Independence Day; Thanksgiving; Christmas Eve & Day. &
Attendance: 140,000 (estimated)
Membership: Full-time Student $20; Discounted Individual $35; Individual $45; Discounted Household $50; Dual $55; Household $65; Sustaining $150; Stegosaurus Club $300; Deinonychus Circle $1,000; Torosaurus Society $2,500; Lifetime $7,500.

WEST ROCK NATURE CENTER, 1080 Wintergreen Ave., New Haven, CT 06515. Mailing Address: c/o Dept. Parks, Recreation & Trees, 720 Edgewood Ave., New Haven, CT 06515. Tel.: 203-946-6559.
Web   *Site:* www.cityofnewhaven.com/parks/parksinformation/westrockpark.asp
Founded: 1946.
Congressional District: 3
Key Personnel: Outdoor Adventure Coord., Martin Torresquintero.
Personnel Profile: Full-Time Paid 3; Part-Time Volunteers 40.
Governing Authority: municipal government. Parent Institution: New Haven Dept. of Parks, Rec. & Trees Affiliated with Connecticut Assoc. of Environmental Educators, Southern Connecticut State University & New Haven Public School System. Tax-exempt.
Institution Type/Description: Nature Center.
Collections: indoor snake collection; native snakes; turtles; insects; reptiles.
Research Fields: Connecticut wildlife.
Facilities: library of books & magazines; nature trails; sledding slope; interpretive building; sundial; organic garden; native reptile display.
Activities: nature study; educational programs; school visitation; New Haven ranger program; volunteer programs; youth volunteer programs; interactive activities.
Hours & Admission Prices: Mon.-Fri. 10-4. No charge; donations accepted. Closed city holidays.
Attendance: 20,000 (estimated)

YALE CENTER FOR BRITISH ART, (M), 1080 Chapel St., New Haven, CT 06510-2302. Mailing Address: P.O. Box 208280, New Haven, CT 06520-8280. Tel.: 203-432-2800; 877-BRIT-ART. Fax: 203-432-4538. Facebook: British Art.
E-mail: ycba.info@yale.edu
Web Site: britishart.yale.edu
Founded: 1977.
Congressional District: 3
Key Personnel: Dir., Amy Meyers; Deputy Dir., Constance Clement; Dep. Dir. Finance & Administration, Rebecca Sender; Assoc. Dir. Advancement & External Affairs, Beth Miller; Cur. Paintings & Sculpture and Operations Mgr., Angus Trumble; Asst. Cur. Paintings & Sculpture and Security Supvr., Cassandra Albinson; Chief Cur. Art Collections & Sr. Cur. Prints & Drawings, Scott Wilcox; Assoc. Cur. Prints & Drawings, Gillian Forrester; Sr. Cur. Rare Books & Manuscripts, Elisabeth Fairman; Cur. Education, Linda Friedlaender; Sr. Paper Conservator, Theresa Fairbanks; Registrar, Timothy Goodhue; Asst. Museum Shop Mgr., Anissa Pellegrino; Sr. Assoc. Communications & Mktg. Mgr., Julienne Richardson; Chief Librarian,

Kraig Binkowski; Chief Conservator, Mark Aronson; Comm. & Visitor Services Coord., Aviva Luria; Operations Mgr., Martin Staffaroni; Security Supvr., Albert Wise, Jr.; Interim Special Events Coord., Caroline Hughes.
Personnel Profile: Full-Time Paid 77; Part-Time Paid 30; Part-Time Volunteers 46; Interns 6.
Governing Authority: university. Parent Institution: Yale University. Tax-exempt: 501(c)(3).
Institution Type/Description: Art Museum.
Collections: British art from 16th-century to present including 1,900 paintings, 20,000 drawings, 30,000 prints & 30,000 books & manuscripts; 100 sculptures.
Major Exhibits: Sculpture by Nicola Hicks, 11/14/13-3/9/14; Fame and Friendship: Pope, Roubiliac, and the Portrait Bust in Eighteenth-Century Britain, 2/20/14-5/19/14; Richard Wilson: and the Transformation of European Landscape Painting, 3/6/14-6/1/14; Art in Focus: Wales, 4/4/14-8/10/14; "Of Green Leaf, Bird, and Flower": Artists' Books and the Natural World, 5/15/14-8/10/14; Bruce Davidson/Paul Caponigro: Two American Photographers in Britain and Ireland, 6/26/14-9/14/14; Victorian Sculpture, 9/11/14-11/30/14; Figures of Empire: Slavery and Portraiture in Eighteenth-Century Atlantic Britain, 10/2/14-12/14/14; Nothcote, 10/2/14-12/14/14.
Research Fields: British art & related British cultural studies.
Facilities: 25,000-vol. non-circulating art reference library specializing in British art from 1550-present; conservation laboratory; 200,000 photo archive of British art worldwide; 200-seat auditorium; classrooms. Museum-related items for sale.
Activities: guided tours; permanent & temporary exhibitions; lectures; films; gallery talks; concerts; symposia; special programs for children; docent program; visiting fellowship program.
Publications: exhibition catalogs; postcards; note cards; slides.
Hours & Admission Prices: Tues.-Sat. 10-5, Sun. 12-5. No charge; donations accepted. Closed New Year's Eve & Day; Memorial Day; Independence Day; Thanksgiving; Christmas Eve & Day. &
Attendance: 100,000 (estimated)
Membership: Free Membership; Friends of British Art Program $1,100 & up.

＊ YALE UNIVERSITY ART GALLERY, (M), 1111 Chapel St., New Haven, CT 06510-2300. Mailing Address: P.O. Box 208271, New Haven, CT 06520-8271. Tel.: 203-432-0600. Fax: 203-432-9523. Facebook: Yale University Art Gallery.
E-mail: artgalleryinfo@yale.edu
Web Site: artgallery.yale.edu
Founded: 1832.
Congressional District: 3
Key Personnel: Dir., Jock Reynolds; Chm. (V), Robert W. Doran; Deputy Dir. Finance & Administration, Jessica Labbe; Deputy Dir. Collections & Education, Pam Franks; Deputy Dir. Museum Resources & Stewardship, Jill Westgard; Dir. Public Rels., Maura Scanlon; Dir. Publications & Editorial, Tiffany Sprague; Chief Conservator, Ian McClure; Cur. American Decorative Arts, Patricia Kane; Cur. Education & Academic Affairs, Kate Ezra; Cur. Asian Art, David Sensabaugh; Cur. African Art, Frederick Lamp; Cur. Coins & Medals, William E. Metcalf; Cur. Prints, Drawings & Photographs, Suzanne Boorsch; Assoc. Cur. Public Education, Jessica Sack; Dir. Exhibitions, Jeffrey Yoshimine; Cur. American Paintings & Sculpture, Helen Cooper; Cur. Early European, Laurence Kanter; Cur. Indo-Pacific Art, Ruth Barnes; Cur. Ancient Art, Susan B. Matheson; Cur. Modern & Contemporary Art, Jennifer Gross; Deputy Dir. Operations & Planning, Carol DeNatale; Coord. Visitor Svcs., Caitlin Zaccaro.
Personnel Profile: Full-Time Paid 139; Part-Time Paid 9; Part-Time Volunteers 25; Interns 96.
Governing Authority: university. Parent Institution: Yale University. Tax-exempt: 501(c)(3).
Institution Type/Description: Art Museum.
Collections: American painting & decorative arts; Greek & Roman art including artifacts from the ancient Roman City of Dura-Europos; Jarves, Griggs, & Rabinowitz collections of early Italian paintings; European, Asian, & African art including the Charles B. Benenson collection of African art; art of the ancient Americas; Societe Anonyme collection of early 20th century European & American art; Impressionist, modern, & contemporary works.
Research Fields: all fields pertaining to collections.
Facilities: Books relating to collections & other museum-related items for sale.
Activities: guided tours; lectures; gallery talks; concerts; permanent & loan exhibitions; family programs; school group tours.
Publications: Yale University Art Gallery Bulletin; interpretive guides to the collection; exhibition catalogues & catalogue raisonnes; membership magazine; calendar of events.
Hours & Admission Prices: July-Aug. Tues.-Fri. 10-5, Sat.-Sun. 11-5; Sept.-June Tues.-Wed. & Fri. 10-5, Thurs. 10-8, Sat.-Sun. 11-5. Tours: no charge. Group Tours: call for fees & information. Closed major holidays. &
Attendance: 105,000 (estimated)

Membership: Student $20; Individual $50; Supporter $75; Contributor $125; Sponsor $250; Benefactor $500; Fellow $1,000; Patron $2,500, $5,000, & $10,000.

YALE UNIVERSITY COLLECTION OF MUSICAL INSTRUMENTS, 15 Hillhouse Ave., New Haven, CT 06511-6823. Mailing Address: P.O. Box 208278, New Haven, CT 06520-8278. Tel.: 203-432-0822. Fax: 203-432-8342.
E-mail: musinst@pantheon.yale.edu
Web Site: www.yale.edu/musicalinstruments
Founded: 1900.
Congressional District: 3
Key Personnel: Interim Dir., William Purvis; Cur., Nicholas Renouf; Cur., Susan E. Thompson.
Governing Authority: university. Tax-exempt: 501(c)(3).
Institution Type/Description: Musical Instruments Museum.
Collections: 16th-18th centuries keyboard instruments; 17th-20th centuries stringed instruments; 15th-20th centuries American & European instruments; bells; mechanical instruments.
Research Fields: musical instruments; organology.
Facilities: 400-vol. library of catalogs, books, brochures, treatises & facsimiles pertaining to the history & identification of musical instruments available for use by approval of university librarian. Publications, recordings, photographs & postcards for sale.
Activities: guided tours; lectures; gallery talks; concerts; formally organized education programs for undergraduate & graduate college students; permanent & temporary exhibitions.
Publications: various catalogs & checklists.
Hours & Admission Prices: Sept.-July Tues.-Fri. 1-4, Sun. 1-5. Closed university recesses, summer & national holidays.
Membership: Student & Senior Citizens over 65 $5; Single $10; Family $20; Regular Associate $25; Family Associate $45; Sustaining $50; Patron $100; Life $500.

New London

CONNECTICUT COLLEGE ARBORETUM, 270 Mohegan Ave., New London, CT 06320-4150. Mailing Address: Campus Box 5201, 270 Mohegan Ave., New London, CT 06320. Tel.: 860-439-5020 & 5060. Fax: 860-439-5482.
E-mail: arbor@conncoll.edu
Web Site: arboretum.conncoll.edu
Founded: 1931.
Congressional District: 2
Key Personnel: Dir., Glenn D. Dreyer; Asst. Dir., Kathy Dame; Arboretum Horticulturist, Leigh S. Knuttel; Groundsperson, Bryan Goulet; Senior Groundsperson, Charles A. McIlwain, III; Staff Asst., Elene Anthopolos; Cur. & Information Resource Mgr., Mary Villa.
Personnel Profile: Full-Time Paid 6; Part-Time Paid 2; Part-Time Volunteers 50.
Governing Authority: college. Parent Institution: Connecticut College. Tax-exempt: 501(c)(3).
Institution Type/Description: Arboretum.
Collections: woody plants native to N.E.U.S.; ornamental trees & shrubs; natural areas.
Research Fields: naturalistic landscaping; terrestrial & wetland ecology; vegetation management with fire and herbicides.
Facilities: nature trails.
Activities: guided tours; lectures; formally organized education programs for adults & undergraduate students; children's activities; demonstrations.
Publications: Arboretum Bulletins & research papers.
Hours & Admission Prices: Daily sunrise-sunset. No charge.
Attendance: 12,000 (estimated)
Membership: Student $10; Individual $30; Organization $40; Sustaining $50; Supporting $100.

GROTON MONUMENT AND MONUMENT HOUSE MUSEUM, Fort Griswold Battlefield State Park, 90 Walbach St., New London, CT 06320. Tel.: 860-445-1729.
Institution Type/Description: History Museum.
Collections: local history; personal artifacts; uniforms; photographs; period furnishings. Historic Structure: granite monument built 1826-1930.
Hours & Admission Prices: Memorial Day to Labor Day daily 9-5. No charge.

HEMPSTED HOUSES, 11 Hempstead St., New London, CT 06320. Tel.: 860-443-7949. Fax: 860-249-4907.
E-mail: hempsted@ctlandmarks.org
Web Site: www.ctlandmarks.org
Key Personnel: Dir., Sheryl Hack.

Governing Authority: private; nonprofit organization. Parent Institution: CT Landmarks, Hartford, CT. Tax-exempt: 501(c)(3).
Institution Type/Description: Historic Houses.
Collections: Hempsted family history; period furnishings; personal artifacts; photographs. Historic Houses: Joshua Hempsted House; Nathaniel Hempsted House.
Research Fields: paranormal investigations.
Facilities: Museum-related items for sale.
Activities: guided tours; workshops; lectures.
Hours & Admission Prices: May-Oct. call for hours; groups by appointment. Adults $7, senior citizens, teachers, & students $6, children 6-18 $4; discount to families; children under 6 no charge.

*   **LYMAN ALLYN ART MUSEUM, (M),** 625 Williams St., New London, CT 06320-4199. Tel.: 860-443-2545. Fax: 860-443-2060.
E-mail: stula@lymanallyn.org
Web Site: www.lymanallyn.org
Formerly: Lyman Allyn Museum of Art at Connecticut College
Founded: 1932.
Congressional District: 2
Key Personnel: Exec. Dir. & Cur., Nancy Stula; Pres., William A. Lieber; Dir. Devel., Nancy Hileman; Asst. to Dir., Ann Wicks.
Personnel Profile: Full-Time Paid 9; Part-Time Paid 6; Part-Time Volunteers 40; Interns 1.
Governing Authority: nonprofit corp. Tax-exempt: 501(c)(3).
Institution Type/Description: Art Museum.
Collections: 10,000 objects of art: contemporary, modern & early American fine arts; American Impressionist paintings; Connecticut decorative arts; 17th-19th century European works on paper. Historic House: The Deshon-Allyn Mansion.
Facilities: 10,000-vol. library of art reference books available by appointment on premises; 78-seat auditorium.
Activities: guided tours; lectures; gallery talks; travel program; docent program; inter-museum loan, permanent & temporary exhibitions; children's art workshops; adult & kids education programs; corporate partnership programs. Museum Sponsors: Monthly Free First Sundays.
Publications: catalogues: The American Collection; Beatrice Cumming 1903-1974; John Day; The Devotion Family; Dolls and Doll Houses and Toys; Charles Ebert; Elizabeth Enders; Lyman Allyn Art Museum Handbook; Mirror of Fashion; Rollie McKenna; New Artist Drawings; New London Silver; A New London Whaler on the California Gold Rush; Painters of Light and Color; The Shakers of Sabbathday Lake; Through the Eyes of Charles Chu; The Barkley L. Hendricks Experience; American Artists Abroad and Their Inspiration; The Vision and Influence of Winslow Ames; James Britton; At Home and Abroad: The Transcendental Landscapes of Christopher Cranch.
Hours & Admission Prices: Tues.-Sat. 10-5, Sun. 1-5. Adults $8, students & seniors $7; AAM, museum members, New London residents, children under 12, Connecticut College faculty, staff & students no charge. Closed major holidays. &
Attendance: 25,000 (accurate)
Membership: Student $25; Individual $40; Family $60; Sustaining $250; Patron $500; Harriet Allyn Society $1,000.

NATHAN HALE SCHOOLHOUSE IN NEW LONDON, 19 Atlantic St., New London, CT 06320. Mailing Address: Connecticut SAR, P.O. Box 411, East Haddam, CT 06423. Tel.: 860-873-3399.
Web Site: www.connecticutsar.org
Founded: 1889.
Governing Authority: Tax-exempt.
Institution Type/Description: Historic Building: housed in the former schoolhouse where Nathan Hale was schoolmaster from 1774-1775.
Collections: local history; Nathan Hale's life & military career; period furnishings.
Hours & Admission Prices: May-Oct. Wed.-Sun. 11-4. No charge; donations accepted.
Attendance: 3,500 (accurate)

NEW LONDON COUNTY HISTORICAL SOCIETY, (M), 11 Blinman St., New London, CT 06320-5677. Tel.: 860-443-1209. Fax: 860-443-1209.
E-mail: info@newlondonhistory.org
Web Site: newlondonhistory.org/
Founded: 1870.
Congressional District: 2
Key Personnel: Exec. Dir., Edward D. Baker; Pres., Nancy H. Steenburg; Librarian, Tricia Royston.

Personnel Profile: Full-Time Paid 1; Part-Time Paid 1; Part-Time Volunteers 45; Interns 3.
Volunteer Hours: 2,000
Operating Expenses: 116,000
Operating Income: 118,000
Governing Authority: private; nonprofit organization. Subsidiary Institution: Shaw Mansion. Tax-exempt: 501(c)(3).
Institution Type/Description: Historic House Museum: housed in the 1756 Shaw Mansion, used as Connecticut's Naval Office during the Revolution.
Collections: Shaw family portraits, furniture; 1,300 manuscripts including early history of New London & inhabitants; decorative arts; textiles; tools; paintings; naval activities during the Revolutionary War; newspapers; whaling; Amistad exhibit documents; Frances M. Caulkins collection.
Major Exhibits: Plunder and Peril: New London During the American Revolution, 10/13-12/14; John Ewen, Jr., Whaling Port Artist, 5/1/14-10/1/14.
Research Fields: local history; genealogy; history of the African American community in Southeastern Connecticut; maritime history; whaling & sealing; Colonial trade with the West Indies; privateering during the American Revolution.
Facilities: 5,000-vol. library of books on genealogy & history; reading room; botanical gardens. Museum-related items for sale.
Activities: guided tours; permanent & temporary exhibitions; lecture series; historic bus tours; History Club for middle school students; special events; first-person interpretive programs. Museum Sponsors: Vintage Baseball Club.
Publications: quarterly newsletter, The NLCHS Newsletter; Diary of Joshua Hempstead, c.1711-1758; Life on a Whaler; Tapestry, A History of Blacks in South East Connecticut; Greetings from New London, A Collection of Early 20th-Century Postcards; Black Roots in Southeastern CT 1650-1900; The Amistad Incident as Reported in the New London Gazette; History of the Amistad Captives; A View from the Sixties: The Black Experience in Southeastern CT; The History of New London, Frances M. Caulkins, 1860; The History of Norwich, Frances M. Caulkins, 1866; New London Goes to War: New London During World War II, 1941-1945; The Rockets' Red Glare: The War of 1812 and Connecticut.
Hours & Admission Prices: Winter: Wed.-Fri. 1-4; Summer: Wed.-Fri. 1-4, Sat. 10-4. Library: by appointment only. Adults $5, senior citizens $4, children under 12 $2; discounts to AAM & NEMA members and groups of 10 or more; members no charge. Closed national holidays; Thanksgiving; Christmas. &
Attendance: 3,100 (estimated)
Membership: Senior $20; Individual & Senior Couples $30; Family $40; Contributing $50; Sustaining $100; Patron $250; Life $1,000

U.S. COAST GUARD MUSEUM, U.S. Coast Guard Academy, 15 Mohegan Ave., New London, CT 06320-4195. Tel.: 860-444-8511. Fax: 860-701-6700.
E-mail: arlyn.s.danielson@uscg.mil
Web Site: www.uscg.mil/hq/cg092/museum
Founded: 1967.
Congressional District: 2
Key Personnel: Cur., Arlyn Danielson.
Personnel Profile: Full-Time Paid 1.
Governing Authority: federal. Parent Institution: U.S. Coast Guard. Subsidiary Institution: U.S. Coast Guard Academy. Tax-exempt.
Institution Type/Description: Maritime Museum.
Collections: lighthouse lenses; rescue equipment; ship & airplane models; paintings & artifacts relating to the U.S. Coast Guard & its predecessors, the Revenue-Cutter service, Lighthouse service & Life-Saving service.
Research Fields: Coast Guard history.
Facilities: 160,000-vol. college undergraduate library, available for use by appointment; military academy.
Activities: small changing exhibits.
Hours & Admission Prices: early Jan. to May & Sept. to mid-Dec. Mon.-Fri. 9-4, Sat. 10-4, Sun. 1-4; Summer: Mon.-Fri. 9-4, 1st & 3rd Sat. 10-4. No charge; donations accepted. Government-issued photo identification needed to enter campus. Closed federal holidays; winter break. &
Attendance: 25,000 (accurate)

New Milford

THE NEW MILFORD HISTORICAL SOCIETY AND MUSEUM, 6 Aspetuck Ave., New Milford, CT 06776. Tel.: 860-354-3069. Fax: 860-210-0263.
E-mail: nmhistorical@gmail.com
Web Site: www.nmhistorical.org
Founded: 1915.
Congressional District: 5

Key Personnel: Pres. Bd. Trustees (V), Joanne Lillis; Cur., Lisa D. Roush; Corresponding Sec., Susan Metcalf.
Personnel Profile: Part-Time Paid 2; Part-Time Volunteers 10.
Governing Authority: board of trustees. Tax-exempt.
Institution Type/Description: Historical Society Museum.
Collections: glass; china; silver; pewter; Wannopee pottery; miniatures; portraits including those by Ralph Earl & Richard Jennys; costumes, samplers, quilts & other textiles; toys, including dolls; furniture; historical & genealogical records; New Milford memorabilia; photographs; archives. Historic Buildings: early 19th-century Knapp House; c.1796 Boardman Store; c.1820 Bank of Litchfield County; c.1843 Hill & Plain Schoolhouse #1.
Research Fields: genealogy; New Milford history.
Activities: tours; lectures; classes; special events.
Publications: books, Edith Newton's New Milford; Howard Peck's New Milford; Voices From The Past; The Courage of Sarah Noble; pamphlet, Two Tales of New Milford; maps; prints of New Milford scenes.
Hours & Admission Prices: Tues.-Thurs. 12-3; other times by appointment. Adults $5; discounts to students; members no charge. Closed legal holidays. ⅁
Membership: Individual $25; Family $50; Sustaining $75; Corporate $95; Life $300.

Newington

AMERICAN RADIO RELAY LEAGUE, 225 Main St., Newington, CT 06111. Tel.: 860-594-0200. Fax: 860-594-0259.
Institution Type/Description: Amateur Radio History.
Collections: Amateur Radio history & equipment; vintage AM station; photographs.
Hours & Admission Prices: Mon.-Fri. 8-5; groups by appointment.

ENOCH KELSEY HOUSE, 1702 Main St., Newington, CT 06111-3938. Tel.: 860-667-0545.
E-mail: NGTNheritage@aol.com
Web Site: www.newingtonhistoricalsociety.org/index.htm
Key Personnel: Dir., Dorothy Abbott.
Governing Authority: Parent Institution: Newington Historical Society & Trust, Inc.
Institution Type/Description: Historic House Museum: home built by Enoch Kelsey & his son David, c.1799.
Collections: period furnishings; personal artifacts.
Hours & Admission Prices: April-Nov. 1st & 3rd Sun. of month 1-3. Closed holidays.

KELLOGG-EDDY HOUSE AND MUSEUM, 679 Willard Ave., Newington, CT 06111-2615. Tel.: 860-666-7118.
E-mail: NGTNheritage@aol.com
Web Site: www.newingtonhistoricalsociety.org/index.htm
Key Personnel: Dir., Dorothy Abbott.
Governing Authority: Parent Institution: Newington Historical Society & Trust, Inc.
Institution Type/Description: Historic House Museum: housed in the former home of General Martin Kellogg, built in 1808.
Collections: period furnishings; personal artifacts.
Hours & Admission Prices: April-Nov. 1st Sun. of month 1-3. Closed holidays.

Niantic

CHILDREN'S MUSEUM OF SOUTHEASTERN CONNECTICUT, 409 Main St., Niantic, CT 06357-3103. Tel.: 860-691-1111. Fax: 860-691-1194.
E-mail: pclaffey@childrensmuseumsect.org
Web Site: childrensmuseumsect.org
Founded: 1992.
Congressional District: 2
Key Personnel: Exec. Dir., Peter Claffey; Pres. (V), Carla Barone; Exhibit Coord., Michael Neville; Education Coord., Marjorie Collins-DellaRocco; Graphic Design Coord., Julie Clements-Reagan.
Personnel Profile. Full-Time Paid 1; Part-Time Paid 12; Part-Time Volunteers 30; Interns 0.
Governing Authority: private; nonprofit organization. Tax-exempt: 501(c)(3).
Institution Type/Description: Children's Museum.
Collections: teaching collections.
Facilities: 5,000 sq. ft. exhibit space; classroom.
Activities: indoor & outdoor hands-on interactive exhibits for children & families; educational programs for children 1-8.
Publications: quarterly newsletter, Building Blocks; annual report.
Hours & Admission Prices: Tues.-Sat. 9:30-4:30, Sun. 12-4:30. Admission $7;

discount to groups of 10 or more; members no charge. Closed Easter, Memorial Day, Independence Day, Labor Day, Thanksgiving, Christmas. ⅁
Attendance: 40,000 (accurate)
Membership: Grandparent $60; Family $75; ACM $100. (military memberships available)

EAST LYME HISTORICAL SOCIETY/THOMAS LEE HOUSE, 228 W. Main St., Niantic, CT 06357. Mailing Address: P.O. Box 112, East Lyme, CT 06333-0112. Tel.: 860-739-9660. Fax: 860-444-6661.
E-mail: info@eastlymehistoricalsociety.org
Web Site: www.eastlymehistoricalsociety.org
Founded: 1894.
Congressional District: 2
Key Personnel: Pres., Norman B. Peck, III; Treas., Sonya Hoisington; Museum Shop Mgr., Elizabeth Kuchta.
Personnel Profile: Part-Time Volunteers 40.
Governing Authority: society. Tax-exempt: 501(c)(3).
Institution Type/Description: Historical Society Museum.
Collections: furnishings; personal artifacts. Historic Buildings: c.1660 Thomas Lee House; Little Boston Schoolhouse c.1734.
Activities: guided tours; lectures; temporary exhibitions; elementary education programs.
Publications: quarterly newsletter.
Hours & Admission Prices: mid-June to Labor Day Tues.-Sun. 1-4; other times by appointment. No charge; donations accepted.
Attendance: 500 (estimated)
Membership: Individual $5; Family $10; Sustaining $25; Life $150.

THOMAS AVERY HOUSE A/K/A SMITH-HARRIS HOUSE, 33 Society Rd., Niantic, CT 06357. Mailing Address: Friends of the Smith-Harris House, P.O. Box 6, Niantic, CT 06357. Tel.: 860-739-0761.
Institution Type/Description: Historic House Museum: housed in the former home of early settlers, Christopher Avery and Nehemiah Smith; built in 1845.
Collections: family history; period furnishings; photographs.
Activities: rental facilities.
Hours & Admission Prices: June-Aug. Fri.-Sun. 12-4; other times by appointment. No charge; donations accepted. Closed Independence Day.

Noank

NOANK HISTORICAL SOCIETY, INC., 17 Sylvan St., Noank, CT 06340-5742. Mailing Address: P.O. Box 9454, Noank, CT 06340-9454. Tel.: 860-536-3021 & 3029.
E-mail: noankhist@sbcglobal.net
Web Site: www.noankhistoricalsociety.org
Founded: 1966.
Congressional District: 2
Key Personnel: Pres., Deborah Bates; Vice Pres., Justin Carmarata; Cur., Mary Anderson; Historian, Arnold Crossman; Treas., Steven Anderson.
Governing Authority: society; nonprofit organization. Subsidiary Institution: Latham/Chester Store Museum, 108 Main St., Noank, CT 06340. Tax-exempt: 501(c)(3).
Institution Type/Description: History Museum: housed in c.1847 mercantile building.
Collections: marine; naval; local history; medical & domestic paraphernalia; art; historical photographs; shipbuilder's tools; exhibition of local fisheries.
Research Fields: Indian; marine; shipbuilding; fishing; histories of old houses; local history; genealogy.
Facilities: Books, early maps, photos & other museum-related items for sale.
Activities: quarterly meetings; group tours; local history program; lectures. Annual Events: Independence Day Art Show & Celebration.
Publications: quarterly newsletter; book, Noank: From the Papers of Claude Chester. Noank: The Ethereal Years; The Log of Downit; Captains B. F. Rathbun of Noank; Noank Celebrating A Maritime Heritage.
Hours & Admission Prices: July-Columbus Day Wed. & Sat.-Sun. 2-5; other times by appointment. No charge; donations accepted.
Attendance: 1,000 (estimated)
Membership: Individual $10; Family $15; Contributing $25 & over.

Norfolk

NORFOLK HISTORICAL MUSEUM, 13 Village Green, Norfolk, CT 06058. Mailing Address: P.O. Box 288, Norfolk, CT 06058-0288. Tel.: 860-542-5761.
E-mail: norfolkhistorical@sbcglobal.net
Web Site: www.norfolkhistoricalsociety.org

Founded: 1960.
Congressional District: 6
Key Personnel: Pres. (V), Barry Webber; 1st Vice Pres., Richard Byrne; 2nd Vice Pres., Eric Anderson; Treas., Susan Dyer; Sec., Sally Vaun.
Personnel Profile: Part-Time Paid 1; Part-Time Volunteers 15.
Governing Authority: nonprofit corporation. Norfolk Historical Society, Inc. Tax-exempt: 501(c)(3).
Institution Type/Description: Historical Society Museum.
Collections: decorative arts; costumes; folklore; maps; manuscripts; artifacts; farm tools; photographs; genealogical material; local history archive including Norfolk area.
Research Fields: early Norfolk history & residents.
Facilities: 200-vol. library of books on local history & genealogy available on premises.
Activities: lectures; permanent & temporary exhibitions.
Publications: pamphlets; annual society newsletter, The Muse.
Hours & Admission Prices: June-Oct. 10 Sat.-Sun. 12-4; Nov.-April 1st Thurs. of month 1-5; other times by appointment. No charge; donations accepted. &
Membership: Family & Friends $10-$25; Supporting $50-$75; Donor $100-$300; Patron $500.

North Haven

NORTH HAVEN HISTORICAL SOCIETY, 27 Broadway, North Haven, CT 06473-2302. Tel.: 203-239-7722.
E-mail: nhhistsoc@gmail.com
Web Site: www.northhavenhistoricalsociety.wordpress.com
Founded: 1957.
Key Personnel: Pres., Robert Iverson; Vice Pres., Steve Nugent; Treas., Walter Brockett; Cur., Gloria Furnival.
Personnel Profile: Part-Time Volunteers 15.
Governing Authority: nonprofit organization. Tax-exempt: 501(c)(3).
Institution Type/Description: History Museum.
Collections: local history & culture; documents, deeds, manuscripts, diaries, journals & account books pertaining to North Haven; genealogical material; Indian artifacts; projectile points; textiles; looms; agricultural items; historic buildings.
Research Fields: local history.
Facilities: library pertaining to basic research & rare books available for research on premises by special arrangement with librarian.
Activities: guided tours; lectures; films; formally organized education programs for children; temporary exhibitions; scholarship fund for North Haven High School senior. Annual Event: Trolley Tour of Town.
Publications: booklets, North Haven in the Revolution; The Quinnipiac; Two to Eight - A Country Childhood When the Century was New; Amidst Cultivated & Pleasant Fields, A Bicentennial History of North Haven, CT; On the Green, The Old Center Cemetery in North Haven, Connecticut 1723-1882.
Hours & Admission Prices: July-Aug. Thurs. 3-6; Sept.-June Tues. & Thurs. 3-6. Closed New Year's Eve & Day; Christmas Eve, Day & week.
Membership: Individual $15; Family $25; Supporting $75; Life Individual $125; Life Family $200.

North Stonington

NORTH STONINGTON HISTORICAL SOCIETY - STEPHEN MAIN HOMESTEAD, 1 Wyassup Rd., North Stonington, CT 06359-1322. Mailing Address: P.O. Box 134, North Stonington, CT 06359. Tel.: 860-535-9448.
E-mail: gchase1@comcast.net
Web Site: nostoningtonhistsoc.homestead.com
Founded: 1970.
Key Personnel: Pres., Frank Eppinger.
Governing Authority: Tax-exempt.
Institution Type/Description: Historical Society Museum: house built in 1781.
Collections: local history & culture; period furnishings; personal artifacts; manuscripts; land records; genealogy.
Activities: demonstrations; special events.
Publications: newsletter; books.
Hours & Admission Prices: Tues. 2-4. No charge; donations accepted. &
Membership: Senior $15; Individual $20; Family $25; Sustaining $50; Corporate $100; Life $500.

Norwalk

ART GALLERY AT NORWALK COMMUNITY COLLEGE, 188 Richards Ave., Norwalk, CT 06854. Tel.: 203-857-7000.
Institution Type/Description: Art Gallery.
Collections: paintings; prints; photography; sculpture.

Hours & Admission Prices: Mon.-Thurs. 9-9, Fri. 9-4, Sat. 9-12. No charge. Closed holidays.

CENTER FOR CONTEMPORARY PRINTMAKING, Mathews Park, 299 West Ave., Norwalk, CT 06850-4002. Tel.: 203-899-7999. Fax: 203-899-7997.
E-mail: info@contemprints.org
Web Site: www.contemprints.org
Founded: 1995.
Congressional District: 4
Key Personnel: Exec. Dir., Julyen Norman.
Personnel Profile: Full-Time Paid 5; Part-Time Paid 1; Part-Time Volunteers 20; Interns 6.
Institution Type/Description: Printmaking Museum.
Collections: prints; printmaking.
Hours & Admission Prices: Mon.-Sat. 9-5, Sun. 12-5. No charge. &
Attendance: 15,000 (estimated)

LOCKWOOD-MATHEWS MANSION MUSEUM, (M), 295 West Ave., Norwalk, CT 06850-4002. Tel.: 203-838-9799. Fax: 203-838-1434.
E-mail: info@lockwoodmathewsmansion.com
Web Site: www.lockwoodmathewsmansion.com
Founded: 1966.
Congressional District: 4
Key Personnel: Exec. Dir., Susan Gilgore; Chm. (V), Patsy Brescia; Museum Shop Mgr., Kathy Gilbertie.
Personnel Profile: Full-Time Paid 2; Part-Time Paid 2; Part-Time Volunteers 60.
Governing Authority: nonprofit organization. Tax-exempt.
Institution Type/Description: Historic House Museum: c.1868 Elm Park Home.
Collections: Victorian furniture; decorative wall paintings; inlaid & carved woodwork; period costumes; 19th-century decorative arts, paintings & archives.
Research Fields: Victorian arts & architecture.
Facilities: Victorian gifts for sale.
Activities: guided tours; lectures; arts festivals; formally organized education programs for children; docent program; traveling exhibitions.
Publications: quarterly newsletter; books, LeGrand Lockwood; The Lockwood Mathews Mansion; Nineteenth Century Architects: Building a Profession; Nineteenth Century Women Photographers: A New Dimension in Leisure; The Tiffanys.
Hours & Admission Prices: Tours: Jan-March by appointment only; April-Dec. Wed.-Sun. 12-4 (last tour at 3 pm). Adults $10, senior citizens $8, youth 8-18 $6; discount to National Preservation Trust members; children under 8 & members no charge. Closed major holidays. &
Attendance: 23,000 (estimated)
Membership: Individual $35; Family $55; Circle Donor $100; Circle Sponsor $250; Circle Patron $500; Circle Benefactor $1,000; Circle Angel $2,500.

THE MARITIME AQUARIUM AT NORWALK, 10 N. Water St., Norwalk, CT 06854-2228. Tel.: 203-852-0700, ext. 2248. Fax: 203-838-5416.
E-mail: marketing@maritimeaquarium.org
Web Site: www.maritimeaquarium.org
Founded: 1988.
Congressional District: 4
Key Personnel: C.E.O. & Pres., Jennifer Herring; Education, Jamie Alonzo; Mktg., Chris Loynd; Dir. Exhibits, Judith Bacal; Museum Shop Mgr., Sergio Munoz.
Personnel Profile: Full-Time Paid 50; Part-Time Paid 50; Part-Time Volunteers 175; Interns 20.
Governing Authority: nonprofit organization. Tax-exempt: 501(c)(3).
Institution Type/Description: Aquarium: housed in 19th-century iron works foundry.
Collections: aquarium of marine life indigenous to Long Island Sound, including harbor seals & sharks; sea turtles; jelly fish; River otters; artifacts of the oystering industry in Norwalk; marine science & culture; sharks & rays touch tank; intertidal animals touch tank.
Research Fields: monitoring physical & chemical water quality parameters, including dissolved oxygen; marine population studies in Long Island Sound; harbor seal census.
Facilities: aquarium; 200-seat restaurant; classrooms; laboratories; 75,000 sq. ft. exhibit space; 310-seat IMAX theatre; 1 research vessel. Environmental education center: classrooms, laboratory, 180-seat school cafeteria. Museum-related items for sale.
Activities: docent program; IMAX films; formal educational programs; guided

tours; ecology cruises; lectures; loan, temporary & traveling exhibitions; mobile vans; interactive exhibits; gallery rental; theatre; conferences; festivals. Annual Events: Holiday Programs; Members Reception; Harbor Seal Symposium.
Publications: quarterly newsletter, The Maritimes; school programs brochure.
Hours & Admission Prices: Aquarium: adults $19.95, seniors 65 & over $17.95, children 12 & under $12.95; children 3 & under & members no charge. Closed Thanksgiving; Christmas. &
Attendance: 500,000 (estimated)
Membership: Individual Plus One $85; Family $110; Family AuPair $130; Family Plus $155; Family Deluxe $275; Friends Society $1,000 & up.

NORWALK HISTORICAL SOCIETY, Mill Hill Historic Park, 2 E. Wall St., Norwalk, CT 06852. Mailing Address: P.O. Box 1640, Norwalk, CT 06852. Tel.: 203-846-0525.
E-mail: info@norwalkhistoricalsociety.org
Web Site: norwalkhistoricalsociety.org
Institution Type/Description: Historical Society Museum.
Collections: local history & culture; period furnishings; personal artifacts; photographs. Historic Buildings: Town House, built in 1835; Gov. Thomas Fitch Law Office, c.1740; Downtown District Schoolhouse, built in 1826.
Facilities: Museum-related items for sale.
Activities: educational programs. Annual Events: 4th of July Bell Ringing Ceremony; Colonial Harvest Festival in November.
Hours & Admission Prices: Call for hours.

SHEFFIELD ISLAND LIGHTHOUSE, Hope Dock (Ferry Svc.), Washington St. & N. Water St., Norwalk, CT 06854. Mailing Address: Norwalk Seaport Association, 132 Water St., 06854, CT 06854. Tel.: 203-838-9444.
Founded: 1978.
Key Personnel: Pres., Irene Dixon
Governing Authority: Tax-exempt.
Institution Type/Description: Historic Lighthouse: housed on Sheffield Island; built in 1868.
Collections: lighthouse; period artifacts.
Hours & Admission Prices: Ferry Service: May-Sept. call for schedule. Adults $22, seniors $20, children 4-12 $12, children 3 & under $5.
Attendance: 5,000 (estimated)
Membership: Senior $25; Senior Family $40; Individual $50; Family $75.

SONO SWITCH TOWER MUSEUM, 77 Washington St., Norwalk, CT 06854-3086. Tel.: 203-246-6958.
Institution Type/Description: Historic Building: built in 1896.
Collections: railroad & switch tower history; railroad memorabilia; period artifacts; photographs.
Hours & Admission Prices: May-Oct. Sat.-Sun. 12-5. No charge; donations accepted.

STEPPING STONES MUSEUM FOR CHILDREN, 303 West Ave., Mathews Park, Norwalk, CT 06850-4002. Tel.: 203-899-0606. Fax: 203-899-0530.
E-mail: information@steppingstonesmuseum.org
Web Site: www.steppingstonesmuseum.org
Founded: 1992.
Key Personnel: Exec. Dir., Rhonda Kiest; Devel., Jennifer Toussaint; Chm. Bd., Sandra Miklave; Dir. Education, Hyla Crane; Dir. Mktg. & Communications, Mary Tobin; Museum Shop Mgr., Dina Berger.
Personnel Profile: Full-Time Paid 45; Part-Time Paid 35; Part-Time Volunteers 63; Interns 1.
Governing Authority: private; nonprofit organization. Tax-exempt: 501(c)(3).
Institution Type/Description: Children's Museum.
Collections: hands-on science, the arts, culture & heritage.
Major Exhibits: Healthyville, 2/2/14-8/14; Conservation Quest, 9/14-12/14.
Research Fields: early childhood development; K-5 learning standards; environmental design; community collaboration.
Facilities: 1,000-vol. library; cafe; 22,000 sq. ft. exhibit space; resource center; gardens. Museum-related items for sale.
Activities: public programs including workshops & performances; school & group tours; traveling workshops; community & youth programs.
Publications: monthly e-newsletter; annual report.
Hours & Admission Prices: Academic Year: Tues.-Sun. 10-5; Summer: daily 10-5. Adults & children $15, seniors $10; children under 1 no charge. Closed New Year's Day; Easter; Thanksgiving; Christmas. &
Attendance: 285,000 (accurate)
Membership: Family $155; Family Plus $205; Friend $400.

Norwich

FAITH TRUMBULL CHAPTER, DAUGHTERS OF THE AMERICAN REVOLUTION, INC., MUSEUM AND CHAPTER HOUSE, 42 Rockwell St., Norwich, CT 06360-3537. Tel.: 860-887-8737.
Founded: 1893.
Congressional District: 2
Key Personnel: Regent, Polly Gunther; Cur., Marianne Vanden Bout.
Personnel Profile: Full-Time Paid 1; Part-Time Volunteers 10.
Governing Authority: nonprofit organization. Parent Institution: The National Society Daughters of the American Revolution. Tax-exempt.
Institution Type/Description: Historical Society Museum.
Collections: furniture; clothing; local portraits; local history; genealogy books. Historic Houses: 1750 Nathaniel Backus House; 1818 Rockwell House.
Research Fields: genealogical tracing; Faith Trumbull DAR Chapter.
Facilities: library of local history books available for use by appointment only.
Activities: guided tours; permanent exhibitions. Museum Sponsors: six area high school scholarships.
Hours & Admission Prices: 2nd & 4th Sun. of month 1-3:30; other times by appointment. No charge; donations accepted. Closed holidays.

THE LEFFINGWELL HOUSE MUSEUM, 348 Washington St., Norwich, CT 06360-2444. Mailing Address: Society of the Founders of Norwich, CT, Inc., P.O. Box 62, Norwich, CT 06360. Tel.: 860-889-9440.
E-mail: info@leffingwellhousemuseum.org
Web Site: www.leffingwellhousemuseum.org
Founded: 1901.
Congressional District: 2
Key Personnel: Pres., Austin Deming.
Personnel Profile: Part-Time Volunteers 20.
Governing Authority: nonprofit. Parent Institution: Society of the Founders of Norwich, CT, Inc. Branch Museums: 1772, Joseph Carpenter Silversmith Shop, 73 E. Town St., Norwich; 1789 E. District School Children's Historical Museum, Norwich. Tax-exempt: 501(c)(3).
Institution Type/Description: Historic House Museum: 1675-1715-1760 The Leffingwell Inn.
Collections: period dolls; Norwich silver; Indian artifacts; genealogical & historical local material; early Norwich colonial silver & pewter; antiques of local origin.
Facilities: 150-vol. library of books pertaining to Norwich historical background; genealogical material pertaining to Norwich families; old deeds; newspapers & documents available for research by properly accredited persons. Books on Norwich history & museum-related items for sale.
Activities: guided tours.
Publications: annual report; books, Samuel Huntington & His Family; Craftsmen & Artists of Norwich; The History of Norwich; A Definitive Index of Caulkins' History; The Autobiography of James L. Smith; Norwich-Century of Growth.
Hours & Admission Prices: May-Oct. Sat.-Sun. 12-4 & also by appointment. Adults $5, students & senior citizens $3, children under 12 $2; discounts to groups of 10 or more; members no charge.
Attendance: 1,500 (estimated)
Membership: Student $10; Senior Citizens (over 65) $15; Single $20; Family $30; Sustaining $60; Patron $150; Life $500.

THE SLATER MEMORIAL MUSEUM - NORWICH FREE ACADEMY, (M), 108 Crescent St., Norwich, CT 06360-3556. Tel.: 860-887-2506 & 425-5560. Fax: 860-885-0379.
E-mail: zoev@nfaschool.org
Web Site: www.slatermuseum.org
Founded: 1886.
Congressional District: 2
Key Personnel: Chm. (V), Richard DesRoches; Dir., Vivian F. Zoe; Museum Educator, Mary Anne Hall; Asst. Dir. & Museum Shop Mgr., Leigh Thomas.
Personnel Profile: Full-Time Paid 3; Part-Time Paid 5; Part-Time Volunteers 30; Interns 5.
Volunteer Hours: 1,800
Operating Expenses: 300,000
Operating Income: 300,000
Governing Authority: nonprofit organization. Parent Institution: Norwich Free Academy Corp. Tax-exempt: 501(c)(3).
Institution Type/Description: Art Museum: housed in an historic building designed by Worcester architect, Stephen Earle c.1886; located on the campus of the Norwich Free Academy.
Collections: fine & decorative art representing 350 years of Norwich history; casts of early Greek, Roman & Renaissance statues; American paintings,

sculpture, graphics & furnishings; Oriental art; African & South Sea Islands art; costumes; Native American & marine items; guns; textiles.

Major Exhibits: 70th Annual CT Artists Juried, 2/14-3/14; Animalia, 6/14-8/14; CT Women Artists National Juried, 8/14-9/14; Renaissance in Pastel, 10/14-11/14; Printmakers' Network of S.E. New England, 11/14-1/15.

Research Fields: John Denison Crocker; Alexander Hamilton Emmons; Ebenezer Tracy; American clock making; Henry Watson Kent; Ellis Ruley; Norwich & Risley Pottery; Joseph Carpenter; Charlotte Fuller Eastman; Ozias Dodge; Irene Weir; Slater, Almy, Ballou families; Helen Marshall.

Facilities: Converse Art Gallery.

Activities: guided tours; lectures; gallery talks; permanent & rotating exhibitions.

Publications: pamphlet, The Slater Memorial Museum; The Slater Museum Cast catalogue; Greek Myths for Young People; quarterly newsletter, The Muse; exhibition catalogues; The Slater's Grand Tour; 20th Century CT Artists; African Art; John Denison Crocker.

Hours & Admission Prices: Tues.-Fri. 9-4, Sat.-Sun. 1-4. Adults $3, seniors & students $2; discounts to NEMA, AAM & ICOM members; members no charge. Closed holidays. &

Attendance: 12,000 (estimated)

Membership: Friends of Slater Museum: Student $10; Senior Citizen $15; Individual $25; Family $35; Contributing $50; Patron $100; Sustaining $200; Benefactor $500.

Oakdale

THE DINOSAUR PLACE AT NATURE'S ART, 1650 Hartford New London Turnpike, Oakdale, CT 06370. Tel.: 860-443-4367. Fax: 860-443-0253.

Institution Type/Description: Mineralogy Museum.

Collections: local history; minerals; fossils.

Facilities: Museum-related items for sale.

Activities: hands-on exhibits; special events; birthday parties.

Hours & Admission Prices: Call for hours.

Old Lyme

* **FLORENCE GRISWOLD MUSEUM,** 96 Lyme St., Old Lyme, CT 06371-1426. Tel.: 860-434-5542. Fax: 860-434-9778 (administrative offices).

E-mail: jeff@flogris.org

Web Site: www.florencegriswoldmuseum.org

Founded: 1936.

Congressional District: 2

Key Personnel: C.E.O. & Dir., Jeffrey W. Andersen; Facilities Mgr., Ted Gaffney; Dir. Devel., Janie Stanley; Dir. Education & Outreach, David Rau; Museum Educator, Julie Riggs; Business Mgr., Therese Kus; Membership Coord., Nathaniel Green; Registrar, Nicole Wholean; Asst. to Dir., Donna Carlson; Cur., Amy Kurtz Lansing; Asst. Cur., Ben Colman; Dir. Mktg., Tammi Amaya Flynn; Dir. Finance, Fred Cote; Devel. Office Coord., Kristin Pishvanov; Visitor Svcs. Coord. & Museum Shop Mgr., Matt Greene; Mktg. Assoc., Cheryl Poirier.

Personnel Profile: Full-Time Paid 14; Part-Time Paid 6; Part-Time Volunteers 375; Interns 2.

Governing Authority: nonprofit. Tax-exempt: 501(c)(3).

Institution Type/Description: Art Museum.

Collections: American paintings; Lyme Art Colony paintings; decorative arts; archives. Historic Houses: c.1794, Huntley-Brown House; William Chadwick Studio; 1817 Florence Griswold House.

Research Fields: regional history; paintings of the Old Lyme Art Colony; art of Connecticut.

Facilities: library & archives on regional history & American art available for use on premises; Hartman Education Center; Rafal Landscape Center; Chadwick Studio; Cafe Flo. Books & museum-related items for sale.

Activities: guided tours; lectures; films; classes.

Publications: catalogues; books, Miss Florence & the Artists of Old Lyme; Hamburg Cove Past and Present; The Lieutenant River; The Lymes' Heritage Cookbook; catalogs, Edward F. Rook American Impressionist; Connecticut and American Impressionism; Clark Voorhees, 1871-1933; Old Lyme: The American Barbizon; Bruce Crane: American Tonalist; The Whites of Waterford; Harry L. Hoffman: A World of Color; En Plein Air: The Art Colonies of East Hampton & Old Lyme, 1880-1930; Thomas W. Nason: New England Virtues; Carved in Wood; portfolio, The Connecticut Impressionists at Old Lyme; Childe Hassam in Connecticut; The Harmony of Nature; The Art and Life of Frank Vincent DuMond; A World Observed: The Art of Everett Longley Warner; Faces of Change: The Art of Ivan Olinsky; A Noble Tradition: American Paintings from the National Arts Club; Lines of Thought: American Works on Paper from a Private Collection; Wilson Henry Irvine and The Poetry of Light; Henry Ward Ranger and The Humanized Landscape; The American Artist in Connecti-

cut, 2002; May Night, Willard Metcalf in Old Lyme, 2005; Visions of Mood: Henry C. White Pastels, 2009; Lyme in Mind: The Clement C. Moore Collection, 2009; Sewell Sillman: Pushing Limits, 2010; With Needle and Brush: Embroidery from the Connecticut Valley, 2011.

Hours & Admission Prices: Tues.-Sat. 10-5, Sun. 1-5. Adults $10, seniors $9, students with ID $8; discounts to NEMA, AAA, ICOM & AAM members; children under 12 & members no charge. &

Attendance: 64,687 (accurate)

Membership: Individual $50; Family $75; Business Sponsor & Supporting $125 & up; Sustaining $250 & up; Patron $500 & up; Leader $1,000 & up; Director's Circle $2,500; President's Circle $5,000 & up.

Old Saybrook

GENERAL WILLIAM HART HOUSE, 350 Main St., Old Saybrook, CT 06475. Mailing Address: P.O. Box 4, Old Saybrook, CT 06475. Tel.: 860-395-1635 & 388-2622.

Institution Type/Description: Historic House Museum: built in 1767. Listed on the National Register of Historic Places.

Collections: Hart family history; period furnishings.

Hours & Admission Prices: June-Aug. daily 1-4; other times by appointment.

Plainville

PLAINVILLE HISTORICAL SOCIETY, 29 Pierce St., Plainville, CT 06062-2207. Tel.: 860-747-6577.

Institution Type/Description: Historical Society Museum.

Collections: local history & culture; letters; manuscripts; personal artifacts; period furnishings; early clothing; photographs.

Activities: educational programs; special events.

Hours & Admission Prices: By appointment.

Pleasant Valley

NATURE MUSEUM, Greenwoods Rd., Pleasant Valley, CT 06063. Mailing Address: P.O. Box 1, Pleasant Valley, CT 06063. Tel.: 860-379-2469.

Web Site: psfnaturemuseum.org

Formerly: Stone Museum

Founded: 1935.

Congressional District: 1

Personnel Profile: Full-Time Paid 1; Part-Time Volunteers 1.

Governing Authority: Parent Institution: State of Connecticut DEEP.

Institution Type/Description: Historic Building: built in 1934.

Collections: local history & culture; mounted wildlife; natural history; photographs; Native American artifacts.

Facilities: nature trails.

Activities: hiking; special events.

Hours & Admission Prices: Memorial Day to June & Sept. to Columbus Day Sun.; July-Aug. Sat.-Sun.

Plymouth

ARTWORKS ART STUDIO & GALLERY, 109 Wilton Rd., Plymouth, CT 06782. Tel.: 860-283-6272. Fax: 860-283-6272.

Institution Type/Description: Art Gallery.

Collections: works by local artists; sculpture; pottery; paintings; pastels.

Activities: classes; birthday parties; summer camps; temporary exhibitions.

Hours & Admission Prices: Mon.-Fri. 9-6, Sat.-Sun. 10-4. No charge.

Portland

RUTH CALLANDER HOUSE MUSEUM - THE PORTLAND HISTORICAL SOCIETY, (M), 492 Main St., Portland, CT 06480-0098. Mailing Address: P.O. Box 98, Portland, CT 06480-0098.

E-mail: portlandhistsoc@yahoo.com

Web Site: portlandhistsoc.com

Founded: 2003.

Institution Type/Description: Historical Society Museum: house built in 1715.

Collections: local history & culture; period furnishings; photographs.

Activities: special events.

Publications: newsletter.

Hours & Admission Prices: 1st Sun. each month 2-4.

Membership: Student 18 & under and Senior 60 & over $10; Individual $15; Family $25; Business Sponsor $50; Patron $100; Lifetime $250.

Ridgefield

✳ THE ALDRICH CONTEMPORARY ART MUSEUM, (M), 258 Main St., Ridgefield, CT 06877-4935. Tel.: 203-438-4519. Fax: 203-438-0198.
E-mail: general@aldrichart.org
Web Site: www.aldrichart.org
Founded: 1964.
Congressional District: 5
Key Personnel: Chm., Mark Goldstein; Dir., Alyson Baker.
Personnel Profile: Full-Time Paid 18; Part-Time Paid 30; Part-Time Volunteers 35; Interns 3.
Governing Authority: nonprofit organization. Tax-exempt: 170(b)(1)(A).
Institution Type/Description: Contemporary Art Museum.
Collections: works by contemporary artists.
Research Fields: Contemporary Art.
Facilities: sculpture garden.
Activities: exhibition related events, tours, lectures, film; nationally recognized education programs; changing exhibitions focusing on contemporary art with an emphasis on emerging & mid-career artists.
Publications: catalog of changing exhibitions; newsletter; exhibition posters.
Hours & Admission Prices: Tues.-Sun. 12-5. Adults $7, senior citizens & college students $4; discount to AAM members; children 18 & under, members and Tues. no charge. Closed New Year's Day; Thanksgiving; Christmas. &

Attendance: 33,000 (estimated)
Membership: Friend $100; Family $125; Partner $500; Donor $1,000; Sponsor $2,500.

KEELER TAVERN MUSEUM, (M), 132 Main St., Ridgefield, CT 06877-4931. Tel.: 203-438-5485 & 431-0815. Fax: 203-438-9953.
E-mail: mconnors@keelertavernmuseum.org
Founded: 1966.
Congressional District: 5
Key Personnel: Pres., Joel Third; 1st Vice Pres., Hilary Micalizzi; Treas., Rhonda Hill; Exec. Dir., Hildegard Grob; Museum Shop Mgr., Margo McEachern.
Personnel Profile: Full-Time Paid 1; Full-Time Volunteers 75; Part-Time Paid 3; Part-Time Volunteers 100.
Governing Authority: society. Parent Institution: Keeler Tavern Preservation Society, Inc. Tax-exempt: 501(c)(3).
Institution Type/Description: Historic Building Museum.
Collections: period tavern furnishings; cannon ball shot by cannon into corner post during battle of Ridgefield still in sight; 18th to 19th-century furnishings, including original family pieces, pewter, woodenware, china, milkglass & costumes; 1890-1930 Joseph Hartmann photographic collection from glass plate negative. Historic Buildings: 1907 summer home of architect Cass Gilbert; 1915 garden house; barn; 1713 tavern & stagecoach inn.
Research Fields: local history.
Facilities: garden; banquet facilities. Gifts & museum-related items for sale.
Activities: guided tours; lectures; temporary exhibitions. Museum Sponsors: Colonial Day; Traditional Christmas Luncheon in Cass Gilbert Garden House in December.
Publications: books, the Diary of Anna M. Resseguie (1851-67); A View From the Inn; The Keeler Family Book.
Hours & Admission Prices: Feb.-Dec. Wed. & Sat.-Sun. 1-4, last tour 3:30; special tours by appointment. Adults $8, senior citizens & children $5; members no charge. Closed New Year's Day; Easter; Memorial Day, Independence Day; Thanksgiving, Christmas.
Attendance: 6,500 (estimated)
Membership: Individual $35; Family $70; Contributing $100; Patron $250; Patriot $500; Innkeeper $1,000; Proprietor $2,000 & up.

RIDGEFIELD HISTORICAL SOCIETY, 4 Sunset Lane, Ridgefield, CT 06877-4643. Tel.: 203-438-5821.
E-mail: ridgefieldhistory@sbcglobal.net
Web Site: www.ridgefieldhistoricalsociety.org
Key Personnel: Pres., Gary Singer
Institution Type/Description: Historical Society Museum.
Collections: Ridgefield history & culture; photographs.
Hours & Admission Prices: Tues.-Thurs. 1-5.

Riverton

GREENWOOD GLASS BLOWING STUDIO GALLERY & SCHOOL, 3 Robertsville Rd., Riverton, CT 06065. Mailing Address: P.O. Box 242, Riverton, NC 06065. Tel.: 860-738-9464.
E-mail: anything@petergreenwood.com
Web Site: www.petergreenwood.com

Founded: 1980.
Institution Type/Description: Art Gallery: housed in historic Union Church; built in 1829.
Collections: Riverton town history; contemporary hand-blown art glass chandeliers; wall sculptures; glass furniture; lighting; goblets; paintings.
Activities: demonstrations; glass-blowing classes; guided tours; workshops.
Hours & Admission Prices: Tues.-Sat. 9-5 by appointment. No charge.
Attendance: 3,000 (estimated)

Rocky Hill

ACADEMY HALL MUSEUM OF THE ROCKY HILL HISTORICAL SOCIETY, INC., 785 Old Main St., Rocky Hill, CT 06067-1519. Mailing Address: P.O. Box 185, Rocky Hill, CT 06067-0185. Tel.: 860-563-6704.
E-mail: info@rockyhillhistory.us
Web Site: rockyhillhistory.wordpress.com
Founded: 1962.
Congressional District: 1
Key Personnel: Pres., Mike Martino.
Personnel Profile: Part-Time Volunteers 15.
Volunteer Hours: 2,600
Operating Expenses: 5,315
Operating Income: 6,942
Governing Authority: society. Tax-exempt: 501(c)(3).
Institution Type/Description: History Museum: housed in 1803 schoolhouse.
Collections: Indian artifacts; domestic implements; 19th-century clothing & accessories; hats; fans; farming equipment; tools; furniture; 500 salts; handmade quilts; early lamps.
Research Fields: local history.
Facilities: 600-vol. library of local & general historical material available in the society's library. Booklets for sale.
Activities: permanent & special exhibitions.
Publications: quarterly newsletter, Rocky Hill Historical Society.
Hours & Admission Prices: June-Sept. Tues. 10-12, 4th Thurs. 7-9, Sat. 12:30-3; Oct.-May Tues. 10-12, Sat. 12:30-3; other times by appointment. No charge; donations accepted.
Attendance: 340 (accurate)
Membership: Individual $15; Family $25; Business & Groups $35; Contributing $100; Life $250.

DINOSAUR STATE PARK, 400 West St., Rocky Hill, CT 06067-3506. Tel.: 860-529-5816. Fax: 860-257-1405.
E-mail: info@dinosaurstatepark.org
Web Site: www.ct.gov/deep/dinosaurstatepark
Founded: 1968.
Congressional District: 1
Key Personnel: Pres. (V), Kathy Kennedy; Vice Pres., Susan Lionberger; Dir., Meg Enkler; Maintainer III, Jeff Thomas; Treas., Jan Locke; Sec., Maryanne Marchinski; Museum Shop Mgr., Carla Roggi.
Personnel Profile: Full-Time Paid 3; Part-Time Paid 9; Part-Time Volunteers 40; Interns 1.
Governing Authority: state. Parent Institution: CT Dept. of Energy and Environmental Protection (DEEP). Subsidiary Institution: Bureau of Outdoor Recreation, State Parks Div. Tax-exempt.
Institution Type/Description: State Park.
Collections: sandstone trackway with 2,000 dinosaur tracks; slabs with tracks; living woody plants of families from Mesozoic Era; geological specimens; herbarium; dioramas; early plants.
Research Fields: dinosaur tracks; local flora & fauna.
Facilities: 1,000-vol. library of geology texts, reprints, bulletins & nature study handbooks available for use on premises by permission of staff; nature trail; picnic area; 100-seat auditorium; classroom; arboretum; butterfly garden; native plant garden. Museum-related items for sale.
Activities: guided tours; temporary & permanent exhibitions; lectures; films; special programs; track casting area.
Publications: quarterly newsletter; Tracks and Trails.
Hours & Admission Prices: Exhibit Center: Tues.-Sun. 9-4:30. Park: daily 9-4:30. Trails: daily 9-4. Exhibit Center: adults 13 & over $6, youth 6-12 $2; children 5 & under no charge. Closed New Year's Day; Thanksgiving; Christmas. &
Attendance: 53,000 (estimated)
Membership: Friends of Dinosaur Park Assoc. Inc.: Individual $20; Family $30; Supporting $50; Coelophysis Club $75; Dilophosaurus Club $150; Corporate $250; Life $400.

Rowayton

ROWAYTON ARTS CENTER, 145 Rowayton Ave., Rowayton, CT 06853-1444. Tel.: 203-866-2744. Fax: 203-866-1123.
E-mail: rowart@snet.net
Web Site: www.rowaytonartscenter.org
Founded: 1960.
Key Personnel: Dir., Debra Randall; Pres., Hu Lindsay; Vice Pres., Jessica Huse.
Personnel Profile: Part-Time Paid 1.
Governing Authority: Tax-exempt.
Institution Type/Description: Art Center.
Collections: paintings; sculpture.
Facilities: Museum-related items for sale.
Activities: adult & children's classes; fundraising events; workshops; demonstrations; receptions. Museum Sponsors: Art in the Park in May; Plein Air Art Walk & Show in June; Holiday Gift Sale in November & December.
Publications: quarterly newsletter.
Hours & Admission Prices: Tues.-Sat. 12-5, Sun. 1-4. No charge; donations accepted. Closed Thanksgiving; Christmas.
Attendance: 700 (estimated)
Membership: Supporting: Individual $30; Family $55; Art Angel $200. Exhibiting: Individual $50; Art Angel $200.

Scotland

D'ELIA ANTIQUE TOOL MUSEUM, 21 Brook Rd., Scotland, CT 06264. Mailing Address: P.O. Box 164, Scotland, CT 06264-0164. Tel.: 860-456-1516.
E-mail: info@deliatoolmuseum.com
Web Site: www.deliatoolmuseum.com
Institution Type/Description: Tool Museum.
Collections: over 1,000 woodworking hand planes; patents; early planemakers.
Hours & Admission Prices: Call for hours.

HUNTINGTON HOMESTEAD, 36 Huntington Rd., Scotland, CT 06264-2209. Mailing Address: P.O. Box 231, Scotland, CT 06264-0231. Tel.: 860-456-8381.
E-mail: info@huntingtonhomestead.org
Web Site: huntingtonhomestead.org
Founded: 1994.
Governing Authority: Parent Institution: Governor Samuel Huntington Trust, Inc., P.O. Box 231, Scotland, CT 06264. Tax-exempt.
Institution Type/Description: Historic House: the birthplace of Samuel Huntington, a signer of the Declaration of Independence.
Collections: period artifacts; photographs; Samuel Huntington signed documents.
Facilities: Museum-related items for sale.
Publications: annual newsletter, The Signature.
Hours & Admission Prices: May-Oct. 1st & 3rd Sat. each month 11-3. No charge; donations accepted.
Membership: Individual $15; Family $20; Citizen $50; Parishioner $100; Lawyer $250; Justice $500; Councilor $1,000; Congressman $5,000; Governor $10,000; President $10,000 & up.

Sharon

SHARON AUDUBON CENTER, 325 Cornwall Bridge Rd., Sharon, CT 06069-2512. Tel.: 860-364-0520. Fax: 860-364-5792.
Web Site: www.audubon.org/local/sanctuary/sharon
Founded: 1961.
Congressional District: 6
Key Personnel: Dir., Scott Heth; Land Mgr., Mike Dudek; Museum Shop Mgr. & Office Mgr., Dawn Osborne; Mgr. Education Programs, Wendy Miller; Wildlife Rehabilitation & Outreach Coord., Erin O'Connell; Caretaker, Dave Paton.
Personnel Profile: Full-Time Paid 6; Part-Time Paid 1; Part-Time Volunteers 10; Interns 2.
Governing Authority: nonprofit. Parent Institution: National Audubon Society, 225 Varick St., 7th Fl., New York, NY 10014. Tax-exempt: 501(c)(3).
Institution Type/Description: Nature Center.
Collections: 1,147 acre area with fields, forest, pond, stream, marsh, trees & plants; wildlife refuge; bird sanctuary; raptor aviaries including 10 species of live birds of prey.
Research Fields: education, woodlot & wildlife management.
Facilities: 900-vol. library of natural history & conservation books available for use on premises; nature center; nine nature trails; trail for handicapped. Books, stationery & museum-related items for sale.
Activities: guided tours; films; hobby workshops for adults; formally organized educational programs; permanent & temporary exhibitions; outreach to local camps; summer day camp for children. Center Sponsors: Maple Sugaring Festival in March; Sharon Audubon Festival in August.
Publications: newsletter, The Otter and the Hawk.
Hours & Admission Prices: Trails: dawn-dusk. Tues.-Sat. 9-5, Sun. 1-5. Adults $3, children under 12 & seniors $1.50; discounts to National Audubon Society members; members no charge. Closed holidays. &
Attendance: 10,000 (estimated)
Membership: Annual $20; National Audubon Society $35.

SHARON HISTORICAL SOCIETY - GAY HOYT HOUSE, (M), 18 Main St., Sharon, CT 06069-2052. Mailing Address: P.O. Box 511, Sharon, CT 06069-0511. Tel.: 860-364-5688.
E-mail: sharonhistoricalsociety@yahoo.com
Web Site: www.sharonhist.org
Founded: 1911.
Key Personnel: Exec. Dir., Elizabeth G. Shapiro; Treas., Allen Reiser; Cur., Rosemary Davis.
Personnel Profile: Part-Time Paid 2; Part-Time Volunteers 80.
Governing Authority: private; nonprofit organization. Tax-exempt: 501(c)(3).
Institution Type/Description: Historical Society Museum: house built in 1775.
Collections: local, regional & state history and culture; decorative arts; period artifacts & furnishings; iron industry; portraits; photographs.
Research Fields: Iron industry in CT.
Facilities: library; 1500 sq. ft. exhibit space. Museum-related items for sale.
Activities: films; formal education programs; guided tours; hobby workshops; lectures; participatory & temporary exhibitions; training programs for professional museum workers. Annual Events: Tree Lighting Open House.
Publications: biannual newsletter, Sharon Archives.
Hours & Admission Prices: Jan. 4-Dec. 20 Wed. & Sat. 10-2, Thurs. & Fri. 10-4; other times by appointment. No charge; donations accepted. Closed New Year's Day; Thanksgiving; Christmas Day & week. &
Attendance: 1,200 (estimated)
Membership: Annual $35; Sustaining $150; Benefactor $300; Sponsor $500; Trustee's Circle $1,000.

Shelton

SHELTON HISTORICAL SOCIETY, 70 Ripton Rd., Shelton, CT 06484. Mailing Address: P.O. Box 2155, Shelton, CT 06484. Tel.: 203-929-7963.
Formerly: Huntington Historical Society
Founded: 1969.
Personnel Profile: Part-Time Paid 3; Part-Time Volunteers 25.
Governing Authority: Tax-exempt.
Institution Type/Description: Historical Society Museum.
Collections: local history & culture; period furnishings; photographs. Historic Buildings: c.1820 Brownson House; c.1860 Wilson barn; 1872 Trap Fall School.
Activities: Annual Event: Father's Day Antique Car Show.
Publications: newsletter.
Hours & Admission Prices: Call for hours.
Attendance: 2,000 (estimated)
Membership: Individual $20; Family $30; Corporate $100.

Sherman

THE SHERMAN HISTORICAL SOCIETY, INC. - NORTHROP HOUSE MUSEUM OF LOCAL HISTORY, 10 Rte. 37 Center, Sherman, CT 06784-1503. Tel.: 860-350-3475 (office) & 354-3083 (store).
E-mail: shermanhistorical@sbcglobal.net
Web Site: www.shermanhistoricalsociety.org
Founded: 1975.
Congressional District: 5
Key Personnel: Pres. (V), Elizabeth Mard; Cur., Gloria Thorne; Museum Shop Mgr., Moira Kelly.
Personnel Profile: Part-Time Volunteers 52.
Governing Authority: Parent Institution: Sherman Historical Society. Tax-exempt: 501(c)(3).
Institution Type/Description: Local History Museum: built in 1829.
Collections: local history & artifacts; period furnishings; photographs. Historic Buildings: Old Store Museum, c.1810; rebuilt Roger Sherman Cobbler Shop, c.1742.
Research Fields: Roger Sherman; local genealogy & history.
Facilities: Museum-related items for sale.
Activities: summer children's program; exhibit openings & previews; band concerts. Museum Sponsors: The Civil War: Sherman Gave a Life! in June; Glowing Pumpkins; Holiday Open House
Publications: newsletter 3 times a year, Sherman Historical Society; book, Sentinel Houses & Barns; books, A History of Sherman; Ned Anderson: Connecticut's Appalachian Trail Blazer - Small Town Renaissance Man.

Hours & Admission Prices: By appointment. No charge; donations accepted. &

Attendance: 2,500 (estimated)
Membership: Household $30; Sponsor $50; Patron $100; Benefactor $250; Life $500.

Simsbury

SIMSBURY HISTORICAL SOCIETY, PHELPS TAVERN MU-SEUM, 800 Hopmeadow St., Simsbury, CT 06070-1825. Mailing Address: P.O. Box 2, Simsbury, CT 06070-0002. Tel.: 860-658-2500. Fax: 860-651-4354.
E-mail: info@simsburyhistory.org
Web Site: www.simsburyhistory.org
Founded: 1911.
Congressional District: 6
Personnel Profile: Full-Time Paid 2; Full-Time Volunteers 4; Part-Time Paid 2; Part-Time Volunteers 120; Interns 4.
Governing Authority: Parent Institution: Simsbury Historical Society Inc. Tax-exempt: 501(c)(3).
Institution Type/Description: History Museum.
Collections: Higley coppers; fuse machinery; sleighs; carriages; artifacts; memorabilia; manuscripts. Historic Houses: 1795 Hendrick Cottage; 1740 First Simsbury School House; replica of 1683 First Meeting House; 1771 Captain Elisha Phelps House, Barn & Outbuildings; mid-19th century Victorian Carriage House; 1876 Simsbury Probate court building.
Research Fields: local & family history; transportation; early manufacturing.
Facilities: 5,000-vol. library and extensive manuscript collection. Commemorative items & publications for sale.
Activities: guided tours; lecture series; historic portrayals; local history programs for children; organized education programs for children; permanent & temporary exhibitions. Annual Event: antique show.
Publications: books, A Spy at Ticonderoga, Newgate From Copper Mine to State Prison; annual calendar; quarterly, Archival; newsletter, The Sign Post.
Hours & Admission Prices: Tours: Tues.-Sat. 12-4. Adults $6, senior citizens over 65 $5, children 6-17 $4; discount to groups & AAA members; members and children 5 & under no charge. Library & Archives: Thurs.-Sat. 12-4. Closed federal holidays.
Attendance: 3,800 (accurate)
Membership: Individual $15; Family $25; Sustaining $40; Corporate $75.

Somers

SOMERS HISTORICAL SOCIETY MUSEUM, 11 Battle St., Somers, CT 06071. Mailing Address: P.O. Box 652, Somers, CT 06071. Tel.: 860-749-6437.
Founded: 1962.
Governing Authority: Tax-exempt.
Institution Type/Description: Historical Society Museum: housed in the former public library; built in 1896.
Collections: local history & culture; period furnishings; personal artifacts; photographs; documents.
Activities: special events.
Publications: quarterly newsletter.
Hours & Admission Prices: Call for hours. No charge; donations accepted.
Membership: $10-$200.

South Glastonbury

WELLES SHIPMAN WARD HOUSE, 972 Main St., South Glastonbury, CT 06073. Mailing Address: P.O. Box 46, Glastonbury, CT 06033. Tel.: 860-633-6890.
E-mail: hsglastonbury@sbcglobal.net
Web Site: www.hsgct.org
Founded: 1963.
Congressional District: 1
Key Personnel: Dir., James Bennett; Pres. (V), Gilbert Tyler.
Personnel Profile: Full-Time Paid 1; Part-Time Paid 2.
Governing Authority: Parent Institution: Historical Society of Glastonbury Inc. Tax-exempt.
Institution Type/Description: Historic House Museum: built c.1755.
Collections: local history & culture; horse-drawn vehicles; 18th-century herb garden; period furnishings; farm equipment.
Activities: special events.
Hours & Admission Prices: mid-June to Sept. Tues. 1-4. Adults $3; members no charge.
Attendance: 3,000 (estimated)

South Norwalk

NORWALK MUSEUM, 41 N. Main St., South Norwalk, CT 06854-2702. Tel.: 203-866-0202. Fax: 203-866-0675.
E-mail: sgunnbromley@norwalkct.org
Web Site: www.norwalkct.org/norwalkmuseum
Key Personnel: Dir., Susan Gunn Bromley
Institution Type/Description: History Museum.
Collections: local history & culture; Norwalk's industry; photographs; oystering; railroads; historic buildings; pottery; quilts.
Facilities: library.
Hours & Admission Prices: Museum: Wed.-Sun. 1-5. Archives: by appointment.

South Windsor

SOUTH WINDSOR HISTORICAL SOCIETY, 771 Ellington Rd., Rte. 74, South Windsor, CT 06074. Mailing Address: P.O. Box 216, South Windsor, CT 06074-0216.
Institution Type/Description: Historical Society Museum.
Collections: local history; period artifacts; books; manuscripts; photographs. Historic Building: 19th-century one-room schoolhouse.
Hours & Admission Prices: April-June & Sept.-Nov. 1st Sun. of month 1-4; July-Aug. Tues. & Thurs. 12-4, 1st Sun. of month 1-4. No charge; donations accepted.

Southbury

AUDUBON CENTER AT BENT OF THE RIVER, 185 E. Flat Hill Rd., Southbury, CT 06488-1151. Tel.: 203-264-5098. Fax: 203-264-6332.
E-mail: bentoftheriver@audubon.org
Web Site: bentoftheriver.audubon.org
Key Personnel: Dir., Leslie Macline Kane.
Governing Authority: Parent Institution: National Audubon Society. Subsidiary Institution: Audubon Connecticut. Tax-exempt.
Institution Type/Description: Wildlife Sanctuary.
Collections: wildlife & their habitats.
Facilities: nature trails.
Activities: educational programs; hiking.
Hours & Admission Prices: Center: Mon.-Fri. 9-5. Trails: daily sunrise to sunset. No charge; donations accepted. &
Attendance: 5,600 (estimated)
Membership: Individual $35.

Southington

BARNES MUSEUM, 85 N. Main St., Southington, CT 06489-2518. Tel.: 860-628-5426. Fax: 860-628-0488.
E-mail: secondom@southington.org
Web Site: barnesmuseum.wordpress.com
Founded: 1973.
Key Personnel: Dir., Marie Secondo.
Personnel Profile: Full-Time Paid 1; Part-Time Paid 1; Part-Time Volunteers 4.
Governing Authority: Tax-exempt.
Institution Type/Description: History Museum.
Collections: period furniture; personal artifacts; diaries; photographs; newspapers; magazines; family Bibles.
Activities: tours.
Hours & Admission Prices: July-Aug. Mon.-Wed. & Fri. 1-5, Thurs. 1-7; Sept.-June Mon.-Wed., Fri., 1st & last Sat. of month 1-5, Thurs. 1-7. Adults $5, seniors $4.
Attendance: 5,500 (estimated)

Stafford Springs

AMERICAN MUSEUM OF AVIATION, 21 Clearview Dr., Stafford Springs, CT 06076. Tel.: 860-208-3095.
E-mail: info@AmericanMuseumofAviation.org
Web Site: www.americanmuseumofaviation.org
Formerly: Propliners of America
Founded: 1998.
Key Personnel: Pres. (V) & Founder, William Bradshaw; Dir. & Chm. (V), J.D. Scroggins; Museum Shop Mgr., H.D. Scroggins.
Personnel Profile: Full-Time Volunteers 2; Part-Time Volunteers 4; Interns 2.
Governing Authority: nonprofit organization. Administrative Offices, Las Vegas, NV. Tax-exempt: 501(c)(3).
Institution Type/Description: Aviation History Museum.
Collections: aviation history & artifacts; photographs; flying & static aircraft

including Douglas DC-7B, Boeing 727, & Piper PA-24-250 Comanche; aircraft cockpits; Douglas C-47A.
Major Exhibits: Boeing 767 Cockpit, 1/14-12/14.
Research Fields: aviation history & archaeology.
Publications: book, Junk Yard Jets.
Hours & Admission Prices: By appointment. No charge; donations accepted. &

Attendance: 420,000 (accurate)

NORTHEAST STATES CIVILIAN CONSERVATION CORPS MUSEUM, 166 Chestnut Hill Rd., Stafford Springs, CT 06076-4007. Tel.: 860-684-3013.
Institution Type/Description: Historic House Museum: housed in the original Civilian Conservation Corps camp building; c.1935.
Collections: corps history; campsite life & work; photographs; equipment; tools.
Hours & Admission Prices: By appointment. No charge. &

Stamford

BARTLETT ARBORETUM & GARDENS, 151 Brookdale Rd., Stamford, CT 06903-4199. Tel.: 203-322-6971, ext. 15. Fax: 203-595-9168.
E-mail: admin@bartlettarboretum.org
Web Site: bartlettarboretum.org
Formerly: Bartlett Arboretum, University of Connecticut
Founded: 1965.
Congressional District: 5
Personnel Profile: Full-Time Paid 8; Part-Time Paid 2; Part-Time Volunteers 100.
Governing Authority: city of Stamford. Parent Institution: Bartlett Arboretum Association. Tax-exempt: 501(c)(3).
Institution Type/Description: Arboretum, Herbarium & Library.
Collections: plant collections; woody trees; shrubs; vines; dwarf & unusual conifers; ericaceous plants; herbarium including plant specimens from 1887 to present; over 3,000 sheets of preserved plants from around the world.
Research Fields: botany; herbarium.
Facilities: Silver Educational Center (June 2011).
Activities: guided tours; lectures; formally organized education programs.
Publications: brochures; seasonal, Bartlett Arboretum Newsletter.
Hours & Admission Prices: Grounds: daily 8:30-sunset. Office: Mon.-Fri. 8:30-4. Adults $6; members no charge. Closed Christmas.
Attendance: 20,000 (estimated)
Membership: Senior $50; Individual $60; Family $75.

MUSEUM OF BLACK WWII HISTORY, Stamford, CT 06906. Mailing Address: 71 Plymouth Rd., Stamford, CT 06906. Tel.: 203-348-6810.
Founded: 2006.
Key Personnel: Founder & Cur., Bruce Bird
Institution Type/Description: History Museum.
Collections: African American military history; photographs; personal artifacts.
Hours & Admission Prices: Feb. 5 to late Nov. Thurs.-Mon. 10-5; other times by appointment. Adults $5, veterans, seniors over 65 & students $3.
Attendance: 350 (estimated)

RICHARD AND HINDA ROSENTHAL GALLERY, Rich Forum, 307 Atlantic St., Stamford, CT 06901-3506. Mailing Address: 61 Atlantic St., Stamford, CT 06901-3506. Tel.: 203-358-2305.
Institution Type/Description: Art Gallery.
Collections: paintings; photographs; sculpture.
Hours & Admission Prices: Call for hours.

SACKLER ART GALLERY, Palace Theatre, 61 Atlantic St., Stamford, CT 06901-2403. Tel.: 203-358-2305.
Institution Type/Description: Art Gallery: housed in a former vaudeville theatre; built in 1927.
Collections: paintings; sculpture; photographs.
Hours & Admission Prices: Call for hours.

SOUNDWATERS COASTAL EDUCATION CENTER, Cove Island Park, 1281 Cove Rd., Stamford, CT 06902-5457. Tel.: 203-323-1978.
Institution Type/Description: Education Center.
Collections: local history; environment; paintings; photographs; freshwater & saltwater fish.
Facilities: aquarium.

Activities: hands-on exhibits; workshops; lectures; educational programs.
Hours & Admission Prices: Memorial Day to Labor Day Tues.-Sat. 10-5.

STAMFORD HISTORICAL SOCIETY, INC., 1508 High Ridge Rd., Stamford, CT 06903-4107. Tel.: 203-322-1565 & 329-1183. Fax: 203-322-1607.
E-mail: administrator@stamfordhistory.org
Web Site: www.stamfordhistory.org
Founded: 1901.
Congressional District: 4
Key Personnel: C.E.O. & Pres., Thomas Zoubek; Chm. (V), Pam Coleman; Administrative Asst., Haideh Molavi; Senior Librarian, Ron Marcus; Museum Shop Mgr., Rosemary Vacca.
Personnel Profile: Part-Time Paid 2; Part-Time Volunteers 40; Interns 1.
Governing Authority: nonprofit organization. Subsidiary Institution: Hoyt Barnum House. Tax-exempt: 170(b)(1)(A).
Institution Type/Description: History & Decorative Arts Museum.
Collections: 17th- to 19th-century Americana; manuscripts; farm implements; needlework; quilts; dolls; household equipment; early pottery & craft tools; archives; costumes; Stamford business & city agency records, 1960-present; Native American & period artifacts; photographs of area people & structures. Historic House: Hoyt-Barnum House, 18th-century farmhouse begun in 1699.
Research Fields: local industrial history; decorative arts; genealogy.
Facilities: 2,500-vol. non-circulating library of books & pamphlets relating to history of Stamford available for use by public.
Activities: guided tours; lectures; permanent & temporary exhibitions; formally organized educational programs.
Publications: books, Stamford Revolutionary War Damage Claims; Springdale Remembered; Stamford - Pictures From The Past; Stamford In The Gilded Age; Fort Stamford - A Concise Study; Elizabeth Clawson: Thou Deserveth To Dye; Stamford Past And Present; Stamford Post Offices And Post Master; History of the Cove in Stamford; Poems on Stone in Stamford, Connecticut; Bibliography of Stamford References. The Civil War Diaries of Noah Webster Hoyt.
Hours & Admission Prices: Museum: Thurs.-Sat. 10-4; Hoyt-Barnum House seasonal & by appointment only. Adults $5, members $2, children $1; discounts to AAM, ICOM & NEMA members. Closed New Year's Day; Thanksgiving; Christmas. &
Attendance: 10,000 (estimated)
Membership: Student & Senior Citizen $25; Individual $35; Family $50; Contributing $100; Supporting $250; Patron $500; Corporate $1,000.

STAMFORD MUSEUM & NATURE CENTER, (M), 39 Scofieldtown Rd., Stamford, CT 06903-4096. Tel.: 203-322-1646. Fax: 203-322-0408.
E-mail: info@stamfordmuseum.org
Web Site: www.stamfordmuseum.org
Founded: 1936.
Congressional District: 4
Key Personnel: Exec. Dir., Melissa H. Mulrooney; Pres., Juanita James; Dir. Finance, William King; Farm Dir., Lauren Satterfield; Cur. Collections, Rosa Portell.
Personnel Profile: Full-Time Paid 20; Part-Time Paid 30; Part-Time Volunteers 35.
Governing Authority: nonprofit organization. Tax-exempt: 501(c)(3).
Institution Type/Description: General Museum.
Collections: agriculture; paintings; sculpture; graphics; astronomy; botany; entomology; geology; herpetology; Native American artifacts, clothing & furnishings; mineralogy; natural history; live domestic & native animals; native wildfowl; cultural artifacts; photography; working farm.
Research Fields: astronomy; variable stars.
Facilities: classrooms; planetarium; observatory; hiking trails.
Activities: formally organized educational programs; permanent & temporary exhibitions; special events. Museum Sponsors: Maple Sugar Sunday; Spring on the Farm; Harvest Spooktacular.
Publications: bimonthly newsletter; quarterly magazine.
Hours & Admission Prices: Mon.-Sat. & holidays 9-5, Sun. 11-5. Adults $8, senior citizens $6, children $4; Stamford residents on Wed., children under 3, members, AAM & ICOM members no charge. Closed New Year's Day; Independence Day; Thanksgiving; Christmas. &
Attendance: 110,000 (accurate)
Membership: Seniors 65 & over $30; Individual $35; Senior Couple $50; Family $80; Family Plus & Contributing $150; Supporting $250; Benefactor $500; Patron $1,000 & up.

UKRAINIAN MUSEUM & LIBRARY, (M), 161 Glenbrook Rd., Stamford, CT 06902-3002. Tel.: 203-323-8866. Fax: 203-357-7681.
E-mail: ukrmulrec@optonline.net
Web Site: ukrainianmuseumlibrary.org
Founded: 1935.
Key Personnel: Dir., John M. Terlecky; Cur., Lubow Wolynetz.
Personnel Profile: Full-Time Paid 2; Full-Time Volunteers 1; Part-Time Paid 2; Part-Time Volunteers 2.
Governing Authority: Library: 39 Clovelly Place, Stamford, CT. Tax-exempt.
Institution Type/Description: Culturally Specific.
Collections: Ukrainian culture & heritage; fine arts; folk arts; religious art; books; periodicals; recordings; photographs; archives; numismatics.
Activities: workshops for children & adults in embroidery; Christmas tree decorating.
Publications: brochure.
Hours & Admission Prices: Wed.-Fri. 1-5, Sat.-Tues. by appointment. No charge; donations accepted. Closed major holidays. &
Attendance: 4,000 (estimated)
Membership: Senior & Student $25; Individual $50; Family & Institution $100; Supporter $500; Patron $1,000; Sponsor $2,500; Benefactor $5,000.

Stonington

CAPTAIN NATHANIEL B. PALMER HOUSE - HOME OF THE DISCOVERER OF ANTARCTICA, 40 Palmer St., Stonington, CT 06378-1014. Mailing Address: The Stonington Historical Society, Inc., P.O. Box 103, Stonington, CT 06378. Tel.: 860-535-8445.
E-mail: director@stoningtonhistory.org
Web Site: www.stoningtonhistory.org/palmer
Founded: 1996.
Congressional District: 2
Key Personnel: Pres. Stonington Historical Society (V), David Purvis; Dir., Mary E. Baker; Treas., Willard Arndt; Museum Shop Mgr., Elizabeth Wood.
Personnel Profile: Full-Time Paid 1; Part-Time Paid 3; Part-Time Volunteers 50; Interns 3.
Governing Authority: Parent Institution: Stonington Historical Society, Inc. Subsidiary Institutions: The Richard Woolworth Library & Research Center; The Old Lighthouse Museum. Tax-exempt.
Institution Type/Description: History Museum.
Collections: memorabilia of Nathaniel's Antarctica discovery; Palmer brothers' personal artifacts; Stonington family portraits, furnishings & artifacts; historic house.
Research Fields: local history; genealogy; maritime.
Facilities: library.
Activities: lectures; programs; special events.
Publications: quarterly newsletter, Footnotes; monographs; biography; history.
Hours & Admission Prices: May-Oct. Wed.-Sun. 1-5. Palmer House & Old Lighthouse Museum: adults $9; members no charge.
Attendance: 2,500 (estimated)
Membership: Individual $35; Family $55; Contributing $100; Sustaining $200; Life $1,000; Life Couple $1,500.

OLD LIGHTHOUSE MUSEUM - STONINGTON HISTORICAL SOCIETY, 7 Water St., Stonington, CT 06378-1422. Mailing Address: P.O. Box 103, Stonington, CT 06378-0103. Tel.: 860-535-1440.
E-mail: director@stoningtonhistory.org
Web Site: www.stoningtonhistory.org
Founded: 1925.
Congressional District: 2
Key Personnel: Exec. Dir. Stonington Historical Society, Mary Beth Baker.
Personnel Profile: Part-Time Paid 6; Part-Time Volunteers 5.
Governing Authority: society. Parent Institution: Stonington Historical Society. Tax-exempt.
Institution Type/Description: Historic Site Museum.
Collections: history; paintings; archives; furniture; 17th-19th century historical artifacts of the town including whaling & War of 1812 relics; Fresnel lighthouse lens; photographs of Long Island Lights; maritime artifacts; children's toys; doll house.
Research Fields: genealogy.
Facilities: 500-vol. library of history & genealogy available by appointment, Call: 860-535-1131; reading room.
Activities: guided tours; lectures; formally organized education programs for children; permanent & temporary exhibitions; climb tower for view of 3 states; weekly walking tours of Stonington Borough.
Publications: quarterly magazine, Historical Footnotes; books, Stonington Graveyards; map of Stonington graveyards; An Hour Walk Through the Borough of Stonington; The Davis Homestead; The Battle of Stonington, War of 1812; Stonington Ice 1874-1947; The Stonington Tragedy.
Hours & Admission Prices: May-Nov. daily 10-5; other times by appointment. Adults $9; discounts to NEMA members; members no charge. Admission includes Capt. Palmer House.
Attendance: 7,500 (accurate)
Membership: Individual $35; Family $55; Contributing $100; Sustaining $200; Benefactor $500; Life $1,000; Dual Life $1,500.

Storrs

THE BALLARD INSTITUTE & MUSEUM OF PUPPETRY, 1 Royce Cir., Storrs, CT 06269. Tel.: 860-486-0339.
E-mail: bimp@uconn.edu
Web Site: www.bimp.uconn.edu
Key Personnel: Dir., Dr. John Bell.
Governing Authority: university.
Institution Type/Description: Puppetry Museum.
Collections: student built puppets; puppets of all types that are hundreds of years old & from around the world.
Activities: theater performances.
Hours & Admission Prices: Call for hours. Suggested Donations: $5. &

CONNECTICUT STATE MUSEUM OF NATURAL HISTORY AND CONNECTICUT ARCHAEOLOGY CENTER, University of Connecticut, Unit 1023, 2019 Hillside Rd., Storrs, CT 06269. Tel.: 860-486-4460. Fax: 860-486-0827.
E-mail: csmnhinfo@uconn.edu
Web Site: www.cac.uconn.edu
Founded: 1982.
Congressional District: 2
Key Personnel: Dir., Leanne Kennedy Harty; Coord. Membership, Emily R.M. Lanz; Coord. Public Information, David C. Colberg; Exhibits Planner, Collin Harty; State Archaeologist, Dr. Nicholas F. Bellantoni; Coord. Programs, Cheri Collins.
Personnel Profile: Full-Time Paid 5; Part-Time Paid 3; Part-Time Volunteers 150.
Governing Authority: college; nonprofit. Parent Institution: University of Connecticut. Subsidiary Institution: University of Connecticut Foundation. Tax-exempt: 501(c)(3).
Institution Type/Description: Natural History Museum.
Collections: Native American artifacts; mammals; birds; reptiles & amphibians; insects; mollusks; minerals; photographs.
Research Fields: biology; anthropology; Connecticut archaeology; ecology.
Facilities: 2 classrooms.
Activities: major events: special events; family programs; guided tours; lectures; education programs for children; workshops; field trips.
Publications: newsletter; activities program; membership brochure; annual report.
Hours & Admission Prices: Wed.-Fri. 10-4. No charge; donations accepted. Closed holidays. &
Attendance: 90,000 (accurate)
Membership: Student & Senior Citizen $25; Individual & Senior Couple $35; Couple $40; Family $45. Donor: Saw-Whet Owl $75; Snow Owl $150; Screech Owl $300; Barn Owl $600.

J. ROBERT DONNELLY HUSKY HERITAGE SPORTS MUSEUM, UConn Alumni Center, 2384 Alumni Dr., Unit-3053, Storrs, CT 06269. Tel.: 888-822-5861.
E-mail: ucaa@uconn.edu
Web Site: www.uconnhuskies.com/trads/museum.html
Formerly: Alumni Center - Husky Heritage Sports Museum
Key Personnel: Exec. Dir., Lisa Lewis; Center Mgr., Julie Sweeney.
Governing Authority: university.
Institution Type/Description: Sports Museum.
Collections: sports memorabilia which reflects more than 100 years of UConn athletic competition; personal artifacts; photographs.
Activities: special events.
Hours & Admission Prices: Mon.-Fri. 8-5. No charge.

MANSFIELD HISTORICAL SOCIETY MUSEUM, 954 Storrs Rd., Storrs, CT 06268-2611. Mailing Address: P.O. Box 145, Storrs, CT 06268-0145. Tel.: 860-429-6575.
E-mail: mansfieldhistorical@snet.net
Web Site: mansfieldct-history.org
Founded: 1957.
Congressional District: 2
Key Personnel: Pres., John Meyers; Dir., Ann Galonska; Treas., Howard Raphaelson.

Personnel Profile: Part-Time Paid 1.
Governing Authority: nonprofit; society. Parent Institution: Mansfield Historical Society. Tax-exempt: 501(c)(3).
Institution Type/Description: History Museum: housed in former Mansfield Town Office building & adjacent Town Hall c.1843.
Collections: furniture; household equipment; tools; farm equipment; industry; textiles; archives; photographs of local houses and people; 18th-19th century costumes of Connecticut town.
Research Fields: early houses in town.
Facilities: 500-vol. library on history, manuscripts & old account books; local genealogy.
Activities: guided tours; permanent & temporary exhibitions.
Publications: Chronology of Mansfield, Connecticut 1702-2002; That Sacred Plant of Paradise; Farming In Mansfield 1690-1955; George Freeman, Miniaturist, 1789-1868. On The Trail of a Legend, Separatist Movement in Mansfield, Ct., 1745-1769; Listen to the Echoes: Early History of Spring Hill; The Mansfield Poor House 1861-1922; World War II on the Homefront: Mansfield (excerpts from the diaries of Edwina Maud Whitney); The Constant Years; Historic Mansfield Center.
Hours & Admission Prices: June-Sept. Thurs. & Sun. 1:30-4:30; tours by appointment. Adults $2; members no charge.
Attendance: 1,000 (estimated)
Membership: Student $10; Individual $20; Family $30; Contributing $50; Small Business $75; Sustaining $100; Patron $150; Corporate $250.

THE WILLIAM BENTON MUSEUM OF ART, (M), University of Connecticut, 245 Glenbrook Rd. U-3140, Storrs, CT 06269-3140. Tel.: 860-486-4520. Fax: 860-486-0234. Facebook: Benton Museum.
E-mail: diane.lewis@uconn.edu
Web Site: www.thebenton.org
Founded: 1966.
Congressional District: 2
Key Personnel: Exec. Dir., Nancy Stula; Business Mgr., Karen Sommer; Registrar, Carla Hill; Asst. Cur. & Academic Project Coord., Lauren Walton; Coord. Membership, Lynn Eriksson; Education Coord., Tracy Lawlor; Public Rels. & Mktg. Coord., Diane Lewis; Preparator, Philip Hollister; Museum Shop Mgr., Jeanne Ahern Mogayzel; Museum Store Assoc., Samantha Smith.
Personnel Profile: Full-Time Paid 4; Part-Time Paid 6; Part-Time Volunteers 25; Interns 2.
Governing Authority: university. Parent Institution: University of Connecticut. Tax-exempt: 501(c)(1).
Institution Type/Description: Art Museum.
Collections: over 6,000 European, American & Asian works in various media from 15th century to present.
Major Exhibits: Persepolis: Word and Image, 1/21/14-3/16/14; Making the Movement Move: Photography, Student Activism, and Civil Rights, 1/21/14-3/14.
Research Fields: 19th-21st century paintings, graphics & photography.
Facilities: coffee shop. Gift items for sale.
Activities: traveling exhibitions; lectures; gallery talks; symposia; interdisciplinary programs; concerts; exhibition-related films; digital media; on-line exhibitions.
Publications: seasonal brochures 3 times a year; exhibition catalogs.
Hours & Admission Prices: Tues.-Fri. 10-4:30, Sat.-Sun. 1-4:30. No charge; donations accepted. Closed between some exhibitions; national holidays. &
Attendance: 38,000 (estimated)
Membership: Student, Senior & Educator $20; Individual $30; Dual & Family $40; Supporting $75; Contributing $125; Sustaining $250; Patron $500; Director's Circle $1,000; Benton Circle $5,000 & up.

Stratford

BOOTHE MEMORIAL PARK AND MUSEUM, 5744 Main St. Putney, Stratford, CT 06614. Mailing Address: P.O. Box 902, Stratford, CT 06615-0902. Tel.: 203-381-2046.
Founded: 1984.
Key Personnel: Dir. Friends of Boothe Park, Bessie Burton; CEO & Pres. (V), Dr. Virginia Harris.
Personnel Profile: Part-Time Volunteers 40; Interns 4.
Volunteer Hours: 6,000
Governing Authority: Parent Institution: Friends of Booth. Subsidiary Institution: Town of Stratford. Tax-exempt.
Institution Type/Description: Americana.
Collections: Boothe family artifacts; baskets; period artifacts. Historic Buildings: homestead building; blacksmith shop; lighthouse; windmill; carriage building shop; Indian artifacts; mineral display.
Facilities: 32 acre park; rose garden.
Activities: picnic grounds.

Publications: biannual newsletter.
Hours & Admission Prices: Museum: June-Oct. 1 Tues.-Fri. 11-1, Sun. 1-4. Grounds: daily. No charge; donations accepted. &
Attendance: 10,000 (estimated)
Membership: Single $10; Family $25.

MERRITT PARKWAY MUSEUM, Ryder's Landing Shopping Ctr., 6580 Main St., Stratford, CT 06614-1605. Mailing Address: Merritt Parkway Conservancy, P.O. Box 17072, Stamford, CT 06907. Tel.: 203-661-3255.
E-mail: jill@merrittparkway.org
Web Site: merrittparkway.org/pages/project_mp_museum.asp#
Founded: 2006.
Institution Type/Description: History Museum.
Collections: attractions & points of interest along the Parkway; building the Merritt video; photographs; archives.
Facilities: archives.
Activities: video.
Hours & Admission Prices: Mon.-Fri. 9-5.

NATIONAL HELICOPTER MUSEUM, INC., 2480 Main St., Stratford, CT 06615-5940. Mailing Address: P.O. Box 775, Stratford, CT 06615-0775. Tel.: 203-375-8857.
Web Site: www.nationalhelicoptermuseum.org
Founded: 1983.
Congressional District: 3
Key Personnel: Dir., Raymond E. Jankowich, M.D.; C.E.O., Frank Falanga; Pres. (V), Ken Pike; Chm. (V) & Museum Shop Mgr., Gale Whittemore.
Personnel Profile: Part-Time Volunteers 15.
Governing Authority: Tax-exempt.
Institution Type/Description: Helicopter Museum.
Collections: helicopter history; photographs; documents; models; videos; Cabin Sikorsky S76; gas turbine T 53 engine; AVCO Lycoming T 55 Cataway gas turbine engine; cabin cockpit S76; Bendix Coaxial Helicopters; Sikorsky Helicopter.
Research Fields: early Connecticut helicopter attempts; Sikorsky helicopters; Bendix helicopter history; Charles Kaman helicopter history.
Hours & Admission Prices: Memorial Day to mid-Oct. Wed.-Sun. 1-4. No charge; donations accepted. &
Attendance: 1,400 (accurate)
Membership: Annual $10; 3-Year $25; Lifetime $100.

PERRY HOUSE VISITORS CENTER, 1128 W. Broad St., Stratford, CT 06615. Tel.: 203-377-3779.
Institution Type/Description: Historic House: built c.1690.
Collections: local history & culture; period furnishings; personal artifacts.
Activities: educational programs; special events.
Hours & Admission Prices: Daily 12-4. No charge.

THE STRATFORD HISTORICAL SOCIETY & CATHARINE B. MITCHELL MUSEUM, (M), 967 Academy Hill, Stratford, CT 06615-0382. Mailing Address: P.O. Box 382, Stratford, CT 06615-0382. Tel.: 203-378-0630. Fax: 203-378-2562.
E-mail: judsonhousestfd@aol.com
Web Site: www.stratfordhistoricalsociety.org
Founded: 1925.
Congressional District: 3
Key Personnel: Pres. (V), Shirley McCormack; Cur., Carol Lovell; Archivist, Gloria Duggan.
Personnel Profile: Full-Time Paid 2; Part-Time Paid 4; Part-Time Volunteers 35.
Governing Authority: society; nonprofit. Parent Institution: Stratford Historical Society. Tax-exempt: 501(c)(3).
Institution Type/Description: Historical Society Museum: housed in 1750 Capt. David Judson Home, located on the original common of 1639.
Collections: archives; decorative arts; baskets; costumes; clothing; glass; china; silver; period Stratford items; toys; dolls; textiles; tools, Revolutionary & Civil Wars; early kitchen items; paintings; photographs; quilts; local artifacts; Native American artifacts; furniture.
Research Fields: local genealogy.
Facilities: 1,300-vol. library; 10,000 sq. ft. exhibit space. Museum-related items for sale.
Activities: guided tours; lectures; films; formally organized education programs; permanent & temporary exhibitions.
Publications: bimonthly newsletter.
Hours & Admission Prices: June-Oct. Wed. & Sun. 12-4. Adults $5, students $2; museum members no charge. &
Attendance: 1,000 (estimated)

Membership: Senior Citizen $15; Adult $20; Corporate $50; Life $200.

Suffield

THE KING HOUSE MUSEUM, 232 S. Main St., Suffield, CT 06078. Mailing Address: P.O. Box 893, Suffield, CT 06078-0893. Tel.: 860-668-5256.
Web Site: www.suffieldhistoricalsociety.org/kinghouse.htm
Founded: 1960.
Congressional District: 2
Key Personnel: Pres. (V), Edward W. Chase, III; Cur., Lester Smith.
Personnel Profile: Part-Time Volunteers 10.
Governing Authority: Parent Institution: Suffield Historical Society. Tax-exempt.
Institution Type/Description: Historic House Museum: built in 1764.
Collections: family & local history; early Connecticut Valley furniture; tobacco industry; early flasks & bottles; Bennington pottery.
Activities: special events.
Publications: newsletter.
Hours & Admission Prices: May-Sept. Wed. & Sat. 1-4; other times by appointment. No charge; donations accepted.
Attendance: 600 (estimated)

PHELPS-HATHEWAY HOUSE, 55 S. Main St., Suffield, CT 06078. Tel.: 860-668-0055. Fax: 860-249-4907.
E-mail: phelps.hatheway@ctlandmarks.org
Web Site: www.ctlandmarks.org
Key Personnel: Dir., Sheryl Hack.
Governing Authority: private; nonprofit organization. Parent Institution: CT Landmarks, Hartford, CT. Tax-exempt: 501(c)(3).
Institution Type/Description: Historic House Museum.
Collections: eighteenth century furniture; period artifacts; personal artifacts.
Activities: special events; educational programs; guided tours; workshops; lectures.
Hours & Admission Prices: May-Oct. call for hours; groups by appointment. Adults $7, senior citizens, teachers & students $6, children 6-18 $4; discounts to groups; members & children under 6 no charge.

Terryville

LOCK MUSEUM OF AMERICA, INC., 230 Main St., Rte. 6, Terryville, CT 06786-5900. Mailing Address: P.O. Box 104, Terryville, CT 06786-0104. Tel.: 860-589-6359. Fax: 860-589-6359 (call first).
E-mail: thomasnsc@aol.com
Web Site: www.lockmuseum.com/
Founded: 1972.
Congressional District: 6
Key Personnel: Pres. & Asst. Cur., Thomas Hennessy, Jr.; Vice Pres., Gilbert Wade; Cur., Thomas F. Hennessy; Sec. & Librarian, Reggie Murawski; Membership Sec., Theresa Kovaleski.
Governing Authority: nonprofit organization. Tax-exempt.
Institution Type/Description: Lock Museum: located on the site of the original offices of the Eagle Lock Co., built in 1859.
Collections: American locks, keys & ornate hardware.
Facilities: 1,000-item library of catalogs, books, financial & sales records of Sargent & Greenleaf Co. & the Eagle Lock Co. from c.1900-1930 available for research on premises; meeting room.
Publications: newsletter.
Hours & Admission Prices: May-Oct. Tues.-Fri. 1:30-4; other times by appointment. Admission $3. &
Attendance: 1,000 (estimated)
Membership: Single $20; Family $25; Company & Organization $100; Individual Life $200; Company Life $1,000.

Thomaston

RAILROAD MUSEUM OF NEW ENGLAND, 242 E. Main St., Thomaston, CT 06787-0400. Mailing Address: P.O. Box 400, Thomaston, CT 06787-0400. Tel.: 860-283-7245. Fax: 860-283-7245.
E-mail: info@rmne.org
Web Site: www.rmne.org
Key Personnel: Pres., Celeste Echlin.
Personnel Profile: Part-Time Volunteers 40.
Governing Authority: Tax-exempt.
Institution Type/Description: History Museum: housed in the New England Thomaston Station; built in 1881.
Collections: railroad heritage, history, & artifacts; New England rolling stock including locomotives, passenger cars, freight cars, & cabooses.

Facilities: Museum-related items for sale.
Activities: train rides; special events.
Hours & Admission Prices: Train Rides: see website for schedule. &
Membership: Individual $35.

Thompson

THOMPSON MUSEUM AT THE ELLEN LARNED MEMO- RIAL BUILDING, Historic Thompson Common, Rte. 193, Thompson, CT 06277. Mailing Address: P.O. Box 47, Thompson, CT 06277. Tel.: 860-923-3776.
E-mail: jiamartino@charter.net
Web Site: www.thompsonhistorical.org
Founded: 1968.
Personnel Profile: Part-Time Volunteers 28.
Governing Authority: Parent Institution: Thompson Historical Society.
Institution Type/Description: Historical Society Museum: housed in a former library building; built in 1902.
Collections: local history & culture; photographs; personal artifacts; period furnishings; Civil War artifacts.
Research Fields: Nipmuc Indians; lithic artifacts.
Publications: quarterly newsletters.
Hours & Admission Prices: May-June & Sept.-Nov. 1st Sat. each month 11-2; other times by appointment. No charge; donations accepted.
Attendance: 350 (estimated)
Membership: Individual $10; Family $15; Contributor $25.

Tolland

DANIEL BENTON HOMESTEAD, Metcalf Rd., Tolland, CT 06084. Mailing Address: P.O. Box 107, Tolland, CT 06084-0107. Tel.: 860-974-1875.
E-mail: tolland.historical@snet.net
Web Site: www.tollandhistorical.org/danielbentonhomestead
Founded: 1969.
Key Personnel: Pres. (V), Stewart R. Joslin; Museum Dir., Gail W. White.
Personnel Profile: Part-Time Volunteers 25.
Governing Authority: society. Tolland Historical Society, Inc., P.O. Box 107, Tolland, CT 06084. Tax-exempt: 501(c)(3).
Institution Type/Description: Historic Building: 1720 house.
Collections: local history; period furnishings.
Facilities: nature trails.
Activities: guided tours; lectures; workshops; formally organized education programs for children; French & Indian War weekend; Revolutionary War Encampment; 18th-century Artisans & Sutlers Market.
Publications: semi-annual newsletter; books, Tolland, An Old Post Road Town; Tolland on the Green.
Hours & Admission Prices: June-Oct. Sun. 1-4; other times by appointment. Donation $2. &
Membership: Single $5; Family $10; Associate $25; Corporate $100.

HICKS-STEARNS FAMILY MUSEUM, 42 Tolland Green, Tolland, CT 06084-3042. Mailing Address: P.O. Box 278, Tolland, CT 06084. Tel.: 860-875-7552.
Founded: 1974.
Key Personnel: Co Dir., Beatrice White-Ramirez; Co Dir., Teresa A. Gerry
Institution Type/Description: Historic House Museum.
Collections: family artifacts; period furnishings.
Facilities: Museum-related items for sale.
Activities: summer concerts. Museum Sponsors: Christmas Open House in December.
Hours & Admission Prices: mid-May to mid-Oct. Sun. & Wed. 1-4; other times by appointment. No charge; donations accepted.

OLD TOLLAND COUNTY COURT HOUSE, 53 Tolland Green, Tolland, CT 06084. Mailing Address: P.O. Box 107, Tolland, CT 06084-0107. Tel.: 860-870-9599. Fax: 860-870-4689.
E-mail: society@tollandhistorical.org
Web Site: www.tollandhistorical.org
Governing Authority: Parent Institution: Tolland Historical Society.
Institution Type/Description: Historic Building: built in 1822.
Collections: local history; period furnishings; photographs; personal artifacts.
Hours & Admission Prices: Call for hours.

TOLLAND HISTORICAL SOCIETY - OLD TOLLAND COUNTY JAIL & MUSEUM, 52 Tolland Green, Tolland, CT 06084. Mailing Address: P.O. Box 107, Tolland, CT 06084-0107. Tel.: 860-870-9599. Fax: 860-870-4689.
E-mail: society@tollandhistorical.org
Web Site: www.tollandhistorical.org
Founded: 1856.
Congressional District: 2
Key Personnel: Dir. & Pres., Kathy Bach; Museum Shop Mgr., Sue Errickson.
Personnel Profile: Part-Time Volunteers 10; Interns 20.
Governing Authority: society. Parent Institution: Tolland Historical Society. Subsidiary Institutions: Old Tolland County Courthouse; Daniel Benton Homestead. Tax-exempt: 501(c)(3).
Institution Type/Description: Historic Society Museum: housed in old jail cell block & jailer's home.
Collections: Native American artifacts; farm implements; 18th, 19th & 20th-century clothing; furniture; early schools & industries.
Activities: guided tours; demonstrations; reenactments. Annual Events: Tolland Antiques Show; Seasonal Exhibits May to October.
Publications: book; quarterly newsletter, Tolland Times; local history publications.
Hours & Admission Prices: mid-May to mid-Oct. Sun. 1-4; other times by appointment. No charge; donations accepted. &
Attendance: 3,030 (accurate)
Membership: Individual $15; Family $25; Business $50; Life $250.

Torrington

ARTWELL GALLERY, 45 Main St., Torrington, CT 06790-5319. Tel.: 860-482-5122. Fax: 860-492-5122.
E-mail: artwell@sbcglobal.net
Web Site: www.artwellgallery.org
Founded: 1995.
Key Personnel: Dir., C.E.O. & Acting Pres., Michael Yurgeles.
Personnel Profile: Full-Time Volunteers 1; Part-Time Paid 1; Part-Time Volunteers 6; Interns 2.
Governing Authority: Tax-exempt.
Institution Type/Description: Art Gallery.
Collections: paintings; photographs; sculpture.
Activities: educational programs; classes; performances; special events.
Hours & Admission Prices: Wed.-Sat. 12-5. No charge; donations accepted.
Attendance: 3,000 (estimated)
Membership: Student $25; Artist $50; Family $75.

TORRINGTON HISTORICAL SOCIETY, INC., (M), 192 Main St., Torrington, CT 06790-5201. Tel.: 860-482-8260.
E-mail: torringtonhistorical@snet.net
Web Site: www.torringtonhistoricalsociety.org
Founded: 1944.
Congressional District: 1
Key Personnel: Exec. Dir., Mark McEachern; Pres. (V), David R. Bennett; Cur., Gail Kruppa; Librarian, Carol Clapp.
Personnel Profile: Full-Time Paid 2; Part-Time Paid 3; Part-Time Volunteers 35; Interns 1.
Governing Authority: society. Tax-exempt: 501(c)(3).
Institution Type/Description: Local History Museum.
Collections: history of Torrington; artifacts & archives; Hendey machine shop artifacts; historic buildings; John Brown birthplace memorial. Historic Building: Hotchkiss-Fyler House, 1900.
Research Fields: historical; genealogical; industrial.
Facilities: 2,000-vol. library of local history & genealogy available for use on premises.
Activities: guided tours; lectures; outreach programs; concerts; children's programs.
Publications: booklets, First Church of Torrington 1741-1841; First Church of Torrington 1841-1942; Torrington Town Meeting 1740-1765; Gleanings from Early Torrington History; The Formative Years, 1740-1852; The Growth Years, 1852-1923; The Annealing Years, 1923-1976; History of Torrington, Conn. (1878) reprint; Erastus Hodges 1781-1847; The John Brown Birthplace; Gertrude Fyler Hotchkiss.
Hours & Admission Prices: House & History Museum: April 15-Oct. Tues.-Sat. 12-4. Library: Tues.-Fri. 1-4. House Museum: adults $5. History Museum: adults $2; discounts to AAM members; members no charge. Closed legal holidays. &
Attendance: 4,500 (estimated)
Membership: Senior Citizens $15; Individual $20; Family $30; Contributing $50; Sustaining $100; Patron $250.

Trumbull

TRUMBULL HISTORICAL SOCIETY MUSEUM AND RESEARCH LIBRARY, 1856 Huntington Tpke., Trumbull, CT 06611. Mailing Address: P.O. Box 312, Trumbull, CT 06611-0312. Tel.: 203-377-6620. Facebook: Trumbull Historical Society, CT.
E-mail: trumbullhistory@gmail.com
Web Site: trumbullhistory.org
Founded: 1964.
Institution Type/Description: Historical Society Museum & Library.
Collections: local history & culture; photographs; period furnishings & clothing; military artifacts.
Facilities: Museum-related items for sale.
Activities: lectures; hikes; summer camps; Annual Event: Mother's Day Tea.
Publications: newsletter, The Gristmiller.
Hours & Admission Prices: 1st & 3rd Sun. of month 2-4. No charge, donations accepted.
Membership: Senior $9; Individual $12; Family $20; Life $100; Business/Corporate $250.

Uncasville

TANTAQUIDGEON INDIAN MUSEUM, (M), Rte. 32, 1819 Norwich-New London Rd., Uncasville, CT 06382-1320. Tel.: 860-848-0594. Fax: 860-862-6025.
E-mail: museum@moheganmail.com
Founded: 1931.
Key Personnel: Exec. Dir., Melissa Tantaquidgeon Zobel; Cur., Stacy Dufresne.
Governing Authority: individual operation.
Institution Type/Description: Indian Museum.
Collections: crafts made by Mohegan & other New England craftsmen of past & present; objects of stone, wood & bone; baskets; bowls; ladles; archaeology; artifacts from Southwestern & Northern Plains Tribes. Historic Church: c.1831 Mohegan Congregational Church.
Hours & Admission Prices: May-Nov. Wed.-Sat. 10-4. No charge. Closed holidays. &

Unionville

UNIONVILLE MUSEUM, 15 School St., Unionville, CT 06085-1029. Tel.: 860-673-2231.
Key Personnel: Pres., Frank Corbeil
Institution Type/Description: History Museum: housed in the restored Andrew Carnegie free public library building, c.1917.
Collections: photographs; letters; advertising items; period clothing; blacksmith tools; paintings.
Activities: Museum Sponsors: annual vintage car parades; suppers.
Publications: historic calendar.
Hours & Admission Prices: Wed. & Sat.-Sun. 2-4. No charge; donations accepted.

Vernon

NEW ENGLAND CIVIL WAR MUSEUM, 14 Park Pl., Vernon Memorial Hall, 2nd Fl, Vernon, CT 06066-3291. Mailing Address: P.O. Box 153, Vernon, CT 06066. Tel.: 860-870-3563.
E-mail: necwm@hotmail.com
Web Site: pages.cthome.net/ne.civilwar.mus
Founded: 1995.
Key Personnel: Exec. Dir., Matt Reardon; Librarian & Cur., Jerry Caroon; Librarian & Cur., Alex Oliphant.
Personnel Profile: Full-Time Volunteers 1; Part-Time Volunteers 2.
Governing Authority: Parent Institution: Alden Skinner Camp #45, Sons of Union Veterans of the Civil War. Tax-exempt.
Institution Type/Description: Civil War Museum.
Collections: Civil War artifacts; personal artifacts of Thomas F. Burpee, Hirst Brothers, & Weston including a sword, belts, cap boxes, holster, photographs & a trumpet; memorial hall.
Hours & Admission Prices: 1st Thurs. of month 3-7, 2nd & 4th Sun. of month 12-3; other times by appointment. No charge; donations accepted. &
Attendance: 1,200 (accurate)

VERNON HISTORICAL SOCIETY MUSEUM, 734 Hartford Turnpike, Vernon, CT 06066-5127. Mailing Address: P.O. Box 2055, Vernon, CT 06066-1455. Tel.: 860-875-4326.
E-mail: vernonhs@sbcglobal.net
Web Site: vhsvernonct.tripod.com
Institution Type/Description: Historical Society Museum.

Collections: local history & culture; paintings; etchings; books; photographs; period artifacts.
Activities: special events.
Hours & Admission Prices: Thurs. & 2nd Sun. of month 2-4. No charge; donations accepted.

Wallingford

WALLINGFORD HISTORICAL SOCIETY, INC., 180 S. Main St., Wallingford, CT 06492-4217. Mailing Address: P.O. Box 73, Wallingford, CT 06492-0073. Tel.: 203-294-1996.
Founded: 1916.
Congressional District: 3
Key Personnel: Pres. (V), Raymond A. Chappell; 1st Vice Pres., Robert Beaumont.
Personnel Profile: Part-Time Volunteers 8.
Governing Authority: society. Tax-exempt: 501(c)(3).
Institution Type/Description: General Museum: housed in 1759 Parsons House.
Collections: costumes; Bibles; scrapbooks; diaries; school books & notebooks.
Research Fields: local history.
Facilities: 200-vol. library of genealogies & newspapers available on premises by permission.
Activities: guided tours; lectures; permanent & temporary exhibitions.
Publications: Wallingford, in the Images of America Series.
Hours & Admission Prices: Memorial Day-Labor Day Sun. 2-4:30; other times by appointment. No charge.
Attendance: 900 (estimated)
Membership: Individual $5; Couple $8; Life $50.

Washington

GUNN MEMORIAL LIBRARY AND MUSEUM, (M), 5 Wykeham Rd., Washington, CT 06793. Mailing Address: P.O. Box 1273, 5 Wykeham Rd., Washington, CT 06793-0273. Tel.: 860-868-7756. Fax: 860-868-7247.
E-mail: gunnmuseum@sbcglobal.net
Web Site: www.gunnlibrary.org
Founded: 1899.
Congressional District: 6
Key Personnel: Exec. Dir., Jean Chapin; Pres., Roger Stikeleather; Pres. (V), Barbara Kohn; Cur., Stephen Bartkus; Asst. Cur., Samantha Heberton.
Personnel Profile: Full-Time Paid 1; Part-Time Paid 1; Part-Time Volunteers 8.
Governing Authority: private; bd. of trustees. Parent Institution: Gunn Memorial Library. Tax-exempt.
Institution Type/Description: General Museum: housed in 1781 building.
Collections: Connecticut history & artifacts; letters; archives; manuscripts; 17th- to 19th-century furnishings; Revolutionary, Civil War, & World War I items; letters of George Washington & Thomas Jefferson; manuscript collections; local photographs; textiles; china & glass; pewter & silver; spinning wheels; dolls & doll houses; 19th- & early 20th-century costumes & portraits; 19th-century farm & carpentry tools.
Research Fields: archives; textiles; glass; doll house & dolls; portraits; photographs.
Facilities: 180-vol. library of history books available for use on premises; genealogical, state & local history reference.
Activities: lectures; gallery talks; loan, permanent & temporary exhibitions.
Publications: Return to Arcadia - The Architecture of Ehrick Rossiter; Rossiter: Country Houses of Washington, Connecticut.
Hours & Admission Prices: Jan.-April Thurs.-Sat. 10-4; May-Dec. Thurs.-Sat. 10-4, Sun. 12-4. No charge; donations accepted. Closed holidays.
Attendance: 3,000 (accurate)

THE INSTITUTE FOR AMERICAN INDIAN STUDIES (IAIS), 38 Curtis Rd., Washington, CT 06793-0260. Mailing Address: P.O. Box 1260, Washington, CT 06793-0260. Tel.: 860-868-0518. Fax: 860-868-1649. Facebook: The Institute For American Indian Studies (IAIS).
E-mail: general@iaismuseum.org
Web Site: iaismuseum.org
Founded: 1975.
Congressional District: 6
Key Personnel: Dir., Elizabeth McCormick; Chm. (V), Dan Sherr; Pres. (V), Ed Sarabia; Archaeologist, Dir. Research & Collections, Dr. Lucianne Lavin; Asst. Cur. Collections, Lisa Piastuch-Temmen; Education Coord. & Dir. Camp, Matthew Barr; Museum Shop Mgr., Christine Peschel.
Personnel Profile: Full-Time Paid 4; Full-Time Volunteers 3; Part-Time Paid 7; Part-Time Volunteers 10; Interns 3.
Governing Authority: nonprofit. Tax-exempt: 501(c)(3).

Institution Type/Description: American Indian Culture & Archaeology Museum.
Collections: prehistoric & historic artifacts primarily from Connecticut & Northeastern United States; ethnographic items from most North American culture areas; contemporary Native art; reconstructed American Indian village; plants the American Indians used in the region over the last 10,000 years.
Research Fields: Southern New England historic & prehistoric American Indian cultural & post white contact culture.
Facilities: 2,000-vol. library available for research for IAIS members, scholars & students with letter from Professor for use on premises only; nature trail; classrooms; arboretum. Books & other gifts for sale.
Activities: guided tours; lectures; films; gallery talks; formally organized education programs for children, adults, college students affiliated with colleges of Hartford, Fairfield, Wesleyan & others; loan, permanent, temporary & traveling exhibitions; school loan service; craft programs; Native American studies program. Museum Sponsors: summer field schools training sessions in archaeology; summer camps & year-round workshops.
Publications: 4 membership newsletters; Occasional Papers; teacher materials; educational pamphlets; books; videos.
Hours & Admission Prices: Mon.-Sat. 10-5, Sun. 12-5. Adults $5, seniors $4.50, children $3; members no charge. Closed New Year's Day; Easter; Labor Day; Memorial Day; Independence Day; Thanksgiving; Christmas. &
Attendance: 60,000
Membership: Senior Citizen $30; Individual $40; Family $55; Contributor $100; Benefactor $250; Sponsor $500; Patron $1,000.

Waterbury

＊ **THE MATTATUCK MUSEUM, (M),** 144 W. Main St., Waterbury, CT 06702-1298. Tel.: 203-753-0381, ext. 10. Fax: 203-756-6283.
E-mail: info@mattatuckmuseum.org
Web Site: www.mattatuckmuseum.org
Formerly: The Mattatuck Museum of the Mattatuck Historical Society
Founded: 1877.
Congressional District: 5
Key Personnel: C.E.O. & Exec. Dir., Robert Burns; Pres. (V), M. Catherine Smith; Cur., Cynthia Roznoy, Ph.D.; Collections Mgr., Suzie Fateh; Dir. Bldg. Svcs., Darryl Dilday; Programming & Education, Marie Galbraith; Finance Mgr., Jan Doughty; Dir. Mktg., Stephanie Harris; Dir. Devel., Eloise Mongillo; Dir. Corp. & Foundation Rels., Stephanie Forrest Yavozturk.
Personnel Profile: Full-Time Paid 10; Part-Time Paid 12; Part-Time Volunteers 200; Interns 5.
Governing Authority: society. Parent Institution: The Mattatuck Historical Society, Inc. Tax-exempt: 501(c)(3).
Institution Type/Description: History & Art Museum: housed in renovated Masonic Temple.
Collections: Connecticut artists collection; 17th-20th century decorative arts; 18th century regional furniture; brass industry; historical material relating to state & local history; colonial; industrial; manuscripts.
Research Fields: Connecticut artists; brass worker's history.
Facilities: library pertaining to local history & Connecticut artists available for use on premises by request; auditorium; cafe. Books, crafts & gift items for sale.
Activities: guided tours; lectures; gallery talks; formally organized educational programs; docent program; permanent, temporary & traveling exhibitions; art classes for children & adults; field trips; performing arts; teacher training.
Publications: annual report; books, Waterbury Pictorial History; Metal, Minds & Machines: Waterbury at Work; Fiddlebacks & Crooked Backs: Elijah Booth & Other Joiners in Newton and Woodbury 1750-1820; Images of Contentment: John Frederick Kensett & the Connecticut Shore; Artist of the Litchfield Hills.
Hours & Admission Prices: Tues.-Sat. 10-5, Sun. 12-5. Adults $5, seniors $4; members no charge. Closed major holidays &
Attendance: 45,000 (accurate)
Membership: Student $10; Seniors $25; Individual $40; Family $60; Sustaining $150; Director's Circle $250; Patron $500; Fellow $1,000; Benefactor $2,500; Kendrick Society $5,000.

TIMEXPO MUSEUM, 175 Union St., Brass Mill Commons, Waterbury, CT 06706-1236. Tel.: 203-755-8463; 800-225-7742. Fax: 203-755-8531.
E-mail: cconti@timexpo.com
Web Site: www.timexpo.com
Founded: 2001.
Congressional District: 5

Key Personnel: Pres., Michael Friend; Dir., Cathy Conti.
Personnel Profile: Full-Time Paid 2; Part-Time Paid 1; Part-Time Volunteers 10.
Governing Authority: Parent Institution: Timexpo Corporation.
Institution Type/Description: History Museum.
Collections: watches & clocks from Waterbury Clock Co. & Waterbury Watch Co., Ingersoll, U.S., Time Corp. & Timex; related letters, photographs & ads; Reed boat from Bolivia; replicas of artifacts from Easter Island, Peru, Egypt, Middle East; videos; interactive exhibits; theories & discoveries of Dr. Thor Heyerdahl, the Norwegian explorer and scientist, with emphasis on ocean navigation by ancient civilizations.
Facilities: 60-seat multi-purpose room; 8,000 sq. ft. exhibit space. Museum-related items for sale.
Hours & Admission Prices: Tues.-Sat. 10-5. Adults $6, seniors $5, children 5-12 $4; discounts to groups with reservation; children under 5 no charge. Closed major holidays. &
Membership: Senior & Teacher $20; Individual $25; Family $60; Patron $150.

Waterford

HARKNESS MEMORIAL STATE PARK, 275 Great Neck Rd., Waterford, CT 06385-3895. Tel.: 860-443-5725. Fax: 860-441-6151.
E-mail: dep.stateparks@ct.gov
Web Site: www.ct.gov/dep
Founded: 1953.
Key Personnel: Park & Recreations Supvr., Mark Darin.
Personnel Profile: Full-Time Paid 4; Part-Time Paid 1.
Governing Authority: state. Connecticut Dept. of Environmental Protection, Division of Parks & Recreation, Hartford, CT 06106.
Institution Type/Description: Park Museum: 1906 Eolia, summer residence of the Edward S. Harkness family. The estate & mansion are located near the confluence of the Thames River & Long Island Sound on the promontory, Goshen Point.
Collections: cutting garden; Italian Garden; Oriental Garden; 42-room mansion; greenhouses; stable; water tower.
Research Fields: virgin shoreline on Long Island Sound; open meadow lands.
Facilities: picnic area.
Activities: picnicking; fishing. Park Sponsors: recreation site for the handicapped.
Publications: brochures.
Hours & Admission Prices: Grounds: daily 8-sunset. In-State cars Mon.-Fri. $5, Sat.-Sun. & holidays $7; out-of-state cars Mon.-Fri. $7, Sat.-Sun. & holidays $10. Season Pass: CT residents $50; Out of State residents $75. Mansion Tours: Memorial to Labor Day Sat.-Sun. & holidays 10-2. No charge; donations accepted. &
Membership: Individual $15; Family $25; Patron $50; Supporter $100; Benefactor $250; Corporate $500; Eolian $1,000.

Watertown

WATERTOWN HISTORICAL SOCIETY, INC., 22 DeForest St., Watertown, CT 06795-2116. Tel.: 860-274-1050.
E-mail: watertownctmuseum@sbcglobal.net
Web Site: www.watertownhistoricalsociety.org
Founded: 1947.
Congressional District: 6
Key Personnel: C.E.O. & Pres., Jeffrey S. Grenier; Vice Pres., Jan J. Guidess; Treas., JoAnn Zanavich; Archivist, John F. Pillis.
Personnel Profile: Part-Time Volunteers 10.
Governing Authority: society. Tax-exempt: 501(c)(3).
Institution Type/Description: Local History Museum: housed in 1846 building.
Collections: articles, utensils, tools & furniture manufactured in Watertown; manuscripts.
Research Fields: town history; genealogies.
Facilities: library of books & files.
Activities: guided tours; lectures; formally organized education programs for children; permanent & temporary exhibitions.
Publications: semiannual newsletter.
Hours & Admission Prices: Wed. 2-4; groups by appointment. No charge; donations accepted. Closed major holidays.
Attendance: 300 (estimated)
Membership: Senior $12; Individual $30; Couple $45; Museum Sponsor $100; Heminway Circle $250; Lifetime $500.

West Hartford

AMERICAN SCHOOL FOR THE DEAF MUSEUM & ARCHIVES, 139 N. Main St., West Hartford, CT 06107. Tel.: 860-570-2304. Fax: 860-570-2301.
E-mail: ed.peltier@asd-1817.org
Web Site: www.asd-1817.org
Key Personnel: Archivist, Edward F. Peltier.
Governing Authority: private school; nonprofit. Parent Institution: American School for the Deaf.
Institution Type/Description: History Museum.
Collections: American deaf heritage & its European antecedents; school archives; rare-books; manuscripts; artifacts created by deaf artists & artisans.
Research Fields: American deaf historical & cultural topics.
Facilities: 1,250-vol. library.
Activities: guided tours; lectures; loan & temporary exhibitions; formal education programs for children; school outreach programs.
Hours & Admission Prices: Mon. & Thurs. 9-3, Fri. 9-2; other times by appointment. No charge; donations accepted. Closed school holidays.
Attendance: 225 (estimated)

ART GALLERY, UNIVERSITY OF SAINT JOSEPH, (M), 1678 Asylum Ave., West Hartford, CT 06117-2791. Tel.: 860-231-5399. Fax: 860-231-5754. Facebook: The Art Gallery, USJ.
E-mail: artgallery@usj.edu
Web Site: www.usj.edu/artgallery
Formerly: Saint Joseph College Art Gallery
Founded: 1937.
Congressional District: 1
Key Personnel: Dir. & Cur., Ann H. Sievers; Collection Mgr. & Registrar, Rochelle L. R. Oakley.
Personnel Profile: Full-Time Paid 2; Part-Time Paid 4; Part-Time Volunteers 4; Interns 3.
Governing Authority: private college; nonprofit. Tax-exempt: 501(c)(3).
Institution Type/Description: Art Museum.
Collections: early 20th-century American paintings; original prints by European & American artists dating from the 15th-century to the present; Japanese woodblock prints; Thomas Nast wood engravings.
Facilities: print study room; 2,150 sq. ft. gallery space.
Activities: formal education programs for undergraduate & graduate college students; temporary exhibitions.
Publications: exhibition checklists.
Hours & Admission Prices: Tues.-Wed. & Fri.-Sat. 11-4, Thurs. 11-7, Sun. 1-4. No charge. Closed national holidays. &
Attendance: 5,000 (estimated)
Membership: Student & Recent Alumni $10; Senior $20; Educator K-12 $25; Individual $35; Family & Household $50; Patron $100; Sponsor $250; Benefactor $500; Director's Circle $1,000; Kelly Associate $10,000.

THE CHILDREN'S MUSEUM, 950 Trout Brook Dr., West Hartford, CT 06119-1492. Tel.: 860-231-2824. Fax: 860-232-0705.
E-mail: info@thechildrensmuseumct.org
Web Site: www.thechildrensmuseumct.org
Formerly: Science Center of Connecticut
Founded: 1927.
Congressional District: 1
Key Personnel: Interim Dir., Don Peterson; Chm. (V), Rick Liftig, D.M.D.; Dir. Guest Rels. & Museum Shop Mgr., Beth Weller; Dir. Education, Sue Carroll; Dir. Travelers Science Dome & Gengras Planetarium, Kristie Mazzoni.
Personnel Profile: Full-Time Paid 21; Part-Time Paid 49; Part-Time Volunteers 80.
Governing Authority: nonprofit corporation. Tax-exempt: 501(c)(3).
Institution Type/Description: Children's Museum.
Collections: fossils; math exhibit; natural history; technology; physical science; hands-on exhibits.
Facilities: planetarium; nature center; wildlife sanctuary.
Activities: birthday parties; rental facilities; summer & vacation science camps; youth group programs; field trips; outreach; nature walks; teacher training; after school programs; evening astronomy programs; science preschool.
Publications: teacher resource guide; annual fund brochure; newsletters; program brochure; e-newsletter.
Hours & Admission Prices: Summer: Mon.-Sat. 10-5, Sun. 11-4; Winter: Tues.-Sat. 9-4, Sun. 11-4. Admission 2-62 $11, senior citizens 63 & over $10; children under 2 no charge. Reciprocal admission program. Closed Easter; Memorial Day; Independence Day; Labor Day; Thanksgiving; Christmas Eve & Day.

Attendance: 210,000 (estimated)
Membership: Caregiver Card $10; One Plus One $75; Family & Grandparent $90; Family Plus $125.

JOSELOFF GALLERY, HARTFORD ART SCHOOL, UNIVERSITY OF HARTFORD, Harry Jack Gray Center, 200 Bloomfield Ave., West Hartford, CT 06117-1545. Tel.: 860-768-4090. Fax: 860-768-5159.
E-mail: joseloff@hartford.edu
Web Site: www.joseloffgallery.org
Founded: 1970.
Congressional District: 1
Key Personnel: Mng. Dir., Lisa Gaumond.
Personnel Profile: Full-Time Paid 1; Part-Time Paid 5.
Governing Authority: university; nonprofit. Parent Institution: University of Hartford/Hartford Art School. Tax-exempt: 501(c)(3).
Institution Type/Description: Art Gallery.
Collections: paintings; photographs; sculpture.
Facilities: 200-seat auditorium; 3,500 sq. ft. exhibit space.
Activities: lectures; opening receptions; informal gallery talks.
Publications: annual calendar of events; catalogues & brochures.
Hours & Admission Prices: Tues.-Fri. 11-4, Sat.-Sun. 12-4. No charge; donations accepted. Closed major university & national holidays. ら
Attendance: 10,000 (estimated)

NEIL'S AMERICAN DREAM MUSEUM, 2021 Albany Ave., West Hartford, CT 06117-2789. Tel.: 860-561-5311.
Key Personnel: Owner & Cur., Neil Sakow
Institution Type/Description: Collectibles Museum.
Collections: toys; sports & cowboy memorabilia; early television collectibles; Americana advertising.
Hours & Admission Prices: By appointment.

*** NOAH WEBSTER HOUSE/WEST HARTFORD HISTORICAL SOCIETY, (M),** 227 S. Main St., West Hartford, CT 06107-3430. Tel.: 860-521-5362. Fax: 860-521-4036.
E-mail: comments@noahwebsterhouse.org
Web Site: www.noahwebsterhouse.org
Formerly: Noah Webster House Museum of West Hartford History
Founded: 1965.
Congressional District: 1
Key Personnel: Interim Exec. Dir., Mary Donohue; Pres. (V), Connie B. Robinson; Archivist & Asst. Cur., Sheila Daley; Coord. Education, Beth Sweeney; Coord. Public Programs, Sarah Mocko St. Germain.
Personnel Profile: Full-Time Paid 4; Part-Time Paid 19; Part-Time Volunteers 80; Interns 2.
Governing Authority: nonprofit organization. Tax-exempt: 501(c)(3).
Institution Type/Description: History Museum: housed in 18th-century Noah Webster House.
Collections: period furnishings & decorative arts of 18th century; 18th to 20th-century period clothing; local & regional manuscripts; Webster publications; paintings & photographs.
Research Fields: West Hartford history; Noah Webster's life & works; Webster family.
Facilities: 800-vol. library of regional & topical history & the life of Noah Webster, available for research by special appointment; 50-seat auditorium; schoolhouse theater. Museum-related items for sale.
Activities: guided tours; special events; lectures; films; gallery talks; outreach educational programs; walking tours; formally organized education programs for children, undergraduate & graduate college students affiliated with museum internships; permanent & temporary exhibitions; cable access monthly TV show; summer camp.
Publications: occasional books & booklets, Noah Webster; From Colonial Parish; filmstrip, A Brief Appreciation of West Hartford; The Blue-backed Speller; filmstrip, Noah Webster: Schoolmaster of America; quarterly newsletter, Spectator.
Hours & Admission Prices: Thurs.-Mon. 1-4. Adults $7, children 6-18 $4; discounts for groups, AAM, NEMA, & GHAHHM members; children 5 & under and members no charge. Closed national holidays. ら
Attendance: 17,500 (accurate)
Membership: Library $10; Student $20; Single $25; Family $40; Sustaining $75.

SARAH WHITMAN HOOKER HOMESTEAD, 1237 New Britain Ave., West Hartford, CT 06110-2404. Mailing Address: c/o 11 Dodge Dr., West Hartford, CT 06107. Tel.: 860-523-5887; 800-475-1233.
E-mail: fransson1701@att.net
Web Site: sarahwhitmanhooker.com

Institution Type/Description: Historic House Museum: built c.1720.
Collections: period furnishings; porcelain; glassware; early clothing.
Hours & Admission Prices: By appointment. Adult $7, children under 13 $3.50.

WESTMOOR PARK EDUCATION CENTER, 119 Flagg Rd., West Hartford, CT 06119. Tel.: 860-561-8260. Fax: 860-236-3815.
E-mail: westmoorpark@westhartford.org
Web Site: westmoorpark.com
Founded: 1974.
Personnel Profile: Full-Time Paid 2; Part-Time Paid 12; Part-Time Volunteers 30.
Governing Authority: city. Parent Institution: Leisure Services Dept.
Institution Type/Description: Park & Education Center.
Collections: environmental, agricultural & horticultural; natural sciences.
Activities: educational programs; scout programs; birthday parties.
Hours & Admission Prices: Office: Mon.-Fri. 9-4:30. Farm: daily 9-4. Grounds: daily dawn to dusk. No charge; donations accepted. ら
Attendance: 105,000 (estimated)

West Haven

SAVIN ROCK MUSEUM, 6 Rock St., West Haven, CT 06516-5846. Mailing Address: 355 Main St., West Haven, CT 06516. Tel.: 203-937-3666.
Web Site: www.savinrockmuseum.com
Key Personnel: Cur., Harold Hartmann
Institution Type/Description: History Museum.
Collections: local history & culture; photographs; period furnishings; personal artifacts.
Hours & Admission Prices: April 5-Sept. 14 Tues. 9am-12pm, Wed. & Fri. 1-4, Sat.-Sun. 4-7; Sept. 17-Dec. 21 Wed. & Fri.-Sun. 1-4. Adults $4, seniors & children under 12 $2.

WARD-HEITMANN HOUSE MUSEUM, 277 Elm St., West Haven, CT 06516. Mailing Address: P.O. Box 573, West Haven, CT 06516. Tel.: 203-937-9823. Fax: 203-937-9823.
E-mail: wardheitmann@sbcglobal.net
Web Site: www.wardheitmann.org
Founded: 1994.
Congressional District: 3
Key Personnel: Chm. (V), Violet Bornemann; Pres. (V), Doreen Breen.
Personnel Profile: Part-Time Volunteers 25.
Governing Authority: Tax-exempt.
Institution Type/Description: Historic House Museum: built c.1684. Listed on the National Register of Historic Places.
Collections: local history; period furnishings; personal artifacts; blacksmith; metalmaking tools; nails; hooks; latches.
Activities: Museum Sponsors: Open House April to December.
Hours & Admission Prices: By appointment. Adults $5; discounts to students; members no charge.
Attendance: 950 (estimated)
Membership: Senior & Student $15; Individual $25; Family $35; Booster $100; Sponsor $250; Patron $500; Benefactor $1,000.

West Simsbury

FLAMIG FARM, 7 Shingle Mill Rd., West Simsbury, CT 06092-2311. Tel.: 860-658-5070.
E-mail: flamigfarm@yahoo.com
Web Site: flamigfarm.com
Institution Type/Description: Farm & Educational Center.
Collections: local food production; decentralized energy production; energy conservation; personal wellness & nutrition.
Facilities: Farm-related items for sale.
Activities: summer camp; pony rides; feed animals; gather eggs; organic gardening; hiking; nature walks; hayrides; rental facilities; farm animal zoo; school tours; demonstrations; educational programs.
Hours & Admission Prices: Daily 9-5.

Westbrook

MILITARY HISTORIANS, HEADQUARTERS & MUSEUM, N. Main St., Westbrook, CT 06498-1944. Tel.: 860-399-9460.
E-mail: military.historians@snet.net
Formerly: The Company of Military Historians Headquarters and Museum
Founded: 2000.
Key Personnel: Admin., Maj. William R. Reid; Cur. Uniforms, Insignia & Photos, Earl Vincent; Librarian, Charles W. Reid.

Personnel Profile: Full-Time Volunteers 2; Part-Time Volunteers 3.
Governing Authority: nonprofit organization. Tax-exempt: 501(c)(3).
Institution Type/Description: Military Uniform Museum.
Collections: military uniforms from 1775 to date; military music & art; military vehicles from World War II to present.
Research Fields: military units of North America.
Facilities: library of books, films, DVDs & photos dealing with the military available for research by special request; reading room.
Activities: guided tours; lectures; films, concerts; permanent exhibitions.
Hours & Admission Prices: Tues.-Thurs. 8-1; other times by appointment. No charge, donations accepted.

STEWART B. MCKINNEY NATIONAL WILDLIFE REFUGE, 733 Old Clinton Rd., Westbrook, CT 06498-1760. Tel.: 860-399-2513. Fax: 860-399-2515.
Web Site: www.fws.gov/northeast/mckinney
Founded: 1972.
Congressional District: 2, 3, & 4
Key Personnel: Mgr., Richard Potvin.
Governing Authority: Parent Institution: U.S. Dept. of the Interior, Fish and Wildlife Service.
Institution Type/Description: Wildlife Refuge.
Collections: wildlife & their habitats; ecology; natural history; geology; cultural history.
Activities: educational programs.
Hours & Admission Prices: Call for hours. No charge.

Weston

THE COLEY HOMESTEAD & BARN MUSEUM, 104 Weston Rd., Weston, CT 06883. Mailing Address: P.O. Box 1092, Weston, CT 06883-0092. Tel.: 203-226-1804.
Founded: 1961.
Congressional District: 5
Key Personnel: Pres. (V), Reg Bowden.
Personnel Profile: Part-Time Volunteers 15.
Governing Authority: society. Parent Institution: The Weston Historical Society. Tax-exempt.
Institution Type/Description: Agriculture Museum: housed in 1835 Old Barn & Victorian house.
Collections: early handicrafts; furnishings; old farm tools; herb garden.
Research Fields: old houses; Weston genealogy.
Facilities: 100-vol. library of old history books.
Activities: guided tours; lectures; formally organized education programs for children; permanent exhibits.
Publications: Weston, the Forging of a Connecticut Town; quarterly newsletter, The Chronicle.
Hours & Admission Prices: Call for hours. No charge; donations accepted.
Attendance: 5,000 (estimated)
Membership: Sustaining $25-$99; Renovator $100-$249; Restorer $250-$399; Preserver $400 & up.

WESTON WOODS INSTITUTE, 389 Newtown Tpke., Weston, CT 06883-1116. Tel.: 203-222-8000 & 454-4005. Fax: 203-222-2263.
E-mail: uryy@sbcglobal.net
Web Site: www.infowestonwoodsinstitute.org
Founded: 1983.
Congressional District: 4
Key Personnel: C.E.O. & Chm. (V), Morton Schindel.
Personnel Profile: Part-Time Paid 1; Part-Time Volunteers 1; Interns 2.
Governing Authority: private; nonprofit organization. Tax-exempt: 501(c)(3).
Institution Type/Description: Children's Museum.
Collections: original animation cels; early mechanical music devices; stereoptical & projection devices.
Research Fields: communication techniques for children.
Facilities: 1,000-vol. library of children's books; 2,500 sq. ft. exhibit space.
Activities: films; formal education programs for children; guided tours; mobile vans; participatory, temporary & traveling exhibitions.
Hours & Admission Prices: By appointment or invitation. No charge; donations accepted.
Attendance: 750 (estimated)

Westport

* **EARTHPLACE - THE NATURE DISCOVERY CENTER,** 10 Woodside Lane, Westport, CT 06880-2322. Mailing Address: P.O. Box 165, Westport, CT 06881-0165. Tel.: 203-227-7253. Fax: 203-227-8909.
E-mail: info@earthplace.org
Web Site: www.earthplace.org
Formerly: Nature Center for Environmental Activities, Inc.
Founded: 1958.
Congressional District: 4
Key Personnel: Interim Exec. Dir., Maire Dalton-Meyer; Pres., Julia Mally.
Personnel Profile: Full-Time Paid 12; Part-Time Paid 21; Part-Time Volunteers 100; Interns 5.
Governing Authority: nonprofit organization. Tax-exempt: 501(c)(3) & 170(b)(1)(A).
Institution Type/Description: Natural History Museum & Environmental Studies Center.
Collections: geology; entomology; ornithology; mineralogy; mammalogy; ichthyology; malacology.
Research Fields: environmental studies; botany; ornithology.
Facilities: nature center; water quality research laboratory; 100-seat auditorium; classrooms; wild animal shelter.
Activities: bird & butterfly demonstration garden; native plant court; guided tours; lectures; films; formally organized educational programs; family local & regional excursions; inter-museum loan; permanent, temporary & traveling exhibitions; water-quality monitoring and research program; nature trails including universal access trail.
Publications: newsletter.
Hours & Admission Prices: Mon.-Sat. 9-5, Sun. 1-4. Adults $7, seniors 62 & over and children 1-12 $5; members no charge. Closed New Year's Day; Easter; Memorial Day; Independence Day; Labor Day; Thanksgiving; Christmas. &
Attendance: 70,000 (estimated)
Membership: Seniors 62 & over $25; Single $35; Family $90; Sponsor $125; Friend $250; Patron $500; Benefactor $1,000.

ROLNICK OBSERVATORY, 182 Bayberry Lane, Westport, CT 06880. Tel.: 203-227-0925.
Institution Type/Description: Observatory.
Collections: 25 inch Newtonian telescope; astronomy.
Activities: classes.
Hours & Admission Prices: Wed.-Thurs. 8pm-10pm. No charge.

WESTPORT ARTS CENTER, 51 Riverside Ave., Westport, CT 06880. Tel.: 203-222-7070.
Web Site: www.westportartscenter.org
Key Personnel: Dir., Nancy Heller; C.E.O., Lance Lundberg
Institution Type/Description: Art Gallery.
Collections: contemporary art; paintings; sculpture.
Activities: educational programs; hands-on activities.
Hours & Admission Prices: Mon.-Fri. 10-4, Sat.-Sun. 12-4. No charge. &
Membership: Individual $45; Household $60; Patron $100-$249; Pacesetter $250-$499; Renaissance $500-$999; Silver Circle $1,000 & up.

WESTPORT HISTORICAL SOCIETY, 25 Avery Place, Westport, CT 06880-3215. Tel.: 203-222-1424. Fax: 203-221-0981.
E-mail: info@westporthistory.org
Web Site: westporthistory.org
Founded: 1899.
Congressional District: 4
Key Personnel: Exec. Dir., Susan Gold; Pres. Westport Historical Society, Susan Wynkoop; Museum Shop Mgr., Olivia Yule.
Personnel Profile: Full-Time Paid 1; Part-Time Paid 3; Part-Time Volunteers 30.
Governing Authority: society; not-for-profit. Tax-exempt: 501(c)(3).
Institution Type/Description: History Museum.
Collections: genealogy; house history; photographs; costumes & textiles; period rooms: parlor, bedroom, kitchen in the Victorian style; Westport history; diorama of historic downtown.
Research Fields: house history; genealogy; costumes; textiles.
Facilities: archives. Gift items for sale.
Activities: formal education programs for children; adult workshops; lectures; temporary exhibitions. Annual Events: Holiday House Tour; Garden Tour.
Publications: bimonthly newsletter; annual report.
Hours & Admission Prices: Mon.-Fri. 10-4, Sat. 12-4. No charge; donations accepted. &
Attendance: 5,000 (estimated)
Membership: Student & Senior Citizen $30; Individual $35; Senior Family $50; Family $60; Patron $100; Minuteman $250; Patron $500; Benefactor $1,000.

Wethersfield

BUTTOLPH-WILLIAMS HOUSE, (M), 249 Broad St., Wethersfield, CT 06109. Mailing Address: 211 Main St., Wethersfield, CT 06109-2339. Tel.: 860-529-0612. Fax: 860-529-0460.
E-mail: info@webb-deane-stevens.org
Governing Authority: Parent Institution: The Antiquarian & Landmarks Society.
Institution Type/Description: Historic House: built c.1711.
Collections: period furnishings; personal artifacts.
Hours & Admission Prices: May-Oct. Mon. & Wed.-Sat. 10-4, Sun. 1-4. Adults $5, seniors 60 & over, student, & children 5-18 $4; members no charge.

ELEANOR BUCK WOLF NATURE CENTER, 156 Prospect St., Wethersfield, CT 06109. Tel.: 860-529-3075.
E-mail: naturecenter@wethersfieldct.com
Institution Type/Description: Nature Center.
Collections: live animals, mammals, birds, & reptiles.
Facilities: library; nature trails. Museum-related items for sale.
Activities: educational programs; hands-on activities.
Hours & Admission Prices: Tues.-Wed. & Fri.-Sat. 10-5, Thurs. 10-7. No charge; donations accepted. &

✳ **THE WEBB-DEANE-STEVENS MUSEUM, (M),** 211 Main St., Wethersfield, CT 06109-2339. Tel.: 860-529-0612. Fax: 860-571-8636.
E-mail: info@webb-deane-stevens.org
Web Site: www.webb-deane-stevens.org
Founded: 1919.
Congressional District: 1
Key Personnel: Pres. (V), Torrey Cooke; Exec. Dir., Charles T. Lyle; Education Coord., Cindy Riccio; Curatorial Asst., Kristen Wands; Curatorial Technician, Sal Cormosino; Rentals, Katie Sullivan; Maintenance Supvr., Richard Agne.
Personnel Profile: Full-Time Paid 1; Part-Time Paid 24; Part-Time Volunteers 40; Interns 1.
Governing Authority: society. Parent Institution: The National Society of The Colonial Dames of America in the State of Connecticut. Tax-exempt: 501(c)(3).
Institution Type/Description: Historic Houses: restored 18th century houses.
Collections: American furniture; decorative arts; ceramics; silver; American needlework; gardens; research library; archives; toys; games; children's items; early 20th-century murals & tea room.
Research Fields: local history; Connecticut Valley furniture & architecture; 18th-century social history; multi-culturalism in 18th century; slavery; colonial revival; women's organizations; American Revolution; child life in the early 18th & early 19th century; Wallace Nutting and the Colonial Revival Movement.
Facilities: 5,000-vol. library of history & decorative arts materials. Museum-related items for sale.
Activities: guided tours; school tours; museum school programs; family programs; academic programs for undergraduate college students affiliated with the greater Hartford Consortium of Colleges; summer internship program; seminars on decorative arts; special events; special interpretive tours. Museum Sponsors: Thanksgiving Holiday Tour; Christmas Holiday Tours.
Publications: newsletter.
Hours & Admission Prices: Jan-March by appointment; April & Nov. Sat. 10-4, Sun. 1-4; May-Oct. Wed.-Mon. 10-4, Sun. 1-4; last tour at 3:30; Nov.-April Sat. 10-4, Sun. 1-4; last tour at 3:30. House: family $25, adults $10, seniors $9, students $5; discounts to AAA & AAM members; members no charge. &
Attendance: 12,500 (accurate)
Membership: Family $20.

WETHERSFIELD HISTORICAL SOCIETY, (M), 150 Main St., Wethersfield, CT 06109-3126. Tel.: 860-529-7656. Fax: 860-563-2609.
E-mail: society@wethhist.org
Web Site: www.wethhist.org
Founded: 1932.
Congressional District: 1
Key Personnel: Interim Dir., Amy Northrop; Pres., Dorene Ciarcia; Vice Pres., Robert L. Fisher; Dir. Mktg. & Events, Margaret Longey.
Personnel Profile: Full-Time Paid 2; Part-Time Paid 6; Part-Time Volunteers 100.
Governing Authority: historical society. Branch Museums: 1804 Old Academy & Library; 1690 The Cove Warehouse; 1793 Capt. James Francis House; 1804 Hurlbut-Dunham House; 1787 Standish House; 1893 Robert Allan Keeney Memorial. Tax-exempt: 501(c)(3).
Institution Type/Description: History Museum.
Collections: local history; historic properties & buildings; tools & crafts; regional furniture; decorative arts; costumes; paintings; wallpaper; marine artifacts. Historic Buildings: 18th century brick Georgian house; 18th century Cove Warehouse.
Research Fields: local history; genealogy.
Facilities: research library; reading room; meeting rooms; lecture halls; visitors center. Museum-related items for sale.
Activities: guided tours; lectures; seminars; docent program; temporary exhibitions; formally organized education programs; daily life programs.
Publications: quarterly newsletter; periodic booklets & books.
Hours & Admission Prices: Old Academy & Research Library: Tues.-Fri. 10-4 & by appointment. No charge. Wethersfield Museum at Keeney Memorial: Tues.-Sat. 10-4, Sun. 1-4. Adults $5; discounts to AAM members; members, children 16 and under & town residents no charge. Hurlbut-Dunham House: mid-May to mid-Oct., Sat. 10-4, Sun. 1-4. Adults $5. Cove Warehouse: mid-May to mid-Oct. Sat. 10-4, Sun. 1-4. Donation: adults $1; children no charge. &
Attendance: 32,500 (estimated)
Membership: Student $10; Senior $20; Individual $30; Senior Citizen Couple 62 & over $35; Family $40; Friend $75; Sponsor $100; Patron $250; Benefactor $500.

Willimantic

CONNECTICUT EASTERN RAILROAD MUSEUM, 55 Bridge St., Willimantic, CT 06226. Mailing Address: P.O. Box 665, Willimantic, CT 06226-0665. Tel.: 860-456-9999.
E-mail: info@cteastrrmuseum.org
Web Site: www.cteastrrmuseum.org
Founded: 1992.
Congressional District: 2
Key Personnel: Pres. (V), Mark F. Granville; Museum Shop Mgr., M. Jean Lambert.
Personnel Profile: Part-Time Volunteers 30.
Governing Authority: nonprofit organization. Parent Institution: Connecticut Eastern Chapter, NRHS, Inc. Tax-exempt: 501(c)(3).
Institution Type/Description: Railroad Museum.
Collections: locomotives; rolling stock; railroad buildings; replica 1850s style pump car. Historic Buildings: roundhouse; turntables; freight house; section house; station house; 1850s 37 ft. reconstructed signal tower.
Publications: newsletter, Ghost Train Journal; annual calendar.
Hours & Admission Prices: May-Oct. Sat.-Sun. 10-4; other times by appointment. Adults $5, seniors 62 & over $4, children 8-12 $1; discounts to groups; members & children under 8 no charge.
Membership: Friend $25; Sponsor $50; Voting Member $59.

WINDHAM TEXTILE AND HISTORY MUSEUM (THE MILL MUSEUM), 411 Main St., Willimantic, CT 06226-3173. Tel.: 860-456-2178.
E-mail: themillmuseum@gmail.com
Web Site: www.millmuseum.org
Founded: 1985.
Congressional District: 2
Key Personnel: Exec. Dir., Brooke Shannon; Pres., Steve Kenton; Educator, Beverly York.
Personnel Profile: Part-Time Paid 2; Part-Time Volunteers 73; Interns 1.
Governing Authority: private; nonprofit organization. Tax-exempt: 501(c)(3).
Institution Type/Description: Textile & History Museum: located in 1877 buildings built as a company store & library for the Willimantic Linen Company.
Collections: material culture of the working families & owners - factory, home life & community life; manuscript collection includes more than 20,000 architectural & engineering drawings, photos and business records.
Research Fields: American, Connecticut & Windham textile industry history; oral history; water power; immigration & migration; mill architecture.
Facilities: 1,850-vol. library; 6,770 sq. ft. exhibit space; field research station. Museum-related items for sale.
Activities: arts festivals; concerts; dance recitals; docent program; films; formal education programs for adults, graduate & undergraduate students affiliated with the University of Connecticut and Eastern Connecticut State University; guided tours; hobby workshops; lectures; loan & temporary exhibitions; rental gallery; school loan service; Girl Scout & Boy Scout badge activities. Annual Events: Meet the Artist; architectural tours; spinning bee; Victorian crafts.
Publications: quarterly newsletter, The Gazette; school curriculum guides; three publications, Willimantic Women: Their Lives and Labors; Willimantic Industry and Community: The Rise & Decline of a Connecticut Textile

City; A Tour of The Oaks (Worker Housing); Mills and Meadows, A Pictorial of Windham & Tolland Counties.
Hours & Admission Prices: Fri.-Sun. 10-4. Adults $7, senior citizens & students $5; discounts to groups, AAM & ICOM members; children & members no charge. Closed New Year's Day; Easter; Thanksgiving; Christmas. &
Attendance: 7,000 (estimated)
Membership: Seniors & Students $10; Individual $15; Family $25; Sustaining $30; Patron $50; Benefactor $1,000; Business $150.

Wilton

WEIR FARM NATIONAL HISTORIC SITE, 735 Nod Hill Rd., Wilton, CT 06897-1309. Tel.: 203-834-1896. Fax: 203-834-2421.
Web Site: www.nps.gov/wefa/
Founded: 1990.
Congressional District: 5
Key Personnel: Supt., Linda Cook.
Personnel Profile: Full-Time Paid 8; Part-Time Paid 3; Part-Time Volunteers 40; Interns 1.
Governing Authority: federal government; nonprofit. Parent Institution: U.S. Dept. of Interior, National Park Service, Northeast Region. Private Partner: Weir Farm Trust. Tax-exempt: 501(c)(3).
Institution Type/Description: Historic Site: comprised of 60 acres of J. Alden Weir's farm (1852-1919) in Branchville, CT. Listed on the National Register of Historic Places.
Collections: focus on J. Alden Weir, American Impressionist painter & owner of the farm from 1882-1919; Weir's art & person artifacts; other artists associated with Weir Farm; furnishings; personal items; archives & manuscript materials relating to Weir's private & professional life. Historic Structures: 3 historic farm houses, two barns, two artists' studios.
Research Fields: historic structures report on all historic buildings; cultural landscape study focusing on the landscape since 1882; historic furnishings report for Weir's house & painting studio & Mahonri Young's sculpture studio; historic painting sites study to locate sites where paintings, drawings & etchings were executed on site.
Facilities: visitor center. Publications about Weir & postcards for sale.
Activities: guided tours; lectures; docent program; formal educational programs; orientation video; computer station of site-related artwork; children's art classes; Resident Artist Programs; art demonstrations; meetings. Annual Event: Jazz in the Garden.
Publications: site bulletin; brochures; exhibit activity guides; quarterly newsletter Weir Farm News; volunteer newsletter.
Hours & Admission Prices: Visitor Center: Wed.-Sun. 9-5; Grounds: daily dawn to dusk. Tours of historic art studios & grounds available, call for schedule. No charge; donations accepted. Closed New Year's Day; Thanksgiving; Christmas.
Attendance: 17,000 (accurate)
Membership: Weir Farm Art Center: Seniors $25; Associate $35; Family $50; Contributor $100; Sponsor $250; Patron $500; Benefactor $1,000.

WILTON HISTORICAL SOCIETY, INC., 224 Danbury Rd., Wilton, CT 06897-6000. Tel.: 203-762-7257. Fax: 203-762-3297.
E-mail: info@wiltonhistorical.org
Web Site: www.wiltonhistorical.org
Founded: 1938.
Congressional District: 4
Key Personnel: Pres., Greg Chann; Vice Pres. Mktg. & Communications, Jeffrey Yates; Museum Shop Mgr., Kate Gluckin; Museum Shop Mgr., Lynda Campbell.
Personnel Profile: Full-Time Paid 3; Part-Time Paid 3; Part-Time Volunteers 80.
Governing Authority: Tax-exempt.
Institution Type/Description: Historical Society Museum.
Collections: 18th to early 20th-century furnishings & accessories; 1740-1910 period rooms; costumes; textile collection; toys; dolls & dollhouses; Norwalk Redware; ceramics; 18th-20th century kitchen equipment; tools. Historic Buildings: c.1739 Betts-Sturges Blackmar House; c.1770 Raymond-Fitch House; 19th century barn; mid 19th century Abbott Barn; working blacksmith shop.
Research Fields: genealogy; local history; survey of 18th & 19th-century houses.
Facilities: research library of history, genealogy & manuscripts pertaining to Wilton history & families. Museum-related items for sale.
Activities: guided house tours; tours to historic sites; lectures; blacksmith shop; educational programs; traveling exhibitions. Museum Sponsors: antiques show; crafts show.
Publications: semiannual newsletter, Eighteenth Century Dwellings In Wilton; Annals of Wilton, Vol. I, II & III; Cannondale: A Connecticut Neighbor-

hood; A Definitive History of Wilton; Wilton Connecticut - Three Centuries of People, Places & Progress.
Hours & Admission Prices: Mon.-Thurs. 10-4:30, Sun. 1-4 for special exhibits & programs. Adults $5; members no charge. &
Attendance: 4,000 (estimated)
Membership: Individual $35; Family $50; Supporter $100; Patron $250.

WOODCOCK NATURE CENTER, 56 Deer Run Rd., Wilton, CT 06897. Tel.: 203-762-7280.
Institution Type/Description: Nature Center.
Collections: natural history; wildlife & their habitats; plants; flowers; trees.
Facilities: Museum-related items for sale.
Hours & Admission Prices: Center: Summer: Mon.-Fri. 9:30-4; Winter: Mon.-Fri. 9:30-4, Sat. call for hours. Trails: daily dawn to dusk. No charge; donations accepted.

Windsor

LUDDY/TAYLOR CONNECTICUT VALLEY TOBACCO MUSEUM, Northwest Park, 135 Lang Rd., Windsor, CT 06095. Tel.: 860-285-1888. Facebook: Connecticut Valley Tobacco Museum.
E-mail: tobaccomuseum@townofwindsorct.com
Founded: 1987.
Key Personnel: Dir., Jay Jackman.
Personnel Profile: Part-Time Paid 1.
Governing Authority: Parent Institution: Connecticut Valley Tobacco Historical Society. Tax-exempt.
Institution Type/Description: History Museum.
Collections: Tobacco Valley history & culture; equipment; photographs; documents; personal artifacts.
Hours & Admission Prices: early March to mid-Dec. Tues.-Thurs. & Sat. 12-4. No charge; donations accepted. &
Attendance: 1,200 (estimated)

MERCY GALLERY, Richmond Art Center - The Loomis Chaffee School, 4 Batchelder Rd., Windsor, CT 06095-3028. Tel.: 860-687-6030 & 6104.
Institution Type/Description: Art Gallery.
Collections: paintings; prints; photographs; sculpture.
Activities: temporary exhibitions.
Hours & Admission Prices: Winter: Mon., Wed. & Fri. 11-4, Tues. & Thurs. 11-4 & 7:45 pm-9:45 pm, Sun. 1-4; other times by appointment.

NORTHWEST PARK & NATURE CENTER, 145 Lang Rd., Windsor, CT 06095. Tel.: 860-285-1886.
Institution Type/Description: Nature Center.
Collections: wildlife & their habitat; natural science.
Facilities: nature trails. Center-related items for sale.
Activities: educational programs; summer camp.
Hours & Admission Prices: Mon.-Sat. 8:30-4:30, Sun. 1-4. No charge. &

OLIVER ELLSWORTH HOMESTEAD AND MUSEUM, 778 Palisado Ave., Windsor, CT 06095-2097. Mailing Address: P.O. Box 791, Windsor, CT 06095-0791. Tel.: 860-688-8717.
Web Site: www.ctdar.org/oeh/index.html
Founded: 1903.
Key Personnel: Chm. (V), Jean E. Kelsey; Pres. (V), Jennie May Rehnberg.
Personnel Profile: Part-Time Volunteers 30.
Governing Authority: Parent Institution: Ellsworth Memorial Association, Inc. Tax-exempt.
Institution Type/Description: Historic House Museum: built in 1781.
Collections: family history; personal artifacts; photographs; portraits; period furnishings.
Activities: special events.
Hours & Admission Prices: May 15-Oct. 15 Fri.-Sat. 12-4, Sun. 1-4; tours by appointment. Adults $4; children 11 & under no charge. Closed major holidays.
Attendance: 1,000 (estimated)

VINTAGE RADIO AND COMMUNICATIONS MUSEUM OF CONNECTICUT, 115 Pierson Lane, Windsor, CT 06095. Mailing Address: P.O. Box 894, Windsor, CT 06095. Tel.: 860-683-2903.
E-mail: radioclctr@aol.com
Web Site: vrcmct.org
Founded: 1990.
Key Personnel: Dir., John C. Ellsworth
Institution Type/Description: History Museum.
Collections: electronic communications history including telephone, telegraph,

mechanical sound recordings; radio & television; early satellite, communications & internet; period furnishings.
Facilities: library.
Activities: rental facilities.
Hours & Admission Prices: Thurs.-Fri. 10-3, Sat. 10-5, Sun. 1-4; groups by appointment. Adults $7, seniors 60 & over $6, students $5; children under 5 no charge.

WINDSOR HISTORICAL SOCIETY, 96 Palisado Ave., Windsor, CT 06095. Tel.: 860-688-3813. Fax: 860-657-1633.
E-mail: info@windsorhistoricalsociety.org
Web Site: www.windsorhistoricalsociety.org
Founded: 1921.
Congressional District: 1
Key Personnel: Dir., Christine Ermenc; Pres (V), John Berky.
Personnel Profile: Full-Time Paid 2; Part-Time Paid 3; Part-Time Volunteers 50.
Governing Authority: nonprofit. Tax-exempt: 501(c)(3).
Institution Type/Description: History Museum.
Collections: local history; manuscripts; period furnishings; personal artifacts; photographs; documents. Historic Buildings: 1758 Strong-Howard; 1767 Dr. Hezekiah Chaffee House.
Facilities: library. Museum-related items for sale.
Activities: research.
Publications: Windsor Historical Society News.
Hours & Admission Prices: Wed.-Sat. 11-4 with historic house tours at 11& 1. Adults $8, seniors & students $6; society members & children under 12 no charge. &
Attendance: 6,471 (accurate)
Membership: Seniors 65 & over $25; Individual $30; Family $40; Donor $100; Patron $250; Life $300; Joint Life $450; Benefactor $500.

Windsor Locks

NEW ENGLAND AIR MUSEUM, Bradley International Airport, 36 Perimeter Rd., Windsor Locks, CT 06096-1069. Tel.: 860-623-3305. Fax: 860-627-2820.
E-mail: staff@neam.org
Web Site: www.neam.org
Founded: 1959.
Congressional District: 1
Key Personnel: Exec. Dir. & C.E.O., Michael P. Speciale; Pres. (V), Scott E. Ashton; Museum Shop Mgr., Gina Maria Alimberti.
Personnel Profile: Full-Time Paid 5; Part-Time Paid 17; Part-Time Volunteers 165.
Governing Authority: nonprofit organization. Owned by Connecticut Aeronautical Historical Association, Inc. Tax-exempt: 501(c)(3).
Institution Type/Description: Aeronautics Museum.
Collections: aircraft; propulsion systems and associated material in aeronautics; manuscripts.
Research Fields: New England aeronautics.
Facilities: 1,000-vol. library of aeronautical, scientific, & transportation books available for use by approval and appointment; indoor & outdoor exhibits featuring aircraft. Aeronautical pamphlets, postcards, photographs, models & museum-related items for sale.
Activities: lectures; inter-museum loan, permanent & temporary exhibitions.
Publications: brochure; pamphlets; Tour Guide; quarterly newsletter.
Hours & Admission Prices: Daily 10-5. Adults $11, senior citizens 65 & over $10, children 4-11 $6; discounts to groups; CAHA members and children 3 & under no charge. Closed New Year's Day; Thanksgiving; Christmas. &
Attendance: 53,058 (accurate)
Membership: CAHA: Individual $60; Family $70; Benefactor $100; Regular Life $1,000.

NODEN-REED FARM MUSEUM, 58 West St., Windsor Locks, CT 06096-1808. Mailing Address: P.O. Box 733, Windsor Locks, CT 06096-0733. Tel.: 860-627-9212.
E-mail: winlocks@aol.com
Founded: 1976.
Key Personnel: Dir., Ruth Bonito; Pres. (V), Joseph Bonito; Cur., James Anderson.
Personnel Profile: Part-Time Volunteers 10.
Governing Authority: Tax-exempt.
Institution Type/Description: History Museum: housed in Victorian farmhouse. Listed on National Register of Historical Places.
Collections: period furnishings; farm tools.
Facilities: nature trails.
Hours & Admission Prices: May-Oct. Sun. 1-5. No charge; donations accepted.

Attendance: 200 (estimated)

Woodbridge

AMITY AND WOODBRIDGE HISTORICAL SOCIETY, 1907 Litchfield Turnpike, Woodbridge, CT 06525. Tel.: 203-387-2823.
E-mail: donaldmenzies@sbcglobal.net
Web Site: www.woodbridgehistory.org
Founded: 1936.
Congressional District: 3
Key Personnel: C.E.O. & Pres., Donald Menzies.
Governing Authority: nonprofit organization. Tax-exempt.
Institution Type/Description: Local History & Historical Society Museum: housed in Thomas Darling House.
Collections: local history; period artifacts.
Activities: guided tours; lectures.
Publications: Amity & Woodbridge Historical Society Journal.
Hours & Admission Prices: By appointment. No charge; donations accepted.
Attendance: 200 (estimated)
Membership: Individual $10; Family $20; Sustaining $50; Life $100.

Woodbury

FLANDERS NATURE CENTER, 5 Church Hill Rd., Woodbury, CT 06798-1718. Tel.: 203-263-3711. Fax: 203-263-2214.
Web Site: www.flandersnaturecenter.org
Founded: 1973.
Key Personnel: Exec. Dir., Arthur S. Milnor
Institution Type/Description: Nature Center.
Collections: wildlife & their habitats.
Facilities: nature trails.
Activities: hiking; educational programs; special events.
Hours & Admission Prices: Daily dawn to dusk. No charge.

THE GLEBE HOUSE MUSEUM AND GERTRUDE JEKYLL GARDEN, 49 Hollow Rd., Woodbury, CT 06798. Mailing Address: P.O. Box 245, Woodbury, CT 06798-0245. Tel.: 203-263-2855. Facebook: Glebe House Museum and Gertrude Jekyll Garden.
E-mail: ghmgjg@snet.net
Web Site: www.theglebehouse.org
Founded: 1923.
Congressional District: 6
Key Personnel: Dir., Judith Kelz; Pres. (V), Carter Booth; Museum Shop Mgr., Rebecca Otten.
Personnel Profile: Full-Time Paid 1; Full-Time Volunteers 1; Part-Time Volunteers 101.
Volunteer Hours: 3,500
Operating Expenses: 100,000
Operating Income: 100,000
Governing Authority: society. Governed by the Seabury Society for the Preservation of the Glebe House. Tax-exempt: 501(c)(3).
Institution Type/Description: Historic Building: c.1750 The Glebe House, birthplace of American Episcopacy.
Collections: original paneling; period furnishings; historical documents relating to Samuel Seabury; birthplace of American Episcopacy; 18th-century prints & painting; books & documents; regional Woodbury 18th century furniture & silver.
Research Fields: Episcopal Church history; state & local history; 18th-century culture; Gertrude Jekyll Gardens & Horticulture; 18th century Woodbury furniture.
Facilities: garden. Photographs, brochures & gift-related items for sale.
Activities: guided tours; formally organized educational programs; lectures, demonstrations, concerts & Symposia; cemetery tours. Museum Sponsors: Fall Colonial Fair; Fall All Hollow's Eve; Holiday Festival.
Publications: brochures & pamphlets; postcards; writing paper; newsletter.
Hours & Admission Prices: May-Oct. Wed.-Sun. 1-4; other times by appointment only. Adults $5, children $2; members no charge.
Attendance: 5,000 (estimated)
Membership: Individual $35; Dual $60; Sponsor $100; Patron $500; Benefactor $1,000.

Woodstock

* **ROSELAND COTTAGE, (M),** 556 Rte. 169, Woodstock, CT 06281-2344. Mailing Address: 141 Cambridge St., Boston, MA 02114. Tel.: 860-928-4074.
Web Site: www.historicnewengland.org
Founded: 1970.

Congressional District: 2
Key Personnel: Site Mgr., Laurie Masciandaro.
Personnel Profile: Full-Time Paid 2; Part-Time Paid 4; Part-Time Volunteers 40.
Governing Authority: society; nonprofit organization. Parent Institution: Historic New England, 141 Cambridge St., Boston, MA 02114. Tel.: 617-227-3956. Tax-exempt: 501(c)(3).
Institution Type/Description: Historic House: 1846 Gothic Revival cottage.
Collections: Gothic Revival furniture; Bowen family possessions; carriage house; outbuildings; parterre garden.
Facilities: garden.
Activities: guided tours; lectures; special events; school & youth programs; functions.
Publications: Historic New England Magazine.
Hours & Admission Prices: June-Oct. 15 Wed.-Sun. 11-4; tours on the hour. Adults $8; members no charge. &
Attendance: 13,112 (accurate)
Membership: National $35; Individual $45; Household $55; Garden & Landscape $75; Contributing and Library & School $100; Young Friends of Historic New England $100-$1,500; Friends of the Library and Archives $125; Historic Homeowner $200; Business & Ogden Codman Design Group $250; Appleton Circle $2,500.

WOODSTOCK HISTORICAL SOCIETY, INC., 523 Rte. 169, Woodstock, CT 06281. Mailing Address: P.O. Box 65, Woodstock, CT 06281-0065. Tel.: 860-928-1035.
E-mail: woodstockhist@att.net
Web Site: www.woodstockhistoricalsociety.org
Founded: 1967.
Congressional District: 3
Key Personnel: Pres. (V), Gail White; Treas., Patricia Simoni.
Personnel Profile: Part-Time Volunteers 20.
Governing Authority: society. Tax-exempt: 170(b)(1)(A).
Institution Type/Description: Historical Society Museum.
Collections: agricultural artifacts; personal artifacts; books & documents; photographs. Historic Buildings: Red-White School; Palmer Hall.
Research Fields: agriculture; architecture; history; genealogy
Facilities: seminar space.
Activities: lectures; field trips; research.
Publications: book, Samuel Sewall & the Town He Named Woodstock; booklet; sketches; Mrs. Dr. Bowen's Bustle & Apple Molasses; From Roxbury Fells to the Eastward Vale: A Journey Through Woodstock, 1686-2011.
Hours & Admission Prices: April-Oct. Sun. 12-4. No charge; donations accepted. Closed major holidays. &
Attendance: 200 (estimated)
Membership: Individual $20; Family $35; Life $250.

DELAWARE

(81 listings)

Bear

WHALE WALLOW NATURE CENTER - LUMS POND STATE PARK, 1068 Howell School Rd., Bear, DE 19701. Tel.: 302-368-6989.
Institution Type/Description: Nature Center.
Collections: local natural history & wildlife; plants; photographs.
Facilities: nature trails.
Activities: educational programs.
Hours & Admission Prices: Park: daily 8am to sunset. Center: Memorial Day to Labor Day daily 12-6.

Bridgeville

BRIDGEVILLE HISTORICAL SOCIETY, 102 S. Williams St., Bridgeville, DE 19933. Mailing Address: P.O. Box 306, Bridgeville, DE 19933-0336. Tel.: 302-337-7600.
Founded: 1976.
Key Personnel: Dir., Mike Collison; Pres., Howard E. Hardesty.
Personnel Profile: Part-Time Volunteers 12.
Governing Authority: Tax-exempt.
Institution Type/Description: Historical Society: building built in 1911. Listed on the National Register of Historic Places.
Collections: local history.
Research Fields: local & regional history.
Publications: 19th century history of Bridgeville, DE (1985, reprint 2005).
Hours & Admission Prices: By appointment. No charge.

Attendance: 312 (accurate)

Christiana

CHRISTIANA FIRE COMPANY MUSEUM, 2 E. Main St., Christiana, DE 19702. Tel.: 302-737-2433.
Institution Type/Description: Fire Company Museum.
Collections: fire company history; photographs; personal artifacts; fire equipment & uniforms; period furnishings.
Hours & Admission Prices: By appointment.

CHRISTIANA HISTORICAL SOCIETY, 49 N. Old Baltimore Pike, Christiana, DE 19702-1637. Tel.: 302-286-6223.
E-mail: kendigger@aol.com
Web Site: www.xtinahs.org/xtinahistoricalsociety.html
Key Personnel: Pres., Ken Baumgardt
Institution Type/Description: Historical Society Museum.
Collections: local history & culture; photographs; personal artifacts.
Hours & Admission Prices: Call for hours.

Claymont

ROBINSON HOUSE, 1 Naamans Rd., Claymont, DE 19703-2701. Tel.: 302-792-0285.
Web Site: www.robinsonhousede.org
Governing Authority: Parent Institution: Claymont Historical Society.
Institution Type/Description: Historic House Museum.
Collections: period furniture.
Facilities: Museum-related items for sale.
Activities: tours. Annual Event: Christmas House Tour in December.
Hours & Admission Prices: By appointment.

Delaware City

FORT DELAWARE SOCIETY, 33 Staff Lane, Ft. Dupont State Park, Delaware City, DE 19706. Mailing Address: P.O. Box 553, Delaware City, DE 19706-0553. Tel.: 302-834-1630. Fax: 302-834-1630.
E-mail: society@fortdelaware.org
Web Site: www.fortdelaware.org
Founded: 1950.
Key Personnel: Pres. (V), David P. Price; Chm. (V), William G. Robelen, IV
Personnel Profile: Part-Time Paid 1.
Governing Authority: state. Division of Parks, affiliated with Fort Delaware Society Bd. of Directors in association with state. Tax-exempt: 170(b)(1)(A).
Institution Type/Description: Military & Park Museum Complex: located in c.1859 former barracks of Civil War Fort Delaware on Pea Patch Island, accessible by boat only.
Collections: materials & artifacts relating to Delaware's role in the Civil War & to the fort on Pea Patch Island; military artifacts; accoutrements; marine items; archaeology finds; gun emplacements; 32 lb. sea coast artillery cannon.
Research Fields: military.
Facilities: picnic area. Books and pamphlets for sale.
Activities: guided tours; films; formally organized education programs for children; permanent & temporary exhibitions; historical interpretations.
Publications: Fort Delaware Notes; quarterly, Fort Delaware Newsletter; books & pamphlets, A Fort Delaware Journal; Confederate Prisoners of War at Fort Delaware; Fort Delaware in the Civil War; Fort Delaware; They Died at Fort Delaware; General M. Jeff Thompson in Fort Delaware; The Story of Fort Delaware; The Chains of Glory; Prison Times; Fire in the Hole.
Hours & Admission Prices: mid-June to Sept. Wed.-Sun. & holidays 10-6. Boat Fee: adults $11, children $6; discounts to senior citizens & active military. &
Attendance: 25,000 (accurate)
Membership: Active $25; Couple or Family $35; Supporting $60; Life $500.

Dover

BIGGS MUSEUM OF AMERICAN ART, (M), 406 Federal St., Dover, DE 19901-3615. Mailing Address: P.O. Box 711, Dover, DE 19903-0711. Tel.: 302-674-2111. Fax: 302-674-5133.
E-mail: ldanko@biggsmuseum.org
Web Site: www.biggsmuseum.org
Formerly: Sewell C. Biggs Museum of American Art
Founded: 1989.
Congressional District: 1

Key Personnel: Dir., Linda A.K. Danko; Chm. (V), Marcia P. Dewitt; Mktg. & Devel. Mgr., Stephanie Fitzpatrick; Cur., Ryan Grover; Education & Programs Mgr., Rebecca Cooper; Museum Mgr., Ellen Arthur.

Personnel Profile: Full-Time Paid 5; Part-Time Paid 4; Part-Time Volunteers 12; Interns 2.

Governing Authority: private; nonprofit association. Tax-exempt.

Institution Type/Description: American Art Museum.

Collections: American fine & decorative arts from 1700-present with emphasis on Delaware & the Delaware Valley including J. Hesselius, the Peale Family, G. Stuart, A. Bierstadt, T. Cole & the Hudson River School, W.M. Chase, F. Schoonover & other Brandywine School Illustrators; Delaware & Philadelphia-region furniture & silver; modern & contemporary Delaware painters.

Major Exhibits: Biggs Picture, 11/1/13-2/23/14.

Research Fields: American fine & decorative arts.

Facilities: 3,000-vol. art library; family archives.

Activities: guided tours; adults & children's activities; lectures; art workshops; crafts; special exhibitions; public celebrations; fundraisers; gala.

Publications: bimonthly, Museletter & calendar; 2-vol. catalogue, The Sewell C. Biggs Collection of American Art; Almost Forgotten: Delaware Women Artists; Delaware Clocks; Greetings from Delaware and Other Artist Communities; Delaware Silver.

Hours & Admission Prices: Tues.-Sat. 9-4:30, Sun. 1:30-4:30. No charge; donations accepted. Closed New Year's Day; Easter; Thanksgiving; Christmas. &

Attendance: 20,000 (estimated)

Membership: Student, Instructor, Military, & Senior $30; Individual $40; Household $50; Family $60.

DELAWARE AGRICULTURAL MUSEUM AND VILLAGE,
866 N. Dupont Hwy., Dover, DE 19901-2012. Tel.: 302-734-1618. Fax: 302-734-0457.

E-mail: damv@verizon.net

Web Site: www.agriculturalmuseum.org

Founded: 1974.

Congressional District: 1

Key Personnel: Pres., Richard Bergold; Dir., Di Rafter; Mgr. Maintenance, Michael Grimes.

Personnel Profile: Full-Time Paid 1; Part-Time Paid 5; Part-Time Volunteers 100.

Governing Authority: nonprofit; bd. of trustees. Tax-exempt: 501(c)(3).

Institution Type/Description: Museum of Rural Life and Agricultural History and Technology.

Collections: artifacts dating from early settlement; manuscript collections; farm equipment; folk art collection; tractors; interpretive poultry & dairy exhibits; horse drawn & transportation equipment; tools; housewares; store furnishings; replica of a 19th century water-powered mill. Historic Structures: c.1890 furnished two-story farm house with summer kitchen; c.1900 barn; c.1880 wagon shed, grainery; c.1860 blacksmith shop; c.1850 one-room schoolhouse; c.1825 cornhouse; c.1820 meathouse; c.1870 country store; c.1860 train station; c.1900 barbershop; c.1858 church.

Research Fields: Delaware farming; general farm life; agricultural technology.

Facilities: 2,000-vol. library pertaining to farming & agriculture available for research; reading room; re-created 1890's village & farmstead, 40,000 sq. ft. exhibit building; 80-seat auditorium. Books, reproductions, craft & other museum-related items for sale.

Activities: guided tours; lectures; youth summer camps; formally organized education programs; intern program; docent program; permanent & temporary exhibits. Special Events: Spring Auction in March; Golf Tournament; Goodies & Goblins; Fall Harvest Festival; Farmer's Christmas in December.

Publications: newsletter, Museum Gazette; brochures; annual report.

Hours & Admission Prices: Tues.-Sat. 10-3. Adults $6, senior citizens 65 & over $4, youth 4-17 $3; discounts for groups, AAA members & Time Travelers; members & children under 3 no charge. Closed New Year's Day; Easter; Thanksgiving; Christmas. &

Attendance: 17,000 (estimated)

Membership: Senior 65 & over $20; Senior Couple & Individual $30; Family $50; Patron $75; Sponsor $150; Business $500; Corporate $1,000; Endowment $5,000.

DELAWARE DIVISION OF HISTORICAL & CULTURAL AFFAIRS, (M),
21 The Green, Dover, DE 19901-3611. Tel.: 302-736-7400. Fax: 302-739-6712.

Web Site: history.delaware.gov

Formerly: Delaware State Museums

Founded: 1931.

Key Personnel: Dir., Timothy Slavin; Deputy Dir., Steve Marz; Cur. Archaeology, Charles H. Fithian; Cur. Collections, Ann M. Baker Horsey; Cur. Education, Madeline Dunn; Cur. Collections Management, Claudia F.

Leister; Landscape Supt., Diane D. Crom; Site Supvr. New Castle, Cynthia Snyder; Site Mgr., Beverly Laing; Site Supvr. Dover, Stacey Johnson; Site Supvr. John Dickinson Plantation, Gloria Henry; Site Supvr. Zwaanendael Museum, Andrea Anderson; Museum Shop Mgr., Bridget Warner.

Personnel Profile: Full-Time Paid 30; Part-Time Paid 45.

Governing Authority: state. Parent Institution: Department of State, Division of Historical & Cultural Affairs. Branch Museums: Archaeology Museum, Museum of Small Town Life, 316 S. Governors Ave., Dover; Old State House, The Green, Dover; State Visitor Center, Federal St., Dover; Johnson Victrola Museum, Corner of Bank Ln. & New St., Dover; Buena Vista State Conference Center, 661 S. DuPont Hwy., New Castle; Old New Castle Courthouse, The Green, New Castle; John Dickinson Plantation, Kitts Hummock Rd., Dover; Octagonal Schoolhouse, Cowgills Corner, Route 9, Dover; Belmont Hall Conference Center, Smyrna; Zwaanendael Museum, Lewes. Monuments: Cooch's Bridge Monument, Old Baltimore Pike, Newark; DeVries Monument, Pilotown Rd., Lewes.

Institution Type/Description: State Historical Agency & Museums: Meeting House Galleries: housed in c.1790 Presbyterian Church now the State Archaeology gallery & 1880 Sunday School Building; Old State House: c.1792; Johnson Victrola Museum; Buena Vista: 1845-47, home of John M. Clayton now used as a conference center; Old New Castle Courthouse: located on the Green laid out by Peter Stuyvesant in 1655; in 1776 the Declaration of Independence & the first Constitution of Delaware were approved; State Visitor Center & Gallery; John Dickinson Plantation, c.1740, historic house, with reconstructed Afro-American tenant house & farm buildings, boyhood home of John Dickinson who wrote, Letters from a Farmer in Pennsylvania, & signed Articles of Confederation & U.S. Constitution; Octagonal Schoolhouse, 1831, one of Delaware's oldest remaining school buildings; Zwaanendael Museum, housed in a 1931 adaptation of the Old Town Hall in Hoorn, Holland. Monuments: Cooch's Bridge Monument commemorating September 3, 1777 skirmish with British; DeVries Monument commemorating landing of the Dutch in 1631, the first settlement in Delaware; c.1770, Belmont Hall Conference Center, home of last President of Delaware.

Collections: lifestyles, government, politics, & decorative arts of Delaware citizens; state archaeological collection; official state portraits. Meeting House Galleries: state archaeology; Main Street Delaware on small towns. Johnson Victrola Museum: phonographs; advertising; records; ephemera; papers of Eldridge Reeves Johnson, founder of the Victor Talking Machine Company. Zwaanendael Museum: early history of Lewes & Sussex County; maritime archaeology of the HMS DeBraak which sank in 1798.

Research Fields: Delaware prehistory, history & culture; legislative, judicial & executive history & material culture; Victor Talking Machine Co. sound archives.

Facilities: 1,000-vol. library on eastern North American prehistory; reading rooms; archaeological laboratory & field research station; visitor center. Museum-related items for sale.

Activities: guided tours; gallery talks; lectures; permanent & temporary exhibitions; inter-museum loan program; formally organized programs for children; museum internship opportunities for undergraduates; archaeological excavation & survey; internship program; in-service training programs for teachers; special educational programs.

Publications: brochures.

Hours & Admission Prices: Buena Vista Conference Center: by appointment. Old New Castle Courthouse: Tues.-Sat. 10-3:30, Sun. 1:30-4:30. Delaware Visitor's Center and Galleries, Delaware Archaeology Museum, Museum of Small Town Life, Johnson Victrola Museum & Historic Old State House: Mon.-Sat. 9-4:30, Sun. 1:30-4:30. John Dickinson Plantation: Jan.-Feb. Tues.-Sat. 10-3:30; March-Dec. Tues.-Sat. 10-3:30, Sun. 1:30-4:30. Zwaanendael: Tues.-Sat. 10-4:30, Sun. 1:30-4:30. No charge; donations accepted. Closed state holidays. For specific information on tour arrangements, call State Visitor Center, Dover, 302-645-1148, or write to State Visitor Center, Federal St., 406 Federal St., Dover, DE 19901. &

Attendance: 100,000 (estimated)

DELAWARE PUBLIC ARCHIVES,
121 Duke of York St., Dover, DE 19901-3638. Tel.: 302-744-5000. Fax: 302-739-2578.

E-mail: archives@state.de.us

Web Site: www.archives.delaware.gov

Founded: 1905.

Key Personnel: Dir., Stephen Marz, C.A.

Governing Authority: Parent Institution: state of Delaware.

Institution Type/Description: Archives.

Collections: history; culture; genealogy; probates; tax assessments; family histories.

Activities: research; tours; genealogy; presentations; workshops. Museum Sponsors: The 1940 Census in April; A Tour of the Delaware Public Archives in May; The Star Spangled Banner Project in June.

Hours & Admission Prices: Mon.-Fri. 8-4:15, 1st Sat. of month 9-4:45. No charge. Closed state holidays. &

DELAWARE STATE POLICE MUSEUM, 1425 N. DuPont Hwy., Dover, DE 19903. Mailing Address: P.O. Box 430, Dover, DE 19903-0430. Tel.: 302-739-7700. Fax: 302-739-7707.
E-mail: dspmuseum@aol.com
Web Site: www.delawaretrooper.com/museum
Key Personnel: Pres., Maj. Raymond Deputy; Vice Pres., Capt. Peter Steil; Treas., Maj. Robert Gouge; Cur., Sgt. John Alstadt
Institution Type/Description: Police Museum.
Collections: criminal investigation methods; highway safety; major crime scenes; police vehicles & equipment; uniforms; weapons.
Activities: tours.
Hours & Admission Prices: Mon.-Fri. 9-3, 3rd Sat. of month 11-3; other times by appointment. No charge.

FIRST STATE HERITAGE PARK WELCOME CENTER AND GALLERIES, First State Heritage Park Welcome Center, 121 Duke of York St., Dover, DE 19901. Tel.: 302-739-5055. Fax: 302-739-3943.
Web Site: history.delaware.gov/
Formerly: Delaware Visitor Center and Galleries
Founded: 1950.
Key Personnel: Dir., Tim Slavin.
Governing Authority: state. Parent Institution: Division of Historical & Cultural Affairs, 21 The Green, Dover, DE 19901.
Institution Type/Description: History Museum & Visitor's Center.
Collections: Delaware history.
Facilities: 500 sq. ft. exhibit space.
Activities: guided tours; lectures; temporary exhibitions; monthly special programs. Annual Event: Old Dover Days.
Publications: local & state information.
Hours & Admission Prices: Mon.-Sat. 8-4:30, Sun. 1:30-4:30, State holidays 9-4:30; groups by appointment. Closed New Year's Day; Easter; Christmas. &

JOHN DICKINSON PLANTATION, 340 Kitts Hummock Rd., Dover, DE 19901-7016. Tel.: 302-739-3277. Fax: 302-739-3173.
Web Site: www.history.delaware.gov
Founded: 1956.
Key Personnel: Supvr., Gloria Henry.
Governing Authority: Parent Institution: State of Delaware. Subsidiary Institution: Division of Historical and Cultural Affairs.
Institution Type/Description: History Museum.
Collections: period history; plantation life; decorative arts; archaeology; architecture; slavery. Historic Buildings: 18th century plantation home; farm buildings.
Activities: demonstrations; tours.
Hours & Admission Prices: Wed.-Sat. 10-3:30, Sun. 1:30-4:30. No charge; donations accepted. Closed state holidays. &

JOHNSON VICTROLA MUSEUM, 375 S. New St., Dover, DE 19901. Mailing Address: First State Heritage Park Welcome Center, 121 Duke of York St., Dover, DE 19901. Tel.: 302-747-5055. Fax: 302-739-3943.
Web Site: history.delaware.gov/museums/jvm/jvm_main.shtml
Founded: 1960.
Key Personnel: Dir., Tim Slavin; Site Supvr., Nena Todd.
Governing Authority: state. Parent Institution: Division of Historical & Cultural Affairs, 21 The Green, Dover, DE 19901.
Institution Type/Description: Technology Museum: Eldridge Reeves Johnson, founder of the Victor Talking Machine Company in 1901, later known as RCA.
Collections: life & work of Eldridge Reeves Johnson; Victor talking machines; phonographs; memorabilia; personal artifacts; paintings; Victor records.
Facilities: 2,000 sq. ft. exhibit space.
Activities: guided tours; monthly family focus programs. Annual Event: Old Dover Day.
Hours & Admission Prices: Wed.-Sat. 9-4:30; groups by appointment. No charge; donations accepted. Closed New Year's Day; Easter; Christmas. Open most state holidays.

THE OLD STATE HOUSE, 25 The Green, Dover, DE 19901. Mailing Address: First State Heritage Park Welcome Center, 121 Duke of York St., Dover, DE 19901. Tel.: 302-739-2438 & 739-5055. Fax: 302-739-3943.
Web Site: history.delaware.gov/museums/sh/sh_main.shtml
Founded: 1976.
Key Personnel: Dir., Tim Slavin; Site Supvr., Nena Todd.

Governing Authority: state. Parent Institution: Division of Historical & Cultural Affairs, 21 The Green, Dover, DE 19901.
Institution Type/Description: Historic Site: housed in the restored capitol building; built in 1792.
Collections: architecture; period furnishings; archival documents.
Research Fields: political records, laws & government.
Facilities: 3,500 sq. ft. exhibit space.
Activities: concerts; guided tours; lectures; monthly family focus programs. Annual Event: Old Dover Days.
Hours & Admission Prices: Tues.-Sat. & state holidays 9-4:30, Sun. 1:30-4:30; groups by appointment. No charge; donations accepted. Closed New Year's Day; Easter; Christmas. &

Dover AFB

AIR MOBILITY COMMAND MUSEUM, (M), 1301 Heritage Rd., Dover AFB, DE 19902-5301. Tel.: 302-677-5938 & 5939. Fax: 302-677-5940.
E-mail: museum@dover.af.mil
Web Site: www.amcmuseum.org
Founded: 1986.
Key Personnel: C.E.O., Michael D. Leister; Pres., Donald Sloan; Operations Mgr., John Taylor; Archivist, Harry Heist; Education, Dick Caldwell; Registrar & Collections Mgr., Deborah Sellars; Museum Shop Mgr., Jim Stewart.
Personnel Profile: Full-Time Paid 2; Full-Time Volunteers 2; Part-Time Paid 3; Part-Time Volunteers 120.
Governing Authority: federal; nonprofit. Parent Institution: USAF Museum. Tax-exempt: 501(c)(3).
Institution Type/Description: Military Museum: housed in a WWII hanger.
Collections: 30 aircraft.
Research Fields: airlift history.
Facilities: 750-vol. library of military aviation; 20,000 sq. ft. exhibit space; 70-seat theater; outdoor air park; commemoration park. Museum-related items for sale.
Activities: films; formal education programs for adults; guided tours; lectures; loan, temporary & traveling exhibitions; training programs for professional museum workers; broadcast/cable programs. Museum Sponsors: Community Appreciation Day.
Publications: quarterly foundation newsletter, Hangar Digest.
Hours & Admission Prices: Tues.-Sun. 9-4. No charge; donations accepted. Closed federal holidays. &
Attendance: 73,000 (accurate)
Membership: Individual $20; Family $30; Patron $100; Aircraft Sponsor & Corporate $200.

Fenwick Island

DISCOVERSEA SHIPWRECK MUSEUM, 708 Coastal Hwy., Fenwick Island, DE 19944. Tel.: 302-539-9366; 888-743-5524. Fax: 302-539-1285.
E-mail: dsmuseum@aol.com
Web Site: www.discoversea.com
Key Personnel: Dir., Dale W. Clifton, Jr.
Institution Type/Description: Maritime History Museum.
Collections: maritime history; shipwreck artifacts.
Activities: lecture; educational programs.
Hours & Admission Prices: June-Aug. daily 11-8; Sept.-May Sat.-Sun. 11-4; groups by appointment. No charge; donations accepted.

FENWICK ISLAND LIGHTHOUSE & MUSEUM, Rt. 54, Fenwick Island, DE 19944. Mailing Address: New Friends of the Fenwick Island Lighthouse, P.O. Box 1001, Selbyville, DE 19975. Tel.: 302-436-8100.
Web Site: www.fenwickislandlighthouse.org
Governing Authority: state.
Institution Type/Description: Historic Lighthouse: built in 1858.
Collections: local & lighthouse history; period artifacts; photographs.
Facilities: Museum-related items for sale.
Hours & Admission Prices: May to Columbus Day: Thurs.-Tues. No charge; donations accepted.

Frederica

BARRATT'S CHAPEL AND MUSEUM, 6362 Bay Rd., Frederica, DE 19946-1505. Tel.: 302-335-5544. Fax: 302-335-5750.
E-mail: barratts@aol.com
Web Site: barrattschapel.org
Founded: 1964.

Congressional District: 33
Key Personnel: Dir., Kenyon Camper; Cur., Barb Duffin; Conference Historian, Rev. Philip Lawtor.
Personnel Profile: Part-Time Paid 2; Part-Time Volunteers 5.
Governing Authority: church denominational group. Parent Institution: Peninsula-Delaware Conference of The United Methodist Church. Tax-exempt.
Institution Type/Description: Religious Museum.
Collections: conference journals; Methodist magazines & artifacts; printed biographies & autobiographies of 18th century preachers; reconstructed 18th-century vestry. Historic Building: 18th century chapel.
Research Fields: Methodist history on Delmarva peninsula.
Facilities: 2,500-vol. library of religious books, conference journals & church records.
Activities: guided tours; lectures; permanent exhibitions.
Publications: New Light on Old Barratts; Cultivating the Methodist Garden.
Hours & Admission Prices: Sat.-Sun. 1:30-4:30; groups & other times by appointment. No charge; donations accepted. &

Attendance: 4,007 (accurate)
Membership: Friends $20.

Georgetown

DELAWARE AVIATION MUSEUM, 21553 Rudder Ln., Georgetown, DE 19947-2016. Tel.: 302-854-0244.
Institution Type/Description: Aviation Museum.
Collections: military aviation history; photographs; WWII & Cold War era artifacts.
Facilities: over 3,000-vol. library.
Hours & Admission Prices: Sat.-Sun. 10-4.

ELSIE WILLIAMS DOLL COLLECTION, 21179 College Dr., Georgetown, DE 19947-0610. Tel.: 302-259-6150. Fax: 302-259-6765.
E-mail: treasures@dtcc.edu
Web Site: www.treasuresofthesea.org/other.html#doll
Founded: 1988.
Key Personnel: Chm. (V), Mark S. Stellini; Pres., Dr. Orlando George; Dir., Robert Hearn.
Personnel Profile: Full-Time Paid 1; Part-Time Paid 1.
Governing Authority: Parent Institution. Delaware Technical & Community College Educational Foundation. Tax-exempt: 501(c)(3).
Institution Type/Description: Doll Museum.
Collections: over 700 dolls including some donated by Elsie Steele Williams, the wife of U.S. Senator John Williams.
Hours & Admission Prices: Mon.-Thurs. 8am-10pm, Fri. 8-4:30, Sat. 9-1. No charge. Closed major holidays; Christmas break. &

MARVEL CARRIAGE MUSEUM, 510 S. Bedford St., Georgetown, DE 19947-1852. Tel.: 302-855-9660.
E-mail: marvelmuseum@juno.com
Web Site: www.marvelmuseum.com
Formerly: Nutter D. Marvel Museum
Key Personnel: Pres., Jim Bowden.
Governing Authority: Parent Institution: Georgetown Historical Society.
Institution Type/Description: Historic Buildings.
Collections: personal artifacts; horse drawn carriages; photographs; newspaper clippings; books; Georgetown history. Historic Buildings: two barrel roof barns; 100 year old church; one room school house; blacksmith shop; two period railroad freight buildings.
Hours & Admission Prices: By appointment.

OLD SUSSEX COUNTY COURTHOUSE, 10 S. Bedford St., Georgetown, DE 19947-1852. Mailing Address: Georgetown Historical Society, 510 S. Bedford St., Georgetown, DE 19947. Tel.: 302-855-9660.
Governing Authority: Parent Institution: Georgetown Historical Society.
Institution Type/Description: Historic Building: built in 1791.
Collections: period furnishings; law & government artifacts.
Hours & Admission Prices: 1st Wed. of month; other times by appointment.

TREASURES OF THE SEA EXHIBIT, Delaware Technical Community College, Owens, 21179 College Dr., Georgetown, DE 19947. Tel.: 302-259-6150. Fax: 302-259-6765.
E-mail: treasures@dtcc.edu
Web Site: www.treasuresofthesea.org
Founded: 1988.
Congressional District: Delaware

Key Personnel: Chm. (V), Mark S. Stellini; Pres., Dr. Orlando George; Dir., Robert Hearn; Education, Registrar & Museum Shop Mgr., Susan Doering; Public Rels., Christine Gillian; Security, William Wood.
Personnel Profile: Part-Time Paid 2; Interns 1.
Governing Authority: college; nonprofit. Parent Institution: Delaware Technical & Community College. Tax-exempt: 501(c)(3).
Institution Type/Description: Maritime Museum.
Collections: shipwreck artifacts: coins, firearms, jewelry, gold & silver bars, cannons; dolls.
Facilities: 75-vol. library of books & video cassettes on shipwreck archaeology available to the public; 10-gallon aquarium; 200-seat auditorium; 40-seat & 200-seat cafeteria; 30-seat theatre; nature trail; picnic area; meeting rooms. Souvenirs & other museum-related items for sale.
Activities: guided tours.
Hours & Admission Prices: Feb. to mid-Dec. Mon.-Tues. 10-4, Fri. 12-4, Sat. 9-1; other times by appointment. Adults $3, senior citizens $2.50, students $1; discounts to WHYY & AAA members; children 4 & under no charge. Closed major holidays; Christmas break. &
Attendance: 2,500 (estimated)

Greenville

SOMERVILLE MANNING GALLERY, Brecks Mill, 2nd Fl., 101 Stone Block Row, Greenville, DE 19807. Tel.: 302-652-0271. Fax: 302-652-1946.
Institution Type/Description: Art Gallery.
Collections: paintings; sculpture; works on paper.
Hours & Admission Prices: Winter: Tues.-Sat. 10-5; Summer: Mon.-Fri. 10-5, Sat. 11-3; other times by appointment.

Harrington

HARRINGTON HISTORICAL SOCIETY, 108-110 Fleming St., Harrington, DE 19952-1145. Tel.: 302-398-3698.
Formerly: Harrington Railroad Museum
Founded: 1978.
Key Personnel: Cur., Jean Miller; Pres. (V), Herman Zeitler
Institution Type/Description: Railroad History Museum.
Collections: railroad & city history; photographs.
Activities: tours. Annual Event: Heritage Day.
Hours & Admission Prices: By appointment. No charge; donations accepted.
Membership: Members $5.

MESSICK AGRICULTURE MUSEUM, 325 Walt Messick Rd., Taylor and Messick, Inc., Harrington, DE 19952-3300. Tel.: 302-398-3729.
Institution Type/Description: Agriculture Museum.
Collections: agriculture history; period John Deere tractors & bicycles.
Hours & Admission Prices: Mon.-Fri. 7:30-4:30. No charge; donations accepted.

Hockessin

DELAWARE NATURE SOCIETY, 3511 Barley Mill Rd., Hockessin, DE 19707-9393. Mailing Address: P.O. Box 700, Hockessin, DE 19707-0700. Tel.: 302-239-2334. Fax: 302-239-2473.
E-mail: dnsinfo@delawarenaturesociety.org
Web Site: www.delawarenaturesociety.org
Founded: 1964.
Key Personnel: Exec. Dir., Michael E. Riska; Pres., Thomas C. Shea, Jr.; Vice Pres., Sharon Struthers; Sec., Richmond Williams; Treas., Clifford H. Hunter.
Personnel Profile: Full-Time Paid 21; Part-Time Paid 58; Part-Time Volunteers 1,000; Interns 6.
Governing Authority: nonprofit organization. Branch Museums: Abbott's Mill Nature Center, 15411 Abbott's Pond Rd., Milford, DE; Ashland Nature Center, Barley Mill & Brackenville Rd., Hockessin, DE; Coverdale Farm, 543 Way Rd., Greenville, DE; DuPont Environmental Education Center, 1400 Delmarva Lane, Wilmington, DE. Tax-exempt: 501(c)(3).
Institution Type/Description: Nature Center.
Collections: books; taxidermy specimens; butterfly house.
Research Fields: environmental.
Facilities: library; nature trails; auditorium.
Activities: teacher workshops; education programs for children, adults, families, schools & special interest groups; internships; Environmental Institution Management Course; environmental issue conferences; environmental advocacy; natural area preservation; native plant demonstration garden. Museum Sponsors: Harvest Moon Festival; Autumn at Abbott's; Native Plant Sale.

Publications: quarterly newsletter; program brochures; Delmarva butterflies field guide; reference book, Birds of Delaware; Amphibians & Reptiles of Delaware.
Hours & Admission Prices: Mon.-Fri. 8:30-4:30, Sat. 9-3. Trail: $2; ANCA & society members no charge. Closed holidays. &
Attendance: 80,000 (estimated)
Membership: College Student & Senior Citizen $30; Individual $40; Household $55; Household Plus $85; Sustaining $125; Supporting $250; Sponsor $500; Director's Circle $1,000.

MT. CUBA CENTER, 3120 Barley Mill Rd., Hockessin, DE 19707. Tel.: 302-239-4244. Fax: 302-239-5366.
E-mail: info@mtcubacenter.org
Web Site: www.mtcubacenter.org
Founded: 2002.
Personnel Profile: Full-Time Paid 36; Part-Time Paid 40; Part-Time Volunteers 20; Interns 7.
Governing Authority: Tax-exempt.
Institution Type/Description: Horticultural Center.
Collections: wildflower gardens; native plants; horticulture books.
Facilities: 589 acre site.
Activities: educational programs; research; conservation; classes; docent tours. Annual Event: Open Day Visitation.
Publications: Trilliums at Mt. Cuba Center: A Visitors Guide; Mt. Cuba Dedicated to the Study of the Piedmont Flora.
Hours & Admission Prices: April 18 to Nov. Fri.-Sat. 10-4; other times by appointment. Adults $6, youth 6-16 $3; discounts to groups; children under 6 no charge.
Attendance: 7,000 (accurate)

Laurel

BALD CYPRESS NATURE CENTER, Trap Pond State Park, 33587 Baldcypress Lane, Laurel, DE 19956-2988. Tel.: 302-875-5163. Fax: 302-875-2697.
Institution Type/Description: Nature Center.
Collections: local history; wildlife; plants; bald cypress trees.
Facilities: nature trails. Gift items for sale.
Activities: hiking; camping; fishing; boating; educational programs.
Hours & Admission Prices: Daily 8 to sunset.

Lewes

THE LEWES HISTORICAL SOCIETY, (M), 110 Shipcarpenter St., Lewes, DE 19958-1210. Tel.: 302-645-7670. Fax: 302-645-2375. Facebook: Lewes Historical Society.
E-mail: info@historiclewes.org
Web Site: www.historiclewes.org
Founded: 1962.
Congressional District: 37
Key Personnel: Dir., Mike DiPaolo; Pres. (V), Russ Allen; Museum Shop Mgr., Ann Hilaman.
Personnel Profile: Full-Time Paid 2; Part-Time Paid 3; Part-Time Volunteers 500; Interns 2.
Governing Authority: society. Tax-exempt.
Institution Type/Description: History Museum.
Collections: local history & culture; period furnishings; 2 Monomoy surfboats; a life-car; beach cart; a Lyle Gun; photographs.
Research Fields: Lewes history, genealogy.
Facilities: Period furnishings & other museum-related items for sale.
Activities: lectures; films; annual craft fair; seminars; guided & self-guided tours. Annual Events: antiques show; art show; Christmas tour of Lewes.
Publications: quarterly newsletter; Pictorial History of Lewes; Victorian Lewes & Her Architecture; Journal of the Lewes Historical Society; Swanendael in New Netherland.
Hours & Admission Prices: May to early Oct. Mon.-Sat. 11-4. Adults $5-$10; members & children under 12 no charge.
Attendance: 60,000 (estimated)
Membership: Individual $25; Family $35; Contributing $50; Century $100; Supporting $250; Life $1,000.

PACKARD REATH GALLERY, 142 Second St., Ste. 2A, Lewes, DE 19958-1396. Tel.: 302-644-7513.
Web Site: www.packardreathgallery.com
Institution Type/Description: Art Gallery.
Collections: works by contemporary & upcoming photographic artists.
Hours & Admission Prices: Daily 11-5.

SEASIDE NATURE CENTER - CAPE HENLOPEN STATE PARK, 15099 Cape Henlopen Dr., Lewes, DE 19958-3153. Tel.: 302-645-8983.
Web Site: www.destateparks.com
Institution Type/Description: Nature Center.
Collections: ocean life; ecology; natural history; photographs.
Facilities: nature trails.
Activities: educational programs; hiking; guided tours.
Hours & Admission Prices: Center: Mon.-Sat. 9-4, Sun. 12-4. Park: daily 8am to sunset.

ZWAANENDAEL MUSEUM, Savannah Rd. & Kings Hwy., Lewes, DE 19958. Mailing Address: 21 The Green, Dover, DE 19901-3611. Tel.: 302-645-1148.
Key Personnel: Site Supvr., Bridget Warner
Institution Type/Description: Historic Building: built in 1931.
Collections: county history; War of 1812; Cape Henlopen Lighthouse artifacts; H.M.S. DeBraak & H.M.S. Severn shipwreck artifacts; political & military history including uniforms & weapons.
Activities: Annual Events: Zwaanendael Heritage Garden Tour in June; Christmas House Tour in December.
Hours & Admission Prices: Tues.-Sat. 10-4:30, Sun. 1:30-4:30. No charge; donations accepted. Closed state holidays.

Middletown

GILBERT W. PERRY JR. CENTER FOR THE ARTS, 51 W. Main St., Middletown, DE 19709. Tel.: 302-449-5396.
Web Site: www.everetttheatre.com
Institution Type/Description: Art Gallery.
Collections: works by local, national & international artists.
Activities: temporary exhibitions; classes.
Hours & Admission Prices: Call for hours.

Milford

ABBOTT'S MILL NATURE CENTER, 15411 Abbott's Pond Rd., Milford, DE 19963-3549. Tel.: 302-422-0847. Fax: 302-422-1849.
E-mail: dnsinfo@delawarenaturesociety.org
Web Site: www.delawarenaturesociety.org/abbotts.html
Key Personnel: Exec. Dir., Michael E. Riska.
Governing Authority: Parent Institution: Delaware Nature Society.
Institution Type/Description: Nature Center.
Collections: wildlife & their habitats; plants; ecology; pine woods. Historic Building: grist mill.
Facilities: nature trails.
Activities: educational programs; classes. Museum Sponsors: Autumn at Abbott's Mill Festival in October.
Hours & Admission Prices: Mon.-Fri. 9-4; other times by appointment.

DUPONT NATURE CENTER, 2992 Lighthouse Rd., Milford, DE 19963. Tel.: 302-422-1329.
Web Site: www.dupontnaturecenter.org
Institution Type/Description: Nature Center.
Collections: wildlife & their habitats.
Activities: educational programs; special events.
Hours & Admission Prices: April-Sept. Tues.-Sun. 10-4; Oct.-March Mon.-Fri. 10-4. No charge; donations accepted. &

MILFORD HISTORICAL SOCIETY, 501 N.W. Front St., Milford, DE 19963-1015. Mailing Address: P.O. Box 352, Milford, DE 19963-0352. Tel.: 302-422-3115.
Formerly: Parson Thorne Mansion
Founded: 1961.
Personnel Profile: Part-Time Volunteers 7.
Governing Authority: Parent Institution: Milford Historical Society.
Institution Type/Description: Historic House Museum: housed in the former home of Milford's founder, Parson Sydenham Thorne; built c.1735. Listed on the National Register of Historic Places.
Collections: family history; period furnishings.
Activities: special events.
Hours & Admission Prices: June 1st Sat. of month; Sept. last Sat. of month; other times by appointment. Admission $2.
Attendance: 1,000
Membership: Individual $10; Joint $15.

MILFORD MUSEUM, 121 S. Walnut St., Milford, DE 19963-1955. Tel.: 302-424-1080.
E-mail: claudia@milforddemuseum.org
Web Site: www.milforddemuseum.org
Founded: 1983.
Congressional District: 1
Key Personnel: Chm., Charles Hammond; Dir., Claudia Leister; Dir. Public Rels., Al Lauckner; Sec., Barbara Jones.
Personnel Profile: Full-Time Paid 1; Part-Time Paid 1; Part-Time Volunteers 20.
Governing Authority: municipal; nonprofit. Parent Institution: City of Milford. Tax-exempt: 501(c)(3).
Institution Type/Description: Local History Museum: housed in c.1910 Federal style brick building.
Collections: local history from 1680-present; store items; costumes; photographs; documents; decorative arts; awards to citizens; shipbuilding artifacts.
Research Fields: local history.
Facilities: 116-vol. library of local history books.
Activities: films; temporary exhibitions; programs pertaining to current exhibits.
Publications: semiannual newsletter, Museum Musings.
Hours & Admission Prices: Tues.-Sat. 10-3:30, Sun. 1-3:30; groups by appointment. No charge; donations accepted. Closed New Year's Day; Easter; Independence Day; Christmas. &
Attendance: 1,500 (estimated)
Membership: Student $5; Individual $10; Family $20; Patron $50; Corporate $100.

Millsboro

MILLSBORO ART LEAGUE, 203 Main St., Millsboro, DE 19966. Tel.: 302-934-6440.
Web Site: www.millsboroartleague.com
Institution Type/Description: Art Gallery.
Collections: works by local & national artists.
Activities: classes; workshops; educational programs.
Hours & Admission Prices: Fri.-Sat. 11-3, Sun. 12-3.

NANTICOKE INDIAN MUSEUM, 27073 John J. Williams Hwy., Millsboro, DE 19966-4642. Tel.: 302-945-7022.
E-mail: nanticok@verizon.net
Personnel Profile: Part-Time Paid 4; Part-Time Volunteers 1.
Institution Type/Description: History Museum.
Collections: history of the Nanticoke Indians; stone artifacts; carvings; pottery; clothing; photographs.
Facilities: library. Museum-related items for sale.
Activities: video presentations. Annual Event: Nanticoke Indian Powwow in Sept.
Hours & Admission Prices: Jan.-March Thurs.-Sat. 10-4, Sun. 12-4; April-Dec. Tues.-Sat. 10-4, Sun. 12-4. Adults $3, children $1.

Milton

MILTON HISTORICAL SOCIETY AND THE LYDIA BLACK CANNON MUSEUM, 210 Union St., Milton, DE 19968-1620. Mailing Address: P.O. Box 112, Milton, DE 19968-0112. Tel.: 302-684-1010.
E-mail: info@historicmilton.org
Web Site: www.historicmilton.org
Founded: 1970.
Key Personnel: Dir., G. Kevin Kelly; Pres., John Bushey.
Personnel Profile: Full-Time Paid 1; Part-Time Paid 1; Part-Time Volunteers 60; Interns 3.
Governing Authority: Tax-exempt.
Institution Type/Description: History Museum: housed in the newly restored 1857 Methodist Church.
Collections: area history & culture; photographs; personal artifacts; shipbuilding tools & records; area agriculture; button press.
Hours & Admission Prices: Wed.-Sat. 11-4; other times by appointment. No charge; donations accepted. Closed federal holidays.
Attendance: 10,150 (accurate)
Membership: Individual $25; Family $35; Supporting $250; Sustaining $500; Partnering $1,000.

PRIME HOOK NATIONAL WILDLIFE REFUGE, 11978 Turkle Pond Rd., Milton, DE 19968-3759. Tel.: 302-684-8419. Fax: 302-684-8504.
E-mail: fw5rw_phnwr@fws.gov
Web Site: www.fws.gov/northeast/primehook
Governing Authority: Parent Institution: U.S. Fish & Wildlife Service.
Institution Type/Description: Wildlife Refuge.
Collections: wildlife & their habitat including birds, mammals, reptiles, amphibian; fish; plants.
Hours & Admission Prices: Refuge: daily sunrise to sunset. Visitor Contact Station & Store: Mon.-Fri. 7:30-4. Visitor Center: April-Nov. Sat.-Sun. 9-4.

New Castle

BELLANCA AIRFIELD MUSEUM AND DELAWARE AVIATION HALL OF FAME, Rte. 273 & Centerpoint Blvd., New Castle, DE 19720. Mailing Address: Friends of Bellanca Airfield, P.O. Box 267, New Castle, DE 19720-0267.
E-mail: contact@friendsofbellanca.org
Web Site: www.friendsofbellanca.org
Institution Type/Description: History Museum: housed in Bellanca airfield, aircraft plant, & service hangar; built in 1928 by aviation pioneer Giuseppe Bellanca & Henry B. duPont
Collections: airfield history; aircraft manufacturing; early airplanes.
Hours & Admission Prices: Daily 11-3. No charge; donations accepted.

NEW CASTLE COURT HOUSE MUSEUM, 211 Delaware St., New Castle, DE 19720-4815. Tel.: 302-323-4453. Fax: 302-323-5319.
E-mail: cynthia.snyder@state.de.us
Web Site: history.delaware.gov
Governing Authority: Parent Institution: Delaware Department of State Division of Historical & Cultural Affairs. Tax-exempt.
Institution Type/Description: Historic Site.
Collections: Delaware history & government; legal history; underground railroad history.
Activities: special events. Museum Sponsors: Day In Old New Castle in May; Separation Day in June; New Castle Christmas in December.
Hours & Admission Prices: Tues.-Sat. 10-3:30, Sun. 1:30-4:30. No charge; donations accepted. Closed state holidays. &
Attendance: 14,000 (accurate)

NEW CASTLE HISTORICAL SOCIETY, (M), 2 E. Fourth St., New Castle, DE 19720-5014. Tel.: 302-322-2794. Fax: 302-322-8923.
E-mail: nchistorical@aol.com
Web Site: www.newcastlehistory.org
Founded: 1934.
Key Personnel: Dir., Michael Connolly; Pres. (V), Richard R. Cooch; Coord. Education, Bruce Dalleo.
Personnel Profile: Part-Time Paid 18; Part-Time Volunteers 120; Interns 1.
Governing Authority: society. Tax-exempt: 501(c)(3).
Institution Type/Description: Historical Society Museum.
Collections: early Dutch furnishings; 18th-century furnishings. Historic Houses: c.1700 Dutch House; 1730 Amstel House.
Research Fields: local history; early Dutch settlement; colonial America.
Activities: traveling exhibit. Museum Sponsors: annual exhibit at Old Library Museum. Annual Event: Historic Preservation Award.
Publications: book, New Castle on the Delaware; Buildings, Books & Blackboards: Education in New Castle 1657 to 1930; Dr. Constance Cooper, 350 Years of New Castle Delaware: Chapter's in a Town's History.
Hours & Admission Prices: April-Dec. Tues.-Sat. 11-4, Sun. 1-4. Amstel House or Dutch House: adults $4, children under 12 $1.50; Combination Tickets: adults $7, children 2-12 $2.50; children under 6, members no charge. Old Library Museum Sat.-Sun. 1-4. no charge. Closed national holidays.
Attendance: 10,500 (accurate)
Membership: Individual $30; Family $50; Business $65; Patron $100; Benefactor $250.

OLD LIBRARY MUSEUM, 40 E. 3rd St., New Castle, DE 19720. Mailing Address: 2 E. 4th St., New Castle, DE 19720-5014. Tel.: 302-322-2794. Fax: 302-322-8923.
E-mail: nchistorical@aol.com
Web Site: www.newcastlehistory.org
Institution Type/Description: Historic Building: housed in a hexagonal brick structure; built in 1892.
Collections: local history & culture; period furnishings; personal artifacts; photographs.
Hours & Admission Prices: May-Dec. Sat.-Sun. 1-4.

READ HOUSE & GARDENS, 42 The Strand, New Castle, DE 19720-4826. Tel.: 302-322-8411. Fax: 302-322-8557.
Web Site: www.dehistory.org
Personnel Profile: Full-Time Paid 2; Part-Time Paid 25; Part-Time Volunteers 10.
Governing Authority: Parent Institution: Delaware Historical Society. Tax-exempt.
Institution Type/Description: Historic House: built in 1801 by George Read Jr., the son of one of Delaware's signers of the Declaration of Independence.
Collections: period furnishings; personal artifacts; photographs; gardens.
Activities: demonstrations; tours; school programs.
Hours & Admission Prices: April-Dec. Wed.-Fri. & Sun. 11-4, Sat. 10-4. Adults $7, senior citizens over 65, military, & students 13-21 $6; discounts to groups; members & children under 6 no charge. Closed major holidays.
Attendance: 12,000 (estimated)

THE VELOCIPEDE MUSEUM, 414 Delaware St., New Castle, DE 19720-5040.
E-mail: thevelocipede@aol.com
Web Site: www.thevelocipede.com
Institution Type/Description: Bicycle Museum.
Collections: early tricycles & bicycles from 1860s to mid 1960s; advertising memorabilia; books; pamphlets; bells; lamps; wheels; signs.
Facilities: library.
Activities: 48" boneshaker interactive exhibit.
Hours & Admission Prices: Sat. 11-5; other times by appointment. No charge.

Newark

CHAMBERS HOUSE NATURE CENTER AT WHITE CLAY CREEK STATE PARK, 1475 Creek Rd., Newark, DE 19711. Mailing Address: 425 Wedgewood Rd., Newark, DE 19711-2123. Tel.: 302-368-6560 & 6900.
Institution Type/Description: Nature Center.
Collections: natural history; photographs; geology.
Facilities: library.
Activities: educational programs; special events; guided tours.
Hours & Admission Prices: May-Oct. Sat.-Sun. 11-4. &

HALE BYRNES HOUSE, 606 Stanton Christiana Rd., Newark, DE 19713-2109. Tel.: 302-737-5792.
Web Site: www.halebyrnes.org
Key Personnel: Resident Property Mgr. & Cur., Ralph Burdick; Resident Property Mgr. & Cur., Kim Burdick.
Governing Authority: Parent Institution: Delaware Division of Historical and Cultural Affairs.
Institution Type/Description: Historic House: housed in the building used as a meeting place for General George Washington & his staff between the Battle of Cooch's Bridge in Delaware and the Battle of Brandywine in Pennsylvania in 1777; built in 1750. Listed on the National Register of Historic Places.
Collections: local history; period furnishings.
Hours & Admission Prices: 1st Wed. of month 12-3; other times by appointment.

IRON HILL MUSEUM, (M), 1355 Old Baltimore Pike, Newark, DE 19702-1110. Tel.: 302-368-5703. Fax: 302-369-4287.
E-mail: info@ironhill-museum.org
Web Site: www.ironhill-museum.org
Founded: 1968.
Key Personnel: Acting Dir., Maureen Zieber; Pres. (V), Michael James; Educational Asst., Cherie Keenan; Bd. Member, Jeffery Gudzune.
Personnel Profile: Full-Time Paid 1; Part-Time Paid 2; Part-Time Volunteers 11.
Governing Authority: Parent Institution: Delaware Academy of Sciences. Tax-exempt.
Institution Type/Description: Natural History Museum.
Collections: human & natural history of Iron Hill region; rocks & minerals; Delaware animals; natural history; earth science; Lenni Lenape Indians; archaeology.
Facilities: nature trail.
Activities: outreach programs; educational programs; teacher workshops; summer camps; public programs; guided tours; Native American program.
Publications: newsletter, News and Views.
Hours & Admission Prices: Tues.-Fri. 9-2, Sat. 12-4; other times by appointment. Adults $2; discounts to ASTC members; members no charge. &
Attendance: 6,000 (estimated)
Membership: Individual $20; Grandfamily $30; Family $40; Patron $100; Lifetime $500.

UNIVERSITY MUSEUMS, UNIVERSITY OF DELAWARE, 30 N. College Ave., 208 Mechanical Hall, Newark, DE 19716. Tel.: 302-831-8037. Fax: 302-831-8057.
E-mail: jat@udel.edu
Web Site: www.udel.edu/museums
Founded: 1978.
Congressional District: 1
Key Personnel: Dir., Janis A. Tomlinson, Ph.D.; Cur. African American Art, Julie L. McGee, Ph.D.; Cur. Education, Ivan D. Henderson; Cur. Collections, Janet Gardner Broske; Cur. Mineralogical Collection, Sharon L. Fitzgerald, Ph.D.; Preparator, Brian Kamen.
Personnel Profile: Full-Time Paid 7; Part-Time Volunteers 1; Interns 1.
Governing Authority: university. Parent Institution: University of Delaware. Branch Museums: Old College Gallery, Old College; Mechanical Hall Gallery; Mineralogical Museum, Penny Hall. Tax-exempt: 501(c)(3).
Institution Type/Description: Old College Gallery: housed in 1832 first major Greek Revival structure in the state. Mechanical Hall: renovated 1898 building.
Collections: Brandywine School; Gertrude Kasebier; African American art; minerals; 20th century American works on paper; Inuit art; pre-Columbian ceramics; survey study collection.
Major Exhibits: Fresh Paint, 2/12/14-6/28/14; Strategy & Structures: The Works of Bob Straight, 2/12/14-6/28/14; Mineralogical Collection, 2/12/14-12/21/14; Permanent Collection, 2/12/14-12/21/14.
Research Fields: multi-disciplinary.
Activities: loan, traveling, permanent & temporary exhibitions.
Publications: exhibit catalogues & brochures.
Hours & Admission Prices: Sept. to May Wed. & Fri.-Sun. 12-5, Thurs. 12-8; Summer: see website for hours. No charge; donations accepted. Closed during installation, university holidays & semester breaks. &
Attendance: 6,000 (accurate)

Odessa

HISTORIC HOUSES OF ODESSA, 109 Main St., Odessa, DE 19730. Mailing Address: P.O. Box 697, Odessa, DE 19730-0697. Tel.: 302-378-4119. Fax: 302-378-4050.
E-mail: info@historicodessa.org
Web Site: www.historicodessa.org
Governing Authority: Parent Institution: Historic Odessa Foundation.
Institution Type/Description: Historic Houses.
Collections: local history; period furnishings; photographs. Historic Buildings: Corbit-Sharp House c.1774; Wilson-Warner House c.1769; Collins-Sharp House c.1700; Brick Hotel c.1822; Odessa Bank c.1853.
Hours & Admission Prices: March-Dec. Thurs.-Sun. 10-4:30. Adults $10, students $8; discounts to members; children under 5 no charge. Closed Easter; Independence Day; Thanksgiving; Christmas Eve & Day.

Port Penn

PORT PENN INTERPRETIVE CENTER, Rte. 9 & Rd. 2, Port Penn, DE 19731. Mailing Address: P.O. Box 170, Delaware City, DE 19706. Tel.: 302-836-2533.
Institution Type/Description: History Museum: housed in a former school-house; built in 1886.
Collections: local history & culture; period hunting decoys; fishing & fur trapping equipment; paintings.
Activities: town walking tour; wetlands hike.
Hours & Admission Prices: Memorial Day to Labor Day Fri.-Sun. 10-4; groups by appointment. No charge; donations accepted.
Attendance: 200 (estimated)

Rehoboth Beach

INDIAN RIVER LIFE-SAVING STATION, 25039 Coastal Hwy., Rehoboth Beach, DE 19971. Tel.: 302-227-6991. Fax: 302-227-6438. Facebook: IRLSS.
Web Site: www.destateparks.com
Personnel Profile: Full-Time Paid 1; Part-Time Paid 2; Interns 2.
Governing Authority: Parent Institution: State of Delaware. Tax-exempt.
Institution Type/Description: Historic Building: built in 1876. Listed on the National Register of Historic Places.
Collections: U.S. Life Saving Service history; photographs.
Facilities: Museum-related items for sale.
Activities: historical programs; nature related programs; boat & kayak tours.
Hours & Admission Prices: Daily 8am to sunset. Adults $4, seniors 62 & up $3, children 6-12 $2; children 5 & under no charge.
Attendance: 22,100

REHOBOTH ART LEAGUE, INC., (M), 12 Dodds Lane, Rehoboth Beach, DE 19971-1668. Tel.: 302-227-8408. Fax: 302-227-4121.
E-mail: info@rehobothartleague.org
Web Site: www.rehobothartleague.org
Founded: 1938.
Congressional District: 7
Personnel Profile: Full-Time Paid 4; Part-Time Paid 5; Part-Time Volunteers 150; Interns 1.
Governing Authority: R.A.L. Board of Trustees. Tax-exempt.
Institution Type/Description: Historic Buildings & Art League: one gallery & studio in c.1740s homestead, two galleries & a studio located on 3 1/2 acres of formal & informal gardens.
Collections: sculpture; paintings; decorative arts; art books. Historic Building: Homestead Museum.
Facilities: classrooms; meeting rooms; studios; gardens.
Activities: art classes; concerts; music programs; art shows for members; shows of work for visiting artists; craft exhibitions; Cottage Tour of Art; bus trips to art museums & theaters; workshops by visiting artists; lectures on art & art history. Museum Sponsors: fine Art Outdoor Show in August; Holiday Fair - Fine Craft Show in November.
Publications: annual brochure; pamphlets, Sand in Your Brush; Story of the Homestead; Doors of Fame.
Hours & Admission Prices: Jan.-March Tues.-Sat. 10-4, Sun. 12-4; April-Dec. Mon.-Sat. 10-4, Sun. 12-4. No charge; donations accepted. Closed New Year's Day; Easter; Thanksgiving; Christmas. &
Attendance: 13,000 (estimated)
Membership: $30-$1,000.

REHOBOTH BEACH MUSEUM, (M), 511 Rehoboth Ave., Rehoboth Beach, DE 19971. Tel.: 302-227-7310.
E-mail: rbhistoricalsociety@verizon.net
Web Site: rehobothbeachmuseum.org
Institution Type/Description: History Museum.
Collections: local history & culture; period furnishings; personal artifacts; photographs; early post cards; maps.
Activities: special events; demonstrations; lectures; educational programs.
Hours & Admission Prices: Memorial Day to Labor Day Mon.-Fri. 10-4, Sat.-Sun. 11-3; Sept.-May Mon. & Thurs.-Fri. 10-4, Sat.-Sun. 11-3.

Seaford

GOVERNOR ROSS PLANTATION, 23669 Ross Station Rd., Seaford, DE 19973-5754. Mailing Address: 203 High St., Seaford, DE 19973-3909. Tel.: 302-628-9500. Fax: 302-628-2984. Facebook: Governor Ross Plantation.
E-mail: seafordsociety@verizon.net
Web Site: www.seafordhistoricalsociety.com
Founded: 1976.
Key Personnel: Pres., Donald E. Allen, Jr.; Treas., John Watson; Public Rels., Ann Nesbitt; Tour Dir. & Museum Shop Mgr., Margaret Alexander.
Personnel Profile: Part-Time Paid 1; Part-Time Volunteers 80.
Governing Authority: private; nonprofit organization. Parent Institution: Seaford Historical Society. Branch Museum: Seaford Museum, 203 High St. Seaford. Tax-exempt: 501(c)(3).
Institution Type/Description: Historic House & Historical Society Museum: c.1860 Italian Villa style home built by Delaware Governor William Ross.
Collections: decorative arts objects appropriate for furnishing a 13-room home; Ross family's personal artifacts; original granary; carriage house; log slave quarters; corn cribs; smokehouse.
Research Fields: slavery & Civil War era in Delaware.
Facilities: library.
Activities: guided tours. Annual Events: Heritage Days during Memorial Day weekend; Victorian Christmas.
Publications: quarterly newsletter, Seaford Historical Society.
Hours & Admission Prices: Sat.-Sun. 1-4; other times by appointment. Adults $5; members & children under 12 no charge. Closed state holidays. &
Attendance: 6,000 (accurate)
Membership: Individual $25; Family $40.

SEAFORD MUSEUM, 203 High St., Seaford, DE 19973-3909. Tel.: 302-628-9828. Fax: 302-628-2984.
E-mail: seafordsociety@verizon.net
Web Site: www.seafordhistoricalsociety.com
Founded: 1997.
Congressional District: 39
Key Personnel: Museum Shop Mgr., Shirley Skinner.
Personnel Profile: Part-Time Paid 3; Part-Time Volunteers 90.

Governing Authority: Parent Institution: Seaford Historical Society. Tax-exempt.
Institution Type/Description: History Museum: housed in the former post office; built in 1930s.
Collections: local history & culture; period furnishings; Native American artifacts; photographs.
Facilities: Museum-related items for sale.
Publications: quarterly newsletter.
Hours & Admission Prices: Thurs.-Sun. 1-4. Adults $5. Closed holidays. &
Attendance: 2,000 (accurate)
Membership: Individual $25; Family $40.

Smyrna

BOMBAY HOOK NATIONAL WILDLIFE REFUGE, 2591 Whitehall Neck Rd., Smyrna, DE 19977-2912. Tel.: 302-653-9345. Fax: 302-653-0684.
E-mail: fw5rw_bhnwr@fws.gov
Web Site: www.fws.gov/northeast/bombayhook
Key Personnel: Mgr., Michael Stroeh
Institution Type/Description: Wildlife Refuge.
Collections: migrating & wintering ducks & geese; mammals.
Facilities: nature trails; picnic area.
Activities: bird watching; hiking; educational programs; workshops.
Hours & Admission Prices: March-May & Sept. to mid-Dec. Mon.-Fri. 8-4, Sat.-Sun. 9-5; mid-Dec. to Feb. Mon.-Fri. 8-4.

SMYRNA MUSEUM, 11 S. Main St., Smyrna, DE 19977-1430. Mailing Address: P.O. Box 335, Smyrna, DE 19977-0335. Tel.: 302-653-1320. Fax: 302-653-8844.
Key Personnel: Pres., Brooks Keen.
Governing Authority: Parent Organization: Duck Creek Historical Society.
Institution Type/Description: History Museum.
Collections: local artifacts; toys & dolls; Native American artifacts; paintings; over 400 pitchers; mid-1700s plank house.
Hours & Admission Prices: Tues., Thurs. & Sat. 10-1; other times by appointment. No charge; donations accepted.

Wilmington

ARDEN CRAFT SHOP MUSEUM, 1807 Millers Rd., Ste. A, Wilmington, DE 19810-4052. Tel.: 302-475-3060.
E-mail: ardencraftshopmuseum@gmail.com
Web Site: www.ardencraftshopmuseum.com
Founded: 2005.
Key Personnel: Chm. (V), Lisa Mullinax.
Personnel Profile: Part-Time Paid 1; Part-Time Volunteers 8; Interns 1.
Institution Type/Description: Art Museum.
Collections: paintings; sculpture; prints; woodcarvings.
Major Exhibits: 10th Anniversary, 10/14-11/14.
Hours & Admission Prices: Sun. 1-3, Wed. 7:30pm-9pm. No charge; donations accepted or requested. &

BELLEVUE HALL MANSION - BELLEVUE STATE PARK, 800 Carr Rd., Wilmington, DE 19809-2163. Tel.: 302-761-6963.
Web Site: www.destateparks.com
Institution Type/Description: Historic House.
Collections: local history & culture; photographs; period furnishings; personal artifacts.
Facilities: nature trails.
Activities: hiking.
Hours & Admission Prices: Park: daily 8am to sunset. Mansion: by appointment.

BRANDYWINE CREEK NATURE CENTER, 41 Adams Dam Rd., Wilmington, DE 19807. Mailing Address: P.O. Box 3782, Greenville, DE 19807-0782. Tel.: 302-655-5740.
Institution Type/Description: Nature Center.
Collections: natural history; photographs.
Facilities: nature trails. Museum-related items for sale.
Activities: educational programs.
Hours & Admission Prices: Daily 8am to sunset.

BRANDYWINE ZOO, 1001 N. Park Dr., Wilmington, DE 19802-3801. Tel.: 302-571-7747. Fax: 302-571-7787.
Web Site: www.brandywinezoo.org
Founded: 1905.
Key Personnel: C.E.O. & Zoo Dir., Nancy M. Falasco; Pres. (V), Greg Ellis;

Cur. Education, Jill Karlson; Asst. Cur. Education, Melody Hendricks; Museum Shop Mgr., Patty Tiano.
Personnel Profile: Full-Time Paid 8; Part-Time Paid 20; Part-Time Volunteers 40; Interns 38.
Governing Authority: state. Subsidiary Institution: Delaware Zoological Society. Tax-exempt.
Institution Type/Description: Zoo.
Collections: North & South America; Temperate Asia.
Facilities: zoological park; classroom. Gift items for sale.
Activities: education programs for school groups; guided tours; outreach programs; teacher workshops; volunteer & docent program; public programs for toddlers & seniors; special events; informal education programs; loan kits; summer camp for children.
Publications: quarterly newsletter, Zoo News; education program guides.
Hours & Admission Prices: Daily 10-4. June-Sept. adults $5, senior citizens $4, children 3-11 $3. Oct.-May adults $4, senior citizens & children 3-11 $2; discount to groups; members & children under 3 no charge. &
Attendance: 90,000 (estimated)
Membership: Individual $25; Senior $35; Household $45; Donor $75; Naturalist $100; Zoo Gooder $250; Animal Enthusiast $500; Rare Bird $1,000.

*** DELAWARE ART MUSEUM, (M),** 2301 Kentmere Pkwy., Wilmington, DE 19806-2096. Tel.: 302-571-9590. Fax: 302-571-0220.
E-mail: info@delart.org
Web Site: www.delart.org
Founded: 1912.
Congressional District: 6
Key Personnel: Acting C.E.O. & C.F.O., Mike Miller; Chief Cur., Margaretta S. Frederick.
Personnel Profile: Full-Time Paid 28; Part-Time Paid 11; Part-Time Volunteers 205; Interns 10.
Volunteer Hours: 11,953
Operating Expenses: 4,911,990
Operating Income: 4,699,491
Governing Authority: nonprofit organization. Tax-exempt: 501(c)(3).
Institution Type/Description: Art Museum.
Collections: The Samuel and Mary R. Bancroft Memorial Collection of British Pre-Raphaelite Art (1848-1915); collection of work by American illustrator Howard Pyle (1853-1911) and other American illustrators from the period 1876-1940; collection of work by American artist John Sloan (1871-1951) and related artists; survey of American art in various media from the early 19th century to the present.
Major Exhibits: Femfolio, 9/13-1/14; American Moderns, 1910-1960: From O'Keefe to Rockwell (T), 10/13-1/14; Blessed are the Peacemakers: Violet Oakley's The Angel of Victory (1941), 2/8/14-5/25/14; Fashion, Circus, Spectacle: Photographs of Scott Heiser, 3/8/14-6/1/14; Retro*active: Performance Art from 1964-1987, 6/14/14-9/21/14; Performance Now (T), 7/12/14-9/21/14; Nature Morte: Platinum Prints by Bruce Katsiff, 10/4/14-1/25/15; From Houdini to Hugo: The Art of Brian Selznick (T), 10/18/14-1/11/15.
Research Fields: American art & illustration; British Pre-Raphaelite art.
Facilities: library; archives; educational facilities; interactive children's learning area; outdoor sculpture garden; cafe. Museum-related items for sale.
Activities: guided tours; lectures; gallery talks; studio art classes; education programs; permanent, temporary & traveling exhibitions.
Publications: quarterly newsletter; studio art classes booklet; summer art camp brochure; school & teacher programs booklet; annual report; E-Bulletin; exhibition catalogs.
Hours & Admission Prices: Wed.-Sat. 10-4, Sun. 12-4. Adults $12, seniors $10, students & youth $6; discounts to AAM members; members, children 6 & under and Sun. no charge. &
Attendance: 70,000 (accurate)
Membership: Basic for One $50; Basic for Two $70; Basic for Three $95; Basic for Foru $125; Sponsor $250; Benefactor $500; John Sloan Society $1,500; Howard Pyle Society $5,000; Rossetti Circle $10,000.

DELAWARE CENTER FOR THE CONTEMPORARY ARTS, (M), 200 S. Madison St., Wilmington, DE 19801-5100. Tel.: 302-656-6466, ext. 7102. Fax: 302-656-6944.
E-mail: info@thedcca.org
Web Site: www.thedcca.org
Founded: 1979.
Congressional District: 1
Key Personnel: Exec. Dir, Maxine Gaiber; Pres. (V), Ron Meick; Treas., Jeff Mitchell; Education Cur., Sarah Ware; Cur., Maiza Hixson; Dir. Mktg. & Public Rels., Sara Teixido; Dir. Special Events, Ashlee Lukoff; Museum Shop Mgr., Marcella Conlin.
Personnel Profile: Full-Time Paid 9; Part-Time Paid 8; Part-Time Volunteers 60; Interns 2.

Governing Authority: private; nonprofit organization. Tax-exempt: 501(c)(3).
Institution Type/Description: Art Museum.
Collections: works by national & regional contemporary artists in all media & crafts.
Major Exhibits: Post Consumed: A Millenial Biennial, 11/13-3/9/14; The Mighty Sparrow, 3/1/14-6/8/14; Magnum Opus: The Alchemical Process of Art, 3/22/14-6/15/14; Street Art, 3/22/14-6/15/14; Mark Stockton, 3/22/14-7/20/14; Scott Kip, 4/19/14-8/17/14; John Moran, 7/5/14-10/26/14; Dawn Hunter, 8/2/14-11/30/14; Hiro Sakaguchi, 11/1/14-4/19/15; Dennis Beach, 11/22/14-5/17/15.
Facilities: 100-seat auditorium; educational facilities; 6,000 sq. ft. exhibit space. Museum-related items for sale.
Activities: concerts; films & video; formal education programs; internships for college students; guided tours; lectures; temporary, participatory & traveling exhibitions. Annual Event: Fundraiser & Art Auction in Fall.
Publications: bimonthly e-newsletter, DCCA News; annual report, DCCA Annual Report; exhibition catalogues.
Hours & Admission Prices: Tues. & Thurs.-Fri. 10-5, Wed. & Sun. 12-5. No charge; donations accepted. Closed major holidays. &
Attendance: 18,000 (estimated)
Membership: Art Student & Senior $40; Individual $45; Household $65; Supporting $125; Sponsor $250; Patron $500; Visionary $1,000.

DELAWARE CHILDREN'S MUSEUM, 550 Justison St., Wilmington, DE 19801-5142. Tel.: 302-654-2340.
Institution Type/Description: Children's Museum.
Collections: hands-on exhibits.
Hours & Admission Prices: Call for hours & admission prices.

DELAWARE DIVISION OF THE ARTS MEZZANINE GALLERY, Carvel State Office Bldg., 820 N. French St., Wilmington, DE 19801. Tel.: 302-577-8278.
Web Site: www.artsdel.org
Institution Type/Description: Art Gallery.
Collections: works by local, national & international artists.
Activities: temporary exhibitions.
Hours & Admission Prices: Mon.-Fri. 8-4:30.

DELAWARE HISTORICAL SOCIETY, (M), 505 Market St., Wilmington, DE 19801-3091. Tel.: 302-655-7161. Fax: 302-655-7844.
E-mail: deinfo@dehistory.org
Web Site: www.dehistory.org
Founded: 1864.
Congressional District: 1
Key Personnel: Exec. Dir., Scott W. Loehr; Pres., Glen C. Gray; Dir. Library, Constance Cooper; Asst. Dir., Michele Anstine; Cur. Education, Antoinette Maccari.
Personnel Profile: Full-Time Paid 15; Part-Time Paid 36; Part-Time Volunteers 25; Interns 4.
Governing Authority: society. Tax-exempt: 501(c)(3).
Institution Type/Description: History Museum.
Collections: documents & artifacts related to Delaware history; Delaware silver; furniture; decorative arts; costumes; paintings; newspapers; maps; photographs; manuscripts. Historic Houses: 1798 Old Town Hall; c.1748-1801 Willingtown Square; 1801 Read House & Gardens.
Research Fields: Delaware history; material culture; architecture; decorative arts.
Facilities: 85,000-vol. library of Delaware history books; 11,000 sq. ft. exhibit space; meeting facilities. Museum-related items for sale.
Activities: guided tours; walking tours of New Castle; lectures; changing exhibitions; formally organized education programs; family discovery room.
Publications: biannual magazine, Delaware history; biannual newsletter; catalogue, Delaware Collections: The Historical Society of Delaware; exhibition catalogues.
Hours & Admission Prices: Library: Mon. 1-9, Tues.-Fri. 9-5. Delaware History Museum: Wed.-Fri. 11-4, Sat. 10-4. Read House: Wed.-Fri. 11-4, Sat. 10-4, Sun. 12-4. Adults $4, senior citizens & children 12-21 $3.50, children 6-12 $2; discounts to groups & AAM members; library & members no charge. Closed Federal holidays. &
Attendance: 60,000 (accurate)
Membership: Student & Institutional $15; Individual $40; Family $55; Contributing $75; Sustaining $100; Sponsor $250; Patron $500; Benefactor $1,000.

DELAWARE MUSEUM OF NATURAL HISTORY, 4840 Kennett Pike, Wilmington, DE 19807-1827. Mailing Address: P.O. Box 3937, Wilmington, DE 19807-0937. Tel.: 302-658-9111. Fax: 302-658-2610.
E-mail: hspruance@delmnh.org
Web Site: www.delmnh.org
Founded: 1957.
Congressional District: 1
Key Personnel: Dir., Halsey Spruance; Pres. Bd., Nina L.S. Burnaford; Dir. Collections & Cur. Ornithology, Dr. Jean Woods; Cur. Mollusks, Dr. Liz Shea; Mgr. Exhibits, Harold Barthold; Controller, Judy Julis; Dir. Devel., Dawn Swartout; Dir. Visitor Svcs., Terri Reed.
Personnel Profile: Full-Time Paid 25; Part-Time Paid 18; Part-Time Volunteers 126.
Governing Authority: nonprofit organization. Tax-exempt: 501(c)(3).
Institution Type/Description: Natural History Museum.
Collections: birds; bird eggs; mollusks; mammals; dinosaurs.
Major Exhibits: Ice Age Imperials (T), 10/13-1/14.
Research Fields: taxonomic ornithology; malacology; mammalogy.
Facilities: 10,000-vol. library of mollusk, bird & other biology manuscripts & archives; discovery room. Molluscan & ornithological research books for sale.
Activities: lectures; films; formally organized educational programs; inter-museum loan; permanent, temporary & school loan exhibits.
Publications: journal, Nemouria & Indo-Pacific Mollusca; books, Exotic Conchology; Living Volutes; Philippine Birds; South Pacific Birds; Woodpeckers of the World; Index Nudibranchia; Ranellidae; Marine Mollusks of Bermuda.
Hours & Admission Prices: Mon.-Sat. 9:30-4:30, Sun. 12-4:30. Adults $9, seniors $8, children 3-17 $7; discounts to groups & AAM members; children under 3 & members no charge. Closed New Year's Day; Independence Day; Thanksgiving; Christmas. &
Attendance: 77,638 (accurate)
Membership: Individual $35; Grandparent $50; Household $60; Sustaining $75; Patron $125; Gold Patron $500; Explorer Society $1,000.

DELAWARE SPORTS MUSEUM AND HALL OF FAME, Frawley Stadium, Entrance on 1st Base Side, 801 Shipyard Dr., Wilmington, DE 19801. Tel.: 302-425-3263. Fax: 302-425-3713.
E-mail: desports@cavtel.net
Web Site: www.desports.org
Founded: 1976.
Key Personnel: Dir., Jon Rafal; Pres. (V), Tom Fort.
Personnel Profile: Full-Time Paid 1.
Governing Authority: Tax-exempt: 501(c)(3).
Institution Type/Description: Sports Museum.
Collections: sports artifacts, memorabilia, equipment, uniforms & photographs; sports history videos from 1860 to present; Hall of Fame including 303 men & women inductees from 1976 to present.
Facilities: 5,000 sq. ft. exhibit space.
Activities: special events. Annual Event: 38th Induction Banquet in May; Hall of Fame Gold Tournament in June.
Publications: annual program, Hall of Fame Induction.
Hours & Admission Prices: April-Oct. Tues.-Sat. 12-5; groups & other times by appointment. Adults $4, seniors over 50 $3, youth 13-19 $2; children 12 & under, Hall of Fame inductees & members no charge. &
Membership: Individual $40.

✳ HAGLEY MUSEUM AND LIBRARY, (M), 298 Buck Rd. E., Wilmington, DE 19807-0630. Mailing Address: P.O. Box 3630, Wilmington, DE 19807-0630. Tel.: 302-658-2400. Fax: 302-658-0568. Facebook: Hagley Museum and Library.
E-mail: dcole@hagley.org
Web Site: www.hagley.org
Founded: 1952.
Congressional District: 1
Key Personnel: Exec. Dir., David A. Cole, Ph.D.; Dir. External Affairs & Devel., Jill A. MacKenzie; Andrew W. Mellon Cur. Prints & Photographs, Jon Williams; Cur. Collections & Exhibitions, Debra Hughes; Dir. Library Svcs., Erik Rau; Dir. Human Resources, Robert Hill; Finance Dir., Jeanne Belk; Chief Cur. Library Collections, Lynn Catanese; Objects Conservator, Ebenezer Kotei; Cur. Interpretation & Visitor Svcs., Lola Russell; Dir. Museum Svcs., Joan Hoge-North; Hagley Shop Mgr. & Buyer, Carole Katchur; Cur. Mechanical Exhibitions, John McCoy; Dir. Facilities, Michael Downs.
Personnel Profile: Full-Time Paid 78; Part-Time Paid 83; Part-Time Volunteers 628.

Governing Authority: nonprofit organization. A division of the Eleutherian Mills-Hagley Foundation, Inc. Tax-exempt: 501(c)(3).
Institution Type/Description: History & Technology Museum: located on the original site of DuPont powder yards c.1802-1921.
Collections: archaeology; technology; patent models; industry; textiles; ceramics; manuscripts, imprints & photographs of the history of business and technology. Historic Buildings: 1803 Eleutherian Mills; 1813 Henry Clay Factory, textile mill; 1817 Brandywine Manufacturer's Sunday School; 1840 Gibbons House; 1858 Millwright/Machine Shop; several 19th-century powder mills.
Research Fields: industrial, economic, social & technological history.
Facilities: research library of manuscript & pictorial collections; rental facilities. Museum-related items for sale.
Activities: guided tours; lectures; education programs for children; formally organized Fellowship education programs for graduate students affiliated with University of Delaware; research grants for visiting scholars; docent program; permanent, temporary & traveling exhibitions; school loan service; narrated bus ride through grounds along Brandywine River.
Publications: guidebooks, Impressions of Hagley; Eleutherian Mills; booklet series, Hagley Museum's Industry In America; catalogues of collections & exhibits; quarterly magazine; annual report; collections guides.
Hours & Admission Prices: Daily 9:30-4:30. Adults $14, seniors & students $10, children 6-14 $5; discounts to AAM, ICOM, AASLH & ASTC members; children under 6 no charge. Closed Thanksgiving; Christmas. &
Attendance: 62,000 (accurate)
Membership: Individual $30; Scholar $35; Staff & Volunteers $50; Household $60; Patron $150-$499; Sponsor $500-$999; Benefactor $1,000 & up.

LOMBARDY HALL, 1611 Concord Pike, Wilmington, DE 19803. Tel.: 302-798-3828.
E-mail: wdmower@verizon.net
Institution Type/Description: Historic House: housed in the former home of Gunning Bedford, Jr., Grand Master of the Grand Lodge of Delaware; 1806-1809.
Collections: local history; Masonic artifacts; period furnishings; personal artifacts; photographs.
Hours & Admission Prices: Dec. 1st Sat.-Sun. 1-4; other times by appointment.

MUSEUM OF BUSINESS HISTORY AND TECHNOLOGY, 1200 Philadelphia Pike, Wilmington, DE 19809-2040. Tel.: 302-798-2100.
Web Site: www.mbht.org
Institution Type/Description: History Museum.
Collections: business technology; inventions; business machines from 1873-1940; 1795 copier; 1820 calculator; early typewriters; books; patents; manuals.
Hours & Admission Prices: By appointment.

NEMOURS MANSION AND GARDENS, 1600 Rockland Rd., Wilmington, DE 19803-3607. Tel.: 800-651-6912. Fax: 302-651-6933.
E-mail: tours@nemours.org
Web Site: www.nemoursmansion.org
Founded: 1977.
Key Personnel: Registrar, Francesca Biella Bonny; Exec. Dir., Grace Gary; Estate Supt., James Solge.
Personnel Profile: Full-Time Paid 21; Part-Time Paid 38.
Governing Authority: nonprofit organization. Parent Institution: The Nemours Foundation. Tax-exempt.
Institution Type/Description: Historic House & Gardens: housed in 1910 Louis XVI-style chateau.
Collections: 14th to 20th-century American, European & Oriental fine & decorative arts; porcelain; paintings; prints; drawings; photographs; sculpture; silver; textiles; tapestries; oriental rugs; furniture; innovative technical equipment.
Research Fields: European fine & decorative arts.
Facilities: formal French gardens.
Activities: guided tours.
Publications: booklet, Nemours Mansion and Gardens.
Hours & Admission Prices: May-Dec. Adults $15; visitors must be at least 12 years old; reservations recommended. Closed Thanksgiving; Christmas. &
Attendance: 20,000 (estimated)

OLD SWEDES HISTORIC SITE, 606 Church St., Wilmington, DE 19801-4421. Tel.: 302-652-5629. Fax: 302-652-8615.
E-mail: admin@oldswedes.org
Web Site: www.oldswedes.org
Formerly: Holy Trinity (Old Swedes) Church & Hendrickson House Museum

Founded: 1947.
Congressional District: 1
Key Personnel: Pres. (V), James A. Bayard, Jr.; Dir., Rebecca L. Wilson.
Personnel Profile: Part-Time Paid 2; Part-Time Volunteers 20.
Governing Authority: nonprofit organization. Tax-exempt: 501(c)(3).
Institution Type/Description: Historic Site.
Collections: early colonial life, Swedish culture; genealogical; burial grounds. Historic Buildings: Andrew Hendrickson house an early 18th century Swedish farmhouse; Old Swedes Church 1698-1699.
Research Fields: Delaware history; colonial history; birth, baptism, marriage & burial records of Old Swede Church.
Facilities: 150-vol. library pertaining to Wilmington history. Gift items for sale.
Activities: tours; ghost tours; concerts; genealogical & historical research.
Publications: annual newsletter, Foundation News.
Hours & Admission Prices: March Wed.-Fri. 1-4, Sat. 10-4; April Dec. Wed.-Sat. 10-4; other times by appointment. Adults $4; discounts to members; children & members no charge. Closed New Year Day; Independence Day; Thanksgiving & day after; Christmas Eve & Day. &
Attendance: 4,000 (estimated)
Membership: Individual $35; Family $50; Contributor $100; Patron $250; Benefactor $500.

ROCKWOOD PARK, 610 Shipley Rd., Wilmington, DE 19809-3609. Tel.: 302-761-4340. Fax: 302-761-4345.
E-mail: pnord@nccde.org
Web Site: www.rockwood.org
Founded: 1976.
Congressional District: 6
Key Personnel: Dir., Philip Nord; Horticulturist, Dena Kirk.
Personnel Profile: Full-Time Paid 2; Part-Time Paid 2; Part-Time Volunteers 25.
Governing Authority: county. Parent Institution: New Castle County. Tax-exempt.
Institution Type/Description: Historic House & Museum Gallery: English Victorian Mansion & grounds.
Collections: American & European turn of the century furnishings; architecture; horticulture; specimen trees; family documents & photographs.
Research Fields: Victorian life.
Facilities: 72 acres of park grounds; picnic area.
Activities: guided tours of mansion & grounds; special events. Annual Events: Partner with Delaware Shakespeare Festival; Holiday Open House.
Publications: Romantic Rockwood.
Hours & Admission Prices: Museum: Wed.-Sun. 10-3. Grounds: daily 7-10. Adults $10, children 2-12 $4; discounts to AAM & ICOM members; children under 2 no charge. Closed major holidays.
Attendance: 15,000 (estimated)

Winterthur

* **WINTERTHUR MUSEUM, GARDEN & LIBRARY, (M),** 5105 Kennett Pike, Winterthur, DE 19735. Tel.: 302-888-4600; 800-448-3883. Fax: 302-888-4820. TDD: 302-888-4907.
E-mail: tourinfo@winterthur.org
Web Site: www.winterthur.org
Formerly: Winterthur Museum & Country Estate
Founded: 1951.
Congressional District: 1
Key Personnel: Dir. & C.E.O., Dr. David P. Roselle; Chm. (V), Rodman Ward, Jr.; C.F.O., Robert Necarsulmer; Dir. Museum Affairs, J. Thomas Savage; Dir. Human Resources, Lisbeth Selsor; Museum Shop Mgr., Ellen Taviano; Dir. Garden & Estate, Christopher Strand; Dir. Collections & Sr. Cur. Textiles, John L. & Marjorie P. McGraw, Linda S. Eaton.
Personnel Profile: Full-Time Paid 171; Part-Time Paid 173; Part-Time Volunteers 700; Interns 12.
Governing Authority: nonprofit organization. Tax-exempt: 501(c)(3).
Institution Type/Description: Decorative Arts & Cultural History Museum: housed in 1839 building with additions in the 1920s, 1930s, 1950s & 1990s.
Collections: American decorative & fine arts from 17th to 20th century; interior architecture; furniture; metalwork; ceramics; glass; textiles; prints, paintings; sculpture; rare books; manuscripts.
Major Exhibits: Costumes of Downton Abbey, 1/5/14-3/14.
Research Fields: decorative arts; American material culture; horticulture; conservation of historic & artistic objects.
Facilities: library of books & periodicals; archives including photographs; 350-seat auditorium; classrooms; restaurant. Museum-related items for sale.
Activities: guided tours; lectures; school tours; outreach educational programs; formally organized education programs; formally organized education programs for graduate students affiliated with the University of Delaware; training programs for professional museum workers; permanent & temporary exhibitions. Museum Sponsors: Point-to-Point Races; annual Delaware Antiques Show; Yuletide at Winterthur.
Publications: Winterthur Portfolio; books, catalogs & monographs on decorative arts; school newsletter; donor newsletter; member calendar.
Hours & Admission Prices: Tues.-Sun. 10-5; call for additional hours during Yuletide. Adults $18, students & seniors $16, children 2-11 $10; discounts to groups by advance arrangement, AAM, ICOM, MAAM & AAMD members; members & children under 2 no charge. Closed Thanksgiving; Christmas. &
Attendance: 110,840 (accurate)
Membership: Individual $60; Dual $80; Family $85; Contributor $150; Patron $250; Benefactor $500; Associate $1,000; Director $2,500.

Yorklyn

MARSHALL STEAM MUSEUM, 3000 Creek Rd., Yorklyn, DE 19736. Mailing Address: Friends of Auburn Heights Inc., P.O. Box 61, Yorklyn, DE 19736-0061. Tel.: 800-349-2134; 302-239-2385.
E-mail: admin@auburnheights.org
Web Site: www.auburnheights.org
Formerly: Auburn Heights Preserve
Founded: 2004.
Key Personnel: Dir., Susan Randolph; Pres. (V), J. Stephen Bryce.
Personnel Profile: Full-Time Paid 2; Part-Time Paid 1; Part-Time Volunteers 80.
Governing Authority: nonprofit organization. Parent Institution: Friends of Auburn Heights Preserve, Inc. Tax-exempt: 501(c)(3).
Institution Type/Description: Historic House: housed in the Marshall family home & carriage house. Listed on the National Register of Historic Places.
Collections: family & automobile history; personal artifacts; period furnishings; early automobiles including Stanley Steamers, Packards, & electric cars; steam train; photographs.
Activities: guided tours; rental facilities; outreach & school programs; Little Tikes Literacy Program.
Publications: quarterly newsletter, Auburn Heights Herald; weekly e-newsletter.
Hours & Admission Prices: Steamin' Days: June-Nov. 1st Sun. each month.
Membership: Individual $35; Dual $45; Family $65.

DISTRICT OF COLUMBIA

(117 listings)

Fort McNair

U.S. ARMY CENTER OF MILITARY HISTORY, MUSEUM DIVISION, (M), 102 Fourth Ave., Fort McNair, DC 20319-5058. Tel.: 202-685-2441. Fax: 202-685-2113.
E-mail: charles.h.cureton.civ@mail.mil
Web Site: www.history.army.mil
Founded: 1946.
Key Personnel: Dir. Army Museums, Dr. Charles H. Cureton.
Personnel Profile: Full-Time Paid 44.
Governing Authority: federal. Parent Institution: Dept. of the Army. Subsidiary Institution: Center of Military History. Tax-exempt.
Institution Type/Description: History Museum and Military Museums.
Collections: Active Army Museums and Museum Activities. Please call for a complete list of museums and other activities.
Research Fields: military history.
Facilities: 1,000-vol. library.
Activities: training & educational programs; programs for soldier/ leaders & professional museum staff; temporary & permanent exhibitions.
Publications: books, The Guide to U.S. Army Museums; Guide for Coordinator of Volunteers in U.S. Army Museums; Certification Inspection Handbook; Professional Assistance Directory; Handbook for Supervisors of Museums; Museum Standing Operating Procedures; Army Museum Information System (AMIS); Army Regulation 870-20; Army Museums, Historical Artifacts, and Art.
Hours & Admission Prices: Library: Mon.-Fri. 9-4. No charge. &

Washington

ADDISON/RIPLEY FINE ART, 1670 Wisconsin Ave., N.W., Washington, DC 20007. Tel.: 202-338-5180. Fax: 202-338-2341.
E-mail: addisonrip@aol.com
Web Site: www.addisonripleyfineart.com
Founded: 1982.
Key Personnel: Owner, Sylvia Ripley; Owner, Christopher Addison; Dir., Romy Silverstein.

Personnel Profile: Full-Time Paid 1; Full-Time Volunteers 1; Part-Time Paid 1; Interns 2.
Volunteer Hours: 240
Operating Expenses: 900,000
Operating Income: 1,100,000
Institution Type/Description: Art Gallery.
Collections: works by contemporary artists; paintings; sculpture; photographs; fine art prints.
Hours & Admission Prices: Tues.-Sat. 11-6; other times by appointment. No charge; donations accepted.
Attendance: 4,000 (estimated)

AFRICAN AMERICAN CIVIL WAR MUSEUM, 1925 Vermont Ave., N.W., Washington, DC 20001. Tel.: 202-667-2667. Fax: 202-667-6771.
E-mail: info@afroamcivilwar.org
Web Site: afroamcivilwar.org
Founded: 2000.
Key Personnel: Exec. Dir. & C.E.O., Dr. Frank Smith, Jr.; Cur., Harold Jones
Governing Authority: Tax-exempt.
Institution Type/Description: History Museum.
Collections: African Americans in the Civil War; period history; photographs; documents; newspaper articles; clothing; uniforms; weapons; Civil Rights.
Facilities: Books & Civil War items for sale.
Activities: special events; educational programs.
Publications: book of names.
Hours & Admission Prices: Mon.-Fri. 10-5, Sat.-Sun. 10-4. No charge; donations accepted.
Attendance: 100,000 (accurate)

THE AMERICAN CENTER OF POLISH CULTURE, 2025 "O" St., N.W., Washington, DC 20036-5913. Tel.: 202-785-2320. Fax: 202-785-2159.
E-mail: assistant@polishcenterdc.org
Web Site: www.polishcenterdc.org
Founded: 1997.
Key Personnel: Chm. (V), Patricia Koproski; Dir., Patricia Whitelaw-Hill.
Personnel Profile: Full-Time Paid 1; Part-Time Paid 3; Part-Time Volunteers 1.
Governing Authority: Tax-exempt.
Institution Type/Description: Cultural History Museum.
Collections: Polish history & cultural heritage; Polish, Jewish & American relations; photographs; paintings; sculpture; decorative arts; textiles; costumes; graphic arts; documentary film.
Research Fields: Polish history & culture.
Facilities: library.
Activities: youth programs; speeches; book presentations; discussions; concerts.
Publications: The Center Line.
Hours & Admission Prices: Mon.-Fri. 10-4. No charge; donations accepted.
Membership: Senior $50; Individual $75; Couple & Family $100.

AMERICAN UNIVERSITY MUSEUM AT THE KATZEN ARTS CENTER, (M), 4400 Massachusetts Ave., N.W., Washington, DC 20016-8003. Tel.: 202-885-1300. Fax: 202-885-1140.
E-mail: aumuseum@american.edu
Web Site: www.american.edu/museum
Formerly: Watkins Gallery, American University
Founded: 2005.
Key Personnel: Dir. & Cur., Jack Rasmussen; Asst. Dir., Katherine MacDiarmid; Chief Preparator & Registrar, Bruce Wick; Asst. Preparator, Zacherey Willis; Sr. Administrative Asst., Samantha Marques-Mordkofsky.
Personnel Profile: Full-Time Paid 5; Part-Time Paid 2; Part-Time Volunteers 55.
Governing Authority: university. Parent Institution: American University. Tax-exempt.
Institution Type/Description: Art Museum.
Collections: 1900-present American & European paintings, prints & drawings.
Facilities: 30,000 sq. ft. exhibit space.
Hours & Admission Prices: Tues.-Sun. 11-4. No charge. &
Membership: Student $15; Individual $45; Family $85; Associate $250; Director $500; Katzen Circle $1,000.

ANACOSTIA COMMUNITY MUSEUM, 1901 Fort Pl., S.E., Washington, DC 20560. Tel.: 202-633-4820. Fax: 202-287-3183. TDD: 202-357-1729.
E-mail: ACMinfo@si.edu
Web Site: anacostia.si.edu

Formerly: Anacostia Museum and Center for African American History & Culture
Founded: 1967.
Key Personnel: Dir., Camille Akeju; Deputy Dir., Sharon Reinckens; Sr. Cur., Portia James; Historian, Gail Lowe.
Personnel Profile: Full-Time Paid 19; Part-Time Volunteers 40; Interns 20.
Governing Authority: Parent Institution: Smithsonian Institution, Washington, DC, which is a nonprofit federally-chartered corporation. Branch Museum: 1901 Fort Pl., S.E., Washington, DC 20560-0004. Tax-exempt: 501(c)(3) & 170(b)(1)(A).
Institution Type/Description: Family & Community History Museum.
Collections: African American social & cultural history; family & community history.
Research Fields: 19th & 20th-century African American history & culture; contemporary African American culture & community life; urban issues.
Facilities: library.
Activities: guided tours; workshops; teacher workshops; performances; educational programs; films. Annual Observances: Martin Luther King Jr. Program.
Publications: exhibit catalogues; newsletter; resource guides; educational materials.
Hours & Admission Prices: Daily 10-5. No charge; donations accepted. Closed Christmas. &
Attendance: 54,000 (accurate)

THE ANN LOEB BRONFMAN GALLERY, 1529 16th St., N.W., Washington, DC 20036. Tel.: 202-518-9400 & 777-3208. Fax: 202-518-9420.
Web Site: washingtondcjcc.org
Key Personnel: Dir., Dafna Steinberg
Institution Type/Description: Art Gallery.
Collections: Jewish culture, heritage & art; paintings; photographs; artifacts; sculpture.
Activities: educational programs.
Hours & Admission Prices: Sun.-Thurs. 10-10, Fri. 10-4. No charge.

ARCHIVES OF AMERICAN ART, SMITHSONIAN INSTITUTION, 750 9th Street, N.W., Ste. 2200, Washington, DC 20001. Mailing Address: P.O. Box 37012, Victor Bldg., Ste. 2200, MRC 937, Washington, DC 20013-7012. Tel.: 202-633-7940. Fax: 202-633-7994.
E-mail: aaaemref@si.edu
Web Site: www.aaa.si.edu
Founded: 1954.
Key Personnel: Dir., John W. Smith; Asst. Dir. Operations, Jody Pettibone; Senior Cur., Elizabeth S. Kirwin; Chief Reference, Marisa Bourgoin; Chief Collections Processing, Barbara J. Aikens; Archives Catalog Database Mgr., Karen B. Weiss; Registrar, Susan Cary.
Personnel Profile: Full-Time Paid 38; Part-Time Paid 9; Part-Time Volunteers 4; Interns 3.
Governing Authority: bureau of the Smithsonian Institution, Washington, DC, which is a nonprofit federally chartered corporation & the board of trustees, Archives of American Art. Regional Centers: New York Center, 1285 Avenue of the Americas, New York, NY 10019; Washington Center, 750 9th St., N.W., Washington, DC 20560; Lawrence A. Fleischman Gallery at the Reynolds Centers, 8th & F Sts., N.W., Washington, DC 20013-7012. Parent Institution: Smithsonian Institution. Tax-exempt: 501(c)(3) & 170(b)(1)(A).
Institution Type/Description: Research Institution.
Collections: the personal & professional records of American artists, dealers, critics, curators & collectors; records of galleries, museums & art societies; oral history material; manuscripts.
Research Fields: American art history.
Facilities: microfilmed archives collection available in research & reading areas of branch offices in: Boston Public Library; New York; San Francisco M.H. deYoung Museum; Washington, DC; Los Angeles, Amon Carter Museum & through inter-library loan.
Activities: inter-museum loans; temporary exhibitions; occasional documentary exhibitions.
Publications: 10-vol. card catalogue of the manuscript collection; 1-vol. card catalogue of the Oral History Collection; quarterly journal, Archives of American Art Journal; books, Collection of Exhibition Catalogues; Directory of Resources; Checklist of the Collection; Arts in America: A Bibliography. guides: Art-related Archival Materials in the Philadelphia Region, 1984-1989 Survey; Inventory of the Records of the National Arts Club, 1898-1960; Art-related Archival Materials in the Chicago Area; Paris: A Guide to Archival Sources for American Art History; A Finding Aid to the Walter Pach Papers; A Finding Aid to the Rockwell Kent papers.
Hours & Admission Prices: Mon.-Fri. 9-5. No charge. Closed Federal holidays. &

Attendance: 555 (estimated)
Membership: Sustaining $65; Associate $125; Sponsor $250; Patron $500; Fellow $1,000; Benefactor $2,500; Chairman's Circle $5,000.

ART MUSEUM OF THE AMERICAS, OAS, (M), 201 18th St., N.W., Washington, DC 20006-5606. Mailing Address: 1889 F St., N.W., Washington, DC 20006-4401. Tel.: 202-458-6016 & 6019. Fax: 202-458-6021.
E-mail: artmus@oas.org
Web Site: www.museum.oas.org
Founded: 1976.
Key Personnel: Coord. Education, Adriana Opsina; Cur. Permanent Collection, Maria Leyva; Cur. Temporary Exhibits, Fabian Goncalves Borrega; Public & Media Rels., Gregory Svitil; Administrative, Charo Marroquin.
Governing Authority: nonprofit. Parent organization: Organization of American States. Tax-exempt.
Institution Type/Description: Latin American Contemporary Art Collection.
Collections: Latin American 20th-century art; paintings; sculpture; graphics; drawings; works by Latin American & Caribbean artists.
Research Fields: Latin American Art.
Facilities: library & archives available for research only on premises.
Activities: guided tours; lectures & seminars; films; inter-museum loan, permanent, temporary & traveling exhibitions; color slides & films on Latin American art & archaeology; documentaries on the lives of the artists of the Americas.
Publications: Audio-visual catalog; exhibition catalogs.
Hours & Admission Prices: Tues.-Sun. 10-5. No charge; donations accepted. Closed major holidays; Good Friday.
Membership: Individual $100; Couple $150.

ARTS CLUB OF WASHINGTON, 2017 I St., N.W., Washington, DC 20006-1804. Tel.: 202-331-7282. Fax: 202-857-3678.
E-mail: membership@artsclubofwashington.org
Web Site: www.artsclubofwashington.org
Founded: 1916.
Key Personnel: Pres. (V), June Hajjar; Vice Pres., Walter Burns; Gen. Mgr., Brennan Hurley; Gallery Mgr., William Owens; Business Mgr., Shelly Gardiner.
Personnel Profile: Full-Time Paid 2; Part-Time Paid 2.
Governing Authority: nonprofit corporation.
Institution Type/Description: Art Gallery: housed in 1802 home of President James Monroe.
Collections: Washington art.
Research Fields: local history.
Activities: promotion of cooperation among artists in all fields; student aid in the creative and performing arts.
Hours & Admission Prices: Sept.-July Tues.-Fri. 10-5, Sat. 10-2. No charge; donations accepted.

B'NAI B'RITH KLUTZNICK NATIONAL JEWISH MUSEUM, 2020 K St., N.W., 7th Fl., Washington, DC 20006-1806. Tel.: 202-857-6647. Fax: 202-857-6601.
E-mail: museum@bnaibrith.org
Web Site: www.bnaibrith.org
Founded: 1957.
Key Personnel: Cheryl Kempler.
Personnel Profile: Full-Time Paid 1.
Governing Authority: denominational group. Parent Institution: B'nai B'rith Henry Monsky Foundation. Tax-exempt: 501(c)(3).
Institution Type/Description: Jewish Heritage & Culture Museum.
Collections: pre-20th century Jewish ceremonial & folk art; Israeli archaeology; Jewish historical documents; contemporary Jewish art; B'nai B'rith artifacts.
Research Fields: Jewish ceremonial folk art; Jewish history; history of B'nai B'rith.
Activities: guided tours; permanent exhibitions; archival research.
Publications: permanent collection, In the Spirit of Tradition.
Hours & Admission Prices: Mon.-Thurs. 12-3 by appointment. No charge; donations accepted. Closed legal & major Jewish holidays. (All information subject to change; call to confirm) ♿

BUREAU OF ENGRAVING & PRINTING, 14th and C Sts., S.W., Washington, DC 20228. Tel.: 202-874-2330; 866-874-2330 (Toll Free).
Key Personnel: Dir., Larry R. Felix
Institution Type/Description: History Museum.

Collections: security documents including U.S. paper currency, U.S. passports, Homeland Security, military ID cards, Immigration & Naturalization certificates.
Activities: view paper currency production.
Hours & Admission Prices: Call for hours.

CIA MUSEUM, (M), Washington, DC 20505. Mailing Address: Central Intelligence Agency, Office of Public Affairs, Washington, DC 20505. Tel.: 703-482-0623. Fax: 703-482-1739.
Web Site: www.cia.gov
Institution Type/Description: Virtual History Museum.
Collections: CIA & national security history; clothing; equipment; weapons; insignia; memorabilia; photographs.
Hours & Admission Prices: Daily.

CHRISTIAN HEURICH HOUSE MUSEUM, (M), 1307 New Hampshire Ave., N.W., Ste. 300, Washington, DC 20036-1537. Tel.: 202-429-1894.
Institution Type/Description: Historic House Museum: housed in the home of local brewer Christian Heurich; built in 1894. Listed on the National Register of Historic Places.
Collections: local history & culture; period furnishings; personal artifacts; photographs.
Hours & Admission Prices: Thurs.-Fri. 11:30 & 1, Sat. 11:30, 1 & 2:30; other times by appointment. Suggested Donation: $5 per person.

＊　**THE CORCORAN GALLERY OF ART, (M),** 17th St. & New York Ave., N.W., Washington, DC 20006. Mailing Address: 500 17th St., N.W., Washington, DC 20006-4899. Tel.: 202-639-1700. Fax: 202-639-1779.
Web Site: www.corcoran.org
Founded: 1869.
Key Personnel: Dir. & Pres., Fred Bollerer; Chm. Bd., Harry F. Hopper; C.O.O., Lauren Garcia; Head Public Rels., Kristin Guiter; Mgr. Mktg., Laith Alnouri; Mgr. Special Events, Allie Gallo; Visitor & Membership Svcs., Kathleen Kane; Registrar, Nancy Swallow.
Personnel Profile: Full-Time Paid 184; Part-Time Paid 39; Part-Time Volunteers 250; Interns 25.
Governing Authority: nonprofit organization. Tax-exempt: 501(c)(3).
Institution Type/Description: Art Museum.
Collections: American paintings & sculptures from 18th-20th century; European paintings, sculptures & decorative arts; American and European drawings, prints & photographs; contemporary art.
Research Fields: William A. Clark collection.
Facilities: 35,000-vol. library of art & art history books available by appointment; archival repository: gallery records including exhibition records & correspondence with American artists available for research by appointment; restaurant. Museum-related items for sale.
Activities: guided tours; lectures; films; gallery talks; docent program; concerts; arts festivals; drama; formally organized education programs; permanent, temporary & traveling exhibitions.
Publications: exhibition catalogs.
Hours & Admission Prices: Wed. & Fri.-Sun. 10-5, Thurs. 10-9. Adults $10, senior citizens 62 & over and students with ID $8; children under 12 & members no charge. Closed New Year's Day; Christmas. ♿
Attendance: 350,000 (estimated)
Membership: Individual $60; Dual $90; Family $100; 1869 Society (additional $65); Supporting $160; Sponsoring $295; Contributing $500.

＊　**DAUGHTERS OF THE AMERICAN REVOLUTION MUSEUM,** 1776 D St., N.W., Washington, DC 20006-5392. Tel.: 202-879-3241. Fax: 202-628-0820.
E-mail: museum@dar.org
Web Site: www.dar.org/museum
Founded: 1890.
Key Personnel: Museum Dir. & Chief Cur., Diane L. Dunkley; Chm. (V), Beverly D. West; Cur. Education, Raina Boyd; Asst. Cur. Education, Kendall Casey; Cur. Collections, Olive Graffam; Cur. Furnishings, Patrick Sheary; Cur. Textiles & Costumes, Alden O'Brien; Collections Mgr., Anne Ruta; Assoc. Registrar, Stephanie Livingston; Museum Shop Mgr., Beverly Cihan.
Personnel Profile: Full-Time Paid 12; Part-Time Paid 1; Part-Time Volunteers 100; Interns 7.
Governing Authority: society. Parent Institution: National Society, Daughters of the American Revolution. Tax-exempt: 501(c)(3).
Institution Type/Description: Decorative Arts & History Museum: housed in 1904 Memorial Continental Hall.
Collections: decorative arts made or used in America in the pre-industrial

period; quilts; coverlets; needlework; costumes; ceramics; glass; silver; pewter; furniture; paintings; miniatures; musical instruments; toys; dolls; Revolutionary War artifacts; 31 American period rooms.
Research Fields: quilts, coverlets & needlework; American material culture, 1700-1840.
Facilities: 2,000-vol. library on decorative arts available for use on premises. Slides, postcards, notecards & handicrafts for sale.
Activities: guided tours; textile identification clinics; lectures; quilt workshops; docent program; inter-museum loan, permanent & temporary exhibitions; organized education programs for children, adults & undergraduate or graduate college students; internship program; costume workshop; summer camps.
Publications: First Flowerings: Early Virginia Quilts; Magnificent Intentions: Decorative Arts of the District of Columbia, 1791-1861; Souvenirs from the Voyage of Life; American Case Furniture 1680-1840: Selections from the DAR Museum Collection; Youth is the Time for Progress: The Importance of American Schoolgirl Art; Preserving the American Spirit at the DAR Museum.
Hours & Admission Prices: Museum: Mon.-Fri. 9:30-4, Sat. 9-5. Period Room Tours: Mon.-Fri. 10-3, Sat. 9-4:30. Call to confirm hours. No charge; donations accepted. Closed major holidays. &
Attendance: 27,000 (estimated)
Membership: Regular $10; Sustaining $25; Sponsor $50; Life & Memorial Tribute $200; Patron $500; Benefactor $5,000.

DC FIRE & EMS MUSEUM, 439 New Jersey Ave., N.W., Washington, DC 20001. Tel.: 202-673-1709; 202-439-1936 (cell).
Key Personnel: Exec. Dir., Walter Gold; Cur. & Museum Mgr., Mark Tennyson
Governing Authority: Tax-exempt.
Institution Type/Description: Firefighting History Museum: housed in the former fire station; built in 1916.
Collections: firefighting history & equipment; books; photographs; hand-made fire bucket.
Facilities: 4,000 sq. ft. exhibition space.
Activities: group tours.
Hours & Admission Prices: Tues.-Sat. by appointment. No charge. &
Attendance: 1,000 (accurate)

DECATUR HOUSE MUSEUM, 748 Jackson Pl., N.W., Washington, DC 20006-4912. Mailing Address: 1610 H St., N.W., Washington, DC 20006-4907. Tel.: 202-842-0920. Fax: 202-842-0030.
E-mail: decatur_house@nthp.org
Web Site: www.decaturhouse.org
Founded: 1956.
Key Personnel: Exec. Dir., Cynthia B. Malinick; Chm. (V), Thomas R. Pickering; Dir. Special Events, Arioth Harrison; Dir. Public Rels. & Mktg., Mame Croze; Museum Shop Mgr., Rosemary Rudd Cohen.
Personnel Profile: Full-Time Paid 6; Part-Time Paid 21; Part-Time Volunteers 3; Interns 4.
Governing Authority: nonprofit organization. Parent Institution: National Trust for Historic Preservation, 1785 Massachusetts Ave., N.W., Washington, DC 20036. Tax-exempt: 501(c)(3).
Institution Type/Description: Historic House Museum: 1819 Commodore Stephen Decatur House, designed by B. Latrobe.
Collections: 18th-, 19th- & 20th-century decorative arts.
Research Fields: residents of house; political & social history of Washington, D.C.; Early 19th Century decorative Arts.
Facilities: rental facilities available. Books & gift items for sale.
Activities: guided tours; lectures; temporary exhibitions; school programs; special events; walking tours.
Publications: Jackson Place Journal.
Hours & Admission Prices: Tues.-Sat. 10-5, Sun. 12-4. Suggested Donation: adult $5. Closed New Year's Day; Thanksgiving; Christmas. &
Attendance: 25,000 (estimated)
Membership: Decatur House Friends: Palladian Society $50; Jackson Place Society $100; Beale Society $250; Decatur Society $500; Lafayette Square Associates $1,000; Latrobe Society $5,000 & up.

DEPARTMENT OF THE TREASURY, Office of the Curator, Rm. 1225, 15th & Pennsylvania N.W., Dept. of the Treasury, Washington, DC 20220. Mailing Address: 1500 Pennsylvania Ave., N.W., Washington, DC 20220-0002. Tel.: 202-622-1250. Fax: 202-622-2294.
E-mail: richard.cote@do.treas.gov
Web Site: www.ustreas.gov/curator
Key Personnel: Cur., Richard Cote; Tour Coord., Mary Edwards; Preservation Specialist, Guy Munsch.
Personnel Profile: Full-Time Paid 4; Part-Time Volunteers 25; Interns 2.

Governing Authority: federal government.
Institution Type/Description: Architecture Museum: 1836-1869 U.S. Treasury Building.
Collections: 19th & 20th-century office furniture; original artwork; portraits of former Secretaries of the Treasury; architectural fragments.
Hours & Admission Prices: Tours: Sat. 9, 9:45, 10:30 & 11:15 by appointment only through your congressional office. &

DIPLOMATIC RECEPTION ROOMS, DEPARTMENT OF STATE, (M), M/FA, Rm. 8213, 2201 C St., N.W., Washington, DC 20520. Tel.: 202-647-1990. Fax: 202-647-3428.
Web Site: www.state.gov/m/drr
Founded: 1961.
Key Personnel: Dir. & Cur., Marcee F. Craighill; Mgr. Collections, Virginia B. Hart; Project Coord., Brianne Brophy; Program Operations Specialist, Jessica A. Wallace.
Governing Authority: Federal. Parent Institution: Dept. of State. Tax-exempt.
Institution Type/Description: National Agency.
Collections: American period furnishings; paintings; decorative period arts.
Research Fields: American furniture; paintings of 18th & 19th century; Chinese Export porcelain; American silver; oriental rugs.
Activities: guided tours.
Publications: Becoming A Nation; Treasures of State.
Hours & Admission Prices: By appointment: Mon.-Fri. 9:30, 10:30, 2:45. Fine Arts Tour: recommended age is 12 & up; photo ID required for adults. No charge; donations accepted. For reservations call or fax 90 days in advance: Department of State, Washington, DC 20520. Tel.: 202-647-3241. Web: https://receptiontours.state.gov. TDD: 202-736-4474; Fax: 202-736-4232. &
Membership: Contributing $500 & up; Sustaining $1,000 & up; Patron $5,000 & up; Sponsors $10,000 & up; Major Sponsors $25,000 & up; Benefactor $50,000 & up; Major Benefactors $100,000 & up; Philanthropists $250,000 & up; Major Philanthropists $500,000 & up; Grand Patrons $1,000,000.

DISTRICT OF COLUMBIA ARTS CENTER, 2438 18th St., N.W., Washington, DC 20009-2004. Tel.: 202-462-7833. Fax: 419-821-9622. Facebook: District of Columbia Arts Center.
E-mail: info@dcartcenter.org
Web Site: www.dcartscenter.org
Founded: 1989.
Key Personnel: Dir., B. Stanley; Chm. (V), Bruce Kogod; Pres. (V), Jay Bothwell.
Personnel Profile: Full-Time Paid 1; Part-Time Paid 6; Part-Time Volunteers 4; Interns 1.
Governing Authority: Tax-exempt.
Institution Type/Description: Visual & Performing Arts Center.
Collections: works by local artists.
Activities: theater performances; special events.
Hours & Admission Prices: Wed.-Sun. 2-7 & during theater performances. No charge.
Attendance: 4,000 (estimated)

＊ **DUMBARTON HOUSE, (M),** 2715 Q St., N.W., Washington, DC 20007-3071. Tel.: 202-337-2288. Fax: 202-337-0348.
E-mail: info@dumbartonhouse.org
Web Site: www.dumbartonhouse.org
Founded: 1932.
Congressional District: 1
Key Personnel: Chm. (V), Anna F. Duff; Pres. (V), Marcy M. Moody; Exec. Dir., Karen L. Daly; Museum Cur., Scott S. Scholz; Dir. Education, Kanani Hoopai.
Personnel Profile: Full-Time Paid 7; Part-Time Paid 30; Part-Time Volunteers 46; Interns 6.
Governing Authority: society. Parent Institution: The National Society of The Colonial Dames of America (NSCDA). Tax-exempt.
Institution Type/Description: Historic House: c.1800.
Collections: federal period furnishings & decorative arts; Nourse & Morris family papers.
Research Fields: Federal period decorative arts; Georgetown, DC history; Washington, DC history; Nourse family; Anthony Morris family; United States Treasury.
Facilities: 120-seat banquet room.
Activities: tours; lectures; special events; concerts; school programs; scout programs; public programs; private events; rental facilities.
Publications: newsletters; calendar; exhibition catalogues.
Hours & Admission Prices: Tues.-Sun. 11-3. Adults $5; discounts to AAA members; AAM, NSCDA & museum members, school groups & students w/ID no charge. Closed Christmas Eve; most national holidays. &
Attendance: 18,000 (estimated)

Membership: Individual $30; Dual $50; Household $75; Supporting $100; Patron $250; Benefactor $500; Sustaining $1,000.

DUMBARTON OAKS RESEARCH LIBRARY & COLLECTION, (M), 1703 32nd St., N.W., Washington, DC 20007-2961. Tel.: 202-339-6414. Fax: 202-339-6419.
E-mail: dumbartonoaks@doaks.org
Web Site: www.doaks.org
Founded: 1940.
Key Personnel: Dir., Prof. Jan Ziolkowski; Dir. Pre-Columbian Studies, Colin McEwan; Dir. Studies in Landscape Architecture, John Beardsley; Dir. Byzantine Studies, Margaret Mullet; Dir. Museum, Gudrun Buehl; Mgr. Image Collection & Fieldwork Archives, Shalimar Fojas White; Mgr. House Collection & Archivist, James Carder; Dir. Gardens & Grounds, Gail Griffin; Museum Shop Mgr., Patti Sheer.
Personnel Profile: Full-Time Paid 100; Part-Time Paid 27; Part-Time Volunteers 41; Interns 20.
Governing Authority: Trustees for Harvard University. Parent Institution: Harvard University. Tax-exempt.
Institution Type/Description: Library & Art Museum.
Collections: Byzantine; Pre-Columbian artifacts; some European & American paintings, sculpture & decorative arts; rare books & manuscripts in garden history & history of landscape architecture.
Major Exhibits: Connecting Collections: Collecting Connections - 50 Years of Pre-Columbian Art at Dumbarton Oaks, 12/12-1/14.
Research Fields: Byzantine; Pre-Columbian; studies in landscape architecture & garden history.
Facilities: 125,150-vol. library on Byzantine research, pre-Columbian studies, & gardens available to accredited scholars by special permission; gardens. Handbooks of the collections & other museum-related items for sale.
Activities: public lectures; permanent exhibitions; conferences; fellowships; concerts.
Publications: handbook, The Byzantine Collection; handbook, The Robert Woods Bliss Collection of Pre-Columbian Art; monographs; conference proceedings; catalogues; guidebook, Dumbarton Oaks: The Collections.
Hours & Admission Prices: Museum: Tues.-Sun. 2-5. No charge. Gardens: April-Oct. Tues.-Sun. 2-6; Nov.-March Tues.-Sun. 2-5. Adults $8, children & senior citizens $5. Closed national holidays. &
Attendance: 38,000 (estimated)

FEDERAL RESERVE BOARD, (M), 20th & C Sts., N.W., Washington, DC 20551. Tel.: 202-452-3778. Fax: 202-736-5680.
E-mail: finearts@frb.gov
Web Site: www.federalreserve.gov/finearts
Founded: 1975.
Key Personnel: Dir., Stephen Bennett Phillips; Chm. (V), Leatrice Eagle; Collections Asst., Rhonda Gray-Young; Fine Arts Program Asst., Nicolette Pisha.
Governing Authority: federal government. Tax-exempt: 170(b)(1)(A).
Institution Type/Description: Art Gallery: housed in a 1935-37 building by Paul Philippe Cret.
Collections: 19th- & 20th-century American and European paintings, prints & works on paper.
Activities: loan exhibitions.
Publications: exhibition brochures.
Hours & Admission Prices: Mon.-Fri. 10-3:30 by appointment. Please call 202-452-3778 or Fax: 202-736-5680. No charge. Closed federal holidays. &

FLASHPOINT, 916 G St., N.W., Washington, DC 20001-4565. Tel.: 202-315-1305. Fax: 202-315-1303.
Web Site: www.culturaldc.org
Key Personnel: Interim Exec. Dir., Travis Bowerman
Institution Type/Description: Art Gallery.
Collections: works by contemporary artists.
Facilities: 75 seat theatre.
Activities: educational programs.
Hours & Admission Prices: Call for hours. &

FOLGER SHAKESPEARE LIBRARY, 201 E. Capitol St., S.E., Washington, DC 20003-1094. Tel.: 202-544-4600. Fax: 202-544-4623.
E-mail: webmaster@folger.edu
Web Site: www.folger.edu
Founded: 1932.
Congressional District: 1
Key Personnel: Dir., Michael Witmore; Chm. (V), Paul T. Ruxin; Dir. Research, Dr. David Schalkwyk; Librarian, Dr. Stephen Enniss; Reference Librarian, Dr. Georgianna Ziegler; Dir. Public Programs, Janet Griffin; Dir.

Devel., Essence Newhoff; Museum Shop Mgr., Barbara Jacoby; Controller, Howard Parks; Head External Rels., Garland Scott.
Personnel Profile: Full-Time Paid 100; Part-Time Volunteers 45.
Governing Authority: Trustees of Amherst College. Tax-exempt: 501(c)(3).
Institution Type/Description: Private Independent Research Library.
Collections: rare books & manuscripts of 15th-18th century on continental & English Renaissance; Shakespeare; theater history from Middle Ages to 20th century; theater memorabilia.
Research Fields: Renaissance; Shakespeare; history of drama.
Facilities: 310,000-vol. library of printed books & 55,000 manuscripts on Renaissance civilization of England & the continent available for use on premises by scholars & graduate students completing doctoral dissertations; reading room; 250-seat Elizabethan theatre. Books, maps, postcards, CDs & DVDs for sale.
Activities: guided tours; public readings & lectures; gallery talks; concerts; seminars; poetry readings; docent program; theatrical performances; drama; education programs; exhibitions.
Publications: magazine, Folger; Shakespeare Quarterly; Folger Library Shakespeare Editions; exhibition catalogs.
Hours & Admission Prices: Library: Mon.-Fri. 8:45-4:45, Sat. 9-12 & 1-4:30. Exhibits Gallery: Mon.-Sat. 10-5, Sun. 12-5. No charge; donations accepted. Closed federal holidays. &
Attendance: 200,000 (estimated)
Membership: Friends of the Folger: For information call 202-675-0359.

FONDO DEL SOL VISUAL ARTS & MEDIA CENTER/EL MUSEO DE CULTURAS Y HERENCIAS AMERICANAS/MOCHA, (M), 2112 R St., N.W., Washington, DC 20008-1932. Tel.: 202-265-9235. Fax: 202-265-1045.
E-mail: fondodelsol@earthlink.net
Web Site: www.fondodelsol.org
Founded: 1973.
Key Personnel: Dir. & Chief Cur., W. Marc Zuver; Co Chm. (Jamaica), A. Michael Auld; Co Chm. (Cuba), Osvaldo Mesa; Co Chm., Dr. Floyd Coleman; Deputy Dir., Education & Website, Alan Urtecho; Dir. Communications & Publications, Sanne Tikjoeb; Asst. Dir. (Peru), Manuel Pereira.
Personnel Profile: Full-Time Paid 1; Part-Time Paid 6; Part-Time Volunteers 2; Interns 2.
Governing Authority: nonprofit. Tax-exempt: 501(c)(3) & 509.
Institution Type/Description: Art Museum & Media Center.
Collections: pre-Columbian & Hispanic Santero collection; contemporary Latino, Chicano & Puerto Rican art.
Research Fields: contemporary Hispanic, Chicano, Puerto Rican, & Latin American art; Native American, Caribbean & Afro American art.
Facilities: film & video library pertaining to Hispanic artists in the U.S.
Activities: guided tours; lectures; films & videos; concerts; arts festivals; TV & radio programs; organized programs for children; internship program; loan, temporary & traveling exhibitions; mobile vans; film & video services for museums & private groups.
Publications: monthly brochures & catalogues for exhibits; monthly program notes for video & media programs.
Hours & Admission Prices: Wed.-Sat. 1-5. Suggested Donation: $3; members & children no charge. Closed major holidays. &
Attendance: 50,000
Membership: Mailing $30; Assistant Sponsor $50; Sponsor $100; Patron $500; Benefactor & Corporate $1,000-$5,000.

FORD'S THEATRE NATIONAL HISTORIC SITE (LINCOLN MUSEUM), (M), 511 10th St., N.W., Washington, DC 20004. Tel.: 202-426-6924. Fax: 202-426-1845. TDD: 202-426-1749.
Web Site: www.nps.gov/foth/
Founded: 1933.
Key Personnel: Site Mgr., Rae Emerson; Asst. Site Mgr., Claudia Anderson; Supt., Kay Fielder.
Personnel Profile: Full-Time Paid 20; Part-Time Volunteers 15.
Governing Authority: federal. Parent Institution: National Park Service. Subsidiary Institution: National Capital Parks-Central, 900 Ohio Dr., S.W., Washington, DC 20242. Tax-exempt.
Institution Type/Description: Historic Site & History Museum.
Collections: Lincoln memorabilia; period furnishings; Oldroyd collection of Lincoln artifacts; John Wilkes Booth gun. Historic Buildings: 1849 Petersen House, where Lincoln died; 1863 Ford's Theatre.
Research Fields: Lincoln's life; his assassination; Civil War period.
Facilities: Books, pamphlets, maps & posters for sale.
Activities: lectures; self-guided tours; permanent exhibitions.
Hours & Admission Prices: Daily 9-5. No charge. Closed Christmas.
Attendance: 1,029,000 (accurate)

* **FREDERICK DOUGLASS NATIONAL HISTORIC SITE,** 1411 W Street, S.E., Washington, DC 20020-4813. Mailing Address: c/o National Parks-East, 1411 W. St., S.E., Washington, DC 20020. Tel.: 202-426-5961. Fax: 202-426-0880. TDD: 202-540-9217.
E-mail: kamal_mcclarin@nps.gov
Web Site: www.cr.nps.gov/csd/exhibits/douglass
Founded: 1916.
Key Personnel: Rgnl. Dir., Peggy O'Dell; Dir., Jon Jarvis; Park Supt., Alex Romero; Cur., Ka'mal McClarin.
Personnel Profile: Full-Time Paid 6; Part-Time Volunteers 3.
Governing Authority: federal. Parent Institution: National Capital Parks-East, National Park Services, U.S. Dept. of the Interior, 1900 Anacostia Dr., S.E., Washington DC. Tax-exempt.
Institution Type/Description: Historic Site.
Collections: films; furnishings, documents & personal artifacts of Frederick Douglass.
Major Exhibits: Douglass & Nature, 2/14; Douglass & Travel, 2/15; Douglass & John Brown, 2/16.
Research Fields: writings, letters, documents related to the activities of Douglass.
Facilities: library of historic records available by appointment; visitor center. Museum-related items for sale.
Activities: guided tours; films; special interpretive programs.
Publications: interpretive pamphlets.
Hours & Admission Prices: April-Oct. daily 9-5; Nov.-March daily 9-4. Adults $3, senior citizens $1.50; children under 6 & school groups with approved waiver no charge. Closed New Year's Day; Thanksgiving; Christmas. &
Attendance: 33,000 (accurate)

* **FREER AND SACKLER GALLERIES OF ART, (M),** 1050 Independence Ave., S.W., Washington, DC 20013-7012. Mailing Address: MRC 707, P.O. Box 37012, Washington, DC 20013-7012. Tel.: 202-633-4880. Fax: 202-357-4911. TDD: 202-786-2374.
E-mail: publicaffairsasia@si.edu
Web Site: www.asia.si.edu
Founded: 1906.
Key Personnel: Dir., Dr. Julian Raby; Exhibitions Coord., Cheryl Sobas; Deputy Dir., Dr. James Ulak; Assoc. Dir. & Cur. Ancient China, Keith Wilson; Assoc. Cur. Chinese Art, Joseph Chang; Cur. Ceramics, Louise Cort; Chief Cur. & Cur. Islamic Art, Dr. Massumeh Farhad; Sr. Assoc. Cur., Ann Yonemura; Assoc. Cur., Lee Glazer; Research Spec. Chinese Literature & History, Stephen Allee; Head Librarian, Reiko Yoshimura; Head Collection Management, Elizabeth Duley; Research Assoc. (Nepalese/Tibetan Art), Dr. Mary Slusser; Mgr. Shops, Peter Musolino; Head, Imaging & Photographic Svcs., John Tsantes; Head Digital Publications, Web & Digital Media, Karen Sasaki; Head Finance & Administration, Pat Kennedy Graham; Dir. External Affairs, Katie Ziglar.
Personnel Profile: Full-Time Paid 141; Part-Time Volunteers 20; Interns 25.
Governing Authority: federal; nonprofit corporation. Parent Institution: Smithsonian Institution, Washington, DC. Tax-exempt: 501(c)(3) & 170(b)(1)(A).
Institution Type/Description: Asian Art Gallery.
Collections: Chinese, Japanese, Korean, Islamic, Ancient Near Eastern art; South & Southeast Asian bronze; jade; sculpture; painting; lacquer; pottery; porcelain; manuscripts; metalwork & glass; works by 19th & early 20th century American artists including James McNeill Whistler; slides; photographs; archives includes Charles Lang Freer papers, Bishop papers, Pope papers, Empress Dowager Cizi photographs, Ernst Herzfeld papers, Myron Bement Smith collection, Dwight William try on papers, Freer-Whistler correspondence.
Research Fields: Asian art history; 19th & early 20th century American art.
Facilities: 80,000-vol. library of books, pamphlets & periodicals half of which are in Chinese & Japanese pertaining to collections available for research in reading room; archives; 300-seat auditorium. Museum-related items for sale.
Activities: guided tours; lectures; gallery talks; permanent & temporary exhibitions; films; concerts; educational programs; after hours events.
Publications: Ars Orientalis; 2 series of monographs, Freer Gallery of Art Oriental Studies; Freer Gallery of Art Occasional Papers; exhibition catalogs; various booklets & pamphlets.
Hours & Admission Prices: Daily 10-5:30. No charge. Closed Christmas. &
Attendance: 723,868 (accurate)
Membership: Friends of the Freer & Sackler Galleries: Patrons' Circle $1,200; Directors' Circle $3,000; Founders' Circle $5,000; Sponsors' Circle $10,000.

GEORGETOWN UNIVERSITY ART COLLECTION, Georgetown University, Healy Hall, Room #107, 3700 O St., N.W., Washington, DC 20057-1174. Mailing Address: Georgetown Univ., Lauinger Library, Special Collections, 5th Fl., 3700 O St., N.W., Washington, DC 20057-1174. Tel.: 202-687-1469. Fax: 202-687-7501.
E-mail: artcollection@georgetown.edu
Web Site: www.library.georgetown.edu/dept/speccoll/guac/
Founded: 1789.
Key Personnel: Cur. Art, LuLen Walker; Asst. Cur., Christen Runge; Head Special Collections, John Buchtel.
Personnel Profile: Full-Time Paid 2; Part-Time Paid 2.
Governing Authority: university. Parent Institution: Georgetown University. Tax-exempt: 501(c)(3).
Institution Type/Description: Art and History Museum: housed in 1879 Healy Hall on the Georgetown University campus.
Collections: historical objects; works by Van Dyck & Luca Giordano; paintings; sculpture; graphics; American portraits; religious objects; decorative arts; period rooms.
Research Fields: historical objects; paintings; sculpture.
Activities: gallery talks; formally organized education programs for undergraduate college students; permanent & temporary exhibitions.
Publications: catalog of the collection.
Hours & Admission Prices: Check website for hours. No charge. &

HERITAGE PRESERVATION, 1012 14th St., N.W., Ste. 1200, Washington, DC 20005-3408. Tel.: 202-233-0800. Fax: 202-233-0807.
E-mail: lreger@heritagepreservation.org
Web Site: www.heritagepreservation.org
Founded: 1979.
Key Personnel: C.E.O. & Pres., Lawrence Reger; Chm. (V), Merv Richard; Treas., Tom Klamer; Vice Pres. Emergency Programs, Lori Foley; Dir. Heritage Health Index II, Lesley Langa; Coord. Conservation Assessment Program, Teresa Martinez.
Personnel Profile: Full-Time Paid 7.
Governing Authority: private; nonprofit organization. Tax-exempt: 501(c)(3).
Institution Type/Description: Heritage Preservation.
Research Fields: Conservation, preservation, emergency preparedness.
Publications: Caring For Your Collections; Caring For Your Historic House; Caring For Your Family Treasures; Heritage Preservation Update; SOS! Update; Capabilities; Field Guide to Emergency Response; The Emergency Response and Salvage Wheel; A Public Trust At Risk: The Heritage Health Index Report on the State of America's Collections.
Hours & Admission Prices: Mon.-Fri. 9-5. Closed New Year's Day; Martin Luther King Jr. Day; Presidents' Day; Memorial Day; Independence Day; Labor Day; Columbus Day; Veterans Day; Thanksgiving; Christmas. &
Membership: Member $150; Supporting $500; Sustaining $1,000; Benefactor $1,001 & up.

* **HILLWOOD ESTATE, MUSEUM & GARDENS, (M),** 4155 Linnean Ave., N.W., Washington, DC 20008-3806. Tel.: 202-686-8500. Fax: 202-966-7846.
E-mail: kphelan@hillwoodmuseum.org
Web Site: www.hillwoodmuseum.org
Founded: 1976.
Key Personnel: C.E.O. & Exec. Dir., Kate Markert; Chm. (V) & Pres. (V) Bd., Ellen MacNeille Charles; C.O.O. & Dir. Interpretation & Visitor Svcs., Angie Dodson; Dir. Collections & Chief Cur., Liana Paredes; Head Exhibitions & Collections Management, Lawrence Waung; Assoc. Cur. Russian & Eastern European Art, Scott Ruby; Cur. American Material Culture & Historian, Estella Chung; Asst. Cur. Costumes & Textiles, Howard Kurtz; Chief Art Librarian & Archives Mgr., Kristen Regina; Head Interpretation, Audra Kelly; Mgr. Youth Audiences, Rebecca Singer; Mgr. Adult Audiences, Erin Lourie; Interpretation Volunteer Mgr., Lisa Leyh; Horticulture Volunteer Coord. & Horticulturist, Bill Johnson; Dir. Mktg. & Communications, Lynn Rossotti; Head Visitor Svcs., Michael Kruelle; Mgr. Visitor Svcs., Arthur Kim; Group Tours & Special Events Coord., Stephen Stuart; Head Merchandising, Lauren Salazar; Dir. Human Resources, Nancy Brown; Dir. Devel., Joan Wetmore; Dir. Finance & Admin. and C.F.O., Madge Minor; Dir. Facilities, Don Rogers; Dir. Horticulture, Brian Barr.
Personnel Profile: Full-Time Paid 63; Part-Time Paid 2; Part-Time Volunteers 277; Interns 1.
Governing Authority: nonprofit organization. Administered by Hillwood Museum & Gardens Foundation. Tax-exempt: 501(c)(3).
Institution Type/Description: Decorative Art Museum: housed in Washington

residence of Marjorie Merriweather Post, founder of Hillwood Museum & Gardens Foundation.

Collections: Russian & French decorative arts; Russian & French books, paintings, sculptures & prints; Russian icons; 18th-century French furniture & porcelain; East Asian works of art; memorabilia & furnishings assembled by Mrs. Post; European paintings & prints; American decorative & fine arts; American material culture; botanical collections; gardens; orchids.

Research Fields: Russian decorative arts; European, fine & decorative arts; East Asian works of art; American decorative arts, paintings, & material culture; 20th-century interior design; 20th-century American history; historic landscape preservation; horticultural sciences.

Facilities: 32,000-vol. library relating to Russian & French decorative arts including rare books, sales catalogues, periodicals & horticultural publications; archives; visitor center; cafe; 13 acres of formal gardens; theater. Museum-related items for sale.

Activities: guided, special interest & audio tours; public programs for adults & children.

Publications: visitor's companion book, Hillwood Museum & Gardens: Marjorie Merriweather Post's Art Collector's Personal Museum; A Taste for Splendor: Russian Imperial & European Treasures from the Hillwood Museum (catalogue of collection); Hillwood Collection Series: Faberge at Hillwood; Russian Icons at Hillwood; Sevres Porcelain at Hillwood; Russian Imperial Porcelain at Hillwood; Russian Glass at Hillwood; French Furniture in the Hillwood Museum Collection; Tradition in Transition: Russian Icons in the Age of the Romanovs; Hillwood: Thirty Years of Collecting 1977-2007; Treasures into Tractors: The Selling of Russia's Cultural Heritage, 1918-1938; Sevres Then and Now: Tradition and Innovation in Porcelain, 1750-2000; Russian Silver in America: Surviving the Melting Pot.

Hours & Admission Prices: Museum: Feb.-Dec. Tues.-Sat. 10-5. Office: Mon.-Fri. 9-5. Estate Donation: adults $15, seniors $12, students $10, children 6-18 $5; discounts to AAM & ICOM members. Closed national holidays. &

Attendance: 64,199 (accurate)

Membership: Individual Supporter $50; Family Supporter $75; Individual Contributor $100; Family Contributor $150; Patron $250; Donor $500; Sponsor $1,000; Collector's Circle & Corporate Neighbor $2,500; Corporate Supporter $5,000; Corporate Sponsor $10,000; Corporate Partner $25,000.

HIRSHHORN MUSEUM AND SCULPTURE GARDEN, SMITHSONIAN INSTITUTION, Seventh St. & Independence Ave., S.W., Washington, DC 20560. Mailing Address: P.O. Box 37012/HMSG MRC 350, Washington, DC 20013-7012. Tel.: 202-633-4674. Fax: 202-786-2682. TDD: 202-357-1729.

E-mail: sawyerd@si.edu

Web Site: hirshhorn.si.edu

Founded: 1966.

Key Personnel: Exec. Dir., Richard Koshalek; Dir. Art & Public Programs, Kerry Brougher; Dir. External Affairs, Gabriel Riera; Program Dir., Milena Kalinovska; Registrar, Barbara Freund; Librarian, Anna Brooke; Chief Conservator, Susan Lake; Chief Exhibits & Design, Al Masino; Security Chief, Adolf Smith; Bldg. Mgr., Fletcher Johnston; Communications & Mktg. Specialist, Erin Baysden; Publications Mgr., Vanessa Mallory; Museum Shop Mgr., Tina Mallett; Dir. Devel., Kevin Chrysler.

Personnel Profile: Full-Time Paid 61; Part-Time Volunteers 100; Interns 33.

Governing Authority: nonprofit organization. Parent Institution: Smithsonian Institution. Tax-exempt.

Institution Type/Description: Art Museum.

Collections: international modern & contemporary art; 19th- & 20th-century sculpture.

Research Fields: 19th- & 20th-century international paintings, sculpture & graphic art; contemporary art.

Facilities: 43,000-vol. library of art books & catalogs available to scholars by appointment on museum premises; 280-seat auditorium.

Activities: guided tours; lectures; films; gallery talks; concerts; education programs for children, adults, undergraduate & graduate college students; docent program; inter-museum loan, permanent, temporary & traveling exhibitions.

Publications: exhibition catalogues & brochures; gallery handouts; seasonal calendars of events.

Hours & Admission Prices: Daily 10-5:30. No charge; donations accepted. Closed Christmas. &

Attendance: 690,544 (accurate)

Membership: Associates' Circle $100-$249; Contributor's Circle $250-$499; Inner Circle $500-$999; Friend's Circle $1,000-$2,499; Curator's Circle $2,500-$4,999; Donor's Circle $5,000-$9,999; Benefactor's Circle $10,000-$24,999; Director's Circle $25,000 & up.

HISTORICAL SOCIETY OF WASHINGTON, DC, 801 K St., N.W., (@ Mt. Vernon Sq.), Washington, DC 20001-3746. Tel.: 202-383-1850 & 249-3955. Fax: 202-417-3823.

E-mail: info@historydc.org

Web Site: www.historydc.org

Key Personnel: Exec. Dir., Sandy Bellamy

Institution Type/Description: Historical Society Museum.

Collections: history; culture; government documents; photographs; books; videos; personal artifacts.

Facilities: library.

Activities: permanent & temporary exhibitions.

Hours & Admission Prices: Library: Mon.-Thurs. by appointment. Exhibit: Mon. & Wed.-Thurs. 10-4. No charge. Closed New Year's Day; Thanksgiving; Christmas.

HOUSE OF THE TEMPLE, 1733 16th St., N.W., Washington, DC 20009-3103. Tel.: 202-232-3579. Fax: 202-464-0487.

E-mail: hcalloway@scottishrite.org

Web Site: www.scottishrite.org

Formerly: The Supreme Council

Founded: 1801.

Key Personnel: C.E.O., Ronald A. Seale; Librarian, Joan Sansbury; Archivist, Art de Hoyos; Asst. Librarian, Larissa Watkins; Cur., Heather K. Calloway; Museum Shop Mgr., Morgan Corr.

Personnel Profile: Full-Time Paid 3; Part-Time Paid 1.

Governing Authority: private; nonprofit organization. Parent Institution: Scottish Rite of Freemasonry. Tax-exempt: 501(c)(3).

Institution Type/Description: Masonic/Fraternal Museum.

Collections: Masonic history with concentration on the Scottish Rite; Burl Ives collection; Albert Pike Americanism.

Research Fields: Freemasonry; fraternal history.

Facilities: 267,000-vol. library.

Activities: guided tours.

Publications: The Scottish Rite Journal; books.

Hours & Admission Prices: Mon.-Thurs. 10-4. Adults 18 & over $8, students 18 & over and seniors 60 & over $3; discounts to groups of 10 or more; members no charge. Closed Federal holidays.

Attendance: 15,000 (estimated)

HOWARD UNIVERSITY GALLERY OF ART, 2455 6th St., N.W., Fine Arts Bldg., Rm. 1025, Washington, DC 20059. Tel.: 202-806-7070. Fax: 202-806-6503.

Web Site: www.art.howard.edu

Founded: 1928.

Key Personnel: Dir., Dr. Tritobia H. Benjamin; Asst. Dir., Scott Baker; Registrar, Eileen Johnston.

Governing Authority: federal. Parent Institution: Howard University. Tax-exempt.

Institution Type/Description: Art Gallery.

Collections: African-American & American painting, sculpture, graphic art; Alain LeRoy Locke African collection; European graphic art; Samuel H. Kress study collection of Italian paintings and sculpture; Irving Gumbel prints.

Research Fields: African-American art.

Activities: inter-museum loan; temporary, permanent & traveling exhibitions.

Publications: catalogs.

Hours & Admission Prices: Winter: Mon.-Fri. 9:30-5, Sun. 12-4. Summer: Mon.-Fri. 9:30-4:30, Sun. call for hours. No charge; donations accepted. Closed national holidays. &

HOWARD UNIVERSITY MUSEUM, MOORLAND SPINGARN RESEARCH CENTER, 500 Howard Pl., N.W., Rm. 129, Washington, DC 20059. Tel.: 202-806-7239. Fax: 202-806-6405.

Web Site: www.founders.howard.edu/moorland-spingarn

Founded: 1914.

Governing Authority: Parent Institution: Howard Univ., 2400 6th St. N.W., Washington DC 20059. Tax-exempt.

Institution Type/Description: Black History Museum.

Collections: African artifacts; historic materials relative to the Black experience.

Research Fields: all areas related to Black history.

Facilities: library & research center containing manuscripts, photographs, tapes, microforms, periodicals & other documents relating to the Black experience, available for use on premises. Museum-related items for sale.

Activities: guided tours; lectures; temporary & permanent exhibitions.

Publications: exhibition catalogs

Hours & Admission Prices: Mon.-Fri. 9-4:30. No charge. &

INDIAN ARTS AND CRAFTS BOARD, 1849 C St., N.W., Rm. 2528, MIB, U.S. Dept. Interior, Washington, DC 20240. Tel.: 202-208-3773. Fax: 202-208-5196.
E-mail: iacb@ios.doi.gov
Web Site: www.iacb.doi.gov
Founded: 1935.
Key Personnel: Dir., Meridith Stanton.
Governing Authority: federal. Affiliated with the U.S. Department of the Interior. Branch Museums: Museum of the Plains Indian, Browning, MT; Southern Plains Indian Museum, Anadarko, Okla.; Sioux Indian Museum, Rapid City, SD. Tax-exempt.
Institution Type/Description: American Indian & Alaska Native Arts Museum.
Collections: contemporary American Indian & Alaska Native arts of the U.S.
Research Fields: pertaining to collection.
Publications: Source Directory of Native American owned & operated arts & crafts businesses; brochures, Know The Law Brochure.
Hours & Admission Prices: Mon.-Fri. 7:45-4:15. No charge. Closed national holidays. &

INTERNATIONAL SPY MUSEUM, (M), 800 F St., N.W., Washington, DC 20004-1505. Tel.: 202-393-7798. Fax: 202-393-7797.
E-mail: aabrell@spymuseum.org
Web Site: www.spymuseum.org
Founded: 2002.
Key Personnel: Exec. Dir., E. Peter Earnest; Media Rels., Amanda Abrell.
Governing Authority: private.
Institution Type/Description: Spy Museum.
Collections: over 600 artifacts; tradecraft, history & contemporary role of espionage; international espionage artifacts; espionage tools & techniques; lipstick pistol; Enigma cipher machine; counterfeit currency; disguised weapons; miniature cameras; concealment devices for weapons; radio transmitters & receivers; dead drops; objects related to specific espionage cases & historic figures such as John Walker, Mata Hari & George Washington; KGB & historic photographs; espionage stories.
Facilities: theater; rental space. Museum-related items for sale.
Activities: audio-visual programs; computer interactives; hands-on components; lectures; tours; book signings; temporary exhibits; special events.
Hours & Admission Prices: Daily 10am. Adults $18, seniors 65 & over $17, children 5-11 $15. Operation Spy: adults 12 & over $14. Operation Spy & Main Exhibit: adults 12 & over $25. Closed New Year's Day; Thanksgiving; Christmas. &
Membership: Spy $65; Spy Partners $95; Spy Family $175; Spy Master $275.

JOHN F. KENNEDY CENTER FOR THE PERFORMING ARTS, 2700 F St., N.W., Washington, DC 20566. Tel.: 202-416-8340. (TTY) 202-416-8524.
Web Site: www.kennedy-center.org
Institution Type/Description: Performing Arts Center & History Museum.
Collections: cultural center history; John F. Kennedy's life & presidency; gifts from more than 60 countries; works of art; personal artifacts; photographs; historic buildings.
Activities: guided tours.
Hours & Admission Prices: Tours: Mon.-Fri. 10-5, Sat.-Sun. 10-1. No charge. &

KENILWORTH PARK AND AQUATIC GARDENS, 1550 Anacostia Ave., N.E., Washington, DC 20019-2028. Mailing Address: 1900 Anacostia Dr., S.E., Washington, DC 20020-6722. Tel.: 202-426-6905. Fax: 202-426-5991.
Web Site: www.nps.gov/keaq/index.htm
Founded: 1882.
Key Personnel: Park Supt., Gayle Hazelwood; Park Ranger, Kate Bucco; Gardener Foreman, Doug Rowley.
Personnel Profile: Full-Time Paid 10; Part-Time Volunteers 32.
Governing Authority: federal. Parent Institution: National Capital Parks-East, National Park Service, U.S. Dept. of Interior. Tax exempt.
Institution Type/Description: Aquatic Garden.
Collections: water plants; tropical & hardy water lilies & lotus.
Research Fields: aquatic ecosystems; tidal marshes; urban river water quality.
Facilities: visitor center.
Activities: guided tours; self-guided walks; birding.
Publications: brochure.
Hours & Admission Prices: Grounds: daily 8-dusk. Aquatic Garden: daily 7-4. Closed New Year's Day; Thanksgiving; Christmas. &
Attendance: 75,000 (estimated)

THE KREEGER MUSEUM, (M), 2401 Foxhall Rd., N.W., Washington, DC 20007-1149. Tel.: 202-337-3050. Fax: 202-337-3051.
E-mail: publicrelations@kreegermuseum.org
Web Site: www.kreegermuseum.org
Founded: 1994.
Key Personnel: Dir., Judy A. Greenberg; Financial Officer, Basil Arendse; Head Mktg. & Public Rels., Membership & Conversation Programs, Derya Samadi; Head Education, Erich Keel; Head Visitor Svcs. & Docent Management, Jane Wilson.
Personnel Profile: Full-Time Paid 8; Full-Time Volunteers 1; Part-Time Paid 1; Part-Time Volunteers 80; Interns 2.
Governing Authority: private; nonprofit foundation. Parent Institution: The David Lloyd Kreeger Foundation. Tax-exempt: 501(c)(3).
Institution Type/Description: Art Museum: housed in the former residence of Carmen & David Lloyd Kreeger; designed by Philip Johnson.
Collections: museum designed by the renowned architect Philip Johnson, showcases the Kreeger's personal art collection of 19th- & 20th-century painting and sculpture, as well as traditional African art. Artists featured: Monet, Picasso, Braque, Miro, Kandinsky, Dubuffet, Moore; contemporary outdoor sculptures.
Research Fields: 19th- & 20th-century European, American, African & Indian art.
Facilities: library.
Activities: docent tours; artists talks; concerts; intern program; master class programs; lecture series; guest speakers; community educational projects; story time for preschool aged children; children's workshops; panel discussions; temporary exhibitions; Conversations at The Kreeger Museum, a program for caregivers or individuals with Alzheimer's disease.
Publications: Gilliam in 3D; Remembering the Present; William Christenberry: Changing Landscape - The Source Revisited; The True Artist is an Amazing Luminous Fountain; The Kreeger Museum Catalogue; Kendell Buster: Inventory of Imagined Places; Tim Rollins & KOS, The Creation; Gene Davis: Interval; The Houses of Philip Johnson; Philip Johnson: Architecture As Art; The Kreeger Museum Catalogue; Dan Steinhilber: Martin Underground.
Hours & Admission Prices: Sept.-July Fri.-Sat. 10-4. Tours: Tues.-Thurs. 10:30 & 1:30 by appointment. Adults $10, seniors 65 & over and student $7; discounts to museum professionals, NARM, ICOM & AAM members; members no charge. Closed New Year's Eve & Day; Martin Luther King Jr. Day; Presidents' Day; Memorial Day; Independence Day; Labor Day; Columbus Day; Thanksgiving & day after; Christmas Eve & Day. &
Attendance: 10,000 (estimated)
Membership: Friend $100; Sustainer $250; Foxhall Circle $500; Patrons Circle $1,000; Director's Circle $5,000; Founders Circle $10,000; Kreeger Circle $25,000.

LAOGAI MUSEUM, 1734 20th St. N.W., Washington, DC 20009. Tel.: 202-408-8300 (office) & 730-9308 (museum). Fax: 202-408-8302.
E-mail: laogai@laogai.org
Web Site: www.laogai.org, www.laogaimuseum.org
Founded: 2008.
Key Personnel: Dir., Harry Wu.
Personnel Profile: Full-Time Paid 8; Part-Time Paid 1.
Governing Authority: Parent Institution: Laogai Research Foundation. Tax-exempt.
Institution Type/Description: History Museum.
Collections: Chinese prison system history & structure; human rights in China; uniforms; photographs; government documents; products manufactured by prisoners; official documents.
Research Fields: human rights in China.
Activities: group tours; film screenings; guest speakers; salon gatherings.
Publications: Laogai in China; Bitter Wind.
Hours & Admission Prices: Mon.-Fri. 10-6, Sat. 10-5. No charge. Closed federal holidays.

LIBRARY OF CONGRESS, 101 Independence Ave., S.E., Washington, DC 20540-0002. Tel.: 202-707-5000 (general) & 8000 (visitor information). Fax: 202-707-1714.
E-mail: libofc@loc.gov
Web Site: www.loc.gov
Founded: 1800.
Key Personnel: Librarian of Congress, James H. Billington; Assoc. Librarian Strategic Initiatives, Laura Campbell; Assoc. Librarian, Library Svcs., Deanna Marcum; Dir. Congressional Research Svcs., Daniel P. Mulhollan; Dir. American Folklife Center, Peggy Bulger; Acting Law Librarian, Donna Scheeder; Register of Copyrights, Copyright Office, Marybeth Peters; Interpretive Programs Officer, William Jacobs; Dir. Publishing Office, W. Ralph Eubanks; Dir. Preservation, Dianne Van Der Reyden; Chief Prints &

Photographs Div., Helen Zinkham; Chief, Packard Campus of the Natl. Audiovisual Conservation Center, Patrick Loughney; Head Retail Mktg. Unit, Anna S. Lee.
Governing Authority: federal. Parent Institution: United States Congress. Tax-exempt: 501(c)(3) Title 26, U.S. Code.
Institution Type/Description: National library with collections housed in the Thomas Jefferson Building, constructed between 1889-1897; the James Madison Memorial Building, constructed between 1970-1981; & the 1939 John Adams Building.
Collections: over 138,000,000 item library of books & pamphlets including Americana, written in 470 languages; bound worldwide newspapers; manuscripts relating to American history & civilization; maps & views; classical to modern music recordings; photographic negatives, prints & slides, including those made by Mathew Brady; prints & drawings, including the Joseph Pennell collection of Whistleriana; motion-picture reels, including Fred Ott's Sneeze, the first to be copyrighted; microfilm; Braille volumes; talking books on records & magnetic tape; caricatures & cartoons; early American history; creating the United States.
Research Fields: research in fields pertaining to legislation for members & committees of Congress; legal research in domestic & international copyright; research for other government agencies & preparation of resulting analyses bibliographies, etc.; technology to improve reading materials for blind & handicapped persons; technology in the preservation of all types of library materials; application of automation to library operations; cataloging, classification & standards for all types of library materials; digitization of library materials.
Facilities: library; Jefferson building: 20,000 sq. ft. exhibit space; reading rooms; auditorium; James Madison Memorial Building: 64-seat projection room. Selected LC publications, facsimiles, music & spoken recordings & other library-related items for sale.
Activities: guided tours; audio tours for major exhibitions; video orientation; national preservation program; lectures & symposia; concerts; dramatic & poetry readings; broadcasts by educational stations in U.S.; inter-library loan; exhibitions, many available for travel to other institutions; permanent exhibitions; coordination of folklife activities for a national constituency; a national program through a network of cooperating libraries to provide reading materials to blind & physically handicapped readers; the digitization of the library's collections of unique American materials.
Publications: biweekly, Library of Congress Information Bulletin; monthly, Calendar of Events in the Library of Congress to addresses within 100 miles of Washington; quarterly, Folklife Center News.
Hours & Admission Prices: Exhibition Halls: Mon.-Fri. 8:30am-9:30pm, Sat. 8:30am-6:30pm. General Reading Rooms: call 201-707-6400 for various hours. No charge. Closed all federal government holidays. &
Attendance: 1,600,000 (estimated)

LIBRARY OF THE SUPREME COUNCIL, 33 DEGREES, 1733 16th St., N.W., Washington, DC 20009-3199. Tel.: 202-232-3579. Fax: 202-464-0487.

E-mail: jsansbury@srmason-sj.org
Web Site: www.srmason-sj.org
Founded: 1888.
Key Personnel: Sovereign Grand Commander, Ronald A. Seale; Cur. & Librarian, Joan Sansbury; Archivist & Historian, Art de Hoyos; Asst. Librarian, Larissa Watkins; Dir. Special Programs, Heather Calloway.
Personnel Profile: Full-Time Paid 3; Part-Time Paid 1.
Governing Authority: Ancient & Accepted Scottish Rite of Freemasonry, Supreme Council.
Institution Type/Description: Masonic & Americanism Museum.
Collections: medals; jewels; patents; art objects; manuscripts; flag carried to the moon by Buzz Aldrin; Masonic book collection; J. Edgar Hoover Collection; Lincoln Collection; Robert Burns Collection; Burl Ives Collection, 1997.
Research Fields: freemasonry; general collection of old non-fiction.
Facilities: 200,000-vol. library of books relating to Masons; religion; history; biography available for research by request, accompanied by Masonic or educational institution reference.
Activities: guided tours.
Publications: magazine, The Scottish Rite Journal.
Hours & Admission Prices: Mon.-Thurs. 8-5. No charge; donations accepted. Closed federal holidays. &
Attendance: 3,600 (estimated)

LILLIAN AND ALBERT SMALL JEWISH MUSEUM, (M), 701 3rd St., N.W., Washington, DC 20001-2624. Mailing Address: 701 4th St., N.W., Washington, DC 20001-2607. Tel.: 202-789-0900. Fax: 202-789-0485.

E-mail: info@jhsgw.org
Web Site: www.jhsgw.org

Founded: 1965.
Key Personnel: Exec. Dir., Laura C. Apelbaum; Pres. (V), Sidney J. Silver.
Personnel Profile: Full-Time Paid 6; Part-Time Paid 1; Part-Time Volunteers 5; Interns 2.
Governing Authority: private; nonprofit organization. Parent Institution: Jewish Historical Society of Greater Washington. Tax-exempt.
Institution Type/Description: Historical Society Museum: housed in the 1876 Old Adas Israel Synagogue, the first structure built in Washington, DC as a Jewish house of worship.
Collections: letters, photographs, documents, records, scrapbooks, diaries, objects & related artifacts which document the Jewish people & the Jewish communities of greater Washington, DC; artifacts related to the origins or historical development of customs, beliefs, activities or material culture as practiced by Jews of the greater Washington, DC area.
Research Fields: communal & social life of Jews in Washington area.
Facilities: 50-vol. library on American Jewish history & Jewish art; historic sanctuary.
Activities: docent program; guided tours; lectures; temporary & traveling exhibitions; video programs of exhibits; internships. Annual Event: meeting of the Jewish Historical Society of Greater Washington.
Publications: annual journal, The Record.
Hours & Admission Prices: Mon., Tues. & Thurs. 1-4. No charge; donations accepted. Closed major Jewish holidays & federal holidays.
Attendance: 3,000 (estimated)
Membership: Individual $36; Family $54; Donor $100; Patron $150; Sponsor $300; Guardian $1,000; Historian $1,800; Heritage $2,500; Legacy $5,000; Benefactor $10,000; President's Club $25,000.

LINCOLN MEMORIAL, W. Potomac Park @ 23rd St., N.W., Washington, DC 20024. Mailing Address: National Mall & Memorial Parks, 900 Ohio Dr., S.W., Washington, DC 20242-0002. Tel.: 202-426-6841; 233-3520.

Web Site: www.nps.org/linc
Founded: 1922.
Key Personnel: Supervisory Park Ranger & Public Affairs Specialist, Toni Braxton
Institution Type/Description: Historic Building: built in 1917 & dedicated in 1922 to honor the 16th president of the United States, Abraham Lincoln.
Collections: 19 ft. marble statue of President Abraham Lincoln; 2 murals; inscriptions of his 2nd inaugural address & the Gettysburg Address.
Facilities: Books & gift items for sale.
Activities: living history programs.
Publications: brochure.
Hours & Admission Prices: Memorial open 24 hours. Park Rangers on duty: daily 9:30am-11:30pm. No charge. &

LINDA K. JORDAN GALLERY, GALLAUDET UNIVERSITY, (M), Washburn Arts Bldg. #127, 800 Florida Ave., N.E., Washington, DC 20002-3600. Tel.: 202-651-5480. Fax: 202-651-5618.

Founded: 2006.
Key Personnel: Chm., Dr. Marguerite Glass
Institution Type/Description: Art Gallery.
Collections: photographs; paintings; drawings; ceramics.
Hours & Admission Prices: Call for hours.

THE LUTHER W. BRADY ART GALLERY, 805 21st St., N.W., Washington, DC 20052-0029. Tel.: 202-994-1525. Fax: 202-994-1632.

E-mail: bradyart@gwu.edu
Web Site: www.gwu.edu/~bradyart
Formerly: The George Washington University Art Gallery
Founded: 1966.
Key Personnel: Dir. University Art Galleries, Lenore D. Miller; Asst. Cur., Olivia Kohler.
Personnel Profile: Full-Time Paid 2; Part-Time Paid 5; Interns 1.
Governing Authority: university. Parent Institution: George Washington University. Tax-exempt: 101(6).
Institution Type/Description: Art Gallery.
Collections: paintings; sculpture; graphic arts & photographs from 18th, 19th & 20th centuries, with special emphasis on American art; W. Lloyd Wright Collection of Washingtoniana; works pertaining to George Washington; U. S. Grant Collection of photographs, documents, prints & newspaper clippings; collection of prints by Joseph Pennell; pre-Columbian artifacts.
Research Fields: art history; documentation of permanent collections.
Facilities: study room.
Activities: graduate & undergraduate programs in museum problems; inter-museum loan & temporary exhibitions; lectures; visual literacy programs for children.
Publications: exhibition catalogs.

Hours & Admission Prices: Tues.-Fri. 10-5. No charge. Closed during school breaks, summer & national holidays. &

Attendance: 3,500 (estimated)

MADAME TUSSAUDS WASHINGTON D.C., 1001 F St., Washington, DC 20004. Tel.: 202-942-7303. Facebook: Madame Tussauds DC.

E-mail: info@madametussaudsdc.com
Web Site: www.madametussaudsdc.com
Institution Type/Description: Wax Museum.
Collections: wax figures including historical & political figures, musicians, sports stars, world leaders, & Hollywood stars.
Major Exhibits: American Icons, 1/14-3/14.
Activities: special events.
Hours & Admission Prices: mid-April to early Sept. Sun.-Fri. 10-6, Sat. 10-8; early Sept. to mid-April Sun.-Thurs. 10-6, Fri.-Sat. 10-8. Adults $20, seniors 60 & over $18, children 3-12 $15; discounts to groups; children 2 & under no charge.

MARIAN KOSHLAND SCIENCE MUSEUM, (M), 525 E. St., N.W., Washington, DC 20001. Mailing Address: The National Academies, 500 5th St., N.W., Washington, DC 20001-2736. Tel.: 202-334-1201. Fax: 202-334-1548.

E-mail: ksm@nas.edu
Web Site: www.koshlandsciencemuseum.org
Founded: 2004.
Key Personnel: Dir., Patrice Legro; Chair, May Barenbaum; Operations & Museum Shop Mgr., Johann Yurgen; Financial & Administrative Officer, Lisa Alston; Deputy Dir., Erika Shugart; Community Outreach, Amy Shaw.
Personnel Profile: Full-Time Paid 7; Part-Time Paid 1; Part-Time Volunteers 22.
Governing Authority: private; nonprofit organization. Parent Institution: National Academy of Sciences. Subsidiary Institution: National Academies, Washington, DC. Tax-exempt: 501(c)(3).
Institution Type/Description: Science Museum.
Collections: science; engineering; medicine.
Research Fields: by requests from Congress or private organizations: agriculture; behavioral & social science; chemistry; earth science; education; health & medicine; physics; technology.
Facilities: 5,000 sq. ft. exhibit space.
Activities: docent program; guided tours; facilitated discussions; participatory & traveling exhibitions; field trip program; evening programs.
Hours & Admission Prices: Wed.-Mon. 10-6. Adults $7, active military & students $5. Closed New Year's Day; Thanksgiving; Christmas. &
Attendance: 30,000 (accurate)

MARY MCLEOD BETHUNE COUNCIL HOUSE NATIONAL HISTORIC SITE, 1318 Vermont Ave., N.W., Washington, DC 20005-3607. Tel.: 202-673-2402. Fax: 202-673-2414.

Web Site: www.nps.gov/mamc
Founded: 1979.
Key Personnel: Supt., Alex Romero; Deputy Supt., Gopaul Noojibail; District Mgr., Joy Kinard; Archivist, Kenneth Chandler; Park Guide, Jamie Euken; Park Guide, Veronica Quiguango; Museum Shop Mgr., Margaret Miles.
Personnel Profile: Full-Time Paid 5; Part-Time Volunteers 1; Interns 26.
Governing Authority: nonprofit. Parent Institution: National Park Service. Tax-exempt.
Institution Type/Description: Historic House.
Collections: photographs; communication artifacts; archives.
Research Fields: Mary McLeod Bethune; African American Women's History.
Facilities: archives.
Activities: programs.
Hours & Admission Prices: Mon.-Sat. 9-5. No charge, donations accepted. Closed New Year's Day; Thanksgiving; Christmas.
Attendance: 8,631 (accurate)

MAURINE LITTLETON GALLERY, 1667 Wisconsin Ave., N.W., Washington, DC 20007. Tel.: 202-333-9307. Fax: 202-342-2004.

E-mail: info@littletongallery.com
Web Site: www.littletongallery.com
Founded: 1984.
Key Personnel: Dir., Harvey K. Littleton
Institution Type/Description: Art Gallery.
Collections: works by contemporary glass artists; sculpture; ceramics.
Activities: temporary exhibitions.
Hours & Admission Prices: Tues.-Sat. 11-6; other times by appointment.

MERIDIAN INTERNATIONAL CENTER, 1624 Crescent Pl., N.W., Washington, DC 20009-4004. Mailing Address: 1630 Crescent Pl., N.W., Washington, DC 20009-4004. Tel.: 202-667-6800. Fax: 202-939-5512.

E-mail: tkharvey@meridian.org
Web Site: www.meridian.org
Formerly: Meridian International Center-Cafritz Galleries
Founded: 2009.
Key Personnel: Pres. & C.E.O., Ambassador Stuart W. Holliday; Chm. (V), Jim Blanchard; Vice Pres. for the Arts, Dr. Curtis Sandberg; Dir. Exhibitions, Terry Harvey.
Personnel Profile: Full-Time Paid 100; Part-Time Volunteers 25; Interns 2.
Governing Authority: nonprofit organization. Tax-exempt: 501(c)(3).
Institution Type/Description: International Arts & Cultural Center.
Collections: French furnishings; over-door paintings; Mortlake tapestry; portraits. Historic Buildings: 1911, White-Meyer House; 1921, Meridian House, both designed by John Russell Pope.
Research Fields: intercultural exchange programs.
Facilities: library; reception & conference facilities; formal gardens; 5,000 sq. ft. exhibit space.
Activities: organizes, presents & circulates international art & historical exhibitions; lectures; films; concerts; center provides programs, orientation & services to international visitors & diplomats.
Publications: newsletter; exhibition catalogues.
Hours & Admission Prices: Wed.-Sun. 2-5. No charge; donations accepted. Closed major holidays. &
Attendance: 25,000 (estimated)
Membership: Individual $100; Family $175.

MID-ATLANTIC ASSOCIATION OF MUSEUMS, 1025 Thomas Jefferson St., N.W., Ste. 500W, Washington, DC 20007-5224. Tel.: 202-452-8040. Fax: 202-833-3636.

E-mail: admin@midatlanticmuseums.org
Web Site: www.midatlanticmuseums.org
Founded: 1947.
Key Personnel: Exec. Dir., Graham Hauck.
Personnel Profile: Full-Time Paid 1; Full-Time Volunteers 2; Part-Time Paid 2; Part-Time Volunteers 1; Interns 1.
Governing Authority: nonprofit organization. Tax-exempt: 501(c)(3).
Institution Type/Description: Museum Service Organization.
Activities: training programs for professional museum workers; workshops. Organization Sponsors: Annual three-day conference.
Publications: quarterly newsletter, Courier; professional publication, Forum; occasional paper series.
Hours & Admission Prices: Mon.-Fri. 9-5. Closed national holidays.
Attendance: 400
Membership: Retired & Student $25; Individual & Individual Affiliate $40; Supporting $60; Contributing $100; Institutional Affiliate $250. Institutional Membership: under 50,000 $50; 50,000-100,000 $75; 100,000-250,000 $100; 250,000-500,000 $150; 500,000-1,000,000 $200; 1,000,000-2,500,000 $250; over 2,500,000 $300.

NATIONAL ACADEMY OF SCIENCES, (M), 2101 Constitution Ave., N.W., Washington, DC 20418-0006. Mailing Address: 500 5th St., N.W., Ste. 1, Washington, DC 20001-2737. Tel.: 202-334-2436. Fax: 202-334-1690.

E-mail: cpnas@nas.edu
Web Site: www.nationalacademies.org/arts
Founded: 1980.
Key Personnel: Dir., Mr. J.D. Talasek; Sr. Program Assoc., Ms. Alana Quinn.
Personnel Profile: Full-Time Paid 2.
Governing Authority: private; nonprofit organization. Tax-exempt: 501(c)(3).
Institution Type/Description: Art & Science Museum.
Collections: art & science; engineering & medicine; paintings; photographs; decorative arts.
Facilities: 650 seat auditorium.
Activities: concerts; guided tours; lectures; temporary & traveling exhibitions. Annual Events: Artist Receptions and Lectures; Classical and Jazz Concerts.
Publications: biannual calendar of events, Exhibitions and Cultural Programs Calendar.
Hours & Admission Prices: Mon.-Fri. 9-5. No charge. Closed New Year's Day; Martin Luther King Jr. Day; Presidents' Day; Memorial Day; Independence Day; Labor Day; Thanksgiving & day after; Christmas Eve & Day.
Attendance: 5,000 (estimated)

✶ **NATIONAL AIR AND SPACE MUSEUM, (M),** Sixth St. & Independence Ave., S.W., Washington, DC 20560. Mailing Address: Office Communications, MRC 321, P.O. Box 37012, Washington, DC 20013-7012. Tel.: 202-633-2370 & 1000. Fax: 202-633-8174. TDD: 202-357-1505.
Web Site: www.nasm.si.edu
Founded: 1946.
Key Personnel: Dir., John R. Dailey; Chm. Center for Earth & Planetary Studies, John Grant; Chm. Space History Dept., Michael Neufeld; Chm. Aeronautics Dept., F. Robert van der Linden; Assoc. Dir. Collections & Research, Peter Jakab; Assoc. Dir. Planning Programs, John Benton; Asst. Dir. Communications, Claire Brown; Dir. Devel., Monecia Taylor; Asst. Dir. Special Events, Linda Hicks; Museum Shop Mgr., Frank Byrne.
Personnel Profile: Full-Time Paid 235; Part-Time Paid 28; Part-Time Volunteers 131; Interns 19.
Governing Authority: federal; nonprofit corporation. Parent Institution: Smithsonian Institution. Subsidiary Institutions: Steven F. Udvar-Hazy Center. Tax-exempt: 501(c)(3) & 170(b)(1)(A).
Institution Type/Description: Aeronautics & Space Museum.
Collections: aeronautical & astronautical items; air & space craft; instruments; equipment; art; uniforms; personal memorabilia; manuscripts.
Research Fields: aviation & aerospace history; flight science technology; earth & planetary studies; contributions of flight to the economy; pioneering efforts of early aviators & astronauts.
Facilities: 40,000-vol. library of material on history of flight & aerospace development; aircraft & space vehicle photographic collection available for inter-library loan, for use on premises & by request; planetarium; theater; reading room; education resource center; cafeteria. Model airplane & space kits, books & other museum-related items for sale.
Activities: guided tours; lectures; films; gallery talks; daily science demonstrations; formally organized education programs for children; docent program or council; inter-museum loan, permanent, temporary & traveling exhibitions.
Publications: periodic series, Studies in Air & Space; Famous Aircraft of the National Air & Space Museum; exhibition booklets; monographs; commemorative books published on special anniversaries; leaflets; catalogs; e-newsletters; official guide to the National Air and Space Museum.
Hours & Admission Prices: Daily 10-5:30. No charge; donations accepted. Closed Christmas. ♿
Attendance: 9,400,000 (estimated)
Membership: Wright Flyer $35; Spirit of St. Louis $50; Glamorous Glennis $100; Friendship $250; Eagle $500; Enterprise $1,000; Voyager Solar System Spacecraft $2,500; North American X-15 $5,000.

THE NATIONAL AQUARIUM, WASHINGTON DC, Department of Commerce Bldg., Rm. B-077, 14th St. & Constitution Ave., N.W., Washington, DC 20230. Tel.: 202-482-2826. Fax: 202-482-4946. Facebook: National Aquarium.
E-mail: dcinfo@aqua.org
Web Site: aqua.org
Founded: 1873.
Key Personnel: Dir. Operations, David Lin; Chm. (V), Tamika Tremaglio; Mgr. Mktg., Emma Connor; Museum Shop Mgr., Tammy Ward.
Personnel Profile: Full-Time Paid 13; Part-Time Paid 5.
Governing Authority: society; nonprofit. Subsidiary Institution: National Aquarium in Baltimore. Tax-exempt.
Institution Type/Description: Aquarium & Fishery Museum.
Collections: approx. 250 species of freshwater & marine animals; 1,500 specimens.
Facilities: Museum-related items for sale.
Activities: educational programs for local school groups.
Publications: quarterly newsletter.
Hours & Admission Prices: Daily 9-5. Adults 12 & over $9.95, seniors 60 & over and military $8.95, children 3-11 $4.95; discounts to groups; members no charge. Closed Thanksgiving; Christmas. ♿
Attendance: 200,000 (estimated)
Membership: Individual $74; Senior $89; Couple $109; Family & Grandparents $159; Family Plus $209; Stingray Club $300; Dolphin Club $500. (Guest Add-On option $50)

NATIONAL ARCHIVES EXPERIENCE, 700 Pennsylvania Ave., N.W., Washington, DC 20408-0002. Mailing Address: Museum Programs, National Archives Bldg., 700 Pennsylvania Ave., Washington, DC 20408. Tel.: 202-357-5000. Fax: 202-357-5926.
E-mail: museumprograms@nara.gov
Web Site: www.archives.gov/nae
Formerly: National Archives and Records Administration
Founded: 1934.

Key Personnel: Acting Archivist, Adrienn Thomas; Exec. Dir. Foundation, Thora Colot; Exec. Dir. Museum Programs, Marvin Pinkert; Dir. Mktg. & Admin., Frank Cordes; Dir. Retail Operations, Chris Derderian
Governing Authority: federal. Maintains National Archives Exhibition Hall, Washington, DC; National Archives at College Park, MD; Franklin D. Roosevelt Library, Hyde Park, NY; Harry S. Truman Library, Independence, MO; Herbert Hoover Library; West Branch, IA; Dwight D. Eisenhower Library, Abilene, KS; John F. Kennedy Library, Boston, MA; Lyndon B. Johnson Library, Austin, TX; Gerald Ford Library, Ann Arbor, MI; Gerald Ford Museum, Grand Rapids, MI; Jimmy Carter Library, Atlanta, GA; Ronald Reagan Library, Simi Valley, CA; Bush Library, College Station, TX; William J. Clinton Presidential Materials Project, Little Rock, AR; 13 Regional Centers. Tax-exempt.
Institution Type/Description: History Museum: housed in the National Archives building; built 1931-1935.
Collections: federal government archives of the United States from 1774 to present including the original Declaration of Independence, the Constitution & the Bill of Rights; photographic records, sound recordings; motion pictures; maps; charts; aerial photographs; textual records; electronic records; the 1297 Magna Carta.
Research Fields: U.S. history & foreign relations.
Facilities: reference library, research rooms; 294-seat theatre; classrooms; snack bar; learning center. Gift items for sale.
Activities: guided tours; conferences; lectures; films; formally organized education programs for children, adults, undergraduate & graduate students; inter-museum loans; permanent, temporary & traveling exhibitions.
Publications: leaflets; facsimiles of documents; exhibit catalogs; quarterly journal, Prologue; annual reports; finding aids; curriculum materials, documentary publications; archives administration publications; presidential library publications; records management handbooks; publications of the Federal Register; monthly calendar of events.
Hours & Admission Prices: Winter: daily 10-5:30; Spring & Summer: daily 10-7. No charge. Research Rooms: Mon. & Wed. 8:45-5, Tues. & Thurs.-Fri. 8:45-9, Sat. 8:45-4:45. Closed Thanksgiving; Christmas. ♿
Attendance: 1,000,000 (estimated)

NATIONAL BUILDING MUSEUM, (M), 401 F St., N.W., Washington, DC 20001-2637. Tel.: 202-272-2448, ext. 3109. Fax: 202-272-2564. Facebook: National Building Museum.
E-mail: brodgers@nbm.org
Web Site: www.nbm.org
Founded: 1980.
Key Personnel: Exec. Dir. & Pres., Chase W. Rynd; Chm. (V), Whayne S. Quin; Sr. Vice Pres. & Cur., G. Martin Moeller, Jr.; Vice Pres. Finance & Admin., Betsy May-Salazar; Vice Pres. Devel., Lauren Harnishfeger; Vice Pres. Education, Scott Kratz; Vice Pres. Exhibitions & Collections, Cathy Crane Frankel; Dir. Mktg. & Communications, Brett Rodgers; Dir. Finance & Controller, Diane Beckham; Dir. Visitor Svcs., Jamee Telford; Dir. Special Events, Chris Frame; Museum Shop Mgr., Michael Higdon; Exec. Asst., Tom Buck.
Personnel Profile: Full-Time Paid 55; Part-Time Paid 48; Part-Time Volunteers 281; Interns 19.
Volunteer Hours: 13,200
Operating Expenses: 8,956,031
Operating Income: 8,608,382
Governing Authority: nonprofit organization. Tax-exempt: 501(c)(3).
Institution Type/Description: Architecture Museum.
Collections: local history & culture; manuscripts; drawings; blueprints; models; photographs; books; artifacts; building & architectural toys.
Major Exhibits: House & Home, 10/13-1/17; Designing for Disaster, 5/14-5/16.
Research Fields: architectural history; architectural projects & trends; design field; urban planning; engineering; construction technology; material technology.
Facilities: cafe/bakery. Museum-related items for sale.
Activities: guided tours; lectures; films; docent program; intern programs for undergraduate & graduate students; day trips/tours; traveling exhibitions; family & outreach programs; rental facilities; youth education.
Publications: monthly calendar of programs; exhibition catalogs; books; annual report.
Hours & Admission Prices: Mon.-Sat. 10-5, Sun. 11-5. Exhibitions: Adults $8, youth 3-17, students with ID and seniors 60 & over $5; discounts to AAM & ICOM members; military & their families between Memorial Day & Labor Day, members and children 2 & under no charge. Building Zone: adults $3. Closed Thanksgiving; Christmas. ♿
Attendance: 525,674 (accurate)
Membership: Student w/ID $40; Senior $50; Individual $60; Dual $80; Family $90; Builder Friend $150; Builder Fellow $250; Builder Contributor $500; Corinthian Friend $1,000-$2,500; Corinthian Fellow $2,500-$5,000; Corinthian Contributor $5,000-$10,000; Corinthian Patron $10,000 & up.

NATIONAL CHILDREN'S MUSEUM, 955 L'Enfant Plaza North, S.W., Ste. 5100, Washington, DC 20024-2103. Tel.: 202-675-4120. Fax: 202-675-4140.
E-mail: info@ncm.museum
Web Site: www.ncm.museum
Formerly: Capital Children's Museum
Founded: 1974.
Key Personnel: Pres., Kathy Dwyer Southern; Chm. (V), S. Ross Hechinger.
Personnel Profile: Full-Time Paid 26.
Governing Authority: nonprofit organization. Tax-exempt: 501(c)(3).
Institution Type/Description: Children's Museum.
Collections: hands-on exhibits.
Publications: quarterly newsletter.
Hours & Admission Prices: Summer: daily 10-7; Labor Day to Memorial Day daily 10-5. Admission $10; members and children one & under no charge. Closed New Year's Day; Thanksgiving; Christmas. &

* **NATIONAL GALLERY OF ART, (M),** 4th St. & Constitution Ave., N.W., Washington, DC 20565. Mailing Address: 2000B S. Club Dr., Landover, MD 20785. Tel.: 202-737-4215. TDD: 202-842-6176.
Web Site: www.nga.gov
Founded: 1937.
Key Personnel: Dir., Earl A. Powell, III; Pres. (V), Victoria P. Sant; Chm. (V), John Wilmerding; Deputy Dir. & Chief Cur., Franklin W. Kelly; Admin., Darrell R. Willson; Acting Treas., Diane Mullis; Sec. Gen. Counsel, Elizabeth A. Croog; Exec. Officer, Devel. & External Affairs, Joseph J. Krakora; Dean, Center for Advanced Study in Visual Arts, Elizabeth Cropper; Cur. Northern Baroque Paintings, Arthur K. Wheelock, Jr.; Cur. Italian Paintings, David Alan Brown; Cur. Northern Renaissance Paintings, John Hand; Cur. Early European Sculpture, Alison Luchs; Cur. Modern & Head, Contemporary Art, Harry Cooper; Sr. Cur. & Head Dept. Photographs, Sarah Greenough; Cur. Special Projects in Modern Art, Ruth Fine; A.W. Mellon Sr. Cur. Prints & Drawings, Andrew C. Robison; Asst. Cur. Old Master Prints, Andaleeb Banta; Cur. & Head Old Master Drawings, Margaret Morgan Grasselli; Cur. & Head Modern Prints & Drawings, Judith Brodie; Cur. & Head Dept., Sculpture & Decorative Arts, Mary Levkoff; Cur. & Head Dept., American & British Paintings, Nancy Anderson; Chief of Conservation, Mervin Richard; Sr. Conservator & Head Painting Conservation, Sarah Fisher; Head Paper Conservation, Kimberly Schenck; Sr. Conservator & Head Object Conservation, Shelley Sturman; Sr. Conservator & Head Textile Conservation, Julia Burke; Head Scientific Research Dept., E. Rene de la Rie; Exec. Librarian, Neal T. Turtell; Chief Library Image Collections, Gregory P.J. Most; Chief Exhibitions, D. Dodge Thompson; Sr. Cur. & Chief Design, Mark Leithauser; Head Education, Lynn Russell; Head Film Programs, Margaret Parsons; Editor in Chief, Publishing Office, Judy Metro; Chief Registrar, Sally Freitag; Chief Imaging & Visual Svcs., Alan Newman; Head Loans & Natl. Lending Svcs., Stephanie T. Belt; Head Digital Imaging Svcs., Robert Grove; Head Visual Svcs., Barbara Bernard; Chief Protocol & Special Events, Carol W. Kelley; Chief Devel. & Corporate Rels. Officer, Christine M. Myers; Chief Press & Public Information Officer, Deborah Ziska; Acting Personnel Officer, Meredith Wiser; Chief Horticulture Svcs., Cynthia Kaufmann; Deputy Chief Admin. Visitor Svcs., Elizabeth Thomas; Chief Gallery Archives, Maygene Daniels; Chief Gallery Shop Div., Ysabel L. Lightner; Special Projects Officer, Pamela Jenkinson.
Personnel Profile: Full-Time Paid 1,055; Part-Time Paid 10; Part-Time Volunteers 513; Interns 20.
Governing Authority: independent establishment of the U.S.; nonprofit organization. Tax-exempt: 170(b)(1)(A); 501(c)(3).
Institution Type/Description: Art Museum.
Collections: European & American painting, sculpture, decorative arts & graphic arts from 13th-21st century; European Old Master paintings; French, Spanish, Italian, American & British 18th- & 19th-century paintings; sculpture from late Middle Ages to present; Renaissance bronzes; Chinese porcelains; medals & plaquettes; modern & contemporary art.
Research Fields: art history; conservation; Center for Advanced Studies in the Visual Arts.
Facilities: library; photographic archives; four restaurants; sculpture garden; education studio. Museum-related items for sale.
Activities: guided tours; illustrated lectures; films; gallery talks; concerts; family activities; docent program; inter-museum loan, permanent & temporary exhibitions; loan teaching services.
Publications: exhibition catalogs; illustrated collection catalogs; catalogue raisones on specific artists; Studies in the History of Art, which includes Conservation Research, symposium papers, and anniversary volumes; the CASVA (Center for Advanced Study in the Visual Arts) annual report; Kress Foundation Studies in the History of European Art; Andrew W. Mellon Lectures in the Fine Arts; National Gallery of Art Bulletin; National

Gallery of Art; annual report; bimonthly calendar of events; quarterly film calendar; quarterly, NGA Kids.
Hours & Admission Prices: Mon.-Sat. 10-5, Sun. 11-6. No charge. &
Attendance: 4,800,000 (estimated)
Membership: Circle: Contributing $1,000; Supporting $2,500; Sustaining $5,000; Patron $10,000.

NATIONAL GEOGRAPHIC MUSEUM, (M), National Geographic Society, 1145 17th St., N.W., Washington, DC 20036-4707. Tel.: 202-857-7700, 7588 & 7000. Fax: 202-857-5864. TDD: 202-857-7198.
E-mail: ngtickets@ngs.org
Web Site: www.ngmuseum.org
Founded: 1930.
Key Personnel: Chm. & C.E.O., Gary E. Knell; Exec. Vice Pres., Mission Programs, Terry D. Garcia; Vice Pres. Public Programs, Gregory McGruder; Vice Pres. Exhibitions, Kathryn Keane; Dir. Museum Operations, Richard McWalters; Cur., Catherine C. Tyson; Dir. Visitor Svcs., Lisa M. Kopp; Art Dir., Museum & Mission Programs, Alan M. Parente; Mgr. Traveling Exhibitions, Abigail Bysshe; Museum Administration Coord., Lauren Ippolito; Museum Shop Mgr., Ellen Tozer.
Personnel Profile: Full-Time Paid 23; Part-Time Paid 7; Part-Time Volunteers 82.
Governing Authority: nonprofit organization. Parent Institution: National Geographic Society. Tax-exempt: 501(c)(3).
Institution Type/Description: General Museum.
Collections: personal artifacts, tools & equipment related to National Geographic Society expeditions & scientific fieldwork; anthropology; archaeology; ethnology; photography; natural history; geology; maps.
Research Fields: work of National Geographic explorers, photographers & scientists.
Facilities: 65,000-vol. library; reading room; auditorium; cafe. Museum-related items for sale.
Activities: changing & traveling exhibitions; guided tours; lectures; films; family festivals; geography education porgram; cultural programs.
Publications: National Geographic Society publications; videos; maps.
Hours & Admission Prices: Daily 10-6. Fee charged for special exhibitions. Closed Christmas. &
Attendance: 160,018 (accurate)

NATIONAL LIBRARY OF EDUCATION, 400 Maryland Ave., S.W., Washington, DC 20202. Tel.: 800-424-1616; 202-205-5015. Fax: 202-401-0547. TTY: 202-205-7561.
E-mail: library@ed.gov
Web Site: ies.ed.gov/ncee/projects/nat_ed_library.asp
Institution Type/Description: Education Library.
Collections: over 60,000 books; journals; microfiche; documents; education legislation.
Activities: research.
Hours & Admission Prices: Mon.-Fri. 9-5. Closed federal holidays.

* **NATIONAL MUSEUM OF AFRICAN ART, SMITHSONIAN INSTITUTION, (M),** 950 Independence Ave., S.W., Washington, DC 20560-0006. Mailing Address: P.O. Box 37012, NMAfA MRC 708, Washington, DC 20013-7012. Tel.: 202-633-4600. Fax: 202-357-4879. TDD: 202-357-4814.
E-mail: nmafaweb@si.edu
Web Site: www.si.edu/nmafa
Founded: 1964.
Key Personnel: Assoc. Dir. Administrative Operations & Facility, Alan Knezevich; Interim Asst. Dir. External Affairs, Dale Mott; Assoc. Dir. Devel., Anita Henri; Cur., Bryna Freyer; Cur., Christine Mullen Kreamer; Asst. Cur., Andrea Nicolls; Cur. Photographic Archives, Amy Staples; Chief Conservator, Stephen Mellor; Registrar, Julie Haifley; Librarian, Janet Stanley.
Personnel Profile: Full-Time Paid 45.
Governing Authority: nonprofit. Bureau of the Smithsonian Institution, Washington, DC. Tax-exempt: 501(c)(3).
Institution Type/Description: Art Museum.
Collections: 7,000 objects of African art, wood, metal, ceramic, ivory & fiber; Eliot Elisofon Photographic Archives of 200,000 color slides, 78,000 black & white negatives & 140,000 ft. of motion picture film & videotape on African art & culture.
Research Fields: African art & culture.
Facilities: 20,000-vol. library of books and periodicals on African arts & culture, providing reference & information service by appointment to the public, by telephone, or by mail; a branch of the Smithsonian Institution Libraries. Books, exhibition catalogues & museum-related items for sale.
Activities: guided tours; lectures; credit courses in African art for area

universities; varied public program department: residency fellowship program administered by the Smithsonian Institution makes research facilities available for advanced scholarly research; outreach programs; demonstrations; performances; films permanent exhibits.
Publications: exhibition catalogs; pamphlets; audiovisual material; instructional material for use by elementary & secondary school teachers.
Hours & Admission Prices: Daily 10-5:30. No charge. Closed Christmas. &
Attendance: 300,000 (estimated)
Membership: Patron $500; Advocate $1,000; Benefactor $2,500; Collector's Society $5,000; Director's Circle $10,000.

*** NATIONAL MUSEUM OF AMERICAN HISTORY, (M),** 1400 Constitution Ave., N.W., Washington, DC 20560. Mailing Address: P.O. Box 37012, Washington, DC 20013-7012. Tel.: 202-633-1000. Fax: 202-633-8053. TTY: 202-633-5285.
Web Site: americanhistory.si.edu
Founded: 1964.
Key Personnel: Dir., Brent D. Glass; Chm. (V), John Rogers; Assoc. Dir. Curatorial, David K. Allison; Assoc. Dir. Management & Administration, Janice G. Lilja; Assoc. Dir. Capital Campaign & External Affairs, Maggie Webster; Assoc. Dir., Judith Gradwohl; Dir. Public Affairs, Melinda Machado; Museum Shop Mgr., Kathy Sklar.
Personnel Profile: Full-Time Paid 200; Part-Time Volunteers 545; Interns 150.
Governing Authority: bureau of the Smithsonian Institution, Washington, DC, a nonprofit federally chartered corporation. Tax-exempt: 501(c)(3) & 170(b)(1)(A).
Institution Type/Description: American History & Technology Museum.
Collections: agriculture & natural resources; armed forces history; ceramics & glass items; community life items; costumes & dress; computers; domestic life; electricity & modern physics; engineering & industry; graphic arts; medical sciences; musical instruments; numismatics; photographic history; physical sciences & mathematics; political history; textiles; transportation; popular culture.
Research Fields: agriculture; architecture; archives; astronomy; business history; ceramics; furnishings; costumes; folk; general; glass; graphic arts; industrial; marine; medical; military; music; manufacturing; mining; photography; political; presidential; preservation methods; naval; numismatic; philatelic; science; technology; textile; transportation.
Facilities: 165,000-vol. library of books & periodicals on American history, technology, science, culture & the military, available for inter-library loan & use on premises; separate laboratory operation; 271-seat auditorium; restaurant. Museum-related items for sale.
Activities: guided tours; lectures; films; gallery talks; concerts; docent program; formally organized education programs for undergraduate & graduate college students; permanent, temporary & traveling exhibitions; hands-on exhibits.
Publications: catalogs; guidebooks; scholarly publications; monographs; pamphlets.
Hours & Admission Prices: Daily 10-5:30. No charge; donations accepted. Closed Christmas. &
Attendance: 4,000,000 (estimated)

NATIONAL MUSEUM OF AMERICAN JEWISH MILITARY HISTORY, 1811 R St., N.W., Washington, DC 20009-1603. Tel.: 202-265-6280. Fax: 202-462-3192.
E-mail: nmajmh@nmajmh.org
Web Site: www.nmajmh.org
Founded: 1958.
Key Personnel: Natl. Exec. Dir., Col. Herb Rosenbleeth; Pres. (V), Norman Rosenshein; Dir. Operations, Larry Richardson.
Personnel Profile: Full-Time Paid 16; Part-Time Paid 1; Part-Time Volunteers 3.
Governing Authority: nonprofit. Parent Institution: Jewish War Veterans of the U.S.A. Tax-exempt: 501(c)(3).
Institution Type/Description: National Museum of American Jewish Military History.
Collections: photos; uniforms; firearms; medals; personal papers; documents; history of JWV organization.
Research Fields: Jewish individuals who participated in the U.S. military.
Facilities: 2,550-vol. library; classrooms; 25,000 sq. ft. exhibit space; rental facilities.
Activities: guided tours; lectures; films; temporary & traveling exhibitions.
Publications: quarterly newsletter; calendar; catalogues, GIs Remember: Liberating the Concentration Camps, Centennial - 100 Years of the JWV; An American, A Sailor and A Jew: The Life and Career of Uriah Phillips Levy USN (1792-1862); Women in the Military: A Jewish Perspective.
Hours & Admission Prices: Mon.-Fri. 9-5, Sun. by appointment only. No charge; donations accepted. Closed federal & Jewish holidays. &
Attendance: 3,000 (estimated)

Membership: Individual $25; Family $36; Patron $50; Sustaining $75; Guardian $100; Life $1,000.

THE NATIONAL MUSEUM OF CATHOLIC ART & LIBRARY, 1500 Massachusetts Ave., N.W., Ste. 122, Washington, DC 20005-1800. Tel.: 917-750-0014; 202-450-5707.
E-mail: info@nmcah.org
Web Site: www.nmcah.org
Founded: 2008.
Congressional District: 15
Key Personnel: Pres. & Exec. Dir., Christina Cox; Chief Cur. Art, Mariavelia Savino.
Personnel Profile: Full-Time Paid 4; Part-Time Paid 4; Part-Time Volunteers 5; Interns 5.
Governing Authority: private; nonprofit. Tax-exempt: 501(c)(3).
Institution Type/Description: Religious Art & Christian History Museum.
Collections: old master & contemporary religious paintings, sculpture, prints; liturgical vestments & objects; photographs.
Research Fields: Catholic History; Liturgical objects & vestments; religious painting, sculpture & decorative arts; contemporary Catholic artists; contribution of Catholicism to the Judeo-Christian tradition in the arts; dialogue between Christianity, other world religions, & the cultures they arise in.
Activities: lectures; Catholic book club; rental facilities; educational programs; guest speakers.
Publications: exhibition catalogues; newsletter; class schedules; quarterly, Arts for the Millennium.
Hours & Admission Prices: Closed for relocation. &
Attendance: 10,000 (estimated)

NATIONAL MUSEUM OF CRIME & PUNISHMENT, 575 7th St., N.W., Ste. 3, Washington, DC 20004-1630. Tel.: 202-393-1099. Facebook: National Museum of Crime and Punishment.
E-mail: media@crimemuseum.org
Web Site: www.crimemuseum.org/
Founded: 2008.
Key Personnel: Dir., Janine Vaccarello.
Governing Authority: Parent Institution: Attraction Concepts.
Institution Type/Description: History Museum.
Collections: history of crime & punishment; photographs; crime solving & fighting.
Facilities: Museum-related items for sale.
Activities: high-speed police chase simulators; simulated FBI shooting range; police lineup; child fingerprinting; autopsy table.
Publications: newsletter, American Police Beat.
Hours & Admission Prices: May 21-Sept. 4 Sun. 10-7, Mon.-Thurs. 9-7, Fri.-Sat. 9-8; Sept. 5-May 20 Sun.-Thurs. 10-7, Fri.-Sat. 10-8; call for holiday hours. Self-Guided Tour: adults 12-59 $21.95, seniors 60 & over, military, law enforcement $16.95, children 5-11 $14.95; members & children under 5 no charge. Audio Tour: adults $26.95, seniors $24.95, children $19.95.Closed Thanksgiving; Christmas. &
Membership: Individual $75; Couple $115; Family $190.

*** NATIONAL MUSEUM OF NATURAL HISTORY, (M),** 10th St. & Constitution Ave., N.W., Washington, DC 20560. Mailing Address: P.O. Box 37012, Washington, DC 20013-7012. Tel.: 202-633-2664. Fax: 202-633-0169. TTY: 202-357-1729.
E-mail: info@si.edu
Web Site: www.mnh.si.edu
Founded: 1846.
Key Personnel: Dir., Dr. Cristian Samper; Chm. (V), Kathryn Fuller; Staff Asst. & Bd. Liaison, Diana Munn; Assoc. Dir. Public Programs, Elizabeth Duggal; Assoc. Dir. Research & Collections, Jonathan Coddington; Assoc. Dir. Operations, Susan Fruchter; Chm. Dept. Paleobiology, Dr. Brian Huber; Chm. Dept. Anthropology, Dr. Mary Jo Arnoldi; Chm. Dept. Entomology, Dr. Ted Schultz; Chm. Dept. Botany, Dr. Warren Wagner; Chm. Dept. Invertebrate Zoology, Jon Norenburg; Public Affairs Mgr., Randall Kremer; Museum Shop Mgr., Betty Stolarun.
Personnel Profile: Full-Time Paid 470; Part-Time Paid 85; Part-Time Volunteers 902; Interns 200.
Governing Authority: nonprofit corporation. Parent Institution: Smithsonian Institution, Washington, DC. Tax-exempt: 501(c)(3) & 170(b)(1)(A).
Institution Type/Description: Natural History Museum.
Collections: anthropology; invertebrate & vertebrate zoology; entomology; botany; fossils; minerals, rocks, gems & meteorites.
Research Fields: anthropology; botany; paleobiology; vertebrate & invertebrate zoology; entomology; mineral sciences including volcanology, meteorites; biodiversity.
Facilities: 250,000-vol. library of books on natural history; restaurant; theater. Books & other museum-related items for sale.

Activities: education programs for children, students & adults; docent program; lectures; films; tours; hands-on exhibits; electronic classroom.

Publications: irregular series, Contributions to: Anthropology, Botany, Earth Sciences, Marine Sciences, Paleobiology, Zoology, Handbook of North American Indians; Quest; calendar of events; annual report; AnthroNotes.

Hours & Admission Prices: Winter: daily 10-5:30; Spring & Summer: call for hours. No charge. Closed Christmas. &

Attendance: 7,421,008 (accurate)

NATIONAL MUSEUM OF THE AMERICAN INDIAN, 4th St. and Independence Ave., S.W., National Mall, Washington, DC 20024. Tel.: 202-633-1000.

Web Site: www.americanindian.si.edu

Key Personnel: Dir., Kevin Grover; Dir. Pub. Affairs, Eileen Maxwell.

Governing Authority: Parent Institution: Smithsonian Institution.

Institution Type/Description: Native American Museum.

Collections: Native American history & culture; wood & stone carvings; masks; painted & quilled hides; clothing & feathered bonnets; pottery & basketry; 18th-century materials from Great Lakes region; C.B. Moore collection; Navajo weavings; works on paper & canvas; Caribbean archaeological objects; ceramics from Costa Rica, Central Mexico & Peru; carved jade from the Olmec & Maya peoples; textiles & gold from Andean cultures; funerary, religious & ceremonial objects; films.

Facilities: theater; cafe; resource center. Museum-related items for sale.

Activities: films; theater; special events.

Publications: magazine, American Indian.

Hours & Admission Prices: Daily 10-5:30; groups of 10 or more by appointment. No charge. Closed Christmas. &

Membership: Golden Prairie Circle $25; Riverbed Circle $35; Everglades Circle $50; Sky Meadows Circle $100; Boundary Waters $250; Desert Sands Circle $500; Director's Council of Friends $1,000, $2,500, $5,000.

✳ NATIONAL MUSEUM OF THE UNITED STATES NAVY, (M), 805 Kidder Breese St., S.E., Washington, DC 20374-5060. Tel.: 202-433-4882. Fax: 202-433-8200. TDD: 202-433-4882.

E-mail: karin.hill@navy.mil

Web Site: www.history.navy.mil

Founded: 1961.

Key Personnel: Dir., Kim Nielsen; Cur., Dr. Edward M. Furgol; Dir. Public Programs, Karin Hill; Museum Shop Mgr., Andre Dyson; Security, James Creasman.

Personnel Profile: Full-Time Paid 11; Part-Time Paid 5; Part-Time Volunteers 18; Interns 10.

Governing Authority: federal. Parent Institution: Dept. of the Navy. Subsidiary Institution: Naval Historical Center. Tax-exempt.

Institution Type/Description: Naval History Museum: housed in the former Breech Mechanism Shop of the old Naval Gun Factory.

Collections: documents; paintings; weapons; photographs; medals; decorations; memorabilia relating to naval personalities; awards; ship models; swords; ordnance; maps; prints; uniforms; flags; scientific instruments; Destroyer Barry (DD-933); 19th & 20th-century guns, cannons submarines, naval artifacts.

Research Fields: U.S. Naval history.

Activities: guided tours; permanent & temporary exhibits pertaining to naval history; educational program; lecture series; musical programs. Museum Sponsors: US Navy in the Korean War exhibit.

Publications: quarterly newsletter, Naval Historical Foundation, Pull Together.

Hours & Admission Prices: April to Labor Day Mon.-Fri. 9-5, Sat.-Sun. 10-5; Sept.-May Mon.-Fri. 9-4, Sat.-Sun. 10-5. No charge; donations accepted. Closed New Year's Day; Thanksgiving; Christmas Eve & Day. &

Attendance: 215,000 (accurate)

Membership: Active $25; Sustaining $50; Associate $75; Fellowship $100; Life $500.

NATIONAL MUSEUM OF WOMEN IN THE ARTS, (M), 1250 New York Ave., N.W., Washington, DC 20005 3970. Tel.: 202-783-5000. Fax: 202-393-3235.

Web Site: www.nmwa.org

Founded: 1981.

Key Personnel: C.E.O. & Dir., Dr. Susan Fisher Sterling; Chm. (V), Wilhelmina Cole Holladay; Pres. (V), Gina Adams; Dir. Education, Deborah Gaston; Chief Cur., Kathryn Wat; Registrar, Catherine Badl; Dir. Library & Research Center, Heather Slavia; Dir. Membership, Christina Knowles; Dir. National Programs, Ilene Gutman; Dir. Retail & Wholesale Operations, Lynda Marks.

Personnel Profile: Full-Time Paid 50; Full-Time Volunteers 129; Part-Time Paid 54; Part-Time Volunteers 175; Interns 20.

Governing Authority: nonprofit organization. Tax-exempt: 501(c)(3).

Institution Type/Description: Art Museum: housed in 1908 second Renaissance revival style structure.

Collections: works by more than 1,000 women artists including paintings, sculptures, pottery, prints, drawings, books, silver, special collections including over 18,500 files on women artists, Irene Rice Pereira's library, 19th & 20th century bookplates & 1,000 unique & limited edition artists' books & photography. Art from the Renaissance to the present.

Major Exhibits: Work by Hand: Hidden Labor & Historical Quilts, 12/20/13-4/27/14; Circa '75: Judy Chicago, 1/17/14-4/13/14; Meret Oppenheim in the Collection of the Nautical Museum of Women in the Arts, 4/25/14-9/14/14; Total Art: Contemporary Video, 6/6/14-10/12/14; Picturing Mary: Woman, Mother, Idea, 12/5/14-4/12/15.

Research Fields: women artists' history & contributions.

Facilities: 18,500-vol. library & research center of books, monographs, exhibition catalogs, & periodicals pertaining to women artists; 200-seat auditorium; 500-seat banquet hall with kitchen; cafe; classrooms. Bookstore with museum-related items for sale.

Activities: guided tours; lectures; films; concerts; symposia; member previews; organized education programs; docent program; loan, temporary, & traveling exhibitions.

Publications: quarterly news magazine; catalogs & brochures.

Hours & Admission Prices: Museum: Mon.-Sat. 10-5, Sun. 12-5. Adults $10, seniors & students w/ID $8; discounts to AAM members; children under 18 & members no charge. Library: Mon.-Fri. 11-2. Closed New Year's Day; Thanksgiving; Christmas. &

Attendance: 126,637 (accurate)

Membership: Supporter $50; Young Professionals Forum $75; Dual Plus $95; Friend $150; Sustainer $275; Benefactor $500; President's Club $1,000; Library Fellow $1,500; Museum Council $2,000; Museum Patron $5,000.

NATIONAL PARK SERVICE, 1201 Eye St., N.W. 2265, Washington, DC 20005-5905. Tel.: 202-354-2000. Fax: 202-371-6757.

E-mail: ronald_wilson@nps.gov

Web Site: www.nps.gov/history/museum/

Founded: 1916.

Key Personnel: Chief Cur., Ronald C. Wilson; Museum Registrar, Kathleen Byrne; Museum Cur., Joan Bacharach; Sr. Cur. Natural History, Greg McDonald; Archivist, John W. Roberts.

Governing Authority: federal. Parent Institution: National Park Service, U.S. Dept. of the Interior, 18th & C Sts., N.W., Washington, DC 20240. Tax-exempt.

Institution Type/Description: National Park Service.

Collections: archeology; ethnology; history; archives; biology; geology; paleontology; 124 million objects, specimens & archival documents in 365 park units.

Research Fields: all areas of collections.

Facilities: library of reference works in museology, history, technology & decorative arts available for use on premises.

Activities: collections management policy & training for National Park System.

Publications: NPS Museum Handbook; Conserve-O-Grams; ANCS & User Manual.

Hours & Admission Prices: Offices: Mon.-Fri. 7:45-4:15. Closed national holidays. &

NATIONAL PARK SERVICE - PEIRCE BARN, 5200 Glover Rd., N.W., Washington, DC 20015. Tel.: 202-895-6222. Fax: 202-895-6230.

E-mail: ricardo_perez@nps.gov

Web Site: www.nps.gov/rocr

Formerly: Peirce Mill and Peirce Barn

Founded: 1936.

Key Personnel: Park Supt., Adrienne Coleman; Park Ranger, Ricardo Perez.

Personnel Profile: Full-Time Paid 1.

Governing Authority: federal. Administered by U.S. Dept. of the Interior, National Park Service, Rock Creek Park.

Institution Type/Description: Historic Building: c.1820 Grist Mill & Peirce Mill located in Rock Creek Park.

Collections: water paddle wheel; Peirce Mill machinery; furniture; records; library. Peirce Barn: The Mills of Rock Creek Valley & The Peirce Estate.

Research Fields: local history.

Facilities: 250-vol. library of books on mill operation, construction & history, and Northwest Washington, D.C. history.

Activities: guided tours; demonstrations; curriculum-based education programs for adults & children.

Publications: brochure, Peirce Mill.

Hours & Admission Prices: Mill & Barn by appointment. Research Facility: by appointment only. Closed federal holidays.

✻ NATIONAL PORTRAIT GALLERY, (M), Eighth & F Sts., N.W., Donald W. Reynolds Center for American Art & Portraiture, Washington, DC 20001. Mailing Address: P.O. Box 37012, Victor Bldg., Ste. 410, MRC-973, Washington, DC 20013-7012. Tel.: 202-633-1000. Fax: 202-633-8243. TDD: 202-633-8229.
E-mail: npgweb@si.edu
Web Site: www.npg.si.edu
Founded: 1962.
Key Personnel: Dir., Kim Sajet; Assoc. Dir. External Affairs & Earned Revenue, Nik Apostolides; Chm., Jack Watson; Librarian, Doug Litts; Conservator, Cindy Lou Molnar; Historian, David Ward; Facilities Mgr., Andrew Smith; Dir. Education Program, Rebecca Kasemeyer; Acting Dir. Devel., Jennifer Renner; Chief Cur., Brandon Fortune; Head Exhibitions, Claire Kelly; Cur. Prints & Drawings, Wendy Wick Reaves; Cur. Photographs, Ann Shumard; Dir. New Media, Deb Sisum; Assoc. Dir. Finance & Operation, Rich Reichley; Chief, Design & Production, Tibor Waldner; Dir., Collections, Informatiion & Research, Linda Thrift; Chief Registrar, John McMahon; registrar for Exhibitions & Loans, Molly Grimsley; Editorial Dir., Dru Dowdy; Museum Shop Mgr., Milissa Ferrari.
Personnel Profile: Full-Time Paid 60; Part-Time Paid 2; Part-Time Volunteers 46; Interns 28.
Governing Authority: nonprofit organization. Parent Institution: Smithsonian Institution. Tax-exempt: 501 (c)(3) & 170(b)(1)(A).
Institution Type/Description: Art & History Museum: housed in c.1840, former U.S. Patent Office building.
Collections: portraits of men & women who have made notable contributions to the history & development of the United States.
Major Exhibits: The Meade Brothers: Pioneers in American Photography, 6/14/13-6/1/14; One Life: Martin Luther King Jr., 6/28/13-6/1/14; Dancing the Dream, 10/4/13-7/13/14; Yousuf Karsh: America Portraits, 11/8/13-11/2/14; Mr. Lincoln's Washington, 12/13/13-1/25/15; American Cool, 2/7/14-9/7/14; Face Value: Portraiture in the Age of Abstraction, 4/18/14-1/11/15; One Life: Grant and Lee, 7/4/14-5/25/15; Portraiture Now: Staging the Self, 8/22/14-4/12/15.
Research Fields: American portraiture; biography; history.
Facilities: 60,000-vol. library of American art, history & biography available for inter-library loan & to qualified persons; field research station; reading room; 346-seat auditorium; classrooms; restaurant. Books, catalogs, reproductions & other museum related items for sale.
Activities: guided tours; lectures; films; gallery talks; formally organized educational programs; docent program; inter-museum loan, permanent & temporary exhibitions; private events.
Publications: catalogs of special exhibitions & various segments of the permanent collection; commissioned works & research.
Hours & Admission Prices: Daily 11:30-7. No charge. Closed Christmas. ♿
Attendance: 1,000,000 (estimated)
Membership: National Portrait Gallery & Smithsonian American Art Museum: Pioneers Circle $1,000; Performers Circle $2,500; Patriots Circle $5,000; Presidents Circle $10,000.

NATIONAL POSTAL MUSEUM, SMITHSONIAN INSTITUTION, 2 Massachusetts Ave., N.E., Washington, DC 20002. Tel.: 202-633-5555. Fax: 202-633-9393. TTY: 202-633-9849.
E-mail: npm@npm.si.edu
Web Site: www.postalmuseum.si.edu.
Founded: 1993.
Key Personnel: Dir., Allen Kane.
Personnel Profile: Full-Time Paid 39; Part-Time Paid 1; Part-Time Volunteers 50; Interns 2.
Governing Authority: federal. Parent Institution: Smithsonian Institution, 900 Jefferson Dr., S.W., Washington, D.C. 20560. Tax-exempt: 501(c)(3).
Institution Type/Description: Postal History Museum: housed in 1914 Old City Post Office Building.
Collections: stamp collection; postal history material pre-dating stamps; postal vehicles; mailboxes; meters; greeting cards; letters & stationery; barcode readings.
Research Fields: postal history; Philately.
Facilities: Museum shop & stamp store.
Activities: hands-on activities; lectures; films.
Publications: books, Reaching Rural America; Great American Post Offices; Mail on the Move; Motorized Mail; Horse-Drawn Mail; Illustrated Guide to the National Postal Museum; Owney: The Mascot of the Railway Service; quarterly member newsletter, En Route.
Hours & Admission Prices: Daily 10-5:30. No charge; donations accepted. Closed Christmas. ♿
Attendance: 500,000 (accurate)
Membership: Annual $25.

NATIONAL SOCIETY OF THE CHILDREN OF THE AMERICAN REVOLUTION MUSEUM, 1776 D St., N.W., Rm. 224, Washington, DC 20006-5303. Tel.: 202-638-3153. Fax: 202-737-3162.
E-mail: hq@nscar.org
Web Site: www.nscar.org
Founded: 1895.
Key Personnel: Sr. National Pres., Lois Schmidt.
Governing Authority: society. Affiliated with the National Society, Daughters of the American Revolution. Tax-exempt: 501(c)(3).
Institution Type/Description: History Museum.
Collections: decorative & applied arts.
Activities: guided tours; inter-museum loan, permanent & temporary exhibitions.
Hours & Admission Prices: Mon.-Fri. 8:30-4, Sat. 9-5. No charge; donations accepted.

NATIONAL TRUST FOR HISTORIC PRESERVATION, (M), 1785 Massachusetts Ave., N.W., Washington, DC 20036-2117. Tel.: 202-588-6000. Fax: 202-588-6038.
E-mail: feedback@nthp.org
Web Site: www.preservationnation.org
Founded: 1949.
Key Personnel: Vice Pres. Stewardship of Historic Sites, Jim Vaughan; Administrative Dir., Stewardship of Historic Sites, Lyn Moriarity; Dir. Museum Collections, Terri Anderson; Dir. Interpretation & Education, Max Van Balgooy; Graham Gund Architect of the Natl. Trust, Barbara Campagna, FAIA, LEED AP; Assoc. Architect, Elizabeth Milnarik; Sr. Archaeologist, Lynne Lewis.
Governing Authority: nonprofit organization. National Trust collection includes 29 historic sites: African Meeting House & Abiel Smith School, Boston, MA; African Meeting House, Nantucket, MA; Belle Grove, Middletown, VA; Brucemore, Cedar Rapids, IA; Chesterwood, Stockbridge, MA; Cliveden, Philadelphia, PA; Cooper-Molera Adobe, Monterey, CA; Decatur House, Washington, DC; Drayton Hall, Charleston, SC; Farnsworth House, Plano, IL; Filoli, Woodside, CA; Frank Lloyd Wright Home & Studio, Oak Park, IL; Frank Lloyd Wright's Pope-Leighey House, Mount Vernon, VA; Frederick C. Robie House, Chicago, IL; Gaylord Building, Lockport, IL; James Madison's Montpelier, Montpelier Station, VA; Kykuit, Tarrytown, NY; Lower East Side Tenement Museum, New York, NY; Lyndhurst, Tarrytown, NY; Oatlands, Leesburg, VA; Shadows-on-the-Teche, New Iberia, LA; Touro Synagogue, Newport, RI; Woodlawn, Mount Vernon, VA; Woodrow Wilson House, Washington, DC; President Lincoln & Soldiers' Home National Monument, Washington, DC; Philip Johnson's Glass House, New Canaan, CT; Villa Finale, San Antonio, TX; Acoma Sky City, Acoma, NM; Hotel de Paris, Georgetown, CO. Tax-exempt: 501(c)(3).
Institution Type/Description: Historical Society.
Collections: decorative & fine arts; local historic artifacts; documents; archaeological artifacts.
Research Fields: American history & culture.
Activities: tours; educational programs; special events.
Publications: bimonthly magazine, Preservation; bimonthly professional journal, Forum, Preservation Information Series, brochures & pamphlets.
Hours & Admission Prices: See individual listings for their hours & admission fees. ♿
Attendance: 809,204 (accurate)
Membership: Individual $20; Family $24; Contributing $50; Sustaining $100; Participating $250; Donor & Corporate $1,000.

NATIONAL WWII MEMORIAL, 17th St., (between Constitution & Independence Aves.), Washington, DC 20006. Mailing Address: National Park Service, 900 Ohio Dr., S.W., Washington, DC 20242-0002. Tel.: 202-619-7222; 800-639-4992.
E-mail: wwiicustserv@americancustomercare.com
Web Site: www.wwiimemorial.com
Institution Type/Description: Military Memorial.
Collections: World War II history.
Activities: special events.
Hours & Admission Prices: Daily 24 hrs. Rangers on duty to answer questions daily 9:30-11:30.

NATIONAL ZOOLOGICAL PARK, SMITHSONIAN INSTITUTION, 3001 Connecticut Ave., N.W., Washington, DC 20008-2598. Tel.: 202-633-4800. Fax: 202-673-4607.
Web Site: www.si.edu/natzoo
Founded: 1889.
Key Personnel: Acting Dir. & Assoc. Dir. Conservation and Science, Dr. Steven Montfort; Chm. (V), Clark Bunting; Assoc. Dir. Communications,

Pamela Baker-Masson; Assoc. Dir. Animal Care, Dr. Don Moore; Assoc. Dir. Education, Joseph Sacco; Chief Management Svcs. Office, Alexander Beim; Project Exec., Debra Nauta-Rodriguez; Sr. Nutritionist, Mike Maslanka; Assoc. Dir. Exhibits & Park Management, Charles Fillah; Head Veterinarian, Dr. Suzan Murray; Head Pathologist, Dr. Timothy Walsh; Chief of Police, Russell Walkowich; Safety Mgr., Donald Williams; Exec. Dir. Friends of the National Zoo (FONZ), Robert Lamb.
Personnel Profile: Full-Time Paid 247; Part-Time Volunteers 1,086.
Governing Authority: federal, nonprofit. Parent Institution: Smithsonian Institution. Tax-exempt: 501(c)(3) & 170(b)(1)(A).
Institution Type/Description: Zoological Park.
Collections: mammals: 131 species, 1,337 individuals; birds: 152 species, 971 individuals; reptiles: 72 species, 482 individuals; amphibians: 19 species, 285 individuals; invertebrates: 110 species; 3,945 individuals.
Research Fields: wildlife ecology & management; animal behavior, breeding, husbandry & conservation; exotic animal medicine & pathology; zoo education; comparative zoology, nutrition, reproductive physiology.
Facilities: 7,000-vol. library of zoological material available for inter-library loan & on premises to qualified scholars; 300-seat auditorium; classrooms; resource room; 163-acre zoological park.
Activities: curriculum guides & teacher resources; cooperative teaching programs with local medical & veterinary school; study grants for students & internships for veterinarians; zoo society, Friends of the National Zoo; series of symposia, seminars & lectures; research fellowships, wildlife management training for officials of developing countries.
Publications: Zoogoer Magazine; Smithsonian New Zoo; Self-Guides for Teachers; How to Zoo: An Educator's Guide to the Smithsonian National Zoo.
Hours & Admission Prices: Grounds: April-Oct. daily 6-8; Nov.-March daily 6-6. Buildings: April-Oct. daily 10-6; Nov.-April daily 10-4:30. No charge; donations accepted. Parking fee $20 maximum. Closed Christmas. &
Attendance: 2,600,000 (estimated)
Membership: Senior Citizen $35; Individual $40; Household $55.

NAVAL HISTORY AND HERITAGE COMMAND, Washington Navy Yard, Navy Art Collection, Bldg. 67, 805 Kidder Breese, S.E., Washington, DC 20374-5060. Tel.: 202-433-3815.
Formerly: Naval Historical Center
Institution Type/Description: History Museum.
Collections: U.S. naval history; over 15,000 paintings, prints, drawings, & sculpture.
Hours & Admission Prices: Mon.-Fri. 9-3:30. Photo ID required for adults to enter base. No charge. Closed Federal holidays.

NEWSEUM, 555 Pennsylvania Ave., N.W., Washington, DC 20001-2114. Tel.: 888-639-7386.
E-mail: info@newseum.org
Web Site: www.newseum.org
Founded: 1997.
Key Personnel: C.E.O., James Duff; Sr. Vice Pres. Broadcasting, Paul Sparrow; Sr. Vice Pres. Devel., Courtney Sorls; Vice Pres. Exhibits, Programs & Media Rels., Cathy Trost; Sr. Vice Pres. Operations, James Thompson.
Personnel Profile: Full-Time Paid 169; Part-Time Volunteers 185.
Governing Authority: private foundation; nonprofit. Parent Institution: Freedom Forum. Tax-exempt: 501(c)(3).
Institution Type/Description: Interactive History & News Museum.
Collections: newspapers; photographs; media technology artifacts & personal artifacts related to major news figures; history of news development & technology.
Research Fields: news history; people; technology; techniques.
Hours & Admission Prices: Daily 9-5. Adults $21.95, seniors $17.95, youth 7-18 $12.95; discounts to AAM members & groups. Closed New Year's Day; Thanksgiving; Christmas. &
Membership: Individual $75; Dual $125; Family $175.

THE OCTAGON MUSEUM, (M), 1799 New York Ave., N.W., Washington, DC 20006-5207. Mailing Address: 1735 New York Ave., N.W., Washington, DC 20006. Tel.: 202-638-3221. Fax: 202-626-7420.
E-mail: ericagees@aia.org
Web Site: www.archfoundation.org/octagon/
Founded: 1942.
Key Personnel: Dir., Erica Rioux Gees.
Personnel Profile: Full-Time Paid 1; Part-Time Paid 8; Part-Time Volunteers 40.
Governing Authority: nonprofit organization. Parent Institution: The AA Legacy. Tax-exempt.

Institution Type/Description: Architecture Museum & Historic House: 1799-1801 Federal townhouse built for Colonel John Tayloe III, based on designs by Dr. William Thornton, served as first temporary White House during the winter of 1814-1815 for President James & Dolley Madison; the Treaty of Ghent was signed there, ending the War of 1812.
Collections: original architectural drawings & photographs; changing exhibits of architecture & allied arts; archaeological artifacts; decorative arts of the period 1800-1828.
Research Fields: architecture; decorative & design arts; Washington history.
Facilities: Gift shop with exhibition catalogs & books for sale.
Activities: guided tours; lectures; inter-museum loan; temporary exhibitions.
Publications: exhibit catalogues; books; research series focusing on the Federal Period through catalogues & exhibitions.
Hours & Admission Prices: Thurs.-Fri. 1-4. Call for information.
Attendance: 1,500 (estimated)

THE OLD STONE HOUSE, 3051 M St., N.W. (Georgetown), Washington, DC 20007-3702. Tel.: 202-426-6851. Fax: 202-895-6230. TDD: 202-426-0125.
Web Site: www.nps.gov/olst
Founded: 1950.
Key Personnel: Asst. Park Supt., Cynthia Cox; Park Ranger, Ron Harvey.
Personnel Profile: Full-Time Paid 3; Part-Time Paid 1; Part-Time Volunteers 3.
Governing Authority: federal. Administered by U.S. Dept. of the Interior, National Park Service. Tax-exempt.
Institution Type/Description: Historic Building: 1765 pre-Revolutionary building located in the Georgetown area of the District of Columbia.
Collections: 18th-century furniture; glassware; cooking ware; kitchen; gardens.
Research Fields: local history.
Facilities: 250-vol. library of books on crafts & history.
Activities: guided tours; formally organized education programs for children, adults & undergraduate college students; permanent & temporary exhibitions.
Publications: brochure, The Old Stone House.
Hours & Admission Prices: Daily 12-5; groups by appointment. No charge; donations accepted. Closed federal holidays. &
Attendance: 70,000 (accurate)

PARISH GALLERY, 1054 31st St., N.W., Washington, DC 20007-6034. Tel.: 202-944-2310.
E-mail: parishgallery@bigplanet.com
Web Site: www.parishgallery.com
Institution Type/Description: Art Gallery.
Collections: paintings; sculpture.
Activities: rental facilities; temporary exhibitions.
Hours & Admission Prices: Tues.-Sat. 12-6; other times by appointment.

*** THE PHILLIPS COLLECTION, (M),** 1600 21st St., N.W., Washington, DC 20009-1090. Tel.: 202-387-2151. Fax: 202-387-2436.
E-mail: info@phillipscollection.org
Web Site: www.phillipscollection.org
Founded: 1921.
Key Personnel: Dir., Dorothy Kosinski; Chm. Bd. Trustees (V), George Vradenburg; Vice Chm., Linda Lichtenberg Kaplan; Vice Chm., Leo E. Zickler; Chief Cur., Eliza Rathbone; C.O.O., Susan J. Nichols; Dir. Center for the Study of Modern Art & Cur.-at-large, Klaus Ottmann; Cur. Modern & Contemporary Art, Vesela Sretenovic; Cur., Elsa Smithgall; Asst. Cur., Sue Behrends Frank; Dir. Budgeting & Reporting, Cheryl Nichols; Installations Mgr., William Koberg; Dir. Music, Caroline Mousset; Chief Registrar, Joseph Holbach; Librarian, Karen Schneider; Dir. Devel., Kara Mullins; Assoc. Registrar, Sarah Anderson; Assoc Registrar Exhibitions, Trish Waters; Mgr. Membership, Jeffrey Petrie; Chief Information Officer, Darci Vanderhoff; Dir. of Education, Suzanne Wright; Dir. Mktg. & Communications, Ann Greer; Dir. Human Resources, Angela Gillespie; Dir. Security & Facilities, Dan Datlow; Museum Shop Mgr., Pete Bernal.
Personnel Profile: Full-Time Paid 70; Part-Time Paid 101; Part-Time Volunteers 87; Interns 18.
Governing Authority: nonprofit organization. Tax-exempt: 501(c)(3).
Institution Type/Description: Art Museum: housed in the former residence of Duncan Phillips family, 1897 Georgian Revival home.
Collections: 19th, 20th & 21st century European & American paintings & sculpture; 3,000 works of art; historic house.
Major Exhibits: Van Gogh Repetitions (T), 10/12/13-1/26/14; John F. Simon, 10/17/13-2/6/14.
Research Fields: 19th- & 20th-century European & American art.
Facilities: 9,000-vol. library of books, periodicals & catalogues, available for

museum members, scholars & researchers by appointment only; auditorium; classroom; meeting facilities; seminar room. Books, catalogues, reproductions & other museum-related items for sale.

Activities: permanent & temporary exhibitions; guided tours; lectures; parent/child workshops; training programs for Art Educators; school tours; gallery talks. Dupont-Kalorama Museums Consortium: Museum Walk Weekend in June; introductory tours; Phillips After 5; Sunday Concerts; Conversations with Artists.

Publications: books, brochures; catalogs of special exhibitions & loan exhibitions; member magazine.

Hours & Admission Prices: Tues.-Wed. & Fri.-Sat. 10-5, Thurs. 10-8:30, Sun. 11-6. Concerts: Oct.-May Sun. 4pm. Adults $12; members no charge. Closed New Year's Day; Independence Day; Thanksgiving; Christmas Eve & Day.

Attendance: 150,000 (estimated)

Membership: Individual $60; Dual & Family $100; Contemporary $125 (for younger members); Associate $175; Friend $300; Contributing Friend $600; Supporting Friend $1,000; Patron $2,000; Galley Patron $3,500; Collection Patron $5,000; Chairman's Circle $10,000; Leadership Circle $25,000; Executive Circle $50,000.

POPE JOHN PAUL II CULTURAL CENTER, 3900 Harewood Rd., N.E., Washington, DC 20017-1505. Tel.: 202-635-5400. Fax: 202-635-5411.

E-mail: info@jp2cc.org
Web Site: www.jp2cc.org
Founded: 2001.
Key Personnel: Dir., Rev. Steven Boguslowski, O.P.; Chm. (V), Archbishop Donald Wuerl; Pres. (V), Cardinal Adam Maida; Museum Shop Mgr., Veronica Osborne.
Personnel Profile: Full-Time Paid 23; Part-Time Paid 30; Part-Time Volunteers 3; Interns 2.
Governing Authority: Tax-exempt.
Institution Type/Description: Cultural Center.
Collections: Catholic Church history; Papal & Polish heritage; photographs; paintings; statues; interactive exhibits.
Facilities: chapel; cafe. Museum-related items for sale.
Activities: children's activities; lectures; movies.
Hours & Admission Prices: Tues. & Thurs.-Sat. 10-5, Sun. 12-5. Suggested Donations: families $15, individuals $5, seniors & students $4; discounts to members. Closed Federal holidays.
Attendance: 29,000 (estimated)
Membership: Individual $35.

PRESIDENT LINCOLN'S COTTAGE, (M), Rock Creek Church Rd., N.W., (at Upshur St. N.W.), Washington, DC 20011. Mailing Address: AFRH-W 558, 3700 N. Capitol St., N.W., Washington, DC 20011-8400. Tel.: 202-829-0436, ext. 31231. Fax: 202-829-0437. Facebook: Lincolns Cottage.

E-mail: lincolnscottage@savingplaces.org
Web Site: www.lincolncottage.org
Founded: 2008.
Key Personnel: Dir., Erin Carlson Mast; Exec. Asst., Zach Klitzman; Assoc. Dir. Devel., John Davison; Mktg. & Membership Coord., Hilary Malson; Assoc. Dir. Programs, Callie Hawkins; Events & Programs Coord., Sahand Miraminy; Sr. Visitor Svcs. Assoc., Jamie Cooper; Preservation Mgr., Jeffrey Larry.
Personnel Profile: Full-Time Paid 7; Part-Time Paid 10; Part-Time Volunteers 3.
Volunteer Hours: 450
Operating Expenses: 1,100,000
Operating Income: 1,200,000
Governing Authority: private; nonprofit organization. Parent Institution: National Trust for Historic Preservation, Washington, DC. Tax-exempt: 501(c)(3).
Institution Type/Description: Historic House: the summer home of Lincoln and his family during his presidency.
Collections: Lincoln history; sculpture; prints; period furnishings.
Major Exhibits: originALs: The Albert Nelson See Diary, Spring 2014.
Research Fields: house; diaries; personal artifacts.
Facilities: educational facilities. Museum-related items for sale.
Activities: formal education programs; guided tours; lectures; permanent, loan & temporary exhibitions; public programs.
Publications: quarterly newsletter, Cottage Courier.
Hours & Admission Prices: Mon.-Sat. 9:30-4:30, Sun. 10:30-4:30. Adults $15, students 6-12 $5; discounts to members, military & groups; children under 6 no charge. Closed New Year's Day; Thanksgiving; Christmas.
Attendance: 26,000 (accurate)

Membership: Dual $50; Sponsor $100; July 22nd Society $250; Presidential Guard $500; Lincoln's Cabinet $1,000; Cottage Benefactor $5,000.

*** THE PRESIDENT WOODROW WILSON HOUSE, (M),** 2340 S St., N.W., Washington, DC 20008-4016. Tel.: 202-387-4062. Fax: 202-483-1466.

E-mail: renholm@woodrowwilsonhouse.org
Web Site: www.woodrowwilsonhouse.org
Founded: 1963.
Key Personnel: Museum Dir., Robert A. Renholm; Chm. (V), Lena Barnes; Cur., Amberly Meli; Cur., Stephanie Daugherty; Mktg. & Events Mgr., Sarah Andrews.
Personnel Profile: Full-Time Paid 3; Part-Time Paid 17; Part-Time Volunteers 45; Interns 2.
Governing Authority: nonprofit. Parent Institution: National Trust for Historic Preservation, 1785 Massachusetts Ave., N.W., Washington, DC 20036. Tax-exempt: 501(c)(3).
Institution Type/Description: Historic House Museum: housed in the former home of President Wilson.
Collections: original furnishings; memorabilia of World War I; early 20th-century decorative arts & political memorabilia; World War I photos & posters; textiles; gifts from various heads of state.
Research Fields: political & social history; interior design.
Facilities: meeting & banquet space; gardens.
Activities: guided tours; school programs; lectures; special events; walking tours.
Publications: newsletter; exhibit catalogues.
Hours & Admission Prices: Tues.-Sun. 10-4. Adults $10, senior citizens $8, students $5; discounts to AAM, ICOM & National Trust members; members & children under 12 no charge. Closed major holidays.
Attendance: 10,023 (accurate)
Membership: Basic $50; Family $100; Patron $250; Heritage $1,000.

PROVISIONS LIBRARY, 5813 Nevada Ave., N.W. #1100, Washington, DC 20015-2547. Tel.: 202-670-7768. Fax: 202-232-1651.

E-mail: provisionslibrary@gmail.com
Web Site: www.provisionslibrary.org
Founded: 2004.
Key Personnel: Exec. Dir. & Cur., Donald Russell; Chm. (V), Ethelbert Miller.
Personnel Profile: Full-Time Paid 2; Full-Time Volunteers 1; Part-Time Paid 1; Part-Time Volunteers 5; Interns 2.
Governing Authority: private; nonprofit organization. Tax-exempt: 501(c)(3).
Institution Type/Description: Library & Art Museum.
Collections: books & periodicals on global social issues.
Research Fields: global social change.
Facilities: library; educational facilities.
Activities: arts festivals; films; formal education programs for adults & college students; guided tours; workshops; lectures; participatory & traveling exhibits; study clubs; training programs for professional museum workers.
Publications: weblog, Signal Fire.
Hours & Admission Prices: Tues.-Fri. 12-5. No charge; donations accepted.
Attendance: 25,000 (estimated)
Membership: Basic $95; Family $150; Sponsor $250 & up.

RAILS-TO-TRAILS CONSERVANCY, The Duke Ellington Building, 2121 Ward Ct., NW, 5th Fl., Washington, DC 20037-1247. Tel.: 202-331-9696. Fax: 202-331-9680.

Web Site: www.railtrails.org
Founded: 1985.
Key Personnel: Pres., Keith Laughlin; Chm. (V), Joe Louis Barrow, Jr.
Personnel Profile: Full-Time Paid 26; Interns 2.
Governing Authority: nonprofit organization. State Chapters: RTC-Michigan, Lansing, MI; RTC-Florida, Tallahassee, FL; RTC-Pennsylvania, Harrisburg, PA; RTC-Ohio, Columbus, OH. Tax-exempt: 501(c)(3).
Institution Type/Description: National Agency.
Collections: files documenting the nation's 1,000 rail-trails open to the public & the 400 in development.
Research Fields: economic benefits to local communities; corridor acquisition & trail development; impacts on adjacent property; technological & legal matters.
Facilities: Publications & trail-related products for sale.
Activities: Annual Events: Regional Seminars; Biennial National Conference.
Publications: quarterly magazine, Rails-To-Trails: A Celebration of Trails and Greenways; annual directory of rail-trails, America's Rail-Trails.
Hours & Admission Prices: Mon.-Fri. 9-5. No charge. Closed federal holidays.

Membership: Individual $18; Family $25; Sustaining $35; Patron $50; Benefactor $100.

* **RENWICK GALLERY OF THE SMITHSONIAN AMERICAN ART MUSEUM, (M),** 1661 Pennsylvania Ave., N.W., (at 17th St.), Washington, DC 20006. Mailing Address: P.O. Box 37012, MRC 510, Washington, DC 20013-7012. Tel.: 202-633-1000.
E-mail: americanartinfo@si.edu
Web Site: www.americanart.si.edu
Formerly: Renwick Gallery of the National Museum of American Art, Smithsonian Institution
Founded: 1972.
Key Personnel: Dir. The Margaret and Terry Stent, Elizabeth Broun; Deputy Dir., Rachel Allen; Chief, Renwick Gallery, Robyn Kennedy; Asst. to Chief, Renwick Gallery, Rebecca Robinson; Deputy Chief Operations, Renwick Gallery, Fern Bleckner; The Charles & Fleur Bresler Cur. American Craft & Decorative Art, Renwick Gallery, Nicholas Bell; Research Asst., Renwick Gallery, Debrah Dunner; Museum Specialist, Renwick Gallery, Marguerite Hergesheimer; Exhibits Specialist, Renwick Gallery, Jim Baxter; Public Affairs Officer, Laura Baptiste; Public Affairs Specialist, Courtney Rothbard; Public Affairs Assoc., Mandy Young.
Governing Authority: Parent Institution: Dept. of the Smithsonian American Art Museum, a unit of the Smithsonian Institution, Washington, DC, a nonprofit federally chartered corporation. Tax-exempt: 501(c)(3) & 170(b)(1)(A).
Institution Type/Description: American Crafts Museum.
Collections: contemporary American crafts.
Research Fields: contemporary & historical American crafts; fellowship program.
Activities: videos; lectures; concerts; craft demonstrations; temporary exhibitions.
Publications: exhibition catalogues; pamphlets; quarterly calendar of events.
Hours & Admission Prices: Daily 10-5:30. No charge. Closed Christmas. &
Attendance: 165,000 (accurate)
Membership: Director's Circle $1,500 & up.

THE RESTAURANT MUSEUM, Washington, DC 20005. Mailing Address: 1330 Massachusetts Ave., Washington, DC 20005. Tel.: 202-479-2572.
E-mail: info@therestaurantmuseum.com
Web Site: therestaurantmuseum.com
Founded: 2010.
Key Personnel: Dir., William Wooby; Public Rels., Jennifer Leupo.
Personnel Profile: Part-Time Volunteers 1.
Governing Authority: private; nonprofit organization.
Institution Type/Description: Virtual History Museum.
Collections: history of restaurants, hotels & the hospitality industry; newspapers; photographs; personal stories.
Hours & Admission Prices: Online museum.

ROCK CREEK NATURE CENTER, 5200 Glover Rd., N.W., Washington, DC 20015-1095. Mailing Address: 3545 Williamsburg Lane, N.W., Washington, DC 20008-1207. Tel.: 202-895-6070. Fax: 202-895-6230. TDD: 202-426-6829.
Web Site: www.nps.gov/rocr
Founded: 1960.
Congressional District: 5
Key Personnel: Asst. Park Supt., Cynthia Cox.
Governing Authority: federal. Parent Institution: National Park Service. Subsidiary Institution: Dept. of Interior. Tax-exempt.
Institution Type/Description: Nature Center.
Collections: botany; entomology; geology; zoology; observation bee hive.
Facilities: reference library of books on natural history in all basic categories: Native Americans, astronomy, conservation and the National Parks available for research on premises by request; auditorium; planetarium. Guides, postcards, charts, books & teaching materials for sale.
Activities: guided nature walks; planetarium shows; animal demonstrations; films; educational programs for school, community & special population groups are regularly scheduled & by reservation; touch tables.
Publications: brochure, Rock Creek Park; Rock Creek Park monthly calendar of events.
Hours & Admission Prices: Wed.-Sun. 9-5. No charge; donations accepted. Closed federal holidays. &
Attendance: 35,000 (accurate)

SECURITIES AND EXCHANGE COMMISSION HISTORICAL SOCIETY, Washington, DC 20004-2514. Mailing Address: 1101 Pennsylvania Ave., N.W., Ste. 600, Washington, DC 20004-2514. Tel.: 202-756-5015. Fax: 202-756-5014.
E-mail: c.rosati@sechistorical.org
Web Site: www.sechistorical.org
Founded: 2002.
Key Personnel: Exec. Dir., Carla L. Rosati, CFRE
Governing Authority: society; nonprofit organization.
Institution Type/Description: Virtual Museum.
Collections: history of the U.S. Securities and Exchange Commission from 1930s to present; securities industry; photographs; oral histories.
Activities: educational programs.
Hours & Admission Prices: Online access only.

SEWALL-BELMONT HOUSE AND MUSEUM, (M), 144 Constitution Ave., N.E., Washington, DC 20002-5608. Tel.: 202-546-1210. Fax: 202-546-3997.
E-mail: info@sewallbelmont.org
Web Site: www.sewallbelmont.org
Founded: 1929.
Congressional District: 2
Key Personnel: Exec. Dir., Page Harrington; Pres., Dianne Chasen Lipsey; Asst. Dir. & Collections Mgr., Jennifer Krafchik; Devel. Mgr., Chitra Panjabi; Public Programs & Outreach Mgr., Elisabeth Crum; Mgr. Events & Facilities, Paul McClellan.
Personnel Profile: Full-Time Paid 6; Part-Time Paid 10; Part-Time Volunteers 10; Interns 4.
Governing Authority: nonprofit organization. Parent Institutions: National Woman's Party; National Capital Park-East, National Park Service, Dept. of Interior. Tax-exempt: 501(c)(3).
Institution Type/Description: Women's History Museum: housed in c.1800 structure later added on, 1800 residence of Albert Gallatin 1801-1813 and 1929-1972 office & home of Alice Paul, founder of the National Woman's Party & author of the ERA.
Collections: fine arts; photographs; banners; scrapbooks; memorabilia of suffrage & equal rights campaign.
Research Fields: Equal Rights & Women's Suffrage Movement.
Facilities: research library; classroom; tented garden.
Activities: guided tours; lectures; films; gallery talks; docent program; permanent exhibitions; volunteer program.
Publications: e-news, Sewall Belmont; Kids Guide to Sewall-Belmont.
Hours & Admission Prices: See website for hours. Tours: $5 per person. Closed holidays. &
Attendance: 15,000 (estimated)
Membership: Annual $35.

* **SMITHSONIAN AMERICAN ART MUSEUM, (M),** 8th & F Streets N.W., Washington, DC 20006. Mailing Address: MRC 970, P.O. Box 37012, Washington, DC 20013-7012. Tel.: 202-633-7970. Fax: 202-633-8535.
E-mail: americanartinfo@si.edu
Web Site: www.americanart.si.edu
Formerly: National Museum of American Art
Founded: 1829.
Key Personnel: Dir. The Margaret and Terry Stent, Elizabeth Broun; Deputy Dir., Rachel M. Allen; Chief Cur., Virginia Mecklenburg; The Fleur & Charles Bresler Cur. of American Craft & Decorative Art, Nicholas R. Bell; Consulting Sr. Cur. Film & Media Arts, John G. Hanhardt; Assoc. Cur. Media Arts, Michael Mansfield; Cur. McEvoy Family Photography, Lisa Hostetler; Cur. Sculpture, Karen Lemmey; The James Dicke Cur. Contemporary Art, Joanna Marsh; Sr. Cur., Eleanor Jones Harvey; Sr. Cur. Graphic Arts, Joann G. Moser; Sr. Cur., William H. Truettner; Special Projects Dir. & Chief Exhibitions, Claire F. Larkin; Chief Devel., Donna Rim; Cur. Latino Art, E. Carmen Ramos; Cur. Folk & Self-Taught Art, Leslie Umberger; Chief External Affairs, Jo Ann Gillula; Registrar, Melissa Kroning; Chm. Lunder Education, Susan Nichols; Chief, Special Events, Emily Chamberlin; Chief Publications, Theresa J. Slowik; Chief Research & Scholars, Christine Hennessey; Chief Conservation, Tiarna Doherty; Chief Renwick Gallery, Robyn L. Kennedy; Administrative Officer, Douglas Wilde; Public Affairs Officer, Laura Baptiste; Public Affairs Assoc., Mandy Young; Public Affairs Specialist, Courtney Rothbard; Mgr. Visitor Svcs., Janet Walker; Web & Social Media Content Mgr., Georgina Goodlander.
Governing Authority: nonprofit federally chartered corporation. Parent Institution: Smithsonian Institution, Washington, DC. Subsidiary Museum: Renwick Gallery, 17th & Pennsylvania Ave. N.W., Washington, DC 20006. Tax-exempt: 501(c)(3) & 170(b)(1)(A).
Institution Type/Description: Art Museum: housed in 1836 Old Patent Office Building.
Collections: American art from the colonial period to the present, including paintings, sculpture, graphics, photography, folk art, and contemporary crafts at the Museum's Renwick Gallery; 19th century landscape painting;

19th & 20th century sculpture; art of the 1930-1940s; African-American art; Latino art; manuscript & clipping files.

Research Fields: all aspects of the collections; American art; active fellowship program.

Facilities: 80,000-vol. library; archives of slides & photographs; restaurant. Museum-related items for sale.

Activities: symposia; lectures; films; concerts; organized education programs; museum intern training; inter-museum loans; permanent, temporary & traveling exhibitions; self-guided & docent-guided tours; family programs; conservation clinics & tours; music performances; monthly jazz performances.

Publications: scholarly journal, American Art; exhibition catalogs; books on American art & artists; pamphlets and booklets on the museum, its collections & programs; quarterly members newsletter & calendar of events.

Hours & Admission Prices: Daily 11:30-7. No charge. Closed Christmas. &

Attendance: 1,300,000 (accurate)

Membership: Director's Circle $1,500 & up.

SMITHSONIAN INSTITUTION, 1000 Jefferson Dr., S.W., Washington, DC 20560-0009. Mailing Address: 1000 Jefferson Dr., S.W., Rm. 369, MRC033, Washington, DC 20560-0009. Tel.: 202-633-2400. Fax: 202-786-2377. TTY: 202-633-5285.

E-mail: liebermane@si.edu

Web Site: www.smithsonian.org

Founded: 1846.

Key Personnel: Acting Under Sec. Science, Scott Miller; Pres., Tom Ott; Dir. Communications & Public Affairs, Evelyn Lieberman.

Personnel Profile: Full-Time Paid 6,250; Part-Time Volunteers 5,500; Interns 900.

Governing Authority: nonprofit independent trust instrumentality of the United States. Museums & Galleries: Anacostia Museum; Cooper-Hewitt, National Design Museum, New York, NY; Freer Gallery of Art; Hirshhorn Museum & Sculpture Garden; National Air & Space Museum; National Museum of African Art; Smithsonian American Art Museum; National Museum of American History; National Museum of the American Indian; National Museum of Natural History; National Portrait Gallery; National Postal Museum; National Zoological Park; Renwick Gallery; Arthur M. Sackler Gallery; Smithsonian Institution Building (the Castle); National Museum of African American History & Culture. Tax-exempt: 501(c)(3) & 170(b)(1)(A).

Institution Type/Description: National Museum.

Collections: natural history specimens; rare books; artworks; science & technology artifacts; rare U.S. & international stamps & coins; U.S. history collections; the national collections (U.S.A.); exotic living animals.

Research Fields: anthropology; biology; ecology; astrophysics; geodesy; meteoritics; space sciences; botany; oceanography; mineral sciences; conservation of natural resources, wildlife, museum object & library materials; the arts; history; museology; exhibit techniques; international, governmental & social history; molecular systematics entomology; vertebrate & invertebrate zoology; paleobiology; artists' papers; Latino Center; American Indian Cultural Resources Center; African American Center; folk life & cultural studies; conservation; tropical rain forests; marine environments; armed forces history; musical instruments; Native American history; archaeology; design; advances in technology; Asian art; African art; American arts & crafts; transportation history; philately; numismatics; political history; Asian Studies Center.

Facilities: 1.5 million-vol. library system including 40,000 rare books, specializing in natural history; horticulture; conservation science; American ethnology; zoology; anthropology; history of astronautics & aeronautics; astrophysics; American history & culture; history of science & technology; philately; American painting, sculpture, portraiture & biography; decorative arts & design; early works of travel & scientific exploration; volcanism; marine systems; tropical biology; Chesapeake Bay area ecology; museology; available for inter-library loan & for use by qualified scholars. Other research facilities include: Astrophysical Observatory; Environmental Research Center; Tropical Research Institute; the Archives of American Art.

Activities: guided tours; lectures; films; gallery talks; concerts; dance performances; arts festivals; workshops; educational activities; special workshops for teachers; Campus on the Mall courses; docent program; training programs for professional museum workers; inter-museum loans; permanent, temporary & traveling exhibitions; biodiversity & other scientific research; study tours; international exchange of scientific publications; scholarly study; virtual tours on the Web; heritage month celebrations. Museum Sponsors: Spring Kite Festival; Summer Smithsonian Folklife Festival; Smithsonian Craft Show.

Publications: monthly magazine, Smithsonian; bimonthly magazine, Air & Space/Smithsonian; books; serials; exhibition catalogs; guides; pamphlets; quarterly research newsletter; bimonthly newsletter about Native American related activities; and multi-volume works on selected subjects, as well as

CD-ROMs, music recordings of remastered originals and web pages on the internet; learned journals.

Hours & Admission Prices: Daily 10-5:30. Visitor Center: daily 8:30-5:30. No charge (except Cooper-Hewitt). Closed Christmas. &

Attendance: 20,100,000 (estimated)

THE SOCIETY OF THE CINCINNATI AT ANDERSON HOUSE, (M), 2118 Massachusetts Ave., N.W., Washington, DC 20008-3640. Tel.: 202-785-2040. Fax: 202-785-0729.

E-mail: admin@societyofthecincinnati.org

Web Site: www.societyofthecincinnati.org

Founded: 1938.

Key Personnel: Exec. Dir., Jack D. Warren, Jr.; Deputy Dir. & Cur., Emily L. Schulz; Dir. Library, Ellen McCallister Clark; Coord. Museum Visitor Svcs., Caren A. Pauley; Museum Collections Mgr., Whitney A. J. Robertson.

Personnel Profile: Full-Time Paid 15; Part-Time Paid 6; Part-Time Volunteers 31; Interns 3.

Governing Authority: nonprofit organization. Parent Institution: The Society of the Cincinnati. Tax-exempt: 501(c)(3).

Institution Type/Description: Historic House Museum: housed in 1902-1905 Anderson House, a beaux arts mansion designed by Little & Browne of Boston.

Collections: Anderson house original furnishings; sculpture; paintings; textiles; ceramics; Asian & European decorative arts; artifacts; manuscripts & books related to the history of the Society of the Cincinnati; armaments, silver, maps, manuscripts, and rare books of the period of the American Revolution; manuscripts, books, and photographs relating to Larz & Isabel Anderson III.

Research Fields: American Revolution; Society of the Cincinnati history; art of war in the 18th century; the lives and collection's of Larz & Isabel Anderson III; Washington history.

Facilities: 45,000-vol. research & reference library on history of the American Revolution, Society of the Cincinnati, Washington DC, and collections; reading room; garden.

Activities: guided tours; lectures; concerts; docent program; volunteer positions available; internships possible; inter-museum loan, permanent & temporary exhibitions.

Publications: biannual journal, Cincinnati Fourteen; Why America is Free; The Insignia of the Society of the Cincinnati; George Rogers Clark lectures; exhibition catalogues.

Hours & Admission Prices: Museum: Tues.-Sat. 1-4. Library: Mon.-Fri. 10-4 by appointment only. No charge; donations accepted. Closed legal holidays & during society meetings. &

Attendance: 9,933 (accurate)

STUDIO GALLERY, 2108 R St., N.W., Washington, DC 20008. Tel.: 202-232-8734.

E-mail: info@studiogallerydc.com

Web Site: www.studiogallerydc.com

Key Personnel: Dir., Idil Periard

Institution Type/Description: Art Gallery.

Collections: works by national & international artists; paintings; sculpture.

Activities: temporary exhibitions.

Hours & Admission Prices: Wed.-Thurs. 1-7, Fri. 1-8, Sat. 1-6.

Attendance: 3,000 (estimated)

THE SUPREME COURT OF THE UNITED STATES, Office of the Curator, One First St., N.E., Washington, DC 20543. Tel.: 202-479-3298 & 3000 (tours); 888-539-4438 (gift shop). Fax: 202-479-2926.

E-mail: curator@supremecourt.gov

Web Site: www.supremecourt.gov

Founded: 1973.

Key Personnel: Cur., Catherine E. Fitts; Assoc. Cur., Matthew Hofstedt; Collections Mgr., Franz Jantzen; Collections Mgr., Devon Westerberg; Mgr. Visitor Programs, Megan Jones; Photographer, Steve Petteway.

Personnel Profile: Full-Time Paid 7; Part-Time Paid 2; Part-Time Volunteers 20; Interns 5.

Governing Authority: federal agency. Tax-exempt.

Institution Type/Description: Historic Agency & Building.

Collections: portraits of all former Justices; marble busts of the Chief Justices & certain Associate Justices; historic photos, etchings & drawings of the Justices & the architecture of the building; memorabilia; archival & manuscript materials on Supreme Court history; 18th & 19th-century American & English furniture & decorative arts; architectural drawings & plans of Supreme Court Building.

Research Fields: lives of the Supreme Court Justices, the traditions of the

Court; architecture & furnishings of Supreme Court Building; administrative history of Supreme Court.

Facilities: library of reference books related to the Supreme Court history available for use in the Curator's Office; cafeteria. Gifts items for sale.

Activities: courtroom lectures; film; permanent & temporary exhibitions.

Publications: booklet, The Supreme Court of the United States; pamphlets, Visitor's Guide to the Supreme Court; Visitor's Guide to Oral Argument.

Hours & Admission Prices: Mon.-Fri. 9-4:30. Courtroom lectures every hour on the half-hour 9:30-3:30 when court is not sitting. No charge. Closed federal holidays. &

Attendance: 300,000 (estimated)

❋　**THE TEXTILE MUSEUM, (M),** 2320 S St., N.W., Washington, DC 20008-4088. Tel.: 202-667-0441. Fax: 202-483-0994.

E-mail: info@textilemuseum.org

Web Site: www.textilemuseum.org

Founded: 1925.

Key Personnel: Pres. Bd. (V), Bruce P. Baganz; C.F.A.O., Doug Maas; Dir. Devel., Eliza Ward; Devel. Mgr., Special Events, Ingrid Faulkerson; Communications & Mktg. Asst., Claire Blaustein; Research Assoc. Western Hemisphere, Ann Rowe; Cur. Eastern Hemisphere, Sumru Krody; Dir. Conservation, Esther Methe; Registrar, Rachel Shabica; Research Assoc. Southeast Asian Textiles, Mattiebelle Gittinger; Assoc. Cur. Eastern Hemisphere, Lee Talbot; Assoc. Conservator, Maria Fusco; Cur. Education, Tom Goehner; Research Assoc. Contemporary Textiles, Rebecca Stevens; Asst. Registrar, Tessa Sabol; Devel. Asst., Sheila Freeman; Librarian, Lydia Fraser; Mgr. Mktg. & Communications, Kathryn Clune; Dir. Facilities & Exhibitions Production, Richard Timpson; Dir. Retail Operations, Chabrina Williams; Exhibition Coord., Emily Robinson; Special Asst. to Dir., Ana Kiss; Education Asst., Hattie Lehman.

Personnel Profile: Full-Time Paid 23; Part-Time Paid 5; Part-Time Volunteers 75; Interns 5.

Governing Authority: nonprofit organization. Tax-exempt: 501(c)(3).

Institution Type/Description: Textile Museum.

Collections: over 19,000 historic & ethnographic handmade textiles; Oriental carpets; pre-Columbian, Peruvian, Islamic, Coptic, Caucasian, southeast & central Asian, indigenous Latin American & North American textiles. Changing exhibitions highlight areas of the permanent collection as well as contemporary art, American quilts & other areas of the textile arts which lie outside the realm of the museum's collection.

Research Fields: preservation; exhibition; study of permanent collections.

Facilities: 20,000-vol. research & reference library of books; periodicals; pamphlets on technical, historical & artistic aspects of rugs & textiles; conservation laboratory. Museum-related items for sale.

Activities: guided tours; lectures; films; gallery talks; docent program; inter-museum loan, temporary & traveling exhibitions; classes in textile techniques; workshops; looking exercises; cultural stories; video; weekly rug & textile appreciation sessions; school & family programs.

Publications: exhibition catalogues; books relating to Museum's collection & activities; quarterly members' magazine.

Hours & Admission Prices: Tues.-Sat. 10-5, Sun. 1-5. Suggested Donation $8. Closed Christmas Eve; federal holidays. &

Attendance: 28,000 (estimated)

Membership: Individual $25; Myers $60; Myers Fellow $125; Supporter $250; Sponsor $500; Patron $1,000.

TOUCHSTONE GALLERY, 901 New York Ave., N.W., Washington, DC 20001-6435. Tel.: 202-347-2787.

E-mail: info@touchstonegallery.com

Web Site: www.touchstonegallery.com

Founded: 1976.

Key Personnel: Dir., Ksenia Grishkova; Pres. (V), Linda Bankero

Institution Type/Description: Art Gallery.

Collections: works by contemporary artists; paintings; mixed media; sculpture.

Major Exhibits: Sculptors, 1/3/14-2/2/14; Leslie Nolan, 2/7/14-3/2/14; Charles St. Charles, 3/7/14-3/30/14; Rosemary Luckett & Shelly Lowenstein, 4/4/14-4/27/14; Fred Tan, 5/2/14-6/1/14; Colleen Sabo, 6/6/14-6/29/14; Mini Solo, 8/1/14-8/29/14; Bill Mould & Pete McCutchen, 9/5/14-9/28/14; Gale Wallar & Judith May, 10/3/14-11/2/14.

Hours & Admission Prices: Wed.-Fri. 11-6, Sat.-Sun. 12-5. No charge. &

Attendance: 15,000 (estimated)

TUDOR PLACE HISTORIC HOUSE & GARDEN, 1644 31st St., N.W., Washington, DC 20007-2924. Tel.: 202-965-0400, ext. 100. Fax: 202-965-0164.

E-mail: info@tudorplace.org

Web Site: tudorplace.org

Founded: 1966.

Congressional District: 1

Key Personnel: Pres. Bd. of Trustees, Timothy B. Matz; Exec. Dir., Leslie Buhler; Mgr. Collections, Fay Winkle; Cur., Erin Kyukendall.

Personnel Profile: Full-Time Paid 10; Part-Time Paid 5; Part-Time Volunteers 5; Interns 10.

Governing Authority: public; nonprofit organization. Parent Institution: Tudor Place Foundation, Inc. Tax-exempt: 501(c)(3).

Institution Type/Description: Historic House & Garden: house designed by Dr. William Thornton, first architect of the U.S. Capitol, completed 1816.

Collections: Washington-Custis-Peter memorabilia; furniture; silver; porcelain; sculpture; paintings; textiles; photographs; manuscripts; household items; historic trees, shrubs & flowers.

Research Fields: Custis-Peter Family; Georgetown and Washington history; Federal period architecture & gardens; Sculpture of Paul Wayland Bartlett, 1865-1925.

Facilities: 5 1/2 acre garden.

Activities: guided tours. Annual Events: Holiday Evenings; Fall Garden Day; Kids Summer History Weeks.

Publications: annual report; quarterly newsletter.

Hours & Admission Prices: Tues.-Sat. 10, 11, 12, 1, 2, 3, Sun. 12, 1, 2, 3. Adults $8; discounts to AAA, NTHP, & AAM members, senior citizens & students; members no charge. Closed New Year's Day; Independence Day; Thanksgiving; Christmas.

Attendance: 17,500 (estimated)

Membership: Individual $45; Family & Dual $80.

U.S. NATIONAL ARBORETUM, 3501 New York Ave., N.E., Washington, DC 20002-1958. Tel.: 202-245-2726. Fax: 202-245-4575.

E-mail: colien.hefferan@ars.usda.gov

Web Site: www.usna.usda.gov

Founded: 1927.

Key Personnel: Dir., Colien Hefferan; Assoc. Dir., Dr. Ramon Jordan; Unit Leader, Education & Visitor Svcs., Nancy Luria; Unit Leader, Gardens, Scott Aker; Research Leader, Floral & Nursery Plants Research Unit, Dr. Margaret Pooler; Administrative & Mktg. Mgr., Deborah Cicala.

Governing Authority: federal. Parent Institution: U.S. Dept. of Agriculture. Subsidiary Institution: Agricultural Research Service. Tax-exempt: 170(c)(1).

Institution Type/Description: Arboretum.

Collections: horticulture; azaleas; rhododendrons; camellias; native plants; ferns; magnolias; holly; perennials; ornamental cherries; crabapples; annuals; Asian plants; ornamental trees & shrubs; herbarium; outdoor museum; national bonsai collection; national herb garden.

Research Fields: test plantings; taxonomy; plant breeding.

Facilities: 10,000-vol. library of books & periodicals on botany & ornamental horticulture available on premises.

Activities: guided tours; lectures; classes; symposia; temporary exhibitions. Annual Events: flower shows.

Publications: U.S. National Arboretum News and Notes (electronic); National Arboretum Contributions.

Hours & Admission Prices: Arboretum: daily 8-5. Visitor Center: Fri.-Mon. 8-4:30. National Bonsai & Penjing Museum Fri.-Mon. 10-4. No charge; donations accepted. &

Attendance: 450,000 (estimated)

❋　**UNITED STATES BOTANIC GARDEN, (M),** 245 First St., S.W., Washington, DC 20024-3201. Tel.: 202-226-8333. Fax: 202-225-1561.

E-mail: usbg@aoc.gov

Web Site: www.usbg.gov

Founded: 1820.

Key Personnel: Acting Dir., Stephen Ayers; Exec. Dir., Holly H. Shimizu; Admin. Officer, Tonda Cave; Safety Officer, Courtney Nicholls; Facility Mgr., John Gallagher; Mgr. Public Programs, Dr. Ari Novy; Horticulture Mgr., James Kaufmann.

Personnel Profile: Full-Time Paid 68; Part-Time Volunteers 184; Interns 4.

Governing Authority: federal. Parent Institution: U.S. Government (Legislative Branch): Joint Committee on the Library; Architect of the Capitol. Tax-exempt.

Institution Type/Description: Botanical Garden.

Collections: tropical, sub-tropical & temperate plants: orchids, cacti & succulents; medicinal & other economic plants; ferns; cycads; bromeliads; epiphytes; carnivorous (insectivorous) plants; rare endangered, threatened plants; native plants (mid-Atlantic U.S.).

Research Fields: plant & culture evaluation.

Facilities: conservatory.

Activities: guided tours by advance appointment; permanent & temporary exhibitions; public classes in botany, horticulture, botanical art & the environment; school programs; teacher workshops; special events; professional conferences.

Publications: assorted horticultural, botanical & environmental information sheets; quarterly, Calendar of Events; book, A Botanic Garden for the Nation; workbook, Junior Botanist; triannual living collections catalog; workbook, Family Field Journal.

Hours & Admission Prices: Conservatory: daily 10-5. No charge. &

Attendance: 1,000,000 (estimated)

UNITED STATES CAPITOL, U.S. Capitol, Architect of the Capitol, Washington, DC 20515. Tel.: 202-228-1222. Fax: 202-228-4602.

E-mail: bwolanin@aoc.gov
Web Site: www.aoc.gov
Founded: 1793.
Key Personnel: Acting Architect of the Capitol, Steven T. Ayers, AIA; Cur., Dr. Barbara A. Wolanin; Photo Branch, Michael Dunn; Registrar, Pamela Violante McConnell.
Governing Authority: federal. Under the direction of the Joint Committee on the Library, U.S. Congress. Branch Facility: U.S. Botanic Garden. Tax-exempt: 501(c)(1).
Institution Type/Description: National Agency & Art Museum.
Collections: 19th & 20th-century paintings, sculpture & decorative arts, including works by Franzoni, Crawford, Rogers, Powers, Roberts, French, Peale, Trumbull, Vanderlyn, Weir, Leutze & Brumidi; restored historic chambers; 100,000 architectural drawings; 70,000 photographs; archives pertaining to the U.S. Capitol art & architecture.
Research Fields: architecture; art; decorative arts; history of the U.S. Capitol complex; history of Washington, D.C.
Facilities: 1,500-vol. specialized reference library & archives pertaining to architecture, art & history of the Capitol, available for consultation on premises; cafeteria & restaurant.
Activities: guided tours; inter-museum loan & permanent exhibitions.
Publications: book, Art in the United States Capitol: A Brief Architecture History; occasional publications.
Hours & Admission Prices: Mon.-Sat. 8:30-4:30. No charge. Closed federal holidays. &

UNITED STATES DEPARTMENT OF THE INTERIOR MUSEUM, (M), 1849 C Street, N.W. Mail Stop 1361, Washington, DC 20240. Tel.: 202-208-4743. Fax: 202-208-1535.

E-mail: diana_ziegler@nbc.gov
Web Site: www.doi.gov/interiormuseum
Founded: 1938.
Key Personnel: Program & Outreach Coord., Diana Ziegler; Registrar, Erin McKeen.
Personnel Profile: Full-Time Paid 3; Interns 2.
Governing Authority: federal. Parent Institution: U.S. Dept. of the Interior, Office of the Secretary Museum Services Branch, Washington, DC 20240. Tax-exempt.
Institution Type/Description: General Museum.
Collections: American Indian, Eskimo, Micronesia, Virgin Island and Guam handicraft & artifacts; rocks; fossils; paintings; sculptures; maps; charts; mapping & surveying equipment; mining equipment; models & dioramas.
Research Fields: land management; North American Indians material culture; federal history.
Activities: lunch time lectures on topics related to the Dept; tours.
Publications: brochure.
Hours & Admission Prices: Closed for renovations. &
Attendance: 30,000 (estimated)

✻ UNITED STATES HOLOCAUST MEMORIAL MUSEUM, (M), 100 Raoul Wallenberg Place, S.W., Washington, DC 20024-2126. Tel.: 202-488-0400. Fax: 202-488-2690. TDD: 202-488-0406.

E-mail: visitorsmail@ushmm.org
Web Site: www.ushmm.org
Founded: 1993.
Key Personnel: Dir. United States Holocaust Memorial Museum, Sara J. Bloomfield; Chm. United States Holocaust Memorial Council, Fred S. Zeidman; Dir. Exhibitions, Steve Goodell; Deputy Dir. Exhibitions, Edward Phillips; Chief of Staff, William Parsons; Cur. Permanent Exhibition, Steven Luckert; Dir. External Affairs, Dara Goldberg; Dir. Center for Advanced Holocaust Studies, Paul Shapiro; Dir. Institutional Stewardship, Diane Saltzman; Sr. Advisor, Arthur S. Berger; Dir. Media Rels., Andrew Hollinger; Dir. Human Resources, Antonio Guzman; Dir. Collections, Michael Grunberger; Dir. Collections Management, Travis Roxlau; Dir. Film & Video, Raye Farr; Chief Conservator, Jane Klinger; Assoc. Cur. Art & Artifacts, Suzy Snyder; Dir. Oral History, Joan Ringelheim; Dir.

Photographic Reference Collection, Judith Cohern; Dir. NIHE, Sarah Ogilvie; Dir. Education, Daniel Napolitano; Dir. Survivor Affairs, Ellen Blalock; Sr. Project Mgr. NIHE, Kristine Donly; Chief Archivist, Henry Mayer; Museum Shop Mgr., Marylou Withem; Dir. Museum Svcs., Paul Garver; Dir. Library, Ronald Coleman; Chief Information Officer Information & Technology, Lawrence Swiader; C.F.O., John Fawsett; Dir. Devel., Jordan Tannenbaum; Dir. Membership, Dana Weinstein.
Personnel Profile: Full-Time Paid 400; Part-Time Paid 10; Part-Time Volunteers 320; Interns 22.
Governing Authority: federal. Tax-exempt: 170(b)(1)(A).
Institution Type/Description: History Museum.
Collections: reflecting the events of the Holocaust of 1933-1945, including artifacts, photographs, films, video, oral histories, books, archives.
Research Fields: Holocaust Studies
Facilities: 55,000-vol. library of Holocaust-related material in 18 languages available for use by the public on the premises only; auditorium; restaurant; resource center; classrooms; 50,000 sq. ft. exhibit space; theatre. Museum-related items for sale.
Activities: research; public programs.
Publications: quarterly, The United States Holocaust Memorial Museum; Journal of Holocaust & Genocide Studies & select papers issued by the Museum's Center for Advanced Holocaust Studies; exhibition catalogues.
Hours & Admission Prices: Daily 10-5:30. No charge; donations accepted. Closed Christmas; Yom Kippur. &
Attendance: 1,373,589 (accurate)
Membership: Associate $25; Member $36; Supporting $100; Sustaining $500; Circle of Remembrance $1,000.

UNITED STATES SENATE COMMISSION ON ART, Rm. S-411, U.S. Capitol Bldg., Washington, DC 20510. Tel.: 202-224-2955. Fax: 202-224-8799.

E-mail: curator@sec.senate.gov
Web Site: www.senate.gov
Founded: 1968.
Key Personnel: Cur., Diane K. Skvarla; Assoc. Cur., Melinda K. Smith; Admin., Scott M. Strong; Museum Specialist, Richard L. Doerner; Registrar, Courtney D. Morfeld; Collections Specialist, Theresa Malanum; Historic Preservation Officer, Kelly Steele; Curatorial Asst., Amy Elizabeth Burton; Staff Asst., Amy Camilleri; Collections Mgr., Deborah Wood.
Personnel Profile: Full-Time Paid 10; Interns 4.
Governing Authority: federal. Parent Institution: U.S. Senate. A commission of the U.S. Senate, Washington, DC 20510. Tax-exempt.
Institution Type/Description: Preservation Projects: c.1850 Old Senate & Old Supreme Court Chambers & other historic areas of the U.S. Capitol Building & exhibits.
Collections: paintings; sculpture; prints; furnishings; documents.
Research Fields: American fine arts & decorative arts; cultural & political history; heritage of U.S. Senate.
Facilities: 250,000-vol. library; food services available.
Activities: guided tours; lectures; internships for undergraduate college students; permanent & temporary exhibitions.
Publications: catalog of collections; guide to exhibits; guides to restored chambers; exhibition posters & catalogs.
Hours & Admission Prices: Daily 9-4:30. No charge. Closed New Year's Day; Thanksgiving; Christmas. &
Attendance: 1,500,000 (estimated)

U.S. CAPITOL HISTORICAL SOCIETY, 200 Maryland Ave., N.E., Washington, DC 20002-5724. Tel.: 202-543-8919; 800-887-9318. Fax: 202-544-8244.

E-mail: uschs@uschs.org
Web Site: www.uschs.org
Founded: 1963.
Key Personnel: Pres., Ronald A. Sarasin, CAE; Vice Pres. Merchandising, Diana Wailes; Dir. Mktg., Mary Hughes; Museum Shop Mgr., Sharron Randolph.
Personnel Profile: Full-Time Paid 35; Full-Time Volunteers 4; Part-Time Volunteers 20; Interns 4.
Governing Authority: Parent Institution: United States Congress. Tax-exempt.
Institution Type/Description: Historical Society Museum.
Collections: history of capitol building & congress; documents; videos.
Research Fields: history of the Capitol & Congress.
Activities: seminars; educational outreach programs.
Publications: We, The People, Washington: Past & Present; Young Person Guide to the Capitol; Understanding Congress; Exploring Capitol Hill; Where the People Speak; A Vision of Freedom; Outstanding Members of Congress Series; The Capitol: Designing and Decorating a National Icon; Capitol For The People - A Poker Guide; Pocket Constitution.

Hours & Admission Prices: Daily 9-4:30. No charge; donations accepted. Closed holidays. &

Attendance: 8,000,000 (estimated)

Membership: Charter $35; Freedom $50; Cornerstone $100; Rotunda $250; Architect of History $500; Capitol Circle $1,000; Benefactor $2,500; Founder $5,000; Brumidi $10,000.

U.S. NAVY MEMORIAL FOUNDATION AND NAVAL HERITAGE CENTER, 701 Pennsylvania Ave., N.W., Suite 123, Washington, DC 20004-2688. Tel.: 202-737-2300, ext. 710 & 725. Fax: 202-737-2308.

E-mail: library@lonesailor.org

Web Site: www.navymemorial.org

Founded: 1991.

Key Personnel: C.E.O. & Pres., VADM John Totushek, USNR (Ret.); Chm. (V), Richard C. Vie; Treas., Edward Walker, USN, (Ret.); Exec. Vice Pres., Cindy McCalip; Cur. & Educator, Mark T. Weber; Registrar, Don Dupuis.

Personnel Profile: Full-Time Paid 13; Full-Time Volunteers 1; Part-Time Paid 6; Part-Time Volunteers 25; Interns 3.

Governing Authority: private; nonprofit organization. Tax-exempt: 501(c)(3).

Institution Type/Description: Naval Museum.

Collections: history of 625,000 sea service veterans; naval art; naval history books; personal accounts; 30,000 photographs; Naval officers who served as US presidents.

Major Exhibits: Year of Military Women, 4/13-3/14.

Research Fields: sea service veterans.

Facilities: 1,500-vol. library; 250-seat auditorium; 250-seat large screen theater; 2,500 sq. ft. exhibit space. Museum-related items for sale.

Activities: arts festival; concerts; docent program; films; guided tours; lectures; rental gallery; loan, traveling & temporary exhibitions; theater. Museum Sponsors: Blessing of the Fleet; Historical Wreath-laying; official Navy events including retirements, re-enlistments & change of command.

Publications: quarterly newsletter, The Lone Sailor.

Hours & Admission Prices: Daily 9:30-5. No charge; donations accepted. Closed New Year's Day; Thanksgiving; Christmas. &

Attendance: 100,000 (estimated)

Membership: $25 & up.

WASHINGTON NATIONAL CATHEDRAL, 3101 Wisconsin Ave., N.W., Washington, DC 20016-5000. Tel.: 202-537-6200. Fax: 202-364-6611.

E-mail: tours@cathedral.org

Web Site: www.nationalcathedral.org

Founded: 1893.

Key Personnel: Bishop, Rt. Rev. John Bryson Chane; Dean, The Very Rev. Samuel Lloyd, III; Dir., Kathleen Cox; Vicar, Rev. Jan N. Cope.

Personnel Profile: Full-Time Paid 90; Part-Time Volunteers 1,200.

Governing Authority: Parent Institution: Protestant Episcopal Cathedral Foundation. Tax-exempt: 509(a)(1).

Institution Type/Description: Religious Institution: housed in a Gothic cathedral.

Collections: religious art & architecture, including stained glass windows, wrought iron, wood & stone carvings, gargoyles & grotesques, needlepoint & tapestries; a stained glass window with embedded moon rock; an altar painting by N.C. Wyeth; garden sculpture of Prodigal Son by Heinz Warneke; bronze sculpture of Abraham Lincoln by Walker Hancock; exterior sculpture by Frederick Hart; exterior fountain & artwork. Nine chapels. Bishop's Garden and Olmsted Woods. 10,650-pipe organ, 10-bell peal, carillon and choirs. Interred remains of Helen Keller, Woodrow Wilson, Adm. George Dewey, Cordell Hull, and others.

Facilities: Bishop's Garden and Olmsted Woods. Gift-related items for sale.

Activities: audio & guided tours; daily worship services; specialty tours; temporary exhibitions; workshops; spirituality speakers; special events. Annual Events: Evensong seasonally; Tour and Tea; Behind the Scenes Tours; Flower Mart in May; Shrove Tuesday Pancake Race; Blessing of the Animals in October.

Publications: brochures; self-guided tours; fact sheets; quarterly magazine, Cathedral Age; Cathedral Voice; books.

Hours & Admission Prices: Winter: Sun.-Fri. 10-5:30, Sat. 10-4; Summer: call for extended hours; groups by appointment. Requested Donation: family $15, adults $5, seniors $3. &

Attendance: 500,000 (accurate)

＊ **THE WHITE HOUSE, (M),** 1600 Pennsylvania Ave., N.W., Washington, DC 20502. Tel.: 202-456-2550. Fax: 202-456-6820.

E-mail: curator@exrwh.gov

Web Site: www.whitehouse.gov

Founded: 1792.

Key Personnel: Cur., William G. Allman; Asst. Cur., Lydia S. Tederick; Asst.

Cur., Melissa C. Naulin; Curatorial Asst., Monica McKiernan; Registrar & Collections Mgr., Donna A. Hayashi Smith.

Governing Authority: federal. Tax-exempt.

Institution Type/Description: Historic House Museum: 1792-1800, The White House.

Collections: American & European furniture, decorative & fine arts from the late 18th, 19th and 20th centuries; collections of Presidential porcelain, silver & glassware; portraits of the Presidents, First Ladies and other national notables; late 18th, 19th and 20th-century paintings and prints; archival & manuscript materials on White House history.

Research Fields: relating to the history of the White House, its furnishings and collections.

Facilities: library of books & other materials relating to all aspects of White House history, the Presidency & White House families.

Activities: tours; permanent & temporary exhibits.

Publications: books, The White House: An Historic Guide; The Presidents of the United States; The First Ladies; The Living White House; The President's House: A History; Art in the White House: A Nation's Pride; The White House: It's Historic Furnishings and First Families; White House Glassware: Two Centuries of Presidential Entertaining; journal, White House History.

Hours & Admission Prices: Please call 202-456-7041 for information. Tours: Tues.-Thurs. 7:30-11, Fri. 7:30-12, Sat. 7:30-1. No charge. &

ZENITH GALLERY, 1429 Iris St., N.W., Washington, DC 20012. Mailing Address: P.O. Box 55295, Washington, DC 20040-5295. Tel.: 202-783-2963.

E-mail: art@zenithgallery.com

Web Site: www.zenithgallery.com

Key Personnel: Dir., Margery Goldberg

Institution Type/Description: Art Gallery.

Collections: works by local & regional artists.

Hours & Admission Prices: By appointment.

FLORIDA

(450 listings)

Anna Maria Island

ANNA MARIA ISLAND HISTORICAL MUSEUM, 402 Pine Ave., Anna Maria Island, FL 34216. Mailing Address: P.O. Box 4315, Anna Maria, FL 34216-4315. Tel.: 941-778-0492.

Web Site: www.amihs.org

Founded: 1992.

Governing Authority: private; nonprofit organization. Parent Institution: Anna Maria Island Historical Society.

Institution Type/Description: History Museum.

Collections: island history & culture; personal artifacts; photographs; videos; shells; fossils; period office equipment; tools.

Activities: educational programs; scholarships.

Hours & Admission Prices: Memorial Day to Labor Day Tues.-Thurs. & Sat. 10-12; Sept.-May Tues.-Thurs. & Sat. 10-3. No charge.

Apalachicola

JOHN GORRIE MUSEUM STATE PARK, 46 6th St. & Avenue D, Apalachicola, FL 32329. Mailing Address: P.O. Box 267, Apalachicola, FL 32329-0267. Tel.: 850-653-9347.

Web Site: baynavigator.com

Founded: 1955.

Congressional District: 2

Key Personnel: Dir. Florida Park Service, Mike Bullock; Park Mgr., Mark Knapke; Ranger, Willie McNair.

Personnel Profile: Full-Time Paid 1.

Governing Authority: state. Parent Institution: St. George Island State Park. Dept. of Environmental Protection, Div. of Recreation and Parks, 3900 Commonwealth Blvd., Tallahassee, FL 32303. Tax-exempt.

Institution Type/Description: History Museum.

Collections: replica of the first ice-making machine invented by Dr. John Gorrie; first refrigeration & air conditioning; history of Apalachicola River & Town.

Activities: tours.

Hours & Admission Prices: Thurs.-Mon. 9-5. Admission $2; children 5 & under no charge. Closed New Year's Day; Thanksgiving; Christmas. &

RANEY HOUSE MUSEUM - APALACHICOLA AREA HISTORICAL SOCIETY, 128 Market St., Apalachicola, FL 32320. Mailing Address: P.O. Box 75, Apalachicola, FL 32329. Tel.: 850-653-4321.
Key Personnel: Pres., Bill Spohrer
Institution Type/Description: Historical Society Museum: housed in the former home of the Raney family; built in 1838.
Collections: local history & culture; period furnishings; personal artifacts; photographs.
Hours & Admission Prices: By appointment.

Apopka

MUSEUM OF APOPKANS, (M), 122 E. 5th St., Apopka, FL 32703-5314. Tel.: 407-703-1707. Fax: 407-703-1773.
E-mail: director@apopkamuseum.org
Web Site: www.apopkamuseum.org
Founded: 1968.
Key Personnel: Mgr., Reba Wilson.
Personnel Profile: Full-Time Paid 1; Part-Time Volunteers 20.
Governing Authority: Parent Institution: Apopka Historical Society. Tax-exempt.
Institution Type/Description: History Museum.
Collections: history of Apopkans; photographs; personal artifacts; Native American artifacts.
Research Fields: NW Orange County history.
Facilities: library; meeting room; rental facilities.
Activities: permanent & temporary exhibitions; film; guest speakers; educational tours; traveling plays for schools; Power Point presentations for schools; membership meetings. Annual Events: Mother's Day Tea; Festival of Trees; Fireball Roberts Memorial Celebration.
Publications: quarterly newsletter, News and Views.
Hours & Admission Prices: Wed.-Thurs. 10-4, Sat. 10-2; other times by appointment. No charge; donations accepted. &
Attendance: 3,000 (estimated)
Membership: Junior $10; Individual $15; Couple $25; Family $35; Business $75; Lifetime Single $200; Lifetime Couple $300.

Avon Park

AVON PARK DEPOT MUSEUM, 3 N. Museum Ave., Avon Park, FL 33825-3153. Tel.: 863-453-3525.
E-mail: apmuseum@centurylink.net
Web Site: www.hsaponline.org
Founded: 1981.
Congressional District: 16
Key Personnel: Dir. Chm. (V), Elaine Levey; Pres. (V), Kathy Johnson.
Personnel Profile: Full-Time Volunteers 1; Part-Time Volunteers 15.
Operating Expenses: 10,000
Operating Income: 10,000
Governing Authority: Parent Institution: City of Avon Park. Subsidiary Institution: Historical Society of Avon Park. Tax-exempt: 501(c)(3).
Institution Type/Description: History Museum.
Collections: Avon Park history & culture; photographs; military; personal artifacts; more than 300 bound newspapers from the 1920s to 2000s.
Major Exhibits: Laundry Suitcase-Mailed, 11/13-1/14; Tin Can Tourists, 12/13-10/14.
Research Fields: military; architecture; old homes.
Facilities: library.
Activities: research; tours; permanent exhibits.
Publications: The First Hundred Years of Avon Park; Avon Park People.
Hours & Admission Prices: Wed.-Fri. 10-3. No charge; donations accepted. &
Attendance: 2,500 (estimated)
Membership: Individual $15; Family $25; Business $50.

SOUTH FLORIDA COMMUNITY COLLEGE MUSEUM OF FLORIDA ART AND CULTURE, 600 W. College Dr., Avon Park, FL 33825-9356. Tel.: 863-784-7240.
E-mail: mofac@southflorida.edu
Web Site: www.mofac.org
Key Personnel: Cur., Mollie Doctrow
Institution Type/Description: Art Museum.
Collections: works by Florida artists.
Hours & Admission Prices: Sept.-May Wed.-Fri. 12:30-4:30. &

Barberville

PIONEER SETTLEMENT FOR THE CREATIVE ARTS, 1776 Lightfoot Ln., Barberville, FL 32105. Mailing Address: P.O. Box 6, Barberville, FL 32105-0006. Tel.: 386-749-2959. Fax: 386-749-2087.
E-mail: info@pioneersettlement.org
Web Site: www.pioneersettlement.org
Founded: 1976.
Key Personnel: Pres. (V), Carl McClancy; Dir. Education, Jewel Tompkins; Coord. Special Events, Dale Barnhart.
Personnel Profile: Full-Time Paid 5; Part-Time Paid 9; Part-Time Volunteers 400.
Institution Type/Description: Cultural Heritage Museum.
Collections: state and regional cultural heritage.
Research Fields: Florida settlement; family histories; early architecture.
Facilities: Museum-related items for sale.
Activities: special events. Annual Events: Folk Art Days in February; Spring Fest & Frolic in April; Summer Folk School in June & July; Harvest Celebration in September; Fall County Jamboree in November; Florida Christmas Remembered in December.
Publications: quarterly newsletter, Pioneer Press.
Hours & Admission Prices: Mon.-Sat. 9-4. Adults $6, children 6-12 $4; members & children under 5 no charge. Closed major holidays. &
Attendance: 65,000
Membership: Student $20; Teacher & Senior Individual $25; Individual $30; Senior Couple $35; Couple $40; Family $50; Patron $100; Small Business $300; Advocate $500; Booster $1,000.

Bartow

POLK COUNTY HISTORICAL MUSEUM, (M), 100 E. Main St., Bartow, FL 33830-4629. Tel.: 863-534-4386. Fax: 863-534-4387.
E-mail: museum@polk-county.net
Web Site: www.polkcountymuseum.org
Founded: 1998.
Congressional District: 12
Key Personnel: Dir., Tom Muir; Museum Asst., Maria Trippe.
Personnel Profile: Full-Time Paid 2; Part-Time Volunteers 12.
Governing Authority: county government; nonprofit. Parent Institution: Polk County Board of County Commissioners. Tax-exempt.
Institution Type/Description: History Museum.
Collections: history of Polk County, Florida from prehistoric to modern times; pioneer life; political history; agriculture; industry; contemporary life of Polk County; general store; hands-on children's exhibits; military artifacts; c.1908 & c.1926 courtrooms.
Facilities: 10,000 sq. ft. exhibit space.
Activities: self-guided tours; participatory exhibits.
Publications: The Historical Marker.
Hours & Admission Prices: Tues.-Fri. 9-5, Sat. 9-3. No charge; donations accepted. Closed New Year's Day; Martin Luther King Jr. Day; Memorial Day; Independence Day; Labor Day; Veterans Day; Thanksgiving weekend; Christmas. &
Attendance: 28,000 (estimated)

Bay Lake

DISNEY'S ANIMAL KINGDOM THEME PARK, 1200 N. Savannah Circle E., Bay Lake, FL 32830. Mailing Address: P.O. Box 10000, Lake Buena Vista, FL 32830-1000. Tel.: 407-939-2468. Fax: 407-939-6240.
Web Site: www.disneyworld.com
Founded: 1998.
Congressional District: 9
Key Personnel: Vice Pres., Jacqueline J. Ogden, Ph.D.; Vice Pres., Michael Colglazier; Administrative Asst., Donna McKiernan; Administrative Asst., Jill Martin; Dir. Animal Opers., Mark R. Penning, BVSc; Dir. Dept. of Animal Health, Scott P. Terrell, D.V.M., Dipl. A.C.V.P.; Dir. Education & Science, Jill Mellen, Ph.D.; Mgr. Animals, Science, & Environment, Heather Eberhart; Sr. Research Biologist, Tamara Bettinger, Ph.D.; Dir. Conservation, Anne Savage, Ph.D.; Cur. Education, Kathy Lehnhardt; Cur. Education, Allyson Atkins; Animal Records & Regulatory Affairs Mgr., Lynn S. McDuffie; Veterinarian, Natalie Mylniczenko, D.V.M., Dipl. A.C.Z.M.; Veterinary Svcs. Operations Mgr., Elizabeth C. Nolan, D.V.M., Dipl. A.C.Z.M.; Veterinary Svcs. Operations Mgr., Don Neiffer, V.M.D., Dipl. A.C.Z.M.; Veterinarian, Greg Fleming, D.V.M., Dipl. A.C.Z.M.; Veterinarian, Deidre Fontenot, D.V.M.; Animal Operations Mgr., Pagani Forest Exploration Trail (R) & Maharajah Jungle Trek (R), Chelle Plasse; Animal Operations Mgr., Savannahs, Joe Christman; Dir. Animal Operations, Matt Hohne; Asst. Animal Operations Mgr., Rafiki's Planet Watch,

Andre J. Daneault; Animal Operations Mgr., Rafiki's Planet Watch, Jerry Brown; Animal Operations Mgr. Disney's Animal Kingdom Lodge, Greg Peccie; Asst. Animal Operations Mgr., Savannahs, Sam Berner; Asst. Animal Operations Mgr., Disney's Animal Lodge, Steve Metzler; Asst. Animal Operations Mgr., Pagani Forest Exploration Trail (R) & Maharajah Jungle Trek (R), Jay Therien; Veterinary Pathologist, Carlos Rodriguez, D.V.M.; Nutritionist, Eduardo V. Valdes, Ph.D.; Gen. Mgr. Park Operations, Tim Sypko; Gen. Mgr. Food & Beverage, Maryann Smith; Dir. Engineering Svcs., Kevin C. Shultz; Mgr. Horticulture Svcs., Wendy Andrew; Animal Operations Mgr., Behavioral Husbandry, Marty MacPhee; Mgr. Human Resources, Park Ops Line of Business, Jeanette Dennis; Gen. Mgr. Merchandise, Robert Kelley.
Governing Authority: Parent Institution: Walt Disney World Resort, a division of Walt Disney Parks and Resorts U.S., Inc., Lake Buena Vista, FL.
Institution Type/Description: Zoological Park.
Collections: 350 species (more than 6,000 specimens) of mammals, birds, reptiles, amphibians, invertebrates & fish; 4 million trees, plants, shrubs, vines, grasses & ferns representing more than 3,000 species.
Research Fields: botany; animal health; applied research related to animal management; reproductive physiology/endocrinology & behavior.
Facilities: library; aquarium; botanical garden; restaurants; classrooms; 500 acre exhibit space; nature center; zoological park. Museum-related items for sale.
Activities: formal education programs; guided tours; lectures; training programs for professional museum workers; children's fossil play maze; wooly mammoth bone dig; guest educational research. Annual Events: Earth Day; Plant Conservation Day; International Migratory Bird Day; Party for the Planet.
Publications: weekly brochures.
Hours & Admission Prices: Summer: 9-8; Winter 9-5. Adults $94.79, children 3-9 $88.40. &

Boca Grande

PORT BOCA GRANDE LIGHTHOUSE & MUSEUM, (M), Gasparilla Island State Park, Boca Grande, FL 33921. Mailing Address: P.O. Box 637, Boca Grande, FL 33921-0637. Tel.: 941-964-0060.
E-mail: barrierislandparkssociety@gmail.com
Web Site: www.barrierislandparkssociety.org
Formerly: Boca Grand Lighthouse Museum and Visitors Center
Founded: 1999.
Key Personnel: Dir., Sharon McKenzie; Pres. (V), Jim Grant; Museum Shop Mgr., Eva Johnson.
Personnel Profile: Full-Time Paid 2; Part-Time Paid 4; Part-Time Volunteers 30.
Governing Authority: Parent Institution: Barrier Island Parks Society, Inc. Tax-exempt: 501(c)(3).
Institution Type/Description: History Museum.
Collections: displays on Native Americas & Spanish, local fishing industry, history of the lighthouse, railroad development, Port Boca Grande, the town of Boca Grande & tarpon fishing.
Publications: newsletter, Tidelines.
Hours & Admission Prices: June-July & Sept.-Oct. Wed.-Sat. 10-4, Sun. 12-4; Nov.-May Mon.-Sat. 10-4, Sun. 12-4. No charge; donations accepted. Closed New Year's Day; Martin Luther King Jr. Day; Easter; Memorial Day; Independence Day; Labor Day; Thanksgiving; Christmas.
Attendance: 22,000 (accurate)
Membership: Student $25; Friend $50; Donor $100; Sponsor $250; Corporate & Business $250 & up; Patron $500; Benefactor $1,000.

Boca Raton

BOCA RATON HISTORICAL SOCIETY, (M), 71 N. Federal Hwy., Boca Raton, FL 33432-3919. Tel.: 561-395-6766, ext. 106. Fax: 561-395-4049.
E-mail: info@bocahistory.org
Web Site: www.bocahistory.org
Founded: 1972.
Congressional District: 23
Key Personnel: Dir., Mary Csar; Pres., James A. Ballerano, Jr.; Cur., Susan Gillis; Educator, Laurie Lynn Jones; Administrative Asst., Jean Scanlon.
Personnel Profile: Full-Time Paid 4; Full-Time Volunteers 1; Part-Time Paid 1; Part-Time Volunteers 150; Interns 2.
Governing Authority: private; nonprofit organization. Tax-exempt: 501(c)(3).
Institution Type/Description: Historical Society Museum.
Collections: Boca Raton history from the late 1890s to modern times; timeline; biographies; map; history of Palm Beach County cities, people & events; restored streamline railcars at F.E.C. Depot; Mizner industry artifacts.
Research Fields: Boca Raton history; historic architecture.

Facilities: research library; 2,000 sq. ft. exhibit space. Museum-related items for sale.
Activities: lectures; loan, temporary & traveling exhibitions; weekly historic tours of city and Boca Raton Resort & Club. Annual Events: Benefit; House Tour; Holiday Gift Show; Wine Festival.
Publications: annual, Spanish River Papers; academic publication on local history.
Hours & Admission Prices: Mon.-Fri. 10-4. Adults $5; members no charge. Closed state holidays. &
Attendance: 25,000 (estimated)
Membership: Young Friend $30; Individual $40; Family $60; Patron $125; Grand Patron $250; Benefactor & Corporate $500.

* **BOCA RATON MUSEUM OF ART, (M),** Mizner Park, 501 Plaza Real, Boca Raton, FL 33432-3982. Tel.: 561-392-2500, ext. 200. Fax: 561-391-6410.
E-mail: info@bocamuseum.org
Web Site: www.bocamuseum.org
Founded: 1950.
Congressional District: 22
Key Personnel: Pres., Dalia Pabon Stiller; Dir., Steven V. Maklansky; Dir. Devel., Christine Mally; Dir. Administration, Roberta Stewart; Dir. Education, Claire Clum; Cur. 20th Century & Contemporary Art, Marisa Pascucci; Cur. Exhibitions & Audience Engagement, Kathleen Goncharov; Asst. to Dir., Valerie Johnson; Dir. Mktg. & Public Rels., Bruce Herman; Dir. Art School, Rebecca Sanders; Registrar, Martin Hanahan; Mgr. Special Events, Kisbel De La Rosa; Dir. Finance, Linda Ursillo; Museum Store Mgr., Laura Toia.
Personnel Profile: Full-Time Paid 29; Part-Time Paid 4; Part-Time Volunteers 500; Interns 8.
Volunteer Hours: 8,149
Operating Expenses: 4,539,492
Operating Income: 4,764,604
Governing Authority: nonprofit organization. Tax-exempt: 501(c)(3).
Institution Type/Description: Art Museum.
Collections: 19th-20th century and contemporary European & American painting, drawing, sculpture & graphics; photography from 1840 to present; The Dr. & Mrs. John J. Mayers Collection of Modern Masters (including Braque, Degas, Demuth, Matisse, Modigliani, Picasso, Prendergast, Seurat and others); The Jean & David Colker Collection of Pre-Columbian Art; West African tribal art of artifacts; Oceanic sculpture, textiles and basketry; Asian sculpture & ceramics; 19th-century English Victorian Doulton-Lambeth stoneware.
Major Exhibits: James Rosenquest's - High Technology and Mysticism: A Meeting Point, 11/23/13-4/6/14; Futurism: Concepts and Imaginings, 1/12/14-3/14; Fascination: The Love Affair between French and Japanese Printmaking, 1/12/14-4/13/14; Pop Culture: Selections from the Weisman Art Foundation, 1/12/14-4/23/14; Investigative Curating: A Mesoamerican Artifact, 4/5/14-8/24/14; Afghan War Rugs: The Modern Art of Central Asia, 5/3/14-7/27/14; Guerra de la Paz: A Matter of Course, 5/3/14-7/27/14; The Art of War: A Collective Experience, 5/3/14-7/27/14.
Research Fields: 20th century European & American modernism, African, Oceanic, Pre-Columbian; photography.
Facilities: 4,000-library of art history; sculpture gardens. Museum-related items for sale.
Activities: guided tours; lectures; films; gallery talks; concerts; study clubs; docent program; traveling exhibitions. Annual Event: Boca Museum Juried Art Festival.
Publications: quarterly newsletter; catalogs for changing exhibitions.
Hours & Admission Prices: Tues. & Thurs.-Fri. 10-5, Wed. 10-9, Sat.-Sun. 12-5. Adults $8, senior citizens $6, students with ID $5; discounts to groups & AAM members; members, children 12 & under & Wed. 5-9 no charge. Admission may change for special exhibits. Closed major holidays. &
Attendance: 200,000 (estimated)
Membership: Student $25; Individual $80; Household $100; Art Enthusiast $150; Art Seeker $300; Art Insider $600; Patron $1,250; Advocate $2,500; Ambassador $5,000; Visionary $10,000.

CHILDREN'S MUSEUM, INC., 498 Crawford Blvd., Boca Raton, FL 33432-3752. Tel.: 561-368-6875. Fax: 561-395-7764.
Founded: 1979.
Congressional District: 14
Key Personnel: Exec. Dir., Poppi Mercier; Pres. (V), Penny Morey.
Personnel Profile: Full-Time Paid 1; Part-Time Paid 6; Part-Time Volunteers 60.
Governing Authority: nonprofit. Tax-exempt: 501(c)(3).
Institution Type/Description: Children's Museum: housed in c.1912 unaltered wooden structure.

Collections: Sophia S. Kuzmick dolls; Korvetz Shell collection; Florida Pioneer Kitchen artifacts; Charles Weiner carousel animals; World of Play toys.
Facilities: 600-vol. library; 1,400 sq. ft. exhibit space; nature & conservation center. Museum-related items for sale.
Activities: docent program; guided tours; temporary, traveling & loan exhibitions; formally organized education programs for children.
Publications: quarterly newsletter, Connections.
Hours & Admission Prices: Tues.-Sat. 9:30-11:45 (reserved for groups), 12-4 (public hours). $3 per person; infants & members no charge. Closed New Year's Day; Memorial Day; Independence Day; Labor Day; Thanksgiving; Christmas. &
Attendance: 50,000 (estimated)
Membership: Individual/Teacher $25; Florida Plus $50; Friend $100; Supporter $250; Patron $500.

CHILDREN'S SCIENCE EXPLORIUM, 300 S. Military Trail, Boca Raton, FL 33486-4302. Tel.: 561-347-3912. Fax: 561-347-3910.
E-mail: explorium@myboca.us
Web Site: www.scienceexplorium.org
Founded: 1998.
Key Personnel: Science Center Cur., Kate Lasher; Coord. Exhibits, Harry Robelen; Visitor Program Specialist, Debbie Blair.
Personnel Profile: Full-Time Paid 4; Part-Time Paid 6; Part-Time Volunteers 75.
Governing Authority: nonprofit. Parent Institution: City of Boca Raton, FL.
Institution Type/Description: Children's Science Museum.
Collections: interactive physical science exhibits.
Major Exhibits: Ocean Bound, 2/14-5/14.
Facilities: 155-seat theatre; demo classroom; 4,000 sq. ft. exhibit space; outdoor science playground; nature trails.
Activities: traveling exhibitions; formal & informal education programs for children; science camps; public programs; special events.
Hours & Admission Prices: Mon.-Fri. 9-6, Sat.-Sun. & holidays 10-5. No charge; $5 donation requested. Closed New Year's Day; Thanksgiving; Christmas. &
Attendance: 167,123 (estimated)

UNIVERSITY GALLERIES, FLORIDA ATLANTIC UNIVERSITY, 777 Glades Rd., Boca Raton, FL 33431-6496. Tel.: 561-297-2966 & 2661.
E-mail: wfaulds@fau.edu
Web Site: www.fau.edu/galleries
Founded: 1970.
Key Personnel: Dir., W. Rod Faulds.
Personnel Profile: Full-Time Paid 2; Part-Time Paid 6; Part-Time Volunteers 20; Interns 4.
Governing Authority: Parent Institution: Florida Atlantic University. Tax-exempt.
Institution Type/Description: Contemporary Art Museums & Galleries.
Collections: Artinian collection of self-portraits by Florida artists; Albert Binny Backus paintings.
Research Fields: contemporary art; art, humanities & popular culture.
Facilities: 5,000 sq. ft. exhibit space.
Activities: docent training for undergraduate students; traveling exhibitions; exhibition-related programs; artists lectures, presentations & residencies to create on-site exhibition projects.
Publications: exhibit catalogues; exhibition announcements; brochures.
Hours & Admission Prices: Ritter Art Gallery & Schmidt Center Gallery: Tues.-Fri. 1-4, Sat. 1-5. No charge; donations accepted. &
Attendance: 15,000 (accurate)
Membership: Individual $30; Family $50.

Bokeelia

KOUCKY GALLERY, 5971 Bay Point Rd, Bokeelia, FL 33922-2860. Tel.: 239-283-4414. Fax: 231-547-2455; 239-283-4495.
E-mail: kouckygallery@gmail.com
Web Site: www.kouckygallery.com
Institution Type/Description: Art Museum.
Collections: fine arts; paintings; sculptures; jewelry.
Hours & Admission Prices: Winter: Mon.-Sat. 9-5; Summer: call for hours.

Bonifay

ROYAL AIR MUSEUM, 1967 Hwy. 173, Bonifay, FL 32425-5635. Tel.: 850-547-9002.
E-mail: jpinto@royalairmuseum.org

Institution Type/Description: Aviation Museum.
Collections: aviation history; aircraft.
Activities: Museum Sponsors: Air Shows.
Hours & Admission Prices: Call for hours.

Bonita Springs

CENTER FOR THE ARTS OF BONITA SPRINGS, (M), 26100 Old 41 Rd., Bonita Springs, FL 34135-8613. Tel.: 239-495-8989. Fax: 239-495-3999.
E-mail: albs@artinusa.com
Web Site: www.artcenterbonita.org
Formerly: Art League of Bonita Springs
Founded: 1959.
Key Personnel: Pres., Susan Bridges.
Personnel Profile: Full-Time Paid 10; Part-Time Paid 5; Part-Time Volunteers 1,102; Interns 25.
Governing Authority: Tax-exempt.
Institution Type/Description: Art Museum.
Collections: paintings; photographs; sculpture.
Activities: educational programs; classes; special events & activities; lectures; festivals; luncheons. Museum Sponsors: Beaux Arts Ball and Auction.
Hours & Admission Prices: Mon.-Fri. 10-4, Sat. 1-5. No charge; donations accepted. Closed New Year's Day; Christmas. &
Attendance: 85,000 (estimated)
Membership: Student $25; Individual $75; Family $100; Business Patron $300; Business Patron Circle $600; Corporate $1,000.

Bowling Green

PAYNES CREEK HISTORIC STATE PARK, 888 Lake Branch Rd., Bowling Green, FL 33834-4078. Tel.: 863-375-4717. Fax: 863-375-4510.
E-mail: jackson.mosley@dep.state.fl.us
Web Site: www.floridastateparks.org
Formerly: Paynes Creek State Historic Site
Founded: 1981.
Congressional District: 10
Key Personnel: Park Mgr., Jacks Mosley; Park Ranger, Ray N. Gilmore; Pres. (V) & Park Ranger, Sam Hale.
Personnel Profile: Full-Time Paid 3; Full-Time Volunteers 6; Part-Time Paid 2; Part-Time Volunteers 10.
Governing Authority: state. Parent Institution: Dept. of Environmental Protection. Tax-exempt.
Institution Type/Description: Historic Site: located near c.1850 Ft. Chokonikla, Seminole Indian War Fort Visitor Center.
Collections: trading post items; Seminole costumes; military uniforms; historic paintings; c.1850 cannon; artifacts from Ft. Chokonikla: bullets, bottles, buttons, utensils & pipes.
Facilities: 65-seat theatre.
Activities: guided tours; organized education programs for children.
Hours & Admission Prices: Daily 8am to sundown. $3 per vehicle. &
Attendance: 31,000 (estimated)

Boynton Beach

SCHOOLHOUSE CHILDREN'S MUSEUM & LEARNING CENTER, 129 E. Ocean Ave., Boynton Beach, FL 33435-4536. Tel.: 561-742-6780. Fax: 561-742-6781.
E-mail: info@schoolhousemuseum.org
Web Site: www.schoolhousemuseum.org
Founded: 1996.
Congressional District: 23
Key Personnel: Dir., Judith Klinek; Pres. (V), Lori Livergood; Museum Shop Mgr., Cheryl Lane.
Personnel Profile: Full-Time Paid 2; Part-Time Paid 8; Part-Time Volunteers 25; Interns 2.
Governing Authority: Parent Institution: Boynton Cultural Centre. Tax-exempt.
Institution Type/Description: Children's Museum.
Collections: hands-on exhibits.
Facilities: classroom; theatre. Gift items for sale.
Activities: birthday parties; special events; holiday workshops; educational programs.
Publications: quarterly newsletter; membership guide; visitor guide.
Hours & Admission Prices: Tues.-Sat. 10-5. Adults $5, grandparents $4.50, children 2-17 $4; discounts to groups of 10 or more & ACM members; members no charge. &
Attendance: 22,600 (accurate)
Membership: Individual $25; Grandparents $45; Family Basic $60; Family Plus $100; Family Ultimate $125.

Bradenton

ARTCENTER MANATEE, 209 9th St., W., Bradenton, FL 34205-8627. Tel.: 941-746-2862. Fax: 941-746-2319.
E-mail: acm@artcentermanatee.org
Web Site: www.artcentermanatee.org
Formerly: Art League of Manatee County
Founded: 1937.
Congressional District: 13
Key Personnel: Dir., Mary Roff; Pres. (V), Linda Vantassell; Museum Shop Mgr., Carla Nierman.
Personnel Profile: Full-Time Paid 4; Part-Time Volunteers 200.
Governing Authority: board of directors; nonprofit organization. Tax-exempt.
Institution Type/Description: Arts Center.
Collections: works by local, regional & national artists.
Facilities: 3,000-vol. library on art, artists, artist's material & techniques available on premises or by special permission; reading room. Handcrafts, art works & gifts for sale.
Activities: lectures; films; gallery talks; arts festivals; workshops; formally organized education programs; instruction in painting, pottery, drawing, sumie & fine crafts; children's programs; temporary exhibits.
Publications: exhibition & class schedules; e-newsletters.
Hours & Admission Prices: Mon. & Fri.-Sat. 9-5, Tues.-Thurs. 9-6. No charge; donations accepted. Closed New Year's Eve & Day; Memorial Day; Independence Day; Labor Day; Thanksgiving; Christmas Eve & Day. &
Attendance: 30,000 (estimated)
Membership: Student $20; Adult $60; Family $80; Friend $125; Patron $250; Benefactor $500; Director's Circle $1,000.

DESOTO NATIONAL MEMORIAL, 75th St., N.W., Bradenton, FL 34209. Mailing Address: P.O. Box 15390, Bradenton, FL 34280-5390. Tel.: 941-792-0458, ext. 105. Fax: 941-792-5094.
E-mail: deso_ranger_activities@nps.gov
Web Site: www.nps.gov/deso/
Founded: 1948.
Congressional District: 13
Key Personnel: Supt., Scott Pardue; Park Ranger, Ben Sims.
Personnel Profile: Full-Time Paid 6; Part-Time Paid 5; Part-Time Volunteers 20.
Governing Authority: federal. Parent Institution: Dept. of the Interior. Affiliated with the National Park Service. Tax-exempt.
Institution Type/Description: Park Museum.
Collections: 16th-century European military artifacts; pre-historic Native American artifacts.
Research Fields: Spanish period of 16th century & related fields.
Facilities: 1,000-vol. library of 16th-century history of the Spanish, Native Americans, Europeans & National Park material available for use on premises; auditorium; 16th-century living history camp; visitor center; nature trails.
Activities: narrative film; permanent exhibits; self-guiding nature trail. Museum Sponsors: living history camp with demonstrations mid-Dec. to mid-April.
Publications: informational brochure.
Hours & Admission Prices: Visitor Center: daily 9-5. History Camp: mid-Dec. to mid-April daily 10-4, call to confirm. No charge; donations accepted. Closed New Year's Day; Thanksgiving; Christmas. &
Attendance: 275,505 (accurate)

MANATEE VILLAGE HISTORICAL PARK, 1404 Manatee Ave. E., Bradenton, FL 34208-1360. Tel.: 941-741-4075. Fax: 941-708-5924.
E-mail: phaedra.rehorn@manateeclerk.com
Web Site: www.manateeclerk.com/historical/manateevillage.aspx
Founded: 1974.
Congressional District: 10
Key Personnel: Suprv., Phaedra Rehorn; Chm., Bill Howard; Museum Shop Mgr., Liz Boling.
Personnel Profile: Full-Time Paid 4; Part-Time Paid 1; Part-Time Volunteers 3.
Governing Authority: county government. Parent Institution: Manatee County Clerk of Circuit Courts. Manatee County Historical Commission (non profit community benefit organization). Tax-exempt: 501(c)(3).
Institution Type/Description: Local History Museum.
Collections: Manatee County history, 1841-1914; period furniture & furnishings; children's hands-on exhibits; Florida barn; smokehouse; sugar kettle & mill; boat works. Historic Buildings: 1860 courthouse; 1887 church; 1912 farm house; 1903 general store; 1908 one-room schoolhouse; 1920s cowhunter bunkhouse; Turpentine Still; blacksmith shop.
Major Exhibits: Stories of Valor, 1/14-1/15.
Research Fields: Manatee County history.

Facilities: 50-vol. library of Manatee County history books, available for use by public; educational facilities.
Activities: guided tours; lectures; organized education programs for children; docent program; participatory, loan, temporary & traveling exhibitions; school loan service; children's hands-on activities. Museum Sponsors: Heritage Days in March; Spirit Voices from Old Manatee in October; Manatee County History Fair in February; Florida Cracker Christmas in December.
Publications: monthly newsletter, The Cabbage Head Gazette.
Hours & Admission Prices: Mon.-Fri. & 2nd & 4th Sat. each month 9-4. No charge; donations accepted. Closed holidays. &
Attendance: 10,000 (estimated)

SOUTH FLORIDA MUSEUM, BISHOP PLANETARIUM & PARKER MANATEE AQUARIUM, (M), 201 10th St., W., Bradenton, FL 34205-8635. Mailing Address: P.O. Box 9265, Bradenton, FL 34206-9265. Tel.: 941-746-4131, ext. 18. Fax: 941-747-2556.
E-mail: bbesio@southfloridamuseum.org
Web Site: www.southfloridamuseum.org
Founded: 1946.
Congressional District: 10
Key Personnel: Dir., Brynne Anne Besio; Chm. & Pres. (V), Tom Breiter; Museum Shop Mgr., Ellen Ferraro.
Personnel Profile: Full-Time Paid 15; Part-Time Paid 14; Part-Time Volunteers 120; Interns 2.
Governing Authority: bd. of directors; nonprofit organization. Tax-exempt: 501(c)(3).
Institution Type/Description: General Museum & Planetarium.
Collections: archaeology; astronomy; live manatee exhibit; geology; history; Indian artifacts; science & natural history; paleontology; medical; ethnology; costumes; manuscripts; transportation; maritime history; decorative arts.
Research Fields: ongoing projects with manatee.
Facilities: educational facilities; hands-on environmental classroom. Educational books & gifts for sale.
Activities: docent program or council; temporary exhibitions; docent led tours for school & adult groups; school programs; camp programs; film programs; science programs; adult public programs; family public programs.
Publications: brochures; e-mail newsletter.
Hours & Admission Prices: Museum: Jan. April & July Mon. Sat. 10 5, Sun. noon-5; May-June & Aug.-Dec. Tues.-Sat. 10-5, Sun. noon-5. Museum, Aquarium & Planetarium: adult $15.95, senior 65 & over $13.95, children 4-12 $11.95; discounts to groups; ASTC & museum members and children under 4 no charge. Closed New Year's Day; Thanksgiving; Christmas. &
Attendance: 85,000 (estimated)
Membership: Individual $50; Family $75; Family Plus $125; Sponsor $250; Patron $500; Benefactor $750.

Bristol

TORREYA STATE PARK, 2576 N.W. Torreya Park Rd., Bristol, FL 32321-2203. Tel.: 850-643-2674. Fax: 850-643-2987.
E-mail: steven.cutshaw@dep.state.fl.us
Web Site: www.floridastateparks.org/torreya
Key Personnel: Park Mgr., Steve Cutshaw.
Personnel Profile: Full-Time Paid 2; Part-Time Paid 1; Part-Time Volunteers 1.
Governing Authority: state. Tax-exempt.
Institution Type/Description: Park & Historic House Museum.
Collections: Historic House: Gregory House built in 1849.
Activities: guided tours.
Hours & Admission Prices: Park: 8-sunset. Tours: Mon.-Fri. 10am, Sat.-Sun. & state holidays 10am, 2pm & 4pm. Adults $3, children 12 & under $2; children under 6 no charge.

Brooksville

HERNANDO HERITAGE MUSEUM, 601 Museum Ct., Brooksville, FL 34601-2631. Tel.: 352-799-0129. Fax: 352-799-4766.
E-mail: info@hernandohistoricalmuseumassoc.com
Web Site: hernandohistoricalmuseumassoc.com/index.htm
Founded: 1980.
Key Personnel: C.E.O., Virginia Jackson; Chm. (V), Virginia Rusk.
Personnel Profile: Part-Time Volunteers 25.
Governing Authority: Parent Institution: Hernando Historical Museum Assoc. Inc.
Institution Type/Description: Historic House Museum: built c.1856. Listed on the National Register of Historic Places.
Collections: personal artifacts; summer kitchen; photographs; school room; war room; doctors room; 1885 train depot; Cracker Country display.

Activities: Living history; annual events: Museum Sponsors: Brooksville Raid Festival in January.
Publications: quarterly newsletters; histories.
Hours & Admission Prices: Tues.-Sat. 12-3. Suggested Donation: adults $5, children $2; students & scouts no charge. &
Attendance: 35,000 (estimated)

Bushnell

DADE BATTLEFIELD HISTORIC STATE PARK, 7200 County Rd. 603, S. Battlefield Dr., Bushnell, FL 33513-3538. Tel.: 352-793-4781. Fax: 352-793-4230.
Web Site: www.floridastateparks.org
Formerly: Dade Battlefield State Historic Site
Founded: 1921.
Congressional District: 6
Key Personnel: Park Mgr., Tracey Stardridge; Pres. (V), Jean Mcnary; Park Ranger, Chuck Wicks; Park Ranger, George Webb.
Personnel Profile: Full-Time Paid 3; Part-Time Paid 1; Part-Time Volunteers 6.
Governing Authority: state. Parent Institution: Dept. of Environmental Protection, Div. of Recreation and Parks, Commonwealth Bldg., 3900 Commonwealth Blvd., Tallahassee, FL 32303. Tax-exempt.
Institution Type/Description: Historic Site and Park Museum: site of a battle between Seminole & US soldiers on Dec. 28, 1835.
Collections: interpretive exhibits; artifacts of the battle; video.
Facilities: picnic area; nature trail; rental hall.
Activities: guided tours, permanent exhibitions, interpretive programs for organized groups can be arranged by contacting site six weeks in advance. Annual Events: WWII Commemorative Day in August; Battle Reenactments in late Dec. & early Jan.
Hours & Admission Prices: Museum: daily 9-5. Grounds: daily 8am to sunset. $3 per vehicle, $1 bicycles & walk-ins. Bus fee $40 or $1 per person. &
Attendance: 30,000 (estimated)
Membership: Dade Battlefield Society: Individual Adult $10. Florida Park Service Annual Permit: Individual $30; Family $60.

Cape Canaveral

AIR FORCE SPACE AND MISSILE HISTORY CENTER, 100 Space Port Way, Cape Canaveral, FL 32920.
Collections: launch history; space-related equipment; photographs.
Facilities: picnic area. Gift items for sale.
Hours & Admission Prices: Tues.-Fri. 9-2, Sat. 9-5, Sun. 12-4. No charge. Closed New Year's Day; Thanksgiving; Christmas.

Cape Coral

CAPE CORAL HISTORICAL MUSEUM, 544 Cultural Park Blvd., Cape Coral, FL 33990-1212. Mailing Address: P.O. Box 150637, Cape Coral, FL 33915-0637. Tel.: 239-772-7037. Fax: 239-573-7518.
E-mail: ccoralmuseum@embarqmail.com
Web Site: www.capecoralhistoricalmuseum.org
Founded: 1986.
Congressional District: 19
Key Personnel: C.E.O. & Cur., Anne C. Cull; Pres. (V), Paul Sanborn; Museum Shop Mgr., Rusty Williamson.
Personnel Profile: Part-Time Paid 1; Part-Time Volunteers 40.
Operating Expenses: 39,000
Operating Income: 39,000
Governing Authority: private; nonprofit organization. Tax-exempt.
Institution Type/Description: History Museum.
Collections: local history & culture; photographs; personal artifacts; memorial rose garden.
Facilities: demonstration native gardens.
Activities: monthly speakers.
Publications: members monthly newsletter.
Hours & Admission Prices: Sept.-June Wed.-Thurs. & Sun. 1-4. Suggested Donation: adults $5, children $2. Closed holidays. &
Attendance: 1,500 (estimated)
Membership: Single $20; Family $25; Corporate $500.

Carrabelle

CAMP GORDON JOHNSTON MUSEUM, City Complex, Gray Ave., 1001 Gray Ave., Carrabelle, FL 32322. Mailing Address: P.O. Box 1334, Carrabelle, FL 32322. Tel.: 850-697-8575.
E-mail: campgordon@fairpoint.net
Web Site: campgordonjohnston.com

Founded: 1998.
Key Personnel: Dir., Linda Minichiello; Chm. (V), Anthony Minichiello.
Personnel Profile: Part-Time Paid 2; Part-Time Volunteers 4.
Governing Authority: Parent Institution: Camp Gordon Johnston Association. Tax-exempt.
Institution Type/Description: World War II Museum.
Collections: camp history; WWII soldier's heritage; photographs; uniforms; military artifacts.
Research Fields: WWII.
Facilities: movie theater. Museum-related items for sale.
Activities: Museum Sponsors: Camp Gordon Johnston Days & Smithsonian National Museum Day.
Publications: The Amphibian.
Hours & Admission Prices: Mon.-Thurs. 1-4, Fri. 12-4, Sat. 10-2. No charge; donations accepted. &
Attendance: 7,000 (accurate)

Cedar Key

CEDAR KEY HISTORICAL SOCIETY MUSEUM, 2nd St. at State Rd. 24, Cedar Key, FL 32625. Mailing Address: P.O. Box 222, Cedar Key, FL 32625-0222. Tel.: 352-543-5549. Facebook: Cedar Key Historical Society Museum.
E-mail: cedarcedar@bellsouth.net
Web Site: cedarkeymuseum.com
Founded: 1979.
Congressional District: 2
Key Personnel: Pres., Ken Young; Vice Pres., George Oakley; Office Mgr., Christie LaVoie; Vice Pres., John Andrews.
Personnel Profile: Part-Time Paid 1; Part-Time Volunteers 20.
Governing Authority: society; nonprofit. Tax-exempt.
Institution Type/Description: Historical Society Museum: housed in c.1871 former private residence.
Collections: from prehistoric to modern times: fossils; shells; pencil & lumber industry; railroad history; maps & charts of early Florida; photographic collection; Indian artifacts & history; commercial fishing industry in Cedar Key.
Facilities: more than 60-vol. library of material on Florida & Cedar Key history; 2,500 sq. ft. exhibit space. Museum-related items for sale.
Publications: semiannual newsletter, The Beacon.
Hours & Admission Prices: Sun.-Fri. 1-4, Sat. 11-5. Adults $2, children over 12 $1; members no charge. Closed Christmas & part of Christmas week. &
Attendance: 9,000
Membership: Student $10; Individual & Family $20; Small Business $25; Supporting $50; Contributing & Corporate Contributing $100; Sustaining $250; Corporate Sustaining & Patron $500; Benefactor & Corporate Benefactor $1,000.

CEDAR KEY MUSEUM STATE PARK, 12231 S.W. 166 Court, Cedar Key, FL 32625-6200. Tel.: 352-543-5350.
Web Site: www.floridastateparks.org
Founded: 1962.
Congressional District: 4
Key Personnel: Dir. Florida Park Svc., Donald Forgione; Park Ranger, Charles Neese; Park Mgr., Kristin Ebersol.
Personnel Profile: Full-Time Paid 1; Part-Time Volunteers 5.
Governing Authority: state. Dept. of Environmental Protection, Div. of Recreation & Parks, Commonwealth Bldg., 3900 Commonwealth Blvd., Tallahassee, FL 32399. Tax-exempt.
Institution Type/Description: State Museum.
Collections: Cedar Key history; shell collection; natural history.
Activities: permanent exhibitions.
Hours & Admission Prices: Thurs.-Mon. 10-5. Admission $2. Closed Christmas. &
Attendance: 20,408 (accurate)

Charlotte Harbor

CHARLOTTE COUNTY HISTORICAL CENTER, (M), 22959 Bayshore Rd., Charlotte Harbor, FL 33980-2000. Tel.: 941-629-7278. Fax: 941-743-3917. Facebook: Charlotte County Historical Center.
E-mail: historicalcenter@charlottefl.com
Web Site: charlottecountyfl.com/historical
Formerly: Florida Adventure Museum of Charlotte County
Founded: 1969.
Congressional District: 72
Key Personnel: Historian, Linda Roberts.
Personnel Profile: Full-Time Paid 2; Part-Time Volunteers 369; Interns 1.

Governing Authority: Parent Institution: Charlotte County Community Services Department. Tax-exempt.
Institution Type/Description: General Museum.
Collections: personal artifacts; archives; local & state history.
Research Fields: Charlotte County & SW Florida history.
Facilities: Gift items for sale.
Activities: self-guided tours; organized education programs; docent program; participatory, temporary & traveling exhibitions; school loan service; annual festival.
Publications: newsletter, The Charlotte Historian.
Hours & Admission Prices: Tues.-Fri. 1-5, Sat. 10-3. Adults $2, children $1; museum, ASTC & CCHC Society members no charge. Closed holidays. &

Attendance: 9,000 (accurate)
Membership: Individual $20; Family & Grandparent $35; Friend $50; Donor $100; Benefactor $500; Angel $1,000.

Chokoloskee

TED SMALLWOOD'S STORE, INC., 360 Mamie St., Chokoloskee, FL 34138. Mailing Address: P.O. Box 310, Chokoloskee, FL 34138-0367. Tel.: 239-695-2989. Fax: 239-695-4454.
E-mail: ishtoholo@ataol.com
Web Site: www.smallwoodstore.com
Founded: 1989.
Key Personnel: Exec. Dir., Ms. Lynn Smallwood McMillin.
Personnel Profile: Full-Time Paid 1; Part-Time Paid 3; Part-Time Volunteers 4.
Governing Authority: private; nonprofit organization. Tax-exempt.
Institution Type/Description: History Museum: housed in 1906 Indian Trading Post.
Collections: Florida history from prehistoric to modern times with emphasis on Chokoloskee.
Activities: killing Mr. Watson reenactment. Museum Sponsors: Seminole Day, 100 Year Celebration.
Hours & Admission Prices: May 2-Nov. Fri.-Tues. 10-4; Dec.-May 1 daily 10-5. Adults $3, senior citizens $2.50; discounts to groups of 10 or more; children under 12 no charge. Closed Thanksgiving; Christmas Eve & Day.
Attendance: 32,000 (estimated)
Membership: Student $5; Senior Citizen $15; Adult $25; Family $40.

Christmas

FORT CHRISTMAS HISTORICAL PARK, 1300 Fort Christmas Rd., Christmas, FL 32709-9427. Tel.: 407-568-4149. Fax: 407-568-9790.
E-mail: cheryl.wasserman@ocfl.net
Web Site: parks.onetgov.net
Founded: 1977.
Key Personnel: Historic Site Supvr., Trudy Trask; Recreation Specialist, Vickie Prewett; Recreation Specialist, Joseph Adams; Recreation Leader, Shirley Truex; Recreation Leader, Cheryl Wasserman; Museum Shop Mgr., Mary Clark; Senior Parks Specialist, Mike Heller; Park Specialist, Jeff Hamilton.
Personnel Profile: Full-Time Paid 7; Full-Time Volunteers 2; Part-Time Paid 2; Part-Time Volunteers 10.
Governing Authority: county. Tax-exempt.
Institution Type/Description: History Museum.
Collections: military artifacts; period furnishings.
Research Fields: family history; architectural history of structures; interviews & oral histories.
Facilities: library; pavilions for outdoor classrooms. Museum-related items for sale.
Activities: formal education programs for children; lectures. Museum Sponsors: Cracker Christmas; Historic Military Encampments; Bluegrass Festival; Pioneer Homecoming.
Hours & Admission Prices: Park: Summer: daily 8-8; Winter: daily 8-6. Museum: Tues.-Sat. 10-5, Sun. 1-5. Pioneer Homes: Tues.-Sat. 10-12 & 1-3:30, Sun. 1-3:30. No charge; donations accepted. Closed holidays. &
Attendance: 123,775 (estimated)
Membership: Individual $10; Family $15; Patron $50, Life $350.

JUNGLE ADVENTURES NATURE ANIMAL PARK, 26205 E. Colonial Dr., Christmas, FL 32709. Tel.: 407-568-2885; 877-424-2867 (Toll Free). Fax: 407-568-0038.
E-mail: jungleadv@aol.com
Web Site: www.jungleadventures.com
Institution Type/Description: Nature Park.
Collections: panthers; bears; wolves; deer; over 200 alligators; native plants; Native American Indian village replica.
Facilities: Park-related items for sale.
Activities: alligator feedings; guided tours.
Hours & Admission Prices: Daily 9:30-5:30. Adults $19.95, seniors 60 & over

$16.95, children 3-11 $12.95; discounts to groups of 15 or more; children under 3 no charge.

Clearwater

CLEARWATER MARINE AQUARIUM, 249 Windward Passage, Clearwater, FL 33767-2244. Tel.: 727-441-1790, ext. 240 & 227. Fax: 727-445-1139.
E-mail: fdame@cmaquarium.org
Web Site: www.seewinter.com
Founded: 1972.
Congressional District: 9
Key Personnel: C.E.O., David Yates; Chm., John Draheim; Vice Chm., Carlen Petersen; Sec., Rosemary Longenecker; Treas., Andy Barwell; Exec. Vice Pres. & C.O.O., Frank Dame.
Personnel Profile: Full-Time Paid 64; Part-Time Paid 105; Part-Time Volunteers 750; Interns 25.
Governing Authority: nonprofit organization. Tax-exempt: 501(c)(3).
Institution Type/Description: Aquarium & Marine Museum.
Collections: specializing in education, marine research & the rescue, rehabilitation & release of injured or sick whales, dolphins, otters & sea turtles.
Research Fields: marine mammal & sea turtle rescue & rehabilitation.
Facilities: library pertaining to sea life; aquarium; educational facilities. Museum-related items for sale.
Activities: films; formal educational programs; guided tours; lectures; temporary exhibitions of your own collections; traveling exhibitions; broadcast programs. Museum Sponsors: Annual Coastal Clean-up.
Publications: quarterly newsletter, Lifelines.
Hours & Admission Prices: Daily 9-6. Adults $19.95, seniors $17.95, children $14.95; discounts to FAA & AAA members & groups; children under 3 & members no charge. Closed Thanksgiving; Christmas. &
Attendance: 750,000 (estimated)
Membership: Senior Citizen $45; Individual $50; Couple $70; Family $99.

MOCCASIN LAKE NATURE PARK, AN ENVIRONMENTAL & ENERGY EDUCATION CENTER, 2750 Park Trail Lane, Clearwater, FL 33759-2602. Tel.: 727-793-2976. Fax: 727-793-2978. TDD: 727-562-4833.
E-mail: cliff.norris@myclearwater.com
Web Site: www.myclearwater.com
Founded: 1982.
Key Personnel: Nature Park Supvr., Cliff Norris; Nature Park Programmer, Lloyd Simmons; Nature Park Support Technician, Dave Weitzel.
Personnel Profile: Full-Time Paid 4; Part-Time Volunteers 135; Interns 1.
Governing Authority: municipal government. Parent Institution: City of Clearwater. Subsidiary Institution: Parks & Recreation Dept. Tax-exempt: 501(c)(3).
Institution Type/Description: Nature Center & Conservation Area; Alternative Energy Demonstration Center.
Collections: plants native to Florida; natural history; alternative/renewable energy systems; permanently injured birds of prey native to Florida; arboretum; live reptiles & amphibian displays; fish tanks; exotic nuisance plants & animals of Florida.
Facilities: library of natural history & energy books; herbarium; 100-seat classroom; educational facilities; nature trail; wildlife preserve. Gift items for sale.
Activities: alternative/renewable energy demonstrations; Families Exploring Nature Club; guided tours; lectures; organized educational programs; docent program; temporary & interactive exhibitions; local chapter meetings of Sierra Club, Audubon & Suncoast Herpetological Society; Science Safari Educational Company nature camps; Native Plant Society.
Publications: quarterly magazine, Clearwater.
Hours & Admission Prices: June-Aug. Mon.-Fri. 9-5; Sept.-May Tues.-Fri. 9-5, Sat. 10-5. Additional hours for special programming. Nonresidents adults $3, Clearwater Residents adults $2; children under 3 no charge. Closed city holidays. &
Attendance: 51,000 (accurate)

Clewiston

* **AH-TAH-THI-KI MUSEUM, (M),** Big Cypress Seminole Indian Reservation, 34725 W. Boundary Rd., Clewiston, FL 33440. Mailing Address: 30290 Josie Billie Hwy., PMB 1009, Clewiston, FL 33440. Tel.: 863-902-1113. Fax: 863-902-1117.
Web Site: www.ahtahthiki.com
Founded: 1989.
Key Personnel: Museum Dir. & THPO, Dr. Paul Backhouse; Operations Mgr., Dr. Annette Snapp; Collections Mgr., Tara Backhouse; Business Mgr., Sonia Perez; Museum Shop Mgr., Rebecca Petrie; Facilities Mgr., Gene Davis.

Personnel Profile: Full-Time Paid 35; Part-Time Paid 1; Part-Time Volunteers 1; Interns 1.

Governing Authority: nonprofit. Parent Institution: Seminole Tribe of Florida, Hollywood, FL.

Institution Type/Description: Tribal Museum.

Collections: Seminole & Southeastern Indian materials with historically related items; Seminole genre paintings by Seminole and non-Indian artists; Seminole war period militaria; botanical park with resident fauna.

Research Fields: Seminole culture & history.

Facilities: 255-vol. library; 8,530 sq. ft. exhibit space; botanical garden; 45-seat large screen theater. Museum-related items for sale.

Activities: guided tours; temporary exhibitions.

Publications: newsletter, Ah-Tah-Thi-Ki Museum News.

Hours & Admission Prices: Daily 9-5. Adults $9, senior citizens and students & military with ID $6; discounts to NARM members & groups; children 4 & under & members no charge. Closed New Year's Day; Labor Day; Veterans Day; Thanksgiving Day; Christmas Day. &

Attendance: 26,793 (accurate)

Membership: Individual $30; Family Clan $60; Osceola Circle $250; Bandolier Bag $500, Chairman's Circle $1,000.

CLEWISTON MUSEUM, INC., 109 Central Ave., Clewiston, FL 33440-3701. Tel.: 863-983-2870.

E-mail: clewistonmuseum@embarqmail.com

Web Site: www.clewistonmuseum.org

Founded: 1984.

Congressional District: 12

Key Personnel: Chm. (V), Brenda Lopez; Dir., Butch Wilson.

Personnel Profile: Full-Time Paid 1.

Governing Authority: nonprofit organization. Parent Institution: Friends of the Museum. Tax-exempt: 501(c)(3).

Institution Type/Description: Heritage & History Museum: housed in c.1928 Clewiston News Building.

Collections: agriculture; Seminole Indians; World War II; community history; fossils; shells; insects; early Native American artifacts.

Facilities: 60-seat theater. Books & crafts for sale.

Activities: traveling exhibitions; films & lectures on the sugar cane & citrus industry of the south Lake Okeechobee area.

Publications: Pioneering in the Everglades by Beardsley.

Hours & Admission Prices: Mon.-Fri. 9-4; other times by appointment. Adults $4, seniors $3, discounts to seniors. Closed major holidays. &

Attendance: 2,300 (estimated)

Membership: Individual $25; Family $35; Sponsor $50; Patron $100; Corporate $250; Corporate Silver $500; Corporate Gold $1,000.

Cocoa

ASTRONAUT MEMORIAL PLANETARIUM AND OBSERVATORY, Brevard Community College, 1519 Clearlake Rd., Cocoa, FL 32922-6598. Tel.: 321-433-7373. Fax: 321-433-7646.

E-mail: leslies@brevardcc.edu

Web Site: www.brevardcc.edu/planet

Founded: 1976.

Congressional District: 11

Key Personnel: Dir., Mark Howard; Assoc. Dir., Suzanne Leslie.

Personnel Profile: Full-Time Paid 3; Part-Time Paid 3; Part-Time Volunteers 20.

Governing Authority: community college; nonprofit. Parent Institution: Brevard Community College.

Institution Type/Description: Astronomy Museum.

Collections: telescope; Minolta Infinium Star Projector.

Research Fields: laser graphics; astrophotography & imagining; U.S., Soviet/Russian/CIS, & all other international astronauts, cosmonauts, and spationautes; advanced exhibit development.

Facilities: planetarium; observatory; theater; classroom.

Activities: films; formal education programs for children and undergraduate & graduate college students; guided tours; participatory & temporary exhibits; portable planetarium; special programs.

Publications: quarterly newsletter, Focal Point.

Hours & Admission Prices: Exhibit Halls: Wed. 1:30-4:30, Fri.-Sat. 6:30pm-10:30pm. Public Shows: Wed. 2 & 3, Fri.-Sat. 7pm, 8pm & 9pm. Observatory: Fri. -Sat. dusk-10pm. Single Show: adults $7, seniors, students, & military $6, children 12 & under $4. Planetarium & Movie: adults $11, seniors, students, & military $9, children 12 & under $7. Planetarium, Movie, & Laser: admission $16. &

Attendance: 40,000 (estimated)

Membership: Single $35; Family $70; Sponsor $120; Corporate $500 & up.

BREVARD MUSEUM OF HISTORY AND NATURAL SCIENCE, 2201 Michigan Ave., Cocoa, FL 32926-5618. Tel.: 321-632-1830 & 1920. Facebook: Brevard Museum of History & Natural Science.

E-mail: bmhs@brevardmuseum.com

Web Site: www.brevardmuseum.com

Founded: 1969.

Congressional District: 29

Key Personnel: C.E.O. & Dir., Nancy Rader; Asst. Dir., Rachel Burris; Pres. (V), Emma Newsham; Museum Shop Mgr. & Receptionist, Philip Burris.

Personnel Profile: Part-Time Paid 3; Part-Time Volunteers 32.

Governing Authority: nonprofit organization. Tax-exempt: 501(c)(3).

Institution Type/Description: History & Natural Science Museum.

Collections: natural science; archaeology; local history; children's discovery room; Windover, 8,000 year old dig exhibit.

Major Exhibits: Space Coast, 1/14-2/14.

Research Fields: anthropology, archaeology, local history, malacology, ornithology, botany, ecology, entomology.

Facilities: 60-seat classroom; laboratory; discovery room; picnic pavilion & nature center with trails.

Activities: school & adult group programming.

Publications: brochures; History of the Cape Canaveral Lighthouse; printed material on special exhibits; cookbook; calendar; bimonthly newsletter, The Dillo's Tale.

Hours & Admission Prices: June-Aug. Thurs.-Sat. 10-4; Sept.-May Tues.-Sat. 10-4. Adults $6, seniors $5.50, college students $5, children 5-16 $4.50; members no charge. Closed New Year's Day; Easter; Thanksgiving; Christmas Eve & Day. &

Membership: Senior $10; Individual $15; Senior Couple & Student $20; Teacher $25; Couple $40; Family $45; Life $500; Corporate $2,500.

FLORIDA HISTORICAL SOCIETY, 435 Brevard Ave., Cocoa, FL 32922-7901. Tel.: 321-690-1971. Fax: 321-690-4388.

E-mail: barbara.west@myfloridahistory.org

Web Site: www.myfloridahistory.org

Founded: 1856.

Key Personnel: Exec. Dir. & Museum Shop Mgr., Benjamin Brotemarkle.

Personnel Profile: Full-Time Paid 5; Part-Time Paid 2; Part-Time Volunteers 20.

Governing Authority: Tax-exempt.

Institution Type/Description: Historical Society Museum.

Collections: local history & culture; photographs; manuscripts.

Research Fields: Florida history.

Facilities: research library.

Publications: Florida Historical Quarterly.

Hours & Admission Prices: Tues.-Sat. 10-4:30. No charge; donations accepted. &

Membership: Student $30; Individual $50; Family & Institutional $75; Founding Fellow $100; Contributing $200; Corporate $500 & up.

Coconut Grove

THE BARNACLE HISTORIC STATE PARK, 3485 Main Hwy., Coconut Grove, FL 33133-5915. Tel.: 305-442-6866. Fax: 305-442-6872.

E-mail: katrina.boler@dep.state.fl.us

Web Site: www.floridastateparks.org/TheBarnacle

Founded: 1973.

Congressional District: 18

Key Personnel: Pres., Alyn Pruett; Park Mgr., Katrina A. Boler; Park Svcs. Specialist, Jessica Cabral.

Personnel Profile: Full-Time Paid 2; Part-Time Paid 3; Part-Time Volunteers 20.

Governing Authority: state; nonprofit. Tax-exempt: 501(c)(3).

Institution Type/Description: Historic House.

Collections: memorabilia of pioneer and yacht designer Commodore Ralph Middleton Munroe & his family; marine artifacts; photographs; cameras; furniture; household items; handmade family quilts; boathouse; tools.

Facilities: Museum-related items for sale.

Activities: arts festivals; concerts; docent program; guided tours; hobby workshops; grounds rental. Museum Sponsors: Regatta (traditional Sailboats); Old-fashioned Independence Day Picnic; Commodore's Birthday Party (volunteer recruit); Firefly Lawn Party (fundraiser); monthly Moonlight Concerts; Old Time Dances; Yoga by the Sea; Washington's Birthday Regatta; Earth Day Celebration; Starlight Movie Classics; Cars & Cigars Father's Day Event; Dog Days of Summer; Up Past Bedtime Children's Movie; Hot Chili Cool Cars.

Publications: quarterly newsletter, The News Packet.

Hours & Admission Prices: Fri.-Mon. 9-5. Tours: 10, 11:30, 1 & 2:30. Group

Tours: Wed.-Thurs. with advanced reservations. Admission $2. Tours: adults $3, children 6-12 $1. Closed New Year's Day; Thanksgiving; Christmas. &

Attendance: 57,157 (estimated)

Membership: Senior Citizen & Student $20; Individual $30; Family $40; Contributing $100; Patron $250; Corporate $500; Life $1,000.

Coral Gables

CORAL GABLES MUSEUM, 285 Aragon Ave., Coral Gables, FL 33134. Tel.: 305-603-8067.

E-mail: info@coralgablesmuseum.org
Web Site: www.coralgablesmuseum.org
Founded: 2003.
Congressional District: 18
Key Personnel: Chm. (V), George Kakouris; Dir., Arva Moore Parks.
Personnel Profile: Full-Time Paid 1; Part-Time Paid 5; Interns 2.
Governing Authority: Tax-exempt.
Institution Type/Description: Civic Arts Museum.
Collections: paintings; photographs; prints; drawings; personal artifacts. Historic Buildings: Phineas Paist Police & Fire Station built in 1936.
Research Fields: architecture, urban design & planning; sustainable development & preservation.
Activities: educational programs; special events; children's camps; lectures; tours; concerts; workshops.
Hours & Admission Prices: Tues.-Thurs. 12-6, Fri. 12-8, Sat. 11-5, Sun. 12-5. Adults $7; member no charge. &
Attendance: 7,000 (estimated)
Membership: Student & Senior $30; Individual $50; Senior Family $60; Dual $70; Family $80; Granada $150.

* **FAIRCHILD TROPICAL BOTANIC GARDEN,** 10901 Old Cutler Rd., Coral Gables, FL 33156-4296. Tel.: 305-667-1651. Fax: 305-661-8953.

E-mail: clewis@fairchildgarden.org
Web Site: www.fairchildgarden.org
Founded: 1938.
Congressional District: 19
Key Personnel: Pres. (V), Bruce W. Greer; Dir., Carl E. Lewis, Ph.D.; Dir. Museum Shop, Erin Fitts.
Personnel Profile: Full-Time Paid 100; Part-Time Paid 19; Part-Time Volunteers 440; Interns 2.
Governing Authority: nonprofit organization. Tax-exempt: 501(c)(3).
Institution Type/Description: Botanical Garden.
Collections: living tropical & sub-tropical plants.
Research Fields: tropical botany & horticulture.
Facilities: 15,000-vol. library of books on tropical botany & horticulture available on premises by appointment. Books for sale.
Activities: guided tours; lectures; classes; plant shows & sales. Museum Sponsors: International Mango Festival; International Orchid Festival; ramble; International Chocolate Festival; Family Harvest Day.
Publications: quarterly magazine, The Tropical Garden.
Hours & Admission Prices: Daily 9:30-4:30. Adults $20, seniors $15, children 6-17 $10; children 5 & under no charge. Closed Christmas. &
Attendance: 146,388 (accurate)
Membership: Student $25; Associate $40; Associate Plus $60; Family & Friends $80; Sustaining $125; Supporting $250; Signature $500; Fellow $1,000-$2,499; Silver Fellow $2,500-$4,999; Gold Fellow $5,000-$9,999; Platinum Fellow $10,000 & up.

* **LOWE ART MUSEUM, UNIVERSITY OF MIAMI, (M),** 1301 Stanford Dr., Coral Gables, FL 33146-2099. Tel.: 305-284-3535. Fax: 305-284-2024.

Web Site: www.lowemuseum.org
Founded: 1950.
Congressional District: 18
Key Personnel: Dir. & Chief Cur., Brian A. Dursum; Adjunct Cur. African Art, Marcilene Wittmer; Adjunct Cur. Renaissance, Perri L. Roberts; Dir. Membership, Yina Balarezo-Badenjki; Asst. Dir., Kara Schneiderman; Assoc. Preparator, Darren Price; Cur. Education, Jodi Sypher; Special Events Coord., Irene Bergmann; Receptionist, Janie Graulich; Head Security, Maria Milhomme; Office Mgr., Lorraine Stassun; Coord. School Programs, Hope Torrents; Registrar, Natasha Cuervo; Asst. to Registrar, Julie Berlin; Preparator, Alessia Lewitt; Communications Specialist, Raymond Mathews; Environmental Svcs. Tech, Diana Mazo-Maza.
Personnel Profile: Full-Time Paid 12; Part-Time Paid 2; Part-Time Volunteers 200; Interns 2.
Governing Authority: Parent Institution: University of Miami. Tax-exempt: 501(c)(3).
Institution Type/Description: Art Museum.

Collections: Samuel H. Kress collection of Renaissance & Baroque Art; Alfred I. Barton collection of North American Indian Art; Samuel K. Lothrop collection of Guatemalan Textiles; pre-Columbian Art; Asian sculpture, bronzes, ceramics & painting; African Art; Art of the Pacific; European & American painting, sculpture & works on paper; Greco-Roman antiquities.
Major Exhibits: Art Lab at the Lowe: The Art of Panama, 5/3/13-4/27/14; Terrestrial Paradises: Imogrey from the Voyages of Captain James Cook, 2/23/13-7/13/14; Pueblo to Pueblo: The Legacy of Southwest Indian Pottery (T), 1/25/14-3/23/14; Art Lab @ the Lowe: Converging Cultures in Spanish Colonial America, 5/9/14-4/26/15; China's Last Empire: The Art and Culture of the Qing Dynasty, 6/21/14-10/19/14; About Materials: Ceramics, 8/2/14-8/20/15; Art in Real Life: Traditional African Art from the Lowe Art Museum, 11/8/14-1/11/15; Transformative Visions: Works by Haitian Artists from the Permanent Collection, 11/8/14-1/11/15.
Research Fields: all fields pertaining to collections.
Facilities: classroom; studio. Paintings, sculptures, books, cards & jewelry for sale.
Activities: guided tours; lectures; films; gallery talks; concerts; arts festivals; family days; formally organized educational programs; educational outreach program; docent program; inter-museum loan; intra-campus lending program; permanent, temporary & traveling exhibitions.
Publications: biannual magazine; Lowe Art Museum magazine; catalogs; brochures.
Hours & Admission Prices: Tues.-Sat. 10-4, Sun. 12-4. Adults $10, senior citizens & students $5; discounts to AAM & ICOM members; children under 12 & members no charge. Closed major holidays. &
Attendance: 41,235 (accurate)
Membership: Academic $35; Individual $50; Dual $75; Family $100; Sustaining $150; Friends of Art $350; Sponsor $500; Patron Friend $1,000; Director's Circle $1,200.

Coral Springs

CORAL SPRINGS MUSEUM OF ART, (M), 2855 Coral Springs Dr., Coral Springs, FL 33065-3825. Tel.: 954-340-5000. Fax: 954-346-4424.

E-mail: ctsaa@coralsprings.org
Web Site: www.csmart.org
Founded: 1997.
Congressional District: 19
Key Personnel: Exec. Dir., Barbara O'Keefe; Pres. (V), Dr. Kerry Kuhn.
Personnel Profile: Full-Time Paid 4; Part-Time Paid 3; Part-Time Volunteers 67.
Governing Authority: private; nonprofit organization. Tax-exempt: 501(c)(3).
Institution Type/Description: Art Museum.
Collections: visual art; contemporary.
Facilities: 15,000 sq. ft. exhibit space; classrooms. Museum-related items for sale.
Activities: formal education programs; guided tours; lectures; temporary exhibitions; artist in residence program; Art Reach; Saturday Chautauqua.
Hours & Admission Prices: Mon.-Sat. 10-5. Adults $6, seniors $5, students $3; Wed. & members no charge. Closed major holidays. &
Attendance: 31,000 (estimated)
Membership: Mondrian $60; Calder $85; Monet $150; Pollock $250; Cezanne $500; Picasso $1,000; Michelangelo $5,000.

Cross Creek

MARJORIE KINNAN RAWLINGS HISTORIC STATE PARK, 18700 S. County Rd. 325, Cross Creek, FL 32640-8403. Tel.: 352-466-3672 & 9273. Fax: 352-466-4743.

E-mail: valerie.rivers@dep.state.fl.us
Web Site: www.floridastateparks.org
Formerly: Marjorie Kinnan Rawlings State Historic Site
Founded: 1970.
Key Personnel: Park Mgr., Valerie Rivers.
Personnel Profile: Full-Time Paid 3; Part-Time Paid 5; Part-Time Volunteers 12; Interns 1.
Governing Authority: state. Dept. of Environmental Protection, Division of Recreation & Parks, 3900 Commonwealth Blvd., Commonwealth Bldg., Tallahassee, FL 32303. Tax-exempt.
Institution Type/Description: Historic House Museum: c.1930's citrus farm & home of Marjorie Kinnan Rawlings, a rambling Cracker farmhouse. A National Historic Landmark.
Collections: furnishings of Marjorie Kinnan Rawlings author of The Yearling, Cross Creek, and other books.
Activities: guided tours.
Hours & Admission Prices: Daily 9-5. $3 per car. House Tours: Oct.-July Thurs.-Sun. 10-4. Adults $3, children 6-12 $2; children under 6 no charge. Closed: Thanksgiving; Christmas. &

Attendance: 25,000
Membership: CSO - Friends of the Marjorie Kinnan Rawlings Farm $20.

Crystal River

COASTAL HERITAGE MUSEUM, 532 Citrus Ave., Crystal River, FL 34428-4017. Tel.: 352-795-1755 & 212-8390. Fax: 352-341-6445.
E-mail: crcoastalmuseum@aol.com
Web Site: coastalheritagemuseum.com
Formerly: The Museum of Citrus County History-Coastal Heritage
Founded: 1986.
Congressional District: 6
Key Personnel: Chm. (V), Sharon Padgett.
Personnel Profile: Part-Time Volunteers 10.
Governing Authority: county. Parent Institution: Citrus County Historical Society. Tax-exempt: 501(c)(3).
Institution Type/Description: Local History Museum: housed in 1939 old City Hall.
Collections: west Citrus County history; archaeological artifacts; furniture; photographs.
Research Fields: Citrus County history & pre-history.
Facilities: archival repository. Museum-related items for sale.
Activities: guided tours; docent program. Annual Event: Tour of Historic Homes in March.
Publications: quarterly newsletter, At Home; brochures; calendar; book, A History of Crystal River, Florida; brochure, Historic Tour Guide of Crystal River, Florida.
Hours & Admission Prices: Aug.-June Tues.-Sat. 10-2. No charge, donations accepted. Closed holidays. &
Attendance: 2,100 (accurate)
Membership: Student $5; Individual $15; Family $22; Business $30; Supporting $100; Life $250; Corporate $500.

CRYSTAL RIVER ARCHAEOLOGICAL STATE PARK, 3400 N. Museum Pt., Crystal River, FL 34428-6207. Tel.: 352-795-3817. Fax: 352-795-6061.
Web Site: floridastateparks.org
Founded: 1965.
Congressional District: 6
Key Personnel: Dir. & Park Mgr., Nicholas D. Robbins; Park Ranger & Museum Shop Mgr., Catherine Wunderlich.
Personnel Profile: Full-Time Paid 3.
Governing Authority: state. Parent Institution: Florida Park Service. Tax-exempt.
Institution Type/Description: State Park Museum.
Collections: archaeology; ancient Florida Indian artifacts.
Research Fields: pre-Columbian Indians.
Facilities: visitor center; museum.
Activities: guided tours for groups.
Publications: brochure.
Hours & Admission Prices: Museum: Thurs.-Mon. 9-5. Park: daily 8am to sundown. Park Entrance: car: $3; motorcycle & pedestrian $2. &
Attendance: 21,000 (estimated)

Dade City

PIONEER FLORIDA MUSEUM ASSOCIATION, INC., (M), 15602 Pioneer Museum Rd., Dade City, FL 33523. Mailing Address: P.O. Box 335, Dade City, FL 33526-0335. Tel.: 352-567-0262. Fax: 352-567-1262.
E-mail: curator@pioneerfloridamuseum.org
Web Site: www.pioneerfloridamuseum.org
Founded: 1961.
Congressional District: 12
Key Personnel: C.E.O. & Pres. (V), Robert Sumner; Dir., Christine Smith; Museum Shop Mgr., Susan Bayes.
Personnel Profile: Full-Time Paid 1; Part-Time Paid 3; Part-Time Volunteers 150.
Governing Authority: bd. of trustees; nonprofit organization, Pioneer Florida Museum Association, Inc. Tax-exempt: 501(c)(3).
Institution Type/Description: Pioneer Museum.
Collections: local pioneer family life & culture including records & personal information; sugar cane patch & grinding exhibit; moon shine still; natural stream; citrus packing house. Historic Buildings: restored school-house, church, depot, C.C. Smith General Store & 1860's home; restored shoe repair shop & two Cypress buildings from Cumner Sons Cypress Saw Mill Company.
Research Fields: pioneer Florida living.
Activities: guided tours; permanent exhibitions; research. Special Events:

Christmas at the Museum; Labor Day, Pioneer Florida Day Celebration; Quilt & Antiques Show & Sale.
Hours & Admission Prices: Tues.-Sat. 10-5; group tours by appointment. Adults $6, seniors 55 & over $5, students 6-18 $2; members and children 5 & under no charge. Closed major holidays. &
Attendance: 20,000 (estimated)
Membership: Individual $25; Family $35; Individual Life $150; Life Couple $250. Mabel Jordan Barn Fund Donation: Pathfinder $100; Settler $250; Pioneer $500; Trailblazer $1,000.

Dania Beach

INTERNATIONAL GAME FISH ASSOCIATION - FISHING HALL OF FAME AND MUSEUM, 300 Gulf Stream Way, Dania Beach, FL 33004-2118. Tel.: 954-922-4212. Fax: 954-924-4220.
E-mail: hq@igfa.org
Web Site: www.igfa.org
Key Personnel: Dir., Ryan Dick; Pres., Rob Kramer.
Governing Authority: Parent Institution: International Game Fish Association. Tax-exempt.
Institution Type/Description: History Museum.
Collections: Hall of Fame inductees; fishing history; angling literature.
Facilities: library. Museum-related items for sale.
Activities: birthday parties; rental facilities; educational programs.
Publications: The International Angler; The International Junior Angler; The World Records Book.
Hours & Admission Prices: Mon.-Sat. 10-6, Sun. 12-6. Adults $8, children $5. Closed Thanksgiving; Christmas. &
Attendance: 70,000
Membership: Junior $20; Regular $40; Family $55.

Davie

BUEHLER PLANETARIUM AND OBSERVATORY, Broward College, 3501 Davie Rd., Davie, FL 33314-1604. Tel.: 954-201-6681. Fax: 954-201-6316.
Web Site: www.iloveplanets.com
Institution Type/Description: Planetarium & Observatory.
Collections: Zeiss M1015 star projector; astronomy.
Activities: star shows; astronomical programs; children's programs. Museum Sponsors: Stories for A Starry Night.
Hours & Admission Prices: Observatory: Wed. & Fri.-Sat. 8pm-10pm. No charge.

FLAMINGO GARDENS, EVERGLADES WILDLIFE SANCTUARY, 3750 Flamingo Rd., Davie, FL 33330-1698. Tel.: 954-473-2955. Fax: 954-473-1738. Facebook: Flamingo Gardens, Everglades Wildlife Sanctuary.
Web Site: www.flamingogardens.org
Founded: 1969.
Key Personnel: Exec. Dir., Stan W. Wood.
Personnel Profile: Full-Time Paid 15; Part-Time Paid 30; Part-Time Volunteers 100.
Governing Authority: Tax-exempt.
Institution Type/Description: Wildlife Sanctuary & Historic Home: housed in the former residence of founders Floyd L. & Jane Wray, built in 1933.
Collections: area natural & cultural heritage; plants; birds; personal artifacts; early South Florida settlers.
Facilities: cafe. Museum-related items for sale.
Activities: tram rides.
Hours & Admission Prices: Daily 9:30-5; ticket booth closes at 4. Adults $18, children 4-11 $10; discounts to seniors, students, military, & groups. Closed Thanksgiving; Christmas. &
Attendance: 125,000 (accurate)
Membership: Individual $50; Single Parent $75; Dual $95; Family $115; Sustaining $245.

OLD DAVIE SCHOOL HISTORICAL MUSEUM, 6650 Griffin Rd., Davie, FL 33314-4331. Tel.: 954-797-1044. Fax: 954-797-1047.
E-mail: director@olddavieschool.org
Web Site: www.olddavieschool.org
Founded: 1984.
Key Personnel: Dir., Leslie Schroeder; Pres., Carrie Weekley; Education, Judy Maxwell.
Personnel Profile: Full-Time Paid 2; Part-Time Paid 1; Part-Time Volunteers 40.
Governing Authority: private; nonprofit organization. Tax-exempt: 501(c)(3).
Institution Type/Description: History Museum: housed in a 1918 two-story

masonry school building designed by August Geiger. Listed on the National Register of Historic Places.

Collections: Western Broward County & Everglades history from prehistoric to modern times; town of Davie history. Historic Buildings: 1912 Viele House; c.1912 Walsh-Osterhoudt home; early pioneer homes.

Facilities: 200-seat auditorium; 2,400 sq. ft. exhibit space. Museum-related items for sale.

Activities: school tours; educational programs; special events; temporary exhibits; docent program; guided tours; theater. Annual Events: Back to School Dinner; Generations of Taste Pioneer Dinner; Old Davie Christmas.

Publications: quarterly newsletter, The Report Card.

Hours & Admission Prices: Tues.-Sat. 10-2. Adults $10, seniors & children $7; discounts to AAM & ICOM members; members no charge. Closed New Year's Day; Memorial Day; Independence Day; Labor Day; Thanksgiving; Christmas. &

Attendance: 14,000 (estimated)

Membership: Individual $25; Family $50; Business $200; Principal's List $500; Honor Roll $1,000.

ROSEMARY DUFFY LARSON GALLERY AT BROWARD COLLEGE, A. HUGH ADAMS CAMPUS, 3501 S.W. Davie Rd., Bldg. 3, Davie, FL 33314-1604. Tel.: 954-201-6984. Fax: 954-201-6518.

E-mail: cgallery@broward.edu

Formerly: Fine Arts Gallery at Broward College, A. Hugh Adams Campus

Founded: 1965.

Key Personnel: Dir., Harumi Abe

Governing Authority: Tax-exempt.

Institution Type/Description: Art Gallery.

Collections: paintings; sculpture; drawings; photographs.

Activities: permanent & temporary exhibitions; receptions.

Hours & Admission Prices: Mon.-Tues. & Thurs.-Fri. 9-3, Wed. 2-8. No charge. &

Attendance: 1,600 (estimated)

∗ **YOUNG AT ART MUSEUM, (M),** 751 S.W. 121st Ave., Ste. 1, Davie, FL 33325-3804. Tel.: 954-424-0085, ext. 23. Fax: 954-370-5057. Facebook: Young At Art Museum.

E-mail: mperrino@youngatartmuseum.org

Web Site: www.youngatartmuseum.org

Founded: 1987.

Congressional District: 20

Key Personnel: Exec. Dir. & C.E.O., Mindy Shrago; Chm. Bd. (V), John Voigt; Dir. Operations, David Tesseo; Dir. Devel., Hannah Hausman; Museum Shop Mgr., Maggie Infantino.

Personnel Profile: Full-Time Paid 40; Part-Time Paid 32; Part-Time Volunteers 825.

Governing Authority: nonprofit organization. Tax-exempt: 501(c)(3).

Institution Type/Description: Art Museum.

Collections: teaching collections; art; paintings; sculpture; hands-on exhibits.

Facilities: Museum-related items for sale.

Activities: formal education programs for children; guided tours; permanent, loan, traveling & participatory exhibitions; art institute studios; ceramic, painting & drawing studios; museum preschool.

Hours & Admission Prices: Mon.-Thurs. 10-5, Fri.-Sat. 10-6, Sun. 11-5. Adults $13, seniors and children 1 & over $12; discounts to groups; members no charge. ACM reciprocal admission to Family Premium & above members. Closed New Year's Day; Easter; Thanksgiving; Christmas. &

Attendance: 225,000 (accurate)

Membership: Grandparents $100; Family $135; Family Premium $150; Family Supporter $250.

Daytona Beach

DAYTONA INTERNATIONAL SPEEDWAY/TOURS, 1801 W. International Speedway Blvd., Daytona Beach, FL 32114-6833. Tel.: 386-681-4078. TTY: 386-681-5093.

E-mail: dkurtz@iscmotorsparts.com

Web Site: www.daytonainternationalspeedway.com

Formerly: Daytona 500 Experience

Founded: 1959.

Institution Type/Description: Racing History Museum.

Collections: Daytona history; stock car champions & their cars; racing history & memorabilia; photographs.

Facilities: IMAX theater. Museum-related items for sale.

Activities: hands-on activities; motion simulator rides; video arcade; tram tour.

Hours & Admission Prices: Daily 10-6; call for extended hours. Adults $23, seniors & children 6-12 $17; children under 5 no charge. Group tours. Closed Thanksgiving; Christmas. &

HALIFAX HISTORICAL MUSEUM & GIFT SHOP, 252 S. Beach St., Daytona Beach, FL 32114-4407. Tel.: 386-255-6976. Fax: 386-255-7605.

E-mail: mail@halifaxhistorical.org

Web Site: www.halifaxhistorical.org

Founded: 1949.

Congressional District: 4

Key Personnel: Dir., Fayn LeVeille; Pres., Ruth Trager; 1st Vice Pres., Walter Snell; 2nd Vice Pres., Dr. Michael Link; 3rd Vice Pres., Warren Trager; Recording Sec., Beth Mindlin; Corresponding Sec., Elizabeth Beecher; Treas., Matt Romanik.

Personnel Profile: Full-Time Paid 2; Full-Time Volunteers 6; Part-Time Volunteers 25; Interns 2.

Governing Authority: society. Parent Institution: Halifax Historical Society. Tax-exempt: 501(c)(3).

Institution Type/Description: Local History Museum.

Collections: Halifax history & culture; Native American artifacts of the area & artifacts from early local plantations; Civil War through WWII military memorabilia; mementos from Pioneer & Victorian families; auto racing history; postcards; period clothing; vintage housewares; Grandma's Attic; film.

Research Fields: early local history; automobile racing; genealogy.

Facilities: 1,500-vol. library of Florida history books, 1,500 rare & out of print books, newspaper files from 1883-present, photographs & vertical files available for research on premises by appointment; 50-seat theatre; research room houses over 10,000 photographs, 5,000 early post cards.

Activities: permanent & temporary exhibitions; educational programs & tours; research facility for genealogists, authors, & home researchers; films on local history & Florida

Publications: semiannual The Halifax Herald; quarterly newsletter to members; books, Daytona Beach and the Halifax River Area; A Brief History of the Halifax Historical Society.

Hours & Admission Prices: Tues.-Fri. 10:30-4:30, Sat. 10-4. Adults $5, children 12 & under $1; members no charge. Thurs. admission by donation. Research: $10 a day; volunteer researcher on staff $20 per hour. Closed major holidays. &

Attendance: 5,000 (estimated)

Membership: Senior $30; Individual $35; Senior Family $40; Family $45, Patron $150,Benefactor $500.

∗ **THE MUSEUM OF ARTS AND SCIENCES, INC. AND CENTER FOR FLORIDA HISTORY, (M),** 352 South Nova Rd., Daytona Beach, FL 32114-4597. Tel.: 386-255-0285. Fax: 386-255-5040.

E-mail: info@moas.org

Web Site: www.moas.org

Founded: 1971.

Congressional District: 7

Key Personnel: Interim Exec. Dir., Deborah B. Allen; Pres. (V), Carol Platig; Grants & Devel. Coord., Jessi Smith; Dir. Membership, Marissa Rodriguez; Dir. Mktg., Rene Adams; Acting Cur. Gary R. Libby Art & Chief Cur., Cynthia Duval; Dir. Operations, Eric Goire; Finance Assoc., Sherman N. Coleman; Sr. Cur. Education & History, Zach Zacharias; Admin. Old St. Augustine Village, Roy Shaffer; Cur. Astronomy, Seth Mayo; Visitor Svcs., Patti Nikolla.

Personnel Profile: Full-Time Paid 12; Part-Time Paid 13; Part-Time Volunteers 100; Interns 4.

Governing Authority: nonprofit. Parent Institution: The Museum of Arts & Sciences. Subsidiary Institutions: Old St. Augustine Village. Tax-exempt: 501(c)(3).

Institution Type/Description: General Museum.

Collections: contemporary Florida paintings, prints & photos by major Florida artists; Pleistocene fossils; regional natural history; Timucuan Indian material from Tick and Stone Island, Florida; Frischer sculpture garden; Cuban paintings of the 18th-20th centuries; African, pre-Columbian, 19th-century European, Central American and Cuban fine & folk art; American fine & decorative arts 1640-1900; European & American works on paper, Oriental fine & decorative arts; Persian miniatures; art; European decorative arts; Indian miniature painting; sloth skeleton; children's hands on exhibits.

Research Fields: Florida contemporary arts & fine arts; Florida history & archaeology; Pleistocene material; Cuban fine arts; American fine and decorative art; French decorative & fine art; Americana Coca Cola.

Facilities: 9,000-vol. library of general reference works available to members on premises; nature trails; planetarium; 268-seat multidisciplinary hall.

Activities: guided tours; lectures; films; marine biology programs; county-wide outreach programs; changing art exhibits; summer art history & science programs; formally organized education programs; planetarium workshops; daily planetarium shows; educational star shows; Florida history programs & activities; changing science exhibitions.

Publications: Arts and Sciences Magazine; yearly monograph; exhibit catalogs; triannual members' bulletin; Cuba: A History in Art in full color; Coast to Coast: The Contemporary Landscape in Florida in full color; A Treasury of American Art: Selections from the American Collection; Great Masters of Cuban Art - Ramos Collection; Reflections: Paintings of Florida 1865-1965.
Hours & Admission Prices: Tues.-Sat. & holidays 9-5, Sun. 11-5. Adults $12.95, seniors & students $10.95, children 6-17 $6.95; discounts to groups, AAM, ICOM, HIA & SEMC members; children 5 & under and members no charge. Closed Thanksgiving; Christmas. &
Attendance: 138,000 (accurate)
Membership: Student $20; Senior Citizen $25; Single $30; Senior Citizen Couple $35; Family $60; Friends of the Museum $125; Renaissance $200-$10,000.

SOUTHEAST MUSEUM OF PHOTOGRAPHY, Daytona State College, 1200 W. International Speedway Blvd., Bldg. 1200, Daytona Beach, FL 32114. Mailing Address: P.O. Box 2811, Daytona Beach, FL 32120-2811. Tel.: 386-506-4475 & 3350. Fax: 386-506-4487. TDD: 386-506-3023.
E-mail: romnesj@daytonastate.edu
Web Site: www.SMPonline.org
Founded: 1979.
Congressional District: 4
Key Personnel: Dir., Kevin R. Miller; Exhibitions Coord., Juliana Romnes; Cur. Education, Christina Katsolis.
Personnel Profile: Full-Time Paid 3; Part-Time Paid 15; Part-Time Volunteers 10; Interns 5.
Governing Authority: public college. Parent Institution: Daytona State College. Tax-exempt.
Institution Type/Description: Photography Museum.
Collections: contemporary & historical photography.
Research Fields: contemporary & historical photography.
Facilities: 1,000-vol. library on museology & photography; 500-seat auditorium; 8,000 sq. ft. exhibit space.
Activities: films; formal education programs for undergraduate college students affiliated with Daytona State College; guided tours; lectures; summer camp; weekend program for ages 8-12.
Publications: occasional catalogs & monographs.
Hours & Admission Prices: Jan. 12-July 30 & Aug. 18-Dec. 16 Tues. & Thurs.-Fri. 11-5, Wed. 11-7, Sat.-Sun. 1-5. No charge; donations accepted. Closed Daytona 500 weekend; Easter; DSC spring break; Independence Day; Thanksgiving & weekend after; Christmas Eve, Day & week. &
Attendance: 60,000 (accurate)

DeLand

AFRICAN AMERICAN MUSEUM OF THE ARTS, 325 S. Clara Ave., DeLand, FL 32720-5884. Mailing Address: P.O. Box 1319, Deland, FL 32721-1319. Tel.: 386-736-4004. Fax: 386-736-4088.
E-mail: art@africanmuseumdeland.org
Web Site: www.africanmuseumdeland.org
Founded: 1994.
Key Personnel: Exec. Dir., Mary Allen; Pres. (V), Jefferson Pendleton.
Personnel Profile: Full-Time Volunteers 1; Part-Time Paid 1; Part-Time Volunteers 3.
Volunteer Hours: 2,000
Operating Income: 30,000
Governing Authority: nonprofit organization. Tax-exempt: 501(c)(3).
Institution Type/Description: Art Museum.
Collections: African American & Caribbean American cultures & art; photographs; paintings; sculptures; masks.
Major Exhibits: Rose Jefferson, 3/12/14-4/19/14; Deland High School IB Art Program, 4/22/14-4/29/14; Joyce Hayes, 5/14/14-6/28/14; AAMA Children Exhibit, 7/5/14; AAMA Permanent Collection, 7/12/14-8/2/14; Kenneth Harris, 8/13/14-9/20/14.
Facilities: amphitheater; cultural park.
Activities: lectures; drama; dance; meetings; reading & creative writing programs; art exhibits; music; Kwanzaa; Juneteenth; after school program; book signing events; jazz festival.
Publications: AAMA brochure.
Hours & Admission Prices: Thurs.-Sat. 10-4. No charge; donations accepted. &
Attendance: 5,000 (estimated)
Membership: Student 6-18 $10; Individual $20; Family $50; Supporter $100; Organization $200; Corporate $250;

DELAND NAVAL AIR STATION MUSEUM, 910 Biscayne Blvd., DeLand, FL 32724-2009. Tel.: 386-738-4149. Fax: 386-738-5405.
E-mail: contact@delandnavalairstation.org
Web Site: www.delandnavalairstation.org
Key Personnel: Exec. Dir., Chris Stubbs.
Governing Authority: private; nonprofit organization.
Institution Type/Description: Military Museum.
Collections: U.S. Naval history; military artifacts; documents; F-14 fighter jet; WWII TBF Avenger torpedo bomber.
Hours & Admission Prices: Tues.-Sat. 12-4. No charge; donations accepted.

FLORIDA MUSEUM OF WOMEN ARTISTS, 100 N. Woodland Blvd., Ste. 1, DeLand, FL 32720-4249. Tel.: 386-873-2976.
E-mail: crecktenwald@floridamuseumforwomenartists.org
Web Site: www.floridamuseumforwomenartists.org
Founded: 2008.
Key Personnel: Exec. Dir., Crystal Recktenwald; Pres. (V), Margaret Schnebly Hodge.
Personnel Profile: Full-Time Paid 1; Part-Time Paid 2; Part-Time Volunteers 3; Interns 4.
Governing Authority: Tax-exempt: 501(c)(3).
Institution Type/Description: Art Museum.
Collections: works by women artists.
Facilities: Museum-related items for sale.
Activities: gallery talks; special events.
Hours & Admission Prices: Tues.-Sat. 11-6. Adults $5; members no charge. Closed major holidays. &
Membership: Student $10; Individual $25; Family $50; Friend $100; Patron $200; Benefactor $500; Sustainer $1,000; Curator's Circle $2,500; Director's Circle $5,000; President's Circle $10,000.

THE GILLESPIE MUSEUM - STETSON UNIVERSITY, 234 E. Michigan Ave., DeLand, FL 32724-3539. Mailing Address: Stetson University, 421 N. Woodland Blvd., Unit 8403, DeLand, FL 32723. Tel.: 386-822-7330. Fax: 386-822-7328.
E-mail: gillespie@stetson.edu
Web Site: www.gillespiemuseum.stetson.edu
Formerly: Gillespie Museum of Minerals, Stetson University
Founded: 1958.
Congressional District: 4
Key Personnel: Dir., Karen Cole, Ph.D.; Cur., Bruce Bradford, Ph.D.; Admin. Asst., Stacy Junkins.
Personnel Profile: Full-Time Paid 1; Part-Time Paid 1; Part-Time Volunteers 10; Interns 10.
Governing Authority: university. Parent Institution: Stetson University. Tax-exempt: 501(c)(3).
Institution Type/Description: Earth & Environmental Science Museum.
Collections: minerals; teaching collection of major rock groups; thematic minerals; environmental science.
Activities: self-guided tours; lectures; formally organized education programs for primary, secondary & college students; permanent exhibitions.
Hours & Admission Prices: Call for hours. Adults $2, seniors & students $1; discounts to groups & ASTC members; members no charge. Closed national & university holidays. &
Attendance: 6,000 (estimated)
Membership: Students $10; Seniors 55 & over $15; Individuals $20; Senior Couples 55 & over $25; Family $35; Corporate $175; Bronze $1,000; Silver $2,500; Gold $5,000; Honorary Curator $10,000.

HENRY A. DELAND HOUSE MUSEUM, 137 W. Michigan Ave., DeLand, FL 32720-3418. Tel.: 386-740-6813. Fax: 386-740-6813.
E-mail: delandhouse@msn.com
Web Site: www.delandhouse.com
Governing Authority: Parent Institution: West Volusia Historical Society.
Institution Type/Description: Historic House: housed in the former home of DeLand's first attorney, Arthur George Hamlin; built in 1886.
Collections: period furnishings; personal artifacts; photographs.
Facilities: Museum-related items for sale.
Hours & Admission Prices: Tues.-Sat. 12-4; other times by appointment.

HOMER AND DOLLY HAND ART CENTER AT STETSON UNIVERSITY, (M), 139 E. Michigan Ave., DeLand, FL 32720. Mailing Address: 421 N. Woodland Blvd., Unit 8423, DeLand, FL 32723. Tel.: 386-822-7270 & 7271.
E-mail: seules@stetson.edu
Web Site: www2.stetson.edu/handartcenter
Key Personnel: Coord., Susanne Eules.
Personnel Profile: Part-Time Paid 1; Part-Time Volunteers 1.

Institution Type/Description: Art Gallery.
Collections: works from the Vera Bluemner Kouba collection & the university's permanent collection; ceramics; paintings; sculpture.
Activities: special events; workshops; demonstrations; artist lectures; educational programs.
Hours & Admission Prices: Academic Year: Mon.-Fri. 11-4. No charge. Closed national holidays; university breaks. &

Attendance: 5,017 (accurate)

MEMORIAL HOSPITAL MUSEUMS, 230 N. Stone St., DeLand, FL 32720-4010. Tel.: 386-740-5800.
Governing Authority: Parent Institution: West Volusia Historical Society. Tax-exempt.
Institution Type/Description: Hospital Museum: listed on the National Register of Historic Buildings.
Collections: medical equipment & history; pharmacy; period electrical systems; personal artifacts; veterans memorabilia.
Hours & Admission Prices: Wed.-Sat. 10-3. No charge; donations accepted. &
Attendance: 500 (estimated)

MUSEUM OF FLORIDA ART, (M), 600 N. Woodland Blvd., DeLand, FL 32720-3447. Tel.: 386-734-4371. Fax: 386-734-7697.
E-mail: dansberger@museumoffloridaart.org
Web Site: www.museumoffloridaart.org
Formerly: The DeLand Museum of Art
Founded: 1951.
Congressional District: 7
Key Personnel: C.E.O., George Bolge; Pres. (V), Judith Thompson; Dir. Finance & Operations, Dorothy Dansberger; Dir. Mktg., Lisa Habermehl; Cur. Art & Exhibition, David Fithian; Cur. Education, Pam Coffman; Dir. Devel., Pattie Pardee; Guest Svcs., Suzi Tanner; Museum Shop Mgr., Jewel Dickson.
Personnel Profile: Full-Time Paid 7; Part-Time Paid 5; Part-Time Volunteers 100; Interns 20.
Governing Authority: nonprofit corporation. Tax-exempt: 501(c)(3).
Institution Type/Description: Art Museum.
Collections: art of Florida; works by Florida artists.
Research Fields: Florida art.
Facilities: meeting rooms; classrooms. Museum-related gifts for sale.
Activities: guided tours; lectures; gallery talks; workshops; formally organized education programs; docent program or council; outreach exhibitions & activities; inter-museum loan; permanent, temporary & traveling exhibitions; travel program.
Publications: monthly newsletter; catalogues; exhibition brochures & programs.
Hours & Admission Prices: Tues.-Sat. 10-4, Sun. 1-4. Adults $5; discounts to AAA, AAM & ICOM members; Southeastern & North American reciprocal membership members, children under 12, members & staff no charge. Closed national holidays. &
Attendance: 30,000 (estimated)
Membership: Student $20; Individual $35; Household $60; Sustaining $150; Patron $500; Leadership Circle $1,000 & up.

ROBERT M. CONRAD RESEARCH AND EDUCATIONAL CENTER, 137 W. Michigan Ave., DeLand, FL 32720-3418. Tel.: 386-740-6813.
Web Site: www.delandhouse.com/conrad.htm
Governing Authority: Parent Institution: West Volusia Historical Society.
Institution Type/Description: History Museum.
Collections: local history; personal artifacts; photographs; newspaper clippings; videos.
Facilities: library.
Activities: research; educational programs.
Hours & Admission Prices: Tues.-Sat. 12-4; other times by appointment.

Deerfield Beach

DEERFIELD BEACH HISTORICAL SOCIETY, 380 E. Hillsboro Blvd., Deerfield Beach, FL 33441-3540. Mailing Address: P.O. Box 755, Deerfield Beach, FL 33443-0755. Tel.: 954-429-0378. Fax: 954-429-0378.
E-mail: carolyn.morris@deerfield-history.org
Web Site: www.deerfield-history.org
Founded: 1973.
Key Personnel: Pres., Bill Muenzenmaier; Vice Pres., Ronald LaVergne; Sec., Catherine Coney; Treas., Dave Noderer; Dir., Carolyn Morris.
Personnel Profile: Part-Time Paid 1; Part-Time Volunteers 12.
Governing Authority: private; nonprofit organization. Subsidiary Institutions: Butler House Museum; 1920 Deerfield School; 1940 Kester Cottage; 1926 Deerfield Beach Elementary School; 1926 Seaboard Airline Railway Station. Tax-exempt: 501(c)(3).
Institution Type/Description: History Museum.
Collections: archives; photographs; oral histories; costumes; period furnishings; artifacts.
Facilities: library. Museum-related items for sale.
Activities: docent training program; internship program; lectures; workshops; guided tours; traveling & permanent exhibits; school programs; historic trail. Annual Events: Pioneer Days; cemetery walk; Breakfast with Santa; History at High Noon; 1/2 day in 1920s classroom; Ice Cream Social; Fall Ball Costume Gala.
Publications: quarterly newsletter; annual lecture series; oral history transcripts; photo essay.
Hours & Admission Prices: Tues.-Thurs. 10-4 by appointment. Suggested Donation: $2; discounts to AAM members. &
Attendance: 500 (estimated)
Membership: Individual $40; Family $50; Professional $100; Sponsor $175; Corporate $200.

SOUTH FLORIDA RAILWAY MUSEUM, 1300 W. Hillsboro Blvd., Deerfield Beach, FL 33442-1716. Tel.: 954-698-6620. Fax: 561-790-4191.
Web Site: www.sfrm.org/
Key Personnel: Pres., Vic Zarzycki; Dir., Richard Bretone
Institution Type/Description: Transportation Museum.
Collections: railroad history; model railroad; historic artifacts.
Hours & Admission Prices: Call for information.

Delray Beach

CASON COTTAGE HOUSE MUSEUM, 5 N.E. 1st St., Delray Beach, FL 33444-3707. Tel.: 561-243-0223 & 243-2577.
E-mail: info@db-hs.org
Governing Authority: Parent Institution: Delray Beach Historical Society.
Institution Type/Description: Historic House Museum: housed in the home of Rev. John R. Cason, a community leader & Methodist minister; built in 1915.
Collections: Florida lifestyle c.1915-1935; personal artifacts; furnishings.
Hours & Admission Prices: Tues.-Fri. 10-4. Adults $4. &

CORNELL MUSEUM OF ART & AMERICAN CULTURE, 51 N. Swinton Ave., Delray Beach, FL 33444-2631. Tel.: 561-243-7922. Fax: 561-243-7018.
E-mail: gadams@delraycenterforthearts.org
Web Site: www.delraycenterforthearts.org
Founded: 1990.
Congressional District: 14
Key Personnel: Dir. Museum, Gloria Rejune Adams; Bd. Pres., Scott Porten; Pres. & C.E.O., Joe Gillie.
Personnel Profile: Full-Time Paid 1; Part-Time Paid 1; Part-Time Volunteers 60.
Governing Authority: nonprofit. Parent Institution: Delray Beach Center for the Arts at Old School Square. Tax-exempt: 501(c)(3).
Institution Type/Description: Art History Museum: housed in 1913 building.
Collections: Black women's achievements against the odds; early 1900s classroom memorabilia. Historic Buildings: 1925 Crest Theatre; 1926 gymnasium.
Major Exhibits: Fascinating World of Golf: Gary Wiren Collection, 11/8/13-4/21/14.
Facilities: educational facilities.
Activities: arts festivals; concerts; docent program; films; guided tours; lectures; loan, temporary & traveling exhibitions; school loan service.
Publications: e-newsletter.
Hours & Admission Prices: June-Oct. Tues.-Sat. 10:30-4:30; Nov.-May Tues.-Sat. 10:30-4:30, Sun. 1-4:30. Adults $10, seniors 65 & over $6, students 13-21 with ID $4, children 4-12 $2; discounts to Florida Trust members; children under 3 & members no charge. Closed New Year's Day; Good Friday; Easter; Memorial Day; Independence Day; Labor Day; Thanksgiving; Christmas Eve & Day. &
Attendance: 50,000 (accurate)
Membership: Basic $40; Friends & Family $75; Honor Role $250; Inner Circle $1,000; Corporate $1,500.

*** THE MORIKAMI MUSEUM AND JAPANESE GARDENS, (M),** 4000 Morikami Park Rd., Delray Beach, FL 33446-2305. Tel.: 561-495-0233. Fax: 561-499-2557.
E-mail: morikami@pbcgov.org
Web Site: www.morikami.org
Founded: 1977.

Congressional District: 14
Key Personnel: Dir. & Sr. Cur., Thomas Gregersen; Trustee Pres., Randal Baker; Park Admin., Bonnie White Lemay; Dir. Advancement, Kimberly Sovinski; Dir. Education, Reiko Nishioka; Membership Mgr., Heather Williams; Volunteer Coord., Sharyn Samuels; Events Mgr., Kizzy Sanchez; Resource Supervisor, Heather Grzybek; School Programs Specialist, Beth Kawazura; Cur. Japanese Art, Susanna Brooks; Cur. Collections, Veljko Dujin; Facility Rental Coord., Alanna Rainey; Museum Shop Mgr., Sallie Chisholm.
Personnel Profile: Full-Time Paid 42; Part-Time Paid 7; Part-Time Volunteers 265.
Governing Authority: county. Parent Institution: Palm Beach County Parks & Recreation Dept., 2700 Sixth Ave. S., Lake Worth, FL 33461. Tax-exempt.
Institution Type/Description: Ethnology Museum specializing in Japanese culture.
Collections: Yamato Colony of Japanese settlers; Japanese fine art, folk art, & ethnological materials; Gulf Stream Bonsai.
Research Fields: history of the Yamato Colony of Japanese farmers.
Facilities: 4,000-vol. library of materials related to Japan, including some in Japanese, available for use on premises by request; 200-acre park with Japanese Gardens; 225-seat theater; picnic facilities.
Activities: guided tours; docent program; permanent, temporary & loan exhibitions; seasonal celebrations; special events. Annual Events: Oshogatsu Japanese New Year; Hatsume Fair; Gala Dinner-Dance; Children's Day; Bon Festival.
Publications: quarterly newsletter & calendar; exhibition catalog.
Hours & Admission Prices: Museum: Tues.-Sun. 10-5. Adults $13, seniors $12, children 6-17 & college students $8; discounts to groups of 15 or more; members & children under 6 no charge. Park: daily sunrise to sundown. No charge. Closed New Year's Day; Easter; Independence Day; Thanksgiving; Christmas. ♧
Attendance: 230,432 (accurate)
Membership: Student $40; Educator Club $5; Individual $60; Dual $90; Netsuke $140; Taiko $280; Samurai $550; Wisdom Ring $1,500.

MUSEUM OF LIFESTYLE & FASHION HISTORY, 322 N.E. 2nd Ave., Delray Beach, FL 33482. Mailing Address: P.O. Box 6127, Delray Beach, FL 33482-6127. Tel.: 561-243-2662. Fax: 561-495-8785.
E-mail: info@mlfhmuseum.org
Web Site: mlfhmuseum.org
Governing Authority: nonprofit organization. Tax-exempt: 501(c)(3).
Institution Type/Description: General Museum.
Collections: hosts exhibits including period fashion designs, popular culture art, decorative arts, interior designs, architecture history; ethnic cultures.
Activities: community outreach programs; special events.
Hours & Admission Prices: Temporarily closed.

SPADY CULTURAL HERITAGE MUSEUM, (M), 170 N.W. 5th Ave., Delray Beach, FL 33444-2653. Tel.: 561-279-8883.
E-mail: cfjones@spadymuseum.org
Web Site: www.spadymuseum.com/
Key Personnel: Exec. Dir., Daisy Fulton; Museum Dir., Charlene Jones; Educator, Brandy Brownlee.
Governing Authority: nonprofit organization.
Institution Type/Description: Cultural Heritage Museum: housed in the former home of Solomon D. Spady, a prominent African American educator & community leader in Delray Beach from 1922-1957.
Collections: black history & culture.
Hours & Admission Prices: Mon.-Fri. 11-4, Sat. by appointment. Adults $5; members no charge.

Doral

MUSEUM OF THE AMERICAS, (M), 2500 N.W. 79th Ave., Ste. 104, Doral, FL 33122-1071. Tel.: 305-599-8088 & 8089.
E-mail: americasmuseum@aol.com
Web Site: www.museumamericas.org
Founded: 1991.
Key Personnel: Dir., Raul M. Oyuela
Institution Type/Description: Art Museum.
Collections: 20th-century Latin American & Caribbean art.
Activities: educational programs.
Hours & Admission Prices: Tues.-Fri. 11-5, Sat. 11-4. No charge.

Dunedin

DUNEDIN FINE ART CENTER, 1143 Michigan Blvd., Dunedin, FL 34698-2799. Tel.: 727-298-3322.
E-mail: gabissett@dfac.org
Web Site: www.dfac.org
Founded: 1974.
Congressional District: 10
Key Personnel: Exec. Dir., George Ann Bissett, C.F.R.E.; Pres. (V), Charles Klein; Museum Shop Mgr., Angelee Carpenter.
Personnel Profile: Full-Time Paid 9; Full-Time Volunteers 2; Part-Time Paid 4; Part-Time Volunteers 75; Interns 1.
Governing Authority: bd. of directors. Tax-exempt: 501(c)(3).
Institution Type/Description: Art Center.
Collections: works by regional, national & international artists.
Major Exhibits: Miniature Art Society of Florida, 1/9/14-2/9/14; Taking Shape, 2/22/14-3/30/14; Altogether Now!, 3/14-4/14; Artistic Discovery, Congressional District 13 High Schools, 4/6/14-4/27/14; The Poetics of Space, 6/14-7/14; In a Dark Time, the Eye Begins to See, 5/23/14-8/17/14; Living Matter (S), 9/5/14-10/20/14; Terra Incognita, 9/5/14-12/23/14.
Activities: lectures; studio classes; workshops; educational & cultural programming.
Publications: quarterly, The Artery.
Hours & Admission Prices: Mon.-Fri. 10-5, Sat. 10-2, Sun. 1-4. Adults $4, senior citizens $3; discounts to AAM & ICOM members; children 2 & under and members no charge; donations accepted. ♧
Attendance: 158,000 (estimated)
Membership: Individual $45; Family $65.

DUNEDIN HISTORICAL SOCIETY & MUSEUM, (M), 349 Main St., Dunedin, FL 34698-5700. Mailing Address: P.O. Box 2393, Dunedin, FL 34697-2393. Tel.: 727-736-1176. Fax: 727-736-4756. Facebook: Dunedin Historical Museum.
E-mail: dunedinhistory@yahoo.com
Web Site: www.dunedinmuseum.org
Founded: 1970.
Congressional District: 5
Key Personnel: Pres., Dave Pauley; Admin. Asst., Lauren Sherbuk; Dir., Vincent Luisi; Museum Shop Mgr., Anita Lanagan.
Personnel Profile: Full-Time Paid 1; Part-Time Paid 3; Part-Time Volunteers 100; Interns 2.
Volunteer Hours: 1,500
Governing Authority: society. Subsidiary Institution: Andrews Memorial Chapel, 1899 San Mateo, Dunedin, FL 34698. Tax-exempt: 501(c)(3).
Institution Type/Description: Local History Museum: housed in 1923 Atlantic Coastline Passenger Station; The Old Freight Warehouse, adjacent to Passenger Station.
Collections: Dunedin and Pinellas photographs & artifacts.
Major Exhibits: Pirates of Florida (T), 1/1/14-3/30/14; Celebrating 50 Years: Stirling Scotland Sisterhood (T), 1/14-4/14; Dunedins Archaeological History, 6/14-9/14.
Research Fields: history of Dunedin area, county & state; documentation for historic preservation; transportation.
Facilities: library of books on history of area, related collections & newspaper articles available for use by appointment only on premises; meeting room. History books & reproductions of documents for sale.
Activities: guided tours; lectures; loan exhibitions; docent program.
Publications: bimonthly newsletter, Dunedin Days; booklet, Historical Highlights-First 100 Years of Settlement; book, Dunedin Thru The Years, 1st & 2nd editions; Arcadia Press - Images of Dunedin; Arcadia Press - Images of Pinellas County.
Hours & Admission Prices: Tues.-Sat. 10-4. Suggested Donation: adults $2; discounts to AAM members; members & children under 12 no charge. Closed legal holidays. ♧
Attendance: 14,500 (accurate)
Membership: Individual $30; Family $45.

Eatonville

ZORA NEALE HURSTON NATIONAL MUSEUM OF FINE ARTS, 227 E. Kennedy Blvd., Eatonville, FL 32751-5303. Tel.: 407-647-3307. Fax: 407-539-2192.
E-mail: info@zorafestival.com
Web Site: zoranealehurstonmuseum.com
Institution Type/Description: Art Museum.
Collections: works by African American artists.
Activities: Annual Event: festival.
Hours & Admission Prices: Mon.-Fri. 9-4; groups by appointment. No charge; donations accepted.

Eglin Air Force Base

AIR FORCE ARMAMENT MUSEUM, 100 Museum St., Eglin Air Force Base, FL 32542-1497. Tel.: 850-651-1808 & 882-4062. Fax: 850-882-3990.
E-mail: info@afarmamentmuseum.com
Web Site: www.afarmamentmuseum.com
Founded: 1985.
Key Personnel: Dir., George Jones; Museum Specialist, Timothy Savoir; Museum Specialist, John "Chuck" Yeager; Operations Dir., Joan Doman.
Personnel Profile: Full-Time Paid 3; Part-Time Paid 3; Part-Time Volunteers 30.
Governing Authority: federal; nonprofit. Parent Institution: Air Force Museum AFB Wright-Patterson, OH. Subsidiary Institution: Eglin AFB, FL. Tax-exempt.
Institution Type/Description: Military Museum.
Collections: armament; gun vault collection; missiles; rockets; aircraft: SR-71, B-17, B-25, F100, P47, P51, F4, F-105, C-47, B-57, F-86, F-89, F101, B-52, F-15, F-104, RB-47, MIG 21, F-111, F-16, A-10, C-130, F-80, T-33, MIG-21, UH-1H, O-2, H-53.
Facilities: 25,000 sq. ft. exhibit space; 60-seat theater. Military & local souvenir items for sale.
Activities: films; docent program; participatory & traveling exhibitions.
Publications: quarterly historical record.
Hours & Admission Prices: Mon.-Sat. 9:30-4:30. No charge; donations accepted. Closed federal holidays. ♿
Attendance: 123,000 (estimated)
Membership: Individual $15.

Ellenton

THE JUDAH P. BENJAMIN CONFEDERATE MEMORIAL AT GAMBLE PLANTATION HISTORIC STATE PARK, 3708 Patten Ave., Ellenton, FL 34222-2152. Tel.: 941-723-4536. Fax: 941-723-4538.
Web Site: www.floridastateparks.org
Founded: 1926.
Congressional District: 13
Key Personnel: Park Mgr., Kevin Kiser.
Personnel Profile: Full-Time Paid 4; Part-Time Volunteers 18.
Governing Authority: state. Parent Institution: Department of Environmental Protection. Tax-exempt.
Institution Type/Description: Historic House: 1840-60 plantation home of Robert Gamble. 1844-1865, Gamble Mansion, the main house of the Gamble sugar plantation; Greek revival vernacular construction.
Collections: antebellum furnishings.
Facilities: picnic shelter; visitor center.
Activities: guided tours.
Publications: brochures.
Hours & Admission Prices: Visitor Center: Thurs.-Mon. 8-5. Park Grounds: daily 8am-sunset. Tours: Thurs.-Mon. 9:30, 10:30 & 1, 2, 3, & 4. Adults $6, children 6-12 $4; under 6 no charge. ♿
Attendance: 56,000 (estimated)

Estero

KORESHAN STATE HISTORIC SITE, 3800 Corkscrew Rd., Estero, FL 33928-1919. Tel.: 239-992-0311.
E-mail: andrew.tetlow@dep.state.fl.us
Web Site: www.floridastateparks.org/koreshan/
Founded: 1961.
Congressional District: 13
Key Personnel: Cur., Andrew Tetlow.
Governing Authority: state; nonprofit. Parent Institution: Florida Park Service. Subsidiary Institution: Koreshan Unity Alliance. Tax-exempt.
Institution Type/Description: Historic Building Complex: 1894-1982 utopian settlement.
Collections: ephemera related to Koreshan members; historic buildings.
Research Fields: communal studies; women's issues; cultural landscape; architecture; music; drama; archaeology.
Facilities: botanical garden.
Activities: docents; special events; weekend guided tours; ghost walks; theatre festival; antique engine show; campfire programs; rental facility; canoe tours.
Publications: brochures; self-guided tour booklet.
Hours & Admission Prices: Daily 8 to sundown. Adults $5 per vehicle; members no charge. ♿
Attendance: 50,000 (estimated)
Membership: Koreshan Unity Alliance, Citizen Support Organization: Student $5; Individual $10; Family $20; Sustaining $25; Sponsor $100; Contributor $500; Patron & Corporate $1,000.

Eustis

EUSTIS HISTORICAL MUSEUM & PRESERVATION SOCI-ETY, INC., 536 N. Bay St., Eustis, FL 32726-3439. Tel.: 352-483-0046. Facebook: Eustis Historical Museum.
E-mail: eustismuseum@gmail.com
Founded: 1983.
Congressional District: 6
Key Personnel: Pres., Danyel Moulden; Treas., John Blankenship; Cur., Jennie Hoon; Historian, Louise Carter; Public Rels., Ethel I. Ryan.
Personnel Profile: Part-Time Paid 1; Part-Time Volunteers 18.
Governing Authority: municipal. Branch Museums: Citrus Museum & Tool Museum. Tax-exempt.
Institution Type/Description: Historic House Museum; housed in the residence of G.D. Clifford, an early settler.
Collections: Florida historical & cultural artifacts; preservation of Florida landmarks; furniture; memorabilia.
Research Fields: history of the area; the citrus industry in central Florida.
Facilities: library; Unity-Bell-Clifford Pavilion available to rent for parties & weddings.
Activities: guided tours; special temporary exhibits; educational programs; meetings. Annual Events: antique appraisal clinic; Christmas in Eustis; Eustis fourth grade history essay contest.
Publications: monthly newsletter; Days of Yesteryear 1875-1911; Eustis, Our Town 1912-1945.
Hours & Admission Prices: Tues.-Fri. 1-5, Sat. 10-3. No charge; donations accepted. Closed New Year's Day; Christmas. ♿
Attendance: 10,000 (estimated)
Membership: Student $10; Individual $20; Family $25; Business $55; Life $500.

LAKE EUSTIS MUSEUM OF ART, 1 W. Orange Ave., Eustis, FL 32726. Tel.: 352-483-2900.
E-mail: lake.eustis.art.museum@gmail.com
Web Site: www.lakeeustisartmuseum.org
Founded: 1994.
Congressional District: 8
Key Personnel: Dir., Richard D. Colvin; Pres. (V), Sam LaPoma.
Personnel Profile: Full-Time Paid 1; Part-Time Volunteers 30; Interns 4.
Governing Authority: Tax-exempt.
Institution Type/Description: Fine Art Museum
Collections: fine art.
Facilities: gallery; classroom.
Activities: special events; art exhibits; classes; workshops; gallery talks; docents.
Publications: newsletter; annual brochure; posters; postcards.
Hours & Admission Prices: Mon.-Fri. 10-4, Sat.-Sun. 12-4. Adults $5; discount to NARM members; members no charge. Closed major holidays. ♿
Attendance: 6,000 (accurate)
Membership: Senior/Student $25; Individual $40; Family $60; Individual NARM $100; Contributor $150; Sustaining $250; Patron $1,000.

Fernandina Beach

AMELIA ISLAND MUSEUM OF HISTORY, 233 S. Third St., Fernandina Beach, FL 32034-4210. Tel.: 904-261-7378. Fax: 904-261-9701.
E-mail: info@ameliamuseum.org
Web Site: www.ameliamuseum.org
Founded: 1986.
Congressional District: 4
Key Personnel: Dir., Phyllis Davis; Bd. Pres., Pat Panella; Asst. Dir. Programs, Alex Buell; Volunteer Coord., Thea Seagraves; Elderhost Coord., Brenda Brubeck; Assoc. Dir. Creative Devel. & Operations, Liz Norris; References & Archives, Teen Peterson; Museum Shop Mgr., Janet Kohler.
Personnel Profile: Full-Time Paid 4; Part-Time Paid 4; Part-Time Volunteers 180.
Governing Authority: nonprofit organization. Tax-exempt: 501(c)(3).
Institution Type/Description: History Museum: housed in c.1937 Nassau County Jail building.
Collections: Nassau County history; over 3000 photos; glassplate negatives; Spanish land grants; 18th-20th century newspapers; textiles; Native American artifacts; 19th century weapons, photography equipment, maps & charts; Victoria era relics; cemetery surveys; city directories; architectural files.
Research Fields: Amelia Island & Nassau County c.2500 B.C.-present.; Civil War; local historic district; Spanish colonial history.
Facilities: library; research facility; archives; 7,000 sq. ft. exhibit space; 100-seat auditorium. Museum & local history-related publications for sale.

Activities: guided tours; Speakers Bureau; student & senior programs; Elderhostel; holiday tours; veterans oral history project; facility rental; lectures.
Publications: monthly & quarterly newsletter; five monographs on local history, Timucuan Indian through early 19th century.
Hours & Admission Prices: Mon.-Sat. 10-4, Sun. 1-4. Museum & Spoken History Tour: 11am & 2pm. Adults $7, students & military $4. Centre Street Historic District Tours: Sept.-June Fri.-Sat.; Summer: call for hours. Admission $10. Ghost Tours: Fri. 6pm. Admission $10. Closed all major holidays. &
Attendance: 35,000 (accurate)
Membership: Individual $50; Family $75; Patron $100; Business Partnership $300.

FORT CLINCH STATE PARK, 2601 Atlantic Ave., Fernandina Beach, FL 32034-2203. Tel.: 904-277-7274. Fax: 904-277-7225.
Web Site: www.floridastateparks.org
Founded: 1935.
Key Personnel: C.E.O., Michael Bullock; Park Mgr., Peter Scalco; Sec. Florida DEP, Michael W. Sole.
Personnel Profile: Full-Time Paid 13; Part-Time Paid 4; Part-Time Volunteers 150.
Governing Authority: state. Parent Institution: Dept. of Environmental Protection, Division of Recreation & Parks, Commonwealth Bldg., 3900 Commonwealth Blvd., Tallahassee, FL 32399. Tax-exempt.
Institution Type/Description: Historic Building: 1864 restored Fort Clinch.
Collections: local history & culture; Civil War era armament.
Facilities: visitor center; nature trails.
Activities: Museum Sponsors: living history demonstrations, life of the 1864 Union soldier at Fort Clinch; Union Garrison Weekend monthly; Confederate Garrison in October.
Publications: quarterly newsletter.
Hours & Admission Prices: Park: daily 8-sundown. $6 per vehicle up to 8 persons. Fort: daily 9-5. $2 per person over 6. Fort: evening programs $3 per person over 6. &
Attendance: 214,630 (estimated)
Membership: Citizen support organization formed Friends of Fort Clinch.

Flagler Beach

BULOW PLANTATION RUINS HISTORIC STATE PARK, 3501 S. Old Kings Rd. S., 9 mi. S.E. of Bunnell State Rd. 5, Flagler Beach, FL 32136-4339. Tel.: 386-517-2084.
Web Site: www.floridastateparks.org/bulowplantation/default.cfm
Founded: 1945.
Congressional District: 4
Key Personnel: Head Ranger, Nicky Makouski; Park Mgr., Benny Woodham.
Personnel Profile: Full-Time Paid 1; Part-Time Paid 2.
Governing Authority: state. Dept. of Environmental Protection, Div. of Recreation & Parks, Commonwealth Bldg., 3900 Commonwealth Blvd., Tallahassee, FL 32303, Tel: 850-488-9872.
Institution Type/Description: History Museum: interpretive center.
Collections: sugar kettle and mill ruins; artifacts & small tools used on plantation; springhouse ruins.
Research Fields: plantation life of early 1800s.
Facilities: nature trail; picnic area with pavilion building; boat ramp; fishing dock.
Activities: permanent exhibitions; occasional tours.
Publications: park brochure.
Hours & Admission Prices: Thurs.-Mon. 9-5. $3 per vehicle, $1 bicycles & walk-ins. Canoe Rentals: $10 hr., $40 day. Pavilion Rental: $30 plus tax. &

Fort Lauderdale

✱ BONNET HOUSE MUSEUM & GARDENS, (M), 900 N. Birch Rd., Fort Lauderdale, FL 33304-3326. Tel.: 954-563-5393. Fax: 954-561-4174. TDD: 954-563-5393.
Web Site: www.bonnethouse.org
Founded: 1987.
Congressional District: 22
Key Personnel: C.E.O., Karen L. Beard; Bd. Chm. (V), Pat Smith; Dir. Devel., Patrick Shavloske; Cur., Stephen Draft; Dir. Education & Volunteer Programs, Linda Schaller; Museum Shop Mgr., Dianne Ennis.
Personnel Profile: Full-Time Paid 9; Part-Time Paid 16; Part-Time Volunteers 313; Interns 1.
Operating Expenses: 1,751,570
Operating Income: 1,949,508
Governing Authority: Bonnet House, Inc. Tax-exempt: 501(c)(3).
Institution Type/Description: Historic House: house built in 1920; property consists of 35 acres from the Atlantic Ocean to the Intracoastal Waterway. Listed on the National Register of Historic Places.

Collections: personal furnishings & collections of artists Frederic C. & Evelyn F. Bartlett; artwork; wood carvings; carousel animals; rare china; orchids; lush tropical gardens.
Facilities: rental facilities; film shoot location. Museum-related items for sale.
Activities: guided tours; lecture series; music series; family days. Annual Event: Orchid Festival.
Publications: Bonnet House Newsletter; Bonnet House: A Legacy of Artistry & Elegance; Bonnet House: The Life & Gift.
Hours & Admission Prices: Tours: Tues.-Sun. 9-4. Adults $20, children 6-12 $16, groups over 15 people $11, grounds only $10; discounts to AAM & AAA members; members & children under 6 no charge. Closed New Year's Day; Thanksgiving; Christmas. &
Attendance: 72,000 (estimated)
Membership: Individual $55; Couples $90; Parrot & Family $125; Monkey $250; Swan $500; Golden Shell $1,000.

BROWARD COUNTY HISTORICAL COMMISSION, 301 S.W. 13th Ave., Fort Lauderdale, FL 33312. Tel.: 954-357-5553. Fax: 954-357-5522.
E-mail: dcunningham@browardlibrary.org
Web Site: www.broward.org/library/history
Founded: 1972.
Congressional District: 14, 15 & 16
Key Personnel: Chm. (V), Betty W. Cobb; Mgr. Library, Peggy D. Davis; Cur., Denyse Cunningham; County Archaeologist, Mathew DeFelice; Administrative Coord., Maria Munoz.
Personnel Profile: Full-Time Paid 3; Part-Time Paid 1; Part-Time Volunteers 1.
Governing Authority: county. Parent Institution: Broward County Board of County Commissioners. Tax-exempt.
Institution Type/Description: Historic Commission.
Collections: books, maps, microfilm, prints, artifacts, documents & photographs related to the exploration, settlement & urbanization of South Florida, particularly Broward County; oral history tapes; manuscripts.
Research Fields: early Broward County & courthouse history; biographies of pioneers.
Facilities: 1,500-vol. library of historical books available for research on premises or for limited loan; reading room; archives; microfilms.
Activities: temporary exhibitions; oral history interviews. Commission Sponsors: seminars; county historical festival; identification of historical sites.
Publications: annual magazine, Broward Legacy.
Hours & Admission Prices: Mon.-Thurs. 8-4:30. No charge; donations accepted. Closed county, state & federal holidays. &
Attendance: 5,000 (estimated)

FORT LAUDERDALE FIRE AND SAFETY MUSEUM, 1022 W. Las Olas Blvd., Fort Lauderdale, FL 33312. Tel.: 954-763-1005.
Institution Type/Description: Firefighting History Museum: housed in Fire Station 3; built in 1927.
Collections: local firefighting history; early firefighting equipment & trucks; photographs; personal artifacts; uniforms.
Facilities: Museum-related items for sale.
Activities: educational programs; fire safety; rental facilities; presentations.
Hours & Admission Prices: Sat. 9 to noon, Sun. 12-4. Programs by appointment. No charge; donations accepted.

FORT LAUDERDALE HISTORICAL SOCIETY, 219 S.W. 2nd Ave., Fort Lauderdale, FL 33301-1825. Tel.: 954-463-4431, ext. 15. Fax: 954-523-6228.
E-mail: executive_director@fortlauderdalehistorycenter.org
Web Site: oldfortlauderdale.org
Founded: 1962.
Congressional District: 12
Key Personnel: Interim Exec. Dir., Wil Trower; Pres. (V), Harry Moon, M.D.; Museum Shop Mgr., Linda Rosen.
Personnel Profile: Full-Time Paid 6; Part-Time Paid 2; Part-Time Volunteers 15.
Governing Authority: society; nonprofit. Parent Institution: Fort Lauderdale Historical Society. Tax-exempt: 501(c)(3).
Institution Type/Description: Historical Society Museum.
Collections: 8,000 three-dimensional artifacts; 350,000 photographs; 5,000 architectural drawings; 2,000 maps & books; 250 community scrapbooks; manuscripts.
Research Fields: local history; historic preservation.
Facilities: 3,000-vol. library of local & Florida history, available by request for use on premises; reading room; archives; photographic archives. Books for sale.
Activities: temporary & permanent exhibitions; historic preservation; educational programs & workshops; lecture series; walking tours.
Publications: membership newsletter, Inn Sider.

Hours & Admission Prices: Museum: Tues.-Sat. 10-5, Sun. 12-5. Archives: Wed. & Fri. 10-4, Sat. 12-4. Adults $10; members no charge. Closed major holidays. &

Attendance: 40,000 (estimated)

Membership: Student & Teacher $35; Individual $50; Family $75; Sustainer $100; Benefactor $250; Steward $500; King's Creek Villager $2,500; Heritage Villager $5,000.

INTERNATIONAL SWIMMING HALL OF FAME, INC., One Hall of Fame Dr., Fort Lauderdale, FL 33316-1694. Tel.: 954-462-6536. Fax: 954-525-4031.

Web Site: www.ishof.org

Founded: 1965.

Congressional District: 15

Key Personnel: C.E.O., Bruce Wigo, Ed.D.; Pres., Richard Korhammer; Chm. (V), Mark Spitz; Cur., Bob Duenkel; Museum Shop Mgr. & Displays Mgr., Laurie Marchwinski.

Personnel Profile: Full-Time Paid 6; Part-Time Paid 4.

Governing Authority: nonprofit organization. Tax-exempt: 501(c)(3).

Institution Type/Description: Aquatic Sports Museum.

Collections: murals, photos, films, books, sculpture & other works of art, devoted primarily to aquatic lore & memorabilia; Olympic medals; swimming; diving; water polo; synchronized swimming; lifesaving; masters & open water swimming; Olympic and historic swimming & bathing attire; 2 Olympic size pools; diving platforms & springboards.

Research Fields: Aquatic field.

Facilities: 200-seat auditorium; 24-seat theater; meeting room. Museum-related items for sale.

Activities: guided tours for groups; films; Hall of Fame Sponsors: International Diving Meet; Annual College Swim Coaches Forum; National and Regional swimming, diving, water polo and synchronized swimming meets.

Publications: quarterly newsletter; induction ceremony books.

Hours & Admission Prices: Daily 9-5. Adults $8, seniors $6, children $4; discounts to groups of 10 or more; children under 12 & members no charge. &

Attendance: 50,000 (estimated)

Membership: Basic Member (U.S.) $35; International Member $50; Contributor/Family Member $100; Competitor Member $250; Champion Member $500. Patrons of Swimming: Century Club Member $1,000; Chairman's Club $5,000; Benefactor's Club $10,000 & up.

* **MUSEUM OF ART/FORT LAUDERDALE, (M),** One E. Las Olas Blvd., Fort Lauderdale, FL 33301-1807. Tel.: 954-525-5500. Fax: 954-524-6011.

Web Site: www.moafl.org

Formerly: Museum of Art, Inc.

Founded: 1958.

Congressional District: 12

Key Personnel: Dir. & Chief Cur., Bonnie Clearwater; Bd. Chm., David W. Horvitz.

Governing Authority: nonprofit organization. Tax-exempt: 501(3)(c).

Institution Type/Description: Art Museum.

Collections: 20th-century European & American art including: Picasso, Dali, Warhol, Mapplethorpe & Stella; William Glackens collection; post-World War II Northern European Expressionist CoBrA art; African, South Pacific, Pre-Columbian, contemporary Cuban & Native American art.

Facilities: 1,000-vol. non-circulating library of art reference books; 256-seat auditorium for meetings & lectures; Miriam & Bernard Peck Sculpture Terrace for weddings & special events; Satellite Learning Center, classrooms for 4th & 5th grades. Museum-related items for sale.

Activities: inter- & intra-museum loan exhibitions; guided gallery tours; lectures; seminars; arts festivals; school coordinated educational programs; acoustiguide; docent training program.

Publications: catalogs for changing exhibitions; gallery guides; newsletters.

Hours & Admission Prices: Mon. Wed. & Fri. Sat. 11-5, Thurs. 11-8, Sun. 12-5. Adults $10, senior citizens 65 & over & military $7, students 13-17 $5; members, children 12 & under, NSU students, faculty & staff no charge. Special exhibit hours & prices may vary. Closed New Year's Day; Independence Day; Thanksgiving; Christmas. &

Attendance: 100,000 (estimated)

Membership: Student/Teacher $40; Individual $60; Couples/Family $100; Museum Enthusiast $150; Academy Supporter $250; Patron $500; Benefactor $1,000; Director's Circle $2,500; Collector's Circle $5,000.

* **MUSEUM OF DISCOVERY AND SCIENCE,** 401 S.W. Second St., Fort Lauderdale, FL 33312-1707. Tel.: 954-467-6637, ext. 311. Fax: 954-467-0046.

E-mail: kcavendish@mods.net

Web Site: www.mods.org

Founded: 1976.

Congressional District: 15

Key Personnel: C.E.O. & Pres., Kim L. Cavendish; Chm. (V), Jon Ferrando; Vice Pres. Finance & C.F.O., Patty Ackerman; Vice Pres. Devel., Patrick Flynn; Bldg. Supt., Ilija Nikolorski; Vice Pres. Mktg. & Communications, Marlene Janetos; Dir. Programs & Exhibits, Joe Cytacki; Museum Shop Mgr., Kevin Stradtner.

Personnel Profile: Full-Time Paid 49; Part-Time Paid 65; Part-Time Volunteers 150; Interns 6.

Operating Expenses: 6,900,000

Operating Income: 7,300,000

Governing Authority: nonprofit organization. Tax-exempt: 501(c)(3).

Institution Type/Description: Science Center.

Collections: live specimens related to Florida ecology; health, physical sciences & space artifacts; fossils; Great Gravity Clock.

Major Exhibits: Goosebumps: The Science of Fear (T), 1/14-8/14; Black Holes: Space Warps & Time Twists (T), 10/14-12/14.

Research Fields: education techniques; ecology; aquariology; corals.

Facilities: educational facilities; IMAX theater; lecture hall; cafe; video/demonstration theater. Museum-related items for sale.

Activities: films; traveling exhibitions; summer camps; snake shows; outreach programs; IMAX films; demonstrations; table talks; school tours; science labs; mall camps; scout programs; camp-ins; sea turtle walks; flight simulators; ride simulator; Youth Advisory Council; Teen Science Track; nanotechnology lectures.

Publications: visitors guide; school services brochure; membership brochure; annual report; quarterly magazine, Explorations.

Hours & Admission Prices: Mon.-Sat. 10-5, Sun. 12-6. Exhibits: adults $14; ASTC & members no charge. IMAX: adults $9; discounts to members & ASTC members. Exhibits & IMAX: adults $19, seniors $18, children $15. &

Attendance: 456,091 (accurate)

Membership: Dual & Grandparent $85; Family $125; Contributing $225; Sponsoring $325.

MY JEWISH DISCOVERY PLACE CHILDREN'S MUSEUM, 6501 W. Sunrise Blvd., Fort Lauderdale, FL 33313-6036. Tel.: 954-792-6700. Fax: 954-792-4839.

E-mail: dgraw@sorefjcc.org

Web Site: sorefjcc.org

Founded: 1979.

Key Personnel: C.E.O. & Exec. Dir., Donald Graw.

Governing Authority: Parent Institution: Soref Jewish Community Center. Tax-exempt.

Institution Type/Description: Children's Museum.

Collections: hands-on exhibits; Jewish culture & history.

Hours & Admission Prices: Tues.-Thurs. 12-4. Admission 2 & over $5; discounts to AAM & ICOM members; one adult accompanied by child no charge.

Attendance: 1,500 (estimated)

Membership: Family with one child $36; Family with 2 children $54; Family with 3 or more children $72.

NAVAL AIR STATION FORT LAUDERDALE MUSEUM, 4000 W. Perimeter Rd., Fort Lauderdale, FL 33315. Tel.: 954-359-4400.

E-mail: allanmcelhiney@yahoo.com

Web Site: www.nasflmuseum.com

Institution Type/Description: History Museum.

Collections: station history; documents; books; naval art; photographs; period artifacts; recreated soldiers barracks.

Facilities: library.

Activities: research.

Hours & Admission Prices: By appointment.

OLD DILLARD MUSEUM, 1009 N.W. 4th St., Fort Lauderdale, FL 33311-8935. Tel.: 754-322-8828. Fax: 754-322-8824.

E-mail: derek.davis@browardschools.com

Web Site: www.broward.k12.Fl.us/olddillardmuseum

Founded: 1991.

Key Personnel: Dir., Derek Davis; Pres. (V), Patricia G. West.

Personnel Profile: Full-Time Paid 3; Part-Time Paid 1.

Governing Authority: Tax-exempt.

Institution Type/Description: History Museum: housed in a former school for Black children.

Collections: African American history, culture, & art; personal artifacts; photographs.

Research Fields: contributions of the Black community in Broward County.

Activities: educational programs; Jazz concerts; field trips; outreach workshops; in-house cultural workshops.
Publications: book, My Soul Is A Witness.
Hours & Admission Prices: Mon.-Fri. 11-4; groups by appointment. No charge; donations accepted.
Attendance: 5,148 (accurate)
Membership: Certificate of Appreciation up to $25; Certificate of Membership up to $35; Membership Certificate $36-$999; Donor $1,000 & up.

PURVIS YOUNG MUSEUM, 725 Progresso Dr., Fort Lauderdale, FL 33304. Tel.: 954-765-0721. Facebook: Purvis Young Collectors; Gallery 721.
E-mail: ltc721@comcast.net
Key Personnel: Owner & Dir., Larry T. Clemons
Institution Type/Description: Art Gallery.
Collections: works by Purvis Young, Howard Finster, Thorton Dial, Mose Tolliver & Sybil Gibson; artifacts about Muhammad Ali, Cassius Clay, Elvis Presley; paintings; drawings; news articles; vintage movie posters; photographs.
Hours & Admission Prices: Tues.-Sat. 2-7. Closed all legal holidays. No charge.
Attendance: 1,500 (estimated)

STRANAHAN HOUSE, INC., 335 S.E. 6th Ave., Fort Lauderdale, FL 33301-2256. Tel.: 954-524-4736. Fax: 954-525-2838.
E-mail: info@stranahanhouse.org
Web Site: www.stranahanhouse.org
Founded: 1981.
Congressional District: 22
Key Personnel: Exec. Dir., April Kirk; Pres., Sandra Casteel; Pres. Elect, Leo Hansen; Sec., Doug Smith; Treas., John Abel; Rental Coord. & Gift Shop Mgr., Donna Betz; Education Dir., Marlene Schotanus; Records & Collection Mgr., Deborah Wood; Weekend & Volunteer Mgr., David Greig; Security, John Della Cerra.
Personnel Profile: Full-Time Paid 2; Part-Time Paid 5; Part-Time Volunteers 35; Interns 2.
Governing Authority: nonprofit organization. Stranahan House Inc. Tax-exempt: 501(c)(3).
Institution Type/Description: Historic Building & Site.
Collections: period furnishings; photographs; personal artifacts.
Facilities: rental venue. Handcrafted Florida based items for sale.
Activities: guided tours; education programs for children; school tours; distance learning; lecture series/classes. Museum Sponsors: River Ghost Tours; Victorian Christmas; Pineapple Jam Fundraiser; Mad Hatter Tea Party; Peter Pan Pirate Party.
Publications: quarterly newsletter, The Trading Post.
Hours & Admission Prices: Guided Tours: daily 1, 2, & 3. Adults $12, seniors $11, students $7; special rates available for large groups (reservations required); members no charge. Closed some holidays.
Attendance: 30,000 (estimated)
Membership: Friend $35; Patron $50; Pioneer $100; Trading Partners $250; Preservation Partner $500; Golden Trader $1,000; Friends of Frank & Ivy $5,000.

TERRAMAR VISITORS CENTER - HUGH TAYLOR BIRCH STATE PARK, 3109 E. Sunrise Blvd., Fort Lauderdale, FL 33304-3313. Tel.: 954-564-4521. Fax: 954-762-3737.
Institution Type/Description: History Museum: housed in the former home of Chicago attorney, Hugh Taylor Birch; built in 1940.
Collections: natural & cultural history; personal artifacts; photographs; period furnishings.
Facilities: 180-acre park; nature trails.
Activities: hiking. canoeing; camping; fishing.
Hours & Admission Prices: Call for hours.

Fort Myers

BOB RAUSCHENBERG GALLERY AT EDISON STATE COLLEGE, 8099 College Pkwy., Fort Myers, FL 33919. Tel.: 239-489-9313. Fax: 239-489-9482.
E-mail: jdellinger@edison.edu
Web Site: bobrauschenberggallery.com
Founded: 1979.
Congressional District: 13
Key Personnel: Dir., Jade Dellinger.
Personnel Profile: Full-Time Paid 2; Part-Time Paid 2; Part-Time Volunteers 26; Interns 2.
Governing Authority: state; college. Parent Institution: Edison College. Tax-exempt.

Institution Type/Description: College Art Gallery.
Collections: works of modern & contemporary artists.
Facilities: library; 194-seat auditorium; classrooms; cafeteria.
Activities: guided tours; films; docent program.
Publications: catalogs from past exhibitions available.
Hours & Admission Prices: Mon.-Fri. 10-4, Sat. 11-3. No charge; donations accepted. Closed holidays. &
Attendance: 10,000 (accurate)

CALUSA NATURE CENTER AND PLANETARIUM, 3450 Ortiz Ave., Fort Myers, FL 33905-7811. Tel.: 239-275-3435. Fax: 239-275-9016.
E-mail: webmaster@calusanature.org
Web Site: www.calusanature.org
Founded: 1970.
Congressional District: 13
Key Personnel: Exec. Dir., Mary Rawl; Pres., Phillip Fowler.
Personnel Profile: Full-Time Paid 5; Part-Time Volunteers 150; Interns 2.
Governing Authority: private; nonprofit organization. Tax-exempt: 501(c)(3).
Institution Type/Description: Nature Center.
Collections: natural history; wildlife; native plants; astronomy.
Facilities: 2,023 sq. ft. educational facilities; 13,700 sq. ft. exhibit space; nature & conservation center; planetarium; 90-seat theater; nature trails; intern apartment; aviary. Gift items for sale.
Activities: films; formal educational programs; guided tours; hobby workshops; lectures; participatory exhibits. Museum Sponsors: laser light shows; astronomy shows; telescope viewing; off-site outreach; field trips. Annual Events: Reptile Day; Bug Day; Creepy Crawlie Fair.
Hours & Admission Prices: Mon.-Sat. 10-5, Sun. 11-5. Adults $10, children 3-12 $5; discount to groups & members, children under 3 no charge. &
Attendance: 60,000 (estimated)
Membership: Individual Senior & Student $25; Individual Adult $30; Family $50.

EDISON & FORD WINTER ESTATES, 2350 McGregor Blvd., Fort Myers, FL 33901-3315. Mailing Address: P.O. Box 2368, Fort Myers, FL 33902-2368. Tel.: 239-334-7419. Fax: 239-461-2688.
E-mail: info@edisonfordwinterestates.org
Web Site: www.edisonfordwinterestates.org
Founded: 1947.
Congressional District: 14
Key Personnel: C.E.O., Chris Pendleton; Chm. (V), Steve Niehaus; C.F.O., Tom Hottovy; Exec. Asst., Corinne Cramsey; Public Rels. & Mktg. Mgr., Lisa Sbuttoni; Museum Store Mgr., Patti Wensel; Cur., Alison Giesen.
Personnel Profile: Full-Time Paid 34; Part-Time Paid 26; Part-Time Volunteers 250; Interns 10.
Volunteer Hours: 28,000
Operating Expenses: 5,168,000
Operating Income: 5,304,000
Governing Authority: nonprofit organization. Operated by Thomas Edison and Henry Ford Winter Estate, Inc. Tax-exempt: 501(c)(3).
Institution Type/Description: History & Science Museum.
Collections: historical & scientific artifacts; Edison's & Ford's inventions; manuscripts; banyan tree extending 400 ft. around the trunk; cycads; prototype Model T; 200 Edison phonographs; memorabilia related to Edison's life; historical plants; period cars; Florida history; 9 historic buildings.
Major Exhibits: Edison and Ford in Florida (T), 1/14-12/14; Holiday Nights, 12/14.
Research Fields: history of technology; social & botanical history; Henry Ford & Edison in Fort Myers, their inventions & business.
Facilities: 500-vol. library; botanical garden. Museum-related items for sale.
Activities: guided & self-guided tours; temporary & permanent exhibitions; education programs; living history programs; elder hostel programs.
Publications: Estates Guide Book; newsletters; recipe book; English & German audio tour devices; French, Spanish, & German print tour material.
Hours & Admission Prices: Daily 9-5. Base: adults $20, children 6-12 $11; Guided: adults $25; Curator Led: adults $40; discounts to AAM, ICOM & AAA members, groups, FL Trust, FL Historic Passport & American Horticulture Society members; members & children under 6 no charge. Closed Thanksgiving; Christmas. &
Attendance: 210,000 (accurate)
Membership: Individual $50; Dual $75; Family $85; Extended Family $175; Contributor $250; Patron $500; Benefactor $1,000; Donor $2,500. Business Affiliate $1,000; Partner $2,000 & up; Alliance $5,000 & up; Council $10,000; Leader $25,000.

FLORIDA GULF COAST UNIVERSITY ART GALLERY, 10501 FGCU Blvd. S., FGCU Library, Fort Myers, FL 32965-6565. Tel.: 239-590-7199. Fax: 239-590-7270.
E-mail: asturdiv@fgcu.edu
Web Site: artgallery.fgcu.edu
Founded: 1997.
Congressional District: 14
Key Personnel: Interim Gallery Dir., Anica Sturdivant; Preparator & Studio Mgr., Stephen R. Coe.
Governing Authority: Parent Institution: Florida Gulf Coast University. Tax-exempt.
Institution Type/Description: Art Museum.
Collections: contemporary art; paintings; sculpture.
Activities: special events; receptions; rental facilities. Annual Event: Annual Student Art Exhibition.
Hours & Admission Prices: Mon.-Fri. 10-4, Sat. 11-2; other times by appointment. No charge; donations accepted. &

IMAGINARIUM HANDS-ON MUSEUM & AQUARIUM, (M), 2000 Cranford Ave., Fort Myers, FL 33916-4006. Tel.: 239-321-7420. Fax: 239-344-5915.
E-mail: imag@cityftmyers.com
Web Site: www.imaginariumfortmyers.com
Founded: 1989.
Congressional District: 14
Key Personnel: Gen. Mgr., Matt Johnson; Business Mgr., Shelby Baucom.
Personnel Profile: Full-Time Paid 9; Part-Time Paid 5; Part-Time Volunteers 18.
Governing Authority: Parent Institution: City of Fort Myers. Subsidiary Institution: Imaginarium Group, Inc. Tax-exempt: 501(c)(3).
Institution Type/Description: Science & Technology Center/Museum: housed in 1938 historic building.
Collections: hands-on exhibits.
Facilities: 100-seat theater; conference center; outdoor pavilion.
Publications: newsletter.
Hours & Admission Prices: Mon.-Sat. 10-5, Sun. 12-5. Adults $12, seniors $10, children 3-12 $8; discounts for ASTC members; museum members & children under 3 no charge. Closed Thanksgiving; Christmas. &
Attendance: 90,000 (accurate)
Membership: Individual $55; Family $75; Extended Family $100; Helping Hand $250.

SOUTHWEST FLORIDA MUSEUM OF HISTORY, (M), 2031 Jackson St., Fort Myers, FL 33901. Tel.: 239-321-7430. Fax: 239-344-5914.
E-mail: museuminfo@cityftmyers.com
Web Site: www.swflmuseumofhistory.com
Formerly: Fort Myers Historical Museum
Founded: 1982.
Congressional District: 14
Key Personnel: Gen. Mgr., Matthew H. Johnson; Business Mgr., Shelby Baucom; Public Rels. & Mktg. Mgr., Helena Finnegan; Education Asst., Pam Miner; Visitors Svc., Chuck Smith; Museum Clerk, Gerri Reaves.
Personnel Profile: Full-Time Paid 4; Part-Time Paid 2; Part-Time Volunteers 10.
Governing Authority: municipal government. Parent Institution: City of Ft. Myers. Tax-exempt: 170(b)(1)(A).
Institution Type/Description: Regional History Museum: housed in 1924 ACL Railroad Depot.
Collections: Calusa & Seminole Indian artifacts; Florida fossils; drawings & photographs of Fort Myers area from 1200B.C.-present; 20-ft. dugout canoe; Ethel Cooper Glass Collection; working switchboard; 10,000 photographs; regional history archives; Paleo Indian plants, animals & typography; Land of Giants: Paleo Florida exhibit; scale models of Historic Buildings: 1874 Jacob Summerlin House; 1856 Fort in Fort Myers; 84 ft. long, 101 ton Pullman private rail car Esperanza built in 1929; Cracker House dwelling built in the late 1800s to early 1900s.
Research Fields: Southwest Florida History.
Facilities: 200-vol. library of Southwest Florida history; educational center; Esperanza railcar & Cracker House. Historical books & related items for sale.
Activities: guided group tours; organized education programs for children; docent program; school loan service; participatory, temporary & traveling exhibitions; audio tour. Museum Sponsors: Escorted Day Trips January to April; Downtown Walking Tours January to April; Authors Evenings.
Publications: brochures; newsletters.
Hours & Admission Prices: Tues.-Sat. 10-5. Adults $9.50, seniors $8.50, students $5; discounts to AAA, AAM & ICOM members; children under 3 & members no charge. Closed all major holidays. &

Attendance: 11,000 (accurate)
Membership: Student $15; Individual $25; Family $45; Donor $75; Benefactor $100; Supporting Donor $500; Sustaining Donor $1,000; Patron Donor $5,000; Founder $10,000.

Fort Pierce

A.E. BACKUS MUSEUM OF ART, (M), 500 N. Indian River Dr., Fort Pierce, FL 34950-3080. Tel.: 772-465-0630. Fax: 772-468-6204.
E-mail: info@backusmuseum.com
Web Site: www.backusmuseum.com
Founded: 1960.
Key Personnel: Dir., Kathleen P. Fredrick.
Personnel Profile: Full-Time Paid 2; Part-Time Paid 9; Part-Time Volunteers 260; Interns 1.
Volunteer Hours: 2,195
Governing Authority: Tax-exempt.
Institution Type/Description: Art Museum.
Collections: works by A.E. Backus, Florida Highwaymen, & Indian River School artists.
Major Exhibits: Pop Rocks - Heroes, Icons and the Mundane (T), 3/21/14-5/2/14; Tribute to Backus - Focus on the Indian River Lagoon, 1/15/14-3/7/14; Through the Eye of the Camera, 5/8/14-7/11/14.
Activities: lectures; trips.
Hours & Admission Prices: Wed.-Sat. 10-4, Sun. 12-4. Adults $2; NARM & museum members, children, students & active military no charge.
Attendance: 28,000 (estimated)
Membership: Individual $50; Couple $75; Family $100; Sustainer $250; Premium $500; Patron $1,000.

HEATHCOTE BOTANICAL GARDENS, INC., 210 Savannah Rd., Fort Pierce, FL 34982-3447. Tel.: 772-464-4672. Fax: 772-464-2676.
E-mail: info@heathcotebotanicalgardens.org
Web Site: www.heathcotebotanicalgardens.org
Founded: 1985.
Congressional District: 16
Key Personnel: Exec. Dir., Cynthia Warren; Pres. & Chm. (V), Cris Adams; Vice Pres., Henry Onitveros; Treas., Thomas Jefferson.
Personnel Profile: Full-Time Paid 3; Part-Time Volunteers 250.
Governing Authority: nonprofit organization. Tax-exempt: 501(c)(3).
Institution Type/Description: Botanical Garden.
Collections: James J. Smith tropical bonsai collection; Japanese garden; subtropical palm walk; native plants; herb garden; trees; shrubs; flowers; Olan Ray Creel endangered Florida bromeliads. Historic Buildings: pioneer house; 1922 home office.
Facilities: botanical garden. Museum-related items for sale.
Activities: guided tours; educational classes; rental facility. Annual Events: Fall Garden Festival; Holiday Nonprofit Market; Artist Days; Bloomin' Art & Plant Sale.
Publications: quarterly newsletter; brochures.
Hours & Admission Prices: May-Oct. Tues.-Sat. 9-5; Nov.-April Tues.-Sat. 9-5, Sun. 1-5. Adults $6, seniors $5, children 6-12 $2; discounts to groups; children under 6, members & AHS reciprocal members no charge. Closed major holidays. &
Attendance: 18,000 (estimated)
Membership: Eco-Explorer $20; Individual $35; Family $50; Contributing $100; Photography $175; Generation Green $250; Green Heart Society $500; Royal Palm Society $1,000 & up.

MANATEE OBSERVATION & EDUCATION CENTER, 480 N. Indian River Dr., Fort Pierce, FL 34950. Tel.: 772-466-1600, ext. 333.
E-mail: manatee@manateecenter.com
Web Site: www.manateecenter.com
Founded: 1996.
Personnel Profile: Full-Time Paid 2; Part-Time Paid 2; Part-Time Volunteers 80.
Institution Type/Description: Environmental Education Center.
Collections: manatees; butterfly garden; hands-on exhibits; wildlife observation area.
Facilities: Exhibit hall; classroom area; Vanishing Mermaid Gift Shop; outdoor observation area; boat tours.
Activities: wildlife observation area; educational programs; special events; lectures.
Hours & Admission Prices: July-Sept. Thurs.-Sat. 10-5; Oct.-June Tues.-Sat. 10-5, Sun. 12-4. Admission $1; members & children under 6 no charge. Closed Easter; Thanksgiving; Christmas.

Attendance: 45,000 (estimated)

NATIONAL NAVY UDT-SEAL MUSEUM, (M), 3300 N. Hwy. A1A, North Hutchinson Island, Fort Pierce, FL 34949-8520. Tel.: 772-595-5845. Fax: 772-595-5847.
E-mail: udtsealm@bellsouth.net
Web Site: www.navysealmuseum.com
Founded: 1985.
Congressional District: 16
Key Personnel: Pres. (V), Willard Snyder; Cur., Ruth McSween; Dir. Mktg. & Media, Rolf Snyder; Education Coord., Suzi Howard; Funds Mgr., Marisa Moffett.
Personnel Profile: Full-Time Paid 4; Full-Time Volunteers 2; Part-Time Paid 2; Part-Time Volunteers 24.
Governing Authority: county; nonprofit. UDT/SEAL Museum Association Inc. Tax-exempt.
Institution Type/Description: Military Museum: located on the 1943-1946, site where the Navy first trained Frog Men (Underwater Demolition Teams).
Collections: 1943-present, equipment used by the Underwater Demolition Teams & SEAL Teams; scuba equipment; weapons; boats; uniforms; photos; explosive devices; dioramas of operations from World War II to Desert Storm; present day Navy SEALs.
Research Fields: various operations performed by UDT/SEALs; other countries with similar programs.
Facilities: library of histories & roster of the teams available to the public; 20-seat theater; 4,000 sq. ft. exhibit space; field research station. Gift items for sale.
Activities: guided tours; lectures; films; theater; organized education programs; temporary & traveling exhibitions. Museum Sponsors: UDT/SEAL reunion; Annual "Muster"-Veterans Day Celebration in November.
Publications: quarterly, Fire in the Hole.
Hours & Admission Prices: Jan.-April Mon.-Sat. 10-4; May-Dec. Tues.-Sat. 10-4, Sun. 12-4. Adults $6, children 6-12 $3; discounts to groups; members & children under 5 no charge. Closed New Year's Eve & Day; Martin Luther King Jr. Day; Presidents' Day; Easter; Memorial Day; Independence Day; Labor Day; Thanksgiving & day after; Christmas Eve & Day. ♿
Attendance: 30,000 (estimated)
Membership: Annual $50; Sponsor $100; Lifetime $800.

ST. LUCIE COUNTY REGIONAL HISTORY CENTER, 414 Seaway Dr., Fort Pierce, FL 34949-3138. Tel.: 772-462-1795. Fax: 772-462-1877.
E-mail: quatraroh@stlucieco.gov
Web Site: www.st-lucie.lib.fl.us/museum
Formerly: St. Lucie County Historical Museum
Founded: 1965.
Congressional District: 10
Key Personnel: Dir., Mark DiMascio; Pres. (V), Nancy Bennett.
Personnel Profile: Full-Time Paid 1; Part-Time Volunteers 65.
Governing Authority: county. St. Lucie County Commission. Tax-exempt.
Institution Type/Description: History Museum.
Collections: Seminole Indian artifacts; Spanish treasure fleet, early settler living; archives; manuscripts; maps; photographs; cattle brands, farm tools & equipment; paintings; firefighting equipment; 1907 Florida cracker house.
Research Fields: local historical sites & history.
Facilities: 1,000-vol. research library of history & genealogy books available for use on premises; Memorial Garden; reading room.
Activities: guided tours; permanent, temporary & traveling exhibits; special exhibits & events.
Publications: book, Pictorial History of St. Lucie County, 1565-1910.
Hours & Admission Prices: Wed.-Sat. 10-4, Sun. 1-4; guided tours by appointment. Adults $4, seniors $3.50, students with ID $2.50, children $1.50. Closed county holidays. ♿
Attendance: 15,000 (accurate)

Fort Walton Beach

CITY OF FORT WALTON BEACH HERITAGE PARK & CULTURAL CENTER, (M), 139 Miracle Strip Pkwy., S.E., Fort Walton Beach, FL 32548-5817. Tel.: 850-833-9595. Fax: 850-833-9675.
E-mail: gmeyer@fwb.org
Web Site: www.fwb.org
Formerly: Indian Temple Mound Museum
Founded: 1962.
Congressional District: 1
Key Personnel: Mgr., Gail Lynn Meyer; Volunteer Coord., Ernie Masters; Coord. Programming, Michael Weech; Coord. Operations, Alicia Gardner.

Personnel Profile: Full-Time Paid 3; Part-Time Paid 1; Part-Time Volunteers 16.
Governing Authority: municipal. Parent Institution: City of Fort Walton Beach, 107 Miracle Strip Pkwy., S.W., Ft. Walton Beach, FL 32548. Tax-exempt.
Institution Type/Description: Heritage Park & Cultural Center: Fort Walton Temple Mound National Historic Landmark.
Collections: lithic & ceramic artifacts of local aboriginal origin; ethnologic & replica items; Indian artifacts; Walton Guard during the Civil War; NW Florida one-room schoolhouse artifacts; postal memorabilia. Historic Buildings: Camp Walton Schoolhouse, c.1912; Garnier Post Office, c.1918.
Research Fields: archaeology & related field notes; anthropology; museology; history.
Facilities: 1,200-vol. library of books, pamphlets & quarterlies on archaeology, museology & history available on premises; reading room. Indian handcrafts, replicas, books, pamphlets & postcards for sale.
Activities: guided tours; lectures; films; gallery talks; formally organized education programs for children; Friends of the Museum; special program by arrangement; inter-museum loan; permanent & temporary exhibitions; school loan service.
Publications: booklets, Excavations at the Fort Walton Temple Mound; The Buck Burial Mound; reprint papers, Florida Anthropologist extracts; Pottery of the Fort Walton Period; Fort Walton Beach Heritage Walk.
Hours & Admission Prices: Indian Temple Mounds Museum: June-Aug. Mon.-Sat. 10-4:30; Sept.-May Mon.-Fri. 12-4:30, Sat. 10-4:30. Historic Buildings: June-Aug. Mon.-Sat. 12-4; Sept.-May Mon.-Sat. 1-3. Adults 18 & over $5, seniors 55 & over & military with ID $4.50, children 4-17 $3; discounts to groups of 12 or more; children under 3 no charge. Closed major holidays. ♿
Attendance: 20,000 (estimated)
Membership: Friends of the Museums: Individual $25; Family $35; Patron $100; Life $1,000.

EMERALD COAST SCIENCE CENTER, (M), 139 Brooks St., Fort Walton Beach, FL 32548-5826. Tel.: 850-664-1261. Fax: 850-664-6862.
E-mail: geninfo@ecscience.org
Web Site: www.ecscience.org
Founded: 1989.
Key Personnel: Exec. Dir., James "Jamie" LaFollette.
Personnel Profile: Full-Time Paid 2; Part-Time Paid 8; Part-Time Volunteers 1.
Governing Authority: Tax-exempt.
Institution Type/Description: Science Center.
Collections: hands-on exhibits.
Activities: picnics; parties.
Hours & Admission Prices: June-Aug. Mon.-Sat. 9-6, Sun. 12-6; Sept.-May Mon.-Fri. 9-2, Sat. 9-4, Sun. 12-4. Adults $5.75, seniors $4.75, children $3.75; discounts to groups of 15 or more; ASTC members and children 2 & under no charge. Closed New Year's Day; Independence Day; Thanksgiving; Christmas.
Attendance: 20,000 (accurate)
Membership: Individual $30; Grandparents $45; Family $55; Friend $80.

GULFARIUM MARINE ADVENTURE PARK, 1010 Miracle Strip Pkwy., S.E., Fort Walton Beach, FL 32548. Tel.: 850-243-9046; 800-247-8575. Fax: 850-244-5809. Facebook; Gulfarium Marine Adventure Park.
E-mail: info@gulfarium.com
Web Site: www.gulfarium.com
Formerly: Florida's Gulfarium
Founded: 1955.
Institution Type/Description: Aquarium.
Collections: over 200 marine mammals including dolphins, sea lions, penguins, otters, & fish; birds, reptiles.
Activities: marine animal shows & interactions; educational programs; camps; special events.
Hours & Admission Prices: Call for hours & admission prices.

Gainesville

ALACHUA COUNTY HISTORIC TRUST: MATHESON MUSEUM, INC., (M), 513 E. University Ave., Gainesville, FL 32601-5451. Tel.: 352-378-2280. Fax: 352-378-1246.
E-mail: info@mathesonmuseum.org
Web Site: www.mathesonmuseum.org
Founded: 1994.
Congressional District: 3
Key Personnel: Exec. Dir., Alicia A. Antone; Pres., Barry Baumstein.
Personnel Profile: Full-Time Paid 3; Part-Time Paid 1; Part-Time Volunteers 20; Interns 1.

Governing Authority: private; nonprofit organization. Tax-exempt: 501(c)(3).
Institution Type/Description: History Museum.
Collections: culture & history of Gainesville, Alachua County & North Central Florida region. Historic House: 1867 historic home.
Major Exhibits: A Timeline through History: Celebrating 20 Years of Historic Presentation with the Matheson Museum, 11/13-12/14; Stetson Kennedy Floriiana & Folk Life Exhibit, 5/14-6/14; Battle of Gainesville Exhibit, 7/14-9/14.
Facilities: 2,300-vol. library of Florida history books; archives; 1,600 sq. ft. exhibit space. Museum-related items for sale.
Activities: lectures; rental gallery; temporary & traveling exhibitions; tours of historic towns & neighborhoods; educational & public programs; docent led tours.
Publications: semi-annual newsletter, The Album.
Hours & Admission Prices: By appointment. Suggested Donation: $5 per person. Closed New Year's Day; Martin Luther King Jr. Day; Easter; Memorial Day; Independence Day; Labor Day; Veterans Day; Thanksgiving & day after; Christmas Day & day after. &
Attendance: 19,000 (estimated)
Membership: Student $15; Senior Individual $25; Individual $30; Family $40; Contributor & Corporate $100; Benefactor $250; Sponsor $500; Patron $1,000; Investor $2,500.

∗ FLORIDA MUSEUM OF NATURAL HISTORY, (M), S.W. 34th St. & Hull Rd., Gainesville, FL 32611. Mailing Address: P.O. Box 117800, Gainesville, FL 32611-7800. Tel.: 352-846-2000 & 392-1721. Fax: 352-392-8783 & 846-0253.
E-mail: museuminfo@flmnh.ufl.edu
Web Site: flmnh.ufl.edu
Formerly: Florida State Museum
Founded: 1917.
Congressional District: 4
Key Personnel: Dir. & Cur. Invertebrate Paleontology, Dr. Douglas S. Jones; Assoc. Dir., Beverly Sensbach; Asst. Dir. Exhibits & Public Programs, Dr. Jaret Daniels; Asst. Dir. Budget & Human Resources, Darlene Novak; Assoc. Dir. Collections & Research and Cur. Mammalogy, Dr. David Reed; Dir. Devel., Josh McCoy; Asst. Dir. Public Rels., Paul Ramey; Assoc. Cur. Informatics, Dr. Reed S. Beaman; Assoc. Cur. Vertebrate Paleontology, Dr. Jonathan I. Bloch; Asst. Cur. Herbarium, Dr. Nico Cellinese; Distinguished Research Cur., Spanish Colonial Archaeology, Dr. Kathleen A. Deagan; Assoc. Scientist & Program Dir. Center for Science Learning, Dr. Betty A. Dunckel; Assoc. Cur. Environmental Archaeology, Dr. Katherine F. Emery; Dir. McGuire Center for Lepidoptera & Biodiversity, Dr. Thomas C. Emmel; Cur. Caribbean Archaeology, Dr. William F. Keegan; Cur. Vertebrate Paleontology, Dr. Bruce J. MacFadden; Cur. Paleobotany, Dr. Steven R. Manchester; Cur. FL Archaeology, Dr. William H. Marquardt; Cur. Latin American Art & Archaeology, Dr. Susan Milbrath; Cur. Lepidoptera, Dr. Jacqueline Y. Miller; Cur. Herpetology, Dr. Max A. Nickerson; Cur. Ichthyology, Dr. Lawrence Page; Cur. Marine Malacology, Dr. Gustav Paulay; Cur. Ornithology, Dr. David W. Steadman; Ordway Professor, Dr. Scott K. Robinson; Distinguished Professor, Molecular Systematics & Evolutionary Genetics, Dr. Pamela S. Soltis; Cur. Malacology, Dr. Fred G. Thompson; Asst. Scientist FL Archaeology, Dr. Karen J. Walker; Cur. Botany & Keeper of the Herbarium, Dr. Norris H. Williams; Asst. Cur. Lepidoptera, Dr. Keith R. Willmott; Asst. Cur. FL Archaeology, Dr. Neill Wallis; Museum Shop Mgr., Stacey Crandall.
Personnel Profile: Full-Time Paid 153; Part-Time Paid 157; Part-Time Volunteers 570; Interns 20.
Governing Authority: university bd. of trustees. Parent Institution: University of Florida. Branch Museums: Randell Research Center, Pineland, FL; McGuire Center for Lepidoptera and Biodiversity. Tax-exempt.
Institution Type/Description: Natural History Museum.
Collections: Southeastern U.S. & Caribbean area mammals; birds; bird eggs; fish; bioacoustics; lepidoptera; herbarium; reptiles; amphibians; invertebrate fossils; vertebrate fossils; marine & freshwater mollusks; environmental archaeology; archaeology; ethnology; fossil plants; pollen; physical anthropology.
Research Fields: paleoecology; ecology; anthropology; archaeology; vertebrate paleontology; lepidoptera; conservation; ethnology; environmental archaeology; herpetology; ornithology; systematics and natural history; zoology; ichthyology; mammalogy; invertebrate paleontology; malacology; botany; zoology; paleobotany; marine malacology; molecular systematics.
Facilities: natural history exhibits; butterfly rainforest; classrooms; research laboratories; collectors shop. Educational materials relating to museum programs for sale.
Activities: guided tours, lectures; films; formally organized education programs; docent program or council; inter-museum loan, permanent, temporary & traveling exhibitions; school loan service. Museum Sponsors: summer program for superior students; travel programs.

Publications: Bullen Monographs; Contributions in Anthropology & History; Natural Science Bulletin; Bulletin of the Allyn Museum of Entomology.
Hours & Admission Prices: Mon.-Sat. 10-5, Sun. 1-5. No charge. Butterfly Rainforest & Special Exhibits: call for admission prices. Closed Thanksgiving; Christmas. &
Attendance: 188,500 (accurate)
Membership: Out-of-Town & Student $25; Individual $50; Dual $75; Family & Group $100; Supporting $150; Explorer $250; Benefactor $500; Fellow $1,000; Patron $2,500; Director's Circle $5,000.

HISTORIC HAILE HOMESTEAD AT KANAPAHA PLANTA-TION, 8500 S.W. Archer Rd., Gainesville, FL 32608. Mailing Address: Historic Haile Homestead, Inc., 4941 S.W. 91st Terrace, Ste. 101, Gainesville, FL 32608-9106. Tel.: 352-336-9096.
E-mail: hailedocent@yahoo.com
Web Site: www.hailehomestead.org
Founded: 1992.
Key Personnel: Pres. (V), Karen A. Kirkman.
Personnel Profile: Full-Time Volunteers 1; Part-Time Volunteers 25.
Governing Authority: Tax-exempt.
Institution Type/Description: Historic House Museum: housed in the home of cotton plantation owner, Thomas Evans Haile & his family, known as Talking Walls; built in 1856. Listed on the National Register of Historic Places.
Collections: Haile family history; photographs; over 12,500 words written on the walls by Haile family & friends; period furnishings.
Activities: Annual Events: Homestead Holidays in December; Candlelight Visits in December.
Hours & Admission Prices: Sat. 10-2, Sun. 12-4; other times by appointment. Adults $5; children under 12 no charge. &
Attendance: 2,000 (estimated)
Membership: Kanapaha Partner $20-$99; Craftsmen's Guild $100-$249; Camden Benefactor's Society $250-$499; Bennet's Circle of Friends $500-$999; Haile Heritage Council $1,000 & up.

KANAPAHA BOTANICAL GARDENS, 4700 S.W. 58th Dr., Gainesville, FL 32608-0808. Tel.: 352-372-4981. Fax: 352-372-5892.
E-mail: kbotanical@gmail.com
Web Site: www.kanapaha.org
Governing Authority: Operated By: North Florida Botanical Society.
Institution Type/Description: Botanical Gardens.
Collections: bamboos; herbs; plants.
Facilities: 62-acrea site. Museum-related items for sale.
Activities: weddings; receptions; conferences; special events.
Hours & Admission Prices: Mon.-Wed. & Fri. 9-5, Sat.-Sun. 9am to dusk or 7pm (whichever comes 1st). Adults $6, children 6-13 $3; discounts to groups; children under 6 no charge.

MORNINGSIDE NATURE CENTER, 3540 E. University Ave., Gainesville, FL 32641-6057. Tel.: 352-334-2170. Fax: 352-334-2248.
Founded: 1972.
Congressional District: 3
Key Personnel: C.E.O., Gary A. Paul; Pres. (V), Penny Weber; Nature Operations Mgr., Steve Phillips; Nature Program Coord., George Chappell.
Personnel Profile: Full-Time Paid 10; Part-Time Paid 10; Part-Time Volunteers 5.
Governing Authority: municipal; nonprofit organization. Parent Institution: City of Gainesville. Subsidiary Institution: Dept. of Recreation & Parks. Tax-exempt.
Institution Type/Description: Nature Center: includes working turn-of-the-century farm & five relocated buildings of historic significance for architecture & time period.
Collections: live representatives of Florida native animal species; taxidermied specimens; animal artifacts; period furnishings & reproductions of turn-of-the-century time period; Timucua Indian family compound.
Facilities: 400-vol. library; educational facilities; nature & conservation center.
Activities: arts festivals; docent program; formal educational programs; guided tours; lectures.
Publications: bimonthly newsletter, The Longleaf Pine.
Hours & Admission Prices: Center: daily 9-5. Living History Farm: Sat. only. Adults $2, children $1. Closed New Year's Day; Easter; Memorial Day; Labor Day; Thanksgiving; Christmas. &
Attendance: 30,000 (estimated)
Membership: Friends of Morningside: Adult $10; Family $20; Contributing $50.

* **SAMUEL P. HARN MUSEUM OF ART, (M),** University of Florida, 3259 Hull Rd., Gainesville, FL 32611-2700. Mailing Address: P.O. Box 112700, Gainesville, FL 32611-2700. Tel.: 352-392-9826. Fax: 352-392-3892.
E-mail: coral@harn.ufl.edu
Web Site: www.harn.ufl.edu
Founded: 1990.
Congressional District: 6
Key Personnel: Dir. & Lecturer, Rebecca Martin Nagy; Dir. Devel., Phyllis Delaney; Cofrin Cur. Asian Art, Jason Steuber; Cur. Modern Art, Dulce Roman; Cur. Contemporary Art, Kerry Oliver-Smith; Dir. Finance & Operations, Mary B. Yawn; Cur. African Art, Susan Cooksey; Registrar, Laura Nemmers; Dir. Mktg. & Public Rels., Tami Wroath; Coord. Membership, Pam Clark; Cur. Academic Programs, Eric Segal; Chief Preparator, Michael Peyton; Cur. Community Programs, Bonnie Bernau; Curatorial & Docent Programs Mgr., Brandi Breslin; Museum Rentals Coord., Vicki Tyson; Volunteer Coord., Dana Szafranski; Museum Shop Mgr., Kathryn Rush.
Personnel Profile: Full-Time Paid 34; Part-Time Paid 28; Part-Time Volunteers 85; Interns 72.
Governing Authority: nonprofit; University of Florida. Tax-exempt: 501(c)(3).
Institution Type/Description: Art Museum.
Collections: modern, contemporary, photography, Asian, African, ancient American & Oceanic art.
Research Fields: 19th & 20th-century American art; African art; Asian art; contemporary art and photography.
Facilities: library; 250-seat auditorium; 32,773 sq. ft. exhibit space; study center; object study room; classrooms; cafe. Museum-related items for sale.
Activities: guided tours; lectures; gallery talks; films & videos; concerts; performances; organized education programs for children, adults, undergraduate & graduate students; docent program; school programs; teacher workshops & summer institutes; training programs for professional museum workers; internships; outreach to senior populations; traveling & temporary exhibitions.
Publications: members' bimonthly newsletter; annual report; exhibition brochures & catalogues.
Hours & Admission Prices: Tues.-Fri. 11-5, Sat. 10-5, Sun. 1-5. No charge; donations accepted. Closed New Year's Day; Martin Luther King Jr. Day; Memorial Day; Independence Day; Labor Day; Veterans Day; Thanksgiving & day after; Christmas. ♿
Attendance: 90,977 (accurate)
Membership: Student $25; Individual $50; Dual $75; Associate $125; Fellow $250; Patron $500. Director's Circle: Sustainer $1,000; Benefactor $2,500. Business & Professional Friends: Contributor $250; Donor $500; Leader $1,000; Executive $2,500.

SANTA FE COLLEGE TEACHING ZOO, 3000 N.W. 83rd St., Gainesville, FL 32606-6200. Tel.: 352-395-5601. Fax: 352-395-7365. Facebook: SF Teaching Zoo.
E-mail: michelle.hagan@sfcollege.edu
Web Site: www.sfcollege.edu/zoo
Formerly: Santa Fe Community College Teaching Zoo
Founded: 1970.
Congressional District: 6
Key Personnel: Zoo Dir., Jonathan Miot; Zoo Cur., Kathleen Coyne-Russell; Advisor, Linda Asbell; Chm., Sture Edvardsson; Pres., Jackson Sasser; Asst. Cur. & Exhibit Mgr., Shawn Jacobs; Asst. Cur. & Mgr. Husbandry Training, Shawntal Abram; Asst. Professor, Joshua Watson; CIT Conservation Education Specialist, Tarah Jacobs.
Personnel Profile: Full-Time Paid 10.
Governing Authority: college. Parent Institution: The Santa Fe College. Tax-exempt.
Institution Type/Description: Zoo.
Collections: representative animal collection used in zoo animal technology, program & zoo.
Research Fields: captive reproductive biology.
Facilities: 100-vol. library of zoological material available for research on premises.
Activities: guided tours; formally organized education programs for undergraduate college students affiliated with Santa Fe College; permanent exhibits.
Hours & Admission Prices: Self-Guided Tours: daily 9-2. Guided Tours: Mon.-Fri. 9:30, 10:30 & 11:30 by appointment. Adults $5, seniors 60 & over, UF students and children 4-12 $4; SF students & staff no charge. Closed Thanksgiving Day Christmas Eve & Christmas Day. ♿
Attendance: 46,110 (accurate)

SANTA FE GALLERY, 3000 N.W. 83rd St., Bldg. M, Rm. 147, Gainesville, FL 32606-6210. Tel.: 352-395-5621. Fax: 352-395-4432.
E-mail: jayne.grant@sfcollege.edu
Web Site: dept.sfcollege.edu/vpa/gallery/schedule.html
Founded: 1978.
Congressional District: 6
Key Personnel: Gallery Mgr., Jayne Grant; Gallery Asst., Daniel Barker.
Personnel Profile: Part-Time Paid 1.
Governing Authority: nonprofit. Parent Institution: Santa Fe Community College. Tax-exempt: 501(c)(3).
Institution Type/Description: Art Gallery.
Collections: paintings; sculpture; photographs.
Facilities: 1,800 sq. ft. exhibit space.
Hours & Admission Prices: Mon.-Fri. 12-4. No charge; donations accepted. Closed all college holidays & breaks. ♿
Attendance: 10,000

THOMAS CENTER GALLERIES, 302 N.E. 6th Ave., Gainesville, FL 32601-5476. Mailing Address: P.O. Box 490, Station 30, Gainesville, FL 32627. Tel.: 352-393-8532. Fax: 352-334-2146.
E-mail: etlingrh@cityofgainesville.org
Web Site: www.gvlculturalaffairs.org/website/programs_events/galleries/galleries.html
Founded: 1979.
Key Personnel: Dir. Parks, Recreation & Cultural Affairs Dept., Steven R. Phillips; Nature & Cultural Mgr., Linda Demetropoulos; Coord. Cultural Affairs Programs, Russell Etling.
Governing Authority: municipal government.
Institution Type/Description: Art Gallery.
Collections: paintings; sculpture; drawings.
Activities: rental space available for workshops; lectures; musical performances; receptions; artist lectures.
Publications: brochures; postcards.
Hours & Admission Prices: Mon.-Fri. 8-5, Sat.-Sun. 1-4. No charge. Closed city holidays; New Year's Day; Independence Day; Thanksgiving; Christmas. ♿
Attendance: 10,000 (estimated)

UNIVERSITY GALLERY, (M), University of Florida, 400 S.W. 13th St., Fine Arts Bldg. B, Gainesville, FL 32611. Mailing Address: P.O. Box 115803, University of Florida, Gainesville, FL 32611-5803. Tel.: 352-273-3041. Fax: 352-846-0266.
E-mail: amyv@ufl.edu
Web Site: www.arts.ufl.edu/galleries
Founded: 1965.
Congressional District: 5
Key Personnel: C.E.O. & Dir., Amy Vigilante.
Personnel Profile: Full-Time Paid 2; Part-Time Paid 7; Interns 2.
Governing Authority: university. Parent Institution: University of Florida, College of Fine Arts, School of Art & Art History. Tax-exempt: 501(c)(3).
Institution Type/Description: Art Gallery.
Collections: master of fine arts student collection; contemporary art.
Research Fields: contemporary art.
Facilities: 30,000-vol. library of books & periodicals on general art & architecture available for inter-library loan.
Activities: lectures; gallery talks; inter-museum loan; temporary & traveling exhibitions.
Publications: exhibitions catalogues.
Hours & Admission Prices: Aug. 24-April 26 Tues.-Wed. & Fri. 10-5, Thurs. 10-7, Sat. 12-4. No charge. Closed academic & national holidays. ♿
Attendance: 8,000 (estimated)
Membership: Individual $25; Family $50; Patron $100 & up; Supporter $250; Donor $500; Title Sponsor $1,000 & up.

Geneva

MUSEUM OF GENEVA HISTORY, (M), 165 First St., Geneva, FL 32732. Mailing Address: Geneva Historical & Genealogical Society, Inc., P.O. Box 91, Geneva, FL 32732-0091.
E-mail: genevahgs@aol.com
Key Personnel: Chm. (V), Mary Jo Martin; Pres. (V), Cindy Simonton.
Governing Authority: Parent Institution: Geneva Historical and Genealogical Society, Inc. Tax-exempt.
Institution Type/Description: History Museum.
Collections: Geneva history; period furnishings; personal artifacts; military uniforms.
Publications: newsletter, Village Crier.

Hours & Admission Prices: Oct.-May 2nd & 4th Sun. of month 2-4; other times by appointment. No charge; donations accepted.
Attendance: 750 (accurate)

Green Cove Springs

MILITARY MUSEUM OF NORTH FLORIDA, One Bunker Ave., Green Cove Springs, FL 32043. Tel.: 904-673-7727.
E-mail: militarymuseumnf@aol.com
Web Site: www.militarymuseumofnorthflorida.com
Founded: 2003.
Congressional District: 6
Key Personnel: Pres., Bud Nelson; Exec. Dir. & Cur., Herb Steigelman; Treas., Cindy Pontrelli; Museum Shop Mgr., Joe Sedlick.
Personnel Profile: Part-Time Volunteers 16.
Volunteer Hours: 3,939
Operating Expenses: 13,280
Operating Income: 14,811
Governing Authority: Tax-exempt.
Institution Type/Description: Military Museum.
Collections: military vehicles & artifacts; photographs; uniforms; weapons; equipment.
Research Fields: military history.
Facilities: library.
Activities: special events; field trips; birthday parties. Museum Sponsors: Armed Forces Day Car Show in May; Memorial Day Ceremony; Railroad Days.
Publications: quarterly newsletter, Mail Call.
Hours & Admission Prices: Wed.-Sat. 10-3, Sun. 12-4. No charge; donations accepted. ♿
Attendance: 1,100 (estimated)

Gulf Breeze

GULF ISLANDS NATIONAL SEASHORE, 1801 Gulf Breeze Pkwy., Gulf Breeze, FL 32563-5000. Tel.: 850-934-2600. Fax: 850-932-9654.
E-mail: guis_information@nps.gov
Web Site: www.nps.gov/guis/
Founded: 1971.
Congressional District: 1
Key Personnel: Supt., Daniel R. Brown; Deputy Supt., Steven McCoy; Chief of Interpretation, Susan Teel; Florida Dist. Interpreter, Stanley Lawhead; Chief Science & Resources Management, Rick Clark; Cultural Resources Program Mgr., David Ogden; Admin. Officer, Dianne Westfaul; Contract Specialist, Evans Ward.
Governing Authority: federal. National Park Services, Dept. of the Interior, Washington, DC (See the listings for Fort Pickens Area-Gulf Islands National Seashore, Santa Rosa Island, FL & William M. Colmer Visitor Center, Ocean Springs, MS for additional information.)
Institution Type/Description: National Park & Historic District: barrier islands stretch from Cat Island in Mississippi, 240 kilometers (160 miles) eastward to the far end of Santa Rosa Island in Florida.
Collections: A portion of the Gulf Islands National Seashore collection was lost due to Hurricane Ivan in 2004 & Hurricane Katrina in 2005. The surviving collection was temporarily transferred to Timucuan Ecological and Historic Preserve in Jacksonville, FL. William M. Colmer Visitor Center with natural history exhibits at Ocean Springs, MS. Historic Military Posts: 1834 Ft. Pickens, FL.; 1844 Ft. Barrancas, FL. (located at Pensacola Naval Air Station); 1859 Advanced Redoubt, FL. (located on board NAS Pensacola;) 1866 Ft. Massachusetts, Ship Island, MS.
Research Fields: history; natural history; archeology.
Facilities: campsites; visitor centers; picnic areas; nature trails; 40-seat auditorium. Books & publications for sale.
Activities: guided tours; films; self guided trails; boat tours; concession boat trip to West Ship Island from Gulfport, MS (April-Oct.) swimming & scuba diving; fishing; formally organized education program for children; volunteer program; permanent exhibitions; talks.
Publications: newsletter, The Gulf Islands Guide; history & natural history pamphlets; folder, Gulf Islands.
Hours & Admission Prices: Entrance fees collected at Perdido Key & Ft. Pickens. All centers & facilities no charge. Closed Thanksgiving, Christmas & New Year's Day. ♿

THE ZOO, 5701 Gulf Breeze Pkwy., Gulf Breeze, FL 32563-9553. Tel.: 850-932-2229. Fax: 850-932-8575.
E-mail: annzoonwf@bellsouth.net
Web Site: www.thezoonorthwestflorida.org
Founded: 1984.
Key Personnel: Exec. Dir., Danyelle Lantz; Gen. Cur., Diane Norris; Dir. Operations, Terry Whitman; Dir. Visitor Svcs., Christi Maeda; Dir. Education, Susan Leveille.
Personnel Profile: Full-Time Paid 30; Part-Time Paid 30; Part-Time Volunteers 25.
Governing Authority: nonprofit. Tax-exempt: 501(c)(3).
Institution Type/Description: Zoo & Botanical Garden.
Collections: over 700 animals; hundreds of birds; flowers; exotic plants; children's zoo.
Facilities: 300-vol. library of zoo & zoo-related books & magazines; 200-seat outdoor amphitheatre; 50-acre zoo; snack bar. Stuffed animals, zoological souvenirs, books, gifts & tee shirts for sale.
Activities: films; docent program; formal educational programs; mobile vans; loan, participatory & traveling exhibitions; broadcast programs; adopt-an-animal program; African photo safaris; zoo camp for kids. Zoo Sponsors: Zoo Lights in December.
Publications: biannual newsletter, The ZOOsletter
Hours & Admission Prices: Summer: daily 9-5; Winter: daily 9-4. Adults $11.50, senior citizens $10.50, children 3-11 $8.25; discounts to groups; children under 3 no charge. Closed Thanksgiving; Christmas. ♿
Attendance: 175,000
Membership: Individual $40; Grandparent & Family $75.

Hollywood

ART AND CULTURE CENTER OF HOLLYWOOD, 1650 Harrison St., Hollywood, FL 33020-6806. Tel.: 954-921-3274. Fax: 954-921-3273.
E-mail: info@artandculturecenter.org
Web Site: www.artandculturecenter.org
Founded: 1978.
Congressional District: 20
Key Personnel: C.E.O. & Exec. Dir., Joy Satterlee; Chm. (V), John Stengel; Bookkeeper, Elizabeth Veszi; Asst. Dir., Susan Rakes; Mktg. & Design Mgr., Alesh Houdek; Technical Dir., Joseph Popejoy; Theater Mgr., Robert Halpern; Dir. Devel., Jeff Rusnak; Visitor Services, Elise Viola; Public Rels. & Membership Mgr., Charmain Yobbi; Cur. Exhibitions, Jane Hart.
Personnel Profile: Full-Time Paid 10; Part-Time Paid 9; Part-Time Volunteers 277; Interns 6.
Volunteer Hours: 4,578
Operating Expenses: 1,225,000
Operating Income: 1,225,000
Governing Authority: private; nonprofit organization. Parent Institution: Hollywood Art and Culture Center, Inc. Tax-exempt: 501(c)(3).
Institution Type/Description: Multi-disciplinary Arts Center
Collections: contemporary paintings & sculpture; ethnographic arts.
Major Exhibits: Abracadabra: Seventh Annual Exhibition and Fundraiser, 1/25/14-3/14/14; Virginia Fifield, 1/25/14-3/16/14; Agustina Woodgate: The Skin Rug Collection, 3/29/14-5/25/14; Johnny Robles, 3/29/14-5/25/14; Nathan Sawaya: The Art of the Brick - In Pieces (T), 6/8/14-8/17/14; Group Show Curated by Robert Adanto- New Orleans Artists' Post-Katrina Work, 6/8/14-8/17/14; Beatriz Monteararo: Soundscapes, 9/6/14-11/2/14; Dave Muller: Too Old to Rock 'n' Roll, 11/15/14-1/18/15; Bhakti Baxter, 11/15/14-1/18/15.
Research Fields: pertaining to the permanent collection & exhibitions.
Facilities: library of art history available for use by special permission; reading room; 500-seat auditorium.
Activities: guided tours; lectures; films; gallery talks; formally organized education programs for children & adults; docent program; permanent exhibitions; summer arts camp; private lessons-voice, guitar, piano; flute.
Publications: catalog.
Hours & Admission Prices: Tues.-Fri. 10-5, Sat.-Sun. 12-4. Adults $7, seniors & students $4; discounts to AAA members & groups of 10 or more; Miami-Dade & Broward teachers, firefighters, police, members and children 3 & under no charge. Closed New Year's Day; Easter; Independence Day; Thanksgiving; Christmas. ♿
Attendance: 61,769 (accurate)
Membership: Discount $30; Individual $50; Family $75; Friend $150; Fellow $250; Patron $500; Benefactor $1,000.

HOLOCAUST DOCUMENTATION & EDUCATION CENTER, INC., 2031 Harrison St., Hollywood, FL 33020-5019. Tel.: 954-929-5690. Fax: 954-929-5635.
E-mail: info@hdec.org
Web Site: www.hdec.org
Founded: 1979.
Congressional District: 17
Key Personnel: Exec. Vice Pres., Rositta E. Kenigsberg; Pres. (V), Harry A. (Hap) Levy; Dir. Educational Outreach, Merle Saferstein; Documentation Dept. Coord., Rita Hofrichter.

Personnel Profile: Full-Time Paid 5; Full-Time Volunteers 2; Part-Time Paid 2; Part-Time Volunteers 350.
Governing Authority: nonprofit organization. Tax-exempt: 501(c)(3).
Institution Type/Description: Holocaust Educational Resource Center.
Collections: video & audio tapes of survivors, liberators & protectors of the Holocaust; World War II & Holocaust memorabilia; paintings.
Research Fields: oral history; prejudice-reduction.
Activities: films; broadcast programs; organized education programs for children, adults & undergraduate & graduate college students; docent program for community exhibits; participatory exhibits; interviewer training; teachers' seminars; visual arts & writing contest for students. Center Sponsors: Holocaust Remembrance; Student Awareness Days; Teachers' Institute on Holocaust Studies; interviews of survivors & eyewitnesses of the Holocaust.
Publications: State of Florida Task Force on Holocaust Education Resource Manuals for grades 4-6; annual newsletter.
Hours & Admission Prices: Mon.-Fri. 9-5. No charge. Closed national & Jewish holidays. &
Membership: Individual $36; Supporting $100; Sustaining $250; Patron $500; Benefactor $1,000; Founder $1,500.

Homeland

HOMELAND HERITAGE PARK, 249 Church Ave., Homeland, FL 33847. Tel.: 863-534-3766. Fax: 863-519-8665.
E-mail: danielgornoski@polk_county.net
Founded: 1985.
Key Personnel: Dir., Gary Hacking; Coord., Daniel Gornoski.
Personnel Profile: Full-Time Paid 1; Part-Time Volunteers 4.
Governing Authority: Parent Institution: Polk County Bd. of County Commissioners.
Institution Type/Description: Historical Park, Sites & Buildings.
Collections: local history & culture; period furnishings; personal artifacts. Historic Buildings: Homeland School, 1878; Old Homeland Methodist Church, 1887; Raulerson House, 1880; English family log cabin & barn, 1888.
Facilities: 5-acre site; rental facilities.
Activities: rental facilities; educational tours K to senior citizens.
Hours & Admission Prices: Mon.-Fri. 8-5, Sat. by appointment. &
Attendance: 15,000 (estimated)

Homestead

BISCAYNE NATIONAL PARK, 9700 S.W. 328th St., Homestead, FL 33033-5634. Tel.: 305-230-1144. Fax: 305-230-1190.
Web Site: www.nps.gov/bisc
Founded: 1968.
Congressional District: 20
Key Personnel: Supt., Mark Lewis; Park Science Coord., Richard Curry; Admin. Officer, Nancy Sanchez; Agent-Florida National Parks & Monument Association & Visitor Center Coord., Bob Showler; Chief Information Officer, Susan Gonshor.
Governing Authority: federal. Parent Institution: Dept. of the Interior, Washington, DC. Tax-exempt.
Institution Type/Description: National Park & Nature Center.
Collections: natural history objects related to the ecology of the park; lobster sanctuary; crabs; seashells; plants; turtles; birds; sponges.
Research Fields: oceanography; marine biology; terrestrial biology; archaeology.
Facilities: 2,280-vol. library of books, journals, reports & documents all pertaining to the history of the park available for research on premises; on-site use or loan of duplicate copies to recognized research organizations; 180,000 acres of tropical marine & subtropical terrestrial natural area; field research station. Publications on natural history for sale.
Activities: guided tours; lectures; formally organized education programs for children; temporary & traveling exhibitions; glass bottom boat; snorkel & scuba tours.
Publications: Seascape, biannual park newsletter.
Hours & Admission Prices: Convoy Point Visitor Center: daily 7-5:30. Dante Fascell Visitor Center: daily 9-5. No charge. Call 305-230-7275 for information. &
Attendance: 500,000 (estimated)

EVERGLADES NATIONAL PARK, 40001 State Rd. 9336, Homestead, FL 33034-6733. Tel.: 305-242-7826 & 7700. Fax: 305-242-7711. TDD: 305-242-7826.
E-mail: nancy_russell@nps.gov
Web Site: www.nps.gov/ever
Founded: 1947.
Congressional District: 20

Key Personnel: Supt., Dan Kimball; Cur., Nancy Russell.
Personnel Profile: Full-Time Paid 2; Part-Time Paid 6; Part-Time Volunteers 1; Interns 3.
Governing Authority: federal. U.S. Dept. of the Interior, National Park Service, Washington, DC 20240. Tax-exempt.
Institution Type/Description: National Park; pre & early settlement, southern Florida.
Collections: herbarium; birds; mammals; fish; insects; corals; tree snails; historical & archaeological artifacts; research archives; 60,000 natural history specimens.
Research Fields: botany; wildlife; ecology; archaeology; history; hydrology.
Facilities: 5,000-vol. library including 4,000 photographs pertaining to natural science available for research on premise; South Florida Research Center Archives; visitor centers: Parachute Key & Flamingo; field research station; 100-seat auditorium. Books on park subjects & other museum related items for sale.
Activities: guided tours; lectures; formally organized education programs for children; permanent & temporary exhibitions.
Publications: biannual park newspaper.
Hours & Admission Prices: Visitor Center's: daily 8-5. Park: $20 per car; senior citizens with Golden Age Passport no charge. &
Attendance: 1,000,000 (estimated)
Membership: Annual Pass $50.

Hutchinson Island

ELLIOTT MUSEUM, (M), 825 NE Ocean Blvd., Hutchinson Island, FL 34996. Tel.: 772-225-1961. Fax: 772-225-2333.
E-mail: info@elliottmuseumfl.org
Web Site: www.elliottmuseumfl.org.
Founded: 1961.
Congressional District: 12
Key Personnel: Chm. (V), Walter Woods; Dir., Jennifer Esler; Dir. Devel., Diane Kimes; Dir. Finance, Amy Martin; Cur., Janel Hendrix.
Personnel Profile: Full-Time Paid 7; Part-Time Paid 5; Part-Time Volunteers 30.
Governing Authority: nonprofit organization. Parent Institution: Historical Society of Martin County. Branch Museum: House of Refuge Museum at Gilbert's Bar. Tax-exempt: 501(c)(3).
Institution Type/Description: General Museum.
Collections: changing exhibit gallery; Wheels of Change; baseball history; art.
Research Fields: Florida & Martin County history.
Facilities: art gallery available for private functions by rental agreement. Gifts & books for sale.
Activities: permanent & temporary exhibitions.
Publications: History of Martin County.
Hours & Admission Prices: Re-opens in March. Adults $12, Seniors $10, children $6, children under 2 no charge. &
Attendance: 50,000 (estimated)
Membership: Student $25; Individual $50; Family $100; Patron $150.

Indian Rocks Beach

GULF BEACH ART CENTER, 1515 Bay Palm Blvd., Indian Rocks Beach, FL 33785-2827. Tel.: 727-596-4331. Fax: 727-596-4331.
E-mail: arts1515@gmail.com
Web Site: beachartcenter.org
Founded: 1978.
Congressional District: 8
Key Personnel: Exec. Dir., Jo Ann Marianne; Pres. (V), Lynda Hamlett.
Personnel Profile: Full-Time Paid 1; Part-Time Paid 1; Part-Time Volunteers 14.
Governing Authority: nonprofit organization. Tax-exempt: 501(c)(3).
Institution Type/Description: Art Gallery & School.
Collections: paintings; sculpture drawings.
Facilities: 300-vol. library of artists & painting material available to members; 1,400 sq. ft. exhibit space; educational facilities.
Activities: arts festivals; workshops; organized education programs for children and adults; temporary exhibitions.
Publications: newsletter.
Hours & Admission Prices: Sept.-June Mon.-Thurs. 9-4, Fri. 9-12. No charge; donations accepted. Closed Thanksgiving; Christmas. &
Membership: Individual $50; Family & Patron $65; Sponsor $250.

Inverness

THE OLD COURTHOUSE HERITAGE MUSEUM, One Courthouse Square, Inverness, FL 34450-4808. Tel.: 352-341-6428. Fax: 352-341-6445.
Formerly: The Museum of Citrus County History-Old Courthouse

Founded: 1985.
Congressional District: 6
Key Personnel: Pres. Historical Society, John Grannan; Pres. Museum, Linda Bega; Vice Pres., Beverly Drinkhouse; Dir. of Historical Resources, Kathy Turner.
Governing Authority: county. Parent Institution: Citrus County Historical Society, Inc. Tax-exempt: 501(c)(3).
Institution Type/Description: History Museum: housed in 1911 Old Courthouse.
Collections: county archives 1885-current; pioneer settlers; photographs; diaries; manuscripts.
Research Fields: pioneer settlement.
Facilities: archival repository available for use by public. Museum-related items & gifts for sale.
Activities: guided tours; docent program.
Publications: bimonthly newsletter, At Home; brochures; calendars; book, Back Home.
Hours & Admission Prices: Museum: Mon.-Fri. 10-4. Historical Resource: Mon.-Fri. 10-5. No charge; donations accepted. &
Attendance: 20,000 (estimated)
Membership: Student (Grades 1-12) $1; Individual $10; Household $15; Organization $20; Life $250; Benefactor $500.

Islamorada

HISTORY OF DIVING MUSEUM, (M), 82990 Overseas Hwy., Islamorada, FL 33036-3600. Tel.: 305-664-9737. Facebook: The History of Diving Museum.
E-mail: info@divingmuseum.org
Web Site: www.divingmuseum.org
Formerly: Florida Keys History of Diving Museum
Founded: 2005.
Congressional District: 18
Key Personnel: Pres. Bd., Dr. Sally E. Bauer; Exec. Dir., Thomas A. Lockyear, II; Mgr. Operations & Dir. Mktg., Courtney Dutton.
Personnel Profile: Full-Time Paid 2; Full-Time Volunteers 2; Part-Time Paid 2; Part-Time Volunteers 12.
Governing Authority: Tax-exempt: 501(c)(3).
Institution Type/Description: Diving Museum.
Collections: diving history; photographs; period artifacts; diving equipment; research documents; sunken treasure; Florida history; books.
Facilities: research library; conference room.
Publications: quarterly newsletter, Bubbles in History.
Hours & Admission Prices: Daily 10-5. Adults $12, children 5-12 $6; discount to AAM & ICOM members; AASLH members, members & children under 5 no charge. Closed New Year's Day; Thanksgiving; Christmas. &
Attendance: 10,000 (estimated)
Membership: Basic $35; Dual $50; Family $150.

Jacksonville

ALEXANDER BREST MUSEUM AND GALLERY, 2800 University Blvd. N., Jacksonville, FL 32211-3321. Tel.: 904-256-7374. Fax: 904-256-7375.
Founded: 1977.
Congressional District: 3
Key Personnel: Dir. & Cur., Prof. Jack Turnock.
Personnel Profile: Full-Time Paid 1; Part-Time Paid 1; Interns 3.
Governing Authority: private university; nonprofit. Parent Institution: Jacksonville University. Tax-exempt.
Institution Type/Description: University Art Museum.
Collections: decorative arts; ivory; Steuben glass; Chinese porcelain & cloisonne; Tiffany glass; Boehm porcelains; Persian carpets; pre-Columbian art & artifacts.
Facilities: 4,250 sq. ft. exhibit space.
Activities: guided tours; lectures; docent program; concerts; temporary exhibits.
Publications: A Guide to the Collection: The Alexander Brest Museum & Gallery.
Hours & Admission Prices: Mon.-Fri. 9-4:30, Sat. 12-5. No charge. Closed Memorial Day; Independence Day; Labor Day; Veterans Day; Thanksgiving; Christmas. &
Attendance: 12,000 (estimated)
Membership: Friends of Fine Arts $35 & up.

*** CUMMER MUSEUM OF ART & GARDENS, (M),** 829 Riverside Ave., Jacksonville, FL 32204-3336. Tel.: 904-356-6857. Fax: 904-353-4101.
Web Site: www.cummer.org
Founded: 1958.

Congressional District: 3
Key Personnel: Dir., Hope McMath; Chm., John Donahoo, III; Chief Cur., Holly Keris; Dir. Finance, Wendy Steve; Registrar, Kristen Zimmerman; Dir. Education, Lynn Norris; Museum Shop Mgr., Susan Tudor.
Personnel Profile: Full-Time Paid 31; Part-Time Paid 6; Part-Time Volunteers 6,205; Interns 24.
Governing Authority: nonprofit organization. A branch of The DeEtte Holden Cummer Museum Foundation. Tax-exempt: 509(a); 4942(j)(3).
Institution Type/Description: Art Museum & Gardens.
Collections: European & American paintings, sculpture; works on paper & decorative arts; early Meissen porcelain; Asian art; ancient art.
Major Exhibits: One Family: Photographs by Vardi Kahana, 10/2/13-2/23/14; Modern Dialect: American Paintings from the John & Susan Horseman Collection (T), 11/13/13-1/26/14; The Art of Empathy, 11/26/13-2/16/14; The Human Figure: Sculptures by Enzo Torcoletti, 9/21/13-930/14; The Prints of William Walmsley, 10/29/13-7/8/14; Our Shared Past, 12/17/13-5/25/14; Our Family Photographs by Vardi Kahana (T), 1/25/14-4/27/14; A Commemoration of the Civil Rights Movement (T), 2/28/14-10/15/14; Collector's Choice: Inside the Hearts and Minds of Regional Collectors, 5/17/14-9/14/14; American Art from the Permanent Collection, 8/5/14-7/15.
Research Fields: early Meissen porcelain; American art.
Facilities: 12,000-vol. library of art books; historic gardens. Art books, catalogs, postcards & art objects for sale.
Activities: guided tours; lectures; concerts; permanent, temporary & traveling exhibitions; interactive educational center; art studios.
Publications: The Wark Collection of Early Meissen Porcelain, 1984; collection Catalogue (2000), The Cummer Museum of Art & Gardens; Vision 2002: The Cummer Contemporary; Passion and Clarity: The Art of Joseph Jeffers Dodge; A Legacy in Bloom: Celebrating a Century of Gardens at The Cummer (2008); Early Meissen Porcelain in the Wark Collection at The Cummer Museum of Art & Gardens (2011); Eugene Savage: The Seminole Paintings (2011); The Art of Empathy: The Mother of Sorrows in Northern Renaissance Art & Devotion David S. Areford (2013).
Hours & Admission Prices: Tues. 10-9, Wed.-Fri. 10-4, Sat. 10-4, Sun. 12-4. Adults $10, senior citizens, military & students $6; discounts to AAM & ICOM members; children 5 & under, members, Tues. 4-9 & Weaver first Sat. each month no charge. Closed major holidays. &
Attendance: 130,667 (accurate)
Membership: Seniors, Military & Educators 10% off any level; Individual $50; Dual $75; Family $100; Associate $250; Sponsor $500; Ponce de Leon $1,000 & up; Benefactor's Circle $5,000; Director's Circle $10,000.

FLORIDA COMMUNITY COLLEGE, KENT CAMPUS MUSEUM/GALLERY, 3939 Roosevelt Blvd., Jacksonville, FL 32205-8946. Tel.: 904-381-3674.
E-mail: Kwarren@fccj.org
Web Site: www.fccj.org
Founded: 1971.
Key Personnel: Dir., Kelly Warren; Support Specialist, Angie Fogle.
Personnel Profile: Part-Time Paid 4.
Governing Authority: college. Parent Institution: Florida Community College at Jacksonville, 501 W. State St., Jacksonville, FL 32202. Tax-exempt.
Institution Type/Description: Art Museum.
Collections: paintings; sculpture.
Facilities: 80-seat auditorium.
Activities: guided tours; lectures; gallery talks; arts festivals; TV programs; temporary & traveling exhibitions.
Publications: Kalliope
Hours & Admission Prices: Mon.-Thurs. 10-4, Fri. 10-3. No charge. Closed holidays. &
Attendance: 1,200 (estimated)

JACKSONVILLE FIRE MUSEUM, 1406 Gator Bowl Blvd., Jacksonville, FL 32202-1310. Tel.: 904-630-0618. Fax: 904-630-4202.
E-mail: taylorw@coj.net
Web Site: www.jacksonvillefiremuseum.com
Founded: 1974.
Congressional District: 3
Key Personnel: Dir., Martin Senterfitt; Cur. & Admin., Wyatt Taylor.
Personnel Profile: Full-Time Paid 1.
Governing Authority: municipal; nonprofit. Tax-exempt: 501(c)(3).
Institution Type/Description: Fire Museum: housed in a historic fire station constructed in 1902. Listed on the National Registered Historical Building.
Collections: antique fire apparatus & artifacts on fire fighting history; history of fire fighting in the city of Jacksonville, Florida.
Facilities: 35-vol. library of fire station log books & accounting records from 1902; 1,200 sq. ft. exhibit space.
Activities: guided tours; fire safety training for school children.
Hours & Admission Prices: Mon.-Fri. 9-4; groups by appointment. No charge; donations accepted.

Attendance: 60,000 (estimated)

JACKSONVILLE MARITIME HERITAGE CENTER, 2 Independent Dr., Ste. 162, Jacksonville, FL 32202-5031. Tel.: 904-355-1101. Fax: 904-355-1106.

E-mail: jaxmarmus@bellsouth.net
Web Site: www.jacksonvillemaritimeheritagecenter.org
Formerly: Jacksonville Maritime Museum Society
Founded: 1985.
Congressional District: 3
Key Personnel: Chm. (V), William C. Sandberg; Dir. & Museum Shop Mgr., Catherine M. Krueger.
Personnel Profile: Part-Time Paid 1; Part-Time Volunteers 10; Interns 2.
Governing Authority: Parent Institution: Jackson Maritime Museum Society. Tax-exempt.
Institution Type/Description: Maritime Museum.
Collections: Jacksonville maritime history; model ships; paintings; photographs; period artifacts.
Publications: monthly electronic newsletter.
Hours & Admission Prices: Tues.-Fri. 11-5; Sat.-Sun. 1-5. No charge; donations accepted. &
Attendance: 7,000 (estimated)
Membership: Individual $25; Family $50; Sustaining $100; Patron $500; Golden Patron $1,000; Corporate $1,500-$2,500.

JACKSONVILLE ZOO AND GARDENS, (M), 370 Zoo Pkwy., Jacksonville, FL 32218-5799. Tel.: 904-757-4463, ext. 210. Fax: 904-757-4315.

E-mail: info@jacksonvillezoo.org
Web Site: www.jacksonvillezoo.org
Formerly: Jacksonville Zoological Gardens
Founded: 1914.
Congressional District: 3
Key Personnel: Exec. Dir., Tony Vecchio; Museum Shop Mgr., Tena Barnidge; Volunteer, Crystal Bienvenue.
Personnel Profile: Full-Time Paid 156; Part-Time Paid 64; Part-Time Volunteers 638; Interns 2.
Governing Authority: nonprofit society. Jacksonville Zoological Society. Tax-exempt: 501(c)(3).
Institution Type/Description: Zoo and Garden.
Collections: mammals; birds; reptiles; amphibians; invertebrates; fish; plants.
Research Fields: field zoology.
Facilities: auditorium; outdoor picnic area; 2 restaurants; 2 snack shops; gardens; meeting facilities. Gift items for sale.
Activities: organized school programs & tours; teacher & family workshops; speakers bureau; docent program or council; sleepovers; after dark programs; night tours; stroller tours; preschool programs; outreach; behind the scenes tour; day camps; after-hours rentals; garden tours.
Publications: quarterly magazine, Wild; member e-newsletter, Animals in Your Box.
Hours & Admission Prices: Daily 9-5. Adults $14.95, seniors $12.95, children $9.95; discounts to groups, military, AAA & zoo members; members & children under 3 no charge. Closed Christmas. &
Attendance: 801,500 (estimated)
Membership: Nanny $40; Individual $45; Couple & Friend $75; One Adult Family $90; Family $115; One Adult Family Plus One $125; Family Plus One $145; Family Plus Two $180.

KINGSLEY PLANTATION, 11676 Palmetto Ave., Jacksonville, FL 32226-2449. Mailing Address: 13165 Mt. Pleasant Rd., Jacksonville, FL 32225-1227. Tel.: 904-251-3537. Fax: 904-251-3577.

E-mail: timu_interpretation@nps.gov
Web Site: www.nps.gov/timu
Founded: 1991.
Congressional District: 4
Key Personnel: Supt., Barbara Goodman; Museum Cur., Anne Lewellen.
Personnel Profile: Full-Time Paid 25; Part-Time Paid 5; Part-Time Volunteers 15.
Governing Authority: federal. Parent Institution: National Park Service. Subsidiary Institution: Timucuan Ecological & Historic Preserve, 13165 Mt. Pleasant Rd., Jacksonville, FL 32225. Tax-exempt.
Institution Type/Description: Historic Site.
Collections: history; archaeological. Historic Buildings: 19th century Kitchen House; Carriage House & slave cabins; 18th century Plantation House.
Research Fields: plantation operations; slavery.
Facilities: interpretive garden. Bookstore.
Activities: guided tours; permanent exhibitions; formal education programs for children.

Publications: book, Junior Ranger Activity Book.
Hours & Admission Prices: Daily 9-5. No charge; donations accepted. Closed New Year's Day; Thanksgiving; Christmas. &
Attendance: 73,500 (estimated)

MANDARIN MUSEUM AND HISTORICAL SOCIETY, (M), 11964 Mandarin Rd., Jacksonville, FL 32223-1339. Mailing Address: P.O. Box 23601, Jacksonville, FL 32241-3601. Tel.: 904-268-0784. Fax: 904-268-0752.

E-mail: mandarinmuseum@bellsouth.net
Web Site: mandarinmuseum.net
Founded: 1992.
Congressional District: 4
Key Personnel: Dir., Anne Morrow; Pres., Sandra Arpen
Governing Authority: Subsidiary Institution: Walter Jones Historical Park, 11964 Mandarin Rd., Jacksonville, FL; Store & Post Office, 12471 Mandarin Rd., Jacksonville, FL. Tax-exempt.
Institution Type/Description: Historic Buildings & Park.
Collections: Mandarin Store & Post Office: period furnishings; local history; Timacua artifacts; Steamboat Maple Leaf artifacts. Walter Jones Historical Park: 1800s homestead; 1875 farmhouse; 1876 barn & sawmill.
Research Fields: historical Florida.
Activities: lectures; educational programs; family events; school groups.
Publications: newsletter; Adventures in Mandarin 1876-1886.
Hours & Admission Prices: Museum: Sat. 9-4. Park: daily 7am-10pm. Store & Post Office: 1st & 3rd Sun. 1-3. No charge; donations accepted. Closed holidays. &
Attendance: 900 (accurate)
Membership: Orange Picker $35; May $50; Maple Leaf $100; Walter Jones Society & Stowe Society $500.

MUSEUM OF CONTEMPORARY ART JACKSONVILLE, (MOCA JACKSONVILLE), 333 N. Laura St., Jacksonville, FL 32202-3505. Tel.: 904-366-6911. Fax: 904-366-6901.

E-mail: info@mocajacksonville.org
Web Site: www.mocajacksonville.org
Formerly: Jacksonville Museum of Modern Art
Founded: 1924.
Congressional District: 3
Key Personnel: Dir., Marcelle Polednik, Ph.D.; Chm. (V), Alan Howard; Cur., Ben Thompson; Assoc. Dir. Education, Allison Galloway; Museum Shop Mgr., Heather Logan.
Personnel Profile: Full-Time Paid 20; Part-Time Paid 24; Interns 8.
Governing Authority: nonprofit. Parent Institution: University of North Florida. Tax-exempt: 501(c)(3).
Institution Type/Description: Contemporary Art Museum.
Collections: emphasis on works from 1960 to present include paintings, printmaking, sculpture, photography; time based media.
Major Exhibits: Unseen Images, Untold Stories: The Lives of LBGT Elders in NE Florida, 12/8/13-1/26/14; Project Atrium: Ingrid Calame, 11/16/13-3/9/14; Material Transformations (T), 1/25/14-4/6/14; FATE (Foundation in Art Theory and Education) Juried Exhibition, 1/28/14-3/2/14; Bede Clark, 3/11/14-5/4/14; NY Times Magazine Photography (T), 4/26/14-8/24/14; Project Atrium: Caroline Lathan-Stiefel, 7/26/14-10/27/14; Get Real: New Directions in Contemporary American Painting (T), 9/13/14-1/4/15; Project Atrium: Ian Johnston, 11/22/14-3/15/15.
Research Fields: contemporary art & art education.
Facilities: 625-vol. library of professional & reference books pertaining to art available for research on premises by application; reading room; exhibit space; 125-seat auditorium; children's interactive area; classroom art facilities: darkroom, printing, ceramics, painting studios. Art, handcrafted items, books & cards for sale.
Activities: school & group tours; docent training programs for special needs individuals, workshops, film, artist talks; academic lectures; performances; community response projects; literacy programs; book club; parent/child workshops; community outreach activities; professional development courses for educators; internships.
Publications: quarterly newsletter.
Hours & Admission Prices: Tues.-Wed. & Fri.-Sat. 11-5, Thurs. 11-9, Sun. 12-5. Adults $8, senior citizens 65 & over, students, military % children $5; UNF students & children under 2 no charge; 1st Wed. of month ArtWalk 5pm-9pm no charge. Closed New Year's Day; Martin Luther King Day; President's Day; Easter; Memorial Day; Independence Day; Labor Day; Veterans Day; Thanksgiving; Christmas. &
Attendance: 78,474 (accurate)
Membership: Individual $50; Dual $75; Family Plus $100; Icon $250; Luminary $500; Avant Garde $1,000; Collectors Circle Individual $1,750; Collectors Circle $3,500.

* **MUSEUM OF SCIENCE & HISTORY OF JACKSON-VILLE, (M),** 1025 Museum Circle, Jacksonville, FL 32207-9053. Tel.: 904-396-6674, ext. 218. Fax: 904-396-7900. TDD: 904-348-3972.
E-mail: admin@themosh.org
Web Site: www.themosh.org
Founded: 1941.
Congressional District: 3 & 4
Key Personnel: Exec. Dir., Maria Hane; Chm., John Magevney; Dir. Finance, Jacquelyn Blanchard; Dir. Education, Christy Turner; Dir. Visitor Svcs. & Facilities, John Rouge.
Personnel Profile: Full-Time Paid 20; Part-Time Paid 15; Part-Time Volunteers 425; Interns 3.
Governing Authority: board of trustees; nonprofit organization. Tax-exempt: 501(c)(3).
Institution Type/Description: Science & History Museum/Planetarium.
Collections: archaeology; ethnology; geology; natural history; natural science; physical science; health science; history; Floridian material culture; decorative arts; costumes; toys, antique dolls.
Research Fields: museum collections, exhibitions & educational programs.
Facilities: galleries of permanent regional history & science exhibitions, physical science interactive & changing exhibitions; classrooms & theaters; 60 ft. dome planetarium; live animal exhibitions; outdoor interpretive courtyard; store collection storage; support spaces.
Activities: tours; lectures; formally organized educational programs; crafts classes; exhibit related trips; science demonstrations; astronomy programs; live animal demonstrations; planetarium programs; day camps; camp-ins; teacher training workshops.
Publications: museum brochure; annual report to members; electronic teachers guide; e-newsletters.
Hours & Admission Prices: Mon.-Thurs. 10-5, Fri. 10-8, Sat. 10-6, Sun. 1-6. Adults $10; members no charge. Closed major holidays. &
Attendance: 145,000 (accurate)
Membership: Individual $50; Dual $60; Family of Three $70; Family of Four $80; Family of Five $90; Family of Six $100; Director's Circle $500; Bartram Society $1,000.

MUSEUM OF SOUTHERN HISTORY, 4304 Herschel St., Jacksonville, FL 32210-2210. Tel.: 904-388-3574. Facebook: Museum of Southern History.
E-mail: shillcw@yahoo.com
Web Site: www.museumsouthernhistory.com
Founded: 1975.
Key Personnel: Chm. Bd., Ben Willingham; Chm. (V), James Shillinglaw.
Personnel Profile: Full-Time Paid 1; Part-Time Volunteers 16.
Governing Authority: private; nonprofit organization. Tax-exempt: 501(c)(3).
Institution Type/Description: General Museum.
Collections: local history & culture; photographs; period artifacts; military weapons; flags.
Research Fields: Black soldiers in Confederate service.
Facilities: 3,000-vol. library; 3,500 sq. ft. exhibit space. Museum-related items for sale.
Activities: guided tours; lectures; loan & participatory exhibits. Annual Event: Battle of Olustee in February.
Publications: quarterly newsletter, The Florida Sandspur.
Hours & Admission Prices: Tues.-Sat. 10-4. Adults $3; senior citizens, students, members & children 16 & under no charge. Closed holidays. &
Attendance: 7,000 (estimated)
Membership: Individual $25; Family $35; Booster $50; Sponsor $100; Benefactor $300.

RITZ THEATRE AND MUSEUM, 829 N. David St., Jacksonville, FL 32202-4734. Tel.: 904-632-5555. Fax: 904-632-5553.
Web Site: www.ritzjacksonville.com
Formerly: Ritz Theatre & LaVilla Museum
Founded: 1999.
Congressional District: 3
Key Personnel: Dir., Carol J. Alexander; Museum Admin., Lydia P. Stewart.
Personnel Profile: Full-Time Paid 7; Full-Time Volunteers 1; Part-Time Paid 2; Part-Time Volunteers 7.
Governing Authority: city. Tax-exempt.
Institution Type/Description: History Museum.
Collections: Jacksonville's African American community history from 1800s to 1960s; furnishings; personal artifacts; photographs.
Research Fields: development of educational institutions, businesses & roles of women in Jacksonville's African American community in the 1800s & 1900s.
Facilities: 11,000 sq. ft. exhibit space; 400-seat theater.

Activities: concerts; dance recitals; docent program; films; guided tours; lectures; loan, temporary & traveling exhibitions; theater. Annual Event: jazz concerts.
Hours & Admission Prices: Tues.-Fri. 10-6, Sat. 10-2. Adults $6, senior citizens, children & students $3; discounts to groups of 25 or more. Closed New Year's Day; Martin Luther King Jr. Day; Presidents' Day; Memorial Day; Independence Day; Labor Day; Veterans Day; Thanksgiving; Christmas. &
Attendance: 6,500 (accurate)

TIMUCUAN ECOLOGICAL AND HISTORIC PRESERVE AND FORT CAROLINE NATIONAL MEMORIAL, 12713 Fort Caroline Rd., Jacksonville, FL 32225-1240. Mailing Address: 13165 Mt. Pleasant Rd., Jacksonville, FL 32225-1227. Tel.: 904-641-7155. Fax: 904-641-3798.
E-mail: timu_interpretation@nps.gov
Web Site: www.nps.gov/timu
Founded: 1953.
Congressional District: 4
Key Personnel: Supt., Barbara Goodman; Chief Interpretation, Brian Loadholtz; Museum Cur., Anne R. Lewellen.
Personnel Profile: Full-Time Paid 25; Part-Time Paid 5; Part-Time Volunteers 15.
Governing Authority: federal; not-for-profit. Parent Institution: National Park Service.
Institution Type/Description: Historic & Natural History Museum.
Collections: furnishings; archaeological remains; Native American culture; colonization of the French colony of La Caroline; plantation period and slavery; Timucuan people exhibits & objects.
Research Fields: plantation operations; slavery; 16th-20th century Native American & European inhabitants; wetland natural history.
Facilities: Books, postcards, field guides & local history publications for sale.
Activities: guided tours; formal education programs for children.
Publications: book, Junior Ranger Activity Book.
Hours & Admission Prices: Daily 9-5. No charge. Closed New Year's Day; Thanksgiving; Christmas. &
Attendance: 1,000,000 (estimated)

Jacksonville Beach

BEACHES MUSEUM & HISTORY PARK, (M), 381 Beach Blvd., Jacksonville Beach, FL 32250-5539. Tel.: 904-241-5657. Fax: 904-241-6243.
E-mail: director@bm-hc.com
Web Site: www.beachesmuseum.org
Founded: 1978.
Key Personnel: Exec. Dir., Dr. Maarten Van De Guchte; Pres. (V), David Wainer.
Personnel Profile: Full-Time Paid 1; Part-Time Paid 4; Part-Time Volunteers 54; Interns 1.
Governing Authority: Parent Institution: Beaches Area Historical Society. Tax-exempt.
Institution Type/Description: History Museum.
Collections: local history & culture; photographs.
Research Fields: first coast beaches in Florida.
Facilities: library; archives.
Activities: lectures; summer camp; educational tours; educational programs. Museum Sponsors: Whistle Talks - Book Authors.
Publications: bimonthly newsletter, Tidings.
Hours & Admission Prices: Tues.-Sat. 10-4:30. Archives: Tues.-Thurs. 10-4:30. Adults $5; discounts to AAM & ICOM members and seniors; members no charge. &
Attendance: 12,000 (accurate)
Membership: Family $50; Sustaining $100; Corporate $250-$500.

Jay

JAY HISTORICAL SOCIETY & MUSEUM, 3946 Hwy. 4, Jay, FL 32565. Mailing Address: P.O. Box 338, Jay, FL 32565-0338. Tel.: 850-675-6122.
Institution Type/Description: Historical Society Museum.
Collections: local history & culture; period furnishings; personal artifacts; photographs.
Hours & Admission Prices: Fri. 10-3.

Jensen Beach

THE CHILDREN'S MUSEUM OF THE TREASURE COAST,
1707 N.E. Indian River Dr., Jensen Beach, FL 34957. Mailing
Address: P.O. Box 2147, Stuart, FL 34995. Tel.: 772-225-7575.
Fax: 772-225-7506.
E-mail: info@childrensmuseumtc.org
Web Site: www.childrensmuseumtc.org
Founded: 2001.
Congressional District: 17
Key Personnel: Dir., Tammy Calabria; Chm. (V), Christine Delvecchio;
Museum Shop Mgr., Olivia Labrador.
Personnel Profile: Full-Time Paid 3; Part-Time Paid 6; Part-Time Volunteers 4.
Governing Authority: Tax-exempt.
Institution Type/Description: Children's Museum.
Collections: hands-on exhibitions.
Activities: educational programs; special events; camps.
Hours & Admission Prices: Mon.-Wed. & Sat. 10-4, Thurs.-Fri. 10-5, Sun.
12-4. Admission $6 per person, children 1-2 yrs. $3; discounts to military;
members & children under one no charge. Closed New Year's Eve & Day;
Easter; Independence Day; Thanksgiving; Christmas Eve & Day. &
Attendance: 50,000 (accurate)
Membership: Grandparent $75; Family $100.

FPL ENERGY ENCOUNTER, 6501 S. Ocean Dr., Jensen Beach,
FL 34957-2041. Tel.: 877-375-4386. Fax: 772-467-7565.
E-mail: energy_encounter@fpl.com
Web Site: www.fpl.com/learning/energy_encounter
Founded: 1990.
Key Personnel: Dir., Vicki Spencer.
Personnel Profile: Full-Time Paid 2; Part-Time Paid 1.
Institution Type/Description: General Museum.
Collections: energy, electricity & nuclear power displays.
Activities: Annual Event: Atomic Energy Merit Badge for Boy Scouts; teacher
workshop - nuclear energy in January.
Hours & Admission Prices: Mon.-Fri. 10-4; groups by appointment. No
charge.
Attendance: 25,000 (accurate)

MARITIME & CLASSIC BOAT MUSEUM, 1707 N.E. Indian
River Dr., Jensen Beach, FL 34957. Mailing Address: P.O. Box
275, Jensen Beach, FL 34957. Tel.: 772-692-1234. Fax: 772-463-
3204.
E-mail: info@mcbmfl.org
Web Site: www.mcbmfl.org
Formerly: Maritime and Yachting Museum of Florida
Founded: 1993.
Key Personnel: Exec. Dir., Jody Clifford.
Personnel Profile: Full-Time Paid 2; Part-Time Paid 3; Part-Time Volunteers
25; Interns 2.
Governing Authority: private; nonprofit organization. Tax-exempt.
Institution Type/Description: Maritime Museum.
Collections: maritime & yachting history; period wooden boats; ship models;
maritime paintings, prints & photographs; navigator's instruments; marine
engines; boatbuilding tools.
Facilities: library.
Activities: educational programs; boat shows; workshops; meetings; boat
restoration; tours; seminars; symposia; lectures.
Publications: newsletter.
Hours & Admission Prices: Mon.-Sat. 10-5, Sun. 1-5. Adults $5, senior
citizens, military & students $4, children $3; children under 5 & members
no charge. Closed major holidays. &
Attendance: 2,500 (estimated)

Juno Beach

LOGGERHEAD MARINELIFE CENTER, 14200 U.S. Hwy. 1,
Loggerhead Park, Juno Beach, FL 33408-1406. Tel.: 561-627-
8280. Fax: 561-627-8305. Facebook: Loggerhead Marinelife Cen-
ter.
E-mail: info@marinelife.org
Web Site: www.marinelife.org
Formerly: Marinelife Center of Juno Beach
Founded: 1983.
Key Personnel: Chm. (V), Brian K. Waxman; Pres. (V) & C.E.O., Jack
Lighton; Chief Conservation Officer, Tommy Cutt; Dir. Finance, Kate
Fratalia; Education Mgr., Rebecca Scarbrough; Communications & Mktg.
Mgr., Tom Longo; Field Operations Mgr., Adrienne McCracken; Data Mgr.,
Sarah Hirsch; Museum Shop Mgr., Barbara Hughes; Programs Coord.,

Hannah Campbell; Hospital Coord., Nicole Montgomery; Science Coord.,
Kerri Allen; Communications & Mktg. Coord., Kat Rumbley; Volunteer
Coord., Caitlin Sampson; Staff Veterinarian, Charles A. Manire, D.V.M.;
Rehabilitation Technician, Victoria Ternullo; Rehabilitation Technician,
Jessica Grishaber; Devel. Asst., Luisa Frasco.
Personnel Profile: Full-Time Paid 17; Part-Time Paid 4; Part-Time Volunteers
300; Interns 3.
Volunteer Hours: 80,000
Operating Expenses: 2,238,452
Operating Income: 2,497,553
Governing Authority: nonprofit organization; board of directors. Tax-exempt:
501(c)(3).
Institution Type/Description: Marine Museum.
Collections: sea turtle specimens; live sea turtles; fish; crab & shell collection;
interactive exhibits.
Research Fields: leatherback sea turtle (Dermochelys coriacea) clutch surviv-
ability; nesting documentation (3 varieties of sea turtle); electronic mapping
of nesting activity; GPS tracking of sea turtle and leatherback audio study.
Facilities: aquarium; educational facilities; 1,500 sq. ft. exhibit space; 2,500
sq. ft. saltwater tank yard; field research station; nature center; zoological
park. Educational items & gifts for sale.
Activities: guided tours; lectures; children's programs; organized educational
programs; docent program; participatory, loan, temporary & traveling
exhibitions.
Publications: print newsletter; brochures; monthly e-newsletter; annual report.
Hours & Admission Prices: Mon.-Sat. 10-5, Sun. 11-5. No charge; donations
accepted. Closed New Year's Day; Easter; Thanksgiving; Christmas. &
Attendance: 215,000 (accurate)

Jupiter

BURT REYNOLDS & FRIENDS MUSEUM, 100 N. U.S. Hwy. #1,
Jupiter, FL 33477-5112. Mailing Address: c/o Burt Reynolds
Institute for Film & Theatre, P.O. Box 264, Jupiter, FL 33468-0264.
Tel.: 561-743-9955. Fax: 561-743-9922.
E-mail: info@brift.org
Web Site: www.brift.org
Key Personnel: Exec. Dir., J. Kelly Dennehy
Institution Type/Description: Celebrity Museum.
Collections: celebrity artifacts; photographs.
Hours & Admission Prices: Thurs.-Sun. 11-4. Adults $5.

HIBEL MUSEUM OF ART, (M), 5353 Parkside Dr., Jupiter, FL
33458-2906. Tel.: 561-622-5560. Fax: 561-622-4881.
E-mail: hibelmuseumjupiter@gmail.com
Web Site: hibelmuseumofart.org
Founded: 1977.
Congressional District: 12
Key Personnel: Gallery Dir., Carol Davis; Chief Cur. Artworks & Exhibitions,
Edna Hibel; Museum Shop Mgr., Adra Farriss.
Personnel Profile: Part-Time Paid 4; Part-Time Volunteers 8.
Governing Authority: nonprofit organization. Parent Institution: Edna Hibel
Art Foundation. Tax-exempt: 501(c)(3).
Institution Type/Description: Art Museum dedicated to the works of Edna
Hibel who specializes in Classical, Impressionist, Figurative & Oriental
styles of painting, drawing, original graphics & sculpture.
Collections: preservation of paintings, sculptures, drawings, graphics, porce-
lain art, crystal art and dolls by Edna Hibel; Oriental & Western antique
furniture; period paperweights; Oriental dolls; snuff bottles.
Research Fields: works and life of Edna Hibel; art history of various cultures.
Facilities: 200-vol. library of cultural & art history; gallery space & function
space for community groups; piano for chamber concerts. Videotapes &
films available & art-related items for sale.
Activities: guided tours; lectures; films; concerts; docent program; temporary,
traveling & loan exhibitions; traveling & lending film program. Museum
Sponsors: seasonal concerts; English Teas December to May.
Publications: educational booklets; catalogs; books; newsletters; brochures;
films and video documentaries.
Hours & Admission Prices: Tues.-Fri. 11-4. No charge; donations accepted.
Closed major holidays. &
Attendance: 6,000 (estimated)
Membership: Friends of the Hibel Museum of Art: Annual $50; Gold One-Year
Family Membership $100.

**LOXAHATCHEE RIVER HISTORICAL SOCIETY - JUPITER
INLET LIGHTHOUSE AND MUSEUM, (M),** 500 Captain
Armour's Way, Jupiter, FL 33469-3508. Tel.: 561-747-8380. Fax:
561-575-3292. Facebook: Jupiter Inlet Lighthouse Museum.
E-mail: visit@jupiterlighthouse.org

Web Site: www.jupiterlighthouse.org
Founded: 1971.
Congressional District: 12
Key Personnel: Bd. Pres., Jim St. Pierre; Vice Chair, Barry Boyce; Pres. & C.E.O., Jamie Stuve; Asst. Dir., Kathleen Glover; Lighthouse Mgr., Chris McKnight.
Personnel Profile: Full-Time Paid 6; Part-Time Paid 4; Part-Time Volunteers 116; Interns 9.
Governing Authority: nonprofit organization. Subsidiary Institution: Jupiter Inlet Lighthouse & Museum. Tax-exempt: 501(c)(3).
Institution Type/Description: General History Museum: lighthouse & historic homes.
Collections: 1860 lighthouse with fresnel lens & related artifacts; regional history; marine & agricultural development; lifesaving station; Seminole Wars; WW II; railroads. Historic Buildings: Jupiter Inlet Oil House; 1892 Tindall House; WW II barracks building; 1800s keepers workshop.
Research Fields: north Palm Beach County; south Martin County; documentation; archaeology.
Facilities: library pertaining to local history for use on premises only by appointment; cafe. Museum-related items for sale.
Activities: guided tours; lecture series; weddings; special events; indoor & outdoor history exhibits; organized educational programs; speakers; permanent exhibitions; sunset & moonrise tours; eco-heritage hike; children's programs.
Publications: newsletters, Update; quarterly periodical; Loxahatchee Currents; booklet, Jupiter Inlet Lighthouse.
Hours & Admission Prices: Jan.-April daily 10-5; May-Dec. Tues.-Sun. 10-5, last lighthouse tour 4pm. Adults $9, children 6-18 & AAM members $5; discounts to groups; children under 5, active military & members no charge. Closed some major holidays. &
Attendance: 58,575 (accurate)
Membership: Individual $35; Companion $50; Family $75; Sustaining $125; Patron $250; Benefactor $500. Lighthouse Keepers $1,000; Captain Yorke Club $2,500; General Meade Circle $5,000. Business: Friend $100; Partner $250.

Key Biscayne

BILL BAGGS CAPE FLORIDA STATE PARK, 1200 S. Crandon Blvd., Key Biscayne, FL 33149-2795. Tel.: 305-361-8779. Fax: 305-365-0003.
Web Site: www.floridastateparks.org
Founded: 1935.
Congressional District: 18
Key Personnel: Park Supt., David Foster.
Personnel Profile: Full-Time Paid 17; Part-Time Paid 3; Part-Time Volunteers 25.
Governing Authority: state. Dept. of Environmental Protection Division of Recreation & Parks, Commonwealth Bldg., 3900 Commonwealth Blvd., Tallahassee, FL 32303. Tax-exempt.
Institution Type/Description: Historic Building: c.1825 restored Lighthouse; reconstruction of Keeper's house.
Collections: native plants; endangered species.
Activities: guided tours; interpretation of the early history of South Florida, Second Seminole War.
Hours & Admission Prices: Park: 8am to sunset. Cape Florida Lighthouse Tours: Thurs.-Mon. 10 & 1; children under 42" not permitted to top of lighthouse. No charge; donations accepted.
Attendance: 700,000 (estimated)
Membership: Florida State Parks Annual Pass (good at all parks): Individual $64.20; Family $128.40.

Key West

AUDUBON HOUSE & TROPICAL GARDENS, (M), 205 Whitehead St., Key West, FL 33040-6522. Tel.: 305-294-2116. Fax: 305-294-4513.
E-mail: audubonhouse@audubonhouse.org
Web Site: www.audubonhouse.org
Founded: 1960.
Congressional District: 19
Key Personnel: Pres. (V), Louis Wolfson, III; Public Rels., Mktg. Dir. & Retail Mgr., Veronique Murphy; Cur. & Museum Mgr., Mary O'Connor.
Personnel Profile: Full-Time Paid 4; Part-Time Paid 5.
Governing Authority: nonprofit organization. Owned & operated by the Mitchell Wolfson Family Foundation, P.O. Box 012440, Miami, FL 33101. Tax-exempt: 501(c)(3).
Institution Type/Description: Historic House Museum & Tropical Gardens: early 19th-century home of Capt. John H. Geiger which commemorates John James Audubon's visit to Key West in 1832.
Collections: original period furniture; original Audubon engravings; the Geiger Archives.
Research Fields: historic home restoration & preservation.
Facilities: botanical garden.
Activities: guided tours; lectures; gallery talks.
Publications: Tropical Garden Catalogue.
Hours & Admission Prices: Daily 9:30-4:30. Adults $12, students $7.50, children 6-12 $5; discounts to military, and AAA, ICOM & AARP members; children under 6 & members no charge.
Attendance: 40,000 (accurate)
Membership: Student $10; Individual $15; Family $25; Donor $100; Contributor $250; Patron $500; Affiliate $1,000; Corporate $1,500.

DONKEY MILK HOUSE, HISTORIC HOME, 613 Eaton St., Key West, FL 33040-6802. Tel.: 305-296-1866. Fax: 305-296-0922.
Founded: 1992.
Key Personnel: C.E.O., Denison Tempel.
Personnel Profile: Full-Time Volunteers 2.
Institution Type/Description: Historic House: located in 1866 10-room mansion; tropical version of Classic Revival architecture containing rare 19th-century interior detailing, including 1890 Italian decorated ceilings; home of U.S. Marshall Williams & family for over 120 years.
Collections: period furniture & furnishings (c.1830-1930) representing items owned by Peter Williams when he built the house & acquisitions over the next 100 years by subsequent generations.
Research Fields: 120 year stewardship of Peter A. Williams & descendants; great fire of 1886, home survived fire due to Marshall Williams deflection with blasts of dynamite.
Facilities: tropical garden.
Activities: garden and part of historical home available for special events.
Hours & Admission Prices: Tours by appointment. Adults $5, senior citizens & members $4, students $2.50; discounts to AAM & ICOM members; children under 12 no charge. &

DRY TORTUGAS NATIONAL PARK, Key West, FL 33041. Mailing Address: P.O. Box 6208, Key West, FL 33041-6208. Tel.: 305-242-7700. Fax: 305-242-7711.
E-mail: ever_reception@nps.gov
Web Site: www.nps.gov/drto
Founded: 1960.
Congressional District: 19
Key Personnel: Public Affairs Officer, Richard Cook.
Governing Authority: federal. National Park Service, Everglades National Park, P.O. Box 279, Homestead, FL 33030. Tel. 305-242-7700.
Institution Type/Description: National Park & Preservation Project: c.1846 Fort Jefferson, third order coastal defense.
Collections: artillery; weaponry; Sooty Tern Rookery; Brown Noddies & frigate birds in their natural habitat.
Research Fields: cultural & natural history.
Facilities: camping; picnic area; snorkeling; fishing; 12-seat auditorium. Reference & marine books for sale.
Activities: visitor interaction computer programming; self-guided trail; intermittent ranger-guided activities.
Publications: brochure.
Hours & Admission Prices: Daily 8-5. Park Entrance: $5. Access to island available from Key West by chartered boat or commercial seaplane. &
Attendance: 95,000

EAST MARTELLO MUSEUM, 3501 S. Roosevelt Blvd., Key West, FL 33040-5209. Mailing Address: 281 Front St., Key West, FL 33040-8313. Tel.: 305-296-3913. Fax: 305-296-6206.
Web Site: www.kwahs.org
Founded: 1951.
Congressional District: 20
Key Personnel: C.E.O. & Exec. Dir., Claudia L. Pennington; Pres., David Harrison Wright; Site Supvr., Jack Holland; Admissions, Michael Buckley; Admissions, Christina Slone.
Personnel Profile: Full-Time Paid 5; Part-Time Paid 4; Part-Time Volunteers 3.
Governing Authority: society; nonprofit organization. Parent Institution: Key West Art & Historical Society. Tax-exempt: 501(c)(3).
Institution Type/Description: Historical Museum and Art Gallery: housed in 1861 brick fort.
Collections: local historic & military artifacts; Mario Sanchez folk art collection; Stanley Papio folk art collections; costumes; WPA paintings; local paintings; period furnishings.
Research Fields: Florida Keys history.
Facilities: Museum-related items for sale.
Activities: monthly art exhibitions; permanent & temporary exhibitions; permanent military displays; tours; educational & cultural programs.

Publications: e-newsletter, The Ghosts of East Martello.
Hours & Admission Prices: Daily 9:30-4:30. Adults $6, seniors 62 & over $5, students & children $3; discounts to local residents & AAA members; children under 5, school tours & members no charge. Closed Christmas. &
Attendance: 38,955 (accurate)
Membership: Student $10; Individual $25; Family $45; Contributing $50; Patron $100; Life $250.

ERNEST HEMINGWAY HOUSE MUSEUM, 907 Whitehead, Key West, FL 33040-7473. Tel.: 305-294-1136. Fax: 305-294-2755.

E-mail: info@hemingwayhome.com
Web Site: www.hemingwayhome.com
Founded: 1964.
Key Personnel: Gen. Mgr., Jacque Sands; C.E.O., Michael A. Morawski; Dir. Events, Dave Gonzales; Museum Shop Mgr., Linda Mendez.
Personnel Profile: Full-Time Paid 22; Part-Time Paid 1.
Governing Authority: individual operation.
Institution Type/Description: Historic House: 1931-1961 Ernest Hemingway Home, built in Spanish Colonial Style of native rock hewn from the grounds.
Collections: furniture; rugs; chandeliers; tile & other mementos from Spain, Africa, France, Cuba and other parts of the world.
Facilities: gardens. Ernest Hemingway books for sale.
Activities: museum & garden facilities available for rent; tours.
Publications: books.
Hours & Admission Prices: Daily 9-5. Adults $12, groups of 12 or more $10, children 6-12 $6; discounts to military, groups & AAA members; children under 5 no charge. &

FORT ZACHARY TAYLOR HISTORIC STATE PARK, 601 Howard England Way, Key West, FL 33040-8396. Tel.: 305-292-6713. Fax: 305-292-6881.

E-mail: david.foster@dep.state.fl.us
Web Site: www.floridastateparks.org/forttaylor
Founded: 1985.
Congressional District: 8
Key Personnel: Park Mgr., David Foster.
Governing Authority: state; nonprofit. Parent Institution: Florida Dept. of Environmental Protection. Subsidiary Institution: Division of Recreation & Parks. Tax-exempt.
Institution Type/Description: History Museum.
Collections: Civil War items; Spanish-American war items; World War I & II items.
Research Fields: Civil War; Spanish-American War; World War I & II.
Facilities: educational facilities; picnic area.
Activities: guided tours; organized education programs for children.
Hours & Admission Prices: Park: 8-sunset. Museum Tours: daily 12-5. Vehicle (up to 8 passengers): $6 per car, $.50 per person, Walk-ins $2.50 per person; children under 6 no charge. &
Attendance: 389,000 (accurate)
Membership: Friends of Fort Taylor $20; Family $30; Business $50.

HARRY S. TRUMAN LITTLE WHITE HOUSE MUSEUM, (M), 111 Front St., Key West, FL 33040-8311. Tel.: 305-294-9911. Fax: 305-294-9988.

E-mail: bwolz@trumanlittlewhitehouse.com
Web Site: www.trumanlittlewhitehouse.com
Founded: 1991.
Congressional District: 20
Key Personnel: Exec. Dir. & C.F.O., Robert J. Wolz; Chm., Christopher Belland; Pres., Edwin Swift; Financial Dir., Benjamin McPherson; Registrar, Robin Kilgo; Public Rels., Piper Smith.
Personnel Profile: Full-Time Paid 12; Part-Time Paid 4; Part-Time Volunteers 22.
Governing Authority: state. Parent Institution: Key West Harry S. Truman Foundation, Inc. Tax-exempt: 501(c)(3).
Institution Type/Description: Historic House: built in 1890 as the Navy Base Commander home, and used by Presidents Taft, Truman, Eisenhower, John F. Kennedy, Carter & Clinton.
Collections: furnishings; video presentations.
Facilities: 100-vol. library of Truman & US Presidency books; 30-seat theater. Museum-related items for sale.
Activities: docent programs; formal education for adults & children; temporary exhibitions; Presidential families program. Annual Event: Educational Conference with Truman Presidential Library.
Publications: newsletter & electronic newsletter.
Hours & Admission Prices: Daily 9-5; last tour begins 4:30 p.m. Adults $15, seniors & military $13, students $5; discounts to AAM, AASLH, ICOM,

Florida Assoc. of Museums, Florida Trust for Historic Preservation, & museum members no charge. &
Attendance: 63,000 (accurate)
Membership: Gift Donations: $25, $50, $75; General: $100, $250, $500, $750, $1,000; Sustaining $1,200, $2,500, $7,500, $10,000, $25,000; Business: $500, $1,000, $1,500, $2,000.

KEY WEST LIGHTHOUSE MUSEUM, 938 Whitehead St., Key West, FL 33040-7423. Tel.: 305-295-6616, ext. 16. Fax: 305-296-6206.

Web Site: www.kwahs.org
Founded: 1966.
Congressional District: 19
Key Personnel: C.E.O. & Dir., Claudia L. Pennington; Pres., David Harrison Wright; Admissions, Gina Hooper; Admissions, Bob Allen; Site Operations, Bob Wandres.
Personnel Profile: Full-Time Paid 20.
Governing Authority: county; federal. Parent Institution: Key West Art & Historical Society, 281 Front St., Key West, FL. Tax-exempt: 501(c)(3).
Institution Type/Description: General Museum: housed in 1887 lighthouse keepers home & 1846 lighthouse museum.
Collections: artifacts relating to the history of the Key West Lighthouse & the lighthouses of the Florida Keys; lenses; photographs; household items; archaeological; models.
Research Fields: lighthouses.
Publications: monthly e-newsletter.
Hours & Admission Prices: Daily & holidays 9:30-4:30. Adults $10, seniors 62 & over $9, children 7-12 $5; discounts to groups, AAA members, & local residents; children under 6 & members no charge. Closed Christmas. &
Attendance: 92,321 (accurate)
Membership: Individual $55; Dual $100; Family $125.

KEY WEST MUSEUM OF ART AND HISTORY, 281 Front St., Key West, FL 33040-8313. Tel.: 305-295-6616, ext. 16. Fax: 305-295-6649.

Web Site: www.kwahs.org
Founded: 1951.
Congressional District: 20
Key Personnel: Exec. Dir., Claudia L. Pennington; Pres., David Harrison Wright; C.F.O., Kathleen Moody; Programs Coord., Gerri Sidoti; Asst. Dir., Michael Gieda; Public Rels., Michael Haskins; Site Supvr., Jack Holland; Admissions, Lynn Dalton; Admissions, Sue Paden.
Personnel Profile: Full-Time Paid 18; Part-Time Paid 6; Part-Time Volunteers 2; Interns 3.
Governing Authority: private; nonprofit organization. Branch Museums: East Martello Museum, Key West, FL; Key West Lighthouse, Key West, FL. Tax-exempt: 501(c)(3).
Institution Type/Description: Art & History Museum: housed in c.1891 Custom House.
Collections: fine art; folk art; historic artifacts; documents; photographs from 1860s to present.
Research Fields: history of Key West & Florida Keys.
Facilities: 2,000-vol. library; 18,000 sq. ft. exhibit space. Museum-related items for sale.
Activities: arts festivals; concerts; guided tours; lectures; loan, traveling & participatory exhibitions; adult programs; children's classes.
Publications: monthly e-newsletter.
Hours & Admission Prices: Daily 9:30-4:30. Adults $7, seniors $6, children & students $5; discount to AAA members & local residents; members & children under 6 no charge. Closed Christmas. &
Attendance: 76,635 (accurate)
Membership: Individual $55; Dual $100; Family $125.

KEY WEST SHIPWRECK TREASURES MUSEUM, 1 Whitehead St., Key West, FL 33040-6634. Tel.: 305-292-8990. Fax: 305-292-1617.

E-mail: shipwreck@historictours.com
Web Site: www.keywestshipwreck.com
Formerly: Key West Shipwreck Historeum Museum
Institution Type/Description: Historic Ship Museum.
Collections: artifacts & memorabilia from the 1985 rediscovery of the wrecked vessel Isaac Allerton.
Hours & Admission Prices: Daily 9:40-5. Adults $14, students, seniors & military $12, children 4-12 $6; children 3 & under no charge. &

KEY WEST TROPICAL FOREST & BOTANICAL GARDEN, 5210 College Rd., Key West, FL 33040-4302. Tel.: 305-296-1504. Fax: 305-296-2242.
E-mail: kwbgs@kwbgs.org
Web Site: www.kwbgs.org
Founded: 1936.
Congressional District: 18
Key Personnel: Gen. Mgr., Misha D. McRae; Pres. (V), Mary Chandler.
Personnel Profile: Full-Time Paid 6; Part-Time Paid 1; Part-Time Volunteers 105; Interns 2.
Governing Authority: private; nonprofit organization. Tax-exempt: 501(c)(3).
Institution Type/Description: Botanical Garden.
Collections: native fragrant plants; tropical fruit trees; native palms; tropical spices; local herbs; flowers.
Facilities: botanical garden; nature center; chapel. Museum-related items for sale.
Activities: docent program; lectures; formal education programs. Annual Events: Migration Mania in March; Cuban Cultural Exposition in April; Key West Art Garden April to June.
Publications: A Complete Guide to Our Museum of Living Collections (2011); A Butterfly's View: Flowering Guide.
Hours & Admission Prices: Daily 10-4. Adults & children $5, seniors $4; members & children under 12 no charge. Closed New Year's Day; Thanksgiving; Christmas. &
Attendance: 25,000 (accurate)
Membership: Student $10; Individual $35; Household Family $50; Patron $120; Ruby $300; Emerald $600; Diamond $1,100.

＊　**MEL FISHER MARITIME HERITAGE SOCIETY,** 200 Greene St., Key West, FL 33040-6516. Tel.: 305-294-2633, ext. 15. Fax: 305-294-5671.
E-mail: info@melfisher.org
Web Site: www.melfisher.org
Founded: 1982.
Congressional District: 20
Key Personnel: C.E.O., Melissa Kendrick; Dir., Rebecca J. Tomlinson, Ph.D.; Chm., Raul Rodriguez; Archivist, Monica Brook; Education Coord., Shannon Burgess; Chief Cur. & Traveling Exhibition Coord., Dylan Kibler; Archaeology, Corey Malcom; Museum Shop Mgr., Yolanda Maloney.
Personnel Profile: Full-Time Paid 25; Full-Time Volunteers 3; Part-Time Paid 13; Part-Time Volunteers 15; Interns 3.
Operating Expenses: 1,713,810
Operating Income: 1,889,421
Governing Authority: society. Not-for-profit. Subsidiary Institution: Key West Turtle Museum, 200 Margaret St., Key West, FL 33040. Tax-exempt: 501(c)(3).
Institution Type/Description: Maritime Museum.
Collections: 17th- & 18th-century coins; decorative arts; ceramics; Slave Trade artifacts; maritime artifacts.
Major Exhibits: Spirits of the Passage (T), 9/13-9/18; Pirates: Mayhem & Madness (T), 9/13-9/18; Last Slave Ships (T), 1/14-1/19; Coins (T), 1/14-1/20.
Research Fields: 15th- to 18th-century shipping & shipbuilding; maritime history of sea turtle fishing industry.
Facilities: 2,200-vol. library of maritime history & Colonial Spanish history available for use by the public; 50-seat theater. Museum-related items for sale.
Activities: formal education programs for children; guided tours; lectures; permanent & traveling exhibitions; elder hostel; summer camps.
Publications: bimonthly newsletter, Navigator; biannual historical magazine, Astrolabe.
Hours & Admission Prices: Daily 8:30-5. Adults $12.50, children $6.25; discounts to AAM & ICOM members, groups, Monroe County residents, & Florida Assoc. of Museums members; members no charge. &
Attendance: 221,344 (accurate)
Membership: Individual $55; Patron $100; Business $500 & up.

THE OLDEST HOUSE & GARDEN MUSEUM, 322 Duval St., Key West, FL 33040. Mailing Address: P.O. Box 689, Key West, FL 33041-0689. Tel.: 305-294-9501. Fax. 305-294-4509.
E-mail: oldisland@comcast.net
Web Site: www.oirf.org
Founded: 1975.
Congressional District: 19
Key Personnel: Admin., Cork Tarplee; Pres., Anthony S. Minore.
Personnel Profile: Full-Time Volunteers 5; Part-Time Paid 3; Part-Time Volunteers 14.
Governing Authority: nonprofit organization. Parent Institution: Old Island Restoration Foundation. Tax-exempt.
Institution Type/Description: Historic House Museum: housed in c.1829 Conch House, home of local sea captain & wrecker.
Collections: 19th-century furniture; Key West history; toys; paintings; documents related to 19th-century wrecking in Key West.
Research Fields: genealogy; local maritime history.
Facilities: gardens.
Activities: guided tours; docent program. Museum Sponsors: garden party.
Publications: A Guide to Historic Key West on the Pelican Path.
Hours & Admission Prices: Mon.-Tues. & Thurs.-Sat. 10-4. No charge; donations accepted. &
Attendance: 4,288 (accurate)
Membership: Old Island Restoration Foundation: Single $50; Couples $95; Family $125; Business $250; Patron $500; Benefactor $1,000; Life $10,000.

RIPLEY'S BELIEVE IT OR NOT!, 108 Duval St., Key West, FL 33040-6506. Tel.: 305-293-9939. Fax: 305-293-9709.
E-mail: museum@ripleyskeywest.com
Web Site: www.ripleys.com
Founded: 1927.
Key Personnel: Mgr., Kevin Hays.
Personnel Profile: Full-Time Paid 10.
Governing Authority: private; profit-making organization. Parent Institution: Ripley Corp., 5720 Major Blvd., Ste. 700, Orlando, FL 32819.
Institution Type/Description: General Museum.
Collections: antiquities & oddities; Asian artifacts; primitive tribal artifacts; human oddities; arcane art.
Facilities: 30-vol. library of Hemingway works available to the public; 30-seat large-screen theater; 10,000 sq. ft. exhibit space. Ripley's Believe It or Not! memorabilia for sale.
Activities: guided tours; theater; participatory & traveling exhibitions.
Hours & Admission Prices: Daily 9:30am-11pm. Adults $14.95, children 5-12 $11.95; children 4 & under no charge. &
Attendance: 86,000 (accurate)

Kissimmee

MUSEUM OF MILITARY HISTORY, (M), 5210 W. 192, Kissimmee, FL 34746. Tel.: 407-507-3894. Fax: 407-507-3894. Facebook: Museum of Military History.
E-mail: rscarmean@embarqmail.com
Web Site: museumofmilitaryhistory.com
Formerly: Veterans Tribute and Museum of Osceola County, Inc.
Founded: 2004.
Key Personnel: C.E.O. & Chm. (V), Donald Smith; Dir. & Museum Shop Mgr., Scott Carmean.
Personnel Profile: Full-Time Paid 3; Part-Time Paid 1; Part-Time Volunteers 25; Interns 3.
Governing Authority: Parent Institution: Veterans Tribute & Museum. Tax-exempt: 501(c)(3).
Institution Type/Description: Military Museum.
Collections: military artifacts, equipment & relics from Civil War to present; photographs; memorabilia.
Research Fields: military history.
Facilities: library & research center.
Hours & Admission Prices: Tues.-Sun. 12-6. Adults $6. &
Attendance: 10,000 (accurate)
Membership: Individual $35; Family $75; Business $100; Corporate $150.

OSCEOLA CENTER FOR THE ARTS, 2411 E. Irlo Bronson Memorial Hwy., Kissimmee, FL 34744-5430. Tel.: 407-846-6257. Fax: 407-846-7902.
E-mail: emoore@ocfta.com
Web Site: www.ocfta.com
Founded: 1964.
Key Personnel: Dir., Ed Moore; Pres. (V), Pete Pace; Museum Shop Mgr., Jules Davidson.
Personnel Profile: Full-Time Paid 7; Part-Time Paid 2; Part-Time Volunteers 20.
Governing Authority: Tax-exempt.
Institution Type/Description: Art Gallery.
Collections: works by local & regional artists.
Facilities: 241-seat theater; 125-seat studio theater.
Activities: visual & performing art classes; special events.
Publications: quarterly event guides.
Hours & Admission Prices: Mon.-Fri. 9-12 & 1-5. Art Gallery: daily. Art Gallery: no charge. Closed national holidays. &
Attendance: 11,000 (estimated)
Membership: Single $100; Family $150; Corporate $250 & up.

PIONEER VILLAGE & MUSEUM - OSCEOLA COUNTY HISTORICAL SOCIETY, 750 N. Bass Rd., Kissimmee, FL 34746. Tel.: 407-396-8644.
Web Site: www.osceolahistory.org
Institution Type/Description: Historical Society Museum.
Collections: pioneer life & history; late-19th century buildings; period furnishings & artifacts.
Activities: special events.
Hours & Admission Prices: Village: Thurs.-Sun. 10-4. Grounds: Tues.-Fri. 9-5. Adults $5, children 6-12 $2; children 5 & under and members no charge.

WHITE 1 FOUNDATION - WWII AIRCRAFT MUSEUM, 233 N. Hoagland Blvd., Kissimmee, FL 34741-4531. Tel.: 727-365-1713; 407-473-3957. Fax: 407-622-9308.
E-mail: white1foundation@yahoo.com
Web Site: www.white1foundation.org
Founded: 1998.
Key Personnel: Dir. & Cur., Dr. M.J. Timken; Devel., Gene Davidson; Public Rels., Melissa Timken; Treas., Peter Lackman; Museum Shop Mgr., G. Murphy, Ph.D.
Personnel Profile: Full-Time Volunteers 4; Part-Time Volunteers 1.
Governing Authority: private; nonprofit organization.
Institution Type/Description: Military History Museum.
Collections: WWII aviation history; aircraft; pilot gear & survival equipment; technical data; Focke Wulf FW 190.
Research Fields: performance of period fighter aircraft to present day; construction techniques.
Facilities: library; educational facilities; 2,000 sq. ft. exhibit space. Museum-related items for sale.
Activities: formal education programs; guided tours; hobby workshops; lectures; participatory, temporary & traveling exhibitions; broadcast programs.
Publications: monthly newsletter email.
Hours & Admission Prices: Mon.-Fri. 9-5, Sat.-Sun. by appointment. No charge; donations accepted. &
Attendance: 5,000 (estimated)
Membership: Individual $40.

LaBelle

LABELLE HERITAGE MUSEUM, 360 N Bridge St., CSR-29 N., LaBelle, FL 33935. Mailing Address: P.O. Box 2846, LaBelle, FL 33975-2846. Tel.: 863-674-0034.
Founded: 1991.
Congressional District: 25
Key Personnel: Pres. (V), Joseph H. Thomas; Museum Shop Mgr., Jeannie Horlacher.
Personnel Profile: Part-Time Paid 32; Part-Time Volunteers 16.
Governing Authority: Tax-exempt.
Institution Type/Description: Historical Museum.
Collections: local history & artifacts; fossils; clothing; family books; furniture; photographs; personal artifacts.
Activities: group tours; special events; Annual events: Old Timers Dinner; Swamp Cabbage Festival in February.
Publications: newsletter.
Hours & Admission Prices: Mon.-Tues., Sat. 9-12, Thurs.-Fri. 1-4; June-Aug. Mon.-Tues. 9-12; other times by appointment; closed all major holidays. Admission $3; members & children under 6 when accompanied by an adult no charge. &
Attendance: 500 (estimated)
Membership: Single $20; Couple $35; Sponsor & Family $50; Patron $100; Corporate $150.

Lake City

LAKE CITY COLUMBIA COUNTY HISTORICAL MUSEUM, INC., 157 S.E. Hernando Ave., Lake City, FL 32025-4428. Mailing Address: P.O. Box 3276, Lake City, FL 32056-3276. Tel.: 386-755-9096. Fax: 904-755-6605.
E-mail: edisto1@alltel.net
Founded: 1984.
Congressional District: 2
Key Personnel: Dir. & Cur., Pat McAlhany; Vice Pres., Paulette Lord; Sec., Sean McMahon, Ph.D.; Education, Margie Stanfield.
Personnel Profile: Full-Time Volunteers 10; Part-Time Volunteers 2.
Governing Authority: private; nonprofit organization. Tax-exempt: 501(c)(3).
Institution Type/Description: History Museum: located in a Historic House.
Collections: display of civil war memorabilia; items relating to history & growth of area; old house furniture.
Research Fields: local family history.

Facilities: 40-vol. library containing civil war references, area history & local family references available to the public; 2,500 sq. ft. exhibit space. Museum-related items for sale.
Activities: guided tours; participatory exhibits.
Hours & Admission Prices: Wed. 10-1. No charge; donations accepted. Closed New Year's Day; Easter; Independence Day; Christmas. &
Attendance: 3,000 (estimated)
Membership: Youth & Senior $5; Individual $10; Family $25.

Lake Wales

BOK TOWER GARDENS, 1151 Tower Blvd., Lake Wales, FL 33853-3470. Tel.: 863-676-1408. Fax: 863-676-6770. Facebook: Bok Tower Gardens.
Web Site: boktowergardens.org
Formerly: Historic Bok Sanctuary
Founded: 1929.
Congressional District: 12
Key Personnel: Pres., David Price; Controller, Steve Jolley; Mktg. & Public Rels. Dir., Brian Ososky; Dir. Retail, Sandra Dent; Dir. Horticulture, Greg Kramer.
Governing Authority: nonprofit organization. Tax-exempt: 501(c)(3).
Institution Type/Description: Botanical Garden.
Collections: cultivated plants; rare & endangered plants; endemic plants; camellias; decorative arts; local historic objects. Historic Building: 1930s Mediterranean-style Mansion.
Research Fields: rare & endangered plant species of the region.
Facilities: 2,600-vol. library of carillon-related material, including 1,000 carillon tapes; 3,000 botanic & historic slides, available to the public on a limited basis; botanical garden; nature walks & nature preserve trail; visitor center; meeting facilities; orientation theatre; cafe. Museum-related items for sale.
Activities: guided tours; lectures; films; indoor & outdoor concerts; organized education programs for children, adults, undergraduate & graduate students in carillon studies; horticulture internships; docent program; participatory & temporary exhibitions; garden wedding ceremonies; facility rentals.
Publications: quarterly newsletter; annual report.
Hours & Admission Prices: Daily 8-6. Adults $12, children 5-12 $3; discount to groups; children under 5, AABGA members & members no charge. &
Attendance: 130,000 (estimated)
Membership: Individual $45; Duo $55; Family $75; Sustainer $100; Donor $150; Patron $300; Sponsor $600; Bok Tower Club $1,200; President's Council $3,000; Chairman's Council $6,000.

LAKE WALES ARTS CENTER, 1099 State Rd. 60 E., Lake Wales, FL 33853-4208. Mailing Address: 228 S. 4th St., Lake Wales, FL 33853-4208. Tel.: 863-676-8426.
E-mail: lwacadmin@verizon.net
Institution Type/Description: Arts Gallery: housed in the Spanish Mission-style building formerly the Holy Spirit Catholic Church; built 1927.
Collections: paintings; sculpture.
Activities: classes; concerts; special events; permanent & temporary exhibits.
Hours & Admission Prices: Mon.-Fri. 9-4. No charge.

LAKE WALES DEPOT MUSEUM & CULTURAL CENTER, 325 S. Scenic Hwy., Lake Wales, FL 33853-3873. Tel.: 863-678-5160. Fax: 863-678-1842.
E-mail: lakewalesdepot@gmail.com
Web Site: www.cityoflakewales.com
Founded: 1976.
Congressional District: 12
Key Personnel: Dir., Mimi Reid Hardman.
Personnel Profile: Part-Time Paid 1; Part-Time Volunteers 5.
Governing Authority: municipal government. Parent Institution: City of Lake Wales. Subsidiary Institution: Seaboard Freight Depot. Branch Museum: Historic Lake Wales Society; Stuart-Dunn-Oliver Historic House Museum; Children's Museum, Lake Wales, FL. Tax-exempt.
Institution Type/Description: Local History Museum: housed in 1928 ACL & Seaboard Coastline Railroad Depot.
Collections: local history artifacts including railroad, turpentine, citrus, cattle exhibits; 1916 Pullman car, 1926 Seaboard Air Line caboose, 1944 engine; photographs and documentary archives; quilt, clothing, toys, Native American & patriotic rotating exhibits. Additional facilities located on the CSX Historic Corridor are the 1916 Seaboard Air Line Freight Depot, the 1920 Children's Museum, & the 1920 Stuart-Dunn-Oliver Historic House Museum, originally a founder's home.
Research Fields: historic building preservation; local history; local black history.
Facilities: 100-vol. library pertaining to local history; documentary & photograph archives; biographical files & manuscripts.

Activities: lectures; dance recitals; theater; hobby workshops; organized education programs; temporary & traveling exhibitions. Museum Sponsors: Local Pioneer Day Festival.
Publications: quarterly newsletter.
Hours & Admission Prices: Mon.-Fri. 9-5, Sat. 10-4. No charge; donations accepted. Closed major holidays. &
Attendance: 20,000 (estimated)
Membership: Historic Lake Wales Society: Student & Associate $10; Individual $25; Family $50; Life $200; Corporate $300.

Lake Worth

LAKE WORTH HISTORICAL MUSEUM, 414 Lake Ave., Lake Worth, FL 33460-3807. Tel.: 561-533-7354. Fax: 561-586-1750.
E-mail: lwlibrary@lakeworth.org
Web Site: www.lakeworth.org
Formerly: Museum of the City of Lake Worth
Founded: 1982.
Congressional District: 14
Key Personnel: Library Svcs. Mgr., Vickie Joslin.
Governing Authority: municipal; nonprofit. Tax-exempt: 509(a)(1).
Institution Type/Description: History Museum.
Collections: Lake Worth history & the surrounding area; photograph collection; period clothing; dishes; tools; office equipment; cameras; kitchen equipment; ethnic displays.
Research Fields: Major General Worth, The Man; 80th Anniversary of the City of Lake Worth; research materials of the area.
Facilities: 500-vol. research library on Florida history available to the public; 3,000 sq. ft. exhibit space; 65-seat auditorium.
Activities: guided tour of museum & city; formal educational programs.
Publications: book, On Lake Worth.
Hours & Admission Prices: Mon.-Wed. 1-4. No charge, donations accepted.

NATIONAL MUSEUM OF POLO AND HALL OF FAME, INC., 9011 Lake Worth Rd., Lake Worth, FL 33467-3617. Tel.: 561-969-3210. Fax: 561-964-8299.
E-mail: polomuseum@att.net
Web Site: www.polomuseum.com
Founded: 1984.
Key Personnel: C.E.O. & Chm. (V), Stephen Orthwein; Exec. Dir. & Vice Pres., George DuPont; Dir. Devel., Brenda Lynn; Pres. (V), Martin S. Cregg.
Personnel Profile: Full-Time Paid 2.
Governing Authority: federal; nonprofit organization. Tax-exempt: 501(c)(3).
Institution Type/Description: Polo Sports History & Hall of Fame.
Collections: books; record books; scrapbooks; polo sporting equipment; trophies; photographs; research material; art work.
Research Fields: polo sports history.
Facilities: 300-vol. library available to the public; 2,000 sports related magazines; 763 sq. ft. exhibit space; field research station. Museum-related items for sale.
Activities: films; docent program; temporary & traveling exhibitions.
Publications: annual newsletter.
Hours & Admission Prices: Jan.-April Mon.-Fri. 10-4, Sat. 10-2; May-Dec. Mon.-Fri. 10-4; other times by appointment. No charge; donations accepted. &

Lakeland

EXPLORATIONS V CHILDREN'S MUSEUM, (M), 109 N. Kentucky Ave., Lakeland, FL 33801-5057. Tel.: 863-687-3869.
E-mail: info@explorationsv.com
Web Site: www.explorationsv.com
Founded: 1991.
Congressional District: 12
Key Personnel: C.E.O., Georgann Carlton.
Personnel Profile: Full-Time Paid 7; Part-Time Paid 8; Part-Time Volunteers 200.
Institution Type/Description: Children's Museum.
Collections: hands-on arts & sciences exhibits.
Facilities: Museum-related items for sale.
Hours & Admission Prices: Mon.-Sat. 9-5:30. Admission $7; discounts to seniors & military; children 2 and under & members no charge. Closed major holidays. &
Attendance: 90,232
Membership: Family $70 & up.

FLORIDA AIR MUSEUM AT SUN'N FUN, 4175 Medulla Rd., Lakeland, FL 33811-1249. Tel.: 863-644-0741. Fax: 863-648-9264.
E-mail: museum@floridaairmuseum.org
Web Site: www.floridaairmuseum.org
Formerly: International Sport Aviation Museum (ISAM)
Founded: 1989.
Key Personnel: Mgr., Lori Bradner; C.E.O., John Leenhouts; Museum Shop Mgr., Susan Highley.
Personnel Profile: Full-Time Paid 4; Part-Time Paid 4; Part-Time Volunteers 25.
Governing Authority: nonprofit organization. Parent Institution: Sun 'n Fun Fly-In, Inc. Tax-exempt: 501(c)(3).
Institution Type/Description: Aviation Museum.
Collections: period aircraft; classic aircraft; experimental aircraft; Howard Hughes aviation archives.
Facilities: 30,000 sq. ft. exhibit space; education programming space. Gift items for sale.
Activities: seminars & workshops; organized education programs for children; temporary exhibitions; summer camps; fly-ins.
Publications: bimonthly newsletter, Sun'n Fun News.
Hours & Admission Prices: Mon.-Fri. 9-5, Sat. 10-4, Sun. 12-4. Contribution: adults $10, senior citizens $8, students $6; ASTC reciprocal program; children 5 & under and members no charge. Closed New Year's Day; Christmas. &
Attendance: 60,000 (estimated)
Membership: Personal $35; Family $65; Personal & Engraved Brick $99.

* 　**POLK MUSEUM OF ART, (M),** 800 E. Palmetto St., Lakeland, FL 33801-5529. Tel.: 863-688-7743. Fax: 863-688-2611.
E-mail: corologas@polkmuseumofart.org
Web Site: www.polkmuseumofart.org
Founded: 1966.
Congressional District: 12
Key Personnel: Exec. Dir., Claire Orologas; Exec. Asst., Palemeschia "Pal" Rivers Powell; Art Cur., Adam Justice; Exhibits Specialist, Gregory Mills; Admin., Terri D'Orsaneo; Operations Mgr., Bill O'Connell; Museum Shop Mgr., Terry Aulisio.
Personnel Profile: Full-Time Paid 12; Part-Time Paid 9; Part-Time Volunteers 200; Interns 2.
Governing Authority: board of trustees; nonprofit organization. Tax-exempt: 501(c)(3).
Institution Type/Description: Art Museum.
Collections: pre-Columbian art; 15th to 18th-century European decorative arts; photography; contemporary American & Florida artists; American paintings and works on paper; 20th-century contemporary Japanese ceramics; Asian decorative arts; contemporary art.
Research Fields: contemporary Florida Art; pre-Columbian art; American Contemporary Art.
Facilities: 37,000 sq. ft. building; 3,500-vol. research library with arts - related volumes, & periodicals; computer graphics teaching laboratory; photography laboratory; ceramics studio; multi-purpose room; teaching auditorium; sculpture garden; studio classrooms. Museum-related items for sale.
Activities: guided tours; lectures; film series; exhibitions; concerts; workshops; formally organized educational programs; docent program; summer program for children; high school for visual arts; permanent, temporary & traveling exhibitions; school loan service. Museum Sponsors: Mayfaire By-The-Lake Art Festival; Festival of Arts and Athletes.
Publications: quarterly newsletter; exhibition catalogues; educational services brochure; traveling exhibition service brochure; annual report; self-guided tour brochure; family gallery guides.
Hours & Admission Prices: Tues.-Sat. 10-5, Sun. 1-5. Adults $5, senior citizens $4; discounts to AAM & ICOM members; members no charge. Closed holidays. &
Attendance: 130,000 (estimated)
Membership: Individual $40; Family & Household $60; Sponsor $100; Advocate $250; Benefactor $500; Patron $1,000; Gold Patron $2,500; Platinum Patron $5,000.

Largo

THE ARMED FORCES HISTORY MUSEUM, (M), 2050 34th Way N., Largo, FL 33771-4094. Tel.: 727-539-8371.
E-mail: info@armedforcesmuseum.com
Web Site: www.armedforcesmuseum.com
Formerly: The Armed Forces Military Museum
Founded: 1996.
Congressional District: 10
Key Personnel: Dir. & C.E.O., John J. Piazza, Sr.; Dir. Mktg., Cindy Bosselmann; Museum Shop Mgr., Janis Beal.

Personnel Profile: Full-Time Paid 10; Part-Time Paid 3; Part-Time Volunteers 40.

Governing Authority: Tax-exempt.

Institution Type/Description: Military & History Museum.

Collections: displays illustrating WWI & II, D-Day landings, the attack on Pearl Harbor, Korea, Vietnam & other conflicts.

Facilities: meeting & conference space.

Activities: tours; military history classes; special events; open houses; civic events; memorial walk; engraved brick program; field trip program for schools.

Publications: bimonthly newsletter.

Hours & Admission Prices: Tues.-Sat. 10-4, Sun. 12-4. Adults $17.95, seniors 65 & over and veterans $14.95, children 4-12 $12.95; discounts to AAA members & groups of 15 or more; members, children 3 & under and active & retired military with ID no charge. Closed major holidays. &

Membership: Individual $35; Dual $50; Family $75; Museum Partner $100; Freedom $250; Preservation $500; Five Star $750.

HERITAGE VILLAGE, (M), 11909-125 St., N., Largo, FL 33774-3611. Tel.: 727-582-2127. Fax: 727-582-2211.

E-mail: ebabb@pinellascounty.org

Web Site: www.pinellascounty.org/heritage

Founded: 1976.

Congressional District: 8

Key Personnel: Pres. Pinelles Cty. Historical Society (V), Rosemarie Kafer; Operations Mgr., Ellen Babb; Museum Interpreter, Paige W. Noel; Museum Shop Mgr., Jane Doyle.

Personnel Profile: Full-Time Paid 3; Part-Time Volunteers 379.

Governing Authority: county. Parent Institution: Pinellas County Government. Subsidiary Institution: Gulf Beaches Historical Museum, 115 10th Ave., St. Pete Beach, FL 33706. Tel.: 727-552-1610. Tax-exempt: 170(b)(1)(A).

Institution Type/Description: Living History Museum.

Collections: over 9,000 photographs of early St. Petersburg, Clearwater & surrounding communities in Pinellas County; audiovisuals; postcards; school & college annuals; books & ledgers; maps & atlases; city directories; newspapers; scrapbooks; census records; clip files; genealogical materials; primitives & home furnishings of early Pinellas Pioneers; early textiles. 28 Historic Buildings: 1907 House of Seven Gables; 1879 Moore House; 1896 Plant-Sumner House; Victorian bandstand; c.1900 gazebo; sugar cane mill; 1911 Lowe Barn; c.1852 Coachman-McMullen Log House; 1876 Boyer Cottage; 1924 Sulphur Springs Depot; 1905 Safety Harbor Church; 1888 Greenwood House; 1890 Safford Pavilion; 1915 Walsingham House; 1915 H.C. Smith store, barber shop, garage & filling station; 1888 Lowe House; 1868 Daniel McMullen House; 1915 Union Academy; Harris School & Firehouse with 1919 LaFrance fire engine; c.1930 Tarpon Springs sponge warehouse.

Research Fields: Pinellas County & Florida history.

Facilities: 3,500-vol. library & archives.

Activities: special events; lecture series; craft demonstrations; fiber arts demonstrations; historic preservation; Pinellas County Historical Society.

Publications: Village Post newsletter.

Hours & Admission Prices: Wed.-Sat. 10-4, Sun. 1-4. No charge; donations accepted. &

Attendance: 200,000 (estimated)

Membership: Pinellas County Historical Society: Junior Docent & College Student $10; Individual $20; Family $35; Contributing $50; Sustaining $100; Patron $500; Benefactor $1,000.

NATIONAL COMEDY HALL OF FAME, (M), 9011 Park Blvd., Ste. 202, Largo, FL 33777-4123. Mailing Address: P.O. Box 20492, St. Petersburg, FL 33742-0492.

E-mail: comedyhall@aol.com

Web Site: comedyhall.com

Key Personnel: Exec. Dir., Tony Belmont

Institution Type/Description: Comedy Museum.

Collections: history of comedy in American theater; comedy legends.

Facilities: library. Museum-related items for sale.

Hours & Admission Prices: Call for hours.

Leesburg

LEESBURG CENTER FOR THE ARTS, 429 W. Magnolia St., Leesburg, FL 34749. Mailing Address: P.O. Box 492857, Leesburg, FL 34749-2857. Tel.: 352-365-0232. Fax: 352-315-1152.

Web Site: www.leesburgcenter4arts.com/index.htm

Governing Authority: Tax-exempt: 501(c)(3).

Institution Type/Description: Art Gallery.

Collections: paintings; photographs; sculpture.

Activities: educational programs. Annual Event: Leesburg Fine Art Festival in March.

Hours & Admission Prices: Call for hours.

Live Oak

SUWANNEE COUNTY HISTORICAL MUSEUM, 208 N. Ohio Ave., Live Oak, FL 32064-2455. Tel.: 386-362-1776.

E-mail: suwanneemuseum@yahoo.com

Web Site: suwanneemuseum.org

Founded: 1981.

Key Personnel: Dir., Randy S. Torrance.

Personnel Profile: Full-Time Paid 1.

Governing Authority: nonprofit. Tax-exempt.

Institution Type/Description: History Museum.

Collections: Suwannee County history; railroad layout; railroad artifacts; telephones; early American kitchen; Timucuan Indian artifacts; post office & farm equipment; photographs.

Activities: special events; seasonal program events & lectures. Museum Sponsors: Civil War Living History Event in February; Chili-Challenge Cook-Off in Spring; Railroad Festival in September; Holiday Open House in December.

Publications: quarterly newsletter, The Suwannee Freight Line.

Hours & Admission Prices: Tues.-Sat. 9-3. No charge; donations accepted.

Membership: Household & Family $10; Corporate & Business $50.

Loxahatchee

LION COUNTRY SAFARI, 2003 Lion Country Safari Rd., Loxahatchee, FL 33470-3977. Tel.: 561-793-1084. Fax: 561-793-9603.

E-mail: sales@lioncountrysafari.com

Web Site: www.lioncountrysafari.com

Founded: 1967.

Key Personnel: Dir. Wildlife, Terry Wolf; Cur. Education, Rhonda Beitmen; Public Rels. & Dir. Mktg., Jennifer Berthume; Coord. Mktg., Esther Sierra

Institution Type/Description: Zoo.

Collections: over 900 animals including giraffe, zebra, chimpanzees, lions, white rhino, antelope, tortoise, birds, reptiles.

Activities: drive-through safari; special events; group tours; animal feedings.

Hours & Admission Prices: Mon.-Fri. 10-5, Sat.-Sun. & holidays 9:30-5:30. Admission 10-64 $26.50, seniors 65 & over $23.50, children 3-9 $19.50; children 2 & under no charge. Parking: $6 per vehicle.

Madison

NORTH FLORIDA COMMUNITY COLLEGE ART GALLERY, 325 N.W. Turner Davis Dr., Ste. A, Madison, FL 32340-1611. Tel.: 850-973-1642. Fax: 850-973-9288.

E-mail: bardenl@nfcc.edu

Founded: 1975.

Congressional District: 2

Key Personnel: Pres., Morris Steen; Dir., William F. Gardner, Jr.

Governing Authority: college. Affiliated with North Florida Community College. Tax-exempt.

Institution Type/Description: College Art Gallery.

Collections: permanent collection.

Facilities: theater; classrooms.

Activities: guided tours; lectures; films; gallery talks; arts festivals; formally organized education programs; permanent exhibitions.

Hours & Admission Prices: Mon.-Fri. 10-12 & 1-3; Sun. special openings. No charge. &

Maitland

ART & HISTORY MUSEUMS - MAITLAND, (M), 231 W. Packwood Ave., Maitland, FL 32751-5596. Tel.: 407-539-2181. Fax: 888-316-5729. Facebook: art & History Museums - Maitland.

E-mail: info@artandhistory.org

Web Site: artandhistory.org

Formerly: Maitland Art Center; Research Studio, Maitland Art Center; and Maitland Historical Society

Founded: 1937.

Congressional District: 7

Key Personnel: C.E.O., Andrea Bailey Cox; Pres. Bd., Valerie Seidel; 1st Vice Pres., Thomas K. Maurer; 2nd Vice Pres., Suzanne Oberholtzer; Treas., Bill Randolph; Sec., Duncan DeWahl; Curator Art & Dir. Education, Rebecca Sexton-Larson; Dir. Mktg., Gretchen Miller Basso; Dir. Devel., Devin Dominguez; Mgr. Operations, Kasey Jones; Rental & Retail Mgr., Brittany Green.

Personnel Profile: Full-Time Paid 7; Part-Time Paid 6; Interns 19.

Governing Authority: nonprofit organization; bd. directors. Parent Institution: Maitland Art & History Association. Subsidiary Institutions: Waterhouse Residence and Carpentry Shop Museums, 820 Lake Lily Dr., Maitland, FL 32751; Telephone Museum and Maitland Historical Museum, 221 W. Packwood Ave., Maitland, FL 32751. Tax-exempt: 501(c)(3).

Institution Type/Description: Art & History Museums.

Collections: works of artists who lived on Art Center grounds since 1937 including Milton Avery & Ralston Crawford; works of founder, Andre Smith; contemporary Florida work; Camera Works photographs: A&H permanent collection; Victorian furnishings; early telephones & tools; local history. Historic Buildings: 1884 Victorian home; 1883 carpentry shop.

Major Exhibits: Environmental Pioneers: The Florida Audubon, 1/9/14-5/4/14; Constructed Landscapes: Works by Jake Fernandez, 1/10/14-2/28/14; Portraits: Birds of Paradise, 1/10/14-2/28/14; Film Stories, 3/14/14-4/20/14; Moving Pictures, 3/14/14-4/20/14; Springtime at the Waterhouse, 3/6/14-5/4/14; Veterans Remembered, 5/15/14-10/5/14; The Waterhouse at War, 5/22/14-8/10/14; Battlefield, 6/6/14-8/10/14.

Research Fields: works of founder, Jules Andre Smith (1880-1959) & those artists who were participants as Bok Fellows at the Center, 1938-1959; early telephones & tools; Victorian Florida; Maitland history.

Facilities: outdoor gardens; chapel; classrooms; artist studios. Books, catalogs, jewelry, art reproductions & other museum-related items for sale.

Activities: history programs; quarterly art classes; special events; research.

Publications: Maitland history book; exhibition catalogs.

Hours & Admission Prices: Art Center Galleries: Tues.-Sun. 11-4. History Museums: Thurs.-Sun. 12-4. Adults $3, seniors 55 & over and children 4-18 $2; children under 3 & members no charge. Closed major holidays. &

Attendance: 55,000 (accurate)

Membership: Student & Senior Citizen $30; Individual Maitland Resident $35; Individual $40; Dual Senior Citizens 55 & over $50; Family & Dual $60; Patron $100; Sustaining $195; Supporting $275.

HOLOCAUST MEMORIAL RESOURCE AND EDUCATION CENTER OF FLORIDA, INC., 851 N. Maitland Ave., Maitland, FL 32751-4461. Tel.: 407-628-0555, ext. 284. Fax: 407-628-1079.

E-mail: info@holocaustedu.org

Web Site: www.holocaustedu.org

Founded: 1983.

Key Personnel: Exec. Dir., Pam Kancher; Chm. (V), Tess Wise; Pres. (V), Randall Ellington; First Vice Pres., Stanley Creel; Second Vice Pres., Jim Shapiro; Sec., Diane Jacobs; Treas., Phillip Senderowitz; Cur., Anita Lam; Archivist, Alice Gamson; Devel., Susan Mitchell; Education, Mitchell Bloomer; Public Rels., Eva Ritt; Admin. Asst., Mary Johnson.

Personnel Profile: Full-Time Paid 4; Part-Time Paid 2; Part-Time Volunteers 13; Interns 1.

Governing Authority: private; nonprofit organization. Tax-exempt: 501(c)(3).

Institution Type/Description: History Museum.

Collections: Holocaust artifacts; original art by professional artists.

Facilities: 5,000-vol. library of books.

Activities: film series; education programs for students; teacher training; courses for adults; traveling & documentary exhibitions. Museum Sponsors: commemorative programs for Kristallnacht & Holocaust; Remembrance Day.

Publications: semiannual newsletter; Holocaust Education Curricula; exhibit & field trip guides.

Hours & Admission Prices: Mon.-Thurs. 9-4, Fri. 9-1, Sun. 1-4. No charge; donations accepted. Closed Jewish holidays; national holidays. &

Attendance: 20,000 (estimated)

Membership: Student $18; Friend $36; Family $100; Patron $150; Supporter $250; Sustainer $500; Benefactor $1,000; Builder $1,500; Founder $5,000.

Marathon

CRANE POINT MUSEUM & NATURE CENTER, 5550 Overseas Hwy., Marathon, FL 33050-2713. Mailing Address: P.O. Box 500536, Marathon, FL 33050-0536. Tel.: 305-743-3900 & 9100 (museum). Fax: 305-743-8172.

E-mail: loretta@cranepoint.net

Web Site: www.cranepoint.net

Founded: 1990.

Congressional District: 20

Key Personnel: Dir. Operations, Loretta Geotis.

Personnel Profile: Full-Time Paid 1; Full-Time Volunteers 0; Part-Time Paid 4; Part-Time Volunteers 20.

Governing Authority: nonprofit organization. Parent Institution: Florida Keys Land & Sea Trust. Tax-exempt: 501(c)(3).

Institution Type/Description: Natural History Museum.

Collections: photos; shipwreck artifacts; dioramas; railroad; nature trail; shells; hardwoods; butterflies; cultural history development of Florida Keys. Historic Adderley Village: restored 1903 Conch home, kitchen house & garden;

Facilities: 63 acre nature trail system. Gift items for sale.

Activities: concerts; docent program; formal education programs for children; guided tours; lectures; participatory exhibits. Museum Sponsors: Summer Day Camp (8-10 yr.); migrating bird festival; Super Science Saturdays (9-11 yr.).

Hours & Admission Prices: Mon.-Sat. 9-5, Sun. 12-5. Adults $12.50, senior citizens 55 & over $11, students 6-12 $8.50; discounts to groups; children under 6 & members no charge. Closed Christmas. &

Attendance: 30,000 (accurate)

Membership: Individual $30; Family $75; Conservator $200; Patron $500; Benefactor $1,000.

Marco Island

MARCO ISLAND HISTORICAL SOCIETY, (M), 180 S. Heathwood Dr., Marco Island, FL 34145-2010. Mailing Address: P.O. Box 2282, Marco Island, FL 34146-2282. Tel.: 239-389-6447. Facebook: marcoislandhistoricalsociety.

Web Site: www.themihs.com

Founded: 1994.

Congressional District: 25

Key Personnel: Pres., Darcie Guerin; Pres. (V), Thomas M. Wagor; Cur., Collections, Austin Bell; Museum Shop Mgr., Lori Wagor.

Governing Authority: Branch Museum: Marco Island Area Assoc. of Realtors, 140 Waterway Dr., Marco Island, FL.

Institution Type/Description: Historical Society Museum.

Collections: local history & culture; photographs.

Facilities: Rose History Auditorium.

Hours & Admission Prices: Tues.-Sat. 9-4. No charge.

Melbourne

BREVARD ZOO, 8225 N. Wickham Rd., Melbourne, FL 32940. Tel.: 321-254-9453. Fax: 321-259-5966.

E-mail: admin@brevardzoo.org

Web Site: www.brevardzoo.org

Institution Type/Description: Zoo.

Collections: over 500 animals including white rhinos, red kangaroos, wallabies, crocodiles, monkeys, & bald eagles; manatees, dolphins.

Activities: petting zoo; train tour; alligator feedings; kayaking; guided tours. Museum Sponsors: Safari Under the Stars; Great Tastes of Suntree; Boo at the Zoo.

Hours & Admission Prices: Daily 9:30-5. Adults $13.50, seniors $12.50, children 2-12 $10; discounts to groups; children under 2 no charge. Closed Thanksgiving; Christmas.

FLORIDA INSTITUTE OF TECHNOLOGY BOTANICAL GARDENS, Florida Tech, 150 W. University Blvd., Melbourne, FL 32901-6982. Tel.: 321-674-8962. Fax: 321-674-7257.

Web Site: www.facilities.fit.edu/botanical_gardens.php

Key Personnel: Dir., John Milbourne

Institution Type/Description: Botanical Gardens.

Collections: palm trees; tropical plants.

Facilities: 30-acres.

Hours & Admission Prices: Daily dawn to dusk.

FOOSANER ART MUSEUM, 1463 Highland Ave., Melbourne, FL 32935-6562. Tel.: 321-674-8916. Fax: 321-242-0798.

E-mail: info@foosanerartmuseum.org

Web Site: www.foosanerartmuseum.org

Formerly: Brevard Art Museum

Founded: 1978.

Congressional District: 15

Key Personnel: Dir. University Museums, Carla Funk; Cur. Exhibitions, Jackie Borsanyi; Administrative Asst. to Dir., Tama Johnson; Mgr. Collections, Jose Marquez; Mgr. Visitor Svcs., Tina Murray.

Personnel Profile: Full-Time Paid 6; Part-Time Paid 4; Part-Time Volunteers 132.

Governing Authority: nonprofit organization. Parent Institution: Florida Institute of Technology. Tax-exempt: 501(c)(3).

Institution Type/Description: Art Museum.

Collections: works by 20th-century women artists; works on paper; contemporary artists; Chase collection of Industrial Design; Ernst Oppler; Central Florida artists; Clyde Butcher.

Facilities: 1,000-vol. library of art reference books & magazines; 5,000 sq. ft. exhibition space; 100-seat auditorium; ceramics & sculpture studio; printmaking, painting & drawing studios. Museum-related items for sale.

Activities: guided tours; lectures; films; gallery talks; concerts; arts festivals; formally organized education programs; docent program; loan & traveling exhibitions.
Publications: select exhibition catalogs; gallery guides; class brochures; e-newsletter.
Hours & Admission Prices: Tues.-Wed. & Fri.-Sat. 10-5, Thurs. 10-7, Sun. 1-5. Adults $5, seniors, children & students $2; members, NARM members & Thurs. no charge. Closed major holidays. ♿
Attendance: 70,000 (estimated)
Membership: Senior & Student $25; Individual $40; Family $60; Patron $400; Director's Circle $500; Renaissance $1,000.

LIBERTY BELL MEMORIAL MUSEUM, 1601 Oak St., Melbourne, FL 32901-4516. Mailing Address: P.O. Box 1776, Melbourne, FL 32902. Tel.: 321-727-1776.
E-mail: susanhonoramerica@cfl.rr.com
Web Site: www.honoramerica.org
Founded: 1985.
Key Personnel: Museum & Educational Programs Dir., Susan Anderson.
Personnel Profile: Part-Time Paid 1.
Institution Type/Description: History Museum.
Collections: full-size replica of the original Liberty Bell; U.S. flags; documents; local history; American War artifacts; model warships & airplanes; weapons; clothing.
Facilities: Museum-related items for sale.
Hours & Admission Prices: Tues.-Fri. 10-4, Sat. 10-2. No charge; donations accepted. ♿
Attendance: 15,000
Membership: Patriot $30; Bronze Patriot $100; Silver Patriot $250; Gold Patriot $500; Platinum Patriot $750; Diamond Patriot $1,000.

Merritt Island

NASA KENNEDY SPACE CENTER, SR 405, Merritt Island, FL 32899. Tel.: 321-867-5000; 866-737-5235.
Institution Type/Description: Space Museum.
Collections: past, present & future space program artifacts; rockets; Space Shuttle Explorer; Astronaut Hall of Fame; NASA's launch headquarters.
Facilities: restaurant; theater. Museum-related items for sale.
Activities: tours; launch programs; NASA programs; interactive space flight simulators.
Hours & Admission Prices: Visitors Center: daily 9-7. Hall of Fame: daily 9-8. Space Center Tour: adults $31, children 3-11 $21. Hall of Fame: adults $17, children 3-11 $13. Closed Christmas; occasional launch days.

Miami

BAY OF PIGS MUSEUM & LIBRARY, 1821 S.W. 9th St., Miami, FL 33135-5101. Tel.: 305-649-4719. Fax: 305-649-8719.
E-mail: bgd2506@gmail.com
Web Site: www.brigada2506.com/museum.htm
Founded: 1979.
Key Personnel: C.E.O. & Museum Shop Mgr., Esteban Bovo; Chm. (V), Felix Rodrigues; Pres. (V), Max Crul; Museum Shop Mgr., Jorge Marquet.
Personnel Profile: Part-Time Paid 2; Part-Time Volunteers 2.
Governing Authority: Parent Institution: Bay of Pigs Veterans Association. Tax-exempt.
Institution Type/Description: History Museum.
Collections: history of the Bay of Pigs; maps; uniforms; arms; organization charts; names of Brigade members; photographs; books; film.
Facilities: library.
Publications: magazine, Giron.
Hours & Admission Prices: Mon.-Fri. 10-9, Sat. 10-2; other times by appointment. No charge; donations accepted. ♿
Attendance: 2,000 (estimated)

BLACK HERITAGE MUSEUM, 15801 S.W. 102nd Ave., Miami, FL 33157-1653. Mailing Address: P.O. Box 570327, Miami, FL 33257-0327. Tel.: 786-287-1157.
E-mail: blkhermu@yahoo.com
Web Site: lets.showmywebsite.com
Founded: 1987.
Key Personnel: Pres., Priscilla G. Stephens Kruize; Financial Dir. & Treas., Gary Roberts; Sec., Eva Cofield.
Governing Authority: nonprofit organization. Tax-exempt: 501(c)(3).
Institution Type/Description: Cultural Arts & History Museum.
Collections: art & artifacts of the Black heritage from around the world; tribal artifacts from Africa & New Guinea; Black Americana; Nigeria's terracotta.
Research Fields: Black heritage & culture; cultural exchanges.

Activities: guided tours; lectures; dance recitals; concerts; permanent & temporary exhibits; special exhibitions.
Publications: newsletter.
Hours & Admission Prices: Call for hours; groups by appointment. No charge; donations accepted. Closed New Year's Day; Easter; Christmas. ♿
Attendance: 4,010
Membership: Active $25; Family $35; Associate $50; Patron $100; Corporate $500.

CORAL CASTLE MUSEUM, 28655 S. Dixie Hwy., Miami, FL 33033. Tel.: 305-248-6345.
Web Site: www.coralcastle.com
Institution Type/Description: Historic Building: built by Edward Leedskalnin from 1923-1951. Listed on the National Register of Historic Places.
Collections: 1,100 tons of carved coral rock.
Activities: rental facilities.
Hours & Admission Prices: Sun.-Thurs. 8-6, Fri.-Sat. 8-8. Adults $9.75, senior citizens 62 & over $6.50, children 7-12 $5; discounts to groups; children under 6 no charge.

DEERING ESTATE AT CUTLER, 16701 S.W. 72nd Ave., Miami, FL 33157-2500. Tel.: 305-235-1668. Fax: 305-254-5866.
Web Site: www.deeringestate.org
Key Personnel: Exec. Dir. Deering Estate Foundation, Mary Pettit
Institution Type/Description: Historic Estate.
Collections: family history; paintings; sculpture; personal artifacts; period furnishings; archaeology; historic buildings.
Activities: classes; programs.
Hours & Admission Prices: Daily 10-5. Adults $12, youth 4-14 $7; discounts to groups; members no charge. Closed Thanksgiving; Christmas. ♿

DORSCH GALLERY, 151 N.W. 24th St., Miami, FL 33127. Tel.: 305-576-1278.
Web Site: dorschgallery.com
Institution Type/Description: Art Gallery.
Collections: sculpture; paintings.
Facilities: 3,600 sq. ft. exhibition space.
Activities: lectures; temporary exhibitions.
Hours & Admission Prices: Tues.-Sat. 12-5.

GALLERY NORTH, MIAMI-DADE COMMUNITY COLLEGE, NORTH CAMPUS, 11380 N.W. 27th Ave., Miami, FL 33167-3418. Tel.: 305-237-1532. Fax: 305-237-1850.
Key Personnel: Dir. Communications, Juan Mendieta
Institution Type/Description: Art Museum.
Collections: paintings; sculpture; photographs.
Activities: lectures.
Hours & Admission Prices: Mon.-Thurs. 9-5, Fri. 9-2. No charge.

GOLD COAST RAILROAD MUSEUM, INC., 12450 S.W. 152nd St., Miami, FL 33177-1402. Tel.: 305-253-0063 & 505-5405; 888-608-7246 (Toll Free). Fax: 305-233-4641.
E-mail: mike196147@yahoo.com
Web Site: www.goldcoast-railroad.org
Founded: 1957.
Congressional District: 18
Key Personnel: C.E.O. & Exec. Dir., Michael D. Hall; Pres., Russell Swain; Vice Pres., Connie Greer; Treas., John McLean; Gift Shop Mgr., Patricia Huffman.
Personnel Profile: Full-Time Paid 6; Full-Time Volunteers 20; Part-Time Paid 6; Part-Time Volunteers 20; Interns 2.
Governing Authority: nonprofit organization. Tax-exempt: 501(c)(3).
Institution Type/Description: Railroad Museum: located on historic NAS RICHMOND, second largest Airship Naval Base World War II.
Collections: Presidential Pullman, Ferdinand Magellan; California Zephyr, Silver Crescent & Silver Stag, FEC Steam Locomotives #113 & 153; FEC Coach Belle Glade & #136; model train area; freight equipment; 7 diesel locomotives & passenger sleepers, coaches, diner & lounge cars; helium tank car; 4 cabooses; cranes. Historic Buildings: Princeton Station; buildings remaining from Richmond Naval Air Station.
Major Exhibits: Presidential Railcar Opening, 1/14-12/14.
Facilities: educational facilities; 52 acres exhibition space. Museum-related items for sale.
Activities: guided tours; training programs for professional museum workers; rental facilities; school summer programs. Annual Events: Dade Heritage Days; Railroad Days; Tropical Agricultural Fiesta in July.
Publications: newsletter, Steam & Steel.
Hours & Admission Prices: Grounds: Tues.-Wed. & Fri. 10-4, Thurs. 10-7,

Sat.-Sun. 11-4. Train Bldg.: Mon.-Fri. 11-2, Sat.-Sun. 11-4. Coach: 2nd weekend each month. Cab Rides: $12. Children's Railroad: $2.50. Coach Rides: adults $8, children 2-11 $6; discounts to local community organization, groups & AAM members; members & children 3 & under no charge. &

Attendance: 120,000 (estimated)
Membership: General $35; Family $50.

HAITIAN HERITAGE MUSEUM, 4141 N.E. 2nd Ave., Ste. 105C, Miami, FL 33137. Tel.: 305-371-5988. Fax: 305-432-3792.
Web Site: www.haitianheritagemuseum.org
Key Personnel: Exec. Dir., Eveline Pierre; Dir. Operations, Serge Rodriguez; Mgr. Education, Sam Joseph
Institution Type/Description: Heritage Museum.
Collections: Haiti's history & culture; personal artifacts; photographs; Haitian art & music.
Facilities: 200-seat theater.
Hours & Admission Prices: Tues.-Fri. 10-5; other times by appointment.

*** HISTORYMIAMI, (M),** 101 W. Flagler St., Miami, FL 33130-1504. Tel.: 305-375-1492. Fax: 305-375-1609. Facebook: HistoryMiami.
E-mail: e.info@historymiami.org
Web Site: www.historymiami.org
Formerly: Historical Museum of Southern Florida
Founded: 1940.
Congressional District: 18
Key Personnel: Interim Pres. & C.E.O., Roxanne Cappello; Chm., Faith Mesnekoff; Cur. Research, Rebecca A. Smith; Chief Cur., Dr. Joanne Hyppolite; Vice Pres. Devel., Emma Heald; Vice Pres. Education, Cecilia Slesnick; Vice Pres. Expansion, Jorge Zamanillo; Vice Pres. Communications, Victoria Cervantes.
Personnel Profile: Full-Time Paid 25; Part-Time Paid 22; Part-Time Volunteers 15; Interns 4.
Governing Authority: society. Parent Institution: Historical Association of Southern Florida. Tax-exempt: 501(c)(3).
Institution Type/Description: History Museum.
Collections: historical artifacts of south Florida; archives & manuscripts; archaeology; maps; photographs; Audubon's Birds of America; recreational powerboats; aviation.
Research Fields: history, folk life & archaeology of southern Florida and Caribbean, West Indies.
Facilities: research center; theater; meeting rooms; sculpture garden. Books & gifts for sale.
Activities: lectures; formally organized educational programs; permanent & temporary exhibitions; city tours; eco-history tours. Annual Event: Miami International Map Fair & Fundraiser.
Publications: annual journal, Tequesta.; annual magazine, HistoryMiami.
Hours & Admission Prices: Tues.-Fri. 10-5, Sat.-Sun. 12-5. Adults $8, seniors & students $7, children 6-12 $5; discounts to AAM & ICOM members; children under 6 & members no charge. Closed New Year's Day; Martin Luther King Jr. Day; Presidents' Day; Memorial Day; Independence Day; Labor Day; Columbus Day; Veteran's Day; Thanksgiving; Christmas. &
Attendance: 94,000 (estimated)
Membership: Senior $40; Individual $50; Senior Dual $55; Dual $65; Family $75; History Buff $125; Trail Blazer $250; Fellow $500.

JUNGLE ISLAND, 1111 Parrot Jungle Trail, Miami, FL 33132-1611. Tel.: 305-400-7000. Fax: 305-400-7290.
E-mail: guestrelations@jungleisland.com
Web Site: www.jungleisland.com
Formerly: Parrot Jungle Island
Founded: 1936.
Congressional District: 19
Key Personnel: Pres., Dr. Bern M. Levine; Education, Krishawana Thornton; Vice Pres. Sales, Andy Juska; Dir. Mktg., Leo Sarmiento; Human Resources, Miglide Garcon; Animal Science, Dr. Jason Chatfield; Operations, Pierre Sasiain; Museum Shop Mgr., Arlene Windess.
Personnel Profile: Full-Time Paid 180; Part-Time Paid 60; Interns 5.
Governing Authority: individual operation. Parent Institution: PJ Birds, Inc.
Institution Type/Description: Aviary & Botanical Garden.
Collections: botanical gardens; exotic plants; birds; reptiles; primates.
Facilities: auditoriums; botanical garden; 200-seat restaurant; meeting rooms; nature & conservation center; 18-acre walk-through park. Gifts & books for sale.
Activities: guided tours; docent program; formal educational programs; lectures; study clubs. Museum Sponsors: plant sales & holiday events.
Hours & Admission Prices: Daily 10-6. Adults $32.95, seniors $30.95,

children $24.95; discounts to groups, AAM members & contracted travel agents. Annual Passes available. &
Attendance: 500,000 (estimated)
Membership: Adult $42.95; Children $30.95.

*** MIAMI ART MUSEUM, (M),** 101 W. Flagler St., Miami, FL 33130-1504. Tel.: 305-375-3000. Fax: 305-375-1725.
E-mail: mamnews@miamiartmuseum.org
Web Site: www.miamiartmuseum.org
Formerly: Center For The Fine Arts
Founded: 1978.
Congressional District: 17
Key Personnel: Chm., Aaron Podhurst; Pres. (V), Gail S. Meyers; Dir., Thomas Collins; Chief Cur., Tobias Ostrander; Cur. Education, Esther "Chipi" Morales; MAM Store Mgr., Michael Balbone.
Personnel Profile: Full-Time Paid 30; Part-Time Paid 29; Part-Time Volunteers 45.
Governing Authority: nonprofit organization. Tax-exempt: 501(c)(3).
Institution Type/Description: Art Museum.
Collections: 20th & 21st centuries international art.
Facilities: auditorium; sculpture court & outdoor plaza; picnic area. Art-related items for sale in MAM store.
Activities: traveling & temporary exhibitions; lectures; films; guided tours available by reservation. Museum Sponsors: free hands-on activities for children and their families - 2nd Sat. of each month.
Publications: e-newsletter, now@mam; Triumph of the Spirit: Carlos Alfonzo, A Survey 1975-1991; Modern Photographs: the Machine, The Body & The City; Jan Dibbets: Perspective Collection; Work in Progress: Herzog & de Meuron's Miami Art Museum; Wifredo Lam at Miami Art Museum.
Hours & Admission Prices: Tues.-Fri. 10-5, Sat.-Sun. 12-5. Adults $8, seniors $4; discounts to AAM, ICOM & AAA members; students with ID, members, children under 12 no charge. &
Attendance: 62,500 (estimated)
Membership: Individual $45; Dual $60; Family $75; Sustaining $125; Contributing $250; Supporting $500; Friends $1,000; Museum Circle $2,500 & up.

MIAMI CHILDREN'S MUSEUM, (M), 980 MacArthur Causeway, Miami, FL 33132-1604. Tel.: 305-373-5437, ext. 100. Fax: 305-373-5431.
E-mail: info@miamichildrensmuseum.org
Web Site: miamichildrensmuseum.org
Founded: 1983.
Congressional District: 19
Key Personnel: Pres. (V), Jeffrey Berkowitz; C.E.O. & Exec. Dir., Deborah Spiegelman; Dir. Exhibits, Mike Neufeld; Dir. Mktg., Raquel Alderman; Dir. Devel., Belissa Alvarez; Museum Shop Mgr., Johanne Bitton.
Personnel Profile: Full-Time Paid 35; Part-Time Paid 100; Part-Time Volunteers 280.
Governing Authority: nonprofit organization. Tax-exempt: 501(c)(3).
Institution Type/Description: Children's Museum.
Collections: hands-on educational exhibits.
Activities: gallery tours; school outreach. resource center; drop-in activities; early childhood series. Museum Sponsors: Children's Film Festival.
Publications: teacher & student educational packets.
Hours & Admission Prices: Daily 10-6. Adults $16, Florida Residents $12; members no charge. Closed Thanksgiving; Christmas. &
Attendance: 399,000 (estimated)
Membership: Grandparent $125; Family & Patron $150; Sponsor $500; Benefactor $1,000.

MIAMI-DADE COLLEGE KENDALL CAMPUS ART GALLERY, (M), Martin and Pat Fine Ctr. for the Arts, 11011 S.W. 104th St., Miami, FL 33176. Mailing Address: MDC Museum and Galleries of Art + Design, 300 N.E. Second Ave., Miami, FL 33132. Tel.: 305-237-2322. Fax: 305-237-2901. TDD: 800-955-8771.
E-mail: lfontana@mdc.edu
Web Site: www.mdc.edu/kendall/art/default.asp
Founded: 1970.
Congressional District: 19
Key Personnel: Acting Dir., Lilia Fontana.
Personnel Profile: Full-Time Paid 1; Part-Time Paid 3.
Governing Authority: college. Tax-exempt: 170(b)(1)(A).
Institution Type/Description: Art Gallery.
Collections: contemporary paintings, prints, sculpture, photography, fibre, video & electronic media; 16th to 19th-century engravings, etchings & lithographs; artists' books.
Research Fields: contemporary, African Diaspora, Latin American & Hispanic art; art of the minorities.

Facilities: 300-seat auditorium; 300-seat restaurant; 3,000 sq. ft. exhibit space.
Activities: guided tours; lectures; films; docent program; training programs for professional museum workers; organized education programs for college students affiliated with Miami-Dade Community College; temporary, loan & traveling exhibitions.
Publications: exhibition catalogues.
Hours & Admission Prices: Sept.-July Mon.-Thurs. 8-7:30, Fri. 8-4, Sat. 9:30-4:30. No charge. &
Attendance: 9,500 (estimated)

* **MIAMI SCIENCE MUSEUM, (M),** 3280 S. Miami Ave., Miami, FL 33129-2899. Tel.: 305-646-4200. Fax: 305-646-4300.
E-mail: gthomas@miamisci.org
Web Site: www.miamisci.org
Founded: 1949.
Congressional District: 18
Key Personnel: Pres. & C.E.O., Gillian Thomas; Chm. (V), Trish Bell; Chm. (V), Dan Bell; C.F.O., Nancy McKee; Vice Pres. Exhibits, Sean Duran; Dir. Wildlife Center, Greta Meally.
Personnel Profile: Full-Time Paid 60; Part-Time Paid 37; Part-Time Volunteers 600.
Governing Authority: county; nonprofit organization. Tax-exempt: 501(c)(3).
Institution Type/Description: Science Museum.
Collections: anthropology; archaeology; astronomy; entomology; ethnology; geology; Florida Indian artifacts; marine; mineralogy; nature center; South Florida birds; dynamic hands on light; sound; optics; chemistry; biology; physics; energy; science; technology; invention; wildlife center with walk-in aviary, insects & reptiles; gallery, The Body in Action, explores wonders of the human body; natural science specimens.
Research Fields: radio; photography; marine life; archaeology; botany; astronomy; meteorology; mineralogy; herpetology; conchology; malacology; ornithology; tropical ecology.
Facilities: 5,000-vol. library of scientific books & material relating to all scientific fields available for inter-library loan & by application; two computer labs; environmental information center; nature center; planetarium; 240-seat theater; 230-seat planetarium; Macintosh Training Center; theater; classrooms; observatory; dark rooms; mini-science theater. Scientific items, items relating to South Florida & other museum-related gifts for sale.
Activities: guided tours; lectures; films; gallery talks; concerts; radio drama; study clubs; hobby workshops; TV & radio programs; formally organized education programs; docent program or council; permanent, temporary & traveling exhibitions; school loan service. Museum Sponsors: Latin-oriented programs; technology training center; teacher training.
Publications: summer camp guide; newsletter, Miamisci.
Hours & Admission Prices: Daily 10-6. Adults $14.95, students & seniors $16, children 3-12 $10.95; discounts to museum members & ASTC members; children under 3 no charge. Closed Thanksgiving; Christmas. &
Attendance: 300,000 (estimated)
Membership: Student & Individual $40; Family $60; Family Plus $80; 21st Century Club $150; Contributor $250; Donor $500; Benefactor $1,000.

MIAMI SEAQUARIUM, 4400 Rickenbacker Causeway, Miami, FL 33149-1095. Tel.: 305-361-5705. Fax: 305-361-6077.
E-mail: cperrina@msq.cc
Web Site: www.miamiseaquarium.com
Founded: 1955.
Congressional District: 15
Key Personnel: Gen. Mgr., Andrew Hertz; Veterinarian, Maya Menchaca, D.V.M.; Financial Dir., Sherryl Moody; Dir. Operations, Charles Gaudio; Museum Shop Mgr., Rosa White.
Governing Authority: business organized for profit. Wometco Enterprises, Inc., 3195 Ponce de Leon, Coral Gables, FL 33134. Tel: 305-529-1400.
Institution Type/Description: Aquarium.
Collections: killer whale; bottlenose dolphins; Pacific white-sided dolphin; Florida manatee; assorted species of shark, sea turtles, tropical & subtropical fishes; sea lions.
Research Fields: study & raising of sea turtle hatchlings; behavioral & nutritional studies concerning Florida manatee husbandry.
Facilities: botanical garden; aquarium; 230-seat cafeteria. Museum-related items for sale.
Activities: occasional lectures; slides; permanent exhibitions; educational activities; swim with the dolphins.
Hours & Admission Prices: Daily 9:30-6. Adults $35.95, children $26.95; discounts to senior citizens, military, AAA & AAM members. Annual Pass: adult $38.90, children $31.90. &
Attendance: 650,000

N'NAMDI CONTEMPORARY MIAMI, 177 N.W. 23rd St., Miami, FL 33127. Tel.: 786-332-4736.
Key Personnel: Dir., Jumaane N'Namdi
Institution Type/Description: Art Gallery.
Collections: works by contemporary artists.
Activities: traveling & temporary exhibitions; special events.
Hours & Admission Prices: Call for hours.

* **THE PATRICIA & PHILLIP FROST ART MUSEUM, (M),** 10975 S.W. 17th St., Miami, FL 33199. Tel.: 305-348-2890. Fax: 305-348-2762.
E-mail: artinfo@fiu.edu
Web Site: www.thefrost.fiu.edu
Formerly: The Art Museum at Florida International University
Founded: 1977.
Congressional District: 16
Key Personnel: Dir., Carol Damian, Ph.D.; Budget & Finance Mgr., Mary Alice Manella; Cur. Collections & Registrar, Debbye Taylor; Cur. Education, Miriam Machado; Administrative Asst., Elisabeth Gonzalez; Membership Coord., Ximena Gallegos; Dir. Devel., Michael Hughes; Asst. Cur., Klaudio Rodriguez; Visitor Svcs. & Events Asst., Jessica Lettsome; Exhib. & Sculpture Park Mgr., Albert Hernandez; Museum Studies Coord., Annette B. Fromm; Asst. Registrar, Sherry Zambrano; New Media Specialist, Raymond Mathews; Security Mgr., Julio Alvarez; Security Guard, Ragan Williams; Security Guard, Luis Tabares.
Personnel Profile: Full-Time Paid 18; Part-Time Paid 10; Part-Time Volunteers 23; Interns 10.
Governing Authority: state. Parent Institution: Florida International University. Tax-exempt: 170(b)(1)(A).
Institution Type/Description: Art Museum.
Collections: European, North & South American paintings, drawings, prints & sculpture; African, Oriental & pre-Columbian artifacts; international outdoor sculpture park, ArtPark.
Research Fields: fields pertaining to collections.
Facilities: library of art books, magazines & catalogues available on premises; sculpture park; auditorium; classrooms.
Activities: museum curated & traveling exhibitions; Critics' Lecture Series; formally organized educational programs; artist walkthroughs; ArtSmart tours. Museum Sponsors: Target Wednesday After Hours.
Publications: newsletter; exhibition catalogues; pamphlets; calendars.
Hours & Admission Prices: Tues.-Sat. 10-5, Sun. 12-5. No charge; donations accepted. Closed holidays. &
Attendance: 112,000 (estimated)
Membership: University Community $35; Individual $75; Family $125; Patron $250; Connoisseur Circle $1,000 & up.

RUBELL FAMILY COLLECTION AND CONTEMPORARY ARTS FOUNDATION, 95 N.W. 29th St., Miami, FL 33127. Tel.: 305-573-6090. Fax: 305-573-6023.
E-mail: info@rfc.museum
Web Site: www.rfc.museum
Key Personnel: Dir., Juan Roselione-Valadez
Institution Type/Description: Art Gallery.
Collections: works by contemporary artists; paintings; sculpture.
Activities: permanent & temporary exhibitions.
Hours & Admission Prices: By appointment. Adults $10; children under 18 $5.

* **VIZCAYA MUSEUM AND GARDENS, (M),** 3251 S. Miami Ave., Miami, FL 33129-2897. Tel.: 305-250-9133, ext. 8452. Fax: 305-285-7137. TDD: 800-955-8771; Facebook: Vizcaya Museum and Gardens.
E-mail: joel.hoffman@vizcaya.org
Web Site: www.vizcaya.org
Founded: 1952.
Congressional District: 18
Key Personnel: Exec. Dir., Joel M. Hoffman; Trust Chm. (V), Rayfield McGhee; Deputy Dir. Finance & Administration, Luis Correa; Asst. to Dir., Kyndal Campbell; Deputy Dir. Advancement, Dennis Fruitt; Deputy Dir. Learning, Ann Loshaw; Deputy Dir. Collections & Curatorial Affairs, Remko Jansonius; Events Dir., Adrienne Kaiser; Learning Programs Mgr., Wendy Wolf; Cur., Gina Wouters; Archivist, Alexander Privee; Grants Mgr., Gina Sacchetti; Chief Horticulturist, Ian Simpkins; Security Chief, Junior Gonzalez; Guiding Programs Mgr., Mark Osterman; Maintenance Supvr., Jim Rustin.
Personnel Profile: Full-Time Paid 50; Part-Time Paid 6; Part-Time Volunteers 70; Interns 5.
Operating Expenses: 5,127,546
Operating Income: 6,058,152

Governing Authority: county. Parent Institution: Miami-Dade County. Tax-exempt.

Institution Type/Description: Historic House: 1916 European-Inspired Villa, Gardens & 11-building village on 50 acres of grounds, formerly the estate of International Harvester Executive James Deering.

Collections: European interiors of 16th- to 19th-century; mixed with American original paneling; murals; ceilings; doors; mantels; mirrors from palaces in Italy, France, Spain, England; historic period furnishings & art objects of marble, bronze, wood, textile, ceramic & ivory; formal gardens including orchids; photography; drawing & document archives; contemporary art.

Research Fields: architecture; decorative arts; gardens.

Activities: guided tours; audio tours; talks; films; concerts; performing arts events; permanent exhibitions; audio tours.

Hours & Admission Prices: Wed.-Mon. 9:30-4:30. Adults $18, seniors $12 students $10, children 6-12 $6; children 5 & under, AAM members and members no charge. Closed Thanksgiving; Christmas. &

Attendance: 192,741 (accurate)

WINGS OVER MIAMI AIR MUSEUM, 14710 S.W. 128 St., Miami, FL 33196-2002. Tel.: 305-233-5197. Fax: 305-232-4134. Facebook: Wings Over Miami.

E-mail: wingsovermiami.suz@bellsouth.net

Web Site: www.wingsovermiami.com

Founded: 2001.

Congressional District: 21

Key Personnel: Pres. (V) & Chm. (V), Suzette Rice; Operations Mgr., Konstantin Oulianov.

Personnel Profile: Full-Time Paid 2; Part-Time Volunteers 100.

Governing Authority: nonprofit organization. Tax-exempt: 501(c)(3).

Institution Type/Description: Aeronautics Museum: located inside Kendall-Tamiami Executive Airport.

Collections: aircraft; engines; instruments; propellers; aviation history of south Florida; airline memorabilia.

Research Fields: restoration of aircraft from beginning of flight to end of World War II era.

Facilities: video booths. Model airplanes, books, pins, patches, bomber jackets & shirts for sale.

Activities: guided tours; facility rental; children's activities; planes flying; special community events.

Publications: quarterly newsletter.

Hours & Admission Prices: Wed.-Sun. 10-5; see website to confirm. Adults $10, senior citizens $7, children 12 & under $6; discounts to groups, AAA, ICOM & AAM members; members no charge. Closed Thanksgiving; Christmas. &

Attendance: 5,500 (estimated)

Membership: Student $15; Associated $20; Individual $30; Family $40.

WOLFSON GALLERIES MIAMI-DADE COLLEGE, 300 N.E. 2nd Ave., Ste. 1365, Miami, FL 33132-2204. Tel.: 305-237-3417. Fax: 305-237-7309,

E-mail: mquiroga@mdcc.edu

Key Personnel: Dir., Mercedes A. Quiroga

Institution Type/Description: Art Museum.

Collections: works by national & international artists.

Hours & Admission Prices: Mon.-Wed. & Fri. 10-4, Thurs. 12-6. No charge. &

ZOO MIAMI, 12400 S.W. 152nd St., One Zoo Blvd., Miami, FL 33177-1402. Tel.: 305-251-0400, ext. 84910. Fax: 305-378-6381.

E-mail: info@zsf.org

Web Site: www.zoomiami.org

Formerly: Miami Metrozoo

Founded: 1980.

Congressional District: 19

Key Personnel: Dir., Eric Stephens; Chm. (V), Rob Hudson; C.E.O. Zoological Society of Florida, Ben Pingree; Zoo Business Mgr., Eric Kaminsky; Admission & Concessions, Julio Mesa; Museum Shop Mgr., Anabella Rodriguez.

Personnel Profile: Full-Time Paid 219; Part-Time Paid 116; Part-Time Volunteers 200; Interns 9.

Governing Authority: county. Parent Institution: Miami Dade County Parks Department. Tax-exempt.

Institution Type/Description: Zoo.

Collections: birds; mammals; ecotherms.

Research Fields: breeding of endangered species, specializing in crocodilians, birds, & ungulates.

Facilities: library; 327-acre exhibit site; amphitheatre; classroom facilities; monorail. Museum-related items for sale.

Activities: wildlife shows; outreach & in-house education programs; field trips;

groups events; party packages; special events; safari cycle rental; guided tram tours; giraffe, pelican & parrot feedings; camel rides; rhino encounter; zookeeper talks.

Publications: newsletter, Keepin' It Wild.

Hours & Admission Prices: Daily 9:30-5:30 (gates close at 4). Adults $15.95, children 3-12 $11.95; children 2 & under, American Zoo and Aquarium Association members no charge. &

Attendance: 810,998 (accurate)

Membership: Zoological Society of Florida: Senior Dual $64; Dual $69; Family $119.

Miami Beach

ARTCENTER/SOUTH FLORIDA, 924-810-800 Lincoln Rd., Miami Beach, FL 33139-2602. Mailing Address: 924 Lincoln Rd,. Ste. 205, Miami Beach, FL 37139. Tel.: 305-674-8278. Fax: 305-674-8772.

E-mail: email@artcentersf.org

Web Site: www.artcentersf.org

Founded: 1984.

Congressional District: 18

Key Personnel: C.E.O. & Exec. Dir., Maria Del Valle; Chm., Kim Kovel; Artistic Dir., Susan Caraballo; Dir. Education, Tammy Key Johnston; Controller, Patricia Leder.

Personnel Profile: Full-Time Paid 8; Part-Time Paid 6; Part-Time Volunteers 20; Interns 2.

Governing Authority: Branch Locations: 800 & 810 Lincoln Rd., Miami Beach, FL. Tax-exempt.

Institution Type/Description: Art Museum.

Collections: works by Miami artists.

Facilities: art education classrooms.

Activities: art programs & classes.

Hours & Admission Prices: Mon.-Thurs. 12-8, Fri.-Sat. 11-10, Sun. 11-9. No charge; donations accepted. &

Attendance: 96,000 (accurate)

Membership: Individual $45; Dual & Family $75; Friend $250; Art Partner $500; Corporate Art Partner $1,000.

*** BASS MUSEUM OF ART, (M),** 2100 Collins Ave., Miami Beach, FL 33139. Tel.: 305-673-7530. Fax: 305-673-7062.

E-mail: info@bassmuseum.org

Web Site: www.bassmuseum.org

Founded: 1963.

Congressional District: 18

Key Personnel: Exec. Dir., Silvia Karman Cubina; Chm. (V) & Pres. (V), George Lindemann; Asst. Dir. Operations, Jean Ortega; Dir. External Affairs, Megan Riley; Dir. Education, Adrienne von Lates; Devel. Assoc., Membership & Volunteer Coord., Denise Wolpert; Registrar & Exhibitions Coord., Chelsea Guerdat; Education Program Coach., Kylee Crook; Chief Preparator & Exhibition Technician, Jan Galliardt; Administrative Asst. to Dir., Leilani Lynch; Admissions Clerk, Gabrielle Peters.

Personnel Profile: Full-Time Paid 10; Part-Time Paid 18; Part-Time Volunteers 50; Interns 5.

Governing Authority: board of trustees. Parent Institution: City of Miami Beach. Tax-exempt: 170(b)(1)(A). Friends of the Bass Museum. Tax-exempt: 501(c)(3).

Institution Type/Description: Art Museum.

Collections: Encyclopedic collection ranging from Old Master paintings, sculpture, textiles; ecclesiastic artifacts; Oriental bronzes; ceramics; decorative arts; 20th-century American graphics, paintings & sculpture; architectural drawings & prints; photography; Latin American & Haitian art.

Major Exhibits: TIME, 12/5/13-3/16/14.

Research Fields: European art & other collections.

Facilities: 7,200 sq. ft. exhibition space. Museum-related items for sale.

Activities: permanent & special exhibitions; lectures; gallery talks; concerts; films; docent program; multidisciplinary symposia; new media projects & presentations.

Publications: collections & special exhibitions catalogues; triannual member magazine.

Hours & Admission Prices: Wed.-Sun. 12-5. General Admission $8, senior citizens & students with valid ID $6; discounts to AAM & ICOM members; Miami Beach residents, members & children under 6 no charge. Call for list of reciprocal memberships. Additional charge for special exhibitions. Closed holidays. &

Attendance: 40,000 (estimated)

Membership: Student $25; Individual $50; Family & Dual $75; Sustaining $125; Contributing $250; Donor $500; Silver Director's Circle $1,000; Gold Director's Circle $2,500; Platinum Director's Circle $5,000.

✳ **JEWISH MUSEUM OF FLORIDA - FIU, (M),** 301 Washington Ave., Miami Beach, FL 33139-6965. Tel.: 305-672-5044. Fax: 305-672-5933. Facebook: Jewish Museum of Florida.
E-mail: director@jewishmuseum.com
Web Site: www.jewishmuseum.com
Founded: 1995.
Congressional District: 18
Key Personnel: Exec. Dir. & Chief Cur., Jo Ann Arnowitz; Pres., Elliot Stone; Founding Exec. Dir., Marcia Jo Zerivitz; Designer, Ira Newman; Education, Chaim Lieberperson; Membership, Nancy Doyle Cohen; Devel., Nancy Rachman; Museum Shop Mgr., Eva Shvedova; Fiscal Administrator & Grants Mgr., Irene Warner; Registrar, Todd Bothel; Coord. Administrative Svcs., Roberta Gordon; Asst. Cur., Jacqueline Goldstein.
Personnel Profile: Full-Time Paid 11; Part-Time Paid 2; Part-Time Volunteers 115; Interns 2.
Governing Authority: private; nonprofit organization. Parent Institution: Florida International University. Tax-exempt: 501(c)(3).
Institution Type/Description: Jewish History of Florida Museum: housed in a 1936 art deco-style building which served as a synagogue with Moorish copper dome & 80 stained-glass windows & 1929 first synagogue on Miami Beach.
Collections: photographs, artifacts, documents & ephemera relating to Florida Jewish history, documented from 1763 to present; Mosaic-Jewish life in Florida.
Major Exhibits: Grocers, Growers & Gefilte Fish: A Gastronomic Look at Florida Jews & Food, 10/13-10/5/14; Graphic Details: Confessional Comics by Jewish Women, 11/13-2/16/14; Cinema Judaica: The War Years, 1939-1949, 3/4/14-8/24/14.
Research Fields: Jewish history of Florida; to document Jewish life in America began in Florida in the 1500s, not in New York in 1654.
Facilities: 150-vol. library of Florida history & general Judaic material; 150-seat auditorium; 400-seat auditorium; educational facilities; 4,000 sq. ft. exhibit space; cafe. Museum-related items for sale.
Activities: concerts; docent program; films; formal educational programs; guided tours; lectures; loan, traveling, temporary & participatory exhibitions; limited FIU Judaic Studies classes.
Publications: quarterly newsletter, TILES; exhibit monographs.
Hours & Admission Prices: Tues.-Sun. 10-5. Adults $6, senior citizens & students $5, children $2.50; discounts to groups, AAM, AAA & PBS members; children under 6, members & Sat. no charge. Closed New Year's Day; Memorial Day; Independence Day; Labor Day; Thanksgiving; Jewish holidays. ♿
Attendance: 47,000 (accurate)
Membership: Senior Individual $30; Individual $36; Senior Couple & Single Parent Family $40; Basic Family $50; Donor $125; Patron $250; Sponsor & Corporate Sponsor $500; Benefactor & Corporate Benefactor $1,000.

MIAMI BEACH BOTANICAL GARDEN, 2000 Convention Center Dr., Miami Beach, FL 33139-1806. Tel.: 305-673-7256. Facebook: Miami Beach Botanical Garden.
Web Site: www.mbgarden.org
Founded: 1962.
Key Personnel: Exec. Dir., Cindy Brown
Institution Type/Description: Botanical Garden.
Collections: palms; Japanese garden; edible garden; wetland garden; native plants; tropical exotics; bromeliads.
Facilities: 2.6 acre garden.
Activities: cultural programs; rental facilities; art exhibitions; workshops; lectures related to green living; community events.
Hours & Admission Prices: Tues.-Sun. 9-5. No charge. ♿
Membership: Student & Senior $25; Individual $35; Family $50.

✳ **THE WOLFSONIAN-FLORIDA INTERNATIONAL UNIVERSITY, (M),** 1001 Washington Ave., Miami Beach, FL 33139-5017. Tel.: 305-531-1001. Fax: 305-531-2133.
E-mail: leffc@fiu.edu
Web Site: www.wolfsonian.org
Founded: 1986.
Congressional District: 18
Key Personnel: Dir., Cathy Leff; Assoc. Dir. Curatorial Affairs & Education, Marianne Lamonaca; Chm. (V), Michael N. Kreitzer; Pres. (V), Ray E. Marchman; Assoc. Dir. Business & Finance, Daniel Nolan, Jr.; Registrar, Kimberly Bergen; Exhibition Designer, Richard Miltner; Asst. Dir. Mktg., Member Rels. & New Media, Ian Rand; Chief Librarian, Francis X. Luca; Museum Shop Mgr., Paola La Rivera.
Personnel Profile: Full-Time Paid 33; Part-Time Paid 5; Part-Time Volunteers 11.
Governing Authority: university. Parent Institution: Florida International University. Tax-exempt: 501(c)(3).

Institution Type/Description: Art, Design, Decorative Arts & Architecture Museum: housed in restored building in the historic Art Deco District.
Collections: over 120,000 art & design objects from 1885-1945 including decorative, design & architectural arts; sculpture; paintings; graphics; industrial design; transportation objects; rare books.
Research Fields: Fellowship Program based in areas of collecting focus.
Facilities: 40,000-vol. library containing rare books & periodicals pertaining to decorative & propaganda arts; 15,000 sq. ft. exhibit space.
Activities: guided tours; lectures; films; concerts; loan, temporary & traveling exhibitions; formal education programs for adults & families; docent program.
Publications: catalogues; books; journal, The Journal of Decorative and Propaganda Arts.
Hours & Admission Prices: Academic Year: Mon.-Tues. & Thurs. and Sat. 12-6, Fri. 12-9; Summer: call for hours. Adults $7, seniors, students & children 6-12 $5; discounts to ICOM members; AAM, ICOM & Wolfsonian members, 6pm-9pm on Fri., children under 6 and students, faculty & staff of state university system of Florida no charge. ♿
Attendance: 35,000 (estimated)
Membership: Artist, Educator, Senior & Student $30; Popular $50; Dual & Family $75; Propagandist $125 & up; Diplomat $250; Ally $500 & up; Patriot $1,000 & up; Futurist $2,500 & up.

WORLD EROTIC ART MUSEUM, (M), 1205 Washington Ave., Miami Beach, FL 33139-4613. Tel.: 305-532-9336. Fax: 305-695-1209.
E-mail: missnaomi@weam.com
Web Site: www.weam.com
Founded: 2005.
Key Personnel: Dir., Pres. (V) & Cur., Naomi Wilzig; Public Rels., Robert G. Harbour; Gen. Mgr. & Security, J.C. Harris; Asst. Cur. & Archivist, Charles G. Haak.
Personnel Profile: Full-Time Paid 5; Part-Time Paid 3.
Governing Authority: municipal.
Institution Type/Description: Art Museum.
Collections: erotic art from Biblical to contemporary.
Facilities: 250-vol. library; 12,000 sq. ft. exhibit space. Museum-related items for sale.
Activities: guided tours; lectures; rental facilities.
Hours & Admission Prices: Mon.-Thurs. 11-10, Fri.-Sun. 11am to midnight; children not admitted. Adults $15, senior citizens $14, students $13.50; discounts to groups of 10 or more; members no charge. Closed New Year's Day; Thanksgiving; Christmas. ♿
Attendance: 5,000 (estimated)
Membership: Annual $100.

Miami Gardens

ST. THOMAS UNIVERSITY LIBRARY, 16401 N.W. 37th Ave., Miami Gardens, FL 33054-6313. Tel.: 305-628-6668. Fax: 305-628-6666.
Web Site: www.stu.edu/Library/tabid/395/Default.aspx
Key Personnel: Dir., L. Bryan Cooper, Ph.D.
Institution Type/Description: Library.
Collections: books; manuscripts; photographs; prints; posters; personal papers; corporate records.
Hours & Admission Prices: Mon.-Thurs. 8am-11pm, Fri. 8-5, Sat. 9-5, Sun. 2-10 by appointment. Closed major holidays.

Micanopy

MICANOPY HISTORICAL SOCIETY MUSEUM, Cholokka Blvd. at Bay St., Micanopy, FL 32667-4112. Mailing Address: P.O. Box 462, Micanopy, FL 32667-0462. Tel.: 352-466-3200. Fax: 352-466-1150.
E-mail: micanopymuseum@aol.com
Web Site: www.afn.org/~micanopy
Founded: 1984.
Congressional District: 6
Key Personnel: Dir., Patricia Crass; C.E.O., Liselotte Hof; Pres. (V), Jean Stream; Museum Shop Mgr., Barbara Lonas.
Personnel Profile: Full-Time Volunteers 1; Part-Time Volunteers 27.
Volunteer Hours: 1,800
Operating Expenses: 10,500
Operating Income: 12,000
Governing Authority: private; nonprofit organization. Parent Institution: Micanopy Historical Society. Tax-exempt.
Institution Type/Description: History Museum.

Collections: Micanopy history; farm & family life from 1860 to present; Seminole portraits; Fort Micanopy records & metal artifacts; Seminole War maps 1836-1839.
Facilities: Museum-related items for sale.
Activities: elderhostels; children's programs for 4th grade & up; history simulation including Micanopy Regulars; talking map.
Publications: museum brochure; newsletter, Micanopy Historical Society; book, Story of Historic Micanopy; walking tour brochure; cookbook, Cherished Recipes of Historic Micanopy (2011).
Hours & Admission Prices: Daily 1-4. Suggested Donation: $2; discounts to AAM members. Closed New Year's Day; Thanksgiving; Christmas. &
Attendance: 5,558 (accurate)
Membership: Individual $10; Family $20; Supporter $50; Contributor $100; Patron $500.

Milton

ARCADIA MILL ARCHAEOLOGICAL SITE, (M), 5709 Mill Pond Lane, Milton, FL 32583-1788. Tel.: 850-626-3084.
E-mail: asams@uwf.edu
Web Site: www.historicpensacola.org/arcadia.cfm
Founded: 2004.
Congressional District: 1
Key Personnel: Site Mgr., Adrianne Sams; Museum Education Coord., Roy Oberto.
Personnel Profile: Full-Time Paid 2; Part-Time Paid 4.
Governing Authority: Parent Institution: West Florida Historic Preservation, Inc./University of West Florida. Subsidiary Institution: T.T. Wentworth; Florida State Museum; Historic Pensacola Village. Tax-exempt: 501(c)(3).
Institution Type/Description: Historic Archaeological Site: housed in a mid-19th century industrial complex.
Collections: archaeological; structural remains.
Research Fields: local historical technology and related people & families.
Facilities: visitors center; nature trail.
Activities: guided tours; archaeological excavations in summer.
Hours & Admission Prices: Tues.-Sat. 10-4. No charge; donations accepted. Closed New Year's Day; Independence Day; Veterans Day; Thanksgiving & day after; day after Christmas. &
Attendance: 15,000 (estimated)

WEST FLORIDA RAILROAD MUSEUM, 5003 Henry St., Milton, FL 32570-6790. Mailing Address: P.O. Box 770, Milton, FL 32572-0770. Tel.: 850-623-3645.
Founded: 1989.
Key Personnel: Pres. (V), Art Tuttle.
Personnel Profile: Part-Time Volunteers 10.
Governing Authority: Tax-exempt.
Institution Type/Description: Railroad History Museum.
Collections: Northwest Florida & South Alabama railroad history.
Hours & Admission Prices: Fri.-Sat. 10-3; other times by appointment. No charge; donations accepted. &
Membership: Annual $30.

Mount Dora

MOUNT DORA CENTER FOR THE ARTS, 138 E. 5th Ave., Mount Dora, FL 32757-5573. Tel.: 352-383-0880. Fax: 352-383-7753.
E-mail: center@mountdoracenterforthearts.org
Web Site: www.mountdoracenterforthearts.org
Founded: 1985.
Congressional District: 5
Key Personnel: Exec. Co.-Chair, Nancy Zinkofsky; Exec. Co-Chair, Elizabeth Miller; Pres. (V), Michell Middleton; Admin. Asst., Krysta Smith.
Personnel Profile: Full-Time Paid 2; Part-Time Paid 1; Part-Time Volunteers 32; Interns 2.
Governing Authority: nonprofit organization. Tax-exempt: 501(c)(3).
Institution Type/Description: Art Center.
Collections: Best of Show pieces from Mount Dora Arts Festival.
Facilities: 700 sq. ft. exhibit space. Gift items for sale.
Activities: arts festivals; formal educational programs; lectures; participatory & temporary exhibitions.
Publications: quarterly, MDCA Newsletter.
Hours & Admission Prices: Mon.-Fri. 10-4, Sat. 10-2. No charge; donations accepted. Closed major holidays. &
Attendance: 35,000 (accurate)
Membership: Student & Senior $20; Individual $25; Senior Couple $30; Family $35; Sponsor $50; Sustaining & Business $100; Associate $250; Patron $500; Benefactor $1,000.

MOUNT DORA HISTORY MUSEUM, 450 Royellou Lane, Mount Dora, FL 32757-5554. Mailing Address: P.O. Box 1166, Mount Dora, FL 32756-1166. Tel.: 352-383-0006.
E-mail: mountdorahistory@gmail.com
Web Site: www.mountdorahistoricalsociety.org
Formerly: Royellou Museum
Founded: 1978.
Key Personnel: Museum Mgr., Carolyn M. Green.
Personnel Profile: Part-Time Paid 1; Part-Time Volunteers 5.
Governing Authority: Parent Institution: Mount Dora Historical Society, Inc. Tax-exempt.
Institution Type/Description: History Museum: housed in the old city jail built in 1923.
Collections: Mount Dora history & photographs; jail cell; personal artifacts; clothing; household artifacts 1880-1950.
Hours & Admission Prices: Tues.-Sun. 1-4. Suggested Donations: adults $2, children $1; children under 5 no charge. Closed Thanksgiving; Christmas. &
Attendance: 2,031 (accurate)
Membership: Student $5; Individual $15; Family $25.

Mulberry

MULBERRY PHOSPHATE MUSEUM, (M), 101 S.E. 1st St., Mulberry, FL 33860-3169. Mailing Address: P.O. Box 707, Mulberry, FL 33860-0707. Tel.: 863-425-2823. Fax: 863-425-0188. Facebook: Mulberry Phosphate Museum.
E-mail: cyoung@cityofmulberryfl.com
Web Site: www.mulberryphosphatemuseum.org
Founded: 1985.
Congressional District: 10
Key Personnel: Dir., Chelsea Young; Museum Asst., Elly Sonth; Museum Asst., Jackie Marcucci.
Personnel Profile: Full-Time Paid 1; Part-Time Paid 2; Part-Time Volunteers 4.
Governing Authority: municipal. Parent Institution: City of Mulberry. Tax-exempt: 501(c)(3).
Institution Type/Description: Natural History Museum: housed in c.1899 Mulberry Train Depot.
Collections: Cenozoic & Bone Valley fossils; Miocene Epoch; local history; minerals; phosphate.
Research Fields: animals living in Florida's Bone Valley region dating back to Miocene Epoch.
Facilities: rental facilities.
Activities: guided tours; lectures; organized educational programs; docent program; loan, temporary exhibitions; school loan service; fossil dig pile.
Hours & Admission Prices: Tues.-Sat. 9-5. No charge; donations accepted. Closed New Year's Day; Memorial Day; Independence Day; Thanksgiving; Christmas. &
Attendance: 9,000 (accurate)

Naples

THE BAKER MUSEUM, ARTIS-NAPLES, (M), 5833 Pelican Bay Blvd., Naples, FL 34108-2740. Tel.: 239-254-2620. Fax: 239-254-2753. Facebook: Artis-Naples.
E-mail: info@artisnaples.org
Web Site: www.artisnaples.org
Formerly: Patty and Jay Baker Naples Museum of Art, Philharmonic Center for the Arts
Founded: 1989.
Congressional District: 13
Key Personnel: Museum Dir. & Chief Cur., Frank Verpoorten; C.E.O. & Pres., Kathleen van Bergen; Chm. (V), Jay Baker; Registrar, Jacqueline Zorn; Preparator, Steven Kravec; Dir. Finance, Renee Neville; Chief Advancement Officer, Mary Deissler; Custome Svc., Elise Guarino; Coord. Special Events & Volunteers, Kelly Rhodes; Cur. Education, Jessica Wozniak; Art Handler, Chris Smith; Exec. Asst. to the Dir. & Chief Cur., Shannon Gallagher.
Personnel Profile: Full-Time Paid 7; Part-Time Paid 1; Part-Time Volunteers 178; Interns 1.
Governing Authority: nonprofit. Parent Institution: Artis-Naples. Tax-exempt: 501(c)(3).
Institution Type/Description: Art Gallery.
Collections: sculpture; painting; American modernism; 20th century Mexican & Latin American art; contemporary art; photography; video; miniatures; glass (chihuly); Mai 68 protest posters.
Major Exhibits: Marcel Duchamp & Family, 1/4/14-4/4/14; Fine Lines: American Drawings from the Brooklyn Museum (T), 1/18/14-4/18/14;

Rediscovering Egypt: The Dahesh Muiseum of Art Collection (T), 1/25/14-4/25/14; Florida Contemporary, 2/1/14-5/1/14; Gods and Heroes: Masterpieces from the EBA (T), 2/19/14-6/19/14; Museum to Scale 1/7 (T), 3/15/14-8/15/14; The Coast and the Sea, 4/19/14-8/19/14; Tallerde Grafica (T), 5/31/14-7/31/14.
Research Fields: on permanent collection objects & subjects of temporary & traveling exhibitions.
Facilities: 200 & 1,221-seat theaters; 22,000 sq. ft. exhibit space.
Activities: permanent, loan & traveling exhibitions; guided tours; docent program; lectures; gallery talks; art trips; performances; formally organized education programs; museum family days.
Publications: catalogues of permanent collections & exhibitions; member newsletter, annual brochure, This Week at the Museum.
Hours & Admission Prices: Sept. 21-July 6 Tues.-Sat. 10-4, Sun. 12-4. Adults $10, students with I.D. $5; discounts to NARM, AAM & ICOM members; children 17 & under & members no charge. Closed New Year's Day; Thanksgiving; Christmas. &
Attendance: 50,000 (estimated)
Membership: Individual $75; Premier $150; Contributing $250; Supporting $500; Sponsor $1,000; Circle $2,500.

COLLIER COUNTY MUSEUM, (M), 3331 Tamiami Trail E., Naples, FL 34112-4901. Tel.: 239-252-8476. Fax: 239-252-8580. Facebook: Collier County Museum.
E-mail: colliermuseums@gmail.com
Web Site: www.colliermuseums.com
Founded: 1978.
Key Personnel: C.E.O., Ron D. Jamro; Cur., Jennifer Guida; Education, Naomi Goren; Volunteer Coord., Mary Margaret Gruszka; Mgr., Lee Mitchell; Mgr., Timothy England; Mgr., Lisa Marciano; Museum Asst., Martha Hutcheson; Museum Asst., Jon Nickerson; Maintenance Specialist, Steve Szokoly; Administrative Asst., Christina Apkarian.
Personnel Profile: Full-Time Paid 13; Part-Time Volunteers 145.
Volunteer Hours: 6,394
Operating Expenses: 1,595,035
Operating Income: 1,744,532
Governing Authority: county. Subsidiary Institutions: Museum of the Everglades, Everglades City, FL; Immokalee Pioneer Museum at Roberts Ranch; Naples Depot Museum, downtown Naples; Marco Island Historical Museum, Marco Island. Tax-exempt: 501(c)(3).
Institution Type/Description: History Museum.
Collections: history & archaeology of Collier County; several restored structures; a steam logging locomotive.
Major Exhibits: Enchantments - The Photographic Adventures of Julian Dimock in SW Florida 1904-1913, 1/2/14-3/29/14; Family Tradition: Tribal Art of Brian and Pedro Zepeda (T), 2/12/14-6/14; Guy La Bree: The Barefoot Artist (T), 2/12/14-6/14; An Artist's Journey - The Detail Within, Paintings by Inez Hudson (T), 4/14-6/14; Photography of Dennis Axler (T), 7/14-9/14; Goodland...A Historic Village, 10/14-12/14.
Research Fields: historic site & structures in Collier County; construction of the Tamiami Trail 1923-1928; Calusa Indians in southwest Florida.
Facilities: 700-vol. non-circulating research library relating to Florida history, Collier County archaeology, Seminole Indians, general American & southern history; botanical garden; educational facilities; 10,000 sq. ft. exhibit space; archaeology laboratory. Museum-related items for sale.
Activities: formal education programs; guided tours; lectures; participatory, temporary & traveling exhibitions. Annual Events: Old Florida Festival. Annual Events: Old Florida Festival; Marjory Stoneman Douglas Festival; USO Show.
Hours & Admission Prices: Mon.-Fri. 9-5. No charge; donations accepted. Closed national & county holidays. &
Attendance: 75,000 (estimated)
Membership: Student & Senior Citizen $15; Individual $25; Family $40; Friend $100; Sponsor $500; Patron $1,000.

CONSERVANCY OF SOUTHWEST FLORIDA NATURE CENTER, 1450 Merrihue Dr., Naples, FL 34102-3449. Tel.: 239-262-0304. Fax: 239-262-0672.
E-mail: info@conservancy.org
Web Site: www.conservancy.org
Founded: 1964.
Key Personnel: C.E.O., Andrew McElwaine
Institution Type/Description: Nature Center.
Collections: owls; hawks; pelicans; bald eagle; injured birds, mammals & reptiles; freshwater & saltwater fish.
Facilities: 21 acres; nature trails. Museum-related items for sale.
Activities: summer camp; school programs; visitor programs; field excursions; canoe & kayak rentals; electric boat tours.
Publications: Conservancy Update; annual report; annual activities book.

Hours & Admission Prices: Call for hours. Adults $12.95; discounts to ANCA members; members no charge. &
Membership: Individual $65 & up.

GOLISANO CHILDREN'S MUSEUM OF NAPLES, 15080 Livingston Rd., Naples, FL 34109. Tel.: 239-514-0084. Fax: 239-260-1616.
E-mail: info@cmon.org
Web Site: www.cmon.org
Founded: 2002.
Key Personnel: C.E.O., Andy Marquart; Chm. (V) & Volunteer Coord., Lindsay Huban; Dir. Institutional Advancement, Sandra Wilson; Dir. Finance, Karysia Demarest.
Personnel Profile: Full-Time Paid 18; Part-Time Paid 29; Part-Time Volunteers 175; Interns 2.
Governing Authority: nonprofit organization. Tax-exempt: 501(c)(3).
Institution Type/Description: Children's Museum.
Collections: hands on participatory exhibits.
Activities: arts festivals; concerts; docent program; guided tours; workshops; lectures; rental gallery; temporary & traveling exhibits.
Publications: quarterly newsletter, C'mon; e-newsletters.
Hours & Admission Prices: Call for hours. Admission $10; discounts to ACM members; children under one no charge. &
Membership: Play $125; Learn $250; Dream $1,000.

HOLOCAUST MUSEUM OF SOUTHWEST FLORIDA, 4760 Tamiami Trail N., Ste. 7, Naples, FL 34103-3065. Tel.: 239-263-9200. Fax: 239-263-9500.
Founded: 2001.
Personnel Profile: Full-Time Paid 4; Part-Time Paid 3; Part-Time Volunteers 150; Interns 1.
Governing Authority: Tax-exempt.
Institution Type/Description: History Museum.
Collections: Holocaust & WWII history & artifacts; survivor documents & stories.
Activities: special events.
Hours & Admission Prices: Tues.-Sun. 1-4. Adults $8; members no charge. Closed public holidays. &
Attendance: 8,300 (accurate)
Membership: $18; $36; $54; $200; $300; $1,000; $1,500. Window of Hope: $1,000-$100,000.

MARIANNE FRIEDLAND GALLERY, 359 Broad Ave. S., Naples, FL 34102. Tel.: 239-262-3484.
E-mail: mfgallery@aol.com
Web Site: www.mariannefriedlandglry.com
Founded: 1991.
Institution Type/Description: Art Gallery.
Collections: works by modern & contemporary artists.
Hours & Admission Prices: Mon.-Sat. 10-5; other times by appointment. No charge.

THE NAPLES ART ASSOCIATION AT THE VON LIEBIG ART CENTER, 585 Park St., Naples, FL 34102-6611. Tel.: 239-262-6517. Fax: 239-262-5404.
E-mail: info@naplesart.org
Web Site: www.naplesart.org
Founded: 1954.
Congressional District: 14
Key Personnel: Dir., Aimee Schlehr; Pres., Stacey Bulloch; Dir. Devel., Maureen Christensen; Cur., Jack O'Brien; Volunteer Coord., Yvonne Gibb; Dir. Education, Callie Spilane; Dir. Festivals, Marianne Megela.
Personnel Profile: Full-Time Paid 9; Part-Time Paid 2; Part-Time Volunteers 600; Interns 60.
Volunteer Hours: 14,000
Operating Expenses: 1,100,000
Operating Income: 1,100,000
Governing Authority: private; nonprofit organization. Parent Institution: Naples Art Association. Tax-exempt: 501(c)(3).
Institution Type/Description: Community Art Center.
Collections: paintings; sculpture; photography; works on paper.
Facilities: 3,000-vol. library; six classrooms/studios; 2,500 sq. ft. exhibit space. Gift items for sale.
Activities: changing exhibitions; education programs; classes; workshops; lectures; facility rental; professional development for working artists. Annual Events: New Year's Festival in January; Naples National Art Festival in February; Nuts About the von Liebig Family Day in February; Downtown Naples Festival of the Arts in March; Goddess Night in April; Fall Festival in November; Art in the Park Nov.-April.

Publications: monthly e-newsletter.
Hours & Admission Prices: Gallery: Mon.-Fri. 10-4; call for seasonal hours. Administration Office: Mon.-Fri. 9-4. No charge; donations accepted. Gift Shop: discounts to NARM members. Closed Easter; Memorial Day; Independence Day; Labor Day; Thanksgiving; Christmas. ♿
Attendance: 50,000 (estimated)
Membership: Individual $75; Family $150; Friends $500; Patron $1,000.

NAPLES BACKYARD HISTORY OLD NAPLES MUSEUM, 1170 Third St., S., Ste. C111, Naples, FL 34102. Mailing Address: P.O. Box 2149, Naples, FL 34106. Tel.: 239-774-2978.
Web Site: www.naplesbackyardhistory.net
Institution Type/Description: History Museum.
Collections: local history & culture; personal artifacts; photographs.
Activities: rental facilities.
Hours & Admission Prices: Thurs. 5-8; other times by appointment.

NAPLES BOTANICAL GARDEN, 4820 Bayshore Dr., Naples, FL 34112-7344. Tel.: 239-643-7275. Fax: 239-649-7306.
E-mail: info@naplesgarden.org
Web Site: www.naplesgarden.org
Founded: 1993.
Congressional District: 14
Key Personnel: Exec. Dir., Brian Holley; Bd. Chm., James A. LaGrippe; C.F.O., Theresa Perkins; Dir. Devel., Phyllis Racine; Mgr. Visitor Svcs., Paula Braida.
Personnel Profile: Full-Time Paid 40; Full-Time Volunteers 1; Part-Time Paid 20; Interns 2.
Governing Authority: Tax-exempt.
Institution Type/Description: Botanical Garden.
Collections: natural science; plants; art; history; gardens.
Major Exhibits: Lego (R) Sculptures in the Garden (T), 2/14-5/11/14.
Facilities: 170 acres; 6 gardens; walking trails.
Activities: dogs in the garden; artists in the garden; yoga; jazz; theatrical performances; guided tours; food events; tree house gardening.
Publications: biweekly email newsletter; seasonal garden magazine.
Hours & Admission Prices: Daily 9-5, Tues. 8-5. Adults $12.95, children 4-14 $7.95; children 3 & under & members no charge. ♿
Attendance: 120,000 (accurate)
Membership: Individual $70; Family $95; Contributing $250; Sustainer $500; Garden Fellow $1,000; Royal Palm Society $1,500, $3,000, $5,000, $10,000.

NAPLES ZOO AT CARIBBEAN GARDENS, 1590 Goodlette-Frank Rd., Naples, FL 34102-5260. Tel.: 239-262-5409.
E-mail: info@napleszoo.org
Web Site: www.napleszoo.com
Governing Authority: nonprofit organization. Tax-exempt: 501(c)(3).
Institution Type/Description: Zoo.
Collections: lions; kangaroos; monkeys; tigers; lions; leopards; spotted hyenas; alligators; snakes.
Facilities: Museum-related items for sale.
Activities: presentations; shows; boat cruise.
Hours & Admission Prices: Daily 9-5. Adults 13-64 $19.95, seniors 65 & over $18.95, children 3-12 $11.95; discounts to active military; members and children 2 & under no charge.

New Port Richey

WEST PASCO HISTORICAL SOCIETY MUSEUM AND LIBRARY, 6431 Circle Blvd., New Port Richey, FL 34652-2360. Tel.: 727-847-0680.
E-mail: info@westpascohistoricalsociety.org
Web Site: westpascohistoricalsociety.org
Founded: 1983.
Congressional District: 5
Key Personnel: C.E.O., David Prace; Pres. (V), Bob Hubach; Treas., Lisa Carey; Museum Shop Mgr., Terresa Stevenson.
Personnel Profile: Part-Time Volunteers 20.
Governing Authority: society. Parent Institution: The West Pasco Historical Society, Inc. Tax-exempt: 501(c)(3).
Institution Type/Description: Historical Society Museum: housed in 1913 two-room schoolhouse.
Collections: Wedgewood china; local history displays; local artifacts; pre-Columbian artifacts; arrowheads; vintage clothing; quilts; period household furnishings; typewriters; Native American clothing; arrowheads; pottery; jewelry; library; paintings.
Research Fields: local history; local photograph collection.
Facilities: 2,000-vol. library of books; newspaper files, available for use by public; classrooms; 2,000 sq. ft. exhibit space. Books & gift items for sale.

Activities: guided tours; lectures; films; arts festivals; docent program; participatory & temporary exhibitions; field trips. Museum Sponsors: fashion show; Christmas Dinner Auction; Rummage Sale.
Publications: quarterly, WPHS Newsletter; books on local history.
Hours & Admission Prices: Fri.-Sat. 1-4. No charge; donations accepted. Closed legal holidays. ♿
Attendance: 3,500 (estimated)
Membership: Student $5; Adult $15; Family $25; Business $50; Life $100.

New Smyrna Beach

ATLANTIC CENTER FOR THE ARTS, INC., 1414 Art Center Ave., New Smyrna Beach, FL 32168-5560. Tel.: 386-427-6975. Fax: 386-427-5669.
E-mail: program@atlanticcenterforthearts.org
Web Site: www.atlanticcenterforthearts.org
Founded: 1977.
Congressional District: 7
Key Personnel: Co-Dir. Community, Nancy Lowden Norman; Co-Dir Residency, Jim Frost; ACA Bd. Chm., Woody Igon; Pres. (V), Patricia Kershner; Treas., Ed Leerdam; Devel. & Public Rels., Kathryn Peterson; Security, Tom Kurtzhals; Museum Shop Mgr., Jill Rox.
Personnel Profile: Full-Time Paid 11; Part-Time Volunteers 150; Interns 2.
Governing Authority: private; nonprofit organization. Tax-exempt: 501(c)(3).
Institution Type/Description: Art Center & Gallery.
Collections: works by former master artists-in-residence.
Facilities: 500-vol. library; 200-seat auditorium; artist studios; 4,000 sq. ft. exhibit space. Museum-related items for sale.
Activities: arts festivals; formal education programs; guided tours; hobby workshops; lectures; temporary & traveling exhibitions.
Publications: yearly annual report, ACA Annual Report; quarterly newsletter, ACA Newsletter.
Hours & Admission Prices: Tues.-Fri. 10-4, Sat. 10-2. No charge; donations accepted. Closed Martin Luther King Jr. Day; Presidents' Day; Memorial Day; Independence Day; Labor Day; Thanksgiving & day after; Christmas week. ♿
Attendance: 5,000 (estimated)
Membership: Artist-in-Residence $25; Individual & Volunteer League $40; Family $70; Art Angel $100; Salon $250; Patron Images & Patron Horsin' Around $1,000; Preservationist $2,500.

Newberry

DUDLEY FARM HISTORIC STATE PARK, 18730 W. Newberry Rd., Newberry, FL 32669-2192. Tel.: 352-472-1142.
E-mail: dudley@gvlhistorichomes.org
Web Site: www.floridastateparks.org/dudleyfarm
Institution Type/Description: Historic Farmstead.
Collections: local history & culture; period furnishings; early tools & equipment; clothing; livestock; 18 historic buildings.
Facilities: visitor center.
Activities: group tours.
Hours & Admission Prices: Wed.-Sun. 9-4. Admission $4.

Niceville

MATTIE KELLY ARTS CENTER GALLERIES AT NORTHWEST FLORIDA STATE COLLEGE, (M), 100 College Blvd., Niceville, FL 32578-1347. Tel.: 850-729-6044. Fax: 850-729-5286. Facebook: Mattie Kelly Arts Center Gallery.
E-mail: artgalleries@nwfsc.edu
Web Site: www.mattiekellyartscenter.org
Formerly: Mattie Kelly Arts Center at Okaloosa-Walton College
Founded: 1997.
Congressional District: 1
Key Personnel: Dir., Jeanette Shires; Gallery Dir., K.C. Williams.
Personnel Profile: Full-Time Paid 1; Part-Time Paid 3; Part-Time Volunteers 32; Interns 2.
Governing Authority: Parent Institution: Northwest Florida State College. Tax exempt.
Institution Type/Description: College Art Museum.
Collections: works by national & international artists.
Facilities: 1,650-seat theater; McIlroy Gallery 1600 sq. ft. exhibition space; Holzhaver Gallery 900 sq. ft. exhibition space.
Activities: education programs; performances.
Hours & Admission Prices: Mon.-Fri. 10-4. No charge. Closed holidays. ♿
Attendance: 15,300 (estimated)
Membership: Member $25; Family $50; Patron $100; Donor $500; Exhibition Sponsor $1,000; Season Benefactor $5,000.

North Miami

* **MUSEUM OF CONTEMPORARY ART (MOCA),** Joan Lehman Bldg., 770 N.E. 125th St., North Miami, FL 33161-5654. Tel.: 305-893-6211. Fax: 305-891-1472.
E-mail: info@mocanomi.org
Web Site: www.mocanomi.org
Founded: 1981.
Congressional District: 10
Key Personnel: Exec. Dir. & Chief Cur., Bonnie Clearwater; Chm. Bd. Dir., Michael Collins; Museum Shop Mgr., Alan Waufle.
Personnel Profile: Full-Time Paid 15; Part-Time Paid 9; Part-Time Volunteers 30; Interns 1.
Governing Authority: nonprofit organization. Tax-exempt: 501(c)(3).
Institution Type/Description: Visual Contemporary Art.
Collections: contemporary art.
Facilities: 23,000 sq. ft. exhibit space.
Activities: guided tours; lectures; films; organized education programs for children organized education & enrichment programs for teens; loan exhibitions; tours to artists' studios; temporary exhibitions of contemporary, national & international artists; monthly jazz concerts; figure drawing classes for adults.
Publications: annual newsletter; catalogs; MOCA'zine teen magazine.
Hours & Admission Prices: MOCA North Miami: Tues. & Thurs.-Sat. 11-5, Wed. 1-9, Sun. 12-5. Adults $5, seniors & students $3; North Miami residents, children under 12 & members no charge. Tues. donations accepted. Closed Thanksgiving; Christmas. &
Attendance: 90,100 (accurate)
Membership: Artist, Student & Educator $30; Individual $50; Family $70; MOCA Shaker $150; Associate $1,000; Benefactor $2,500; International Collector Circle $5,000.

North Miami Beach

ANCIENT SPANISH MONASTERY, (M), 16711 W. Dixie Hwy., North Miami Beach, FL 33160-3714. Tel.: 305-945-1461. Fax: 305-945-6986.
E-mail: info@spanishmonastery.com
Web Site: www.spanishmonastery.com
Founded: 1952.
Key Personnel: Exec. Admin. & C.E.O., Dr. Gregory Mansfield; Museum Shop Mgr., Tania Witten; Parish Sec., Mayten Battalle.
Personnel Profile: Full-Time Paid 6; Full-Time Volunteers 2; Part-Time Paid 2; Part-Time Volunteers 20; Interns 1.
Governing Authority: church. Parent Institution: Episcopal Diocese of S.E. Florida. Tax-exempt.
Institution Type/Description: Art Museum: housed in The Cloister and Refectory, reconstruction of monastery, built in 1141 in Segovia, Spain, with original stones brought to United States by William Randolph Hearst.
Collections: paintings; sculpture; historical artifacts of religious nature; historical architectural artifacts.
Facilities: Religious articles, books & related items for sale.
Activities: guided tours; arts festivals; drama; inter-museum loan & permanent exhibitions.
Hours & Admission Prices: Mon.-Sat. 9-5, Sun. 12-5. Adults 12-55 $8, senior citizens 55 & over $4, students 12 & under $1; discounts to AAM & ICOM members. &
Attendance: 100,000 (estimated)

North Palm Beach

JOHN D. MACARTHUR BEACH STATE PARK & NATURE CENTER, 10900 Jack Nicklaus Dr., North Palm Beach, FL 33408-3440. Tel.: 561-624-6952 & 6950.
E-mail: friends@macarthurbeach.org
Web Site: www.macarthurbeach.org
Institution Type/Description: Park & Nature Center.
Collections: aquariums; video; natural history.
Facilities: picnic area. Museum-related items for sale.
Activities: recreational activities; video; education programs.
Hours & Admission Prices: Nature Center: daily 9-5. Park: daily 8 to sunset. Entrance Fee: $5 per vehicle. &

Ocala

* **APPLETON MUSEUM OF ART, (M),** 4333 E. Silver Springs Blvd., Ocala, FL 34470-5001. Tel.: 352-291-4455. Fax: 352-291-4460.
E-mail: ormej@cf.edu
Web Site: www.appletonmuseum.org

Founded: 1986.
Congressional District: 6
Key Personnel: Dir., Cindi Morrison; Coord. Finance Svcs., Kathleen Balboni; Cur., Ruth Grim; Mgr. Membership & Events, Colleen Harper; Registrar, David Reutter; Coord. Facilities, Russell Days; Staff Asst. III, Joyce Orme.
Personnel Profile: Full-Time Paid 17; Part-Time Paid 27; Part-Time Volunteers 100; Interns 2.
Volunteer Hours: 3,500
Operating Expenses: 1,790,000
Operating Income: 1,800,000
Governing Authority: nonprofit organization. Parent Institution: College of Central Florida. Tax-exempt.
Institution Type/Description: Fine Arts Museum.
Collections: fine arts, including American, European, Pre-Columbian, African, Asian, Islamic & decorative arts; contemporary art; Florida artists antiquities.
Research Fields: all items pertaining to collections.
Facilities: art reference library; 250-seat auditorium; conference rooms; 39,000 sq. ft. exhibit space. Museum-related items for sale.
Activities: guided tours; lectures; films; concerts; organized education programs for children, adults, undergraduate & graduate college students; docent program; loan, temporary & traveling exhibitions; workshops.
Publications: calendar of events; quarterly newsletter.
Hours & Admission Prices: Tues.-Sat. 10-5, Sun. 12-5. Adults $6, seniors 55 & over and students over 19 $4, youth 10-18 $3; children 9 & under, military & their family and members no charge. Reciprocal memberships with other museums. Closed New Year's Day; Thanksgiving; Christmas. &
Attendance: 50,000 (accurate)
Membership: Students & Educator $15; Senior $25; Individual $30; Dual Senior $40; Dual Family $50; Director's Circle $100-$499, $500-$999, $1,000-$2,499, $2,500-$4,999, $5,000 & up.

CF WEBBER CENTER GALLERY, 3001 S.W. College Rd., Ocala, FL 34474-4415. Tel.: 352-873-5809. Fax: 352-873-5886.
E-mail: bjornh@cf.edu
Web Site: www.cf.edu
Founded: 1985.
Congressional District: 6
Key Personnel: Dir. Visual & Performing, Dr. Jennifer Fryns; Exhibits Coord., Heather Bjorn.
Personnel Profile: Part-Time Paid 4; Part-Time Volunteers 1.
Governing Authority: Parent Institution: College of Central Florida. Tax-exempt.
Institution Type/Description: Art Museum.
Collections: paintings; photographs; prints; drawings; graphic arts; sculpture.
Major Exhibits: A Permanent Mark: Highlights of CF's Permanent Collection, 1/9/14-2/14/14; Art History 101: Icons of Western Art (T), 2/20/14-3/28/14; 2014 CF Student Art & Exhibition, Imprints Magazine, 4/9/14-5/2/14; Summer Spotlight XVII: Visual Artists' Society, 5/22/14-6/27/14; Nature's Best Photography (Sites) (T), 8/2/14-10/26/14; Visual Artists' Society: Best of the Season, 11/14-12/14; 19th Annual Trains at the Holidays, 12/15/14-12/30/14.
Facilities: 2,000 sq. ft. exhibit space; conference room; outside patio.
Activities: videos; films; seminars; lectures; poetry readings; story-tellings; musical offerings; temporary exhibitions.
Hours & Admission Prices: Mon.-Fri. 10-4. No charge; donations accepted. Closed college holidays. &
Attendance: 9,000 (estimated)
Membership: Webber Gallery Patron Society $25.

THE DISCOVERY CENTER, 701 N.E. Sanchez Ave., Ocala, FL 34470. Tel.: 352-401-3900. Fax: 352-368-5514.
E-mail: discovery@ocalafl.org
Web Site: www.mydiscoverycenter.org
Formerly: Discovery Science and Outdoor Center
Founded: 1993.
Congressional District: 5
Key Personnel: Supvr., Suzanne Shuffitt; Asst., Mary Keck.
Personnel Profile: Full-Time Paid 1; Part-Time Paid 4; Part-Time Volunteers 200.
Governing Authority: nonprofit organization. Parent Institution: City of Ocala. Tax-exempt.
Institution Type/Description: Science Center.
Collections: hands-on science exhibits; children's art.
Facilities: 35 acre park setting; children's playground.
Activities: science & educational programs; participatory, loan & traveling exhibitions; explainer program.
Publications: Science Matters.

Hours & Admission Prices: Office: Mon.-Fri. 9-5. Programs: by appointment. &

Attendance: 25,000 (estimated)
Membership: Individual & Home School $30; Family $50.

DON GARLITS MUSEUM OF DRAG RACING INC., 13700 S.W. 16th Ave., I-75, Exit 341, Ocala, FL 34473-3970. Tel.: 352-245-8661; 877-271-3278 (toll-free). Fax: 352-245-6895. Facebook: Don Garlits Museum of Drag Racing Inc.
E-mail: donna@garlits.com
Web Site: www.garlits.com
Founded: 1976.
Congressional District: 6
Key Personnel: C.E.O. & Chm., Donald G. Garlits; Gen. Mgr. & C.F.O., Donna Garlits; Treas., Gay Lyn Capitano; Dir., Carl Schiefer; Dir., Chris Karamesines; Dir., Steve Ellis; Dir. Legal, George Albright; Museum Shop Mgr., Chris Bupus; Admin. Asst., Marlene Martian.
Personnel Profile: Full-Time Paid 6; Part-Time Paid 2; Part-Time Volunteers 4.
Governing Authority: board of directors. Tax-exempt: 501(c)(3).
Institution Type/Description: Drag Racing & Antique Cars & Tools Museum.
Collections: drag racing from late 1940s-present; race cars; engine exhibits; early parts; pictures; model exhibits; period cars.
Research Fields: drag racing.
Facilities: 5000-vol. library pertaining to drag racing; video theater. Auto-related items for sale.
Activities: guided tours; films; organized education programs; loan & temporary exhibitions.
Publications: membership newsletter.
Hours & Admission Prices: Daily 9-5. Adults $15, senior citizens & students $13, children $6; discounts to groups & AAA members; children under 5 no charge. Closed Thanksgiving; Christmas. &
Attendance: 75,000 (estimated)
Membership: Annual $25.

SILVER RIVER MUSEUM & ENVIRONMENTAL EDUCATION CENTER, 1445 N.E. 58th Ave., Ocala, FL 34470-1189. Tel.: 352-236-5401. Fax: 352-236-7142.
E-mail: scott.mitchell@marion.k12.fl.us
Web Site: www.silverrivermuseum.com
Founded: 1991.
Congressional District: 6
Key Personnel: Dir., Cur. & Public Rels., Scott Mitchell; Financial Dir., Archivist, Registrar & Museum Shop Mgr., Janet Scantlan; Education, Dorothy Lorch.
Personnel Profile: Full-Time Paid 4; Part-Time Paid 6; Part-Time Volunteers 25.
Governing Authority: public school district; nonprofit organization. Parent Institution: Marion County School System. Tax-exempt.
Institution Type/Description: History & Natural History of Florida Museum.
Collections: history; natural history; archaeology; paleontology; geology.
Research Fields: Florida natural & cultural history.
Facilities: research library; educational facilities. Museum-related items for sale.
Activities: educational program; docent program.
Hours & Admission Prices: Student Tours: Mon.-Fri. General Public: Sat.-Sun. 9-5. Adults $2; children 6 & under no charge. &
Attendance: 20,000 (estimated)

Ochopee

BIG CYPRESS NATIONAL PRESERVE, Oasis Visitors Center, 52105 Tamiami Trail E., Ochopee, FL 34141. Mailing Address: 33100 Tamiami Trail, E., Ochopee, FL 34141. Tel.: 239-695-1201 & 2000. Fax: 239-695-3901.
Web Site: nps.gov/bicy
Founded: 1974.
Congressional District: 12, 19 & 16
Key Personnel: Supt., Pedro Ramos.
Governing Authority: federal. Parent Institution: National Park Services, Dept. of the Interior, Washington, DC.
Institution Type/Description: National Park & Visitor Center.
Collections: archeological findings stored at the Southeast Archeological Center, Tallahassee, FL; 2,400 sq. miles consisting of sandy islands of slash pine; mixed hardwood hammocks; wet prairies; marshes; estuarine mangrove forests; airplants; bromeliads; orchids; animals in natural habitat.
Research Fields: natural sciences; south Florida environment.
Facilities: small resource library.
Hours & Admission Prices: Oasis Visitor Center: daily 9-4:30. No charge. Closed Christmas. &

Orange Park

MUSEUM OF GREAT CHRISTIAN PREACHERS & MINISTRIES, 2941 Greenridge Rd., Orange Park, FL 32073-6411. Tel.: 904-375-0047.
Founded: 2000.
Key Personnel: Dir., Minister Donald L. Crutch; Chm. (V), Shaun Harper; Pres. (V), Lisa A. Crutch; Devel. & Cur., Samuel Alston; Education & Public Rels., Joni Meyers; Treas., Patricia A. Strader; Registrar & Museum Shop Mgr., Marcus Debnam.
Personnel Profile: Full-Time Paid 1; Full-Time Volunteers 6; Interns 2.
Governing Authority: private; nonprofit organization. Tax-exempt: 501(c)(3).
Institution Type/Description: Religious Art Museum.
Collections: lives & travel of Christian leaders; wax figures of Christian leaders; photographs.
Research Fields: religions; Bible.
Activities: concerts. Annual Event: Gospel Music Concert.
Hours & Admission Prices: Oct.-Aug. Mon.-Fri. 10-6, Sat. 9-12. Adults $7, senior citizens, students & children $3. Closed New Year's Eve & Day; Martin Luther King Jr. Day; Lincoln's Birthday; Washington's Birthday; Good Friday; Memorial Day; Independence Day; Thanksgiving & day before; Christmas Eve & Day.
Attendance: 400

Orlando

ANITA S. WOOTEN GALLERY AT VALENCIA COMMUNITY COLLEGE EAST CAMPUS, 701 N. Econlockhatchee Trail, Orlando, FL 32825-6404. Mailing Address: P.O. Box 3028, Orlando, FL 32802-3028. Tel.: 407-582-2298. Fax: 407-582-8917.
Web Site: www.valenciacc.edu/gallery
Founded: 1967.
Congressional District: 5
Key Personnel: Dir., Jackie Otto-Miller.
Personnel Profile: Part-Time Paid 2.
Governing Authority: college. Parent Institution: Valencia Community College. Tax-exempt: 501(c)(3).
Institution Type/Description: Art Museum.
Collections: contemporary art; Florida art; small works.
Facilities: 30-vol. library of artists books available for inter-library loan; educational facilities; 1,200 sq. ft. exhibit space.
Activities: guided tours; lectures; films; concerts; dance recitals; theatre; broadcast programs; organized education programs for adults & undergraduate or graduate college students; temporary, loan & traveling exhibitions. Museum Sponsors: Small Works Competition.
Publications: exhibition catalogs.
Hours & Admission Prices: Summer: Mon.-Thurs. 8:30-4:30, Fri. 8:30am-12pm; Winter: Mon.-Fri. 8:30-4:30. No charge. &
Attendance: 18,000 (estimated)

HARRY P. LEU GARDENS, 1920 N. Forest Ave., Orlando, FL 32803-1537. Tel.: 407-246-2620. Fax: 407-246-2849.
E-mail: robert.bowden@cityoforlando.net
Web Site: www.leugardens.org
Founded: 1961.
Congressional District: 3
Key Personnel: Dir., Robert E. Bowden; Membership Coord., Lynn Williams; Mktg., Tracy Micciche; Museum Shop Mgr., Miriam Maldanado.
Personnel Profile: Full-Time Paid 20; Part-Time Paid 20; Part-Time Volunteers 186.
Governing Authority: municipal; board of trustees; nonprofit. Parent Institution: City of Orlando. Tax-exempt: 501(c)(3).
Institution Type/Description: Botanical Garden.
Collections: camellias; roses; citrus; palms; cycads; bromeliads.
Facilities: library of horticulture books available to the public; classroom; lecture hall; meeting rooms; archives; herbarium. Museum-related items for sale.
Activities: concerts; docent program; formal educational programs; guided tours; hobby workshops; lectures. Annual Events: Spring & Fall Moonstrolls; International Christmas, plant sales, fall gardening festival.
Publications: quarterly newsletter, Garden View; quarterly class schedule, Leu Gardens Classes & Workshops.
Hours & Admission Prices: Daily 9-5. Adults $7, children grades K-12 $2; discounts to groups & American Horticultural Society; Mon., museum members & members of AABGA RAP program no charge. Closed Christmas. &
Attendance: 135,000 (accurate)
Membership: Senior $30; Individual $35; Household $45; Contributing $60; Sustaining $125; Orchid $300; Rose $500; Camellia $1,000.

MENNELLO MUSEUM OF AMERICAN ART, (M), 900 E. Princeton St., Orlando, FL 32803-1437. Tel.: 407-246-4278. Fax: 407-246-4329.
E-mail: mennellomuseum@cityoforlando.net
Web Site: www.mennellomuseum.com
Founded: 1998.
Congressional District: 8
Key Personnel: C.E.O., Frank Holt; Chm. (V), David Cross; Office & Museum Shop Mgr., Kim Robinson.
Personnel Profile: Full-Time Paid 2; Part-Time Paid 3; Part-Time Volunteers 16; Interns 2.
Operating Expenses: 525,000
Governing Authority: municipal; not-for-profit. Parent Institution: City of Orlando. Tax-exempt: 501(c)(3).
Institution Type/Description: American Art Museum.
Collections: paintings by Earl Cunningham; American Folk art; early 20th-century paintings.
Major Exhibits: Southwestern Allure: The Art of the Santa Fe Art Colony (T), 1/14-4/14; Mingled Visions: The North American Indian By Edward S. Curtis (T), 5/14-10/14; Southwest Pottery, 5/14-10/14; George Catlin's American Buffalo (T), 10/14-1/15; The Taos Society of Artists, 10/14-1/15.
Research Fields: identification & research on American Folk artists; Earl Cunningham & early 20th century American modernism.
Facilities: library; 3,511 sq. ft. exhibit space; lakeside grounds. Museum-related items for sale.
Activities: docent program; guided tours; lectures; loan & traveling exhibitions.
Publications: member's newsletter; exhibit catalogs; American Folk Art Masters; Earl Cunningham: Dreams Realized; Earl Cunningham's America, The Paintings of George Bellows; John Sloan: Gloucester & Santa Fe; Style & Grace; The Michael & Marilyn Mennello Collection of American Art.
Hours & Admission Prices: Tues.-Sat. 10:30-4:30, Sun. 12-4:30. Adults $5, seniors $4, students $1; discounts to SARM, SERM & AAM members; active military, members & children under 12 no charge. Closed major holidays. &
Attendance: 27,000 (accurate)
Membership: Student & Senior $15; Individual $35; Family $50; Sponsor $100; Patron $500; Benefactor $1,000.

✳ ORANGE COUNTY REGIONAL HISTORY CENTER, (M), 65 E. Central Blvd., Orlando, FL 32801-2401. Tel.: 407-836-8500. Fax: 407-836-8550.
E-mail: sara.vanarsdel@ocfl.net
Web Site: www.thehistorycenter.org
Formerly: Orange County Historical Museum
Founded: 1957.
Congressional District: 5 & 9
Key Personnel: Exec. Dir., Sara Van Arsdel; Chm. (V), Johanna Clark; Pres. (V), Tom McAleavy; Museum Shop Mgr., Henry Nauman.
Personnel Profile: Full-Time Paid 29; Part-Time Paid 4; Part-Time Volunteers 125; Interns 5.
Governing Authority: county government. Parent Institution: Orange County Historical Society, Inc. Affiliated with the Board of County Commissioners. Subsidiary Institution: Historical Society of Central Florida, Inc. Tax-exempt: 501(c)(3).
Institution Type/Description: Local Central Florida History & American History (12,000 years ago to the present).
Collections: permanent & special exhibits, tracing central Florida history; national & international exhibits on historical subjects; furniture collections; Indian artifacts; Seminoles; Pioneers; early Florida industries including: cattle, citrus, aviation, transportation & tourism; Heritage Square park & plaza; restored 1927 Courtroom B; community exhibits.
Research Fields: Floridiana; oral history; citrus industry; tourism.
Facilities: 5,000-vol. library; theater; photo archives; old city directories; telephone books; maps; ledgers; scrapbooks; yearbooks; postcards; archival material.
Activities: educational programs for children, students & adults; rental facility; social events; seminars; lectures; family fun events; special exhibits; tours; hands-on activities; camps.
Publications: newsletter; brochures, educational pamphlets; quarterly magazine.
Hours & Admission Prices: Mon.-Sat. 10-5, Sun. 12-5. Adults $9, seniors 60 & over $7, children 3-12 $6; discounts to AAM members & Orange County employees; children under 3, members & regional Florida teachers no charge. Closed county holidays. &
Attendance: 100,000 (accurate)
Membership: Senior Citizen & Student $40; Individual $50; Senior Family $65; Family $75; Contributing Patron $125; Sustaining Patron $250; Director's Circle Patron $500; Special Patron $1,000; Legacy Patron $1,500; 1892 Cornerstone Patron $1,892; History Maker Patron $2,500.

✳ ORLANDO MUSEUM OF ART, (M), 2416 N. Mills Ave., Orlando, FL 32803-1483. Tel.: 407-896-4231. Fax: 407-896-9920.
E-mail: info@omart.org
Web Site: www.omart.org
Founded: 1924.
Congressional District: 8
Key Personnel: Dir., Glen Gentele; Chm. Bd., Ted R. Brown, Esq.; Chief Operations, Alex Garcia; Interim Public Rels. & Mktg. Mgr., Randall Ross; Cur. Education, Jane Ferry; Cur., Hansen Mulford; Museum Shop Mgr., Maryann Keane; Finance & Admin. Dir., Dana Dougherty; Registrar, Andrea Long.
Personnel Profile: Full-Time Paid 27; Part-Time Paid 7; Part-Time Volunteers 656; Interns 6.
Governing Authority: nonprofit organization. Tax-exempt: 501(c)(3).
Institution Type/Description: Art Museum.
Collections: contemporary art; 19th & early 20th century American Art; art of the ancient Americas; African Art.
Major Exhibits: Living in Style: African Art of Everyday Life from the Collection of William D. & Norma Canelas Roth, 1/14-12/14; Aztec to Zapotec II: Selections from the Art of the Ancient Americas Collection, 1/14-12/14; Rembrandt, Rubens, Gainsborough and the Golden Age of Painting in Europe (T), 1/25/14-5/25/14; Acquisition Trust: Celebrating Thirty Years of Collecting at the Orlando Museum, 10/11/14-11/2/14.
Research Fields: 18th-, 19th- & 20th-century American art; art of the Ancient Americas; African art; contemporary art.
Facilities: 10,922-vol. reference library; 250-seat auditorium; classrooms; meeting rooms; full catering kitchen.
Activities: permanent & temporary exhibitions; adult & children's programming; guided tours; educator workshops; lectures; art classes; facility rental program; corporate lease programs.
Publications: newsletters; exhibition catalogs; e-calendar; collection guides; pamphlets.
Hours & Admission Prices: Tues.-Fri. 10-4, Sat.-Sun. 12-4. Adults $8, seniors 65 & over, college students, & military $7, children 4-17 $5; discount to groups of 10 or more & AAM members; children 3 & under and members no charge. Closed national holidays. &
Attendance: 93,202 (accurate)
Membership: Student $40; Individual $55; Dual & Family $80; Contributing $125; Supporting $250; Sustaining $500; Ambassador $1,000-$25,000. (Seniors $10 discount on all levels).

✳ ORLANDO SCIENCE CENTER, INC., (M), 777 E. Princeton St., Orlando, FL 32803-1291. Tel.: 407-514-2000. Fax: 407-514-2277.
E-mail: info@osc.org
Web Site: www.osc.org
Founded: 1959.
Congressional District: 8
Key Personnel: C.E.O. & Pres., Dr. Brian Tonner; Chm. (V), Theodore Bradford; Vice Pres. Finance & C.F.O., John Mallozzi; Vice Pres. & C.O.O., JoAnn Newman; Dir. Public Rels., Jeff Stanford; Vice Pres. Participant Experience, Kim Hunter.
Personnel Profile: Full-Time Paid 70; Part-Time Paid 50; Part-Time Volunteers 200; Interns 10.
Governing Authority: nonprofit organization. Tax-exempt: 501(c)(3) & 509(a).
Institution Type/Description: Science & Technology Center.
Collections: interactive exhibits focusing on physical, natural, space & health sciences; applied technology-aerospace, lasers, simulation & entertainment; career; early childhood; CineDome: 15/70 film, Digistar II planetarium; Observatory: 10-inch refractor telescope.
Research Fields: educational research; visitor studies.
Facilities: 300-seat combination Iwerks large film format theater & planetarium; astronomical observatory; meeting rooms; 250-seat theater; restaurant. Books, science games, educational & science-related items for sale.
Activities: classes & courses; volunteer training program; formally organized education programs for pre-K through undergraduate college students, teachers & adults; large-scale demonstrations & science theater; permanent & temporary exhibitions; planetarium shows; educational tours; workshop.
Publications: magazine, SCOPE.
Hours & Admission Prices: Thurs.-Tues. 10-5. Adults $17, students & seniors 55 & over $16, children 3-11 $12; discounts to active military, ASTC members, reciprocal participants & corporate sponsor employees; members no charge. Closed Thanksgiving; Christmas. &
Attendance: 390,000 (accurate)
Membership: Individual $80; Couple $95; Family $110.

RIPLEY'S BELIEVE IT OR NOT! MUSEUM, 8201 International Dr., Orlando, FL 32819-9326. Tel.: 407-363-4418 & 351-5803. Fax: 407-345-0803.
E-mail: haggadone@ripleys.com
Web Site: www.ripleys.com/orlando
Founded: 1992.
Institution Type/Description: General Museum.
Collections: antiquities & oddities from around the world & throughout time; personal artifacts; folk art.
Facilities: Museum-related items for sale.
Activities: self-guided tours.
Hours & Admission Prices: Daily 9am to midnight. Adults $19.99, children 4-12 $12.99; discounts to groups of 10 or more, AAM & ICOM members. &

TERRACE GALLERY-CITY OF ORLANDO, City Hall, 400 S. Orange Ave., Orlando, FL 32801-3360. Mailing Address: City Hall, P.O. Box 4990, Orlando, FL 32802-4990. Tel.: 407-246-4279. Fax: 407-246-3434.
E-mail: paul.wenzel@cityoforlando.net
Web Site: www.cityoforlando.net/arts
Founded: 1992.
Congressional District: 8
Key Personnel: Cur. & Public Art Coord., Paul F. Wenzel.
Personnel Profile: Full-Time Paid 2; Interns 1.
Governing Authority: municipal. Parent Institution: City of Orlando. Branch Gallery: Harry P. Leu Gardens, Garden House Gallery, 1920 N. Forest Ave., Orlando, FL. Tax-exempt: 501(c)(3).
Institution Type/Description: Art Gallery.
Collections: contemporary Florida artists; 850 works in all media.
Research Fields: Florida art.
Facilities: 3,000 sq. ft. exhibit space.
Activities: guided tours; lectures; loan, temporary & traveling exhibitions.
Publications: exhibit catalogs.
Hours & Admission Prices: Terrace Gallery: Mon.-Fri. 8-9, Sat.-Sun. 12-5. No charge. Mayor's Gallery: Mon.-Fri. 8-5. No charge. Garden House Gallery: daily 9-5. Adults $5, children $1. &
Attendance: 35,000 (estimated)

UNIVERSITY OF CENTRAL FLORIDA ART GALLERY, Visual Arts Bldg. #51, Orlando, FL 32816. Mailing Address: P.O. Box 161342, Orlando, FL 32816-1342. Tel.: 407-823-5470 & 3161. Fax: 407-823-6470. Facebook: UCF Art Gallery.
E-mail: gallery@ucf.edu
Founded: 1974.
Congressional District: 24
Key Personnel: Dir., Diane Daugherty.
Personnel Profile: Full-Time Paid 2; Part-Time Paid 2; Part-Time Volunteers 10.
Governing Authority: public university. Parent Institution: University of Central Florida.
Institution Type/Description: Art Museum.
Collections: paintings; sculpture; photography.
Activities: guided tours; special events.
Hours & Admission Prices: Mon.-Fri. 10-5. No charge.
Attendance: 7,000 (estimated)

WYCLIFFE DISCOVERY CENTER, 11221 John Wycliffe Blvd., Orlando, FL 32832-7013. Mailing Address: P.O. Box 628200, Orlando, FL 32862-8200. Tel.: 407-852-3626. Fax: 407-852-3781.
Web Site: www.wycliffe.org/wordspring
Formerly: WordSpring Discovery Center
Founded: 2002.
Congressional District: 24
Key Personnel: Dir., Patricia Cox; Mgr. Events, Brian Shaffer; Mgr. Education Program, Dorcas Winfrey; Museum Shop Mgr., Tim Holloran.
Personnel Profile: Full-Time Paid 6; Part-Time Volunteers 8.
Governing Authority: private; nonprofit organization. Parent Institution: Wycliffe Bible Translators. Tax-exempt: 501(c)(3).
Institution Type/Description: Religious Museum.
Collections: Bible translations; ethnic costumes from around the world; audio samples of the world's languages.
Research Fields: linguistics; literacy.
Facilities: 500-vol. library of Bibles in different languages; 5,300 sq. ft. exhibit space; auditorium; cafe; 35-seat theater. Museum-related items for sale.
Activities: films; guided tours; lectures; participatory exhibits; theater.
Hours & Admission Prices: Mon.-Fri. 9-4, Sat. & holidays call for hours.

Adults $8, seniors $7, children grades 1-12 $6; discounts to groups; children under 6 no charge. &
Attendance: 15,000 (accurate)
Membership: Children $12; Senior $14; Adult $16.

Ormond Beach

ORMOND MEMORIAL ART MUSEUM & GARDEN, (M), 78 E. Granada Blvd., Ormond Beach, FL 32176-6534. Tel.: 386-676-3347. Fax: 386-676-3244.
E-mail: omam78e@aol.com
Web Site: www.ormondartmuseum.org
Founded: 1946.
Congressional District: 4
Key Personnel: Dir., Susan Tucker; Chm. (V) & Pres. (V), Theresa Gryer; Pres. (V), Mary Greenlees; Vice Pres., Patti Alexander; Treas., Alexis Lenssen; Cur. Education, Barbara Saunders; Cur. Education, Regina Stengel; Cur. Education, Linda King; Administrative Coord., Vanessa Elliott.
Personnel Profile: Full-Time Paid 3; Part-Time Paid 3; Part-Time Volunteers 110.
Governing Authority: private; nonprofit. Tax-exempt: 501(c)(3).
Institution Type/Description: Art Museum & Botanical Garden: the museum is located in a four-acre botanical garden.
Collections: symbolic paintings by Malcolm Fraser.
Facilities: botanical garden; 4,500 sq. ft. exhibit space.
Activities: arts festivals; docent program; formal education program for children; guided tours; lectures; loan, temporary & traveling exhibitions. Annual Events: Starry Starry Night in January; holiday exhibitions & receptions; Presidents Day School Alternative in February.
Publications: quarterly newsletter.
Hours & Admission Prices: Mon.-Fri. 10-4, Sat.-Sun. 12-4. Donation: adults $2; discounts to AAM members; members, seniors 60 & over and children no charge. Closed New Year's Day; Good Friday; Easter; Memorial Day; Independence Day; Labor Day; Thanksgiving; Christmas. &
Attendance: 18,000 (estimated)
Membership: Retiree & Individual $25; Family $35. Heritage Club: Sponsor $85; Benefactor $150; Patron $500.

Osprey

* **HISTORIC SPANISH POINT, (M),** 337 N. Tamiami Trail, Osprey, FL 34229-8911. Mailing Address: P.O. Box 846, Osprey, FL 34229-0846. Tel.: 941-966-5214. Fax: 941-966-1355.
E-mail: info@historicspanishpoint.org
Web Site: historicspanishpoint.org
Founded: 1980.
Congressional District: 13
Key Personnel: C.E.O. & Exec. Dir., John D. Mason; Dir. Education, Kara Pallin; Site Horticulturist, Nancy Paul.
Personnel Profile: Full-Time Paid 8; Part-Time Paid 4; Part-Time Volunteers 200.
Governing Authority: nonprofit organization. Parent Institution: Gulf Coast Heritage Association Inc. Tax-exempt: 501(c)(3).
Institution Type/Description: Historic Site.
Collections: late 19th-century pioneer artifacts; prehistoric Indian artifacts; shell tools; pottery; textiles.
Research Fields: local & regional history; archaeology; environmental studies.
Facilities: visitors center, nature trails; 500 sq. ft. exhibit gallery space; pioneer buildings; archaeological exhibition; picnic area. Museum-related items for sale.
Activities: guided tours; lectures; organized education programs; docent program; temporary exhibitions.
Publications: quarterly newsletter, Vision.
Hours & Admission Prices: Mon.-Sat. 9-5, Sun. 12-5. Adults $12, seniors $10, children $5; children under 5, NARM & members no charge. Closed New Year's Day; Easter; Thanksgiving; Christmas. &
Attendance: 30,000 (estimated)
Membership: Individual $40; Family $65; Donor $125; Settler $250; Heritage Club $500 & up.

Palatka

RAVINE GARDENS STATE PARK, 1600 Twigg St., Palatka, FL 32177-5637. Tel.: 386-329-3721. Fax: 386-329-3718.
E-mail: alfred.bea@dep.state.fl.us
Web Site: www.floridastateparks.org/ravinegardens/default.cfm
Founded: 1933.
Congressional District: 7
Key Personnel: Park Mgr., Nathan Sommons.
Personnel Profile: Full-Time Paid 10; Part-Time Paid 3; Part-Time Volunteers 12.

Governing Authority: state. Tax-exempt.
Institution Type/Description: Botanical Garden.
Collections: plants; flowers; trees.
Facilities: 240-seat auditorium; 4 meeting rooms.
Activities: guided tours; guided trail walks; garden tours; wagon tours.
Publications: Park & Trail Guides.
Hours & Admission Prices: Daily 8-sunset. Cars with up to 8 people $5, motorcycles & single occupant cars $4, bicycles & walk-ins $2; children under 6 no charge. Annual park passes available. &
Attendance: 145,963 (accurate)
Membership: The Friends of Ravine Gardens (CSO) Citizen Support Organization: Student $5; Individual $10; Family $20; Corporation $30.

Palm Beach

* **FLAGLER MUSEUM, (M),** Cocoanut Row & Whitehall Way, Palm Beach, FL 33480. Mailing Address: P.O. Box 969, Palm Beach, FL 33480-0969. Tel.: 561-655-2833. Fax: 561-655-2826.
E-mail: mail@flaglermuseum.us
Web Site: www.flaglermuseum.us
Formerly: Henry Morrison Flagler Museum
Founded: 1959.
Congressional District: 14
Key Personnel: Exec. Dir., John Michael Blades; Business Mgr., Donovan Owen; Chief Cur., Tracy Kamerer; Dir. Member Svcs., Sarah Brutschy; Dir. Public Affairs, David Carson; Facility Mgr., William Fallacaro; Dir. Education, Allison Goff.
Personnel Profile: Full-Time Paid 37; Part-Time Paid 14; Part-Time Volunteers 90.
Governing Authority: nonprofit organization. Tax-exempt: 501(c)(3).
Institution Type/Description: Historic House: Whitehall, 1902 Gilded Age estate of Henry Morrison Flagler.
Collections: furnishings original to Flagler's Gilded Age estate and other period objects.
Research Fields: America's Gilded Age; Henry Flagler's role in the development of Florida; Florida history.
Facilities: library & archives by appointment.
Activities: guided tours; lectures; concerts; docent program; permanent & temporary exhibitions; special events.
Publications: video, The Wilderness Shall Bloom; newsletter, Inside Whitehall; exhibit catalogues, Palm Beach Panorama; Henry M. Flagler's Paintings - The Taste of a Gilded Age Collector; Romance of the Season: Five Florida East Coast Hotel Novelettes from 1904 (facsimile reproduction); guidebook, Flagler Museum: An Illustrated Guide; exhibit catalogue, A Society of Painters: Flagler's St. Augustine Art Colony; video/DVD, Whitehall: A National Historic Landmark; exhibit catalogue, Tiffany at the World's Columbian Exposition.
Hours & Admission Prices: Tues.-Sat. 10-5, Sun. 12-5. Adults $18, youth 13-18 $10, children 6-12 $3; discount to groups; members & AAM members no charge. Closed New Year's Day; Thanksgiving; Christmas. &
Attendance: 85,000 (accurate)
Membership: Individual $75; Family $125; Sustaining $225; Sponsor $500; Patron $1,000 Benefactor $2,500; Associate $5,000.

* **THE SOCIETY OF THE FOUR ARTS, (M),** 2 Four Arts Plaza, Palm Beach, FL 33480-4102. Tel.: 561-655-7227. Fax: 561-655-7233.
Web Site: www.fourarts.org
Founded: 1936.
Congressional District: 11
Key Personnel: Pres., Ervin S. Duggan; Chm. Bd., Patrick Henry; Exec. Vice Pres., Nancy Mato.
Personnel Profile: Full-Time Paid 30; Part-Time Paid 7; Part-Time Volunteers 3.
Volunteer Hours: 500
Operating Expenses: 5,600,000
Operating Income: 5,900,000
Governing Authority: corporation. Parent Institution: The Society of the Four Arts. Tax-exempt: 501(c)(3) & 170(b)(1)(A).
Institution Type/Description: General Museum.
Collections: paintings; sculptures; Japanese & Chinese porcelain; Japanese & Chinese textiles; shells.
Major Exhibits: Illustrating Words: The Wondrous Fantasy World of Robert L. Forbes, Poet and Ronald Searle, Artist, 11/13-6/15; The Coast and the Sea: Marine and Maritime Art from the New York Historical Society (T), 1/25/14-3/9/14; Illuminating the Word: The Saint John's Bible (T), 3/22/14-4/23/14; Light In the Desert: Photographs from the Monastery of Christ in the Desert by Tony O'Brien (T), 3/22/14-4/23/14.
Research Fields: paintings; sculpture.
Facilities: 60,000-vol. library of general books available for inter-library loan

& on premises; children's library; reading room; sculpture garden; botanical garden; 700-seat auditorium.
Activities: lectures; films; concerts; dance & opera; formally organized education programs for children; inter-museum loan & traveling exhibitions.
Publications: catalogs of exhibitions; triannual calendar; annual schedule of events.
Hours & Admission Prices: Gallery & Auditorium: Dec. to mid-April Mon.-Sat. 10-5, Sun. 1-5. Exhibits: $5 per person. Metropolitan Operas: $27 per person. National Theatre of London: $25 per person. Evening Concerts: $40-$45 per person. Sunday Concerts: $20 per person. Films: $5 per person. Exhibition Great Art on Screen $12 per person. Closed New Year's Day; Easter; Thanksgiving; Christmas. Children's Library: May-July & Sept.-Oct. Mon.-Fri. 10-4:45; Nov.-April Mon.-Fri. 10-4:45, Sat. 10-12:45. Closed New Year's Eve & Day; Presidents' Day; Memorial Day; Independence Day; Thanksgiving & two days after; Christmas Eve & Day. Adult Library: June-Oct. Mon.-Fri. 10-5; Nov.-May Mon.-Fri. 10-5, Sat. 10-1. Closed New Year's Eve & Day; Presidents' Day; Memorial Day; Independence Day; Thanksgiving & two days after; Christmas Eve & Day. Dixon Education Bldg.: Mon.-Fri. 10-5. Closed New Year's Eve & Day; Presidents' Day; Memorial Day; Independence Day; Thanksgiving & two days after; Christmas Eve & Day. Sculpture & Botanical Gardens: daily 10-5 (weather permitting). Closed major holidays. Discounts to AAM members & groups. &
Attendance: 111,464 (estimated)
Membership: Individual $1,300.

Palm Coast

FLORIDA AGRICULTURAL MUSEUM, 7900 Old Kings Rd., Palm Coast, FL 32137-8285. Tel.: 386-446-7630. Fax: 386-446-7631.
E-mail: info@myagmuseum.com
Web Site: www.myagmuseum.com
Founded: 1983.
Congressional District: 7
Key Personnel: C.E.O., Bruce J. Piatek; Chm. (V), Tom Torrence; Trustee, Doyle Conner; Trustee, Ben Hill Griffin, III; Trustee, Brenda Tucker Boyd; Trustee, Michael Kenney; Trustee, William Livingston; Trustee, Louis Parrish; Trustee, Clark Bailey; Trustee, Frank Ford; Trustee, Rudy Bradley.
Personnel Profile: Full-Time Paid 3; Part-Time Paid 4; Part-Time Volunteers 45; Interns 2.
Governing Authority: public; nonprofit. Tax-exempt: 501(c)(3).
Institution Type/Description: Agriculture Museum.
Collections: period artifacts; specimens; historical documents & heirlooms representing the history of Florida agriculture; heritage livestock breeds including endangered Florida Cracker horses and cattle. Historic Buildings: 1890s farmstead & outbuildings; 1900s dry goods store & commissary.
Research Fields: Florida agricultural history; Florida shortline railroads; rural Florida architecture & lifeways; heritage livestock breeds.
Facilities: 10,000-vol. library of historic & contemporary agriculture; 460 acres of grounds.
Activities: guided tours; lectures; monthly community projects.
Hours & Admission Prices: Wed.-Sun. 9-5. Adults $8, children 5-12 $6; members & children under 5 no charge. &
Attendance: 20,000 (estimated)
Membership: Farmhand $30; Ma & Pa $45; Farm Family $60; Farmer $125-$249; Rancher $250-$499; Planter $500-$999; Cattle King $1,000 & up; Director's Posse $5,000 & up; Chairman's Council $10,000 & up.

THE LILLYWHITE FAMILY MUSEUM, Niblick's Reach, Hammock Dunes, 4 Riviera Pl., Palm Coast, FL 32137-2270. Tel.: 386-446-3679. Fax: 386-446-3679.
E-mail: jwlillywhite@mac.com
Web Site: www.thelillywhitefamilymuseum.com
Founded: 2007.
Congressional District: 7
Key Personnel: Dir. & Cur., John W. Lillywhite; Chm. (V), Elisabet G. Wauters-Lillywhite.
Personnel Profile: Full-Time Volunteers 2.
Governing Authority: private.
Institution Type/Description: Family Sports History Museum.
Collections: family sports memorabilia; sporting goods; photographs; illustrations; ephemera; books; family business of manufacturing & retailing of sporting goods from 1844 to present.
Research Fields: life & times of the Victorian Lillywhite Family of sportsmen & business entrepreneurs.
Facilities: 300-vol. library; 300 sq. ft. exhibit space.
Activities: traveling exhibitions.

Hours & Admission Prices: Mon.-Fri. 9-1 by appointment only. No charge; donations accepted. Closed holidays.
Attendance: 300 (estimated)

WASHINGTON OAKS GARDENS STATE PARK, 6400 N. Ocean Shore Blvd., Palm Coast, FL 32137-2415. Tel.: 386-446-6780. Fax: 386-446-6781.
E-mail: matgeo@mail.state.fl.us
Web Site: www.floridastateparks.org/washingtonoaks
Founded: 1970.
Key Personnel: Park Mgr., Douglas Carter; Asst. Mgr., Renee Paolini.
Personnel Profile: Full-Time Paid 10; Part-Time Paid 2; Part-Time Volunteers 10.
Governing Authority: state. Department of Environmental Protection, Commonwealth Bldg., 3900 Commonwealth Blvd., Tallahassee, FL 32303.
Institution Type/Description: State Gardens.
Collections: botanical.
Publications: park brochures.
Hours & Admission Prices: Park: daily 8-sundown. Vehicle (up to 8 people) $5, Single Occupant Vehicle$4, Pedestrians & Bicyclists $1; children 5 & under no charge.

Palm Harbor

PALM HARBOR MUSEUM, 2043 Curlew Rd., Palm Harbor, FL 34683-6820. Tel.: 727-724-3054. Facebook: Palm Harbor Museum.
Web Site: palmharbormuseum.com
Formerly: North Pinellas Historical Museum
Founded: 1997.
Congressional District: 10
Key Personnel: Exec. Consultant, Colin Bissett.
Personnel Profile: Part-Time Volunteers 12.
Governing Authority: private; nonprofit organization. Parent Institution: Palm Harbor Historical Society, Inc. Tax-exempt: 501(c)(3).
Institution Type/Description: History Museum.
Collections: local & regional history; photographs; audiovisuals; personal artifacts.
Facilities: library. Museum-related items for sale.
Activities: guided tours; temporary exhibitions; broadcast programs; luncheons for groups of 20 or more by appointment. Annual Events: Pioneer Days; English High Tea.
Hours & Admission Prices: Mon. & Sat. 10-4; groups of 10 or more by appointment. No charge; donations accepted. Closed New Year's Day; Thanksgiving; Christmas. &
Attendance: 900 (estimated)
Membership: Individual $25; Family $40; Business $100; Contributing $250; Patron $500; Benefactor $1,000; History Angel $2,500.

Panama City

GULF COAST COMMUNITY COLLEGE ART GALLERY, 5230 W. Hwy. 98, Panama City, FL 32401-1058. Tel.: 850-872-3887. Fax: 850-872-3836.
E-mail: tmarinuzzi@gulfcoast.edu
Institution Type/Description: Art Gallery.
Collections: paintings; sculpture; photographs.
Hours & Admission Prices: Mon.-Fri. 8-4. No charge.

SCIENCE AND DISCOVERY CENTER OF NORTHWEST FLORIDA, 308 Airport Rd., Panama City, FL 32405-4610. Tel.: 850-769-6128. Fax: 850-769-6129.
Web Site: scienceanddiscoverycenter.org
Formerly: The Junior Museum of Bay County
Founded: 1969.
Congressional District: 1
Key Personnel: Exec. Dir., Rae Cotton; Dir. Education, Mickey Busby; Admin., Sarah Sapp.
Personnel Profile: Full-Time Paid 4; Part-Time Paid 2; Part-Time Volunteers 120.
Governing Authority: nonprofit organization. Tax exempt: 501(c)(3).
Institution Type/Description: Children's Museum.
Collections: pioneer Florida log structures; grist mill; Bay County historical material; agriculture implements; nature trail; science & nature hands-on exhibits.
Research Fields: local history; natural sciences.
Facilities: 80-seat auditorium; classrooms; nature trail. Museum-related items for sale.
Activities: self-guided tours; children summer classes; weekend programs; permanent & temporary exhibits.

Publications: brochures; newsletter, The Adventures.
Hours & Admission Prices: Tues.-Sat. 10-5, Sun. 12-5. Adults $7, children, seniors & military $6; discounts to groups; teachers & members no charge. Closed major holidays. &
Attendance: 27,000 (accurate)
Membership: Individual $35; Family & Grandparents $60; Enhanced Family $125; Patron $250; Benefactor $500; Sustaining $1,000.

VISUAL ARTS CENTER OF NORTHWEST FLORIDA, 19 E. 4th St., Panama City, FL 32401-3106. Tel.: 850-769-4451. Fax: 850-785-9248.
E-mail: visualartscenterofnwfla@comcast.net
Web Site: www.vac.org.cn
Founded: 1988.
Key Personnel: Exec. Dir., Ellen Killough; Operations Mgr., Stacey Dyer; Exhibits & Education, Jayson Kretzer; Volunteer Coord., Bonnie Rodenburg.
Personnel Profile: Full-Time Paid 2; Part-Time Paid 2; Part-Time Volunteers 36.
Governing Authority: nonprofit organization. Parent Institution: Panama Art Association. Tax-exempt: 501(c)(3).
Institution Type/Description: Art Gallery; 1925 former city courthouse, jail & fire station.
Collections: permanent collections focus on artists of northwest Florida; paintings in watercolor, oil, acrylics, pen & ink; drawings; photographs; sculpture.
Facilities: large main gallery; smaller Higby gallery; art studios; children's hands-on interactive gallery. Museum-related items for sale.
Activities: lecture series; workshops; tours.
Publications: newsletter, Images.
Hours & Admission Prices: Tues. & Thurs. 10-8, Wed. & Fri.-Sat. 10-6, Sun. 1-5. No charge; donations accepted. Closed New Year's Eve & Day; Christmas. &
Attendance: 10,078 (accurate)
Membership: Student, Senior Citizen 55 & over and Educator $15; Senior Couple 55 & over $25; Individual $35; Family $60; Group (10 or more) $75; O'Keefe $100l; Picasso $250; Monet $500; Renoir $1,000; Van Gogh $2,500; Rembrandt $5,000; Michelangelo $10,000.

Panama City Beach

MAN IN THE SEA MUSEUM, 17314 Panama City Beach Pkwy., Panama City Beach, FL 32413-6038. Tel.: 850-235-4101. Fax: 850-235-4101.
E-mail: momits@bellsouth.net
Web Site: www.maninthesea.org
Formerly: The Museum of Man In the Sea, Inc.
Founded: 1980.
Congressional District: 1
Key Personnel: Mgr., Leslie Baker; Pres. (V), Michael A. Zinszer; Education & Museum Shop Mgr., Sarah Hough.
Personnel Profile: Full-Time Paid 2; Part-Time Paid 1; Part-Time Volunteers 36.
Governing Authority: not-for-profit organization. Parent Institution: Institute of Diving, Inc., 17413 Panama City Beach Pkwy., Panama City Beach, FL 32413. Tax-exempt: 501(c)(3).
Institution Type/Description: History & Science Museum.
Collections: how man can live, work & play underwater; SCUBA; commercial & military diving equipment; underwater habitats; diving bells; submersibles; armored diving suits; recreational diving equipment; history of diving.
Research Fields: commercial; military & recreational diving equipment; underwater habitats; artificial reefs; marine pollution; marine archaeology; history of diving.
Facilities: 300-vol. library. Books, jewelry & T-shirts for sale.
Activities: guided tours; lectures; loan & participatory exhibits; school loan service.
Publications: quarterly, Institute of Diving Newsletter.
Hours & Admission Prices: Tues.-Sun. 10-4. Admission $5; members & children under 6 no charge. Closed New Year's Day; Thanksgiving; Christmas. &
Attendance: 30,080 (accurate)
Membership: Annual $35; Three-Year Individual $90; Lifetime $1,000.

Parrish

FLORIDA RAILROAD MUSEUM, INC., U.S. 301, 12210 83rd St. E., Parrish, FL 34219. Mailing Address: P.O. Box 355, Parrish, FL 34219-0355. Tel.: 941-776-0906. Fax: 941-917-0081.
Web Site: www.frrm.org

Formerly: Florida Gulf Coast Railroad Museum
Founded: 1983.
Congressional District: 11
Key Personnel: Pres. (V), Steven Wonderly; Sec., Dr. Edward Dunham; Treas., William Maddock.
Personnel Profile: Full-Time Paid 1; Full-Time Volunteers 30; Part-Time Volunteers 30; Interns 5.
Governing Authority: board of directors. Tax-exempt.
Institution Type/Description: Railroad Museum.
Collections: concentration on railroad history from 1890s to present with emphasis on Florida railroad history & the economic history of the railroad industry; steam engine #12; operating railroad; railroad tools.
Research Fields: economic impact of the railroads on Florida history.
Activities: guided tours; docent program; formal education program for adults; operation of a branch line passenger train. Annual Events: WWII & Civil War Reenactments; Halloween specials; Hole in Head Gang Train Robbers; Christmas Train.
Publications: members newsletter, Tracks; e-newsletter.
Hours & Admission Prices: Train rides: Sat.-Sun. 11 & 2. Adults $12, children $8, extra charge for special events, please visit website for special event dates; members no charge. Closed New Year's Day; Easter; Christmas.
Attendance: 45,000 (estimated)
Membership: Junior $20; Individual $40; Supporting $50; Family $100; Sponsor $300; Trainmaster $500; Sustaining $1,000.

Patrick Air Force Base

AIR FORCE SPACE AND MISSILE MUSEUM, 191 Museum Cir., Patrick Air Force Base, FL 32925-2535. Tel.: 321-853-9171.
Founded: 1966.
Institution Type/Description: Space Museum.
Collections: history of rocketry & space flight; Launch Complex 26; Launch Complex 5/6; missiles; rockets; space-related equipment; photographs; personal artifacts.
Facilities: library. Gift items for sale.
Activities: group tours.
Hours & Admission Prices: Call for hours.

Pembroke Pines

THE ART GALLERY, BROWARD COMMUNITY COLLEGE SOUTH CAMPUS, 7200 Hollywood Blvd., Pembroke Pines, FL 33024-7225. Tel.: 954-210-8895. Fax: 954-201-8934.
E-mail: kbelan@broward.edu
Web Site: www.broward.edu/locations/south/artgallery
Key Personnel: Dir., Dr. Kyra Belan
Institution Type/Description: Art Gallery.
Collections: paintings; photographs; sculpture.
Hours & Admission Prices: Mon.-Fri. 10-2. No charge.

Pensacola

ANNA LAMAR SWITZER CENTER FOR VISUAL ARTS, Pensacola State College, 1000 College Blvd., Pensacola, FL 32504-8910. Tel.: 850-484-2554 & 2550. Fax: 850-484-2564.
E-mail: vspencer@pensacolastate.org
Web Site: www.pensacolastate.edu/visarts
Formerly: Visual Arts Gallery
Founded: 1970.
Congressional District: 3
Key Personnel: Dir., Vivian L. Spencer.
Personnel Profile: Full-Time Paid 2; Part-Time Paid 6; Interns 7.
Governing Authority: college. Parent Institution: Pensacola State College. Tax-exempt.
Institution Type/Description: College Art Museum.
Collections: contemporary drawings; prints; paintings; photography; crafts; sculpture.
Facilities: studio complex.
Activities: guided tours; formally organized education programs for college students; loan & temporary exhibitions.
Publications: brochures; catalogs; announcements; posters.
Hours & Admission Prices: Sept.-May Mon.-Thurs. 8-9, Fri. 8-3:30. No charge. Closed holidays. ♿
Attendance: 58,000 (accurate)
Membership: daVinci $250; Michelangelo $500; Medici $1,000.

HISTORIC PENSACOLA VILLAGE, (M), 120 Church St., Pensacola, FL 32502-5941. Mailing Address: P.O. Box 12866, Pensacola, FL 32591-2866. Tel.: 850-595-5985, ext. 100. Fax: 850-595-5989. Facebook: Historic Pensacola Village.
E-mail: roverton@uwf.edu
Web Site: www.historicpensacola.org
Founded: 1987.
Congressional District: 1
Key Personnel: Interim Exec. Dir., Malinda Horton; Bd. Pres. (V), Jerry Maygarden; C.O.O., Robert Overton; Special Events Coord., Casey Campbell; Registrar, Carolyn Prime; Archivist, Jacquelyn Wilson; Dir. Education, Sheyna Marcey; Cur. Exhibits, Gale Messerschmidt; Educator, Jim McMillen; Living History Coord., Ryan Arvay; Historic Preservationist, Ross Pristera; Museum Shop Mgr., Wendi Davis.
Personnel Profile: Full-Time Paid 17; Part-Time Paid 8; Part-Time Volunteers 30; Interns 2.
Governing Authority: Parent Institution: University of West Florida, West Florida Historic Preservation, Inc. Branch Institutions: T.T. Wentworth Jr., FL State Museum; Pensacola Historical Society Resource Center; Pensacola Children's Museum; Arcadia Mill Archaeological Stie, Milton, FL. Tax-exempt: 501(c)(3).
Institution Type/Description: Historic Houses & Buildings.
Collections: T.T. Wentworth Jr. collection of 150,000 Southeastern U.S. historical artifacts and archives, primarily 19th & 20th century: paintings, lithographs, photographs, documents, weapons, hardware, household items, furniture, textiles, farming equipment, broadsides and newspapers, transportation vehicles; marine artifacts; archaeology; doll houses; local Civil War artifacts & documents. Historic Houses: 1803 Lavalle House; 1805 Julee Cottage; 1810 Walton House; 1879 Moreno Cottage; 1871 Dorr House; c.1835 Barkley House; 1903 L & N Marine Terminal; 1930 Piney Woods sawmill; 1888 Christ Church parish school; 1888 Lear house; 1832 Old Christ Church; 1876 Pfeiffer house; 1870 & 1890 warehouses (museums of commerce & industry); 1888 Manuel Barrios Cottage; 1880 McMillan House; 1885 Shotgun House.
Research Fields: regional historical, architectural & archaeological research; maritime & industrial history.
Facilities: 50,000 sq. ft. exhibit space in eleven museum buildings; high-tech classroom/conference room. Museum-related items for sale.
Activities: guided tours; inter-museum loan, permanent & temporary exhibitions; living history program; rental facilities.
Hours & Admission Prices: Village: Tues.-Sat. 10-4. Adults $6, children $3; discounts to senior citizens, AAM & ICOM members, groups over 20 & military personnel; members no charge. Closed New Year's Eve & Day; Martin Luther King Jr. Day; Memorial Day; Independence Day; Labor Day; Veterans Day; Thanksgiving & day after; Christmas Eve, Day & week. ♿
Attendance: 13,351 (accurate)
Membership: Senior & Student $20; Individual $35; Couple $45; Family $60; Patron $100; Supporter $250; Grand Benefactor $500.

✱ **NATIONAL NAVAL AVIATION MUSEUM,** 1750 Radford Blvd., Ste. C, Pensacola, FL 32508-5402. Tel.: 850-452-3604. Fax: 850-452-3296.
E-mail: museuminfo.navalaviation@mchsi.com
Web Site: navalaviationmuseum.org
Founded: 1963.
Congressional District: 1
Key Personnel: Dir., Capt. Robert Rasmussen, USN (Ret); Found. Pres. & C.E.O., VADM (Ret) G.L. Hoewing, USN (Ret); Admin., John Carro; Dir. Mktg., Shelley Ragsdale; Flight Deck Museum Shop Mgr., Belinda Bentley; Deputy Dir. & Cur., Robert Macon; Education Coord., Sam Shilling; IMAX Operations, Fred Geiger; C.F.O., Steve Flint; MIS, Gary Petersen.
Personnel Profile: Full-Time Paid 27; Part-Time Volunteers 405.
Governing Authority: federal government. Parent Institution: United States Navy. Tax-exempt: 501(c)(3).
Institution Type/Description: Naval Aviation Museum.
Collections: history of naval aviation; photographs; charts; diagrams; period artifacts; scale models; over 150 full size aircraft; models & aviation art.
Research Fields: aeronautics.
Facilities: library; cafe. Museum-related items for sale.
Activities: guided tours; historical movies; audiovisual exhibits; permanent & temporary exhibitions; IMAX; flight simulators; flight adventure deck.
Publications: exhibit & program brochures; Foundation magazine; newsletter, Fly By.
Hours & Admission Prices: Daily 9-5. No charge; donations accepted. Closed New Year's Day; Thanksgiving; Christmas. ♿
Attendance: 814,000 (estimated)
Membership: One Year $35; Two Year $65; Three Year $90; Benefactor's Circle (Lifetime) $1,000.

PENSACOLA CHILDREN'S MUSEUM, (M), 115 E. Zaragoza St., Pensacola, FL 32502. Mailing Address: P.O. Box 12866, Pensacola, FL 32591-2866. Tel.: 850-595-1559. Fax: 850-595-5989. Facebook: Pensacola Children's Museum.
E-mail: roverton@uwf.edu
Web Site: www.historicpensacola.org
Founded: 2012.
Congressional District: 1
Key Personnel: Interim Exec. Dir., Malinda Horton; Bd. Pres. (V), Jerry Maygarden; C.O.O., Robert Overton; Registrar, Carolyn Prime; Dir. Education, Sheyna Marcey; Cur. Exhibits, Gale Messerschmidt; Educator, Jim McMillen; Museum Shop Mgr., Wendi Davis.
Personnel Profile: Full-Time Paid 17; Part-Time Volunteers 30.
Governing Authority: Parent Institution: University of West Florida, West Florida Historic Preservation, Inc. Branch Museums: T.T. Wentworth Jr., FL State Museum; Historica Pensacola Village; Pensacola Historical Society; Arcadia Mill Archaeological Site, Milton, FL.
Institution Type/Description: Children's Museum.
Collections: local historical artifacts representing various cultures, occupations, & historical eras.
Facilities: rental facilities. Museum-related items for sale.
Activities: permanent & temporary exhibitions; special events; tours; parties; traveling trunks.
Hours & Admission Prices: Tues.-Sat. 10-4. Admission $3; children under one & members no charge. Closed New Year's Day; Martin Luther King Jr. Day; Memorial Day; Independence Day; Labor Day; Veterans Day; Thanksgiving & day after; Christmas.
Attendance: 18,939
Membership: Senior & Student $20; Individual $35; Couple $45; Family $60; Patron $100; Supporter $250; Grand Benefactor $500.

PENSACOLA HISTORICAL SOCIETY, (M), 110 Church St., Pensacola, FL 32502-5962. Mailing Address: P.O. Box 12866, Pensacola, FL 32591-2866. Tel.: 850-595-5840. Fax: 850-595-5989. Facebook: Pensacola Historical Society.
E-mail: roverton@uwf.edu
Web Site: historicpensacola.org
Founded: 1933.
Congressional District: 1
Key Personnel: Interim Exec. Dir., Malinda Horton; Bd. Pres. (V), Jerry Maygarden; C.O.O., Robert Overton; Historic Preservationist, Ross Pristera; Chief Cur., B. Lynne Robertson; Registrar, Carolyn Prime; Archivist, Jacki Wilson; Museum Shop Mgr., Wendi Davis.
Personnel Profile: Full-Time Paid 17; Part-Time Paid 1; Part-Time Volunteers 30; Interns 2.
Governing Authority: Parent Institution: West Florida Historic Preservation, Inc. Branch Institutions: T.T. Wentworth, Jr.; Florida State Museum; Historic Pensacola Village; Pensacola Children's Museum; Arcadia Mill Archaeological Site, Milton, FL. Tax-exempt: 501(c)(3).
Institution Type/Description: Resource Center & Archives.
Collections: local history & artifacts; clothing; local family silver; 19th-century men's & women's accessories; American Indian artifacts; local bottles; 19th-century household items; Viola A. Blount fine art glass; the Carpenter glass negative collection; Escambia County & Pensacola history including archaeological materials; library collection: Indian, maritime, multicultural & military themes; manuscripts; paintings; photographs; postcards; personal papers; maps; early furnishings; T.T. Wentworth Jr. artifacts.
Research Fields: history of Pensacola, Escambia County & northwest Florida; history & genealogy of inhabitants; history of 19th-century businesses; area brick manufacture 18th-20th centuries.
Facilities: 2,000-vol. library of material relevant to the history of Pensacola, Escambia County & extreme northwest Florida; technical reference section; oral history section; microfilm; microfiche; tape recordings; 1780-1910 censuses; reading area; maps; 60,000 photographs & negatives; archives. Books, publications, postcards, stationery & map reprints for sale.
Activities: lectures; ghost tours.
Publications: booklets; magazine; Pensacola History Illustrated: A Journal of Pensacola and Northwest Florida History; bimonthly newsletter, Pensacola History Today.
Hours & Admission Prices: Resource Center: Tues.-Fri. 10-4. Resource Center: $5.50 per day; members no charge. Closed state & university holidays. &
Attendance: 7,404 (accurate)
Membership: Student, Senior or Out of Towner $20; Individual $35; Couple $45; Family $60; Patron $100; Supporter $250; Grand Benefactor $500.

PENSACOLA LIGHTHOUSE & MUSEUM, (M), 2081 Radford Blvd., Pensacola, FL 32508. Tel.: 850-393-1561.
Institution Type/Description: Historic Building: lighthouse built in 1869. Listed on the National Register of Historic Places.
Collections: local history & culture; keeper's quarters; personal artifacts; period furnishings; photographs.
Facilities: Museum-related items for sale.
Activities: educational programs; lectures; sunset tours; special events; rental facilities.
Hours & Admission Prices: Mon.-Sat. 10:30-5:30, Sun. 12:30-5:30. Adults $5, children 7-12, active military and seniors 65 & over $3.

*** PENSACOLA MUSEUM OF ART, (M),** 407 S. Jefferson St., Pensacola, FL 32502-5901. Tel.: 850-432-6247. Fax: 850-469-1532. Facebook: Pensacola Museum of Art.
E-mail: frontdesk@pensacolamuseum.org
Web Site: www.pensacolamuseum.org
Founded: 1954.
Congressional District: 1
Key Personnel: Exec. Dir., Jodi Gup; Pres. (V), Betty Roberts; Dir. Finance & Admin., Justin Gates; Dir. Mktg., Stacy Kendall; Visitors Svcs. & Education Coord., Cortlandt Glover; Exec. Asst. & Coord. Membership, Kate Moloney; Assoc. Cur., Leah Griffin; Coord. Education, Patrick Jennings; Coord. Devel., Kate Sutley; Registrar & Preparator, Nick Christopher; Media Coord., Stephanie Kress.
Personnel Profile: Full-Time Paid 4; Part-Time Paid 1; Part-Time Volunteers 120.
Governing Authority: Pensacola Museum of Art, Inc. Tax-exempt: 501(c)(3).
Institution Type/Description: Art Museum: housed in 1908 old city jail.
Collections: glass, paintings & graphics with emphasis on American contemporary artists.
Major Exhibits: Art of the Brick (T), 5/14-8/14; Adventures in Aft (T), 8/14-10/14.
Facilities: 1,500-vol. library of art reference material available for inter-library loan or use by responsible persons or any institution; 100-seat auditorium; classrooms. Museum-related items for sale.
Activities: guided tours; lectures; films; formally organized educational programs; loan, permanent & traveling exhibitions.
Publications: newsletter; catalogues for major exhibitions.
Hours & Admission Prices: Tues.-Fri. 10-5, Sat.-Sun. 12-5. Adults $5, students, military & seniors $3; children 5 & under, Tues., and members no charge. Closed national holidays. &
Attendance: 80,000 (estimated)
Membership: Student $20; Individual $50; Family $75; PMA Contemporaries $100; Sustaining $250; Patron $500; Benefactor $1,000; Corporate $250-$10,000.

T.T. WENTWORTH, JR. FLORIDA STATE MUSEUM, (M), 330 S. Jefferson St., Pensacola, FL 32502-5943. Mailing Address: P.O. Box 12866, Pensacola, FL 32591-2866. Tel.: 850-595-5985, ext. 100. Fax: 850-595-5989. Facebook: T.T. Wentworth, Jr. Florida State Museum.
E-mail: roverton@uwf.edu
Web Site: www.historicpensacola.org
Founded: 1988.
Congressional District: 1
Key Personnel: Interim Exec. Dir., Malinda Horton; Bd. Pres. (V), Jerry Maygarden; C.O.O., Robert Overton; Registrar, Carolyn Prime; Cur. Exhibits, Gale Messerschmidt; Dir. Education, Sheyna Marcey; Museum Shop Mgr., Wendy Davis.
Personnel Profile: Full-Time Paid 17; Part-Time Paid 8; Part-Time Volunteers 30; Interns 2.
Governing Authority: Parent Institution: University of West Florida, West Florida Historic Preservation, Inc., Pensacola, Fl. Branch Institutions: Pensacola Historical Society; Pensacola Children's Museum; Arcadia Mill Archaeological Site, Milton, FL; Historic Pensacola Village, Pensacola, Fl. Tax-exempt: 501(c)(3).
Institution Type/Description: Historic Building: housed in the former Pensacola City Hall building.
Collections: local history; period artifacts; photographs; artifacts from local land & underwater sites.
Research Fields: local & regional history.
Facilities: 7,000 sq. ft. exhibit space. Bookstore & museum-related items for sale.
Activities: formal education programs for University of West Florida students; guided tours; inter-museum loan, participatory & temporary exhibitions. Annual Event: Open House in June & October.
Hours & Admission Prices: Tues.-Sat. 10-4. No charge; donations accepted.

Closed New Year's Day; Martin Luther King Jr. Day; Memorial Day; Independence Day; Labor Day; Veterans Day; Thanksgiving & day after; Christmas. &

Attendance: 22,916 (accurate)

Membership: Student & Senior $20; Individual $35; Couple $45; Family $60; Patron $100; Supporter $250; Grand Benefactor $500.

UNIVERSITY OF WEST FLORIDA ART GALLERY, 11000 University Pkwy., Bldg. 82, Pensacola, FL 32514-5750. Tel.: 850-474-2696. Fax: 850-474-2043.

E-mail: artgallery@uwf.edu

Web Site: uwf.edu/art

Founded: 1970.

Congressional District: 1

Key Personnel: Gallery Dir., Nick Croghan.

Personnel Profile: Full-Time Paid 1; Part-Time Paid 1; Part-Time Volunteers 6; Interns 5.

Governing Authority: university; state. Parent Institution: University of West Florida. Tax-exempt.

Institution Type/Description: University Art Gallery.

Collections: contemporary paintings; photographs.

Activities: guided tours; visiting artists & scholar's lecture series; films; gallery talks; concerts; loan, curated, temporary & traveling exhibitions.

Publications: catalogues; flyers; posters.

Hours & Admission Prices: Mon.-Fri. 10-5, during school year. No charge, donations accepted. Closed Easter; Memorial Day; Independence Day; Veterans Day; Thanksgiving. &

Attendance: 14,000 (accurate)

WEST FLORIDA HISTORIC PRESERVATION, INC., (M), 120 Church St., Pensacola, FL 32502-5941. Mailing Address: P.O. Box 12866, Pensacola, FL 32591-2866. Tel.: 850-595-5985. Fax: 850-595-5989. Facebook: West Florida Historic Preservation, Inc.

E-mail: roverton@uwf.edu

Web Site: www.historicpensacola.org

Formerly: Historic Pensacola Preservation Board

Founded: 1967.

Congressional District: 1

Key Personnel: Interim Exec. Dir., Malinda Horton; Bd. Pres. (V), Jerry Maygarden; Museum Shop Mgr., Wendi Davis; C.O.O., Robert Overton; Historic Preservationist, Ross Pristera; Cur. Exhibits, Gale Messerschmidt; Registrar, Carolyn Prime; Archivist, Jacquelyn Wilson; Dir. Education, Sheyna Marcey; Educator, Jim McMillen; Coord. Special Events, Casey Campbell; Living History Coord., Ryan Arvay.

Personnel Profile: Full-Time Paid 17; Part-Time Paid 10; Part-Time Volunteers 30; Interns 2.

Governing Authority: Parent Institution: University of West Florida. Subsidiary Institution: Pensacola Historical Society. Branch Museums: Pensacola Children's Museum; Historic Pensacola Village; T.T. Wentworth, Jr. Florida State Museum; Arcadia Mill Archaeological Site, Milton, FL. Tax-exempt: 501(c)(3).

Institution Type/Description: Preservation Society.

Collections: archives; local history; photographs; historic houses.

Research Fields: local & regional history; genealogy.

Facilities: research library & archives.

Activities: research; permanent & temporary exhibits; formal education programs for Univ. of West FL students; guided tours; rental facilities; lecture series; living history programs & demonstrations; summer camp.

Publications: semi-annual magazine; bimonthly newsletter; brochures.

Hours & Admission Prices: Offices & Archives: Mon.-Fri. 8-4:30. Research: Tues.-Fri. Research: $5.50 per day; discounts to AAM & ICOM members; members no charge. Closed New Year's Eve & Day; Martin Luther King Jr. Day; Memorial Day; Independence Day; Labor Day; Veterans Day; Thanksgiving & day after; Christmas Eve, Day & week. &

Attendance: 49,707 (accurate)

Membership: Student, Senior, & Out of Town $20; Individual $35; Couple $45; Family $60; Patron $100; Supporter $250; Grand Benefactor $500. Time only members: 80 hrs volunteer time annually.

Perry

FOREST CAPITAL STATE MUSEUM, S. U.S. Hwy. 19, 204 Forest Park Dr, Perry, FL 32348-6320. Tel.: 850-584-3227. Fax: 850-584-3488.

Web Site: www.floridastateparks.org

Founded: 1973.

Key Personnel: Park Svcs. Specialist, Lindsay West.

Personnel Profile: Full-Time Paid 1; Part-Time Paid 2.

Governing Authority: state. Tax-exempt: 47-04-023952-52C/2.

Institution Type/Description: Logging & Lumber Museum: Timber Industry Museum & Forestry including c.1865 Pioneer-Cracker Homestead.

Collections: exhibits interpreting turpentine production, virgin forest cutting, modern forest practices; indigenous tree species; turn-of-the-century furnishings; turn-of-the-century interpretive homestead area.

Facilities: picnic area.

Activities: guided tours. Museum Sponsors: Florida Forest Festival in October.

Publications: State Parks Guide.

Hours & Admission Prices: Thurs.-Mon. 9-5. Adults $2; children 5 & under & Florida public school groups no charge. Closed New Year's Day; Thanksgiving; Christmas. &

Attendance: 7,000 (estimated)

Pine Island Center

MUSEUM OF THE ISLANDS, 5728 Sesame, Pine Island Center, FL 33956. Mailing Address: P.O. Box 305, Saint James City, FL 33956-0305. Tel.: 239-283-1525.

E-mail: info@museumoftheislands.com

Web Site: www.museumoftheislands.com

Founded: 1990.

Congressional District: 14

Key Personnel: Pres. (V), Sharon Traylor; Museum Shop Mgr., Barbara "Bobbie" Mahaffey.

Personnel Profile: Part-Time Volunteers 20.

Governing Authority: Tax-exempt.

Institution Type/Description: General Museum.

Collections: Pine Island history; seashells; photographs; portraits; dolls; period household artifacts; fishing gear.

Hours & Admission Prices: May-Oct. Tues., Thurs. & Sat. 11-3; Nov.-April Tues.-Sat. 11-3, Sun. 1-4; groups by appointment. Adults $2, children $1. Closed Easter; Thanksgiving; Christmas. &

Attendance: 8,000 (estimated)

Membership: Individual $10; Family $15; Club $25; Sponsor $50; Life $100.

Plant City

DINOSAUR WORLD, 5145 Harvey Tew Rd., Plant City, FL 33565. Tel.: 813-717-9865. Fax: 813-707-9776.

Institution Type/Description: Natural History Museum.

Collections: over 150 life size dinosaurs; dinosaur eggs; raptor claws.

Facilities: theatre; picnic area. Museum-related items for sale.

Activities: educational programs; classes; fossil dig; outreach programs; scout programs; special events; birthday parties.

Hours & Admission Prices: Feb.-Nov. daily 9-6; Dec.-Jan. daily 9-5. Adults $12.75, seniors over 60 $10.75, children 3-12 $9.75; discounts to groups & military dependents; active military no charge.

EAST HILLSBOROUGH HISTORICAL SOCIETY, INC., 605 N. Collins St., Plant City, FL 33563-3321. Tel.: 813-757-9226.

Formerly: 1914 Plant City High School Community Center

Founded: 1974.

Key Personnel: Pres. (V) & Archivist, Shelby Bender; Vice Pres., Roy Wise.

Personnel Profile: Part-Time Paid 2; Part-Time Volunteers 20.

Governing Authority: private; nonprofit organization. Tax-exempt: 501(c)(3).

Institution Type/Description: Pioneer Heritage Museum: housed in c.1914 high school bldg.

Collections: local pioneer family artifacts.

Facilities: 1,000-vol. library; 500-seat auditorium; 8,500 sq. ft. exhibit space.

Activities: formal education programs for adults; guided tours; hobby workshops; genealogical workshops. Annual Events: Pioneer Day; Christmas Candlelight Tour of Homes in Historic District.

Publications: Plant City: It's Origin & History; Vintage Postcards: Plant City; Images of America: Plant City; Postcards of America: Plant City's Berry Seasons.

Hours & Admission Prices: Pioneer Museum: call for hours. Quintilla Geer Bruton Archives Center: Tues. 10-5, Wed.-Sat. 1-5, evenings by appointment. No charge; donations accepted. &

Attendance: 10,000 (estimated)

Membership: Individual $15; Family & Business $25; Lifetime $100; Patron $1,000.

LOOKOUT MOUNTAIN WILD ANIMAL PARK, 10627 Paul Buchman Hwy., Plant City, FL 33565-7213.

E-mail: doc@lookoutmtnzoo.com

Web Site: lookoutmtnzoo.com

Governing Authority: nonprofit organization.

Institution Type/Description: Zoo.

Collections: animals from around the world including monkeys, parrots, llamas, Bengal tigers; black bear; buffalo; lynx.

Activities: feeding, training & playing with animals.
Hours & Admission Prices: Call for hours.

PLANT CITY PHOTO ARCHIVES & HISTORY CENTER, 106
S. Evers St., Plant City, FL 33563. Tel.: 813-754-1578.
E-mail: info@plantcityphotoarchives.org
Web Site: www.plantcityphotoarchives.org
Founded: 2000.
Key Personnel: Dir., Gilbert V. Gott; Pres. (V), Edward M. Verner.
Personnel Profile: Full-Time Paid 2; Part-Time Paid 1; Part-Time Volunteers 6;
Interns 1.
Governing Authority: Parent Institution: Plant City Photo Archives, Inc.
Tax-exempt.
Institution Type/Description: History Museum.
Collections: photographs; community documents & papers.
Research Fields: west central Florida people, places & events.
Facilities: library; 1,000 sq. ft. exhibition space.
Activities: research.
Publications: select monographs.
Hours & Admission Prices: Tues.-Sat. 10-4. No charge; donations accepted.
Closed major holidays. &
Attendance: 5,000 (estimated)
Membership: Basic $25; Friend $50; Patron $100; Founder $250; Trustee $500
& up.

Plantation

PLANTATION HISTORICAL MUSEUM, 511 N. Fig Tree Lane,
Plantation, FL 33317-1849. Tel.: 954-797-2722. Fax: 954-797-2717.
E-mail: jfeeney@plantation.org
Web Site: www.plantation.org/Museum/index.html
Founded: 1975.
Congressional District: 20
Key Personnel: Cur., Shirley Schuler; Office Mgr., Rosemary Schafer; Asst.
Cur., John Feeney.
Personnel Profile: Part-Time Paid 3; Part-Time Volunteers 15.
Governing Authority: municipal; society; nonprofit organization. Parent Insti-
tution: City of Plantation, FL. Tax-exempt: 501(c)(3).
Institution Type/Description: Local History Museum.
Collections: 20th century artifacts; fire fighting equipment; 1956 GMC front-
end pumper Fire Engine No. 1; Nature: Everglades Diorama.
Facilities: Locally handmade items for sale.
Activities: guided tours; hobby workshops; temporary & traveling exhibitions;
oral history program of pioneer residents.
Publications: monthly newsletter; city quarterly; county quarterly.
Hours & Admission Prices: Tues., Thurs. & Sat. 10-12 & 1-4:30, Fri. 12-4:30;
groups by appointment on Tues. & Thurs. No charge; donations accepted.
Closed national holidays. &
Attendance: 16,000 (estimated)
Membership: Individual $15; Family $25; Life $150.

SCHACKNOW MUSEUM OF FINE ARTS, 7080 Northwest 4th
St., Plantation, FL 33317-2201. Tel.: 954-583-5551. Fax: 954-583-5557.
E-mail: schacknowmuseum1@bellsouth.net
Web Site: www.smofa.com
Founded: 2000.
Key Personnel: Founder & Dir., Max Schacknow; Museum Shop Mgr., Olga
Avellone.
Governing Authority: Tax-exempt.
Institution Type/Description: Art Museum.
Collections: paintings; photography; sculpture.
Hours & Admission Prices: Tues.-Sat. 10-5. Adults $5, seniors & children over
12 $3; discounts to AAM members & members of other museums; members
no charge. &
Membership: Individual $75, Couple $95; 4 or More $100.

Point Washington

EDEN STATE GARDENS AND MANSION, 181 Eden Garden
Rd., Point Washington, FL 32459. Tel.: 850-267-8320.
Key Personnel: Pres. (V), Maryjo Morris.
Personnel Profile: Part-Time Volunteers 6.
Volunteer Hours: 12,000
Operating Expenses: 66,000
Operating Income: 82,000
Institution Type/Description: Historic House: housed in the former home of
William Henry Wesley, founder of a timber company; built in 1890.

Collections: family & local history; period artifacts; Louis XVI furniture;
personal artifacts; photographs; gardens.
Facilities: rental facilities; gardens. Museum-related items for sale.
Activities: rental facilities.
Hours & Admission Prices: Tours: Thurs.-Mon. Adults $4. &
Attendance: 60,000 (estimated)

Polk City

AMERICAN WATER SKI EDUCATIONAL FOUNDATION,
1251 Holy Cow Rd., Polk City, FL 33868-8200. Tel.: 863-324-2472. Fax: 863-324-3996.
E-mail: awsefhalloffame@cs.com
Web Site: waterskihalloffame.com
Founded: 1968.
Key Personnel: Exec. Dir., Carole Lowe; Chm., Jim Grew; Pres., Paul Chapin;
Treas., Mark Harvat.
Personnel Profile: Full-Time Paid 1.
Governing Authority: nonprofit organization. Subsidiary Museum: Water Ski
Museum & Hall of Fame, Polk City, FL. Tax-exempt: 501(c)(3).
Institution Type/Description: Waterski Sports Museum.
Collections: first water skis, 1922; show ski costumes; water ski ropes &
handles; photographs; archives; memorabilia.
Facilities: library; resource center.
Activities: films; formal education programs; participatory exhibits; annual
ceremony; lake events.
Publications: newsletter, Water Ski Hall of Fame.
Hours & Admission Prices: Mon.-Fri. 10-5. Adults $5, seniors $4, children
6-12 $3; discounts to AAM members; members and children 5 & under no
charge. Closed New Year's Day; Martin Luther King Jr. Day; Memorial
Day; Independence Day; Thanksgiving; Christmas. &
Attendance: 10,000 (estimated)
Membership: Regular $25; Supporting $100; Sponsor $250; Patron $500;
Benefactor $1,000; Founder $50,000.

FANTASY OF FLIGHT, 1400 Broadway Blvd., S.E., Polk City, FL
33868-9109. Tel.: 863-984-3500.
Web Site: www.fantasyofflight.com
Key Personnel: Owner, Kermit Weeks
Institution Type/Description: Aircraft Museum.
Collections: aircraft from early flight to 1950s
Facilities: Museum-related items for sale.
Activities: biplane & hot air balloon rides.
Hours & Admission Prices: Daily 10-5. Adults $28.95, children 6-15 $14.95;
discounts to groups & military; children 4 & under no charge. Closed
Thanksgiving; Christmas.

Pompano Beach

ELY EDUCATIONAL MUSEUM, 1500 N.W. 6th Ave., Pompano
Beach, FL 33060. Mailing Address: c/o Pompano Beach Historical
Society, P.O. Box 154, Pompano Beach, FL 33061. Tel.: 954-781-2256.
Institution Type/Description: Historic House Museum: housed in the former
home of Blanche and Joseph Ely.
Collections: Ely family history & culture; education in the black community;
personal artifacts; period furnishings; photographs.
Hours & Admission Prices: Call for hours.

**KESTER COTTAGES - POMPANO BEACH HISTORICAL
SOCIETY,** 220 N.E. 3rd Ave., Pompano Beach, FL 33060.
Mailing Address: P.O. Box 154, Pompano Beach, FL 33061. Tel.:
954-782-3015.
E-mail: info@pampanohistory.com
Institution Type/Description: History Museum: housed in cottages built in the
1930s.
Collections: local history & culture; period furnishings; personal artifacts;
photographs.
Hours & Admission Prices: Call for hours.

OLD POMPANO FIRE STATION, 219 N.E. 4th Ave., Pompano
Beach, FL 33060. Mailing Address: P.O. Box 154, Pompano Beach,
FL 33061. Tel.: 954-782-3015.
Institution Type/Description: Historic Building: housed in the city's first fire
station; built in 1925.
Collections: fire fighting history & equipment; early fire engines; photographs;
personal artifacts.
Hours & Admission Prices: Call for hours.

SAMPLE-MCDOUGALD HOUSE, 450 N.E. 10th St., Pompano Beach, FL 33060. Mailing Address: P.O. Box 1599, Pompano Beach, FL 33061. Tel.: 954-946-9700.
Web Site: www.samplemcdougaldhouse.com
Key Personnel: Exec. Dir., Dan Hobby
Institution Type/Description: Historic House Museum: built in 1916. Listed on the National Register of Historic Places.
Collections: Sample family history; period furnishings; personal artifacts; photographs.
Hours & Admission Prices: Call for hours.

Ponce Inlet

MARINE SCIENCE CENTER, 100 Lighthouse Dr., Ponce Inlet, FL 32127-7325. Tel.: 386-304-5545.
Institution Type/Description: Science Center.
Collections: marine life including sea turtles & native fish; mangroves; shells; artificial reefs; dune habitats.
Facilities: 5,000 gallon aquarium; classroom. Center-related items for sale.
Activities: observation area; sea turtle & bird rehabilitation area; group tours; educational programs; summer camp.
Hours & Admission Prices: Tues.-Sat. 10-4, Sun. 12-4. Adults 13 & over $5, seniors $4, youth 3-12 $2; children under 3 no charge. Closed New Year's Day; Martin Luther King Jr. Day; Thanksgiving; Christmas.

PONCE DELEON INLET LIGHTHOUSE PRESERVATION ASSOCIATION, INC., (M), 4931 S. Peninsula Dr., Ponce Inlet, FL 32127-7301. Tel,: 386-761-1821. Fax: 386-761-3121.
E-mail: lighthouse@ponceinlet.org
Web Site: www.ponceinlet.org
Founded: 1972.
Congressional District: 4
Key Personnel: Exec. Dir., Ed Gunn; Pres., Tami Lewis; Dir. Operations, Mike Bennett; Cur., Ellen Henry; Program Mgr., Mary Wontzel.
Personnel Profile: Full-Time Paid 11; Part-Time Paid 16; Part-Time Volunteers 31.
Governing Authority: nonprofit organization. Tax-exempt: 501(c)(3).
Institution Type/Description: History Museum, Historic Site & National Historic Landmark.
Collections: local history & lighthouse keepers' lives. Historic Structures: c.1887 Lighthouse Tower; 3 Keepers' Houses; 4 out-buildings; Lens exhibit building.
Facilities: 10-acre grounds. Museum-related items for sale.
Activities: guided tours; lectures; video theater.
Publications: quarterly newsletter; brochure.
Hours & Admission Prices: Memorial Day to Labor Day daily 10-9; Sept.-May daily 10-6. Adults $5, children $1.50; members no charge. Closed Christmas Day.
Attendance: 145,600 (accurate)
Membership: General $20; Family $40.

Ponte Vedra Beach

MARIAL MUSEUM OF SACRED ART, The Gallery at Christ Episcopal Church, 400 San Juan Dr., Ponte Vedra Beach, FL 32082-2838. Tel.: 904-285-6127.
Institution Type/Description: Art Gallery.
Collections: modern & contemporary religious art.
Hours & Admission Prices: Call for hours.

Port St. Joe

CONSTITUTION CONVENTION MUSEUM STATE PARK, 200 Allen Memorial Way, Port St. Joe, FL 32456-2342. Tel.: 850-229-8029; 850-227-1327. Fax: 850-227-1488.
E-mail: mark.knapke@dep.state.fl.us
Web Site: www.floridastateparks.org/constitutionconvention/
Formerly: Constitution Convention State Museum
Founded: 1955.
Congressional District: 2
Key Personnel: Park Mgr., Mark Knapke; Asst. Mgr., Danny Kemp; Park Ranger, William Wilkinson.
Personnel Profile: Full-Time Paid 1; Part-Time Paid 1.
Governing Authority: state. Parent Institution to: St. Joseph Peninsula State Park. Tax-exempt.
Institution Type/Description: History Museum: site of the Constitution Convention for the Territory of FL.
Collections: Old St. Joseph history & artifacts.
Activities: guided tours; permanent exhibitions.

Hours & Admission Prices: Visitor Center: Thurs.-Mon. 9-12 & 1-5. Adults $2; children 4 & under no charge. Closed New Year's Day; Thanksgiving; Christmas.
Attendance: 3,500 (estimated)

Punta Gorda

MILITARY HERITAGE MUSEUM, (M), 1200 W. Retta Esplanade, Unit 48, Punta Gorda, FL 33950-5325. Tel.: 941-575-9002.
E-mail: info@freedomisntfree.org
Web Site: www.freedomisntfree.org
Founded: 2001.
Congressional District: 16
Key Personnel: Exec. Dir., Kim Lovejoy; Chm. (V), Bill Minerich.
Personnel Profile: Full-Time Paid 2; Part-Time Volunteers 62.
Volunteer Hours: 7,607
Governing Authority: nonprofit organization. Tax-exempt: 501(c)(3).
Institution Type/Description: Military Heritage Museum.
Collections: military heritage, history, & artifacts; photographs; personal artifacts; war memorabilia.
Hours & Admission Prices: Summer: Mon.-Sat. 10-6, Sun. 12-5; Winter: call for extended hours. No charge; donations accepted.
Attendance: 43,851 (accurate)
Membership: Student $15; Adult $25; Corporate $250; Life $500.

Quincy

GADSDEN ARTS CENTER, 13 N. Madison, Quincy, FL 32351-2409. Tel.: 850-875-4866. Fax: 850-627-8606.
E-mail: grace@gadsdenarts.org
Web Site: www.gadsdenarts.org
Founded: 1994.
Congressional District: 2
Key Personnel: Exec. Dir., Grace R. Maloy; Chm. (V) & Pres. (V), Dawn McMillan; Cur., Angela Barry; Museum Shop Mgr., Becky Reep.
Personnel Profile: Full-Time Paid 3; Part-Time Paid 2; Part-Time Volunteers 100; Interns 6.
Governing Authority: Parent Institution: Gadsden Arts, Inc. Tax-exempt: 501(c)(3).
Institution Type/Description: Fine Arts Museum.
Collections: paintings; sculpture; American art including southern vernacular & contemporary regional.
Major Exhibits: New Visions of Florida Landscape, 1/14-3/14; Owen Gray Landscapes, 1/14-3/14; GAC Artist's Guild - Rotating Exhibitions, 1/14-12/14; Masterworks from Wesleyan College (T), 3/14-5/14; Chris Nolan: Outside the Box, 3/14-5/14; Hot Wax & Hot Glass, 6/14-8/14; Russell Bellamy Sculpture, 6/14-8/14; Michael Harrell & Wendy Durand, 8/14-10/14; Edward Babcock: Divine Essence, 8/14-10/14; 26th Art in Gadsden Juried Exhibition, 10/14-12/14.
Facilities: studios.
Activities: guided tours; art classes; workshops; lectures; summer camps; after school art programs; school-based programs.
Publications: monthly e-newsletter, GAC Quarterly News.
Hours & Admission Prices: Tues.-Sat. 1-5. Adults $1; discounts to AAM, ROAM & NARM members; members & children no charge. Closed holidays.
Attendance: 18,000 (estimated)
Membership: Student $10-$25; Individual $25-$50; Family $45-$100; NARM & Artists Guild $100; Corporate $250-$10,000.

Safety Harbor

SAFETY HARBOR MUSEUM OF REGIONAL HISTORY, 329 Bayshore Blvd. S., Safety Harbor, FL 34695-4053. Tel.: 727-726-1668. Fax: 727-725-9938.
E-mail: info@safetyharbormuseum.org
Web Site: www.safetyharbormuseum.org
Founded: 1970.
Congressional District: 9
Key Personnel: Pres., Les Griffith; Dir. Operations, Bobbie Davidson; Dir. Exhibits, Ronald Fekete; Dir. Education, Robert S. Anderson.
Personnel Profile: Full-Time Paid 1; Part-Time Paid 2; Part-Time Volunteers 30; Interns 2.
Governing Authority: private; nonprofit organization. Tax-exempt: 501(c)(3).
Institution Type/Description: Florida & Regional History Museum.
Collections: dioramas depicting key events in Florida history; fossils; displays & artifacts pertaining to the pre-history of the Tocobaga Indians, paleo & archaic periods; the recorded history of the Spanish encounter; pioneer era of Odet Philippe; Heritage Gallery collection of Safety Harbor memorabilia.

Research Fields: Florida archaeology; Florida history; history of Safety Harbor area.
Activities: guided tours; lectures; public school tour with lecture.
Publications: newsletters; brochure, Tocobaga Fishing Industry in Tampa Bay; biography, Odet Philippe.
Hours & Admission Prices: Wed. & Fri. 10-4, Thurs. 10-7, Sat. 10-2, Sun. 1-4; other times by appointment. Adults $4, seniors $3; discounts to AAA members; children under 7 & members no charge. Closed major holidays. &
Attendance: 8,783 (accurate)
Membership: Individual $25; Family $50; Supporter $100; Sponsor $250; Corporate $500.

Saint Augustine

CASTILLO DE SAN MARCOS NATIONAL MONUMENT, One S. Castillo Dr., Saint Augustine, FL 32084-3252. Tel.: 904-829-6506. Fax: 904-823-9388.
Web Site: www.nps.gov/casa
Founded: 1935.
Congressional District: 4
Key Personnel: Supt., Gordon J. Wilson; Eastern National Bookstore, Bruce Harris.
Governing Authority: federal. Parent Institution: National Park Service, Dept. of the Interior, Washington, DC 20240. Tax-exempt.
Institution Type/Description: Park Museum: housed in 1672-95 restored Spanish Castillo de San Marcos.
Collections: Indian, Spanish & English pottery shards; artillery; manuscripts.
Research Fields: St. Augustine fort.
Activities: guided tours; lectures; films; permanent exhibitions; living history demonstrations.
Publications: folder, Castillo de San Marcos, Florida; handbook, The Building of Castillo de San Marcos.
Hours & Admission Prices: Daily 8:45-4:45. Adults $6; children under 16 no charge. Closed Christmas. &

CRISP-ELLERT ART MUSEUM - FLAGLER COLLEGE, (M), 48 Sevilla St., Saint Augustine, FL 32084. Mailing Address: 74 King St., Saint Augustine, FL 32084. Tel.: 904-826-8530.
E-mail: crispellert@flagler.edu
Web Site: www.flagler.edu/crispellert
Founded. 2007.
Key Personnel: Dir., Julie Dickover
Institution Type/Description: Art Museum.
Collections: works by contemporary artists.
Hours & Admission Prices: Academic Year: Mon.-Fri. 10-4; Summer: Tues.-Fri. 10-4. No charge. &

DE MESA-SANCHEZ HOUSE, 43 St. George St., Saint Augustine, FL 32085. Mailing Address: c/o Colonial Quarter, LLC, 12 S. Castillo Dr., Saint Augustine, FL 32084. Tel.: 904-819-1444.
Institution Type/Description: Historic House Museum.
Collections: local history & culture; period furnishings; personal artifacts; photographs.
Activities: guided tours.
Hours & Admission Prices: Tours: Sun. 11:30, 1, 2:30, 3:30, Mon.-Sat. 11, 1, 2:30, 3:30. Adults $5, seniors 62 & over $4, children 6-17 $2.

FIRST LIGHT MARITIME SOCIETY - ST. AUGUSTINE LIGHTHOUSE & MUSEUM, INC., (M), 81 Lighthouse Ave., Saint Augustine, FL 32080-4650. Tel.: 904-829-0745. Fax: 904-808-1248.
E-mail: info@staugustinelighthouse.org
Web Site: www.staugustinelighthouse.org
Founded: 1988.
Congressional District: 4
Key Personnel: Exec. Dir., Kathy Allen Fleming; Chm. (V), Thersa Floyd; Deputy Dir., Rick Cain; Dir. Collections, Interpretations, Brenda Swann; Dir. Sales, Lee Capitano; Volunteer Coord., Loni Wellman; Dir. Museum Advancement, Michele Adams; Museum Shop Mgr., Sam Andrews.
Personnel Profile. Full-Time Paid 25; Part-Time Paid 11; Part Time Volunteers 206.
Volunteer Hours: 8,246
Operating Expenses: 2,900,000
Operating Income: 3,000,000
Governing Authority: nonprofit. Tax-exempt.
Institution Type/Description: Maritime & History Museum: National Register site.
Collections: lightkeepers' implements; historic lens on loan from U.S. Coast Guard.

Research Fields: lighthouse history & preservation; maritime archaeology; shrimping history.
Facilities: rental gallery.
Activities: Museum Sponsors: Lighthouse Festival in March; Luminary Light in December.
Publications: member's newsletter, The Spyglass.
Hours & Admission Prices: Winter: daily 9-6; Summer: call for extended hours. Adults $9, seniors 60 & over $7.50, children 12 & under $7; discounts to AAM members & St. Johns County School District students; retired, active military & members no charge. Closed Thanksgiving; Christmas Eve & Day. &
Attendance: 204,000 (accurate)
Membership: Keeper $30; Keeper's Family $50; Family Plus $100; Guardians $100 for 5 years; Heritage Club $25 for 5 years; Legacy Circle $500 for 5 years; Founding $1,000 for 5 years.

FORT MATANZAS NATIONAL MONUMENT, 8635 A1A S., Unit A, Saint Augustine, FL 32080-8400. Tel.: 904-829-6506, ext. 227 (Headquarters). Fax: 904-471-7605.
E-mail: linda_chandler@nps.gov
Web Site: www.nps.gov/foma
Founded: 1935.
Congressional District: 4
Key Personnel: District Ranger, Andrew Rich; Park Ranger, Linda Chandler; Bookstore Mgr., Bruce Harris; Supt., Gordon Wilson.
Personnel Profile: Full-Time Paid 3; Part-Time Paid 3; Part-Time Volunteers 5.
Governing Authority: federal. Parent Institution: National Park Service, Dept. of the Interior, Washington, DC 20240. Museum Shop: Eastern National Parks & Monuments Association. Tax-exempt.
Institution Type/Description: Park Museum: Site of first European battle for control of New World.
Collections: Historic Building: 1742 Fort Matanzas, Spanish fort.
Research Fields: all aspects related to the defense at Matanzas Inlet as accessory to the fort at St. Augustine.
Facilities: nature trail; picnic area. Publications & postcards for sale.
Activities: orientation video; permanent exhibitions; guided tours; free ferry service to Fort & living history demonstrations; school group lectures; nature trails.
Publications: brochure in English & Spanish; self-guiding map to Ft. Manzas N.M.
Hours & Admission Prices: Daily 9-5:30. Ferry: 9:30-4:30. No charge for admission. Cooperative association with National Park Service. Closed Christmas. &
Attendance: 700,000 (accurate)

GOVERNMENT HOUSE MUSEUM, 48 King St., Saint Augustine, FL 32085. Mailing Address: P.O. Box 210, Saint Augustine, FL 32085-0210. Tel.: 904-825-5034.
E-mail: staugustineinfo@admin.ufl.edu
Web Site: www.staugustine.ufl.edu/govHouse.html
Governing Authority: Parent Institution: University of Florida.
Institution Type/Description: Historic House Museum.
Collections: local history & culture; period furnishings; personal artifacts; archaeological artifacts; Native American canoe; colonial weapons; gold & silver from Spanish shipwrecks; photographs.
Activities: rental facilities; Spanish & British reenactment ceremonies.
Hours & Admission Prices: Temporarily closed.

LIGHTNER MUSEUM, (M), City Hall, Museum Complex, 75 King St., Saint Augustine, FL 32084. Mailing Address: P.O. Box 334, Saint Augustine, FL 32085-0334. Tel.: 904-824-2874. Fax: 904-824-2712.
E-mail: info@lightnermuseum.org
Web Site: www.lightnermuseum.org
Founded: 1948.
Congressional District: 4
Key Personnel: C.E.O., Robert W. Harper, III; Chm. (V), David C. Drysdale; Cur., Barry W. Myers, Jr.; Visitor Svcs., Jenifer Jordan; Business Mgr., Angela Blankenship; Museum Shop Mgr., Janice Phelan.
Personnel Profile: Full-Time Paid 6; Part-Time Paid 7; Part-Time Volunteers 64; Interns 2.
Governing Authority: municipal; nonprofit organization. Tax-exempt: 501(c)(3).
Institution Type/Description: General Museum: housed in 1887 Alcazar Hotel.
Collections: 19th-century fine & decorative arts & material culture, featuring American & European glass, ceramics, metal work & furniture; natural science, industry & anthropology collection.
Research Fields: 19th-century decorative arts material culture.

Facilities: 12,000-vol. reference library; permanent & changing exhibition galleries; meeting room; restaurant; courtyard garden. Museum-related items for sale.
Activities: demonstrations of 19th century mechanical musical instruments; lectures; concerts; seminars.
Publications: exhibition catalogs; reproduction postcards; guide books; multilingual guides; Lost Colony - The Artists of Saint Augustine, 1930-1950.
Hours & Admission Prices: Daily 9-5. Adults $10, active military with ID $6; college students with ID & youth 12-18 $5 students $5; discount to groups, military & AAM members; children under 12 with adult no charge. Closed Christmas. &
Attendance: 133,424 (accurate)

OLD FLORIDA MUSEUM/FORT MENENDEZ, 259 San Marco Ave., Saint Augustine, FL 32084-1628. Tel.: 800-813-3208; 904-824-8874.
E-mail: info@oldfloridamuseum.com
Web Site: www.oldfloridamuseum.com
Founded: 1996.
Key Personnel: Dir., William G. Pitzalis
Institution Type/Description: History Museum.
Collections: hands-on exhibits; Florida history.
Hours & Admission Prices: Daily 10-5. Call for admission prices. &

OLDEST HOUSE MUSEUM COMPLEX, 14 St. Francis St., Saint Augustine, FL 32084-5047. Mailing Address: 271 Charlotte St., St. Augustine, FL 32084-5033. Tel.: 904-824-2872. Fax: 904-824-2569.
E-mail: sahsdirector@bellsouth.net
Web Site: www.oldesthouse.org
Founded: 1883.
Congressional District: 7
Key Personnel: C.E.O., Dr. Susan R. Parker; Pres. (V), Dr. Kathleen Deagan; Office Mgr., Magen Wilson.
Personnel Profile: Full-Time Paid 6; Part-Time Paid 15; Part-Time Volunteers 27.
Governing Authority: society. Parent Institution: St. Augustine Historical Society, 271 Charlotte St., Saint Augustine, FL. Tax-exempt: 501(c)(3).
Institution Type/Description: Historic Houses, Military Museum & Ornamental Garden.
Collections: Spanish, English & American antiques, 1727-present; St. Augustine history items; interpretive exhibits including tools used by the Spanish & English; manuscripts; maps & photographs.
Research Fields: St. Augustine history: domestic, social & military.
Facilities: library; archives.
Activities: tours; school programs; elder hostels; lecture series.
Publications: annual journal, El Escribano; semiannual newsletter, East Florida Gazette; books, The Oldest City; St. Augustine: A Saga of Survival; The Houses of St. Augustine; Awakening of St. Augustine; St. Augustine Historical Society News.
Hours & Admission Prices: Daily 9-5. Adults $8, seniors $7, students & adult tours $4; discounts to groups & military; members & children under 6 no charge. Closed Easter; Thanksgiving; Christmas. &
Attendance: 40,000 (estimated)
Membership: Student $20; Individual $35; Family $50; Contributor $100; Supporter $250; Benefactor $500; Guardian $1,000. Organizations & Firms: Associate $100; Promoter $250; Sustainer $500; Developer $1,000; Patron $2,500.

OLDEST STORE MUSEUM, 167 San Marco Ave., Saint Augustine, FL 32084-3269. Tel.: 904-829-9729.
Key Personnel: Museum Shop Mgr., John Stabley
Institution Type/Description: History Museum.
Collections: over 100,000 period artifacts.
Facilities: Museum-related items for sale.
Hours & Admission Prices: June-Aug. Mon.-Sat. 9-5, Sun. 10-5; Sept.-May Mon.-Sat. 9-5, Sun. 12-5. Adults $5, children $1.50.

THE PENA-PECK HOUSE MUSEUM, 143 St. George St., Saint Augustine, FL 32084-3642. Tel.: 904-829-5064. Fax: 904-829-3898.
E-mail: lthom2007@aol.com
Founded: 1932.
Congressional District: 4
Key Personnel: Pres. (V), Karen Jurgensen; Museum Shop Mgr., Val Rosselear.
Personnel Profile: Full-Time Paid 1; Part-Time Paid 2; Part-Time Volunteers 60.

Governing Authority: municipal; private; nonprofit organization. The Woman's Exchange. Tax-exempt.
Institution Type/Description: History Museum: Housed in ante bellum home of Dr. Seth & Sarah Lay Peck. Built in 1870 with original furnishings dating back to 1837.
Collections: home & furnishings of a family who came to St. Augustine from Connecticut in the early 1830s.
Research Fields: Peck-Burt records.
Facilities: Museum-related items for sale.
Activities: guided tours; candlelight evening tours by appointment.
Publications: book, The Treasurer's House.
Hours & Admission Prices: Tours: daily 12:30-4. Shop: Mon.-Sat. 10-5, Sun. 12-5. No charge; donations accepted. Closed most major holidays.
Attendance: 5,000 (estimated)

PONCE DE LEON'S FOUNTAIN OF YOUTH ARCHAEO-LOGICAL PARK, 11 Magnolia Ave., Saint Augustine, FL 32084. Tel.: 800-356-8222; 904-829-3168. Fax: 904-829-1529.
E-mail: fountain@aug.com
Web Site: www.fountainofyouthflorida.com
Founded: 1867.
Key Personnel: Dir. & C.E.O., John W. Fraser; Chm. (V), Kit Keating; Museum Shop Mgr., Leah Collier.
Personnel Profile: Full-Time Paid 25; Part-Time Paid 4; Part-Time Volunteers 20.
Governing Authority: Parent Institution: Fountain of Youth Properties, Inc.
Institution Type/Description: History Museum.
Collections: local history & culture; period furnishings; personal artifacts; historic buildings.
Facilities: planetarium; nature trails; landmark spring & cross; Native American Timucuan interpretive village; 600 ft. long observation trail & platform over marsh; 35 ft. tall Spanish watchtower; reconstruction of Nombre de Dios Mission; 3,000 sq. ft. covered pavillion; cafe.
Activities: live cannon firings daily; Native American interactive living history; celestial navigation show; Spanish history globe show; drink from spring.
Hours & Admission Prices: Daily 9-5. Adults $12, seniors over 60 $11, children 6-12 $8; discounts to groups, active military & AAA members; children under 6 no charge. Closed Christmas. &
Attendance: 179,000 (accurate)

POTTER'S WAX MUSEUM, 31 Orange St., Saint Augustine, FL 32084-3634. Tel.: 904-829-9056.
E-mail: potterswaxmuseum@yahoo.com
Web Site: www.potterswax.com
Founded: 1948.
Congressional District: 7
Key Personnel: Dir., Kimber Ponce.
Personnel Profile: Full-Time Paid 2; Part-Time Paid 6.
Institution Type/Description: Wax Museum.
Collections: 150 wax figures including presidents, athletes, entertainers, authors, artists, inventors, scientists & explorers.
Activities: tours.
Hours & Admission Prices: Sun.-Thurs. 10-5, Fri.-Sat. 10-7. Adults $10, seniors 55 & over $9, children 6-12 $7; discount to local, military & AAA members; children 5 & under no charge.
Attendance: 60,000 (estimated)

RIPLEY'S BELIEVE IT OR NOT! MUSEUM, 19 San Marco Ave., Saint Augustine, FL 32084-3278. Tel.: 904-824-1606.
Web Site: staugustine.ripleys.com
Key Personnel: Mgr., Ed Shaffer
Institution Type/Description: General Museum.
Collections: over 800 antiquities & oddities from around the world; personal artifacts.
Hours & Admission Prices: Daily 9-8. Adults 12 & over $14.99, senior citizens 55 & over $12.26, children 5-11 $7.99; children under 5 no charge.

ST. AUGUSTINE ALLIGATOR FARM, 999 Anastasia Blvd., Saint Augustine, FL 32080-4619. Tel.: 904-824-3337, ext. 10. Fax: 904-829-6677.
E-mail: jbrueggen1@aol.com
Web Site: www.alligatorfarm.com
Founded: 1893.
Key Personnel: Dir., John Brueggen; Gift Shop Mgr., Natalie Hurtado.
Personnel Profile: Full-Time Paid 27; Part-Time Paid 3.
Institution Type/Description: Zoological Park.
Collections: 23 species of crocodiles from around the world; tropical birds.
Hours & Admission Prices: June-Aug. daily 9-6; Sept.-May daily 9-5. Adults

$21.95, children 3-11 $10.95; discounts to seniors, groups, AAM, ICOM & AAA members; children under 3 & members no charge. &
Attendance: 213,000 (estimated)
Membership: Individual $49.95; Photo; $59.95; Grandparent $69.95; Family $79.95.

ST. AUGUSTINE ART ASSOCIATION, (M), 22 Marine St., Saint Augustine, FL 32084-4438. Tel.: 904-824-2310. Fax: 904-824-0716.

E-mail: info@staaa.org
Web Site: www.staaa.org
Founded: 1924.
Governing Authority: Tax-exempt.
Institution Type/Description: Art Museum.
Collections: paintings; photographs; period artifacts.
Facilities: 3,000 sq. ft. exhibition space; rental facilities.
Activities: lectures; classes; children's programs & camps; special events; master artist workshops & demonstrations; walking tours.
Publications: newsletter.
Hours & Admission Prices: Tues.-Sat. 12-4, Sun. 2-5. No charge; donations accepted. Closed holidays. &
Membership: Student $25; Individual $50; Family $65; Patron $250; Life $1,000.

ST. AUGUSTINE PIRATE & TREASURE MUSEUM, 12 S. Castillo Dr., Saint Augustine, FL 32084. Tel.: 877-467-5863.

E-mail: sknott@piratesoul.com
Web Site: www.piratesoul.com
Formerly: Pirate Soul Museum
Founded: 2005.
Key Personnel: Exec. Dir., Cindy Stavely; Museum Shop Mgr., Tracy King
Institution Type/Description: History Museum.
Collections: pirate artifacts; Captain Kidd; pirate treasure chest.
Hours & Admission Prices: Daily 9-8. Adults $13.95, children $7.95. &

ST. PHOTIOS GREEK ORTHODOX NATIONAL SHRINE, 41 St. George St., Saint Augustine, FL 32084-3607. Mailing Address: P.O. Box 1960, St. Augustine, FL 32085. Tel.: 904-829-8205; 800-222-6727. Fax: 904-829-8707. Facebook: St. Photios Shrine.

E-mail: info@stphotios.com
Web Site: www.stphotios.com
Founded: 1982.
Congressional District: 4
Key Personnel: Dir., Polexeni M. Hillier; Chm. (V), Archbishop Demetrios; Pres. (V), Metropolitan Alexios of Atlanta; 1st Vice Pres. (V), Dr. Manuel Tissura; Museum Shop Mgr., Jenny Harker; Admin., Brenda Gale; Museum Shop Mgr., Eugenia Mercado.
Personnel Profile: Full-Time Paid 2; Part-Time Paid 2; Part-Time Volunteers 14; Interns 3.
Volunteer Hours: 1,096
Operating Expenses: 240,000
Operating Income: 240,000
Governing Authority: nonprofit organization. Parent Institution: Greek Orthodox Archdiocese of America. Tax-exempt: 501(c)(3).
Institution Type/Description: History Museum: housed in 1740 Avero House. Listed on the National Register of Historic Places.
Collections: 1768 Greek landing with Minorcans, Italians, Corsicans & Greeks; frescoes; icons; artifacts reflecting the Hellenic Heritage & teachings of the Orthodox Christian Faith; early life of Greeks in America; photographs; historical documents; relics of 18 saints of the early church.
Major Exhibits: Oils of Greece (T), 1/13-12/14; Priests of the Early Church in America (T), 1/15-12/15.
Research Fields: Greeks in America.
Facilities: 300-vol. library of Greek-related material available to the public; 25-seat video theater. Greek items & items relating to Greek Orthodoxy for sale.
Activities: lectures; films; loan exhibitions. Museum Sponsors: St. Photios Feast Day; Annual St. Photios National Shrine Pilgrimage in February; House of Worship Tours in February; Lecture Series in February, June & October; Renewal retreat in April; Greek Landing Day in June.
Publications: annual newsletter, Friends.
Hours & Admission Prices: Mon.-Sat. 9-5, Sun. 12-6. No charge; donations accepted. Closed New Year's Day; Greek Orthodox Good Friday; Independence Day; Thanksgiving; Christmas. &
Attendance: 100,000 (estimated)
Membership: Friend $50; Supporter $100; Sustainer $250; Patron $500; Benefactor $1,000; Wall-of-Tribute $2,000.

WORLD GOLF HALL OF FAME & MUSEUM, (M), One World Golf Pl., Saint Augustine, FL 32092-2724. Tel.: 904-940-4000. Fax: 904-940-4391. Facebook: Golf Hall of Fame.

E-mail: info@wghof.org
Web Site: www.worldgolfhalloffame.org
Founded: 1998.
Congressional District: 7
Key Personnel: Sr. Vice Pres. & C.O.O., Jack Peter; Dir. Museum Operations, Brodie Waters; Vice Pres. Hall of Fame, Bruce Lahti; Dir. Mktg., Mary Altman; Dir. Communications, Travis Hill; Dir. Exhibits, Andy Hunold; Museum Shop Mgr., Jody Sutton.
Personnel Profile: Full-Time Paid 30; Part-Time Paid 25; Part-Time Volunteers 150.
Governing Authority: nonprofit organization. Parent Institution: World Golf Foundation. Tax-exempt: 501(c)(3).
Institution Type/Description: Sports Museum.
Collections: golf clubs & equipment used by famous golfers; paintings; ceramics; trophies; medals; sculpture; photographs; period clubs, balls & other golf memorabilia; clothing; jewelry.
Research Fields: database on Hall of Fame members.
Facilities: 300-seat IMAX theater; 18-hole putting course; cafe; walk of champions. Museum-related items for sale.
Activities: interactive exhibits including golf simulator & putting surfaces; audio tour; mini theaters.
Publications: annual, World Golf Hall of Fame; annual visitor guide, World Golf Hall of Fame.
Hours & Admission Prices: Mon.-Sat. 10-6, Sun. 12-6. Adults $19.50, seniors & students $17.50, children 5-12 $9; members & children under 5 no charge. IMAX: adults $8, seniors & students $7, children 3-12 $5. Closed Thanksgiving; Christmas. &
Attendance: 200,000 (estimated)
Membership: 6 month $35; Individual $50; Associate $125; Sustaining $250; Patron $500; Benefactor $1,000; Ambassador $2,500; Legends $5,000; Champions $10,000.

XIMENEZ-FATIO HOUSE MUSEUM, 20 Aviles St., Saint Augustine, FL 32084-4442. Mailing Address: 28 Cadiz St., Saint Augustine, FL 32084. Tel.: 904-829-3575. Fax: 904-829-3445.

Web Site: www.ximenezfatiohouse.org
Governing Authority: Parent Institution: The National Society of The Colonial Dames of America in the State of Florida
Institution Type/Description: Historic House Museum.
Collections: local history & culture; period furnishings; personal artifacts.
Hours & Admission Prices: Wed.-Sat. 11-4. Family $15, adults $6, senior, students 6-17 & military $4. &

Saint George Island

ST. GEORGE ISLAND VISITOR CENTER AND LIGHTHOUSE MUSEUM, 2 E. Gulf Beach Dr., Saint George Island, FL 32328-2883. Tel.: 888-927-7744.

Web Site: www.seestgeorgeisland.com
Institution Type/Description: Maritime History Museum.
Collections: local history & culture; lighthouse.
Hours & Admission Prices: Visitor Center: Fri.-Wed. 10-5. Lighthouse Tours: Mon.-Wed. 9-12 & 1-3, Sat. 9-1, Sun. 1-3. Adults $5, children 16 & under $3; Lighthouse Assn. members & children 6 and under no charge.

Saint Petersburg

∗ THE DALI MUSEUM, One Dali Blvd., Saint Petersburg, FL 33701. Tel.: 727-823-3767. Fax: 727-894-6068.

E-mail: info@thedali.org
Web Site: salvadordalimuseum.org
Formerly: Salvador Dali Museum
Founded: 1954.
Congressional District: 8
Key Personnel: Dir. & C.E.O., Dr. Charles Henri Hine; Chm. (V), Eleanor Morse; Pres. (V), Thomas James; Deputy Dir. Collections, Joan R. Kropf; Dir. Mktg., Kathy White; Cur. Exhibitions, William Jeffett; Museum Shop Mgr., Nancy McCue.
Personnel Profile: Full-Time Paid 30; Part-Time Paid 6; Part-Time Volunteers 200; Interns 4.
Governing Authority: nonprofit organization. Parent Institution: Salvador Dali Institute. Tax-exempt: 509(a).
Institution Type/Description: Art Museum.
Collections: Salvador Dali oils, drawings, watercolors, graphics & sculpture.
Research Fields: Salvador Dali; surrealism.
Facilities: 5,000-vol. library of books by or about Dali available for use by private appointment.

Activities: lectures; permanent & traveling exhibitions; children's programs; film program; music programs.
Publications: books, Dali Draftsmanship Catalog; Dali Primer; Dali, A Panorama of his Art, 1974; Dali in Public Museum Collections, 1974; Poetic Homage to Dali, 1973; Dali-Picasso, A Study in their Similarities & Differences, 1973; Notes on the Paintings, Student Edition, 1973; The Dali Adventure, A Photo Album, 1973; Tragic Myth of Millett's Angelus by Dali; Dali in the Nude (Reprint); Passions of Salvador Dali; Dali's Animal Crackers; Dali, A Collection, 1972; exhibition catalogues, Dante Divine Comedy; Surrealist Drawings; Isidro Clott Sculpture catalog; Horses; Crucifixion (Corpus Hypercubus); Dali's Graphic Art; Kenny Scharf, Pop Surrealist; Dali, The Early Years; Andy Warhol at the Dali; Dali by Design; Man Ray's Paris Portraits 1921-1939; Surrealism in America During the 1930s & 1940s; Masson, Masterpieces of Surrealism; James Rosenquist: Paintings; James Rosenquist: Selects; A Disarming Beauty: The Venus de Milo in 20th Century Art; Salvador Dali: The Salvador Dali Museum Collection; Dali Objects/Dali Fetishes: Love and Death; Dali and Two French Writers; Dali and Miro c.1928; Persistence & Memory: New Critical Perspectives on Dali at the Centennial; Joan Fontcuberta: Imaginary Gardens; Jordi Colomer: Arabian Stars; Pollock To Pop: America's Brush with Dali.
Hours & Admission Prices: Mon.-Wed. & Fri.-Sat. 10-5:30, Thurs. 10-8, Sun. 12-5:30. Adults $21, senior citizens 65 & over and students 18 & over $15, children 6-12 $7; discounts to AAA, AAM & ICOM members, Hospitality Industry Association (H.I.A.) and Florida Attraction members with proper I.D.; children 5 & under and members no charge. Closed Thanksgiving; Christmas. &
Attendance: 210,000 (estimated)
Membership: Student & Seniors $35; Individual $40; Zodiac Individual $60; Family $70; Zodiac Family $90. NARM: Individual $100; Family $300; Dadaist $250; Surrealist $500.

FLORIDA CRAFTSMEN GALLERY, 501 Central Ave., Saint Petersburg, FL 33701. Tel.: 727-821-7391. Fax: 727-822-4294.
E-mail: info@floridacraftsmen.net
Web Site: www.floridacraftsmen.net
Key Personnel: Exec. Dir., Diane Shelly.
Governing Authority: nonprofit organization.
Institution Type/Description: Art Gallery.
Collections: works by Florida artists.
Activities: workshops; educational programs.
Hours & Admission Prices: Mon.-Sat. 10-5:30.

∗　**FLORIDA HOLOCAUST MUSEUM, (M),** 55 5th St., S., Saint Petersburg, FL 33701-4146. Tel.: 727-820-0100; 800-960-7448. Fax: 727-821-8435. Facebook: Florida Holocaust Museum.
E-mail: eblankenship@flholocaustmuseum.org
Web Site: www.flholocaustmuseum.org
Founded: 1989.
Congressional District: 10
Key Personnel: Exec. Dir., Elizabeth Gelman; Cur. Exhibitions & Collections, Erin Blankenship.
Personnel Profile: Full-Time Paid 9; Part-Time Paid 7; Part-Time Volunteers 270; Interns 5.
Governing Authority: private; nonprofit organization. Tax-exempt: 501(c)(3).
Institution Type/Description: History Museum.
Collections: principle concentration on items related to the time of the Holocaust; additional items of Judaica, artwork concerning the Holocaust, genocide & the human condition; mixed media; traveling art & cultural exhibits.
Major Exhibits: Holocaust Images by D.K. Lubarsky (T), 10/12/13-1/26/14; Reflections of a Survivor: Artwork by Michael Smuss (T), 12/21/13-3/16/14; Fire in My Heart: The Story of Hannah Senesh (T), 1/11/14-4/27/14; Fragments: Portraits of Survivors (T), 2/21/14; Whoever Saves A Life (T), 4/14-6/14.
Facilities: 7,000-vol. library on the Holocaust, human rights & other genocides; 12,000 sq. ft. exhibit space; Teaching Trunks for classrooms K-12. Museum-related items for sale.
Activities: docent & teacher training; guided tours; education programs & outreach; traveling art exhibits; Survivor services & video testimony; self-guided audio tours. Annual Events: Kristallnacht Night of Broken Glass Commemoration; To Life Award Dinner; Yom HaShoah (Day of Remembrance) Commemoration; Summer Institutes for Teachers K-16; Student Awareness Days; Anne Frank Humanitarian Award.
Publications: exhibit catalogues; curricula.
Hours & Admission Prices: Daily 10-5 (last admission at 3:30); call for additional hours. Adults $16, senior citizens 65 & over $14, college students $10, students under 18 $8; active military, children 6 & under and members

no charge. Closed New Year's Day; Martin Luther King Jr. Day; Rosh Hashanah; Yom Kippur; Thanksgiving; Christmas. &
Attendance: 100,000 (accurate)
Membership: Student & Educator $25; Individual $45; Dual (2 adults) $65; Family $85; Circle of Friends $275; Circle of History $500; Circle of Heritage $1,000; Circle of Hope $3,000; Circle of Tolerance $5,000; Circle of Equality $10,000.

GREAT EXPLORATIONS CHILDREN'S MUSEUM, (M), 1925 Fourth St. N., Saint Petersburg, FL 33704-4307. Tel.: 727-821-8992. Fax: 727-823-7287. Facebook: Great Ex Kids.
E-mail: ahowell@greatex.org
Web Site: www.greatex.org
Founded: 1986.
Congressional District: 10
Key Personnel: Exec. Dir., Angeline Howell; Deputy Dir., Alan Kahle; Finance Mgr., Laurel Ginn; Chm. (V), David Hood.
Personnel Profile: Full-Time Paid 18; Part-Time Paid 11; Part-Time Volunteers 300.
Governing Authority: private; nonprofit organization. Tax-exempt.
Institution Type/Description: Children's Museum.
Collections: science, art & history interactive exhibits; teaching collection for outreach & demonstrations (snakes, spiders, scorpions, millipedes, beetles); sound; sailboats; racecars; roleplay; baby garden; air pressure; electromagnetism; climbing wall.
Major Exhibits: The Wonderful Wizard of Oz (T), 12/10-12/15.
Facilities: 1926 Mediterranean Revival building.
Activities: daily preschool; youth leadership program; afterschool program; summer & holiday camps; educational demonstrations & outreach shows; workshops; special events; art & music therapy outreaches.
Publications: e-newsletter, Great News; education guide.
Hours & Admission Prices: Mon.-Sat. 10-4:30, Sun. 12-4:30; see website for additional fall hours. Admission 2-54 $10; discounts for ASTC members & reciprocal members with ACM; seniors 55 & over no charge. &
Attendance: 165,000 (accurate)
Membership: Basic $100; Great $150; Explorer $250; Lifetime $1,500.

MOREAN ARTS CENTER, 719 Central Ave., Saint Petersburg, FL 33701-3627. Tel.: 727-822-7872. Fax: 727-821-0516.
E-mail: amanda@moreanartscenter.org
Web Site: www.moreanartscenter.org
Formerly: The Arts Center
Founded: 1964.
Key Personnel: Exec. Dir., Katee Tully; Cur., Amanda Cooper.
Governing Authority: Tax-exempt.
Institution Type/Description: Art Museum.
Collections: works by contemporary artists.
Activities: art shows.
Hours & Admission Prices: Mon.-Sat. 9-6, Sun. 12-6. Adults $8, seniors 65 & over $6, students & children $5; children 5 & under no charge. Closed Thanksgiving; Christmas.

∗　**MUSEUM OF FINE ARTS, ST. PETERSBURG, (M),** 255 Beach Dr., N.E., Saint Petersburg, FL 33701-3498. Tel.: 727-896-2667. Fax: 727-894-4638. Facebook: MFAS St. Pete.
Web Site: www.fine-arts.org
Founded: 1961.
Congressional District: 13
Key Personnel: Bd. Chm. & Pres., Dr. Edward A. Amley; Dir., Dr. Kent Lydecker; Assoc. Dir. Advancement, Don Howe; Hazal & William Hough Chief Cur., Dr. Jennifer Hardin; Dir. Devel., Daryl DeBerry; Mgr. Mktg. & Events, Allison Canfield; Registrar, Louise Reeves; Cur. Public Programs, Anna Alexander Glenn; Assoc. Cur. Public Programs, Mary Szaroleta; Asst. Cur. of Art after 1950, Katherine Pill; Curatorial Asst., Sabrina Hughes; Coord. Curatorial Affairs, Bridget Bryson; Financial Officer, Diana Dillon; Sr. Art Preparator & Photographer, Thomas U. Gessler; Installations, Dimitri Lykoudis; Dir. Public Rels., David O. Connelly; Asst. Visitor Svcs. Mgr., Billy Summer; Mgr. Museum Store & Visitor Svcs., Audrie Ranon; Asst. Store Mgr., Jeff Eversole; Coord. Membership & Mktg., Bailey Nicholas; Devel. Assoc., Amanda Bonanno; Asst. to Dir., Vicki Sofranko; Mgr. Bldg., J.P. Fatseas.
Personnel Profile: Full-Time Paid 23; Part-Time Paid 11; Part-Time Volunteers 600.
Volunteer Hours: 20,730
Operating Expenses: 3,340,399
Operating Income: 3,346,603
Governing Authority: board of trustees; nonprofit organization. Tax-exempt: 501(c)(3).
Institution Type/Description: Art Museum.

Collections: American & European paintings, drawings, prints, sculpture & photographs; Ancient Greek & Roman, pre-Columbian, Native American, African & Asian Art; decorative arts; sculpture garden; gallery of Steuben glass; new media; video art; ceramics.
Major Exhibits: Recent Acquisitions: Prints, Drawings, and Photographs, 10/13-3/2/14; New Mexico and the Arts of Enchantment featuring the Raymond James Financial Collection, 1/18/14-5/11/14; My Generation: Young Chinese Artists, 6/7/14-9/28/14; Jaime Myeth's Portrait of Rudolph Nureyev: Images of the Dancer from the Brandywine Art Museum, 9/14-12/14.
Facilities: 27,000-vol. library of art reference books; 225-seat theatre & recital hall; cafe; gardens; sculpture garden; glass conservatory. Museum-related items for sale.
Activities: docent tours; lectures; films; gallery talks; family days; concerts; formally organized educational programs; docent program; inter-museum loan; permanent, temporary & traveling exhibitions; workshops; classes; tours and educational materials for people with physical & emotional challenges; curriculum collaboration with public schools.
Publications: quarterly newsletter, Mosaic; exhibition catalogues & brochures; educational materials for children & adults.
Hours & Admission Prices: Mon.-Wed. & Fri.-Sat. 10-5, Thurs. 10-8, Sun. 12-5. Adults $17, senior citizens 65 & over $15; Thurs. after 5pm $5; discounts to students, groups of 10 or more & AAM members; MFA members and children 6 & under no charge. Closed Thanksgiving; Christmas. &
Attendance: 129,474 (accurate)
Membership: Scholar (Student & Educator) $35; Individual $60; Family $115; Friend $250; Benefactor $500; Sustainer $1,200; Founder $10,000; Director $25,000. Support Group: Friends of Decorative Arts $20; Friends of Photography $30; Marly Music Society $75; Collector's Circle $500.

THE PIER AQUARIUM, 800 Second Ave., N.E., Ste. 2001, Saint Petersburg, FL 33701-3503. Tel.: 727-895-7437. Fax: 727-894-1212.
E-mail: info@pieraquarium.org
Web Site: www.pieraquarium.org
Founded: 1988.
Congressional District: 10
Key Personnel: Pres. & C.E.O., E. Howard Rutherford; Chm. (V), Mark Luther, Ph.D.; Museum Shop Buyer, Laurie Zakaroff
Personnel Profile: Full-Time Paid 5; Part-Time Paid 7; Part-Time Volunteers 50; Interns 4.
Governing Authority: Tax-exempt.
Institution Type/Description: Aquarium.
Collections: marine environments from around the world; Tampa Bay touch tank.
Facilities: 2,000 sq. ft. aquarium; marine laboratory.
Activities: touch tank; fish feeding. Museum Sponsors: Fish Head Ball in October.
Publications: Fresh Fish Quarterly.
Hours & Admission Prices: Mon.-Sat. 10-8, Sun. 12-6. Adults $5, students 7 & over and seniors 65 & over $4; children under 6 & members no charge. Tampa Bay Touch Tank: daily 1-4. Shark Feeding: daily 3pm. &
Attendance: 65,000 (accurate)
Membership: Student, Senior, Volunteer & Teacher $20; Adult $25; Couple $40; Family $50 ($5 each additional).

RANSOM VISUAL ARTS CENTER, Eckerd College, 4200 54th Ave. S., Saint Petersburg, FL 33711-4744. Tel.: 727-864-8340 & 8342.
Web Site: www.eckerd.edu/tour/index.php?f=arts3
Key Personnel: Dir., Arthur Skinner
Institution Type/Description: Art Center.
Collections: works by students & local artists.
Facilities: 900 sq. ft. exhibit space.
Hours & Admission Prices: Mon.-Fri. 10-4:30. No charge.

ST. PETERSBURG MUSEUM OF HISTORY, 335 Second Ave., N.E., Saint Petersburg, FL 33701-3501. Tel.: 727-894-1052, ext. 207. Fax: 727-823-7276.
E-mail: george.banez@stpetemuseumofhistory.org
Web Site: www.spmoh.org
Founded: 1920.
Congressional District: 6
Key Personnel: Exec. Dir., Joel Cohen; Pres. (V), Harold Heller; Vice Pres., Andrew Hayes; Co-Vice Pres., Joe Griner; Dir. Education & Outreach, Nevin Sitler; Archivist, Ann Wikoff; Dir. Events & Mktg., James Parrish.

Personnel Profile: Full-Time Paid 2; Part-Time Paid 6; Part-Time Volunteers 30; Interns 2.
Governing Authority: nonprofit organization. Parent Institution: St. Petersburg Historical Society. Tax-exempt: 501(c)(3).
Institution Type/Description: History Museum.
Collections: exhibits reflect St. Petersburg; the Pinellas Peninsula and Florida history.
Research Fields: St. Petersburg & Florida history; Pinellas Peninsula.
Facilities: learning/teaching center; archival center. Museum-related items for sale.
Activities: gallery tours; focused educational programs; lectures; exhibitions; outreach programs; evening rentals.
Publications: quarterly newsletter, The Sea Breeze.
Hours & Admission Prices: Wed.-Sat. 10-5, Sun. 1-5. Adults $12; discounts to students, seniors & teachers; members no charge &
Attendance: 35,000 (accurate)
Membership: Student $20; Individual $35; Family $50; Florida Settler $100; Archivist $500; Collector $1,000.

THE SCIENCE CENTER OF PINELLAS COUNTY, 7701 22nd Ave., N., Saint Petersburg, FL 33710-3899. Tel.: 727-384-0027. Fax: 727-343-5729.
E-mail: info@steic.org
Web Site: www.steic.org
Founded: 1959.
Congressional District: 10
Key Personnel: Chm., Mike Mikurak; Dir., Joseph S. Cuenco.
Personnel Profile: Full-Time Paid 10; Part-Time Paid 14; Part-Time Volunteers 21.
Governing Authority: private; nonprofit organization. Tax-exempt: 501(c)(3).
Institution Type/Description: Science Center.
Collections: anatomy; anthropology; aquarium; archaeology; astronomy; botany; entomology; geology; herpetology; marine; medicine; natural history; paleontology; zoology; 600-gallon touch-tank; cyber security certification; robotics; forensics; astrophotography; mousetrap physics.
Research Fields: all fields of science.
Facilities: science classrooms; planetarium with Minolta's Mediaglobe projector; 170-seat auditorium; computer facilities; observatory with Meade 16 inch LX200 telescope.
Activities: guided tours; lectures; films; formally organized educational programs; permanent exhibitions.
Publications: workshop brochures; newsletter, Flash.
Hours & Admission Prices: Mon.-Fri. 9-4, Sat. special events. Adults $5; discounts to ASTC & FAM members; members no charge. Closed New Year's Day; Thanksgiving; Christmas. &
Attendance: 52,000 (estimated)
Membership: Seniors $25; Individual $40; Family $75.

TED WILLIAMS MUSEUM AND HITTERS HALL OF FAME, One Tropicana Dr., Saint Petersburg, FL 33705-1703. Tel.: 888-326-7297.
E-mail: info@tedwilliamsmuseum.com
Web Site: www.tedwilliamsmuseum.com
Key Personnel: Exec. Dir., Dave McCarthy; Deputy Dir., John Papelbon.
Governing Authority: nonprofit. Tax-exempt: 501(c)(3).
Institution Type/Description: Sports Museum.
Collections: history of baseball; Ted Williams' life & career; sculpture; photographs; personal artifacts; Hitters Hall of Fame.
Facilities: Museum-related items for sale.
Activities: tours; dinners; special events. Annual Event: Museum Induction Ceremony.
Hours & Admission Prices: During Tampa Bay Devil Rays games for game ticket purchasers.

Sanderson

OLUSTEE BATTLEFIELD HISTORIC STATE PARK, 5890 Battlefield Trail Rd., Sanderson, FL 32087. Tel.: 386-758-0400.
E-mail: olusteecso@yahoo.com
Web Site: www.battleofolustee.org
Founded: 1909.
Key Personnel: Park Mgr., Benjamin Faure; Park Ranger, Francis J. Loughran.
Governing Authority: state. Dept. of Natural Resources, Div. of Recreation & Parks, 3900 Commonwealth Blvd., Commonwealth Bldg., Tallahassee, FL 32303.
Institution Type/Description: State Historic Site & Military Museum.
Collections: military artifacts; battlefield monument.
Facilities: nature trails.
Activities: permanent exhibitions; interpretive programs.
Publications: brochures.

Hours & Admission Prices: Park: Daily: 8-5. Visitor Center: daily 9-5. No charge; donations accepted; (except for special events). Closed Thanksgiving; Christmas. &

Sanford

CENTRAL FLORIDA ZOO & BOTANICAL GARDENS, 3755 N.W. Hwy. 17-92, Sanford, FL 32771. Mailing Address: P.O. Box 470309, Lake Monroe, FL 32747-0309. Tel.: 407-323-4450, ext. 100. Fax: 407-321-0900.
E-mail: information@centralfloridazoo.org
Web Site: www.centralfloridazoo.org
Founded: 1971.
Congressional District: 7
Key Personnel: C.E.O., Joe Montisano; Chm. (V), Rob Panepinto; C.F.O., Chuck Grimes; Vice Pres. Communications & Community Resources, Shonna Green; Vice Pres. Education, Stephanie Williams; Vice Pres. Operations, David Tetzlaff.
Personnel Profile: Full-Time Paid 37; Part-Time Paid 32; Part-Time Volunteers 100.
Volunteer Hours: 20,000
Operating Expenses: 3,500,000
Operating Income: 3,500,000
Governing Authority: private; nonprofit. Tax-exempt: 501(c)(3).
Institution Type/Description: Zoo.
Collections: over 566 specimens (194 species) of mammals, birds, reptiles, amphibians & invertebrates; emphasis focuses on American Zoo and Aquarium (AZA) cooperative conservation programs for rare & endangered species; species emphasized include neotropical primates, small felids, soft-billed birds, large lizard species and tortoises.
Research Fields: behavioral, nutritional & reproductive biology; focal species include black howler monkey, black-footed cat, clouded leopard, wreathed hornbill and Grand Cayman Island rock iguana.
Facilities: 700-vol. library on natural sciences & ecology; educational facilities; nature/conservation center. Shirts, books & zoo-related items for sale.
Activities: docent program; guided tours; college credit internships; lectures; mobile vans; educational programs; community outreach programs; on-site special events. Annual Events: Black Tie on the Wild Side; Zoo Boo Bash; Hippity Hop Adventure; Birthday with the Big Boys.
Publications: quarterly members' newsletter, ZooViews; annual report.
Hours & Admission Prices: Daily 9-5. Adults $14.95, seniors $12.95, children 3-12 $10.95; discounts to groups, military & AAA members; children under 2 & members no charge. Closed Thanksgiving; Christmas. &
Attendance: 288,000 (accurate)
Membership: Single $55; Senior Plus $69; Single Plus $79; Family & Grandparent $89; Premier Family $140; Keeper's Circle $250; Give & Get Wild $500; Conservationist $1,000.

MUSEUM OF SEMINOLE COUNTY HISTORY, (M), 300 Bush Blvd., Sanford, FL 32773-6135. Tel.: 407-665-2489. Fax: 407-665-5220.
E-mail: knelson@seminolecountyfl.gov
Web Site: www.seminolecountyfl.gov/leisure/museum/
Founded: 1983.
Congressional District: 5
Key Personnel: Museum Coord., Kim Nelson.
Personnel Profile: Part-Time Paid 1; Part-Time Volunteers 1.
Governing Authority: county; nonprofit. Parent Institution: Seminole County Board of County Commissioners. Tax-exempt.
Institution Type/Description: History Museum.
Collections: artifacts from Civil War; Seminole wars; modern times; documents & photos; exhibits on the history of Seminole County & Central Florida.
Major Exhibits: Alligators: Dragons in Paradise (T), 6/14-8/30/14.
Facilities: 200-vol. library on Florida & regional history available to the public; meeting room; 3,500 sq. ft. exhibit space; 173,500 sq. ft. (bldg 1); 2,400 sq. ft. (bldg. 2).
Activities: docent program; guided tours; lectures; loan & temporary exhibitions.
Publications: The Early Days of Seminole County; Touring Seminole County.
Hours & Admission Prices: Tues.-Fri. 1-5, Sat. 9-1. Adults $3, children 4-18 $1; members & children under 4 no charge. Closed New Year's Day; Martin Luther King Jr. Day; Memorial Day; Independence Day; Labor Day; Veterans Day; Thanksgiving & day after; Christmas. &
Attendance: 3,000 (accurate)
Membership: Single $15; Family $20; Patron $50; Life $250.

SANFORD MUSEUM, 520 E. First St., Sanford, FL 32771-1410. Mailing Address: P.O. Box 1788, Sanford, FL 32772-1788. Tel.: 407-688-5198. Fax: 407-688-5125.
E-mail: clarkea@sanfordfl.gov
Formerly: Henry Shelton Sanford Memorial Library & Museum
Founded: 1957.
Congressional District: 5
Key Personnel: Cur., Alicia Clarke.
Personnel Profile: Full-Time Paid 2; Part-Time Paid 3; Part-Time Volunteers 6.
Governing Authority: municipal; nonprofit. Parent Institution: City of Sanford. Tax-exempt.
Institution Type/Description: Museum, Research Library & Archives.
Collections: photos; culture; local baseball history; library of 19th-century legal, diplomatic & government books, classics & periodicals; 1820-1890 portrait collection; 1823-1891 papers of H.S. Sanford; 19th-century decorative arts; local history.
Research Fields: local history; Henry S. Sanford; local architectural history & African-American history; diplomatic service; Congo; civil war; Florida; citrus.
Facilities: 2,700-vol. library, 55,000 papers & 149 microfilm; available to the public, microfilm available for inter-library loan; meeting room.
Activities: guided tours; lectures; temporary exhibitions.
Hours & Admission Prices: Tues.-Fri. 11-4, Sat. 1-4 & by appointment; No charge; donations accepted. Closed federal holidays; day after Thanksgiving. &
Attendance: 3,000 (estimated)
Membership: Affiliated Historical Society: Student $5; Individual $15; Family $25; Patron $100; Corporate $500.

Sanibel

＊　THE BAILEY-MATTHEWS SHELL MUSEUM, (M), 3075 Sanibel-Captiva Rd., Sanibel, FL 33957-3111. Mailing Address: P.O. Box 1580, Sanibel, FL 33957-1580. Tel.: 239-395-2233; 888-679-6450 (toll free). Fax: 239-395-6706.
E-mail: dhipschman@shellmuseum.org
Web Site: www.shellmuseum.org
Founded: 1986.
Congressional District: 13
Key Personnel: Dir., Dorrie Hipschman; Cur., Jose H. Leal, Ph.D.; Museum Store Mgr., Gretchen Falk.
Personnel Profile: Full-Time Paid 5; Part-Time Paid 5; Part-Time Volunteers 110.
Governing Authority: nonprofit organization. Parent Institution: The Shell Museum & Educational Foundation, Inc. Tax-exempt: 501(c)(3).
Institution Type/Description: Natural History Museum.
Collections: mollusks; archives; malacological library.
Research Fields: malacology.
Facilities: 6,000-vol. library of popular to monographic works & scientific journals on mollusks; 150-seat auditorium; 5,000 sq. ft. exhibit space. Museum-related items, except shells, for sale.
Activities: video on shell life; Children's Learning Lab Center.
Publications: quarterly newsletter; books, Seashells Photo Postcards; Pliocene Mollusca of Southern Florida; Edge of the Fossil Sea; The Nautilus, Idea to Reality.
Hours & Admission Prices: Daily 10-5. Adults $9, youths 5-16 $5; discounts to AAM members; children 4 and under & members no charge. Closed major holidays. &
Attendance: 50,000 (estimated)
Membership: Periwinkle $50; Wentletrap $70; Golden Olive $100; Lion's Paw $250; Junonia $500; Angel Wing $1,000 & up.

SANIBEL-CAPTIVA CONSERVATION FOUNDATION, INC., 3333 Sanibel-Captiva Rd., Sanibel, FL 33957-3100. Mailing Address: P.O. Box 839, Sanibel, FL 33957-0839. Tel.: 239-472-2329. Fax: 239-472-6421.
E-mail: sccf@sccf.org
Web Site: www.sccf.org
Founded: 1967.
Congressional District: 14
Key Personnel: Exec. Dir., A. Erick Lindblad; Pres. (V), Bill Fenniman; Business Mgr., Wendy Cerdan; Legacy Funds Coord., Cheryl Giattini; Dir. Education, Kristie J. Seaman Anders; Landscaping for Wildlife Educator, Dee Century-Serage; Native Plant Nursery Mgr., Jenny Evans; Research Scientist, Dr. Eric Milbrandt; Member Rels. Dir., Marti Bryant; Dir. Wildlife Habitat Mgmt., Brad Smith.
Personnel Profile: Full-Time Paid 22; Part-Time Paid 1; Part-Time Volunteers 300; Interns 4.
Governing Authority: nonprofit organization. Tax-exempt: 501(c)(3).
Institution Type/Description: Land Trust; Nature Center.

Collections: native plants of Florida; island butterflies; herbarium.
Research Fields: ecology of gopher tortoises on Sanibel; surface water management & eradication of exotic pest plants; sea turtle research & monitoring; restoration ecology; shorebird monitoring.
Facilities: 1,000-vol. library of barrier island flora & fauna material, available to members only; 60-seat auditorium; educational facilities; 1,300 sq. ft. exhibit space; field research station; nature center; native plant nursery; nature trails. Nature oriented items for sale.
Activities: docent program; formal educational programs; intern program for undergraduate or graduate college students; guided tours; lectures; participatory exhibits. Annual Events: Great Island Pickup; Beach Clean-up; Open House; Earth Day.
Publications: monthly newsletter, Update; book, Growing Native; book, A Natural Course; annual report.
Hours & Admission Prices: June-Sept. Mon.-Fri. 8:30-3; May-Oct. Mon.-Fri. 8:30-4; Dec.-April Mon.-Fri. 8:30-4, Sat. 10-3. Native Plant Nursery: May-Nov. Mon.-Fri. 8:30-5; Dec.-April Mon.-Fri. 8:30-5, Sat. 10-3. Adults $3; members & children under 17 no charge. Closed New Year's Day; Memorial Day; Independence Day; Labor Day; Thanksgiving; Christmas. ♿
Attendance: 15,000 (estimated)
Membership: Individual $25; Family $50; Corporate $100.

Sarasota

ART CENTER SARASOTA, INC, 707 N. Tamiami Tr., Sarasota, FL 34236-4050. Tel.: 941-365-2032. Fax: 941-366-0585.
E-mail: artsarasota@aol.com
Web Site: www.artsarasota.org
Formerly: Sarasota Visual Arts Center
Founded: 1926.
Congressional District: 8
Key Personnel: Exec. Dir., Fayanne Hayes; Pres., Kathleen McDonald.
Personnel Profile: Full-Time Paid 2; Full-Time Volunteers 10; Part-Time Paid 7; Part-Time Volunteers 30; Interns 2.
Governing Authority: nonprofit organization. Tax-exempt: 501(c)(3).
Institution Type/Description: Art Gallery.
Collections: paintings; sculpture; graphics; photography.
Facilities: approx. 200-vol. library of art reference material available on premises.
Activities: guided tours; lectures; gallery talks; arts festivals; painting demonstrations; workshops; life-sketch groups; arts & crafts demonstrations; permanent & temporary exhibitions. Gallery Sponsors: The Sarasota County Public Schools Art Show; Annual Fund Drive for PBS T.V. WUSF as a collection point.
Publications: book, Sarasota Art Association Yearbook; monthly news bulletin; workshop brochures.
Hours & Admission Prices: Tues.-Sat. 10-4. Suggested Donation: adults $3. Closed on all major holidays. ♿
Attendance: 15,000 (estimated)
Membership: Single $50; Family & Sponsor $75; Associate $200; Patron $500; Life & Angel $1,500 minimum.

CROWLEY MUSEUM & NATURE CENTER, 16405 Myakka Rd., Sarasota, FL 34240-9192. Tel.: 941-322-1000. Fax: 941-322-1000.
E-mail: info@cmncfl.org
Web Site: www.cmncfl.org
Founded: 1974.
Congressional District: 13
Key Personnel: Chm. Bd., Bill Cowdright; Program Dir., Laney Poire.
Personnel Profile: Full-Time Paid 2; Part-Time Paid 2; Part-Time Volunteers 25.
Governing Authority: nonprofit. Tax-exempt: 501(c)(3).
Institution Type/Description: History Museum, Historic Building & Nature Center.
Collections: Florida pioneer lifestyles from late 1800s to early 1900s; blacksmith shop; replica homestead cabin; Victorian era home artifacts; general store; post office; Florida pioneer tools. Historic Buildings: 1892 Cracker House; sugar cane mill; sugar shack.
Research Fields: Crowley family; local Myakka area & Florida pioneers; Tatum Family.
Facilities: picnic pavilion; nature trails. Museum-related items for sale.
Activities: self-guided tours; lectures; workshops; organized educational programs; docent program; participatory exhibits. Museum Sponsors: Southwest Florida Heritage Festival in January.
Publications: e-newsletter.
Hours & Admission Prices: June-Sept. Fri.-Sun. 8-2; Oct.-May Thurs.-Sun. 10-4. Adults $8, children 5-12 $3; members no charge. Closed New Year's Day; Independence Day; Thanksgiving; Christmas. ♿

Attendance: 6,100 (accurate)
Membership: Student $15; Individual $30; Family $40; Patron $100; Donor $500.

✱ **JOHN AND MABLE RINGLING MUSEUM OF ART, (M),** 5401 Bay Shore Rd., Sarasota, FL 34243-2161. Tel.: 941-359-5700. Fax: 941-359-7704. TDD: 941-359-5700.
E-mail: info@ringling.org
Web Site: www.ringling.org
Founded: 1927.
Congressional District: 13
Key Personnel: Exec. Dir., Steve High; Chm. (V), Clifford L. Walters, III; C.F.O. & Assoc. Dir., Jennifer Price; Cur. Circus Museum & Archivist, Deborah Walk; Head Librarian, Linda McKee; Chief Conservator, Michelle Scalera; Deputy Dir., Grady Enlow; Chief Mktg. & Communications, Pam Fendt; Chief Security & Safety, Mitch Ladewski; Chief Facilities, Steven Dunay; Mgr. Human Resources, Cindy Clenney; Mgr. Collections, Francoise Hack; Assoc. Dir. Exhibitions & Programs, Dwight Currie; Cur. European Art, Dr. Virginia Brilliant; Cur. Modern & Contemporary Art, Dr. Matthew McLendon; Museum Shop Mgr., Rachel Allen; Keeper of Ca' d'zen Mansion, Ron McCarty; Coord. Exhibitions, Donn Roll.
Personnel Profile: Full-Time Paid 126; Part-Time Paid 115; Part-Time Volunteers 725; Interns 10.
Governing Authority: state. Parent Institution: Florida State University. Tax-exempt.
Institution Type/Description: Art Museum & Estate.
Collections: Art Museum: Old Master paintings from the Renaissance through the 19th century including Venetian Baroque and Rubens; American paintings, sculpture, prints, drawings & photographs from 12th to 21st centuries; Asian art; decorative arts; archaeological material from Cyprus & the ancient Mediterranean. Circus Museum: circus memorabilia; wagons; costumes; miniatures; posters. Tibbals Learning Center: a miniature circus. Historic Asolo Theater: late 18th-century Italian theater. Ca'd'Zan Mansion: John & Mable Ringling's furnished 1920s residence; gardens.
Major Exhibits: Icons of Style (T), 10/13-1/14.
Research Fields: Baroque art; circus history; 20th-century art; Chinese ceramics.
Facilities: 70,000-vol. library; archives; gardens; 260-seat theater; classrooms; restaurant. Books, prints, decorative objects, jewelry, furnishings, statuary & other museum-related items for sale.
Activities: guided tours; lectures; gallery talks; education programs; inter-museum loan, permanent, temporary & traveling exhibitions; volunteer & members events; daily theater performances.
Publications: quarterly, Ringling Museum Newsletter; books, The John & Mable Ringling Museum of Art; Museum Once Forgotten: The Rebirth of The John and Mable Ringling Museum; John Ringling, Dreamer, Builder, Collector; The John and Mable Ringling Museum of Art Guide to the Collections; The Circus in Miniature: The Howard Bros. Circus Model; exhibition catalogs; catalogue, The Italian Paintings Before 1800, Italian Collection, Great Paintings from the John & Mable Ringling Museum of Art; catalogue, The Flemish & Dutch Paintings: 1400-1900 Ringling Collection.
Hours & Admission Prices: Daily 10-5:30. Adults $25, seniors 65 & over $20, students & children 6-17 $10; discounts to groups & AAM members; members & children under 6 no charge. Call to confirm. ♿
Attendance: 360,000 (estimated)
Membership: Friend $75; Associate $100; Contributor $175; Sponsor $500; Colleague $1,000; Patron $5,000.

✱ **THE MARIE SELBY BOTANICAL GARDENS, INC.,** 811 S. Palm Ave., Sarasota, FL 34236-7995. Tel.: 941-366-5731. Fax: 941-366-9807.
E-mail: marketing@selby.org
Web Site: www.selby.org
Founded: 1973.
Congressional District: 8
Key Personnel: C.E.O., Tom Buchter; Dir. Mktg., Grace Carlson; Museum Shop Mgr., Amy Sullivan.
Personnel Profile: Full-Time Paid 37; Part-Time Paid 14; Part-Time Volunteers 600; Interns 4.
Governing Authority: nonprofit organization. Tax-exempt: 501(c)(3).
Institution Type/Description: Arboretum & Botanical Garden: original building & gardens of Marie & William Selby.
Collections: herbarium of over 60,000 mounted & accessioned specimens in orchidaceae, gesneriaceae & bromeliaceae; tropical epiphytic plants; live plant collection of 8,000 documented, largely wild-collected epiphytes, including 4,500 orchidaceae.
Research Fields: taxonomical research related to epiphytic plants; canopy biology.
Facilities: 5,000-vol. library of books primarily epiphytic botany & less

extensive in general taxonomic botany, tropical display house; outdoor gardens; 75-seat auditorium. Plants, books & gifts associated with plants for sale.

Activities: guided tours; lectures; films; concerts; arts festivals; docent program or council; formally organized education programs for undergraduate & graduate college students; changing exhibits; elementary school science program.

Publications: annual journal, Selbyana; membership e-newsletter, Tropical Dispatch.

Hours & Admission Prices: Daily 10-5. Adults $17, children 6-11 $6; children 5 & under and members no charge. Closed Christmas. &

Attendance: 200,000 (accurate)

Membership: Gardens Friend $60; Gardens Family $90; Contributing $125; Sustaining $250; Sponsor $500; Stewards of the Earth $1,000.

MOTE MARINE LABORATORY/AQUARIUM, 1600 Ken Thompson Pkwy., Sarasota, FL 34236-1096. Tel.: 941-388-4441, ext. 332. Fax: 941-388-4312. Facebook: Mote Marine Laboratory/Aquarium.

E-mail: info@mote.org
Web Site: www.mote.org
Formerly: Cape Haze Marine Laboratory
Founded: 1955.
Congressional District: 13
Key Personnel: Pres., Dr. Michael P. Crosby; Pres. (V), Sue Stolberg; Chm. Bd., Eugene H. Beckstein; Administrative Dir., Dena J. Smith; Library & Archive Dir., Susan Stover; Cur., Dan Bebak; Archivist, Erin Mahaney; Grants, Ellen Vandernoot; Public Rels., Nadine Slimak; Security, Earl Stockton; Gift Shop Mgr., Joyce Gaffney.
Personnel Profile: Full-Time Paid 192; Part-Time Paid 2; Part-Time Volunteers 1,665; Interns 159.
Volunteer Hours: 220,773
Operating Expenses: 19,200,000
Governing Authority: nonprofit organization. Parent Institution: Mote Marine Laboratory. Tax-exempt: 501(c)(3).
Institution Type/Description: Marine Laboratory Museum.
Collections: marine animals; papers of MML; herbarium; archive collections on the history of scientific development in 20th century US; photographic materials; historical research papers; Archival Collections: documents, records, photographs relating to the environmental history of SW Florida dating 1920s to present.
Major Exhibits: Survivors: Extreme Adaptions, 2/1/14-9/14/14.
Research Fields: marine biology; environmental assessment; aquatic chemistry; coastal engineering; physical oceanography; marine biomedicine & immunology; coastal ecology; marine fisheries; aquaculture; sharks; sea turtles; sea mammals.
Facilities: 26,000-vol. library automated with marine-related research available for inter-library loan; 25,000 sq. ft. exhibit space; 385-seat auditorium; marine educational resource center; field research station; nature center. Marine-related items for sale.
Activities: docent program; films; formal education programs; guided tours; mobile exhibit. Annual Events: Dinners; Lecture Series; Oceanic Evening; Party on the Pass; Sea Turtle Beach Fun Run.
Publications: quarterly magazine, Mote Magazine; Mote Technical Reports; Collected Papers from Mote Marine Laboratory.
Hours & Admission Prices: Daily 10-5. Adults $19, Senior (over 65) $18, children 4-12 $14, children 0-3 & members no charge. &
Attendance: 352,000 (accurate)
Membership: Loggerhead $60; Dolphin Pair $75; Manatee Family $100; Jellyfish Family $150; Seahorse Herd $300; Sea Stars $500; Angelfish $1,000.

RINGLING COLLEGE OF ART AND DESIGN, SELBY GALLERY, (M), 2700 N. Tamiami Trail, Sarasota, FL 34234-5895. Tel.: 941-359-7563. Fax: 941-309-1969.

E-mail: selby@ringling.edu
Web Site: www.ringling.edu/selbygallery; selbygallery.org
Founded: 1986.
Congressional District: 13
Key Personnel: Dir., Kevin Dean; Asst. Dir., Laura Avery; Gallery Asst., Tim Jaeger.
Personnel Profile: Full-Time Paid 3; Part-Time Paid 3; Interns 1.
Governing Authority: private college; nonprofit. Parent Institution: Ringling College of Art and Design, Sarasota. Tax-exempt: 501(c)(3).
Institution Type/Description: Art Museum.
Collections: 20th-century prints.
Major Exhibits: Glass and Charcoal: The Art of Kathleen Elliot and Huguette Despault May (T), 1/6/14-2/12/14; Revelations on the World As It Is: Adrianne Colburn and Christina Seely, 2/21/14-4/5/14; Annual Ringling College Student Exhibition, 4/11/14-4/23/14; Annual Ringling College

Senior Thesis Exhibition, 5/2/14-5/6/14; Annual Community Exhibition, 5/7/14-5/30/14; Frank Rampolla, 7/18/14-8/8/14; Contemporary Graphic Novels & Punk Immortalized, 8/15-9/20/14; Annual Ringling College Faculty Exhibition, 9/26/14-10/24/14.
Facilities: 150-seat auditorium; 3,000 sq. ft. exhibit space.
Activities: lectures; gallery talks; panel discussions; outreach.
Publications: catalogues for select exhibitions.
Hours & Admission Prices: Mon. & Wed.-Sat. 10-4, Tues. 10-7. No charge; donations accepted. Closed school holidays. &
Attendance: 28,000 (accurate)

SARASOTA CLASSIC CAR MUSEUM, 5500 N. Tamiami Trail, Sarasota, FL 34243-2199. Tel.: 941-355-6228.

E-mail: info@sarasotacarmuseum.org
Web Site: www.sarasotacarmuseum.org
Founded: 1953.
Key Personnel: Pres., Martin Godbey.
Personnel Profile: Full-Time Paid 1; Part-Time Paid 12; Part-Time Volunteers 60; Interns 2.
Governing Authority: private; nonprofit organization. Tax-exempt: 501(c)(3).
Institution Type/Description: Classic Car Museum.
Collections: John & Mable Ringling automobile collection; horseless carriages; automotive history from war & post-war periods to the present; vintage, antique & classic automobiles; antique game arcade.
Facilities: library; 60,000 sq. ft. exhibit space. Museum-related items for sale.
Activities: docent program; formal education programs for children; guided tours; rental gallery; study clubs. Museum Sponsors: Antique Car Auction; Soap Box Derby.
Publications: quarterly newsletter, Friends of Classic Cars.
Hours & Admission Prices: Daily 9-6. Adults 13-61 $8.50, seniors 62 & over $7.50, children 6-12 $6.50; discounts to groups; children 5 & under no charge. Closed Christmas. &
Attendance: 42,000 (estimated)
Membership: Children $16; Senior Citizens $18; Adults $20.

THE TURNER MUSEUM AND THOMAS MORAN GALLERIES, 930 N. Tamiami Trail, Suite 807, Sarasota, FL 34236-4070. Tel.: 941-365-1649.

E-mail: turnermuseum@gmail.com
Web Site: www.turnermuseum.org
Founded: 1973.
Congressional District: 13
Key Personnel: C.E.O., Pres. (V) & Dir., Douglass Montrose-Graem; Pres. (V) & Museum Shop Mgr., Isis Graham; Bd. Sec., Katherine Vaggalis.
Personnel Profile: Full-Time Volunteers 3; Part-Time Volunteers 55; Interns 2.
Governing Authority: nonprofit organization. Tax-exempt: 501(c)(3).
Institution Type/Description: Art Center.
Collections: works of J.M.W. Turner & Thomas Moran including engravings, ranging from preliminary etchings to reprints, watercolors, touched proofs & drawings.
Research Fields: J.M.W. Turner; Thomas Moran; Hokusai; Renoir.
Facilities: 1,000-vol. library of material relating to J.M.W. Turner & Thomas Moran, available for inter-library loan. Prints, books & other material relating to the artists for sale.
Activities: guided tours; lectures; gallery talks; live concerts combined with dinners/buffets; TV & radio programs; permanent, temporary & traveling exhibitions; monthly meetings; appraisals. Breakfasts, Luncheons, Duchess' High Teas (English Style) & candle-lit dinners served by appointment.
Publications: catalogue, Turner & Moran; Turner's Cosmic Optimism; Turner's Angels; Turner's Rainbows; Turner's Children; Turner's Powerful Allegories.
Hours & Admission Prices: By appointment. Adults $15; discounts to seniors and AAM & ICOM members. &
Attendance: 100,000 (estimated)
Membership: Children $1; Student $5; Junior $10; Family $25; Senior $50; Associate $100; Life Candidate $250; Business $500; Life $1,000; Corporate $5,000.

Sebring

CHILDREN'S MUSEUM OF THE HIGHLANDS, 219 N. Ridgewood Dr., Sebring, FL 33870-7204. Mailing Address: P.O. Box 1243, Sebring, FL 33871-1243. Tel.: 863-385-5437.

E-mail: linda@childrensmuseumhighlands.com
Web Site: www.childrensmuseumhighlands.com
Founded: 1990.
Congressional District: 16
Key Personnel: Dir., Linda Crowder.
Governing Authority: nonprofit organization.
Institution Type/Description: Children's Museum.

Collections: hands-on exhibits.
Activities: field trips; birthday parties; programs & special events; summer classes.
Hours & Admission Prices: Tues.-Wed. & Fri.-Sat. 10-5, Thurs. 10-8. Admission $3, Thurs. after 5 $1; members no charge.
Attendance: 20,000 (estimated)

HIGHLANDS ART LEAGUE, 351 W. Center Ave., Sebring, FL 33870-3109. Tel.: 863-385-6682. Fax: 863-385-6611.
E-mail: info@highlandsartleague.com
Web Site: highlandsartleague.org/
Formerly: Highland Museum Of The Arts
Founded: 1969.
Key Personnel: Dir., Debbie Kendrick; Pres. (V), Jeri Wohl; Mgr., Susan James.
Personnel Profile: Part-Time Paid 2; Part-Time Volunteers 15.
Governing Authority: private; nonprofit organization. Parent Institution: Highlands Art League, Inc. Tax-exempt: 501(c)(3).
Institution Type/Description: Art Museum: located in the Alan Altvater Civic Complex.
Collections: paintings; photographs; sculpture.
Activities: competitions; receptions; permanent exhibitions.
Publications: newsletter, Artistically Speaking.
Hours & Admission Prices: Mon.-Fri. 10-2. No charge; donations accepted. Closed major holidays.
Attendance: 1,500 (estimated)
Membership: Individual $50; Family $75.

HIGHLANDS HAMMOCK STATE PARK/CIVILIAN CONSERVATION CORPS MUSEUM, 5931 Hammock Rd., Sebring, FL 33872-7408. Tel.: 863-386-6094. Fax: 863-386-6095.
E-mail: dorothy.l.harris@dep.state.fl.us
Web Site: www.floridastateparks.org
Founded: 1994.
Congressional District: 10
Key Personnel: Park Mgr., Steven Dale; Asst. Park Mgr., Brian Pinson; Park Specialist, Dorothy Harris.
Personnel Profile: Part-Time Volunteers 30.
Governing Authority: state. Dept. of Environmental Protection, Div. of Recreation & Parks, 3900 Commonwealth Blvd., Commonwealth Bldg., Tallahassee, FL 32303. Tax-exempt.
Institution Type/Description: Park Museum: housed in 1930s building constructed of heavy native timbers, lumber cut & fabricated on site by the Civilian Conservation Corps.
Collections: CCC memorabilia; photographs; tools; printed documents; life in the 1930s & 1940s; Florida land boom; bust; Great Depression; civilian conservation corps; World War II.
Facilities: nature trails; bicycle paths; campgrounds.
Activities: permanent, nature & CCC era exhibitions; guided tours; seasonal campfire programs; bicycle rentals. Annual Event: CCC Era Festival in November.
Publications: interpretive booklet for the nature trails; checklist of birds & other vertebrates; technical listing of plants available for the serious student.
Hours & Admission Prices: Daily 8-sundown. Park entrance fee: $6 per vehicle (2-8 passengers), $4 per vehicle (1 passenger), pedestrians & bicycles $2. Call ranger station for tour bus rates. Entrance to museum is included in regular entrance fees. Museum hours: daily 9-4. Annual State Park Entrance Pass available. &
Attendance: 15,000 (estimated)

SEBRING HISTORICAL SOCIETY, 321 W. Center Ave., Sebring, FL 33870. Tel.: 863-471-2522.
Institution Type/Description: Historical Society Museum.
Collections: local history; books; photographs.
Facilities: archives.
Hours & Admission Prices: Mon.-Thurs. 9:30-5

Seminole

PANAMA CANAL MUSEUM, 7985 113th St., Ste. 100, Seminole, FL 33772-4785. Tel.: 727-394-9338. Fax: 727-394-2737.
E-mail: pankee@aol.com
Web Site: www.panamacanalmuseum.org
Key Personnel: Contact, Elizabeth Neily
Institution Type/Description: History Museum.
Collections: Panama Canal history; documents; photographs.
Hours & Admission Prices: Mon.-Sat. 10-4; other times by appointment. Admission $3.

St. Marks

ST. MARKS NATIONAL WILDLIFE REFUGE VISITOR CENTER, 1255 Lighthouse Rd., St. Marks, FL 32355. Mailing Address: P.O. Box 68, St. Marks, FL 32355-0068. Tel.: 850-925-6121. Fax: 850-925-6930. Facebook: SMSVNWRS.
E-mail: saintmarks@fws.gov
Web Site: www.fws.gov/saintmarks
Founded: 1931.
Congressional District: 2
Key Personnel: Refuge Mgr., Terry Peacock; Chm. (V), David Moody; Pres. (V), Betsy Kellenberger; Museum Shop Mgr., Betty Hamilton.
Personnel Profile: Full-Time Paid 23; Part-Time Volunteers 150; Interns 2.
Governing Authority: Parent Institution: U.S. Dept. of Interior Fish and Wildlife Service. Tax-exempt.
Institution Type/Description: Wildlife Refuge.
Collections: wildlife & their habitat. Historic Building: St. Marks Lighthouse, built 1832.
Research Fields: fire; longleaf pine; endangered species; whooping cranes; monarch butterflies.
Facilities: nature trails; visitor center; observation towers.
Activities: wildlife observation; educational programs; tours.
Publications: The Eagle's Eye.
Hours & Admission Prices: Refuge: daily sunrise to sunset. Visitor Center: Mon.-Fri. 8-4, Sat.-Sun. 10-5. Refuge: $5 per car; Federal Recreation Fee passes accepted. Visitor Center: no charge. Closed Thanksgiving; Christmas. &
Attendance: 275,000 (estimated)

SAN MARCOS DE APALACHE HISTORIC STATE PARK, 148 Old Fort Rd., St. Marks, FL 32355-0027. Mailing Address: 3600 Indian Mounds Rd., Tallahassee, FL 32303-2300. Tel.: 850-922-6007. Fax: 850-488-0366.
Web Site: www.floridastateparks.org
Founded: 1964.
Congressional District: 2
Key Personnel: Park Mgr., Rob Lacy; Administrative Asst., Leeanne Zimmerman.
Personnel Profile: Full-Time Paid 1.
Governing Authority: state. Dept. of Environmental Protection, Div. of Recreation and Parks, 3900 Commonwealth Blvd., Tallahassee, FL 32303. Tax-exempt.
Institution Type/Description: State History Museum: museum built on the foundation of the Civil War-era Marine Hospital.
Collections: archaeology; military; Indian artifacts; fort ruins.
Facilities: military cemetery.
Activities: interpretive video; guided tours; self-guided trail.
Publications: site specific brochure.
Hours & Admission Prices: Thurs.-Mon. 9-5. Adults $2; children under 6 no charge. Closed New Year's Day; Thanksgiving; Christmas. &
Attendance: 15,000 (estimated)

Stuart

FLORIDA OCEANOGRAPHIC COASTAL CENTER, 890 N.E. Ocean Blvd., Hutchinson Island, Stuart, FL 34996-1627. Tel.: 772-225-0505.
Web Site: www.floridaocean.org
Formerly: Coastal Science Center at Hutchinson Island
Founded: 1964.
Key Personnel: Exec. Dir., Mark Perry.
Governing Authority: Tax-exempt: 501(c)(3).
Institution Type/Description: Maritime Museum.
Collections: marine artifacts; aquariums; replica of Saballariid reef.
Research Fields: oyster reef restoration & monitoring; habitat restoration.
Facilities: Museum-related items for sale.
Activities: videos; education; research; restoration; advocacy.
Hours & Admission Prices: Mon.-Sat. 10-5, Sun. 12-4. Adults $12, children 3-12 $6; discounts to active military; children under 3 & members no charge. Closed New Year's Day; Easter; Thanksgiving; Christmas. &
Attendance: 50,000 (estimated)
Membership: Individual $75; Family $125; Patron $250; Benefactor $500; Business Partner & President's Club $1,000.

HOUSE OF REFUGE MUSEUM, (M), 301 S.E. MacArthur Blvd., Stuart, FL 34996. Tel.: 772-225-1875. Fax: 772-225-2333.
E-mail: info@elliottmuseumfl.org
Web Site: elliottmuseumfl.org/houseofrefuge/index.html
Formerly: Gilbert's Bar House of Refuge

Founded: 1875.
Congressional District: 12
Key Personnel: Keeper, Barbara Dewhirst.
Personnel Profile: Full-Time Paid 3; Part-Time Paid 2; Part-Time Volunteers 20.
Governing Authority: society. Parent Institution: Historical Society of Martin County, Stuart. Tax-exempt.
Institution Type/Description: Historic House Museum: four room furnished house representing the period 1890-1904. National Register of Historic Places.
Collections: archives; life saving equipment.
Facilities: Gift items for sale.
Activities: special events.
Publications: History of Martin County.
Hours & Admission Prices: Mon.-Sat. 10-4, Sun. 1-4. Adults $6, children 5-12 $3; children under 5 no charge. Closed major holidays. &
Attendance: 17,000
Membership: Individual $35; Family $50; Sponsor $75; Patron $125; Car Club $150.

STUART HERITAGE MUSEUM, 161 S.W. Flagler Ave., Stuart, FL 34994-2139. Tel.: 772-220-4600. Fax: 772-781-3416.
E-mail: stuartheritage1@yahoo.com
Founded: 1988.
Key Personnel: Exec. Dir., Mary Walton Jones; Mgr., Betty Hardwick.
Personnel Profile: Part-Time Paid 1; Part-Time Volunteers 12.
Governing Authority: Tax-exempt: 501(c)(3).
Institution Type/Description: Local County History Museum: built in 1901.
Collections: 10,000 period artifacts including photographs, furnishings, period clothing, county artifacts, maps & books.
Activities: monthly meeting.
Publications: monthly members newsletter; Martin County, Our Heritage; Stuart Heritage Museum Activity Book; Stuart Heritage Cookbook; Stuart Heritage Commemorative Album.
Hours & Admission Prices: Daily 10-3. No charge; donations accepted. Closed New Year's Day; Thanksgiving; Christmas. &
Attendance: 4,995 (accurate)
Membership: Single $10; Family $20; Sponsor $50; Sustaining $100; Lifetime $500.

Tallahassee

ALFRED B. MACLAY GARDENS STATE PARK, 3540 Thomasville Rd., Tallahassee, FL 32309-3413. Tel.: 850-487-4556 & 4115. Fax: 850-487-8808.
E-mail: ginger.nichols@dep.state.fl.us
Web Site: www.floridastateparks.org/maclay
Founded: 1953.
Congressional District: 2
Key Personnel: Park Mgr., Beth Weidner.
Personnel Profile: Full-Time Paid 11; Part-Time Paid 5; Part-Time Volunteers 207; Interns 1.
Governing Authority: state. Parent Institution: Florida Dept. of Environmental Protection. Subsidiary Institution: Division of Recreation & Parks. Tax-exempt.
Institution Type/Description: Historic House Museum & Gardens: c.1909 Maclay House.
Collections: gardens containing plant life native to the Florida Panhandle; camellias; azaleas; dogwoods; mixed hardwood & pine forest; historic house.
Facilities: library; 28 acres of ornamental gardens; 17-acre recreation area; nature trail; rental picnic pavilion; banquet facilities; 2.8 mile hiking, bicycle & horse trail. Museum-related items for sale.
Activities: guided tours; lectures; organized education programs; docent program. Gardens' Sponsors: Tour of Gardens; Moon Over Maclay; Jazz Concert.
Publications: Maclay Gardens: a seasonal newsletter provided by Friends of Maclay Gardens.
Hours & Admission Prices: Maclay House & Gardens: daily 9-5. Park: 8am to sundown. House & Gardens: Jan.-April adults $6, children under 12 $3. Park & Gardens: May-Dec. $6 per car (8 people); park fee includes gardens; bicyclers, walkers, extra persons in vehicle $2 per person; members 1st Sat. of each month no charge. &
Attendance: 159,000 (accurate)
Membership: Gardener $25; Garden Family $50; Patron $100; Sponsor $250; Donor $500.

CHALLENGER LEARNING CENTER, 200 S. Duval St., Tallahassee, FL 32301. Tel.: 850645-7796. Fax: 850-645-7784.
Institution Type/Description: Science Center.
Collections: space science; hands-on exhibitions.
Facilities: IMAX 3D theatre.
Activities: educational programs; birthday parties; camps; space mission simulator; special events.
Hours & Admission Prices: Call for hours & admission prices.

FLORIDA ASSOCIATION OF MUSEUMS, 459 Cedar Hill Rd, Tallahassee, FL 32312-1046. Mailing Address: P.O. Box 10951, Tallahassee, FL 32302-2951. Tel.: 850-222-6028. Fax: 850-222-6112.
E-mail: Fam@flamuseums.org
Web Site: www.flamuseums.org
Founded: 1986.
Congressional District: 2
Key Personnel: Exec. Dir., Malinda Horton; Pres. (V), Daniel E. Stetson.
Personnel Profile: Full-Time Paid 1; Part-Time Paid 2.
Governing Authority: nonprofit state association. Tax-exempt.
Institution Type/Description: State Museum Association.
Publications: FAM Annual Directory.
Hours & Admission Prices: Call for hours.
Membership: Individual $30; Vendor & Affiliate $75; Institutional $250 or .1% of annual operating budget.

* **FLORIDA STATE UNIVERSITY MUSEUM OF FINE ARTS, (M),** 530 W. Call St., 250 Fine Arts Bldg., Tallahassee, FL 32306-1140. Mailing Address: P.O. Box 3061140, Tallahassee, FL 32306-1140. Tel.: 850-644-6836 & 1254. Fax: 850-644-7229.
E-mail: apalladinocraig@fsu.edu
Web Site: www.mofa.fsu.edu
Founded: 1950.
Congressional District: 2
Key Personnel: C.E.O. & Dir., Allys Palladino-Craig.
Personnel Profile: Full-Time Paid 5; Part-Time Paid 4; Part-Time Volunteers 30; Interns 6.
Volunteer Hours: 4,946
Operating Expenses: 468,857
Operating Income: 468,857
Governing Authority: state. Parent Institution: Florida State University. Subsidiary Institution: College of Visual Art, Theatre & Dance. Tax-exempt.
Institution Type/Description: Art Museum.
Collections: The Victor & Mary Carter Collection of Peruvian Art, European Art, Asian Art & contemporary art; glass; graphics; photography; the Mary Lewis American Basketry Collection.
Research Fields: contemporary art & art historical topics, subject to guest curatorial publications, Thematic.
Facilities: lecture room; lounge; sculpture garden.
Activities: lectures; gallery talks; formally organized education programs for graduate students; inter-museum loan, temporary & traveling exhibitions; performing arts events; theatre; music & dance; community outreach program.
Publications: brochures; exhibitions catalogues; art history journal, Athanor; curatorial publications, Thematic.
Hours & Admission Prices: May-Aug. Mon.-Fri. 9-4; Sept.-April Mon.-Fri. 9-4, Sat.-Sun. 1-4. No charge; donations accepted. Closed university holidays. &
Attendance: 65,112 (accurate)
Membership: Artist's League, Student & Senior Citizen $15; Individual $25; Family $35; Supporting $50; Century Club $100; Business $250; Westcott Society $500; Benefactor $1,000; Angel $1,000 & up.

FLORIDA'S HISTORIC CAPITOL, (M), 400 S. Monroe St., Tallahassee, FL 32399-6536. Tel.: 850-487-1902. Fax: 850-410-2233.
E-mail: info@flhistoriccapitol.gov
Web Site: www.flhistoriccapitol.gov
Key Personnel: Cur., John B. Phelps.
Institution Type/Description: History Museum.
Collections: local history & culture relating to Florida's legislative history; oral papers; photographs.
Hours & Admission Prices: Mon.-Fri. 9-4:30, Sat. 10-4:30, Sun. & holidays 12-4:30. No charge; donations accepted. Closed Thanksgiving; Christmas.

FOSTER-TANNER FINE ARTS GALLERY, Florida A&M University, Foster Tanner Arts Bldg., 1630 Pinder Dr., Tallahassee, FL 32307. Tel.: 850-599-3161. Fax: 850-599-8761.
E-mail: harriswiltsher@famu.edu
Key Personnel: Dir., Harris Wiltsher
Institution Type/Description: Art Museum.
Collections: African American & Native American artists.
Hours & Admission Prices: Jan. 3-Dec. 13 Mon.-Fri. 11-5. No charge; donations accepted.

GOODWOOD MUSEUM AND GARDENS, (M), 1600 Miccosukee Rd., Tallahassee, FL 32308-5166. Tel.: 850-877-4202. Fax: 850-877-3090.
E-mail: goodwood@goodwoodmuseum.org
Web Site: www.goodwoodmuseum.org
Founded: 1990.
Congressional District: 2
Key Personnel: Exec. Dir., Andy McLeod; Chm. (V), Rick Barnett
Institution Type/Description: Historic House: built c.1840.
Collections: period furnishings; personal artifacts.
Facilities: 16 acres. Museum-related items for sale.
Hours & Admission Prices: House: Mon.-Fri. 10-4, Sat. 10-2; groups by appointment. Gardens: Mon.-Fri. 9-5, Sat. 10-2. House: admission $6; children under 3 & members no charge. Gardens: no charge. Closed New Year's Eve & Day; Thanksgiving; Christmas Eve, Day & week. &
Membership: Individual $35; Family $55; Contributor $100; Sponsor $250; Benefactor $500; Patron $1,000.

*** KNOTT HOUSE MUSEUM, (M),** 301 E. Park Ave., Tallahassee, FL 32301-1513. Tel.: 850-922-2459 & 245-6400. Fax: 850-413-7261. Facebook: Knott House Museum.
E-mail: beatrice.cotellis@dos.myflorida.com
Web Site: www.museumoffloridahistory.com/about/sites/knott/about.cfm
Founded: 1992.
Congressional District: 2
Key Personnel: Site Mgr., Beatrice Cotellis; Educator, Ana Perez; Educator, Sarah Shaw.
Personnel Profile: Full-Time Paid 1; Part-Time Paid 2; Part-Time Volunteers 30; Interns 1.
Governing Authority: state. Parent Institution: Museum of Florida History, R.A. Gray Bldg., 500 S. Bronough St., Tallahassee, FL 32399-0250. Tax-exempt: 501(c)(3).
Institution Type/Description: History Museum: housed in 1843 colonial revival house.
Collections: original furnishings; personal effects; books; cards; poems written by Mrs. Knott; documents, receipts & correspondence interpreting life in Tallahassee during 1928-1941.
Major Exhibits: Early African American Influences on the Knott House, 2/13-2/15; Knott House in Territorial Period, 9/13-6/14.
Research Fields: women's history; 1930s in Florida; 1930 Florida gardens & leisure activities; Florida politics; African American history.
Facilities: classroom; 380 sq. ft. exhibit space. Museum-related items for sale.
Activities: art event; concerts; films; formal education programs; guided tours; participatory & temporary exhibitions; theater; teenage workshop; poetry readings; swing dance; walking tours; scout programs; lectures; adult poetry workshops. Museum Sponsors: Emancipation Day.
Publications: Tales of Tallahassee, Twice Told and Untold.
Hours & Admission Prices: Sept.-July Museum: Tues.-Sat. 10-4. Tours: Wed.-Fri. 1, 2 & 3, Sat. on the hour from 10-3. Museum: no charge; donations accepted. Group Tours: $1 per person. Closed Thanksgiving; Christmas. &
Attendance: 5,000 (estimated)
Membership: Student $15; Senior & Teacher $25; Individual $35; Family $60.

LAKE JACKSON MOUNDS ARCHAEOLOGICAL STATE PARK, 3600 Indian Mounds Rd., Tallahassee, FL 32303-2300. Tel.: 850-922-6007. Fax. 850-488-0366.
Web Site: www.floridastateparks.org
Founded: 1970.
Congressional District: 2
Key Personnel: Park Mgr., Rob Lacy; Administrative Asst., Leeanne Zimmerman.
Personnel Profile: Full-Time Paid 1.
Governing Authority: state; nonprofit; Parent Institution: State of Florida, Dept. of Environmental Protection. Division of Recreation & Parks. Tax-exempt.
Institution Type/Description: Archaeological Site: 1200-1500 ceremonial center of the Fort Walton period.
Collections: local history & culture; 1800s grist mill.

Facilities: library of printed references to artifact types, animal & plant species available to the public.
Activities: guided tours; self-guided trail; lectures; Powerpoint.
Publications: site specific brochure.
Hours & Admission Prices: Daily 8-sunset. Cars: $3; pedestrians & bikers $2; donations accepted. &
Attendance: 16,464 (estimated)

LEMOYNE CENTER FOR THE VISUAL ARTS, 125 N. Gadsden St., Tallahassee, FL 32301-1507. Tel.: 850-222-8800. Fax: 850-224-2714.
E-mail: director@lemoyne.org
Web Site: www.lemoyne.org
Formerly: LeMoyne Art Foundation, Inc.
Founded: 1964.
Congressional District: 2
Key Personnel: Exec. Dir., Hillary Brett; Pres. (V), Kelly Dozier; Vice Pres., Eva B. Armstrong; Volunteer Coord. & Museum Shop Mgr., Betty Bayes Lessinger; Dir. Education, Anna Myers; Education Coord., Amanda Wilke; Education Coord., Jennifer Infinger; Cur., Lesley Marchessault; Events Coord., Sheri Sanderson.
Personnel Profile: Full-Time Paid 3; Part-Time Paid 6; Part-Time Volunteers 50; Interns 2.
Governing Authority: nonprofit organization. Tax-exempt: 501(c)(3).
Institution Type/Description: Art Center: housed in 1852 wooden structure used as hospital during Civil War.
Collections: contemporary art; sculpture garden.
Research Fields: history; collect books & prints related to LeMoyne.
Facilities: educational facility. Gift items for sale.
Activities: guided tours; lectures; films; gallery talks; rental gallery; TV & radio programs; formally organized education programs; docent program or council, inter-museum loan, permanent, temporary & traveling exhibitions.
Publications: newsletter to members; exhibition catalogs.
Hours & Admission Prices: Tues.-Sat. 10-5. Adults $2; discounts to AAM & ICOM members; students & members no charge. Fees charged only for special exhibits. Closed major holidays. &
Attendance: 115,000 (estimated)
Membership: Regular: Individual $30; Family $50; Patron $120; Donor $250; Fellow $500; Benefactor $1,000; Life $5,000. Business: Patron $250; Contributing $500; Donor $1,000; Fellow $2,500; Benefactor $5,000; Life $10,000.

THE MARY BROGAN MUSEUM OF ART AND SCIENCE, 350 S. Duval St., Tallahassee, FL 32301-1711. Tel.: 850-513-0700, ext. 236. Fax: 850-513-0143.
E-mail: thanson@thebrogan.org
Web Site: www.thebrogan.org
Formerly: Odyssey Science Center/Museum of Art/Tallahassee
Founded: 1990.
Congressional District: 2
Key Personnel: C.E.O., Chucha Barber; C.O.O., Trish Hanson; Mgr. Public Rels., Kallen Lunt; Dir. Art Collections & Registrar, Michelle Smith Grindberg.
Personnel Profile: Full-Time Paid 7; Part-Time Paid 10; Part-Time Volunteers 200; Interns 8.
Governing Authority: Tax-exempt.
Institution Type/Description: Art & Science Museum.
Collections: visual arts; hands-on science exhibits.
Facilities: classrooms; 6,500 sq. ft. exhibit space. Museum-related items for sale.
Activities: guided tours; lectures; docent program; participatory & traveling exhibitions; field trips; birthday parties; summer camp; rental facilities.
Publications: catalogues; newsletters; annual report.
Hours & Admission Prices: Mon.-Sat. 10-5, Sun. 1-5. Adults $10, children 3-17, students, senior citizens 60 & over and military $5; discounts to groups, AAA & AAM members; children 2 & under, ASTC & museum members no charge. &
Attendance: 85,000 (accurate)
Membership: College $25; Individual $30; Membership for Two $40; Grandparents $50, Family $60, Plaza Passport $150, Visionary $350, Champion $1,200. Smithsonian Benefits can be added to any level for $20.

MISSION SAN LUIS, 2100 W. Tennessee St., Tallahassee, FL 32304-3119. Tel.: 850-245-6406.
E-mail: info@missionsanluis.org
Web Site: www.missionsanluis.org
Founded: 1983.
Congressional District: 2

Key Personnel: Dir., Robert S. Blount; Chm. (V), Vern Williams; Museum Shop Mgr., Janelle Willingham.
Personnel Profile: Full-Time Paid 23; Part-Time Paid 2; Part-Time Volunteers 70; Interns 19.
Volunteer Hours: 9,157
Operating Expenses: 1,544,037
Operating Income: 10,637,414
Governing Authority: Parent Institution: Florida Dept. of State. Subsidiary Institution: Friends of Mission San Luis, Inc. Tax-exempt.
Institution Type/Description: History Museum.
Collections: re-created Apalachee Indian village & Spanish structures; period furnishings; personal artifacts; Spanish Colonial religious art.
Research Fields: 1656-1704 history & archaeology of Mission San Luis.
Facilities: classrooms; theater; rental facilities; nature trail; visitor center. Museum-related items for sale.
Activities: orientation video; education programs & activities; demonstrations; rental facilities.
Publications: newsletter, La Pluma.
Hours & Admission Prices: Tues.-Sun. 10-4. Adults $5, seniors 65 & over $3, children 6-17 $2; children under six, active military & members no charge. Closed New Year's Day; Easter; Independence Day; Thanksgiving; Christmas Eve & Day. &
Attendance: 42,414 (accurate)
Membership: Student $10; Senior & Teacher $20; Individual $25; Senior Family & Teacher Family $30; Family & Household $40; Donor $100; Patron $300; Benefactor $500.

***   MUSEUM OF FLORIDA HISTORY, (M),** 500 S. Bronough St., Tallahassee, FL 32399-0250. Tel.: 850-245-6400. Fax: 850-245-6433. Facebook: Museum of Florida History.
E-mail: jeana.brunson@dos.myflorida.com
Web Site: www.museumoffloridahistory.com
Founded: 1967.
Congressional District: 2
Key Personnel: Bureau Chief, Dr. Jeana Brunson; Cur. Historic Sites & Education, K.C. Smith; Cur. Research & Collections, Lea Ellen Thornton; Cur. Design & Fabrication, Drew Ericson; Devel. & Financial, Elyse Cornelison; Museum Shop Mgr., Paige Breshike.
Personnel Profile: Full-Time Paid 22; Part-Time Paid 8; Part-Time Volunteers 100.
Governing Authority: state. Parent Institution: Florida Dept. of State. Tax-exempt: 501(c)(3).
Institution Type/Description: History Museum.
Collections: historical material reflecting the history & culture of the State, including archaeological materials, Spanish numismatics & maritime objects; artifacts from citrus industry & steamboat era; early tourist memorabilia. Historic Building: 1843 restored Knott House Museum.
Research Fields: museology; Florida history; archaeology; Florida government.
Facilities: 49,000 sq. ft. exhibit space; 240-seat auditorium; cafe. Museum-related gifts for sale.
Activities: permanent, temporary & traveling exhibitions; formally organized education programs; statewide services; traveling exhibit program rental. Museum Sponsors: State History Fair.
Publications: newsletter, Historically Speaking; site brochures, education guide; Heritage Education program guides.
Hours & Admission Prices: Mon.-Fri. 9-4:30, third Thurs. of month 9-8, Sat. 10-4:30, Sun. & holidays 12-4:30. No charge; donations accepted. Closed Thanksgiving; Christmas. &
Attendance: 62,862 (accurate)
Membership: Student $15; Senior & Teacher $25; Individual $35; Senior Family $50; Family $60.

RILEY HOUSE MUSEUM OF AFRICAN AMERICAN HISTORY & CULTURE, (M), 419 E. Jefferson St., Tallahassee, FL 32301-1817. Tel.: 850-681-7881. Fax: 850-386-4368.
E-mail: staff@rileymuseum.org
Web Site: www.rileymuseum.org; www.faahph.com
Founded: 1996.
Congressional District: 2
Key Personnel: Exec. Dir. & C.E.O., Althemese Barnes; Chm. (V), Dr. David Jackson; Chm., Gwendolyn Spencer; Program Coord., Maggie Lewis Butler; Mktg., Publicity Dir. & Museum Shop Mgr., Marion McGee.
Personnel Profile: Full-Time Paid 3; Full-Time Volunteers 2; Part-Time Paid 2; Part-Time Volunteers 38; Interns 2.
Governing Authority: private; nonprofit organization. Parent Institution: John G. Riley Foundation Inc. Tax-exempt.

Institution Type/Description: Historic House Museum: housed in home of John Gilmore Riley, first African American principal of Lincoln High School in Tallahassee, Florida; built in 1890.
Collections: grassroots African American Florida history; black abolitionists papers; American Missionary Society papers; Thelma Thurston Gorham collection; Black history books.
Research Fields: African American education in Florida 1865-1968; tenant farming & midwifery history in northwest Florida 1865-1950s; African American burial sites; Black abolitionist & American Missionary Society.
Facilities: 900 sq. ft. exhibit space; visitor center. Museum-related items for sale.
Activities: formal educational programs; guided tours; loan, temporary, traveling & participatory exhibitions; school loan service; training programs for professional museum workers; broadcast & cable programs. Museum Sponsors: Members Reception; Statewide Conference; Emancipation Activity; Holiday Event.
Publications: quarterly newsletter, Out of the Past... A Noble Witness; Paths To Freedom; Black education teacher's guide and DVD, Milestone Memories.
Hours & Admission Prices: Jan. 3-Dec. 21 Mon.-Thurs. 10-4, Fri.-Sat. 10-2; groups by appointment. Adults $2, members $1.50. Closed New Year's Day; Martin Luther King, Jr. Day; Easter weekend; Memorial Day; Independence Day; Veterans Day; Thanksgiving & day before. &
Attendance: 8,002 (accurate)
Membership: Individual $30; Contributor $100; Benefactor $250; Founder $1,000; Corporate $2,500.

TALLAHASSEE AUTOMOBILE MUSEUM, 6800 Mahan Dr., Tallahassee, FL 32308-1402. Mailing Address: P.O. Box 120, Hosford, FL 32334-0120. Tel.: 850-942-0137. Fax: 850-576-8500.
Web Site: www.tacm.com
Formerly: Antique Car Museum
Founded: 1996.
Governing Authority: Tax-exempt.
Institution Type/Description: Automobile Museum.
Collections: period automobiles; collectibles; Smoky Mountain trains; Indian artifacts; outboard motors; fishing lures; cash registers; calculators; golf memorabilia; baby rattles; beaded purses.
Hours & Admission Prices: Mon.-Fri. 8-5, Sat. 10-5, Sun. 12-5; Thanksgiving & Christmas by appointment. Adults $16, 2 adults $13.50, students $10.75, children 5-9 $7.50; discounts to groups. Train Museum: $6 per person. &
Attendance: 7,509 (accurate)

***   TALLAHASSEE MUSEUM, (M),** 3945 Museum Dr., Tallahassee, FL 32310-6325. Tel.: 850-575-8684. Fax: 850-574-8243.
E-mail: rdaws@tallahasseemuseum.org
Web Site: www.tallahasseemuseum.org
Formerly: Tallahassee Museum of History and Natural Science
Founded: 1957.
Congressional District: 2
Key Personnel: C.E.O. & Exec. Dir., Russell S. Daws; Pres. (V), Michael Stehlik; Facilities Mgr., Mike Sullivan; Dir. Finance & Museum Shop Mgr., Theresa Davis; Chief Cur., Linda Deaton; Cur. Animals, Mike Jones; Education Dir., Jennifer Golden.
Personnel Profile: Full-Time Paid 16; Part-Time Paid 63; Part-Time Volunteers 545; Interns 8.
Governing Authority: nonprofit organization. Tax-exempt: 501(c)(3).
Institution Type/Description: General Museum.
Collections: 19th-century regional history & natural history; agricultural implements; household artifacts; textiles; clothing; native Florida wildlife. Historic Buildings: 1840s plantation house; African-American church & schoolhouse; caboose; 1880s farmstead.
Research Fields: regional history related to restored buildings; participants in Red Wolf captive breeding program; Florida Panther Recovery Program.
Facilities: natural habitat zoo; nature trails; discovery center; food service; meeting room. Museum-related items for sale.
Activities: formally organized educational programs; traveling, permanent & temporary exhibitions; school loan service; preschool; 19th-century crafts & skills demonstrations. Museum Sponsors: Arts & Crafts Festival; Music Festivals; Folklife Festival; summer camp.
Publications: bimonthly, Museum News.
Hours & Admission Prices: Mon.-Sat. 9-5, Sun. 12:30-5. Adults $9, seniors & college students $8.50, children $6; discounts to AAM & FAM members; members & children 3 & under no charge. Closed New Year's Day; Thanksgiving; Christmas Eve & Day. &
Attendance: 100,000 (accurate)
Membership: Student $25; Basic $40; Dual $50; Family $60; Contributing $100-$174; Habitat Club $175-$349.

Tampa

AMERICAN VICTORY MARINERS MEMORIAL & MUSEUM SHIP, 705 Channelside Dr., Tampa, FL 33602-5600. Tel.: 813-228-8766.
E-mail: marketing@americanvictory.org
Web Site: www.americanvictory.org
Founded: 1945.
Key Personnel: Dir., Bill Kuzmick
Institution Type/Description: Maritime History Museum: housed on a 1940s-era merchant cargo ship.
Collections: maritime culture & industry; photographs; medals & documents; personal artifacts; navigational equipment; weaponry; hands-on exhibits.
Activities: ship tours; hands-on exhibits; memberships; special events; private event rentals; educational programs.
Hours & Admission Prices: Mon. 12-5; Tues.-Sat. 10-5, Sun. 12-5. Adults $10, veterans & seniors 65+ $8, children $4; children 3 & under no charge.
Attendance: 60,000 (estimated)
Membership: $25-$250

BIG CAT RESCUE, 12802 Easy St., Tampa, FL 33625-3702. Tel.: 813-920-4130. Fax: 866-571-4523.
E-mail: info@bigcatrescue.org
Web Site: www.bigcatrescue.org
Institution Type/Description: Wildlife Refuge.
Collections: over 150 big cats including jaguars, lions, tigers, cougars, panthers; bobcats, & leopards.
Activities: educational programs. Annual Event: Fur Ball in October.
Hours & Admission Prices: Call for hours & admission prices.

CRACKER COUNTRY, (M), 4800 N. Hwy. 301, Tampa, FL 33680. Mailing Address: P.O. Box 11766, Tampa, FL 33680-1766. Tel.: 813-627-4225. Fax: 813-740-3518. TDD: 813-621-7821.
Web Site: www.crackercountry.org
Founded: 1979.
Congressional District: 9
Key Personnel: Mgr., Cindy Horton; Museum Programs Supvr., Jennifer Becker; Facilities Maintenance Cur., Ron Tarlton; Cultural Commerce Supvr., Linda Mahoney.
Personnel Profile: Full Time Paid 4; Part-Time Volunteers 150; Interns 3
Governing Authority: state. Parent Institution: Florida State Fair Authority. Tax-exempt.
Institution Type/Description: Historic Village & Museum: 12 turn-of-the-century buildings, 1870-1912, located on the Florida State Fairgrounds in a section known as Cracker Country.
Collections: all of the buildings are furnished with period furnishings; old Lionel train collection in Depot; collection of oil portraits of Florida's Governors beginning with Andrew Jackson; rural Florida late 1800s, early 1900s.
Facilities: Museum-related items for sale.
Activities: education programs for children; grade school tours; docent program. Annual Events: Florida State Fair in February; public openings selected Saturdays throughout the year.
Publications: brochure, Cracker Country; monthly newsletter, Traditions.
Hours & Admission Prices: Grade School Tours: March-May & Sept.-Dec. Tues.-Fri. Admission $5; school tours: adults & students $7; Saturday Public Openings: adults $7, children 6-11 $6, seniors & children 5 & under no charge.
Attendance: 300,000 (estimated)

FLORIDA AQUARIUM, 701 Channelside Dr., Tampa, FL 33602-5614. Tel.: 813-273-4000.
Web Site: www.flaquarium.org
Institution Type/Description: Aquarium.
Collections: over 20,000 aquatic plants & animals.
Facilities: cafe. Museum-related items for sale.
Activities: Eco-tour dolphin encounter; boat rides.
Hours & Admission Prices: Daily 9:30-5. Adults $21.95, seniors 60 & over $18.95, children under 12 $16.95; children 2 & under no charge. Closed Thanksgiving; Christmas.

FLORIDA MUSEUM OF PHOTOGRAPHIC ARTS, 400 N. Ashley Dr., C200, Tampa, FL 33602. Tel.: 813-221-2222.
E-mail: museummanager@fmopa.org
Web Site: www.fmopa.org
Formerly: Tampa Gallery of Photographic Art
Founded: 2001.
Congressional District: 11
Key Personnel: Dir., Jane Simon; Chm., Doug Bartley; Chm. (V), Carol Gaynor; Museum Shop Mgr., Natalie Sparkman.
Personnel Profile: Full-Time Paid 2; Part-Time Paid 3; Part-Time Volunteers 50; Interns 8.
Governing Authority: Tax-exempt.
Institution Type/Description: Art Museum.
Collections: works by national & international photographic artists.
Major Exhibits: Tampa's Past, 11/13-2/14; David Hilliard, 4/3/14-6/15/14; Permanent Collection, 6/19/14-8/17/14; New Visions, 8/21/14-11/2/14.
Activities: children's literacy through photography program; photography classes & workshops.
Hours & Admission Prices: Tues.-Thurs. & Sat. 10-5, Fri. 10-8, Sun. 12-5. Adults $10, students, military & seniors $8; discounts to local educational institutions & NARM members; members no charge. Closed New Year's Day; Independence Day; Thanksgiving; Christmas.
Attendance: 7,500 (estimated)
Membership: Student & Senior $25; Individual $45; Family $85; Founder $250; Silver $500; Gold $1,000; Platinum $2,500.

GLAZER CHILDREN'S MUSEUM, 110 W. Gasparilla Plaza, Tampa, FL 33602. Tel.: 813-443-3861. Fax: 813-443-3841. Facebook: Glazer Children's Museum.
E-mail: info@glazermuseum.org
Web Site: www.glazermuseum.org
Founded: 2010.
Key Personnel: Pres. & C.E.O., Al Najjar; Mgr. Mktg. & Communications, MaryJane Craig
Institution Type/Description: Children's Museum.
Collections: hands-on exhibits.
Activities: special events; rental facilities; birthday parties.
Hours & Admission Prices: Mon.-Fri. 10-5, Sat. 10-6, Sun. 1-6. Adults $15, military & seniors $12.50, children $9.50; children under one no charge. Closed New Year's Day; Easter; Thanksgiving; Christmas.
Membership: Family $100; Premium $150; Deluxe $300.

✽ **HENRY B. PLANT MUSEUM, (M),** 401 W. Kennedy Blvd., Tampa, FL 33606-1450. Tel.: 813-258-7301. Fax: 813-258-7272.
E-mail: czinober@ut.edu
Web Site: www.plantmuseum.com
Founded: 1933.
Congressional District: 7
Key Personnel: C.E.O., Cynthia Gandee Zinober; Cur. & Registrar, Susan Carter; Operations & Membership, Scott Waltz; Museum Store Mgr., Jenny Rutz; Museum Rels., Sally Shifke; Curatorial Asst., Nora Armstrong; Cur. Education, Heather Brown.
Personnel Profile: Full-Time Paid 7; Part-Time Paid 2; Part-Time Volunteers 85; Interns 1.
Governing Authority: municipal. Parent Institution: The University of Tampa. Tax-exempt: 501(c)(3).
Institution Type/Description: History & Decorative Arts Museum: housed in 1891 Tampa Bay Hotel.
Collections: original furnishings & objects of the Gilded Age from 1891 Tampa Bay Hotel; Plant System railroad; steamship & hotel artifacts; Spanish-American War collection.
Major Exhibits: Gasparilla: A Tampa Tradition, 1/14-2/14.
Research Fields: decorative arts; The Tampa Bay Hotel; local history; Plant System railroads, steamships & hotels; early Florida tourist industry; Spanish-American War.
Facilities: library of reference materials relating to the collection available for use on the premises.
Activities: guided tours; outside lectures; school programs; special events; music programs; monthly antique evaluation clinics; audio wand tour. Museum Sponsors: Victorian Christmas Stroll. Theater: Upstairs/Downstairs at the Tampa Bay Hotel.
Publications: booklets, An Architectural Guide to the Tampa Bay Hotel, A Walking Tour of the Gardens, Moments in Time: The Tampa Bay Hotel, Its History & Glory 1891-1931; video, The Tampa Bay Hotel: Florida's First Magic Kingdom; video, Dateline Tampa 1898: Florida and the Spanish American War, narrated by General H. Norman Schwarzkopf. Henry Bradley Plant: The 19th century King of Florida; If Our Hotel Could Talk; Maggie and Max at the Museum.
Hours & Admission Prices: Tues.-Sat. 10-5, Sun. 12-5. Adults $10; discounts to AAM members & SE Museums; members no charge. Closed Thanksgiving; Christmas.
Attendance: 55,000 (accurate)
Membership: Friend $50; Family $75; Patron $250; Veranda Circle $500; Crescent Club $1,000; Henry's Private Club $2,500; 1891 Landmark Society $5,000.

✳ **MUSEUM OF SCIENCE & INDUSTRY, (M),** 4801 E. Fowler Ave., Tampa, FL 33617-2099. Tel.: 813-987-6000. Fax: 813-987-6310.
E-mail: shannon.herbon@mosi.org
Web Site: www.mosi.org
Founded: 1955.
Congressional District: 7
Key Personnel: Pres., Wit Ostrenko; Sr. Vice Pres., Vicki Ahrens; Vice Pres. Exhibits, Dave Conley; Vice Pres. Education, Anthonette Carregal; Vice Pres. Facilities, Donald D. Toeller; Vice Pres. Finance, Kathy Prossick; Vice Pres. Research & Institutional Devel., Dr. Judith L Lombana; Vice Pres. Mktg. & Corp. Rels., Tanya Vomacka; Vice Pres. Human Resources, Kelly Currington; Volunteer Coord., Joel Bates; Media Rels. Specialist, Shannon Herbon; Museum Shop Mgr., Laurie Pinna.
Personnel Profile: Full-Time Paid 61; Part-Time Paid 117; Part-Time Volunteers 308.
Governing Authority: county. Parent Institution: Museum of Science & Industry Foundation. Tax-exempt.
Institution Type/Description: Science, Industry & Technology Center.
Collections: science; technology; industry; human history; butterfly habitat.
Research Fields: relative to collections & geographic area.
Facilities: 74-acre site; 340-seat IMAX dome theater; 400,000 sq. ft. exhibit area; 400-seat auditorium; planetarium; 30-acre natural habitat; 120-seat science demonstration theater; learning center; classrooms; Head Start Center; 40,000 sq. ft. children's science center; cafe; lab. Museum-related science items for sale.
Activities: permanent & changing exhibits; science demonstrations; classes & group programs; community outreach activities; Head Start program; summer camps.
Publications: newsletter; annual report; magazine, MOSI Matters; Summer Science Camp catalog, guidebook to group programs.
Hours & Admission Prices: Mon.-Fri. 9-5, Sat.-Sun. 9-6. Adults 13-59 $21.95, seniors 60 & over $19.95, children 2-12 $17.95 (combo-MOSI, IMAX & MOSI); discounts for AAM, ASTC, & FDEP members; members no charge. Call to confirm. ♿
Attendance: 611,365 (accurate)
Membership: Basic $45, $79, $99; Plus IMAX & Plus Rides $60, $109, $154; Premium $75, $139, $199.

SCARFONE/HARTLEY GALLERY, 310 N. Blvd., Tampa, FL 33606-1403. Mailing Address: 401 W. Kennedy, Tampa, FL 33606-1450. Tel.: 813-253-6217. Fax: 813-258-7497.
E-mail: dcowden@ut.edu
Web Site: www.ut.edu
Founded: 1977.
Congressional District: 7
Key Personnel: Dir., Dorothy C. Cowden.
Personnel Profile: Full-Time Paid 1; Part-Time Paid 10; Part-Time Volunteers 20; Interns 2.
Governing Authority: university. University of Tampa, 401 W. Kennedy, Tampa, FL 33606. Tax-exempt: 501(c)(3).
Institution Type/Description: Fine Arts Gallery.
Collections: contemporary 3-D & 2-D sculpture; prints; paintings; drawings; ceramics.
Major Exhibits: Sedrick Huckaby, 1/27/14-2/22/14; Alex Kanersky, 2/28/14-3/28/14.
Facilities: teaching gallery; lecture facility.
Activities: guided tours; lectures; films; concerts; dance performances; arts festivals; drama; training programs for professional museum workers; visual art exhibitions.
Publications: brochures for events & special exhibitions; catalogs for special events.
Hours & Admission Prices: Aug.-May Tues.-Fri. 10-4, Sat. 1-4. No charge; donations accepted. Closed national holidays. ♿
Attendance: 12,000 (estimated)
Membership: Friends of the Gallery: General $35; Patron $125-$1,000.

TAMPA BAY HISTORY CENTER, INC., (M), 801 Old Water St., Tampa, FL 33602-5418. Tel.: 813-228-0097. Fax: 813-223-7021.
E-mail: info@tampabayhistorycenter.org
Web Site: www.tampabayhistorycenter.org
Founded: 1989.
Congressional District: 7
Key Personnel: Chm., Paul Whiting, Jr.; Pres. & C.E.O., C.J. Roberts; Dir. Devel., Lisa Richardson; Cur., Rodney Kite Powell; Cur. Education, Julie Matus; Dir. Mktg., Manny Leto; Mgr. Collections, Malerie Dorman; Accountant, Maria Steijlen; Mgr. Admin. Svcs., Judy Miller; Museum Shop Mgr., Krissy Kite-Powell; Assoc. Dir. Advancement, Andrea Gallagher.

Personnel Profile: Full-Time Paid 15; Part-Time Paid 5; Part-Time Volunteers 80; Interns 3.
Volunteer Hours: 5,500
Governing Authority: private; nonprofit. Tax-exempt: 501(c)(3).
Institution Type/Description: History Museum.
Collections: Florida, Hillsborough County & the Tampa Bay region from prehistoric era to present.
Major Exhibits: Charting the Land of Flowers: 500 Years of Florida Maps, 9/13-2/14.
Research Fields: history, archaeology & multicultural heritage of Hillsborough County and the Tampa Bay region.
Facilities: 4,000-vol. library of regional history books; 17,000 sq. ft. exhibit space; educational facilities.
Activities: lectures; guided tours; formal educational programs.
Publications: quarterly newsletter, Cotanchobee; Tampa Bay Frontier Series (12 vol.); journal, Tampa Bay History.
Hours & Admission Prices: Daily 10-5. Adults $12.95; discounts to FAM, AAM & ICOM members; members no charge. Closed Thanksgiving; Christmas. ♿
Attendance: 75,575 (estimated)
Membership: Student & Teacher (Associate) $30; Individual $55; Companion $75; Family $85; Supporter $150; Sponsor $275; Patron $500; Benefactor $1,000; Founder $12,500 ($2,500 per year for 5 years).

TAMPA FIREFIGHTERS MUSEUM, 720 Zack St., Tampa, FL 33602. Tel.: 813-964-6862.
Institution Type/Description: Firefighting History Museum: housed in the former fire headquarters of the Tampa Fire Department; built in 1911.
Collections: local firefighting history & equipment; personal artifacts; photographs; uniforms; hands-on exhibitions; firefighter memorial.
Facilities: Museum-related items for sale.
Activities: educational programs; rental facilities; training workshops; hands-on exhibits.
Hours & Admission Prices: Tues.-Sun. 10-2. ♿

✳ **TAMPA MUSEUM OF ART,** 120 Gasparilla Plaza, Tampa, FL 33602. Tel.: 813-274-8130. Fax: 813-274-8732.
Web Site: www.tampamuseum.org
Founded: 1979.
Congressional District: 7
Key Personnel: Chm. (V), Peter Hepner; Exec. Dir., Todd D. Smith; Dir. External Affairs, Leslie Langford; Richard E. Perry Cur. Greek & Roman Art & Chief Cur., Seth D. Pevnick.
Personnel Profile: Full-Time Paid 13; Part-Time Paid 12; Part-Time Volunteers 125.
Governing Authority: bd. of trustees. Parent Institution: Tampa Museum of Art, Inc. Subsidiary Institution: Tampa Museum of Art Foundation. Tax-exempt.
Institution Type/Description: Art Museum.
Collections: 19th-century to contemporary paintings, sculpture & works on paper; Greek & Roman antiquities; C. Paul Jennewein collection of sculpture; photography.
Major Exhibits: Graphic Studio: Uncommon Practice at USF, 2/1/14-5/18/14; My Generation: Young Chinese Artists (T), 6/7/14-9/28/14; Poseiden and the Sea: Myth, Cult & Daily Life (T), 6/14/14-11/30/14.
Research Fields: 19th- to 20th-century American art; Greek and Roman antiquities, 2500 B.C.-300 A.D.
Activities: community outreach programs; special lectures; monthly events; workshops & programs.
Publications: member newsletter; exhibition catalogues for major TMA-organized exhibitions; gallery guides for some smaller exhibitions; educational collateral.
Hours & Admission Prices: Mon.-Thurs. 11-7, Fri. 11-8, Sat.-Sun. 11-5. Adults $10-12; members no charge. Closed major holidays. ♿
Attendance: 85,000 (estimated)
Membership: Student $35; Individual $50; Dual $70; Household $80; Sustainer $125; Patron $250; Silver Patron $500; Gold Patron $1,000; Benefactor $2,500. Director's Circle $5,000 & up; President's Circle $10,000 & up; Philanthropist $20,000.

TAMPA'S LOWRY PARK ZOO, 1101 W. Sligh Ave., Tampa, FL 33604-5958. Tel.: 813-935-8552. Fax: 813-935-9486.
E-mail: information@lowryparkzoo.com
Web Site: www.lowryparkzoo.com
Founded: 1957.
Congressional District: 11
Key Personnel: Chm., Robert Thomas; Exec. Dir. & C.E.O., Craig Pugh; Museum Shop Buyer & Mgr., Kim Gefre.
Personnel Profile: Full-Time Paid 142; Part-Time Paid 179; Part-Time Volunteers 5,987; Interns 6.

Volunteer Hours: 81,000
Operating Expenses: 15,574
Operating Income: 16,402
Governing Authority: society; nonprofit. Parent Institution: Lowry Park Zoo-logical Society of Tampa, Inc. Tax-exempt: 501(c)(3).
Institution Type/Description: Zoo.
Collections: over 1,700 animals & their habitats.
Major Exhibits: Sea Lion Splash (T), 3/14-5/14.
Facilities: library; children's zoo; aquarium; botanical garden; cafeteria; environmental education center; 2 amphitheaters; nature & conservation center; event pavilion. Museum-related items for sale.
Activities: arts festivals; concerts; docent program; formal educational pro-grams; lectures; study clubs. Museum Sponsors: Annual Karamu, ZooFari; Summer Safari Night; ZooBoo; Wild Wonderland; Safari Africa; WaZoo; Fiesta Zoo.
Publications: newsletter quarterly; triannual newsletter, Zoo Chatter; quarterly education program guide; annual report.
Hours & Admission Prices: Daily 9:30-5. Adults $24.95, senior citizens $22.95, children $19.95; discounts to AZA member & groups; members no charge. Closed Thanksgiving; Christmas. &
Attendance: 1,100,000 (estimated)
Membership: Guest Pass $30; Individual $60; Dual $95; Single Parent Family $130; Family $160; Family Plus $210.

UNIVERSITY OF SOUTH FLORIDA BOTANICAL GARDEN,
4202 E. Fowler Ave., NES 107, Tampa, FL 33620-9951. Tel.: 813-974-2329. Fax: 813-974-4808.
E-mail: lwalker@cas.usf.edu
Web Site: www.cas.usf.edu/garden
Key Personnel: Dir., Laurie Walker; Coord. Special Events, Kim Hutton; Mgr. Plant Shop, Valerie Nye; Sr. Agricultural Asst., David Ropp
Institution Type/Description: Botanical Garden.
Collections: plants.
Facilities: Museum-related items for sale.
Activities: research; education programs.
Hours & Admission Prices: Mon.-Fri. 9-5, Sat. 9-4, Sun. 12-4. Adults $5. Closed major holidays.

* UNIVERSITY OF SOUTH FLORIDA CONTEMPORARY ART MUSEUM, (M), 3821 USF Holly Dr., Tampa, FL 33620. Mailing Address: 4202 E. Fowler Ave., CAM101, Tampa, FL 33620-7360. Tel.: 813-974-4133. Fax: 813-974-5130.
E-mail: caminfo@arts.usf.edu
Web Site: www.usfcam.usf.edu/CAM/cam_about.html
Founded: 1968.
Congressional District: 7
Key Personnel: Dir., Margaret A. Miller; Assoc. Dir., Alexa Favata; Collec-tions Cur., Peter Foe; Exhibitions Supvr., Tony Palms; Dir. Public Art, Vincent Ahern; Security, David Waterman; Program Asst., Victoria Billig; New Media Cur., Don Michael Fuller; Business Coord. & Museum Shop Mgr., Randall West.
Personnel Profile: Full-Time Paid 11; Part-Time Paid 3; Part-Time Volunteers 3; Interns 6.
Governing Authority: university. Parent Institution: University of South Florida. Tax-exempt.
Institution Type/Description: Contemporary Art Museum.
Collections: contemporary graphics; photography; public art commissions.
Research Fields: graphics; public art.
Facilities: Museum-related items for sale.
Activities: guided tours; lectures; films; gallery talks; formally organized education programs for undergraduate & graduate college students affiliated with University of South Florida; inter-museum loan, temporary & travel-ing exhibitions; art bank exhibition loan service.
Publications: exhibition catalogs; newsletters; posters; public art videos.
Hours & Admission Prices: Museum: Mon.-Fri. 10-5, Sat. 1-4. No charge. Closed university holidays. &
Attendance: 58,000 (estimated)
Membership: Student & Senior $15; Individual $25; Dual (2) $45; Benefactor $250; Patron $1,500; Corporate $2,500-$100,000.

YBOR CITY MUSEUM STATE PARK, 1818 Ninth Ave. E., Tampa, FL 33605-3818. Tel.: 813-247-6323. Fax: 813-233-3343.
E-mail: douglas.kinder@dep.state.fl.us
Web Site: www.ybormuseum.org
Formerly: Ybor City State Museum
Founded: 1980.
Key Personnel: Dir., Chantal Hevia; Park Mgr., Kimberlee Tennille; Asst. Park Mgr., Bobby Tootacker; Pres. (V), Rich Simmons; Park Ranger, Nancy Garrison; Park Svcs. Specialist, Alex Kinder; Museum Shop Mgr., Cookie Ginex.

Personnel Profile: Full-Time Paid 2; Part-Time Volunteers 38; Interns 1.
Governing Authority: state. Parent Institution: Dept. of Environmental Protec-tion, Div. of Recreation & Parks. Tax-exempt.
Institution Type/Description: Industrial Museum: 1923 Bakery Building; original brick commercial ovens. Complex also contains c.1895 restored & furnished house, originally rented to workers.
Collections: exhibits & artifacts interpreting the cigar industry in Tampa's Latin District, Ybor City; cigar-making tools.
Research Fields: cigar industry in Ybor City; various Latin ethnic groups.
Facilities: 2,800 sq. ft. exhibit space; garden available for rental. Museum-related items for sale.
Activities: docent program; museum/LaCasita tours; cigar rolling demonstra-tions.
Hours & Admission Prices: Daily 9-5. Adults $4; children under 6 no charge. LaCasita: daily 10-3 depending on docent availability. Closed New Year's; Thanksgiving; Christmas. &
Attendance: 29,714 (estimated)
Membership: Senior & Student $25; Museum Friend $35; Museum Family $50; Y.M. Ybor Patron $100; LaCasita Society $250; Latin Quarter Club $500; 1886 Founders Circle $1,000.

Tarpon Springs

* LEEPA-RATTNER MUSEUM OF ART AT ST. PETERS-BURG COLLEGE, (M), 600 Klosterman Rd., Tarpon Springs, FL 34689-1299. Mailing Address: P.O. Box 1545, Tarpon Springs, FL 34688-1545. Tel.: 727-712-5762. Fax: 727-712-5223.
E-mail: lrma@spcollege.edu
Web Site: www.spcollege.edu/museum
Founded: 2002.
Congressional District: 9
Key Personnel: Dir., Ann Larsen; Chm. Bd. Dirs., William Schumacher; Membership, Michele Schnerdenbach; Cur., Lynn Whitelaw; Museum Shop Mgr., Lynn F. Pierson.
Personnel Profile: Full-Time Paid 6; Part-Time Paid 6; Part-Time Volunteers 75.
Governing Authority: Parent Institution: St. Petersburg College Bd. of Trust-ees. Subsidiary Institution: Leepa-Rattner Museum of Art, Inc. Bd. of Directors. Tax-exempt.
Institution Type/Description: Art Museum.
Collections: paintings; photographs; sculpture.
Research Fields: mid-20th & 21st-century modern & contemporary American & European art.
Facilities: library & archives.
Hours & Admission Prices: Tues.-Wed. & Sat. 10-5, Fri. 10-4, Thurs. 10-8, Sun. 1-5. Adults $6, seniors $4; discounts to AAM members; children, students, members, military & Sun. no charge. &
Attendance: 35,000 (accurate)
Membership: Student, Educator, SPC Faculty & Staff $20; Individual $40; Dual $60; Family $65; Friend $125; Contributor $250; Advocate $500; Sustainer $1,000.

SAFFORD HOUSE HISTORIC HOUSE MUSEUM, 23 Parkin Court, Tarpon Springs, FL 34689-3235. Tel.: 727-937-1130. Fax: 727-938-2429.
E-mail: tbucuvalas@ctsfl.us
Web Site: www.tarponarts.org/info_venues_safford.html
Founded: 1995.
Congressional District: 9
Key Personnel: Cur., Tina Bucuvalas.
Personnel Profile: Full-Time Paid 1; Part-Time Volunteers 10.
Governing Authority: municipal. Parent Institution: City of Tarpon Springs, P.O. Box 5004, Tarpon Springs, FL 34688. Tax-exempt.
Institution Type/Description: Historic House Museum: home of Anson P.K. Safford from 1883-1891.
Collections: Safford family history; life in west central Florida during late 19th century; period furnishings.
Research Fields: Safford family; late Victorian life in Florida; late 19th century medical practice.
Facilities: 20-vol. library. Museum-related items for sale.
Activities: concerts; docent program; guided tours. Annual Events: Holiday Candlelight Tour in December.
Hours & Admission Prices: Wed. & Fri. 11-3; other times by appointment. Adults $3; discounts to Florida Association of Museums; children no charge. Closed New Year's Day; Memorial Day; Independence Day; Labor Day; Thanksgiving & day after; Christmas. &
Attendance: 986 (accurate)

TARPON SPRINGS AREA HISTORICAL SOCIETY - DEPOT MUSEUM, 160 E. Tarpon Ave., Tarpon Springs, FL 34689-3452. Tel.: 727-943-4624.
E-mail: tarpon.historical@verizon.net
Founded: 1976.
Congressional District: 9
Key Personnel: Pres., Cynthia Tarapani; Treas., Phyllis Kolianos; Cur., Tina Bucavolous, Ph.D.; Museum Shop Mgr., Margaret Rayner.
Personnel Profile: Part-Time Paid 1; Part-Time Volunteers 35.
Governing Authority: private; nonprofit organization. Tax-exempt: 501(c)(3).
Institution Type/Description: Historical Society Museum.
Collections: Tarpon Springs artifacts & archives from mid 19th century to present.
Research Fields: oral town histories; impact of railroad on the town.
Facilities: 47-vol. library. Museum-related items for sale.
Activities: docent program; films; lectures. Museum Sponsors: Historic House Tour. Annual Events: Cemetery Tour; Remembrance Tea.
Publications: monthly newsletter, Newsletter.
Hours & Admission Prices: Tues.-Sat. 11:30-4. No charge; donations accepted. Closed New Year's Day; Memorial Day; Independence Day; Thanksgiving; Christmas. &
Attendance: 4,670 (accurate)
Membership: Individual $20; Family $30; Sustaining $50; Contributing $100; Life $200.

TARPON SPRINGS CULTURAL CENTER, 101 S. Pinellas Ave., Tarpon Springs, FL 34689-3631. Tel.: 727-942-5605. Fax: 727-938-2429.
E-mail: bpoteat1@ci.tarpon-springs.fl.us
Founded: 1987.
Congressional District: 9
Key Personnel: Dir., Dr. Kathleen Monahan; Mktg., Gen Haley; Cur., Tina Bucuvalas, Ph.D.; Box Office Mgr., Lisa Cobb; Site Mgr., Billie Poteat.
Personnel Profile: Full-Time Paid 4; Part-Time Paid 2; Part-Time Volunteers 40.
Governing Authority: municipal government. Parent Institution: City of Tarpon Springs. Tax-exempt.
Institution Type/Description: Cultural Center: housed in a c.1915 city hall, national register listed building.
Collections: concentration on historical, artistic & ethnographic material relating to Tarpon Springs; photos.
Research Fields: Greek-American culture in Tarpon Springs; local & regional history.
Facilities: 35-vol. library; 90-seat theater. Florida folk crafts for sale.
Activities: concerts; dance recitals; films; formal education programs for children; guided tours; workshops; lectures; loan, participatory, temporary & traveling exhibitions; theater.
Hours & Admission Prices: Mon.-Fri. 9-4, Sat. 12-4. No charge; donations accepted. Closed New Year's Day; Independence Day; Veterans Day; Thanksgiving; Christmas. &
Attendance: 6,726 (accurate)
Membership: Senior $30; Individual $35; Senior Family $40; Family $45; Bronze $65; Silver $135; Gold $195; Platinum $295; Ruby $575; Diamond $1,295.

TARPON SPRINGS HERITAGE MUSEUM, 100 Beekman Lane, Tarpon Springs, FL 34689-3555. Mailing Address: P.O. Box 5004, Tarpon Springs, FL 34688-5004. Tel.: 727-937-0686 & 0699. Fax: 727-937-0657.
E-mail: info@tarponarts.org
Web Site: www.tarponarts.org
Founded: 2001.
Congressional District: 9
Key Personnel: Dir., Dr. Kathleen Monahan; Public Rels., Gen Haley; Cur., Judy LeGath.
Personnel Profile: Full-Time Paid 3; Part-Time Paid 1; Part-Time Volunteers 1.
Governing Authority: municipal. Parent Institution: City of Tarpon Springs. Tax-exempt.
Institution Type/Description: History Museum.
Collections: period photographs; Greek American history; Christopher Still murals.
Hours & Admission Prices: Mon.-Fri. 10-4. Adults $3; children no charge. Closed public holidays. &
Attendance: 2,377 (accurate)

Tavares

LAKE COUNTY HISTORICAL SOCIETY AND MUSEUM, Lake County Historic Courthouse, 317 W. Main St., Tavares, FL 32778-3813. Mailing Address: P.O. Box 7800, Tavares, FL 32778-7800. Tel.: 352-343-9890. Fax: 352-343-9814.
E-mail: lakecounty1887@yahoo.com
Founded: 1954.
Key Personnel: Dir., Cur. & Office Mgr., Lavonda Morris.
Personnel Profile: Part-Time Paid 1; Part-Time Volunteers 10.
Governing Authority: municipal. Tax-exempt.
Institution Type/Description: County Museum.
Collections: history of Lake County; Native American artifacts; Spanish relics.
Activities: hands-on exhibits.
Publications: Lake County Florida: A Pictorial History; History Fast Facts.
Hours & Admission Prices: Mon.-Fri. 8:30-5. No charge. &
Attendance: 5,000 (estimated)
Membership: Student $5; Individual $15; Family $20; Corporation $125; Lifetime $200.

Tequesta

LIGHTHOUSE ARTCENTER, Gallery Square North, 373 Tequesta Dr., Tequesta, FL 33469-3027. Tel.: 561-746-3101. Fax: 561-746-3241.
Web Site: www.lighthousearts.org
Formerly: Lighthouse Center for the Arts
Founded: 1965.
Key Personnel: Exec. Dir., Katie Deits.
Personnel Profile: Full-Time Paid 6; Part-Time Paid 2; Part-Time Volunteers 45; Interns 1.
Governing Authority: board of directors. Tax-exempt.
Institution Type/Description: Art Gallery.
Collections: paintings; photographs; sculpture.
Facilities: Museum-related items for sale.
Activities: art classes; lectures; permanent & temporary exhibitions; jazz concerts; luncheon series.
Publications: newsletters.
Hours & Admission Prices: Mon.-Fri. 10-4, Sat. 10-2. Adults $5; members no charge. &
Attendance: 50,000 (estimated)
Membership: Individual $75; Family $100.

Titusville

AMERICAN POLICE HALL OF FAME AND MUSEUM, 6350 Horizon Dr., Titusville, FL 32780-8002. Tel.: 321-264-0911. Fax: 321-264-0033.
E-mail: barrys@aphf.org
Web Site: aphf.org
Founded: 1960.
Key Personnel: C.E.O., Donna Shepherd; C.F.O., Debra Chitwood; Pres. (V), Jack Rinchich; Exec. Dir., Barry Shepherd; Museum Shop Mgr., Lori Shepherd.
Personnel Profile: Full-Time Paid 24; Part-Time Paid 5; Part-Time Volunteers 4.
Governing Authority: Parent Institution: National Association of Chiefs of Police. Tax-exempt.
Institution Type/Description: Police Museum.
Collections: Hall of Fame inductees; American law enforcement history; weapons; photographs; memorials; gangsters; future of law enforcement; cars; motorcycles.
Activities: indoor firing range; air tours.
Publications: Police Family News; The Chief of Police.
Hours & Admission Prices: Daily 10-6. Adults $13, members $9, seniors, military, & children 4-12 $8; discounts to AAA members & military; law enforcement & family survivors no charge. Closed New Year's Day; Thanksgiving; Christmas. &
Attendance: 50,000 (accurate)

NORTH BREVARD HISTORICAL MUSEUM, 301 S. Washington Ave., Titusville, FL 32796-3539. Mailing Address: P.O. Box 5265, Titusville, FL 32783-5265. Tel.: 321-269-3658.
Web Site: www.nbbd.com/godo/history
Founded: 1989.
Key Personnel: Pres. (V), Betty J. Mattingly.
Personnel Profile: Part-Time Volunteers 75.
Governing Authority: Operated by Historical Society of North Brevard. Tax-exempt.

Institution Type/Description: Historical Museum.
Collections: personal artifacts; photographs; records; clothing; local history.
Activities: group tours.
Publications: newsletter.
Hours & Admission Prices: Tues.-Sat. 10-3. No charge; donations accepted.
Attendance: 2,500 (estimated)
Membership: Individual $15; Family $25; Associate & Business $50; Lifetime $500.

SPACE WALK OF FAME MUSEUM, 4 Main St., Titusville, FL 32796. Tel.: 321-264-0434. Fax: 321-264-0767. Facebook: Space Walk of Fame Museum.
E-mail: info@spacewalkoffame.com
Web Site: www.spacewalkoffame.com
Founded: 1996.
Key Personnel: Pres. (V), Charlie Mars; Museum Shop Mgr., Lori Fore.
Governing Authority: Tax-exempt.
Institution Type/Description: Space History Museum.
Collections: space program history; photographs; personal artifacts; astronaut memorials & monuments.
Activities: special events.
Hours & Admission Prices: Mon.-Sat. 10-5. No charge; donations accepted.
Attendance: 2,000 (estimated)
Membership: Mercury $100-$499; Gemini $500-$999; Apollo $1,000 & up.

UNITED STATES ASTRONAUT HALL OF FAME, 6225 Vectorspace Blvd., Titusville, FL 32780-8040. Tel.: 321-455-7000 & 269-6100.
Web Site: www.kennedyspacecenter.com
Institution Type/Description: Hall of Fame.
Collections: astronaut hall of fame; personal artifacts; space suits & equipment; photographs.
Hours & Admission Prices: Daily 10-6:30. Adults $17, children 3-11 $13.

VALIANT AIR COMMAND WARBIRD AIR MUSEUM, 6600 Tico Rd., Titusville, FL 32780-8009. Tel.: 321-268-1941. Fax: 321-268-5969.
E-mail: vacwarbirds@bellsouth.net
Web Site: www.vacwarbirds.org
Founded: 1977.
Key Personnel: Commander, Lloyd Morris.
Governing Authority: nonprofit. Tax-exempt: 501(c)(3).
Institution Type/Description: Aviation History Museum.
Collections: aviation history; military aircraft.
Facilities: 30,000 sq. ft. exhibit space. Museum-related items for sale.
Activities: tours; airshow.
Publications: newsletter.
Hours & Admission Prices: Daily 9-5. Adults $12, senior citizens & military $10, children 5-12 $5; discounts to groups. Closed New Year's Day; Thanksgiving; Christmas.
Attendance: 60,000 (estimated)
Membership: Single $75; Family $100; Lifetime $500, $750 & $1,000.

Valparaiso

HERITAGE MUSEUM OF NORTHWEST FLORIDA, 115 Westview Ave., Valparaiso, FL 32580-1387. Tel.: 850-678-2615. Fax: 850-678-4547. Facebook: Heritage Museum of Northwest Florida.
E-mail: heritagemuseum@co.okaloosa.fl.us
Web Site: heritage-museum.org
Founded: 1971.
Congressional District: 1
Key Personnel: Museum Mgr., Gina Marini; Chm. (V), Gordon King.
Personnel Profile: Full-Time Paid 1; Part-Time Paid 1; Part-Time Volunteers 25; Interns 1.
Governing Authority: nonprofit; society. Parent Institution: Heritage Museum Association, Inc. Tax-exempt: 501(c)(3).
Institution Type/Description: History Museum.
Collections: Northwest Florida history from prehistoric times to mid-20th century.
Research Fields: genealogy; agricultural history; turpentine industry; lumbering; railroading & fishing.
Facilities: 1,500-vol. library; classrooms. Handcrafted items for sale.
Activities: guided tours; lectures; gallery talks; heritage crafts classes; formally organized education programs for children; docent program or council; permanent & temporary exhibitions.
Publications: newsletter; monographs on local history.

Hours & Admission Prices: Tues.-Sat. 10-4. Adults $5, seniors & military $4, children $3; members & children under 4 no charge. Closed legal holidays. &
Attendance: 17,000 (accurate)
Membership: Senior Citizen & Student $15; Individual $35; Family $50; Advocate $100; Sustaining $250; Benefactor $500; Heritage Circle $1,000 & up. Business memberships available.

Venice

VENICE ART CENTER, 390 Nokomis Ave. S., Venice, FL 34285-2416. Tel.: 941-485-7136. Fax: 941-484-4361.
E-mail: info@veniceartcenter.com
Web Site: www.veniceartcenter.com
Key Personnel: Exec. Dir., Mary Morris
Institution Type/Description: Art Center.
Collections: paintings; photographs; sculpture.
Facilities: classrooms.
Hours & Admission Prices: Summer: Mon.-Fri. 9-5; Winter: Mon.-Sat. 9-5.

Vero Beach

THE HERITAGE CENTER AND INDIAN RIVER CITRUS MUSEUM, (M), 2140 14th Ave., Vero Beach, FL 32960-3432. Tel.: 772-770-2263. Fax: 772-770-2131.
E-mail: veroheritage@bellsouth.net
Web Site: www.veroheritage.org
Founded: 1991.
Congressional District: 8
Key Personnel: Dir., Rebecca A. Rickey; Pres. (V), Elizabeth Graves Bass; Museum Shop Mgr., Barbara Walsh.
Personnel Profile: Full-Time Paid 1; Part-Time Paid 6; Part-Time Volunteers 40.
Governing Authority: Tax-exempt: 501(c)(3).
Institution Type/Description: History Museum.
Collections: citrus industry history; period tools & artifacts; photographs.
Facilities: Museum-related items for sale.
Hours & Admission Prices: June-Aug. Tues.-Fri. 10-3; Sept.-May Tues.-Fri. 10-4. No charge; donations accepted. Closed New Year's Eve & Day; Christmas Day & week. &
Attendance: 3,000 (estimated)
Membership: Individual $25; Family $35; Corporate $75; Trailblazer $125; Benefactor $250; Homesteader $1,000.

MCKEE BOTANICAL GARDEN, (M), 350 U.S. Hwy. 1, Vero Beach, FL 32962-2906. Tel.: 772-794-0601. Fax: 772-794-0602.
E-mail: info@mckeegarden.org
Web Site: www.mckeegarden.org
Founded: 1932.
Congressional District: 15
Key Personnel: Pres., Barbara Holmen McKenna; Dir., Christine Hobart; Museum Shop Mgr., Gail Galbraith.
Personnel Profile: Full-Time Paid 6; Part-Time Paid 8; Part-Time Volunteers 125.
Volunteer Hours: 4,000
Operating Expenses: 1,125,000
Operating Income: 1,125,000
Governing Authority: private; nonprofit organization. Tax-exempt: 501(c)(3).
Institution Type/Description: Botanical Garden: listed on the National Register of Historic Places.
Collections: plants; trees; flowers.
Major Exhibits: Seward Johnson Sculpture Exhibition (T), 1/3/14-4/27/14.
Facilities: botanical garden. Museum-related items for sale.
Activities: formal educational programs for adults; lectures; seasonal events & activities.
Publications: newsletter 3 times a year, News from McKee Botanical Garden.
Hours & Admission Prices: Tues.-Sat. 10-5, Sun. 12-5. Adults $12, seniors $11, children $5; reciprocal admission to American Horticultural Society; members no charge. Closed major holidays. &
Attendance: 35,000 (accurate)
Membership: Individual $35; Family $50; Sustaining $100; Patron $500; Associate $1,000; Gatekeeper $1,500.

MCLARTY TREASURE MUSEUM, 13180 Highway N. 1A, Part of the Sebastian Inlet State Park, Vero Beach, FL 32963-9400. Tel.: 772-589-2147. Fax: 321-984-4854.
Web Site: www.myflorida.com
Founded: 1970.
Congressional District: 16

Key Personnel: Dir., Ed Perry; Pres. (V), Fred Marshall; Park Mgr., Ronald N. Johns; Gift Shop Mgr., Barbara Kmack.
Personnel Profile: Full-Time Paid 1; Full-Time Volunteers 14; Part-Time Volunteers 20.
Governing Authority: state. Dept. of Environmental Protection (DEP), Div. of Recreation & Parks, 3900 Commonwealth Blvd., Tallahassee, FL 32303. Tax-exempt.
Institution Type/Description: History and Film Museum: built on 1715 site of Spanish salvage campsite adjacent to shipwreck in Atlantic Ocean.
Collections: a portion of the State's share of the treasures & artifacts (worked metals, ceramics, glass) of the Spanish Fleet Shipwrecks of 1715.
Facilities: Books of local historical information for sale.
Activities: DVD presentation of A&E The Queen's Jewels & the 1715 Fleet; artifacts on display; observation deck overlooking shipwreck sites.
Publications: brochures.
Hours & Admission Prices: Daily 10-4:30; last film showing 3:15. Admission $2 per person; children under 6 no charge. &

Attendance: 14,690 (accurate)

✱　**VERO BEACH MUSEUM OF ART, INC., (M),** 3001 Riverside Park Dr., Vero Beach, FL 32963-1874. Tel.: 772-231-0707. Fax: 772-231-0938. Facebook: Vero Beach Museum of Art.
E-mail: info@verobeachmuseum.org
Web Site: www.verobeachmuseum.org
Founded: 1979.
Congressional District: 12
Key Personnel: Chm. Bd., Peter M. Thompson; Exec. Dir., Lucinda H. Gedeon; Dir. Devel., Robyn P. Orzel; Dir. Mktg. & Communications, Sophie Bentham Wood; Dir. Education, J. Marshall Adams; Cur. Collections & Exhibitions, Jay Williams; Registrar, J'Laine K. Newcombe; Coord. Public Rels., Joe Ellis; Dir. Security & Technical Svcs., John Janssen; Asst. to Exec. Dir., Bonnie Wetherell; Museum Shop Mgr., Jo Anne Miller.
Personnel Profile: Full-Time Paid 26; Part-Time Paid 61; Part-Time Volunteers 519.
Governing Authority: nonprofit organization. Tax-exempt: 501(c)(3).
Institution Type/Description: Art Museum.
Collections: 20th-century American art; 21st-century American & International art.
Research Fields: American art.
Facilities: 5,000-vol. library pertaining to general art history, materials & techniques available for reference; 250-seat auditorium; museum art school including sculpture foundry; sculpture park; lecture hall; cafe. Museum-related items for sale.
Activities: multi-disciplinary art instructional studios; museum school; temporary & permanent exhibitions; art festivals; guided tours; lectures; films; concerts.
Publications: exhibition catalogues; annual report; quarterly magazine; public programs brochure; class schedule; e-newsletter.
Hours & Admission Prices: Memorial Day to Labor Day Tues.-Sat. 10-4:30, Sun. 1-4:30; Sept.-May Mon.-Sat. 10-4:30, Sun. 1-4:30. Adults $10; discounts to groups and AAM & ICOM members; members and children 17 & under no charge. Closed New Year's Day; Easter; Memorial Day; Independence Day; Labor Day; Thanksgiving; Christmas. &

Attendance: 75,000 (accurate)
Membership: Individual $45; Family $60; Benefactor $150; Donor $375; Business Patron $500; Patron $625; Chairman's Club $1,500; Corporate Partners Club $2,000; Director's Silver Society $2,500; Director's Gold Society $5,000; Director's Platinum Society $10,000; Director's Diamond Society $25,000.

Weirsdale

GRAND OAKS RESORT & MUSEUM, (M), 3000 Marion County Rd., Weirsdale, FL 32195-5168. Tel.: 352-750-5500. Fax: 352-750-3342. Facebook: Grand Oaks Resort & Museum.
E-mail: laureen@thegrandoaks.com
Web Site: www.thegrandoaks.com
Formerly: Florida Carriage Museum and Resort
Founded: 1995.
Key Personnel: Dir., C.E.O. & Chm., Thomas Warriner; Business Mgr., Laureen Oliver.
Governing Authority: Parent Institution: Grand Oaks Operating LLC.
Institution Type/Description: Transportation Museum.
Collections: over 150 European & American carriages; period horse-related artifacts; horse-drawn fire equipment; period vehicles.
Activities: Annual Events: Carriage & Horse Festival; Pleasure Driving Competition; CDE Competitions.
Hours & Admission Prices: Tues.-Sat. 10-4, Sun. 12-4; groups by appointment. Adults $11, students 5-18 $5; children 4 & under no charge. Closed holidays. &

Attendance: 10,000 (estimated)
Membership: Individual $35.

West Palm Beach

THE ARMORY ART CENTER, 1700 Parker Ave., West Palm Beach, FL 33401-7042. Tel.: 561-832-1776. Fax: 561-832-0191.
Web Site: www.armoryart.org
Formerly: The Robert & Mary Montgomery Armory Art Center
Founded: 1987.
Congressional District: 22
Key Personnel: C.E.O., Sandra B. Coombs; Chm., James Swope; Pres., Stephen Rabb; Dir. Mktg., Kati Erickson; Dir. Education, Talya Lerman.
Personnel Profile: Full-Time Paid 11; Part-Time Paid 60; Part-Time Volunteers 20; Interns 6.
Governing Authority: private; nonprofit organization. Tax-exempt: 501(c)(3).
Institution Type/Description: Art Center: housed in Historic Palm Beach County National Guard Armory.
Collections: works by contemporary artists.
Major Exhibits: Hughette Despault & Kathleen Eliott (T), 2/21/14-3/22/14; Armory Faculty Exhibition, 2/21/14-3/22/14; All Student Showcase, 3/28/14-5/3/14; Artists in Residence Exhibition, 3/28/14-5/3/14.
Facilities: 5,200 sq. ft. exhibit space; studio classrooms.
Activities: traveling exhibits; formal educational programs; guided tours; lectures; art classes; film series; demonstrations.
Publications: weekly newsletter, 8 week school catalogue; exhibition catalogues.
Hours & Admission Prices: Mon.-Sat. 9-4. Receptions: no charge. Closed New Year's Day; Memorial Day; Independence Day; Labor Day; Thanksgiving; Christmas. &

Attendance: 25,000 (accurate)
Membership: Youth $35; Individual $75; Family $100; Patron $500; Studio Guild $750; Gallery Patron $1,000.

HISTORICAL SOCIETY OF PALM BEACH COUNTY, (M), 300 N. Dixie Hwy., West Palm Beach, FL 33401-4605. Mailing Address: P.O. Box 4364, West Palm Beach, FL 33402-4364. Tel.: 561-832-4164. Fax: 561-832-7965.
E-mail: info@historicalsocietypbc.org
Web Site: www.historicalsocietypbc.org
Founded: 1937.
Congressional District: 22
Key Personnel: C.E.O. & Pres., Jeremy W. Johnson; Chm., Grier Pressly, III; Dir. Research & Archives and Chief Cur., Debi Murray; Cur. Collections & Exhibits, Steven F. Erdmann; Cur. Education, Tony Marconi; Research & Curatorial Asst., Nick Golubov; Membership Assoc., Carol Elder; Office Mgr., Sharon Poss; Mktg. & Special Events Coord., Jillian Markwith; Communications & Advancement Specialist, Melissa Sullivan.
Personnel Profile: Full-Time Paid 7; Part-Time Paid 2; Part-Time Volunteers 60; Interns 1.
Governing Authority: nonprofit organization. Subsidiary Institution: The Richard & Pat Johnson Palm Beach County History Museum. Tax-exempt: 501(c)(3).
Institution Type/Description: History Museum.
Collections: furniture; American Indian artifacts; sheet music; maps; vertical clipping file; photographic archive; Addison Mizner furniture & architectural drawings; Treanor & Fatio original architectural drawings & manuscript collection; microfilm of early Palm Beach County newspapers.
Major Exhibits: People of the Water, 9/3/13-6/28/14.
Research Fields: Palm Beach County; South Florida; Caribbean.
Facilities: 3,000-vol. library of books, newspapers & pamphlets of the history of the Palm Beach area & on the state of Florida available on premises; research room.
Activities: lectures; traveling exhibits; collaborations with other museums; fund-raising dinner. Historical Society Sponsors: Young Friends Events.
Publications: Tustenegee.
Hours & Admission Prices: Museum: Tues.-Sat. 10-5. No charge; donations accepted. Research by appointment. Research Fee $20; discounts to AAM & ICOM members; members no charge. &

Attendance: 60,000 (accurate)
Membership: Individual $50; Family $75; Barefoot Mailman $125; Mizner Circle $250; Flager Circle $500; Pioneer Circle $1,000; Benefactor $2,500.

MOUNTS BOTANICAL GARDEN, 531 N. Military Trail, West Palm Beach, FL 33415-1311. Tel.: 561-233-1757. Fax: 561-233-1723. Facebook: Mounts Botanical Garden.
Web Site: www.mounts.org
Founded: 1954.
Key Personnel: Dir., Allen Sistrunk; Pres. (V), Mike Zimmerman
Institution Type/Description: Botanical Garden.

Collections: tropical & subtropical plants from six continents; Florida's plants, trees, tropical fruit, herbs, citrus & palms.
Facilities: Museum-related items for sale.
Publications: quarterly magazine, The Leaflet.
Hours & Admission Prices: Mon.-Sat. 8-4, Sun. 12-4. Suggested Donation: $5. Closed New Year's Day; Thanksgiving; Christmas Eve & Day.
Attendance: 60,000 (estimated)
Membership: Individual $40; Family & Friends $60; Donor $100; Sustaining $250; Patron $500; Gardening Angel $1,000.

* **NORTON MUSEUM OF ART, (M),** 1451 S. Olive Ave., West Palm Beach, FL 33401-7162. Tel.: 561-832-5196. Fax: 561-659-4689.
E-mail: museum@norton.org
Web Site: www.norton.org
Founded: 1941.
Congressional District: 22
Key Personnel: Exec. Dir. & C.E.O., Hope Alswang; C.F.O., Lucy Bukowski; Co Chm., Harry Howell; Co Chm., Janine Mayville; Deputy Dir., James Brayton Hall; Human Resources Mgr., Jane Wattick; Chm. Curatorial Affairs, Cheryl Brutvan; Cur. Photography, Timothy Wride; Deputy Dir. Devel., Holly Davis; Cur. American Art, Ellen E. Roberts; Cur. Education, Glenn Tomlinson; Cur. Chinese Art, Laurie Barnes; Registrar, Pamela Parry; Assoc. Cur. Education, Carole Gutterman; Dir. Communications, Scott Benarde; Museum Shop Mgr., Heather Blades.
Personnel Profile: Full-Time Paid 68; Part-Time Paid 8; Part-Time Volunteers 400; Interns 4.
Governing Authority: nonprofit corporation. Tax-exempt: 501(c)(3).
Institution Type/Description: Art Museum.
Collections: 19th & 20th century American paintings & sculpture; 15th - 20th century European paintings & sculpture; Chinese jades, bronzes, ceramics & Buddhist sculpture; contemporary art after 1970; photography.
Major Exhibits: The Four Princely Gentlemen: Plum Blossoms, Orchids, Bamboo, and Chrysanthemums, 11/14/13-1/26/14; Phyllida Barlow: HOARD, 12/3/13-2/23/14; The Polaroid Years: Instant Photography and Experimentation, 12/19/13-3/23/14; David Webb: Society's Jeweler, 1/16/14-4/13/14; To Jane, Love Andy: Warhol's First Superstar, 2/2/14-5/25/14; Qing Chic: Chinese Textiles from the 19th to early 20th Century, 2/6/14-5/4/14; Industrial Sublime: Modernism and the Transformation of New York's Rivers, 1900-1940, 3/20/14-6/22/14.
Research Fields: 3
Facilities: 4,000-vol. library of art history; reference materials; art periodicals; 112,500 sq. ft. exhibit space; 33 galleries; 3 gardens; education wing; 200-seat auditorium; cafe. Museum-related items for sale.
Activities: guided tours; lectures; films; gallery talks; concerts; formally organized educational programs; docent program; permanent, temporary & traveling exhibitions; teacher enrichment programs; school tours; community outreach; special events.
Publications: catalogs of the permanent collections; calendars; exhibition catalogs; posters; postcards of major works in permanent collection; member newsletter, annual report; books.
Hours & Admission Prices: Tues.-Wed. & Fri.-Sat. 10-5, Thurs. 10-9, Sun. 11-5. Adults $12, students w/valid ID $5; discounts to AAM members; children 12 & under & members no charge. No additional charge for special exhibitions. Closed major holidays. &
Attendance: 100,000 (accurate)
Membership: Student $25; Individual $70; Household $100; Supporter $125; Contributor $250; Young Friends $250 & $750; Patron $1,000; Founders' Circle $1,750-$10,000.

PALM BEACH PHOTOGRAPHIC CENTRE, 415 Clematis St., West Palm Beach, FL 33401-5319. Tel.: 561-276-9797. Fax: 561-276-1932.
E-mail: info@workshop.org
Web Site: www.workshop.org
Key Personnel: Exec. Dir., Fatima NeJame; Mng. Dir., Art NeJame
Institution Type/Description: Photography Museum.
Collections: photographs.
Activities: photography classes; community programs; workshops; educational activities; seminars.
Hours & Admission Prices: Mon.-Sat. 9-6.

PALM BEACH ZOO, 1301 Summit Blvd., West Palm Beach, FL 33405-3098. Tel.: 561-547-9453. Fax: 561-585-6085.
E-mail: info@palmbeachzoo.org
Web Site: www.palmbeachzoo.org
Founded: 1969.
Congressional District: 23
Key Personnel: Pres., Luis Fernandez; Dir. Living Collections, Keith Lovett; C.O.O., W. Garrett Hambuechen; C.F.O., Kathleen Breland; Dir. Mktg., Claudia Harden; Dir. Education, Kristen Cytacki; Museum Shop Mgr., Lyn Monnette.
Personnel Profile: Full-Time Paid 79; Full-Time Volunteers 24; Part-Time Paid 21; Part-Time Volunteers 100; Interns 13.
Governing Authority: society; nonprofit Parent Institution: Zoological Society of the Palm Beaches. Tax-exempt: 501(c)(3).
Institution Type/Description: Zoo.
Collections: over 1,700 animals from around the world.
Research Fields: husbandry & breeding of endangered species.
Facilities: 5,000-vol. library for staff use; botanical garden; 23 acres exhibit space. Museum-related items for sale.
Activities: volunteer & docent program; formal education programs for children; lecture & presentation series; wildlife rescue & rehabilitation; interactive fountain.
Publications: monthly e-newsletter, Zoo Bytes; quarterly magazine, Palm Beach Zoo.
Hours & Admission Prices: Daily 9-5. Adults $12.95, senior citizens $9.95, children 3-12 $8.95; discounts to groups, AAA & AZA members; children under 3 & zoo members no charge. Closed Thanksgiving; Christmas. &
Attendance: 250,000 (estimated)
Membership: Senior $55; Individual Plus $60; Family $85; Friends of the Zoo $150; Big Cat Society $1,000.

SOUTH FLORIDA SCIENCE CENTER & AQUARIUM, INC., 4801 Dreher Trail N., West Palm Beach, FL 33405-3017. Tel.: 561-832-1988. Fax: 561-833-0551.
E-mail: media@sfsm.org
Web Site: www.sfsciencecenter.org
Formerly: South Florida Science Museum
Founded: 1959.
Congressional District: 11
Key Personnel: C.E.O., Lew Crampton; Chm. Bd., Matthew Lorentzen; C.O.O., Kate Arrizza; Dir. Devel., Marcy Hoffman; Dir. Events, Kristina Holt; Dir. Operations, Jeff Gourdouze; Aquarium Curator, Rebecca Shearer; Museum Shop Mgr., Lila Klix.
Personnel Profile: Full-Time Paid 25; Part-Time Paid 10; Part-Time Volunteers 50.
Governing Authority: nonprofit organization Tax-exempt: 501(c)(3); 170(b)(1)(a)(vi).
Institution Type/Description: Science Center.
Collections: natural history; paleontology; archaeology; astronomy; geology; concology.
Major Exhibits: Titanic (T), 11/13-4/14.
Research Fields: shark tagging; coral surveys; aquatic plants; aquaculture.
Facilities: planetarium; discovery room; observatory; aquarium; auditorium; exhibit hall; outdoor science trails; Egypt gallery.
Activities: lectures; films; formally organized educational programs; traveling, permanent & temporary exhibits; science theater; teacher in-services.
Publications: annual report; Teacher's Guide; magazine.
Hours & Admission Prices: Mon.-Fri. 9-5, Sat.-Sun. 10-6. Adults $13.50, seniors $12, children 3-12 $10; discounts to groups, AAM, AAA & ASTC members; children under 3 & members no charge. Call for information on Planetarium & Laser Light shows. Closed Thanksgiving; Christmas. &
Attendance: 117,000 (accurate)
Membership: Individual $80; Family $100; Family Plus $150.

YESTERYEAR VILLAGE, 9067 Southern Blvd., South Florida Fairgrounds, West Palm Beach, FL 33411-3625. Mailing Address: P.O. Box 210367, West Palm Beach, FL 33421-0367. Tel.: 561-790-5232; 800-640-3247. Fax: 561-753-2124.
Founded: 1990.
Congressional District: 23
Key Personnel: C.E.O. & Dir., Richard Vymlatil; Devel., Vicki Chouris; Chm. (V), Dare Peterson; Chm. (V), Dan Hulen; Public Rels., John Picano; Treas., Matt Wallsmith; Security, Ray Gregory.
Personnel Profile: Full-Time Paid 2; Part-Time Volunteers 576.
Governing Authority: Tax-exempt.
Institution Type/Description: History Museum.
Collections: 30 restored buildings including blacksmith shop; general store; pineapple processing plant; Riddle house.
Facilities: Museum-related items for sale.
Activities: education programs; birthday parties; rental facility; school outings.
Hours & Admission Prices: Tues.-Sun. 10-5. Adults $6, seniors $5, children $4; children under 5 no charge. &

White Springs

STEPHEN FOSTER FOLK CULTURE CENTER STATE PARK, 11016 Lillian Saunders Dr., White Springs, FL 32096-0435. Mailing Address: P.O. Drawer G, White Springs, FL 32096-0435. Tel.: 386-397-2733. Fax: 386-397-4262. www.stephenfostercso.org.
E-mail: andrea.thomas@dep.state.fl.us
Web Site: www.floridastateparks.org/stephenfoster
Founded: 1939.
Congressional District: 4
Key Personnel: CSO Pres. (V), Carol Stob; Asst. Park Mgr., Michelle Waterman; Museum Docent, Pat Cromer; Gift Shop Mgr., Susan Conley.
Personnel Profile: Full-Time Paid 15; Part-Time Volunteers 80.
Governing Authority: state. Parent Institution: Dept. of Environmental Protection, Div. of Recreation and Parks, 3900 Commonwealth Blvd., Commonwealth Bldg., Tallahassee, FL 32303. Tax-exempt: 170(b)(1)(A).
Institution Type/Description: Folk Culture Museum: located on the Suwannee River at White Springs, FL.
Collections: Howard Chandler Christy paintings; musical instruments; minstrel show material; manuscripts; moving dioramas depicting Stephen Foster's folk songs & desk; 97 bell carillon tower; world's largest carillon bell system.
Research Fields: life & works of Stephen C. Foster; the Folk Arts; Suwannee River Life & History; Florida folklife.
Facilities: Florida crafts & museum-related gifts for sale.
Activities: guided tours; concerts; cultural festivals; formally organized education programs for children; workshops; craft demonstrations permanent exhibitions; Jeanie auditions & scholarship program; Christmas programs. Museum Sponsors: Stephen Foster Day in January; Florida Folk Festival.
Publications: books, Suwannee River & Biography of Stephen C. Foster; American Troubadour; single sheets, Biography of Stephen C. Foster.
Hours & Admission Prices: Daily 8-sunset. $5 per vehicle (up to 8 people), $2 per person over 8; motorcoach rates available. &
Attendance: 100,000 (estimated)
Membership: Annual fee $20.

Winter Garden

CENTRAL FLORIDA RAILROAD MUSEUM, 101 S. Boyd St., Winter Garden, FL 34787-3500. Mailing Address: P.O. Box 770567, Winter Garden, FL 34777-0567. Tel.: 407-656-0559.
E-mail: info@cfrhs.org
Web Site: www.cfrhs.org/
Founded: 1981.
Congressional District: 8
Key Personnel: Pres. (V), Phil Cross; Cur., Ken Murdock; Museum Shop Mgr., Irv Lipscomb.
Personnel Profile: Part-Time Paid 1; Part-Time Volunteers 18.
Governing Authority: Parent Institution: Winter Garden Heritage Foundation. Subsidiary Institution: Central Florida Railway Historical Society. Tax-exempt.
Institution Type/Description: Historical Railroad Museum.
Collections: railroading history in Central Florida; railroad dining car china; 1938 Fairmont motor car; Clinchfield caboose; railroad industry artifacts.
Activities: Field trips.
Publications: monthly newsletter, The Flatwheel; Outline History of Central Florida Railroads.
Hours & Admission Prices: Daily 1-5. No charge; donations accepted.
Attendance: 6,825 (accurate)
Membership: Society Family $10; Society Student $15; Society Friend $25; Society Individual $45; Society Century $100; Bronze Corporate $250; Gold Corporate $500; Diamond Corporate $750; Platinum Corporate $1,000.

WINTER GARDEN HERITAGE MUSEUM, 1 N. Main St., Winter Garden, FL 34787-2824. Mailing Address: P.O. Box 770657, Winter Garden, FL 34777-0657. Tel.: 407-656-3244. Fax: 407-656-0110.
E-mail: museum@wghf.org
Web Site: www.wghf.org
Founded: 1998.
Congressional District: 10
Key Personnel: Dir., Kay Cappleman; Chm. (V), Lori Gibson.
Personnel Profile: Full-Time Paid 1; Part-Time Paid 6; Part-Time Volunteers 248; Interns 2.
Volunteer Hours: 6,000
Operating Expenses: 190,000
Operating Income: 189,255
Governing Authority: Parent Institution: Winter Garden Heritage Foundation.

Subsidiary Institution: Central Florida Railroad Museum, 101 S. Boyd St., Winter Garden, FL. Tax-exempt.
Institution Type/Description: Heritage History Museum.
Collections: Winter Garden history; architecture; photographs; Native American artifacts; citrus labels.
Major Exhibits: Shipping Gold: the Packing Houses of West Orange County (T), 2/14-4/14; Orange County's Most Popular Sheriff: The Life and Legacy of Dave Starr (T), 11/14-12/14.
Activities: Annual Event: Music Fest in October.
Publications: Heritage Windows.
Hours & Admission Prices: Daily 1-5. No charge; donations accepted.
Attendance: 20,000 (estimated)
Membership: Silver Partnership: Senior & Student $35; Partner $50; Corporate $500. Gold Partnership: Senior & Student $50; Partner $75; Corporate $750. Diamond Partnership: Senior & Student $75; Partner $100; Corporate $1,000.

Winter Park

ALBIN POLASEK MUSEUM AND SCULPTURE GARDENS, 633 Osceola Ave., Winter Park, FL 32789-4429. Tel.: 407-647-6294. Fax: 407-647-0410.
Web Site: www.polasek.org/
Founded: 1961.
Institution Type/Description: Art Museum.
Collections: sculptures; paintings.
Facilities: gardens.
Activities: weddings; special events.
Hours & Admission Prices: Sept.-June Tues.-Sat. 10-4, Sun. 1-4; groups by appointment. Adults $5, seniors $4, students $3; members & children under 12 no charge. &
Membership: Individual $40; Student/Teacher/Senior 60 & over $30; Family or Dual $60; Participating $100. Friends of the Polasek: Contributing $200; Supporting $350; Sustaining $500; Corporate & Benefactor $1,000; Grand Benefactor $2,500; Preservationist $5,000; Lifetime $25,000 & up.

THE CHARLES HOSMER MORSE MUSEUM OF AMERICAN ART, (M), 445 N. Park Ave., Winter Park, FL 32789-3212. Tel.: 407-645-5311 & 5316. Fax: 407-647-1284.
Web Site: www.morsemuseum.org
Founded: 1942.
Congressional District: 9
Key Personnel: Dir., Dr. Laurence J. Ruggiero; Cur. & Collections Mgr., Jennifer Thalheimer; Dir. Public Affairs, Catherine Hinman; Bldg. Mgr., Tom Mobley; Museum Shop Mgr., Ava Maxwell; Cur., Donna Climenhage; Cur. Education, Betsy Peters.
Personnel Profile: Full-Time Paid 24; Part-Time Paid 12; Part-Time Volunteers 62.
Volunteer Hours: 1,200
Governing Authority: nonprofit organization. Parent Institution: Charles Hosmer Morse Foundation, 445 N. Park Ave., Winter Park, FL 32789. Tax-exempt.
Institution Type/Description: American Art Museum.
Collections: windows, lamps, blown glass, jewelry, pottery & paintings by Louis Comfort Tiffany & other turn-of-the-century masters; late 19th and early 20th-century American paintings; art pottery; exhibits include Tiffany's 1893 Chapel for the World's Columbian Exposition, windows, furniture and other objects designed by Tiffany for his personal use & from his Long Island home, Laurelton Hall.
Research Fields: American art 1850s-1930s.
Facilities: galleries. Museum-related items for sale.
Activities: guided docent tours; lectures; interpretive programs.
Publications: American Art Pottery; The Tiffany Chapel at the Morse Museum.
Hours & Admission Prices: May-Oct. Tues.-Sat. 9:30-4, Sun. 1-4; Nov.-April Tues.-Thurs. & Sat. 9:30-4, Fri. 9:30-8, Sun. 1-4; guided tours by appointment. Adults $5, seniors $4, students $1; discounts to museum professionals; children 12 and under, Fri. 4-8 & members no charge. Closed New Year's Day; Memorial Day; Labor Day; Thanksgiving; Christmas. &
Attendance: 79,426 (accurate)
Membership: Student & Teacher $5; Individual $20; Family $30; Contributing $50; Benefactor $100; Sustaining $1,000.

CREALDE SCHOOL OF ART - ALICE & WILLIAM JENKINS GALLERY, 600 St. Andrews Blvd., Winter Park, FL 32792. Tel.: 407-671-1886. Fax: 407-671-0311.
E-mail: rberrie@crealde.org
Web Site: www.crealde.org/galleries.html
Founded: 1975.
Institution Type/Description: Art Gallery.
Collections: works by national & international artists; paintings; sculpture.

Activities: classes; special events. Museum Sponsors: Annual Juried Student Show.
Hours & Admission Prices: Mon.-Thurs. 9-4, Fri.-Sat. 9-1. No charge; donations accepted. &

*** THE GEORGE D. AND HARRIET W. CORNELL FINE ARTS MUSEUM, (M),** Rollins College, 1000 Holt Ave., Winter Park, FL 32789-4499. Tel.: 407-646-2526. Fax: 407-646-2524. Facebook: Cornell Fine Arts Museum.
E-mail: stodd@rollins.edu
Web Site: www.rollins.edu/cfam
Founded: 1978.
Congressional District: 8
Key Personnel: Dir. Bruce A. Beal, Ena Heller, Ph.D.; Chm. (V), John Giaimo; Registrar, Leslie Cone; Cur., Jonathan F. Walz, Ph.D.; Cur. Education, Daniel Powell; Exec. Asst., Sandy Todd; Donor & Guest Rels. Liaison, Dana Thomas; Guest Svcs., Louise Buyo; Security Officer, Brian Glenn.
Personnel Profile: Full-Time Paid 5; Part-Time Paid 2; Part-Time Volunteers 96; Interns 1.
Governing Authority: college. Parent Institution: Rollins College. Tax-exempt: 501(c)(3).
Institution Type/Description: Art Museum.
Collections: 5,000 objects from the 14th to 21st centuries: European paintings of the Renaissance through contemporary; American & European paintings; bronzes; sculptures; prints; drawings; posters; photographs; Native American art; Kress Collection; Smith Watch Key Collection; decorative arts: furniture, glass, ceramics.
Research Fields: American & European art.
Facilities: college art museum.
Activities: temporary, loan & traveling exhibitions; lectures; films; workshops; guided tours.
Publications: permanent collection handbook, Treasures of the Cornell Fine Arts Museum; newsletter; exhibition catalogues; museum brochures; post cards; exhibition schedule brochures; posters; education brochure.
Hours & Admission Prices: Tues.-Fri. 10-4, Sat.-Sun. 12-5. No charge; donations accepted. Closed major holidays. &
Attendance: 25,000 (estimated)
Membership: Individual $30; Basic $60; Contributing $100; Patron $200; Salon $500; Benefactor $1,000 & up.

HANNIBAL SQUARE HERITAGE CENTER, 642 W. New England Ave., Winter Park, FL 32789. Tel.: 407-539-2680.
Web Site: www.crealde.org/heritage_center_gallery.html
Institution Type/Description: Art Gallery.
Collections: photographs; oral histories.
Activities: educational programs; special events.
Hours & Admission Prices: Tues.-Thurs. 12-4, Fri. 12-5, Sat. 10-2. No charge.

SHOWALTER HUGHES COMMUNITY GALLERY, 600 St. Andrews Blvd., Winter Park, FL 32792. Tel.: 407-671-1886.
Web Site: www.crealde.org
Founded: 1975.
Institution Type/Description: Art Gallery.
Collections: works by students, faculty & emerging artists.
Activities: outreach programs; lectures; temporary exhibitions.
Hours & Admission Prices: Mon.-Thurs. 9-4, Fri.-Sat. 9-1. No charge, donations accepted.

Zephyrhills

ZEPHYRHILLS DEPOT MUSEUM, 39110 South Ave., Zephyrhills, FL 33542-5255. Tel.: 813-780-0067.
Founded: 1997.
Key Personnel: Dir., Vicki S. Elkins.
Governing Authority: city.
Institution Type/Description: History Museum: housed in the restored 1927 Atlantic Coast Line Railroad Depot.
Collections: history of trains, local area, & culture; personal artifacts; photographs; school memorabilia; railroad artifacts.
Facilities: Museum-related items for sale.
Hours & Admission Prices: Thurs.-Fri. 9-1, Sat. 9-12. No charge. &
Attendance: 1,200 (estimated)

Zolfo Springs

CRACKER TRAIL MUSEUM, 2822 Museum Dr., Zolfo Springs, FL 33890-9433. Tel.: 863-735-0119.
E-mail: judith.george@hardeecounty.net
Web Site: www.hardeecounty.net
Founded: 1967.
Congressional District: 13
Key Personnel: Cur., Judith George; Supt. Bldg. & Grounds, Daniel Weeks.
Personnel Profile: Full-Time Paid 1; Part-Time Volunteers 4.
Governing Authority: county. Parent Institution: Hardee County BoCC. Tax-exempt.
Institution Type/Description: Park Museum.
Collections: aeronautics; agriculture; anthropology; archaeology; archives; paintings; decorative arts; circus costumes; history; Indian artifacts; industrial; medical; natural history; textiles; transportation; old train; old-time blacksmith shop.
Major Exhibits: Pioneer Park Days, 2/26/14-3/2/14.
Facilities: animal refuge; campgrounds; playground; closed & open air building & pavilion rentals.
Activities: guided tours specializing in educational groups. Annual Event: Pioneer Park Days.
Hours & Admission Prices: Mon.-Fri. 9-5. Adults $2; children under 5 no charge. &
Attendance: 6,809 (accurate)
Membership: Individual $20.

GEORGIA

(349 listings)

Albany

ALBANY CIVIL RIGHTS INSTITUTE, 326 Whitney Ave., Albany, GA 31701-2861. Mailing Address: P.O. Box 6036, Albany, GA 31706-6036. Tel.: 229-432-1698. Fax: 229-432-2150.
E-mail: iturner@acrmm.org
Web Site: www.albanycivilrightsinstitute.org
Formerly: Albany Civil Rights Movement Museum at Old Mt. Zion Church
Founded: 1994.
Congressional District: 2
Key Personnel: Pres. (V), Charlie Crapps; Admin., Irene L. Turner.
Personnel Profile: Part-Time Paid 6; Part-Time Volunteers 10; Interns 1.
Governing Authority: private; nonprofit organization. Tax-exempt: 501(c)(3).
Institution Type/Description: History Museum.
Collections: photographs; artifacts; archives; Civil Rights; African American history.
Research Fields: civil rights movement in Albany & southwest Georgia; history of Mount Zion Baptist Church in Albany GA; the significance of music in the civil rights movement.
Facilities: library; archives.
Activities: temporary & permanent exhibitions; monthly lecture series; monthly performances of ACRI Freedom Singers.
Hours & Admission Prices: Tues.-Sat. 10-4. Adults $6, senior citizens, students 5th-12th grade, military & college students $5, children 1st-4th grade $3, pre-school $2; discounts to groups of 20 or more; members & children under 4 no charge. Closed holidays. &
Attendance: 6,500 (accurate)
Membership: Student & Senior Citizen $25; Individual $45; Family $65; Nonprofit Organization $325; Corporation & Business $625.

*** ALBANY MUSEUM OF ART,** 311 Meadowlark Dr., Albany, GA 31707-5704. Tel.: 229-439-8400. Fax: 229-439-1332.
E-mail: info@albanymuseum.com
Web Site: www.albanymuseum.com
Founded: 1964.
Congressional District: 2
Key Personnel: Interim Exec. Dir., Karen Kemp; Chm. (V), Banks Margeson; Cur., Merritt Giles.
Personnel Profile: Full-Time Paid 4; Full-Time Volunteers 2; Part-Time Paid 4; Part-Time Volunteers 300.
Governing Authority: nonprofit organization. Tax-exempt: 501(c)(3).
Institution Type/Description: Art Museum.
Collections: African art; 16th- to 17th-century European drawings; American paintings & works on paper since 1945; American & European decorative arts; folk art; Southern art.
Research Fields: 20th century American art; American & European paintings & graphics; African objects.
Facilities: classrooms; auditorium; preparatory space; hands-on children's gallery.
Activities: guided tours; lectures; curriculum & lesson plan packets for teachers; art classes; children's membership programs; birthday parties; children's art fair; auction; education programs; docent program; loan, permanent, temporary & traveling exhibitions; community loan service; outreach programs.
Publications: bimonthly newsletter, ARTalk.

Hours & Admission Prices: Tues.-Sat. 10-5. No charge. Closed major holidays.
&

Attendance: 20,000 (estimated)

Membership: Student/Military Family $30; Family/Individual $50; Supporting $100; Individual Patron $150; Patron Couple $250; Benefactor $500.

CHEHAW WILD ANIMAL PARK, 105 Chehaw Park Rd., Albany, GA 31701-1260. Tel.: 229-430-5275. Fax: 229-430-3035.

E-mail: info@chehaw.org

Web Site: www.chehaw.org

Institution Type/Description: Zoo.

Collections: over 212 animals representing more than 55 species; African black rhino; American bald eagle; lemurs; meerkats; zebras; cheetahs; alligators; tortoise.

Facilities: 100-seat amphitheater; nature trails; picnic areas. Gift items for sale.

Activities: educational programs; special events; train rides; alligator feedings; cheetah runs.

Hours & Admission Prices: Park: adults $2, seniors 62 & over, military and children 4-12 $1.Park & Zoo: adults 13-61 $8.25, seniors 62 & over $7.25, military & children 4-12 $5.25; children under 4 no charge. Train: $3; children under 2 no charge.

Membership: Single $35; Dual $50; Family $75.

FLINT RIVER QUARIUM, 101 Pine Ave., Albany, GA 31701-2593. Mailing Address: 117 Pine Ave., Albany, GA 31701-2593. Tel.: 229-639-2650. Fax: 229-639-2707.

Web Site: www.flintriverquarium.com

Founded: 2004.

Key Personnel: Dir., Scott W. Loehr; Chm. (V), Emily McAfee; Devel. & Public Rels., Wendy Bellacomo; Education, Melissa Martin; Cur., Richard Brown; Accounts Mgr., Vonda Hancock; Operations Mgr., Kathy Batson; Imagination Theater Mgr., Vashion Milledge; Membership & Guest Svcs. Mgr., Vicki Churchman; Aquarist, Kelly Putnam; Aquarist, Melissa Scott; Aquarist, Amanda Margraves; Museum Shop Mgr., Claudia Durham.

Personnel Profile: Full-Time Paid 18; Part-Time Paid 10; Part-Time Volunteers 92; Interns 2.

Governing Authority: private; nonprofit organization. Tax-exempt: 501(c)(3).

Institution Type/Description: Natural Science Museum.

Collections: local flora & fauna; aquatic animals.

Facilities: aquarium; aviary; educational facilities; 28,000 sq. ft. exhibit space; 103-seat large format theater. Gift items for sale.

Activities: docent program; films; guided tours; participatory exhibits; rental gallery; theater.

Publications: quarterly newsletter, The Blue Hole Banner.

Hours & Admission Prices: Mon.-Fri. 9-5, Sat. 10-6, Sun. 1-5. Adults $9, senior citizens $8, children $6, students $4.50; discounts to groups of 15 or more; members no charge. Closed Thanksgiving; Christmas. &

Attendance: 66,285 (accurate)

Membership: Individual $49; Family Plus $79; Family Adventure $99; Friend $149; Contributor $349; Blue Hole Society $1,000-$10,000.

THRONATEESKA HERITAGE FOUNDATION, INC., 100 W. Roosevelt Ave., Albany, GA 31701-2325. Tel.: 229-432-6955. Fax: 229-435-1572.

E-mail: info@heritagecenter.org

Web Site: www.heritagecenter.org

Founded: 1974.

Congressional District: 2

Key Personnel: C.E.O. & Exec. Dir., Tommy Gregors; Pres. (V), Anne Stokes; Mgr. Operations, Cheryl Jones.

Personnel Profile: Full-Time Paid 6; Part-Time Paid 5; Part-Time Volunteers 6; Interns 2.

Governing Authority: nonprofit corporation. Tax-exempt.

Institution Type/Description: History Museum.

Collections: costumes; pioneer tools; rocks; minerals; shells; historic artifacts; Indian artifacts; Nelson Tift furnishings; antique carriages; Confederate War artifacts; local artifacts of the history of Southwest Georgia area; railroad cars with model train exhibit; 1911 steam locomotive; manuscripts. Historic Buildings: 1913 Union Station; 1857 depot; 1919 Railway Express Bldg.

Research Fields: planetarium; local history.

Facilities: archives; planetarium; community meeting room. Wetherbee Planetarium: 40' diameter high definition full dome screen.

Activities: guided tours; formally organized education programs; temporary exhibits.

Publications: weekly science newsletter, Word From the Wetherbee; monthly astronomy outlook & calendar.

Hours & Admission Prices: Museum: Thurs.-Sat. 10-4; other times by appointment. No charge. Planetarium: Thurs.-Fri. 10:30-11:30, Sat. 1, 2, 3.

Admission $3.50; members no charge. Closed New Year's Day; Martin Luther King Jr. Day; Memorial Day; Labor Day; Thanksgiving; Christmas.
&

Attendance: 25,418 (accurate)

Membership: Conductor $25; Station Master $35; Signalman $50; Brakeman $100; Steward $250; Engineer $500; Locomotive $1,000.

Alpharetta

MANSELL HOUSE AND GARDENS, 1835 Old Milton Pkwy., Alpharetta, GA 30009. Tel.: 770-475-4663.

E-mail: info@alpharettahistoricalsociety.org

Governing Authority: nonprofit organization. Parent Institution: Alpharetta Historical Society. Tax-exempt: 501(c)(3).

Institution Type/Description: Historic House Museum: built in 1912.

Collections: Alpharetta & old Milton County history; genealogy records; period furnishings; photographs.

Facilities: rental facilities. Gift items for sale.

Activities: rental facilities; research.

Hours & Admission Prices: Mon. 11-2, Wed. & Fri. 10-2; other times by appointment.

Americus

RYLANDER THEATER, 310 W. Lamar St., Americus, GA 31709-3543. Mailing Address: P.O. Box 864, Americus, GA 31709-0864. Tel.: 912-931-0001. Fax: 912-928-2466.

Web Site: www.rylander.org

Key Personnel: Gen. Dir., Norman S. Easterbrook

Institution Type/Description: Historic Building: restored working theatre built in 1921.

Collections: art exhibits; classic movies.

Facilities: theater.

Activities: guided tours; stage performances.

Hours & Admission Prices: Tours: by appointment. Performances: call for hours. Closed New Year's Day & day after; Memorial Day; Independence Day; Thanksgiving & day after; Christmas Eve, Day & day after.

Andersonville

*** ANDERSONVILLE NATIONAL HISTORIC SITE, (M),** 496 Cemetery Rd., Andersonville, GA 31711-4040. Tel.: 229-924-0343. Fax: 229-924-1086.

Web Site: www.nps.gov/ande

Founded: 1971.

Congressional District: 2

Key Personnel: Supt., Brad Bennett; Chief of Interpretation & Education, Eric Leonard; Museum Shop Mgr., David Stroup.

Personnel Profile: Full-Time Paid 15; Part-Time Paid 1; Part-Time Volunteers 147; Interns 2.

Governing Authority: federal. Parent Institution: National Park Service, Dept. of the Interior. Tax-exempt: 170(b)(1)(A).

Institution Type/Description: National Park & Historic Site: Civil War P.O.W. camp.

Collections: items related to the Andersonville prison; items related to P.O.W. camps; Civil War & later conflicts; 1890-1910 monuments & sculptures. Historic Buildings: 1877 Cemetery Sexton's residence; 1908 Cemetery chapel.

Research Fields: P.O.W.'s in Andersonville; Civil War; P.O.W.'s in other wars.

Facilities: 3,000-vol. library, official records of the War of the Rebellion, general Civil War, Civil War prisons, Andersonville prison & later P.O.W. camps & woman's relief corps journals available for research on premises only. Books, postcards & slides for sale.

Activities: guided tours; lectures; slide shows; permanent & temporary exhibitions.

Publications: brochure; bulletin, P.O.W. Site.

Hours & Admission Prices: Visitors Center: daily 8:30-5. Park: daily 8-5. No charge. Visitor Center: closed New Year's Day; Thanksgiving; Christmas. Grounds open. &

Attendance: 150,000 (estimated)

Membership: Friends of Andersonville: Senior Citizen $10; Individual $15; Family $25; Corporate & Business $100.

Athens

CHURCH-WADDEL-BRUMBY HOUSE MUSEUM & ATHENS WELCOME CENTER, 280 E. Dougherty St., Athens, GA 30601-2611. Tel.: 706-353-1820. Fax: 706-353-1770.

E-mail: athenswc@negia.net

Web Site: www.athenswelcomecenter.com
Founded: 1968.
Congressional District: 10
Key Personnel: Dir. & Museum Shop Mgr., Evelyn Reece; C.E.O., Amy Kissane; Pres. (V), Jan Levinson.
Personnel Profile: Full-Time Paid 1; Part-Time Paid 2; Part-Time Volunteers 10; Interns 2.
Volunteer Hours: 950
Governing Authority: nonprofit organization. Administered by Athens-Clarke Heritage Foundation, Firehall No. 2, 489 Prince Ave., Tel.: 706-353-1801, Fax: 706-552-0753. Tax exempt: 501(c)(3).
Institution Type/Description: Historic House: housed in c.1820 Church-Waddel-Brumby House.
Collections: photographs; period furnishings.
Research Fields: local history.
Facilities: Gift items for sale.
Activities: daily guided tours & self-guided tours.
Hours & Admission Prices: Mon.-Sat. 10-5, Sun. 12-5. No charge; donations accepted. Closed New Year's Day; Thanksgiving; Christmas. No charge; donations accepted. &
Attendance: 13,000 (accurate)

CIRCLE GALLERY/COLLEGE OF ENVIRONMENT & DESIGN - UNIVERSITY OF GEORGIA, G14 Caldwell Hall, Athens, GA 30602. Mailing Address: 609 Caldwell Hall, Athens, GA 30602. Tel.: 706-542-8292. Fax: 706-542-4485.
E-mail: rds@uga.edu
Web Site: www.sed.uga.edu/gallery
Formerly: SED Gallery/School of Environmental Design - University of Georgia
Founded: 1993.
Congressional District: 11
Key Personnel: C.E.O., Rene D. Shoemaker.
Personnel Profile: Full-Time Paid 1; Part-Time Paid 7; Interns 1.
Governing Authority: public university; nonprofit organization. Parent Institution: University of Georgia. Subsidiary Institution: College of Environment & Design.
Institution Type/Description: General Museum.
Collections: architecture; landscape architecture; historic preservation; art.
Facilities: 6,000-vol. library.
Activities: lectures, loan & temporary exhibitions; receptions.
Hours & Admission Prices: Mon.-Fri. 8:30-6. No charge. Closed holidays. &
Attendance: 2,500 (estimated)

* **GEORGIA MUSEUM OF ART, UNIVERSITY OF GEORGIA, (M),** 90 Carlton St., Athens, GA 30602-6719. Tel.: 706-542-4662. Fax: 706-542-1051. Facebook: Georgia Museum of Art.
E-mail: mlachow@uga.edu
Web Site: www.georgiamuseum.org
Founded: 1945.
Congressional District: 10
Key Personnel: Dir., William U. Eiland; Deputy Dir., Annelies Mondi; Chm. (V), B. Heyward Allen, Jr.; Pres. Friends of the Museum, Karen Prasse; Coord. Public Rels., Michael Lachowski; Chief Preparator, Todd Rivers; Editor, Hillary Brown; Cur. Education, Carissa Dicindio; Assoc. Cur. Education, Callan Steinmann; Business Mgr., Lisa Conley; Art Handler, Larry Forte; Museum Shop Mgr., Amy Miller; Cur. Pierre Daura, Lynn Boland; Cur. Decorative Arts, Dale Couch.
Personnel Profile: Full-Time Paid 30; Part-Time Paid 5; Part-Time Volunteers 150; Interns 14.
Governing Authority: state. Parent Institution: University of Georgia. Subsidiary Institution: Friends of the Museum. Tax-exempt: 170(B)(1)(A).
Institution Type/Description: Art Museum.
Collections: 19th- to 20th-century American painting; Kress Study Collection of Italian Renaissance paintings; Larry D. and Brenda A. Thompson Collection of African American art; European, American and Japanese prints & drawings; contemporary American photographs; decorative arts; folk art.
Major Exhibits: Art Interrupted: Advancing American Art and the Politics of Cultural Diplomacy (T), 1/25/14-4/20/14; Rugs of the Caucuses, 1/30/14-4/27/14; Lamar Dodd School of Art MFA Candidates Exhibition 2014, 4/12/14-5/4/14; Carroll Cloar Prints Show (T), 5/17/14-8/10/14; Women, Art, and Social Change: The Newcomb Pottery Enterprise (T), 5/17/14-8/31/14; Picturing America: Signature Works from the Westmoreland Museum of American Art (T), 6/14/14-8/24/14; Tomata du Plenty, 8/30/14-11/23/14; Elephant 6 Collective Exhibitions, 9/13/14-11/16/14; The History of the American Band: Art, Instruments, and Ephemerae from the Collection of George Foreman, 11/1/14-1/25/15; Not Ready to Make Nice: Guerilla Girls in the Art World & Beyond (T), 12/6/14-3/1/15.

Research Fields: American paintings, drawings & prints; old master & European modern prints.
Facilities: library of art history books; 16,000 sq. ft. gallery space; 200-seat auditorium; studio classroom; 3 classrooms; outdoor sculpture garden; study center; A-V theater; cafe; teacher resource center. Museum-related items for sale.
Activities: guided tours; docent program; gallery talks for children & adults; internships; educational programs for children, adults, families, senior citizens & university students; lectures; statewide outreach, temporary & traveling exhibitions; regular film series.
Publications: Georgia Museum of Art Bulletin; quarterly newsletter; exhibition catalogs & brochures; gallery notes; docent newsletter; calendar.
Hours & Admission Prices: Tues.-Wed. & Fri.-Sat. 10-5, Thurs. 10-9, Sun. 1-5. No charge; donations accepted. &
Attendance: 85,000 (accurate)
Membership: Student $20; Senior $25; Individual $35; Senior Couple $45; Family /Couple $65; Contributing $100; Donating $250; Sustaining $500; Benefactor $5,000.

GEORGIA MUSEUM OF NATURAL HISTORY, Natural History Bldg., University of Georgia, Athens, GA 30602-1882. Tel.: 706-542-1663. Fax: 706-542-3920.
E-mail: musinfo@uga.edu
Web Site: museum.nhm.uga.edu
Founded: 1977.
Congressional District: 10
Key Personnel: Dir. & Cur. Zoology, Dr. B. Freeman; Pres. (V), Jean Porter; Cur. Zooarchaeology, Dr. E. Reitz; Cur. Mineralogy, Dr. P. Schroeder; Cur. Archaeology, Dr. D.J. Hally; Cur. Education & Outreach, Dr. C. Hoffman; Cur. Botany, Dr. D. Giannasi; Cur. Botany, Dr. W. Zomlefer; Cur. Entomology, Dr. J. McHugh; Cur. Economic Geology, Dr. D. Crowe; Cur. Mycology, Dr. R. Hanlin.
Personnel Profile: Full-Time Paid 17; Part-Time Paid 12; Part-Time Volunteers 3; Interns 26.
Governing Authority: university. Parent Institution: University of Georgia. Tax-exempt.
Institution Type/Description: Natural History Museum.
Collections: archaeology; botany; entomology; economic geology; mycology; paleontology; zooarchaeology; palynology; mineralogy; mammalogy; herpetology; ornithology; ichthyology.
Research Fields: anthropology; archaeology; botany; entomology; geology; mycology; zoology.
Facilities: laboratories; classrooms.
Activities: educational outreach programs; science box program; tours; special exhibits; collections available to qualified researchers.
Hours & Admission Prices: By appointment. No charge. Closed holidays. &
Attendance: 10,000 (estimated)
Membership: Student & Senior Citizens $15; Individual $30; Family $40.

LYNDON HOUSE ARTS CENTER, 293 Hoyt St., Athens, GA 30601-2648. Tel.: 706-613-3623. Fax: 706-613-3627.
Web Site: lyndonhouseartsfoundation.wordpress.com
Founded: 1973.
Congressional District: 10
Key Personnel: Dir., Claire Benson; Cur., Nancy Lukasiewicz; Pres. (V) Lyndon House Arts Foundation, Lou Kudon; Museum Shop Mgr., Celia Brooks.
Personnel Profile: Full-Time Paid 4; Part-Time Paid 1; Part-Time Volunteers 500.
Volunteer Hours: 5,000
Operating Expenses: 500,000
Operating Income: 500,000
Governing Authority: municipal. A facility of Athens-Clarke County Dept. of Leisure Services. Tax-exempt.
Institution Type/Description: Art Center: c.1850 Ware-Lyndon House, originally built by Dr. Edward R. Ware & later sold to Dr. Edward Smith Lyndon.
Collections: decorative arts; furnishings; personal artifacts; historic house.
Major Exhibits: 39th Juried Exhibition, 2/14; S.E. Pastel Society International Juried Exhibition, 5/14; Three Dimensions of Art in Georgia, 8/14; Full House, 11/14.
Facilities: library; studio classrooms; meeting room; historic house museum; galleries; gallery shop; children's wing.
Activities: special events; guided tours; lectures; gallery talks; arts festivals; meetings & workshops for local art organizations; formally organized education & awareness programs for children, adults, specialized groups & general public; art courses in photography; clay painting, printmaking, weaving & youth art; stained glass, internships for students at the University of Georgia.

Publications: periodic catalogs; newsletter.
Hours & Admission Prices: Tues. & Thurs. 12-9, Wed. & Fri.-Sat. 9-5. No charge; donations accepted. Closed Thanksgiving; Christmas. &
Attendance: 70,000 (accurate)

THE STATE BOTANICAL GARDEN OF GEORGIA, 2450 S. Milledge Ave., Athens, GA 30605-1674. Tel.: 706-542-1244. Fax: 706-542-3091.
E-mail: garden@uga.edu
Web Site: www.botgarden.uga.edu
Founded: 1968.
Congressional District: 10
Key Personnel: Dir., Wilf Nicholls; Pres. (V) Friends of the Garden, Mike Sikes; Chm. (V) Advisory Bd., Betsy Ellison; Dir. Research, Dr. James Affolter; Horticulturist, Jeannette Coplin; Dir. Education, Anne Shenk; Visitor Center Mgr., William Tonks; Information Specialist, Lisa Kennedy; Museum Shop Mgr., Susan Cooper; Plant Conservation Coord., Jennifer Ceska; Office Mgr., Jason Burdette; Administrative Sec., Beverly Morton; Coord. (V), Andrea Fischer.
Personnel Profile: Full-Time Paid 28; Part-Time Paid 10; Part-Time Volunteers 150; Interns 1.
Governing Authority: university. Parent Institution: University of Georgia. Subsidiary Institution: Friends of the Garden. Tax-exempt.
Institution Type/Description: University Botanical Garden.
Collections: woody plant collections; native flora; native & adapted trees & shrubs including shade & ornamental trees.
Research Fields: horticulture; botany; natural history; ecology; plant conservation.
Facilities: 140-seat chapel; 125-seat auditorium; classrooms; nature trails; visitor center & conservatory of tropical plants; conference facilities for up to 300 people. Museum-related items for sale.
Activities: lectures; films; concerts; workshops; trail walks; temporary exhibition of botanical art; guided tours.
Publications: quarterly, Garden News.
Hours & Admission Prices: Grounds: April-Sept. daily 8-8; Oct.-March daily 8-6. Visitor Center: Tues.-Sat. 9-4:30, Sun. 11:30-4:30. Cafe: Tues.-Fri. 11-2, Sat.-Sun. 11:30-3. No charge; donations accepted. Closed University holidays. &
Attendance: 200,000 (estimated)
Membership: Student $15; Senior Citizen $20; Nonprofit Organization $25; Senior Couple $30; Individual $35; Family $45; Sponsor $100; Donor $250; Contributor $500; Patron $1,000; Ambassador $2,500; Life $5,000.

T.R.R. COBB HOUSE, 175 Hill St., Athens, GA 30601. Tel.: 706-369-3513. Fax: 706-354-1054.
Key Personnel: Cur., Sam Thomas
Institution Type/Description: Historic House Museum: housed in the former home of lawyer & Confederate Army officer Thomas Cobb.
Collections: life & career of Thomas Cobb; period furnishings; photographs; personal artifacts.
Hours & Admission Prices: Tues.-Sat. 10-4. Suggested Donation: adults $2; children & students no charge. Closed holidays.

TAYLOR-GRADY HOUSE, 634 Prince Ave., Athens, GA 30601-2453. Tel.: 706-549-8688. Fax: 706-613-0860. facebook.com/taylorgradyhouse.
E-mail: jlathens@aol.com
Web Site: www.taylorgradyhouse.com
Founded: 1968.
Congressional District: 10
Key Personnel: Dir., Sarah Doughton.
Personnel Profile: Full-Time Paid 3.
Governing Authority: nonprofit organization. Junior League of Athens, Inc. Tax-exempt.
Institution Type/Description: Historic House Museum: 1845 Henry W. Grady Home.
Collections: period furnishings.
Activities: guided tours; rental gallery.
Publications: pamphlet.
Hours & Admission Prices: Mon., Wed. & Fri. 9-3, Tues. & Thurs. 9-1. Suggested Donation $3. Closed holidays. &
Attendance: 5,000 (estimated)

Atlanta

THE APEX MUSEUM, 135 Auburn Ave., N.E., Atlanta, GA 30303-2567. Tel.: 404-523-2739. Fax: 404-523-3248 (call first).
E-mail: apexmuseum@aol.com
Web Site: www.apexmuseum.org

Founded: 1978.
Congressional District: 5
Key Personnel: Pres., Dan Moore, Sr.; Gallery & Tour Coord., Michele Mitchell.
Personnel Profile: Full-Time Paid 4; Part-Time Volunteers 40; Interns 4.
Governing Authority: nonprofit organization. Parent Institution: Collections of Life & Heritage, Inc. Tax-exempt: 501(c)(3).
Institution Type/Description: History Museum & Building: 1910 John Wesley Dobbs Building, former School Book Depository; entity of the Sweet Auburn historic Freedom Walk.
Collections: Sankoya wood & brass artifacts; collection represents 10 West African countries; permanent African-American art collection which includes pieces by nationally renowned artists.
Facilities: 7,500 sq. ft. exhibit space; 50-seat Trolley Theatre. Museum-related items for sale.
Activities: guided tours; lectures; films; theatre; rental gallery; loan & temporary exhibitions; school loan service; docent program; underground railroad trek; tribute to the ancestors.
Publications: bimonthly, The APEX Times.
Hours & Admission Prices: Tues.-Sat. 10-5. Adults $6, senior citizens 55 & over and students $5; discounts to groups; members & children under 4 no charge. &
Attendance: 65,000 (accurate)
Membership: Student & Senior Citizen $15; Individual Adults $30; Family of 4 $50; Patron $100; Donor $250; Life $500.

AMERICAN BAPTIST HISTORICAL SOCIETY, 3001 Mercer University Dr., Atlanta, GA 30341-4115. Tel.: 678-547-6680.
E-mail: abhs@abhsarchives.org
Web Site: abhsarchives.org/index.shtml
Founded: 1853.
Congressional District: 34
Key Personnel: Exec. Dir., Dr. Deborah Bigham-Van Broekhoven; Communications & Reader Svcs., Clarence Brown, Jr.
Governing Authority: church. Affiliated with American Baptist Churches in United States Valley Forge, PA. 19481. Tax-exempt: 501(c)(3).
Institution Type/Description: Research Library.
Collections: archives; manuscripts; original records; books; pamphlets; photographs; paintings.
Research Fields: Baptist history; American religion.
Facilities: 60,000-vol. library of books pertaining to Baptistiana, archives & manuscripts collections available for use under staff supervision on premises.
Publications: quarterly magazine, American Baptist Quarterly; newsletter, The Primary Source.
Hours & Admission Prices: By appointment. &
Membership: Basic $25; Subscribers $40.

ATLANTA BOTANICAL GARDEN, 1345 Piedmont Ave., N.E., Atlanta, GA 30309-3366. Tel.: 404-876-5859. Fax: 404-876-7472.
E-mail: info@atlantabotanicalgarden.org
Web Site: www.atlantabotanicalgarden.org
Founded: 1976.
Congressional District: 5
Key Personnel: Pres. & C.E.O., Mary Pat Matheson; Supt., Dorothy Chapman Fuqua Conservatory, Ron Determann; Mktg., Sabina Carr; Mktg. Vice Pres., Mildred Pinnell; Horticulture Vice Pres., Josh Todd; Museum Shop Mgr., Kathleen Guy; Public Rels. Mgr., Danny Flanders; Visitor Svcs. Mgr., Jason Diem.
Personnel Profile: Full-Time Paid 80; Part-Time Paid 25; Part-Time Volunteers 600; Interns 2.
Governing Authority: nonprofit organization. Tax-exempt.
Institution Type/Description: Arboretum & Botanical Garden.
Collections: 15 acres of Georgia's native plant material; 15 acres of annual, perennial & woody plants; tropical & desert collections in Fuqua Conservatory; Fuqua Orchid Center with high elevation orchid house; tropical orchid display house; children's garden; Canopy Walk suspension bridge; edible garden.
Major Exhibits: Orchid Daze, 2/14-4/14; Imaginary Words, 5/14-10/14; Garden Lights, Holiday Nights, 11/14-1/15.
Research Fields: native & tropical plant conservation programs; conservation breeding programs involving poison & tree frogs; micro-propagation techniques for endangered species.
Facilities: 5,000-vol. library of botany, horticulture & taxonomy books available for research to members & visitors on premises; orchid reference library; tissue culture lab; glass conservatory; classrooms; workshop; conservation greenhouses. Museum-related items for sale.
Activities: guided tours; lectures; adult & children's classes; seasonal children's camps; entertainment program for children in Children's garden;

films; art shows; plant society exhibits; docent program; permanent exhibitions; off premises garden tours; festivals; social functions.
Publications: member newsletter, Clippings; 3 volunteer newsletter, Digging In; annual report.
Hours & Admission Prices: April Tues.-Sun. 9-7; May-Oct. Tues.-Wed. & Fri.-Sun. 9-7, Thurs. 9am-10pm; Nov.-March Tues.-Sun. 9-5. Adults $18.95, children 3-12 $12.95; members & children under 3 no charge. Closed New Year's Day; Thanksgiving; Christmas. &
Attendance: 425,000 (accurate)
Membership: Individual $69; Dual $89; Family $99; Family Plus $129; Contributing $179; Supporting $329; Donor $500; Director's $1,000; Arbor Circle $2,500; Magnolia Circle $5,000; Orchid Circle $10,000.

ATLANTA CONTEMPORARY ART CENTER, 535 Means St., N.W., Atlanta, GA 30318-5729. Tel.: 404-688-1970. Fax: 404-577-5856.
E-mail: info@thecontemporary.org
Web Site: www.thecontemporary.org
Formerly: Nexus Contemporary Art Center
Founded: 1973.
Congressional District: 5
Key Personnel: Bd. Pres., Tim Schrager; Mng. Dir., Stacie Lindner; Devel. Dir., Lane Conville-Canney; Artistic Dir., Stuart Horodner.
Personnel Profile: Full-Time Paid 4; Part-Time Paid 5; Part-Time Volunteers 4; Interns 25.
Governing Authority: nonprofit organization. Tax-exempt: 501(c)(3).
Institution Type/Description: Contemporary Art Gallery
Collections: works by contemporary artists.
Research Fields: contemporary art; strong emphasis on socio-political contemporary.
Facilities: 6,000 sq. ft. exhibit space; educational facilities.
Activities: talks; lectures; panel discussions; workshops; screenings; tours; opening receptions; open studio events.
Publications: exhibition catalogues; seasonal programming brochures.
Hours & Admission Prices: Tues.-Wed. & Fri.-Sat. 11-5, Thurs. 11-8, Sun. 12-5. Adults $8, students & seniors $5; Thurs. & members no charge. &
Attendance: 7,000 (estimated)
Membership: Student $15; Artist $25; Individual $40; Dual $65; Supporter $125; Patron Artist $200; Enthusiast $250; Advocate $500; Leader $1,000; Champion $2,500; Visionary $5,000.

THE ATLANTA CYCLORAMA, 800-C Cherokee Ave., S.E., Atlanta, GA 30315-1470. Tel.: 404-658-7625. Fax: 404-658-7045.
Web Site: www.atlantacyclorama.org
Founded: 1898.
Congressional District: 5
Key Personnel: Dir., Camille R. Love; Bookstore Mgr., Beverly D. Williams; Mktg. & Public Rels., Yakingma L. Robinson.
Personnel Profile: Full-Time Paid 6.
Governing Authority: city. Parent Institution: Atlanta Department of Parks, Recreation & Cultural Affairs. Tax-exempt.
Institution Type/Description: Art & Civil War Museum: located in c.1921 structure, housing a 42-ft. high x 356 ft. in circumference, c.1886 Cyclorama that depicts the July 22, 1864 Battle of Atlanta.
Collections: Civil War artifacts.
Research Fields: cyclorama art form; Civil War history.
Facilities: visitor center; available for rental. Books & museum-related items for sale.
Activities: lectures; gallery talks; film festivals; living history; art exhibits.
Publications: brochure; fact sheet; book, The Battle of Atlanta-The Cyclorama.
Hours & Admission Prices: Tues.-Sat. 9:15-4:30. Adults 13 & over $10, senior citizens 65 & over and children 4-12 $8; children 3 & under no charge. &
Attendance: 175,000

✳ **ATLANTA HISTORY CENTER, (M),** 130 W. Paces Ferry Rd., N.W., Atlanta, GA 30305-1380. Tel.: 404-814-4000. Fax: 404-814-2041. TDD: 404-814-4000; Facebook: Atlanta History Center.
E-mail: hhardwick@atlantahistorycenter.com
Web Site: www.atlantahistorycenter.com
Formerly: Atlanta Historical Society
Founded: 1926.
Congressional District: 5
Key Personnel: C.E.O. & Pres., Sheffield Hale; Exec. Vice Pres., Michael Rose; Vice Pres. Operations, Sean Thorndike; Vice Pres. Public Programs, Kate Whitman; Vice Pres. Devel., Cheri Snyder; Vice Pres. Capital Projects, Jackson McQuigg; Vice Pres. Mktg., Hillary Hardwick; Museum Shop Mgr., Michael Mims.
Personnel Profile: Full-Time Paid 58; Part-Time Paid 66; Part-Time Volunteers 358; Interns 20.

Governing Authority: board of trustees. Parent Institution: Atlanta Historical Society, Inc. Subsidiary Institution: Margaret Mitchell House. Tax-exempt: 501(c)(3).
Institution Type/Description: History Museum.
Collections: history of Atlanta, the Civil War & African-American history; manuscripts; photographs; maps; books; architectural drawings; Atlanta newspapers; Margaret Mitchell memorabilia; early 19th-century furniture & tools; extensive costume & textile collection; folklife collection; Civil War artifacts; decorative arts; gardens; Coca-Cola artifacts; Indian artifacts; 1996 Olympic Games; Shutze decorative arts. Historic Houses: 1845 Tullie Smith Farm; 1890 Victorian Playhouse; 1928 Swan House; 1920 Margaret Mitchell House.
Research Fields: local & state history; regional, Civil War, Black, urban & suburban histories; folklife & decorative arts.
Facilities: cafe; 33-acre property; 22-acre gardens. Museum-related items for sale.
Activities: permanent & changing exhibitions; workshops; tours; lectures; seminars; festivals; formally organized programs; teacher training classes; active auxiliary groups for crafts; current affairs & educational support.
Hours & Admission Prices: Museum: Mon.-Sat. 10-5:30, Sun. 12-5:30. Kenan Research Center: Wed.-Sat. 10-5. Adults $16.50, students 13 & over and seniors 65 & over $13, youth 4-12 $11; discounts to groups, ICOM & AAM members; members and children 3 & under no charge. Additional charge for some special events. Admission included access to Margaret Mitchell House. Closed New Year's Day; Thanksgiving; Christmas Eve & Day. &
Attendance: 235,000 (accurate)
Membership: Individual $50; Family & Dual $75; Sustaining $125; Sponsor $250; Patron $500; Director's Roundtable $1,000.

BRAVES MUSEUM & HALL OF FAME/TURNER FIELD TOURS, 755 Hank Aaron Dr., Atlanta, GA 30315-1120. Tel.: 404-614-2311. Fax: 404-614-1423.
Web Site: braves.com/tours
Founded: 1997.
Key Personnel: Dir., Carolyn Serra.
Personnel Profile: Full-Time Paid 2; Part-Time Paid 12.
Institution Type/Description: Sports Museum.
Collections: more than 700 Braves artifacts & photographs.
Hours & Admission Prices: April-Sept. Mon.-Sat. 9-3, Sun. 1-3; Oct.-March Mon.-Sat. 10-2. Non-game Days: adults $5. Game Days: adults $2. Turner Field Tours & Braves Museum: adults $17, junior 10-13 $13, child 3-9 $9. &

CALLANWOLDE FINE ARTS CENTER, 980 Briarcliff Rd., N.E., Atlanta, GA 30306-2650. Tel.: 404-872-5338. Fax: 404-872-5175.
E-mail: info@callanwolde.org
Web Site: www.callanwolde.org
Founded: 1973.
Congressional District: 4
Key Personnel: Interim Exec. Dir., Jerry Poole; Interim Exec. Dir., Tommie Nichols; Pres. (V), Andrew Keenan; Gallery Dir., Laurie Allan; Dir. Arts Education, Casey Parsons Bagby; Chief Registrar, Robin L. Van Horne.
Personnel Profile: Full-Time Paid 15; Part-Time Paid 5; Part-Time Volunteers 50.
Governing Authority: nonprofit organization; The Callanwolde Foundation, Inc. Tax-exempt: 501(c)(3).
Institution Type/Description: Arts Center: housed in 1917-1920 Callanwolde, home of Charles Howard Candler, son of Asa Candler of Coca-Cola fame.
Collections: personal artifacts; 1920s Gothic mansion.
Facilities: classrooms; rental gallery. Paintings, prints, sculpture, woven items, pottery & jewelry for sale.
Activities: guided tours; concerts; dance recitals; arts festivals; drama; formally organized education programs; docent program or council; traveling exhibitions; workshops.
Publications: quarterly, Classes Publication; quarterly newsletter, Callanwolde Fine Art Center.
Hours & Admission Prices: Mon.-Fri. 10-8, Sat. 10-3. No charge. Closed legal holidays. &
Attendance: 25,000 (estimated)
Membership: Student & Senior Citizen $15; Individual $25; Family $50; Sustaining $125; Contributor $250; Patron $500; Bronze $1,500-$3,499; Silver $3,500; Gold $5,000; Platinum $7,500; Diamond $10,000.

CENTENNIAL OLYMPIC PARK VISITOR CENTER, 265 Park Ave. West, N.W., Atlanta, GA 30313-1591. Mailing Address: c/o Georgia World Congress Center, 285 Andrew Young International Blvd., NW, Atlanta, GA 30313-1591. Tel.: 404-223-4412 & 222-7275. Fax: 404-223-4499.
E-mail: info@centennialpark.com

Web Site: www.centennialpark.com
Key Personnel: Asst. Gen. Mgr., Joe Skopitz; Prog. Coord., Greg Knight; Visitor Center & Volunteer Coord., Jennifer Tinker; Public Rels. Specialist, Christina Skowronski; Exec. Asst., Lisa Stock
Institution Type/Description: Memorial Park & Visitor Center.
Collections: 1996 Olympic history & memorial; Fountain of Rings; 500,000 commemorative bricks; Quilt of Remembrance honoring the 111 people injured in the explosion; eternal light shines in memory of the victim of the explosion.
Facilities: 21 acre park; visitor center.
Activities: audio walking tour; community events; festivals; fundraisers; private events. Center Sponsors: Independence Day Celebration; Wednesday Wind Down Concert Series; Fourth Saturday Family fun Days.
Hours & Admission Prices: Daily call for hours. No charge.
Attendance: 3,000,000 (estimated)

CENTER FOR PUPPETRY ARTS, (M), 1404 Spring St., N.W., Atlanta, GA 30309-2820. Tel.: 404-873-3089. Fax: 404-873-9907.
E-mail: info@puppet.org
Web Site: www.puppet.org
Founded: 1978.
Congressional District: 4
Key Personnel: Exec. Dir., Vincent Anthony; Chm. (V), John Chandler; Administration, Lisa Rhodes; Production, Kristin Jarvis; Museum Shop Mgr., Julie Coenson.
Personnel Profile: Full-Time Paid 37; Part-Time Paid 25; Part-Time Volunteers 32.
Governing Authority: nonprofit organization. Tax-exempt: 501(c)(3).
Institution Type/Description: Puppetry Museum.
Collections: global collections of puppets; toy & poster collection; collection of books, videos, DVDs & films on puppetry.
Research Fields: puppetry.
Facilities: 2,000-vol. library of books & periodicals, including 950 videotapes, available to the public; 170-seat theater; 350-seat theater; 4,000 sq. ft. exhibit space; educational facilities. Puppets, books & clothing for sale.
Activities: guided tours; lectures; theater; organized education programs; docent program; participatory, loan, temporary & traveling exhibitions; films.
Publications: monthly e-newsletter, Center News; exhibit catalogs.
Hours & Admission Prices: Tues.-Fri. 9-3, Sat. 10-5, Sun. 12-5. Museum: admission $8.25; discounts to groups, AAM & ICOM members; children under 2 & members no charge. Museum, Show & Workshop: adults $16, members $9. Closed New Year's Day; Easter; Memorial Day; Independence Day; Labor Day; Thanksgiving; Christmas. &
Attendance: 65,000 (estimated)
Membership: Individual $45-$74; Dual & Family $75-$149; Sponsor $150-$249; Sustainer $250-$499; Benefactor $500-$999; Impressario $1,000-$2,499.

CHATTAHOOCHEE RIVER NATIONAL RECREATION AREA, 1978 Island Ford Pkwy., Atlanta, GA 30350-3432. Tel.: 678-538-1200 & 1280. Fax: 770-399-8087. Facebook: Chattahoochee River National Recreation Area.
Web Site: www.nps.gov/chat
Founded: 1978.
Congressional District: 5, 6 & 9
Key Personnel: Supt., Bill Cox; Chief Park Ranger, Scott M. Pfeninger; Chief Visitor Svcs., Rudy Evenson.
Personnel Profile: Full-Time Paid 32; Part-Time Volunteers 35; Interns 6.
Volunteer Hours: 40,000
Operating Income: 3,100,000
Governing Authority: federal. Parent Institution: National Park Service, S.E. Region, 75 Spring St., S.W., Atlanta, GA 30303. Tax-exempt.
Institution Type/Description: Natural & Cultural Area.
Collections: prehistoric village sites, rock shelters & fish weirs; historic Indian village sites & Civil War remains. Historic Structures: 1850 Allenbrook; 1850 ruins of Akers Mill-Marietta Paper Mill; 1840 Ivy (Laurel) Mill Ruins.
Research Fields: cultural resource inventory; archaeology; historic resources; natural resource base inventory.
Facilities: visitor contact areas; trails; open activity fields; picnic areas; concessions; cooperative association operations; collective research documents on natural & cultural history.
Activities: recreational & educational opportunities; guided walks; lectures; canoe, kayak & raft rentals; shuttle service; skill instructional clinics; horseback riding; 3-mile fitness trail with exercise stations; 50-mile hiking trails
Publications: brochures; trail maps.
Hours & Admission Prices: Daily 9-5. No charge. Entrance Fee: daily $3, annual $25. Closed Christmas. &
Attendance: 3,200,000 (accurate)

CLARK ATLANTA UNIVERSITY ART GALLERIES, Trevor Arnett Hall, 2nd Fl., Atlanta, GA 30314. Mailing Address: 223 James P. Brawley Dr., S.W., Atlanta, GA 30314-4358. Tel.: 404-880-8671. Fax: 404-880-6968.
E-mail: tdunkley@cau.edu
Web Site: www.cau.edu/academics_art_galleries.aspx
Founded: 1942.
Congressional District: 5
Key Personnel: Dir., Tina Dunkley; Asst. Cur., Cynthia Ham.
Personnel Profile: Full-Time Paid 2; Part-Time Volunteers 8; Interns 2.
Governing Authority: Parent Institution: Clark Atlanta University. Tax-exempt.
Institution Type/Description: Art Galleries.
Collections: African-American art from 1942-1970; American & World art.
Research Fields: African diaspora art; African American art.
Activities: lectures; guided tours; films; gallery talks; permanent & temporary traveling exhibits; formal education program for children.
Publications: In The eye of the Muses; selections from the Clark Atlanta University art collection.
Hours & Admission Prices: Tues.-Fri. 11-4. No charge; donations accepted. &
Attendance: 10,000 (estimated)
Membership: Friends $25-$1,000.

DAVID J. SENCER CDC MUSEUM, (M), 1600 Clifton Rd., N.E., (at CDC Pkwy. -MS A14), Atlanta, GA 30329-4018. Tel.: 404-639-0830. Fax: 404-639-0834. Facebook: CDC Museum.
E-mail: museum@cdc.gov
Web Site: www.cdc.gov/museum
Formerly: CDC/Global Health Odyssey Museum
Founded: 1996.
Congressional District: 4
Key Personnel: Dir., Judy M. Gantt; Collections Mgr., Mary Hilpertshauser; Dir. Education, Lynda Flage; Cur., Louise E. Shaw; Instructional Technologist, Alex Rogers.
Personnel Profile: Full-Time Paid 5; Part-Time Volunteers 50.
Governing Authority: federal; nonprofit. Parent Institution: Centers for Disease Control & Prevention. Tax-exempt.
Institution Type/Description: Medical Museum.
Collections: the history of Public Health Service & Centers for Disease Control & Prevention with emphasis on benefits to nation and world of CDC's work: protection of health & safety; credible information, and promotion of health in all stages of life.
Major Exhibits: Health is a Human Right, 9/13-1/14.
Facilities: 150-seat cafeteria; 50-seat cafe. Museum-related items for sale.
Activities: guided tours; video presentations. Museum Sponsors: Yearly Disease Detective Camp.
Hours & Admission Prices: Mon.-Wed. & Fri. 9-5, Thurs. 9-7. No charge. Closed federal holidays &
Attendance: 90,000 (accurate)

DECORATIVE ARTS COLLECTION, INC. MUSEUM OF DECORATIVE PAINTING, (M), 650 Hamilton Ave., S.E., Ste. M, Atlanta, GA 30312. Mailing Address: P.O. Box 18028, Atlanta, GA 30316-0028. Tel.: 404-627-3662.
E-mail: dac@decorativeartscollection.org
Web Site: www.decorativeartscollection
Founded: 1982.
Key Personnel: Dir., Andy Jones; Chm. (V), Peggy Harris; Coord. Art Collection, Bette Marken.
Personnel Profile: Part-Time Paid 2; Part-Time Volunteers 20.
Governing Authority: Tax-exempt.
Institution Type/Description: Art Museum.
Collections: historic & contemporary decorative painting; early American folk art from around the world.
Research Fields: appraisal, historical value of each piece as well as monetary worth.
Facilities: 1,375-vol. library on decorative art; classroom. Gift items for sale.
Activities: guided tours; loan, temporary & traveling exhibitions.
Publications: exhibition catalogues, Fathers of American Decorative Painting; Sentimental Collection of Roses; Celebrate St. Nicholas; The Book of Painted Quilts.
Hours & Admission Prices: Mon.-Fri. 11-4; other times by appointment. No charge; donations accepted. &
Attendance: 600 (estimated)
Membership: Friend $40; Contributor $100; Sponsor $250; Patron $500; Benefactor $1,000; Silver Benefactor $2,500; Gold Benefactor $5,000.

DELTA AIR TRANSPORT HERITAGE MUSEUM, (M), Delta World Headquarters, 1060 Delta Blvd., B-914, Atlanta, GA 30354. Tel.: 404-715-7886 (Office) & 773-1219 (Store). Fax: 404-715-2037.
E-mail: museum.delta@delta.com
Web Site: www.deltamuseum.org
Founded: 1995.
Key Personnel: Chm. (V), Harold Bevis; Dir., Tiffany Meng; Mgr. Archives, Marie Force; Museum Shop Mgr., Judy Bean.
Personnel Profile: Full-Time Paid 4; Part-Time Paid 1; Part-Time Volunteers 25; Interns 1.
Governing Authority: Parent Institution: Delta Air Lines. Tax-exempt.
Institution Type/Description: Corporate History Museum.
Collections: Delta history; aircraft; airline uniforms; photographs; films; 2 1940s aircraft maintenance hangars; corporate records.
Research Fields: Delta Air Lines history.
Hours & Admission Prices: Mon.-Thurs. 9-4 by appointment. Requested Donation: $5.
Attendance: 25,000 (estimated)

DEWBERRY GALLERY OF SCAD, 1545 Peachtree St., Ste. 225, Atlanta, GA 30309. Tel.: 404-815-2931.
Web Site: www.scad.edu/exhibitions/galleries
Institution Type/Description: Art Gallery.
Collections: works by student artists.
Activities: temporary exhibitions.
Hours & Admission Prices: Tues.-Fri. 1-6. No charge.

FEDERAL RESERVE BANK OF ATLANTA, VISITOR'S CENTER AND MONETARY MUSEUM, 1000 Peachtree St., N.E., Atlanta, GA 30309-4470. Tel.: 404-498-7874. Fax: 404-498-8050.
E-mail: amy.hennessy@atl.frb.org
Web Site: www.frbatlanta.org/about/tours/museum.cfm
Formerly: Atlanta Visitor's Center and Monetary Museum
Founded: 2001.
Congressional District: 5
Key Personnel: Dir., Amy Hennessy.
Governing Authority: Parent Institution: Federal Reserve Bank of Atlanta.
Institution Type/Description: Money Museum.
Collections: artifacts & memorabilia pertaining to the history of money; structure & function of the Federal Reserve System; personal finance; how the economy works.
Publications: Econ South (general public); Extra Credit (for teachers).
Hours & Admission Prices: Tours: Mon.-Fri. 9-4. No charge.
Attendance: 20,000 (accurate)

* **FERNBANK MUSEUM OF NATURAL HISTORY, (M),** 767 Clifton Rd., N.E., Atlanta, GA 30307-1274. Tel.: 404-929-6300 & 6400 (tickets). Fax: 404-929-6405 & 6406.
Web Site: www.fernbankmuseum.org
Founded: 1992.
Congressional District: 4
Key Personnel: C.E.O. & Pres., Susan E. Neugent; Bd. Chm., Hampton Morris; Vice Pres. Devel., Leslie Marlowe; Vice Pres. Education, Christine Bean; Museum Shop Mgr., Linda Gerber.
Personnel Profile: Full-Time Paid 77; Part-Time Paid 30; Part-Time Volunteers 350; Interns 6.
Governing Authority: private; nonprofit organization. Parent Institution: Fernbank, Inc. Tax-exempt: 501(c)(3).
Institution Type/Description: Natural History Museum.
Collections: geology; paleontology; zoology; anthropology; archaeology; ethnography.
Research Fields: archaeology.
Facilities: 181-seat auditorium; 150-seat cafe; 60,000 sq. ft. exhibit space; 315-seat IMAX theatre; children's discovery rooms; facility rentals for private events.
Activities: lectures; loan, temporary, participatory & traveling exhibitions; volunteer program; summer camp; summer archaeology educational/research program; school programs; after-school program in partnership with Atlanta Public Schools; Urban Watch Atlanta; children's programming; special exhibitions; Martinis & IMAX(R) theatre; facilities rental.
Hours & Admission Prices: Mon.-Sat. 10-5, Sun. 12-5. Museum: adults $15, senior citizens & students $14, children 12 & under $13; discount to groups & ASTC members; members no charge. IMAX: adults $13, seniors & students $12, children $11, members $8. Closed Thanksgiving; Christmas.
Attendance: 402,412 (accurate)

Membership: Individual $60; Family & Dual $85; Contributing $135; Family Advantage $150; Patron Circle $250; Benefactor's Circle $500.

FERNBANK SCIENCE CENTER, 156 Heaton Park Dr., N.E., Atlanta, GA 30307-1398. Tel.: 678-874-7102. Fax: 678-874-7110.
E-mail: fernbank@fernbank.edu
Web Site: www.fsc.fernbank.edu
Founded: 1967.
Congressional District: 4
Key Personnel: Dir., Douglas J. Hrabe.
Personnel Profile: Full-Time Paid 40; Part-Time Volunteers 30.
Governing Authority: public school district. Parent Institution: DeKalb County (GA) Bd. of Education. Tax-exempt: 501(c)(3).
Institution Type/Description: Science Museum, Planetarium & Observatory.
Collections: skins; hides; entomology; ornithology; herbarium; geology; skeletal; gems; minerals; birds; mammals; insects.
Research Fields: life & physical sciences with emphasis on astronomy & ecology.
Facilities: 26,000-vol. library of biological & physical science books for reference only; planetarium; classrooms; forest; observatory; greenhouses; slide sets.
Activities: classes in nature & science; guided tours; formally organized education programs; formally organized education programs for undergraduate college students affiliated with Emory University; permanent, traveling & temporary exhibitions; children's programs; public programs for adults & children. Center Sponsors: classes for the handicapped.
Publications: Primarily for Understanding Science; Just for Understanding Science; Ready for Understanding Science.
Hours & Admission Prices: Mon.-Wed. 12-5, Thurs.-Fri. 12-10, Sat. 10-5; groups by appointment. Museum: no charge. Planetarium: adults $5, students & seniors $4; discounts to AAM & ICOM members; members no charge.
Attendance: 300,000 (estimated)

FOX THEATRE, 660 Peachtree St., N.E., Atlanta, GA 30308-1929. Tel.: 404-881-2100 & 688-3353 (Tours). Fax: 404-872-2972.
Governing Authority: nonprofit organization. Parent Institution: Atlanta Landmarks.
Institution Type/Description: Historic Building: built c.1920. A National Historic Landmark.
Collections: theatre history & architecture; films; performances.
Facilities: 4,678 auditorium. Museum-related items for sale.
Activities: performances; special events; outreach & educational programs; workshops; summer film series.
Hours & Admission Prices: Tours by appointment: Mon. & Wed.-Thurs. 10am, Sat. 10am & 11am. Adults $10, students & seniors $5.

THE GALLERY AT CHASTAIN ARTS CENTER, 135 W. Wieuca Rd., N.W., Atlanta, GA 30342. Tel.: 404-252-2927. Fax: 404-851-1270.
Web Site: www.ocaatlanta.com
Institution Type/Description: Art Gallery.
Collections: paintings; drawings; sculpture.
Activities: temporary exhibitions.
Hours & Admission Prices: Mon.-Fri. 9:30-5, Sat. 10-3.

GALLERY SEE - SAVANNAH COLLEGE OF ART AND DESIGN, 1600 Peachtree St., Bldg. C, 4th Fl., Atlanta, GA 30309. Tel.: 404-815-2931.
Web Site: www.scad.edu/exhibitions/galleries
Institution Type/Description: Art Gallery.
Collections: paintings; illustrations; drawings.
Activities: temporary exhibitions.
Hours & Admission Prices: Mon.-Fri. 8:30-5:30. No charge.

GALLERY 1600 - SCAD ATLANTA, 1600 Peachtree St., Bldg. A, 2nd Fl., Atlanta, GA 30309. Tel.: 404-815-2931.
Web Site: www.scad.edu/exhibitions/galleries
Institution Type/Description: Art Gallery.
Collections: paintings; sculpture; drawings.
Activities: temporary exhibitions.
Hours & Admission Prices: Mon.-Fri. 8:30-5:30. No charge.

GEORGIA AQUARIUM, 225 Baker St., Atlanta, GA 30313-1809. Tel.: 404-581-4000. Facebook: Georgia Aquarium.
Web Site: www.georgiaaquarium.org
Founded: 2005.
Congressional District: 5

Key Personnel: Pres. & C.O.O., David Kimmel; Chm. (V), Bernie Marcus.
Personnel Profile: Full-Time Paid 347; Part-Time Paid 272; Part-Time Volunteers 1,700; Interns 15.
Governing Authority: Tax-exempt.
Institution Type/Description: Aquarium.
Collections: aquatic animals representing 600 species from around the world; photographs; videos.
Major Exhibits: Sea Monsters Revealed: Aquatic Bodies (T), 9/13-12/14.
Research Fields: whale sharks; belugas; sea turtles; coral; dolphins; manta ray; sea otters.
Facilities: theater; cafe; ballroom; catering. Museum-related items for sale.
Activities: special events; tours; rental facilities; birthday parties; school programs; sleepovers; scuba & swim programs.
Publications: Bringing the Ocean to Atlanta.
Hours & Admission Prices: Sun.-Fri. 10-5, Sat. 9-6; see website for additional hours. Peak: adults $34.95, seniors 65 & over $30.95, children 3-12 $28.95; discounts to AZA members, groups & military. Off Peak: adults $29.95, seniors $25.95, children $23.95. Annual Pass: adults $65, seniors $54, children $48. &
Attendance: 2,200,000 (accurate)

GEORGIA CAPITOL MUSEUM, 206 Washington St., Atlanta, GA 30334. Mailing Address: 2 Martin Luther King Dr., Ste. 820, Atlanta, GA 30334-9000. Tel.: 404-656-2846; 463-4536 (tours). Fax: 404-657-3801.
E-mail: tfrilingos@sos.state.ga.us
Web Site: www.sos.georgia.gov/archives/state_capitol
Founded: 1895.
Congressional District: 4 & 5
Key Personnel: Mgr., Timothy Frilingos.
Personnel Profile: Full-Time Paid 3; Interns 1.
Governing Authority: state. Parent Institution: State of Georgia. Tax-exempt.
Institution Type/Description: General Museum.
Collections: Georgia history; the Capitol building; portraits; sculpture; historic flags.
Activities: permanent exhibitions.
Publications: State Capitol pamphlets.
Hours & Admission Prices: Mon.-Fri. 8-5 by appointment. No charge. Closed legal holidays. &
Attendance: 60,000 (estimated)

GEORGIA STATE UNIVERSITY SCHOOL OF ART & DE-SIGN GALLERY, 10 Peachtree Center Ave., Atlanta, GA 30303-3003. Mailing Address: P.O. Box 4107, Atlanta, GA 30302-4107. Tel.: 404-413-5221. Fax: 404-651-1779.
E-mail: artgallery@gsu.edu
Founded: 1970.
Congressional District: 5
Key Personnel: Dir., Waduda Muhummad; Cur. Visual Resources, Ann England.
Personnel Profile: Full-Time Paid 1; Part-Time Paid 2; Part-Time Volunteers 2; Interns 2.
Governing Authority: university. Parent Institution: Georgia State University. Subsidiary Institution: School of Art & Design. Tax-exempt: 170(b)(1)(A).
Institution Type/Description: University Art Gallery.
Collections: American contemporary paintings, prints, photographs & crafts.
Research Fields: contemporary art issues.
Facilities: 3,000-vol. library of books, 100,000 slides, catalogues & specialized art publications available for research by special permission; reading room; 400-seat auditorium; classrooms; 400-seat cafeteria.
Activities: lectures; films; artist talks; symposiums; special programs; loan, temporary & traveling exhibitions.
Hours & Admission Prices: Call for hours. No charge. Closed school holidays; New Year's Day; Independence Day; Labor Day; Thanksgiving. &
Attendance: 6,000 (estimated)

GOVERNOR'S MANSION, 391 W. Paces Ferry Rd., N.W., Atlanta, GA 30305-1001. Tel.: 404-261-1776. Fax: 4042318621.
E-mail: mansionevents@gov.state.ga.us
Web Site: www.mansion.georgia.gov
Key Personnel: Dir., Joy Forth
Institution Type/Description: Historic Home: home of current Governor Nathan Deal.
Collections: 19th century neoclassical furnishings; paintings; porcelain; personal artifacts; photographs.
Hours & Admission Prices: Tours: Tues.-Thurs. 10-11:30 am; call or visit website for additional winter hours. No charge.

HAMMONDS HOUSE MUSEUM, 503 Peeples St., S.W., Atlanta, GA 30313-1815. Tel.: 404-612-0500.
E-mail: myrna.anderson@fultoncountyga.gov
Web Site: hammondshouse.org
Key Personnel: Exec. Dir., Myrna Anderson-Fuller; Cur., Kevin Sipp; Communications Mgr. & Exec. Admin., Y'na Snipes; Facilities Mgr./Security, Byron Simmons
Institution Type/Description: Art Museum: former home of late Dr. Otis Thrash Hammonds, built around 1872.
Collections: more than 350 art works from the mid-19th century; Haitian paintings, African sculptures & masks.
Hours & Admission Prices: Tues.-Fri. 10-6, Sat.-Sun. 1-5. Adults $4, senior citizens, students & children $2; members no charge. Closed national holidays.

THE HERNDON HOME, 587 University Pl., N.W., Atlanta, GA 30314-4126. Tel.: 404-581-9813. Fax: 404-588-0239.
E-mail: hhinfo@herndonhome.org
Web Site: www.herndonhome.org
Institution Type/Description: Historic Home: former home of the Alonzo Herndon family, built in 1910. A National Historic Landmark.
Collections: family history; personal artifacts; period furnishings.
Hours & Admission Prices: Tours: Tues. & Thurs. 10-4, Sat. by appointment.

* **HIGH MUSEUM OF ART, (M),** 1280 Peachtree St., N.E., Atlanta, GA 30309-3549. Tel.: 404-733-HIGH. Fax: 404-733-4450.
E-mail: highmuseum@woodruffcenter.org
Web Site: www.high.org
Founded: 1905.
Congressional District: 5
Key Personnel: Dir., Michael E. Shapiro; C.O.O., Philip Verre; C.F.O., Rhonda Matheison; Dir. Mktg. & Museum Advancement, Kristen M. Delaney; Mgr. Exhibitions, Amy Simon; Dir. Collections & Exhibitions, David Brenneman; Wieland Family Cur. Modern & Contemporary Art, Michael Rooks; Cur. Fred & Rita Richman African Art, Carol Thompson; Margaret and Terry Stent Cur. American Art, Stephanie Heydt; Eleanor McDonald Storza Dir. Education, Virginia Shearer; Head Museum Interpretation, Julia Forbes; Head School Programs, Lisa Hooten; Exhibitions Designer, Jim Waters; Assoc. Dir. Devel., Matthew Tanner; Mgr. Foundation & Government Relations, Susan Aspinwall; Mgr. Corporate & Exhibitions Support, Anika Madden; Mgr. Individual Support, Ruth Richardson Kelly; Mgr. Wine Auction, Cate Candler Singerman; Mgr. Public Rels., Marci Tate; Registrar, Frances Francis; Mgr. Creative Svcs., Angela Jaeger; Controller, Amy Arant; Mgr. Facilities & Logistics, Kevin Streiter; Head Retail Operations, Sylvia Roberts; Chief Security, Stanley Gray.
Personnel Profile: Full-Time Paid 254; Part-Time Paid 5; Part-Time Volunteers 1,000.
Governing Authority: nonprofit organization. Parent Institution: Robert W. Woodruff Arts Center, Inc. Tax-exempt: 501(c)(3).
Institution Type/Description: Art Museum.
Collections: 19th-20th century American art; European paintings & decorative art; African art; modern & contemporary art; photography; Southern artists; folk & self-taught art.
Major Exhibits: The Louvre and the Taileries Garden (T), 10/13-1/14; Bangles to Benches Contemporary Jewelry & Design, 10/13-6/14; Go West! Art of the American Frontier from the Buffalo Bill Historical Center, 11/13-4/14; The Bunnen Collection of Photography, 1/14-2/14; Dream Cars: Innovative Design, Visionary Ideas, 5/14-9/14; Wynn Bullock: Revelations (T), 6/14-1/15.
Research Fields: permanent collection; exhibition programs.
Facilities: reading room; 226-seat auditorium; classrooms. Museum-related items for sale.
Activities: guided tours; lectures; gallery talks; concerts; dance recitals; arts festivals; study clubs; formally organized educational programs; docent program or council; inter-museum loan, temporary, traveling & permanent exhibitions. Museum Sponsors: community outreach program; Friday Jazz monthly.
Publications: monthly calendars; exhibition catalogues; newsletter; periodic institutional reports.
Hours & Admission Prices: Tues.-Wed. & Fri.-Sat. 10-5, Thurs. 10-8, Sun. 12-5. Adults $19.50, senior citizens & college students $16.50; children 6-17 $12, discounts to groups of 10 or more; children under 6 & members no charge. Closed New Year's Day; Martin Luther King, Jr. Day; Independence Day; Labor Day; Thanksgiving; Christmas. Friday Jazz 3rd Fri. of month 5-10. &
Attendance: 400,000 (estimated)
Membership: Student $35; Senior & Educator $55; Associate $60; Individual $65; Senior Dual, Educator Dual, & Educator Family Dual $85; Dual & Family $95; Young Professional & Contributing $150.

HISTORIC OAKLAND CEMETERY, Historic Oakland Foundation, 248 Oakland Ave., S.E., Atlanta, GA 30312-2220. Tel.: 404-688-2107. Fax: 404-658-6092.
E-mail: oaklandcemetery@mindspring.com
Web Site: www.oaklandcemetery.com
Founded: 1850.
Key Personnel: Exec. Dir., David S. Moore; Chm. (V), Alfred Kennedy; Museum Shop Mgr., Sally Smith.
Personnel Profile: Full-Time Paid 4; Part-Time Paid 7; Part-Time Volunteers 115.
Governing Authority: private; nonprofit organization. Parent Institution: City of Atlanta. Tax-exempt.
Institution Type/Description: Historical Society: office is located in the 1899 Bell Tower building (Norman).
Collections: Victorian funerary sculpture, monuments & architecture; Jewish, African-American & Confederate sections.
Activities: gardening; tours; picnicking; jogging; photography; landscape painting.
Publications: members newsletter, The Oakland Herald.
Hours & Admission Prices: Office: Mon.-Fri. 9-5. Tours: Dec.-March Sat.-Sun. 2pm; other times by appointment. Cemetery no charge. Walking Tours: family $26, adults $10, seniors, students & children $5; members and children 6 & under no charge. &
Attendance: 65,000 (estimated)
Membership: Individual $40; Dual-Family $65; Angels $150; Out in the Rain Fountain Friends $300; Obelisk Overseers $500; Bell Tower Club $1,000; Lion of Atlanta League $2,500; Grand Slam Society $5,000; 1850 Society $10,000.

IMAGINE IT! THE CHILDREN'S MUSEUM OF ATLANTA, 275 Centennial Olympic Park Dr., N.W., Atlanta, GA 30313-1827. Tel.: 404-659-5437. Fax: 404-223-3675.
E-mail: askme@childrensmuseumatlanta.org
Web Site: www.imagineit-cma.org
Founded: 2003.
Key Personnel: Exec. Dir., Jane Turner; Museum Shop Mgr., Melissa Boggan
Institution Type/Description: Children's Museum.
Collections: hands-on exhibits.
Hours & Admission Prices: Mon.-Tues. & Thurs.-Fri. 10-4, Sat.-Sun. 10-5. Admission 1 & over $12.75; children under one & members no charge. Closed Thanksgiving; Christmas.
Attendance: 207,000 (estimated)
Membership: $75; $125; $250; $500.

JIMMY CARTER LIBRARY AND MUSEUM, (M), 441 Freedom Pkwy., Atlanta, GA 30307-1497. Tel.: 404-865-7100. Fax: 404-865-7102. Facebook: Jimmy Carter Library & Museum.
E-mail: carter.library@nara.gov
Web Site: www.jimmycarterlibrary.gov
Founded: 1986.
Congressional District: 5
Key Personnel: Deputy Dir., David Stanhope; Cur., Sylvia Mansour Naguib; Museum Shop Mgr., James E. Stewart.
Personnel Profile: Full-Time Paid 32; Part-Time Paid 1; Part-Time Volunteers 39; Interns 1.
Governing Authority: federal. A unit of the National Archives and Records Administration, Washington, DC 20408. Tax-exempt: 170(b)(1)(A).
Institution Type/Description: Presidential Library.
Collections: personal papers; government records; still photographs; motion picture films; audio & video tapes; sound recordings; head of state gifts; gifts from private citizens; political campaign items; personal & family memorabilia.
Research Fields: life, times, career & presidential administration of President Carter.
Facilities: research archives of materials from the Carter White House, including textual & audiovisual resources; research room; 25,000 sq. ft. exhibit space; two auditoriums & seminar room. Museum-related items for sale.
Activities: guided tours; lectures; films; permanent, temporary & traveling exhibitions; organized education programs for children, adults, undergraduate & graduate students.
Publications: list of holdings; general information brochure; museum store catalog.
Hours & Admission Prices: Mon.-Sat. 9-4:45, Sun. 12-4:45. Adults $8, senior citizens, students & military $6; discounts to AAM members; children under 16 & members no charge. Closed New Year's Day; Thanksgiving; Christmas. &
Attendance: 82,757 (accurate)

Membership: Friends of the Jimmy Carter Library: Individual $25; Family $40; Associate $75; Sustaining $150; Patron $400.

✻ **THE MARGARET MITCHELL HOUSE, (M),** 990 Peachtree St., Atlanta, GA 30309-3901. Tel.: 404-249-7015. Fax: 404-249-7118. Facebook: Margaret Mitchell House.
E-mail: info@margaretmitchellhosue.com
Web Site: www.margaretmitchellhouse.com
Founded: 1990.
Key Personnel: Exec. Dir., Michael Rose; C.E.O. & Pres., Sheffield Hale; Chm. (V), Bill Shearer; Museum Shop Mgr., Michael Mims.
Governing Authority: Parent Institution: Atlanta Historical Center. Tax-exempt.
Institution Type/Description: Historic House: the former home of Margaret Mitchell where she wrote the Pulitzer prize-winning novel, Gone With The Wind.
Collections: Mitchell's apartment in 1899 house restored to its original condition; Gone With the Wind movie memorabilia; Southern literature.
Facilities: visitor center. Museum-related items for sale.
Activities: individual & group tours; fundraising & promotional events; book signings; author programs; adult & youth writing workshops.
Publications: brochures.
Hours & Admission Prices: Mon.-Sat. 10-5:30, Sun. 12-5:30. Adults $13, seniors 65 & over and students 13-18 with ID $10, children 4-12 $8.50; discounts for online purchases. Center & House: adults $22, seniors 65 & over and students 13-18 $17.50, youth 4-12 $12; discounts for online purchases. Closed New Year's Day; Thanksgiving; Christmas Eve & Day. &
Membership: Individual $50; Dual $65; Family $75; Sustaining $125; Sponsor $250; Patron $500; Roundtable $1,000.

MARTIN LUTHER KING, JR. CENTER FOR NONVIOLENT SOCIAL CHANGE, INC., 449 Auburn Ave., N.E., Atlanta, GA 30312-1503. Tel.: 404-526-8900. Fax: 404-526-8932.
E-mail: information@thekingcenter.org
Web Site: www.thekingcenter.org
Founded: 1968.
Congressional District: 5
Key Personnel: C.E.O. & Co Pres., Martin L. King, III; Exec. Dir., Tim L. Richardson; Museum Shop Mgr., Juanita Robinson.
Personnel Profile: Full-Time Paid 20.
Governing Authority: nonprofit organization. Tax-exempt.
Institution Type/Description: History Museum, Educational Center & Archives: located at the Martin Luther King, Jr. National Historic Site.
Collections: King family's furnishings & personal effects; artwork; manuscripts; memorabilia; works of art executed by artists in memory of Dr. King. Historic House: 1895 Dr. Martin Luther King birthplace. Historic Site: tomb of Dr. King.
Research Fields: life of Dr. King; American Civil Rights Movement.
Facilities: 5,000-vol. library; archival & museum materials pertaining to Dr. King, Black history & civil rights movement; reading room; reflecting pool; chapel; screening room. Books, recordings & other museum-related items for sale.
Activities: guided tour; permanent & temporary exhibitions.
Publications: brochures.
Hours & Admission Prices: Summer: Mon.-Fri. 9-6; Winter: Mon.-Fri. 9-5. No charge; donations accepted. Closed legal holidays. &
Attendance: 1,000,000

MARTIN LUTHER KING, JR. NATIONAL HISTORIC SITE AND PRESERVATION DISTRICT, 450 Auburn Ave., N.E., Atlanta, GA 30312-1504. Tel.: 404-331-5190. Fax: 404-730-3112.
E-mail: judy_forte@nps.gov
Web Site: www.nps.gov/malu
Founded: 1980.
Congressional District: 5
Key Personnel: Supt., Judy Forte; Administrative Officer, Tonya Perkins; Chief Interpretation, Robert Pazkee; Chief Ranger, Clark Moore.
Personnel Profile: Full-Time Paid 34; Part-Time Volunteers 30; Interns 1.
Governing Authority: federal. National Park Service. Tax-exempt.
Institution Type/Description: Historic Site & District: neighborhood in which Dr. Martin Luther King, Jr. grew up, includes birthplace, boyhood home, church & gravesite.
Collections: historic photos; local & oral history; historic furnishings & buildings; African-American history.
Research Fields: African-American history; Civil Rights Movement; Dr. Martin Luther King, Jr.
Facilities: library.

Activities: interpretive programs; talks on the community & the Civil Rights Movement; film presentation; tours of Dr. King's birthplace; walking tour of historic site.
Publications: catalog of historic structures; general management plan; book, Sweet Auburn-The Thriving Hub of Black Atlanta.
Hours & Admission Prices: Tours: mid-June to mid-Aug. daily 9-5. No charge; donations accepted. Closed New Year's Day; Thanksgiving; Christmas. &
Attendance: 900,000 (accurate)

✳ **MICHAEL C. CARLOS MUSEUM, (M),** Emory University, 571 S. Kilgo Cir., Atlanta, GA 30322. Tel.: 404-727-4282 & 0573. Fax: 404-727-4292. TDD: 404-727-8017.
E-mail: priyanka.sinha@emory.edu
Web Site: www.carlos.emory.edu
Founded: 1920.
Congressional District: 4
Key Personnel: Dir., Bonnie Speed; Assoc. Dir., Catherine Howett Smith; Co Chm. (V), Ed Snow; Co Chm. (V), Dirk Brown; Cur. Egyptian, Nubian & Near Eastern Art, Dr. Peter Lacovara; Cur. Greek & Roman Art, Dr. Jasper Gaunt; Cur. Ancient American Art, Dr. Rebecca Rollins Stone; Assoc. Cur. African Art, Jessica Stephenson; Assoc. Cur. Works on Paper, Margaret Shufeldt; Dir. of Education, Elizabeth Hornor; Dir. Exhibitions & Collections, Nancy Roberts; Coord. Patron Rels., Gail Habif; Conservator, Renee Stein; Mgr. Budget & Personnel, Travis Myers; Public Rels. & Mktg., Priyanka Sinha; Registrar, Todd Lamkin; Museum Shop Mgr., Mark Burell.
Personnel Profile: Full-Time Paid 30; Full-Time Volunteers 72; Part-Time Paid 8; Part-Time Volunteers 100; Interns 6.
Governing Authority: Parent Institution: Emory University. Tax-exempt: 501(c)(3).
Institution Type/Description: Art & Archaeology Museum.
Collections: fine art works on paper, photographs & painting: Mediterranean, Egyptian, Near Eastern, Mesoamerican, Central & South American, North American, ethnographic; African, Oceanic & Asian.
Major Exhibits: Romare Bearden: A Black Odyssey, 12/14/13-3/9/14; Mirroring the Saints: The Jesuit Wierix Collection from the Krijtberg, Amsterdam, Spring 2014.
Activities: special exhibitions; lectures; docent-led tours; gallery talks; films; storytelling; camps; concerts.
Publications: exhibition catalogs; newsletter; educational materials; catalogue, Surrealist Vision & Technique; handbook, Michael C. Carlos Museum.
Hours & Admission Prices: Tues.-Fri. 10-4, Sat. 10-5, Sun. 12-5. Suggested Donation: adults $8, students, seniors & children 6-17 $6; discounts to AAM & ICOM members; members no charge. &
Attendance: 100,000 (estimated)
Membership: Teacher & University student, faculty or staff $30; Individual $40; Dual $60; Family $75; Doric $150; Ionic $250; Corinthian $500; Curator's Council $1,000; Collector's Council $2,500; Director's Council $5,000; Carlos Partnership $10,000.

MOUNT WILSON OBSERVATORY & MUSEUM, c/o CHARA, Georgia State Univ., One Park Pl., S. #720, Atlanta, GA 30303. Mailing Address: P.O. Box 1909, Atlanta, GA 30301-1909. Tel.: 404-413-5484. Fax: 404-413-5481.
Web Site: www.mtwilson.edu/index.php
Key Personnel: Dir., Dr. Harold McAlister; Deputy Dir., Dr. Arthur Vaughan
Institution Type/Description: Science Museum.
Collections: astronomy; photographs.
Facilities: observatory; 256-seat auditorium.
Activities: summer lecture series.
Hours & Admission Prices: April-Nov. daily 10-4; weather permitting.

THE MUSEUM OF CONTEMPORARY ART OF GEORGIA (MOCA GA), 75 Bennett St., #A2, Atlanta, GA 30309-1275. Tel.: 404-367-8700. Fax: 404-367-1477. Facebook: MOCAGA.
E-mail: info@mocaga.org
Web Site: www.mocaga.org
Founded: 2000.
Congressional District: 5
Key Personnel: C.E.O. & Dir., Annette Cone-Skelton; Chm. (V), Philip Babb.
Personnel Profile: Full-Time Paid 5; Part-Time Paid 2; Part-Time Volunteers 35; Interns 6.
Volunteer Hours: 2,500
Governing Authority: private; nonprofit organization. Tax-exempt: 501(c)(3).
Institution Type/Description: Art Museum.
Collections: permanent art collection of works by Georgia artists including paintings, prints, photographs, sculpture, video & historical archives on the visual arts of Georgia; archives of Georgian artists & institutions.
Research Fields: contemporary Georgia artists.

Facilities: exhibition galleries; art library; archives; education & resource center.
Activities: workshops; lectures; temporary exhibitions; films; guided tours. Annual Events: ArtMerge; MOCA Gala Art Auction Fundraiser; Movers & Shakers; Off-the-Wall Pin Up Show & Sale; Cafe Moca; Spooktacular Halloween Party.
Publications: exhibition catalogues.
Hours & Admission Prices: MOCA GA: Tues.-Sat. 10-5. Adults $5, seniors & students $1; active military & members no charge. Closed New Year's Day; Martin Luther King Jr. Day; Memorial Day; Independence Day; Labor Day; Thanksgiving; Christmas. &
Attendance: 10,000 (estimated)
Membership: Artist $25; Individual $35; Dual $55; Family $60; Friend $100; Contributor $250; Patron $500; Supporting Patron $1,000; Benefactor $2,500; Supporting Benefactor $5,000; Founder's Circle $10,000 & up.

MUSEUM OF DESIGN ATLANTA, 1315 Peachtree St., Atlanta, GA 30309. Tel.: 404-979-6455. Fax: 404-521-9311.
E-mail: info@museumofdesign.org
Web Site: www.museumofdesign.org
Formerly: Atlanta International Museum of Art and Design
Founded: 1989.
Congressional District: 5
Key Personnel: Exec. Dir., Brenda Galina, Ph.D.; Assoc. Dir., Laura Flusche, Ph.D.; Chm. (V), Bruce McEvoy.
Personnel Profile: Full-Time Paid 5; Part-Time Paid 2; Part-Time Volunteers 10; Interns 2.
Governing Authority: nonprofit. Affiliated with the Washington based Smithsonian Institution. Tax-exempt.
Institution Type/Description: Design Museum.
Collections: architecture, industrial & product design.
Facilities: Museum-related items for sale.
Activities: lectures; workshops; film screenings; home tours; summer camp.
Publications: e-newsletter.
Hours & Admission Prices: Tues.-Sat. 11-5. Adults $10; members no charge. &
Attendance: 15,000 (estimated)
Membership: Individual $35; Dual $55; Family $75; Journeyman $100; Artisan $250; Craftsman $500; Master $1,000.

OGLETHORPE UNIVERSITY MUSEUM OF ART, (M), 4484 Peachtree Rd., N.E., Atlanta, GA 30319-2797. Tel.: 404-364-8555. Fax: 404-364-8556. Facebook: Oglethorpe University Museum of Art.
E-mail: epeterson1@oglethorpe.edu
Web Site: museum.oglethorpe.edu
Founded: 1993.
Key Personnel: Dir., Elizabeth Peterson; Chm. (V), William Shropshire, Ph.D.; Collections Mgr., John Daniel Tilford.
Personnel Profile: Full-Time Paid 1; Part-Time Paid 10; Interns 1.
Governing Authority: private university; not-for-profit. Parent Institution: Oglethorpe University. Tax-exempt: 501(c)(3).
Institution Type/Description: University Museum.
Collections: focus on representational & figurative artwork which is often international, spiritual & metaphysical.
Facilities: 3,500 sq. ft. exhibit space. Museum-related items for sale.
Activities: guided tours; lectures; docent program; music recitals.
Publications: catalogs.
Hours & Admission Prices: Tues.-Sun. 12-5. Adults $5; groups of 10 or more $3 per person; children under 12 & members no charge. Closed major holidays; University holidays; Christmas-New Year's week. &
Attendance: 8,000 (accurate)
Membership: Senior $20; Individual $35; Family $55; Patron $100 and up.

PARKS, RECREATION & HISTORIC SITES DIVISION, GEORGIA DEPT. OF NATURAL RESOURCES, 2 Martin Luther King Jr. Dr. E., Ste. 1352, Atlanta, GA 30334-9000. Tel.: 404-656-2770, ext. 5335. Fax: 404-651-5871.
Web Site: www.gastateparks.org
Founded: 1925.
Congressional District: 5
Key Personnel: Commissioner, Chris Clark; Dir., Becky Kelley; Chief Operations, Walley Woods; Interpretation, Dr. Debbie Wallsmith.
Governing Authority: state. Parent Institution: State of Georgia. Subsidiary Institution: Dept. of Natural Resources. Branch Museums: Vann House, Chatsworth; Historic Traveler's Rest, Toccoa; New Echota, Calhoun; Etowah Mounds Archaeological Area, Cartersville; Fort McAllister, Richmond Hill; Fort King George, Darien; Dahlonega Courthouse Gold Museum, Dahlonega; New Lapham-Patterson House, Thomasville, Jarrell

Plantation, Juliette; Wormsloe, Savannah; Robert Toombs House, Washington; Fort Morris Historic Site, Midway; Alexander H. Stephens Memorial; Liberty Hall & the Confederate Museum, Crawfordville; Hofwyl-Broadfield Plantation, Darien; Sweetwater Creek Park; Stephen C. Foster State Park Museum, Fargo; Panola Mountain State Park Museum; Providence Canyon State Park Museum, Lumpkin; Elijah Clark State Park Museum, Lincolnton; Pickett's Mill Battlefield, Dallas; Indian Springs Museum, Indian Springs; Kolomoki Mounds Museum, Blakely; Hamburg State Park Museum, Mitchell; Georgia Veterans Memorial, Cordele; Fort Yargo State Park, Winder; F.D. Roosevelt's Little White House, Warm Springs; Jefferson Davis Memorial Historic Site. Tax-exempt.
Institution Type/Description: State Historic Agency.
Collections: general history; archaeology; marine; preservation project; Indian artifacts; mineralogy; medicine; military; politics; domestic; historical markers; agriculture; geological; fossils.
Research Fields: archaeology; general history; Georgia natural resources; marine; preservation; politics; agriculture; Georgia military history.
Facilities: 600-vol. library of research books available on premises. Publications & postcards for sale.
Activities: guided tours; gallery talks; permanent exhibitions; audiovisual programs.
Publications: brochures.
Hours & Admission Prices: Mon.-Fri. 8-6, Sat.-Sun. 9-5. Adults $1.50, children 6-17 $.75; discounts to handicapped veterans; children under 6 no charge. Office: Mon.-Fri. 8-4:30. Roosevelt's Little White House: adults $4, children 6-18 $2; children 5 & under no charge. Parking $2. Closed Thanksgiving; Christmas. &
Attendance: 2,356,821 (accurate)

PHOTOGRAPHIC INVESTMENTS GALLERY, 3977 Briarcliff Rd., N.E., Atlanta, GA 30345-2647. Tel.: 404-320-1012. Fax: 404-320-3465.
E-mail: ecsymmes@aol.com
Founded: 1979.
Key Personnel: Dir., Edwin C. Symmes, Jr.
Governing Authority: individual operation.
Institution Type/Description: Photography Art Gallery.
Collections: 19th-century photography representing all processes & images worldwide; 19th-century & earlier Oriental Scrolls.
Facilities: original 19th-century photography & Oriental original paintings for sale.
Activities: guided tours; gallery talks; loan, permanent, temporary & traveling exhibitions.
Publications: catalogs.
Hours & Admission Prices: Daily by appointment. No charge. Closed legal holidays.

RHODES HALL, 1516 Peachtree St., N.W., Atlanta, GA 30309-2908. Tel.: 404-885-7800. Fax: 404-875-2205.
E-mail: events@georgiatrust.org
Web Site: www.georgiatrust.org/historic_sites/rhodes_hall.htm
Founded: 1904.
Institution Type/Description: Historic House: former home of Rhodes Furniture founder, Amos Rhodes.
Collections: period artifacts & memorabilia.
Publications: Rambler.
Hours & Admission Prices: Tues. 11-3, Sat. 10-2; tours on the hour. Behind-the-Scenes Tour: adults $7; discounts to AAM members. 1st Floor Tour: adults $5; discounts to senior citizens, students & children 6-12; Georgia Trust members & children under 6 no charge. Closed major holidays. &
Attendance: 7,500 (estimated)
Membership: See website for information.

ROBERT C. WILLIAMS PAPER MUSEUM, (M), Institute of Paper Science & Technology, 500 10th St., N.W., Atlanta, GA 30332. Mailing Address: Institute of Paper Science & Technology, Mail Code 0620, Georgia Tech, Atlanta, GA 30332-0620. Tel.: 404-894-7840. Fax: 404-894-4778.
E-mail: teri.williams@ipst.gatech.edu
Web Site: www.ipst.gatech.edu/amp
Formerly: Robert C. Williams American Museum of Papermaking at Georgia Tech
Founded: 1936.
Congressional District: 5
Key Personnel: C.E.O. & Museum Dir., Teri Williams; Program Coord., Juan Chevere; Cur. Education, Virginia Howell.
Personnel Profile: Full-Time Paid 3; Part-Time Paid 2; Interns 7.
Governing Authority: Parent Institution: Georgia Institute of Technology. Tax-exempt: 501(c)(3).
Institution Type/Description: History, Art & Technology Museum.
Collections: artifacts pertaining to the evolution of papermaking from 200 BC to modern times technology; included are watermarks, paper molds, wood blocks, prints, photographs, parchment & vellum; historic books; early papermaking machines; pre-paper artifacts.
Research Fields: pulp, paper & related science & technology; history papermaking & recycling; art of paper.
Facilities: 2,000-vol. library pertaining to the making of paper by hand & machine available for use on premises.
Activities: traveling exhibitions; papermaking for kids; video products.
Publications: newsletter; brochures; 75th Anniversary Book; From Appleton to Atlanta - The Institutes First 75 Years.
Hours & Admission Prices: Mon.-Fri. 9-5. No charge; donations requested. Closed holidays. &
Attendance: 25,000 (estimated)
Membership: Student $30; Individual $50; Family $60; Company & Institution $120; Patron $250; Sustaining $500; Life Membership $1,000.

THE SALVATION ARMY SOUTHERN HISTORICAL CENTER & MUSEUM, 1032 Metropolitan Pkwy., S.W., Atlanta, GA 30310-3488. Tel.: 404-752-7578 & 753-4166. Fax: 404-753-1932.
E-mail: historical_center@uss.salvationarmy.org
Web Site: www.salvationarmyhistory.org
Founded: 1986.
Congressional District: 5
Key Personnel: Dir. & Archivist, Michael Nagy; Museum & Archival Asst., Andrea Troxclair.
Personnel Profile: Full-Time Paid 2.
Governing Authority: denominational group; nonprofit organization. Parent Institution: The Salvation Army-USA Southern Territory. Subsidiary Institution: Evangeline Booth College. Tax-exempt.
Institution Type/Description: Religious Museum.
Collections: Salvation Army's movement from England to the southern U.S.; development & history in the south; photos; artifacts; microfilm; costumes; films; video & audio interviews; personal papers; musical instruments; music; flags.
Research Fields: history & methodology of various aspects of the Salvation Army's work in the southern states.
Facilities: Hicks Memorial Library, 20,000-vol. library of secular, religious & Salvation Army material available for inter-library loan & to the public; educational facilities; 4,670 sq. ft. exhibit space; theatres; research center; archives.
Activities: guided tours; lectures; seminars.
Publications: occasional brochures.
Hours & Admission Prices: By appointment. No charge; donations accepted. Closed New Year's Day; Good Friday; Easter; Memorial Day; Independence Day; Labor Day; Thanksgiving; Christmas. &
Attendance: 621 (accurate)

SAVANNAH COLLEGE OF ART & DESIGN-ATLANTA GALLERIES, Woodruff Arts Center, 1280 Peachtree St., N.E., Atlanta, GA 30309. Mailing Address: P.O. Box 3146, Savannah, GA 31402-3146. Tel.: 404-815-2931.
Web Site: www.acagallery.org
Formerly: Atlanta College of Art
Founded: 1905.
Congressional District: 5
Key Personnel: Chm. (V), John W. Spiegel; Pres., Paula S. Wallace; Vice Pres. Business Affairs, Timothy Spaeth; Asst. Dir. Conf. & Exhibitions, Katy Barnes.
Personnel Profile: Full-Time Paid 68; Part-Time Paid 95.
Governing Authority: college. Parent Institution: R.W. Woodruff Arts Center. Tax-exempt.
Institution Type/Description: Art Gallery.
Collections: artists' books.
Research Fields: issues related to fine art & design.
Facilities: 30,000-vol. library of fine arts books available for inter-library loan during regular library hours; studios; classrooms. Art supplies for sale.
Activities: guided tours; lectures; films; traveling exhibitions; granting of BFA degree; continuing education courses.
Publications: catalogues for BFA degrees program & continuing education courses; annual news report magazine; exhibition brochures.
Hours & Admission Prices: Tues.-Thurs. 11-5, Fri. 11-8, Sat.-Sun. 12-5. No charge; donations accepted. Closed holidays. &
Attendance: 12,000 (estimated)

SOUTH ARTS, 1800 Peachtree St., N.W., Ste. 808, Atlanta, GA 30309-2512. Tel.: 404-874-7244. Fax: 404-873-2148. TDD: 404-876-6240.
Web Site: www.southarts.org
Founded: 1975.
Congressional District: 5
Key Personnel: Exec. Dir., Gerri Combs; Chm. Bd. Dirs., Todd Lowe; Traditional Arts & ADA Programs, Teresa Hollingsworth; Contemporary Arts, Katy Malone.
Personnel Profile: Full-Time Paid 12; Part-Time Paid 2; Part-Time Volunteers 2; Interns 3.
Governing Authority: private; nonprofit organization. Tax-exempt: 501(c)(3).
Institution Type/Description: A regional arts agency dedicated to providing leadership & support to affect positive change in the arts throughout the south.
Activities: folk arts & southern culture traveling exhibitions.
Publications: Directory of Southern Visions Traveling Exhibits.
Hours & Admission Prices: Mon.-Fri. 9-5. No charge. &
Attendance: 100,000 (estimated)

SPELMAN COLLEGE MUSEUM OF FINE ART, (M), 350 Spelman Lane, S.W., Atlanta, GA 30314-4399. Mailing Address: Box 1526, Atlanta, GA 30314-4399. Tel.: 404-270-5607. Fax: 404-270-5980.
E-mail: museum@spelman.edu
Web Site: www.museum.spelman.edu
Founded: 1996.
Congressional District: 5
Key Personnel: Dir., Andrea D. Barnwell; Cur. Collections, Anne Collins Smith.
Personnel Profile: Full-Time Paid 3; Part-Time Paid 1; Part-Time Volunteers 54; Interns 4.
Governing Authority: private college. Parent Institution: Spelman College. Tax-exempt: 501(c)(3).
Institution Type/Description: Art Museum.
Collections: 19th & 20th century American, African American & European works; African sculpture; textiles; crafts; works by & about women.
Research Fields: contemporary African American.
Facilities: 500-vol. library of art books not available to the public; conservation/restoration lab; video viewing room. Museum-related items for sale.
Activities: formal education programs for undergraduate &r graduate college students; Directed Studies course for art majors: museum preparation & exhibition installation; temporary & traveling exhibitions.
Hours & Admission Prices: Tues.-Fri. 10-4, Sat. 12-4. Suggested Donation: $3 per person Closed Good Friday; all federal holidays & Spelman College breaks. &
Attendance: 7,000 (estimated)
Membership: Atlanta University Center Faculty & Staff & Students $25; Individual $35; Alumnae $50; Dual/Family $60; Contributor $100; Bronze $500; Silver $750; Gold $1,000; Platinum $3,000; Director's Circle $5,000.

SPRUILL GALLERY, 4681 Ashford-Dunwoody Rd., Atlanta, GA 30338. Tel.: 770-394-4019. Fax: 770-394-6179. www.spuillgallery.blogspot.com.
E-mail: gallery@spruillarts.org
Web Site: www.spruillarts.org
Formerly: Spruill Center for the Arts Gallery
Founded: 1975.
Congressional District: 4
Key Personnel: C.E.O., Robert Kinsey; Chm. (V), Beth Saxe; Gallery Mgr. & Exhibition Coord., Jennifer Price.
Personnel Profile: Full-Time Paid 7; Part-Time Paid 2; Part-Time Volunteers 25; Interns 1.
Governing Authority: private; nonprofit organization. Parent Institution: Spruill Center for the Arts, 5339 Chamblee Dunwoody Rd., Atlanta, GA 30338. Tax-exempt: 501(c)(3).
Institution Type/Description: Art Museum.
Collections: paintings; sculpture; photographs; drawings.
Major Exhibits: Spruill Arts Student & Faculty Juried Exhibition, 6/14-8/14; Atlanta Celebrates Photography 2014, 9/14-10/14; Holiday Artists Market, 11/14-12/14.
Facilities: 2,000 sq. ft. exhibit space. Museum-related items for sale.
Activities: lectures; workshops; arts festivals; formal education programs; guided tours; hobby workshops; participatory exhibits. Museum Sponsors: Members' Party; Fall Festival; Holiday Artists Market; Sidewalk Art Sales; Opening Receptions.
Publications: quarterly catalog, Catalog of Courses; monthly newsletter.
Hours & Admission Prices: Jan. 12-Oct. 31 Tues.-Sat. 11-5; Nov. 1-Dec. 24

Tues.-Sun. 10-6. No charge; donations accepted. Closed Memorial Day; Independence Day; Labor Day; Thanksgiving. &
Attendance: 25,000 (estimated)
Membership: Senior $20; Individual $35; Household $70.

TROIS GALLERY - SCAD ATLANTA, 1600 Peachtree St., Bldg. A, 4th Fl., Atlanta, GA 30309. Tel.: 404-815-2931.
Web Site: www.scad.edu/exhibitions/galleries
Institution Type/Description: Art Gallery.
Collections: paintings; sculpture; drawings.
Activities: temporary exhibitions.
Hours & Admission Prices: Mon.-Fri. 8:30-5:30. No charge.

VSA ARTS OF GEORGIA - ARTS FOR ALL GALLERY, The Healey Bldg., 57 Forsyth St., N.W., Ste. R-1, Atlanta, GA 30303-2226. Tel.: 404-221-1270. Fax: 404-221-1984.
E-mail: info@vsaartsga.org
Web Site: www.vsaartsga.org
Founded: 1974.
Key Personnel: Exec. Dir., Elizabeth Labbe-Webb, M.B.A.
Governing Authority: Tax-exempt.
Institution Type/Description: Art Gallery.
Collections: works by area artists who may be disabled or economically disadvantaged.
Activities: special programs & events.
Hours & Admission Prices: Mon.-Fri. 10-4. No charge. &
Attendance: 6,400 (estimated)

THE WILLIAM BREMAN JEWISH HERITAGE MUSEUM, (M), 1440 Spring St., N.W., Atlanta, GA 30309-2832. Tel.: 678-222-3700. Fax: 404-881-4009.
E-mail: dschcndowich@thebreman.org
Web Site: www.thebreman.org
Formerly: The Breman Jewish Heritage & Holocaust Museum
Founded: 1996.
Congressional District: 5
Key Personnel: Dir., Aaron Berger; Pres., Tom Asher; Archivist & Registrar, Jeremy Katz; Coord. Special Exhibitions & Programs, Ghila Sanders; Weinberg Center for Holocaust Educ., Dr. Lili Baxter; Public Rels. & Mktg., David Schcndowich; Museum Shop Mgr., Rachel Katz.
Personnel Profile: Full-Time Paid 15; Full-Time Volunteers 3; Part-Time Paid 1; Part-Time Volunteers 150; Interns 2.
Governing Authority: private; nonprofit organization. Parent Institution: Jewish Federation of Greater Atlanta, 1440 Spring St., Atlanta, GA 30309.
Institution Type/Description: Jewish Heritage Museum.
Collections: concentration on Georgia Jewish history from 1845 to the present; Holocaust history with special emphasis on the experiences of Holocaust survivors who have made new lives in Atlanta; Georgia Jewish history.
Major Exhibits: Return to Rich's: The Story Behind the Store, 11/17/13-5/26/14.
Research Fields: Atlanta Jewish history; Holocaust history; Georgia Jewish history; Alabama Jewish History.
Facilities: library of southern Jewish history, Holocaust, genealogy and general Judaica books, available to the public; 8,000 sq. ft. exhibit space; 200-seat auditorium; educational facilities; 500-seat theatre. Museum-related items for sale.
Activities: docent programs; films; formal education programs for adults; guided tours; lectures; school loan service; temporary exhibitions; theatre; children & family hands-on workshops; Holocaust survivors speakers bureau.
Publications: exhibition catalogues, Creating Community: The Jews of Atlanta from 1845 to the Present; Absence of Humanity; The Holocaust Years; Historical Survey: 150 Years of Creating Community; catalogues, Seeking Justice: The Leo Frank Case Revisited; ZAP! POW! BAM! The Superhero, The Golden Age of Comic Books, 1938-1950.
Hours & Admission Prices: Mon.-Thurs. 10-5, Fri. 10-3, Sun. 1-5. Adults $10, senior citizens 62 & over $7, students $5, children 3-6 $2; discount to groups; children under 3 & members no charge. Closed New Year's Day; Independence Day; Thanksgiving; Jewish Holy Days. &
Attendance: 39,000 (estimated)
Membership: Student & Senior $36; Individual $50; Family $75; Donor $150; Patron $250; Sponsor $500; Benefactor $1,000; Chai $1,800; President's Council $5,000.

WORLD OF COCA-COLA PAVILION, 121 Baker St., N.W., Atlanta, GA 30313-1807. Tel.: 404-676-5151; 800-676-2653.
Web Site: www.worldofcoca-cola.com
Institution Type/Description: History Museum.

Collections: Coke history & memorabilia; Dr. John Pemberton's handwritten formula book; photographs.
Facilities: theater. Museum-related items for sale.
Activities: sample soft drinks from around the world.
Hours & Admission Prices: Call for hours. Adults $16, seniors 65 & over $14, youth 3-12 $12; children 2 & under no charge. Closed Thanksgiving; Christmas.

WREN'S NEST, 1050 Ralph David Abernathy Blvd., S.W., Atlanta, GA 30310-1812. Tel.: 404-753-7735, ext. 1. Fax: 404-753-8535.
E-mail: info@wrensnestonline.com
Web Site: www.wrensnestonline.com
Founded: 1909.
Congressional District: 5
Key Personnel: Dir., Sue Gilman; Pres. (V), Lain Shakespeare.
Personnel Profile: Full-Time Paid 1; Part-Time Paid 4; Part-Time Volunteers 40; Interns 1.
Governing Authority: nonprofit. Parent Institution: Joel Chandler Harris Association Inc. Tax-exempt.
Institution Type/Description: Historic House: 1881 home of Joel Chandler Harris, creator of Uncle Remus & chronicler of stories about Br'er Rabbit.
Collections: original furnishings belonging to the Harris family; personal artifacts of Mr. Harris including his typewriter, hat, umbrella & rolltop desk which he used while employed at the Atlanta Constitution and where the Uncle Remus stories were introduced; photographs & memorabilia related to Harris, his family & the Uncle Remus stories.
Research Fields: literature, history; 19th century homes & furnishings; African-American Folktales; Turn-of-the-Century Journalism; Victorian furnishings & accessories.
Facilities: gardens; picnic grounds. Books & other museum related items for sale.
Activities: guided tours for adults & children; storytelling programs; permanent exhibits; special events.
Publications: periodic newsletters; brochures; pamphlets; books of stories & poems written by area children.
Hours & Admission Prices: Tues.-Sat. 10-2:30. Storytelling: Sat. 1pm. Adults $8, senior citizens & teens $7, children 4-12 $5; discounts for groups & AAM members; members no charge. Closed major holidays. &
Attendance: 9,850 (estimated)
Membership: Individual & Family $35; Br'er Fox $100; Br'er Bear $250; Br'er Rabbit $500; Br'er Patch Patron $1,000 & up.

ZOO ATLANTA, 800 Cherokee Ave., S.E., Ste. A, Atlanta, GA 30315-1470. Tel.: 404-624-5600.
E-mail: info@zooatlanta.org
Web Site: www.zooatlanta.org
Founded: 1889.
Congressional District: 5
Key Personnel: Pres. & C.E.O., Dr. David Allen; C.F.O., Kathy Heagney Williams; Vice Pres. Veterinary Svcs., Maria Crane, D.V.M.; Vice Pres. & C.O.O., Blythe Randolph; Sr. Vice Pres. Operations, Cary Burgess; Sr. Vice Pres. Animal Programs & Science, Dwight Lawson; Vice Pres. Mktg. & Sales, Marcus Margerum; Museum Shop Mgr., Greg Cain.
Personnel Profile: Full-Time Paid 172; Part-Time Paid 61; Part-Time Volunteers 275.
Governing Authority: nonprofit organization. Tax-exempt.
Institution Type/Description: Zoo.
Collections: zoological collection.
Research Fields: primatology; herpetology; animal behavior.
Activities: summer classes; animal demonstrations; tours; educational programs.
Publications: guidebook; members' magazine, ZOOMagazine.
Hours & Admission Prices: April-Oct. Mon.-Fri. 9:30-4:30, Sat.-Sun. 9:30-5:30; Nov.-March Mon.-Fri. 9:30-4:30. Adults $18.99, seniors $14.99, children 3-11 $13.99; discounts to groups & reciprocal members; members & children under 3 no charge. Closed Thanksgiving; Christmas. &
Attendance: 710,000 (accurate)
Membership: Friend of Zoo Atlanta: Individual $59; Family $99; Serengeti $150; Patron $200; Keeper $300; Curator $500; (add a guest $20).

Augusta

✳ **AUGUSTA MUSEUM OF HISTORY, (M),** 560 Reynolds St., Augusta, GA 30901-1430. Tel.: 706-722-8454. Fax: 706-724-5192.
E-mail: amh@augustamuseum.org
Web Site: www.augustamuseum.org
Formerly: Augusta Richmond County Museum
Founded: 1937.
Congressional District: 10
Key Personnel: Dir., Nancy J. Glaser; Pres. (V), Maj. Gen. (Ret. USAF) Perry

M. Smith; Operations Mgr., Kristie Linn; Exhibit Technician, Larry Graham; Registrar, Amanda Klaus; Visitor Svcs., W. Keith Bates; Bldgs. & Grounds, Larry Taylor.
Personnel Profile: Full-Time Paid 5; Part-Time Paid 3; Part-Time Volunteers 50; Interns 2.
Governing Authority: board of trustees. Branch Museum: Ezekiel Harris House, 1822 Broad St., Augusta. Tax-exempt: 501(c)(3).
Institution Type/Description: History Museum presenting the past of Augusta, Georgia and its environs.
Collections: local & military history; regional archaeology; railroading artifacts; textiles; geology; costumes; early American artifacts. Historic House: 1797 Ezekiel Harris House.
Research Fields: local and regional history.
Facilities: research library & archives; classrooms; theatre.
Activities: guided tours; lectures; special events; permanent & traveling exhibitions; film series; adult & family programs.
Publications: magazine.
Hours & Admission Prices: Museum: Thurs.-Sat. 10-5, Sun. 1-5. House: Tues.-Fri. by appointment, Sat. 10-5. Museum: adults $4, seniors 65 & over $3, children 6-18 $2; discounts to AAM members; children 5 & under and members no charge. House: adults $2, children $1; children under 5 no charge. Regional & national reciprocal membership programs. &
Attendance: 35,000 (estimated)
Membership: Individual $30; Family & Dual $60; Friend $125; Contributor $250; Patron $500; Benefactor $1,000; Oglethorpe Society $10,000.

BOYHOOD HOME OF PRESIDENT WOODROW WILSON, (M), 419 Seventh St., Augusta, GA 30901-2317. Mailing Address: P.O. Box 37, Augusta, GA 30903-0037. Tel.: 706-722-9828. Fax: 706-724-3083.
E-mail: erick@historicaugusta.org
Web Site: www.wilsonboyhoodhome.org
Founded: 1991.
Congressional District: 12
Key Personnel: Exec. Dir., Erick D. Montgomery; Pres. (V), Paul G. King; Museum Shop Mgr., Stephanie Herzberg.
Personnel Profile: Full-Time Paid 3; Part-Time Paid 4; Part-Time Volunteers 15.
Governing Authority: private; nonprofit organization. Parent Institution: Historic Augusta, Inc., Augusta, GA. Tax-exempt: 501(c)(3).
Institution Type/Description: Historic House Museum: childhood home of President Woodrow Wilson 1860-1870.
Collections: original & period furniture; detached kitchen & service building; carriage house; 1860 decorative arts.
Research Fields: Woodrow Wilson; mid-19th century Augusta, GA; mid-19th century domestic decorative arts.
Facilities: 200-vol. library of Woodrow Wilson biographical material; rental space available. Museum-related items for sale.
Activities: docent program; guided tours; costumed interpretation. Annual Event: Woodrow Wilson Lecture; Holiday Tours.
Publications: book, Thomas Woodrow Wilson: Family Ties and Southern Perspectives.
Hours & Admission Prices: Thurs.-Sat. 10-4 on the hour. Adults $5, senior citizens $4, students $3; discounts to groups. Closed New Year's Day; Thanksgiving; Christmas. &
Attendance: 3,500 (estimated)
Membership: Student $20; Individual & Nonprofit $50; Family & Double $75; Sustainer $100; Contributor $150; Donor $300; Patron $500; Benefactor $1,000; Heritage $2,500; Landmark $5,000; Corporate Diamond $10,000.

GERTRUDE HERBERT INSTITUTE OF ART, 506 Telfair St., Augusta, GA 30901-2310. Tel.: 706-722-5495. Fax: 706-722-3670.
E-mail: ghia@ghia.org
Web Site: ghia.org
Founded: 1937.
Congressional District: 10
Key Personnel: Exec. Dir., Rebekah Henry; Chm. (V), Janice Williams Whiting.
Personnel Profile: Full-Time Paid 3; Part-Time Paid 3; Part-Time Volunteers 40.
Governing Authority: board of trustees. Tax-exempt: 101(6).
Institution Type/Description: Art Museum.
Collections: paintings; graphics; monthly changing exhibitions. Historic House: 1818 Ware's Folly.
Facilities: painting & drawing ateliers; sculptor's studio; potter's studio with kiln; education annex: photography studio, darkroom, printmaking studio.
Activities: lectures; formally organized education programs; exhibitions; outreach programs for artists.
Publications: information brochure; quarterly newsletter.
Hours & Admission Prices: Tues.-Fri. 10-5, Sat. by appointment. No charge;

donations accepted. Closed holidays; Independence Day; Thanksgiving; Christmas Eve to New Year's Day. &

Attendance: 20,000 (estimated)

Membership: Student (18 & under) & Senior Citizen (65 & up) $20; Artist & Individual $30; Family $50; Donor $100; Sponsor $300; Patron $500; Benefactor $1,000.

LUCY CRAFT LANEY MUSEUM OF BLACK HISTORY AND CONFERENCE CENTER, (M), 1116 Phillips St., Augusta, GA 30901-2724. Tel.: 706-724-3576. Fax: 706-724-3576.

E-mail: info@lucycraftlaneymuseum.com

Web Site: www.lucycraftlaneymuseum.com

Founded: 1991.

Congressional District: 125

Key Personnel: Exec. Dir., Christine Miller-Betts.

Personnel Profile: Full-Time Paid 3; Part-Time Paid 1; Part-Time Volunteers 10.

Governing Authority: private; nonprofit organization. Parent Institution: Delta House, Inc. Tax-exempt 501(c)(3).

Institution Type/Description: Art History Museum.

Collections: African Americans from the Augusta & Savannah River area who have made tremendous contributions to the areas of education, religion, science, sports, entertainment & medicine.

Research Fields: history of Augusta; local African American history.

Facilities: conference & children's center.

Activities: films; formal education programs for children; guided tours; lectures; temporary exhibitions; theater performances; art classes. Museum Sponsors: story telling & senior luncheons 1st Wed. August to June.

Publications: quarterly, Newsletter; brochures.

Hours & Admission Prices: Tues.-Fri. 9-5, Sat. 10-4; other times by appointment. Adults $5, senior citizens $3, students & children $2; discounts to military & seniors. Closed holidays. &

Attendance: 9,000 (estimated)

Membership: Friend $25-$99; Sponsor $100-$249; Supported $250-$499; Sustainer $500-$999.

MEADOW GARDEN - THE HISTORIC FARM HOME OF GEORGE WALTON, 1320 Independence Dr., Augusta, GA 30901-1038. Tel.: 706-724-4174.

E-mail: meadowgarden@att.net

Web Site: www.historicmeadowgarden.org

Founded: 1900.

Congressional District: 10

Key Personnel: Mgr., Susan Jackson; Chm. (V), Virginia Nicholson; State Regent, Virginia Lingelbach.

Personnel Profile: Part-Time Paid 1; Part-Time Volunteers 40.

Governing Authority: Parent Institution: Georgia State Society, National Society Daughters of the American Revolution. Tax-exempt.

Institution Type/Description: Historic House: 1792-1804 residence of George Walton, youngest Georgia signer of the Declaration of Independence.

Collections: George Walton's personal documents & memorabilia; 18th-19th century American & English furnishings; porcelains; paintings; early household equipment.

Activities: guided tours.

Hours & Admission Prices: Mon.-Fri. 10-4, Sat. by appointment. Adults $5, seniors $4, students K-12 $1; discount to military and groups of 10 & over. &

Attendance: 1,368 (accurate)

Membership: Friends & Supporters $10-$49; Patrons $50 & up; Society $100; Sons & Daughters of Liberty $250; Octavia's Legacy $500; George Walton Society $1,000; Meadow Garden Preservation Society $5,000.

MORRIS MUSEUM OF ART, One Tenth St., Augusta, GA 30901-1134. Tel.: 706-724-7501. Fax: 706-724-7612.

E-mail: kgrogan@themorris.org

Web Site: www.themorris.org

Founded: 1985.

Congressional District: 10

Key Personnel: Exec. Dir. & Cur., Kevin Grogan; Chm. (V), William S. Morris, III; Dir. Devel., Phyllis Giddens; Cur. Education, Michelle Schulte; Asst. Cur. Education, Matt Porter; Creative Dir., Todd Beasley; Mktg. & Public Rels. Dir., Nicole McLeod; Coord. Membership Svcs., Blake Leverett; Registrar, Mindy Gales; Preparator & Exhibition Designer, Dwayne Clark; Senior Guard, Frank Lozito; Finance Officer, Louis Gangarosa; Archivist & Librarian, Cary Wilkins; Curatorial Asst., Julia Bruton; Special Events Coord., Lauren Land; Museum Store Mgr., Barbara Morphy; Office Mgr., Brenda Hall; Mgr. Museum Programs, Jessica Martindale; Consulting Editor, Keith Claussen.

Personnel Profile: Full-Time Paid 16; Part-Time Paid 6; Part-Time Volunteers 50; Interns 4.

Governing Authority: Tax-exempt: 501(c)(3).

Institution Type/Description: Art Museum.

Collections: works of art by artists of the American South, Federal period to the present.

Major Exhibits: The Worlds of Hunt Slovem, 12/7/13-2/23/14; King Snake Press: A 15th Anniversary Overview, 1/11/14-3/9/14; Blues Haiku and New Monotypes by Phil Garrett, 1/21/14-3/14; Soldier Artist: Conrad Wise Chapman, 3/15/14-5/25/14; Drawings and Watercolors by James Calvert Smith, 4/14-6/22/14; This Happy Land: Paintings by William Entrekin, 5/31/14-8/17/14; Art of the African Diaspora: The Collection of Jonathan Green and Richard Weedman, 6/14-9/28/14; Drawings by Philip Morseberger, 6/28/14-9/7/14; Sculpture by Anita Huffington, 6/28/14-9/7/14; Oh! Augusta: Photographs by William Greiner, 8/23/14-11/2/14.

Research Fields: Art and artists of the American South, colonial era to the present.

Facilities: 18,000-vol. library; 120-seat auditorium; 18,000 sq. ft. exhibit space. Museum-related items for sale.

Activities: concerts; docent program; guided tours; lectures; loan, temporary & traveling exhibitions. Annual Events: gala.

Publications: newsletter; monographs; exhibition catalogues; gallery guides.

Hours & Admission Prices: Tues.-Sat. 10-5, Sun. 12:30-5. Adults $5, senior citizens, military & students $3; discounts to groups, AAM, ICOM, SEMC, GAMG & Southeastern Art Museum Directors' Forum Members (SEAMD); children under 12, members & Sunday no charge. Closed New Year's Day; Easter; Independence Day; Thanksgiving; Christmas. &

Attendance: 36,000 (accurate)

Membership: Student $15; National, Teacher & Docent $30; Individual $40; Family $50; Supporter $100; Donor $250; Patron $500; Benefactor $1,000; Museum Society $2,500; Director's League $5,000; Chairman's Circle $10,000.

SACRED HEART CULTURAL CENTER, 1301 Greene St., Augusta, GA 30901. Tel.: 706-826-4700.

Key Personnel: Exec. Dir., Sandra Fenstermacher; Rental Coord., Rachel Gregory; Museum Shop Mgr., Judy Evans

Institution Type/Description: Cultural Center.

Collections: local history & culture; paintings; photographs; personal artifacts.

Activities: guided tours.

Hours & Admission Prices: Mon.-Fri. 9-5; other times by appointment.

1797 EZEKIEL HARRIS HOUSE, 1822 Broad St., Augusta, GA 30904-3918. Mailing Address: 560 Reynolds St., Augusta, GA 30901. Tel.: 706-722-8454. Fax: 706-724-5192.

E-mail: amh@augustamuseum.org

Web Site: www.augustamuseum.org/harris.htm

Founded: 1965.

Congressional District: 10

Key Personnel: Dir., Nancy J. Glaser; Pres. (V), Maj. Gen. (Ret. USAF) Perry M. Smith.

Personnel Profile: Part-Time Paid 2; Part-Time Volunteers 3.

Governing Authority: Augusta-Richmond County. Tax-exempt.

Institution Type/Description: Historic House: 1797 Ezekiel Harris House.

Collections: furnishings.

Research Fields: colonial; Revolutionary War.

Activities: guided tours; gallery talks; permanent exhibitions; living history re-enactments.

Publications: brochure.

Hours & Admission Prices: Tues.-Fri. 10-4 by appointment, Sat. 10-5. Adults $2, children 6-18 $1; discounts to AAM members; members & children 5 and under no charge.

Attendance: 2,000 (accurate)

Membership: Individual $30; Family $60; Friend $125; Contributor $250; Patron $500; Benefactor $1,000.

Bainbridge

FIREHOUSE CENTER & GALLERY, 119 W. Water St., Bainbridge, GA 39817-3693. Mailing Address: P.O. Box 35, Bainbridge, GA 39818-0035. Tel.: 229-243-1010.

Institution Type/Description: Art Museum.

Collections: paintings; sculpture; photographs.

Activities: special events.

Hours & Admission Prices: Mon.-Fri. 12-4, Sat.-Sun. 1-5; groups by appointment.

Blairsville

MISTY MOUNTAIN TRAIN MUSEUM, 16 Misty Mountain Lane, Blairsville, GA 30512. Tel.: 706-745-9819.
Web Site: http://mistymountainmodelrailroad.com/
Institution Type/Description: Model Railroad Museum.
Collections: 3,400 sq. ft. O-gauge model train layout.
Hours & Admission Prices: Tours: Jan.-April Mon., Wed. & Sat. 2 pm; May-Dec. Mon., Wed. & Fri.-Sat. 2 pm; groups by appointment. Donation: adults $5; children under 17 no charge.

UNION COUNTY HISTORICAL SOCIETY MUSEUM, 3 Town Square, Blairsville, GA 30512. Mailing Address: P.O. Box 35, Blairsville, GA 30514-0035. Tel.: 706-745-5493. Fax: 706-781-1899.
E-mail: history1@windstream.net
Web Site: unioncountyhistory.org
Founded: 1976.
Congressional District: 9
Key Personnel: Pres. (V), William Akins; Vice Pres., Ed Reed; Treas., Carolyn Jarrard; Admin., Edie Rich; Museum Shop Mgr., Frances Partin.
Personnel Profile: Part-Time Paid 3; Part-Time Volunteers 40.
Governing Authority: private; nonprofit organization. Parent Institution: Union Co. Historical Society. Tax-exempt: 501(c)(3).
Institution Type/Description: Historical Society Museum: housed in old Union County Courthouse.
Collections: concentration on Union County and North Georgia history, 19th century to present. The holdings include the 1899 historic courthouse, a 1906 home and a 1860 log cabin as well as Margarita Morgan miniatures.
Facilities: 86-vol. library of genealogical materials.
Activities: guided tours; temporary exhibits. Annual Events: Mountain Gospel Music Convention 2nd weekend in May; Mountain Marketplace Heritage Festival in September; Bluegrass Festival 4th weekend in September; Halloween on the Square; Christmas on the Square.
Publications: quarterly e-newsletter.
Hours & Admission Prices: May-Nov. Mon.-Sat. 10-4; Dec.-April Mon.-Fri. 10-4. Suggested Donations: adults $2, students $1; members no charge. Closed New Year's Day; Martin Luther King Jr. Day; Labor Day; Christmas. &
Attendance: 10,000 (estimated)
Membership: Individual $20; Family $30; Sustaining $100; Business $100-$250; Lifetime $350-$500.

Blakely

KOLOMOKI MOUNDS STATE PARK MUSEUM, Off U.S. Hwy. 27, follow signs for park, 205 Indian Mounds Rd., Blakely, GA 39823-4460. Tel.: 229-724-2150. Fax: 229-724-2152.
E-mail: kolomoki_park@dnr.state.ga.us
Web Site: www.gastateparks.org
Founded: 1938.
Congressional District: 2
Key Personnel: Park Mgr., Matt Bruner; Interpreter & Museum Shop Mgr., Billy Adams.
Personnel Profile: Full-Time Paid 6; Part-Time Paid 2; Part-Time Volunteers 4.
Governing Authority: state. Administered by Parks, Recreation & Historic Sites Div., Georgia Dept. of Natural Resources, 205 Butler St., S.E., Atlanta, GA 30334.
Institution Type/Description: Historic Site: 13th-century Indian burial mound & village site.
Collections: archaeology; Indian artifacts; excavated mound; Mississippian mound complex.
Research Fields: Weeden Island period aboriginal occupation.
Facilities: nature trails.
Activities: guided tours; lectures; camping; swimming pool; playground.
Publications: report, Excavations at Kolomoki.
Hours & Admission Prices: Museum: daily 8-5. Historic Sites: adults $5, seniors & children 6-17 $4, children under 6 $1; discounts to groups. Closed New Year's Day; Thanksgiving; Christmas. &
Attendance: 20,000 (accurate)
Membership: Friends of Georgia State Parks & Historic Site $50.

Brunswick

HOFWYL-BROADFIELD PLANTATION STATE HISTORIC SITE, 5556 U.S. Hwy. 17 N., Brunswick, GA 31525-4651. Tel.: 912-264-7333. Fax: 912-262-3346.
Web Site: www.gastateparks.org/info/hofwyl/
Founded: 1974.
Congressional District: 1
Key Personnel: Site Mgr., Bill Giles; Park Ranger (Interpretation), Faye Cowart; Park Ranger (Interpretation), Andy Beckman.
Personnel Profile: Full-Time Paid 5; Part-Time Volunteers 3.
Governing Authority: state. Administered by Parks, Recreation & Historic Sites Div., Dept. of Natural Resources. Tax-exempt.
Institution Type/Description: Historic House: c.1850 Hofwyl-Broadfield Plantation.
Collections: dairy equipment; rice tools; furnishings from the period 1790-1972.
Research Fields: rice culture; dairying; Georgia Coastal Life; Gullah/Geechee culture.
Facilities: trails; visitor center.
Activities: guided tours; audiovisual programs. Annual Events: Black History Program in February; Annual Plantation Christmas Program in December.
Publications: trail guide; postcards; brochures.
Hours & Admission Prices: Thurs.-Sat. 9-5. Adults $6.50, senior citizens 62 & over $6, youth 6-18 $3.75; discount to groups. Closed New Years Day; Thanksgiving; Christmas. &
Attendance: 25,000 (accurate)

Buckhead

STEFFEN THOMAS MUSEUM OF ART, 4200 Bethany Rd., Buckhead, GA 30625-1729. Tel.: 706-342-7557. Fax: 706-342-4348.
E-mail: info@steffenthomas.org
Web Site: www.steffenthomas.org
Founded: 1998.
Key Personnel: Acting Dir. & Arts Outreach Program Coord., Lisa Conner; Pres., Nancy Vaughan; Vice Pres., Preston Small; Sec. & Treas., Betty Straw Brown; Office Administration, P. Tommany; Visitor Svcs., Sadie Carter; Visitor Svcs., Ashley Myers.
Personnel Profile: Full-Time Paid 1; Full-Time Volunteers 1; Part-Time Paid 3.
Governing Authority: private; nonprofit organization. Tax-exempt: 501(c)(3).
Institution Type/Description: Visual Arts Museum.
Collections: works by Steffen Thomas; sculpture; paintings; mosaics; furniture; works on paper.
Facilities: 100-vol. library; 8,000 sq. ft. exhibit space. Museum-related items for sale.
Activities: seminars; workshops; formal education programs; guided tours; loan, temporary & traveling exhibitions. Annual Events: Outdoor Spring Festival; Founder's Day in summer; Wine Tasting & Art Auctions in fall; Literary Arts Program in winter.
Publications: quarterly newsletter, STMA News.
Hours & Admission Prices: Tues.-Sat. 11-4. Adults $5, senior citizens & students $3; members & children under 6 no charge. Closed New Year's Day; Martin Luther King, Jr. Day; Memorial Day; Independence Day; Labor Day; Thanksgiving Day; Christmas. &
Attendance: 5,000 (accurate)
Membership: Student $15; Senior $20; Individual $25; Family $50; Supporting $100; Sustaining $150; Partner $250 & up.

Calhoun

NEW ECHOTA STATE HISTORIC SITE, 1211 Chatsworth Hwy., N.E., Calhoun, GA 30701. Tel.: 706-624-1321. Fax: 706-624-1324.
Web Site: www.gastateparks.org
Formerly: New Echota State Historical Society
Congressional District: 9
Key Personnel: Supt., David Gomez.
Personnel Profile: Full-Time Paid 1.
Governing Authority: state. Parent Institution: Parks, Recreation & Historic Sites Div., Dept. of Natural Resources, 205 Butler St., S.E., Atlanta, GA 30334.
Institution Type/Description: Preservation Project: 1825 capital town of Cherokee Nation.
Collections: Historic Buildings: 1828 Rev. Samuel Worcester's mission; Vann tavern; replica of Cherokee Council House & Cherokee Phoenix print shop; courthouse.
Research Fields: Cherokee genealogy; Trail of Tears; architecture.
Facilities: Publications & postcards for sale.
Activities: guided tours; gallery talks; permanent exhibitions.
Hours & Admission Prices: Thurs.-Sat. 9-5. Self-Guided Tours: Thurs.-Sat. Adults $6.50, seniors 62 & over $6, youth 6-17 $5; children under 6 $2; discount to AAA & AAM members and groups. Closed New Year's Day; Thanksgiving; Christmas. &
Attendance: 20,000 (estimated)

ROLAND HAYES MUSEUM, 212 S. Wall St., Calhoun, GA 30701-2499. Tel.: 706-629-2599. Fax: 706-602-2599.
Web Site: harrisartscenter.com/events/rolandhayes/tabid/67/default.aspx
Key Personnel: Interim Dir. Harris Arts Center, Toni Molleson
Institution Type/Description: History Museum.
Collections: personal artifacts & memorabilia from Hayes' life; Hayes' piano; photographs.
Activities: rental facilities; tours. Museum Sponsors: Roland Hayes Birthday Celebration in June.
Hours & Admission Prices: Mon. 10-6, Tues.-Thurs. 10-4, Fri.-Sat. 10-2. No charge.

Canton

CHEROKEE COUNTY HISTORY MUSEUM & VISITOR CEN-TER, (M), 100 North St., Ste. 140, Canton, GA 30114. Mailing Address: P.O. Box 1287, Canton, GA 30169. Tel.: 770-345-3288. Fax: 770-345-3289.
E-mail: sjoyner@rockbarn.org
Web Site: www.rockbarn.org
Formerly: Historic Cherokee County Courthouse
Institution Type/Description: Historical Society Museum.
Collections: local history & culture; photographs; period artifacts; county jail.
Activities: rental facilities.
Hours & Admission Prices: Call for hours. No charge; donations accepted.

Cartersville

BARTOW HISTORY MUSEUM, 4 E. Church St., Cartersville, GA 30120-3331. Tel.: 770-382-3818. Fax: 770-383-9314.
Web Site: www.bartowhistorymuseum.org
Founded: 1987.
Congressional District: 7
Key Personnel: Dir., Trey Gaines; Education Coord., Charity Chastain; Museum Shop Mgr., Elaine Popham.
Personnel Profile: Full-Time Paid 4; Part-Time Paid 4; Part-Time Volunteers 8; Interns 1.
Governing Authority: Parent Institution: Georgia Museums Inc. Tax-exempt: 501(c)(3).
Institution Type/Description: History Museum.
Collections: personal artifacts; farming tools; furnishings; documents/archives; photographs; textiles.
Research Fields: Bartow County history; northwest Georgia.
Facilities: 10,000 sq. ft. exhibit space. Museum-related items for sale.
Activities: guided tours; lectures; broadcast programs; temporary & loan exhibits; workshops; children's programs; summer day camp; staff development workshops for teachers.
Publications: Evolution of a Potter; The General, The Great Locomotive Dispute; Architecture of Bartow County; quarterly newsletter, Recollections.
Hours & Admission Prices: Mon.-Sat. 10-5. Adults $5.50, seniors & students $4.50; active military & members no charge. Southeastern Reciprocal Membership Program. Closed major holidays. &
Attendance: 10,575 (accurate)
Membership: Individual $30; Family $60; Contributor $250; Patron $500; Curator $1,000.

BOOTH WESTERN ART MUSEUM, 501 N. Museum Dr., Cartersville, GA 30120-3272. Mailing Address: P.O. Box 3070, Cartersville, GA 30120-1702. Tel.: 770-387-1300. Fax: 770-387-1319.
Web Site: www.boothmuseum.org
Founded: 2000.
Congressional District: 11
Key Personnel: Dir., Seth Hopkins; Devel., Matt Jarrard; Education, Lisa Wheeler; Mktg., Tom Shinall; Treas., Cathy Lee Eckert; Registrar, Nikki Morris; Cur., Jeff Donaldson; Librarian & Archivist, Liz Gentry; Museum Shop Mgr., Macra Adair; Security, Ken Wade.
Personnel Profile: Full-Time Paid 40; Part-Time Paid 9; Part-Time Volunteers 77; Interns 1.
Governing Authority: private; nonprofit organization. Parent Institution: Georgia Museums, Inc., Cartersville, GA. Tax-exempt: 501(c)(3).
Institution Type/Description: Art Museum.
Collections: Western American art & culture; Presidential letters & portraits; Western movie posters; contemporary Civil War art; Western illustration.
Research Fields: American West art; presidential history; Civil War.
Facilities: 20,000-vol. library; 55,000 sq. ft. exhibit space; 140-seat theatre; cafe. Museum-related items for sale.
Activities: lectures; demonstrations; discussions; film screenings; family events; arts festivals; concerts; docent program; guided tours; loan, participatory, traveling & temporary exhibitions; theater.

Publications: quarterly newsletter, The Booth Bulletin.
Hours & Admission Prices: Tues.-Wed. & Fri.-Sat. 10-5, Thurs. 10-8, Sun. 1-5. Adults $10, seniors 65 & over $8, students $7; discounts to groups of 15 or more; military, children 12 & under and members no charge. &
Attendance: 39,789 (accurate)
Membership: Individual $50; Family and Me & My Grandkids $95; Friend $150; Museum Package $200; Contributor $250; Patron $500; Curator $1,000; Collector's Circle $2,500; Director's Circle $5,000.

ETOWAH INDIAN MOUNDS HISTORICAL SITE, 813 Indian Mounds Rd., S.W., Cartersville, GA 30120-6415. Tel.: 770-387-3747. Fax: 770-387-3972. Facebook: Etowah Indian Mounds.
E-mail: etowah_mounds@dnr.state.ga.us
Web Site: www.gastateparks.org/info/etowah
Founded: 1953.
Congressional District: 7
Key Personnel: Cur., Keith Bailey.
Personnel Profile: Full-Time Paid 1; Part-Time Paid 1.
Governing Authority: state. Affiliated with Parks, Recreation & Historic Sites Div., Dept. of Natural Resources, Suite 1352, 205 Butler St., S.E., Atlanta, GA 30334.
Institution Type/Description: Archaeology Museum.
Collections: archaeological excavations of prehistoric American Indian center.
Research Fields: archaeology.
Facilities: American Indian books & postcards for sale.
Activities: guided tours; gallery talks; permanent exhibitions; flint chipping; basket weaving. Annual Event: Torchlight Tours in spring; Hay Rides in fall.
Hours & Admission Prices: Wed.-Sat. 9-5, Sun. 2-5:30. Adults $5.50, seniors $4.50, youth $4; discount to school groups of 15 or more; children under 5 no charge. Closed New Year's Day; Thanksgiving; Christmas. &
Attendance: 50,000 (accurate)
Membership: Individual $50; Family $75; Supporting $125; Patron $500; Trustee $5,000.

EUHARLEE COVERED BRIDGE AND HISTORIC MUSEUM, 118 Covered Bridge Rd., Cartersville, GA 30120. Mailing Address: 30 Burge's Mill Rd., Euharlee, GA 30145. Tel.: 770-607-2017.
Founded: 1997.
Congressional District: 7
Key Personnel: Chm. (V), Jean G. Cowart.
Governing Authority: Parent Institution: City of Euharlee. Tax-exempt.
Institution Type/Description: Historic Site: bridge built in 1886 by Washington W. King, a black contractor. Listed on the National Register of Historic Places.
Collections: local history; photographs; personal artifacts.
Activities: seasonal events; ghost walk.
Hours & Admission Prices: Museum: Wed.-Sat. 10-4, Sun. 2-4. No charge; donations accepted. &
Attendance: 2,000 (accurate)
Membership: Annual $12.

HISTORIC DEPOT AT FRIENDSHIP PLAZA, One Friendship Plaza, Cartersville, GA 30120-3570. Mailing Address: P.O. Box 200397, Cartersville, GA 30120. Tel.: 770-387-1357; 800-733-2280.
Institution Type/Description: Historic Building: built in 1854.
Collections: local history; personal artifacts; photographs.
Hours & Admission Prices: Mon.-Fri. 9-5, Sat. 11-2, Sun. 1:30-4:30. &

ROSE LAWN MUSEUM, 224 W. Cherokee Ave., Cartersville, GA 30120-3004. Tel.: 770-387-5162. Fax: 770-386-1527.
E-mail: roselawnga@comcast.net
Web Site: www.roselawnmuseum.com
Founded: 1973.
Congressional District: 7
Key Personnel: Dir., Jane Drew; C.E.O. & Commissioner, Clarence Brown.
Personnel Profile: Full-Time Paid 1; Part-Time Paid 3; Part-Time Volunteers 10.
Governing Authority: county; nonprofit. Parent Institution: Bartow County. Tax-exempt: 170(b)(1)(A).
Institution Type/Description: Historic House: c.1880 Victorian mansion, former home of evangelist Samuel Porter Jones.
Collections: furniture; farm implements; documents; clothing & costumes; silver; memorabilia belonging to Samuel Porter Jones; documents & memorabilia belonging to Rebecca Latimer Felton, first woman U.S. Senator; historic garden.
Research Fields: architecture; ministerial.

Facilities: reception facilities; garden. Antiques & museum-related items for sale.

Activities: guided tours; concerts; arts festivals; Tea & Tour.

Publications: books, History of Bartow County (reprint); Thunderbolts (reprint); Life and Sayings of Sam P. Jones; Quit Your Meanness; Laughter in the Amen Corner.

Hours & Admission Prices: Tues.-Fri. 10-12 & 1-5; other times by appointment. Adults $5, students $2. Closed holidays. &

Attendance: 10,000 (accurate)

TELLUS SCIENCE MUSEUM, (M), 100 Tellus Dr., Cartersville, GA 30120. Mailing Address: P.O. Box 3663, Cartersville, GA 30120-1712. Tel.: 770-606-5700. Fax: 770-386-0600.

E-mail: joes@tellusmuseum.org

Web Site: www.tellusmuseum.org

Formerly: Tellus: Northwest Georgia Science Museum

Founded: 2009.

Congressional District: 11

Key Personnel: Dir., Jose Santamaria.

Governing Authority: private; nonprofit organization. Parent Institution: Georgia Museums Inc. Tax-exempt: 501(c)(3).

Institution Type/Description: Science Museum.

Collections: minerals; fossils; gems; mining artifacts; rocks; Georgia mining heritage; books & magazines; period vehicles; space artifacts.

Research Fields: Georgia minerals, geology & mining; astronomy; transportation technology.

Facilities: planetarium; 32,000 sq. ft. exhibit space; lecture hall; classrooms; cafe; banquet facilities. Museum-related items for sale.

Hours & Admission Prices: Daily 10-5. Adults $14, seniors 65 & over $10, children 3-17 & students with ID $10; active military & members no charge. Planetarium show $3.50. &

Attendance: 200,000 (estimated)

Membership: Individual $55; Student & Educator $35; Family or Me and My Grand Kids $95; Friend $150; Museum Package $200; Patron $500; Curator $1,000; Collector $2,500; Director $5,000 & up.

Cedartown

THE POLK COUNTY HISTORICAL SOCIETY, 205 N. College St., Cedartown, GA 30125. Mailing Address: P.O. Box 203, Cedartown, GA 30125-0203. Tel.: 770-749-0073.

Founded: 1974.

Congressional District: 6

Key Personnel: Pres. (V), Tom Lowe; Museum Shop Mgr., Ann White.

Personnel Profile: Part-Time Volunteers 3.

Governing Authority: society; nonprofit organization. Tax-exempt: 501(c)(3).

Institution Type/Description: Historical Society Museum: housed in 1921 former Hawke's Children's Library.

Collections: local artifacts.

Research Fields: artifacts & local history of Polk County; genealogy.

Facilities: display area; 100-seat meeting room.

Activities: guided tours; art festivals; permanent & temporary exhibitions. Annual Event: Tour of Homes.

Publications: quarterly newsletter.

Hours & Admission Prices: Wed. 1:30-4, last Sun. of month 1-5. No charge; donations accepted.

Attendance: 300 (estimated)

Membership: Single $15; Couple $20; Family $25; Corporate $100.

Chatsworth

CHIEF VANN HOUSE HISTORIC SITE, 82 Hwy. 225 N., Chatsworth, GA 30705-6331. Tel.: 706-695-2598. Fax: 706-517-4255.

E-mail: vann_house_park@dnr.state.ga.us

Web Site: www.gastateparks.org

Founded: 1952.

Congressional District: 10

Key Personnel: Pres. & Chm. (V), Carolyn Luffman; Co-Chm., Jan McNeil; Museum Shop Mgr. & Site Supvr., Jeff Stancil; Seasonal, Tim Howard; Seasonal, Ethan Calhoun; Seasonal, Andrea Hart.

Personnel Profile: Full-Time Paid 1; Part-Time Paid 3; Part-Time Volunteers 30.

Governing Authority: state. Parent Institution: Georgia Dept. of Natural Resources. Subsidiary Institution: Parks & Historic Sites Div., Floyd Towers E., 205 Butler St., S.E., Atlanta, GA 30334. Tax-exempt.

Institution Type/Description: Historic House Museum: housed c.1804 Vann House.

Collections: furniture; personal items.

Research Fields: Vann family genealogy; Cherokee Indian history.

Facilities: nature trail. Books for sale.

Activities: guided tours; permanent exhibitions.

Publications: Murray County's Indian Heritage; If the Vann House Could Speak; The Vann House Speaks Again.

Hours & Admission Prices: Thurs.-Sat. 9-5. Adults $6, seniors $5.50, youth $4, school groups $3.50; children 5 & under and members no charge. Closed New Year's Day; Thanksgiving; Christmas. &

Attendance: 12,000 (accurate)

Membership: Individual $5; Family $10; Vann Partner $15-$25; Cherokee Trader $26-$50; Scots Brigade $51-$75; Road Builders $76-$100; Mission Associates $101-$200; Braves $201-$500; Chief's Council $501-$1,000; Principal People $1,000 and up.

FORT MOUNTAIN STATE PARK, 181 Fort Mountain Park Rd., Chatsworth, GA 30705-6669. Tel.: 706-422-1932. Fax: 706-422-1930.

Web Site: www.gastateparks.org

Founded: 1938.

Congressional District: 9

Key Personnel: Supt., Brian Ensley.

Personnel Profile: Full-Time Paid 9; Part-Time Paid 3; Part-Time Volunteers 5.

Governing Authority: state. Parent Institution: State of Georgia, Parks Recreation & Historic Sites Div., Dept. of Natural Resources. Tax-exempt.

Institution Type/Description: State Park.

Collections: pre-historic stone wall.

Research Fields: Woodland period; ceremonial sites; fault zone geology.

Facilities: nature trails; tower overlook.

Activities: seasonal guided walks & naturalist programs; camping.

Publications: brochure.

Hours & Admission Prices: Park: daily 7am-10pm. Office: daily 8-5. Parking $5.

Chickamauga

WALKER COUNTY GEORGIA REGIONAL HERITAGE MUSEUM, 100 Gordon St., Chickamauga, GA 30707-1453. Tel.: 706-375-4488 & 3289.

Formerly: Walker County Regional Heritage & Model Train Museum

Founded: 1996.

Key Personnel: Dir., Kathie M. Ufford.

Personnel Profile: Full-Time Paid 1; Part-Time Volunteers 3.

Institution Type/Description: Historic Building: housed in a former train depot; built in 1890.

Collections: local history; war memorabilia; Native American artifacts; furniture; period guns; Lionel Old Gauge model trains; Civil War history & artifacts.

Facilities: Museum-related items for sale.

Activities: summer train excursions.

Hours & Admission Prices: Mon.-Sat. 10-4. Admission: $2.

Clarkesville

THE MAULDIN HOUSE, MILLINERY SHOP AND BIG HOLLY CABIN, 458 Jefferson St., Clarkesville, GA 30523. Mailing Address: P.O. Box 21, Clarkesville, GA 30523. Tel.: 706-754-2220. Fax: 706-754-2231.

E-mail: mbhorton@clarkesvillega.com

Web Site: www.clarksvillega.com

Congressional District: 10

Governing Authority: Parent Institution: City of Clarkesville.

Institution Type/Description: History Museum.

Collections: local history & culture; period furnishings & clothes; hats; millinery artifacts. Historic Buildings: late 1800s Mauldin house; 1822 Big Holly log cabin; late 1800s millinery shop.

Hours & Admission Prices: Mon.-Fri. 8-5; other times by appointment. No charge; donations accepted. &

Attendance: 6,000 (estimated)

Cleveland

WHITE COUNTY HISTORICAL SOCIETY, White County Court House, Cleveland, GA 30528. Mailing Address: P.O. Box 1139, Cleveland, GA 30528-0022. Tel.: 706-865-3225.

Web Site: www.whitecountyhistoricalsociety.com

Founded: 1965.

Congressional District: 9

Key Personnel: Pres., Mark S. Johnson; Vice Pres., Emory Jones; Treas., Emily Freeman; Museum Shop Mgr., Norma Holeman.

Personnel Profile: Part-Time Paid 2.

Governing Authority: membership. Parent Institution: White County Historical. Tax-exempt.
Institution Type/Description: Local History Museum.
Collections: old newspapers; Civil War documents; diaries & letters; historic local pictures; historic papers of happenings & events in county. Historic Building: 1859-60 Courthouse; historic covered bridge in Sauter/Nacoochee Valley.
Research Fields: oral history; folklore.
Activities: guided tours; lectures; gallery talks; hobby workshops; permanent & temporary exhibitions.
Publications: monthly newsletter.
Hours & Admission Prices: Thurs.-Sat. 10-3. No charge; donations accepted. &
Attendance: 800
Membership: Individual $15; Family $20; Life $150; Family Life $200.

Columbus

COCA-COLA SPACE SCIENCE CENTER, 701 Front Ave., Columbus, GA 31901-2925. Tel.: 706-649-1470. Fax: 706-649-1478.
E-mail: info@ccssc.org
Web Site: www.ccssc.org
Founded: 1996.
Congressional District: 2
Key Personnel: Exec. Dir., Dr. Shawn Cruzen; Asst. Dir., Mary Johnson; Dir. Challenger Learning Center, Scott Norman; Dir. Planetarium, Lance Tankersley; Coord. Visitor Svcs., Dutch Cummings.
Personnel Profile: Full-Time Paid 10; Part-Time Paid 10; Interns 2.
Governing Authority: Parent Institution: Columbus State University. Tax-exempt.
Institution Type/Description: Science Center.
Collections: science; mathematics; technology.
Research Fields: astronomy.
Facilities: Mead Observatory; Omnisphere Theatre; Challenger Center.
Activities: group talks; K-12 activities; public astronomy nights.
Hours & Admission Prices: Mon.-Fri. 10-4, Sat. 10:30-8. Adults $6, military & seniors $5, children $4, CSU ID card holders $3; discounts to ASTC members. &
Attendance: 35,000 (estimated)
Membership: Astronaut $40; Mercury $100; Gemini $250; Apollo $500.

＊ **THE COLUMBUS MUSEUM, (M),** 1251 Wynnton Rd., Columbus, GA 31906-2899. Tel.: 706-748-2562. Fax: 706-748-2570.
E-mail: information@columbusmuseum.com
Web Site: www.columbusmuseum.com
Founded: 1953.
Congressional District: 3
Key Personnel: Dir., Charles T. Butler; Asst. to Dir., Patricia Butts; Pres. Bd. of Trustees (V), Wade H. (Trip) Tomlinson, III; Graphic Designer, Marcolm Tatum; Dir. Collections & Exhibitions, Kristen Miller Zohn; Cur. History, Rebecca Bush; Registrar, Aimee Brooks; Asst. Registrar, Mellda Alexander; Exhibitions Coord., Katie Coakley; Art Handler, Chris Land; Art Preparator, Leslie Shirah; Coord. Community Programs, Kaci Norman; Coord. Youth & Family Programs, Jessamy South; Cur. Education, Abbie Edens; Dir. Devel., Donna Atkins; Mgr. Membership, Anna Bradley; Social Media & Online Coord., Whitney Lackey; Mktg. & Media Mgr., Ashley Bice; Devel. Programs Asst., Laura Narr; Deputy Dir. Operations, Kimberly Beck; Information Asst., Mary Goff; Museum Shop Mgr. & Volunteer Coord., Jennifer Blomqvist; Weekend Receptionist, Logan Arrowood; Security Chief, Rick McGowan; Security Officer, Alfred Johnson; Security Officer, Gayle Solomon Kittrell; Building Engineer, Ted Sanchez.
Personnel Profile: Full-Time Paid 29; Part-Time Paid 7; Part-Time Volunteers 500; Interns 3.
Volunteer Hours: 4,500
Operating Expenses: 2,938,032
Operating Income: 5,478,430
Governing Authority: nonprofit corporation; board of trustees. Tax-exempt: 501(c)(3).
Institution Type/Description: Art & History Museum.
Collections: American paintings; sculpture; drawings; archaeology; decorative arts; southern history; ethnology; costumes; folk art; crafts.
Major Exhibits: Resistance: Art by Najee, 9/28/13-2/10/14; Annotations: Wynnton & Midtown Postcards, 10/26/13-3/6/14; Painting With Fire: Works by Betsy Eby (T), 10/27/13-2/23/14; Feel the Heat: Contemporary Glass Sculpture, 11/24/13-3/23/14; Greetings From Midtown: Historic Postcard Vignettes, 12/22/13-6/15/14; Shalom Y'all: The Valley's Jewish Heritage, 2/20/14-7/13/14; William Beckman Drawings: A Retrospective, 1967-2013 (T), 5/18/14-9/7/14; Beth Van Hoesen: Printmaker, 7/14-12/14; Obsolete Objects, 7/6/14-6/7/15; Leaving Mississippi - Reflections on Heroes & Folklore: Works by Najee Dorsey, 8/21/14-1/4/15; Two Repub-

lics, 10/14-1/15; American Paintings from the Collection of Wesleyan College (T), 12/14-1/25/15.
Research Fields: American art; regional history & culture.
Facilities: library of art, art reference books, special exhibition catalogs, collection catalogs & periodicals available for research on premises.
Activities: guided tours; films; gallery talks; formally organized education programs; inter-museum loan, permanent, temporary & traveling exhibitions.
Publications: quarterly newsletter; exhibition catalogs; annual report.
Hours & Admission Prices: Tues.-Wed. & Fri.-Sat. 10-5, Thurs. 10-8, Sun. 1-5. No charge; donations accepted. Closed legal holidays. &
Attendance: 65,000 (estimated)
Membership: Civic $30; Supporting $50; Contemporaries $50; Contributing $150; Patron $500; Master Circle $1,000; Collector Circle $2,500; Director Circle $5,000.

HISTORIC COLUMBUS FOUNDATION, INC., 1440 Second Ave., Columbus, GA 31901-2124. Mailing Address: P.O. Box 5312, Columbus, GA 31906-0312. Tel.: 706-322-0756. Fax: 706-576-4760.
E-mail: hcfinc@historiccolumbus.com
Web Site: www.historiccolumbus.com
Founded: 1966.
Congressional District: 3
Key Personnel: Bd. Chm., George G. Flowers; Pres., Jack B. Key, III; Exec. Dir., Elizabeth K. Barker; Membership, Jane Etheridge; Dir. Mktg., Carroll Hudson; Office Mgr., Debbie Lipscomb; Dir. Planning & Devel., Justin Krieg; Dir. Cultural Outreach, Ridley Stallings.
Personnel Profile: Full-Time Paid 5; Part-Time Paid 2.
Governing Authority: nonprofit organization. Tax-exempt.
Institution Type/Description: Historic Foundation: five house museums including c.1870 first brick house in original residential part of city.
Collections: early 1800s-late Victorian furniture & furnishings; Coca-Cola collectibles & memorabilia. Historic Houses: 1860 The Rankin House; 1828 The Walker-Peters-Langdon House; 1870 700 Broadway House; 1840 Pemberton House; 1840s Farm House; c.1820 log cabin.
Research Fields: Columbus' history; Blacks in the history of Columbus.
Facilities: interpretive center. Gift items for sale.
Activities: guided tours; lectures. Foundation Sponsors: heritage balls; trips.
Publications: quarterly, HCF Newsletter; Images; Our Town: An Introduction to the History of Columbus, Georgia; Architectural Styles of Our Town: Col. 6A; Heritage Park - The Industrial Heritage of Columbus, GA.
Hours & Admission Prices: Office: Mon.-Fri. 9-5. Tour of Houses: Wed.-Sat. 2. Adults: $5, students $1; members no charge.
Attendance: 3,000 (estimated)
Membership: Individual $40; Family $60; Business $100; Friend $125; Patron $250; Landmark $400; Corporate $500; Sponsor $600; Benefactor $1,000; Heritage $2,500.

NATIONAL CIVIL WAR NAVAL MUSEUM AT PORT COLUMBUS, 1002 Victory Dr., Columbus, GA 31901-3429. Tel.: 706-327-9798. Fax: 706-324-7225.
E-mail: cwnavy@portcolumbus.org
Web Site: www.portcolumbus.org
Founded: 1962.
Key Personnel: C.E.O. & Exec. Dir., Bruce Smith; Museum Shop Mgr., Susan Egram.
Personnel Profile: Full-Time Paid 4; Full-Time Volunteers 7; Part-Time Paid 3; Part-Time Volunteers 14.
Governing Authority: Consolidated Govt. Columbus & Muscogee County, Ga. Tax-exempt.
Institution Type/Description: Naval Museum.
Collections: Civil War naval operations history; Confederate naval gunboats; CSS Jackson & CSS Chattahoochee on display; exhibits pertaining to U.S. & Confederate Navies.
Research Fields: design & construction of American Civil War era warships.
Facilities: Publications & museum-related items for sale.
Activities: guided tours; permanent exhibitions.
Publications: newsletter, Port City Ram.
Hours & Admission Prices: Daily 9-5. Adults $6.50, active military and seniors 65 & over $5.50, students $5; members no charge. Closed Christmas. &
Attendance: 28,000 (accurate)
Membership: Shipmate $25; Lieutenant $50; Commander $100; Captain $250; Commodore $500; Admiral $1,000.

NATIONAL INFANTRY MUSEUM, 3800 S. Lumpkin Rd., Columbus, GA 31903. Mailing Address: 1775 Legacy Way, Columbus, GA 31903. Tel.: 706-545-2958. Fax: 706-545-5158.
E-mail: zachary.f.hanner@conus.army.mil

Web Site: www.infantry.army.mil/museum
Founded: 1959.
Congressional District: 3
Key Personnel: Cur. & Dir., Z. Frank Hanner; Registrar, Edward Annable.
Personnel Profile: Full-Time Paid 12; Part-Time Volunteers 50.
Governing Authority: U.S. Army. Parent Institution: Center of Military History, Washington, DC. Tax-exempt.
Institution Type/Description: Military History Museum.
Collections: militaria; weapons; military art; photograph archives; firearms including experimental U.S. items; presidential collection.
Facilities: 6,000-vol. library of Army manuals, books on infantry weapons, uniforms, equipment & history; 250,000 photos; 100-seat auditorium.
Activities: permanent & temporary exhibitions.
Hours & Admission Prices: Mon.-Sat. 9-5, Sun. 11-5. No charge; donations accepted. &
Attendance: 77,493 (accurate)
Membership: Friend of the Infantry Museum $50. National Infantry Foundation, P.O. Box 2823, Columbus, GA 31902-2823.

Cordele

GEORGIA VETERANS MEMORIAL MUSEUM, 2459-A Hwy. 280 W., Cordele, GA 31015-9511. Tel.: 229-276-2371. Fax: 229-276-2711.
Founded: 1962.
Congressional District: 2
Key Personnel: Mgr., Randell Meeks; Interpretive Ranger, Mike Goodwin.
Personnel Profile: Full-Time Paid 1.
Governing Authority: state. Administered by the Parks, Recreation and Historic Sites Div., Georgia Dept. of Natural Resources, 270 Washington St. S.W., Atlanta, GA 30334. Tax-exempt.
Institution Type/Description: Military Museum.
Collections: historic aircraft; fighting vehicles; uniforms; weapons; accoutrements.
Activities: self-guided tours; permanent exhibits.
Hours & Admission Prices: Daily 8-5. Museum: no charge; donations accepted. Park: $5 per vehicle. Closed Thanksgiving & Christmas. &
Attendance: 3,403 (accurate)

Cornelia

LOUDERMILK BOARDING HOUSE MUSEUM & EVERY-THING ELVIS MUSEUM, 271 Foreacre St., Cornelia, GA 30531-3659. Tel.: 706-778-2001. Facebook: Loudermilk Boarding House Museum.
E-mail: elvisqueen@windstream.net
Founded: 1999.
Key Personnel: Dir. & Museum Shop Mgr., Joni Mabe
Institution Type/Description: History Museum: housed in the former home of Robert Loudermilk and his wife, Phanettia Henderson; built in 1908.
Collections: Loudermilk family history; personal artifacts; period furnishings; photographs; Elvis Presley memorabilia.
Hours & Admission Prices: Fri.-Sat. 10-5. Admission $10; children under 6 no charge.

Crawfordville

CONFEDERATE MUSEUM, 456 Alexander St., Crawfordville, GA 30631-2903. Mailing Address: P.O. Box 310, Crawfordville, GA 30631-0310. Tel.: 706-456-2221 (Museum) & 2602 (Park Office). Fax: 706-456-2396.
Founded: 1952.
Congressional District: 10
Key Personnel: Dir., Andre McLenton.
Personnel Profile: Full-Time Paid 1; Part-Time Volunteers 1.
Governing Authority: state. Administered by Parks, Recreation & Historic Sites Div., Georgia Dept. of Natural Resources, Floyd Tower E., Suite 1352, 205 Butler St., S.E., Atlanta, GA. Tax-exempt.
Institution Type/Description: History Museum.
Collections: arms & memorabilia of Civil War; furnishings; servants quarters & outbuildings. Historic House: 1875 Liberty Hall, home of Alexander H. Stephens.
Research Fields: political history of Civil War; life of A. H. Stephens.
Facilities: Gift items for sale.
Activities: permanent exhibitions; guided tours; audiovisual.
Hours & Admission Prices: Tues.-Sun. 10-5. Adults $4, senior citizens $3.50, children 6-18 $2; discounts to groups; children 5 & under no charge. Closed New Year's Day; Thanksgiving; Christmas. &
Attendance: 92,000

Dahlonega

DAHLONEGA GOLD MUSEUM STATE HISTORIC SITE, #1 Public Square, Dahlonega, GA 30533-1210. Tel.: 706-864-2257. Fax: 706-864-8370.
E-mail: dahlonega@dnr.state.ga.us
Web Site: www.gastateparks.org
Formerly: Historic Lumpkin County Courthouse
Founded: 1966.
Congressional District: 9
Key Personnel: Site Mgr., David Foot; Interpretive Ranger, Teresa Krummel; Interpretive Ranger, Robin Glass.
Personnel Profile: Full-Time Paid 5; Part-Time Paid 2; Part-Time Volunteers 2.
Governing Authority: state. Affiliated with Georgia Dept. of Natural Resources, State Parks & Historic Sites, Floyd Towers E., 205 Butler St., S.E., Atlanta, GA 30334. Tax-exempt.
Institution Type/Description: Historic Building: housed in c.1836 Lumpkin County Courthouse.
Collections: Dahlonega's Gold Rush history; native Georgian gold & Dahlonega Mint coins; mining equipment; photographs.
Research Fields: Georgia gold mining history.
Facilities: Mining publications & other museum related items for sale.
Activities: guided tours; permanent exhibitions; 27-minute film.
Publications: book, History of Georgia's Gold Museum.
Hours & Admission Prices: Mon.-Sat. 9-5, Sun. 10-5. Adults $7, senior citizens 62 & over $6.50, children 6-17 $4.50, children under 6 $2 ; discount to groups with reservation. Annual Pass available. Closed New Year's Day; Thanksgiving; Christmas. &
Attendance: 60,000 (accurate)

Dallas

PAULDING COUNTY HISTORICAL SOCIETY & MUSEUM, 295 N. Johnston St., Dallas, GA 30132-3603. Mailing Address: P.O. Box 333, Dallas, GA 30132. Tel.: 770-505-3485.
Institution Type/Description: Historical Society Museum.
Collections: local history & culture; Native American artifacts; gems; minerals; clothing; photographs; Civil War artifacts; WWI & II artifacts; period furnishings.
Hours & Admission Prices: Thurs. 12-4. Adults $2, children $1.

Dalton

CREATIVE ARTS GUILD, 520 W. Waugh St., Dalton, GA 30720-3474. Mailing Address: P.O. Box 1485, Dalton, GA 30722-1485. Tel.: 706-278-0168. Fax: 706-278-6996.
E-mail: terryt@creativeartsguild.org
Web Site: www.creativeartsguild.org
Founded: 1963.
Congressional District: 9
Key Personnel: Chm. (V), Cindy McCreery; Guild Mgr., Leanne Lawson.
Personnel Profile: Full-Time Paid 4; Part-Time Paid 20; Part-Time Volunteers 1.
Governing Authority: nonprofit organization. Tax-exempt.
Institution Type/Description: Art Commission & Gallery.
Collections: various types of art media.
Facilities: classrooms; galleries; studios.
Activities: guided tours; lectures; concerts; dance recitals; arts festivals; drama; docent programs; permanent, temporary & traveling exhibitions; performing arts events; intercultural activities.
Publications: monthly bulletin, CAG; monthly class brochures.
Hours & Admission Prices: Mon.-Thurs. 9-6, Fri. 9-4:30. No charge; donations accepted. Closed New Year's Day; Memorial Day; Independence Day; Labor Day; Thanksgiving; Christmas. &
Attendance: 200,000
Membership: Senior Citizen $30; Standard $45; Family $50; Art Lovers & Friends of the Arts $100; Growth Partner $250; Founder's Circle $1,000.

PRATERS MILL FOUNDATION, 5845 Georgia Hwy. 2, Dalton, GA 30721-1282. Mailing Address: P.O. Drawer H, Varnell, GA 30756-1008. Tel.: 706-694-6455. Fax: 706-694-8413. Facebook: Praters Mill.
E-mail: info@pratersmill.org
Web Site: www.PratersMill.org
Founded: 1971.
Congressional District: 9
Key Personnel: Pres. (V), Judy Alderman; Treas., Melanie Chapman; Dir., Sherry Sexton.
Personnel Profile: Part-Time Paid 1.

Governing Authority: private; nonprofit organization. Subsidiary Institution: Varnell Heritage Center. Tax-exempt.
Institution Type/Description: Historic Site Museum: housed in c.1855 grist mill & c.1898 Prater's Country Store.
Collections: horse-drawn farm implements.
Research Fields: Native American history of area; pioneer families, gristmills, music, dance, crafts, traditions, cotton gin, country store & agriculture.
Facilities: library; nature trail; park; outdoor shed.
Activities: arts & crafts festivals; guided tours; temporary exhibitions. Museum Sponsors: Prater's Mill Foundation.
Publications: brochures; books; historical reviews.
Hours & Admission Prices: Grounds: sunrise to sunset. No charge. Group Tours: by appointment. Adults $8. Country Fair: adults $7; discounts to groups; children under 12 no charge.
Attendance: 10,000 (estimated)

WHITFIELD-MURRAY HISTORY CENTER & ARCHIVES, 715 Chattanooga Ave., Dalton, GA 30720-8800. Mailing Address: P.O. Box 6180, Dalton, GA 30722-6180. Tel.: 706-278-0217. Facebook: Whitfield-Murray History Center & Archives.
E-mail: wmhs@optilink.us
Web Site: whitfield-murrayhistoricalsociety.org
Formerly: Crown Gardens and Archives
Founded: 1976.
Congressional District: 9
Key Personnel: Exec. Dir., Jennifer Detweiler; Pres., Ellen Thompson.
Personnel Profile: Full-Time Paid 1; Part-Time Volunteers 10.
Governing Authority: society; nonprofit organization. Whitfield-Murray Historical Society. Tax-exempt: 501(c)(3).
Institution Type/Description: Preservation Project & Historic Building: c.1890 Crown Cotton Mill office building located in the Crown Mill Historic District; 1848 Blunt House; 1908 Wright Hotel-Chatsworth; 1840 John Hamilton House-Chatsworth Depot.
Collections: handmade tufted bedspreads; machines for making spreads; material; textiles; period furnishings; artifacts pertaining to the Black community; Sims Collection of hand-carved wooden objects; c.1890 furniture from the Loveman home; records of Mills; Whitfield Co. Census Indexes; Murray Co. Census Indexes; Civil War artifacts; antique apothecary jars-Hamilton Mountain and Dug Gap Battle Park; Civil War sites.
Research Fields: genealogy for the North Georgia area.
Facilities: 500-vol. library Robert Loveman Room; genealogical research material available for research on premises; archives; reading room; meeting room; banquet facilities; Indian & Civil War display. Books, recipe books & other museum-related items for sale.
Activities: exhibits; historical society meetings.
Publications: magazine, The quarterly of Whitfield-Murray Historical Society; books, Official History of Whitfield County, Georgia; Murray County Heritage.
Hours & Admission Prices: Tues.-Fri. 10-5, Sat. 9-1. Crown Gardens: no charge; donations accepted. Hamilton House & Blunt House: $5; discount to groups. Closed New Year's Day; Independence Day week; Thanksgiving; Christmas week. &
Attendance: 7,000 (estimated)
Membership: Student $10; Individual $20; Family $30.

Darien

FORT KING GEORGE STATE HISTORIC SITE, 302 McIntosh Rd., S.E., Darien, GA 31305. Mailing Address: P.O. Box 711, Darien, GA 31305-0711. Tel.: 912-437-4770. Fax: 912-437-5479. Facebook: Fort K. George.
E-mail: ftkgeo@darientel.net
Web Site: www.gastateparks.org/fortkinggeorge
Founded: 1961.
Congressional District: 1
Key Personnel: Historic Site Supt., Steven Smith; Pres. (V), Walter T. Pitts; GA State Parks & Historic Sites Ambassador, Adam W. Young; Interpreter, Jason Baker; Museum Shop Mgr., Anabela Tortorell.
Personnel Profile: Full-Time Paid 3; Part-Time Paid 1; Part-Time Volunteers 37.
Governing Authority: state. Administered by Parks, Recreation & Historic Sites Div., Dept. of Natural Resources, 2 Martin Luther King Dr., S.E., Atlanta, GA 30334. Tax-exempt.
Institution Type/Description: History Museum & Historic Site.
Collections: Aboriginal & Spanish artifacts; reproductions of uniforms, weapons & accoutrements of British garrison.
Research Fields: coastal Georgia Indians; European exploration & colonization of the Southeast; early Georgia history; Darien & McIntosh county; history of early sawmilling in Georgia.
Facilities: Publications, postcards & gifts for sale,

Activities: tours of reconstructed 18th-century fort & blockhouse; permanent & changing exhibits; interpretive programs; slide presentations.
Publications: Fort King George-Step One to Statehood; Garrison Newsletter.
Hours & Admission Prices: Tues.-Sun. 9-5. Adults $7, senior citizens $6.50, youth $4; discount to groups of 25 or more. Closed New Year's Day; Thanksgiving; Christmas. &
Attendance: 48,000 (estimated)

Decatur

DALTON GALLERIES, Agnes Scott College, Dana Fine Arts Bldg., 141 E. College Ave., Decatur, GA 30030-5361. Tel.: 404-471-5361. Fax: 404-471-5369.
E-mail: daltongallery@agnesscott.edu
Web Site: daltongallery.agnesscott.edu
Founded: 1957.
Personnel Profile: Full-Time Paid 1; Part-Time Paid 1.
Governing Authority: college; nonprofit organization. Tax-exempt: 170(b)(1)(A).
Institution Type/Description: Art Gallery.
Collections: Harry L. Dalton collection; Steffen Thomas collection; Ferdinand Warren collection; Clifford Clarke collection.
Facilities: theater; classrooms.
Activities: drama; permanent & temporary exhibitions.
Hours & Admission Prices: During academic school year: Mon.-Fri. 10-4:30, Sat.-Sun.12-4. No charge. &
Attendance: 1,500 (estimated)

DEKALB HISTORY CENTER MUSEUM, 101 E. Court Sq., Decatur, GA 30030-2544. Tel.: 404-373-1088. Fax: 404-373-8287.
E-mail: info@dekalbhistory.org
Web Site: www.dekalbhistory.org
Formerly: DeKalb Historical Society Museum
Founded: 1947.
Congressional District: 4
Key Personnel: Exec. Dir., Melissa Forgey; Treas., John Hewitt.
Personnel Profile: Full-Time Paid 5; Full-Time Volunteers 50; Part-Time Paid 2; Part-Time Volunteers 100; Interns 3.
Governing Authority: society; nonprofit organization. Tax-exempt: 501(c)(3).
Institution Type/Description: History Museum: housed in the Old Courthouse on Decatur Square.
Collections: seven exhibit rooms; county-wide history; manuscripts. Historic Buildings: 1830-40 Benjamin Swanton House; 1825 Thomas-Barber Cabin; 1822 John Biffle Cabin.
Research Fields: local history.
Facilities: 1,500-vol. library on historical subjects & private collections of classics available for research on premises only; reading room. Books & photographs for sale.
Activities: guided tours; lectures; films; oral history programs; formally organized education programs; loan & permanent exhibitions.
Publications: monthly newsletter, DeKalb Historical Society; books, The History of DeKalb County 1822-1900, The Story of Decatur, Life in Dixie During the War; postcards of DeKalb; tour guides.
Hours & Admission Prices: Museum: call for hours. Swanton House & Biffle Cabin: open by appointment. Closed county & national holidays. &
Attendance: 3,600 (estimated)
Membership: Student, Teacher & Senior $25; Individual $30; Family $50; Patron $75; Sustaining $125.

Demorest

JOHNNY MIZE ATHLETIC CENTER AND MUSEUM, Piedmont College, 165 Central Ave., Demorest, GA 30535. Tel.: 706-778-3000.
Institution Type/Description: Sports Museum.
Collections: life & career of Johnny Mize; baseball artifacts; photographs; personal artifacts.
Hours & Admission Prices: Mon.-Fri. 6 am-8 pm, Sat. 8-3.

MASON-SCHARFENSTEIN MUSEUM OF ART - PIEDMONT COLLEGE, (M), 165 Central Ave., Demorest, GA 30535. Tel.: 800-277-7020.
E-mail: museum@piedmont.edu
Web Site: www.piedmont.edu/art/gallery.html
Institution Type/Description: Art Gallery.
Collections: paintings & sculptures by local, national & international artists.
Activities: special events.
Hours & Admission Prices: Call for hours.

Douglas

HERITAGE STATION MUSEUM, 219 W. Ward St., Douglas, GA 31533-3501. Tel.: 912-389-3461. Fax: 912-389-3446. Facebook; City of Douglas.
E-mail: info@cityofdouglas.com
Founded: 1999.
Institution Type/Description: History Museum.
Collections: railroad history; Georgia's cultural heritage; Native American; business & industry; education; medical; early 1900s clothing; photography.
Facilities: library. Museum-related items for sale.
Activities: rental facilities; group tours; school groups; temporary exhibits.
Hours & Admission Prices: Thurs. & Fri. 10-4. $1.

WORLD WAR II FLIGHT TRAINING MUSEUM, 3 Airport Cir., Douglas, GA 31534. Tel.: 912-383-9111.
E-mail: douglas63rd@windstream.net
Web Site: www.wwiiflighttraining.org
Institution Type/Description: Military History Museum.
Collections: flight training history; replica 1942 barracks including bunks, desks, & chairs; 63rd Flying Training Detachment history; war training schools; WWII bombers & fighter aircraft; survival gear; pilot & aircrew combat equipment.
Hours & Admission Prices: Fri.-Sat. 11-4.

Douglasville

CULTURAL ARTS COUNCIL OF DOUGLASVILLE - DOUGLAS COUNTY, 8652 Campbellton St., Douglasville, GA 30134-1825. Mailing Address: P.O. Box 2018, Douglasville, GA 30133-2018. Tel.: 770-949-2787. Fax: 770-949-5788.
E-mail: cultureom@earthlink.net
Web Site: www.artsdouglas.org
Founded: 1986.
Congressional District: 13
Key Personnel: Exec. Dir., Laura Lieberman.
Personnel Profile: Full-Time Paid 2; Part-Time Paid 1; Part-Time Volunteers 30; Interns 2.
Governing Authority: Parent Institution: Cultural Arts Council of Douglasville/Douglas County, Inc. Tax-exempt.
Institution Type/Description: Cultural Arts Center: housed in the historic Roberts/Mozley house, built in 1901. Listed on the National Register of Historic Places.
Collections: works by local & regional artists including sculptures & paintings.
Research Fields: local artists from Georgia.
Activities: classes; rental facilities; special events; meetings; performances.
Publications: quarterly newsletters; annual, Atlanta Celebrates Photography & Festival Guide; National Juried Fine Arts Exhibition; annual report.
Hours & Admission Prices: Jan. 3-Dec. 21 Mon.-Fri. 9-5. No charge; donations accepted. &
Attendance: 15,000 (accurate)
Membership: Student & Senior Citizen $10; Individual $20; Family $30; Friend $50; Patron & Business $100; Sponsor $500; Angel $1,000.

DOUGLAS COUNTY MUSEUM OF HISTORY AND ART, 6754 W. Broad St., Old Douglas County Courthouse, Douglasville, GA 30134-1711. Tel.: 770-949-4090.
E-mail: info@douglascountymuseum.com
Web Site: douglascountymuseum.com
Institution Type/Description: History Museum: housed in the Old Douglas County Courthouse.
Collections: mid-20th century artifacts; personal artifacts; cocktail shakers; school lunchboxes; children's phonographs; Coca Cola memorabilia.
Hours & Admission Prices: Call for hours.

Dublin

DUBLIN-LAURENS MUSEUM, 311 Academy Ave., Dublin, GA 31021-5219. Mailing Address: P.O. Box 1461, Dublin, GA 31040-1461. Tel.: 478-272-9242.
E-mail: museum@laurenshistory.org
Founded: 1979.
Congressional District: 8
Key Personnel: Dir., Scott Thompson, Sr.; Mgr., Betty Page.
Personnel Profile: Part-Time Paid 1; Part-Time Volunteers 15.
Governing Authority: society; nonprofit. Parent Institution: The Laurens County Historical Society. Tax-exempt.
Institution Type/Description: Historical Society Museum: housed in 1904 restored Carnegie Library.
Collections: textiles; photographs; farm tools & implements; art memorabilia; genealogy; Indian artifacts.
Research Fields: local history.
Facilities: 2,000 sq. ft. exhibit space.
Activities: guided tours; lectures; TV programs; permanent & temporary exhibitions; field trips.
Publications: quarterly newsletter, Laurens Co. Historical Society.
Hours & Admission Prices: Tues.-Fri. 1-4:30; other times by appointment. No charge; donations accepted. &
Attendance: 4,000 (estimated)
Membership: Individual $15; Family $20; Patron, Institutional, School Libraries & Museums $25; Sustaining $50; Sponsor $100; Benefactor $200.

Duluth

GWINNETT COUNTY PARKS & RECREATION - MCDANIEL FARM PARK, 3251 McDaniel Rd., Duluth, GA 30096-4605. Mailing Address: 75 Langley Dr., Lawrenceville, GA 30045-6936. Tel.: 770-814-4920. Fax: 770-814-4922.
E-mail: mark.patterson@gwinnettcounty.com
Web Site: www.gwinnettcounty.com
Formerly: Lanier Museum of Natural History
Founded: 2005.
Congressional District: 9
Key Personnel: Dir., Dr. Mark A. Patterson.
Personnel Profile: Full-Time Paid 1; Part-Time Paid 4; Part-Time Volunteers 6; Interns 2.
Governing Authority: county; nonprofit organization. Gwinnett County Parks & Recreation, 75 Langley Dr., Lawrenceville, GA 30045. Tax-exempt: 501(c)(3) & 701(b)(1)(A).
Institution Type/Description: Historic Site.
Collections: period artifacts; birds; mammals; fish; butterflies & insects; rocks; minerals; fossils; shells; snakes; hands-on exhibits; journals; historic buildings.
Research Fields: natural history; science; cultural & historical local history.
Facilities: reference center.
Activities: guided tours; films; study & hobby workshops; formally organized education programs; docent program; changing exhibit areas.
Publications: quarterly journal, The Eyrie.
Hours & Admission Prices: Fri.-Sat. 10-4, Sun. 12-4. Admission $1; discounts to AAM & ICOM members; members & Gwinnett History Museum members no charge.
Attendance: 40,000 (estimated)
Membership: Student, Educator, Senior Citizen 55 & over $10; Individual $15; Families $25.

JACQUELINE CASEY HUDGENS CENTER FOR THE ARTS, 6400 Sugarloaf Pkwy., Bldg. 300, Duluth, GA 30097-7419. Tel.: 770-623-6002. Fax: 770-623-3555.
E-mail: info@thehudgens.org
Web Site: www.thehudgens.org
Formerly: Gwinnett Fine Arts Center
Founded: 1981.
Congressional District: 10
Key Personnel: Bd. Chm., Stan Hall; Exec. Dir., Teresa Osborn; Dir. Mktg., Kelly Olson; Dir. Education, Angela Nichols.
Personnel Profile: Full-Time Paid 2; Part-Time Paid 6; Part-Time Volunteers 250; Interns 1.
Governing Authority: nonprofit. Parent Institution: Gwinnett Council for the Arts. Tax-exempt: 501(c)(3).
Institution Type/Description: Art Museum.
Collections: works by noted Georgia artists in all medias & styles since 1920; also works by Picasso, Kandinsky, Litchtenstein, Miro & Rauschenberg.
Facilities: botanical & sculpture garden; educational facilities; 6,000 sq. ft. exhibit space; The Glass Pyramid. Art items for sale.
Activities: arts festivals; concerts; formal education programs; guided tours; lectures; loan, temporary & traveling exhibitions; puppet theater; black box theater.
Publications: quarterly & monthly newsletters
Hours & Admission Prices: Tues.-Sat. 10-5. Adults $5, seniors & students $3; children 2 & under no charge. Closed New Year's Day; Martin Luther King Jr. Day; Memorial Day; Independence Day; Labor Day; Thanksgiving; Christmas. &
Attendance: 45,000 (estimated)
Membership: Student/Artist/Senior $20; Individual $30; Family $50; Contributing $100; Sponsor $300.

SOUTHEASTERN RAILWAY MUSEUM, 3595 Buford Hwy., Duluth, GA 30096. Mailing Address: P.O. Box 1267, Duluth, GA 30096-0023. Tel.: 770-476-2013. Fax: 770-573-3754.
E-mail: admin@southeasternrailwaymuseum.org
Web Site: www.srmduluth.org
Founded: 1968.
Congressional District: 10
Key Personnel: Admin., Randy Pirkle; Pres. (V), John Pollock; Operations Mgr., Dale Grice; Preservation Mgr. & Librarian, Nick Whitehouse; Museum Shop Mgr., Lallie Morris.
Personnel Profile: Part-Time Paid 2; Part-Time Volunteers 50.
Governing Authority: society; nonprofit organization. Parent Institution: National Railway Historical Society. Affiliated with the Atlanta Chapter of the National Railway Historical Society, P.O. Box 1267, Duluth, GA 30096. Tax-exempt: 501(c)(3).
Institution Type/Description: Railway Museum.
Collections: over 90 pieces of railroad rolling stock including Pullmans, steam locomotives & wooden freight cars; business records; blueprints; maps; books; magazines; trade journals & house organs; newspaper articles.
Facilities: 4,000-vol. library of railroad-related history available for research by appointment; over 30 pieces of rolling stock open for touring. Railroad-related gifts for sale.
Activities: special events; restored caboose rides on steam & diesel 1940s locomotives.
Publications: quarterly newsletter, Milepost 613.
Hours & Admission Prices: Jan.-March Sat. 10-5; April-Dec. Thurs.-Sat. 10-5. Adults $8, seniors 65 & over $6, children 2-12 $4; children under 2 no charge. &
Attendance: 15,150 (accurate)
Membership: Youth $20; Senior $35; Adult $45; Family $90.

East Point

EAST POINT HISTORICAL SOCIETY, 1685 Norman Berry Dr., East Point, GA 30344. Mailing Address: P.O. Box 90675, East Point, GA 30364-0675. Tel.: 404-767-4656.
E-mail: ephistoricalsociety@gmail.com
Web Site: www.eastpoinths.org
Governing Authority: nonprofit organization.
Institution Type/Description: Historical Society Museum.
Collections: local history & culture; period furnishings; photographs.
Activities: Annual Event: Holiday Open House in December.
Publications: newsletter.
Hours & Admission Prices: Thurs. 1-4, Sat. 11-3. No charge.

Eatonton

UNCLE REMUS MUSEUM, 214 Oak St., Eatonton, GA 31024. Mailing Address: P.O. Box 3184, Eatonton, GA 31024-3184. Tel.: 706-485-6856.
Founded: 1963.
Congressional District: 11
Key Personnel: C.E.O., J. Marshal; Dir. & Museum Shop Mgr., Lanelle Frost; Pres. (V), Mona Betzel.
Personnel Profile: Full-Time Paid 5.
Governing Authority: nonprofit organization.
Institution Type/Description: History Museum: housed in a former slave cabin, c.1820.
Collections: furniture; personal artifacts; agriculture & wood working; books of Joel Chandler Harris, author of The Uncle Remus stories; Civil War; 3 slave cabins; G.C. Harris writings & life; tools; Brer Rabbit; Brer Fox; shadow boxes with carved figures.
Facilities: Volumes of the Uncle Remus stories & related items for sale.
Activities: guided tours; lectures; permanent exhibitions.
Hours & Admission Prices: March to mid-Nov. Mon.-Sat. 10-5, Sun. 2-5; mid-Nov. to Feb. Mon. & Wed.-Sat. 10-5. Adults $2, children 8 & under $1. Closed New Year's Day; Mother's Day; Independence Day; Christmas. &
Attendance: 12,000 (estimated)
Membership: Individual $1; Life $50.

Elberton

ELBERTON GRANITE MUSEUM & EXHIBIT, 1 Granite Plaza, Elberton, GA 30635. Mailing Address: P.O. Box 640, Elberton, GA 30635-0640. Tel.: 706-283-2551. Fax: 706-283-6380.
E-mail: granite@egaonline.com
Web Site: www.egaonline.com/egaassociation/museum
Founded: 1981.
Congressional District: 10

Key Personnel: Exec. Vice Pres. & Dir., Doyle "Doy" G. Johnson.
Personnel Profile: Part-Time Paid 3.
Governing Authority: trade association. Affiliated with Elberton Granite Assn., Inc., P.O. Box 640, Elberton, GA 30635. Tax-exempt.
Institution Type/Description: Granite Museum.
Collections: historical exhibits; artifacts; educational displays; granite monuments.
Facilities: classrooms.
Activities: guided tours for groups by prior arrangement; lectures; films.
Hours & Admission Prices: Mon.-Sat. 2-5; call for additional hours. No charge. Closed holidays.
Attendance: 3,885 (accurate)

Ellijay

GILMER ARTS AND HERITAGE ASSOCIATION, 207 Dalton St., Ellijay, GA 30540-9000. Tel.: 706-635-5605. Fax: 706-636-5606.
E-mail: gaha@ellijay.com
Web Site: www.gilmerarts.org
Key Personnel: Pres., John Mahan.
Personnel Profile: Full-Time Paid 1.
Governing Authority: tax-exempt.
Institution Type/Description: Art Association, Heritage.
Collections: works by regional & national artists; Gilmer County artifacts.
Publications: newsletter.
Hours & Admission Prices: Mon.-Fri. 10-5. No charge.
Attendance: 500 (estimated)

Fairburn

OLD CAMPBELL COUNTY MUSEUM, Intersection of E. Broad St. & Cole St., Fairburn, GA 30213. Mailing Address: c/o OCCHS, P.O. Box 463, Fairburn, GA 30213-0342. Tel.: 770-548-9181.
E-mail: info@oldcampbellcountyhistoricalsociety.com
Web Site: www.oldcampbellcountyhistoricalsociety.com/homepage.html
Founded: 1971.
Key Personnel: Dir., Nancy Cornell; Pres. (V), Stan Jones
Institution Type/Description: Historical Society Museum.
Collections: artifacts & memorabilia pertaining to Campbell County.
Hours & Admission Prices: Tues. 11-4; other times by appointment. No charge; donations accepted. &
Attendance: 1,000 (estimated)
Membership: Student $5; Individual $17; Member & Spouse $27; Business $50; Life $250.

Fairmount

SUNRISE PLANETARIUM & SCIENCE MUSEUM, 1427 Slate Mine Rd., Fairmount, GA 30139-2835. Tel.: 706-337-2775.
Institution Type/Description: Planetarium & Science Museum.
Collections: anthropology; archaeology; astronomy; entomology; ethnology; geology.
Hours & Admission Prices: Oct.-May 1 Sun. 3; other times by appointment. No charge.

Fargo

STEPHEN C. FOSTER STATE PARK, 17515 Hwy. 177, Fargo, GA 31631-5004. Tel.: 912-637-5274. Fax: 912-637-5587.
Web Site: www.gastateparks.org/info/scfoster
Founded: 1954.
Congressional District: 8
Key Personnel: Mgr., Travis Griffin.
Personnel Profile: Full-Time Paid 12.
Governing Authority: state. Parent Institution: Georgia Dept. of Natural Resources. Administered by: State Parks & Historic Sites. Tax-exempt.
Institution Type/Description: State Park Museum.
Collections: natural history of Okefenokee Swamp; lumbering; turpenting.
Research Fields: natural & cultural history of Okefenokee Swamp.
Facilities: canoe, boat & walking trails; boardwalk; camping; cottages.
Activities: tours; boat tours & educational programs.
Publications: Trail Guide; postcards.
Hours & Admission Prices: Park: 7am-10pm. Suwannee River Center: Wed.-Sun. 8-5. Office: Fall & Winter: daily 8-5; Spring & Summer: daily 7-6. National Park Pass: $5 per vehicle. Boat tours: adults $10, children $6; children under 4 no charge. &
Attendance: 65,000 (estimated)

Fayetteville

HOLLIDAY-DORSEY-FIFE MUSEUM, 140 Lanier Ave., W., Fayetteville, GA 30214-1606. Tel.: 770-716-5332. Fax: 770-460-3906.
E-mail: bloisehill@fayetteville-ga.gov
Web Site: hdfhouse.com
Founded: 2003.
Congressional District: 3
Key Personnel: House Mgr. & Museum Shop Mgr., John Lynch; Dir., Brian Wismer; Chm. (V), Debi Riddle.
Personnel Profile: Full-Time Paid 1; Part-Time Volunteers 4.
Governing Authority: Parent Institution: Fayetteville Downtown Dev. Authority. Tax-exempt.
Institution Type/Description: History Museum.
Collections: historic artifacts & treasures of Fayette County, including Gone with the Wind memorabilia; Lower Creek Indian arrowheads & drill tips; relics & documents from the War Between the States.
Activities: Museum Sponsors: Living History Days third weekend in May; Cemetery Spirit Walk in October; Victorian Christmas first weekend in December.
Hours & Admission Prices: Thurs.-Sat. 10-3. Adults $5, senior citizens $4.
Attendance: 500 (estimated)

Fitzgerald

BLUE AND GRAY MUSEUM, 116 N. Johnston St., Fitzgerald, GA 31750-2476. Tel.: 229-426-5069; 800-386-4642. Fax: 229-426-5069.
E-mail: bgmuseum@mchsi.com
Founded: 1961.
Key Personnel: Dir., Al Strom; Asst. Dir., Mary Spicer; Dir. Tourism, Alesia Biggers.
Personnel Profile: Part-Time Paid 23; Part-Time Volunteers 2.
Governing Authority: Tax-exempt.
Institution Type/Description: History Museum.
Collections: history of Fitzgerald; photographs; household artifacts; film documentary; life of General Raymond Davis, USMC (Ret.); period clothing; china; glassware; cooking utensils, American Civil War artifacts; oral histories; African American history; Southern Cross of Honor medal; Medal of Honor.
Activities: film.
Hours & Admission Prices: Tues.-Sat. 10-4, Sun. 1-5. Adults $3, students $1; discounts to AAM members, seniors & groups of 10 or morel; members no charge. Closed New Year's Day; Memorial Day; Independence Day; Labor Day; Thanksgiving & day after; Christmas. &
Attendance: 1,396 (estimated)
Membership: Individual $25; Couple $35; Business Club & Family $50; Patron $100; Sustaining Patron $250; Benefactor $500.

Flovilla

INDIAN SPRINGS STATE PARK MUSEUM, Hwy. 42, 5 mi. S. of Jackson, Flovilla, GA 30216. Mailing Address: 678 Lake Clark Rd., Flovilla, GA 30216-2309. Tel.: 770-504-2277. Fax: 770-504-2178.
Web Site: www.georgiastateparks.org
Founded: 1825.
Congressional District: 3
Key Personnel: Mgr., Ken Lalumiere.
Personnel Profile: Full-Time Paid 6; Part-Time Paid 1; Part-Time Volunteers 2.
Governing Authority: state. Administered by Parks, Recreation & Historic Sites Div., Georgia Dept. of Natural Resources, Floyd Towers E., 205 Butler St., Atlanta, GA 30334. Tax-exempt: 170(b)(1)(A).
Institution Type/Description: History Museum.
Collections: photographs of local historical hotels, Civilian Corps., Hoard Mullis Amusement Park; pictures of local Creek Indians; treaties; replica of traditional Indian apparel; pottery; items that reflect stages of Indian civilization.
Research Fields: Creek Indians in Georgia.
Facilities: cottages; group camp facilities; fishing; picnicking; camping; beach; pedal boats; fishing boats; mini-golf. Gift items for sale.
Hours & Admission Prices: Memorial Day-Labor Day Sat.-Sun. 12-4. Museum: no charge. Parking $3. &
Attendance: 400,000 (accurate)

Flowery Branch

FLOWERY BRANCH DEPOT MUSEUM, 5517 Main St., Flowery Branch, GA 30542. Tel.: 770-967-6472.
Institution Type/Description: History Museum: housed in the former Flowery Branch train depot; built in 1890.
Collections: local history; photographs; railroad artifacts.
Activities: rental facilities.
Hours & Admission Prices: Sat. 11-3.

Forest Park

NATIONAL MUSEUM OF COMMERCIAL AVIATION, 5442 Frontage Rd., Ste. 110, Forest Park, GA 30297. Tel.: 404-675-9266.
E-mail: info@nationalaviationmuseum.com
Web Site: www.nationalaviationmuseum.com
Institution Type/Description: Aviation Museum.
Collections: commercial aviation history & artifacts; timetables & ticket jackets; 200 early uniforms; model airplanes; period aviation toys; in-flight serving ware dating back to 1930s; travel posters; aircraft components; Eastern Airlines Martin 404 cockpit; Gate Gourmet catering truck; Tug tractor.
Facilities: library; archives.
Activities: group & school tours; rental facilities.
Hours & Admission Prices: Wed.-Sat. 10-4; other times by appointment. Adults $4, children 4 & up $2.

Fort Gaines

SUTTONS CORNER FRONTIER STORE MUSEUM, 115 S. Washington St., Fort Gaines, GA 39851. Mailing Address: P.O. Box 787, Fort Gaines, GA 39851-0787. Tel.: 229-221-2502.
E-mail: karen@suttonscorner.org
Key Personnel: Cur., Karen Klear.
Personnel Profile: Part-Time Volunteers 5.
Governing Authority: Parent Institution: Fort Gaines Historical Society. Tax-exempt.
Institution Type/Description: History Museum; housed in a former country store; built c.1850.
Collections: local history & culture; period artifacts; wooden cash registers; grist mill; photographs; personal artifacts.
Hours & Admission Prices: Thurs.-Sat. 10-3 by appointment. No charge; donations accepted. Closed Thanksgiving; Christmas. &
Attendance: 300 (estimated)

Fort Gordon

U.S. ARMY SIGNAL CORPS MUSEUM, 504 Chamberlain Ave., Bldg. 29807, Fort Gordon, GA 30905-5735. Tel.: 706-791-2818 & 3856. Fax: 706-791-6069.
E-mail: atzh-pom-m@gordon.army.mil
Web Site: www.gordon.army.mil/ocos/Museum/
Founded: 1965.
Congressional District: 10
Key Personnel: Dir., Robert Anzuoni.
Personnel Profile: Full-Time Paid 3.
Governing Authority: federal government; U.S. Army. Parent Institution: Center of Military History. Tax-exempt.
Institution Type/Description: Military/Science & Technology Museum.
Collections: U.S. Army signal corps history; c.1860-present signal flags; weapons; equipment; communication, science & technology; Signal Corps; uniforms & swords; 10th Armored Division artifacts.
Research Fields: Signal Corps equipment.
Facilities: 10,000-vol. library pertaining to Signal Corps equipment available to the public by appointment.
Activities: guided tours; films; training programs for professional museum workers; temporary & traveling exhibitions; school loan service; science & history classes for military and K-12; living history demonstrations.
Publications: brochures, Code Talker; Story of Wig-Wag; Signaling Souls; The Korean War; Signal Camp of Instruction; Hello Girls; V-Mail; Junior Signalier; Army Values.
Hours & Admission Prices: Tues.-Fri. 8-4, Sat. 10-4. No charge; donations accepted. Closed federal holidays. &
Attendance: 34,000 (accurate)

Fort Oglethorpe

CHICKAMAUGA AND CHATTANOOGA NATIONAL MILITARY PARK, 3370 LaFayette Rd., Fort Oglethorpe, GA 30742. Mailing Address: P.O. Box 2128, Fort Oglethorpe, GA 30742-0128. Tel.: 706-866-9241.
Web Site: www.nps.gov/chch
Founded: 1890.
Congressional District: 9
Personnel Profile: Full-Time Paid 24; Part-Time Paid 8; Part-Time Volunteers 40; Interns 1.
Governing Authority: federal. Parent Institution: U.S. Government Dept. of Interior. Subsidiary Institution: National Park Service. Tax-exempt.
Institution Type/Description: Military Museum.
Collections: Civil War artifacts; Fuller gun collection of American military shoulder arms. Historic Houses: 1866 Cravens House; 1850s & 1860s Brotherton, Snodgrass & Kelley Cabins.
Research Fields: Civil War History.
Facilities: 2,000-vol. library of books on Civil War history available on premises by appointment; 150-seat auditorium at Chickamauga.
Activities: seasonal guided tours & living history demonstrations; lectures; films & slides.
Publications: park brochure; trail map; self-guiding tour pamphlet.
Hours & Admission Prices: Daily 8:30-5. No charge; donations accepted. Closed Christmas. &
Attendance: 900,000 (accurate)

6TH CAVALRY MUSEUM, #6 Barnhardt Circle, Fort Oglethorpe, GA 30742-3646. Mailing Address: P.O. Box 2011, Fort Oglethorpe, GA 30742-0011. Tel.: 706-861-2860.
E-mail: info@6thcavalrymuseum.com
Web Site: www.6thcavalrymuseum.com
Founded: 1981.
Key Personnel: Dir., Christine McKeever; Chm. (V), Kyle Russell.
Personnel Profile: Full-Time Paid 1; Part-Time Volunteers 1.
Governing Authority: Tax-exempt.
Institution Type/Description: Military History Museum: listed on the National Register of Historic Places.
Collections: military artifacts; uniforms; weapons; photographs; Patton Tank and Cobra Gunship Helicopter.
Facilities: 6,500 sq. ft. exhibit space.
Hours & Admission Prices: Tues.-Sat. 9-4. Families $10, adults $3, students & seniors $2; discounts to groups; children under 5 no charge.
Attendance: 5,000 (estimated)
Membership: Family $25.

Fort Stewart

FORT STEWART MUSEUM, 2022 Frank Cochran Dr., Bldg. T904, Fort Stewart, GA 31314-4936. Tel.: 912-767-7885. Fax: 912-767-2121.
E-mail: walter.meeks@stewart.army.mil
Web Site: www.stewart.army.mil/about/history.asp
Founded: 1977.
Congressional District: 1
Key Personnel: Dir., Walter W. Meeks, III
Governing Authority: federal. Part of Army Museum System. Parent Institution: U.S. Army Center of Military History, Attn: DAMH-MD, 1099 14th St. N.W., Washington, DC 20005-3402. Tel. 202-761-5373. Tax-exempt.
Institution Type/Description: Military Museum.
Collections: focuses on the history of 3rd Infantry Division, 24th Infantry Division & Fort Stewart follows history of units & the post Word War I, World War II, Korea & Desert Storm.
Research Fields: history, lineage & honors relating to the 3rd Infantry Division, 24th Infantry Division & Fort Stewart; the history of Fort Stewart; technical data on weapons, equipment & vehicles.
Facilities: 185-vol. library of hardback books, 91 paperback manuals, photo file of several hundred pieces on the history of division & its units, post, U.S. military history, general military history, historical data on arms & equipment available for research on premises & by special arrangement.
Activities: guided tours; lectures; gallery talks; permanent & temporary exhibitions; staff rides.
Publications: brochure, 3rd Division History & Fort Stewart.
Hours & Admission Prices: Temporarily closed for renovations. &
Attendance: 45,036 (accurate)

Fort Valley

A.L. FETTERMAN EDUCATIONAL MUSEUM, Massee Lane Gardens, 100 Massee Lane, Fort Valley, GA 31030-6974. Tel.: 478-967-2358. Fax: 478-967-2083.
E-mail: crichard@americancamellias.org
Web Site: www.americancamellias.org
Founded: 1989.
Congressional District: 8
Key Personnel: Exec. Dir., Celeste Richard; Pres., Don Bergamini; Museum Shop Mgr., Stefanie Turner.
Governing Authority: society; nonprofit. Parent Institution: American Camellia Society. Tax-exempt: 501(c)(3).
Institution Type/Description: Library & Porcelain Museum.
Collections: two museum buildings with rare books, original paintings, prints; porcelain sculptures from Boehm, Connoisseur, Bronn, Cybis, Royal Worchester, Dorothy Doughty, Gunther Granget, Hummel & other art objects.
Research Fields: camellia cultivars & breeding.
Facilities: 1,500-vol. library of rare & current books relating to horticulture, with emphasis on camellias; 170-seat auditorium; botanical garden; banquet facilities available. Museum-related items for sale.
Activities: guided tours; lectures; organized education programs for adults; docent program; environmental day camp for children 6-11.
Publications: quarterly magazine, The Camellia Journal; annual, The Camellia Yearbook.
Hours & Admission Prices: Tues.-Sat. 10-4:30, Sun. 1-4:30. Adults $5, seniors 65 & over $4; ACS members & children under 12 no charge. Closed national holidays. &
Attendance: 22,500 (estimated)
Membership: Single $25; Joint or Family $27.50; Sustaining $50; Patron $125; Single Life $500; Joint Life $550.

MASSEE LANE GARDENS, HOME OF AMERICAN CAMELLIA SOCIETY, 100 Massee Lane, Fort Valley, GA 31030-6974. Tel.: 478-967-2358. Fax: 478-967-2083.
E-mail: crichard@americancamellias.org
Web Site: www.americancamellias.org
Founded: 1945.
Congressional District: 8
Key Personnel: Pres., Judge Roger Vinson; Operations Mgr., Celeste Richard; Museum Shop Mgr., Leisa Dortch.
Personnel Profile: Full-Time Paid 5; Part-Time Paid 1; Part-Time Volunteers 12.
Governing Authority: society; nonprofit organization. Parent Institution: American Camellia Society. Branch Museums: A. L. Fetterman Educational Museum; Stevens-Taylor Gallery. Tax-exempt: 501(c)(3).
Institution Type/Description: Horticultural Society.
Collections: rare books; original paintings & prints; Boehm porcelains; Boehm; Connoisseur; Cybis; Bronn; Royal Copenhagen; Hummel; gardens; S.E. United States camellias & plants; porcelains.
Research Fields: horticulture.
Facilities: 5,000-vol. library of books on horticulture available for research on site; botanical garden; greenhouse; rose garden; Japanese garden; reading room; environmental garden. Museum-related items for sale.
Activities: lectures; gallery talks; workshops.
Publications: quarterly magazine, The Camellia Journal; book, The American Camellia Yearbook.
Hours & Admission Prices: Tues.-Sat. 10-4:30, Sun. 1-4:30. Adults $5; discounts to senior citizens, groups, AAA, AAM & ICOM members; children under 12 & members no charge. &
Attendance: 35,000 (estimated)
Membership: Single $25; Joint $27.50; Patron $100; Life-Single $500; Life-Double $550.

Gainesville

BRENAU UNIVERSITY GALLERIES, BRENAU UNIVERSITY, (M), 500 Washington St., S.E., Gainesville, GA 30501-3697. Tel.: 770-534-6263 (gallery). Fax: 770-538-4599.
E-mail: gallery@brenau.edu
Web Site: www.brenau.edu
Founded: 1985.
Congressional District: 9
Key Personnel: Dir. & Cur., Vanessa Grubbs; University Pres., Ed Schrader; Dir. Art & Design Dept., Mary Beth Looney.
Personnel Profile: Full-Time Paid 2; Part-Time Paid 6; Part-Time Volunteers 9; Interns 1.
Governing Authority: private university; nonprofit. Parent Institution: Brenau University. Tax-exempt: 501(c)(3).

Institution Type/Description: University Art Gallery: three gallery spaces, 2 housed in adjacent buildings, one is in the main floor of the Simmons Visual Arts Center, a recently renovated 1914 structure; the other gallery space is outside the balcony area of Brenau's c.1880s restored Pearce Auditorium. The third is in the new John S. Burd Center for the Performing Arts.

Collections: American artists, particularly women, pop artists & historical styles; paintings; drawings; prints; sculpture; photos; craft works: glass, metals, clay & fiber.

Research Fields: pop art movement of the 1960s; women's art; 18th-19th century American/European painting.

Facilities: Simmons Visual Arts Center; the Presidents Gallery; the Leo Castelli Art Gallery; 700-seat auditorium; 300-seat theatre; studios; lecture & recital halls. Museum-related items for sale.

Activities: docent program; formal education programs; interdisciplinary events; workshops; guided tours; lectures; loan, participatory, temporary & traveling exhibitions; training program for professional museum workers through the Arts Management degree program. University Fine Arts & Humanities Department Sponsors: Firespark, a summer program in the arts for high school students.

Publications: exhibition catalogues & brochures.

Hours & Admission Prices: Simmons Visual Arts Center & President's Gallery: Tues.-Fri. 10-4, Sun. 2-5. Leo Costell Art Gallery: Tues.-Fri. 1-4. No charge; donations accepted. Closed Easter; spring break (early March); Memorial Day; Independence Day; Labor Day; Thanksgiving week; Christmas break. &

Attendance: 26,000 (estimated)

Membership: Individual $25; Family $50; Patron $100; Sponsor $250; Benefactor $500; Director's Circle $1,000; Corporate $5,000.

ELACHEE NATURE SCIENCE CENTER, 2125 Elachee Dr., Gainesville, GA 30504-7158. Tel.: 770-535-1976. Fax: 770-535-2302.

E-mail: elachee@elachee.org
Web Site: www.elachee.org
Founded: 1976.
Congressional District: 9
Key Personnel: Pres. & C.E.O., Andrea Timpone; Pres., R.K. Whitehead; Education, Peter Gordon; Devel. & Public Rels., Lavon Callahan; Museum Shop Mgr., Judy Stock.
Personnel Profile: Full-Time Paid 8; Part-Time Paid 25; Part-Time Volunteers 120.
Governing Authority: private; nonprofit organization. Tax-exempt: 501(c)(3).
Institution Type/Description: Nature Science Center.
Collections: native amphibians; reptiles.
Research Fields: archaeology; water quality; plant & animal biodiversity inventory.
Facilities: museum center; hiking trails; picnic area. Gift items for sale.
Activities: Southern Association of Colleges and Schools (SACS) accredited education programs for children; public programs; summer & school-break camps.
Hours & Admission Prices: April-Nov. Mon.-Sat. 10-5; Dec.-March Mon.-Fri. 10-3. Adults $5, children 2-12 $3; discounts to ANCA members; members & children under 2 no charge. Closed New Year's Day; Thanksgiving; Christmas. &
Attendance: 70,000 (estimated)
Membership: Individual $30; Family $45; Friend $100; Donor $250; Patron $500; Grand Patron $1,000.

INTERACTIVE NEIGHBORHOOD FOR KIDS, 999 Chestnut St., S.E., Ste. 11, Gainesville, GA 30501-6962. Tel.: 770-536-1900. Fax: 770-536-5459. Facebook: INK for Kids.

E-mail: info@inkfun.org
Web Site: www.inkfun.org
Founded: 2002.
Congressional District: 9
Key Personnel: Exec. Dir., Sheri Hooper; Chm. (V), Alan Wayne
Institution Type/Description: Children's Museum.
Collections: hands-on exhibitions.
Activities: birthday parties; rental facilities; special events; educational programs.
Hours & Admission Prices: Mon.-Sat. 10-5. Admission $8; discounts Sun. 1-5. &
Attendance: 70,000 (accurate)
Membership: Individual $99.

NORTHEAST GEORGIA HISTORY CENTER AT BRENAU UNIVERSITY, (M), 322 Academy St., N.E., Gainesville, GA 30501. Mailing Address: P.O. Box 1451, Gainesville, GA 30503-1451. Tel.: 770-297-5900. Fax: 770-297-5933.

Web Site: www.negahc.org
Formerly: Georgia Mountains History Museum at Brenau University
Founded: 1981.
Congressional District: 9
Key Personnel: C.E.O., Glen Kyle; Pres. (V), Dr. Patricia Burd.
Personnel Profile: Full-Time Paid 2; Part-Time Paid 1; Part-Time Volunteers 100; Interns 2.
Governing Authority: private; nonprofit organization. Branch Museums: Chief White Path, Gainesville, GA. Tax-exempt: 501(c)(3).
Institution Type/Description: History Museum.
Collections: textile; poultry; medical; African-American; Indian; firefighting tools & equipment; 19th-century American arts & crafts; Gen. James Longstreet memorabilia; Ed Dodd & Mark Trail exhibit.
Facilities: library; 8,000 sq. ft. exhibit space; American Freedom Garden. Gift items for sale.
Activities: docent program; guided tours; traveling exhibitions. Annual Events: Taste of History luncheon; Spring Estate sale.
Publications: monthly newsletter.
Hours & Admission Prices: Tues.-Sat. 10-4. Adults $5, senior citizens, students & children $3; discounts to AAA members; members no charge. Closed New Year's Day; Independence Day; Christmas. &
Attendance: 6,000 (accurate)
Membership: Student $15; Individual $35; Family $70; Sponsor $100; Contributor $250; Curator $500; Friend $1,000. Partnership Levels: $1,500; $2,500; $5,000.

QUINLAN VISUAL ARTS CENTER, 514 Green St., N.E., Gainesville, GA 30501-3314. Tel.: 770-536-2575. Fax: 678-343-2738. Facebook: Quinlan Visual Arts Center.

E-mail: info@qvac.org
Web Site: www.qvac.org
Founded: 1946.
Congressional District: 10
Key Personnel: Exec. Dir., Amanda McClure; Asst. Dir., Paula E. Lindner; Admin., Margaret Tingley.
Personnel Profile: Full-Time Paid 2; Part-Time Paid 1; Part-Time Volunteers 100.
Governing Authority: Parent Institution: Quinlan Arts Inc., Board of Trustees. Tax-exempt.
Institution Type/Description: Arts Center.
Collections: paintings; sculpture.
Facilities: classrooms. Museum-related items for sale.
Activities: educational programs; summer art camps; workshops. Annual Event: 35th Annual Gala Fine Art Auction February to March.
Publications: biannual class schedule; newsletter; workshop brochure.
Hours & Admission Prices: Mon.-Fri. 9-5, Sat. 10-4. No charge; donations accepted. Closed Memorial Day; Independence Day; Labor Day; Thanksgiving; Christmas. &
Attendance: 9,000 (estimated)
Membership: Senior Citizen, Student, Teacher & Artist $50; Individual $55; Family $100; Contributing Patron $150; Donor Patron $300; Sustaining Patron $500; Medallion Associate $1,000; Medallion Patron $2,500; Medallion Benefactor $5,000.

Glennville

TATTNALL MUSEUM, 211 S. Tillman St., Glennville, GA 30427-1737. Mailing Address: P.O. Box 607, Glennville, GA 30427-0607. Tel.: 912-654-5276. Fax: 912-538-3156.
Key Personnel: Museum Bd. Chm., Daine Bazemore.
Governing Authority: Parent Institution: city of Glennville.
Institution Type/Description: Art, Science & History Museum.
Collections: art; science; local history.
Hours & Admission Prices: Mon.-Thurs. 8am-9pm. Closed national holidays.

Grovetown

GROVETOWN MUSEUM, 106 W. Robinson Ave., Grovetown, GA 30813. Tel.: 706-863-1867.
Web Site: www.choosecolumbiacounty.com
Institution Type/Description: History Museum.
Collections: local history & culture; period furnishings; photographs; personal artifacts.
Hours & Admission Prices: Thurs.-Fri. 10-4, Sat. 1-4.

Harlem

LAUREL & HARDY MUSEUM, 250 N. Louisville St., Harlem, GA 30814. Mailing Address: P.O. Box 99, Harlem, GA 30814. Tel.: 888-288-9108; 706-556-0401.
E-mail: museum@harlemga.org
Founded: 2002.
Institution Type/Description: History Museum: birthplace of Oliver Hardy.
Collections: life & career history of Laurel & Hardy; photographs; movies; personal artifacts; posters; memorabilia.
Activities: movies.
Hours & Admission Prices: Tues.-Sat. 10-4.

Hartwell

HART COUNTY HISTORICAL MUSEUM - TEASLEY-HOLLAND HOUSE, 31 E. Howell St., Hartwell, GA 30643. Tel.: 706-376-6330.
Institution Type/Description: History Museum: housed in the former home of Isham Asbury Teasley; built in 1880.
Collections: local history & culture; period furnishings; photographs.
Hours & Admission Prices: Mon.-Fri. 8:30-5. No charge.

Hawkinsville

HAWKINSVILLE/PULASKI COUNTY ARTS COUNCIL, 100 Lumpkin St., Hawkinsville, GA 31036-1518. Mailing Address: 42 Lumpkin St., Hawkinsville, GA 31036. Tel.: 912-783-1884. Fax: 912-783-2333.
E-mail: ArtsCouncil@cstel.net
Web Site: www.hawkinsvilleoperahouse.com
Key Personnel: Office Mgr., Julianna Stewart
Institution Type/Description: Art Gallery.
Collections: works by Georgia artists.
Activities: cultural & educational activities; tours.
Hours & Admission Prices: Tours: Mon.-Fri. 10-4.

Hiawassee

BRASSTOWN BALD VISITOR CENTER, 2941 State Hwy. 180, Hiawassee, GA 30546. Mailing Address: 2042 Hwy. 515 W., Blairsville, GA 30512-3773. Tel.: 706-745-6928 & 896-2556. Fax: 706-745-7494.
Web Site: www.fs.fed.us/conf
Founded: 1967.
Congressional District: 9
Key Personnel: Resource Asst., Alison Koopman; Recreation Program Mgr., Valencia Morris.
Personnel Profile: Part-Time Paid 4; Part-Time Volunteers 4.
Governing Authority: federal. U.S. Dept. of Agriculture Forest Service, Southern Region.
Institution Type/Description: Park Museum: located within the Chattahoochee National Forest.
Collections: natural history; exhibits pertaining to Man & the Mountain.
Facilities: rooftop observation deck; mountaintop theater; hiking trails; picnic area. Museum-related items for sale.
Activities: video programs.
Publications: brochure.
Hours & Admission Prices: mid-April to May Sat.-Sun. 10-5; Memorial Day to Veterans Day daily 10-5. Parking & Shuttle Ride: 16 & over $5; discounts to seniors, Golden Age Pass & Interagency Access Pass holders. Center: No charge; donations accepted. &
Attendance: 75,000 (estimated)

Jefferson

CRAWFORD W. LONG MUSEUM, 28 College St., Jefferson, GA 30549-1036. Tel.: 706-367-5307.
E-mail: info@crawfordlong.org
Web Site: www.crawfordlong.org
Founded: 1957.
Congressional District: 9
Key Personnel: Administrative Mgr., Vicki Starnes; Consultant, Lesa Campbell; Pres. (V), Roxane Rose; Museum Shop Mgr., Laura Collier.
Personnel Profile: Part-Time Paid 3; Part-Time Volunteers 3.
Governing Authority: municipal; City of Jefferson, GA.
Institution Type/Description: Medical Museum, History Museum & Historic Building & Site.
Collections: 19th-century medical artifacts; personal effects of Dr. C.W. Long; history of the discovery & development of anesthesia; documents; photographs; 1840s doctor's office & apothecary shop; 19th-century general store.
Research Fields: Dr. C.W. Long; history of anesthesia; medicine; local history.
Activities: guided tours; temporary exhibits; lectures; workshops; quarterly lunch & learns; annual historic cemetery tours.
Hours & Admission Prices: Tues.-Fri. 10-5, Sat. 10-4. Adults $5, senior citizens $4, students $3; discounts to military; children 5 & under no charge. Closed on major holidays. &
Attendance: 2,178 (estimated)
Membership: Student $20; Individual $25; Family $35; Corporate $100; Patron $200; Benefactor $1000 & up.

Jekyll Island

THE GEORGIA SEA TURTLE CENTER, 214 Stable Rd., Jekyll Island, GA 31527-0844. Tel.: 912-635-4444.
E-mail: georgiaseaturtlecenter@jekyllisland.com
Web Site: www.georgiaseaturtlecenter.org
Founded: 2007.
Congressional District: 1
Key Personnel: Dir. & Veterinarian, Terry Norton; Coord. Research, Kimberly Andrews; Coord. Education, Katie Higgins; Educator, Kristen Lee; Educator, Kira Sterns.
Governing Authority: Parent Institution: Jekyll Island Authority. Tax-exempt.
Institution Type/Description: Marine Museum.
Collections: exhibits on sea turtle conservation & rehabilitation.
Hours & Admission Prices: March-Nov. Mon. 10-2, Tues.-Sun. 9-5; Dec.-Feb. Tues.-Sun. 9-5. Adults 13 & over $7, senior citizens 65 & over $6, children 4-12 $5; children 3 & under no charge. Closed New Year's Day; Christmas Eve & Day.

JEKYLL ISLAND MUSEUM, 100 Stable Rd., Jekyll Island, GA 31527-0870. Mailing Address: 381 Riverview Dr., Jekyll Island, GA 31527-0874. Tel.: 912-635-4036. Fax: 912-635-4420.
E-mail: jhunter@jekyllisland.com
Web Site: www.jekyllisland.com
Founded: 1954.
Congressional District: 1
Key Personnel: Dir., John S. Hunter; Cur., Gretchen Greminger; Programming, Andrea Marroquin; Tours, Shirley Martin.
Personnel Profile: Full-Time Paid 16; Part-Time Paid 10; Part-Time Volunteers 40; Interns 7.
Governing Authority: State Park Authority; Jekyll Island Authority. Subsidiary Institution: Jekyll Island Museum Associates. Tax-exempt.
Institution Type/Description: Historic Site & Preservation Project.
Collections: furnishings; documents; photographs; clothing & memorabilia; Tiffany stained-glass window. Historic Houses: 1896 Moss Cottage, former Struthers-Macy home; 1907 Goodyear Cottage, former F. H. Goodyear home: 1897 Club House; 1917 Crane Cottage, former R.T. Crane, Jr. home; 1891 Hollybourne Cottage, former C.S. Maurice home; 1927 Villa Ospo, former W. Jennings home; 1904 Cherokee Cottage, former Shrady-James home; 1900 Mistletoe Cottage, former Porter-Claflin home; 1904 Faith Chapel; 1900 service facilities including Club Dock; 1930 Club indoor Tennis court; 1890 infirmary; Baker Crane Stable; boat engineer's house & servant quarters; 1742 tabby ruins of Major Horton House; 1884 DuBignon Cottage; 1898 Visitor Center, former Jekyll Island Club Stables; 1892 Indian Mound, former McKay-W. Rockefeller home; 1928 Villa Marianna, former F. M. Gould home.
Research Fields: The Jekyll Club era, 1886-1942; Colonial Georgia, 1500-1776; decorative art; natural history, barrier island.
Facilities: 120-seat theater; document & photograph archives. Museum-related books & gifts for sale.
Activities: formally organized educational programs; volunteer program; permanent & temporary exhibits; internships; guided tours.
Publications: The Jekyll Island Club Historic District, 100 Years; Jekyll Island Club Historic District Walking Tour.
Hours & Admission Prices: Daily 9-5. Passport to the Century Tour: two period furniture restored cottages. Adults $10, students 6-18 $6; children under 6 no charge. Call for tour hours. Closed New Year's Day; Christmas. &
Attendance: 37,246 (accurate)
Membership: Friends of Historic Jekyll Island, Inc: Individual $25; Family $50; Preservation Partner $100; Business & Sponsor $250; Patron $500; Benefactor $1,000.

Johns Creek

AUTREY MILL NATURE PRESERVE & HERITAGE CENTER, 9770 Autrey Mill Rd., Johns Creek, GA 30022-7168. Tel.: 678-366-3511. Fax: 678-366-3512.
E-mail: events@autreymill.org
Web Site: autreymill.org
Key Personnel: Dir., Ben Team; Heritage & Events, Claudette Lopez; Office Asst., Michael Hanff.
Personnel Profile: Full-Time Paid 1; Part-Time Paid 2; Interns 4.
Institution Type/Description: Nature Preserve.
Collections: local history & culture; early tenant farmlife artifacts including irons, beds, kitchen & laundry equipment; general store inventory & ledgers (1929-1945); Native American artifacts including projectile points, spearheads, & arrowheads; live amphibians, reptiles & insects; farming equipment; chapel. Historic Buildings: Summerour House, c.1880; Old Warsaw Church, c.1860; Green Country Store; 1800s Tenant Farmhouse; 1942 Program Barn; 1860s tenant house.
Hours & Admission Prices: Park: daily 8am to dusk. Visitor Center: Mon.-Sat. 10-4. No charge; donations accepted.
Attendance: 12,000 (estimated)
Membership: Senior $25; Individual $35; Family $50.

Jonesboro

ROAD TO TARA MUSEUM, 104 N. Main St., Jonesboro, GA 30236-8315. Tel.: 770-478-4800; 800-662-7829. Fax: 770-478-1888.
E-mail: info@visitscarlett.com
Web Site: visitscarlett.com/roadtotaramuseum.html
Founded: 2000.
Key Personnel: Dir., Frenda Turner; Museum Shop Mgr., Julie Bustamante; Chm. (V), Linda Summerlin.
Personnel Profile: Full-Time Paid 4; Part-Time Paid 8.
Governing Authority: Parent Institution: Clayton County Tourism Authority. Tax-exempt.
Institution Type/Description: History Museum: housed in 1867 Historic Train Depot.
Collections: Civil War & Gone with the Wind memorabilia.
Hours & Admission Prices: Mon.-Fri. 8:30-5:30, Sat. 10-4. Adults $7, senior citizens & children $6.
Attendance: 13,000 (accurate)

STATELY OAKS PLANTATION, (M), 100 Carriage Lane, Jonesboro, GA 30236. Mailing Address: P.O. Box 922, Jonesboro, GA 30237-0922. Tel.: 770-473-0197. Fax: 770-473-9855.
E-mail: statelyoaks@historicaljonesboro.org
Web Site: historicaljonesboro.org
Founded: 1972.
Congressional District: 13
Key Personnel: Pres. (V), Barbara Emert.
Personnel Profile: Full-Time Paid 1; Part-Time Paid 4; Part-Time Volunteers 35; Interns 2.
Volunteer Hours: 8,000
Governing Authority: Parent Institution: Historical Jonesboro/Clayton County, Inc. Tax-exempt.
Institution Type/Description: Historic House.
Collections: period artifacts & memorabilia.
Major Exhibits: Victorian Valentines, 2/14.
Research Fields: Battle of Jonesboro; Gone with the Wind; Victorian costumes.
Facilities: 1839 Greek revival house; 1839 log kitchen; 1894 country store; 1900s tenant cabin; 1900s Bethel house classroom; Creek Indian village.
Activities: school tours.
Hours & Admission Prices: Mon.-Sat. 10-4. Adults $12, senior citizens 55 & over and military with ID $9, children under 11 & retired military $6.
Attendance: 5,000 (estimated)
Membership: Individual $10; Family $75.

Juliette

JARRELL PLANTATION GEORGIA STATE HISTORIC SITE, 711 Jarrell Plantation Rd., Juliette, GA 31046-2525. Tel.: 478-986-5172. Fax: 478-986-5919.
E-mail: jarrell.plantation.park@dnr.state.ga.us
Web Site: gastateparks.org
Founded: 1974.
Congressional District: 8
Key Personnel: Dir., Ken Lalumiere; Ranger, Bretta Perkins.

Personnel Profile: Full-Time Paid 2; Part-Time Volunteers 60.
Volunteer Hours: 860
Operating Income: 27,300
Governing Authority: state. Parent Institution: Georgia Dept. of Natural Resources, 2 Martin Luther King, Jr. Dr., S.E., Ste. 1352, Atlanta, GA 30334. Subsidiary Institution: Parks, Recreation & Visitor Sites Div. Tax-exempt.
Institution Type/Description: Living Farm Historic Site.
Collections: tools; furnishings; clothing; implements; grist mill; cotton gin; planer mill; boiler & steam engines; syrup evaporators; cane mill; saw mill; blacksmith forge.
Research Fields: Georgia agriculture & farm life, 1840-1960; Georgia history; Antebellum South; tenant farmer era.
Facilities: visitor center; barns; outbuildings; two dwellings. Publications & brochures for sale.
Activities: seasonal special events.
Hours & Admission Prices: Thurs.-Sat. 9-5. Admission: $4-$6.50; discount to groups & Friends of Georgia State Parks & Historic Sites members; children under 6 no charge. Closed New Year's Day; Thanksgiving; Christmas.
Attendance: 5,000 (estimated)
Membership: Friends of Jarrell Plantation & Friends of Georgia State Parks & Historic Sites $50 & up.

Kennesaw

BERNARD A. ZUCKERMAN MUSEUM OF ART, 1000 Chastain Rd., #3104, Kennesaw, GA 30144-5591. Tel.: 770-499-3223. Fax: 770-499-3345. Facebook: Zuckerman Museum.
E-mail: zma@kennesaw.edu
Web Site: zuckerman.kennesaw.edu
Formerly: Kennesaw State University Art Museum & Galleries
Congressional District: 11
Key Personnel: Dir., Justin Rabideau.
Personnel Profile: Full-Time Paid 5; Part-Time Paid 10; Part-Time Volunteers 3; Interns 3.
Governing Authority: Parent Institution: Kennesaw State University. Tax-exempt.
Institution Type/Description: Art Museum.
Collections: contemporary & historical works of art with focus on American artists; modern & contemporary works on paper; Southern Graphics Council International Print & Archive collection.
Major Exhibits: See Through Walls, 1/14-3/14; 2013-14 Walthall Fellows, 4/14-7/14; Rowland Ricketts: The Substance of Color, 7/14-10/14; Here Say: Imaging the South, 7/14-10/14.
Activities: demonstrations; lectures; permanent & temporary exhibitions.
Hours & Admission Prices: Tues.-Thurs. & Sat. 11-4. No charge; donations accepted. Closed university & public holidays.

KENNESAW MOUNTAIN NATIONAL BATTLEFIELD PARK, 900 Kennesaw Mountain Dr., Kennesaw, GA 30152-4854. Tel.: 770-427-4686. Fax: 770-528-8398. Facebook: Kennesaw Mountain National Battlefield Park.
Web Site: www.nps.gov/kemo
Founded: 1935.
Congressional District: 11
Key Personnel: Park Supt., Nancy Walther; Chief Interpreter & Resource Mgr., Anthony Winegar; Cur., Amanda K. Corman; Historian, W.R. Johnson; Museum Shop Mgr., Dan Beard.
Governing Authority: federal. Affiliated with National Park Service, U.S. Dept. of Interior. Tax-exempt.
Institution Type/Description: Civil War History Museum: located on the site of a Civil War battlefield.
Collections: dress & weapons of Civil War soldiers. Historic House: 1836 Kolb Farm.
Research Fields: Atlanta Campaign of 1864.
Facilities: 1,200-vol. non-circulating library of Civil War books, papers & diaries available on premises; 98-seat auditorium; trails.
Activities: self-guided tours; auto tour; hiking tour; auto-visual program; Civil War soldier living history demonstrations spring to fall; Sat. afternoon lectures spring & fall.
Publications: park brochure; site bulletins.
Hours & Admission Prices: Park: March-Oct. daily 7:30am-8pm; Nov.-Feb. daily 7:30-6. Visitor Center: daily 9-5. Center: no charge; donations accepted. Shuttle Service Fee: adults 12 & over $3, children 6-11 $1.50; America the Beautiful - The National Parks & Federal Recreational Land Pass holders & children under 6 no charge. Visitor Center: closed New Year's Day; Thanksgiving; Christmas.
Attendance: 1,900,000 (estimated)

MUSEUM OF HISTORY AND HOLOCAUST EDUCATION, **(M),** Kennesaw State Univ. Center, 3333 Busbee Dr., Kennesaw, GA 30144. Mailing Address: 1000 Chastain Rd., MD 3308, Kennesaw, GA 30144-5588. Tel.: 678-797-2083. Fax: 770-420-4432.
Web Site: www.kennesaw.edu/historymuseum
Institution Type/Description: History Museum.
Collections: WWII & Holocaust history; photographs.
Activities: educational programs.
Hours & Admission Prices: Mon.-Fri. 10-5; groups by appointment. Closed during university holidays & breaks.

SMITH-GILBERT GARDENS, 2382 Pine Mountain Rd., Kennesaw, GA 30152. Tel.: 770-919-0248.
Key Personnel: Dir. Operations, Susan Schroeder
Institution Type/Description: Botanical Garden.
Collections: over 3,000 species of plants; sculptures. Historic House: Hiram Butler Home, 1880.
Facilities: banquet facilities.
Activities: special events; rental facilities; educational programs; classes.
Hours & Admission Prices: Tues.-Sat. 9-4; other times by appointment. Adults $7, seniors & active military $6, children 6-12 $5; children 5 & under no charge.

THE SOUTHERN MUSEUM OF CIVIL WAR AND LOCOMOTIVE HISTORY, (M), 2829 Cherokee St., Kennesaw, GA 30144-2823. Tel.: 770-427-2117. Fax: 770-421-8485.
E-mail: rbanz@kennesaw.ga.gov
Web Site: www.southernmuseum.org
Formerly: Kennesaw Civil War Museum
Founded: 1972.
Congressional District: 16
Key Personnel: Exec. Dir., Dr. Richard Banz; Cur., Jonathan Scott; Archivist, Sallie Loy; Dir. Operations, Dena Bush.
Personnel Profile: Full-Time Paid 10; Part-Time Paid 2; Part-Time Volunteers 27.
Governing Authority: municipal. Parent Institution: Smithsonian Institution. Tax-exempt.
Institution Type/Description: Civil War & Train Museum: located on the site where the Great Locomotive chase began.
Collections: Civil War memorabilia; history of Georgia railroads; graphic presentation of the Andrews Raid; locomotive history; replica of a 1910 locomotive manufacturing facility which is belt driven with all machinery, production line, pattern shop, offices, forge & laboratories; 2 locomotives on assembly line. Historic Structure: c.1875 railroad depot.
Facilities: library; archives; theater; education center. Gift items for sale.
Activities: guided tours; lectures; films; special events.
Publications: brochures; quarterly newsletter.
Hours & Admission Prices: Mon.-Sat. 9:30-5, Sun. 11-6. Adults $7.50, senior citizens $6.50, children 4-12 $5.50; members & children under 3 no charge. Closed Easter; Thanksgiving; Christmas Eve & Day. &
Attendance: 50,000 (estimated)
Membership: Individual $50; Family Plus Smithsonian $75; Sustaining $150; Patron $250; Sponsor $500; Benefactor $1,000.

Kingsland

KINGSLAND DEPOT, 200 E. King Ave., Kingsland, GA 31548. Tel.: 912-673-1890.
Institution Type/Description: Historic Building & Welcome Center: housed in a former railroad depot; built c.1920. The OWN Network & BBC America rented the depot to be their Lovetown USA Headquarters from Feb. 14 to March 14, 2012. Oprah Winfrey filmed the opening for Remembering Whitney: The Oprah Interview in the Kingsland Depot.
Collections: local history & culture; period furnishings; photographs; personal artifacts.
Hours & Admission Prices: Mon.-Fri. 8-5.

Kingston

KINGSTON WOMAN'S HISTORY CLUB MUSEUM, 13 E. Main St., N.W., Kingston, GA 30145-2307. Mailing Address: P.O. Box 261, Kingston, GA 30145-0261. Tel.: 770-546-3116.
Governing Authority: private; nonprofit corporation. Tax-exempt: 501(c)(3).
Institution Type/Description: History Museum.
Collections: local history & culture; scrapbooks; photographs.
Hours & Admission Prices: Sat.-Sun. 1-4; other times by appointment. No charge; donations accepted.
Membership: Associate $15; Active $25.

LaFayette

THE MARSH HOUSE, (M), 308 N. Main St., LaFayette, GA 30728-2422. Mailing Address: P.O. Box 722, LaFayette, GA 30728-0722. Tel.: 706-638-5187. Facebook: The Marsh House of Lafayette.
Web Site: marshhouseoflafayette.com
Founded: 2002.
Congressional District: 9
Key Personnel: Pres., Evelle Dana; Recording Sec., Dr. David P. Boyle; House Mgr., Marjorie Craig; Events Coord., Mary Smitherman; Docent Training, Jennie Chandler.
Personnel Profile: Part-Time Paid 1.
Governing Authority: Parent Institution: Walker County Historical Society. Tax-exempt.
Institution Type/Description: Historic House: built in 1836.
Collections: local history & culture; period artifacts; furnishings.
Publications: quarterly, Walker County Historical Society.
Hours & Admission Prices: Tours by appointment. Adults $5. &
Attendance: 500 (estimated)
Membership: General $10.

LaGrange

HILLS AND DALES ESTATE, 1916 Hills and Dales Dr., LaGrange, GA 30240-2958. Mailing Address: P.O. Box 790, LaGrange, GA 30241-0014. Tel.: 706-882-3242. Fax: 706-882-3464. Facebook: Hills and Dales Estate.
E-mail: cwood@hillsanddales.org
Web Site: www.hillsanddales.org
Founded: 1998.
Congressional District: 7
Key Personnel: Exec. Dir., Carleton Wood; Pres. Foundation, H. Speer Burdette, III; Treas. Foundation, Esther S. Rainey; Museum Shop Mgr., Carrie Mills.
Personnel Profile: Full-Time Paid 12; Part-Time Paid 9; Part-Time Volunteers 1.
Governing Authority: private; nonprofit organization. Parent Institution: Fuller E. Callaway Foundation. Tax-exempt: 501(c)(3).
Institution Type/Description: Historic House Museum: housed in the former home of textile magnate Fuller E. Callaway Sr. and his family.
Collections: decorative arts & artifacts related to the Fuller E. Callaway family; Ferrell Gardens including boxwood plantings, fountains, herb garden & greenhouse.
Research Fields: history of Hills and Dales Estate and Ferrell Gardens.
Facilities: 35-acre estate; 1,500 sq. ft. exhibit space; theater; gardens. Gift-related items for sale.
Activities: formal education programs; guided tours of house & garden.
Publications: biannual newsletter.
Hours & Admission Prices: March-June Tues.-Sat. 10-6, Sun. 1-6; July-Feb. Tues.-Sat. 10-5. Adults $15, students $7. Closed major holidays. &
Attendance: 10,000 (accurate)
Membership: Individual $30; Family $45.

LAGRANGE ART MUSEUM, 112 Lafayette Pkwy., LaGrange, GA 30240-3209. Tel.: 706-882-3267. Fax: 706-882-2878.
E-mail: art@lagrangeartmuseum.org
Web Site: www.lagrangeartmuseum.org
Formerly: Chattahoochee Valley Art Museum
Founded: 1963.
Congressional District: 3
Personnel Profile: Full-Time Paid 2; Part-Time Paid 2; Part-Time Volunteers 2.
Governing Authority: nonprofit organization. Tax-exempt: 501(c)(3).
Institution Type/Description: Art Museum: housed in 1892 Victorian structure, originally a county jail.
Collections: contemporary American works of art.
Research Fields: art history.
Facilities: classrooms; art resource center.
Activities: guided tours; lectures; gallery talks; workshops; formally organized education programs; temporary exhibitions. Museum Sponsors: Creative Youth Art League - Youth Art Competition; LaGrange National Biennial Competition; For the Love of the Arts - Arts Festival.
Publications: Artsplace, quarterly schedule of classes & activity; newsletter; exhibition catalogues.
Hours & Admission Prices: Tues.-Fri. 9-5, Sat. 11-5, Mon. by appointment. Suggested Donation: adults $5; Troup County residents no charge. Closed New Year's Day; Memorial Day; Independence Day; Thanksgiving; Christmas. &
Attendance: 3,582 (accurate)

Membership: Apprentice $25; Journeyman $50; Artisan $85; Family $100; Craftsman $250; Master $500; Curator's Circle $1,000; Sustainer Level $2,500.

LAMAR DODD ART CENTER, LAGRANGE COLLEGE, 302 Forrest Ave., LaGrange, GA 30240. Mailing Address: 601 Broad St., LaGrange, GA 30240-2955. Tel.: 706-880-8211. Fax: 706-880-8007.
E-mail: dmarrin@lagrange.edu
Web Site: lagrange.edu/academics/art/lamar.dodd.htm
Founded: 1982.
Congressional District: 3
Key Personnel: Dir., John Lawrence; College Comptroller, Marty Pirrman.
Personnel Profile: Full-Time Paid 5.
Governing Authority: private college; nonprofit. Parent Institution: LaGrange College. Tax-exempt: 501(c)(3).
Institution Type/Description: College Art Museum.
Collections: Southwestern American Indian; Lamar Dodd paintings & drawings; 20th-century photography; Christ-Janer prints.
Activities: guided tours; LaGrange National juried art competition.
Publications: exhibition catalogues.
Hours & Admission Prices: Jan.-May & Sept.-Nov. Mon.-Fri. 8:30-4:30. No charge. &

Attendance: 20,000 (estimated)

LEGACY MUSEUM ON MAIN: A HISTORY MUSEUM FOR WEST GEORGIA, 136 Main St., LaGrange, GA 30240-3218. Mailing Address: P.O. Box 1051, LaGrange, GA 30241-0019. Tel.: 706-884-1828. Fax: 706-884-1840. Facebook: Troup County Archives.
E-mail: info@trouparchives.org
Web Site: www.trouparchives.org
Founded: 2008.
Congressional District: 8
Key Personnel: Exec. Dir., Kaye L. Minchew; Archivist, Shannon Gavin-Harris; Historian, F.C. Johnson, III
Personnel Profile: Full-Time Paid 3; Part-Time Paid 4; Part-Time Volunteers 6; Interns 1.
Governing Authority: private; nonprofit organization. Parent Institution: Troup County Historical Society. Subsidiary Institution: Troup County Archives. Tax-exempt: 501(c)(3).
Institution Type/Description: History Museum.
Collections: concentration on West Georgia & East Alabama history from prehistoric to modern times including Creek Indians; early settlers; first railroads; founding of educational institutions.
Facilities: library; educational facilities. Museum-related items for sale.
Activities: guided tours; lectures; temporary exhibitions.
Publications: quarterly newsletter, Troup County Historical Society & Archives.
Hours & Admission Prices: Mon.-Fri. 9-5, Sat. 10-4. No charge; donations accepted. Closed federal holidays. &
Attendance: 4,680 (accurate)
Membership: Builder $25; Sponsor $50; Sustaining $100; Founder $200; Patriot $300; Senator $500; Governor $1,000; Marquis $5,000.

Lawrenceville

GWINNETT HISTORICAL SOCIETY, 185 Crogan St., Lawrenceville, GA 30046-0261. Mailing Address: P.O. Box 261, Lawrenceville, GA 30046-0261. Tel.: 770-822-5174.
E-mail: ghs@gwinnetths.org
Web Site: www.gwinnetths.org
Founded: 1966.
Congressional District: 9
Key Personnel: Pres. (V) & C.E.O., Vicki Watkins; Trustee, Joye Quinn; Librarian, Harriett Nicholls; Corresponding Sec., Paula Wall McGee; Recording Sec., Rachel Bronnum.
Personnel Profile: Full-Time Volunteers 18; Part-Time Volunteers 30.
Governing Authority: society, nonprofit organization. Tax-exempt.
Institution Type/Description: Historical Society Museum.
Collections: photographs; farm tools; county newspapers 1872-1964; 1853 U.S. Navy sword & scabbard; family history files; cemetery card file; 120-vol. official records of the War of the Rebellion; manuscripts; photo albums; documents; Civil War letters.
Research Fields: genealogy; local history; historic preservation.
Facilities: 2,000-vol. library pertaining to local history, genealogy & court records available for research on premises only; reading room; slide show Vanishing Gwinnett available for rent to schools & civic clubs. Museum-related items for sale.

Activities: lectures; changing monthly displays; formally organized education tours. Museum Sponsors: Winn Festival held at Elisha Winn House to raise funds for restoration of house.
Publications: quarterly publication, Gwinnett Historical Society Quarterly; The Heritage; books on local history.
Hours & Admission Prices: Mon.- Fri. 10-2. No charge; donations accepted. &
Attendance: 1,600 (estimated)
Membership: Student $10; Regular $25; Family $35; Lifetime $500.

GWINNETT HISTORY MUSEUM, Lawrenceville Female Seminary Bldg., 455 S. Perry St., S.W., Lawrenceville, GA 30045-4836. Tel.: 770-822-5178. Fax: 770-237-5612.
E-mail: gwinnetthistorymuseum@gwinnettcounty.com
Founded: 1974.
Congressional District: 4
Key Personnel: Dir., Jennifer Collins; Outreach Coord., Kim Elmore; Outreach Coord., Jillian Waters Griffin.
Personnel Profile: Full-Time Paid 1; Part-Time Paid 2.
Governing Authority: county; nonprofit. Parent Institution: Gwinnett County. Gwinnett County Community Svcs., Parks & Recreation Dept. Tax-exempt.
Institution Type/Description: History Museum: housed on the second floor of the c.1854 Lawrenceville Female Seminary, an all-girls school. Listed on the National Register of Historic Places.
Collections: artifacts related to the history of Gwinnett County & its people; exhibits include Farmlife, Textiles, Old School Days, Wartime Gwinnett, Shape-Note Singing and Old-Time Music.
Research Fields: shape-note singing; old time music; family history; Gwinnett County, GA.
Facilities: library; 1,090 sq. ft. exhibit space. Books of local historical interest, notecards, 19th-century toys, games, Civil War posters, coloring books & museum-related items for sale.
Activities: formal education programs; guided tours; hobby workshops; lectures; outreach programs to local schools. Annual Event: folk music festival. Museum Sponsors: three monthly organizational meetings; coffeehouse nights monthly.
Publications: quarterly newsletter, The Gwinnett Muse; monthly coffeehouse newsletter, Java Jottings.
Hours & Admission Prices: Mon.-Thurs. 10-4, Sat. by appointment. Adults $1; discounts to AAM members; members no charge. Closed New Year's Day; Memorial Day; Independence Day; Labor Day; Christmas; federal holidays.
Attendance: 12,000 (estimated)
Membership: Senior & Student $8; Individual $15; Nonprofit Organization $20 & up; Family $25; Pioneer Circle $100; Female Seminary Circle $200; Button Gwinnett Circle $500; Corporate Sponsor $1,000 & up.

GWINNETT VETERANS MEMORIAL MUSEUM, 185 Crogan St., Rm. 118, Lawrenceville, GA 30045. Mailing Address: P.O. Box 166, Snellville, GA 30078. Tel.: 770-985-0901 & 921-1326.
E-mail: topchief@earthlink.net
Web Site: vetmemorialmuseum.tripod.com
Governing Authority: nonprofit organization.
Institution Type/Description: Military History Museum: housed in the historic Gwinnett County Courthouse.
Collections: history & artifacts of American Veterans from the Revolutionary War to present; personal artifacts; photographs; military artifacts & memorabilia; uniforms.
Hours & Admission Prices: Mon.-Fri. 10-4, Sat. 10-2. No charge; donations accepted.

Leslie

GEORGIA RURAL TELEPHONE MUSEUM, 135 Bailey Ave., Leslie, GA 31764-2601. Mailing Address: P.O. Box 187, Leslie, GA 31764-0187. Tel.: 229-874-4786.
Key Personnel: C.E.O., Tommy C. Smith.
Governing Authority: nonprofit organization.
Institution Type/Description: Telephone Museum: housed in renovated 1920s cotton warehouse.
Collections: history of communications; 1876 to 21st century telephones; telephone memorabilia.
Hours & Admission Prices: Mon.-Fri. 9-12 & 1-3:30. Adults $5, seniors $4, children $2; discounts to groups. Closed Memorial Day; Independence Day; Labor Day; Thanksgiving & day after; Christmas Eve & Day.

Lilburn

BAPS SHRI SWAMINARAYAN MANDIR, 460 Rockbridge Rd., N.W., Lilburn, GA 30047. Tel.: 678-906-2277. Fax: 678-906-2984.
E-mail: info.atlanta@usa.baps.org

Web Site: atlanta.baps.org
Founded: 2007.
Institution Type/Description: Religious Museum.
Collections: Hindu history, culture & religion; religious artifacts; paintings; sculpture.
Activities: audio tours; group tours.
Hours & Admission Prices: Mandir: daily 9-6; groups by appointment. Sacred Shrines: daily 9-10:30, 11:15-12, & 4-6. No charge.

Lincolnton

ELIJAH CLARK MEMORIAL MUSEUM, 2959 McCormick Hwy., Lincolnton, GA 30817-3909. Tel.: 706-359-3458. Fax: 706-359-5856.
E-mail: eclark@g-net.net
Web Site: gastateparks.org/info/elijah
Founded: 1961.
Congressional District: 10
Key Personnel: Dir., Nelson S. Noble.
Governing Authority: state. Administered by Parks, Recreation & Historic Sites Div., Georgia Dept. of Natural Resources, Floyd Towers E. 205 Butler St. S.E., Atlanta, GA 30334. Tax-exempt.
Institution Type/Description: State Park Museum: housed in replica of the house of Elijah Clark.
Collections: archives; uniforms; history artifacts of 1770s; reconstructed log house & outbuildings.
Research Fields: colonial weaponry & life styles.
Activities: guided tours.
Hours & Admission Prices: April-Nov. Sat.-Sun. 9-5; other times by appointment. No charge.

Lithia Springs

SWEETWATER CREEK STATE PARK, 1750 Mount Vernon Rd., Lithia Springs, GA 30122-3501. Mailing Address: P.O. Box 816, Lithia Springs, GA 30122-0816. Tel.: 770-732-5871. Fax: 770-732-5874.
E-mail: sweetwater_creek.park@dnr.state.ga.us
Web Site: gastateparks.org
Founded: 1972.
Congressional District: 6
Key Personnel: Park Mgr., Brad Ballard.
Personnel Profile: Full-Time Paid 5; Part-Time Paid 7; Part-Time Volunteers 9; Interns 1.
Governing Authority: state. Administered by the Parks, Recreation & Historic Sites Div., Georgia Dept. of Natural Resources, 270 Washington St. S.W., Atlanta, GA 30334. Tax-exempt.
Institution Type/Description: State Park: located at the site of the ruins of an 1849 textile mill burned in Gen. Sherman's Atlanta Campaign.
Collections: local & natural history; wildlife; historic building; Native American artifacts & petroglyph.
Facilities: nature trails; visitor's center; picnic shelter; conference room. Gift items for sale.
Activities: self-guided & guided tours; permanent exhibits; boating; canoeing; hiking; picnicking; fishing; geocaching.
Publications: brochure.
Hours & Admission Prices: Park: daily 7am to sunset. Visitor Center & Museum: daily 9-5. No charge. Parking: $5. &
Attendance: 200,000 (estimated)

Lumpkin

BEDINGFIELD INN MUSEUM, Cotton St. on the Square, Lumpkin, GA 31815. Mailing Address: P.O. Box 818, Lumpkin, GA 31815-0818. Tel.: 229-838-6419. Fax: 229-838-6134.
Web Site: www.bedingfieldinn.org
Founded: 1965.
Congressional District: 2
Key Personnel: Pres., Rhoda Averett; Vice Pres., Joanne Bowles; Treas., Frank Heard.
Personnel Profile: Full-Time Volunteers 20; Part-Time Volunteers 20.
Governing Authority: nonprofit organization. Parent Institution: Stewart County Historical Commission. Tax-exempt: 501(c)(3).
Institution Type/Description: Historic Building Museum: 1836 Stagecoach Inn.
Collections: decorative arts.
Facilities: Books for sale.
Activities: guided tours; permanent exhibitions. Museum Sponsors: Candlelight Tour; Fair on the Square in October.
Publications: books, The Bedingfield Inn Cookbook; Inventory of Early Stewart County Furniture.

Hours & Admission Prices: By appointment. Adult $5, children $2; members no charge.
Attendance: 500 (estimated)
Membership: Individual $25; Family $48; Business $100.

HISTORIC WESTVILLE, INC., (M), 9294 Singer Pond Rd., Lumpkin, GA 31815. Mailing Address: P.O. Box 1850, Lumpkin, GA 31815-1850. Tel.: 229-838-6310. Fax: 229-838-4000. Facebook: Historic Westville.
E-mail: info@westville.org
Web Site: www.westville.org
Formerly: Westville Historic Handicrafts
Founded: 1966.
Congressional District: 2
Key Personnel: Exec. Dir., Leo J. Goodsell; Chm. (V), Mike Bunn; Dir. Interpretation, Michelle Alexander; Museum Shop Mgr., Betty Shannon; Mgr. Restaurant, Shirley Platt; Office Mgr., Veronica Wiese; Devel. Dir., Alma P. Hubbard.
Personnel Profile: Full-Time Paid 8; Part-Time Paid 4; Part-Time Volunteers 932.
Volunteer Hours: 6,397
Governing Authority: nonprofit organization. Parent Institution: Historic Westville, Inc. Tax-exempt: 501(c)(3).
Institution Type/Description: Historic Village Museum: thirty-four buildings & houses c.1850.
Collections: early 19th-century decorative arts; 1825-1860 appropriate Georgia landscaping & gardens. 34 Historic Houses & Structures dating from 1800-1864: 1850 Randle-Morton Store; 1832 Stewart County Academy; 1842 Grimes-Feagin House; 1843 McDonald House; 1836 Cabinet Shop; 1838 Shoemakers Shop; 1838 Singer House; 1845 Doctor's Office; 1854 Chattahoochee County Courthouse; 1851 Climax Presbyterian Church; 1831 Bryan-Worthington House; 1840 Bagley Gin House; Log Cabins; Patterson-Marrett Farmhouse; Mule Barn; 1827 Wells House; 1850 Blacksmith Shop; c.1850 Carriage Shelter; Adam's Store; 1836 Lawson House; 1840 Yellow Creek Camp Meeting Tabernacle; 1840 Moye House; Pottery Pug Mill; 1878 Damascus Methodist Church; c.1850s Caproni Outhouse.
Research Fields: celebrated events of 1850; history of the 1826 land cession of Georgia; 19th-century interior painting in the Deep South; genealogy of West Georgia; Indian history of West Georgia; Stewart County African-American heritage.
Facilities: 1,500-vol. library available for research on premises; reading room; 300-seat outdoor auditorium. Handcrafted items & other museum-related items for sale.
Activities: self-guided tours; formally organized education programs for adults & students; permanent exhibitions; seasonal special events; school programs. Museum Sponsors: Spring Festival; Creek Indian War Reenactment; Independence Day Celebration; Labor Day Festivities; Fall Harvest Days; Christmas Time at Westville.
Publications: e-newsletter & newsletter, The Westville Mirror; books, The Groaning Board; Pot Liquor, Magic and Mystery of Westville.
Hours & Admission Prices: Thurs.-Sat. 10-5. Adults $10, senior citizens, college students & military $8, children grades K-12 $5; discounts to groups, AAM & ALHFAM members; members & Pre-K students no charge. Closed New Year's Eve & Day; Thanksgiving; Christmas. &
Attendance: 13,000 (estimated)
Membership: Individual $25; Builder $50; Sustainer $100; Patron $250; Westville Society $500; Millennium Society $1,000.

Mableton

MABLE HOUSE COMPLEX, 5239 Floyd Rd., Mableton, GA 30126. Tel.: 770-819-3285 & 7765.
Institution Type/Description: Historic House Museum: housed in the former home of Robert Mable; built in 1843. Listed on the National Register of Historic Places.
Collections: local history & culture; period furnishings; photographs.
Activities: guided tours; rental facilities. Annual Event: Storytelling Festival in fall.
Hours & Admission Prices: June-Sept. Thurs. 8:30-12:30.

Macon

THE ALLMAN BROTHERS BAND MUSEUM AT THE BIG HOUSE, 2321 Vineville Ave., Macon, GA 31208. Mailing Address: The Big House Foundation, P.O. Box 4291, Macon, GA 31208-4291. Tel.: 478-741-5551.
E-mail: info@thebighousemuseum.org
Web Site: www.thebighousemuseum.org
Founded: 2010.
Key Personnel: Dir., Lisa C. McLendon; Chm. (V), James M. Wells, III

Personnel Profile: Full-Time Paid 2; Part-Time Volunteers 30.
Governing Authority: Parent Institution: The Big House Foundation, Inc. Tax-exempt.
Institution Type/Description: Band History Museum.
Collections: Band history, music & awards; guitars; basses; drums; keyboards; clothing; posters; photographs; band memorabilia; personal artifacts; audio & video recordings.
Activities: special events.
Hours & Admission Prices: Thurs.-Sun. 11-6. Adults $8, students, seniors & military $6, children under 6 & members no charge. &
Attendance: 5,000 (estimated)
Membership: $50-$50,000.

THE CANNONBALL HOUSE, (M), 856 Mulberry St., Macon, GA 31201-6755. Tel.: 478-745-5982. Fax: 478-745-5944.
E-mail: cbhinfo@yahoo.com
Web Site: www.cannonballhouse.org
Founded: 1963.
Congressional District: 8
Key Personnel: Dir., Earl Colvin; Pres. (V), Cheryl Aultman.
Personnel Profile: Full-Time Paid 1; Part-Time Paid 3; Part-Time Volunteers 3.
Governing Authority: Parent Institution: Friends of the Cannonball House, Inc. Tax-exempt.
Institution Type/Description: History Museum: c.1853.
Collections: period furnishings & clothing; personal artifacts; quilts; Civil War uniforms, equipment & firearms; flags; servants quarters; Alpha Delta Pi & Phi Mu furnishings.
Activities: monthly programs.
Publications: biannual newsletter.
Hours & Admission Prices: Summer: Mon.-Sat. 10-5, Sun. by appointment; Winter: Mon.-Sat. 11-5. Adults $6, military and senior citizens 65 & over $5, students $3; discounts to AAA members & groups of 15 or more; children 6 & under and members no charge. Closed most major holidays.
Attendance: 10,000 (estimated)
Membership: Individual $35; Family $75; Patron $150; Grand Patron $250; Donor $500; Benefactor $1,000.

GEORGIA MUSIC HALL OF FAME, 200 Martin Luther King Jr. Blvd., Macon, GA 31201-3490. Mailing Address: 75 5th St. N.W., Ste 1200, Atlanta, GA 30308-1020. Tel.: 888-GA-ROCKS; 478-751-3334. Fax: 478-751-3100.
E-mail: ccreekmore@georgia.org
Web Site: www.georgiamusic.org
Founded: 1996.
Congressional District: 8
Key Personnel: Dir., Lisa Love; Chm., Eugene C. Dunwody, Jr.; Cur., Joseph Johnson; Music Store Coord., Melvina Spence; Public Rels. & Event Mgr., Katie Roberts; Asst. Cur. & Educator, Kristin Veline; Visitor Svcs. Mgr., Mary Stansfield; Visitor Svcs. Coord. & Volunteer Coord., Steven Fulbright; Administrative Operations Coord., Cissie Creekmore; Mgr. Facilities, A.B. Goel.
Personnel Profile: Full-Time Paid 9; Part-Time Paid 6; Part-Time Volunteers 4; Interns 4.
Governing Authority: state. Parent Institution: Georgia Department of Economic Development. Tax-exempt: 501(c)(3).
Institution Type/Description: Music Museum.
Collections: history of Georgia music from pre-colonial to present.
Research Fields: Georgia musical artists; Georgia instrument makers; Georgia music industry leaders & personalities; Georgia musical folkways & related international traditions.
Facilities: 1,100-vol. library; banquet room; archives; reading room; 15,000 sq. ft. exhibit space. Museum-related items for sale.
Activities: concerts; exhibit openings; volunteer & intern programs; guided tours; school tours; hobby & educational workshops; lectures; films; rental facilities; participatory & temporary exhibitions.
Publications: magazine, Georgia Music.
Hours & Admission Prices: Tues.-Sat. 9-5. Adults $8, senior citizens, students, tour & family groups $6, children $3.50; discounts to AAA members; members no charge. Closed New Year's Day; Thanksgiving; Christmas & day after. &
Attendance: 45,000 (estimated)
Membership: Solo $30; Duet $50; The Band (family) $75; Manager $100; Agent (patron) $250; Promoter $500.

GEORGIA SPORTS HALL OF FAME, (M), 301 Cherry St., Macon, GA 31201-3398. Mailing Address: P.O. Box 4644, Macon, GA 31208-4644. Tel.: 478-752-1585, ext. 100. Fax: 478-752-1587.
Web Site: www.gshf.org

Founded: 1999.
Congressional District: 8
Key Personnel: Dir. & Devel., Ben Sapp; Chm., Paul Holmes, Jr.; Treas., Gwen Arrington; Museum Shop Mgr., Candice Barca; Museum Shop Sr. Assoc., Courtney Freeman; Rental Coord., Eric Thomas.
Personnel Profile: Full-Time Paid 4; Part-Time Paid 2; Interns 1.
Governing Authority: state; nonprofit.
Institution Type/Description: State Agency.
Collections: history of sports in Georgia; sports heritage of Georgia's athletes.
Facilities: library; 14,000 sq. ft. exhibit space; 205-seat auditorium; large screen theater. Museum-related items for sale.
Activities: dance recitals; films; guided tours; lectures; participatory exhibits; rental gallery; temporary & traveling exhibitions; theater.
Publications: quarterly newsletter, Sports Report.
Hours & Admission Prices: Tues.-Sat. 9-5. Adults $8, senior citizens, students & military with ID $6, children $3.50; discounts to groups & AAA members; members no charge. Closed New Year's Day; Easter; Thanksgiving; Christmas & day after. &
Attendance: 13,560 (accurate)
Membership: Individual $25; Friend $50; Contributor $100; Benefactor $250; Patron $500; Grant Patron $1,000.

HAY HOUSE MUSEUM, (M), 934 Georgia Ave., Macon, GA 31201-6708. Tel.: 478-742-8155. Fax: 478-745-4277.
Web Site: www.georgiatrust.org; www.hayhouse.org
Founded: 1977.
Congressional District: 8
Personnel Profile: Full-Time Paid 3; Part-Time Paid 12; Part-Time Volunteers 50; Interns 2.
Governing Authority: nonprofit organization. Parent Institution: The Georgia Trust for Historic Preservation. Tax-exempt: 501(c)(3).
Institution Type/Description: Historic House: 1855-59 Italian Renaissance Revival Mansion.
Collections: period furnishings; porcelains; decorative arts relative to house & region.
Research Fields: architecture; decorative arts; regional history.
Facilities: rental facilities.
Activities: guided tours; lectures; education programs for students; docent program; rental facilities. Annual Event: Cherry Blossom Festival in March; Macon Gardens, Mansions & Moonlight Home & Garden Tour in May; Christmas at Hay House November-December.
Publications: bimonthly newsletter, The Rambler, from the Georgia Trust for Historic Preservation.
Hours & Admission Prices: Jan.-Feb. & July-Aug. Tues.-Sat. 10-4; March-June & Sept.-Nov. Tues.-Sat. 10-4, Sun. 1-4; Dec. daily 10-4; last tour at 3pm. Regular Season: adults $9, senior citizens & military $8, students $5; discounts to AAA members; children under 6 & members no charge. Christmas Season: adults $11, seniors & military $10, students $7. Closed major holidays. &
Attendance: 16,752 (accurate)

HISTORIC MACON FOUNDATION, INC. - SIDNEY LANIER COTTAGE HOUSE MUSEUM, Sidney Lanier Cottage, 935 High St., Macon, GA 31201. Mailing Address: P.O. Box 13358, Macon, GA 31208-3358. Tel.: 478-743-3851 & 742-5084.
Web Site: www.historicmacon.org
Formerly: Middle Georgia Historical Society, Inc.
Founded: 1964.
Congressional District: 8
Key Personnel: Exec. Dir., Josh Rogers; Dir. SLC, Janis I. Haley; Dir. Finance, Cantey Ayres.
Personnel Profile: Full-Time Paid 5; Part-Time Paid 5; Part-Time Volunteers 6; Interns 1.
Governing Authority: bd. of trustees. Tax-exempt: 501(c)(3).
Institution Type/Description: Historic Site: c.1840 birthplace of poet Sidney Lanier.
Collections: documents & photographs; renovation & restoration of poet Sidney Lanier's birthplace; memorabilia of Sidney Lanier; books; musical instruments.
Facilities: archives.
Activities: guided house tours; lectures; poetry readings; trips.
Publications: quarterly bulletin; Butler's History of Macon & Central Georgia; Macon's Architectural & Historic Heritage; Macon, A Pictorial History; Our Town Macon Coloring Guide for children; Macon Sets A Fine Table; Eneas Africanus; Living Macon Style; DVDs of Sidney Laniers Life, Rosehill Cemetery & Indian Historical Tour; Historic Macon.
Hours & Admission Prices: Thurs.-Sat. 10-4; groups by appointment. Adults $5, seniors $4, children 6-18 $3; discounts to AAA members, groups of 10 or more & military; members no charge. Closed major holidays. &
Attendance: 10,000 (estimated)

Membership: Student $20; Individual $40; Family $60; Patron $150; Benefactor $250; Supporting $400; Sustaining $600; Historic Macon Club $1,000.

*   **MUSEUM OF ARTS AND SCIENCES, (M),** 4182 Forsyth Rd., Macon, GA 31210-4869. Tel.: 478-477-3232. Fax: 478-477-3251.
E-mail: lfisher@masmacon.com
Web Site: www.masmacon.org
Founded: 1956.
Congressional District: 8
Key Personnel: Dir., Susan Welsh; Operations Dir., Lisa Gant Fisher; Bd. Pres., Charlotte Bogle; Cur. Education, Susan Mays; Cur. Science, Paul Fisher; Museum Shop Mgr., Beth Fisher.
Personnel Profile: Full-Time Paid 12; Part-Time Paid 12; Part-Time Volunteers 232.
Volunteer Hours: 4,835
Governing Authority: private; nonprofit corporation. Tax-exempt: 501(c)(3).
Institution Type/Description: General Museum.
Collections: art, science & humanities artifacts; archaeological artifacts; exotic moths & butterflies; zyghoriza fossil; rocks & minerals; living zoological specimens; toys; paintings; drawings; prints & sculpture by regional & American artists; ceramics; hands-on exhibitions. Historic House: c.1928 The Kingfisher Cabin.
Major Exhibits: The Lost Mural of Ellis Island, 1/14-3/14; Quilts, 1/14-5/14; Jennifer Crenshaw-Textile Artist, 1/14-5/14; Judy Gales - Fiber Art, 1/14-5/14; Construction, 2/14-3/14; Emerging Artists, 3/14-6/14; Wetmore Ships, 5/14-9/14; Black Holes (T), 6/14-8/14; Abrasion Holography, 6/14-9/14; Festival of Trees/Celebrate, 11/14-12/14.
Research Fields: art, science & history of clay.
Facilities: 3 galleries featuring changing exhibits in art, science & humanities; 3-story Discovery House (interactive); 2-story habitat for live collections; planetarium; observatory; auditorium; classroom; nature trails; caboose; cabin; store carries imported crafts, original art/ceramics, books; kits, jewelry & unique gifts.
Activities: guided tours; lectures; films; gallery talks; classes; field trips; hobby workshops; formally organized education programs; docent program or council; permanent, temporary & traveling exhibitions; school loan service.
Publications: quarterly newsletter, Museum Muse; program guide; exhibit catalogues; gallery guides.
Hours & Admission Prices: Museum: Tues.-Sat. 10-5, Sun. 1-5. Adults $10, seniors (62+) $8; students w/ID $7, children 3-17 $5, children under 3 no charge; discounts to military & Smithsonian affiliates; members no charge. Planetarium Programs: Tues.-Fri. 11:30 & 4, Sat. 11:30, 2 & 4, Sun. 2 & 4. Mini-Zoo Programs: Sun. & Tues.-Fri. 3 p.m., Sat. 1 & 3. Closed New Year's Day; Easter; Memorial Day; Independence Day; Labor Day; Thanksgiving; Christmas Eve & Day. ♿
Attendance: 68,783 (accurate)
Membership: Student $20; Teacher & Senior $25; Individual $35; Family $65; Friend $150; Patron $300; Benefactor $500; President's Roundtable $1,000.

OCMULGEE NATIONAL MONUMENT, 1207 Emery Hwy., Macon, GA 31217-4320. Tel.: 478-752-8257, ext. 210. Fax: 478-752-8259.
E-mail: ocmu_superintendent@nps.gov
Web Site: www.nps.gov/ocmu
Founded: 1936.
Congressional District: 8
Key Personnel: Supt., James David; Museum Shop Mgr., Lisa Lemon.
Personnel Profile: Full-Time Paid 11; Part-Time Volunteers 358; Interns 1.
Governing Authority: federal. Affiliated with National Park Service, U.S. Dept. of Interior, Washington, DC. Tax-exempt: 501(c)(3).
Institution Type/Description: Park Museum.
Collections: Indian artifacts; archaeology representing six culture levels covering 10,000 years; anthropology; history; ethnology; British Colonial Trading Post; Lamar Type Site. Historic Building: 900-1100 A.D. Earthlodge.
Research Fields: southeast archaeology; Ocmulgee site history; ten prehistoric Mississippian style mounds.
Facilities: 2,000-vol. library of books on anthropology, archaeology & history of southeastern United States available on premises to authorized archaeologists, historians & students; discovery lab; nature trails; picnic area. Publications & American Indian crafts for sale.
Activities: lectures; formally organized education programs for children & college students in discovery lab; permanent exhibitions; native craft demonstrations; school loan service; driving & self-guiding trails.
Hours & Admission Prices: Daily 9-5. No charge; donations accepted. Closed New Year's Day; Christmas. ♿
Attendance: 170,000 (accurate)
Membership: Cooperating Association Memberships available. Student &

Senior Citizen $10; Individual $15; Family $25; Contributing $50; Sustaining $100; Patron $250; Corporate $500.

TUBMAN AFRICAN-AMERICAN MUSEUM, 340 Walnut St., Macon, GA 31201-0515. Mailing Address: P.O. Box 6671, Macon, GA 31208-6671. Tel.: 478-743-8544. Fax: 478-743-9063.
E-mail: guestservices@tubmanmuseum.com
Web Site: www.tubmanmuseum.com
Founded: 1982.
Congressional District: 3
Key Personnel: Bd. Chm., Tom Sands; Bd. Vice Chm., Morris Butler; Exec. Dir., Andy Ambrose; C.F.O., Barrie Miller-Howard; IT & Office Mgr., Adra Dudley; Dir. Exhibitions, Jeff Bruce; Dir. Education, Tonya Parker Ontlay; Guest Services Coord., Patricia Stephens; Programs Mgr., Shandaleia Shepard.
Personnel Profile: Full-Time Paid 8; Part-Time Paid 2; Part-Time Volunteers 100; Interns 2.
Governing Authority: nonprofit.
Institution Type/Description: History Museum.
Collections: African-American art, history & culture.
Facilities: 1,000-vol. library.
Publications: six catalogs annually.
Hours & Admission Prices: Tues.-Fri. 9-5, Sat. 11-5. Adults $6, seniors & military with ID $5, children $4; discounts to AAM & ICOM members; museum members no charge. Closed New Year's Day; Memorial Day; Independence Day; Thanksgiving; Christmas. ♿
Attendance: 57,372 (accurate)
Membership: Senior Citizen & Student $15; Individual $25; Family $50; Friend $100; Patron $250; Grand Patron $500; Benefactor $1,000; Grand Benefactor $2,500.

Madison

MADISON-MORGAN CULTURAL CENTER, 434 S. Main St., Madison, GA 30650-1640. Tel.: 706-342-4743. Fax: 706-342-1154.
E-mail: info@mmcc-arts.org
Web Site: www.mmcc-arts.org
Founded: 1976.
Congressional District: 10
Key Personnel: Mng. Dir., Ruth Bracewell; Chm., Steve Schaefer; Vice Chm., Bob Hughes; Dir. Devel. & Membership, Amy Knight; Dir. Mktg., Theresa Dickinson; Performance Dir., Rob Seymour; Dir. Visitor Svcs. & Box Office, Deanna Lamar.
Personnel Profile: Full-Time Paid 6; Part-Time Paid 3; Part-Time Volunteers 50.
Governing Authority: nonprofit organization. Parent Institution: Morgan County Foundation Inc. Tax-exempt: 501(c)(3).
Institution Type/Description: Multi-disciplinary Cultural Center: housed in the 1895 Madison Graded School, in Madison Historic District.
Collections: artifacts, photographs, papers pertaining to the history of Madison, Morgan County & the Piedmont region of Georgia; decorative arts; costumes; architectural fragments; tools; household utensils; Civil War materials; school related artifacts & furnishings; history & local education exhibits; restored classroom.
Research Fields: Madison & Morgan County history & architecture; arts & crafts decorative arts.
Facilities: 395-seat auditorium; member's room. Literature pertaining to Madison & Morgan County, exhibit catalogs & gift items for sale.
Activities: guided tours; lectures; theatre; concerts; demonstrations; formally organized education programs; docent programs; loan, permanent & traveling exhibitions; residencies; workshops; classes; seminars. Center Sponsors: annual four-day arts festival.
Publications: quarterly newsletter, Madison-Morgan Cultural Center; periodic catalogs of specific exhibitions.
Hours & Admission Prices: Tues.-Sat. 10-5, Sun. 2-5. Adults $3, seniors $2.50, students $2; discounts to groups of 20 or more; children under 6, museum & AAM members no charge. Closed New Year's Day; Memorial Day; Independence Day; Labor Day; Thanksgiving; Christmas Eve & Day. ♿
Attendance: 20,000 (estimated)
Membership: Students $10; Individual $20; School $25; Friend & Family $30; Civic Club or Group $60; Sustainer $60-$99; Contributor $100-$249; Associate $250-$499; Patron $500-$999; Benefactor $1,000 and up.

MADISON MUSEUM OF FINE ART, (M), 300 Hancock St., Madison, GA 30650-1305. Mailing Address: P.O. Box 814, Madison, GA 30650. Tel.: 706-485-4530.
E-mail: mbechtell@prodigy.net
Web Site: madisonmuseum.org
Founded: 2003.

Key Personnel: C.E.O., Pres. (V) & Dir., Michele L. Bechtell; Sec., Sean Gallagher; Treas., Henry C. Ransom.
Personnel Profile: Full-Time Volunteers 1; Part-Time Volunteers 10.
Governing Authority: private; nonprofit organization. Tax-exempt: 501(c)(3).
Institution Type/Description: Visual Art History Museum.
Collections: paintings; sculptures.
Facilities: 60-seat auditorium; botanical garden; educational facilities; 5,000 sq. ft. exhibit space; 60-seat theater. Museum-related items for sale.
Activities: concerts; docent program; films; arts festivals; lectures; loan & traveling exhibitions; theater; training programs; broadcast programs. Annual Events: Spring Art Walk; Twelfth Night Renaissance Supper.
Publications: quarterly newsletter.
Hours & Admission Prices: Mon.-Sat. 1-5. No charge; donations accepted. &
Attendance: 5,000 (estimated)
Membership: Individual $35; Family $50; Friend $150; Contributing $250; Corporate $500; Patron $1,000; Director's Circle $2,500.

MORGAN COUNTY AFRICAN AMERICAN MUSEUM, 156 Academy St., Madison, GA 30650-1202. Tel.: 706-342-9191.
Institution Type/Description: History Museum.
Collections: African American history & culture; personal artifacts; period furnishings; photographs.
Activities: special events.
Hours & Admission Prices: Tues.-Sat. 10-4. Adults $5, students $3.

ROGERS HOUSE & ROSE COTTAGE, 179 E. Jefferson, Madison, GA 30650-1361. Mailing Address: c/o City of Madison, P.O. Box 32, Madison, GA 30650. Tel.: 706-343-0190.
Institution Type/Description: Historic Houses.
Collections: local history & culture; period furnishing; personal artifacts; photographs.
Hours & Admission Prices: Mon.-Sat. 10-4:30, Sun. 1:30-4:30.

Marietta

BRUMBY HALL & GARDENS, 500 Powder Springs St., Marietta, GA 30064. Tel.: 770-427-2500.
Institution Type/Description: Historic House Museum: housed in the former home of Colonel Arnoldus V. Brumby; built in 1851.
Collections: local history; period furnishings; personal artifacts; photographs; English gardens including topiary, boxwood, rose, perennial & knot gardens.
Facilities: banquet facilities.
Activities: special events; rental facilities.
Hours & Admission Prices: By appointment.

COBB COUNTY YOUTH MUSEUM, (M), 649 Cheatham Hill Dr., Marietta, GA 30064-5512. Mailing Address: P.O. Box 78, Marietta, GA 30061-0078. Tel.: 770-427-2563. Fax: 770-427-1060.
E-mail: youthmuseum@aol.com
Web Site: www.theyouthmuseum.org
Founded: 1964.
Congressional District: 6
Key Personnel: Pres., Caroline Whaley; Pres. Elect, Susan Potter; Dir., Anita S. Barton; Administrative Asst. & Museum Shop Mgr., Eleanor Watson.
Personnel Profile: Full-Time Paid 1; Part-Time Paid 7; Part-Time Volunteers 3.
Governing Authority: nonprofit organization. Tax-exempt.
Institution Type/Description: Youth Museum.
Collections: jet trainer; caboose; street car stop; canoe.
Major Exhibits: America's Pathways to Independence, 1/13-12/14.
Activities: guided tours; docent program; puppet shows; speakers bureau; living history exhibits; summer program. Annual Event: BBQ & Silent Auction.
Hours & Admission Prices: Tours: Sept.-May Mon.-Fri. 9:30-1:30 by appointment. Admission $10. Closed school holidays. &
Attendance: 13,817 (accurate)

GONE WITH THE WIND MOVIE MUSEUM, 18 Whitlock Ave., Marietta, GA 30064-2346. Tel.: 770-794-5576.
E-mail: csutherland@mariettaga.gov
Web Site: www.mariettaga.gov/gwtw/
Key Personnel: Dir., Connie Sutherland
Institution Type/Description: Movie Museum.
Collections: original costumes; posters; books; props; sketches; photographs.
Facilities: Museum-related items for sale.
Activities: video.
Hours & Admission Prices: Mon.-Sat. 10-5. Adults $7, seniors & students $6; discounts to groups of 15 or more.

MARIETTA/COBB MUSEUM OF ART, 30 Atlanta St., S.E., Marietta, GA 30060-1975. Tel.: 770-528-1444. Fax: 770-528-1440.
E-mail: smacaulaymcma@bellsouth.net
Web Site: www.mariettacobbartmuseum.org
Founded: 1989.
Congressional District: 7
Key Personnel: Dir., Sally Macaulay; Chm., Ray Worden; Dir. Operations, Jennifer Fox; Dir. Education, Emily Ryals.
Personnel Profile: Full-Time Paid 3; Part-Time Paid 1; Part-Time Volunteers 50.
Governing Authority: nonprofit organization. Tax-exempt: 501(c)(3).
Institution Type/Description: Fine Art Museum: housed in 1909 Greek Revival Post Office Building.
Collections: 19th & 20th century American art; paintings; prints; drawings; watercolors; pastels; Ash Can School works; contemporary art from 1980s-present; Soho, New York.
Research Fields: 19th & 20th century art.
Facilities: educational facilities.
Activities: docent program; formal education programs for adults; guided tours; workshops; lectures; temporary exhibitions of museum's American collections; loan, participatory & traveling exhibitions; weddings; meetings. Museum Sponsors: Annual Children's Art Festival; Art Camp.
Publications: newsletter, MCMA Chronicle.
Hours & Admission Prices: Tues.-Fri. 11-5, Sat. 11-4. Adults $5, students & senior citizens $3; discounts to groups, AAA, AAM & ICOM members; members & children under 6 no charge. Closed major holidays. &
Attendance: 75,000 (estimated)
Membership: Student & Senior $25; Individual $35; Family Dual $50; Supporting $100; Contributing $250; Patron $500; Sustaining $1,000; Benefactor $5,000; Grand Benefactor $10,000.

MARIETTA FIRE MUSEUM, 112 Haynes St., Fire Station #1, Marietta, GA 30060-1973. Tel.: 770-794-5491.
Web Site: www.marietta.gov/departments/emergency/fire/museum.aspx
Institution Type/Description: Fire-Fighting Museum.
Collections: late 1800s photographs; early fire helmets; helmets from around the world; period fire-fighting equipment, hoses, nozzles & bells; 1879 Silsby Steamer; 1921 American LaFrance pumper; 1929 Seagrave pumper; 1949 Pirsch ladder truck; 1952 Chevrolet panel truck.
Hours & Admission Prices: Mon.-Fri. 8-5, Sat.-Sun. by appointment. No charge.

MARIETTA MUSEUM OF HISTORY, (M), 1 Depot St., Ste. 200, Marietta, GA 30060-1905. Tel.: 770-794-5710. Fax: 770-794-5733.
E-mail: info@mariettahistory.org
Web Site: www.mariettahistory.org
Founded: 1996.
Key Personnel: C.E.O. & Founder, Dan Cox; Exec. Dir., Jan Galt.
Personnel Profile: Full-Time Paid 4; Part-Time Paid 2; Part-Time Volunteers 30; Interns 2.
Governing Authority: Tax-exempt.
Institution Type/Description: History Museum.
Collections: Marietta & Cobb County Georgia history.
Hours & Admission Prices: Mon.-Sat. 10-4. Adults $5, seniors & students $3; members & children under 6 no charge. Closed New Year's Day; Independence Day; Thanksgiving; Christmas. &
Attendance: 10,049 (accurate)
Membership: Bronze $30; Silver $50; Gold $75; Platinum $150; Diamond $500.

THE ROOT HOUSE MUSEUM, N. Marietta Loop & Polk St., Marietta, GA 30060. Mailing Address: 145 Denmead St., Marietta, GA 30060-1934. Tel.: 770-426-4982. Fax: 770-499-9540.
E-mail: clhs2@bellsouth.net
Web Site: cobblandmarks.com
Founded: 1972.
Key Personnel: Dir., Daryl Barksdale; Chm., Skip Harper; Cur., Mary Ellen Higginbotham.
Personnel Profile: Part-Time Paid 5; Part-Time Volunteers 33.
Governing Authority: Parent Institution: Cobb Landmarks & Historical Society, Inc. Tax-exempt.
Institution Type/Description: Historic House Museum: built c.1845.
Collections: Historic Building: c.1845 William Root house.
Facilities: Museum-related items for sale.
Publications: monthly newsletter, Landmarker; book, The 1st 100 Years of Cobb County; articles, Historic Highlights of Cobb County; Cobb County in the 20th Century.

Hours & Admission Prices: Wed.-Sat. 11-4. Adults $4, senior citizens $3, children $2; members no charge. &

Attendance: 6,000 (estimated)

Membership: Student & Institutional $20; Annual $45; Patron $100; Business $150; Donor $ 250; Silver Sponsor $500; Gold Sponsor $1,000; Platinum Sponsor $2,500

McDonough

THE BROWN HOUSE - THE GENEALOGICAL SOCIETY OF HENRY & CLAYTON COUNTIES, INC., 71 Macon St., McDonough, GA 30253. Mailing Address: P.O. Box 1296, McDonough, GA 30253-1296. Tel.: 770-954-1456.

E-mail: genealsoc@bellsouth.net

Institution Type/Description: Genealogical Society: housed in the former home of Revolutionary War soldier, Andrew Brown, c.1820. Listed on the National Register of Historic Places.

Collections: county, state, immigration, international & genealogical records.

Research Fields: genealogy.

Facilities: library.

Activities: genealogical research; classes.

Hours & Admission Prices: Jan. to mid-Dec. Wed. & Fri. 10-3. Closed holidays.

HISTORICAL VETERANS MUSEUM & HISTORIC PARK, 101 Lake Dow Rd., McDonough, GA 30252. Tel.: 770-288-7300.

Institution Type/Description: Military History Museum.

Collections: military history; uniforms; military memorabilia from WWI to present; military vehicles; log cabin; cookhouse; schoolhouse; blacksmith shop; 1934 steam engine locomotive; Veterans Wall of Honor.

Facilities: library.

Hours & Admission Prices: Museum: Mon.-Sat. 10-3; other times by appointment. Park Tours: by appointment. No charge; donations accepted.

Midway

DORCESTER ACADEMY MUSEUM OF AFRICAN-AMERICAN HISTORY, 8787 E. Oglethorpe Hwy., Midway, GA 31320. Mailing Address: P.O. Box 51, Midway, GA 31320-0051. Tel.: 912-884-2347.

E-mail: doracad1@coastalnow.net

Web Site: dorchesteracademy.com

Founded: 1871.

Congressional District: 1

Key Personnel: Dir., Deborah Robinson; C.E.O., William Austin.

Personnel Profile: Part-Time Volunteers 21.

Governing Authority: Parent Institution: Historic Dorchester Academy. Tax-exempt.

Institution Type/Description: History Museum.

Collections: local history; culture; period artifacts; farming tools.

Hours & Admission Prices: Tues.-Fri. 11-2, Sat.-Sun. 2-4. No charge.

Attendance: 300 (estimated)

FORT MORRIS STATE HISTORIC SITE, 2559 Fort Morris Rd., Midway, GA 31320-6205. Tel.: 912-884-5999. Fax: 912-884-5285.

E-mail: fortmorris@coastalnow.net

Web Site: gastateparks.org

Founded: 1978.

Congressional District: 1

Key Personnel: Site Mgr., Arthur C. Edgar, Jr.

Personnel Profile: Full-Time Paid 1; Part-Time Volunteers 15.

Governing Authority: state. Administered by Parks Recreation & Historic Sites Div. Georgia Dept. of Natural Resources. Tax-exempt.

Institution Type/Description: Military Museum.

Collections: 1814 remains of revolutionary earthwork.

Research Fields: Colonial American revolution; Civil War; Georgia history.

Facilities: visitor Center. Postcards, gifts & books for sale.

Activities: guided tours; monthly programs; summer youth activities; military encampments; living history demonstrations, cannon firings; film.

Hours & Admission Prices: Thurs.-Sat. & holidays 9-5. Adults $4.50, senior citizens $4, youth 6-18 $3; children under 5 no charge. Group Tours: adults $3.25, youth 6-18 $2.50, children 5 & under $1. Closed New Year's Day; Thanksgiving; Christmas. &

Attendance: 13,000 (estimated)

MIDWAY MUSEUM, INC., 491 N. Coastal Hwy., Midway, GA 31320. Mailing Address: P.O. Box 195, Midway, GA 31320-0195. Tel.: 912-884-5837. Fax: 912-884-5837.

E-mail: midwaymuseum@yahoo.com

Web Site: themidwaymuseum.org

Founded: 1957.

Congressional District: 1

Key Personnel: Chm. (V), Tina Scott.

Personnel Profile: Part-Time Paid 1; Part-Time Volunteers 10.

Governing Authority: society. Tax-exempt.

Institution Type/Description: History Museum.

Collections: 1700s-1800s furnishings.

Research Fields: local genealogy & history.

Facilities: Museum-related items for sale.

Activities: guided tours; permanent exhibitions.

Publications: brochures.

Hours & Admission Prices: Tues.-Sat. 10-4; groups by appointment. Adults $6, senior citizens $5, children $3; discounts to groups & DAC members; children under 6 no charge. Closed holidays.

Attendance: 50,000 (estimated)

Membership: Students, Military UDC & DAC $15; Individual $25; Family $50; Gwinnett $100; Hall $500; Stewart $1,000.

SEABROOK VILLAGE, 660 Trade Hill Rd., Midway, GA 31320-6215. Tel.: 912-884-7008. Fax: 912-884-3046.

E-mail: seabrookvillage@yahoo.com

Web Site: seabrookvillage.org

Founded: 1994.

Key Personnel: Administrative Mgr., Florence Tate-Roberts; Pres. (V), Jolonda Greene.

Personnel Profile: Part-Time Paid 1; Part-Time Volunteers 25.

Governing Authority: Tax-exempt.

Institution Type/Description: History Museum.

Collections: open air African-American village.

Hours & Admission Prices: Tues.-Sat. 10-2. Self-Guided Tours: adults $5, children under 12 $3; discounts to active military, military reserve, AASLH, Coastal Museum Assoc., groups & AAM members; members & children under 6 no charge. Closed holidays. &

Attendance: 2,500 (estimated)

Membership: Seniors & Students $5; Adults $15; Family $30; Business $50-$250; Supporting $100.

Milledgeville

BLACKBRIDGE GALLERY - GEORGIA COLLEGE & STATE UNIVERSITY DEPARTMENT OF ART, (M), 209 Blackbridge Hall, Milledgeville, GA 31061. Mailing Address: Campus Box 94, Milledgeville, GA 31061. Tel.: 478-445-4572.

Web Site: www.gcsu.edu/art/blackbridgegallery.htm

Key Personnel: Dir., Carlos M. Herrera

Institution Type/Description: Art Gallery.

Collections: photographs; paintings.

Hours & Admission Prices: Mon.-Fri. 9-5.

GEORGIA COLLEGE AND STATE UNIVERSITY MUSEUM, 221 N. Clarke St., Milledgeville, GA 31061. Mailing Address: Campus Box 43, Milledgeville, GA 31061. Tel.: 478-445-4391. Fax: 478-445-6847.

E-mail: museum@gcsu.edu

Web Site: library.gcsu.edu/museum

Formerly: Museum & Archives of Georgia Education

Founded: 1975.

Congressional District: 11

Key Personnel: Cur., Shannon Morris; Dean, Dr. Rachel Schipper.

Personnel Profile: Full-Time Paid 1; Interns 1.

Governing Authority: university. Parent Institution: Georgia College & State University. Tax-exempt.

Institution Type/Description: History Museum.

Collections: Flannery O'Connor; Georgia College and State University history.

Research Fields: historical records of schools in the state; comparison of textbooks.

Facilities: 3,000-vol. library of textbooks prior to 1950; records of school systems; biographical information on computers of state's retired teachers available for research by special request; reading room.

Activities: guided tours; lectures; school & community outreach programs; permanent & temporary exhibitions.

Hours & Admission Prices: Mon.-Sat. 10-4; other times by appointment. No charge; donations accepted. &

Attendance: 2,500 (estimated)

GEORGIA COLLEGE NATURAL HISTORY MUSEUM AND PLANETARIUM, (M), Herty Hall, 1st Fl., W. Montgomery & N. Wilkinson Sts., Milledgeville, GA 31061. Mailing Address: Dept. of Biological & Environmental Sciences, Campus Box 081, Milledgeville, GA 31061. Tel.: 478-445-2395.
E-mail: nhm@gcsu.edu
Web Site: www.gcsu.edu/nhm
Founded: 2004.
Institution Type/Description: Natural History Museum & Planetarium.
Collections: earth sciences; paleontology.
Facilities: 2,500 sq. ft. exhibition space.
Activities: Planetarium: educational programs & shows.
Hours & Admission Prices: Mon.-Fri. 8-4, 1st Sat. each month 10-4. Closed holidays.

GEORGIA'S OLD CAPITAL MUSEUM, 201 E. Greene St., Milledgeville, GA 31061. Mailing Address: P.O. Box 1177, Milledgeville, GA 31059-1177. Tel.: 478-453-1803. Fax: 478-453-4813.
E-mail: info@oldcapitalmuseum.org
Web Site: www.oldcapitalmuseum.org
Founded: 1993.
Key Personnel: Exec. Dir., Dr. Amy Wright.
Personnel Profile: Full-Time Paid 1; Full-Time Volunteers 1; Part-Time Paid 2.
Governing Authority: private; nonprofit organization. Tax-exempt: 501(c)(3).
Institution Type/Description: History Museum.
Collections: area history.
Research Fields: cultural history.
Facilities: research library; 6,000 sq. ft. exhibit space. Museum-related items for sale.
Activities: formal education programs for adults; guided tours; lectures.
Hours & Admission Prices: Tues.-Fri. 10-4, Sat. 12-4. No charge; donations accepted. Closed legal holidays.
Attendance: 9,500 (accurate)
Membership: Student $15; Senior Citizen $20; Individual $25; Family $40; Organization $50; Friend $100; Business $150; Patron $250; Corporate Donor $500.

JOHN MARLOR ARTS CENTER, 201 N. Wayne St., Milledgeville, GA 31061-3437. Tel.: 478-452-3950. Fax: 478-452-5321.
E-mail: alliedarts@windstream.net
Web Site: milledgevillealliedarts.com
Founded: 1977.
Congressional District: 12
Key Personnel: Exec. Dir., Randy Cannon.
Personnel Profile: Full-Time Paid 2; Part-Time Paid 1.
Governing Authority: Tax-exempt.
Institution Type/Description: Art Center: housed in home built by John Marlor in 1830.
Collections: paintings.
Hours & Admission Prices: Mon.-Fri. 9-4:30. No charge; donations accepted.
Attendance: 1,500 (estimated)

OLD GOVERNOR'S MANSION - GEORGIA COLLEGE & STATE UNIVERSITY, 231 W. Hancock St., Milledgeville, GA 31061. Tel.: 478-445-4545.
Key Personnel: Dir., James C. Turner
Institution Type/Description: Historic Building: housed in the residence of Georgia's Governors; built in 1839.
Collections: local history & culture; period furnishings; personal artifacts; photographs.
Activities: school tours.
Hours & Admission Prices: Tues.-Sat. 10-4, Sun. 2-4; groups by appointment. Adults $10, senior citizens $7, students $2; children under 6 and GCSU faculty, staff & students no charge. Closed holidays.

STETSON-SANFORD HOUSE, 601 W. Hancock St., Milledgeville, GA 31061-3215. Mailing Address: 201 E. Green St., Milledgeville, GA 31061-3519. Tel.: 800-653-1804. Fax: 478-453-4813.
E-mail: sally@oldcapitalmuseum.org
Web Site: www.oldcapitalmuseum.org
Governing Authority: Parent Institution: Georgia's Old Capital Museum Society. Tax-exempt.
Institution Type/Description: Historic House Museum: housed in a former hotel built for George T. Brown; purchased by merchant Daniel B. Stetson in 1857 whose daughter Elizabeth married Judge Daniel B. Sanford, Clerk of the Secession Convention.
Collections: family & local history; period furnishings; photographs.

Hours & Admission Prices: House: Thurs.-Sat. with CVB Trolley Tour; other times by appointment. Museum: Tues.-Fri. 10-4, Sat. 12-4. Museum: adults $5.50, seniors $4.50, students $2.25. Closed major holidays.
Membership: Student & Senior $20; Individual $25; Family $40.

Mitchell

HAMBURG STATE PARK MUSEUM, 6071 Hamburg State Park Rd., Mitchell, GA 30820-2999. Mailing Address: P.O. Box 310, Crawfordville, GA 30631-0310. Tel.: 478-552-2393; 800-864-7275.
Web Site: gastateparks.org/info/hamburg
Founded: 1968.
Congressional District: 10
Key Personnel: Park Ranger, Earvin Cordry.
Personnel Profile: Full-Time Paid 5; Part-Time Paid 1; Part-Time Volunteers 2.
Governing Authority: state. Parent Institution: Georgia Dept. of Natural Resources, Parks & Historic Sites Division, 2 Martin Luther King Jr. Dr., S.E., Atlanta, GA 30334. Tax-exempt.
Institution Type/Description: Industrial Museum: housed in 1920 water turbine powered gin & milling complex.
Collections: farm tools; ginning equipment; milling machinery; corn shellers; animal drawn farm equipment.
Facilities: Corn meal for sale.
Activities: self-guided tours; milling demonstrations; permanent exhibits; guided & self guided tours.
Hours & Admission Prices: Park: daily 7am-10pm. Museum: daily 8-5. No charge. Parking: $3; Wed. no charge. Closed Thanksgiving; Christmas. &
Attendance: 3,500 (estimated)

Moreland

LEWIS GRIZZARD MUSEUM, 2769 US Hwy. 29 S., Moreland, GA 30259-2407. Tel.: 800-8COWETA.
Web Site: explorecoweta.com/lewisgrizzardmuseum.html
Institution Type/Description: History Museum.
Collections: personal artifacts & memorabilia of syndicated newspaper columnist & author Lewis Grizzard.
Hours & Admission Prices: Call for hours.

OLD MILL MUSEUM, Downtown Moreland Sq., Moreland, GA 30259. Mailing Address: P.O. Box 128, Moreland, GA 30259-0128. Tel.: 770-254-2627. Fax: 770-254-2628.
Institution Type/Description: History Museum.
Collections: local history; period farming equipment; WWII artifacts.
Hours & Admission Prices: Sat.-Sun. 1-4; other times by appointment. No charge; donations accepted.

Moultrie

COLQUITT COUNTY ARTS CENTER, 401 7th Ave., S.W., Moultrie, GA 31768-4633. Tel.: 912-985-1922. Fax: 912-890-6746.
E-mail: info@colquittcountyarts.com
Key Personnel: Exec. Dir., Jeffery Ophime
Institution Type/Description: Arts Center: housed in the former Moultrie High School.
Collections: works by national & international artists.
Activities: educational programs; special events. Annual Events: theater productions.
Hours & Admission Prices: Mon.-Fri. 10-5:30, Sat. 10-2.

MUSEUM OF COLQUITT COUNTY HISTORY, 4th Ave. & 5th St., S.E., Moultrie, GA 31788. Mailing Address: P.O. Box 86, Moultrie, GA 31776-0086. Tel.: 229-890-1626.
E-mail: olreb@moultriega.net
Institution Type/Description: History Museum.
Collections: local history; personal artifacts; photographs; period furnishings.
Activities: special events.
Hours & Admission Prices: Fri.-Sat. 10-5, Sun. 2-5; other times by appointment. No charge; donations accepted.

Mountain City

THE FOXFIRE MUSEUM & HERITAGE CENTER, 98 Foxfire Lane, Mountain City, GA 30562. Mailing Address: P.O. Box 541, Mountain City, GA 30562-0541. Tel.: 706-746-5828. Fax: 706-746-5829. Facebook: The Foxfire Museum & Heritage Center.
E-mail: foxfire@foxfire.org
Web Site: www.foxfire.org
Founded: 1966.
Congressional District: 9
Key Personnel: Dir. & Pres., Ann Moore; Chm. (V), Hunter Moorman; Museum Shop Mgr., Paulette Carpenter.
Personnel Profile: Full-Time Paid 3; Part-Time Paid 2; Part-Time Volunteers 16.
Governing Authority: nonprofit organization. Parent Institution: The Foxfire Fund, Inc. Tax-exempt.
Institution Type/Description: Appalachian Historic Buildings.
Collections: period artifacts; over 20 historic log cabins, some dating back to the early 1800s; pottery; folk art. Buildings: chapel; blacksmith shop; mule barn; wagon shed; single-room home; gristmill; smokehouse.
Facilities: Museum-related items for sale.
Activities: self-guided & guided tours; workshops.
Publications: Foxfire book series (1-12); Foxfire 40 Years; Aunt Arie, A Foxfire Portrait; Appalachian Cookery; Foxfire Book of Toys and Games; A Foxfire Christmas; Foxfire Magazine; book, Foxfire Book of Woodstove Cookery; book, Singin', Praisin', Raisin': Foxfires 45th Anniversary.
Hours & Admission Prices: Mon.-Sat. 8:30-4:30; groups & guided tours by appointment. Self-Guided Tour: adults 11 & up $6, children 7-10 $3; discounts to AAA & AARP members; children 6 & under no charge. Closed Thanksgiving; Christmas. &
Attendance: 13,500 (accurate)

Nelson

PICKENS COUNTY MARBLE MUSEUM, 1985 Kennesaw Ave., Nelson, GA 30151. Tel.: 770-735-2211.
Institution Type/Description: History Museum.
Collections: local marble mining history; tools & equipment.
Activities: marble quarry tours.
Hours & Admission Prices: Mon.-Fri. 9-4. No charge; donations accepted.

Newnan

NEWNAN-COWETA HISTORICAL SOCIETY - MALE ACADEMY MUSEUM, 30 Temple Ave., Corner of Temple Ave. & College St., Newnan, GA 30263-2066. Tel.: 770-251-0207. Fax: 770-683-0208.
E-mail: nchs@newnanbiz.net
Key Personnel: Admin., Tom Redwine; Asst. Admin., Carlisle Young
Institution Type/Description: Historical Society Museum.
Collections: area cultural, historical & architectural heritage.
Activities: special events.
Hours & Admission Prices: Tues.-Sat. 10-12 & 1-3, Sun. 2-5. Adults $3, children under 12 $1.

Norcross

BLUE MOON CYCLE'S BMW & EUROPEAN MOTORCYCLE MUSEUM, 752 W. Peachtree St., Norcross, GA 30071-1866. Tel.: 770-447-6945. Fax: 770-447-5798. Facebook: Blue Moon Cycle.
E-mail: generalinfo@bluemooncycle.com
Web Site: www.bluemooncycle.com
Founded: 1990.
Institution Type/Description: Motorcycle Museum.
Collections: early European motorcycles.
Hours & Admission Prices: Tues.-Fri. 9-6, Sat. 9-5. No charge.

Omaha

PROVIDENCE CANYON STATE PARK, 218 Florence Rd., Omaha, GA 31821-2218. Tel.: 229-838-4244; 800-864-7275 (reservations). Fax: 229-838-6735.
Web Site: www.gastateparks.org
Founded: 1971.
Congressional District: 2
Key Personnel: Mgr., Joy L. Joyner; Interpreter, Sherry Stephens; Interpreter, Thomas Wilson.
Personnel Profile: Full-Time Paid 5; Part-Time Paid 2.
Governing Authority: state. Administered by Parks, Recreation & Historic Sites Div., Dept. of Natural Resources. Tax-exempt.
Institution Type/Description: State Park Natural History Museum.
Collections: natural history; erosion.
Research Fields: endangered wildflowers; erosion; natural history.
Facilities: nature trails; interpretive center; picnic shelter; group shelter.
Activities: guided tours; self-guided tours; wildflower audio-visual program; hiking trails; backpacking; pioneer campground for organized groups.
Hours & Admission Prices: Park: mid-Sept. to mid-April daily 7-6; mid-April to mid-Sept. 7am-9pm. Office: daily 8-5. Buses: more than 30 passengers $50; Vans: 13-30 passenger $20. Parking: $3 per vehicle; Wed. no charge. &
Attendance: 90,000 (accurate)

Palmetto

PALMETTO TRAIN DEPOT MUSEUM, 549 Main St., Palmetto, GA 30268-1241. Tel.: 770-463-3377.
Institution Type/Description: Historic Building: housed in a former train depot; built in early 1900s.
Collections: local history & culture; period furnishings & toys; early 1900 baskets.
Hours & Admission Prices: Tues. & Thurs. 10-2. No charge.

Peachtree City

CAF DIXIE WING MUSEUM, 1200 Echo Ct., Peachtree City, GA 30269. Tel.: 678-364-1110.
Web Site: dw.squawk1200.net
Founded: 1986.
Congressional District: 5
Key Personnel: Dir. & Museum Shop Mgr., Billy G. Baldwin
Governing Authority: Parent Institution: Commemorative Air Force. Tax-exempt.
Institution Type/Description: Military Museum.
Collections: WWII & Korea warplanes; military artifacts & memorabilia; uniforms; weapons; newspapers; aircraft engines; model warplanes; WWII toys.
Activities: special events. Museum Sponsors: Annual Air Show; WWII Day.
Hours & Admission Prices: Tues., Thurs., & Sat. 9-4; other times by appointment. No charge; donations accepted. Closed major holidays. &
Attendance: 1,100 (estimated)

Pine Mountain

CALLAWAY GARDENS, Hwy. 18, Pine Mountain, GA 31822-2000. Mailing Address: P.O. Box 2000, Pine Mountain, GA 31822-2000. Tel.: 706-663-2281. Fax: 706-663-5004.
E-mail: info@ callawaygardens.com
Web Site: www.callawaygardens.com
Founded: 1952.
Congressional District: 3
Key Personnel: C.E.O., Edward C. Callaway; Pres., C. Robert McElerey; Dir. Retail, Julie Hicks; Dir. Devel., John P. Byrne; Dir. Historic Preservation, Michael A. Anderson; Museum Shop Mgr., Kevin Bridges.
Personnel Profile: Full-Time Paid 117; Part-Time Paid 49; Part-Time Volunteers 700; Interns 6.
Governing Authority: nonprofit organization. Parent Institution: Ida Cason Callaway Foundation. Subsidiary Institution: Callaway Gardens Resort Inc. Tax-exempt: 501(c)(3).
Institution Type/Description: Arboretum/Botanical Garden.
Collections: southeastern U.S. native plants. Historic House: 1799 Log Cabin.
Research Fields: applied horticulture.
Facilities: botanical garden; theater; classrooms; butterfly center; horticultural center; discovery center. Gift items for sale.
Activities: guided tours; lectures; films; concerts; art festivals; formally organized educational programs for children, adults, undergraduate & graduate college students; summer intern program for students in horticulture & natural history; slide lending library.
Publications: newsletter, Nature Naturally; newsletter, Foundation.
Hours & Admission Prices: Labor Day to March 13 daily 9-5; March 19 to Sept. daily 9-6. Fantasy in Lights 5-10. Adults $15, children 6-12 $7.50; discounts to groups & AABGA members; members and children 5 & under no charge. &
Attendance: 1,000,000 (accurate)
Membership: Child 6-12 $15; Senior $25; Adult 13 & up $30; Large Family Price Cap $100; Deluxe Adult $105; Deluxe Adult for Two $160; Azalea $250; Magnolia $500; Dogwood $750.

Plains

JIMMY CARTER NATIONAL HISTORIC SITE, 300 N. Bond St., Plains, GA 31780-5562. Tel.: 229-824-4104. Fax: 229-824-3441.
Web Site: www.nps.gov/jica
Key Personnel: Supt., Gary Ingram; Cur., Kate Funk
Institution Type/Description: Presidential Museums.
Collections: Plains Depot: presidential campaign memorabilia. Plains High School: furnished classroom; principal's office; auditorium; Carters' lives; political & business careers, education, family, religion & post-presidency; audiovisual tour of Carters' house. Boyhood Farm: restored 1937 farm including Carters' boyhood home & commissary.
Facilities: trails.
Activities: audiovisual tour; walking trails.
Hours & Admission Prices: Plains High School Museum: daily 9-5. Plains Depot - 1976 Carter Presidential Campaign Headquarters Museum: daily 9-4:30. Jimmy Carter Boyhood Farm Museum: daily 10-5. No charge. Closed New Year's Day; Thanksgiving; Christmas. &

Pooler

MIGHTY EIGHTH AIR FORCE MUSEUM, 175 Bourne Ave., Pooler, GA 31322-9516. Mailing Address: P.O. Box 1992, Savannah, GA 31402-1992. Tel.: 912-748-8888. Fax: 912-748-0209.
E-mail: development@mightyeighth.org
Web Site: www.mightyeighth.org
Founded: 1996.
Congressional District: 12
Key Personnel: C.E.O. & Pres., Henry Skipper; Chm. (V), Dr. William Cathcart; Dir. Devel., Brenda Elmgren; Dir. Oral Histories, Vivian Rogers-Price; Dir. Education, Heather Theis; Mktg. & Public Rels., Mandy Livingston; Dir. Finance, Pam Sconyers; Security & Operations, Bruce Johnson; Museum Store Mgr., Felice Stelljes; Registrar, Jean Prescott; Memorial Gardens, Peggy Harden; Advanced Coord., Susan Eiseman.
Personnel Profile: Full-Time Paid 24; Part-Time Volunteers 102; Interns 3.
Governing Authority: private; nonprofit organization. Tax-exempt: 501(c)(3).
Institution Type/Description: Military History Museum.
Collections: 8th Air Force history from WWII to present; military & veterans artifacts; vehicles & aircraft; archives & iconographic collections; oral histories.
Research Fields: 8th Air Force history; American air power and the science of flight.
Facilities: 10,000-vol. library of books; 125-seat restaurant; 37,000 sq. ft. exhibit space; 4 theaters; memorial garden; chapel; educational facilities; rental facilities. Museum-related items for sale.
Activities: docent program; formal education programs for children; guided tours; lectures; rental gallery; theater; art exhibits; workshops; loan, participatory, temporary & traveling exhibitions; school loan service; summer camp. Annual Events: Museum Anniversary Celebration; Golf Tournament; Warbirds Ball and Gala.
Publications: quarterly newsletter, Mighty Eighth Air Force Museum News.
Hours & Admission Prices: Daily 9-5. Adults $10, senior citizens & students $9, children $6; members and children 5 & under no charge. Closed New Year's Day; Easter; Thanksgiving; Christmas. &
Attendance: 84,222 (accurate)
Membership: Veteran, Student & Heir of the Eighth $25; Individual $35; Family $50; Honor Guard $100; Squadron Leader $500; Wing Commander $1,000.

Powder Springs

SEVEN SPRINGS HISTORICAL SOCIETY, 3901 Brownsville Rd., Powder Springs, GA 30127. Mailing Address: 4484 Marietta St., P.O. Box 46, Powder Springs, GA 30127. Tel.: 678-567-5611.
Institution Type/Description: Historical Society Museum.
Collections: local history & culture; photographs; period furnishings; personal artifacts.
Hours & Admission Prices: Wed. 10 to noon, Sat. 12-4, Sun. 1-4.

Riceboro

GEECHEE KUNDA CULTURAL ARTS CENTER AND MUSEUM, 622 Ways Temple Rd., Riceboro, GA 31323-4307. Tel.: 912-884-4440.
Institution Type/Description: History Museum.
Collections: African American history & culture; personal artifacts; period furnishings; photographs.

Facilities: Museum-related items for sale.
Hours & Admission Prices: Tues.-Sat. 10-4. No charge; donations accepted. Closed Federal holidays.

Richmond Hill

FORT MCALLISTER, 3894 Fort McAllister Rd., Richmond Hill, GA 31324-4862. Tel.: 912-727-2339. Fax: 912-727-3614.
E-mail: ftmcallr@coastalnow.net
Web Site: www.fortmcallister.org
Founded: 1958.
Congressional District: 1
Key Personnel: Park Mgr., Daniel Brown.
Personnel Profile: Full-Time Paid 5.
Governing Authority: state. Parent Institution: Parks, Recreation & Historic Sites Div., Georgia Dept. of Natural Resources, 270 Washington St., S.W., Atlanta, GA 30334. Tax-exempt.
Institution Type/Description: Preservation Project: located on the site of 1861 Confederate fort.
Collections: restored earthworks; military.
Research Fields: development of artillery; static defenses; ironclad ships.
Facilities: camp sites; nature trails. Publications & postcards for sale.
Activities: guided tours; permanent exhibitions; living history demonstrations during summer.
Hours & Admission Prices: Daily 8-5. Adults $5, seniors $4.50, children $3.50; discounts to groups; children under 5 no charge. Closed Thanksgiving; Christmas. &
Attendance: 140,000 (estimated)
Membership: Friends of GA $35.

THE RICHMOND HILL HISTORICAL SOCIETY AND MUSEUM, 11460 Ford Ave., Richmond Hill, GA 31324. Mailing Address: P.O. Box 381, Richmond Hill, GA 31324. Tel.: 912-756-3697.
Institution Type/Description: Historical Society Museum: housed in the former Henry Ford Kindergarten.
Collections: local history & culture; period furnishings; personal artifacts; photographs.
Hours & Admission Prices: Mon. & Thurs.-Sat. 10-3. No charge. Closed holidays.

Rincon

GEORGIA SALZBURGER SOCIETY MUSEUM, 2980 Ebenezer Rd., Rincon, GA 31326-3716. Tel.: 912-826-5629.
E-mail: info@georgiasalzburgers.com
Web Site: www.georgiasalzburgers.com
Founded: 1925.
Key Personnel: Pres., Keith Zeigler; Treas., Janice Moody; Cur., Martha Zeigler.
Personnel Profile: Part-Time Volunteers 10.
Governing Authority: society. Tax-exempt: 501(c)(3).
Institution Type/Description: General Museum.
Collections: tools; furniture; letters; books; Bibles; deeds; maps; records of early settlers or their descendants before or just after the Civil War; Indian artifacts; Boltzius Monument. Historic Structure: 1769 Jerusalem Lutheran church; frame cottage.
Research Fields: genealogy.
Activities: tours; meetings.
Publications: newsletters, Salzburger Genealogy; Salzburger Cook Book; book, The Salzburgers & Their Descendants; genealogy books, Seckinger, Exley Helmle.
Hours & Admission Prices: Wed. & Sat.-Sun. 3-5; other times by appointment. No charge; donations accepted. &
Attendance: 950 (estimated)
Membership: Regular $25; Life $250.

Ringgold

OLD STONE CHURCH MUSEUM, GA Hwy. #2, Ringgold, GA 30736. Mailing Address: CCHS P.O. Box 113, Ringgold, GA 30736-0113. Tel.: 706-935-5232.
Institution Type/Description: Historic Building: housed in a former church built c.1849.
Collections: local history; Civil War artifacts; personal artifacts.
Hours & Admission Prices: Thurs.-Sun. 1-5. No charge; donations accepted.

Rome

CHIEFTAINS MUSEUM/MAJOR RIDGE HOME, (M), 501 Riverside Pkwy., Rome, GA 30161-0373. Mailing Address: P.O. Box 373, Rome, GA 30162-0373. Tel.: 706-291-9494. Fax: 706-291-2410.
E-mail: chmuseum@bellsouth.net
Web Site: www.chieftainsmuseum.org
Founded: 1969.
Congressional District: 7
Key Personnel: Pres. (V), Ruth Demeter; Program Coord. & Museum Shop Mgr., Debby Brown.
Personnel Profile: Full-Time Paid 1; Part-Time Paid 1; Part-Time Volunteers 25.
Governing Authority: nonprofit organization. Parent Institution: Chieftains Museum, Inc. Tax-exempt.
Institution Type/Description: History Museum: housed in 1790 log cabin, expanded in 1828, to a plantation house belonging to Cherokee leader Major Ridge.
Collections: items from Archaic Indian occupation to present time; trading post artifacts from archaeological dig; period furniture; costumes; photographs; paintings.
Research Fields: Ridge family history, Cherokee history & culture, architecture, archaeology.
Activities: guided tours; gallery talks; docent programs; occasional TV programs; loan, permanent & temporary exhibitions; school tours; Cherokee history & culture programs.
Publications: newsletter.
Hours & Admission Prices: Wed.-Sat. 10-5. Adults $5, senior citizens 60 & over $3, children $2; discounts to AAM members, groups & school tours; members & Cherokee Nation registered members no charge. Closed major holidays. &
Attendance: 6,000 (accurate)
Membership: Individual $35; Family $50; Supporter $75; Friend $100; Sustainer $250; Patron $500; Benefactor $1,000; Chieftains $2,500.

OAK HILL AND THE MARTHA BERRY MUSEUM, (M), 24 Veterans Memorial Hwy., Rome, GA 30161. Mailing Address: P.O. Box 490189, Mount Berry, GA 30149-0189. Tel.: 706-368-6789; 800-220-5504. Fax: 706-368-6787.
E-mail: oakhill@berry.edu
Web Site: www.berry.edu/oakhill
Founded: 1972.
Congressional District: 11
Key Personnel: Asst. Dir., Rebecca Roberts; Coord. Mktg., Patrice Shannon; Mgr. Operations & Museum Shop Mgr., Cheli Rouse; Ground Mgr., Kristin McNully; Horticulturalist, Heather Miller; Museum Asst., Rebecca Henry.
Personnel Profile: Full-Time Paid 3; Part-Time Paid 9.
Governing Authority: private college. Parent Institution: Berry College, 2277 Martha Berry Hwy., Mt. Berry, GA. Tax-exempt: 501(c)(3).
Institution Type/Description: Historic House Museum: home and gardens of Martha Berry, Berry College founder. History Museum: maps the evolution of Berry College from an industrial and agricultural school to a liberal arts college.
Collections: period furnishings; china; silver; books; artifacts belonging to Martha Berry and her school; paintings; Italian & American arts.
Facilities: 48-seat theater; public garden. Museum-related items for sale.
Activities: guided tours; lectures; temporary exhibitions. Museum Sponsors: Garden Events; Christmas Festival.
Publications: newsletter, The Oak Hill Gardener.
Hours & Admission Prices: Mon.-Sat. 10-5. Adults $5, children $3, tours & family group rates $4 per person; discounts to AAA members & senior citizens. Closed New Year's Day, Martin Luther King Jr. Day; Easter; Memorial Day; Independence Day; Labor Day; Christmas week. &
Attendance: 15,000 (estimated)

ROME AREA HISTORY MUSEUM, 305 Broad St., Rome, GA 30161-3005. Tel.: 706-235-8051. Fax: 706-235-6631.
E-mail: lbarba@romehistorymuseum.com
Web Site: www.romehistorymuseum.com
Founded: 1996.
Key Personnel: Dir., Leigh Barba; Chm. (V), Gardner Wright; Pres. Bd., Janet Byington; Archivist, Russell McClanahan; Exec. Asst., Donna Shaw.
Personnel Profile: Full-Time Paid 1; Part-Time Paid 2; Part-Time Volunteers 8; Interns 2.
Governing Authority: Tax-exempt.
Institution Type/Description: History Museum.
Collections: area history; Native Americans; early settlers through 20th century; culture; industries; documents; photographs; personal artifacts.

Activities: monthly lecture series; monthly afternoon tea; school programs.
Hours & Admission Prices: Wed.-Fri. 10-4, Sat. 10-1. No charge. Closed New Year's Day; Independence Day; Labor Day; Thanksgiving; Christmas Eve & Day. &
Attendance: 12,407 (accurate)
Membership: Student $15; Individual $30; Family $50; Patron $100.

Rossville

THE CHIEF JOHN ROSS HOUSE ASSOC., 200 E. Lake Ave., Rossville, GA 30741. Mailing Address: 826 Chickamauga Ave., Rossville, GA 30741-1407. Tel.: 706-861-3954 & 866-5171. Fax: 706-861-3967.
Founded: 1797.
Congressional District: 7
Key Personnel: C.E.O. & Pres., W. Larry Rose.
Personnel Profile: Part-Time Paid 2.
Governing Authority: nonprofit organization. John Ross House Assoc. Tax-exempt.
Institution Type/Description: Historic House: 1797 two-story log house of Chief John Ross.
Collections: Cherokee alphabet; arrowheads; pictures; letters; furniture; rugs & linens.
Research Fields: Cherokee Indian records.
Activities: guided tours; arts festivals; permanent exhibitions; special events.
Publications: brochures.
Hours & Admission Prices: June-Aug. 15 Thurs.-Sat. 10-2; groups of 12 or more at other times by appointment. Adults $2; children under 12 no charge.
Attendance: 130 (accurate)
Membership: Individual & Family $25; Corporate $50; Chieftain $100 & up.

Roswell

ARCHIBALD SMITH PLANTATION HOME, 935 Alpharetta St., Roswell, GA 30075-3827. Tel.: 770-641-3978. Fax: 770-641-3974.
E-mail: cdouglas@roswellgov.com
Founded: 1845.
Congressional District: 6
Key Personnel: Dir., Chuck Douglas.
Personnel Profile: Full-Time Paid 1; Part-Time Paid 5; Part-Time Volunteers 50; Interns 2.
Governing Authority: Parent Institution: City of Roswell, Ga. Tax-exempt.
Institution Type/Description: History Museum.
Collections: personal artifacts; period furnishings; outbuildings; servants quarters; smoke house; corn crib; carriage house.
Hours & Admission Prices: Mon.-Sat. 10-3, Sun. 1-3. Adults $8, seniors 65 & over $7, students & children 6-18 $6; discounts to AAA members; members & children under 6 no charge. Closed major holidays. &
Attendance: 9,000 (accurate)

BARRINGTON HALL, 535 Barrington Dr., Roswell, GA 30075. Tel.: 678-639-7500; 770-225-2457.
Governing Authority: city.
Institution Type/Description: Historic House Museum: housed in the former home of Roswell co-founder, Barrington King, c.1842. Listed on the National Register of Historic Places.
Collections: local history & culture; period furnishings; personal artifacts; photographs.
Activities: group tours.
Hours & Admission Prices: Mon.-Sat. 10-3, Sun. 1-3. Adults $8, seniors 65 & over $7, children 6-12 $6; discounts to groups & AAA members; children under 6 no charge. Closed National holidays.

BULLOCH HALL, 180 Bulloch Ave., Roswell, GA 30075-4420. Mailing Address: P.O. Box 1309, Roswell, GA 30077-1309. Tel.: 770-992-1731 & 1951. Fax: 770-587-1840.
E-mail: info@bullochhall.org
Web Site: www.bullochhall.org
Founded: 1978.
Congressional District: 5
Key Personnel: Dir., Pam Billingsley; Chm. (V), Bill W. Gray; Vice Chm., Tom Campbell; Asst. Site Coord., Janice Metzler.
Personnel Profile: Full-Time Paid 1; Part-Time Paid 8; Part-Time Volunteers 40.
Governing Authority: municipal. Tax-exempt.
Institution Type/Description: Historic House: c.1839 Antebellum Greek Revival House & Cottage.
Collections: period furnishings; family photos; manuscripts.
Research Fields: Bulloch family; local history.

Facilities: archives.
Activities: guided tours; lectures; docent program & council; craft guilds; heritage craft classes.
Publications: quarterly newsletter, The Hallmark.
Hours & Admission Prices: Mon.-Sat. 10-3, Sun. 1-3; tours on the hour. Adults $8, seniors $7, children 6-18 $6; members & children under 6 no charge. Closed New Year's Eve & Day; Martin Luther King Jr. Day; Easter; Memorial Day; Independence Day; Labor Day; Thanksgiving; Christmas Eve & Day. &
Attendance: 11,000 (estimated)
Membership: Individual $25; Family $35; Business $50; Patron $100; Donor-Patron $500; Benefactor $1,000.

CHATTAHOOCHEE NATURE CENTER, 9135 Willeo Rd., Roswell, GA 30075-4723. Tel.: 770-992-2055.
E-mail: marketing@chattnaturecenter.org
Institution Type/Description: Nature Center.
Collections: reptiles; amphibians; birds of prey; native & endangered plants.
Facilities: Museum-related items for sale.
Activities: adult programs; birthday parties; scout groups; special events.
Hours & Admission Prices: Mon.-Sat. 9-5, Sun. 12-5. Adults $8, seniors $6, children 3-12 $5; children 2 & under no charge. Closed New Year's Day; Thanksgiving; Christmas.

ROSWELL FIRE & RESCUE MUSEUM, 1002 Alpharetta St., Roswell, GA 30075-3661. Mailing Address: 1810 Hembree Rd., Alpharetta, GA 30009. Tel.: 770-641-3730. Fax: 770-641-3843.
E-mail: fire@roswellgov.com
Web Site: www.roswellgov.com/index.aspx?NID=208
Founded: 1986.
Key Personnel: Fire Chief, Ricky Spencer.
Personnel Profile: Full-Time Paid 18; Part-Time Paid 140.
Institution Type/Description: Fire-Fighting Museum.
Collections: local history & culture; fire-related photographs; Atlanta's fire history; 1947 Ford American LaFrance Pumper; World Trade Center steel.
Hours & Admission Prices: Call for hours. No charge. &
Attendance: 500 (estimated)

ROSWELL VISUAL ARTS CENTER & GALLERY, 10495 Woodstock Rd., Roswell, GA 30075 2941. Mailing Address: 38 Hill St., Ste. 100, Roswell, GA 30075. Tel.: 770-594-6122.
E-mail: rrpd@ci.roswell.ga.us
Web Site: www.roswellgov.com
Key Personnel: Arts Supvr., Jan Gibbons.
Governing Authority: Branch: Roswell Art Center West, 1355 Woodstock Rd., Roswell, GA 30075. Tel.: 770-641-3990.
Institution Type/Description: Art Gallery.
Collections: works by regional, local & international artists; drawings; paintings; printmaking; sculpture; ceramics; fiber; photography; mixed media.
Facilities: Museum-related items for sale.
Activities: special events; classes.
Hours & Admission Prices: Mon.-Fri. 9:30-6, Sat. 9-1. Roswell Art Center West: Wed.-Sat. 10-6. No charge. &

TEACHING MUSEUM NORTH, 793 Mimosa Blvd., Roswell, GA 30075. Tel.: 770-552-6339.
Founded: 1988.
Institution Type/Description: History Museum.
Collections: local history & culture; period furnishings; photographs; hands-on exhibitions; period furnishings; courtroom furnishings & artifacts; Anne Frank memorabilia; log cabin replica c.1880; drawings; presidential artifacts; clothing; government documents.
Activities: hands-on exhibitions; educational programs; special events; scout programs; research. Museum Sponsors: Black History Celebration; Saturday Science.
Hours & Admission Prices: Mon.-Fri. 8-4; other times by appointment.

Royston

TY COBB MUSEUM, 461 Cook St., Royston, GA 30662-4003. Tel.: 706-245-1825. Fax: 706-245-1831. Facebook: Ty Cobb Museum.
E-mail: julie.ridgway@tycobbhealthcare.com
Web Site: www.tycobbmuseum.org
Founded: 1998.
Congressional District: 9
Key Personnel: Dir., Julie Ridgway; C.E.O., Greg Hearn; Chm. (V), Randy Moore; Museum Shop Mgr., Sharri Hobbs.

Personnel Profile: Full-Time Paid 1; Part-Time Paid 3; Part-Time Volunteers 1.
Governing Authority: Parent Institution: Ty Cobb Healthcare System. Tax-exempt.
Institution Type/Description: Sports Museum.
Collections: baseball & Ty Cobb history; art; memorabilia; books; video; film; personal artifacts; photographs.
Facilities: theater.
Activities: special events.
Hours & Admission Prices: Mon.-Fri. 9-4, Sat. 10-4; call to confirm. Adults $5, seniors 62 & over $4, students $3; discounts to groups of 10 or more and AAM & ICOM members; children under 5 & active military no charge. &
Attendance: 3,700 (accurate)
Membership: Support Club $1-$99; 367 Club $100 & up; Century Club $1,000 & up; Crown Club $5,000 & up.

Saint Marys

CUMBERLAND ISLAND NATIONAL SEASHORE, 129 Osborne St., Saint Marys, GA 31558-8416. Mailing Address: 101 Wheeler St., Saint Marys, GA 31558-8421. Tel.: 912-882-4336 & 4335. Fax: 912-882-6284.
E-mail: john_a_mitchell@nps.gov
Web Site: www.nps.gov/cuis
Founded: 1972.
Congressional District: 1
Key Personnel: Supt., Fred Boyles; Cur., John A. Mitchell.
Governing Authority: federal. Parent Institution: National Park Service. Tax-exempt.
Institution Type/Description: Historical Museum.
Collections: Historic Site: 2000 B.C. American Indian village site & Dungeness ruins. Historic Houses: c.1880-1910 Plum Orchard Mansion; Ice House Museum; outbuildings.
Major Exhibits: Forgotten Invasion, 1/12-1/15.
Research Fields: historic research appropriate to Cumberland Island; natural history; sea turtle research; bird tagging; family genealogy, history of island & region, natural history, pre-historic life-ways, architecture; majority of museum is in storage & needs special advance admission for research purposes; U.S. history & the War of 1812.
Facilities: natural history education center; visitor center.
Activities: guided tours; lectures; films; formally organized educational programs for children, adults, undergraduate & graduate students.
Publications: park brochure.
Hours & Admission Prices: Office: daily 8-4:30. Ferry: March-Nov. daily 9 am & 11:45 am; Dec.-Feb. Mon. & Thurs.-Sun. 9 am & 11:45 am. Round Trip: adults $20, senior citizens 65 & over $18; children 12 & under $12. Day use fee $4 per person; call office for reservations 912-673-7747. &
Attendance: 40,000 (estimated)

ORANGE HALL, 311 Osborne St., Saint Marys, GA 31568. Tel.: 912-576-3644.
Institution Type/Description: Historic House Museum: c.1820.
Collections: local history & culture; period furnishings; photographs; personal artifacts.
Activities: tours.
Hours & Admission Prices: Daily call for hours.

ST. MARYS FILM MUSEUM, 111 Osborne St., Saint Marys, GA 31558. Tel.: 912-576-5488.
Web Site: www.coastalgeorgiafilm.org
Institution Type/Description: Film Museum.
Collections: film history; films; photographs.
Activities: self-guided tours.
Hours & Admission Prices: Mon.-Sat. 9-5.

ST. MARYS SUBMARINE MUSEUM, 102 St. Marys St. W., Saint Marys, GA 31558-4945. Tel.: 912-882-2782. Fax: 912-882-2748.
E-mail: submus@tds.net
Web Site: www.stmaryssubmuseum.com
Founded: 1995.
Key Personnel: Pres., Doug Cooper; Museum Shop Mgr., John Crouse.
Personnel Profile: Full-Time Paid 1; Part-Time Paid 2.
Governing Authority: Tax-exempt.
Institution Type/Description: Military & Submarine Museum.
Collections: submarine artifacts, history & memorabilia; photographs.
Research Fields: quarterly newsletter; submarine history.
Facilities: Gift items for sale.
Activities: submarine periscope look out. Museum Sponsors: U.S. Subvets WWII Annual Memorial Service.

Publications: quarterly newsletter; quarterly Georgia submarine veterans newsletter; submarine history CD.
Hours & Admission Prices: Tues.-Sat. 10-4, Sun. 1-5. Adults 19-62 $4, senior citizens 63-99 & military $3, children 6-18 $2; children under 6 & members no charge. Closed New Year's Eve, Day & week; Easter; Thanksgiving; Christmas week. &
Attendance: 10,000 (estimated)
Membership: Children $5; Individual $15; Family $25

Saint Simons Island

THE ARTHUR J. MOORE METHODIST MUSEUM, 100 Arthur Moore Dr., Saint Simons Island, GA 31522. Mailing Address: P.O. Box 24081, Saint Simons Island, GA 31522-7081. Tel.: 912-638-4050. Fax: 912-638-9050.
E-mail: mmarchive@mooremuseum.org
Web Site: www.mooremethodistmuseum.org
Founded: 1965.
Congressional District: 1
Key Personnel: Dir., Judi Fergus; Chm. (V), Becky Bridges; Pres. (V), Rev. Dave Hanson; Museum Shop Mgr., Dianne Smoot.
Personnel Profile: Full-Time Paid 2; Part-Time Volunteers 10.
Governing Authority: church. Parent Institution: South Georgia Annual Conference of the U.M.C. Affiliated with the South Georgia Methodist Conference Epworth by the Sea, Inc. Tax-exempt.
Institution Type/Description: Religious Museum: near the site of Oglethorpe, John and Charles Wesley's activities in 1736.
Collections: John & Charles Wesley memorabilia, letters, land grants; Nativities from around the world.
Research Fields: history of Methodist churches.
Facilities: 6,000-vol. library of religious material available for research; reading room.
Activities: guided tours; docent program or council; temporary exhibitions.
Publications: brochure, Historical Highlights.
Hours & Admission Prices: Tues.-Sat. 10-4. Adults $5; children 5 & under no charge. Closed New Year's Day; Independence Day; Thanksgiving; Christmas. &
Attendance: 10,000 (accurate)
Membership: Basic $35 & up; Life $500.

*** COASTAL GEORGIA HISTORICAL SOCIETY - ST. SIMONS ISLAND LIGHTHOUSE MUSEUM, THE MARITIME CENTER AT THE HISTORIC COAST GUARD STATION, (M),** 610 Beachview Dr., Saint Simons Island, GA 31522-4821. Mailing Address: P.O. Box 21136, Saint Simons Island, GA 31522-0636. Tel.: 912-638-4666. Fax: 912-638-6609.
E-mail: admin@saintsimonslighthouse.org
Web Site: saintsimonslighthouse.org
Formerly: A.W. Jones Heritage Center and St. Simons Island Lighthouse Museum - Coastal Georgia Historical Society
Founded: 1965.
Congressional District: 1
Key Personnel: Pres., Anne Stembler; Dir., Sherri Jones; Public Rels., Leigh Ann Stroud; Cur., Ellen Rogers; Museum Shop Mgr., Curt Smith.
Personnel Profile: Full-Time Paid 6; Part-Time Paid 8; Part-Time Volunteers 100.
Operating Income: 950,000
Governing Authority: nonprofit organization. St. Simons Island Lighthouse Museum, Beachview Dr. & 12th St., Saint Simons Island, GA. Tax-exempt: 501(c)(3).
Institution Type/Description: Local History Museum; Lighthouse Museum.
Collections: artifacts; books; journals; manuscripts; photographs; slides; graphic arts; audio-visual material of the coastal Georgia area from 1788-present. Historical Structures: c.1872 lighthouse; 1890 oil house; 1872 lightkeepers' dwelling & archives complex.
Major Exhibits: Saint Simons Island: Its History and Lighthouses, 11/13-3/14.
Research Fields: all aspects of coastal history, crafts & lifestyles from the Spanish period to the present.
Facilities: 500-vol. library on coastal history available for use on premises. Books & museum-related items for sale.
Activities: guided tours for groups; lectures; docent program; exhibitions; craft demonstrations; historical excursions; seminars.
Publications: quarterly newsletter.
Hours & Admission Prices: Mon.-Sat. 10-5, Sun. 1:30-5. Adults $10, children 6-11 $5; discounts to groups; military, children under 6 & members no charge. (Admission prices include all museums & venues). Closed New Year's Day; Easter; Thanksgiving; Christmas Eve & Day. &
Attendance: 100,000 (accurate)
Membership: Individual $50; Members Guild $75; Family $100; Curator $150; Lamplighter $300; Assistant Keeper $500; Keeper of the Light $1,000.

FORT FREDERICA NATIONAL MONUMENT, 6516 Frederica Rd., Saint Simons Island, GA 31522. Tel.: 912-638-3639. Fax: 912-634-5357.
E-mail: denise_spear@nps.gov
Web Site: www.nps.gov/fofr
Founded: 1936.
Congressional District: 1
Personnel Profile: Full-Time Paid 11; Part-Time Paid 2; Part-Time Volunteers 35.
Governing Authority: Parent Institution: Dept. of the Interior. Tax-exempt.
Institution Type/Description: Park Museum.
Collections: study collection of artifacts found at Fort Frederica and town of Frederica. Historic Sites: 1736-1758 ruins of English barracks, fort and house foundations.
Research Fields: Frederica archeological reports; Frederica's early settlers.
Facilities: library; auditorium; visitor center. Historical books & historical reproductions for sale.
Activities: talks; permanent & temporary exhibitions; 25-min. film on Frederica history; tours; living history; children's program; Wayside exhibits. Museum Sponsors: Frederica Festival; Theatrical digital tour.
Publications: pamphlet, Fort Frederica National Monument; booklet, Fort Frederica Color Book; books, The Bloody Summer of 1742: A Colonial Boy's Journal; Voyage to Georgia; The First Families of Frederica; Phoebe's Secret Diary: Daily Life and First Romance of a Colonial Girl.
Hours & Admission Prices: Visitor Center: daily 9-5. Adults $3, annual park pass $10; children under 16, Golden Age & Access Pass holders no charge. Closed Christmas. &
Attendance: 292,505 (accurate)

THE GLYNN ART ASSOCIATION, INC., 529 Beachview Dr., Saint Simons Island, GA 31522-4705. Tel.: 912-638-8770. Fax: 912-634-2787.
E-mail: glynnart@bellsouth.net
Web Site: www.glynnart.org
Founded: 1953.
Congressional District: 1
Key Personnel: Pres. (V), Keith Crusan; Vice Pres., Joyce Ledingham.
Personnel Profile: Full-Time Paid 1; Part-Time Paid 1; Part-Time Volunteers 40.
Governing Authority: nonprofit organization. Glynn Art Association, Inc. dba Coastal Alliance for the Arts. Tax-exempt: 509(a)(2), 501(c)(3).
Institution Type/Description: Art Gallery.
Collections: series, Yesterday's Golden Isles, early primitive paintings of the area; pottery; works represent 160 artists.
Facilities: library of general art books available for research upon request; reading room; classrooms. Museum replicas, arts & craft items for sale.
Activities: guided tours; lectures; films; gallery talks; concerts; dance recitals; arts festivals; drama; study clubs; TV programs; formally organized education programs; docent program; traveling exhibitions.
Publications: newsletter; Sapelo.
Hours & Admission Prices: Tues.-Sat. 9-5. No charge; donations accepted. Closed New Year's Day; Independence Day; Labor Day; Thanksgiving; Christmas. &
Attendance: 10,000 (estimated)
Membership: Individual $25; Family $35; Donor $50; Patron $75; Sponsor $125; Friend $250; Benefactor $500 & up.

*** MARITIME CENTER - COASTAL GEORGIA HISTORICAL SOCIETY, (M),** 4201 First St., Saint Simons Island, GA 31522-3902. Mailing Address: P.O. Box 21136, Saint Simons Island, GA 31522-0636. Tel.: 912-638-4666. Fax: 912-638-6609.
E-mail: ssi1872@comcast.net
Formerly: Maritime Museum at Historic Coast Guard Station
Founded: 1965.
Congressional District: 1
Key Personnel: Dir., Sherri Jones; Pres., Jonathan Raclin; Public Rels., Jerri Hager; Cur., Ellen Rogers; Museum Shop Mgr., Curt Smith.
Personnel Profile: Full-Time Paid 5; Part-Time Paid 8; Part-Time Volunteers 100.
Governing Authority: private; nonprofit organization. Parent Institution: Coastal GA Historical Museum. Tax-exempt: 501(c)(3).
Institution Type/Description: Coast Guard & Natural History Museum.
Collections: Coast Guard & military history; coastal Georgia's natural history; military artifacts; structures; photographs; personal artifacts.
Facilities: outdoor classroom; nature center. Museum-related items for sale.
Hours & Admission Prices: Mon.-Sat. 10-5, Sun. 1:30-5. Adults $10, children 6-11 $5; military, children under 6 & members no charge. (Admission prices include all museums & venues) Closed New Year's Day; Easter; Thanksgiving; Christmas Eve & Day. &

Attendance: 40,000 (estimated)
Membership: Individual $50; Members Guild $75; Family $100; Curator $150; Lamplighter $300; Assistant Keeper $500; Keeper of the Light $1,000.

MILDRED HUIE MUSEUM, 1819 Frederica Rd., Saint Simons Island, GA 31522. Mailing Address: P.O. Box 30841, Sea Island, GA 31561. Tel.: 912-638-3057.
E-mail: medhouse@bellsouth.net
Founded: 2001.
Key Personnel: Dir., Dayna Caldwell, M.F.A.; C.E.O., Mildred Huie Wilcox
Institution Type/Description: Art Museum: housed in the Mediterranean House; built in 1929.
Collections: Mildred Huie's life, career & memorabilia; paintings; sculpture; sketches; plantation models; books.
Hours & Admission Prices: Fri.-Sun. 1-5; other times by appointment. No charge; donations accepted.

Sandersville

THE BROWN HOUSE MUSEUM, 268 N. Harris St., Sandersville, GA 31082. Mailing Address: P.O. Box 6088, Sandersville, GA 31082-6088. Tel.: 478-552-1965 & 2963.
E-mail: fcveal@yahoo.com
Founded: 1999.
Congressional District: 10
Key Personnel: Dir., Frances C. Veal; Pres. Society, Layne Kilchens.
Personnel Profile: Full-Time Volunteers 18; Part-Time Paid 8; Part-Time Volunteers 16.
Governing Authority: Parent Institution: Washington County Historical Society.
Institution Type/Description: History Museum: housed in c.1850 house, headquarters for General Sherman in 1864.
Collections: Washington County history; pottery; Civil War artifacts; schools of Wash Co; old city cemetery.
Hours & Admission Prices: Tues. & Thurs-Fri. 2-5, Sat. 10-3; other times by appointment. No charge; donations accepted. Closed holidays. &
Attendance: 1,000 (estimated)

THE CHARLES EDWARD CHOATE EXHIBIT, 131 W. Haynes St., Ste. B, Sandersville, GA 31082-1737. Tel.: 478-552-3288. Fax: 478-552-1449.
Web Site: www.washingtoncounty-ga.com
Founded: 1996.
Congressional District: 12
Governing Authority: Parent Institution: Washington County Chamber of Commerce.
Institution Type/Description: Art Museum.
Collections: Sterling Everett paintings; photographs.
Activities: self guided tours.
Hours & Admission Prices: Mon.-Fri. 9-12 & 1-5. No charge. Closed holidays. &
Attendance: 50 (estimated)

GENEALOGY RESEARCH CENTER AND OLD JAIL MUSEUM, 129 Jones St., Sandersville, GA 31082-1768. Mailing Address: P.O. Box 6088, Sandersville, GA 31082-6088. Tel.: 478-552-6965.
E-mail: genealogyresearch@att.net
Web Site: www.rootsweb.com/gawashin/genweb/
Formerly: Washington County Museum
Founded: 1978.
Congressional District: 10
Key Personnel: Dir. & Chm. (V), Loretta Cato; C.E.O., Ronald Jackson.
Personnel Profile: Part-Time Paid 4; Part-Time Volunteers 20.
Governing Authority: private; nonprofit organization. Branch Museums: The Brown House Museum, 268 N. Harris St., Sandersville, GA 31082. Tel. 912-552-1965. Genealogy Research Center and Museum, 129 Jones St., Sandersville, GA 31082. Tel. 912-552-6965. Tax-exempt.
Institution Type/Description: History Museum: housed in c.1891 sheriffs' house with attached jail.
Collections: Washington County history; genealogy.
Research Fields: genealogy.
Publications: quarterly newsletter.
Hours & Admission Prices: Tues. & Thurs.-Fri. 2-5, St. 10-3. No charge; donations accepted. Closed holidays. &
Attendance: 400 (estimated)
Membership: Adult $12; Family $15.

Sandy Springs

ABERNATHY ARTS CENTER, 254 Johnson Ferry Rd., N.W., Sandy Springs, GA 30328. Tel.: 404-613-6172. Fax: 404-303-6135.
Institution Type/Description: Art Gallery.
Collections: works by students & faculty.
Facilities: classrooms.
Activities: classes; temporary exhibitions.
Hours & Admission Prices: Tues.-Fri. 9-5, Sat. 9-2.

GEORGIA COMMISSION ON THE HOLOCAUST, 5920 Roswell Rd., Ste. 209, Sandy Springs, GA 30328. Tel.: 770-206-1558.
Web Site: www.holocaust.georgia.gov
Institution Type/Description: History Museum.
Collections: life of Anne Frank; Holocaust history; over 600 photographs; period furnishings.
Activities: video.
Hours & Admission Prices: Tues.-Thurs. 10-4, Fri. 10-2, Sat.-Sun. 12-4. No charge.

SANDY SPRINGS HISTORIC SITE & MUSEUM, 6075 Sandy Springs Cir., Sandy Springs, GA 30328-3841. Mailing Address: c/o Heritage Sandy Springs, 6110 Bluestone Rd., Sandy Springs, GA 30328. Tel.: 404-851-9111. Fax: 404-851-9807.
Key Personnel: Exec. Dir., Carol Thompson
Institution Type/Description: Historic House: housed in the Williams-Payne house; built in 1869.
Collections: period artifacts; photographs; books; oral histories; maps; manuscripts.
Hours & Admission Prices: House: by appointment. Park: daily dawn to 8pm. No charge.

Sautee

SAUTEE-NACOOCHEE CENTER, 283 Hwy. 255 N., Sautee, GA 30571-2606. Mailing Address: P.O. Box 460, Sautee, GA 30571-0460. Tel.: 706-878-3300. Fax: 706-878-1395.
Web Site: www.snca.org
Formerly: Sautee-Nacoochee Community Association
Founded: 1982.
Key Personnel: C.E.O. & Dir., Kathy Blandin, Ph.D.; Chm., Bob Prim; Cur., Sam Schultz; Cur. Folk Pottery, John Burrison; Dir. Folk Pottery Museum, Chris Brooks.
Personnel Profile: Full-Time Paid 3; Part-Time Paid 8; Part-Time Volunteers 312; Interns 2.
Governing Authority: private; nonprofit organization. Parent Institution: Sautee Nacoochee Community Association. Branch Museums: History Museum of Sautee-Nacoochee; The Center Gallery. Tax-exempt.
Institution Type/Description: Art & History Museum.
Collections: paintings; pottery; glass; jewelry.
Research Fields: folk pottery.
Hours & Admission Prices: Office: Mon.-Fri. 9-5. Museum & Gallery: Sat. 10-5, Sun. 1-5. Adults $5; members no charge. &
Attendance: 22,300 (estimated)
Membership: Student $28; Senior Individual $38; Individual $53; Senior Family $58; Family $83; Patron $150.

Savannah

ALEXANDER HALL GALLERY - SAVANNAH COLLEGE OF ART AND DESIGN, 668 Indian St., Savannah, GA 31401-1105. Tel.: 912-525-4727. Fax: 912-525-4952.
Web Site: www.scad.edu/exhibitions/galleries
Institution Type/Description: Art Gallery.
Collections: paintings; photographs; drawings; sculpture.
Activities: temporary exhibitions.
Hours & Admission Prices: Mon.-Fri. 9-5. No charge.

ANDREW LOW HOUSE, (M), 329 Abercorn St., Savannah, GA 31401-4634. Tel.: 912-233-6854. Fax: 912-233-9239.
E-mail: sbohlin@andrewlowhouse.com
Web Site: www.andrewlowhouse.com
Founded: 1927.
Congressional District: 1
Key Personnel: Chm., Karen Powell; Pres., Mrs. Richard Dudley Williams; Dir., Stephen Bohlin; Registrar, Jessica Mumford.
Personnel Profile: Full-Time Paid 1; Part-Time Paid 18; Part-Time Volunteers 60.

Governing Authority: private; nonprofit organization. Governing Authority: National Society of the Colonial Dames in the State of GA. Tax-exempt: 501(c)(3).
Institution Type/Description: Historic Site: housed in the former home of Andrew Low, a wealthy Savannah cotton factor.
Collections: Andrew Low history; 19th-century decorative arts, 1800-1850.
Research Fields: building archaeology.
Facilities: 1,000-vol. library; botanical garden. Museum-related items for sale.
Activities: guided tours.
Publications: annual report, The Low Exchange.
Hours & Admission Prices: mid-Jan. to Dec. Mon.-Sat. 10-4, Sun. 12-4. Adults $10, students & children $6; discounts to groups of 10 or more & AAA members; members no charge. Closed Labor Day; Thanksgiving; Christmas.
Attendance: 24,701 (accurate)

ARCHIVES MUSEUM, TEMPLE MICKVE ISRAEL, 20 E. Garden St. on Monterey Square, Savannah, GA 31401. Mailing Address: P.O. Box 816, Savannah, GA 31402-0816. Tel.: 912-233-1547. Fax: 912-233-3086.
E-mail: lauree@mickveisrael.org
Web Site: www.mickveisrael.org
Founded: 1733.
Congressional District: 12
Key Personnel: Rabbi, Arnold Belzer; Pres., Steve Gordon; Chm. (V), Eileen Lobel; Vice Pres., Brian Markowitz; Gift Shop Mgr., Glenda McNew; Temple Admin., Anne Maner; Asst. Dir. & Museum Committee Member, Lauree San Juan.
Personnel Profile: Full-Time Paid 5; Part-Time Volunteers 10.
Governing Authority: nonprofit organization. Tax-exempt.
Institution Type/Description: Religious Museum: housed in c.1876 Gothic style, Congregation Mickve Israel Synagogue.
Collections: 18th-20th century artifacts of Jewish life.
Research Fields: archival records for research on Jews in Savannah, Georgia & the South from 1733-present day.
Facilities: 20,000-vol. library pertaining to Jewish studies available for research by special permission. Ceremonial objects, books & jewelry for sale.
Activities: guided tours; permanent exhibitions; heritage tours.
Publications: biweekly event newsletter, Contact.
Hours & Admission Prices: Mon.-Fri. 10-1 & 2-4, call to confirm. Adults $5; children under 12 & members no charge. Closed holidays. &

Attendance: 8,000 (estimated)

DAVENPORT HOUSE MUSEUM, 324 E. State St., Savannah, GA 31401-3411. Tel.: 912-236-8097. Fax: 912-233-7938. Facebook: Davenport House Museum.
E-mail: info@davenporthousemuseum.org
Web Site: davenporthousemuseum.org
Founded: 1955.
Congressional District: 12
Key Personnel: Pres. (V), Brooke Wilford; C.E.O. Historic Savannah Foundation, Daniel Carey; Dir., Jamie Credle; Asst. Dir., Jeff Freeman; Volunteer Coord., Dottie Kraft; Museum Shop Mgr., Ben Head; Maintenance Technician, Raleigh Marcell.
Personnel Profile: Full-Time Paid 2; Part-Time Paid 15; Part-Time Volunteers 75; Interns 3.
Governing Authority: board of trustees. Parent Institution: Historic Savannah Foundation. Tax-exempt: 501(c)(3).
Institution Type/Description: Historic House Museum.
Collections: period furniture based on the 1828 probate court inventory; Federal furniture; Kennedy pharmacy.
Research Fields: paint analysis; historic preservation; 19th century decorative arts; heritage education; antebellum social history.
Facilities: garden; meeting facilities. Museum-related items for sale.
Activities: guided tours; docent program; junior interpreter program; living history; holiday evening tours; Madeira traditions experience; tea program.
Publications: books: Savannah; Historic Savannah Survey Book; The Davenport House Museum: Savannah's Beacon of Preservation; tour book; monthly volunteer newsletter; semiannual newsletter, Friends; DVDs - Isaiah Davenport: Portrait of a Master Builder.
Hours & Admission Prices: Mon.-Sat. 10-4, Sun. 1-4; tours on the half hour; special and group tours available upon request. Adults $8, students 6-18 $5; discount to museum employees & volunteers, AAA, Coastal Museum Assoc. members & National Trust Partner Place members; members & children under 6 no charge. Closed New Year's Day; St. Patrick's Day; Easter; Independence Day; Thanksgiving; Christmas.
Attendance: 36,155 (accurate)
Membership: Annual Giving Levels: Brick Mason $1-$49; Carpenter $50-

$149; Master Builder $150-$249; Alderman $250-$599; Fire Warden $600-$999; McKinnon Circle $1,000.

FAHM HALL GALLERY - SAVANNAH COLLEGE OF ART AND DESIGN, 1 Fahm St., Savannah, GA 31401-1140. Tel.: 912-525-4727.
Institution Type/Description: Art Gallery.
Collections: sculpture; paintings.
Activities: temporary exhibitions.
Hours & Admission Prices: Mon.-Fri. 9-5. No charge.

FIRST AFRICAN BAPTIST CHURCH & MUSEUM, 23 Montgomery St., Savannah, GA 31401-2429. Tel.: 912-233-6597. Fax: 912-234-7950.
Institution Type/Description: Religious Museum.
Collections: church history; 18th century artifacts & memorabilia; photographs; personal & religious artifacts.
Facilities: Museum-related items for sale.
Activities: guided tours.
Hours & Admission Prices: Mon.-Fri. 10-4; other times by appointment. Adults $8, seniors & students $6. &
Attendance: 10,000 (estimated)

FLANNERY O'CONNOR CHILDHOOD HOME, 207 E. Charlton St., Savannah, GA 31401-4605. Tel.: 912-233-6014.
E-mail: flanneryoconnorhome@gmail.com
Web Site: www.flanneryoconnorhome.org
Founded: 1989.
Congressional District: 12
Key Personnel: Pres. (V), Helen Borrello; Resident Mgr., Toby Aldrich.
Personnel Profile: Part-Time Paid 1.
Governing Authority: Tax-exempt.
Institution Type/Description: Historic House: housed in the childhood home of author, Flannery O'Connor.
Collections: family history; photographs; personal artifacts; period furnishings.
Activities: workshop. Museum Sponsors: Literary Events in Fall & Spring; Ursrey Memorial Lecture in Fall.
Publications: newsletter, Friends of Flannery.
Hours & Admission Prices: Fri.-Wed. 1-4. Adults $6, students $5; members and youth 15 & under no charge. Closed major holidays.
Attendance: 2,349 (accurate)
Membership: Individual $25; Family $50; Patron $100; Peacock Circle $500.

FORT PULASKI NATIONAL MONUMENT, U.S. Hwy. 80 E., Savannah, GA 31410. Mailing Address: P.O. Box 30757, Savannah, GA 31410-0757. Tel.: 912-786-5787. Fax: 912-786-6023.
Web Site: www.nps.gov/fopu/
Founded: 1924.
Congressional District: 1
Key Personnel: Supt., Randall Wester.
Personnel Profile: Full-Time Paid 17; Part-Time Paid 8; Part-Time Volunteers 30.
Governing Authority: federal. National Park Service, Dept. of the Interior, Washington, DC 20240. Tax-exempt.
Institution Type/Description: History Museum & Park.
Collections: Civil War cannon and projectiles; uniform accessories; personal effects of soldiers; bottle collection; c.1862 period & replica room furnishings; c.1890 battery Horace Hambright.
Research Fields: garrison life of Federal and Confederate soldiers; heavy seacoast artillery; seacoast defenses of the U.S.
Facilities: 1,200-vol. library of general reference books & specialized research on Civil War, cultural & natural resource management, 1934-1940, historic preservation of Cockspur Island, Coastal Georgia & South Carolina on premises only with two weeks advance reservation; visitor center; history & nature trails; picnic area. Handbooks on various historic sites, natural history field guides, post cards & mementoes relating to Fort Pulaski for sale.
Activities: tours; lectures; costumed presentations; permanent & temporary exhibitions.
Publications: guidebooks; pamphlets.
Hours & Admission Prices: Labor Day-Memorial Day daily 9-5; call for extended summer hours. Adult 16 & over $5; children 15 & under no charge. Golden Age, Eagle & National Park Pass accepted. Closed Thanksgiving; Christmas. &
Attendance: 400,000 (estimated)

GEORGIA HISTORICAL SOCIETY, 501 Whitaker St., Savannah, GA 31401-4889. Tel.: 912-651-2125. Fax: 912-651-2831.
E-mail: ghs@georgiahistory.com
Web Site: www.georgiahistory.com
Founded: 1839.
Congressional District: 12
Key Personnel: C.E.O. & Pres., Dr. W. Todd Groce; Chm., Robert L. Brown, III; Exec. Vice Pres., Laura Garcia-Culler; Dir. Library & Archives, Lynette Stoudt; Sr. Historian, Dr. Stan Deaton.
Personnel Profile: Full-Time Paid 16; Part-Time Paid 8; Part-Time Volunteers 19; Interns 5.
Governing Authority: society. Tax-exempt: 501(c)(3).
Institution Type/Description: Historical Society Museum: housed in 1874-1875 building designed by Detlief Lienau.
Collections: Georgia history & culture; books; manuscripts; maps; photographs; prints; newspapers; paintings; portraits; period artifacts.
Research Fields: 18th, 19th & 20th-century Georgia manuscripts; southern history; genealogy.
Facilities: 25,000-vol. library of books on Georgia & American history; reading room.
Activities: lectures; temporary exhibitions; historical marker program; technical services & programming for local historical societies & museums; teacher institutes; educational programs for school children.
Publications: quarterly magazine, Georgia Historical Quarterly; occasional volume, Collections of the Georgia Historical Society; quarterly newsmagazine, Georgia History Today.
Hours & Admission Prices: Wed.-Fri. 12-5, 1st & 3rd Sat. each month 10-5. Adults $5; members no charge. Closed national & state holidays. &
Attendance: 35,000 (accurate)
Membership: Student $35; Individual $55; Family $65; Library $85; Sponsor $100; Benefactor $250; Corporate & Sustainer $500; John MacPherson Berrien Circle $1,000; William Brown Hodgson Circle $2,500; 1839 Society $5,000; James Edward Oglethorpe Circle $10,000.

GEORGIA STATE RAILROAD MUSEUM, 655 Louisville Rd., Savannah, GA 31401. Tel.: 912-651-6823.
Web Site: www.chsgeorgia.org/railroad-museum.html
Institution Type/Description: Railroad History Museum: A National Historic Landmark.
Collections: early railcars & rolling stock; handcar.
Activities: train rides; educational programs; guided tours; turntable & blacksmithing demonstrations.
Hours & Admission Prices: Call for hours.

GIRL SCOUT FIRST HEADQUARTERS MUSEUM, 330 Drayton St., Savannah, GA 31401-4433. Tel.: 888-223-3883.
Institution Type/Description: Girl Scout History Museum: housed in the former carriage house to the Andrew Low House, built in 1848. Girl Scout founder, Juliette Low converted building into Girl Scout Headquarters in 1912.
Collections: Girl Scout history from 1912 to present.
Facilities: Museum-related items for sale.
Activities: educational programs; patch program.
Hours & Admission Prices: Mon.-Tues. & Thurs.-Sat. 10-4.

GREEN-MELDRIM HOUSE MUSEUM, 14 W. Macon St., Savannah, GA 13401. Tel.: 912-233-3845.
Institution Type/Description: Historic House Museum: housed in the former home of Charles Green who offered General William T. Sherman the use of his home as headquarters during the Civil War in 1864.
Collections: local history; period furnishings; military artifacts; photographs.
Hours & Admission Prices: Tues. & Thurs.-Fri. 10-4, Sat. 10-1. Adults $7, children $2.

GUTSTEIN GALLERY - SAVANNAH COLLEGE OF ART AND DESIGN, 201 E. Broughton St., Savannah, GA 31401-3401. Mailing Address: P.O. Box 3146, Savannah, GA 31402-3146. Tel.: 912-525-4735. Fax: 912-525-4952.
E-mail: exhibitions@scad.edu
Web Site: www.scad.edu
Founded: 1979.
Key Personnel: Exec. Dir., Laurie Ann Farrell.
Personnel Profile: Full-Time Paid 18; Part-Time Paid 5.
Governing Authority: Parent Institution: Savannah College of Art and Design (SCAD). Tax-exempt.
Institution Type/Description: Art Museum.
Collections: work of prominent artists & emerging Savannah College of Art and Design alumni.

Publications: exhibition catalogs.
Hours & Admission Prices: Mon.-Fri. 10-6, Sat. 12-5. No charge; donations accepted. &
Attendance: 25,000 (estimated)

HALL STREET GALLERY - SAVANNAH COLLEGE OF ART AND DESIGN, 212 W. Hall St., Savannah, GA 31402. Tel.: 912-525-4727.
Institution Type/Description: Art Gallery.
Collections: paintings; drawings.
Activities: temporary exhibitions.
Hours & Admission Prices: Mon.-Fri. 10-5. No charge.

∗ **JEPSON CENTER,** 207 W. York St., Savannah, GA 31401. Tel.: 912-790-8800.
Institution Type/Description: Contemporary Art Center.
Collections: works by contemporary artists.
Facilities: 220-seat auditorium; 7,500 sq. ft. exhibition space; education studios.
Activities: temporary & traveling exhibitions; educational programs.
Hours & Admission Prices: Sun.-Mon. 12-5, Tues.-Wed. & Fri.-Sat. 10-5, Thurs. 10-8. Adults $20, seniors 65 & over and military $18, student K-College $5; discounts to AAA members; members & children under 5 no charge. Closed New Year's Day; Martin Luther King, Jr. Day; St. Patrick's Day; Easter; Labor Day; Thanksgiving; Christmas. &

JULIETTE GORDON LOW BIRTHPLACE, (M), 10 E. Oglethorpe Ave., Savannah, GA 31401-3707. Tel.: 912-233-4501. Fax: 912-233-4659. TDD: 912-233-4501.
E-mail: info@juliettegordonlowbirthplace.org
Web Site: www.girlscouts.org/birthplace
Founded: 1956.
Congressional District: 1
Key Personnel: C.E.O. & Dir., Fran Powell Harold; Cur., Sherry Lang; Program Mgr., Katherine Keena; Museum Shop Mgr., Linda LeFurgy.
Personnel Profile: Full-Time Paid 12; Part-Time Paid 55; Part-Time Volunteers 100; Interns 2.
Governing Authority: nonprofit organization. Parent Institution: Girl Scouts of the U.S.A., 420 5th Ave., New York, NY 10018. Tax-exempt.
Institution Type/Description: Historic House Museum: 1818-1821 Wayne-Gordon House.
Collections: 19th-century memorabilia, art & furniture of Juliette Gordon Low & the Gordon family; Girl Scouts of the U.S.A.
Research Fields: Juliette Gordon Low; history of Wayne-Gordon House; founding & early years, 1912-1915, of Girl Scouts of the U.S.A.; 19th-century lifestyles.
Facilities: Museum-related items for sale.
Activities: guided tours; formally organized education programs for Girl Scouts, school groups & others; permanent & temporary exhibitions; special events.
Publications: Birthplace Bound.
Hours & Admission Prices: mid-Jan. to Feb. & Nov.-Dec. Mon.-Tues. & Thurs.-Sat. 10-4, Sun. 11-4; March-Oct. Mon.-Sat. 10-4, Sun. 11-4. Adults $8; Girl Scout adults & students 6-20 $7, Girl Scouts 6-18 $6; children 5 & under no charge. Closed New Year's Day; St. Patrick's Day; Easter; Thanksgiving; Christmas Eve & Day. &
Attendance: 65,000 (accurate)
Membership: Girl Scouts of U.S.A.: Annual $10; the Birthplace Circle of Friends: Annual Green Level $50-$249; Silver Level $250-$499; Gold Level $500-$999; Platinum Level $1,000 & up.

KING-TISDELL COTTAGE - MUSEUM OF BLACK HISTORY, 514 E. Huntingdon St., Savannah, GA 31401-5115. Mailing Address: Beach Institute, 502 E. Harris St., Savannah, GA 31401. Tel.: 912-234-8000. Fax: 912-234-8001.
E-mail: kingtisdell@bellsouth.net
Web Site: www.kingtisdell.org
Institution Type/Description: Historic House Museum: housed in an 1896 Victorian cottage.
Collections: African American history & culture; photographs; personal artifacts; period furnishings.
Hours & Admission Prices: Mon.-Fri. 12-5, Sat.-Sun. 1-4. Adults $1.50, children $.75; discounts to groups. Closed holidays.

MASSIE SCHOOL, 207 E. Gordon St., Savannah, GA 31401-5003. Tel.: 912-395-5070. Fax: 912-201-5224.
E-mail: candy.lowe@sccpss.com
Web Site: www.massieschool.com

Formerly: Massie Heritage Center
Congressional District: 1
Key Personnel: Dir., Candy Lowe; Pres. (V), Emma Adler; Museum Shop Mgr., Sandee Lipsitz.
Personnel Profile: Full-Time Paid 4; Part-Time Paid 6; Part-Time Volunteers 6.
Governing Authority: Parent Institution: Savannah-Chatham County Public Schools. Tax-exempt.
Institution Type/Description: Heritage Center.
Collections: local history & culture; city model; photographs; documents; personal artifacts.
Facilities: classrooms. Museum-related items for sale.
Activities: educational programs; guided tours.
Hours & Admission Prices: Mon.-Fri. 9-4; groups by appointment. Guided Tours: adults $8, children 5-12 $3; members no charge. Self-Guided Tours: adults $5, children 5-12 $3; discounts to NTHP members; children under 4 & members no charge. Closed holidays. &

MAY POETTER GALLERY, 342 Bull St., Savannah, GA 31401. Tel.: 912-525-4948.
Web Site: www.scadexhibitions.com
Institution Type/Description: Art Gallery.
Collections: paintings; sculpture.
Activities: special events; lectures.
Hours & Admission Prices: Mon.-Fri. 9-5.

MERCER WILLIAMS HOUSE MUSEUM, 429 Bull St., Monterey Square, Savannah, GA 31401. Tel.: 912-236-6352; 877-430-6352 (toll free). Fax: 912-238-3993.
Institution Type/Description: Historic House Museum: housed in the former home of John Wilder; built c.1868.
Collections: local history & culture; period furnishing; personal artifacts; photographs; drawings; portraits.
Facilities: Gift items for sale.
Hours & Admission Prices: Mon.-Sat. 10:30-4:10, Sun. 12-4. Adults $12.50, students $8.

OATLAND ISLAND WILDLIFE CENTER, 711 Sandtown Rd., Savannah, GA 31410-1019. Tel.: 912-395-1212 & 1500. Fax: 912-898-3983.
Web Site: www.oatlandisland.org
Formerly: Oatland Island Education Center
Founded: 1974.
Congressional District: 1
Key Personnel: Dir., Heather Merbs; Education, Pam Keener; Education, Annie Quinting; Education, Max McKelvey; Administrative Sec., Shirley Calhoun; Naturalist, Pam Hewatt.
Personnel Profile: Full-Time Paid 12; Part-Time Paid 3; Part-Time Volunteers 50; Interns 2.
Governing Authority: public school district. Parent Institution: Chatham-Savannah Board of Education, 208 Bull St., Savannah, GA 31406. Tel.: 912-395-5600. Tax-exempt.
Institution Type/Description: Environmental Education Nature Center.
Collections: wild animals indigenous to state of Georgia; Phillips barn; Martin Cane mill. Historic Houses: 1835 Delk-Dawson House; 1835 Wayne County Cabin.
Facilities: zoological park; nature center; auditorium; classrooms.
Activities: guided tours; formally organized education programs for children, adults & undergraduate college students.
Publications: brochure; trail guide; program guide; newsletter.
Hours & Admission Prices: Daily 10-4. Adults $5, seniors, children 4-17 & military $3; members no charge. Closed New Year's Day; Thanksgiving; Christmas Eve & Day. &
Attendance: 67,080 (accurate)
Membership: Individual $25; Family $35; Donor $125; Patron $250.

OLD FORT JACKSON, 1 Ft. Jackson Rd., Savannah, GA 31404-1039. Tel.: 912-232-3945. Fax: 912-236-5126.
E-mail: oldfortjackson@chsgeorgia.org
Web Site: www.chsgeorgia.org/jackson
Founded: 1975.
Congressional District: 1
Key Personnel: Site Mgr., Marty Liebschner.
Personnel Profile: Full-Time Paid 6; Part-Time Paid 4; Part-Time Volunteers 25.
Governing Authority: board of directors. Parent Institution: Coastal Heritage Society. Tax-exempt.
Institution Type/Description: American Military Museum.

Collections: artifacts relating to the military history of Savannah including the American Revolution, the War of 1812 & the Civil War.
Research Fields: U.S. military in Savannah, 1800-1877; history of Deptford Tract where Fort Jackson now stands; history of construction technology; Savannah Naval Forces & CSS Georgia.
Facilities: picnic area. Museum-related items for sale.
Activities: daily self-guided tours with canon firing demonstrations & uniformed interpreters; holiday programs; special school & adult programs.
Publications: quarterly newsletter; book, Tide Craft.
Hours & Admission Prices: Daily 9-5. Adults $4.25, students, senior citizens & military (past & present) $3.75; discounts to AAA & CHS members; children 6 & under no charge. Closed New Year's Day; Thanksgiving; Christmas. &
Attendance: 75,000 (estimated)
Membership: Student $15; Basic $25; Patron $50; Contributing $100; Sustaining $250; Sponsor $500; Turnbull $1,000.

OWENS-THOMAS HOUSE, 124 Abercorn St., Savannah, GA 31401-3732. Mailing Address: P.O. Box 10081, Savannah, GA 31412-0281. Tel.: 912-233-9743.
Institution Type/Description: Historic House Museum: housed in a home designed by architect, William Jay; c.1819.
Collections: local history & culture; period furnishings; personal artifacts; photographs; decorative arts from 1750-1830.
Hours & Admission Prices: Tues.-Sat. 10-5, Sun. 2-5. Closed New Year's Day; Martin Luther King Jr. Day; St. Patrick's Day; Easter; Labor Day; Thanksgiving; Christmas.

PEI LING CHAN GALLERY - SAVANNAH COLLEGE OF ART AND DESIGN, 322 & 324 Martin Luther King Jr. Blvd., Savannah, GA 31401. Mailing Address: P.O. Box 3146, Savannah, GA 31402-3146. Tel.: 912-525-4727.
Web Site: www.scad.edu/exhibitions/galleries
Institution Type/Description: Art Gallery.
Collections: paintings; sculpture.
Activities: temporary exhibitions.
Hours & Admission Prices: Mon.-Fri. 10-5:30. No charge.

PINNACLE GALLERY - SAVANNAH COLLEGE OF ART AND DESIGN, 320 E. Liberty St., Savannah, GA 31401. Tel.: 912-525-4950.
Web Site: www.scadexhibitions.com
Institution Type/Description: Art Gallery.
Collections: paintings; sculpture.
Activities: lectures; special events.
Hours & Admission Prices: Mon.-Fri. 9-5:30, Sat. 10-5, Sun. 1-4.

RALPH MARK GILBERT CIVIL RIGHTS MUSEUM, 460 Martin Luther King, Jr. Blvd., Savannah, GA 31401-4800. Tel.: 912-231-8900. Fax: 912-234-2577.
E-mail: nancyeand@yahoo.com
Web Site: www.sip.armstrong.edu/CivilRightsMuseum/Civilindex.html
Institution Type/Description: History Museum: named in honor of the late Dr. Ralph Mark Gilbert, father of Savannah's civil rights movement & NAACP leader.
Collections: civil rights history; photographs.
Hours & Admission Prices: Tues.-Sat. 9-5. Museum Tours: adults $8, senior citizens 65 & over $6, students $4; discounts to groups of 10 or more; members no charge.

ROUNDHOUSE RAILROAD MUSEUM, 601 W. Harris St., Savannah, GA 31401-3193. Tel.: 912-651-6823. Fax: 912-651-3194.
E-mail: roundhouse@chsgeorgia.org
Institution Type/Description: Railroad Museum: A National Historic Landmark.
Collections: railroad history; seven historic structures; locomotives; rolling stock; model trains; operating turntable.
Hours & Admission Prices: Daily 9-5. Call for admission prices.

SAVANNAH BOTANICAL GARDENS, 1388 Eisenhower Dr., Savannah, GA 31406. Tel.: 912-355-3883.
E-mail: sacgc@att.net
Web Site: www.savannahbotanical.org
Governing Authority: Parent Institution: Savannah Area Council of Garden Clubs, Inc.
Institution Type/Description: Botanical Garden.
Collections: plants; trees; flowers. Historic Building: Reinhard House.
Facilities: nature trails; amphitheater; banquet facilities.

Activities: nature trails; guided & self-guided tours; rental facilities.
Hours & Admission Prices: Mon.-Sat. 8-8, Sun. 8am-8:45pm. Garden: fee may be charged for groups of 10 or more.

SAVANNAH HISTORY MUSEUM, 303 Martin Luther King Jr. Blvd., Savannah, GA 31401-4217. Tel.: 912-651-6825 & 238-1779.
Web Site: www.chsgeorgia.org/shm/
Governing Authority: nonprofit organization. Tax-exempt: 501(c)(3).
Institution Type/Description: History Museum: housed in the old Central of Georgia Railway passenger shed, built 1850-1860s.
Collections: Savannah's history from 1733 to present; Revolutionary War Battle of Savannah; dugout canoes; 19th & 20th-century women's fashions; Forest Gump's bench; weapons; military uniforms; Savannah's railway history.
Facilities: theater. Museum-related items for sale.
Activities: film presentations; programs.
Hours & Admission Prices: Mon.-Fri. 8:30-5, Sat.-Sun. 9-5. Call for admission prices.

SAVANNAH - OGEECHEE CANAL MUSEUM & NATURE CENTER, 681 Fort Argyle Rd., Savannah, GA 31419-9239. Tel.: 912-748-8068.
E-mail: info@savannahogeecheecanalsociety.com
Web Site: www.savannahogeecheecanal society.org
Founded: 1993.
Key Personnel: Pres. (V), Steve Elkins; Museum Shop Mgr., Chica Arndt.
Personnel Profile: Part-Time Paid 4; Part-Time Volunteers 4.
Institution Type/Description: History Museum & Nature Center.
Collections: canal history; Civil War artifacts; early household items; personal artifacts; period furnishings; photographs.
Facilities: nature trails.
Activities: guided tours; special events; educational programs.
Hours & Admission Prices: Call for hours. Adults $2, children 5-12 $1; members no charge.
Attendance: 2,500 (estimated)
Membership: $20-$40.

SCAD MUSEUM OF ART, 601 Turner Blvd., Savannah, GA 31401-4242. Mailing Address: P.O. Box 3146, Savannah, GA 31402-3146. Tel.: 912-525-7191. Facebook: SCAD Museum of Art.
E-mail: scadmoa@scad.edu
Web Site: www.scadmoa.org
Founded: 2002.
Congressional District: 12
Key Personnel: Mng. Dir., Kimberly Shreve.
Personnel Profile: Full-Time Paid 5; Part-Time Volunteers 1; Interns 2.
Governing Authority: college. Parent Institution: Savannah College of Art and Design. Subsidiary Institution: Earle W. Newton Center for British and American Studies; Walter O. Evans Center for African American Studies. Tax-exempt.
Institution Type/Description: Art Museum.
Collections: paintings including 17th-19th century British & colonial American portraits; prints from the Newton collection; photographs including the 19th-20th century Rhoades collection; African American art from the Evans collection; 20th-21st century costumes including the C.Z. Guest collection.
Research Fields: British & American art; African-American art.
Activities: temporary & permanent exhibitions; group tours; gallery talks; lectures; special events.
Hours & Admission Prices: Call for hours. Adults $10, senior & military $8; members no charge. &
Attendance: 45,000 (estimated)
Membership: Teacher, Student & Alumni $30; Individual $50; Family & Dual $70.

SHIPS OF THE SEA MARITIME MUSEUM/WILLIAM SCARBROUGH HOUSE AND GARDENS, (M), 41 Martin Luther King Blvd., Savannah, GA 31401-2435. Tel.: 912-232-1511. Fax: 912-234-7363.
E-mail: contact@shipsofthesea.org
Web Site: www.shipsofthesea.org
Founded: 1966.
Key Personnel: Dir., Tony Pizzo; Cur. Exhibits & Education, Wendy Melton; Mgr. Communication & Events, Michelle Riley.
Personnel Profile: Full-Time Paid 3; Part-Time Paid 10; Part-Time Volunteers 1.
Governing Authority: private; nonprofit organization. Tax-exempt: 501(c)(3).
Institution Type/Description: Maritime Museum.

Collections: 18th-19th century large scale ship models, paintings & maritime antiques; navigational instruments; seafaring artifacts; video presentations.
Facilities: botanical garden; educational facilities. Museum-related items for sale.
Activities: formal educational programs for children; guided tours; lectures; instrument-making; boatbuilding.
Publications: book, Flotsam and Jetsam; William Scarbrough's House; Savannah Line; Savannah & The Civil War at Sea.
Hours & Admission Prices: Tues.-Sun. 10-5 (last admission 4:15). Family $20, adults $8.50, senior citizens, students, AAA, & military $6.50; discounts to AAM members & groups of 10 or more. Closed New Year's Eve & Day; St. Patrick's Day; Easter; Thanksgiving; Christmas Eve & Day. &
Attendance: 20,000 (estimated)

*   **TELFAIR MUSEUM OF ART, (M),** Telfair Academy of Arts & Sciences, 121 Barnard St., Savannah, GA 31401. Mailing Address: P.O. Box 10081, Savannah, GA 31412-0281. Tel.: 912-232-1177, ext. 16. Fax: 912-232-6954.
E-mail: hadaways@telfair.org
Web Site: www.telfair.org
Founded: 1875.
Congressional District: 1
Key Personnel: Dir. & C.E.O., Lisa N. Grove; Chm. (V) & Pres. (V), T. Mills Fleming; Registrar, Jessica M. Estes; Designer & Preparator, Milutin Pavlovic; Sr. Cur. Education, Harry H. DeLorme; Admin., Sandra S. Hadaway; Cur. Owens-Thomas House & Decorative Arts, Tania J. Sammons; Asst. Cur., Elizabeth Moore; Cur., Courtney McNeil; Dir. Devel. & Communications, Kristin Boylston; Mgr. Corporate & Foundation Rels., Laura C. Dixon; Mgr. Donor Rels., Catherine Renner; Data Lease Mgr., Jessie Kotarski; Education Resource Specialist, Martha Mythlo; Designer & Preparator, Heath Ritch; Dir. Retail Operations, Lisa Ocampo; Graphic Designer, Holly Akkerman; Mgr. Events, Jessica Denmark; Exec. Asst. to Dir., Kelly Balenzano; Dir. Cultural Diversity & Access, Vaughnette Goode-Walker; Lead Interpreter, Paulette Thompson.
Personnel Profile: Full-Time Paid 38; Part-Time Paid 46; Part-Time Volunteers 900; Interns 10.
Governing Authority: board of trustees; nonprofit. Parent Institution: Telfair Museum of Art, Inc. Sites: Telfair Academy of Arts & Sciences; Jepson Center for the Arts, 207 W. York St., Savannah, GA; Owens-Thomas House, 124 Abercom St., Savannah, GA. Tax-exempt: 501(c)(3).
Institution Type/Description: Art Museum & Historic House: housed in 1819 Regency mansion.
Collections: Regency mansion contains many Telfair family objects from late 18th & early 19th century, including pieces commissioned from Duncan Phyfe and Thomas Cooke; Savannah-made silver; portraits. Art museum wing contains 18th-20th century American & European paintings; prints; drawings; sculptures; porcelain; costumes; collection of works by Kahlil Gibran; decorative arts; sculpture; prints; drawings; graphic arts; African American art.
Research Fields: pertaining to the collections: American painting, decorative arts; architecture.
Facilities: Jepson Center for the Arts; educational facilities. Museum-related items for sale.
Activities: docent tours; guided tours; special guided tours in French, Italian & Spanish available upon request; lectures; films; gallery talks; concerts; school & docent programs; education programs; permanent & temporary exhibitions.
Publications: books, Christopher P.H. Murphy: A Retrospective; The Octagon Room: Classical Savannah: Fine & Decorative Arts 1800-1840; Looking Back: Art in Savannah 1900-1960; Nostrums for Fashionable Entertainments: Dining in Georgia, 1800-1850.
Hours & Admission Prices: Museum of Art: Sun. 1-5, Mon. 12-5, Tues.-Sat. 10-5. Jepson Center: Sun. 1-5, Mon. 12-5, Tues.-Wed. & Fri.-Sat. 10-5, Thurs. 10-8. Family $40, adults $20, senior citizens & military $18, K to college students $5; discounts to AAA, AAM & ICOM members; members no charge. NARM Reciprocal Museum Program. Closed holidays. &
Attendance: 177,611 (accurate)
Membership: Senior, Student, Artist & Teacher $35; Individual $45; Senior & Dual $65; Family $75; Friend $150; Donor $500; Patron $1,000; Grand Patron $1,500; Sponsor $2,500; Benefactor $5,000; President's Council $10,000.

THE UGA MARINE EXTENSION SERVICE - MARINE EDUCATION CENTER & AQUARIUM, 30 Ocean Science Circle, Savannah, GA 31411. Tel.: 912-598-2496. Fax: 912-598-2302.
E-mail: fish@uga.edu
Web Site: www.marex.uga.edu/aquarium
Founded: 1972.
Key Personnel: Dir., Randall Walker; Assoc. Dir., Anne Lindsay.
Governing Authority: Parent Institution: University of Georgia.

Institution Type/Description: Aquarium.

Collections: organism typical of various habitats found along the coast, the tidal creeks of salt marshes, the ocean beaches, & the open waters of the continental shelf; 17 tanks housing over 150 live animals representing 50 species of fish, turtles, & invertebrates found along the Georgia coast.

Activities: educational programs; workshops; public programs; summer camps.

Publications: quarterly newsletter, MECA Minute.

Hours & Admission Prices: Mon.-Fri. 9-4, Sat. 10-5. Adults 13 & over $6, seniors 55 & over, military and children 3-12 $3; children under 2 no charge. Season Passes available. &

Attendance: 15,000 (estimated)

WORMSLOE STATE HISTORIC SITE, 7601 Skidaway Rd., Savannah, GA 31406-6449. Tel.: 912-353-3023. Fax: 912-353-3023.

E-mail: wormsloe@bellsouth.net

Web Site: www.gastateparks.org

Founded: 1973.

Congressional District: 1

Key Personnel: Mgr., Chris Floyd; Interpretive Ranger, Lane Harris; Museum Shop Mgr., Kathy Morris.

Personnel Profile: Full-Time Paid 4; Part-Time Paid 2; Part-Time Volunteers 15.

Governing Authority: state. Parent Institution: Parks, Recreation & Historic Sites Div., Dept. of Natural Resources. Tax-exempt.

Institution Type/Description: Historic Site: ruins of 1739 fortified house.

Collections: archaeological artifacts from period c.1733-1850.

Research Fields: Georgia cultural & natural history.

Facilities: interpretive center; audiovisual room; living history area.

Activities: guided & self-guided touring walks; audiovisual; special programs.

Hours & Admission Prices: Tues.-Sun. 9-5; groups by appointment. Adults $5, senior citizens & tour groups of 15 or more $4.50, children 6-17 $3.50, youth groups of 15 or more $3; children 5 & under no charge. Closed New Year's Day; Thanksgiving; Christmas. &

Attendance: 100,000 (estimated)

Membership: Child $10; Adult $15; Family $30.

Senoia

THE BUGGY SHOP MUSEUM, 74 Main St., Senoia, GA 30276. Tel.: 770-253-1018.

E-mail: buggyshopsenoia@aol.com

Founded: 1995.

Institution Type/Description: History Museum.

Collections: local history & culture; early cars & buggies including Model T & Model A; musical instruments; jewelry; doctor's tools; early tractor; sewing & washing machines; Coca-Cola memorabilia; quilts; clothing; WWII books; Scripto pens; early cameras; plows & farm equipment.

Hours & Admission Prices: April-Oct. 3rd Sat.-Sun. each month 1-4. No charge; donations accepted.

Smyrna

SMYRNA MUSEUM OF HISTORY, 2861 Atlanta Rd., S.E., Smyrna, GA 30080-3657. Tel.: 770-435-7549 & 431-2858.

E-mail: smyrnamuse@aol.com

Web Site: smyrnahistory.org

Founded: 1985.

Congressional District: 13

Key Personnel: Dir., Harold Smith.

Governing Authority: city. Subsidiary Institution: Smyrna Historical Society and City of Smyrna.

Institution Type/Description: Historical Society Museum.

Collections: photographs; publications; historical & genealogical research materials.

Publications: bimonthly, Lives and Times; Historic Smyrna - An Illustrated History; Smyrna Memorial Cemetery.

Hours & Admission Prices: Tues.-Sat. 10-4; other times by appointment. No charge; donations accepted. &

Membership: Individual $20; Family $25; Donor $50; Sponsor $100.

Springfield

HISTORIC EFFINGHAM SOCIETY - THE EFFINGHAM MUSEUM AND LIVING HISTORY SITE, 1002 Pine St., Springfield, GA 31329. Mailing Address: P.O. Box 999, Springfield, GA 31329. Tel.: 912-754-2170.

E-mail: histeffingham@aol.com

Web Site: www.historiceffinghamsociety.org

Formerly: Effingham Old Jail House Museum

Founded: 1994.

Congressional District: 1

Key Personnel: Pres. & Dir., Norma Jean Morgan; Vice Pres., Walter Gnann; Museum Shop Mgr., Beverly Poole.

Personnel Profile: Part-Time Paid 2; Part-Time Volunteers 15.

Governing Authority: Parent Institution: Historic Effingham Society. Subsidiary Institution: Effingham Museum. Tax-exempt.

Institution Type/Description: History Museum & Living History Site.

Collections: local history & culture; period furnishings; personal artifacts; photographs; Native American artifacts; weapons. Historic Buildings: 1800s summer house; 1790s barn; outbuildings; railroad depot.

Facilities: library of local history & genealogy records.

Activities: research. Annual Events: Old Effingham Days Festival in April; Old Fashioned Christmas in December.

Publications: River to River, The History of Effingham County, GA; John Adam Truetlen, First Constitutional Governor of Georgia; Images of America: Effingham County; War Stories & School Day Incidents for the Children; Those Gallant Georgians Who Served in the War Between the States; Effingham County Cemeteries, 1734-2001; Growing Up on the Farm; The Elizabeth Waller Jones Story.

Hours & Admission Prices: Tues.-Fri. 9-3, Sat.-Sun. by appointment. Adults $5, seniors & military $4, students 6-18 $3; members & children under 6 no charge. &

Attendance: 750 (accurate)

Membership: Individual $20; Family $30; Business $150.

Statesboro

* **GEORGIA SOUTHERN UNIVERSITY MUSEUM, (M),** 2142 Southern Dr., Statesboro, GA 30458. Mailing Address: P.O. Box 8061, Statesboro, GA 30460-1000. Tel.: 912-478-5444. Fax: 912-478-0729.

E-mail: btharp@georgiasouthern.edu

Web Site: www.georgiasouthern.edu/museum

Founded: 1980.

Congressional District: 1

Key Personnel: Dir., Dr. Brent W. Tharp; Asst. Dir., Debbie Gleason; Administrative Sec., LaMaudice Holmes; Cur. Education, Ruby Ashley; Cur. Paleontology, Dr. Kathlyn Smith.

Personnel Profile: Full-Time Paid 4; Part-Time Paid 15; Part-Time Volunteers 5.

Governing Authority: university; nonprofit. Parent Institution: Georgia Southern University. Tax-exempt.

Institution Type/Description: University Museum.

Collections: natural, cultural & geological history; fossil skeletons; 26-foot Mosasaur; 20-foot Middle-Eocene archaeocete whale; fossil vertebrates of Coastal Georgia; Southeastern Indians; fossil oysters; fish; Wiss cutlery, scissors & shears; dolphin & Bryde's whale skeletons.

Major Exhibits: Life as a Dinosaur, 10/13-1/26/14; Victory From Within: The American Prisioner of War Experience (T), 2/10/14-6/1/14; The Mad Scientist's Laboratory, 6/16/14-1/25/15.

Research Fields: paleontology; geology; anthropology; technology.

Activities: guided tours; lectures; films; broadcast programs; organized education programs adults, children & undergraduate or graduate college students; participatory, temporary & traveling exhibitions; science & social studies kit program & teacher training.

Publications: annual brochure, Georgia Southern Museum; occasional papers.

Hours & Admission Prices: Tues.-Fri. 9-5, Sat.-Sun. 2-5. Admission $2; discounts to AAM, Coastal Museum Assn. & Ga. Assn. of Museums & Galleries members; GSU students, members and children 3 & under no charge. Closed university holidays. &

Attendance: 20,000 (estimated)

Membership: Individual $30; Family $50; Curator $100; Patron $250; Director $500; Monasaur Society $1,000.

Stockbridge

PANOLA MOUNTAIN STATE CONSERVATION PARK, (M), 2600 Hwy. 155, S.W., Stockbridge, GA 30281-5250. Tel.: 770-389-7801. Fax: 770-389-7925.

E-mail: panola_mountain@mail.dnr.state.ga.us

Web Site: www.gastateparks.org/info/panolamt

Founded: 1974.

Congressional District: 6

Key Personnel: Resource Mgr., Jody Rice.

Personnel Profile: Full-Time Paid 6; Part-Time Paid 2; Part-Time Volunteers 4.

Governing Authority: state. Parks, Recreation & Historic Sites Div., Dept. of Natural Resources. Tax-exempt.

Institution Type/Description: State Conservation Park.

Collections: granite monadnock; granite outcrop ecology; butterflies; moths; skippers; live reptiles; observation beehive; mounted mammal collection; live bats.
Research Fields: field ecology.
Facilities: visitor center; trails; scenic vistas; picnic area; playground.
Activities: teacher workshops; guided walks; audiovisual programs; environmental education programs.
Publications: newsletter.
Hours & Admission Prices: Park: Sept. 15-April 14 daily 7-6; April 15-Sept. 14 daily 7am-9pm. Interpretive Center: daily 8:30-5. $3 per vehicle; vans $20. Annual Pass: $30. &
Attendance: 201,000 (accurate)

Stone Mountain

ART STATION, 5384 Manor Dr., Stone Mountain, GA 30083-3067. Mailing Address: P.O. Box 1998, Stone Mountain, GA 30086. Tel.: 770-469-1105.
E-mail: info@artstation.org
Web Site: www.artstation.org
Founded: 1986.
Key Personnel: Founder, Pres. & Artistic Dir., David Thomas
Institution Type/Description: Art Museum.
Collections: photographs; paintings.
Facilities: classrooms; studios; theater; rental facilities. Museum-related items for sale.
Activities: programs; performances; classes.
Hours & Admission Prices: Tues.-Fri. 10-5, Sat. 10-3. No charge; donations accepted. &
Attendance: 5,000 (estimated)

GEORGIA'S STONE MOUNTAIN PARK, Hwy. 78, Stone Mountain, GA 30086. Mailing Address: P.O. Box 778, Stone Mountain, GA 30086-0778. Tel.: 770-498-5690. Fax: 770-498-5735. TDD: 770-498-5702 (available through switchboard).
E-mail: smpmarketing@stonemountainpark.com
Web Site: www.stonemountainpark.com
Founded: 1958.
Congressional District: 55
Key Personnel: Gen. Mgr., Gerald Rakestraw; Coord. Mktg., Mike Duchock.
Personnel Profile: Full-Time Paid 18; Interns 1.
Governing Authority: Parent Institution: Silver Dollar City. state. Tax-exempt: 501(c)(3).
Institution Type/Description: General Museum.
Collections: Indian artifacts; Civil War; preservation projects; Revolutionary War Colonial; early automobiles; musical instruments; 19th-century band organs; c.1830 Georgia-made furniture; 18th- & 19th-century English & American furniture. Historic Houses: 1790 Thornton House; 1845 Kingston House; 1845 Dickey House; 1845 Clayton House; 1869 grist mill; 1892 covered bridge.
Research Fields: Indian artifacts; geology; flora.
Facilities: botanical garden; nature trails.
Activities: school services; educational materials; guided tours; films; lectures. Park Sponsors: Yellow Daisy Festival and special events.
Hours & Admission Prices: Visitors Center: daily 8:30-5:30. One Day Adventure Pass: adults 12 & up $26, senior & military $23, children 3-11 $21. Closed Christmas. &
Attendance: 6,000,000 (accurate)

Tallapoosa

WEST GEORGIA MUSEUM OF TALLAPOOSA, 185 Mann St., Tallapoosa, GA 30176. Mailing Address: P.O. Box 725, Tallapoosa, GA 30176-0725. Tel.: 770-574-3125.
Founded: 1990.
Key Personnel: Dir., Mildred McElroy; Chm. (V), Bud Jones.
Personnel Profile: Full-Time Paid 1.
Governing Authority: Parent Institution: city of Tallapoosa.
Institution Type/Description: History Museum.
Collections: local history & culture; period furnishings; photographs; 30ft. Tyrannosaurus Rex; horse-drawn buggy; restored 1923 Ford peddlers wagon; personal artifacts.
Publications: museum newsletter.
Hours & Admission Prices: Tues.-Fri. 9-4, Sat. 9-5. Adults $2, children $1.
Attendance: 2,000 (estimated)
Membership: Museum League $25.

Thomasville

BIRDSONG NATURE CENTER, 2106 Meridian Rd., Thomasville, GA 31792-0417. Tel.: 229-377-4408; 800-953-BIRD (2473). Fax: 229-377-8723.
E-mail: birdsong@birdsongnaturecenter.org
Web Site: www.birdsongnaturecenter.org
Founded: 1986.
Congressional District: 2
Key Personnel: Exec. Dir., Kathleen D. Brady; Pres. (V), Robert Pando; Chm. (V), Bailey White.
Personnel Profile: Full-Time Paid 3; Part-Time Paid 8; Part-Time Volunteers 100.
Governing Authority: private; nonprofit organization. Tax-exempt: 501(c)(3).
Institution Type/Description: Nature Center: former plantation; farmhouse & 1880s barn.
Collections: widely varying habitats; plants & wildlife; field, pine & hardwood forests; ponds; swamps.
Research Fields: bluebird breeding study; plant & animal inventory; prescribed burning; ecological management.
Facilities: nature center. Books, birdfeeders & museum-related items for sale.
Activities: guided tours; lectures; formal education programs for adults, children & tailored programs for undergraduate & graduate students. Annual Events: Fall Festival; Winter Solstice Celebration; Butterfly Festival; Nature Writers Event; ongoing seasonal programs on weekends.
Publications: bimonthly newsletter & calendar of events, Birdsong Nature Center.
Hours & Admission Prices: Wed. & Fri.-Sat. 9-5, Sun. 1-5; groups by appointment. Adults $5, children $2.50; discount to groups; Association of Nature Center Administrators members no charge. Closed New Year's Day; Christmas.
Attendance: 3,700 (accurate)
Membership: Individual Friend $25; Family Friends $35; Friends of the Cardinal $50; Friends of the Chicadee $100; Friends of the Bluebird $500; Birdsong Naturalist $1,000; Corporate Friends $100-$1,000.

JACK HADLEY BLACK HISTORY MUSEUM, (M), 214 Alexander St., Thomasville, GA 31792-4996. Tel.: 229-226-5029. Fax: 229-226-5084.
E-mail: jackhadleyblackhistorymuseum@rose.net
Web Site: www.jackhadleyblackhistorymuseum.com
Governing Authority: Parent Institution: The Jack Hadley Black History Memorabilia, Inc. Tax-exempt: 501(c)(3).
Institution Type/Description: History Museum.
Collections: local African American history & culture; period furnishings; personal artifacts; photographs; documents; prints; posters; books; prints.
Hours & Admission Prices: Tues.-Sat. 10-5. Adults $5, children & college students $3. Closed New Year's Day; Thanksgiving; Christmas Eve & Day.

LAPHAM-PATTERSON HOUSE, 626 N. Dawson St., Thomasville, GA 31792-4449. Mailing Address: Thomas County Historical Society, 725 N. Dawson St., Thomasville, GA 31792. Tel.: 229-226-7664.
Founded: 1974.
Congressional District: 2
Key Personnel: C.E.O. & Cur., Cheryl Walters; Interpretive Ranger, Voncile Jones; Chm. (V), John Wood.
Personnel Profile: Full-Time Paid 2; Part-Time Paid 2; Part-Time Volunteers 14.
Institution Type/Description: Historic House: c.1884 Lapham-Patterson House, Victorian home.
Collections: furnishings; period rooms.
Research Fields: Victorian architecture.
Facilities: library.
Activities: guided tours; permanent & temporary exhibitions; community programs.
Publications: post cards; books.
Hours & Admission Prices: Temporarily closed. &
Attendance: 5,000 (accurate)
Membership: Individual $10; Family $15; Business $25; Corporate $250; Sustaining $500; Benefactor $1,000.

PEBBLE HILL PLANTATION, 1251 U.S. 319 S., Tallahassee Rd., Thomasville, GA 31792. Mailing Address: P.O. Box 830, Thomasville, GA 31799-0830. Tel.: 229-226-2344. Fax: 229-227-0095.
E-mail: wwhite@pebblehill.com
Web Site: www.pebblehill.com
Founded: 1983.

Congressional District: 2
Key Personnel: Chm. (V), Warren Bicknell; Gen. Mgr., Wallace Goodman.
Personnel Profile: Full-Time Paid 15; Part-Time Paid 12; Part-Time Volunteers 5; Interns 1.
Governing Authority: nonprofit organization. Parent Institution: Pebble Hill Foundation, Inc.
Institution Type/Description: Historic Site, Buildings & Art Gallery.
Collections: 18th- to 19th-century furniture; 18th- to 20th-century sporting art; photograph collections; decorative arts.
Major Exhibits: Works by Ogden Pleissner, 10/13-4/14.
Research Fields: support research; site & family research related to interpretation; decorative arts; English/American art collection; photograph collection.
Facilities: 3,600-vol. library of books. Gift items for sale.
Activities: guided tours; docent program; temporary exhibitions; special events; walking tour with informative signage.
Hours & Admission Prices: Tues.-Sat. 10-5, Sun. 1-5; last tour of the Main House begins at 4. Gate: adults $5, children 2-12 $2. House: adults $10, children 6-12 $4; children under 6 not admitted; discounts to groups of 18 & up with advance reservations. Closed New Year's Day; Thanksgiving; Christmas Eve & Day. &
Attendance: 20,000 (estimated)
Membership: Yearly membership $30.

POWER OF THE PAST AVIATION MUSEUM, 882 Airport Rd., Thomasville, GA 31757. Mailing Address: 432 Colton Ave, Thomasville, GA 31792. Tel.: 229-226-3010 & 403-4888.
E-mail: ljdekle@rose.net
Web Site: powerofthepast.org
Founded: 1975.
Congressional District: 11
Key Personnel: Dir., James Dekle; Chm. (V), John Dekle.
Personnel Profile: Part-Time Volunteers 7.
Institution Type/Description: Aviation History Museum.
Collections: aircraft including 1931 RNF WACO; 1928 Curtis Wright Travel Air 2000; airplane engines & propellers; signs; nostalgia material including Amelia Earhart & Charles Lindbergh
Activities: Annual Event: Thomasville Fly-In in October.
Hours & Admission Prices: Sun. 2-6; other times by appointment. No charge; donations accepted.
Attendance: 1,000 (estimated)

THOMAS COUNTY MUSEUM OF HISTORY, 725 N. Dawson St., Thomasville, GA 31792-4452. Tel.: 229-226-7664. Fax: 229-226-7466.
E-mail: history@rose.net
Web Site: www.thomascountyhistory.org
Governing Authority: Parent Institution: Thomas County Historical Society.
Institution Type/Description: History Museum.
Collections: local history & artifacts; period furnishings; personal artifacts; photographs.
Activities: special events.
Hours & Admission Prices: Sept. to mid-Aug. Mon.-Sat. 10-12 & 2-5; other times by appointment. Adults $5, students $1; discounts to groups; members no charge. Closed New Year's Eve & Day; Independence Day; Thanksgiving; Christmas Eve & Day.

THOMASVILLE CULTURAL CENTER, INC., 600 E. Washington St., Thomasville, GA 31792-4648. Mailing Address: P.O. Box 2177, Thomasville, GA 31799-2177. Tel.: 229-226-0588. Fax: 229-226-0599.
E-mail: info@thomasvilleculturalcenter.org
Web Site: www.thomasvilleculturalcenter.org
Founded: 1978.
Congressional District: 2
Key Personnel: C.E.O. & Exec. Dir., Tricia Collins; Pres. (V), Peggy Rich; Dir. Devel., Susan O'Neal; Dir. Finance, Kelly Swan; Dir. Education & Outreach, Mary Oglesby; Coord. Visual Arts, Amy Wheeler; Member Svcs. & Communications Coord., Rachael Fink; Dir. PWAF, Sharlene Celaya Cannon; Asst. Dir. PWAF, Holly Jarvis; Mgr. Bldg. & Grounds, Herbert Brinson; Administrative Asst., Casie Vela.
Personnel Profile: Full-Time Paid 10; Part-Time Paid 1; Part-Time Volunteers 197; Interns 1.
Governing Authority: nonprofit organization. Tax-exempt: 501(c)(3) and 170(b)(1)(A).
Institution Type/Description: Cultural Center: housed in 1915 East Side School.

Collections: American and European paintings; wildlife art; over 800 ceramic objects; sculpture.
Facilities: 550-seat auditorium; instructional studio; educational facilities. Artwork on consignment & gifts of local interest for sale.
Activities: temporary & permanent exhibits; concerts; lectures; music & visual art classes for children & adults; guided tours; annual arts festivals.
Publications: newsletter, Plantation Wildlife; Arts Festival Magazine.
Hours & Admission Prices: Office: Tues.-Fri. 9-5. Gallery: Tues.-Fri. 9-5, Sat. 1-5. No charge; donations accepted. Closed New Year's Eve & Day; Martin Luther King Jr. Day; Easter; Memorial Day; Independence Day; Labor Day; Thanksgiving; Christmas Eve & Day. &
Attendance: 63,781 (estimated)
Membership: Individual $25-$50; Household $50-$100; Business $500; Artist's Guild (additional $10 at any level).

Thomson

HICKORY HILL, 502 Hickory Hill Dr., Thomson, GA 30824-7655. Tel.: 706-595-7777; 877-595-9777 (toll free). Fax: 706-595-7177. Facebook: Hickory Hill House Museum.
Web Site: www.hickory-hill.org
Founded: 2004.
Congressional District: 9
Key Personnel: Pres., Tad Brown; Chm. (V), Byron Attridge; Cur., Michelle Zupan.
Personnel Profile: Full-Time Paid 2; Part-Time Volunteers 1; Interns 1.
Operating Expenses: 285,000
Operating Income: 285,000
Governing Authority: private; nonprofit organization. Parent Institution: The Watson Brown Foundation. Tax-exempt: 501(c)(3).
Institution Type/Description: Historic House Museum: housed in the home of Thomas E. Watson.
Collections: personal artifacts; clothing; furniture; documents; photographs; 12 historic buildings; gardens.
Research Fields: Tom Watson's publishing plant, The Jeffersonian; Thomas Watson; populism.
Facilities: library; 256 acres; walking trails; gardens. Museum-related items for sale.
Activities: formal education programs; guided tours; scholar's forums; public history event. Annual Events: Tom Watson Watermelon Festival; Dig History Archaeology Camp; Sticks & Stones.
Publications: quarterly newsletter, Legacy; biannual, reprints of Tom Watson's books.
Hours & Admission Prices: Mon.-Fri. 10-5, Sat. by appointment. Adults $3, senior citizens $2, children $1; discounts AASLH, AAM, ICOM members & groups. Closed major holidays; Hickory Hill Forum. &
Attendance: 4,800 (estimated)

MAC ON MAIN ART GALLERY AND STUDIO, 107 Main St., Thomson, GA 30824. Tel.: 706-699-1804.
Institution Type/Description: Art Gallery.
Collections: works by local & regional artists.
Hours & Admission Prices: Tues.-Thurs. 11-5, Fri. 10-4.

MCDUFFIE MUSEUM, 121 Main St., Thomson, GA 30824. Tel.: 706-595-9923.
E-mail: info@mcduffiemuseum.com
Web Site: www.mcduffiemuseum.com
Institution Type/Description: History Museum.
Collections: local history & culture; period furnishings; personal artifacts; photographs.
Hours & Admission Prices: Tues.-Sat. 12-5.

Tifton

GEORGIA MUSEUM OF AGRICULTURE & HISTORIC VILLAGE AT ABAC, Interstate 75 Exit 63B at 8th St., Tifton, GA 31793. Mailing Address: 1392 Whiddon Mill Rd., Tifton, GA 31793-7800. Tel.: 229-391-5205; 800-767-1875. Fax: 229-391-5201. Facebook: ABAC GMA.
E-mail: gboone@abac.edu
Web Site: www.abac.edu/museum
Formerly: Agrirama, Georgia's Living History Museum & Village
Founded: 1972.
Congressional District: 2
Key Personnel: Dir., Paul Willis; Mktg., Garrett Boone; Cur., Dr. Lisa Lishman; Maintenance & Restoration, David King.
Personnel Profile: Full-Time Paid 8; Part-Time Paid 12; Part-Time Volunteers 4.

Governing Authority: state. Parent Institution: University System at Georgia. Subsidiary Institution: Abraham Baldwin Agricultural College. Tax-exempt.
Institution Type/Description: 19th-Century Living History Museum.
Collections: agriculture equipment; printing & typesetting equipment; furniture & furnishings of the period; naval stores implements; medical & dental equipment. Historic Buildings: 1896 farmhouse; 1886 log cabin; 1895 one-room school; 1890 sawmill; 1879 grist mill; 1885 printing office; 1899 railroad depot; 1887 doctor's office; 1882 church; 1879 commissary. Historic Reconstructions: 1890 turpentine still & cooper's shed; c.1890 cotton gin; c.1890 drug store; c.1900 variety works; 1877 log house; 1887 Victorian house.
Major Exhibits: Back Roads of GA - Photography, 10/13-1/14.
Research Fields: 18th, 19th & early 20th century rural & agricultural history.
Facilities: 250-seat conference facility; full service kitchen; snack bar; picnic area; outdoor theatre. Gift items for sale.
Activities: tours; workshops; expositions; permanent & temporary exhibits; entertainment; fishing lake.
Publications: monthly newsletter, ABAC Focus; historic newspaper, Georgia Recorder.
Hours & Admission Prices: Tues.-Sat. 9-4:30. Tues.-Fri. adults $7, senior citizens $6, children 5-16 $4; Sat. adults $10, senior citizens $8, children 5-16 $5; discounts for groups of 20 or more; children under 5 no charge. Closed New Year's Day; Thanksgiving; Christmas week. &
Attendance: 40,000 (accurate)
Membership: Individual $25; Family $50; Family Friend or Contributing $100; Family Patron $150; Family Donor or Sustaining $250; Life $1,000. Business: Associate $500; Supporter $1,000; Partner $2,500; Benefactor $5,000.

Toccoa

CURRAHEE MILITARY MUSEUM, 160 N. Alexander St., Toccoa, GA 30577. Tel.: 706-282-5055.
Institution Type/Description: Military History Museum.
Collections: military history & artifacts; personal artifacts; photographs; documents; Civil War, WWI & WWII artifacts.
Activities: special events.
Hours & Admission Prices: Mon.-Sat. 10-4, Sun. 1-4. Closed New Year's Eve & Day; Christmas Eve & Day.

Tunnel Hill

WESTERN & ATLANTIC TUNNEL AND MUSEUM, 215 Clisby Austin Rd., Tunnel Hill, GA 30755. Mailing Address: P.O. Box 6177, Dalton, GA 30722-6177. Tel.: 706-876-1571.
E-mail: tscalf@visitdaltonga.com
Formerly: Tunnel Hill Heritage Center Museum & Historic W & A Railroad Tunnel
Founded: 2003.
Congressional District: 9
Key Personnel: Dir., Thomas Scalf; Pres. (V), Janet Cochran.
Personnel Profile: Full-Time Paid 2; Part-Time Paid 4.
Volunteer Hours: 150
Governing Authority: Parent Institution: Tunnel Hill Historical Foundation. Subsidiary Dalton Convention & Visitors Bureau. Tax-exempt.
Institution Type/Description: History Museum.
Collections: history & heritage of Tunnel Hill; Native American; early settlers; 1850 railroad restoration; Civil War artifacts; personal artifacts; period furnishings.
Activities: school groups; tours.
Hours & Admission Prices: Mon.-Sat. 9-5. Adults $8, children under 12 $5; groups of 20 or more $5. Closed New Year's Day; Thanksgiving; Christmas Eve. &
Attendance: 7,000 (accurate)
Membership: Individual $20; Family $25.

Tybee Island

TYBEE ISLAND LIGHT STATION AND TYBEE MUSEUM, 30 Meddin Dr., Tybee Island, GA 31328-9733. Mailing Address: P.O. Box 366, Tybee Island, GA 31328-0366. Tel.: 912-786-5801. Fax: 912-786-6538.
E-mail: tybeelh@bellsouth.net
Web Site: www.tybeelighthouse.org
Formerly: Tybee Museum and Lighthouse
Founded: 1960.
Congressional District: 1
Key Personnel: Dir., Cullen Chambers; Pres. (V), Kevin Sofa; Dir. Operations, Sarah Jones; Museum Shop Mgr., Candy Carter.
Personnel Profile: Full-Time Paid 13; Part-Time Paid 3; Part-Time Volunteers 56.

Governing Authority: board of directors. Parent Institution: Tybee Island Historical Society. Affiliated with Tybee Museum Association. Tax-exempt.
Institution Type/Description: History Museum: housed in old Spanish-American War Coastal Defense Battery; c.1867 Tybee Island lighthouse & cottages.
Collections: guns & pistols; period dolls; railroad memorabilia; military artifacts; lighthouse artifacts.
Research Fields: U.S. coastal defense 1897-1946; U.S. Lighthouse Service 1789-1939.
Activities: tours; lectures by appointment only.
Publications: books, Historic Tybee Island, Tybee Island Recipe Book, Fort Screven 1897-1945.
Hours & Admission Prices: Wed.-Mon. 9-5:30. Adults $9, seniors, military & children 6-17 $7; children under 6 no charge. Last ticket sold at 4:30. Closed New Year's Day; St. Patrick's Day; Thanksgiving; Christmas.
Attendance: 100,000 (accurate)
Membership: Seniors & Students $35; Single $45; Couple $60; Household $75; 2nd Asst. Keepers $125; 1st Asst. Keepers $150; Head Keepers $250; Business $250; Lifetime $1,000.

Valdosta

THE CRESCENT, VALDOSTA GARDEN CENTER, INC., 904 N. Patterson St., Valdosta, GA 31601-4531. Mailing Address: P.O. Box 2423, Valdosta, GA 31604-2423. Tel.: 229-244-6747. Fax: 912-242-1005.
Founded: 1951.
Congressional District: 2
Key Personnel: Pres., Betty Becton.
Governing Authority: nonprofit. Parent Institution: The Garden Center, Inc. Tax-exempt.
Institution Type/Description: Nature Center & Historic House: 1898 mansion, home of former U.S. Sen. William S. West.
Collections: period furniture; Day Lily, perennial & annuals gardens.
Research Fields: restoration; landscaping & flower arranging.
Facilities: 200-vol. library of books pertaining to gardening, landscaping & flower arranging available to view on premises; reading room; botanical garden; garden; field research station; 200-seat auditorium. Museum-related items for sale.
Activities: guided tours; lectures; films; study clubs; formally organized education programs. Museum Sponsors: Christmas Open House; Flower Show; Antique Shows.
Publications: yearbook.
Hours & Admission Prices: Mon.-Fri. 2-5; other times by appointment when not rented. No charge; donations accepted. &
Attendance: 4,286 (accurate)

LOWNDES COUNTY HISTORICAL SOCIETY AND MUSEUM, (M), 305 W. Central Ave., Valdosta, GA 31601-5404. Mailing Address: P.O. Box 56, Valdosta, GA 31603-0056. Tel.: 229-247-4780. Fax: 229-247-2840.
E-mail: history@valdostamuseum.com
Web Site: www.valdostamuseum.org
Founded: 1967.
Congressional District: 1
Key Personnel: Exec. Dir., Donald O. Davis; Pres., Patsy Giles; Financial Dir., Redden Hart.
Personnel Profile: Full-Time Paid 1; Part-Time Paid 3; Part-Time Volunteers 10; Interns 1.
Governing Authority: private; nonprofit organization. Tax-exempt: 501(c)(3).
Institution Type/Description: History Museum: housed in 1913 Carnegie Library.
Collections: artifacts, documents & photographs concerning Lowndes County Georgia; Doc Holliday exhibit.
Facilities: 500-vol. library of city directories, phone books, genealogical research, History of Georgia & Counties; 3,000 sq. ft. exhibit space.
Activities: formal education programs for children; guided tours; lectures. Museum Sponsors: History 100 Dinner, annual fundraiser.
Publications: monthly newsletter, Yesterday & Today; Doc Holliday Book; Way Back When Vol. I, II, III; Pines and Pioneers, A History of Lowndes County, Georgia 1825-1900.
Hours & Admission Prices: Mon.-Fri. 10-5, Sat. 10-2. No charge; donations accepted. Closed New Year's Day; Sat. before Easter; Memorial Day; Independence Day; Labor Day; Thanksgiving; Christmas to New Year's Day. &
Attendance: 26,000 (accurate)
Membership: Student $10; Annual $25; Family $30; Organization & Business $50; Contributing $100; Patron $250; Life $1,000.

VALDOSTA STATE UNIVERSITY FINE ARTS GALLERY, Fine Arts Bldg., Rm 107, 1500 N. Patterson St., Valdosta, GA 31698. Tel.: 229-333-5835. Fax: 229-259-5121.
E-mail: apearce@valdosta.edu
Web Site: valdosta.edu/art
Founded: 1906.
Congressional District: 8
Key Personnel: Gallery Dir., Julie Bowland.
Personnel Profile: Full-Time Paid 1; Part-Time Volunteers 25.
Governing Authority: board of regents; college. Affiliated with Valdosta State College. Tax-exempt.
Institution Type/Description: Art Gallery.
Collections: The Lamar Dodd Collections.
Facilities: library; planetarium; reading room; auditorium; theater; classrooms.
Activities: guided tours; lectures; films; gallery talks; concerts; docent program or council; formally organized education programs for undergraduate college students.
Publications: exhibition catalogs; posters; postcards.
Hours & Admission Prices: Mon.-Thurs. 10-4, Fri. 10-3, Sat.-Sun, evenings open for performing events. No charge; donations accepted. Closed school holidays. &
Attendance: 10,000

Vidalia

ALTAMA MUSEUM OF ART & HISTORY, 611 Jackson St., Vidalia, GA 30474-4721. Mailing Address: P.O. Box 33, Vidalia, GA 30475-0033. Tel.: 912-537-1911. Facebook: 280 Art Guild.
E-mail: altama@bellsouth.net
Key Personnel: Dir., Rebecca Smelser.
Personnel Profile: Part-Time Paid 1.
Institution Type/Description: Art Museum: housed in 1911 Brazell House.
Collections: porcelain; prints; wood sculptures; Girl Scout memorabilia; paintings; local history; Staffordshire porcelain; antique furniture; vintage clothing.
Activities: art celebration for kids; workshops.
Hours & Admission Prices: Jan.-June & Sept.-Nov. Mon.-Tues. & Thurs.-Fri. 11-4. No charge; donations accepted. &
Membership: Individual $20; Family $30; Sponsor $50; Patron $100; Benefactor $250; Angel $500.

Warm Springs

ART IN MOTION VINTAGE MOTORCYCLE MUSEUM, 78 Lil Sturgis St., Warm Springs, GA 31830. Mailing Address: 31 Red Bud Trail, Newman, GA 30263. Tel.: 770-502-0028; 678-296-3326.
E-mail: presto434343@yahoo.com
Web Site: prestonopportunities.com
Founded: 2008.
Key Personnel: Dir., Preston Evans.
Personnel Profile: Part-Time Paid 3.
Governing Authority: Subsidiary Institution: Foto Excursion Museum and Photography Studio, Warm Springs, GA.
Institution Type/Description: Motorcycle Museum.
Collections: motorcycles; jukeboxes; slot machines; early advertising.
Activities: special events.
Hours & Admission Prices: Fri.-Sat. 11-6, Sun. 1-6; other times by appointment. Adults $5-$12.
Attendance: 5,000 (estimated)

ROOSEVELT'S LITTLE WHITE HOUSE STATE HISTORIC SITE, 401 Little White House Rd., Warm Springs, GA 31830-2157. Tel.: 706-655-5870. Fax: 706-655-5872.
Web Site: www.gastateparks.org
Founded: 1946.
Congressional District: 10
Key Personnel: Site Mgr., Robin Glass; Asst. Mgr., Mary F. Thrash; Museum Shop Mgr., Diane Crane.
Personnel Profile: Full-Time Paid 19; Part-Time Paid 1; Part-Time Volunteers 3.
Governing Authority: state. Parent Institution: Georgia Department of Natural Resources. Tax-exempt.
Institution Type/Description: Historic Buildings Museum: 1932 Georgia home of Pres. Roosevelt, where he died April 12, 1945.
Collections: personal items; gifts to Pres. Roosevelt from individuals, states and foreign countries; personal & official correspondence.
Research Fields: geology; prehistory & history of Warm Springs; historical significance of Franklin D. Roosevelt.

Facilities: research library; theater. Museum-related gifts for sale.
Activities: tours; films; permanent exhibitions.
Hours & Admission Prices: Daily 9-4:45. Adults $7, seniors over 62 $6, children 6-18 $4; discount to groups of 15 or more. Closed New Year's Day; Thanksgiving; Christmas. &
Attendance: 100,000 (estimated)

Warner Robins

***   **MUSEUM OF AVIATION AT ROBINS AIR FORCE BASE, GA,** 1942 Heritage Blvd., Robins AFB, Warner Robins, GA 31099. Tel.: 478-926-6870. Fax: 478-926-5566.
E-mail: kenneth.emery@robins.af.mil
Web Site: www.museumofaviation.org
Founded: 1984.
Key Personnel: Dir., Kenneth Emery; Pres. & C.O.O. Museum Foundation, Pat Bartness; Chm. Bd. Directors Museum Foundation, Carolyn Crayton; Cur., Mike Rowland; Retail Shops Mgr., Sarah Parker.
Personnel Profile: Full-Time Paid 28; Part-Time Paid 29; Part-Time Volunteers 83.
Governing Authority: nonprofit organization. Parent Institution: U.S. Air Force. Subsidiary Institution: Robins Air Force Base, GA. Tax-exempt: 501(c)(3).
Institution Type/Description: Aviation Museum.
Collections: World War II-present, aviation memorabilia; over 100 historic aircraft; missiles.
Facilities: library; 250-seat auditorium; 40-seat cafeteria; 200,000 sq. ft. exhibit space; 51 acre site. Aviation-related items for sale.
Activities: docent program; films; guided tours; Georgia Youth Science & Technology Center; 6 axis motion simulator. Museum Sponsors: Young Astronaut Day; Georgia Aviation Hall of Fame induction.
Publications: History of Robins Air Force Base; book, God is My Co-Pilot - 1943.
Hours & Admission Prices: Daily 9-5. No charge; donations accepted. Closed New Year's Day; Easter; Thanksgiving; Christmas. &
Attendance: 540,000 (estimated)
Membership: Students 18 & under $10; Individual $35; Silver Eagle Individual $100; Silver Eagle Club & Business $150; Gold Eagle Individual $500; Sustaining $1,000.

Warrenton

MUSEUM OF CULTURAL HERITAGE, 46 S. Norwood St., Warrenton, GA 30828. Tel.: 706-465-9604.
Institution Type/Description: History Museum: housed in the former East Warrenton Depot.
Collections: local history & culture; period furnishings; photographs; personal artifacts.
Hours & Admission Prices: Mon.-Fri. 10-4. No charge.

Warthen

WARTHEN OLD JAIL, 7616 Hwy. 15 N., Warthen, GA 31094. Tel.: 478-552-3288.
Web Site: www.washingtoncountyga.com
Institution Type/Description: Historic Building: housed in a former wooden jail. Listed on the National Register of Historic Places.
Collections: local history & culture; period furnishings.
Hours & Admission Prices: Mon.-Fri. 9-5. No charge.

Washington

CALLAWAY PLANTATION, 2160 Lexington Rd., Washington, GA 30673-3310. Tel.: 706-678-7060. Fax: 706-678-7060.
E-mail: callaway@washingtongeorgia.net
Founded: 1960.
Key Personnel: Dir., David Van Hart; Museum Shop Mgr., Olivia Jackson.
Personnel Profile: Full-Time Paid 2; Part-Time Paid 2; Part-Time Volunteers 5.
Governing Authority: Parent Institution: City of Washington, GA. Tax-exempt.
Institution Type/Description: Historic House Museum.
Collections: local history & culture; period furnishings; personal artifacts; photographs; farming equipment; historic buildings.
Activities: special events. Annual Events: Spring Tour in April; Mule Day Festival in October; Christmas Tour in December.
Hours & Admission Prices: Tues.-Sat. 10-5, Sun. by appointment. Call for admission prices. Closed major holidays. &
Attendance: 6,000 (estimated)

ROBERT TOOMBS HOUSE, 216 E. Robert Toombs Ave., Washington, GA 30673-2037. Tel.: 706-678-2226.
E-mail: roberttoombshouse16@live.com
Web Site: www.gastateparks.org
Founded: 1982.
Congressional District: 10
Key Personnel: Site Mgr., Marcia Campbell.
Personnel Profile: Full-Time Paid 1; Part-Time Volunteers 3.
Governing Authority: Parent Institution: Wilkes County Board of Commissioners. Tax-exempt.
Institution Type/Description: Historic House Museum.
Collections: historic house; outbuildings; books; furniture & furnishings from the period c.1840-1900; Toombs family artifacts.
Research Fields: life of Toombs; Civil War; politics; Reconstruction Era after the Civil War.
Activities: tours; school programs; living history programs.
Hours & Admission Prices: Tues.-Sat. 10-4. Adults 13 & over $5, children 6-12 $3, children 3-5 $1; discount to groups; children under 3 no charge. Closed New Year's Day; Thanksgiving; Christmas.
Attendance: 2,600 (estimated)

WASHINGTON HISTORICAL MUSEUM, 308 E. Robert Toombs Ave., Washington, GA 30673-2038. Tel.: 706-678-2105. Fax: 706-678-3752.
E-mail: historical@washingtonwilkes.org
Founded: 1959.
Congressional District: 10
Key Personnel: Coord., Stephanie Macchia.
Personnel Profile: Full-Time Paid 1; Part-Time Paid 4; Part-Time Volunteers 15.
Governing Authority: municipal. Parent Institution: City of Washington, GA. Subsidiary Institution: Washington-Wilkes Historical Foundation. Tax-exempt.
Institution Type/Description: Historical Museum: 1836 Barnett-Slaton House, built by Albert Gallatin Semmes.
Collections: Confederate & Wilkes County history.
Research Fields: local Civil War & Wilkes County history.
Facilities: Music, publications & postcards for sale.
Activities: guided tours; permanent exhibitions.
Publications: books relating to Washington-Wilkes.
Hours & Admission Prices: Tues.-Sat. 10-5; groups by appointment. Adults $3; discount to groups of 15 or more; children under 12 no charge. Closed major holidays. &

Attendance: 2,400 (accurate)

Watkinsville

EAGLE TAVERN MUSEUM, 26 N. Main St., Watkinsville, GA 30677. Mailing Address: 21 N. Main St., Watkinsville, GA 30677-2064. Tel.: 706-769-5197. Fax: 706-310-1682.
E-mail: pholcomb@oconee.ga.us
Web Site: www.visitoconee.com
Founded: 1966.
Congressional District: 13
Key Personnel: Dir. Tourism & Cur., Peggy Holcomb.
Personnel Profile: Full-Time Paid 1; Part-Time Volunteers 5; Interns 1.
Governing Authority: county. Oconee County Bd. of Commissioners. Tax-exempt.
Institution Type/Description: Historic Building: tavern stage coach stop.
Collections: late 18th-19th century furnishings.
Activities: guided tours; permanent exhibitions.
Publications: brochures pertaining to the state of Georgia & Eagle Tavern.
Hours & Admission Prices: Tours: Tues.-Fri. 10-4. Adults & seniors $2, children 6-15 $1; children under 6 no charge. &
Attendance: 7,000 (accurate)

Waycross

OKEFENOKEE HERITAGE CENTER, 1460 N. Augusta Ave., Waycross, GA 31503-4954. Tel.: 912-285-4260. Fax: 912-283-2858. Facebook: Okenfenokee Heritage Center.
E-mail: sbean@mediastreamus.net
Web Site: www.okefenokeeheritagecenter.com
Founded: 1975.
Congressional District: 8
Key Personnel: Dir. & Cur., Steven Bean; Museum Shop Mgr., Betty Callahan.
Personnel Profile: Full-Time Paid 1; Part-Time Paid 2.
Governing Authority: nonprofit organization. Tax-exempt: 501(c)(3).
Institution Type/Description: History Museum/Art Center.

Collections: regional art & history of Okefenokee Swamp & areas surrounding the swamp; renovated 1912 train, including steam engine & tender, 1 baggage car, 1 baggage/postal car, passenger car & caboose; Native American artifacts from 1200 AD to 1840. Historic Buildings: 1900s Depot; c.1832 General Thomas Hilliard House; late 1800s print shop.
Research Fields: local history; arts; oral history.
Facilities: classrooms; conference room. Gift items for sale.
Activities: tours; art classes; art shows; performances; special events.
Publications: newsletter; exhibition catalogues; brochures.
Hours & Admission Prices: Tues.-Sat. 9-2. Adults $7, children 6-18 $5; children 5 & under no charge. Closed New Year's Day; Thanksgiving; Christmas. &
Attendance: 13,000 (estimated)
Membership: Individual $30; Family $50; Heritage Club $200; Business $300; Corporate $500.

SOUTHERN FOREST WORLD, 1440 N. Augusta Ave., Waycross, GA 31503-4954. Tel.: 912-285-4056. Fax: 912-285-4056.
E-mail: southernforestworld@gmail.com
Web Site: www.southernforestworld.org
Founded: 1981.
Congressional District: 8
Key Personnel: Exec. Dir., Reba Smith, Ed.D.; Pres. (V), Barry Deas; Museum Shop Mgr., Dan King.
Personnel Profile: Full-Time Paid 1; Part-Time Paid 1; Part-Time Volunteers 10.
Governing Authority: nonprofit organization. Tax-exempt: 501(c)(3).
Institution Type/Description: Forestry Museum.
Collections: artifacts & specimens relating to the history & development of forestry in the South, including a 38' tall model of a Loblolly Pine; a cross-section of the nation's largest Slash Pine; 1905 steam powered logging locomotive; naval stores tools & cups; giant Cypress; working scale model of a turpentine still; 22' fire tower; 1900 logging cart; NASA astronaut space uniform.
Research Fields: forestry.
Facilities: permanent exhibitions. Gift items for sale.
Activities: youth programs; tree plantings; association seminars.
Hours & Admission Prices: Tues.-Fri. 9-5. Groups call for appointment. Adults 19 & above $5, students 6-17 $4; children 5 & under & members no charge. Closed New Year's Day; Easter; Independence Day; Labor Day, Thanksgiving; Christmas. &
Attendance: 6,000 (estimated)
Membership: Individual $20; Family $30; Donor $50; Sponsor $100; Sustaining $250; Friend $500; Patron $1,000; Business & Corporate $100-$10,000.

Waynesboro

BURKE COUNTY MUSEUM, 536 Liberty St., Waynesboro, GA 30830. Mailing Address: 628 Myrick St., Waynesboro, GA 30830. Tel.: 706-437-9557.
Key Personnel: Cur., Robert L. Hammond
Institution Type/Description: History Museum.
Collections: county history & culture; photographs; personal artifacts.
Hours & Admission Prices: Mon.-Fri. 8-4, Sat.-Sun. & holidays by appointment.

Williamson

CANDLER FIELD MUSEUM - PEACH STATE AERODROME, 349 Jonathan's Roost Rd., Williamson, GA 30292. Tel.: 770-467-9490.
Key Personnel: Owner, Ron Alexander.
Governing Authority: nonprofit organization.
Institution Type/Description: Airport History Museum.
Collections: airport history; airplanes; clothing; cars; photographs.
Activities: special events. Annual Event: Vintage Day.
Hours & Admission Prices: Tues.-Sun.

Winder

BARROW COUNTY HISTORICAL SOCIETY AND MUSEUM, 94 E. Athens St., Winder, GA 30680. Mailing Address: P.O. Box 277, Winder, GA 30680. Tel.: 770-307-1183.
Institution Type/Description: Historical Society Museum: housed in the former Barrow County jail; built in 1915. Listed on the National Register of Historic Places.
Collections: local history; period artifacts; hanging tower; 3 jail cells; Senator Richard B. Russell's life & career; photographs; WW I & II uniforms; early barber chair; corn sheller; early tools; period school & church furnishings.

Facilities: archives.
Activities: research.
Hours & Admission Prices: Mon.-Fri. 1-4.

FORT YARGO STATE PARK, 210 S. Broad St., Winder, GA 30680-2059. Tel.: 770-867-3489. Fax: 770-867-7517.
E-mail: fort_yargo_park@dnr.state.ga.us
Web Site: www.gastateparks.org
Founded: 1954.
Congressional District: 9
Key Personnel: Supt., Eric Bentley.
Governing Authority: state. Administered by the Parks, Recreation and Historic Sites Div., Georgia Dept. of Natural Resources, Floyd Tower E., Ste. 1352, 2 Martin Luther King, Jr. Dr., Atlanta, GA 30334. Tax-exempt.
Institution Type/Description: Historic Building: restored blockhouse used during the Creek Indian Wars.
Collections: Historic House: 1792 Timber Blockhouse.
Research Fields: Georgia frontier.
Facilities: picnic & camping facilities.
Activities: conducted tours.
Hours & Admission Prices: Daily 7am-10pm. Old Fort Tours: by appointment. Park Pass: daily $5 per vehicle; Annual $50 per vehicle, senior citizens $25.
Attendance: 400,000

Winterville

CARTER-COILE COUNTRY DOCTORS MUSEUM, 111 Marigold Ln., Winterville, GA 30683. Mailing Address: P.O. Box 306, Winterville, GA 30683-0306. Tel.: 706-742-8600. Fax: 706-742-5476.
E-mail: winterville@charter.net
Web Site: www.cityofwinterville.com/doctor_museum.html
Founded: 1971.
Congressional District: 11
Key Personnel: Municipal Clerk, Wendy Martin.
Governing Authority: nonprofit organization. Parent Institution: City of Winterville. Tax-exempt.
Institution Type/Description: Medical Museum: housed in 1874 frame building used as an office by Dr. Warren Carter & Dr. Frank Coile.
Collections: medical equipment, furnishings, instruments, books, & special anatomy exhibits received from medical & other health personnel chiefly used in Clarke, GA & adjacent counties in the late 1800s & early 1900s & from Loree Florence, first women graduate of University of Georgia Medical School, 1926.
Research Fields: Winterville area physicians & their families; architecture.
Activities: guided tours; permanent exhibitions.
Hours & Admission Prices: Temporary closed. Historians & researchers by appointment. Call for special event openings.

Woodbine

WOODBINE INTERNATIONAL FIRE MUSEUM, 100 Bedell Ave., Woodbine, GA 31569-0058. Tel.: 912-576-5351.
Founded: 1991.
Key Personnel: Dir., Jodie G. Briese; Cur., Robert J. Briese.
Personnel Profile: Part-Time Volunteers 2.
Governing Authority: private; nonprofit organization.
Institution Type/Description: General Museum.
Collections: fire fighting equipment from Revolution to present from all over the world.
Research Fields: fire history & disasters.
Facilities: 300-vol. library.
Activities: guided tours; lectures.
Hours & Admission Prices: Mon.-Fri. 9:30-4, Sun. 11:30-4. No charge; donations accepted. Closed most major holidays.
Attendance: 2,700 (estimated)

Woodstock

AIR ACRES MUSEUM, 2115 Jep Wheeler Rd., Woodstock, GA 30188-6520. Tel.: 770-517-6090.
Institution Type/Description: Military Aircraft Museum.
Collections: period military aircraft.
Activities: Annual Event: Low Country Broil.
Hours & Admission Prices: Tues.-Sat.

HAWAII

(94 listings)

Captain Cook

AMY B.H. GREENWELL ETHNOBOTANICAL GARDEN, 82-6160 Mamalahoa Hwy., Captain Cook, HI 96704. Mailing Address: P.O. Box 1053, Captain Cook, HI 96704-1053. Tel.: 808-323-3318. Fax: 808-323-2394.
Web Site: www.bishopmuseum.org/greenwell
Founded: 1974.
Congressional District: 2
Key Personnel: Mgr., Peter Van Dyke; Cur., Brian Kiyabu.
Personnel Profile: Full-Time Paid 5; Part-Time Paid 1; Part-Time Volunteers 20.
Governing Authority: private; nonprofit organization. Parent Institution: Bishop Museum, Honolulu, HI. Tax-exempt: 501(c)(3).
Institution Type/Description: Botanical Garden.
Collections: native plants; Hawaiian crops.
Facilities: library; botanical garden; archaeology site. Museum-related items for sale.
Activities: docent program; formal education programs; guided tours; hobby workshops; lectures; school outreach program. Annual Events: Horticultural Festival; Seed Exchange; Arbor Day.
Hours & Admission Prices: Tues.-Sun. 9-4. Adults $7, members $5; discounts to AAM members. Closed New Year's Day; Presidents' Day; Memorial Day; King Kamehameha Day; Independence Day; Labor Day; Thanksgiving & day after; Christmas.
Attendance: 12,040 (estimated)
Membership: Garden Friend $30; Garden Family $45.

KONA COFFEE LIVING HISTORY FARM, 82-6199 Mamalahoa Hwy., Captain Cook, HI 96704. Mailing Address: P.O. Box 398, Captain Cook, HI 96704. Tel.: 808-323-2006. Fax: 808-323-2398.
E-mail: coffeefarm@konahistorical.org
Web Site: www.konahistorical.org
Key Personnel: Dir., Joy Holland.
Governing Authority: private; nonprofit organization. Tax-exempt: 501(c)(3).
Institution Type/Description: Living History Farm: established in 1913. Listed on the National Register of Historic Places.
Collections: local history & culture.
Hours & Admission Prices: Mon.-Fri. 10-2; last tour begins at 1pm. Adults $15, senior citizens $13, children $5; members no charge.
Membership: Individual $35; Family $55; Partner $100; Business $250; Associate $500; Patron $1,000.

Ewa

HAWAIIAN RAILWAY SOCIETY, 91-1001 Renton Rd., Ewa, HI 96706. Mailing Address: P.O. Box 60369, Ewa, HI 96706. Tel.: 808-681-5461. Fax: 808-681-4860.
E-mail: info@hawaiianrailway.com
Web Site: hawaiianrailway.com
Institution Type/Description: Railroad History Museum.
Collections: railroading history; steam locomotives; freight cars; parlor cars.
Facilities: Museum-related items for sale.
Activities: train rides.
Hours & Admission Prices: Train Rides: Sun. 1 & 3; other times by appointment. Adults $12, children 2-12 and seniors 62 & over $8; children under 2 no charge. Closed Christmas Eve & Day.

Fort DeRussy

U.S. ARMY MUSEUM OF HAWAII, Battery Randolph, Kalia Rd., Fort DeRussy, HI 96815. Mailing Address: P.O. Box 8064, Honolulu, HI 96830-0064. Tel.: 808-942-0318. Fax: 808-438-2819.
Web Site: www.hiarmymuseumsoc.org
Founded: 1976.
Congressional District: 1
Key Personnel: Dir. & Cur., Judith Bowman; Museum Shop Mgr., Mr. Sheldon Tyau.
Personnel Profile: Full-Time Paid 3; Part-Time Volunteers 22.
Governing Authority: federal; nonprofit. Parent Institution: U.S. Army Garrison, Hawaii. Subsidiary Institution: Hawaii Army Museum Society. Tax-exempt: 501(c)(3).
Institution Type/Description: Military History Museum: housed in Battery Randolph, a Taft Period coast artillery battery; located at Fort DeRussy, one of the earliest military posts established by the U.S. Army in Hawaii.

Collections: military materials, domestic & foreign, associated with the history of the U.S. Army in the Pacific; Hawaiian military history.
Research Fields: military history of Hawaii & the role played by Hawaii & Hawaiians in the nation's defense.
Facilities: permanent exhibitions.
Activities: research assistance; school & group tours. Annual Event: Living History Day.
Hours & Admission Prices: Tues.-Sun. 9-5. No charge; donations accepted. &
Attendance: 100,000 (estimated)

Haleiwa

WAIMEA VALLEY BOTANICAL GARDEN, (I), 59-864 Kamehameha Hwy., Haleiwa, HI 96712-9406. Tel.: 808-638-7766. Fax: 808-638-7776.
E-mail: jhoh@waimeavalley.net
Web Site: waimeavalley.net
Formerly: Waimea Arboretum and Botanical Garden; Waimea Valley Audubon Center
Founded: 1973.
Congressional District: 2
Key Personnel: Dir., Richard Pezzulo; Chm. (V), Hoku Haiku; Botanical Mgr., Josephine Hoh; Museum Shop Mgr., Gail Cabalce.
Personnel Profile: Full-Time Paid 12; Interns 1.
Governing Authority: Parent Institution: Hi'ilei Aloha LLC. Subsidiary Institution: Hi'ipaka LLC. Tax-exempt.
Institution Type/Description: Living Plant Museum, Arboretum & Botanical Garden.
Collections: living plant collection with scientific field data, arranged by genera, family or geographical region; collections of ancient Hawaiian strains of economic plants; seed collection; endangered bird species; herbarium collections of plants & seedlings; flora of the island ecosystems.
Research Fields: propagation & preservation of rare & endangered tropical & sub-tropical plants; cultivation techniques of a wide range of plant species; excavation of historical sites; Hawaiian ethnobotany section, medicinal, food & useful plants; recreated Hawaiian living area.
Facilities: 2,000-vol. library & herbarium of books available upon request on premises only; botanical garden; nature center; 400-seat restaurant. Museum-related items for sale.
Activities: guided tours; formally organized education programs for children.
Hours & Admission Prices: Daily 9-5. Out-of-State Residents: adults $15, children 4-12 & seniors $7.50. Hawaii Residents: adults $10, children 4-12 & seniors $5. Military: $10. Closed New Year's Day; Thanksgiving; Christmas. &
Attendance: 250,000 (estimated)
Membership: Annual Passes: Individual $50; Family 2 adults 6 children up to 18 yrs. of age $100; Senior 60 & over $750; Lifetime $1,000.

Hana, Maui

HANA CULTURAL CENTER, 4974 Uakea Rd., Hana, Maui, HI 96713. Mailing Address: P.O. Box 27, Hana, Maui, HI 96713-0027. Tel.: 808-248-8622. Fax: 808-248-7898.
E-mail: mail@hanaculturalcenter.org
Web Site: www.hanaculturalcenter.org
Founded: 1971.
Key Personnel: Gen. Operations Mgr., Meiling Hoopai; Pres., Esse Sinenci; Vice Pres., Malia Henderson; Chm. (V), Jackie Kahula; Program Mgr., Leinaala Estrella; Receptionist, Sydney Shamblin.
Personnel Profile: Full-Time Paid 1; Full-Time Volunteers 1; Part-Time Paid 3; Part-Time Volunteers 2.
Governing Authority: private; nonprofit organization. Tax-exempt: 501(c)(3).
Institution Type/Description: Cultural Center.
Collections: Hawaiian articles.
Facilities: library; botanical garden. Museum-related items for sale.
Activities: guided tours; hobby workshops. Museum Sponsors: craft demonstrations June-August.
Publications: annual newsletter, brochures.
Hours & Admission Prices: Mon.-Fri. 10-4. Adults $3; members no charge. Closed New Year's Day; Easter; Thanksgiving; Christmas. &
Attendance: 26,510 (estimated)
Membership: Annual $25; Ohana $60; Life $250.

Hanalei

WAIOLI MISSION HOUSE MUSEUM, 4050 Nawiliwili Rd., Hanalei, HI 96714. Mailing Address: c/o Waioli Corporation, P.O. Box 1631, Lihue, HI 96766-5631. Tel.: 808-245-3202. Fax: 808-245-7988.
Key Personnel: Caretaker, Barbara Kennedy; Caretaker, Roger Kennedy.

Governing Authority: Parent Institution: Waioli Corp., Lihue, HI.
Institution Type/Description: Historic House: housed in an 1837 missionary house. Listed on the National Register of Historic Places.
Collections: period furnishings; personal artifacts.
Hours & Admission Prices: Guided Tours: Tues., Thurs., Sat. 9-3.

Hawaii Volcanoes National Park

HAWAII VOLCANOES NATIONAL PARK, KILAUEA VISITOR CENTER, Headquarters Bldg. #1, Crater Rim Dr., Hawaii Volcanoes National Park, HI 96718. Mailing Address: P.O. Box 52, Hawaii National Park, HI 96718-0052. Tel.: 808-985-6000. Fax: 808-985-6004.
Web Site: www.nps.gov/havo/
Founded: 1916.
Congressional District: 2
Key Personnel: Park Superintendent, Cindy Orlando.
Governing Authority: federal. Dept. of the Interior, National Park Service.
Institution Type/Description: Visitor Center.
Collections: Hawaii artifact; natural history material; paintings; geological specimens; birds; archaeology; anthropology; ethnology; photographs.
Research Fields: Hawaiiana; park natural history.
Facilities: 2,500-vol. library available to on-site researchers by prearrangement; nature & conservation center; 210-seat auditorium. Books for sale.
Activities: guided tours; lectures; films; formally organized educational programs; temporary exhibitions.
Publications: Hawaii Natural History Association.
Hours & Admission Prices: Park open 24 hours. Kilauea Visitor Center: daily 7:45-5. Jaggar Museum: daily 8:30-7:30. 7-day vehicle entrance fee $10, 7-day individual entrance fee $5; Martin Luther King, Jr. Day, National Park Week, Hawai'i Volcanoes National Park Cultural Festival, National Park Service Day, National Public Lands Day & Veterans Day Weekend no charge. &
Attendance: 1,500,000 (estimated)
Membership: Annual Pass $25.

Hilo

EAST HAWAII CULTURAL CENTER, 141 Kalakaua St., Hilo, HI 96720-2807. Tel.: 808-961-5711.
E-mail: arts@ehcc.org
Web Site: www.ehcc.org
Key Personnel: Acting Chm., Kay Yokoyama; Exec. Dir., Dennis Taniguchi
Institution Type/Description: Art Gallery.
Collections: works by local, national & international artists.
Facilities: theater. Museum-related items for sale.
Activities: performances; meetings.
Hours & Admission Prices: Call for hours. Gallery: Mon.-Sat. 10-4. No charge; donations accepted.

IMILOA ASTRONOMY CENTER, 600 Imiloa Place, Hilo, HI 96720. Tel.: 808-969-9703 & 9700. Fax: 808-969-9748.
E-mail: info@imiloahawaii.org
Web Site: www.imiloahawaii.org
Institution Type/Description: Astronomy Museum.
Collections: earth; planets; stars; galaxy; universe.
Activities: educational programs; special events.
Hours & Admission Prices: Tues.-Sun. 9-5. Closed New Year's Day; Thanksgiving; Christmas.

✻ **LYMAN MUSEUM, (M),** 276 Haili St., Hilo, HI 96720-2978. Tel.: 808-935-5021. Fax: 808-969-7685. Facebook: Lyman Museum.
E-mail: info@lymanmuseum.org
Web Site: www.lymanmuseum.org
Founded: 1931.
Congressional District: 2
Key Personnel: Pres., Chm. (V) & Exec. Dir., Richard Henderson.
Personnel Profile: Full-Time Paid 10; Part-Time Paid 2; Part-Time Volunteers 20; Interns 2.
Governing Authority: nonprofit organization. Tax-exempt: 501(c)(3).
Institution Type/Description: General Museum: depicting the natural & cultural history of Hawaii.
Collections: early 19th- & 20th-century Hawaiian artists; Hawaiian & missionary artifacts; native & immigrant people of Hawaii and local culture; World Class minerals; manuscripts; worldwide seashells; Hawaii land shells; Flora & Fauna; natural history of Hawaii including its volcanic origins. Historic Mission House: built in 1839 by missionaries from New England.

Research Fields: Hawaiiana & local history.

Facilities: 3,500-vol. library on Hawaiiana available for research on premises; 20,000 photo library of Old Hawaii; 200 blueprints; charts, daguerreotypes; glassplates; journals; letters; newsletters; 660 New England newspapers; 390 prints & maps; 508 rare newspapers; 2,500 ephemera collection. Gift items for sale.

Activities: guided tours; permanent and special exhibitions; lectures; workshops; elderhostel/road scholar programs.

Publications: books, The Lymans of Hilo; Hilo 1825-1925: A Century of Paintings & Drawings; The Private Japanese Hospital: An Unique Social Phenomenon on Hawaii, 1907-1960; Japanese Painting, Calligraphy & Lacquer; A Record of the Descendants of David Belden Lyman & Sarah Joiner Lyman of Hawaii 1832-1933; The Lymans of Hawaii Island. Educational Outreach Booklets: 'Ohe Kapala: Bamboo Printing; Kane, Kanaloa, Ku & Lono: Na Akua Nui O Hawai'i, (Four Major Gods of Hawaii); Early Hawaiian Musical Instruments; Ku'i I Ke Kalo: Pounding Taro; Gourds in the Garden at Work and in the Writings (of Old); Na Mea Nui O Na Ali'i Hawai'i: Symbols of Royalty. Museum newsletter; booklet, Lyman Mission House; books, Sarah Joiner Lyman of Hawaii: Her Own Story; Crystals of the Lyman Museum.

Hours & Admission Prices: Mon.-Sat. 10-4:30. Adults $10, senior citizens $8, children 6-17 $3; discounts to AAA, AASLH & Hawaii Museums Association members; members no charge. Closed New Year's Day; Memorial Day; Independence Day; Labor Day; Thanksgiving; Christmas. &

Attendance: 14,284 (accurate)

Membership: Student $10; Out of State $15; Individual $30; Dual & Family $55; Dual Family & Guests $75; Steward $120; Conservator $250; Heritage Partner $500; David & Sarah Lyman Society $1,000.

NANI MAU GARDENS, 421 Makalika St., Hilo, HI 96720-5899. Tel.: 808-959-3500. Fax: 808-959-3501.

E-mail: corp@hottours.us

Web Site: www.nanimaugardens.com

Institution Type/Description: Gardens.

Collections: tropical flowers & plants; orchids; palms & tropical fruit orchards.

Facilities: restaurant. Museum-related items for sale.

Activities: garden tours.

Hours & Admission Prices: Daily 9-4:30. Gardens: adults $10, children 4-10 $5. Garden & Tram Tour: adults $17, children 4-10 $10.

PACIFIC TSUNAMI MUSEUM, 130 Kamehameha Ave., Hilo, HI 96720-2833. Mailing Address: P.O. Box 806, Hilo, HI 96720. Tel.: 808-935-0926. Fax: 808-935-0842. Facebook: Pacific Tsunami Museum.

E-mail: tsunami@tsunami.org

Web Site: www.tsunami.org

Founded: 1994.

Congressional District: 2

Key Personnel: Exec. Dir., Marlene Murray; Pres., Jim D. Wilson; Treas., Jill Jacunski; Cur. & Archivist, Barbara J. Muffler; Administrative Asst., Colleen DeSa.

Personnel Profile: Full-Time Paid 3; Part-Time Paid 2; Part-Time Volunteers 18.

Governing Authority: private; nonprofit organization. Tax-exempt: 501(c)(3).

Institution Type/Description: Natural History Museum.

Collections: tsunami photographs; 1946 tsunami in Hilo.

Facilities: classrooms; 7,250 sq. ft. exhibit space. Museum-related items for sale.

Activities: guided tours; lectures; participatory exhibits; docent program; formal education programs for children. Annual Event: Tsunami Story Festival.

Publications: biannual newsletter; Driving & Walking Tour of Historical Tsunami Sties in East Hawaii; Tsunami Education: A Blueprint for Coastal Communities.

Hours & Admission Prices: Mon.-Sat. 9-4. Adults $8, senior citizens & Kama'aina $7, children 6-17 $4; members and children 3 & under no charge. Closed New Year's Day; Independence Day; Thanksgiving; Christmas Eve & Day. &

Attendance: 17,194 (accurate)

Membership: Individual $15; Family $35; Supporting $50; Associate $100; Affiliate $250; Patron $500; Supporter $1,000; Benefactor $2,000.

PANAEWA RAINFOREST ZOO, Mamaki St., (one mile off Hwy. 11), Hilo, HI 96720. Mailing Address: Friends of the Zoo, P.O. Box 738, Kea'au, HI 96749-0738. Tel.: 808-959-9233. Fax: 808-961-8411.

E-mail: foz@hilozoo.com

Web Site: www.hilozoo.com

Institution Type/Description: Zoo.

Collections: 80 animal species including the endangered Nene (Hawaii State Bird); white Bengal Tiger.

Facilities: 12 acres. Museum-related items for sale.

Activities: tiger feeding daily.

Hours & Admission Prices: Daily 9-4. Petting Zoo: Sat. 1:30-2:30. No charge. Closed New Year's Day; Christmas.

WAILOA ARTS & CULTURAL CENTER, 200 Piopio St., Wailoa State Park, Hilo, HI 96720. Tel.: 808-933-0416. Fax: 808-933-0417.

E-mail: wailoa@yahoo.com

Founded: 1967.

Key Personnel: C.E.O. & Dir., Codie M. King.

Personnel Profile: Full-Time Paid 1; Part-Time Volunteers 18.

Governing Authority: Parent Institution: State of Hawaii. Subsidiary Institution: Department Land & Natural Resources, State Parks Division.

Institution Type/Description: Art Museum.

Collections: works by local artists; Big Island history & culture.

Facilities: library; visitors center.

Activities: demonstrations; seminars; workshops; classes; school tour & outreach programs; films; concerts; live performances.

Hours & Admission Prices: Mon.-Tues. & Thurs.-Fri. 8:30-4:30, Wed. 12-4:30. No charge; donations accepted. Closed holidays. &

Attendance: 30,000 (estimated)

Honolulu

BATTLESHIP MISSOURI MEMORIAL, Historic Ford Island, Pearl Harbor, 63 Cowpens St., Honolulu, HI 96818-5006. Mailing Address: USS Missouri Memorial Assoc., Inc., P.O. Box 879, Aiea, HI 96701. Tel.: 808-455-1600; 877-644-4896 (toll free).

E-mail: mightymo@ussmissouri.org

Web Site: www.ussmissouri.com

Key Personnel: Dir., Peter Berg

Institution Type/Description: Military Museum: housed on the USS Missouri, built in 1941.

Collections: battleship history; military artifacts & equipment; personal artifacts; photographs.

Facilities: Museum-related items for sale.

Activities: school group tours.

Hours & Admission Prices: June-Aug. daily 8-5; Sept.-May daily 8-4. Adults $22, children $11. Closed New Year's Day; Thanksgiving; Christmas.

BERNICE PAUAHI BISHOP HERITAGE CENTER, THE KAMEHAMEHA SCHOOLS, 1887 Makuakane St., Honolulu, HI 96817-1800. Tel.: 808-842-8635. Fax: 808-842-8603.

E-mail: heritagecenter@ksbe.edu

Web Site: apps.ksbe.edu/heritagecenter/aloha

Founded: 1988.

Key Personnel: Cur., Nuulani Atkins

Institution Type/Description: Heritage Center: school founded by Bernice Pauahi Bishop & her husband, Charles Reed Bishop.

Collections: Bishop's furniture & personal artifacts.

Hours & Admission Prices: Mon.-Fri. 9-2 by appointment.

*** BISHOP MUSEUM, (M),** 1525 Bernice St., Honolulu, HI 96817-2704. Tel.: 808-847-3511. Fax: 808-841-8968. Facebook: Bishop Museum.

E-mail: museum@bishopmuseum.org

Web Site: www.bishopmuseum.org

Founded: 1889.

Congressional District: 19

Key Personnel: Chm. Bd., Allison Holt Gendreau; C.E.O. & Pres., Blair D. Collis; Dir. Strategic Initiatives, Elizabeth Tatar, Ph.D.; Dir. Cultural Collections, Betty Lou Kam; Dir. Human Resources, Anna Scott; Dir. Bldgs. & Grounds, Special Events & Exhibits, Larry Schmitt; Dir. Visitor Experience, Michael Shanahan; Museum Shop Mgr., Maria Young.

Personnel Profile: Full-Time Paid 135; Part-Time Paid 58; Part-Time Volunteers 524; Interns 11.

Governing Authority: private charitable corporation. Subsidiary Institution: Hawaii Maritime Center. Tax-exempt: 501(c)(3), 170(b)(1)(A), 509(a)(1).

Institution Type/Description: Cultural & Natural History Museum.

Collections: culture & natural history of Hawaii & the Pacific; anthropology; archaeology; archives; botany; entomology; ethnology; geology; herpetology; ichthyology; malacology; zoology; herbarium; philatelic; photographs; invertebrate zoology; mammalogy; ornithology; maps; motion picture film; art on paper & canvas; living ethnobotanical; pamphlets; manuscripts.

Research Fields: anthropology; archaeology; botany; entomology; ethnobotany; history; malacology; vertebrate & invertebrate zoology.
Facilities: 100,000-vol. library; 55,000 sq. ft. exhibit space; 12-acre campus on Oahu; 15-acre gardens on Hawaii Island; planetarium; 70-seat theatre. Books, handicrafts, jewelry, reproductions & video tapes for sale.
Activities: educational classes; school group visits; lectures; films; docent program; changing & traveling exhibitions; volunteer program; special events; planetarium programs; observatory telescope viewing; excursions. Museum Sponsors: Family Sundays.
Publications: four bulletins, anthropology, botany, entomology, zoology; occasional papers; monograph, Insects of Micronesia; handbooks, WAU Ecology; Indo-Pacific Fishes; special publications; miscellaneous publications; calendars; guidebooks; newsletter.
Hours & Admission Prices: Wed.-Mon. 9-5. Adult non-resident $17.95, senior citizens over 65 $14.95, adult resident $10.95, youth 4-12 $8.95; discounts for senior citizens, active military & AAM members; members & children under 4 no charge. Library & Archives: appointments granted & scheduled based on staff availability. Closed Christmas. &
Attendance: 327,967 (accurate)
Membership: Friend & Senior $35; Dual $50; Family & Friends $65; Patron $100; Benefactor $250; Visionary $500.

THE CONTEMPORARY MUSEUM AT FIRST HAWAIIAN CENTER, 999 Bishop St., Honolulu, HI 96813-4423. Mailing Address: 2411 Makiki Heights Dr., Honolulu, HI 96822-2547. Tel.: 808-526-1322. Fax: 808-536-5973.

E-mail: itully@tcmhi.org
Web Site: www.tcmhi.org
Founded: 1996.
Congressional District: 1
Key Personnel: Dir., Allison Wong; Pres.(V), Violet Loo; Volunteer & Special Events Coord., Sheryl Kramer; Deputy Dir. & Chief Cur., James Jensen; Cur. Exhibitions, Inger Tully; Cafe Mgr. & Mgr. Retail Operations, Bob Madison; Coord. Membership, Nicole Higa; Dir. Devel., Joan Yanagihara; Bldg. & Grounds Mgr., Garry Ka'aihue; Chief Preparator, John Koga; Cur. Education, Aaron Padilla; Deputy Dir. Finance & Operations, John Talkington; Asst. to Dir., Gordon Wong; Registrar, Tae Kitakata; Mgr. Cafe Kitchen, Scott Sakaguchi; Security Mgr., Michael Chock.
Personnel Profile: Full-Time Paid 15; Part-Time Paid 10; Part-Time Volunteers 200; Interns 3.
Governing Authority: public; nonprofit organization. Parent Institution: The Contemporary Museum Board of Trustees. Tax-exempt: 501(c)(3).
Institution Type/Description: Art Museum.
Collections: art of local artists who live and/or work in the state of Hawaii, or were born & moved from the island, or who came to the island and produced work in the state.
Facilities: 3,500 sq. ft. in the main banking hall & designated gallery at The First Hawaiian Center building, downtown Honolulu.
Activities: guided tours; gallery talks; temporary exhibitions; one man & group shows by contemporary artists.
Publications: exhibition brochures; newsletter; catalogues.
Hours & Admission Prices: Mon.-Thurs. 8:30-4, Fri. 8:30-6. Adults $8. Reciprocal memberships. Closed bank holidays. &
Attendance: 85,216 (estimated)
Membership: Student $15; Nonresident $20; Individual $45; Dual $75; Sponsor $125; Associate $250; Partner $500; Contemporary Circle & Patron $1,000; President's Circle & Leader $2,500; Director's Circle & Benefactor $5,000; Champion's Circle $10,000.

DORIS DUKE FOUNDATION FOR ISLAMIC ART - SHANGRI LA, (M), 4055 Papu Circle, Honolulu, HI 96816. Tel.: 808-734-1941. Fax: 808-732-4361.

Web Site: www.shangrilahawaii.org
Congressional District: 1
Key Personnel: Exec. Dir., Deborah Pope.
Personnel Profile: Full-Time Paid 22; Part-Time Paid 6; Part-Time Volunteers 1; Interns 2.
Governing Authority: Parent Institution: Doris Duke Charitable Foundation. Tax-exempt.
Institution Type/Description: Art & History Museum: housed in the seasonal home of Doris Duke, built in 1937.
Collections: Islamic art & culture; period furnishings; religious works of art; 17th-19th centuries decorative arts.
Research Fields: Islamic art & culture.
Facilities: 5-acre waterfront property.
Activities: guided tours; lectures; symposia; scholarly research; educational programs and field studies; permanent & temporary exhibitions.
Publications: Doris Duke's Shangri La.
Hours & Admission Prices: Wed.-Sat. by appointment. Out-of-State $25; In-State $20. &

Attendance: 13,312 (accurate)

EAST-WEST CENTER, 1601 East-West Rd., Honolulu, HI 96848-1601. Tel.: 808-944-7584. Fax: 808-944-7376.

E-mail: ewcinfo@eastwestcenter.org
Web Site: www.eastwestcenter.org
Founded: 1961.
Key Personnel: Pres. & C.E.O., Charles Morrison; Arts Coord., William Feltz; Cur., Michael Schuster, Ph.D.
Personnel Profile: Full-Time Paid 3; Part-Time Volunteers 10; Interns 1.
Governing Authority: nonprofit. Tax-exempt.
Institution Type/Description: General Museum.
Collections: art, ethnography & history exhibits.
Hours & Admission Prices: Mon.-Fri. 8-5, Sun. 12-4. No charge; donations accepted. Closed federal holidays. &
Attendance: 6,000

442ND VETERANS CLUB ARCHIVES, 933 Wiliwili St., Honolulu, HI 96826. Mailing Address: 933 Wiliwili St., Apt. 102, Honolulu, HI 96826-2766. Tel.: 808-945-0032 & 949-7997. Fax: 808-949-1539.

E-mail: 442archives@hawaiiantel.net
Formerly: 442nd Veterans Archives & Learning Center
Key Personnel: Pres., William Thompson.
Personnel Profile: Part-Time Paid 2.
Governing Authority: archives & exhibits of the 442nd combat team.
Institution Type/Description: Military Museum.
Collections: personal artifacts.
Hours & Admission Prices: Mon.-Fri. 8-3; call to confirm. No charge; donations accepted.

HAROLD L. LYON ARBORETUM, 3860 Manoa Rd., Honolulu, HI 96822-1198. Tel.: 808-988-0456. Fax: 808-988-0462.

E-mail: lyonarb@hawaii.edu
Web Site: www.hawaii.edu/lyonarboretum
Founded: 1918.
Congressional District: 1
Key Personnel: Dir., Christopher Dunn, Ph.D.; Pres. Lyon Arboretum Assoc., Mrs. Emmy Seymour; Junior Researcher, Nelli Sugii; Horticulturalist, Elizabeth Huppman; Research Assoc. II, Karen Shigematsu; Research Assoc. IV, Ray Baker; Educational Specialist, Jill Laughlin; Research Technician, Carol Nakamura; Maintenance Worker I, Kenneth Seamon; Arborist, Leon Marcus; Sec., Tokiko Murakami; Research Affiliate, D.W. Arthur Whistler; Research Affiliate, Dr. George Staples, III; Research Affiliate, Dr. John Moody; Research Affiliate, Dr. Rich Criley.
Personnel Profile: Full-Time Paid 10; Part-Time Volunteers 350.
Governing Authority: state; not-for-profit organization. Parent Institution: University of Hawaii. Tax-exempt: 501(c)(3).
Institution Type/Description: Botanical Garden & Arboretum.
Collections: native & endemic plants of Hawaii & tropics; Hawaiian, southeast Asian & Pacific ethnobotanical plants; germplasm of ornamental & economic plants.
Research Fields: taxonomy of native & endemic plants of Hawaii; taxonomy of tropical plants; ethnobotany of Pacific, Southeast Asian & Asian peoples; propagation & horticulture of tropical plants; propagation by micropropagation; culture; ecosystem restoration; rare Hawaiian plants conservation & biology; seed storage & conservation.
Facilities: botanical garden; classrooms; herbarium; laboratory; craft center; micropropagation facilities. Museum-related items for sale.
Activities: guided tours; hiking trails; lectures; docent program; temporary exhibits; formally organized education programs for children, adults, undergraduate & graduate college students affiliated with the University of Hawaii & private universities in Hawaii.
Publications: occasional periodical, Lyonia; annual periodical, Harold L. Lyon Arboretum Lecture; educational series; occasional pamphlets; books through Univ. of Hawaii Press; quarterly newsletter, The Kukui Leaf
Hours & Admission Prices: Mon.-Fri. 9-4, Sat. 9-3; other times by appointment. Suggested Donation: $5 per person. Closed holidays. &
Attendance: 30,000 (estimated)
Membership: Student $10; Senior Citizen-Kupuna 65 & over $20; Individual $25; Family (one household) $35; Kukui $50; Niu $100; Koa $250; 'Ohi'a $1,000.

HAWAII CHILDREN'S DISCOVERY CENTER, 111 Ohe St., Honolulu, HI 96813-5517. Tel.: 808-524-5437. Fax: 808-524-5400.

E-mail: info@discoverycenterhawaii.org
Web Site: www.discoverycenterhawaii.org
Founded: 1985.

Congressional District: 1
Key Personnel: Chm. (V), Loretta Yajima; Treas., Arthur Tokin; Pres., Liane Usher.
Personnel Profile: Full-Time Paid 8; Full-Time Volunteers 1; Part-Time Paid 8; Part-Time Volunteers 50.
Governing Authority: not-for-profit organization. Tax-exempt: 501(c)(3).
Institution Type/Description: Children's Museum.
Collections: hands-on exhibits.
Facilities: 45,000 sq. ft. exhibit space.
Activities: outreach activities including a children's fun run & other special events; spring, summer & winter day camps for children; weekly toddler classes & other special events.
Publications: quarterly newsletter; annual report; membership brochures.
Hours & Admission Prices: Tues.-Fri. 9-1, Sat.-Sun. 10-3. Admission $10; children under one & members no charge. Closed New Year's Day; Easter; Labor Day week; Thanksgiving; Christmas. &
Membership: Family $100 & up.

HAWAII MARITIME CENTER, 1525 Bernice St., Honolulu, HI 96817-2704. Tel.: 808-523-6151. Fax: 808-536-1519.
Web Site: www.bishopmuseum.org
Founded: 1988.
Key Personnel: C.E.O., Tim Johns; Dir. Sales & Event Planning, Linda Chalk.
Personnel Profile: Full-Time Paid 3; Part-Time Paid 2; Part-Time Volunteers 32.
Governing Authority: private; nonprofit. Parent Institution: Bishop Museum. Tax-exempt: 501(c)(3).
Institution Type/Description: Maritime Museum.
Collections: focus on maritime history of Hawaii from the time of the arrival of the Polynesians to the present; history; maritime commerce; recreational activities; historic vessel.
Hours & Admission Prices: Daily 9-5. Adults $7.50, children 6-17 $4.50; discounts to active duty military, tour groups & local residents; children under 6 & members no charge. Closed Christmas Day. &
Attendance: 68,400 (accurate)
Membership: Individual $35; Family $45; Navigator & Mate $60; Captain $100; Commodore $250.

HAWAII PACIFIC UNIVERSITY GALLERY, 1166 Fort St. Mall, Ste. 200, Honolulu, HI 96813-2785. Tel.: 808-544-0228. Fax: 808-544-1424.
E-mail: jishaque@hpu.edu
Web Site: www.hpu.edu
Founded: 1983.
Congressional District: 1
Key Personnel: Pres., Geoffrey Bannister, Ph.D.; Chm. (V), Michael J. Chun, Ph.D.; Dir., Sanit Khewhok.
Personnel Profile: Part-Time Paid 1; Part-Time Volunteers 10.
Governing Authority: private; nonprofit organization. Parent Institution: Hawaii Pacific University. Tax-exempt.
Institution Type/Description: Art Gallery.
Collections: paintings; sculpture; photographs.
Facilities: 2,000 sq. ft. exhibit space.
Activities: Annual Events: six exhibitions per academic year August to June.
Hours & Admission Prices: Mon.-Sat. 8-5. No charge. Closed New Year's Eve & Day; Thanksgiving; Christmas. &
Attendance: 7,000 (estimated)

HAWAII STATE CAPITOL, 415 S. Beretania St., Honolulu, HI 96813-2477. Mailing Address: 415 S. Beretania St., Rm 415, Honolulu, HI 96813-2407. Tel.: 808-586-0178 & 0146. Fax: 808-586-0046.
Web Site: governor.hawaii.gov/about/capitol-tour-information/
Institution Type/Description: History Museum.
Collections: Hawaii history; photographs.
Activities: group & school tours.
Hours & Admission Prices: Mon.-Fri. 9-3:30. Guided Tours: Tues.-Fri. 1. No charge. Closed holidays.

HAWAII STATE FOUNDATION ON CULTURE & THE ARTS AND HAWAII STATE ART MUSEUM, 250 S. Hotel St., 2nd Fl., Honolulu, HI 96813-2831. Tel.: 808-586-0300. Fax: 808-586-0308.
E-mail: ken.hamilton@hawaii.gov
Web Site: www.hawaii.gov/sfca
Founded: 1965.
Congressional District: 1
Key Personnel: Exec. Dir., Ronald Yamakawa.

Personnel Profile: Full-Time Paid 20; Part-Time Volunteers 15.
Governing Authority: The State of Hawaii, Department of Accounting & General Services. Tax-exempt.
Institution Type/Description: Art Museum.
Collections: over 5,000 works by Hawaiian contemporary artists from 1967 to present.
Research Fields: historical records of ethnic organizations in Hawaii; traditional Hawaiian arts & music.
Facilities: cafe; multipurpose room. Museum-related items for sale.
Activities: permanent & temporary exhibitions; guided tours; lectures; concerts; monthly family events; training programs for arts education & museum professionals; grants & funding program.
Publications: monthly newsletter; annual report.
Hours & Admission Prices: Offices: Mon.-Fri. 7:45-4:30. Museum: Tues.-Sat. 10-4; first Fri. of every month 5-9. No charge. Closed state holidays. &
Attendance: 30,000
Membership: Student & Senior $20; Friend $45; Family $100; Contributor $250; Patron $500; Benefactor $1,000 & up.

HAWAIIAN HISTORICAL SOCIETY, 560 Kawaiaha'o St., Honolulu, HI 96813-5023. Tel.: 808-537-6271. Fax: 808-537-6271.
E-mail: hhsbarb@lava.net
Web Site: www.hawaiianhistory.org
Founded: 1892.
Key Personnel: Administrative Dir. and Librarian, Barbara Dunn.
Governing Authority: nonprofit organization.
Institution Type/Description: Historical Society Museum.
Collections: historical documents; newspapers; photographs; manuscripts.
Facilities: library.
Activities: research; history conferences; lectures; special events; membership meetings. Annual Event: Open House in December.
Publications: book, The Hawaiian Journal of History; newsletter, Na Mea Kahiko.
Hours & Admission Prices: Tues.-Fri. 10-4. No charge.
Membership: Student $20; Senior $30; Individual $40; Family $50.

* **HAWAIIAN MISSION HOUSES HISTORIC SITE AND ARCHIVES, (M),** 553 S. King St., Honolulu, HI 96813-3002. Tel.: 808-447-3910. Fax: 808-545-2280.
E-mail: info@missionhouses.org
Web Site: www.missionhouses.org
Formerly: Mission Houses Museum
Founded: 1920.
Congressional District: 1
Key Personnel: Exec. Dir., Thomas A. Woods; Pres., Bonnie F. Rice; Head Librarian, John Barker; Dir. Devel. & Society Rels. Dir., Mary Ann Lentz; Cur. Programs, Mike Smola; Cur. Object Collections, Alana Cole Faber; Volunteer Coord., Marcia L. Timboy; Museum Shop Mgr., Dianne Ching; Office Mgr., Lisa Solomine; Accountant, Gabriela Bonilla.
Personnel Profile: Full-Time Paid 10; Part-Time Paid 2; Part-Time Volunteers 40.
Governing Authority: nonprofit organization. Parent Institution: Hawaiian Mission Children's Society.
Institution Type/Description: Historic House Museum.
Collections: missionary & Polynesian artifacts including domestic artifacts & household furnishings; archival collections of missionary & Hawaiian church records; Hawaiian language materials; working replica of Ramage printing press; manuscripts. Historic Houses: 1821 frame mission house; 1831 Chamberlain House; 1841 printing office.
Research Fields: Protestant mission in Hawaii & the Pacific; 19th-century local history & cultural contact; daily life & material culture in 19th-century Hawaii; missionary-Hawaiian trade & exchange; archaeology.
Facilities: 12,000-vol. research library. Museum-related gifts for sale.
Activities: walking tours of historic Honolulu; guided tours; seasonal family programs; living history programs; permanent & temporary exhibitions; workshops & lectures; educational programs for primary, secondary & college level students; quarterly community festivals; premier juried holiday craft fair in the islands.
Publications: books, Missionary Album; Voyages to Hawaii Before 1860; The Hawaii Journals of the New England Missionaries 1813-1894; A Guide to the Holdings of the HMCS Library; The Journals of Cochran Forbes c.1831-1864; Mission Houses Museum Guidebook, 2001; Engraved at Lahainaluna.
Hours & Admission Prices: Tues.-Sat. 10-4. Adults $10, Kama'aina, military & seniors $8, students 6 to college $6, school tours $2-$5 per student; discount to AAM members; members and children 5 & under no charge. Closed New Year's Day; Easter; Thanksgiving; Christmas.
Attendance: 22,000 (estimated)

Membership: Student $25; Individual $40; Family $100; Friend $125; Contributor $250; Patron $500; Silver Circle $1,000; Gold Circle $2,500; Platinum Circle $5,000; Diamond $10,000.

HONOLULU BOTANICAL GARDENS, 50 N. Vineyard Blvd., Honolulu, HI 96817-3759. Tel.: 808-522-7060. Fax: 808-522-7050.
E-mail: hbg@honolulu.gov
Web Site: www.honolulu.gov/parks/hbg
Founded: 1931.
Key Personnel: Dir., Winifred Singeo; Education, Joyce Spoehr; Orchid Horticulturist, Scot Mitamura; Horticulturist, Joshlyn Sand; Parks Grounds Supvr., Derrick Miyasaki; Botanist, Naomi Fenstenmacher.
Personnel Profile: Full-Time Paid 30; Part-Time Paid 7; Part-Time Volunteers 100.
Governing Authority: Parent Institution: Dept. of Parks & Recreation. Branch Museums: Foster Botanical Garden, Honolulu, HI; Wahiawa Botanical Garden, Wahiawa, HI; Ho'omaluhia Botanical Garden, Kaneohe, HI; Koko Crater Botanical Garden, Honolulu, HI; Lili'uokalani Botanical Garden, Honolulu, HI.
Institution Type/Description: Botanical Gardens.
Collections: Hawaiian, Polynesian & tropical plants; hybrids; orchids.
Research Fields: cooperator: chemical & biological controls of invasive flora and fauna.
Facilities: botanical garden; visitor centers; educational facilities. Museum-related items for sale.
Activities: arts festivals; concerts; docent program; guided tours; lectures. Annual Events: Midsummer Night's Gleam; Summer Twilight Concerts.
Hours & Admission Prices: Foster, Ho'omaluhia & Wahiawa daily 9-4. Lili'uokalani daily 7-5. Koko Crater daily sunrise to sunset. Foster: adults $5, students 6-12 $3; children 5 & under no charge. Ho'omaluhia, Wahiawa, Lili'uokalani & Koko Crater no charge. Closed New Year's Day; Christmas. &
Attendance: 204,998 (estimated)

* **HONOLULU MUSEUM OF ART, (M),** 900 S. Beretania St., Honolulu, HI 96814-1495. Tel.: 808-532-8700. Fax: 808-532-8787.
Web Site: www.honolulumuseum.org
Formerly: Honolulu Academy of Arts
Founded: 1922.
Congressional District: 1
Key Personnel: Dir., Stephan Jost; Chm. (V), Lynne Johnson; Deputy Dir., Allison Wong; Head Finance, Wei Robertson; Acting Dir. Devel., Jessica Welch; Cur. Asian Art, Shawn Eichman; Mgr. Textiles, Sara Oka; Cur. European & American Art, Theresa Papanikolas; Dir. Tour & Docent Programs, Betsy Robb; Dir. Honolulu Museum of Art School, Vince Hazen; Cur. Film, Abigail Algar; Registrar, Collections, Pauline Sugino; Registrar, Exhibitions, Cynthia Low; Visitor & Volunteer Svcs., Vicki Reisner; Head Operational Svcs., Robert White; Museum Shop Mgr., Kathee Hoover; Cafe Mgr., Mike Nevin; Librarian, Sachiko Kawai'ae'a.
Personnel Profile: Full-Time Paid 110; Part-Time Paid 45; Part-Time Volunteers 400; Interns 4.
Governing Authority: nonprofit organization. Tax-exempt: 501(c)(3).
Institution Type/Description: Art Museum: housed in c.1927 building designed by Bertram G. Goodhue Associates.
Collections: art from ancient to modern times including painting, sculpture, ceramics, bronzes, lacquer & furniture from China, Japan, India, Indonesia, Korea, the Philippines, & Southeast Asia; James A. Michener Japanese prints; European & American painting, prints, sculpture & decorative arts; Kress collection of Italian Renaissance painting; traditional arts of Oceania, the Americas & Africa; traditional & modern Hawaiian art; Islamic art; contemporary art.
Major Exhibits: Georgia O'Keeffe and Ansel Adams: The Hawai'i Pictures (T), 7/13-1/14.
Research Fields: Asian, European and American art.
Facilities: 45,000-vol. library of art books available for use by members & scholars; reading room; 24,000 sq. ft. Art Center Studio Program; 300-seat auditorium; classrooms; restaurant; Mediterranean & Asian gardens. Art books, reproductions, museum replicas, slides, cards, & locally made artwork for sale.
Activities: permanent & temporary exhibitions; guided tours; lectures; films; gallery talks; art classes for children & adults; concerts; dance recitals; arts festivals; formally organized education programs; inter-museum loan, school loan service.
Publications: exhibition catalogs; books; pamphlets; members' magazine.
Hours & Admission Prices: Tues.-Sat. 10-4:30, Sun. 1-5. Adults $10, children 4-17 $5; discounts to AAM & ICOM members; children under 3, members & first Wed. of each month no charge. Closed New Year's Day; Independence Day; Labor Day; Thanksgiving; Christmas. &
Attendance: 250,000 (estimated)

Membership: Student $20; National & International $40; Individual $55; Family $95; Subscriber $150; Contributor $350; Sponsor $700; Guardian $1,500; Benefactor $2,500; Patron $5,000; Director's Circle $10,000.

HONOLULU MUSEUM OF ART SPALDING HOUSE, 2411 Makiki Heights Dr., Honolulu, HI 96822-2547. Tel.: 808-237-5214. Fax: 808-536-5973.
E-mail: info@honolulumuseum.org
Web Site: www.honolulumuseum.org
Formerly: The Contemporary Museum
Founded: 1961.
Congressional District: 1
Key Personnel: Art Dir., Stephan Jost; Pres. (V), Lynne Johnson; Cur. Education, Aaron Padilla; Dir. Facilities, Sheryl Kramer; Administrative Asst., Kelly Meskin; Chef, Susan Lai Hipp; Operations, Garry Ka'aihue.
Personnel Profile: Part-Time Volunteers 50.
Governing Authority: public; nonprofit organization. Tax-exempt: 501(c)(3).
Institution Type/Description: Art Museum.
Collections: paintings; sculpture; photographs.
Major Exhibits: Social Studies Through Art, 10/13-2/14.
Facilities: 3.5-acre site with gardens & outdoor sculpture; cafe. Museum-related items for sale.
Activities: guided tours; gallery talks; permanent & temporary exhibitions; workshops.
Hours & Admission Prices: Makiki Museum: Tues.-Sat. 10-4, Sun. 12-4. Adults $10, children 4-17 $5; discounts to AAM, Western Museum Group & North American Reciprocal Membership Program members; children under 3, members & third Thurs. of month no charge. Closed major holidays. &
Attendance: 25,000 (estimated)

HONOLULU POLICE DEPARTMENT LAW ENFORCEMENT MUSEUM, 801 S. Beretania St., Honolulu, HI 96813-5790. Tel.: 808-723-3475. Fax: 808-529-3960.
E-mail: hpd@honolulupd.org
Web Site: www.honolulupd.org/community/index.php?page=museum
Founded: 1984.
Governing Authority: Parent Institution: Honolulu Police Department.
Institution Type/Description: History Museum.
Collections: department history; Harley Davidson police motorcycle; uniforms; early 911 switchboard; Wall of Chiefs; documents; photos; weapons; restraint devices; police call box; a 1904 arrest log; other items that document the 19th & 20th century Hawaii social & urban history.
Hours & Admission Prices: Mon.-Fri. 9-3:30. No Charge. Closed state & federal holidays. &
Attendance: 8,100 (estimated)

HONOLULU SURFING MUSEUM, 2300 Kalakaua Ave., Honolulu, HI 96815. Tel.: 808-791-1200. Fax: 808-791-1250.
E-mail: info@honolulusurfmuseum.com
Web Site: www.honolulusurfmuseum.com
Founded: 2009.
Key Personnel: Cur., Mark Fragale.
Institution Type/Description: Surfing History Museum.
Collections: history of surfing; surfboards; photographs; personal artifacts; early ukuleles; guitars; paintings; movie posters.
Facilities: restaurant. Museum-related items for sale.
Hours & Admission Prices: Daily 11am. No charge. &

HONOLULU ZOO, 151 Kapahulu Ave., Honolulu, HI 96815-4096. Tel.: 808-971-7171 & 7174. Fax: 808-971-7173.
Web Site: www.honzoosoc.org
Founded: 1947.
Congressional District: 1
Key Personnel: Dir., Jeff Mahon, PhD.; Asst. Dir., Baird Fleming, D.V.M.; Volunteer Coord., Barbara Thacker; Administrative Asst., Lois Yoshikawa; Museum Shop Mgr., Kim Borges.
Personnel Profile: Full-Time Paid 74; Part-Time Paid 8; Part-Time Volunteers 256.
Governing Authority: municipal; county. Parent Institution: City & County of Honolulu. Subsidiary Institution: Dept. of Enterprise Services. Tax-exempt.
Institution Type/Description: Zoo.
Collections: live animals & plants from Africa, Asia, Oceania & Central/South America.
Research Fields: wild animal husbandry; propagation.
Facilities: 200-vol. zoological library available for staff and volunteer use only; children's zoo; education pavilion.
Activities: guided tours; docent program; summer concert series; adoption

program; elephant encounters; keeper talks; educational programs. Museum Sponsors: Vacation Adventures for children; Zoo by Moonlight; Breakfast with a Keeper.
Publications: quarterly newsletter, Honolulu Zoological Society.
Hours & Admission Prices: Daily 9-4:30. Adults $8, children 6-12 $1; children 5 & under no charge. Residents & military stationed in Hawaii with local ID: adults $4, children 6-12 $1. &
Attendance: 667,981 (accurate)
Membership: Senior Citizen & Student $15; Individual $20; Family $25 (basic) & $35 (premium); Supporting $75; Patron $125; Benefactor $175.

IOLANI PALACE, (M), 364 S. King St., Honolulu, HI 96813-2900. Mailing Address: P.O. Box 2259, Honolulu, HI 96804-2259. Tel.: 808-522-0822. Fax: 808-532-1051.
E-mail: info@iolanipalace.org
Web Site: www.iolanipalace.org
Founded: 1966.
Congressional District: 1
Key Personnel: Exec. Dir., Kippen de Alba Chu; Pres., Ronald K. Jarrett; Cur., Heather Diamond; Museum Shop Mgr., Chandra Watson.
Personnel Profile: Full-Time Paid 21; Part-Time Paid 13; Part-Time Volunteers 119.
Volunteer Hours: 13,739
Operating Expenses: 2,025,260
Operating Income: 2,128,390
Governing Authority: state. Tax-exempt: 501(c)(3).
Institution Type/Description: Historic Building: Iolani Palace c.1882 erected as state residence for Hawaii's last king, Kalakaua; Iolani Barracks c.1871 housed Royal Guard; Coronation Pavilion c.1883; 11-acre grounds with many original plantings.
Collections: furnishings & artifacts of the Hawaiian Royal Family from the period of the late 19th-century Hawaiian Monarchy, 1882-1893.
Research Fields: Hawaiian Monarchy period; decorative arts; local history.
Facilities: Museum-related items for sale.
Activities: guided tours of Iolani Palace; lectures.
Publications: quarterly newsletter; books, Iolani Palace Guidebook; Iolani Palace Cookbook; The Queen's Quilt.
Hours & Admission Prices: Guided Tours: Tues.-Thurs. 9-10, Fri.-Sat. 9-11:15; reservations required. Adults $21.75, children 5-12 $6; children 4 & under not admitted. Audio Tours: Mon. 9-4, Tues.-Thurs. 10:30-4, Fri.-Sat. 12-4. Adults $14.75, children 5-12 $6; children under 4 not admitted. Galleries: Mon.-Sat. 9:30-5. Adults $7, children 5-12 $3; children 4 & under welcome; residents with ID, second Sun. of month no charge. Closed New Year's Day; Martin Luther King's Day; Presidents Day; Memorial Day; Veterans Day; Thanksgiving; Christmas. &
Attendance: 115,562 (accurate)
Membership: Na Keiki 12 & under $10; Na Haumana (full-time students) $15; Na Kukui (active palace volunteers) $20; Na Kupuna (Seniors 65 & up) $25; Na Kakoo (Individuals) $35; Nonprofit Organizations $50; Na Ohana (Family) $100; Na Koa (Soldiers) $250; Na Kahu (Attendants) $500; Chamberlain's Circle $1,000; King's Royal Order $2,500; Kaulana Na Pua (Life) $5,000.

JAPANESE CULTURAL CENTER OF HAWAII, 2454 S. Beretania St., Honolulu, HI 96826-1524. Tel.: 808-945-7633. Fax: 808-944-1123.
E-mail: info@jcch.com
Web Site: www.jcch.com
Founded: 1987.
Congressional District: 1
Key Personnel: Exec. Dir., C.E.O. & Pres., Carole Hayashino; Museum Shop Mgr., Christy Takamune.
Personnel Profile: Full-Time Paid 10; Part-Time Paid 3; Part-Time Volunteers 200.
Governing Authority: private; nonprofit. Tax-exempt: 501(c)(3).
Institution Type/Description: History Museum.
Collections: Japanese-American history in Hawaii from 1885 to contemporary times; 8,000 books; periodicals; pamphlets; newspapers; literature on the Japanese in Hawaii & Japan; Japanese on the mainland.
Facilities: 8,000-vol. library on Japanese American history & Japanese culture; education facilities; 5,000 sq. ft. exhibit space. Museum-related items for sale.
Activities: educational panels; forums; lectures. Annual Events: New Year's Ohana Festival; Shi Chi Go San: Keiki Kimono Dressing/Annual Fundraising Gala; Kodomo no Hi: Keiki Fun Fest.
Publications: newsletter, Legacies.
Hours & Admission Prices: Gallery & Gift Shop: Mon.-Sat. 10-4. Adults $7; discounts to senior citizens & school groups; members no charge. Parking validated. &

Attendance: 15,000 (estimated)
Membership: Student $15; Regular $35; Family $50; Individual & Nonprofit Organization $100; Family Sustaining & Supporting $250; Premier I $500; Imperial $1,000.

JOHN YOUNG MUSEUM OF ART, Univ. of Hawaii at Manoa, Krauss Hall, 2500 Dole St., Honolulu, HI 96822-2349. Tel.: 808-956-3634.
Web Site: www.outreach.hawaii.edu/jymuseum
Institution Type/Description: Art Museum.
Collections: Asian, Pacific & African art; art books.
Facilities: library.
Hours & Admission Prices: Call for hours. No charge; donations accepted.

KING KAMEHAMEHA V JUDICIARY HISTORY CENTER, (M), 417 S. King St., Honolulu, HI 96813-2943. Tel.: 808-539-4999. Fax: 808-539-4996.
E-mail: matt@jhchawaii.net
Web Site: www.jhchawaii.net
Founded: 1989.
Congressional District: 1
Key Personnel: Exec. Dir., Matt Mattice.
Personnel Profile: Full-Time Paid 4; Part-Time Volunteers 11.
Governing Authority: executive board; nonprofit. Tax-exempt.
Institution Type/Description: History Museum: housed in Ali'iolani Hale (1874), the Hawaii Supreme Court Building.
Collections: reproductions of artwork; court memorabilia; jury chair; dioramas; restored 1913 courtroom; artifacts, manuscripts; documents relating to Hawaii's legal & judicial history.
Research Fields: translation of 19-century court minute books from Hawaiian to English; oral history of jurists; biographical information on 19th- to 21st-century lawyers & judges; statistical information on 19th-century court cases.
Facilities: 45-seat theater; 5,000 sq. ft. exhibit space.
Activities: guided tours; theater; organized education programs for students; teacher workshops; docent program; temporary exhibitions; exhibits interpreting the transition from Hawaiian law to Western law; Hawaii under martial law.
Publications: semiannual newsletter, Kaulike; books, Kaahumanu, Molder of Change; Ali'iolani Hale, A Sentinel in Time; Curriculum Guide; Trial of a Queen: 1895 Military Tribunal; Judges in the Classroom; Hawaii Under Martial Law, 1941-1944; Annual Report.
Hours & Admission Prices: Mon.-Fri. 8-4; group & school tours by appointment. No charge; donations accepted. Closed state & federal holidays. &
Attendance: 47,715 (estimated)
Membership: Student & Senior Citizen $15; Individual $25; Family $50; Contributing $100; Corporate $300; Associate $500; Patron $1,000.

MANOA HERITAGE CENTER, (M), 2859 Manoa Rd., Honolulu, HI 96822-1752. Tel.: 808-988-1287.
E-mail: manoaheritagecenter@hawaiiantel.net
Web Site: www.manoaheritagecenter.org
Key Personnel: Educ. Dir., Margo Vitelli.
Governing Authority: nonprofit organization.
Institution Type/Description: History Museum: listed on the National Register of Historic Places.
Collections: stone heiau/sacred site; native plants. Historic House: Ku'ali'i.
Hours & Admission Prices: Tues.-Sat. by appointment. Adults $7; seniors & military $4; children & students no charge.
Membership: Century Circle $1,000.

MOANALUA GARDENS FOUNDATION, 1352 Pineapple Place, Honolulu, HI 96819-1796. Tel.: 808-839-5334. Fax: 808-839-3658.
E-mail: moanaluaf001@hawaii.rr.com
Web Site: moanaluagardensfoundation.org
Founded: 1970.
Congressional District: 1
Key Personnel: Exec. Dir., Alexander Jamile.
Personnel Profile: Full-Time Paid 1; Full-Time Volunteers 1; Part-Time Paid 1; Part-Time Volunteers 125.
Governing Authority: nonprofit organization. Tax-exempt: 501(c)(3).
Institution Type/Description: Historic Foundation.
Collections: native & ethnic flora & fauna.
Research Fields: Hawaiian natural & cultural history.
Facilities: 3,000-vol. library of books, videotapes, & maps; reading room; nature center.
Activities: guided tours; lectures; formally organized educational program; field trips; hula festival; teacher training; service trips.

Publications: annual newsletter, Na Makani O Moanalua; workbooks; field-site guides; curricula manuals; videotape.
Hours & Admission Prices: By reservation: Mon.-Tues. & Fri. 8-4, Wed.-Thurs. 8:30-1. Fees for guided tours, lectures & trips. Requested donations for Prince Lot Hula Festival. Closed state holidays.
Membership: Individual $30; Family $40; Friend $100; Supporter $250; Sustainer $1,000; Steward $5,000; Protector $10,000.

PACIFIC AVIATION MUSEUM - PEARL HARBOR, (M), Hangar 37, Ford Island, 319 Lexington Blvd., Honolulu, HI 96818-5004. Tel.: 808-441-1000. Fax: 808-441-1019.

E-mail: reservations@pacificaviationmuseum.org
Web Site: www.pacificaviationmuseum.org
Founded: 1997.
Key Personnel: Exec. Dir., Kenneth H. DeHoff, Jr.; Chm. (V), Adm. Ronald Hays; Pres. (V), Clinton Churchill; Treas., Michael L. Olsen; Dir. Operations, Scotty Scott; Dir. Devel., Carol Arnott; Dir. Mktg. & Business Devel., Anne Murata; Dir. Education, Shauna Tonkin, Ph.D.; Cur., Burl Burlingame.
Governing Authority: private; nonprofit organization. Tax-exempt.
Institution Type/Description: Military Aviation Museum.
Collections: U.S. military aviation in the Pacific from 1913 to present with emphasis on WWII, Korea & Vietnam.
Research Fields: WWII aircraft crash sites in Pacific region.
Facilities: 600-vol. library; aerospace education center; 217,000 sq. ft. exhibit space; 140-seat large screen theater; food court. Museum-related items for sale.
Activities: formal education programs; lectures; theater; tours. Annual Events: Annual Donor Dinner; member events.
Publications: bimonthly newsletter, Pacific Aviation Museum Newsletter.
Hours & Admission Prices: Daily 9-5. Adults $20, children 4-12 $10; discounts to military, Kama'aina & school groups; members & children under 4 no charge. Closed New Year's Day; Thanksgiving; Christmas. &
Attendance: 150,000
Membership: Individual $50; Family $100; Sponsor $500; Advocate $1,000; Patron $5,000.

QUEEN EMMA SUMMER PALACE, 2913 Pali Hwy., Honolulu, HI 96817-1417. Tel.: 808-595-3167. Fax: 808-595-4395.

E-mail: info@daughtersofhawaii.org
Web Site: www.daughtersofhawaii.org
Founded: 1903.
Congressional District: 1
Key Personnel: Regent, Dale Bachman; Museum Shop Mgr., Marty Sczesny; Public Rels., Kelly Ota.
Personnel Profile: Full-Time Paid 13; Part-Time Paid 4; Part-Time Volunteers 50.
Governing Authority: society. Parent Institution: Daughters of Hawaii. Branch Museum: 1837-38, Hulihee Palace, 75-5718 Alii Dr., Kailua-Kona, HI 96740. Tax-exempt: 501(c)(3).
Institution Type/Description: Historic House Museum: c.1847 former home of Queen Emma & King Kamehameha IV, located on site where King Kamehameha I assembled his army, 1795 prior to conquering the island of Oahu & bringing the islands of the Hawaiian group under one head.
Collections: household furnishings & personal effects of Queen Emma & her family; period pieces; portraits; photographs; Hawaiian artifacts; tapa; feather work; Hawaiian quilts.
Research Fields: history of the collection; era presented; ancient Hawaiian history Queen Emma's era; history of Hawaiian quilts.
Facilities: meeting room with stage & kitchen; terraced area. Museum-related items for sale.
Activities: guided tours; special exhibits; demonstrations of dance, arts, crafts of ancient Hawaii; audiovisual programs; Hawaiian crafts; school programs; craft classes; choral group; language; hula classes.
Publications: Daughters of Hawai'i publications: Amos Starr Cooke & Juliette Montague Cooke 'Emalani; Hawaiian Furniture & Hawaii's Cabinetmakers; Liholiho & Emma; Memories of Majesty; Na Lani Kaumaka-A Century of Historic Preservation; Queen Emma & the Bishop; Reflections of Royalty; Stories of Long Ago; Treasures of the Hawaiian Kingdom (English & Japanese versions); Dining with the Daughters: Second Helping. Videos: Hulihe'e Palace-Treasure of the Hawaiian Kingdom; Queen Emma Summer Palace-A Royal Retreat.
Hours & Admission Prices: Daily 9-4. Adults $6, senior citizens $4, youth 17 & under $1; discounts to groups; members no charge. Closed holidays. &
Attendance: 14,144 (estimated)
Membership: Calabash Cousins: Individual $40; Family $60; Patron $110; Life $800; Family Life $1,200. Daughters of Hawaii: Junior $20; Adult $45; Sustaining $100; Life $1,000.

ROYAL MAUSOLEUM STATE MONUMENT, 2261 Nuuanu Ave., Honolulu, HI 96817-1713. Tel.: 808-536-7602.

Key Personnel: Cur., William Maioho
Institution Type/Description: Historic Building: burial site of the Kamehameha & Kalahaua royal families; built in 1865.
Collections: royal family history; burial grounds; chapel.
Hours & Admission Prices: Mon.-Fri. 8-4; guided tours by appointment. No charge.

UNIVERSITY OF HAWAI'I ART GALLERY, (M), 2535 McCarthy Mall, Honolulu, HI 96822-2233. Tel.: 808-956-6888. Fax: 808-956-9659.

E-mail: gallery@hawaii.edu
Web Site: www.hawaii.edu/artgallery
Founded: 1976.
Key Personnel: Dir., Rod Bengston; Assoc. Dir., Sharon Tasaka; Exhibition Design Asst., Wayne Kawamoto.
Personnel Profile: Full-Time Paid 3; Part-Time Paid 14; Part-Time Volunteers 20; Interns 4.
Governing Authority: university. Parent Institutions: University of Hawai'i at Manoa; Dept. of Art & Art History. Tax-exempt: 170(b)(1)(A).
Institution Type/Description: Art Museum.
Collections: works by national & international artists.
Major Exhibits: Graduate MFA Thesis Exhibition, 1/26/14-2/14/14; YUAN - An Exhibition by Beili Liu, 3/9/14-4/11/14; Bachelor FIne Arts Exhibition 2014, 4/27/14-5/16/14.
Research Fields: Asian & Pacific art history.
Facilities: 300-seat auditorium; classrooms; two galleries.
Activities: guided tours; lectures; films; gallery talks; arts festivals; traveling & temporary exhibitions.
Publications: Pranas Domsaitis; The Art of Korea; Egyptian Antiquities from the Charles Pankow Collection; A Tradition of Excellence; Shoebox Sculpture Exhibition catalogues; Glass: Another View; First Impressions: Japanese Prints of Foreigners; Symbol and Surrogate: The Picture Within; The Art of Asian Costume; Sum of the Parts; Naked Truths; Huc Luquien's Hawaii: Prints 1918-1950; Crossings '97: France/Hawaii; Jose Guadalupe Posada: My Mexico; Crossing 2003: Korea/Hawaii; Making Connections: Treasures from the University of Hawaii Libraries; Painting with Threads: The Art of Japanese Embroidery; Reconstructing Memories; Excelling the Work of Heaven: Personal Adornment from China; Writing With Thread: Traditional Textiles of Southwest Chinese Minorities; exhibition catalogues; brochures.
Hours & Admission Prices: late Aug. to mid-May Mon.-Fri. 10-4, Sun. 12-4. No charge; donations accepted. Closed holidays. &
Attendance: 50,000 (estimated)

USS BOWFIN SUBMARINE MUSEUM & PARK, 11 Arizona Memorial Dr., Honolulu, HI 96818-3104. Tel.: 808-423-1341. Fax: 808-422-5201.

E-mail: info@bowfin.org
Web Site: www.bowfin.org
Founded: 1978.
Congressional District: 1
Key Personnel: C.E.O. & Pres., Gerald Hofwolt.
Personnel Profile: Full-Time Paid 46; Part-Time Paid 3; Part-Time Volunteers 1.
Governing Authority: nonprofit association. Parent Institution: Pacific Fleet Submarine Memorial Association. Tax-exempt.
Institution Type/Description: Military Museum.
Collections: WWII submarine, USS Bowfin (SS-287); submarine related artifacts; assorted missile & torpedo displays; WWII patrol reports; files on submarines; files include foreign & domestic submarines & a variety of related topics; publications on submarines; A-1 Polaris missile & A-3 Polaris missile; periscope viewing structure; USS Parche conning tower; Japanese one-man suicide torpedo, Kaiten; C3 Poseidon missile.
Research Fields: submarines; rescue & salvage; deep sea diving; WWII naval history.
Facilities: 2,000 vol. research library; 10,000 item archives; WWII Waterfront Memorial; park.
Activities: self-guided tours; digital audio tours; permanent exhibitions.
Publications: semi-annual newsletter; book, On Eternal Patrol.
Hours & Admission Prices: Daily 7-5. Submarine & Museum: adults $10, military, seniors & Kama'aina $7, children 4-12 $4; Museum Only: adults $5, children $3; members & children under 4 no charge in museum (not permitted on submarine for safety reasons). Closed New Year's Day; Thanksgiving; Christmas.
Attendance: 222,448 (accurate)

WAIKIKI AQUARIUM, 2777 Kalakaua Ave., Honolulu, HI 96815-4027. Tel.: 808-923-9741. Fax: 808-923-1771. Facebook: Waikiki Aquarium.
E-mail: director@waquarium.org
Web Site: www.waquarium.org
Founded: 1904.
Congressional District: 1
Key Personnel: Dir., Dr. Andrew Rossiter; Chm. (V), Dr. Chuck Kelley; Dir. Operations, Jerry Crow; Museum Shop Mgr., Indahwati Soediamto.
Personnel Profile: Full-Time Paid 33; Part-Time Paid 22; Part-Time Volunteers 200.
Governing Authority: state. Parent Institution: University of Hawaii. Tax-exempt: 501(c)(3).
Institution Type/Description: Marine Aquarium.
Collections: Hawaiian & Western tropical Pacific Ocean marine fauna, flora: living fishes, corals, jellyfish, molluscs, crustaceans, echinoderms, Hawaiian monk seals, Hawaiian coastal plants; influence of the ocean on early Hawaiian culture.
Research Fields: husbandry, behavior of aquarium specimens; nautilus husbandry, reproduction, propagation; husbandry, propagation, conservation of corals; Hawaiian monk seal behavior, digestive efficiency and interactions with humans; ID, treatment, prevention of diseases of captive marine life; propagation and biology of Pacific giant clams.
Facilities: rental facilities. Books & other marine-related items for sale.
Activities: formally organized education programs for school groups (pre-K to college); enrichment programs for families, children, adults; permanent exhibitions; study tours; lecture series; rental facilities.
Publications: quarterly newsletter, Kilo I'a.
Hours & Admission Prices: Daily 9-5. Adults $9, Hawaii residents, military & seniors over 60 $6, youths 13-17 $4, juniors 5-12 $2; children 4 & under, schools, approved institutions & FOWA members no charge. Closed Christmas. &
Attendance: 290,479 (accurate)
Membership: Friends of Waikiki Aquarium: Plus One $25 (add one person to existing membership); Senior $30; Individual $40; Grandparents & Family $60; Family Plus $85. Corporate Membership: Coral Reef Circle $1,000; Nautilus Circle $2,500; Mano Circle $5,000; Hawaiian Monk Seal Circle $10,000.

WASHINGTON PLACE, 320 S. Beretania St., Honolulu, HI 96813-2420. Tel.: 808-586-0248. Fax: 808-586-0790.
E-mail: corinne.chun@hawaii.gov
Key Personnel: Dir., Cameron Heen; Cur., Corinne Chun Fujimoto.
Personnel Profile: Full-Time Paid 3.
Governing Authority: Parent Institution: Office of the Governor/State of Hawaii. Subsidiary Institution: Washington Place Foundation.
Institution Type/Description: Historic House: built in 1847 by Capt. John Dominis; home of Queen Lili'uokalani, Hawaii's last monarch & daughter in-law of Dominis.
Collections: Dominis period furnishings; the Queen's personal artifacts including her koa wood piano & Victorian furnishings.
Hours & Admission Prices: Tours: Thurs. by appointment. No charge; donations accepted. Closed holidays. &
Attendance: 9,000 (estimated)

WORLD WAR II VALOR IN THE PACIFIC NATIONAL MEMORIAL, 1 Arizona Memorial Pl., Honolulu, HI 96818. Tel.: 808-422-3300. Fax: 808-483-8608.
Web Site: nps.gov/valr
Formerly: USS Arizona Memorial
Founded: 1980.
Key Personnel: Superintendent, Paul DePrey.
Governing Authority: federal. Dept. of the Interior, National Park Service, Washington, DC & the Arizona Memorial Museum Association.
Institution Type/Description: Park Visitor Center & Military Museum; Historic Site & World War II Memorial Museum.
Collections: American & Japanese preparations for Pearl Harbor amidst two years of World War prior to U.S. entry; history of war time Hawaii & the attacks on Oahu's military sites; salvage & repair of ships; administrative history of the development of the memorial; History WWII Pacific through battle of Midway; USS Arizona (BB-39) and its crew; Pearl Harbor survivors & facilities.
Research Fields: martial law & wartime conditions in Hawaii; internment camps; pre war preparations & World War II in the Pacific through the Battle of Midway; military & civilian history; USS Arizona (BB-39); ships and military units on Oahu on Dec. 7, 1941.
Facilities: research library relating to Pearl Harbor history & World War II available for use on premises only. Books, video tape programs & other museum-related items for sale.

Activities: film presentation; shuttle boat to the Memorial; formally organized education programs for children; permanent & temporary exhibitions.
Publications: catalog; brochures.
Hours & Admission Prices: Pearl Harbor Visitor Center: daily 7-4:30; USS Arizona Tours: daily 8-3. No charge; donations accepted. Closed New Year's Day; Thanksgiving Day; Christmas Day. &
Attendance: 1,420,000
Membership: Brass $25; Copper $50; Bronze $100; Silver $250; Gold $500.

Kahului

MAUI ARTS AND CULTURAL CENTER, One Cameron Way, Kahului, HI 96732-1137. Tel.: 808-242-2787. Fax: 808-244-4665.
E-mail: info@mauiarts.org
Web Site: www.mauiarts.org
Key Personnel: Pres. & C.E.O., Karen A. Fischer; Gallery Dir., Neida Bangerter
Institution Type/Description: Art Gallery.
Collections: national & international artists; paintings; sculpture.
Activities: cultural programs.
Hours & Admission Prices: Mon.-Fri. 9-5. No charge.

Kailua-Kona

ASTRONAUT ELLISON S. ONIZUKA SPACE CENTER, Keahole-Kona International Airport, Kailua-Kona, HI 96745. Mailing Address: P.O. Box 833, Kailua-Kona, HI 96745-0833. Tel.: 808-329-3441. Fax: 808-326-9752.
E-mail: tashima@aloha.net
Key Personnel: Cur., Nancy Tashima
Institution Type/Description: Space Museum: dedicated to the memory of Hawaii's first astronaut who perished in the 1986 Challenger space shuttle disaster.
Collections: personal artifacts; moon rock; Apollo 13 space suit; gravity well; interactive rock-propulsion exhibit.
Facilities: library; 45-seat theater. Museum-related items for sale.
Activities: interactive exhibits.
Hours & Admission Prices: Daily 8:30-4:30. Adults & seniors $3, children 12 & under and students $1. Closed New Year's Day; Thanksgiving; Christmas.

DAUGHTERS OF HAWAII - HULIHEE PALACE, 75-5718 Alii Dr., Kailua-Kona, HI 96740-1702. Tel.: 808-329-1877 & 9555. Fax: 808-329-1321.
E-mail: hp1838@huliheepalace-museum.org
Web Site: www.daughtersofhawaii.org
Formerly: Hulihee Palace
Founded: 1928.
Congressional District: 2
Key Personnel: HP Docent Coord./Publicity & Mktg., Casey Ballao; Sr. Docent, Jolee Chip; Office Mgr., Anita Okimoto; Co-Chair, Lolly Davis; Co-Chair, Lisa Twiggs-Smith; Museum Shop Mgr., Shirley Finan.
Personnel Profile: Full-Time Paid 3; Part-Time Paid 3; Part-Time Volunteers 6.
Governing Authority: society. Parent Institution: Daughters of Hawaii. Tax-exempt.
Institution Type/Description: Historic House Museum: c.1838 vacation residence for Hawaiian royalty, built by Gov. Kuakini, recently restored to the Kalakaua period.
Collections: Hawaiian artifacts; household furnishings; portraits; furniture; tapa; featherwork; Hawaiian quilts.
Research Fields: era presented; ancient Hawaiian artifacts.
Facilities: meeting room with kitchen. Museum-related items for sale.
Activities: guided tours; pageants & concerts by the sea; hula demonstrations; crafts.
Publications: Treasures of the Hawaiian Kingdom; books, Hawaiian Furniture & Hawaii's Cabinet; Stories of Long Ago; pamphlet, Emma & Liholiho.
Hours & Admission Prices: Museum: Mon.-Sat. 9-4; Gift Shop: Mon.-Sat. 9:30-4. Adults $8, senior citizens & Kama'aina $6, students 18 & under $1. Guided tours available upon request at additional cost. Closed holidays. &
Attendance: 26,000 (accurate)
Membership: Support Group $30; Individual $45.

Kalaheo

NATIONAL TROPICAL BOTANICAL GARDEN, 3530 Papalina Rd., Kalaheo, HI 96741-9599. Tel.: 808-332-7324, ext. 203. Fax: 808-332-9765.
E-mail: administration@ntbg.org
Web Site: www.ntbg.org

Founded: 1964.
Congressional District: 2
Key Personnel: C.E.O. & Dir., Chipper Wichman; Chm., Merrill L. Magowan; Museum Shop Mgr., Gwen Silva.
Personnel Profile: Full-Time Paid 94; Part-Time Paid 21; Part-Time Volunteers 250; Interns 5.
Volunteer Hours: 19,412
Operating Expenses: 10,533,623
Operating Income: 10,881,279
Governing Authority: board of trustees; nonprofit. Parent Institution: Organization Headquarters & main research center, Lawai, HI. Garden Sites: McBryde Garden, Allerton Garden, Lawai, HI; Limahuli Garden & Preserve, Haena, HI; Kahanu Garden, Hana, HI; Preserves, Big Island of HI; The Kampong, Coconut Grove, FL. Tax-exempt: 501(c)(3).
Institution Type/Description: Tropical Botanical Garden.
Collections: 15,000-vol research library including the Loy McCandless Marks Botanical Library collection of books; thousands of living plants including rare & endangered species; plants of medicinal & nutritional value; tropical fruits; spices, palms; flowering trees; erythrinas; gingers; breadfruit, coconut; taro; bananas; vanilla; tropical ornamentals; ethnobotanical plants; ancient taro terraces; natural preserves; herbarium: 56,000 specimens, including native naturalized & cultivated plants of Kauai & other Hawaiian islands; vouchers of NTBG plants; 6,500 botanical prints; 17,000 photographic slides; culturally significant structures, objects & plants; sculptured gardens; water features. Allerton Gardens: former estate of Robert Allerton, managed by NTBG; McBryde Garden: scientific & conservation collections; Limahuli Garden: native Hawaiian plants, ancient taro terraces; Kahanu Garden: Polynesian ethnobotanical plants (including world's largest breadfruit collection), Piilanihale Heiau, Historical Landmark; The Kampong: former home horticulturist David Fairchild, tropical fruit cultivars, register historic places.
Research Fields: tropical plants; conservation of rare & endangered species; systematics & taxonomy, restoration ecology, evolutionary & floristic studies on flora of Hawaii & Polynesia; flora of Fiji; flora of the Marquesas Islands & Samoa; nutrition (breadfruit cultivars) medicinal & ethnobotanical plants of Polynesia; plants important to medicine, nutrition, conservation, horticulture & preservation; Florida: plants of the New and Old World tropics; tropical fruit cultivars; Pacific Islands.
Facilities: Headquarters: conservation center; botanical research center; lecture hall. Lawai Valley, Limahuli, Kahanu & Kampong visitor centers. Museum-related items for sale.
Activities: lectures; workshops; educational tours; graduate & post-graduate courses; cooperation education with State Dept. of Education & University of Hawaii; in-school curriculum focusing on native plants; active collaborative relationship with Hawaii botanical & horticultural organizations, focusing on conservation of native plants; Florida horticultural classes University-level, horticultural organizations, tours, botanical drawing classes, ethnobotanic activities.
Publications: regional floras of Hawaii & Fiji; Polynesian medicinal plants manual; taxonomic revisions & monographs; an annotated botanical bibliography: Allertonia (scientific journal); Hawaii: A Natural History; members magazine; pamphlets & brochures; occasional papers; evolutionary studies.
Hours & Admission Prices: Reservations required. Allerton (guided), McBryde (self guided), call 808-742-2623; Limahuli Garden (guided & self guided), call 808-826-1053; Kahanu Garden (self guided), call 808-248-8912; The Kampong (guided) call 305-442-7169. Call for admission fees. Closed holidays. ⟁
Attendance: 96,000 (estimated)
Membership: Individual $50; Dual $85; Contributing $150; Sustaining $250; Supporting $500; Fellow $1,500; Sponsoring $3,000; Patron $5,000; Benefactor $10,000; Chairman's Circle $20,000.

Kalaupapa

KALAUPAPA NATIONAL HISTORICAL PARK, 7 Puahi St., Kalaupapa, HI 96742. Mailing Address: P.O. Box 2222, Kalaupapa, HI 96742-0040. Tel.: 808-567-6802. Fax: 808-567-6729.
Web Site: www.nps.gov/kala/gov
Founded: 1980.
Congressional District: 2
Key Personnel: Supt., Erika Stein Espaniola; Cur., T. Scott Williams; Museum Technician, Kirk Dietz; Museum Technician, Kellie M. Ellis.
Personnel Profile: Full-Time Paid 3.
Governing Authority: federal. Parent Institution: National Park Service, U.S. Dept. of the Interior. Tax-exempt.
Institution Type/Description: National Park: site of the 1886-present Molokai Island leprosy settlement.
Collections: history artifacts; areas related to early Hawaiian settlement; scenic & geologic resources; habitats for rare & endangered species.

Research Fields: natural sciences; historical archaeology; early 20th-century historical Hawaiian architecture.
Hours & Admission Prices: Park: Mon.-Fri. 7-3:30. Tours: Mon.-Sat. Call Damien Tours, 808-567-6171, for tour of the park.

Kamuela

ANNA RANCH HERITAGE CENTER, 65-1480 Kawaihae Rd., Kamuela, HI 96743-8554. Tel.: 808-885-4426.
E-mail: info@annaranch.org
Web Site: www.annaranch.org
Key Personnel: Programs Mgr., Maka Wiggins.
Governing Authority: Parent Institution: Anna Perry-Fiske Charitable Trust. Tax-exempt.
Institution Type/Description: Historic House: former home of Anna Leialoha Lindsey Perry-Fiske. Listed on the Hawaii State Register of Historic Places.
Collections: pa'u costumes, hats, boots & saddles; fine china; Koa furniture.
Facilities: rental facilities
Activities: blacksmith & saddle maker demonstrations.
Hours & Admission Prices: Tues.-Fri. 10-3; guided tours at 10 & 1. Admission $10; discount to Kama'aina residents. Closed most major holidays.
Attendance: 850 (estimated)

PARKER RANCH HISTORIC HOMES & GARDENS, 66-1304 Mamalahoa Highway, Kamuela, HI 96743-8433. Tel.: 808-885-7311. Fax: 808-885-5602.
E-mail: info@parkerranch.com
Web Site: www.parkerranch.com
Founded: 1847.
Key Personnel: C.E.O., Neil T. Kuyper.
Governing Authority: Parent Institution: Parker Ranch Foundation Trust/Parker Ranch Inc.
Institution Type/Description: Historic Homes: housed in Mana Hale, the home of founder John Palmer Parker and Puuopelu, the Parker Hawaiian Victorian Estate from 1879 to 1992.
Collections: Mana Hale: native koa wood interiors; handmade furniture; Hawaiian quilts. Puuopelu: period furnishings; French impressionist art; Asia figurines; gardens.
Activities: rental facilities; tours.
Hours & Admission Prices: Self Guided Tours: Mon.-Fri. 8-4. No charge. ⟁
Attendance: 200

Kaneohe

GALLERY IOLANI AT WINDWARD COMMUNITY COLLEGE, 45-720 Keaahala Rd., Kaneohe, HI 96744. Tel.: 808-236-9155.
Key Personnel: Dir., Antoinette Martin
Institution Type/Description: Art Gallery.
Collections: paintings; posters; drawings.
Hours & Admission Prices: Mon.-Tues. 1-5 & 6-8, Wed.-Fri. & Sun. 1-5.

SENATOR FONG'S PLANTATION & GARDENS, 47-285 Pulama Rd., Kaneohe, HI 96744-5026. Tel.: 808-239-6775. Fax: 808-239-6469.
E-mail: info@fonggarden.com
Web Site: www.fonggarden.net
Key Personnel: Mgr., Patsy Fong
Institution Type/Description: Plantation & Garden Museum.
Collections: palms; fruit & nut trees; ferns; flowers.
Facilities: 725-acre garden. Museum-related items for sale.
Activities: walking tours; weddings; receptions; parties; luncheons; lei classes.
Hours & Admission Prices: Sun.-Fri. 10-2:30. Tours: 10:30am & 1pm. Adults $14.50, children 5-12 $9; discounts to military, AAA members & groups. Closed New Year's Day; Christmas.

Kapolei

HAWAII MUSEUM OF FLYING DBA NAVAL AIR MUSEUM BARBERS POINT, Bldg. 1792, Midway Rd., Kalaeloa Airport, Kapolei, HI 96707. Mailing Address: P.O. Box 75253, Kapolei, HI 96707-0253. Tel.: 808-682-3982. Fax: 808-682-3983.
E-mail: info@nambarberspoint.org
Web Site: www.nambarberspoint.org
Founded: 1999.
Governing Authority: Tax-exempt: 501(c)(3).
Institution Type/Description: Military History Museum.
Collections: naval aviation history; Navy fighter & attack jets; airplanes; helicopters; ground vehicles & equipment.

Hours & Admission Prices: By appointment. Adults $7, children under 18 $5.

Kealakekua

KONA HISTORICAL SOCIETY - H.N. GREENWELL STORE,
81-6551 Mamalahoa Hwy., Kealakekua, HI 96750. Mailing Address: Kona Historical Society, P.O. Box 398, Captain Cook, HI 96704-0398. Tel.: 808-323-3222. Fax: 808-323-2398.
E-mail: khs@konahistorical.org
Web Site: www.konahistorical.org
Founded: 1976.
Key Personnel: Exec. Dir., Joy Holland; Pres. (V), Allen Wall
Governing Authority: nonprofit organization. Tax-exempt.
Institution Type/Description: Historical Society Museum: housed in a restored general store; built c.1875. Listed on the National Register of Historic Places.
Collections: local history & culture; period general store merchandise.
Facilities: Kona Coffee Living History Farm.
Activities: interpretive programs.
Publications: book, Guide to Old Kona; map, Kailua Village Walking Tour; brochure, The Kona Districts.
Hours & Admission Prices: Mon. & Thurs. 10-2. H.N. Greenwell Store: adults $7, children 5-12 $3; discounts to seniors, military & AAA members. &
Attendance: 6,000
Membership: Single $35; Family $55.

Kekaha

KOKEE NATURAL HISTORY MUSEUM, Kokee State Park, Kekaha, HI 96752. Mailing Address: P.O. Box 100, Kekaha, Kauai, HI 96752-0100. Tel.: 808-335-9975. Fax: 808-335-6131.
E-mail: kokeemuseum@earthlink.net
Web Site: www.kokee.org
Founded: 1952.
Key Personnel: Dir., Marsha Erickson.
Personnel Profile: Full-Time Paid 4; Part-Time Paid 44.
Governing Authority: Parent Institution: Hiu O Laka. Tax-exempt.
Institution Type/Description: Natural History Museum.
Collections: local flora, fauna, & natural history; shells; Hawaiian artifacts; photographs.
Hours & Admission Prices: Daily 10-4. No charge; donations accepted. &
Attendance: 95,595 (estimated)

Kilauea

KILAUEA POINT NATIONAL WILDLIFE REFUGE, Kilauea Point, Kilauea, HI 96754. Mailing Address: P.O. Box 1128, Kilauea, HI 96754-1128. Tel.: 808-828-1413. Fax: 808-828-1414.
Institution Type/Description: Wildlife Refuge.
Collections: seabirds; native plants; marine mammals; lighthouse.
Facilities: nature trails.
Activities: hiking; hands-on environmental programs.
Hours & Admission Prices: Daily 10-4.

Kohala

PUUKOHOLA HEIAU NATIONAL HISTORIC SITE, 62-3601 Kawaihae Rd., Kohala, HI 96743-9720. Mailing Address: 62-3601 Kawaihae Rd., Kamuela, HI 96743-9720. Tel.: 808-882-7218. Fax: 808-882-1215.
Web Site: www.nps.gov/puhe
Founded: 1972.
Congressional District: 2
Key Personnel: Supt., Daniel Kawaiaea.
Personnel Profile: Full-Time Paid 9; Full-Time Volunteers 9; Part-Time Paid 4.
Governing Authority: federal. Administered by National Park Service, U.S. Dept. of the Interior. Tax-exempt.
Institution Type/Description: Park Museum: ruins of Puukohola Heiau, Temple on the Hill of the Puukohola Whale, war temple built 1790-1791 by King Kamehameha the Great.
Collections: archaeological excavations & findings.
Facilities: visitor center.
Activities: hiking trails. Annual Events: National Park Week in April; Asian/Pacific American Heritage Day in May; Hawaiian Flag Day in July; Hawaiian Cultural Festival in August.
Publications: park information folder.
Hours & Admission Prices: Daily 7:30-5. No charge; donations accepted. &
Attendance: 256,639 (accurate)

Kualapua

MOLOKAI MUSEUM AND CULTURAL CENTER, Kala'e Hwy., Kualapua, HI 96757. Mailing Address: P.O. Box 269, Kualapua, HI 96757-0269. Tel.: 808-567-6436. Fax: 808-567-6624.
Key Personnel: Exec. Dir., Noelani Keliikipi
Institution Type/Description: Historic Site: housed on the site of the former R.W. Meyer Sugar Mill, built in 1878. Listed on the National Register of Historic Places.
Collections: restored sugar mill; sugar cane maker; photographs.
Facilities: Museum-related items for sale.
Activities: guided hikes; seminars; rental facilities; dance performances.
Hours & Admission Prices: Mon.-Sat. 10-2. Adults $2.50, students 5-18 $1. Closed New Year's Day; Thanksgiving; Christmas.

Kurtistown

FUKU-BONSAI CULTURAL CENTER & HAWAII STATE BONSAI REPOSITORY, 17-656 Olaa Rd., Kurtistown, HI 96760. Mailing Address: P.O. Box 6000, Kurtistown, HI 96760-6000. Tel.: 808-982-9880. Fax: 808-982-9883.
Web Site: www.fukubonsai.com
Institution Type/Description: Nature Center.
Collections: Hawaiian, Japanese, & Chinese bonsai; cultural artifacts; sculpture.
Hours & Admission Prices: Mon.-Sat. 8-4. No charge; donations accepted.

Lahaina

LAHAINA RESTORATION FOUNDATION, 120 Dickenson, Lahaina, HI 96761-1224. Tel.: 808-661-3262. Fax: 808-661-9309.
E-mail: info@lahainarestoration.org
Web Site: www.lahainarestoration.org
Founded: 1962.
Congressional District: 2
Key Personnel: Exec. Dir., Theo Morrison; Pres., David Allaire.
Personnel Profile: Full-Time Paid 7; Part-Time Paid 9; Part-Time Volunteers 4.
Governing Authority: nonprofit organization. Tax-exempt: 501(c)(3).
Institution Type/Description: History Museums.
Collections: marine; archives; medical; history; architecture; period furniture; art; Hawaiian stone artifacts & tools; manuscripts; microfilm; 19th-century ship logs & journals; Maui's Humpback whales & the undersea world of Hawaii; Chinese artifacts; working Taoist Altar; Lahaina life during the Hawaiian Monarchy 1820-1893; plantation era artifacts. Historical Houses; 1835 Baldwin Home; 1834 Masters Reading Room; 1837 Hale Pai Printing House; 1900 Wo Hing Chinese Temple; Hale Paahao Prison; Hale Aloha Church.
Research Fields: local history; ethnic & anthropological studies.
Facilities: 500-vol. library of early Hawaiiana available on premises by request; reading room. Books, microfilms, artifact reproductions & prints for sale.
Activities: guided tours; lectures; films; permanent & temporary exhibitions; films. Museum Sponsors: ethnic & cultural development & preservation programs for children.
Publications: monthly newsletter; walking tour map.
Hours & Admission Prices: Daily 10-4. Adults $7; members & children no charge. Closed Thanksgiving; Christmas. &
Attendance: 240,000 (estimated)
Membership: Individual $25; Ohana $35; Supporting $75; Business $100; Historian $250; Preservationist $500; Patron $1,000; Heritage Partner $2,500; Heritage Leader $5,000.

Lahaina, Maui

WHALERS VILLAGE MUSEUM, 2435 Ka'anapali Pkwy., Bldg. H-6, Lahaina, Maui, HI 96761-1980. Tel.: 808-661-4567.
E-mail: info@whalersvillage.com
Web Site: www.whalersvillage.com/museum/museum.htm
Institution Type/Description: Whaling Museum.
Collections: Lahaina's whaling era, 1825-1860; photo murals; recreated whaling ship forecastle; period ornaments & utensils made from whale ivory and bone; 19th century scrimshaw.
Facilities: Museum-related items for sale.
Activities: self-guided audio tours; short videos.
Hours & Admission Prices: Daily 9am-10pm. No charge.

Laie

POLYNESIAN CULTURAL CENTER, 55-370 Kamehameha Hwy., Laie, HI 96762-1113. Tel.: 808-293-3005. Fax: 808-293-3022.
E-mail: internetrez@polynesia.com
Web Site: www.polynesia.com
Institution Type/Description: Cultural Center.
Collections: people & cultures of Polynesia including Hawaii, Samoa, Aotearoa, Fiji, the Marquesas, Tahiti, and Tonga.
Facilities: IMAX theater. Museum-related items for sale.
Activities: shows.
Hours & Admission Prices: Mon.-Sat. 12:30-6:30. Call for admission prices.

Lanai City

LANAI CULTURE & HERITAGE CENTER, 730 Lanai Ave., Ste. 118, Lanai City, HI 96763. Mailing Address: P.O. Box 631500, Lanai City, HI 96763-1305. Tel.: 808-565-7177.
E-mail: mikala@lanaichc.org
Web Site: www.lanaichc.org
Institution Type/Description: History Museum.
Collections: local heritage & culture; photographs; ranching; plantation history.
Activities: research.
Hours & Admission Prices: Mon.-Fri. 8:30-3:30. Sat. 9-1; groups by appointment. No charge.

Laupahoehoe

LAUPAHOEHOE TRAIN MUSEUM, 36-2377 Mamalahoa Hwy., Laupahoehoe, HI 96764. Mailing Address: P.O. Box 358, Laupahoehoe, HI 96764-0358. Tel.: 808-962-6300. Fax: 808-963-6957.
E-mail: thetrainmuseum@yahoo.com
Web Site: www.thetrainmuseum.com
Founded: 1998.
Key Personnel: Treas., Doug Connors.
Governing Authority: Tax exempt.
Institution Type/Description: History Museum.
Collections: islands history of railroads, plantations & local cultures; photographs; video.
Facilities: Museum-related items for sale.
Hours & Admission Prices: Thurs.-Sun. 10-5; other times & groups by appointment. Family $15, adults $6, seniors $5, children $3; discounts to AAM members. Closed holidays.
Attendance: 7,000 (estimated)
Membership: Grand Dancer $25; Brakeman $40; Fireman $100; Conductor $500; Engineer $1,000.

Lihue

GROVE FARM MUSEUM, 4050 Nawiliwili Rd., Lihue, HI 96766. Mailing Address: P.O. Box 1631, Lihue, Kauai, HI 96766. Tel.: 808-245-3202. Fax: 808-245-7988.
E-mail: tours@grovefarm.org
Web Site: grovefarm.org
Key Personnel: Dir., Robert J. Schleck; Cur., Moises Madayag
Institution Type/Description: Sugar Plantation: listed on the National Register of Historic Places.
Collections: sugar plantation history; period furnishings; historic buildings.
Activities: guided tours.
Hours & Admission Prices: Mon. & Wed.-Thurs. 10am & 1pm by appointment. Adults $20, children 5-12 $10.
Membership: Kupuna $35; Palaka $40; Ohana $70; Kaipu $120; Cane Fire $250; Sugar Mill $500; Sisters $1,000; G.N. $2,500.

KAUAI HISTORICAL SOCIETY, 4396 Rice St., Ste. 101, Lihue, HI 96766-1371. Mailing Address: P.O. Box 1778, Lihue, HI 96766-5778. Tel.: 808-245-3373. Fax: 808-245-8693.
E-mail: info@kauaihistoricalsociety.org
Web Site: www.kauaihistoricalsociety.org
Founded: 1914.
Key Personnel: Dir., Mary Requilnan.
Personnel Profile: Full-Time Paid 1; Part-Time Paid 1; Part-Time Volunteers 12.
Governing Authority: nonprofit organization. Tax-exempt.
Institution Type/Description: Historical Society Museum & Archive.
Collections: period furniture; art; manuscripts; maps; photographs.
Facilities: library.
Activities: guided & school tours; permanent exhibits; historical programs; classes; workshops; performances; monthly lectures. Annual Event: The Royal Pa'ina in April.
Hours & Admission Prices: Office: Mon.-Fri. 8-4. Archives: by appointment. No charge; donations accepted.
Membership: Senior & Student $25; Individual $35; Family $50; Small Business $100 & up; Supporting $125; Contributing $250; Patron & Corporation $500.

KAUAI MUSEUM, 4428 Rice St., Lihue, HI 96766-1338. Tel.: 808-245-6931. Fax: 808-245-6864.
E-mail: director@kauaimuseum.org
Web Site: kauaimuseum.org
Founded: 1960.
Congressional District: 2
Key Personnel: Exec. Dir., Education & Art Mezzanine Gallery and Museum Shop Mgr., Jane Gray; Pres. (V), Holbrook Goodale; Cur., Chris Faye.
Personnel Profile: Full-Time Paid 3; Part-Time Paid 7; Part-Time Volunteers 70.
Governing Authority: nonprofit corporation. Tax-exempt: 501(c)(3)
Institution Type/Description: History & Art Museum: housed in early 20th-century Wilcox building designed by Hart Wood.
Collections: Hawaiiana collection with particular emphasis on items dealing with the island of Kauai; Kauai photographs; school art exhibits; ethnic & heritage displays; paintings; dolls; textiles; WWII memorabilia; furnishings; weapons; scrimshaw.
Research Fields: Hawaiiana, particularly the Island of Kauai.
Facilities: 1,000-vol. library of books; pamphlets; photographs dealing with the history of the Island of Kauai available on premises. Hawaiiana books Niihau shell leis prints & handicrafts from the South Pacific for sale.
Activities: guided tours; films; arts festivals; permanent & temporary exhibitions. Museum Sponsors: Filipino Fiesta in January; Chinese New Year in February; Chinese Rice Farming in March; Kaua'i Merrie Monarch Maidens in April; Annual May Day Lei Contest in May; Ukulele Festival in June; 5th Annual Japanese Cultural Festival in July; Floating Lantern Traditions in August; Na Mele O Kaua'i in September; Traditions and Games of Makahiki in October; Lonoikamakahiki in November; Aloha Kalikimaka in December.
Publications: Hawaiian Quilting on Kauai; Early Kauai Hospitality 1820-1920: A Family Book of Receipts; Amelia: a novel of midcentury Hawaii; Moki Learns to Count: a children's book; Kauai, The Separate Kingdom; Kauai Museum Quilt Collection.
Hours & Admission Prices: Mon.-Sat. 10-5. Adults $10, seniors 65 & over $8, youths 13-17 $6, children 6-12 $2; discounts for members' guests, Western Museum Assoc. & Hawaii Museum members; 1st Sat. each month, members and children 5 & under no charge. Closed New Year's Day; Independence Day; Memorial Day; Labor Day; Thanksgiving; Christmas.
Attendance: 25,175 (estimated)
Membership: Individual $25; Family $45; Contributing $75; Sustaining $100; Patron $250; Corporate $500 & up.

KAUAI VETERANS MUSEUM, 3215 Kapule Hwy., Lihue, HI 96766. Tel.: 808-246-1135.
E-mail: kumuseum@gmail.com
Web Site: kauaiveteranscenter.com
Institution Type/Description: Military Museum.
Collections: military history & artifacts; photographs; personal artifacts; military weapons & uniforms.
Hours & Admission Prices: Mon.-Fri. 10-2; guided tours by appointment. No charge; donations accepted. Closed holidays.

KILOHANA PLANTATION, 3-2087 Kaumualii Hwy., Lihue, HI 96766-9505. Mailing Address: P.O. Box 3121, Lihue, HI 96766-6121. Tel.: 808-245-5608. Fax: 808-245-7818.
E-mail: kilohana@hawaiilink.net
Web Site: www.kauaikilohana.com
Founded: 1986.
Institution Type/Description: Historic House: housed in the former home of sugar baron, Gaylord Parke Wilcox; built in 1936.
Collections: period furnishings; personal artifacts.
Hours & Admission Prices: Mon.-Sat. 10-9:30, Sun. 9:30-3. No charge.

Makawao

HUI NO'EAU VISUAL ARTS CENTER, 2841 Baldwin Ave., Makawao, HI 96768-9642. Tel.: 808-572-6560. Fax: 808-572-2750.
E-mail: info@huinoeau.com
Web Site: huinoeau.com
Founded: 1934.
Key Personnel: Exec. Dir., Caroline Killhour; Sr. Devel. Officer, Shay Belisle; Registrar, Jessica Hoecker; Sr. Program Mgr., Anne-Marie Forsythe; Programs & Exhibitions Coord., Lana Coryell.
Governing Authority: private; nonprofit organization.
Institution Type/Description: Visual Arts Center: housed in 1917 Mediterranean-style home.
Collections: artifacts of Harry & Ethel Baldwin & their daughter Francis Cameron.
Facilities: library of historical references of art history, art magazines & books available to members; botanical garden; educational facilities; lecture & slide room. Museum-related items for sale.
Activities: docent program; guided tours; lectures. Museum Sponsors: Christmas House; Art Affair; Open House.
Publications: quarterly newsletter, Hui News.
Hours & Admission Prices: Mon.-Sat. 10-4. No charge; donations accepted. &
Attendance: 30,000 (estimated)
Membership: Individual $45-$74; Family $75-$124; Sponsor $125-$249; Supporter $250-$499; Patron $500-$999; Benefactor $1,000-$2,499; President's Circle $2,500-$4,999; Director's Circle $5,000 & up.

Paia

MAUI CRAFTS GUILD, 120 Hana Hwy., Paia, HI 96779. Mailing Address: P.O. Box 790609, Paia, HI 96779-0609. Tel.: 808-579-9697. Fax: 808-579-8694.
E-mail: info@mauicraftsguild.com
Web Site: www.mauicraftsguild.com
Institution Type/Description: Art Museum.
Collections: works by local artists including ceramics, sculpture, prints, textiles, photographs, baskets.
Hours & Admission Prices: Daily 10-6.

Papaikou

HAWAII TROPICAL BOTANICAL GARDEN, 27-717 Old Mamalahoa Hwy., Papaikou, HI 96781-7746. Mailing Address: P.O. Box 80, Papaikou, HI 96781-0080. Tel.: 808-964-5233. Fax: 808-964-1338. Facebook: Hawaii Tropical Botanical Garden.
E-mail: htbg@ilhawaii.net
Web Site: www.hawaiigarden.com
Founded: 1978.
Congressional District: 2
Key Personnel: Dir., David C. Tan; Pres., Pauline Lutkenhouse.
Governing Authority: nonprofit. Tax-exempt: 501(c)(3).
Institution Type/Description: Botanical Garden.
Collections: tropical & subtropical plants; palms; heliconias; gingers; bromeliads.
Facilities: 40-acres.
Hours & Admission Prices: Garden: daily 9-5. Gift Shop: daily 8:30-5. Adults $15, children 6-16 $5; children under 6 no charge. Closed New Year's Day; Thanksgiving; Christmas.
Attendance: 80,000 (estimated)
Membership: Garden Patron $1,000; Life $2,500; Tree of Life $5,000; Garden Benefactor $10,000.

Puunene

ALEXANDER & BALDWIN SUGAR MUSEUM, (M), 3957 Hansen Rd., Puunene, HI 96784. Mailing Address: P.O. Box 125, Puunene, HI 96784-0125. Tel.: 808-871-8058. Fax: 808-871-4321.
E-mail: sugarmus@maui.net
Web Site: sugarmuseum.com
Founded: 1980.
Key Personnel: Dir., Roslyn Lightfoot; Pres. (V), Douglas Sheehan; Contact, Holly Buland; Museum Shop Mgr., Lani Medeiros.
Personnel Profile: Full-Time Paid 2; Part-Time Paid 5; Part-Time Volunteers 20.
Governing Authority: private; nonprofit organization. Tax-exempt.
Institution Type/Description: Agriculture Museum: housed in 1902 former sugar factory superintendent's residence, located across the way from Hawaii's largest sugar mill.
Collections: focus on sugar industry & plantation life on the island of Maui from 1850s to 1950s; emphasis on native cultures of various ethnic groups brought to Hawaii as labor.
Research Fields: sugar industry on island of Maui, 1850-1950; Hawaiian plantation life.
Facilities: 1,800 sq. ft. exhibit space. Hawaiian sugar-based food products, local craft items & other museum-related items for sale.
Activities: docent programs; educational programs for children; traveling exhibits.
Publications: newsletter.
Hours & Admission Prices: Daily 9:30-4:30. Adults $7, children $2. Closed New Year's Day; Easter; Thanksgiving; Christmas. &
Attendance: 35,000 (estimated)

Schofield Barracks

TROPIC LIGHTNING MUSEUM, Bldg. #361, Waianae Ave., Schofield Barracks, HI 96857-5000. Mailing Address: Directorate of Plans, Training, Mobilization & Security, 745 Wright Ave., Bldg. 107, Wheeler Army Airfield, Schofield Barracks, HI 96857-5026. Tel.: 808-655-0438.
Founded: 1958.
Key Personnel: Cur., Kathleen Ramsden
Institution Type/Description: Military History Museum.
Collections: history of the 25th Infantry Division from 1941 to present; Wheeler Army Airfield; military equipment & uniforms; artillery guns; military vehicles; Schofield Barracks.
Hours & Admission Prices: Tues.-Sat. 10-4. No charge; donations accepted. Closed holidays. &
Attendance: 14,000 (estimated)

Volcano

VOLCANO ART CENTER GALLERY, 19-4074 Old Volcano Rd., Volcano, HI 96785. Mailing Address: P.O. Box 129, Volcano, HI 96785-0129. Tel.: 808-967-7565. Fax: 808-967-7511.
E-mail: gallery@volcanoartcenter.org
Web Site: www.volcanoartcenter.org
Founded: 1974.
Congressional District: 2
Key Personnel: C.E.O., Tanya Aynessazian; Gallery Mgr., Shelby Smith; Bd. Chm. (V), Linda Pratt; Bd. Treas., Mike Miyahira; Education Coord., Marsha Hee; Office Mgr. & Bookkeeper, Jane Buchholz.
Personnel Profile: Full-Time Paid 10; Part-Time Paid 9; Part-Time Volunteers 300; Interns 1.
Governing Authority: not-for-profit organization. Offices: 19-4074 Old Volcano Rd., Volcano, HI 96785. Tax-exempt: 501(c)(3).
Institution Type/Description: Cultural Center & Gallery: housed in a hotel built in 1877.
Collections: paintings; photographs; prints; ceramics; wood; glass; jewelry.
Facilities: 300-seat theatre; 3 classrooms; 7.4 acre rainforest site with nature trails.
Activities: concerts; dance recitals; guided tours; workshops; lectures; theatre. Museum Sponsors: May Day is Lei Day; Christmas in the Country; Love the Arts.
Publications: quarterly, Art Beat Magazine.
Hours & Admission Prices: Daily 9-5. No charge; donations accepted. Closed Christmas. &
Attendance: 100,000 (estimated)
Membership: Students & Seniors $20; Individual $40; Dual & Family $50; Contributor $100; Supporting $250; Sponsor $500; Patron $1,000; Benefactor $2,500.

Wailuku

HAWAII NATURE CENTER, 875 Iao Valley Rd., Wailuki, HI 96793. Mailing Address: 2131 Makiki Heights Dr., Honolulu, HI 96822. Tel.: 808-955-0100; 244-6500.
E-mail: info@hawaiinaturecenter.org
Web Site: www.hawaiinaturecenter.org
Founded: 1981.
Key Personnel: Dir. Operations, Casey Carmichael; Dir. Philanthropy, Kathryn A. Currier
Institution Type/Description: Interactive Nature Museum.
Collections: island environment, culture, & history; wildlife; plants.
Hours & Admission Prices: Mon.-Fri. 12-4, Sat.-Sun. 11-3. Grounds no charge. Closed New Year's Day; Thanksgiving; Christmas.
Attendance: 15,000 (estimated)
Membership: Family $50.

Wailuku

HAWAII NATURE CENTER, MAUI, 875 Iao Valley Rd., Wailuku, HI 96793-3009. Tel.: 808-244-6525.
E-mail: bookmaui@hawaiinaturecenter.org
Web Site: www.hawaiinaturecenter.org
Founded: 1992.
Key Personnel: Museum Shop Mgr., Dee Dee Santos
Institution Type/Description: Nature Center.
Collections: island environment, culture, & history; wildlife; plants.
Activities: educational programs; school group tours; hiking.
Publications: quarterly newsletter; brochures.
Hours & Admission Prices: Daily 10-4. Adults $6; members no charge.
Membership: Individual $25; Family $50; Contributing $250; Sustaining $500; Steward $1,000.

MAUI HISTORICAL SOCIETY - BAILEY HOUSE MUSEUM, 2375A Main St., Wailuku, HI 96793-1661. Tel.: 808-244-3326.
E-mail: baileyhousemuseum@clearwire.net
Web Site: www.mauimuseum.org
Formerly: Hale Hoikeike
Founded: 1951.
Congressional District: 2
Key Personnel: Exec. Dir., Nicole McMullen
Personnel Profile: Full-Time Paid 2; Part-Time Paid 2; Part-Time Volunteers 20; Interns 2.
Governing Authority: nonprofit organization. Parent Institution: Maui Historical Society. Tax-exempt: 501(c)(3) & 170(b)(1)(A).
Institution Type/Description: Historic House: housed on an ancient Hawaiian site.
Collections: Maui & Hawaiian history & culture; native plant garden; native Hawaiian artifacts of stone, wood, shell, & feathers; furniture; missionary era clothing; paintings by E. H. Bailey; Kaho'olawe Island artifacts; photographs.
Research Fields: history & archaeology of Maui County; ancient Hawaiian history & missionary era.
Facilities: Museum-related items and books, clothing, jewelry, cards, stationery & Hawaiian crafts for sale.
Activities: guided tours; lectures; inter-museum loan, permanent & temporary exhibitions.
Publications: Index to the Maui News; Maui, A Guide to Resources; journal, Imi Ike.
Hours & Admission Prices: Mon.-Sat. 10-4. Adults $7, seniors $5, children 6-12 $2; discounts to seniors 60 & over; active military, members & travel professionals no charge. Closed holidays.
Attendance: 11,000 (estimated)
Membership: Student & Senior Citizen $10 discount for each level; Individual $35; Family $50; $100; $250; $500; $1,000; $3,000.

MAUI OCEAN CENTER, 192 Maalaea Rd., Wailuku, HI 96793-5931. Tel.: 808-270-7000. Fax: 808-270-7070. Facebook: Maui Ocean Center.
E-mail: info@mauioceancenter.com
Web Site: www.mauioceancenter.com
Founded: 1998.
Key Personnel: Gen. Mgr., Kate Zolezzi; Controller, Bridget Reardon; Education, Erin Iberg; Sales & Mktg., Lori Mellenbruch; Cur., John Gorman; Museum Shop Mgr., Tapani Vuori.
Governing Authority: Parent Institution: Coral World International.
Institution Type/Description: Aquarium.
Collections: Hawaiian marine life & cultural artifacts; living coral; reef fish; invertebrates; sharks; stingrays; sea turtles.
Facilities: aquarium; 200-seat restaurant. Museum-related items for sale.
Activities: lectures. Annual Events: Earth Day; World Ocean Day; Shark Week.
Hours & Admission Prices: July-Aug. daily 9-6; Sept.-May daily 9-5. Adults $25.50, senior citizens 65 & over $22.50, children 3-12 $18.50; discounts to groups; members no charge. &
Attendance: 350,000 (estimated)
Membership: Children 3-12 $42; Senior 65 & over $50; Adult $55; Family $165.

MAUI OKINAWA CULTURAL CENTER, 688 Nukuwai Place, Wailuku, HI 96793-1340. Mailing Address: P.O. Box 1884, Wailuku, HI 96793-6884. Tel.: 808-242-1560. Fax: 808-242-5952.
Institution Type/Description: Art Museum.
Collections: Okinawan culture; pottery; textiles; lacquerware; bingata; calligraphy.
Hours & Admission Prices: Mon.-Fri. 8:30-11:30; other times by appointment. No charge; donations accepted.

Waimanalo

SEA LIFE PARK HAWAII, 41-202 Kalanianaole Hwy., Ste. 7, Waimanalo, HI 96795-1897. Tel.: 866-393-5158 (Toll Free).
Institution Type/Description: Nature Center.
Collections: marine mammals including dolphins, sea lions, rays, & penguins.
Facilities: theater; restaurant. Gift items for sale.
Activities: interactive programs; swim with dolphins.
Hours & Admission Prices: Daily 10:30-5. Adult $29.99, children 3-11 $19.

Waimea

FAYE MUSEUM, 9400 Kaumualii Hwy., Waimea, HI 96796. Mailing Address: c/o Waimea Plantation Cottages, P.O. Box 367, Waimea, HI 96796. Tel.: 808-338-1625.
E-mail: waimeasugar@hawaiilink.net
Key Personnel: Cur., Chris Faye
Institution Type/Description: History Museum: a pioneer sugar planter in West Kauai, H.P. Faye helped form Kekaha Sugar, incorporated in 1898.
Collections: local history; sugar industry; personal artifacts; photographs.
Hours & Admission Prices: Open Year Round: Mon.-Fri. 9-5. No charge.

GALLERY WEST, 9400 Kaumualii Hwy., Waimea, HI 96796. Mailing Address: c/o Waimea Plantation Cottages, P.O. Box 367, Waimea, HI 96796. Tel.: 808-338-1625 & 2340.
Key Personnel: Mgr., Kathleen Miguel
Institution Type/Description: Art Gallery.
Collections: works by local artists.
Hours & Admission Prices: Call for hours.

WAIMEA SUGAR MILL CAMP MUSEUM, 9400 Kaumualii Hwy., Waimea, HI 96796. Mailing Address: c/o Waimea Plantation Cottages, P.O. Box 367, Waimea, HI 96796. Tel.: 808-337-1005. Fax: 808-337-9449.
Institution Type/Description: History Museum.
Collections: local history & culture.
Hours & Admission Prices: Tues., Thurs. & Sat. 9am by appointment.

WEST KAUAI TECHNOLOGY & VISITOR CENTER, 9565 Kaumualii Hwy., Waimea, HI 96796. Mailing Address: c/o West Kauai Business & Professional Association, P.O. Box 903, Waimea, HI 96796. Tel.: 808-338-1332.
Institution Type/Description: History Museum.
Collections: area culture & history; science; technology.
Activities: multimedia presentations; demonstrations; walking tour; special events.
Hours & Admission Prices: Mon.-Fri. 9:30-5.

Waipahu

HAWAII OKINAWA CENTER, 94-587 Ukee St., Waipahu, HI 96797-4214. Tel.: 808-676-5400. Fax: 808-676-7811.
Institution Type/Description: Cultural Center.
Collections: local history & culture; Okinawa crafts including pottery, doll making, & fabrics; early plantation & immigration.
Hours & Admission Prices: Mon.-Fri. 8:30-5. Closed New Year's Day; Presidents' Day; Memorial Day; Independence Day; Labor Day; Thanksgiving; Christmas.

HAWAII'S PLANTATION VILLAGE, 94-695 Waipahu St., Waipahu, HI 96797-2601. Tel.: 808-677-0110. Fax: 808-676-6727.
E-mail: hpv.waipahu@hawaiiantel.net
Web Site: www.hawaiiplantationvillage.org
Founded: 1973.
Congressional District: 2
Key Personnel: Pres., Faith P. Evans; 1st Vice Pres., Deanna Espinas; 2nd Vice Pres., Loretta Pang; 3rd Vice Pres., Richard Oshiro; Sec., Robert Castro; Treas., Glenn Ifuku.
Personnel Profile: Full Time Paid 6; Part-Time Paid 18; Part-Time Volunteers 120.
Governing Authority: private; nonprofit organization. Tax-exempt: 501(c)(3).
Institution Type/Description: General Museum.
Collections: 1900-1940 Hawaii plantation family personal artifacts; ethnic & cultural materials related to the 8 major ethnic groups to work on the plantations.
Research Fields: plantation history; Oahu sugar records.
Facilities: 200-vol. library of Hawaii plantation & ethnic history books; 100 sq. ft. exhibit space; educational facilities. Museum-related items for sale.

Activities: docent program; guided tours; lectures; temporary exhibitions; ethnic heritage events.

Publications: quarterly newsletter, Plantation News.

Hours & Admission Prices: Office: Mon.-Sat. 8-4:30. Guided Tours: Mon.-Sat. 10-2. Adults $13, seniors 62 & over $10, Kama'aina & military $7, youth 4-11 $5. Closed New Year's Day; Good Friday; Memorial Day; Independence Day; Labor Day; Thanksgiving; Christmas. &

Attendance: 60,000 (estimated)

Membership: Youth $10; Student $15; Double Seniors $20; Individual $25; Family $40; Family Plus $55; Hoe Hana $100-$249; Wai Hana $250-$499; Luna $500-$749; Manager $750-$999; Sustainer $1,000 & up.

IDAHO

(126 listings)

Almo

CITY OF ROCKS NATIONAL RESERVE, 3035 Elba Almo Rd., Almo, ID 83312. Mailing Address: P.O. Box 169, Almo, ID 83312-0169. Tel.: 208-824-5910.

Founded: 1988.

Congressional District: 2

Governing Authority: Parent Institution: NPS/Idaho Dept. of Parks & Recreation.

Institution Type/Description: Park Museum.

Collections: local history; geology; flowers; trees; plants.

Facilities: visitor center.

Hours & Admission Prices: Park: daily. Visitor Center: April 17-Oct. 23 daily 8-4:30; Oct. 24-April 11 Mon.-Fri. 8-4:30. No charge; donations accepted. &

American Falls

MASSACRE ROCKS STATE PARK, 3592 Park Lane, American Falls, ID 83211-5556. Tel.: 208-548-2672. Fax: 208-548-2671.

E-mail: mas@idpr.state.id.us

Web Site: parksandrecreation.idaho.gov/parks/massacrerocks.aspx

Founded: 1967.

Congressional District: 2

Key Personnel: Park Mgr., L. Max Newlin.

Personnel Profile: Full-Time Paid 3.

Governing Authority: state. Tax-exempt.

Institution Type/Description: Park Museum.

Collections: Indian artifacts; natural history; pioneer & Oregon Trail history.

Research Fields: Oregon Trail history.

Facilities: information center.

Activities: guided nature walks; evening campfire programs; Oregon Trail tours.

Publications: Geology of Massacre Rocks; Jane A. Gould Journal; The Oregon Trail Revisited; Historic Sites along the Oregon Trail; maps.

Hours & Admission Prices: Visitor Center: May 15-Sept. 15 daily 7:30am-9pm. Park: Summer: daily 8:30-8:30; Sept.-June daily 7-3:30. $4 per vehicle. &

Attendance: 50,000

Arco

CRATERS OF THE MOON NATIONAL MONUMENT AND PRESERVE, 18 mi. S.W. of Arco on Hwy. 26, Arco, ID 83213. Mailing Address: P.O. Box 29, Arco, ID 83213-0029. Tel.: 208-527-1300. Fax: 208-527-3073.

Web Site: www.nps.gov/crmo

Formerly: Craters of the Moon National Monument

Founded: 1924.

Congressional District: 2

Key Personnel: Supt., Dan Buckley.

Personnel Profile: Full-Time Paid 1; Part-Time Paid 1.

Governing Authority: federal. Parent Institution: National Park Service. Tax-exempt.

Institution Type/Description: Natural History Museum.

Collections: local geology; history; herbarium; manuscripts.

Facilities: 500-vol. library on natural science. Books for sale.

Activities: guided tours; campfires during summertime; films; permanent exhibitions.

Publications: Around the Loop; Geological Map of Craters of the Moon; pocket flower guide; Unearthly Landscape, Craters of the Moon-official handbook (published by the supporting Natural History Association).

Hours & Admission Prices: Memorial Day-Labor Day daily 8-6; Sept.-May

daily 8-4:30. Visitor Center: no charge; donations accepted. Park: $8 per vehicle. Closed winter holidays. &

Attendance: 217,000 (accurate)

Athol

FARRAGUT STATE PARK, 13550 E. Hwy. 54, Athol, ID 83801. Tel.: 208-683-2425.

Institution Type/Description: State Park.

Collections: natural history; wildlife; local history.

Facilities: visitor center; naval training center. Museum-related items for sale.

Activities: hiking; biking; camping; guided walks; educational programs; model airplane flyer's field.

Hours & Admission Prices: Daily.

Atlanta

ATLANTA HISTORICAL SOCIETY, Middle Fork Rd., Atlanta, ID 83601. Mailing Address: P.O. Box 53, Atlanta, ID 83601.

Governing Authority: nonprofit. Tax-exempt: 501(c)(3).

Institution Type/Description: Historical Society Museum.

Collections: local history; period artifacts; photographs; pioneer life. Historic Buildings; 1910 jail; c.1870 cabin.

Hours & Admission Prices: Summer: daily 10-9; other times by appointment. No charge.

Blackfoot

BLACKFOOT COMMUNITY CENTER'S CHILDREN'S MUSEUM OF IDAHO, 157 W. Sexton, Blackfoot, ID 83221. Tel.: 208-785-8022.

E-mail: thecenter@cableone.net

Web Site: blackfootcommunitycenter.blogspot.com

Key Personnel: Dir., Ashlee Howell

Institution Type/Description: Children's Museum.

Collections: hands-on exhibitions.

Activities: interactive exhibitions; educational programs.

Hours & Admission Prices: Call for hours.

IDAHO POTATO MUSEUM, 130 N.W. Main St., Blackfoot, ID 83221-0801. Mailing Address: P.O. Box 801, Blackfoot, ID 83221-0801. Tel.: 208-785-2517. Fax: 208-785-7974.

E-mail: info@idahopotatomuseum.com

Web Site: idahopotatomuseum.com

Founded: 1984.

Congressional District: 2

Key Personnel: Dir., Tish Dahmen; Pres. (V), Jonathan Harris; Museum Shop Mgr., Jesica Smith.

Personnel Profile: Full-Time Paid 2; Part-Time Paid 1.

Governing Authority: Parent Institution: Idaho Potato Museum Board of Directors. Tax-exempt.

Institution Type/Description: Agriculture Museum.

Collections: potato industry history; period farming equipment.

Facilities: Museum-related items for sale.

Activities: video.

Hours & Admission Prices: April-Sept. Mon.-Sat. 9:30-5; Oct.-March Mon.-Fri. 9:30-3. Adults $3, seniors 55 & over $2.50, children 6-12 $1; discounts to groups, military & AAA members.

Attendance: 6,000 (estimated)

Boise

BASQUE MUSEUM & CULTURAL CENTER, 611 Grove St., Boise, ID 83702-5971. Tel.: 208-343-2671. Fax: 208-336-4801.

E-mail: info@basquemuseum.com

Web Site: www.basquemuseum.com

Founded: 1985.

Key Personnel: Dir., Patty Miller; Cur., Michael J. Vogt

Institution Type/Description: Culture & History Museum.

Collections: area history & culture; photographs; personal artifacts.

Facilities: library. Museum-related items for sale.

Activities: tours; special events; educational programs.

Hours & Admission Prices: Tues.-Fri. 10-4, Sat. 11-3. Adults $5, students & seniors $4, children 6-12 $3; children 5 & under and members no charge. Closed holidays.

Membership: Seniors $25; Individual $35.

✱ BOISE ART MUSEUM, (M), 670 Julia Davis Dr., Boise, ID 83702-7646. Tel.: 208-345-8330, ext. 10. Fax: 208-345-2247. Facebook: Boise Art Museum.
E-mail: info@boiseartmuseum.org
Web Site: www.boiseartmuseum.org
Founded: 1937.
Congressional District: 1
Key Personnel: Exec. Dir. & C.E.O., Melanie Fales; Pres. Bd. of Trustees, Nicole Snyder; Devel. Specialist, Erin Kennedy; Cur. Art, Sandy Harthorn; Registrar, Kathy Bettis; Cur. Education, Terra Feast; Curatorial Asst., Catherine Rakow; Museum Store Mgr., Melissa Swafford; Preparator, Todd Newman; Financial Mgr., Mary Schaefer; Asst. Preparator, Dave Darraugh; Museum Resources Coord., Hana Van Huffel.
Personnel Profile: Full-Time Paid 14; Part-Time Paid 15; Part-Time Volunteers 280; Interns 8.
Governing Authority: nonprofit organization. Parent Institution: Boise Art Museum, Inc. Tax-exempt.
Institution Type/Description: General Art Museum.
Collections: 20-century American Art; Northwest artists; international ceramics; international prints & photographs; James Castle collection; Asian & African objects & adornments.
Major Exhibits: 2013 Idaho Triennial, 11/16/13-4/27/14.
Facilities: 10,000-vol. art reference library available on premises by permission; 3,000 exhibition catalogs; reading room; 34,800 sq. ft. building; 2,800 sq. ft. indoor sculpture court; 16,000 sq. ft. gallery space. Crafts by Northwest artists, books & postcards for sale in the museum store.
Activities: children's programming; guided tours; lectures; films; gallery talks; performances; educational programs; docent program; temporary & traveling exhibitions; annual fundraising events.
Publications: quarterly, calendar; 2007 Idaho Triennial Catalog; exhibition catalogs: 100 Years of Idaho Art; James Barsness; Fabricated Nature; Jack Dollhausen: A 30 Year Start; Gary Hill: Language Willing; John Grade: Sculpture and Drawings; James Castle: Drawings, Constructions and Books; Collection of the Boise Art Museum; Twice Removed: Deborah Oropallo; Unraveled: Hildur Bjarnadottir; American Art: Wilfred Davis Fletcher Collection; Scott Fife: Big Trouble; Kendall Buster: New Growth; Lead Pencil Studio: After; Devorah Sperber: Threads of Perception; 2010 Idaho Triennial; Stephen Knapp: Lightpaintings; Mike Rathbun: The Situation He Found Himself In; In the Abstract: Wilfred Davis Fletcher Collection.
Hours & Admission Prices: Tues.-Sat. 10-5 (1st Thurs. 10-9), Sun. 12-5. Adults $5, college students with ID & senior citizens $3, children grades K-12 $1; discounts to AAM members, museum professionals & 1st Thurs. of month; members & children under 6 no charge. Closed New Year's Day; Martin Luther King Jr. Day; Presidents' Day; Memorial Day; Independence Day; Labor Day; Columbus Day; Thanksgiving; Christmas. &
Attendance: 49,343 (accurate)
Membership: Student & Senior Citizen $35; Individual $45; Family $60; Advocate $125; Contributor $250; Sustainer $500; Patron $1,000; Benefactor's Circle $2,500 & up.

THE DISCOVERY CENTER OF IDAHO, 131 Myrtle St., Boise, ID 83702-7652. Tel.: 208-343-9895. Fax: 208-343-0105.
E-mail: dcifilter@gmail.com
Web Site: www.scidaho.org
Founded: 1986.
Congressional District: 2
Key Personnel: Exec. Dir., Janine Boire; Pres. (V), Mark Solon; Dir. Exhibits, Bill Molina; Exhibit Builder, Mike Twitchell; Business Mgr., Jane Baird; Senior Sec. & Museum Shop Mgr., Joanne Beall; Dir. Public Programs, Susan Dittus; Coord. Education, Kris Allison; Dir. Personnel, Sally Stivison-Dunne.
Personnel Profile: Full-Time Paid 9; Full-Time Volunteers 150; Part-Time Paid 2; Part-Time Volunteers 200.
Governing Authority: not-for-profit organization. Tax-exempt: 501(c)(3).
Institution Type/Description: Science Museum.
Collections: hands-on exhibits including science, math, & technology.
Facilities: educational facilities; portable, inflatable planetarium. Scientific tools, toys, games, books & charts for sale.
Activities: docent program; formal education programs for children; lectures; loan, participatory & traveling exhibitions; teacher workshops for credit; speaker series.
Publications: quarterly newsletter, Discovery News.
Hours & Admission Prices: Winter: Tues.-Thurs. 9-5, Fri. 9-7, Sat. 10-5, Sun. 12-5; Summer: Mon.-Thurs. 9-5, Fri. 9-7, Sat. 10-5, Sun. 12-5. Adults 13 & over $6.50, senior citizens 60 & over $5.50, children 3-17 $4; discounts to groups & ASTC affiliate members; members and children 2 & under no charge. Closed New Year's Day; Easter; Thanksgiving; Christmas. &
Attendance: 98,700 (accurate)

Membership: Individual $25; Grandparent $40; Family & Grandparent Plus $50; Family Plus $60; Adventurer $100; Sir Isaac Newton Society $250 & up.

IDAHO AQUARIUM, 64 N. Cole Rd., Boise, ID 83704. Tel.: 208-375-1932.
E-mail: idahoaquarium@gmail.com
Web Site: www.idahoaquarium.net
Key Personnel: Dir., Ammon Covino.
Governing Authority: nonprofit organization. Tax-exempt: 501(c)(3).
Institution Type/Description: Aquarium.
Collections: over 250 different species of animals and marine life; sharks; rays; turtles; octopus; fish.
Activities: educational programs; guided tours; feed the sharks & birds; birthday parties; puppet show.
Hours & Admission Prices: Daily 10-8. Adults $9, seniors & military $7, children under 12 $6; discounts to groups; children 2 & under no charge.

IDAHO BLACK HISTORY MUSEUM, 508 Julia Davis Dr., Boise, ID 83702-7694. Tel.: 208-433-0017.
E-mail: museum@ibhm.org
Web Site: www.ibhm.org
Founded: 1995.
Congressional District: 2
Key Personnel: Pres., Phillip Thompson
Institution Type/Description: History Museum.
Collections: history & culture of Idaho blacks.
Activities: lectures; workshops; musical presentations; permanent exhibitions.
Publications: newsletter.
Hours & Admission Prices: Sat.-Sun. 11-4; groups by appointment. No charge; donations accepted. &
Membership: Student & Senior $15; Individual $30; Family $50; Friend $100; Associate $250; Patron $500; President's Circle $1,000; Corporate $2,500; Benefactor $5,000.

IDAHO BOTANICAL GARDEN, 2355 Old Penitentiary Rd., Boise, ID 83712. Tel.: 208-343-8649. Fax: 208-343-3601.
E-mail: info@idahobotanicalgarden.org
Web Site: www.idahobotanicalgarden.org
Founded: 1984.
Congressional District: 2
Key Personnel: Exec. Dir., Julia Rundberg; Pres. (V), Dena Shipton; Educator, Elizabeth Dickey; Head Horticulture, Toby Mancini; Gardener, Rebecca Needles.
Personnel Profile: Full-Time Paid 16; Full-Time Volunteers 1; Part-Time Paid 15; Part-Time Volunteers 585.
Governing Authority: private; nonprofit organization. Tax-exempt: 501(c)(3).
Institution Type/Description: Botanical Garden.
Collections: Gardens: herb; rose; peony; native plants; Sacajawea monument; Lewis & Clark.
Facilities: library; botanical garden. Museum-related items for sale.
Activities: weddings; community events; school tours; special events; concerts; docent tours; formal education programs; guided tours; hobby workshops; lectures. Annual Events: plant sale; Great Garden Escape in summer; Bug Day; Outlaw Field Music Concerts; Fall Harvest Festival; Winter Garden aGlow.
Publications: quarterly newsletter, Garden Thymes.
Hours & Admission Prices: Summer: Mon. & Wed.-Thurs. 9-5, Tues. & Fri. 9-9, Sat.-Sun. 10-6. Winter: Mon.-Fri. 9-5. Adults $5, seniors & children 6-12 $3; discounts to groups of 10 or more; members no charge. &
Attendance: 120,000 (accurate)
Membership: Individual $35; Individual Plus $40; Family $50; Family Plus $60; Contributing $75; Sustaining $150; Sponsoring $300; Patron $500; Perennial Society $1,000.

IDAHO MILITARY HISTORY MUSEUM, 4692 W. Harvard St., Boise, ID 83705. Mailing Address: 4040 W. Guard St., Boise, ID 83705-5004. Tel.: 208-272-4841. Facebook: Idaho Military History Museum.
E-mail: kswanson@imd.idaho.gov
Web Site: museum.mil.idaho.gov
Founded: 1995.
Key Personnel: Chm. (V), Russ Trebby; Dir. & Museum Shop Mgr., Kenneth Swanson.
Personnel Profile: Full-Time Paid 1; Part-Time Volunteers 20; Interns 4.
Governing Authority: nonprofit organization. Parent Institution: Idaho Military Historical Society. Tax-exempt: 501(c)(3).
Institution Type/Description: Military History Museum.

Collections: area military history & artifacts; Idaho National Guard; WWII; Vietnam War; photographs; 1857 Napoleon civil War era cannon; armored vehicles; Korean War; WWI; Medal of Honor; Navy; Marines' Air Force; Air National Guard; Gowen Field; Paul Gowen.
Research Fields: Idaho military history.
Facilities: library; archives.
Activities: school & group tours.
Publications: newsletter, Pass In Review.
Hours & Admission Prices: Tues.-Sat. 12-4; other times by appointment. No charge; donations accepted. Closed New Year's Day; Easter; Thanksgiving; Christmas.
Attendance: 4,500 (estimated)
Membership: Individual: Student & Associate $10; Senior (60 & over) $15; General $25; Lifetime $375. Organizational: Bronze $50; Silver $100; Gold $250; Platinum $500.

IDAHO MUSEUM OF MINING AND GEOLOGY, 2455 Old Penitentiary Rd., Boise, ID 83712-8254. Tel.: 208-368-9876.
Web Site: www.idahomuseum.org
Founded: 1989.
Personnel Profile: Part-Time Volunteers 60.
Governing Authority: private; nonprofit organization. Tax-exempt: 501(c)(3).
Institution Type/Description: Mining & Geology Museum.
Collections: displays of Idaho's geologic features; historic photographs & artifacts from 19th-20th century Idaho mining towns; mineral collection.
Activities: educational earth science activities; school tours & gold panning demos; lecture series; field trips.
Hours & Admission Prices: April-Oct. Wed.-Sun. 12-5. No charge; donations accepted. &
Attendance: 10,000 (estimated)
Membership: Junior Geologist $5; Individual $25; Family $45.

IDAHO STATE CAPITOL, 700 W. Jefferson, Boise, ID 83720-0002. Mailing Address: P.O. Box 83720, Boise, ID 83720. Tel.: 208-334-2100.
Institution Type/Description: Historic Building: built in 1886.
Collections: local & state government, history & political leaders; official portraits; personal artifacts; period furnishings; paintings; photographs; architecture; flags; statues.
Activities: guided tours.
Hours & Admission Prices: During Legislative Sessions: Mon.-Fri. 6am-10pm, Sat.-Sun. 9-5; Between Legislative Sessions: Mon.-Fri. 6-6, Sat.-Sun. 9-5.

✱ **IDAHO STATE HISTORICAL MUSEUM, (M),** 610 N. Julia Davis Dr., Boise, ID 83702-7646. Tel.: 208-334-2120. Fax: 208-334-4059.
E-mail: jody.ochoa@ishs.idaho.gov
Web Site: www.idahohistory.net/museum.html
Founded: 1881.
Congressional District: 2
Key Personnel: Admin., Museum & Historic Sites, Jody Ochoa; Dir. Idaho State Historical Society, Steve Guerber; Cur., Joe Toluse; Museum Shop Mgr., Karen Scheider.
Personnel Profile: Full-Time Paid 8; Part-Time Paid 6; Part-Time Volunteers 30; Interns 3.
Governing Authority: state. Parent Institution: Idaho State Historical Society. Tax-exempt: 501(c)(3).
Institution Type/Description: History Museum.
Collections: historical objects pertaining to Idaho & Pacific Northwest history including mining, ranching, timber industry, transportation & the Oregon Trail; Indian artifacts; Chinese artifacts; manuscripts; archives; maps; photographs; books; newspapers.
Research Fields: history of Idaho & Pacific Northwest.
Facilities: research library available for on premises use; 13,000 sq. ft. exhibit space; auditorium; classroom. Gift items for sale.
Activities: guided tours; lectures; films; workshops; TV & radio programs; formally organized education programs; training programs for museum workers & college students; inter-museum loans; permanent, temporary & traveling exhibitions; school loan service.
Publications: quarterly magazine, Idaho Yesterdays; children's magazine, Prospector; quarterly newsletter, Mountain Light; miscellaneous short research papers on Idaho history.
Hours & Admission Prices: May-Sept. Tues.-Sat. 9-5, Sun. 1-5; Oct.-April Tues.-Fri. 9-5, Sat. 11-5. Adults $5, seniors $4, children & students with ID $3; discounts to AAM members; members & children under 6 no charge. Call for group rates. Closed New Year's Day; Thanksgiving; Christmas. &
Attendance: 190,000 (accurate)
Membership: Full-time Student & Senior Citizen (65 & over) $15; Senior

Couple $25; Individual $35; Family $50; Contributing $75; Friend $150; Sponsoring $300; Patron $500; Benefactor $1,000; Director's Club $5,000.

MORRISON KNUDSEN NATURE CENTER, IDAHO DEPARTMENT OF FISH & GAME, 600 S. Walnut St., Boise, ID 83712-7729. Tel.: 208-334-2225. Fax: 208-287-2905.
E-mail: david.cannamela@idfg.idaho.gov
Web Site: http://fishandgame.idaho.gov/public/education/?getpage=234
Founded: 1990.
Key Personnel: Supt., David Cannamela.
Personnel Profile: Full-Time Paid 3; Part-Time Paid 5; Part-Time Volunteers 400; Interns 2.
Governing Authority: Parent Institution: State of Idaho - Department of Fish & Game. Tax-exempt.
Institution Type/Description: Nature Center.
Collections: local history; wildlife including fish, birds, waterfowl, & mammals; plants; trees; flowers.
Facilities: 4.6 acre fish & wildlife habitat and trails.
Activities: guided tours; educational programs; open to the public self guided.
Hours & Admission Prices: Park: daily sunrise to sunset. Visitors Center: Tues.-Fri. 9-5, Sat.-Sun. 11-5. No charge; donations accepted. &
Attendance: 150,000 (estimated)

OLD IDAHO PENITENTIARY STATE HISTORIC SITE, 2445 Old Penitentiary Rd., Boise, ID 83712-8254. Tel.: 208-334-2844. Fax: 208-334-3225.
Web Site: www.idahohistory.net/oldpen.html
Governing Authority: Parent Institution: Idaho State Historical Society. Tax-exempt.
Institution Type/Description: Historic Building Museum: built in 1870, the penitentiary received over 13,000 convicts while in operation for more than a century. Listed on the National Register of Historic Places.
Collections: prison history; notorious inmates; daily prison life; Solitary Confinement; Death Row; the Gallows; period weapons.
Activities: education programs.
Hours & Admission Prices: Memorial Day to Labor Day daily 10-5; Sept.-May daily 12-5. Adults $5, seniors $4, children 6-12 $3; discounts for groups of 10 or more; children under 6 no charge. Closed state holidays.
Attendance: 23,539 (accurate)

WORLD CENTER FOR BIRDS OF PREY, 5668 W. Flying Hawk Lane, Boise, ID 83709-7289. Tel.: 208-362-8687. Fax: 208-362-2376.
E-mail: tpf@peregrinefund.org
Web Site: www.peregrinefund.org
Formerly: Velma Morrison Interpretive Center
Founded: 1984.
Key Personnel: Dir., Jack Cafferty; Museum Shop Mgr., Nick Piccono.
Personnel Profile: Full-Time Paid 4; Full-Time Volunteers 1; Part-Time Paid 1; Part-Time Volunteers 85.
Governing Authority: Parent Institution: The Peregrine Fund. Tax-exempt.
Institution Type/Description: Nature Center.
Collections: birds of prey.
Activities: raptor presentations; hands-on exhibits; falconry tours; educational programs.
Hours & Admission Prices: March-Oct. daily 9-5; Nov.-Feb. Tues.-Sun. 10-4. Adults 17 & over $7, seniors 62 & over $6, children 4-16 $5; members & children under 4 no charge. &
Attendance: 35,000 (accurate)
Membership: Individual $25; Family $50; Family Plus $100.

WORLD SPORTS HUMANITARIAN HALL OF FAME, 1910 University Dr., Boise, ID 83725. Mailing Address: P.O. Box 9324, Boise, ID 83707-3324. Tel.: 206-262-7301. Fax: 208-343-0831. Facebook: World Sports Humanitarian Hall of Fame.
Founded: 1994.
Key Personnel: Pres., Rick Frisch
Institution Type/Description: Hall of Fame.
Collections: history of world-class; humanitarian athletes; photographs; personal artifacts; Hall of Fame Inductees.
Hours & Admission Prices: Mon.-Fri. 9-4. No charge.

ZOO BOISE, 355 Julia Davis Dr., Boise, ID 83702-7670. Tel.: 208-608-7735. Fax: 208-384-4059. TDD: 208-384-4240.
E-mail: zooboise@cityofboise.org
Web Site: www.zooboise.org
Founded: 1916.
Key Personnel: Dir., Steve Burns; Registrar, Corinne Roberts.

Governing Authority: Parent Institution: City of Boise. Subsidiary Institution: Friends of Zoo Boise. Tax-exempt.
Institution Type/Description: Zoo.
Collections: over 200 animals representing over 100 species.
Major Exhibits: Butterflies in Bloom, 6/14-9/1/14.
Facilities: cafe. Museum-related items for sale.
Hours & Admission Prices: Daily 10-5. May-Sept. adults $10, seniors 62 & over $8, children 3-11 $7; children 2 & under no charge; Oct.-April adults $7, senior citizens 62 & over $4.50, children 3-11 $4.25; discount on Thurs.; children under 4 no charge. Closed New Year's Day; Thanksgiving; Christmas. &
Attendance: 329,084 (accurate)
Membership: Individual $45; Couple $65; Grandparent $75; Household $90; Household Plus $110; Contributing $125. Pride of the Zoo: Level 1 $300; Level 2 $500; Level 3 $1,000.

Bonners Ferry

BOUNDARY COUNTY HISTORICAL SOCIETY AND MUSEUM, 7229 Main St., Bonners Ferry, ID 83805. Mailing Address: P.O. Box 808, Bonners Ferry, ID 83805-0808. Tel.: 208-267-7720.
Founded: 1969.
Congressional District: 1
Key Personnel: Pres., Jill Nystrom; Sec., Cal Russell; Treas., Stephanie Tucker; Cur. & Collections Mgr., Sue Kemmis.
Personnel Profile: Part-Time Paid 1; Part-Time Volunteers 20.
Governing Authority: nonprofit organization.
Institution Type/Description: History Museum.
Collections: local history; photographs; period tools; dishes; clothing.
Publications: History of Boundary County, Vol. 1; cookbook, Family Favorites.
Hours & Admission Prices: May-Sept. Tues.-Sat. 10-4; Oct.-April Fri.-Sat. 10-4. Admission $2, family $5. &
Attendance: 3,000 (estimated)
Membership: Senior Individual $10; Individual & Senior Couple $15; Basic Family $25; Donor $100; Sponsor $500; Patron $1,000.

KOOTENAI NATIONAL WILDLIFE REFUGE, 287 Westside Rd., Bonners Ferry, ID 83805-5172. Tel.: 208-267-3888. Fax: 208-267-5570.
Key Personnel: Mgr., Dianna Ellis
Institution Type/Description: Wildlife Refuge.
Collections: over 300 species of wildlife.
Hours & Admission Prices: Refuge: daily dawn to dusk. Office: Mon.-Fri. 7:30-4.
Attendance: 20,000

Burley

CASSIA COUNTY MUSEUM, E. Main & Hiland Ave., Burley, ID 83318. Mailing Address: P.O. Box 331, Burley, ID 83318-0331. Tel.: 208-678-7172.
E-mail: cassiamuseum@cassiacounty.org
Founded: 1972.
Key Personnel: Pres., Rod Smith; Financial Dir. & Treas., Joel Robins; Cur., Valerie Bowen.
Personnel Profile: Full-Time Paid 1; Part-Time Volunteers 5.
Institution Type/Description: History Museum.
Collections: schoolhouse; general store; cabin; train car; farm machinery; county covered wagon.
Activities: guided tours; Museum Sponsors: History Alive Days for county 4th graders.
Hours & Admission Prices: April-Oct. Tues.-Sat. 10-5. No charge; donations accepted. Closed Independence Day. &
Attendance: 1,750 (estimated)
Membership: Individual $10; Family $25.

Caldwell

AEROPLANES OVER IDAHO MUSEUM, 5017A Aviation Way, Caldwell, ID 83605. Tel.: 208-455-1708.
E-mail: info@aeroplanesoveridaho.org
Web Site: www.aeroplanesoveridaho.org
Founded: 2003.
Governing Authority: nonprofit organization.
Institution Type/Description: Aviation History Museum.
Collections: 27 aircraft from 1940 to present including 1945 Yak3U, AN-2Colt, & Skyhawk A-4.
Activities: airplane rides.

Hours & Admission Prices: Daily. No charge; donations accepted.

CITY OF CALDWELL VAN SLYKE MUSEUM, Caldwell Municipal Park, Harrison St., Caldwell, ID 83605. Mailing Address: 411 Blaine St., Caldwell, ID 83605-3619. Tel.: 208-455-3011. Fax: 208-455-3003.
E-mail: smiller@ci.caldwell.id.us
Web Site: www.cityofcaldwell.com
Formerly: Kiwanis Van Slyke Museum Foundation Inc.
Founded: 1958.
Congressional District: 10
Key Personnel: Chm. City Hall, Susan Miller.
Personnel Profile: Part-Time Volunteers 5.
Governing Authority: nonprofit organization. Parent Institution: Caldwell, ID Kiwanis Club, P.O. Box 925, Caldwell 83605.
Institution Type/Description: Park Museum & Visitors Center.
Collections: farm implements; agriculture; history. Historic Houses: 1864 McKenzie log cabin; 1864 Johnston Brothers log cabin.
Research Fields: Pioneer tools & equipment used to convert an arid sagebrush land to the now highly productive agriculture community of today.
Facilities: fenced-in compound in city park.
Activities: guided tours.
Hours & Admission Prices: By appointment only. No charge; donations accepted. &
Attendance: 350 (estimated)

ORMA J. SMITH MUSEUM OF NATURAL HISTORY, (M), The College of Idaho, 2112 Cleveland Blvd., Caldwell, ID 83605-4432. Tel.: 208-459-5507.
E-mail: bclark@collegeofidaho.edu
Web Site: www.collegeofidaho.edu/campus/community/museum
Founded: 1976.
Congressional District: 1
Key Personnel: Dir., Adjunct Prof. of Biology, Cur. Invertebrates, & Bd. Member, William H. Clark; Research Assoc., Dr. Paul Blom; Cur. Lepidoptera, Dr. Paul Castrovillo; Archaeologist, Cur. Ethnology, Janet L. Summers Duffy; Cur. Paleontology, Howard Emry; Research Assoc., Mammology & Bd. Member, Dr. Sean Farley; Cur. Paleontology, Research Assoc., Paleobotany, Dr. Patrick F. Fields; Cur. Entomology, Dr. Alan R. Gillogly; Cur. Mollusca, Stephen J. Lysne; Cur. Ethnography, Bill Nance; Asst. to Dir., Research Assoc., James Pike; Cur. Entomology, Dr. James K. Ryan; Cur. Entomology, Dr. Craig R. Baird; Assoc. Cur. Mollusca, Richard A. Salisbury; Cur. Fossil Fishes, Dr. Gerald R. Smith; Facilities & Bd. Member, Leland Thames; Asst. Prof. Biology, Assoc. Cur. Fishes & Bd. Member, Dr. Chris Walser; Gen. Entomology, Dr. David Ward; Research Assoc., Librarian & Cur. Invertebrates, Jerry Wood; Prof. Biology, Cur. Mammals, Biology Dept. Coord. of Museum Affairs & Bd. Member, Dr. Eric Yensen; Cur. Fishes, Dr. Donald W. Zaroban; Bd. Member & Research Assoc. Entomology, Dr. Ron Bitner.
Personnel Profile: Part-Time Paid 1; Part-Time Volunteers 40; Interns 4.
Governing Authority: Parent Institution: The College of Idaho. Tax-exempt.
Institution Type/Description: Natural History Museum.
Collections: entomology; anthropology; Native American artifacts; paintings; mammals; fossils; natural history; mollusca including snails & bivalves; birds; reptiles & amphibians; books; photographs.
Research Fields: entomology (arthropods of Baja California, Mexico).
Facilities: research library.
Activities: education programs; scientific workshops; fundraisers. Museum Sponsors: Monthly Museum Volunteer Workdays; BioBlitz in summer; 4th of July Butterfly Count; Bug Day in August.
Publications: museum brochures; staff & curator's research publications.
Hours & Admission Prices: By appointment. No charge. &
Attendance: 3,000 (accurate)

OUR MEMORIES INDIAN CREEK MUSEUM, 1122 Main St., Caldwell, ID 83605. Mailing Address: c/o Canyon County Historical Society, P.O. Box 595, Nampa, ID 83651. Tel.: 208-467-7611.
Web Site: www.canyoncountyhistory.com/ourmemories
Personnel Profile: Part-Time Volunteers 6.
Governing Authority: Parent Institution: Canyon Historical Society.
Institution Type/Description: History Museum.
Collections: local history & culture; period furnishings; personal artifacts; photographs.
Activities: special events.
Publications: newsletter.
Hours & Admission Prices: Tues. & Fri. 11-4; other times by appointment. Adults $3; members no charge.
Attendance: 1,000 (estimated)

Membership: Individual $15; Couple $20; Family $25; Contributing $30; Sustaining $60; Sponsor $125.

ROSENTHAL GALLERY OF ART, The College of Idaho, 2112 Cleveland Blvd., Caldwell, ID 83605-4432. Tel.: 208-459-5321 & 5209. Fax: 208-459-5885.
E-mail: gclaassen@collegeofidaho.edu
Founded: 1891.
Key Personnel: Dir., Garth Claassen.
Governing Authority: college. Parent Institution: College of Idaho. Tax-exempt: 501(c)(3).
Institution Type/Description: Art Exhibition Gallery.
Collections: prints; paintings; sculptures.
Activities: guided tours; lectures; films; gallery talks; concerts; dance; poetry readings; inter-museum loan; permanent, temporary & traveling exhibitions.
Publications: annual brochures on exhibits.
Hours & Admission Prices: Mon.-Fri. 10-5; Sat.-Sun. by appointment. No charge; donations accepted. Closed academic holidays & breaks. ♿
Attendance: 1,500

WHITTENBERGER PLANETARIUM, Boone Science Hall, The College of Idaho, 2112 Cleveland Blvd., Caldwell, ID 83605. Tel.: 208-459-5211. Fax: 208-459-5414.
E-mail: atruksa@collegeofidaho.edu
Web Site: www.collegeofidaho.edu/planetarium
Founded: 1970.
Key Personnel: Dir. & Museum Shop Mgr., Amy Truksa; Administrative Asst., Kinga Britschgi.
Personnel Profile: Part-Time Paid 2; Part-Time Volunteers 2.
Governing Authority: Parent Institution: The College of Idaho.
Institution Type/Description: Planetarium.
Collections: astronomy; geological.
Activities: fieldtrips; public shows; Astronomy Day activities.
Hours & Admission Prices: Call for hours. Adults $4, students $2; discount to groups.
Attendance: 3,000 (estimated)

Cambridge

CAMBRIDGE MUSEUM, 15 N. Superior, Cambridge, ID 83610. Mailing Address: P.O. Box 35, Cambridge, ID 83610-0035. Tel.: 208-257-3485.
E-mail: shansen@ctcweb.net
Founded: 1983.
Key Personnel: Dir., Sandra Hansen; Museum Shop Mgr., Bo Thorsen.
Personnel Profile: Part-Time Volunteers 7.
Volunteer Hours: 1,240
Operating Expenses: 3,500
Operating Income: 4,000
Governing Authority: Parent Institution: City of Cambridge. Tax-exempt.
Institution Type/Description: History Museum.
Collections: local history; geological; farming; Native American; pioneer artifacts.
Major Exhibits: Vintage Tractor, 1/14-12/14; 1960's Clothing, 1/14-12/14.
Research Fields: local family history.
Facilities: genealogy & research library.
Activities: school tours.
Publications: book, A Saga & A History Revisited; They Came Before Us, Vol. 1; They Came Before Us, Vol. 2.
Hours & Admission Prices: June-Aug. Wed.-Sat. 10-4; call to confirm. No charge; donations accepted. ♿
Attendance: 850 (accurate)

Cataldo

COEUR D'ALENES OLD MISSION STATE PARK, 31732 S. Mission Rd., Cataldo, ID 83810. Mailing Address: P.O. Box 30, Cataldo, ID 83810-0030. Tel.: 208-682-3814. Fax: 208-682-4032.
E-mail: old@idpr.idaho.gov
Web Site: www.parksandrecreation.idaho.gov
Key Personnel: Park Mgr., Kathleen Durfee.
Governing Authority: state; not-for-profit organization. Tax-exempt.
Institution Type/Description: Historic Building: built by the Coeur d'Alene Tribe & Jesuit missionaries in 1850.
Collections: structure & furnishings; tools & equipment; religious artifacts; artworks; cultural artifacts.

Hours & Admission Prices: April-Oct. 9-5, Nov.-March 10-3. Entrance fee $5 per vehicle. Closed New Year's Day; Thanksgiving; Christmas. ♿
Attendance: 90,000 (estimated)

Challis

NORTH CUSTER MUSEUM, 1201 E. Main Ave., Challis, ID 83226. Mailing Address: P.O. Box 776, Challis, ID 83226-0776. Tel.: 208-879-2846.
Founded: 1998.
Congressional District: 2
Key Personnel: Dir., Marion McDaniel; Pres. (V), John Rose.
Personnel Profile: Part-Time Paid 1; Part-Time Volunteers 4.
Volunteer Hours: 50
Operating Expenses: 4,727
Operating Income: 602
Governing Authority: Parent Institution: North Custer Historical Society. Tax-exempt.
Institution Type/Description: History Museum.
Collections: local history & culture; personal artifacts; photographs.
Publications: annual newsletter.
Hours & Admission Prices: Memorial Day to Labor Day Sat.-Sun. 10-4. No charge; donations accepted. ♿
Attendance: 525 (estimated)
Membership: Individual $5; Family $10; Business $25.

Coeur d'Alene

THE ART SPIRIT GALLERY, 415 Sherman Ave., Coeur d'Alene, ID 83814-2728. Tel.: 208-765-6006.
E-mail: steve@theartspiritgallery.com
Web Site: www.theartspiritgallery.com
Founded: 1997.
Key Personnel: Dir., Steven J. Gibbs.
Personnel Profile: Full-Time Paid 1; Part-Time Paid 1.
Institution Type/Description: Art Gallery.
Collections: paintings; drawings; sculpture; pottery.
Facilities: 2,500 sq. ft. exhibition space.
Activities: permanent & temporary exhibitions; receptions.
Hours & Admission Prices: Tues.-Sat. 11-6. No charge. ♿

MUSEUM OF NORTH IDAHO, (M), 115 N.W. Blvd., Coeur d'Alene, ID 83814-2798. Mailing Address: P.O. Box 812, Coeur d'Alene, ID 83816-0812. Tel.: 208-664-3448.
E-mail: dd@museumni.org
Web Site: www.museumni.org
Founded: 1968.
Congressional District: 1
Key Personnel: Dir., Dorothy Dahlgren; Pres., Mike Dolan.
Personnel Profile: Full-Time Paid 1; Part-Time Volunteers 30.
Governing Authority: nonprofit organization. Branch Museum: Fort Sherman Museum, Coeur D'Alene. Tax-exempt: 501(c)(3).
Institution Type/Description: Local History Museum.
Collections: exhibits of Coeur d'Alene region's history from early explorers to present; logging, mining, & transportation materials; Forest Service; textiles; firearms; Native American artifacts; photographs.
Facilities: library pertaining to history of Coeur d'Alene Region. Museum-related items for sale.
Activities: traveling exhibitions; school loan service.
Publications: newsletter, Museum of North Idaho Newsletter; books: In All the West No Place Like This; North Fork of the Coeur d'Alene River: White Pine Route; Inland Empire Electric Line; Spokane International Railway; COE: The First Sixty Years; Wildflowers of the Inland Northwest; The Milwaukee Road In Idaho: A Guide to Sites & Locations; Up the Swiftwater: A Pictorial History of the Colorful St. Joe Country; The Milwaukee Road Olympian, A Ride to Remember; Swiftwater People; From Hell to Heaven: Death Related Mining Accidents in North Idaho; Lookout Cookbook; The Milwaukee Road's Western Extension: The Building of a Transcontinental Railroad.
Hours & Admission Prices: Museum of North Idaho: April-Oct. Tues.-Sat. 11-5. Library: by appointment. Adults $3, children 6-16 $1; discounts to AAM, ICOM & AASLH members; members no charge. Closed Independence Day. ♿
Attendance: 5,500 (accurate)
Membership: Amelia Wheaton $25; Flyer $50; Idaho $100; Georgie Oakes $100 & up.

Coolin

PRIEST LAKE STATE PARK, 314 Indian Creek Park Rd., Coolin, ID 83821-9769. Tel.: 208-443-2200. Fax: 208-443-3893.
Institution Type/Description: State Park.
Collections: natural history; wildlife.
Facilities: nature trails.
Activities: camping; fishing; hiking; educational programs; guided walks.
Hours & Admission Prices: Daily.

Cottonwood

HISTORICAL MUSEUM AT ST. GERTRUDE, (M), 465 Keuterville Rd., Cottonwood, ID 83522-5183. Tel.: 208-962-2050. Fax: 208-962-2059.
E-mail: museum@stgertrudes.org
Web Site: www.historicalmuseumatstgertrude.org
Formerly: St. Gertrude's Museum
Founded: 1931.
Congressional District: 1
Key Personnel: Dir., Museum Shop Mgr. & Dir. Displays, Dr. Sam Couch; C.E.O., Sister Clarissa Goeckner; Museum Technician, Shirley Gehring.
Personnel Profile: Full-Time Paid 2; Part-Time Paid 1; Part-Time Volunteers 15.
Governing Authority: nonprofit organization. Parent Institutions: Monastery of St. Gertrude; Idaho Corp. of Benedictine Sisters. Subsidiary Institution: St. Gertrude's Museum. Tax-exempt: 501(c)(3).
Institution Type/Description: History Museum.
Collections: local history; Indian artifacts; old books; textiles; paintings; chinaware; tools; mineralogical & biological displays; dated medical equipment; war accoutrements; turn of the century music makers; cultural items from various foreign countries; Winifred Rhoads Emmanuel Collection (Asian art); Polly Bemis Collection; Buckskin Bill Collection; photographs.
Research Fields: local Idaho history.
Facilities: library containing maps, manuscripts & books on local history & mineralogy.
Activities: guided tours for groups; lectures on local history. Museum Sponsors: Raspberry Festival & quilt show in August.
Publications: Pioneer Days in Idaho County, Volumes I & II; Idaho Chinese Lore; Polly Bemis; coloring book, Old Idaho; museum newsletter, Rediscover; museum journal, Echoes of the Past.
Hours & Admission Prices: Tues.-Sat. 9:30-4:30. Adults $5; discounts to AAM members; members no charge. &
Attendance: 7,000 (estimated)
Membership: Individual $20; Couple $30; Family $40; Associate $80; Lifetime $1,000.

Council

COUNCIL VALLEY MUSEUM, Galena St., Council, ID 83612. Mailing Address: P.O. Box 252, Council, ID 83612-0252. Tel.: 208-253-4582.
Founded: 1972.
Congressional District: 1
Key Personnel: Chm. (V), Dale Fisk.
Personnel Profile: Part-Time Volunteers 4.
Governing Authority: city. Tax-exempt.
Institution Type/Description: History Museum.
Collections: local history; photographs; personal artifacts.
Hours & Admission Prices: Memorial Day to Labor Day Tues.-Sat. 10-4, Sun. 1-4. &
Attendance: 1,000 (estimated)

Donnelly

VALLEY COUNTY MUSEUM, 13131 Farm to Market Rd., Donnelly, ID 83615. Mailing Address: P.O. Box 444, Donnelly, ID 83615-0444. Tel.: 208-325-8628. Facebook: Historic Roseberry.
E-mail: info@historicroseberry.com
Web Site: www.historicroseberry.com
Formerly: Long Valley Museum
Founded: 1973.
Congressional District: 8
Key Personnel: Acting Pres. (V), Barbara Kwader; Museum Shop Mgr., Ann McQuade; Volunteer Coord., Katie Morgan; Groundskeeper, Thomas Schmidt.
Personnel Profile: Part-Time Paid 2.
Governing Authority: Parent Institution: Long Valley Preservation Society. Tax-exempt.

Institution Type/Description: Historical Museum.
Collections: local history & culture; photographs; historic buildings; early 1900s town site & artifacts; period farm equipment; Native American tools.
Research Fields: genealogy; social history.
Facilities: research center.
Activities: research; rental facilities; concerts. Museum Sponsors: Music Festival; Arts & Crafts Show; Ice Cream Social; Pioneer Picnic.
Publications: newsletter.
Hours & Admission Prices: May & Oct. Sun. 1-5; June-Sept. Sat.-Sun. 1-5; other times by appointment. No charge; donations accepted. &
Attendance: 10,500 (estimated)
Membership: Senior $10; Individual $15; Family $25; Life $250; Patron $1,000; Benefactor $2,000.

Dubois

HERITAGE HALL MUSEUM, 110 S. Reynolds, Dubois, ID 83423. Mailing Address: P.O. Box 253, Dubois, ID 83423-0253.
Founded: 1969.
Key Personnel: Chm. (V), Conni Owen; Pres. (V), Barbara Kidd.
Governing Authority: city. Tax-exempt.
Institution Type/Description: Historical Society Museum.
Collections: local history & culture; photographs; personal artifacts.
Facilities: Museum-related items for sale.
Hours & Admission Prices: Memorial Day to Labor Day Fri.-Sat. 2-6; other times by appointment. No charge; donations accepted.
Attendance: 300 (estimated)

Elk River

ELK RIVER HISTORICAL MUSEUM, Community Center, Second & Main, Elk River, ID 83827. Mailing Address: P.O. Box 63, Elk River, ID 83827-0063. Tel.: 208-826-3219.
E-mail: k_mathome@moscow.com
Key Personnel: Dir. & Pres. (V), Deanna L. Kreisher; Sec. & Museum Shop Mgr., Keith Lunders; Museum Shop Mgr., Marge Lunders.
Volunteer Hours: 6
Governing Authority: Tax-exempt.
Institution Type/Description: History Museum.
Collections: local history & culture; photographs; news articles; period furnishings; personal artifacts; early logging equipment including a donkey logger, loading jammer, logging truck, & marion.
Hours & Admission Prices: Summer: Sat. 10 to noon; other times by appointment. No charge; donations accepted. &
Attendance: 110 (accurate)
Membership: Individual $5; Lifetime $25.

Emmett

GEM COUNTY HISTORICAL VILLAGE MUSEUM, 501 E. First St., Emmett, ID 83617-3005. Tel.: 208-365-9530 & 4340.
E-mail: gemcohs@bigskytel.com
Web Site: www.gemcohs.org
Formerly: Gem County Historical Society and Museum
Founded: 1974.
Key Personnel: Pres. Bd. (V), Kathleen Derig; Dir. Museum Shop Mgr., Meg Davis.
Personnel Profile: Part-Time Paid 1; Part-Time Volunteers 4.
Governing Authority: Parent Institution: Gem County Historical Society.
Institution Type/Description: Historical Society Museum.
Collections: area history; Native American artifacts; trapping; mining; transportation; lumber; military; firearms; photographs; personal artifacts; medical; ranching.
Research Fields: Gem County history.
Activities: tours; meetings; special programs. Annual Events: Fundraising Chuckwagon Supper in September; Historical Reenactment in October.
Publications: quarterly newsletter.
Hours & Admission Prices: Wed.-Sat. 1-5; tours & other times by appointment. No charge; donations accepted. &
Attendance: 1,000 (accurate)
Membership: Individual $10; Family $15; Benefactor $25.

Filer

TWIN FALLS COUNTY HISTORICAL MUSEUM, 21337A Hwy. 30, Filer, ID 83328-5513. Tel.: 208-736-4675. Fax: 208-736-4675.
E-mail: tfcountymuseum@msn.com
Founded: 1957.

Congressional District: 24
Key Personnel: Pres., Ron Yates; Vice Pres., Steve Westphal; Admin., Mychel Matthews.
Personnel Profile: Part-Time Volunteers 8.
Governing Authority: nonprofit organization. Tax-exempt: 501(c)(3).
Institution Type/Description: History Museum & Visitor Center: housed in the former Union School built in 1914.
Collections: period clothing; phonographs; musical instruments; schoolhouse display; pictures; sewing machines; wagons; gas pumps; farm machinery; washing machines; period rooms; steam engine; picture gallery; doctors equipment; blacksmith shop.
Research Fields: local history; agriculture; family history.
Activities: guided tours; programs; events.
Publications: newsletter, The Union School Report Card.
Hours & Admission Prices: Tues.-Sat. 10-5. No charge; donations accepted. Closed holidays. &
Attendance: 1,600 (accurate)
Membership: Student & Senior Citizens $10; Adult $15; Family $25; Nonprofit $50; Centennial $100; Sustaining & Lifetime $500; Millenial $1,000.

Fort Hall

SHOSHONE-BANNOCK TRIBAL MUSEUM, Simplot Rd., Fort Hall, ID 83203. Mailing Address: Box 306, Fort Hall, ID 83203-0306. Tel.: 208-237-9791. Fax: 208-237-0797.
Web Site: www.sbtribes.com
Founded: 1993.
Key Personnel: Mgr., Coord. & Museum Shop Mgr., Rosemary Devinney.
Governing Authority: Parent Institution: Shoshone-Bannock Tribes.
Institution Type/Description: Tribal Museum.
Collections: Shoshone-Bannock history; culture; art; ceremonial clothing; photographs; Wrensted collections 1895-1912.
Facilities: Museum-related items for sale.
Activities: tours.
Hours & Admission Prices: June-Aug. daily 9:30-5; Sept.-May Mon.-Fri. 9:30-5. Adults $3.50, youth 6-17 $2; Native American Indians with Tribal ID no charge. Closed Federal & Tribal holidays.
Attendance: 3,000 (estimated)

Glenns Ferry

GLENNS FERRY HISTORICAL MUSEUM, 201 W. Cleveland Ave., Glenns Ferry, ID 83623. Tel.: 208-366-2192.
Institution Type/Description: Historic Building: housed in a former school; built in 1909. Listed on the National Register of Historic Places.
Collections: local history & culture; photographs; period furnishings; personal artifacts.
Activities: permanent & temporary exhibits.
Hours & Admission Prices: June-Sept. Sat.-Sun. 12-5; other times by appointment. No charge; donations accepted.

Gooding

GOODING COUNTY HISTORICAL SOCIETY MUSEUM, 273 Euskadi Lane, Gooding, ID 83330. Mailing Address: P.O. Box 580, Gooding, ID 83330-0580. Tel.: 208-934-5318. Fax: 208-934-4885.
E-mail: gchstoponis@yahoo.com
Web Site: goodingcountyhistoricalsociety.shutterfly.com
Founded: 1971.
Key Personnel: Pres. (V), Ilene Rounsfell; Dir., Sharon Cheney; Cur., Coy Jones
Governing Authority: Tax-exempt.
Institution Type/Description: Historical Society Museum.
Collections: local history; photographs; personal artifacts; furnishings.
Facilities: Museum-related items for sale.
Activities: workshops; educational programs; research; rental facilities.
Publications: newsletter, Timepiece.
Hours & Admission Prices: By appointment. No charge; donations accepted. &
Attendance: 150 (estimated)
Membership: Single $15; Family $25; Lifetime $500.

Grangeville

BICENTENNIAL HISTORICAL MUSEUM, 305 N. College, Grangeville, ID 83530. Mailing Address: P.O. Box 212, Grangeville, ID 83530-0212. Tel.: 208-983-2104 & 2277.
E-mail: info@bicentennialmuseum.com
Institution Type/Description: History Museum.

Collections: local history & culture; photographs; period furnishings; personal artifacts; clothing.
Hours & Admission Prices: June-Sept. Wed. & Fri. 1-5; other times by appointment. No charge; donations accepted.

Hagerman

HAGERMAN FOSSIL BEDS NATIONAL MONUMENT, 221 N. State St., Hagerman, ID 83332. Mailing Address: P.O. Box 570, Hagerman, ID 83332-0570. Tel.: 208-837-4793. Fax: 208-837-4857.
Institution Type/Description: History Museum.
Collections: horse fossils; local history & culture; photographs.
Hours & Admission Prices: mid-June to Labor Day daily 9-5; Sept. to mid-June Thurs.-Mon. 9-5. Closed New Year's Day; Thanksgiving; Christmas.

HAGERMAN VALLEY HISTORICAL SOCIETY, Hagerman State St., 100 S. State St., Hagerman, ID 83332. Mailing Address: P.O. Box 86, Hagerman, ID 83332-0086. Tel.: 208-837-6288.
Key Personnel: Pres. (V), Orval Vader; Chm. (V), Leroy Jazwick.
Personnel Profile: Part-Time Volunteers 15.
Volunteer Hours: 292
Operating Expenses: 1,500
Operating Income: 2,000
Institution Type/Description: Historical Society Museum.
Collections: local history; photographs; fossilized Hagerman horse replica.
Hours & Admission Prices: April-Oct. Wed.-Sun. 1-4. No charge; donations accepted.
Attendance: 500 (estimated)
Membership: Single $15; Couple $25.

Hailey

BLAINE COUNTY HISTORICAL MUSEUM, 218 N. Main St., Hailey, ID 83333. Mailing Address: P.O. Box 124, Hailey, ID 83333. Tel.: 208-788-1801 & 4210. Fax: 208-788-1801.
E-mail: bcmuseum@mindspring.com
Web Site: bchistoricalmuseum.org
Founded: 1964.
Congressional District: 2
Key Personnel: Dir., Teddie Daley; Pres. (V), Robert MacLeod.
Personnel Profile: Full-Time Volunteers 12; Part-Time Paid 3; Part-Time Volunteers 12.
Governing Authority: independent. Parent Institution: Blaine County. Subsidiary Institution: City of Hailey, ID. Tax-exempt: 501(c)(3).
Institution Type/Description: History Museum: housed in an 1880 old armory & social center.
Collections: 5,000 political buttons & items donated by Joseph W. Fuld; replica of a mine tunnel; kitchen; living room; school; office; sewing room displays; antiques; first Idaho switchboard telephone; Ezra Pound exhibit, Poetry & Politics; photographs; biographical & historical items and data from Mallory Collection; journals dating to mid-1800s; legal documents; cameras; some military uniforms from first and second world wars.
Research Fields: Hailey, birthplace of Ezra Pound; genealogy; history.
Facilities: library.
Activities: Annual Events: Open House; Living History Day; Blaine County Heritage Court.
Publications: newsletter.
Hours & Admission Prices: Memorial Day weekend to Oct. daily 11-5. No charge; donations accepted. &
Attendance: 1,500 (accurate)

Harrison

CRANE HISTORICAL SOCIETY MUSEUM, 201 Coeur d'Alene Ave., Harrison, ID 83833. Mailing Address: P.O. Box 152, Harrison, ID 83833-0152. Tel.: 208-689-3111.
E-mail: cranehistsoc@gmail.com
Web Site: www.cranehistoricalsociety.org
Founded: 1984.
Congressional District: 5
Key Personnel: Pres. (V), Berti Arnzen.
Personnel Profile: Part-Time Volunteers 8.
Volunteer Hours: 250
Operating Expenses: 2,860
Operating Income: 4,560
Governing Authority: Tax-exempt.
Institution Type/Description: Historical Society Museum.

Collections: local history & culture; photographs; period furnishings; personal artifacts; 1920 town jail; logging; saw mill; early 1900s equipment; oral histories.
Hours & Admission Prices: Memorial Day to Labor Day Sat.-Sun. & holidays 12-4; other times by appointment. No charge; donations accepted.
Attendance: 701 (accurate)
Membership: Single $10; Sawyer $25; Foreman $50; Mill Owner $75; Life $250.

Idaho City

BOISE BASIN MUSEUM, 503 Montgomery St., Idaho City, ID 83631. Mailing Address: P.O. Box 358, Idaho City, ID 83631-0358. Tel.: 208-392-9551. Fax: 208-392-9905.
Institution Type/Description: History Museum.
Collections: local history; photographs; personal artifacts; Gold Rush; pioneer life.
Facilities: Museum-related items for sale.
Activities: rental facilities; tours.
Hours & Admission Prices: May & Sept. Sat.-Sun. 11-4; Memorial Day to Labor Day daily 11-4; tours by appointment. Adults $2, seniors & students $1.50; children under 6 no charge.

IDAHO CITY VISITOR CENTER, 100 Main St., Ste. A, Idaho City, ID 83631. Mailing Address: P.O. Box 350, Idaho City, ID 83631-0350. Tel.: 208-392-6040. Fax: 208-392-9512.
Institution Type/Description: Visitor Center.
Collections: local history & culture; photographs.
Hours & Admission Prices: Summer: daily 9-5; Winter: call for hours.

Idaho Falls

THE ART MUSEUM OF EASTERN IDAHO, (M), 300 S. Capital Ave., Idaho Falls, ID 83402-3952. Tel.: 208-524-7777. Fax: 208-529-6666.
E-mail: info@theartmuseum.org
Web Site: www.theartmuseum.org
Formerly: Eagle Rock Art Museum and Education Center
Founded: 2002.
Congressional District: 2
Key Personnel: Exec. Dir., Miyai Abe Griggs; Administrative Asst., Jessica Hull Livesay; Dir. Education, Alexa Stanger; Office Asst., Myrta Zietz.
Personnel Profile: Full-Time Paid 2; Part-Time Paid 2; Part-Time Volunteers 30; Interns 1.
Governing Authority: Tax-exempt.
Institution Type/Description: Art Museum.
Collections: works by Idaho artists including paintings; sculpture & photographs.
Facilities: Museum-related items for sale.
Activities: summer art camps; art classes for children & adults; art workshops; art education outreach in public schools; family art days; open studios.
Publications: monthly newsletter, MuseNews.
Hours & Admission Prices: Tues.-Sat. 11-5. Adults $4, youth 6-18 $2; children under 6 no charge. North American reciprocal museums.
Attendance: 30,000 (estimated)
Membership: $35-$1,000.

COLLECTORS' CORNER MUSEUM, 900 John Adams Pkwy., Idaho Falls, ID 83401-4049. Tel.: 208-528-9900.
Web Site: idahofallsarts.org
Founded: 2003.
Congressional District: 2
Key Personnel: Co Dir. & Museum Shop Mgr., Nida Gyorfy; Co Dir., Jim Gyorfy.
Personnel Profile: Full-Time Volunteers 2; Part-Time Volunteers 10.
Institution Type/Description: Collectors Museum.
Collections: local history & culture; personal artifacts; period furnishings; photographs; dolls; tools.
Research Fields: hobbies
Activities: group & school tours; special events; club affiliations.
Publications: bi-annual newsletter.
Hours & Admission Prices: Tues.-Sat. 10-5. Adults $5, senior citizens $4, students $3; children under 3 no charge. Closed holidays.
Attendance: 1,900 (accurate)

LDS TEMPLE VISITOR CENTER, 1000 Memorial Dr., Idaho Falls, ID 83402. Tel.: 208-523-4504.
Key Personnel: Dir., Gary J. Higley.

Governing Authority: Parent Institution: Church of Jesus Christ of Latter Day Saints.
Institution Type/Description: Religious Museum.
Collections: temple history; local culture; religious artifacts & furnishings; paintings; photographs; films.
Hours & Admission Prices: Center: daily 9-9. Temple Grounds: Summer daily. Temple interior closed to public. No charge.
Attendance: 50,000 (estimated)

MUSEUM OF IDAHO, 200 N. Eastern Ave., Idaho Falls, ID 83402-4029. Tel.: 208-522-1400. Fax: 208-524-5060. Facebook: Museum of Idaho.
E-mail: marketinqasst@museumofidaho.org
Web Site: www.museumofidaho.org
Founded: 1985.
Key Personnel: Exec. Dir., David Pennock, Ph.D.; Dir. Devel., Nick Gailey; Dir. Education, Sunny Katseangs; Acting Dir. Business Affairs, Allison Ball; Dir. Exhibitions, Rod Hansen; Acting Dir. Mktg., Laura Cooley; Business Asst., Susan Van Orden; Devel. Asst., Nicki McDaniel; Facilities Dir., Greg Stoddard.
Personnel Profile: Full-Time Paid 9; Part-Time Paid 1; Part-Time Volunteers 250; Interns 1.
Governing Authority: private; nonprofit organization. Parent Institution: Bonneville County Historical Society. Tax-exempt.
Institution Type/Description: History Museum.
Collections: regional natural history; cultural history; personal artifacts; furnishing; archaeological.
Major Exhibits: Olde Fashioned Christmas & Winter Festivals, 12/13-1/14; Race to the End of the Earth (T), 1/14-9/14.
Facilities: Museum-related items for sale.
Activities: educational programs.
Publications: annual report; quarterly newsletter.
Hours & Admission Prices: Mon.-Tues. 9-8, Wed.-Sat. 9-5. Adults $8, senior citizens 62 & over $7, youth 4-17 $6, family $25; discounts to groups, AAM members & military; members & children under 4 no charge.
Attendance: 100,000 (accurate)
Membership: Senior & Student $25; Adult $30; Dual $55; Family $100.

TAUTPHAUS PARK ZOO, 2725 Carnival Way, Idaho Falls, ID 83402. Mailing Address: P.O. Box 50220, Idaho Falls, ID 83405-0220. Tel.: 208-612-8552. Fax: 208-528-6256.
E-mail: ifzoo@idahofallszoo.org
Web Site: www.idahofallszoo.org
Founded: 1935.
Key Personnel: Supt., Beth Rich.
Personnel Profile: Full-Time Paid 12; Part-Time Paid 14; Part-Time Volunteers 60; Interns 3.
Governing Authority: Parent Institution: City of Idaho Falls. Subsidiary Institution: Parks & Recreation Division.
Institution Type/Description: Zoo.
Collections: over 400 animals.
Activities: educational programs.
Hours & Admission Prices: mid-April to Sept. daily 9-4; Memorial Day to Labor Day daily 9-5. Adults 13 & over $7, seniors 62 & over $5.50, children 4-12 $4; children 3 & under no charge.
Attendance: 125,000 (estimated)
Membership: Individual $30; Dual $40; General $65.

WILLARD ARTS CENTER/CARR GALLERY, 498 A St., Idaho Falls, ID 83402-3617. Tel.: 208-522-0471.
E-mail: bnewton@idahofallsarts.org
Web Site: www.idahofallsarts.org
Key Personnel: Interim Exec. Dir. & Operations Mgr., Jill Barnes; Carr Gallery Cur., Nathan Barnes; Technical Dir., Brad Higbee; Visual Arts. & Educational Dir., Catherine Smith; Dir. Devel., Gaylene Verdoorn
Institution Type/Description: Art Gallery.
Collections: artwork by local artists.
Hours & Admission Prices: Carr Gallery: Mon.-Fri. 11-5, Sat. 10-4. No charge.

Island Park

JOHNNY SACK'S CABIN, Big Springs, Hwy. 20, Island Park, ID 83429. Mailing Address: Ashton/Island Park Ranger District, Box 858, Ashton, ID 83420. Tel.: 208-652-7442.
Governing Authority: Parent Institution: U.S. Forest Service.
Institution Type/Description: Historic House: housed in the hand-built former home of German immigrant, Johnny Sack; built in 1939. Listed on the National Register of Historic Places.

Collections: Johnny Sack's life & history; personal artifacts; handmade furnishings.
Hours & Admission Prices: mid-June to Labor Day daily 10-4. No charge; donations accepted.

Jerome

JEROME COUNTY HISTORICAL SOCIETY, INC. AND IDAHO FARM & RANCH MUSEUM, 212 1st Ave. E., Jerome, ID 83338-2325. Mailing Address: P.O. Box 50, Jerome, ID 83338-0050. Tel.: 208-324-5641. Fax: 208-324-7694.
E-mail: info@historicaljeromecounty.com
Web Site: www.historicaljeromecounty.com
Founded: 1981.
Congressional District: 2
Key Personnel: C.E.O., Dale Ross; Treas., Shonna Fraser; Cur., Marguerite Roberson.
Personnel Profile: Part-Time Paid 1; Part-Time Volunteers 10.
Governing Authority: private; nonprofit organization. Tax-exempt: 501(c)(3).
Institution Type/Description: General Museum.
Collections: early farm machinery; historic structures; agricultural artifacts; history of Jerome County including World War II relocation camp; archaeological artifacts.
Facilities: library.
Activities: Annual Event: Live History Day in September.
Publications: monthly newsletter.
Hours & Admission Prices: Museum: Jan.-March Thurs.-Sat. 1-4; April-Dec. Tues.-Sat. 1-5. Idaho Farm & Ranch Museum by appointment (under construction). No charge; donations accepted. Closed legal holidays. &
Attendance: 750 (estimated)
Membership: Regular $15; Family $25; Supporting $75; Life $500.

LAND OF THE YANKEE FORK HISTORIC ASSOCIATION MUSEUM, 249 N. 250 W., Jerome, ID 83338. Mailing Address: c/o Land of the Yankee Fork Historic Association, P.O. Box 57, Stanley, ID 83278. Tel.: 208-410-5168.
E-mail: musicratdog@gmail.com
Formerly: Custer Museum
Founded: 1961.
Congressional District: 26
Key Personnel: District Forest Ranger, Russ Camper; Pres. (V), Dan Fansler; Museum Shop Mgr., Zora Fansler.
Personnel Profile: Full-Time Volunteers 4; Part-Time Paid 4; Part-Time Volunteers 6.
Governing Authority: federal. Parent Institution: U.S. Department of Agriculture, Forest Service. Subsidiary Institution: Idaho Parks & Recreation. Tax-exempt: 501(c)(3).
Institution Type/Description: History Museum: located on site of Custer Gold Mining Town.
Collections: gold rush mining equipment; household artifacts; Chinese personal items pertaining to mining; photographs; freighting equipment; school books & supplies. Historic Buildings: 1900 Custer School House; 1890 Custer Saloon & Doctor's Office; Brockman Cabin; McKenzie House.
Facilities: 40-vol. library of school books, hotel ledgers, 1880-1900 business records available on premises.
Activities: printed guide tours; lectures; permanent exhibitions; slide shows; gold panning. Annual Event: Custer Day Celebration in July.
Publications: pamphlets, Custer, A Walking Tour of Custer Idaho; Land of the Yankee Fork Historic Area.
Hours & Admission Prices: Memorial Day weekend to Labor Day daily 10-5. No charge; donations accepted. &
Attendance: 10,000 (estimated)
Membership: Individual $5; Family $7.50; Lifetime $100.

Kellogg

CRYSTAL GOLD MINE MUSEUM, 51931 Silver Valley Rd., Kellogg, ID 83837. Mailing Address: P.O. Box 510, Kellogg, ID 83837-0510. Tel.: 208-783-4653. Fax: 208-783-4653.
Web Site: www.goldmine-idaho.com
Founded: 1997.
Institution Type/Description: Natural History Museum: housed in an 1880s underground gold mine.
Collections: gold mine history; gold mining.
Facilities: Museum-related items for sale.
Activities: guided underground tour; seasonal gold-panning.
Hours & Admission Prices: May-Sept. daily 9-6; Oct.-April daily 10-4. Adults $12, senior citizens $11, children 4-17 $8.50; children under 4 no charge.

SHOSHONE COUNTY MINING AND SMELTING MUSEUM DBA THE STAFF HOUSE MUSEUM, 820 McKinley Ave., Kellogg, ID 83837-2525. Mailing Address: P.O. Box 783, Kellogg, ID 83837-0783. Tel.: 208-786-4141.
Web Site: www.staffhousemuseum.com
Founded: 1986.
Key Personnel: Pres. (V), Marlene Martin; Museum Shop Mgr., Becky Powers.
Personnel Profile: Full-Time Volunteers 1; Part-Time Paid 3; Part-Time Volunteers 3.
Governing Authority: bd. of directors. Tax-exempt.
Institution Type/Description: Mining & Smelting Museum: housed in the historic home of Stanley A. Easton, manager of Bunker Hill & Sullivan Mining & Concentrating Company; later converted to a residence for single Bunker Hill staff members.
Collections: mining; smelting; historic & cultural artifacts related to the history & industry of the area; scouting memorabilia.
Facilities: Museum-related items for sale.
Publications: biannual newsletter.
Hours & Admission Prices: May-Sept. daily 10-6; other times by appointment. Suggested: adults $4, seniors 55 & over $3, children 6-18 $1; discounts to groups; members & children under 6 no charge.
Attendance: 2,600 (accurate)
Membership: Single $15; Couple $20; Family $25; Business $30-$100; Golden $200; Life $500.

Ketchum

GAIL SEVERN GALLERY, 400 First Ave. N., Ketchum, ID 83340. Mailing Address: P.O. Box 1679, Ketchum, ID 83340. Tel.: 208-726-5079. Fax: 208-726-5092.
E-mail: info@gailseverngallery.com
Web Site: www.gailseverngallery.com
Institution Type/Description: Art Gallery.
Collections: paintings; sculpture.
Hours & Admission Prices: Mon.-Sat. 9-6, Sun. 12-6.

KETCHUM/SUN VALLEY HISTORICAL SOCIETY, 180 1st Ave. E., Ketchum, ID 83340. Mailing Address: P.O. Box 2746, Ketchum, ID 83340-2746. Tel.: 208-726-8118.
Web Site: www.ksvhistoricalsociety.org
Founded: 1989.
Institution Type/Description: Historical Society Museum.
Collections: local history & culture; photographs; personal artifacts.
Facilities: Museum-related items for sale.
Hours & Admission Prices: Mon.-Fri. 12-4, Sat. 1-4. Adults $5; children under 18 & members no charge; donations accepted.
Attendance: 750 (estimated)
Membership: Individual $40; Friend $75; Family $100; Sponsor $125; Lifetime Membership $200; Family Lifetime Membership $300.

* **SUN VALLEY CENTER FOR THE ARTS, (M),** 191 5th St. E., Ketchum, ID 83340. Mailing Address: P.O. Box 656, Sun Valley, ID 83353-0656. Tel.: 208-726-9491. Fax: 208-726-2344.
E-mail: info@sunvalleycenter.org
Web Site: www.sunvalleycenter.org
Founded: 1971.
Key Personnel: Exec. Dir., Sally Boettger; Pres. (V), Tod Hamachek; Vice Pres., Shelley Williams; Artistic Dir., Kristin Poole.
Personnel Profile: Full-Time Paid 17; Part-Time Paid 1.
Governing Authority: Subsidiary Institution: Company of Fools. Tax-exempt.
Institution Type/Description: Art Museum.
Collections: works by regional & national artists.
Activities: music performances; lectures; classes; youth education outreach into schools; temporary exhibitions; theater performances.
Hours & Admission Prices: Summer: Mon.-Fri. 9-5, Sat. 11-5; Winter: Mon.-Fri. 9-5. No charge. &
Membership: Friend $50; Family $75; Sustainer $100; Advocate $250; Associate $500; Patron $1,000; Benefactor $2,500; Curator $5,000; Director $10,000.

Kooskia

LOCHSA HISTORICAL RANGER STATION, US Hwy. 12 Mile Marker 121.5, Kooskia, ID 83539. Mailing Address: 502 Lowry St., Kooskia, ID 83539. Tel.: 208-926-4274. Fax: 208-926-6450.
Personnel Profile: Part-Time Volunteers 30.
Institution Type/Description: Historic Building: housed in a former forest service ranger station used during the 1920s & 1930s.

Collections: local history; period furnishings; personal artifacts; photographs.
Facilities: nature trails.
Activities: hiking.
Hours & Admission Prices: Memorial Day to Labor Day daily 9-5. No charge; donations accepted.

Lava Hot Springs

SOUTH BANNOCK COUNTY HISTORICAL CENTER, (M), 110 E. Main St., Lava Hot Springs, ID 83246. Mailing Address: P.O. Box 387, Lava Hot Springs, ID 83246-0387. Tel.: 208-776-5254.
Founded: 1980.
Congressional District: 2
Key Personnel: C.E.O. & Pres., Kim Harris; Dir., Cathy Sher; Treas., Janet Berreth; Cur. & Clerk, Mary Avery; Cur. & Clerk, Deb Bower; Museum Shop Mgr., Debbie Fagnant; Cur. & Clerk, Betty Sikkenga; Custodian, Ken Sikkenga.
Personnel Profile: Part-Time Paid 6; Part-Time Volunteers 4.
Governing Authority: society; board of directors. Tax-exempt: 501(c)(3).
Institution Type/Description: Historical Society.
Collections: artifacts; photographs; documents; family histories; Native American.
Research Fields: local, area & state history; genealogical; individual & family history.
Facilities: research center.
Activities: temporary & permanent exhibits; slide presentations; walking tours upon request.
Publications: walking tour guide; bimonthly newsletter; biographical sketches, Charley, The Virginian; Bob Dempsey-The Greatest Trader on the Immigrant Trail; Trails, Trappers, Trains & Travelers; A Century of Transition 1890-1990.
Hours & Admission Prices: Daily 12-5; other times by appointment. No charge; donations accepted. &
Attendance: 16,123 (accurate)
Membership: Single $12; Couple $20; Business $30.

Lewiston

LEWIS-CLARK STATE COLLEGE CENTER FOR ARTS & HISTORY, 500 8th Ave., Lewiston, ID 83501-2698. Mailing Address: 415 Main St., Lewiston, ID 83501-1821. Tel.: 208-792-2243. Fax: 208-792-2850.
Web Site: www.lcsc.edu/museum/contactus/default.htm
Institution Type/Description: Art & History Museum: housed in the former Vollmer Great Bargain Store designed by architect, Kirtland Cutter; built in 1884. Listed on the National Register of Historic Places.
Collections: local history; visual arts.
Activities: special events.
Hours & Admission Prices: Tues.-Sat. 11-4.

NEZ PERCE COUNTY HISTORICAL SOCIETY AND MUSEUM, 0306 Third St., Lewiston, ID 83501-1860. Tel.: 208-743-2535.
E-mail: npcdirector1@cableone.net
Web Site: www.npchistsoc.org
Formerly: Luna House Museum
Founded: 1960.
Congressional District: 1
Key Personnel: Pres. (V), Richard Riggs; Registrar, Lora Feucht.
Personnel Profile: Full-Time Paid 1; Part-Time Paid 2; Part-Time Volunteers 1.
Governing Authority: board of directors; Nez Perce County Historical Society. Tax-exempt: 501(c)(3).
Institution Type/Description: History Museum.
Collections: artifacts & photographs depicting Nez Perce County history from 1800 to present. Historic Building: Heritage House
Research Fields: Nez Perce County, ID.
Facilities: library; archives.
Activities: temporary exhibits; educational programs; special lectures.
Publications: The Golden Age, journal of the Nez Perce County Historical Society (published semiannually); A Walking Tour of Historical Downtown Lewiston; annual historical photographic calendar; Lewiston Country, An Armchair History.
Hours & Admission Prices: March-Dec. Tues.-Sat. 10-4. Adults $4, seniors 60 & up $3, students 7-17 $2; children 6 & under & members no charge. Closed major federal holidays. &
Attendance: 5,000 (accurate)
Membership: Student $20; Member $30-$49; Friend $50-$99; Sponsor $100-$499; Patron $500-$999.

Mackay

LOST RIVER MUSEUM, 312 Capitol St., Mackay, ID 83251. Mailing Address: P.O. Box 572, Mackay, ID 83251. Tel.: 208-588-3148.
Founded: 1975.
Key Personnel: Dir., Earl Lockie; Pres. (V), Lana Pehrson
Governing Authority: Parent institution: South Custer Historical Society. Tax-exempt.
Institution Type/Description: History Museum: housed in a former church; built c.1900.
Collections: local history & culture; mining equipment & implements; railroad memorabilia; early printing equipment; personal artifacts; period clothing; household utensils; photographs; 70 yrs. of hard copy newspapers.
Activities: Museum Sponsors: Ice Cream Social in July; Christmas Bazaar.
Hours & Admission Prices: Memorial Day to Sept. Sat.-Sun. 1-5; other times by appointment. &
Attendance: 500 (estimated)

Malad

ONEIDA COUNTY PIONEER MUSEUM, 27 Bannock St., Malad, ID 83252-1240. Tel.: 208-766-4847.
Web Site: www.maladidaho.org/museum/museum.htm
Institution Type/Description: History Museum: built in 1914.
Collections: period artifacts & furnishings.
Hours & Admission Prices: Tues.-Sat. 1-5; other times by appointment. No charge; donations accepted.

McCall

CENTRAL IDAHO HISTORICAL MUSEUM, 1001 State St., McCall, ID 83638-3705. Tel.: 208-634-4497.
E-mail: cihmuseum@gmail.com
Web Site: centralidahohistoricalmuseum.com
Formerly: Central Idaho Cultural Center
Founded: 1992.
Key Personnel: Pres. (V), Marlee Wilcomb
Governing Authority: Tax-exempt.
Institution Type/Description: Historic Site: built in 1937 by the Civilian Conservation Corps. Listed on the National Register of Historic Places.
Collections: local history; personal artifacts; period furnishings; photographs; Civil Conservation Corp. statue. Historic Buildings: fire warden's house; crew quarters; machine shop; pump house.
Research Fields: McCall area history; Civil Conservation Corp. in Idaho.
Publications: annual newsletter.
Hours & Admission Prices: Memorial Day to Sept. 15 Wed.-Sat. 12-3:30; Winter: by appointment. House Tours: adults $3. &
Attendance: 1,250 (estimated)
Membership: $25; $50; $100 & up.

Melba

CELEBRATION PARK, 6530 Hot Spot Ln., Melba, ID 83641-5275. Tel.: 208-495-2745.
Founded: 1990.
Governing Authority: Parent Institution: Canyon County, ID.
Institution Type/Description: Park Museum.
Collections: local history; archaeology; Native American art & artifacts; photographs.
Facilities: visitor center.
Hours & Admission Prices: Daily 9-4. Closed major holidays.

Montpelier

NATIONAL OREGON/CALIFORNIA TRAIL CENTER, 320 N. 4th St., Montpelier, ID 83254-1256. Mailing Address: P.O. Box 323, Montpelier, ID 83254-0323. Tel.: 208-847-3800; 866-847-3800 (toll free). Fax: 208-847-1863.
E-mail: info@oregontrailcenter.org
Web Site: www.oregontrailcenter.org
Founded: 1997.
Key Personnel: Dir., Becky Smith; Pres. (V), Al Harrison; Vice Pres., Steve Allred; Sec. & Treas., Jene Parker; Museum Shop Mgr., Cindy Raymond.
Personnel Profile: Full-Time Paid 2; Part-Time Paid 20; Part-Time Volunteers 2.
Governing Authority: Tax-exempt.
Institution Type/Description: History Museum.
Collections: pioneer history & heritage; early settlers; paintings.

Facilities: Museum-related items for sale.
Activities: simulated wagon train.
Hours & Admission Prices: May-Sept. Sun.-Thurs. 9-5, Fri.-Sat. 9-6; Oct. Tues.-Sat. 10-3; Nov.-April Mon.-Thurs. 10-3 by reservation. Tours every 1/2 hour. Discounts to AAM & ICOM members.
Attendance: 48,000 (accurate)
Membership: Valued Pioneer $1,000; Outrider $5,000; Trail Boss $10,000; Wagon Master $20,000.

Moscow

THE APPALOOSA MUSEUM AND HERITAGE CENTER, 2720 W. Pullman Rd., Moscow, ID 83843-4024. Tel.: 208-882-5578, ext. 279. Fax: 208-882-8150.
E-mail: museum@appaloosa.com
Web Site: www.appaloosamuseum.org
Founded: 1973.
Congressional District: 1
Key Personnel: Pres. (V), King Rockhill.
Personnel Profile: Full-Time Paid 1; Part-Time Paid 2; Part-Time Volunteers 5; Interns 1.
Governing Authority: board of directors; nonprofit organization. Associated with the Appaloosa Horse Club, Inc., P.O. Box 8403, Moscow 83843. Tax-exempt.
Institution Type/Description: History Museum.
Collections: equipment, photographs & art work relative to the history of the Appaloosa horse & the Nez Perce Indians who bred the Appaloosa horse.
Research Fields: the Appaloosa horse in European & U.S. history.
Facilities: Museum-related items for sale.
Activities: guided tours; lectures. Museum Sponsors: Annual Holiday Open House; Annual Trail Ride.
Publications: quarterly newsletter.
Hours & Admission Prices: Mon.-Fri. 10-5, Sat. 10-4. Suggested Donation: adults $2. Closed legal holidays.
Attendance: 5,000 (estimated)
Membership: The Spurs $25; Roping Club $50; Bridle Club $100; Saddle Club $500; Patron $1,000; Benefactor $5,000.

IDAHO FOREST FIRE MUSEUM, 310 N. Main St., Moscow, ID 83843-2629. Tel.: 208-882-4767.
Founded: 1997.
Institution Type/Description: Firefighting History Museum.
Collections: forest fire history; Idaho's 1910 forest fire; photographs; Smoky the Bear memorabilia.
Hours & Admission Prices: Jan.-May Mon.-Fri. 9-4; June-Dec. Mon.-Fri. 9-5, Sat. 9-4. No charge.

MCCONNELL MANSION, 110 S. Adams, Moscow, ID 83843-2829. Mailing Address: 327 E. Second St., Moscow, ID 83843-2819. Tel.: 208-882-1004. Fax: 208-882-0759.
E-mail: lchslibrary@latah.id.us
Web Site: users.moscow.com/lchs
Founded: 1968.
Congressional District: 1
Key Personnel: Dir., Daniel Crandall; Pres. (V), Brian Magelky; Cur., Ann Catt.
Personnel Profile: Part-Time Paid 2; Part-Time Volunteers 30.
Governing Authority: board of trustees. Latah County Historical Society. Tax-exempt.
Institution Type/Description: Local History Museum: housed in 1886 Governor McConnell Mansion.
Collections: concentration on the history of Latah County, Idaho, including interpretation of lifestyles in rural area Pacific Northwest from 1870-present; archives; photographs; oral history transcriptions.
Research Fields: lumber industry; railroads; oral history; local businesses & families; social history; conservation; historic preservation & restoration; agricultural history.
Facilities: research facilities.
Activities: guided tours; school tours; classroom instruction; speaker's series; workshops; interpretive exhibits; ice cream social; slide programs; hands-on exhibits & artifacts; treasure hunt for children; educational program.
Publications: books on local history; annual journal, Latah Legacy; newsletter; exhibit brochures.
Hours & Admission Prices: May-Sept. Tues.-Sat. 1-5; Oct.-April Tues.-Sat. 1-4; other times by appointment. Suggested Donation: family $4, adults $2; members & children under 16 no charge.
Attendance: 3,500 (estimated)
Membership: Individual $25; Cultivator $35; Prospector $50; Builder $100; Engineer $250; Founder $500 & up.

UNIVERSITY OF IDAHO PRICHARD ART GALLERY, 414 S. Main St., Moscow, ID 83843-2916. Mailing Address: P.O. Box 444405, Moscow, ID 83844-4405. Tel.: 208-885-3586. Fax: 208-885-3622.
E-mail: rrowley@uidaho.edu
Key Personnel: Dir., Roger Rowley
Institution Type/Description: Art Gallery.
Collections: paintings; sculpture.
Activities: lectures; presentations.
Hours & Admission Prices: Winter: Tues.-Sat. 10-8, Sun. 10-6; Summer: Tues.-Fri. 1-7, Sat. 9-4. No charge.

Mountain Home

BRUNEAU DUNES OBSERVATORY, 27608 Sand Dunes Rd., Mountain Home, ID 83647. Tel.: 208-366-7919. Fax: 208-366-2844.
E-mail: bru@idpr.idaho.gov
Institution Type/Description: Observatory & Visitor Center.
Collections: Observatory: 25 inch telescope. Visitor Center: wildlife & the dunes; birds of prey; insects; fossils.
Facilities: visitor center. Gift items for sale.
Activities: audiovisual orientation program; bird watching.
Hours & Admission Prices: Observatory: April to mid-Oct. Fri.-Sat. one hour before sundown to midnight. Park: $5 per vehicle. Observatory: $3 per person; children 5 & under no charge.

MOUNTAIN HOME HISTORICAL MUSEUM, 180 S. 3rd E., Mountain Home, ID 83647-3019. Tel.: 208-587-6847. Facebook: Mountain Home Historical Society.
E-mail: director@mountainhomemuseum.com
Web Site: www.mountainhomemuseum.com
Key Personnel: Dir., Debbie Shoemaker; Pres. (V), Marilyn Landers
Institution Type/Description: Historic Building: housed in the former Carnegie Public Library; built in 1908. Listed on the National Register of Historic Places.
Collections: local history; cultural heritage; mining; agriculture; railroad memorabilia; Native American artifacts; WWI & WWII.
Hours & Admission Prices: Tues.-Fri. 10-4. No charge; donations accepted. Closed New Year's Day; Martin Luther King Jr. Day; Presidents' Day; Memorial Day; Independence Day; Labor Day; Columbus Day; Veterans Day; Thanksgiving; Christmas.
Membership: Individual $15; Family $25; Business $100; Lifetime $500.

Mullan

CAPTAIN JOHN MULLAN MUSEUM, 229 Earle St., Mullan, ID 83846. Mailing Address: P.O. Box 675, Mullan, ID 83846-0675. Tel.: 208-744-1155 (June-Aug.) & 1557 (Sept.-May).
Founded: 1985.
Congressional District: 1
Key Personnel: Pres. & Chm. (V), Butch Jacobson.
Personnel Profile: Part-Time Volunteers 60.
Governing Authority: Parent Institution: Mullan Historical Society. Tax-exempt.
Institution Type/Description: History Museum: housed in the old Liberty Theater.
Collections: Mullan history; period furnishings; clothing; photographs; newspapers; mining artifacts; sports memorabilia; logging equipment; drugstore artifacts.
Research Fields: area history; genealogy.
Hours & Admission Prices: June-Aug. 10-4; other times by appointment. No charge; donations accepted. Closed holidays.
Attendance: 850
Membership: Yearly $20; Sustaining $25.

Murphy

OWYHEE COUNTY HISTORICAL MUSEUM, 17085 Basey St., Murphy, ID 83650. Mailing Address: P.O. Box 67, Murphy, ID 83650-0067. Tel.: 208-495-2319. Fax: 208-495-9824.
E-mail: administration@owyheemuseum.org
Web Site: www.owyheemuseum.org
Founded: 1960.
Congressional District: 1
Key Personnel: Exec. Dir., Joe Demshar; Pres. (V), Meril Ebbers; Office Mgr., Vivian Good; Librarian/Archivist, Joan Bachman; Museum Shop Mgr., Neva Miller.
Personnel Profile: Full-Time Paid 1; Part-Time Paid 2; Part-Time Volunteers 31; Interns 1.

Governing Authority: society. Parent Institution: Owyhee County Historical Society. Subsidiary Institution: Owyhee County Museum. Tax-exempt.
Institution Type/Description: Historical Society Museum.
Collections: schoolhouse; Owyhee County historical items; agriculture; Indian artifacts; general; geology; archives; mining; oral history; Owyhee county artist; Owyhee County artifacts; Norman Reich bit and spur collection.
Research Fields: agriculture; Indian artifacts; archives; mining; bits & spurs.
Facilities: Gifts & books for sale.
Activities: guided tours; lectures; field trips; temporary & permanent exhibits; activities & lectures for local school children.
Publications: annual magazine, Owyhee Outpost; Owyhee County Blue Book 1898; Sketches of Owyhee County; Outpost: A Journal of Owyhee County History.
Hours & Admission Prices: Tues.-Sat. 10-4. Adults $2; discounts to AAM & ICOM members & Owyhee County Residence; members no charge. Closed holidays. &
Attendance: 14,321 (estimated)
Membership: Senior & Student $15; Individual $25; Family $35; Supporting $50; Associate $250; Life $500.

Nampa

CANYON COUNTY HISTORICAL SOCIETY & MUSEUM, 1200 Front St., Nampa, ID 83651-3931. Mailing Address: P.O. Box 595, Nampa, ID 83653-0595. Tel.: 208-467-7611.
E-mail: info@canyoncountyhistory.com
Web Site: www.canyoncountyhistory.com
Founded: 1976.
Personnel Profile: Part-Time Volunteers 4.
Governing Authority: Branch Museum: Our Memories Indian Creek Museum, 1122 Main St., Caldwell, ID 83605. Tax-exempt: 501(c)(3).
Institution Type/Description: Historical Society Museum.
Collections: area history; period artifacts including cider press & cast iron stove; bottles; 1915 Edison phonograph; caboose phone & tools. Our Memories Museum: local history; c.1920s dental office; c.1940s operating room; period artifacts.
Facilities: Museum-related items for sale.
Activities: tours; research; special events. Museum Sponsors: Fund-raiser Tea & Style Show in March; Christmas Open House in December.
Publications: quarterly newsletter, Rivers, Rails, and Trails.
Hours & Admission Prices: Depot: May-Oct. Thurs.-Fri. 11-4, Sat. 10-2; Nov.-April Thurs.-Fri. 11-4, Sat. 11-2. Our Memories Indian Creek Museum: Tues. & Fri. 11-4; other times by appointment. Adults $3, youth 13-18 $2, seniors & children 6-12 $1; members no charge.
Attendance: 5,000 (estimated)
Membership: Individual $10; Couple $15; Family $18; Contributing $25; Sustaining $50; Sponsor $100; Business $250; Corporate $500; Founder $1,000.

DEER FLAT NATIONAL WILDLIFE REFUGE, 13751 Upper Embankment Rd., Nampa, ID 83686-8046. Tel.: 208-467-9278. Fax: 208-467-1019.
Founded: 1909.
Congressional District: 1
Key Personnel: Acting Mgr., Stan Culling.
Personnel Profile: Full-Time Paid 5.
Institution Type/Description: Wildlife Refuge.
Collections: wildlife & their habitats; birds; fish; plants.
Facilities: nature trails; visitor center.
Activities: hiking; hands-on activities; viewing platforms & blind; environmental education.
Hours & Admission Prices: Park: daily daylight hours. Visitor Center: Mon.-Fri. 8-4, Sat. 10-4. No charge. Closed Federal holidays. &
Attendance: 170,000 (estimated)

WARHAWK AIR MUSEUM, 201 Municipal Dr., Nampa, ID 83687-8582. Tel.: 208-465-6446. Fax: 208-465-6232.
E-mail: admin@warhawkairmuseum.org
Web Site: warhawkairmuseum.org
Founded: 1989.
Congressional District: 1
Key Personnel: Exec. Dir., Sue Paul; Pres. (V), John Paul; Museum Shop Mgr., Tammy John; Administrative Asst., Heather Mullins.
Personnel Profile: Full-Time Paid 1; Full-Time Volunteers 1; Part-Time Paid 2; Part-Time Volunteers 30.
Governing Authority: Tax-exempt.
Institution Type/Description: Military Museum.
Collections: WWI & WWII artifacts; WWII fighter airplanes including Curtiss P-40 & P-51C Mustang; 1940s survival gear & equipment; NASA; Cold War Era - Korea & Vietnam aircraft including Huey helicopter, F-86 jet, MIGS 17 & 21, O-1 Bird airplane and F-104 jet; personal veterans' memorabilia.
Facilities: library. Museum-related items for sale.
Activities: special events; tours.
Publications: newsletter, Flight Line.
Hours & Admission Prices: Tues.-Sat. 10-5, Sun. 11-4. Adults $10, senior citizens & military with ID $8, children 5-12 $4; discount to AAA members; members no charge. &
Attendance: 23,500 (estimated)
Membership: Individual $35; Family $60; Sponsorship $250; Sponsorship II $500, $1,000, $5,000 & $10,000.

Oakley

OAKLEY VALLEY HISTORICAL MUSEUM, 140 W. Main St., Oakley, ID 83346. Mailing Address: P.O. Box 239, Oakley, ID 83346-0239. Tel.: 208-862-7890.
Founded: 2000.
Congressional District: 2
Key Personnel: Pres., Robert Fehlman.
Personnel Profile: Part-Time Volunteers 25.
Governing Authority: Parent Institution: City of Oakley. Tax-exempt.
Institution Type/Description: History Museum: listed on the National Register of Historic Places.
Collections: area history; historic buildings.
Hours & Admission Prices: Summer: Fri.-Sat. 1-4; other times by appointment. No charge; donations accepted.
Attendance: 1,000 (estimated)
Membership: Individual $20; Couple $30.

Orofino

CLEARWATER HISTORICAL MUSEUM, 315 College Ave., Orofino, ID 83544. Mailing Address: P.O. Box 1454, Orofino, ID 83544-1454. Tel.: 208-476-5033.
E-mail: chmuseum@frontier.com
Web Site: www.clearwatermuseum.org
Founded: 1960.
Congressional District: 1
Key Personnel: Pres., Nick Albers; Vice Pres., Mike McHone; Sec., Donna Heieren; Treas., Martha Dempsey; Museum Dir., Bernice Pullen.
Personnel Profile: Part-Time Paid 1; Part-Time Volunteers 2.
Governing Authority: state & county. Parent Institution: Clearwater Historical Society. Tax-exempt.
Institution Type/Description: History Museum.
Collections: books; Nez Perce Indian artifacts; photographs; logging equipment used with horsepower; tapes of oral history; county artifacts; clippings from newspapers; Lewis-Clark histories & early mining days in Pierce City, Idaho.
Research Fields: history of Clearwater County & early family records.
Facilities: library of Idaho & County history, taped oral history & photograph collection; reading room. Local history books for sale.
Activities: guided tours; group tours; permanent & temporary exhibitions.
Publications: quarterly newspaper.
Hours & Admission Prices: June-Sept. Tues.-Sat. 12:30-5:30; Oct.-May Tues.-Sat. 1:30-4:30. No charge; donations accepted. Closed holidays. &
Attendance: 1,500 (estimated)
Membership: Individual $12; Life $100.

Paris

PARIS TABERNACLE HISTORICAL SITE, 109 S. Main St., Paris, ID 83261. Tel.: 208-945-2686.
Governing Authority: Parent Institution: Church of Jesus Christ of Latter-Day Saints. Tax-exempt.
Institution Type/Description: Historic Building: housed in the Church of Jesus Christ of Latter Day Saints; built by Mormon pioneers in 1889. Listed on the National Register of Historic Places.
Collections: local history & culture; religious artifacts & furnishings; paintings.
Activities: organ recitals.
Hours & Admission Prices: Guided Tours: Memorial Day to Labor Day daily 9:30-5:30. No charge.
Attendance: 9,000 (accurate)

Parma

OLD FORT BOISE REPLICA AND MUSEUM, 20847 Old Fort Boise Rd., Parma, ID 83660. Mailing Address: P.O. Box 942, Parma, ID 83660-0942. Tel.: 208-722-5138.
Institution Type/Description: History Museum.
Collections: local history & culture; photographs; period furnishings; personal artifacts.
Hours & Admission Prices: June-Aug. Fri.-Sun. 1-3; other times by appointment. No charge; donations accepted. ♿
Attendance: 450 (estimated)

Payette

PAYETTE COUNTY MUSEUM, 90 S. 9th St., Payette, ID 83661. Mailing Address: P.O. Box 696, Payette, ID 83661-0696. Tel.: 208-642-4883.
E-mail: payettemuseum@qwestoffice.net
Founded: 1973.
Governing Authority: Parent Institution: Payette County Historical Society.
Institution Type/Description: History Museum: housed in a former church.
Collections: local history & culture; period furnishings; 1861 Confederate Civil War Cannon barrel.
Hours & Admission Prices: Wed.-Sat. 12-4; other times by appointment. No charge; donations accepted. ♿

Pierce

J. HOWARD BRADBURY MEMORIAL LOGGING MUSEUM, 101 S. Main St., Pierce, ID 83546. Mailing Address: P.O. Box 378, Weippe, ID 83553-0378. Tel.: 208-464-2531 & 435-4670.
Institution Type/Description: Logging History Museum.
Collections: logging industry history; photographs; tools.
Hours & Admission Prices: mid-June to mid-Oct. Fri.-Sat. 12-4; other times by appointment.

Pocatello

BANNOCK COUNTY HISTORICAL MUSEUM, 3000 Alvord Loop (Upper Level of Ross Park), Pocatello, ID 83201. Mailing Address: P.O. Box 253, Pocatello, ID 83204-0253. Tel.: 208-233-0434.
E-mail: bancohismus@gmail.com
Web Site: www.bchm-id.org
Founded: 1963.
Congressional District: 2
Key Personnel: Society Pres., Robert Myers; Dir. & Cur., Tabatha Butler; Sec., Fred Evans.
Personnel Profile: Full-Time Paid 1; Part-Time Paid 2; Part-Time Volunteers 3.
Governing Authority: society. Parent Institution: Bannock County Historical Society. Tax-exempt: 501(c)(3).
Institution Type/Description: History Museum.
Collections: Wrensted collection; Indian photographs; Peake collection; Larsen collection; railroad photographs; farm equipment; general store; medical history.
Activities: guided tours; permanent & temporary exhibitions; games for kids.
Publications: annual newsletter.
Hours & Admission Prices: Memorial Day to Labor Day daily 10-6; Sept.-May Tues.-Sat. 10-4; other times by appointment. Winter: adults $2, children 6-12 $1; Summer: adults $4, seniors $3, children $2; children under 6 & members no charge. (includes admission to Fort Hall replica). Closed legal holidays. ♿
Attendance: 3,000 (estimated)
Membership: Individual $10.

FORT HALL REPLICA, 3002 Alvord Loop, Pocatello, ID 83201. Mailing Address: Pocatello Parks & Recreation, 144 Wilson Ave., Pocatello, ID 83201. Tel.: 208-234-1795. Fax: 208-234-6578.
Web Site: www.forthall.net
Founded: 1963.
Congressional District: 2
Key Personnel: Dir. Pocatello Park & Recreation, Jerry Sepich; Chm. (V) Ft. Hall Replica Commission, Jacques Alvord.
Personnel Profile: Part-Time Paid 2.
Governing Authority: Parent Institution: City of Pocatello, Parks & Recreation Dept. Tax-exempt.
Institution Type/Description: History Museum: housed in a replica of the historic facility that served pioneers along the Oregon Trail.
Collections: Oregon Trail history; pioneer life & culture; covered wagon; tepee; period furnishings & artifacts; photographs.

Hours & Admission Prices: Memorial Day to Labor Day Mon.-Sat. 10-6, Sun. 1-5. Adults $4, seniors $3, youth 3-17 $2; discounts to school & large groups.
Attendance: 2,500 (estimated)

✱ **IDAHO MUSEUM OF NATURAL HISTORY, (M),** 5th & Dillon, Pocatello, ID 83209. Mailing Address: 921 S. 8th Ave., Stop 8096, Pocatello, ID 83209-0002. Tel.: 208-282-3168. Fax: 208-282-5893.
E-mail: imnh@isu.edu
Web Site: imnh.isu.edu
Founded: 1934.
Congressional District: 2
Key Personnel: Dir. & Cur. Anthropology, Dr. Herbert Maschner; Store Mgr., Bill Angle; Earth Sciences Cur., Dr. Leif Tapanila; Education Resource Coord. & Education Cur., Kelly Pokorny; Life Sciences Cur., Dr. Rick Williams; Southeast Idaho Repository Mgr., Amy Commendador; Registrar, Curt Schmitz; Office Admin., Mary Moses.
Personnel Profile: Full-Time Paid 8; Part-Time Paid 16; Part-Time Volunteers 20; Interns 3.
Governing Authority: state. Administrative Authority: Idaho State University. Tax-exempt: 170(b)(1)(A).
Institution Type/Description: Natural History Museum.
Collections: Great Basin collections in archaeology & ethnology; vertebrate fossils from the Intermountain West, specializing in Pleistocene; Crabtree flintworking; manuscripts & photographs of north Rocky Mountains region; major herbarium holdings; modern ostial; modern vertebrate; ornithology; mammalogy.
Research Fields: archaeology; botany; ethnography; herpetology; ornithology; recent vertebrates; historical photographs & documents; vertebrate paleontology; pertaining to the Great Basin & northern Rocky Mountains.
Activities: guided gallery tours; films; formally organized education programs; adult lectures; classes; field trips; temporary & permanent exhibitions; field work in anthropology & paleontology; summer programs for children & adults; self-guided ISU Tree Walk; Discovery Room; ScienceTrek sleepover; Natural History Academy; Pint-sized Science Academy; Forays into the Field; summer science snack.
Publications: Tebiwa, Journal of the Idaho Museum of Natural History; special publications; occasional papers series; booklets & pamphlets.
Hours & Admission Prices: Tues.-Sat. 11-5. Adults $5, youth under 18 $1. Closed holidays. ♿
Attendance: 8,829 (accurate)
Membership: Individual $35; Household $50; Sustaining $200.

MIND'S EYE GALLERY - IDAHO STATE UNIVERSITY, Rendezvous Complex, 921 S. 8th Ave., Pocatello, ID 83209. Tel.: 208-282-3451.
Institution Type/Description: Art Gallery.
Collections: photographs; drawings.
Hours & Admission Prices: Mon.-Fri. 8-5.

THE MUSEUM OF CLEAN, 711 S. Second Ave., Pocatello, ID 83201-6520. Mailing Address: P.O. Box 700, Pocatello, ID 83201. Tel.: 208-232-3535 & 236-6906.
E-mail: donaslett@varsityfs.com
Web Site: museumofclean.com
Formerly: Don Aslett's Cleaning Museum
Founded: 2007.
Congressional District: 2
Key Personnel: C.E.O., Don Aslett.
Personnel Profile: Full-Time Paid 1.
Governing Authority: Parent Institution: Don Aslett Clean World Foundation. Tax-exempt: 501(c)(3).
Institution Type/Description: History Museum.
Collections: cleaning techniques, products, implements & machines.
Research Fields: environmental protection history; home living.
Hours & Admission Prices: Call for hours. Adults $5. ♿
Attendance: 5,000 (estimated)

POCATELLO ART CENTER, GALLERY AND SCHOOL, 444 N. Main, Pocatello, ID 83204-5071. Tel.: 208-232-0970.
E-mail: pocartctr@ida.net
Web Site: www.pocatelloartctr.org
Key Personnel: Pres. (V), Carolyn Purnell
Institution Type/Description: Art Gallery.
Collections: paintings; porcelain; photographs; china.
Activities: classes; workshops; temporary & permanent exhibitions. Museum Sponsors: 1st Friday's Art Walk & Sale.

Hours & Admission Prices: Mon.-Fri. No charge; donations accepted. Closed Federal holidays.
Membership: Individual $25.

POCATELLO ZOO, 3101 Ave. of the Chiefs, Pocatello, ID 83204-2135. Tel.: 208-234-6264.
E-mail: sransom@pocatellor.us
Web Site: www.pocatellozoo.org
Key Personnel: Zoo Dir., Scott Ransom; Cur. Education, Bonnie Jakubos
Institution Type/Description: Zoo.
Collections: wildlife native to North America's Intermountain West.
Hours & Admission Prices: April 15-April 30 Sat.-Sun. 9-5; May-June 15 daily 9-5; June 16 to Labor Day daily 10-6; Sept.-Oct. Sat.-Sun. 10-4. Adults 12-59 $4.25, seniors 60 & over $3, children 3-11 $2.25; children 2 & under no charge.

THE TRANSITION GALLERY - IDAHO STATE UNIVER-SITY, Earl R. Pond Student Union, 921 S. 8th Ave., Pocatello, ID 83209. Tel.: 208-282-3451.
Institution Type/Description: Art Gallery.
Collections: paintings; sculpture.
Hours & Admission Prices: Mon.-Fri. 10-8.

Priest Lake

PRIEST LAKE MUSEUM, 38 W. Lakeshore Dr., Priest Lake, ID 83856. Mailing Address: P.O. Box 44, Coolin, ID 83821-0044. Tel.: 208-443-2676.
E-mail: priestlakemuseum@gmail.com
Web Site: www.plmuseum.org
Founded: 1990.
Key Personnel: Pres. (V), Tom Weitz; Museum Shop Mgr., Kay Coykendall.
Personnel Profile: Part-Time Volunteers 60; Interns 1.
Volunteer Hours: 1,500
Operating Expenses: 9,850
Operating Income: 10,050
Governing Authority: Tax-exempt.
Institution Type/Description: Historic Building Museum: housed in a former forest ranger cabin; c.1930s.
Collections: local history & culture; period furnishings; personal artifacts; photographs.
Major Exhibits: Prohibition & Moonshining at Priest Lake, 6/14-9/14.
Research Fields: Priest Lake; Idaho area history.
Activities: summer outings to historical sites.
Publications: Pioneer Voices; DVD.
Hours & Admission Prices: Memorial Day to Labor Day Tues.-Sun. 10-4. No charge; donations accepted. &
Attendance: 3,500 (accurate)
Membership: $10-$500.

Priest River

KEYSER HOUSE TIMBER MUSEUM, 301 Montgomery St., Priest River, ID 83856. Mailing Address: Priest River Chamber of Commerce, P.O. Box 929, Priest River, ID 83856-0929. Tel.: 208-448-2721.
Institution Type/Description: Historic House Museum: housed in the former home of Henry and Elizabeth Keyser; built in 1895.
Collections: local history & culture; pioneer life; Keyser family artifacts; timber industry; period furnishings; photographs; quilts; tools.
Hours & Admission Prices: Call for hours.

Rexburg

LEGACY FLIGHT MUSEUM, Rexburg Airport, 400 Airport Rd., Rexburg, ID 83440. Mailing Address: P.O. Box 122, Rexburg, ID 83440-0122. Tel.: 208-359-5905. Fax: 208-356-7989.
E-mail: legacyflightmuseum@yahoo.com
Web Site: www.legacyflightmuseum.com
Founded: 2004.
Key Personnel: Dir. & Pres. (V), John Bagley; Chm. (V), Glenn Embree; Museum Shop Mgr., George Howard.
Personnel Profile: Part-Time Volunteers 10.
Governing Authority: Parent Institution: City of Rexburg. Tax-exempt.
Institution Type/Description: Military History Museum.
Collections: military aircraft history; personal artifacts; photographs.
Research Fields: aviation.
Facilities: museum/hangar floor leased for receptions & company parties.

Activities: flight lessons.
Hours & Admission Prices: Memorial Day to Labor Day Mon.-Sat. 9-5; Sept.-May Sat. 9-5. Adults $6; discounts to veterans & BYU-Idaho students.
Attendance: 1,000 (estimated)

TETON FLOOD MUSEUM, 51 N. Center St., Rexburg, ID 83440-1539. Tel.: 208-359-3063. Fax: 208-359-3063.
E-mail: jillspencer2@yahoo.com
Key Personnel: Dir., Cur. & Museum Shop Mgr., Jill Spencer.
Personnel Profile: Part-Time Paid 1; Part-Time Volunteers 6.
Governing Authority: city. Tax-exempt.
Institution Type/Description: History Museum: the disaster of the Teton Dam collapse on June 5, 1976.
Collections: personal artifacts; video of actual dam break; photographs; pioneer; western artifacts; war memorabilia.
Activities: tour of the actual dam; video of the break; temporary & permanent exhibitions.
Hours & Admission Prices: May-Sept. Mon. 10-7, Tues.-Sat. 10-5; Oct.-April Mon. 11-7, Tues.-Fri. 11-5. Adults $2, youth 12-18 $1, children under 12 $.50. &
Attendance: 5,000 (estimated)

UPPER SNAKE RIVER VALLEY HISTORICAL SOCIETY, 51 N. Center St., Rexburg, ID 83440-1539. Mailing Address: P.O. Box 244, Rexburg, ID 83440-0244. Tel.: 208-356-9100.
E-mail: usrvhistsoc@hotmail.com
Web Site: rexburghistoricalsociety.com
Founded: 1965.
Congressional District: 2
Key Personnel: C.E.O., Dir. & Cur., Louis Clements; Pres., Harvey Jackman; Vice Pres., Harold Forbush.
Personnel Profile: Full-Time Volunteers 1; Part-Time Volunteers 20.
Governing Authority: society. Branch Museums: Fremont County Museum, St. Anthony, ID; Teton Valley Museum, Driggs, ID; Jefferson County Museum, Rigby, ID; Bonnville County Museum, Idaho Falls, ID; Teton Flood Museum, Rexburg, ID. Tax-exempt: 170(b)(1)(A).
Institution Type/Description: Local Historic Museum.
Collections: emphasis on Idaho & local history for teaching & information purposes; Teton flood collection; guns; agriculture; period artifacts; manuscripts.
Research Fields: Idaho history; local history; Teton flood.
Facilities: 1,800-vol. library of Idaho & local history available for inter-library loan & for research; reading room. Western history books for sale.
Activities: guided tours; lectures; films; gallery talks; TV & radio programs; formally organized education programs for children; inter-museum loan, permanent, temporary & traveling exhibitions. Institute Sponsors: History Fair, an annual gathering of artifact displays from over the state.
Hours & Admission Prices: Memorial Day to Labor Day Mon. 10-7, Tues.-Sat. 10-5; Winter: Mon. 11-7, Tues.-Fri. 11-5. Adults $2, children 12-18 $1; children under 12 no charge. Closed national holidays. &
Attendance: 10,129 (accurate)

Rigby

JEFFERSON COUNTY HISTORICAL SOCIETY AND FARN-SWORTH TV PIONEER MUSEUM, 118 West 1st South, Rigby, ID 83442-1303. Mailing Address: P.O. Box 284, Rigby, ID 83442-0284. Tel.: 208-745-8423.
Web Site: www.blacksmithinn.com/museum.html
Founded: 1975.
Congressional District: 2
Key Personnel: Pres. (V), Gary Spaulding; Sec., Kathryn McLain; Treas., Linda Kite; Cur., Glena R. Smith.
Personnel Profile: Full-Time Volunteers 5; Part-Time Volunteers 28.
Governing Authority: Parent Institution: Jefferson County Historical Society. Tax-exempt.
Institution Type/Description: History Museum.
Collections: area history; Philo Farnsworth, inventor with over 300 international patents; photographs; Native American.
Activities: tours.
Hours & Admission Prices: Tues.-Sat. 1-5. Requested Donations: adults $2, children 6-17 $1; discounts to veterans; children under 6 no charge. &
Attendance: 1,500 (estimated)
Membership: Jefferson County Historical Society $10; Lifetime $100.

Rupert

MINIDOKA COUNTY HISTORICAL MUSEUM, 99 E. Baseline Rd., Rupert, ID 83350. Mailing Address: Box 21, Rupert, ID 83350-0021. Tel.: 208-436-0336.
E-mail: mchs@pmt.org
Web Site: minidokamuseum.weebly.com
Founded: 1970.
Congressional District: 2
Key Personnel: Pres., Gus Bryngelson; Museum Shop Mgr., Ginger Cooper.
Personnel Profile: Part-Time Paid 2; Part-Time Volunteers 10.
Operating Expenses: 31,605
Operating Income: 32,747
Governing Authority: Tax-exempt.
Institution Type/Description: History Museum.
Collections: local history; Native American artifacts; early 1900s marble soda fountain; over 600 bottles & jars; horse drawn farm equipment; Russell steam engine tractor; wooden wheeled carts; 1906 railroad depot history; photographs; Ice Age bones; newspapers from 1905 to 2008; 1940 fire engine; early 1900 wooden fire carts; 1906 HPPR Depot.
Facilities: research library.
Activities: programs at monthly meetings. Museum Sponsors: member appreciation picnic; Thanksgiving dinner; Christmas party.
Publications: monthly newsletter.
Hours & Admission Prices: Call for hours. No charge; donations accepted. &
Attendance: 900 (accurate)
Membership: Youth 13-18 $4; Single $10; Couple $15; Lifetime $100.

MINIDOKA NATIONAL WILDLIFE REFUGE, Rte. 4, 961 E. Minidoka Dam, Rupert, ID 83350-9414. Tel.: 208-436-3589.
Institution Type/Description: Wildlife Refuge.
Collections: wildlife including swans; white pelicans; songbirds; mule deer; beaver; muskrat; coyote.
Facilities: nature trails.
Activities: hiking.
Hours & Admission Prices: Call for hours.

Saint Maries

HISTORICAL HUGHES HOUSE, 538 Main Ave., Saint Maries, ID 83861. Tel.: 208-245-1501.
Institution Type/Description: Historic House Museum: housed in a former men's club and later a doctor's office; built in 1902.
Collections: local history & culture; photographs; period furnishings; personal artifacts.
Hours & Admission Prices: mid-May to Sept. Wed.-Sun. 12-4; call to confirm. No charge; donations accepted.
Attendance: 300 (estimated)

Salmon

LEMHI COUNTY HISTORICAL MUSEUM, 210 Main, Salmon, ID 83467-4111. Tel.: 208-756-3342.
E-mail: llemhi@centurytel.net
Web Site: www.lemhicountymuseum.org
Founded: 1963.
Congressional District: 2
Key Personnel: Pres. (V), Hope Benedict; Vice Pres., Laurie Rugierro; Sec., Fredde Haworth; Cur., Clair Wiley.
Personnel Profile: Part-Time Paid 2; Part-Time Volunteers 15.
Governing Authority: society; nonprofit organization. Tax-exempt.
Institution Type/Description: History Museum.
Collections: local history & culture; Indian artifacts; mining; ranching; Lemhi Shoshoni artifacts; Asian artifacts including lost-wax bronzes & Ming Dynasty vases.
Publications: map, Lewis & Clark Route through Pass to Bitterroot; Madame Charbonneau.
Hours & Admission Prices: May 15 to mid-Oct. Mon.-Sat. 10-5. Admission 12 & over $2; children under 12 no charge. &
Attendance: 5,500 (estimated)
Membership: Individual $10.

SACAJAWEA INTERPRETIVE, CULTURAL & EDUCATIONAL CENTER, (M), 110 Lewis and Clark, Salmon, ID 83467-5340. Mailing Address: 200 Main St., Salmon, ID 83467-4111. Tel.: 208-756-1188.
E-mail: jlb.sac@gmail.com
Web Site: www.sacajaweacenter.org
Founded: 2003.
Congressional District: 35
Key Personnel: Dir., Judy Barkley.
Personnel Profile: Part-Time Paid 4; Part-Time Volunteers 40.
Governing Authority: Parent Institution: city of Salmon. Tax-exempt.
Institution Type/Description: History Museum.
Collections: Sacajawea's life & cultural history; Sacajawea monument; wildlife; tipi encampment.
Facilities: learning center; visitor center; outdoor amphitheater; picnic areas; garden; interpretive trail; nature trail; low ropes challenge course.
Activities: special events; demonstrations; interpretive presentations; hands-on primitive living skills; classes; outdoor & environmental education; concerts; interactive educational exhibits.
Hours & Admission Prices: Park: daily. Visitor Center: May-Oct. Family $12, adult $5. &

Sandpoint

BIRD AVIATION MUSEUM & INVENTION CENTER, Bird Ranch Rd., Sandpoint, ID 83864. Mailing Address: P.O. Box 817, Sandpoint, ID 83864. Tel.: 208-255-4321. Fax: 208-255-7630.
Key Personnel: Dir., Rachel Riddle Schwam
Institution Type/Description: Aviation History Museum.
Collections: history of aviation, aviators, & innovators; photographs; air planes; Hall of Fame inductees.
Facilities: cafe. Museum-related items for sale.
Activities: special events; rental facilities.
Hours & Admission Prices: May 17 to Oct. 8 Mon.-Sat. 8-4; other times by appointment. No charge.

BONNER COUNTY HISTORICAL MUSEUM, (M), 611 S. Ella Ave., Sandpoint, ID 83864-1100. Tel.: 208-263-2344.
E-mail: bchs@frontier.com
Web Site: www.bonnercountyhistory.org
Founded: 1972.
Congressional District: 1
Key Personnel: Dir., Olivia Luther; Pres., Kathy Osborne.
Personnel Profile: Part-Time Paid 1; Part-Time Volunteers 25.
Governing Authority: society; nonprofit organization. Parent Institution: Bonner County Historical Society, Inc. Tax-exempt.
Institution Type/Description: Local History Museum.
Collections: local history; early Americana; geology; American Indian artifacts; forest history; logging; sawmilling; oral history; photographs; railroads.
Research Fields: historical site survey; local history.
Facilities: research library; photo display area; arboretum.
Activities: oral history program; historical site survey program; field trips; evening historical programs.
Publications: membership newsletter; Beautiful Bonner Vol. I & Vol. II.
Hours & Admission Prices: Tues.-Fri. 10-4. Adults $3, students $1; discounts to AAM members; members & children under 6 no charge. &
Attendance: 5,000 (estimated)
Membership: Senior & Student $15; Individual $20; Family $30.

Shoshone

SHOSHONE INDIAN ICE CAVES, 1561 N. Hwy. 75, Shoshone, ID 83352-5246. Tel.: 208-886-2058.
Institution Type/Description: History Museum.
Collections: local history; geology; lava ice case; Native American artifacts; minerals; gems.
Activities: guided tours.
Hours & Admission Prices: May-Sept. daily 8am-7:15pm. Adults $10, seniors $8, children 4-12 $6; discounts to groups; children 3 & under no charge.

Spalding

NEZ PERCE NATIONAL HISTORICAL PARK, 39063 U.S. Hwy. 95, Spalding, ID 83540-9715. Mailing Address: P.O. Box 237, Wisdom, MT 59761-0237. Tel.: 208-843-7001. Fax: 208-843-7003.
E-mail: bob_chenoweth@nps.gov
Web Site: www.nps.gov/nepe
Founded: 1965.
Congressional District: 1
Key Personnel: Supt., Steve Black; Cur., Bob Chenoweth; Museum Shop Mgr., Mary Lou Tiede; Archivist & Library Technician, Robert Applegate; Museum Technician, Linda J. Paisano.
Personnel Profile: Full-Time Paid 27; Part-Time Paid 1; Part-Time Volunteers 2; Interns 1.

Governing Authority: federal. Parent Institution: National Park Service, Dept. of Interior, Washington DC 20240. Park Interpretive Shelters at Kamiah & White Bird, ID. Tax-exempt.
Institution Type/Description: Park Museum: located on a site of prehistoric occupation & early mission.
Collections: Nez Perce ethnological; prehistoric lithics; Western Americana; local pioneer & historical documents & memorabilia; 4,000 photos on the Nez Perce Indians & the local region; Big Hole National ethnographic, historic & archeological related to the battlefield; Lake Roosevelt National Recreation area archeological & historical items associated with Fort Spokane; City of Rock National Reserve archeological material.
Research Fields: Nez Perce Indian culture; early mission activity; local historical events pertaining to park sites & westward American expansion.
Facilities: 2,500-vol. library of books about Nez Perce Indian & Pacific Northwest history available at research center by appointment. Books, postcards & handcrafts for sale.
Activities: Nez Perce Indian craft demonstrations & cultural programs each summer; slide & film programs; guided tours; special events.
Publications: interpretive & information brochures; site bulletins.
Hours & Admission Prices: Winter: daily 8-4:30; Summer: daily 8-5:30. No charge. Closed New Year's Day; Thanksgiving; Christmas. &
Attendance: 152,393 (accurate)

Twin Falls

***   HERRETT CENTER FOR ARTS & SCIENCE, FAULKNER PLANETARIUM AND CENTENNIAL OBSERVATORY, (M),** 315 Falls Ave., College of Southern Idaho, Twin Falls, ID 83301-3367. Mailing Address: P.O. Box 1238, Twin Falls, ID 83303-1238. Tel.: 208-732-6655. Fax: 208-736-4712.
E-mail: herrett@csi.edu
Web Site: herrett.csi.edu/
Founded: 1952.
Congressional District: 2
Key Personnel: Dir., Teri Fattig; Mgr. Collections, Facilities Specialist & Special Events Coord., Sarah Ngakoutou; Exhibits Mgr. & Graphic Designer, Joey Heck; Art Gallery Mgr., Milica Popovic; Facility, Display & Planetarium Technician, Robert (Nick) Peterson; Office Mgr. & Museum Shop Mgr., Carolyn Browning; Planetarium Mgr., Rick Greenawald; Mktg. Coord, Doug Maughan; Education Facilitator, Darcy Thornborrow; Planetarium Production Specialist, Chris Anderson.
Personnel Profile: Full-Time Paid 7; Part-Time Paid 4; Part-Time Volunteers 4.
Governing Authority: Junior College District. Parent Institution: College of Southern Idaho. Tax-exempt.
Institution Type/Description: Arts & Science Museum, Planetarium & Observatory.
Collections: Americas & European archaeology; natural history; modern paintings & sculptures.
Research Fields: Lithic technology.
Facilities: 5,000-vol. library of books available to students for research; art gallery; planetarium (Digistar II); observatory.
Activities: presentations for classes of students; permanent, temporary & traveling exhibitions.
Hours & Admission Prices: Tues. & Fri. 9:30-9, Wed.-Thurs. 9:30-4:30, Sat. 1-9. Planetarium: call for show times. Planetarium: adults $4.50, senior $3.50, students $2.50; discounts to AAM members; children under 2 no charge. Museum: no charge. Observatory: call for hours. Closed holidays. &
Attendance: 51,000 (estimated)

Wallace

NORTHERN PACIFIC DEPOT RAILROAD MUSEUM, 219 6th St., Wallace, ID 83873-2283. Mailing Address: P.O. Box 469, Wallace, ID 83873-0469. Tel.: 208-752-0111. Fax: 208-753-9361.
E-mail: northernpacificdepot@cebridge.net
Founded: 1986.
Key Personnel: Museum Shop Mgr., Diane Van Broeke.
Governing Authority: Tax-exempt.
Institution Type/Description: Railroad Museum: listed on the National Register of Historical Places.
Collections: depot & railroading history; 13 ft. glass map.
Activities: Annual Event: Classic Car Show & Festival in May.
Hours & Admission Prices: April & Oct. 1-Oct. 15 Mon.-Sat. 10-3; May & Sept. daily 9-5; July-Aug. daily 9-7. Family $6, adults $2.50, children 6-16 $1.
Attendance: 8,500 (estimated)

OASIS BORDELLO MUSEUM, 605 Cedar St., Wallace, ID 83873. Mailing Address: P.O. Box 928, Osburn, ID 83849. Tel.: 208-753-0801.
Founded: 1993.
Key Personnel: Owner, Michelle Mayfield
Institution Type/Description: History Museum: housed in a former brothel.
Collections: local history; furnishings; early wine press; personal artifacts.
Facilities: visitor center. Gift items for sale.
Activities: guided tours.
Hours & Admission Prices: May-Oct. Mon.-Sat. 10-5, Sun. 11-3. Adults $5.

SIERRA SILVER MINE TOUR, 420 5th St., Wallace, ID 83873-2212. Tel.: 208-752-5151. Facebook: Sierra Silver Mine.
E-mail: info@silverminetour.org
Web Site: www.silverminetour.org
Founded: 1982.
Institution Type/Description: Geology Museum.
Collections: local history; mining equipment & tools; geology.
Activities: guided tours; demonstrations.
Hours & Admission Prices: May & Sept. daily 10-2; June-Aug. daily 10-4. Adults $12, seniors 60 & over $11, children 4-16 $8.50; discounts to groups; children under 4 no charge.
Attendance: 10,000

WALLACE DISTRICT MINING MUSEUM, 509 Bank St., Wallace, ID 83873-2224. Mailing Address: P.O. Box 469, Wallace, ID 83873-0469. Tel.: 208-556-1592. Fax: 208-556-1592.
E-mail: info@wallaceminingmuseum.org
Web Site: www.wallaceminingmuseum.org
Founded: 1956.
Congressional District: 1
Key Personnel: Exec. Dir., Jim McReynolds; Pres., Dale Lavigne; Sec., Dennis O'Brien; Museum Shop Mgr., Peggy McReynolds.
Personnel Profile: Full-Time Paid 2; Part-Time Paid 4; Part-Time Volunteers 2.
Governing Authority: nonprofit. Tax-exempt.
Institution Type/Description: Mining Museum.
Collections: mining exhibits; 3D Sunshine Mine model; timbered mine.
Research Fields: Various records, maps, reports, documents, photographs, statistics for the Coeur d'Alene Mining District.
Facilities: Brochures, membership stock certificates & other museum-related items for sale.
Activities: permanent & temporary exhibitions.
Publications: brochure; booklet, Gold Strikes & Silver Linings.
Hours & Admission Prices: See website for hours & admission prices. &
Attendance: 49,000 (estimated)
Membership: Sustaining $20; Contributing $50; Patron $100; Life $500.

Weippe

WEIPPE DISCOVERY CENTER, 204 Wood St., Weippe, ID 83553. Mailing Address: P.O. Box 435, Weippe, ID 83553-0435. Tel.: 208-435-4058. Fax: 208-435-4374.
E-mail: discovery@weippe.com
Web Site: www.weippediscoverycenter.com
Founded: 2003.
Congressional District: 1
Key Personnel: Dir., Terri Summerfield
Institution Type/Description: History Museum.
Collections: Lewis & Clark history; painted murals; life-size map of trail & campsites; Nez Perce artifacts.
Hours & Admission Prices: Mon. & Thurs.-Fri. 10-5, Tues.-Wed. 10-7, Sat. 10-1. No charge; donations accepted. &
Attendance: 600

WEIPPE HILLTOP HERITAGE MUSEUM, 105 N. 1st St. E., Weippe, ID 83553. Mailing Address: P.O. Box 279, Weippe, ID 83553-0279. Tel.: 208-435-4200.
Founded: 1999.
Congressional District: 1
Key Personnel: Pres. (V), Everet Martin; Dir., Lorna D. Hubler.
Personnel Profile: Part-Time Volunteers 1.
Governing Authority: Tax-exempt.
Institution Type/Description: History Museum.
Collections: local history & culture; personal artifacts; photographs; period furnishings.
Publications: children's book, The Weippe Story.
Hours & Admission Prices: Wed. 10-4; other times by appointment. No charge; donations accepted. &

Attendance: 600 (estimated)
Membership: Individual $5; Family $20; Business $50.

Weiser

SNAKE RIVER HERITAGE CENTER, 2295 Paddock Ave., Weiser, ID 83672-1195. Mailing Address: P.O. Box 307, Weiser, ID 83672-0307. Tel.: 208-549-0205. Facebook: Weiser Museum.
E-mail: info@weisermuseum.com
Web Site: www.weisermuseum.com
Founded: 1962.
Congressional District: 1
Key Personnel: Pres. (V), Lynn Isaacson.
Personnel Profile: Part-Time Paid 1; Part-Time Volunteers 12.
Governing Authority: private; nonprofit organization. Tax-exempt: 170(b)(1)(A).
Institution Type/Description: Cultural Center & Museum: housed in 1899-1933 preparatory school.
Collections: local history; music; costumes; folklore; glass; geology; Holocaust history; automotive garages & history; alumni memorabilia of former institute students. Historic Buildings: 1880 Harper Cabin; 1923 Hooker Hall.
Research Fields: Weiser, Washington County & regional history.
Facilities: newspapers, pamphlets, letters, & photographs available for research on premises; 300-seat auditorium. Museum-related gifts for sale.
Activities: guided tours; drama; demonstrations; historical slide show presentation.
Publications: annual newsletter.
Hours & Admission Prices: Memorial Day-Labor Day Thurs.-Sat. 1-4; tours by appointment. No charge; donations accepted.
Attendance: 1,000 (estimated)
Membership: Family & Business $25; Lifetime $300.

Winchester

WOLF EDUCATION & RESEARCH CENTER, 518 Joseph Ave., Winchester, ID 83555. Mailing Address: P.O. Box 217, Winchester, ID 83555. Tel.: 888-422-1110, ext. 1.
Key Personnel: Dir., Chris Anderson
Institution Type/Description: Nature Center.
Collections: wolves & their habitats; natural artifacts; photographs.
Facilities: 300-acre sanctuary. Museum-related items for sale.
Activities: educational programs; tours.
Hours & Admission Prices: May & Sept. Self-Guided Tours: Sat.-Sun. 9-4; Guided Tours: call for appointment. Memorial Day to Labor Day Self-Guided Tours: daily 9-5; Guided Tours: daily 7:30-7; other times by appointment.
Attendance: 3,400 (estimated)

ILLINOIS
(493 listings)

Aledo

ESSLEY-NOBLE MUSEUM: MERCER COUNTY HISTORI-CAL SOCIETY, 1406 S.E. 2nd Ave., Aledo, IL 61231-2504. Mailing Address: P.O. Box 269, Aledo, IL 61231-0269. Tel.: 309-582-2280.
E-mail: mcmuseum@frontier.com
Web Site: www.mchsil.org
Founded: 1959.
Congressional District: 17
Key Personnel: Pres., Bill Bertrand; Cur., Veda Meriwether.
Personnel Profile: Part-Time Paid 1; Part-Time Volunteers 15.
Governing Authority: society; nonprofit organization. Parent Institution: Mercer County Historical Society. Tax-exempt: 501(c)(3).
Institution Type/Description: Historical & Preservation Society: housed in the Essley-Noble Memorial Building.
Collections: local history; c.1900 kitchen; pearl button cutting lathe; old farming tools; Civil War items; Lincoln desk; 1880 parlor room; original Mercer county court house flag; showcase of early American toys; New Boston Millstone; Roosevelt Military academy; photographs; newspapers on microfilm; one room country school; genealogy; machine shed with farming tools & implements; William & Vashti College artifacts.
Research Fields: genealogy; local history & lore; county census-microfilm newspapers.
Facilities: 8-vol. library of Mercer County DAR cemetery records & various county genealogy material available for research on premises.

Activities: guided tours; permanent & temporary exhibitions; spring & fall meeting open to public with program. Annual Event: ice cream social open to public in August.
Publications: quarterly newsletter.
Hours & Admission Prices: April-Oct. Wed. & Sat.-Sun. 1-5; Nov.-March Sat. 12-4; school & groups by appointment. No charge; donations accepted.
Attendance: 1,200 (estimated)
Membership: Individual $20; Life $300.

Alton

ALTON MUSEUM OF HISTORY AND ART, INC., 2809 College Ave., Alton, IL 62002-4743. Tel.: 618-462-2763.
E-mail: altonmuseum@gmail.com
Web Site: altonmuseum.com
Founded: 1971.
Congressional District: 12
Key Personnel: Pres., Dr. Norman Showers; 1st Vice Pres., Brian Combs; 2nd Vice Pres., Patti Culp; Immediate Past Pres. & Office Mgr., Charlene Johnson; Koenig House Hostess & Museum Shop Mgr., Lois Lobbig; Founder, Pres. Emeritus & Member-At-Large, Charlene Gill.
Personnel Profile: Full-Time Volunteers 1; Part-Time Volunteers 65; Interns 1.
Governing Authority: nonprofit organization. Historic House: Koenig House, 829 E. 4th St., Alton, IL 62002. Tax-exempt: 501(c)(3).
Institution Type/Description: History & Art Museum: c.1832 educational building & museum.
Collections: river memorabilia; local paintings, sculptures, industry display; artifacts from local railroads, industries & business places; model trains; photographs; Freedom collection, including Elijah Lovejoy, Abraham Lincoln & Lyman Trumbull; Lovejoy print shop; underground railroad; Robert Wadlow, the world's tallest recorded man; fine arts; Black pioneers. Historic Building: 1887 Koenig House.
Research Fields: Alton Civil War prison; Black Pioneers; historic characters of region; underground railroad; Elijah Lovejoy; early education & transportation.
Facilities: 2,500 sq. ft. exhibit space. Art prints of local artists, books & other museum-related items for sale.
Activities: guided tours; lectures; arts festivals; organized educational programs; training programs for professional museum workers; participatory & temporary exhibitions; film; Koenig House tours; heirloom evaluation days; annual membership meeting. Museum Sponsors: Friday Night at the Museum monthly; Miles Davis Jazz Celebration in May; Annual Guardian of History Award; Margaret Davis Weber Award; antique auction; Alton Landing Festival; Christmas Tea.
Publications: brochure; annual report; historic reprints; quarterly newsletter.
Hours & Admission Prices: Wed.-Sat. 10-4, Sun. 1-4. Adults $5, children $1; discounts to seniors on Wed.; members no charge. Closed New Year's Day; Easter; Independence Day; Thanksgiving; Christmas.
Attendance: 11,000 (accurate)
Membership: Youth $5; Senior $20; Active (Single) $25; Family $30; History Loner $40; Patron of the Arts $50; Business $100; Life $1,000; Benefactor $5,000.

Amboy

AMBOY DEPOT MUSEUM, 50 S. East Ave., Amboy, IL 61310. Mailing Address: P.O. Box 108, Amboy, IL 61310. Tel.: 815-857-4700.
E-mail: information@amboydepotmuseum.org
Web Site: www.amboydepotmuseum.org
Founded: 1973.
Congressional District: 14
Key Personnel: Chm. (V), Jack Dempsey; Museum Shop Mgr., Carol Biester.
Personnel Profile: Part-Time Volunteers 28.
Governing Authority: Parent Institution: Amboy Depot Commission. Tax-exempt.
Institution Type/Description: Historic Building Museum: housed in a former depot and division headquarters of the Illinois Central Railroad. Listed on the National Register of Historic Places.
Collections: Amboy & the Illinois Central Railroad history; period artifacts & furnishings; photographs.
Activities: community events. Museum Sponsors: Depot Days in August.
Publications: quarterly membership newsletter.
Hours & Admission Prices: Sun. & Thurs. 1-4, Fri.-Sat. 10-4. No charge; donations accepted.
Attendance: 2,436 (accurate)
Membership: Individual $10; Family $20; Gold $100.

Antioch

LAKES REGION HISTORICAL SOCIETY ARCHIVE CENTER, 965 Main St., Antioch, IL 60002. Tel.: 847-395-4912.
E-mail: info@lakesregionhistory.org
Web Site: lakesregionhistory.org
Institution Type/Description: Historical Society Museum.
Collections: local history; birth & marriage records; cemetery & funeral home records; census & church records; county biographies; diaries; ledgers; family histories; high school yearbooks; maps; photographs; newspapers; telephone books; scrapbooks.
Activities: research.
Hours & Admission Prices: 1st Sat. each month; other times by appointment.

LAKES REGION HISTORICAL SOCIETY MUSEUM, 817 Main St., Antioch, IL 60002. Mailing Address: c/o Lakes Region Historical Society, 965 Main St., Antioch, IL 60002. Tel.: 847-395-4912.
E-mail: info@lakesregionhistory.org
Web Site: lakesregionhistory.org
Institution Type/Description: Historical Society Museum: housed in a former school house; built in 1892.
Collections: local history & culture; period furnishings; photographs; personal artifacts.
Hours & Admission Prices: March-Dec. Sat. 11-3. No charge; donations accepted.

LAKES REGIONAL HISTORICAL SOCIETY MEETING HOUSE, 977 Main St., Antioch, IL 60002. Tel.: 847-395-4912.
E-mail: info@lakesregionhistory.org
Web Site: lakesregionhistory.org
Institution Type/Description: Historical Society Museum: housed in a former church; built in 1863.
Collections: local history; period furnishings; photographs; military artifacts; Studebaker buggy; sleigh.
Activities: rental facilities.
Hours & Admission Prices: 1st Sat. each month; other times by appointment. No charge; donations accepted.

Arcola

ILLINOIS AMISH INTERPRETIVE CENTER, 125 N. County Rd. 425E, Arcola, IL 61910-3756. Tel.: 888-452-6474. Fax: 217-268-4810.
E-mail: amishcenter@consolidated.net
Institution Type/Description: History Museum.
Collections: Amish culture & history; Arthur Amish settlement; personal artifacts.
Facilities: Museum-related items for sale.
Activities: tours.
Hours & Admission Prices: May-Oct. Wed.-Sun 10-5, adult $10, senior 60 & over $8, children (5-12) $6. Large groups year-round by reservation.

Arlington Heights

ARLINGTON HEIGHTS HISTORICAL MUSEUM, (M), 110 W. Fremont St., Arlington Heights, IL 60004-5912. Tel.: 847-255-1225. Fax: 847-255-1570.
Web Site: www.ahmuseum.org
Founded: 1957.
Congressional District: 10th & 12th
Key Personnel: Pres., Warren Wahl; Museum Admin., Kristina Christie; Devel. Coord., Carol Frieburg; Cur., Mickey Horndasch; Asst. Coord., Teri Ozawa.
Personnel Profile: Full-Time Paid 1; Part-Time Paid 9; Part-Time Volunteers 100.
Governing Authority: nonprofit organization. Parent Institution: Arlington Heights Historical Society. Subsidiary Institutions: Arlington Heights Park District; Village of Arlington Heights. Tax-exempt: 501(c)(3).
Institution Type/Description: History Museum.
Collections: local history 1836-present. Heritage Gallery: temporary exhibits. Historic Buildings: c.1836 replica log cabin; c.1882 Muller house; c.1908 Banta house; c.1900 coach house; c.1906 soda pop factory building.
Research Fields: local history.
Facilities: library of clippings, photographs & archival material relating to Arlington Heights available for use by public. Heritage Gallery: museum-related items for sale.
Activities: guided tours; formally organized educational programs for school & scout children; permanent & temporary exhibitions; exhibit receptions & programs; craft workshops; family programs; special events.

Publications: bimonthly newsletter, The Dunton Post.
Hours & Admission Prices: Tours: Sat.-Sun. 2 & 3; other times & groups by appointment. Adults $4, children $2; discount to AAM members; members no charge. Heritage Gallery (current exhibit): Fri.-Sun. 1:30-4:30. No charge; donations accepted. Closed national holidays.
Attendance: 14,000 (estimated)
Membership: Individual $25; Family $45; Nonprofit $75; Patron $100; Individual Life $500. Business Memberships available.

Athens

ABRAHAM LINCOLN LONG NINE MUSEUM, 200 S. Main St., Athens, IL 62613. Tel.: 217-636-8755.
Web Site: www.abrahamlincolnlongninemuseum.com
Founded: 1972.
Congressional District: 18
Key Personnel: Dir., John R. Eden
Institution Type/Description: History Museum: built by Colonel Matthew Rogers in 1832. Listed on the National Register of Historic Places.
Collections: Lincoln history; paintings; period artifacts.
Publications: local walking & driving tour maps.
Hours & Admission Prices: June-Sept. 1 Tues.-Sat. 1-5. No charge; donations accepted.

Atlanta

ATLANTA LIBRARY & MUSEUM, 100 S.E. Race St., Atlanta, IL 61723-7526. Mailing Address: P.O. Box 568, Atlanta, IL 61723. Tel.: 217-648-2003.
Institution Type/Description: Historic Building: built in 1908. Listed on the National Register of Historic Places.
Collections: local history; period high school photographs, plaques, & trophies; tools; personal artifacts.
Hours & Admission Prices: Tues.-Wed. & Fri. 12:30-4:30, Thurs. 12:30-8, Sat. 9-1.

Aurora

AURORA HISTORICAL SOCIETY, 305 Cedar St., Aurora, IL 60507. Mailing Address: P.O. Box 905, Aurora, IL 60507. Tel.: 630-897-9029.
E-mail: ahs@aurorahistory.net
Web Site: aurorahistory.net
Founded: 1906.
Key Personnel: Dir., John Jaros.
Personnel Profile: Full-Time Paid 2; Part-Time Paid 2; Part-Time Volunteers 70.
Governing Authority: Subsidiary Institutions: Tanner House Museum, Cedar St., Aurora, IL; David L. Pierce Art & History Center, 20 E. Downer Pl., Aurora, IL. Tax-exempt.
Institution Type/Description: History Museum.
Collections: local history from 1830s to present; letters; legal documents; books; maps; photographs. Historic Building: 1857 house.
Activities: research.
Hours & Admission Prices: Tues. & Thurs. by appointment. Adults $5; members no charge.
Membership: Student & Senior $20; Individual $30; Family $40; Patron $125; Benefactor $250; Lifetime $500.

AURORA REGIONAL FIRE MUSEUM, (M), New York Ave. & Broadway, Aurora, IL 60507. Mailing Address: P.O. Box 1782, Aurora, IL 60507-1782. Tel.: 630-892-1572.
E-mail: arfminfo@aol.com
Web Site: www.auroraregionalfiremuseum.org
Founded: 1990.
Key Personnel: Mgr., Deborah Davis; Cur., David Lewis.
Personnel Profile: Full-Time Paid 2; Part-Time Volunteers 12.
Governing Authority: private; nonprofit organization. Tax-exempt: 501(c)(3).
Institution Type/Description: Firefighting Museum: housed in the 1894 Old Central Fire Station.
Collections: firefighting history in Aurora & surrounding communities.
Facilities: library; 50-seat auditorium; 15,000 sq. ft. exhibit space; theater. Museum-related items for sale.
Activities: films; formal education programs; lectures. Annual Event: Fire Engine Muster.
Publications: quarterly newsletter, Fire Museum News.
Hours & Admission Prices: Thurs.-Sat. 1-4; groups by appointment. Adults $5, children 12 & under $3; discounts to AAM members. Closed all major holidays.

Attendance: 5,500 (estimated)

BLACKBERRY FARM PIONEER VILLAGE, 100 S. Barnes Rd., Aurora, IL 60506-8118. Tel.: 630-892-1550. Fax: 630-892-1597.
E-mail: ssmith@fvpd.net
Web Site: www.foxvalleyparkdistrict.org
Founded: 1969.
Congressional District: 15
Key Personnel: Exec. Dir., Nancy McCaul; Mgr., Sandy Smith; Bd. Pres., Theodia Gillespie; Supvr., Laureen Baumgartner.
Personnel Profile: Full-Time Paid 2; Part-Time Paid 40; Part-Time Volunteers 3; Interns 1.
Governing Authority: nonprofit organization. Parent Institution: Fox Valley Park District, 712 S. River St., Aurora, IL 60507. Tel.: 708-897-0516. Branch Museum: Red Oak Nature Center, Batavia, IL. Tax-exempt.
Institution Type/Description: Village Museum.
Collections: restored carriages; 19th-century farm equipment; contents of 11 stores; decorative arts. Historic Buildings: stave silo; late 19th-century Big Rock train station; 1840s post and beam house built by John Wagner, Aurora, IL.
Research Fields: agricultural history of northern Illinois; c.1830-1910 one-room school in the late 19th century.
Facilities: gardens; classroom; food service. Gift items for sale.
Activities: guided tours; arts festivals; craft workshops; daily craft demonstrations; formally organized educational programs; permanent & temporary exhibitions.
Publications: brochure; quarterly activity guide.
Hours & Admission Prices: May-Aug. Mon.-Fri. 9:30-3:30, Sat.-Sun. 11-5; Sept. Sat.-Sun. 11-5; Oct. call for hours. Adults $7, children $6; discounts to members. &
Attendance: 72,500 (accurate)
Membership: Fox Valley Park District: Resident $25; Nonresident $30.

DAVID L. PIERCE ART & HISTORY CENTER, 20 E. Downer Place, Aurora, IL 60505-3302. Mailing Address: 44 E. Downer Place, Aurora, IL 60505-3302. Tel.: 630-906-0654 & 0650. Fax: 630-906-0657.
E-mail: rchurch@aurora-il.org
Web Site: www.aurora-il.org
Founded: 1996.
Congressional District: 14
Key Personnel: Chm., Chris Hoban; Dir. & Cur., Rena J. Church.
Personnel Profile: Full-Time Paid 2; Part-Time Paid 3.
Governing Authority: municipal; nonprofit. Parent Institution: City of Aurora. Tax-exempt.
Institution Type/Description: Art Commission: located in Stolp Island Historic District.
Collections: multi-dimensional in a variety of medium including oil, watercolor, intaglio, photography, as well as wooden & acrylic sculpture.
Activities: commission of major public pieces. Annual Events: photography competition & show; juried fine arts show; youth art program in summer.
Hours & Admission Prices: Wed.-Fri. 12-4. No charge. &
Attendance: 9,000 (accurate)

SCHINGOETHE CENTER FOR NATIVE AMERICAN CULTURES, (M), Dunham Hall - Lower Level, 1400 Marseillaise Place (corner Marseillaise & Randall Rd.), Aurora, IL 60506-4892. Mailing Address: c/o Aurora University, 347 S. Gladstone Ave., Aurora, IL 60506-4892. Tel.: 630-844-7843. Fax: 630-844-6529.
E-mail: museum@aurora.edu
Web Site: www.aurora.edu/museum
Founded: 1989.
Congressional District: 14
Key Personnel: Exec. Dir. & Chief Cur., Meg Bero; Aurora Univ. Pres., Rebecca Sherrick; Finance Dir., Beth Reisenweber; Asst. Cur., Dave Spencer.
Personnel Profile: Full-Time Paid 3; Part-Time Paid 4; Part-Time Volunteers 12.
Governing Authority: private university; not-for-profit. Parent Institution: Aurora University. Tax-exempt.
Institution Type/Description: Native American Museum.
Collections: Native American art; cultural materials of North, Central & South American peoples from prehistoric times to the present.
Research Fields: cultural history; visual arts; intercultural contact; pedagogy; material culture studies.
Facilities: 3 exhibit galleries; research library of Native American art & culture, Native American periodicals, rare books, newspapers of Native American origin & pamphlet file available to the public; educational facilities through the university.

Activities: docent program; films; formal education programs for children & undergraduate or graduate college students; guided tours; lectures; loan & temporary exhibitions; school loan service; art & craft presentations by Native Americans; in-service programs for teachers. Annual Events: Native American film series.
Hours & Admission Prices: Academic Year: Tues. 10-7, Wed.-Fri. 10-4, Sun. 1-4; Summer: Tues.-Fri. 10-4. Suggested Donation: adults $3, students & seniors $2, children under 12 $1. Closed university holidays. &
Attendance: 9,912 (accurate)
Membership: Student $15; Individual $25; Family $45; Patron & Contributor $100; Benefactor $500.

SCITECH HANDS ON MUSEUM, 18 W. Benton, Aurora, IL 60506-6013. Tel.: 630-859-3434, ext. 214. Fax: 630-859-8692.
E-mail: operations@scitechmuseum.org
Web Site: www.scitechmuseum.org
Formerly: SciTech Hands On Science Center
Founded: 1988.
Congressional District: 14
Key Personnel: Exec. Dir., Arlene S. Hawks; Operations Dir., Camille Coller.
Personnel Profile: Full-Time Paid 6; Part-Time Paid 6; Part-Time Volunteers 10; Interns 1.
Governing Authority: nonprofit organization. Tax-exempt: 501(c)(3).
Institution Type/Description: Science & Technology Museum: housed in historic former Aurora Post Office Building.
Collections: science & technology hands on exhibits on weather, light & color, heat, mathematics, sound & music, magnets & electricity, physics, chemistry, biology, astronomy & nuclear physics.
Facilities: outdoor science park; Monkey Business Cafe; Discovery Academy - Preschool & Kindergarten. Science-related materials for sale.
Activities: informal educational programs for children; field trips; camps; overnights; parties.
Hours & Admission Prices: Tues.-Sat. 10-3. Adults $8, seniors 60 & over $7, children 4-17 $6; discount to ASTC members; children 3 & under no charge. Closed New Year's Eve; Memorial Day; Independence Day; Thanksgiving; Christmas. &
Attendance: 52,000 (accurate)
Membership: Single $60; Family $75; Business Member $300; Life Member $1,500.

THE WILLIAM TANNER HOUSE MUSEUM - AURORA HISTORICAL SOCIETY, Cedar & Oak Sts., Aurora, IL 60506. Mailing Address: P.O. Box 905, Aurora, IL 60507-0905. Tel.: 630-906-0650. Fax: 630-906-0657.
E-mail: jjshanahan@aurorahistory.net
Web Site: www.aurorahistoricalsociety.org
Founded: 1906.
Congressional District: 4
Key Personnel: Exec. Dir., John R. Jaros; Pres., Jessica Byrne; Sr. Cur., Dennis Buck; Museum Shop Mgr., Jacqueline Shanahan.
Personnel Profile: Full-Time Paid 3; Part-Time Paid 4; Part-Time Volunteers 45.
Governing Authority: society; nonprofit organization. Tax-exempt.
Institution Type/Description: Local History Museum: housed in 1857 mansion. Listed on the National Register of Historic Places.
Collections: Indian and pioneer artifacts; mastodon bones; 19th & early 20th-century tools & household items; Victorian furnishings; timepieces; musical instruments; local photographs, books, documents; vehicles; locally manufactured goods; railroad items.
Research Fields: local history & genealogy.
Facilities: library; research room.
Activities: guided group tours; slide-lecture presentations; tour groups. Museum Sponsors: Fourth of July Bell-Ringing Ceremony; Christmas Open House.
Publications: booklets on history of Aurora; bimonthly newsletter; photos of museum.
Hours & Admission Prices: Tanner House: seasonal Mon. & Sat. 12-3. Art & History Center: Wed.-Sun. 12-4. Adults $4, students & seniors $2; members no charge.
Attendance: 10,000 (accurate)
Membership: Student & Senior Citizen $15; Individual $25; Family $35; Patron $100; Benefactor $250; Life $500; Business & Organization: $100, $250, $500 & $1,000.

Barrington

BARRINGTON HISTORY MUSEUM, 212 W. Main St., Barrington, IL 60010-3011. Tel.: 847-381-1730. Fax: 847-381-1766.
Web Site: bahsil.org
Founded: 1969.

Congressional District: 12
Key Personnel: C.E.O. & Pres., Michael J. Harkins.
Governing Authority: nonprofit organization. Tax-exempt: 501(c)(3).
Institution Type/Description: History Museum: housed in two Folk Victorian Houses.
Collections: early farm tools & equipment; blacksmith & harness making tools; clothing & furniture; memorabilia from community & surrounding areas; Barrington area history including photos & maps; blacksmith shop.
Research Fields: historic sites in area; history of Barrington & Cuba Townships.
Facilities: meeting room.
Activities: guided tours; permanent & changing exhibitions; school programs; slide lectures; community programs; genealogy series.
Publications: monthly newsletter; annual report.
Hours & Admission Prices: Tues.-Fri. 10-4, Sat. 10-1. Tours by appointment. Blacksmith Shop: Sat. 12-4. &
Membership: Senior Citizen $10; Senior Citizen Couple $15; Individual $25; Family $35; Business & Organization $50; Annual Contributing $100; Patron $250; Benefactor $500; Annual Friend $1,000.

CRABTREE NATURE CENTER, 3 Stover Rd., Barrington, IL 60010-5342. Tel.: 847-381-6592.
Web Site: fpdcc.com
Founded: 1971.
Congressional District: 8
Governing Authority: Parent Institution: Forest Preserve District of Cook County. Tax-exempt.
Institution Type/Description: Nature Center.
Collections: wildlife & their habitats; native plants.
Facilities: nature trails.
Activities: group talks; educational programs.
Publications: monthly program guide; trail maps.
Hours & Admission Prices: March to late Oct. Sat.-Thurs. 9-4:30; late Oct. to Feb. Sat.-Thurs. 9-3:30. No charge; donations accepted. Closed New Year's Day; Thanksgiving; Christmas. &

Bartlett

BARTLETT DEPOT MUSEUM, 100 W. Railroad Ave., Bartlett, IL 60103. Mailing Address: 2228 Main St., Bartlett, IL 60103. Tel.: 630-837-0800.
Web Site: www.villageofbartlettmuseums.org
Governing Authority: Parent Institution: Village of Bartlett Museums. Branch Museum: Bartlett History Museum, 228 S. Main St., Bartlett, IL 60103.
Institution Type/Description: Railroad History Museum: housed in a former railroad depot; built in 1873.
Collections: depot & railroading history; period furnishings; photographs.
Hours & Admission Prices: Tues. 12-7, Thurs. 12-6, 1st Sat. each month 9am to noon. No charge. &

BARTLETT HISTORY MUSEUM, 228 S. Main St., Bartlett, IL 60103. Tel.: 630-837-0800.
Web Site: www.villageofbartlettmuseums.org
Key Personnel: Dir., Pam Rohleder.
Personnel Profile: Full-Time Paid 1.
Governing Authority: Parent Institution: Village of Bartlett Museums. Branch Museum: Bartlett Depot Museum, 100 W. Railroad Ave., Bartlett, IL 06103.
Institution Type/Description: History Museum.
Collections: local history & culture; period furnishings; photographs; personal artifacts.
Hours & Admission Prices: Mon.-Fri. 8:30-4:30, Sat. 9-12. No charge. &

Batavia

BATAVIA DEPOT MUSEUM, (M), 155 Houston St., Batavia, IL 60510-1924. Mailing Address: P.O. Box 14, Batavia, IL 60510. Tel.: 630-406-5274. Fax: 630-593-5202.
E-mail: carlah@bataviaparks.org
Web Site: bataviahistoricalsociety.org
Founded: 1960.
Congressional District: 14
Key Personnel: Dir., Carla Hill; Cur., Chris Winter.
Governing Authority: society, municipal; nonprofit organization. Parent Institution: Batavia Historical Society & the Batavia Park District. Tax-exempt: 501(c)(3).
Institution Type/Description: Historical Society Museum: housed in 1854 Batavia Depot, one of the oldest railroad station on the Burlington Line.
Collections: local history; clothing; tools; railroad; windmills; pioneer; genealogical; photographs of local historic houses; Mary Todd Lincoln display; manuscripts.

Research Fields: local genealogy; windmills; Newton Wagons; Mary Todd Lincoln.
Facilities: library of historical books & manuscripts.
Activities: guided tours; permanent & temporary exhibitions; lectures.
Publications: quarterly circular, Newsletter of Historical Society; books, Batavia Past & Present; Historical Batavia; Little Town In A Big Woods (grade school level).
Hours & Admission Prices: March-Thanksgiving Mon., Wed., Fri.-Sun. 2-4; other times by appointment. No charge; donations accepted. &
Membership: Junior $2; Classroom $5; Individual $10; Joint/Family $15; Individual Life $100; Family & Business or Institution Life $150.

LEDERMAN SCIENCE EDUCATION CENTER AT FERMI-LAB, Fermilab, Kirk Rd. & Pine St., Batavia, IL 60510. Mailing Address: Fermilab MS 777, Box 500, Batavia, IL 60510. Tel.: 630-840-8258. Fax: 630-840-2500.
E-mail: edreg@fnal.gov
Web Site: ed.fnal.gov/office/index.shtml
Institution Type/Description: Science Center.
Collections: hands-on mathematics & science exhibitions.
Facilities: laboratories. Museum-related items for sale.
Activities: hands-on exhibitions; educational programs; group tours; scout programs; special events; guided tours; science adventures. Center Sponsors: Ask-a-Scientist.
Hours & Admission Prices: Mon.-Fri. 8:30-4:30, Sat. 9-3; groups by appointment.

RED OAK NATURE CENTER, 2343 S. River St., Batavia, IL 60510-9664. Tel.: 630-897-1808. Fax: 630-264-0056.
E-mail: redoak@fvpd.net
Web Site: www.foxvalleyparkdistrict.org
Founded: 1978.
Personnel Profile: Full-Time Paid 1; Part-Time Paid 4.
Governing Authority: Parent Institution: Fox Valley Park District.
Institution Type/Description: Nature Center.
Collections: hands-on exhibits; natural history; wildlife & their habitats; plants.
Facilities: nature trails.
Activities: educational programs; family programs; camps; special events.
Hours & Admission Prices: Mon.-Fri. 9-4:30, Sat.-Sun. 10-3. No charge; donations accepted.
Attendance: 23,000 (estimated)

Belleville

LABOR AND INDUSTRY MUSEUM, 123 N. Church St., Belleville, IL 62220-1418. Tel.: 618-222-9430.
E-mail: laborandindustry@yahoo.com
Web Site: www.laborandindustrymuseum.org
Founded: 1996.
Key Personnel: Chm. (V), Pat Schmeder; Coord. Collections, Judy Belleville.
Governing Authority: Tax-exempt.
Institution Type/Description: History Museum.
Collections: Illinois' labor & industry history; photographs.
Facilities: archives; learning center.
Hours & Admission Prices: Sat. 10-2; other times by appointment.

ST. CLAIR COUNTY HISTORICAL SOCIETY, 701 E. Washington St., Belleville, IL 62220-3846. Tel.: 618-234-0600. Fax: 618-234-3060.
E-mail: stcchs@att.net
Founded: 1905.
Congressional District: 57
Key Personnel: Pres., Douglas Gain; Treas., Kenneth Lickenbrook; Cur., William Shannon; Museum Shop Mgr. & Office Admin., Norma Walker.
Personnel Profile: Full-Time Paid 1; Part-Time Paid 1; Part-Time Volunteers 50.
Volunteer Hours: 1,000
Operating Expenses: 60,000
Operating Income: 40,000
Governing Authority: society; board of directors. Branch Museum: 1830, The Kunz House, 602 Fulton St., Belleville, IL 62221. Tax-exempt.
Institution Type/Description: Historical Museum: housed in 1866, Victorian home.
Collections: period furniture; costumes; tools; memorabilia; manuscript collections.
Major Exhibits: Early Medicine & Tools, 10/13-6/14; Chronicles Local Woman's Journey to Japan, 10/13-9/14.
Research Fields: historical records of St. Clair County.

Facilities: research library; meeting room.
Activities: guided tours; permanent exhibitions; landmark & historic site awards; lectures. Annual Events: Annual Fashion Promenade and Luncheon; Candlelight House Tour in December.
Publications: monthly newsletter; annual journal.
Hours & Admission Prices: Mon.-Fri. 10-2; other times by appointment. Adults $2, children $1; discounts to AAM & AAA members; members no charge. Closed national holidays. &
Attendance: 1,500 (estimated)
Membership: Student $5; Individual $25; Family $40; Business $75; Life $500.

WILLIAM & FLORENCE SCHMIDT ART CENTER, (M), 2500 Carlyle Ave., Belleville, IL 62221-5859. Tel.: 618-222-5278; 800-222-5131, ext. 5278.
Web Site: swic.edu/sac
Founded: 2002.
Congressional District: 12
Key Personnel: Exec. Dir., Libby Reuter; Asst. Cur. & Registrar, Christina Cosio; Cur. & Facility Coord., Nicole Dutton; Asst. Cur., Jessica Mannisi
Institution Type/Description: Art Center.
Collections: works by contemporary artists.
Hours & Admission Prices: Academic Year: Mon.-Wed. 9-5, Thurs. 9-8, Fri. 10-5, Sat. 11-5; Summer: call for hours. No charge.

Bellflower

BELLFLOWER HISTORICAL AND GENEALOGICAL SOCIETY, 210 N. Latcha St., Bellflower, IL 61724. Mailing Address: P.O. Box 140, Bellflower, IL 61724-0140. Tel.: 309-722-3458.
Founded: 1976.
Congressional District: 22
Key Personnel: Pres. (V), Phyllis Kumler.
Personnel Profile: Part-Time Volunteers 6.
Governing Authority: nonprofit. Tax-exempt.
Institution Type/Description: Historic Building: early 1900 Illinois Central Railroad Depot.
Collections: local memorabilia.
Activities: guided tours; permanent & temporary exhibitions.
Publications: Highlights.
Hours & Admission Prices: By appointment. No charge; donations accepted.
Attendance: 128 (accurate)
Membership: Single $2; Couple $3; Family $4.

Belvidere

BOONE COUNTY HISTORICAL MUSEUM, 314 S. State, Belvidere, IL 61008-3609. Tel.: 815-544-8391. Fax: 815-547-1691.
E-mail: info@bchmuseum.org
Web Site: bchmuseum.org
Founded: 1968.
Congressional District: 16
Key Personnel: Dir., Mary Hale; Pres., Carol Rowe; Research, Lonna Bentley.
Personnel Profile: Full-Time Paid 2; Part-Time Paid 1; Part-Time Volunteers 25.
Governing Authority: nonprofit organization. Subsidiary Institution: Leader Building 1890 Collection. Library: 121 E. Locust, Belvidere, IL 61008. Tel: 815-544-2580. Tax-exempt: 501(c)(3).
Institution Type/Description: Historical Society Museum.
Collections: Civil War; sewing machines; washing machines; 19th-century Duxstad Log House; Newton natural history of Boone County; Chrysler automobiles; manuscripts; Native American arrow-heads, hatchets, axes; farm tools; archival files with photographs, letters & records of Boone County; photos; genealogy; pioneer artifacts; toys; vindex; Judi Ford (Miss America 1969) exhibit; 1906 Eldridge Runabout, artifacts from the National Sewing Machine Company; vintage clothing from 1830s to present.
Research Fields: pertaining to collections, people, houses, land, genealogy, & events.
Facilities: 100-vol. library of history pertaining to Boone County for use on premises only; chapel. Books for sale.
Activities: guided tours; lectures; films; genealogy, maps; weddings.
Publications: books, Landmarks, The Story of Boone County; Then & Now; quarterly society newsletter; reprint, 95th Illinois Regiment.
Hours & Admission Prices: Museum: Tues.-Fri. 9-5, Sat. 9-3; other times by appointment. Library: Tues.-Fri. 9-3. No charge; donations accepted. &
Attendance: 4,000 (estimated)
Membership: Senior 55 & over and Student $10; Individual $25; Grandpass $35; Family $50; Organizations & Small Business $100; Business (21 plus employees) $250.

Bement

BRYANT COTTAGE, 146 E. Wilson Ave., Bement, IL 61813-1250. Mailing Address: P.O. Box 41, Bement, IL 61813-0041. Tel.: 217-678-8184.
E-mail: hpa.bryantcottage@illinois.gov
Web Site: www.bement.com/bryant.htm
Founded: 1925.
Congressional District: 15
Key Personnel: Site Mgr., Marilyn L. Ayers.
Personnel Profile: Full-Time Paid 1.
Governing Authority: Parent Institution: Illinois Historic Preservation Agency. Tax-exempt: 501(c)(3).
Institution Type/Description: Historic House Museum: 1856 Bryant Cottage where the Lincoln-Douglas Debates were verbally agreed to be part of the 1858 Senate campaigns.
Collections: original & period furnishings.
Research Fields: 1850s lifestyle; local history; Lincoln & Douglas.
Activities: guided tours; lectures; permanent exhibitions; school program.
Publications: Historic Illinois; brochures.
Hours & Admission Prices: Call for hours. Donations accepted. Closed New Year's Day; Martin Luther King Jr. Day; President's Day; Thanksgiving; Christmas. &

Berwyn

BERWYN ROUTE 66 MUSEUM, 7003 W. Ogden Ave., Berwyn, IL 60402. Tel.: 708-484-9349. Fax: 708-484-1015.
E-mail: info@berwynrt66museum.org
Web Site: www.berwynrt66museum.org
Founded: 2010.
Congressional District: 3
Key Personnel: Dir., Jon Fey; Museum Shop Mgr., Myles Slaughter.
Personnel Profile: Full-Time Volunteers 2; Part-Time Paid 6.
Institution Type/Description: History Museum.
Collections: Route 66 history; photographs; period artifacts; local & regional art.
Hours & Admission Prices: Mon., Wed. & Fri. 1-4.
Attendance: 3,550 (estimated)

Bishop Hill

BISHOP HILL HERITAGE MUSEUM, (M), Steeple Building, 103 N. Bishop Hill St., Bishop Hill, IL 61419. Mailing Address: P.O. Box 92, Bishop Hill, IL 61419-0092. Tel.: 309-927-3899. Fax: 309-927-3010.
E-mail: bhha@mymctc.net
Web Site: www.bishophillheritage.org
Founded: 1962.
Congressional District: 17
Key Personnel: Dir., Todd L. DeDecher; Museum Shop Mgr., Marie Watson.
Personnel Profile: Full-Time Paid 1; Part-Time Paid 5; Part-Time Volunteers 30.
Governing Authority: nonprofit organization. Parent Institution: Bishop Hill Heritage Association. Tax-exempt: 501(c)(3).
Institution Type/Description: History Museum: housed in 1854 Steeple Building, located in Bishop Hill, a Swedish communal settlement founded in 1846 by religious dissenters.
Collections: artifacts of Bishop Hill Colony; manuscript collections; photograph collections. Historic Buildings: 1857 blacksmith shop; 1853 Bishop Hill Colony store; 1854 steeple; 1855 dairy building.
Major Exhibits: From Generation to Generation: Folk Arts of Illinois (T), 6/1-9/1/14.
Research Fields: history of Swedish immigration; history of Bishop Hill Colony.
Facilities: 400-vol. library of archival materials on Bishop Hill Colony & Swedish immigration available in Steeple building by appointment; reading room. Swedish imports, folk & designer crafts for sale.
Activities: lectures; craft workshops; demonstrations of traditional folk crafts; formally organized education programs for adults; Bishop Hill documentary; costume guided tours.
Publications: biannual news bulletin.
Hours & Admission Prices: Jan.-March daily 10-4; April-Oct. Sun. 12-5, Mon.-Sat. 10-5. No charge; donations accepted. &
Attendance: 90,000 (estimated)
Membership: Friend $5-$34.99; Premium $35-$99; Sustaining $100-$249; Benefactor $250 & up.

BISHOP HILL STATE HISTORIC SITE, 304 S. Bishop Hill St., Bishop Hill, IL 61419-0104. Mailing Address: P.O. Box 104, Bishop Hill, IL 61419-0104. Tel.: 309-927-3345. Fax: 309-927-3343.
E-mail: bishophill@winco.net
Web Site: www.bishophill.com
Founded: 1946.
Congressional District: 17
Key Personnel: Site Mgr., Martha Downey.
Personnel Profile: Full-Time Paid 5; Part-Time Volunteers 30.
Governing Authority: state. Parent Institution: State of Illinois, Illinois Historic Preservation Agency, Division of Historic Sites, Old State Capitol, Springfield, IL 62701. Tax-exempt.
Institution Type/Description: History Museum: complex consists of 1848 Colony Church, located on the site of a Swedish communal settlement, and 1850s Bjorklund Hotel.
Collections: paintings pertaining to Bishop Hill colony by Olof Krans; artifacts & furniture from Bishop Hill colony during 1846-1861. Historic Building: 1852 Bjorklund Hotel operated by the Bishop Hill colony; 1988 Bishop Hill Museum.
Research Fields: restoration & immigrant history 1840-1860; Illinois history 1845-1870; Illinois Folk Art.
Facilities: picnic area.
Activities: guided tours; permanent exhibitions.
Hours & Admission Prices: March-Oct. Wed.-Sun. 9-5, Nov.-Feb. Wed.-Sun. 9-4. Suggested Donation: family $10, adults $4, children 17 & under $2. Closed New Year's Day; Martin Luther King Jr. Day; Presidents' Day; Columbus Day; Veterans Day; Thanksgiving; Christmas. &
Attendance: 79,381 (accurate)

Blandinsville

BLANDIN HOUSE MUSEUM, 215 S. Chestnut St., Blandinsville, IL 61420. Mailing Address: P.O. Box 21, Blandinsville, IL 61420-9169. Tel.: 309-652-3673.
Founded: 1970.
Key Personnel: Pres. (V), Karen Moore
Institution Type/Description: History Museum.
Collections: local history & culture; period artifacts; photographs.
Hours & Admission Prices: Mon.-Fri. call for hours. No charge; donations accepted.
Membership: Annual $10.

Bloomington

THE DAVID DAVIS MANSION, 1000 Monroe Dr., Bloomington, IL 61701-3333. Tel.: 309-828-1084. Fax: 309-828-3493. Facebook: David Davis Mansion.
E-mail: davismansion@yahoo.com
Web Site: www.davismansion.org
Founded: 1960.
Congressional District: 15
Key Personnel: Site Mgr., Marcia Young; Asst. Mgr., Jeannie Riordan; Cur., Jeff Saulsbery; Museum Shop Mgr., Bekah Litchfield.
Personnel Profile: Full-Time Paid 3; Part-Time Paid 2; Part-Time Volunteers 80; Interns 4.
Governing Authority: Parent Institution: Illinois Historic Preservation Agency. Tax-exempt.
Institution Type/Description: Historic House: 1872 David Davis' Second Empire Italianate brick mansion. David Davis was the campaign manager for Abraham Lincoln who appointed him as a judge of the U.S. Supreme Court.
Collections: furnishings; decorative arts; textiles; 19th-century mechanical systems; barn artifacts.
Research Fields: 19th century cultural, social & political history; decorative arts; architectural history.
Facilities: visitor center; garden.
Activities: guided tours; special events; educational programs.
Publications: David Davis Mansion Newsletter; Clover Lawn Souvenir booklet; Sarah's Garden; Managing Clover Lawn: A Guide to the Kitchen of Sarah Davis and the Life That Filled It.
Hours & Admission Prices: Wed.-Sun. 9-4. No charge; donations accepted. Closed New Year's Day; Thanksgiving; Christmas; state holidays. &
Attendance: 60,000 (estimated)
Membership: Individual $35; Family $50.

MCLEAN COUNTY ARTS CENTER, 601 N. East St., Bloomington, IL 61701-3094. Tel.: 309-829-0011. Fax: 309-829-4928.
E-mail: info@mcac.org
Web Site: www.mcac.org
Founded: 1922.
Congressional District: 15
Key Personnel: C.E.O. & Exec. Dir., Douglas Johnson; Cur., Alison Hatcher; Education Coord., Rob Fifield.
Personnel Profile: Full-Time Paid 2; Part-Time Paid 1; Part-Time Volunteers 8; Interns 4.
Governing Authority: nonprofit organization. Tax-exempt.
Institution Type/Description: Arts Center.
Collections: permanent collection of McLean County Art Association.
Facilities: sales & rental gallery; rotating exhibits. Art works for sale.
Activities: gallery talks; permanent & temporary collections; educational program; monthly shows; trips to various museums.
Publications: quarterly newsletter, ART'S ALIVE!
Hours & Admission Prices: Tues. 10-7, Wed.-Fri. 10-5, Sat. 12-4. No charge; donations accepted. Closed national holidays. &
Attendance: 12,548 (accurate)
Membership: Individual $30; Family $50; Sponsor $100; Patron $250; Benefactor $500; Major Benefactor $1,000.

* **MCLEAN COUNTY MUSEUM OF HISTORY, (M),** 200 N. Main, Bloomington, IL 61701-3912. Tel.: 309-827-0428. Fax: 309-827-0100.
E-mail: mcmh@mchistory.org
Web Site: www.mchistory.org
Founded: 1892.
Congressional District: 11
Key Personnel: Exec. Dir., Greg Koos; Pres., Rob Fozzin; Treas., John Killion; Mgr. Devel. Operations, Graham Cowger; Cur., Susan Hartzold; Librarian & Archivist, William Kemp; Dir. Mktg., Jeff Woodord; Dir. Education, Candace Summers; Registrar, Terri Clemens; Coord. Education Programs, Rachael Kramp; Volunteer Coord., Mary Anne Scheirman.
Personnel Profile: Full-Time Paid 8; Part-Time Paid 5; Part-Time Volunteers 240; Interns 8.
Governing Authority: society; nonprofit. Parent Institution: McLean County Historical Society. Tax-exempt: 501(c)(3).
Institution Type/Description: Historic Building: 1904 courthouse.
Collections: McLean County from prehistoric to present; textiles; costumes; settlement period collections; Civil War artifacts; 19th-century material culture; archives.
Major Exhibits: Fiesta! A Celebration of Mexican Popular Arts, 1/13 8/14; Greening of the Prairie: Irish Immigration & Settlement in McLean County, 1/13-8/9/14.
Research Fields: local history.
Facilities: 12,000-vol. library pertaining to local history & genealogy; archives; discovery room. Books for sale.
Activities: guided tours; docent programs; organized education programs for children & undergraduate or graduate college students affiliated with Illinois State University & Illinois Wesleyan University; research services; permanent, traveling & temporary exhibit.
Publications: books, The Illustrated History of McLean County; The Heart of the Cornbelt: An Illustrated History of Corn Farming in McLean County, Illinois; Places of Pride: The Work & Photography of Clara Brian; A Matter of Life & Death: Health Illness & Medicine in McLean County; Prairie Roots: A Guide to Illinois Genealogy; History of African-Americans in McLean County, Illinois 1835-1975; Irish Immigrants in McLean County, Illinois; It Is Begun: The Pantagraph Reports The Civil War; Lincoln's Bloomington and Normal Illinois: A Tour Narrated by Abraham Lincoln; Gifts to the Prairie: The Work of Pioneer Nurserymen and the Art of the Prestele Family; Come & Get It! The McDonaldization & The Disappearance of Local Food from a Central Illinois Community.
Hours & Admission Prices: Mon. & Wed.-Sat. 10-5, Tues. 10-9. Adults $5, senior citizens $4; discounts to Time Travelers and AAM & ICOM members; students & society members no charge. Closed New Year's Eve & Day; Memorial Day; Independence Day, Labor Day; Thanksgiving; Christmas Eve & Day. &
Attendance: 27,257 (accurate)
Membership: Student/Senior $15; Primary $30; Allin $50; Davis $100; Fell $250; Fifer $500; Stevenson $750; Lincoln $1,000.

MILLER PARK ZOO, 1020 S. Morris Ave., Bloomington, IL 61701-6307. Tel.: 309-434-2250. Fax: 309-434-2823. Facebook: Miller Park Zoo.
E-mail: jtetzloff@cityblm.org
Web Site: www.millerparkzoo.org
Founded: 1891.
Key Personnel: Supt., Jay Tetzloff; Cur., Jonathan Reding; Education Instructor, Shannon Reedy; Museum Shop Mgr., Jennifer Rogers.
Personnel Profile: Full-Time Paid 9; Part-Time Paid 5.

Volunteer Hours: 16,668
Operating Expenses: 1,217,949
Operating Income: 572,809
Governing Authority: city. Tax-exempt.
Institution Type/Description: Zoo.
Collections: big cats; Wallaby Walk-About & Rain Forest exhibits; children's zoo; bald eagles; wolves; owls; animals of Asia.
Facilities: zoo lab.
Hours & Admission Prices: Zoo: daily 9:30-4:30. Adults 13-59 $5.95, senior citizens 60 & over and children 3-12 $3.95; members & children under 3 no charge. Closed Thanksgiving; Christmas. &
Attendance: 97,006 (accurate)
Membership: Individual $35; Extended $35; Joint $50; Family & Grandparent $60.

PRAIRIE AVIATION MUSEUM, 2929 E. Empire St., Bloomington, IL 61704-5452. Tel.: 309-663-7632. Fax: 309-663-8411.
E-mail: info@prairieaviationmuseum.org
Web Site: www.prairieaviationmuseum.org
Founded: 1984.
Key Personnel: Pres. (V), John Eckley; Museum Shop Mgr., Frank Thompson.
Personnel Profile: Part-Time Volunteers 100.
Governing Authority: Tax-exempt.
Institution Type/Description: Aviation History Museum.
Collections: aviation history; aircraft; personal artifacts; photographs.
Activities: special events.
Hours & Admission Prices: Summer: Tues.-Sat. 11-4, Sun. 12-4; Winter: Thurs.-Sat. 11-4, Sun. 12-4. Adults $5, children 6-11 $2; children 5 & under and members no charge. Closed New Year's Day; Thanksgiving; Christmas.
Attendance: 3,000 (estimated)
Membership: Individual $30.

Bolingbrook

ILLINOIS AVIATION MUSEUM, 130 S. Clow International Pkwy., Bolingbrook, IL 60440. Tel.: 630-771-1937. Fax: 630-771-9063.
E-mail: info@illinoisaviationmuseum.org
Web Site: www.illinoisaviationmuseum.org
Institution Type/Description: Aviation History Museum.
Collections: aircraft including Long EZ, T-2, Nieuport 11, Nieuport 12, Pazmany, T-33, Fokker E-3; Russian flight suit.
Activities: educational programs; special events.
Hours & Admission Prices: Sat. 10-2; other times by appointment.

Bourbonnais

EXPLORATION STATION...A CHILDREN'S MUSEUM, 459 N. Kennedy Dr., Bourbonnais, IL 60914-1970. Tel.: 815-933-9905. Fax: 815-928-6054.
E-mail: sarahw@btpd.org
Web Site: www.exploration-station.org
Founded: 1987.
Congressional District: 11
Key Personnel: Dir., Hollice Clark, III; Gen. Mgr., Sarah C. Winkle; Outreach Coord., Tammy Anderson; Office Mgr., Cathy Gagnon; Education & Community Rels. Coord, Katie Gildin.
Personnel Profile: Full-Time Paid 2; Part-Time Paid 15; Part-Time Volunteers 75.
Governing Authority: nonprofit. Bourbonnais Township Park District. Tax-exempt.
Institution Type/Description: Children's Museum.
Collections: aircraft & NASA replicas.
Activities: guided tours; lectures; participatory & traveling exhibitions; study clubs; storytelling; art & science workshops; workshops for toddlers; Sat. specials; hands-on interactive exhibits for children. Annual Events: A Night in Sleepy Hollow.
Publications: quarterly newsletter; seasonal brochure.
Hours & Admission Prices: April-Aug. Mon., Wed. & Fri.-Sat. 10-5, Tues. & Thurs. 10-5:30; Sept.-March Tues. & Thurs. 10-5:30, Wed. & Fri.-Sat. 10-5. Adults & children $6, senior citizens $5; discounts to members & Bourbonnais Twp. residents. &
Attendance: 39,000 (accurate)
Membership: Basic $50; Family $65-$85; Supporter $110.

Brookfield

BROOKFIELD HISTORICAL SOCIETY, 8820-1/2 Brookfield Ave., Brookfield, IL 60513-1670. Tel.: 708-485-3420.
Institution Type/Description: Historic Building: housed in the former Grossdale Railroad Station; built in 1889.
Collections: local history & culture; period furnishings; photographs; personal artifacts.
Hours & Admission Prices: 2nd & 4th Sun. of the month 1-4.

CHICAGO ZOOLOGICAL SOCIETY/BROOKFIELD ZOO, (M), 3300 Golf Rd., Brookfield, IL 60513-1060. Tel.: 708-688-8400. Fax: 708-485-3532. TDD: 708-485-0360; Facebook Brookfield Zoo.
E-mail: webmaster@czs.org
Web Site: www.czs.org
Founded: 1921.
Congressional District: 3
Key Personnel: Chm. (V), W. C. Kunkler; Women's Bd. Pres., Diane Dygert; C.E.O. & Pres., Stuart D. Strahl, Ph.D.; Sr. Vice Pres. Finance & Administration, Ken Kaduk; Sr. Vice Pres. Institutional Advancement, Cindy Zeigler; Sr. Vice Pres. Conservation, Education, & Training, Alejandro Grajal, Ph.D.; Sr. Vice Pres. Collections & Animal Care, William Zeigler.
Personnel Profile: Full-Time Paid 404; Part-Time Paid 1,019; Part-Time Volunteers 710; Interns 46.
Governing Authority: Managed by Chicago Zoological Society & owned by Cook County Forest Preserve District. Tax-exempt: 501(c)(3).
Institution Type/Description: Zoo.
Collections: butterflies; scorpion; spider; salamander; frogs; tortoises; lizards; snakes; crocodile; chameleon; ostrich; emu; cassowary; heron; egret; ibis; stork; trumpeter swan; waterfowl; Humboldt penguin; stork; roadrunner; Andean condor; bald eagle; hawks; owls; peacock; tern; dove; touraco; trumpeter hornbill; green-winged macaw; Mitchell's cockatoo; lorikeet; woodpecker; kookaburra; Micronesian kingfisher; Bali mynah; red bird-of-paradise; tanagers; echidna; wombat; kangaroo; shrew; flying fox; bats; lemurs; capuchin; mangabey; mandrill; colobus; tamarins; gibbon; orangutan; gorilla; sloth; anteater; mole-rat; porcupine; bottle-nosed dolphin; wolves; fox; African wild dog; sloth bear; brown bear; polar bear; otters; binturong; mongoose; meerkat; lion; caracal; leopard; tiger; seal; sea lion; aardvark; rock hyrax; zebra; Przewalski's wild horse; tapir; rhinoceros; pygmy hippopotamus; warthog; red river hog; camel; giraffe; okapi; bison; duiker; addax; waterbuck; klipspringer; domesticated animals.
Research Fields: animal behavior; behavioral endocrinology; wildlife endocrinology; animal welfare science; ecological & evolutionary processes; molecular genetics; population genetics; population biology (restricted and/or captive populations); conservation biology; contraception & reproduction in animals; aging in animals; animal physiology; environmental physiology (UV light research); status of free-ranging populations; amphibian decline; marine mammal research; conservation medicine; disease dynamics & patterns; pathology; conservation psychology; visitor & other audience research.
Facilities: 10,000-vol. library of books & 210 scientific journals on biology, animal medicine, zoology, pathology, ecology, behavioral science & photography available for inter-library loan & use on premises; aquarium; restaurants; banquet facility; picnic area. Science books & animal art novelties for sale.
Activities: outdoor interactive game; carousel; college experiential learning opportunities; community outreach; docent & volunteer programs; educational programs; inter-museum loan & permanent exhibitions; outreach programs for underserved audiences; playgrounds; school of environmental education for high school students; Share the Care; special events; Splash Pad; subscription programs; tram tours; wild encounters; zoo camp; Zoo Chats from Memorial Day to Labor Day.
Publications: quarterly member magazine, Gateways; biweekly member email; opt-in animal email; annual report; social media.
Hours & Admission Prices: Memorial Day-Labor Day Mon.-Sat. 9:30-6, Sun. 9:30-7:30; Sept.-May daily 10-5. Adults $12, senior citizens 65 & over and children 3-11 $10.50; members, active military, reservists, Jan.-Feb. Tues., Thurs. & Sat.-Sun. & Oct.-Dec. Tues. & Thurs. no charge. Parking: buses $12, cars $10. Additional fees for shows & special areas. &
Attendance: 2,158,185 (accurate)
Membership: Senior $58; Individual $60; One Plus $69; Family $97; Family Plus $115; Supporting $225; Sustaining $450.

Buffalo Grove

RAUPP MEMORIAL MUSEUM, (M), 901 Dunham Lane, Buffalo Grove, IL 60089. Tel.: 847-459-2318. Fax: 847-459-3148.
Institution Type/Description: History Museum.
Collections: local history; period artifacts; photographs.
Hours & Admission Prices: Sun. 1-4, Mon.-Thurs. 11-4:30.

Byron

BYRON MUSEUM OF HISTORY, 110 N. Union St., Byron, IL 61010. Mailing Address: P.O. Box 186, Byron, IL 61010. Tel.: 815-234-5031. Fax: 815-234-4447.
E-mail: info@byronmuseum.org
Web Site: byronmuseum.org
Founded: 1990.
Key Personnel: Dir., Kira Halvey
Institution Type/Description: History Museum.
Collections: local history & culture; period artifacts; period furnishings; photographs; costumes; textiles; historic buildings.
Facilities: Museum-related items for sale.
Hours & Admission Prices: Wed.-Sat. 10-3; other times by appointment. No charge; donations accepted. Closed New Year's Day; Thanksgiving & day after; Christmas Eve & Day.

Cahokia

CAHOKIA COURTHOUSE STATE HISTORIC SITE, 107 Elm St., Cahokia, IL 62206-1014. Tel.: 618-332-1782. Fax: 618-332-1737.
Founded: 1940.
Congressional District: 12
Key Personnel: Site Supt., Molly McKenzie.
Personnel Profile: Full-Time Paid 3; Part-Time Paid 2; Part-Time Volunteers 60; Interns 1.
Governing Authority: state, nonprofit. Parent Institution: State of Illinois, Illinois Historic Preservation Agency, Historic Sites Div. Old State Capitol, Springfield, IL 62706. Tax-exempt: 501(c)(3).
Institution Type/Description: History Museum: housed in 1737 residence.
Collections: artifacts; furniture. Historic Buildings: 1810 Jarrot Mansion; 1799 Holy Family Church; 1790 Martin-Boismenue House.
Research Fields: local history; French colonial architecture; territorial state government; American government in the old Northwest Territory; French colonial & American frontier way of life.
Facilities: visitors center.
Activities: guided tours; lectures; permanent exhibitions; special events; living history programs.
Publications: bimonthly newsletter, Historic Illinois.
Hours & Admission Prices: Tues.-Sat. 9-5. No charge. Jarrot Mansion, Martin-Boismenue, Holy Family Church: by appointment. Closed major holidays. ♿
Attendance: 25,600 (estimated)

GREATER ST. LOUIS AIR & SPACE MUSEUM, (M), 2300 Vector Dr., Cahokia, IL 62206. Tel.: 618-332-3664.
E-mail: info@airandspacemuseum.org
Web Site: www.airandspacemuseum.org
Founded: 2005.
Governing Authority: nonprofit organization. Tax-exempt: 501(c)(3).
Institution Type/Description: Aviation History Museum.
Collections: aviation history; early air & space artifacts; photographs.
Research Fields: St. Louis area aviation and aerospace history; Lambert - St. Louis International Airport historical documents.
Activities: education programs; instrument flight training; link trainer training courses.
Publications: book, St. Louis Aviation.
Hours & Admission Prices: Call for hours. No charge; donations accepted.
Membership: Students and Seniors 65 & over $30; Individual $40; Supportive $75; Sustaining $150; Patron $250; Corporate $500; Life $1,000.

Cairo

CAIRO CUSTOMS HOUSE MUSEUM, 1400 Washington Ave., Cairo, IL 62914-1870. Tel.: 618-734-9632.
Institution Type/Description: Historic Building: housed in the former United States Custom House; built from 1867-1872.
Collections: local history; Civil War memorabilia; replica of the U.S.S. Cairo gunboat; 1865 Cairo Fire Dept. hand-operated pumper; 1937 flood in Cairo; period furnishings.

Hours & Admission Prices: Tues.-Fri. 10-12 & 1-3. No charge.

MAGNOLIA MANOR, 2700 Washington Ave., Cairo, IL 62914-1458. Mailing Address: P.O. Box 286, Cairo, IL 62914-0286. Tel.: 618-734-0201. Fax: 618-734-0201.
Founded: 1952.
Congressional District: 24
Key Personnel: C.E.O. & Pres., Charles McGinness; Vice Pres., Elizabeth Morin; Cur. & Museum Shop Mgr., Tim Slapinski.
Personnel Profile: Full-Time Paid 1; Part-Time Volunteers 25.
Governing Authority: nonprofit organization. Owned & operated by Cairo Historical Association. Tax-exempt: 501(c)(3).
Institution Type/Description: Historic House Museum: 1869 Magnolia Manor.
Collections: 19th-century furnishings; Cairo history; Civil War items.
Activities: guided tours. Museum Sponsors: Victorian Christmas.
Publications: cookbook; history book.
Hours & Admission Prices: Summer: Mon.-Sat. 9-4:30, Sun. 1-4:30. Adults $8, children $2. Closed New Year's Eve & Day; Easter; Independence Day; Thanksgiving; Christmas Eve, Day & week.
Attendance: 2,500 (estimated)
Membership: Individual $20.

Carbondale

THE SCIENCE CENTER, University Mall, 1237 E. Main Space 1048, Carbondale, IL 62901-5830. Tel.: 618-529-5431. Fax: 618-457-2195.
E-mail: thesciencecenter@hotmail.com
Web Site: yoursciencecenter.org
Founded: 1994.
Congressional District: 12
Key Personnel: Exec. Dir., Chris Walls; Pres. (V), Roxanne Conley.
Personnel Profile: Full-Time Paid 4; Part-Time Paid 2; Part-Time Volunteers 20; Interns 4.
Governing Authority: private; nonprofit organization. Tax-exempt: 501(c)(3).
Institution Type/Description: Children's Science Museum.
Collections: hands-on interactive exhibits based on scientific concepts for children 3-13 & their parents.
Facilities: educational facilities; 2,200 sq. ft. exhibit space. Museum-related items for sale.
Activities: docent program; education programs for children; loan exhibitions; summer camps; camp-ins.
Publications: quarterly newsletter, Discover Us.
Hours & Admission Prices: Wed.-Thurs. 11-5, Fri. 11-6, Sat. 10-6, Sun. 12-5. Admission $3.50; discount to ASTC members; children under 4 & members no charge. Closed Easter; Thanksgiving; Christmas. ♿
Attendance: 15,000 (estimated)
Membership: Family $50; Lifetime $1,000.

✳ UNIVERSITY MUSEUM, (M), Southern Illinois University Carbondale, 1000 Faner Dr., MC 4508, Carbondale, IL 62901-4328. Tel.: 618-453-5388. Fax: 618-453-7409.
E-mail: museum@siu.edu
Web Site: www.museum.siu.edu
Founded: 1874.
Congressional District: 22
Key Personnel: Dir. Museum & Museum Studies, Dona R. Bachman; Dir. Museum Education, Robert DeHoet; Cur. Anthropology, Susannah Munson; Cur. Exhibits, Nathaniel Steinbrink; Adjunct Cur. Geology, Harvey Henson; Registrar, Eric Jones.
Personnel Profile: Full-Time Paid 5; Part-Time Paid 24; Part-Time Volunteers 75; Interns 8.
Governing Authority: state; university. Parent Institution: Southern Illinois University. Tax-exempt: 501(c)(3).
Institution Type/Description: General Museum.
Collections: European & American paintings, drawings, prints from 13th to 20th century with emphasis on 19th & 20th centuries; 20th-century photography; 20th-century sculpture, metals, ceramics; musical instruments; extensive Oceanic collection; natural history; archaeology; geology; decorative arts; costumes; various Asiatic holdings; Southern Illinois history; Native American artifacts; WPA artifacts.
Major Exhibits: Cast in Carbondale, 1/14-5/15; Master Artists from the Collection, 1/14-5/15.
Research Fields: history; geology; anthropology; archaeology; ethnology; museology; textiles; costumes; oral history; folklore; ethnomusicology.
Facilities: reference library; auditorium; laboratories; computer lab; galleries; sculpture garden; Japanese garden. Museum-related items for sale.
Activities: guided tours; lectures; films; inter-museum loans; temporary &

traveling exhibitions; public outreach programs; volunteer program; museum studies program; museum student group.

Publications: catalogs for special exhibits; annual report; scholarly reports; museum newsletter.

Hours & Admission Prices: Tues.-Fri. 10-4, Sat. 1-4; pre-arranged group tours available. No charge; donations accepted. Closed university & national holidays. &

Attendance: 15,000 (estimated)

Membership: Patron $100; Gold Patron $500; Corporate Patron $1,000.

Carlinville

MACOUPIN COUNTY HISTORICAL SOCIETY - ANDERSON MANSION, 920 Breckenridge St., Carlinville, IL 62626. Mailing Address: P.O. Box 432, Carlinville, IL 62626. Tel.: 217-854-2141.

E-mail: info@carlinvillechamber.com

Web Site: www.macsociety.org

Founded: 1973.

Institution Type/Description: Historical Society Museum: housed in the former home of John Anderson; built in 1883.

Collections: local history & culture; period furnishings; personal artifacts; photographs. Historic Buildings: one-room schoolhouse; blacksmith shop; church; wash house; granary.

Activities: special events. Museum Sponsors: Spring Festival in May; Fall Festival in September.

Hours & Admission Prices: April-Nov. Wed. 10-2. No charge; donations accepted.

Membership: Individual $10; Lifetime $150.

Carmi

WHITE COUNTY HISTORICAL SOCIETY, 203 N. Church St., Carmi, IL 62821. Mailing Address: P.O. Box 121, Carmi, IL 62821-0121. Tel.: 618-382-8425.

E-mail: w.c.h.s.-genealogy@hotmail.com

Web Site: www.rootsweb.com/~ilwcohs

Founded: 1957.

Congressional District: 19

Key Personnel: Chm. (V), Lecta Hortin; Pres. (V), Paula Pierson.

Personnel Profile: Part-Time Paid 1; Part-Time Volunteers 40.

Governing Authority: society. Subsidiary Institution: Mary Smith Fay Genealogy Library. Branch Museums: Robinson-Stewart House, Main Cross St., Carmi, IL 62821; L. Haas Store Museum, E. Main St., Carmi, IL; Ratcliff Inn Museum, E. Main St., Carmi, IL; Matsel Cabin, Robinson St., Carmi, IL. Tax-exempt.

Institution Type/Description: History Museum.

Collections: letters written by U.S. Senator John M. Robinson (1792-1843) & other pioneers; pioneer cemetery. Historic Buildings: 1814 Robinson-Stewart House; 1828 Ratcliff Inn; 1896 L. Haas Store; c.1850-1870 Matsel cabin; log house.

Research Fields: genealogy.

Facilities: library of genealogical materials.

Activities: guided tours; lectures; films; permanent exhibitions; dinner meetings with programs for membership & public.

Publications: members newspaper, White County Historian; Heritage Houses of White Co, IL; A Sketch of Crossville, IL in the Early Part of the 20th Century; Centennial History of Crossville and Phillips Township; books, Marriages from White County Illinois Vols. I-IV; Cemeteries of White County Illinois, Vols. I-II; Hamilton County Illinois Cemeteries & Probate Index; Early Land Grants of White County Illinois; White County Illinois Wills, 1816-1916; Index to White County Illinois Court Papers Vols. I-II; Original Land Grants, White County Illinois.

Hours & Admission Prices: Ratcliff Inn; Robinson-Stewart House; L. Haas Store Museum; Matsel Cabin; Genealogy Library: Tues.-Fri. 11:30-4:30; other times by appointment, contact Mary Smith Fay Library 618-382-8425. No charge; donation accepted. Closed national holidays.

Attendance: 1,500 (estimated)

Membership: Individual $25; Family $35; Sustaining $50; Patron $100; Corporate $500; Life $1,000.

Carterville

JOHN A. LOGAN COLLEGE MUSEUM, (M), 700 Logan College Rd., Carterville, IL 62918-2501. Tel.: 618-985-2828, ext. 8287. Fax: 618-985-2248. TDD: 618-985-2752.

E-mail: museum@jalc.edu

Web Site: www.jalc.edu/museum/index.html

Founded: 1983.

Congressional District: 12

Key Personnel: Dir., Adrienne Barkley Giffin

Governing Authority: public college. Parent Institution: John A. Logan College. Tax-exempt.

Institution Type/Description: College Museum.

Collections: contemporary regional art & craft; wildlife art; historical memorabilia relating to the life of General John A. Logan. Historic Building: c.1860 one-room schoolhouse.

Activities: one room school, spring & fall sessions.

Hours & Admission Prices: During regular college hours: Mon.-Fri. 8-9, Sat. 9-5. No charge. &

Attendance: 40,000 (estimated)

Carthage

HANCOCK COUNTY HISTORICAL SOCIETY, 306 Walnut, Carthage, IL 62321.

Founded: 1968.

Key Personnel: Pres., Keith Bruns; Vice Pres., Ned Casady; Sec., Janet Holtman; Treas., Susan Cheney; Office Mgr., Frank Burkett.

Personnel Profile: Part-Time Volunteers 10.

Governing Authority: society; nonprofit organization. Tax-exempt.

Institution Type/Description: Historical Society: housed in Kibbe Museum at Carthage site.

Collections: personal artifacts; history books; genealogy books; census microfilm; area newspaper microfilm; obituary books; DAR books; military records & books.

Research Fields: local history; genealogy.

Facilities: 250-vol. library of history books, atlases, directories, bound newspapers & genealogical material available for research on premises; 1830-1950, genealogical card index of early Hancock County families.

Activities: guided tours; lectures; radio programs; formally organized educational programs.

Publications: books, Pictorial History of Hancock County Illinois; Families of Hancock County Illinois, A Biographical History; I'll Be Seeing You World War II 1941-1945; Tour Guides; Granddads Automobiles. Reprint of 1874 Atlas.

Hours & Admission Prices: Mon.-Fri. 9-3. No charge; donations accepted. Closed holidays. &

Attendance: 1,000 (estimated)

Membership: Active & Family $10; Life $125.

Champaign

CHAMPAIGN COUNTY HISTORICAL MUSEUM, (M), 102 E. University Ave., Champaign, IL 61820-4111. Tel.: 217-356-1010. Fax: 217-356-1478.

E-mail: director@champaignmuseum.org

Web Site: www.champaignmuseum.org

Founded: 1972.

Congressional District: 15

Key Personnel: Co Pres. (V), Hal Balbach; Co Pres. (V) & Treas., Susanne Wood; Sec., Sandy Roberts.

Personnel Profile: Part-Time Volunteers 5.

Governing Authority: nonprofit organization. Tax-exempt: 501(c)(3).

Institution Type/Description: County History Museum: housed in the Cattle Bank; built in 1857. Listed on the National Register of Historic Places.

Collections: period furnishings & furniture; documents; clothing; textiles; books; Champaign County history.

Research Fields: Champaign County history, including oral history; information on the built environment, the natural environment, agriculture & urban and rural community developments.

Facilities: 500-vol. library of books available for use on premises by appointment; reading room; classrooms. Books & gift items in keeping with historical theme for sale.

Activities: guided tours of local historic properties; lectures; concerts; temporary & traveling exhibitions; formally organized programs for children.

Publications: quarterly newsletter, Champaign County Historical Museum Newsletter; tour guide, Historic Sites in Champaign-Urbana; tour booklet, Historic Sites in Champaign County; booklets, Grass Roots Preservation; J.O. Cunningham; Views of the City of Champaign and the University of Illinois, 2009; essays, Historical Geography of Champaign County; brochure, Champion County Historical Museum.

Hours & Admission Prices: Sat.-Sun. 12-5; other times by appointment; groups by appointment. No charge; donations accepted.

Attendance: 2,400 (estimated)

Membership: Student $10; Individual $35; Family $50; Sustaining $100; Donor $200; Patron $500.

✻ **KRANNERT ART MUSEUM AND KINKEAD PAVILION,** **(M),** 500 E. Peabody Dr., University of Illinois, Champaign, IL 61820-6913. Tel.: 217-333-1861. Fax: 217-333-0883.
E-mail: kam@illinois.edu
Web Site: www.kam.illinois.edu
Founded: 1961.
Congressional District: 19
Key Personnel: Dir., Kathleen Harleman; Dir. Mktg. & Publications, Diane Schumacher; Dir. Education, Anne Sautman; Cur., Robert LaFrance; Cur., Allyson Purpura; Collections Mgr., Kimberly Sissons; Assoc. Dir., Claudia Corlett-Stahl; Registrar, Christine Saniat; Office Admin., Chris Schaede; Exhibition Designer, Eric Lemme; Exhibition Designer, Walter Wilson; Security Chief, David Holzner; Coord. Education Center, Virginia Erickson.
Personnel Profile: Full-Time Paid 18; Part-Time Paid 13; Part-Time Volunteers 155; Interns 3.
Governing Authority: university. Parent Institution: University of Illinois at Urbana-Champaign. Tax-exempt: 501(c)(3) & 170(c).
Institution Type/Description: Art Museum.
Collections: Krannert Art Museum's collection of over 10,000 works of art represents the cultures of Africa, Asia, Europe and the Americas. In addition to the permanent collection, the Museum offers outstanding temporary exhibitions throughout the year.
Major Exhibits: Auto-Graphics: Victor Ekpuk, 1/14-4/14; Not Ready to Make Nice: Guerilla Girls, 1/14-4/14; Yoko Inoue, 1/14-5/14.
Research Fields: European & American art.
Facilities: auditorium; lecture room; cafe.
Activities: guided visits; lectures; gallery discussions; formally organized education programs; instructional materials loaned; inter-museum loan; permanent, temporary & traveling exhibitions; concerts & dance performances.
Publications: exhibition catalogs; brochures; biannual newsletter.
Hours & Admission Prices: Tues.-Wed. & Fri.-Sat. 9-5, Thurs. 9-9, Sun. 2-5. No charge; donations accepted. Closed national holidays. ♿
Attendance: 138,156 (accurate)
Membership: Student $15; Individual $45; Family & Dual $80; Benefactor $150; Patron $500; Director's Circle $1,000.

ORPHEUM CHILDREN'S SCIENCE MUSEUM, 346 N. Neil St., Champaign, IL 61820-3614. Tel.: 217-352-5895. Fax: 217-352-8160.
E-mail: orpheumkids@gmail.com
Web Site: www.orpheumkids.com
Founded: 1994.
Congressional District: 15
Key Personnel: C.E.O., Carolyn Knepp; Acting Dir., Zoe Stinson; Pres., James Quisenberry; Museum Shop Mgr., Susan Kitson.
Personnel Profile: Part-Time Paid 5; Part-Time Volunteers 113.
Governing Authority: Tax-exempt.
Institution Type/Description: Children's Museum.
Collections: hands-on science exhibits.
Activities: school field trips; summer science camps; special events. Museum Sponsors: Weekend Wizards.
Publications: quarterly newsletter.
Hours & Admission Prices: Mon.-Fri. 10-4, Sat.-Sun. 1-5; other times by appointment. Adults & children $5, seniors $4; discounts to ACM & ASTC members; members no charge. Closed New Year's Day; Easter; Christmas. ♿
Attendance: 13,658 (accurate)
Membership: Child $20; Family & Sponsor $40; Family Plan $60; Super Family $100; Supporting $250.

PARKLAND COLLEGE ART GALLERY, 2400 W. Bradley Ave., Champaign, IL 61821-1899. Tel.: 217-351-2485 & 2200. Fax: 217-373-3899.
E-mail: lcostello@parkland.edu
Web Site: www.parkland.edu/gallery
Founded: 1980.
Congressional District: 15
Key Personnel: Dir., Lisa Costello; Chm. (V), Prof. Chris Berti; Coord Exhibits Coord., Paula McCarty; Coord. Collections, Laura O'Donnell.
Personnel Profile: Full-Time Paid 1; Part-Time Paid 7; Part-Time Volunteers 35.
Governing Authority: college; nonprofit. Parent Institution: Parkland College. Tax-exempt.
Institution Type/Description: Art Gallery.
Collections: contemporary art by regional & national artists: drawings; paintings; photography; crafts; sculpture; ethnic arts.
Activities: lectures; loan & traveling exhibitions; TV & radio programs. Annual

Events: Art Faculty Exhibit; National Drawing Invitational; Art Student Exhibit; High School Art Seminar. Biennial Event: Ceramics Invitational.
Publications: national biennial ceramics catalogue.
Hours & Admission Prices: Academic Year: Mon.-Thurs. 10-7, Fri. 10-3, Sat. 12-2. Summer: Mon.-Thurs. 10-7. No charge; donations accepted. Closed college & major holidays. ♿
Attendance: 10,000 (estimated)
Membership: Friend $25; Patron $50; Champion $50; Fellow $250; Guardian $500.

SOUSA ARCHIVES AND CENTER FOR AMERICAN MUSIC, Harding Band Bldg., Second Fl., 1103 S. Sixth St., Univ. of Illinois at Urbana-Champaign, Champaign, IL 61820. Tel.: 217-244-9309. Fax: 217-333-2868.
E-mail: sousa@library.illinois.edu
Web Site: sousaarchives.org
Formerly: Sousa Archives for Band Research
Founded: 1994.
Congressional District: 15
Key Personnel: Archivist Music & Fine Arts, Scott Schwartz.
Personnel Profile: Full-Time Paid 1; Part-Time Paid 7; Part-Time Volunteers 16.
Governing Authority: Parent Institution: University of Illinois. Tax-exempt.
Institution Type/Description: American Music Museum.
Collections: original music compositions & arrangements by John Philip Sousa; 20th century electronic & avant-garde music; ethnomusicological research papers; period band uniforms; musical instruments; sound recording devices; local Urbana-Champaign; Illinois popular music groups.
Major Exhibits: The James Bond Theme: Music to Live, Die & Love Another Day, 4/11/13-3/14/14; How the Sousa Band Music Library Came to the University of Illinois, 6/1/13-6/2/14; A Musical Life: The Travels of Otto Mesloh, 7/1/13-6/30/14; Traveling America's Early Highways: A Strothkamp Family Road Trip, 8/30/13-7/7/14; Celebrating the Harding-Hindsley-Begian-Keene Band Legacy at the University of Illinois, 9/9/13-8/18/14; Live from the Crossroads: A Snapshot of Champaign-Urbana's Local Music Scene 1980-2000, 10/7/13-9/29/14.
Research Fields: musicology; ethnomusicology; computer music; John Philip Sousa; American band music.
Activities: Museum sponsors "One Community Together" programming for Urbana's Sweet Corn Festival in August & the University of Illinois' American Music Month celebration in November.
Hours & Admission Prices: Mon.-Fri. 8:30-12 & 1-5. No charge; donations accepted. ♿
Attendance: 3,606 (accurate)

WILLIAM M. STAERKEL PLANETARIUM, Parkland College, 2400 W. Bradley Ave., Champaign, IL 61821-1806. Tel.: 217-351-2568. Fax: 217-373-3809.
E-mail: dleake@parkland.edu
Web Site: www.parkland.edu/planetarium
Founded: 1987.
Key Personnel: Dir., David C. Leake.
Personnel Profile: Full-Time Paid 3; Part-Time Paid 1.
Governing Authority: Parent Institution: Parkland College. Tax-exempt.
Institution Type/Description: Planetarium.
Collections: Carl Zeiss M1015 Star Projector.
Activities: special events; rental facilities; sky shows; children's programs; films.
Publications: 3 annual Schedules; newsletter, Friends.
Hours & Admission Prices: Aug.-May call for hours. Adults $5, children, seniors, & students $4; members no charge.
Attendance: 30,000 (estimated)
Membership: Parkland Student, Staff & Seniors $25; Individual %30; Family (students/staff) $45; Family $50.

Charleston

✻ **TARBLE ARTS CENTER, EASTERN ILLINOIS UNIVERSITY, (M),** 2010 9th St., Charleston, IL 61920. Mailing Address: 600 Lincoln Ave., Charleston, IL 61920-3011. Tel.: 217-581-2787. Fax: 217-581-7138.
E-mail: tarble@eiu.edu
Web Site: www.eiu.edu/~tarble
Founded: 1982.
Congressional District: 15
Key Personnel: Dir., Michael Watts; Pres., Dorothy Bennett; Asst. Dir., Michael Schuetz; Dean, Dr. Bonnie Irwin; Cur. Education, Kit Morice.
Personnel Profile: Full-Time Paid 4; Part-Time Paid 6; Part-Time Volunteers 50.

Governing Authority: university. Parent Institution: Eastern Illinois University. Tax-exempt: 501(c)(3).
Institution Type/Description: University Art Center.
Collections: late 20th century Illinois folk arts; contemporary U.S. works on paper; American Scene prints; Paul T. Sargent paintings.
Major Exhibits: Joe Meiser Installation, 1/2/14-2/16/14; Kate Brooks: In the Light of Darkness, 1/11/14-2/23/14; Dab Devening: KIOSK, 1/17/14-3/9/14; Generation to Generation: Folk Arts of Illinois, 4/8/14-6/29/14; Ansel Adams: Master Works, 8/17/14-10/19/14; Michael Aurbach: Sculpture, 1/17/15-2/18/15.
Research Fields: eastern, central & southern Illinois 20th century folk arts; Paul T. Sargent; Robert M. Root; Charles Turzak.
Facilities: 8,800 sq. ft. exhibition space; meeting room; outdoor sculpture court. Gift items for sale.
Activities: changing exhibitions; visiting artists & scholars; lectures & residencies; tours, enrichment & educational programs; classes; workshops; demonstrations; chamber music series; film & video; special events; docent program. Annual Event: Celebration: A Festival of the Arts in April.
Publications: monthly e-newsletter; semesterly exhibitions & activities listing; exhibition catalogues, brochures & announcement cards.
Hours & Admission Prices: late Aug. to mid-May Tues.-Fri. 10-5, Sat. 10-4, Sun. 1-4; mid-May to mid-Aug. Tues.-Sat. 10-4, Sun. 1-4. No charge; donations accepted. Closed major holidays. ♿
Attendance: 16,000 (estimated)
Membership: Friend $50; Sponsor $100; Patron $250; Benefactor $500; Associate $1,000.

Chatham

CHATHAM RAILROAD MUSEUM, 100 N. State St., Chatham, IL 62629-1350. Tel.: 217-483-7792.
Web Site: www.chathamrailroadmuseum.org
Founded: 1999.
Key Personnel: Pres. (V), William Shannon.
Personnel Profile: Part-Time Volunteers 12.
Institution Type/Description: Railroad Museum: housed in 1902 era depot.
Collections: history of area & national railroads; passenger & freight railroad transportation from mid-1800s.
Hours & Admission Prices: 2nd & 4th Sun. 2-4. No charge; donations accepted. Closed holidays.

Chester

RANDOLPH COUNTY ARCHIVES AND MUSEUM, 1 Taylor St., Chester, IL 62233-1970. Tel.: 618-826-2667. Fax: 618-826-3363.
Founded: 1795.
Congressional District: 12
Key Personnel: Dir., Emily Lyons.
Personnel Profile: Part-Time Paid 1; Part-Time Volunteers 8.
Governing Authority: Parent Institution: Randolph County. Tax-exempt.
Institution Type/Description: History Museum.
Collections: county history; photographs; documents.
Hours & Admission Prices: Mon.-Fri. 9-4; other times by appointment. No charge; donations accepted. Closed holidays.
Attendance: 800 (accurate)

Chicago

A. PHILIP RANDOLPH PULLMAN PORTER MUSEUM, (M), 10406 S. Maryland Ave., Chicago, IL 60628-3090. Mailing Address: P.O. Box 6276, Chicago, IL 60680-6276. Tel.: 773-928-3935. Fax: 773-928-8372.
Web Site: www.aphiliprandolphmuseum.com
Key Personnel: Dir., Lyn Hughes.
Governing Authority: nonprofit organization. Tax-exempt: 501(c)(3).
Institution Type/Description: African American History Museum.
Collections: A. Philip Randolph's life & history; labor history of African-Americans in America.
Activities: Museum Sponsors: Black History Events in February.
Hours & Admission Prices: April-Dec. 1 Thurs. 1-4, Fri.-Sat. 11-4; other times by appointment. General admission $5.

✱ **ADLER PLANETARIUM & ASTRONOMY MUSEUM, (M),** 1300 S. Lake Shore Dr., Chicago, IL 60605-2489. Tel.: 312-922-7827. Fax: 312-322-9909. TDD: 312-322-0995.
E-mail: museum@adlerplanetarium.org
Web Site: www.adlerplanetarium.org
Founded: 1930.

Congressional District: 7
Key Personnel: Chm. (V), Bryan Cressey; Pres., Dr. Paul Knappenberger, Jr.; Exec. Vice Pres. & C.O.O., Margaret A. Marek; Vice Pres. Research, Dr. Lucy Fortson; Dir. Astronomy, Dr. Geza Gyuk; Sr. Vice Pres. Finance & Admin. & C.F.O., Michael Lo Presti; Collections Mgr., Devon Pyle-Vowles; Dir. Museum Communications, Molly O'Connell; Dir. Foundation & Government Rels., Paula Pergament; Vice Pres. External Affairs, Charles L. Katzenmeyer; Dir. Campaign, Ginevra Ranney; Cur. Emerita, Marjorie K. Webster; Dir. Exhibits & Theaters, Susan Harrison; Vice Pres. Exhibits & Programs, Dr. Susan Wagner; Dir. Astronomy History, Dr. Marvin Bolt; Dir. Operations, William J. Wilhelm; Dir. Human Resources, Marguerite E. Dawson; Cur., Dr. Bruce Stephenson; Dir. Education, Karen Carney; Dir. Corporate Rels., Ave Costa; Dir. Sales & Private Events, Julie Bishop; Dir. Information Systems, Kenneth P. Kobus; Museum Shop Mgr., Linda Stucky.
Personnel Profile: Full-Time Paid 104; Part-Time Paid 107; Part-Time Volunteers 155; Interns 5.
Governing Authority: independent; nonprofit corporation. Tax-exempt: 501(c)(3).
Institution Type/Description: Planetarium, History & Science Museum.
Collections: early scientific instruments in astronomy; time-keeping devices; navigation & engineering; rare books dealing with astronomy & related sciences; photographs related to scientific instruments; space exploration & related artifacts.
Research Fields: astronomy; astrophysics; history of science.
Facilities: 193-seat Star Rider Theater; 280-seat Zeiss Planetarium; 280-seat KROC Universe theatre; classrooms; restaurant; 18-unit computer classroom; distance learning studio.
Activities: family programs; demos; private events; public lectures; formal & informal education programs; permanent & temporary exhibitions; monthly public observing sessions with video hook-up to 20-inch computer-controlled Cassegrain telescope with charge-coupled device. Planetarium Sponsors: Astro-Science Workshop for high ability high school students; astronomy outreach programs with local schools & community groups; astronomy-related travel programs to global locations.
Publications: quarterly member's newsletter; gallery guides.
Hours & Admission Prices: Mon.-Fri. 9:30-4:30, Sat.-Sun. 10-4:30. Adults $10, seniors over 65 $8, children 3-14 $6; discounts to Chicago residents, AAM & ICOM members; members no charge. Closed Thanksgiving; Christmas. ♿
Attendance: 400,000 (estimated)
Membership: Student & Senior $45; Individual $60; Family $80; Mercury $150; Venus $250; Mars $500; Shepard Society $1,000.

ARC GALLERY (ARTISTS, RESIDENTS OF CHICAGO), 2156 N. Damen Ave., Apt. 2, Chicago, IL 60647-6482. Tel.: 312-733-2787. Fax: 312-733-2787.
E-mail: arcgallery@yahoo.com
Web Site: www.arcgallery.org
Founded: 1973.
Key Personnel: Administrative Dir., Brooke Demos; Mng. Dir., Charlotte Segal; Treas., Mirjana Ugrinov; Co-Treas., Deva Suckerman; Co-Treas., Melanie Adcock.
Personnel Profile: Full-Time Volunteers 15; Part-Time Paid 1; Part-Time Volunteers 20; Interns 2.
Governing Authority: not-for-profit organization. Tax-exempt: 501(c)(3).
Institution Type/Description: Art Gallery.
Collections: works by contemporary Chicago artists.
Research Fields: emerging Chicago artists; women's issues.
Facilities: media room; special events gallery.
Activities: lectures. Annual Events: Members & Affiliates Show; panel discussions.
Publications: biannual Raw Space catalog; quarterly newsletter; members catalog; book show catalog; Poetic Dialogue exhibition catalog.
Hours & Admission Prices: Wed.-Sat. 12-6, Sun. 12-4. No charge; donations requested. Closed all major holidays. ♿
Attendance: 6,200 (estimated)
Membership: Arc Angel Artist members $25 per year, yearly juried exhibitions fees waived; Artist members who operate non-profit space $60 per month.

✱ **THE ART INSTITUTE OF CHICAGO, (M),** 111 S. Michigan Ave., Chicago, IL 60603-6492. Tel.: 312-443-3600. Fax: 312-443-0849. TDD: 312-443-3890.
Web Site: artinstituteofchicago.org
Founded: 1879.
Congressional District: 1
Key Personnel: Chm. (V), Thomas J. Pritzker; Dir. & Pres., James Cuno; Exec. Vice Pres., Julia E. Getzels; Exec. Vice Pres. & CFO, Eric Anyah; Chm. African & Amerindian Art, Richard F. Townsend; Cur. African Art,

Kathleen Bickford-Berzock; Chm. American Arts, Judith A. Barter; Cur. American Arts, Sarah E. Kelly; Assoc. Cur. American Art, Ellen E. Roberts; Acting Chm. Architecture & Design, Zoe J. Ryan; Asst. Cur. Architecture & Design, Alison Fisher; Cur. Ancient Art, Karen Manchester; Cur. Indian & Islamic Art, Madhuvanti Ghose; Assoc. Cur. Chinese Art, Elinor L. Pearlstein; Assoc. Cur. Japanese Art, Janice Katz; Cur. Fellow Asian Art, Yuka Kadoi; Asst. Cur. Ancient Art, Mary C. Greuel; Chm. Contemporary Art, James Rondeau; Asst. Cur. Contemporary Art, Lisa B. Dorin; Chm. European Decorative Arts, Christopher P. Monkhouse; Cur. European Decorative Arts, Ghenete Zelleke; Chm. Med-Mod European Paint & Sculpture and Chm. Prints & Drawings, Douglas Druick; Cur. Modern Art, Stephanie D'Alessandro; Cur. European Painting Before 1750, Martha Wolff; Cur. European Painting, Gloria Groom; Assoc. Cur. Painting & Sculpture before 1750, Eve Straussman-Pflanzer; Asst. Cur. Medieval Art, Christina M. Nielsen; Chm. Photography, Matthew S. Witkovsky; Conservator, Photography, Douglas G. Severson; Assoc. Cur. Photography, Elizabeth Siegel; Assoc. Cur. Photography, Katherine A. Bussard; Cur. Prints & Drawings, Suzanne McCullagh; Cur. Prints & Drawings, Martha Tedeschi; Conservator Prints & Drawings, Harriet Stratis; Cur. Prints & Drawings, Mark Pascale; Cur. Prints & Drawings, Peter Zegers; Asst. Cur. Textiles, Odile V. Joassin; Conservator, Textiles, Lauren K. Chang; Chm. Textiles, Daniel S. Walker; Exec. Dir., Conservation, Frank Zuccari; Sr. Conservator, Objects, Barbara Hall; Conservator, Objects, Suzanne Schnepp; Conservator, Paintings, Kristin Lister; Conservator, Paintings, Faye Wrubel; Assoc. Conservator, Objects, Emily D. Heye; Exec. Dir. Libraries, Jack P. Brown; Archivist, Bart Ryckbosch; Exec. Dir. Museum Education, Robert Eskridge; Dir. Adult Programs, Admin. & Interpretive Media Museum Education, David Stark; Dir. Interpretive Exhibitions & Family Programs Museum Education, Jean Sousa; Exec. Dir. Publications, Robert V. Sharp; Dir. Publications, Sarah E. Guernsey; C.O.O., David A. Thurm; Vice Pres., Auxiliary Operations, Elizabeth Grainer; Exec. Dir. Museum Registration, Patricia Loiko; Vice Pres. Collection Management & Technology, Samuel Quigley; Sr. Registrar Loans & Exhibitions, Darrell Green; Sr. Registrar Permanent Collections, Sally-Ann Felgenhauer; Registrar Permanent Collection, Gregory Tschann; Vice Pres. Exhibitions & Museum Administration, Dorothy Schroeder; Vice Pres. Museum Finances, Jeanne M. Ladd; Assoc. Vice Pres. Physical Plant, William D. Caddick; Assoc. Vice Pres. Protection Svcs., Michelle Lehman Jenness; Vice Pres. Mktg., Carrie Heinonen; Dir. Graphic Design, Jeffrey Wonderland; Assoc. Dir. Business & Civic Rels., George J. Martin; Vice Pres. Human Resources, Michael Nicolai; Vice Pres. Information Svcs. & CIO, Eugene B. Adams, Jr
Personnel Profile: Full-Time Paid 530; Part-Time Paid 20; Part-Time Volunteers 720; Interns 60.
Governing Authority: nonprofit organization. Branch Institution: School of the Art Institute, Columbus Dr. & Jackson Blvd. Tax-exempt: 501(c)(3).
Institution Type/Description: Art Museum.
Collections: all periods of European & American painting; sculpture; prints; drawings; decorative arts; textiles; Chinese, Japanese, Indian & Middle Eastern art; European medieval art; classical art; photography; African, pre-Hispanic & Native American art; architectural drawings & fragments; arms & armor.
Research Fields: pertaining to collections.
Facilities: 341,679-vol. library of monographs, periodicals, exhibition catalogs, 490,000 slide images, photographs, color prints, architectural drawings, plans, manuscripts available for inter-library loan, to university faculty, staff of other museums & for use on premises by request; reading room; 4 auditoriums; theater; restaurants; education center. Books, reproductions, slides, postcards, jewelry, original decorative items & games for sale.
Activities: public lectures; tours; subscription programs; formally organized education programs for children, families & adults; outreach programs, workshops for teachers & paraprofessionals; permanent, temporary & traveling exhibitions; volunteer programs in all areas; subscription programs.
Publications: The Art Institute of Chicago Annual Report; exhibition catalogues.
Hours & Admission Prices: Thurs.-Fri. 10:30-8, Sat.-Wed. 10:30-5. Adults $18, children, students & senior citizens 65 & over $12; children under 14 & members no charge. Closed Thanksgiving; Christmas. &
Attendance: 1,800,000 (estimated)
Membership: Student $40; E-Member $60; Member $80; Member Plus $125; Premium Member $175.

AVERILL AND BERNARD LEVITON A + D GALLERY, 619 S. Wabash Ave., Chicago, IL 60605. Tel.: 312-369-8687. Fax: 312-369-8009.
Web Site: www.colum.edu/adgallery
Key Personnel: Gallery Dir., Jennifer Murray; Asst. Dir., Julianna Cuevas; Preparator, Megan Ross.

Governing Authority: Parent Institution: Columbia College Chicago. Tax-exempt.
Institution Type/Description: Art Gallery.
Collections: works by emerging & established artists.
Activities: educational programs.
Hours & Admission Prices: Tues.-Wed. & Fri.-Sat. 11-5, Thurs. 11-8. No charge. &
Attendance: 3,500 (estimated)

BALZEKAS MUSEUM OF LITHUANIAN CULTURE, (M), 6500 S. Pulaski Rd., Chicago, IL 60629-5136. Tel.: 773-582-6500. Fax: 773-582-5133.
E-mail: info@balzekasmuseum.org
Web Site: ww.balzekasmuseum.org
Founded: 1966.
Congressional District: 23
Key Personnel: Chm. & Pres. (V), Stanley Balzekas, Jr.; Exec. Dir., Rita Janz; Genealogist & Editor, Karile Vaitkute; Head Library, Robert Balzekas; Dir. Periodicals Collection, Irene Norbut; Librarian, Irena Pumputiene; Mgr. Membership & Collections, Regina Vasiliauskiene; Dir. Intl. Programs, Rasa Rudzykte; Museum Shop Mgr., Ruth Akelaityte; Office Mgr., Barbara Howley.
Personnel Profile: Full-Time Paid 7; Full-Time Volunteers 5; Part-Time Paid 5; Part-Time Volunteers 5.
Volunteer Hours: 1,000
Governing Authority: nonprofit organization. Tax-exempt: 501(c)(3).
Institution Type/Description: Lithuanian Culture Museum.
Collections: Lithuanian memorabilia & art; numismatics; philately; Lithuanian hagiology; rare books & maps; textiles; armor & period weapons; genealogy; Lithuanian art; folklore; ethnology; costumes; amber exhibit; photography; audiovisual, cartography; art archives.
Research Fields: Lithuanian & Eastern European history & folklore, Lithuanian immigration to the U.S.; genealogy; Lithuanian art.
Facilities: 75,000-vol. library of books written in all languages pertaining to Lithuania & 50,000-item archive on Eastern European history; manuscript division; 1,600 periodical titles; numismatics, armor, Center for the Study of U.S. Presidents & antique weapons available for inter-library loan & for use on premises. Amber, coins, stamps, folk art & miscellaneous items for sale.
Activities: guided tours; lectures; films; gallery talks; hobby workshops; discussion panel; formally organized educational programs; docent program or council; inter-museum loan, permanent, temporary & traveling exhibitions; school loan service. Museum Sponsors: folk art & art classes.
Publications: quarterly newsletter, Lithuanian Museum Review; genealogy.
Hours & Admission Prices: Daily 10-4. Adults $9, senior citizens & students $7, children 12 & under $3; discounts to AAM & ICOM members; members no charge. Closed New Year's Day; Easter; Christmas. &
Attendance: 44,000 (estimated)
Membership: Individual $35; Family $40; Supporting $50; Genealogy $75; Patron $100; Organization $250; Benefactor $500; Sponsor $1,000; Life $5,000.

BRONZEVILLE CHILDREN'S MUSEUM, 9301 S. Stony Island Ave., Chicago, IL 60617-3644. Tel.: 773-721-9301. Fax: 773-721-9303.
E-mail: bcm@bronzevillechildrensmuseum.org
Web Site: www.bronzevillechildrensmuseum.com
Key Personnel: Pres., Peggy A. Montes
Institution Type/Description: Children's Museum.
Collections: hands-on exhibits.
Facilities: Museum-related items for sale.
Hours & Admission Prices: Tues.-Sat. 10-4. Admission $5.

CAMBODIAN AMERICAN HERITAGE MUSEUM AND KILLING FIELDS MEMORIAL, 2831 W. Lawrence Ave., Chicago, IL 60625-3619. Tel.: 773-878-7090. Fax: 773-878-5299.
Key Personnel: Dir., Charles Daas; Museum Archivist, Ty Tim
Institution Type/Description: Cambodian Heritage Museum.
Collections: Cambodian history & culture; photographs.
Hours & Admission Prices: Museum & Memorial: Mon.-Fri. 10-4, Sat. by appointment. Community Center: Mon.-Fri. 9-5. No charge, donations accepted.

CHARNLEY-PERSKY HOUSE MUSEUM, (M), 1365 N. Astor St., Chicago, IL 60610-2144. Tel.: 312-915-0105 & 573-1365. Fax: 312-573-1141.
E-mail: psaliga@sah.org
Web Site: www.sah.org
Founded: 1998.

Key Personnel: C.E.O., Pauline Saliga; Pres. (V), Barry Bergdoll.
Personnel Profile: Part-Time Paid 1; Part-Time Volunteers 15.
Governing Authority: nonprofit educational organization. Parent Institution: Society of Architectural Historians. Tax-exempt.
Institution Type/Description: Historic House Museum: 1891-92 Charnley-Persky house designed by Louis Sullivan & Frank Lloyd Wright.
Collections: historic house.
Activities: tours.
Hours & Admission Prices: Tour of House: Wed. & Sat. 12 noon; groups by appointment. Adults $10, seniors $8, children 5-12 $5; IL teachers, members & Wed. no charge.
Attendance: 1,354 (accurate)

✳ **THE CHICAGO ACADEMY OF SCIENCES AND ITS PEGGY NOTEBAERT NATURE MUSEUM, (M),** 2430 N. Cannon Dr., Chicago, IL 60614-2874. Tel.: 773-755-5100. Fax: 773-755-5199.

Web Site: naturemuseum.org
Founded: 1857.
Congressional District: 5
Key Personnel: Pres. Emeritus & Academy Counselor, Dr. Paul G. Heltne; Pres. & C.E.O., Deborah Lahey; Chm. (V), Susan Whiting; Dir. Finance & Administration and C.F.O., Sharon Walton; Vice Pres. Education, Rafael Rosa; Cur. Biology and Vice Pres. Conservation & Research, Dr. Douglas Taron; Vice Pres. Exhibits & Museum Experience, Alvaro Ramos; Vice Pres. External Affairs, Marc Miller.
Personnel Profile: Full-Time Paid 52; Part-Time Paid 11; Part-Time Volunteers 225; Interns 29.
Volunteer Hours: 11,843
Operating Expenses: 7,066,486
Operating Income: 6,934,586
Governing Authority: nonprofit organization. Parent Institution: Chicago Academy of Sciences. Tax-exempt: 501(c)(3).
Institution Type/Description: Natural Science Museum.
Collections: birds; mammals; herpetology; geology; paleontology; malacology; botany; entomology; zoology; photography.
Major Exhibits: Animal Secrets, 10/13-1/14; Emerge & Unfurl: Photography of Ilze Arajs, 2/8/14-5/11/14; Nature's Struggle: Survival & Extinction, 3/22/14-9/21/14; Endangered Species: Print Project/Jenny Kendler and Molly Schafer, 4/22/14-9/21/14; Prairie Splendor: Photography of Frank Mayfield, 5/24/14-9/24/14; Lost Bird Project: Sculptures of Todd McGrain, 6/21/14-6/21/15; Faces of the Southern ocean, 9/6/14-11/14; Rainforest Adventure, 10/18/14-1/18/15.
Research Fields: natural sciences; effective role of museums in teacher professional development; ecology of urban wildlife; conservation biology of North American butterflies; headstart program for Blanding's turtle; director of the Illinois Butterfly Monitoring Network; Director of Project Squirrel.
Facilities: 25,000 sq. ft. exhibit space; 2,700 sq. ft. walk-in butterfly vivarium; 6.2 acres of botanical communities; children's gallery; meeting rooms; cafe. Museum-related items for sale.
Activities: youth summer camps; volunteer & intern opportunities; science & nature lectures; early childhood classes; onsite workshops & classes; teacher professional development workshops; permanent & temporary exhibitions; public programs; school & community science, technology & nature outreach; teen college preparedness & job skill training.
Publications: quarterly newsletter; annual report; Nature Museum News; Nature Museum Program Calendar.
Hours & Admission Prices: Mon.-Fri. 9-5, Sat.-Sun. 10-5. Adults $9, seniors & students $7, children 3-12 $6; discounts to groups and AAM, ICOM & ASTC members; suggested donation days for Illinois residents on Thurs. Closed May 2, Thanksgiving & Christmas. ♿
Attendance: 278,984 (accurate)
Membership: Individual $42; Senior $47; Senior Family $55; Family $60; Premier $130.

CHICAGO ARCHITECTURE FOUNDATION, 224 S. Michigan Ave., Chicago, IL 60604. Tel.: 312-922-3432, ext. 245. Fax: 312-922-2607.

E-mail: losmond@architecture.org
Web Site: www.architecture.org
Founded: 1966.
Congressional District: 7
Key Personnel: Pres. & C.E.O., Lynn Osmond; Chm., John Pintozzi; C.F.O., Asron Andersen; Vice Pres. Devel., Jennifer Van Valkenburg; Vice Pres. Tours & Retail, Michael Malak.
Personnel Profile: Full-Time Paid 55; Part-Time Paid 13; Part-Time Volunteers 470; Interns 7.
Governing Authority: nonprofit foundation. Tax-exempt: 501(c)(3).
Institution Type/Description: Architecture Museum & Center.

Collections: sculptures; architecture.
Major Exhibits: Chicago: City of Big Data, 3/14-9/15; Chicago Women in Architecture 40th Anniversary, 5/14-8/14.
Research Fields: Chicago architecture; urban planning & design, sustainability & infrastructure.
Facilities: Chicago Architecture Center; ArcelorMittal Design Studio; lecture facility; permanent exhibition space; education studios; Shop & Tour Orientation.
Activities: architectural tours of Chicago; lectures; permanent & temporary exhibits; docent training; teacher training, in-school & community youth programs, Newhouse Architecture Program.
Publications: annual tour announcements; books, A Walk Through Graceland Cemetery, A View From the River; tour brochure; A.I.A. guide to Chicago; quarterly newsletter, InSites; The Architecture Handbook: A Student Guide to Understanding Buildings; Schoolyards to Skylines - Teaching with Chicago's Amazing Architecture; Chicago Architecture: 1885 to Today; annual report.
Hours & Admission Prices: Mon.-Sat. 9-7, Sun. 9-6. Tours: Bus $40; River Cruise Mon.-Fri. $28, Sat.-Sun. $32; Walking $15. Closed New Year's Day; Memorial Day; Independence Day; Labor Day; Thanksgiving & day after; Christmas. ♿
Attendance: 488,000 (estimated)
Membership: Student $35; Senior & National $40; International $47; Individual $55; Senior Household & Household National $60; Household International $67.

CHICAGO CHILDREN'S MUSEUM, Navy Pier, 700 E. Grand Ave., Suite 127, Chicago, IL 60611-3577. Tel.: 312-527-1000. Fax: 312-527-9082.

Web Site: www.chicagochildrensmuseum.org
Founded: 1982.
Congressional District: 7
Key Personnel: Chm. Bd., Bob Barnett; Pres. & C.E.O., Jennifer Farrington; Vice Pres. Experience Devel. & Educational Programming, Natalie Bortoli; C.F.O., Dave Susalla; Vice Pres. Human Resources, Catherine Patyk; Volunteer & Intern Resources Coord., Sarah Williams; Assoc. Vice Pres. Exhibit & Bldg. Operations, Peter Williams.
Personnel Profile: Full-Time Paid 55; Part-Time Paid 28; Part-Time Volunteers 103; Interns 8.
Governing Authority: nonprofit organization. Tax-exempt: 501(c)(3).
Institution Type/Description: Children's Museum.
Collections: hands-on exhibits.
Major Exhibits: Forts, 1/14-5/14; Circus Zirkus, 5/14-9/14; Shadowwood, 9/14-11/14; Snow Much Fun, 11/14-1/15.
Facilities: 30,000 sq. ft. exhibit space. Gift items for sale.
Activities: hands-on permanent, temporary & traveling exhibits for children & their families; school & group, family travel & trunk shows; teacher & parent training; community outreach programs; paid programming; free family night; membership activities; performances. Special Events: annual meeting; Family Benefit; museum sponsored events.
Publications: monthly e-newsletter.
Hours & Admission Prices: Thurs. 10-8, Fri.-Wed. 10-5. Admission $14; children under one, Thurs. 5pm-8pm & members no charge. Closed Thanksgiving; Christmas. ♿
Attendance: 467,000 (accurate)
Membership: Family $115; Explorers $145; Corporate $5,000 & up.

CHICAGO CULTURAL CENTER, 78 E. Washington St., Chicago, IL 60602-4801. Tel.: 312-744-6630. Fax: 312-744-2089. TDD: 312-744-2947.

E-mail: culture@cityofchicago.org
Web Site: www.chicagoculturalcenter.org
Founded: 1897.
Congressional District: 7
Key Personnel: Commissioner, Chicago Dept. Cultural Affairs, Lois Weisberg; Deputy Commissioner Visual Arts, Gregory Knight; Cur. Exhibitions, Lanny Silverman; Exhibitions Designer & Preparator, Greg Lunceford.
Personnel Profile: Full-Time Paid 59; Part-Time Volunteers 150.
Governing Authority: Parent Institution: City of Chicago. Subsidiary Institution: Dept. of Cultural Affairs. Tax-exempt.
Institution Type/Description: Art Museum & Cultural Center.
Collections: paintings; photographs; sculpture.
Research Fields: modern & contemporary art.
Facilities: 2 theaters; 2 concert halls; civic reception halls & performing arts areas; dance studio; Chicago Office of Tourism Visitor Information Center; cafe. Museum-related items for sale.
Activities: changing exhibitions on wide range of subjects: contemporary art; crafts; cultural traditions; musical concerts; dance; theater; lectures; film

series; special exhibits & programs for children; literary arts program; group tours; building tours.

Publications: exhibition brochures; catalogs & educational brochures related to special exhibitions; performing arts brochures; self-guided tour of the Chicago Cultural Center brochure with map; book, The People's Palace: The Story of the Chicago Cultural Center; monthly calendar; building brochure; public art guides; tourism brochures.

Hours & Admission Prices: Cultural Center: Mon.-Thurs. 8-7, Fri. 8-6, Sat. 9-6, Sun. 10-6. No charge. Closed holidays. &

Attendance: 875,000 (estimated)

Membership: MOSAIC Membership: Student $25; Senior Individual $40; Individual $55; Senior Household $60; Household $75; Gem $125; Pillar $250; Tiffany $500; Luminary $1,000.

✻   CHICAGO HISTORY MUSEUM, (M), 1601 N. Clark St., Chicago, IL 60614-6038. Tel.: 312-642-4600. Fax: 312-266-2077. TDD: 800-526-0857.

Web Site: www.chicagohistory.org
Formerly: Chicago Historical Society
Founded: 1856.
Congressional District: 9
Key Personnel: Dir., C.E.O. & Pres., Gary T. Johnson; Exec. Vice Pres. & Chief Historian, Russell Lewis; Chm. (V), T. Bondurant French; Vice Pres. Interpretation & Education, Phyllis Rabineau; Sr. Cur., Olivia Mahoney; Dir. Corporate Events, Barb Siska; Dir. Research & Access, Ellen Keith; Dir. Education, The Elizabeth F. Cheney, D. Lynn McRainey; Dir. Institutional Advancement, Randy Adamsick; Vice Pres. Finance & C.F.O., Cheryl Obermeyer; Dir. Exhibitions, Tamara Biggs; Dir. Properties, John Yelen; Dir. Accounting, Leigh Stevenson; Dir. Technology, Don Pasqualini; Dir. Human Resources, Diane Ohi; Dir. Print & Multimedia Publications, Rosemary Adams; Dir. Visitor Svcs., Virginia Fitzgerald; Dir. Collections, Andrew W. Mellon, Kathleen Plourd; Dir. Curatorial Affairs, John Russick; Cur., Jill Austin; Cur., Joy Bivins; Dir. Museum Shop, Jennifer Vlna.
Personnel Profile: Full-Time Paid 81; Part-Time Paid 15; Part-Time Volunteers 111; Interns 12.
Governing Authority: nonprofit organization. Parent Institution: Chicago Historical Society. Tax-exempt: 501(c)(3).
Institution Type/Description: History Museum.
Collections: American history, 1760-1865 including: Revolutionary War; westward expansion; Civil War; Lincoln. Chicago & Illinois history 1690-present includes: fur trade; railroad; industrialization; urban culture & society; archives; manuscripts; architectural drawings; models; fragments; paintings; sculpture; decorative & industrial arts; costumes; prints; photographs.
Major Exhibits: Thailand - World's Columbian Exposition, 9/2/13-3/4/14; Railroad Portraits, 11/16/13-9/7/15.
Research Fields: Chicago & Illinois history; urban history; post-1945 Chicago history; Lincoln; Civil War; race & ethnicity, fashion; politics; neighborhoods & communities; labor; cultural.
Facilities: research center; auditorium; meeting rooms; rental facility; outdoor plaza. Postcards, books, pamphlets, colored slides & reproductions of photos for sale.
Activities: guided tours; lectures; films; gallery talks; concerts; study clubs; docent program; inter-museum loan, permanent, temporary & traveling exhibitions.
Publications: quarterly journal, Chicago History; quarterly newsletter; calendar; catalogues; monographs related to research fields; encyclopedia of Chicago (online version).
Hours & Admission Prices: Museum: Mon.-Sat. 9:30-4:30, Sun. 12-5. Research Center: Sept.-Dec. Tues.-Fri. 1-4:30, Sat. 10-4:30. Audio Tours: adults $14, senior citizens 65 & over and students 13-22 $12; discounts for groups of 10 or more, AAM & ICOM members; DuSable Museum & the National Museum of Mexican Art members, museum members, children 12 and under no charge. Research Center: annual $15, daily $5; members no charge. Closed New Year's Eve & Day; Thanksgiving & day before; Christmas Eve, Day & week &
Attendance: 2,068,008 (accurate)
Membership: Student & Senior 65 and over $45; Individual $50; Student & Senior Family $55; Family $60.

CHICAGO MARITIME MUSEUM, (M), 310 S. Racine Ave., Ste. 6N, Chicago, IL 60607-2841. Tel.: 312-421-9096. Fax: 312-850-1994.
E-mail: chimaritime@gmail.com
Web Site: chicagomaritimesociety.org
Formerly: Chicago Maritime Society
Founded: 1982.
Congressional District: 7

Key Personnel: Pres. (V), Gerald H. Thomas; Museum Shop Mgr., Don Glasell.
Personnel Profile: Full-Time Volunteers 1; Part-Time Paid 2; Part-Time Volunteers 11; Interns 1.
Governing Authority: Parent Institution: Chicago Maritime Society. Tax-exempt.
Institution Type/Description: Maritime Museum.
Collections: maritime history & artifacts; photographs; personal artifacts.
Facilities: Models of ships & museum-related items for sale.
Activities: outreach programs; research; educational programs.
Publications: books, From Lumber Hookers to the Hooligan Fleet, A Treasury of Chicago Maritime History; Images of America - Maritime Chicago; Schooner Passage; Chicago Maritime - An Illustrated History.
Hours & Admission Prices: Call for hours. No charge; donations accepted. &
Attendance: 300 (estimated)
Membership: Regular $35; Supporting $100; Sponsor $1,000; Director $2,000; Patron $5,000.

CHICAGO PUBLIC LIBRARY, 400 S. State St., Harold Washington Library Center, Special Collections & Preservation Div., Chicago, IL 60605-1216. Mailing Address: 9N-10, Special Collections & Preservation Div., 400 S. State St., Chicago, IL 60605. Tel.: 312-747-4883 & 4875.
E-mail: eholland@chipublib.org
Web Site: www.chipublib.org
Founded: 1872.
Congressional District: 7
Key Personnel: Commissioner, Brian Bannon.
Personnel Profile: Full-Time Paid 9; Part-Time Paid 1.
Governing Authority: municipal. Parent Institution: City of Chicago. Tax-exempt: 501(c)(3).
Institution Type/Description: History Museum: general exhibition site housed in the Chicago Public Library, Harold Washington Library Center.
Collections: artifacts; books; manuscripts; photographs; graphics; art; archives.
Major Exhibits: Vivian Maier: Out of the Shadows (T), 3/29/14-9/28/14.
Research Fields: Civil War; Chicago literary, cultural & social history; book arts.
Facilities: 18,000-vol. library pertaining to the Civil War & Chicago history; 390-seat auditorium, three exhibition spaces, special collections reading room.
Activities: lectures; films; concerts; organized educational programs; temporary & traveling exhibitions.
Publications: brochures; bibliographies & exhibitions catalogs.
Hours & Admission Prices: Library: Mon.-Thurs. 9-9, Fri.-Sat. 9-5, Sun. 1-5. Special Collections: Mon.-Tues. 12-6, Fri.-Sat. 12-4. No charge. &
Attendance: 182,489 (estimated)

CHICAGO STATE UNIVERSITY, PRESIDENT'S GALLERY, 9501 S. King Dr., Cook Admin. Bldg., 3rd Fl., Chicago, IL 60628-1598. Tel.: 773-995-3905.
Key Personnel: Cur., Joyce Owens Anderson
Institution Type/Description: Art Gallery.
Collections: paintings; photographs.
Activities: lectures; demonstrations; workshops; special events.
Hours & Admission Prices: Mon.-Fri. 8:30-5. No charge.

CHINESE-AMERICAN MUSEUM OF CHICAGO, 238 W. 23rd St., Chicago, IL 60616-1904. Tel.: 312-949-1000. Fax: 312-949-1001.
Web Site: www.ccamuseum.org
Founded: 2005.
Key Personnel: Pres. (V), Dr. Kim K. Tee.
Governing Authority: Parent Institution: Chinatown Museum Foundation, Chicago. Tax-exempt.
Institution Type/Description: Chinese-American Museum.
Collections: local history & culture; documents; photographs; period artifacts; clothing; paintings; ceramics.
Major Exhibits: Great Wall to the Great Lakes, 2/10-2/16.
Facilities: Museum-related items for sale.
Activities: lectures; educational programs.
Publications: newsletter.
Hours & Admission Prices: Fri. 9:30-1:30, Sat.-Sun. 10-5. Suggested Donation: adults $2, seniors & students $1; members no charge.
Membership: Student $5; Individual $25; Family $50; Corporate $150; Honorary $200

CITY GALLERY AT THE HISTORIC WATER TOWER, 806 N. Michigan Ave., Chicago, IL 60611-2103. Mailing Address: Department of Cultural Affairs, 78 E. Washington St., Chicago, IL 60602-4801. Tel.: 312-744-2400 & 742-0808.
Institution Type/Description: Photography Gallery: housed in the historic Chicago Water Tower; built in 1869. Listed on the National Register of Historic Places.
Collections: Chicago-themed photographs by Chicago photographers.
Hours & Admission Prices: Mon.-Sat. 10-6:30, Sun. 10-5. No charge.

* **CLARKE HOUSE MUSEUM, (M),** 1827 S. Indiana Ave., Chicago, IL 60616-1308. Mailing Address: Glessner House Museum, 1800 S. Prairie Ave., Chicago, IL 60616-1320. Tel.: 312-745-0040. Fax: 312-745-0077.
E-mail: info@clarkehousemuseum.org
Web Site: www.clarkehousemuseum.org
Founded: 1984.
Congressional District: 7
Key Personnel: Cur., Rebecca LaBarre; Liaison from City of Chicago Dept. of Cultural Affairs & Special Events, Julie Burros.
Personnel Profile: Full-Time Paid 1; Part-Time Volunteers 60; Interns 1.
Governing Authority: municipal government. Owned by the city of Chicago. Operated by Glessner House Museum. Furnished by The National Society of Colonial Dames of America in The State of Illinois. Tax-exempt: 501(c)(3).
Institution Type/Description: Historic House Museum: 1836 Greek Revival building.
Collections: period rooms; domestic life in Chicago, 1836-1860; 19th-century decorative arts; early urban development of Chicago.
Research Fields: early Chicago; domestic technology.
Facilities: 300-vol. library of books on 19th-century history, decorative arts & architecture.
Activities: guided tours; lectures; organized education programs for children; docent program; candlelight holiday tours.
Hours & Admission Prices: Guided Tours (start at Glessner House Museum): Wed.-Sun. 12 & 2. Adults $10, senior citizens & students $9, children 5-12 $6; discounts to AAM members; children under 5 & Wed. no charge. Clarke & Glessner Combination: adults $15, seniors & students $12, children 5-12 $8; discounts to AAM, AAA, National Trust & Public Broadcasting members. Closed New Year's Day; Easter; Memorial Day; Independence Day; Labor Day; Thanksgiving; Christmas Eve & Day. &
Attendance: 6,000 (estimated)

COLUMBIA COLLEGE CHICAGO CENTER FOR THE BOOK & PAPER ARTS, (M), 1104 S. Wabash, 2nd Fl., Chicago, IL 60605-2334. Tel.: 312-369-6630. Fax: 312-369-8082.
E-mail: bookandpaper@colum.edu
Web Site: www.colum.edu/book_and_paper
Key Personnel: Dir., Dr. Steve Woodall
Institution Type/Description: Art Gallery.
Collections: papermaking studio; letterpress studio; hot stampers; Vandercook and Chandler & Price presses; over 800 drawers of type.
Activities: lectures; special events.
Hours & Admission Prices: Gallery: Mon.-Fri. 10-6. Office: Mon.-Fri. 9:30-5. No charge.

DEPAUL ART MUSEUM, 935 W. Fullerton Ave., Chicago, IL 60614. Tel.: 773-325-7506 & 7593. Fax: 773-325-4506.
E-mail: artmuseum@depaul.edu
Web Site: depaul.edu/museum
Formerly: DePaul University Art Museum
Founded: 1987.
Congressional District: 9
Key Personnel: Dir., Louise Lincoln; Asst. Dir., Laura Fatemi; Asst. Cur., Gregory Harris; Administrative Asst., Alison Kleiman.
Personnel Profile: Full-Time Paid 4; Part-Time Paid 10; Part-Time Volunteers 2; Interns 2.
Governing Authority: private university; nonprofit. Parent Institution: DePaul University. Tax-exempt: 170(b)(1)(A).
Institution Type/Description: University Art Museum.
Collections: paintings; photographs; sculpture; prints; African sculpture.
Major Exhibits: The Sochi Project, 1/14-3/14.
Activities: temporary, traveling & loan exhibitions; lectures.
Publications: gallery notes; exhibition catalogs.
Hours & Admission Prices: Mon.-Thurs. & Sat. 11-5, Fri. 11-7, Sat.-Sun. 12-5. No charge; donations accepted. Closed university holidays. &
Attendance: 18,500 (accurate)
Membership: $50-$1,000.

DUSABLE MUSEUM OF AFRICAN-AMERICAN HISTORY, INC., 740 E. 56th Place, Chicago, IL 60637-1495. Tel.: 773-947-0600, ext. 246. Fax: 773-947-0677.
E-mail: cbethea@dusablemuseum.org
Web Site: www.dusablemuseum.org
Founded: 1961.
Congressional District: 1
Key Personnel: C.E.O. & Pres., Carol L. Adams, Ph.D.; C.O.O., Charles E. Bethea; Chm. (V), Cheryl Blackwell Bryson; Dir. Education, Sandra Gaither; Dir. Special Events, Tracey Williams; Dir. Visitor Svcs., Jomo Cheatham; Museum Shop Mgr., Lillian Roberts.
Personnel Profile: Full-Time Paid 23; Part-Time Paid 21; Part-Time Volunteers 40; Interns 12.
Governing Authority: nonprofit organization. Tax-exempt: 501(c)(3).
Institution Type/Description: History Museum.
Collections: archives; African, Afro-American art; memorabilia; historical artifacts; sculpture; photographs; books.
Research Fields: African-American art history; Chicago & Illinois history.
Facilities: approx. 6,000-vol. library of African & Afro-American books, tapes, phonograph records, pictures, slides, biographical & vertical files available for use in library; reading room; auditorium; exhibition galleries. Curios, sculpture, prints, artifacts & publications for sale.
Activities: guided tours; lectures; traveling exhibitions; classes; film program; annual school oratorical; essay contests; after-school program.
Publications: Heritage Calendar; book of poems, What Shall I Tell My Children Who Are Black; books, Black Power in Old Alabama; The Birth & The Building of the Dusable Museum; Figures in Black History; Poetry of Prison.
Hours & Admission Prices: Mon.-Sat. 10-5, Sun. 12-5; groups by appointment. Adults $10, students & senior citizens $7, children 6-11 $3; discounts to AAM & ICOM members, Chicago residents & police, military; children under 5, members, Museums in the Park staff & Sun. no charge. Closed New Year's Day; Easter; Independence Day; Thanksgiving; Christmas. &
Attendance: 180,000 (accurate)
Membership: Seniors & Students $15; General $25; Family $35; Annual Sponsor $100; Patron $500 & up; Annual Sustainer $1,000.

EDGEWATER HISTORICAL SOCIETY, 5358 N. Ashland Ave., Chicago, IL 60640. Tel.: 773-506-5358.
Web Site: www.edgewaterhistory.org
Institution Type/Description: History Museum.
Collections: local history & culture; period furnishings; personal artifacts; photographs; books.
Activities: group tours; special events.
Hours & Admission Prices: Sat.-Sun. 1-4. Closed New Year's Eve & Day; Christmas Eve & Day.

FASHION STUDY COLLECTION, Columbia College Chicago, 618 S. Michigan Ave., 8th Fl., Chicago, IL 60605-1901. Mailing Address: 600 S. Michigan Ave., Chicago, IL 60605. Tel.: 312-369-6283. Fax: 312-369-8422.
E-mail: jwayneguite@colum.edu
Web Site: www.colum.edu/fashion_collection
Formerly: Fashion Columbia Study Collection
Founded: 1989.
Key Personnel: Collections Mgr., Jacqueline WayneGuite.
Governing Authority: Parent Institution: Columbia College Chicago. Tax-exempt.
Institution Type/Description: Fashion Collection.
Collections: 6,000 items of dress including European, American & Japanese fashion designers; period artifacts; ethnic clothing.
Research Fields: fashion; costume; dress; textiles; ethnography; history.
Facilities: research center.
Hours & Admission Prices: Research Center: Mon.-Thurs. 10-5. Appointments: Mon.-Thurs. 9-5, Fri. 9-3. No charge; donations accepted. &

* **FIELD MUSEUM OF NATURAL HISTORY, (M),** 1400 S. Lake Shore Dr., Chicago, IL 60605-2496. Tel.: 312-922-9410. Fax: 312-922-0741. TDD: 312-341-9299.
Web Site: www.fieldmuseum.org
Founded: 1893.
Congressional District: 7
Key Personnel: C.E.O. & Pres., Richard W. Lariviere, PhD; Bd. Chm. (V), John Rowe; Exec. Vice Pres., J.W. Croft; Vice Pres. Institutional Advancement, Charles Katzenmeyer; Vice Pres. Admin., Shawn VanDerziel; Sr. Vice Pres. Science & Education, Debra Moskovits; Sr. Vice Pres. Museum Enterprises, Laura M. Sadler.

Personnel Profile: Full-Time Paid 460; Part-Time Paid 32; Part-Time Volunteers 450; Interns 10.
Governing Authority: board of trustees; nonprofit organization. Tax-exempt: 501(c)(3) & 170(b)(1)(A).
Institution Type/Description: Natural History Museum.
Collections: anatomy; anthropology; archaeology; archives; botany; costumes; entomology; ethnology; geology; herbarium; herpetology; Indian; mineralogy; natural history; paleontology; science; textiles; zoology.
Research Fields: archaeology; botany; geology; zoology; anthropology; environmental & conservation programs.
Facilities: 250,000-vol. library of natural history material available for inter-library loan & use on premises; full scale replica of a Pawnee Earth Lodge; Place for Wonder, natural & cultural artifacts; reading room; theater; classrooms; restaurant. Gift items, books & cards for sale.
Activities: guided tours; lectures; films; performances; demonstrations; adult education courses; formally organized education programs for children & graduate students affiliated with Northwestern University, University of Chicago & University of Illinois at Chicago, Northern Illinois University; one-day environmental field trips for adults & families; museum-wide volunteer program; docent program or council; training programs for professional museum workers; permanent, temporary & special exhibitions; school loan service.
Publications: monographs, Fieldiana Anthropology; Fieldiana Botany; Fieldiana Geology; Fieldiana Zoology; bimonthly periodical, In the Field.
Hours & Admission Prices: Daily 9-5. Adults $15, senior citizens 65 & over and students with ID $12, children 3-11 $10; discounts to members, Chicago residents, AAM & ICOM members, military personnel, individual teachers, children under 3 & groups of 10 or more with appointment. All Access Pass: adults $30, student & senior $25, children 3-11 $20. Closed Christmas. &
Membership: Student & Senior $70; Individual $80; Family $110.

FIRE MUSEUM OF GREATER CHICAGO, (M), 5218 S. Western, Chicago, IL 60609. Mailing Address: 517 Senon Dr., Lemont, IL 60439-4093. Tel.: 877-225-7491.

Web Site: www.firemuseumofgreaterchicago.org
Founded: 1997.
Personnel Profile: Part-Time Volunteers 20.
Governing Authority: Tax-exempt.
Institution Type/Description: Fire History Museum.
Collections: Chicago area fire & firefighting history; photographs; firefighting equipment; helmets; badges; uniforms; toys.
Research Fields: Chicago firefighters.
Facilities: library.
Publications: quarterly newsletter.
Hours & Admission Prices: Open House: Jan.-Nov. 4th Sat. each month, call for hours; groups by appointment. No charge; donations accepted. &
Membership: Basic $30; Founding $100; Corporate $1,000.

FREDERICK C. ROBIE HOUSE, 5757 S. Woodlawn Ave., Chicago, IL 60637-1698. Mailing Address: 209 S. LaSalle St., Ste. 118, Chicago, IL 60604. Tel.: 312-994-4000 & 4040. Fax: 773-324-6099.

E-mail: info@flwright.org
Web Site: flwright.org
Founded: 1996.
Congressional District: 1
Key Personnel: C.E.O. & Pres., Celeste Adams; Chm., Graham Rarity.
Personnel Profile: Full-Time Paid 30; Part-Time Paid 20; Part-Time Volunteers 75.
Governing Authority: private; nonprofit organization. Parent Institution: Frank Lloyd Wright Trust, Chicago, IL. Tax-exempt: 501(c)(3).
Institution Type/Description: Historic House Museum: 3-story residence designed by Frank Lloyd Wright, c.1908.
Collections: structures, furniture & decorative arts designed by Frank Lloyd Wright; Japanese art & decorative materials collected by Wright; drawings; prairie style architecture.
Research Fields: Frank Lloyd Wright's life & work from 1889-1916.
Facilities: 5,040 sq. ft. exhibit space. Museum-related items for sale.
Activities: docent program; formal education programs; guided tours; lectures; neighborhood walk; special events. Annual Events: Wright Plus Housewalk in May.
Publications: quarterly newsletter, Wright Angles; books, The Oak Park Home & Studio of Frank Lloyd Wright; Frank Lloyd Wright and the Prairie; Building a Legacy: The Restoration of Frank Lloyd Wright's Home and Studio; Frank Lloyd Wright's Fifty Views of Japan; In Wright's Shadow: Artists and Architects in the Oak Park Studio; Hometown Architect: The Complete Buildings of Frank Lloyd Wright in Oak Park and River Forest; The Wright Family Library.

Hours & Admission Prices: Tours: Thurs.-Mon. 11-3; groups by appointment. Book Shop: Thurs.-Mon. 9-4. Adults $15, students and seniors 65 & over $12; discount to members; children under 3 no charge. Tours: no charge to Frank Lloyd Wright Preservation Trust members. Closed New Year's Day; Thanksgiving; Christmas Eve & Day.
Attendance: 33,000 (estimated)
Membership: Individual $50; Family $65; Prairie Society $125; Octagon Society $250; Inglenook Society $500; Skylight Society $1,000.

GALLERY 400 - UNIVERSITY OF ILLINOIS AT CHICAGO, (M), Art and Design Hall, 1st Fl., 400 S. Peoria St. (MC 034), Chicago, IL 60607. Tel.: 312-996-6114. Fax: 312-355-3444.

E-mail: gallery400@uic.edu
Web Site: gallery400.uic.edu
Founded: 1983.
Congressional District: 7
Key Personnel: Dir., Lorelei Stewart.
Personnel Profile: Full-Time Paid 2; Part-Time Paid 5; Interns 9.
Governing Authority: Parent Institution: State of Illinois. Subsidiary Institution: University of IL at Chicago. Tax-exempt.
Institution Type/Description: Art Gallery.
Collections: works by local & national artists.
Activities: lectures; film screenings; performances.
Hours & Admission Prices: Tues.-Fri. 10-6, Sat. 12-6; other times by appointment. No charge. &
Attendance: 6,702

GARFIELD PARK CONSERVATORY, 300 N. Central Park Ave., Chicago, IL 60624-1945. Tel.: 312-746-5100. Fax: 773-638-1777.

E-mail: donorservices@garfieldpark.org
Web Site: www.garfieldconservatory.org
Founded: 1908.
Congressional District: 7
Key Personnel: Dir. Conservatories, Mary Eysenbach; Gen. Foreman, Miguel del Valle; Foreman, Matthew Barrett; Foreman, Thomas Costanza; Foreman, Koch Unni.
Personnel Profile: Full-Time Paid 21; Part-Time Paid 3; Part-Time Volunteers 120.
Governing Authority: municipal; nonprofit. Parent Institution: Chicago Park District, 541 N. Fairbanks, Chicago, IL 60611. Subsidiary Institution: Garfield Park Conservatory Alliance. Tax-exempt.
Institution Type/Description: Horticultural Conservatory.
Collections: botanical collection; tropicals; deserts; urban demonstration gardens; palms; aroids; ferns; cycads; perennials; children's garden; sensory garden.
Facilities: botanical garden; greenhouse; demonstration garden; children's garden.
Activities: education programs; workshops; tours; permanent & temporary exhibitions; summer gardens; family programs. Museum Sponsors: Holiday Show; Spring Show; Tropical Show.
Publications: member magazine, Chicago Greenscapes; brochures; maps.
Hours & Admission Prices: Wed. 9-8, Thurs.-Tues. 9-5. No charge; donations accepted. &
Attendance: 160,000 (accurate)
Membership: Seedling $25; Flower $50; Fern $100; Palm $250; Director $500; Jensen Club $1,000 & up.

✴ GLESSNER HOUSE MUSEUM, (M), 1800 S. Prairie Ave., Chicago, IL 60616-1320. Tel.: 312-326-1480. Fax: 312-326-1397.

E-mail: glessnerhouse@sbcglobal.net
Web Site: www.glessnerhouse.org
Founded: 1994.
Congressional District: 7
Key Personnel: Exec. Dir., William Tyre; Pres., Rolf Achilles; Museum Shop Mgr., Gwen Carrion.
Personnel Profile: Full-Time Paid 4; Part-Time Paid 1; Part-Time Volunteers 60; Interns 4.
Volunteer Hours: 3,200
Operating Expenses: 349,255
Operating Income: 352,731
Governing Authority: nonprofit organization. Tax-exempt: 501(c)(3).
Institution Type/Description: Historic House: 1887 home designed by H.H. Richardson.
Collections: English arts & crafts; aesthetic movement furniture; decorative arts; period furniture; glass; ceramics; engravings; books; photographs.
Research Fields: Arts & Crafts Movement; Aesthetic Movement; H.H. Richardson. Glessner family; Chicago architecture; Prairie Avenue Historic District; late 19th century, urban home designed by HH Richardson on Prairie Ave.

Facilities: event rental space.
Activities: guided tours; lectures; educational programs; docent program.
Publications: membership newsletter, The Glessner Journal.
Hours & Admission Prices: Tour Center: Wed.-Sun. 11:30-4. Guided Tours: 1 & 3. Clarke House Tours: 12 & 2. Adults $10, senior citizens & students $9, children 5-12 $6; discounts to AAM, AAA, National Trust & Public Broadcasting members; children under 5, members & Wed. no charge. Glessner & Clarke House: adults $15, seniors & students $12, children 5-12 $8. Closed New Year's Day; Easter; Memorial Day; Independence Day; Labor Day; Thanksgiving; Christmas Eve & Day.
Attendance: 10,000 (estimated)
Membership: Student & Senior $30; Individual $40; Family $50; Friend $100 & up; Preservationist $250 & up; Curator $500 & up; Prairie Avenue Society $1,000 & up; Decorative Arts Society $2,500 & up; Richardson Society $5,000 & up.

HISTORIC PULLMAN FOUNDATION, Visitor Center & Museum, 11141 S. Cottage Grove Ave., Chicago, IL 60628-4614. Mailing Address: 614 E. 113th St., Chicago, IL 60628-5100. Tel.: 773-785-8181, 3828 (Tours) & 8901 (Visitor Center). Fax: 773-785-8182. Facebook: Historic Pullman Foundation.
E-mail: foundation@pullmanil.org
Web Site: www.pullmanil.org
Founded: 1973.
Congressional District: 2
Key Personnel: Pres. (V), Michael Shymanski; Vice Pres. & Treas. (V), Cynthia Martin-McMahon.
Governing Authority: nonprofit organization. Tax-exempt: 501(c)(3).
Institution Type/Description: Historic District: 1880-84 built by Pullman's Palace Car Co.; first planned model industrial community.
Collections: Historic Buildings: 1881 Hotel Florence; 1881 The Historic Pullman Center; 1892 The Pullman Market Hall; 1883 Arcade Site & Pullman Visitor Center; 1882-5 various North Pullman Row Houses.
Research Fields: Pullman's history.
Facilities: visitor center; archives. Museum-related items for sale.
Activities: guided tours; year-round educational tour program. Museum Sponsors: annual House tour in October; Historic Preservation Programs Bus Tour; Candlelight House Walk & Buffet Dinner in Dec.
Publications: quarterly newsletter, Update.
Hours & Admission Prices: Visitor Center: Tues.-Sun. 11-3. Adults $5, seniors & students under 14 $4. Tours: May-Oct. first Sun. of month 1:30; groups by appointment. Adults $7, seniors $5, students $4; children 12 & under no charge when accompanied by an adult. Closed national holidays. &
Attendance: 475,000 (estimated)
Membership: General $20; Associate $30; Supporting $50; Friend of the Foundation $100; Donor $500; Pullman Patron $1,000.

HYDE PARK ART CENTER, 5020 S. Cornell Ave., Chicago, IL 60615-3016. Tel.: 773-324-5520. Fax: 773-324-6641.
E-mail: info@hydeparkart.org
Web Site: hydeparkart.org
Founded: 1939.
Congressional District: 1
Key Personnel: Exec. Dir., Chuck Thurow; Chm. (V), Lawrence Furnstahl.
Personnel Profile: Full-Time Paid 9; Part-Time Paid 40; Part-Time Volunteers 65; Interns 12.
Governing Authority: nonprofit. Tax-exempt: 501(c)(3).
Institution Type/Description: Art Gallery & School.
Collections: Chicago-rooted, national & international artists.
Facilities: classrooms; ceramics studio; art resource center.
Activities: education programs for children & adults with a hands-on focus; guided gallery tours; one day workshops; panel discussions; temporary exhibitions; community outreach program; artist & curator talks; family days; monthly creative events. Annual Events: benefit auction; student & faculty show; special benefits.
Publications: quarterly newsletter, News; quarterly school calendar; occasional exhibition catalogues.
Hours & Admission Prices: Mon.-Thurs. 9-8, Fri.-Sat. 9-5, Sun. 12-5. No charge; donations accepted. &
Attendance: 45,000 (estimated)
Membership: Student, Senior Citizen & Artist $30; Individual $45; Family $55; Sponsor $100; Patron $300; Benefactor $500; Ruth's Circle $1,000; Creative Leader $5,000.

✻ ILLINOIS STATE MUSEUM, CHICAGO GALLERY, 100 W. Randolph, Ste. 2-100, Chicago, IL 60601-3921. Tel.: 312-814-5322. Fax: 312-814-3471.
Web Site: www.museum.state.il.us/ismsites/chicago/

Formerly: State of Illinois Art Gallery
Founded: 1985.
Congressional District: 7
Key Personnel: Dir., Kent Smith; Assoc. Cur. & Gallery Admin., Jane Stevens; Asst. Cur., Douglas Stapleton.
Personnel Profile: Full-Time Paid 2; Part-Time Paid 1; Part-Time Volunteers 2; Interns 2.
Governing Authority: state. Parent Institution: Illinois State Museum. Tax-exempt.
Institution Type/Description: Art Gallery.
Collections: historic & contemporary Illinois art including paintings, drawings, printmaking, video, ceramics, photography & textiles.
Research Fields: Illinois artists.
Facilities: 3,200 sq. ft. exhibition space.
Activities: guided tours; lectures; organized educational programs; changing exhibitions.
Publications: brochures; catalogs; handouts.
Hours & Admission Prices: Mon.-Fri. 9-5. No charge. Closed national holidays. &
Attendance: 41,762 (accurate)
Membership: Artist $20; Individual $35; Family $50; Contributing $100; Sustaining $300; Life $500.

INDO-AMERICAN HERITAGE MUSEUM, 6328 N. California Ave., Chicago, IL 60659.
E-mail: info@iahmuseum.org
Web Site: www.iahmuseum.org
Founded: 2008.
Congressional District: 9
Institution Type/Description: Heritage Museum.
Collections: Indian cultural heritage; immigrant history; photographs; personal artifacts.
Activities: educational programs & activities; special events.
Hours & Admission Prices: Mon.-Sat. 10-5 by appointment.

INSTITUTE OF PUERTO RICAN ARTS & CULTURE, 3015 W. Division St., Chicago, IL 60622. Tel.: 773-486-8345. Fax: 773-486-8806.
E-mail: info@iprac.org
Web Site: www.iprac.org
Founded: 2001.
Key Personnel: Exec. Dir., Jose Lopez
Institution Type/Description: Art Gallery.
Collections: works by Puerto Rican artists.
Facilities: Gallery-related items for sale.
Activities: lectures; workshops; film series; cafe; classrooms; theater. Annual Event: Fine Arts & Crafts Festival.
Hours & Admission Prices: Call for hours. No charge; donations accepted.
Attendance: 25,000 (estimated)

INTERNATIONAL MUSEUM OF SURGICAL SCIENCE, 1524 N. Lake Shore Dr., Chicago, IL 60610-1651. Tel.: 312-642-6502, ext. 3130. Fax: 312-642-9516.
E-mail: leonard@imss.org
Web Site: www.imss.org
Founded: 1953.
Congressional District: 7
Key Personnel: Pres. (V), Raymond Dieter, M.D.; Cur., Lindsey Thieman; Museum Shop Mgr., Lynnea Smith.
Personnel Profile: Full-Time Paid 2; Part-Time Paid 1; Part-Time Volunteers 10; Interns 2.
Governing Authority: nonprofit organization. Parent Institution: International College of Surgeons, 1516 N. Lake Shore Dr., Chicago, IL 60610. Tax-exempt: 501(c)(3).
Institution Type/Description: Medical Museum: housed in 1917 Howard Van Doren Shaw Mansion.
Collections: medical collection showing the growth & perfection of many surgical specialties such as obstetrics & gynecology, orthopedics, urology, radiology, X-ray, heart research & acupuncture; manuscript collections; extensive collection of medical artifacts & art.
Facilities: 2,000-vol. library: books dealing with various fields of medicine; 80-seat auditorium.
Activities: self-guided tours; lectures; loan, permanent, temporary & traveling exhibitions.
Publications: quarterly newsletter, MuseLetter.
Hours & Admission Prices: May-Sept. Tues.-Sun. 10-4; Oct.-April Tues.-Sat. 10-4. Adults $10, senior citizens & students $6; discounts to AAM & ASTC members; members & Tues. no charge. Closed New Year's Day; Independence Day; Thanksgiving; Christmas. &

Attendance: 15,000 (accurate)
Membership: Student $15; Individual $25; Family $50; Director's Club $100; Chairman's Club $500; Founder's Club $1,000. Corporate & Business Membership: Sponsor $100; Director's Circle $1,000; Chairman's Circle $5,000; Founder's Circle $10,000.

INTUIT: THE CENTER FOR INTUITIVE AND OUTSIDER ART, 756 N. Milwaukee Ave., Chicago, IL 60642-5939. Tel.: 312-243-9088. Fax: 312-243-9089.

E-mail: intuit@art.org
Web Site: www.art.org
Founded: 1991.
Congressional District: 5
Key Personnel: C.E.O. & Dir., Joel Mangers; Devel. & Membership, Chris Renton; Education, Carol Ng-He; Pres. (V), Ralph Concepcion; Public Rels. & Museum Shop Mgr., Heather Holbus; Treas., Gary Zickel; Registrar, Aza Quinn-Brauner; Cur., Jan Petry.
Personnel Profile: Full-Time Paid 3; Full-Time Volunteers 5; Part-Time Paid 2; Part-Time Volunteers 2; Interns 2.
Governing Authority: private; nonprofit organization. Tax-exempt: 501(c)(3).
Institution Type/Description: Art Museum.
Collections: art brut; non-traditional folk art; self-taught art; visionary art.
Major Exhibits: Mad Musee, 9/13-1/14.
Research Fields: art & artists from the genres of art brut, non-traditional folk art, self-taught art, and/or visionary art; art environments, art & psychiatry, and the role of Chicago in the movement of folk & outsider art in America.
Facilities: library.
Activities: 75-seat auditorium; 1,300 sq. ft. exhibit space; films; formal education programs; guided tours; lectures; loan, temporary & traveling exhibitions. Museum-related items for sale. Annual Event: The Visionary Ball.
Publications: annual newsletter, The Outsider.
Hours & Admission Prices: Tues.-Wed. & Fri.-Sat. 11-5, Thurs. 11-7:30. Adults $5; members & children under 12 no charge. Closed national holidays. &
Attendance: 5,000 (estimated)
Membership: Student $25; Individual $40; Family & Partner $60; Associate $100-$249; Enthusiast $250-$499; Patron $500-$1,199; Benefactor $1,200-$2,499; Leadership Circle $2,500-$3,499; Visionary Circle $3,500 & up.

IRISH AMERICAN HERITAGE CENTER, 4626 N. Knox Ave., Chicago, IL 60630-4035. Tel.: 773-282-7035. Fax: 773-282-0380.

E-mail: info@irishahc.org
Web Site: irish-american.org
Founded: 1976.
Congressional District: 10
Key Personnel: Pres. (V), John Gorski; Museum Shop Mgr., Tom Boyle.
Personnel Profile: Full-Time Paid 2; Part-Time Paid 2; Part-Time Volunteers 200.
Institution Type/Description: Heritage Center.
Collections: Irish history & culture; furniture; personal artifacts; Irish lace; late 19th century piano; Belleek Parian china; period maps.
Facilities: library; archives; ballroom
Activities: special events; temporary exhibits.
Publications: monthly newsletter; Irish American News.
Hours & Admission Prices: Wed.-Sat. by appointment. No charge; donations accepted. &
Attendance: 9,000,000 (estimated)
Membership: Individual $30; Family $40.

JAMES P. FITZGIBBONS HISTORICAL MUSEUM, 9801 Ave. G, Calumet Park Fieldhouse, Chicago, IL 60617. Tel.: 312-721-7948.

Key Personnel: Pres. (V), Bernard Janecki.
Governing Authority: Parent Institution: Southeast Chicago Historical Society.
Institution Type/Description: Historical Society Museum.
Collections: local history & culture; personal artifacts; photographs.
Hours & Admission Prices: Thurs. 1-4, 1st Sun. of month 12-3.

JANE ADDAMS HULL-HOUSE MUSEUM, UNIVERSITY OF ILLINOIS CHICAGO, (M), 800 S. Halsted, (M/C 051), Chicago, IL 60607-7017. Tel.: 312-413-5353. Fax: 312-413-2092.

E-mail: jahh@uic.edu
Web Site: www.uic.edu/jaddams/hull/
Founded: 1967.
Congressional District: 7
Key Personnel: Dir., Lisa Yun Lee; Facilities Mgr. & Head Docent, Dan Portincaso.
Personnel Profile: Full-Time Paid 4; Part-Time Paid 6.

Governing Authority: university. Parent Institution: University of Illinois at Chicago. Tax-exempt.
Institution Type/Description: Historic House Site: 1856 Hull Mansion occupied by Jane Addams in 1889; serving as the first settlement building of Hull House complex; 1907 Resident's Dining Hall.
Collections: structures; paintings; photographs; documents; artifacts; memorabilia; furniture; paintings & other artwork; books; papers, letters, clipping & manuscripts relating to history of Hull House, Jane Addams & the surrounding neighborhood.
Research Fields: history of settlement house movement in Chicago; life of Jane Addams; local history; history of social welfare reform; immigration & ethnic history.
Activities: guided tours; lectures; permanent & temporary exhibitions; audio-visual programs.
Publications: brochure.
Hours & Admission Prices: Tues.-Fri. 10-4, Sun. 12-4. No charge. Closed major holidays.
Attendance: 13,000 (accurate)

JOHN G. SHEDD AQUARIUM, (M), 1200 S. Lake Shore Dr., Chicago, IL 60605-2490. Tel.: 312-939-2438. Fax: 312-939-8069.

E-mail: contactus@sheddaquarium.org
Web Site: www.sheddaquarium.org
Founded: 1924.
Congressional District: 7
Key Personnel: Pres. & C.E.O., Ted A. Beattie; Chm. Bd., Sarah Nava Garvey; C.O.O., Michael Delfini; Vice Pres. Facilities, Robert Wengel; Vice Pres. Legislation & Regulation, Jim Robinett; Vice Pres. Devel., Sandy Marek; Sr. Vice Pres. Global Field Expeditions, Cheryl Mell; Sr. Vice Pres. Human Resources, Nancy Anschel; Exec. Vice Pres. & C.F.O., Joyce Simon; Exec. Vice Pres. Mktg., Guest Experiences & Sales, Amy Ritter Cowen; Exec. Vice Pres. Animal Care & Training, Ken Ramirez; Exec. Vice Pres. External Affairs & Communs., Roger Germann; Museum Gift Store Dir., Alan Criss.
Personnel Profile: Full-Time Paid 270; Part-Time Paid 115; Part-Time Volunteers 750.
Governing Authority: nonprofit organization. Parent Institution: Shedd Aquarium Society. Tax-exempt: 501(c)(3).
Institution Type/Description: Aquarium.
Collections: 32,500 animals representing more than 1,500 species; invertebrates; fish; birds; reptiles; amphibians; mammals; 90,000-gallon Caribbean exhibit; Oceanarium: features whales, dolphins, sea otters, sea lions & penguins in re-creations of natural habitats; Amazon Rising features 14 floor-to-skylight habitats; Indo-Pacific shark & coral reef exhibit features shark, coral, & other marine life; Wild Reef includes 25 sharks & live coral.
Research Fields: marine mammals; coral reefs; wetlands; tropical marine fish propagation; snails; iguanas.
Facilities: 15,000-vol. library; 400 periodicals; auditorium; restaurant; cafeteria; classrooms; amphitheater. Educational materials & gift items for sale.
Activities: educational programs & materials for school groups; college credit courses; evening & weekend classes for adults & children; special exhibits on aquatic topics presented semiannually; nature films; lectures; performances; aquarium orientation; volunteer training programs; marine mammal shows; Caribbean feeding; natural history trips.
Publications: WaterShedd; annual report.
Hours & Admission Prices: Memorial Day to Labor Day daily 9-6; Sept.-May Mon.-Fri. 9-5, Sat.-Sun. 9-6. Total Experience Pass (Waters of the World, Amazon Rising, Wild Reef, Oceanarium, Polar Play Zone, & one 4-D Experience): adults $28.95, youth $21.95. Shedd Pass (Waters of the World, Amazon Rising, Wild Reef, Oceanarium, & Polar Play Zone): adults $24.95, youth $17.95. Aquarium only: adults $8, youth $6; members no charge. Closed Christmas. &
Attendance: 2,000,000 (estimated)
Membership: Family $175; Family Plus $190; Associate $250-$499; Sponsor $500-$999; Ambassador $1,000.

LATVIAN FOLK ART MUSEUM, 4146 N. Elston Ave., Chicago, IL 60618-1828. Tel.: 773-588-2085. Fax: 773-588-3405. Facebook: Latvian Folk Art Museum.

E-mail: latvianfolkart@gmail.com
Founded: 1978.
Key Personnel: Vice Pres., Dace Kezbers; Pres. (V), Mara Jauntirans.
Personnel Profile: Part-Time Volunteers 2.
Governing Authority: Tax-exempt.
Institution Type/Description: Folk Art Museum.
Collections: textiles; ceremonial costumes; musical instruments.
Major Exhibits: Displaced to this Place-Baltic DP Experiences After WW II (T), 1/14-12/14.
Hours & Admission Prices: By appointment. No charge; donations accepted. Closed holidays.

LINCOLN PARK CONSERVATORY, 2391 N. Stockton Dr., Chicago, IL 60614-3419. Tel.: 312-742-7736 & 746-5995. Fax: 312-742-5619.
Web Site: www.chicagoparkdistrict.com
Founded: 1890.
Congressional District: 9
Key Personnel: Dir. Conservatories, Mary Eysenbach; Gen. Foreman, Miguel del Valle; Foreman, Don Fuller; Foreman, Rose Bialis; Horticulturalist, Steve Meyer.
Personnel Profile: Full-Time Paid 14; Part-Time Paid 2; Part-Time Volunteers 250.
Governing Authority: municipal. Parent Institution: Chicago Park District, 541 N. Fairbanks, Chicago, IL 60611. Tax-exempt.
Institution Type/Description: Horticultural Conservatory: housed in 1892 Lincoln Park Conservatory.
Collections: botanical collection; cycads; fern grotto; epiphytes including orchids and bromeliads; tropicals including palms, fruit trees, shrubs and ground covers; orchids.
Facilities: conservatory; outdoor gardens; greenhouses.
Activities: annual flower shows; outdoor summer gardens.
Publications: brochures; on-site map.
Hours & Admission Prices: Daily 9-5; guided tours by appointment. No charge; donations accepted. &

Attendance: 460,000 (estimated)

LINCOLN PARK ZOOLOGICAL GARDENS, (M), 2001 N. Clark St., Chicago, IL 60614-4757. Mailing Address: P.O. Box 14903, Chicago, IL 60614-0903. Tel.: 312-742-2000. Fax: 312-742-2137.
E-mail: lpz@lpzoo.org
Web Site: www.lpzoo.org
Founded: 1868.
Congressional District: 5
Key Personnel: Zoo Dir., Pres. & C.E.O., Kevin J. Bell; Bd. Chair, John Ettelson; Sr. Dir. Huruis Center, Leah Melber; Sr. Vice Pres. Operations, Troy Baresel; Sr. Vice Pres. Capital & Programmatic Planning, Steven Thompson, Ph.D.; Vice Pres. Devel., Christine Zrinsky; Vice Pres. Human Rels. & Administration, Linda Leadbitter; Dir. Education, Allison Price; Vice Pres. Conservation, Lisa Faust; Vice Pres. Communications & Public Affairs, Marybeth Johnson; Cur. Large Mammals, Mark Kamhout; Cur. Small Mammals & Reptiles, Diane Mulkerin; Vice Pres. Animal Care, Megan Ross, Ph.D.; Cur. Birds, Sunny Nelson; Cur. Primates, Maureen Leahy; Gen. Cur., Dave Bernier; Veterinarian, Kathryn Gamble, D.V.M.; Museum Shop Mgr., Marla Molinelli.
Personnel Profile: Full-Time Paid 220; Part-Time Paid 118; Part-Time Volunteers 370; Interns 15.
Governing Authority: society. Parent Institution: Lincoln Park Zoological Society. Tax-exempt: 501(c)(3).
Institution Type/Description: Zoo.
Collections: 1,275 specimens of mammals, birds, reptiles, amphibians, invertebrates.
Research Fields: zoo medicine; animal nutrition; reproduction; comparative pathology; primatology; small population biology; epidemiology; field conservation; small population management; endocrinology; human/animal interactions in urban settings.
Facilities: 49 acres of exotic animal exhibits & educational opportunities.
Activities: volunteer programs on grounds; formally organized education programs for children, adults, students & teachers (K-12); website programs.
Publications: quarterly, Lincoln Park Zoo Magazine; quarterly, public programs mailer; brochure, group sales; catalogue, fall & winter/spring programs.
Hours & Admission Prices: Winter: daily 9-5; Summer: Mon.-Fri. 9-6, Sat.-Sun. & holidays 9-7. No charge. &

Attendance: 3,500,000 (estimated)
Membership: Individual $65; Household $90; Safari $175; Curators' Circle $365; Explorers' Circle $500; Ecologists' Circle 1,000; Conservators' Circle $1,500.

THE LITHUANIAN MUSEUM, 5600 S. Claremont Ave., Chicago, IL 60636-1039. Tel.: 773-434-4545. Fax: 773-434-9363.
E-mail: info@lithuanianresearch.org
Web Site: www.lithuanianresearch.org
Founded: 1989.
Congressional District: 3
Key Personnel: Pres., Augstinas Idzelis, Ph.D., J.D.; Dir. & Dir. Medical Museum, Skirmante Miglinas; Financial Dir. & Chm. (V), Dr. Robert Vitas, Ph.D.; Public Rels., Kristina Lapienyte; Dir. Technical Svcs., Thomas R. Miglinas

Governing Authority: nonprofit organization. Parent Institution: The Lithuanian Research & Studies Center, Inc. Subsidiary Museum: Lithuanian Museum of Medicine. Tax-exempt: 501(c)(3).
Institution Type/Description: Lithuanian History Museum.
Collections: arts & crafts; history; militaria; music; medicine; coins; stamps; biography; posters; costumes; textiles; crosses; art; graphics; amber jewelry & objects.
Research Fields: Lithuanian art, culture & history; Lithuanian immigration to the U.S.
Facilities: 100,000-vol. library of books & periodicals available for use by public; educational facilities; 2,000 sq. ft. exhibit space.
Activities: guided tours; loan, temporary & traveling exhibitions.
Publications: books, Lithuanian-American Medical Directory; Lithuania & the United States: The Establishment of State Relations; Lithuanica Collections in European Research Libraries: A Bibliography; The Samogitian Crusade; journal reprint, Varpas, 1889-1905; The Baltic Crusade, second revised & enlarged edition; Samogitian Crusade; Tannenberg and After; Livonian Crusade.
Hours & Admission Prices: Wed.-Fri. 9-3; other times by appointment. No charge; donations accepted.
Attendance: 6,000 (estimated)
Membership: Individual $50; Supporter $100; Honorary $500; Perpetual $1,000.

* **LOYOLA UNIVERSITY MUSEUM OF ART, (M),** 820 N. Michigan Ave., Chicago, IL 60611-2147. Tel.: 312-915-7600. Fax: 312-915-6388. Facebook: LUMA Chicago.
E-mail: luma@luc.edu
Web Site: www.luc.edu/luma
Founded: 2005.
Congressional District: 42
Key Personnel: Chm. (V), Matthew Pattilo, Esq.; Dir. Cultural Affairs & Devel., Pamela E. Ambrose; Devel., Ann Fruland; Cur., Jonathan Canning; Cur. Education, Ann M. Meehan; Head Preparator, Tim Duncan; Administrative Mgr. & Museum Shop Mgr., Guadalupe Pastenes; Registrar, Lisa Marshall; Events Coord., Mary Arhondonis; Graphic Designer, Jessica Kleoppel; Information Asst., Noreen Jones.
Personnel Profile: Full-Time Paid 10; Part-Time Paid 18; Part-Time Volunteers 30; Interns 9.
Operating Expenses: 1,164,000
Operating Income: 1,076,000
Governing Authority: private university. Parent Institution: Loyola University Chicago. Tax-exempt: 501(c)(3).
Institution Type/Description: Art Museum.
Collections: Martin D'Arcy, S.J. Collection of Medieval, Renaissance & Baroque art; LUMA Collection of modern & contemporary art.
Major Exhibits: Art & Faith of the Creche: The Collection of James and Emilia Govan (T), 11/13-1/14; Crossings & Dwellings, 714-11/14; Edward Gore: Elegant Enigmas, 2/18/14-6/2/14.
Facilities: library of art & architecture books; 25,000 sq. ft. exhibit space; lecture hall; children's gallery. Museum-related items for sale.
Activities: concerts; docent program; films; formal education programs; guided tours; lectures; loan, traveling & temporary exhibitions; internships; special events.
Publications: quarterly members' magazine, The LUMANARY; exhibition catalogs.
Hours & Admission Prices: Tues. 11-8, Wed.-Sun. 11-6. Adults $8, senior citizens $6, students under 25 with ID $2; discounts to groups, AAM, AAMUG NARM & ICOM members; military families, children under 18, members of the clergy, museum professionals, LUC Staff, faculty, students & Tues. no charge. Closed New Year's Eve & Day; Good Friday; Easter; Independence Day; Thanksgiving; Christmas Eve & Day. &

Attendance: 15,000 (estimated)
Membership: Individual $75; Family $125; Supporter $250; Contributor $500; Sustainer $1,000; Patron $5,000; Benefactor $10,000.

MAYA POLSKY GALLERY, 215 W. Superior St., Chicago, IL 60654-3528. Tel.: 312-440-0055. Fax: 312-440-0501.
E-mail: info@mayapolskygallery.com
Institution Type/Description: Art Gallery.
Collections: paintings; sculpture.
Hours & Admission Prices: Tues.-Fri. 10-5, Sat. 11-5.

MUSEUM OF BROADCAST COMMUNICATIONS, 360 N. State St., Chicago, IL 60654-5411. Tel.: 312-245-8200. Fax: 312-245-8207.
Institution Type/Description: Broadcast History Museum.
Collections: American radio & television history.
Facilities: cafe. Museum-related items for sale.

Activities: view vintage shows; tape a newscast in the television studio.
Hours & Admission Prices: Temporarily closed for relocation.

✷ MUSEUM OF CONTEMPORARY ART CHICAGO, (M), 220 E. Chicago Ave., Chicago, IL 60611-2644. Tel.: 312-280-2660. Fax: 312-799-3510. TDD: 312-397-4006.
E-mail: webmaster@mcachicago.org
Web Site: www.mcachicago.org
Founded: 1967.
Congressional District: 7
Key Personnel: Dir. Pritzker, Madeleine Grynsztejn; Chm., Mary Ittelson; Deputy Dir., Janet Alberti; C.F.O., Peter Walton; Dir. Admin., Helen Dunbeck; Dir. Collections & Exhibitions, Jennifer Draffen; James W. Alsdorf Chief Cur., Michael Darling; Dir. Facilities Management, Don Meckley; Dir. Performance Programs, Peter Taub; Dir. Retail Operations, Mark Millmore; Dir. Education, Erika Hanner; Dir. Media, Karla Loring; Dir. Devel., Lisa Key; Dir. Security, Eddie Sallie.
Personnel Profile: Full-Time Paid 99; Part-Time Paid 70; Part-Time Volunteers 200; Interns 85.
Governing Authority: private; nonprofit organization; board of trustees. Tax-exempt: 501(c)(3).
Institution Type/Description: Art Museum & Center.
Collections: post-World War II painting; sculpture; graphics; photography; mixed media; artists' conceptual works; artists' books; videos.
Facilities: 16,000-vol. library; artist files; catalogues; educational facilities; video orientation space; cafe. Books & gift items for sale.
Activities: temporary exhibitions & rotating installations of the permanent collection of contemporary art & photography; film series; performance art; lecture series; tours; educational outreach program; seminars; teacher workshops; children's art programs.
Publications: exhibit catalogs; gallery guides; quarterly calendar of events; museum brochure.
Hours & Admission Prices: Tues. 10-8, Wed.-Sun. 10-5. Adults $12, students with ID & seniors $7; discounts to AAM & ICOM members; members, military, Tues. and children 12 & under no charge. Closed New Year's Day; Thanksgiving; Christmas. ♿
Attendance: 275,000 (estimated)
Membership: Student & Out of Town $30; Senior $40; Individual $60; Household $75.

✷ MUSEUM OF CONTEMPORARY PHOTOGRAPHY, CO-LUMBIA COLLEGE CHICAGO, (M), 600 S. Michigan Ave., Chicago, IL 60605-1900. Tel.: 312-663-5554. Fax: 312-369-8067.
E-mail: mocp@colum.edu
Web Site: www.mocp.org
Founded: 1976.
Congressional District: 7
Key Personnel: C.E.O., Alan Turner; Pres., Warrick L. Carter; Chm. (V), Bill Hood; Dir., Natasha Egan; Assoc. Dir. & Cur., Karen Irvine; Head Operations, Stephanie Conaway.
Personnel Profile: Full-Time Paid 7; Part-Time Paid 4; Part-Time Volunteers 1; Interns 12.
Governing Authority: college; nonprofit organization. Parent Institution: Columbia College Chicago. Tax-exempt: 501(c)(3).
Institution Type/Description: Photography Museum.
Collections: works of contemporary American photographers, including Zeke Berman, Ruth Bernhard, Dawoud Bey, Marsha Burns, Harry Callahan, Carl Chiarenza, Larry Clark, Linda Connor, Eileen Cowin, Barbara Crane, Louise Dahl-Wolfe, Bruce Davidson, Roy DeCarava, Elliott Erwitt, Walker Evans, Lee Friedlander, Emmet Gowin, Robert Heinecken, Barbara Kasten, Mark Klett, Dorothea Lange, Danny Lyon, Susan Meiselas, Ray Metzker, Nicholas Nixon, Anne Noggle, Irving Penn, Aaron Siskind, Mike & Doug Starn, Ruth Thorne-Thomsen, Jerry N. Uelsmann, James Van Der Zee, Carrie Mae Weems, William Wegman, Minor White, Garry Winogrand, Joel-Peter Witkin.
Research Fields: contemporary photography.
Facilities: classrooms; print study room; research & viewing center.
Activities: guided tours; lectures & symposia; traveling exhibitions originated by the museum; video, computer & participatory programs; publications; membership benefits; fine print program.
Publications: annual exhibition catalogues.
Hours & Admission Prices: Mon.-Wed. & Fri.-Sat. 10-5, Thurs. 10-8, Sun. 12-5. No charge; donations accepted. Closed New Year's Eve & Day; Martin Luther King Jr. Day; Memorial Day; Independence Day; Labor Day; Thanksgiving weekend; Christmas Eve, Day & week. ♿
Attendance: 50,000 (estimated)
Membership: Student $20 Individual $35; Dual $50; Group Photo $100; Scholar $300; Collector $500; Director's Circle $1,000.

✷ MUSEUM OF SCIENCE AND INDUSTRY, (M), 5700 S. Lake Shore Dr., Chicago, IL 60637-2003. Tel.: 773-684-1414. Fax: 773-684-7141. TDD: 773-684-3323.
Web Site: www.msichicago.org
Founded: 1926.
Congressional District: 1
Key Personnel: Pres. & C.E.O., David R. Mosena; Asst. to Pres., Eileen M. Cabrera; Chm., Robert S. Morrison; Vice Pres. Exhibits & Collections, Kurt Haunfelner; Vice Pres. Finance & Administration, Bob Fisher; Vice Pres. Capital Campaign, Shannon Alexander; Vice Pres. Mktg. & Public Rels., Valerie Waller; Vice Pres. Education & Guest Svcs., Andrea Ingram; Dir. Temporary Exhibits, Anne Rashford; Dir. Business Operations, Andy Zakrajsek; Dir. Membership, Susan Rawls.
Personnel Profile: Full-Time Paid 287; Part-Time Paid 73; Part-Time Volunteers 605.
Governing Authority: nonprofit organization. Tax-exempt: 501(c)(3).
Institution Type/Description: Science & Technology Museum: housed in classic Greek structure constructed as the Palace of Fine Arts for the World's Fair Columbian Exposition of 1893 in Chicago, located on the site of the Exposition in Jackson Park.
Collections: U-505 submarine; hands-on exhibits; aviation; scientific principles, technological applications & social implications; industry; Spirit of America; Empire 999; Pioneer Zephyr; Apollo 8 capsule.
Facilities: restaurants. Books & other museum-related items for sale.
Activities: guided tours; lectures; films; children & family classes; science fairs; arts festivals; inter-museum loan, permanent, temporary & traveling exhibitions; live science experiences; teacher development programs; student learning labs & video conferencing programs; group tours.
Publications: e-newsletters; member magazine, Momentum; teacher education guides; annual education calendars.
Hours & Admission Prices: Summer: Mon.-Sat. 9:30-5:30, Sun. 11-5:30; Winter: Mon.-Sat. 9:30-4, Sun. 11-4. Adults $13, seniors $12, children 3-11 $9; discounts to Chicago residents, and AAM & ICOM members; Illinois teachers, active military, police & fire personnel, members & children under 3 no charge. Omnimax Theater: adults $8, seniors $7, children 3-11 $6; children under 3 (on adult's lap) no charge. Combination tickets available. Closed Christmas. ♿
Attendance: 1,675,109 (accurate)
Membership: Associate Senior & Student $55; Bachelors Individual $70; Masters Family $105; Doctorate Family $175.

NATIONAL HELLENIC MUSEUM, (M), 333 S. Halsted St., Chicago, IL 60661-5415. Tel.: 312-655-1234. Fax: 312-655-1221.
E-mail: info@hellenicmuseum.org
Web Site: www.nationalhellenicmuseum.org
Formerly: Hellenic Museum & Cultural Center
Founded: 1983.
Congressional District: 7
Key Personnel: Dir., Connie Mourtoupalas; Dir. External Affairs, Toula Georgakopoulas; Collections Mgr. & Registrar, Christopher Helms; Dir. Museum Experience & Cur., Bethany Fleming; Accounting Asst., Naomi Janovec-Easley; Dir. Devel., Tula Gogolak; Dir. Mktg., Amelia Dellos; Dir. Operations, Kevin Miller; Museum Educator, Elise Freed-Brown; Mgr. Library & Oral History, John Anagnostopoulos; Membership & Volunteer Coord., Hannah Imber; Museum Shop Mgr., Michelle Diakatos.
Personnel Profile: Full-Time Paid 15; Part-Time Paid 2; Part-Time Volunteers 50; Interns 6.
Governing Authority: nonprofit. Tax-exempt.
Institution Type/Description: Greek History, Culture & Art Museum.
Collections: Greek history, culture, art, & immigration; life of contemporary Greek-American; period artifacts.
Research Fields: Greek & Greek-American history; immigration; Greek language & Greek publications.
Facilities: classroom; special events hall.
Activities: children's programs & tours; lectures; public programs; seminars; symposia.
Publications: newsletter; brochures.
Hours & Admission Prices: Mon., Wed. & Fri. 10-5, Thurs. 10-8, Sat.-Sun. 11-5. Adults $10; members no charge. ♿
Attendance: 10,000 (estimated)
Membership: Senior $25; Individual $50; Family $100; Fellow $250; Patron $500; Benefactor $1,000.

NATIONAL ITALIAN AMERICAN SPORTS HALL OF FAME, 1431 W. Taylor St., Chicago, IL 60607-4625. Tel.: 312-226-5566. Fax: 312-226-5678.
E-mail: george@niashf.org
Web Site: www.niashf.org/
Founded: 1977.

Key Personnel: Founder & Chm., George Randazzo.
Personnel Profile: Full-Time Paid 6; Full-Time Volunteers 1; Part-Time Paid 3; Interns 1.
Governing Authority: nonprofit. Tax-exempt.
Institution Type/Description: Sports Museum.
Collections: history & heritage of Italian Americans in sports.
Facilities: rental facilities. Museum-related items for sale.
Activities: special events; boxing matches; guest speakers. Annual Event: NIASHF Induction and Awards Ceremony.
Publications: quarterly magazine, Red, White and Green; quarterly newsletter, NIASHF News.
Hours & Admission Prices: Mon.-Fri. 9-5, Sat.-Sun. 11-4; other times by appointment. Adults $5, seniors, children & groups over 15 $3. &

✳ **NATIONAL MUSEUM OF MEXICAN ART,** 1852 W. 19th St., Chicago, IL 60608-2706. Tel.: 312-738-1503. Fax: 312-738-9740.
E-mail: carlos@nationalmuseumofmexicanart.org
Web Site: www.nationalmuseumofmexicanart.org
Formerly: Mexican Fine Arts Center Museum
Founded: 1982.
Congressional District: 1
Key Personnel: C.E.O., Pres. & Founder, Carlos Tortolero; Vice Pres., Juana Guzman; Vice Pres. & C.F.O., Martin Sandoval; Chm., Martin R. Castro; Cur. Permanent Collections, Rebecca D. Meyers; Dir. Devel., Randy Adamsick; Chief Cur. & Dir. Visual Arts, Cesareo Moreno; Performing Arts Dir., Jorge Valdivia; Museum Shop Mgr., Raquel Rios.
Personnel Profile: Full-Time Paid 38; Part-Time Paid 35; Part-Time Volunteers 32; Interns 11.
Governing Authority: nonprofit organization. Tax-exempt: 501(c)(3).
Institution Type/Description: Art Museum.
Collections: Mexican art, folk art & culture as it manifests itself inside and outside of Mexico; photography; graphic arts; contemporary art.
Facilities: 48,000 sq. ft. exhibit space; educational facilities; 11,000 sq. ft. building off-site which houses a youth museum and WRTE-FM Radio Arte, a youth run radio station. Gift items for sale.
Activities: guided tours; lectures; arts festivals; organized education programs for children; participatory, loan, temporary & traveling exhibitions; performing arts events.
Publications: catalogues.
Hours & Admission Prices: Tues.-Sun. 10-5. No charge; donations accepted. Closed major holidays. &
Attendance: 168,000 (accurate)
Membership: Students & Senior Citizens $20; Individual $35; Family $55.

NATIONAL PUBLIC HOUSING MUSEUM, 750 S. Halsted St., Ste. 843 MC 117, Chicago, IL 60607. Tel.: 312-996-0738. Fax: 312-996-0708.
E-mail: info@nphm.org
Web Site: www.nphm.org
Founded: 2007.
Congressional District: 7
Key Personnel: Dir., Keith L. Magee.
Personnel Profile: Full-Time Paid 4; Part-Time Paid 3; Part-Time Volunteers 15; Interns 3.
Governing Authority: private; nonprofit organization. Tax-exempt: 501(c)(3).
Institution Type/Description: History Museum.
Collections: local history & culture; oral histories of public housing residents; photographs; personal artifacts.
Research Fields: affordable housing; public housing planning & design; public history; social history of public housing.
Activities: formal education programs for adults & children; lectures; participatory, temporary & traveling exhibitions; theater; study clubs. Annual Event: Afternoon of Good Times Benefit.
Hours & Admission Prices: Opening 2014. No charge; donations accepted.
Membership: Senior & Student $15; Associate $25; Contributor $50; Friend $100; Supporter $250.

THE NATIONAL SOCCER HALL OF FAME, 1801 S. Prairie Ave., Chicago, IL 60616-1319. Tel.: 607-432-3351. Fax: 607-432-8429.
E-mail: info@soccerhall.org
Web Site: www.soccerhall.org
Founded: 1981.
Congressional District: 25
Key Personnel: Pres. & C.O.O., Stephen H. Baumann; Chm. (V), Douglas B. Willies; Vice Pres. Business Devel. & Operations, Jonathan Ullman; Dir.

Devel., Kathryn Dailey; Dir. Museum & Archives, Jack Huckel; Museum Shop Mgr., Mary Myers.
Personnel Profile: Full-Time Paid 8; Full-Time Volunteers 1; Part-Time Paid 20; Part-Time Volunteers 20; Interns 5.
Governing Authority: nonprofit organization. Tax-exempt: 501(c)(3).
Institution Type/Description: Sports History Museum.
Collections: soccer history; uniforms; balls; trophies; league records; archives; videotapes; photographs; media guides; films; paintings.
Research Fields: history of soccer.
Facilities: 13,500-vol. library pertaining to soccer; 40,000 sq. ft. exhibit space. Museum-related items for sale.
Activities: temporary exhibitions; organized education programs for undergraduate or graduate college students affiliated with State University of New York at Oneonta. Annual Events: Soccer Tournaments & Clinics; Hall of Fame Induction; Soccer Camps.
Publications: newsletter, Hall of Famer; online newsletter.
Hours & Admission Prices: Call for hours. Adults $12.50, students $9.50, senior citizens & veterans $8.50, youth 6-12 $7.50; discounts to military, AAM, NSCAA, NISOA, AYSO card members; children under 6 no charge. &
Attendance: 44,000 (estimated)

NATIONAL VETERANS ART MUSEUM, 4041 N. Milwaukee Ave., Chicago, IL 60641-1834. Tel.: 312-326-0270. Fax: 312-326-9767.
E-mail: info@nvvam.org
Web Site: www.nvvam.org
Formerly: National Vietnam Veterans Art Museum
Founded: 1981.
Key Personnel: Gen. Mgr. & Museum Shop Mgr., Ted Stanuga; Dir., Levi Moore; Chm. (V), Chief Cur. & Head Art Committee, Michael Helbing; Cur. & Project Devel., Aaron Hughes; Dir. Education & Outreach, Jennifer Komorowski; Museum Shop Mgr., Susan Zieliwski.
Personnel Profile: Part-Time Paid 6; Part-Time Volunteers 6; Interns 1.
Institution Type/Description: Military Art Museum.
Collections: paintings; photographs; sculpture.
Facilities: cafe. Museum-related items for sale.
Publications: newsletter, Artifacts.
Hours & Admission Prices: Tues.-Sat. 10-5. Adults $10, students $7; discounts to groups & active military; members no charge. &
Attendance: 8,112 (estimated)
Membership: Individual $50; Family $100.

ORIENTAL INSTITUTE MUSEUM, UNIVERSITY OF CHI-CAGO, (M), 1155 E. 58th St., Chicago, IL 60637-1569. Tel.: 773-702-9520. Fax: 773-702-9853.
E-mail: oi-museum@uchicago.edu
Web Site: oi.uchicago.edu
Founded: 1894.
Congressional District: 1
Key Personnel: Dir., Gil Stein; Head Museum Education & Pub. Programs, Catherine Kenyon; Conservator, Laura D'Alessandro; Archivist, John Larson; Registrar & Cur., Helen McDonald; Gift Shop Mgr., Denise Browning; Asst. Conservator, Alison Whyte; Security Supvr., Jason Barcus.
Personnel Profile: Full-Time Paid 12; Part-Time Paid 22; Part-Time Volunteers 107; Interns 1.
Governing Authority: university. Parent Institution: University of Chicago, 5801 Ellis Ave. Tax-exempt: 501(c)(3).
Institution Type/Description: Archaeology, Ancient History & Art Museum.
Collections: art & archaeology of the Ancient Near East, Egypt, Iraq, Iran, Israel, Syria, Cyprus, Palestine, Turkey, Nubia, Early Christian, Islamic material; manuscript collections.
Research Fields: Ancient Near Eastern archaeology & history, sculpture, decorative arts; Ancient Near Eastern languages; Islamic languages & civilization.
Facilities: 25,000-vol. research library; archives of Oriental Institute field expeditions; 65,000 photos; 8,600 slides of the ancient near East; 275-seat auditorium; classrooms. Books & other museum-related items for sale.
Activities: guided tours; lectures; films; formally organized education programs for adults, children, undergraduate & graduate students affiliated with University of Chicago; docent program; inter-museum loan, permanent & temporary exhibitions.
Publications: guide; occasional pamphlets; exhibition catalogs; museum brochure.
Hours & Admission Prices: Tues. & Thurs.-Sat. 10-6, Wed. 10-8:30, Sun. 12-6. No charge; donations accepted. &
Attendance: 60,113 (accurate)
Membership: Students $20 p.a. (USA only); Seniors, UC Faculty or Staff & Long-distance members $40 p.a.; Basic $50 p.a.; Sustaining $75 p.a.;

Supporting $100 p.a.; Contributing $250 p.a.; Sponsoring $500 p.a.; Breasted Patron $1,000 p.a.; Director's Circle $2,500 p.a.

PACKER SCHOPF GALLERY, 942 W. Lake St., Chicago, IL 60607. Tel.: 312-226-8984.
E-mail: packer@packergallery.com
Institution Type/Description: Art Gallery.
Collections: works by contemporary artists.
Hours & Admission Prices: Tues.-Sat. 11-5:30.

THE PALETTE & CHISEL, 1012 N. Dearborn, Chicago, IL 60610-2804. Tel.: 312-642-4400. Fax: 312-642-4317.
E-mail: fineart1012@sbcglobal.net
Web Site: www.paletteandchisel.org
Founded: 1895.
Key Personnel: Exec. Dir., William Ewers; Pres. (V), Dayle Sazonoff; Vice Pres., Leslie Outten; Treas., Christine Sauser; Corporate Sec., Christine Jones; Corresponding Sec., Barbara Graefen.
Personnel Profile: Full-Time Paid 1; Part-Time Paid 2; Part-Time Volunteers 2.
Governing Authority: nonprofit organization. Tax-exempt: 501(c)(3).
Institution Type/Description: Art Academy: housed in c.1875 double bay Italianate mansion.
Collections: club archives 1895-present; c.1900-present artwork from members.
Facilities: library of donated art books; painting & sculpture classrooms.
Activities: arts festivals; formal education programs for adults; guided tours; lectures; temporary exhibitions. Annual Events: auction; juried shows; workshops & lectures; all day modeling marathons on New Year's Day, Labor Day & Memorial Day.
Publications: quarterly newsletter, The Cowbell.
Hours & Admission Prices: Mon.-Fri. 1-5; other times by appointment. No charge for exhibitions; fees for classes; workshops & lectures; discounts for members. Office closed New Year's Day; Memorial Day; Independence Day; Labor Day; Thanksgiving; Christmas Eve & Day.
Attendance: 750 (estimated)
Membership: Patron $50, $125, $1,000; Non-Resident Artist (Basic) $45; Non-Resident Artist (Workshop) $150; Resident Artist $360.

THE PAPERWEIGHT MUSEUM, 410 S. Michigan Ave., Chicago, IL 60605. Tel.: 831-427-1177. Fax: 831-427-0111.
E-mail: lselman@paperweight.com
Web Site: www.theglassgallery.com
Institution Type/Description: Paperweight Museum.
Collections: paperweights.
Hours & Admission Prices: Mon.-Fri. 9-5, Sat. 10-4, Sun. by appointment.

PERIMETER GALLERY, 210 W. Superior St., Chicago, IL 60654. Tel.: 312-266-9473. Fax: 312-266-7984.
E-mail: perimeterchicago@perimetergallery.com
Web Site: perimetergallery.com
Founded: 1982.
Key Personnel: Dir., Frank Paluch; Asst. Dir., Scott Ashley; Registrar, Holly Sabin
Institution Type/Description: Art Gallery.
Collections: works of contemporary art including paintings; sculpture & works on paper.
Hours & Admission Prices: Tues.-Sat. 10:30-5:30. No charge.

POLISH MUSEUM OF AMERICA, 984 N. Milwaukee Ave., Chicago, IL 60642-4101. Tel.: 773-384-3352 & 3731. Fax: 773-384-3799.
E-mail: pma@polishmuseumofamerica.org
Web Site: www.polishmuseumofamerica.org
Founded: 1935.
Congressional District: 4
Key Personnel: Chm., Joseph A. Drobot, Jr.; Dir., Mr. Jan M. Lorys; Pres. (V), Maria Ciesla; Archivist, Halina Misterka; Sec., Geraldine Balut Coleman; Treas., Mitchelle Kmiec; Librarian, Malgorzata Kot; Museum Shop Mgr., Mary Jane Robles.
Personnel Profile: Full-Time Paid 6; Part-Time Paid 6; Part-Time Volunteers 200; Interns 2.
Governing Authority: nonprofit. Parent Institution: Polish Roman Catholic Union. Subsidiary Institution: PRCUA. Tax-exempt.
Institution Type/Description: Ethnic Museum.
Collections: 1,400 originals of Polish & Polish American artists insignia & drawings of Polish kings; costumes; religious artifacts; Polish military memorabilia; memorabilia of T. Kosciuszko; Kossak paintings; memorabilia of Paderewski; folk art; sculptures; manuscripts; periodicals; medals; coins; large part of Polish pavilion from the 1939 New York World's Fair.
Research Fields: Polish & Polish American culture.
Facilities: 62,000-vol. library of publications pertaining to Poland, by American-Polish authors & by Polish-American publishing firms; slides, strips, maps, photos & phono records available for research on the premises only.
Activities: art contests; guided tours; rotating exhibitions; Polish classical concerts; lectures.
Publications: quarterly periodical, Polish Museum of America; newsletter.
Hours & Admission Prices: Museum: Fri.-Wed. 11-4. Library: Mon.-Tues. & Fri.-Sat. 10-4, Wed. 1-7. Donation Requested: adults $5, students & senior citizens $4, children under 12 $3. &
Attendance: 4,593 (accurate)
Membership: Individual $25; Family $40; Contributing $50; Supporting $100; Sustaining $500; Lifetime $1,000.

THE PRITZKER MILITARY LIBRARY, 104 S. Michigan Ave., Chicago, IL 60603. Tel.: 312-374-9333. Fax: 312-374-9314.
E-mail: info@pritzkermilitarylibrary.org
Web Site: www.pritzkermilitarylibrary.org
Founded: 2003.
Key Personnel: Pres. & C.E.O., Kenneth Clarke; Chm. (V), Col. J.N. Pritzker, IL ARNG (Ret.); Chief Librarian, Theresa A.R. Embrey, MLIS
Personnel Profile: Full-Time Paid 20; Part-Time Paid 50; Interns 6.
Governing Authority: Tax-exempt.
Institution Type/Description: Library & Gallery.
Collections: books, videos, posters, uniforms, rare coins, medals & stamps relating to military history; ship models; military equipment.
Facilities: library.
Publications: quarterly publication, Frontline.
Hours & Admission Prices: Tues.-Wed. & Fri.-Sat. 10-4, Thurs. 10-7. Adults $5; members & active duty military with ID no charge. &
Membership: Student $50; Educator & Active Military $95; Associate $125-$199; Household $200-$399; Household Plus $400-$999. Founder's Club: Member $1,000-$2,499; Bronze $2,500-$4,999; Silver $5,000-$9,999; Gold $10,000-$24,999; Platinum $25,000-$49,999; Diamond $50,000.

THE RENAISSANCE SOCIETY AT THE UNIVERSITY OF CHICAGO, 5811 S. Ellis Ave., Rm. 418, Chicago, IL 60637-1404. Tel.: 773-702-8670. Fax: 773-702-9669
E-mail: haberman@uchicago.edu
Web Site: www.renaissancesociety.org
Founded: 1915.
Congressional District: 1
Key Personnel: Dir. & Cur., Susanne Ghez; Pres., Greg Cameron; Dir. Devel., Lori Bartman; Dir. Education & Assoc. Cur., Hamza Walker; Devel. Assoc., Zachary Kaplan; Registrar & Dir. Publications, Karen Reimer; Preparator, Robert Bain; Dir. Mktg., Mia Ruyter; Mktg. Assoc., Yuri Stone; Office Mgr., Lise Haberman.
Personnel Profile: Full-Time Paid 7; Part-Time Paid 2; Part-Time Volunteers 2; Interns 3.
Governing Authority: board of directors. 5811 Ellis Ave., Chicago, IL 60637. Tax-exempt.
Institution Type/Description: Contemporary Art Museum.
Collections: works by contemporary artists.
Activities: guided tours; lectures; films; gallery talks; concerts; temporary & traveling exhibitions.
Publications: catalogs; checklists.
Hours & Admission Prices: Sept.-June Tues.-Fri. 10-5, Sat.-Sun. 12-5. No charge. Closed national holidays. &
Attendance: 20,000 (estimated)
Membership: Member $50; Friend $100; Supporter $500; Associate $1,000; Patron $2,500; Benefactor $5,000; Director's Circle $10,000.

RIDGE HISTORICAL SOCIETY - GRAVER-DRISCOLL HOUSE, 10621 S. Seeley Ave., Chicago, IL 60643-2618. Tel.: 773-881-1675. Facebook: Ridge Historical Society.
E-mail: ridgehistory@hotmail.com
Web Site: www.ridgehistoricalsociety.org
Institution Type/Description: Historical Society Museum. housed in the former home of the Herbert Spencer Graver family; built in 1922.
Collections: local & family history; personal artifacts; period furnishings; photographs.
Facilities: library.
Activities: research; special events; rental facilities; educational programs; lectures.
Hours & Admission Prices: By appointment; call for hours.
Membership: Student under 18 $10; Individual $25; Family $50; Contributor $100; Patron $150; Guarantor $500.

ROY BOYD GALLERY, 739 N. Wells St., Chicago, IL 60654. Tel.: 312-642-1606. Fax: 312-642-2143.
E-mail: info@royboydgallery.com
Web Site: www.royboydgallery.com
Institution Type/Description: Art Gallery.
Collections: paintings; works on paper; sculpture.
Activities: temporary exhibitions.
Hours & Admission Prices: Tues.-Sat. 10-5:30; other times by appointment.

SACKRIDER MUSEUM OF HANDBAGS, (M), Chicago, IL 60647. Mailing Address: 2452 N. Richmond St., Chicago, IL 60647-2618. Tel.: 312-330-7601.
E-mail: info@thesack.org
Web Site: thesack.org
Key Personnel: Founding Dir. & Pres., Jill Brady
Institution Type/Description: Virtual Handbag Museum.
Collections: handbags, purses, chatelaines, reticules, minaudieres, & travel suitcases from 1890 to present.
Hours & Admission Prices: Virtual Museum.

SCHNEIDER GALLERY, 230 W. Superior St., Chicago, IL 60654. Tel.: 312-988-4033. Fax: 312-440-9256.
E-mail: schneidergalleryinfo@gmail.com
Key Personnel: Dir., Martha Schneider; Asst. Dir., Rose Licavoli
Institution Type/Description: Art Gallery.
Collections: photographs.
Hours & Admission Prices: Tues.-Fri. 10:30-5, Sat. 11-5.

SCHOOL OF THE ART INSTITUTE OF CHICAGO, BETTY RYMER GALLERY, 280 S. Columbus Dr., Chicago, IL 60603. Tel.: 312-443-3703. Fax: 312-443-1493.
E-mail: exhibitions-saic@saic.edu
Web Site: www.artic.edu/exhibitions
Institution Type/Description: Art Gallery.
Collections: works by established & emerging artists.
Hours & Admission Prices: Tues.-Sat. 11-6.

SCHOOL OF THE ART INSTITUTE OF CHICAGO, SULLIVAN GALLERIES, 33 S. State St., 7th Fl., Chicago, IL 60603-2809. Tel.: 312-629-6635. Fax: 312-629-6636.
E-mail: exhibitions-saic@saic.edu
Web Site: www.saic.edu/exhibitions
Formerly: School of the Art Institute of Chicago, Gallery 2
Founded: 1984.
Key Personnel: Exec. Dir., Mary Jane Jacob.
Personnel Profile: Full-Time Paid 7; Part-Time Paid 1; Interns 30.
Governing Authority: private college; nonprofit. Subsidiary Institution: Art Institute of Chicago, IL. Tax-exempt: 501(c)(3).
Institution Type/Description: College Art Gallery.
Collections: works of art in all media.
Activities: temporary exhibitions; performances; lectures; screenings by SAIC students, faculty & professional artists.
Hours & Admission Prices: Tues.-Sat. 11-6. No charge. Closed holidays. &
Attendance: 97,098 (accurate)

✳ SMART MUSEUM OF ART, (M), University of Chicago, 5550 S. Greenwood Ave., Chicago, IL 60637-1506. Tel.: 773-702-0200. Fax: 773-702-3121.
E-mail: smart-museum@uchicago.edu
Web Site: smartmuseum.uchicago.edu
Founded: 1974.
Congressional District: 1
Key Personnel: Dir., Anthony G. Hirschel; Deputy Dir. & Chief Cur., Stephanie Smith; Senior Cur., Richard A. Born; Cur. & Assoc. Dir. Academic Initiatives, Anne Leonard; Mgr. Tour & Teacher Initiatives, Lisa Davis; Mgr. Devel. Operations, Jennifer Ruehl; Dir. Finance & Admin., Peg Liput; Chief Preparator, Rudy Bernal; Asst. Registrar, Sara Patrello; Preparator, Ray Klemchuck; Assoc. Dir. Communications, C.J. Lind; Asst. Dir. Hospitality & Special Events, Sarah Polachek; Security & Facilities Mgr., Todd Hengsteler; Security Supvr., Paul Bryan; Assoc. Cur. Contemporary Art, Jessica Moss; Mgr. Devel. Communications, Kate Nardin; Business Mgr., Joyce Norman; Exec. Asst. for Leadership Support, Cindy Hansen; Consulting Cur., Wu Hung; Head Registrar, Angela Steinmetz; Asst. Registrar, Sara Hindmarch; Study Room Supvr., Alice Kain; Dir. Education & Interpretation, Michael Christiano; Dir. Devel. & External Relations, Warren Davis; Assoc. Programs Dir., Erik Peterson; Cafe & Gift Shop Mgr., Kate Kelly; Exec. Asst. Program Support, Sarah Mendelsohn.
Personnel Profile: Full-Time Paid 26; Interns 12.

Governing Authority: university. Parent Institution: University of Chicago, 5801 S. Ellis Ave., Chicago. Tax-exempt: 501(c)(3).
Institution Type/Description: University Art Museum.
Collections: painting, sculpture, decorative arts, drawings, prints, photography & East Asian art ranging from ancient to modern.
Research Fields: related to the permanent collection; special exhibitions.
Facilities: Exhibition catalogues, scholarly monographs in art & art history, posters, slides, postcards & stationery items for sale.
Activities: guided tours; lectures; gallery talks; symposia; colloquia; educational programs; permanent collection; special exhibitions.
Publications: Smart Museum Bulletin; exhibition catalogues; collection handbook.
Hours & Admission Prices: Tues., Wed., Fri.-Sun. 10-5, Thurs. 10-8. No charge; donations accepted. Closed holidays. &
Attendance: 70,183 (accurate)
Membership: University of Chicago Students, Faculty & Staff $35; Smart Partner $50; Smart Partner Plus $150, $500, $1,000, $2,500 & up.

SMITH MUSEUM OF STAINED GLASS WINDOWS, Navy Pier, 600 E. Grand Ave., Chicago, IL 60611-3419. Tel.: 312-595-7437.
Web Site: www.navypier.com
Key Personnel: Cur., Rolf Achilles
Institution Type/Description: Stained Glass Museum.
Collections: over 150 stained glass windows including secular & religious art.
Hours & Admission Prices: See website for hours. No charge.

SPERTUS MUSEUM, 610 S. Michigan Ave., Chicago, IL 60605-1901. Tel.: 312-322-1747 & 1700. Fax: 312-922-3934.
E-mail: museum@spertus.edu
Web Site: www.spertus.edu
Founded: 1968.
Congressional District: 20
Key Personnel: Pres. & C.E.O., Howard A. Sulkin, Ph.D.; Dir., Rhoda Rosen; Registrar & Collections Mgr., Arielle Weininger; Asst. Registrar, Tom Gengler; Sr. Cur., Staci Boris; Sr. Judaica Cur., Felicitas Heimann-Jelinek; Educator, Amanda Friedeman; Dir. Design, Mark Akgulian; Designer, Tracy Kostenbader; Design Asst., Tony Doyle; Exhibition Coord., Sheila Cronin; Museum Shop Mgr., Alana Aldort; Asst. Cur., Sarah Nelson; Asst. Cur. Judaica, Ilana Segal.
Personnel Profile: Full-Time Paid 11; Part-Time Paid 10; Part-Time Volunteers 30; Interns 2.
Governing Authority: college. Parent Institution: Spertus Institute of Jewish Studies. Tax-exempt: 501(c)(3).
Institution Type/Description: Cultural, Art & Children's Museum.
Collections: archives; archaeology; art library; fine arts; Jewish ceremonial arts; ethnology; folklore; contemporary art.
Research Fields: Judaica & pertinent fine arts; Jewish culture & civilization.
Facilities: 1,000-vol. library of books, catalogs & folios relating to Jewish art, graphics & historic material available for research by public; theater; rental facilities; classrooms; cafe. Museum-related items for sale.
Activities: guided tours; lectures; gallery talks; films; musical programs; workshops; docent programs; family programs; community-based outreach programs; formally organized education programs for graduate students; inter-museum loan, permanent & temporary exhibitions.
Publications: exhibition catalogue, The New Authentics.
Hours & Admission Prices: Sun.-Wed. 10-5, Thurs. 10-6, Fri. 10-3. Adults $7, students & seniors $5; children under 5, Tues. 10-12 & Thurs. 2-6 no charge. Closed Jewish & national holidays. &
Attendance: 50,000 (accurate)
Membership: SpertusNet $30-$55; Senior $40-$70; Basic $50-$90; Household $65-$120; Associate $150; Fellow $250; Patron $500; Benefactor $1,000; Angel $5,000.

STEPHEN A. DOUGLAS TOMB, 636 E. 35th St., Chicago, IL 60616-4196. Tel.: 312-225-2620. Fax: 312-225-7855.
Founded: 1865.
Congressional District: 1
Key Personnel: Site Supt., Michael Carson.
Governing Authority: state. Parent Institution: Illinois Historic Preservation Agency. Tel.: 312-814-2070. Tax-exempt.
Institution Type/Description: Historic Site: 1866-1881 Stephen A. Douglas Tomb designed by Leonard W. Volk. Listed on the National Register of Historic Places.
Collections: monument; sculpture; Senator Douglas' life history & tomb.
Activities: guided tours; lectures.
Publications: bimonthly, newsletter, Historic Illinois.
Hours & Admission Prices: Wed.-Sun. 9-5. No charge. Closed New Year's Day; Thanksgiving; Christmas. &
Attendance: 15,000 (estimated)

SWEDISH AMERICAN HISTORICAL SOCIETY, 3225 W. Foster Ave., Chicago, IL 60625-4823. Mailing Address: 3225 W. Foster Ave., Box 48, Chicago, IL 60625-4823. Tel.: 773-538-5722.
E-mail: info@swedishamericanhist.org
Web Site: www.swedishamericanhist.org
Founded: 1948.
Congressional District: 5
Key Personnel: Chm. (V), Kevin Proescholdt; Pres. (V), Philip J. Anderson.
Personnel Profile: Part-Time Paid 1.
Governing Authority: society. Tax-exempt: 501(c)(3).
Institution Type/Description: Historical Society Archives.
Collections: books; periodicals; manuscripts; microfilm; oral histories.
Research Fields: archives; religion; education; journalism; fine arts; government; science; business; industry from 1840 to present day of Swedes in the United States.
Facilities: 3,000-vol. library of ethnic history available on premises. Books for sale.
Activities: annual & spring meetings; tours.
Publications: journal; The Swedish American Historical Quarterly; book.
Hours & Admission Prices: Mon.-Fri. by appointment. No charge. Closed legal holidays. &
Membership: Student $10; Annual $25; Sustaining $50; Donor $100; Benefactor $250; Life $1,000.

SWEDISH AMERICAN MUSEUM, (M), 5211 N. Clark St., Chicago, IL 60640-2101. Tel.: 773-728-8111. Fax: 773-728-8870.
E-mail: museum@samac.org
Web Site: www.swedishamericanmuseum.org
Founded: 1976.
Congressional District: 9
Key Personnel: Exec. Dir., Karin Moen Abercrombie; Pres., Joan Papadopoulous; Museum Shop Mgr., Ann Cutler.
Personnel Profile: Full-Time Paid 5; Part-Time Paid 1; Part-Time Volunteers 54; Interns 5.
Governing Authority: nonprofit organization. Tax-exempt: 170(b)(1)(A).
Institution Type/Description: Swedish History & Immigration Museum.
Collections: paintings, tools & artifacts describing the emigration of the Swedish people to the United States; special displays featuring famous Swedes in this country; furniture & woodcarvings by Swedish Americans; models; 1910 organ.
Research Fields: Swedish pioneer life; contributions of Swedes to development of America.
Facilities: classroom. Museum-related items for sale.
Activities: formally organized educational programs; loan, permanent & traveling exhibitions; concerts; lectures.
Publications: newsletter, Flaggan.
Hours & Admission Prices: Tues.-Fri. 10-4, Sat.-Sun. 11-4. Adults $4, seniors, students & children $3, family $10; members no charge. Children's Museum of Immigration: Tues.-Fri. 1-4, Sat.-Sun. 11-4. &
Attendance: 45,000 (estimated)
Membership: Student & Senior $15; Senior Couple $25; Individual $35; Family $50; Nonprofit Organization & Sustaining $75; Sandburg $100-$249; Linnaeus $250-$520; 521 Club $521-$999; Three Crowns $1,000.

UKRAINIAN INSTITUTE OF MODERN ART, 2320 W. Chicago Ave., Chicago, IL 60622-4722. Tel.: 773-227-5522.
E-mail: info@uima-chicago.org
Web Site: www.uima-chicago.org
Founded: 1971.
Congressional District: 7
Key Personnel: Chm., Dr. Paul Nadzikewycz; Pres., Orysia Cardoso; Cur., Stanislav Grezdo.
Personnel Profile: Part-Time Paid 3; Part-Time Volunteers 10; Interns 2.
Governing Authority: Tax-exempt.
Institution Type/Description: Art Museum.
Collections: works by Chicago artists & Ukrainian sculptors & painters.
Major Exhibits: Glass & Ceramic, 12/6/13-1/14; Themed Group Exhibit, 2/14-3/14; Three Chicago Artists, 4/14-5/14; Morris Barazani (retrospective), 6/14-7/14; Themed Exhibit, 8/14-9/14; Chicago Artist Month, 10/14-11/14; Winter Exhibit, 12/14.
Activities: film screenings; music recitals; literary events; concerts; lectures; cultural events.
Publications: exhibit catalogs.
Hours & Admission Prices: Wed.-Sun. 12-4. Suggested Donation: $5. &
Attendance: 3,000 (estimated)
Membership: Individual $45; Family $60; Associate $100-249; Sponsor $250-499; Benefactor $500 & up.

UKRAINIAN NATIONAL MUSEUM, INC., 2249 W. Superior St., Chicago, IL 60612-1327. Tel.: 312-421-8020. Fax: 773-772-2883.
E-mail: hankewych@msn.com
Web Site: www.ukrainiannationalmuseum.org
Founded: 1952.
Congressional District: 7
Key Personnel: Pres. Bd. Directors, Jaroslaw J. Hankewych; Cur., Maria Klimchak; Admin., Anna Chychula.
Personnel Profile: Full-Time Paid 2; Full-Time Volunteers 1; Part-Time Paid 1; Part-Time Volunteers 25.
Governing Authority: nonprofit organization. Tax-exempt: 501(c)(3).
Institution Type/Description: Folk Art Museum.
Collections: Ukrainian folk art; textiles; woodworking; embroideries; beadwork; weaving exhibits; historical exhibits; Ukrainian artists paintings & sculptures; local & national archives relating to Ukrainian subjects; organizational archives; rare book collection; photos; music; history of Ukrainian immigration to America, particularly Chicago.
Facilities: 16,320-vol. Ukrainian & East European subjects available for public use.
Activities: guided tours; lectures; hobby workshops; organized education programs for children; loan & temporary exhibitions.
Publications: quarterly newsletter, Museum Chronicle.
Hours & Admission Prices: Thurs.-Sun. 11-4, Mon.-Wed. by appointment only. Adults $5, students & children $2; tours require advanced notice. Closed New Year's Day; Easter; Independence Day; Labor Day; Thanksgiving; Christmas. &
Attendance: 5,623 (accurate)
Membership: Seniors $15; Regular $25; Supporting $100; Benefactors $500; Founders $1,000; Patrons $5,000.

WESTERN EXHIBITIONS, 845 W. Washington Blvd., 2nd Fl., Chicago, IL 60607. Tel.: 312-480-8390.
E-mail: scott@westernexhibitions.com
Web Site: www.westernexhibitions.com
Founded: 2004.
Key Personnel: Dir., Scott Speh.
Personnel Profile: Full-Time Paid 1; Interns 2.
Institution Type/Description: Art Gallery.
Collections: works by contemporary artists; paintings; sculpture.
Hours & Admission Prices: Wed.-Sat. 11-6. No charge. &

WOMAN MADE GALLERY, 685 N. Milwaukee Ave., Chicago, IL 60642-8021. Tel.: 312-738-0400. Fax: 312-738-0404.
E-mail: gallery@womanmade.org
Web Site: www.womanmade.org
Founded: 1992.
Congressional District: 7
Key Personnel: Exec. Dir., Beate Minkovski; Gallery Coord., Ruby Thorkelson.
Personnel Profile: Full-Time Paid 1; Part-Time Paid 1; Part-Time Volunteers 40; Interns 10.
Governing Authority: Tax exempt.
Institution Type/Description: Art Gallery.
Collections: works by women artists.
Facilities: Museum-related items for sale.
Activities: rental facilities; special events.
Publications: quarterly e-newsletter.
Hours & Admission Prices: Wed.-Fri. 12-7, Sat.-Sun. 12-4. No charge.
Attendance: 6,000 (estimated)
Membership: Basic $40; Advanced $100; Online Registry $200.

Chicago Heights

UNION STREET GALLERY, 1527 Otto Blvd., Chicago Heights, IL 60411-3442. Tel.: 708-754-2601. Facebook: Union Street Gallery.
E-mail: unionstreetart@gmail.com
Web Site: www.unionstreetgallery.org
Founded: 1995.
Key Personnel: Dir., Jessica Freudenberg-Segal; Pres. (V), Michael "Pete" Petrouski; Museum Shop Mgr., Erica Lessie.
Personnel Profile: Full-Time Paid 1; Part-Time Paid 1; Part-Time Volunteers 35.
Volunteer Hours: 933
Operating Expenses: 100,698
Operating Income: 141,352
Governing Authority: Tax-exempt.
Institution Type/Description: Art Gallery.

Collections: works by regional & national artists; paintings; photographs; sculpture; print drawing; graphic arts.
Major Exhibits: Earth and Sky, 1/15/14-2/8/14; Fused, 1/15/14-2/8/14; National Juried Exhibit, 2/26/14-3/29/14; Photography Exhibit From Gallery 7, 4/19/14-5/10/14; Regional Juried Exhibit, 4/19/14-5/10/14; Homewood Flossmore High School, 5/14/14-5/17/14; Insider Art, 6/18/14-7/19/14; Chicago Pastel Painters Exhibit, 8/6/14-9/6/14; National Juried Exhibit, 9/24/14-10/25/14; Twelfth Month, 11/12/14-12/27/14.
Activities: special events. Annual Events: Union Street Gallery Pop-Up Thrift Shop; Garden Walk.
Hours & Admission Prices: Wed.-Thurs. 12-5, Fri. 12-6 & 6-9 (receptions), Sat. 11-4. No charge; donations accepted. Closed New Year's Day; Easter; Thanksgiving; Christmas. &
Attendance: 3,000 (estimated)
Membership: Student & Senior $20; Individual $25; Artist $35; Family $45; Patron $100-$499; Benefactor $500-$999.

Chillicothe

CHILLICOTHE AREA HISTORY MUSEUM, 723 N. Fourth St., Chillicothe, IL 61523. Mailing Address: P.O. Box 181, Chillicothe, IL 61523-0181. Tel.: 309-274-9076.
Founded: 1971.
Congressional District: 18
Personnel Profile: Part-Time Volunteers 5.
Operating Expenses: 7,300
Operating Income: 9,700
Governing Authority: Parent Institution: Chillicothe Historical Society, Inc. Tax-exempt.
Institution Type/Description: History Museum.
Collections: local history & culture; period furnishings personal artifacts; photographs. Historic House: 1928 cottage.
Hours & Admission Prices: Wed. & 1st Sun. each month 1-4. Adults $3, children 6-12 $2; prices include admission to all 3 locations if visited on the same day.
Attendance: 1,100 (estimated)

ROCK ISLAND DEPOT MUSEUM, Cedar St. & 3rd St., Chillicothe, IL 61523. Mailing Address: P.O. Box 181, Chillicothe, IL 61523-0181. Tel.: 309-274-9076.
Web Site: www.chillicothehistorical.org
Governing Authority: Parent Institution: Chillicothe Historical Society.
Institution Type/Description: Historic Depot Museum: built in 1889.
Collections: railroad history & artifacts; period furnishings; personal artifacts; photographs; 1929 Santa Fe caboose.
Hours & Admission Prices: Wed. & 1st Sun. each month 1-4. Adults $3, children 6-12 $2; prices include admission to all 3 locations if visited on the same day.
Attendance: 1,100 (estimated)

Cicero

HAWTHORNE WORKS MUSEUM, Morton College, 3801 S. Central Ave., Cicero, IL 60804-4300. Tel.: 708-656-8000, ext. 2322. Fax: 708-656-3297.
E-mail: jennifer.butler@morton.edu
Web Site: www.morton.edu/museum/index.html
Founded: 2008.
Key Personnel: Pres., Dr. Leslie Navarro; Dir., Jennifer Butler.
Personnel Profile: Full-Time Paid 1; Full-Time Volunteers 1; Part-Time Paid 1; Part-Time Volunteers 1.
Governing Authority: Parent Institution: Morton College, Cicero, IL. Tax-exempt.
Institution Type/Description: Local Industrial History Museum.
Collections: local industrial history; 20th-century American history; photographs; communications artifacts; papers; Western Electric's Hawthorne Works, 1906-1986.
Research Fields: telephone & communication history; computer technology; local industry; immigrants; women's history.
Activities: classes; community outreach.
Hours & Admission Prices: Call for hours. No charge. Closed Martin Luther King Jr. Day; Presidents' Day; Pulaski's Birthday; spring & winter break; Memorial Day; Columbus Day; Thanksgiving & weekend after. &
Attendance: 1,000 (estimated)

Clinton

C.H. MOORE HOMESTEAD, 219 E. Woodlawn St., Clinton, IL 61727-1052. Tel.: 217-935-6066. Fax: 217-935-0553.
E-mail: chmoore123@yahoo.com
Web Site: chmoorehomestead.org
Founded: 1967.
Congressional District: 21
Key Personnel: C.E.O. & House Chm., Steve Perring; Cur. Farm & Railroad Museum, Robert McMath; Treas., Connie Durbin.
Personnel Profile: Full-Time Paid 2; Part-Time Paid 2.
Governing Authority: nonprofit organization. Parent Institution: De Witt County Museum Association. Tax-exempt: 501(c)(3).
Institution Type/Description: General Museum: housed in c.1867 C.H. Moore home, remodeled & expanded in 1876 & 1887.
Collections: clothing; dolls; agriculture; railroad; period furnishings.
Facilities: Gift items for sale.
Activities: guided tours; lectures; permanent & temporary exhibitions. Museum Sponsors: apple & pork festival; ice cream social; quilt show; Christmas tea.
Publications: booklet, The Homestead; annual newsletter.
Hours & Admission Prices: April-Dec. Tues.-Sat. 10-5, Sun. 1-5. Adults $3, children 12-18 $1; children under 12 no charge.
Attendance: 5,000 (estimated)
Membership: Individual $15; Family $20; Business $30; Supporting $50. DeWitt County: Landmark $100; Patron $500; Settler $1,000; Pioneer $10,000.

Coal Valley

NIABI ZOO, 12908 Niabi Zoo Rd., Coal Valley, IL 61240-9467. Tel.: 309-799-3482 & 5107. Fax: 309-799-5761.
E-mail: ask@niabizoo.com
Web Site: www.niabizoo.com
Founded: 1964.
Key Personnel: Dir., Marc Heinzman; Pres. (V), John Ferrell; Asst. Dir. & Dir. Education, Sharon Freedman; Registrar, Carla Friedland.
Personnel Profile: Full-Time Paid 20; Part-Time Paid 20; Part-Time Volunteers 100; Interns 12.
Governing Authority: county. Tax-exempt: 501(c)(3).
Institution Type/Description: Zoo.
Collections: animals from North America, Asia, Africa & Australia.
Facilities: snack bar; educational facilities; zoological park. Museum-related items for sale.
Activities: formal education programs for children; train rides; endangered species carrousel. Annual Events: Elephants' Birthday; Zoofari Ball; Boo at the Zoo.
Publications: quarterly newsletter, Zoonooz.
Hours & Admission Prices: Spring & Summer: daily 9:30-5; Fall: Mon.-Fri. 11-4, Sat.-Sun. 9:30-5; Winter: Sat.-Sun. 11-4. Adults $6, senior citizens $5.50, children $5; members no charge. &
Attendance: 250,000 (estimated)
Membership: Children $5, Senior Citizen & Student $25; Adult $30, additional Adult $15.

Cobden

UNION COUNTY HISTORICAL AND GENEALOGICAL SOCIETY, 117 N. Appleknocker Dr., Cobden, IL 62920. Mailing Address: P.O. Box 93, Cobden, IL 62920. Tel.: 618-893-2865.
Institution Type/Description: Historical Society Museum.
Collections: local history & culture; period furnishings; photographs; personal artifacts.
Hours & Admission Prices: March to mid-Dec. Sat.-Sun. 1-5; other times by appointment. No charge; donations accepted.

Collinsville

CAHOKIA MOUNDS STATE HISTORIC SITE, 30 Ramey St., Collinsville, IL 62234-7617. Tel.: 618-346-5160. Fax: 618-346-5162.
E-mail: cahokia.mounds@sbcglobal.net
Web Site: www.cahokiamounds.org
Founded: 1930.
Congressional District: 12
Key Personnel: C.E.O. Cahokia Mounds Museum Society, Lori Belknap; Pres. (V) Cahokia Mounds Museum Society, Bill Meister; Site Mgr., Mark Esarey; Asst. Site Mgr., Matt Migalla; Asst. Site Mgr. & Dir. Public Rels., William Iseminger; Site Interpreter, Marilyn Harvey; Museum Shop Mgr.

Cahokia Mounds Museum Society, Linda Krieg; Site Technician, Kevin Fernandez; Site Technician, Gene Stratmann; Site Technician, Joseph Seago.
Personnel Profile: Full-Time Paid 7; Part-Time Paid 5; Part-Time Volunteers 90; Interns 2.
Governing Authority: state. Parent Institution: Illinois Historic Preservation Agency, Old State Capitol, Springfield, IL 62701. Tel.: 217-785-1584. Tax-exempt: 501(c)(3).
Institution Type/Description: Archaeology Museum & Site: 800-1500 A.D., site of largest prehistoric Indian city in North America.
Collections: prehistoric Indian artifacts.
Research Fields: archaeology.
Facilities: 2,000-vol. library of books on archaeology, anthropology & American Indians available for research on premises; 150-seat orientation theatre; 265-seat auditorium. Museum-related items for sale.
Activities: guided tours; lectures; slides; formally organized educational programs; permanent & traveling exhibitions; school loan service. Special Events: Kid's Day & Contemporary Indian Art Show; Archaeology Day; Indian Market.
Publications: quarterly newsletter, Cahokian; books, Cahokia City of the Sun; Journey to Cahokia.
Hours & Admission Prices: May-Oct. daily 9-5; Nov.-April Wed.-Sun. 9-5. Suggested Donations: families $15, adults $7, seniors $5, students $2. Closed most holidays. ♿
Attendance: 330,000 (estimated)
Membership: Senior & Student $30; Individual $40; Family $50; Contributor $100; Donor $500; Patron $1,000; Corporate $10,000 & up.

Crystal Lake

COLONEL PALMER HOUSE, 660 E. Terra Cotta Ave., Crystal Lake, IL 60014. Mailing Address: One E. Crystal Lake Ave., Crystal Lake, IL 60014. Tel.: 815-477-5873.
E-mail: palmerhouse@crystallakeparks.org
Web Site: www.crystallakeparks.org
Founded: 1998.
Congressional District: 12
Key Personnel: Facility Mgr., Mary Ott.
Personnel Profile: Part-Time Paid 1.
Governing Authority: Parent Institution: Crystal Lake Park District. Tax-exempt.
Institution Type/Description: Historic House Museum: built in 1858.
Collections: local history & culture; period furnishings; personal artifacts; photographs.
Facilities: archives.
Activities: special events; research; scout programs.
Hours & Admission Prices: Tues. 3-7, Thurs. 11-4, Fri. 10-4. No charge; donations accepted.

Cypress

ILLINOIS DEPARTMENT OF NATURAL RESOURCES, CACHE RIVER STATE NATURAL AREA, BARKHAUSEN CACHE RIVER WETLANDS CENTER, 8885 State Rte. 37 S., Cypress, IL 62923-2323. Mailing Address: 930 Sunflower Lane, Belknap, IL 62908. Tel.: 618-657-2064. Fax: 618-657-2065.
E-mail: molie.oliver@illinois.gov
Web Site: www.dnr.illinois.gov
Founded: 2005.
Congressional District: 19
Governing Authority: state. Parent Institution: Illinois Department of Natural Resources. Subsidiary Institution: Cache River State Natural Area. Tax-exempt.
Institution Type/Description: Wetlands Center.
Collections: cultural & natural history; flowers; plants; trees; wildlife.
Facilities: 2,000 sq. ft. exhibit space; nature trails; visitor's center.
Activities: wildlife viewing area; orientation video; hiking.
Hours & Admission Prices: Wed.-Sun. 9-4. No charge. Closed New Year's Day; Lincoln's Birthday; Veterans Day; Thanksgiving & day after; Christmas. ♿
Attendance: 15,000 (estimated)

Danville

VERMILION COUNTY MUSEUM SOCIETY, 116 N. Gilbert St., Danville, IL 61832-8506. Tel.: 217-442-2922. Fax: 217-442-2001.
E-mail: vermilioncounty@att.net
Web Site: www.vermilioncountymuseum.org
Founded: 1964.
Congressional District: 15
Key Personnel: Pres. (V), Donald Richter; Dir. & Museum Shop Mgr., Susan Richter.
Personnel Profile: Full-Time Paid 2; Part-Time Paid 2; Part-Time Volunteers 37.
Volunteer Hours: 3,306
Operating Expenses: 124,713
Operating Income: 118,509
Governing Authority: society. Tax-exempt: 501(c)(3).
Institution Type/Description: History Museum: housed in 1855 Dr. Fithian Home, doctor's residence, often visited by Abraham Lincoln.
Collections: costumes; paintings; sculpture; graphics; decorative arts; herbarium; natural history; 1857 Mann's Chapel; 1860s Doctor Office; 1890s Parlor, Lincoln Room; 1920 Dental Office. New Museum Building contains recreations of: coal mine tunnel; main street Danville c.1929; Lincoln/Lamon law office; one room school.
Major Exhibits: Lincoln in Illinois - Photographs, 2/14; Midwest Heritage Quilt Show, 7/14; Spirit of Vermilion County, 10/14.
Research Fields: costumes; natural history; paintings; sculpture; graphics; decorative arts; herbarium; county history.
Facilities: 300-vol. library of material on Lincoln, Civil War & history of area; reading room. Gifts for sale.
Activities: guided tours; lectures; arts festivals; permanent, temporary & traveling exhibitions. Annual Event: 31st Annual Midwest Heritage Quilt Show in July.
Publications: quarterly magazine, Heritage of Vermilion County.
Hours & Admission Prices: Tues.-Sat. 10-5. Adults: $2.50 one site, $4 two sites, children 13-17 $1; discounts to AAA members & Time Travelers; members & children under 13 no charge. Closed Independence Day; Thanksgiving; Christmas.
Attendance: 5,324 (accurate)
Membership: Student & Senior Citizen, Organizational & Contributing $15; Individual $20; Family $25; Sustaining $50; Patron $75; Business $100; Life $450.

VERMILION COUNTY WAR MUSEUM, (M), 307 N. Vermilion, Danville, IL 61832-4769. Tel.: 217-431-0034. Facebook: Vermillion County War Museum.
E-mail: vcwm@comcast.net
Web Site: vcwm.org
Founded: 1997.
Congressional District: 15
Key Personnel: Pres. (V), J. Kouzmanoff.
Personnel Profile: Part-Time Volunteers 30; Interns 3.
Volunteer Hours: 3,246
Governing Authority: Tax-exempt.
Institution Type/Description: Military History Museum.
Collections: military history from the Revolutionary War to present.
Hours & Admission Prices: Tues.-Fri. 12-3, Sat. 10-4. Adults $2. ♿
Membership: Individual $15; Family $25; Sponsor $100; Lifetime $250.

DeKalb

THE ANTHROPOLOGY MUSEUM, (M), Northern Illinois University, Department of Anthropology, DeKalb, IL 60115. Tel.: 815-753-2520. Fax: 815-753-7027.
E-mail: jkirker@niu.edu
Web Site: www.niu.edu/anthro_museum
Founded: 1964.
Congressional District: 14
Key Personnel: Dir., Jennifer Kirker Priest, M.A.; Cur., Laura McDowell Hopper, M.A.
Personnel Profile: Full-Time Paid 2; Part-Time Paid 10; Part-Time Volunteers 10; Interns 2.
Governing Authority: university. Parent Institution: Northern Illinois University. Tax-exempt: 170(b)(1)(A).
Institution Type/Description: Anthropology & Archaeology Museum.
Collections: ethnographic materials from Plains, Southwest & Northwest coasts & eastern North America, Mexico, South America, Pacific Islands, Southeast Asia, Africa & Greece; North America archaeology material.
Research Fields: archaeology; ethnography; physical anthropology.
Facilities: separate lab operation; classrooms.
Activities: guided tours; lectures; gallery talks; permanent & temporary exhibitions; outreach program.
Publications: Archaeological Site Reports.
Hours & Admission Prices: Tues.-Fri. 10-4, Sat. 12-4. No charge; donations accepted. Closed school holidays. ♿
Attendance: 2,000 (estimated)

BARB CITY MOTORCYCLE MUSEUM, Pierce Harley-Davidson, 969 Peace Rd., DeKalb, IL 60115. Tel.: 815-756-4558.
Institution Type/Description: Motorcycle Museum.
Collections: over 90 motorcycles from 1918 to present; 1903 motorcycle replica.
Hours & Admission Prices: Mon.-Tues. & Thurs. 9-6, Wed. 9-5, Fri. 9-7, Sat. 9-3, Sun. 10-2.

ELLWOOD HOUSE MUSEUM, (M), 509 N. First St., DeKalb, IL 60115-3232. Tel.: 815-756-4609. Fax: 815-756-4645.
E-mail: ellwoodhouse@tbcnet.com
Web Site: www.ellwoodhouse.org
Founded: 1965.
Congressional District: 14
Key Personnel: Pres. (V), Jill Olson; Dir., Gerald J. Brauer; Cur. Education & Exhibits, Rebecca Nickels; Coord. Visitor & Volunteer Svcs., Donna Gable.
Personnel Profile: Full-Time Paid 2; Part-Time Paid 2; Part-Time Volunteers 50; Interns 1.
Governing Authority: nonprofit corporation. Parent Institution: Ellwood House Association in conjunction with DeKalb Park District. Tax-exempt: 501(c)(3).
Institution Type/Description: Historic House Museum: housed in 1879 Victorian mansion Ellwood House.
Collections: Victorian furnishings; figural Staffordshire; costume collections; barbed wire history; horse drawn vehicles; buggies; cutters; private collections of books belonging to the Ellwood family. Historic Structures: c.1890 Victorian Play House; c.1905 Museum House; c.1880 Stone Water Tower.
Research Fields: barbed wire industry; Ellwood Family history; DeKalb County history; decorative arts.
Facilities: education & visitor center; 120-seat meeting room; park; gardens; wildflower study area. Museum-related items for sale.
Activities: guided tours; permanent & temporary exhibitions; docent program; special seasonal events; school outreach programs. Museum Sponsors: Art Fair; Ice Cream Social; Victorian Christmas; Decorative Arts Lecture Series.
Publications: quarterly bulletin, Ellwood House Herald; booklets, Ellwood House: An Estate of the Gilded Age; brochures.
Hours & Admission Prices: Museum: March to mid-Dec. Tues.-Sun. 1-5. Tours: Tues.-Fri. 1 & 3, Sat.-Sun. 1, 2 & 3; other times by appointment. Visitors Center: call for hours. Adults $8, youth 6-17 $3; discounts to AAM members; members no charge. &
Attendance: 8,000 (estimated)
Membership: Adult $25; Family $35; Contributing $50; Sustaining $100.

JACK OLSON GALLERY - NORTHERN ILLINOIS UNIVERSITY, (M), School of Art, 200 Visual Arts Bldg., DeKalb, IL 60115. Tel.: 815-753-4521. Fax: 815-753-7701.
E-mail: pvanael@niu.edu
Web Site: www.olsongallery.niu.edu
Key Personnel: Gallery Coord., Peter Van Ael
Institution Type/Description: Art Gallery.
Collections: works by students & faculty.
Activities: lectures.
Hours & Admission Prices: Sept. to early May Mon.-Thurs. 10-4, Fri. 10am to noon. No charge; donations accepted. &

NIU ART MUSEUM, Northern Illinois University, Altgeld Hall, DeKalb, IL 60115-2825. Mailing Address: NIU Art Museum, 1425 W. Lincoln Hwy., DeKalb, IL 60115. Tel.: 815-753-1936. Fax: 815-753-7897.
E-mail: jburke2@niu.edu
Web Site: www.niu.edu/artmuseum
Founded: 1970.
Congressional District: 14
Key Personnel: Dir., Jo Burke; Asst. Dir., Peter Olson; Coord. Mktg. & Education, Heather Green.
Personnel Profile: Full-Time Paid 2; Part-Time Paid 6; Interns 1.
Governing Authority: public university. Parent Institution: Northern Illinois University. Tax-exempt.
Institution Type/Description: University Art Museum.
Collections: contemporary works on paper; original prints, drawings & photographs; Burmese art.
Major Exhibits: NIU School of Art Faculty, 1/7/14-2/15/14; Hoarding, Amassing & Excess, 3/25/14-5/23/14; Collecting (T), 4/3/14-5/23/14; Dressing Difference: Fashion, Gender & Ethnicity in British Burma (T), 8/26/14-11/14/14; Sexuality & Gender Portrayal in the Arts, 8/26/14-11/14/14.
Research Fields: Burmese art; contiguous southeast Asian countries; contemporary art.

Facilities: 4,000 sq. ft. exhibit space.
Activities: formal education programs for undergraduate or graduate students affiliated with Northern Illinois Univ.; guided tours; lectures; workshops for teachers; loan, temporary & traveling exhibitions; training programs for professional museum workers; school tours; programs in schools.
Publications: exhibition catalogs.
Hours & Admission Prices: Sept. to May Tues.-Fri. 10-5, Sat. 12-4; group tours by appointment. No charge; donations accepted. &
Attendance: 9,000 (estimated)
Membership: Student $10; Senior $15; Individual $25; Dual $40; Sponsor $100; Patron $250.

Decatur

BIRKS MUSEUM, Millikin University, 1184 W. Main, Decatur, IL 62522-2039. Tel.: 217-424-6337. Fax: 217-424-3992.
E-mail: ewalker@mail.millikin.edu
Web Site: www.millikin.edu/academics/cfa/Pages/BirksMuseum.aspx
Founded: 1981.
Congressional District: 20
Key Personnel: Cur., Ed Walker.
Personnel Profile: Part-Time Paid 1; Interns 2.
Governing Authority: college; nonprofit. Parent Institution: Millikin University. Tax-exempt: 501(c)(3).
Institution Type/Description: China & Glass Museum.
Collections: 17th- to 20th-century ceramics & glass from Europe, America & China; American & European decorative arts; Millikin University memorabilia.
Facilities: 160-vol. library pertaining to ceramics, glass & decorative arts available for research on premise only.
Activities: guided tours; lectures; organized education programs for children, adults & undergraduate or graduate students affiliated with Millikin University; loan, temporary & traveling exhibitions.
Publications: brochure, The Birks Museum.
Hours & Admission Prices: Daily during the school year 1-4. No charge; donations accepted.
Attendance: 1,700 (estimated)

CHILDREN'S MUSEUM OF ILLINOIS, 55 S. County Club Rd., Decatur, IL 62521-4470. Tel.: 217-423-5437. Fax: 217-423-5455. Facebook: CMOFIL.
E-mail: info@cmofil.org
Web Site: www.cmofil.org
Founded: 1990.
Key Personnel: Exec. Dir., Nicole Bateman.
Personnel Profile: Full-Time Paid 2; Part-Time Paid 16.
Governing Authority: Tax-exempt.
Institution Type/Description: Children's Museum.
Collections: hands-on exhibits.
Facilities: Museum-related items for sale.
Activities: special events; educational series.
Hours & Admission Prices: May-Aug. Mon.-Fri. 9:30-4:30, Sat. 10-5, Sun. 1-5; Sept.-April Tues.-Fri. 9:30-4:30, Sat. 10-5, Sun. 1-5. Admission $5; members no charge. Closed major holidays. &
Attendance: 60,000 (accurate)
Membership: Family, Grandparent & Childcare Provider $75; Fun Family $135; Support $250; Life $1,000.

GALLERY 510 ARTS GUILD, 160 E. Main St., Ste. 100, Decatur, IL 62523-1283. Tel.: 217-422-1509. Fax: 217-475-1509.
E-mail: info@gallery510.org
Web Site: gallery510.org
Key Personnel: Exec. Dir., Barbara J. Dove.
Governing Authority: nonprofit organization.
Institution Type/Description: Art Gallery.
Collections: works by local artists.
Facilities: Museum-related items for sale.
Hours & Admission Prices: Tues.-Fri. 11-5, Sat. 11-3.
Membership: Friend up to $49; Contributor $50-$149; Sponsor $150-$249; Patron $250-$499; Benefactor $500 & up.

HIERONYMUS MUELLER MUSEUM, 420 W. Eldorado St., Decatur, IL 62522-2189. Tel.: 217-423-6161.
E-mail: muellermuseum@aol.com
Key Personnel: Mgr., Mike Deatherage
Institution Type/Description: History Museum.
Collections: photographs; early Mueller family & company history.
Hours & Admission Prices: Thurs.-Sat. 1-4. Adults $2, children under 17 $1.50.

KIRKLAND FINE ARTS CENTER, 1184 W. Main St., Decatur, IL 62522-2039. Tel.: 217-424-6253. Fax: 217-424-3993.
E-mail: kfac@Millikin.edu
Web Site: www.millikin.edu/kirkland/pages/default.aspx
Formerly: Perkinson Gallery
Founded: 1969.
Key Personnel: Dir., Jan Traughber.
Governing Authority: university. Parent Institution: Millikin University, 1184 W. Main St., Decatur, IL 62522. Tax-exempt.
Institution Type/Description: Art Museum.
Collections: 19th- & 20th-century art: graphics, decorative art, paintings, sculpture.
Facilities: 1,903-seat auditorium; 269-seat auditorium 167-seat auditorium; art galleries; lecture halls; rehearsal halls.
Activities: concerts; dance recitals; arts festivals; drama; formally organized education programs for undergraduate college students affiliated with Millikin University; invitational exhibitions.
Hours & Admission Prices: Mon.-Fri. during the school year 8-5. Closed university holidays. &
Attendance: 20,000
Membership: Collaborator: $50-$99; Partner: 100-$249; Promoter $250-$499; Sponsor $500-$999; Producer $1,000-$1,999.

MACON COUNTY HISTORY MUSEUM, 5580 N. Fork Rd., Decatur, IL 62521-1859. Tel.: 217-422-4919. Fax: 217-422-4773.
E-mail: info@mchsdecatur.org
Web Site: www.mchsdecatur.org
Founded: 1979.
Congressional District: 13
Key Personnel: Exec. Dir., Nathan Pierce; Pres. (V), Karen Anderson.
Personnel Profile: Full-Time Paid 1; Part-Time Paid 1; Part-Time Volunteers 60.
Governing Authority: society; Macon County Historical Society; nonprofit organization. Tax-exempt: 501(c)(3).
Institution Type/Description: Historical Society Museum.
Collections: 1820-present, historical artifacts from Macon County & Central Illinois including photographs, tools, clothing, late Victorian & early 20th-century mass-produced decorative art; farm & kitchen paraphernalia; Lincoln's life & career; Civil War; Grand Army of the Republic. Historic Buildings: 1855 log home; 1863 Salem Rural School; 1870s Gazebo; 1830s Lincoln Courthouse; The Lincoln Connection.
Research Fields: historical research in Decatur, Macon County & Central Illinois.
Facilities: 3,000-vol. library.
Activities: guided tours; school program; permanent & temporary exhibitions.
Publications: bimonthly newsletter.
Hours & Admission Prices: Tues.-Sat. & 4th Sun. 1-4. Adults $2; discounts to members. &
Attendance: 5,000 (estimated)
Membership: Regular $25; Couple $40; Family $50; Donor $100; Patron $500; Benefactor $1,000; Founder $5,000.

SCOVILL ZOO, 71 S. Country Club Rd., Decatur, IL 62521-4470. Tel.: 217-421-7435. Fax: 217-422-7330. Facebook: Scovill Zoo.
Web Site: www.decatur-parks.org/scovill-zoo/
Key Personnel: Dir., Dave Webster; Asst. Dir., Ken Frye.
Personnel Profile: Full-Time Paid 11; Part-Time Paid 25; Part-Time Volunteers 60; Interns 2.
Institution Type/Description: Zoo.
Collections: zoological.
Facilities: Museum-related items for sale.
Activities: petting zoo.
Hours & Admission Prices: Spring: Mon.-Fri. 10-5, Sat.-Sun. 10-7; Summer: daily 10-7. Adults $5.50, seniors 65 & over $4.50, children 2-12 $3.50; children under 2 no charge. Z.O. & O Express Train: adults $2.50, seniors 65 & over and children 2-12 $2.50; children under 2 no charge. Carousel Ride. $2. &
Attendance: 98,000 (accurate)
Membership: Individual $30; Family $50.

Des Plaines

DES PLAINES HISTORY CENTER, (M), 781 Pearson St., Des Plaines, IL 60016-4506. Tel.: 847-391-5399. Fax: 847-297-4741.
E-mail: contact@desplaineshistory.org
Web Site: desplaineshistory.org
Formerly: Des Plaines Historical Museum
Founded: 1967.
Congressional District: 9

Key Personnel: Dir., Shari Caine; Pres., Elizabeth Makelim.
Personnel Profile: Full-Time Paid 1; Part-Time Paid 2; Part-Time Volunteers 75.
Governing Authority: society. Parent Institution: Des Plaines Historical Society. Tax-exempt: 501(c)(3).
Institution Type/Description: Local History Museum.
Collections: local history; archives; photographs.
Research Fields: history of city of Des Plaines & Maine Township, Cook County, Illinois.
Facilities: library of monographs, biographies, histories, manuscripts, documents, transparencies, 10,000 photographs; history and visitor center. Post cards, stationery, Historical Society related items & hand-made items for sale.
Activities: public programs for adults, children and families; special events; community partnerships; guided tours; exhibits.
Publications: newsletter, Cobweb; Greetings From Des Plaines: A Community History Through Postcards.
Hours & Admission Prices: Tues.-Fri. 10-4, see website to confirm; other times by appointment. No charge; donations accepted. &
Attendance: 5,000 (estimated)
Membership: Senior & student $15; Individual $25; Household $35; Business/Organization $100.

KOEHNLINE MUSEUM OF ART, Oakton Community College, 1600 E. Golf Rd., Des Plaines, IL 60016-1234. Tel.: 847-635-2633. Fax: 847-635-1764.
E-mail: nharpaz@oakton.edu
Web Site: www.oakton.edu/museum
Key Personnel: Mgr. & Cur., Nathan Harpaz
Institution Type/Description: Art Museum.
Collections: contemporary art; sculpture park.
Hours & Admission Prices: June-Aug. Mon.-Thurs. 10-7; Sept.-May Mon.-Fri. 10-6, Sat. 11-4.

MCDONALD'S #1 STORE MUSEUM, 400 N. Lee St., Des Plaines, IL 60016-4610. Mailing Address: McDonald's Corp., 2111 McDonald's Dr., Oak Brook, IL 60523. Tel.: 847-297-5022.
Web Site: www.mcdonalds.com/corp/about/museum_info.html
Institution Type/Description: General Museum: a recreation of the first McDonald's Restaurant opened in Des Plaines, Illinois by founder, Ray Kroc, on April 15, 1955.
Collections: Speedee road sign; period equipment; mannequins representing the all male crew dressed in the 1955 uniform; photographs; early advertising.
Activities: video presentation.
Hours & Admission Prices: Memorial Day to Labor Day. No charge.

Dixon

JOHN DEERE HISTORIC SITE, 8334 S. Clinton St., Main, Grand Detour, Dixon, IL 61021-9499. Tel.: 815-652-4551. Fax: 815-652-3835.
E-mail: holstbrian@johndeere.com
Web Site: johndeereattractions.com
Founded: 1953.
Congressional District: 16
Key Personnel: Dir. & Pres., Mara Sovey; Site Mgr., Brian Holst.
Personnel Profile: Full-Time Paid 1; Part-Time Paid 10.
Governing Authority: company organized for profit, The John Deere Foundation, Deere & Co., John Deere Rd., Moline, IL 61265. Tax-exempt.
Institution Type/Description: Historic Site: 1836 Deere House.
Collections: archaeology exhibit; blacksmith shop; Deere house.
Facilities: Museum-related items for sale.
Activities: guided tours
Hours & Admission Prices: May-Oct. Wed.-Sun. 9-5; Winter: tours by appointment only. Adults $5; children under 12 no charge. &
Attendance: 22,000 (estimated)

RONALD REAGAN BOYHOOD HOME, (M), 816 S. Hennepin Ave., Dixon, IL 61021-3646. Mailing Address: Ronald Reagan Home Foundation, P.O. Box 816, Dixon, IL 61021-0816. Tel.: 815-288-5176. Fax: 815-288-3642.
E-mail: info@reaganhome.org
Web Site: www.reaganhome.org
Founded: 1980.
Congressional District: 14
Key Personnel: Exec. Dir., Brandi Langner; Pres. (V), Jeffrey A. Lovett.

Personnel Profile: Full-Time Paid 1; Part-Time Paid 2; Part-Time Volunteers 50.
Governing Authority: private; nonprofit organization. Tax-exempt.
Institution Type/Description: Historic House: boyhood home of 40th U.S. President Ronald Reagan.
Collections: 1920s furnishings.
Facilities: library available to scholars & researchers; visitors center. Museum-related items for sale.
Activities: guided tours.
Publications: quarterly newsletter, News From Home.
Hours & Admission Prices: April-Oct. Mon.-Sat. 10-4, Sun. 1-4. Adults 12 & over $5; members no charge. Closed Easter.
Attendance: 8,712 (accurate)
Membership: Friends of Ronald Reagan Boyhood Home: Lifeguard League $40-$124; Sportscaster Society $125-$249; Actor's Guild $250-$499; Governor's Group $500-$999; President's Circle $1,000-$4,999; World Leaders Alliance $5,000 & up.

Downers Grove

THE DOWNERS GROVE PARK DISTRICT MUSEUM, (M), 831 Maple Ave., Downers Grove, IL 60515-4904. Tel.: 630-963-1309. Fax: 630-963-0496.
E-mail: cchristensen@dgparks.org
Web Site: www.dgparks.org (click museum)
Founded: 1969.
Congressional District: 13
Key Personnel: Pres. Bd. of Park Commissioners, Robert Gelwicks; Vice Pres. Bd. of Park Commissioners, Ron Smith; Sec. Bd. of Park Commissioners, Katheryn Engel-Accettura; Treas. Bd. of Park Commissioners, Cathy Mahoney; Commissioner Bd., Janet Barr; District Admin., Dan Cermak; Dir. Recreation, Sandy Divon; Museum Supvr., Christa Christensen; Sec., David Docekal; Weekend Mgr., Carol Wandschneider; Asst. Weekend Mgr., David Roberts; Mgr. Collections, Sarah Ebel; Custodian, Karen Kurey.
Personnel Profile: Full-Time Paid 1; Part-Time Paid 5; Part-Time Volunteers 40.
Governing Authority: Parent Institution: Downers Grove Park District. Tax-exempt: 501(a).
Institution Type/Description: Local History.
Collections: period furniture; decorative arts; textiles; vintage clothing; toys; tools; home paraphernalia; photographs; personal documents.
Research Fields: historic homes; genealogy; local history.
Activities: organized education programs; docent program; temporary exhibitions. Museum Sponsors: special events.
Publications: monthly newsletter, The Plank.
Hours & Admission Prices: Public Tours: Sun.-Fri. 1-3. Group Tours: Mon.-Fri. 9-1 & 3-4:30 by appointment only. Office & Research: Mon.-Fri. 8:30-4:30. No charge; donations accepted. Closed holidays. &
Attendance: 2,800 (estimated)

Dundee

DUNDEE TOWNSHIP HISTORICAL SOCIETY, INC., 426 Highland Ave., Dundee, IL 60118-1225. Mailing Address: 437 Highland Ave., Dundee, IL 60118-1277. Tel.: 847-428-6996.
E-mail: dths@sbcglobal.net
Web Site: www.northstarnet.org/dukhome/DTHS
Founded: 1964.
Congressional District: 33
Key Personnel: Pres. (V), Marge Edwards; Museum Co-Chm., Nancy Wendt; Museum Co-Chm., Jack Wendt; Programs Chm. & Library Volunteer, Marge Edwards; Library Volunteer, Mary Lamp; Library Volunteer, Lana Graf; Library Volunteer, Connie Kashub.
Governing Authority: society. Tax-exempt: 501(c)(3).
Institution Type/Description: Historical Society.
Collections: local history; manuscript collections; exhibits highlighting Allen Pinkerton; exhibit featuring local industries & early settlers.
Major Exhibits: Dundee & The Civil War, 3/11-1/14.
Research Fields: local history; genealogy.
Facilities: 1,000-vol. library of historical & biographical books, Illinois legislative records 1824-present, township census, cemetery records, genealogies, photographs, newspaper files on microfilm & magazines available for research by appointment; meeting room. Museum-related items for sale.
Activities: historical bus tours with slides for students, Scouts, & organizations available by appointment; permanent & semi-annual changing exhibits. Society Sponsors: bus tour in April; house walk in September; cemetery walk in October; Dickens in Dundee in December.
Publications: bimonthly society newsletter, The Informer; book, Dundee Township, 1835-1985.
Hours & Admission Prices: Wed. & Sun. 2-4; other times by appointment. Adults $1; members donations accepted.

Attendance: 1,000 (estimated)
Membership: Senior Citizens & Youth $7.50; Individual $10; Couple & Family $17.50; Sustaining Individual & Family $40; Business & Organization $60; Life $250.

Dunlap

WHEELS O' TIME MUSEUM, 1710 W. Woodside Dr., Dunlap, IL 61525. Mailing Address: P.O. Box 9636, Peoria, IL 61612-9636. Tel.: 309-243-9020. Fax: 309-243-5616. Facebook: Wheels O' Time Museum.
E-mail: wotmuseum@aol.com
Web Site: wheelsotime.org
Founded: 1977.
Congressional District: 19
Key Personnel: Pres., Gary O. Bragg; Treas., Fred Roland; Public Rels., Bobbie Rice; Museum Shop Mgr., Janice M. Bragg.
Personnel Profile: Part-Time Paid 2; Part-Time Volunteers 40.
Governing Authority: company organized for profit. Parent Institution: Wheels O' Time, Inc.
Institution Type/Description: Transportation & Mechanical Museum.
Collections: airplanes; vintage & classic automobiles; barber shop with singing quartet; circus; clocks; gas & steam engines; fire engines & equipment; grandma's kitchen; military; tools; toys; tractors; model & full scale trains; washing machines; historical artifacts; early Ford V-8 display; 1931 Ahrens Fox fire truck.
Facilities: 2,500-vol. library; 30,000 sq. ft. exhibit space. Gift items for sale.
Activities: guided tours; loan exhibitions; temporary exhibitions of your own collections; school group tours.
Publications: brochures; quarterly newsletter.
Hours & Admission Prices: May-Oct. Wed.-Sun. 12-5; other times by appointment. Call for admission prices; discounts to AAA members, groups over 20 & handicapped. &
Attendance: 7,000 (estimated)

East Alton

NATIONAL GREAT RIVERS MUSEUM, 2 Lock & Dam Way, East Alton, IL 62024-2406. Mailing Address: P.O. Box 337, Alton, IL 62002-0337. Tel.: 618-462-6979, ext. 7807. Fax: 618-462-7650.
E-mail: angela.n.smith@usace.army.mil
Web Site: www.mvs.usace.army.mil/rivers
Founded: 2000.
Key Personnel: Dir., Angela N. Smith.
Personnel Profile: Full-Time Paid 3; Part-Time Paid 4; Part-Time Volunteers 25.
Governing Authority: federal government; nonprofit.
Institution Type/Description: Science & History Museum.
Collections: history of the Mississippi River; history and workings of the locks & dams system.
Facilities: 7,000 sq. ft. exhibit space; educational facilities; aquarium; 100-seat theater. Museum-related items for sale.
Activities: formal educational programs.
Hours & Admission Prices: Daily 9-5; groups of 10 or more by appointment. Tour of Lock & Dam at 10, 1 & 3. No charge; donations accepted. Closed New Year's Day; Thanksgiving; Christmas. &
Attendance: 90,000 (accurate)

East Saint Louis

KATHERINE DUNHAM DYNAMIC MUSEUM, 1005 Pennsylvania Ave., East Saint Louis, IL 62201-1407. Mailing Address: P.O. Box 6, East Saint Louis, IL 62202-0006. Tel.: 618-874-8560. Fax: 618-874-8562.
E-mail: kdcahgloria@sbcglobal.net
Web Site: www.kdcah.com
Founded: 1978.
Key Personnel: Chm. (V), Riley Owens; Museum Shop Mgr., Gloria Atkins.
Personnel Profile: Part-Time Paid 2; Part-Time Volunteers 1; Interns 1.
Governing Authority: Parent Institution: Katherine Dunham Center for Arts & Humanities. Tax-exempt.
Institution Type/Description: General Museum.
Collections: Katherine Dunham's writings, films, & works of visual arts; over 250 African & Caribbean art objects from around the world; tapestries; paintings; sculpture; musical instruments; ceremonial costumes; photographs; programs; awards & mementos from Miss Dunham's career; furniture; instruments.
Facilities: Gift items for sale.
Activities: seminar; training program for children.

Hours & Admission Prices: By appointment. Private Tours: $100; discounts to groups.
Attendance: 500 (estimated)

Edwardsville

COLONEL BENJAMIN STEPHENSON HOUSE, 409 S. Buchanan, Edwardsville, IL 62025. Mailing Address: P.O. Box 754, Edwardsville, IL 62025. Tel.: 618-692-1818. Fax: 618-692-6418.
E-mail: stephensonhouse@sbcglobal.net
Web Site: www.stephensonhouse.org
Founded: 2000.
Congressional District: 19
Key Personnel: Dir., RoxAnn Raisner
Institution Type/Description: Historic House Museum: built in 1820.
Collections: Stephenson family history; life during the 1820s; period furnishings; early gardening.
Activities: educational programs; field trips; seminars; lectures; special events.
Hours & Admission Prices: Jan.-Feb. Sat. 10-4, Sun. 12-4; March-Dec. Thurs.-Sat. 10-4, Sun. 12-4.

MADISON COUNTY HISTORICAL MUSEUM & ARCHIVAL LIBRARY, (M), 715 N. Main St., Edwardsville, IL 62025-1111. Tel.: 618-656-7562 (museum) & 7569 (library). Fax: 618-659-3457.
E-mail: mchm@co.madison.il.us
Founded: 1924.
Congressional District: 21
Key Personnel: MCHS Bd. Pres (V), Gary Denue; MCHS Bd. Vice Pres., Miriam Burns; Dir. Museum & Library, Suzanne Dietrich; Cur. Objects & Textiles, Jennifer Walta; Mgr. Operations, Mary Westerhold; Archivist, LaVerne Bloemker; Archivist, Carol Frisse.
Personnel Profile: Full-Time Paid 1; Part-Time Paid 5; Part-Time Volunteers 20.
Governing Authority: society & county government. Parent Institution: Madison County Historical Society & Madison County Government. Subsidiary Institution: Friends of the Madison County Museum. Tax-exempt: 501(c)(3).
Institution Type/Description: Historical Society Museum: housed in 1836 John H. Weir home. Listed on the National Register of Historic Places.
Collections: Native American Indian & pioneer artifacts; books; manuscripts; furniture; clothing; tools; documents; maps; eight-room restored home; land grants.
Major Exhibits: A History of Surveying, 11/13-2/14.
Research Fields: genealogy; local history; industries; historic houses.
Facilities: 2,500-vol. library on Illinois & Madison County history; 40-seat meeting room. Books, gift items for sale.
Activities: school & scout field trips; guided tours; lectures; permanent & changing seasonal exhibits of historic costumes, quilts, needlework, Christmas decorations & highlights of Madison County history.
Publications: biannual, Museum Progress Report; bimonthly newsletter.
Hours & Admission Prices: Wed.-Fri. 9-4, Sun. 1-4. No charge; donations recommended. Closed holidays. &
Attendance: 2,420 (accurate)
Membership: Student $15; Individual $35; Family $50; James Madison $100; Edward Coles $250; Elijah Lovejoy $500; John Weir M.D. $1,000.

THE UNIVERSITY MUSEUM, Southern Illinois University at Edwardsville, Edwardsville, IL 62026. Mailing Address: Box 1150, Southern Illinois University at Edwardsville, Edwardsville, IL 62026. Tel.: 618-650-2996. Fax: 618-650-2995.
E-mail: ebarnet@siue.edu
Founded: 1959.
Congressional District: 19
Key Personnel: Dir., Eric B. Barnett; Museum Exhibit Designer, Jerry L. Fahey.
Personnel Profile: Full-Time Paid 3; Part-Time Paid 4.
Governing Authority: state; university. Parent Institution: Southern Illinois University Edwardsville. Tax-exempt.
Institution Type/Description: General Museum.
Collections: Stroup Korean Pottery Collection; Milton K. & Doris T. Harrington Collection; MFA at Edwardsville Collection; Nelson-Wagner Collection; Olin Collection; Sullivan Ornament Collection; University Drawing Collection; Nause & Graf Collection; Anthropology Teaching Museum; Folklore Archive; musical instruments.
Research Fields: architectural ornament restoration.
Facilities: classrooms.
Activities: guided tours; lectures; gallery talks; arts festivals; formally organized education programs for undergraduate & graduate students; training programs for professional museum workers; loan, permanent, temporary & traveling exhibitions.
Publications: catalogue, Louis H. Sullivan Architectural Ornament Collection.
Hours & Admission Prices: Mon.-Fri. 9-4:30. No charge; donations accepted. &
Attendance: 30,000 (estimated)
Membership: Senior & Student $20; Basic $35.

Elburn

TOWN AND COUNTRY PUBLIC LIBRARY, 320 E. North St., Elburn, IL 60119. Tel.: 630-365-2244. Fax: 630-365-2358.
E-mail: library@elburn.lib.il.us
Web Site: www.elburn.lib.il.us
Key Personnel: Dir., Mary Lynn Alms
Institution Type/Description: Library.
Collections: books; magazines; newspapers; photographs; Elburn Historical Society's municipal records, school records, probate records, letters, journals, autograph books, & physical artifacts.
Hours & Admission Prices: Mon.-Thurs. 9-9, Fri.-Sat. 9-5, Sun. 1-5. No charge.

Elgin

ELGIN AREA HISTORICAL SOCIETY, (M), 360 Park St., Elgin, IL 60120-4455. Tel.: 847-742-4248. Fax: 847-931-6199.
E-mail: elginhistory@foxvalley.net
Web Site: www.elginhistory.org
Founded: 1961.
Congressional District: 14
Key Personnel: Dir., Elizabeth Marston; Pres. (V), George Rowe; Museum Shop Mgr., Melanie Burki; Treas., Bill Briska; Registrar, Beth Nawara; Researcher, David Siegenthaler; Educator, Lucy Elliott.
Personnel Profile: Part-Time Paid 3; Part-Time Volunteers 75; Interns 1.
Governing Authority: private; nonprofit organization. Tax-exempt: 501(c)(3).
Institution Type/Description: History Museum: housed in c.1856 Greek Revival landmark building.
Collections: local city history; Elgin National Watch Company; Elgin photos & images; Elgin road races; local architecture; Courier News photo negatives, 1936-1983; Elgin businesses.
Facilities: 450-vol. library; 3,000 sq. ft. exhibit space. Museum-related items for sale.
Activities: docent program; films; formal education programs; guided tours; lectures; radio programs; loan, temporary & participatory exhibitions. Annual Events: Cemetery Walk; Holiday Tea; Annual Benefit.
Publications: bimonthly newsletter, Crackerbarrel.
Hours & Admission Prices: Wed.-Sat. 12-4. Adults $3, students $1; discounts to AAM, ICOM, AASLH & AMM members; children 6 & under & members no charge. Closed New Year's Day; Independence Day; Thanksgiving; Christmas Eve & Day. &
Attendance: 5,000 (estimated)
Membership: Student $10; Adult $30; Family $45; Sponsor $75; Century $100; Century Plus $150; Donor $500; Life $1,500.

ELGIN FIRE BARN NO. 5 MUSEUM, 533 St. Charles St., Elgin, IL 60120. Tel.: 847-697-6242. Facebook: Elgin Fire Museum.
E-mail: efbn5m@firehousemail.com
Web Site: elginfiremuseum.weebly.com
Founded: 1991.
Key Personnel: Vice Pres., Cathy Hemmings.
Personnel Profile: Part-Time Volunteers 15.
Institution Type/Description: Firefighting History Museum: housed in the city's former fire barn; built in 1903.
Collections: firefighting history, tools & equipment; personal artifacts; photographs; early fire trucks; uniforms; badges; period furnishings.
Hours & Admission Prices: Feb. 3-Dec. 14 Sat.-Sun. 12-4; call to confirm. Adults $2, children 5 & over $1; members no charge.
Membership: Child $5; Individual $15; Family $25; Scout Troop $30; Business $100; Corporate $500.

ELGIN PUBLIC MUSEUM, (M), 225 Grand Blvd., Elgin, IL 60120-4278. Tel.: 847-741-6655. Fax: 847-931-6787.
E-mail: epm@cityofelgin.org
Web Site: www.elginpublicmuseum.org
Founded: 1904.
Congressional District: 12
Key Personnel: Exec. Dir., Margaret (Peggie) Stromberg; Pres. (V), Marty Kellams; Education Coord., Sara Russell.
Personnel Profile: Part-Time Paid 4; Part-Time Volunteers 20.

Governing Authority: nonprofit organization. Tax-exempt: 501(c)(3).
Institution Type/Description: Natural History Museum: housed in 1907, Neo-Classical building with 30 ft. sky-lighted ceiling.
Collections: natural history specimens; botany; zoology; geology; anthropology; paleontology; endangered & extinct species including passenger pigeon, California condor, eagles, owls, hawks & skeletal material of the Irish deer; Native American objects; ornithology collection; interpretive collections including Illinois wildlife, fish of Illinois, butterflies, insects & fossils.
Research Fields: experiential teaching materials; natural history education for youth & cross-cultural understanding.
Facilities: 500-vol. natural history research library available on a limited basis to qualified researchers & students; 5,600 sq. ft. exhibit space.
Activities: guided tours; lectures; hobby workshops; organized educational programs; participatory, loan, temporary, permanent & traveling exhibitions; school loan service.
Publications: bimonthly newsletter, EPM News; booklets.
Hours & Admission Prices: June-Sept. 1 Tues.-Sat. 12-4; Sept. 2-May Sat.-Sun. 12-4; school programs by reservation. No charge; donations accepted. Closed most major holidays. &
Attendance: 20,000 (estimated)
Membership: Student & Senior Citizen $15; Individual $25; Family $35; Contributing $50; Sustaining $75; Patron $150.

Elk Grove Village

ELK GROVE HISTORICAL MUSEUM, (M), 399 Biesterfield Rd., Elk Grove Village, IL 60007-3381. Tel.: 847-439-3994 & 690-1440.
E-mail: sdenninger@elkgroveparks.org
Web Site: www.elkgroveparks.org
Formerly: The Schuette Biermann Farmhouse Museum
Founded: 1975.
Key Personnel: Dir. & Cur., Sandy Denninger; Chm., Cliff Schultz.
Personnel Profile: Part-Time Paid 3; Part-Time Volunteers 6; Interns 1.
Governing Authority: municipal government; nonprofit. Parent Institution: Elk Grove Park Dist., 499 Biesterfield Rd., Elk Grove Village, IL 60007. Tax-exempt: 170(b)(1)(A).
Institution Type/Description: Farmhouse Museum.
Collections: Elk Grove Township & village history from Native American to present; mid 19th century farm life; early artifacts from 1850-1900; reproduction of 1910 one-room schoolhouse. Historic Buildings: 1856 2-story vernacular frame house; outbuildings; 1880s horse barn.
Major Exhibits: The Green Scale (T), 3/14-8/14.
Research Fields: farm families 1830s to present; buildings; pictures.
Facilities: classroom; 1,200 sq. ft. exhibit space. Museum-related items for sale.
Activities: hobby workshops; lectures; participatory exhibits; docent program; formal education programs for children; guided tours; badge-related programs for scout groups. Annual Events: Spring Planting in May; Ice Cream Social in July; Pioneer Days in September; Cemetery Program in October; Native American Event in November.
Publications: quarterly newsletter/calendar, The Historian.
Hours & Admission Prices: June-Aug. Tues.-Fri. 12-5, Sat. 11-2; Sept.-May Wed. & Fri. 2:30-5:30, Sat. 11-2; other times by appointment. Adults $1; members no charge. Closed major holidays.
Attendance: 4,500 (estimated)
Membership: Student & Senior $8; Adult $15; Family $20.

Ellis Grove

PIERRE MENARD HOME STATE HISTORIC SITE, 4230 Kaskaskia St., Ellis Grove, IL 62241-1718. Tel.: 618-859-3031; 618-2847230. Fax: 618-859-3031.
Web Site: www.illinois.gov/ihpa/Experience/Sites/Southwest/Pages/Pierre-Menard.aspx
Founded: 1927.
Personnel Profile: Full-Time Paid 1; Part-Time Paid 1; Part-Time Volunteers 4.
Governing Authority: state. Parent Institution: Illinois Historic Preservation Agency, 313 S. Sixth St., Springfield, IL 62701. Tax-exempt: 501(c)(3).
Institution Type/Description: Historic House Museum: c.1815-1820 home of Pierre Menard, the first Lt. Governor of Illinois and U.S. Agent of Indian Affairs.
Collections: original furniture & artifacts; furnishings of the period; separate kitchen; smokehouse; springhouse.
Research Fields: political & business history; Menard family history; the French in Illinois.
Facilities: picnic areas.
Activities: orientation video; guided tours; special events.
Publications: bimonthly newsletter, Historic Illinois.

Hours & Admission Prices: May-Oct. Wed.-Sun. 9-5 or by appointment. No charge; donations requested. Closed New Year's Day; Martin Luther King Jr. Day; Thanksgiving; Christmas. &
Attendance: 6,000 (accurate)

Elmhurst

AMERICAN MOVIE PALACE MUSEUM, 152 N. York Rd., 2nd Fl., Elmhurst, IL 60126. Tel.: 630-782-1800.
Web Site: www.historictheatres.org
Institution Type/Description: Theater History Museum.
Collections: American theater history; large scale-model of Chicago's 1927 Avalon Theatre; photographs; books; blueprints; memorabilia.
Activities: videos; special events.
Hours & Admission Prices: Tues.-Fri. 9-4, 3rd Sat. of month 9:30-1:30; groups by appointment. No charge; donations accepted.

ELMHURST ART MUSEUM, (M), 150 S. Cottage Hill Ave., Elmhurst, IL 60126-3329. Tel.: 630-834-0202. Fax: 630-834-0234.
E-mail: info@elmhurstartmuseum.org
Web Site: www.elmhurstartmuseum.org
Founded: 1981.
Congressional District: 6
Key Personnel: Dir., Phyllis O'Neill; Asst. Dir., Stephanie Grow; Controller, Heather Pastore; Museum Shop Mgr., Josephine Keane.
Personnel Profile: Full-Time Paid 6; Part-Time Paid 2; Part-Time Volunteers 20; Interns 2.
Governing Authority: private; not-for-profit organization. Tax-exempt: 501(c)(3).
Institution Type/Description: Art Museum.
Collections: artwork by regional American and European artists - 20th century.
Activities: guided tours; lectures; participatory & traveling exhibits; broadcast programs of exhibits. Annual Event: Art in the Park Fine Arts Festival.
Publications: quarterly newsletter, Start with Art!
Hours & Admission Prices: Tues.-Thurs. & Sat. 10-5, Fri. 10-8. Adults $7, seniors & students $5; children under 12 & Fri. no charge. Closed national holidays. &
Attendance: 35,000 (accurate)
Membership: Individual $40; Family $55; Century $100.

ELMHURST HISTORICAL MUSEUM, (M), 120 E. Park Ave., Elmhurst, IL 60126-3420. Tel.: 630-833-1457. Fax: 630-833-1326. Facebook: Elmhurst Historical Museum.
E-mail: ehm@elmhurst.org
Web Site: www.elmhursthistory.org
Founded: 1952.
Congressional District: 6
Key Personnel: Dir., Brian F. Bergheger; Pres. Foundation, Gene Evans; Cur. Collections, Nancy Wilson; Cur. Exhibits, Lance Tawzer.
Personnel Profile: Full-Time Paid 3; Part-Time Paid 5; Part-Time Volunteers 65.
Governing Authority: municipal. Parent Institution: City of Elmhurst, IL. Tax-exempt.
Institution Type/Description: History Museum: housed in 1892 Glos Mansion.
Collections: local history artifacts; manuscripts; literature on local history; photographs; slides; history of town & citizens; U.S. & state census; local newspapers.
Major Exhibits: Sandburg in Elmhurst, 11/13-4/20/14; Cubs vs. Sox: Chicago's Civil War, 5/16/14-9/28/14; New Exhibition on History of Elmhurst, Fall 2014; Let Children Be Children (T), 1/9/15-4/5/15.
Research Fields: German-American settlement, architecture; 19th- to 20th-century Midwest history; suburbanization.
Facilities: over 500-vol. library of local history & museum administration; genealogy, archival, manuscript & iconographic holdings.
Activities: school services; lectures; performances; guided tours; workshops; special events.
Publications: books, Elmhurst Prairie to Tree Town; Old Elmhurst; Elmhurst: Origin of Names; Elmhurst: Trails From Yesterday; Elmhurst: Scenes from Yesterday; Visionary: An Elmhurst Retrospective.
Hours & Admission Prices: Tues.-Sun. 1-5. No charge; donations accepted. &
Attendance: 15,000 (accurate)
Membership: Student $20; Senior $30; Individual $50; Family $75; Corporate/Business & Supporting $180.

LIZZADRO MUSEUM OF LAPIDARY ART, (M), 220 Cottage Hill Ave., Elmhurst, IL 60126-3351. Tel.: 630-833-1616.
E-mail: info@lizzadromuseum.org
Web Site: www.lizzadromuseum.org
Founded: 1962.

Congressional District: 6
Key Personnel: Exec. Dir., John S. Lizzadro; Dir., Dorothy J. Asher; Museum Shop Mgr., Laura McCall.
Personnel Profile: Full-Time Paid 3; Part-Time Paid 6; Part-Time Volunteers 3; Interns 1.
Volunteer Hours: 200
Operating Expenses: 440,536
Operating Income: 455,772
Governing Authority: nonprofit organization. Tax-exempt: 501(c)(3).
Institution Type/Description: Lapidary Arts Museum.
Collections: lapidary art; jade; ivory; hard-stone carvings; gemstones; gem materials; crystals; mineral specimens; mineralogy; paleontology; geology; archaeology.
Major Exhibits: Smithsonian Jewelry 20th Century, 10/13-4/6/14.
Research Fields: areas pertaining to collections.
Facilities: 1,000-vol. library pertaining to lapidary arts, history of jades, archaeology, paleontology, mineralogy & geology available for research by members only; 100-seat auditorium. Stone carvings, minerals, fossils & jewelry items with natural stones for sale.
Activities: guided tours; lectures; films; gallery talks; formally organized educational programs; permanent, temporary & traveling exhibitions; field trips.
Publications: quarterly newsletter, Lizzadro Museum; book, The Lizzadro Collection-Chinese Jades & Other Hard Stone Carvings.
Hours & Admission Prices: Tues.-Sat. 10-5, Sun. 1-5. Adults $5, senior citizens $4, students $3, children 7-12 $2; discounts to AAA members; children under 7, museum members, active members of the armed forces & Fridays no charge. Closed New Year's Day; Easter; Independence Day; Thanksgiving; Christmas. &
Attendance: 29,491 (accurate)
Membership: Annual $30; Two-year $55; Sustaining $100; Life $1,000; Benefactor $2,500 & up.

THEATRE HISTORICAL SOCIETY OF AMERICA, (M), 152 N. York St., 2nd Fl., Elmhurst, IL 60126-2806. Tel.: 630-782-1800. Fax: 630-782-1802.
E-mail: execdirector@historictheatres.org
Web Site: www.historictheatres.org
Founded: 1969.
Key Personnel: Exec. Dir., Richard Fosbrink; Dir. Archives, Kathy McLeister; Pres. (V), Karen Colizzi Noonan.
Personnel Profile: Full-Time Paid 3; Part-Time Paid 1.
Governing Authority: Tax-exempt.
Institution Type/Description: Historical Society Museum: housed in the York Theatre; built in 1924.
Collections: history on 17,000 theatres; photographs; documents; early artifacts; American theatre architecture & history.
Major Exhibits: Theatres of the Tonys, 5/13-5/14.
Facilities: archives.
Publications: quarterly journal, Marquee; quarterly newsletter; annual magazine.
Hours & Admission Prices: Tues.-Fri. 9-4 by appointment. No charge; donations accepted.
Attendance: 1,500 (estimated)

Elsah

VILLAGE OF ELSAH MUSEUM, 26 LaSalle, Elsah, IL 62028. Mailing Address: P.O. Box 28, Elsah, IL 62028-0028. Tel.: 618-374-1059. Fax: 618-374-2625.
E-mail: historicelsah@gmail.com
Web Site: www.elsah.org
Founded: 1978.
Congressional District: 17
Key Personnel: Sec., Jane Pfeifer.
Personnel Profile: Part-Time Paid 2; Part-Time Volunteers 4.
Governing Authority: municipal. Parent Institution: Village of Elsah. Subsidiary Institution: Historic Elsah Foundation. Tax-exempt.
Institution Type/Description: History Museum: housed in 1887 village hall.
Collections: local historical memorabilia including archaeological, geological, photographic, social & industrial components.
Major Exhibits: Village of Elsah Museum Annual Photography Exhibit, 4/5/14-8/2/14.
Research Fields: local history.
Facilities: 40-vol. library of local history & Illinois periodicals available for research on consultation with the director. Gift items for sale.
Activities: permanent exhibitions. Annual Event: Local Area Photography Contest April-June.
Publications: biannual newsletter; guide book, Elsah History.

Hours & Admission Prices: April-Oct. Fri.-Sun. 1-4. No charge; donations accepted. &
Attendance: 1,200 (accurate)

Eureka

BURGESS HALL ART GALLERY - EUREKA COLLEGE, 300 E. College Ave., Burgess Hall, 3rd Fl., Eureka, IL 61530-1500. Tel.: 309-467-6866.
E-mail: redge@eureka.edu
Web Site: www.eureka.edu/arts/visual/exhibitions
Congressional District: 18
Key Personnel: Dir., Prof. Rhea Edge.
Governing Authority: Parent Institution: Eureka College.
Institution Type/Description: Art Gallery.
Collections: works by contemporary artists & the art archives.
Hours & Admission Prices: Mon.-Fri. 8-5; other times by appointment. No charge. &

THE RONALD W. REAGAN SOCIETY OF EUREKA COLLEGE, 300 E. College Ave., Eureka, IL 61530. Tel.: 309-467-6477.
E-mail: reagan@eureka.edu
Key Personnel: Cur., Dr. Brian Sajko
Institution Type/Description: History Museum.
Collections: over 10,000 personal & public artifacts of Ronald Reagan from his student days at Eureka, his movie & television career, as Governor of California, his campaign for presidency, & his two terms in office.
Facilities: garden.
Hours & Admission Prices: late Aug. to May Mon.-Fri. 8-8, Sat. 10-6, Sun. 12-8; Summer: Mon.-Fri. 8-4, Sat. 10-2. No charge; donations accepted. Closed holidays.

Evanston

DITTMAR MEMORIAL GALLERY - NORTHWESTERN UNIVERSITY, Norris University Center, 1999 S. Campus Dr., Evanston, IL 60208. Tel.: 847-491-2348. Fax: 847-491-4333.
E-mail: dittmargallery@northwestern.edu
Institution Type/Description: Art Gallery.
Collections: sculpture; paintings; photographs; prints.
Hours & Admission Prices: Daily 10-10.

EVANSTON ART CENTER, 2603 Sheridan Rd., Evanston, IL 60201-1776. Tel.: 847-475-5300. Fax: 847-475-5330.
Web Site: www.evanstonartcenter.org
Founded: 1929.
Congressional District: 9
Key Personnel: Exec. Dir., Alan Leder; Pres. (V), Harold Bauer.
Personnel Profile: Full-Time Paid 7; Full-Time Volunteers 50; Part-Time Paid 50; Part-Time Volunteers 5; Interns 4.
Governing Authority: nonprofit organization. Tax-exempt: 501(c)(3).
Institution Type/Description: Contemporary Art Center.
Collections: works by regional & national artists.
Research Fields: emerging regional & contemporary artists; painting; sculpture & installation.
Facilities: studios; classrooms.
Activities: art classes; workshops; demonstrations; lectures; performances; youth outreach programs; outdoor sculpture installations; exhibitions program; curated exhibitions.
Publications: Art Center schedule of classes brochures; exhibition catalogs; quarterly newsletter, Concentrics; annual report.
Hours & Admission Prices: Mon.-Thurs. 10am-10pm, Fri.-Sat. 10-4, Sun. 1-4. Suggested donation: $3. Closed legal holidays. &
Attendance: 25,000 (estimated)
Membership: Senior Citizen $25; Student $30; Individual $35; Family $40; Contributing $50; Sponsor $100; Contemporary Circle $250; President's Roundtable $500; Patron $1,000; Benefactors $2,000; Sustaining Benefactors $5,000.

EVANSTON ENVIRONMENTAL ASSOCIATION/EVANSTON ECOLOGY CENTER, 2024 McCormick Blvd., Evanston, IL 60201-3055. Tel.: 847-448-8256. Fax: 847-448-8805.
E-mail: ecologycenter@cityofevanston.org
Web Site: www.laddarboretum.org
Founded: 1975.
Congressional District: 9
Key Personnel: Ecology Center Coord., Linda Lutz; Environmental Educator,

Karen Taira; Environmental Education, Ellen Fierer; Coord. Garden, Becky Kass; Office Mgr., Elizabeth Cullen.

Personnel Profile: Full-Time Paid 4; Part-Time Paid 4; Part-Time Volunteers 16.

Governing Authority: nonprofit organization. Parent Institution: City of Evanston. Branch Museums: Ladd Arboretum & Ecology Center, 2024 McCormick, Evanston, IL 60201; Grosse Point Lighthouse, Nature Center, Central & Sheridan Rd., Evanston, IL. Tax-exempt: 501(c)(3).

Institution Type/Description: Environmental Education Center & Arboretum.

Collections: natural history of region; history of Grosse Pointe Lighthouse; native wildflowers; alternative energy demonstration projects. Historic Building: 1874, Grosse Pointe Lighthouse Station.

Research Fields: natural landscaping; interpretive services; greenhouses; environmental studies.

Facilities: 450-vol. library of books relating to general public interest & environmental concerns available for research on premises; botanical garden; solar greenhouse; classrooms; nature center; arboretum; demonstration energy systems. Booklets & pamphlets relating to environmental concerns for sale.

Activities: guided tours; lectures; films; workshops; formally organized educational programs; docent program; permanent exhibitions.

Publications: quarterly newsletter, Futures.

Hours & Admission Prices: Summer: Mon.-Fri. 9-4:30; Labor Day-Memorial Day Mon.-Sat. 9-4:30. No charge. Closed national holidays; Memorial Day; Independence Day weekend; Labor Day weekend. &

Attendance: 8,000 (estimated)

Membership: Individual $25; Family $50; Nonprofit, School & Contributing $100; Sustaining $250; Life $500; Corporate $1,000.

EVANSTON HISTORICAL CENTER AND CHARLES GATES DAWES HOUSE, (M), 225 Greenwood St., Evanston, IL 60201-4713. Tel.: 847-475-3410. Fax: 847-475-3599.

E-mail: evanstonhs@northwestern.edu

Web Site: www.evanstonhistorycenter.org

Founded: 1898.

Congressional District: 10

Key Personnel: Pres. (V), Robert Barr.

Personnel Profile: Full-Time Paid 2; Part-Time Paid 5; Part-Time Volunteers 45; Interns 1.

Governing Authority: private; nonprofit organization. Tax-exempt: 501(c)(3).

Institution Type/Description: Historic House Museum: restored 1894 home of former Vice Pres. & Nobel laureate Dawes: 28-room mansion. National Historic Landmark.

Collections: Gen. Charles G. Dawes memorabilia; historical Evanston artifacts; archival material; costumes; photograph & newspaper archives; 1920s furnished rooms.

Research Fields: genealogical; Evanston history & architecture.

Facilities: 1.8 acre lakefront property.

Activities: tours of house; permanent & rotating exhibits; lectures; holiday events; special events. Annual Events: Mother's Day House Walk; Fall Ice Cream Social.

Publications: biannual membership newsletter, TimeLines.

Hours & Admission Prices: Tours: Fri.-Sun. 1-5. Tours: adults $5, students & seniors $3; discounts to AAM members; members no charge. Research: Tues., Thurs., Sat. 1-4 $5; members & students no charge. Closed legal holidays. &

Attendance: 6,000 (accurate)

Membership: Individual $30; Family & Dual $45; Sustainer $100; Patron $250; Currey Club Conservator $500; Curry Club Circle $1,000.

FRANCES WILLARD HOUSE MUSEUM, 1730 Chicago Ave., Evanston, IL 60201-4502. Tel.: 847-328-7500.

Web Site: www.wctu.org/house.html

Formerly: The Willard House (WCTU Museum)

Founded: 1900.

Congressional District: 10

Key Personnel: Nat. Pres., Woman's Christian Temperance Union, Rita Kaye Wert.

Governing Authority: nonprofit organization. Parent Institution: Woman's Christian Temperance Union. Subsidiary Institution: Frances Willard Historical Assoc. Tax-exempt: 501(c)(3).

Institution Type/Description: Historic House: home of Frances E. Willard (1865-1898).

Collections: period furniture; books; costumes; pictures; maps; items from around the world gathered by Frances Willard.

Activities: guided tours.

Hours & Admission Prices: 1st & 3rd Sun. of each month 1-4; other times by appointment only. Adults $10, children 12 & under $5; discounts to AAM & ICOM members; Woman's Christian Temperance Union & Frances Willard Historical Association members no charge.

Attendance: 400 (estimated)

Membership: .

LEVERE MEMORIAL TEMPLE, 1856 Sheridan Rd., Evanston, IL 60201-3837. Mailing Address: P.O. Box 1856, Evanston, IL 60204-9918. Tel.: 847-475-1856; 800-233-1856. Fax: 847-475-2250. Facebook: Levere Memorial Temple.

Web Site: www.sae.net

Founded: 1929.

Congressional District: 9

Key Personnel: Dir., Blaine Ayers; Pres. (V), Ed Fuller, Jr.

Personnel Profile: Part-Time Paid 1.

Governing Authority: nonprofit organization. Parent Institution: Sigma Alpha Epsilon Foundation. Tax-exempt: 501(c)(3).

Institution Type/Description: Preservation Society.

Collections: books; letters; fraternity jewelry & artifacts; early fraternity life & customs; manuscripts; Tiffany stained glass windows depicting fraternity & U.S.A. history; Fraternity Hall of Fame; restored Tiffany sketches, 70 chapel stained glass window templates; murals.

Facilities: library; 325-seat chapel.

Activities: guided tours; permanent exhibitions.

Publications: quarterly magazine for fraternity members, The Record; The Phoenix.

Hours & Admission Prices: Memorial Day to Labor Day Mon.-Thurs. 9-4:30, Fri. 9 to noon, Sat.-Sun. by appointment; Sept.-May Mon.-Fri. 9-4:30, Sat.-Sun. by appointment only. No charge; donations accepted. Closed New Year's Eve & Day; Presidents' Day; Easter; Memorial Day; Independence Day; Labor Day; Thanksgiving & day after; Christmas Day & week.

Attendance: 3,000 (estimated)

* MARY AND LEIGH BLOCK MUSEUM OF ART, NORTHWESTERN UNIVERSITY, (M), 40 Arts Circle Dr., Evanston, IL 60208-2410. Tel.: 847-491-4001. Fax: 847-491-2261.

E-mail: block-museum@northwestern.edu

Web Site: www.blockmuseum.northwestern.edu

Founded: 1980.

Congressional District: 9

Key Personnel: Dir., Lisa G. Corrin; Chm., Christine O. Robb; Dir. Devel., Helen Hilken; Interim Business Admin., Rita Shorts; Dir. Education Programs, Judy Koon; Grants Mgr., Nicole Druckman; Communications Mgr., Burke Patten; Collections & Exhibitions Asst., Elizabeth Wolf; Block Cinema, Mimi Brody; Sr. Cur., Debora Wood; Assoc. Cur., Corinne Granof; Registrar, Kristina Bottomley; Mgr. Exhibitions & Facilities, Dan Silverstein; Asst. to Dir., Emily Forsgren; Security Coord., James Foster; Security Asst., Aaron Chatman.

Personnel Profile: Full-Time Paid 13; Part-Time Paid 83; Part-Time Volunteers 88; Interns 4.

Governing Authority: nonprofit; university. Northwestern University, 633 Clark St., Evanston, IL 60208. Tax-exempt: 501(c)(3) & 170(b)(1)(A).

Institution Type/Description: University Art Museum.

Collections: 15th-21st century prints, drawings, & photographs; architectural renderings by Walter Burley and Marion Mahony Griffin; textiles by Theo Leffman; modernist artist including Arp, Moore, Miro, Hepworth & more in outdoor sculpture garden.

Research Fields: history of prints; sculpture; photography.

Activities: lectures & symposia with artists and scholars; film series; guided exhibition tours; gallery talks; education programs for schools, children & families; classes and resources for graduate & undergraduate students affiliated with Northwestern University.

Publications: exhibition catalogues & guides; quarterly newsletter, Around the Block; quarterly exhibition & programming calendar; annual report.

Hours & Admission Prices: Tues. & Sat.-Sun. 10-5, Wed.-Fri. 10-8. No charge; donations accepted. &

Attendance: 39,000 (accurate)

Membership: Student $15; Individual $40; Family $60; Patron $100; Sustainer $250; Benefactor $500; Block Circle $1,000.

MITCHELL MUSEUM OF THE AMERICAN INDIAN, (M), 3001 Central St., Evanston, IL 60201-1102. Tel.: 847-475-1030.

E-mail: kmcdonald@mitchellmuseum.org

Web Site: www.mitchellmuseum.org

Founded: 1977.

Congressional District: 9

Key Personnel: Exec. Dir., Kathleen McDonald; Cur. Exhibitions & Collections, Melissa Halverson.

Personnel Profile: Full-Time Paid 2; Part-Time Paid 3; Part-Time Volunteers 30.

Governing Authority: nonprofit organization. Tax-exempt.

Institution Type/Description: Native American Museum.

Collections: 10,000 North American Indian artifacts from cultures of the Woodlands, Plains, Arctic, Southwest & Northwest coast
Major Exhibits: Storytellers, 1/14-12/14; Childhood in Indian Country, 9/14-8/15.
Research Fields: American Indians of the U.S. & Canada.
Facilities: 3,000-vol. library. Museum-related items for sale.
Activities: guided tours; lectures; performances; workshops; school loan boxes; hands-on exhibits; family programs; permanent & temporary exhibits.
Publications: brochure; newsletter.
Hours & Admission Prices: Tues.-Wed. & Fri.-Sat. 10-5, Thurs. 10-8, Sun. 12-4. Adults $5, students, children & senior citizens $3; discounts to AAM, ICOM, AAA & AASLH members; tribal citizens no charge. Closed New Year's Eve & Day; Independence Day; Thanksgiving; Christmas Eve & Day. &
Attendance: 10,000 (estimated)
Membership: Individual $35; Family & Partners $50; Travelers Pass $100.

NOYES CULTURAL ARTS CENTER, 927 Noyes St., Ste. 100, Evanston, IL 60201-6205. Tel.: 847-448-8260.
Institution Type/Description: Art Gallery.
Collections: paintings; sculpture.
Activities: theatrical performances; special events.
Hours & Admission Prices: Mon.-Sat. 10-7, Sun. 10-6. No charge. &

THE PREHISTORIC LIFE MUSEUM, 704 Main St., Evanston, IL 60202-1702. Tel.: 847-866-7374. Fax: 847-866-6854.
E-mail: rockshop1@att.net
Web Site: www.davesrockshop.com
Founded: 1970.
Institution Type/Description: Paleontology Museum.
Collections: dinosaur footprints; fossils; bones; dinosaur egg; insects; cave bear skeleton.
Facilities: Museum-related items for sale.
Hours & Admission Prices: Mon.-Tues. & Thurs.-Fri. 10:30-5:30, Sat. 10-5. No charge.
Attendance: 6,000 (estimated)

SHOREFRONT LEGACY CENTER, 2010 Dewey Ave., 205, Evanston, IL 60201. Mailing Address: P.O. Box 1894, Evanston, IL 60204. Tel.: 847-864-7467. Facebook: Shorefront.
E-mail: shorefront@me.com
Web Site: www.shorefrontlegacy.org
Founded: 2002.
Congressional District: 9
Key Personnel: Exec. Dir., Dino Robinson.
Personnel Profile: Part-Time Volunteers 1.
Governing Authority: Tax-exempt: 501(c)(3).
Institution Type/Description: History Museum.
Collections: Black history & culture; photographs.
Research Fields: local African American culture.
Facilities: library; research center.
Activities: educational programs; special events; outreach; temporary exhibits; lectures.
Publications: Shorefront Journal; Gatherings; Through the Eyes of Us.
Hours & Admission Prices: Sat. 9-2; other times by appointment. No charge; donations accepted. Closed major holidays. &
Attendance: 300 (estimated)
Membership: Basic $25; Family $50; Organization $100; Corporate $500.

Fairbury

FAIRBURY ECHOES MUSEUM, 126 W. Locust, Fairbury, IL 61739-1549. Tel.: 815-692-2191.
E-mail: museum1976@yahoo.com
Founded: 1979.
Key Personnel: Pres. (V), Diane Pawlowski
Governing Authority: Tax-exempt.
Institution Type/Description: History Museum.
Collections: history of Fairbury & local area; photographs; personal artifacts.
Publications: quarterly newsletter.
Hours & Admission Prices: Thurs.-Fri. 1-4:30, Sat. 9-11:30. No charge; donations accepted.

Fairfield

WAYNE COUNTY HISTORICAL SOCIETY'S HANNA HOUSE MUSEUM, 101 E. Center, Fairfield, IL 62837-2101. Mailing Address: 300 S.E. 2nd St., Fairfield, IL 62837-2127.
Formerly: Hanna House Museum

Founded: 1953.
Key Personnel: Pres. (V), Judith Puckett; Museum Mgr., Jami Roethe; Museum Mgr., Niki Roethe.
Personnel Profile: Part-Time Paid 2; Part-Time Volunteers 12.
Governing Authority: Parent Institution: Wayne County Historical Society. Tax-exempt.
Institution Type/Description: Local History Museum.
Collections: local history including prohibition-era gangland figures; oil boom; medicine; funeral equipment; education; military; sports; Lions' memorabilia.
Research Fields: Wayne County.
Facilities: historic house; archives.
Publications: bimonthly membership newsletter.
Hours & Admission Prices: April-Oct. Sat. 10-2. No charge; donations accepted. Closed holidays.
Attendance: 500 (estimated)
Membership: Annual $15; Lifetime $300.

Farmer City

FARMER CITY GENEALOGICAL AND HISTORICAL SOCIETY MUSEUM, 220 S. Main St., Farmer City, IL 61842. Mailing Address: P.O. Box 173, Farmer City, IL 61842. Tel.: 309-928-9411.
Institution Type/Description: Historical Society Museum.
Collections: local history & culture; period furnishings; personal artifacts; photographs.
Publications: quarterly newsletter, Mirror.
Hours & Admission Prices: Mon., Wed. & Fri. 9-12.

Frankfort

FRANKFORT FIRE DEPARTMENT MUSEUM, 333 W. Nebraska St., Frankfort, IL 60423. Tel.: 815-469-1700.
Institution Type/Description: Firefighting History Museum.
Collections: local firefighting history & equipment; 1905 hose cart; personal artifacts; photographs; uniforms; early fire trucks.
Hours & Admission Prices: Call for hours.

KIDSWORK CHILDREN'S MUSEUM, 11 S. White St., Frankfort, IL 60423. Tel.: 815-469-1199.
Institution Type/Description: Children's Museum.
Collections: hands-on exhibitions.
Activities: birthday parties; special events; educational programs; summer camps.
Hours & Admission Prices: Tues.-Sat. 9-4, Sun. 11-4. Admission $6, seniors $5.

Freeport

FREEPORT ART MUSEUM, (M), 121 N. Harlem Ave., Ste. B, Freeport, IL 61032-3845. Tel.: 815-235-9755. Fax: 815-235-6015.
E-mail: director@freeportartmuseum.org
Web Site: www.freeportartmuseum.com
Founded: 1975.
Congressional District: 16
Key Personnel: C.E.O. & Dir., Jessica J. Caddell; Pres. (V), Vanessa Hughes; Dir. Education, Barry Treu; Visitor Svcs. & Business Mgr., Carrie Baxter.
Personnel Profile: Full-Time Paid 3; Part-Time Volunteers 55; Interns 2.
Governing Authority: private; nonprofit organization. Tax-exempt: 501(c)(3).
Institution Type/Description: Art Museum.
Collections: European artifacts including 19th century paintings, sculptures, prints & Florentine mosaics from the W.T. Rawleigh collection; Near & Far East artifacts including jewelry; glass; mummy case & other artifacts from Ancient Egypt, China & S.E. Asia; Native American artifacts including Sioux beadwork, Pueblo pottery (including pieces by Nampeyo & Martinez), Kachina dolls & pre-Columbian artifacts; Oceanic art including masks, ceremonial pieces, Indonesian textiles, pottery & musical instruments; works by contemporary artists such as Richard Boschulte, Dan Edler, Winn Jones, Andrew Langoussis, Michael Johnson, Jeanette Sloan, Roland Poska & Duane Smith; student art.
Research Fields: late 1800s & early 1900s paintings, ethnographic art; Egyptian, Oceanic, Greek & Roman antiquities; Native American artifacts.
Facilities: library; classrooms; meeting space.
Activities: guided tours; inter-museum loan & permanent exhibitions; area arts development programs-school artists in residence; extension exhibits; dance school, instructional classes; performances; concerts; poetry readings; festivals; family programs.
Publications: newsletter; catalogs.
Hours & Admission Prices: Tues.-Fri. 10-5, Sat. 12-5. No charge; donations accepted. Closed holidays. &

Attendance: 11,000 (accurate)
Membership: Family $40; Furst $80; Rilling $125; Parvin $250; Dedrick $500; 121 Club $1,000; Georgia Sales $2,500; Rawleigh $5,000.

HIGHLAND COMMUNITY COLLEGE ARBORETUM, 2998 Pearl City Rd., Freeport, IL 61032-9341. Tel.: 815-235-6121. Fax: 815-235-6130. TDD: 815-235-9584.
E-mail: pete.willging@highland.edu
Founded: 1970.
Congressional District: 16
Key Personnel: Pres. (V), Joe Kanofsky; Business Officer, Pam Schleich; Dir. Public Rels., Pete Willging; Dir. Physical Plant & Maintenance, Rich Eads.
Personnel Profile: Full-Time Paid 2; Part-Time Paid 4.
Governing Authority: college. Parent Institution: Highland Community College Foundation. Tax-exempt: 170(b)(1)(A).
Institution Type/Description: Arboretum.
Collections: 3,000 plantings representing over 300 different species of ground coverings, trees & shrubs.
Research Fields: hican & pecan cultivation.
Facilities: library; 150-seat auditorium; 400-seat theater; classrooms; cafeteria. Textbooks, instructional supplies, school & art supplies for sale.
Activities: guided tours.
Hours & Admission Prices: Daily 7-7. No charge; donations accepted. Closed legal holidays. &
Attendance: 5,000 (estimated)

SILVERCREEK MUSEUM, (M), 2954 S. Walnut Rd., Freeport, IL 61032-9528. Tel.: 815-235-7329 & 2198.
E-mail: jkayklever@frontier.com
Web Site: www.thefreeportshow.com
Founded: 1988.
Congressional District: 12
Key Personnel: Pres. (V), Larry Buttel; Vice Pres., Mary Seefeldt-Swanson; Sec., Siri McMahon; Treas., Kris McNames; Dir., Jerry Klever; Dir., Ida DeBoer; Dir., Ed Keech.
Personnel Profile: Part-Time Volunteers 50.
Governing Authority: nonprofit. Parent Institution: Stephenson County Antique Engine Club. Tax-exempt.
Institution Type/Description: General Museum: housed in 1906 county poor farm.
Collections: 450 pieces of red wing pottery; W.T. Rawleigh Collection; brass; agricultural implements; art.
Activities: guided tours; temporary exhibitions; antique steam train rides. Annual Events: Pancake Supper & Bake Sale in April & November; Ice Cream Social in May; Holiday Treat & Cookies Sale in December.
Publications: member newsletters 3-4 per year.
Hours & Admission Prices: Memorial Day to mid-Oct. 11-4. Adults $4; discounts to AAM members. &
Attendance: 5,000 (estimated)
Membership: Stephenson County Antique Engine Club: Family $25.

STEPHENSON COUNTY HISTORICAL SOCIETY, (M), 1440 S. Carroll Ave., Freeport, IL 61032-6530. Tel.: 815-232-8419. Fax: 815-297-0313. Facebook: Stephenson County Historical Society.
E-mail: director@stephcohs.org
Web Site: stephcohs.org
Founded: 1944.
Congressional District: 16
Key Personnel: C.E.O., Edward F. Finch; Pres. (V), Connie Sorn; Museum Shop Mgr., Brigitte Rayhorn.
Personnel Profile: Full-Time Paid 1; Part-Time Paid 3; Part-Time Volunteers 22.
Volunteer Hours: 325
Operating Expenses: 64,295
Operating Income: 65,909
Governing Authority: society. Tax-exempt: 501(c)(3).
Institution Type/Description: History Museum: housed in 1857 house built by Oscar & Malvina Taylor, listed on National Register of Historic Sites.
Collections: Lincoln-Douglas debate material; industrial items connected with county industries; 1843 log cabin; arboretum; Jane Addams collection; arcade toys; Henney Motor Co., W.T. Rawleigh Co. & Stephens Motor Co. items in the horsepower age 1850-1910. Historic Building: 1900-1915 one-room schoolhouse.
Facilities: 100-vol. library of local history available on premises by request; arboretum.
Activities: guided tours; local history lecture series in fall; American history lecture series in winter; classic film series in fall & winter; permanent & temporary exhibitions.
Publications: catalogs, Arcade Toy Company (reprinted); Henney Buggy Company (reprinted); The Early Black Settlers of Stephenson County; pictorial companion, A Hundred Year Journal; bimonthly newsletter, The Museum Times.
Hours & Admission Prices: Wed.-Sun. 12-4; groups of 10 or more by appointment. Adults $8, children 6-12 $4; members no charge. &
Attendance: 7,800 (accurate)
Membership: Individual $25; Family $35; Tutty $50; Addams $100; Lincoln $250; Taylor $500.

Galena

GALENA-JO DAVIESS COUNTY HISTORICAL SOCIETY & MUSEUM, 211 S. Bench St., Galena, IL 61036-2203. Tel.: 815-777-9129. Fax: 815-777-9131.
E-mail: info@galenahistorymuseum.org
Web Site: www.galenahistorymuseum.org
Founded: 1938.
Congressional District: 17
Key Personnel: Pres. (V), Steve Coates; Exec. Dir., Nancy Breed.
Personnel Profile: Full-Time Paid 2; Part-Time Paid 6; Part-Time Volunteers 25.
Governing Authority: nonprofit organization. Subsidiary Institution: Old Blacksmith Shop. Tax-exempt: 501(c)(3).
Institution Type/Description: Historical Society Museum: housed in 1858 Daniel A. Barrows house.
Collections: regional artifacts & manuscripts of the upper Mississippi lead mine region; 1830s mine shaft; 19th-century mining tools; farm implements; costumes; industries in Galena; Civil War artifacts. Historic Building: 1858 U.S. Grant Leather Store; geology.
Research Fields: oral history; lead mining; steamboating; architecture; U.S. Grant; Civil War; upper Mississippi River.
Facilities: 1,000-vol. library of photographs, manuscripts, newspapers & publications relating to the area available for research on premises by appointment. Museum-related items for sale.
Activities: guided tours; lectures; gallery talks; formally organized educational programs; permanent & temporary exhibitions; school loan service; tours of Galena's historic district.
Publications: posters; books; exhibit catalogs; quarterly newsletter.
Hours & Admission Prices: Daily 9-4:30. Family $20; adults $8, seniors 65 & over $7, youth 10-18 $6; children under 10 & members no charge. Closed New Year's Eve & Day; Easter; Thanksgiving; Christmas Eve & Day.
Attendance: 13,945 (accurate)
Membership: Student & Military $15; Senior 65 & over $25; Adult $30; Senior Couple 65 & over $35; Family $50; Business $100; Lead Miner $150; Blacksmith $250; Steamboat Captain $500; Galena General $1,000.

OLD MARKET HOUSE STATE HISTORIC SITE, Market Square, Galena, IL 61036. Mailing Address: 307 Decatur St., P.O. Box 333, Galena, IL 61036-0333. Tel.: 815-777-3310. Fax: 815-777-3310.
E-mail: granthome@granthome.com
Founded: 1947.
Congressional District: 16
Key Personnel: Site Mgr., Terry J. Miller.
Personnel Profile: Full-Time Paid 5; Part-Time Paid 7; Part-Time Volunteers 15.
Governing Authority: state. Illinois Historic Preservation Agency, Div. of Historic Sites, Old State Capitol, Springfield, IL 62701. Tax-exempt.
Institution Type/Description: Historic House Museum: c.1845 Greek Revival-style Market House.
Collections: Galena artifacts; President Ulysses S. Grant artifacts.
Research Fields: history; architecture.
Facilities: meeting hall; market square.
Activities: guided tours; lectures; inter-museum loan exhibitions; permanent exhibitions. Museum Sponsors: weekly Galena farmers' market May to October.
Publications: bimonthly newsletter, Historic Illinois; brochure, Market House.
Hours & Admission Prices: Wed.-Sun. 9-12 & 1-5; Winter: 10-12 & 1-4. Donations suggested.
Attendance: 32,395 (accurate)

U.S. GRANT'S HOME STATE HISTORIC SITE, 500 Bouthillier St., Galena, IL 61036-2704. Mailing Address: 307 Decatur St., P.O. Box 333, Galena, IL 61036-0333. Tel.: 815-777-3310. Fax: 815-777-3310.
E-mail: granthome@granthome.com
Web Site: www.granthome.com
Founded: 1932.
Congressional District: 16

Key Personnel: Site Mgr., Terry J. Miller.

Personnel Profile: Full-Time Paid 5; Part-Time Paid 7; Part-Time Volunteers 15.

Governing Authority: state. Parent Institution: Illinois Historic Preservation Agency, Div. of Historic Sites, Old State Capitol, Springfield, IL 62701. Tax-exempt.

Institution Type/Description: Historic House Museum: 1860 Italianate bracketed style house presented to Gen. Grant in 1865.

Collections: Grant mementos; Victorian furnishings; pictorial exhibit on Grant's life & Galena. Historic Buildings: c.1850 Greek Revival row house; 1844-1860 Elihu B. Washburne House State Historic Site; two 1840s log houses; 1891 frame house; 1838 frame house; c.1838 log house.

Research Fields: U.S. Grant; Galena & regional history.

Facilities: 300-vol. library of material on Grant & local history, 1828-1931 Galena newspapers.

Activities: permanent exhibitions. Special Event: Lamplight Tour in December.

Publications: bimonthly newsletter, Historic Illinois; brochure, Grant's Home.

Hours & Admission Prices: March-Oct. Wed.-Sun. 9-4:45; Nov.-Feb. Wed.-Sun. 9-4. Suggested Donation: adults $4, children under 18 $2. Closed New Year's Day; Martin Luther King Jr. Day; Presidents' Day; Election Day; Veterans Day; Thanksgiving; Christmas. &

Attendance: 84,770 (accurate)

Galesburg

CARL SANDBURG STATE HISTORIC SITE, 313 E. 3rd St., Galesburg, IL 61401-6021. Mailing Address: P.O. Box 106, Bishop Hill, IL 61419. Tel.: 309-342-2361. Fax: 309-342-2141.

E-mail: carl@sandburg.org

Web Site: www.sandburg.org

Founded: 1945.

Congressional District: 17

Key Personnel: Site Supt., Martha Downey.

Personnel Profile: Part-Time Paid 1.

Governing Authority: state. Parent Institution: Illinois Historic Preservation Agency.

Institution Type/Description: Historic Home Museum: c.1870 immigrant railroad worker's cottage.

Collections: period furniture; family pieces of the Sandburgs; Sandburg publications & memorabilia; memorial garden. Historic Site: Remembrance Rock where Sandburg's ashes were buried; 1858 house used as Visitor Center.

Facilities: garden; gravesite.

Activities: guided tours; permanent exhibits; video presentations.

Publications: newsletter, Inklings and Idlings.

Hours & Admission Prices: May-Oct. Thurs.-Sun. 9-5. Suggested Donations: family $10, adults $4, children 17 & under $2. &

Attendance: 1,000 (estimated)

DISCOVERY DEPOT CHILDREN'S MUSEUM, 128 S. Chambers St., Galesburg, IL 61401-4966. Tel.: 309-344-8876.

E-mail: discoverydepot@grics.net

Web Site: www.discoverydepot.org

Governing Authority: nonprofit organization.

Institution Type/Description: Children's Museum.

Collections: hands-on exhibits.

Facilities: Museum-related items for sale.

Activities: birthday parties; special programs.

Hours & Admission Prices: July-Aug. Mon.-Sat. 10-5, 3rd Fri. of month 10-7, Sun. 1-5; Sept.-June Tues.-Sat. 10-5, 3rd Fri. of month 10-7, Sun. 1-5. Admission $3.50.

GALESBURG CIVIC ART CENTER, 114 E. Main St., Galesburg, IL 61401-4601. Tel.: 309-342-7415.

E-mail: info@galesburgarts.org

Web Site: www.galesburgarts.org

Founded: 1923.

Congressional District: 17

Key Personnel: C.E.O., Heather L. Norman; Bd. Pres., Jim Straub; Office Mgr., Lynn Miller.

Personnel Profile: Part-Time Paid 2; Part-Time Volunteers 100.

Governing Authority: nonprofit organization. Parent Institution: Galesburg Civic Art League. Tax-exempt.

Institution Type/Description: Art Gallery & Civic Center: housed in c.1897 building.

Collections: W.P.A. prints including Thomas Hart Benton, Raphael Soyer, M. Albright; regional landscape paintings, including Harold Gregor; paintings from national competitions since 1965; works by Joe Bujnowski, J. Butler, H. Gerardia & Salvador Dali, Clare Smith.

Research Fields: W.P.A. prints; regional central & central western Illinois.

Facilities: Museum-related items for sale.

Activities: lectures; loan & temporary exhibitions; educational programs. Annual Events: Galex National Exhibit; Art-in-the-Park fair; Studios Midwest Residency Program; The Black Earth Film Festival.

Publications: Artifacts Newsbulletin; brochures.

Hours & Admission Prices: Tues.-Fri. 10:30-4:30, Sat. 10:30-3; tours by appointment. No charge; donations accepted. Closed all holidays.

Attendance: 8,000 (estimated)

Membership: Senior & Student $20; Individual $40; Family $60; Patron-of-the-Arts $150; Muse $300; Art Champion $500; Benefactor $1,000 & up.

GALESBURG RAILROAD MUSEUM, 211 S. Seminary St., Galesburg, IL 61401-4955. Tel.: 309-342-9400.

Web Site: www.galesburgrailroadmuseum.org

Founded: 1980.

Congressional District: 17

Key Personnel: Pres., Jim Clayton; Museum Shop Mgr., Annette Godsil.

Personnel Profile: Part-Time Paid 4.

Governing Authority: Tax-exempt.

Institution Type/Description: History Museum.

Collections: railroad history & artifacts; memorabilia; 1930 passenger train; Pullman parlor car; caboose; tools; photographs; uniforms; utensils; books; magazines; china; Railway Express combination mail/baggage car.

Hours & Admission Prices: April-Sept. Tues.-Sat. 10-4, Sun. 12-4. Adults $5, children 8-16 $3; children 7 & under and members no charge. Closed holidays. &

Attendance: 16,000 (estimated)

Membership: Individual $20; Family $30; Lifetime $300; Lifetime Couple $400.

ILLINOIS CITIZEN SOLDIER MUSEUM, 1001 Michigan Ave., Galesburg, IL 61401-6481. Tel.: 309-342-1181.

E-mail: vfwpost2257@centurylink.net

Founded: 1988.

Congressional District: 17

Key Personnel: Bldg. Mgr. & Museum Shop Mgr., Jim Verheyen; Pres. (V) & Treas., Mike Lummis.

Personnel Profile: Full-Time Volunteers 1; Part-Time Volunteers 5.

Governing Authority: veterans organization; nonprofit. Parent Institution: V.F.W. 2257. Tax-exempt.

Institution Type/Description: Military Museum: housed in Admiral James Stockdale building, home of Veterans of Foreign Wars Post 2257.

Collections: Admiral James Stockdale memorabilia; artifacts & military memorabilia of Illinois service men and women from Spanish American War to present.

Facilities: 50-vol. library on miscellaneous military subjects; educational facilities.

Activities: guided tours; lectures; loan, participatory & temporary exhibitions; school loan service; study clubs; broadcast programs.

Hours & Admission Prices: Mon.-Fri. 9-1:30, Sat. 9-4. No charge; donations accepted. Closed New Year's Day; Christmas; federal holidays. &

Attendance: 1,000 (estimated)

NATIONAL RAILROAD HALL OF FAME, 200 E. Main St., Galesburg, IL 61401-4707. Tel.: 309-345-4634.

E-mail: info@nrrhof.org

Web Site: www.nrrhof.org

Key Personnel: Exec. Dir., Julie King

Institution Type/Description: Railroad Museum.

Collections: railroad history; Hall of Fame inductees.

Hours & Admission Prices: By appointment. No charge.

Geneva

FABYAN VILLA MUSEUM & JAPANESE GARDEN, 1511 S. Batavia Ave., Geneva, IL 60134. Mailing Address: P.O. Box 903, St. Charles, IL 60174-0903. Tel.: 630-377-6424. Fax: 630-377-6424.

E-mail: fabyanvilla@ppfv.org

Web Site: www.ppfv.org/fabyan.htm

Formerly: Fabyan Villa Museum, Dutch Windmill & Japanese Gardens

Founded: 1941.

Congressional District: 14

Key Personnel: Dir., Lynn Dransoff; Museum Asst., Hannah Walters.

Personnel Profile: Part-Time Paid 3; Part-Time Volunteers 23.

Governing Authority: nonprofit. Kane County. Parent Institution: Forest Preserve District of Kane County. Tax-exempt.

Institution Type/Description: Historic House & Gardens: c.1907 home of Col.

George Fabyan, re-designed by Frank Lloyd Wright, located in forest preserve, land formerly the estate of Col. Fabyan.

Collections: Asian aesthetic & cultural artifacts; historic photographs; Frank Lloyd Wright architecture & mission-style furniture; natural history specimens; scientific equipment; WWI documents & cryptology apparatus; Silvio Silvestri sculptures; Japanese garden. Historic Structures: 19th-century Dutch windmill; lighthouse.

Research Fields: local history; military history.

Facilities: gardens.

Activities: guided tours; lectures; slide show; kids programs; adult education programs; concerts.

Publications: The Preservation Advocate.

Hours & Admission Prices: Museum: May & Sept. to mid-Oct. Wed. & Sat.-Sun. 1-4:30; June-Aug. Wed.-Thurs. 1-4, Sat.-Sun. 1-4:30. Japanese Gardens: mid-May to mid-Oct. Sun. & Wed. 1-4. Adults $2, seniors & children $1.

Attendance: 6,075 (accurate)

Membership: Individual $25; Family $35.

Glen Ellyn

STACY'S TAVERN MUSEUM AND GLEN ELLYN HISTORICAL SOCIETY, (M), 800 N. Main St., Glen Ellyn, IL 60137. Mailing Address: P.O. Box 283, Glen Ellyn, IL 60138-0283. Tel.: 630-469-1867, ext. 101.

E-mail: director@gehs.org

Web Site: www.gehs.org

Founded: 1969.

Congressional District: 6

Key Personnel: Exec. Dir., Jane Shupert-Arick; Pres. (V), Ruth Wright; Museum Shop Mgr., Jennifer Porter.

Personnel Profile: Part-Time Paid 2; Part-Time Volunteers 90.

Governing Authority: society; nonprofit organization. Affiliated with the Glen Ellyn Historical Society & the Village of Glen Ellyn. Tax-exempt: 501(c)(3).

Institution Type/Description: Historic House: 1846 Moses Stacy House & Inn; local Glen Ellyn Historical Exhibition & Archives.

Collections: pre-1850 American country furnishings; photographs; Glen Ellyn historical & archival collections.

Research Fields: local history; genealogy.

Facilities: library available for research on premises. Gift items & local history books for sale.

Activities: guided tours; lectures; films; arts festivals; docent program or council; temporary & traveling exhibitions.

Publications: quarterly newsletter; book series: Stories from Glen Ellyn's Past (Vol. I, 2009)

Hours & Admission Prices: Wed. & Sun. 1:30-4:30; other times by appointment. No charge; donations accepted; special tours by arrangement. Closed New Year's Eve & Day; Easter; Mother's Day; Independence Day; Thanksgiving; Christmas. &

Attendance: 3,500 (accurate)

Membership: Senior $20; Individual $25; Family $40; Patron $100.

Glencoe

*** CHICAGO BOTANIC GARDEN, (M),** 1000 Lake Cook Rd., Glencoe, IL 60022-1168. Tel.: 847-835-5440, ext. 0. Fax: 847-835-4484. TDD: 847-835-0790.

E-mail: jmccaffrey@chicagobotanic.org

Web Site: www.chicagobotanic.org

Founded: 1965.

Congressional District: 10

Key Personnel: Chm. (V), Robert F. Finke; Pres. & C.E.O., Sophia Siskel; Exec. Vice Pres. & Dir., Kris S. Jarantoski; Exec. Vice Pres. & C.F.O., Thomas J. Nissly; Vice Pres. Visitor Experience & Business Devel., Harriet Resnick; Vice Pres. Mktg. & Devel., James F. Boudreau; Negaunee Foundation Vice Pres., Science, Dr. Gregory Mueller; Vice Pres. Human Resources, Aida Giglio; Vice Pres. Government Affairs, Ginny Hotaling; Vice Pres. Education & Community Programs, Patsy Benveniste; Dir. Community Gardening, Angela Mason; Dir. Horticulture, Timothy D. Johnson; Dir. Medard & Elizabeth Welch Dir. Plant and Science Conservation, Kayri Havens, Ph.D.; Dir. Ornamental Plant Research, James R. Ault, Ph.D.; Dir. Living Plant Documentation, Boyce Tankersley; Dir. Restoration Ecology, Woman's Bd. Cur. Aquatic Plant & Urban Lake Studies, Robert Kirschner, B.S.; Dir. Foundation & Government Rels., Melissa Matterson; Dir. Visitor Programs & Events, Jodi Zombolo; Assoc. Vice Pres. Devel., Steve Ball; Dir. Planned & Major Gifts, Patricia M. Shanahan; Devel. Officer, Vivienne Jones; Dir. Lenhardt Library, Leora Siegel; Dir. Visitor Operations, Darren Bochat; Dir. Volunteer Svcs., Judith M. Cashen; Dir. Finance & Information Systems, Rob Pollack; Dir. Maintenance, Gregory L. Detlie.

Personnel Profile: Full-Time Paid 234; Part-Time Paid 25; Part-Time Volunteers 1,300; Interns 51.

Volunteer Hours: 112,000

Operating Expenses: 29,930,000

Operating Income: 30,819,000

Governing Authority: managed by the Chicago Horticultural Society. Parent Institution: Forest Preserves of Cook County. Tax-exempt.

Institution Type/Description: Botanic Garden.

Collections: 2.7 million living plants; 9,871 taxa including prairie, woodland & riparian, aquatic native plants, bulbs, edible plants, dwarf conifers, tropical & desert plants, annuals, perennials, roses, ornamental grasses, trees, shrubs, vines & ground covers. Scientific: National Tallgrass Prairie Seed Bank; herbarium. Fine & Decorative Arts: porcelains; prints; paintings; photographs; sculpture; oriental decorative arts; 26 display gardens; 4 natural areas.

Major Exhibits: Healing Plants: Illustrated Herbals, 11/13-2/9/14; The Orchid Show, 2/15/14-3/16/14; Exotic Orchids: Orchestrated in Prints, 2/15/14-5/11/14; Antiques & Garden Fair, 4/11/14-4/13/14; Moth Exhibition, 4/19/14-8/17/14; Model Railroad Garden, 5/10/14-10/26/14; Moku Hanga: The Art of Japanese Woodblock Printing, 5/16/14-8/10/14; Ex Libris: Bookplates Through the Ages, 8/15/14-11/9/14; Succulents: Featuring Redoute Masterpieces, 11/14/14-2/8/15; Wonderland Express, 11/28/14-1/4/15.

Research Fields: Horticulture: environmental horticulture; ornamental plant breeding; plant evaluation & introduction; Chicagoland Grows(R); Plant Science and Conservation: invasive plant science and policy, plant biology; economic botany; plant systematics; plant conservation; habitat fragmentation; climate change; land management; restoration genetics; rare plant restoration; Project Budburst; seed banking; restoration ecology; lake, prairie, river and woodland ecosystems; soil ecology.

Facilities: library; 200-seat auditorium; classrooms; herbarium; events pavilion; cafe. Gifts items for sale.

Activities: guided tram & walking tours; lectures; carillon concerts; formally organized education programs; narrated tram rides; workshops; demonstrations; art exhibits; plant society shows; seasonal plant sales; cultural events; certificate program; bachelors, masters & doctoral programs; urban agriculture programs; after school & camp programs; scout programs; wellness & lifestyle programs; horticultural therapy; public festivals.

Publications: Member Magazine; brochures; Plant Evaluation Notes; educational publications; Bonsai: A Patient Art (published by Yale Press); monthly e-newsletters to garden members & Presidents' Circle members.

Hours & Admission Prices: Memorial Day to Labor Day daily 7am-9pm. Garden: no charge. Parking: $25 per car; society reciprocal members no charge. Grand Tram Tours: April-Oct. Mon.-Fri. 10-4, Sat.-Sun. 10-5. Bright Encounters Tours: Mon.-Fri. 10:15-3:15, Sat.-Sun. 10:15-5:15. Adults $6, seniors $5, children 3-12 $4; discounts to members and groups of 15 & over. &

Attendance: 955,000 (accurate)

Membership: Garden $85; Garden Plus $115; Garden Two Year $160; Garden Plus Two Years $220.

GLENCOE HISTORICAL SOCIETY AT THE EKLUND HISTORY CENTER & GARDEN, 375 Park Ave., Glencoe, IL 60022-1551. Tel.: 847-835-0040.

E-mail: info@glencoehistoricalsociety.org

Web Site: glencoehistoricalsociety.org

Founded: 1937.

Congressional District: 10

Key Personnel: Pres. (V), John Carothers.

Personnel Profile: Part-Time Volunteers 35.

Governing Authority: Tax-exempt.

Institution Type/Description: Historical Society Museum.

Collections: items pertaining to the Village of Glencoe.

Hours & Admission Prices: Museum: Jan.-July & Sept.-Dec. Wed. 10-3, Sun. 1-4. No charge; donations accepted.

Membership: Annual $50 & up.

Glenview

GLENVIEW HISTORY CENTER, 1121 Waukegan Rd., Glenview, IL 60025-3036. Tel.: 847-724-2235. Fax: 847-724-2235.

E-mail: berdaw@juno.com

Web Site: www.glenviewhistory.org

Formerly: Glenview Area Historical Society

Founded: 1965.

Congressional District: 10

Key Personnel: Pres., Mary A. Long; Librarian, Beverly Dawson.

Personnel Profile: Part-Time Volunteers 25.

Governing Authority: society. Tax-exempt: 501(c)(3).

Institution Type/Description: Historical Society Museum: housed in original 1864 farm house.

Collections: local history; clothing; furniture & furnishings; pictures; papers; Naval Air Station Glenview photographs & written material.
Research Fields: local history.
Facilities: 100-vol. library containing books on local history, available for research by prior arrangement; reading room.
Activities: guided tours; illustrated lectures; docent program.
Publications: G.A.H.S. newsletter.
Hours & Admission Prices: Museum: Sun. 1-4; other times by appointment. Library: Tues. 1-4. No charge; donations accepted. Closed holidays.
Attendance: 750 (estimated)
Membership: Individual $25; Business $25-$50; Household $50; Contributing Life $400.

THE GROVE, 1421 Milwaukee Ave., Glenview, IL 60025-1436. Tel.: 847-299-6096. Fax: 847-299-0571.
Web Site: www.thegroveglenview.org
Founded: 1976.
Congressional District: 10
Key Personnel: Dir., Stephan Swanson; Museum Shop Mgr., Kris VanVoorhis.
Personnel Profile: Full-Time Paid 10; Part-Time Paid 27; Part-Time Volunteers 53; Interns 1.
Governing Authority: municipal. Affiliated with the Glenview Park District, 1930 Prairie St., Glenview, IL 60025. Tel. 847-724-5670. Tax-exempt.
Institution Type/Description: Natural History Museum: housed in 1856 Kennicott House, site of original grounds.
Collections: Kennicott family papers; Donald Culross Peattie books; manuscripts; functioning blacksmith shop.
Research Fields: botanical gardens; natural history research; ecological studies.
Facilities: 400-vol. library of Kennicott Family history available for research on premises; botanical garden; nature conservation center; field research station; classrooms; Redfield Cultural Center; Interpretive Center.
Activities: guided tours; lectures; formally organized education programs; docent program; permanent exhibitions.
Publications: monthly newsletter, The Rustlings.
Hours & Admission Prices: Mon.-Fri. 8-4:30, Sat.-Sun. 9-5. No charge; donations accepted. Closed New Year's Day; Independence Day; Christmas. ♿
Attendance: 75,000 (estimated)
Membership: Individual $25; Family $40; Organization & Business $100; Contributing $250; Sustaining $500; Life $1,000.

ILLINOIS PGA GOLF HALL OF FAME MUSEUM, 2901 W. Lake Ave., Glenview, IL 60026-1264. Tel.: 847-724-7272 & 729-5700, ext. 100.
Institution Type/Description: Golf History Museum.
Collections: golf history; Hall of Fame inductees; golf memorabilia & equipment; photographs.
Activities: hands-on exhibits.
Hours & Admission Prices: Daily. No charge.

KOHL CHILDREN'S MUSEUM OF GREATER CHICAGO, 2100 Patriot Blvd., Glenview, IL 60026-8018. Tel.: 847-832-6600. Fax: 847-724-6469. Facebook: Kohl Children's Museum.
E-mail: info@kohlchildrensmuseum.org
Web Site: www.kohlchildrensmuseum.org
Founded: 1985.
Congressional District: 29
Key Personnel: C.E.O. & Pres., Sheridan Turner; Chm., Paul Sutenbach; Controller, Howard Fox; Vice Pres. Programs, Stephanie Bynum; Vice Pres. Devel., Brook McNulty; Sr. Dir. Public Rels. & Mktg., Dave Judy; Vice Pres. Business Affairs, Bill Sanders; Museum Shop Mgr., Norma Fernandez.
Personnel Profile: Full-Time Paid 35; Part-Time Paid 35; Part-Time Volunteers 200; Interns 5.
Governing Authority: nonprofit organization. Tax-exempt: 501(c)(3).
Institution Type/Description: Children's Museum.
Collections: 16 interactive hands-on exhibits including a child-sized supermarket; hands-on house; music; art studio; 2-acre outdoor exhibit area; sculpture trail.
Major Exhibits: Japan and Nature: Spirits of the Seasons, 1/14-7/14.
Activities: daily activities; Focus Field Trips; in-school programs; early childhood connections.
Publications: monthly calendar of activities.
Hours & Admission Prices: Mon. 9:30-12, Tues.-Sat. 9:30-5, Sun. 12-5. Adults & children $9.50, senior citizens $8.50; discounts ACM reciprocal program members; children under one no charge. Closed New Year's Day; Independence Day; Labor Day; Thanksgiving; Christmas Eve & Day. ♿
Attendance: 345,475 (accurate)

Membership: Kids Club Grand $100; Kids Club $110; Kids Club Plus $130.

NAVAL AIR STATION GLENVIEW MUSEUM, 2040 Lehigh Ave., Glenview, IL 60026-1619. Mailing Address: c/o Glenview Hangar One Foundation, P.O. Box 198, Glenview, IL 60028-0198. Tel.: 847-657-0000. Facebook: Naval Air Station Glenview Museum.
E-mail: wam51@comcast.net
Web Site: www.thehangarone.org
Key Personnel: Pres., Bill Marquardt
Institution Type/Description: Naval Air Museum.
Collections: naval aviation history; models of aircraft; photographs; military uniforms & equipment.
Hours & Admission Prices: Sat. 10-5, Sun. 12-5; other times by appointment. Closed major holidays.
Membership: Student/Senior $25; Individual $50; Family $75; Sponsor $150; Cornerstone $500.

Granite City

OLD SIX MILE HISTORICAL SOCIETY AND OLD SIX MILE MUSEUM, 3279 Maryville Rd., Granite City, IL 62040. Mailing Address: P.O. Box 483, Granite City, IL 62040. Tel.: 618-731-3101.
E-mail: sixmilehistorical@gmail.com
Web Site: oldsixmile.wordpress.com
Institution Type/Description: Historical Society Museum.
Collections: local history & culture; period furnishings; personal artifacts; photographs.
Hours & Admission Prices: May-Oct. Sun. 1-4; groups by appointment. No charge.

Grayslake

ROBERT T. WRIGHT COMMUNITY GALLERY OF ART, Library, College of Lake County, 19351 W. Washington St., Grayslake, IL 60030-1148. Tel.: 847-543-2240.
E-mail: sjones@clcillinois.edu
Web Site: gallery.clcillinois.edu/
Founded: 1981.
Congressional District: 8
Key Personnel: Dir., Steven Jones; Museum Shop Mgr., Rachel Wolfe.
Personnel Profile: Full-Time Paid 1; Part-Time Paid 2.
Governing Authority: Parent Institution: College of Lake County. Tax-exempt.
Institution Type/Description: Art Gallery.
Collections: works by local, national & international artists; outdoor sculptures.
Publications: exhibition brochures & catalogs.
Hours & Admission Prices: Mon.-Thurs. 8am-9pm, Fri. 8-4:30, Sat. 9-4:30, Sun. 1-5. No charge; donations accepted. ♿
Attendance: 8,000 (estimated)
Membership: Student $15; Friend of Gallery & Artist $30.

Great Lakes

GREAT LAKES NAVAL MUSEUM, (M), Bldg. 42, 610 Farragut Ave., Great Lakes, IL 60088-3607. Tel.: 847-688-3154. Fax: 847-688-3169. Facebook: Great Lakes Naval Museum.
E-mail: glnm@navy.mil
Web Site: www.history.navy.mil/glnm
Founded: 1991.
Congressional District: 10
Key Personnel: Dir., Jennifer Searcy.
Personnel Profile: Full-Time Paid 4.
Governing Authority: Parent Institution: Naval History and Heritage Command
Institution Type/Description: Military Museum.
Collections: United States Navy history; Navy boot camp.
Hours & Admission Prices: Wed.-Thurs. & Sat. 1-5, Fri. 11-5. No charge. ♿
Attendance: 40,000 (accurate)

Greenup

CUMBERLAND COUNTY HISTORIC AND GENEALOGICAL SOCIETY, 216 Cumberland St., Greenup, IL 62428. Mailing Address: P.O. Box 582, Greenup, IL 62428-0582. Tel.: 217-923-9306.
E-mail: historic@rr1.net
Governing Authority: nonprofit organization. Tax-exempt: 501(c)(3).

Institution Type/Description: Historical Society Museum: housed in the Johnson Building. Listed on the National Register of Historic Places.
Collections: local history & culture; period furnishings; personal artifacts; early toys; photographs. Historic Building: Greenup Depot.
Hours & Admission Prices: Mon.-Sat. 10-4, Sun. 12-4. No charge; donations accepted. Closed Easter; Thanksgiving; Christmas.

Greenville

HOILES-DAVIS MUSEUM, 318 W. Winter St., Greenville, IL 62246-1722. Mailing Address: P.O. Box 376, Greenville, IL 62246-0376. Tel.: 618-664-1590.
Founded: 1955.
Congressional District: 55
Key Personnel: Pres., Sharon Grimes; Vice Pres., Kathy Brewer; Treas., John S. Coleman; Historian, Kevin Kaegy; Sec., Jane Hopkins.
Personnel Profile: Interns 1.
Governing Authority: society. Parent Institution: Bond County Historical Society. Tax-exempt.
Institution Type/Description: Historical Society Museum.
Collections: personal artifacts; books; newspaper articles; period furnishings; Bond County memorabilia from 1818-1960; WWI posters; vintage business items; one room school display; cemetery & original voters records of Bond County.
Activities: temporary exhibitions.
Publications: pamphlet, History of Bond County.
Hours & Admission Prices: Sat. 1-3, Sun. 2-4. No charge; donations accepted.
Attendance: 700 (estimated)
Membership: $15.

Gridley

GRIDLEY TELEPHONE MUSEUM, 318 N. Center St., Gridley, IL 61744. Mailing Address: P.O. Box 370, Gridley, IL 61744. Tel.: 309-747-2284 & 4118 (Tours).
Web Site: www.telephonemuseumofgridley.org
Institution Type/Description: Telephone Museum.
Collections: telephones; records; equipment; switchboards; period furnishings.
Hours & Admission Prices: Tues.-Fri. 1:30-3:30, Sat. by appointment. No charge; donations accepted.

Gurnee

MOTHER RUDD HOME MUSEUM, 4690 Old Grand Ave., Gurnee, IL 60031. Mailing Address: P.O. Box 84, Gurnee, IL 60031-0084. Tel.: 847-263-9540.
E-mail: info@motherrudd.org
Web Site: motherrudd.org
Institution Type/Description: Historical Society Museum: housed in a former inn; built in 1841.
Collections: local history & culture; period furnishings; photographs; personal artifacts.
Activities: research.
Hours & Admission Prices: By appointment. No charge; donations accepted.
Membership: Individual $7; Family $12; Friend $25; Patron $50.

Hanna City

WILDLIFE PRAIRIE STATE PARK, 3826 N. Taylor Rd., Hanna City, IL 61536-9467. Tel.: 309-676-0998. Fax: 309-676-7783.
E-mail: jwilliamson@wildlifeprairiestatepark.org
Web Site: www.wildlifeprairiestatepark.org
Founded: 1978.
Congressional District: 18
Key Personnel: Exec. Dir., Jeanne Williamson; Operations Mgr., Mike McKim; Cur. Animals, Brenda Henon; Mktg. & Public Rels. Coord., Keri Budde; Controller, Becki Salmon; Security, Alan Burgett.
Personnel Profile: Full-Time Paid 10; Part-Time Paid 7; Part-Time Volunteers 300; Interns 3.
Governing Authority: nonprofit organization. Parent Institution: Friends of Wildlife Prairie, 3826 N. Taylor Rd., Hanna City, IL 61536. Tax-exempt.
Institution Type/Description: Wildlife Park.
Collections: animals & plants native to the prairie; pioneer farmstead.
Research Fields: animal behavior.
Facilities: botanical garden; zoological park; nature & conservation center; 650-seat capacity banquet hall; catering facilities; overnight lodging; train ride; pioneer farmstead. Gift items for sale.
Activities: guided tours; lectures; films; formally organized education pro-grams for children, adults & undergraduate college students; docent program; train ride; fishing; hiking; mountain biking.
Publications: members newsletter, Outlook; facilities guide; tourism & lodging brochures; animal guide.
Hours & Admission Prices: mid-March to April & Oct. to mid-Dec. 9-4:30; May-Sept. 9-6:30. Adults $7, children 4-12 $5; discounts to groups; members and children 3 & under no charge.
Attendance: 150,000 (accurate)
Membership: Individual $40; Family & Grandparent $50; Keepers Club $250-$499; Explorers Club $500-$999; Naturalist Club $1,000-$2,499; Director's Club $2,500-$4,999; Benefactor's Club $5,000-$9,999; Rutherford Society $10,000 & up.

Harrisburg

SALINE CREEK PIONEER VILLAGE & POOR HOUSE MUSEUM, 1600 S. Feazel St., Harrisburg, IL 62946. Tel.: 618-253-7342.
Key Personnel: Pres. (V), Mary E. Shackleford.
Volunteer Hours: 100
Governing Authority: Tax-exempt.
Institution Type/Description: Historic Village.
Collections: local history & culture. Historic Buildings: 1877 Poor house; Blockhouse; saddle bag cabin; barn with threshing floor; post office; school; Quaker Church; jail; cabin.
Activities: annual events: Christmas tree lighting; Blessing of County; Life on the Illinois Frontier (May).
Hours & Admission Prices: Tours: Tues.-Sat. 2pm; other times by appointment. Adults $5.
Attendance: 200 (estimated)

Harvard

GREATER HARVARD AREA HISTORICAL SOCIETY, 308 N. Hart Blvd., Harvard, IL 60033-3018. Mailing Address: P.O. Box 505, Harvard, IL 60033-0505. Tel.: 815-943-6141.
Founded: 1977.
Congressional District: 33
Key Personnel: C.E.O., Chm. (V) & Pres., Brian Schultz; Cur., Elaine Fiducci; Sec., Elzora Stoxen; Treas., Jim Finke.
Personnel Profile: Part-Time Volunteers 12.
Governing Authority: municipal. Tax-exempt.
Institution Type/Description: History Museum & Historic Building.
Collections: military uniforms & artifacts; G.A.R. records; Woman's Relief Corps records; country school records; 1890-1950s clothing; WMCW original radio broadcasting equipment; original patents, financial records, catalogs from Starline Company; milk testing equipment; Milk Day memorabilia; toys; early small farm equipment; city tax & Justice of Peace records; early kitchen & household furnishings; period drugstore & doctors' equipment; model schoolroom with textbooks; early hospital records; Future Farmers of America high school records & farm tools; Eldridge Ayer Burbank original & print collection.
Research Fields: genealogical & local history.
Facilities: local history books, newspapers, documents available for research.
Activities: temporary exhibits; school visitations; programs for membership & the public; class reunions.
Publications: membership letters; brochure, G.H.A. Historical Society; historic walking tour guide; reprinted Harvard History Book; curators' report.
Hours & Admission Prices: May-Oct. Wed. 9:30-12, Thurs. 10-12; Sun. 1:30-4; or by appointment, call 815-943-3561. No charge; donations accepted.
Attendance: 800 (estimated)
Membership: Individual $15; Family of 3 or more $18; Organizations & Business $25; Sustaining $30; Annual Contributor $100 & up; Life $150.

Heyworth

SIMPKINS MILITARY HISTORY MUSEUM, 605 E. Cole St., Heyworth, IL 61745. Mailing Address: P. O. Box 336, Heyworth, IL 61745-0336. Tel.: 309-473-3989.
E-mail: simpkinsmuseum@gmail.com
Web Site: www.simpkinsmuseum.com
Formerly: Simpkins War Museum
Founded: 1979.
Congressional District: 11
Key Personnel: Dir., Gary A. Simpkins; Chm. (V), Carol J. Simpkins.
Personnel Profile: Part-Time Volunteers 5; Interns 1.
Institution Type/Description: Military Museum.
Collections: military artifacts from the Civil War, Spanish-American War,

WWI & WWII, Korean War, Vietnam War, Cold War, & Desert Storm; photographs; personal artifacts; 34 star American flag; military uniforms & weapons.
Hours & Admission Prices: Tues., Thurs. & Sat.-Sun. 1-5. No charge; donations accepted. Closed legal holidays except Veterans Day. &
Attendance: 1,250 (estimated)

Highland Park

THE ART CENTER - HIGHLAND PARK (TAC), 1957 Sheridan Rd., Highland Park, IL 60035-2540. Tel.: 847-432-1888. Fax: 847-432-9106. Facebook: The Art Center HP.
E-mail: info@theartcenterhp.org
Web Site: www.theartcenterhp.org
Founded: 1960.
Key Personnel: Dir., Gabrielle Rousso; Pres. (V), Jonathan Plotkin.
Governing Authority: Tax-exempt.
Institution Type/Description: Art Gallery.
Collections: paintings; sculpture; photographs.
Activities: classes; special events; summer camp.
Hours & Admission Prices: Mon.-Sat. 9-4:30; other times by appointment. No charge.
Membership: Individual $45; Family $75; Affiliate $250.

HIGHLAND PARK HISTORICAL SOCIETY, 326 Central Ave., Highland Park, IL 60035-2611. Tel.: 847-432-7090.
E-mail: hphistorical@sbcglobal.net
Web Site: www.highlandparkhistory.com
Founded: 1966.
Congressional District: 10
Key Personnel: Pres. (V), V. Robert Rotering; Archivist, Nancy Webster.
Personnel Profile: Full-Time Paid 1; Full-Time Volunteers 1; Part-Time Paid 2; Part-Time Volunteers 50.
Governing Authority: society. Branch Museums: Jean Butz James Museum, 326 Central Ave., Highland Park, IL; Stupey Log Cabin, 1700 Block of St. Johns Ave., Highland Park, IL; Walt Durbahn Tool Museum, 326 Central Ave., Highland Park, IL; Bob Robinson Bandstand, Laurel & Prospect, Highland Park, IL. Tax-exempt: 170(b)(1)(A).
Institution Type/Description: Local History Museum: housed in 1871 brick Victorian House.
Collections: local history; slides; period artifacts; tools; Ravinia Festival history. Historic House: 1847 Stupey cabin.
Research Fields: local history; politics & government; early photography; transportation; club movements; women's suffrage; education; community music.
Facilities: library of books on local history available on premises by appointment.
Activities: guided tours; lectures; permanent & temporary exhibitions; meetings.
Publications: bimonthly newsletter; quarterly, Lamplighter.
Hours & Admission Prices: Feb. 15-Dec. 15 Wed.-Fri. 1-4, Sun. 2-4. Cabin: spring & summer only. Occasional special exhibit fee charged. Closed holidays. &
Attendance: 2,300 (estimated)
Membership: Individual $40; Dual & Family $65; Friend $100; Supporting $250; Heritage $500, $1,000, $2,000.

WALTER E. HELLER NATURE CENTER, 2821 Ridge Rd., Highland Park, IL 60035-1533. Mailing Address: 636 Ridge Rd., Highland Park, IL 60035-4361. Tel.: 847-433-6901 & 831-3810. Fax: 847-433-8856.
E-mail: heller@pdhp.org
Web Site: www.hellernaturecenter.org
Founded: 1980.
Congressional District: 10
Key Personnel: Site Mgr., Jeff Smith; Naturalist, Leah Holloway; Naturalist, Jessica Reyes.
Governing Authority: municipal. Parent Institution: Park District of Highland Park. Tax-exempt.
Institution Type/Description: Nature Center & Preserve.
Collections: living botanical collections; wildflowers.
Research Fields: prairie vegetation cultivation.
Facilities: library of natural history, environmental education, plant & animal guides & Illinois animal life available for research on premises; nature center; classrooms; 100 acre site; 3 miles of trails.
Activities: guided tours; lectures; environmental & outdoor educational programs.
Publications: quarterly newsletter.

Hours & Admission Prices: Mon.-Fri. 8:30-5, Sat.-Sun. 9-3. No charge; donations accepted. Closed national holidays. &
Attendance: 17,000 (estimated)

Hinsdale

HINSDALE HISTORICAL SOCIETY, (M), 15 S. Clay St., Hinsdale, IL 60521-3244. Mailing Address: P.O. Box 336, Hinsdale, IL 60522-0336. Tel.: 630-789-2600.
E-mail: info@hinsdalehistory.org
Web Site: www.hinsdalehistory.org
Founded: 1975.
Key Personnel: Pres. (V), Cindy Klima; Archivist, Melissa D'Landro.
Personnel Profile: Part-Time Paid 3; Part-Time Volunteers 100; Interns 3.
Governing Authority: board of trustees; nonprofit organization. Subsidiary Institution: Hinsdale History Museum; Immanuel Hall; Roger & Ruth Anderson Architecture Center; R. Harold Zook Home & Studio. Tax-exempt.
Institution Type/Description: Historical Society Museum: housed in a fully restored 1874 middle class residence.
Collections: concentration on Victorian era & local history with special emphasis on the Village of Hinsdale.
Research Fields: residential construction dates; village structures; local architecture.
Facilities: 70-vol. library; garden. Museum-related items for sale.
Activities: preservation workshops; children's & adult educational programs; guided tours; temporary exhibits; lectures.
Hours & Admission Prices: Fri.-Sat. 10-2; other times by appointment. No charge; donations accepted.
Attendance: 600 (estimated)
Membership: Individual $25; Family $40; Corporate $50; Friend $100; Lifetime $500.

ROBERT CROWN CENTER FOR HEALTH EDUCATION, 21 Salt Creek Lane, Hinsdale, IL 60521-2902. Tel.: 630-325-1900. Fax: 630-325-3970.
E-mail: rcche@robertcrown.org
Web Site: www.robertcrown.org
Founded: 1958.
Congressional District: 13
Key Personnel: C.E.O., Kathleen M. Burke; Chm. (V), Ross Forbes; Mgr. Chicago Campus, Tierra Johnson; Mgr. Aurora Campus, Althea Motley; Dir. Strategic Mktg., Jon Scoles.
Personnel Profile: Full-Time Paid 15; Part-Time Paid 5; Part-Time Volunteers 86; Interns 1.
Governing Authority: nonprofit. Tax-exempt: 501(c)(3).
Institution Type/Description: Health Education Center.
Collections: human body; health; healthy living.
Facilities: education facilities.
Activities: general health education programs; family living & sex education; drug abuse prevention; summer camps; evening programs; classroom & outreach programs.
Publications: puberty education booklets; reality check newsletters; annual report.
Hours & Admission Prices: School Groups: mid-Sept. to mid-June Mon.-Fri. 9-2:30 by appointment. Admission $5; teachers no charge. Outreach: $7 per person. Closed national holidays. &
Attendance: 120,000 (accurate)

Ingleside

FOX LAKE-GRANT TOWNSHIP AREA HISTORICAL SOCIETY, 411 Washington St., Ingleside, IL 60041. Mailing Address: P.O. Box 224, Ingleside, IL 60041. Tel.: 847-587-0544.
E-mail: flgranthall@yahoo.com
Congressional District: 8
Key Personnel: Pres. (V), Nancy Kubalanza.
Governing Authority: Tax-exempt.
Institution Type/Description: Historical Society Museum.
Collections: local history & culture; period furnishings; personal artifacts; photographs; scrapbooks.
Publications: paper, Hometown Heritage.
Hours & Admission Prices: Feb.-Dec. 1st & 3rd Sun. of the month 1-4. No charge; donations accepted. &
Attendance: 800 (estimated)
Membership: Individual $10; Family $25, $50, $100.

Jacksonville

DAVID STRAWN ART GALLERY, (M), 331 W. College Ave., Jacksonville, IL 62650-2474. Mailing Address: P.O. Box 1213, Jacksonville, IL 62651-1213. Tel.: 217-243-9390.
Web Site: www.strawnartgallery.org
Founded: 1873.
Congressional District: 18
Key Personnel: Pres., Addie Coultas; Dir., Kelly M. Gross.
Personnel Profile: Part-Time Paid 3; Part-Time Volunteers 100.
Governing Authority: society. Parent Institution: The Art Association of Jacksonville.
Institution Type/Description: Art Gallery: housed in 1915 former home of Dr. David Strawn.
Collections: early Mississippi Indian pottery; Miriam Cougar Allen collection of antique & collectible dolls; Charles Prentice Thompson classical collection.
Activities: formally organized education programs.
Publications: pamphlet, Pre-Columbian pottery; books, Strawn Family and Home History; The Art Association of Jacksonville History.
Hours & Admission Prices: Sept.-May Tues.-Sat. 4-6, Sun. 1-3. No charge. Closed holidays. &
Attendance: 5,974 (estimated)
Membership: Student & Associate (lives more then 50 miles away) $20; Single $35; Family $50; Sustaining $100; Individual Life $600; Couple Life $1,000.

Joliet

JOLIET AREA HISTORICAL MUSEUM, (M), 204 N. Ottawa St., Joliet, IL 60432-4007. Tel.: 815-723-5201. Fax: 815-723-9039.
E-mail: t.contos@jolietmuseum.org
Web Site: www.jolietmuseum.org
Formerly: Joliet Area Historical Society
Founded: 1999.
Congressional District: 4
Key Personnel: Exec. Dir., Anthony B. Contos; Museum Shop Mgr., Elaine Stonich.
Personnel Profile: Full-Time Paid 5; Part-Time Paid 3; Part-Time Volunteers 50; Interns 1.
Governing Authority: nonprofit. Tax-exempt: 501(c)(3).
Institution Type/Description: Local History Museum.
Collections: Joliet industries, businesses, schools & citizens; photos; advertising items.
Facilities: orientation theater; 800 sq. ft. exhibit space.
Activities: lectures; temporary exhibitions; public programs; bus tours; guided tours; fundraisers.
Publications: quarterly newsletter, Legacy.
Hours & Admission Prices: Tues.-Sat. 10-5, Sun. 12-5. Adults $6, senior citizens 60 & over $5, youth 4-17 $3; discounts to AAM members; members no charge. Closed New Year's Eve & Day; Good Friday; Easter; Memorial Day; Independence Day; Labor Day; Thanksgiving; Christmas Eve & Day. &
Attendance: 25,000 (accurate)
Membership: Senior $20; Senior Family, Student & Teacher $25; Individual $35; Family $45.

LAURA A. SPRAGUE ART GALLERY, Joliet Junior College Main Campus, 1215 Houbolt Rd., Spicer-Brown Hall, Joliet, IL 60431-8938. Tel.: 815-280-2423 & 2223.
Web Site: www.jjc.edu/dept/finearts/lauraspraguegallery.htm
Key Personnel: Gallery Dir., Joe Milosevich
Institution Type/Description: Art Gallery.
Collections: paintings.
Hours & Admission Prices: Mon.-Fri. 8-8.

RIALTO SQUARE THEATRE, 15 E. Van Buren St., Joliet, IL 60432-4211. Tel.: 815-726-7171.
Institution Type/Description: Historic Building Museum: housed in a former movie theatre; built in 1926.
Collections: theatre history & culture; photographs; period furnishings; personal artifacts.
Facilities: theatre.
Activities: rental facilities; tours; children's programs; performances; shows; concerts.
Hours & Admission Prices: Tours: Tues. 1:30; other times by appointment. $5 per person.

SLOVENIAN WOMEN'S UNION HERITAGE MUSEUM, 431 N. Chicago St., Joliet, IL 60432-1785. Tel.: 815-727-1926. Fax: 815-723-0670.
E-mail: swvhome@swva.org
Web Site: www.swua.org
Institution Type/Description: Heritage Museum.
Collections: Slovenian history & culture; period clothing; musical instruments; religious articles; photographs; art; furniture.
Facilities: 1,000-vol. library.
Hours & Admission Prices: Mon.-Fri. 10-2. No charge; donations accepted.

Kampsville

CENTER FOR AMERICAN ARCHEOLOGY, Hwy. 100, Kampsville, IL 62053. Mailing Address: P.O. Box 366, Kampsville, IL 62053-0366. Tel.: 618-653-4316. Fax: 618-653-4232.
E-mail: caa@caa-archeology.org
Web Site: www.caa-archeology.org
Formerly: Kampsville Archeological Museum
Founded: 1970.
Congressional District: 13
Key Personnel: Pres. (V), Jane E. Buikstra, Ph.D.; Museum Shop Mgr., Dr. Carol Colaninno.
Personnel Profile: Full-Time Paid 10; Part-Time Paid 3; Interns 3.
Governing Authority: nonprofit organization; board of directors. Tax-exempt: 501(c)(3).
Institution Type/Description: Archaeology Museum.
Collections: archaeological materials representing cultures dating back to 10,000 BC.
Research Fields: archaeology research.
Facilities: archaeology library of books available for research with special permission from director; research station; separate laboratory operation.
Activities: guided tours; lectures; formally organized education programs for children, adults, undergraduate & graduate students.
Publications: archaeology books; Research Series; Technical Report Series.
Hours & Admission Prices: May-Oct. Tues.-Fri. 10-5, Sat. 10-4, Sun. 12-4. No charge; donations accepted. Fee for tours. &
Attendance: 3,000 (estimated)
Membership: $25; $45; $50; $75; $100; $250; $500; $1,000 & up.

Kankakee

KANKAKEE COUNTY MUSEUM, 801 S. 8th Ave., Kankakee, IL 60901-4744. Tel.: 815-932-5279. Fax: 815-932-5204.
Web Site: www.kankakeecountymuseum.com
Formerly: Kankakee County Historical Society Museum
Founded: 1906.
Congressional District: 11
Key Personnel: Dir., Connie Licon; Pres. Bd. (V), Bob de Oliveira; Coord. Research, Jorie Walters; Museum Asst., Camille Grosso; Cur. Exhibits, Katey Moore; Cur. Collections, Nick Grad; Coord. Education, Sarah Schoon; Office Mgr., Julie White.
Personnel Profile: Full-Time Paid 2; Part-Time Paid 5; Part-Time Volunteers 50; Interns 1.
Governing Authority: nonprofit organization; board of directors. Subsidiary Institution: Kankakee County Historical Society. Tax-exempt.
Institution Type/Description: County Historical Museum.
Collections: Native American artifacts; household objects; firearms; textiles; framed objects; decorative arts; agricultural items; toys; political item tools; edge weapons; approximately 75 plaster & marble studies of the sculptor George Grey Barnard; 1855 home of Dr. A.L. Small; boyhood home of Illinois Gov. Len Small 1921-1929; 1904 one room schoolhouse.
Research Fields: genealogy.
Facilities: 5,000-vol. library of history & genealogy. Traditional gift items & books on county history for sale.
Activities: guided tour groups; lectures; slide programs; interpretive programs; research for fee; traveling & special exhibits by appointment; lecture series; temporary exhibits. Special Events: Rhubarb Fest in May; Gallery of Trees Christmas event in December.
Publications: brochure, Kankakee County Historical Society Museum; museum newsletter.
Hours & Admission Prices: Tues.-Thurs. 10-4, Sat. 1-4. Adults $3, seniors & students $2, children 15 & under $1; members & Tues. no charge. Closed holidays. &
Attendance: 25,000 (estimated)
Membership: Annual $20; Family $40; Business & Professional $75; Life $250.

WILLOWHAVEN INTERPRETIVE CENTER, 1451 N. 4000 E Rd., Kankakee, IL 60901. Mailing Address: Bourbonnais Township Park District, 456 N. Kennedy Dr., Bourbonnais, IL 60914. Tel.: 815-933-9905, ext. 472.
E-mail: nicole@btpd.org
Key Personnel: Facility Mgr., Nicole Jenkins
Institution Type/Description: Natural History Museum & Aquarium.
Collections: nature; wildlife; conservation; natural history; live animals.
Facilities: aquarium.
Activities: children's reading area; microscope lab table.
Hours & Admission Prices: Call for hours.

Kenilworth

THE KENILWORTH HISTORICAL SOCIETY, 415 Kenilworth Ave., Kenilworth, IL 60043-1134. Tel.: 847-251-2565. Fax: 847-251-2565.
E-mail: kenilworthhistory@sbcglobal.net
Web Site: www.kenilworthhistory.org
Founded: 1922.
Congressional District: 10
Key Personnel: Pres., Carol M. Schulz; Vice Pres., Sheila Mitchell; Treas., Peter Tyor; Sec., Rachel Noel; Cur., Melinda F. Kwedar.
Personnel Profile: Part-Time Paid 3; Part-Time Volunteers 15.
Governing Authority: nonprofit. Tax-exempt: 501(c)(3).
Institution Type/Description: Historical Society Museum.
Collections: displays materials related to village & residents; house histories; costumes; local documents & artifacts; heirloom needlepoint rug; photographs; architectural & personal archival materials related to architect George W. Maher and other Kenilworth families & organizations.
Research Fields: histories of homes; biographical material of Kenilworth residents since 1882.
Facilities: library pertaining to books about Kenilworth & its residents for in-house use.
Activities: organized education programs for children; temporary exhibitions. Annual Events: Christmas Open House; Memorial Day Open House.
Publications: Kenilworth Tree Stories; George Washington Maher in Kenilworth; Joseph Sears & His Kenilworth; Jens Jensen in Kenilworth.
Hours & Admission Prices: Mon. 9-4:30, Thurs. 9-12; other times by appointment. No charge; donations accepted. &
Attendance: 750 (estimated)
Membership: Regular $35; Sustaining $100; Contributor $200; Life $1,000; Life Benefactor $5,000.

Knoxville

KNOX COUNTY MUSEUM, c/o City Hall, 33 Public Square, Knoxville, IL 61448-1378. Tel.: 309-289-2814. Fax: 309-289-8825.
E-mail: peg@leag.biz
Web Site: www.kville.org
Founded: 1953.
Congressional District: 47
Key Personnel: C.E.O. & Pres. (V), Peg Bivens; Cur., Sally Hutchcroft.
Personnel Profile: Part-Time Paid 1; Part-Time Volunteers 1.
Governing Authority: Affiliated with Knox County Historical Sites, Inc.
Institution Type/Description: Historical Society Museum.
Collections: Abingdon pottery. Historic Buildings: 1839 Knox County Court House; 1845 Knox County Jail; 1832 John G. Sanburn Log Cabin.
Facilities: Gift items for sale.
Activities: guided tours.
Hours & Admission Prices: Year-round Tues.-Sat. 10-2; May-Sept. Sun. 2-4; Oct.-April tours by appointment; closed Sun. No charge, donations accepted. &
Attendance: 1,000 (estimated)
Membership: Annual $10; Lifetime $100.

La Grange

LA GRANGE AREA HISTORICAL SOCIETY, (M), 444 S. LaGrange Rd., La Grange, IL 60525-2448. Tel.: 708-482-4248. Fax: 708-482-4248. Facebook: Le Grange Area Historical Society.
E-mail: lagrangehistory@sbcglobal.net
Web Site: lagrangehistory.org
Founded: 1972.
Key Personnel: Pres. (V), Karen Lynch
Institution Type/Description: Historical Society Museum: housed in the Vial House, built in 1874.
Collections: local history & culture; period furnishings; personal artifacts; photographs; costumes.

Hours & Admission Prices: Open House: last Sun. each month 1-4. Research: Wed. 9:30-12:30; other times by appointment. No charge; donations accepted. &

LaFox

GARFIELD FARM MUSEUM, 3N016 Garfield Rd., LaFox, IL 60147-0403. Mailing Address: P.O. Box 403, LaFox, IL 60147-0403. Tel.: 630-584-8485. Fax: 630-584-8522.
E-mail: info@garfieldfarm.org
Web Site: garfieldfarm.org
Founded: 1977.
Congressional District: 14
Key Personnel: Dir., Jerome Martin Johnson; Dir. Museum Operations, William Wolcott; Asst. Site Mgr., Joseph Coleman; Pres. C.H.A.L. Inc., Helen Bauer; Pres. G.H.S. Inc., Susan Lloyd; Treas. G.H.S. Inc., Marty Germann.
Personnel Profile: Full-Time Paid 4; Part-Time Paid 1; Part-Time Volunteers 200.
Governing Authority: nonprofit. Parent Institution: Campton Historic Agricultural Land, Inc. Subsidiary Institution: Garfield Heritage Society Inc. Tax-exempt: 501(c)(3).
Institution Type/Description: Living Historic Farm & Inn Museum.
Collections: over 2,000 19th-century documents, diaries & artifacts of the Garfield and Mighell families; 600 photographs; family furnishings & tools; manuscripts. Historic Structures: 1842 hay barn; 1846 inn; 1849 horse barn; 1890s granary; 1906 dairy barn; reconstructed chicken house; 1840s Burr family farmhouse; 1859 Edward Garfield/Mongerson Brothers Farmstead.
Research Fields: 1830-1850, Northern Illinois settlement, farm, tavern & rural life history.
Facilities: 370 acres farmland with prairie.
Activities: guided tours; lectures; rental gallery; docent program; organized education programs for children, undergraduate & graduate college students; training programs for professional museum workers. Museum Sponsors: Fall Festival in October.
Publications: biannual brochure; newsletter, Fairfield-Campton Crier; booklet, Garfield Farm: Setting the Benchmark; Prairie Messenger.
Hours & Admission Prices: June-Sept. Wed. & Sun. 1-4; other times by appointment only. Adults $3, youth groups under 13 $2; members no charge. Closed Thanksgiving; Christmas. &
Attendance: 10,000 (estimated)
Membership: Individual $20; Family $30; Patron $75; Commercial $100; Life $1,000.

LaSalle

HEGELER CARUS MANSION, 1307 7th St., LaSalle, IL 61301. Tel.: 815-224-6543. Fax: 815-224-5801. Facebook: Hegeler Carus Mansion.
E-mail: execdirector@hegelercarus.org
Web Site: www.hegelercarus.org
Institution Type/Description: Historic Home Museum: built in 1874.
Collections: Hegeler & Carus family history; period furnishings; personal artifacts.
Activities: guided tours; lectures; school groups; rental facilities.
Hours & Admission Prices: Wed.-Sun. 12, 1, 2, & 3; groups by appointment. Suggested Donation: adults $10, students K-12 $5; discounts to groups of 20 or more. Closed New Year's Eve & Day; Easter; Memorial Day; Independence Day; Labor Day; Thanksgiving; Christmas Eve & Day. &

Lake Forest

LAKE FOREST-LAKE BLUFF HISTORICAL SOCIETY, (M), 361 E. Westminster, Lake Forest, IL 60045-2255. Tel.: 847-234-5253.
E-mail: info@lflbhistory.org
Web Site: www.lflbhistory.org
Founded: 1972.
Congressional District: 10
Key Personnel: Exec. Dir., Janice Hack; Cur., Laurie Stein.
Personnel Profile: Part-Time Paid 3; Part-Time Volunteers 75; Interns 1.
Governing Authority: private; nonprofit organization. Tax-exempt: 501(c)(3).
Institution Type/Description: Historical Society Museum.
Collections: concentration on Lake Forest & Lake Bluff history.
Research Fields: Lake Forest; Lake Bluff; estate architects.
Facilities: 1,000-vol. library; 1,000 sq. ft. exhibit space. Museum-related items for sale.
Activities: lectures; temporary exhibitions; tours.
Publications: quarterly newsletter, The Heritage Timepiece; Lake Forest Day: 100 Years of Celebration; Walter Frazier: Homes of Chicago's North Shore 1940-1970.

Hours & Admission Prices: Tues.-Thurs. 10-4, Fri. by appointment, Sun. 1-4. No charge. Closed major holidays.
Attendance: 1,500 (estimated)
Membership: Individual $40; Family $65; Silver $100; Gold $250; Platinum $500; Museum $1,000.

Lerna

LINCOLN LOG CABIN STATE HISTORIC SITE, (M), 402 S. Lincoln Hwy. Rd, Lerna, IL 62440. Tel.: 217-345-1845. Fax: 217-345-6472. Facebook: Lincoln Log Cabin.
E-mail: hpa.lincolnlog@illinois.gov
Web Site: www.lincolnlogcabin.org
Founded: 1936.
Congressional District: 19
Key Personnel: Site Mgr., Matthew Mittelstaedt; Museum Shop Mgr., Susan Colgrove.
Personnel Profile: Full-Time Paid 9; Part-Time Paid 15; Part-Time Volunteers 200.
Governing Authority: state. Affiliated with the State of Illinois Historic Preservation Agency, Old State Capitol, Springfield, IL 62701. Tax-exempt: 501(c)(3).
Institution Type/Description: Historic Houses & Site: reconstructed cabin and farm of Thomas and Sara Lincoln on original site.
Collections: period furnishings.
Research Fields: Lincoln family; Pioneer farm life.
Facilities: visitor center; picnic areas.
Activities: guided tours; permanent exhibitions. Museum Sponsors: living history in summer months.
Publications: bimonthly, newsletter, Historic Illinois; site brochure.
Hours & Admission Prices: Daily 9-5. No charge; donations accepted. Closed New Year's Day; Thanksgiving; Christmas. &
Attendance: 126,000 (accurate)
Membership: $25; $50; $110; $250; $500; $1,000; $2,500.

REUBEN MOORE HOME STATE HISTORIC SITE, 400 S. Lincoln Hwy. Rd., Lerna, IL 62440-2840. Mailing Address: 400 S. Lincoln Hwy. Rd., P.O. Box 100, Lerna, IL 62440-0100. Tel.: 217-345-1845. Fax: 217-345-6472. Facebook: Lincoln Log Cabin.
E-mail: hpa.lincolnlog@illinois.gov
Web Site: www.lincolnlogcabin.org
Founded: 1929.
Congressional District: 19
Key Personnel: Site Supt., Matthew Mittelstaedt.
Personnel Profile: Part-Time Paid 2.
Governing Authority: state. A part of the State of Illinois, Illinois Historic Preservation Agency, Old State Capitol, Springfield, IL 62701. Tax-exempt: 501(c)(3).
Institution Type/Description: Historic House Museum: mid-19th century home of Matilda Moore, stepsister of Abraham Lincoln; site of Lincoln's last visit with his stepmother & family before leaving to assume the presidency in 1861.
Collections: period furnishings.
Research Fields: 1850s-70s rural village life in east central Illinois.
Activities: permanent exhibitions.
Publications: bimonthly newsletter, Historic Illinois; brochure.
Hours & Admission Prices: June-Aug. Wed.-Sun.; Sept.-May by appointment only. No charge. Closed New Year's Day; Thanksgiving; Christmas. &
Attendance: 4,326 (accurate)

Lewistown

* **DICKSON MOUNDS MUSEUM,** 10956 N. Dickson Mounds Rd., Lewistown, IL 61542-9112. Tel.: 309-547-3721. Fax: 309-547-3189. TDD: 217-782-9175.
E-mail: wiant@museum.state.il.us
Web Site: www.museum.state.il.us/ismsites/dickson/
Founded: 1927.
Congressional District: 17
Key Personnel: Dir. Dickson Mounds Museum, Dr. Michael D. Wiant; Assoc. Cur., Dr. Michael Conner; Asst. Cur., Christa Christensen; Asst. Cur., Alan D. Harn; Exhibits Preparator, Kelvin Sampson; Office Mgr., Kim Dunnigan.
Personnel Profile: Full-Time Paid 10; Part-Time Paid 1; Part-Time Volunteers 5.
Governing Authority: state. Parent Institution: Illinois State Museum, 502 S. Spring St., Springfield, IL 62706. Tax-exempt.
Institution Type/Description: Anthropology Museum & Site.
Collections: archaeological materials from West Central Illinois; Mississippian & Middle Woodland sites on grounds; American Indian cultures of Illinois;

Paleo-Indian to historic periods. Historic Buildings: 1839 school; 1850 Toll House & 1907 school.
Research Fields: archaeology; anthropology; local & state history.
Facilities: discovery center; resource center; archaeology laboratories & sites; picnic grounds; auditorium; outdoor stage; coffee shop. Museum-related items for sale.
Activities: guided tours; formally organized educational programs; workshops; demonstrations; performances; conferences; festivals; participatory programs; permanent & special exhibits; audiovisual programs.
Publications: quarterly newsletter, Fieldnotes; Dickson Mounds Museum Anthropological Studies.
Hours & Admission Prices: Daily 8:30-5. No charge; donations accepted. Closed New Year's Day; Thanksgiving; Christmas. &
Attendance: 40,103 (accurate)
Membership: Individual $35; Family $50; Contributing $100; Sustaining $300.

Libertyville

LIBERTYVILLE-MUNDELEIN HISTORICAL SOCIETY, INC., 413 N. Milwaukee Ave., Libertyville, IL 60048-2247. Tel.: 847-362-2330. Fax: 847-362-0006.
Founded: 1955.
Congressional District: 10
Key Personnel: Chm. Acquisitions Committee, Beverly Schar.
Personnel Profile: Part-Time Volunteers 35.
Governing Authority: nonprofit organization. Parent Institution: Cook Memorial Public Library District. Tax-exempt: 501(c)(3).
Institution Type/Description: History Museum: housed in 1878 Ansel B. Cook home. Listed on The National Register of Historic Places.
Collections: pioneer artifacts; Civil War items; historical books, pictures, maps & papers from pioneer families of the area.
Research Fields: history & artifacts of the Libertyville-Mundelein & Lake County areas.
Facilities: 500-vol. library of papers, pamphlets & books on local history available for use on premises to members & their guests; reading room.
Activities: guided tours of Victorian home. Museum Sponsors: Victorian period Christmas.
Publications: monthly newsletter.
Hours & Admission Prices: Summer: Sun. 2-4; Winter: call Faith Sage 847-362-3992 for appointment. Adults $2, students & seniors $1. &
Attendance: 1,100 (accurate)
Membership: Junior under 18 & Senior Citizens over 65 $5; Individual $8; Family $15; Contributor $25; Sponsor $50; Life $100.

Lincoln

HERITAGE IN FLIGHT MUSEUM, Logan County Airport, 1351 Airport Rd., Lincoln, IL 62656-5444. Tel.: 217-732-3333.
Institution Type/Description: History Museum: housed in a former WWII barracks from Camp Ellis.
Collections: aviation & military history; aircraft from helicopters to carrier jets; military & personal artifacts; photographs.
Hours & Admission Prices: Sat. 9-12 by appointment.

LINCOLN HERITAGE MUSEUM, 300 Keokuk St., Lincoln, IL 62656-1699. Tel.: 217-732-3155, ext. 295.
E-mail: rkeller@lincolncollege.edu
Web Site: www.lincolncollege.edu/museum
Formerly: Lincoln College Museum
Founded: 1942.
Congressional District: 18
Key Personnel: Dir. & Cur., Ron Keller; Asst. Dir. & Cur., Anne Suttles.
Personnel Profile: Full-Time Paid 1; Part-Time Volunteers 3.
Governing Authority: Parent Institution: Lincoln College. Tax-exempt.
Institution Type/Description: History Museum.
Collections: Lincoln's life, heritage & history; personal artifacts; Presidential artifacts; photographs; period furnishings.
Facilities: Museum-related items for sale.
Activities: educational outreach; temporary exhibits; special programs.
Publications: newsletter, The Lincoln.
Hours & Admission Prices: Mon.-Fri. 9-4, Sat. 1-4. No charge; donations accepted. Closed Federal holidays except Lincoln's Birthday. &
Attendance: 4,000 (estimated)

LOGAN COUNTY GENEALOGICAL & HISTORICAL SOCIETY, 114 N. Chicago St., Lincoln, IL 62656. Tel.: 217-732-3200. www.rootsweb.ancestry.com/~illcghs/.
E-mail: lcghs1@hotmail.com
Web Site: www.logancoil-genhist.org

Founded: 1979.
Personnel Profile: Part-Time Volunteers 12.
Institution Type/Description: Historical Society Museum.
Collections: local history; books; family histories; birth, marriage, & death records; photographs.
Facilities: library.
Activities: research.
Publications: quarterly, Roots & Branches.
Hours & Admission Prices: Tues.-Fri. 11-4, Sat. 10-1; call to confirm hours. No charge; donations accepted.
Attendance: 1,500 (accurate)
Membership: Single $15; Dual $20.

POSTVILLE COURTHOUSE STATE HISTORIC SITE, 914 Fifth St., Lincoln, IL 62656-2308. Mailing Address: P.O. Box 355, Lincoln, IL 62656-0355. Tel.: 217-732-8930 & 8687.
Founded: 1953.
Personnel Profile: Part-Time Paid 1.
Governing Authority: state. Parent Institution: Illinois Historic Preservation Agency, Old State Capitol, Springfield, IL 62701. Tax-exempt: 501(c)(3).
Institution Type/Description: History Site/Museum.
Collections: replica of the first Logan County Illinois Court House; part of old 8th Judicial Circuit from 1839-48 where Abraham Lincoln practiced law; period furnishings including Lincoln's rocker; Mary Lincoln's garden; well.
Research Fields: Lincoln; the Illinois 8th Judicial Circuit; local history.
Facilities: garden.
Activities: guided tours; lectures; organized tours for schoolchildren & tour buses. Special Events: 1800s Craft Fair; Abe Lincoln's Birthday; Christmas Open House; Quilt Show.
Hours & Admission Prices: Tues.-Sat. 12-4; other times by appointment. No charge; donations accepted. Closed New Year's Day; Lincoln's Birthday; Veterans Day; Independence Day; Thanksgiving; Christmas.
Attendance: 2,600 (accurate)

Lisle

JURICA-SUCHY NATURE MUSEUM, (M), Benedictine University, 5700 College Rd., Lisle, IL 60532-2851. Tel.: 630-829-6545. Fax: 630-829-6547.
E-mail. TSUCHY@ben.edu
Web Site: alt.ben.edu/resources/J_Museum/index.htm
Formerly: Jurica Nature Museum
Founded: 1970.
Congressional District: 14
Key Personnel: Pres., Dr. William Carroll; Chief Cur., Rev. Theodore D. Suchy, OSB
Personnel Profile: Part-Time Paid 5; Part-Time Volunteers 2.
Governing Authority: college. Parent Institution: Benedictine University. Tax-exempt.
Institution Type/Description: Natural History Museum.
Collections: invertebrate; vertebrate; mammals, reptiles, fish, insects, birds; plants; skeletons of fossils; bird eggs; corals; crustaceans; minerals; American Indian artifacts; large mounted animals.
Facilities: library; reading room; classrooms; laboratory.
Activities: guided tours; formally organized education programs for elementary, high school & college students; permanent & temporary exhibitions.
Hours & Admission Prices: Mon.-Fri. 1-5, Sun. 2-4; groups by advance arrangement. No charge; donations accepted. Closed school vacations. &
Attendance: 6,000 (accurate)

✳ THE MORTON ARBORETUM, (M), 4100 Illinois Rte. 53, Lisle, IL 60532-1293. Tel.: 630-968-0074. Fax: 630-719-2433.
E-mail: trees@mortonarb.org
Web Site: www.mortonarb.org
Founded: 1922.
Congressional District: 13
Key Personnel: Pres. & C.E.O., Dr. Gerard T. Donnelly; Volunteer Coord., Kristin Sabatino; Museum Shop Mgr., Jacqueline Fucilla.
Personnel Profile. Full-Time Paid 150; Part-Time Paid 100; Part-Time Volunteers 1,000; Interns 2.
Volunteer Hours: 62,000
Operating Expenses: 27,000,000
Operating Income: 27,000,000
Governing Authority: nonprofit organization. Tax-exempt: 501(c)(3).
Institution Type/Description: Arboretum.
Collections: living collections including 190,000 plant specimens representing 4,200 kinds of plants; trees, shrubs & woody plants arranged by taxonomic, geographic, special habitat, horticultural, rare & endangered collections; herbarium included 189,000 dried plant specimens; library includes 39,000 botanical books & periodicals, and works of art.
Research Fields: tree health; tree improvement; woodland conservation.
Facilities: library; education center; visitor center; children's garden; restaurant; rental facilities; store; 1,700 acre site.
Activities: education programs for youth, family, schools & adults; special events & exhibits; guided tours; theater-hikes; 5K run; plant sale; bicycling; self-guided visits over 16 miles of trails & 9 miles of roads. Annual Events: Arbor Day Celebration; Fall Color Festival.
Publications: members newsletter; education program; plant information brochures; members e-mail; annual report (web-based).
Hours & Admission Prices: Grounds: daily 7am to sunset. Buildings: call for hours. Adults $12, seniors $11, children 2-17 $9; discounts on Wed; children under 2 & members no charge. &
Attendance: 808,542 (accurate)
Membership: Morton I $55; Morton II $80; Family $115; Morton Plus $145; Friend $250; Partner $500; Thornhill Society $1,000 & up.

THE MUSEUMS AT LISLE STATION PARK, 921 School St., Lisle, IL 60532-1951. Tel.: 630-968-0499.
E-mail: museum@lisleparkdistrict.org
Web Site: www.lisleparkdistrict.org
Founded: 1978.
Congressional District: 13
Key Personnel: Cur., Brian Failing.
Personnel Profile: Part-Time Paid 1; Part-Time Volunteers 70; Interns 2.
Governing Authority: regional government. Parent Institution: Lisle Park District, 1925 Ohio St., Lisle 60532. Tax-exempt.
Institution Type/Description: History Museum.
Collections: focuses on the history of the Lisle community from the settlement period of the 1830s-present; history of the CB&Q railroad in Lisle. Historic Structures: 1874 CB&Q depot with living quarters; 1881 CB&Q caboose; 1830s Beaubien Tavern; 1850s Greek Revival farmhouse; 19th century style blacksmith shop.
Facilities: 1,200 sq. ft. exhibit space; facility rental available. Handcrafted items for sale.
Activities: formal educational programs; guided tours; lectures; temporary & traveling exhibitions. Annual Events: Lisle Depot Days in September; Chicago History Author Series during Fall, Lights of Lisle & Once Upon A Christmas in December.
Publications: seasonal newsletter, Lisle Heritage Society Newsletter.
Hours & Admission Prices: June-Sept. Tues. 1-8, Thurs. 1-4, Sat. 10-4; Sept.-May Sat. 1-4. No charge; donations accepted. &
Attendance: 8,000 (accurate)

Lockport

THE GAYLORD BUILDING - NATIONAL TRUST HISTORIC SITE, 200 W. 8th St., Lockport, IL 60441-2878. Tel.: 815-838-9400. Fax: 815-838-9449.
E-mail: info@gaylordbuilding.org
Web Site: www.gaylordbuilding.org
Founded: 1987.
Congressional District: 13
Key Personnel: Dir., Mark S. Harmon, M.A.
Personnel Profile: Full-Time Paid 2; Part-Time Paid 2; Part-Time Volunteers 20; Interns 1.
Governing Authority: Parent Institution: National Trust for Historic Preservation. Tax-exempt: 501(c)(3).
Institution Type/Description: Historic Building: housed in the former building that stored construction materials for the Illinois & Michigan Canal; built in 1838. A Historic Site.
Collections: local history & culture; period artifacts; canal history; photographs.
Facilities: visitor center; restaurant.
Activities: special events.
Hours & Admission Prices: Mon.-Sat. 10-5, Sun. 11-5. No charge; donations accepted; Closed major holidays. &
Attendance: 87,000 (estimated)

✳ ILLINOIS STATE MUSEUM/LOCKPORT GALLERY, 201 W. 10th St., Ste. 1 W, Lockport, IL 60441-4039. Tel.: 815-838-7400. Fax: 815-838-7448.
E-mail: jjaskowiak@museum.state.il.us
Web Site: www.museum.state.il.us/ismsites/lockport/
Founded: 1987.
Congressional District: 13
Key Personnel: Dir., Jennifer Jaskowiak.
Personnel Profile: Full-Time Paid 2; Part-Time Volunteers 3.
Governing Authority: Parent Institution: Illinois State Museum.

Institution Type/Description: Art Museum.
Collections: fine, decorative & ethnographic arts of Illinois.
Major Exhibits: Mise en Place: Sketches, Notes, Journals and Other Preparations, 11/13-4/14.
Research Fields: art & artists of Illinois.
Hours & Admission Prices: Mon.-Fri. 9-5, Sun. 12-5. No charge; donations accepted. Closed state holidays. &
Attendance: 16,500 (estimated)
Membership: Individual $35; Family $50; Contributing $100; Sustaining $300.

WILL COUNTY HISTORICAL SOCIETY, 803 S. State St., Lockport, IL 60441-3433. Tel.: 815-838-5080. Fax: 815-838-4547.
E-mail: sandyvasko@willcountyhistoricalsociety.org
Web Site: www.willcountyhistoricalsociety.org
Formerly: Illinois and Michigan Canal Commissioners Office Museum
Founded: 1964.
Congressional District: 4
Key Personnel: Exec. Dir. & Pres. (V), Sandy Vasko.
Personnel Profile: Full-Time Volunteers 1; Part-Time Volunteers 35; Interns 1.
Governing Authority: nonprofit organization. Tax-exempt: 501(c)(3).
Institution Type/Description: History Museum.
Collections: Will County history including governmental record books & original documents from the Illinois & Michigan Canal. Historic Building: I & M Canal Office, built 1837.
Research Fields: history & genealogy.
Activities: guided tours; permanent & temporary exhibitions; school, scout & family programs.
Publications: bimonthly newsletter, Rudder; quarterly journal, Quarterly.
Hours & Admission Prices: Museum: mid-Feb. to mid-Nov. Tues.-Sun. 12-4. Suggested Donations: adults $3, seniors $2, children $1; members no charge. Closed major holidays.
Attendance: 4,500 (estimated)
Membership: Student, Military & Seniors $7.50; Senior Plus $10; Individual $15; Family $25; Organization $40; Business $50; Contributing $100; Sustaining $500 & up.

Lombard

LILACIA PARK-LOMBARD PARK DISTRICT, 227 W. Parkside Ave., Lombard, IL 60148-2592. Tel.: 630-627-1281. Fax: 630-627-1286.
E-mail: info@lombardparks.com
Web Site: www.lombardparks.com
Congressional District: 6
Key Personnel: Museum Shop Mgr., Jackie Brzezinski.
Personnel Profile: Full-Time Paid 45; Part-Time Paid 200; Part-Time Volunteers 125.
Governing Authority: municipal. Tax-exempt.
Institution Type/Description: Horticultural Park.
Collections: 800 lilacs; 6 acres of lilac plantings. Historic House: 1888 Coach House.
Research Fields: botanical.
Facilities: botanical garden; information center. Gift items for sale.
Activities: children's entertainment; concerts; lilac planting & care seminars; family events; Lilac Queen coronation; tour program.
Publications: annual brochures, Lilac Time in Lombard.
Hours & Admission Prices: Daily 9-6. First two weeks of May (Lilac Time). No charge; donations accepted. &
Attendance: 10,500 (estimated)

LOMBARD HISTORICAL SOCIETY - VICTORIAN COTTAGE MUSEUM, (M), 23 W. Maple, Lombard, IL 60148-2512. Tel.: 630-629-1885. Fax: 630-629-9927.
Web Site: lombardhistory.org
Founded: 1970.
Key Personnel: Dir., Jeanne Schultz Angel; Pres. (V), Corinne Flemm; Treas., Stephanie Zabela; Membership, Diane Zeigler.
Personnel Profile: Full-Time Paid 1; Part-Time Paid 3; Part-Time Volunteers 25; Interns 2.
Governing Authority: municipal. Parent Institution: Lombard Historical Society. Tax-exempt.
Institution Type/Description: Local History Museum: housed in a historic home; features lifestyle of a middle class family in Lombard during the 1870s.
Collections: pre-1880 artifacts; Victorian crafts; photographs, artifacts & archives of Lombard history.
Research Fields: Victorian decorative arts; Lombard history.
Facilities: 525-vol. local history, archives & library available to the public by appointment.

Activities: guided tours; films; docent program; temporary exhibitions; walking tours; education program. Museum Sponsors: Holiday Programs, Ice Cream Social.
Publications: quarterly newsletter.
Hours & Admission Prices: Jan. to mid-Dec. Wed. & Sun. 1-4; other times by appointment. No charge; donations accepted. &
Attendance: 7,500 (accurate)
Membership: Students $5; Senior Individual $7.50; Senior Family $12.50; Individual $15; Family $25; Commercial $40; Supporting Individual $50; Supporting Commercial $100; Life $150; Life Family $200.

SHELDON PECK HOMESTEAD, 355 E. Parkside, Lombard, IL 60148-2776. Mailing Address: Lombard Historical Society, 23 W. Maple, Lombard, IL 60148. Tel.: 630-629-1885. Fax: 630-629-9927.
E-mail: lombardhistory@att.net
Web Site: lombardhistory.org
Founded: 1999.
Key Personnel: Dir., Jeanne Schultz Angel; Pres., Corinne Flemm; Treas., Stephanie Zabela; Museum Shop Mng., Diane Zeigler.
Personnel Profile: Full-Time Paid 1; Part-Time Paid 1; Part-Time Volunteers 65; Interns 1.
Governing Authority: private; nonprofit organization. Parent Institution: Lombard Historical Society, 23 W. Maple, Lombard, IL 60148. Tax-exempt: 501(c)(3).
Institution Type/Description: Historic House Museum: housed in an 1840s restored farmhouse; underground railroad site.
Collections: reproduction artwork of Sheldon Peck; 19th century furniture.
Research Fields: underground railroad.
Facilities: Museum-related items for sale.
Activities: docent program; guided tours; hobby workshops; lectures; loan & temporary exhibitions; broadcast programs. Annual Event: ice cream social in summer.
Hours & Admission Prices: Tues., Thurs. & Sun. 1-4. No charge; donations accepted. Closed holidays. &
Attendance: 7,500 (accurate)
Membership: Student $5; Senior Individual $7.50; Senior Family $12.50; Individual & Organization $15; Family $25; Commercial $40; Supporting $50; Supporting Commercial $100; Life $150; Life Family $200.

Macomb

WESTERN ILLINOIS UNIVERSITY ART GALLERY, (M), 1 University Circle, Macomb, IL 61455-1390. Tel.: 309-298-1587.
E-mail: AM-Hayes-Hawkinson@wiu.edu
Web Site: www.wiu.edu/artgallery
Founded: 1899.
Congressional District: 17
Key Personnel: Dir., Ann Marie Hayes-Hawkinson.
Personnel Profile: Full-Time Paid 1; Part-Time Paid 2; Part-Time Volunteers 10.
Governing Authority: state. Parent Institution: Western Illinois University. Tax-exempt: 501(c)(3).
Institution Type/Description: Art Museum.
Collections: contemporary graphics; paintings; sculpture; WPA graphics & paintings.
Research Fields: WPA; contemporary prints & drawings.
Facilities: two galleries.
Activities: gallery talks & presentations; traveling exhibitions; exhibitions organized by university faculty; visiting artist lecture series; docent gallery talks & presentations.
Publications: exhibition catalogues.
Hours & Admission Prices: School Year Tues.-Fri. 9-4, Sat. 1-4. No charge. Closed university vacations. &
Attendance: 6,958 (accurate)

Mahomet

MABERY GELVIN BOTANICAL GARDEN, 109 S. Lake of the Woods Rd., Mahomet, IL 61853. Mailing Address: P.O. Box 1040, Mahomet, IL 61853. Tel.: 217-586-4389 & 4630. Fax: 217-586-6852.
E-mail: hq@ccfpd.org
Institution Type/Description: Botanical Garden.
Collections: plants; trees; flowers.
Activities: rental facilities.
Hours & Admission Prices: Daily 7 am to sundown. No charge.

✻ **MUSEUM OF THE GRAND PRAIRIE, (M),** 600 N. Lombard, Mahomet, IL 61853. Mailing Address: P.O. Box 1040, Mahomet, IL 61853-1040. Tel.: 217-586-2612. Fax: 217-586-3491.
Web Site: www.ccfpd.org
Formerly: Early American Museum
Founded: 1967.
Congressional District: 15
Key Personnel: Dir., Cheryl Kennedy; Asst. Dir., Barbara Oehlschlaeger-Garvey.
Personnel Profile: Full-Time Paid 3; Part-Time Paid 5; Part-Time Volunteers 100; Interns 1.
Governing Authority: county. Parent Institution: Champaign County Forest Preserve. Tax-exempt: 170(b)(1)(A).
Institution Type/Description: History Museum.
Collections: tools, implements, furniture, lighting devices & household items dating from c.1850-1920; one-room schoolhouse; town hall.
Facilities: reference library; discovery room. Gift items for sale.
Activities: lectures; events; changing displays; formally organized educational programs; permanent exhibitions.
Publications: brochures; newsletters.
Hours & Admission Prices: March-May & Sept.-Dec. daily 1-5; June-Aug. Mon.-Sat. 10-5, Sun. 1-5. Fee for special programs only. Closed Easter; Christmas Eve & Day. &

Attendance: 14,000 (estimated)

Makanda

GIANT CITY STATE PARK, 235 Giant City Rd., Makanda, IL 62958-3207. Tel.: 618-457-4836.
Volunteer Hours: 1,000
Institution Type/Description: Park Museum.
Collections: natural & cultural history of the park; geology; plants; wildlife; film; Native American artifacts.
Facilities: nature trails; visitor center; observation deck. Museum-related items for sale.
Activities: film; hiking; camping.
Hours & Admission Prices: Call for hours. No charge.
Attendance: 14,715 (accurate)

Marion

CRAB ORCHARD NATIONAL WILDLIFE REFUGE, 8588 Rte. 148, Marion, IL 62959-5822. Tel.: 618-997-3344.
Institution Type/Description: Wildlife Refuge.
Collections: migratory waterfowl; fish; endangered wildlife.
Facilities: nature trails; visitor center.
Activities: hiking trails; picnicking; wildlife observation; youth camps.
Hours & Admission Prices: Visitor Center: daily 8-4:30. Admission: Daily: $2 per vehicle; Weekly: $5 per vehicle.

WILLIAMSON COUNTY HISTORICAL SOCIETY, 105 S. Van Buren, Marion, IL 62959-2509. Tel.: 618-997-5863 (Thurs. 9-3).
Web Site: www.thewchs.net
Founded: 1976.
Key Personnel: Pres. (V), Bob Jackson; Cur., Mary Jean DeMattei; Membership Chm., Dolores Thetford.
Personnel Profile: Part-Time Volunteers 10; Interns 1.
Governing Authority: nonprofit organization. Tax-exempt.
Institution Type/Description: County Historical Society Museum: housed in 1913 former county jail & sheriff's home.
Collections: genealogy records; military; bank; drug store; jail cell; grocery store; Native American room with arrowheads, stone tools & other artifacts.
Research Fields: history; genealogy.
Activities: programs during meetings of society; tours.
Publications: quarterly for members, Footprints in Williamson County; area history subject books.
Hours & Admission Prices: March to late Nov. Thurs. 9-3, third Thurs. of the month 9-6, Sat. 10-1; other times by appointment. Research Library: March-Nov. Requested Tour Donation: adults $2, children under 12 $1; discounts to students; members no charge. Closed Thanksgiving & day after.
Attendance: 600 (estimated)
Membership: Annually $20.

Marshall

CLARK COUNTY MUSEUM, 502 S. Fourth St., Marshall, IL 62441. Mailing Address: P.O. Box 207, Marshall, IL 62441-0207.
E-mail: publishdc@gmail.com

Founded: 1969.
Congressional District: 22
Key Personnel: Dir. & Pres. (V), Dwight Connelly.
Personnel Profile: Part-Time Volunteers 12.
Governing Authority: society. Parent Institution: the Clark County Historical Society. Tax-exempt: 501(c)(3).
Institution Type/Description: Local History Museum.
Collections: articles relating to Clark County history; china; furniture; photos & documents; primitive kitchen equipment; quilts; hand tools; microfilm copies of local paper 1870-1986.
Facilities: genealogical library south of courthouse for family research.
Activities: guided tours; special exhibits; marking historical sites.
Publications: periodic, Clark County Museum Newsletter.
Hours & Admission Prices: May-Oct. Sun. 2-4; other times by appointment. No charge; donations accepted.
Attendance: 269 (accurate)
Membership: Annual $10.

Maywood

WEST TOWN MUSEUM OF CULTURAL HISTORY, 104 S. 5th Ave., Maywood, IL 60153-1308. Tel.: 708-343-3554. Fax: 708-343-3557.
Key Personnel: Pres., Northica H. Stone
Institution Type/Description: Cultural History Museum.
Collections: Maywood history & culture; Underground Railroad history.
Hours & Admission Prices: Wed. 9:30-3; other times by appointment. Adults $5, children under 13 $3.

Mendota

BREAKING THE PRAIRIE MUSEUM, 684 8th St., Mendota, IL 61342-0433. Mailing Address: P.O. Box 433, Mendota, IL 61342-0433. Tel.: 815-539-3373.
E-mail: mmhsmuseum@yahoo.com
Web Site: www.mendotamuseums.org
Institution Type/Description: History Museum.
Collections: local history & agriculture; replica of Mathesius Brothers' barn; country chapel; early tractors; Scheidenhelm & Schaller buggy; farm equipment & tools; restored pump organ. Historic Structure: mid-1800s Elgin-Hummer windmill
Facilities: rental facilities.
Activities: tours.
Hours & Admission Prices: Daily call for hours.

HUME-CARNEGIE MUSEUM, 901 Washington St., Mendota, IL 61342. Mailing Address: P.O. Box 433, Mendota, IL 61342. Tel.: 815-539-3373.
Institution Type/Description: History Museum: housed in the Andrew Carnegie Library; c.1904.
Collections: local history, heritage & culture; early business & commerce; personal artifacts; photographs; organs; medical artifacts.
Facilities: Museum-related items for sale.
Activities: educational programs.
Hours & Admission Prices: March-Nov. Sat.-Sun. 1-4. Suggested Donation: adults $2, student $1. &

UNION DEPOT RAILROAD MUSEUM, 783 Main St., Mendota, IL 61342. Mailing Address: P.O. Box 433, Mendota, IL 61342-0433. Tel.: 815-539-3373.
E-mail: mmhsmuseum@yahoo.com
Web Site: www.mendotamuseums.org/udrr.htm
Institution Type/Description: Railroad Museum.
Collections: railroad history; HO scale mode of 1940s Mendota. Rolling Stock: 1911 CB&Q wooden caboose; 1923 CB&Q steam locomotive & tender; 1938 Milwaukee Road passenger car; 1949 Golden State Limited diner.
Activities: interactive exhibitions.
Hours & Admission Prices: Wed. & Sat.-Sun. 12-4. Adults $2, students $1.

Metamora

METAMORA COURTHOUSE HISTORIC SITE, 113 E. Partridge, Metamora, IL 61548-7021. Mailing Address: P.O. Box 628, Metamora, IL 61548. Tel.: 309-367-4470. Facebook: Metamora Courthouse Historic Site.
E-mail: metcourt@mtco.com
Web Site: www.villageofmetamora.com/?hiscourt
Founded: 1845.

Congressional District: 15
Key Personnel: Cur., Jean Myers.
Personnel Profile: Full-Time Paid 1; Part-Time Volunteers 6.
Governing Authority: state. Tax-exempt: 501(c)(3).
Institution Type/Description: History Museum: housed in a restored 1845 Greek Revival courthouse located in the 8th Judicial Circuit that Abraham Lincoln traveled as a circuit lawyer.
Collections: pioneer items used in Woodford County during the Lincoln era; 8th Judicial Circuit exhibit.
Activities: guided tours.
Publications: bimonthly newsletter, Historic Illinois; brochure.
Hours & Admission Prices: March-Oct. Tues.-Sat. 1-5; Nov.-Feb. Tues.-Sat. 12-4; groups by appointment. No charge; donations accepted. Closed New Year's Day; Martin Luther King Jr. Day; Presidents' Day; Veterans Day; Election Day; Thanksgiving; Christmas. &
Attendance: 6,869 (accurate)

Metropolis

THE ALLARD HOUSE AND MORRIS TOY MUSEUM, 21 Henry Rd., Metropolis, IL 62960-2939. Tel.: 270-988-3591.
Institution Type/Description: Historic House & Toy Museum.
Collections: local history; period furnishings; toys from 1800s to present.
Hours & Admission Prices: Call for hours. Admission: $2 per person.

FORT MASSAC STATE PARK AND HISTORIC SITE, 1308 E. 5th St., Metropolis, IL 62960-2380. Tel.: 618-524-9321 & 4712. Fax: 618-524-9321.
E-mail: dnr.r5parks@illinois.gov
Founded: 1908.
Congressional District: 59
Key Personnel: Site Supt., Terry Johnson.
Personnel Profile: Full-Time Paid 4; Part-Time Volunteers 32.
Governing Authority: state. Parent Institution: Illinois Dept. of Natural Resources, One Natural Resource Way, Springfield, IL 62702. Tax-exempt: 501(c)(3).
Institution Type/Description: History Museum: located on the site of 1756-1814, Military Post.
Collections: artifacts & reconstructed period buildings.
Research Fields: U.S. military history 1790-1820.
Activities: guided tours; lectures.
Publications: site brochure; calendar of events.
Hours & Admission Prices: Daily by appointment. No charge. Closed Thanksgiving; Christmas. &
Attendance: 1,604,115 (accurate)

Moline

DEERE & COMPANY, ADMINISTRATIVE CENTER, One John Deere Place, Moline, IL 61265-8098. Tel.: 800-765-9588.
Web Site: www.deere.com/en_us/attractions/worldhq/index.html
Founded: 1837.
Key Personnel: Dir. Community Affairs, James H. Collins; Mgr. Library Svcs., Betty S. Hagberg; Archivist, Vicki L. Eller.
Governing Authority: profit-making organization.
Institution Type/Description: History Museum.
Collections: agriculture; archives; industrial; archaeology.
Research Fields: agriculture; archives.
Facilities: 25,000-vol. library of agricultural research & reference books, magazines, equipment catalogs & advertising literature available for use by appointment on the premises.
Activities: guided tours; permanent exhibitions.
Hours & Admission Prices: Daily 9-5. No charge. Closed national holidays.

JOHN DEERE PAVILION, 1400 River Dr., Moline, IL 61265. Tel.: 309-765-1000.
Governing Authority: Parent Institution: Deere & Company.
Institution Type/Description: Agricultural History Museum.
Collections: company & equipment history; interactive displays; children's exhibitions.
Activities: rental facilities.
Hours & Admission Prices: Mon.-Fri. 9-5, Sat. 10-5, Sun. 12-4. No charge. Closed Good Friday; Memorial Day; Independence Day; Labor Day.

ROCK ISLAND COUNTY HISTORICAL SOCIETY, 822 11th Ave., Moline, IL 61265-1221. Tel.: 309-764-8590. Fax: 309-764-4748.
E-mail: richs@netexpress.net
Web Site: www.richs.cc

Founded: 1905.
Congressional District: 17
Key Personnel: Pres., Jack Michalski; Chm. (V), Orin Rockhold.
Personnel Profile: Full-Time Volunteers 1; Part-Time Volunteers 50.
Governing Authority: society. Tax-exempt.
Institution Type/Description: Historical Society Museum: housed in c.1877 Victorian three story house.
Collections: folklore; military; agriculture; costumes; Indian artifacts; carriage house.
Research Fields: local history of Rock Island County.
Facilities: 10,000-vol. library of books, leaflets, clippings, papers, diaries available for research on premises.
Activities: guided tours; lectures; Jr. Historian Jamboree-4th & 5th grade selected students workshop. Museum Sponsors: Open House.
Publications: quarterly newsletter.
Hours & Admission Prices: House Museum & Carriage House: by appointment. Library: Wed.-Sat. 9-4. Adults $5; discounts to AAM & ICOM members; members no charge. &
Attendance: 5,000 (estimated)
Membership: Household & Business $30; Contributing $50-$99; Patron $100-$499; President's Circle $500 & up.

Monmouth

BUCHANAN CENTER FOR THE ARTS, 64 Public Square, Monmouth, IL 61462-1756. Tel.: 309-734-3033. Fax: 309-734-3554.
E-mail: bca2@frontiernet.net
Web Site: bcaarts.org
Founded: 1990.
Congressional District: 17
Key Personnel: Dir., Susan Twomey; Museum Shop Mgr., Karen Gillen.
Personnel Profile: Full-Time Paid 2.
Governing Authority: nonprofit. Tax-exempt.
Institution Type/Description: Art Museum & Arts Agency.
Collections: furnishings; personal artifacts; drawings; paintings; decorative arts; costumes; textiles.
Facilities: 2400 sq. ft. gallery & studio. Museum-related items for sale.
Activities: classes; workshops.
Publications: Artscoop.
Hours & Admission Prices: Tues.-Fri. 10-5, Sat. 10-2. No charge; donations accepted. Closed major holidays. &
Attendance: 12,170 (accurate)
Membership: Friend $15 & up; Family & Organization Member $30; Grantor & Partner $50 & up; Sponsor & Associate $100 & up; Sustainer & Principal $250 & up; Benefactor & Advocate $500 & up; Patron $1,000 & up; Guarantor $2,500 & up; Founder $5,000 & up.

WYATT EARP BIRTHPLACE HOME, 406 S. 3rd St., Monmouth, IL 61462. Tel.: 309-734-6771.
E-mail: wyattearpbirthp@aol.com
Web Site: www.earpmorgan.com/wyattearpbirthplacewebsite.html
Founded: 1986.
Congressional District: 47
Key Personnel: Pres. (V), Mark Novak; Chm. (V) & Museum Shop Mgr., Audie Durand.
Personnel Profile: Full-Time Volunteers 1; Part-Time Volunteers 20.
Governing Authority: nonprofit organization. Parent Institution: Wyatt Earp Birthplace, Inc. Tax-exempt: 501(c)(3).
Institution Type/Description: Historic House: built c.1841. Listed on the National Register of Historic Places.
Collections: photographs; news items; 1840-1880 furnishings; memorabilia of Wyatt Earp; genealogy; books.
Research Fields: life of Wyatt Earp.
Facilities: library; period furnished rooms. Books & museum-related items for sale.
Activities: guided tours; film & slide shows; reenactments; portrayals. Museum Sponsors: Birthday Celebration including TV's Wyatt Earp - Hugh O'Brian; talks by Earp authors & researchers.
Publications: brochures; annual newsletter; books, The Wyatt Earp Birthplace: A Review; Part I; Part II; Wyatt Earp, Native Son.
Hours & Admission Prices: By appointment. Tours: adults $3, children 6-12 $1; discounts to AAM & ICOM members; members no charge. &
Attendance: 500 (accurate)
Membership: Local Student $1; Individual $5; Supporter $10; Sponsor $25, $50, $75, $100, $250, $500, $750, $1,000.

Monticello

ALLERTON PARK AND RETREAT CENTER, 515 Old Timber Rd., Monticello, IL 61856-8279. Tel.: 217-333-3287 & 762-7011. Fax: 217-762-3742.
Institution Type/Description: Historic House & Park.
Collections: family history; personal artifacts; period furnishings; photographs.
Facilities: visitor center.
Hours & Admission Prices: Visitor Center: Staffed: April-Nov. Sat.-Sun. 9-5; Unstaffed: Dec.-March daily 9-5. Closed New Year's Day; Thanksgiving; Christmas.

MONTICELLO RAILWAY MUSEUM, 992 Iron Horse Pl., Monticello, IL 61856. Mailing Address: P.O. Box 401, Monticello, IL 61856-0401. Tel.: 217-762-9011; 877-762-9011.
E-mail: info@mrym.org
Web Site: www.mrym.org
Founded: 1966.
Congressional District: 15
Key Personnel: Supt. Locomotives, Kent McClure; Pres. (V) & Museum Shop Mgr., Donna McClure; Vice Pres., Bill Crisp; Chm. (V), T.O. Grant; Sec., Derek Kouzmanoff; Gen. Mgr., Sylvester Keller; Treas., Doug Butzow.
Personnel Profile: Full-Time Volunteers 2; Part-Time Volunteers 40.
Governing Authority: nonprofit organization. Tax-exempt: 501(c)(3).
Institution Type/Description: Railway Museum.
Collections: late 1890-1950 steam locomotives; diesel locomotives; freight & passenger cars. Historic Buildings: restored c.1920 Illinois Central Depot; 1899 Wabash Depot.
Facilities: Gift shop with railroad-related items for sale.
Activities: train rides May-Oct. Sat.-Sun. & holidays; Throw Momma on the Train; Father's Day; monthly steam weekend; Polar Express(TM); railroad days; ghost train; Throttle Time.
Publications: The Second Section; Iron Horse Times; Yella Board.
Hours & Admission Prices: May-Oct. Sat.-Sun. Museum: no charge. Train Rides: adults $10, senior citizens $8, youth 2-12 $6; discount to Illinois Central Railroad employees, AAA members & groups of 20 or more; children under 2 & members no charge. Write for schedule of events. Fares may differ for special events. &
Attendance: 22,000 (accurate)
Membership: Youth $20; Adult $35.

PIATT COUNTY MUSEUM, (M), 1 Heritage Ln., Monticello, IL 61856. Tel.: 217-762-4731.
Web Site: www.piattmuseum.org
Founded: 1965.
Congressional District: 19
Key Personnel: Pres. (V), T.J. Shambaugh, IV; Trustee, Lorin I. Nevling, Ph.D.; Dir. Devel., Peg Bargon.
Personnel Profile: Part-Time Volunteers 15.
Governing Authority: nonprofit organization. Tax-exempt: 501(c)(3).
Institution Type/Description: General Museum.
Collections: Fine arts; Civil War relics; model railroad & miniature city; WWI to WWII relics; agriculture.
Activities: Annual Event: Barn Tour in October.
Publications: newsletters.
Hours & Admission Prices: Call for hours.
Attendance: 600 (estimated)
Membership: Individual Patron $25; Family $35; Organization $75.

Morris

GRUNDY COUNTY HISTORICAL SOCIETY, 510 W. Illinois Ave., Morris, IL 60450. Tel.: 815-942-4880. Facebook: Museum of the Grundy County Historical Society - Illinois.
E-mail: grundyhistory@sbcglobal.net
Web Site: www.grundycountyhs.org
Founded: 1923.
Personnel Profile: Part-Time Paid 2; Part-Time Volunteers 8.
Operating Income: 30,000
Institution Type/Description: Historical Society Museum.
Collections: local history & culture; photographs; WWI & II memorabilia; Civil War artifacts; EJ & E railroad caboose.
Facilities: Museum-related items for sale.
Hours & Admission Prices: Thurs.-Sat. 10-3; groups by appointment. No charge; donations accepted.
Attendance: 1,600
Membership: Individual $20; Family $25; Contributing & Business $50; Lifetime $200; Endowment $300.

Morrison

MORRISON'S HERITAGE MUSEUM, 202 E. Lincoln Way, Morrison, IL 61270-2825. Mailing Address: P.O. Box 1, Morrison, IL 61270. Tel.: 815-772-8889 & 3084.
Founded: 1976.
Key Personnel: Pres., Harvey Zuidema.
Personnel Profile: Part-Time Volunteers 29.
Institution Type/Description: History Museum.
Collections: manufacturing history; military artifacts; religious; American Indian; train models.
Facilities: Museum-related items for sale.
Activities: monthly programs.
Publications: quarterly newsletter.
Hours & Admission Prices: April-Nov. Fri.-Sun. 1-4; other times by appointment. No charge; donations accepted. &
Attendance: 1,200 (estimated)
Membership: Single $10; Family $20; Business & Organization $25.

Morton Grove

MORTON GROVE HISTORICAL MUSEUM/HAUPT-YEHL HOUSE, 6240 Dempster St., Morton Grove, IL 60053. Mailing Address: 6834 N. Dempster, Morton Grove, IL 60053-2631. Tel.: 847-965-0203. Fax: 847-965-7484.
E-mail: mbusch@mortongroveparks.com
Web Site: www.mortongroveparks.com
Founded: 1970.
Congressional District: 9
Key Personnel: Pres. Morton Grove Historical Society, Donna Hedrick; Cur., Mary Busch.
Personnel Profile: Part-Time Paid 3; Part-Time Volunteers 25.
Governing Authority: nonprofit. Parent Institution: Morton Grove Park District. Tax-exempt: 501(c)(3).
Institution Type/Description: Historical Society Museum: housed in c.1888 farm home.
Collections: photographs; local memorabilia & artifacts; furniture & furnishings.
Research Fields: maps; printed material on local history; oral histories.
Activities: guided tours.
Publications: bimonthly newsletter, Historical Highlights.
Hours & Admission Prices: Sun. 2-4, Wed. 1-3; other times tours of 10 persons or more by appointment. No charge; donations accepted. Closed national holidays.
Attendance: 2,234 (accurate)
Membership: Student $1; Senior $5; Individual $7.50; Family $15; Sponsor $25; Individual Life $50; Family Life $100.

Mount Carmel

WABASH COUNTY MUSEUM, 320 N. Market St., Mount Carmel, IL 62863. Mailing Address: P.O. Box 512, Mount Carmel, IL 62863-0512. Tel.: 618-262-8774.
E-mail: ctdant@frontier.com
Web Site: www.museum.wabash.il.us
Founded: 1992.
Key Personnel: Pres. (V), Claudia Dant.
Personnel Profile: Part-Time Paid 1; Part-Time Volunteers 70.
Governing Authority: Parent Institution: Wabash County Museum District. Tax-exempt.
Institution Type/Description: History Museum.
Collections: local history & culture; photographs; period furnishings; personal artifacts.
Research Fields: local history.
Facilities: library.
Activities: research.
Hours & Admission Prices: Tues., Thurs. & Sun. 2-5. No charge; donations accepted. Closed Easter; Thanksgiving; Christmas. &
Attendance: 2,000 (estimated)

Mount Prospect

MOUNT PROSPECT HISTORICAL SOCIETY MUSEUMS, 101 S. Maple, Mount Prospect, IL 60056-3229. Tel.: 847-392-9006. Fax: 847-577-9660. Facebook: Mount Prospect Historical Society.
E-mail: info@mtphistory.org
Web Site: www.yourcentralschool.org
Founded: 1968.
Congressional District: 12

Key Personnel: Pres. (V), Frank Corry; Exec. Dir., Lindsay Rice; Financial Dir., Chad Busse; Administrative Asst., Cindy Brok.
Personnel Profile: Part-Time Paid 3; Part-Time Volunteers 63; Interns 1.
Governing Authority: nonprofit. Parent Institution: Dietrich Friedrichs House, 101 S. Maple, Mount Prospect, IL 60056. Tax-exempt: 501(c)(3).
Institution Type/Description: Local History Museum: housed in a c.1906 home.
Collections: early 20th century domestic life in Mt. Prospect; exhibits relating to the history of Mt. Prospect; decorative arts; photographs; costumes; archives.
Research Fields: rural town life; history of Mt. Prospect; history of Illinois schooling; economic development; 20th-century suburban development; immigration history.
Facilities: library of local history materials; education center. Books, historic afghans & stationery items for sale.
Activities: guided tours; lectures; docent program; temporary exhibitions; educational programs; cemetery walk. Annual Events: House Walk; Stadtmitte history crawl.
Publications: quarterly newsletter; History Book of Mt. Prospect; Mt. Prospect: Where Town & Country Met; Images of America: Mount Prospect; Lost Mount Prospect.
Hours & Admission Prices: Tues.-Thurs. 10-3:30 by appointment. No charge; donations accepted. Closed New Year's Day; Independence Day; Thanksgiving; Christmas. &

Attendance: 5,000 (estimated)
Membership: Senior $20; Single $25; Family $30; Sponsor $75; Benefactor $150; Patron $250; Life $1,000.

Mount Pulaski

MOUNT PULASKI COURTHOUSE STATE HISTORIC SITE, 113 S. Washington St., Mount Pulaski, IL 62548. Mailing Address: 113 S. Washington St., P.O. Box 171, Mount Pulaski, IL 62548-0171. Tel.: 217-792-3919.
Founded: 1936.
Congressional District: 18
Personnel Profile: Part-Time Paid 1; Part-Time Volunteers 16.
Governing Authority: state. Parent Institution: Illinois Historic Preservation Agency, Old State Capitol, Springfield, IL 62701. Tax-exempt: 501(c)(3).
Institution Type/Description: Historic Building: 1848 Greek Revival County Court House; part of Illinois 8th Judicial Circuit where Abraham Lincoln practiced law.
Collections: furnishings.
Research Fields: history of area; Abraham Lincoln & the 8th Judicial Circuit.
Activities: guided tours; school programs.
Publications: bimonthly newsletter, Historic Illinois; brochure.
Hours & Admission Prices: Tues.-Sat. 12-4. No charge; donations accepted. Closed most holidays.
Attendance: 1,150 (estimated)

MOUNT PULASKI HISTORICAL SOCIETY MUSEUM, 104 E. Cooke St., Mount Pulaski, IL 62548. Mailing Address: 102-104 E. Cooke St., Mount Pulaski, IL 62548-0181. Tel.: 217-792-5430.
E-mail: sueeddie@frontier.com
Founded: 1995.
Key Personnel: Pres. (V), Sue Schaffenacker
Governing Authority: Tax-exempt.
Institution Type/Description: Historical Society Museum.
Collections: local history, culture, & business; Abraham Lincoln's life & family; personal artifacts; period furnishings; photographs; bank equipment & furnishings.
Facilities: library. Museum-related items for sale.
Publications: quarterly newsletter.
Hours & Admission Prices: April-Nov. Wed.-Fri. 12-4, Sat. by appointment. No charge; donations accepted. Closed New Year's Day; Martin Luther King's Birthday; Presidents' Day; Veterans Day; General Election Day; Thanksgiving; Christmas. &
Attendance: 300 (accurate)
Membership: Individual $15; Couple $25.

Mount Vernon

CEDARHURST CENTER FOR THE ARTS, 2600 Richview Rd., Mount Vernon, IL 62864. Mailing Address: P.O. Box 923, Mount Vernon, IL 62864-0019. Tel.: 618-242-1236. Fax: 618-242-9530.
E-mail: mitchellmuseum@cedarhurst.org
Web Site: cedarhurst.org

Formerly: Mitchell Museum at Cedarhurst
Founded: 1973.
Congressional District: 22
Key Personnel: Exec. Dir., Sharon Bradham; Craft Fair Coord., Linda Wheeler; Dir. Visual Arts, Rusty Freeman; C.F.O., Heather Owens; Dir. Communications, Sarah Sledge; Dir. Devel., Hillary Settle; Dir. Education, Jennifer Sarver; Dir. Operations, Greg Hilliard; Museum Shop Mgr. & Historian, Sarah Lou Bicknell.
Personnel Profile: Full-Time Paid 16; Part-Time Paid 2; Part-Time Volunteers 25.
Governing Authority: nonprofit organization. Parent Institution: John R. & Eleanor R. Mitchell Foundation. Tax-exempt: 501(c)(3).
Institution Type/Description: Art Museum & Sculpture Park.
Collections: late 19th & early 20th-century American paintings; contemporary sculpture & art from the Midwest.
Research Fields: 19th & 20th-century American painting; contemporary sculpture & art from the Midwest.
Facilities: 1,900-vol. library; nature trails; bird sanctuary; children's gallery; 90-acre sculpture park.
Activities: guided tours; sculpture walks; chamber music series; dinner theater; children's theater; summer concerts; docent program; temporary, traveling & juried exhibitions; art classes in the Shrode Art Center. Museum Sponsors: Annual Cedarhurst Craft Fair in September; Collector's Club.
Publications: newsletter; brochures; exhibition catalogues; annual report.
Hours & Admission Prices: Tues.-Sat. 10-5, Sun. 1-5. No charge; donations accepted. Closed national holidays. &
Attendance: 50,000 (estimated)
Membership: Individual $50; Family $75; Patron $150; Sponsor $250; Benefactor $500; Guarantor $1,000.

SHRODE ART CENTER, 2600 Richview Rd., Mount Vernon, IL 68264. Mailing Address: P.O. Box 923, Mount Vernon, IL 68264. Tel.: 618-242-1236, ext. 249.
E-mail: carrie@cedarhurst.org
Web Site: www.cedarhurst.org
Founded: 1973.
Key Personnel: Dir., Carrie Gibbs.
Governing Authority: Parent Institution: Cedarhurst Center for the Arts. Tax-exempt.
Institution Type/Description: Art Gallery.
Collections: quilts; photographs; sculpture; ceramics; stained glass; painting; drawings.
Facilities: library; classrooms.
Activities: classes; workshops; temporary exhibitions.
Hours & Admission Prices: Tues.-Sat. 10-5, Sun. 1-5. Closed national holidays.

Murphysboro

GENERAL JOHN A. LOGAN MUSEUM, (M), 1613 Edith St., Murphysboro, IL 62966-2542. Mailing Address: P.O. Box 563, Murphysboro, IL 62966-0563. Tel.: 618-684-3455. Fax: 618-684-3569.
E-mail: johnaloganmuseum@globaleyes.net
Web Site: www.loganmuseum.org
Founded: 1989.
Congressional District: 12
Key Personnel: Dir., P. Michael Jones; Pres. (V), Michael J. McNerney.
Personnel Profile: Full-Time Volunteers 1; Part-Time Paid 4; Part-Time Volunteers 10.
Governing Authority: Tax-exempt.
Institution Type/Description: Military History Museum.
Collections: life of General John A. Logan; 19th & early 20th century American history; Civil War; personal artifacts; photographs.
Hours & Admission Prices: June-Aug. Tues.-Sat. 10-4, Sun. 1-4; Sept.-May Tues.-Sun. 1-4; groups by appointment. No charge; donations accepted. Closed New Year's Eve; Christmas Eve; holidays. &
Attendance: 3,500 (accurate)

JACKSON COUNTY HISTORICAL SOCIETY, 1616 Edith St., Murphysboro, IL 62966. Tel.: 618-684-6989.
E-mail: jchs@verizon.net
Institution Type/Description: Historical Society Museum.
Collections: local history & culture; photographs; period furnishings; personal artifacts.
Hours & Admission Prices: Wed. & Fri. 12-3, Thurs. 12-3 & 6:30-9.

Naperville

DUPAGE CHILDREN'S MUSEUM, 301 N. Washington St., Naperville, IL 60540-4537. Tel.: 630-637-8000. Fax: 630-637-1276. Facebook; DePage Children's Museum.
E-mail: admin@dupagechildrensmuseum.org
Web Site: www.dupagechildrensmuseum.org
Founded: 1987.
Congressional District: 13
Institution Type/Description: Children's Museum.
Collections: hands-on exhibits.
Facilities: cafe.
Activities: special programs; birthday parties; rental facilities; exhibit activities, science & art residencies; Third Thursday; Tiny Great Performances (TM).
Publications: DCM newsletter online; the DCM blog: www.thechildrensmuseumblog.blog-city.com; eNews Club.
Hours & Admission Prices: Mon. 9-1, Tues.-Thurs. 9-4, Third Thurs. of month & Fri. 9-8, Sat. 9-5, Sun. 12-5. Admission $11, seniors 60 & over $10, discounts to ACM members; members & children under one no charge. Closed New Year's Eve & Day; Easter; Memorial Day; Independence Day; Labor Day; Thanksgiving; Christmas Eve & Day. (Rates & hours subject to change, check website). &
Attendance: 325,144 (accurate)
Membership: Child $90; Family & Grandparent $110; Family Plus $130; Family Premier $150; Children's Circle $250; Benefactor $1,000.

✱　**NAPER SETTLEMENT, (M),** 523 S. Webster St., Naperville, IL 60540-6517. Tel.: 630-420-6010. Fax: 630-305-4044.
E-mail: towncrier@naperville.il.us
Web Site: www.napersettlement.com
Founded: 1969.
Congressional District: 41
Key Personnel: Chm., Sally Pentecost; Vice Pres. Museum Svcs., Debbie Grinnell; Chief Cur., Louise Howard; Vice Pres. Organization Resources, Harriet Pistorio; Vice Pres. Devel. & Mktg., Kevin Burns; Dir. Learning Experiences, Nancy Smith; Dir. Buildings & Grounds, Sharon Bennett Hinkle; Dir. Financial Svcs., Christine Jepsen.
Personnel Profile: Full-Time Paid 28; Part-Time Paid 14; Part-Time Volunteers 1,300; Interns 2.
Governing Authority: municipal, City of Naperville. Parent Institution: Naperville Heritage Society. Tax-exempt.
Institution Type/Description: Historic Museum.
Collections: furnishings; costumes; dolls; glass; manuscripts; china; folk paintings; interactive displays; historical artifacts. Historic Buildings: 1883 Martin Mitchell Mansion & Carriage House; 1864 Century Memorial Chapel; c.1840s Halfway House; Conestoga Wagon; c.1860s Fire House; late 19th century Stonecarver's Shop; 1833 Paw Paw Post Office; 1841 Meeting House; c.1880s working Print Shop; c.1860 recreated Blacksmith Shop; c.1840 recreated one-room schoolhouse; c.1832 recreated fort; c.1840s log cabin.
Research Fields: history of Naperville, including as it relates to history in Midwest; genealogy.
Facilities: research library; 12-acre grounds; gardens; rental facilities.
Activities: guided & self-guided tours; school programs; adult & children's programs; lecture series; geocaching; group outings; volunteer programs; special events; permanent & temporary exhibitions. Annual Events: Civil War Days; Maple Sugaring Days; Naper Nights Concert Series; Dinner on the Town; Oktoberfest; History Speaks Lecture Series; All Hallows Eve: Village of Fear.
Publications: magazine, Treasures; brochures; calendar of events; annual report; Memories of the Past - Sharing Moments Through History; Tidbits of the Past, A Fun Collection of Historic Brainteasers; postcard & notecard series of current and historic photographs.
Hours & Admission Prices: Summer: April-Oct. Tues.-Sat. 10-4, Sun. 1-4. Adults $12, senior citizens $10, youth 4-12 $8; members no charge. Audio Tour: $3; geocache GPS unit $1; discounts to AAA members. Winter: Nov.-March Tues.-Fri. 10-4. Adults $5.25, senior citizens $4.75, youths 4-12 $4, audio tour included; discounts to AAA members; members no charge. Closed New Year's Day; Thanksgiving & day after; Christmas. &
Attendance: 118,000 (estimated)
Membership: Individual $50; Family $100; Sustaining Member $100-$500.

Nauvoo

JOSEPH SMITH HISTORIC SITE, 865 Water St., Nauvoo, IL 62354. Mailing Address: P.O. Box 338, Nauvoo, IL 62354-0338. Tel.: 217-453-2246. Fax: 217-453-6416.
E-mail: jshs@frontiernet.net
Web Site: cofchrist.org/js/
Founded: 1918.
Congressional District: 19
Key Personnel: C.E.O., Joyce A. Shireman, Ph.D.
Personnel Profile: Full-Time Paid 4; Full-Time Volunteers 12; Part-Time Paid 1; Part-Time Volunteers 10; Interns 10.
Governing Authority: church. Parent Institution: Community of Christ. Subsidiary Institution: Historic Sites Division. Tax-exempt: 501(c)(3).
Institution Type/Description: Historic Sites & Buildings: 1803 & 1840 Joseph Smith Homestead; 1842 Joseph Smith Red Brick Store; 1843 Joseph Smith Mansion House.
Collections: original 19th-century furnishings, structures & recovered artifacts.
Research Fields: historic archaeology; Mormon history.
Facilities: library. Museum-related items for sale.
Activities: guided tours; summer kitchen-activities of daily living-1840s: cooking; carding wood; spinning candle making; herb garden; games.
Hours & Admission Prices: May-Oct. Mon.-Sat. 9-5, Sun. 1-5; Nov.-Dec. Tues.-Sat. 10-4. Summer Kitchen: Memorial Day to mid-Aug. daily 10-4. Guided Tour: $2 per person. &
Attendance: 90,000 (estimated)
Membership: $120, $300, $420 & $600; Heritage Club $1,000.

NAUVOO HISTORICAL SOCIETY MUSEUM, 980 S. Bluff St., Nauvoo, IL 62354. Mailing Address: P.O. Box 426, Nauvoo, IL 62354-0426. Tel.: 217-453-2512. Fax: 217-453-2512.
Founded: 1953.
Congressional District: 17
Key Personnel: Chm. & Pres. (V), Mary Reed; Site Supt., Regan Ramsey; Maintenance Supvr., Michael Locke; Site Tech, Mike Middendorf.
Personnel Profile: Part-Time Paid 3; Part-Time Volunteers 3.
Governing Authority: state; nonprofit organization. Affiliated with State of Illinois, Div. of Natural Resources, 600 N. Grand Ave. W, Springfield, IL 62706. Tax-exempt.
Institution Type/Description: Historical Society Museum.
Collections: geology; history; American Indian artifacts; 1850 wine cellar & grape arbor.
Research Fields: Icarian period.
Facilities: wine cellar; 2 acre grape arbor.
Activities: guided tours; permanent exhibitions.
Publications: bimonthly newsletter, Historic Illinois; brochure.
Hours & Admission Prices: mid-May to mid-Oct. daily 1-5. No charge; donations accepted. &
Attendance: 5,000 (estimated)
Membership: Individual $2; Family $5.

Newton

NEWTON PUBLIC LIBRARY AND MUSEUM, 100 S. Vanburen, Newton, IL 62448-1559. Tel.: 618-783-8141. Fax: 618-783-8149.
Founded: 1928.
Key Personnel: Head Librarian, Connie Davidson.
Personnel Profile: Full-Time Paid 2; Part-Time Paid 2.
Governing Authority: municipal. Tax-exempt.
Institution Type/Description: Local History Museum.
Collections: military; early household wares; Indian artifacts; farm implements; uniforms; early German books & Bibles.
Facilities: 14,435-vol. library of reference material, children's books, adult fiction & genealogical material available for research on premises.
Activities: art exhibits; summer reading programs for children; adult book club.
Hours & Admission Prices: Mon., Wed. & Fri. 10-5, Tues. & Thurs. 10-7, Sat. 10-1. No charge. Closed national holidays. &
Attendance: 23,000 (estimated)

Nokomis

BOTTOMLEY-RUFFING-SCHALK BASEBALL MUSEUM, 121 W. State St., Nokomis, IL 62075-1658. Mailing Address: P.O. Box 75, Nokomis, IL 62075-0075. Tel.: 217-563-8807. Fax: 217-324-6616.
E-mail: info@brsmuseum.org
Web Site: brsmuseum.org
Founded: 1981.
Key Personnel: Treas., Steve Johnson
Institution Type/Description: Sports Museum.
Collections: Bottomley, Ruffing & Schalk personal artifacts & career history; baseball memorabilia.
Publications: bimonthly, The Bullpen.
Hours & Admission Prices: Mon. & Wed.-Sat. 9am-11am, Tues. 9am-11am & 6pm-9pm; other times by appointment. No charge; donations accepted. &
Attendance: 700 (estimated)
Membership: Individual $10.

Normal

CHILDREN'S DISCOVERY MUSEUM, 101 E. Beaufort, Normal, IL 61761-3026. Tel.: 309-433-3444. Fax: 309-451-3614.
E-mail: museum@normal.org
Web Site: childrensdiscoverymuseum.net
Founded: 1994.
Congressional District: 15
Key Personnel: Museum Mgr., Sheila Riley; Chm. (V), Dr. Monica Noraian; Education Coord., Bethany Thomas; Volunteer & Membership Coord., Shelly Hanover.
Personnel Profile: Full-Time Paid 10; Part-Time Paid 40; Part-Time Volunteers 800; Interns 5.
Governing Authority: municipal. Parent Institution: Town of Normal. Tax-exempt.
Institution Type/Description: Children's Museum.
Collections: hands-on exhibits; agriculture; environmental; imagination & creative play; art; climbing structure.
Research Fields: parent & child interaction in a cultural and educational setting.
Facilities: classrooms; birthday party rooms; 12,000 sq. ft. exhibit space; hands on exhibits; Luckey Climber; vending machines. Educational, developmentally appropriate toys, games, books & other items for sale.
Activities: formal education programs for children; participatory exhibits; education outreach.
Publications: quarterly newsletter; seasonal flyers; educators guide.
Hours & Admission Prices: Jan.-May & Sept.-Dec. Tues.-Wed. & Sat. 9-5, Thurs.-Fri. 9-8; June-Aug. Mon.-Wed. & Sat. 9-5, Thurs.-Fri. 9-8. Admission $7, field trips $4; ASTC, ACM, members & children under 2 no charge. Closed New Year's Day; Easter; Labor Day; Thanksgiving; Christmas. &
Attendance: 143,000 (accurate)
Membership: Basic Family $95; Plus Family $130; Grandparent $125.

NORMAL EDITIONS WORKSHOP - ILLINOIS STATE UNIVERSITY, 5620 School of Art, Normal, IL 61790-5620. Tel.: 309-438-7530. Fax: 309-438-2215.
E-mail: normaleditionsworkshop@ilstu.edu
Web Site: www.cfa.ilstu.edu/normal_editions
Founded: 1976.
Congressional District: 13
Key Personnel: Dir., Richard D. Finch; Assoc. Dir., Veda M. Rives.
Governing Authority: Parent Institution: Illinois State University.
Institution Type/Description: Printmaking.
Collections: works by emerging & established artists.
Research Fields: lithography; intaglio; woodcut & letterpress printmaking processes.
Facilities: research facility with 2,000 exhibit areas.
Activities: temporary exhibitions.
Publications: Richard D. Finch & Veda M. Rives, Ed. Marks from the Matrix, Collaborative Limited Edition Prints 1976-2006, Normal Editions Workshop: Normal, Illinois, 2007; Richard D. Finch & Veda M. Rives, Ed. Rudy Pozzatti: The Twelve Labors of Hercules, Normal Editions Workshop: Normal, Illinois, 2011; Veda M. rives, Ed. A Community of Printmakers: A Gift Portfolio Honoring James D. Butler; Normal Editions Workshop: Normal, Illinois,2012.
Hours & Admission Prices: Mon.-Thurs. 9-5; other times by appointment. No charge. &

UNIVERSITY GALLERIES OF ILLINOIS STATE UNIVERSITY, 110 Center for Visual Arts, Normal, IL 61790. Mailing Address: P.O. Box 5600, Normal, IL 61790-5600. Tel.: 309-438-5487. Facebook: University Galleries of Illinois State University.
E-mail: gallery@ilstu.edu
Web Site: www.cfa.ilstu.edu/galleries
Key Personnel: Dir., Barry Blinderman; Cur. Exhibitions, Kendra Paitz; Cur. & Interpretive Programs Coord., Tony Preston-Schreck; Registrar, Gabriel Johnson.
Personnel Profile: Full-Time Paid 3.
Governing Authority: Parent Institution: Illinois State University. Tax-exempt.
Institution Type/Description: Art Gallery.
Collections: works by contemporary artists; paintings; sculpture; photographs.
Major Exhibits: Carrie Schneider: Burning House, 1/14/14-2/16/14; Stanya Kahn, 1/14/14-2/16/14; Anne Boyden Varnot: W/hole, 2/25/14-3/13/14; Regan Golden: Ground Swell, 2/25/14-3/13/14; Winding On and Through the Lands, 2/25/14-4/6/14; Brian Gillis: His Room, 4/15/14-5/11/14; Student Annual, 4/15/14-5/11/14; Walter Robinson, 10/14-12/14.
Publications: books, Jason Lazarus: Your Time is Gonna Come; Melanie Schiff: Sun Land; Nadia Hotait.
Hours & Admission Prices: Academic Year: Tues. 9:30-7, Wed.-Fri. 9:30-4:30, Sat.-Mon. 12-4; Summer: Mon. & Wed.-Sat. 12-4, Tues. 12-7. &

North Chicago

FEET FIRST: THE SCHOLL STORY, ROSALIND FRANKLIN UNIVERSITY OF MEDICINE AND SCIENCE, 3333 Green Bay Rd., North Chicago, IL 60064-3037. Tel.: 847-578-8417. Fax: 847-578-8643.
E-mail: kelly.reiss@rosalindfranklin.edu
Web Site: www.rosalindfranklin.edu
Founded: 1993.
Congressional District: 10
Key Personnel: C.E.O., K. Michael Welch; Chm., Ruth Rothstein.
Personnel Profile: Full-Time Paid 1.
Governing Authority: private university; nonprofit. Parent Institution: Rosalind Franklin University of Medicine & Science. Tax-exempt.
Institution Type/Description: Podiatry Museum: housed in the Dr. William M. Scholl College of Podiatric Medicine.
Collections: photographs; period shoes; memorabilia from life of Dr. William Scholl; catalogs; shoe fluoroscope; exhibits to educate the public about the foot.
Research Fields: life of Dr. Scholl and his company; history of the institution.
Activities: guided tours for schools & groups.
Publications: Dr. Scholl: Foot Doctor to the World; Podiatric Medicine and the Dr. William M. Scholl College; Rosalind Franklin University of Medicine & Science: A Centennial View.
Hours & Admission Prices: Daily 9-4. No charge; donations accepted. &
Attendance: 5,000 (estimated)

O'Fallon

O'FALLON HISTORICAL SOCIETY, 101 W. State St., O'Fallon, IL 62269-0344. Mailing Address: P.O. Box 344, O'Fallon, IL 62269-0344. Tel.: 618-624-8409.
E-mail: info@ofallonhistory.net
Web Site: ofallonhistory.net
Governing Authority: nonprofit organization. Tax-exempt: 501(c)(3).
Institution Type/Description: Historical Society Museum: housed in the former First National Bank; built in 1904.
Collections: local history & culture; period furnishings; personal artifacts; photographs.
Hours & Admission Prices: Wed. & Fri.-Sat. 1-4; other times by appointment.

Oak Brook

GRAUE MILL AND MUSEUM, (M), 3800 York Rd., Oak Brook, IL 60523-2738. Tel.: 630-655-2090 & 920-9720. Fax: 630-920-9721.
E-mail: info@grauemill.org
Web Site: www.grauemill.org
Founded: 1950.
Congressional District: 13
Key Personnel: Exec. Dir., Whitney Templeton; Pres. (V), George Mueller.
Personnel Profile: Full-Time Paid 1; Part-Time Paid 23; Part-Time Volunteers 50.
Governing Authority: nonprofit corporation. Tax-exempt.
Institution Type/Description: History Museum: housed in 1852 restored waterwheel gristmill.
Collections: Civil War; farm implements & vehicles; dolls; period furnishings; artifacts. Historic Building: c.1859 Miller's Home.
Facilities: Museum-related items for sale.
Activities: spinning & weaving demonstrations.
Publications: brochures, The Story of The Old Graue Mill; Field Trip Guide; children's book, Discover Graue Mill.
Hours & Admission Prices: mid-April to mid-Nov. Tues.-Sun. 10-4:30. Adults $4.50, senior citizens $4, children 3-15 $2; discounts to groups of 20 or more; members & children under 3 no charge. &
Attendance: 40,000 (estimated)
Membership: Individual $30; Family $50; Sustaining $100-$249; Sponsor $250-$499; Benefactor $500-$999; Foundation: 1,000.

MAYSLAKE PEABODY ESTATE, (M), 1717 W. 31st St., Oak Brook, IL 60523-1701. Tel.: 630-206-9588. Fax: 630-850-2362. Facebook: Mayslake Peabody Estate.
E-mail: jfowers@dupageforest.com
Web Site: www.mayslakepeabody.com
Congressional District: 6
Personnel Profile: Full-Time Paid 4; Part-Time Paid 0; Part-Time Volunteers 42.
Governing Authority: Parent Institution: Forest Preserve District DuPage County. Tax-exempt.

Institution Type/Description: Historic House Museum: housed in the former home of Francis Stuyvesant Peabody, a coal baron & national figure in Democratic politics; built in 1921.

Collections: Francis Stuyvesant Peabody's life & family history; personal artifacts; period furnishings; photographs.

Activities: special events; theatrical performances; cultural events; educational programs; rental facilities; exhibits.

Publications: newsletter four times per year.

Hours & Admission Prices: Guided Tours: mid-Jan. to mid-Dec. Wed. 11 & 12:30, Sat. 9:30, 10, 11, & 11:30; other times by appointment. $5 per person.

Attendance: 120,000 (estimated)

Membership: Monkey $100-$249; Piper $250-$499; Angel $500-$999; Dragon $1,000-$2,499; Pilgrim $2,500-$4,999; Gryphon $5,000 & up.

Oak Lawn

CHILDREN'S MUSEUM IN OAK LAWN, 5100 Museum Dr., Oak Lawn, IL 60453-7005. Tel.: 708-423-6709. Fax: 708-423-6723.

E-mail: general.information@cmoaklawn.org

Web Site: www.cmoaklawn.org

Founded: 2001.

Key Personnel: Dir., Adam Woodworth; Pres. (V), Cathy Cepican.

Personnel Profile: Full-Time Paid 3; Part-Time Paid 2; Part-Time Volunteers 40; Interns 1.

Governing Authority: Tax-exempt.

Institution Type/Description: Children's Museum.

Collections: hands-on exhibits.

Activities: birthday parties; special events; educational programs.

Hours & Admission Prices: Wed. 8:30-3, Thurs.-Sat. 10-3. Adults 17 & over $3.50, children 1-16 $3. Closed Christmas. &

Attendance: 20,000

Membership: Child $25; Family $55; Grand $65; All-in-the-Family $100.

Oak Park

✳ FRANK LLOYD WRIGHT HOME AND STUDIO, (M), 951 Chicago Ave., Oak Park, IL 60302-2007. Mailing Address: 209 S. LaSalle St., Ste. 118, Chicago, IL 60604. Tel.: 312-994-4000.

E-mail: info@flwright.org

Web Site: flwright.org

Founded: 1974.

Congressional District: 7

Key Personnel: C.E.O. & Pres., Celeste Adams; Chm. Bd., Graham Rarity.

Personnel Profile: Full-Time Paid 30; Part-Time Paid 20; Part-Time Volunteers 325.

Governing Authority: private; nonprofit organization. Parent Institution: Frank Lloyd Wright Trust, Chicago, IL. Tax-exempt: 501(c)(3).

Institution Type/Description: Historic House Museum: 1889-1909 residence & office of Frank Lloyd Wright; birthplace of prairie-style architecture.

Collections: structures, furniture & decorative arts designed by Frank Lloyd Wright; Japanese Art & decorative materials collected by Wright; drawings; prairie style architecture.

Research Fields: Frank Lloyd Wright's life & work, 1889-1916.

Facilities: library; 8,223 sq. ft. exhibit space. Museum-related items for sale.

Activities: docent program; formal education programs; guided tours; lectures; walking tours; bus tours; workshops. Annual Event: Wright Plus Housewalk in May.

Publications: quarterly newsletter, Wright Angles; books, The Oak Park Home and Studio of Frank Lloyd Wright; Frank Lloyd Wright and the Prairie; Building a Legacy: The Restoration of Frank Lloyd Wright's Home and Studio; Frank Lloyd Wright's Fifty Views of Japan; In Wright's Shadow: Artists and Architects in the Oak Park Studio; Hometown Architect: The Complete Buildings of Frank Lloyd Wright in Oak Park and River Forest; The Wright Family Library.

Hours & Admission Prices: Tours: Memorial Day to Labor Day daily 10-4; Sept.-May daily 11-4; groups by appointment, call 312-994-4040. Book Shop: daily 9-5. Adults $15, students and seniors 65 & over $12; children 3 & under no charge. Closed New Year's Day; Thanksgiving; Christmas Eve & Day.

Attendance: 64,000 (accurate)

Membership: Individual $50; Family $65; Prairie Society $125; Octagon Society $250; Inglenook Society $500; Skylight Society $1,000.

THE HEMINGWAY MUSEUM AND THE ERNEST HEMINGWAY BIRTHPLACE, 200 N. Oak Park Ave., Oak Park, IL 60302-2128. Mailing Address: P.O. Box 2222, Oak Park, IL 60303-2222. Tel.: 708-848-2222. Fax: 708-386-2952.

E-mail: ehfop@sbcglobal.net

Web Site: www.ehfop.org

Founded: 1990.

Congressional District: 7

Key Personnel: Chm. (V), Allan Baldwin; Vice Chm., Virginia R. Cassin; Archivist, Barbara Ballinger; Museum Shop Mgr., Conni Irwin.

Personnel Profile: Part-Time Paid 2; Part-Time Volunteers 200; Interns 1.

Governing Authority: private; nonprofit organization. The Ernest Hemingway Birthplace, 339 N. Oak Park Ave., Oak Park, IL 60302. Tax-exempt: 501(c)(3).

Institution Type/Description: History Museum: housed in the birthplace of Ernest Hemingway.

Collections: Hemingway Museum: personal artifacts including high school papers & letters, studies in world languages, movie posters. Birthplace: Queen Ann Victorian style house contains artifacts depicting the life & development of Ernest Hemingway.

Research Fields: Hemingway's Oak Park years.

Facilities: archives; public lecture area. Museum-related items for sale.

Activities: arts festivals; docent program; films; guided tours; lectures; loan, temporary & traveling exhibitions; study clubs; broadcast programs. Annual Events: Hemingway birthday colloquist & lecture; Hemingway Birthplace Boxing Day.

Publications: biannual member newsletter, The Hemingway Despatch.

Hours & Admission Prices: Sun.-Fri. 1-5, Sat. 10-5. Adults $10, senior citizens & students $8; discounts to AAA members & tour groups of 10 or more; members & children 5 & under no charge. Closed New Year's Day; Martin Luther King Jr. Day; Easter; Memorial Day; Independence Day; Labor Day; Thanksgiving; Christmas. &

Attendance: 6,976 (estimated)

Membership: Student & Senior Citizen $30; Individual $45; Family $60.

HISTORICAL SOCIETY OF OAK PARK & RIVER FOREST, 217 Home Ave., Oak Park, IL 60302-3101. Mailing Address: P.O. Box 771, Oak Park, IL 60303-0771. Tel.: 708-848-6755. Fax: 708-848-0246.

E-mail: oprfhistorian@sbcglobal.net

Web Site: oprfhistory.org

Founded: 1968.

Congressional District: 7

Key Personnel: Exec. Dir., Frank Lipo; Pres. (V), Jan Novak Dressel; Pres. (V), Mary Ann Porucznik; Treas., Jim Taglia; Public Rels., Jean Guarino.

Personnel Profile: Full-Time Paid 1; Part-Time Paid 1; Part-Time Volunteers 30; Interns 1.

Governing Authority: nonprofit organization. Tax-exempt: 501(c)(3).

Institution Type/Description: Historical Society Museum: housed in 1897 Prairie style mansion.

Collections: Oak Park & River Forest history, 1830-present; exhibit on Edgar Rice Burroughs, creator of Tarzan; costumes; photographs; books; archives; artifacts.

Research Fields: local history.

Facilities: 800-vol. library of local history books & local newspapers; 700 sq. ft. exhibit space. Postcards & books on local sites & history for sale.

Activities: formal education programs for children; lectures; temporary exhibitions.

Publications: quarterly newsletter, Village Yesteryears; book, Ernest Hemingway as Recalled by his High School Contemporaries; Guidebook to Forest Home Cemetery; Grace Wilson Trout biography.

Hours & Admission Prices: Museum: Thurs.-Sun. 12:30-3:30; group tours by appointment. Office: Tues. & Thurs. 1-5. Adults $10, children 18 & under $3; Fri. & members no charge. Closed New Year's Day; Easter; Christmas.

Attendance: 5,000 (accurate)

Membership: Senior Citizen $25; Individual $35; Family & Patron $50; Life $1,000.

THE OAK PARK CONSERVATORY, 615 Garfield St., Oak Park, IL 60304-2001. Tel.: 708-386-4700. Fax: 708-386-3221.

Web Site: www.oakparkparks.com

Founded: 1929.

Congressional District: 3

Key Personnel: Mgr., Henrietta Yardley.

Personnel Profile: Full-Time Paid 3; Part-Time Paid 5; Part-Time Volunteers 30.

Governing Authority: municipal. Parent Institution: Park District of Oak Park, 218 Madison St., Oak Park, 60302. Tel.: 708-383-0002. Subsidiary Institution: Friends of the Oak Park Conservatory.

Institution Type/Description: Conservatory.

Collections: cacti & succulent collection; exotic tropical plants; subtropical plants; seasonal flora; culinary, medicinal, scented & dye herbs; botanical art.

Facilities: library of horticultural publications, books & catalogs available for research by special request; conservatory center; kitchenette for rent; prairie garden with native Illinois species; herb garden; classrooms.

Activities: guided tours; lectures; films; hobby workshops; radio programs; formally organized education programs for children, adults & undergraduate college students; training programs; permanent & temporary exhibitions; plant clinic.

Publications: brochure; quarterly members newsletter, Conservatory Conversations; monthly e-news, What's Blooming.

Hours & Admission Prices: Mon. 2-4, Tues.-Sun. 10-4, holidays 10-3. Suggested Donations: adult $2, children $1. &

Attendance: 34,105 (accurate)

Membership: Individual $25; Family & Dual $40; Associate $75; Premier $150; Sustaining $300.

THE SUBURBAN, 125 N. Harvey Ave., Oak Park, IL 60302. Tel.: 708-305-2657.

E-mail: bkmgcar@comcast.net

Web Site: www.thesuburban.org

Institution Type/Description: Art Gallery.

Collections: paintings; sculpture.

Hours & Admission Prices: Call for hours.

WONDER WORKS, 6445 W. North Ave., Oak Park, IL 60302-1009. Tel.: 708-383-4815.

E-mail: info@wonder-works.org

Web Site: www.wonder-works.org

Institution Type/Description: Children's Museum.

Collections: hands-on exhibits.

Activities: birthday parties; educational programs.

Hours & Admission Prices: Mon. & Wed.-Sat. 10-5, Sun. 12-5. Admission $6; children under one & members no charge. Closed New Year's Day; Memorial Day; Independence Day; Labor Day; Thanksgiving; Christmas.

Oakbrook Terrace

LAKE VIEW NATURE CENTER, 17W063 Hodges Rd., Oakbrook Terrace, IL 60181-4505. Tel.: 630-941-8747. Fax: 630-941-3558.

E-mail: lvnc@obtpd.org

Web Site: www.obtpd.org

Founded: 1994.

Congressional District: 6

Key Personnel: Dir., Liane Knight.

Personnel Profile: Full-Time Paid 1; Part-Time Paid 5; Part-Time Volunteers 4.

Governing Authority: municipal. Parent Institution: Oakbrook Terrace Park District. Tax-exempt.

Institution Type/Description: Nature Center.

Collections: regional flora, fauna & natural history.

Facilities: 800-vol. library; native botanical garden; nature center; 1,500 sq. ft. exhibit space.

Activities: formal environmental education programs; participatory exhibits. Annual Event: Spring Celebration in May; Fall Open House in November.

Publications: newsletter, Lake View's Nature News.

Hours & Admission Prices: Mon.-Fri. 9-4, Sat.-Sun. 12-4. No charge; donations accepted. Closed 1st Mon. each month; New Year's Eve & Day; Good Friday; Easter weekend; Memorial Day; Independence Day; Labor Day; Thanksgiving & day after; Christmas Eve & Day. &

Attendance: 20,000 (accurate)

Oglesby

STARVED ROCK STATE PARK, 2678 E. 873rd Rd., Oglesby, IL 61348. Mailing Address: P.O. Box 509, Utica, IL 61373-0509. Tel.: 815-667-4906 & 5356. Fax: 815-667-5354.

E-mail: starvedrockv@ivnet.com

Web Site: dnr.state.il.us

Founded: 1911.

Congressional District: 15

Key Personnel: Supt., Tom Levy; Site Interpreter, Tobias Miller.

Governing Authority: state. A part of Illinois Dept. of Natural Resources, Div. of Land Management, 1 Natural Resources Way, Springfield, IL 62702. Tax-exempt.

Institution Type/Description: Historic Park & Museum: located on the sites of former Indian village of Illinois Indians; 1673-1760 French occupation and 1683 French Fort St. Louis.

Collections: cultural & natural history; archeological; geological.

Research Fields: Indian & French history; biology.

Facilities: interpretive center, lodge with overnight accommodations & dining room; 15 miles of hiking trails; seasonal concessions.

Activities: guided tours; lectures; films; formally organized educational programs; permanent exhibitions; camping.

Publications: brochure, Starved Rock State Park.

Hours & Admission Prices: Park: daily 5am-9pm. Interpretive Center: daily 9-4. No charge; donations accepted. &

Attendance: 216,869 (accurate)

Ottawa

OTTAWA HISTORICAL AND SCOUTING HERITAGE MUSEUM, (M), 1100 Canal St., Ottawa, IL 61350-4940. Mailing Address: P.O. Box 2241, Ottawa, IL 61350-6841. Tel.: 815-431-9353.

E-mail: scouter07@hotmail.com

Web Site: www.ottawascoutingmuseum.org

Founded: 1992.

Congressional District: 11

Key Personnel: C.E.O., Cur. & Public Rels., Mollie Perrot; Pres. & Museum Shop Mgr., Christine Hasty; Financial Dir., Kathy Hite; Archivist, Bruno Polli; Security, Steve Perrot.

Personnel Profile: Full-Time Volunteers 1; Part-Time Volunteers 30.

Governing Authority: private; nonprofit. Tax-exempt: 501(c)(3).

Institution Type/Description: Local History & Scouting Heritage Museum.

Collections: Ottawa history & scouting heritage; local history; photographs; personal artifacts; books; clothing; camping gear; uniforms; badges; patches; awards; medals; handbooks; Native American artifacts; W.D. Boyce display; literature dating from the early days of scouting to the present; Baden-Powell.

Research Fields: Native American culture, nature & conservation as they relate to scouting; Ottawa, IL history.

Facilities: 2,800-vol. library; 60-seat auditorium; nature center; 2,700 sq. ft. exhibition space. Museum-related items for sale.

Activities: docent program; hobby workshops; temporary exhibitions; adult programs. Annual Events: monthly Preservation of Ottawa History group meetings.

Publications: quarterly newsletter, Memory Lane.

Hours & Admission Prices: Thurs.-Mon. 10-4. Adults $3, students & children $2; discount to groups of 10 or more; members no charge. Closed New Year's Eve & Day; Easter; Independence Day; Labor Day; Thanksgiving; Christmas Eve & Day. &

Attendance: 3,000 (accurate)

Membership: Youth $3; Adult $10; Family $15; Contributing $25; Patron $50; Benefactor $100; Guarantor $1,000.

REDDICK MANSION AND GARDENS, 100 W. Lafayette St., Ottawa, IL 61350. Tel.: 815-433-6100.

E-mail: contact@reddickmansion.org

Web Site: reddickmansion.org

Founded: 1974.

Congressional District: 11

Key Personnel: Pres. (V), Diane Sanders; Museum Shop Mgr., Hedy Bernholdt.

Personnel Profile: Part-Time Volunteers 7.

Governing Authority: Tax-exempt.

Institution Type/Description: Historic House Museum: housed in the former home of businessman & politician, William Reddick; built in 1856.

Collections: Reddick family history; period furnishings; personal artifacts; photographs.

Activities: special events. Museum Sponsors: Vintage Clothing of Civil War Era in April; Themed Afternoon Tea in May; Graden Walk in June; Ice Cream Social in Garden in August; Kids Event in September; Victorian Christmas in November; Dinner & Christmas Concert in December; Gingerbread House Contest.

Hours & Admission Prices: Mansion Tours: Mon. & Wed.-Sat. 11-3, Sun. 11-2. Adults $5; discounts to AAM members; members no charge.

Attendance: 3,000 (accurate)

Membership: Individual $20; Family $30; Corporate $100.

Paris

BICENTENNIAL ART CENTER & MUSEUM, 132 S. Central Ave., Paris, IL 61944-1729. Tel.: 217-466-8130. Fax: 217-466-8130.

E-mail: parisartcenter@frontier.com

Web Site: www.parisartcenter.com

Founded: 1975.

Congressional District: 19

Key Personnel: Dir., Susan Stafford.

Personnel Profile: Part-Time Paid 1; Part-Time Volunteers 100.

Governing Authority: nonprofit. Tax-exempt: 501(c)(3).

Institution Type/Description: Art Museum.

Collections: paintings in oils & watercolors; drawings; sculpture; photography.

Facilities: 240-vol. library available for public use; classrooms.

Activities: formal education programs; classes & workshops; visual arts; participatory exhibits; monthly exhibits of professional artists. Museum Sponsors: County Schools Art Show; Fall Art Exhibit; Town & Country, Photography Show; annual Paint Illinois Exhibit (juried-open to all IL artists).
Publications: bimonthly newsletter.
Hours & Admission Prices: Tues.-Fri. 10-4. No charge. Closed New Year's Eve & Day; Easter; Independence Day; Thanksgiving & day after; Christmas Eve & Day. &
Attendance: 5,456 (accurate)
Membership: Student $5; Individual $20; Family $25-$49; Sponsor $50-$99; Benefactor $100-$150; Sustaining $200-$500; Life $1,000 & up.

EDGAR COUNTY HISTORICAL MUSEUM, 408 N. Main, Paris, IL 61944-1549. Tel.: 217-463-5305.
Founded: 1969.
Congressional District: 53
Key Personnel: Pres. (V), Kay Wolfe.
Personnel Profile: Part-Time Volunteers 24.
Governing Authority: society. Affiliated with Edgar County Historical Society, 408 N. Main St., Paris, IL 61944-1549. Tax-exempt: 170(b)(1)(A).
Institution Type/Description: Historical Society Museum: housed in 1876 Arthur House. Listed on the National Register of Historic Places.
Collections: artifacts from Edgar County 19th-century family life; natural history; period clothing & furniture.
Research Fields: genealogical & historical.
Facilities: display areas apart from museum.
Activities: guided tours; lectures; films; docent program or council; field trips; permanent & temporary exhibitions.
Publications: quarterly, Edgar County Historical Society Newsletter.
Hours & Admission Prices: Wed.-Fri. 9-4. No charge. &
Attendance: 1,200 (estimated)
Membership: Single $10; Family $15; Individual Life $100; Couple Life $150.

Park Forest

TALL GRASS ARTS ASSOCIATION, 367 Artist Walk, Park Forest, IL 60466-2059. Tel.: 708-748-3377. Fax: 708-748-9132.
E-mail: tallgrass367@sbcglobal.net
Web Site: www.tallgrassarts.org
Founded: 1955.
Key Personnel: Exec. Dir., Margaret Donohue; Chm. (V), Janet Muchnik; Museum Shop Mgr., Gisele Perrault.
Personnel Profile: Part-Time Paid 2.
Governing Authority: Tax-exempt.
Institution Type/Description: Art Gallery.
Collections: paintings; fine art.
Facilities: Museum-related items for sale.
Activities: art classes. Museum Sponsors: 57th Annual Art Fair.
Publications: newsletter.
Hours & Admission Prices: Tues.-Sat. 11-4. No charge; donations accepted. &
Attendance: 10,000 (estimated)
Membership: $40, $75; $150; $250.

Park Ridge

BRICKTON ART CENTER, 306 Busse Hwy., Park Ridge, IL 60068-3251. Tel.: 847-823-6611. Fax: 847-823-6622.
Institution Type/Description: Art Gallery.
Collections: paintings; sculpture.
Activities: classes; lectures; educational programs.
Hours & Admission Prices: Mon.-Thurs. 10-5, Fri.-Sat. 10-4.

WOOD LIBRARY-MUSEUM OF ANESTHESIOLOGY, 520 N. Northwest Hwy., Park Ridge, IL 60068-2538. Tel.: 847-825-5586. Fax: 847-825-2085.
E-mail: wlm@asahq.org
Web Site: www.woodlibrarymuseum.org
Founded: 1950.
Key Personnel: Pres., Mary Ellen Warner, M.D.; Librarian & Dir., Karen Bieterman, MLIS, Asst. Librarian, Teresa Jimenez, MS LIS; Cur., Hon. George S. Bause, M.D.; Museum Registrar, Judith Robins, M.A.; Library Asst., Margaret Jenkins; Archivist, Felicia Reilly, MALS
Personnel Profile: Full-Time Paid 5; Full-Time Volunteers 1; Part-Time Volunteers 1.
Governing Authority: American Society of Anesthesiologists. Tax-exempt: 501(c)(3).
Institution Type/Description: Medical Museum.
Collections: medical & anesthesiology history; anesthesiology equipment; manuscripts; video oral history; current anesthesia books & journals.

Research Fields: medical & anesthesiology history.
Facilities: 13,000-vol. library of anesthesiology available for use on the premises; reading room.
Activities: annual Wood Library-Museum Fellowships; annual historical lectureship; quadrennial election of historian-laureate.
Publications: history of anesthesiology & education in anesthesiology publications; Living History of Anesthesiology (videotaped interviews).
Hours & Admission Prices: By appointment only. No charge. Closed holidays. &
Attendance: 200 (estimated)
Membership: Friends $40; 3 Year Friend $100; Friends for Life $500.

Paxton

FORD COUNTY HISTORICAL SOCIETY, 145 S. Market St., Paxton, IL 60957-1284. Mailing Address: P.O. Box 115, Paxton, IL 60957-0115. Tel.: 217-379-3723.
E-mail: fordcohistsoc@hotmail.com
Web Site: www.rootsweb.com/~ilford
Founded: 1967.
Congressional District: 15
Key Personnel: Chm. (V), Cynthia Swanson; Chm. (V), Judith Ondercho; Pres. (V), Judith Jepsen-Popel.
Personnel Profile: Part-Time Volunteers 4.
Governing Authority: society; nonprofit organization. Parent Institution: Ford County Historical Society. Tax-exempt: 501(c)(3).
Institution Type/Description: Historical Museum: housed in historic water tower.
Collections: historic flags; Native American artifacts; historic dresses & uniforms.
Research Fields: local history.
Facilities: library.
Activities: tours.
Publications: Ford County Histories; Ford County History 1985; book, A Ticket to the Best, a socioeconomic history of Paxton, IL 1989; bimonthly newsletter, Ford County Historical; book reprint, Remembrances of a Pioneer; atlases and biographical records for 1876, 1884, 1892, 1901 & 1916; Remembrances of A Pioneer - Autobiography of Jane Patton.
Hours & Admission Prices: Mon.-Fri. 9-5, Sun. 2-4 and by appointment. No charge; donations accepted. Closed national holidays. &
Attendance: 3,000 (estimated)
Membership: Regular $10; Life $100.

Pekin

TAZEWELL COUNTY GENEALOGICAL & HISTORICAL SOCIETY LIBRARY, Ehrlicher Research Center, 719 N. Eleventh St., Pekin, IL 61555. Mailing Address: P.O. Box 312, Pekin, IL 61555-0312. Tel.: 309-477-3044.
E-mail: tcghs@tcghs.org
Web Site: www.tcghs.org
Founded: 1978.
Key Personnel: Pres. (V), Dorinda Shannon.
Personnel Profile: Part-Time Volunteers 30.
Operating Expenses: 17,065
Operating Income: 17,095
Governing Authority: Tax-exempt.
Institution Type/Description: Historical Society Museum.
Collections: books; photographs; genealogy; period artifacts; clothing; diaries; school records.
Research Fields: family history; local history.
Activities: research.
Publications: monthly newsletter, The Monthly.
Hours & Admission Prices: Library: Mon. & Thurs.-Fri. 9-1, Tues. 9-1 & 7-9, Wed. 9-4:30, Sun. 2-4:30; call to confirm. No charge; donations accepted. Closed holidays and holiday weekends.
Attendance: 2,000 (estimated)
Membership: Student $18; Individual $20; Family, Canada & Foreign $23; Life $180; Life Family $255.

TAZEWELL COUNTY MUSEUM, (M), 2950 Court St., Pekin, IL 61554. Tel.: 309-347-8375 & 346-1889.
Governing Authority: nonprofit organization.
Institution Type/Description: History Museum.
Collections: local history & culture; personal artifacts; photographs; period furnishings.
Activities: special events.
Hours & Admission Prices: Mon. & Wed. 10-2, Sat. 10-12.

Peoria

CATERPILLAR VISITORS CENTER, 110 S.W. Washington St., Peoria, IL 61602. Tel.: 309-675-0606.
E-mail: caterpillarvisitorscenter@cat.com
Web Site: www.caterpillar.com/visitors-center
Institution Type/Description: Company History Museum.
Collections: Caterpillar company history & vehicles.
Facilities: Gift items for sale.
Activities: school group tours; hands-on exhibitions; video.
Hours & Admission Prices: Mon.-Sat. 10-5. Adults $7, seniors 55 & over, veterans and active military $6, military 55 & over $5; discounts to groups; children 12 & under no charge. Closed holidays. &

CONTEMPORARY ART CENTER, 305 S.W. Water St., Peoria, IL 61602-1425. Tel.: 309-674-6822.
E-mail: artcentr@mtco.com
Web Site: www.peoriacac.org
Institution Type/Description: Art Gallery.
Collections: paintings; sculpture.
Hours & Admission Prices: Tues.-Sat. 11-5, Fri. 11-8. No charge. &

* **LAKEVIEW MUSEUM OF ARTS AND SCIENCES, (M),** 1125 W. Lake Ave., Peoria, IL 61614-5985. Tel.: 309-686-7000. Fax: 309-686-0280.
E-mail: kathleen@lakeview-museum.org
Web Site: www.lakeview-museum.org
Founded: 1965.
Congressional District: 18
Key Personnel: Pres. & C.E.O., James J. Richerson; Chm. (V), Jim Vergon; Vice Pres. Education, Sheldon Schafer; Vice Pres. Collections, Kristan H. McKinsey; Vice Pres. Devel., Nikki Cole; Vice Pres. Communications, Kathleen Woith; Dir. Exhibitions, Cory Tibbits; Dir. Education, Ann Schmitt; Dir. Preschool, Sherry Woessner; Museum Store Mgr., Linda Gouvia.
Personnel Profile: Full-Time Paid 21; Part-Time Paid 19; Interns 5.
Governing Authority: nonprofit organization. Tax-exempt: 501(c)(3).
Institution Type/Description: General Museum.
Collections: 18th-20th century European & American paintings, sculpture, works on paper, decorative arts; Illinois folk art & duck decoys; Western African, Pre-Columbian, Native American, Asian art & artifacts; natural sciences collections including entomological, mineral & fossil collections.
Research Fields: Illinois related artists; collection areas.
Facilities: research library; permanent exhibition area; three changing galleries; planetarium; art & science education wing; enriched preschool; performing arts auditorium; children's discovery center; school loan center; sculpture garden; community solar system model. Illinois Folk Art Gallery. Art objects & used books for sale.
Activities: guided tours; lectures; gallery talks; video; public planetarium shows; workshops & demonstrations; educational programs for children & adults in the arts, sciences & humanities; art, science & ethnic activities; permanent & temporary exhibitions; recycled book sales; school tours; picture person program; school loans.
Publications: monthly newsletter.
Hours & Admission Prices: Museum: Tues.-Sat. 10-4, Sun. 12-4. Adults $6, seniors 60 & over $5, children 3-17 $4; AAM members, members, children 3 & under no charge. Planetarium: adults $4, senior citizens & children 4-17 $3.50; discounts to AAM members. Combo tickets available. Closed major holidays. &
Attendance: 139,867 (accurate)
Membership: College Student $28; Individual $40; Grandpass & Family $60; Patron $100; Renaissance $1,000.

LUTHY BOTANICAL GARDEN, 2520 N. Prospect Rd., Peoria, IL 61603-2126. Tel.: 309-686-3362. Fax: 309-685-6240.
Web Site: peoriaparks.org
Formerly: George L. Luthy Memorial Botanical Garden
Founded: 1951.
Congressional District: 18
Key Personnel: Dir., Bonnie Noble; Business Officer, Jan Budzynski; Mgr., Bob Streitmatter; Museum Shop Mgr., Mary Mulay.
Personnel Profile: Full-Time Paid 4; Part-Time Paid 15; Part-Time Volunteers 63; Interns 1.
Governing Authority: municipal. Branch of Peoria Park District. Tax-exempt.
Institution Type/Description: Botanical Garden.
Collections: tropical plants; orchids; woodland garden; rose garden; herb garden; perennial garden; numerous woody ornamentals; spring & fall borders; Hosta Glade.

Facilities: botanical garden; conservatory. Small plants, books & supplies for sale.
Activities: guided tours; lectures; concerts; art shows; study clubs; hobby workshops; formally organized educational programs; permanent & temporary exhibitions; landscape consulting.
Publications: quarterly newsletter, Leaves From The Garden.
Hours & Admission Prices: Houses: Tues.-Sat. 10-5, Sun. 12-5. Adults $2.50; members no charge. Closed New Year's Day; Thanksgiving; Christmas. &
Attendance: 135,000 (estimated)
Membership: Individual $20; Family $30; Contributing $50; Donor $100; Patron & Corporate $250 & up.

THE PEORIA ART GUILD, 203 Harrison St., Peoria, IL 61602-1536. Tel.: 309-637-2787, ext. 2.
Web Site: www.peoriaartguild.org
Key Personnel: Dir., Beth Reusch
Institution Type/Description: Contemporary Art Museum.
Collections: contemporary art.
Facilities: Museum-related items for sale.
Activities: studio school. Annual Event: Fine Art Fair.
Hours & Admission Prices: Mon.-Thurs. 10-6, Fri.-Sat. 10-5.

PEORIA HISTORICAL SOCIETY, 611 S.W. Washington St., Ste. A, Peoria, IL 61602-5105. Tel.: 309-674-1921. Fax: 309-674-1882. Facebook: Peoria Historical Society.
E-mail: adminphs@peoriahistoricalsociety.org
Web Site: www.peoriahistoricalsociety.org
Founded: 1934.
Congressional District: 18
Key Personnel: Pres., Mark Johnson; Vice Pres., Deborah Dougherty; Exec. Dir., Walter C. Ruppman; Dir. Collections, Robert Killion; Museum Shop Mgr., Kathy Dallinger.
Personnel Profile: Part-Time Paid 3; Part-Time Volunteers 100; Interns 6.
Governing Authority: incorporated by State of Illinois. Subsidiary Institution: Flanagan House & Pettengill-Morron House. Tax-exempt: 501(c)(3).
Institution Type/Description: Historical Society: housed in c.1837 Flanagan House & 1868 Pettengill-Morron House.
Collections: Pettengill-Morron House, c.1868: furniture; silver; china; glassware; paintings; Oriental rugs; jewelry. Flanagan House, c.1837: carpenter shop with tools; pre-Civil War bedroom; period kitchen; local history artifacts.
Research Fields: local history of industry, recreation, architecture; oral histories; ethnic influences, including Native American, French, African American.
Facilities: library of histories, portrait and biographical albums, journals, legal documents & manuscripts, available for research on the premises; Bradley University Library, available to general public for research; reading room.
Activities: guided tours; lectures; arts festivals; temporary exhibitions; historic bus tours of Peoria. Museum Sponsors: historical lectures; school programs; Christmas season open house and tour.
Publications: bimonthly newsletter, Timeline.
Hours & Admission Prices: Open House: 1st Sun each month 1-4; other times by appointment. Adults $7, children 12 & under $3; members no charge. Closed holidays.
Attendance: 1,700 (estimated)
Membership: Student $10; Individual $25; Family $35; Pimiteoui $50; Fritz Triebel $100; Charles Duryea $250; Captain Henry Detweiller $500; Belle Reynolds $1,000; Jean Baptiste Du Sable $2,500; Father Jacques Marquette $5,000.

PEORIA RIVERFRONT MUSEUM, 222 S.W. Washington St., Peoria, IL 61602. Tel.: 309-686-7000.
Web Site: www.peoriariverfrontmuseum.org
Institution Type/Description: History Museum & Planetarium.
Collections: local history & culture; period furnishings; photographs; hands-on exhibitions.
Facilities: theater; planetarium.
Activities: special events; interactive exhibitions.
Hours & Admission Prices: Mon.-Thurs. & Sat. 10-5, Fri. 10-8, Sun. 12-5. Galleries & Planetarium: adults $11, seniors 60 & over $10, youth 3-17 $9. Movies: additional fee. Combo tickets available. Closed New Year's Day; Easter; Thanksgiving; Christmas.

PEORIA ZOO AT GLEN OAK PARK, 2218 N. Prospect Rd., Peoria, IL 61603-2126. Tel.: 309-686-3365. Fax: 309-685-6240.
E-mail: info@peoriazoo.org
Web Site: www.peoriazoo.org
Founded: 1955.

Congressional District: 47 & 93
Key Personnel: Dir., Yvonne Strode; Exec. Dir. Parks, Bonnie Noble; Museum Shop Mgr., Mary Mulay.
Personnel Profile: Full-Time Paid 18; Part-Time Paid 32; Part-Time Volunteers 60; Interns 9.
Governing Authority: municipal. Parent Institution: Peoria Park District. Tax-exempt.
Institution Type/Description: Zoo.
Collections: over 300 animals including mammals, birds, reptiles, amphibians, fish, & invertebrates representing 100 species.
Facilities: Gift items for sale.
Activities: guided tours; lectures; formally organized educational programs; docent program or council; traveling exhibitions; workshops; rental facilities.
Publications: newsletter, Zoo Tales.
Hours & Admission Prices: Daily 10-5. Adults $9, seniors $8, children 3-12 $5.25; PZS, AZA, other zoo members and children 2 & under no charge. Closed New Year's Eve & Day; Thanksgiving; Christmas Eve & Day. &
Attendance: 163,281 (accurate)
Membership: Individual $30; Joint $60; Grandparent & Family $85; Grandparent & Family Plus $105; Booster $150; Benefactor $300; Patron $500; Pride $1,000.

Petersburg

LINCOLN'S NEW SALEM STATE HISTORIC SITE, 15588 History Lane, Petersburg, IL 62675-6010. Tel.: 217-632-4000. Fax: 217-632-4010.
Web Site: www.lincolnsnewsalem.com
Founded: 1917.
Key Personnel: Site Supt., Tim Guinan; Pres., Al Grosboll; Account Clerk, Glen Baum; Museum Shop Mgr., Donna Hitchcock.
Personnel Profile: Full-Time Paid 10; Part-Time Volunteers 350.
Governing Authority: state. State of Illinois, Historic Preservation Agency, R.R. 1, Petersburg, IL 62675. Subsidiary Institution: New Salem Lincoln League. Tax-exempt: 501(c)(3).
Institution Type/Description: Village Museum: site of New Salem Village, in the 1830s where Lincoln lived as a young man. Structures are reconstructed.
Collections: twelve timber houses; Rutledge Tavern; 10 shops; stores; industries; school; period artifacts; furnishings; farm tools.
Research Fields: Lincoln & New Salem; local history; pioneer community life; agriculture in 1830s.
Facilities: picnic grounds; campgrounds. Museum-related items for sale.
Activities: self-guided tours; staffed cabins; permanent exhibitions; summer outdoor dramas.
Publications: bimonthly newsletter; brochures.
Hours & Admission Prices: March-April 15 & Sept.-Oct. Wed.-Sun. 9-5; April 16 to Labor Day daily 9-5; Nov.-Feb. Wed.-Sun. 8-4. No charge; donations accepted. Closed New Year's Day; Thanksgiving; Christmas. &
Attendance: 400,000 (estimated)
Membership: Annual $15.

STARHILL FOREST ARBORETUM, 12000 Boy Scout Tr., Petersburg, IL 62675-6034. Tel.: 217-632-3685. Fax: 217-632-3685.
E-mail: guy@starhillforest.com
Web Site: www.starhillforest.com
Founded: 1976.
Congressional District: 18
Key Personnel: Owner, Guy Sternberg; Owner, Edie Sternberg.
Personnel Profile: Full-Time Paid 1.
Governing Authority: Parent Institution: Illinois College. Tax-exempt.
Institution Type/Description: Arboretum & Botanical Gardens.
Collections: living plant collections. natural forms & native species of woody arborescent plants; national oak collection for the North American Plant Collections Consortium.
Research Fields: ornamental horticulture; forestry; land architecture; ecology.
Facilities: 1,000 vol. reference library pertaining to horticulture & ecology for use on location only; botanical garden; field research station.
Activities: guided tours; lectures.
Publications: reports; brochures.
Hours & Admission Prices: By appointment only. No charge; donations accepted. &
Attendance: 200 (estimated)

Pontiac

CATHERINE V. YOST MUSEUM & ARTS CENTER, 298 W. Water, Pontiac, IL 61764-1757. Mailing Address: 115 W. Howard, Pontiac, IL 61764-1819. Tel.: 815-844-6574 & 5847 (Pontiac tourism).
Key Personnel: Cur., Carol Gardner
Institution Type/Description: Historic House Museum: housed in Queen Anne-style home.
Collections: Yost family furnishings & personal artifacts.
Hours & Admission Prices: May-Dec. by appointment. Donations accepted.

THE INTERNATIONAL WALLDOG MURAL & SIGN ART MUSEUM, 217 N. Mill St., Pontiac, IL 61764. Tel.: 815-842-1848.
E-mail: kristen@muralmuseum.com
Founded: 2010.
Institution Type/Description: Art Museum.
Collections: history of early sign painters; outdoor mural & wall advertising.
Hours & Admission Prices: April-May & Aug.-Oct. Mon.-Fri. 9-5, Sat.-Sun. 10-4; June-July Mon.-Thurs. 9-5, Fri. 9-9, Sat.-Sun. 10-4; Nov.-March daily 10-4. No charge; donations accepted.

THE JONES HOUSE, 314 E. Madison St., Pontiac, IL 61764. Mailing Address: 115 W. Howard St., Pontiac, IL 61764. Tel.: 815-844-5847; 800-835-2055.
Institution Type/Description: Historic House Museum: housed in the former home of Henry C. Jones; built in 1857. Listed on the National Register of Historic Places.
Collections: local history; period furnishings.
Hours & Admission Prices: By appointment.

LIVINGSTON COUNTY WAR MUSEUM & DAL ESTES EDUCATION CENTER, 321 N. Main St., Pontiac, IL 61764-1929. Tel.: 815-842-0301.
Web Site: www.warmuseum.us
Founded: 2004.
Congressional District: 15
Key Personnel: Pres., Jack Murphy.
Personnel Profile: Full-Time Volunteers 6; Part-Time Volunteers 20.
Governing Authority: bd. dirs.
Institution Type/Description: Military History Museum.
Collections: WWI to present; personal artifacts; uniforms.
Publications: quarterly newsletter.
Hours & Admission Prices: Tues.-Sat. 10-4, Sun. 12-4; other times by appointment. No charge. &
Attendance: 10,000
Membership: Annual $20; Life $100.

PONTIAC OAKLAND AUTOMOBILE MUSEUM, 205 N. Mill St., Pontiac, IL 61764. Tel.: 815-842-2345. Facebook: Pontiac Oakland Automobile Museum.
E-mail: info@pontiacoaklandmuseum.org
Web Site: www.pontiacoaklandmuseum.org
Founded: 2010.
Key Personnel: Dir., Tim Dye.
Personnel Profile: Full-Time Paid 2.
Governing Authority: Tax-exempt.
Institution Type/Description: Automobile Museum.
Collections: early automobiles; dealer artifacts & signs; over 2,000 oil cans; maps; brochures; drawings; service manuals.
Facilities: library.
Publications: quarterly, Friends of Pontiac.
Hours & Admission Prices: April-Oct. daily 9-5; Nov.-March daily 10-4. No charge; donations accepted. Closed New Year's Day; Easter; Thanksgiving; Christmas. &
Attendance: 20,000 (accurate)
Membership: Basic: Senior 60 & over $20; Individual $25; Senior Dual $30; Family $40. Premium: Chieftain $100; Super Chief $150; Star Chief $250. Lifetime: 60 & over $400; 40-59 $500; 39 & over $600. Corp. & Club: Sponsor-A-Car $400 (annual); Sponsor-A-Display $2,500 (one time).

ROUTE 66 ASSOCIATION HALL OF FAME & MUSEUM, 110 W. Howard St., Pontiac, IL 61764-1820.
Institution Type/Description: History Museum.

Collections: Route 66 history; photographs; plaques; personal artifacts; Hall of Fame inductees.
Hours & Admission Prices: Summer: Mon.-Fri. 9-5, Sat.-Sun. 10-4; Winter: Mon.-Fri. 11-3, Sat.-Sun. 10-4.

Poplar Grove

VINTAGE WINGS AND WHEELS, 5151 Orth Rd., Ste. A-1, Poplar Grove, IL 61065. Tel.: 815-547-3115. Fax: 815-547-3117.
E-mail: vintagemuseum@gmail.com
Web Site: poplargrovewingsandwheels.com
Institution Type/Description: Transportation Museum.
Collections: transportation history; early aircraft; personal artifacts.
Hours & Admission Prices: May-Sept. Mon.-Fri. 11-3, 1st Sat. each month 10-2, 2nd Sun. each month 9-1.

Prairie du Rocher

FORT DE CHARTRES STATE HISTORIC SITE & MUSEUM, Fort de Chartres Historic Site, 1350 State Rt. 155, Prairie du Rocher, IL 62277. Tel.: 618-284-7230. Fax: 618-284-7230.
E-mail: ftdchart@htc.net
Founded: 1913.
Congressional District: 22
Key Personnel: Site Mgr., Darrell Duensing; Asst. Mgr., Dennis Thomas.
Personnel Profile: Full-Time Paid 3; Part-Time Paid 4.
Governing Authority: Owned & operated by the State of Illinois Historic Preservation Agency, Old State Capitol, Springfield, IL 62701. Affiliated with Illinois Dept. of Conservation, Bureau of Land & Historic Sites, 405 E. Washington, Springfield, IL 62706. Tax-exempt: 501(c)(3).
Institution Type/Description: State Park Museum & History Museum: located on the original site of the French Fort de Chartres built in 1753.
Collections: historical texts; archaeology; military; tools; weapons; utensils; artifacts; reconstructed period buildings. Historic Building: 1753 powder magazine.
Research Fields: archaeology; French, colonial & military life in North America during the 18th century.
Facilities: 85-vol. library of historical text of the 1720-1772 period available on premises; over 500-vol. non-lending research library open to public with microfilm & microfiche available.
Activities: self-guided tours; lectures; visitor center; permanent exhibitions. Museum Sponsors: annual Traders' Rendezvous; 18th-century kid's weekend; French & Indian War Assemblage.
Publications: brochure.
Hours & Admission Prices: Wed.-Sun. 9-5. No charge; donations accepted. Closed New Year's Day; Thanksgiving; Christmas. &
Attendance: 65,000 (estimated)

Princeton

BUREAU COUNTY HISTORICAL SOCIETY MUSEUM, 109 Park Ave., W., Princeton, IL 61356-1927. Tel.: 815-875-2184.
E-mail: bchsmuseum@yahoo.com
Web Site: www.bureaucountymuseum.com
Founded: 1911.
Congressional District: 17
Key Personnel: Pres., Kathryn Cartwright; Dir., Pam Lange.
Personnel Profile: Full-Time Paid 1; Part-Time Paid 4.
Governing Authority: nonprofit. Tax-exempt.
Institution Type/Description: Historical Society Museum.
Collections: local lore; family information & photographs of early residents.
Research Fields: genealogy, local county history.
Facilities: 500-vol. library of local history available on premises. Locally made & published books for sale.
Activities: guided tours; permanent exhibitions.
Publications: newsletter.
Hours & Admission Prices: March-Dec. Wed.-Sat. 1-5; group tours by advance reservation. Requested Donation: adults $3, children $1. Closed Easter; Mother's Day; Independence Day; Labor Day; Thanksgiving.
Attendance: 3,358 (accurate)
Membership: Annual $15; Life $150.

Quincy

ALL WARS MUSEUM, Illinois Veterans Home, 1707 N. 12th St., Quincy, IL 62301-1355. Tel.: 217-222-8641, ext. 380. Fax: 217-222-9621.
Key Personnel: Dir., Rick Gengenbacher; Cur., Bob Craig
Institution Type/Description: Military History Museum.

Collections: U.S. military history from the Revolutionary War to present; military artifacts; photographs; personal artifacts; M60 tank; M5 Stuart Tank; weapons; WWII-Korean War ambulance; jeeps.
Hours & Admission Prices: March-Dec. 7 Tues.-Sat. 9-12 & 1-4, Sun. 1-4. No charge; donations accepted.

THE GARDNER MUSEUM OF ARCHITECTURE & DESIGN, 332 Maine St., Quincy, IL 62301-3929. Tel.: 217-224-6873. Fax: 217-224-0006.
E-mail: gardnermuseum@sbcglobal.net
Web Site: www.gardnermuseumarchitecture.org
Founded: 1974.
Congressional District: 20
Key Personnel: Pres., Susan Deege; Museum Coord., Vicki Ebbing.
Personnel Profile: Full-Time Paid 2; Part-Time Paid 1; Part-Time Volunteers 30.
Governing Authority: nonprofit. Tax-exempt.
Institution Type/Description: Architecture & Design Museum: housed in 1889, Richardsonian Romanesque style Old Public Library.
Collections: architectural artifacts from Quincy & Adams County; stained glass; architectural drawings; historic photographs; files on historic buildings in Quincy & Adams County; stone sculpture yard.
Research Fields: regional history; regional architecture; state & local architects; the building arts.
Facilities: 600-vol. library pertaining to American architecture, design and architectural preservation.
Activities: guided tours; lectures; permanent, temporary & traveling exhibitions; living history program; walking tours; children's programming.
Publications: bimonthly newsletter; local and regional tour guides; exhibit brochures & catalogues; museum brochures.
Hours & Admission Prices: Gallery & Store: Wed.-Sun. 1-4. Offices & Library: Mon.-Fri. 9-4. Adults $3, children $1.50; members no charge. &
Attendance: 5,000 (estimated)
Membership: Individual $20; Family $35; Participating $50; Contributing $100; Supporting $200; Sustaining $500; Benefactor $1,000.

HISTORICAL SOCIETY OF QUINCY AND ADAMS COUNTY, 425 S. 12th St., Quincy, IL 62301-4303. Tel.: 217-222-1835. Fax: 217-222-8212.
E-mail: hsqac@sbcglobal.net
Web Site: www.adamscohistory.org
Founded: 1896.
Congressional District: 17
Key Personnel: Exec. Officer, Reg Ankrom; Pres. (V), Chuck Radel; Treas., John Johannes; Museum Shop Mgr., Joe Winkelmann.
Personnel Profile: Part-Time Paid 5; Part-Time Volunteers 65; Interns 1.
Governing Authority: private; nonprofit organization. Tax-exempt: 501(c)(3).
Institution Type/Description: Historic House Museum: 1835 home of former Governor of Illinois, John Wood, also founder of Quincy.
Collections: items relating to the history of Quincy & Adams County.
Research Fields: local & state history.
Facilities: 1,200-vol. library of history of the Midwest especially Illinois & local area; 80-seat meeting facility; 3,000 sq. ft. exhibit space. Museum-related items for sale.
Activities: guided tours; lectures; temporary exhibits; educational programs; community outreach. Annual Event: Christmas Candlelight tours of Gov. John Wood Mansion.
Hours & Admission Prices: Tours: April-Oct. Tues.-Sat. 10-2. Office: Mon.-Fri. 10-2. Adults $4, children $2; members no charge. Time Traveler program. Closed holidays.
Attendance: 5,500 (estimated)
Membership: Individual $25; Pioneer $50; Builder $100; John Wood Society $500; Mayor $1,000; Quartermaster $5,000; Governor $10,000; Founder $25,000.

QUINCY ART CENTER, (M), 1515 Jersey St., Quincy, IL 62301-4250. Tel.: 217-223-5900. Fax: 217-223-6950. Facebook: Quincy Art Center.
E-mail: jnelson@quincyartcenter.org
Web Site: quincyartcenter.org
Founded: 1923.
Congressional District: 20
Key Personnel: Exec. Dir. & Cur., Julie D. Nelson; Pres. (V), Bruce Broemmel; Chm. (V), Dan Selby.
Personnel Profile: Full-Time Paid 2; Part-Time Paid 4; Part-Time Volunteers 96; Interns 4.
Volunteer Hours: 4,334
Operating Expenses: 250,000
Operating Income: 250,000

Governing Authority: nonprofit organization. Tax-exempt: 501(c)(3).
Institution Type/Description: Art Museum: housed in 1887 carriage house designed by Joseph Lyman Silsbee, mentor of Frank Lloyd Wright.
Collections: paintings; prints; drawings; photographs.
Major Exhibits: Mel Watkin: Fallen Trees / Mark Rospenda: Unbreakable, 1/31/14-3/9/14; Tin: Recycled Enchantment, 1/31/14-3/9/14; 40th Annual High School Student Art Competition, 3/28/14-5/11/14; Barb Fedeler: Charcoal Lands, 3/28/14-5/11/14; The World of Cameron Fuller / Jenny Chi: The Natural State, 5/23/14-6/27/14; 5th Annual Members Exhibit / Permanent Collection Choices, 7/11/14-8/3/14; Watercolor NOW!, 8/15/14-9/21/14; Mary S. Oakley - Lee Lindsay Artists Showcase, 11/21/14-1/18/15.
Facilities: classroom & studio space including printmaking shop & ceramics studio.
Activities: temporary exhibits; one annual & one bi-annual juried competition, one serves artists within a 50 mile radius, the other artists residing in Illinois, Iowa, Missouri & Indiana; annual High School Competitive Exhibit; class & workshop program for children & adults; guided tours; tour & hands-on program with area schools; artist demonstration; gallery talks; community outreach program, smArt Kids: Art Mentor Program.
Publications: calendar of exhibits; exhibition announcements; juried exhibit catalog; Mary S. Oakley Area Artists Showcase brochure; class & workshop flyers; high school competition catalog.
Hours & Admission Prices: Mon.-Fri. 9-4, Sat.-Sun. 1-4. No charge; donations accepted. &
Attendance: 21,900 (estimated)
Membership: Senior Citizen $25; Individual $30; Household $40, $75, $100, $150, $250, $500 & up; Business $150, $250, $500, $1,000 & up.

THE QUINCY MUSEUM, 1601 Maine St., Quincy, IL 62301-4264. Tel.: 217-224-7669.
E-mail: quinmu1@adams.net
Web Site: thequincymuseum.com
Founded: 1965.
Congressional District: 20
Key Personnel: Exec. Dir., Barbara Wilkinson; Pres., Richard Hopkins; 1st Vice Pres., Joseph Mays; Volunteer Coord., Sandra Huddleston; Collections, Jane Huelsmeyer.
Personnel Profile: Full-Time Paid 1; Part-Time Paid 6; Part-Time Volunteers 144; Interns 1.
Governing Authority: nonprofit. Parent Institution: The Quincy Museum, Inc. Tax-exempt: 501(c)(3).
Institution Type/Description: Natural History Museum.
Collections: local history; fossils; minerals; oil paintings; Victorian era furnishings; firearms from Revolution to World War II; American Indian artifacts; natural history artifacts; dinosaurs; archaeological; ethnographical; decorative arts.
Research Fields: natural history of the Mississippi River Valley; archaeology; Victorian life.
Facilities: 1,500-vol. library.
Activities: education programming, lectures, field trips, hands-on activities for children.
Publications: monthly newsletter.
Hours & Admission Prices: Tues.-Sun. 1-5. Adults $4, children 5-18 & college students $2; discounts to AAM members; members & children under 5 no charge. &
Attendance: 8,500 (accurate)
Membership: Individual $20; Family $35; Patron $100; Business $150 & up; Friend $500; Life $5,000.

WORLD AEROSPACE MUSEUM, 1645 Hwy. 104, Quincy, IL 62305. Tel.: 217-885-3143.
Institution Type/Description: Aerospace History Museum.
Collections: aircraft from soviet block countries.
Hours & Admission Prices: Mon.-Fri. 10-4 by appointment.

Rantoul

CHANUTE AIR MUSEUM, (M), 1011 Pacesetter Dr., Rantoul, IL 61866-3672. Tel.: 217-893-1613. Fax: 217-892-5774. Facebook: Chanute Air Museum.
E-mail: curator@aeromuseum.org
Web Site: www.aeromuseum.org
Formerly: Octave Chanute Aerospace Museum
Founded: 1992.
Congressional District: 15
Key Personnel: Pres., Nancy Kobel; Cur., Mark Hanson.
Personnel Profile: Full-Time Paid 1; Part-Time Paid 7; Part-Time Volunteers 43.

Governing Authority: private; nonprofit organization. Parent Institution: Octave Chanute Aerospace Heritage Foundation. Tax-exempt.
Institution Type/Description: Aerospace Museum.
Collections: Chanute AFB history; 30 fighter, bomber & training aircraft used by USAAF, USAF & Navy; replicas of JN4D Jenny & Chanute Glider; 3000 artifacts related to air flight, USAAF, USAF & Navy history; ICBM training silos; 100,000 photographs, slides & negatives; Illinois aviation history.
Research Fields: Chanute Air Force Base history from 1917-1993; Illinois aviation; aerospace history; military aviation; 99th Pursuit Squadron/Tuskegee Airmen.
Facilities: research library available by appointment; 126,000 sq. ft. exhibit space.
Activities: formally organized education programs for grade school students; guided tours; summer camps.
Publications: quarterly newsletter; brochure; web site listings.
Hours & Admission Prices: April-Oct. Mon.-Sat. 10-5, Sun. 12-5; Nov.-March call for hours. Adults $10, senior citizens & retired military $8, students K-12 & college students w/ID $5; discounts to groups; children 4 & under and members no charge. Closed New Year's Day; Easter; Thanksgiving; Christmas. &
Attendance: 17,000 (estimated)
Membership: Military, Student, Educator & Senior $30; Individual $35; Family $50; Business Crew Chief $100; Patron $250; Family Lifetime $1,000. Business: Pilot $500; Flight Commander $1,000; Operations Officer $2,000; Squadron Commander $3,000; Group Commander $4,000; Wing Commander $5,000.

River Grove

CERNAN EARTH & SPACE CENTER, Triton College, 2000 N. 5th Ave., River Grove, IL 60171-1907. Tel.: 708-456-0300, ext. 3372. Fax: 708-583-3153.
E-mail: cernan@triton.edu
Web Site: www.triton.edu/cernan
Founded: 1974.
Congressional District: 6
Key Personnel: Dir., Bart Benjamin; Production Asst., Dan Troiani; Financial & Membership Coord., Karen Pieranunzi; Technician, Joe Schultz.
Personnel Profile: Full-Time Paid 3; Part-Time Paid 9; Part-Time Volunteers 8.
Governing Authority: college. Parent Institution: Triton College. Tax-exempt: 501(c)(3).
Institution Type/Description: Planetarium.
Collections: meteorites; space hardware of historical interest.
Research Fields: educational uses of media.
Facilities: planetarium; classrooms. Gift items for sale.
Activities: lectures; films; concerts; laser light show; loan & permanent exhibitions.
Publications: bimonthly newsletter; teacher guides; promotional literature.
Hours & Admission Prices: Mon.-Thurs. 9-5, Fri. 9-1 & 6:30-9:30, Sat. 6pm-9:30pm, Sun. 1:30-4:30. Earth and Sky Show: adults $8, children & senior citizens $4. Laser Show: adults $10, children & senior citizens $5. Closed major holidays. &
Attendance: 14,500 (accurate)
Membership: Mars $40; Venus $50; Saturn $75; Jupiter $100.

Riverside

RIVERSIDE HISTORICAL MUSEUM, (I), 10 Pine Ave., Riverside, IL 60546-2264. Mailing Address: 27 Riverside Rd., Riverside, IL 60546-2264. Tel.: 708-447-2542.
E-mail: history@riverside.il.us
Web Site: www.riversidemuseum.net
Founded: 1975.
Key Personnel: Chm., Judith Cizck.
Personnel Profile: Full Time Volunteers 7.
Volunteer Hours: 600
Governing Authority: Parent Institution: Village of Riverside. Tax-exempt.
Institution Type/Description: History Museum.
Collections: local history & culture; photographs; maps; manuscripts; architectural drawings; surveys; books; audiovisual tapes; Native American artifacts.
Facilities: archives.
Activities: lectures; educational outreach.
Hours & Admission Prices: Sat. 10-2; other times by appointment. No charge; donations accepted. &
Attendance: 5,000
Membership: Individual $35.

Rochelle

ROCHELLE FIRE DEPARTMENT MUSEUM, 401 5th Ave., Rochelle, IL 61068. Tel.: 815-562-2122.
Institution Type/Description: Firefighting History Museum.
Collections: local firefighting history & equipment; personal artifacts; photographs; 1922 Seagrave fire truck.
Hours & Admission Prices: Call for hours.

Rock Island

AUGUSTANA COLLEGE ART MUSEUM, 7th Ave. & 38th St., Art & Art History Dept., Rock Island, IL 61201-2296. Mailing Address: 639 38th St., Rock Island, IL 61201-2273. Tel.: 309-794-7231. Fax: 309-794-7678.
E-mail: sherrymaurer@augustana.edu
Web Site: www.augustana.edu/arts/artmuseum
Founded: 1983.
Congressional District: 17
Key Personnel: C.E.O., Steven Bahls; Dir., Sherry C. Maurer; Devel. & Membership, Lynn Jackson; Preparator & Registrar, Dana Densberger; Public Rels., Scott Cason.
Personnel Profile: Full-Time Paid 1; Part-Time Paid 8; Part-Time Volunteers 2.
Governing Authority: college. Parent Institution: Augustana College. Tax-exempt: 501(c)(3).
Institution Type/Description: Art Museum.
Collections: Swedish-American art; 19th & 20th-century works; Inkstands from Schonstedt collection; Olson-Brandelle North American Indian art.
Research Fields: Swedish-American art; Southwestern Native American Indian art; inkwells
Facilities: 3,310 sq. ft. exhibit space.
Activities: guided tours; lectures; films; concerts; organized education programs for adults & undergraduate college students; loan, temporary & traveling exhibitions.
Publications: annual museum calendar.
Hours & Admission Prices: Sept.-May Tues.-Sat. 12-4. No charge; donations accepted. Closed college holidays. &
Attendance: 43,708 (accurate)

BLACK HAWK STATE HISTORIC SITE: HAUBERG INDIAN MUSEUM, 1510 46th Ave., Rock Island, IL 61201-6853. Tel.: 309-788-9536. Fax: 309-788-9865.
E-mail: haubergmuseum@aol.com
Web Site: www.blackhawkpark.org/
Founded: 1927.
Congressional District: 19
Key Personnel: Site Supt., Scott Roman; Museum Dir., Elizabeth Carvey-Stewart.
Personnel Profile: Full-Time Paid 4; Part-Time Volunteers 4.
Governing Authority: state; nonprofit. Parent Institution: Illinois Historic Preservation Agency, Old State Capital, Springfield, IL 62701. Tax-exempt.
Institution Type/Description: State Historic Site Museum: 1740-1831 site of the main villages of the Sauk & Fox Nations.
Collections: dioramas containing life-size figures of Sauk & Fox Indians; depicting daily life in an Indian village; Sauk & Fox material culture; 1933-1935 Civilian Conservation Corps camp photographs & video.
Research Fields: Sauk & Fox Indians; site research.
Facilities: lodge with banquet & meeting rooms.
Activities: guided tours; lectures; changing exhibitions; seminars.
Publications: bimonthly newsletter, Historic Illinois; brochure.
Hours & Admission Prices: March-Oct. daily 9-12 & 1-5; Nov.-Feb. daily 9-12 & 1-4. No charge; donations accepted. Closed New Year's Day; Thanksgiving; Christmas. &
Attendance: 32,344 (accurate)

COLONEL DAVENPORT HISTORICAL FOUNDATION, Hillman St. and Mississippi River, Rock Island Arsenal, Rock Island, IL 61201. Mailing Address: P.O. Box 4603, Rock Island, IL 61204-4603. Tel.: 309-786-7336.
E-mail: coloneldavenport1833@hotmail.com
Web Site: www.davenporthouse.org
Founded: 1978.
Congressional District: 17
Key Personnel: Pres. (V), Judy Tumbleson; Membership, Suzanne Hoskins; Museum Shop Mgr., Ginny Bauersfeld.
Personnel Profile: Part-Time Volunteers 50.
Governing Authority: society; nonprofit. Tax-exempt: 501(c)(3).
Institution Type/Description: Historical House Museum: George Davenport Home.

Collections: period furniture; personal artifacts.
Research Fields: Col. George Davenport; Native Americans, regional history.
Activities: guided tours; lectures; films; organized educational programs.
Publications: quarterly newsletter, The Cornucopia; periodic historical monographs.
Hours & Admission Prices: May-Oct. Thurs.-Sun. 12-4; Nov.-April by appointment. Families $10, adults $5, students & children $3; senior citizens $3; discounts to YPN Hot Spot, AAA, CDHF & WQPT members. &
Attendance: 1,500 (estimated)
Membership: Student $15; Senior Citizen $20; Single $25; Family $30; Patron $50; Sustaining $100; Benefactor $500; Life $2,000.

FRYXELL GEOLOGY MUSEUM, (M), Swenson Hall of Geosciences, Augustana College, 38th St., Rock Island, IL 61201-2296. Tel.: 309-794-7318. Fax: 309-794-7564.
E-mail: susanwolf@augustana.edu
Web Site: http://www.augustana.edu/fryxellmuseum
Founded: 1929.
Congressional District: 19
Key Personnel: Dir., William R. Hammer.
Personnel Profile: Part-Time Paid 2; Part-Time Volunteers 1.
Governing Authority: college. Parent Institution: Dept. of Geology, Augustana College, Rock Island, IL.
Institution Type/Description: Geology Museum.
Collections: paleontology; fossils; minerals; geology; remote sensing exhibits.
Research Fields: Triassic & Jurassic Antarctic vertebrates.
Facilities: college library is available for museum use.
Activities: guided tours; formally organized education programs; permanent & temporary exhibitions.
Publications: booklet, Fryxell Geology Museum; teacher's tour guide, Discovery Trip to the Fryxell Geology Museum.
Hours & Admission Prices: Sept.-May Mon.-Fri. 8-4:30, Sat.-Sun. 1-4. No charge. Closed holidays. &
Attendance: 8,000 (estimated)

JOHN DEERE PLANETARIUM, Augustana College, 639 38th St., Rock Island, IL 61201-2210. Tel.: 309-794-7327, 7000 & 7318. Fax: 309-794-7564.
E-mail: leecarkner@augustana.edu
Web Site: helios.augustana.edu/astronomy
Founded: 1969.
Congressional District: 17
Key Personnel: Dir., Lee Carkner; Sec., Gail Parsons.
Personnel Profile: Part-Time Paid 2; Part-Time Volunteers 1.
Governing Authority: university. Parent Institution: Augustana College, 639-38 St. Tax-exempt.
Institution Type/Description: Planetarium.
Collections: meteorites; space materials.
Research Fields: meteorites.
Facilities: planetarium; 200-seat auditorium; classrooms.
Activities: lectures; films; formally organized education programs for elementary through college students; permanent exhibitions.
Hours & Admission Prices: Gallery: Sept.-May Mon.-Fri. 8-4. Planetarium: by reservation only. No charge. &
Attendance: 6,000 (estimated)

QUAD CITY ARTS, 1715 2nd Ave., Rock Island, IL 61201. Tel.: 309-793-1213. Fax: 309-793-1265.
E-mail: info@quadcityarts.com
Web Site: www.quadcityarts.com
Key Personnel: Exec. Dir., Carmen Darland
Institution Type/Description: Art Gallery.
Collections: works by local & regional artists; paintings; sculpture; photographs.
Hours & Admission Prices: Gallery: Tues.-Fri. 10-5, Sat. 11-5; other times by appointment. Office: Mon.-Fri. 9-5. Closed New Year's Day; Martin Luther King Jr. Day; Good Friday; Memorial Day; Independence Day; Labor Day; Thanksgiving & day after; Christmas Eve.

QUAD CITY BOTANICAL CENTER, 2525 Fourth Ave., Rock Island, IL 61201. Tel.: 309-794-0991. Fax: 309-794-1572.
E-mail: jenkins@qcgardens.com
Web Site: www.qcgardens.com
Key Personnel: Exec. Dir., Ami Jenkins; Dir. Education, Gary Koeller; Accounting Supvr., Gretchen A. Hampsey; Mktg. & Guest Svcs., Beth Peters-Carpenter.
Governing Authority: nonprofit organization. Tax-exempt: 501(c)(3).

Institution Type/Description: Botanical Garden.
Collections: plants; flowers; day lilies; mums; ornamental grasses; butterfly garden.
Facilities: library; banquet facilities. Gift items for sale.
Activities: special events; educational programs; guided tours; banquet facilities.
Hours & Admission Prices: April-Oct. Mon. & Wed.-Sat. 10-5, Tues. 10-8, Sun. 12-5; Nov.-March Mon.-Sat. 10-4, Sun. 12-4. Adults $6, seniors 60 & over and military $5, youth 6-15 $4, children 2-5 $2; members and children 2 & under no charge. Closed Thanksgiving; Christmas.

ROCK ISLAND ARSENAL MUSEUM, Bldg. 60, Rock Island Arsenal, Rock Island, IL 61299-5000. Mailing Address: 1 Rock Island Arsenal, Attn: IMRI-PLT, Rock Island, IL 61299-5000. Tel.: 309-782-5021. Fax: 309-782-3598.
E-mail: kris.g.leinicke.civ@mail.mil
Formerly: John M. Browing Memorial Museum
Founded: 1905.
Congressional District: 19
Key Personnel: Dir., Kris G. Leinicke; Cur. Collections, William E. Johnson; Registrar, Jodie Creen Wesemann; Cur. Education, Jodean Rousey Murdock.
Personnel Profile: Full-Time Paid 4; Part-Time Volunteers 36.
Volunteer Hours: 2,124
Governing Authority: federal. Parent Institution: U.S. Army Center of Military History. Subsidiary Institution: Rock Island Arsenal Historical Society. Tax-exempt.
Institution Type/Description: Military Museum.
Collections: weapons; equipment.
Research Fields: history of Arsenal Island; military production & development.
Facilities: resource center; theater. Books & museum-related items for sale.
Activities: lectures.
Publications: books, The Spanish - American War: Its Impact on the Rock Island Arsenal; A Short History of the Rock Island Prison Barracks; An Illustrated History of the Rock Island Arsenal and Arsenal Island.
Hours & Admission Prices: Tues.-Sat. noon-4. No charge; donations accepted. Closed major holidays. &
Attendance: 20,202 (accurate)

Rockford

ANDERSON JAPANESE GARDENS, 318 Spring Creek Rd., Rockford, IL 61107-1035. Tel.: 815-229-9390.
Institution Type/Description: Japanese Gardens.
Collections: Japanese plants, flowers, & trees; sculpture.
Facilities: 14-acre site; restaurant; visitors center; nature trails. Museum-related items for sale.
Activities: tours.
Hours & Admission Prices: May-Nov. 3 Mon.-Fri. 9-6, Sat. 9-4, Sun. 10-4. Adults $8, seniors 62 & over $7, students $6; children 5 & under no charge.

BURPEE MUSEUM OF NATURAL HISTORY, (M), 737 N. Main St., Rockford, IL 61103-6966. Tel.: 815-965-3433. Fax: 815-965-2703. Facebook: Burpee Museum.
E-mail: maureen.mall@burpee.org
Web Site: www.burpee.org
Founded: 1942.
Congressional District: 16
Key Personnel: Chm. (V), Patricia M. Gomez; Exec. Dir., Maureen Maul; Dir. Education, Sheila Rawlings; Collection Mgr., Scott Williams; Mktg. Mgr., MacKenna Atteberry; Guest Experience Mgr., Hillary Parks.
Personnel Profile: Full-Time Paid 12; Part-Time Paid 5; Part-Time Volunteers 45.
Volunteer Hours: 1,885
Governing Authority: nonprofit corporation. Harry & Della Burpee Museum Association. Parent Institution: Rockford Park District. Tax-exempt: 501(c)(3).
Institution Type/Description: Natural History Museum: housed in 1893 Victorian mansion, the Fletcher Barnes home.
Collections: natural history; zoology; geology; science; paleontology; mineralogy; archaeology; anthropology; American Indian; herpetology; herbarium; biology.
Major Exhibits: Megalodon: The Largest Shark That Ever Lived, 2/1/14-4/14.
Research Fields: invertebrate paleontology; vertebrate paleontology.
Facilities: 500-vol. library of reference books, journals & files of publications pertaining to local natural history available on premises or by special

arrangements; 42,000 sq. ft. Robert H. Solem wing; picnic area; meeting areas. Books, booklets, specimens, rocks, minerals, shells & miniature models for sale.
Activities: guided tours; lectures; films; study clubs; permanent & temporary exhibitions; natural history classes; field trips; birthday parties; environmental education programs; vertebrate paleontology field expeditions to Montana & Utah; virtual field trips; family fossil field trips in Rockford area; home school classes; scout programs.
Publications: triannual newsletter.
Hours & Admission Prices: Daily 10-5. Adults $8, children (4-12) 47, children 3 & under no charge; Tues. are donation days; discounts to ASTC members; members no charge. Closed New Year's Day; Easter Sunday, Thanksgiving Day, Christmas Eve & Christmas Day. &
Attendance: 42,297 (accurate)
Membership: General Membership $70.

DISCOVERY CENTER MUSEUM, 711 N. Main St., Rockford, IL 61103-7204. Tel.: 815-963-6769. Fax: 815-968-0164.
E-mail: sarahw@discoverycentermuseum.org
Web Site: www.discoverycentermuseum.org
Founded: 1981.
Congressional District: 16
Key Personnel: Pres. Bd., Joel Huotari; Exec. Dir., Sarah Wolf; Exhibits Coord., Bruce Quast; Mktg. Mgr., Ann Marie Walker; Operations Mgr., Nancy Blank; Dir. Education & Programs, Corinne Sosso; Education Specialist, Jessica Williams; Assoc. Dir., Mike Rathbun; Dir. Devel., Lynn Momberger; Dir. Early Childhood Education, Gloria Svanda; Museum Shop Mgr., Joyce Mazzola.
Personnel Profile: Full-Time Paid 10; Part-Time Paid 18; Part-Time Volunteers 308.
Governing Authority: nonprofit organization. Tax-exempt: 501(c)(3).
Institution Type/Description: Children's Science Museum.
Collections: hands-on exhibits; children's garden.
Facilities: 400-vol. library of children's science books, available to the public; planetarium; 5,000 sq. ft. auditorium; classrooms; 23,000 sq. ft. exhibit space; 8,000 sq. ft. outdoor science park. Museum-related items for sale.
Activities: guided tours; formally organized education programs for children; participatory & traveling exhibitions; outreach programs; working replica of TV news studio.
Publications: quarterly newsletter, Discover.
Hours & Admission Prices: Daily 10-5. Admission $8; discount to ASTC & ACM member; members no charge. Closed New Year's Day; Easter; Thanksgiving; Christmas. &
Attendance: 134,492 (accurate)
Membership: Grand Pass & Family $70; Super Pass $110.

THE ERLANDER HOME MUSEUM, 404 S. 3rd St., Rockford, IL 61104-2013. Mailing Address: P.O. Box 5443, Rockford, IL 61125-0443. Tel.: 815-963-5559. Facebook: Swedish Historical Society.
E-mail: museum@swedishhistorical.org
Web Site: www.swedishhistorical.org
Founded: 1951.
Congressional District: 16
Personnel Profile: Part-Time Volunteers 100.
Governing Authority: Parent Institute: Swedish Historical Society. Tax-exempt: 501(c)(3).
Institution Type/Description: Historic House Museum.
Collections: Swedish-American cultural history; Rockford-made furniture; Charlotte Weibull dolls; Swedish immigration to northern Illinois; area Swedish heritage.
Activities: special events; cultural programs. Museum Sponsors: National Kubb Tournament in May; Midsummer Festival in June; Lucia in December.
Publications: Nyheter.
Hours & Admission Prices: March-Dec. Sun. 1-3; other times by appointment only with 72 hours advanced notice. Adults $5; members no charge. Closed major holidays.
Attendance: 5,000 (estimated)
Membership: Student 18-25 $15; Senior Individual 62 & over $20; Senior Couple $25; Individual $30; Couple $35; Family $40; Sustaining Individual $100; Patron $250; Erlander $500; Lifetime $1,000; Stockholm $2,000.

ETHNIC HERITAGE MUSEUM, 1129 S. Main St., Rockford, IL 61101. Mailing Address: P.O. Box 382, Rockford, IL 61105-0382. Tel.: 815-962-7402 & 877-2287. Fax: 815-962-7402.
E-mail: ehm1129@comcast.net
Web Site: www.ethnicheritagemuseum.org

Founded: 1989.
Congressional District: 16
Key Personnel: C.E.O., Pres. & Dir. (V), Sue Lewandowski; Education & Vice Pres. Devel., Lynell Cannell; Treas., Joanne Baylis; Sec., Anna Delgado.
Personnel Profile: Full-Time Volunteers 4; Part-Time Volunteers 15; Interns 1.
Governing Authority: private; nonprofit organization. Tax-exempt: 501(c)(3).
Institution Type/Description: History Museum.
Collections: 1st settlers of the southwest banks of the Rock River which include African American, Irish, Polish, Italian, Lithuanian & Latino cultures; photographs; folk art; costumes; oral histories.
Major Exhibits: Celebrate Woman of the Year, 2/14-3/14; Italian Weddings (gowns, linens etc.), 4/14-11/9/14; Celebrate Man of the Year, 6/14-7/14; Day of the Dead Ceremony, 11/1/14; Christmas Holiday Traditions Open House, 11/15/14-12/22/14.
Research Fields: The Crusader, African American newspaper, 1953-1973; oral history of family settlers; The South Rockford News: 1932-1942; Italian history; genealogy.
Facilities: library; 1,500 sq. ft. exhibit space; computer lab. Museum-related items for sale.
Activities: arts festivals; docent program; guided tours; hobby workshops; lectures; participatory & temporary exhibits; girl scouts program; new Lockwood Gallery addition; Head Start program-City Source; genealogy research & services. Annual Events: Black History Events in February; Celebrating Rockford's Women in March; Italian, Polish & Lithuanian Folk Art workshop; fundraiser; Holiday & Christmas Event in December.
Publications: quarterly newsletter.
Hours & Admission Prices: Feb.-Dec. Sun. 2-4; other times by appointment. Special Events: family $10, individual $5; discounts to AAM & ICOM members; members no charge. Closed New Year's Day & day after; Mother's Day; Memorial Day; Father's Day; Independence Day; Labor Day; Christmas Day & day after. &
Attendance: 1,900 (estimated)
Membership: Individual $15; Family $25; Trail Blazers $50 & up; Settlers $100; Pioneer $250; Discoverer $500.

KLEHM ARBORETUM & BOTANIC GARDEN, 2715 S. Main St., Rockford, IL 61102-3925. Tel.: 815-965-8146. Fax: 815-965-5914.
E-mail: driggs@klehm.org
Web Site: www.klehm.org
Formerly: Northern Illinois Botanical Society
Founded: 1989.
Key Personnel: Dir., Daniel Riggs; Pres. (V), Alise Howlett.
Personnel Profile: Full-Time Paid 5; Part-Time Paid 3.
Governing Authority: Tax-exempt.
Institution Type/Description: Arboretum & Botanical Garden.
Collections: fountain garden; children's garden; butterfly garden; 5 demonstration gardens; peony garden; hosta garden; prehistoric garden; rhododendron & azalea dell; magnolia collection; conifers collections; over 400 species of trees.
Facilities: library; classrooms; education center.
Activities: guided tours; horticultural programs & classes; concerts; rental facilities.
Hours & Admission Prices: April-Oct. daily 9-4; June-Aug. Fri. 9-8, Sat.-Thurs. 9-4; Nov.-March Tues.-Sat. 9-4. Adults $6, seniors 65 & over, children and students under 18 $3; members no charge. Closed New Year's Eve & Day; Thanksgiving; Christmas Eve & Day. &
Attendance: 35,622 (accurate)
Membership: Senior $30; Individual $40; Senior Family $50; Family $65. Call for corporate memberships.

LAURENT HOUSE MUSEUM, 4646 Spring Brook Rd., Rockford, IL 61114-6362. Tel.: 815-877-2952.
E-mail: info@laurenthouse.com
Web Site: www.laurenthouse.com
Institution Type/Description: Historic House Museum: designed by Frank Lloyd Wright; built in 1949.
Collections: local history; period furnishings; personal artifacts; photographs.
Hours & Admission Prices: By appointment. Adults $20, seniors 65 & over, military and students $15; discounts to groups. &

MIDWAY VILLAGE MUSEUM, 6799 Guilford Rd., Rockford, IL 61107-2613. Tel.: 815-397-9112. Fax: 815-397-9156.
E-mail: info@midwayvillage.com
Web Site: www.midwayvillage.com
Founded: 1970.
Congressional District: 16
Key Personnel: Chm., Jan Jones; Pres., David Byrnes; Administrative Asst., Deb Nau; Coord. Special Events, Jessica MacDonald; Mgr. Site & Operations, Shawn Baxter; Mgr. Customer Svc., Fran Hogan; Cur. Collections, Laura Furman; Cur. Education, Mark Herman; Asst. Cur., Regina Gorman; Educator, Lydia Cassinelli; Volunteer Coord., Gina Moseley; Business Mgr., Nancy McIntosh; Garden Historian, Tari Rowland; Dir. Mktg., Lonna Converso; Dir. Devel., Deb Nai.
Personnel Profile: Full-Time Paid 9; Part-Time Paid 29; Part-Time Volunteers 370.
Governing Authority: nonprofit. Midway Village. Branch Museum: Old Dolls' House Museum. Tax-exempt: 170(b)(1)(A).
Institution Type/Description: History Village Museum: 26 historic turn-of-the-century buildings located on 137 acres.
Collections: Camp Grant military artifacts; industrial building with machines & display of local industries; artifacts & research materials relevant to local history; plumbing shop; hospital; hotel; period artifacts; gardens. Historic Buildings: 1874, Harlem Town Hall; 1892, Holcomb Bank; 1870 farmhouse; 1902 Stone School; 1880 Church; 1900s blacksmith shop; 1860s Breckenridge House; 1855 jail; hardware store; general store; millhouse; fire station; print shop; police station; 1850, 1902 barns; 1910 barn silo; 1922 water tower; 1838 Marsh house; 1890s Pepper house; 1910 Ralston house; 1840s Mowry Brown house; 1860 Chamberlain Hotel.
Research Fields: exhibit projects.
Facilities: library; classrooms; banquet facilities. Museum-related items for sale.
Activities: guided tours; 1890 school enrichment program for children; formally organized education programs for children, adults & college students affiliated with Rock Valley College; lecture series; field trips; rental facilities; reenactments; docent program or council; Visions of the Past living history program; Summer Camps for Kids. Special Events: All Hallows Eve; Baseball & Barbershop; Bikes & Kites; Sock Monkey Madness Festival; World War II reenactment; Model Train Show; Vintage Baseball play by 1850s rules; Chautauqua; Scarecrow Harvest Festival; Holiday Traditions; Mystery Dinners; Victorian Wedding; Victorian Teas.
Publications: quarterly newsletter, Artifacts; Goldenrod Series of books for use in schoolhouse.
Hours & Admission Prices: Museum Center, Mill House & Old Dolls' House Museum: Jan.-April & Sept.-Dec. Tues.-Fri. 10-4, Sat. 10:30-4, May-Aug. Tues.-Fri. 10-4, Sat.-Sun. 10:30-4. Village: May Thurs.-Sun. 11-4; June-Aug. Tues.-Sun. 11-4; Sept.-April by appointment. Office: Mon.-Fri. 8:30-5. Adults $6, children 3-17 $4; discounts to senior citizens & AAM members; members no charge. Special Events: adults $6, children 3-17 $4; members no charge. Closed major holidays. &
Attendance: 306,929 (accurate)
Membership: Senior $25; Individual $30; Senior Family $50; Family $55; Kent Society $500-$999; 1834 Society $1,000 & up.

ROCKFORD ART MUSEUM, (M), 711 N. Main St., Rockford, IL 61103-7204. Tel.: 815-968-2787. Fax: 815-316-2179.
E-mail: staff@rockfordartmuseum.org
Web Site: www.rockfordartmuseum.org
Founded: 1913.
Congressional District: 16
Key Personnel: Pres., Lisa Lindman; Exec. Dir., Linda Dennis; Office Mgr., Nancy Sauer; Coord. Education, Stacey Sauer; Dir. Education, Elizabeth Dailing; Coord. Communications, Dave Dixon; Community Rels. Coord., Sarah McNamara; Registrar, Jeremiah Blankenbaker; Customer Svc. Coord., Carrie Johnson.
Personnel Profile: Full-Time Paid 8; Part-Time Paid 2; Part-Time Volunteers 450.
Governing Authority: nonprofit organization. Tax-exempt: 501(c)(3).
Institution Type/Description: Art Museum.
Collections: American masters; American photography; contemporary glass art; modern & contemporary art; outsider art; history of art slide collection.
Research Fields: 19th- & 20th-century American art; photography; glass; outsider art.
Facilities: library of books & periodicals dealing with all aspects of the visual arts available for use by appointment; 17,000 sq. ft. exhibit space; classrooms; auditorium. Museum-related items for sale.
Activities: guided tours; educational programs; permanent, temporary & traveling exhibitions; lectures; special openings. Museum Sponsors: annual Greenwich Village Art Fair; Young Artist Show; Art in the Garden and Evergreen Ball; fundraising galas.
Publications: exhibition announcements; brochures; exhibition catalogs; quarterly, members publication; educator's guide, Year in Review.
Hours & Admission Prices: Mon.-10-5, Sun. 12-5. Adults $6, seniors & students $3; children under 12, members & Tues. no charge. Closed New Year's Day; Thanksgiving; Christmas. &
Attendance: 42,380 (accurate)
Membership: Student $20; Individual $35; Family $50; Contributor $100; Benefactor $250; Sustainer $500; Patrons' Circle $1,000.

ROCKFORD COLLEGE ART GALLERY/CLARK ARTS CENTER, 5050 E. State St., Rockford, IL 61108-2393. Tel.: 815-226-4105. Fax: 815-394-5167.
E-mail: cwatters@rockford.edu
Founded: 1847.
Congressional District: 16
Key Personnel: Dir., Carey Watters.
Governing Authority: Rockford College. Tax-exempt: 501(c)(3).
Institution Type/Description: College Art Gallery.
Collections: 20th century paintings; prints; photography; ceramics; drawings; installations, assemblage; ethnographic art; prints by modern & contemporary masters.
Facilities: 1,400 sq. ft. exhibit space.
Activities: guided tours; lectures; organized education programs for college students; loan, temporary & traveling exhibitions.
Publications: art exhibition catalogs: Hollis Sigler, Breast Cancer Journal: Walking with the Ghosts of My Grandmothers; The New Woman in Chicago, 1910-1945: Paintings from Illinois Collections; No Feathers: Manifest Destiny Indicted; Prodigal Daughter; Studio Faculty, Rockford College Art Dept.; To Print Magnificently: Early Modern & Contemporary prints from the Rockford College Collection; Judy Chicago: Thinking About Trees.
Hours & Admission Prices: Sept.-May Tues. 12-3, Wed. 6pm-8pm, Thurs.-Sun. 3-6. No charge; donations accepted. &

TINKER SWISS COTTAGE MUSEUM, 411 Kent St., Rockford, IL 61102-2915. Tel.: 815-964-2424. TDD: 815-963-3323; Facebook: Tinker Swiss Cottage Museum.
E-mail: info@tinkercottage.com
Web Site: www.tinkercottage.com
Founded: 1943.
Congressional District: 16
Key Personnel: Dir., Steve Litteral.
Personnel Profile: Full-Time Paid 3; Part-Time Paid 1; Part-Time Volunteers 76.
Governing Authority: not-for-profit; volunteer board of trustees; Rockford Park district. Tax-exempt: 990(a) & 501(c)(3).
Institution Type/Description: Historic House Museum: 1865 Swiss-style home built by Robert H. Tinker.
Collections: original Tinker household furnishings, art, manuscripts & ephemera; reconstructed 1891 suspension bridge.
Research Fields: Robert Tinker diaries, personal letters, records & correspondence; history of Rockford, IL.
Facilities: education/visitor center; rose garden. Museum-related items for sale.
Activities: guided tours. Museum Sponsors: paranormal tours; murder mystery dinners; Victorian Christmas guided tours; workshops; programs for schools, seniors, families & civic groups.
Publications: brochure; quarterly newsletter, Tinker Topics; general brochure; grounds tour brochure; journals of museum founder Robert Tinker.
Hours & Admission Prices: Guided tours: Tues.-Sun. 1 & 3; closed major holidays. Adults $7, senior citizens 65 & up $6, students $5; discounts to Time Traveler, AASLH & AAA, members; members & children 5 & under no charge. &
Attendance: 12,000 (estimated)
Membership: See website for membership information.

Rockton

MACKTOWN, A LIVING HISTORY EDUCATION CENTER, 2221 Freeport Rd., Rockton, IL 61072-1817. Mailing Address: P.O. Box 566, Rockton, IL 61072-0566. Tel.: 815-624-4200. Fax: 815-624-4200.
E-mail: macktown1@verizon.net
Web Site: www.macktownlivinghistory.com
Founded: 1952.
Congressional District: 16
Key Personnel: Pres. (V) Macktown, A Living History Education Center, Ray Ferguson; Pres. Rockton Historical Society, Marilyn Mohring.
Personnel Profile: Full-Time Volunteers 2; Part-Time Volunteers 30.
Governing Authority: county; society. Parent Institution: Winnebago County Forest Preserve District. Affiliated with Rockton Township Historical Society. Tax-exempt.
Institution Type/Description: Historic Houses: 1839 Stephen Mack House, two-story farm house, home to one of the first white settlers in Winnebago County; 1846 Whitman Trading Post, limestone building.
Collections: period artifacts & furnishings; cradle made by Stephen Mack; early wooden woodworking tools; 1830s heritage garden.
Research Fields: new archaeological dig.
Facilities: garden.

Activities: school programs; permanent exhibitions; archaeology digs; tours; cultural programs; workshops; special programs & events. Museum Sponsors: pre-1850s reenactment in April; Antiques Roadshow; Mother-Daughter Tea; 1830s Christmas Breakfast.
Publications: semiannual newsletter.
Hours & Admission Prices: Fri.-Sat. 10-2, Sun. 12-4. No charge; donations accepted.
Attendance: 6,500 (estimated)
Membership: Pioneer $25; Frontiersman $50; Trader $100; Corporate $500; Lifetime $1,000.

Romeoville

ISLE A LA CACHE MUSEUM, 501 E. Romeo Rd., Romeoville, IL 60446-1538. Tel.: 815-886-1467.
Web Site: www.fpdwc.org/isle.cfm
Founded: 1983.
Personnel Profile: Full-Time Paid 5; Part-Time Volunteers 15.
Governing Authority: Parent Institution: Forest Preserve District at Will County. Tax-exempt.
Institution Type/Description: History Museum.
Collections: exhibits of the French fur trade; birch bark canoe; Native American artifacts including metal, beads & cloth.
Hours & Admission Prices: Tues.-Sat. 10-4, Sun. 12-4. &

Roscoe

HISTORIC AUTO ATTRACTIONS, 13825 Metric Dr., Roscoe, IL 61073-7607. Tel.: 815-389-9999. Fax: 800-779-6461.
E-mail: museum@historicautoattractions.com
Web Site: www.historicautoattractions.com
Founded: 2001.
Congressional District: 16
Key Personnel: Dir., Wayne Lensing; Treas., Cathy Ellis.
Personnel Profile: Full-Time Paid 1.
Governing Authority: private.
Institution Type/Description: History Museum.
Collections: presidential & world leaders' cars & limousines; Kennedy artifacts; WWII history; Hollywood, crime & Western memorabilia; racing & television artifacts.
Research Fields: presidential history; American political history in relation to the White House.
Facilities: Museum-related items for sale.
Activities: Annual Events: Lensing Autumn Classic Car Show; Annual Motorcycle Show.
Hours & Admission Prices: Memorial Day to Labor Day Tues.-Sat. 10-5, Sun. 11-4; Sept.-Nov. Sat. 10-5, Sun. 11-4. Adults $12, senior citizens $10, students 6-15 $7; children under 6 no charge. Season Pass: $35. &
Attendance: 10,000 (estimated)

Roselle

ROSELLE HISTORY MUSEUM, 102 S. Prospect St., Roselle, IL 60172. Mailing Address: 39 E. Elm St., Roselle, IL 60172. Tel.: 630-351-5300.
E-mail: rosellehistory@sbcglobal.net
Web Site: www.rosellehistory.com
Formerly: Roselle Historical Society
Founded: 1975.
Key Personnel: Raymond Hitzemann.
Personnel Profile: Part-Time Paid 1.
Governing Authority: Tax-exempt.
Institution Type/Description: History Museum.
Collections: local history & culture; photographs; documents. Historic Buildings: 1874 Sumner house; 1878 Richter house; Coach house.
Publications: quarterly members' newsletter.
Hours & Admission Prices: Sun. 2-4, other times by appointment. No charge; donations accepted.
Attendance: 500 (estimated)
Membership: Individual $25.

Rosemont

DONALD E. STEPHENS MUSEUM OF HUMMELS, 9511 Higgins Rd., Rosemont, IL 60018. Tel.: 847-692-4000.
E-mail: stephensmuseum@rosemont.com
Web Site: www.stephenshummelmuseum.com
Institution Type/Description: General Museum.
Collections: over 1,000 M.I. Hummel figurines & ANRI woodcarvings.
Facilities: Museum-related items for sale.

Activities: tours.
Hours & Admission Prices: Mon.-Fri. 10-4, Sat. 9-3; groups by appointment. No charge.

Rosiclare

THE AMERICAN FLUORITE MUSEUM, Main St., Rosiclare, IL 62982. Mailing Address: P.O. Box 755, Rosiclare, IL 62982. Tel.: 618-285-3513.
Institution Type/Description: Mineral Museum.
Collections: fluorspar mining industry & tools; minerals; photographs.
Hours & Admission Prices: March-Dec. Thurs.-Fri. & Sun. 1-4, Sat. 10-4; tours by appointment. Adults $3, children 6-12 $1.

Roxana

WOOD RIVER REFINERY HISTORY MUSEUM, 900 S. Central, Roxana, IL 62084. Mailing Address: P.O. Box 76, Roxana, IL 62084-0076. Tel.: 618-255-3718. Fax: 618-2552552.
Governing Authority: nonprofit organization.
Institution Type/Description: History Museum.
Collections: refinery history; period artifacts; early cars; photographs.
Hours & Admission Prices: Wed.-Thurs. 10-4.

Rushville

SCHUYLER JAIL MUSEUM, 200 S. Congress St., Rushville, IL 62681-1410. Tel.: 217-322-6975.
Founded: 1968.
Key Personnel: Pres., Lillian Hoover; Cur., Maryjane Busby.
Governing Authority: nonprofit organization. Parent Institution: Schuyler-Brown Historical & Genealogical Society, Congress & Madison Sts., Rushville, IL 62681. Tax-exempt.
Institution Type/Description: Historical Society Museum: housed in 1857-58 Schuyler Jail.
Collections: local memorabilia of Schuyler County; genealogy.
Research Fields: genealogy.
Facilities: library of genealogical books available for members of society; reading room.
Activities: guided tours.
Publications: quarterly magazine, The Schuylerite.
Hours & Admission Prices: April-Nov. daily 1-5; Dec.-March Sat.-Sun. 1-4. No charge. Closed major holidays. &
Membership: Individual $15.

Russell

RUSSELL MILITARY MUSEUM, 43363 Old Hwy. 41, Russell, IL 60075. Tel.: 847-395-7020. Fax: 847-395-7025. Facebook: Russell Military Museum.
E-mail: ksonday@db3mail.com
Web Site: www.russellmilitarymuseum.com
Formerly: Kenosha Military Museum
Founded: 1986.
Key Personnel: Dir., Mark Sonday; Museum Shop Mgr., Joyce Sonday.
Governing Authority: Tax-exempt.
Institution Type/Description: Military Museum.
Collections: Ch-54 Skycranes; C-130 Hercules; WWI French howitzer; WWII Higgins boat; M-4 Sherman tanks; M-48 bridge layer; M-41 Walker Bulldog tank; Vietnam era river patrol boat; Huey helicopters; experimental hovercraft; Hind; Cobra; MWRAP.
Facilities: Museum-related items for sale.
Activities: auction; birthday parties; school, scout & group specials.
Hours & Admission Prices: See website for hours. Adults 13 & over $10, children & seniors $5; children 2 & under no charge. &
Attendance: 20,000 (estimated)

Saint Charles

BEITH HOUSE MUSEUM, 8 W. Indiana St., Saint Charles, IL 60174-2829. Mailing Address: c/o PPFV, P.O. Box 903, Saint Charles, IL 60174-0903. Tel.: 630-377-6424. Fax: 630-377-6424.
E-mail: info@ppfv.org
Web Site: www.ppfv.org/beith.htm
Key Personnel: Dir., Elizabeth Safanda
Institution Type/Description: Historic House Museum: housed in the home of Kane County stone mason, William Beith, built in 1850.
Collections: period furnishings; personal artifacts; photographs; archaeological remains.

Facilities: library of historical research & conservation resources.
Activities: Museum Sponsors: Old House Mysteries; Archaeology Dig; Is He or Isn't She Challenge.
Publications: preservation partners of the Fox Valley's Advocate.
Hours & Admission Prices: May-Oct. Tues. 1-4; other times by appointment. Suggested Donations: $1 per person, $3 per family.

DURANT HOUSE MUSEUM, 37W370 Dean St., Saint Charles, IL 60175. Mailing Address: P.O. Box 903, Saint Charles, IL 60174-0903. Tel.: 630-377-6424. Fax: 630-377-6424.
E-mail: info@ppfv.org
Web Site: www.ppfv.org/durant.htm
Institution Type/Description: Historic House Museum: farmstead built in 1843.
Collections: period furnishings; personal artifacts; period tools & equipment.
Activities: seasonal programs; educational programs by appointment.
Publications: preservation partners of the Fox Valley's Advocate.
Hours & Admission Prices: June-Aug. Thurs. & Sun. 1-4; Sept.-Oct. Sun. 1-4. Suggested Donation: adults $2, children $1.

ST. CHARLES HERITAGE CENTER, (M), 215 E. Main St., Saint Charles, IL 60174-2040. Tel.: 630-584-6967. Fax: 630-584-6077.
E-mail: info@stcmuseum.org
Web Site: www.stcmuseum.org
Founded: 1933.
Key Personnel: Pres., Laura Haule; Registrar, Julie Bunke.
Personnel Profile: Full-Time Paid 2; Part-Time Paid 1; Part-Time Volunteers 40; Interns 3.
Governing Authority: society; nonprofit. Subsidiary Institution: Dunham-Hunt Museum. Tax-exempt.
Institution Type/Description: Historical Museum: housed in 1920s era gas station & a c.1840 historic house.
Collections: local history from Native American occupation to present; local history archive; Percheron horse archive materials; 1841 Sheldon Peck painting.
Research Fields: local history.
Facilities: library; 2,500 sq. ft. exhibit space.
Activities: guided tours; lectures; organized educational programs; permanent & temporary exhibitions; architectural quest; living history events.
Publications: quarterly newsletter, The Charlemagne Newsletter; booklets, Celebrating History-A Pictorial Essay of St. Charles; Reflections of St. Charles; The Potawatomi, A Native American Legacy; St. Charles In The Civil War; City Lights.
Hours & Admission Prices: Tues.-Sat. 10-4. No charge; donations accepted. Closed most holidays. &
Attendance: 6,000 (estimated)
Membership: St. Charles Junior Historical Society Age 9-14 $10; Individual $25; Family $40; Patron $125; Corporate $250; Sponsor $500; Benefactor $1,000.

Sandwich

SANDWICH HISTORICAL SOCIETY/STONE MILL MUSEUM, (M), 315 E. Railroad, Sandwich, IL 60548-2241. Mailing Address: P.O. Box 82, Sandwich, IL 60548-0082. Tel.: 815-786-2513.
Founded: 1969.
Congressional District: 14
Key Personnel: Pres. (V), Pat Clapper.
Personnel Profile: Part-Time Volunteers 20.
Governing Authority: society; nonprofit. Parent Institution: Sandwich Historical Society. Tax-exempt.
Institution Type/Description: Historical Society Museum: housed in an 1856 mill.
Collections: Indian artifacts; barbwire collection; local artifacts; manuscript collection; blacksmith tools; oldtime fire equipment; toys; military memorabilia; patent models; kitchen; dental parlor; bedroom; parlor; children's playroom clothes closet; farm machines; tools; c.1909 Regal touring car; early equipment manufactured in Sandwich; local artifacts.
Research Fields: Indian artifacts; farm machinery; carpenter & blacksmith tools.
Facilities: library of newspapers from Sandwich used for genealogy, reading room. Gifts for sale.
Activities: guided tours.
Hours & Admission Prices: April-Oct. Sun. 1-4; other times by appointment. No charge; donations accepted. Closed holidays.
Attendance: 1,500 (estimated)
Membership: Individual $10; Family $20.

Schaumburg

SPRING VALLEY NATURE CENTER & HERITAGE FARM, 1111 E. Schaumburg Rd., Schaumburg, IL 60194-3648. Tel.: 847-985-2100. Fax: 847-985-9692.
Web Site: www.parkfun.com
Formerly: Spring Valley Nature Sanctuary & Volkening Heritage Farm
Founded: 1983.
Key Personnel: Exec. Dir., Jean Schlinkmann; Conservation Svcs. Mgr., David Brooks.
Personnel Profile: Full-Time Paid 10; Part-Time Paid 20; Part-Time Volunteers 250.
Governing Authority: municipal; nonprofit. Parent Institution: Schaumburg Park District. Tax-exempt.
Institution Type/Description: Natural History Museum & Sanctuary; Living History Farm.
Collections: 1848 house; c.1875 dairy barn; c.1880 summer kitchen & smokehouse.
Research Fields: taxonomical study.
Facilities: 1,000-vol. library for public use; 8,500 sq. ft. earth-sheltered, passive solar interpretive center; 135 acres total; 3.5 miles handicapped accessible trails; native habitat greenhouse; Discovery Niches; meeting room. Honey, maple syrup, nature books, pamphlets, animal puppets & bird feeders for sale.
Activities: guided tours; lectures; films; study clubs; hobby workshops; organized education programs; docent program; participatory exhibits; outreach environmental education program. Museum Sponsors: Maple Syrup Fair; Autumn Pioneer Festival; Christmas In the Valley; Backyards for Nature Fair; Halloween Ghost Jaunt; Junior Girl Scout Badge Workshops.
Publications: bimonthly newsletter, Natural Enquirer; brochures, Spring Valley Nature Sanctuary; trail map; self-guiding trail book; Group Venture Program; Step Into the Wild Volunteer Program.
Hours & Admission Prices: Grounds: April-Oct. daily 8-8; Nov.-March daily 8-5. Visitors Center: daily 9-5. Farm Site: April-Oct. Tues.-Fri. 9-2, Sat.-Sun. 10-4; special events & groups by appointment. No charge; donations accepted. Closed New Year's Day; Thanksgiving; Christmas. &
Attendance: 85,000 (estimated)
Membership: Spring Valley Nature Club: Student & Senior Citizen $5; Family $10; Contributing $15; Supporting $25; Life $100; Sponsoring $1,000.

Shabbona

SHABBONA-LEE-ROLLO HISTORICAL MUSEUM, 119 W. Comanche, Shabbona, IL 60550-0334. Mailing Address: P.O. Box 334, Shabbona, IL 60550-0334. Tel.: 815-824-2597.
E-mail: slrmuseum@frontierl.com
Founded: 1992.
Key Personnel: Cur., Ada Gallagher
Institution Type/Description: History Museum.
Collections: history; genealogy; family history documents; census records; newspaper microfilm; plat books; publications; photographs.
Hours & Admission Prices: Tues. & Thurs. 9-11:30am & 12:30-3, Sat. 9-11am. No charge; donations accepted.
Membership: Individual $10; Family $15; Contributing $35; Life $250.

Shelbyville

PAM AND BOB BOARMAN CHEVY BELAIR MUSEUM, 224 W. Main St., Shelbyville, IL 62565-1614. Tel.: 217-774-4919.
Institution Type/Description: Car Museum.
Collections: 10 early Chevy's; 50s memorabilia; Chevy clocks; neon signs; early jukebox; 1933 Indy Pace car.
Hours & Admission Prices: Mon.-Fri. 8-7, Sat. 8-6.

Skokie

ILLINOIS HOLOCAUST MUSEUM AND EDUCATION CENTER, 9603 Woods Dr., Skokie, IL 60077. Tel.: 847-967-4800. Fax: 847-967-4801.
E-mail: info@ilhmec.org
Web Site: www.ilholocaustmuseum.org
Formerly: Holocaust Memorial Foundation of Illinois, Inc.
Founded: 2009.
Key Personnel: Exec. Dir., Richard S. Hirschhaut; Deputy Dir., Evette L. Simon; Dir. Finance & Accounting, Eric Schwager; Dir. Devel., Ken Cooper; Dir. Mktg. & Communications, Karen Goodman Minter; Dir. Training & Public Programs, Noreen Brand; Dir. Educational Outreach & Genocide Initiatives, Kelley H. Szany; Chief Cur. Collections & Exhibitions, Arielle Weininger.
Personnel Profile: Full-Time Paid 30; Part-Time Paid 2; Part-Time Volunteers 294; Interns 26.
Governing Authority: Tax-exempt.
Institution Type/Description: History Museum.
Collections: Holocaust history; photographs; personal artifacts; documents; Make a Difference! The Harvey L. Miller Family Youth Exhibition; Legacy of Absence Gallery.
Major Exhibits: Keep Calm and Carry On: Textiles on the Home Front in WWII Britain (T), 9/20/13-1/14; Ruth Gruber: Photojournalist (T), 2/14-6/14; Charlotte Salomon: Life? or Theater? (T), 6/14-9/14; RACE: Are We So Different? (T), 10/14-1/15.
Activities: hands-on exhibits; educational programs; special events.
Hours & Admission Prices: Mon.-Wed. & Fri. 10-5, Thurs. 10-8, Sat.-Sun. 11-4. Adults $12, seniors 65 & over and students 12-22 $8, children 5-11 $6; discounts to military members; member no charge.
Attendance: 75,180 (accurate)
Membership: Holocaust Survivor $25; Individual $40; Family $75; Supporter $250; Sponsor $500; Patron $1,000.

SKOKIE HERITAGE MUSEUM & LOG CABIN, 8031 Floral Ave., Skokie, IL 60077-3605. Tel.: 847-674-1500, ext. 3000; 677-6672.
E-mail: ajhanson@skokieparks.org
Web Site: www.skokieparks.org
Founded: 1992.
Congressional District: 9
Key Personnel: Facility Mgr., Amanda Hanson.
Personnel Profile: Full-Time Paid 1; Part-Time Paid 3; Part-Time Volunteers 3.
Governing Authority: Parent Institution: Skokie Park District. Subsidiary Institution: Skokie Historical Society. Tax-exempt.
Institution Type/Description: History Museum.
Collections: period furnishings; documents; memorabilia; fire department equipment. Historic Buildings: Engine House; 1847 log cabin.
Activities: classes; educational school programs; special events.
Hours & Admission Prices: Museum: Thurs.-Fri. 12-4, Sat.-Sun. 10-2. Log Cabin: by appointment. No charge; donations accepted. &
Attendance: 4,000 (estimated)

South Elgin

FOX RIVER TROLLEY MUSEUM/FOX RIVER TROLLEY ASSOCIATION, 361 S. La Fox (IL Rte. 31), South Elgin, IL 60177. Mailing Address: P.O. Box 315, South Elgin, IL 60177-0315. Tel.: 847-697-4676.
E-mail: info@foxtrolley.org
Web Site: www.foxtrolley.org
Founded: 1958.
Congressional District: 14
Key Personnel: Pres. (V), Edward J. Konecki; Public Rels., Don MacBean; Museum Shop Mgr., Laura Taylor.
Personnel Profile: Part-Time Volunteers 300.
Governing Authority: private; nonprofit organization.
Institution Type/Description: Trolley Museum.
Collections: 25 cars & 2 diesel locomotives; maintain & operate 2.5 miles of trolley track.
Research Fields: interurban & street cars of the Greater Chicago area.
Facilities: Museum-related items for sale.
Activities: hobby workshops. Museum Sponsors: Caboose Day in Spring & Fall; Trolley Fest in late August; Pumpkin & Haunted Trolley in October.
Publications: quarterly newsletter, Fox River Lines; annual visitor's guide, Trolley Times.
Hours & Admission Prices: May 12-June 29 & Sept.-Oct. Sun. 11-5, call for additional hours; June 29 to Labor Day Sat.-Sun. 11-5. One Ride: adult $4, senior 65 & up $3, child 3-11 $2, each additional ride $1. All day ticket: $8 per person; discount to groups of 20 or more.
Attendance: 10,500 (accurate)
Membership: Associate $25; Family $40.

South Holland

SOUTH HOLLAND HISTORICAL SOCIETY, 16250 Wausau Ave., South Holland, IL 60473-2199. Mailing Address: Box 48, South Holland, IL 60473-0048. Tel.: 708-596-2722.
Founded: 1969.
Congressional District: 3
Key Personnel: Pres. (V), Robin Scheldberg; Vice Pres., Bill Paarlberg; Sec., Barbara O'Donnell; Treas., Joyce Becker; Acting Cur., Edward Smith.
Personnel Profile: Part-Time Volunteers 20.
Governing Authority: society; nonprofit organization. Tax-exempt.

Institution Type/Description: Historical Society Museum: housed in Village Library.
Collections: clothing; jewelry; books; tools; military; local memorabilia; diorama of South Holland in the 1920s. Historic Houses: 1869 Paarlberg Farm; 1858 Van Oostenbrugge Homestead.
Research Fields: local family histories.
Facilities: library of church books & newspaper clippings available for research on premises.
Activities: guided tours.
Publications: newsletter, The Onionskin.
Hours & Admission Prices: Sat. 1-4. Adults $1, children $.50; members no charge. &
Attendance: 300 (estimated)
Membership: Regular $10; Business $25; Life $100.

Springfield

ABRAHAM LINCOLN PRESIDENTIAL LIBRARY & MUSEUM, (M), 212 N. 6th St., Springfield, IL 62701-1004. Tel.: 1-800-610-2094.
E-mail: HPA.info@illinois.gov
Web Site: www.alplm.org
Founded: 2005.
Congressional District: 18
Key Personnel: Exec. Dir., Eileen Mackevich; Deputy Dir., David Blanchette; Cur. Lincoln, James Cornelius; Dir. Library Svcs., Kathryn M. Harris; ALPLM Foundation, Rene Brethorst; Dir. Guest Svcs., Clare Thorpe; Volunteer Svcs. Coord., Jeremy Carrell; Museum Shop Mgr., Amy Miller.
Personnel Profile: Full-Time Paid 90; Part-Time Volunteers 500; Interns 10.
Governing Authority: Parent Institution: State of Illinois. Subsidiary Institution: Illinois Historic Preservation. Tax-exempt.
Institution Type/Description: Historical Library & Museum.
Collections: Abraham Lincoln; Illinois history; genealogy; Illinois newspapers; Civil War; books; maps; manuscripts; photographs; Lincolniana.
Facilities: library; restaurant; visitor center. Museum-related items for sale.
Activities: Union Square Park: outdoor performances & events.
Publications: Foundation newsletter, Four Score and Seven.
Hours & Admission Prices: Library: Mon.-Fri. 9-5. Museum: daily 9-5. Research: call for appointment. Adults $12, students & seniors 62 & over $9, children 5-15 $6; discounts to military, group & AAA members; children under 5 no charge. Closed New Year's Day; Thanksgiving; Christmas. &
Attendance: 350,000 (accurate)
Membership: Basic $70; Family $85; Family Plus $110; Circuit Rider $300; Railsplitter $500; Presidents' Cabinet $1,000.

AIR COMBAT MUSEUM, 835 S. Airport Rd., Springfield, IL 62707. Tel.: 217-522-2181.
Institution Type/Description: Military Museum.
Collections: military aircraft including North American P-51D Mustang; Vought F4U-5N Corsair; Ryan PT-22 Recruit.
Hours & Admission Prices: Mon.-Fri. 9-12 & 1-4; guided tours by appointment. Closed national holidays.

THE DANA-THOMAS HOUSE STATE HISTORIC SITE, 301 E. Lawrence Ave., Springfield, IL 62703-2232. Tel.: 217-782-6776 & 6773. Fax: 217-788-9450.
E-mail: dthf@warpnet.net
Web Site: dana-thomas.org
Founded: 1981.
Congressional District: 20
Key Personnel: Foundation Exec. Dir., Regina Albanese.
Personnel Profile: Full-Time Paid 3; Part-Time Paid 6; Part-Time Volunteers 110; Interns 1.
Governing Authority: state. Parent Institution: The Illinois Historic Preservation Agency. Tax-exempt: 501(c)(3).
Institution Type/Description: Historic House Museum: 1902 Frank Lloyd Wright prairie period house.
Collections: original Wright-designed oak furniture; zinc-camed art glass; historic photographs; family documents; decorative arts; archival building documents.
Research Fields: turn-of-the-century arts & crafts; household arts; period architecture.
Facilities: 250-vol. library pertaining to Frank Lloyd Wright & architectural material available to the public; 45-seat orientation auditorium. Bookshop with Dana House, Frank Lloyd Wright & architecture-related material available.
Activities: guided tours; lectures; films; concerts; organized educational programs; docent program.

Publications: quarterly, Dana-Thomas House Foundation Newsletter, Wright in Springfield; monthly, Volunteer Interpreter Newsletter.
Hours & Admission Prices: Wed.-Sun. 9-5; last tour at 4. Suggested Donation: adults $5, children 3-17 $3; school classes no charge. Closed some state holidays.
Attendance: 40,000 (accurate)
Membership: The Dana-Thomas House Foundation: Supporter $35; Contributor $50; Donor $75-$125; Sponsor $250; Patron $500; Lifetime Individual $2,500.

DAUGHTERS OF UNION VETERANS OF THE CIVIL WAR MUSEUM, 503 S. Walnut, Springfield, IL 62704-1932. Mailing Address: P.O. Box 211, Springfield, IL 62705-0211. Tel.: 217-544-0616.
Institution Type/Description: History Museum.
Collections: local history; records of the Union & Confederate Armies; medals; photographs; drums; books; Civil War uniforms; personal artifacts.
Hours & Admission Prices: Mon.-Fri. 9-12 & 1-4; other times by appointment.

GRAND ARMY OF THE REPUBLIC MEMORIAL MUSEUM, 629 S. 7th St., Springfield, IL 62703-1636. Tel.: 217-522-4373.
Governing Authority: Parent Institution: Woman's Relief Corps Auxiliary.
Institution Type/Description: History Museum.
Collections: Civil War Union history & memorabilia; personal artifacts; documents; photographs.
Hours & Admission Prices: March-Dec. Tues.-Sat. 10-4. No charge; donations accepted.

HENSON ROBINSON ZOO, 1100 E. Lake Dr., Springfield, IL 62712-5536. Tel.: 217-585-1821.
Web Site: www.hensonrobinsonzoo.org
Founded: 1970.
Key Personnel: Dir., Jackie Peeler; Special Activites Coord., Kim Alexander; Education Cur., Emily McEvoy.
Personnel Profile: Full-Time Paid 16; Part-Time Paid 15; Part-Time Volunteers 80.
Governing Authority: municipal; not-for-profit. Parent Institution: Springfield Park District. Tax-exempt: 501(c)(3).
Institution Type/Description: Zoo.
Collections: zoological.
Research Fields: reproductive biology with lemurs & felines; endangered species survival programs for lemurs, tamarins, cheetahs & red wolves.
Facilities: 125-vol. zoological related library; zoological park. Plush animals & zoo-related items for sale.
Activities: guided tours. Annual Events: Spring Opening Festival; Fall Closing Festival; Safe Halloween at Zoo.
Publications: quarterly newsletter, Cougar Chronicle.
Hours & Admission Prices: April-Oct. Mon.-Fri. 10-5, Sat.-Sun. 10-6; Nov.-Dec. daily 10-4; Jan.-Feb. Sat.-Sun. 10-4; March daily 10-4. Adults $5.25, seniors over 62 $3.75, children 3-12 $3.50; members and children 2 & under no charge. &
Attendance: 103,546 (accurate)
Membership: Individual $30; Family $50; Donor $75; Patron $100; President $250; Life $500.

ILLINOIS EXECUTIVE MANSION, 410 E. Jackson St., Springfield, IL 62701. Tel.: 217-782-6450.
Institution Type/Description: Historic Building: housed in the home of the Illinois Governor; built in 1855. Listed on the National Register of Historic Places.
Collections: Illinois history; period furnishings; personal artifacts; photographs; paintings.
Activities: guided tours.
Hours & Admission Prices: Tours: Tues. & Thurs. 9:30-11:30 & 2-3:15, Sat. 9:30-11.

ILLINOIS FIRE MUSEUM, State Fairgrounds, Main & Central Aves., Springfield, IL 62794. Mailing Address: P.O. Box 19427, Springfield, IL 62794-9427. Tel.: 217-524-8754.
Institution Type/Description: Fire-Fighting Museum: housed in the former fire house; built in 1857.
Collections: firefighting history & memorabilia; fire equipment & trucks; 1857 horse drawn hand pumper; 1939 Diamond T Pumper; fire department patches; Firefighting Medal of Honor; uniforms; call boxes; nozzles; pike poles; fire extinguishers; photographs.
Hours & Admission Prices: By appointment.

ILLINOIS STATE MILITARY MUSEUM, Dept. of Military Affairs, 1301 N. MacArthur, Springfield, IL 62702-2317. Tel.: 217-761-3910. Fax: 217-761-3709.
Founded: 1878.
Congressional District: 18
Key Personnel: Dir., Stewart Reeve; Cur., William Lear.
Personnel Profile: Full-Time Paid 2; Part-Time Volunteers 6; Interns 1.
Governing Authority: Parent Institution: Illinois Dept. of Military Affairs.
Institution Type/Description: Military History Museum.
Collections: military vehicles, aircraft, weapons, uniforms & equipment; flags; photographs; documents.
Facilities: library.
Activities: outreach programs.
Hours & Admission Prices: Tues.-Sat. 1-4:30; other times by appointment. No charge; donations accepted.
Attendance: 10,000 (accurate)

* **ILLINOIS STATE MUSEUM,** 502 S. Spring St., Springfield, IL 62706-5000. Tel.: 217-782-7387. Fax: 217-782-1254. TTY: 217-782-9175.
E-mail: webmaster@museum.state.il.us
Web Site: www.museum.state.il.us
Founded: 1877.
Congressional District: 20
Key Personnel: Museum Dir., Dr. Bonnie W. Styles; Bd. Chm. Illinois State Museum Board, Guerry Suggs; Assoc. Dir., Karen A. Witter; Dir. Resource Allocation, C.F.O. & Human Resources Dir., Charlotte A. Montgomery, CPA; Dir. Art, Jim Zimmer; Asst. Dir. Art, Robert Sill; Exhibits Chief, Paul Stromdahl; Exhibits Prep., Paul Countryman; Education Chair, Beth Shea; Assoc. Cur. Education, Nina Walthall; Mary Ann MacLean Educator, Elizabeth Bazan; Anthropology Chair, Dr. Terry Martin; Cur. Anthropology, Dr. Robert Warren; Cur. Anthropology, Dr. Jonathan Reyman; Geology Chair, Dr. Jeffrey J. Saunders; Asst. Cur. Geology, Dr. Chris Widga; Botany Chair, Dr. Eric Grimm; Cur. Botany, Dr. Hong Qian; Zoology Chair, Dr. Everett D. Cashatt; Asst. Cur. Zoology, Dr. Meredith Mahoney; Asst. Cur. Zoology, H. David Bohlen; Librarian, Pat Burg; Assoc. Cur. Technology, Dr. Erich Schroeder; Dickson Mounds Museum Dir., Dr. Michael Wiant; Assoc. Cur. Anthropology Dickson Mounds Museum, Dr. Michael Conner; Asst. Cur. Anthropology Dickson Mounds Museum, Alan Harn; Asst. Cur. Education Dickson Mounds Museum, Christa Christensen; Illinois Artisans Program Dir., Carolyn Patterson; Admin. Illinois State Museum Chicago Gallery, Jane Stevens; Asst. Cur. Art Chicago Gallery, Doug Stapleton; Southern Illinois Art & Artisans Center, Mary Lou Galloway; Assoc. Cur. Southern Illinois Art Gallery, Debra Tayes; Museum Shop Mgr., Cheryl Staley.
Personnel Profile: Full-Time Paid 78; Part-Time Paid 30; Part-Time Volunteers 116; Interns 8.
Governing Authority: Illinois state museum bd. Parent Institution: Illinois Department of Natural Resources. Branch Museums: Dickson Mounds Museum, Lewistown, IL 61542; Illinois State Museum Chicago Gallery, Chicago, IL 60601; Illinois State Museum Lockport Gallery, Lockport, IL 60441; Southern Illinois Art & Artisans Center, Whittington, IL 62897. Tax-exempt.
Institution Type/Description: Natural History, Anthropology, History & Art Museum.
Collections: anthropology; archaeology; botany; geology; zoology; entomology; herbarium; herpetology; natural history; paleontology; fine art; decorative arts; art history; Illinois history.
Research Fields: anthropology; archaeology; botany; ethnography; geology; paleobotany; paleoecology; paleontology; vertebrate paleontology; zoology; arts & history in Illinois.
Facilities: 10,000-vol. library of primarily scientific reference books & reports pertaining to the collections, publications & reprints by museum staff; laboratory center for study of prehistoric culture & environments.
Activities: guided tours; films; lectures; gallery talks; formally organized education programs; inter-museum loan, permanent, temporary, & traveling exhibitions; docent program; archaeological & paleontological field expeditions.
Publications: quarterly magazine, The Living Museum; quarterly periodical, Transactions of the Illinois State Academy of Science; quarterly periodical, Impressions; occasional popular & scientific publications.
Hours & Admission Prices: Mon.-Sat. 8:30-5, Sun. 12-5. No charge; donations accepted. Closed New Year's Day; Thanksgiving; Christmas. ♿
Attendance: 374,122 (accurate)
Membership: Individual $35; Family $50; Contributing $100; Sustaining $300.

ILLINOIS STATE POLICE HERITAGE FOUNDATION MUSEUM, 4000 N. Peoria Rd., Springfield, IL 62702-1033. Tel.: 217-525-1922.
Founded: 2000.
Congressional District: 18
Key Personnel: Pres. (V), Aaron Cross; Museum Shop Mgr., H. Mueller.
Personnel Profile: Full-Time Volunteers 1; Part-Time Volunteers 21.
Governing Authority: Tax-exempt.
Institution Type/Description: Police History Museum.
Collections: automobiles; police equipment & uniforms; photographs.
Publications: newsletter; brochure.
Hours & Admission Prices: Thurs. & Sat. 10-2. Suggested Donation: $1. Closed holidays. ♿
Attendance: 475 (estimated)

KOREAN WAR NATIONAL MUSEUM, 9 S. Old State Capitol Plaza, Springfield, IL 62701-1510. Mailing Address: 47 W. Division St., Box 1950, Chicago, IL 60610. Tel.: 888-419-5053.
E-mail: kwnm@kwnm.org
Web Site: www.kwnm.org
Founded: 1997.
Congressional District: 15
Key Personnel: Pres. & C.E.O., Anthony Enrietto; Dir. Administration, Mary Spurr.
Personnel Profile: Full-Time Paid 3; Part-Time Paid 1; Part-Time Volunteers 15.
Governing Authority: private; nonprofit organization. Tax-exempt: 501(c)(3).
Institution Type/Description: Military & War Museum.
Collections: Korean War memorabilia; photographs; documents; books; artifacts.
Research Fields: unit histories; Korean War; casualty documentation.
Facilities: library of books related to Korean War.
Activities: lectures.
Publications: newsletter, The Forgotten Voices; brochure.
Hours & Admission Prices: Tues.-Sat. 9-5, Sun. 1-5. Office: Mon.-Fri. 8-5. Suggested donation $3. Closed major holidays. ♿
Attendance: 20,000 (estimated)
Membership: Veteran $25; Veteran Family $30; General Public $35; General Public Family $40; Corporate Civic $50; Sponsor $100; Life $300.

LINCOLN DEPOT, 930 Monroe St., Springfield, IL 62701-1612.
Institution Type/Description: Historic Building: housed in the former depot from which Lincoln left Springfield to start his inaugural journey to Washington DC. on Feb. 11, 1861.
Collections: Lincoln history; personal artifacts; period furnishings; photographs.
Hours & Admission Prices: Call for hours.

LINCOLN HERDON LAW OFFICES STATE HISTORIC SITE, 112 N. Sixth St., Springfield, IL 62701. Mailing Address: Old State Capitol Complex, Springfield, IL 62701. Tel.: 217-785-7289. Fax: 217-557-0282.
Web Site: www.illinoishistory.gov
Founded: 1970.
Congressional District: 20
Key Personnel: Mgr., Justin Blanford; Foundation Chm., Sandy Pecori; Volunteer Coord., Sandy Temple; Museum Shop Mgr., Ron Hohman; Museum Shop Mgr., Dana Hohman.
Personnel Profile: Full-Time Paid 8; Full-Time Volunteers 50; Part-Time Paid 8; Part-Time Volunteers 50.
Governing Authority: state agency. Parent Institution: Illinois Historic Preservation Agency. Tax-exempt.
Institution Type/Description: State Historic Site: located in 1840-1841 three-story commercial Greek Revival building built by Springfield merchant Seth M. Tinsley across from the 1839 Illinois Statehouse, now the Old State Capital Historic Site.
Collections: 19th-century law office & Federal Court Complex furnishings; 1841-1849 recreated Post Office; Lincoln's legal career historic site.
Facilities: Museum-related items for sale.
Activities: annual Illinois Statehood Day - Dec. 3rd.
Hours & Admission Prices: Call for hours. Requested Donations: adults $2, children $1. Closed New Year's Eve & Day; Martin Luther King Jr. Day; Presidents' Day; Veterans' Day; Thanksgiving; Christmas; any general election day. ♿
Attendance: 30,000 (accurate)
Membership: Old State Capitol Complex Foundation (Support Group) $15; $40; $100; $250; $500.

LINCOLN HOME NATIONAL HISTORIC SITE, (M), 413 S. 8th St., Springfield, IL 62701-1905. Tel.: 217-492-4241, ext. 221. Fax: 217-492-4673. Facebook: Lincoln Home National Historic Site; TDD: 217-492-4244.
E-mail: liho_superintendent@nps.gov
Web Site: www.nps.gov/liho
Founded: 1972.
Congressional District: 20
Key Personnel: Park Supt., Dale K. Phillips; Cur., Susan M. Haake; Historian, Timothy P. Townsend; Chief Operations, Laura Gundrum; Museum Shop Mgr., David Mull.
Personnel Profile: Full-Time Paid 30; Part-Time Paid 25; Part-Time Volunteers 125.
Volunteer Hours: 7,495
Operating Expenses: 2,600,000
Operating Income: 2,700,000
Governing Authority: federal. Parent Institution: National Park Service, Dept. of the Interior, Washington, DC. Tax-exempt.
Institution Type/Description: Historic District & Historic House Museum: Home of Abraham Lincoln, 16th President of the United States.
Collections: house furnishings; artifacts relating to & associated with the Lincoln family; architecture; archeology; archives.
Major Exhibits: Wish You Were Here, 10/13-1/14; Council of War, 10/13-2/14; Archeology in the Neighborhood, 1/14-12/14; Julius Rosenwald Project, 2/14-7/14.
Research Fields: historic structure report; historic landscape plan; historic furnishings plan.
Facilities: visitor center. Publications & museum-related items for sale.
Activities: guided tours & walks; dramatizations; lectures; films; audio-visual programs; formally organized educational programs; costumed interpretation programs; street theater.
Publications: brochure; information sheet; books.
Hours & Admission Prices: Daily 8:30-5. tickets to entrance available on first-come, first serve basis at Visitor's Center. No charge; donations accepted. Closed New Year's Day; Thanksgiving; Christmas. &
Attendance: 231,593 (accurate)

LINCOLN MEMORIAL GARDEN AND NATURE CENTER, 2301 E. Lake Dr., Springfield, IL 62712-8908. Tel.: 217-529-1111. Fax: 217-529-0134.
E-mail: lmg2301@comcast.net
Web Site: www.lincolnmemorialgarden.org
Founded: 1935.
Congressional District: 20
Key Personnel: Exec. Dir., Jim Matheis; Pres., Barbara Rogers; Education Dir. & Naturalist, Betsy Irwin; Museum Shop Mgr., Michele Cannon.
Personnel Profile: Full-Time Paid 3; Part-Time Paid 7; Part-Time Volunteers 100; Interns 1.
Governing Authority: nonprofit organization. Tax-exempt: 501(c)(3).
Institution Type/Description: Arboretum & Nature Center.
Collections: living flora & fauna native or naturalized to central Illinois.
Research Fields: prairie propagation; prairie & native Illinois landscape restoration; protection & propagation of endangered species; maintaining Jens Jensen designed landscape.
Facilities: 300-vol. library of books on plant & animal taxonomy, general information & natural science interpretation available for research; botanical garden; nature center; 110 acres; 5 miles of trails. Hand crafted nature-related items & science books for sale.
Activities: guided tours; lectures; formally organized education programs for children, adults, undergraduate & graduate college students affiliated with University of Illinois at Springfield; temporary exhibitions.
Publications: newsletter, Nature Center News.
Hours & Admission Prices: Tues.-Sat. 10-4, Sun. 1-4. No charge; donations accepted. Building: closed Easter; Independence Day; Thanksgiving; Christmas week. &
Attendance: 50,000 (estimated)
Membership: Senior Citizen $30; Individual $40; Family & Business $75; Railsplitter $150.

LINCOLN TOMB STATE HISTORIC SITE, Oak Ridge Cemetery, Springfield, IL 62702. Mailing Address: 1500 Monument Ave., Springfield, IL 62702-2500. Tel.: 217-782-2717. Fax: 217-524-3738.
Web Site: tombtour.org
Founded: 1874.
Congressional District: 20
Key Personnel: Site Mgr., Candy Knox; Site Service Specialist, Mikle Siere.
Personnel Profile: Full-Time Paid 2; Part-Time Paid 1; Part-Time Volunteers 5.

Governing Authority: state. A part of the State of Illinois, Illinois Historic Preservation Agency, Old State Capitol, Springfield, IL 62701. Tax-exempt.
Institution Type/Description: Historic Site: 1874, the tomb of Abraham Lincoln.
Collections: c.1865, public vault where Lincoln's body first lay; burial place of Abraham Lincoln, Mary Todd Lincoln, Edward Baker Lincoln, William Wallace Lincoln, Thomas Lincoln.
Research Fields: President Lincoln's funeral train; funeral services; public reaction.
Activities: oral presentations; flag ceremonies; memorial services; veterans and fraternal organization ceremonies; Boy Scout Pilgrimage. Site Sponsors: Tuesday evening formal flag lowering ceremonies & troop inspections by the 114th Regiment of the Illinois Volunteer Infantry June-August.
Publications: bimonthly newsletter, Historic Illinois; brochure.
Hours & Admission Prices: March-April & Sept.-Oct. Tues.-Sat. 9-5; May to Labor Day daily 9-5; Nov.-Feb. Tues.-Sat. 9-4. No charge; donations accepted. Closed New Year's Day; Martin Luther King Jr. Day; Presidents' Day; Veterans' Day; General Election Day; Thanksgiving; Christmas. &
Attendance: 345,000 (accurate)
Membership: Lincoln Monument Association: Annual $20.

MUSEUM OF FUNERAL CUSTOMS, 215 S. Grand Ave. W., Springfield, IL 62704-3838. Tel.: 217-544-3480. Fax: 217-544-3484.
E-mail: funeralmuseum@ifda.org
Web Site: www.funeralmuseum.org
Founded: 2000.
Congressional District: 18
Key Personnel: Dir., Jon N. Austin; Pres. (V), Paula Staab Polk.
Personnel Profile: Full-Time Paid 2; Part-Time Paid 1; Part-Time Volunteers 7; Interns 2.
Governing Authority: private; nonprofit organization. Tax-exempt: 501(c)(3).
Institution Type/Description: History Museum.
Collections: American funeral & mourning customs; history of undertaking & funeral directing from 1840 to present; art & science of embalming; funeral rite, coffin & casket styles; memorialization; interment; grief & mourning.
Research Fields: U.S. Patents related to undertaking.
Facilities: 1,000-vol. library; 3,600 sq. ft. exhibit space. Museum-related items for sale.
Activities: guided tours; lectures; loan & temporary exhibitions. Annual Events: Symposium in October; poetry reading in March.
Publications: seven themed brochures related to funeral history and mourning.
Hours & Admission Prices: Tues.-Sat. 10-4, Sun. 1-4. Adults $4, senior citizens $3, students $2; discounts to AAM members; members and children 5 & under no charge. Closed New Year's Day; Independence Day; Thanksgiving & day after; Christmas. &
Attendance: 7,685 (accurate)

***　OLD STATE CAPITOL, (M),** Sixth at Adams Sts., 1 Old State Capitol Plaza, Springfield, IL 62701. Tel.: 217-785-7960. Fax: 217-557-0282. TDD: 217-524-7128.
E-mail: justin.blanford@illinois.gov
Web Site: www.illinois.gov/ihpa/
Founded: 1969.
Congressional District: 20
Key Personnel: Site Mgr., Justin A. Blanford.
Governing Authority: state. Parent Institution: Illinois Historic Preservation Agency: Subsidiary Institution: Historic Sites Division. Tax-exempt.
Institution Type/Description: Historic Building: 1839-1876, Illinois's fifth Statehouse.
Collections: period & original furnishings.
Research Fields: Lincolniana.
Facilities: Museum-related items for sale.
Activities: guided tours; costumed interpreters on Fri.-Sat.; first person interpretation & third person. Special Events: candlelight tours; statehood day in Dec.; naturalization ceremonies; Holocaust Observance; youth outreach.
Publications: Old State Capitol brochure; elementary school study guide; quarterly newsletter.
Hours & Admission Prices: Seasonal hours, call for confirmation. No charge; donations requested. Closed New Year's Day; Martin Luther King Jr. Day; Presidents Day; Labor Day; Veterans Day; Thanksgiving; Christmas. &
Attendance: 108,000 (accurate)
Membership: Old State Capitol Foundation (Support Group) $15; $40; $100; $250; $500.

THE PEARSON MUSEUM, 801 N. Rutledge, Springfield, IL 62702-4910. Mailing Address: S.I.U. School Medicine, P.O. Box 19635, Springfield, IL 62794-9635. Tel.: 217-545-8017 & 4261. Fax: 217-545-7903.
E-mail: probertson@siumed.edu
Web Site: www.siumed.edu/medhum/pearson
Founded: 1974.
Congressional District: 20
Key Personnel: Dir. & Devel. Dir., Phillip V. Davis, Ph.D.; Registrar, Allona Beasley Mitchell; Business Mgr., Patricia Robertson, M.A.
Personnel Profile: Part-Time Paid 3.
Governing Authority: university. Parent Institution: Southern Illinois University. Subsidiary Institution: School of Medicine. Tax-exempt: 501(c)(3).
Institution Type/Description: University Medical Science Museum.
Collections: medical, dental & pharmaceutical artifacts; graphic art; photographs; slides.
Research Fields: 19th & early 20th-century medicine, dentistry & pharmacy practice in the upper Mississippi River basin area.
Facilities: auditorium; classrooms; teaching theatre; laboratory.
Activities: guided tours; permanent, temporary & traveling exhibitions.
Hours & Admission Prices: Tues. 8:30-4:30; group tours by appointment. No charge; donations accepted. &
Attendance: 6,000 (estimated)
Membership: Friends of the Pearson Museum $50 & up.

SHEA'S GAS STATION MUSEUM, 2075 Peoria Rd., Springfield, IL 62702-1837. Tel.: 217-522-0475.
Institution Type/Description: History Museum.
Collections: station history; gas pumps; signs; photographs; service station memorabilia; personal artifacts.
Hours & Admission Prices: Tues.-Sat. 7-4; other times by appointment. Admission $2.

SPRINGFIELD ART ASSOCIATION, 700 N. Fourth St., Springfield, IL 62702-5232. Tel.: 217-523-2631. Fax: 217-523-3866.
E-mail: office@springfieldart.org
Web Site: www.springfieldart.org
Founded: 1913.
Congressional District: 17
Key Personnel: Coord. Education, Erin Swendsen; Education Coord., Katie Rasmussen; Exec. Dir., Betsy Dollar; Dir. Library, Jan Arnold; Librarian, Mark Jenkins; Cur. Collections, Erika Holst; Office Mgr., Megan DeMaris.
Personnel Profile: Full-Time Paid 4; Part-Time Paid 2; Part-Time Volunteers 150; Interns 3.
Governing Authority: nonprofit organization. Tax-exempt: 501(c)(3).
Institution Type/Description: Historic Home: c.1833.
Collections: 19th & 20th-century paintings; eight rooms of 19th-century & early American furniture.
Major Exhibits: Holiday Hall Sale, 11/11-12/17.
Research Fields: Asian collection; George Peter Alexander Healy, portrait painter.
Facilities: art library; classrooms.
Activities: guided tours; lectures; films; gallery talks; arts festivals; formal organized educational programs for children, adults & undergraduate college students; permanent & temporary exhibitions; art outreach program with local schools.
Publications: quarterly bulletin; exhibitions & collections catalogues.
Hours & Admission Prices: House Tours: Tues.-Sat. 11-2. Library: Mon.-Fri. 9-5, Sat. 10-3. Gallery: Mon.-Fri. 10-5, Sat. 10-3. No charge; donations accepted. Closed New Year's Day; Presidents' Day; Columbus Day; Labor Day; Thanksgiving & day after; Christmas Eve, Day & week. &
Attendance: 5,000 (estimated)
Membership: College Student $35; Art Educator $40; Individual $55; Family & Dual $75; Patron $150; Sustainer $250; Benefactor $500; Thomas Condell Circle $750; Edwards Place Society $1,000 & up.

STATE OF ILLINOIS-HISTORIC PRESERVATION AGENCY, HISTORIC SITES DIVISION, 313 S. 6th St., Springfield, IL 62701-1805. Tel.: 217-785-7930. Fax: 217-785-8117.
E mail: hpa.info@illinois.gov
Web Site: www.illinoishistory.gov
Founded: 1985.
Congressional District: 20
Key Personnel: Dir., Amy Martin; Chm. (V), Sunny Fischer; Supt. Historic Sites, Alyson Grady; Capital Projects, Jane Rhetta; Exhibits, Steve Leonard; Historian, Mark Johnson; Conservator, Malcolm Brown.
Personnel Profile: Full-Time Paid 62; Part-Time Volunteers 500.
Governing Authority: state. Branch Museums: Albany Mounds; Bishop Hill; Black Hawk; Bryant Cottage; Buel House; Cahokia Courthouse; Cahokia Mounds; Campbell's Island Memorial; Carl Sandburg Birthplace; Dana Thomas House; David Davis Mansion; Douglas Tomb; Emerald Mound; Fort de Chartres; Fort Kaskaskia; Governor Bond Memorial; Governor Coles Memorial; Governor Horner Memorial; Grand Village of the Illinois; Halfway Tavern; Hofmann Tower; Illinois Vietnam Veterans Memorial; Jarrot Mansion; Jubilee College; Kaskaskia Bell Memorial; Kincaid Mounds; Korean War Memorial; Lewis and Clark Memorial; Lincoln-Herndon Law Offices; Lincoln Log Cabin; Lincoln Monument Memorial; Lincoln Tomb; Lincoln Trail Memorial; Lincoln's New Salem; Lovejoy Memorial; Martin/Boismenue House; Metamora Courthouse; Moore Home; Mt. Pulaski Courthouse; Norwegian Settlers Memorial; Old Market House; Old State Capitol; Pierre Menard Home; Postville Courthouse; Pullman; Rose Hotel; Shawneetown Bank; U.S. Grant Home; Vachel Lindsay Home; Vandalia Statehouse; Washburne House; Wild Bill Hickok Memorial; Apple River Fort. Tax-exempt.
Institution Type/Description: State Agency.
Collections: period furnishings; buildings; art; artifacts; primary source materials.
Research Fields: state & local history.
Activities: guided tours; lectures; restoration; interpretation; special events; audiovisual presentations; living history; outreach interpretive programs.
Publications: brochures for individual sites; bimonthly newsletter, Journal of Illinois History.
Hours & Admission Prices: Call for hours & admission information. No charge; donations accepted. &
Attendance: 2,400,000 (accurate)

TRUTTER MUSEUM, 5250 Shepherd Rd., Springfield, IL 62794. Tel.: 217-786-2784.
Institution Type/Description: History Museum.
Collections: Trutter family art, artifacts & cultural items from around the world.
Activities: special events; workshops; lectures.
Hours & Admission Prices: mid-Aug. to mid-May Tues.-Thurs. 10-3; other times by appointment.

VACHEL LINDSAY HOME STATE HISTORIC SITE, 603 S. Fifth St., Springfield, IL 62703-1604. Tel.: 217-524-0901. Fax: 217-557-0282.
E-mail: hpa.vachellindsay@illinois.gov
Web Site: www.illinois-history.gov & www.vachellindsay.org
Founded: 1946.
Congressional District: 20
Key Personnel: Mgr. Old State Capitol Complex, Justin Blandford; Admin., Jennie Battles.
Personnel Profile: Full-Time Paid 1; Part-Time Volunteers 10.
Governing Authority: nonprofit organization. Parent Institution: Illinois Historic Preservation Agency. Tax-exempt.
Institution Type/Description: Historic House: 1846 home of Vachel Lindsay.
Collections: drawings; letters; manuscripts; books; paintings; sculpture; videotapes.
Facilities: library of poetry, prose & records of the poet. Museum-related items for sale.
Activities: lectures; gallery talks; resource & referral service; poetry & music programs.
Publications: book, The Little Turtle; book, DVD, Look Into Your Heart: The Challenge of Vachel Lindsay, available through The Vachel Lindsay Association.
Hours & Admission Prices: Tours: Tues.-Sat. 10-4; groups of 10 or more by appointment. Suggested Donations: adults $4, students $2.
Attendance: 4,978 (estimated)
Membership: The Vachel Lindsay Association: Individual $25; Family $35; Patron $50; Sustaining $100; Benefactor $250; Life $2,500.

WASHINGTON PARK BOTANICAL GARDEN, (M), 1740 W. Fayette, Springfield, IL 62704-2356. Tel.: 217-546-4116. Fax: 217-546-0257.
E-mail: chad@springfieldparks.org
Web Site: www.springfieldparks.org
Founded: 1972
Congressional District: 20
Key Personnel: Interim Exec. Dir., Derek Harms; Supt. Botanical Garden, Chad Scaife; Park Bd. Pres., Leslie Sgro.
Personnel Profile: Full-Time Paid 4; Part-Time Paid 5; Part-Time Volunteers 5.
Governing Authority: municipal. Parent Institution: Springfield Park District. Tax-exempt.
Institution Type/Description: Botanical Garden.
Collections: tropical plants; daylilies; roses; iris; cacti; trees; perennials; rockery garden; monocot garden.

Major Exhibits: Spring Floral Show, 3/22/14-4/6/14; Historic Photography Display, 8/16/14-8/31/2014; Winter Holiday Floral Show, 12/6/14-12/21/14.

Facilities: 162-vol. library pertaining to horticulture & botany; conservatory; educational facilities; gardens.

Activities: guided tours; lectures; organized educational programs; plant sales. Annual Events: seasonal shows; Prairie State Orchid Society Show in February; Sangamon Valley Iris Society - Iris Show in May; Central Illinois Daylily Society Show in June.

Publications: quarterly newsletter, The Washington Park Botanical Garden News.

Hours & Admission Prices: Mon.-Fri. 12-4, Sat.-Sun. 12-5; tours by appointment. No charge; donations accepted. Closed state holidays. &

Attendance: 80,000 (estimated)

Membership: Individual $11; Family $14; Benefactor $25 & up.

Sterling

STERLING-ROCK FALLS HISTORICAL SOCIETY MUSEUM, 1005 E. 3rd. St., Sterling, IL 61081-2813. Mailing Address: 1005 E. Third St., P.O. Box 65, Sterling, IL 61081-2813. Tel.: 815-622-6215.

E-mail: srfhs@comcast.net

Web Site: www.svonline.net/~srfhs

Founded: 1959.

Congressional District: 15

Key Personnel: Dir. & Cur., Terence Buckaloo; Pres. (V), David B. Lowe.

Personnel Profile: Full-Time Paid 1; Part-Time Volunteers 20; Interns 1.

Governing Authority: nonprofit. Parent Institution: Sterling-Rock Falls Historical Society. Tax-exempt: 501(c)(3).

Institution Type/Description: Historical Society Museum.

Collections: local general history; Indian & pioneer heritage; rare books; military; archives; paintings; industrial.

Research Fields: local history; Indian & pioneer heritage; genealogy; historical structures.

Activities: guided tours; gallery talks; formally organized education programs for children.

Hours & Admission Prices: Tues., Thurs. & Sat. 10-12 & 1-4, Sun. 1-5. No charge; donations accepted. &

Attendance: 2,700 (accurate)

Membership: Individual $15; Family $25; Sustaining $50; Business $100; Life $300.

Stockton

STOCKTON HERITAGE MUSEUM, 107 W. Front Ave., Stockton, IL 61085-1317. Mailing Address: P.O. Box 93, Stockton, IL 61085-0093. Tel.: 815-947-9144. Facebook: Stockton Heritage Museum.

Congressional District: 17

Key Personnel: Dir. & Cur., Bobbi Reagan; Pres. (V), Melody Heidenreich.

Personnel Profile: Part-Time Volunteers 28.

Governing Authority: Tax-exempt: 501(c)(3).

Institution Type/Description: History Museum.

Collections: local history; Kraft Cheese; Chicago Great Western Railroad.

Publications: quarterly newsletter.

Hours & Admission Prices: May-Dec. Sat. 9-1. No charge; donations accepted. &

Attendance: 450 (estimated)

Membership: Students $5; Seniors $10; Couple $20; Family $25.

Stone Park

ITALIAN CULTURAL CENTER OF CASA ITALIA, 1621 N. 39th Ave., Stone Park, IL 60165-1186. Tel.: 708-345-3842. Fax: 708-345-3891.

Web Site: www.casaitaliachicago.net

Key Personnel: Chm., Leonardo S. DeFranco; Dir. Exhibits, Josetta Mentesana Weber

Institution Type/Description: Italian Heritage Museum.

Collections: contemporary art; local history.

Facilities: library.

Hours & Admission Prices: Mon.-Fri. 9-4 by appointment.

Sugar Grove

AIR CLASSIC INC. MUSEUM OF AVIATION, 43W624 U.S. 30, Sugar Grove, IL 60554. Tel.: 630-466-0888.

Web Site: www.airclassicsmuseum.org

Congressional District: 14

Key Personnel: Pres., Mike Luman; Vice Pres., Lawrence Matt

Institution Type/Description: Aviation & Military Museum.

Collections: planes from the Korean War II to Desert Storm.

Hours & Admission Prices: early April to late Nov. Sat.-Sun. 10-3. Group Tours: Tues.-Fri. Adults $8, seniors & children 5-16 $5; children under 5 no charge.

Attendance: 3,000 (estimated)

Membership: Student $15; Individual $30; Family $45.

Sullivan

MOULTRIE COUNTY HERITAGE CENTER, 117 E. Harrison St., Sullivan, IL 61951. Mailing Address: P.O. Box 588, Sullivan, IL 61951-0588. Tel.: 217-728-4085.

Governing Authority: Parent Institution: Moultrie County Historical and Genealogical Society.

Institution Type/Description: History Museum.

Collections: local history & culture; period furnishings; personal artifacts; photographs.

Facilities: library.

Activities: research; annual events.

Hours & Admission Prices: Mon. & Sat. 1-5; other times by appointment.

Sycamore

MIDWEST MUSEUM OF NATURAL HISTORY, 425 W. State St., Sycamore, IL 60178-1410. Tel.: 815-895-9777. Fax: 815-899-2552.

E-mail: director@mmnh.org

Web Site: www.mmnh.org

Founded: 2005.

Congressional District: 14

Key Personnel: Exec. Dir., Molly Holman

Institution Type/Description: Natural History Museum.

Collections: mounted & live animals from around the world.

Facilities: classroom; cafeteria.

Activities: hands-on exhibits; school group programs.

Hours & Admission Prices: Tues.-Sat. 10-5, Sun. 12-5. Adults $6, seniors & children $5; discounts to groups of 20 or more; children under 2 no charge. &

Membership: Single Parent Family $55; Family & Grandparent $65.

SYCAMORE HISTORY MUSEUM, 1730 N. Main St., Sycamore, IL 60178. Mailing Address: P.O. Box 502, Sycamore, IL 60178-0502. Tel.: 815-895-5762.

E-mail: info@sycamorehistory.org

Web Site: sycamorehistory.org

Formerly: Sycamore Historical Society & Museum

Founded: 1999.

Key Personnel: Dir., Michelle Donahoe; Pres., Jim Lyon

Governing Authority: Tax-exempt.

Institution Type/Description: History Museum.

Collections: Sycamore's history & culture; photographs.

Research Fields: local history.

Activities: walking tours in summer; school groups; scouts. Museum Sponsors: monthly Brown Bag Lunch; Cemetery Walk in October.

Publications: newsletter.

Hours & Admission Prices: Tues.-Sat. 10-3 by appointment. Adults $5; members & children under 14 no charge. &

Attendance: 5,000 (estimated)

Membership: Individual $30; Family $50.

Tampico

RONALD REAGAN BIRTHPLACE & MUSEUM, 111 S. Main St., Tampico, IL 61283. Mailing Address: P.O. Box 344, Tampico, IL 61283. Tel.: 815-438-2130.

E-mail: reaganbirthplace@thewisp.net

Founded: 1976.

Congressional District: 17

Key Personnel: Dir. (V), Joan Johnson.

Personnel Profile: Part-Time Volunteers 16; Interns 1.

Volunteer Hours: 1,900

Institution Type/Description: Historic Building: housed in the birthplace of Ronald Reagan; born Feb. 6, 1911.

Collections: Reagan family history; photographs; personal artifacts.

Facilities: Museum-related items for sale.

Activities: special events; Ronald Reagan Birthday Party, open house annually.

Hours & Admission Prices: April-Oct. Mon.-Sat. 10-4, Sun. 1-4; other times by appointment. No charge; donations accepted. Closed Easter; Mother's Day.
Attendance: 2,500 (estimated)

TAMPICO AREA HISTORICAL SOCIETY, 119 S. Main St., Tampico, IL 61283. Mailing Address: P.O. Box 154, Tampico, IL 61283. Tel.: 815-438-7581.
E-mail: tampicohistoricalsociety@gmail.com
Web Site: www.tampicohistoricalsociety.com
Founded: 1990.
Congressional District: 17
Key Personnel: Pres. (V), Joan Johnson.
Personnel Profile: Part-Time Volunteers 7.
Governing Authority: Tax-exempt.
Institution Type/Description: Historical Society Museum.
Collections: local history & culture; period artifacts; photographs; family histories.
Research Fields: genealogy.
Activities: research.
Hours & Admission Prices: Open by appointment. No charge; donations accepted. &
Attendance: 500 (estimated)
Membership: Single $10; Family $15; Lifetime $50.

Teutopolis

TEUTOPOLIS MONASTERY MUSEUM, Rte. 40, & S. Garrott St., Teutopolis, IL 62467-1161. Tel.: 217-857-3586.
Founded: 1975.
Congressional District: 54
Key Personnel: Pres. (V), Ray Vahling; Vice Pres., Carol Hoelscher; Treas., Henry Hawickhorst; Sec. & Helping Tour Chm., Joyce Vahling; Helping Tour Chm., Eleanor Gebben.
Personnel Profile: Part-Time Volunteers 15.
Governing Authority: church; nonprofit organization. St. Francis Church. Tax-exempt.
Institution Type/Description: Local History Museum: housed in 1862 Franciscan Novitiate.
Collections: furniture used by the Franciscan order; religious items & books; early pioneer artifacts; family portraits; clocks; archives.
Facilities: library of religious material; Teutopolis history; books in 21 languages. Gift items & books for sale.
Activities: guided tours.
Hours & Admission Prices: April-Nov. first Sun. of month 12:30-4; tours by appointment. Adults $3, children $1.
Attendance: 500 (estimated)

Tinley Park

TINLEY PARK HISTORICAL SOCIETY, (M), 6727 W. 174th St., Tinley Park, IL 60477-3529. Mailing Address: P.O. Box 325, Tinley Park, IL 60477-0325. Tel.: 708-429-4210. Fax: 708-444-5099. TDD: 708-444-5000.
E-mail: bbettenh@tinleypark.org
Founded: 1974.
Congressional District: 3
Key Personnel: Chm. (V), Brad L. Bettenhausen; Treas., Marian Block.
Personnel Profile: Part-Time Volunteers 8.
Governing Authority: private; nonprofit organization. Tax-exempt: 501(c)(3).
Institution Type/Description: Historical Society Museum: housed in an 1884 frame church in Prairie Gothic style.
Collections: history of the Village of Tinley Park & surrounding area integral to its growth & development; ethnic heritage (German) of community & Rock Island Railroad; newspapers; local genealogy.
Research Fields: history of community until 1940.
Facilities: research library; 2,200 sq. ft. exhibit space; 175-person capacity church/meeting room; 90-person capacity kitchen/meeting room.
Activities: docent program; guided tours; temporary exhibitions.
Publications: quarterly newsletter, New Bremen News.
Hours & Admission Prices: Wed. 10-2; other times by appointment. No charge; donations accepted. Closed federal & state holidays. &
Membership: Senior & Student $15; Adult $25; Institutional & Business $50; Supporting $100; Life $250; Founders Circle $500.

Toulon

STARK COUNTY HISTORICAL SOCIETY, 318 W. Jefferson St., Toulon, IL 61483. Tel.: 309-286-3104.
Institution Type/Description: Historical Society Museum.
Collections: local history & culture; period furnishings; personal artifacts; photographs.
Hours & Admission Prices: By appointment.
Membership: Life $25.

Tremont

TREMONT MUSEUM & HISTORICAL SOCIETY, Madison & S. Sampson Sts., Tremont, IL 61568. Mailing Address: P.O. Box 738, Tremont, IL 61568. Tel.: 309-925-5262.
Founded: 1987.
Personnel Profile: Part-Time Volunteers 30; Interns 1.
Governing Authority: Tax-exempt: 501(c)(3).
Institution Type/Description: History Museum.
Collections: Tremont history & culture; period furnishings; personal artifacts; photographs; Keystone Steel & Wire Co. history; Morton pottery.
Publications: bimonthly, Tremont Tales and Trails.
Hours & Admission Prices: 2nd Sun. each month 2-4; other times by appointment. Adults $1; children & members no charge.
Attendance: 400 (estimated)
Membership: Single $10; Family $15; Life $150.

Union

DONLEY'S WILD WEST TOWN MUSEUM, 8512 S. Union Rd., Union, IL 60180-9661. Tel.: 815-923-9000. Fax: 815-923-2253.
Web Site: www.wildwesttown.com
Founded: 1975.
Institution Type/Description: History Museum.
Collections: Wild West history & artifacts; cowboy memorabilia including wooly chaps, gun belts, guns, spurs, lariats, & boots; death masks; handcuffs; badges; mining tools & equipment; Civil War; period furnishings including music boxes & phonographs.
Hours & Admission Prices: April-May 24 & Sept. 5-Oct. Sat.-Sun. 10-6; May 25-Sept. 4 daily 10-6. Adults $15; children 2 & under no charge. (museum included in Wild West Town admission).

ILLINOIS RAILWAY MUSEUM, (M), 7000 Olson Rd., Union, IL 60180. Mailing Address: P.O. Box 427, Union, IL 60180-0427. Tel.: 815-923-4391, ext. 404. Fax: 815-923-2006.
E-mail: nkallas@irm.org
Web Site: www.irm.org
Founded: 1953.
Congressional District: 12
Key Personnel: Exec. Dir., Nick Kallas; Pres. (V), Joe Stupar; Museum Shop Mgr., Tom Blodgett.
Personnel Profile: Full-Time Paid 4; Full-Time Volunteers 5; Part-Time Paid 3; Part-Time Volunteers 75.
Governing Authority: nonprofit organization. Tax-exempt: 501(c)(3).
Institution Type/Description: Railway Museum: housed in c.1851 Marengo, Illinois rail depot.
Collections: steam, diesel & electric locomotives; passenger cars; streetcars; interurban cars; rapid transit cars; electric work equipment; box motors & freight; trolley buses; freight cars; motor buses; steam railroad work equipment; 1910 Spaulding interlocking tower; railroad neon signs. Historic Buildings: 1851 railroad station; 1910 Chicago L Station; O'Mahoney Roadside Diner.
Research Fields: railway technology.
Facilities: 5,000-vol. library of technical railway materials available by request. Books & rail-oriented items for sale.
Activities: films; workshops; inter-museum loan, permanent & temporary exhibitions; rides.
Publications: quarterly newsletter, Rail and Wire.
Hours & Admission Prices: Memorial Day-Labor Day daily 10-4; Spring & Fall Sat.-Sun. 10-5. Admission & rides: adults $14, children $10; discounts to AAM members & groups; members no charge.
Attendance: 65,470 (accurate)
Membership: Associate $40; Family $65.

MCHENRY COUNTY HISTORICAL SOCIETY AND MUSEUM, (M), 6422 Main St., Union, IL 60180. Mailing Address: Box 434, Union, IL 60180-0434. Tel.: 815-923-2267. Fax: 815-923-2271.
E-mail: info@mchsonline.org

Web Site: www.mchsonline.org
Founded: 1963.
Congressional District: 8 & 16
Key Personnel: C.E.O., Kurt Begalka; Pres. (V), Bob Frenz; Vice Pres., Molly Walsh; Museum Shop Mgr., Nancy Roozee; Coord. Exhibits & Collection, Kira Halvey.
Personnel Profile: Full-Time Paid 3; Part-Time Volunteers 300.
Governing Authority: society. Tax-exempt: 501(c)(3).
Institution Type/Description: Historical Society Museum.
Collections: papers; records of early settlers & the history of McHenry County. Historic Houses: 1847 log cabin; 1867 limestone one-room school house on original site; 1885 Town Hall; 1895 one-room schoolhouse; 1898 country church; 1949 modern tourist cabin; traveling history museum bus, The James.
Research Fields: local history; local genealogy.
Facilities: 2,000-vol. library of original land grants, early textbooks, newspapers, biography files, business, governmental & organization records, photographs, diaries & maps available for research.
Activities: area meetings; special lecture & history programs; formally organized educational programs & tours for adults & students; historic plaquing program; traveling history bus.
Publications: quarterly newsletter, The Tracer and Society's Page.
Hours & Admission Prices: May & Oct. Tues.-Fri. & Sun. 1-4; June-Sept. Tues.-Fri. & Sun. 1-4. Adults $5, children & senior citizens 60 and over $3; discounts to AAM members & those who bring nonperishable goods; veterans & members no charge. Closed holidays. &
Attendance: 5,500 (estimated)
Membership: Seniors & Students $10; Individual $20; Organizations $25; Couple & Family (includes children in school) $30; Business $100; Life $500.

University Park

NATHAN MANILOW SCULPTURE PARK, GOVERNORS STATE UNIVERSITY, 1 University Pkwy., University Park, IL 60484-3165. Tel.: 708-534-4486. Fax: 708-534-8959. TDD: 708-534-8650.
E-mail: g-bates@govst.edu
Web Site: www.govst.edu/sculpture
Founded: 1969.
Congressional District: 11
Key Personnel: Dir. & Cur., Geoffrey Bates; Chm. (V), Jacqueline Lewis.
Personnel Profile: Full-Time Paid 1; Part-Time Volunteers 16.
Governing Authority: public university. Parent Institution: Governors State University. Subsidiary Institution: Governors State University Foundation. Tax-exempt.
Institution Type/Description: Sculpture Park.
Collections: over 20 works of sculpture.
Facilities: library.
Activities: guided group tours; educational programming; special events.
Publications: brochure; catalog; newsletter.
Hours & Admission Prices: Dawn to dusk. No charge; donations accepted. Guided Tours: Adults $5, students $3.50.
Attendance: 15,000 (estimated)
Membership: Prairie Associate $40; Grassroots Supporter $100; Premier Partner $500; Sculptor's Forum $1,000; Collector's Circle $5,000; Visionary $10,000.

Urbana

✴ **SPURLOCK MUSEUM, UNIVERSITY OF ILLINOIS AT URBANA-CHAMPAIGN, (M),** 600 S. Gregory St., Urbana, IL 61801-3759. Tel.: 217-333-2360 & 244-3355. Fax: 217-244-9419.
E-mail: ksheahan@illinois.edu
Web Site: www.spurlock.illinois.edu
Founded: 1911.
Congressional District: 19
Key Personnel: Dir., Wayne Pitard; Bd. Pres., Robin Fossun; Collections Mgr., Christa Deacy-Quinn; Asst. Collections Mgr., John Holton; Coord. Collections, Melissa Sotelo; Registrar, Jennifer White; Asst. Registrar, Amy Heggemeyer; Dir. Education, Tandy Lacy; Asst. Dir. Education, Kim Sheahan; Coord. Education, Beth Watkins; Coord. Learning Center, Julia Robinson; Coord. Education Program, Brook Taylor; Information Technology, Jack Thomas; Business Mgr., Karen Flesher; Coord. Special Events, Brian Cudiamat; Cur. Asia, Chiou-Peng TzeHuey; Cur. Africa, Mahir Saul; Cur. East Asia, Kai Wing Chow; Cur. Oceania, Janet Keller; Cur. South America, Norman Whitten; Cur. East Asia, Yu Wang; Security Supvr., Cipriano Martinez; Lead Guard, Thomas Yu; Guard, Michael Albert; Guard, Larry Booth; Guard, Robert Krickus; Guard, Gary Higgs.
Personnel Profile: Full-Time Paid 19; Part-Time Paid 50; Part-Time Volunteers 24.

Governing Authority: university. Affiliated with the University of Illinois at Urbana-Champaign. Tax-exempt.
Institution Type/Description: World Cultures Museum.
Collections: archaeology; art; history; numismatics; glass; pottery; bronzes; textiles; cast reproductions; classical & medieval & European sculpture; anthropology; American Indian; cuneiform tablets; Canelos Quichua ceramics; Amazonian bark cloth.
Research Fields: cuneiform tablets; ancient pottery; Merovingian jewelry; cylinder seals; masks; storytelling.
Facilities: auditorium; reception facilities available; educational resource center; multipurpose learning center.
Activities: guided tours; lectures; gallery talks; permanent & temporary exhibitions; special events; dance; theatre; outreach programs; storytelling.
Publications: annual magazine, Spurlock; newsletter.
Hours & Admission Prices: Tues. 12-5, Wed.-Fri. 9-5, Sat. 10-4, Sun. 12-4. No charge; donations accepted. Closed university holidays. &
Attendance: 19,000 (estimated)
Membership: Museum Friends: Individual $25; Family $35; Sustaining $50; Donor $100; Sponsor $500; Patron $1,000; Benefactor $5,000; Founder $10,000.

WANDELL SCULPTURE GARDEN, Meadowbrook Park, Vine St., Urbana, IL 61801. Mailing Address: Urbana Park District, 303 W. University Ave., Urbana, IL 61801. Tel.: 217-367-1536. Fax: 217-367-1391.
Institution Type/Description: Sculpture Garden.
Collections: oak, stainless steal, bronze, & concrete sculptures.
Hours & Admission Prices: Daily dawn to dusk.

Utica

LASALLE COUNTY HISTORICAL SOCIETY MUSEUM, (M), 101 E. Canal, Utica, IL 61373. Mailing Address: P.O. Box 278, Utica, IL 61373-0278. Tel.: 815-667-4861. Fax: 815-310-7613. Facebook: LaSalle County Historical Society Museum.
E-mail: lchsmuseum@gmail.com
Web Site: lasallecountymuseum.org
Founded: 1907.
Congressional District: 11
Key Personnel: Pres. (V), Doug Holland.
Personnel Profile: Part-Time Paid 6; Part-Time Volunteers 100; Interns 4.
Governing Authority: society. Parent Institution: LaSalle County Historical Society; CB&Q #4978 engine; herb & prairie plant gardens. Branch Historical Sites: Artesian Well House, Ottawa; blacksmith shop, Utica; Aitken School House (one-room); barn. Tax-exempt: 501(c)(3).
Institution Type/Description: History Museum: housed in 1848 pre-Civil War stone warehouse along the Illinois Michigan canal.
Collections: American Indian artifacts; furnishings of early pioneer homes; clothing; farm tools; pioneer implements; Lincoln carriage.
Facilities: approx. 800-vol. library of books on the history of LaSalle County available for use on premises; reading room. Museum-related items for sale.
Activities: guided tours; lectures; radio programs; permanent & temporary exhibitions. Museum Sponsors: land cruises; Burgoo Festival.
Publications: bulletin, The Society Story; booklet, Focus on the Past; booklet, The First Kaskaskia; reprints of old LaSalle County books; Pioneers, Powwows & Prairie Playgrounds; Remembering the Rural One Room Schools of LaSalle County.
Hours & Admission Prices: Wed.-Fri. 10-4, Sat.-Sun. 12-4; other times by appointment. No charge; donations accepted. Closed major holidays. &
Attendance: 12,000 (estimated)
Membership: Individual $25; Family $30; Nonprofit $50; Organization $100; Life $250; Historian $500; Benefactor $1,000; Patron $5,000.

Vandalia

THE LITTLE BRICK HOUSE, 621 Saint Clair St., Vandalia, IL 62471. Mailing Address: Vandalia Historical Society, Inc., 105 S. 4th, Vandalia, IL 62471-2809. Tel.: 618-283-4866.
Founded: 1960.
Congressional District: 55
Key Personnel: Dir., Dale Timmermann.
Governing Authority: individual operation. Parent Institution: Vandalia Historical Society, Inc.
Institution Type/Description: Historic House Museum: 1840-1860 James W. Berry property.
Collections: furniture; china; portraits; engravings; original buildings from 1825-1895; manuscripts by Vandalia authors; pictures of James Hall's family; books of early settlers; autographed books by Vandalia authors; sketches of Capital day leaders; Lincoln memorabilia.

Research Fields: frontier history; Abraham Lincoln; James Hall; capital leaders; 1820s-1840s culture.
Facilities: 100-vol. library of early Illinois & frontier period history available for research only. Antiques, books on Illinois history & other museum related items for sale.
Activities: guided tours; lectures; temporary exhibitions.
Publications: books, Seven Stories; European Journey; James Hall of Lincoln's Frontier World; Vandalia: Wilderness Capital of Lincoln's Land.
Hours & Admission Prices: By appointment only. Suggested Donations: adults $3, children 12 yrs. & under $1.
Attendance: 150 (estimated)

VANDALIA STATEHOUSE STATE HISTORIC SITE, 315 W. Gallatin St., Vandalia, IL 62471-2820. Tel.: 618-283-1161.
E-mail: hpa.vandalia@illinois.gov
Founded: 1836.
Congressional District: 19
Key Personnel: Dir. IHPA, Jan Grimes; Site Supt., Mary Cole.
Personnel Profile: Full-Time Paid 2.
Governing Authority: state. Affiliated with Illinois Historic Preservation Agency, Old State Capitol, Springfield, IL 62701. Tax-exempt.
Institution Type/Description: Historic Site: 1836 Vandalia Statehouse is the oldest Capitol building in the state of Illinois.
Collections: period furnishings.
Activities: guided tours; lectures; craft demonstrations. Museum Sponsors: Abraham Lincoln Birthday Celebration in February; Grande Levee Festival in September; Christmas Open House in December.
Publications: bimonthly newsletter, Historic Illinois.
Hours & Admission Prices: March-Oct. Tues.-Sat. 9-5; Nov.-Feb. Tues.-Sat. 9-4. Suggested Donations: adults $4, children $2. Closed New Year's Day; Thanksgiving; Christmas. ♿
Attendance: 35,000 (accurate)

Vernon Hills

LOYOLA AT CUNEO MANSION & GARDENS, (M), 1350 N. Milwaukee, Vernon Hills, IL 60061-1540. Tel.: 847-362-3042 & 3054. Fax: 847-362-4130.
E-mail: mdavi13@luc.edu
Web Site: www.luc.edu/cuneo
Formerly: Cuneo Mansion & Gardens
Founded: 1991.
Congressional District: 10
Key Personnel: C.E.O. & Pres., Rev. Garanzini; Exec. Dir., Kevin Ginty.
Personnel Profile: Full-Time Paid 8; Part-Time Volunteers 40.
Governing Authority: bd. of directors; nonprofit. Parent Institution: Loyola University Chicago. Tax-exempt.
Institution Type/Description: Historic House & Garden: 1914 Cuneo Mansion.
Collections: Renaissance paintings; 17th-century tapestries; period oriental rugs & furnishings.
Major Exhibits: Hawthorn Mellody Farms, 4/14-12/14.
Facilities: garden; conservatory; banquet facilities; 3,000 sq. ft. exhibit space.
Activities: arts festivals; concerts; docent program; organized education programs for adults; guided tours; lectures.
Publications: newsletter, CuneoGram.
Hours & Admission Prices: Feb.-Dec. Fri.-Sun. 11-4. Adults $10, seniors & students with ID $9; members no charge. Closed major holidays. ♿
Attendance: 30,000 (estimated)
Membership: Individual $45; Family $75; Director's Circle $150; Charter $250.

Versailles

VERSAILLES AREA GENEALOGICAL & HISTORICAL SOCIETY, 113 W. First St., Versailles, IL 62378. Mailing Address: P.O. Box 92, Versailles, IL 62337. Tel.: 217-225-3401.
E-mail: vaghs83@yahoo.com
Web Site: vaghs.tripod.com
Founded: 1983.
Institution Type/Description: Historical Society Museum.
Collections: local history & culture; period furnishings; photographs; personal artifacts.
Activities: research.
Hours & Admission Prices: April-Oct. Mon., Wed. & Fri. 1:30-5; Nov.-March Wed. & Fri. 1:30-5; other times by appointment. No charge; donations accepted.
Attendance: 200 (estimated)

Villa Park

VILLA PARK HISTORICAL MUSEUM, 220 S. Villa Ave., Villa Park, IL 60181. Tel.: 630-941-0223. Facebook: Villa Park Historical Museum.
Web Site: www.vphistorical society.com
Founded: 1976.
Congressional District: 6
Key Personnel: Pres. (V), Carol Marcus.
Personnel Profile: Part-Time Paid 2; Part-Time Volunteers 10.
Governing Authority: Tax-exempt.
Institution Type/Description: Historical Museum.
Collections: local history & culture; period furnishings; personal artifacts; photographs.
Major Exhibits: Villa Park 100th anniversary, 4/14-12/14; Chicago Aurora & Elgin Railroad, 4/14-12/14; Ovaltine, 4/14-12/14; 1930's Kitchen, 4/14-12/14; Avertising Souvenirs-Local Business, 4/14-12/14.
Facilities: former train station listed on National Register.
Activities: special events.
Publications: Whistlestop, newsletter.
Hours & Admission Prices: Tues.-Fri. 1-5, Sat.-Sun. 10-4. No charge; donations accepted.
Attendance: 5,089 (accurate)
Membership: Students & Seniors $10; Adults $15; Family $25; Corporate $30.

Volo

VOLO AUTO MUSEUM ATTRACTION, 27582 W. Volo Village Rd., Volo, IL 60073-9613. Tel.: 815-385-3644. Fax: 815-385-0703.
E-mail: brian@volocars.com
Web Site: www.volocars.com
Formerly: Volo Antique Auto Museum and Village
Founded: 1961.
Key Personnel: C.E.O. & Owner, Greg Grams; Dir., Brien Grams; Devel. & Museum Shop Mgr., Myra Grams.
Personnel Profile: Full-Time Paid 20; Part-Time Paid 2.
Governing Authority: company organized for profit.
Institution Type/Description: Automobile Museum.
Collections: over 300 collector automobiles including famous TV & movie cars; military artifacts.
Facilities: restaurant. Gift items for sale.
Activities: trolley tours; kiddie rides; train rides; combat zone.
Hours & Admission Prices: Daily 10-5. Winter: adults $11.95, senior citizens $9.95, children 5-12 $6.95; members no charge. Summer: adults $13.95, senior citizens $11.95, children 5-12 $8.95. Closed Easter; Thanksgiving; Christmas. ♿
Attendance: 500,000 (estimated)
Membership: Single $50; Family $85; Lifetime $500.

Washington

WASHINGTON HISTORICAL SOCIETY, 105 Zinser Place, Washington, IL 61571. Mailing Address: P.O. Box 54, Washington, IL 61571. Tel.: 309-444-4793. Facebook: Washington Historical Society.
Web Site: www.washington-historical-society.org
Key Personnel: Pres., Bev Riggins; Vice Pres., Sue Freeberg.
Personnel Profile: Part-Time Volunteers 6.
Governing Authority: Tax-exempt.
Institution Type/Description: Historical Society Museum: housed in the former medical office of Dr. Harley Zinser; built in 1916.
Collections: local history & culture; period furnishings; medical equipment; personal artifacts; photographs.
Hours & Admission Prices: March-Nov. Thurs.-Sat. 11-2; other times by appointment. No charge; donations accepted.
Attendance: 150 (estimated)

Watseka

IROQUOIS COUNTY HISTORICAL SOCIETY, 103 W. Cherry St., Watseka, IL 60970-1524. Tel.: 815-432-2215. Fax: 815-432-2215.
E-mail: ichs2215@mchsi.com
Web Site: www.iroquoiscountyhistoricalsociety.com
Founded: 1967.
Congressional District: 15
Key Personnel: Pres., Rolland Light; Vice Pres., Jean Hiles; Sec., Marilyn Wilken; Treas., Bob Ficke; Office Mgr., Judy Ficke.
Personnel Profile: Full-Time Paid 1; Part-Time Paid 4; Part-Time Volunteers 5.

Governing Authority: society; nonprofit. Subsidiary Institution: Iroquois Co. Genealogical Society. Tax-exempt: 501(c)(3).

Institution Type/Description: General Museum: housed in 1866 Old Iroquois County Courthouse.

Collections: art; archaeology; geology; mineralogy; medical; apothecary; natural history; agriculture; antiques; architecture; costumes; furniture; guns; hobbies; horology; industrial; lapidary; numismatics; musical instruments; toys; dolls; military quilts.

Research Fields: genealogy; historic sites.

Facilities: library of county records, historical books & old school books available for use under the supervision of genealogy society members; reading room; 145-seat auditorium. Novelties, t-shirts, sweatshirts & books for sale.

Activities: guided tours; concerts; rental gallery; permanent, temporary & traveling exhibitions; garden walk & garden faire. Museum Sponsors: Golden Wedding anniversary celebration for senior citizens; Christmas Tree Lane; Harvest Daze.

Publications: quarterly, Iroquois County Historical Society Newsletter; quarterly magazine, Iroquois Stalker; numerous centennial histories of towns in the county plus hardbound history books. Write for complete listing.

Hours & Admission Prices: Mon.-Fri. 10-4; other times by appointment. Suggested Donations: adults $2, children $.50; members no charge. Closed major holidays. &

Attendance: 5,000 (estimated)

Membership: Individual $15; Family $40; Life $150; Commercial Life $300.

Wauconda

*** LAKE COUNTY DISCOVERY MUSEUM, (M),** Rte. 176, Fairfield Rd. & Lakewood Forest Preserve, Wauconda, IL 60084. Mailing Address: 27277 N. Forest Preserve Dr., Wauconda, IL 60084-2016. Tel.: 847-968-3400. Fax: 847-526-0024.

E-mail: lcmuseum@lcfpd.org

Web Site: www.lakecountydiscoverymuseum.org

Founded: 1976.

Congressional District: 12

Key Personnel: C.E.O., Thomas E. Hahn; Pres., Ann Maine; Chm., Craig Taylor; Pres. (V), Dennis Leopold; Dir., Katherine Hamilton-Smith; Devel. Officer, Rebekah Snyder; Collections Coord., Diana Dretske; Cultural Resources Mgr., Andrew Osborne; Mgr. Historical Resources, Christine Pyle; Mgr. Exhibits, Steve Furnett; Imaging & Licensing Specialist, Heather Johnson; Museum Shop Mgr., Alicia Fullerton.

Personnel Profile: Full-Time Paid 19; Part-Time Paid 3; Part-Time Volunteers 101; Interns 3.

Governing Authority: county; nonprofit. Parent Institution: Lake County Forest Preserve District. Tax-exempt.

Institution Type/Description: General Museum.

Collections: decorative arts; clothing; vehicles; household & farming equipment; regional history archives: documents; photographs; diaries; ledgers; Curt Teich Postcard Archives: view & advertising cards; photographic prints & negatives, 1898-1974; Lake County history in relation to Chicago; history & significance of postcards.

Research Fields: regional history; history of postcards; Civil War history (96th Illinois volunteer regiment); Civil War homefront.

Facilities: 1,500-vol. library pertaining to local history for use on premises through inter-library loan; archives; 6,000 sq. ft. exhibit gallery; education space.

Activities: guided tours; lectures; films; workshops; organized education programs; volunteer program; organized intern program for undergraduate or graduate college students; participatory exhibits.

Publications: quarterly, Image File.

Hours & Admission Prices: Mon.-Sat. 10-4:30, Sun. 1-4:30. Adults $6, students $2.50; discounts to AAM members; members & pre-schoolers no charge. Closed New Year's Day; Thanksgiving; Christmas Eve & Day. &

Attendance: 70,000 (accurate)

Membership: Friends of the Lake County Discovery Museum & Curt Teich Archives: Individual $30; Family $45; Individual 2 years $55; Family Plus $75; Family 2 years $85; Discovery Circle $150; Mastodon Club $500.

WAUCONDA TOWNSHIP HISTORICAL SOCIETY, 711 Main St., Wauconda, IL 60084. Mailing Address: P.O. Box 256, Wauconda, IL 60084-0256. Tel.: 847-526-9303.

Founded: 1973.

Congressional District: 8

Key Personnel: Pres., Lynn McAlister; Treas., Roberta Francisco; Archivist, Registrar & Sec., Dale Buttolph.

Personnel Profile: Part-Time Volunteers 20.

Governing Authority: society; nonprofit organization. Affiliated with Wauconda Township Historical Society. Tax-exempt: 170(b)(1)(A).

Institution Type/Description: Historical Society Museum: housed in c.1840 brick farm home of Andrew C. Cook.

Collections: farm equipment of area; toys; quilts; furnishings of 1860s; dolls; Civil War cavalry equipment; books; photographs; fans; cameras; china; kitchen utensils & cow bells.

Research Fields: genealogy & history of Wauconda Township.

Facilities: library of books ranging from cookbooks to poetry available for research on premises. Museum-related items for sale.

Activities: guided tours; lectures; films; permanent & temporary exhibitions; open houses.

Publications: brochure, Andrew C. Cook Residence; Memory Books, interviews with long-time residents of the area.

Hours & Admission Prices: May-Sept. Sun. 1-4; other times by appointment. No charge; donations accepted. &

Attendance: 150 (estimated)

Membership: Student $2; Individual $10; Life $150.

Waukegan

WARBIRD HERITAGE FOUNDATION & MUSEUM, 3000 Corporate Dr., Waukegan, IL 60087. Tel.: 847-244-8701. Fax: 847-244-8703.

Web Site: www.warbirdheritagefoundation.org

Governing Authority: nonprofit organization. Tax-exempt: 501(c)(3).

Institution Type/Description: Military Aircraft Museum.

Collections: early military aircraft.

Activities: special events.

Hours & Admission Prices: By appointment.

WAUKEGAN HISTORY MUSEUM, WAUKEGAN HISTORICAL SOCIETY, 1917 N. Sheridan Rd., Bowen Park, Waukegan, IL 60087-5131. Tel.: 847-336-1859 & 360-4749. Fax: 847-662-6190.

E-mail: museum@waukeganhistorical.org

Web Site: www.waukeganhistorical.org

Formerly: Haines Museum, Waukegan Historical Society

Founded: 1968.

Congressional District: 31

Key Personnel: Pres. (V), Dennis Moisio; Supvr., Ty Rohrer; Librarian, Beverly Millard.

Personnel Profile: Full-Time Paid 1; Part-Time Volunteers 20; Interns 1.

Governing Authority: nonprofit organization. Subsidiary Institution: Waukegan Park District. Tax-exempt.

Institution Type/Description: Local History Museum.

Collections: artifacts of the Waukegan area; John Raymond Memorial Research Library; Lincoln Room; Civil War artifacts; furnishings; clothing; more than 8,000 photos of local & county historic landmarks; Indian artifacts; wax cylinders.

Major Exhibits: Back to School, 10/12/13-6/7/14.

Research Fields: genealogy; city of Waukegan; near north historical district; Waukegan landmarks; restoration of historic buildings; library resource center.

Facilities: library of material on Waukegan & Lake County history available for use on premises; reading room.

Activities: speakers bureau, slides of old Waukegan; walking, bicycle & automobile tours; lectures; school tours; inter-museum loan, permanent & temporary exhibitions. Museum Sponsors: annual tour of homes; restoration of historic buildings; Oakwood Cemetery Walk.

Publications: quarterly newsletter; brochure on museum; brochure, Near North Historic District; Waukegan's Legacy, Our Landmarks; Waukegan: A History.

Hours & Admission Prices: Tues. & Thurs. 10-4, Sat. 1-4; other times by appointment. No Charge; donations accepted. Closed all major holidays. &

Attendance: 1,500 (estimated)

Membership: Student & Senior Citizen $7; Individual $15; Family $25; Patron $50; Individual Lifetime $100; Couple Lifetime $150; Corporate $500.

West Chicago

KLINE CREEK FARM, 1N600 County Farm Rd., West Chicago, IL 60185. Mailing Address: P.O. Box 5000, Wheaton, IL 60189-5000. Tel.: 630-876-5900. Fax: 630-293-9421.

E-mail: kcf@dupageforest.com

Web Site: www.dupageforest.com

Founded: 1989.

Congressional District: 6

Key Personnel: Pres., D. (Dewey) Pierotti, Jr.; Supvr., Keith R. McClow; Coord. Collection, Carol Nardbrook; Staff Asst., Sue Clark; Heritage Interpreter, Kate Garret; Heritage Interpreter, Dennis Buck; Heritage Interpreter, Wayne Hill; Heritage Interpreter, Carmen Guerrero.

Personnel Profile: Full-Time Paid 5; Part-Time Paid 5; Part-Time Volunteers 63; Interns 5.

Governing Authority: county; nonprofit. Parent Institution: Forest Preserve District, DuPage County, Inc. Tax-exempt: Illinois State Statute 96.5.

Institution Type/Description: Living History Museum: re-creation of a turn-of-the-century farm in northeast Illinois.

Collections: 1890s farm equipment & household furnishings; purebred livestock. Historic Buildings: barn; house; summer kitchen; ice house; smoke house.

Research Fields: 1890s rural life; social customs; turn-of-the-century building methods; Victorian era interior decorating; farming methods & equipment; period clothing; purebred livestock.

Facilities: picnic areas.

Activities: guided tours; organized programs; farming demonstrations. Museum Sponsors: Ice Harvest; Maple Sugaring; Country Fair; Memorial Day Remembered; Holiday on the Farm.

Hours & Admission Prices: Thurs.-Mon. 9-5. No charge; donations accepted. Closed New Year's Eve & Day; Independence Day; Thanksgiving; Christmas Eve & Day. &

Attendance: 74,000 (estimated)

KRUSE HOUSE MUSEUM, 527 Main St., West Chicago, IL 60185-2842. Mailing Address: P.O. Box 246, West Chicago, IL 60186-0246. Tel.: 630-231-0564 & 2329.

E-mail: krusehouse@comcast.net

Web Site: www.krusehousemuseum.org

Founded: 1976.

Congressional District: 14

Key Personnel: Pres. (V), Lance Conkright; Museum Shop Mgr., Donna Orlandini.

Personnel Profile: Part-Time Volunteers 30.

Governing Authority: society. West Chicago Historical Society. Tax-exempt.

Institution Type/Description: Historical Society Museum: housed in 1917 Kruse House.

Collections: c.1920 household items & costumes belonging to railroad family.

Facilities: Museum & gift-related items for sale.

Activities: guided tours; special exhibits.

Publications: bimonthly newsletter.

Hours & Admission Prices: May-Sept. Sat. 11-3; tours by appointment. No charge; donations accepted.

Attendance: 520 (estimated)

Membership: Individual $15; Family $25; Organization $50; Individual Life $150; Family Life & Organization Life $250.

WEST CHICAGO CITY MUSEUM, (M), 132 Main St., West Chicago, IL 60185-2835. Tel.: 630-231-3376 & 293-2266. Fax: 630-293-2943.

E-mail: museum@westchicago.org

Web Site: www.westchicago.org

Founded: 1976.

Congressional District: 14

Key Personnel: Dir., Curator, Sara Phalen; Archivist, Sally DeFauw.

Personnel Profile: Part-Time Paid 2; Part-Time Volunteers 30.

Governing Authority: municipal. Parent Institution: City of West Chicago. Subsidiary Institution: CB & Q Rail Depot. Tax-exempt.

Institution Type/Description: History Museum: housed in the former Town Hall & Fire Station; Chicago, Burlington & Quincy Station; c.1860. Listed on the National Register of Historic Places.

Collections: C&NW, CB&Q, CA&E Railroad; woodworking, 1880-1920; domestic tools & tools from various crafts; farm items; documents & photographs pertaining to local history & Illinois railroad history; genealogy files; manuscripts. Historic Railroad Park: mid 19th century CB&Q Station.

Research Fields: Illinois railroad history; local history & genealogy; John W. Gates; tools.

Facilities: 200-vol. library of books & maps on local & Illinois railroad history available by appointment; microfilm reels of local newspapers; genealogical material for 11,000 former residents.

Activities: guided tours; lectures; formally organized educational programs for schools; school loan service; A/V presentations.

Publications: History of West Chicago; Historic Homes of West Chicago; History of an Old Railroad Town; Tales Tombstones Tell, Stirring Up History.

Hours & Admission Prices: Jan.-March Wed.-Fri. 12-4; April-Dec. Wed.-Sat. 12-4; other times by appointment. Research: by appointment. No charge; donations accepted. &

Attendance: 3,000 (estimated)

Membership: Student $10; Senior $15; Individual $20; Senior Family $25; Household & Nonprofit Organization $35; Business $40; Life $500.

West Frankfort

FRANKFORT AREA HISTORICAL MUSEUM, 2000 E. St. Louis St., West Frankfort, IL 62896-1647. Tel.: 618-932-6159.

Founded: 1972.

Congressional District: 25

Key Personnel: Pres., Mary Ellen Maragni; Vice Pres. & Museum Shop Mgr., Winona Harris; Sec., Sylvia Tharp; Treas., Ervin Thomas; Asst. Dir. & Cur. Veteran's Museum, Jim Hall; Librarian, Shirley Payne; Asst. Dir., Dean Tharp.

Personnel Profile: Full-Time Volunteers 20; Part-Time Volunteers 40.

Governing Authority: society; nonprofit organization. Tax-exempt: 501(c)(3).

Institution Type/Description: Historical Building: housed in a former school building; built c.1916.

Collections: Southern Illinois history; farm utensils; tools; furnishings; toys; literature; war memorabilia; food preparation & preserving; methods of holiday celebration; one room school; coal mining. Historic Building: early 1900s home.

Research Fields: local history; genealogy.

Facilities: 10,000-vol. library of rare books on genealogy, local history & town records for research on premises only; maps; newspapers; microfilms; slides; tapes; periodicals; 60-capacity auditorium; educational facilities; tea room. Homemade crafts & other museum-related items for sale.

Activities: guided tours; lectures; films; hobby workshops; TV & radio programs; docent program; organized education programs children, adults & undergraduate or graduate college students affiliated with Southern Illinois University, John A. Logan College & Rend Lake College; training programs for museum workers; participatory, loan, temporary & traveling exhibitions; school loan service. Museum Sponsors: Victorian Spring Luncheon; Ladies Spring Luncheon; Autumn High Tea; area Quilt Show in July; area Flea Market in September; Veterans' Recognition in November; Holiday House in December.

Publications: books, History of West Frankfort; Cooking With Kindness Vol. 1 & 2; brochures; Franklin County History Book.

Hours & Admission Prices: Wed.-Thurs. 9-3. No charge; donations accepted. Closed New Year's Day; Easter; Independence Day; Thanksgiving; Christmas. &

Attendance: 9,000 (estimated)

Membership: Friend $10; Patron $20; Associate $30; Partner $50; Executive $100; Benefactor $1,000.

VETERANS DEPOT MUSEUM, 101 W. Main St., West Frankfort, IL 62896-2317. Mailing Address: 2000 E. St. Louis, West Frankfort, IL 62896-1647.

Key Personnel: Dir., Robert Rogers; Chm. (V), Dwight Tharp.

Personnel Profile: Part-Time Volunteers 4.

Governing Authority: Subsidiary Institution: Frankfort Area Historical Museum. Tax-exempt.

Institution Type/Description: Military Museum.

Collections: Civil War, WWI, WWII, Korean War, Vietnam War & Gulf War memorabilia; books; videos; WW radio equipment; military memorabilia.

Activities: guided tours; loan exhibitions. Annual Event: Veterans Day Ceremonies.

Hours & Admission Prices: March-Nov. Sun. 1-4; Dec.-Feb. by appointment. No charge; donations accepted. Closed national holidays.

Attendance: 400 (estimated)

Western Springs

WESTERN SPRINGS HISTORICAL SOCIETY, 4211 Grand Ave., Western Springs, IL 60558-1435. Mailing Address: P.O. Box 139, Western Springs, IL 60558-0139. Tel.: 708-246-9230.

Founded: 1967.

Congressional District: 4

Key Personnel: Pres., Allyson Zak; Vice Pres., Kimberly Knake; Cur. Res. & Artifacts, John Devona.

Personnel Profile: Part-Time Volunteers 12.

Governing Authority: society; nonprofit organization. Tax-exempt.

Institution Type/Description: General Museum: housed in 1892 Old Water Tower.

Collections: history of Western Springs; records; books; manuscripts; maps; photographs.

Research Fields: history; genealogy; government; education.

Facilities: library of research aids, books, genealogy aids, classified local interests & history available for research on premises.

Activities: guided tours; films; training programs; permanent & temporary exhibitions. Museum Sponsors: Preservation Awards (homes).

Publications: books, Western Springs: A Centennial History of the Village, 1886-1986; 19th-Century Houses in Western Springs; Western Springs: The Story of Village Leadership; Walking tour of Western Springs.

Hours & Admission Prices: Call for hours. No charge.
Attendance: 1,000 (estimated)
Membership: Individual $25; Family $40; Business & Sustaining $50; Patron $100; Life $500.

Wheaton

BILLY GRAHAM CENTER MUSEUM, 500 E. College Ave., Wheaton, IL 60187-5534. Tel.: 630-752-5909, ext. 0. Fax: 630-752-5916.
E-mail: BGCMus@wheaton.edu
Web Site: www.billygrahamcenter.org/museum
Founded: 1975.
Congressional District: 14
Key Personnel: Museum Coord., Christian Sawyer; Dir. Resources, Paul Ericksen.
Personnel Profile: Full-Time Paid 3; Part-Time Paid 4.
Governing Authority: college. Affiliated with Wheaton College. Tax-exempt: 501(c)(3).
Institution Type/Description: Religious Museum.
Collections: materials related to the history of evangelism, revival & missions in America; contemporary Christian art.
Research Fields: American church history; emphasis on evangelism, revival & missions.
Facilities: 65,000-vol. library of books & 130,000 microforms relating to missions, evangelism, church history & general theology available for research through Wheaton College; reading room; 500-seat auditorium; classrooms. Books for sale.
Publications: booklet, The Collections at the Billy Graham Center; Researching Modern Evangelicalism, A Guide to the Holdings of the Billy Graham Center, With Information on Other Collections.
Hours & Admission Prices: Mon.-Sat. 9:30-5:30, Sun. 1-5. Suggested Donations: adults $7, seniors & students $4, children 12 & under $2. Closed New Year's Eve & Day; Christmas Eve, Day & week. ♿

CENTER FOR HISTORY, 315 W. Front St., 2nd Fl., Wheaton, IL 60187-5015. Mailing Address: P.O. Box 373, Wheaton, IL 60187-0373. Tel.: 630-871-6601.
E-mail: info@wheatonhistory.com
Web Site: www.wheatonhistory.org
Formerly: Wheaton History Center
Founded: 1986.
Key Personnel: Pres. & C.E.O., Alberta Adamson, CFRE; Chm., Ed Ewoldt.
Personnel Profile: Full-Time Paid 1; Part-Time Volunteers 5.
Governing Authority: nonprofit organization. Parent Institution: Wheaton Historic Preservation Council. Branch Museum: 606 N. Main St., Wheaton, IL 60187. Tax-exempt: 501(c)(3).
Institution Type/Description: History Museum.
Collections: golf collection including artifacts, publications & archives; slave artifacts & documents; oral histories; WWII; S.S. Eastland; local history; period costumes.
Research Fields: Civil War; WWII; slavery; abolition; architecture; S.S. Eastland; Wheaton history; golf.
Activities: Community Outreach; facility rental; interpretation; lectures; research library & archives; school-based curriculum. Museum Sponsors: Housewalk; Architecture Tours; S.S. Eastland Reunion.
Publications: Recollections of the War of the Rebellion.
Hours & Admission Prices: Mon.-Sat. 10-5; tours, programs & research by appointment. Admission $5; discounts to AAM members; members and children 8 & under no charge. Closed New Year's Day; Easter; Memorial Day; Labor Day; Thanksgiving; Christmas. ♿
Attendance: 9,000 (estimated)
Membership: Individual $30; Family $50; Corporate $100.

COSLEY ZOO, 1356 N. Gary Ave., Wheaton, IL 60187. Tel.: 630-665-5534. Fax: 630-260-6408. Facebook: Cosley Zoo.
E-mail: cosleyzoo@wheatonparks.org
Web Site: cosleyzoo.org
Founded: 1974.
Congressional District: 6
Key Personnel: Dir., Susan Wahlgren; Pres. (V), Arthur Pape.
Personnel Profile: Full-Time Paid 6; Part-Time Paid 25; Part-Time Volunteers 30; Interns 12.
Governing Authority: Parent Institution: Wheaton Park District. Tax-exempt.
Institution Type/Description: Zoo.
Collections: over 200 animals representing more than 70 species; amphibians; invertebrates; birds; mammals; reptiles; fish.
Facilities: Gift items for sale.
Activities: educational programs.

Publications: quarterly newsletter. Cosley Tails.
Hours & Admission Prices: Jan.-March & Nov. daily 9-4; April-Oct. daily 9-5; late Nov.-Dec. daily 9-9; groups by appointment. Adults $5, seniors 55 & over $4; members, youth 17 & under, AZA members and Wheatong Park District residents no charge. Closed New Year's Day; Thanksgiving; Christmas. ♿
Attendance: 140,000 (accurate)
Membership: Individual $39; Family $59; Family Plus One $79-$99.

DUPAGE COUNTY HISTORICAL MUSEUM, (M), 102 E. Wesley St., Wheaton, IL 60187-5321. Tel.: 630-510-4941. Fax: 630-665-5880.
Web Site: www.dupagemuseum.com
Founded: 1965.
Congressional District: 14
Key Personnel: Dir., Mike Benard; Cur., Sara Arnas; Educator, Sara Buttita.
Personnel Profile: Full-Time Paid 2; Part-Time Paid 3; Part-Time Volunteers 8; Interns 2.
Governing Authority: Parent Institution: Wheaton Park District and DuPage County. Subsidiary Institution: Du Page County Museum Association. Tax-exempt.
Institution Type/Description: History Museum.
Collections: 19th, 20th & 21st-century material culture & archives pertaining to DuPage County; Colonial Coverlet Guild collection; costumes; model railroad.
Major Exhibits: Fashion Accessories in Vogue...and Out, 1/25/14-8/24/14; 175th Anniversary of DuPage County, 9/13/14-8/3/14.
Research Fields: county history; early families; home history research; textiles; 19th, 20th & 21st-century material culture; genealogy.
Facilities: 2,000-vol. library of county & local histories; city directories & archives available for research on premises. Local history books for sale.
Activities: monthly family programs; organized education programs for children; adult lectures & seminars; permanent & changing exhibits; research services; interactive exhibits; landscaped HO gauge model railroad depicting Chicago & Northwestern, Chicago, Aurora & Elgin, and Burlington Northern railroads, operates 3rd & 5th Sat. 1:30-3:30.
Publications: brochures; newsletter; quarterly calendar of events.
Hours & Admission Prices: Mon.-Fri. 8:30-4:30, Sat.-Sun. 12-4. No charge; donations accepted. Closed New Year's Eve & Day; Memorial Day; Independence Day; Labor Day; Thanksgiving; Christmas Eve & Day. ♿
Attendance: 8,669 (accurate)

THE FIRST DIVISION MUSEUM AT CANTIGNY, (M), 1s151 Winfield Rd., Wheaton, IL 60189-3353. Tel.: 630-260-8185. Fax: 630-260-9298.
E-mail: info@firstdivisionmuseum.org
Web Site: www.FirstDivisionMuseum.org
Founded: 1960.
Congressional District: 14
Key Personnel: Exec. Dir., Paul H. Herbert, Ph.D.; Dir. Museum Operations, Keith R. Gill; Dir. Research Center, Eric Gillespie; Mgr. Public Programs, Gayln Piper; Dir. Publications, Steve Hawkins; Cur. Collections, Bill Brewster; Asst. Cur. Collections, Chris Zielinski; Mgr. Exhibits, Teri Bianchi; Museum Educator, Melissa Neumann; Collections Mgr., Shana Keil; Librarian, Tracy Cirar; Research Historian, Andrew Woods; Graphic Designer & Editor, Dave Blake; Registrar, John Maniatis; Archivist, Kate Kleiderman; Media Design Specialist, Christie Walsh.
Personnel Profile: Full-Time Paid 20; Part-Time Paid 2; Part-Time Volunteers 51; Interns 5.
Governing Authority: nonprofit; self-financing; Cantigny First Division Foundation. Parent Institution: Robert R. McCormick Foundation. Tax-exempt: 501(c)(3).
Institution Type/Description: Military Museum: located on the grounds of Cantigny Park.
Collections: story of 1st Infantry Div. from World War I to present; archives; outdoor display of military vehicles.
Research Fields: 1st Infantry Division; 1917-present; American military history; freedom of the press; publishing & journalism history; Chicago area history.
Facilities: 12,000-vol. library of books; 41,000 photographs; 73,000 documents & manuscripts; 15,000 artifacts pertaining to U.S. military history.
Activities: guided tours; lectures; audio-visual programs; educational programs; temporary exhibits; patriotic events; conferences & symposia.
Publications: Cantigny Military History Series, books; Bridgehead Sentinel, 1st Division Veterans' newspaper; conference reports; historical monographs; memoirs; bibliographies; videos; brochures.
Hours & Admission Prices: Feb. Fri.-Sun. 10-4; March-April & Nov.-Dec.

Tues.-Sun. 10-4; May-Oct. Tues.-Sun. 10-5. Gardens & Grounds: dawn-dusk. $5 per car parking fee & road use Tues.-Sun. Closed New Year's Day; Thanksgiving; Christmas. &
Attendance: 145,000 (accurate)

ROBERT R. MCCORMICK MUSEUM AT CANTIGNY, (M), 1 S. 151 Winfield Rd., Wheaton, IL 60189-3353. Tel.: 630-260-8163. Fax: 630-260-8160. Facebook: Robert R. McCormick Museum.
E-mail: mccormickmuseum@cantigny.org
Web Site: www.cantigny.org
Founded: 1955.
Congressional District: 14
Key Personnel: Dir., Diane Gutenkauf; C.E.O., David Hiller; Asst. Dir., William Buhlig; Exec. Dir. Cantigny Foundation, Matt Lafond; Tour Coord., Jeff Anderson; Museum Shop Mgr., Alicia Anderson.
Personnel Profile: Full-Time Paid 3; Part-Time Paid 8; Part-Time Volunteers 10.
Governing Authority: private; self-financing, nonprofit organization. Parent Institution: Cantigny Foundation. Subsidiary Institution: The First Division Museum. Tax-exempt: 501(c)(3).
Institution Type/Description: Historic House: 1896, country home of Joseph Medill, editor of the Chicago Tribune built by architect C.A. Coolidge; house enlarged in 1932 for Col. Robert R. McCormick (grandson of Medill) editor & publisher of the Chicago Tribune.
Collections: furnishings; books & memorabilia of Col. Robert McCormick.
Facilities: 5,000-vol. library of books & documents; movie theater; gardens; small picnic area.
Activities: guided tours; lectures; chamber music; patriotic celebrations; outdoor concerts; scout camp grounds.
Publications: Brochure.
Hours & Admission Prices: Feb. Fri.-Sun. 10-4; March-April & Nov.-Dec. Tues.-Sun. 10-4; May-Oct. Tues.-Sun. 10-5. Annual Pass: $60; Bus: $80. Grounds: daily 7am to sunset. Closed New Year's Eve & Day; Thanksgiving & day after; Christmas Eve & Day. &
Attendance: 55,342 (accurate)

White Hall

GREGORY HOUSE MUSEUM, Rte. 1, White Hall, IL 62092. Mailing Address: RR1, Box 94, White Hall, IL 62092-0094. Tel.: 217-374-6715.
Founded: 1993.
Key Personnel: Cur., Emily B. Esarey
Institution Type/Description: Historic House Museum.
Collections: Gregory family history & personal artifacts; local agricultural, political & community history; paperweights.
Activities: Museum Sponsors: Greene County Days in September.
Hours & Admission Prices: By appointment. No charge; donations accepted. Closed Thanksgiving; Christmas. &
Attendance: 150 (estimated)

Whittington

SOUTHERN ILLINOIS ART & ARTISANS CENTER, 14967 Gun Creek Trail, Whittington, IL 62897-1000. Tel.: 618-629-2220. Fax: 618-629-2704.
E-mail: mgalloway@museum.state.il.us
Web Site: www.museum.state.il.us
Founded: 1990.
Key Personnel: Dir., Mary Lou Galloway; Cur., Debra Tayes; Museum Shop Mgr., Romaula Coleman.
Personnel Profile: Full-Time Paid 6; Part-Time Paid 5.
Governing Authority: state; nonprofit. Parent Institution: Illinois State Museum, 502 S. Spring St., Springfield, IL 62706. Tax-exempt: 501(c)(3).
Institution Type/Description: Art & Artisans Center.
Collections: works by members of the Illinois Artisans Program.
Facilities: visitors center. Museum-related items for sale.
Activities: arts festivals; guided tours; craft workshops; lectures; temporary & traveling exhibitions; artisan demonstrations & workshops; outdoor festivals & special events. Annual Events: Children's Festival; Art & Wine Festival; Illinois Art & Fine Craft.
Publications: annual brochure, Calendar of Events & Programs.
Hours & Admission Prices: Daily 9-5. No charge. Closed New Year's Day; Easter; Thanksgiving; Christmas. &
Attendance: 39,741
Membership: Artist & Student $20; Individual $35; Family $50; Contributing $100; Life $500.

Williamsville

DIE CAST AUTO SALES, 117 N. Elm St., Williamsville, IL 62693-7503. Tel.: 217-566-3898.
Institution Type/Description: History Museum: housed in a former 1930s service station.
Collections: die-cast cars; Coca Cola collectibles; Route 66 artifacts.
Hours & Admission Prices: Call for hours.

Wilmette

WILMETTE HISTORICAL MUSEUM, (M), 609 Ridge Rd., Wilmette, IL 60091-2441. Tel.: 847-853-7666. Fax: 847-853-7706.
E-mail: museum@wilmette.com
Web Site: www.wilmettehistory.org
Founded: 1949.
Congressional District: 10
Key Personnel: Dir., Kathy Hussey-Arntson.
Personnel Profile: Full-Time Paid 1; Part-Time Paid 2; Part-Time Volunteers 55.
Governing Authority: municipal. Parent Institution: Village of Wilmette. Subsidiary Institution: Wilmette Historical Society. Tax-exempt: 501(c)(3).
Institution Type/Description: Local History Museum.
Collections: local history archives; costumes; history.
Research Fields: pertaining to collection.
Facilities: library; archives including photographs relating to Wilmette, Gross Point & vicinity available for research.
Activities: guided tours; lectures; formally organized education programs; temporary exhibitions.
Publications: quarterly newsletter.
Hours & Admission Prices: Sun.-Thurs. 1-4:30. No charge. Closed national holidays. &
Attendance: 6,000 (estimated)

Winnetka

WINNETKA HISTORICAL SOCIETY, (M), Museum & Research Center, 411 Linden, Winnetka, IL 60093. Mailing Address: P.O. Box 365, Winnetka, IL 60093-0365. Tel.: 847-446-0001. Fax: 847-501-3221.
E-mail: winnetka411@comcast.net
Web Site: www.winnetkahistory.org
Founded: 1932.
Congressional District: 10
Key Personnel: Pres., Nan Greenough; Exec. Dir., Patti Van Cleave; Cur., Katherine Macica; Cur. Costume, Elizabeth Carlson.
Personnel Profile: Part-Time Paid 3; Part-Time Volunteers 15; Interns 1.
Governing Authority: nonprofit. Branch Museum: Schmidt-Burnham Log House, 1140 Willow Rd., Winnetka, IL 60093. Tax-exempt: 501(c)(3).
Institution Type/Description: Local History Museum and Archives.
Collections: history of Winnetka & surrounding area; photographs; costumes; paintings; documents; artifacts. Historic House: c.1837 log house.
Research Fields: local, Chicago & Illinois history; authors from Winnetka.
Facilities: 900-vol. library of general & local history available to the public; reception area. Museum-related items for sale.
Activities: guided tours; lectures; educational programs for children & adults. Museum Sponsors: parties; benefit.
Publications: book, Winnetka Architecture: Where Past is Present; semiannual newspaper Winnetka Historical Society Gazette; DVD, Winnetka Story.
Hours & Admission Prices: Call for hours. Museum & participants of Time Travelers: no charge. Log House: call for admission prices. &
Attendance: 1,200 (estimated)
Membership: Family Pass (2 adults & children under 18) $50.

Zion

PLATEN PRESS PRINTING MUSEUM, 3051 Sheridan Rd., Zion, IL 60099-3243. Tel.: 847-746-8170.
Institution Type/Description: Printing Museum.
Collections: printing; bindery equipment; addressograph machines; metal casting; letterpress; bookbinding.
Hours & Admission Prices: By appointment.

ZION HISTORICAL SOCIETY, 1300 Shiloh Blvd., Zion, IL 60099-2622. Tel.: 847-746-2427 & 872-4566.
E-mail: tr91752@sbcglobal.net
Web Site: www.zionhs.com
Founded: 1967.
Congressional District: 10

Key Personnel: Pres., Carol Ruesch; Vice Pres., Lorna Yates.
Personnel Profile: Part-Time Volunteers 12.
Governing Authority: society. Tax-exempt: 501(c)(3).
Institution Type/Description: Historic House Museum: 1902 Shiloh House, the residence of the founder of the city of Zion.
Collections: original furnishings: antiques; religious artifacts; manuscripts.
Facilities: library of religious & historical books available for use on premises. Gift items & local crafts for sale.
Activities: guided tours; lectures; films; inter-museum loan, permanent & temporary exhibitions.
Hours & Admission Prices: Memorial Day-Labor Day Sun. 2-5; other times by appointment. Adults $5, children $2.
Attendance: 150 (estimated)
Membership: Individual $7.50; Family $15; Contributing $25; Sustaining $35; Life $150.

INDIANA

(321 listings)

Albion

OLD JAIL MUSEUM, 215 W. Main St., Albion, IN 46701-1115. Mailing Address: P.O. Box 152, Albion, IN 46701-0152. Tel.: 260-740-8932.
Web Site: www.rootsweb.ancestry.com/~innchs/index.html
Founded: 1968.
Congressional District: 4
Key Personnel: Vice Pres., Bill Landon; Dir., Richard Recker; Dir., Carol Kirsch; Pres., Bill Shultz; Bd. Directors, Mary Stolte; Treas., Judy Richter; Dir. & Museum Shop Mgr., Margaret Ott; Bd. Directors, Sondra Luke; Sec., Sarah Knopp.
Personnel Profile: Part-Time Volunteers 30.
Volunteer Hours: 325
Operating Expenses: 10,137
Operating Income: 9,619
Governing Authority: society; nonprofit organization. Parent Institution: Noble County Historical Society, Inc. Tax-exempt.
Institution Type/Description: Historic Building: 1876 Noble County Old Jail & sheriff's residence.
Collections: furnishings; artifacts; documents.
Facilities: library of old books & magazines. Gift items for sale.
Activities: guided tours; permanent exhibitions.
Publications: quarterly letter, Pioneer Echoes.
Hours & Admission Prices: Memorial Day to 3rd week in Sept. Sat. 1:30-4:30; group tours by appointment. Adults $3, school children $1; members no charge.
Attendance: 340 (accurate)
Membership: Student $1.50; Individual $10; Life $100.

Alexandria

ALEXANDRIA MONROE TOWNSHIP HISTORICAL SOCI-ETY, 313 N. Harrison St., Alexandria, IN 46001. Tel.: 765-724-2993.
Institution Type/Description: Historical Society Museum.
Collections: local history & culture; period furnishings; personal artifacts; photographs.
Hours & Admission Prices: Call for hours.

Anderson

THE ANDERSON CENTER FOR THE ARTS, (M), 32 W. Tenth St., Anderson, IN 46016-1409. Mailing Address: P.O. Box 1218, Anderson, IN 46015-1218. Tel.: 765-649-1248. Fax: 765-649-0199.
E-mail: info.taca@sbcglobal.net
Web Site: www.andersonart.org
Formerly: Anderson Fine Arts Center
Founded: 1966.
Congressional District: 2
Key Personnel: Dir., Deborah McBratney-Stapleton; Pres., Harry Carter; Administrative Asst., Cheryl Mitchell.
Personnel Profile: Full-Time Paid 4; Part-Time Paid 4; Part-Time Volunteers 30; Interns 3.
Volunteer Hours: 500
Operating Expenses: 344,000
Operating Income: 280,000
Governing Authority: private; nonprofit organization. Parent Institution: Anderson Fine Arts Foundation, Inc. Tax-exempt: 501(c)(3).

Institution Type/Description: Fine Arts Center.
Collections: paintings, sculpture, drawings and prints by Midwestern and Indiana artists & 20th-century American artists; crafts; pottery; metalsmithing.
Research Fields: American art.
Facilities: classroom; conference & meeting space.
Activities: guided tours; lectures; films; festivals; art classes; workshops; organized education programs; docent program; temporary & traveling exhibitions; outreach programs.
Hours & Admission Prices: Tues.-Fri. 12-5, Sat. 10-5, Sun. 2-5. Families $5, adults $2, seniors $1.50, children & students $1; discounts to AAM & ICOM members; Tues., 1st. Sun. of each month, members & children under 4 no charge. Closed national holidays. &
Attendance: 35,000 (estimated)
Membership: Student K-12 $10; Individual $40; Family $60; Patron $125; Sustaining $250; Benefactor $500; Master $1,000 & up. Corporate: Friend $50; Patron $125; Sustaining $250; Sponsor $500; Master $1,000 & up.

GRUENEWALD HISTORIC HOUSE, 626 Main St., Anderson, IN 46016-1514. Tel.: 765-648-6875.
Web Site: www.gruenewaldhouse.com
Founded: 1976.
Congressional District: 6
Key Personnel: Dir., Jean Whitsell-Sherman.
Personnel Profile: Part-Time Paid 1; Part-Time Volunteers 12.
Governing Authority: Tax-exempt.
Institution Type/Description: Historic House Museum.
Collections: personal artifacts; 1890s Victorian-style furnishings; gardens.
Activities: Museum Sponsors: History Awareness Series.
Publications: members' newsletters.
Hours & Admission Prices: April-Dec. Mon., Wed. & Fri. 10-3, Sat. 12-3. Adults $5; discounts to AAA members; students & members no charge. Closed Thanksgiving; Christmas. &
Attendance: 750 (estimated)
Membership: Individual $35; Family $50; Corporate $100.

GUSTAV JEENINGA MUSEUM OF BIBLE AND NEAR EAST-ERN STUDIES, (M), Theology Bldg., 1123 Anderson University Blvd., Anderson, IN 46012-3495. Mailing Address: 1100 E. 5th St., Anderson, IN 46012-3462. Tel.: 765-641-4526. Fax: 765-641-3005.
E-mail: dlneidert@anderson.edu
Web Site: www.anderson.edu/campus/museum/index.html
Founded: 1963.
Congressional District: 6
Key Personnel: Dir., David Neidert.
Governing Authority: university. Parent Institution: Anderson University, Inc. Tax-exempt: 501(c)(3).
Institution Type/Description: Archaeological Museum.
Collections: archaeological objects related to biblical & Near Eastern studies.
Activities: formally organized education programs for undergraduate & graduate students.
Publications: biannual newsletters, Illumination.
Hours & Admission Prices: Mon.-Fri. 9-5. No charge. &
Attendance: 2,000 (estimated)
Membership: Active $5; Family $10; Contributing $25; Sustaining $50.

JESSIE C. WILSON ART GALLERIES, Anderson University, Krannert Fine Arts Center, 1100 E. 5th St., Anderson, IN 46012-3462. Tel.: 765-641-4322. Facebook: Jessie C. Wilson Galleries - Anderson University.
Web Site: anderson.edu/academics/art-and-design/wilson
Key Personnel: Dir., Tai Lipan; Office Mgr., Robyn Davis
Institution Type/Description: Art Gallery.
Collections: over 140 works by Warner Sallman.
Hours & Admission Prices: Mon.-Fri. 9-4. No charge.

Angola

TRINE UNIVERSITY, GENERAL LEWIS B. HERSHEY MU-SEUM, 1 University Ave., Angola, IN 46703-1764. Tel.: 260-665-4162 & 4100. Fax: 260-665-4283.
E-mail: library@trine.edu
Web Site: www.trine.edu
Founded: 1970.
Congressional District: 4
Key Personnel: C.E.O. & Cur., Dr. Earl D. Brooks, II
Personnel Profile: Part-Time Volunteers 1.

Governing Authority: university. Parent Institution: Trine University. Tax-exempt: 170(b)(1)(A).
Institution Type/Description: Military Museum.
Collections: memorabilia of General Lewis B. Hershey.
Facilities: classrooms.
Hours & Admission Prices: Mon.-Fri. 8-4:30; other times upon request. No charge; donations accepted. Closed national holidays. &

Auburn

*** AUBURN AUTOMOTIVE HERITAGE, INC. DBA AUBURN CORD DUESENBERG AUTOMOBILE MUSEUM, (M),** 1600 S. Wayne St., Auburn, IN 46706-3509. Mailing Address: P.O. Box 271, Auburn, IN 46706-0271. Tel.: 260-925-1444. Fax: 260-925-6266.
E-mail: info@automobilemuseum.org
Web Site: automobilemuseum.org
Founded: 1973.
Congressional District: 3
Key Personnel: Chm. (V), Michael Eikenberry; Exec. Dir., Laura Brinkman; C.O.O., Kendra Klink; Cur., Aaron Warkentin; Museum Store Mgr., Karen Grogg.
Personnel Profile: Full-Time Paid 11; Part-Time Paid 25; Part-Time Volunteers 120.
Governing Authority: nonprofit organization. Parent Institution: Auburn Automotive Heritage, Inc. Tax-exempt: 501(c)(3).
Institution Type/Description: Transportation Museum: located in original 1930 Administration Building of the Auburn Automobile Co. Listed on the National Register of Historic Places; a National Historic Landmark.
Collections: over 120 period & classic cars; photographs; literature & memorabilia on cars manufactured in or associated with the City of Auburn & their impact on American culture.
Research Fields: automobile manufacturing operations based in Auburn, IN, from 1900 to 1937.
Facilities: archives of original photographs, sales manuals, technical data, & news articles related to automobiles built in Auburn, IN, available for inspection & research; catering & banquet facilities. Museum & automotive-related items for sale.
Activities: lectures; films; archives.
Publications: semiannual newsletter, The Accelerator; annual report; It's A Duesy.
Hours & Admission Prices: Daily 10-7 Mon.-Fri., 10-5 Sat. & Sun. Last admission 1 hour before closing. Adults $12.50, students $7.50; discounts to groups, AAM, ICOM & AAA members or any member of a recognized car club; children under 5 & members no charge. Closed New Year's Day; Thanksgiving; Christmas. &
Attendance: 47,283 (accurate)
Membership: Individual $45; Senior $50; Family $60; Grandparent $60; Patron $100; Sustaining $250; Life $1,500.

HOOSIER AIR MUSEUM, 2822 Cty. Rd. 62, Auburn, IN 46706. Mailing Address: P.O. Box 87, Auburn, IN 46706. Tel.: 260-927-0443.
Web Site: www.hoosierairmuseum.org
Congressional District: 3
Key Personnel: Pres. (V), Richard Dodge; Museum Shop Mgr., Larry Stone.
Personnel Profile: Part-Time Volunteers 20.
Governing Authority: nonprofit organization. Parent Institution: Hoosier Air Museum. Tax-exempt: 501(c)(3).
Institution Type/Description: Military History Museum.
Collections: aviation history & memorabilia; aircraft; radial, in-line & jet engines; model airplanes; photographs.
Major Exhibits: Vintage Aircrafts, 3/14-12/14.
Facilities: Museum-related items for sale.
Activities: rental facilities.
Hours & Admission Prices: mid-March to mid-Dec. Wed.-Sun. 10-4; groups by appointment. Adults $5, students 12-18 $4; active military in uniform, students and children 11 & under no charge.
Attendance: 4,400 (estimated)

KRUSE AUTOMOTIVE & CARRIAGE MUSEUM, 5634 County Road 11A, Auburn, IN 46706. Tel.: 260-927-9144. Fax: 260-927-8043. Facebook: National Military History Center.
E-mail: info@dvkfoundation.org
Institution Type/Description: Transportation Museum.
Collections: cars including a Duesenberg, Indy race car, Carl Casper custom cars, the Batmobile; British Royal Carriages; Wells Fargo Stage Coach; TV & movie cars.
Hours & Admission Prices: Daily 9-5. Adults $10, seniors 55 & over $8,

children 7-12 $6, active military $4; discounts to groups; WWII veterans & children under 7 no charge. Closed New Year's Day; Thanksgiving; Christmas.

NATIONAL AUTOMOTIVE & TRUCK MUSEUM OF THE UNITED STATES, INC. (NATMUS), 1000 Gordon M. Buehrig Place, Auburn, IN 46706-3525. Tel.: 260-925-9100. Fax: 260-925-8695.
E-mail: info@natmus.com
Web Site: www.natmus.org
Founded: 1988.
Congressional District: 3
Key Personnel: Pres. (V), John Pontius; Exec. Dir., Donald Grogg; Business Mgr. & Museum Shop Mgr., Audra Wilcoxson.
Personnel Profile: Full-Time Paid 1; Part-Time Paid 4; Part-Time Volunteers 143.
Volunteer Hours: 2,478
Operating Expenses: 242,032
Operating Income: 528,502
Governing Authority: not-for-profit organization. Tax-exempt: 501(c)(3).
Institution Type/Description: Automotive & Truck Museum: housed in c.1923 former Service Building & 1928 former L29 Cord/Experimental Building of the Auburn Automobile Company.
Collections: automobiles; trucks; models; toys; automobilia; automotive library; gas & oil pumps; pedal cars.
Research Fields: automotive, truck & business history; automotive and truck-related toys & models.
Facilities: 1,500-vol. library of automotive & truck literature; 112,000 sq. ft. exhibit space; meeting room.
Activities: guided tours; loan exhibitions.
Publications: quarterly newsletter; National Automotive & Truck Newsletter.
Hours & Admission Prices: Daily 9-5. Adults $8, children 6-12 $4; discount to AAA members; children 5 & under and NATMUS members no charge. Closed New Year's Day; Thanksgiving; Christmas. &
Attendance: 10,848 (accurate)
Membership: Student & Senior Citizen $25; Family $30; Contributing $100; Sustaining $250; Classic $500; Life $1,000; Corporation $1,500.

NATIONAL MILITARY HISTORY CENTER, 5634 County Rd. 11A, Auburn, IN 46706. Mailing Address: P.O. Box 1, Auburn, IN 46706. Tel.: 260-927-9144. Fax: 260-927-8043.
Web Site: www.militaryhistorycenter.org
Formerly: World War II Victory Museum
Founded: 2003.
Key Personnel: Chm. (V), Dir. Education & Volunteer Coord., Emily Disbro; Pres. (V), Dean V. Kruse; Mgr. Operations & Museum Shop Mgr., Tammy Hantz; Property Mgr., Mike Phares.
Personnel Profile: Full-Time Paid 4; Part-Time Paid 7; Part-Time Volunteers 125; Interns 4.
Governing Authority: nonprofit. Parent Institution: Dean V. Kruse Foundation, Inc. Tax-exempt.
Institution Type/Description: Military Museum.
Collections: WWII history; military artifacts; uniforms; equipment; personal artifacts; photographs; vehicles; weapons.
Facilities: library; rental facilities. Museum-related items for sale.
Activities: Annual Events: Memorial Day Celebration; Veterans Day Celebration; History Fest; Festival of Trees; A Night with Santa.
Publications: newsletter, The Salute.
Hours & Admission Prices: Mon.-Sat. 9-5, Sun. 1-5. Adults $10, seniors 55 & over $8, children 7-12 $6; discounts to groups; children under 7 & WWII veterans no charge. &
Attendance: 50,000 (estimated)
Membership: Senior $45; Individual $50; Family $100; Corporate $500; Lifetime $2,000.

Aurora

HILLFOREST HOUSE MUSEUM, 213 Fifth St., Aurora, IN 47001-1211. Mailing Address: P.O. Box 127, Aurora, IN 47001-0127. Tel.: 812-926-0087. Fax: 812-926-1075.
E-mail: hillforest@embarqmail.com
Web Site: www.hillforest.org
Founded: 1956.
Congressional District: 9
Key Personnel: Pres. (V), Richard Strzynski; Dir., Cindy Schuette; Volunteer Coord., Suzanne Ullrich.
Personnel Profile: Full-Time Paid 1; Part-Time Paid 3; Part-Time Volunteers 30.
Governing Authority: nonprofit organization. Parent Institution: Hillforest Historical Foundation, Inc. Tax-exempt: 501(c)(3).

Institution Type/Description: Victorian House Museum: 1852-91 Hillforest, Victorian Ohio River Valley mansion.
Collections: household furnishings & cultural items; archival materials; genealogical information.
Research Fields: regional & local history; local industry; Thomas Gaff & family.
Facilities: food service. Museum-related items for sale.
Activities: guided tours; lectures; docent program; historical & cultural programs; formally organized education programs for docents; educational outreach programs; youth day camps.
Publications: quarterly newsletter; annual calendar of events; brochure; cookbook; souvenir booklet; annual report.
Hours & Admission Prices: April-Dec. Tues.-Sun. 1-5. Adults $5, students 7-13 $3; discounts to groups & AAA members; children 6 & under and members no charge. Closed major holidays. &
Attendance: 3,000 (estimated)
Membership: Junior Historian $5; Individual $25; Family $40; Contributor $60; Sponsor $125; Patron $300; Benefactor $500; Life $5,000.

Batesville

BATESVILLE AREA HISTORICAL SOCIETY MUSEUM, 15 W. George St., Batesville, IN 47006. Tel.: 812-932-0999.
E-mail: bahs@etczone.com
Web Site: batesvilleareahistoricalsociety.org
Founded: 1999.
Congressional District: 9
Key Personnel: Pres. (V), Jean Struewing
Governing Authority: Tax-exempt.
Institution Type/Description: Historical Society Museum.
Collections: local history & culture; period furnishings; period artifacts; photographs.
Hours & Admission Prices: April - Sept: Tues. & Thurs. 9-3, Fri.-Sat. 9 to noon; Oct - March: Thurs 9-3, Sat 9 to noon. No charge, donations accepted &
Attendance: 1,104 (estimated)
Membership: $15 per year.

Battle Ground

HISTORIC PROPHETSTOWN, 3549 Prophetstown Tr., Battle Ground, IN 47920-7018. Mailing Address: P.O. Box 331, Battle Ground, IN 47920-0331. Tel.: 765-567-4700. Fax: 765-567-4736.
E-mail: reservations@prophetstown.org
Web Site: www.prophetstown.org
Formerly: The Museums At Prophetstown
Founded: 1996.
Congressional District: 7
Key Personnel: C.O.O., Dris Abraham; Pres., Robert Kennedy.
Personnel Profile: Full-Time Paid 3; Part-Time Paid 6; Part-Time Volunteers 20.
Governing Authority: bd. of directors. Tax-exempt: 501(c)(3).
Institution Type/Description: Agriculture Museum.
Collections: agriculture & horse powered farming.
Research Fields: agriculture & lifestyles.
Activities: field trips programs; supper programs; educational programs; draft horse clinics. Special Events: Plow Days; Planting Days; Wheat Harvest; Threshing Show; Corn Husking; Country Fair; Barn Dance.
Publications: newsletter, Prophetstown Dispatch.
Hours & Admission Prices: April-Oct. daily 8-5; Nov.-March Mon.-Fri. 9-4. State Park entry fee: cars $6; members no charge. &
Attendance: 35,000 (estimated)
Membership: Individual $45; Family $60.

TIPPECANOE BATTLEFIELD MUSEUM & PARK, 200 Battle Ground Ave., Battle Ground, IN 47920-7026. Mailing Address: c/o Tippecanoe County Hist. Assoc., 1001 South St., Lafayette, IN 47901. Tel.: 765-567-2147. Fax: 765-567-2149.
Web Site: www.tippecanoehistory.org
Founded: 1972.
Congressional District: 7
Key Personnel: Dir., Kathy Atwell; Museum Shop Mgr., Rick Conwell.
Personnel Profile: Full-Time Paid 1; Part-Time Paid 1; Part-Time Volunteers 6.
Governing Authority: private; nonprofit. Parent Institution: Tippecanoe County Historical Association. Tax-exempt: 501(c)(3).
Institution Type/Description: Park Museum & Visitor Center, site of Nov. 7, 1811 Battle of Tippecanoe. National Historic Site.
Collections: material pertaining to the Battle of Tippecanoe & the election of 1840; prehistoric Native artifacts.
Research Fields: Wabash Valley; United Methodist Church history, re: frontier

era, 1850 active campground years; Battle of Tippecanoe; Tecumseh; Wm. H. Harrison; Prophetstown.
Facilities: retreat center; chapel; picnic grounds. Gifts & books for sale.
Activities: guided tours; lectures; education programs for children; living history interpretation.
Publications: newsletter; history & nature trail brochures.
Hours & Admission Prices: Thurs.-Tues. 10-5, call for winter hours. Adults $5, seniors, AAM & AAA members $4, children & students $2; members no charge. Closed New Year's Day; Easter; Thanksgiving; Christmas. &
Attendance: 35,000 (accurate)
Membership: Senior Individual $30; Individual $35; Senior Couple $40; Family $50; Patron $100; Sustaining $250; Treasure $500.

TIPPECANOE COUNTY HISTORICAL ASSOCIATION, 200 Battle Ground Ave., Battle Ground, IN 47920. Mailing Address: 1001 South St., Lafayette, IN 47901-1571. Tel.: 765-476-8411. Fax: 765-476-8414.
Web Site: www.tcha.mus.in.us
Formerly: Tippecanoe County Historical Museum
Founded: 1925.
Congressional District: 2
Personnel Profile: Full-Time Paid 2; Part-Time Paid 4; Part-Time Volunteers 20; Interns 1.
Governing Authority: society. Branch Museums: Battle of Tippecanoe Museum; Fort Ouiatenon; McCollough Archives & Research Center. Tax-exempt: 501(c)(3).
Institution Type/Description: House Museum.
Collections: pre-contact Native American artifacts & archaeology collections from the sites of Fort Ouiatenon & the Battle of Tippecanoe; personal artifacts including manuscripts, household goods & toys; agricultural, commercial, industrial, educational, social & military history of the county & region; newspapers; artworks & manuscripts by Indiana artist George Winter; local documents, photographs, negatives; county governmental records.
Research Fields: genealogy; 18th-century artifacts; local history; French Colonial archaeology; Tippecanoe Battlefield archaeology.
Facilities: 4,000-vol. library of books on history of Tippecanoe County & Indiana available in association's library across the street. Books & handcrafted items for sale.
Publications: newsletter, TCHA News; leaflet series, Tippecanoe Tales; Books, Tippecanoe County Cartoonist; Historical Map of Tippecanoe County; Battle of Tippecanoe: Conflict of Cultures; Ouabache Potpourri; 100 Years of the Courthouse; The House That Moses Fowler Built; 1894 Bird's Eye View, Lafayette; Old Lafayette, 1825-1854; Old Lafayette, Vol. II, 1854-1876; Grist Mills of Tippecanoe County.
Hours & Admission Prices: Temporarily closed. &
Attendance: 150,000 (estimated)
Membership: Basic $35; Sustaining $250.

Bedford

LAWRENCE COUNTY HISTORICAL AND GENEALOGICAL SOCIETY, 929 15th St., Bedford, IN 47421-3813. Tel.: 812-278-8575. Fax: 812-278-8583.
E-mail: lchgs@hpcisp.com
Web Site: www.lawrencecountyhistory.org
Founded: 1928.
Congressional District: 9
Key Personnel: Pres., Rowena Cross-Najafi; Dir. Library, Joyce Shepherd; Museum Shop Mgr., Kenneth White.
Personnel Profile: Full-Time Paid 2; Part-Time Paid 1; Part-Time Volunteers 45.
Governing Authority: county. Tax-exempt.
Institution Type/Description: General Museum.
Collections: pioneer items; Civil War artifacts; genealogy; books & literature on Lawrence County.
Research Fields: genealogy; local history.
Facilities: research library; meeting room. Museum-related items for sale.
Activities: guided tours; lectures; films; audiovisual programs; formally organized education programs for children. Museum Sponsors: Duke Energy World of Discovery program for children.
Publications: newsletter; quarterly, Seedling Patch.
Hours & Admission Prices: Tues.-Fri. 9-4, Sat. 9-3. No charge; donations accepted. &
Attendance: 3,500 (accurate)
Membership: Student $5; Individual $15; Couple $20; Family $25; Sustaining $100-$249; Contributor $250-$499; Patron $500.

Berne

SWISS HERITAGE VILLAGE & MUSEUM, 1200 Swiss Way, Berne, IN 46711. Mailing Address: P.O. Box 88, Berne, IN 46711-0088. Tel.: 260-589-8007. Facebook: Swiss Heritage Village & Museum.
E-mail: debbyn@swissheritage.org
Web Site: www.swissheritage.org
Founded: 1985.
Key Personnel: Exec. Dir., Debby Neuenschwander
Institution Type/Description: Historic Buildings: late 1800s & early 1900s village.
Collections: local history, culture & heritage; period furnishings; photographs.
Activities: educational programs.
Hours & Admission Prices: Call for hours. Adults $6, seniors 65 & over $5, children 6-18 $3; children under 6 no charge.

Beverly Shores

BEVERLY SHORES HISTORY MUSEUM & ART GALLERY, INC., 525 Broadway, Beverly Shores, IN 46301. Mailing Address: P.O. Box 305, Beverly Shores, IN 46301-0305. Tel.: 219-878-1517.
E-mail: nichols-nancy@comcast.net
Founded: 1998.
Key Personnel: Pres. (V), Nancy Nichols.
Personnel Profile: Part-Time Volunteers 1.
Governing Authority: Tax-exempt.
Institution Type/Description: History Museum.
Collections: local history & culture; photographs; period furnishings; personal artifacts; paintings; maps.
Major Exhibits: Beginnings, 5/14-6/14; The Work of the Shirley Heinze Land Trust, 7/14-8/14; Century of Progress Homes, 9/14-10/14.
Facilities: Museum-related items for sale.
Hours & Admission Prices: May-Oct. Fri.-Sun. 11-3 & Mon. Memorial Day-Labor Day 11-3. No charge; donations accepted. &
Attendance: 1,000 (estimated)

Bloomington

GRUNWALD GALLERY OF ART, INDIANA UNIVERSITY, (M), 1201 E. 7th St., Bloomington, IN 47405-5501. Tel.: 812-855-8490. Fax: 812-855-7498.
E-mail: grunwald@indiana.edu
Web Site: www.indiana.edu/~grunwald
Formerly: School of Fine Arts Gallery, Indiana University
Founded: 1987.
Congressional District: 8
Key Personnel: Dir., Betsy Stirratt.
Personnel Profile: Full-Time Paid 2; Part-Time Paid 10; Part-Time Volunteers 2.
Governing Authority: public university; nonprofit. Tax-exempt: 170(b)(1)(A).
Institution Type/Description: Art Gallery.
Collections: paintings; sculpture; photographs.
Activities: national & regional artist exhibitions; student exhibitions.
Publications: exhibition catalogs
Hours & Admission Prices: Tues.-Sat. 12-4. No charge; donations accepted. Closed spring break; Thanksgiving; Christmas. &
Attendance: 20,000 (estimated)
Membership: Student $35; Basic $100; FAB $500; Deluxe $1,000; Primo $5,000.

✻ INDIANA UNIVERSITY ART MUSEUM, 1133 E. Seventh St., Bloomington, IN 47405-7509. Tel.: 812-855-5445. Fax: 812-855-1023.
E-mail: iuam@indiana.edu
Web Site: www.artmuseum.iu.edu
Founded: 1941.
Congressional District: 7
Key Personnel: Dir., Adelheid M. Gealt; Chm. (V), Tony Morarec; Assoc. Dir. Editorial Svcs., Linda Baden; Assoc. Dir. Devel., Jeremy Hatch; Assoc. Dir. Curatorial Svcs., Diane Pelrine; Assoc. Dir. Administration, David Tanner; Museum Shop Mgr., Murat Candiler.
Personnel Profile: Full-Time Paid 38; Part-Time Paid 46; Part-Time Volunteers 64; Interns 10.
Governing Authority: state. Parent Institution: Indiana University. Tax-exempt: 501(c)(3).
Institution Type/Description: Art Museum.
Collections: early Egyptian, Greek, Roman sculpture; vases; jewelry; coins & glass; 14th through 21st-century European & American paintings, sculpture, prints, drawings, photography, decorative arts; African, Oceanic, pre-Columbian, Japanese, Chinese, S.E. Asian paintings, sculpture, prints, ceramics, decorative arts.
Research Fields: ancient art; African art; Renaissance decorative arts; modernism; German expressionism; American regionalism.
Facilities: library; laboratories; cafe. Museum-related items for sale.
Activities: inter-museum loan, permanent, temporary & traveling exhibitions; gallery talks; museum education tours by arrangement; concerts.
Publications: exhibition catalogs; guide to the collections; occasional papers; annual reports; monthly calendars.
Hours & Admission Prices: Tues.-Sat. 10-5, Sun. 12-5. No charge; donations accepted. Closed New Year's Day; Memorial Day; Independence Day; Labor Day; Thanksgiving; Christmas. &
Attendance: 40,883 (accurate)

THE KINSEY INSTITUTE GALLERY, Morrison Hall, Rm. 313, 1165 E. Third St., Bloomington, IN 47405. Tel.: 812-855-7686. Fax: 812-855-8277.
E-mail: kinsey@indiana.edu
Web Site: www.kinseyinstitute.org
Founded: 1990.
Governing Authority: Parent Institute: Indiana University. Tax-exempt: 501(c)(3).
Institution Type/Description: Health Museum.
Collections: sexual health; gender; reproduction; photographs; paintings; artifacts.
Research Fields: sexual health & behavior.
Facilities: library.
Activities: guided & self-guided tours; permanent & temporary exhibitions; special events; lectures.
Publications: newsletter.
Hours & Admission Prices: Mon.-Fri. 1:30-5; other times by appointment. Children under 18 not admitted without parent. No charge. Closed major holidays. &
Attendance: 4,000 (estimated)

✻ MATHERS MUSEUM OF WORLD CULTURES, (M), 416 N. Indiana Ave., Bloomington, IN 47408-3742. Tel.: 812-855-6873. Fax: 812-855-0205.
E-mail: mathers@indiana.edu
Web Site: www.mathers.indiana.edu
Formerly: William Hammond Mathers Museum
Founded: 1963.
Congressional District: 9
Key Personnel: Asst. Dir., Judith Kirk; Registrar, Theresa Harley-Wilson; Conservator, Judith Sylvester; Business Mgr. & Museum Shop Mgr., Sandra Warren; Cur. Collections, Ellen Sieber; Cur. Education, Sarah Hatcher; Exhibits Co-Cur., Elaine Gaul; Security Coord., Kelly Wherley; Exhibits Co-Cur., Matthew Sieber.
Personnel Profile: Full-Time Paid 8; Part-Time Paid 2; Part-Time Volunteers 30.
Governing Authority: Indiana University. Tax-exempt: 170(b)(1)(A).
Institution Type/Description: General Museum.
Collections: archaeological; ethnological; historical collections from North America, Latin America, Europe, Africa, Asia & Oceania.
Research Fields: anthropology; folklore & history.
Facilities: 1,000-vol. library relating to collections available on premises; classrooms.
Activities: guided tours; lectures; films; gallery talks; docent program; discovery kits; formally organized education programs for undergraduate & graduate students affiliated with Indiana University; rotating exhibits.
Publications: Wm. Hammond Mathers Museum Annual Report; Monograph Series: Occasional Papers; Interpreting Our Past; Photographs as Research Documents.
Hours & Admission Prices: Tues.-Fri. 9-4:30, Sat.-Sun. 1-4:30. No charge; donations accepted. Closed national holidays. &
Attendance: 35,000 (estimated)

MONROE COUNTY HISTORICAL SOCIETY, (M), 202 E. Sixth St., Bloomington, IN 47408-3518. Tel.: 812-332-2517. Fax: 812-355-5593. Facebook: Monroe County History Center.
E-mail: director@monroehistory.org
Web Site: www.monroehistory.org
Founded: 1980.
Congressional District: 8
Key Personnel: Dir., Helmut Hentschel; Pres. (V), Laura Newton; Cur., Rebecca Vaughn; Asst. Dir., Hillary Detty; Education, Angi St. Clair; Museum Shop Mgr., Mary Lee Deckard.

Personnel Profile: Full-Time Paid 6; Part-Time Paid 2; Part-Time Volunteers 100.
Volunteer Hours: 4,951
Operating Expenses: 357,717
Operating Income: 376,906
Governing Authority: society; nonprofit organization. Parent Institution: Monroe County Historical Society. Tax-exempt.
Institution Type/Description: Historical Society & History Museum: housed in the former Carnegie Library Building.
Collections: objects & documents related to the history of Monroe County; natural history specimens; tools; clothing; items relating to limestone & local industries; household items.
Research Fields: local history; material culture; folklore; limestone industry; architectural history.
Facilities: genealogy library. Gift & museum-items for sale.
Activities: guided tours; docent program; lectures; workshops; outreach program; permanent & temporary exhibitions.
Publications: newsletter; books; local history & cemetery records.
Hours & Admission Prices: Tues.-Sat. 10-4. Adults $2, children 6-18 $1; discounts to AAM, AAA & AASLH members; members no charge. Closed major holidays. &
Attendance: 12,000 (accurate)
Membership: Basic $35; Family $60; Sustaining $100; Contributing $250; Patron $500; Corporate: Level 1 $100-$249; Level 2 $250-$499; Level 3 $500-$999; Level 4 $1,000 & up.

THE SAGE COLLECTION, 2805 E. 10th St., Ste. 140, Bloomington, IN 47405. Tel.: 812-855-4627. Fax: 812-855-4627.
E-mail: ksrichar@indiana.edu
Web Site: www.indiana.edu/~sagecoll/index.html
Formerly: Elizabeth Sage Historic Costume Collection
Founded: 1935.
Congressional District: 7
Key Personnel: Exec. Dir., Nelda M. Christ; Cur., Kathleen L. Rowold; Asst. Cur., Kelly Richardson.
Personnel Profile: Full-Time Paid 1; Part-Time Paid 1; Part-Time Volunteers 4; Interns 1.
Governing Authority: university. Parent Institution: Indiana University. Tax-exempt: 170(b)(1)(A).
Institution Type/Description: Textile & Costume Museum.
Collections: 19th & 20th-century Western European & American clothing; accessories.
Research Fields: 19th & 20th-century fashions; systematic methods of categorizing & identification of clothing items; by appointment only.
Facilities: 100-vol. library pertaining to fashion plates & advertising material related to 19th & 20th-century fashion available for research on premises by appointment only.
Activities: research & tours of storage by appointment only; off-site lectures & films; formally organized education program for graduate & undergraduate college students; temporary exhibits.
Publications: semiannual newsletter, Historic Costume News; brochure.
Hours & Admission Prices: By appointment only. No charge; donations accepted.
Membership: Friends of Elizabeth Sage: Associate Annual $25; Patron Annual $50; Contributor Annual $100; Life Single $500; Memorial Single $1,000; Corporate $1,000 & up.

WONDERLAB MUSEUM OF SCIENCE, HEALTH AND TECHNOLOGY, 308 W. Fourth St., Bloomington, IN 47404-5120. Tel.: 812-337-1337. Fax: 812-330-1337. Facebook: www.facebook.com/WonderLab.Museum.
E-mail: writeus@wonderlab.org
Web Site: www.wonderlab.org/
Founded: 1995.
Congressional District: 9
Key Personnel: Dir., Catherine Olmer; Volunteer Dir., Jeanne Gunning; Museum Shop Mgr., Colleen Couper.
Personnel Profile: Full-Time Paid 10; Part-Time Paid 10; Part-Time Volunteers 968.
Governing Authority: private; nonprofit organization. Tax-exempt: 501(c)(3).
Institution Type/Description: Children's Museum.
Collections: hands-on exhibits.
Facilities: Museum-related items for sale.
Activities: special events.
Hours & Admission Prices: Tues.-Sat. 9:30-5, Sun. 1-5, 1st Fri. of month 9:30-8:30. General admission $7; discounts to groups of 10 or more with reservation; children under one, members & ASTC Passport members no charge. &
Attendance: 89,000 (accurate)

Membership: Dual $59; Basic Family $93; Grandparent $112; Premier Family $132.

WYLIE HOUSE MUSEUM, 307 E. 2nd St., Bloomington, IN 47401-4799. Mailing Address: 317 E. 2nd St., Bloomington, IN 47401-4701. Tel.: 812-855-6224.
E-mail: libwylie@indiana.edu
Web Site: www.indiana.edu/~libwylie
Founded: 1960.
Key Personnel: Dir., Jo Burgess; Outdoor Interpreter, Sherry Wise; Cur. Education, Bridget Edwards.
Personnel Profile: Full-Time Paid 3; Part-Time Volunteers 12; Interns 1.
Governing Authority: public university; nonprofit. Parent Institute: Indiana University. Tax-exempt: 501(c)(3).
Institution Type/Description: Historic House Museum: housed in 1835 home of Indiana University's first president.
Collections: a few original Wylie furnishings; personal artifacts; garden containing 19th-century varieties of flowers, herbs & vegetables which are raised for seeds to sell.
Research Fields: daily life activities & furnishing of 1840s; Andrew Wylie biography; Theophilus Wylie Family.
Facilities: 175-vol. library of books available for use on premises; botanical garden; 4,000 sq. ft. exhibit space. Museum-related items for sale.
Activities: concerts; docent program; guided tours. Annual Events: Spring Seed Sale in March; Candlelight Tour in December.
Publications: semiannual newsletter, Wylie House.
Hours & Admission Prices: March-Nov. Tues.-Sat. 10-2. No charge; donations accepted. Closed major holidays.
Attendance: 2,000 (accurate)

Bluffton

WELLS COUNTY HISTORICAL MUSEUM, 420 W. Market St., Bluffton, IN 46714. Mailing Address: P.O. Box 143, Bluffton, IN 46714-0143. Tel.: 260-824-9956.
E-mail: jcsturgeon@adamswells.com
Web Site: www.wchs-museum.org
Founded: 1935.
Congressional District: 4
Key Personnel: Pres., James Sturgeon; Vice Pres., Connie Brubaker; Sec., Marcia Hotopp; Treas., Greg Waters.
Governing Authority: society. Parent Institution: Wells County Historical Society. Tax-exempt.
Institution Type/Description: Local History Museum.
Collections: items pertaining to Wells County history, culture & industry.
Research Fields: County site & structure survey; genealogy; family history.
Facilities: library of material on Wells County history available for use on premises; reading room.
Activities: guided tours; permanent exhibitions.
Publications: quarterly newsletters; Wells County History; Architectural Atlas of Wells County, Indiana; Wells County, Indiana - A Pictorial History 1999; Wells County Towns & Townships.
Hours & Admission Prices: April-Oct. Sun. 1-4; June-Aug. Wed. 1-4. No charge; donations accepted.
Attendance: 610 (estimated)
Membership: Student $5; Individual $10; Family $15; Patron $25; Life $150.

Boonville

WARRICK COUNTY MUSEUM, INC., 217 S. First St., Boonville, IN 47601-1701. Tel.: 812-897-3100. Fax: 812-897-6104.
E-mail: wcmuseum@aol.com
Web Site: www.warrickcountymuseum.org
Founded: 1976.
Congressional District: 8
Key Personnel: Pres., Connie Barnhill; Treas., Jeffrey Byrne; Sec., Colleen Talley.
Personnel Profile: Part-Time Volunteers 15.
Governing Authority: private; nonprofit. Tax-exempt.
Institution Type/Description: History Museum: located in 1901 Ella Williams School, in historic district.
Collections: life in Warrick County from the early 1800s.
Research Fields: Black Heritage underground railroad (Freedom Trails) sites.
Facilities: 5,000 sq. ft. exhibit space.
Activities: guided tours; temporary & loan exhibitions. Annual Events: Boo in Boonville; Christmas in Boon-Village.
Publications: Warrick County Sites, Scenes and Citizens; Warrick County Atlases 1880 and 1899; The Keith Murder Trial; Gold Fillings.
Hours & Admission Prices: March-Dec. Tues.-Wed. 1-4; tours by appointment. Adults $2. Closed holidays. &

Attendance: 1,250 (accurate)
Membership: Student & Senior Citizen $10; Individual $15; Family $30; Business $50; Hemenway $100; Lincoln $250; Hoover $500; Gold $1,000.

Bristol

BONNEYVILLE MILL, 53373 County Rd. 131, Bristol, IN 46507. Mailing Address: 211 W. Lincoln Ave., Goshen, IN 46526-3218. Tel.: 574-825-9324.
E-mail: info@elkhartcountyparks.org
Web　　　　　　　　　　　　　　　　　　　　　　　　　　*Site:* www.elkhartcountyparks.org/properties_locations/bonneyville_mill.htm
Institution Type/Description: History Museum: housed in mid-1830s mill.
Collections: working mill grinds corn, wheat, rye & buckwheat.
Facilities: Museum-related items for sale.
Hours & Admission Prices: May-Oct. daily 10-5; guided tours by appointment. No charge.

ELKHART COUNTY HISTORICAL MUSEUM, (M), 304 W. Vistula St. (St. Rd. 120), Bristol, IN 46507. Mailing Address: P.O. Box 434, Bristol, IN 46507. Tel.: 574-848-4322. Fax: 574-848-5703. TDD: 574-535-6420.
E-mail: museum@elkhartcountyparks.org
Web Site: elkhartcountyhistory.org
Founded: 1896.
Congressional District: 3
Key Personnel: Mgr., Matthew Schuld; Park Dir., Larry Neff.
Personnel Profile: Full-Time Paid 4; Part-Time Paid 2; Part-Time Volunteers 30.
Governing Authority: county. Parent Institution: Elkhart County Park Dept. Subsidiary Institution: Elkhart County Historical Society. Tax-exempt.
Institution Type/Description: General Museum: 1st consolidated school in Elkhart County; listed on the National Register.
Collections: agriculture; railroad history; home furnishings; apparel; one-room school; military artifacts & Native American artifacts.
Research Fields: Elkhart County history & genealogy.
Facilities: 1,500-vol. library of books; archives.
Activities: guided tours; lectures. Historical Society Programs: A Crystal Ball.
Publications: newsletter.
Hours & Admission Prices: Feb.-Dec. 1 Tues.-Fri. 10-4, Sun. 1-5. No charge; donations accepted. Closed federal holidays. &
Attendance: 5,500 (estimated)
Membership: Individual $10; Family $15; Lifetime Individual $150.

Brook

GEORGE ADE MEMORIAL ASSOCIATION INC., Hwy. #16, Brook, IN 47922. Mailing Address: P.O. Box 221, Brook, IN 47922-0221. Tel.: 219-275-0895.
Founded: 1961.
Congressional District: 5
Key Personnel: Cur., Richard Gerts.
Governing Authority: nonprofit organization. Tax-exempt.
Institution Type/Description: Historic House Museum: home of George Ade, humorist, author & playwright.
Collections: objects of art & interest collected by George Ade during his trips at the turn of the century; furniture; mementos.
Activities: guided tours; permanent exhibitions; social functions.
Hours & Admission Prices: Call for hours. No charge; donations accepted. &

Brookville

FRANKLIN COUNTY SEMINARY AND MUSEUM, 5th & Mill St., Brookville, IN 47012. Mailing Address: P.O. Box 342, Brookville, IN 47012-0342. Tel.: 765-647-5182.
E-mail: beneker@verizon.net
Founded: 1969.
Congressional District: 9
Key Personnel: Pres., Franklin County Historical Society, Pamela Beneker; Museum Shop Mgr., Martha Shea.
Personnel Profile: Part-Time Volunteers 5.
Governing Authority: nonprofit organization. Parent Institution: The Franklin County Historical Society. Tax-exempt: 170(b)(1)(A).
Institution Type/Description: Local History Museum: housed in c.1828-30 original Franklin County Seminary building.
Collections: records & artifacts of early Franklin County. Historic Houses: 1812 Little Cedar Grove Baptist Church; 1820-21 Old Brick Church & cemetery.
Hours & Admission Prices: By appointment only. No charge; donations accepted.

Attendance: 250 (estimated)
Membership: Society $8.

Buckskin

HENAGER ”MEMORIES AND NOSTALGIA“ MUSEUM AND NATIONAL VETERANS MEMORIAL, Hwy. 57, Buckskin, IN 47613. Mailing Address: 8837 S. State Rd. 57, Elberfeld, IN 47613-8445. Tel.: 812-795-2230 & 2237. Facebook: Henager Museum.
E-mail: nationalveteransmemorial@gmail.com
Web Site: www.henagermuseum.com; www.nationalveteransmemorial.org
Founded: 1996.
Congressional District: 8
Key Personnel: Chm. (V), James G. Henager.
Governing Authority: Tax-exempt.
Institution Type/Description: History & Military Museum: the National Veterans Memorial headquarters.
Collections: pop culture; WWI & II, Korean and Vietnam veterans; Roy Rodgers; the Beatles; Smokey Bear history & collectibles; woodworking art & history; Santa's workshop; scouting; automotive heritage; military artifacts including flags, mess kits, medals, headgear & photographs; veterans memorabilia; TV movies; heroes & legends; American music; veterans memorial.
Hours & Admission Prices: June-Aug. Mon.-Thurs. 8-7, Fri. 8-5, Sat. 8-4; Sept.-May Mon.-Fri. 8-5, Sat. 8-4. Adults $6, children $3; discounts to members, OLSR members & AAA members.
Membership: Individual $25; Family $65.

Cambridge City

HUDDLESTON FARMHOUSE MUSEUM, 838 National Rd., Mt. Auburn, Cambridge City, IN 47327. Mailing Address: P.O. Box 284, Cambridge City, IN 47327-0284. Tel.: 765-478-3172. Fax: 765-478-3410.
E-mail: huddleston@historiclandmarks.org
Web Site: www.historiclandmarks.org
Founded: 1977.
Congressional District: 2
Key Personnel: Museum Dir., Karen Trent; Pres., Marsh Davis, Program Assoc., James Orr.
Personnel Profile: Part-Time Paid 2; Part-Time Volunteers 12.
Governing Authority: nonprofit organization. Parent Institution: Historic Landmarks Foundation of Indiana, 340 W. Michigan, Indianapolis, IN 46202. Tax-exempt: 501(c)(3).
Institution Type/Description: Historic Building: c.1840 federal style brick, 3-story farmhouse.
Collections: farmhouse; outbuildings complex; 1840-1860 period furnishings; farm implements; Huddleston family records; history of travelers & residents along National Road in Indiana; Quaker history.
Research Fields: architectural history; local history; preservation technology; National Road history; middle 19th-century life.
Facilities: 400-vol. preservation resource library; architectural slide & photo collection (eastern Indiana); meeting facilities.
Activities: tours daily; seasonal special events; school programs; group tours; summer camps for children; hearth dinners; lectures; Civil War encampment.
Hours & Admission Prices: Call for hours. &
Attendance: 3,633 (accurate)
Membership: Active $20; Contributing & Nonprofit Organization $30; Sponsoring $50; Sustaining $100; Fellow $250; Patron $500; Landmark Member $1,000.

MUSEUM OF OVERBECK ART POTTERY, Cambridge City Public Library, 600 W. Main St., Cambridge City, IN 47327-1117. Tel.: 765-478-3335. Fax: 765-478-6144.
E-mail: ccitypl@yahoo.com
Web Site: www.cclib.lib.in.us
Key Personnel: Dir., Jill King; Dir. Library, Vicki Molok.
Personnel Profile: Part-Time Volunteers 1.
Volunteer Hours: 200
Governing Authority: bd. of directors. Parent Institution: Cambridge City Public Library. Tax-exempt.
Institution Type/Description: Art Museum.
Collections: Overbeck sisters' art.
Hours & Admission Prices: Mon.-Sat. 10-12 & 2-5. No charge; donations accepted. &
Attendance: 100 (estimated)

Carmel

CARMEL CLAY HISTORICAL SOCIETY - THE MONON DEPOT MUSEUM, 211 First St., S.W., Carmel, IN 46032. Tel.: 317-846-7117.
E-mail: carmelclayhistory@yahoo.com
Web Site: www.carmelclayhistory.org
Founded: 1975.
Key Personnel: Dir., Katherine Dill; Pres. (V), Jennifer Hershberger; Museum Shop Mgr., Susan Bock.
Personnel Profile: Part-Time Paid 1; Interns 1.
Governing Authority: Tax-exempt.
Institution Type/Description: Historical Society Museum: depot built in 1883.
Collections: local history & culture; personal artifacts; photographs; railroad artifacts.
Activities: scout programs; public programs; children's activities; educational programs.
Hours & Admission Prices: Museum: April-Oct. Fri.-Sat. 1-4, Sun. 2-4. Archives: Tues.-Wed. 8:30-3. No charge; donations accepted. &
Attendance: 10,000 (estimated)
Membership: Regular $25; Family $40; Life $500.

EVAN LURIE FINE ART GALLERY, 30 W. Main St., Carmel, IN 46032. Tel.: 317-844-8400. Fax: 317-844-8460.
E-mail: info@evanluriegallery.com
Key Personnel: Dir., Evan Lurie
Institution Type/Description: Art Gallery.
Collections: paintings; sculpture; photographs.
Activities: temporary exhibitions.
Hours & Admission Prices: Mon.-Fri. 11-5, Sat.-Sun. 12-4.

THE MUSEUM OF MINIATURE HOUSES AND OTHER COL-LECTIONS, INC., (M), 111 E. Main St., Carmel, IN 46032-1823. Tel.: 317-575-9466. Fax: 317-575-0240.
E-mail: mmhaoc@aol.com
Web Site: www.museumofminiatures.org
Founded: 1991.
Key Personnel: Dir., Suzanne Moffett; Treas. & Museum Shop Mgr., Suzanne H. Landshof.
Personnel Profile: Full-Time Volunteers 3; Part-Time Paid 1; Part-Time Volunteers 36.
Governing Authority: private; nonprofit organization. Tax-exempt: 501(c)(3).
Institution Type/Description: Decorative Arts Museum.
Collections: miniatures; dollhouses.
Facilities: library. Museum-related items for sale.
Activities: guided tours; hobby workshops; special events.
Hours & Admission Prices: Wed.-Sat. 11-4, Sun. 1-4. Adults $5, children under 10 $3; discounts to AAM members; members no charge. Closed New Year's Day; Easter; Memorial Day; Independence Day; Labor Day; Thanksgiving; Christmas. &
Attendance: 6,200 (accurate)
Membership: Friend $20; Family $35; Patron $100; Donor $250; Benefactor $1,000.

NATIONAL ASSOCIATION OF MINIATURE ENTHUSIASTS, 130 N. Rangeline Rd., Carmel, IN 46032-1743. Mailing Address: P.O. Box 69, Carmel, IN 46082-0069. Tel.: 317-571-8094. Fax: 317-571-8105.
E-mail: name@miniatures.org
Web Site: www.miniatures.org
Founded: 1972.
Congressional District: 10
Key Personnel: Pres., Karen Barone; Program Coord., Toni Cochran.
Personnel Profile: Full-Time Paid 3; Part-Time Paid 3.
Governing Authority: private; nonprofit organization.
Institution Type/Description: Arts & Crafts/Hobby Museum.
Collections: miniatures.
Facilities: 110-vol. library of NAME publications, videos & slide programs; 1,000 sq. ft. exhibit space. Logo merchandise & NAME items for sale.
Activities: guided tours; hobby workshops; loan & participatory exhibits; arts festivals; study clubs. Museum Sponsors: national convention; annual business meeting; regional conventions.
Publications: quarterly, Miniature Gazette.
Hours & Admission Prices: Mon.-Fri. 9-4, Sat.-Sun. by appointment only. No charge; donations accepted. Closed New Year's Eve & Day; Labor Day; Thanksgiving & day after; Christmas Eve, Day & week. &
Attendance: 500 (estimated)
Membership: Additional Family Members $10; Individual $25; Foreign $30; Shop Membership $100.

Cedar Lake

LAKE OF THE RED CEDARS MUSEUM, 7408 Constitution Ave., Cedar Lake, IN 46303-9186. Mailing Address: P.O. Box 421, Cedar Lake, IN 46303-0421. Tel.: 219-374-6157. Fax: 219-374-6157.
Web Site: cedarlakehistory.org
Founded: 1977.
Congressional District: 1
Key Personnel: C.E.O. & Pres. (V), James Laud; Dir., Anne Zimmerman.
Personnel Profile: Part-Time Volunteers 20.
Governing Authority: nonprofit organization. Parent Institution: Cedar Lake Historical Association Inc. Tax-exempt: 501(c)(3).
Institution Type/Description: Local History Museum: housed in 1920s 60-room hotel.
Collections: Cedar Lake history & the surrounding area; Chicago, IL history as it pertains to Cedar Lake history; prehistoric-modern artifacts; domestic collections; photographs; industrial artifacts; farming artifacts; archives.
Research Fields: ice industry; Mr. Samuel Bartlett.
Facilities: 8,150 sq. ft. exhibit space. Gift items for sale.
Activities: docent program; films; guided tours; hobby workshops; lectures; temporary & participatory exhibits. Museum Sponsors: Ice Cream Social; Dinner in October.
Publications: monthly newsletter.
Hours & Admission Prices: May-Oct. Thurs.-Sun. 1-4; other times by special arrangement for groups. Adults $2, children $1. &
Attendance: 1,000 (estimated)
Membership: Senior & Student $5; Senior Couple $7; Educator $8; Adult $10; Family $20; Patron & Sustaining $25; Benefactor $100.

Charlestown

THOMAS DOWNS HOUSE, 1045 Main St., Charlestown, IN 47111-1223. Mailing Address: 8524 State Rd. 403, Charlestown, IN 47111. Tel.: 812-256-5777.
E-mail: hart2827@sbcglobal.net
Congressional District: 9
Key Personnel: Cur., Donna Hart; Dir., Chm. (V) & Museum Shop Mgr., Sue Koetter.
Governing Authority: Parent Institution: Clark's Grant Historical Society. Tax-exempt.
Institution Type/Description: History Museum.
Collections: period furnishings.
Hours & Admission Prices: Call for hours. No charge; donations accepted.

Chesterton

CHESTERTON ART CENTER, 115 S. 4th St., Chesterton, IN 46304-2344. Tel.: 219-926-4711.
E-mail: gallery@chestertonart.com
Web Site: www.chestertonart.com
Founded: 1960.
Key Personnel: Dir., Judy Gregurich; Chm. (V), Joni Clem; Pres. (V), John Mullin.
Personnel Profile: Part-Time Paid 3; Part-Time Volunteers 100.
Governing Authority: Parent Institution: Association of Artists & Craftsmen of Porter County. Tax-exempt.
Institution Type/Description: Art Gallery.
Collections: works by regional artists.
Activities: classes; special events; art fair.
Publications: monthly newsletter.
Hours & Admission Prices: Mon.-Fri. 11-4, Sat.-Sun. 1-4. No charge; donations accepted. &
Attendance: 3,000 (estimated)
Membership: Individual $25; Family $35; Patron $50; Benefactor $100; Sponsor $1,000.

WESTCHESTER TOWNSHIP HISTORY MUSEUM, 700 W. Porter Ave., Chesterton, IN 46304-2205. Tel.: 219-983-9715. Fax: 219-926-6424.
E-mail: museum@wpl.lib.in.us
Web Site: www.wpl.lib.in.us/museum
Founded: 1998.
Congressional District: 1
Key Personnel: Cur., Serena Sutliff.
Personnel Profile: Full-Time Paid 1; Part-Time Paid 4; Part-Time Volunteers 12.
Governing Authority: Parent Institution: Westchester Public Library. Tax-exempt.

Institution Type/Description: History Museum.
Collections: local history & culture; photographs; personal artifacts.
Hours & Admission Prices: Wed.-Sun. 1-5; other times by appointment. No charge.
Attendance: 4,650 (accurate)

Clarksville

FALLS OF THE OHIO STATE PARK, (M), 201 W. Riverside Dr., Clarksville, IN 47129-3148. Tel.: 812-280-9970, ext. 400. Fax: 812-280-7110. Facebook: Falls of the Ohio State Park.
E-mail: park@fallsoftheohio.org
Web Site: www.fallsoftheohio.org
Founded: 1994.
Congressional District: 9
Key Personnel: Park Mgr., Lucas Green; Registrar & Interpretive Naturalist, Alan Goldstein; Interpretive Mgr., Kelley Morgan; Interpretive Naturalist, Jeremy Beavin.
Personnel Profile: Full-Time Paid 8; Part-Time Paid 3; Part-Time Volunteers 104.
Governing Authority: state; nonprofit. Tax-exempt.
Institution Type/Description: Historic Site: State Park includes the George Rogers Clark homesite.
Collections: 200 acre Middle Devonian fossil beds in the Ohio River, known since the 18th-century. Historic House: log cabin representing George Rogers Clark's home, 1803-1809.
Research Fields: new methods of paleontology education for students & teachers; Silurian & Devonian paleontology & stratigraphy; pre-history at the Falls of the Ohio, bird populations.
Facilities: 1,000-vol. library of books related to ornithology, geology, paleontology, Ohio River history & life sciences; 2 marine & 1 freshwater aquariums; classroom; 2,000 sq. ft. exhibit space; nature/conservation center; 122-seat theater. Fossils, minerals, jewelry, books, T-shirts & gift items with nature themes for sale.
Activities: docent program; films; formal education programs for teachers, children & college students; guided hikes; lectures; temporary exhibitions; theater; orientations for teachers bringing their classes to the park. Annual Events: Raptor Day; Earth Day; Archaeology Day; Family Fun Fair; Earth Discovery Day; Cabin Fever Festival; George Rogers Clark Weekend.
Publications: quarterly, Friends of the Falls Volunteer Newsletter; semiannual, Falls of the Ohio Newsletter; brochures about park-related themes, fossils, wildlife & trails; Educator's Handbook (most available through website).
Hours & Admission Prices: Mon.-Sat. 9-5, Sun. 1-5. Adults $5, children 3-18 $2; Indiana State Park Pass & Golden Hoosier Pass members and children under 3 no charge. Parking: $2 per day. Closed Thanksgiving; Christmas. &
Attendance: 25,544 (accurate)
Membership: Friend $25; Associate $50; Sponsor $100; Patron $200. Corporate: $500; $1,000; $ 2,500; $5,000.

Columbia City

WHITLEY COUNTY HISTORICAL MUSEUM, 108 W. Jefferson St., Columbia City, IN 46725-1744. Tel.: 260-244-6372. Facebook: Whitley County Historical Museum.
E-mail: wcmuseum@whitleymuseum.com
Web Site: www.whitleymuseum.com
Founded: 1963.
Congressional District: 4
Key Personnel: Dir., Dani Tippmann; Cur., Susan Richey.
Governing Authority: state. Parent Institution: Whitley Co. Historical Society. Subsidiary Institution: Whitley County. Tax-exempt: 501(c)(3).
Institution Type/Description: Historic Home & History Museum: housed in c.1875-1908 Vice Pres. Thomas R. Marshall home.
Collections: artifacts of prehistory & early history of Whitley County; early courthouse records; furniture belonging to Thomas R. Marshall; Marshall memorabilia; natural history displays.
Research Fields: genealogy; archaeology; pioneer.
Facilities: genealogy library; educational center.
Activities: children's programming. Museum Sponsors: Whitley County Historical Society; Old Settlers week special events; pioneer crafts; amateur archaeology association; nursing home programs; Christmas series; special Sunday displays & programs; spring & fall lecture series.
Publications: quarterly, Whitley County Historical Bulletin.
Hours & Admission Prices: Tues.-Thurs. 9-5, Fri. 9-12; other times by appointment. No charge; donations accepted. &
Attendance: 9,400 (estimated)
Membership: Personal Donor $50; Lloyd C. Douglas Level: $100; Simon J. Peabody Level $250; Ralph F. Gates Level $500; Mihsihkinaahkwa (Chief Little Turtle) Level $1,000; Thomas R. Marshall Level $5,000 & up.

Columbus

ATTERBURY-BAKALAR AIR MUSEUM, 4742 Ray Boll Blvd., Columbus, IN 47203. Tel.: 812-372-4356.
E-mail: abmuseum@att.net
Web Site: atterburybakalarairmuseum.org
Institution Type/Description: Military History Museum.
Collections: military history including WWII, Korean War, Cuban & Vietnam conflicts; photographs; military equipment & aircraft; The Indiana Aviation Hall of Fame; chapel.
Hours & Admission Prices: By appointment. No charge; donations accepted. &

BARTHOLOMEW COUNTY HISTORICAL SOCIETY, 524 Third St., Columbus, IN 47201-6724. Tel.: 812-372-3541. Fax: 812-372-3113.
E-mail: info@bartholomewhistory.org
Web Site: www.bartholomewhistory.org
Founded: 1921.
Congressional District: 2
Key Personnel: Exec. Dir., Julie Hughes; Pres. (V), Amy Kaiser.
Personnel Profile: Full-Time Paid 2; Part-Time Paid 1; Part-Time Volunteers 40.
Governing Authority: private corp. Tax-exempt: 501(c)(3).
Institution Type/Description: Local History Museum.
Collections: archives; Bartholomew County; manuscripts. Historic Houses: c.1867 McEwen-Samuels-Marr Home; c.1870 Henry Breeding Farm Home.
Research Fields: local history; genealogy services.
Facilities: 600-vol. library of manuscripts & material relating to Bartholomew County available for research on premises.
Activities: permanent & temporary exhibits; special programs for children; public events; heritage arts workshops; quarterly meetings. Annual Event: Reeves Pancake Breakfast.
Publications: newsletter; county history; 1879, Atlas & reprints.
Hours & Admission Prices: Tues.-Fri. 9-4; other times by appointment. No charge; donations accepted.
Attendance: 4,000 (estimated)
Membership: Curator $50; Preserver $100; Heritage Builder $250; Founder (Life) $1,000.

KIDSCOMMONS CHILDREN'S MUSEUM, 309 Washington St., Columbus, IN 47201. Tel.: 812-378-3046. Fax: 812-373-9085.
Web Site: www.kidscommons.org
Founded: 2005.
Institution Type/Description: Children's Museum.
Collections: hands-on exhibitions.
Hours & Admission Prices: Tues.-Sat. 10-5, Sun. 1-5.

Connersville

FAYETTE COUNTY HISTORICAL MUSEUM, 103 S. Vine St., Connersville, IN 47331-2649. Mailing Address: P.O. Box 197, Connersville, IN 47331. Tel.: 765-825-5325.
Congressional District: 6
Key Personnel: Pres. (V) & Museum Shop Mgr., Paulette Hayes.
Personnel Profile: Part-Time Volunteers 12.
Governing Authority: Parent Institution: Historic Connersville, Inc. Tax-exempt.
Institution Type/Description: History Museum.
Collections: local history; Civil War artifacts; photographs; personal artifacts; 4 locally-made automobiles.
Activities: monthly membership programs.
Publications: Pictorial History of Connersville; The Lexington Automobile.
Hours & Admission Prices: March-Dec. Thurs. & Sun. 1-4; other times by appointment. No charge. &
Attendance: 400
Membership: Single $15; Couple $25.

WHITEWATER VALLEY RAILROAD MUSEUM, 455 Market St., Connersville, IN 47331-2073. Mailing Address: P.O. Box 406, Connersville, IN 47331-0406. Tel.: 765-825-2054. Fax: 765-825-4550.
E-mail: officemanager@whitewatervalleyrr.org
Web Site: www.whitewatervalleyrr.org
Founded: 1972.
Key Personnel: Chm. (V), Ryan Scott; Pres. (V) & Museum Shop Mgr., John R. Hillman.

Personnel Profile: Full-Time Paid 1; Part-Time Volunteers 80.

Governing Authority: Branch Museum: Train Depot, 455 Market St., Connersville, IN 47331. Tax-exempt.

Institution Type/Description: Railroad Museum.

Collections: diesel locomotives; open window coaches; local history; railroad history, transportation & equipment.

Activities: 19 mile scenic train rides; educational programs.

Hours & Admission Prices: May-Oct. daily 9-5. Train Rides (round trip): adults $22, children 2-12 $14.

Attendance: 40,000 (estimated)

Membership: Annual $25; Life $200.

Corydon

CORYDON CAPITOL HISTORIC SITE, (M), 126 E. Walnut St, Corydon, IN 47112-1516. Tel.: 812-738-4890. Fax: 812-738-4904.

E-mail: corydoncapitol@seidata.com

Founded: 1930.

Congressional District: 9

Key Personnel: Cultural Admin., Bec Riley; Senior Interpreter, Nancy Snyder.

Personnel Profile: Full-Time Paid 3; Part-Time Paid 5.

Governing Authority: state. Parent Institution: Indiana Dept. of Natural Resources. Subsidiary Institution: Museum & Historic sites. Tax-exempt.

Institution Type/Description: Historic Site: 1816, first State Capitol Building; Gov. William Hendricks headquarters 1822-1825.

Collections: period furnishings.

Research Fields: early Indiana government.

Activities: guided tours.

Hours & Admission Prices: April to mid-Dec. Tues.-Sat. 9-5, Sun. 1-5; mid-Dec. to March call for hours. Adults $3.50, seniors $3, children $2, students $1; discounts to groups; children under 3 no charge. Closed New Year's Day; Easter; Veterans Day; Columbus Day; Election Day; Thanksgiving; Christmas.

Attendance: 87,800 (estimated)

Crawfordsville

CARNEGIE MUSEUM OF MONTGOMERY COUNTY, 222 S. Washington St., Crawfordsville, IN 47933-2444. Mailing Address: 205 S. Washington St., Crawfordsville, IN 47933. Tel.: 765-362-4618 & 4622. Facebook: Carnegie Museum of Montgomery County.

E-mail: carnegie@cdpl.lib.in.us

Web Site: www.cdpl.lib.in.us/carnegie

Founded: 2007.

Key Personnel: Dir., Catherine Burkhart.

Personnel Profile: Full-Time Paid 1; Part-Time Paid 3; Part-Time Volunteers 15; Interns 1.

Governing Authority: Parent Institution: Crawfordsville District Public Library. Tax-exempt.

Institution Type/Description: Interdisciplinary Museum: housed in the former Crawfordsville Public Library; built in 1902.

Collections: local history & culture; paintings; sculpture; science; business & industry; military; sports & pop culture; literature.

Publications: newsletter, Mosaic.

Hours & Admission Prices: Wed.-Sat. 10-5; other times by appointment. No charge; donations accepted. &

Attendance: 4,850 (accurate)

Membership: Individual $15; Household $25.

GENERAL LEW WALLACE STUDY & MUSEUM, (M), 200 Wallace Ave., Crawfordsville, IN 47933-2546. Mailing Address: P.O. Box 662, Crawfordsville, IN 47933-0662. Tel.: 765-362-5769. Fax: 765-362-5769.

E-mail: study@ben-hur.com

Web Site: www.ben-hur.com

Formerly: Ben-Hur Museum (Gen. Lew Wallace Study)

Founded: 1905.

Congressional District: 7

Key Personnel: Dir., Larry Paarlberg; Pres., Helen Hudson; Assoc. Dir., Amanda Wesselmann; Visitor Svcs. Rep., Kara Edie; Grounds Mgr., Deb King.

Personnel Profile: Full-Time Paid 2; Part-Time Paid 2; Part-Time Volunteers 30; Interns 1.

Governing Authority: municipal. Parent Institution: City of Crawfordsville Park & Rec. Dept. Subsidiary Institution: Lew Wallace Study Preservation Society. Tax-exempt.

Institution Type/Description: History Museum.

Collections: personal effects of Gen. Wallace, including artwork & inventions;

items relating to the movies & Broadway play of Ben-Hur; Civil War & World War I items.

Research Fields: local history; Civil War; World War I; Ben-Hur movies & Broadway play; American literature; political history.

Facilities: 3 1/2-acre grounds; picnic area. Museum-related items for sale.

Activities: guided & self-guided tours; temporary & permanent exhibitions; video presentation; structured educational programming.

Publications: brochures; newsletter.

Hours & Admission Prices: Feb. to mid-Dec. Wed.-Sat. 10-5, Sun. 1-5; groups by appointment. Adults $3, students $1; discounts to AAM, AAA & AASLH members; LWS Preservation Society members & children under 6 no charge. Closed holidays. &

Attendance: 7,000 (accurate)

Membership: Lieutenant $25; Major General $50; Governor $100; Ambassador $250; Ben-Hur Club $500.

MONTGOMERY COUNTY HISTORICAL SOCIETY AT LANE PLACE, 212 S. Water St., Crawfordsville, IN 47933-2535. Mailing Address: P.O. Box 127, 212 S. Water St., Crawfordsville, IN 47933-0127. Tel.: 765-362-3416. Facebook: Montgomery County Historical Society.

E-mail: info@lane-mchs.org

Web Site: www.lane-mchs.org

Founded: 1912.

Congressional District: 7

Key Personnel: Asst. Dir., Ann Harvey; Cur., Jennifer Rigsby.

Personnel Profile: Part-Time Paid 3; Part-Time Volunteers 26.

Governing Authority: society. Parent Institution: Montgomery County Historical Society. Tax-exempt.

Institution Type/Description: Historic House: c.1845 home of Henry S. Lane, governor & senator of Indiana.

Collections: oil paintings; china; dolls; period furnishings; Lincoln memorabilia. Historic Building: log cabin, former home of John Allen Speed, Mayor of Crawfordsville.

Major Exhibits: A Stitch in Time: Quilts Presented by the Montgomery County Historical Society, 3/14-12/15/14.

Research Fields: Abraham Lincoln; birth of Republican Party.

Activities: guided tours; temporary exhibitions; special & seasonal events.

Publications: annual local history book.

Hours & Admission Prices: March to mid-Dec. Tues. 1-4:30, Wed.-Sun. 10-4:30. Adults $5, children under 12 $3; members no charge. Closed all holidays.

Attendance: 2,000 (estimated)

Membership: Individual $15; Family $20.

ROPKEY ARMOR MUSEUM, 5649 E. 150 N., Crawfordsville, IN 47933. Tel.: 765-794-0238.

E-mail: info@ropkeyarmormuseum.com

Web Site: ropkeyarmormuseum.com

Institution Type/Description: Military History Museum.

Collections: military history, heritage, vehicles, equipment, & weapons; M3A1 scout car; WWII ear Shermans; M109 155mm self-propelled howitzer.

Hours & Admission Prices: March-Dec. Mon.-Fri. 10:30-4:30; other times by appointment. No charge; donations accepted.

THE ROTARY JAIL MUSEUM, (M), 225 N. Washington, Crawfordsville, IN 47933-1737. Mailing Address: Montgomery County Cultural Foundation, Inc., P.O. Box 771, Crawfordsville, IN 47933-0771. Tel.: 765-362-5222. Fax: 765-362-5222.

E-mail: oldjailmuseum@accelplus.net

Web Site: www.oldjailmuseum.net

Formerly: The Old Jail Museum

Founded: 1975.

Congressional District: 7

Key Personnel: Dir., Tamara Hemmerlein; Pres. (V), Carrie Sosbe.

Personnel Profile: Part-Time Paid 3; Part-Time Volunteers 15.

Governing Authority: nonprofit organization. Parent Institution: Montgomery County Cultural Foundation, Inc. Tax-exempt: 501(c)(3).

Institution Type/Description: Historic Building: 1882 Old Jail only remaining rotating circular jail still in working condition.

Collections: Native American artifacts, clothing, furniture, books, toys, textiles, photos, tools; history & culture of Montgomery County, Indiana & the Midwest as a cultural whole.

Research Fields: relating to the exhibits.

Facilities: 111-vol. library available for use upon request; three exhibit rooms in former living quarters of the sheriff. Museum-related items for sale.

Activities: guided tours; docent program; temporary exhibitions; adult & youth art classes.

Publications: quarterly newsletter.

Hours & Admission Prices: March to mid-Dec. Wed.-Sat. 10-5, Sun. 1-5. Adults $3; members no charge. Closed holidays. &
Attendance: 5,000 (estimated)
Membership: Individual $10; Family $15; Patron $25; Contributing $50; Sustaining $100; Benefactor $250.

Crown Point

LAKE COUNTY HISTORICAL MUSEUM, Old Lake Courthouse, Courthouse Sq., Ste. 202, Crown Point, IN 46307. Tel.: 219-662-3975.
Institution Type/Description: History Museum.
Collections: local history & culture; photographs; period furnishings & artifacts; farm implements; clothing; musical instruments; personal artifacts.
Hours & Admission Prices: May-Oct. Thurs.-Sat. 1-4. Adults $1, children $.50.

Dale

DR. TED'S MUSICAL MARVELS, 11896 S. U.S. 231, Dale, IN 47523. Mailing Address: 911 Hickory Dr., Huntingburg, IN 47542. Tel.: 812-937-4250. Fax: 812-937-4250.
E-mail: mmarvels@psci.net
Web Site: drted.com
Formerly: House of Mechanical Music Machines
Founded: 1984.
Congressional District: 9
Key Personnel: Dir., Theodore Waflart, M.D.; Pres. (V), Mary Kay Waflart; Museum Shop Mgr., Millie Schum.
Personnel Profile: Part-Time Volunteers 4.
Governing Authority: private.
Institution Type/Description: Musical Instrument Museum.
Collections: self-playing musical instruments from 1800s to mid-1900s.
Hours & Admission Prices: By appointment for groups of 15 or more. &
Attendance: 1,000 (estimated)

Dana

ERNIE PYLE STATE HISTORIC SITE, 120 W. Briarwood Ave., Dana, IN 47847-0338. Mailing Address: P.O. Box 81, Vincennes, IN 47591-0081. Tel.: 765-665-3633. Fax: 765-665-9312.
E-mail: erniepyle@abcs.com
Founded: 1976.
Congressional District: 7
Key Personnel: Tour Guide, Janice Duncan; Tour Guide, Joanie Rumple.
Personnel Profile: Full-Time Paid 1; Part-Time Paid 2.
Governing Authority: state. Branch of the Indiana State Museum System, Div. of the Dept. of Natural Resources, Indiana State Museum, 650 W. Washington St., Indianapolis, IN 46204. Tax-exempt.
Institution Type/Description: Visitor Center & Historic House.
Collections: Visitor Center: two World War II-era Quonset huts; WWII uniforms, weapons & gear; Ernie Pyle memorabilia; photographs. Historic House: 1851 farmhouse, birthplace of noted WWII journalist Ernie Pyle was born in 1900.
Research Fields: life & writings of Ernie Pyle; World War II.
Facilities: visitor center. Museum-related items for sale.
Activities: audio & video stations; guided tours; permanent exhibits.
Hours & Admission Prices: May 3-Nov. 17 Thurs.-Sat. 10-5, Sun. 1-5. Closed Easter; Columbus Day; Election Day; Thanksgiving.
Attendance: 8,000 (estimated)

Danville

HENDRICKS COUNTY HISTORICAL MUSEUM, 170 S. Washington St., Danville, IN 46122. Mailing Address: P.O. Box 226, Danville, IN 46122. Tel.: 317-718-6158.
E-mail: museum@co.hendricks.in.us
Web Site: hendrickscountyhistoricalmuseum.org
Founded: 1976.
Key Personnel: Pres. (V), Gail Tharp
Governing Authority: Tax-exempt.
Institution Type/Description: History Museum.
Collections: local history & culture; period furnishings; personal artifacts; photographs; Central Normal College & Canterbury College artifacts from 1878-1951.
Activities: Annual Event: Civil War Heritage Days - Hendricks County in June.
Publications: newsletter, 3 times per year.
Hours & Admission Prices: Sat. 11-3; other times by appointment. No charge; donations accepted.

Attendance: 1,800 (estimated)
Membership: Home $25; Settler $100; Pioneer $250; Trailblazer $500; Founder $1,000 & up.

Decatur

ADAMS COUNTY HISTORICAL MUSEUM, 420 W. Monroe St., Decatur, IN 46733-1622. Mailing Address: P.O. Box 262, Decatur, IN 46733-0262. Tel.: 260-724-3493.
Web Site: www.rootsweb.ancestry.com/~inadams/
Founded: 1965.
Congressional District: 10
Key Personnel: Pres., Herb Myers.
Governing Authority: society; nonprofit organization. Parent Institution: Adams County Historical Society. Tax-exempt: 501(c)(3).
Institution Type/Description: Historical Society Museum: housed in c.1903 Charles Dugan Home.
Collections: local historical items.
Activities: guided tours; lectures.
Publications: newsletter, The Trumpeter; The History of Adams Co., Indiana, Vol. I & II.
Hours & Admission Prices: June-Sept. Sun. 1-4; other times by appointment only. No charge; donations accepted.
Attendance: 3,000 (estimated)

Delphi

CARROLL COUNTY HISTORICAL SOCIETY MUSEUM, 101 W. Main St., Ground Fl., Court House, Delphi, IN 46923-1574. Mailing Address: P.O. Box 277, Delphi, IN 46923-0277. Tel.: 765-564-3152.
E-mail: carrollcountyhistoricalsociety@ffni.com
Web Site: www.carrollcountymuseum.org
Founded: 1924.
Key Personnel: Dir., Mark Smith; Pres., Steve Mulligan.
Personnel Profile: Full-Time Volunteers 1; Part-Time Volunteers 8.
Governing Authority: Parent Institution: Carroll County Historical Society. Tax-exempt.
Institution Type/Description: History Museum & Genealogy Library.
Collections: local history; cultural heritage; Conestoga wagon; period tools; dolls; paintings; genealogy.
Activities: research.
Publications: quarterly newsletter; Carroll County Indiana Legacy, 1824-2005.
Hours & Admission Prices: Tues. & Thurs.-Fri. 9-4; other times by appointment. No charge; donations accepted. Closed holidays. &
Attendance: 6,500 (estimated)
Membership: Junior $5; Single $20; Family $25; Business Club & Group $50; Corporation $100; Lifetime $300.

Dugger

DUGGER COAL MUSEUM, 1080 S. Section St., Dugger, IN 47848. Mailing Address: P.O. Box 501, Dugger, IN 47848-0501.
Founded: 1980.
Key Personnel: Pres. (V), Beth Secrest; Cur., Martha Marlow.
Personnel Profile: Part-Time Volunteers 4.
Governing Authority: private; nonprofit organization. Tax-exempt.
Institution Type/Description: Coal Mine Museum.
Collections: history & artifacts from deep coal mines & strip mines; manuals; equipment models; photographs; news clippings; miners tools; miniature electric drugline model; working model of a mine tipple; clothing; protective gear; hard hats; soft caps; all varieties of miners' lights; old mining reports; paychecks & old letters by & about miners; operators control station for an 8900 dragline.
Facilities: Museum-related items for sale.
Activities: Museum Sponsors: Alumni Banquet in September; Coal Festival in September.
Hours & Admission Prices: Coal Festival: Sept. daily 9-9; other times by appointment. No charge; donations accepted.
Attendance: 500 (estimated)
Membership: Individual $4; Lifetime $100.

Dunkirk

THE GLASS MUSEUM, 309 S. Franklin, Dunkirk, IN 47336-1209. Tel.: 765-768-6872. Fax: 765-768-6894.
E-mail: glassmuseumdunkirk@gmail.com
Web Site: www.dunkirkpubliclibrary.com
Founded: 1979.

Congressional District: 4
Key Personnel: Pres. (V), Sandy Rogers; Exec. Dir., Ailesia Franklin; Cur., Blake Watson.
Personnel Profile: Part-Time Paid 2; Part-Time Volunteers 10.
Governing Authority: nonprofit organization. Parent Institution: Dunkirk Public Library. Tax-exempt: 501(c)(3).
Institution Type/Description: Glass Museum: housed next to the Dunkirk library.
Collections: 8,500 pieces of glassware; Depression; Vaseline; Custard; Albany; hand blown glass; hand painted machine-pressed; handcrafted glassware; glass bottles; glass canes; Indiana glass; patterns & items from Indiana & 115 other glass factories from around the world; cup plates Cambridge glass; Fenton glass.
Activities: guided tours; lectures; rotating, permanent & temporary exhibits. Annual Event: Glass Days Festival in June.
Publications: brochures; annual newsletter.
Hours & Admission Prices: May-Oct. Tues.-Fri. 10-4, Sat. 10-2; Sun. by appointment only. Adults $2; discounts to AAA members & Enjoy Indiana cardholders; children no charge. Closed Memorial Day; Independence Day; Labor Day. &
Attendance: 1,200 (accurate)
Membership: Friend Club $5.

Edinburgh

CAMP ATTERBURY MUSEUM, Welcome Ctr., Hospital Rd., Edinburgh, IN 46124-5000. Mailing Address: P.O. Box 5000, Camp Atterbury, Edinburgh, IN 46124-5000. Tel.: 812-526-1744.
E-mail: int-cajmtcpao@ng.army.mil
Web Site: www.campatterbury.in.ng.mil
Founded: 1998.
Key Personnel: Artifact Responsible Officer, Maj. Lisa Kopczynski.
Personnel Profile: Full-Time Paid 1; Part-Time Volunteers 10.
Governing Authority: Tax-exempt: 501(c)(3).
Institution Type/Description: Military Museum.
Collections: photographs; WWII artifacts; military equipment; POW Chapel.
Activities: veteran activities. Annual Events: POW reunion; former land owners picnic; Memorial Ceremony.
Hours & Admission Prices: Mon.-Fri. 8:30-4, Sat. 1-4. No charge; donations accepted. Closed federal holidays. &
Attendance: 1,500 (estimated)

Elkhart

MIDWEST MUSEUM OF AMERICAN ART, (M), 429 S. Main St., Elkhart, IN 46516-3210. Mailing Address: P.O. Box 1812, Elkhart, IN 46515-1812. Tel.: 574-293-6660. Fax: 574-293-6660 (call first). Facebook: Midwest Museum of American Art.
E-mail: mdwstmsmam@aol.com
Web Site: www.midwestmuseum.us
Founded: 1978.
Congressional District: 2
Key Personnel: Chm. (V), Dr. Rick Burns; Dir., Jane Burns; Cur., Brian Byrn; Museum Shop Mgr., Gertrude Basquin.
Personnel Profile: Full-Time Paid 1; Full-Time Volunteers 1; Part-Time Paid 1; Part-Time Volunteers 54; Interns 1.
Volunteer Hours: 970
Operating Expenses: 276,000
Operating Income: 276,000
Governing Authority: nonprofit organization. Tax-exempt: 501(c)(3).
Institution Type/Description: Art Museum.
Collections: American Impressionists, Regionalists & Abstract Expression Illustrators; Chicago & Pop Art; prints, drawings, sculpture & photography; works by Robert Reid, Charles Hawthorne, Earnest Lawson, John Singer Sargent, Grandma Moses, Grant Wood & Thomas Hart Benton.
Major Exhibits: Anthony Droege: A 30-Year Survey of Figures, Still Life & Landscape, 12/13/13-2/23/14; Youth Art, 3/14; Gardens, Rivers & Lakes: The American Landscapes (T), 4/4/14-7/13/14; 35 Years Best of Show Regional (T), 7/18/14-10/5/14; 36th Elkhart Juried Regional, 10/3/14-12/7/14; Barn Again - Gwen Gutwein (T), 12/12/14-3/1/14.
Research Fields: American folk painting & American Art Movements.
Facilities: Museum-related items for sale.
Activities: guided tours; lectures; films; gallery talks; concerts; formally organized education programs; docent program; permanent, temporary, loan & traveling exhibitions.
Publications: catalogue, Panorama of American Art; bulletin, Midwest Museum; bimonthly, Art News; catalogues, Midwest Photo 80; Midwest Photo 81; catalogue, Midwest Museum of American Art Permanent Collection.
Hours & Admission Prices: Tues.-Fri. 10-4, Sat.-Sun. 1-4. Adults $5, senior citizens $3, students $2; discounts to AAM, AAA & Mobil Guide members;

Sun. & members no charge. Closed New Year's Day; Memorial Day; Independence Day; Labor Day; Thanksgiving; Christmas. &
Attendance: 20,000 (accurate)
Membership: Student & Senior $10; Individual $20; Family $60; Junior Society $100; Society of Associates $150; Patrons $250; Benefactor $500; Sustainer $1,000; Guarantor $2,000; Endower $5,000.

NATIONAL NEW YORK CENTRAL RAILROAD MUSEUM, 721 S. Main St., Elkhart, IN 46516-3112. Mailing Address: 229 S. Second St., Elkhart, IN 46516. Tel.: 574-294-3001. Fax: 574-295-9434.
E-mail: info@nycrrmuseum.org
Web Site: www.nycrrmuseum.org
Founded: 1987.
Congressional District: 2
Key Personnel: Exec. Dir., Robin Hume.
Personnel Profile: Full-Time Paid 1; Part-Time Paid 3; Part-Time Volunteers 15; Interns 1.
Governing Authority: municipal government; nonprofit. Parent Institution: City of Elkhart. Tax-exempt.
Institution Type/Description: Transportation Museum: housed in c.1885-1906 New York Central Freighthouse complex.
Collections: the New York Central railroad, its predecessor & successor lines and their relationship to Elkhart, Indiana; National Railroad Adjustment Board annals; Engine House crew books; rules & regulations for railroad operation; steam & diesel engine equipment; personal papers & correspondence; O gauge layout.
Research Fields: immigrants & the railroad; written histories of ex-New York Central employees; New York Central & it's employees impact on Elkhart.
Facilities: 345-vol. library; 6,000 sq. ft. exhibit space. Railroad-oriented items for sale.
Activities: docent program; guided tours; lectures; temporary & traveling exhibitions; 2 ft. gauge train tour of rolling stock.
Publications: quarterly newsletter.
Hours & Admission Prices: Tues.-Sat. 10-5, Sun. 12-4. Adults $5, senior citizens & children $4. Closed major holidays. &
Attendance: 10,000 (estimated)
Membership: Single $25; Family $40.

RV/MH HERITAGE FOUNDATION, INC., 21565 Executive Pkwy., Elkhart, IN 46514-9693. Tel.: 574-293-2344; 800-378-8694. Fax: 574-293-3466.
E-mail: rvmhhall@aol.com
Web Site: rvmhhallofffame.org
Founded: 1972.
Congressional District: 3
Key Personnel: Pres., Carl A. Ehry; Chm. (V), Bill Garpow.
Personnel Profile: Full-Time Paid 2; Part-Time Paid 2; Part-Time Volunteers 20.
Governing Authority: private; nonprofit organization. Subsidiary Institution: RV/MH Hall of Fame. Tax-exempt: 501(c)(3).
Institution Type/Description: Industrial Museum.
Collections: publications, images & records of the recreational vehicle and manufactured housing industries; travel trailers, motorized units, manufactured equipment & supplies of the industries.
Research Fields: advent, growth & development of the recreational vehicle and manufactured housing industries; manufacturing technology, techniques & distribution; retailing; recreational & practical use of industry's product.
Facilities: 10,000-vol. library of industry & consumer publications; educational facilities; 14,000 sq. ft. exhibit space.
Activities: guided tours.
Publications: Heritage Happenings.
Hours & Admission Prices: Mon.-Fri. 9-5, Sat. 9-3; other times by appointment. Adults $8, seniors $6, youth 6-16 $3; discounts to groups. &
Attendance: 30,000 (estimated)
Membership: Bronze Patron $250; Silver Patron $500; Gold Patron $1,000; Star Patron $2,500.

RUTHMERE MUSEUMS CAMPUS, 302 E. Beardsley Ave., Elkhart, IN 46514-2719. Tel.: 574-264-0330; 888-287-7696. Fax: 574-266-0474.
Web Site: www.ruthmere.org
Founded: 1973.
Congressional District: 2
Key Personnel: Exec. Dir., Bill Firstenberger; Cur., Jennifer Johns; Asst. Dir., Bob Edel; Communications Dir., Laura Horst; Collections Mgr., Joy Olsen; Wedding & Hospitality Coord., Annett Kozak; Bldg. & Grounds Mgr., Ronald Wolschlager.

Personnel Profile: Full-Time Paid 4; Part-Time Paid 6; Part-Time Volunteers 50; Interns 1.

Governing Authority: bd. of trustees; nonprofit organization. Tax-exempt: 501(c)(3).

Institution Type/Description: Historic House Museum: c.1910.

Collections: Louis XV revival furniture; Dresden & Sevres porcelain; Wedgwood pottery; art glass by Steuben & Tiffany Studios; sculpture by A. Rodin, C. Claudel, W.O. Partridge; paintings by Samuel F.B. Morse, Alma-Tadema & other 19th-century American and European artists & craftsmen.

Research Fields: architect E. Hill Turnock; American Midwestern material culture 1910-1945.

Facilities: 1,771-vol. reference library of American decorative arts & domestic architecture.

Activities: guided tours; public programs; group tours; concerts; lectures.

Publications: quarterly newsletter; semi-annual Ruthmere Record; monthly e-newsletter.

Hours & Admission Prices: April-Dec. Ruthmere Tours: Tues.-Sat. 10-4, Sun. 1-4; tours on the hour; last tour at 3pm. Adults $10, students $4; discounts to AAM & Blue Start Museum members & groups; members no charge. Havilah Beardsley House & Historic District Tour: Tues.-Sun. 1-4. Adults $5, students $2; discounts to AAM & Blue Start Museum members; members no charge. Campus Day Pass: adults $13, students $5, family $40 (5 visitors); children under 5 no charge.

Attendance: 12,000 (estimated)

Membership: Individual $50; Family $75; Patron $100; Bronze Patron $250; Silver Patron $500; Gold Patron $1,000; Rose Gold Patron $2,500; Platinum Patron $5,000.

Evansville

*** ANGEL MOUNDS STATE HISTORIC SITE, (M),** 8215 Pollack Ave., Evansville, IN 47715-6231. Tel.: 812-853-3956. Fax: 812-858-7686.

E-mail: angelmoundsshs@indianamuseum.org

Web Site: www.angelmounds.org

Founded: 1946.

Congressional District: 8

Key Personnel: C.E.O., Historic Site Cur. & Site Mgr., Mike Linderman; Pres. (V), Jennifer Gudorf; Sectional Archaeology Program Developer, Haley Tallman; Museum Shop Mgr., Patrick Thomas.

Personnel Profile: Full-Time Paid 5; Part-Time Paid 1; Part-Time Volunteers 30; Interns 2.

Governing Authority: state. Parent Institution: Indiana State Museums & Historic Sites. Tax-exempt.

Institution Type/Description: Archaeology Museum and Pre-Historic Site: Middle Mississippian Indian Site & WPA facility.

Collections: Middle Mississippian Indian artifacts; Angel Mounds' excavation photographs; reconstructed village.

Research Fields: archaeology of Middle Mississippian Indians.

Facilities: 70-seat auditorium.

Activities: orientation video; permanent & temporary exhibits; lectures; tours; organized education programs for adults & students; special events.

Publications: historic site brochure & walking tour map.

Hours & Admission Prices: Tues.-Sat. 9-5, Sun. 1-5. Adults $5, senior citizens $4, children $2, school groups $1; members no charge.

Attendance: 49,000 (estimated)

Membership: Friends Group: Individual $10; Family $15; Contributor $25; Supporting $50; Patron $100; Corporate $250.

EVANSVILLE AFRICAN AMERICAN MUSEUM, 579 S. Garvin, Evansville, IN 47713. Mailing Address: P.O. Box 3124, Evansville, IN 47731-3124. Tel.: 812-423-5188. Fax: 812-426-1885.

E-mail: eaamuseum@sbcglobal.net

Web Site: www.evansvilleaamuseum.com

Founded: 1997.

Congressional District: 8

Key Personnel: Exec. Dir., Lu Porter; Pres. (V), Harold W. Calloway; Museum Shop Mgr., Nancy McClure.

Personnel Profile: Full-Time Paid 2; Part-Time Paid 3; Part-Time Volunteers 12.

Governing Authority: Tax-exempt.

Institution Type/Description: History Museum: housed in one of the nation's first housing projects, Lincoln Gardens; built in 1938.

Collections: African American culture & history; personal artifacts; photographs.

Major Exhibits: Let Freedom Resound: The Abolitionist Movement in Indiana, Ohio, and Kentucky, 1/7/14-2/1/14; Freedom Rides, 2/10/14-3/10/14; Laboring Women: African American Midwifery, 5/6/14-8/30/14; American Mosaic: The History of a Multicultural America, 6/10/14-9/27/14; 20 and Odd...: Slavery in America, 10/1/14-1/3/15.

Activities: Museum Sponsors: Fabulous First Fridays monthly.

Hours & Admission Prices: Tues.-Sat. 10-5. Adults $5, children $3; members no charge. Closed New Year's Day; Martin Luther King, Jr. Day; Memorial Day; Independence Day; Labor Day; Thanksgiving; Christmas.

Attendance: 4,018 (estimated)

Membership: Individual $25; Family $75; Sustaining $100; Not-for-Profit & Sustaining Family $250; Corporation (under 50 employees) $500; Lifetime & Corporation (over 50 employees) $1,500.

*** EVANSVILLE MUSEUM OF ARTS, HISTORY & SCIENCE, (M),** 411 S.E. Riverside Dr., Evansville, IN 47713-1098. Mailing Address: P.O. Box 3435, Evansville, IN 47733. Tel.: 812-425-2406. Fax: 812-421-7509.

E-mail: info@emuseum.org

Web Site: www.emuseum.org

Founded: 1926.

Congressional District: 8

Key Personnel: Interim Dir. & Cur. Collections, Mary McNamee Bower; Pres., Sharon Walker; Registrar, Liz Fuhrman Bragg; Cur. History, Thomas R. Lonnberg; Dir. Koch Science Center & Planetarium, Mitchell Luman; Dir. Devel., Susan Washburn; Dir. Membership & Mktg., Josh Gilmore; Cur. Education, Karen Malone.

Personnel Profile: Full-Time Paid 16; Full-Time Volunteers 1; Part-Time Paid 18; Part-Time Volunteers 50; Interns 10.

Governing Authority: nonprofit organization. Tax-exempt: 501(c)(3).

Institution Type/Description: Arts, History & Science Museum.

Collections: painting; sculpture; prints & drawings; decorative arts; transportation; arms & armor; steam locomotive & cars; science displays; anthropology.

Facilities: 3,500-vol. library & archives related to art, history & science; planetarium; classroom; science center. Gift items for sale

Activities: guided tours; lectures; special programs; gallery talks; musicals; formally organized education programs; docent program; student volunteer program; demonstrations; inter-museum loans; permanent, temporary & traveling exhibitions; camp-ins.

Publications: members magazine, The COPIA; books: Architectural Heritage of Evansville; The Eye & the Heart: Watercolors of John Stuart Ingle; Simplicity, A Grace: Jacob Maentel in Indiana; A Charmed Vision: The Art of Carolyn Plochmann; Art School: A Homage to the Masters, Paintings by George Deem; T.C. Steele: An American Master of Light; catalogs of special exhibits; biennial catalogs of Mid-States Craft Exhibition & Mid-States Art Exhibition.

Hours & Admission Prices: Tues.-Sat. 11-5, Sun. 12-5. Adults $7; ASTC & museum members no charge. Call for theater prices. Closed New Year's Day; Easter; Independence Day; Thanksgiving; Christmas Eve & Day.

Attendance: 72,500 (accurate)

Membership: Contributor $50-$124; Patron $125-$249; Donor $250-$499; Director's Associate $500-$999; President's Circle $1,000-$2,499; Founder's Society $2,500 & up.

KOCH FAMILY CHILDREN'S MUSEUM OF EVANSVILLE, 22 S.E. 5th St., Evansville, IN 47708-1604. Mailing Address: P.O. Box 122, Evansville, IN 47701-0122. Tel.: 812-464-2663. Fax: 812-477-4339.

E-mail: info@cmoekids.org

Web Site: www.cmoekids.org

Formerly: Hands On Discovery

Founded: 1994.

Key Personnel: Exec. Dir., Stephanie Terry; Pres. (V), Kate Miller.

Personnel Profile: Full-Time Paid 7; Part-Time Paid 19; Interns 3.

Governing Authority: nonprofit organization. Tax-exempt: 501(c)(3).

Institution Type/Description: Children's Museum.

Collections: hands-on exhibits.

Hours & Admission Prices: Tues.-Sat. 10-5, Sun. 12-5. Admission 18 months & over $7; discount to ACM Reciprocal members; members no charge.

Attendance: 51,839 (accurate)

Membership: Family $99; Family Plus $129; Grandparents $139.

MESKER PARK ZOO & BOTANIC GARDEN, 1545 Mesker Park Dr., Evansville, IN 47720-8206. Tel.: 812-435-6143. Fax: 812-435-6140.

E-mail: zooservices@meskerparkzoo.com

Web Site: www.meskerparkzoo.com

Founded: 1929.

Congressional District: 8

Key Personnel: Exec. Dir., Amos D. Morris, Jr.; Dir. Operations, Erik Beck;

Visitor Svcs. Cur., Stephanie Sanderson; Education Cur., Diana Barber; Animal Cur., Susan Lindsey, Ph.D.; Registrar, Dana Duke; Botanic Cur., Paul Bouseman.

Personnel Profile: Full-Time Paid 35; Part-Time Paid 42; Part-Time Volunteers 450.

Governing Authority: municipal. A branch of Evansville Park Dept. & Recreation Dept. Civic Center. Evansville Zoological Society. Tax-exempt.

Institution Type/Description: Zoo.

Collections: animal exhibits; 200 species of mammals, birds, reptiles; over 650 specimens.

Facilities: 500-vol. library of zoological books available for use on premises; 50 acre grounds; children's zoo. Museum-related items for sale.

Activities: lectures; permanent exhibitions; narrated safari train tour; outreach; summer camp; Kinderfun.

Publications: quarterly magazine; quarterly members newsletter, Mesker Messenger.

Hours & Admission Prices: Daily 9-5. Adults $8.50, children 3-12 $7.50; discounts to AAZPA & other zoo society members; members & children 2 & under no charge. &

Attendance: 140,019 (accurate)

Membership: Senior $29; Individual $34; Couple $49; Family $59; (additional adult $12).

REITZ HOME MUSEUM, 224 S.E. First St., Evansville, IN 47713-1002. Mailing Address: P.O. Box 1322, Evansville, IN 47706-1322. Tel.: 812-426-1871. Fax: 812-426-2179. Facebook: Reitz Home Museum.

E-mail: information@reitzhome.com

Web Site: www.reitzhome.com

Founded: 1974.

Congressional District: 8

Key Personnel: Dir., Matt Rowe; Cur., Pam Guthrie.

Personnel Profile: Full-Time Paid 3; Part-Time Paid 1; Part-Time Volunteers 250; Interns 2.

Governing Authority: private; nonprofit organization. Parent Institution: The Reitz Home Preservation Society. Tax-exempt: 501(c)(3).

Institution Type/Description: Victorian Historic House: located in the Riverside Historic District, the 1871 house was built by hardwood lumber magnate John Augustus Reitz.

Collections: Victorian furniture & decorative arts from the 1871-1920 period.

Research Fields: original wall coverings & house history.

Facilities: educational facilities; 18,255 sq. ft. exhibit space. Museum-related items for sale.

Activities: docent program; formal education programs; guided tours; participatory exhibits. Annual Events: Wine Down to the Weekend(s); Bring the Derby Home; Mystery Event; Wine Dive; Victorian Christmas.

Publications: quarterly newsletter.

Hours & Admission Prices: Tues.-Sat. 11-3:30, Sun. 1-3:30; last tour 2:30. Adults $7.50, students $2.50, children 12 & under $1.50; discounts available to groups, schools & AAM members; Reitz Home members no charge. &

Attendance: 12,000 (accurate)

Membership: Reitz Home Guild $15; Friend $50-$124; Contributor $125-$249; Donor $250-$499; Patron $500-$999; Benefactor $1,000 & up.

WESSELMAN WOODS NATURE PRESERVE, 551 N. Boeke Rd., Evansville, IN 47711-5923. Tel.: 812-479-0771. Fax: 812-479-7573.

E-mail: info@wesselmannaturesociety.org

Web Site: www.wesselmannaturesociety.org

Formerly: Wesselman Park Nature Center

Founded: 1973.

Congressional District: 8

Key Personnel: Dir., John Scott Foster; Pres. (V), Luke Yaeger.

Personnel Profile: Full-Time Paid 3; Part-Time Paid 8; Interns 5.

Governing Authority: Parent Institution: Wesselman Nature Society, Inc.

Institution Type/Description: Nature Center.

Collections: old growth forest & bottomland hardwood forest; habitats; wildlife.

Research Fields: biological.

Facilities: nature center; bird viewing area; 6 miles of trails.

Activities: special events; educational programs; school, scout & camp programs. Annual Events: Maple Sugarbush Festival in March; Harvest Festival in October; Wandering Owl Wine Trail in Fall.

Publications: quarterly newsletter, Old Growth Gazette.

Hours & Admission Prices: Tues.-Sun. 8-6. No charge; donations accepted. &

Attendance: 40,000 (estimated)

Fairmount

FAIRMOUNT HISTORICAL MUSEUM & GIFT SHOP, 203 E. Washington St., Fairmount, IN 46928-1700. Mailing Address: P.O. Box 92, Fairmount, IN 46928-0092. Tel.: 765-948-4555. Fax: 765-948-4259.

Web Site: www.jamesdeanartifacts.com/index.php

Founded: 1975.

Congressional District: 5

Key Personnel: Fairmount Historian, Ann Warr; Vice Pres., Gale Hikade; Chm. (V), Robert McManaman.

Governing Authority: Tax-exempt.

Institution Type/Description: History Museum.

Collections: local history; photographs; personal artifacts of James Dean & his family; Jim Davis, creator of Garfield.

Activities: Museum sponsors: Fairmount Museum Day Remembering: James Dean in September

Publications: annual membership newsletter.

Hours & Admission Prices: March-Nov. Mon.-Sat. 10-5, Sun. 12-5; other times by appointment. No charge; donations accepted. Closed Easter; Thanksgiving. &

Attendance: 7,000 (accurate)

Membership: Annual $6; 6 Years $30; 10 Years $50; Life $100.

Fishers

*   **CONNER PRAIRIE INTERACTIVE HISTORY PARK, (M),** 13400 Allisonville Rd., Fishers, IN 46038-4499. Tel.: 317-776-6000; 800-966-1836. Fax: 317-776-6014.

E-mail: rosenthal@connerprairie.org

Web Site: www.connerprairie.org

Founded: 1964.

Congressional District: 5

Key Personnel: Pres. & C.E.O., Ellen Rosenthal; Chm. Museum Bd., Tim Hassinger; Chm. Foundation Bd., Berkley Duck; C.F.O., Kyle Wenger; C.O.O., Ken Bubp; Vice Pres. Exhibits, Programs, Facilities, Cathy Ferree; Deputy Dir. Museum Experience, Tim Crumrin; Mgr. Public Rels., Lynelle Mellady; Vice Pres. Devel., Cameron McGuire; Vice Pres. Institutional Advancement, Dan Folta; Dir. Mktg. Communications, Amy Ahlersmeyer; Dir. Human Resources, Susan Johnson; Dir. Facilities, John Spicklemire; Dir. Festivals & Special Events, Kelly Backus; Museum Shop Mgr., Elaine Molin; Dir. Corporate & Foundation Rels., Joe Hammer.

Personnel Profile: Full-Time Paid 89; Part-Time Paid 238; Part-Time Volunteers 488.

Governing Authority: Tax-exempt: 501(c)(3).

Institution Type/Description: Interactive History Park & Museum.

Collections: microfilm of relevant magazines & newspapers; pioneer artifacts; 1859 balloon voyage & related artifacts; early aviation history; modern balloon flight. Historic Buildings: 1836 rural settlement with houses, store, blacksmith shop, pottery, carpenter shop, inn, schoolhouse, barns & other outbuildings; 1823 William Conner home; barn & supporting structures; loomhouse; 1863 Civil War area; covered bridge; 1816 Lenape Indian Village.

Research Fields: Indiana & midwestern history, science, technology & environment.

Facilities: 3,500-vol. library of material relating to rural living, crafts, local & Indiana history and science available for research on premises; welcome center; picnic pavilions; restaurant; 800 acres; amphitheater; rental facilities. Park-related items for sale.

Activities: interactive exhibits; organized education programs; permanent exhibitions; special programs; costumed staff doing seasonal tasks; period craft programs; weddings; harvests; agricultural fairs; Underground Railroad program; Civil War era programs & activities; play & learning area for young children. Annual Events: Hearthside Suppers January to March; Civil War Days in May; Fourth of July Celebration; County Fair in September; Headless Horseman Fall Festival; 1880s Agricultural Fair; Candlelight Tours in December.

Publications: member magazine, Closer Look.

Hours & Admission Prices: April Thurs.-Sat. 10-5, Sun. 11-5; May-Sept. Tues.-Sat. 10-5, Sun. 11-5; Oct. Wed.-Sat. 10-5, Sun. 11-5; Nov.-March call for special programs & hours. Adults $14, senior citizens 65 & over $13, youth 2-12 $9; discounts to AAM & ICOM members; children 2 & under and members no charge. Closed Easter; Thanksgiving; Christmas Eve & Day. &

Attendance: 315,455 (accurate)

Membership: Individual $45; Family & Grandparent $65; Family & Grandparent Plus $95; Heritage $100-$249; Prairietown Citizen $250-$499; Golden Eagle $500-$999; Conner Society $1,000; Conner Society Silver $2,000; Conner Society Gold $5,000 & up.

Fort Wayne

AFRICAN-AMERICAN HISTORICAL MUSEUM, 436 E. Douglas Ave., Fort Wayne, IN 46802-3539. Tel.: 260-420-0765. Fax: 260-426-9773.
E-mail: fwaahm@aol.com
Web Site: www.african-americanfw.com
Founded: 1999.
Congressional District: 4
Key Personnel: Dir., Hana L. Stith; Pres. (V), Rubin L. Brown; Museum Shop Mgr., Patsy Brewer
Governing Authority: Tax-exempt.
Institution Type/Description: History Museum.
Collections: African-American history & culture; photographs; personal artifacts; furnishings.
Hours & Admission Prices: Tues.-Fri. 9-1, Sat. 12-4, Sun. by appointment. Adults $5, children $3; members no charge. Closed holidays.
Attendance: 3,000 (accurate)
Membership: Under 16 $5; Senior Citizen $10; Adult $25; Family $40; Contributor & Organization $100; Bronze $500; Sustaining $500 & up; Silver $1,000; Gold $5,000.

ARTLINK, 300 E. Main St., Fort Wayne, IN 46802. Tel.: 260-424-7195. Fax: 260-424-8453.
Web Site: www.artlinkfw.com
Founded: 1978.
Key Personnel: Exec. Dir., Deb Washler; Bd. Pres., Bill Shewman; Education, Rebecca Stockert.
Personnel Profile: Full-Time Paid 2; Part-Time Paid 2.
Governing Authority: private; nonprofit organization. Tax-exempt: 501(c)(3).
Institution Type/Description: Art Gallery.
Collections: works by regional & national artists.
Facilities: 3,000 sq. ft. exhibit space.
Activities: workshops. Annual Events: National Print Exhibition; Wet Paint Auction.
Publications: quarterly newsletter, Genre.
Hours & Admission Prices: Tues.-Fri. 10-5, Sat. 12-6, Sun. 12-5. No charge; donations accepted. Closed New Year's Day; Memorial Day; Independence Day; Labor Day; Thanksgiving; Christmas Eve & Day. &
Attendance: 16,278 (estimated)
Membership: Student $10; Senior $15; Individual Artist $25; Individual $30; Family & Household $40; Contributor $75; Patron $150; Benefactor $300.

CATHEDRAL MUSEUM, Archbishop Noll Catholic Center, 915 S. Clinton St., Fort Wayne, IN 46801. Mailing Address: P.O. Box 390, Fort Wayne, IN 46801. Tel.: 260-422-4611, ext. 3337. Fax: 260-744-1972.
E-mail: stpeter1872@frontier.com
Web Site: www.diocesefwsb.org
Founded: 1980.
Congressional District: 3
Key Personnel: Dir., Rev. Phillip A. Widmann; Chm. (V), Margaret Venderley.
Personnel Profile: Part-Time Volunteers 25.
Governing Authority: Parent Institution: Diocese of Fort Wayne, South Bend. Tax-exempt.
Institution Type/Description: Catholic Church Museum.
Collections: northern Indiana Catholic church history; wood carvings; sacred vessels; church vestments; paintings; religious artifacts, relics; books.
Facilities: Books & gift items for sale.
Activities: special events.
Publications: newsletter.
Hours & Admission Prices: Tues.-Fri. 10-2; other times by appointment. No charge; donations accepted. Parking: no charge. &
Attendance: 6,500 (accurate)

FOELLINGER-FREIMANN BOTANICAL CONSERVATORY, 1100 S. Calhoun St., Fort Wayne, IN 46802-3007. Tel.: 260-427-6440. Fax: 260-427-6450.
Web Site: www.botanicalconservatory.org
Founded: 1983.
Key Personnel: Dir., Mitch Sheppard; Museum Shop Mgr., Jane Ford; Business Devel., Linda Miller.
Personnel Profile: Full-Time Paid 11; Part-Time Paid 6; Part-Time Volunteers 200.
Governing Authority: city. Parent Institution: Fort Wayne Parks and Recreation Department. Tax-exempt.
Institution Type/Description: Botanical Garden.
Collections: orchids; palm trees; cacti; tropical specimens; midwest trees & shrubs.

Facilities: Gift items for sale.
Activities: field trips; group tours; special events; horticultural workshops; plant sales; plant shows; plant swaps; art displays; summer day camp; garden weddings.
Publications: event brochure; quarterly program guide; volunteer newsletter.
Hours & Admission Prices: Tues.-Wed. & Fri.-Sat. 10-5, Thurs. 10-8, Sun. 12-4. Adults $5, children 3-17 $3; discount to AAA members & AHS reciprocal program members; members and children 2 & under no charge. Closed New Year's Day; Christmas. &
Attendance: 70,000 (accurate)
Membership: Add A Guest $15; Individual $35; Family $50; Friend $75; Booster $125; Patron $250; Advocate $500; Manager's Circle $1,000 & up.

FORT WAYNE CHILDREN'S ZOO, (M), 3411 Sherman Blvd., Fort Wayne, IN 46808-1594. Tel.: 260-427-6800. Fax: 260-427-6820. Kids Zoo.
E-mail: zoo@kidszoo.org
Web Site: www.kidszoo.org
Founded: 1965.
Congressional District: 3
Key Personnel: Dir., Jim Anderson; Animal Cur., Mark Weldon; Education, Cheryl Piropato; Devel., Kelly Hagerman; Veterinarian, Joe Smith; Operations, Jim McGowin; Finance, Ann Barker.
Personnel Profile: Full-Time Paid 60; Part-Time Paid 100; Part-Time Volunteers 400; Interns 15.
Volunteer Hours: 32,000
Operating Expenses: 5,800,000
Operating Income: 5,800,000
Governing Authority: nonprofit. Subsidiary Institution: Fort Wayne Zoological Society. Tax-exempt: 501(c)(3).
Institution Type/Description: Zoo.
Collections: African, Australian, & Indonesian animals; Indo-pacific fish & sharks.
Research Fields: Australian animals; native rattlesnakes.
Facilities: 2,000-vol. library of materials on natural history; food service available. Gift items for sale.
Activities: formal education programs; guided tours; lectures; mobile vans; participatory exhibits.
Publications: magazine, Zoo to You; newsletters, Zoo to You Extra; Kangaroo Pouch; z-mail electronic newsletter.
Hours & Admission Prices: late April to mid-Oct. daily 9-5. Adults $13.50, seniors $10.50, children 2-14 $8.50; discounts to AZA members & reciprocal zoos; children under 2 no charge. &
Attendance: 545,000 (accurate)
Membership: Add-A-Guest $30; Individual $75; Single Parent $86; Family & Grandparent $99; Basic Plus $170; Safari Club $250; Director's Circle $500; King of the Jungle $1,000.

FORT WAYNE FIREFIGHTERS MUSEUM, 226 W. Washington Blvd., Fort Wayne, IN 46802-3021. Tel.: 260-426-0051. Facebook: Fort Wayne Firefighters Museum.
E-mail: fwfmuseum@netscape.net
Web Site: www.fortwaynefiremuseum.com
Founded: 1974.
Key Personnel: Pres. (V), Joel Degitz.
Governing Authority: private; nonprofit organization. Tax-exempt: 501(c)(3).
Institution Type/Description: Fire-Fighting Museum: housed in an 1893 structure.
Collections: history of the Fort Wayne Fire Department dating back to 1800s; apparatus; equipment; patch collection.
Facilities: 200-vol. library. Museum-related items for sale.
Activities: guided tours; formal education for adults & children; temporary exhibitions.
Hours & Admission Prices: Mon.-Tues. & Thurs.-Fri. 10-4, Sat. 10-3. Adults $4, senior citizens & students $3; discounts to groups with reservations; members and children 5 & under no charge. Closed New Year's Day; Memorial Day; Independence Day; Labor Day; Thanksgiving; Christmas.
Attendance: 3,113 (accurate)

✱ **FORT WAYNE MUSEUM OF ART, (M),** 311 E. Main St., Fort Wayne, IN 46802-1997. Tel.: 260-422-6467. Fax: 260-422-1374.
E-mail: mail@fwmoa.org
Web Site: www.fwmoa.org
Founded: 1922.
Congressional District: 4
Key Personnel: Pres., Leonard Helfrich; Vice Pres., Ben Eisbart; Business Mgr. & Museum Shop Mgr., Lon R. Braun; Dir., Charles A. Shepard, III;

Dir. External Affairs, Linda Dykhuizen; Cur. Children & Family Education, Max Meyer; Cur. Exhibitions & Programs, Sarah Aubrey; Registrar, Leah Reeder.

Personnel Profile: Full-Time Paid 18; Part-Time Paid 1; Part-Time Volunteers 200.

Governing Authority: nonprofit organization. Tax-exempt: 501(c)(3).

Institution Type/Description: Art Museum.

Collections: 19th- & 20th-century American prints & paintings; sculpture; works on paper; contemporary art.

Research Fields: 20th-century contemporary art.

Facilities: library; auditorium; classroom. Museum-related items for sale.

Activities: guided tours; gallery talks; films; lectures; education programs.

Publications: trimonthly newsletters; catalogs for special exhibits; biennial report.

Hours & Admission Prices: Tues.-Sat. 10-5, Sun. 12-5. Adults $5; discounts to AAM members; members, Wed. & 1st Sun. of each month no charge. Closed New Year's Day; Easter; Memorial Day; Independence Day; Labor Day; Thanksgiving; Christmas. &

Attendance: 80,000 (estimated)

Membership: Individual $35; Dual $50; Family $65; Donors: Bronze $100-$249; Silver $250-$499; Gold $500-$749; Platinum $750-$999; Director's Circle: Founder $1,000-$1,499; Sustainer $1,500-$2,499; Ambassador $2,500-$4,999; Visionary $5,000 & up.

GREATER FORT WAYNE AVIATION MUSEUM, Fort Wayne International Airport, Lt. Paul Baer Terminal, 3801 W. Ferguson Rd., Fort Wayne, IN 46809-3142. Mailing Address: P.O. Box 9573, Fort Wayne, IN 46899. Tel.: 260-478-7146.

E-mail: rogerfortwayne@aol.com

Web Site: www.fwairport.com/air-museum.aspx

Founded: 1984.

Congressional District: 3

Key Personnel: Pres. (V), Barbara Kraegel; Cur., Roger Myers.

Personnel Profile: Part-Time Volunteers 12.

Governing Authority: Tax-exempt.

Institution Type/Description: Aviation Museum.

Collections: area aviation history; military, commercial & general aviation; photographs; personal artifacts.

Facilities: 6,000 sq. ft. exhibit space.

Activities: group tours; scholarship granted to an Allen County high school senior.

Publications: quarterly newsletter, Contact.

Hours & Admission Prices: By appointment. No charge; donations accepted. &

Attendance: 3,000 (estimated)

Membership: Annual $20.

THE HISTORY CENTER, (M), 302 E. Berry St., Fort Wayne, IN 46802-2708. Tel.: 260-426-2882. Fax: 260-424-4419. Facebook: FW History Center.

E-mail: histsociety@fwhistorycenter.com

Web Site: www.fwhistorycenter.com

Formerly: Allen County-Fort Wayne Historical Society Museum

Founded: 1921.

Congressional District: 3

Key Personnel: Pres. (V), Alan Grinsfelder; Exec. Dir., Todd Maxwell Pelfrey; Cur., Walter Font; Community & Project Coord., Robert Nern; Museum Shop Mgr., Kathy Baker; Exhibitor, Randy Elliott; Program Coord., Leanna Harney.

Personnel Profile: Full-Time Paid 4; Part-Time Paid 6.

Governing Authority: nonprofit organization. Parent Institution: Allen County-Fort Wayne Historical Society. Subsidiary Institution: Old City Hall Museum & Chief Richardville House, 5705 Bluffton Rd., Fort Wayne, IN. Tax-exempt: 501(c)(3).

Institution Type/Description: History Museum: housed in Fort Wayne City Hall; c.1893.

Collections: manuscripts & archives pertaining to Fort Wayne, Allen County & northeastern Indiana; 19th & 20th century clothing; c.1790-1820 military artifacts; paintings; toys; china; glass; maps; tools; Indian artifacts; industrial products & equipment; law enforcement artifacts. Historic House: Chief Richardville House c.1827; a National Historic Landmark.

Major Exhibits: Prohibition from IHS, 4/14.

Research Fields: 19th & 20th century local history.

Facilities: library of Allen County history, manuscripts covering Fort Wayne history, 200 rolls of National Archives microfilm on military & Indian affairs & 25,000 photographs of locality available for use by public on premises. Books on local & Indiana history & historically oriented items for sale.

Activities: guided tours; lectures; formally organized education programs;

docent program on permanent & temporary exhibitions; interactive exhibits; Native American study groups.

Publications: magazine, Old Fort News; brochures; newsletter.

Hours & Admission Prices: Mon.-Fri. 10-5, Sat. & 1st Sun. of month 12-5. Adults $5, seniors & students $3; members & children 5 and under no charge. Chief Richardville House: May-Nov. 1st Sat. of month 1-4. Adults $7, senior citizens & students $5; members no charge. New Year's Day; Memorial Day; Independence Day; Labor Day; Thanksgiving; Christmas. &

Attendance: 41,167 (estimated)

Membership: Senior $30; Senior Couple & Individual $35; Family $50; Hamilton $100; Anthony Wayne $250; Kekionga $500; Chief Richardville $1,000.

MACEDONIAN MUSEUM, 124 W. Wayne St., Ste. 204, Fort Wayne, IN 46802-2505. Tel.: 260-422-5900. Fax: 260-422-1348.

E-mail: mtfw@macedonian.org

Web Site: www.macedonian.org

Founded: 1986.

Congressional District: 3

Key Personnel: Pres. (V), Fred Meanchoff; Museum Shop Mgr., Lois Eubank.

Personnel Profile: Full-Time Paid 1; Part-Time Volunteers 3.

Governing Authority: Parent Institution: Macedonian Patriotic Organization. Tax-exempt.

Institution Type/Description: History Museum.

Collections: Macedonian history, art, religion & culture; costumes; personal artifacts; photographs.

Facilities: Museum-related items for sale.

Activities: special events; educational programs.

Publications: Macedonian Tribune.

Hours & Admission Prices: Tues.-Fri. 11-1:30; other times by appointment. Adults $3, youth 13-19 $2; members and children 12 & under no charge. Closed Federal holidays; Christmas week; day after Thanksgiving; snow emergencies.

Attendance: 150 (estimated)

Membership: Senior & Student $20; Individual $25; Family $45; St. Naum $50; St. Clement $100; St. Cyril $250; St. Methodius $500; Mother Teresa $1,000.

SCIENCE CENTRAL, 1950 N. Clinton St., Fort Wayne, IN 46805-4049. Tel.: 260-424-2400. Fax: 260-422-2899.

Web Site: www.sciencecentral.org

Founded: 1995.

Congressional District: 3

Key Personnel: Exec. Dir., Martin S. Fisher.

Personnel Profile: Full-Time Paid 9; Part-Time Paid 53; Part-Time Volunteers 186; Interns 11.

Governing Authority: Tax-exempt.

Institution Type/Description: Science Museum.

Collections: hands-on science, math & technology exhibits.

Facilities: classrooms; demonstration theater; cafeteria.

Activities: field trips; outreach; summer, spring, & winter science camps; teacher workshops; scout programs; birthday parties; temporary exhibitions.

Publications: member newsletter.

Hours & Admission Prices: June-Aug. Tues.-Sat. 10-5, Sun. 12-5; Sept.-May Wed.-Fri. 10-4, Sat. 10-5, Sun. 12-5. Adults $8; discounts to seniors & AAA members; members, active military and children under 2 no charge. Closed New Year's Day; Easter; Memorial Day; Independence Day; Labor Day; Thanksgiving; Christmas Eve & Day. &

Attendance: 74,170 (accurate)

Membership: Basic $65; Deluxe $100.

SWINNEY HOMESTEAD, 1424 W. Jefferson Blvd., Fort Wayne, IN 46802-4111. Tel.: 260-424-7212.

Web Site: www.settlersinc.org

Founded: 1971.

Congressional District: 3

Key Personnel: Pres. (V), Linda H. Huge; Museum Shop Mgr., Kristine Conner.

Personnel Profile: Part-Time Volunteers 25.

Governing Authority: bd. of directors. Tax-exempt.

Institution Type/Description: Historic House Museum: housed in the former home of Thomas & Lucy Swinney, built in 1844. Listed on the National Register of Historic Places.

Collections: period artifacts & furnishings. Historic Building: 1849 log house.

Facilities: herb garden.

Activities: open houses; living history program; in-school activities; Hand Arts Series; special events.

Publications: The Broadside.
Hours & Admission Prices: Groups by appointment. Homestead: no charge; donations accepted. Hands Art Series: $60 for 6 sessions. &
Attendance: 2,032 (estimated)
Membership: General $30; Hand-Arts-Series $60.

Fountain City

LEVI COFFIN HOUSE, (M), 113 U.S. 27 North, Fountain City, IN 47341. Mailing Address: Box 77, Fountain City, IN 47341-0077. Tel.: 765-847-2432. Fax: 765-847-2498.
E-mail: manager@levicoffin.org
Web Site: www.waynet.org
Founded: 1967.
Congressional District: 2
Key Personnel: Pres. (V), Janice McGuire; Museum Shop Mgr., Saundra Jackson.
Personnel Profile: Part-Time Volunteers 30.
Governing Authority: society; nonprofit organization. Parent Institution: Levi Coffin House Association through an operating agreement with the Indiana Sate Museum and Historic Sites, 650 W. Washington St., Indianapolis, IN 46204. Tax-exempt.
Institution Type/Description: Historic House: 1839 Levi Coffin House.
Collections: furnishings & furniture of the period.
Activities: guided tours; volunteer program.
Publications: History Booklet.
Hours & Admission Prices: June-Aug. Tues.-Sat. 1-4; Sept.-Oct. Sat. 1-4; school groups by appointment only. Adults $2, students & school groups $1; members no charge. Closed Independence Day.
Attendance: 5,000 (estimated)
Membership: Student $10; Individual $20; Family $30; Contributing $50; Patron $100; Corporate $200.

Frankfort

CLINTON COUNTY MUSEUM, 301 E. Clinton St., Frankfort, IN 46041-1908. Tel.: 765-659-2030 & 4079. Fax: 765-654-7773.
E-mail: cchsm@geetel.net
Web Site: www.cchsm.org; www.oldstoney.org
Founded: 1980.
Congressional District: 5
Key Personnel: Pres. Bd., Dr. Mark Griffith; Vice Pres., Neil Conner; C.E.O. & Dir., Nancy Hart; Treas., Steve Beets; Sec., Joe Palmer; Bd. Member, Dr. Roger Robison; Bd. Member, Audrey Branagin; Bd. Member, Gloria Ponton; Bd. Member & Newsletter Editor, Donna Harmon; Archivist & Librarian, Joan Cox Bohm; Clinton County Historian, James Miller; Sec., Fay Gilbert.
Personnel Profile: Full-Time Volunteers 12; Part-Time Volunteers 16.
Governing Authority: nonprofit organization. Branch of Clinton County Historical Society. Tax exempt: 501(c)(3).
Institution Type/Description: Local History Museum: housed in c.1892 Old Stoney, former Frankfort High School building.
Collections: 19th-century costumes; journals; manuscripts; rare books; novels & classic literature; photographs; post cards; letters; cut glass; period artifacts; quilts; agricultural tools; transportation objects; 1890s classroom interior objects; local art; medical tools; ornithology; general store; parlor; kitchen; bedroom & military displays following all wars; performance recordings & journals from tours of Freddie Shaffer's All Girl Band; Zerna Sharp's personal papers & photos; Pop Stairs' flight instructor papers, maps & photos; blacksmith shop; outhouse; train room; church; 1905 dental office; archives 1830-1925. Historic Building: 1859 log cabin.
Research Fields: local history; oral history; genealogy; business & industry; schools; post offices; transportation & communication; Freddie Shaffer's All Girl Band, USO tours 1940-1953, members of band were Clinton County natives until the last 2 years of the band.
Facilities: 1,000-vol. library of Indiana history, local high school yearbooks & city directories available for research in presence of museum volunteers on premises by appointment; reading room. Publications & museum-related items for sale.
Activities: guided tours; docent program or council; loan, permanent, temporary & traveling exhibitions; monthly DAR meetings. Museum Sponsors. America Blooms, We Partner With Main Street; Annual Armed Forces Day Program; Clinton County Gem in March, May, August & October; Christmas Open House.
Publications: quarterly newsletter, Clinton County Historical Society & Museum News; books, Dick & Jane, and the Lady Responsible for Their Being; Zerna Sharp, A Clinton County Native; Historical Notes; Pictorial History; Life in the Early 1900's; Civil War Soldiers from County; When Basketball was King & Everett Case; Pictorial - Clinton Co.; I Ran Away with an All Girl Band.

Hours & Admission Prices: Mon.-Fri. 9-4, 3rd Sat. of month 1-4. No charge; donations accepted. Closed holidays. &
Attendance: 1,563 (accurate)
Membership: General $20; Life $200; Business $500; Industry $1,000.

Franklin

JOHNSON COUNTY MUSEUM OF HISTORY, (M), 135 N. Main St., Franklin, IN 46131-1720. Tel.: 317-346-4500. Fax: 317-736-5451.
E-mail: bcundiff@co.johnson.in.us
Web Site: www.johnsoncountymuseum.org
Founded: 1931.
Congressional District: 2
Key Personnel: Pres., Lyman Snyder; Dir., Brenna Cundiff; Cur., Anna Musun-Miller; Librarian, Linda Talley.
Personnel Profile: Full-Time Paid 2; Part-Time Paid 3; Part-Time Volunteers 40; Interns 3.
Governing Authority: society. Subsidiary Institution: Johnson County Historical Society. Tax-exempt.
Institution Type/Description: History Museum.
Collections: textiles; costumes; furniture; decorative accessories; tools; artifacts; period rooms include Pioneer, Civil War, & Victorian artifacts; 1940s & 1950s artifacts; military; works by local artists.
Facilities: library of genealogy & public records area available for research.
Activities: guided tours; lectures; formally organized education programs for children; research.
Publications: quarterly newsletter, Nostalgia News.
Hours & Admission Prices: Mon.-Fri. 9-4, Sat. 10-3. Donation Requested $2. &
Attendance: 10,000 (accurate)
Membership: Student $10; Individual $20; Family $35; Life $500; Family Life $750.

French Lick

BODY REFLECTIONS SALON & HAIR MUSEUM, 448 S. Maple St., French Lick, IN 47432-1083. Tel.: 812-936-7008.
Founded: 1985.
Congressional District: 62
Key Personnel: Dir., Tony Kendall
Institution Type/Description: Hair Museum.
Collections: hair styling history; cosmetology industry; period razors, shears, combs, hairbrushes, hair tonics, curling irons & permanent wave machines; hair; 1800s hair wreath; period hair dryers; bottles; photographs; magazines; hair jewelry.
Activities: Annual Event: Fall Hair & Trade Show.
Hours & Admission Prices: Tues.-Fri. 9-7, Sat. 9-2. No charge.

FRENCH LICK WEST BADEN MUSEUM, (M), 469 S. Maple St., French Lick, IN 47432. Mailing Address: P.O. Box 250, French Lick, IN 47432. Tel.: 812-936-3592.
E-mail: flwbmuseum@gmail.com
Founded: 2011.
Institution Type/Description: History Museum.
Collections: local history from 1900-1930; miniature Hagenbeck-Wallace Circus; local famous artists, politicians, horsemen & sports figures.
Hours & Admission Prices: Tues.-Sat. 10-4. Adults $5; members no charge. &

INDIANA RAILWAY MUSEUM, 8594 W. State Rd. 56, French Lick, IN 47432. Mailing Address: P.O. Box 150, French Lick, IN 47432-0150. Tel.: 800-74-TRAIN. Fax: 812-936-2904.
E-mail: infoirm@indianarailwaymuseum.org
Web Site: www.indianarailwaymuseum.org
Founded: 1961.
Personnel Profile: Full-Time Paid 7; Part-Time Paid 4; Part-Time Volunteers 40.
Governing Authority: Tax-exempt.
Institution Type/Description: Railway Museum.
Collections: railroad history; photographs.
Activities: train rides.
Hours & Admission Prices: Trains: March-Dec. see website for schedule. Adults $16, children 2-11 $8; children under 2 no charge. &
Attendance: 47,668 (accurate)
Membership: Individual $20; Family $60.

Garrett

GARRETT HISTORICAL MUSEUM, Heritage Park, 300 N. Randolph St., Garrett, IN 46738. Mailing Address: P.O. Box 225, Garrett, IN 46738-0225. Tel.: 260-357-5575 & 4812.
E-mail: jmohre@mchsi.com
Web Site: garretthistoricalsociety.org
Founded: 1971.
Congressional District: 4
Key Personnel: Pres. & C.E.O., John Mohre; Cur., Cleo Talley; Museum Shop Mgr., Katrina Custer.
Personnel Profile: Part-Time Volunteers 10.
Governing Authority: nonprofit organization. Tax-exempt.
Institution Type/Description: History Museum.
Collections: railroad memorabilia; local history; caboose; 3 railroad signals; HO scale model railroad; railway post office car; 1875 B & O freight house; diesel locomotive.
Research Fields: railroad history; local history.
Facilities: railroad research library. Photos of old engines, post cards & museum-related items for sale.
Activities: guided tours.
Publications: quarterly newsletter, The Dispatcher.
Hours & Admission Prices: March-Dec. Sat.-Sun. 1-4; Winter: by appointment. No charge; donations accepted. &
Attendance: 4,400 (estimated)
Membership: Single $10; Family $25; Sponsor $50; Lifetime $500.

Geneva

LIMBERLOST STATE HISTORIC SITE, (M), 200 E. 6th St., Geneva, IN 46740-1004. Mailing Address: P.O. Box 356, Geneva, IN 46740-0356. Tel.: 260-368-7428. Fax: 260-368-7007. Facebook: Friends of the Limberlost.
E-mail: limberlostshs@indianamuseum.org
Web Site: www.indianamuseum.org/limberlost
Founded: 1947.
Congressional District: 6
Key Personnel: Pres. (V), Dave Cramer; Cur. & Site Mgr., Randy Lehman; Rgnl. Ecologist, Ken Brunswick; Museum Shop Mgr., Fran Austin.
Personnel Profile: Full-Time Paid 2; Part-Time Paid 4; Part-Time Volunteers 4.
Governing Authority: state. Parent Institution: Indiana State Museums & Historic Sites, 650 W. Washington St., Indianapolis, IN 46204. Branch of Indiana State Museum System. Tax-exempt.
Institution Type/Description: Historic House: Home of Gene Stratton-Porter 1895-1913, author & naturalist.
Collections: period furnishings; memorabilia of Gene Stratton Porter; Loblolly Marsh Wetland Preserve.
Research Fields: the works of Gene Stratton-Porter.
Facilities: Museum-related items for sale.
Activities: guided tours; special programming; outreach; tours of former wetland, a bird sanctuary; school programming.
Publications: pamphlet on Gene Stratton-Porter; newsletter; calendar of events.
Hours & Admission Prices: Tues.-Sat. 9-5, Sun. 1-5; other times by appointment. Adults $5, seniors $4, children $2; members of other Indiana State Historic Sites & state museum members no charge. Closed national holidays. &
Attendance: 6,000 (accurate)
Membership: Senior Citizen & Limited Income $10; Individual $20; Family $35; Business $50; Lifetime $1,000.

Greencastle

PUTNAM COUNTY MUSEUM, 1105 N. Jackson St., Greencastle, IN 46135-1072. Tel.: 765-653-8419.
E-mail: tanis@putnamcountymuseum.org
Web Site: putnamcountymuseum.org
Founded: 2000.
Key Personnel: Exec. Dir., Tanis Monday; Asst. Dir., Sarah Myers; Pres. Bd., Therese Cunningham.
Personnel Profile: Full-Time Paid 1; Part-Time Paid 1; Part-Time Volunteers 30; Interns 6.
Governing Authority: private; nonprofit organization. Tax-exempt: 501(c)(3).
Institution Type/Description: History Museum.
Collections: Putnam County history; natural & cultural artifacts; hands-on exhibits.
Facilities: 3,000 sq. ft. exhibit space.
Activities: guided tours; participatory & temporary exhibitions; presentations; formal education programs for children; DePauw University winter course. Museum Sponsors: Annual Dinner & Meeting; Evening At The Elms.

Publications: Our Past, Their Present: Historical Essays on Putnam County, IN; Peeler Pottery: A Retrospective; newsletter, The Arch.
Hours & Admission Prices: Tues.-Fri. 1-4, Sat. 10-2; other times by appointment. No charge; donations accepted. Closed Memorial Day; Independence Day; Labor Day; Thanksgiving; Christmas. &
Attendance: 3,000 (accurate)
Membership: Student $10; Collectible $25-$49; Keepsake $50-$99; Artifact $100-$249; Antique $250-$499; Treasure $500-$999; Heirloom $1,000-$4,999; Partner $5,000 & up.

RICHARD E. PEELER ART CENTER, (M), 10 W. Hanna St., Greencastle, IN 46135-1911. Mailing Address: DePauw University, 10 W. Hanna St., Greencastle, IN 46135. Tel.: 765-658-4336. Fax: 765-658-6552.
E-mail: craighadley@depauw.edu
Web Site: www.depauw.edu/arts/galleries
Founded: 2002.
Congressional District: 7
Key Personnel: Dir. & Cur. University Galleries, Museums & Collections, Kaytie Johnson; Registrar Univ. Exhibitions & Collections, Christie Anderson.
Personnel Profile: Full-Time Paid 2; Part-Time Paid 3.
Governing Authority: private university. Parent Institution: DePauw University. Tax-exempt: 501(c)(3).
Institution Type/Description: Art Museum.
Collections: contemporary art; Japanese & Tibetan art; American Impressionism.
Facilities: 8,000 sq. ft. exhibit space; 90-seat auditorium; educational facilities.
Activities: lectures; loan, traveling, temporary & participatory exhibits.
Publications: Mind Storm: Contemporary American Folk Art from the Arient Family Collection; Sally Heller: Material Minutiae; 2005 DePauw Biennial: Contemporary Art in the Midwest; Skirting the Line: Conceptual Drawing; Chuck Ramirez: Deeply Superficial.
Hours & Admission Prices: Tues.-Fri. 10-4, Sat. 11-5, Sun. 1-5. No charge. Closed university holidays & breaks. &
Attendance: 5,000 (estimated)

Greenfield

JAMES WHITCOMB RILEY BIRTHPLACE & MUSEUM, Riley Home, 250 W. Main St., Greenfield, IN 46140. Mailing Address: Greenfield Parks and Recreation, 280 N. Apple St., Greenfield, IN 46140-2656. Tel.: 317-462-8539 & 477-4340 (Park Office). Fax: 317-477-4341.
E-mail: parks_rec@greenfieldin.org
Web Site: www.greenfieldin.org
Founded: 1937.
Congressional District: 10
Personnel Profile: Part-Time Paid 10; Part-Time Volunteers 16.
Governing Authority: municipal. Parent Institution: City of Greenfield. Subsidiary Institution: Riley Old Home Society. Tax-exempt.
Institution Type/Description: Historic House & Museum: housed in 1849 birthplace of James Whitcomb Riley.
Collections: furniture; relics; manuscripts; period books; paintings; china; crafts; furniture made by Reuben Riley; John A. Riley poems; James Whitcomb Riley mementos; scrapbooks; local history; newspapers; yearbooks; Will Vawter paintings.
Research Fields: costumes; home decorating of period 1849-1864; literature; local & county history.
Facilities: library. Books, reproductions of John Singer Sargent portrait of Riley & other museum-related items.
Activities: guided tours; permanent exhibitions; special programs. Museum Sponsors: Little Orphan Annie; Pixy Magic Gardens; Happy Birthday to Riley; Annual Riley Days celebration in October; Christmas at the Riley Home.
Publications: brochures.
Hours & Admission Prices: April-Oct. Tues.-Sat. 11-4. House: adults $4, seniors $3.50, children 6-17 $1.50, school groups $1; children under 5 no charge. &
Attendance: 4,118 (accurate)

OLD LOG JAIL AND CHAPEL MUSEUMS, Rte. 40 & Apple St., Greenfield, IN 46140. Mailing Address: P.O. Box 375, Greenfield, IN 46140-0375. Tel.: 317-462-7780 & 0631.
Founded: 1966.
Congressional District: 6
Key Personnel: C.E.O. & Pres. (V), Greg Roland; Cur. & Museum Shop Mgr., Jim Arthur.
Personnel Profile: Part-Time Paid 2; Part-Time Volunteers 9.

Governing Authority: society. Parent Institution: Hancock County Historical Society, P.O. Box 375, Greenfield, IN 46140. Tax-exempt.
Institution Type/Description: Historic Buildings: 1853 Log Jail; 1856 Wooden Chapel.
Collections: Tom's Indian arrowhead collection; coverlets of the mid 1800s; pictures of local family, business & agriculture of 1870-1917; Knoop-Bodkin collection of Indian artifacts. Historic Building: 1856 Philadelphia Church.
Research Fields: jails of the mid-1800s.
Facilities: 50-vol. library. Museum-related items for sale.
Activities: guided tours; lectures; temporary exhibitions; school loan service.
Publications: newsletter, Log Chain.
Hours & Admission Prices: April-Oct. Sat.-Sun. 1-5; other times by appointment. Adults $2, children $1.
Attendance: 800
Membership: Junior $1; Adult $10; Life $100.

Greensburg

DECATUR COUNTY HISTORICAL SOCIETY, 222 N. Franklin St., Greensburg, IN 47240. Mailing Address: P.O. Box 163, Greensburg, IN 47240. Tel.: 812-663-2764.
E-mail: dechissoc@etczone.com
Web Site: www.decaturcountyhistory.org
Institution Type/Description: Historical Society Museum.
Collections: local history & culture; period furnishings; personal artifacts; photographs.
Hours & Admission Prices: Jan.-March Tues. & Thurs. 10-2; April-Dec. Tues., Thurs. & Sat. 10-2, Sun. 1-4.

Greentown

GREENTOWN GLASS MUSEUM, INC., 112 N. Meridian, Greentown, IN 46936-0161. Mailing Address: P.O. Box 161, Greentown, IN 46936-0161. Tel.: 765-628-6206.
Web Site: www.greentownglass.org
Founded: 1969.
Congressional District: 5
Key Personnel: Pres. (V), Merrill Swisher; Vice Pres., Jeffrey Martin; Sec., Sally Mower; Head Cur., Norma Jean David; Treas., Sharon Oldaker.
Personnel Profile: Part-Time Paid 5; Part-Time Volunteers 18.
Governing Authority: nonprofit organization. Parent Institution: Greentown Glass Museum. Tax-exempt: 501(c)(3).
Institution Type/Description: Glass Museum.
Collections: glassware manufactured by the Indiana Tumbler & Goblet Co. of Greentown, 1894-1903; Chocolate glass from other factories poured by Jacob Rosenthal, developer of Chocolate glass at Greentown; tools & materials used in the making of glassware; Holly Amber (Golden Agate), Chocolate & Nile Green Glassware collections.
Research Fields: history of the town at the time of the glassworks; history of the glassworks & the people working there; glassware poured at the Indiana Tumbler & Goblet Co.
Facilities: library of material relating to the Indiana Tumbler & Goblet Co. & the glassware manufactured there. Books & museum-related items for sale.
Activities: guided tours; lectures; films; arts festivals; temporary exhibitions. Annual Event: Fundraiser in April.
Publications: brochures.
Hours & Admission Prices: March-May 14 & Nov. Sat.-Sun. 1-4; May 15-Oct. Wed.-Fri. 10-12 & 1-4, Sat.-Sun. 1-4. No charge; donations accepted. Closed Easter. &
Attendance: 2,000 (estimated)
Membership: Share $25.

Hagerstown

HAGERSTOWN MUSEUM AND ARTS PLACE, 96 1/2 E. Main, Hagerstown, IN 47346-1213. Mailing Address: P.O. Box 126, Hagerstown, IN 47346-0126. Tel.: 765-489-4005. Fax: 765-489-4005.
E-mail: tom@hagerstownmuseum.comcastbiz.net
Formerly: Nettle Creek Cultural Center
Founded: 1973.
Key Personnel: Dir., Tom Butters.
Personnel Profile: Part-Time Paid 1; Part-Time Volunteers 25.
Governing Authority: Parent Institution: Historic Hagerstown, Inc. Tax-exempt.
Institution Type/Description: Historic Building: built in 1880.
Collections: local history & culture; murals; industry; works by local artists.
Facilities: Works by area artists for sale.
Activities: art & violin lessons.

Hours & Admission Prices: March 2-Dec. 24 Wed.-Sat. 1-6; groups & other times by appointment. No charge; donations accepted. Closed major holidays.
Membership: Student $15; Individual $20; Family 25; Business $30; Friend $50-$100; Patron $100-$300; Supporter $301-$500; Benefactor $501 & up.

WILBUR WRIGHT BIRTHPLACE & MUSEUM, 1525 N. Co. Rd. 750 E., Hagerstown, IN 47346. Tel.: 765-332-2495.
E-mail: wilbur@nltc.net
Web Site: www.wwbirthplace.com
Founded: 1929.
Congressional District: 2
Key Personnel: Chm. (V), Pat MaLott; Vice Chm., Sylvia Ward; Treas., Susan Thornburg; Museum Shop Mgr., Thornton McKay.
Personnel Profile: Full-Time Paid 1; Part-Time Paid 1; Part-Time Volunteers 30.
Governing Authority: Wilbur Wright Birthplace Preservation Society. Tax-exempt.
Institution Type/Description: Historic House: c.1867 re-created birthplace of Wilbur Wright.
Collections: period furniture; replica of 1903 Wright Flyer; F-84F jet fighter plane; local history, Brethren Church associations with The Wright Family; Wright smokehouse; replica birthplace house of Wilbur Wright.
Facilities: picnic area; meeting room; welcome center. Museum-related items for sale.
Activities: tours.
Publications: newsletter.
Hours & Admission Prices: House & Museum: April-Oct. Tues.-Sat. 10-5, Sun. 1-5. Family $10, adults $4, seniors $3, students $2; members no charge. &
Attendance: 10,000 (estimated)
Membership: Single $20; Family $25; Corporation $100; Lifetime $1,000.

Hammond

JOHN DILLINGER MUSEUM, 7770 Corinne Dr., Hammond, IN 46323. Tel.: 219-989-7979. Fax: 219-989-7777.
E-mail: info@dillingermuseum.com
Web Site: www.dillingermuseum.com
Founded: 1999.
Key Personnel: Dir., Spero Batistatos; Chm. (V), Vic DeMeyer.
Governing Authority: Parent Institution: South Shore Convention & Visitors Authority.
Institution Type/Description: History Museum.
Collections: life of John Dillinger & other criminals during the 1930s; photographs; early 20th century crime fighting technology; life-size wax figures; personal artifacts.
Facilities: Museum-related items for sale.
Hours & Admission Prices: Daily 10-4. Adults $4, seniors $3, children 6-12 $2; discounts to active law enforcement & military personnel with valid ID; children 5 & under no charge. &

SUZANNE G. LONG LOCAL HISTORY ROOM, 564 State St., Hammond, IN 46320-1532. Tel.: 219-931-5100, ext. 306. Fax: 219-931-3474. TDD: 219-852-2232.
E-mail: lytler@hammond.lib.in.us
Web Site: www.hammond.lib.in.us
Formerly: Hammond Historical Society
Founded: 1960.
Congressional District: 1
Key Personnel: Dir., Rene Greenleaf; Librarian, Richard Lytle; Public Information Coord., Linda Swisher.
Governing Authority: nonprofit organization. Affiliated with Hammond Public Library. Tax-exempt: 501(c)(3).
Institution Type/Description: Historical Society Museum.
Collections: books; photographs; tapes on Hammond & Calumet region history; video cassettes; scrapbooks; maps; documents.
Research Fields: regional & local history; early settlement; business; education; architectural history
Facilities: 250-vol. library of books, newspaper negatives & pamphlets, including 150 cassettes of programs & tours; photocopies of print material by request; reading room.
Activities: temporary exhibitions.
Publications: 8 issues, newsletter, Hammond Historical Society.
Hours & Admission Prices: Tues. & Thurs.-Fri. 1-5, Wed. 1-9, Sat. 9-5. No charge. &
Attendance: 1,500
Membership: Individual $10; Life $100.

Hartford City

BLACKFORD COUNTY HISTORICAL SOCIETY, 321 N. High St., Hartford City, IN 47348. Mailing Address: P.O. Box 264, Hartford City, IN 47348-0264.
Governing Authority: Tax-exempt.
Institution Type/Description: Historical Society Museum.
Collections: local history & culture; photographs; personal artifacts; period furnishings.
Publications: newsletter.
Hours & Admission Prices: April-Nov. Sun. 1-4. No charge; donations accepted. &
Attendance: 200 (estimated)
Membership: Individual $10.

Hebron

RAILROAD DEPOT MUSEUM, 125 N. Main St., Hebron, IN 46341. Mailing Address: P.O. Box 679, Hebron, IN 46341. Tel.: 219-996-3192.
Founded: 1994.
Congressional District: 4
Institution Type/Description: History Museum: housed in a former depot; built in 1868.
Collections: local history; railroad artifacts & memorabilia; period furnishings; photographs; local newspapers; uniforms.
Hours & Admission Prices: By appointment. No charge; donations accepted. &
Membership: Individual $10.

Hobart

HOBART HISTORICAL SOCIETY MUSEUM, 706 E. Fourth St., Hobart, IN 46342-4411. Mailing Address: P.O. Box 24, Hobart, IN 46342-0024. Tel.: 219-942-0970.
Founded: 1968.
Congressional District: 1
Key Personnel: Pres., Elin Christianson; Business Officer, Robert Green.
Personnel Profile: Part-Time Volunteers 12.
Governing Authority: nonprofit organization. Parent Institution: Hobart Historical Society, Inc. Tax-exempt.
Institution Type/Description: Historical Society Museum: housed in c.1914-1915 Carnegie Library Building. Listed on the National Register of Historic Places.
Collections: local history & culture. Ballantyne Gallery: wheelwright & woodworking tools; a replica of blacksmith shop & print shop; agricultural implements.
Research Fields: local history; local genealogy.
Facilities: 1,500-vol. library on the Calumet area and Hobart history available for inter-library loan or on-site research; c.1891 to present microfilms of local newspapers; 1840-1920 microfilm of Hobart Township census; reading room.
Activities: guided tours; lectures; films; gallery talks; permanent, temporary, loan & traveling exhibitions; school loan service.
Publications: quarterly newsletter, Hobart History News; books, George Earle & Family of Hobart Indiana; Along the Route: History of Hobart Post Office; The Old Settlers Cemetery (1975); The Nine Day Wonder and The One Month Doodlebug (2002); Growing Up in Hobart (1994); Hobart Memories (1996); Lake County Communities Past & Present (2001); periodical newspaper, Hobart History Advocate; Hobart's Historic Buildings (2002).
Hours & Admission Prices: Sat. 10-12; other times by appointment. No charge; donations accepted.
Attendance: 1,500 (accurate)
Membership: Senior Citizens $1; Personal $5; Contributing $25; Institutional $50; Century Club $100.

WOOD'S HISTORIC GRIST MILL, 9410 Old Lincoln Hwy., Hobart, IN 46342-7049. Mailing Address: Lake County Parks & Recreation Dept., 8411 E. Lincoln Hwy., Crown Point, IN 46307. Tel.: 219-947-1958. Fax: 219-947-7105.
E-mail: info@lakecountyparks.com
Web Site: www.lakecountyparks.com/activities/deep_river/woods_mill.html
Founded: 1979.
Key Personnel: C.E.O., Robert J. Nickovich; Supt. Business Devel., James Basala; Supt. Visitor Svcs., Sandra Basala; Supt. Park Opers., Robert Studebaker; Supt. Park Svcs., David Kwolek
Institution Type/Description: Historic Building: grist mill built late 1800s.
Collections: local history; period artifacts; photographs.

Facilities: nature trails.
Activities: demonstrations; hiking; picnicking.
Publications: Pathfinder.
Hours & Admission Prices: May-Oct. Mon.-Thurs. 10-4, Sat. 10-5, Sun. 12-5. No charge; donations accepted. Closed Memorial Day, Independence Day; Labor Day; Columbus Day. &

Huntington

HUNTINGTON COUNTY HISTORICAL SOCIETY MUSEUM, 315 Court St., Huntington, IN 46750-2862. Tel.: 260-356-7264.
E-mail: huntingtonhistoricalmuseum@gmail.com
Web Site: www.huntingtonhistoricalmuseum.com
Founded: 1932.
Congressional District: 5
Key Personnel: Dir. & Museum Shop Mgr., Sarah Schmidt; Pres., Patti Sovers.
Personnel Profile: Part-Time Paid 1; Part-Time Volunteers 25.
Governing Authority: society. Tax-exempt.
Institution Type/Description: Historical Society Museum.
Collections: county items; Indian stones & arrowheads.
Facilities: 8,500 sq. ft. exhibit space.
Activities: permanent exhibitions.
Publications: bimonthly newsletter.
Hours & Admission Prices: Feb.-Dec. Wed.-Fri. 10-4, Sat. 1-4. Adults $2; members no charge. Closed Independence Day; Thanksgiving; Christmas. &
Attendance: 3,500 (estimated)
Membership: Student $12; Individual $18; Family $24; Sustaining $50; Life $250; Corporate $350.

ROBERT E. WILSON GALLERY, Huntington University, 2303 College Ave., Huntington, IN 46750-1237. Tel.: 260-356-6000.
E-mail: mduffer@huntington.edu
Web Site: www.huntington.edu/Wilson-Gallery/
Key Personnel: Dir., Melissa Duffer.
Governing Authority: Tax-exempt.
Institution Type/Description: Art Museum.
Collections: contemporary paintings, prints & sculptures.
Hours & Admission Prices: Mon.-Fri. 9-5 and before & after all Merillat Centre for the Arts performances; other times by appointment. No charge. &
Attendance: 3,000

SHEETS WILDLIFE MUSEUM, 200 Safari Trail, Huntington, IN 46750-8049. Tel.: 260-356-9453.
E-mail: sheetswildlife@gmail.com
Web Site: sheetswildlifemuseum.com
Founded: 2005.
Key Personnel: Dir., Shirley M. Schug.
Personnel Profile: Full-Time Paid 1.
Governing Authority: Tax-exempt.
Institution Type/Description: Wildlife Museum.
Collections: animals & fish from around the world; shell art.
Facilities: theater. Museum-related items for sale.
Publications: quarterly, Museum Messenger.
Hours & Admission Prices: Tues.-Sat. 10-4. Adults $4, seniors & children $2; discounts to groups; children under 4 no charge. &
Attendance: 1,327 (accurate)

U.S. VICE PRESIDENTIAL MUSEUM AT THE DAN QUAYLE CENTER, 815 Warren St., Huntington, IN 46750-2151. Mailing Address: P.O. Box 856, Huntington, IN 46750-0856. Tel.: 260-356-6356. Fax: 260-356-1455.
E-mail: info@quaylemuseum.org
Web Site: www.quaylemuseum.org
Formerly: The Dan Quayle Center & Museum
Founded: 1993.
Congressional District: 4
Key Personnel: Exec. Dir., Daniel Johns.
Governing Authority: private; nonprofit organization. Tax-exempt.
Institution Type/Description: History Center: housed in 1919 former Christian Scientist Church; focus on the Vice Presidency & its challenges.
Collections: life of Dan Quayle from birth until the end of his vice presidency of the United States; history of four other Vice Presidents from Indiana, and the artifacts relating to all U.S. Vice Presidents.
Research Fields: history of the U.S. Vice Presidency; Vice Presidents from Indiana; Dan Quayle's life.
Facilities: 4,000 sq. ft. exhibit center.
Activities: seminars; lectures.

Publications: quarterly newsletter.
Hours & Admission Prices: Mon.-Fri. 9:30-4:30. Adults $3, children 7-17 $1; children under 6 & members no charge. Closed major holidays. &
Attendance: 6,000 (accurate)
Membership: Grass Roots $30; Contributing $50; Sustaining $100; Supporting $250; Dan's Circle $1,000; Builder's Club $2,500; Founder's Club $5,000.

Indianapolis

*** BENJAMIN HARRISON PRESIDENTIAL SITE, (M),** 1230 N. Delaware St., Indianapolis, IN 46202-2531. Tel.: 317-631-1888. Fax: 317-632-5488.
E-mail: harrison@bhpsite.org
Web Site: www.bhpsite.org
Formerly: President Benjamin Harrison Home
Founded: 1966.
Congressional District: 11
Key Personnel: Pres. & C.E.O., Phyllis Geeslin; Education, Roger Hardig; Education, David Pleiss; Cur., Jennifer Capps; Dir., Melissa Calahan; Financial Admin., Margaret Sallee; Dir. Devel., Erin Trisler; Events, Stacy Clark.
Personnel Profile: Full-Time Paid 7; Part-Time Paid 3; Part-Time Volunteers 65; Interns 2.
Governing Authority: nonprofit corporation. Tax-exempt: 501(c)(3).
Institution Type/Description: Historic House Museum: c.1874-75 President Benjamin Harrison home.
Collections: furniture; furnishings; books; election campaign items; inaugural Bible & other memorabilia; gifts to the president; law library; rare books; personal papers; women suffrage materials.
Research Fields: sponsored Harrison biographies by Fr. Harry J. Sievers, S. J.; materials on Benjamin Harrison, his grandfather, Wm. H. Harrison, Mary Lord Dimmick Harrison & Caroline Scott Harrison; Belva Lockwood; Arthur Jordan.
Facilities: library of books, general literature & court records of the Harrison collection available for use on premises; meeting & banquet rooms. Gift shop with postcards, books, pamphlets & notepaper, reproduction White House china for sale.
Activities: guided tours; docent program; educational programs & special events; changing exhibitions; Live from Delaware Street. Museum Sponsors: Legacy of Leadership in February.
Publications: quarterly newsletter, The Statesman.
Hours & Admission Prices: late Jan. to May & Aug.-Dec. Mon.-Sat. 10-3:30; June-July Mon.-Sat. 10-3:30, Sun. 12-3:30. Adults $10, seniors $8, students & college students $5; discounts for groups, AAM & AAA members; members no charge. Closed New Year's Day; Easter; 500 Race Weekend; Memorial Day; Labor Day; Thanksgiving; Christmas Eve & Day. &
Attendance: 26,000 (accurate)
Membership: Voters (Individuals) $25; Electors (Family) $35; Delegates $50; Representatives $100; Senators $250; Justices $500; Cabinet $1,000 & up.

BROAD RIPPLE GALLERY, 714 E. 65th St., Indianapolis, IN 46220-1610. Tel.: 317-253-5340. Fax: 317-253-5468.
E-mail: info@hoosiersalon.org
Web Site: www.hoosiersalon.org
Formerly: Hoosier Salon Gallery
Founded: 1925.
Congressional District: 10
Key Personnel: Exec. Dir., Donnae Dole.
Personnel Profile: Full-Time Paid 1; Part-Time Paid 2; Part-Time Volunteers 30.
Governing Authority: nonprofit organization. Parent Institution: Hoosier Salon Patrons Association, Inc. Tax-exempt: 501(c)(3).
Institution Type/Description: Art Gallery.
Collections: paintings; sculpture; prints.
Research Fields: Hoosier (Indiana) artists.
Activities: gallery talks; permanent, temporary & traveling exhibitions; community events. Gallery Sponsors: annual juried exhibition of living Indiana artists; artist seminar; spring & fall community tours; celebration of Hoosier artists.
Publications: quarterly, Hoosier Salon Newsletter; book, A Grand Tradition: The Art and Artists of the Hoosier Salon, 1925-1990.
Hours & Admission Prices: Tues.-Fri. 11-5, Sat. 11-3; other times by appointment. No charge. &
Attendance: 35,000 (estimated)
Membership: Artist $30; Patron $65; Business Patron, Contributing & Sustaining Business $100; Sustaining Patron $250; Corporate Sponsor $500; Lifetime Patron $2,000.

*** THE CHILDREN'S MUSEUM OF INDIANAPOLIS, (M),** 3000 N. Meridian St., Indianapolis, IN 46208-4716. Mailing Address: P.O. Box 3000, Indianapolis, IN 46206-3000. Tel.: 317-334-3322. Fax: 317-921-4019. TDD: 317-920-2020.
E-mail: customerservice@childrensmuseum.org
Web Site: www.childrensmuseum.org
Founded: 1925.
Congressional District: 10
Key Personnel: Chm. Trustees (V), Yvonne Shaheen; Pres. (V), Jeffrey H. Patchen; Vice Pres. Finance, Karen Kennelly; Dir. Education, David Cassady; Dir. Community Svcs., Janet Boston; Dir. Investments Management, John Grogan; Dir. Health, Safety & Security, Clarence Taylor; Museum Shop Mgr., Robert Tate.
Personnel Profile: Full-Time Paid 180; Part-Time Paid 200; Part-Time Volunteers 1,575.
Governing Authority: nonprofit corporation. Tax-exempt: 501(c)(3).
Institution Type/Description: Children's Museum.
Collections: archaeology; ethnology; geology; history; natural history; paleontology; transportation with exhibits on railroading; natural science; physical science; prehistory; ancient Egypt; leisure time pursuits; operating Dentzel carousel; operating toy train layout; early toy trains; model trains; toys; dolls; folk art; objects from around the world; seasonal displays.
Research Fields: how children learn in informal environments.
Facilities: 4,000-vol. library related to collections for use on premises & inter-library loan; resource center; 130-seat planetarium; 350-seat theater; orientation classroom; meeting rooms; restaurant; theater. Gift items for sale.
Activities: guided tours; lectures; films; permanent & temporary exhibits; planetarium programs; gallery talks & demonstrations; science & computer festivals; drama; formally organized education programs for children & adults; craft classes; nature field trips; nature walks; performing arts including concerts, theater, dance & puppetry; mobile performing stage; science demonstrations; teacher workshops; Scout badge classes; hands-on exhibits.
Publications: monthly newsletter; quarterly report; souvenir booklet; annual report; gallery guides; teachers' guides; docent training manuals; pamphlets, museum history.
Hours & Admission Prices: March to Labor Day daily 10-5; Sept.-Feb. Tues.-Sun. 10-5. Adults $13.50, seniors $12.50, children 2-17 $8.50; 1st Thurs. of month 4-8 & members no charge. Closed Easter; Thanksgiving; Christmas. &
Attendance: 1,200,000 (accurate)
Membership: Individual $40; Family $55; Grandparents $60; Family Plus $75.

CRISPUS ATTUCKS CENTER MUSEUM, 1140 Dr. Martin Luther King Jr. St., Indianapolis, IN 46202-2221. Tel.: 317-226-2430. Fax: 317-226-4611.
E-mail: museumca@ips.k12.us
Web Site: www.crispusattucksmuseum.ips.k12.in.us/
Founded: 1998.
Key Personnel: Dir., Pat Payne; Cur., Robert Chester.
Personnel Profile: Full-Time Paid 2; Part-Time Paid 1; Part-Time Volunteers 5.
Institution Type/Description: History Museum.
Collections: African American history; school history; Tuskegee Airmen; Crispus Attucks High School memorabilia.
Hours & Admission Prices: Mon.-Fri. 9-5, Sat. 12-5; guided tours by appointment. Adults $5, senior citizens 65 & over, college students & youth 6-13 $2; discount to groups of 15 & over.
Attendance: 5,000 (estimated)

*** EITELJORG MUSEUM OF AMERICAN INDIANS AND WESTERN ART, (M),** 500 W. Washington, Indianapolis, IN 46204-2775. Tel.: 317-636-9378, ext. 0. Fax: 317-264-1724.
Web Site: www.eiteljorg.org
Founded: 1985.
Key Personnel: Pres. & C.E.O., John Vanausdall; Chm., Betsey Harvey; Vice Pres. & Chief Cur. Officer, James H. Nottage; Vice Pres. Devel., Susie Maxwell; Vice Pres. Administration & C.F.O., Susan Lewis; Vice Pres. Public Programs & Visitor Svcs., Martha Hill; Vice Pres. Facilities & Security, Jim Fulton; Dir. Communications & Mktg., Tamara Winfrey Harris; Cur. Contemporary Art, Jennifer Complo; Dir. Collections, Amy McKune; Gund Cur. Western Art History & Culture, Suzan Campbell; Museum Shop Mgr., Judy Kirkwood.
Personnel Profile: Full-Time Paid 51; Part-Time Paid 19; Part-Time Volunteers 806; Interns 9.
Governing Authority: nonprofit organization. Tax-exempt: 501(c)(3).
Institution Type/Description: Art Museum.
Collections: paintings, drawings, & bronzes relating to the American West; Native American art & cultural artifacts.

Research Fields: Native American cultures; artists & artwork of the West.
Facilities: library of printed material; 120-seat theater. Gift items for sale.
Activities: guided tours; lectures; films; organized education programs; docent program; loan, temporary & traveling exhibitions; school loan service; artist in-residence program; rental facilities.
Publications: quarterly newsletter; exhibition catalogs.
Hours & Admission Prices: Mon.-Sat. 10-5, Sun. 12-5; groups by appointment. Adults $8, senior citizens 65 & over $7, children 5-17 & full-time students $5; discount to AAM members, military & groups; members, IUPUI students, and children 4 & under no charge. Closed New Year's Day; Thanksgiving; Christmas Eve & Day. &
Attendance: 113,447 (accurate)
Membership: Teacher Plus Guest $45; Individual Plus Guest $50; Dual & Teacher Family $55; Family & Grandparent $60; Contributing $100; Sustaining $250; Patron $500; Eagle $1,500; Golden Eagle $2,500; Presidents Society $5,000.

EMIL A. BLACKMORE MUSEUM OF THE AMERICAN LEGION, 700 N. Pennsylvania St., Indianapolis, IN 46204-1129. Mailing Address: P.O. Box 1055, Indianapolis, IN 46206-1055. Tel.: 317-630-1200. Fax: 317-630-1241.
E-mail: library@legion.org
Web Site: www.legion.org
Founded: 1967.
Key Personnel: Natl. Adjutant, Daniel S. Wheeler; Dir. Library & Museum, Howard Trace.
Governing Authority: nonprofit organization. Parent Institution: The American Legion. Tax-exempt: 501(c)(19).
Institution Type/Description: Military History Museum.
Collections: 20th-century U.S. military history & American Legion development.
Hours & Admission Prices: Mon.-Fri. 8-4. No charge. Closed New Year's Eve & Day; Martin Luther King Jr. Day; Presidents' Day; Good Friday; Memorial Day; Independence Day; Labor Day; Veterans Day; Thanksgiving & day after; Christmas Eve & Day. No charge for admission. &

THE ENDOWMENT FUND OF THE PHI KAPPA PSI FRATERNITY-HERITAGE HALL, 5395 Emerson Way, Indianapolis, IN 46226-1415. Tel.: 317-632-1852. Fax: 317-637-1898.
E-mail: info@pkpfoundation.org
Web Site: www.phikappapsi.com
Founded: 1978.
Congressional District: 10
Key Personnel: Exec. Dir., Shawn M. Collingsworth; Chm., Wayne W. Wilson; Pres., Paul R. Wineman.
Personnel Profile: Full-Time Paid 2; Part-Time Volunteers 1.
Governing Authority: society; nonprofit. Affiliated Organization: Phi Kappa Psi Fraternity. Tax-exempt.
Institution Type/Description: General Museum: housed in c.1876 two-story brick Italianate home.
Collections: decorative arts; period furniture.
Facilities: 1,200-vol. library.
Activities: guided tours.
Publications: quarterly magazine, The Shield.
Hours & Admission Prices: Mon.-Fri. 1-4. No charge. Closed national holidays.
Attendance: 750 (estimated)

FRANKLIN TOWNSHIP HISTORICAL SOCIETY, 6510 S. Franklin Rd., Indianapolis, IN 46239. Mailing Address: P.O. Box 39015, Indianapolis, IN 46239. Tel.: 317-862-8822.
E-mail: kimgada2004@yahoo.com
Founded: 1975.
Governing Authority: private; nonprofit organization. Tax-exempt: 501(c)(3).
Institution Type/Description: Historical Society Museum: housed in a former church; built in 1871. Listed on the National Register of Historic Buildings.
Collections: local history & culture; personal artifacts; period furnishings; photographs.
Facilities: 200-vol. library.
Activities: education programs; school loan service; temporary exhibitions. Annual Events: Quilt Shows; Antique Show & Tell; dinner meeting.
Publications: quarterly newsletter.
Hours & Admission Prices: March-Oct. 1st Sat. & 3rd Sun. 1-4; other times by appointment. No charge; donations accepted. &
Attendance: 300 (estimated)
Membership: Individual $10; Sustaining $25; Lifetime $100.

FREETOWN VILLAGE LIVING HISTORY MUSEUM, 625 Indiana Ave., Indianapolis, IN 46202-3133. Mailing Address: P.O. Box 1041, Indianapolis, IN 46206-1041. Tel.: 317-631-1870. Fax: 317-631-0224.
E-mail: freetown_info@freetownvillage.org
Web Site: www.freetown.org
Founded: 1982.
Congressional District: 7
Key Personnel: Dir., Ophelia Wellington; Program Dir., Mirriam A. Umar
Institution Type/Description: History Museum.
Collections: Indiana's African American history & culture.
Activities: programs; workshops; special events.
Publications: The Freetown Villager.
Hours & Admission Prices: Mon.-Fri. 10-5. No charge; donations accepted. Closed major holidays. &
Attendance: 17,500 (estimated)
Membership: Individual $15; Family $35; Village Elder $100; Corporate Elder $500; Heritage Builder $1,000.

HERRON GALLERIES, 735 W. New York St., Indianapolis, IN 46202. Tel.: 317-278-9419. Fax: 317-278-9435. Facebook: Herron Galleries.
E-mail: katzp@iupui.edu/galleries
Web Site: herron.iupui.edu
Founded: 1979.
Key Personnel: Dir., Paula Katz.
Personnel Profile: Full-Time Paid 2; Part-Time Paid 10; Interns 1.
Governing Authority: Parent Institution: Herron School of Art & Design. Tax-exempt.
Institution Type/Description: Art Gallery.
Collections: works by local, national & international artists; performance & installation art.
Major Exhibits: Ossuary (T), 1/14-2/14; Juvenile in Justice (T), 3/14-4/14.
Research Fields: contemporary art.
Facilities: 4,500 sq. ft. exhibition space.
Activities: visiting artist lectures; scholars; films; panel talks.
Hours & Admission Prices: Mon.-Tues. & Thurs.-Sat. 10-5, Wed. 10-8. No charge; donations accepted. &
Attendance: 25,000 (estimated)

HOOK'S HISTORICAL DRUG STORE AND PHARMACY MUSEUM, Indiana State Fairgrounds, 1202 E. 38th St., Indianapolis, IN 46205-2807. Tel.: 317-924-1503.
Web Site: www.hooksmuseum.org
Key Personnel: Chm., Bob Hunt; Pres., Tabitha Cross.
Personnel Profile: Part-Time Paid 5; Part-Time Volunteers 15.
Governing Authority: Parent Institution: Greenfield Museum Initiative.
Institution Type/Description: History Museum: built in 1849.
Collections: American pharmacy artifacts; soda fountain memorabilia; period furnishings.
Hours & Admission Prices: Call for hours. No charge; donations accepted.
Attendance: 50,000 (estimated)

HOOSIER SALON PATRONS ASSOCIATION GALLERIES, 714 E. 65th St., Indianapolis, IN 46220. Tel.: 317-253-5340. Fax: 317-253-5468.
E-mail: info@hoosiersalon.org
Web Site: www.hoosiersalon.org
Institution Type/Description: Art Gallery.
Collections: paintings; sculpture.
Activities: temporary exhibitions.
Hours & Admission Prices: Call for hours.

IUPUI CULTURAL ARTS GALLERY, 420 University Blvd., Ste. 278, Indianapolis, IN 46202-5147. Tel.: 317-278-8511. Fax: 317-278-0828.
E-mail: hayesjoe@iupui.edu
Web Site: www.iupui.edu/~cagcc/
Founded: 2008.
Congressional District: 7
Key Personnel: Mgr., Jes Shepard.
Governing Authority: public university. Parent Institution: Indiana University Purdue University Indianapolis.
Institution Type/Description: Art Museum.
Collections: paintings; sculpture.
Hours & Admission Prices: Mon.-Sat. 10-7, Sun. 1-7. No charge. &
Attendance: 18,000 (estimated)

*** IMA - INDIANAPOLIS MUSEUM OF ART, (M),** 4000 Michigan Rd., Indianapolis, IN 46208-3326. Tel.: 317-923-1331. Fax: 317-931-1978. Facebook: IMA Indianapolis Museum of Art.
E-mail: ima@imamuseum.org
Web Site: www.imamuseum.org
Founded: 1883.
Congressional District: 7
Key Personnel: Mel & Bren Simon Dir. & C.E.O., Dr. Charles L. Venable; Chm., June McCormack; Vice Chm., Thomas Hiatt; Vice Chm., Rick Johnson; Vice Chm., Kathleen Postlethwait; C.O.O., Nicholas Cameron; Deputy Dir. Public Affairs, Katie Zarich; Deputy Dir. Environmental & Historic Preservation, The Ruth Lilly, Mark Zelonis; Dir. Historic Resources, Bradley Brooks; Dir. Human Resources, Laura McGrew; C.F.O., Jennifer Bartenbach; Wood-Pulliam Distinguished Sr. Cur., Ellen W. Lee; Cur. Asian Art, Dr. John T. Teramoto; Sr. Cur. Contemporary Art, Dr. Lisa Freiman; Assoc. Cur. Contemporary Art, Sarah Green; Assoc. Cur. Prints, Drawings & Photographs, Annette Schlagenhauff; Cur. Prints, Drawings & Photographs, Martin F. Krause, Jr.; Cur. Textile & Fashion Arts, Niloo Imami-Paydar; Cur. Paintings & Sculpture Before 1800, Ronda J. Kasl; Dir. Audience Engagement, Preston Bautista; Mgr. Exhibitions, Kayla Tackett; Deputy Dir. Collections & Exhibitions, Katie Haigh; Dir. Security, Martin Whitfield; Sr. Cur. Design Arts, R. Craig Miller; Dir. Retail Svcs., Jenny Geiger.
Personnel Profile: Full-Time Paid 208; Part-Time Paid 134; Part-Time Volunteers 493; Interns 21.
Governing Authority: nonprofit. Subsidiary Institutions: Miller House and Garden, Columbus, IN; 100 Acres: The Virginia B. Fairbanks Art & Nature Park, Indianapolis, IN. Tax-exempt: 501(c)(3).
Institution Type/Description: Art Museum & Historic Site: Lilly House - former home of J.K. Lilly, Jr., the late Indianapolis businessman, collector & philanthropist. National Historic Landmark.
Collections: European & American paintings & sculpture; African, American, Asian, European, contemporary & decorative art, including paintings, sculpture, prints, drawings, photographs, textiles & costumes; Asian art; textiles; costumes; decorative arts; pre-Columbian art; Clowes Fund collection; J.M.W. Turner collection; the Holliday collection of Neo-Impressionist art; Eiteljorg collection of African & South Pacific Art; manuscripts; Josefowitz collection of Gauguin and the School of Pont-Aven. Japanese Edo-period paintings. Oldfields - Lilly House & Gardens: 22-room mansion on a 26-acre estate. 100 Acres: The Virginia B. Fairbanks Art & Nature Park - contemporary sculptures.
Facilities: reference library; educational resource center; gardens; 175-seat lecture hall & auditorium; educational resource center; restaurant; 500-seat special events pavilion; 600-seat theater; conservation science laboratory. Museum-related items for sale.
Activities: guided tours; lectures; films; gallery talks; concerts; arts festivals; formally organized education programs; docent program or council; permanent, temporary & traveling exhibitions.
Publications: monthly email newsletter; catalogs of permanent collection & exhibitions; brochures; quarterly member magazine.
Hours & Admission Prices: Gardens & Grounds: daily dawn to dusk. Museum & House: Tues.-Wed. & Sat. 11-5, Thurs.-Fri. 11-9, Sun. 12-5. Fee for special exhibitions only. Closed New Year's Day; Thanksgiving; Christmas. &
Attendance: 414,000 (accurate)
Membership: Student $35; Individual $55; Dual & Family $75; Associate $125; Advocate $250; Patron $500.

INDIANA HISTORICAL SOCIETY, (M), Eugene and Marilyn Glick Indiana History Center, 450 W. Ohio St., Indianapolis, IN 46202-3269. Tel.: 317-232-1882; 800-447-1830. Fax: 317-234-0079. TDD: 317-233-6615.
E-mail: welcome@indianahistory.org
Web Site: www.indianahistory.org
Founded: 1830.
Key Personnel: C.E.O. & Pres., John A. Herbst; Chm. (V), Jerry Semler; Exec. Vice Pres., Stephen L. Cox; Sr. Dir. Collections, Steve Haller; Mgr. Media Rels., Amy Lamb; Vice Pres. Devel. & Membership, Andrew Halter; Vice Pres. Business & Operations, Jeff Matsuoka; Sr. Dir. Human Resources, April Kerber; Vice Pres. Mktg. & Public Rels., Jeanne Scheets; Controller, Kathleen A. Grothe, CPA; Sr. Dir. Conservation, Ramona Duncan-Huse; Sr. Dir. Special Events & Visitor Svcs., Natalie Palmer; Sr. Dir. Public Programs, Trina Nelson Thomas; Museum Shop Mgr., Phil Janes.
Personnel Profile: Full-Time Paid 65; Part-Time Paid 35; Part-Time Volunteers 213; Interns 6.
Governing Authority: nonprofit organization. Tax-exempt: 501(c)(3).
Institution Type/Description: Historical Society & Archives.
Collections: history of Indiana & the Old Northwest Territory; archives; photographs; manuscripts; books; printed items; artifacts.

Major Exhibits: You Are There 1939: Healing Bodies, Changing Minds, 10/12-1/14; You Are There 1913: A City Under Water, 10/12-2/14.
Research Fields: Indiana & Old Northwest history.
Facilities: library; 4,500 sq. ft. exhibit space; 300-seat theatre; cafe; conservation & microfilming labs; classrooms; music room. Museum-related items for sale.
Activities: Indiana Experience; lectures; educational programs; arts & cultural programs; loan, temporary & traveling exhibitions; field services; Jr. Historical Society programs; concerts; performances; workshops. Museum Sponsors: history conferences.
Publications: bimonthly newsletter; quarterly popular history magazine, Traces of Indiana and Midwestern History; semi-annual, The Hoosier Genealogist: Connections; various special interest newsletters; books.
Hours & Admission Prices: Tues.-Sat. 10-5. Adults $7, seniors $6.50; discount to AAM & ICOM members; members and active military & their families no charge. Library: no charge. Closed New Year's Day; Thanksgiving; Christmas. &
Attendance: 101,100 (estimated)
Membership: Student $20; Senior $40; Individual $50; Household $65; Sustaining $100; Benefactor $250; History Patron $500; Council of 1816 $1,000.

INDIANA LANDMARKS, 1201 Central Ave., Indianapolis, IN 46202-2656. Mailing Address: 1201 Central Ave., Indianapolis, IN 46202-2656. Tel.: 317-639-4534. Fax: 317-639-6734.
E-mail: info@indianalandmarks.org
Web Site: www.indianalandmarks.org
Formerly: Historic Landmarks Foundation of Indiana
Founded: 1960.
Congressional District: 10
Key Personnel: Pres., Marsh Davis; Exec. Vice Pres., Tina Connor; Honorary Chm., Randall T. Shepard; Chm., Tim Shelley; Dir. Devel., Sharon Gamble; Vice Pres. Preservation Svcs., Mark Dollase; Dir. Southern Rgnl. Office, Greg Sekula; Vice Pres., Mary Burger; Dir. Northern Rgnl. Office, Todd Zieger; Dir. Eastern Rgnl. Office, J.P. Hall; Dir. Western Rgnl. Office, Tommy Kleckner; Dir. Morris-Butler House, Gwendolen Raley; Dir. Library & Education, Suzanne Stanis.
Personnel Profile: Full-Time Paid 39; Part-Time Paid 8; Part-Time Volunteers 200; Interns 3.
Governing Authority: nonprofit organization. Branch Museums: 1865 Morris-Butler House, Indianapolis; 1840 Huddleston Farmhouse, Cambridge City; Veraestau Historic Site, Aurora. Tax-exempt: 501(c)(3).
Institution Type/Description: Historic Houses.
Collections: period furnishings; paintings by early Indiana artists.
Research Fields: nineteenth-century art, architecture & culture; architecture of Indiana; restoration & preservation techniques; community conservation; historic buildings survey.
Facilities: 2,000-vol. library on architecture, architectural history, restoration & renovation techniques, journals & periodicals, over 13,000 slides, black & white photo archives, historic buildings survey material, press clippings & archives on historic Indiana buildings available for use on premises by members; reading room.
Activities: guided tours; lectures; films; slide talks; formally organized education programs; docent program; permanent & temporary exhibitions; statewide revolving loan fund for restoration activities; preservation and design consultation; publications; resource centers; architectural survey; restoration projects; operation of museums; grants programs.
Publications: bimonthly magazine, Indiana Preservationist; technical leaflets; preservation bulletins; museum brochures; special publications.
Hours & Admission Prices: Office: Mon.-Fri. 8:30-5. No charge; donations accepted. Closed legal holidays. &
Membership: Student & Senior $20; Individual $35; Nonprofit Organization $40; Household $50; Portico $100; Business $125; Cornerstone $250; Newel $500; Pillar $1,000; Cornice $2,500; Pinnacle $5,000.

INDIANA MEDICAL HISTORY MUSEUM, 3045 W. Vermont St., Indianapolis, IN 46222-4943. Tel.: 317-635-7329. Fax: 317-635-7349.
E-mail: edenharter@imhm.org
Web Site: www.imhm.org
Founded: 1969.
Congressional District: 10
Key Personnel: Exec. Dir., Mary Ellen Hennessey Nottage; Pres. (V), Guy Hansen.
Governing Authority: nonprofit organization. Tax-exempt: 501(c)(3).
Institution Type/Description: Medical Museum: housed in c.1896 pathology laboratory.
Collections: over 15,000 medical artifacts & health care artifacts which include

surgical & dental equipment; diagnostic instruments; nursing uniforms; pharmaceutical bottles; quack devices; health care history.
Research Fields: late 19th century & early 20th-century medicine in Indiana.
Facilities: 3,000-vol. library pertaining to late 19th & early 20th-century medicine.
Activities: guided tours; presentations; temporary exhibits; educational programs; research; rental facilities.
Publications: quarterly newsletter.
Hours & Admission Prices: Guided Tours: Thurs.-Sat. 10-4, on the hour, last tour at 3. Adults $7, university students $5, students 18 & under $3; discounts to AAM members; members & children under 6 no charge. Closed New Year's Day; Independence Day; Thanksgiving Day, day after and weekend; Christmas. &
Attendance: 8,000 (estimated)
Membership: Basic $35; Supporting $50; Contributing $100; Central State Circle $500; Edenharter Circle $1,000.

✱   INDIANA STATE MUSEUM & HISTORIC SITES, (M), 650 W. Washington St., Indianapolis, IN 46204-2185. Tel.: 317-232-1637. Fax: 317-232-7090. TDD: 317-234-2447; Facebook: Indiana State Museum.
E-mail: museumcommunication@indianamuseum.org
Web Site: www.indianamuseum.org
Founded: 1869.
Congressional District: 1-10
Key Personnel: C.E.O. & Pres., Tom King; Dir. & C.O.O., Kathleen McLary; Chm. Bd., William A. Browne, Jr.; Dir. Education, Colleen Smyth; Vice Pres. Mktg. & Devel., Santina Sullivan; Chief Cur. Cultural History, Dale Ogden; Chief Cur. Natural History, Ron Richards; Cur. Historic Archaeology, Bill Wepler; Cur. Geology, Margaret Fisherkeller; Cur. Biology, Damon Lowe; Cur. Fine Arts, Mark Ruschman; Cur. Social History, Mary Jane Teeters-Eichacker; Vice Pres. Arts & Culture, Susannah Koerber; Cur. Agriculture, Industry & Technology, Todd Stockwell; Mgr. New Media, Leslie Lorance; Mgr. Facility & Security, Ron Tolan; Vice Pres. State Historic Sites, Bruce Beesley; Museum Shop Mgr., Susan Dyar.
Personnel Profile: Full-Time Paid 110; Full-Time Volunteers 1,230; Part-Time Paid 25; Part-Time Volunteers 200; Interns 10.
Governing Authority: state; bd. of trustees. Parent Institution: State of Indiana. Branch Museums: Angel Mounds, Evansville; 1813 Corydon Capitol, Corydon; Culbertson Mansion, New Albany; 1844 Lanier Mansion, Madison; 1895 Limberlost, Geneva; c.1820 New Harmony; 1914 Gene Stratton Porter Home, Rome City; 1910 T.C. Steele Home & Studio, Nashville; c.1840 Whitewater Canal, Metamora; 1839 Levi Coffin Home, Fountain City; Vincennes, c.1810-1838; Indiana State Museum & Historic Sites, Indianapolis. (Check individual listings for hours & further information). Tax-exempt.
Institution Type/Description: General Museum.
Collections: period furnishings; art; science; culture; artifacts; natural history specimens; music; political; primary source material; prints; drawings; graphic arts; sculpture.
Major Exhibits: Fearless Furniture, 10/5/13-5/27/14; Indiana's Ice Age Giants: The Mystery of Mammoths and Mastodons, 11/13-7/14.
Research Fields: Indiana natural & cultural history; art.
Facilities: orientation theatre; auditorium; meeting rooms; two restaurants; IMAX theater. Museum-related items for sale.
Activities: lectures; restoration; slide, film & tape presentations; formally organized education programs; special events; festivals; special traveling exhibits. Museum Sponsors: Sankofa Black Heritage Festival; 10th Annual Indiana Art Fair in February; Pinewood Derby in April; Artworks Asian American Festival of the Arts in May; Juneteenth Festival in June; Latino Festival of the Arts in September; GeoFest in October; Day of the Dead in October; Family New Year's Eve in December.
Publications: brochures for individual sites; member publications.
Hours & Admission Prices: Mon.-Sat. 10-5, Sun. 11-5. Adults $9.50; ASTC members & members no charge. Closed Thanksgiving; Christmas. &
Attendance: 325,000 (accurate)
Membership: Individual $55; Family & Grandparent $70; Patron $100.

INDIANA STATE POLICE MUSEUM AND EDUCATION CENTER, 8660 E. 21st St., Indianapolis, IN 46219-2562. Tel.: 317-899-8293. Fax: 317-899-8289.
E-mail: isp@isp.in.gov
Web Site: www.in.gov/isp
Formerly: Indiana State Police Youth Education and Historical Center
Founded: 1993.
Congressional District: 5
Key Personnel: Dir., F/Sgt. Brian Olehy; Museum Shop Mgr., Sgt. Ray Poole.
Governing Authority: Parent Institution: Indiana State Police Youth Education Museum and Memorial Fund. Tax-exempt.
Institution Type/Description: History Museum.

Collections: police vehicles & firearms; photographs; personal artifacts; uniforms; cars; equipment.
Activities: guided tours; talks with prior reservation.
Hours & Admission Prices: Tours: Mon.-Fri. 9-3; other times by appointment. No charge. Closed state holidays. &
Attendance: 35,000 (estimated)

INDIANA WAR MEMORIALS, 431 N. Meridian St., Indianapolis, IN 46204-1711. Tel.: 317-232-7615. Fax: 317-233-4285.
E-mail: iwm@iwm.in.gov
Web Site: www.in.gov/iwm/
Founded: 1927.
Congressional District: 10
Key Personnel: Pres. Commission, Carol Mutter; Exec. Dir., J. Stewart Goodwin; Administrative Dir., Nina Gaither; Museum Dir., Ethan Wright; Museum Specialist Collections, Donna M. Schmink; Museum Specialist Database, Chase Brazel; Museum Specialist Tours, Ryan Krenzke; Physical Plant Dir., Steve Stiegelmeyer.
Personnel Profile: Full-Time Paid 15; Part-Time Paid 4; Part-Time Volunteers 2.
Governing Authority: state commission. Indiana War Memorials Commission: maintains: Indiana War Memorials; Soldiers & Sailors Monument, Colonel Eli Lilly Civil War Museum, Monument Circle, USS Indianapolis Memorial. Tax-exempt.
Institution Type/Description: Historic Commission & Military Museum: housed in 1927 Indiana War Memorial.
Collections: uniforms, weapons, battle flags, pictures, helicopter, jeep, missile, cannon, from Revolutionary War to Operation Enduring Freedom.
Facilities: library of books, pictures & combat maps available for research on premises; auditorium; USS Indianapolis Radio Room.
Activities: permanent exhibitions.
Hours & Admission Prices: Indiana War Memorial Shrine Room & Military Museum: Wed.-Sun. 9-5. Soldiers & Sailors Monument, Colonel Eli Lilly Civil War Museum, Observation Deck & Gift Shop Fri.-Sun. 10:30-5:30. No charge; donations accepted. &
Attendance: 250,000 (accurate)

INDIANAPOLIS ART CENTER, 820 E. 67th St., Indianapolis, IN 46220-1199. Tel.: 317-255-2464. Fax: 317-254-0486.
E-mail: info@indplsartcenter.org
Web Site: www.indplsartcenter.org
Founded: 1934.
Congressional District: 6
Key Personnel: Pres. & C.E.O., Carter Wolf; Chm. (V), Tanya Stuart Overdorf; Office Mgr., Jennifer Collins; Dir. Operations, Pamela Rosenberg; Dir. Devel., Kelly Lamb; Gallery Dir. & Artist Svcs. Mgr., Patrick Flaherty; Dir. Outreach, Laura Alvarado; Vice Pres. & Dir. Programs, David S. Thomas; Education Assoc., Breiana Satchwell; Dir. Special Events, Iris Dillon; Dir. Mktg., Lisa DeHayes; Dir. Finance, Doug Halman; Business Mgr., Alisia Morales; Facilities Mgr., Brett Sommers.
Personnel Profile: Full-Time Paid 22; Part-Time Paid 6; Part-Time Volunteers 4.
Governing Authority: nonprofit. Tax-exempt: 501(c)(3).
Institution Type/Description: Studio Art Teaching Center & Exhibitions.
Collections: contemporary visual arts.
Research Fields: Indiana & 250 mi. radius of artists.
Facilities: 1,100-vol. library dedicated to arts related subjects, slides & videos of over 2,000 regional artists; 13 studio classrooms; 224 seat auditorium; 40,000 sq. ft. exhibition space; resource center; wildflower & sculpture garden.
Activities: over 300 different art classes; art education for all ages & skill levels; community arts resource center; local & regional artist exhibitions; programming for culturally diverse audiences including lectures, concerts & films; fine art day camp for children; art tours; outreach programs to youth in underserved areas.
Publications: newsletter; schedule of events & classes; occasional exhibition catalogues & posters; monthly event postcard.
Hours & Admission Prices: Mon.-Fri. 9am-10pm, Sat. 9-6, Sun. 12-6; hours may vary when classes are not in session. No charge; donations accepted. Closed New Year's Day; Memorial Day; Independence Day; Labor Day; Thanksgiving; Christmas. &
Attendance: 324,451 (estimated)
Membership: Student & Senior Citizen $30; Individual $45; Family $60.

INDIANAPOLIS FIREFIGHTERS MUSEUM & HISTORICAL SOCIETY, 748 Massachusetts Ave., Indianapolis, IN 46204-1609. Tel.: 317-262-5161. Fax: 317-262-5163.
Founded: 1996.
Congressional District: 7

Governing Authority: Parent Institution: Indianapolis Professional Firefighters Local 416. Tax-exempt.
Institution Type/Description: Fire-Fighting Museum.
Collections: Indiana fire department & firefighting history; 1921 Stutz engine; 1921 Stutz ladder truck; horse drawn steam pumper; hose cart; badges; photographs.
Hours & Admission Prices: Mon.-Fri. 8:30-3. No charge; donations accepted. Closed holidays. &
Attendance: 5,000 (estimated)
Membership: Firefighter $26; Captain (Sponsor) $100; Fire Chief (Corporate) $220.

INDIANAPOLIS MOTOR SPEEDWAY HALL OF FAME MU-SEUM, (M), 4790 W. 16th St., Indianapolis, IN 46222-2573. Mailing Address: P.O. Box 24518, Speedway, IN 46224-0152. Tel.: 317-492-6784. Fax: 317-492-6449.
E-mail: ebireley@brickyard.com
Web Site: www.indianapolismotorspeedway.com
Founded: 1956.
Congressional District: 11
Key Personnel: Dir., Ellen K. Bireley.
Personnel Profile: Full-Time Paid 9; Part-Time Paid 26; Interns 1.
Governing Authority: nonprofit organization. Tax-exempt: 501(c)(3).
Institution Type/Description: Transportation Museum.
Collections: racing-related cars; period & classic passenger cars; racing & automotive memorabilia; art; trophy collection.
Facilities: library of books on motor racing, antique & classic cars available for research by special request. Gift items for sale.
Activities: permanent & temporary exhibitions.
Hours & Admission Prices: March-Oct. daily 9-5; Nov.-Feb. daily 10-4; call for additional hours. Museum: adults $5, youth 6-15 $3; children 6 & under no charge. Bus ride around track when not in use: adults $5, youth 6-15 $3; discounts to AAM members; children 6 & under no charge. Closed Thanksgiving; Christmas. &
Attendance: 350,000 (estimated)

INDIANAPOLIS MUSEUM OF CONTEMPORARY ART, 1043 Virginia Ave., Ste. 5, Indianapolis, IN 46203-1761. Tel.: 317-634-6622. Fax: 317-634-1977.
Web Site: www.indymoca.org
Founded: 2001.
Key Personnel: Dir., Shauta Marsh; Pres., Brandon Judkin; Vice Pres., Jennifer Boehm; Education, Elizabeth Mix.
Personnel Profile: Full-Time Paid 2; Part-Time Paid 1; Part-Time Volunteers 100.
Governing Authority: nonprofit. Tax-exempt: 501(c)(3).
Institution Type/Description: Art Museum.
Collections: works by contemporary artists.
Facilities: 1,200 sq. ft. exhibit space. Museum-related items for sale.
Activities: concerts; films; formal education programs; lectures; participatory & temporary exhibitions.
Hours & Admission Prices: Thurs.-Sat. 11-6, Sun. 12-3. No charge. Closed holidays.
Attendance: 25,000 (accurate)
Membership: Individual $30; Dual $50; Family $70; Sustaining $100; Friend $150.

* **INDIANAPOLIS ZOO,** 1200 W. Washington St., Indianapolis, IN 46222-0309. Tel.: 317-630-2001. Fax: 317-630-5153.
E-mail: info@indyzoo.com
Web Site: www.indianapoliszoo.com
Formerly: Indianapolis Zoological Society
Founded: 1964.
Congressional District: 11
Key Personnel: C.E.O. & Pres., Michael I. Crowther; Chm. (V), Alan Cohen; Deputy Dir. & Senior Vice Pres., Paul Grayson; Senior Vice Pres. External Rels., Karen Burns; Vice Pres. Operations, Tim Savona; Vice Pres. Internal Rels., Mary Jane Bennett; Vice Pres. Life Sciences, Rob Shumaker.
Personnel Profile: Full-Time Paid 215; Part-Time Paid 28; Part-Time Volunteers 616; Interns 40.
Governing Authority: nonprofit organization. Subsidiary Institution: Indianapolis Zoo & White River Gardens. Tax-exempt: 170(b)(1)(A); 501(c)3.
Institution Type/Description: Zoo, Aquarium, & Botanical Garden.
Collections: live animals & plants from around the world.
Research Fields: reproduction in endangered & threatened species & nutritional studies.
Facilities: zoological park; 200-seat auditorium-theater; classrooms; concession stand; restaurant; aquarium; botanical garden with conservatory. Gift items for sale.

Activities: lectures; gallery talks; concerts; TV & radio programs; formally organized education programs for children, adults, undergraduate & graduate college students affiliated with Indiana University, Butler University, Ball State University, University of Indianapolis, Anderson College; docent program or council; training programs; permanent & temporary exhibitions; day camps; pre-school programs; volunteer council; cooperative programming with other agencies; train, pony, coaster & carousel rides; animal demonstrations in performance area; special events for families & children.
Publications: biennial members magazine, Indianapolis Zoo Magazine; biannual donor magazine, Inside the Zoo.
Hours & Admission Prices: Jan. to mid-March & Nov. Wed.-Sun. 9-4; mid-March to May & Sept.-Oct. Mon.-Thurs. 9-4, Fri.-Sun. 9-5; June-Aug. Mon.-Thurs. 9-5, Fri.-Sun. 9-6; Dec. Wed.-Sun. 12-9. Adults $16.95, seniors $15.95, youth $11.95; members no charge. Closed New Year's Day; Thanksgiving; Christmas Eve & Day. &
Attendance: 1,013,730 (accurate)
Membership: Individual Plus One $105; Family & Grandparents $125; Family Plus 2 & Grandparent Plus 2 $165.

J.I. HOLCOMB OBSERVATORY AND PLANETARIUM, 4600 Sunset Ave., Indianapolis, IN 46208-3443. Tel.: 317-940-8333.
E-mail: holcombobservatory@butler.edu
Web Site: www.butler.edu/holcomb-observatory
Founded: 1954.
Key Personnel: Dir., Dr. Brian Murphy; Assoc. Dir., Richard Brown.
Personnel Profile: Full-Time Paid 1; Part-Time Paid 1.
Governing Authority: university. Parent Institution: Butler University. Subsidiary Institution: Dept. of Physics & Astronomy. Tax-exempt: 501(c)(3).
Institution Type/Description: Planetarium & Observatory.
Collections: astronomical tables; journals & books on astronomy & astrophysics.
Research Fields: astronomy.
Facilities: 250-vol. library of astronomical tables, research papers & ephemerides available for inter-library loan & by arrangement with Butler University Library; planetarium; classrooms.
Activities: guided tours; lectures; films; formally organized education programs for children & undergraduate college students.
Hours & Admission Prices: June-July Fri.-Sat. 9:15 & 10:15; Sept.-May Fri.-Sat. 7 & 8:15 pm. Adults $3, senior citizens & children over 4 $2.

JAMES WHITCOMB RILEY MUSEUM HOME, 528 Lockerbie St., Indianapolis, IN 46202-3617. Tel.: 317-631-5885. Fax: 317-955-0619.
E-mail: rileyhome@rileykids.org
Founded: 1922.
Key Personnel: Dir., Jim Obergfell; C.E.O., Kevin O'Keefe.
Personnel Profile: Full-Time Paid 3; Part-Time Paid 1.
Governing Authority: philanthropic organization. Parent Institution: Riley Children's Foundation. Tax-exempt: 501(c)(3).
Institution Type/Description: Historic House Museum: 1872 home of James Witcomb Riley.
Collections: art works; books; furnishings.
Facilities: 1,000-vol. library of general books. Books by James Whitcomb Riley & other memorabilia for sale.
Activities: guided tours.
Hours & Admission Prices: Tues.-Sat. 10-3:30. Adults $4, students $1; children under 7 no charge. Closed major holidays.
Attendance: 7,500 (accurate)

MILITARY LIBRARY AND MUSEUM, 10801 N. College Ave., Indianapolis, IN 46280. Mailing Address: c/o Clay Township Trustee, 10701 N. College Ave., Indianapolis, IN 46280. Tel.: 317-582-0507.
Institution Type/Description: Military History Museum.
Collections: local history; genealogy; archives; photographs.
Facilities: library.
Activities: research.
Hours & Admission Prices: Call for hours.

MORRIS-BUTLER HOUSE, (M), 1204 N. Park Ave., Indianapolis, IN 46202-2638. Tel.: 317-636-5409. Fax: 317-636-2630.
E-mail: morris-butler@indianalandmarks.org
Web Site: www.indianalandmarks.org
Founded: 1969.
Congressional District: 10
Key Personnel: Chm. (V) Advisory Committee, Bradley Brooks; Pres., Marsh Davis; Dir., Gwendolen Raley; Program Coord., Kelly Gascoine.

Personnel Profile: Full-Time Paid 2; Part-Time Paid 1; Part-Time Volunteers 45; Interns 2.
Governing Authority: nonprofit organization. Parent Institution: Indiana Landmarks. Tax-exempt: 501(c)(3).
Institution Type/Description: Historic House: house built in 1865.
Collections: c.1865-1900 Victorian decorative arts; theme rooms.
Research Fields: decorative arts; local & social history; interior preservation.
Facilities: meeting room; event space.
Activities: lectures; workshops; special events; exhibits.
Publications: brochures; exhibit-related booklets.
Hours & Admission Prices: Call for availability & pricing. &
Attendance: 4,000 (accurate)
Membership: Senior 60 & over $20; Individual $35; Nonprofit Organization $40; Household $50; Portico $100; Business $125; Founder's Club $250-$500; President's Circle $1,000-$2,500; Chairman's Council $5,000 & up.

NATIONAL ART MUSEUM OF SPORT, INC., (M), P.O. Box 441155, Indianapolis, IN 46244. Tel.: 317-931-8600. Facebook: National Art Museum of Sport.
E-mail: evarner@nationalartmuseumofsport.org
Web Site: www.nationalartmuseumofsport.org
Founded: 1959.
Congressional District: 7
Key Personnel: Chm. (V), Patrick T. Perrolla; Exec. Dir., Elizabeth C. Varner.
Personnel Profile: Full-Time Paid 1; Part-Time Volunteers 10.
Governing Authority: nonprofit corporation. Indiana charter, board of governors with an operating board of governors. Tax-exempt: 501(c)(3).
Institution Type/Description: Sports Art Museum.
Collections: paintings; sculpture; graphics; photographs of sporting subjects.
Research Fields: Sport art & artists who incorporated sporting subjects into their work; photography & sport history; social issues & influences on sporting art.
Facilities: library pertaining to sporting art & cataloged works of sports world-wide; 10,500 sq. ft. exhibit space.
Activities: docent guided tours; permanent & temporary exhibitions; collaborative projects with other agencies.
Publications: exhibition catalogs; newsletter, Score Board.
Hours & Admission Prices: Mon.-Fri. 8-5. No charge; donations accepted. Closed major holidays. &
Attendance: 136,000 (estimated)
Membership: Friend $45; Benefactor $250; Germaine G. Glidden Society $1,000; 21st Century Society $2,000.

NCAA HALL OF CHAMPIONS, One NCAA Plaza, 700 W. Washington, Indianapolis, IN 46204-2710. Tel.: 800-735-NCAA (toll free); 317-916-HALL (local). Fax: 317-916-4254.
E-mail: hocmail@ncaa.org
Web Site: ncaahallofchampions.org
Key Personnel: Dir., Mike King; Asst. Dir., Kelly Dodds; Public & Media Rels., Gail Dent
Institution Type/Description: Sports Museum.
Collections: photographs; sports memorabilia.
Major Exhibits: Celebrating Historically Black Colleges and Universities, 10/15/13-5/1/14.
Facilities: theater. Museum-related items for sale.
Hours & Admission Prices: Tues.-Sat. 10-5, Sun. 12-5. Adults $5, senior citizens 65 & over and youth 6-18 $3; children under 6 no charge. &
Attendance: 52,163 (accurate)

RHYTHM! DISCOVERY CENTER, 110 W. Washington St., Ste. A, Indianapolis, IN 46204-3423. Tel.: 317-275-9030. Fax: 317-974-4499. Facebook: Rhythm! Discovery Center.
E-mail: rhythm@pas.org
Web Site: www.rhythmdiscoverycenter.org
Formerly: Percussive Arts Society Museum
Founded: 1990.
Congressional District: 7
Key Personnel: C.E.O., Jeff Hartsough; Pres. (V), John R. Beck; Museum Shop Mgr., Erin Jeter.
Personnel Profile: Full-Time Paid 8; Part-Time Paid 6; Interns 1.
Governing Authority: Parent Institution: Percussive Arts Society. Tax-exempt.
Institution Type/Description: Musical Instruments Museum.
Collections: percussion instruments; books & scores; recorded music; historical & modern use of drums & percussion around the world; hands-on exhibits.
Major Exhibits: Drums from the Circle City - The Leedy Drum Company, 5/13-5/14; DRUMset: Driving the Beat of American Music, 11/13-11/15.
Facilities: library. Gift items for sale.

Activities: concerts; clinics; classes; labs; workshops; panels & presentations.
Publications: Percussive Arts Society: Percussion News; Percussive Notes.
Hours & Admission Prices: Mon.-Sat. 10-5, Sun. 12-5. Adults $9, senior citizens $8, children & students $6; children 5 & under and members no charge. Closed New Year's Day; Thanksgiving; Christmas. &
Attendance: 10,000 (estimated)
Membership: $35-$60.

Jamestown

JACKSON TOWNSHIP HISTORICAL SOCIETY, 41 W. Main St., Jamestown, IN 46147. Mailing Address: P.O. Box 297, Jamestown, IN 46147-0297. Tel.: 765-676-5891.
Founded: 2001.
Governing Authority: Tax-exempt.
Institution Type/Description: Historical Society Museum.
Collections: local history & culture; personal artifacts; period furnishings; photographs; printing machines; linotypes.
Hours & Admission Prices: Call for hours. No charge; donations accepted. &
Attendance: 100 (estimated)
Membership: Individual $20.

Jasper

DUBOIS COUNTY MUSEUM, 2704 N. Newton St., Jasper, IN 47546. Tel.: 812-634-7733.
E-mail: jim.hagedors@duboiscountymuseum.org
Web Site: duboiscountymuseum.org
Key Personnel: Dir. Programs & Exhibits, Janet Kluemper; Pres. (V), Jim Hagedorn
Institution Type/Description: History Museum.
Collections: local history & culture; period furnishings; personal artifacts; photographs; military artifacts; sports memorabilia; Native American artifacts; dollhouses; miniatures; early farm equipment; woodworking tools.
Activities: special events & programs; workshops.
Hours & Admission Prices: Tues.-Fri. 10-2, Sat. 10-4, Sun. 1-4. No charge; donations accepted. &
Attendance: 12,000 (estimated)

INDIANA BASEBALL HALL OF FAME, Vincennes Univ. - Jasper Ruxer Student Ctr., 851 College Ave., Jasper, IN 47546. Mailing Address: 1436 Leopold St., Jasper, IN 47546-2117. Tel.: 812-482-2262. Fax: 812-482-1982.
E-mail: rajahoward@psci.net
Web Site: indbaseballhalloffame.org
Founded: 1977.
Congressional District: 63
Governing Authority: not for profit. Tax-exempt: 501(c)(3).
Institution Type/Description: Hall of Fame.
Collections: Hall of Fame inductees including pro-players, high school, college, & pro coaches and managers; photographs; personal artifacts; baseballs; gloves; uniforms; bats.
Activities: interactive championship highlights; hands-on baseball simulation.
Hours & Admission Prices: mid-May to mid-Aug. daily 11-3; mid-Aug. to mid-May Thurs.-Sun. 11-3. Adults 13 & over $4, children 5-12 $3, senior citizens 60 & over $2; children 4 & under no charge. &

KREMPP GALLERY, 951 College Ave., Jasper, IN 47546-9382. Tel.: 812-482-3070. Fax: 812-634-6997.
Web Site: www.jasperarts.org
Founded: 1975.
Congressional District: 9
Key Personnel: Dir., Kit Miracle.
Personnel Profile: Full-Time Paid 4; Part-Time Paid 7; Part-Time Volunteers 300; Interns 2.
Governing Authority: city. Tax-exempt.
Institution Type/Description: Art Center.
Collections: paintings.
Major Exhibits: Elzbieta Bittner - Tapestries, 1/3/14-1/31/14; Judi Krew - Pastels & Sculptures, 2/5/14-2/28/14; Youth Art Month, 3/4/14-3/31/14; Chad Hartwig - Ceramics, 4/2/14-4/28/14; James Henderson - Photographs, 4/2/14-4/28/14; Linda Stephen - Origami, 5/1/14-5/30/14; Abby Laux - Painting, 6/3/14-6/30/14; Dubois County Art Guild, 6/14-7/14; 21st Annual Juried Exhibit, 9/14; Chris Leib - Paintings, 10/14; Russell May - Paintings, 10/14; Carrie Ann Schumacher - Sculpture, 11/14.
Activities: lecture; workshop; art classes.
Publications: gallery cards; bi-annual newsletter; annual brochure.
Hours & Admission Prices: Mon.-Wed. & Fri. 10-5, Thurs. 10-7, Sun. 12-3. No charge; donations accepted. &

Attendance: 40,000 (estimated)

Jeffersonville

HOWARD STEAMBOAT MUSEUM AND MANSION, 1101 E. Market St., Jeffersonville, IN 47130-4333. Mailing Address: P.O. Box 606, Jeffersonville, IN 47131-0606. Tel.: 812-283-3728. Fax: 812-283-6049.
E-mail: hsmsteam@aol.com
Web Site: www.steamboatmuseum.org
Formerly: Howard Steamboat Museum, Inc.
Founded: 1958.
Congressional District: 9
Key Personnel: Pres. (V), Rick Madden; Admin. & Museum Shop Mgr., Yvonne Knight; Cur. Collections, Keith Norrington.
Personnel Profile: Full-Time Paid 1; Part-Time Paid 6; Part-Time Volunteers 5; Interns 1.
Governing Authority: nonprofit organization; bd. of directors. Parent Institution: Clark County Historical Society. Tax-exempt: 501(c)(3).
Institution Type/Description: Historic House Museum: 1894 home of Edmonds J. Howard, son of James E. Howard, founder of the Howard Shipyards 1834.
Collections: original 1893 furnishings; steamboat artifacts & models; photographs; half-breadth models; tools.
Research Fields: steamboat architecture & design; Ohio River history.
Facilities: Steamboat prints, books & museum-related items for sale.
Activities: guided tours; lectures; permanent exhibitions. Annual Event: Spring Festival: A Victorian Chautauqua in May.
Publications: brochures; semiannual newsletter; catalogs for Fall Into Art exhibitions; book, Scenes From Memory; booklet, 57.1 ft. The 1937 Flood Remembered.
Hours & Admission Prices: Tues.-Sat. 10-4, Sun. 1-4. Adults $6; discounts to groups, students, seniors, military, AARP, AAA & AAM members; under 6 & members no charge. Closed most holidays. &
Attendance: 10,000 (estimated)
Membership: Annual Personal Memberships: Student or Senior Citizen 65 & over $10; Individual $15; Family $25; Contributor $26-$99; Friend $100-$199; Benefactor $200-$499; James E. & Loretta Howard Society $500 & up.

Kendallville

MID-AMERICA WINDMILL MUSEUM, 732 S. Allen Chapel Rd., Kendallville, IN 46755-3220. Mailing Address: P.O. Box 5048, Kendallville, IN 46755-5048. Tel.: 260-347-2334.
E-mail: hobshill@ligter.com
Web Site: www.midamericawindmillmuseum.org
Founded: 1991.
Key Personnel: Dir. & Chm. (V), Sara Hobson; Pres. (V), Kevin Kelham; Museum Shop Mgr., JoAnn Burke.
Personnel Profile: Part-Time Paid 1; Part-Time Volunteers 50.
Governing Authority: Parent Institution: Kendallville Historical Society. Tax-exempt.
Institution Type/Description: Technology Museum.
Collections: windmills; history of wind power & windmills; photographs.
Facilities: library.
Activities: rental facilities; weddings; special events; video.
Publications: Windmill Clipper.
Hours & Admission Prices: April-Nov. Tues.-Fri. 10-4, Sat. 10-5, Sun. 1-4. Adults $4, senior citizens $3.50, students $1.50; children under 6 no charge.
Membership: Annual $20; Business $50; Life $600.

Knox

STARKE COUNTY HISTORICAL SOCIETY, 401 S. Main St., Knox, IN 46534. Tel.: 574-772-5393.
E-mail: jimshilling@gmail.com
Web Site: www.starkehistory.com
Founded: 1974.
Personnel Profile: Part-Time Volunteers 25.
Institution Type/Description: Historical Society Museum.
Collections: local history; agriculture; military artifacts; period furnishings; photographs; toys; personal artifacts.
Activities: Annual Events: July Ice Cream Social; Christmas Open House.
Publications: annual newsletter.
Hours & Admission Prices: Tues.-Fri. 12-4. No charge; donations accepted.
Attendance: 900 (estimated)
Membership: Single $10; Family $15.

Kokomo

AUTOMOTIVE HERITAGE MUSEUM, 1500 N. Reed Rd., U.S. 31 N., Kokomo, IN 46901-2592. Tel.: 765-454-9999. Fax: 765-454-9956.
E-mail: jsgkphone@yahoo.com
Web Site: www.kokomoautomotivemuseum.org
Key Personnel: Gen. Mgr., James Parsons
Institution Type/Description: Automobile Museum.
Collections: over 100 period vehicles from 1895 to 1970s.
Activities: facility rental.
Hours & Admission Prices: Tues.-Sun. 10-4; groups by appointment. Adults $5, seniors $4, children 7-14 $2.

ELWOOD HAYNES MUSEUM, 1915 S. Webster St., Kokomo, IN 46902-2040. Tel.: 765-456-7500. Fax: 765-456-7577.
E-mail: kfrazer@cityofkokomo.org
Founded: 1967.
Congressional District: 5
Key Personnel: Cur., Kay J. Frazer; Cur., Pete Kelley.
Personnel Profile: Full-Time Paid 1; Part-Time Paid 1.
Governing Authority: municipal; operated by the city Park Dept. Tax-exempt.
Institution Type/Description: Industrial Museum: housed in 1915 home of Elwood Haynes.
Collections: automobiles, stainless steel & stellite invented by Elwood Haynes; industrial products of the city of Kokomo; sculptures; photographs; period clothing; personal artifacts.
Facilities: 40-seat auditorium.
Activities: guided tours; films; formally organized education programs for children; permanent exhibitions.
Publications: books, Elwood Haynes 1857-1925; The Complete Motorist; Alloys & Automobiles, the Life of Elwood Haynes.
Hours & Admission Prices: Tues.-Sat. 11-4, Sun. 1-4. No charge; donations accepted. Closed most holidays. &
Attendance: 4,000 (estimated)

HOWARD COUNTY HISTORICAL MUSEUM, (M), 1200 W. Sycamore St., Kokomo, IN 46901-4386. Tel.: 765-452-4314. Fax: 765-452-4581.
E-mail: dave.broman@howardcountymuseum.org
Web Site: howardcountymuseum.org
Founded: 1916.
Congressional District: 5
Key Personnel: Exec. Dir., Dave Broman.
Personnel Profile: Full-Time Paid 3; Part-Time Paid 5; Part-Time Volunteers 100; Interns 2.
Governing Authority: nonprofit organization. Parent Institution: Howard County Historical Society. Subsidiary Institution: Howard County Museum. Tax-exempt.
Institution Type/Description: History Museum & Historic House.
Collections: Howard County history; ethnological; archaeology; military; agriculture; glass & manufacturing. Historic Houses: Seiberling Mansion; c.1890 Elliott House.
Research Fields: Howard County history; Howard County inventions & manufacturing.
Facilities: research library; rental facilities.
Activities: children's educational programs; tour bus travelers activity. Museum Sponsors: Christmas at the Seiberling.
Publications: newsletter, Museum Hi-Lites; Howard County History In The Mail, Monroe Seiberling's Mansion.
Hours & Admission Prices: Feb.-Dec. Tues.-Sun. 1-4. Adults $4; discounts to seniors tour groups; members no charge. Closed national holidays. &
Attendance: 10,000 (estimated)
Membership: Individual $25; Family $30; Patron $50; Seiberling $100; Benefactor $250; Leadership $500; Founder's Circle $1,000.

INDIANA UNIVERSITY KOKOMO ART GALLERY, 2300 S. Washington St., Kokomo, IN 46902-3557. Tel.: 765-455-9523.
E-mail: gallery2@iuk.edu
Web Site: www.iuk.edu/admin-services/gallery/
Key Personnel: Gallery Dir., Susan Skoczen-Southard, M.F.A.
Institution Type/Description: Art Gallery.
Collections: works by local, regional, national & international artists; new media.
Activities: lectures; workshops; demonstrations; special events.
Hours & Admission Prices: Tues. & Thurs. 10-4, Wed. 10-8. No charge.

KAA ART CENTER, 525 W. Ricketts St., Kokomo, IN 46902-2029. Tel.: 765-457-9480.
Web Site: www.kokomoartassociation.org
Formerly: Kokomo Art Center
Founded: 1927.
Key Personnel: Chm. (V) & Volunteer Coord., Elaine Wanke; Pres. (V), Cheryl Sullivan; Bd. Member, Colette Inderhees
Governing Authority: Sponsored By: Kokomo Parks Dept. Subsidiary Institution: Artworks Gallery, 210 N. Main St., Kokomo, IN 46901. Tax-exempt.
Institution Type/Description: Art Center.
Collections: Hoosier art & artists; Norman Rockwell signed prints from Tom Sawyer book; American Indian art.
Activities: classes; workshops; seminars; summer kid's art camp. Museum Sponsors: Annual Spring Show in May; Hippensteel High School Award Winner Exhibit in June.
Publications: bimonthly newsletter, Art Strokes; annual directory.
Hours & Admission Prices: House Tours: Feb.-Nov. Tues.-Sat. 1-4. No charge. Closed Labor Day; Memorial Day; Independence Day; Good Friday; Easter weekend; Thanksgiving. &
Attendance: 400 (estimated)
Membership: Student $10; Single $25; Lifetime $1,000.

La Porte

HESSTON STEAM MUSEUM, 1201 E. 1000 N., La Porte, IN 46350-8642. Tel.: 219-778-2783.
Web Site: www.hesston.org
Governing Authority: nonprofit. Parent Institution: LaPorte County Historical Steam Society. Tax-exempt: 501(c)(3).
Institution Type/Description: History Museum.
Collections: steam locomotives; steam powered saw mill; 92 ton railroad steam crane; electric power plant; steam traction engines.
Activities: Museum Sponsors: Hesston Steam & Power Show in September; Halloween Ghost Train in October; Santa's Candy Cane Express in December.
Hours & Admission Prices: Museum: Sat.-Sun. & holidays 11:30-5. No charge. Train Rides: Memorial Day to Labor Day Sat.-Sun. & holidays 12-5. Adults $5, children over 3 $3; children under 3 no charge.

LA PORTE COUNTY HISTORICAL SOCIETY MUSEUM, 2405 Indiana Ave., Ste. 1, La Porte, IN 46350-6063. Tel.: 219-324-6767. Fax: 219-324-9029.
E-mail: info@laportecountyhistory.org
Web Site: www.laportecountyhistory.org
Founded: 1906.
Congressional District: 13
Key Personnel: Pres. (V), Arnold Bass; Cur., Susie Richter; Museum Shop Mgr. & Asst. Cur., Janet Sikorski.
Personnel Profile: Full-Time Paid 1; Part-Time Paid 2; Part-Time Volunteers 10.
Governing Authority: society. Tax-exempt.
Institution Type/Description: History Museum.
Collections: local history; Jones collection of firearms; manuscripts; 35 Kesling automobiles; 14 period rooms.
Research Fields: LaPorte County history; genealogical research.
Facilities: research library; meeting room.
Activities: guided tours; lectures; permanent exhibitions. Museum Sponsors: Annual Civil War Reenactment in June; Annual Classic Car Show in July.
Publications: quarterly newsletter, Old Letter; book, The Belle Gunness Story; pamphlets on La Porte County history; reprints of county history & plat books; LaPorte and Its Environs.
Hours & Admission Prices: Tues.-Sat. 10-4:30. Out-of-county adults $6, out-of-county seniors 60 & over $5, children 12-17 & in-county adults $3; members, children under 12 & Time Travelers no charge. Closed national holidays. &
Attendance: 7,634 (accurate)
Membership: Student $2.50; Active $10; Sustaining $20; Patron $35.

LaGrange

LAGRANGE COUNTY HISTORICAL SOCIETY, INC., 109 S. High St., LaGrange, IN 46761. Mailing Address: P.O. Box 134, LaGrange, IN 46761-0134. Tel.: 260-463-3763; 350-8561 (cell).
E-mail: blmccoy2@yahoo.com
Founded: 1966.
Congressional District: 4
Key Personnel: Pres. (V), Izara Miller.
Governing Authority: private; nonprofit. Tax-exempt.
Institution Type/Description: Historical Society Museum.

Collections: historical papers; local artifacts. Historic Structure: log cabin; period clothing.
Research Fields: local history; genealogy.
Facilities: library of local histories, pamphlets, newspapers, clippings, pictures, cemetery records, war records, family records, & old & abandoned school records available for research by appointment.
Activities: formally organized education programs for adults; log cabin tours; picture & paper research.
Publications: History of LaGrange County, 1936; 1882, LaGrange County Histories; cemeteries of LaGrange County Indiana, 1832-1982; My Town, Your Town-LaGrange, 1836-1986.
Hours & Admission Prices: By appointment only. No charge; donations accepted.
Attendance: 500 (estimated)
Membership: Annual Single $10; Couple $20; Family $25; Contributing $50; Life $250.

Lafayette

*** ART MUSEUM OF GREATER LAFAYETTE, (M),** 102 S. 10th St., Lafayette, IN 47905-1173. Tel.: 765-742-1128. Fax: 765-742-1120.
E-mail: ksmith@artlafayette.org
Web Site: www.artlafayette.org
Formerly: Greater Lafayette Museum
Founded: 1909.
Congressional District: 7
Key Personnel: Exec. Dir., Kendall Smith, II; Museum Admin., Glenda McClatchey; Cur. Collections & Exhibitions, Michael Atwell.
Personnel Profile: Full-Time Paid 2; Part-Time Paid 4; Part-Time Volunteers 100; Interns 2.
Governing Authority: nonprofit organization. Tax-exempt: 501(c)(3).
Institution Type/Description: Art Museum.
Collections: 19th-21st century American paintings, prints & drawings; historical and contemporary Indiana art; Latin American works; American art pottery.
Research Fields: Indiana art.
Facilities: 3,000 sq. ft. exhibition gallery; classrooms. Museum-related items for sale.
Activities: guided tours of exhibitions & studios; national & regional temporary exhibits; children's activities; lecture series; education program; studio art classes.
Publications: quarterly, newsletter; class schedules.
Hours & Admission Prices: Daily 11-4. No charge; donation accepted. Closed New Year's Day; Martin Luther King. Jr. Day; Presidents' Day; Memorial Day; Independence Day; Labor Day; Thanksgiving; Christmas Eve & Day. &
Attendance: 15,000 (accurate)
Membership: Student $20; Individuals $40; Family $60; Friend $120; Advocate $250; Patron $500; Sustainer $1,000; Benefactor $2,500; Founder $5,000; Director's Circle $10,000.

COLUMBIAN PARK ZOO, 1915 Scott St., Lafayette, IN 47904-2929. Tel.: 765-807-1540. Fax: 765-807-1547.
E-mail: claufman@city.lafayette.in.us
Web Site: www.lafayette.in.gov/zoo
Founded: 1908.
Congressional District: 7
Key Personnel: Zoo Dir. & Zoo Gift Shop, Claudine Laufman; Asst. Zoo Dir., Dana Rhodes.
Personnel Profile: Full-Time Paid 7; Part-Time Paid 6; Part-Time Volunteers 27; Interns 2.
Governing Authority: municipal. Branch of Lafayette Board of Parks & Recreation, 1915 Scott St., Lafayette, IN 47904. Tel.: 765-807-1500. Tax-exempt: 501(c)(3).
Institution Type/Description: Zoo.
Collections: exotic mammals; birds; reptiles; native wildlife.
Facilities: zoological park. Gift items for sale.
Activities: formally organized education programs for children; docent program or council; traveling exhibitions.
Publications: program guide, Funformation.
Hours & Admission Prices: Summer: daily 10-4:30. &
Attendance: 75,000 (estimated)
Membership: Individual $25; Family $40; Patron $50; Associate $75; Sustaining $100; Benefactor $250; Zoo Club $500; Zoo Fellow $1,000.

IMAGINATION STATION, 600 N. 4th St., Lafayette, IN 47901. Tel.: 765-420-7780. Fax: 765-420-8260.
E-mail: info@imagination-station.org

Web Site: www.imagination-station.org
Key Personnel: Operations Mgr., Patty Janssen
Institution Type/Description: Children's Museum.
Collections: hands-on science, space & technology exhibitions.
Activities: demonstrations; classes; educational programs; workshops; story time.
Hours & Admission Prices: Tues. 10-1 & 5:30-8, Sat. 9-2. Children 3-16 $4, adults 17 & up $2; children 2 & under no charge.
Membership: Indiana Resident $60; Out of State $75.

Lawrenceburg

DEARBORN COUNTY HISTORICAL SOCIETY, 508 W. High St., Lawrenceburg, IN 47025. Tel.: 812-537-4075.
E-mail: deahistory@embarqmail.com
Web Site: www.rootsweb.ancestry.com~indbchs
Founded: 1984.
Key Personnel: Pres. (V), Charles Whiting.
Personnel Profile: Part-Time Paid 1.
Operating Expenses: 39,748
Operating Income: 41,769
Governing Authority: Tax-exempt.
Institution Type/Description: Historical Society Museum.
Collections: local history & culture; photographs; period furnishings; personal artifacts; documents.
Research Fields: genealogy.
Activities: educational programs.
Publications: bimonthly, Doorway to History.
Hours & Admission Prices: Mon.-Fri. 1-4; other times by appointment. No charge; donations accepted. Closed holidays. &
Attendance: 150 (estimated)
Membership: Student $3; Individual $10; Family $20; Organization $30; Sustaining $50; Business & Industry $75.

Lebanon

CRAGUN HOUSE, 404 W. Main St., Lebanon, IN 46052. Mailing Address: P.O. Box 141, Lebanon, IN 46052-0140. Tel.: 765-483-9414.
Governing Authority: Parent Institution: Boone County Historical Society.
Institution Type/Description: Historic House: built in 1893.
Collections: local history & culture; period furnishings; personal artifacts; photographs.
Facilities: rental facilities.
Activities: research.
Hours & Admission Prices: Mon.-Tues. 1-4; other times by appointment.

Liberty

UNION COUNTY HISTORICAL MUSEUM, 488 S. CR 100 W., Liberty, IN 47353. Mailing Address: P.O. Box 143, Liberty, IN 47353-0143. Tel.: 765-458-8928.
E-mail: jk2083@frontier.com
Founded: 1929.
Congressional District: 5
Key Personnel: Pres., Virginia Bostick.
Personnel Profile: Part-Time Volunteers 5.
Governing Authority: Parent Institution: Union County Historical Society. Tax-exempt.
Institution Type/Description: History Museum.
Collections: local history & culture; photographs; personal artifacts.
Publications: annual newsletter.
Hours & Admission Prices: Call for hours. No charge; donations accepted. &
Attendance: 200 (estimated)
Membership: Individual $5; Family $10.

Ligonier

LIGONIER HISTORICAL MUSEUM, 503 S. Main St., Ligonier, IN 46767. Mailing Address: 300 S. Main St., Ligonier, IN 46767. Tel.: 260-894-7580; 260-894-4511. Facebook: Ligonier Historical Society.
Web Site: www.ligoniertemple.blogspot.com
Institution Type/Description: History Museum.
Collections: Ligonier's history; Jewish artifacts.
Hours & Admission Prices: May-Nov. Tues. & Sat.-Sun. 1-4; other times by appointment. No charge.

STONE'S TAVERN, State Rd. 5 & U.S. 33, Ligonier, IN 46767-9603. Mailing Address: 1588 N. 650 W., Kimmell, IN 46760. Tel.: 574-529-3693.
E-mail: stonestrace1964@hotmail.com
Web Site: www.stonestrace.com
Founded: 1964.
Congressional District: 4
Key Personnel: Pres., Jim Hossler.
Governing Authority: society; nonprofit organization. Tax-exempt: 501(c)(3).
Institution Type/Description: Historic Building Museum: 1839 Stone's Tavern.
Collections: archives.
Activities: permanent exhibitions. Museum Sponsors: Pioneer Crafts Festival in September.
Hours & Admission Prices: June-Sept. Sun. afternoons; other times by appointment. No charge; donations accepted. Sept. Crafts Festival Sat.-Sun. after Labor Day 10-5. Adults $5; children under 12 no charge.
Attendance: 5,000 (estimated)
Membership: Individual $3; Family $5.

Lincoln City

LINCOLN BOYHOOD NATIONAL MEMORIAL, 3027 E. South St., Lincoln City, IN 47552. Mailing Address: P.O. Box 1816, Lincoln City, IN 47552-1816. Tel.: 812-937-4541. Fax: 812-937-9929. TDD: 812-937-4541.
E-mail: libo_superintendent@nps.gov
Web Site: www.nps.gov/libo
Founded: 1962.
Congressional District: 9
Key Personnel: Chief of Interpretation & Resource Management, Mike Capps; Supt., Kendell Thompson.
Personnel Profile: Full-Time Paid 11; Part-Time Paid 7; Part-Time Volunteers 3.
Governing Authority: federal. U.S. Dept. of Interior, National Park Service. Tax-exempt.
Institution Type/Description: Park History Museum.
Collections: Lincoln-related books; pioneer artifacts; Lincoln Living Historical Farm; reproduction of pioneer period cultural items. Historic Sites: Memorial Building; Nancy Hanks Lincoln Gravesite; Cabin Site Memorial; Trail of Twelve Stones; Lincoln Spring.
Research Fields: pioneer history of southern Indiana; Lincoln 1816-1830; Lincoln & Hanks genealogy.
Facilities: 1,000-vol. library of books on subjects related to Lincoln, family history, Spencer county, state & pioneer life available for research on premises; nature trails; auditorium; visitor center. Museum-related items for sale.
Activities: guided tours; self-guided walks; historical interpretation; research; living history; environmental educational programs; loaning of various Lincoln films; on-site school group programs; off-site interpretive services.
Publications: brochures; teachers' packets.
Hours & Admission Prices: Daily 8-5. Family $5, adults $3. Closed New Year's Day; Thanksgiving; Christmas. &
Attendance: 150,000 (accurate)
Membership: Lifetime Golden Access Passport no charge; Lifetime Nationwide Golden Age Passport 62 & over or Lincoln Boyhood Pass $10; Nationwide Golden Eagle Passport $50.

LINCOLN STATE PARK & COLONEL JONES HOME, Hwy. 162, Lincoln City, IN 47552. Mailing Address: Lincoln State Park, P.O. Box 216, Lincoln City, IN 47552-0216. Tel.: 812-937-4710 & 2802. Fax: 812-937-4833.
E-mail: mcrews@dnr.in.gov
Web Site: www.in.gov/dnr/parklake/parks/lincoln.html
Formerly: Colonel William Jones State Historic Site/Lincoln State Park & Colonel Jones Home
Founded: 1976.
Congressional District: 8
Key Personnel: Site Mgr., Michael Crews.
Personnel Profile: Full-Time Paid 1; Part-Time Paid 3; Part-Time Volunteers 6.
Governing Authority: state. Parent Institution: Indiana Department of Natural Resources. Subsidiary Institution: Division of Parks and Reservoirs. Tax-exempt.
Institution Type/Description: Historic House: c.1834 Colonel William Jones House & property.
Collections: house furnishings; papers & mementoes relating to house history.
Research Fields: Colonel William Jones family history; relationship of Colonel Jones & Abraham Lincoln.
Facilities: 100 acres of forests.
Activities: guided tours by docents in costume; annual festival.

Publications: folder, house history; Friends' group newsletter.
Hours & Admission Prices: May 10-Oct. Sat.-Sun. 11-4; other times by appointment. Suggested Donation: adults $2.
Attendance: 2,000 (accurate)

Linden

LINDEN DEPOT MUSEUM, 520 N. Main St., Linden, IN 47955-0061. Mailing Address: P.O. Box 154, Linden, IN 47955-0154. Tel.: 765-339-7245; 877-643-6371. Fax: 765-339-4896.
E-mail: information@lindendepot.com
Web Site: lindendepot.com
Founded: 1986.
Congressional District: 4
Key Personnel: Pres., Bob Strew; Vice Pres., Jim Davis; Treas., Joe Weaver; Sec. & Dir. Devel., Bob Straw.
Personnel Profile: Part-Time Paid 5; Part-Time Volunteers 12; Interns 1.
Governing Authority: nonprofit organization. Parent Institution: Historic Linden, Inc. Subsidiary Institution: Railway Heritage Trust. Tax-exempt: 501(c)(3).
Institution Type/Description: Railroad Museum & Railroad History Resource Center.
Collections: memorabilia & records of railroads, 1800s-present.
Research Fields: Montgomery Co. (IN) railroads; Monon Railroad; Cloverleaf Railroad (Frankfort, IN-St. Louis, MO)
Facilities: 500-vol. library of railroad company records, available for use by the public & inter-library loan.
Activities: lectures; participatory, loan, temporary & traveling exhibitions; school loan service; annual awards series.
Publications: semiannual newsletter, Railway Heritage News; annual report, Annotated Activity Annual.
Hours & Admission Prices: May-Sept. Wed.-Sun. 1-5. Adults $2, children 13-18 $1, children 6-12 $.50; children under 6 no charge.
Membership: Railway Heritage Associate $10.

Logansport

CASS COUNTY HISTORICAL SOCIETY, 1004 E. Market St., Logansport, IN 46947-3560. Tel.: 574-753-3866. Fax: 574-722-9267.
E-mail: cchistoricalsoc@frontier.com
Web Site: www.casshistory.com
Founded: 1907.
Congressional District: 5
Key Personnel: Pres., Donald Snyder; Vice Pres., Harry Rodkey; Sec., Becky Rivers; Treas., Linda Lantz.
Personnel Profile: Full-Time Paid 1; Part-Time Volunteers 19.
Governing Authority: county. Tax-exempt: 501(c)(3).
Institution Type/Description: Historical Society: housed in 1853 Jerolaman-Long House.
Collections: lustreware; glass; natural history; oral history transcriptions; regional art; Civil War room; log cabin; 1920 Revere car; genealogy; Native American artifacts; fossils; two Geode & Mastodon jaw bones.
Research Fields: Civil War; Indians; Cass County history; genealogy.
Facilities: 5,000-vol. library of Civil War & general history books available for use on premises; reading room.
Activities: guided tours; lectures.
Publications: newsletter; book of early photographs, Where Two Rivers Meet; oral history transcriptions, Airing Cass County: Memories of the Old Times; History of Cass County IN 1913-2002.
Hours & Admission Prices: Jan.-Feb. by appointment; March-Dec. Tues.-Sat. 1-5. No charge; donations accepted. Closed legal holidays.
Attendance: 2,500 (estimated)
Membership: Junior $10; General $25-$49; Sustaining $50-$99; Friend of the Museum $100-$499; Benefactor $500-$999; History Patron $1,000 & up.

COLE CLOTHING MUSEUM, 900 E. Broadway, Logansport, IN 46947-3162. Tel.: 574-753-4058.
Founded: 1992.
Congressional District: 2
Institution Type/Description: Clothing Museum.
Collections: period clothing from late 1800s to 1970s.
Activities: tea & style shows.
Hours & Admission Prices: April-Dec. Tues. & Thurs.-Fri. 1-4; other times by appointment. Adults $3, children $1.50.

Madison

HISTORIC MADISON, INC., 500 West St., Madison, IN 47250-3399. Tel.: 812-265-2967.
E-mail: hmi@historicmadisoninc.com
Web Site: www.historicmadisoninc.com
Founded: 1960.
Congressional District: 9
Key Personnel: Exec. Dir. & Pres., John M. Staicer.
Personnel Profile: Full-Time Paid 4; Part-Time Paid 4; Part-Time Volunteers 100.
Governing Authority: nonprofit organization. Subsidiary Institution: Historic Madison Foundation Inc. Branch Museums: 1818 Jeremiah Sullivan House; 1843 Dr. Wm. Hutchings Office & Hospital; The Talbott-Hyatt Pioneer Garden; 1835 HMI Auditorium; 1850 Francis Costigan House; 19th century Saddletree Factory; 1839 St. Michael The Archangel Church; 1850 AME Church. Tax-exempt: 501(c)(3).
Institution Type/Description: Preservation Project and Historic House Museum.
Collections: house furnishings, original & period; office & hospital furnishings of early 19th-century; woodworking & metalworking tools; transportation equipment & artifacts.
Research Fields: 19th-century architecture; 19th-century floriculture, horticulture, handcrafts; period house furnishings; urban development; industrial history - saddletree manufacturing; underground railroad.
Activities: guided tours; lectures; gallery talks; docent program or council; permanent exhibitions.
Publications: brochures; booklets, Jeremiah Sullivan House; A Horse & Buggy Doctor in Southern Indiana; Madison and the Garber Family: A Community and Its Newspaper 1837-1992; book, The Early Architecture of Madison, Indiana; video: Remembering Madison, 1961.
Hours & Admission Prices: Sullivan House & Dr. Wm. Hutchings Office: May-Oct. Sun. & Tues.-Thurs. 1-4:30, Mon. & Sat. 10-4:30. Francis Costigan House: May-Oct. Mon. 10-4:30, Sat.-Sun. 1-4:30. Schroeder Saddletree Factory Museum: May-Oct. Mon. 10-4:30, Sat.-Sun. 1-4:30. Adults $3; students & members no charge.
Membership: Annual $10; Family $15; Participating $25; Sustaining $50; Living Endowment $100; Patron $250; Benefactor $1,000.

JEFFERSON COUNTY HISTORICAL SOCIETY, INC., 615 W. First St., Madison, IN 47250-3731. Tel.: 812-265-2335. Fax: 812-273-5023.
E-mail: info@jchshc.org
Web Site: www.jchshc.org
Founded: 1850.
Congressional District: 9
Key Personnel: Dir., John Nyberg; Museum Shop Mgr., Diana Hand.
Personnel Profile: Full-Time Paid 1; Part-Time Paid 3; Part-Time Volunteers 80; Interns 2.
Governing Authority: nonprofit organization; society. Tax-exempt: 501(c)(3).
Institution Type/Description: County History Museum & Madison Railroad Station: built in 1895.
Collections: artifacts reflecting history of the region from prehistory to 1980s; textiles; GAR memorabilia; county & city archives; caboose.
Research Fields: Ohio valley history.
Facilities: Museum-related items for sale.
Activities: guided tours; lectures; arts festivals; organized education programs for adults; docent program; participatory, loan, traveling & temporary exhibitions; rental facilities. Society Sponsors: Madison in Bloom garden tour; Chautauqua of the Arts.
Publications: quarterly newsletter, Composite Columns; Beloved Madison, a 300 color photo book of walking tours of historic district.
Hours & Admission Prices: April 2-Dec. 22 Mon.-Sat. 10-4:30. Adults $4, discounts to members. &
Attendance: 18,000 (accurate)
Membership: Individual $25; Family $35; Sustaining $50; Sponsor $100; Donor $250; Patron $500; Corporate Benefactor $1,000.

*** LANIER MANSION STATE HISTORIC SITE, (M),** 601 W. First St., Madison, IN 47250-3731. Tel.: 812-625-3526. Fax: 812-265-3501.
E-mail: laniermansionshs@indianamuseum.org
Web Site: www.indianamuseum.org/explore/lanier-mansion
Formerly: Lanier State Historic Site
Founded: 1925.
Congressional District: 6
Key Personnel: Site Mgr., Gerry Reilly; Historic Sites Program Mgr., Anne Fairchild; Site Administrator, Phyllis Stephens.

Governing Authority: state. Parent Institution: Indiana State Museum and Historic Sites Corporation. Tax-exempt.
Institution Type/Description: Historic House and Site: 1844 Greek revival home of J.F.D. Lanier.
Collections: period furnishings; early to mid-Victorian Southern Indiana life.
Research Fields: 1817-1850 Indiana history.
Facilities: Museum-related items for sale.
Activities: guided tours; lectures; educational programs; special events.
Hours & Admission Prices: Daily 9-5; last tour begins at 4. Adults $5, seniors 60 & over $4, children 3-12 $2; discount to school groups; members & children under 3 no charge. Closed most state holidays.
Attendance: 10,162 (accurate)
Membership: Indiana State Museum and Historic Sites: Individual $57; Family & Grandparent $72; Patron $102.

SCHOFIELD HOUSE, 217 W. Second, Madison, IN 47250-3722. Mailing Address: P.O. Box 243, Madison, IN 47250-0243. Tel.: 812-265-4759.
Institution Type/Description: Historic Tavern House: c.1816.
Collections: period furnishings; personal artifacts.
Hours & Admission Prices: April-Nov. Mon.-Sat. 9:30-4, Sun. 12:30-4. Adult $3; children no charge.

Marengo

MARENGO CAVE, 400 E. State Rd. 64, Marengo, IN 47140. Mailing Address: P.O. Box 217, Marengo, IN 47140-0217. Tel.: 888-702-2837; 812-365-2705.
Web Site: www.marengocave.com
Key Personnel: Owner, Gary Roberson
Institution Type/Description: Natural History Museum.
Collections: natural history.
Facilities: cafe. Museum-related items for sale.
Activities: camping cabins; walking tours; pan for gemstones; cave simulator.
Hours & Admission Prices: Memorial Day to Labor Day Mon.-Fri. 9-6, Sat.-Sun. 9-6:30; Sept.-May daily 9-5. Adults $13-$20.50, children $7-$9; children 3 & under no charge. Closed Thanksgiving; Christmas.

Marion

MARION PUBLIC LIBRARY & MUSEUM, (M), Carnegie Bldg., 600 S. Washington St., Marion, IN 46953-1963. Tel.: 765-668-2900, ext. 150. Fax: 765-668-2911. TDD: 765-668-2907.
E-mail: jfelton@marion.lib.in.us
Web Site: www.marion.lib.in.us
Founded: 1884.
Congressional District: 5
Key Personnel: Dir., Mary Eckerle; Head of Indiana History & Genealogy Svcs., Rhonda Stoffer; Cur., June Felton.
Personnel Profile: Full-Time Paid 1; Part-Time Volunteers 1.
Governing Authority: public library district; not-for-profit municipal corporation. Parent Institution: Marion Public Library. Tax-exempt.
Institution Type/Description: General Museum & History Museum: housed in renovated 1902 Carnegie library building.
Collections: concentration on the history of Marion & Grant County.
Research Fields: history & genealogy of Marion and Grant County.
Facilities: 4,000-vol. library of local history & genealogy; auditorium.
Activities: guided tours; docent program. Annual Events: Quilt Show in July; Annual Quilt Show in July; Archaeology/Paleontology Day in September.
Publications: bimonthly newsletter, Special Edition; e-newsletter.
Hours & Admission Prices: Summer: Mon., Wed. & Fri. 9-5:30, Tues. & Thurs. 9-8, Sat. 9-5; Sept.-May Mon., Wed. & Fri. 9-5:30, Sat. 9-5, Sun. 1-4. No charge; donations accepted. Closed New Year's Eve & Day; Presidents' Day; Good Friday; Easter; Memorial Day; Independence Day; Labor Day; Thanksgiving; Christmas Eve & Day.
Attendance: 8,000 (estimated)

THE QUILTERS HALL OF FAME, INC., 926 S. Washington St., Marion, IN 46953-1969. Mailing Address: P.O. Box 681, Marion, IN 46952-0681. Tel.: 765-664-9333. Fax: 765-664-9333. Facebook: The Quilters Hall of Fame.
E-mail: quiltershalloffame@sbcglobal.net
Web Site: quiltershalloffame.net
Founded: 1979.
Congressional District: 20
Key Personnel: Bd. Pres. & Public Rels., Kathy Boxell; Treas., Arlan Christ; Museum Mgr., Deb Geyer; Collections, Dale Drake.
Personnel Profile: Full-Time Paid 1; Part-Time Volunteers 40.
Governing Authority: nonprofit organization. Tax-exempt.
Institution Type/Description: History Museum: housed in the former home of Marie D. Webster, a nationally known quilt designer of the early 20th century.
Collections: works & accomplishments of Hall of Fame inductees including quilts & patterns; quilting history.
Facilities: library; garden.
Activities: guided tours; workshops; lectures; temporary exhibits. Annual Event: Annual Celebration in July.
Publications: biannual newsletter; Celebration Program Booklet.
Hours & Admission Prices: April-Dec. Thurs.-Sat. 10-3. Adults $4, seniors & Students $3; members no charge. Closed major holidays.
Attendance: 3,000 (estimated)
Membership: Senior $20; Individual $30; Associate $50; Donor $75; Sponsor & Quilt Guild $100; Patron $250; Benefactor $500.

Mentone

LAWRENCE D. BELL AIRCRAFT MUSEUM INC., 210 S. Oak St., Mentone, IN 46539. Mailing Address: P.O. Box 411, Mentone, IN 46539-0411. Tel.: 574-353-7113.
E-mail: bilinda2@frontier.com
Founded: 1976.
Key Personnel: Dir. & Pres. (V), Tim Whetstone; Treas. & Museum Shop Mgr., Mary Boggs.
Personnel Profile: Full-Time Volunteers 10.
Governing Authority: private; nonprofit organization. Tax-exempt: 501(c)(3).
Institution Type/Description: Aviation History Museum.
Collections: aviation history & artifacts from WWII, Korean & Vietnam era; aircraft; Lawrence D. Bell's personal memorabilia & artifacts; 3 helicopters; uniforms & equipment; restored Vietnam-era Huey 801.
Major Exhibits: Bryan Pyle, 12/20/13-12/30/14.
Research Fields: aviation & Bell Corp. history.
Facilities: 230-vol. library; 30-seat theater. Museum-related items for sale.
Activities: school presentations; guided tours. Annual Event: Helicopter Fly-In.
Publications: annual newsletter, Echoes of the Bellringer.
Hours & Admission Prices: Summer: Sun. 1-5; other times by appointment. No charge; donations accepted.
Attendance: 1,500 (estimated)
Membership: Student $5; Individual $40; Family $50; Corporation $250, $500, $1,000.

Merrillville

MERRILLVILLE COMMUNITY PLANETARIUM, Clifford Pierce Middle School, 199 E. 70th Ave., Merrillville, IN 46410-3679. Tel.: 219-650-5486. Fax: 219-650-5470.
E-mail: info@mcpstars.org
Web Site: www.mcpstars.org
Founded: 1973.
Key Personnel: Planetarium Dir., Gregg Williams; Gift Shop Mgr., Pam Powell; Show Presenter, Linda Charnetzky; Show Presenter, Pam Gower; Show Presenter, Ruth Drapeau.
Personnel Profile: Full-Time Paid 1; Part-Time Paid 4; Part-Time Volunteers 40.
Governing Authority: Parent Institution: Merrillville Community School Corp. Tax-exempt.
Institution Type/Description: Planetarium.
Collections: Spitz model 512 star projector; Sky-Skan DigitalSky system.
Facilities: 64-seat theater. Museum-related items for sale.
Activities: group programs; public shows; private group shows.
Hours & Admission Prices: School Days: 7:30-3:30; private shows by appointment. Adults $3, children $2.
Attendance: 29,000 (accurate)

Metamora

WHITEWATER CANAL STATE HISTORIC SITE, (M), 19083 Clayborn St., Metamora, IN 47030. Mailing Address: P.O. Box 88, Metamora, IN 47030-0088. Tel.: 765-647-6512. Fax: 765-647-2734.
E-mail: wwcshs@indianamuseum.org
Web Site: www.indianastatemuseum.org/whitewater
Founded: 1948.
Congressional District: 2
Key Personnel: Historic Site Cur., Jay Dishman.
Personnel Profile: Full-Time Paid 6; Part-Time Paid 5.
Governing Authority: state. Parent Institution: Indiana State Museum & Historic Sites, Inc., 650 W. Washington St., Indianapolis, IN 46204. Tax-exempt.
Institution Type/Description: Historic Site & Buildings.

Collections: milling machinery; early transportation & industrial development; early tools. Historic Structures: 1890 operative grist mill, c.1840 restored working canal locks, boat & aqueduct.

Research Fields: Whitewater Canal History; general canal history; local & regional history; grist mill history.

Facilities: picnic grounds; hiking trails.

Activities: milling operations; horse-drawn boat trip along the canal; guided tours.

Hours & Admission Prices: Grist Mill: Wed.-Sun. 9-5. No charge; donations accepted. Ben Franklin III Canal Boat: adults 54, seniors 60-89 $4, children 3-12 $2; discounts to school groups; over 90 & children under 3 no charge. Closed New Year's Day; Easter; Thanksgiving; Christmas. &

Attendance: 131,896 (accurate)

Michigan City

BARKER MANSION, 631 Washington St., Michigan City, IN 46360-3419. Tel.: 219-873-1520. Fax: 219-873-1520.
E-mail: c_zubler@comcast.net
Web Site: emichigancity.org
Key Personnel: Dir., Cecelia Zubler.
Governing Authority: city. Tax-exempt.
Institution Type/Description: Historic House Museum: housed in the former home of John H. Barker owner of Haskell & Barker Railroad Car Company, later known as Pullman-Standard.
Collections: books; paintings; family artifacts; furnishings; portraits.
Hours & Admission Prices: June-Oct. Mon.-Fri. 10, 11:30 & 1, Sat.-Sun. 12 & 2; Nov.-May Mon.-Fri. 10, 11:30 & 1. Adults $5, children 4-18 $2; children 3 & under no charge. Closed holidays. &
Attendance: 10,000 (accurate)

GREAT LAKES MUSEUM OF MILITARY HISTORY, 360 Dunes Plaza, W., U.S. Hwy. 20, Michigan City, IN 46360-7342. Mailing Address: 350 Menke Rd., Trail Creek, IN 46360-6521. Tel.: 219-872-2702; 800-726-5912.
E-mail: info@militaryhistorymuseum.org
Web Site: www.militaryhistorymuseum.org/
Founded: 1993.
Key Personnel: C.E.O., Terrye Mansfield; Chm. (V), Ruth Mokrycki.
Personnel Profile: Part-Time Paid 2; Part-Time Volunteers 1.
Governing Authority: nonprofit organization. Tax-exempt: 501(c)(3).
Institution Type/Description: Military Museum.
Collections: military heritage; military memorabilia from the Revolutionary War to present.
Activities: Annual Events: black-tie event honoring veterans; Military Vehicle Show
Publications: newsletter, The Bugler
Hours & Admission Prices: Memorial Day to Labor Day Tues.-Fri. 9-4, Sat. 10-4, Sun. 12-4; Sept.-May Tues.-Fri. 9-4, Sat. 10-4. Adults $3, seniors & veterans $2, children 8-18 $1; discounts to AIM members; children under 8 & active military no charge. Closed New Year's Eve & Day; Thanksgiving; Christmas Eve & Day. &
Attendance: 2,500 (estimated)

LUBEZNIK CENTER FOR THE ARTS, (M), 101 W. 2nd St., Michigan City, IN 46360-3228. Tel.: 219-874-4900. Fax: 219-872-6829.
E-mail: artinfo@lubeznikcenter.org
Web Site: www.lubeznikcenter.org
Formerly: John G. Blank Center for the Arts
Founded: 1977.
Congressional District: 3
Key Personnel: Exec. Dir., Carolyn Saxton; Pres. (V), Rachel Saxon; Dir. Exhibits & Education, Janet Bloch; Cur. Exhibitions, Carol Ann Brown; Museum Shop Mgr., Esther Guncheon.
Personnel Profile: Full-Time Paid 5; Part-Time Paid 6; Part-Time Volunteers 125; Interns 2.
Governing Authority: nonprofit. Tax-exempt: 501(c)(3).
Institution Type/Description: Art Museum.
Collections: contemporary graphics, paintings, ceramics; photographs.
Major Exhibits: Citizen-Soldier-Citizen, 11/13-2/14; Theater of Conflict, 11/13-2/14; Embroidered Archetypes, 3/14-5/14; What Is Left Unspoken, 3/14-5/14; Materials Possessions, 5/14-7/14; Invasive Species, 8/14-10/14; Juried Call - Up Cycling, 8/14-10/14; Comics & Codes, 10/14-1/15; Anime All The Way, 10/14-1/15; Bill Eddy, 10/14-1/15.
Research Fields: pertaining to collection.
Facilities: Museum-related items for sale.
Activities: art classes; tours; special presentations; facility rentals. Museum

Sponsors: Annual Lakefront Art Festival in August; Annual Gala in June; First Friday Cultural Events.
Publications: annual report; exhibit catalogues.
Hours & Admission Prices: June-Aug. Mon.-Wed. & Fri. 10-5, Thurs. 2-7, Sat.-Sun. 11-4; Sept.-May Mon.-Fri. 10-5, Sat.-Sun. 11-4. Suggested Donation: $3 per person; discounts to groups & active military; members no charge. Closed New Year's Eve & Day; Martin Luther King Jr. Day; Independence Day; Memorial Day; Labor Day; Thanksgiving; Christmas. &
Attendance: 25,000 (estimated)
Membership: Senior Individual $30; Individual $36; Family $60; Sustaining $120; Advocate $240; Art Makers $500 & up.

OLD LIGHTHOUSE MUSEUM, 100 Heisman Harbor Rd., Washington Park, Michigan City, IN 46360. Mailing Address: P.O. Box 512, Michigan City, IN 46361-0512. Tel.: 219-872-6133.
E-mail: mchistorical@att.net
Web Site: www.oldlighthousemuseum.org
Founded: 1973.
Congressional District: 1
Key Personnel: Dir. & Museum Shop Mgr., Laura Shields.
Personnel Profile: Part-Time Paid 1; Part-Time Volunteers 30.
Governing Authority: society. Parent Institution: Michigan City Historical Society, Inc. Tax-exempt: 501(c)(3).
Institution Type/Description: Maritime & History Museum: housed in 1858 keepers dwelling.
Collections: lighthouse service artifacts; boat builder's tools; shipwreck artifacts; local Indian artifacts; local historical articles.
Research Fields: local history; maritime.
Facilities: 300-vol. library of material on county & area history, lakefront & shipping available for use in museum library by appointment. Postcards, historical maps, pamphlets & books for sale.
Activities: guided tours; lectures; docent council; permanent & temporary exhibitions.
Publications: quarterly newsletter, Old Lighthouse Museum News; booklets, Abijah Bigelow, Revolutionary Soldier, History of Michigan City, the Life of a Town; Great Lakes' First Submarine, L.D. Phillips; Fool Killer; U.S. Life Saving/Coast Guard 1889-1989; Michigan City's First Hundred Years; Little Bit of History Series; #1-Tribute to G.C. Calvert - Michigan City History; #2-History of the Trail Creek Region; #3-Two Speeches of the Mayor Martin T. Krueger; #4-Memories of Early Michigan City; #5-Memories of the Michigan City Lighthouse; #6-Eddyville & Eddyville school.
Hours & Admission Prices: April-Nov. Tues.-Sun. 1-4; other times by appointment. Adults and children 14 & over $5, children under 14 $2; members no charge. Closed New Year's Day; Good Friday; Easter; Memorial Day; Independence Day; Thanksgiving; Christmas.
Attendance: 5,000 (estimated)
Membership: Individual $8; Family $10; Patron $25; Sponsor $50; Life $150.

WASHINGTON PARK ZOO, 115 Lakeshore Dr., Michigan City, IN 46360-3256. Tel.: 219-873-1510. Fax: 219-873-1540. Facebook: Washington Park Zoo.
E-mail: jmartinez@washingtonparkzoo.com
Web Site: www.washingtonparkzoo.com
Formerly: Washington Park Zoological Gardens
Founded: 1928.
Key Personnel: Zoo Dir., Johnny Martinez; Asst. Dir., Jamie LeBlanc-Huss; Cur., Elizabeth Emerick; Office Mgr., Shawne Sheldon.
Governing Authority: municipal. Tax-exempt.
Institution Type/Description: Zoological Gardens & Nature Center.
Collections: 35 species mammals, 91 specimens; 25 species birds, 59 specimens; 23 species reptiles, 37 specimens; Children's Castle; Petting Barn.
Facilities: zoological park. Zoological-related items for sale.
Activities: lectures; films; study clubs; TV & radio programs; formally organized education programs; docent program; loan, permanent, temporary & traveling exhibitions; school loan service.
Publications: quarterly newsletter.
Hours & Admission Prices: April 1-Memorial Day & Labor Day-Oct. 31 daily 10-4, Memorial Day to Labor Day daily 10-5. Adults $6.50, seniors $4.50, children 3-11 $5.50; discounts reciprocal with member zoos; children 2 & under no charge. &
Attendance: 63,000 (accurate)
Membership: Individual $40; Family & Grandparent $60; Family Plus $70; Zookeeper $125; Business Member $250; Business Associate $500; Business Partner $1,000; Business Patron $2,500; Business Benefactor $5,000.

Middletown

VERA'S LITTLE RED DOLLHOUSE MUSEUM, 4385 W. County Rd. 850, N., Middletown, IN 47356-9462. Tel.: 765-533-3453.
Institution Type/Description: Doll Museum.
Collections: dolls; stuffed animals.
Hours & Admission Prices: Call for hours. No charge; donations accepted. &

Mishawaka

HANNAH LINDAHL CHILDREN'S MUSEUM, (M), 1402 S. Main St., Mishawaka, IN 46544-5241. Tel.: 574-254-4540. Fax: 574-254-4585.
E-mail: director@hlcm.org
Web Site: www.hlcm.org
Founded: 1946.
Congressional District: 3
Key Personnel: Pres. (V), Dave Eisen; Dir. & Cur., Lexie Schroeder.
Personnel Profile: Full-Time Paid 1; Part-Time Volunteers 30; Interns 5.
Governing Authority: public school district. Parent Institution: School City of Mishawaka. Tax-exempt: 170(b)(1).
Institution Type/Description: Local History & Children's Museum.
Collections: Native American artifacts; culture & history of Mishawaka; ethnic articles & period clothing; archaeology finds; military items; tools; books; natural history displays; Japanese house & gardens; airplane model collection; c.1800, Mishawaka village; Survive Alive House teaching fire safety; student projects.
Research Fields: history & ethnic cultures of Mishawaka.
Facilities: Gift items for sale.
Activities: guided tours; demonstrations; inter-school & inter-museum loans; education kits; historical bus tours.
Publications: quarterly newsletter.
Hours & Admission Prices: June Tues.-Thurs. 10-2; Sept.-May Tues.-Fri. 9-4. Adults $3, senior citizens 60 & over, students 5-17 $2, children 2-4 $1; discounts to WNIT & ACM reciprocal members; members no charge. Closed school holidays. &
Attendance: 5,500 (estimated)
Membership: First Time Family $10; Friend $20; Donor $35; Benefactor $50; Sponsor $75; Reciprocal $100; Corporate $150.

MISHAWAKA SPORTS MUSEUM, 109 Lincoln Way E., Mishawaka, IN 46544-2016. Tel.: 574-257-0039.
Founded: 1995.
Key Personnel: Museum Shop Mgr., Larry La Cluyse
Institution Type/Description: Sports Museum.
Collections: early 1900s sports equipment; programs; autographs; period artifacts; photographs; magazines; books; Notre Dame memorabilia.
Hours & Admission Prices: Mon.-Fri. 11-5. No charge; donations accepted. Closed holidays.

OTIS BOWEN MUSEUM & ARCHIVES, 1001 Bethel Cir., Mishawaka, IN 46545-2232. Tel.: 574-257-3347. Fax: 574-257-3499.
Web Site: www.bethelcollege.edu/library/archives
Founded: 1983.
Congressional District: 2
Key Personnel: Dir., Dr. Clyde Root.
Governing Authority: Parent Institution: Bethel College. Tax-exempt.
Institution Type/Description: History Museum.
Collections: Gov. Otis Bowen's personal artifacts & documents; photographs.
Hours & Admission Prices: By appointment. No charge. &

Mitchell

PIONEER VILLAGE & VIRGIL GRISSOM MEMORIAL AT SPRING MILL STATE PARK, 3333 State Rd. 60 E., Mitchell, IN 47446. Mailing Address: P.O. Box 376, Mitchell, IN 47446-0376. Tel.: 812-849-3534. Fax: 812-849-5249.
E-mail: springmillstatepark@dnr.in.gov
Web Site: www.in.gov/dnr/parklake/2968.htm
Formerly: Spring Mill State Park Pioneer Village & Grissom Memorial
Founded: 1927.
Congressional District: 8
Key Personnel: Chm. (V), Coletta Prewitt; Property Mgr., Mark Young; Asst. Property Mgr., Jon Winne; Interpretive Naturalist, Brad Gilley.
Personnel Profile: Full-Time Paid 16; Part-Time Paid 47; Part-Time Volunteers 162.

Governing Authority: state. Parent Institution: Dept. of Natural Resources. Div. of State Parks, 402 W. Washington, Rm. W298, Indianapolis, IN 46204. Tax-exempt.
Institution Type/Description: Village Museum: housed in 1817 Grist Mill, & other restored buildings, located on the site of a flourishing pioneer village in the 1800s.
Collections: over 1,000 artifacts from the 1800s period of the village; early pioneer garden; Gemini III space capsule & suite; pioneer cemetery; mementoes of the life of Virgil I. (Gus) Grissom, pioneer in America's exploration of outer space & the second American astronaut to go into space.
Research Fields: botanical; history of the area.
Facilities: nature center; picnic areas; nature trails.
Activities: guided tours; candlelight tours; cave boat rides; craft demonstrations; naturalist programs; special events; fishing; mountain bikes; hayrides.
Publications: Gristmill Gazette; The Village That Slept Awhile; Pioneer Garden; Village coloring book; garden coloring book.
Hours & Admission Prices: Call for hours. Out-of-State Residents $7 per vehicle; Indiana Residents $5 per vehicle. Closed New Year's Day; Thanksgiving; Christmas. &
Attendance: 670,000 (estimated)
Membership: Indiana Residents $36 (annual entrance pass); Out-of-State Residents $46 (annual entrance pass).

Monticello

WHITE COUNTY HISTORICAL SOCIETY MUSEUM, 101 S. Bluff St., Monticello, IN 47960-2308. Tel.: 574-583-3998.
E-mail: wcmuseum@centurylink.net
Web Site: www.rootsweb.ancestry.com/~inwchs/
Founded: 1911.
Congressional District: 2
Key Personnel: Dir. & Museum Shop Mgr., Judith Baker Elliott; Pres., Kean MacOwan; Vice Pres., John Baum.
Personnel Profile: Part-Time Paid 1; Part-Time Volunteers 7.
Governing Authority: nonprofit organization. Tax-exempt: 501(c)(3).
Institution Type/Description: Historical Society Museum.
Collections: people & history of White County, Indiana.
Research Fields: family history; White County history.
Facilities: library available for use on premises.
Activities: guided tours; lectures; formally organized education programs for children; permanent & temporary exhibitions; research.
Publications: newsletter four times per year, Museum Musings.
Hours & Admission Prices: Wed.-Fri. 10-4; other times by appointment. No charge; donations accepted. Closed Christmas Eve & Day; national holidays.
Attendance: 1,000 (estimated)
Membership: Single $15; Family $25; Business $100; Life $150, Business Life $500.

Mooresville

ACADEMY OF HOOSIER HERITAGE, 250 N. Monroe St., Fl. 2, Mooresville, IN 46158-1551. Tel.: 317-831-9001.
E-mail: jrkylelee@comcast.net
Web Site: www.academymuseum.org
Formerly: Academy Building Museum
Founded: 2000.
Key Personnel: Dir., Julie Kyle-Lee; Pres. (V), William Kirk.
Personnel Profile: Part-Time Paid 1; Part-Time Volunteers 15; Interns 1.
Institution Type/Description: Local History Museum: housed in a former school built in 1861. Listed on the National Register of Historic Places.
Collections: community heritage & history; photographs; personal artifacts; period furnishings; local & state school history; one-room school; veterans memorial.
Major Exhibits: Toys Through the Years, 12/13-1/14; Cards from the Heart: The Ruth Sears Collection, 2/14; Memorial Salute to Veterans of the Past, 5/14-6/14; Salute to Veterans, 11/14.
Research Fields: school history; one-room schools; Civil War; local 1800s history.
Facilities: community meeting room.
Activities: special events; temporary & permanent exhibits; hands-on activities; school & educational programs. Annual Events: Civil War Days; Victorian Christmas; Quilt Show.
Hours & Admission Prices: Wed. 2-7, 2nd Sat. each month 11-3; other times by appointment. No charge; donations accepted.
Membership: Students and Seniors 60 & over $15; Individual $20; Family $45; Settler $75; Founder $100; Trailblazer $300.

Muncie

ACADEMY OF MODEL AERONAUTICS/NATIONAL MODEL AVIATION MUSEUM, 5151 E. Memorial Dr., Muncie, IN 47302-9252. Tel.: 765-287-1256. Fax: 765-281-7904. Facebook: Academy of Model Aeronautics/National Model Aviation Museum.
E-mail: museum@modelaircraft.org
Web Site: www.modelaircraft.org/museum/museuminfo.aspx
Founded: 1936.
Congressional District: 2
Key Personnel: Pres. (V), Dave Mathewson; Archivist, Jackie Shalberg; Collectioins Mgr., Maria VanVreede; Museum Dir., Michael Smith; Education Specialist, Emily Loy; Museum Shop Mgr., Wendy Neal.
Personnel Profile: Full-Time Paid 4; Part-Time Paid 1; Part-Time Volunteers 12; Interns 2.
Governing Authority: private; nonprofit organization. Parent Institution: Academy of Model Aeronautics. Tax-exempt.
Institution Type/Description: Aeronautics Museum.
Collections: concentration on the developmental history of flying model/miniature aircraft; plans; photographs.
Research Fields: aviation & model aviation history.
Facilities: library & archive of books & magazines covering all aspects of aviation & model aviation; classroom; theater. Clothing, accessories, books, videotapes & items related to model aviation for sale.
Activities: docent program; guided tours; educational make-and-fly programming.
Publications: monthly magazine, Model Aviation; quarterly newsletter, Cloud Nine.
Hours & Admission Prices: April 1-Sept. 30 Mon.-Fri. 8-4:30; Oct. 1- March 31 Tues.-Sat. 10-4; call for holiday hours; Adults $3, children 6-17 $1.50; discounts to group of 10 or more; members no charge. &
Attendance: 5,709 (accurate)
Membership: Patron $25; Supporter $100; Life $1,000.

*** DAVID OWSLEY MUSEUM OF ART BALL STATE UNIVERSITY,** 2101 W. Riverside Ave., Muncie, IN 47306. Tel.: 765-285-5242 & 5270. Fax: 765-285-4003.
E-mail: artmuseum@bsu.edu
Web Site: www.bsu.edu/artmuseum
Formerly: Ball State University Museum of Art
Founded: 1936.
Congressional District: 10
Key Personnel: Interim Dir., Carl Schafer; Cur. Education, Tania Said; Preparator & Exhibition Designer, Randy Salway; Admin., Christine Lussier.
Personnel Profile: Full-Time Paid 4; Part-Time Paid 18; Part-Time Volunteers 40; Interns 2.
Governing Authority: state. Parent Institution: Ball State University. Tax-exempt: 170(b)(1)(A).
Institution Type/Description: Art Museum.
Collections: Italian Renaissance art; 17th- to 19th-century European art; 19th- & 20th-century American art; Chinese art; Indian art; African art; art of ancient cultures; Mesoamerican, Pacific Island & Japanese art.
Research Fields: pertaining to collections.
Activities: guided tours for children & adults; lectures; gallery talks; inter-museum loan, permanent, temporary & traveling exhibitions.
Publications: catalogs for special exhibitions; annual report; newsletter; Ball State University Museum of Art at 75: The Museum and a History of its Collection.
Hours & Admission Prices: Mon.-Fri. 9-4:30, Sat.-Sun. 1:30-4:30. No charge; donations accepted. Closed holidays. &
Attendance: 30,000 (estimated)
Membership: Student $15; Individual $25; Family & Dual $50; Sponsor $100-$249; Patron $250-$499; Sustaining $500-$999; Benefactor $1,000-$1,999; Philanthropist $2,000 & up; Connoisseur, accumulated giving of $50,000 & up.

MINNETRISTA, (M), 1200 N. Minnetrista Pkwy., Muncie, IN 47303-2925. Tel.: 765-282-4848. Fax: 765-741-5110. Facebook: Minnetrista.
E-mail: info@minnetrista.net
Web Site: www.minnetrista.net
Formerly: Minnetrista Cultural Center
Founded: 1988.
Congressional District: 2
Key Personnel: Pres. & C.E.O., Betty Brewer; Chm. (V), James P. Borgmann; Vice Pres. Visitor Experience, Rebecca Holmquist; C.F.O., Phillip Dunn; Vice Pres. Philanthropy, Bob Scott; Dir. Collections, Karen Vincent; Dir.

Education & Experience, George Buss; Dir. Operations, Rob Keisling; Retail Mgr., Molly Harty.
Personnel Profile: Full-Time Paid 39; Part-Time Paid 27; Part-Time Volunteers 193; Interns 13.
Volunteer Hours: 12,815
Operating Expenses: 4,230,289
Operating Income: 4,207,525
Governing Authority: nonprofit. Subsidiary Institution: Minnetrista Cultural Foundation, Inc. Tax-exempt: 501(c)(3).
Institution Type/Description: General Museum.
Collections: artifacts documenting the heritage of East Central Indiana including documents, manuscripts, photographs, fruit jars, industrial products, clothing, quilts & coverlets, furniture, fine & decorative arts, military, toys, dolls, games, transportation; historic & contemporary gardens; 4 historic buildings.
Major Exhibits: Weird & Wonderful: 25 Reasons to Love Our Community, 10/25/13-3/30/14; Casey's Art Studio, 1/11/14-2/9/14; Annual Juried Art Shows Sale, 2/22/14-5/4/14; Glass 2014, 4/19/14-7/20/14; Open Space: Art About the Land (T), 8/30/14-9/28/14; Water's Extreme Journey (T), 8/30/14-1/14/15; Painting Indiana III: A Dignity of Place (T), 10/5/14-11/30/14.
Research Fields: history, environment, art, & industry of east central Indiana.
Facilities: 2,500-vol. library of local history materials; 12,500 sq. ft. exhibit space; 4 historic buildings; educational facilities; 21 acres of gardens; 6.3 acre nature area. Museum-related items for sale.
Activities: guided tours; lectures; films; concerts; arts festivals; study clubs; organized educational programs; docent program; participatory, loan, traveling & temporary exhibits; seasonal events; museum theatre & outreach.
Publications: triannual, Columns Magazine.
Hours & Admission Prices: Mon.-Sat. 9-5:30, Sun. 11-5:30. Admission $5; discounts to groups, APGA, AAA & AAM members; members no charge; donations accepted. Gift Shop: The Orchard Shop at Minnetrista: Mon.-Sat. 9-5:30, Sun. 11-5:30. Closed New Year's Day; Easter; Thanksgiving; Christmas. &
Attendance: 114,900 (estimated)
Membership: Individual $30; Dual $50; Family $70; Premier $125; Maplewood Circle $250; Nebosham Circle $500; Oakhurst Circle $1,000.

THE MOORE-YOUSE HOME MUSEUM, 122 E. Washington St., Muncie, IN 47305-1734. Mailing Address: 120 E. Washington St., Muncie, IN 47303-1734. Tel.: 765-282-1550.
E-mail: museum@the-dchs.org
Web Site: www.the-dchs.org/moore-youse_home_museum.htm
Founded: 1925.
Congressional District: 6
Key Personnel: Dir., Mary Maxon; C.E.O., Jack Maxon.
Governing Authority: Parent Institution: Delaware County Historical Society, Inc. Tax-exempt.
Institution Type/Description: Historic House Museum.
Collections: Muncie history; furnishings; paintings; documents; photographs.
Publications: quarterly newsletter.
Hours & Admission Prices: March-Nov. 1st Sun. of month 1-4; other times by appointment. Adults $3; members no charge.
Membership: Individual $20; Family $30; Corporate $100.

MUNCIE CHILDREN'S MUSEUM, 515 S. High St., Muncie, IN 47305-2376. Mailing Address: P.O. Box 544, Muncie, IN 47308-0544. Tel.: 765-286-1660. Fax: 765-286-1662.
E-mail: museum@munciemuseum.com
Web Site: www.munciechildrensmuseum.com
Founded: 1977.
Congressional District: 2
Key Personnel: Dir. Operations, Kynda Rinker; Chm. (V), James Borgmann; Pres. (V), Patrick Burkey.
Personnel Profile: Full-Time Paid 3; Part-Time Paid 9; Part-Time Volunteers 200.
Governing Authority: nonprofit organization. Tax-exempt: 501(c)(3).
Institution Type/Description: Children's Museum.
Collections: hands-on exhibits.
Facilities: 3 education classrooms. Educational items for sale.
Activities: guided tours; lectures; concerts; workshops; formally organized educational programs; docent program; permanent, temporary & loan exhibitions.
Publications: newsletter.
Hours & Admission Prices: Tues.-Sat. 10-5, Sun. 1-5. Adults $6; discounts to ASTC members. Closed major holidays. &
Attendance: 37,850 (accurate)
Membership: Family $70; Extended Family $100.

Munster

MUNSTER HISTORY MUSEUM, 1154 Ridge Rd., Munster, IN 46321. Mailing Address: Munster Historical Society, Museum Committee, 1005 Ridge Rd., Munster, IN 46321. Tel.: 219-836-6530. Fax: 219-838-3296.
E-mail: info@munsterhistory.org
Web Site: munsterhistory.org
Institution Type/Description: History Museum.
Collections: local history & culture; period furnishings; personal artifacts; photographs.
Hours & Admission Prices: 1st & 3rd Sat. each month 10-1; groups by appointment.

SOUTH SHORE ARTS, 1040 Ridge Rd., Munster, IN 46321-1876. Tel.: 219-836-1839, ext. 100. Fax: 219-836-1863.
E-mail: kelly@southshoreartsonline.org
Web Site: www.southshoreartsonline.org
Formerly: Northern Indiana Arts Association
Founded: 1969.
Congressional District: 1
Key Personnel: Exec. Dir., John Cain; Pres., Michael Luongo; Administrative Asst., Kelly Freeman; Dir. Finance & Administration, Christine McCabe; Dir. Exhibitions, Jillian Van Volkenburgh; Dir. Mktg. & Devel., Tricia Hernandez; Education Coord., Kimberly McKinley; Special Projects Dir., Donna Catalano; Museum Shop Mgr., Jackie Wickland.
Governing Authority: nonprofit. Tax-exempt: 501(c)(3).
Institution Type/Description: Art Association Museum.
Collections: works by regional artists.
Facilities: library of reference, visual art & arts management material available to the public for use on premises; 370-seat auditorium; 5,000 sq. ft. exhibit space; educational facilities. Art museum-related gift items for sale.
Activities: guided tours; lectures; concerts; arts festivals; theater; organized education programs; docent program. Museum Sponsors: Salon Show; Elementary, Junior High & High School Art Shows; regional, national & international exhibits.
Publications: quarterly newsletter.
Hours & Admission Prices: Mon.-Fri. 10-5, Sat. 10-4, Sun. 12-4. Adults $3, students $2; members no charge. &
Membership: Student $25; Artist $40; Individual $45; Family $65; Patron $125-$249, Sponsor $250-$499; Benefactor $500-$999; Life $1,000.

Nappanee

AMISH ACRES, 1600 W. Market St., Nappanee, IN 46550. Tel.: 800-800-4942; 574-773-4188. Fax: 574-773-4180.
E-mail: amishacres@amishacres.com
Web Site: www.amishacres.com
Founded: 1968.
Congressional District: 3
Institution Type/Description: Historic Farm Museum.
Collections: local history & culture; period furnishings & equipment; personal artifacts. Historic Buildings: farmhouse; two room house; barn; wagon shed; ice house; cider mill; one room schoolhouse.
Hours & Admission Prices: May-Oct. Tues.-Sat. 10-7, Sun. 10-5. &

Nashville

BROWN COUNTY ART GALLERY AND MUSEUM, #1 Artist Dr., Nashville, IN 47448-8101. Mailing Address: P.O. Box 443, Nashville, IN 47448-0443. Tel.: 812-988-4609.
E-mail: brncagal@att.net
Web Site: www.browncountyartgallery.org
Founded: 1926.
Congressional District: 9
Key Personnel: Pres., Lyn Letsinger-Miller; Dir., Jeanne Bennett; Vice Pres., Cheryl Eyed.
Personnel Profile: Full-Time Paid 1; Full-Time Volunteers 17; Part-Time Paid 2; Part-Time Volunteers 12.
Governing Authority: Parent Institution: Brown County Art Gallery Foundation. Tax-exempt.
Institution Type/Description: Art Association Gallery.
Collections: oil paintings & pastels by Glen Cooper Henshaw; memorial collection of artwork by prominent early Indiana artists & deceased artist members; paintings; etchings; prints; exhibits of contemporary artist members.
Activities: gallery talks; study clubs; annual exhibit; school & youth program; classes in fine & performing arts. Annual Events: Spring Patron Art Show in May; Indiana Heritage Arts Show in June; Neophytes Art Show in August; Brown County Artists Today in September; Collector's Showcase in October; Christmas Art Show & Sale in November-December.
Publications: annual catalogue.
Hours & Admission Prices: Mon.-Sat. 10-5, Sun. 12-5. No charge; donations accepted. Closed New Year's Day; Thanksgiving; Christmas. &
Attendance: 7,500 (estimated)
Membership: Student $10; Single $25; Family $30; Patron & Group $40; Century Club & Corporate $100; Life $1,000.

BROWN COUNTY HISTORY CENTER, 46 E. Gould St., Nashville, IN 47448. Mailing Address: Box 668, Nashville, IN 47448-0668. Tel.: 812-988-2899.
E-mail: julia@browncountyhistorycenter.org
Web Site: browncountyhistorycenter.org
Formerly: Brown County Historical Society Pioneer Village
Founded: 1957.
Congressional District: 9
Key Personnel: Dir. & Museum Shop Mgr., Julia Pearson; Pres. (V), Ivan Lancaster; Archives, Rhonda Dunn.
Personnel Profile: Part-Time Paid 1; Part-Time Volunteers 60.
Governing Authority: society; nonprofit. Parent Institution: Brown County Historical Society. Subsidiary Institution: Pioneer Women's Club. Affiliated with the Indiana State Historical Society, Indianapolis, IN. Tax-exempt.
Institution Type/Description: Historical Society Museum Village.
Collections: furnishings & furniture; completely equipped early doctor's office; weaving & spinning collections; pioneer artifacts. Historic Structures: log cabin; barn; log jail; doctor's office; medical & dental items; blacksmith shop.
Research Fields: archaeology of Brown County; historical research.
Facilities: library of genealogical material & archives available for use by permission of archivist or genealogist on premises only.
Activities: guided tours; temporary exhibitions; spinning; weaving; blacksmith; woodworking; fireplace cooking; candle dipping; fabric dyeing.
Publications: quarterly newsletter; annual genealogy records; History of County Reprints.
Hours & Admission Prices: May-Oct. Sat.-Sun. 1-4:30.
Attendance: 6,000 (estimated)
Membership: Student $5; Annual $20; Sustainer $250; Business $100; Life $1,000.

✵ **T.C. STEELE STATE HISTORIC SITE, (M),** 4220 T.C. Steele Rd., Nashville, IN 47448-9586. Tel.: 812-988-2785. Fax: 812-988-8457.
E-mail: tcsteeleshs@indianamuseum.org
Web Site: www.tcsteele.org
Founded: 1945.
Congressional District: 9
Key Personnel: Historic Site Mgr., Andrea Smith DeTarnowsky.
Personnel Profile: Full-Time Paid 3; Part-Time Paid 3; Part-Time Volunteers 50.
Governing Authority: state. Parent Institution: Indiana State Museum & Historic Sites Corporation, 650 W. Washington St., Indianapolis, IN 46204. Tax-exempt.
Institution Type/Description: Historic Site.
Collections: Steele paintings; period furnishings; books; fine & decorative art. Historic Buildings: c.1907 T.C. Steele Home; c.1916 Studio.
Facilities: 5 hiking trails; over 3 acres of gardens; state nature preserve.
Activities: self-guided & guided tours; nature trails.
Hours & Admission Prices: Tues.-Sat. 9-5, Sun. 1-5. Adults $3.50, seniors 65 & over $3, children 12 & under $2; discounts to groups, AAA & National Trust for Historic Preservation members; Friends Group & Indiana State Museum Society members no charge. Closed most holidays. &
Attendance: 10,340 (accurate)
Membership: Individual $25; Family $40; Supporting $100; Sponsor $200; Patron $500; Lifetime $1,000.

New Albany

CARNEGIE CENTER FOR ART & HISTORY, (M), 201 E. Spring St., New Albany, IN 47150-3422. Tel.: 812-944-7336. Fax: 812-981-3544. Facebook: Carnegie Center for Art and History.
E-mail: snewkirk@carnegiecenter.org
Web Site: carnegiecenter.org
Founded: 1971.
Congressional District: 9
Key Personnel: Dir., Sally Newkirk; Cur., Karen Gillenwater; Dir. Mktg. & Outreach, Laura Wilkins; Public Rels. Assoc., Delesha Thomas.
Personnel Profile: Full-Time Paid 4; Part-Time Volunteers 15.
Governing Authority: nonprofit. NA-FC Public Library. Tax-exempt.

Institution Type/Description: History Museum & Art Gallery: housed in 1904 Carnegie Library.
Collections: Underground Railroad history; local history & culture; works by regional artists; Civil War nurse, Lucy Higgs Nichols.
Major Exhibits: The Potential in Everything: Albertus Gorman and R. Michael Wimmer, 1/24/14-4/5/14; Floyd County Secondary Schools Art Show & Competition, 4/12/14-4/26/14; Form Not Function: Quilt Art at the Carnegie, 5/9/14-7/12/14; Penny Sisto, 10/24/14-1/10/15.
Activities: tours, lectures; art & craft workshops for adults & children; New Albany Public Art Project. Annual Event: National Juried Exhibition of Contemporary Quilt Art.
Hours & Admission Prices: Tues.-Sat. 10-5:30. No charge. Closed major holidays. &
Attendance: 19,000 (accurate)
Membership: Patron $30; Curator $50; Director $100; John Cody Society $250; Scribner Society $500; Carnegie Society $1,000.

CULBERTSON MANSION STATE HISTORIC SITE, (M), 914 E. Main St., New Albany, IN 47150-5841. Tel.: 812-944-9600. Fax: 812-949-6134.
E-mail: culbertsonmansionshs@dnr.in.gov
Web Site: www.indianamuseum.org/shs
Founded: 1976.
Congressional District: 9
Key Personnel: Pres. (V), Eleene Metcalf; Historic Site Cur. & Museum Shop Mgr., Jamie Mauch.
Personnel Profile: Full-Time Paid 2; Part-Time Paid 2; Part-Time Volunteers 110.
Governing Authority: state. Branch of Indiana State Museum System, Div. of the Dept. of Natural Resources Indiana, 650 West Washington St., Indianapolis, IN 46204. Tax-exempt.
Institution Type/Description: Historic House: 1867 W.S. Culbertson Home, a 25-room Victorian mansion.
Collections: period furnishings.
Research Fields: mid-Victorian & Southern Indiana life.
Activities: guided tours; lectures.
Publications: semi-annually, The Culbertson Newsletter.
Hours & Admission Prices: April to mid-Dec. Tues.-Sat. 9-5, Sun. 1-5. Adults $3.50, seniors $3, children under 12 $2; discounts to groups with appointment; children 3 & under no charge. Closed Election Day; Veterans Day; Thanksgiving, Christmas. &
Attendance: 20,000 (accurate)
Membership: Student & Seniors 55 & over $5; Individual $10; Family $20; Sponsor $25; Patron $50; Benefactor $100; Life $1,000.

New Carlisle

HISTORIC NEW CARLISLE, INC. LOCAL HISTORY MUSEUM, 304 E. Michigan St., New Carlisle, IN 46552. Mailing Address: P.O. Box 107, New Carlisle, IN 46552-0107. Tel.: 574-654-3897.
E-mail: historicnc@townofnewcarlisle.com
Web Site: www.historicnewcarlisle.org/local historymuseum.htm
Founded: 1989.
Key Personnel: Dir., Dana Groves; Pres. (V), Marcy Kauffman.
Personnel Profile: Full-Time Paid 1.
Governing Authority: Tax-exempt.
Institution Type/Description: History Museum.
Collections: local history & culture; period furnishings; personal artifacts; photographs; postcards; newspapers; New Carlisle High School yearbooks.
Publications: bimonthly newsletter.
Hours & Admission Prices: By appointment. No charge, donations accepted.
&
Membership: Senior $10; Individual $15; Family $20; Business $50; Corporate $100; Benefactor $500.

New Castle

ART ASSOCIATION OF HENRY COUNTY, INC., 218 S. 15th St., New Castle, IN 47362-3201. Mailing Address: P.O. Box 842, New Castle, IN 47362-0842. Tel.: 765-529-2634. Facebook: Art Association of Henry County.
E-mail: ncartcenter@yahoo.com
Web Site: henrycountyarts.comeze.com
Founded: 1965.
Key Personnel: Pres., Steve Weidert; Dir., Vicky White.
Personnel Profile: Full-Time Paid 1; Part-Time Paid 1.
Institution Type/Description: Art Association & Gallery.
Collections: paintings; photography; local, county & state artists.

Major Exhibits: Henry Co. Elementary Exhibit, 1/19/14-2/2/14; Art from the Heart: Places & People We Love, 2/9/14-4/6/14; Miniatures, 2/9/14-4/6/14; Henry Co. Student Show, 4/13/14-5/11/14; Henry Co. Historical Society, 5/18/14-6/22/14; Raintree Camera Club, 6/29/14-7/27/14; Rose Show, 8/3/14-8/31/14; 49th Annual Hall Show, 9/14/14-10/19/14; Open Space: About the Land (T), 10/26/14-11/23/14; Starving Artist, 11/29/14-12/7/14.
Facilities: arts park & pavillion; gallery; annex; rental facilities.
Activities: open studios; children's classes; workshops; summer day camp; ceramics.
Publications: bimonthly newsletter.
Hours & Admission Prices: Tues.-Fri. 9-4, Sun. 1-4. No charge; donations accepted. &
Attendance: 2,000 (estimated)
Membership: Student $15; Individual $25; Family & Club $40; Supporting $100; Sustaining $200; Benefactor $300; Corporate $500.

HENRY COUNTY HISTORICAL SOCIETY, 606 S. 14th St., New Castle, IN 47362-3339. Tel.: 765-529-4028.
E-mail: henrycountyhistoricalsociety@gmail.com
Web Site: www.henrycountyhs.org
Founded: 1887.
Congressional District: 6
Key Personnel: Pres., Gene Ingram; Exec. Dir., Elizabeth Edstene.
Personnel Profile: Part-Time Paid 2; Part-Time Volunteers 10.
Governing Authority: county. Tax-exempt.
Institution Type/Description: Historic House: 1870 home of Civil War General William Grose. Listed on the National Register of Historic Places.
Collections: archives; furniture; tools; china; glass; silver; county cemetery records; natural history; Indian relics; dolls.
Research Fields: genealogical; local history.
Facilities: 1,000-vol. library of books on local, state & Civil War history & biographical & genealogical materials available for use on premises; reading room.
Activities: guided tours; permanent exhibitions.
Publications: newsletter, Historicalog.
Hours & Admission Prices: Wed.-Sat. 1-4:30. Adults $2, students $1; children & members no charge. Closed holidays. &
Attendance: 1,200 (estimated)
Membership: Individual $20; Family $30.

INDIANA BASKETBALL HALL OF FAME, 408 Trojan Lane, New Castle, IN 47362. Mailing Address: One Hall of Fame Ct., New Castle, IN 47362-2941. Tel.: 765-529-1891. Fax: 765-529-0273.
E-mail: info@hoopshall.com
Web Site: www.hoopshall.com
Founded: 1962.
Congressional District: 10
Key Personnel: Exec. Dir., Chris May; Pres. (V), Marvin Tudor; Gift Shop Mgr., Becky Beavers.
Personnel Profile: Full-Time Paid 3; Part-Time Paid 3; Part-Time Volunteers 100.
Governing Authority: nonprofit. Tax-exempt: 501(c)(3).
Institution Type/Description: Sports Museum.
Collections: sports equipment; uniforms; photos; films; videos; newspapers.
Research Fields: sports history; history of Indiana high schools.
Facilities: 250-vol. library containing information on rules, strategy & history monographs; reading room; theater.
Activities: tours; tournaments; awards banquet.
Publications: quarterly newsletter, IBHF News; quarterly magazine, Indiana Basketball History.
Hours & Admission Prices: Sun. 1-5, Mon.-Sat. 10-5. Adults $5; discounts to AAM members; members no charge. Closed major holidays. &
Attendance: 10,000 (accurate)
Membership: Regular Member $100; Benefactor $250; Patron $500; Lifetime $3,000.

New Harmony

HISTORIC NEW HARMONY, (M), 603 West St., New Harmony, IN 47631. Mailing Address: P.O. Box 579, New Harmony, IN 47631-0579. Tel.: 812-682-4488. Fax: 812-682-4313.
E-mail: harmony@usi.edu
Web Site: www.usi.edu/hnh
Founded: 1974.
Congressional District: 8
Key Personnel: Dir. Historic New Harmony, Connie Weinzapfel; Collections Mgr., Amanda Bryden; Collections Asst., Heather Baldus; Community

Engagement Mgr., Missy Parkison; Visitor Svcs. Coord., Melissa Williams; Administrative Asst., Christine Crews; Museum Shop Buyer, Sara Rhoades.
Personnel Profile: Full-Time Paid 9; Part-Time Paid 37; Part-Time Volunteers 6; Interns 1.
Governing Authority: nonprofit organization. Parent Institution: Univ. of Southern Indiana. Subsidiary Institution: USI/New Harmony Foundation; Atheneum/Visitors Center, 401 N. Arthur St., New Harmony, IN 47631. Tax-exempt: 501(c)(3).
Institution Type/Description: History Museum & Preservation Project: located on the site of two early utopian experiments-George Rapp's 1814-1825 Harmony Society & Welsh-born social reformer & industrialist Robert Owen's New Harmony; it is also the site of the early headquarters of the U.S. Geological Survey; it provided many of the earliest collections of the Smithsonian Institution.
Collections: 1814-1834 house museums of the Harmonist & Owen periods; geological & natural science collections of the earliest geological surveys; early theatre; manuscripts; 81 hand-colored Maximilian-Bodmer lithographs. Historic Buildings: approx. 12 c.1814-1824 structures; 20 c.1830-1920 structures.
Research Fields: geology; archaeology; decorative arts; communal studies; architecture.
Facilities: archives; 250 & 427-seat auditorium; 200-seat theater; visitor's center. Books, prints & museum-related items for sale.
Activities: guided tours; films on history; gallery talks; concerts; drama; workshops; educational programs; dance recitals; arts festivals; drama; formally organized education programs for adults & undergraduate college students; permanent & traveling exhibitions.
Publications: biannual newsletter.
Hours & Admission Prices: March 15-Dec. 30 daily 9:30-5; Dec. 30-March 14 groups by appointment. Family $25, adults $12, children 7-17 $5; discounts to groups, AAM, ICOM & AAA members. &
Attendance: 21,183 (accurate)
Membership: Donor Categories: Historian $25; Educator $50; Naturalist $100; Preservationist $250; Philanthropist $500; Golden Raintree $1,000; Door of Promise $2,000.

NEW HARMONY GALLERY OF CONTEMPORARY ART, 506 Main St., New Harmony, IN 47631. Mailing Address: P.O. Box 627, New Harmony, IN 47631-0627. Tel.: 812-682-3156. Fax: 812-682-3870.
E-mail: skrhoades@usi.edu
Web Site: www.nhgallery.com
Founded: 1975.
Congressional District: 8
Key Personnel: Asst. Dir., Sara Rhoades.
Personnel Profile: Full-Time Paid 1; Part-Time Paid 2.
Governing Authority: nonprofit organization. University of Southern Indiana. Tax-exempt: 501(c)(3).
Institution Type/Description: Contemporary Art Gallery.
Collections: changing exhibits; contemporary art.
Activities: lectures; workshops; monthly exhibitions.
Publications: catalogs.
Hours & Admission Prices: Jan.-March Tues.-Sat. 10-5; April-Dec. Tues.-Sat. 10-5, Sun. 12-4. No charge. &
Attendance: 40,000 (estimated)
Membership: Student $20; Individual $25; Family $40; Associate $100; Donor $200; Benefactor $500; Patron of the Arts $1,000.

WORKING MEN'S INSTITUTE, 407 W. Tavern St., New Harmony, IN 47631. Mailing Address: P.O. Box 368, New Harmony, IN 47631-0368. Tel.: 812-682-4806.
Web Site: www.workingmensinstitute.org
Formerly: New Harmony Working Men's Institute
Founded: 1838.
Congressional District: 8
Key Personnel: Dir., Stephen Cochran.
Personnel Profile: Full-Time Paid 1; Part-Time Paid 2; Part-Time Volunteers 4.
Governing Authority: Tax-exempt. 501(c)(3).
Institution Type/Description: 19th Century Museum & Library.
Collections: natural history; local history; portraits; books; textiles; education; teaching tools.
Research Fields: utopian & communal studies; Civil War.
Publications: book, New Harmony Story.
Hours & Admission Prices: Tues.-Sat. 10-4. No charge; donations accepted. Closed New Year's Eve & Day; Easter; Independence Day; Thanksgiving; Christmas Eve & Day.
Attendance: 8,500 (accurate)
Membership: Individual & Family $15; Sponsor $25; Donor $50; Patron $100; Corporation & Benefactor $250 & up.

Noblesville

HAMILTON COUNTY MUSEUM OF HISTORY, 810 Conner St., Noblesville, IN 46060. Mailing Address: P.O. Box 397, Noblesville, IN 46061. Tel.: 317-770-0775. Fax: 317-770-0775.
E-mail: hamiltoncomuseum@att.net
Web Site: hamiltoncoinhs.org
Founded: 1963.
Key Personnel: Dir., Diane Nevitt.
Personnel Profile: Part-Time Paid 1.
Governing Authority: Parent Institution: Hamilton County Historical Society. Tax-exempt.
Institution Type/Description: History Museum: housed in the Old Hamilton County Sheriff's Residence and Jail; built in 1875.
Collections: county history & culture; period furnishings; photographs; personal artifacts.
Publications: newsletter.
Hours & Admission Prices: Thurs.-Sat. 10-4; other times by appointment. No charge; donations accepted. &
Attendance: 2,550 (accurate)
Membership: Students & Senior Citizen $10; Regular $15; Sustaining $50; Life $200.

INDIANA TRANSPORTATION MUSEUM, 701 Cicero Rd., Forest Park, Noblesville, IN 46060. Mailing Address: P.O. Box 83, Noblesville, IN 46061-0083. Tel.: 317-773-6000 & 776-7887. Fax: 317-770-1902.
E-mail: info@itm.org
Web Site: www.itm.org
Founded: 1960.
Congressional District: 5
Key Personnel: Chm., Jeffrey Kehler; Dir. Rail Operations, Paul Meister; Treas., John McNichol; Sec., Craig Presuer.
Personnel Profile: Part-Time Paid 1; Part-Time Volunteers 200.
Governing Authority: nonprofit organization. Tax-exempt: 501(c)(3).
Institution Type/Description: Rail Transportation and Technology Museum: office housed in 1930 railroad station from Hobbs, IN.
Collections: early 20th century railway & trolley equipment; steam, diesel & electric railway equipment, timetables, photographs, blueprints, other transportation artifacts.
Research Fields: Industrial History & Technology.
Facilities: library of railroad material; picnic area. Museum-related items for sale.
Activities: permanent exhibitions; guided tours daily by appointment; steam and diesel train excursions to various cities; machinery demonstration area.
Hours & Admission Prices: April-Oct. Sat.-Sun. 11-4. Adults $3, children 3-12 $2; children 2 & under and members no charge. Excursion fares: call for information.
Attendance: 40,000 (estimated)
Membership: Individual $20; Family $35; Sustaining $50; Contributing $100; 20th Century $250; Patron $500; Benefactor $1,000.

North Judson

HOOSIER VALLEY RAILROAD MUSEUM, 507 Mulberry St., North Judson, IN 46366-0075. Mailing Address: P.O. Box 75, North Judson, IN 46366-0075. Tel.: 574-896-3950.
E-mail: questions@hoosiervalley.org
Web Site: www.hoosiervalley.org
Institution Type/Description: Railroad Museum.
Collections: railroad history, equipment & artifacts; photographs; C&O No. 2789 steam locomotive; over 30 pieces of rolling stock.
Facilities: Museum-related items for sale.
Activities: train rides; educational programs; special events.
Hours & Admission Prices: Museum: Sat. 9-4. No charge.

North Manchester

MANCHESTER COLLEGE ART COLLECTION, 604 E. College Ave., North Manchester, IN 46962. Tel.: 260-982-5000.
Institution Type/Description: Art Gallery.
Collections: paintings; drawings; sculpture.
Hours & Admission Prices: Call for hours.

NORTH MANCHESTER HISTORICAL SOCIETY, 124 E. Main St., North Manchester, IN 46962. Mailing Address: P.O. Box 361, North Manchester, IN 46962. Tel.: 260-982-0672. Facebook: North Manchester Historical Society.
E-mail: nmhistory@cinergymetro.net

Web Site: www.nmanchesterhistory.org
Founded: 1972.
Key Personnel: Pres. (V), Mary L. Chrastil.
Personnel Profile: Full-Time Volunteers 3; Part-Time Volunteers 40.
Governing Authority: Subsidiary Institution: North Manchester Center for History. Tax-exempt.
Institution Type/Description: Historical Society Museum: housed in the former Oppenheim's Department Store and the Thomas Marshall House.
Collections: local history & culture; period furnishings; personal artifacts; photographs; manuscripts; antique agricultural artifacts.
Facilities: library.
Activities: special events; educational programs.
Publications: quarterly newsletter.
Hours & Admission Prices: Call for hours. Adults $3; discounts to Time Traveler members; members no charge. &
Attendance: 5,000 (estimated)
Membership: Individual $30; Couples $50; Family & Grandparents $60; Sustaining $75; Supporting $100.

North Vernon

HAYDEN HISTORICAL MUSEUM, 6715 W. County Rd. 20 S., North Vernon, IN 47265. Mailing Address: P.O. Box 58, Hayden, IN 47245. Tel.: 812-346-8212.
Institution Type/Description: History Museum: housed in a former chicken house; built in 1949.
Collections: local history & culture; period furnishings; military artifacts; photographs.
Hours & Admission Prices: Memorial Day to Labor Day Sun. 1-4, Wed. 4-8; Sept.-May Mon. 4-5:30, Wed. 4-8; other times by appointment.

Notre Dame

MUSEUM OF BIODIVERSITY AND GREENE-NIEUWLAND HERBARIUM, Dept. of Biological Sciences, Univ. of Notre Dame, Notre Dame, IN 46556-0369. Tel.: 574-631-6684. Fax: 574-631-7413.
E-mail: Barbara.J.Hellenthal.2@nd.edu
Founded: 1876.
Congressional District: 3
Key Personnel: Dir. Museum of Biodiversity, Ronald A. Hellenthal; Guest Dir. Herbarium, Richard J. Jensen; Cur., Barbara J. Hellenthal.
Personnel Profile: Full-Time Paid 1; Part-Time Volunteers 1.
Governing Authority: university. Parent Institution: University of Notre Dame. Tax-exempt.
Institution Type/Description: Herbarium Museum.
Collections: 600,000 specimens; Edward Lee Greene collection includes 5,000 type of specimens from western North American plants; Nieuwland collection from around the world.
Research Fields: plant taxonomy; plant geography; ecology; entomology; parasitology; vertebrate biology.
Activities: formally organized education programs for undergraduate & graduate college students; inter-museum loan for research; Greene Herbarium searchable database of 67,000 specimens.
Publications: American Midland Naturalist.
Hours & Admission Prices: By appointment only. No charge. &
Attendance: 2,000 (accurate)

❋ THE SNITE MUSEUM OF ART, UNIVERSITY OF NOTRE DAME, (M), University of Notre Dame, 100 M. Krause Cir., Notre Dame, IN 46556-0368. Mailing Address: P.O. Box 368, Notre Dame, IN 46556-0368. Tel.: 574-631-5466. Fax: 574-631-8501.
Web Site: sniteartmuseum.nd.edu
Founded: 1842.
Congressional District: 3
Key Personnel: Dir., Charles R. Loving; Assoc. Dir., Ann M. Knoll; Cur. Ethnographic Arts, Douglas E. Bradley; Cur. European Art, Cheryl K. Snay, Ph.D.; Mktg. & Public Affairs Specialist, Gina Costa; Exhibition Designer, John Phegley; Cur. Education, Public Programs, Sarah Martin; Exhibit Coord., Ramiro Rodriguez; Registrar, Rebeka Ceravolo; Coord. Friends of the Snite Museum, Heidi Williams.
Personnel Profile: Full-Time Paid 17; Part-Time Paid 20; Part-Time Volunteers 100; Interns 4.
Governing Authority: religious order; Congregation of Holy Cross. Parent Institution: University of Notre Dame. Tax-exempt: 501(c)(3).
Institution Type/Description: Art Museum.
Collections: Italian Renaissance paintings and sculpture; pre-Columbian sculptures, textiles; ceramics; 15th-20th century drawings and prints; Kress study collection; 17th & 18th century Italian, Dutch, Flemish and English paintings; 19th century French oil sketches, drawings, & paintings; 20th century paintings, photography & sculpture; paintings; sculpture; furniture; ceramics; glass; Mestrovic sculpture.
Research Fields: history & provenance of paintings; sculpture; drawings; prints; photographs; objects of art; pre-Columbian, Native American, Old Master drawings; French oil sketches.
Facilities: catalogue & reference library.
Activities: guided tours; lectures; films; gallery talks; concerts; formally organized education programs; permanent, temporary & traveling exhibitions; symposia.
Publications: newsletter, Calendar of Events; catalogs, Victor Higgins: An American Master; Master Drawings from the Reilly Collection; White Swan: Crow Indian, Warrior and Painter; Ivan Mestrovic; Rembrandt Etchings; Selected Works from the Snite Museum of Art; University of Notre Dame Master Drawings: The Wisdom-Reilly Collection; Taos Artists and Their Patrons, 1898-1950; A Gift of Light: Photographs in the Janos Scholz Collections; Passages of Light and Time: George Rickey's Life in Motion; Darkness and Light: Death and Beauty in Photography; Para la Gente: Art, Politics and Cultural Identity of the Taller de Grafica Popular; Eclectic Antiquity: The Classical Collection of the Snite Museum of Art; Parallel Currents: Highlights of the Ricardo Pau-Llosa Collection of Latin American Art; handbook, Selected Works Collection; Nineteenth Century French Drawings.
Hours & Admission Prices: Tues.-Wed. 10-4, Thurs.-Sat. 10-5, Sun. 1-5. No charge; donations accepted. Closed major holidays. &
Attendance: 49,959 (accurate)
Membership: Senior Citizen $25; Active $40; Family $60; Sustaining $100; Supporting $250; Patron $500; Benefactor $750; Donor $1,000; Director's Circle $5,000; Premium $10,000.

Ogden Dunes

HISTORICAL SOCIETY OF OGDEN DUNES, 115 Hillcrest Rd. #101, Ogden Dunes, IN 46368.
Institution Type/Description: Historical Society Museum.
Collections: local history & culture; period furnishings; personal artifacts; photographs.
Hours & Admission Prices: 3rd Sun. of month 4-5:30.

Parker City

ME'S ZOO, 12441 W. County Rd. 300 S., Parker City, IN 47368. Tel.: 765-468-8559. Fax: 765-468-9014.
Web Site: www.meszoo.com
Institution Type/Description: Zoo.
Collections: over 300 animals.
Facilities: Museum-related items for sale.
Activities: petting zoo. Museum Sponsors: Holiday Lights in November & December.
Hours & Admission Prices: April 14-Sept. Tues.-Thurs. & Sat. 10-5, Sun. 12-5; Nov.-Dec. call for hours. Adults $6.50, children 1-12 $5.50; discounts on Wed. &

Peru

CIRCUS CITY FESTIVAL MUSEUM, 154 N. Broadway, Peru, IN 46970-2234. Tel.: 765-472-3918. Fax: 765-472-2826.
E-mail: perucirc@perucircus.com
Web Site: www.perucircus.com
Founded: 1959.
Key Personnel: Exec. Vice Pres., Michelle Boswell; Vice Pres. Museum & Exhibits, Timothy Bessignano; Vice Pres. Festival, Kevin Gallahan; Exec. Sec. & Office Mgr., Linda Cawood; Pres. Circus & Recording Sec., Sandy Ploss.
Personnel Profile: Full-Time Paid 1; Part-Time Volunteers 1,000.
Governing Authority: nonprofit organization. Parent Institution: Circus City Festival, Inc. Tax-exempt: 501(c)(3).
Institution Type/Description: Circus Museum.
Collections: circus history; circus lithographs; photographs; uniforms & costumes of famous performers; trapeze; rigging; harness; guns; mouthpieces & other such items; furniture from The Wallace Circus Train, The Ben Wallace Home, The Mugivan Home; wild animal cage used at the Old Circus Winter Quarters; miniature circus wagons & scale circuses; paintings; 1894 Sicilian donkey cart.
Facilities: Gifts, circus & other museum-related items for sale.
Activities: guided tours; lectures; films.
Publications: newsletter; program book.
Hours & Admission Prices: July call for hours; Aug.-June Mon.-Fri. 9-1 & 2-4. No charge; donations accepted. Closed holidays. &
Attendance: 13,000 (estimated)

Membership: Youth $5; Annual & Adult Member $10; Circus Family Membership $20; Contributing $25; Sustaining $50; Century Club $100; Donor $250; Patron $500; Life $1,000.

GRISSOM AIR MUSEUM, (I), 1000 W. Hoosier Blvd., Peru, IN 46970-3723. Tel.: 765-689-8011; 574-398-1451.
E-mail: director@grissomairmuseum.com
Web Site: www.grissomairmuseum.com
Founded: 1981.
Congressional District: 4
Key Personnel: Dir. & Museum Shop Mgr., Jim Price; Chm. (V), Chris Birk.
Personnel Profile: Full-Time Paid 1; Part-Time Paid 1; Part-Time Volunteers 35.
Governing Authority: private; nonprofit. Tax-exempt: 501(c)(3).
Institution Type/Description: Military & Aviation Museum: located adjacent to Grissom Air Reserve Base. An Indiana State Historic Site.
Collections: aircraft; equipment; memorabilia ranging from World War II to the present; B-17 Flying Fortress; B-58 nuclear bomber; aviation art; 25 aircraft; 5 story SAC era alert tower; 5 aircraft cockpits.
Major Exhibits: D-21 Drone, 4/12-12/15.
Research Fields: history of the aircraft, personnel & organizations related to the adjacent Air Base, which initially opened as Bunker Hill Air Station in 1942.
Facilities: 1,200-vol. library of technical aircraft manuals & aviation-related magazines; 40-seat meeting room; 6,000 sq. ft. exhibit space; 20 acres of outdoor displays. Gift items for sale.
Activities: aviation art shows; model shows; guest speakers; guided tours; loan & temporary exhibitions; 5 story alert tower; 5 open cockpits in exhibit hall; 1-2 open aircraft on Saturdays. Museum Sponsors: Fly Market in May; Annual Car Show in August; Festival of Flight; Wheels & Wings Motorcycle Show; Armed Forces Day; Memorial Day; Independence Day; Patriots Day; Veterans Day.
Publications: quarterly newsletter, Grissom Air Museum Times.
Hours & Admission Prices: Check website for hours. Group tours available at reduced price. &
Attendance: 15,000 (estimated)
Membership: Wingman $35-$99; Century $100-$249; Co Pilot $250-$499; Pilot $500-$999; Eagle $1,000 & up.

INTERNATIONAL CIRCUS HALL OF FAME & MUSEUM, 3076 E. Circus Lane, Peru, IN 46970-7133. Tel.: 800-771-0241.
Institution Type/Description: Circus History Museum.
Collections: circus life, history & professionals; posters; wagons; models; miniature 1934 Hagenbeck Wallace Circus replica; Hall of Fame; photographs.
Hours & Admission Prices: May-June & Aug.-Oct. Mon.-Fri. 10-4; July Mon.-Sat. 10-4, Sun. 12-4.

MIAMI COUNTY MUSEUM, (M), 51 N. Broadway, Peru, IN 46970-2237. Tel.: 765-473-9183. Fax: 765-473-3880. Facebook: Miami County Museum.
E-mail: admin@mcmuseum.org
Web Site: www.miamicountyhistory.org
Founded: 1916.
Congressional District: 5
Key Personnel: Pres., Gary Hawley; Cur., Elise Kordis; Collections Mgr. & Registrar, Katie Ebeling; Administrative Asst. & Museum Shop Mgr., Marge Urnik; Archivist, Angelyn Hellman.
Personnel Profile: Full-Time Paid 3; Full-Time Volunteers 0; Part-Time Paid 1; Part-Time Volunteers 57.
Governing Authority: society. Miami County Historical Society, 51 N. Broadway, Peru, IN 46970. Tax-exempt.
Institution Type/Description: General Museum.
Collections: local history items; Miami & American Indian artifacts; pioneer items; circus memorabilia; transportation vehicles; archives & research; microfilm; Victorian rooms & stores; Cole Porter display & his 1955 Fleetwood Cadillac.
Research Fields: local history; railroad; Miami Indian; Cole Porter; Civil War; circus; Wabash & Erie Canal; local history, genealogy
Facilities: library of local newspapers 1837-1938 available for research. Museum-related items for sale.
Activities: tours; lectures; temporary & permanent exhibitions; outreach programs.
Publications: monthly bulletin (newsletter); books & leaflets pertaining to Miami County history; 1996 Pictorial History Book of Miami County.
Hours & Admission Prices: Tues.-Sat. 9-5. Suggested donation: $3. Closed holidays. &
Attendance: 7,000 (estimated)

Membership: General $25-$99; Friend $100-$400; Benefactor $500-$999; Patron $1,000-$5,000; Founder's Circle $5,000 & up.

Plymouth

MARSHALL COUNTY HISTORICAL MUSEUM, 123 N. Michigan St., Plymouth, IN 46563-2132. Tel.: 574-936-2306. Fax: 574-936-9306. Facebook: Marshall County Museum Historic Crossroads Center.
E-mail: mchistory@mchistoricalsociety.org
Web Site: www.mchistoricalsociety.org
Founded: 1957.
Congressional District: 13
Key Personnel: Dir. & C.E.O., Linda Rippy; Bd. Pres. (V), Dr. Ronald Liechty.
Personnel Profile: Full-Time Paid 2; Part-Time Paid 4; Part-Time Volunteers 40.
Governing Authority: nonprofit organization. Parent Institution: Marshall County Historical Society, Inc., Plymouth, IN. Tax-exempt: 501(c)(3).
Institution Type/Description: History Museum.
Collections: local county history artifacts; tools; furniture; household items; local genealogy; photos; Indian artifacts; clothing; books.
Research Fields: local history; family genealogy.
Facilities: 150-vol. library of books on local history & genealogy, also publications & newspapers available for use on premises.
Activities: guided tours; lectures; permanent & temporary exhibitions.
Publications: quarterly newsletter; quarterly journal.
Hours & Admission Prices: Tues.-Sat. 10-4. No charge; donations accepted. Closed county holidays. &
Attendance: 5,000 (estimated)
Membership: Annual $25; Annual Business $50; Five Year $100; Ten Year $200.

Portage

PORTAGE COMMUNITY HISTORICAL SOCIETY, 5250 U.S. Hwy. 6, Portage, IN 46368. Mailing Address: P.O. Box 305, Portage, IN 46368-0305. Tel.: 219-762-8349. Facebook: Portage Community Historical Society (Indiana).
E-mail: pchs.1@frontier.com
Web Site: www.geocities.com/portagehistorical society
Formerly: Al Goin Historical Museum
Founded: 1988.
Key Personnel: Pres. (V), Valerie Roach; Newsletter, Kathy Heckman.
Personnel Profile: Part-Time Volunteers 25; Interns 1.
Operating Expenses: 10,500
Operating Income: 11,000
Governing Authority: Subsidiary Institution: Portage City Park Dept. Tax-exempt.
Institution Type/Description: Local History.
Collections: city history; tools & farm equipment; Indian artifacts; school history; early settlers; period fire truck. Historic Buildings: farmhouse; barn; stagecoach.
Activities: genealogy research. Museum Sponsors: Historical Festival in June; Christmas Open House in December.
Publications: bimonthly newsletter; books, Township History, 2003; Historic Homes, 2005; Historic Homes (2012).
Hours & Admission Prices: March to early Dec. Fri.-Sat. 11-3; call to confirm. No charge; donations accepted. Closed Independence Day. &
Attendance: 2,500 (estimated)
Membership: Individual $18; Family $48; Business $58; Lifetime $109.

Porter

INDIANA DUNES NATIONAL LAKESHORE, 1100 N. Mineral Springs Rd., Porter, IN 46304-1225. Tel.: 219-926-7561, ext. 225. Fax: 219-926-7561.
Web Site: www.nps.gov/indu
Founded: 1970.
Congressional District: 2
Key Personnel: Supt., Costa Dillon; Asst. Supt., Garry Traynham.
Personnel Profile: Full Time Paid 130; Part Time Paid 55; Part Time Volunteers 595.
Governing Authority: federal. Parent Institution: National Park Service, Dept. of the Interior, Washington DC. Tax-exempt.
Institution Type/Description: Park Visitor Center.
Collections: prehistoric cultural collection; fur trade items; display of native plants & animals of the region; 19th-century farming & settlement; natural history. Historic Structures: 19th-century, Bailly Homestead; 19th-century Chellberg Farm; 1933 World's Fair Houses.
Research Fields: air & water quality; plant & animal ecology.

Facilities: 1,600-vol. park library & study collection available for research; two visitor centers; herbarium. Books & items related to park themes for sale.

Activities: guided tours; interpretive programs; audiovisual; formally organized educational programs; overnight educational programs.

Publications: English & Spanish brochure; guides; handbooks.

Hours & Admission Prices: Park: daily. $4 parking fee at West Beach. Visitor Center: daily 8-5, call for extended summer hours. No charge; donations accepted. &

Attendance: 1,834,435 (estimated)

Portland

JAY COUNTY HISTORICAL SOCIETY MUSEUM AND GENEALOGY LIBRARY, (M), 903 E. Main St., Portland, IN 47371. Tel.: 260-726-7168. Fax: 260-726-7178.

E-mail: research@jaycountyhistory.org
Web Site: www.jaycountyhistory.org
Personnel Profile: Part-Time Paid 2; Part-Time Volunteers 50.
Governing Authority: Tax-exempt.
Institution Type/Description: History Museum & Library.
Collections: local history & culture; period furnishings; photographs; personal artifacts; books; genealogy.
Activities: special events. Museum Sponsors: Heritage Festival in fall.
Publications: Jay County Journal.
Hours & Admission Prices: Mon.-Fri. 10-4. No charge; donations accepted. &
Membership: Annual $15; $25; $50; $100; $250; Life $300; $500.

MUSEUM OF THE SOLDIER, 510 E. Arch St., Portland, IN 47371-1525. Mailing Address: P.O. Box 518, Portland, IN 47371-0518. Tel.: 260-726-2967. Facebook: Museum of the Soldier.

E-mail: museum@bright.net
Web Site: www.museumofthesoldier.com/
Key Personnel: Bd. Mem., Matt Simmons; Bd. Mem., Jim Waechter.
Governing Authority: nonprofit organization. Tax-exempt: 501(c)(3).
Institution Type/Description: Military Museum.
Collections: military uniforms, weapons & equipment; personal artifacts.
Activities: lectures; banquets; temporary & permanent exhibits.
Hours & Admission Prices: 1st & 3rd Sat.-Sun. 12-5; other times by appointment. No charge; donations accepted.
Membership: Individual $15; Family $30; Patron $50-$499; Corporate $100 & up; Organization $500; Benefactor $500 & up.

Rensselaer

LILIAN FENDIG ART GALLERY, 301 N. Van Rensselaer St., Rensselaer, IN 47978-2630. Tel.: 219-866-5278. Fax: 219-866-5278.

E-mail: prairie@rhsi.tv
Founded: 1993.
Congressional District: 4
Key Personnel: Pres., John Groppe.
Personnel Profile: Part-Time Paid 1; Part-Time Volunteers 22.
Governing Authority: Parent Institution: Prairie Arts Council, Inc. Tax-exempt.
Institution Type/Description: Art Gallery.
Collections: paintings; archives; ceramics; fused glass.
Facilities: archives.
Activities: Museum Sponsors: ARTCAMP; Junior Art Club; Taste of Rensselaer; Holiday Art Show & Sale.
Publications: exhibition brochures; catalogs; newsletters; annual report.
Hours & Admission Prices: Mon.-Fri. 2-4. No charge; donations accepted. &
Attendance: 12,000 (estimated)
Membership: Student $10; Individual $25; Family $35; Sponsor & Organization $50; Patron $100-$249; Sustaining $250-499; Benefactor $500-$999.

Richmond

AGNES AND ABRAM GAAR FOUNDATION - GAAR HOUSE MUSEUM, 2593 Pleasant View Rd., Richmond, IN 47374-2050. Mailing Address: 1623 N. St. Rd. 227, Richmond, IN 47374. Tel.: 765-966-1262 & 7184.

Web Site: www.thegaarhouse.com
Formerly: Gaar Mansion and Farm Museum
Founded: 1974.
Congressional District: 6
Governing Authority: Tax-exempt.
Institution Type/Description: Historic Building: housed in the home of Abram Gaar & his wife Agnes; built in 1876. Listed on the National Register of Historic Places.

Collections: period furnishings; personal artifacts; paintings.
Activities: rental facilities for weddings & special occasions.
Hours & Admission Prices: June-Aug. 1st & 3rd Sun. 1-4; Dec. 1st 3 Sun. 1-4; tours by appointment. Adults $5, children 18 & under $2.

HAYES ARBORETUM, 801 Elks Rd., Richmond, IN 47374-2526. Tel.: 765-962-3745. Fax: 765-966-1931.

E-mail: stephenhayes13@yahoo.com
Web Site: www.hayesarboretum.org
Founded: 1959.
Congressional District: 10
Key Personnel: C.E.O., Stephen H. Hayes; Operations Supvr., Stephen H. Hayes, Jr., LEED, AP
Personnel Profile: Full-Time Paid 2; Part-Time Paid 1; Part-Time Volunteers 20.
Governing Authority: nonprofit. The Stanley W. Hayes Research Foundation Inc., P.O. Box 1404, Richmond, IN 47374. Tax-exempt: 501(c)(3), 4942(J)(3).
Institution Type/Description: Arboretum & Nature Center.
Collections: 10-acre native woody plant preserve, including one specimen of each species of native woody plant indigenous to the Whitewater Drainage Basin of Indiana & Ohio, incorporating 10 counties of Indiana & 4 counties in Ohio.
Research Fields: arboricultural; botanical; taxonomical.
Facilities: science related library available for on-site use; arboretum; nature & conservation center; reading room; 100-seat auditorium; classrooms.
Activities: guided tours; children's summer classes; cooperative programs with Ball State, Miami University & Earlham College; permanent & temporary exhibitions.
Publications: brochures, Welcome; Auto Tour Guide.
Hours & Admission Prices: March-Oct. Tues.-Sat. 9-5. No charge; donations accepted. &
Attendance: 10,000 (estimated)

INDIANA FOOTBALL HALL OF FAME, (M), 815 N. A St., Richmond, IN 47374-3119. Mailing Address: P.O. Box 40, Richmond, IN 47375-0040. Tel.: 765-966-2235. Fax: 765-966-5700.

E-mail: contact@indiana-football.org
Web Site: www.indiana-football.org
Founded: 1973.
Congressional District: 89
Key Personnel: Dir., Lou Ann Moore, Ph.D.; Pres. (V), Gerry Keesling.
Personnel Profile: Part-Time Paid 2; Part-Time Volunteers 3.
Governing Authority: Tax-exempt.
Institution Type/Description: Sports Museum.
Collections: Hall of Fame inductee photographs; football history & equipment; team photographs; high school helmets; scholarship honor wall; uniforms; plaques; biographies; football library.
Activities: Museum Sponsors: East-West IFHOF Football Classic in June. HOF Golf Classic in July Inductions; 6 annual scholarships to graduating high school senior scholar athletes.
Publications: newsletters; brochure.
Hours & Admission Prices: mid-Jan. thru Dec. Tues.-Sat. 11-5; other times by appointment. No charge; donations appreciated. Closed major National holidays.

JOSEPH MOORE MUSEUM, Earlham College, 801 National Rd. W., Richmond, IN 47374. Tel.: 765-983-1303. Fax: 765-983-1497. Facebook: Joseph Moore Museum.

E-mail: lernere@earlham.edu
Web Site: www.earlham.edu/joseph-moore-museum
Founded: 1887.
Congressional District: 10
Key Personnel: Dir., Dr. Heather R.L. Lerner; Coord. Educational Outreach, Carol Stocksdale; Mgr. Collections, Dr. Ann-Eliza Lewis.
Personnel Profile: Full-Time Paid 1; Part-Time Paid 3; Part-Time Volunteers 2; Interns 20.
Governing Authority: college. Parent Institution: Earlham College. Tax-exempt: 501(c)(3).
Institution Type/Description: Natural History Museum.
Collections: skins of birds & mammals; dried & liquid-preserved insects; live & preserved reptiles; vertebrate & invertebrate fossils; Indian artifacts; natural history.
Research Fields: mammalogy; ornithology; entomology; paleoecology; herpetology; paleontology.
Facilities: research library; planetarium. Museum-related items for sale.
Activities: guided tours; permanent & temporary exhibitions; natural history workshops.

Hours & Admission Prices: Sun.-Mon., Wed. & Fri. 1-5; tours by appointment only. No charge; donations accepted. ♿
Attendance: 4,500 (estimated)

LEEDS GALLERY - EARLHAM COLLEGE, 801 National Rd. W., Richmond, IN 47374-4095. Tel.: 765-983-1410.
Institution Type/Description: Art Gallery.
Collections: paintings; photographs; sculpture; drawings; ceramics.
Hours & Admission Prices: Mon.-Fri. 9-8, Sat.-Sun. 1-8.

MODEL T MUSEUM, 309 N. 8th St., Richmond, IN 47374. Mailing Address: P.O. Box 126, Centerville, IN 47330-0126. Tel.: 765-488-0026. Fax: 765-855-3428. Facebook: Model T Museum.
E-mail: admin@mtfca.com
Web Site: www.mtfca.com
Founded: 2007.
Key Personnel: C.E.O., Jay Klehfoth; Museum Shop Mgr., Barbara Klehfoth.
Personnel Profile: Full-Time Paid 2; Part-Time Paid 4; Part-Time Volunteers 30.
Governing Authority: Parent Institution: MTFCA. Tax-exempt.
Institution Type/Description: Automobile Museum.
Collections: Model T automobiles; photographs.
Publications: The Vintage Ford published 6 times per year by MTFCA; 7 restoration manuals; 40 restoration DVDs.
Hours & Admission Prices: Tues.-Sun. 10-5. Adults $3. ♿
Attendance: 10,000 (estimated)
Membership: Family $40.

RICHMOND ART MUSEUM, (M), 350 Hub Etchison Pkwy., Richmond, IN 47374-5339. Mailing Address: P.O. Box 816, Richmond, IN 47375. Tel.: 765-966-0256. Fax: 765-973-3738.
E-mail: shaun@richmondartmuseum.org
Web Site: www.richmondartmuseum.org
Formerly: Art Association of Richmond
Founded: 1898.
Congressional District: 2
Key Personnel: Exec. Dir., Shaun Dingwerth; Pres., Barry Jonston.
Personnel Profile: Full-Time Paid 2; Part-Time Paid 2; Part-Time Volunteers 75; Interns 2.
Governing Authority: board of trustees; nonprofit organization. Tax-exempt.
Institution Type/Description: Art Museum.
Collections: American paintings; sculpture; graphics; decorative arts; Indiana artists; Richmond School; Hoosier group art pottery; Overbeck.
Research Fields: art education; Indiana & Richmond artists.
Facilities: 1,200-vol. library of art available on premises; archives; reading room; 525-seat auditorium.
Activities: guided tours; lectures; films; gallery talks; art classes for children; education programs; docent program or council; inter-museum loan, permanent, temporary & traveling exhibitions; school loan service; juried professional & amateur artists competition.
Publications: newsletter, 1898-1978 Art in Richmond; Richmond Art Museum History and Permanent Collection.
Hours & Admission Prices: Tues.-Fri. 10-4, Sun. 1-4. No charge; donations accepted. Closed national & school holidays. ♿
Attendance: 20,000 (estimated)
Membership: Individual $35; Family $50; Supporting $75; Sustaining $100; Friends of the Museum $250; Sustaining Business $300; Museum Advocate & Patron & Contributing Business $500; Benefactor & Corporate Sponsor $1,000; Donor $1,500.

RONALD GALLERY - EARLHAM COLLEGE, 801 National Rd. W., Richmond, IN 47374-4095. Tel.: 765-983-1410.
Key Personnel: Cur., Julia May
Institution Type/Description: Art Gallery.
Collections: over 3,000 paintings, prints & sculptures.
Hours & Admission Prices: Mon.-Thurs. 8am to midnight, Fri. 8am-10pm, Sat. 10-10, Sun. noon to midnight.

WAYNE COUNTY HISTORICAL MUSEUM, (M), 1150 N. A St., Richmond, IN 47374-3298. Tel.: 765-962-5756. Fax: 765-939-0909.
E-mail: director@wchmuseum.org
Web Site: www.waynecountyhistoricalmuseum.org
Founded: 1930.
Congressional District: 10
Key Personnel: C.E.O. & Museum Shop Mgr., James Harlan; Pres. (V), A.J. Sickmann.

Personnel Profile: Full-Time Paid 2; Part-Time Paid 3; Part-Time Volunteers 60.
Governing Authority: society; nonprofit organization. Affiliated with Wayne County Historical Society. Tax-exempt: 501(c)(3).
Institution Type/Description: General Museum: housed in 1864 Hicksite Friends Meeting House.
Collections: china; glass; silver; apothecary; agriculture; architecture; firefighting equipment; guns; hobbies; musical instruments; toys and dolls; transportation; Wooten desk; 1929 Davis plane; early Richmond-made autos; C. Francis Jenkins (radiovision) collection; Starr-Gennett jazz history; Gaar-Scott steam engines; 3,000 yr. old Egyptian Mummy & related artifacts; textile collection - clothing, quilts, coverlets. Historic Houses: 1823 Soloman Dickinson Log House.
Facilities: 500-vol. library of county histories, scrapbooks & newspapers available for use on premises. Historical reproductions & museum-related items for sale.
Activities: guided tours; lectures; arts festivals; hobby workshops; docent program or council; permanent exhibitions.
Publications: quarterly newsletter.
Hours & Admission Prices: Jan.-Feb. Mon.-Fri. 9-4, Sat. 1-4; March-Dec. Mon.-Fri. 9-4, Sat.-Sun. 1-4. Adults $7, students 6-18 $4; discounts to groups & AAA members; children under 6 & members no charge. Closed national holidays. ♿
Attendance: 12,000 (estimated)
Membership: Student & Individual $30; Family $50; Contributing $100; Patron $150; Business $250; Life $1,000; Corporate Sponsorship $2,500.

Rising Sun

OHIO COUNTY HISTORICAL SOCIETY, INC. & MUSEUM, 212 S. Walnut, Rising Sun, IN 47040. Tel.: 812-438-4915. Facebook: Ohio County Museum.
Web Site: www.ohiocountymuseum.org
Founded: 1972.
Key Personnel: Dir. & Museum Shop Mgr., Clifford Wm. Thies; Pres. (V), Lane Siekman.
Personnel Profile: Full-Time Paid 1; Part-Time Paid 1.
Governing Authority: Parent Institution: Ohio County Historical Society. Tax-exempt.
Institution Type/Description: Historical Society Museum.
Collections: local history & culture; period furnishings; personal artifacts; photographs.
Research Fields: genealogy.
Activities: educational programs. Museum Sponsors: Quilt Fest in March.
Hours & Admission Prices: Call for hours. No charge; donations accepted. ♿
Attendance: 1,500 (estimated)
Membership: Student $5; Senior 55 & over $10; Individual $20; Family $25; Director's Circle $50.

Rochester

FULTON COUNTY HISTORICAL SOCIETY MUSEUM, 37 E. 375 N., Rochester, IN 46975-9718. Tel.: 574-223-4436. Fax: 574-224-4436.
E-mail: fchs@rtcol.com
Web Site: www.fultoncountyhistory.org
Founded: 1963.
Congressional District: 5
Key Personnel: Pres. Emerita, Shirley Willard; Dir., Melinda Clinger; Pres., Fred Oden, Jr.; Treas., Lola Riddle.
Personnel Profile: Full-Time Paid 1; Part-Time Paid 3; Part-Time Volunteers 90; Interns 1.
Governing Authority: nonprofit organization. Tax-exempt: 501(c)(3).
Institution Type/Description: Living History Village & History Museum: 35-acre village including 14 buildings.
Collections: furniture; farm equipment; photos; tools; literature; manuscripts; period artifacts; Elmo Lincoln's first Tarzan films & posters; reconstructed 1924 round barn which houses farm machinery; 1971 caboose from Norfolk & Western; boxcar; 160 ft. track. Historic Structures: Loyal, Indiana; 1832 William Polke House/Stagecoach Inn; 1874 Rochester Depot; 1860s Pioneer Woman's Log Cabin; 1910 Dr. Shafer's office; 1920 general store; 1915 small red round barn/granary; replica blacksmith shop, 1920s print shop, windmill, 1920s footbridge made by Rochester Bridge Co.; 1940 round chicken house; 1912 jail; 1900 cider mill; 1961 octagonal corn crib; 1930s railcar & garage.
Research Fields: genealogy; local history; Trail of Death removal of Potowatomi Indians in 1838.
Facilities: 1,500-vol. library of old school texts, histories, manuscripts, county newspapers 1858 to present, Civil War diaries, Bibles & newspapers available for use on premises. Handcrafts, Indian jewelry & local history books for sale.

Activities: guided tours; lectures; films; permanent & temporary exhibitions; school loan service. Museum Sponsors: Living History Fair in March; Redbud Trail Rendezvous in April; Historical Power Show in June; Potawatomi Trail of Death from Indiana to Kansas; Trail of Courage Living History Festival in September; Toy Show in October; Haunted Woods Walk in October; Farmers Market November to Dec.; Christmas Crafts & Open House in December.

Publications: magazine, Fulton County Images; books, Fulton County Folks, Volumes 1 & 2; Historical Atlas of Fulton County 1883 reprint; Home Folks-Tales of Old Settlers of Fulton County 1910 reprint; Wagoner Family History, 1941 reprint; 1935 Shields Genealogy reprint; semiannual genealogical newsletter, Fulton County Folk Finder; semiannual newsletter, Potawatomi Trail of Death Association; semiannual newsletter, Fulton County Historical Power Association News; Historical Society Newsletter; book, Potawatomi Trail of Death 1838 Indiana to Kansas, Fulton County Folks Vol. 1 & 2.

Hours & Admission Prices: Mon.-Sat. 9-5. No charge; donations accepted. Closed holidays.

Attendance: 35,000 (accurate)

Membership: Individual $20; Family $30; Sponsor $50; Club $100; Corporation $250; Benefactor $500; Life $750.

Rockport

LINCOLN PIONEER VILLAGE & MUSEUM, 928 Fairground Dr., Rockport, IN 47635. Tel.: 812-649-9147.

Institution Type/Description: History Museum.

Collections: 14 Lincoln-era replica cabins including a Pioneer Schoolhouse, Lincoln Homestead Cabin, the Old Pigeon Baptist Church; period furnishings; personal artifacts; photographs.

Hours & Admission Prices: Call for hours.

Rockville

BILLIE CREEK VILLAGE, INC., 65 S. Billie Creek Rd., Rockville, IN 47872. Mailing Address: P.O. Box 357, Rockville, IN 47872-0357. Tel.: 765-569-0252. Fax: 765-569-3582.

E-mail: billiecreekvillage@billiecreekvillage.org

Web Site: www.billiecreekvillage.org

Founded: 1968.

Key Personnel: Dir. & C.E.O., Charles Cooper; Museum Shop Mgr., Sue Cooper.

Governing Authority: Tax-exempt.

Institution Type/Description: History Museum.

Collections: recreated turn-of-the-century village & farmstead; 38 buildings.

Hours & Admission Prices: Village: Wed.-Sat. 10-4, Sun. 12-4. General Store: Mon.-Sat. 10-4, Sun. 12-4. Admission $5; children under 6 no charge. Special Events: $7.

PARKE COUNTY HISTORICAL MUSEUM, 503 W. Ohio St., Rockville, IN 47872. Mailing Address: P.O. Box 332, Rockville, IN 47872. Tel.: 765-569-2223.

E-mail: info@parkecountyhistoricalsociety.org

Web Site: www.parkecountyhistoricalsociety.org

Governing Authority: Parent Institution: Parke County Historical Society.

Institution Type/Description: History Museum: housed in a former seminary; later used as an armory during the Civil War; built in 1839.

Collections: local history & culture; period furnishings & clothing; photographs; personal artifacts; early musical instruments.

Activities: special events.

Hours & Admission Prices: Wed.-Sun. 1-5.

Rome City

GENE STRATTON-PORTER STATE HISTORIC SITE, (M), 1205 Pleasant Point, Rome City, IN 46784-9644. Tel.: 260-854-3790. Fax: 260-854-9102. Facebook: Gene Stratton-Porter State Historic Site.

E-mail: genestrattonportershs@indianamuseum.org

Web Site: indianamuseum.org

Founded: 1946.

Congressional District: 4

Personnel Profile: Full-Time Paid 2; Part-Time Paid 3; Part-Time Volunteers 25.

Governing Authority: state. Parent Institution: Indiana State Museum & Historic Sites, Corp., 650 W. Washington St., Indianapolis, IN 46204. Tax-exempt.

Institution Type/Description: Historic House: c.1920 home of Gene Stratton Porter.

Collections: period furnishings; Gene Stratton Porter photographs & memorabilia.

Facilities: 125-acres of grounds; 3 miles of walking trails & scenic paths; formal gardens; arboretum; picnic area; bird sanctuary.

Activities: guided tours; lectures.

Publications: newsletter, Gene Stratton-Porter Memorial Society.

Hours & Admission Prices: Carriage House Visitor Center: Tues.-Sat. 10-5, Sun. 1-5. No charge. Cabin Tours: Tues.-Sat. 10-4, Sun. 1-4. Adults seniors 55 & over $4, children 12 & under $2; discounts to groups. Clo New Year's Day; Memorial Day; Independence Day; Labor Day; Thangiving; Christmas.

Attendance: 60,000 (estimated)

Membership: Single $10; Family $15; Sustaining $20; Nonprofit $25; Busin $50; Life Single $100; Husband & Wife $150.

Rushville

RUSH COUNTY HISTORICAL SOCIETY MUSEUM, 619 Perkins St., Rushville, IN 46173. Mailing Address: P.O. Box 30 Rushville, IN 46173. Tel.: 765-932-2492.

E-mail: rchs1@frontier.com

Web Site: rushcountyhistory.org

Institution Type/Description: Historical Society Museum.

Collections: local history & culture; period furnishings; personal artifac historic buildings; photographs.

Hours & Admission Prices: March-Nov. Mon. & Thurs. call for hours. charge; donations accepted.

Salem

STEVENS MUSEUM, 307 E. Market St., Salem, IN 47167-211 Tel.: 812-883-6495.

E-mail: jhc@blueriver.net

Web Site: www.stevensmuseum.com

Founded: 1897.

Congressional District: 9

Key Personnel: Pres. Bd., Danny Newby; Dir., Willie Harlen; Sec., Katheri Simpson; Treas., Clara Marie Burns; Museum Shop Mgr., Kathy Wade.

Personnel Profile: Full-Time Paid 2; Part-Time Paid 2; Part-Time Voluntee 10.

Governing Authority: bd. of directors. Parent Institution: Washington Coun Historical Society. Subsidiary Institution: John Hay Center. Tax-exemp 501(c)(3).

Institution Type/Description: Local History, 1824 John Hay birthplace, Pione Village & Depot Museum.

Collections: manuscripts; newspapers; period furnishings; coverlets & quil furniture; farm tools; educational; church; costumes; dentistry; medic pioneer; Indians; wars; guns; dishes; textiles.

Research Fields: genealogy; history; wars; school; cemetery; period furnitu & furnishings; church.

Facilities: 5,000-vol. genealogy library available on premises; reading roor 125-seat auditorium.

Activities: guided tours; lectures; permanent & temporary exhibitions.

Publications: quarterly newsletter.

Hours & Admission Prices: Tues.-Sat. 9-5. Genealogy research: Tues.-Sat. 9- Museum: adults $2-$3; members & children under 6 no charge. Library: $ Closed New Year's Day; Easter; Mother's Day; Memorial Day; Father Day; Independence Day; Labor Day; Thanksgiving; Christmas.

Attendance: 30,000 (estimated)

Membership: Individual $15; Life $200; Couple Life $250.

Santa Claus

SANTA CLAUS MUSEUM AND VILLAGE, 69 N. State Rd. 245, Santa Claus, IN 47579. Mailing Address: P.O. Box 1, Santa Claus, IN 47579. Tel.: 812-544-2434.

E-mail: elf@santaclausmuseum.org

Web Site: www.santaclausmuseum.org

Founded: 2006.

Key Personnel: Dir., Emily Thompson.

Governing Authority: Tax-exempt.

Institution Type/Description: History Museum.

Collections: local history; Santa Claus memorabilia; books; photographs; early toys, furnishings & posters; news articles; 1935 Santa statue. Historic Buildings: c.1880 church; c.1856 post office.

Activities: write & receive a letter from Santa; special events.

Hours & Admission Prices: Call for hours. No charge; requested donation of $2 per person & $5 per family.

Attendance: 9,000 (accurate)

Membership: Individual $25; Couple $35; Family $40; Bronze $100; Copper $250; Silver $500; Gold $1,000; Platinum $2,000.

Scottsburg

SCOTT COUNTY HERITAGE CENTER AND MUSEUM, 1050 S. Main St., Scottsburg, IN 47170-6663. Mailing Address: P.O. Box 122, Scottsburg, IN 47170-0122. Tel.: 812-752-1050.
Governing Authority: Parent Institution: Preservation Alliance, Inc. Tax-exempt.
Institution Type/Description: History Museum.
Collections: local history & culture; personal artifacts; photographs.
Hours & Admission Prices: Mon.-Fri. 9-5, Sat. 9-1. No charge. &

Seymour

FREEMAN ARMY AIRFIELD MUSEUM, 1025 A Ave. - Freeman Field, Seymour, IN 47274. Tel.: 812-522-2031.
E-mail: lbothe@comcast.net
Web Site: www.freemanfield.org/images/data/museum.htm
Institution Type/Description: Military Museum.
Collections: airfield history; military artifacts; personal artifacts; photographs; uniforms.
Hours & Admission Prices: Sat. 9am to noon; other times by appointment. No charge; donations accepted.

Shelbyville

LOUIS H. & LENA FIRN GROVER MUSEUM, 52 W. Broadway, Shelbyville, IN 46176-1256. Tel.: 317-392-4634.
E-mail: director@grovermuseum.org
Web Site: www.grovermuseum.org
Founded: 1980.
Congressional District: 2
Key Personnel: Dir., Candace Miller; Pres. (V), Gregg Steele; Museum Shop Mgr., Debbie Ewing; Staff Asst., Teresa Tungate.
Personnel Profile: Full-Time Paid 2; Part-Time Volunteers 10.
Governing Authority: society. Parent Institution: Shelby County Historical Society. Tax-exempt: 501(c)(3).
Institution Type/Description: Historical Society Museum.
Collections: agricultural implements; medical; photographs; documents; tools; toys; textiles; model railroad - trains & buildings; street scene - 30 buildings and artifacts from 1900-1910; Shelby County archaeological artifacts. Historical House: 1820 Thomas Hendricks Boyhood Home.
Research Fields: local history topics.
Facilities: 100-seat auditorium; classrooms. Books, pictures & postcards for sale.
Activities: formally organized education programs for adults; temporary exhibitions; heritage skills classes. Annual Events: Quilt Show & Luncheon; Bear Paw Quilt Guild; Christmas Tree in November & December.
Publications: quarterly newsletter, Echoes of Old Shelby.
Hours & Admission Prices: Tues.-Sat. 9-4. No charge; donations accepted. Closed holidays. &
Attendance: 7,800 (accurate)
Membership: Single $20; Family $30; Sustaining $40.

Sheridan

SHERIDAN HISTORICAL SOCIETY MUSEUM, 308 S. Main St., Sheridan, IN 46069. Tel.: 317-758-5054.
E-mail: sheridanhistorical@sbcglobal.net
Web Site: www.sheridanhistoricalsociety.com
Founded: 1969.
Congressional District: 5
Key Personnel: C.E.O., Jim Pickett
Personnel Profile: Part-Time Volunteers 20.
Institution Type/Description: Historical Society Museum.
Collections: local history & culture; period furnishings; personal artifacts; photographs; pioneer life.
Research Fields: genealogy.
Activities: Museum Sponsors: Boxley Lecture Series in February; Sheridan Fireside Tales in June; Sheridan Blue Grass Fever in July.
Publications: Touchstones in History.
Hours & Admission Prices: Call for hours. No charge; donations accepted.
Attendance: 400 (estimated)
Membership: Annual $20; Family $35; Life $110.

Shipshewana

HOSTETLER'S HUDSON AUTO MUSEUM, Shipshewana Town Center, 760 S. Van Buren St., Shipshewana, IN 46565. Mailing Address: P.O. Box 486, Shipshewana, IN 46565-0486. Tel.: 260-768-3021. Fax: 260-768-3023.
Web Site: www.hostetlershudsons.com
Founded: 2007.
Key Personnel: Mgr. Operations & Museum Shop, JR. Hostetler.
Personnel Profile: Part-Time Volunteers 20.
Governing Authority: town. Subsidiary Institution: Shipshewana Car Museum, Inc. Tax-exempt.
Institution Type/Description: Auto Museum.
Collections: early Hudson-produced cars, trucks, & utility vehicles, 1909-1956.
Facilities: 1,000-seat convention center.
Activities: rental facilities.
Hours & Admission Prices: Memorial Day to Labor Day Mon.-Tues. 9-8, Wed.-Sun. 9-5; Sept.-May Mon.-Sat. 9-5; groups by appointment. Adults $8, seniors 60 & over $7, students 6-17 $4; discounts to groups; members & children under 5 no charge.
Attendance: 8,000 (estimated)
Membership: Volunteer & Senior $20; Individual $30; Family $50; Contributor $100; Sustaining $250; Visionary $500; Corporate $1,000; Life $1,500.

Shoals

MARTIN COUNTY HISTORICAL SOCIETY COUNTY MUSEUM, 220 Capitol, Shoals, IN 47581-0564. Mailing Address: P.O. Box 564, Shoals, IN 47581-0564. Tel.: 812-247-1133.
E-mail: historical@frontier.com
Founded: 1937.
Key Personnel: Dir., Pres. (V) MCHS & Museum Shop Mgr., Jim Marshall.
Personnel Profile: Part-Time Volunteers 7.
Volunteer Hours: 400
Operating Expenses: 3,500
Operating Income: 5,000
Governing Authority: Tax-exempt.
Institution Type/Description: History Museum.
Collections: period artifacts; tax records; assessment records; marriages; school records; scrapbooks; photographs.
Publications: quarterly newsletters.
Hours & Admission Prices: May-Oct. Mon., Wed. & Fri. 10-4. No charge; donations accepted. &
Attendance: 500 (estimated)
Membership: Individual $10; Family $25.

South Bend

CENTER FOR HISTORY, (M), 808 W. Washington, South Bend, IN 46601-1439. Tel.: 574-235-9664. Fax: 574-235-9059.
E-mail: director@centerforhistory.org
Web Site: www.centerforhistory.org
Formerly: Northern Indiana Center for History
Founded: 1867.
Congressional District: 2
Key Personnel: C.E.O. & Exec. Dir., Randy W. Ray; Pres. (V), Mark D. Noeldner; Cur., David S. Bainbridge; Dir. Visitor Svcs., Ken Cencelewski; Dir. Education, Travis Childs; Dir. Devel., Amanda Miller; Dir. Mktg., Marilyn Thompson; Dir. Facilities & Grounds, Tom Rapach; Registrar, Kristi Dunn; Finance Mgr., Marilyn Jurgonski.
Personnel Profile: Full-Time Paid 19; Part-Time Paid 15; Part-Time Volunteers 125.
Governing Authority: nonprofit organization. Parent Institution: Northern Indiana Historical Society, Inc. Tax-exempt: 501(c)(3).
Institution Type/Description: History Museum & Children's Interactive Museum.
Collections: local history & culture from the prehistoric era to the present; pioneer; Native American; works of art; decorative art; military; industrial; costume; toys; dolls; games; transportation; ethnic; political; textile; sports; All-American Girls Professional Baseball League artifacts; tools; Archives includes fur trading journals, regional newspapers from 1830-1960s, court documents & records from St. Joseph County, genealogical manuscripts, diaries, letters, bound volumes on regional & Indiana history, photographs, sheet music, military, ethnic.
Research Fields: prehistoric era to present including pioneer, Native American, works of art, decorative art, military, industrial, costume, toys, dolls, games,

transportation, ethnic, political, textile, sports; All-American Girls Professional Baseball League; fur trading; genealogy & history of St. Joseph County & state of Indiana.

Facilities: research library; auditorium; amphitheatre. Museum-related items for sale.

Activities: self-guided tours; guided tours; lectures; classes; workshops; docent program; films; historical archaeology programs; weekly newspaper article; off-site exhibitions; treasure hunt; experiential classroom kits; living history programs; historical competitions; publications; children's hands-on history exhibits.

Publications: newsletters, volunteer corps newsletter; historical calendar; books; pamphlets; reprints.

Hours & Admission Prices: Mon.-Sat. 10-5, Sun. 12-5. Adults $8, seniors $6.50, children $5; discounts to staff of other museums & AAM members; members & children under 6 no charge. &

Attendance: 48,250 (accurate)

Membership: Student & Senior $30; Individual $40; Family $60; Patron & Sustainer $125; Sponsor $250.

COLLEGE FOOTBALL HALL OF FAME, 111 S. Saint Joseph St., South Bend, IN 46601-1901. Tel.: 800-440-FAME. Fax: 574-235-5720.

E-mail: lisa.klunder@collegefootball.org
Web Site: www.collegefootball.org
Founded: 1951.
Key Personnel: Exec. Dir, Lisa Klunder; Pres., Steve Hatchell; Chm. (V), Archie Manning; Dir. Mktg., Katie Berrettini; Dir. Sales & Special Events, Ashley Burbin; Public Rels., Kim Bugos; Financial Svcs. Coord., Patti Glascoe; AV Coord., Richard Allen; Collections Mgr., Kent Stephens; Dir. Operations, Mark Maurer; Museum Shop Mgr., Laura Holaway.
Personnel Profile: Full-Time Paid 9; Part-Time Paid 7; Part-Time Volunteers 126; Interns 7.
Governing Authority: company. Parent Institution: National Football Foundation & College Football Hall of Fame, 22 Maple Ave., Morristown, NJ 07960. Tax-exempt.
Institution Type/Description: Sports Museum.
Collections: football history, equipment & memorabilia; photographs; manuscripts; personal artifacts; Hall of Fame inductees.
Facilities: 230-seat auditorium; theater; 70-seat restaurant; 400-seat banquet facility. Museum-related gifts for sale.
Activities: films; loan, permanent & temporary exhibitions; photograph reproduction service; luncheons; autograph sessions; Festival-film transfer.
Hours & Admission Prices: Memorial Day to late Nov. Mon.-Thurs. 10-5, Fri.-Sat. 9-6, Sun. 9-5, Thanksgiving to Memorial Day daily 10-5. Adults (non-residents) $12, St. Joseph County residents $11, students & senior citizens $10, children 5-12 $7; discounts to groups of 20 or more, ICOM & AAA members; children under 4 no charge. Closed New Year's Day; Easter; Thanksgiving; Christmas. &
Attendance: 80,000 (estimated)

COPSHAHOLM - THE OLIVER MANSION, Center for History, 808 W. Washington, South Bend, IN 46601-1439. Tel.: 574-235-9664. Fax: 574-235-9059.

E-mail: director@centerforhistory.org
Web Site: www.centerforhistory.org
Formerly: Copshaholm House Museum & Historic Oliver Gardens
Founded: 1990.
Congressional District: 3
Key Personnel: C.E.O. & Dir., Randy W. Ray; Pres. (V), Mark Noeldner; Cur., Kristie Erickson; Dir. Mktg., Marilyn Thompson; Dir. Facilities & Grounds, Tom Rapach; Registrar, Kristi Dunn; Dir. Education, Travis Childs; Visitor Svcs. Dir., Ken Cencelewski; Dir. Devel., Amanda Miller; Finance Mgr., Marilyn Jurgonski.
Personnel Profile: Full-Time Paid 18; Part-Time Paid 14; Part-Time Volunteers 120.
Governing Authority: nonprofit organization. Parent Institution: Center for History. Tax-exempt.
Institution Type/Description: Historic House; 1895-1896 38-room Oliver Family mansion.
Collections: original mid-17th to mid-20th century furnishings; furniture; decorative arts; household objects; works of art; costumes; archival collection includes business ledgers & papers of South Bend Iron Works, Oliver Chilled Plow Works, Oliver Farm Equipment Company, personal papers, letters, diaries, photos, blueprints & plans for Copshaholm & gardens.
Research Fields: history of America's industrial midwest in late 19th & early 20th centuries; the Oliver Family & companies; historic American gardens, architecture, decorative arts & household lifestyles.
Facilities: visitor center. Museum-related items for sale.

Activities: guided tours; lectures; classes; workshops; docent program; student designed tours.
Publications: newsletter; docent manual; publications related to Copshaholm.
Hours & Admission Prices: Mon.-Sat. 10-5, Sun. 12-5; last tour leaves at 2. Adults $8, seniors $6.50, children $5; children under 5, staff of other museums, AAM, & NIHS members no charge. Closed major holidays.
Attendance: 25,000 (accurate)
Membership: Student & Senior $30; Individual $40; Family $60; Sustainer $125; Sponsor $250.

HEALTHWORKS! KIDS' MUSEUM, (M), Memorial Leighton Healthplex, 111 W. Jefferson St., Ste. 200, South Bend, IN 46601-1993. Tel.: 574-647-5437. Fax: 574-239-6459.

E-mail: jsimmons@memorialsb.org
Web Site: www.healthworkskids.org
Founded: 1999.
Institution Type/Description: Children's Health Museum.
Collections: hands-on exhibits.
Activities: special events; school programs.
Hours & Admission Prices: Tues.-Fri. 9-4, Sat.-Sun. 12-4. Admission $5; members, ACM & children under 2 no charge. Closed New Year's Day; Memorial Day; Independence Day; Labor Day; Thanksgiving; Christmas Eve & Day. &
Attendance: 70,000 (estimated)
Membership: Best Buddy $50; Sensational Sidekick $75; Awesome Amigo $100; Precious Pal $250; Fantastic Friend $500; Spectacular Sponsor $1,000.

POTAWATOMI ZOO, 500 S. Greenlawn Ave., South Bend, IN 46615-1341. Mailing Address: P.O. Box 1764, South Bend, IN 46634. Tel.: 574-288-4639. Fax: 574-289-3776.

E-mail: info@potawatomizoo.org
Web Site: potawatomizoo.org
Founded: 1902.
Congressional District: 3
Key Personnel: Zoo Dir., Terry DeRosa; Society Dir., Marcy Dean; Veterinarian, Dr. Carol Bradford; Office Mgr., Pat Fenters.
Personnel Profile: Full-Time Paid 20; Part-Time Paid 12; Part-Time Volunteers 35; Interns 3.
Governing Authority: municipal. Parent Institution: City of South Bend, South Bend, IN 46617. Tel.: 574-284-9401. Tax-exempt: 501(c)(3).
Institution Type/Description: Zoo.
Collections: animals from around the world; mammals; birds; reptiles.
Facilities: 23-acre zoological park; butterfly house; classroom; concessions. Museum-related items for sale.
Activities: guided tours; educational classes; camps; special events; zoo train.
Publications: newsletter, UPROAR.
Hours & Admission Prices: April to late Nov. daily 10-5. Adults $7.50, children & seniors $5.50; discount to school groups; children 2 & under, AZA, & zoo society members no charge. &
Attendance: 185,000 (accurate)
Membership: Senior Citizen $25; Individual $35; Grandparents $55; Family $60; Patron $100; Contributor $250.

* SOUTH BEND MUSEUM OF ART, 120 S. St. Joseph St., South Bend, IN 46601-1902. Tel.: 574-235-9102. Fax: 574-235-5782.

E-mail: info@southbendart.com
Web Site: southbendart.org
Formerly: South Bend Regional Museum of Art
Founded: 1947.
Congressional District: 3
Key Personnel: Exec. Dir., Susan R. Visser; Pres. Bd. Trustees (V), Marika Smith; Cur. Collections, Kim Hoffmann; Assoc. Cur. & Registrar, Mark Rospenda; Cur. Education, Jessica Lentych Loyd; Dir. Mktg., Peg Luecke; Business Mgr., Bill Seybold; Devel. Mgr. & Membership Coord., Claudia Maskowski.
Personnel Profile: Full-Time Paid 8; Part-Time Paid 11; Part-Time Volunteers 40; Interns 3.
Governing Authority: private. Tax-exempt: 501(c)(3).
Institution Type/Description: Art Museum.
Collections: historic & contemporary Midwestern artists; 19th-21st century American paintings & works on paper.
Research Fields: historic & contemporary American and Indiana art.
Facilities: classroom; special children's studio; activity rooms for films, lectures & meetings. Museum-related items for sale.

Activities: lectures; films; gallery talks; formally organized education programs; studio art program for adults & children; docent training; inter-museum loan; family programs; traveling exhibitions; scholarships to talented and underprivileged students. Museum Sponsors: outdoor jazz & sculpture events, Meet Me On the Island; Day of the Dead Celebration.

Publications: monthly e-newsletter; exhibition announcements; checklists; catalogues; class brochures.

Hours & Admission Prices: Office: Mon.-Fri. 9-5. Galleries & Retail: Wed.-Sun. 12-5. Suggested Donation: adults $5; members no charge. Closed major holidays. ♿

Attendance: 60,000 (estimated)

Membership: Students & Senior Citizens $30; Individual $40; Household $60; Sustaining $100; Supporting $250; Donor $500; Benefactor $1,000.

✻ STUDEBAKER NATIONAL MUSEUM, INC., (M), 201 S. Chapin St., South Bend, IN 46601-2521. Tel.: 574-235-9714; 888-391-5600. Fax: 574-235-5522.

E-mail: info@studebakermuseum.org
Web Site: www.studebakermuseum.org
Founded: 1977.
Congressional District: 2
Key Personnel: Exec. Dir., Rebecca Bonham; Pres. Bd. Trustees, Mike Kendzicky; Business Mgr., Bonnie Oswald; Archivist, Andrew Beckman; Mgr. Mktg. & Asst. Dir., Jo McCoy; Cur., Drew Vande Wielle; Museum Shop Asst., Susan Boocher; Exhibit Designer & Facilities Mgr., Don Filley; Education & Public Programs, Courtney Bogunia.
Personnel Profile: Full-Time Paid 7; Part-Time Paid 6; Part-Time Volunteers 54.
Governing Authority: bd. of trustees; nonprofit corporation. Tax-exempt: 501(c)(3).
Institution Type/Description: Automobile Museum.
Collections: 1824-1966 Studebaker historic vehicles including carriages, wagons & automobiles; products manufactured in South Bend & Mishawaka, Indiana from 1830 to the present.
Research Fields: South Bend area industrial & commercial history; Studebaker & Packard records of production & history.
Facilities: archives. Museum-related items for sale.
Activities: guided tours; lectures; docent program; education programs; permanent, temporary & traveling exhibits; inter-museum loans.
Publications: quarterly newsletter; annual report; visitor brochures.
Hours & Admission Prices: Mon.-Sat. 10-5, Sun. 12-5. Archives: Mon.-Thurs. 10-4; other times by appointment. Museum: adults $8, senior citizens over 60 & student over 18 $6.50, children 6-18 $5; discounts to groups and AAM & AAA members; members & children under 6 no charge. Campus: adults $12, seniors over 60 & student over 18 $10, youth 6-18 $7; discounts to groups; members no charge. Museum: closed New Year's Day; Easter; Thanksgiving; Christmas Eve & Day. ♿
Attendance: 39,428 (accurate)
Membership: Museum: Senior Citizen over 60 $30; Individual $40; Dual $50; Family $60. Campus: Senior over 60 $45; Individual $60; Dual $75; Family $90. Corporate $1,000 & up.

Syracuse

SYRACUSE-WAWASEE HISTORICAL MUSEUM, 1013 N. Long Dr., Syracuse, IN 46567-1060. Tel.: 574-457-3599.

E-mail: director@syracusemuseum.org
Web Site: www.syracusemuseum.org
Institution Type/Description: History Museum.
Collections: local history & culture; photographs; period artifacts.
Activities: educational programs.
Hours & Admission Prices: Summer: Wed.-Sat. 1-5; Winter: Thurs.-Fri. 1-5; other times by appointment. No charge; donations accepted.
Attendance: 2,000 (estimated)

Terre Haute

CANDLES HOLOCAUST MUSEUM AND EDUCATION CENTER, 1532 S. Third St., Terre Haute, IN 47802-1012. Tel.: 812-234-7881. Fax: 812-478-2824.

E-mail: info@candlesholocaustmuseum.org
Web Site: www.candlesholocaustmuseum.org
Founded: 1984.
Key Personnel: Dir., Kiel Majewski; Chm. (V), Peggy Tierney; Vice Chair, Tanna Srulouitch; Museum Coord., Dorothy Chambers; Operations Dir., Nicole Sconce.
Personnel Profile: Full-Time Paid 3; Part-Time Paid 6; Part-Time Volunteers 1.
Volunteer Hours: 500
Operating Expenses: 446,902
Operating Income: 452,221

Governing Authority: Tax-exempt.
Institution Type/Description: Holocaust Museum.
Collections: Holocaust history; Mengele twins; photographs.
Activities: lectures.
Hours & Admission Prices: Tues.-Fri. 10-4, Sat. 1-4. No charge; donations accepted. Closed New Year's Day; Independence Day; Thanksgiving & day after; Christmas. ♿
Attendance: 10,000 (estimated)
Membership: Individual $18; Students & Seniors $9.

CLABBER GIRL MUSEUM, 900 Wabash Ave., Terre Haute, IN 47807-3208. Mailing Address: P.O. Box 150, Terre Haute, IN 47808-0150. Tel.: 812-232-9446. Fax: 812-478-7181.

E-mail: mmorgan@clabbergirl.com
Web Site: www.myclabbergirl.com
Founded: 2003.
Key Personnel: Cur., Meegan Morgan; Museum Shop Mgr., Lisa Yowell
Institution Type/Description: Historic Building: housed in the former home of Herman Hulman & his sons; built by them in 1892 and where they founded their family wholesale grocery business.
Collections: Hulman family history; wholesale grocery business; history of baking; history of automobile racing & wheels in motion.
Facilities: Museum-related items for sale.
Activities: culinary tour packages, Rex Roasting Co. (coffee roasting company); interactive culinary classes.
Publications: quarterly newsletter.
Hours & Admission Prices: Mon.-Fri. 10-6, Sat. 9-3; guided tours by appointment. No charge. Closed holidays. ♿

EUGENE V. DEBS HOME, 451 N. Eighth St., Terre Haute, IN 47807-3006. Mailing Address: P.O. Box 9454, Terre Haute, IN 47808-9454. Tel.: 812-232-2163 & 237-3443.

E-mail: charles.king@indstate.edu
Web Site: debsfoundation.org
Founded: 1962.
Congressional District: 8
Key Personnel: Exec. Vice Pres., Noel Beasley; Treas., Mick Love; Dir. Museum & Cur., Karon Brown; Sec., Charles King.
Personnel Profile: Full-Time Paid 1.
Governing Authority: nonprofit organization. Parent Institution: Eugene V. Debs Foundation. Tax-exempt.
Institution Type/Description: Historic House: c.1894 home of Eugene V. Debs.
Collections: period furnishings; mementos of Debs' labor activism and his campaigns for the presidency; historical photographs.
Research Fields: labor; socialism; social reform.
Activities: guided tours; permanent exhibitions.
Publications: semi-annual newsletter.
Hours & Admission Prices: Tues.-Sat. 1-4:30. No charge; donations accepted. Closed national holidays; New Year's Eve & Day; Christmas Eve, Day & week.
Attendance: 1,170 (accurate)
Membership: Student & Limited Income $10; Regular $25; Supporting $50; Sustaining $100; Life $250.

NATIVE AMERICAN MUSEUM, 5170 E. Poplar St., Dobbs Park, Terre Haute, IN 47803-9313. Tel.: 812-877-6007.

E-mail: jane.creedon@terrehaute.in.gov
Web Site: www.terrehaute.in.gov
Founded: 1994.
Congressional District: 7
Key Personnel: Cur., Jane Creedon.
Personnel Profile: Full-Time Paid 1; Part-Time Paid 4.
Governing Authority: municipal; nonprofit. Tax-exempt.
Institution Type/Description: Museum of Native American Cultures.
Collections: Indiana history from prehistoric to modern times with special emphasis on the native tribes of western Great Lakes, Wabash River Valley & Ohio's Lower River Valley; wigwam; longhouse; tools & materials for cooking; clothing; basketry; weapons; dolls.
Facilities: 320-vol. library of books on native peoples culture; 2,600 sq. ft. exhibit space; botanical garden; educational facilities.
Activities: guided tours; hobby workshops; lectures; loan & temporary exhibitions. Annual Events: Buffalo Chip Throwing Contest; Archaeology Week; Museum Day.
Hours & Admission Prices: Tues.-Sat. 9-5, Sun. 12-5. No charge; donations accepted. Closed all legal holidays. ♿
Attendance: 13,000 (estimated)

PAUL DRESSER MEMORIAL BIRTHPLACE, First & Farrington Sts., Terre Haute, IN 47802. Mailing Address: 1411 S. 6th St., Terre Haute, IN 47802. Tel.: 812-235-9717. Fax: 812-235-9717.
E-mail: vchs@joink.com
Founded: 1967.
Congressional District: 7
Key Personnel: Exec. Dir., Marylee Hagan; Asst. Dir., Barbara Carney.
Personnel Profile: Full-Time Paid 1; Part-Time Paid 1; Part-Time Volunteers 1.
Governing Authority: nonprofit organization. Vigo County Historical Society. Branch Museums: Historical Museum of the Wabash Valley, Terre Haute, IN. Tax-exempt: 501(c)(3).
Institution Type/Description: Historic House Museum: c.1850 Paul Dresser Birthplace.
Collections: furniture; household items; period artifacts.
Research Fields: local history.
Activities: tour groups.
Publications: quarterly pamphlets, Leaves of Thyme.
Hours & Admission Prices: May-Sept. Sun. 1-4. Museum: Tues.-Sun. 1-4; other times by appointment, weather permitting. No charge; donations accepted. Closed holidays.
Attendance: 15,000 (estimated)
Membership: Individual $30; Household $50; Patron $100; Sustaining $150; Business Sponsor $500; Life $1,000.

*   **SWOPE ART MUSEUM, (M),** 25 S. 7th St., Terre Haute, IN 47807-3692. Tel.: 812-238-1676. Fax: 812-238-1677.
E-mail: info@swope.org
Web Site: www.swope.org
Founded: 1942.
Congressional District: 7
Key Personnel: Exec. Dir., Marianne Richter; Pres. Bd. Mgrs. (V), Rick Shagley; Pres. Bd. Overseers, Jonathan Ford; Mgr. Communications, Kristi Finley; Cur., Elizabeth Petrulis; Mgr. Collection, Jenna Lanman.
Personnel Profile: Full-Time Paid 4; Part-Time Paid 7; Part-Time Volunteers 50; Interns 1.
Governing Authority: nonprofit organization. Tax-exempt: 501(c)(3).
Institution Type/Description: Art Museum: housed in 1901 Renaissance Revival building.
Collections: 19th- to 20th-century American paintings, sculpture & works on paper.
Major Exhibits: Flotsam and Jetsam, 11/13-1/25/14.
Research Fields: 19th & 20th Century American art.
Activities: guided tours; lectures; gallery talks; art classes; formally organized education programs; docent program; inter-museum loan; permanent, temporary & traveling exhibitions.
Publications: descriptive brochure; semiannual newsletter; special exhibition catalogs; class schedule; family guides; books Swope Art Museum: Selected Works from the Collection.
Hours & Admission Prices: Tues.-Fri. 10-5, Sat. 12-5; call for additional hours. No charge; donations accepted. Closed national holidays.
Attendance: 13,000 (accurate)
Membership: Individual $40; Dual & Family $65; N.A.R.M. $125; Sponsor $250; Patron $500; Benefactor $1,000; Director's Circle $2,500.

TERRE HAUTE CHILDREN'S MUSEUM, 727 Wabash Ave., Terre Haute, IN 47807-3203. Tel.: 812-235-5548.
E-mail: info@terrehautechildrensmuseum.com
Web Site: www.terrehautechildrensmuseum.com
Founded: 1988.
Institution Type/Description: Children's Museum.
Collections: hands-on exhibits.
Hours & Admission Prices: Tues.-Thurs. 10-6, Fri. 10-8, Sat. 10-5, Sun. 12-5. Admission $7; children under 2 no charge.
Membership: Family $125.

UNIVERSITY ART GALLERY, INDIANA STATE UNIVERSITY, Center for Performing & Fine Arts, N. 7th & Chestnut St., Terre Haute, IN 47809. Mailing Address: Dept. of Art, FA 108, Terre Haute, IN 47809. Tel.: 812-237-3720 & 3787. Fax: 812-237-4359.
E-mail: mvandenberg@isugw.indstate.edu
Web Site: www.indstate.edu/artgallery
Founded: 1939.
Congressional District: 7
Key Personnel: Interim University Cur., Crystal Vicars-Pugh; Chm., Alden Cavanaugh; Pres., Dan Bradley.
Personnel Profile: Full-Time Paid 1; Part-Time Paid 6; Part-Time Volunteers 2; Interns 1.
Governing Authority: university. Parent Institution: Indiana State University. Tax-exempt.
Institution Type/Description: Art Gallery.
Collections: paintings; sculpture; prints; ceramics.
Facilities: 2,600 sq. ft. exhibition space.
Activities: regional, national & international contemporary art exhibitions; educational programs for undergraduate & graduate studies.
Publications: 2 selected catalogs for special exhibits per year.
Hours & Admission Prices: Mon.-Wed. & Fri. 11-4, Thurs. 11-8. No charge. Closed holidays.
Attendance: 18,000 (estimated)

VIGO COUNTY HISTORICAL MUSEUM, 1411 S. 6th St., Terre Haute, IN 47802-1114. Tel.: 812-235-9717. Fax: 812-235-4998.
E-mail: vchs@joink.com
Founded: 1958.
Congressional District: 7
Key Personnel: C.E.O. & Exec. Dir., Marylee Hagan; Pres., Gary Greiner; Asst. Dir., Barbara Carney.
Personnel Profile: Full-Time Paid 1; Part-Time Paid 2; Part-Time Volunteers 25.
Governing Authority: society; owned & operated by Vigo County Historical Society, Inc. Branch Museums: Paul Dresser Memorial Birthplace. Tax-exempt: 501(c)(3).
Institution Type/Description: General Museum: housed in 1868 brick two story Italianate building.
Collections: textiles; archaeology; costumes; military; restored Victorian rooms & country store; schoolroom; dressmakers shop; Coca Cola bottle memorabilia.
Research Fields: local history.
Facilities: library; 50-seat meeting room. Museum-related items for sale.
Activities: guided tours; lectures; films; formally organized education programs; docent program; permanent & temporary exhibitions.
Publications: Four annual pamphlets, Leaves of Thyme.
Hours & Admission Prices: Tues.-Sun. 1-4. No charge; donations accepted. Closed national holidays.
Attendance: 15,000 (estimated)
Membership: Individual $30; Household $50; Patron $100; Sustaining $150; Business Sponsor $500; Life $1,000.

WABASH VALLEY RAILROADERS MUSEUM, 1316 Plum St., Terre Haute, IN 47801. Mailing Address: The Haley Tower Historical & Technical Society, P.O. Box 10291, Terre Haute, IN 47801. Tel.: 812-238-9958.
Institution Type/Description: History Museum.
Collections: railroad history & artifacts; photographs; documents; hands-on exhibitions.
Hours & Admission Prices: May-Oct. Sat.-Sun. 11-4; other times by appointment.

Thorntown

THORNTOWN HERITAGE MUSEUM, 124 W. Main St., Thorntown, IN 46071-1128. Mailing Address: 124 N. Market St., Thorntown, IN 46071-1144. Tel.: 765-436-7348. Fax: 765-436-7011.
E-mail: kniemeyer@thorntown.lib.in.us
Founded: 1973.
Congressional District: 4
Key Personnel: Pres. (V), John Gillan.
Personnel Profile: Part-Time Volunteers 21.
Governing Authority: Parent Institution: Thorntown Public Library. Tax-exempt.
Institution Type/Description: History Museum.
Collections: local history & culture; photographs; period furnishings; personal artifacts; violins; stringed instruments. Museum Sponsors: 1863-67 brick house.
Activities: tours; ice cream social; educational programs.
Hours & Admission Prices: May-Sept. Fri. 4-7, Sat. 12-4, Sun. 1-4. No charge; donations accepted.
Attendance: 450 (estimated)
Membership: Annual $15.

Tipton

TIPTON COUNTY HERITAGE CENTER, (M), 323 W. South St., Tipton, IN 46072-2068. Tel.: 765-675-5828.
Founded: 1921.
Key Personnel: Dir., Jill Curnutt-Howerton; Pres. (V), Donald Manlove.

Personnel Profile: Part-Time Volunteers 8.
Governing Authority: Tax-exempt.
Institution Type/Description: History Museum.
Collections: local history & culture; period documents; photographs.
Publications: quarterly newsletter.
Hours & Admission Prices: Tues.-Sat. 1-5. No charge.
Attendance: 400 (estimated)
Membership: Individual $10.

Union City

ART ASSOCIATION OF RANDOLPH COUNTY, INC., 115 N.
Howard, Union City, IN 47390-1435. Tel.: 765-964-7227. Fax:
765-964-4569. Facebook: Arts Depot.
E-mail: info@artsdepot.org
Web Site: www.artsdepot.org
Founded: 1955.
Key Personnel: Exec. Dir., Vicki Vardaman; Pres. (V), Lyn Orr.
Personnel Profile: Full-Time Paid 1.
Governing Authority: private; nonprofit organization. Tax-exempt: 501(c)(3).
Institution Type/Description: Art Museum.
Collections: local artists from east central Indiana; dolls.
Major Exhibits: Mildred Whitesel, 1/14/14-2/20/14; Whitewater Artists Guild,
2/25/14-3/27/14; 60th Annual Art Show, 4/8/14-4/24/14.
Facilities: 250-vol. library; 80-seat auditorium; educational facilities; restored
train depot.
Activities: formal education programs; concerts; guided tours; hobby work-
shops; lectures; loan & participatory exhibits; rental gallery; dinners.
Annual Events: Art Show; Photograph Show; Quilt Show.
Publications: monthly newsletter, Artracks.
Hours & Admission Prices: Tues.-Fri. 10-4; other times by appointment.
Museum: no charge; donations accepted. Special events may require an
admission donation. Closed major holidays. &
Attendance: 5,000 (estimated)
Membership: Student $10; Individual $25; Family $40; Friend of Art $100 &
up; Patron $250 & up; Partner $500 & up; Angel $1,000 & up.

Valparaiso

BRAUER MUSEUM OF ART, (M), Valparaiso University Center
for the Arts, Valparaiso, IN 46383-6349. Mailing Address: Val-
paraiso University Center for the Arts, 1709 Chapel Dr., Valparaiso,
IN 46383-4519. Tel.: 219-464-5365. Fax: 219-464-5244.
E-mail: gregg.hertzlieb@valpo.edu
Web Site: www.valpo.edu/artmuseum
Founded: 1953.
Congressional District: 5
Key Personnel: Dir. & Cur., Gregg Hertzlieb; Assoc. Cur. & Registrar, Gloria
Ruff.
Personnel Profile: Full-Time Paid 2; Part-Time Volunteers 8; Interns 4.
Governing Authority: university. Tax-exempt: 501(c)(3).
Institution Type/Description: Art Museum.
Collections: 19th, 20th, & 21st century American Art; paintings, drawings,
sketchbooks & archival material by Junius Sloan; world religious art.
Research Fields: American & religious art.
Facilities: Arts Center: theatre; recital hall; arts instructional spaces.
Activities: lectures; permanent, temporary & traveling exhibitions; guided
tours; gallery talks; inter-museum loans; virtual exhibitions available on
web site; receptions.
Publications: Exhibition brochures; Inaugural booklet.
Hours & Admission Prices: Academic year: Tues. & Thurs.-Fri. 10-5, Wed.
10-8:30, Sat.-Sun. 12-5. Academic recess & summer: Tues.-Sun. 12-5. No
charge; donations accepted. Closed New Year's Eve & Day, Good Friday;
Easter; Independence Day; Thanksgiving; Christmas Eve & Day. &
Attendance: 10,000 (estimated)
Membership: Apprentice $25; Craftsman $50; Artist $100; Mentor $250;
Master $500; Curator $1,000.

THE MEMORIAL OPERA HOUSE, 104 Indiana Ave., Valparaiso,
IN 46383-5603. Tel.: 219-548-9137.
Key Personnel: Dir., Brian Schafer.
Governing Authority: Parent Institution: Porter County.
Institution Type/Description: Historic Building: built in 1893. Listed on the
National Register of Historic Places.
Collections: theater history; period furnishings; photographs.
Facilities: 364-seat theatre.
Activities: performances; concerts; films; special events; tours.
Hours & Admission Prices: Mon.-Sat. 10-2; groups by appointment. No
charge; donations accepted.

PORTER COUNTY MUSEUM OF HISTORY, (M), 153 Franklin
St., Valparaiso, IN 46383-5631. Tel.: 219-465-3595. Facebook:
Porter County Museum of History.
E-mail: info@pocomuse.org
Web Site: pocomuse.org
Formerly: Old Jail Museum
Founded: 1916.
Congressional District: 1
Key Personnel: Chm., Joanne Urschel; Exec. Dir., Kevin Matthew Pazour.
Personnel Profile: Full-Time Paid 1; Part-Time Volunteers 14; Interns 2.
Governing Authority: county government. Parent Institution: Porter County
Heritage Corporation.
Institution Type/Description: Local History Museum.
Collections: mastodon bones found in county; relics from pioneer families &
Indians; period photographs.
Major Exhibits: Disasters, Catastrophes & Epidemics, 9/13/13-3/11/14.
Activities: guided tours; education programs for children.
Publications: newsletters.
Hours & Admission Prices: Wed.-Sat. 9-5. No charge: donations accepted.
Closed some national holidays.
Attendance: 4,000 (estimated)
Membership: Student $15; Basic $20; Individual Plus One $30; Family $40;
Museum SupPorter $120; Heritage Circle $240.

Veedersburg

FOUNTAIN COUNTY WAR MUSEUM, INC., 116 E. First St.,
Veedersburg, IN 47987-1402. Tel.: 765-376-6474.
E-mail: auters69@yahoo.com
Founded: 1995.
Congressional District: 7
Key Personnel: Dir. & Pres. (V), Archie Campbell; Chm. (V), Tye Auter;
Treas., Leroy Lindquist.
Personnel Profile: Part-Time Volunteers 8.
Governing Authority: private; nonprofit organization. Tax-exempt.
Institution Type/Description: War Museum.
Collections: military artifacts; personal artifacts; photographs.
Facilities: library.
Activities: guided tours; lectures; temporary exhibitions. Annual Event: festi-
val.
Hours & Admission Prices: Mon.-Fri. 10-2, Sat. 10-4; other times by
appointment. No charge; donations accepted. &
Attendance: 500 (estimated)

Vernon

JENNINGS COUNTY HISTORICAL SOCIETY, 134 E. Brown
St., Vernon, IN 47282. Mailing Address: P.O. Box 335, Vernon, IN
47282. Tel.: 812-346-8989.
E-mail: hector1838@frontier.com
Web Site: jenningscounty.org
Founded: 1961.
Key Personnel: Pres. (V), Chris Asher; Museum Shop Mgr., Wanda Kay
Wright.
Personnel Profile: Part-Time Paid 1.
Governing Authority: Tax-exempt.
Institution Type/Description: Historical Society Museum: housed in the North
American House; built in 1838.
Collections: local history & culture; period furnishings; personal artifacts;
photographs.
Activities: festivals; school tours.
Hours & Admission Prices: Wed.-Fri. 11-4 and during festivals. No charge;
donations accepted. &
Attendance: 15,000 (estimated)
Membership: Family $25; Supporter $50-$99; Benefactor & Business $100-
$999; Championship $1,000 & up.

Versailles

**RIPLEY COUNTY, INDIANA, HISTORICAL SOCIETY MU-
SEUM,** Water & Main, Versailles, IN 47042. Mailing Address: Box
525, Versailles, IN 47042 0525.
Founded: 1930.
Congressional District: 9
Key Personnel: Dir. & Treas., Owen Menchhofer; Pres. & Acting Cur., Cheryl
Welch.
Personnel Profile: Part-Time Volunteers 23.
Governing Authority: society; bd. of directors. Tax-exempt.
Institution Type/Description: Historical Society Museum.
Collections: Civil War artifacts; Native American artifacts; books; primitive
tools; costumes & textiles.

Research Fields: cemeteries; genealogies; local history.
Facilities: archives.
Activities: permanent exhibitions. Museum Sponsors: Pumpkin Show.
Publications: cemetery records; quarterly bulletin; Ripley County History, 1818-1988, Vol. I.
Hours & Admission Prices: Museum: Memorial Day to Labor Day Sun. 2-4. Library: Mon.-Fri. 1-4. No charge; donations accepted. Library Research: nonmembers $5; members no charge. &
Attendance: 1,000 (estimated)
Membership: Individual $15.

Vevay

LIFE ON THE OHIO RIVER HISTORY MUSEUM, 208 E. Market St., Vevay, IN 47043-1233.
Founded: 2004.
Congressional District: 9
Key Personnel: Dir. & Coord. Collections, Martha Bladen; Pres. (V), Sundra Whitham; Treas., Anita Danner; Sec., Joyce Benbow.
Personnel Profile: Full-Time Paid 1; Part-Time Paid 3; Part-Time Volunteers 5; Interns 1.
Governing Authority: Parent Institution: Switzerland County Historical Society. Tax-exempt.
Institution Type/Description: History Museum.
Collections: river history; steamboat models; pilot wheel; photos; documents.
Research Fields: Ohio River history.
Facilities: genealogy library.
Activities: riverboat cruises during the Swiss Wine Festival in August.
Publications: newsletter, Grapevine; postcards.
Hours & Admission Prices: April-Oct. daily 10-4; Nov.-March daily 12-4. No charge. Closed Christmas Eve & Day. &
Attendance: 4,200 (accurate)
Membership: Basic $10; Contributing $20; Supporting $30; Sustaining $50; Life $200.

SWITZERLAND COUNTY HISTORICAL MUSEUM, 210 E. Main, Vevay, IN 47043. Mailing Address: P.O. Box 201, Vevay, IN 47043-0201. Tel.: 812-427-3560.
E-mail: swcomuseums@embarqmail.com
Web Site: www.switzcomuseums.org
Founded: 1924.
Congressional District: 9
Key Personnel: Dir., Martha Bladen; Pres. (V), Sundra Whitham; Recording Sec., Joyce Benbow.
Personnel Profile: Full-Time Paid 1; Part-Time Paid 3; Interns 1.
Governing Authority: Parent Institution: Switzerland County Historical Society. Tax-exempt.
Institution Type/Description: County History Museum.
Collections: costumes; pioneer relics; documents; Indian artifacts; textiles & quilts; dolls; first piano in Indiana. Historic Building: 1860 Presbyterian Church located in Vevay.
Research Fields: genealogy.
Facilities: genealogy library.
Activities: monthly programs; summer garden tour; storyfest; Christmas tour of homes; permanent & temporary exhibits.
Publications: newsletter, The Grapevine; Harraman's History of Switzerland County; The American Vine-Dresser's Guide; The Vevay Cook Book.
Hours & Admission Prices: April-Oct. daily 10-5; Nov.-March daily 12-4. Adults $3; members no charge. Closed Easter; Christmas Eve & Day. &
Attendance: 4,200 (accurate)
Membership: Individual $10; Family $15; Contributing $20; Business & Supporting $30; Sustaining $50; Lifetime $200.

Vincennes

GEORGE ROGERS CLARK NATIONAL HISTORICAL PARK, 401 S. Second St., Vincennes, IN 47591-1001. Tel.: 812-882-1776, ext. 110. Fax: 812-882-7270.
E-mail: gero_ranger_activities@nps.gov
Web Site: www.nps.gov/gero
Founded: 1967.
Congressional District: 8
Key Personnel: Supt., Dale K. Phillips.
Personnel Profile: Full-Time Paid 11.
Governing Authority: federal. Dept. of the Interior, National Park Service. Tax-exempt.
Institution Type/Description: Park Museum.
Collections: British & American frontier weapons, uniforms, clothing, accoutrements; French Canadian household items; flat work; maps, diagrams & interpretive texts; oil murals; bronze statue of George Rogers Clark.

Research Fields: American Revolutionary War in the West; Old Northwest Territory.
Facilities: 600-vol. library of books, journals, maps on Revolutionary War in the West, early settlement of Old Northwest & evolution from territory status to statehood.
Activities: self-guided tours; movie; rifle & musket demonstrations; museum talks.
Publications: handbook.
Hours & Admission Prices: Daily 9-5. Adults 17 & over $3. Closed New Year's Day; Thanksgiving; Christmas. &
Attendance: 129,950 (accurate)

INDIANA MILITARY MUSEUM INC., 715 S. 6th St., Vincennes, IN 47591-8922. Mailing Address: P.O. Box 977, Vincennes, IN 47591-8922. Tel.: 812-882-1941.
Web Site: www.indymilitarymuseum.com
Founded: 1983.
Key Personnel: Dir. & Pres. (V), Jim R. Osborne; Volunteer Coord., Kassie Roush.
Personnel Profile: Part-Time Volunteers 70.
Governing Authority: Tax-exempt.
Institution Type/Description: Military Museum.
Collections: military history, artifacts, memorabilia, vehicles, aircraft, artillery, uniforms & equipment from the Civil War to Desert Storm.
Hours & Admission Prices: Daily 10-4. Adults $5, children $3; discounts to groups, seniors & veterans; members no charge. Closed New Year's Day; Thanksgiving; Christmas. &
Attendance: 5,000 (accurate)
Membership: Single $20; Family $50; Life $500.

MICHEL BROUILLET HOUSE & MUSEUM, 509 N. 1st St., Vincennes, IN 47591-1401. Mailing Address: P.O. Box 1979, Vincennes, IN 47591. Tel.: 812-882-7422. Fax: 812-882-0928.
Founded: 1975.
Congressional District: 8
Key Personnel: Chm. Bd., Jimmy Morrison; Cur., Richard Day; Pres., P.R. Sweeney; Mgr. Vincennes Historic Sites, Bruce Beesley.
Governing Authority: nonprofit. Parent Institution: The Old Northwest Corporation. Tax-exempt: 501(c)(3).
Institution Type/Description: History Museum: housed in 1806 restored French pioneer home.
Collections: furnishings; fur traders; Stone Age implements; Indian weapons; domestic utensils; dioramas; fossils.
Research Fields: early French settlement of Vincennes; prehistoric period in Wabash Valley.
Activities: lectures; guided tours; special shows.
Publications: monthly newsletter, Old Northwest Corporation Newsletter.
Hours & Admission Prices: By appointment. Adults $2, children $1.
Membership: Individual $10; Family $15; Business $25; Corporate $100.

OLD CATHEDRAL LIBRARY & MUSEUM, 205 Church St., Vincennes, IN 47591-1133. Tel.: 812-882-5638.
Founded: 1794.
Congressional District: 8
Key Personnel: Dir., John Schipp.
Personnel Profile: Part-Time Volunteers 6.
Governing Authority: Parent Institution: St. Francis Xavier Church. Tax-exempt.
Institution Type/Description: Religious Museum.
Collections: books; church records; Simon Brute de Remur collection; St. Francis Xavier Church, Vincennes records 1749 to present; theology, Bibles & Biblical works; historical civil & religious works.
Facilities: over 10,000-vol. library.
Hours & Admission Prices: June to mid-Aug. Mon.-Fri. 1-4; other times by appointment. Adults $1, children 12 & under $.50.

OLD FRENCH HOUSE & INDIAN MUSEUM, 1st & Seminary St., Vincennes, IN 47591. Mailing Address: The Old Northwest Corp., P.O. Box 1979, Vincennes, IN 47591. Tel.: 812-882-7742; 800-886-6443.
E-mail: vincennesshes@nr.in.gov
Key Personnel: Cur., Richard Day.
Governing Authority: nonprofit organization. Parent Institution: The Old Northwest Corp., P.O. Box 1979, Vincennes, IN 47591.
Institution Type/Description: History Museum: housed in the former home of Michel Brouillet; c.1806.
Collections: local history & culture; period furnishings; personal artifacts; fur trade.

Hours & Admission Prices: Tours: Mon.-Sat. 2:30 pm; other times by appointment. Adults $2, students $1.

RED SKELTON MUSEUM AND EDUCATION CENTER, 20 Red Skelton Blvd., Vincennes, IN 47591. Mailing Address: Vincennes University, 1002 N. 1st St., DC-38, Vincennes, IN 47591. Tel.: 812-888-4850.
Web Site: www.redskeltonmuseum.org
Institution Type/Description: History Museum.
Collections: Red Skelton's life & career; personal artifacts; photographs; costumes; awards; paintings. Historic House: Red Skelton's boyhood home.
Activities: clown school. Museum Sponsors: Red Skelton Festival in June.
Hours & Admission Prices: Call for hours.

VINCENNES STATE HISTORIC SITE, INDIANA TERRI-TORY, (M), First & Harrison Sts., Vincennes, IN 47591. Mailing Address: P.O. Box 81, Vincennes, IN 47591-0081. Tel.: 812-882-7422. Fax: 812-882-0928.
Web Site: www.state.in.us/ism/sites/vincennes/
Founded: 1949.
Congressional District: 8
Key Personnel: Cur., Bruce A. Beesley; Asst. Cur., Richard Day.
Personnel Profile: Full-Time Paid 4; Part-Time Volunteers 5.
Governing Authority: state. Parent Institution: Indiana Dept. of Natural Resources. Subsidiary Institution: Indiana State Museum. Tax-exempt.
Institution Type/Description: Historic Site: c.1805 Territory Capitol Building.
Collections: period furnishings; printing equipment; replica 1804 print shop; replica 1801 schoolhouse.
Activities: guided tours; lectures.
Hours & Admission Prices: Mon.-Sat. 9-5. Adults $3.50, seniors $3, children $2. Closed state holidays. &
Attendance: 33,000 (accurate)

VINCENNES STATE HISTORIC SITE: VINCENNES BRANCH OF STATE BANK OF INDIANA, (M), 114 N. Second St., Vincennes, IN 47591-1217. Mailing Address: P.O. Box 81, Vincennes, IN 47591-0081. Tel.: 812-882-7422. Fax: 812-882-0928.
E-mail: rday@indianamuseum.org
Web Site: www.state.in.us/ism/sites/vincennes/
Founded: 1838.
Congressional District: 8
Key Personnel: Historic Site Cur., Bruce Beesley; Asst. Cur., Richard Day.
Personnel Profile: Full-Time Paid 4; Part-Time Volunteers 5.
Governing Authority: government. State of Indiana. Subsidiary Institution: Ind. Dept. of Natural Resources. Tax-exempt.
Institution Type/Description: History Museum: housed in the former state bank building; built in 1838.
Collections: local history & culture; period artifacts.
Activities: tours/interpretation.
Publications: newsletters.
Hours & Admission Prices: By appointment. Adults $3.50. &
Attendance: 5,000 (estimated)

WILLIAM HENRY HARRISON MANSION, GROUSELAND, 3 W. Scott St., Vincennes, IN 47591-1433. Tel.: 812-882-2096. Fax: 812-882-7626. Facebook: Grouseland Foundation.
E-mail: grouseland@sbcglobal.net
Web Site: grouselandfoundation.org
Founded: 1911.
Congressional District: 8
Key Personnel: Pres., James Corridan; Exec. Dir., Lisa Ice-Jones; Treas., Nancy Lawless; Cur., Dennis Latta.
Personnel Profile: Full-Time Paid 1; Part-Time Paid 4; Part-Time Volunteers 25.
Governing Authority: private; nonprofit organization. Parent Institution: Grouseland Foundation, Inc. Tax-exempt: 501(c)(3).
Institution Type/Description: History Museum: Presidential Home. A National Historic Landmark.
Collections: the life of William Henry Harrison from his birth in Virginia to his death in the White House; personal & military artifacts; early 19th century furnishings; household tools.
Research Fields: William Henry Harrison genealogy & artifacts; Indiana Territory; War of 1812; American policy & culture.
Facilities: Museum-related items for sale.
Activities: guided tours; temporary exhibitions; living history events. Annual Events: dinner in April; Candlelight tours in May; Peony sales (fall);
Hours & Admission Prices: Jan.-Feb. daily 11-4; March-Dec. daily 10-5.

Adults $5, seniors $4, students & children $3. Closed New Year's Day; Easter; Thanksgiving; Christmas Eve & Day.
Attendance: 10,500 (accurate)
Membership: Student $25; Individual $40; Family $75; Grouseland Patron $100; Governor's Society $250; President's Society $500; Harrison's Cabinet $1,000.

Wabash

CHARLEY CREEK GARDENS, 551 N. Miami St., Wabash, IN 46992. Mailing Address: P.O. Box 454, Wabash, IN 46992-0454. Tel.: 260-563-1020.
Key Personnel: Dir., Kelly Smith
Institution Type/Description: Gardens.
Collections: native & foreign plants; horticulture.
Facilities: gardens.
Activities: educational programs; research; guided tours; rental facilities.
Hours & Admission Prices: Gardens: daily dawn to dusk. Center: by appointment.

DR. JAMES FORD HISTORIC HOME, 177 W. Hill St., Wabash, IN 46992-3049. Mailing Address: P.O. Box 454, Wabash, IN 46992-0454. Tel.: 260-563-8686.
E-mail: director@jamesfordmuseum.org
Web Site: jamesfordmuseum.org
Founded: 2005.
Key Personnel: Dir., J. Tyler Hardwork.
Personnel Profile: Full-Time Paid 2.
Governing Authority: Parent Institution: Charley Creek Foundation.
Institution Type/Description: Historic House Museum: housed in a 19th century physician's home; built 1841.
Collections: mid- to late 19th century furnishings; 19th century surgeon's office.
Research Fields: Ford family.
Facilities: library; gardens. Museum-related items for sale.
Activities: special events; lectures; recitals; festivals.
Publications: newsletter, Faith, Family, Posterity.
Hours & Admission Prices: Fri. 12-5, Sat.-Sun. 10-5; groups by appointment. Adults $3, children 14 & under $2.
Attendance: 3,000

HONEYWELL CENTER, 275 W. Market St., Wabash, IN 46992-3057. Tel.: 260-563-1102. Fax: 260-563-0873.
Web Site: www.honeywellcenter.org
Key Personnel: Exec. Dir., Tod Minnich; Gallery Mgr., Bonnie McKee
Institution Type/Description: Art Gallery.
Collections: works by local & national artists; sculpture.
Facilities: 1,500-seat theater.
Activities: traveling exhibitions.
Hours & Admission Prices: Daily 9am-10pm. No charge.

WABASH COUNTY HISTORICAL MUSEUM, (M), 36 E. Market St., Wabash, IN 46992-3124. Tel.: 260-563-9070. Fax: 260-569-9070.
E-mail: director@wabashmuseum.org
Web Site: www.wabashmuseum.org
Key Personnel: Exec. Dir., Mitch Figert; Chm. (V), Lee Ann George; Assoc. Dir., Emily Perkins; Dir. Operations, Brian Haupert.
Personnel Profile: Full-Time Paid 3; Part-Time Paid 1; Part-Time Volunteers 30.
Governing Authority: Tax-exempt.
Institution Type/Description: History Museum.
Collections: county history & culture; photographs.
Facilities: archives; 32-seat theater.
Hours & Admission Prices: Tues.-Sat. 10-4; Christmas Day 2-4. Adults $5, children 6-12 and seniors 60 & over $3; discounts to groups; members, museum associates and children 5 & under no charge. Closed major holidays. &
Attendance: 6,000 (estimated)
Membership: Individual $25; Grandparent $40; Family $45.

Wakarusa

BIRD'S EYE VIEW MUSEUM OF MINIATURES, 325 S. Elkhart St., Wakarusa, IN 46573-9727. Mailing Address: c/o Wakarusa Historical Society, P.O. Box 2, Wakarusa, IN 46573. Tel.: 574-862-2367.
Institution Type/Description: History Museum.

Collections: 215 miniature models of Indiana landmarks.
Activities: guided tours.
Hours & Admission Prices: Mon.-Fri. 8-5, Sat. 8am to noon. No charge; donations accepted.

Warsaw

KOSCIUSKO COUNTY HISTORICAL SOCIETY, 121 N. Indiana St., Warsaw, IN 46581. Mailing Address: P.O. Box 1071, Warsaw, IN 46581-1071. Tel.: 574-269-1078.
E-mail: director@kosciuskohistory.com
Web Site: www.kosciuskohistory.com
Key Personnel: Pres., Jerry Frush; Museum Dir., Sally Hogan.
Governing Authority: nonprofit organization.
Institution Type/Description: Historical Society Museum.
Collections: local history & culture; photographs.
Activities: special programs.
Publications: newsletter.
Hours & Admission Prices: Wed.-Sat. 10-4, Sat. 10-4. Closed major holidays.
Membership: Family $25; Contributing $50; Coporate $200; Lifetime $450.

Washington

DAVIESS COUNTY HISTORICAL SOCIETY MUSEUM, 212 1/2 E. Main St., Washington, IN 47501. Mailing Address: P.O. Box 2341, Washington, IN 47501-0981. Tel.: 812-444-9360. Fax: 812-257-0301.
E-mail: dchistory@sbcglobal.net
Web Site: www.daviesscountyhistory.net
Formerly: Daviess County Museum
Founded: 1966.
Congressional District: 8
Key Personnel: Dir., Vincent A. Sellers; Pres. (V), Dean Dorrell.
Personnel Profile: Full-Time Paid 1; Full-Time Volunteers 1; Part-Time Volunteers 10; Interns 1.
Governing Authority: Tax-exempt: 501(c)(3).
Institution Type/Description: History Museum.
Collections: local history & culture; photographs; personal artifacts; coal; farming; Amish; B&O railroad; Wabash & Erie Canal; funeral practices from 1776 to present.
Facilities: meeting facilities.
Activities: quarterly programs.
Publications: quarterly newsletter; Facebook.
Hours & Admission Prices: Mon.-Fri. 10-4, Sat. 9-12. Adults $2; members no charge. &
Attendance: 1,500 (estimated)
Membership: Student $10; Basic $20; Family $35; Sustaining $100; Benefactor $250.

Waynetown

MID'TOWN MUSEUM, 102 E. Washington St., Waynetown, IN 47990. Tel.: 765-376-1728 & 275-2328.
E-mail: jokokojo@tctc.com
Web Site: midtownnac.tripod.com
Institution Type/Description: History Museum.
Collections: local history & culture; period furnishings; personal artifacts; clothing; photographs.
Facilities: Museum-related items for sale.
Hours & Admission Prices: Tues.-Fri. 9-5, Sat. 9-2. Adults $2, seniors $1.50, children $1; discounts to groups.

West Lafayette

THE ARTHUR & KRIEBEL HERBARIA, PURDUE UNIVERSITY, 915 W. State St., Dept. of Botany & Plant Pathology, West Lafayette, IN 47907-2054. Fax: 765-494-0363.
E-mail: herbaria@purdue.edu
Web Site: www.btny.purdue.edu/Herbaria/herbaria.html
Founded: 1873.
Congressional District: 4
Key Personnel: Dir., Gregory Shaner.
Personnel Profile: Part-Time Paid 1.
Governing Authority: university. Parent Institution: Purdue University. Subsidiary Institution: Dept. Botany & Plant Pathology.
Institution Type/Description: Herbarium.
Collections: rust plant parasitic fungi; plants; fungi.
Research Fields: taxonomy & ecology of plant parasitic fungi, mainly rust fungi worldwide; plants & mushrooms of Indiana.

Facilities: lab.
Activities: formally organized education programs for graduate, undergraduate & grade school students; public exhibitions; talks & excursions.
Hours & Admission Prices: Mon.-Fri. 9-5. No charge. Closed state & national holidays. &
Attendance: 150 (estimated)

PURDUE UNIVERSITY GALLERIES, Yue-Kong Pao Hall, 552 W. Wood St., West Lafayette, IN 47907-2002. Tel.: 765-494-3061. Fax: 765-496-2817.
E-mail: galleries@purdue.edu
Web Site: www.purdue.edu/galleries
Founded: 1978.
Congressional District: 7
Key Personnel: Dir., Craig Martin; Asst. Dir., Michal Hathaway; Asst. to Dir., Mary Ann Anderson.
Personnel Profile: Full-Time Paid 3; Part-Time Paid 20; Part-Time Volunteers 6.
Governing Authority: university. Parent Institution: Purdue University. Tax-exempt: 501(c)(3).
Institution Type/Description: University Art Gallery.
Collections: More than 5,200 objects are in the permanent collection: 3,000 historic to contemporary prints, photographs, and works on paper; Art of the Americas, including Native American baskets, Pre-Columbian textiles and ceramics. Paintings, drawings and prints by the Mexican Modernists.
Research Fields: pertaining to collections.
Activities: guided tours; lectures; films; loan, temporary & traveling exhibitions; Art Cart, education outreach activity; Living Graphic Novel, educational outreach activity.
Publications: newsletter.
Hours & Admission Prices: Mon.-Wed. & Fri.-Sat. 10-5, Thurs. 10-8, Sun. 1-5. No charge. Closed university holidays. &
Attendance: 24,797 (accurate)

Westfield

WESTFIELD WASHINGTON HISTORICAL SOCIETY & MUSEUM, 145 S. Union St., Westfield, IN 46074. Mailing Address: P.O. Box 103, Westfield, IN 46074. Tel.: 317-804-5365.
Web Site: wwwhs.museum.com
Founded: 2008.
Congressional District: 5
Institution Type/Description: Historical Society Museum.
Collections: local & regional history; Underground Railroad; Quakers; Orphan Train; local business & industry; photographs; personal artifacts; period furnishings.
Publications: bi-monthly newsletter.
Hours & Admission Prices: Sat. 11-3, call to confirm.

Whiting

WHITING-ROBERTSDALE HISTORICAL MUSEUM, 1610 119th St., Whiting, IN 46394-1702. Tel.: 219-659-1432.
Institution Type/Description: History Museum.
Collections: local history & culture; photographs; personal artifacts.
Hours & Admission Prices: Tues.-Wed. & Sat. 1-4.

Winchester

RANDOLPH COUNTY HISTORICAL MUSEUM, 416 S. Meridian, Winchester, IN 47394-2028. Tel.: 317-584-1334.
E-mail: rchsin2@comcast.net
Web Site: www.randolphcountyindianahistoricalsociety.org
Founded: 1968.
Congressional District: 2
Key Personnel: Pres. (V), Marjorie Birtwhistle; Vice Pres., Saundra Jackson; Treas., Krista Hayes; Sec. & Museum Shop Mgr., Monisa Wisener.
Personnel Profile: Full-Time Volunteers 1; Part-Time Volunteers 5; Interns 1.
Governing Authority: society; nonprofit. Parent Institution: Randolph County Historical Society. Tax-exempt: 170(b)(1)(A).
Institution Type/Description: Genealogy & History Museum: housed in 1858 brick home built by Carey Goodrich.
Collections: clothes; dishes; photos; tools; books; furniture & furnishings of the period; genealogical materials; military artifacts; Native American; toys; advertising; newspapers from 1858; 1903 buggy.
Facilities: research library of genealogical & historical information. Museum-related items for sale.
Activities: guided tours; lectures; permanent exhibitions; research materials including genealogy, school records, history books & maps & newspapers from 1858-2000; rental facilities.

Publications: newsletter; books, 1818-1990 Randolph County History; reprint, 1882 History Randolph County; 1895 & 1918 Atlas (Randolph County).
Hours & Admission Prices: Mon.-Fri. 1:30-6; other times by appointment. No charge; donations accepted. ♿
Attendance: 750 (estimated)
Membership: Family $15.

Winona Lake

THE BILLY SUNDAY HOME AND THE RENEKER MUSEUM OF WINONA HISTORY, Westminster Hall, Grace College, Winona Lake, IN 46590-1062. Mailing Address: Westminster Hall, Grace College, 200 Seminary Dr., Winona Lake, IN 46590. Tel.: 574-372-5193.
Web Site: www.villageatwinona.com
Key Personnel: Dir., Bill Firstenberger.
Personnel Profile: Full-Time Paid 1; Part-Time Volunteers 50; Interns 2.
Governing Authority: Parent Institution: Village at Winona. Tax-exempt.
Institution Type/Description: Historic Site: housed in former home of professional baseball player turned evangelist, Billy Sunday.
Collections: personal artifacts; furnishings; paintings; needlework; folk art; needlecrafts; baseball memorabilia.
Research Fields: Billy Sunday; art & crafts design.
Facilities: visitors center.
Hours & Admission Prices: Tues.-Sat. 10-5. Visitors Center: donations accepted. Sunday Home Tour: adults $4; discounts to groups with appointment; children 12 & under no charge. Closed New Year's Day; Good Friday; Thanksgiving; Christmas. ♿
Attendance: 7,000 (estimated)

Wolcott

HISTORIC WOLCOTT HOUSE, 500 N. Range St., Wolcott, IN 47995-8276. Mailing Address: Anson Wolcott Historical Society, P.O. Box 242, Wolcott, IN 47995. Tel.: 219-279-2951.
Key Personnel: Pres., Ann Cain
Institution Type/Description: Historic House Museum.
Collections: period furnishings.
Hours & Admission Prices: By appointment. No charge; donations accepted.

Wolf Lake

LUCKEY HOSPITAL MUSEUM, US 33 & SR 109, Wolf Lake, IN 46796. Mailing Address: P.O. Box 143, Wolf Lake, IN 46796. Tel.: 260-635-2490.
E-mail: info@luckeyhospitalmuseum.org
Web Site: www.luckeyhospitalmuseum.org
Founded: 2004.
Key Personnel: Chm. (V), Sandra Huntsman; Pres. (V), Hal Stump
Governing Authority: Tax-exempt.
Institution Type/Description: History Museum.
Collections: hospital history; 1930s nurse's caps & uniforms; surgery room; early medical instruments & equipment; iron lung; early baby bed; fully furnished doctors lounge, patient room, labor room & delivery room.
Publications: quarterly newsletter to members.
Hours & Admission Prices: Memorial Day to Labor Day Wed. & Sat. 10-2; groups & other times by appointment. Adults $5, K-12 $3; members & pre-school students no charge.
Attendance: 500 (accurate)
Membership: Bronze $10; Silver $25; Gold $50; Platinum $100; Diamond $1,000.

Zionsville

ANTIQUE FAN MUSEUM, 10983 Bennett Pkwy., Zionsville, IN 46077. Tel.: 888-567-2055 (Toll Free); 317-733-4113. Fax: 317-733-4162.
Institution Type/Description: History Museum.
Collections: over 450 period ceiling & desk fans dating back to the 1880s; hand fans; memorabilia.
Hours & Admission Prices: Call for hours.

P.H. SULLIVAN MUSEUM AND GENEALOGY LIBRARY, 225 W. Hawthorne St., Zionsville, IN 46077-1620. Tel.: 317-873-4900.
Web Site: www.sullivanmunce.org
Founded: 1973.
Congressional District: 6
Key Personnel: Pres. Bd. Dir. (V), Kelly Masoncup; Exec. Dir., Cynthia Young; Pres. Museum Women's Guild (V), Judi Potts.
Personnel Profile: Full-Time Paid 2; Part-Time Volunteers 30.
Governing Authority: nonprofit organization. Parent Institution: P.H. Sullivan Foundation. Branch Museums: Munce Art Center. Tax-exempt: 501(c)(3).
Institution Type/Description: Local History Museum.
Collections: Boone County & Indiana history.
Research Fields: United States genealogy.
Facilities: 4,500-vol. library of material on Boone County, Indiana & other states available on premises; reading room; permanent & temporary exhibitions.
Activities: lectures; workshops; temporary exhibitions; artist-in-residence program.
Hours & Admission Prices: Tues.-Sat. 10-4. No charge; donations accepted. Closed New Year's Day; Thanksgiving; Christmas. ♿
Attendance: 7,000 (estimated)
Membership: $25; $50; $100; $250; $500; $1,000.

SULLIVANMUNCE CULTURAL CENTER, 205-225 W. Hawthorne St., Zionsville, IN 46077-1620. Tel.: 317-873-4900.
Web Site: www.sullivanmunce.org
Formerly: Munce Art Center
Founded: 1981.
Congressional District: 6
Key Personnel: Exec. Dir., Cynthia Young; Pres., Kelly Masoncup.
Personnel Profile: Full-Time Paid 2.
Governing Authority: nonprofit organization. Parent Institution: P.H. Sullivan Foundation. Tax-exempt: 501(c)(3).
Institution Type/Description: History and The Arts.
Collections: local & early Indiana art.
Facilities: library of art books & genealogical records; archives; classrooms.
Activities: classes; temporary exhibitions.
Publications: quarterly, class schedule; tri-annual, newsletter.
Hours & Admission Prices: Tues.-Sat. 10-4. No charge; donations accepted. Closed holidays. ♿
Attendance: 11,000 (estimated)
Membership: $25 & up.

IOWA

(369 listings)

Ackley

ACKLEY HERITAGE CENTER, 120 State St., Ackley, IA 50601-1545. Tel.: 641-847-2201.
E-mail: ackleyhc@mchsi.com
Web Site: www.ackleyheritagecenter.com
Founded: 1987.
Key Personnel: Chm. (V), Marie Arends; Pres. (V), Duan Sudtelgte
Institution Type/Description: History Museum.
Collections: local history & culture; photographs; personal artifacts.
Hours & Admission Prices: Tues. & Thurs.-Fri. 1:30-4:30; other times by appointment. No charge; donations accepted. ♿
Attendance: 850 (estimated)

Adel

ADEL HISTORICAL MUSEUM, 1129 Main St., Adel, IA 50003-1424. Mailing Address: 1204 Locust St., Adel, IA 50003. Tel.: 515-993-1032. Fax: 515-993-3384.
E-mail: donprice555@msn.com
Web Site: www.adeliowa.org
Key Personnel: Dir. (V), Jan Price.
Governing Authority: city. Tax-exempt.
Institution Type/Description: History Museum: housed in a former schoolhouse; built in 1857.
Collections: local history; period furnishings; photographs; personal artifacts.
Hours & Admission Prices: mid-May to mid-Sept. Sat. 10-4; other times by appointment. No charge; donations accepted.
Attendance: 400 (estimated)

Albert City

ALBERT CITY HISTORICAL MUSEUM, 212 2nd St., N., Albert City, IA 50510-1210. Mailing Address: Box 431, Albert City, IA 50510. Tel.: 712-843-5684.
Founded: 1974.
Key Personnel: Pres. (V), Dick Aronson.
Operating Expenses: 13,165
Operating Income: 13,945

Governing Authority: Tax-exempt.
Institution Type/Description: History Museum.
Collections: military room; grocery store; tool shop; early 1900s cars, music boxes, glassware figurines, railroad memorabilia; medical instruments; period furnishings. Historic Buildings: 1900s home; 1875 schoolhouse.
Hours & Admission Prices: Sun. 2-5; other times by appointment. Adults $5; members no charge.
Attendance: 200 (estimated)

Algona

CAMP ALGONA POW MUSEUM, 114 S. Thorington St., Algona, IA 50511-2616. Mailing Address: P.O Box 174, Algona, IA 50511-0174. Tel.: 515-395-2267 & 295-3719.
E-mail: yocumcampalgona@netamumail.com
Web Site: www.pwcamp.algona.org
Founded: 2001.
Congressional District: 4
Key Personnel: Pres. & Chm. (V), Richard Schiek; Vice Pres., Jerry Yocum; Sec. & Treas., Don Hansen.
Volunteer Hours: 500
Operating Expenses: 58,000
Operating Income: 60,000
Governing Authority: Tax-exempt.
Institution Type/Description: History Museum: former camp for German POW's from 1944-1946.
Collections: weapons; World War II uniforms & artifacts; poetry; paintings; woodcarvings.
Hours & Admission Prices: April-Dec. Sat.-Sun. 1-4; other times by appointment. Adults $3. &
Attendance: 2,000 (accurate)
Membership: Individual $10.

PHARMACISTS MUTUAL COMPANIES MUSEUM, 808 Hwy. 18 W., Algona, IA 50511. Tel.: 515-295-2461; 800-247-5930.
E-mail: info@phmic.com
Web Site: www.phmic.com
Institution Type/Description: History Museum.
Collections: replica of 20-century drugstore; early pharmacy tools, bottles & remedies; perfumes.
Hours & Admission Prices: Daily call for hours. &

Allison

BUTLER COUNTY HISTORICAL SOCIETY, 219 1/2 S. Main St., Allison, IA 50602-9507. Mailing Address: Box 14, Allison, IA 50602-0014. Tel.: 319-267-2255.
E-mail: djpoppen@netins.net
Formerly: Butler County Historical Museum-Little Yellow Schoolhouse
Founded: 1956.
Congressional District: 3
Key Personnel: Pres., Anita Hardy; Vice Pres., Ruth Haan; Sec. & Treas., Judi Poppen.
Personnel Profile: Part-Time Volunteers 10.
Governing Authority: nonprofit. Branch Museum: Little Yellow School House; Log Cabin; Hall of Fame; Butler County Museum; Anna Pals House. Tax-exempt.
Institution Type/Description: Historical Society Museum: residing on the Butler County Fairgrounds.
Collections: agriculture; history; preservation project. Historic Buildings: 1888 Little Yellow Schoolhouse - one room country school; Hall of Fame includes accomplishments of Butler County residents; 1850s Log Cabin reproduction; restored home of Anna Pals.
Facilities: 200-vol. library.
Activities: guided tours; permanent & temporary exhibitions.
Publications: booklet, History of the Little Yellow School House; book, Hall of Fame, 1976-2011.
Hours & Admission Prices: June-Sept. by appointment. No charge; donations accepted. &
Attendance: 850 (estimated)
Membership: Individual $10; Lifetime $100.

Amana

AMANA HERITAGE MUSEUM, 705 44th Ave., Amana, IA 52203-0081. Mailing Address: P.O. Box 81, Amana, IA 52203-0081. Tel.: 319-622-3567. Fax: 319-622-6481.
E-mail: amanaheritage@southslope.net
Web Site: www.amanaheritage.org

Formerly: Museum of Amana History
Founded: 1969.
Congressional District: 3
Key Personnel: Dir., Lanny Haldy; Pres. (V), Allyn Neubauer; Cur., Kelly Duffy.
Personnel Profile: Full-Time Paid 2; Part-Time Paid 20.
Governing Authority: nonprofit organization. Parent Institution: Amana Heritage Society. Tax-exempt: 501(c)(3).
Institution Type/Description: Historical Society Museum: housed in 1865 & 1870 buildings.
Collections: artifacts pertaining to Amana (Community of True Inspiration) history; crafts; costumes; decorative arts; archives; photographs; furnishings; textiles.
Research Fields: local history; history of Community of True Inspiration; German immigration; Pietism.
Facilities: library; 85-seat auditorium; 5,000 sq. ft. exhibit space. Books for sale.
Activities: guided tours; lectures; organized education programs for children; temporary exhibitions; guided walking tours of village of Amana.
Publications: quarterly newsletter.
Hours & Admission Prices: April-Oct. 30 Mon.-Sat. 10-5, Sun. 12-5. Adults $7; discounts to groups; children, students & members no charge. &
Attendance: 25,000 (accurate)
Membership: Contributing $25; Silver Heritage $75; Gold Heritage $200.

Ames

✱ BRUNNIER ART MUSEUM, (M), University Museums, Iowa State University, 290 Scheman Bldg., Ames, IA 50011. Tel.: 515-294-3342. Fax: 515-294-3342.
Web Site: www.museums.iastate.edu
Founded: 1975.
Key Personnel: Dir. & Chief Cur., Lynette L. Pohlman; Collections Mgmt. & Communications Coord., Allison Sheridan; Interpretive Specialist, Dorothy Witter; Devel. Sec., Sue Olson; Educator, Visual Literacy & Learning, Nancy Girard; Administrative Specialist, Angela Shippy.
Personnel Profile: Full-Time Paid 4; Part-Time Paid 2; Interns 12.
Governing Authority: state. Parent Institution: University Museums, Iowa State University. Tax-exempt.
Institution Type/Description: Decorative & Fine Arts Museum.
Collections: decorative arts including glass, dolls, ceramics, jade.
Research Fields: decorative arts; glass; ceramics.
Facilities: library of reference books available for research on premises; reading room. Catalogues & slides for sale.
Activities: guided tours; gallery talks; docent program; permanent, temporary & traveling exhibitions. Annual Events: Brunnier in Bloom.
Publications: bimonthly e-newsletter, Interpretations.
Hours & Admission Prices: Tues.-Fri. 11-4, Sat.-Sun.1-4. No charge; donations accepted. Closed Memorial Day; Independence Day; Labor Day; Thanksgiving; Christmas; university holidays. &
Attendance: 15,000 (estimated)
Membership: Household $50; Sustained $100; Pacesetter $250; Director's Guild $500; Benefactor $1,000; Patron $2,500; Curator's Associate (Lifetime) $10,000; Curators $15,000.

CHRISTIAN PETERSEN ART MUSEUM, Morrill Hall, Iowa State University, Ames, IA 50011. Mailing Address: University Museums, Iowa State University, 290 Scheman Bldg., Ames, IA 50011. Tel.: 515-294-9500.
Web Site: www.museums.iastate.edu
Founded: 2007.
Congressional District: 4
Key Personnel: Dir. & Chief Cur., Lynette Pohlman; Collections Mgmt. & Communications Coord., Allison Sheridan; Interpretive Specialist, Dorothy Witter; Devel. Sec., Sue Olson; Educator, Visual Literacy & Learning, Nancy Girard; Administrative Specialist, Angela Shippy.
Personnel Profile: Full-Time Paid 4; Part-Time Paid 2; Interns 13.
Governing Authority: state; nonprofit. Parent Institution: University Museums, Iowa State University, Ames, IA. Tax-exempt: 501(c)(3).
Institution Type/Description: Art Museum.
Collections: 20th century & contemporary art; works by Christian Petersen.
Research Fields: American regionalism; campus public art.
Facilities: 200-vol. library; 5,060 sq. ft. exhibit space; classroom.
Activities: films; formal education programs for college students; guided tours; lectures; temporary & traveling exhibitions.
Publications: bimonthly e-newsletter, e-Interpretations.
Hours & Admission Prices: Mon.-Fri. 11-4. No charge; donations accepted. Closed major holidays; semester breaks. &
Attendance: 15,000 (estimated)
Membership: Household $50; Sustained $100; Pacesetter $250; Director's

Guild $500; Benefactor $1,000; Patron $2,500; Curator's Associate (Lifetime) $10,000; Curators $15,000.

✱ FARM HOUSE MUSEUM, Farm House Ln., Iowa State University, Ames, IA 50011. Mailing Address: University Museums, Iowa State University, 290 Scheman Bldg., Ames, IA 50011. Tel.: 515-294-7426. Fax: 515-294-3342.
Web Site: www.museums.iastate.edu
Founded: 1974.
Key Personnel: Dir. & Chief Cur., Lynette Pohlman; Collections Mgmt. & Communications Coord., Allison Juull; Interpretive Specialist, Amanda Hall; Devel Sec., Sue Olson; Educator, Visual Literacy & Learning, Nancy Gebhart; Administrative Specialist & Museum Shop Mgr., Angela Shippy.
Governing Authority: state; nonprofit. Parent Institution: University Museums, Iowa State University. Tax-exempt.
Institution Type/Description: Historic Site: c.1860-1864 The Farm House.
Collections: c.1860-1910 decorative arts; period furnishings.
Research Fields: historical objects; Iowa & Iowa State University history.
Facilities: 4,080 sq. ft. exhibit space.
Activities: guided tours; gallery talks; docent program; loan, permanent, temporary & traveling exhibitions. Annual Event: Haunted Iowa State.
Publications: quarterly newsletter, News From University Museums.
Hours & Admission Prices: Mon.-Fri. 12-4 by appointment only. No charge; donations accepted. Closed university holidays; semester breaks. ♿
Attendance: 6,000 (estimated)
Membership: College Student $10; Household $50; Sustained $100; Pacesetter $250; Director's Guild $500; Benefactor $1,000; Patron $2,500; Curator's Associates (Lifetime) $10,000; Curators $15,000.

OCTAGON CENTER FOR THE ARTS, 427 Douglas Ave., Ames, IA 50010-6281. Tel.: 515-232-5331. Fax: 515-232-5088.
E-mail: gallery@octagonarts.org
Web Site: www.octagonarts.org
Founded: 1966.
Congressional District: 4
Key Personnel: Acting Exec. Dir. & Cur., Heather Johnson; Pres., Mark Peterson; Vice Pres., Alan Johnson; Octagon Shop Mgr., Ruth Wiedemeier; Educ. Dir., Beth Weninger.
Personnel Profile: Full-Time Paid 6; Part-Time Paid 8; Part-Time Volunteers 75; Interns 3.
Governing Authority: nonprofit organization. Tax-exempt: 501(c)(3).
Institution Type/Description: Art Center.
Collections: traditional & contemporary international masks; Feinberg Mask collection.
Research Fields: traditional & contemporary crafts.
Facilities: 150-seat auditorium; educational facilities. Art work for sale.
Activities: guided tours; lectures; concerts; arts festivals; theater; organized education programs; loan, temporary, & traveling exhibitions; exhibitions geared towards children; internships for undergraduate & graduate college students; outreach programming to schools, organizations & agencies. Museum Sponsors: Annual national juried show.
Publications: monthly, The Octagon Center for the Arts Newsletter; exhibition catalogues; quarterly, class schedules.
Hours & Admission Prices: Gallery: Tues.-Fri. 10-5, Sat.-Sun. 1-5. Office: Mon.-Fri. 8-5. Shop: Mon.-Wed. & Fri. Sat. 10-5, Thurs. 10-8. Gallery: no charge; donation suggested. Fees for classes & special events. Closed major holidays. ♿
Attendance: 30,000 (estimated)
Membership: Student $15; Senior Citizen $25; Individual $35; Family $50; Sustaining $75-$200; Founder's Club $300; Sponsorships $300 & up.

REIMAN GARDENS - IOWA STATE UNIVERSITY, 1407 University Blvd., Ames, IA 50011. Tel.: 515-294-2710. Fax: 515-294-4817.
Key Personnel: Cur., Nathan Brockman
Institution Type/Description: Garden.
Collections: native plants, flowers & trees; butterflies.
Facilities: Gift items for sale.
Activities: educational programs; special events.
Hours & Admission Prices: Gardens: June-Sept. 6 daily 9-6; Sept. 7-May daily 9-4:30. Butterfly Wing: daily 9-4:30. Adults 18-64 $8, seniors 65 & over $7, youth 4-17 $4; children 3 & under, members & ISU students no charge. Closed New Year's Day; Thanksgiving; Christmas.

STORY COUNTY CONSERVATION CENTER, McFarland Park, 56461 180th St., Ames, IA 50010-9451. Tel.: 515-232-2516. Fax: 515-232-6989.
E-mail: conservation@storycounty.com
Web Site: www.storycountyconservation.org
Institution Type/Description: Conservation Center.
Collections: environmental conservation; mounted wildlife; photographs.
Activities: rental facilities; educational programs; trails.
Hours & Admission Prices: April-Oct. Mon.-Fri. 8:30-4:30, Sun. 2-5. No charge.

Anamosa

ANAMOSA STATE PENITENTIARY MUSEUM, 406 N. High St., Anamosa, IA 52205-1199. Mailing Address: P.O. Box 144, Anamosa, IA 52205-0144. Tel.: 319-462-2386.
E-mail: aspmuseum@mchsi.com
Web Site: www.asphistory.com/museum
Founded: 2001.
Key Personnel: Dir., Richard Snavely; Chm. (V), Don Folkerts; Museum Shop Mgr., Jan Pearson.
Personnel Profile: Part-Time Volunteers 15.
Governing Authority: Tax-exempt: 501(c)(3).
Institution Type/Description: Penitentiary Museum.
Collections: Iowa's 132 year prison history; prison life; photographs; cell replica; video.
Facilities: Museum-related items for sale.
Activities: video.
Hours & Admission Prices: Memorial Day to Oct. Fri.-Mon. 12-4; other times by appointment. Adults $3; children, students & members no charge. ♿
Attendance: 1,700 (accurate)
Membership: Annual $10; Lifetime $125.

GRANT WOOD ART GALLERY, 124 E. Main St., Anamosa, IA 52205-1879. Tel.: 319-462-4267.
E-mail: grantwoodartgallery@grantwoodgallery.org
Web Site: www.grantwoodartgallery.org
Founded: 1973.
Key Personnel: Chm. (V) & Museum Shop Mgr., Jon D. Hatcher.
Personnel Profile: Part-Time Volunteers 25.
Governing Authority: Tax-exempt.
Institution Type/Description: Art Gallery.
Collections: Grant Wood's life & paintings; prints; photography; murals.
Facilities: Museum-related items for sale.
Hours & Admission Prices: Jan.-March Sun. 1-4; April-Dec. Mon.-Sat. 10-4, Sun. 1-4. No charge. ♿

NATIONAL MOTORCYCLE MUSEUM, 102 Chamber Dr., Anamosa, IA 52205-1806. Mailing Address: P.O. Box 405, Anamosa, IA 52205-0405. Tel.: 319-462-3925. Fax: 319-462-3982.
E-mail: museum@nationalmcmuseum.org
Web Site: www.nationalmcmuseum.org
Founded: 1989.
Personnel Profile: Full-Time Paid 4; Part-Time Paid 4.
Governing Authority: nonprofit. Tax-exempt.
Institution Type/Description: Motorcycle Museum.
Collections: motorcycle industry history; vintage bikes; photographs; Hall of Fame.
Facilities: Museum-related items for sale.
Hours & Admission Prices: April-Oct. daily 9-5; Nov.-March Mon.-Sat. 9-5, Sun. 10-4. Adults $8, seniors $7; children 12 & under no charge. ♿
Attendance: 18,000 (accurate)
Membership: Annual $50; Sustaining $100; Lifetime $500.

Ankeny

ANKENY AREA HISTORICAL SOCIETY MUSEUM, 301 S.W. Third St., Ankeny, IA 50023. Mailing Address: P.O. Box 1111, Ankeny, IA 50021-0973. Tel.: 515-965-5795.
E-mail: history@ankenyhistorical.org
Web Site: ankenyhistorical.org
Founded: 1988.
Congressional District: 3
Key Personnel: Dir., Rosemary Hutton Taylor; Pres. (V), Ron Sampson.
Personnel Profile: Part-Time Paid 1.
Governing Authority: Tax-exempt.
Institution Type/Description: Historical Society Museum.
Collections: local history & culture; personal artifacts; period furnishings; photographs; military artifacts. Historic Buildings: 1905 farm house & barn.
Research Fields: northern Polk County, Iowa; Ankeny, Des Moines Ordnance Plant - WWII.
Activities: guided tours. Annual Event: Spring Luncheon; Summerfest BBQ;

Halloween Uptown; Christmas Cookie Sale; Uptown Ankeny Christmas Event in December.
Publications: newsletter.
Hours & Admission Prices: April-Nov. 1st Sun. of month 2-4; other times by appointment. No charge; donations accepted. &
Attendance: 1,200 (estimated)
Membership: Individual $15; Family $30; Business, Institution & Professional $40; Extraordinary Support $50.

ANKENY ART CENTER, 1520 S.W. Ordnance Rd., Ankeny, IA 50023-2510. Tel.: 515-965-0940. Fax: 515-963-1009.
E-mail: ankenyarts@ankenyartcenter.com
Web Site: www.ankenyartcenter.com
Founded: 1981.
Key Personnel: Dir., Barb Vaske.
Personnel Profile: Part-Time Paid 2; Interns 1.
Governing Authority: nonprofit organization. Parent Institution: Ankeny Friends of the Arts. Tax-exempt.
Institution Type/Description: Art Center.
Collections: works by Iowa artists.
Activities: classrooms; outdoor events; education programs; special events; summer art camps.
Hours & Admission Prices: Tues.-Wed. & Fri. 9-1, Thurs. 4-7, Sat. 9-12. No charge.
Attendance: 5,000 (estimated)

IOWA AVIATION HERITAGE MUSEUM, 3704 S.E. Convenience Blvd., Ankeny, IA 50021. Tel.: 515-964-4515.
E-mail: mlcallison@myway.com
Web Site: iowaaviationheritagemuseum.webs.com
Institution Type/Description: Aviation History Museum.
Collections: military aviation history; models; aircraft; personal artifacts; photographs.
Hours & Admission Prices: Tues. & Thurs. 10-2, Sat. 10-4; other times by appointment.

Armstrong

ARMSTRONG HERITAGE MUSEUM, 425 Sixth St., Armstrong, IA 50514. Mailing Address: 401 C Ave., Armstrong, IA 50514. Tel.: 712-868-3593.
E-mail: cggc@ringtelco.com
Web Site: www.armstrongiowa.net/libmus.php
Founded: 1989.
Key Personnel: Dir., Paula Dyer
Governing Authority: Tax-exempt.
Institution Type/Description: History Museum.
Collections: local history & culture.
Hours & Admission Prices: By appointment. No charge; donations accepted.
Attendance: 350 (estimated)
Membership: Annual $10; Lifetime $50.

Arnolds Park

ABBIE GARDNER STATE SITE, 34 Monument Dr., Arnolds Park, IA 51331. Mailing Address: P.O. Box 74, Arnolds Park, IA 51331-0074. Tel.: 712-332-7248. Fax: 515-282-0502.
Founded: 1856.
Congressional District: 6
Key Personnel: Dir., Mike Koppert
Institution Type/Description: Historic House Museum: housed in log cabin built in 1856; located on the site of the Spirit Lake Massacre in 1857. Listed on the National Register of Historic Places.
Collections: period furnishings; personal artifacts.
Hours & Admission Prices: Memorial Day to Labor Day daily 12-4; Sept. Sat. 12-4. No charge; donations accepted. &
Attendance: 15,000

IOWA GREAT LAKES MARITIME MUSEUM, 243 W. Broadway, Arnolds Park, IA 51331-7779. Mailing Address: P.O. Box 726, Arnolds Park, IA 51331-0726. Tel.: 712-332-5264. Fax: 712-332-7714.
Web Site: www.arnoldspark.com
Founded: 1987.
Key Personnel: Gen. Mgr., Scott Pyle; Cur., Mary Kennedy.
Personnel Profile: Part-Time Paid 1; Part-Time Volunteers 10.
Governing Authority: private; nonprofit organization. Tax-exempt: 501(c)(3).
Institution Type/Description: Maritime Museum.

Collections: wooden boats from the Iowa Great Lakes area of Northwest Iowa; personal artifacts; photographs.
Facilities: 1,850-vol. library; 6,000 sq. ft. exhibit space; 150-seat theater.
Activities: films; formal education programs for adults; guided tours; lectures; temporary exhibitions; theater.
Publications: biannual newsletter, The Steam Whistle.
Hours & Admission Prices: Daily 10am, closing time varies. No charge; donations accepted. Closed Easter; Thanksgiving; Christmas. &
Attendance: 75,000 (estimated)
Membership: Captain $35; Rear Admiral $75; Commander $150; Fleet $250; Wish List $1,000.

IOWA ROCK 'N ROLL MUSIC ASSOCIATION MUSEUM, 91 Lake St., Arnolds Park, IA 51331. Mailing Address: P.O. Box 557, Arnolds Park, IA 51331-0557. Tel.: 712-332-6540.
E-mail: jhardypedersen@iowarocknroll.com
Web Site: iowarocknroll.com
Founded: 1997.
Governing Authority: nonprofit organization. Tax-exempt.
Institution Type/Description: Music Museum: housed on the site of the Roof Garden Ballroom.
Collections: rock 'n roll history; Iowa music, entertainers, radio stations, ballrooms & promoters; photographs; recording studio.
Publications: monthly newsletter.
Hours & Admission Prices: mid-May to mid-Sept. daily 11-7; mid-Sept. to mid-May Thurs.-Sat. 12-4. Adults 13 & over $1; IRRMA members and children 12 & under no charge. Closed New Year's Day; Veterans Day; Thanksgiving; Christmas. &
Attendance: 5,317 (estimated)
Membership: Opening Act $40; Headliner $100; Corporate $200; Gold Record $550; Super Star $700.

Ashton

DEBOER GROCERY MUSEUM AND LITTLE HOUSE MUSEUM, 320 Third St., Ashton, IA 51232. Tel.: 712-724-6239.
E-mail: chonkomp@hotmail.com
Founded: 1996.
Congressional District: 5
Governing Authority: Parent Institution: Osceola County Historical Society. Tax-exempt.
Institution Type/Description: History Museum.
Collections: original grocery store shelving; fruit & vegetable counters; walk-in meat cooler; scales; early cash register.
Hours & Admission Prices: Memorial Day-Labor Day 1st & 3rd Sun. 10-2; other times by appointment. No charge; donations accepted.
Attendance: 200 (estimated)
Membership: Individual $15; Family $25.

Aurora

RICHARDSON-JAKWAY HISTORIC SITE, 2791 136th St., Aurora, IA 50607. Mailing Address: 1883 125th St., Hazleton, IA 50641-9695. Tel.: 319-636-2617. Fax: 319-636-2624.
E-mail: bccbdan@iowatelecom.net
Web Site: www.buchanancountyparks.com
Formerly: Fontana Interpretive Nature Center
Key Personnel: Dir., Dan Cohen
Institution Type/Description: Historic House Museum.
Collections: furnishings; historic house.
Activities: educational programs. Museum Sponsors: Old Time Meals in May.
Hours & Admission Prices: By appointment. No charge; donations accepted.

Avoca

FARMALL-LAND USA MUSEUM, 2101 N. LaVista Heights Rd., Avoca, IA 51521. Tel.: 712-307-6806 & 343-6354.
Institution Type/Description: History Museum.
Collections: tractors; threshing machine; fire truck; dolls.
Facilities: Museum-related items for sale.
Hours & Admission Prices: April-Oct. Tues.-Sat. 10-5, Sun. 12-5; other times by appointment. Adults $8, students $5, children 5-12 $3; children under 5 no charge.

SWEET VALE OF AVOCA MUSEUM, 504 N. Elm St., Avoca, IA 51521-3521. Tel.: 712-343-2477.
Institution Type/Description: History Museum.
Collections: local history & culture; over 75 mounted animals; Danish plates.
Hours & Admission Prices: Memorial Day to Labor Day daily 1-4.

Beaman

BEAMAN HERITAGE CENTER, 223 Main St., Beaman, IA 50609. Mailing Address: P.O. Box 135, Beaman, IA 50609-0135. Tel.: 641-366-2912. Fax: 641-366-3141.
E-mail: bcmlib@heartofiowa.net
Web Site: library.beaman.lib.ia.us
Founded: 1996.
Key Personnel: Dir., LaVonne Sternhagen.
Personnel Profile: Part-Time Paid 1.
Governing Authority: Parent Institution: Beaman Community Memorial Library. Tax-exempt.
Institution Type/Description: History Museum.
Collections: local history, heritage & art; personal artifacts; photographs; period furnishings.
Activities: Museum Sponsors: Memorial Day celebration.
Hours & Admission Prices: Mon.-Fri. 2-6, Sat. 10 to noon; other times by appointment. No charge; donations accepted. &
Attendance: 75 (accurate)

Bedford

TAYLOR COUNTY HISTORICAL MUSEUM AND ROUND BARN, 1001 Pollock Blvd., Bedford, IA 50833. Tel.: 712-523-2041.
E-mail: taylorcomuseum@frontiernet.net
Institution Type/Description: History Museum.
Collections: local history & culture; period furnishings; personal artifacts; photographs. Historic Buildings: round barn; log cabin; rural schoolhouse; depot; print shop; bank building; chapel.
Hours & Admission Prices: Call for hours. &

Bellevue

THE YOUNG MUSEUM, 406 N. Riverview, Bellevue, IA 52031. Tel.: 563-872-4456.
E-mail: dcephoto@iowatelecom.net
Institution Type/Description: Historic House Museum.
Collections: local history & culture; period furnishings; personal artifacts; photographs.
Hours & Admission Prices: May-Oct. Sat. 1-5; tours by appointment.

Belmond

JENISON-MEACHAM MEMORIAL ART CENTER AND MUSEUM, 1179 Taylor Ave., Belmond, IA 50421-7568. Tel.: 641-444-3557 & 4635.
Institution Type/Description: Art & History Museum.
Collections: paintings; murals; period furnishings; personal artifacts; photographs. Historic Buildings: 1900s house, barn, schoolhouse, & "Andy's Maytag House.
Activities: Annual Event: Antique Power and Craft Show in August.
Hours & Admission Prices: May-Nov. Sat.-Sun. 1:30-4:30. No charge. &

Bettendorf

* **FAMILY MUSEUM, (M),** 2900 Learning Campus Dr., Bettendorf, IA 52722-7710. Tel.: 563-344-4106. Fax: 563-344-4164. Facebook: Family Museum.
E-mail: familymuseum@bettendorf.org
Web Site: www.familymuseum.org
Founded: 1974.
Congressional District: 1
Key Personnel: Program Mgr., Kim Kidwell; Exhibits Mgr., Tom Stanger; Coord. Group Svcs., Julie Klein; Coord. Events, Becky Ortner; Coord. Public Rels., Elly Gerdts; Business & Community Rels. Mgr., Jeff Reiter; Volunteer Coord., Nina Bouxsein; Dance Coord., Jessica Halfhill; Museum Shop Mgr., Caroline O'Sullivan-Jens.
Personnel Profile: Full-Time Paid 12; Part-Time Paid 11; Part-Time Volunteers 240; Interns 3.
Volunteer Hours: 6,000
Operating Expenses: 2,000,000
Operating Income: 2,000,000
Governing Authority: municipal. Parent Institution: City of Bettendorf. Tax-exempt.
Institution Type/Description: Children's Museum.
Collections: hands-on exhibits.
Major Exhibits: Dora the Explorer, Summer 2014 (T).
Facilities: multi-purpose room; meeting room; theatre.

Activities: guided tours; permanent & temporary exhibitions; classes; special events; birthday parties.
Publications: brochure; class catalogs; weekly member e-mail.
Hours & Admission Prices: Memorial Day-Labor Day: Mon.-Sat. 9-5, Sun. 12-5; Sept.-May: Mon.-Thurs. 9-8, Fri.-Sat. 9-5, Sun. 12-5. Admission 2-59 $7, seniors 60 & over $4, children $4; discounts to ASTC & ACM members; children under one & members no charge. Closed major holidays. &
Attendance: 96,000 (estimated)
Membership: Household $70 plus $10 each additional person; Grandparent $70 plus $10 each grandchild; ACM Plus $125; Triple $195.

Bloomfield

FINDLEY HOUSE - DAVIS COUNTY HISTORICAL SOCIETY, 205 S. Dodge St., Bloomfield, IA 52537. Tel.: 641-664-1855.
Congressional District: 2
Key Personnel: Chm. (V), Nancy Clancy
Governing Authority: Tax-exempt.
Institution Type/Description: Historical Society Museum: housed in the former home of Civil War physician, Dr. William Findley; built in 1844.
Collections: local history & culture; period furnishings; personal artifacts; photographs.
Hours & Admission Prices: Memorial Day to Aug. Sat. 1-4; other times by appointment. No charge; donations accepted.

Boone

BOONE COUNTY HISTORICAL CENTER, 602 Story St., Boone, IA 50036-2832. Tel.: 515-432-1907.
E-mail: director@boonecountyhistory.org
Web Site: www.boonecountyhistory.org
Founded: 1990.
Congressional District: 4
Key Personnel: Dir., Pamela Schwartz; Chm., Lee McNair; Sec., Janet Tait; Treas., Judy Russell.
Personnel Profile: Full-Time Paid 1; Part-Time Paid 1; Part-Time Volunteers 35.
Governing Authority: nonprofit. Parent Institution: Boone County Historical Society, 602 Story St., Boone, IA 50036. Tax-exempt: 501(c)(3).
Institution Type/Description: Cultural Center & Museum: housed in 1907 Masonic Temple.
Collections: Boone County & Iowa's natural, military, domestic, agricultural & industrial history.
Research Fields: local history; Iowa history.
Facilities: archival material available to the public; 150-seat auditorium; classrooms; conference & meeting rooms.
Activities: cultural festivals; concerts; docent programs; films; guided tours; formal education program for children; loan, temporary & traveling exhibitions; genealogical & historical research.
Publications: biannual historical journal, Trail Tales; biannual newsletter, History Page.
Hours & Admission Prices: Mon.-Fri. 1-4; other times by appointment. Adults $3; children & members no charge. Closed New Year's Day; Easter; Thanksgiving; Christmas. &
Attendance: 5,150 (accurate)
Membership: Individual & Family $40; Business & Organizational $100; Life $600.

IOWA RAILROAD HISTORICAL SOCIETY, 225 10th St., Boone, IA 50036-0603. Mailing Address: P.O. Box 603, Boone, IA 50036-0603. Tel.: 515-432-4249, ext. 14; 800-626-0319. Fax: 515-432-4253.
E-mail: bandsvrr@tdsi.net
Web Site: www.scenic-valleyrr.com
Founded: 1983.
Congressional District: 5
Key Personnel: C.E.O., Fenner Stevenson; Pres. (V), Alan Schroeder; Museum Dir., Mike Wendel; Financial Dir., Dale Mount; Museum Shop Mgr., Jim Barkwill.
Personnel Profile: Full Time Paid 11; Part Time Paid 3; Part Time Volunteers 200.
Governing Authority: private; nonprofit organization. Parent Institution: Iowa Railroad Historical Society. Subsidiary Institution: Boone & Scenic Valley Railroad. Tax-exempt: 501(c)(3).
Institution Type/Description: Historical Railroad Museum.
Collections: full-sized railroad cars & engines; switch keys; photographs; railroad history.
Research Fields: history of railroading in Iowa.
Facilities: 9,000 sq. ft. exhibit space. Railroad-related items for sale.

Activities: train rides; guided tours; participatory & temporary exhibitions. Annual Events: Pufferbilly Days; day out with Thomas event; Pumpkin Trains; Santa Express(TM) Trains.
Publications: quarterly newsletter, Keeping Track.
Hours & Admission Prices: Memorial Day to Oct. Mon.-Fri. 8:30-4:30, Sat.-Sun. 10-6; Nov.-May Mon.-Fri. 8:30-4:30. Historical Society & Museum: adults $20, children $7; Museum: adults $8, children $3. Closed New Year's Day; Thanksgiving; Christmas. &
Attendance: 50,799 (accurate)
Membership: Individual $25; Bronze $125; Silver $250; Gold $500; Platinum $1,000.

KATE SHELLEY RAILROAD MUSEUM, 1198 - 232nd St., Boone, IA 50036-7118. Mailing Address: 602 Story St., Boone, IA 50036-2832. Tel.: 515-432-1907.
E-mail: director@boonecountyhistory.org
Web Site: www.boonecountyhistory.org
Founded: 1976.
Congressional District: 4
Key Personnel: Dir., Pamela Schwartz; Chm., Lee McNair; Treas., Judy Russell; Sec., Janet Tait.
Personnel Profile: Full-Time Paid 1; Part-Time Paid 1.
Governing Authority: nonprofit. Parent Institution: Boone County Historical Society, 602 Story St., Boone 50036. Tax-exempt: 501(c)(3).
Institution Type/Description: Historic Site & Railroad Depot.
Collections: artifacts relating to railroad heroine Kate Shelley's life; recreated 19th century railroad passenger station; working telegraph system; railroad memorabilia.
Activities: guided tours; video presentation on the Kate Shelly story.
Publications: biannual journal, Trail Tales; biannual newsletter, History Page.
Hours & Admission Prices: Museum: by appointment. Grounds & Trails: daily dawn to dusk. No charge for general admission; bus tours: $1 per person.
Attendance: 2,000 (accurate)
Membership: Boone County Historical Society $40; Business & Organizational $100; Life $600.

MAMIE DOUD EISENHOWER BIRTHPLACE, 709 Carroll St., Boone, IA 50036. Mailing Address: 602 Story St., Boone, IA 50036-2832. Tel.: 515-432-1907.
E-mail: director@boonecountyhistory.org
Web Site: www.boonecountyhistory.org
Founded: 1970.
Congressional District: 5
Key Personnel: Exec. Dir., Pamela Schwartz; Pres. (V), Lee McNair.
Personnel Profile: Full-Time Paid 1; Part-Time Volunteers 50.
Governing Authority: nonprofit organization. Tax-exempt: 501(c)(3).
Institution Type/Description: Historic House: c.1890 Mamie Doud Eisenhower birthplace.
Collections: chronology of Mamie Doud Eisenhower in memorabilia; photographs; newspapers; campaign & inaugural items; gifts from Mrs. Eisenhower; Doud & Carlson family history; manuscripts; period furniture & furnishings; books; video & audio tapes; magazines; political cartoons; awards to Mamie Doud Eisenhower. Historic Automobiles: 1962 Plymouth Valiant once owned by Mamie Doud Eisenhower; 1949 Chrysler Windsor.
Research Fields: Doud, Carlson & Eisenhower family history; Boone County history.
Facilities: 2,000-vol. library of books, magazines, newspapers, photographs, papers & letters of Mamie Doud Eisenhower, Dwight D. Eisenhower & their families available for research on premises & by special arrangement; reading room. Books, publications & other museum-related articles for sale.
Activities: guided tours; lectures; docent program or council; loan, permanent & temporary exhibitions; slide programs.
Publications: brochures; books; newspapers; newsletter.
Hours & Admission Prices: Mon.-Sat. 10-5, Sun. 1:30-5. Adults $5; children 17 & under no charge.
Attendance: 6,000 (estimated)
Membership: Boone County Historical Society: Individual $40; Business & Organizational $100; Lifetime $600.

Britt

HANCOCK COUNTY AGRICULTURAL MUSEUM, 2210 Jewel Ave., Britt, IA 50423-8584. Mailing Address: 2090 James Ave., Britt, IA 50423-8549. Tel.: 641-843-4362.
Founded: 2000.
Congressional District: 4
Key Personnel: Pres. (V) & Museum Shop Mgr., Darrell C. Schaper.
Personnel Profile: Part-Time Volunteers 40.
Volunteer Hours: 460

Operating Expenses: 3,765
Operating Income: 5,152
Governing Authority: Tax-exempt.
Institution Type/Description: Agriculture Museum.
Collections: local history; period farm equipment & machinery from 1840s to 1960s; personal artifacts.
Activities: special events; middle school tours; oat threshing demonstration. Annual Event: Hancock County Fair Tractor Tour.
Hours & Admission Prices: By appointment. No charge; donations accepted. &
Attendance: 1,450 (estimated)
Membership: Lifetime $100.

HOBO MUSEUM, 51 Main Ave. S., Britt, IA 50423-1664. Mailing Address: P.O. Box 143, Britt, IA 50423. Tel.: 641-843-9104.
E-mail: lindah@hobo.com
Web Site: hobo.com
Key Personnel: Pres. Hobo Foundation & Mgr., Linda Hughes.
Personnel Profile: Part-Time Paid 2; Part-Time Volunteers 4.
Governing Authority: Tax-exempt.
Institution Type/Description: History Museum.
Collections: hobo history & way of life; photographs; video; railroad memorabilia; steamtrain collection; personal belongings of Maury Graham, P.N. Kid & Hardknock Kid.
Activities: group tours. Annual Event: National Hobo Convention in August.
Publications: quarterly newsletter.
Hours & Admission Prices: June-Aug. 15 Mon.-Fri. 10-5; other times by appointment. Admission $3. &
Attendance: 1,200 (estimated)
Membership: Friends $15.

Burlington

ART GUILD OF BURLINGTON, INC., 301 Jefferson St., Burlington, IA 52601-5333. Tel.: 319-754-8069. Fax: 319-754-4731.
E-mail: arts4living@aol.com
Web Site: www.artguildofburlington.org
Founded: 1966.
Congressional District: 3
Key Personnel: Exec. Dir., Ann Distelhorst; Education, Lillian Rubin.
Personnel Profile: Part-Time Paid 2; Interns 1.
Governing Authority: private; nonprofit organization. Tax-exempt: 501(c)(3).
Institution Type/Description: Art Gallery.
Collections: fine art & crafts by regional artists.
Facilities: library; classroom; 175-seat auditorium; 2,400 sq. ft. exhibit space; sculpture garden. Museum-related items for sale.
Activities: arts festivals; concerts; dance recitals; films; formal education program; guided tours; hobby workshops; lectures; temporary, loan & traveling exhibitions. Annual Events: member exhibit; Snake Alley Art Fair in June.
Publications: monthly newsletter; annual report.
Hours & Admission Prices: No charge; donations accepted. &
Attendance: 10,000 (estimated)
Membership: Student $15; Individual $30; Family $50; Sustaining $100; Patron $250.

HAWKEYE LOG CABIN, c/o Des Moines County Historical Society, 501 N. Fourth St., Burlington, IA 52601. Tel.: 319-752-7449.
E-mail: info@dmchs.org
Founded: 1971.
Congressional District: 1
Key Personnel: Chm. (V), Randy Bloomberg; Pres., Lyle Magneson; Treas., Terri Dowell; Exec. Sec., Debra Olson.
Personnel Profile: Part-Time Paid 3; Part-Time Volunteers 25.
Governing Authority: nonprofit. Parent Institution: Des Moines Co. Historical Society. Branch Museums: Phelps House; The Apple Trees Museum. Tax-exempt: 509(A)(54); 509(A)(87).
Institution Type/Description: History Museum; Historic House: 1909 log cabin.
Collections: pioneer tools & furnishings.
Facilities: Gift items for sale.
Activities: guided tours.
Publications: monthly newsletter; annual report; yearbook.
Hours & Admission Prices: May-Sept. Sat.-Sun. 1:30-4:30. Admission $2 for special arranged tours; call 319-753-5981. No charge; donations accepted. &
Attendance: 1,654 (accurate)
Membership: Individual $15 & up; Family $30 & up; Sponsor $50 & up;

Benefactor $100 & up; ShoQuoQuon Society $500 & up; Grimes-Salter Society $1,000 & up.

PHELPS HOUSE, 521 Columbia, Burlington, IA 52601-5117. Tel.: 319-753-5880.
E-mail: dmchs.jhunt@yahoo.com
Founded: 1974.
Congressional District: 1
Key Personnel: Pres. (V), Lyle Magneson; Vice Pres., Randy Bloom Berg; Treas., Terri Dowell; Dir., Angela Beeken; Publications & Public Rels. Dir., Debra Olson.
Personnel Profile: Part-Time Paid 3; Part-Time Volunteers 20.
Governing Authority: nonprofit. Parent Institution: Des Moines County Historical Society. Branch Museums: Log Cabin Museum; The Des Moines County Heritage Center, 501 N. 4th St., Burlington, IA. Tax-exempt: 504(A)(54); 504(A)(87).
Institution Type/Description: Historical Preservation of Historic House: 1851 Victorian Mansion.
Collections: period furniture & furnishings; medical museum artifacts.
Activities: group guided tours by appointment. Annual Event: Victorian Christmas Open House in December.
Publications: Membership/Program Yearbook; monthly newsletter; annual report; town meeting publicity.
Hours & Admission Prices: May-Oct. Sat.-Sun. 1:30-4:30; group tours by appointment. Adults $3; members and children 12 & under no charge.
Attendance: 1,300 (accurate)
Membership: Individual $15; Family & Grandparents $30; Sponsor $50; Benefactor $100; Sho Quo Quon Society $500; Grimes-Saller Society $1,000. Corporate: Contributing $50; Supporting $75; Sustaining $100.

Burr Oak

LAURA INGALLS WILDER PARK & MUSEUM, INC., 3603 236th Ave., Burr Oak, IA 52101-7889. Tel.: 563-735-5916. Fax: 563-735-5464.
E-mail: museum@lauraingallswilder.us
Web Site: www.lauraingallswilder.us
Founded: 1976.
Congressional District: 3
Key Personnel: Dir. & Museum Shop Mgr., Barb Olson; Pres. Bd., Dawn Austin; Vice Pres., Tammy Bjork; Treas., Tessa Flak.
Personnel Profile: Part-Time Paid 7; Part-Time Volunteers 6.
Governing Authority: nonprofit organization. Tax-exempt: 501(c)(3).
Institution Type/Description: Park & Museum.
Collections: restored hotel; furniture of the 1800 period; Laura Ingalls Wilder's personal artifacts; 2 historic buildings.
Facilities: picnic shelter.
Activities: guided tours; hands-on activities; educational programming.
Publications: book, Laura Ingalls Wilder: The Iowa Story.
Hours & Admission Prices: April-May & Sept.-Oct. Mon.-Sat. 10-4, Sun. 12-4; Memorial Day to Labor Day Mon.-Sat. 9-5, Sun. 12-4. Adult $8, children 6-17 $6; members no charge. &
Attendance: 6,400 (accurate)
Membership: Half Pint $20; Trailblazer $30; Family $50; Settler $100; Homesteader $200; Pioneer $350.

Carroll

SWAN LAKE FARMSTEAD MUSEUM, 22676 Swan Lake Dr., Carroll, IA 51401-9153. Tel.: 712-792-4614. Fax: 712-792-8078.
E-mail: info@carrollcountyconservation.com
Web Site: www.carrollcountyconservation.com
Institution Type/Description: Farm Museum.
Collections: agricultural machinery; farm animals; conservation; mounted animals.
Hours & Admission Prices: May-Oct. daily 5 am-10:30 pm. No charge.

Cascade

TRI-COUNTY HISTORICAL SOCIETY MUSEUM, 608 2nd Ave., S.W., Cascade, IA 52033. Tel.: 563-852-3371.
Institution Type/Description: Historical Society Museum.
Collections: local history & culture; period furnishings; personal artifacts; photographs; Baseball Hall of Famer, Urban "Red" Faber; railroad artifacts; box car replica; military artifacts & uniforms.
Hours & Admission Prices: May-Oct. 1st Sun. 1-4; other times by appointment.

Cedar Falls

CEDAR FALLS HISTORICAL SOCIETY, 308 W. 3rd St., Cedar Falls, IA 50613-2745. Tel.: 319-266-5149. Fax: 319-268-1812.
E-mail: cfhistory@cfu.net
Web Site: www.cfhistory.org
Founded: 1962.
Congressional District: 1
Key Personnel: Dir., Karen Smith.
Governing Authority: society. Branch Museums: Victorian Home & Carriage House Museum, 308 W. 3rd St., Cedar Falls, IA; Ice House & Little Red School House, First & Clay Sts., Cedar Falls, IA 50613; William J. Lenoir Model Train Collection, 308 W. 3rd St., Cedar Falls, IA; Behrens-Rapp Service Station Museum, 1st & Clay Sts., Cedar Falls, IA. Tax-exempt.
Institution Type/Description: Local History Museums.
Collections: clothes; taped interviews; local historic photographs; aviator John H. Livingston trophies & files; novelist Bess Streeter Aldrich books; ice industry; local archives; William Lenoir model trains; agriculture-related items; genealogy & cemetery records; horse drawn transportation artifacts.
Major Exhibits: Before William Sturgis, 2/14-12/14.
Research Fields: local history; natural ice industry; late 19th-century agriculture.
Facilities: reading room.
Activities: guided tours; films; gallery talks; permanent & temporary exhibitions; open houses; newspaper column.
Hours & Admission Prices: Victorian Home: Tues.-Sat. 10-4, Sun. 1-4. Ice House Museum: May-Oct. Sat. 10-4, Sun. 1-4. Adults $5; children 12 & under no charge. Little Red Schoolhouse: May: Sat.-Sun. 1-4; June-Aug.: Wed. & Sat.-Sun. 1-4; Sept.: Sat.-Sun. 1-4. &
Membership: Individual $35; Family $75.

HARTMAN RESERVE NATURE CENTER, 657 Reserve Dr., Cedar Falls, IA 50613-4723. Tel.: 319-277-2187. Fax: 319-277-4420.
E-mail: hartmanreserve@co.black-hawk.ia.us
Web Site: www.hartmanreserve.org
Founded: 1976.
Congressional District: 3
Key Personnel: Dir., Ed Gruenwald; Pres., Greg Greco; Program Coord., Chris Anderson; Dir. Devel., Anne Duncan.
Personnel Profile: Full-Time Paid 6; Part-Time Paid 3; Interns 4.
Governing Authority: county. Black Hawk County Conservation Board, 2410 W. Lone Tree Rd., Cedar Falls, IA 50613. Tax-exempt.
Institution Type/Description: Nature Center.
Collections: insects; mammal skins; animal parts; mounted birds & mammals; amphibians.
Research Fields: ecological monitoring of woodland.
Facilities: 150-vol. library on nature study, environmental education & nature-related periodicals; nature & conservation center; classroom; nature trails.
Activities: guided tours; lectures; organized education programs; school field trips; in-service training; teacher workshops; special programs for organizations & other groups; temporary exhibitions.
Publications: quarterly newsletter; annual informational brochures.
Hours & Admission Prices: March-May & Sept.-Oct. Mon.-Fri. 8-4:30, Sun. 1-5; June-Aug. & Nov.-Feb. Mon.-Fri. 8-4:30. No charge; donations accepted. Fees charged for specific programs. &
Attendance: 51,000 (estimated)
Membership: Student $10; Senior $20; Individual $25; Family $35; Contributor $50; Developer $100; Sustainer $250; Champion $500; Founder $1,000; Life $2,500.

ICE HOUSE MUSEUM, Franklin St., (at W. 1st St.), Cedar Falls, IA 50613. Mailing Address: Cedar Falls Historical Society, 308 W. 3rd St., Cedar Falls, IA 50613-2745. Tel.: 319-266-5149.
E-mail: cfhistory@cfu.net
Web Site: www.cfhistory.org
Key Personnel: Dir., Karen Smith.
Governing Authority: Parent Institution: Cedar Falls Historical Society. Tax-exempt.
Institution Type/Description: History Museum: listed on the National Register of Historic Places.
Collections: cutting, harvesting, storing, selling & uses of natural ice; tools & cutting implements; photographs; ice-boxes; ice wagon; horse drawn vehicles.
Hours & Admission Prices: May-Oct. Wed. & Sat. 10-4, Sun. 1-4. Adults $5; children 12 & under no charge. &

JAMES & MERYL HEARST CENTER FOR THE ARTS AND HEARST SCULPTURE GARDEN, 304 W. Seerley Blvd., Cedar Falls, IA 50613-4050. Tel.: 319-273-8641. Fax: 319-273-8659.
E-mail: mary.huber@cedarfalls.com
Web Site: www.hearstartscenter.com
Founded: 1989.
Congressional District: 2
Key Personnel: Pres. Bd., Joni Krejchi; Dir., Mary Huber; Cur., Emily Drennan; Svcs. Coord., Gail LeFlore; Devel. Coord., Vicki Simpson; Museum Shop Mgr., Abby Haigh.
Personnel Profile: Full-Time Paid 2; Part-Time Paid 16; Part-Time Volunteers 74; Interns 3.
Governing Authority: municipal; nonprofit. Parent Institution: City of Cedar Falls. Subsidiary Institution: Friends of the Hearst. Tax-exempt.
Institution Type/Description: Art Museum, Sculpture Garden & Community Public Art Collections.
Collections: two & three dimensional works by regional artists; book illustrations by Gary Kelley; special collections; public art; sculpture garden; children's book illustration.
Research Fields: regional artists.
Facilities: auditorium; three classrooms; meeting rooms; sculpture garden.
Activities: guided tours; lectures; concerts; dance recitals; arts festival; study clubs; organized education programs; loan exhibitions; internship; temporary & traveling exhibitions. Museum Sponsors: activities at Sturgis Falls Festival; College Hill Art Festival; Cedar Trails Festival & Main Street Art Fair and Festival.
Publications: exhibition catalogs; quarterly brochure; volume of poetry, Selected Poems by James Hearst; volume of photographs, Platinum Scenes: Photographs by Irving Herman; The Complete Poetry of James Hearst.
Hours & Admission Prices: Tues. & Thurs. 8-9, Wed. & Fri. 8-5, Sat.-Sun. 1-4. No charge; donations accepted. Closed New Year's Day; Easter weekend; Memorial Day; Independence Day; Labor Day weekend; Thanksgiving weekend; Christmas.
Attendance: 56,200 (estimated)
Membership: Student & Senior $20; Individual $25; Family $35; Supporter $50; Sponsor $100; Benefactor $500; Collector $1,000.

LITTLE RED SCHOOL HOUSE MUSEUM, 1 W. 1st St., Cedar Falls, IA 50613. Mailing Address: Cedar Falls Historical Society, 308 W. 3rd St., Cedar Falls, IA 50613. Tel.: 319-266-5149.
E-mail: cfhistory@cfu.net
Web Site: www.cfhistory.org
Formerly: "R" Little Red School House Museum
Governing Authority: Parent Institution: Cedar Falls Historical Society. Tax-exempt.
Institution Type/Description: Historic Building Museum: built in 1909.
Collections: local history; period furnishings; books; bell tower & bell.
Activities: educational programs; research.
Hours & Admission Prices: May: Sat.-Sun. 1-4; June-Aug.:Wed., Sat. & Sun. 1-4; Sept.: Sat.-Sun. 1-4.

UNI GALLERY OF ART, UNIVERSITY OF NORTHERN IOWA, 1601 W. 27th St., 104 Kamerick Art Bldg., Cedar Falls, IA 50614-0362. Tel.: 319-273-3095. Fax: 319-273-7333.
E-mail: galleryofart@uni.edu
Web Site: www.uni.edu/artdept/gallery/
Founded: 1978.
Congressional District: 3
Key Personnel: Head Dept. Art, Jeffery Byrd; Dir. Gallery, Darrell Taylor.
Personnel Profile: Full-Time Paid 1; Part-Time Paid 11; Part-Time Volunteers 5.
Governing Authority: university. Parent Institution: University of Northern Iowa, Cedar Falls, IA 50614. Tax-exempt: 170(b)(1)(A).
Institution Type/Description: Art Gallery.
Collections: 20th century American & European; contemporary artists.
Major Exhibits: Avant-garde in Art and Music, 1/13/14-5/10/14; Organic Abstraction, 9/29/14-10/25/14; Vertigo A-GoGo, 12/4/14.
Research Fields: 20th century European & American Art.
Activities: guided tours; lectures; films; gallery talks; formally organized education programs for children & undergraduate college students; permanent, temporary & traveling exhibitions.
Publications: Contemporary Chicago Painters; De Kooning 1969-78; Standards by Allan Kaprow; Reuben Nakian; Leda & The Swan; Art for Public Spaces; University of Northern Iowa Dept. Art Faculty Exhibition-1982; 75th Anniversary Alumni Invitational-1982; 1985 Art Faculty Exhibition; The Contemporary American Potter: Recent Vessels; Jose de Creeft 1884-1982, works from the Collection of Nina de Creeft Ward & William de Creeft; David Delafield: A Retrospective of An Iowa Artist; Philip Pearlstein Painting to Watercolors; Art Nouveau Glass & Pottery from the

Syracuse University Art Collections; Juane Quick-to-see Smith & George Longfish: Personal Symbols; Walter Dusenbery: Classical Echoes; Born In Iowa: 29 Artists; John Page: A Retrospective Exhibition in Three Parts; Magic Silver Catalog; exhibition catalogs, A Question of Faith (juror: Eleanor Heartney), Residue of Silence (essay: Matthew Baigell), Figured Ceramics; Rie Hachiyanagi: A Retrospective; David Delafield Retrospective; Mary Snyder Behrens: New Work; Transformations in the Nervepool: The Rituals & Zoacodes of Ebon Fisher; Creating Our World: Russian/American Children's Art Exhibition; Justice Illuminated: The Art of Arthur Szyk; Quiet Village: Recent Works by Michael Krueger and Jenny Schmid; George Longfish - A Retrospective; Dean and Gunnar Schwarz: Pottery Form and Inherent Expression; Marguerite Wilderhain; World Views: Photographs by Tina Barney, Linda Conner, Andrew Moore, and JoAnn Verburg; New Polyphonies: Contemporary Art from Portugal; Highlights from the Collection; Body Prop - New Works by Nick Dong, Erica Duffy, Lauren Kalman, and Deb Todd Wheeler; Frje Echeverria: Four Decades of Working Beside Students; Love Me or Die: Cat Chow.
Hours & Admission Prices: Mon.-Thurs. 10-7, Fri.-Sat. 12-5 & by appointment. No charge. Closed holidays; when classes are not in session.
Attendance: 12,000 (accurate)

UNIVERSITY OF NORTHERN IOWA MUSEUMS & COLLECTIONS - MARSHALL CENTER ONE-ROOM SCHOOL, 3219 Hudson Rd., Cedar Falls, IA 50614-0199. Tel.: 319-273-2188. Fax: 319-273-6924.
E-mail: doris.mitchell@uni.edu
Web Site: www.uni.edu/museum/mcs/index.html
Key Personnel: Dir., Dr. Sue Grosboll
Institution Type/Description: Historic Building Museum.
Collections: period furnishings & artifacts.
Hours & Admission Prices: By appointment. No charge. Closed holidays.

UNIVERSITY OF NORTHERN IOWA MUSEUMS & COLLECTIONS - UNIVERSITY MUSEUM, (M), 3219 Hudson Rd., Cedar Falls, IA 50614-0199. Tel.: 319-273-2188. Fax: 319-273-6924.
E-mail: doris.mitchell@uni.edu
Web Site: www.uni.edu/museum
Founded: 1892.
Congressional District: 1
Key Personnel: Dir., Katherine Martin; Sec., Doris Mitchell.
Personnel Profile: Full-Time Paid 1; Part-Time Volunteers 10; Interns 3.
Governing Authority: state. Parent Institution: University of Northern Iowa. Tax-exempt: 170(c)(1).
Institution Type/Description: Natural & Human History Museum.
Collections: minerals; rocks; fossils; mammals; birds; reptiles; fish; marine invertebrates; ethnological pieces from Africa, Asia & the Americas; university & rural education history. Historic Building: Historic School House.
Facilities: classroom; meeting room.
Activities: guided tours; lectures; club meetings; internships; research; traveling trunks.
Hours & Admission Prices: University Museum: Mon.-Fri. 9-4:30. Marshall Center School: by appointment. No charge; donations accepted. Closed holidays.
Attendance: 15,500 (accurate)
Membership: Student $20; Explorer $35; Naturalist $60; Researcher $125; Conservator $250; Curator $500; Director $1,000.

Cedar Rapids

AFRICAN AMERICAN HISTORICAL MUSEUM & CULTURAL CENTER OF IOWA, 55 12th Ave., S.E., Cedar Rapids, IA 52401-2202. Mailing Address: P.O. Box 1626, Cedar Rapids, IA 52406-1626. Tel.: 319-862-2101. Fax: 319-862-2105.
E-mail: information@blackiowa.org
Web Site: blackiowa.org
Founded: 1994.
Key Personnel: Exec. Dir., Thomas Moore.
Personnel Profile: Full-Time Paid 10; Part-Time Paid 3; Interns 1.
Governing Authority: private; nonprofit organization. Tax-exempt: 501(c)(3).
Institution Type/Description: History Museum.
Collections: African American history of Iowa.
Facilities: library; educational facilities; 4,200 sq. ft. exhibit space; celebration hall. Museum-related items for sale.
Activities: docent program; formal education programs; guided tours; lectures; loan, participatory, temporary & traveling exhibitions. Annual Events: Juneteenth; golf outings; banquet; Kwanza.
Publications: quarterly newsletter, Griot.

Hours & Admission Prices: Mon.-Sat. 10-4. Adults $4, children $2.50; discounts to groups; members no charge. Closed New Year's Day; Martin Luther King Jr. Day; Memorial Day; Independence Day; Labor Day; Thanksgiving; Christmas. &

Attendance: 10,772 (accurate)

Membership: Youth $5; Individual $25; Family $50; Golden $100; Century $200; Corporate $250 & up.

BRUCEMORE, 2160 Linden Dr., S.E., Cedar Rapids, IA 52403-1748. Tel.: 319-362-7375. Fax: 319-362-9481.

E-mail: mail@brucemore.org

Web Site: www.brucemore.org

Founded: 1981.

Congressional District: 1

Key Personnel: Exec. Dir., David Janssen; Pres., Brenda Duello; Bldg. & Grounds Supt., Roger Johnson; Asst. Dir., Maura Pilcher; Head Gardener, David Morton; Museum Shop Mgr., Kaycie Schatz; Asst. Gardener, David Morton; Mgr. Interpretations & Collections, Jessica Peel-Austin; Accountant, Kelly Costello.

Personnel Profile: Full-Time Paid 10; Part-Time Paid 7; Part-Time Volunteers 325.

Governing Authority: nonprofit organization. Parent Institution: National Trust for Historic Preservation, 1785 Massachusetts Ave., N.W., Washington, DC 20036. Tax-exempt: 501(c)(3).

Institution Type/Description: Historic Site & Community Cultural Center.

Collections: furnishings, archives, archaeology. Historic Structure: c.1886 mansion.

Research Fields: family history; Midwest social history.

Facilities: 3,000-vol. library; archives; formal gardens; orchard; 26 acre estate.

Activities: guided tours; lectures; concerts; docent program or council; outdoor theater. Museum Sponsors: Garden & Landscape Show.

Publications: quarterly newsletter.

Hours & Admission Prices: Tours: March-Dec. Tues.-Sat. 10-3, Sun. 12-3; other times by appointment. Adults $7, students $3; members, Brucemore & National Trust members no charge. &

Attendance: 45,400 (estimated)

Membership: Individual $35; Grandparent & Household $50; Patron $100; Donor $250; Benefactor $500; Trustees Club $1,000; The George and Irene Douglas Circle $2,500; The Howard and Margaret Hall Heritage Club $5,000.

THE CARL & MARY KOEHLER HISTORY CENTER, (M), 615 First Ave., S.E., Cedar Rapids, IA 52401-1315. Tel.: 319-362-1501. Fax: 319-362-6790.

E-mail: history@historycenter.org

Web Site: www.historycenter.org

Formerly: Linn County Historical Society

Founded: 1969.

Congressional District: 1

Key Personnel: Exec. Dir., Caitlin Treece; Pres. (V), Dr. Adam Ebert.

Personnel Profile: Full-Time Paid 3; Part-Time Paid 5; Part-Time Volunteers 65; Interns 3.

Governing Authority: private; nonprofit organization. Parent Institution: Linn County Historical Society. Tax-exempt: 501(c)(3).

Institution Type/Description: History Museum.

Collections: Linn County, Iowa history from prehistoric-present; history of Cedar Rapids.

Research Fields: business, social & cultural history of Linn County, prehistoric-1859; history of Cedar Rapids during the 1865 period & at the turn-of-the-century, 1960.

Facilities: library; archives; research facilities; 5,000 sq. ft. exhibit space.

Activities: guided tours; adult & children's programs.

Publications: quarterly newsletter, Linn County Time Lines.

Hours & Admission Prices: Tues.-Sat. 10-4. No charge; donations requested. Closed New Year's Eve & Day; Memorial Day; Labor Day; Thanksgiving & day after; Christmas Eve & Day. &

Attendance: 8,000 (accurate)

Membership: Individual $40; Family $50; Contributor $100-$499; Benefactor $500 & up.

✳ **CEDAR RAPIDS MUSEUM OF ART, (M),** 410 Third Ave, S.E., Cedar Rapids, IA 52401-1620. Tel.: 319-366-7503. Fax: 319-366-4111.

E-mail: info@crma.org

Web Site: www.crma.org

Founded: 1905.

Congressional District: 2

Key Personnel: Devel., Joanne Wzontek; C.F.O. & Business Mgr., Deanna Clemens Pedersen; Preparator, Judy Frauenholtz; Cur., Sean Ulmer; Coord.

Communications, Cindy Motsinger; Dir. Education, Erin Thomas; Special Events Coord., Beth Roof; Bldg. Supvr., Carlis Faurot; Museum Shop Mgr., Casey Dunagan.

Personnel Profile: Full-Time Paid 8; Part-Time Paid 11; Part-Time Volunteers 224; Interns 5.

Operating Expenses: 1,228,035

Operating Income: 1,527,154

Governing Authority: nonprofit corporation. Tax-exempt: 501(c)(3).

Institution Type/Description: Art Museum.

Collections: 20th century modern and regionalist paintings; works by Grant Wood, Marvin Cone, James Swann, Malvina Hoffman, Bertha Jaques, Mauricio Lasansky; 19th & 20th century sculpture, paintings, prints, photographs; ancient Roman portrait busts.

Major Exhibits: Conger Metcalf, 1/18/14-5/11/14; Papier Francais: French Works on Paper, 2/15/14-5/25/14; Marvin Cone on My Mind: The Ceramics of Dean Schwarz, 3/15/14-11/2/14; Carl Van Vechten: Photographer to the Stars, 5/24/14-9/7/14; Grant Wood: American Impressionist, 6/14/14-9/21/14.

Research Fields: local regionalist artists.

Facilities: classrooms; auditorium.

Activities: guided tours; lectures; films; gallery talks; concerts; Grant Wood Studio tours; workshops; temporary & traveling exhibitions.

Publications: newsletters; catalogs; selected exhibitions.

Hours & Admission Prices: Tues.-Wed., Fri. & Sun. 12-4, Thurs. 12-8, Sat. 10-4. Adults $5, students & senior citizens $4; members; children 18 & under, members & North American Reciprocal Museum Program members no charge. Closed national holidays. &

Attendance: 31,344 (estimated)

Membership: Student, Educator & Senior Citizen $30; Individual & Senior Couple $40; Family $60; Patron $125; Benefactor $250; Conewood Society $500; Turner Society $1,000 & up.

COE COLLEGE ART GALLERIES, 1220 1st Ave., N.E., Cedar Rapids, IA 52402-5092. Tel.: 319-399-8647. Fax: 319-399-8557.

Key Personnel: Dir., Mariah Dekkenger

Institution Type/Description: Art Gallery.

Collections: paintings; photographs; sculpture.

Hours & Admission Prices: Daily 3-5. No charge.

IOWA MASONIC LIBRARY AND MUSEUM, 813 1st Ave., S.E., Cedar Rapids, IA 52402-5001. Mailing Address: P.O. Box 279, Cedar Rapids, IA 52406-0279. Tel.: 319-365-1438. Fax: 319-365-1439.

E-mail: librarian@gl-iowa.org

Web Site: www.grandlodgeofiowa.com

Founded: 1844.

Congressional District: 1

Key Personnel: Grand Sec. & Librarian, William R. Crawford; Asst. Librarian, William R. Kreuger.

Personnel Profile: Full-Time Paid 1; Part-Time Volunteers 10.

Governing Authority: nonprofit organization. Affiliated with Grand Lodge of Iowa, A.F. & A.M., The Free Masons' Lodges in Iowa. Tax-exempt.

Institution Type/Description: Library & History Museum.

Collections: Masonic collection; Swab collection; manuscripts.

Research Fields: Free masonry; Iowa history; religion.

Facilities: 100,000-vol. library of Masonic history, Iowa history, biography, poetry & literature, Burnisiana & Lincolniana; microfilms of Cedar Rapids newspapers available for inter-library loan; reading room.

Activities: guided tours; lectures; films; permanent exhibitions.

Publications: Grand Lodge Bulletin.

Hours & Admission Prices: Mon.-Fri. 8-12 & 1-5. No charge; donations accepted. Closed national holidays.

Attendance: 1,300 (accurate)

✳ **NATIONAL CZECH & SLOVAK MUSEUM & LIBRARY, (M),** 1400 Inspiration Place S.W., Cedar Rapids, IA 52404. Tel.: 319-362-8500. Fax: 319-363-2209.

Web Site: www.ncsml.org

Founded: 1974.

Congressional District: 2

Key Personnel: C.E.O. & Pres., Gail Naughton; Chm. (V), Lu Barton; Cur., Stefanie Kohn; Librarian, David Muhlena; Dir. Education, Programs & Visitor Svcs., Janet L. Stoffer; Vice Pres. Devel., Jason Wright; Museum Store Mgr., catherine Otto.

Personnel Profile: Full-Time Paid 20; Part-Time Paid 4; Part-Time Volunteers 200.

Governing Authority: nonprofit organization. Tax-exempt: 501(c)(3).

Institution Type/Description: Heritage Center; Ethnic Museum; Czech and Slovak history and culture. Historic Building: 1880-1900, restored Czech immigrant home.
Collections: folk costumes from the Czech Republic & Slovakia; glass; ceramics; porcelain; ethnic dolls; handwork; wood-carved items; painting & prints; maps & graphic materials; folk, decorative & fine arts; farm tools & implements.
Major Exhibits: Beauty of the Czechoslovak Republic, 7/13-4/15; Spiritual Dimension in Czech Printmaking (T), 11/13-2/14; Written in the Czech Landscape (T), 12/13-3/14; Traditional Customs of the Czech Republic, 3/14-1/15; PHOS (T), 4/14-9/14; Prague Through the Lens of the Secret Police, 10/14-1/15.
Research Fields: Czech and Slovak history and culture; 20th century Cold War emigre oral history project.
Facilities: 20,000-vol. library: Czech & Slovak history & culture. Ethnic glass, ceramics, books & instructional booklets, music & language tapes for sale.
Activities: guided tours; out-reach programs; arts & crafts demonstration; special exhibits; educational programs.
Publications: newsletter, MOST; magazine, Slovo.
Hours & Admission Prices: Mon.-Sat. 9:30-4, Sun. 12-4. Adults $10, seniors $9, active military $5, students 14 & up $5, youth 6-13 $3, children 5 & under & members no charge. Closed New Year's Day; Thanksgiving; Christmas. &
Attendance: 34,000 (accurate)
Membership: Individual $35; Family $45; Contributing $100; Sustaining $250.

SCIENCE STATION, Lindale Mall, 444 1st Ave., N.E., Ste. 200, Cedar Rapids, IA 52402-3250. Tel.: 319-363-4629. Fax: 319-366-4590.
E-mail: john@sciencestation.org
Web Site: www.sciencestation.org
Founded: 1986.
Congressional District: 1
Key Personnel: Exec. Dir., John Swanson; Pres. (V), Todd Bergen; Education Coord., Stacey Brooks; Business Mgr., Terri Breheny.
Personnel Profile: Full-Time Paid 9; Part-Time Paid 20; Part-Time Volunteers 50.
Governing Authority: nonprofit organization. Tax-exempt: 501(c)(3).
Institution Type/Description: Science Museum: housed in c.1917 Fire Station building.
Collections: hands-on science exhibits emphasizing light & perception, sound, motion & energy.
Facilities: classrooms; IMAX theater. Science kits & laboratory supplies for sale.
Activities: films; organized education programs; participatory & traveling exhibitions.
Hours & Admission Prices: Temporarily located at Lindale Mall, 4444 First Ave., NE, Cedar Rapids, IA. June-Aug. 30. Mon.-Sat. 10-5, Sun. 12-5; Aug. 31-May Tues.-Sat. 10-5, Sun. 12-5. Family $10, adults & preschoolers $3; members no charge. Closed major holidays. &
Attendance: 90,000 (accurate)
Membership: Dual $60; Family $80.

Centerville

APPANOOSE COUNTY HISTORICAL & COAL MINING MU-SEUM, 100 W. Maple St., Centerville, IA 52544-2211. Tel.: 641-856-8040. Facebook: Appanoose County Historical & Coal Mining Museum.
E-mail: appanoosehistory@yahoo.com
Web Site: www.appunoosehistory.com
Institution Type/Description: History Museum: housed in the former Centerville Post Office; built 1903.
Collections: local history; military & theatrical artifacts; post office equipment; cameras; household artifacts; farm tools & equipment; coal mining; local transportation; photographs.
Hours & Admission Prices: April & Sept.-Oct. Mon.-Fri. 1-5; Memorial Day to Labor Day Mon.-Fri. 1-5, Sat. 10-2; Nov.-March Wed.-Fri. 1-4. Adults $4, students $1; members no charge. Closed New Year's Eve & Day; Memorial Day; Independence Day; Labor Day; Thanksgiving.
Membership: Single $15; Couple $27.

Chariton

LUCAS COUNTY HISTORICAL SOCIETY MUSEUM, 123 N. 17th St., Chariton, IA 50049-1618. Mailing Address: P.O. Box 807, Chariton, IA 50049. Tel.: 641-774-4464. Facebook: Lucas County Museum.
E-mail: lchs@iowatelecom.net

Web Site: www.lucascountyhistoricalsociety.blogspot.com/
Founded: 1965.
Congressional District: 27
Key Personnel: Pres. (V), Frank Myers
Governing Authority: Tax-exempt.
Institution Type/Description: History Museum.
Collections: local history & culture; period furnishings; personal artifacts.
Publications: newsletter.
Hours & Admission Prices: Memorial Day to Oct. 1 Tues.-Sat. 1-4; other times by appointment. No charge.
Attendance: 1,000 (estimated)
Membership: Individual $5.

Charles City

CARRIE CHAPMAN CATT GIRLHOOD HOME AND MU-SEUM, 2379 Timber Ave., Charles City, IA 50616-8979. Mailing Address: P.O. Box 33, Charles City, IA 50616. Tel.: 641-228-3336.
E-mail: visit@catt.org
Web Site: catt.org
Founded: 1991.
Key Personnel: Pres. (V), Susan McDonnell.
Personnel Profile: Part-Time Volunteers 50.
Governing Authority: Tax-exempt.
Institution Type/Description: Historic House: housed in the childhood home of Carrie Chapman Catt, coordinator of the woman suffrage movement & political strategist. Listed on the National Register of Historic Places.
Collections: Carrie Chapman Catt's life & career; photographs; period furnishings; personal artifacts.
Hours & Admission Prices: Memorial Day to Labor Day Mon.-Sat. 10-4, Sun. 12-4; other times by appointment. No charge; donations accepted.
Attendance: 700 (estimated)

CHARLES CITY ARTS CENTER, 301 N. Jackson St., Charles City, IA 50616-2006. Tel.: 641-228-6284.
Institution Type/Description: Art Center.
Collections: works by regional artists.
Activities: classes; special events.
Hours & Admission Prices: Wed.-Thurs. 1-9, Fri.-Sat. 1-5. No charge.

FLOYD COUNTY HISTORICAL MUSEUM, 500 Gilbert St., Charles City, IA 50616-2738. Tel.: 641-228-1099. Fax: 641-228-1157.
E-mail: fchs@fiai.net
Web Site: www.floydcountymuseum.org
Founded: 1961.
Congressional District: 2
Key Personnel: Co-Pres. (V), Jody Flint; Co-Pres. (V), Tony Lessin; Dir., Mary Ann Townsend; Office Asst. & Receptionist, Elaine Mead.
Personnel Profile: Full-Time Paid 1; Part-Time Paid 2; Part-Time Volunteers 20.
Governing Authority: society; board of trustees. Parent Institution: Floyd County Historical Society. Tax-exempt.
Institution Type/Description: History Museum.
Collections: archaeology; medical; pharmacology; military; 1850-1900 period rooms; model railroad club; quilts; archives; Hart-Parr, Oliver & White tractors; horse drawn conveyances; 1910 Cretors Popcorn Wagon; Salsbury's Veterinary Laboratories; 1856 Mutchlar log cabin; 1873 Legel's Drugstore.
Research Fields: Hart-Parr & Oliver Tractor company archives; agricultural equipment; Carrie Lane Chapman Catt; Floyd County history.
Facilities: library of 1858-1937 newspaper files, documents & history books available for use on premises. Museum-related items for sale.
Activities: guided & unguided tours; films; permanent exhibitions; short term exhibits.
Publications: quarterly newsletter, Floyd County Heritage; Hart-Parr, Oliver & White Farm Equipment manuals.
Hours & Admission Prices: June-Aug. Mon.-Fri. 9-4:30, Sat.-Sun. 1-4; Sept.-May Mon.-Fri. 9-4:30. Adults 13 & over $5, children 6-12 $3; members & children under 10 no charge. Closed New Year's Day; Thanksgiving; Christmas. &
Attendance: 5,500 (accurate)
Membership: Single $10; Family $20.

MOONEY ART COLLECTION - CHARLES CITY LIBRARY, 106 Milwaukee Mall, Charles City, IA 50616. Tel.: 641-257-6319. Fax: 641-257-6325.
Institution Type/Description: Library & Art Gallery.
Collections: engravings, etchings, & prints by artists including Rembrandt, Gauguin, & Picasso.

Hours & Admission Prices: May-Sept. Mon.-Thurs. 10-8, Fri. 10-5, Sat. 1-5; Labor Day to Memorial Day Mon.-Thurs. 10-8, Fri. 10-5, Sat.-Sun. 1-5. No charge. &

Cherokee

*** SANFORD MUSEUM AND PLANETARIUM, (M),** 117 E. Willow St., Cherokee, IA 51012-1854. Tel.: 712-225-3922.
E-mail: sanfordmuseum@iowatelecom.net
Web Site: sanfordmuseum.org
Founded: 1951.
Congressional District: 4
Key Personnel: Dir., Linda Burkhart; Asst. Dir., Michele Deiber Kumm; Educator & Museum Shop Manager, Kerisa Pingel; Cur. Archaeology, Jason Titcomb.
Personnel Profile: Full-Time Paid 4; Part-Time Paid 4.
Governing Authority: Tiel Sanford Memorial Fund. Tax-exempt: 501(c)(3).
Institution Type/Description: General Museum.
Collections: archaeology; history; geology; paleontology; zoology; ethnology; archives.
Research Fields: archaeology; history; paleontology.
Facilities: 5,000-vol. library of research materials in archaeology, geology & general works available for inter-library loan & on request; reading room. Reproductions, books, rocks, fossils, shells & collecting equipment for sale.
Activities: guided tours; lectures; films; gallery talks; drama; formally organized education programs; inter-museum loan, permanent, temporary & traveling exhibitions; planetarium demonstrations; monthly astronomy night programs (observation of night sky, weather permitting); monthly art displays; annual archaeology field school.
Publications: newsletter, Northwest Chapter Iowa Archaeological Society.
Hours & Admission Prices: Museum: Mon.-Fri. 9-5, Sat.-Sun. 12-5. No charge. Planetarium Programs: last Sun. of month 2 p.m.; other times by appointment. Closed major holidays. &
Attendance: 20,000 (estimated)
Membership: Sanford Museum Association Membership: Contributing $10; Sustaining $25; Associate $50; Patron $100.

Clarinda

GLENN MILLER BIRTHPLACE HOME, 601 S. Glenn Miller Ave., Clarinda, IA 51632-2657. Mailing Address: P.O. Box 61, Clarinda, IA 51632. Tel.: 712-542-2461. Fax: 712-542-2461.
E-mail: gmbs@heartland.net
Web Site: www.glennmiller.org
Institution Type/Description: Historic House Museum: housed in the birthplace of band leader, Glenn Miller.
Collections: Glenn Miller's life & career; personal artifacts; period furnishings; photographs.
Activities: special events.
Hours & Admission Prices: May-Oct. Tues.-Sun. 1-5.

GLENN MILLER BIRTHPLACE MUSEUM, 122 W. Clark St., Clarinda, IA 51632. Mailing Address: P.O. Box 61, Clarinda, IA 51632. Tel.: 712-542-2461.
E-mail: gmbs@heartland.net
Web Site: www.glennmiller.org
Governing Authority: Parent Institution: The Glenn Miller Birthplace Society.
Institution Type/Description: History Museum.
Collections: Glenn Miller's life & career; personal artifacts; photographs; period furnishings.
Activities: Annual Event: Glenn Miller Festival in June.
Hours & Admission Prices: Tues.-Sun. 1-5; other times by appointment. Adults $6, seniors 62 & over $5; children 12 & under no charge. Closed holidays.

NODAWAY VALLEY HISTORICAL MUSEUM, 1600 S. 16th St., Clarinda, IA 51632. Mailing Address: P.O. Box 393, Clarinda, IA 51632-0393. Tel.: 712-542-3073.
E-mail: nvm@iowatelecom.net
Web Site: nodawayvalleymuseum.org
Institution Type/Description: History Museum.
Collections: local history & culture; period furnishings; personal artifacts; photographs; local business & industry; farming; schools; transportation; agriculture; 3H & 4 H development.
Activities: Annual Event: Anniversary Heritage Day Festival in June.
Hours & Admission Prices: Tues. 9-4, Wed.-Sun. 1-4; groups by appointment.

Clarion

4-H SCHOOLHOUSE MUSEUM, 1st Ave. & Central Ave. W., Clarion, IA 50525. Mailing Address: 302 S. Main St., Clarion, IA 50525. Tel.: 515-532-3453. Fax: 515-532-2511.
E-mail: clchamb@goldfieldaccess.net
Web Site: www.clarion-iowa.com
Founded: 1955.
Congressional District: 6
Key Personnel: Chm. (V), Yvonne Stevens.
Governing Authority: municipal; nonprofit organization. Parent Institution: City of Clarion. Subsidiary Institution: 4-H Museum.
Institution Type/Description: History Museum: birthplace of the 4-H emblem.
Collections: 4-H memorabilia; history of 4-H movement; early 1900s school artifacts; clothing.
Facilities: 225-vol. library.
Activities: guided tours.
Hours & Admission Prices: June-Aug. Tues. & Thurs. 1-4, Sat. 9-12; other times by appointment. No charge; donations accepted.
Attendance: 1,200 (accurate)

HEARTLAND MUSEUM, Hwy. 3 W. & 9th St., S.W., Clarion, IA 50525. Mailing Address: P.O. Box 652, Clarion, IA 50525-0652. Tel.: 515-602-6000. Fax: 515-532-2396.
E-mail: hmfclarionia@hotmail.com
Web Site: www.heartlandmuseum.org
Founded: 1999.
Congressional District: 5
Key Personnel: Pres. (V), George Boyington; Museum Shop Mgr., Normajene Callier.
Personnel Profile: Full-Time Volunteers 5; Part-Time Volunteers 8.
Volunteer Hours: 2,000
Governing Authority: bd. of directors. Tax-exempt.
Institution Type/Description: History Museum.
Collections: agricultural machinery including farm tractors & construction machinery; horse drawn wagons, carriages, & sleighs; toys; local history & cultures; hats; international teddy bears; Victorian Era streetscapes; 1930s doctor's office.
Research Fields: genealogy.
Facilities: Museum-related items for sale.
Activities: fundraising activities. Museum Sponsors: Trial Reenactments in July; Festival of Trees in December.
Publications: brochure, Heartland News; newsletters, Teddy Bear; Hat Parlor.
Hours & Admission Prices: May-Sept. Sat. 1:30-4; other times by appointment. Adults $6, children 11 & under $3; discounts to groups of 25 or more. &
Attendance: 4,100 (estimated)
Membership: Child $10; Senior $16; Adults $18; Family $35; Friend $100.

Clear Lake

CLEAR LAKE ARTS CENTER, 17 S. 4th St., Clear Lake, IA 50428-1816. Tel.: 641-357-1998.
E-mail: clac@netins.net
Institution Type/Description: Art Gallery.
Collections: 2-D & 3-D art.
Activities: temporary exhibits.
Hours & Admission Prices: Tues.-Sat. 10-5. No charge.

CLEAR LAKE FIRE MUSEUM, 112 N. 6th St., Clear Lake, IA 50428. Mailing Address: c/o Clear Lake Fire Dept., 711 2nd Ave. N., Claer Lake, IA 50428. Tel.: 641-357-2613.
Institution Type/Description: Firefighting History Museum.
Collections: firefighting history & equipment; 1924 Ahrens-Fox pumper truck; 1883 hand-pulled hose cart; early fire extinguishers; newspaper clippings; photographs; fire bell; firefighters memorial.
Hours & Admission Prices: Memorial Day to Labor Day Sat.-Sun. 1-4. No charge; donations accepted.

PIONEER MUSEUM & HISTORICAL SOCIETY OF NORTH IOWA, 9184 G 265th St., Clear Lake, IA 50428-8507. Mailing Address: P.O. Box 421, Mason City, IA 50402-0421. Tel.: 641-423-1258.
Founded: 1964.
Congressional District: 3
Key Personnel: Pres., Richard Peterson; Dir., Kay Ingersoll; Sec., Suzanne Kisner; Treas., Paul Pirkl.
Personnel Profile: Part-Time Paid 2; Part-Time Volunteers 25.
Governing Authority: society; nonprofit organization. Tax-exempt.

Institution Type/Description: Historical Society Museum.
Collections: local historical items; Indian artifacts; furniture & furnishings; fossil collection; handmade implements; old dolls & toys; hand-drawn fire engine; milk wagon; harp. 1911 Colby Car made in Mason City. 1906 Ford Model N; 1923 Model T Ford; 1929 Model A Ford Sedan; Barbie doll collection; Regina Orchestra Corona. Historic Buildings: 1900 schoolhouse; 1856 log cabin.
Facilities: Museum-related items for sale.
Activities: guided tours; permanent & temporary exhibitions.
Publications: Memories of Old Cerro Gordo: First Person and Contemporary Tales; Gone But Not Forgotten Wheelwood, IA; The Dougherty Fighting Irish; Rocky Hill - Undated History of Cerro Gordo County
Hours & Admission Prices: May-Sept. Tues.-Sun. 1-5. Adults $3, children $1; members no charge. Season Pass: family $20, adult $10. &
Attendance: 3,000 (estimated)
Membership: Individual $10; Family $20.

Clermont

MONTAUK, 26223 Harding Rd., Clermont, IA 52135-8600. Mailing Address: P.O. Box 372, Clermont, IA 52135-0372. Tel.: 563-423-7173. Fax: 563-423-7378.
E-mail: montauk@acegroup.cc
Web Site: www.iowahistory.org/historic-sites/montauk/index.html
Founded: 1968.
Congressional District: 2
Key Personnel: Acting Mgr., Wade Schott.
Personnel Profile: Full-Time Paid 4; Part-Time Paid 6; Part-Time Volunteers 2.
Governing Authority: state. State Historical Society of Iowa. Dept. of Cultural Affairs, Capitol Complex, Des Moines, IA 50319. Tel. 515-281-7650. Branch Museum: Clermont Museum. Tax-exempt.
Institution Type/Description: Historic House: 1874 home of William Larrabee, Iowa's 12th Governor.
Collections: household furnishings, clothing & memorabilia of the Larrabee family; small art; natural science; manuscripts; history collections of the Clermont Museum. Historic Building: 1857 Union Sunday School.
Research Fields: Victorian furnishings; Iowa political history.
Activities: guided tours; concerts; temporary exhibitions; special events.
Publications: Iowa Heritage Illustrated.
Hours & Admission Prices: Memorial Day to Labor Day daily 12-4; Sept.-Oct. Fri.-Sun. 12-4. No charge; donations accepted. &
Attendance: 4,500 (accurate)

Clinton

BICKELHAUPT ARBORETUM, 340 S. 14th St., Clinton, IA 52732-5432. Tel.: 563-242-4771.
E-mail: margo.hansen@bickelhaupt.org
Web Site: www.bickelhaupt.org
Founded: 1970.
Congressional District: 2
Key Personnel: Dir. Programs, Margo Hansen.
Personnel Profile: Full-Time Paid 2; Part-Time Volunteers 68.
Governing Authority: nonprofit organization. Tax-exempt: 501(c)(3).
Institution Type/Description: Arboretum.
Collections: woody plants; perennial herbaceous plants; eastern Iowa & northwestern Illinois native plants.
Facilities: 600-vol. library of books & periodicals pertaining to horticulture available for inter-library loan; 65-seat education center; 70-seat outdoor amphitheatre; indoor plant conservancy.
Activities: guided tours; lectures; formally organized education programs.
Publications: annual report.
Hours & Admission Prices: Daily dawn-dusk. No charge; donations accepted. &
Attendance: 31,000 (estimated)

CLINTON COUNTY HISTORICAL SOCIETY, 601 S. 1st St., Clinton, IA 52732-4118. Mailing Address: P.O. Box 2435, Clinton, IA 52733-2435. Tel.: 563-242-1201.
Founded: 1965.
Congressional District: 1
Key Personnel: Pres., Don Dethmann; Treas., Janice Hansen.
Personnel Profile: Part-Time Volunteers 20.
Governing Authority: private; nonprofit organization. Tax-exempt: 501(c)(3).
Institution Type/Description: Historical Society Museum.
Collections: Clinton County, city & area history.
Facilities: library; 5,000 sq. ft. exhibit space.
Activities: guided tours; lectures.
Publications: bimonthly members newsletter.

Hours & Admission Prices: Wed. & Sun. 1-4; other times by appointment. No charge. &
Attendance: 2,000 (estimated)
Membership: Individual $10; Family $15; Supporting $25; Patron $100.

FELIX ADLER CHILDREN'S DISCOVERY CENTER, 332 8th Ave. S., Clinton, IA 52732. Tel.: 563-243-3600.
E-mail: discoverycenter@qwestoffice.net
Web Site: adlerdiscoverycenter.org
Founded: 1993.
Institution Type/Description: Children's Museum.
Collections: hands-on exhibitions.
Facilities: Museum-related items for sale.
Activities: field trips; educational programs; special events; birthday parties.
Hours & Admission Prices: Wed.-Sat. 10-4, Sun. 1-4. Admission $4, seniors 65 & over $3; children under 2 & members no charge.

FELIX ADLER MEMORIAL ASSOCIATION, INC., Felix Adler Children's Discovery Center, 332 8th Ave. S., Clinton, IA 52732-5666. Tel.: 563-243-3600. Fax: 563-243-3600.
E-mail: discoverycenter@qwestoffice.net
Formerly: Children's Discovery Center
Founded: 1993.
Congressional District: 1
Key Personnel: Dir., Chm. (V) & Museum Shop Mgr., Theo Smith.
Personnel Profile: Full-Time Volunteers 2; Part-Time Paid 1; Part-Time Volunteers 30.
Governing Authority: Subsidiary Institution: Felix Adler Children's Discovery Center. Tax-exempt.
Institution Type/Description: Children's Museum.
Collections: hands-on exhibits.
Activities: special events; educational programs; miniature golf course; early out reading program; after school mentor program. Museum Sponsors: Felix Adler Day Celebration in June.
Publications: newsletter; school event flyers.
Hours & Admission Prices: Wed.-Sat. 10-4, Sun. 1-4. Adults $4, senior citizens 65 & over $3; discounts to AAM & ICOM members; members & children under 2 no charge. &
Attendance: 250 (estimated)
Membership: Individual $30; Grandparent $40; Family $50; Reciprocal $100.

Colfax

TRAINLAND U.S.A., 3135 Hwy. 117 N., Colfax, IA 50054-7534. Tel.: 515-674-3813. Fax: 515-674-3813.
E-mail: red@trainlandusa.com
Web Site: www.trainlandusa.com
Founded: 1981.
Congressional District: 4
Key Personnel: Pres. (V), Leland Atwood; Museum Shop Mgr., Judy Smith-Atwood.
Personnel Profile: Part-Time Paid 6.
Governing Authority: private.
Institution Type/Description: Toy Museum.
Collections: 1916-1987 Lionel toy trains & accessories; steam & diesel railroads. Historic Building: 1850s CNW depot.
Facilities: 4,400 sq. ft. exhibit space. Train-related items for sale.
Activities: guided tours; lectures; hobby workshops; organized educational programs for children.
Hours & Admission Prices: Memorial Day-Labor Day daily 10-6. Adults $7.50, senior citizens over 55 $7, children 2-12 $5; discounts to groups of 15 or more; children under 2 no charge. (prices subject to change) &
Attendance: 15,000 (estimated)

Coralville

ANTIQUE CAR MUSEUM OF IOWA, 860 Quarry Rd., Coralville, IA 52241-2226. Tel.: 319-354-3310. Fax: 319-354-3310.
E-mail: info@acmoi.com
Web Site: www.acmoi.com
Founded: 2006.
Key Personnel: Pres., Dean Oakes.
Personnel Profile: Full-Time Volunteers 1; Part-Time Paid 2; Part-Time Volunteers 12.
Governing Authority: bd. of directors. Tax-exempt: 501(c)(3).
Institution Type/Description: Car Museum.
Collections: 90 automobiles from 1899 to present.
Facilities: 28,000 sq. ft. exhibit space.
Activities: guided tours.

Hours & Admission Prices: Tues.-Sat. 10-5, Sun. 12-5. Adults $5. &
Attendance: 6,300 (estimated)

THE IOWA CHILDREN'S MUSEUM, (M), 1451 Coral Ridge Ave., Ste. 502A, Coralville, IA 52241-2804. Tel.: 319-625-6255.
Web Site: www.theicm.org/
Key Personnel: Exec. Dir., Deb Dunkhase; Dir. Devel. & Mktg., Fran Jensen; Dir. Visitor Experience, Jordan Hougham; Coord. Special Programs, Julie Thomas; Exhibit Fabricator, Leonid Stepanov.
Personnel Profile: Full-Time Paid 5; Full-Time Volunteers 1; Part-Time Paid 15; Part-Time Volunteers 125; Interns 3.
Governing Authority: nonprofit organization. Tax-exempt: 501(c)(3).
Institution Type/Description: Children's Museum.
Collections: hands-on exhibits.
Facilities: Museum-related items for sale.
Activities: make your own video; interactive exhibits; birthday parties; field trips; special programs.
Hours & Admission Prices: June to mid-Sept. Mon.-Thurs. & Sat. 10-6, Fri. 10-8, Sun. 11-6; Sept.-May Tues.-Thurs. & Sat. 10-6, Fri. 10-8, Sun. 11-6. Admission $6, seniors 60 & over $5; discounts to groups of 10 or more; children under 1 & members no charge. Closed major holidays.
Attendance: 90,000 (accurate)
Membership: Family Circle $90; Family Circle Plus $100; Grandparent Circle $125.

JOHNSON COUNTY HISTORICAL SOCIETY, (M), 860 Quarry, Coralville, IA 52241-2226. Mailing Address: P.O. Box 5081, Coralville, IA 52241-5081. Tel.: 319-351-5738. Fax: 319-351-5310.
E-mail: questions@johnsoncountyhistory.org
Web Site: www.johnsoncountyhistory.org
Formerly: Johnson County Heritage Museum
Founded: 1973.
Congressional District: 2
Key Personnel: Pres. (V), Elaine Haddy; Chm. (V), Steve Weeber.
Personnel Profile: Full-Time Paid 2; Part-Time Paid 5; Part-Time Volunteers 80.
Governing Authority: society. Tax-exempt: 501(c)(3).
Institution Type/Description: Historical Society Museum: located in 1876 two-story school.
Collections: photographs; local advertising; tools; glassware; textiles; costumes; books; toys; furnishings.
Research Fields: Johnson County history.
Facilities: Museum-related items for sale.
Activities: guided tours; lectures; organized education programs; docent program; loan & temporary exhibitions.
Publications: brochure, Heritage Museum; bimonthly newsletter; exhibit-related catalogues.
Hours & Admission Prices: Tues.-Sat. 10-5, Sun. 12-5. Adults $5; members no charge. Bus tours welcome. Closed major holidays.
Attendance: 6,200 (estimated)
Membership: Sustaining Individual $25; Household $35.

Corning

AMERICA'S FRENCH ICARIAN VILLAGE, 710 Davis Ave., Ste. 1, Corning, IA 50841. Tel.: 641-322-4717.
E-mail: lcaria@frontiernet.net
Web Site: www.icaria.net
Formerly: Icaria Museum and Research Library
Congressional District: 3
Personnel Profile: Part-Time Paid 1; Part-Time Volunteers 20.
Governing Authority: Parent Institution: French Icarian Colony Foundation. Tax-exempt.
Institution Type/Description: History Museum.
Collections: Icaria history & culture; genealogy; photographs; drawings. Historic Buildings: 1860 one-room school; 1878 refectory.
Activities: special events; educational programs.
Hours & Admission Prices: Tues.-Fri. 10-3; other times by appointment. Donation requested.

CORNING CENTER FOR THE FINE ARTS, 706 Davis Ave., Corning, IA 50841-1451. Tel.: 641-322-4549.
Institution Type/Description: Art Gallery.
Collections: paintings; sculpture; photographs.
Activities: Artist in Residency program.
Hours & Admission Prices: Wed.-Fri. 10-5, Sat. 10-4. No charge.

JOHNNY CARSON BIRTHPLACE HOME MUSEUM, 500 13th St., Corning, IA 50841-1106. Mailing Address: 701 Davis Ave., Corning, IA 50841-1418. Tel.: 641-322-3212.
Institution Type/Description: Historic House Museum: housed in the birthplace of Johnny Carson, born Oct. 23, 1925.
Collections: Johnny Carson's life & career; personal artifacts; period furnishings; photographs.
Activities: guided tours.
Hours & Admission Prices: By appointment.

Correctionville

CORRECTIONVILLE MUSEUM, Fifth and Driftwood Sts., Correctionville, IA 51016-7732. Mailing Address: Rural Woodbury County Historical Society, P.O. Box 255, Correctionville, IA 51016-0255. Tel.: 712-372-4791.
E-mail: cville@ruralwaves.us
Founded: 1997.
Key Personnel: Pres. (V), Marjorie E. Hoppe; Treas., Sonya Kostan.
Personnel Profile: Part-Time Volunteers 8.
Governing Authority: Parent Institution: Rural Woodbury County Historical Society. Tax-exempt.
Institution Type/Description: History Museum: housed in old Merchants State Bank.
Collections: local history & culture.
Hours & Admission Prices: Memorial Day-Labor Day Sat. 10-1, Sun. 2-4. No charge; donations accepted.
Attendance: 130 (estimated)

Corydon

PRAIRIE TRAILS MUSEUM OF WAYNE COUNTY, Hwy. 2 E., Corydon, IA 50060. Mailing Address: P.O. Box 104, Corydon, IA 50060-0104. Tel.: 641-872-2211. Fax: 641-872-2211.
E-mail: ptmuseum@grm.net
Web Site: www.prairietrailsmuseum.org
Founded: 1942.
Congressional District: 4
Key Personnel: Pres. (V), Hal Greenlee; Mgr., Brenda DeVore.
Personnel Profile: Part-Time Paid 1; Part-Time Volunteers 30.
Governing Authority: society. Parent Institution: Wayne County Historical Society. Tax-exempt.
Institution Type/Description: General Museum.
Collections: agriculture; paintings; Native American artifacts; natural history; machinery; cars; Mormon Trail; family & household artifacts.
Facilities: library of genealogical research books.
Activities: Museum Sponsors: Freedom Ring in July; Pioneer Festival in October.
Publications: quarterly newsletter, Wayne County Historical Society Newsletter.
Hours & Admission Prices: April-May & Sept.-Oct. Mon.-Fri. 1-5; June-Aug. Mon.-Fri. 10-5, Sat.-Sun. 1-5; call for additional hours. Family $15, adults $5, college $3, students 7th-12th grade $2, children K-6th grade $1; members with card no charge. Tour: $4. &
Attendance: 2,500 (accurate)
Membership: Individual $10; Life $100.

Council Bluffs

GREAT PLAINS WING MUSEUM, 16803 McCandless Rd., Council Bluffs, IA 51503. Tel.: 712-322-2435.
E-mail: nlswede@netins.net
Web Site: www.greatplainswing.org
Founded: 1988.
Congressional District: 3
Institution Type/Description: Military History Museum.
Collections: over 1,600 military & homefront artifacts from the U.S., Germany, Britain, & Japan; photographs; models; personal artifacts; flying P-51 Mustang & AF6.
Activities: special events.
Hours & Admission Prices: Wed. 6pm-9pm, Sat. 9-4, Sun. 12-4. No charge. &
Attendance: 2,500 (estimated)

HISTORIC GENERAL DODGE HOUSE, 605 3rd St., Council Bluffs, IA 51503-6614. Mailing Address: 621 3rd St., Council Bluffs, IA 51503-6614. Tel.: 712-322-2406 & 3504. Fax: 712-322-3504.
E-mail: generaldodgehouse@windstream.net
Web Site: www.dodgehouse.org

Founded: 1964.
Congressional District: 5
Key Personnel: Exec. Dir., Kori L. Nelson; Pres., chuck Hansen; Museum Shop Mgr., Cathy Born.
Personnel Profile: Full-Time Paid 1; Part-Time Paid 5; Part-Time Volunteers 65.
Governing Authority: municipal; nonprofit. Tax-exempt.
Institution Type/Description: Historic House: 1869 Victorian 14-room home of General Grenville M. Dodge. Victorian Arts, RR history, Civil War history.
Collections: Victorian furnishings; 1869-1916 artifacts.
Research Fields: Victorian period costumes & furniture; Dodge's involvement in the Civil War & Iowa 4th Infantry troop; early railroad construction, especially the Union Pacific.
Facilities: Museum-related items for sale.
Activities: audio guided tours; films; docent program; organized educational programs for students; special annual events.
Publications: quarterly newsletter.
Hours & Admission Prices: Feb.-Dec. Tues.-Sat. 10-5, Sun. 1-5. Adults $7, seniors 62 & over $5, children 6-16 $3; discounts to groups of 20 or more & AAA members; children under 6 & members no charge. Closed some holidays.
Attendance: 11,000 (accurate)
Membership: Individual $25; Dual $40; Family $55; Sustainer $75; Century $100; Donor Circle $250. Business memberships available.

HISTORIC SQUIRREL CAGE JAIL MUSEUM, 226 Pearl St., Council Bluffs, IA 51503. Mailing Address: P.O. Box 2, Council Bluffs, IA 51502-0002. Tel.: 712-323-2509.
E-mail: info@thehistoricalsociety.org
Web Site: thehistoricalsociety.org
Key Personnel: Dir., Carla Borgaila; Museum Shop Mgr., Ed Ritchie.
Personnel Profile: Full-Time Paid 2; Part-Time Paid 2; Part-Time Volunteers 4.
Governing Authority: Parent Institution: Historical Society of Pottawattamie County. Tax-exempt.
Institution Type/Description: Historic Building Museum: housed in the former county jail; built in 1885. Listed on the National Register of Historic Places.
Collections: local history; revolving jail; period artifacts.
Publications: bimonthly members' newsletter, The Historical Society of Pottawattamie County.
Hours & Admission Prices: Feb.-Dec. Tues.-Sat. 10-4, Sun. 1-4. Adults $7, seniors 60 & over $6, children 6-12 $5; discounts to AAM & AAA members & groups; children 5 & under and members no charge. Closed Jan. & major holidays.
Attendance: 15,000 (estimated)
Membership: Single $20; Family $35.

IOWA SCHOOL FOR THE DEAF MUSEUM, 3501 Harry Langdon Blvd., Council Bluffs, IA 51503. Tel.: 712-366-0571. Fax: 712-366-3218. TTY: 712-366-0571.
E-mail: cangeroth@iowaschoolforthedeaf.org
Web Site: www.iowaschoolforthedeaf.org/lmc-and-museum
Institution Type/Description: History Museum.
Collections: school history; deaf culture; sign language development; electronic aids for the deaf; photographs.
Hours & Admission Prices: By appointment.

RAILS WEST RAILROAD MUSEUM, 16th Ave. & S. Main St., Council Bluffs, IA 51503. Mailing Address: P.O. Box 2, Council Bluffs, IA 51502-0002. Tel.: 712-323-5182.
E-mail: info@thehistoricalsociety.org
Web Site: thehistoricalsociety.org
Key Personnel: Dir., Carla Borgaila; Museum Shop Mgr., Ed Ritchie.
Personnel Profile: Full-Time Paid 2; Part-Time Paid 2; Part-Time Volunteers 3.
Governing Authority: Tax-exempt.
Institution Type/Description: Historic Building Museum: housed in the former Chicago Rock Island and Pacific Railroad Passenger Depot; built in 1899.
Collections: local history & culture; railroad artifacts; rolling stock; period furnishings; photographs; model railroad.
Publications: bi-monthly members newsletter, The Historical Society of Pottawattamie County Newsletter.
Hours & Admission Prices: Feb.-Dec. Tues.-Sat. 10-4, Sun. 1-4; other times by appointment. Adults $7, seniors 60 & over $6, children 6-16 $5; discounts to AAA members & groups of 15 or more; members and children 5 & under no charge. Closed Jan. & major holidays.
Attendance: 15,000 (estimated)
Membership: Single $20; Family $35.

UNION PACIFIC RAILROAD MUSEUM, (M), 200 Pearl St., Council Bluffs, IA 51503-0825. Tel.: 712-329-8307. Fax: 712-323-4973.
E-mail: palabount@up.com
Web Site: www.uprrmuseum.org
Founded: 1921.
Congressional District: 5
Key Personnel: Mgr. Operations & Facilities, Beth Lindquist; Mgr. Collections & Outreach, Patricia LaBounty; Communication & Volunteer Rels. Mgr., Abby Cape.
Personnel Profile: Full-Time Paid 1; Part-Time Volunteers 40.
Governing Authority: private; nonprofit. Parent Institution: Union Pacific Railroad. Subsidiary Institution: Friends of the UPRR Museum. Tax-exempt: 501(c)(3).
Institution Type/Description: History Museum.
Collections: western railroad history to modern technology; American history 1850s to present; artifacts, archives & photographs.
Facilities: 700-vol. library; 18,000 sq. ft. exhibit space; theater. Museum-related items for sale.
Activities: temporary & participatory exhibits.
Publications: quarterly newsletter, Golden Spike.
Hours & Admission Prices: Tues.-Sat. 10-4. No charge; donations accepted. Closed New Years Eve & Day; Presidents Day; Memorial Day; Independence Day; Labor Day; Thanksgiving; Christmas Eve & Day. &
Attendance: 24,000 (accurate)
Membership: Surveyor $35; Grader $60; Gandy Dancer $100; Track Foreman $250; Chief Engineer $500; Superintendent $1,000.

WESTERN HISTORIC TRAILS CENTER, 3434 Richard Downing Ave., Council Bluffs, IA 51501-7962. Tel.: 712-366-4900. Fax: 712-366-5080.
E-mail: teressa.swand@iowa.gov
Web Site: www.iowahistory.org/historic-sites
Institution Type/Description: History Center.
Collections: Lewis & Clark; Mormon Pioneer; Oregon & California trails; sculptures; hands-on exhibits; photographs; video; Native American artifacts; early & present-day travel.
Activities: educational programs; hands-on exhibits; film.
Hours & Admission Prices: April-Oct. daily 9-5; Nov.-March Tues.-Sun. 9-5. No charge.

Creston

UNION COUNTY HISTORICAL COMPLEX, McKinley Park, 116 W. Adams St., Creston, IA 50801. Tel.: 641-782-8220.
Web Site: unioncountyiowatourism.com/sites.html
Founded: 1966.
Key Personnel: Dir., Mark Huff.
Personnel Profile: Part-Time Paid 4; Part-Time Volunteers 25.
Governing Authority: nonprofit organization. Tax-exempt.
Institution Type/Description: Village Museum.
Collections: Historic Buildings: schoolhouse; depot; log cabin; barn; church; country store (replica); railroad watch tower; railroad caboose & house; barber shop; leather & blacksmith shop; 2 machine sheds.
Facilities: library of books available for research by appointment.
Activities: guided tours by appointment only; rental facilities; school groups.
Hours & Admission Prices: June to Labor Day daily 1-5. No charge; donations accepted.
Attendance: 1,000 (estimated)
Membership: Regular $1; Senior Citizen Life $20; Life $25.

Dakota City

HUMBOLDT COUNTY HISTORICAL ASSOCIATION MUSEUM, 905 1st Ave., N., Dakota City, IA 50529-5134. Mailing Address: P.O. Box 162, Humbolt, IA 50548-0162. Tel.: 515-332-5280.
E-mail: hcha@goldfieldaccess.net
Web Site: www.humboldtiowahistory.org
Founded: 1962.
Congressional District: 6
Key Personnel: Dir., Sandra Back; Pres. (V), Carolyn Logan.
Personnel Profile: Part-Time Paid 1.
Governing Authority: nonprofit organization. Tax-exempt: 501(c)(3).
Institution Type/Description: County Historical Museum.
Collections: local history & culture; Native American artifacts; manuscripts. Historic Structures: 1879 Mill Farm house; 1883 Norway No. 6 District School; 1883 Hardy Methodist Church; red barn; Rutland Jail; 1875 chickenhouse; log cabin; kettle shed.

Research Fields: state & local county history.
Facilities: 100-vol. library of Humboldt County & Iowa history books available for use by special arrangement. Handmade articles for sale.
Activities: guided tours; lectures; permanent & temporary exhibitions; meeting programs throughout the year; tapes of interviews of old residents. Museum Sponsors: special displays & demonstrations during the summer.
Publications: book, History of the City of Humboldt-First 100 Years; Gotch: Biography of World's Champion Wrestler, Frank Gotch.
Hours & Admission Prices: June-Sept. Mon.-Tues. & Thurs.-Sat. 10-4, Sun. 1:30-4:30; special tours by appointment. General admission: $5.
Attendance: 2,000 (estimated)
Membership: Annual $10; Lifetime $200.

Dallas Center

BRENTON ARBORETUM, 25141 260th St., Dallas Center, IA 50063-8336. Tel.: 515-992-4211. Fax: 515-992-3303.
E-mail: info@thebrentonarboretum.org
Web Site: thebrentonarboretum.org
Founded: 1997.
Key Personnel: Dir., Lynn Kuhn.
Personnel Profile: Full-Time Paid 2; Part-Time Paid 1; Part-Time Volunteers 5.
Governing Authority: Tax-exempt.
Institution Type/Description: Arboretum.
Collections: landscape management & conservation; native wildflowers; over 2,600 trees & shrubs.
Major Exhibits: Vanishing Acts (T), 6/1-10/14.
Research Fields: Kentucky coffee tree.
Facilities: library; multi-purpose room; nature trails; pavilion.
Activities: guided tours; educational programs; classes; special events; research; demonstrations; walking trails.
Publications: newsletter.
Hours & Admission Prices: Tues.-Sun. 9am to sunset; groups by appointment. No charge, donations accepted. &
Attendance: 1,150 (estimated)
Membership: Individual $20; Family $30; Organization $50; Supporting $100; Sustaining $250; Enduring $500.

Davenport

* **FIGGE ART MUSEUM, (M),** 225 W. 2nd St., Davenport, IA 52801-1804. Tel.: 563-326-7804. Fax: 563-326-7876.
E-mail: clerk@figgeartmuseum.org
Web Site: figgeartmuseum.org
Formerly: Davenport Museum of Art
Founded: 1925.
Congressional District: 1
Key Personnel: Pres. (V), Andrew J. Butler; Dir., Tim Schiffer; C.F.O., Todd Woeber; Cur. Education, Melissa Hueting; Assoc. Cur., Dr. Rima Girnius; Registrar, Andrew Wallace; Outreach Coord., Laura Dunn; Dir. Museum Svcs., Jennifer Brooke; Devel. & Membership, Susan Horan; Museum Shop Mgr., Chris Sweeney.
Personnel Profile: Full-Time Paid 16; Part-Time Paid 6; Part-Time Volunteers 113; Interns 13.
Governing Authority: private. Tax-exempt.
Institution Type/Description: Art Museum.
Collections: American, European, Haitian, Mexican Colonial & Asian art; Grant Wood & the Regionalists.
Research Fields: Haitian art of the native tradition; Mexican-Colonial Art; American Art.
Facilities: 6,000-vol. library of art reference; auditorium; classrooms. Museum-related items for sale.
Activities: guided tours; lectures; gallery talks; concerts; arts festivals; formally organized education programs for children, adults & undergraduate college students; docent program or council; inter-museum loan, permanent & temporary exhibitions.
Publications: quarterly newsletter; invitations; annual report; books, exhibition catalogues.
Hours & Admission Prices: Tues.-Wed. & Fri.-Sat. 10-5, Thurs. 10-9, Sun. 12-5. Adults $7; discount to North American Reciprocal Member Veterans; members no charge. &
Attendance: 60,000 (estimated)
Membership: Senior, Student & Educator $40; Individual $50; Household $75.

GERMAN AMERICAN HERITAGE CENTER, 712 W. Second St., Davenport, IA 52802-1410. Tel.: 563-322-8844. Fax: 563-322-2687.
E-mail: admin@gahc.org
Web Site: www.gahc.org
Founded: 1995.

Congressional District: 1
Key Personnel: Dir., Janet Brown-Lowe; Chm. (V), Cal Werner; Museum Shop Mgr., Joan Finkenhoefer.
Personnel Profile: Full-Time Paid 2; Part-Time Paid 3; Part-Time Volunteers 60.
Governing Authority: Tax-exempt.
Institution Type/Description: Cultural Heritage Museum.
Collections: German American cultural heritage; photographs; personal artifacts.
Facilities: 100-seat banquet facility.
Activities: special events & programs.
Publications: quarterly newsletter, Infoblatt.
Hours & Admission Prices: Tues.-Sat. 10-4, Sun. 12-4; other times by appointment. Group tours $25 min. hostess fee, adults $5, senior citizens $4, children 5-17 $3; children under 5 & members no charge. &
Attendance: 8,000 (estimated)
Membership: Single $35; 2 Adult Household $45; Family $55 or 2 Grandparents & Grandchildren $55; Life $1,000; 2x Life $1,500.

PALMER MUSEUM OF CHIROPRACTIC HISTORY, 1000 Brady St., Davenport, IA 52803. Tel.: 563-884-5404.
Founded: 1997.
Key Personnel: Dir., Alana Callender; Education, Dr. Roger Hynes; Public Rels., Julie Arnold.
Personnel Profile: Full-Time Paid 2; Part-Time Paid 1.
Governing Authority: private college. Tax-exempt.
Institution Type/Description: History Museum.
Collections: chiropractic history from 1895 to present; chiropractic table & technique tools.
Activities: guided tours; loan exhibitions.
Publications: chiropractic history.
Hours & Admission Prices: Jan. 4-Dec. 21 Mon.-Fri. 8-4, Sat.-Sun. by appointment. Tours: $5 per person; discounts to AAM members.
Attendance: 400 (estimated)

* **PUTNAM MUSEUM OF HISTORY & NATURAL SCIENCE, (M),** 1717 W. 12th St., Davenport, IA 52804-3597. Tel.: 563-324-1933 & 1054. Fax: 563-324-6638.
E-mail: museum@putnam.org
Web Site: www.putnam.org
Founded: 1867.
Congressional District: 1
Key Personnel: C.E.O. & Pres., Kim Findlay; Chm., Dana Waterman; C.F.O., Kim Nickels; Chief Cur., Eunice Schlichting; Cur. History, Christina Kastell; Dir. Education, Donna Murray; Cur. Natural Science, Christine Chandler; Dir. Theatre Operations, Dean K. Fick; Dir. Mktg., Lori Arguello; Dir. Human Resources & Administration, Drue Curry; Dir. Facilities, Mike Murphy; Dir. Visitor Svcs., Beth Knaack; Retail Sales Mgr., Sue Folwell.
Personnel Profile: Full-Time Paid 24; Part-Time Paid 14; Part-Time Volunteers 101; Interns 21.
Governing Authority: nonprofit corporation. Tax-exempt: 501(c)(3).
Institution Type/Description: History, Natural Science & Anthropology Museum.
Collections: regional history with emphasis on the Quad cities of Iowa & Illinois: geology; ethnology; archaeology; botany; paleontology; anthropology; zoology; archives & decorative arts.
Research Fields: history & natural history of the Quad Cities of Iowa & Illinois.
Facilities: 5,000-vol. library of history, natural history & fine art books available for use by special request; classrooms; Heritage Theatre (R); scilab; 300-seat auditorium; discovery room; meeting rooms; 270-seat IMAX(R) 3D Theatre. Books, publications & gifts related to the museum for sale.
Activities: permanent & changing exhibits; IMAX films, lectures; concerts; classes; family workshops; school tours; school & community outreach programs; special events; World Adventure Travelogue Series.
Publications: brochures; proceedings of the Davenport Academy of Sciences; Vascular Plants of Scott & Muscatine Counties; The Indian Mounds at Albany, Illinois; Prehistoric Moundbuilders of the Mississippi Valley; Watching the River, Walking the Land: A Natural History of Eastern Iowa and Western Illinois; The Nazca Pottery of Ancient Peru
Hours & Admission Prices: Mon.-Sat. 10-5, Sun. 12-5. Museum: $4-$6; discounts to members. IMAX: $6.50-$8.50; discounts to members. IMAX & Museum: $8-$10; discount to members. Closed Thanksgiving; Christmas. &
Attendance: 160,000 (estimated)
Membership: Individual $35; Adventurer $40; Duo $60; Grand & Family $80; IMAX $125; Triple $190.

VANDER VEER BOTANICAL PARK, 215 W. Central Park Ave., (Btw. Brady & Harrison Sts.), Davenport, IA 52803. Mailing Address: 214 W. Central Park Ave., Davenport, IA 52803-1503. Tel.: 563-326-7818.
Web Site: www.cityofdavenportiowa.com
Institution Type/Description: Botanical Garden.
Collections: plants & flowers including rose garden, Hosta Glade; seasonal exhibits.
Activities: temporary exhibitions.
Hours & Admission Prices: Park: daily sunrise to sunset. The Conservatory and Park Store: Tues.-Sun. 10-4. Adults $1, children 15 & under no charge.
Attendance: 60,000 (estimated)

DeWitt

CENTRAL COMMUNITY HISTORICAL SOCIETY MUSEUM, 628 6th Ave., DeWitt, IA 52742. Tel.: 563-659-3686.
E-mail: ahsoenk@gmail.com
Web Site: centralcommunityhistoricalsociety.webs.com
Founded: 1977.
Key Personnel: Dir. & Pres. (V), Ann Soenksen.
Personnel Profile: Part-Time Volunteers 12.
Governing Authority: nonprofit organization. Tax-exempt.
Institution Type/Description: Historical Society Museum.
Collections: local history & culture; county records; military artifacts; household items; period furnishings; personal artifacts; photographs; early telephones, typewriters & cameras; clothing; sewing equipment; farm machinery & equipment; fire equipment; 1930s fire engine.
Hours & Admission Prices: Mon. 8:30 am-11:30 am, Thurs. 8-10:30, Sun. 1-4. No charge; donations accepted. Closed holidays. &
Attendance: 800 (estimated)
Membership: Individual $10; Family 18; Society Friend $50; Patron $100.

Decorah

FINE ARTS COLLECTION, LUTHER COLLEGE, 700 College Dr., Luther College Library, Decorah, IA 52101-1041. Tel.: 563-387-1328 & 1300. Fax: 563-387-1132.
E-mail: elli03@luther.edu
Web Site: finearts.luther.edu
Congressional District: 4
Key Personnel: Cur., Kate Elliott; Gallery Coord., David Kamm.
Personnel Profile: Full-Time Paid 1; Part-Time Paid 1.
Governing Authority: private college; nonprofit. Parent Institution: Luther College. Tax-exempt: 501(c)(3).
Institution Type/Description: Art Museum.
Collections: contemporary & historic prints, paintings & sculpture; Gerhard Marcks drawings & prints; Marguerite Wildenhain pottery & drawings; pre-Columbian pottery; Inuit sculpture; Scandinavian immigrant paintings.
Research Fields: Marguerite Wildenhain; Gerhard Marcks; pre-Columbian Panamanian ceramics; Inuit sculpture.
Facilities: 2,265 sq. ft. exhibit space.
Activities: student, temporary & traveling exhibits; gallery talks; scholarly research & education.
Publications: exhibition catalogs; brochures; flyers.
Hours & Admission Prices: Sept.-June Mon.-Fri. 8-10, Sat. 9-5, Sun. 12-10. No charge. Closed college holidays. &
Attendance: 10,000

GEOLOGY COLLECTION, LUTHER COLLEGE, 700 College Dr., Decorah, IA 52101-1045. Tel.: 563-387-1508.
E-mail: youngjea@luther.edu
Web Site: geology.luther.edu
Key Personnel: Cur., Jean Young
Institution Type/Description: Geology Museum.
Collections: minerals; rocks; fossils.
Hours & Admission Prices: By appointment. No charge.

LUTHER COLLEGE ETHNOGRAPHIC AND ARCHAEOLOGICAL COLLECTIONS, 700 College Dr., Anthropology Lab, Decorah, IA 52101-1041. Tel.: 563-387-2156.
E-mail: landch01@luther.edu
Web Site: anthrophology.luther.edu
Founded: 1969.
Congressional District: 4
Personnel Profile: Full-Time Paid 1; Part-Time Paid 12.
Governing Authority: Parent Institution: Luther College. Tax-exempt.
Institution Type/Description: History Museum.
Collections: North American & Midwest archaeology; Native American,
Inupiat, African, Asian ethnographic artifacts; period Greek & Roman coins & numismatic collections; paper currency.
Hours & Admission Prices: Call for hours.

THE PORTER HOUSE MUSEUM, (M), 401 W. Broadway St., Decorah, IA 52101. Mailing Address: P.O. Box 115, Decorah, IA 52101-0115. Tel.: 563-382-8465.
E-mail: porterhousemuseum@gmail.com
Web Site: porterhousemuseum.org
Founded: 1966.
Key Personnel: Chm. (V), Mildred Kjome; Dir., Emily Mineart.
Personnel Profile: Full-Time Paid 1.
Governing Authority: Tax-exempt.
Institution Type/Description: Historic House Museum: built in 1867.
Collections: local history & culture; period furnishings; personal artifacts; photographs; A.F. Porter's butterflies from around the world; journals; early 20th century natural history collection; Bert's unique butterfly art.
Activities: special events.
Hours & Admission Prices: June-Aug. Mon.-Sat. 10-4, Sun. 1-4, Sept.-Oct. Fri.-Sun. 1-4; off-season and school & bus groups by appointment. Adults $5, seniors over 65 $4, children 6-17 $3; children under 6, school groups & members no charge.
Membership: Friend $25; Patron $50; Corporate $50 & up; Supporting $100 & up.

* **VESTERHEIM NORWEGIAN-AMERICAN MUSEUM, (M),** 523 W. Water St., Decorah, IA 52101-1733. Mailing Address: P.O. Box 379, Decorah, IA 52101-0379. Tel.: 563-382-9681. Fax: 563-382-8828. Facebook: Vesterheim Norwegian-American Museum.
E-mail: info@vesterheim.org
Web Site: www.vesterheim.org
Founded: 1877.
Congressional District: 4
Key Personnel: Chm., Kate Martinson; Exec. Dir., Steve Johnson; Dir. Admin., Marcia McKelvey; Coord. Tours to Norway, Michelle Whitehill; Volunteer Coord., Martha Griesheimer; Membership Mgr., Peggy Sersland; Chief Cur., Laurann Gilbertson; Registrar & Archivist, Jennifer Johnston Kovarik; Editor, Charlie Langton; Office Asst., Jocelyn Bruening; Technology Specialist, Faust Gertz; Devel. Asst., Stephanie Johnson; Museum Shop Mgr., Ken Koop.
Personnel Profile: Full-Time Paid 10; Part-Time Paid 13; Part-Time Volunteers 253; Interns 3.
Volunteer Hours: 3,343
Operating Expenses: 1,961,224
Operating Income: 2,242,312
Governing Authority: nonprofit organization. Tax-exempt: 501(c)(3).
Institution Type/Description: Ethnic Museum.
Collections: Norwegian, Norwegian immigrant & Norwegian-American artifacts; folk art; fine art; furnishings; tools & technology; clothing & personal artifacts; archives. Historic buildings; 1880 log schoolhouse; 1860 drying house; 1860 waterpower grain mill; 1851 grist mill; 1863 stone church; 1860 Norwegian house; 1850-1929 immigrant farmstead; 1855 Stovewood house; 1877 hotel; 1852 pioneer immigrant log home; 1859 pioneer immigrant log home; 1859 frame general store; 1879 northern plains frame house; 1905 northern plains frame church.
Major Exhibits: Four From the North (T), 10/13-4/14; Daughters of Norway, 10/13-4/14; Flora Metamorphicae (T), 12/13-11/14; Favorite Things, 1/14-4/14; Syttende Mai (T), 5/14-4/15; National Exhibition of Folk Art in Norwegian Tradition, 6/14-7/14.
Research Fields: Norwegian American material culture.
Facilities: 10,000-vol. research library; reading room; classrooms. Museum-related items for sale.
Activities: guided tours; lectures; arts festivals; folk-art classes; formally organized education programs for adults & youth; inter-museum loan, permanent & temporary exhibitions. Museum Sponsors: special undergraduate internships for Luther College students; Nordic Fest; national juried exhibitions in traditional Norwegian arts & crafts; folk art tours in Norway.
Publications: newsletters; semiannual magazine; specialty newsletters, Rosemaling; books & pamphlets on Norwegian and Norwegian-American textiles, painting, woodcarving, silversmithing.
Hours & Admission Prices: May-Oct. daily 9-5.; Nov.-April daily 10-4. Adults $10, seniors $8, children 6-18 $5; discounts for families & groups; first Thurs. of month, children 6 & under and members no charge. Closed New Year's Day; Easter; Thanksgiving; Christmas. &
Attendance: 15,727 (accurate)

Membership: Senior $30; Individual $35; Senior Family $40; Family $50; International $60; Friend $125; Supporter $250; Sponsor $500; Sustaining Fellow $1,000.

Denison

DONNA REED HERITAGE MUSEUM, 1305 Broadway, Denison, IA 51442-1923. Tel.: 712-263-3334. Fax: 712-263-8026.
E-mail: info@donnareed.org
Web Site: www.donnareed.org
Institution Type/Description: History Museum: housed in a former German opera house; built in 1914.
Collections: Donna Reed's life & career; personal artifacts; photographs; awards & honors; movie memorabilia; videos.
Activities: guided tour.
Hours & Admission Prices: Mon.-Fri. 10-2; other times by appointment. No charge; donations accepted.

W.A. MCHENRY HOUSE, 1428 First Ave. N., Denison, IA 51442-1402. Mailing Address: P.O. Box 741, Denison, IA 51442. Tel.: 712-674-3750.
Founded: 1967.
Key Personnel: Pres. (V), Connie Volkman; Museum Tours, Nancy Bliesman.
Personnel Profile: Part-Time Volunteers 12.
Governing Authority: Parent Institution: Crawford County Historical Society. Tax-exempt.
Institution Type/Description: Historic House Museum.
Collections: local history & culture; period furnishings; personal artifacts; photographs military.
Hours & Admission Prices: Memorial Day to July & Sept.-Oct. Sat.-Sun. 1-4. Adults $3.
Attendance: 500 (estimated)
Membership: Single $10; Family $25; Corporate $100.

Des Moines

ANDERSON GALLERY - DRAKE UNIVERSITY, Harmon Fine Arts Center, 25th St. & Carpenter Ave., Des Moines, IA 50311. Tel.: 515-271-1994. Fax: 515-271-2558.
E-mail: heatherskeens@drake.edu
Web Site: www.drake.edu/andersongallery
Founded: 1996.
Key Personnel: Dir., Heather Skeens
Institution Type/Description: Art Gallery.
Collections: paintings; sculpture.
Activities: lectures; workshops.
Hours & Admission Prices: Tues.-Wed. & Fri.-Sun. 12-4, Thurs. 12-8; other times by appointment. No charge.

BLANK PARK ZOO, 7401 S.W. 9th St., Des Moines, IA 50315-6667. Tel.: 515-285-4722. Fax: 515-974-2590.
E-mail: info@blankparkzoo.com
Web Site: www.blankparkzoo.com
Founded: 1963.
Congressional District: 3
Key Personnel: Dir. Operations, Anne Shimerdla; C.E.O., Mark Vukovich; Dir. Animal Care & Conservation, Kevin Drees; Dir. Devel., Azure Christensen; Public Rels., Ryan Bickel.
Personnel Profile: Full-Time Paid 35; Part-Time Paid 17; Part-Time Volunteers 90; Interns 5.
Governing Authority: Parent Institution: Blank Park Zoo Foundation. Tax-exempt.
Institution Type/Description: Zoo.
Collections: 104 animal species.
Facilities: library of books related to animals; cafeteria with outdoor seating. Animal-related items for sale.
Activities: mobile vans; day camp; annual events.
Publications: quarterly newsletter, Zootracks.
Hours & Admission Prices: May-Sept. daily 9-5; Oct.-April daily 10-4. Adults $11, senior citizens $9, children 3-12 $6; members and children 2 & under no charge. AZA reciprocal zoos. &
Attendance: 413,000 (accurate)
Membership: One Plus One $69; Family & Grandparent $89; Family Plus $99.

* **DES MOINES ART CENTER, (M),** 4700 Grand Ave., Des Moines, IA 50312-2099. Tel.: 515-277-4405. Fax: 515-271-0357.
E-mail: cdoolittle@desmoinesartcenter.org
Web Site: www.desmoinesartcenter.org
Founded: 1948.

Congressional District: 66
Key Personnel: Pres. Bd., James Habbell; Dir., Jeff Fleming; Dir. Studio Programs, Peggy Leonardo; Registrar, Rose Wood; Protection Svc. Dir., Michael O'Neal; Dir. Finance, Cheryl Larkin; Museum Shop Mgr., Sarah Jane Shimasaki.
Personnel Profile: Full-Time Paid 60; Part-Time Paid 40; Part-Time Volunteers 100.
Governing Authority: nonprofit organization. Parent Institution: Edmunson Art Foundation. Tax-exempt: 501(c)(3).
Institution Type/Description: Art Museum.
Collections: late 19th-21st-century European & American painting & sculpture; prints; African Art.
Facilities: 14,000-vol. research library; 220-seat auditorium; classrooms; studios; restaurant. Museum-related items for sale.
Activities: inter-museum loan; permanent, temporary & traveling exhibitions; lectures; gallery talks; performing arts programs; film programs; docent tours; educational programs; musical entertainment; guided tours. Museum Sponsors: Contemporary Collectors; Art Noir; Print Club; Arts After Hours monthly.
Publications: exhibition catalogs; bulletin, News; gallery handouts; program brochures; gallery guides; Visitor's Guide; class schedules.
Hours & Admission Prices: Tues.-Wed. & Fri.-Sat. 11-4, Thurs. 11-9, Sun. 12-4. No charge; donations accepted. Closed New Year's Eve & Day; Independence Day; Thanksgiving; Christmas. &
Attendance: 377,062 (accurate)
Membership: Individual $35; Household $50; Contributor $100; Sustainer $500; Patron $1,000.

FORT DES MOINES MUSEUM AND EDUCATION CENTER, 75 E. Army Post Rd., Des Moines, IA 50315-5866. Tel.: 888-828-FORT; 515-282-8060.
Key Personnel: Exec. Dir., Joe Nolte; Museum Shop Mgr., Diana Wheeland
Institution Type/Description: History Museum.
Collections: African American men of WWI; Women's Army Auxiliary Corps; photographs; period art.
Activities: facilities rental; educational programs; weddings; receptions; meetings.
Hours & Admission Prices: Mon.-Sat. 10-4. Adults $2; members no charge. &
Attendance: 10,000 (accurate)
Membership: Individual $30; Family $60; Bronze $100; Silver $250; Gold $500.

GREATER DES MOINES BOTANICAL CENTER, 909 Robert D. Ray Dr., Des Moines, IA 50309-2897. Tel.: 515-323-6290. Fax: 515-243-2631.
E-mail: info@dmbotanicalgarden.com
Web Site: www.dmbotanicalgarden.com
Formerly: Des Moines Botanical Center
Founded: 1979.
Congressional District: 4
Key Personnel: Pres. & C.E.O., Stephanie Jutila; Horticulture Mgr., Kelly Noms; Guest Svcs. & Facility Mgr., Megan Luna.
Personnel Profile: Full-Time Paid 15; Part-Time Paid 7; Part-Time Volunteers 210.
Governing Authority: municipal. Department of Parks & Recreation, 3226 University, Des Moines, IA 50311. Parent Institution: City of Des Moines. Tax-exempt.
Institution Type/Description: Botanical Garden.
Collections: 2,250 tropical & semi-tropical plants; arid indoor; outdoor cactus; succulent garden; dwarf conifer collection; twenty seven Bonsai; medicinal, culinary & fragrance herb garden; butterfly garden; perennial garden; rock garden.
Research Fields: horticulture; ornamental horticulture.
Facilities: 1,000-vol. library of botanical, horticultural books & periodicals available for research on premises; botanical garden; garden cafe. Small plants, horticultural books & other museum-related items for sale.
Activities: guided tours; lectures; hobby workshops; formally organized education programs; temporary exhibitions.
Publications: quarterly, Botanical Center Newsletter; yearly events calendar; plant information fact sheets.
Hours & Admission Prices: Daily 10-5. Adults $4, senior citizens & children $3; discount to groups of 10 or more, AABGA, American Association of Botanical Gardens, American Horticultural Society & Arboretum members; members & children under 3 no charge. Closed New Year's Day; Memorial Day, Independence Day, Labor Day, Thanksgiving; Christmas. &
Attendance: 200,000 (estimated)
Membership: Senior Citizen & Student $20; Individual $30; One Plus One, Family & Grandparent $40; Cultivators: Seed $100; Sprout $250; Bud $500 & up; Flower $1,000 & up; Garden $2,500 & up.

HERITAGE ART GALLERY, 111 Court Ave., Des Moines, IA 50309. Tel.: 515-286-2242.
Web Site: www.heritagegallery.org
Institution Type/Description: Art Gallery.
Collections: photographs; paintings; sculpture.
Hours & Admission Prices: Mon.-Fri. 11-4:30. No charge.

HOYT SHERMAN PLACE, 1501 Woodland Ave., Des Moines, IA 50309-3283. Tel.: 515-244-0507. Fax: 515-237-3582.
Web Site: www.hoytsherman.org
Founded: 1907.
Congressional District: 4
Key Personnel: Exec. Dir., Carol Pollock.
Personnel Profile: Full-Time Paid 8; Part-Time Paid 1; Part-Time Volunteers 163.
Governing Authority: nonprofit. Parent Institute: Hoyt Sherman Place Foundation. Tax-exempt: 501(c)(3).
Institution Type/Description: Art Museum Complex: comprised of 1877 House; 1907 Art Museum; 1923 Theater.
Collections: 17th- to 19th-century paintings; statuary; antique furniture; 16th century carved Swiss cabinets; decorative arts; artifacts.
Facilities: meeting rooms; 1,250-seat theater.
Activities: guided tours; lectures; docent program; participatory exhibits.
Publications: quarterly members newsletter.
Hours & Admission Prices: Mon.-Fri. 9-4. No charge; donations accepted. Guided Tours: $2 per person. &
Attendance: 90,000 (estimated)
Membership: Supporting Cast $50; Leading Roles $150; Director $300; Producer $600; Visionary $1,250 & up.

IOWA DEPARTMENT OF NATURAL RESOURCES, 502 E. 9th St., Des Moines, IA 50319-0034. Tel.: 515-281-5918. Fax: 515-281-6794. TDD: 515-242-5967.
Web Site: www.iowadnr.gov
Founded: 1935.
Key Personnel: Dir., Richard Leopold.
Governing Authority: state. Tax-exempt.
Institution Type/Description: State Park Museum.
Collections: Museum: Iowa Civilian Conservation Corps Museum At Backbone State Park, Delaware Co. Historic Houses: 1847 Ft. Atkinson grounds & museum; 1850 Wildcat Den State Park, 1847 Pine Creek old grist mill, one-room school building; 1951 Cedar Rock, home, boat house & grounds designed by Frank Lloyd Wright.
Publications: monthly, Iowa Conservationist.
Hours & Admission Prices: Museum: May-Oct. Sat.-Sun. 11-5. Wildcat Den Mill & School: park & grounds open year-round, buildings open only for special events. Fort Atkinson: mid-May to Oct. Tues-Sun. 1-5. Cedar Rock: May-Oct. Tues.-Sun. 11-5. No charge. &
Attendance: 35,000

IOWA HALL OF PRIDE, 330 Park St., Des Moines, IA 50309-1701. Tel.: 515-280-8969. Fax: 515-280-3211.
E-mail: jack@iowahallofpride.com
Web Site: www.iowahallofpride.com
Key Personnel: Dir., Jack Lashier; Education Coord., Shelly Johnson
Institution Type/Description: History Museum.
Collections: state history; Iowa's sports legends, movie stars, scientists & heroes; art.
Facilities: theater.
Activities: special events; birthday parties; facility rental; school & scout groups.
Hours & Admission Prices: Mon.-Sat. 9:30-4:30, Sun. by appointment. Adults $5, out of state students $4; discounts to groups; Iowa children K-12 no charge. Closed major holidays. &

POLK COUNTY HERITAGE GALLERY, Polk County Office Bldg., 111 Court Ave., Des Moines, IA 50309-2218. Tel.: 515-286-2242. Fax: 515-286-3082. TDD: 515-286-2003.
E-mail: info@heritagegallery.org
Web Site: www.heritagegallery.org
Founded: 1980.
Congressional District: 7
Key Personnel: Pres. (V), Tom Green.
Personnel Profile: Full-Time Volunteers 1; Part-Time Volunteers 20.
Governing Authority: Parent Institution: Polk County Heritage Gallery Board. Tax-exempt.
Institution Type/Description: Art Gallery: housed in 1908 post office. Listed on the National Register of Historic Places.

Collections: local history & culture; period furnishings; personal artifacts; photographs.
Research Fields: Postal History.
Activities: changing exhibits.
Hours & Admission Prices: Mon.-Fri. 11-4:30. No charge; donations accepted. Closed legal holidays. &
Attendance: 4,000 (accurate)

SALISBURY HOUSE & GARDENS, (M), 4025 Tonawanda Dr., Des Moines, IA 50312-2909. Tel.: 515-274-1777. Fax: 515-274-0184.
E-mail: contactus@salisburyhouse.org
Web Site: www.salisburyhouse.org
Founded: 1993.
Congressional District: 4
Key Personnel: Exec. Dir., Mark J. Heppner; Chm. (V), Mike Simonson.
Personnel Profile: Full-Time Paid 8; Part-Time Paid 2; Part-Time Volunteers 60.
Governing Authority: Tax-exempt: 501(c)(3).
Institution Type/Description: Historic House Museum & Gardens: built 1923-28.
Collections: 14th to 20th-century furnishings & tapestries; manuscript collections; paintings; rare books.
Facilities: 2,100-vol. library of rare books available for use on premises. Brochures & postcards for sale.
Activities: guided tours; annual education & public programs; rental facilities; special events.
Publications: quarterly newsletter, Beneath the Rafters.
Hours & Admission Prices: Tours: March-Dec. Tues.-Fri. 1 & 2:30. Adults $7, seniors $6, children 6-12 $3; discounts to members & school groups sponsored by members. Closed most holidays. &
Attendance: 30,000 (estimated)
Membership: Individual $35; Family $50; Friend $100; Contributor $250; Patron $500; Benefactor $1,000.

*   **SCIENCE CENTER OF IOWA, (M),** 401 W. Martin Luther King Jr. Pkwy., Des Moines, IA 50309-4776. Tel.: 515-274-6868, ext. 222. Fax: 515-274-3404.
E-mail: info@sciowa.org
Web Site: www.sciowa.org
Founded: 1965.
Congressional District: 4
Key Personnel: Pres. & C.E.O., Curt Simmons; Dir. Retail Operations, Kent Maahs.
Personnel Profile: Full-Time Paid 45; Part-Time Paid 45; Part-Time Volunteers 10; Interns 14.
Governing Authority: nonprofit organization. Tax-exempt: 501(c)(3).
Institution Type/Description: Science & Technology Museum.
Collections: rocks, minerals & fossils of Iowa; native Iowa animals; native animals; interactive exhibits.
Facilities: 110-seat auditorium; classrooms; preschool; IMAX Dome Theater; Star Theater with walk through planetarium; John Deere Adventure Theater; cafe. Telescopes, science kits, science books for sale.
Activities: demonstration; lectures; films; hobby workshops; formally organized education programs; statewide outreach program; docent program; temporary & traveling exhibitions; volunteer programs; small discoveries children 7 & under; live performances.
Hours & Admission Prices: Mon.-Sat. 10-5, Sun. 12-5. Science Center: adults $11, senior citizens $10, children 2-12 $7; members no charge. IMAX: adults $8-$13, senior citizens 65 & over $6-$11, children $5-$10. Closed Easter; Thanksgiving; Christmas. &
Attendance: 350,000 (accurate)
Membership: SCI: One Plus One $70; Family & Grand Pass $90; Family Plus $120. Combination memberships available with the Blank Park Zoo & Living History Farms.

STATE HISTORICAL MUSEUM OF IOWA, 600 E. Locust St., Des Moines, IA 50319-1006. Tel.: 515-281-5111. Fax: 515-282-0502. TDD: 515-242-5147.
E-mail: susan.kloewer@iowa.gov
Web Site: www.iowahistory.org
Formerly: State Historical Society of Iowa
Founded: 1892.
Congressional District: 4
Key Personnel: State Cur. & Historical Sites Coord., Jerome Thompson; Dir. Iowa Dept. Cultural Affairs, Mary Cownie; Museum Dir., Susan Kloewer; Registrar, Jodi Evans; Exhibits Mgr., Andrew Harrington; Museum Cur., Leo Landis; Conservator, Pete Sixbey.

Governing Authority: state. Affiliated with State Historical Society of Iowa, Historical Division. Parent Institution: Iowa Dept. of Cultural Affairs. Subsidiary Institution: Museum Bureau. Tax-exempt.
Institution Type/Description: History Museum.
Collections: archaeology; ornithology; geology; paleontology; neontology; decorative arts; crafts; clothing; textiles; agricultural; tools & equipment; medical equipment; transportation; politics; social history.
Research Fields: textiles; furniture; military; social history.
Facilities: library of history, genealogy, manuscripts, state archives, Iowa census records, & newspapers.
Activities: guided tours.
Publications: Iowa Heritage Illustrated; Annals of Iowa; Iowa Historian.
Hours & Admission Prices: Mon.-Sat. 9-4:30, Sun. 12-4:30. No charge; donations accepted. Closed state holidays. ♿
Attendance: 51,000 (estimated)
Membership: Basic $50; Heritage Circle $100.

TERRACE HILL HISTORIC SITE AND GOVERNOR'S MANSION, 2300 Grand Ave., Des Moines, IA 50312-5308. Tel.: 515-242-5841.
E-mail: barbara.filer@iowa.gov
Web Site: www.terracehilliowa.org
Founded: 1971.
Key Personnel: Admin., Barbara Filer; Horticulturist, Montgomery Lovell; Coord. Communications & Programs, Meredith Sillau; Maintenance, Mike Miner.
Personnel Profile: Full-Time Paid 5; Part-Time Volunteers 80.
Governing Authority: state government. Subsidiary Institution: Terrace Hill Foundation. Tax-exempt.
Institution Type/Description: Historic Site: restored to the opulent Victorian lifestyle of the late 1880s-early 1900s, a mansion, carriage house & gardens are situated on 9 acres.
Collections: concentration on the Victorian era, late 1880s to early 1900s.
Facilities: 20-vol. library of Victorian information relating to decorating, gardening & entertaining; research & reference materials; educational facilities; 15,000 sq. ft. exhibit space; formal gardens. Museum-related items for sale.
Activities: docent program; internships; guided tours; lectures. Annual Events: Jazz in July; Holly & Ivy Holiday Tour.
Publications: quarterly newsletter, Terrace Hill; annual magazine, Terrace Hill; books, Little Man with the Long Shadow; Our Governor's Mansions; cookbook, Fresh from Terrace Hill.
Hours & Admission Prices: March-Dec. Tues.-Sat. 10-1:30. Tours at 10:30, 11:30, 12:30 & 1:30. Adults $5, children 6-12 $2. Terrace Hill Society members one free admission per year. Closed state holidays; New Year's Eve; Christmas Eve. ♿
Attendance: 10,703 (accurate)
Membership: Individual $15; Household $25; Supporting $50; Patron $100; Benefactor $250; Preservation Club $500 & up.

THE WALLACE CENTERS OF IOWA - WALLACE HOUSE FOUNDATION, 756 16th St., Des Moines, IA 50314-1601. Tel.: 515-243-7063.
E-mail: info@wallace.org
Web Site: www.wallace.org
Founded: 1988.
Congressional District: 3
Governing Authority: Tax-exempt.
Institution Type/Description: Historic House Museum: built in 1883.
Collections: Wallace family artifacts & photographs; period furnishings.
Hours & Admission Prices: Tues.-Fri. 9-2; other times by appointment. No charge; donations accepted.
Attendance: 1,000 (estimated)
Membership: Learner $25; Grown $50; Believer $100; Thinker $250; Innovator $500; Dreamer $1,000.

Diagonal

DIAGONAL PRINTING MUSEUM, 100 E. 1st St., Diagonal, IA 50845. Tel.: 641-734-5540.
E-mail: ringgoldtourism@gmail.com
Institution Type/Description: History Museum.
Collections: printing history; printing press & Linotype; period artifacts; genealogy.
Hours & Admission Prices: Memorial Day to Labor Day call for hours. ♿

Dow City

DOW HOUSE HISTORIC SITE, 513 S. Prince St., Dow City, IA 51528. Mailing Address: Crawford County Conservation Board, 1202 Broadway, Courthouse 1st Fl., Denison, IA 51442. Tel.: 712-263-2748. Fax: 712-263-3352.
Web Site: www.crawfordcountyconservationboard.com/historic.html
Founded: 1974.
Key Personnel: Dir., Lance Nelson; Chm. (V), Mary Eiten; Pres. (V), Neal Moeller.
Personnel Profile: Part-Time Volunteers 3.
Governing Authority: Parent Institution: Crawford County Conservation Board. Tax-exempt.
Institution Type/Description: Historic House: housed in the home of businessman, Simon E. Dow; built in 1872. Listed on the National Register of Historic Places.
Collections: Dow family history; personal artifacts; period furnishings; photographs; 1870 era farm equipment.
Activities: Museum Sponsors: Vintage Christmas Tours.
Hours & Admission Prices: Memorial Day to Labor Day Sat.-Sun. 12-4. Adults $2, children 14 & under $1.
Attendance: 500 (estimated)

Dows

DOWS DEPOT WELCOME CENTER, 1896 Railroad St., Dows, IA 50071. Mailing Address: P.O. Box 287, Dows, IA 50071-0287. Tel.: 515-852-3595.
Web Site: www.dowsiowa.com/welcomecenter.html
Institution Type/Description: Railroad Museum: Wright County's first depot built in 1896. Listed on the National Register of Historical Places.
Collections: railroad memorabilia; local period artifacts.
Facilities: Museum-related items for sale.
Hours & Admission Prices: Mon.-Fri. 9:30-4:30, Sun. 1-4. No Charge. Closed New Year's Day; Easter; Thanksgiving; Christmas. ♿
Attendance: 4,000 (estimated)

1880 ONE-ROOM SCHOOLHOUSE, 201 E. Ellsworth, Dows, IA 50071. Tel.: 515-852-3595.
E-mail: dowsdepot@fbx.com
Web Site: dowsiowa.com
Institution Type/Description: Historic Building & Museum.
Collections: local history; period furnishings; school artifacts; schoolbooks; potbellied stove; desks; dunce stool.
Hours & Admission Prices: Call for hours.

EVANS LITTLE PRAIRIE HOUSE, 201 W. Railroad St., Dows, IA 50071. Tel.: 515-852-3595. Fax: 515-852-4326.
Web Site: www.dowsiowa.com
Institution Type/Description: Historic House Museum: built late 1800s.
Collections: local history; period furnishings.
Hours & Admission Prices: Mon.-Sat. 9-5, Sun. 12-5. No charge.

QUASDORF BLACKSMITH AND WAGON MUSEUM, Railroad St., Dows, IA 50071. Mailing Address: P.O. Box 312, Train St. at Depot, Dows, IA 50071. Tel.: 515-852-3595. Fax: 515-852-4326.
Web Site: www.dowsiowa.com/quasdorfmuseum.html
Institution Type/Description: Historic House Museum: built in 1899. Listed on the National Register of Historic Places.
Collections: machines; tools; belt-driven & electric welding equipment; wagon wheels; blacksmithing artifacts; books.
Hours & Admission Prices: Mon.-Sat. 9-5, Sun. 12-5. Call for hours. No charge.

Dubuque

DUBUQUE ARBORETUM & BOTANICAL GARDENS, 3800 Arboretum Dr., Dubuque, IA 52001-1040. Tel.: 563-556-2100. Fax: 563-556-2443.
E-mail: dabg@qwestoffice.net
Web Site: www.dubuquearboretum.com
Founded: 1980.
Congressional District: 2
Key Personnel: Pres. (V), Jack Frick; Financial Dir. & Museum Shop Mgr., Ms. Barbara Barton; Horticulturist, Mr. Jim Schwarz; Devel., Mr. Lloyd Streief; Rosarian, Mr. Marlyn Bausman.
Personnel Profile: Full-Time Volunteers 1; Part-Time Volunteers 360.
Governing Authority: private; nonprofit organization. Parent Institution: Dubuque Arboretum, Inc. Tax-exempt: 501(c)(3).

Institution Type/Description: Arboretum & Botanical Garden.
Collections: All-American rose selections: hybrid, miniature, shrub & old garden roses; All-American annuals & perennials; Dwarf Conifer collections: Bill Walter collection, Hermsen collection; formal herb garden; Hostas: over 13,000 plants, 850 varieties; woodland wildflowers; water & shade gardens; prairie wildflowers & grasses; ornamental tree & shrub collections: viburnam, azalea, lilacs; displays: iris, peony, lily, day lily, chrysanthemum, dahlia, cactus; Japanese garden.
Research Fields: Japanese garden.
Facilities: 2,000-vol. library; botanical garden; educational facilities. Gift items for sale.
Activities: concerts; dance recitals; formal education programs; guided tours; hobby workshops; lectures; participatory exhibits; rental gallery; school loan service.
Publications: quarterly newsletter, Groundcover; yearly program, Programs & Events; yearly music program, Music in the Gardens; rose festival program.
Hours & Admission Prices: Arbor Day-Oct. 8am to sunset; Nov.-April Mon.-Fri. 9-5, Sat. 9-1. Gift Shop & Library: 9-6. No charge; donations accepted. Closed New Year's Day; Thanksgiving; Christmas. ⅷ
Attendance: 100,000 (estimated)
Membership: Individual $25; Family $50; Supporting $100.

✳ **DUBUQUE MUSEUM OF ART, (M),** 701 Locust St., Dubuque, IA 52001-6817. Tel.: 563-557-1851. Fax: 563-557-7826.
E-mail: info@dbqart.com
Web Site: www.dbqart.com
Founded: 1874.
Congressional District: 1
Key Personnel: Exec. Dir., Mark D. Wahlert; Deputy Dir., Diane Sass; Pres. (V), Christopher Wand; Dir. Education, Margaret Buhr; Mgr. Collections & Exhibitions, Stacy Gage.
Personnel Profile: Full-Time Paid 4; Part-Time Paid 4; Part-Time Volunteers 25; Interns 3.
Governing Authority: nonprofit. Tax-exempt: 501(c)(3).
Institution Type/Description: Art Museum
Collections: state & national art.
Facilities: classrooms.
Activities: guided tours; lectures; gallery talks; concerts; arts festivals; formally organized education programs; temporary exhibitions. Museum Sponsors: bimonthly exhibits.
Publications: quarterly newsletter.
Hours & Admission Prices: Tues-Fri. 10-5, Sat.-Sun. 1-4. No charge; donations accepted. ⅷ
Attendance: 8,527 (accurate)
Membership: Student $15; Individual & Senior Citizens $40; Family $60; Partners $100; Donors $250; Benefactors $500; Director's Circle $1,000; Visionaries $2,000 & up.

✳ **MATHIAS HAM HOUSE HISTORIC SITE,** 2241 Lincoln Ave., Dubuque, IA 52001-1424. Mailing Address: 350 E. 3rd St., Dubuque, IA 52001-2302. Tel.: 800-226-3369; 563-557-9545 (local). Fax: 563-583-1241.
E-mail: info@rivermuseum.com
Web Site: www.rivermuseum.com
Founded: 1964.
Congressional District: 2
Key Personnel: Pres. & C.E.O., Jerry A. Enzler; Vice Pres. Operations, Alan Stache; Historical Cur., Cristin Waterbury; Vice Pres. Organizational Devel., Ginger Sakas; Dir. Mktg., John Sutter; Mgr. Sales & Mktg., Nate Breitsprecker; Museum Shop Mgr., Mary Jo Gothard; Coord. Mktg., Emily Graves; Education, Mark Wayner; Registrar, Tish Boyer.
Personnel Profile: Full-Time Paid 1; Part-Time Paid 8; Part-Time Volunteers 100.
Governing Authority: private; nonprofit society. Parent Institution: Dubuque County Historical Society. Subsidiary Institutions: National Mississippi River Museum & Aquarium; Old Jail Museum. Tax-exempt: 501(c)(3).
Institution Type/Description: Historical Society Museum.
Collections: local history. Historic Buildings: c.1856 restored Victorian Mansion; c.1833 log cabin; c.1883 one-room schoolhouse; c.1840 granary.
Research Fields: local history; utilitarian & decorative arts.
Facilities: Museum-related gift items for sale.
Activities: guided tours; walking tours; preservation of buildings; architectural program for county schools; permanent, temporary & traveling exhibitions.
Hours & Admission Prices: Memorial Day to Labor Day Wed.-Sun. 11-4. Adults $5, children 7-17 $3.50; discounts to museum & Time Travelers members no charge.
Attendance: 1,250 (estimated)
Membership: Supporter: Individual $65; Family & Grandparents $125. Na-

tional Rivers Hall of Fame: Pilot $175; Chief Pilot $250; Captain $500; Commodore & Corporate $1,000.

NATIONAL MISSISSIPPI RIVER MUSEUM & AQUARIUM, (M), 350 E. 3rd St., Dubuque, IA 52001-2302. Tel.: 800-226-3369; 563-557-9545. Fax: 563-583-1241. Facebook; River Museum.
E-mail: info@rivermuseum.com
Web Site: www.rivermuseum.com
Founded: 1950.
Congressional District: 2
Key Personnel: Pres. & C.E.O., Jerry A. Enzler; Pres. & Chm. (V), Tim Butler; Vice Pres. Organizational Devel., Ginger Sakas; Devel. Coord., Vicky Sutter; Cur., Cristin Waterbury; Vice Pres. Operations, Alan Stache; Project Mgr., John Oglesby; Sales Mgr., Nate Breitsprecker; Membership, Donor & Youth Group Mgr., Melissa Wersinger; Office Mgr., Marilyn Snyder; Exhibit Design, Wayne McDermott; Mktg. Dir., John Sutter; Registrar, Tish Boyer; Communications Editor, Amanda Dolter; Museum Shop Mgr., Mary Jo Gothard.
Personnel Profile: Full-Time Paid 39; Part-Time Paid 22; Part-Time Volunteers 450; Interns 4.
Governing Authority: private; nonprofit society. Parent Institution: Dubuque County Historical Society. Subsidiary Institutions: Mathias Ham House Historic Site; Old Jail Museum. Tax-exempt: 501(c)(3).
Institution Type/Description: Marine Museum & Aquarium.
Collections: upper Mississippi River & Rivers of America; canoes; flatboats; steamboats; artifacts of Upper Mississippi River Valley; steam engines; towboating; river history; National Rivers Hall of Fame. Historic Structures: 1934 riverboat, sidewheeler William M. Black, 1940 towboat Logsdon; 1901 renovated railroad freight house; blacksmith shop.
Research Fields: river history; especially Upper Mississippi River Valley; amphibian conservation.
Facilities: 18,000 item library & archives on river history; 3D/4D theatre; National Rivers Hall of Fame; 12 aquariums; Mississippi River Center; National River Center; outdoor connecting Mississippi Plaza. Cafe. Museum-related items for sale.
Activities: guided tours; school programs; publications; lectures; theater; participatory & traveling exhibits; rental facilities; films.
Publications: walking tours; monthly member e-newsletter; annual report; Kids Club.
Hours & Admission Prices: Memorial Day to Labor Day daily 9-6; Sept.-Oct. daily 9-5; Nov. to May daily 10-5. Adults $15, seniors 65 & over $13, youth 3-17 $10; discounts to AAA members; members & children under 3 no charge. AZA Reciprocal Program and Time Travelers Reciprocal Program. Closed Thanksgiving; Christmas. ⅷ
Attendance: 180,000 (estimated)
Membership: Supporter: Individual $65; Family & Grandparents $125. National Rivers Hall of Fame: Pilot $175; Chief Pilot $250; Captain $500; Commodore & Corporate $1,000.

OLD JAIL MUSEUM, 8th & Central Ave., Dubuque, IA 52001. Mailing Address: 350 E. Third St., Dubuque, IA 52001. Tel.: 563-557-9545. Fax: 563-583-1241.
E-mail: info@rivermuseum.com
Web Site: www.rivermuseum.com/features_historicsites_jail.cfm
Founded: 2005.
Congressional District: 2
Key Personnel: Pres. & C.E.O., Jerry Enzler; Vice Pres. Organizational Devel., Ginger Sakas; Vice Pres. Operations, Alan Stache; Education, Mark Wagner; Dir. Mktg. & Sales, John Sutter; Coord. Mktg., Emily Graves; Registrar, Tish Boyer; Cur., Cristin Waterbury; Museum Shop Mgr., Mary Jo Gothard.
Personnel Profile: Full-Time Volunteers 1; Part-Time Paid 8; Part-Time Volunteers 100.
Governing Authority: private; nonprofit organization. Parent Institution: Dubuque County Historical Society, 350 E. 3rd St., Dubuque, IA 52001. Tax-exempt: 501(c)(3).
Institution Type/Description: Historic Building: housed in the former Dubuque County Jail; built in 1857. Listed on the National Register of Historic Places & a National Landmark Building.
Collections: county history; personal artifacts; photographs; period furnishings.
Research Fields: local history
Facilities: Museum-related items for sale.
Activities: guided tours; loan & temporary exhibitions.
Hours & Admission Prices: Memorial Day to Labor Day Wed.-Sun. 11-4. Adults $5, children 7-17 $3.50; discounts to members; children 6 & under no charge.
Attendance: 1,525 (estimated)
Membership: Supporter: Individual $65; Family & Grandparent $125. National

River Hall of Fame: Pilot $175; Chief Pilot $250; Captain $500; Commodore & Corporate $1,000.

Dunlap

MCLEAN MUSEUM AND DOUGAL HOUSE, 1211 Iowa Ave., Dunlap, IA 51529-1538. Tel.: 712-643-5721 & 5908.
E-mail: dunlapia@loganet.net
Founded: 1988.
Governing Authority: Parent Institution: Dunlap Historical Society.
Institution Type/Description: History Museum.
Collections: local history & culture; photographs; period furnishings; personal artifacts; plat books; assessor's records; local industry; Booster Buck. Historic Buildings: 1879 brick church; 1870 barn.
Activities: Annual Event: Barn Festival in July.
Hours & Admission Prices: Summer: Sat.-Sun. call for hours; other times by appointment.

Dyersville

DYER-BOTSFORD VICTORIAN HOUSE AND DOLL MUSEUM, 331 1st Ave., Dyersville, IA 52040. Tel.: 563-875-2414.
E-mail: dyersvillehs@iowatelecom.net
Web Site: dyersvillehistory.com
Institution Type/Description: Historic House & Doll Museum: housed in an 1850 Victorian home.
Collections: local history & culture; period furnishings; personal artifacts; over 2,000 dolls; hand-carved miniature circus; replica of 1850 castle; German leather Christmas tree; German, Polish & Czech ornaments.
Hours & Admission Prices: May to early Nov. Mon.-Fri. 10-4, Sat.-Sun. 1-4.

FIELD OF DREAMS MOVIE SITE, 28995 Lansing Rd., Dyersville, IA 52040-8005. Tel.: 888-875-8404. Fax: 563-875-7253.
E-mail: fodbetty@aol.com
Web Site: www.fodmoviessite.com
Founded: 1989.
Institution Type/Description: Movie Site: Iowa farm used for the filming of Field of Dreams.
Collections: baseball field.
Facilities: concession stand. Museum-related items for sale.
Activities: play ball on baseball field.
Publications: brochures.
Hours & Admission Prices: April-Nov. daily 9-6. No charge; donations accepted. &
Attendance: 60,000 (estimated)

NATIONAL FARM TOY MUSEUM, 1110 16th Ave. Ct., S.E., Dyersville, IA 52040-2374. Tel.: 563-875-2727. Fax: 563-875-8467.
E-mail: farmtoys@dyersville.com
Web Site: www.national/farmtoymuseum.com
Founded: 1986.
Key Personnel: Exec. Dir., Jacque Rahe; Chm. (V), Dave Bell; Membership Coord., Amanda Schwartz; Asst. Mgr. & Museum Shop Mgr., Ellen Hunt.
Personnel Profile: Full-Time Paid 2; Part-Time Paid 12.
Governing Authority: Parent Institution: Dyersville Industries, Inc. Tax-exempt: 501(c)(3).
Institution Type/Description: Farm Toy.
Collections: farm toys; tractors; trucks; life-size toy tractors; toy pedal tractors; toy manufacturers; collectors toys.
Hours & Admission Prices: Daily 8-6. Adults $5, seniors 65 & over $4, youth 6-17 $3; discounts to AAA & IMA members; children 5 & under no charge. Closed New Year's Day; Easter; Thanksgiving; Christmas. &
Attendance: 15,000 (estimated)
Membership: Individual $25; Family $50; Enthusiast $100; Collector $150.

Eldon

AMERICAN GOTHIC HOUSE CENTER, 300 American Gothic St., Eldon, IA 52554-9654. Tel.: 641-652-3352. Fax: 641-652-3352. Facebook: American Gothic House.
E-mail: theamericangothichouse@gmail.com
Web Site: www.americangothichouse.net
Founded: 2007.
Congressional District: 2
Key Personnel: Pres. (V), Steve Siegel; Admin., Holly Berg.
Personnel Profile: Full-Time Paid 1; Part-Time Paid 1; Part-Time Volunteers 30; Interns 1.
Operating Expenses: 107,951

Operating Income: 63,282
Governing Authority: county. Tax-exempt.
Institution Type/Description: Historic House: housed next to the home used as the backdrop of Grant Wood's 1930 painting; built in 1882. Listed on the National Register of Historic Places.
Collections: Grant Wood's life & art; American Gothic history & parodies; video of Grant Wood.
Facilities: Museum-related items for sale.
Activities: take costumed photos in front of the house; video about Grant Wood; educational programs; rental facilities; children's activities.
Publications: e-newsletter.
Hours & Admission Prices: May-Sept. Sun.-Mon. 1-4, Tues.-Sat. 10-5; Oct.-April Tues.-Fri. 10-4, Sat.-Mon. 1-4. No charge; donations accepted. Closed New Year's Day; Martin Luther King Jr. Day; Presidents' Day; Veterans Day; Thanksgiving & day after; Christmas Eve & Day. &
Attendance: 15,000 (estimated)

Eldora

ELDORA WELCOME CENTER AND RAILROAD MUSEUM, 1215 Park St., Eldora, IA 50627. Mailing Address: 1442 Washington, Eldora, IA 50627-1633. Tel.: 641-939-3241. Fax: 641-939-7555.
E-mail: eldoraecondeu@heartofiowa.net
Web Site: www.eldoraiowa.com
Institution Type/Description: Visitors Center & History Museum.
Collections: local history; railroad memorabilia.
Publications: brochures.
Hours & Admission Prices: Memorial Day to Oct. 1 Sat. 12-4, Sun. 1-5; other times by appointment. No charge.

HARDIN COUNTY FARM MUSEUM, 203 Washington St., Eldora, IA 50627. Mailing Address: P.O. Box 41, Eldora, IA 50627-0041. Tel.: 641-939-7107.
E-mail: farmmuseum@heartofiowa.net
Web Site: eldoraiowa.com/pages/farm-museum
Founded: 1998.
Congressional District: 4
Key Personnel: Pres. (V), Jason Reinertson.
Personnel Profile: Part-Time Volunteers 7.
Governing Authority: Tax-exempt.
Institution Type/Description: Farm Museum.
Collections: farm history & equipment; agriculture industry; farm life; round-roofed barn; farm buildings.
Activities: special events. Museum Sponsors: tractor pulls; threshing days; craft & antique shows; barn dance.
Publications: newsletter.
Hours & Admission Prices: By appointment. No charge; donations accepted.
Attendance: 1,505 (accurate)
Membership: Single $10; Family $15.

HARDIN COUNTY HISTORICAL SOCIETY, 1603 Washington St., Eldora, IA 50627. Mailing Address: P.O. Box 187, Eldora, IA 50627-0187. Tel.: 641-939-5137.
E-mail: dlbabcock@heartofiowa.net
Web Site: hardincountyhistoricalsociety.org
Founded: 1972.
Key Personnel: Chm. (V) & Pres. (V), Leola Babcock; Cur., John Farmer.
Personnel Profile: Part-Time Volunteers 9.
Governing Authority: Tax-exempt.
Institution Type/Description: Historical Society Museum.
Collections: local history & culture; period furnishings & artifacts; photographs; documents; books; records.
Publications: newsletter, The Hitching Post.
Hours & Admission Prices: May-Oct. 3rd Sun. of the month 1-4; other times by appointment. No charge; donations accepted. &
Attendance: 1,000 (estimated)
Membership: Single $10; Family $15.

Elgin

GILBERTSON NATURE CENTER, 22580 A Ave., Elgin, IA 52141-9567. Tel.: 563-426-5740.
Institution Type/Description: Nature Center.
Collections: wildlife & their habitats; hands-on exhibits. Historic Buildings: Mavis & Conner Dummermuth Building including period tools & artifacts; Hart Dummermuth House including period furnishings.
Activities: petting zoo.
Hours & Admission Prices: Center: Memorial Day to Labor Day Wed.-Sun. 11-7; other times by appointment. Petting Zoo: May by appointment.

MAVIS & CONNER DUMMERMUTH HISTORICAL BUILD-ING AND HART DUMMERMUTH HISTORICAL HOUSE, 22580 A Ave., Elgin, IA 52141. Tel.: 563-426-5740.
E-mail: gncfccb@alpinecom.net
Web Site: www.elginiowa.org/GILBERTSON.html
Institution Type/Description: Historic Buildings.
Collections: farm & home history & artifacts; period furnishings; photographs; personal artifacts.
Hours & Admission Prices: mid-May to Labor Day call for hours. &

Elk Horn

BEDSTEMOR'S HOUSE, 2105 College St., Elk Horn, IA 51531-8005. Mailing Address: 2212 Washington St., Elk Horn, IA 51531. Tel.: 800-759-9192. Fax: 712-764-7002.
E-mail: registrar@danishmuseum.org
Web Site: www.danishmuseum.org
Founded: 1990.
Key Personnel: Exec. Dir., John Mark Nielson.
Personnel Profile: Part-Time Paid 3; Part-Time Volunteers 1.
Governing Authority: Parent Institution: The Danish Immigrant Museum. Tax-exempt.
Institution Type/Description: Historic House Museum.
Collections: Danish heritage & culture; local history; period furnishings; personal artifacts.
Activities: weekend programs, games & crafts demonstrations. Annual Events: Tivoli Fest in May; Julefest in November.
Hours & Admission Prices: May 15-Sept. 15 daily 1-4. Adults $5; discounts to AAM members; members no charge.
Attendance: 700 (accurate)

THE DANISH IMMIGRANT MUSEUM, (M), 2212 Washington St., Elk Horn, IA 51531-2116. Tel.: 712-764-7001. Fax: 712-764-7002.
E-mail: info@danishmuseum.org
Web Site: www.danishmuseum.org
Founded: 1983.
Congressional District: 4
Key Personnel: Exec. Dir., John Mark Nielsen; Pres. (V), Marc Petersen; Treas., John Molgaard; Museum Shop Mgr., Joni Soe-Butts.
Personnel Profile: Full-Time Paid 9; Part-Time Paid 4; Part-Time Volunteers 100.
Governing Authority: private; nonprofit organization. Tax-exempt: 501(c)(3).
Institution Type/Description: Danish Immigrant Museum.
Collections: 35,000 artifacts documenting the life, culture & diversity of Danish-American immigrants to North America & their descendants; period furnishings. Historic House: Bedstemor's turn-of-the-century home.
Research Fields: Danish American material culture; transformations of Danish customs & traditions in an American setting.
Facilities: 13,000-vol. library; 10,000 sq. ft. exhibit space. Gift items for sale.
Publications: The America Letter.
Hours & Admission Prices: Museum: Mon.-Fri. 9-5, Sat. 10-5, Sun. 12-5. Bedstemor's House: May 15-Sept. 15 daily 1-4. Family History Center: May-Oct. Tues.-Wed. & Sat. 9-5; Nov.-April Tues.-Wed. & Fri. 10-4. Adults $5, children 8-17 $2; members no charge. Closed New Year's Day; Easter; Thanksgiving; Christmas. &
Attendance: 10,000 (estimated)
Membership: Active $30; National $50; Contributing & Business/Organization $100; Sustaining $250; Sponsoring $500; Benefactor $1,000; Patron $2,500 & up.

Elkader

CARTER HOUSE MUSEUM, 101 S.E. High St., Elkader, IA 52043-0444. Mailing Address: c/o Elkader Historical Society, P.O. Box 444, Elkader, IA 52043-0444. Tel.: 563-245-1573.
E-mail: carterhousemuseum@alpinecom.net
Web Site: www.carterhousemuseum.com
Founded: 1983.
Key Personnel: Dir. & Pres. (V), Betty Buchholz.
Personnel Profile: Part-Time Volunteers 10.
Volunteer Hours: 2,200
Operating Expenses: 13,240
Operating Income: 8,656
Governing Authority: Tax-exempt.
Institution Type/Description: Historic House Museum: built in 1855.
Collections: local history & culture; period furnishings; clothing; personal artifacts; military uniforms; farm tools; early fire-fighting equipment; early medical supplies & equipment; 19th century drug store supplies; fossils.

Major Exhibits: Old 1928 Glass Negatives, 5/14-9/14.
Research Fields: local history of ethnic areas.
Facilities: 18-room Carter House; large room for exhibits & research.
Activities: holiday exhibits & entertainment.
Hours & Admission Prices: Memorial Day to Labor Day Sat. 10-4, Sun. 12-4; groups by appointment. Adults $4, senior citizens $3. &
Attendance: 500 (estimated)

OSBORNE VISITOR, WELCOME AND NATURE CENTER, 29862 Osborne Rd., Elkader, IA 52043. Tel.: 563-245-1516. Fax: 563-245-2222.
E-mail: cccb@claytoncountyia.gov
Web Site: claytoncountyconservation.org
Personnel Profile: Full-Time Paid 6; Part-Time Paid 3; Part-Time Volunteers 50; Interns 5.
Governing Authority: Parent Institution: Clayton County.
Institution Type/Description: Visitor Center & Nature Center.
Collections: wildlife & their habitats; conservation; arboretum; pioneer village.
Facilities: nature center; wild animal park; nature trails; classrooms. Museum-related items for sale.
Activities: special events. Annual Events: Heritage Days in October.
Publications: newsletter; Osborne Oracle; Milling Around.
Hours & Admission Prices: mid-April to mid-Oct. Mon.-Sat. 8-4, Sun. 12-4; mid-Oct. to mid-April Mon.-Sat. 8-4. No charge; donations accepted.
Attendance: 8,063 (accurate)

Emmetsburg

VICTORIAN ON MAIN, 1703 Main St., Emmetsburg, IA 50536-1653. Tel.: 712-852-3781.
Institution Type/Description: Historic House: built in 1883.
Collections: period furnishings.
Hours & Admission Prices: Sun. 1-4; other times by appointment.

Estherville

H.G. ALBEE MEMORIAL MUSEUM, 1720 Third Ave., S., Estherville, IA 51334. Mailing Address: P.O. Box 101, Estherville, IA 51334-0101. Tel.: 712-362-2750.
Founded: 1964.
Congressional District: 6
Key Personnel: Pres., Mildred Bryan; Sec., Mary Gray; Treas., Mike Maloney.
Personnel Profile: Part-Time Volunteers 60.
Governing Authority: nonprofit organization. Parent Institution: Emmet County Historical Society, Inc. Branch Museums: Restored Country School; Country Church. Tax-exempt: 501(c)(3).
Institution Type/Description: Historical Society Museum.
Collections: general collection pertaining to Emmet County & country schools; horse drawn farm machinery; country church; blacksmith shop.
Research Fields: genealogy.
Facilities: 20-vol. library of photo books with captions, brief history & special events from each township in the county available for use on premises.
Activities: guided tours; permanent & temporary exhibitions.
Publications: Emmet County History, Vol. III.
Hours & Admission Prices: June-Aug. daily 2-5; other times by appointment. No charge; donations accepted. &
Attendance: 1,200 (accurate)
Membership: Individual $5; Life $50.

Exira

AUDUBON COUNTY HISTORICAL SOCIETY, Courthouse Museum, Washington St., Exira, IA 50076. Mailing Address: 1745 160th St., Audubon, IA 50025-7483. Tel.: 712-563-3984.
Founded: 1960.
Congressional District: 5
Key Personnel: Pres., Carma Hutchins; Vice Chm. (V), William Roth.
Personnel Profile: Part-Time Paid 1; Part-Time Volunteers 20.
Governing Authority: society. Branch Museums: Nathaniel Hamlin Museum, Hwy. 71, Audubon, IA 50025; Old Court House Museum, Exira, IA. Tax-exempt.
Institution Type/Description: History Museum.
Collections: Indian history; 1958 Audubon County flood memorabilia; painting; 100 ft. painted mural.
Research Fields: development of hybrid corn.
Facilities: 50-vol. library of county maps & other items available for research by special appointment.
Activities: guided tours; permanent & temporary exhibitions. Museum Sponsors: Fall Festival in September.

Publications: brochure; book, Mudroad Mac.
Hours & Admission Prices: Memorial Day-Labor Day Sun. 1:30-4:30. Adults $3; discounts to members; students no charge. &
Attendance: 1,500 (estimated)
Membership: Annual $2; Life $10.

Fairfield

CARNEGIE HISTORICAL MUSEUM, 112 S. Court, Fairfield, IA 52556. Mailing Address: P.O. Box 502, Fairfield, IA 52556-0009. Tel.: 641-472-6343.
E-mail: carnegie@lisco.com
Web Site: www.fairfieldmuseum.com
Founded: 1870.
Congressional District: 3
Key Personnel: Pres., Gene Luedtke; Bd. Member, Jaimie Johnston; Bd. Member, Mandy Mellum; Dir., Mark Shafer; Treas., Keith Dimmitt.
Personnel Profile: Part-Time Paid 1.
Governing Authority: municipal. Tax-exempt.
Institution Type/Description: General Museum: housed in Carnegie library building.
Collections: Indian artifacts; Iowa pioneer artifacts; period furnishings; mounted birds; mammals; rocks; period drug store; Civil War artifacts; Roman antiquities; Parsons College collection; international ethnic curios; dolls.
Activities: guided tours; gallery talks. Museum Sponsors: 1st Friday Art Walk.
Hours & Admission Prices: Tues., Thurs. & Sat. 1-4; other times by appointment. First Friday Art Walk: 6pm-9pm. No charge; donations accepted. &
Membership: Students $5; Individual $10; Family $15; Contributing & Business $25; Sustaining $50; Patron $100; Life $200.

FAIRFIELD ART ASSOCIATION, 200 N. Main, Fairfield, IA 52556-2835. Tel.: 641-472-2000. Fax: 641-472-7890.
E-mail: info@fairfieldacc.com
Web Site: www.fairfieldacc.com/artgallery.html
Founded: 1964.
Key Personnel: Pres. & Volunteer Dir., Suzan Kessel.
Governing Authority: nonprofit organization. Tax-exempt.
Institution Type/Description: Arts & Convention Center.
Collections: paintings & sculptures by Iowa & Midwest artists.
Facilities: classrooms; gallery.
Activities: guided tours; lectures; films; gallery talks; arts festivals; formally organized education programs; temporary & traveling exhibitions.
Publications: monthly newsletter, Fairfield Art Association Newsletter.
Hours & Admission Prices: Mon.-Fri. 9-5; call for additional hours. No charge; donations accepted. &
Attendance: 1,000 (estimated)
Membership: Student $5; Individual $10; Family $15; Contributing $25; Friends $50; Patron $100.

Forest City

HERITAGE PARK OF NORTH IOWA - WINNEBAGO HISTORICAL SOCIETY, 336 Clark St. N., Forest City, IA 50436. Tel.: 641-585-4332.
E-mail: jjnoulman@wctatel.net
Web Site: heritageparkofnorthiowa.com
Founded: 1999.
Congressional District: 4
Key Personnel: Chm. (V), Jim Oulman; Chm. (V), Wyndham Sellers
Governing Authority: Parent Institution: Winnebago Historical Society. Tax-exempt.
Institution Type/Description: Regional History Museum.
Collections: farm; equipment; transportation.
Hours & Admission Prices: May-Oct. Sun. 1-4; other times by appointment. Events: $2-$6.
Attendance: 5,000 (estimated)
Membership: Individual $25; Family $50; Family $300; Corporate $100; Founder $1,000.

MANSION MUSEUM - WINNEBAGO HISTORICAL SOCIETY, 336 Clark St. N., Forest City, IA 50436. Mailing Address: P.O. Box 27, Forest City, IA 50436. Tel.: 641-581-3283.
E-mail: jnoulman@wctatel.net
Founded: 1977.
Congressional District: 4
Key Personnel: Chm. (V) & Pres. (V), Riley Lewis.
Personnel Profile: Part-Time Volunteers 15; Interns 2.

Governing Authority: Tax-exempt.
Institution Type/Description: Historical Society Museum: built c.1900.
Collections: local history & culture; period furnishings; personal artifacts; photographs.
Hours & Admission Prices: May-Oct. Sun. 1-4; other times by appointment. Adults $2; members no charge.
Attendance: 500 (estimated)
Membership: Individual $25; Family $50; Family Life $300; Corporate $100; Founder $1,000.

Fort Dodge

BLANDEN MEMORIAL ART MUSEUM, 920 Third Ave., S., Fort Dodge, IA 50501-4723. Tel.: 515-573-2316. Fax: 515-573-2317.
E-mail: pkay@blanden.org
Web Site: www.blanden.org
Founded: 1930.
Congressional District: 5
Key Personnel: Dir., Margaret A. Skove; Pres., Dr. Kenneth Adams; Business Office, Pam Kay; Educator, Linda Flaherty; Maintenance & Security, Mark Jessen.
Personnel Profile: Full-Time Paid 3; Part-Time Paid 2; Part-Time Volunteers 22.
Governing Authority: municipal. Parent Institution: City of Fort Dodge. Subsidiary Institution: Blanden Charitable Foundation. Tax-exempt: 501(c)(3).
Institution Type/Description: Art Museum.
Collections: American & European paintings & sculpture; 16th- to 21st-century prints; drawings; 17th- to 21st-century Asian art; 19th- & 20th-century tribal art.
Major Exhibits: Mexican Folk Art (T), 10/13-2/14.
Research Fields: 19th & 20th-century European & American painting, prints & sculpture.
Facilities: 4,000-vol. non-circulating fine arts library available to public; studio art classrooms. Cards; jewelry; hand-blown glass & pottery items for sale.
Activities: permanent & traveling exhibitions; gallery talks; artist's lectures; guided tours; education program; community outreach program; concerts; outreach art education program; artist-in residence workshops; music & vocal performances.
Publications: triannual color magazine; exhibition catalogues; 76 Years of Collecting: Blanden Art Museum.
Hours & Admission Prices: Tues.-Sat. 11-5. No charge; donations accepted. Closed holidays. &
Attendance: 13,421 (accurate)
Membership: Blanden Charitable Foundation: Individual $25; Household $40; Contributing & Business $100-$249; Patron $250-$499; Benefactor $500; Conservator $1,000.

FORT MUSEUM, S. Kenyon & Museum Rd., Fort Dodge, IA 50501. Tel.: 515-573-4231. Fax: 515-573-4231.
Institution Type/Description: History Museum.
Collections: local history; period furnishings; personal artifacts; photographs; statue.
Activities: guided tours.
Hours & Admission Prices: mid-April to mid-Oct. Mon.-Sat. 9-5, Sun. 11-5. Adults $6, students 6-18 $3; children 5 & under no charge.

Fort Madison

NORTH LEE COUNTY HISTORICAL SOCIETY AND SANTA FE DEPOT MUSEUM COMPLEX, 814 10th St. & Ave. H, Fort Madison, IA 52627-0285. Mailing Address: Box 285, Fort Madison, IA 52627-0285. Tel.: 319-372-7661.
E-mail: nlchs@iowatelecom.net
Founded: 1962.
Congressional District: 2
Key Personnel: Pres., Andy Andrews.
Personnel Profile: Part-Time Paid 2; Part-Time Volunteers 6.
Governing Authority: board of trustees. Subsidiary Institution: North Lee County Historic Center. Tax-exempt 501(c)(3).
Institution Type/Description: Historic Center: located in 1910 Santa Fe Railroad depot (National Register of Historic Buildings in historic district); caboose; 1870s country school; Brush College; 1993 Flood museum; Louis Koch Gallery of Historic Paintings; furniture.
Collections: Indian artifacts; early Fort Madison artifacts; Sheaffer Pen Co.; Prison; Mississippi River; Santa Fe Railway; Silsby Fire Engine Pumper; Civil War; Louis Koch collection.
Research Fields: local history; artifacts; railroad history; firefighting lore; Sheaffer Fountain Pen history; Iowa territorial & pioneer days.

Activities: guided tours; formally organized education programs for children &
  schools. Museum Sponsors: annual flea market in September; car show in
  May.
Publications: yearly newsletter; museum brochure.
Hours & Admission Prices: Memorial Day-Labor Day Mon.-Sat. 10-4, Sun.
  12-4; Sept.-Oct. Fri.-Sat. 10-2; other times by appointment. Adults $2,
  children $1; discounts to groups of 15 or more; members & children under
  6 no charge. Brush College: by appointment only. &
Attendance: 4,000 (accurate)
Membership: Students $5; Adults $10; Immediate Family $25; Sponsor $30;
  Patron $40; Business $50.

SHEAFFER PEN MUSEUM, 627 Ave. G, Fort Madison, IA 52627.
  Tel.: 319-372-1674.
Governing Authority: nonprofit organization. Tax-exempt: 501(c)(3).
Institution Type/Description: History Museum.
Collections: W.A. Sheaffer Pen Company history, artifacts & memorabilia;
  local history; science; art; writing instruments.
Hours & Admission Prices: Mon. & Fri.-Sat. 10-2.
Membership: Prelude $25; Nostalgia $75; Triumph $150.

Garnavillo

GARNAVILLO HISTORICAL MUSEUM, 205 N. Washington,
  Garnavillo, IA 52049-7220. Mailing Address: P.O. Box 371,
  Garnavillo, IA 52049-0371. Tel.: 563-964-2607.
E-mail: garnavillohistoricalsociety@gmail.com
Founded: 1965.
Congressional District: 2
Key Personnel: Pres., Barbara Leitgen; Dir., Tracy Werges; Dir., Richard
  Owings; Dir., Lanny Kuehl.
Personnel Profile: Full-Time Volunteers 4; Part-Time Volunteers 25.
Volunteer Hours: 500
Operating Expenses: 10,000
Operating Income: 10,000
Governing Authority: society. Tax-exempt: 501(c)(3).
Institution Type/Description: General Museum: housed in 1866 church.
Collections: archives; archaeology; anthropology; manuscripts; extensive ge-
  nealogy file; general; history; Indian artifacts; restored furnished Log
  Cabin; 1860 Lodge Hall; history of the surrounding community with
  spotlight on prior residents; written documentation of the life & times of a
  local prominent business man & entrepeneur.
Research Fields: archaeology; local history; genealogy.
Facilities: 7,400 card reference file containing names & historical references;
  newspapers dating from 1800 available for research on the premises.
Activities: permanent exhibitions.
Publications: 1836-1876 history of Garnavillo; Garnavillo, Gem of the Prairie.
Hours & Admission Prices: Memorial Day-Labor Day Sat.-Sun. & holidays
  1-4. No charge; donations accepted.
Attendance: 700 (estimated)
Membership: Annual $5; Life $25.

George

GEORGE BICENTENNIAL MUSEUM, 204 E. Michigan Ave.,
  George, IA 51237. Mailing Address: 210 N. Washington St.,
  George, IA 51237. Tel.: 712-475-3612 (Office) & 2581 (Chm.).
Founded: 1976.
Key Personnel: Chm. (V), Helen Fiihr.
Personnel Profile: Full-Time Volunteers 7; Part-Time Volunteers 47.
Governing Authority: city. Tax-exempt.
Institution Type/Description: History Museum.
Collections: pioneer life & history; farm tools & equipment; household
  artifacts; musical instruments; military uniforms.
Hours & Admission Prices: June-Aug. Sun. 2-4; July 4th 9-4; other times by
  appointment. No charge; donations accepted. &
Attendance: 100 (estimated)

Gladbrook

MATCHSTICK MARVELS TOURIST CENTER & MUSEUM,
  319 Second St., Gladbrook, IA 50635-7718. Tel.: 641-473-2410.
E-mail: glbktheater@iowatelecom.net
Web Site: www.matchstickmarvels.com
Institution Type/Description: General Museum.
Collections: matchstick models of people & structures including buildings,
  aircraft & boats; model drawings & plans; tools & equipment; video.
Activities: video.
Hours & Admission Prices: April-Nov. daily 1-5; groups by appointment.
  Adults $3, children 5-12 $1; children under 5 no charge.

Glenwood

MILLS COUNTY HISTORICAL SOCIETY AND MUSEUM, 2
  Lake Dr., Glenwood, IA 51534. Mailing Address: P.O. Box 255,
  Glenwood, IA 51534-0255. Tel.: 712-527-9221 & 5038.
E-mail: carriemerritt1925@msn.com
Founded: 1959.
Congressional District: 4
Key Personnel: C.E.O. & Pres., Carrie Merrit.
Personnel Profile: Full-Time Volunteers 88; Part-Time Volunteers 92.
Governing Authority: society. Tax-exempt.
Institution Type/Description: Local History Museum.
Collections: machinery hall; dolls; guns; household furniture; glass & china;
  pre-historic Indian artifacts; period furnishings; c.1900 metal toys; pioneer
  artifacts; Burlington train caboose. Historic Buildings: 1881 country
  schoolhouse; farm cottage.
Facilities: Indian earth lodge.
Activities: guided tours; permanent exhibitions; performances.
Publications: newsletter 10 issues per year.
Hours & Admission Prices: May-Sept. Sat.-Sun. 1-4; other times by appoint-
  ment. Pay what you wish.
Attendance: 2,015 (estimated)
Membership: Annual $5; Life $25.

Grafton

GRAFTON HERITAGE DEPOT MUSEUM, County Rds A39 &
  S62, Grafton, IA 50440. Mailing Address: Worth County Historical
  Society, 917 Central Ave., Northwood, IA 50459-1525. Tel.:
  641-748-2337.
Institution Type/Description: Historic Building: built in 1878. Listed on the
  National Register of Historic Places.
Collections: local history; period furnishings; photographs; depot agent's
  office; freight room; waiting room.
Activities: guided tours.
Hours & Admission Prices: Memorial Day to Labor Day Sun. 1:30-5; groups
  by appointment.

Greenfield

ADAIR COUNTY HERITAGE MUSEUM, 2393 S. Lakeview Dr.,
  Hwy. 92 W., Greenfield, IA 50849. Mailing Address: P.O. Box 214,
  Greenfield, IA 50849-0214. Tel.: 641-743-2232.
Web Site: www.adaircountymuseum.com
Founded: 1963.
Congressional District: 3
Key Personnel: Pres. (V), Earl Carroll.
Personnel Profile: Part-Time Paid 1.
Volunteer Hours: 800
Operating Expenses: 8,230
Operating Income: 7,848
Governing Authority: Tax-exempt.
Institution Type/Description: History Museum.
Collections: local history & culture; photographs; horse-drawn farm machin-
  ery & equipment; early furnishings; medical & military artifacts; 1880-1940
  household items & tools; 1880-1920 quilts. Historic Buildings: 1893
  church; 1872 George A. Wilson's birthplace home; 1910 railroad depot;
  country schoolhouse; city barn.
Research Fields: county history.
Facilities: library.
Activities: Annual Events: Chicken and Biscuit Dinner in June; County Fiar Ice
  Cream Social in July; Country School Experience in September; Soup
  Supper in November.
Hours & Admission Prices: May-Oct. 1 daily 1-4; other times by appointment.
  Adults $3, children 5-12 $1.50; children under 5 no charge. &
Attendance: 200
Membership: Individual $10; Family $25; Lifetime $200.

IOWA AVIATION MUSEUM, Greenfield Municipal Airport, 2251
  Airport Rd., Greenfield, IA 50849. Mailing Address: P.O. Box 31,
  Greenfield, IA 50849-0031. Tel.: 641-343-7184. Fax: 641-343-
  7184. Facebook: Iowa Aviation Museum.
E-mail: aviation@iowatelecom.net
Web Site: www.flyingmuseum.com
Formerly: Iowa Aviation Preservation Center
Founded: 1990.
Congressional District: 3
Key Personnel: Pres., Greg Schildberg; Vice Pres. & Treas., Dick Westbrook.

Personnel Profile: Full-Time Paid 1; Part-Time Paid 2; Part-Time Volunteers 30.
Governing Authority: nonprofit organization. Parent Institution: Antique Preservation Association. Tax-exempt: 501(c)(3).
Institution Type/Description: Aviation Museum.
Collections: Iowa aviation history; photographs; memorabilia; 1928-1946 airplanes; books & magazines.
Facilities: 1,500-vol. library of aviation books; 7,500 sq. ft. exhibit space. Museum-related items for sale.
Activities: guided tours; traveling exhibitions; pilot training courses; antique plane rides; public events. Annual Events: Hall of Fame banquet; Fly-In in summer & winter.
Publications: quarterly, APA Newsletter.
Hours & Admission Prices: May-Sept. Mon.-Sat. 10-5, Sun. 1-5; Oct.-April Mon.-Fri. 10-5, Sat.-Sun. 1-5. Adults $5.50, seniors $3.50, children 5-12 $2.50; discounts to groups of 20 or more; members & children 4 and under no charge. Closed New Year's Day; Easter; Thanksgiving; Christmas Eve & Day. &
Attendance: 3,000 (estimated)
Membership: Individual $40.

Grinnell

FAULCONER GALLERY AT GRINNELL COLLEGE, (M), 1108 Park St., Grinnell, IA 50112-1643. Tel.: 641-269-4660. Fax: 641-269-4626.
E-mail: Strongdj@Grinnell.edu
Web Site: www.grinnell.edu/faulconergallery
Formerly: Faulconer Gallery
Founded: 1999.
Congressional District: 1
Key Personnel: Dir., Lesley Wright; Assoc. Dir. & Cur. Exhibitions, Daniel Strong; Cur. Collections, Kay Wilson; Cur. Academic & Community Outreach, Tilly Woodward; Exhibition Designer, Milton Severe; Admin. Support Asst., Constance Gause.
Personnel Profile: Full-Time Paid 6; Part-Time Paid 19; Interns 3.
Governing Authority: private college. Parent Institution: Grinnell College. Tax-exempt: 501(c)(3).
Institution Type/Description: College Art Museum.
Collections: c.1300 to present, art on paper, African ceramics.
Facilities: 7,500 sq. ft. gallery designed by Cesar Pelli; print & drawing study room.
Activities: temporary exhibitions; lectures; musical performances.
Publications: Sandy Skoglund: Raining Popcorn; Layers of Brazilian Art; John Wilson: A Retrospective; William Kentridge Prints; Scandinavian Photography 1: Sweden; Scandinavian Photography 2: Denmark; Austin Thomas: Perches and Drawings; Hin: The Quiet Beauty of Japanese Bamboo Art; Frank Breuer Photographs; Where Are You From: Contemporary Portuguese Art; Angela Strassheim: Left Behind; Works in Progress: Prints from Wildwood Press.
Hours & Admission Prices: Daily 11-5. No charge; donations accepted. Closed holidays. &
Attendance: 15,000 (estimated)

GRINNELL HISTORICAL MUSEUM, 1125 Broad St., Grinnell, IA 50112. Mailing Address: 903 16th Ave., Grinnell, IA 50112-1106. Tel.: 641-236-7827.
E-mail: ghmuseum@iowatelecom.net
Web Site: www.grinnellhistoricalmuseum.org
Founded: 1954.
Key Personnel: Pres., Howard McDonough; Sec., Michele Parslow; Treas., Vera Cousins.
Personnel Profile: Part-Time Volunteers 30.
Governing Authority: private; nonprofit organization. Tax-exempt: 501(c)(3).
Institution Type/Description: History Museum: housed in an 1895-1896 home.
Collections: history of Grinnell; period artifacts; photographs.
Research Fields: genealogical research.
Facilities: 2,000 sq. ft. exhibit space.
Activities: guided tours.
Publications: annual newsletter.
Hours & Admission Prices: March-May & Sept.-Dec. Sat.-Sun. 2-4; June-Aug. Tues.-Sun. 2-4; groups by appointment. No charge; donations accepted. Closed New Year's Day; Independence Day; Christmas.
Attendance: 800 (estimated)

SPAULDING CENTER FOR TRANSPORTATION - HOME OF THE IOWA TRANSPORTATION MUSEUM, 829 Spring St., Grinnell, IA 50112-2043. Mailing Address: 927 4th Ave., Grinnell, IA 50112. Tel.: 641-236-9860. Fax: 641-236-2626.
Web Site: www.iowatransportationmuseum.com
Founded: 2012.
Key Personnel: Exec. Dir., Charles Brooke.
Governing Authority: Tax-exempt.
Institution Type/Description: Transportation Museum.
Collections: Iowa's transportation history.
Activities: special events; rental facilities.
Hours & Admission Prices: Call for hours. No charge; donations accepted. &

Griswold

CASS COUNTY HISTORICAL SOCIETY MUSEUM, 412 Main St., Griswold, IA 51535-5800. Mailing Address: P.O. Box 254, Griswald, IA 51535. Tel.: 712-778-5040 & 2191.
Web Site: www.casscountymuseumiowa.com
Institution Type/Description: Historic Building: Listed on the National Register of Historic Places.
Collections: local history & culture; period furnishings; personal artifacts; photographs; clothing.
Hours & Admission Prices: May-Dec. Wed.-Sun. 1-4; other times by appointment. No charge; donations accepted. Closed holidays. &

Grundy Center

HERBERT QUICK SCHOOLHOUSE, Hwy. 175 E., Grundy Center, IA 50638. Mailing Address: 705 F Ave., Grundy Center, IA 50638. Tel.: 319-825-3838.
E-mail: gcdall@gcmuni.net
Institution Type/Description: Historic Building: housed in the former school of author Herbert Quick.
Collections: local history; Herbert Quick's life & career; personal artifacts; period furnishings.
Hours & Admission Prices: By appointment. No charge.

Guttenberg

LOCKMASTER'S HOUSE HERITAGE MUSEUM, Lock & Dam Lane, Guttenberg, IA 52052. Mailing Address: Guttenberg Heritage Society, P.O. Box 701, Guttenberg, IA 52052-0701. Tel.: 319-252-1531.
Institution Type/Description: Historic Building: housed in the former lockmaster's house. Listed on the National Register of Historic Places.
Collections: local history; period furnishings; photographs; personal artifacts.
Hours & Admission Prices: Memorial Day to mid-Oct. Tues.-Sun. 12-4.

Hampton

REA POWER PLANT MUSEUM, 1450 110th St., Hampton, IA 50441. Mailing Address: P.O. Box 114, Hampton, IA 50441-0442. Tel.: 641-456-5777.
Institution Type/Description: History Museum.
Collections: Iowa's electric power history; farming.
Hours & Admission Prices: By appointment. No charge.

Harlan

SHELBY COUNTY HISTORICAL MUSEUM, 1805 Morse Ave., Harlan, IA 51537-2042. Tel.: 712-755-2437. Facebook: Shelby County Historical Museum.
E-mail: shelbyco.museum@gmail.com
Founded: 1964.
Congressional District: 5
Personnel Profile: Full-Time Paid 1; Part-Time Paid 1.
Institution Type/Description: History Museum.
Collections: local history & culture; early firefighting equipment; period furnishings; clothing; books; historic buildings.
Hours & Admission Prices: June-Aug. Mon.-Fri. 8-4, Sun. 2-4; other times by appointment. No charge.
Membership: Individual $20; Individual Lifetime $100.

Haverhill

MATTHEW EDEL BLACKSMITH SHOP, 214 1st St., Haverhill, IA 50120. Mailing Address: Historical Society of Marshall County, P.O. Box 304, Marshalltown, IA 50158. Tel.: 641-752-6664.
Institution Type/Description: History Museum: housed in the former blacksmith shop & 1st home of Matthew Edel which he purchased in 1883. Listed on the National Register of Historic Places.
Collections: Matthew Edel's family history & career; period furnishings; personal artifacts; photographs.
Activities: blacksmithing demonstrations; guided tours.
Hours & Admission Prices: Memorial Day to Labor Day daily 12-4; groups by appointment. No charge.

Hawarden

CALLIOPE VILLAGE, 19th St. & Ave. E, Hawarden, IA 51023. Mailing Address: City of Hawarden, 1150 Central Ave., Hawarden, IA 50123-1815. Tel.: 712-551-2403.
E-mail: happ@cityofhawarden.com
Institution Type/Description: Village History Museum.
Collections: local history & culture; personal artifacts; period furnishings; photographs; church; log house. Historic Buildings: harness shop; 1872 school; medical building; bank; jail; 1878 cabin; barn; 1882 depot; 1886 Carlson home; 1881 general store; Shoemaker Museum.
Hours & Admission Prices: Memorial Day to Labor Day Sun. 1-4.

Homestead

HOMESTEAD STORE MUSEUM, 4430 V St., Homestead, IA 52236. Tel.: 319-622-3567.
Web Site: www.amanaheritage.org
Institution Type/Description: Historic Building Museum: housed in a former general store.
Collections: local history, culture & industry; carpet weaving; bookbinding; printing; tinsmithing; woolen mill industry; photographs; personal artifacts; period furnishings.
Hours & Admission Prices: May-June 9 & mid-Aug. to Oct. Sat 10-5; June 10 to mid-Aug. Mon.-Fri. 11-5, Sat. 10-5.

Honey Creek

HITCHCOCK NATURE CENTER, 27792 Ski Hill Loop, Honey Creek, IA 51542-4398. Tel.: 712-545-3283.
Founded: 1991.
Congressional District: 5
Key Personnel: C.E.O., Mark Shoemaker; Program Mgr., Tina Popson.
Personnel Profile: Full-Time Paid 7; Part-Time Paid 10; Part-Time Volunteers 50.
Governing Authority: county. Parent Institution: Pottawattamie County Conservation board.
Institution Type/Description: Nature Center.
Collections: plants; flowers; wildlife & their habitats.
Facilities: nature trails.
Activities: observation tower; educational programs; hiking trails; wildlife viewing
Publications: newsletter.
Hours & Admission Prices: Gallery: March-Nov. Tues.-Sat. 9-5, Sun. 1-5; Dec.-Feb. Fri.-Sat. 9-5. Park: daily 6 am-10 pm. Park: $2 per vehicle. &
Attendance: 36,000 (estimated)

Hopkinton

DELAWARE COUNTY HISTORICAL MUSEUM COMPLEX, College Square Hwy. 38, Hopkinton, IA 52237. Tel.: 563-926-2639.
Institution Type/Description: Historic Buildings: listed on the National Register of Historic Places.
Collections: local history; period furnishings & artifacts; photographs; historic buildings.
Hours & Admission Prices: June-Sept. Tues.-Sun. 1-4. Adults $3.

Independence

WAPSIPINICON MILL MUSEUM, 100 1st St. W., Independence, IA 50644-2601. Mailing Address: P.O. Box 321, Independence, IA 50644-0321. Tel.: 319-334-4616. Facebook: Wapsi Mill.
E-mail: leannekay@indytel.com
Web Site: www.buchanancountyhistory.com

Founded: 1976.
Governing Authority: Tax-exempt.
Institution Type/Description: Historic Building: housed in an 1870s grain mill. Listed on the National Register of Historic Places.
Collections: 1870s grain milling process; mill history.
Hours & Admission Prices: mid-May to mid-Sept. Tues.-Sun. 12-4; other times by appointment. No charge; donations accepted. Group Tours: $3 per person. &
Attendance: 3,000 (estimated)
Membership: Individual $10; Lifetime $100.

Indianola

NATIONAL BALLOON MUSEUM AND US BALLOONING HALL OF FAME, 1601 N. Jefferson Way, Indianola, IA 50125-1484. Mailing Address: P.O. Box 149, Indianola, IA 50125-0149. Tel.: 515-961-3714.
E-mail: museum@nationalballoonmuseum.com
Web Site: www.nationalballoonmuseum.com
Formerly: National Balloon Museum and Hall of Fame
Founded: 1973.
Key Personnel: Chm. (V), Linda Nicholson; Pres. (V), Denis Freischmeyer; Cur., Becky Wigeland; Museum Shop Mgr., Betty Crawford
Governing Authority: Tax-exempt.
Institution Type/Description: Balloon Museum & Hall of Fame; aviation research library.
Collections: ballooning & its history; ballooning hall of famers; photographs; balloon gondolas & memorabilia; gas balloons.
Facilities: aviation research library. Museum-related items for sale.
Activities: ballooning events; tour of balloon manufacturer. Museum Sponsors: Balloon Fashion Flare & Brunch in January; US Ballooning Hall of Fame in August.
Publications: book, Indianola; Ballooning Capital of Iowa; newsletter; History of National Balloon Museum.
Hours & Admission Prices: Feb.-April & Dec. 27-Dec. 30 daily 1-4; May-Dec. 23 Mon.-Fri. 9-4, Sat. 10-4, Sun. 1-4; other times by appointment. Adults $3; members no charge. Tour of Balloon Manufacturer: Adults $5, children $1. Closed holidays. &
Attendance: 4,700 (estimated)
Membership: Balloon Watcher $25; Museum Chaser $50; Museum Supporter/Family $100; Museum Sponsor/Corporate $250; Friend of the Museum $500; Museum Patron $1,000; Life Patron $2,500 or $500 per year.

SIMPSON COLLEGE/FARNHAM GALLERIES, 701 N. C St., Mary Berry Hall, 3rd Fl., Indianola, IA 50125-1202. Tel.: 515-961-1761; 800-362-2454. Fax: 515-961-1498.
E-mail: jnostra@simpson.edu
Web Site: www.simpson.edu/art/gallery
Founded: 1982.
Congressional District: 5
Key Personnel: Dir. & Chair, Art Dept., Justin Nostrala; Vice Pres. Business & Finance, Ken Birkenholtz.
Personnel Profile: Full-Time Paid 1; Part-Time Paid 2.
Governing Authority: college; nonprofit. Parent Institution: Simpson College. Tax-exempt.
Institution Type/Description: Art Gallery.
Collections: paintings; sculpture; photographs.
Activities: formal education programs for college students; traveling exhibitions.
Hours & Admission Prices: Mon.-Fri. 8:30-4:30; other times by appointment. No charge; donations accepted. &
Attendance: 300 (estimated)

WARREN COUNTY HISTORICAL SOCIETY MUSEUM, 1400 W. 2nd Ave., Indianola, IA 50125. Mailing Address: P.O. Box 256, Indianola, IA 50125-0256. Tel.: 515-961-8085.
Institution Type/Description: Historical Society Museum.
Collections: local history & culture; period furnishings; personal artifacts; pottery; tools.
Hours & Admission Prices: Thurs. 9-4, Sat. 9am-12pm; other times by appointment.

Iowa City

OLD CAPITOL MUSEUM, The University of Iowa, 21 Old Capitol, Iowa City, IA 52242. Tel.: 319-335-0548. Fax: 319-353-2982.
E-mail: shalla-wilson@uiowa.edu

Web Site: www.uiowa.edu/~oldcap/
Founded: 1976.
Congressional District: 3
Key Personnel: Dir., Pamela Trimpe; Cur., Shalla Wilson.
Personnel Profile: Full-Time Paid 2; Part-Time Paid 8; Part-Time Volunteers 35.
Governing Authority: university. Affiliated with The University of Iowa. Tax-exempt.
Institution Type/Description: Historic Building: Old Capitol, Iowa's last territorial capitol from 1842-46; first state capitol with the admission of Iowa to the Union in 1846.
Collections: furniture & furnishings of the period; original library collection.
Research Fields: pertaining to the structure & its contents & history.
Facilities: library.
Activities: self-guided tours; guided tours; discovery center; Iowa Humanities Gallery.
Publications: folders, Old Capitol; Old Capitol Tour Guide; booklet, Old Capitol.
Hours & Admission Prices: Tues.-Wed. & Fri. 10-3, Thurs. & Sat. 10-5, Sun. 1-5. No charge; donations accepted. Closed national holidays; Dec. 23-Jan. 13. &
Attendance: 30,000 (accurate)

PLUM GROVE HISTORIC HOME, 1030 Carroll St., Iowa City, IA 52240-4601. Mailing Address: Johnson County Historical Society Museum, 860 Quarry Rd., Coralville, IA 52241-2421. Tel.: 319-337-6846 & 351-5738. Fax: 319-351-5310.
E-mail: questions@johnsoncountyhistory.org
Web Site: www.johnsoncountyhistory.org
Founded: 1944.
Congressional District: 1
Key Personnel: C.E.O., Alexandra Drehman; Chm. (V), Steve Weeber.
Personnel Profile: Full-Time Paid 1; Part-Time Paid 2; Part-Time Volunteers 10.
Governing Authority: state. Parent Institution: State Historical Society of Iowa. Subsidiary Institution: Johnson County Historical Society. Tax-exempt.
Institution Type/Description: Historic Building: 1844 home of Robert Lucas, first governor of Territory of Iowa.
Collections: period furnishings.
Research Fields: anthropology; politics; architecture; early life; geography.
Activities: guided tours; formally organized education programs.
Publications: brochures.
Hours & Admission Prices: Memorial Day to Labor Day Wed.-Sun. 1-5; Sept.-Oct. Sat.-Sun. 1-5. No charge; donations accepted. Closed Independence Day. &
Attendance: 2,000 (accurate)

PROJECT ART-UNIVERSITY OF IOWA AND CLINICS, 200 Hawkins Dr., Iowa City, IA 52242-1009. Tel.: 319-353-6417. Fax: 319-384-8141.
E-mail: adrienne-drapkin@uiowa.edu
Web Site: uihealthcare.com/projectart
Founded: 1978.
Congressional District: 1
Key Personnel: Dir., Adrienne Drapkin.
Personnel Profile: Full-Time Paid 2; Part-Time Paid 2; Part-Time Volunteers 8.
Governing Authority: state government. Parent Institution: University of Iowa Hospitals and Clinics, 200 Hawkins Dr., Iowa City, IA. Tax-exempt.
Institution Type/Description: Art Museum.
Collections: original American art; contemporary Iowa and Midwest artists; Native American art in all media both historic and modern; contemporary glass by internationally recognized artists; weavings from early 20th-century Turkey.
Research Fields: fine art related to health issues, the body; Iowa based art; contemporary photography, glass.
Facilities: 200-seat restaurant; 2,000 sq. ft. exhibit space.
Activities: concerts; lectures; art instruction for patients; summer concert series. Annual Event: Festivals of World Cultures.
Hours & Admission Prices: Mon.-Fri. 8-5; permanent collection is on view 24 hours a day in the public corridors, lobbies & waiting rooms. No charge. &
Attendance: 40,000 (estimated)

UNIVERSITY OF IOWA ATHLETICS HALL OF FAME AND MUSEUM, 2425 Prairie Meadow Dr., Iowa City, IA 52242. Tel.: 319-384-1031.
Institution Type/Description: Sports Museum.
Collections: Hawkeye history & memorabilia; personal artifacts; photographs; sports equipment & uniforms; trophies.
Hours & Admission Prices: Call for hours. &

UNIVERSITY OF IOWA HOSPITALS & CLINICS MEDICAL MUSEUM, 200 Hawkins Dr. 8014 RCP, Iowa City, IA 52242-1009. Mailing Address: UIHC Medical Museum, Iowa City, IA 52242. Tel.: 319-353-6417; 356-7106. Fax: 319-384-8141.
E-mail: adrienne-drapkin@uiowa.edu
Web Site: uihealthcare.com/medmuseum
Founded: 1989.
Key Personnel: Dir., Adrienne Drapkin.
Personnel Profile: Full-Time Paid 1.
Governing Authority: public university; nonprofit. Parent Institution: University of Iowa. Subsidiary Institution: University of Iowa Hospitals & Clinics. Tax-exempt: 501(c)(3).
Institution Type/Description: Medical Museum.
Collections: instruments; artifacts; photographs; books; documents; uniforms; medical & healthcare history.
Research Fields: instrumentation; medical history; current medical procedures & technologies.
Facilities: 125-vol. library of 1850-1950s medical texts; College of Medicine notebooks & documentation; historical hospital records; 1,900 sq. ft. exhibit space.
Activities: guided tours; lectures; films; participatory & temporary exhibitions.
Publications: exhibition brochures & catalogues.
Hours & Admission Prices: Mon.-Fri. 8-5, Sat.-Sun. 1-4. No charge. Closed New Year's Day; Thanksgiving; Christmas. &
Attendance: 50,000 (estimated)

*　**UNIVERSITY OF IOWA MUSEUM OF ART, (M),** 1375 Hwy. 1 W., 1840 Studio Arts, Iowa City, IA 52242-1789. Tel.: 319-335-1727. Fax: 319-335-3677.
E-mail: uima@uiowa.edu
Web Site: uima.uiowa.edu
Founded: 1967.
Congressional District: 1
Key Personnel: Dir., Sean O'Harrow, Ph.D.; Chief Cur., Kathleen Edwards; Cur. Education, Dale William Fisher; Mgr. Exhibitions & Collections, Jeff Martin; Cur. African & Non-Western Art, Catherine Hale.
Personnel Profile: Full-Time Paid 10; Part-Time Paid 9; Part-Time Volunteers 165.
Governing Authority: state. Parent Institution: University of Iowa. Tax-exempt: 501(c)(3).
Institution Type/Description: Art Museum.
Collections: paintings; prints & drawings; African art; pre-Columbian art; metalwork & jewelry; photography; sculpture.
Activities: guided tours; lectures; educational programs; permanent & traveling exhibitions; special exhibitions; docent program; study tours; the Elliott Society.
Publications: exhibition catalogs; magazine.
Hours & Admission Prices: Temporarily Locations: UIMA & Black Box Theater: Tues.-Wed. & Fri. 10-5, Thurs. 10-9, Sat.-Sun. 12-5; Old Capitol Museum: Tues.-Wed. & Fri.-Sat. 10-5, Thurs. 10-8, Sun. 1-5; Figge Art Museum: Tues.-Sat. 10-5, Thurs. 10-9, Sun. 12-5. No charge; donations accepted. &
Attendance: 40,000 (estimated)
Membership: Contributor $25; Benefactor $100; Elliott Society $150; Curator's Circle $250; Sponsor $500; Director's Circle $1,000; Patron $5,000.

UNIVERSITY OF IOWA MUSEUM OF NATURAL HISTORY, (M), 11 Macbride Hall, Iowa City, IA 52242-1322. Tel.: 319-335-0606. Fax: 319-335-3677. Facebook: Iowa Natural History.
E-mail: uimnh@uiowa.edu
Web Site: www.uiowa.edu/mnh
Founded: 1858.
Congressional District: 3
Key Personnel: Dir., John Logsdon; Assoc. Dir., Trina Roberts; Education & Outreach Coord. & Museum Shop Mgr., Sarah Horgen; Collections Mgr., Cindy Opitz; Exhibits Mgr., Byron Preston.
Personnel Profile: Full-Time Paid 5; Part-Time Paid 8; Part-Time Volunteers 28; Interns 2.
Governing Authority: university. Parent Institution: University of Iowa. Tax-exempt: 501(c)(3).
Institution Type/Description: Natural History Museum.
Collections: vertebrate & invertebrate research collections from around the world including North American birds & mammals; marine & freshwater invertebrates & insects; ethnographic materials from the North American Arctic & Midwest regions, the Philippines, Africa, & New Zealand; local geology, archaeology & ecology; Laysan Island Cyclorama.
Research Fields: midwest archaeology; paleontology; botany; mammalogy; ornithology; environmental studies.

Facilities: reference library of natural history subjects; 18,000 sq. ft. exhibit space. Postcards, books & collection-related items for sale.

Activities: guided tours; public lecture series; formally organized programs for children; school curriculum programming; formally organized museum studies program for undergraduate & graduate students affiliated with University of Iowa; training programs for professional museum workers; inter-museum loan, permanent & temporary exhibitions; school loan service.

Publications: series, University of Iowa Studies in Natural History; tour guides & brochures; ASC Featured Institution reprints; school curriculum materials.

Hours & Admission Prices: Tues.-Wed. & Fri.-Sat. 10-5, Thurs. 10-8, Sun. 1-5. No charge; donations accepted. Closed national holidays. &

Attendance: 30,000 (estimated)

Iowa Falls

CALKINS NATURE AREA/FIELD MUSEUM, 18335 135th St., Iowa Falls, IA 50126-8512. Tel.: 641-648-9878. Fax: 641-648-9878.

E-mail: calkinsnatureareahccb@gmail.com

Founded: 1890.

Congressional District: 4

Key Personnel: Exec. Dir., Wes Wiese; Chm. (V), William H. Schmidt.

Personnel Profile: Full-Time Paid 2; Part-Time Paid 3; Part-Time Volunteers 20.

Governing Authority: college. Parent Institution: Hardin County Conservation Board. Subsidiary Institution: Ellsworth College. Tax-exempt.

Institution Type/Description: Natural Science & History Museum.

Collections: mounted bird & mammal specimens; Indian artifacts; eggs from 197 species of birds; fossils & minerals; shells; insects; other invertebrates.

Facilities: library; 100-seat multipurpose room; classroom.

Activities: guided tours; lectures; educational programs; permanent & temporary exhibits.

Publications: newsletter, Calkins Nature Notes; book, The History of the Ellsworth College Museum.

Hours & Admission Prices: Interpretive Center: Mon.-Fri. 8-4. No charge; donations accepted. &

Attendance: 4,700

Membership: Individual & Family $25; Donor $50; Supporter $100; Sustainer $200; Benefactor $500.

DOW HOUSE - IOWA FALLS HISTORICAL SOCIETY, 519 Stevens St., Iowa Falls, IA 50126. Mailing Address: P.O. Box 364, Iowa Falls, IA 50126. Tel.: 641-648-4017 & 3606.

Institution Type/Description: Historic House Museum: built in 1909.

Collections: local history & culture; period furnishings & clothing; photographs; Earl Kuns' point collection.

Hours & Admission Prices: Summer: Tues.-Thurs. 1-4; other times by appointment.

ILLINOIS CENTRAL DEPOT MUSEUM, Rocksylvania Ave., Iowa Falls, IA 50126. Mailing Address: P.O. Box 364, Iowa Falls, IA 50126. Tel.: 641-648-5900 & 2849.

Institution Type/Description: Historic Building: depot built in 1904.

Collections: local history; period furnishings; railroad artifacts; photographs.

Hours & Admission Prices: By appointment.

IOWA FALLS HISTORICAL MUSEUM, 520 Rocksylvania Ave., Carnegie Ellsworth Library Bldg., Iowa Falls, IA 50126. Mailing Address: P.O. Box 364, Iowa Falls, IA 50126. Tel.: 641-648-4017.

Institution Type/Description: History Museum.

Collections: local history & culture; replica of early downtown; Bill Riley memorabilia; international dolls.

Hours & Admission Prices: June-Sept. Thurs. 1-3; Oct.-May 1st Sun. each month 1-3. &

PAT CLARK ART COLLECTION GALLERY, 520 Rocksylvania Ave., Iowa Falls, IA 50126. Tel.: 641-648-8576.

Institution Type/Description: Art Gallery.

Collections: paintings; sculpture.

Hours & Admission Prices: Mon.-Fri. 9-12 & 1-4; guided tours by appointment. No charge.

Jefferson

GREENE COUNTY HISTORICAL SOCIETY MUSEUM, 219 E. Lincolnway, Jefferson, IA 50129-2205. Tel.: 515-386-8544.

Governing Authority: Tax-exempt.

Institution Type/Description: Historical Society Museum.

Collections: local history & culture; Native American artifacts; WWII memorabilia; early transportations; photographs; period furnishings.

Hours & Admission Prices: May-Sept. Wed. 1-5, Sat. 9 to noon; other times by appointment. No charge. &

Membership: Single $15; Family $20.

JEFFERSON TELEPHONE MUSEUM, 105 W. Harrison, Jefferson, IA 50129-2105. Mailing Address: P.O. Box 269, Jefferson, IA 50129-0269. Tel.: 515-386-4141. Fax: 515-386-2600.

E-mail: jtcobob@netius.net

Web Site: www.jeffersontelecom.com

Founded: 1958.

Key Personnel: Gen. Mgr., James Daubendiek.

Governing Authority: company; nonprofit.

Institution Type/Description: Technology Museum.

Collections: telephones & related telephone equipment; history & evolution of telephony.

Activities: guided tours.

Hours & Admission Prices: Mon.-Fri. 9-4. No charge. Closed major holidays.

Attendance: 100 (estimated)

Johnston

IOWA GOLD STAR MILITARY MUSEUM, 7105 N.W. 70th Ave., Johnston, IA 50131-1824. Tel.: 515-252-4531. Fax: 515-727-3107.

E-mail: goldstarmuseum@iowa.gov

Web Site: www.iowanationalguard.com

Founded: 1985.

Congressional District: 3

Key Personnel: Dir., Sherrie Cobert; Chm. (V), Bob Holliday.

Personnel Profile: Full-Time Paid 2; Part-Time Paid 1; Part-Time Volunteers 80; Interns 1.

Governing Authority: Tax-exempt.

Institution Type/Description: Military Museum.

Collections: Iowa veterans; Iowa military history; photographs; personal artifacts.

Facilities: conference area.

Activities: group tours.

Hours & Admission Prices: Mon.-Fri. 8:30-4:30, Sat. 10-4. Closed holidays. No charge; donations accepted. &

Attendance: 21,121 (accurate)

Kalona

KALONA HISTORICAL VILLAGE, 715 D Ave., Kalona, IA 52247. Tel.: 319-656-3232.

E-mail: kalonatours@kctc

Web Site: www.kalonaiowa.org

Institution Type/Description: Historic Buildings: 1800s village including 13 historic buildings.

Collections: local history & culture; period furnishings; personal artifacts; photographs. Historic Buildings: Snyder log home & Wahl house.

Hours & Admission Prices: April-Oct. Mon.-Sat. 10-5; Nov.-March Mon.-Sat. 11-3.

Kellogg

KELLOGG HISTORICAL SOCIETY, 218 High St., Kellogg, IA 50135. Mailing Address: P.O. Box 295, Kellogg, IA 50135-0295.

Founded: 1980.

Congressional District: 3

Key Personnel: Pres. (V), David Faircloth; Financial Dir., Jenna Shine.

Personnel Profile: Part-Time Volunteers 9.

Governing Authority: private; nonprofit organization. Tax-exempt: 501(c)(3).

Institution Type/Description: Historical Society Museum.

Collections: 1,100 pitchers; 150 nativity sets; period artifacts; farm-related items; military; dinner pails. Historic Buildings: two-story agricultural building; restored country church & school; furnished factory & bank; machine shed & blacksmith shop with equipment & tools.

Facilities: library of genealogy & cemetery records. Museum-related items for sale.

Activities: guided tours. Historical Society Sponsors: biennial 2 day Down Home Christmas Celebration in December.

Publications: quarterly newsletter, The Kellogg Enterprise.
Hours & Admission Prices: Memorial Day weekend to Sept. 1 Mon.-Fri. 9-4, Sun. 1:30-5, Sat. by appointment. Donations requested. Closed all holidays. &
Attendance: 1,000 (estimated)
Membership: Annual $15.

Kensett

KENSETT COMMUNITY CHURCH MUSEUM, Second St., Kensett, IA 50448. Mailing Address: Worth County Historical Society, 917 Central Ave., Northwood, IA 50459-1525. Tel.: 641-324-1180.
Governing Authority: Parent Institution: Worth County Historical Society. Tax-exempt.
Institution Type/Description: Historic Building: built in 1899. Listed on the National Register of Historic Places.
Collections: local history; construction & membership documents; period furnishings.
Activities: special events; community meetings.
Hours & Admission Prices: June-Aug. Sun. 2-4; call for additional hours. No charge; donations accepted.
Attendance: 200 (estimated)

Keokuk

KEOKUK RIVER MUSEUM, 101 Mississippi Dr., Keokuk, IA 52632. Mailing Address: P.O. Box 400, Keokuk, IA 52632-0400. Tel.: 319-524-4765. Fax: 319-524-2642. geomverity.org.
E-mail: chuckpietscher@cityofkeokuk.org
Web Site: keokukiowatourism.org
Founded: 1962.
Congressional District: 1
Key Personnel: Chm. (V), Charles R. Pietscher.
Personnel Profile: Part-Time Volunteers 5.
Governing Authority: municipal. Parent Institution: City of Keokuk. Tax-exempt: 170(b)(1)(A).
Institution Type/Description: Maritime Museum: housed in 1927 George M. Verity Mississippi River Steamboat.
Collections: river transportation; marine.
Research Fields: Upper Mississippi river boats.
Facilities: Museum-related items for sale.
Activities: Civil War re-enactment on hospital boat.
Hours & Admission Prices: April-Oct. Mon. & Thurs.-Fri. 9-12 & 4-6, Sat.-Sun. 9-6. Adults $4, children $1.50; special group rates for 12 or more.
Attendance: 6,000 (accurate)

SAMUEL F. MILLER HOUSE & MUSEUM, 318 N. 5th St., Keokuk, IA 52632. Tel.: 319-524-7283.
Governing Authority: Parent Institution: Lee County Historical Society.
Institution Type/Description: Historic House Museum: built in 1859.
Collections: local history; period furnishings; personal artifacts; photographs.
Hours & Admission Prices: Memorial Day to Labor Day Fri.-Sun. 1-4.

Keosauqua

VAN BUREN COUNTY HISTORICAL SOCIETY, (M), 1st. St., Keosauqua, IA 52565. Mailing Address: P.O. Box 236, Keosauqua, IA 52565-0236. Tel.: 319-293-3088.
E-mail: dorissecor@netins.net
Web Site: www.villagesofvanburen.com
Founded: 1960.
Congressional District: 1
Key Personnel: Pres., Marvin Danneil; Sec., Doris Secor; Treas., Joe Stump.
Personnel Profile: Part-Time Volunteers 12; Interns 1.
Governing Authority: society. Tax-exempt: 501(c)(3).
Institution Type/Description: Historical House Museum.
Collections: local history & culture; guns; military equipment; personal artifacts; photographs; period furnishings. Historical Buildings: c.1845 Pearson House; 1855 Aunty Green Hotel, Bonaparte; restored log cabins, Selma & Keosauqua; country schoolhouse, sheds; c.1870 Twombly Building.
Facilities: 40-vol. library of maps; atlases; 100 rolls microfilm; census; county newspapers; documents; scrapbooks & books on local history & genealogies available for use on premises. Museum-related items for sale.
Activities: guided tours; formally organized education programs for children; permanent exhibitions.
Publications: books, History of Bonaparte, Iowa; History of Selma-Douds; History of Milton; History of Stockport; History of Birmingham; History of Cantril; Four Seasons-Life on a Pioneer Van Buren County Farm.

Hours & Admission Prices: Society: June to mid-Oct. Sun. & holidays 1-4; other times by appointment. No charge; donations accepted. Pearson House Memorial Day to Labor Day Sun. 1-4; other times by appointment. Suggested Donation: adults $2; students $1. Closed holidays.
Attendance: 325 (accurate)
Membership: Van Buren Historical Society: Individual $10.

Knoxville

NATIONAL SPRINT CAR HALL OF FAME & MUSEUM, One Sprint Capital Pl., Knoxville, IA 50138. Mailing Address: P.O. Box 542, Knoxville, IA 50138-0542. Tel.: 641-842-6176. Fax: 641-842-6177.
E-mail: bbaker@sprintcarhof.com
Web Site: www.sprintcarhof.com
Founded: 1986.
Congressional District: 3
Key Personnel: Exec. Dir., Bob Baker; Pres. (V), Jeff Savage; Cur., Tom Schmeh; Devel., Doug Lockin; Mktg. Mgr., David Herrmann; Education, Lori Demoss; Registrar & Museum Shop Mgr., Sharon Spaur; Security, Gary Van Waardhuizen.
Personnel Profile: Full-Time Paid 5; Part-Time Paid 4; Part-Time Volunteers 25.
Governing Authority: private; nonprofit organization. Tax-exempt: 501(c)(3).
Institution Type/Description: Sports Museum.
Collections: history of sprint car racing from early 1900s; open wheel racers to super-modified & big car racing to its present form of winged & non-winged sprint cars.
Facilities: library; 18,000 sq. ft. exhibit space; 40-seat theater; suites & skyboxes. Museum-related items for sale.
Activities: films; guided tours; hobby workshops; mobile vans; theater; traveling exhibitions. Annual Events: Induction Banquet; golf tourneys; auctions; film festival; autograph sessions; trade show; exclusive exhibits.
Publications: bimonthly newsletter, Hallmarks; annual program, NSCHOF Induction Program.
Hours & Admission Prices: Mon.-Fri. 8-5, Sat.-Sun. 12-5. Adults $4; discounts to AAA & ISHA members; members no charge. Closed New Year's Day; Easter; Thanksgiving; Christmas. &
Attendance: 11,508 (accurate)
Membership: Supporter $25; International $40; Family $50; Team $100; Friend of the Hall $500; Patron $1,000.

Lake City

CENTRAL SCHOOL PRESERVATION - CENTRAL SCHOOL MUSEUM, 211 S. Center St., Lake City, IA 51449-2003. Tel.: 712-464-8639.
Institution Type/Description: History Museum: housed in a restored 1884 schoolhouse. Listed on the National Historic Register.
Collections: local history & culture; photographs; period furnishings.
Hours & Admission Prices: Daily 9:30-11:30 & 1-5; other times by appointment.
Attendance: 900 (estimated)

Lake Mills

COUNTRY SCHOOL MUSEUM, 308 N. Mill St., Lake Mills, IA 50450-1221. Mailing Address: c/o Lake Mills Area Historical Society, 202 S. Lincoln St., Lake Mills, IA 50450. Tel.: 641-592-5253. Fax: 641-592-5252.
E-mail: lmcdc@wctatel.net
Web Site: www.lakemillsiowa.com
Institution Type/Description: Historic Buildings: housed in a former rural country school used from 1936-1954.
Collections: local history; period school furnishings; books; player piano.
Hours & Admission Prices: Memorial Day to Labor Day Sun. 2-4; other times by appointment. No charge.

Lamoni

LIBERTY HALL HISTORIC CENTER - HOME OF JOSEPH SMITH III, 1138 W. Main St., Lamoni, IA 50140-1273. Tel.: 641-784-6133.
E-mail: libhall@grm.net
Web Site: www.libhall.net
Founded: 1976.
Congressional District: 5
Key Personnel: C.E.O., Lach Mackay; Pres., Jeff Naylor; Museum Site Dir., Steve Smith.

Personnel Profile: Full-Time Volunteers 5; Part-Time Volunteers 18.
Governing Authority: nonprofit. Parent Institution: Community of Christ. Tax-exempt: 501(c)(3).
Institution Type/Description: Historic House: 1881 building; 1876 schoolhouse.
Collections: period furnishings, textiles, clothing; quilt collections; emphasis on Community of Christ.
Research Fields: biographic research; archaeological dig.
Activities: lectures; organized education programs for children, adults & undergraduate or graduate college students affiliated with Graceland University; bird walks; full moon walks; history events. Annual Events: teas; Christmas Festival.
Publications: quarterly, Restoration Trail Forum; semi-annual, Community of Christ Historic Sites Foundation Newsletter.
Hours & Admission Prices: March-Dec. 20 Tues.-Sat. 10-4, Sun. 1:30-4. No charge, donations accepted; group tours by appointment, call 641-784-6133.
Attendance: 2,500 (estimated)

Lansing

RIVER HISTORY MUSEUM, 50 S. Front St., Lansing, IA 52151. Tel.: 563-538-4641.
Institution Type/Description: History Museum.
Collections: local history & culture; period furnishings; personal artifacts; photographs; early commercial fishing equipment; boat motors; paintings; memorabilia.
Activities: Annual Event: Lansings's Fish Days in August.
Hours & Admission Prices: Summer: Sat. 1-4; other times by appointment. No charge; donations accepted.

Latimer

LATIMER HISTORICAL MUSEUM, 108 S. Akir St., Latimer, IA 50452-7593. Tel.: 641-579-6452. Fax: 641-579-6508.
E-mail: latimer@iowaconnect.com
Web Site: www.latimeriowa.net
Institution Type/Description: History Museum.
Collections: local history; photographs; military artifacts; early toys; fire equipment.
Hours & Admission Prices: Mon.-Fri. 8-3, Sat. by appointment. No charge.

Laurens

POCAHONTAS COUNTY IOWA HISTORICAL SOCIETY MUSEUM, 271 N. 3rd St., Laurens, IA 50554-1274. Mailing Address: P.O. Box 148, Laurens, IA 50554-0148. Tel.: 712-841-2577.
Founded: 1977.
Congressional District: 4
Key Personnel: C.E.O., Ann Beneke; Chm. (V) & Pres. (V), Marcia Leu; Financial Dir., Kristy Mather; Cur., Joseph Sobotka; Archivist, Jane Kirchner.
Personnel Profile: Part-Time Volunteers 11.
Governing Authority: society; nonprofit. Parent Institution: Pocahontas County Iowa Historical Society. Tax-exempt.
Institution Type/Description: Historical Society Museum: housed in Carnegie library.
Collections: medical instruments & records of early local physicians; Native American artifacts; World War I uniforms; surveying instruments; farm tools; stereoscope & slides; household items; furnishings; school trophies; costumes; toys; Civil War books. Historic Site: c.1900, Wiegert Farm; two-story farmhouse; barn; smokehouse; hoghouse; machinery building; one-room schoolhouse; church; garden.
Research Fields: county farmsteads.
Facilities: 2,500 sq. ft. exhibit space. Museum-related items for sale.
Activities: school class trips; guided tours; living history farm. Annual Event: farm festival in August.
Publications: quarterly newsletter; annual membership book.
Hours & Admission Prices: Memorial Day-Oct. every other Sun. 2-5, call for reservations. No charge; donations accepted.
Attendance: 650 (estimated)
Membership: Annual $2.50; Life $25.

Le Claire

BUFFALO BILL MUSEUM OF LE CLAIRE, IOWA, INC., 199 N. Front St., Le Claire, IA 52753-7713. Mailing Address: P.O. Box 284, Le Claire, IA 52753-0284. Tel.: 563-289-5580 & 4603.
E-mail: buffalobill@lowatekcom.net
Web Site: buffalobillmuseumleclaire.com
Founded: 1957.
Congressional District: 1
Key Personnel: Pres., Debbie Smith; Exec. Dir., Robert Schiffke; Vice Pres., Jim Fletcher; Sec., Connie Carlott; Treas., Janet Willman; Museum Shop Mgr., Terri Applegate.
Personnel Profile: Part-Time Paid 1; Part-Time Volunteers 25.
Governing Authority: nonprofit organization; board of trustees. Tax-exempt: 170(b)(1)(A).
Institution Type/Description: Preservation Project.
Collections: Buffalo Bill Cody memorabilia; riverboat memorabilia; steamboat models; 1869 wooden hall coal fired work boat on the Mississippi River; Lone Star Steamboat; Indian & pioneer artifacts; airplane flight recorders & automobile safety models of Prof. James Ryan II; manuscripts; tax records; rapids pilots; Scott County & Le Claire records.
Research Fields: Cody family; rapids pilots; pioneer families.
Facilities: library. Gift items for sale.
Activities: guided tours; lectures; permanent exhibitions.
Publications: Le Claire, Iowa, A Mississippi River Town; National Historic District Guide to the City of Le Claire, Iowa; newsletter, River Pilot's Pier.
Hours & Admission Prices: Mon.-Sat. 9-5, Sun. 12-5; Winter: Mon.-Sat. 9-4, Sun. 12-4. Adults $5, students & children 5-15 $1; discounts to tour groups of 10 or more, seniors 65 & over, military and AAM members; scout groups, school groups & members no charge. Closed New Year's Day; Thanksgiving; Christmas.
Attendance: 15,000 (accurate)
Membership: Single $10; Couple $15; Family $20.

Le Mars

BLUE BUNNY ICE CREAM PARLOR & MUSEUM, Wells Dairy, Inc., 115 Central Ave. N.W., Le Mars, IA 51031. Tel.: 712-546-4522.
E-mail: dlsusemihl@bluebunny.com
Web Site: www.bluebunny.com/parlor
Formerly: Ice Cream Capital of the World Visitor Center
Institution Type/Description: General Museum: home of Blue Bunny Ice Cream.
Collections: history of ice cream & Wells Dairy; photographs; interactive exhibitions.
Hours & Admission Prices: April-Sept. Mon.-Sat. 9am-10pm, Sun. 12-10; Oct. March Mon.-Sat. 10-9, Sun. 12-9. No charge.

PLYMOUTH COUNTY HISTORICAL MUSEUM, 335 1st Ave., S.W., Le Mars, IA 51031-2000. Tel.: 712-546-7002.
E-mail: pchmuseum@gmail.com
Web Site: plymouthcountymuseum.homestead.com/museum.html
Key Personnel: Admin., Judy Bowman; Registrar & Exhibit Mgr., Mary Holub
Institution Type/Description: History Museum. Listed on the National Register of Historic Places.
Collections: local history & culture relating to Plymouth County.
Major Exhibits: Museum Manger, 11/13-1/14.
Hours & Admission Prices: Tues.-Sun. 1-5. No charge; donations accepted.
Membership: Single $20; Couple $40; Patron $50; Century $100; Benefactor $250; Heritage $500; Pioneer $1,000.

Leon

DECATUR COUNTY MUSEUM, 111 N. Main, Leon, IA 50144. Tel.: 641-446-4841.
E-mail: chet.redman@greatwesternbank.com
Institution Type/Description: History Museum.
Collections: local history & culture; period furnishings; general strore; blacksmith shop; personal artifacts; photographs.
Hours & Admission Prices: Memorial Day to Labor Day call for hours.

Lewis

HITCHCOCK HOUSE, 63788 567th Lane, Lewis, IA 51544-5137. Mailing Address: 1609 Lomas Circle, Atlantic, IA 50022-2734. Tel.: 712-769-2323.
Key Personnel: Chm. (V), Floyd Pearce; Museum Shop Mgr., Sandy Fairbairn.

Personnel Profile: Part-Time Volunteers 2; Interns 30.
Governing Authority: Parent Institution: State of Iowa Dept. of Natural Resources. Tax-exempt.
Institution Type/Description: Historic House Museum: housed in the former home of Rev. George Hitchcock; built in 1856. A National Historic Landmark.
Collections: local history & culture; Underground Railroad; period furnishings.
Activities: temporary exhibitions; special events.
Publications: newsletter.
Hours & Admission Prices: May-Sept. Tues.-Sun. 1-5. High School & up $5; bus tour facilitators & bus drivers no charge.
Attendance: 2,250 (accurate)
Membership: $10; $25; $50; $100.

Lockridge

JOHNNY CLOCK MUSEUM, 711 W. Main St., Lockridge, IA 52635. Tel.: 319-696-3711.
Key Personnel: Owner, John McClain; Owner, Pat McClain
Institution Type/Description: Clock Museum.
Collections: hand-carved clocks.
Hours & Admission Prices: Call for hours.

Logan

MUSEUM OF RELIGIOUS ARTS, 2697 Niagara Trail, Logan, IA 51546-0122. Tel.: 712-644-3888.
E-mail: museum@loganet.net
Web Site: www.mrarts.org
Founded: 1995.
Key Personnel: Pres. (V), Kris Haase; Dir., Rhonda McHugh.
Personnel Profile: Full-Time Paid 2; Part-Time Paid 3; Part-Time Volunteers 17.
Governing Authority: nonprofit organization. Tax-exempt: 501(c)(3).
Institution Type/Description: Religious Museum.
Collections: religious history, arts, tradition & culture; statues; Biblical scenes; illuminate films; 44 life-size wax figures.
Facilities: library; chapel; hospitality room; meeting room; theater. Museum-related items for sale.
Activities: chapel rental; tours; films.
Hours & Admission Prices: Tues.-Sat. 10-4, Sun. 12-4. Adults $5; discounts to children groups, and AAA & AAM members. Closed New Year's Day, Easter, Memorial Day, Independence Day, Labor Day, Thanksgiving & Christmas. &
Attendance: 12,000 (estimated)
Membership: Single $25; Family $40.

Lone Tree

LONE TREE HISTORICAL MUSEUM, 203 S. Devoe, Lone Tree, IA 52755. Tel.: 319-338-0104.
E-mail: daleandardath@mchsi.com
Institution Type/Description: History Museum.
Collections: local history & culture; period furnishings; photographs; country store; barbershop; farm equipment & tools; miniature 1900s farm diorama.
Hours & Admission Prices: May to mid-Oct. call for hours. &

Long Grove

DAN NAGLE WALNUT GROVE PIONEER VILLAGE, 18817 290th St., Long Grove, IA 52756-9615. Tel.: 563-328-3283.
E-mail: conservation@scottcountyiowa.com
Web Site: www.scottcountyiowa.com/conservation
Founded: 1980.
Key Personnel: Dir., Chm. & Pres., Deborah Leistikow.
Personnel Profile: Full-Time Paid 1; Part-Time Paid 9; Part-Time Volunteers 125.
Governing Authority: Parent Institution: Scott County Conservation. Tax-exempt.
Institution Type/Description: Historic Buildings.
Collections: 18 Historic Buildings: c.1890 Walnut Grove Bank; c.1870 Donahue Train Depot; farm machinery building; barber shop; Bison Saloon; blacksmith shop; carpenter shop; doctor's office; Donahue Train Depot; firehouse; Keppy & Nagle General Store; school; soda fountain shop; telephone office.
Hours & Admission Prices: April-Oct. daily 9-6. No charge; donations accepted.
Attendance: 10,000 (estimated)

Lucas

JOHN L. LEWIS MINING & LABOR MUSEUM, 102 Division St., Lucas, IA 50151-7700. Mailing Address: P.O. Box 3, Lucas, IA 50151-0003. Tel.: 641-766-6831. Fax: 641-766-6831.
E-mail: danallen@coalmininglabormuseum.com
Web Site: www.coalmininglabormuseum.com
Founded: 1990.
Key Personnel: Pres. (V), Earl Seymour.
Volunteer Hours: 920
Governing Authority: nonprofit organization. Tax-exempt.
Institution Type/Description: Mining & Labor Museum.
Collections: coal mining & labor history; photographs; mine workers memorabilia; period coal mining equipment, tools & documents; John L. Lewis' personal artifacts; family history.
Facilities: theater.
Publications: annual newsletter.
Hours & Admission Prices: April 15-Oct. 15 Mon.-Sat. 9-3; groups by appointment. Adults $2, students 11 & over $1; children under 10 no charge. Closed holidays. &
Attendance: 1,000 (estimated)
Membership: Individual $5.

Lynnville

WAGAMAN MILL/MUSEUM, 200 East St., Lynnville, IA 50153. Mailing Address: P.O. Box 113, Lynnville, IA 50153. Tel.: 641-792-9780. Fax: 641-791-9976.
E-mail: conservation@co.jasper.ia.us
Web Site: www.co.jasper.ia.us/conservation
Founded: 1997.
Congressional District: 98
Key Personnel: Treas., Garnet Gertsma Walt Van Maanen.
Personnel Profile: Part-Time Volunteers 10.
Governing Authority: Parent Institution: Lynnville Historical Society. Tax-exempt.
Institution Type/Description: Historic Building: built in 1848. Listed on the National Register of Historic Places.
Collections: mill history; photographs; early area settlers; period equipment & furnishings.
Publications: quarterly newsletter.
Hours & Admission Prices: Memorial Day to Sept. Sun. 1-4; groups by appointment. No charge; donations accepted. Groups: $1 per person. &
Attendance: 538
Membership: Individual $10.

Macedonia

STEMPEL BIRD COLLECTION, 311 Main St., Macedonia, IA 51549-3002. Tel.: 712-486-2323.
Institution Type/Description: General Museum.
Collections: over 300 mounted birds.
Hours & Admission Prices: April-Nov. 1 Mon.-Fri. 9-3; other times by appointment. No charge. Closed holidays. &

Madrid

MADRID HISTORICAL MUSEUM, 109 W. Second St., Madrid, IA 50156-1339. Mailing Address: P.O. Box 105, Madrid, IA 50156. Tel.: 515-795-3249.
E-mail: madridmuseum@windstream.net
Web Site: www.madridiamuseum.com
Formerly: Clay Castle Museum & Madrid Historical Museum
Congressional District: 5
Key Personnel: Dir., Earl Check; Pres. (V), Roger Peterson.
Governing Authority: Parent Institution: Madrid Historical Society. Tax-exempt: 501(c)(3).
Institution Type/Description: Historical Society Museum.
Collections: over 1,000 dolls; toys; period furnishings; clothing; early bank vaults; replica coal mine; mining artifacts; Swedish artifacts.
Hours & Admission Prices: Call for hours. No charge; donations accepted. &
Attendance: 500 (estimated)
Membership: Individual $10.

Manning

MANNING HAUSBARN HERITAGE PARK, 12196 311th St., Manning, IA 51455. Tel.: 712-655-3131. Fax: 712-655-2941.
E-mail: heritag@mmctsu.com

Web Site: www.germanhausbarn.com
Founded: 1999.
Congressional District: 5
Key Personnel: Dir., Freda Dammann; Chm. (V), Warren Puck.
Personnel Profile: Full-Time Paid 1; Part-Time Volunteers 60.
Volunteer Hours: 1,600
Operating Expenses: 90,028
Operating Income: 101,136
Governing Authority: nonprofit organization. Parent Institution: Manning Community Foundation. Tax-exempt: 501(c)(3).
Institution Type/Description: Historic Building: housed in a dwelling that consists of family living quarters & areas for livestock and farm equipment; built in 1660 in Schleswig-Holstein Germany, dismantled and reconstructed here in 1999.
Collections: German history & culture; period furnishings; photographs. Historic Buildings: Leet/Hassler Farmstead, c.1910; Trinity Church, c.1901.
Publications: quarterly newsletter.
Hours & Admission Prices: May-Oct. Mon.-Sat. 10-4, Sun. 1-4. Adults $6.
Attendance: 5,000 (estimated)
Membership: Traditional $30; Supporting $60; Historic $120; Foundation $500.

Maquoketa

CLINTON ENGINES MUSEUM, 605 E. Maple St., Maquoketa, IA 52060. Mailing Address: P.O. Box 1245, Maquoketa, IA 52060-1245. Tel.: 563-652-1803. Fax: 563-652-1803.
E-mail: museum@jciahs.com
Web Site: www.jciahs.com
Founded: 2008.
Congressional District: 2
Key Personnel: C.E.O. & Chm., David Stockham; Dir., Bonnie W. Mitchell; Museum Shop Mgr., Marcella Heneke.
Personnel Profile: Full-Time Paid 1; Part-Time Paid 3; Part-Time Volunteers 100.
Governing Authority: Parent Institution: Jackson County Historical Society. Tax-exempt.
Institution Type/Description: History Museum.
Collections: corporation history & artifacts; gasoline engines; hands-on exhibits.
Facilities: research library; meeting rooms; audiovisual center.
Activities: tours.
Publications: Clinton Engines - 1950-1959, The Don Thomas Years; Clinton Engines 1959 Until the End; The Maquoketa I Remember; Maquoketa, One of a Kind.
Hours & Admission Prices: Tues.-Fri. 10-4, Sat.-Sun. 12-4. Adults $5; discounts to Time Travelers; members no charge. Closed major holidays. &
Attendance: 550 (estimated)
Membership: Single $20; Family $30; Contributor $50.

JACKSON COUNTY HISTORY MUSEUM & MACHINE SHED, 1212 E. Quarry, Fairgrounds, Maquoketa, IA 52060. Mailing Address: P.O. Box 1245, Maquoketa, IA 52060-1245. Tel.: 563-652-5020. Fax: 563-652-5020.
E-mail: museum@jciahs.com
Web Site: www.jciahs.com
Formerly: Jackson County Historical Museum & Research Library
Founded: 1964.
Congressional District: 2
Key Personnel: C.E.O. & Pres. (V), David Stockham; Dir., Bonnie W. Mitchell; Museum Shop Mgr., Marcella Heneke.
Personnel Profile: Part-Time Paid 6; Part-Time Volunteers 100.
Governing Authority: historical society board. Parent Institution: Jackson County Historical Society. Subsidiary Institution: Clinton Engines Corporation. Iowa Orphan Train Association Headquarters. Tax-exempt: 501(c)(3).
Institution Type/Description: History Museum.
Collections: horse-drawn farm machinery; dolls; books; township history; genealogical research material; crafts; spinning; quilting; carpentry tools; transportation; small farm tools; local paintings; country school; printing shop; business & professional; postal; log cabin; firearms; 1914 Case Steam Engine; reproduction of a McCormack Reaper; military artifacts; early cameras & radios; period tractors & agricultural equipment; general store; textiles; wildlife; phones; photographs; lime kilns.
Research Fields: genealogy & township histories of Jackson county; textile information; railroad; lime kilns; archeological; architectural; agricultural.
Activities: school, club & tourist tours; tours of local attractions; open houses & demonstrations. Museum Sponsors: Brown Bag Lunch every Tuesday; monthly programs on various topics; garage sale second week of June;

Pioneer Day, first Sunday in October; Heritage Dinner, first Sunday in November; Holiday Basket Auction, second Tuesday evening in December.
Publications: quarterly historical society newsletter, Timelines; local interest history books, Ghost Towns of Jackson County; Old Creameries; Agriculture; Country Schools; Cemetery Tours; Couple Dozen in All; Vigilantes; Clinton Engines - 1950-1959, The Don Thomas Years; The Maquoketa I Remember; Maquoketa, One of a Kind.
Hours & Admission Prices: Tues.-Fri. 10-4, Sat.-Sun. 12-4. Adults $5; discount to Time Travelers; members no charge. Closed major holidays. &
Attendance: 6,000 (estimated)
Membership: Individual $20; Family $30; Contributor $50.

OLD CITY HALL ART GALLERY, 121 S. Olive, Maquoketa, IA 52060. Tel.: 563-652-3405.
E-mail: wefrantzen@yahoo.com
Web Site: oldcityhallgallery.com
Institution Type/Description: Art Gallery: housed in the former town's fire station; built in 1901.
Collections: oil paintings by Rose Frantzen; drawings & paintings by Charles Morris; photography by Wayne Frantzen; pottery by Sylvia March.
Hours & Admission Prices: Daily 11-7; call to confirm.

Marion

GRANGER HOUSE MUSEUM, 970 Tenth St., Marion, IA 52302-3572. Mailing Address: P.O. Box 753, Marion, IA 52302-0753. Tel.: 319-377-6672.
E-mail: grangerhouse@marionhistoricalsociety.org
Web Site: marionmuseums.org
Founded: 1973.
Key Personnel: Dir., Kathy Wilson; Pres. (V), Mark Morgan.
Personnel Profile: Part-Time Paid 1.
Governing Authority: Parent Institution: Marion Historical Museum Inc., d/b/a Marion Historical Society. Tax-exempt.
Institution Type/Description: Historic House & Site.
Collections: Victorian household items & furnishings; 19th & early 20th century agricultural & transportation artifacts.
Research Fields: local history.
Activities: tours; programs; special events.
Hours & Admission Prices: May-Aug. Thurs.-Sun. 1-4; Sept.-Dec. Sat.-Sun. 1-4; other times by appointment; additional hours for special events & programs. Adults $5, students 6-18 $2; society members no charge. Closed New Year's Eve & Day; Christmas & week after.
Attendance: 1,200 (estimated)
Membership: Individual $30; Family $50.

MARION HERITAGE CENTER, 590 Tenth St., Marion, IA 52302-4409. Mailing Address: P.O. Box 753, Marion, IA 52302-0753. Tel.: 319-477-6376.
E-mail: marionheritage@juno.com
Founded: 2000.
Congressional District: 1
Key Personnel: Dir., Lynette Brenzel; Marion Historical Society Pres., Mark Morgan.
Personnel Profile: Part-Time Paid 1.
Governing Authority: municipal. Parent Institution: Marion Historical Society. Tax-exempt.
Institution Type/Description: History Museum; changing exhibits 2 times per year, 3 months art & 6 months local historical topic.
Collections: local history.
Major Exhibits: Annual Quilt Walk, 8/17/13-4/26/14; Farm Progress, 9/7/13-4/26/14; Art by Your Friends & Neighbors, 5/17/14-7/27/14.
Research Fields: business of Marion; Francis Marion; local history.
Facilities: Museum-related items for sale.
Activities: tours; special events; programs; genealogy research.
Hours & Admission Prices: Wed.-Sun. 1-4. Adults $5, children $2; members & Blue Star military no charge. &
Attendance: 2,000 (estimated)
Membership: Individual $30; Family $50; Sustaining $100.

Marshalltown

CENTRAL IOWA ART ASSOCIATION, Fisher Community Center, Marshalltown, IA 50158. Mailing Address: 709 S. Center St., Ste. #1, Marshalltown, IA 50158-2876. Tel.: 641-753-9013.
E-mail: ciaa@iowatelecom.net
Web Site: www.centraliowaartassociation.org
Founded: 1946.
Congressional District: 3

Key Personnel: Pres. (V), William Flowers; Museum Shop Mgr., Jeanne Newton-Schoborg.
Governing Authority: nonprofit organization. Tax-exempt: 501(c)(3).
Institution Type/Description: Impressionist & Ceramics Museum.
Collections: Sisley, Van Dongen, Bonnard, Signac, Henner, Vlaminck, Minaux, Buffet, Utrillo, Degas, Monticelli, Vuillard, Mattisse, Cassatt, Pissarro, Fisher Gallery, ceramics study collection.
Facilities: 500-vol. library of art books available for research to members; studio. Paintings, prints, ceramics, art supplies & jewelry for sale.
Activities: lectures; films; gallery talks; arts festivals; hobby workshops; rental gallery; formally organized education programs; permanent & temporary exhibitions; school loan service; classes & workshops; field trips; local artists gallery. Museum Sponsors: monthly art exhibits.
Publications: monthly print & electronic newsletter.
Hours & Admission Prices: Mon.-Fri. 11-5. No charge; donations accepted. Closed holidays. &
Attendance: 4,000 (estimated)
Membership: Military Veterans $15; Senior Citizen 55 & over, High School & College $25; Individual $35; Family $55; Scholarship Fund $100; Arts Angel $200; Painter's Silver $400; Painter's Gold $600; Painter's Platinum $1,000.

FISHER ART GALLERY, 709 S. Center St., Marshalltown, IA 50158-2876. Tel.: 515-753-9013.
E-mail: ciaa@iowatelecom.net
Governing Authority: Parent Institution: Fisher Foundation. Tax-exempt.
Institution Type/Description: Art Gallery.
Collections: works by regional & local artists; ceramics; sculpture; 19 French impressionist & post impressionist paintings.
Hours & Admission Prices: mid-April to mid-Oct. daily 11-5; mid-Oct. to mid-April Mon.-Fri. 11-5. No charge; donations accepted. &
Attendance: 500 (estimated)

HISTORICAL SOCIETY OF MARSHALL COUNTY, 202 E. Church St., Marshalltown, IA 50158-2943. Tel.: 641-752-6664.
Formerly: Marshall County Historical Museum
Founded: 1908.
Institution Type/Description: History Museum.
Collections: local history & culture; geology; archaeology; business & industry; medicine; dentistry; military; period furnishings; government.
Hours & Admission Prices: Tues.-Thurs. 10-4. No charge; donations accepted.

Mason City

* **CHARLES H. MACNIDER MUSEUM, (M),** 303 2nd St., S.E., Mason City, IA 50401-3988. Tel.: 641-421-3666. Fax: 641-422-9612.
E-mail: eblanchard@masoncity.net
Web Site: www.macniderart.org
Founded: 1964.
Congressional District: 4
Key Personnel: Dir., Edith M. Blanchard; Pres., Jay Hansen; Coord. Education, Linda Willeke; Registrar & Curatorial Asst., Mara Linskey; Museum Shop Mgr., Jennifer Klein.
Personnel Profile: Full-Time Paid 7; Part-Time Volunteers 50.
Governing Authority: municipal. Parent Institution: City of Mason City, IA. Tax-exempt: 170(b)(1)(A).
Institution Type/Description: American Art Museum.
Collections: American and Iowa art including paintings, prints, drawings & pottery; ceramics; puppets, marionettes & related items created & collected by the late master puppeteer Bil Baird.
Research Fields: American, Iowa, & regional art.
Facilities: 1,500-vol. library of books on art history, techniques & biographical with emphasis on American art available for use on premises; reading room; classrooms. Paintings, drawings, pottery, jewelry, prints & sculpture for sale.
Activities: guided tours; lectures; films; gallery talks; concerts; arts festivals; drama; rental gallery; formally organized education programs; docent program; permanent, temporary & traveling exhibitions.
Publications: quarterly newsletter; exhibit catalogs; annual report.
Hours & Admission Prices: Tues.-Wed. & Fri.-Sat. 9-5, Thurs. 9-9. No charge; donations accepted. Closed national holidays. &
Attendance: 20,043 (accurate)
Membership: Student $25; Individual $30 & up; Household $40 & up; Bronze $100 & up; Silver $250 & up; Gold $500 & up; Platinum $1,000 & up.

FRANK LLOYD WRIGHT STOCKMAN HOUSE AND ARCHITECTURAL INTERPRETIVE CENTER, 530 First St., N.E., Mason City, IA 50401-3534. Mailing Address: P.O. Box 565, Mason City, IA 50402. Tel.: 641-423-1923.
E-mail: robertandbonniemccoy@gmail.com
Web Site: www.stockmanhouse.org
Founded: 1990.
Key Personnel: Chm. (V), Judy Wagner; Dir. Operations, Joanne Hardinger; Museum Shop Mgr., Kathy Kinsey.
Personnel Profile: Part-Time Paid 1; Part-Time Volunteers 30.
Governing Authority: Tax exempt: 501(c)(3).
Institution Type/Description: Historic House Museum.
Collections: local history; reproductions of period Frank Lloyd Wright furnishings; architectural models of historic prairie school buildings.
Facilities: auditorium; classroom. Museum-related items for sale.
Activities: tours; architecture & design workshops; lecture series; educational programs; orientation video.
Publications: quarterly newsletter.
Hours & Admission Prices: Tours begin at the Architectural Center, 520 1st St., N.E. House: mid-May-Oct. regularly scheduled tours; Nov.-April tours only scheduled by appointment (hours & tour days & times subject to change based on time of year). Adults 10, teens 12-17 $5, children 11 & under $2. Center no charge, donations accepted. &
Attendance: 2,500 (estimated)
Membership: Individual $25; Dual $40; Corporate $100.

KINNEY PIONEER MUSEUM, Municipal Airport, Hwy. 122 W., Mason City, IA 50401. Tel.: 641-423-1258.
Institution Type/Description: History Museum.
Collections: local history & culture; Iowa frontier life; personal artifacts; early furnishings. Historic Buildings: one-room school; log cabin; jail; blacksmith shop.
Hours & Admission Prices: May-Sept. Tues.-Sun. 1-5; groups & other times by appointment. Adults $3, children under 12 $1; discounts to groups.

LIME CREEK NATURE CENTER, 3501 Lime Creek Rd., Mason City, IA 50401-9256. Tel.: 515-423-5309. Fax: 641-423-1566.
E-mail: tvonehw@co.cerro-gordo.ia.us
Web Site: www.co.cerro-gordo.ia.us
Founded: 1984.
Congressional District: 4
Key Personnel: Dir., Fred Heinz; Chm. (V), Paul Hertzel.
Personnel Profile: Full-Time Paid 1; Part-Time Volunteers 45; Interns 1.
Governing Authority: Parent Institution: Cerro Gordo County Conservation Board. Tax-exempt.
Institution Type/Description: Nature Center.
Collections: wildlife & their habitats; mounted animals & birds; native plants.
Facilities: nature trails.
Activities: public programs; school community outreach programs.
Publications: In Nature's Care.
Hours & Admission Prices: Summer: Mon.-Fri. 7:30-4, Sat. 9-5, Sun. 1-5; Winter: Mon.-Fri. 7:30-4, Sat. 9-5, Sun. 1-4. No charge; donations accepted. &
Attendance: 35,000 (estimated)

MEREDITH WILLSON BOYHOOD HOME, 314 S. Pennsylvania Ave., Mason City, IA 50401-3913. Tel.: 641-424-2852; 866-228-6262 (Toll Free).
Web Site: www.themusicmansquare.org
Institution Type/Description: Historic House Museum: housed in the birthplace & boyhood home of Meredith Willson, songwriter & playwright of The Music Man.
Collections: family memorabilia; period furnishings; personal artifacts.
Hours & Admission Prices: Tues.-Sun. 1-5; other times by appointment. Adults $6, students $3. Closed holidays.

Maxwell

THE COMMUNITY HISTORICAL SOCIETY, Main St., Maxwell, IA 50161. Tel.: 515-382-4085.
E-mail: bamddswans@gmail.com
Founded: 1964.
Congressional District: 4
Key Personnel: Pres. (V), Robert Swanson; Vice Pres., Jerry White; Historian, Mrs. Mildred McIntosh.
Personnel Profile: Part-Time Volunteers 10.
Governing Authority: society. Tax-exempt: 501(c)(3).
Institution Type/Description: General Museum.

Collections: agriculture; costumes; history; Indian artifacts; children's museum; transportation; textiles.
Facilities: 600-vol. library of Iowa history books available for use on premises; archives; reading room.
Activities: permanent, temporary & traveling exhibitions.
Publications: Maxwell Centennial History
Hours & Admission Prices: April-Sept. Sun., holidays & by appointment. No charge; donations accepted. ♿
Attendance: 1,350 (accurate)
Membership: Annual $5; Life & Memorials $25.

Melcher

MELCHER-DALLAS COAL MINING AND HERITAGE MUSEUM, 101 N.E. Center, Melcher, IA 50163. Mailing Address: P.O. Box 412, Melcher, IA 50163-0412. Tel.: 641-947-5651 & 891-7438.
E-mail: gramspop_1@iowatelecom.net
Founded: 1996.
Key Personnel: Dir., Sandra Haug.
Volunteer Hours: 40
Operating Income: 500
Governing Authority: Tax-exempt.
Institution Type/Description: Mining Museum.
Collections: mining history & tools; blacksmith shop; 1900s garden; period furnishings.
Hours & Admission Prices: Memorial Day to Labor Day Sat.-Sun. 1-4. No charge. ♿
Attendance: 225 (estimated)

Middle Amana

COMMUNAL KITCHEN AND COOPERSHOP MUSEUM, 1003 26th Ave., Middle Amana, IA 52203. Mailing Address: Amana Heritage Society, P.O. Box 81, Amana, IA 52203-0081. Tel.: 319-622-3567.
Web Site: www.amanaheritage.org
Institution Type/Description: History Museum.
Collections: period kitchen artifacts; cooper trade tools; restored 1930s communal kitchen.
Hours & Admission Prices: Call for hours. Adults $4, children 8-17 $1; children 7 & under no charge.

Milford

CLARK MUSEUM OF OKOBOJI AREA AND IOWA HISTORY, 2151 213th Ave., Milford, IA 51351-7200. Tel.: 712-338-2147.
E-mail: ijclark1@msm.com
Web Site: clarkmuseum.com
Founded: 1984.
Governing Authority: Tax-exempt.
Institution Type/Description: History Museum.
Collections: local & state history and culture; historic photographs; period farm equipment; early furnishings, restaurant & lodging memorabilia; personal artifacts; period advertising; signs; horse-drawn school bus; gas engines; calvary wheels.
Hours & Admission Prices: April-Oct. Tues.-Sat. 10-6, Sun. 11-6. No charge; donations accepted.

Minburn

THE VOAS NATURE AREA/VOAS MUSEUM, 1930 Lexington Rd., Minburn, IA 50167-8148. Mailing Address: 14581 K Avenue, Perry, IA 50220-6379. Tel.: 515-465-3577. Fax: 515-465-3579.
Founded: 1991.
Congressional District: 4
Key Personnel: C.E.O., Mike Wallace; Cur., Archivist & Education, Pete Malmberg.
Personnel Profile: Full-Time Paid 2; Part-Time Volunteers 10.
Governing Authority: county. Parent Institution: Dallas County Conservation, 1477 K Ave., Perry, IA 50220. Tax-exempt.
Institution Type/Description: Geological Museum.
Collections: rocks; fossils; minerals; rare native elements; quartz specimens.
Facilities: 800 acre recreation & conservation area; natural resource center; wetland restoration area; woodland & trails; 2,000 sq. ft. exhibit space; botanical garden.
Activities: formal education programs; guided tours.
Publications: triannual newsletter, Raccoon River Greenbelt Newsletter.

Hours & Admission Prices: May-Oct. Sat.-Sun. 1-4 when volunteers are available; other times by appointment; Nov.-April by appointment only. No charge; donations accepted.
Attendance: 1,000 (estimated)

Missouri Valley

DESOTO NATIONAL WILDLIFE REFUGE, 1434 316th Lane, Missouri Valley, IA 51555-7033. Tel.: 712-642-4121.
E-mail: desoto@fws.gov
Web Site: midwest.fws.gov/desoto
Institution Type/Description: Wildlife Refuge.
Collections: waterfowl; birds; wildlife; Missouri River Basin history; ecology; natural history.
Facilities: 8,358 acre refuge.
Activities: environmental education & interpretive programs; wildlife observation area.
Hours & Admission Prices: Refuge: sunrise to sunset. Visitor Center: daily 9-4:30. Buses & Vans: $20-$30; Cars: $3. Closed federal holidays.

HARRISON COUNTY HISTORICAL VILLAGE, 2931 Monroe Ave., Missouri Valley, IA 51555. Tel.: 712-642-2114. Fax: 712-642-2114.
E-mail: welcome@harrisoncountyparks.org
Institution Type/Description: History Museum & Welcome Center.
Collections: local history & culture; period furnishings; personal artifacts. Historic Buildings: log cabin; general store; school.
Facilities: Museum-related items for sale.
Activities: special events; educational programs.
Hours & Admission Prices: Village: mid-April to Nov. daily. Welcome Center: Mon.-Sat. 9-5, Sun. 12-5. No charge; donations accepted. Closed New Year's Day; Easter; Thanksgiving; Christmas.

STEAMBOAT BERTRAND COLLECTION, DeSoto National Wildlife Refuge, 1434 316th Lane, Missouri Valley, IA 51555-7033. Tel.: 712-388-4800; 402-359-1299. Fax: 712-388-4872. Facebook: Desoto Boyer Chute.
E-mail: dean_knudsen@fws.gov
Web Site: refuges.fws.gov/refuge/Desoto/wildlife_and_habitat/steamboat_bertrand.html
Founded: 1969.
Congressional District: 4
Key Personnel: Refuge Mgr., Tom Cox; Cur., Dean Knudsen.
Personnel Profile: Full-Time Paid 2; Part-Time Volunteers 1.
Governing Authority: federal. Parent Institution: U.S. Fish & Wildlife Service, Dept. of Interior, Washington, DC. Tax-exempt.
Institution Type/Description: Park Museum Center & Historic Site: excavation of the 1865 steamboat Bertrand.
Collections: excavated items from the Bertrand; 1865 sunken steamboat cargo; tools; hardware; textiles; clothing & shoes; armaments; patent medicine; bottled & canned food; housewares; mining, lumbering, farming & building supplies.
Research Fields: steamboat construction; mercantile packing & shipping; tools; hardware; clothing; textiles; food preservation; bottle & can manufacture; bitters; wines; Missouri River transportation; Civil War-era civilian frontier material culture.
Facilities: 1,300-vol. library pertaining to cultural history, western river navigation, steamboating & nature available for research on premises by appointment; nature & conservation center; 80-seat auditorium; theater; visitor center. Historical & nature books, prints & craft items for sale.
Activities: films; formally organized education programs for children; volunteer program; loan & permanent exhibitions.
Publications: guidebook; books, The Steamboat Bertrand; The Bertrand Stores.
Hours & Admission Prices: Daily 9-4:30. $3 per vehicle. Bus: 20 people or less $20; over 20 people $30. Federal Golden Age Passport, Federal Golden Access Passport, Federal Golden Eagle Passport, Federal Duck Stamp no charge. Closed federal holidays except Memorial Day, Independence Day & Labor Day. ♿
Attendance: 60,000 (accurate)
Membership: Desoto Refuge Annual Pass $15.

WISECUP FARM MUSEUM, 1200 Canal St., Missouri Valley, IA 51555. Mailing Address: 1772 305th St., Missouri Valley, IA 51555. Tel.: 402-689-4002 & 1984. Fax: 712-642-4232.
E-mail: cjwcup@live.com
Web Site: wisecupfarmmuseum.com
Founded: 2000.
Key Personnel: Dir., Bd. Member & Museum Shop Mgr., Charles Wisecup;

Bd. Member, Jeff Synder; Bd. Member, Paul Lane; Bd. Member, Dennis Smith; Bd. Member, Julie Wisecup; Bd. Member, Mary Hansen.
Personnel Profile: Full-Time Volunteers 2; Part-Time Volunteers 1.
Governing Authority: Tax-exempt.
Institution Type/Description: Farm History Museum.
Collections: farm history; farm equipment including 25 tractors; buggies; a sleigh; wooden wagon; 1946 GMC truck; 1961 fire truck; 1900's dress shop; barber shop; Dr. office; general store; survey office; toy shop; 1800's-1900's household artifacts; tools; one room county school; church; native grass plot; log cabin.
Hours & Admission Prices: Memorial Day to Sept. Tues. & Sun. 1-5, Sat. 9-5. No charge; donations accepted. &
Attendance: 800 (estimated)

Montezuma

POWESHIEK COUNTY HISTORICAL & GENEALOGICAL SOCIETY, 200 S. 3rd St., Montezuma, IA 50171. Mailing Address: P.O. Box 280, Montezuma, IA 50171. Tel.: 641-623-3322.
E-mail: leon429@zumatel.net
Founded: 1978.
Congressional District: 3
Key Personnel: Pres. (V), Sue Eichhorn; Dir., Joan Ehrig; Dir., Carol Klein; Dir., Sandy Cooper; Dir., Bev Creps; Dir., Pat Smith.
Personnel Profile: Part-Time Volunteers 7.
Governing Authority: Tax-exempt.
Institution Type/Description: History Museum.
Collections: local history & culture; period furnishings; personal artifacts; photographs; jail.
Facilities: library.
Activities: research.
Publications: quarterly newsletter, The Searcher.
Hours & Admission Prices: Mon. & Thurs. 9-4, Sat. 9-2. No charge; donations accepted. Closed holidays. &
Attendance: 675 (estimated)
Membership: Annual $10; Lifetime $200.

Moravia

MORAVIA WABASH DEPOT MUSEUM COMPLEX, 800 W. North St., Moravia, IA 52571. Mailing Address: P.O. Box 216, Moravia, IA 52571. Tel.: 641-724-3736.
E-mail: judy@iowatelecom.net
Volunteer Hours: 1,200
Operating Expenses: 250
Operating Income: 2,000
Governing Authority: Parent Institution: Moravia Historical Society. Tax-exempt.
Institution Type/Description: History Museum: depot built by the Wabash Railroad in the early 1900s. Listed on the National Register of Historic Places.
Collections: depot & railroad history; photographs; signal board; early 1900s section car.
Hours & Admission Prices: May-Oct. 1st weekend each month. No charge; donations accepted. &
Attendance: 500 (estimated)
Membership: Annual $10.

Morrison

GRUNDY COUNTY HERITAGE MUSEUM, 204 Fourth St., Morrison, IA 50657. Tel.: 319-345-2688. Fax: 319-345-2688.
E-mail: gccb@gccourthouse.org
Web Site: www.grundycounty.org
Key Personnel: Dir., Kevin Williams; Museum Shop Mgr., Sue Eckhoff.
Personnel Profile: Full-Time Paid 3; Part-Time Paid 1.
Governing Authority: Parent Institution: Grundy County Conservation Board. Subsidiary Institution: Historical Collections of Grundy County. Tax-exempt.
Institution Type/Description: History Museum.
Collections: over 250 mounted animals; railroad artifacts; photographs; period furnishings.
Hours & Admission Prices: Tues.-Thurs. 8-4:30. No charge; donations accepted. &
Attendance: 3,000 (estimated)

Moulton

MOULTON HISTORICAL SOCIETY MUSEUM, Hwy. 202, Moulton, IA 52572. Tel.: 641-642-3684.
Institution Type/Description: Historical Society Museum.
Collections: local history & culture; railroad artifacts; rural schoolhouse; period furnishings.
Hours & Admission Prices: Memorial Day to Labor Day Sun. 1-4; tours by appointment.

Mount Pleasant

HARLAN-LINCOLN HOUSE, 101 W. Broad St., Mount Pleasant, IA 52641-1337. Mailing Address: Iowa Wesleyan College, 601 N. Main St., Mount Pleasant, IA 52641. Tel.: 319-385-6215. Fax: 319-385-6324.
E-mail: iwcarch@iwc.edu
Web Site: iwc.edu
Founded: 1959.
Congressional District: 1
Key Personnel: C.E.O., Lynn Ellsworth; Chm. (V), Elizabeth Garrels.
Personnel Profile: Part-Time Paid 1; Part-Time Volunteers 15.
Governing Authority: college. Affiliated with Iowa Wesleyan College, N. Main St. Tax-exempt: 501(c)(3).
Institution Type/Description: Historic House: housed in retirement home of U.S. Senator James Harlan (1876-1899); the summer home of the Robert Todd Lincoln Family (1876-1907).
Collections: original furnishings and memorabilia of the Harlan & Lincoln families; objects of same period.
Activities: guided tours.
Publications: newsletter, Friends of the Harlan-Lincoln House.
Hours & Admission Prices: By appointment only. Adults $3; members no charge. &
Attendance: 1,000 (accurate)
Membership: Individual $30; Family $50; Ambassador $100-$249; Cabinet $250-$499; Senator $500-$999; Harlan Society $1,000-$4,999.

MIDWEST OLD SETTLERS & THRESHERS ASSOCIATION, INC., 405 E. Threshers Rd., Mount Pleasant, IA 52641-2584. Tel.: 319-385-8937. Fax: 319-385-0563.
E-mail: info@oldthreshers.org
Web Site: www.oldthreshers.org
Founded: 1950.
Congressional District: 1
Key Personnel: C.E.O., Lennis Moore; Pres. (V), Chris Heaton; Cur. Theatre, Dr. Mike Kramme; Pub. Rels. Coord., Terry McWilliams; Museum Shop Mgr., Linda Dovenspike.
Personnel Profile: Full-Time Paid 7; Full-Time Volunteers 4; Part-Time Paid 2; Part-Time Volunteers 497; Interns 4.
Governing Authority: nonprofit organization. Parent Institution: Midwest Old Threshers. Branch Museum: Museum of Repertoire Americana, Mt. Pleasant, IA. Tax-exempt: 501(c)(3).
Institution Type/Description: Agricultural History Museum.
Collections: steam traction engines, stationary steam, agricultural implements; transportation; folk theatre; antiques. Historic Buildings: authentic railroad depot; narrow gauge railroad; trolleys; log houses; country church; schoolhouse; barber shop; bandstand.
Research Fields: agricultural history; Chautauqua & repertoire theater.
Facilities: research library; theater. Museum-related items for sale.
Activities: guided tours; lectures; films; TV & radio programs; formally organized education programs; permanent, temporary & traveling exhibitions; repertoire theater; docent program. Museum Sponsors: Bussey Doll Convention; Printers' Fair; Midwest Haunted Rails; Thrashers' House of Terror.
Publications: quarterly newspaper, Threshers Chaff; annual, Threshers Review.
Hours & Admission Prices: Office: Mon. Fri. 8 5. Museum: call for hours. Adults $5; children 14 & under no charge. Theatre: $3 per person. &
Attendance: 5,127 (estimated)
Membership: Annual $20.

Mount Vernon

PETER PAUL LUCE GALLERY, MCWETHY HALL, CORNELL COLLEGE, 600 1st St., S.W., Mount Vernon, IA 52314-1098. Tel.: 319-895-4491. Fax: 319-895-4519.
E-mail: scoleman@cornellcollege.edu
Web Site: www.cornellcollege.edu
Formerly: Armstrong Gallery, Cornell College

Founded: 1853.
Congressional District: 2
Key Personnel: Pres., Jonathan Brand; Dept. Chm., Tony Plant; Dean, Joseph Dieker; Business Officer, Katie Green; Coord. Exhibitions, Susan Coleman; Dir. Public Information, DeeAnn Rexroat.
Personnel Profile: Full-Time Paid 1; Interns 2.
Governing Authority: college. Parent Institution: Cornell College. Tax-exempt.
Institution Type/Description: College Art Gallery.
Collections: Sonnenschein collection of baroque drawings; drawings & prints of Thomas Nast; Whiting collection of Phoenician glass; Karel Appel painting; Roy Lichtenstein lithograph prints; Grant Wood lithographs; Charles Atherton Cumming's paintings; Richard Anuskiewicz; Larry Rivers painting; Henry A. Mills paintings.
Major Exhibits: Jocelyn Chateauvent, Paper Installation, 1/19/14-3/2/14; Senior Art Shows, 3/14-5/7/14; Cornell Studio Art Faculty Exhibition, 9/14-10/14; Gillian Pederson-Krag, Paintings & Etchings, 11/14-12/14.
Activities: lectures; formally organized education programs for undergraduate college students; temporary & traveling exhibitions.
Publications: catalog, Doug Hanson Retrospective (2012).
Hours & Admission Prices: Academic Year: Mon.-Fri. 9-4, Sun. 2-4; Summer: by appointment only. No charge. Closed school holidays. &
Attendance: 5,000 (estimated)

Muscatine

* **MUSCATINE ART CENTER, (M),** 1314 Mulberry Ave., Muscatine, IA 52761-3429. Tel.: 563-263-8282. Fax: 563-263-4702. Facebook: Muscatine Art Center.
E-mail: art@muscatineiowa.gov
Web Site: www.muscatineartcenter.org
Founded: 1965.
Congressional District: 1
Key Personnel: Dir., Melanie K. Alexander; Program Coord., Katy Doherty; Registrar, Virginia Cooper; Office Coord., Lynn Bartenhagen.
Personnel Profile: Full-Time Paid 3; Part-Time Paid 5; Part-Time Volunteers 40.
Governing Authority: municipal. Tax-exempt: 501(c)(3).
Institution Type/Description: Art Center: housed in 1908 Edwardian style Musser Mansion & 1976 Stanley Gallery.
Collections: 19th & 20th century American art; paintings; sculpture; prints; drawings; oriental rugs; textiles; decorative arts; historical artifacts of Iowa & Muscatine; antique glass paperweights; children's toys.
Research Fields: related to art collection, emphasizing the Great River Collection of works documenting the Mississippi River.
Facilities: 1,000-vol. library of art history, period furnishings, decorative arts, available for research on premises only; 100-seat auditorium; classrooms; gallery.
Activities: guided tours; lectures; films; gallery talks; concerts; formally organized education programs; docent program; permanent & temporary exhibitions.
Publications: catalogues of exhibitions.
Hours & Admission Prices: Tues.-Wed. & Fri. 10-5, Thurs. 10-7, Sat.-Sun. 1-5. No charge; donations accepted. Closed national holidays. &
Attendance: 32,000 (accurate)
Membership: Individual $30; Family $50; Contributing $100; Sustaining $500; Benefactor $1,000.

MUSCATINE HISTORY & INDUSTRY CENTER, 117 W. Second St., Muscatine, IA 52761-3714. Tel.: 563-263-1052. Facebook: Muscatine History and Industry Center.
E-mail: muscatinehistory@machlink.com
Web Site: www.muscatinehistory.org
Founded: 1987.
Key Personnel: Exec. Dir., Mary Wildermuth; Pres., Steve Forbes.
Personnel Profile: Full-Time Paid 1; Part-Time Paid 2; Part-Time Volunteers 15; Interns 1.
Governing Authority: private; nonprofit organization. Parent Institution: Historic Muscatine, Inc., 117 W. 2nd St., Muscatine, IA 52761. Tax-exempt: 501(c)(3).
Institution Type/Description: History Museum.
Collections: 1890-1960s pearl button memorabilia & machinery; history of the freshwater pearl button industry in Muscatine, IA; local business artifacts.
Research Fields: clamming history; history of button manufacturing.
Facilities: Museum-related items for sale.
Activities: guided tours; corporate events.
Hours & Admission Prices: Tues.-Sat. 10-4. Suggested Donation: adults $5, students $1. &
Attendance: 10,000 (accurate)
Membership: Senior $25; Individual $35; Household $50; Corporate $500 & up.

PINE CREEK GRIST MILL, Wildcat Den State Park, 1884 Wildcat Den Rd., Muscatine, IA 52761-9479. Tel.: 319-263-4337. Fax: 319-264-8329.
E-mail: jim.ohl@dnr.state.ia.us
Web Site: www.pinecreekgristmill.com
Institution Type/Description: Historic Building: built in 1848 by Benjamin Nye. Listed on the National Register of Historic Places.
Collections: structure; 19th-century milling industry.
Activities: educational programs for students.
Hours & Admission Prices: May & Sept. 26-Oct. 11 Sat.-Sun. 12:30-4:30; June-Sept. 20 Wed.-Sun. 12:30-4:30

Nashua

CHICKASAW COUNTY HISTORICAL SOCIETY BRADFORD PIONEER VILLAGE MUSEUM, 2729 Cheyenne Ave., Nashua, IA 50658-9611. Tel.: 641-435-2567.
E-mail: chickasawhist@gmail.com
Founded: 1953.
Congressional District: 2
Key Personnel: Pres. & Museum Shop Mgr., Leatha Springer; Vice Pres., Ruth Rosauer; Sec. & Treas., Barb Cairns.
Governing Authority: nonprofit organization. Tax-exempt: 170(b)(1)(A).
Institution Type/Description: General Museum: located on site of 1859 original Bradford village.
Collections: pioneer farm machinery; Indian artifacts; arts & crafts; clothing; Victorian cottage & furniture; Dr. Pitts Medical Office; depot & railroad museum; caboose & railroad track; 2 log homes; country store; laundry; blacksmith shop; agriculture building; heritage house; toy shop; country school; early jail.
Research Fields: genealogy.
Facilities: old school texts & medical books available for use by appointment. Gifts & handicrafts for sale.
Activities: guided tours; lectures; permanent exhibitions.
Publications: pamphlet; annual report; newsletter; guide book tours.
Hours & Admission Prices: May-Oct. Mon.-Sat. 9-5, Sun. 12-5; groups by appointment only. Adults $5, children K-12 $3; members no charge.
Attendance: 3,000 (estimated)
Membership: Annual $20.

New London

DOVER HISTORICAL MUSEUM, 213 W. Main St., New London, IA 52645-1337. Tel.: 877-468-7700.
Web Site: www.dovermuseum.org
Founded: 1994.
Key Personnel: Pres., Gwen Moore.
Personnel Profile: Part-Time Volunteers 20.
Governing Authority: Tax-exempt.
Institution Type/Description: Historic Building: listed on the National Register of Historic Places.
Collections: local history; genealogy; rocks & minerals; quilts; early businesses; Masons & Eastern Star; one-room school; agriculture; military; railroad artifacts. Historic Building: 1870 railroad depot.
Activities: demonstrations for children.
Publications: triannual newsletter.
Hours & Admission Prices: May-Nov. Sat.-Sun. 1-4. No charge; donations accepted. &
Attendance: 1,700 (estimated)
Membership: Student $1; Individual $10; Family $15; Friend $50-$99; Sustaining $100-$249; Supporting $250-$499; Benefactor $500-$999; Sponsor $1,000 & up.

Newton

JASPER COUNTY MUSEUM, 1700 S. 15th Ave. W., Newton, IA 50208-4321. Tel.: 641-792-9118.
E-mail: jascomus@iowatelecom.net
Web Site: www.jaspercountymuseum.net
Founded: 1979.
Key Personnel: Dir., Audrey C. Rex; Pres. (V), Linda K. Wormley.
Governing Authority: Tax-exempt.
Institution Type/Description: History Museum.
Collections: local & coal mining history; advertising agency industry history; Maytag washing machine history; farm tools & equipment; period furnishings; washing machines; posters; historic building.
Activities: county historical tours. Museum Sponsors: Brown Bay Lunch & Learns; Newton Chamber of Commerce Breakfast in April; Maytag Collectors Annual Meeting; Jasper County cemetery board meeting; Youth Day; Historical Cemetery Walk; Family Heritage Day; Christmas Open House.

Publications: quarterly newsletter, Jasper Journal.
Hours & Admission Prices: By appointment. Adults over 19 $5, children under 19 $2. &

Attendance: 600 (estimated)
Membership: Individual $10; Family $20.

NEWTON ARBORETUM AND BOTANICAL GARDEN, 3000 N. 4th Ave. E, Newton, IA 50208-8745. Tel.: 641-791-3021.
Institution Type/Description: Arboretum & Botanical Garden.
Collections: trees; plants; flowers; gardens.
Hours & Admission Prices: Daily daylight hours. Learning Center: Mon.-Fri. 8-4. &

Northwood

GLADYS PIXLEY MEMORIAL LOG HOUSE, Central Ave. & 4th St., Northwood, IA 50459. Mailing Address: Worth County Historical Society, 917 Central Ave., Northwood, IA 50459-1525. Tel.: 641-324-1180.
Institution Type/Description: Historic Building: housed in the former home of James Randall; built in 1858.
Collections: local history; period furnishings; spinning wheel; bentwood cradle; spool bed; early household utensils & artifacts.
Hours & Admission Prices: Memorial Day to Labor Day Sun. 2-4. No charge; donations accepted.
Membership: Individual $10; Lifetime $100.

MACHINERY MUSEUM, Central Ave. & 4th St., Northwood, IA 50459. Mailing Address: Worth County Historical Society, 917 Central Ave., Northwood, IA 50459-1525. Tel.: 641-324-1180.
Institution Type/Description: Historic Building: housed on the former site of the Charles Wordall Saw Mill, the A.J. Dwelle Grist Mill, and the Ice House.
Collections: local history; period machinery & equipment; horse mower; threshing machine; 1886 racing sulky; Hoosier shovel seeder; walking plow; gang plow; oats binder; corn binder.
Hours & Admission Prices: Memorial Day to Labor Day Sun. 2-4. No charge; donations accepted.
Membership: Individual $10; Lifetime $100.

MAIN MUSEUM - WORTH COUNTY HISTORICAL SOCI-ETY, 917 Central Ave., Northwood, IA 50459-1525. Tel.: 641-324-1180.
Institution Type/Description: Historical Society Museum: housed in the former Worth County Courthouse; built in 1879.
Collections: local history & culture; photographs; period furnishings; personal artifacts; early clothing; quilts.
Hours & Admission Prices: Memorial Day to Labor Day Sun. 2-4. No charge; donations accepted. &
Membership: Individual $10; Lifetime $100.

OLD CREAMERY MUSEUM, Central Ave. & 4th St., Northwood, IA 50459. Mailing Address: Worth County Historical Society, 917 Central Ave., Northwood, IA 50459-1525. Tel.: 641-324-1180.
Institution Type/Description: Historic Building: built in 1892.
Collections: local history; period furnishings; rug loom; churn; scythe; equipment; tools; cream separators; swill cart; chain pump; print shop.
Hours & Admission Prices: Memorial Day to Labor Day Sun. 2-4. No charge; donations accepted.
Membership: Individual $10; Lifetime $100.

SWENSRUD SCHOOL MUSEUM, Central Ave. & 10th St., Northwood, IA 50459. Mailing Address: Worth County Historical Society, 917 Central Ave., Northwood, IA 50459. Tel.: 641-324-1180.
Institution Type/Description: Historic Building: housed in a former school built by William G. Stott in 1874.
Collections: local history; period furnishings; desks; pot-bellied stove; lunch pails; early school books.
Hours & Admission Prices: Memorial Day to Labor Day Sun. 2-4. No charge; donations accepted.
Membership: Individual $10; Lifetime $100.

Norway

IOWA BASEBALL MUSEUM OF NORWAY, 112 E. Railroad St., Norway, IA 52318. Tel.: 319-721-6288.
Web Site: www.norwaybaseballmuseum.com
Institution Type/Description: Sports Museum.
Collections: baseball history; championship trophies; local & national major league players; photographs; personal artifacts; baseball memorabilia.
Hours & Admission Prices: Tues., Thurs. & Sat. 1-4.

Oakland

NISHNA HERITAGE MUSEUM, 117 N. Main St., Oakland, IA 51560. Tel.: 712-482-6802.
Web Site: www.nishnaheritagemuseum.com
Founded: 1975.
Congressional District: 5
Key Personnel: Pres., Gayle Perkins; Finance Dir., Wilson Pechacek; Trustee, Jo Kates.
Personnel Profile: Full-Time Volunteers 1; Part-Time Volunteers 8.
Governing Authority: society; nonprofit organization. Parent Institution: Oakland Historical Society. Tax-exempt: 501(c)(3).
Institution Type/Description: Heritage Museum: housed in 1905 general store; 1907 hardware store.
Collections: thimbles; sewing machines; washing machines; keys; buttons; buckles; ladies combs; irons; clothing; fruit jars; lighting fixtures; ice making machinery; furnishings; local artifacts; household items; old conveyances; soda fountain, ice cream table & chairs; dolls; model Conestoga wagon. Historic Building: 1907 hardware store; archaeology bones of 1 to 64 million years ago; history of printing artifacts; salesman models.
Research Fields: c.1900 artifacts.
Activities: guided tours; lectures; organized education programs for children.
Publications: weekly columns; newspaper.
Hours & Admission Prices: Mon.-Fri. 11-3. Adults $5. &
Attendance: 500 (estimated)
Membership: Annual $10.

Odebolt

ODEBOLT HISTORICAL MUSEUM, 2nd & Maple Sts., Odebolt, IA 51458. Mailing Address: 3181 Fox Ave., Odebolt, IA 51458-0196. Tel.: 712-668-2264 & 2766.
E-mail: cklarson@netins.net
Web Site: www.odebolt.net/museum.html
Key Personnel: Pres. & Cur., Kathy Larson; Vice Pres., Alice Hemphill; Sec., Mary Schroeder; Treas., Renae Babcock
Institution Type/Description: History Museum.
Collections: military artifacts; household items; farming tools & equipment; Buffalo Bill's buffalo robe; photographs.
Hours & Admission Prices: Memorial Day & Odebolt Creek Days in June; other times by appointment. No charge; donations accepted.
Attendance: 344 (accurate)

Oelwein

OELWEIN AREA HISTORICAL SOCIETY MUSEUM, 900 2nd Ave., S.E., Oelwein, IA 50662-3055. Mailing Address: P.O. Box 445, Oelwein, IA 50662-0445. Tel.: 319-283-4203.
Institution Type/Description: Historical Society Museum.
Collections: local history & culture; photographs; personal artifacts.
Hours & Admission Prices: June-Sept. Sun. 1-4; Oct.-May by appointment. Adults $2, students $1; discounts to groups; children under 12 & members no charge. &
Attendance: 750 (estimated)
Membership: Individual $10.

Ogden

HICKORY GROVE RURAL SCHOOL MUSEUM, Baltin Chapel Complex, Junction of E 41 & J Ave., Ogden, IA 50212. Mailing Address: 602 Story St., Boone, IA 50036-2832. Tel.: 515-432-1907.
E-mail: director@boonecountyhistory.org
Web Site: www.boonecountyhistory.org
Founded: 1972.
Congressional District: 4
Key Personnel: Exec. Dir., Pamela Schwartz; Pres. (V), Lee McNair; Sec., Janet Tait; Treas., Judy Russell.

Personnel Profile: Full-Time Paid 1; Part-Time Volunteers 10.
Governing Authority: nonprofit organization. Parent Institution: Boone County Historical Society, 602 Story St., Boone, IA 50036. Tax-exempt: 501(c)(3).
Institution Type/Description: Historic House & Museum: housed in an 1889 restored rural school.
Collections: original double desks; stage curtain; pot-bellied stove; angle lamps; pump organ; photographs; documents; teacher's & children's costumes.
Research Fields: rural school history of Boone County & Iowa.
Facilities: 200-vol. library of rural school textbooks; adjacent park with fishing, golfing & camping available.
Activities: guided tours; organized educational programs for children; video tape presentation.
Publications: biannual journal, Trail Tales; biannual newsletter, History Page.
Hours & Admission Prices: By appointment. Adults $1, children 17 & under no charge.
Attendance: 400 (estimated)
Membership: Boone County Historical Society: Individual & Family $40; Business & Organization $100; Life $600.

Okoboji

THE HIGGINS MUSEUM, (M), 1507 Sanborn Ave., P.O. Box 258, Okoboji, IA 51355. Mailing Address: P.O. Box 1, Boone, IA 50036. Tel.: 712-332-5859. Fax: 712-332-5859.
E-mail: ladams@thehigginsmuseum.org
Web Site: www.thehigginsmuseum.org
Founded: 1978.
Congressional District: 5
Key Personnel: Bd. Pres. & Chm. (V), Dean Oakes; Cur., Larry Adams.
Personnel Profile: Full-Time Paid 1; Part-Time Paid 1; Part-Time Volunteers 5.
Governing Authority: private; nonprofit organization.
Institution Type/Description: History & Numismatics Museum.
Collections: national bank notes from 1863-1935; concentrating on Iowa & adjoining states (MN, NE, WI, IL, MO, SD) with representation from all issuing states; 20,000 real photos, turn of the century Iowa postcards; security printing; large collection of bank directories & comptroller of the currency annual reports.
Research Fields: national banks in Iowa, Minnesota, South Dakota, Nebraska, Missouri & other states.
Facilities: 2,500-vol. library of numismatics; 7,000 sq. ft. exhibit space. Museum-related items for sale.
Activities: guided tours; temporary exhibitions. Annual Event: coin, paper money & post card show in August.
Hours & Admission Prices: Memorial Day Weekend to Labor Day Tues.-Sun. 11-5:30. No charge; donations accepted. &
Attendance: 950 (estimated)

PEARSON LAKES ART CENTER, 2201 Hwy. 71, Okoboji, IA 51355-0255. Mailing Address: P.O. Box 255, Okoboji, IA 51355-0255. Tel.: 712-332-7013. Fax: 712-332-7014.
E-mail: info@lakesart.org
Web Site: www.lakesart.org
Founded: 1965.
Key Personnel: C.E.O., Tom Tourville; Dir. Visual Arts, Danielle Clouse; Dir. Performing Art, Rachelle Fratzke; Dir. Education, Holly Zinn.
Personnel Profile: Full-Time Paid 3; Part-Time Paid 4; Interns 1.
Governing Authority: Tax-exempt.
Institution Type/Description: Art Center.
Collections: hands-on exhibits; works by international & national artists.
Activities: festivals; readings; educational programs; exhibit openings; musical & theatrical events.
Publications: annual season catalog.
Hours & Admission Prices: June-Aug. Mon.-Wed. & Fri.-Sat. 10-4, Thurs. 10-9, Sun. 12-3; Sept.-May Tues.-Wed. & Fri.-Sat. 10-4, Thurs. 10-9. No charge; donations accepted. &
Attendance: 20,000 (accurate)
Membership: Single $45; Family $65; Supporter $150; Patron $250; Associate $500; Benefactor $1,000; Corporate $5,000.

Onawa

MONONA COUNTY HISTORICAL MUSEUM, 47 12th St., Onawa, IA 51040. Mailing Address: Box 382, Onawa, IA 51040-0382. Tel.: 712-423-3452.
E-mail: jrobbins@longlines.com
Founded: 1992.
Key Personnel: Pres., Jim Robbins.
Personnel Profile: Part-Time Volunteers 15.

Governing Authority: Parent Institution: Loess Hills Historical Society. Tax-exempt.
Institution Type/Description: History Museum: birthplace of the Eskimo Pie.
Collections: Eskimo Pie company history; dipping machine & equipment; farm implements; period furnishings.
Hours & Admission Prices: Memorial Day to Labor Day Sat.-Sun. 1-4:30; other times by appointment. No charge; donations accepted. &
Attendance: 1,000 (estimated)
Membership: Individual $5; Business $15; Sustaining $25; Patron $50.

MONONA COUNTY VETERAN'S MEMORIAL MUSEUM, 203 13th St., Onawa, IA 51040. Mailing Address: P.O. Box 418, Onawa, IA 51040-0418. Tel.: 712-423-2411. Facebook: Monona County Veteran's Memorial Museum.
E-mail: democrat@longlines.com
Web Site: www.webspawner.com/users/mononacountyvmm
Founded: 2000.
Key Personnel: Dir. & Cur., William Wonder.
Personnel Profile: Full-Time Volunteers 2; Part-Time Volunteers 10.
Volunteer Hours: 800
Operating Expenses: 11,000
Operating Income: 11,000
Governing Authority: city. Parent Institution: Onawa Community Foundation. Subsidiary Institution: Monona County Veteran's Board. Tax-exempt.
Institution Type/Description: Military Museum.
Collections: M60 armored tank; A7D Corsair II fighter jet; Vietnam ear UH-1 Huey helicopter; photographs; uniforms; personal artifacts; 105 Howitzer; '42 Ford Jeep; Danforth ship's anchor; military weapons; 1942 Diamond T Half Track; M3 Half Track; 1917 horse-drawn ammunition & machine gun cart; B-26 Marauder fuselage section.
Facilities: library; multimedia center.
Activities: monthly legion meetings; military-related programs; movies.
Publications: brochures.
Hours & Admission Prices: May-Sept. Sat.-Sun. 1-4; other times by appointment. No charge; donations accepted. Bus Tours: $3 per person. &
Attendance: 2,750 (accurate)

Orient

HENRY A. WALLACE COUNTRY LIFE CENTER, 2773 290th St., Orient, IA 50858. Mailing Address: P.O. Box 363, Greenfield, IA 50849-0363. Tel.: 641-337-5019.
E-mail: haw@mddc.com
Web Site: www.henryawallacecenter.com
Key Personnel: Dir., Diane Weiland
Institution Type/Description: History Museum: housed on the birthplace farmstead of Henry A. Wallace, born Oct. 7, 1888.
Collections: Wallace's life & career; personal artifacts; period furnishings; photographs; writings.
Facilities: nature trails.
Activities: special events; rental facilities.
Hours & Admission Prices: Self-Guided Tours: daily. Guided Tours: Summer daily 10-5. Donations requested.

Osage

CEDAR VALLEY MEMORIES, 1 1/2 Mile W. Hwy. 9, Osage, IA 50461. Mailing Address: Cedar Valley Memories c/o Mitchell County Historical Society, P.O. Box 51, Osage, IA 50461-0051. Tel.: 641-732-1269.
Governing Authority: county. Parent Institution: Mitchell County Historical Society. Tax exempt: 501(c)(3).
Institution Type/Description: History Museum.
Collections: 5 vintage steam engines including 1922 32 H.P. Advance-Rumley, 1912 Reeves 40-140 Cross Compound, 1878 Blumentrit, two cylinder; first gas running car built in Osage, 1901; agricultural items.
Activities: Museum Sponsors: Cedar Valley Memories Power Show in August.
Hours & Admission Prices: Memorial Day to Labor Day Sat.-Sun. 1-4; other times by appointment. No charge.

MILTON R. OWEN NATURE CENTER, 18793 Hwy. 9, Osage, IA 50461. Tel.: 641-732-5204. Fax: 641-732-1138. Facebook: Mitchell County Conservation Board.
E-mail: mccb@osage.net
Web Site: www.mitchellcountyconservation.com
Formerly: Mitchell County Nature Center
Founded: 1998.
Governing Authority: Tax-exempt.
Institution Type/Description: Nature Center.

Collections: wildlife specimens; photographs.
Facilities: rentable space; full kitchen; flush restrooms.
Activities: nature observation; bird feeding observation; camping; cross-country skiing; hiking; biking; equestrian trail; kayak; canoe; archery; hunting; fishing; photography; trumpeter swan enclosure; children's play area & activities; education displays; historical displays; environmental education programs.
Publications: quarterly newsletter, The Tributary.
Hours & Admission Prices: Tues.-Fri. 9-4, Sun. 1-4. No charge. &

MITCHELL COUNTY HISTORICAL MUSEUM, 809 Sawyer Dr., Ste. 2, Osage, IA 50461. Tel.: 641-832-2574.
Founded: 1965.
Congressional District: 14
Key Personnel: Chm. & Dir., Ellen Elsbury; Pres., Kurt Meyer; Genealogy Research & Library Coord., Jerry Fisk.
Personnel Profile: Part-Time Paid 2; Part-Time Volunteers 125; Interns 2.
Governing Authority: county. Parent Institution: Mitchell County Historical Society. Tax-exempt: 501(c)(3).
Institution Type/Description: History Museum.
Collections: clothing; guns; tools; household items; books; portraits; documents; mail wagon, covered wagon; musical instruments; medical instruments; spinning wheels; original beauty shop equipment; Indian artifacts; Hamlin Garland.
Research Fields: genealogy; cemeteries of the county; county history; Century Farms rural school records; war veterans.
Facilities: library; reading room; children's activity area. Museum-related items for sale.
Activities: guided tours; lectures; cultural events; rental facilities; class reunions; gallery tours; local artists' exhibits; children's activity area.
Publications: cookbook, Heritage from the Kitchen; History of David, Iowa; Osage History 150 Years; New Haven, Iowa; History of Stacyville; History of Little Cedar, Iowa; History of Riceville, Iowa; DVDs, Mitchell County Memories; Cedar Valley Seminary Museum; Living Art of Erik Budd.
Hours & Admission Prices: Daily 1-5; other times by appointment. No charge; donations accepted. Closed Easter; Christmas. &
Attendance: 10,000 (accurate)
Membership: Mitchell County Historical Society: Individual $15; Family $25.

Osceola

CLARKE COUNTY MUSEUM, 1030 S. Main, Osceola, IA 50213. Tel.: 641-342-3027.
E-mail: keefe@iowatelecom.net
Institution Type/Description: History Museum.
Collections: local history & culture; period furnishings; personal artifacts; photographs; country schoolhouse; early equipment. Historic Building: log cabin.
Hours & Admission Prices: May-Sept. call for hours. No charge. &

Oskaloosa

NELSON PIONEER FARM AND MUSEUM, (M), 2211 Nelson Lane, Oskaloosa, IA 52577-9609. Mailing Address: Mahaska County Historical Society, P.O. Box 578, Oskaloosa, IA 52577-0578. Tel.: 641-672-2989. Facebook: Nelson Pioneer Farm - Mahaska County Historical Society.
E-mail: curator@nelsonpioneer.org
Web Site: www.nelsonpioneer.org
Founded: 1942.
Key Personnel: Dir., Kelley Halbert; Pres., Jay Fox.
Personnel Profile: Full-Time Paid 1; Part-Time Paid 18; Part-Time Volunteers 2.
Volunteer Hours: 274
Operating Expenses: 68,000
Operating Income: 75,000
Governing Authority: society; nonprofit educational institution. Owned and operated by Mahaska County Historical Society. Tax-exempt.
Institution Type/Description: General Museum: housed on an 1844 homestead.
Collections: agriculture; Indian artifacts; archaeology; farm machinery exhibit. Historic Houses: 1853, Prine Schoolhouse; 1861, Littler Log Cabin; Spring Creek Voting House; 1915 W.L. Mott & Son General Store; 1865 Buffalo farm scale & Scale House; 1864 Coal Creek Friends meeting house; Wright, Iowa Post Office; Kalbach Lumberyard first office; blacksmith shop; Hopewell building; 1853 Daniel & Margaret Nelson Home & 1856 Nelson barn.
Research Fields: agriculture.
Facilities: 1,000-vol. library of books, scrap books, folders & clippings available for use on premises. Gift items for sale.

Activities: guided tours; permanent exhibitions; monthly events & dances. Museum Sponsors: Fall Festival Day in September.
Publications: quarterly newsletter, News Bulletin.
Hours & Admission Prices: mid-May to mid-Oct. Tues.-Sat. 10-4; bus tours by special arrangement. Adults $7, students 5-16 $2; children under 5 & society members no charge. &
Attendance: 3,537 (accurate)
Membership: Individual $10; Family $18.

Ottumwa

AIRPOWER MUSEUM INC., 22001 Bluegrass Rd., Ottumwa, IA 52501-8569. Tel.: 641-938-2773. Fax: 641-938-2093.
E-mail: antiqueairfield@sirisonline.com
Web Site: www.antiqueairfield.com
Founded: 1965.
Congressional District: 3
Key Personnel: Chm. & C.E.O., Robert L. Taylor; Treas., Brent Taylor; Graphics Editor, Cindy Reis.
Governing Authority: nonprofit organization. Tax-exempt: 501(c)(3).
Institution Type/Description: Aeronautics Museum.
Collections: period airplanes; model airplanes; aviators clothing; manuscripts; medals; trophies; 57 full-size aircraft; 150 aircraft models; 50 aircraft engines; miscellaneous items related to aviation history.
Research Fields: aviation history; historical technical information.
Facilities: library of books on aviation history; reading room. Books & gift items for sale.
Activities: loan & permanent exhibitions.
Publications: quarterly magazine, Airpower Museum (APM); bulletin.
Hours & Admission Prices: Mon.-Fri. 9-5, Sat. 10-5, Sun. 1-5. No charge; donations accepted. Closed New Year's Day; Independence Day; Labor Day; Thanksgiving; Christmas. &
Attendance: 6,500 (estimated)

WAPELLO COUNTY HISTORICAL MUSEUM, 210 W. Main, Ottumwa, IA 52501-2500. Tel.: 641-682-8676. Fax: 641-682-8676.
E-mail: wchs@pcsia.net
Web Site: www.wapellocountymuseum.com
Founded: 1959.
Congressional District: 1
Key Personnel: Pres., Registrar & Coord., James C. Johnson; Vice Pres., Carol Hoffman; Sec., Mary Ellen Schmitz.
Personnel Profile: Part-Time Paid 1; Part-Time Volunteers 37.
Governing Authority: society. Parent Institution: Wapello County Historical Society. Tax-exempt.
Institution Type/Description: History Museum.
Collections: artifacts from all aspects of life in Wapello County from prehistory to present; records of early industries; old blacksmith, carpenter & stonemason tools; detailed scale model of 1890-1891 Ottumwa Coal Palace Industrial Exhibit Hall; old kitchen utensils; early furniture; costumes; telecommunications; Dain/Deere.
Research Fields: local & county history; early Iowa Indian Territory.
Facilities: 200-vol. library of state, county & local history, available for use on premises.
Activities: guided tours; lectures; permanent exhibitions; special events.
Publications: quarterly newsletter; local history brochures & pamphlets.
Hours & Admission Prices: Tues.-Fri. 10-4, Sat. 12-4. Adults $3, children under 12 $1; scheduled student tours & members no charge. &
Attendance: 1,423 (accurate)
Membership: Individual $15; Family $25; Life $250; Corporate $500.

Panora

GUTHRIE COUNTY HISTORICAL VILLAGE, 206 W. South St., Panora, IA 50216-1015. Tel.: 641-755-2989. Fax: 641-755-4066. Facebook: Guthrie County Historical Village.
E-mail: gchv@netins.net
Web Site: www.panora.org/museum
Founded: 1968.
Key Personnel: Dir., Kristine Jorgensen.
Governing Authority: Parent Institution: Guthrie County. Subsidiary Institution: Guthrie County Historical Village Foundation. Tax-exempt.
Institution Type/Description: Historical Village.
Collections: local history & culture; period furnishings; personal artifacts. Historic Buildings: 1851 log house; c.1900 depot; 1913 church; schoolhouses 1800s-1950s; blacksmith shop; general store; print shop; 1878 law office.
Hours & Admission Prices: Tues.-Fri. 10-4:30, Sat. 1-4:30; other times by appointment. Adults $2, children 6-17 $1; children 5 & under no charge. &
Attendance: 2,500 (estimated)

Membership: Individual & Individual Gift $20; Family & Family Gift $30; Corporate $75-$150.

Parkersburg

PARKERSBURG HISTORICAL HOME, 401 5th St., Parkersburg, IA 50665. Mailing Address: P.O. Box 142, Parkersburg, IA 50665. Tel.: 319-346-1461.
Key Personnel: Pres. (V), Becky Thorne
Institution Type/Description: Historic House Museum: built in 1895. Listed on the National Register of Historic Places.
Collections: local history & culture; period furnishings.
Hours & Admission Prices: Memorial Day to Labor Day Sun. 1-4. Adults $5.
Membership: Annual $10.

Pella

PELLA HISTORICAL VILLAGE, (M), 507 Franklin St., Pella, IA 50219-1671. Mailing Address: P.O. Box 145, Pella, IA 50219-0145. Tel.: 641-628-4311. Fax: 515-628-9192.
E-mail: pellatuliptime@iowatelecom.net
Web Site: www.pellatuliptime.com
Founded: 1965.
Key Personnel: Pres. (V), Mike Morgan.
Personnel Profile: Full-Time Paid 2; Part-Time Paid 6.
Governing Authority: society. Parent Institution: Pella Historical Society. Tax-exempt: 501(c)(3).
Institution Type/Description: Ethnic (Dutch) Museum Complex: located in 20 historic buildings.
Collections: archives; ethnology; industrial; preservation project; authentic Dutch costumes; Dutch bakery; complete set of newspapers printed in Pella. Historic Buildings: 1851 Wyatt Earp boyhood home; 1843 pioneer log cabin; 1853 Van Spankeren store; 1874 Amsterdam school; 1850 Dutch windmill.
Research Fields: archives; ethnology; industrial; preservation project.
Facilities: Items from the Netherlands for sale.
Activities: guided tours; films; arts festivals.
Publications: annual brochure, Tulip Time in Pella; quarterly newsletter, Historic Village Newsletter.
Hours & Admission Prices: March-Dec. Mon.-Sat. 9-4. Adults $8, K-12 $2; members no charge. Closed national holidays. &
Attendance: 20,000 (estimated)
Membership: Individual $35; Silver $55; Family $65; Gold $105.

SCHOLTE HOUSE MUSEUM, 728 Washington, Pella, IA 50219-1523. Tel.: 641-628-3684.
E-mail: scholtehouse@windstream.net
Web Site: scholtehouse.com
Founded: 1982.
Key Personnel: Dir., Beverly J. Graves.
Personnel Profile: Part-Time Volunteers 14.
Governing Authority: private; nonprofit organization. Parent Institution: Pella Historical Society. Tax-exempt.
Institution Type/Description: Historic House Museum: home of the founding father of the town of Pella, Dominie Hendrik Pieter Scholte c.1847.
Collections: furnishings and art belonging to Dominie Scholte and his family including clothing, letters, china, silverware, bedding (knitted, crocheted, quilts).
Facilities: library of books written by Dominie Scholte. Museum-related items for sale.
Activities: concerts; teas; self-guided tours; workshops for volunteers. Museum Sponsors: Children's Fun Day of 1800s Activities; Christmas Tea.
Publications: brochures; quarterly newsletter.
Hours & Admission Prices: March-Dec. Mon.-Sat. 1-4; other times by appointment. Adults $5, students & children $2; Pella district & Central College students, members no charge. Windmill Historical Village & Scholte House: adults $13. Closed New Year's Day; Easter; Memorial Day; Independence Day; Labor Day; Thanksgiving; Christmas Eve & Day. &
Attendance: 3,000 (estimated)

VERMEER MUSEUM & GLOBAL PAVILION, 2110 Vermeer Rd. E., Pella, IA 50219. Tel.: 641-621-7017.
E-mail: pavilion@vermeer.com
Institution Type/Description: History Museum.
Collections: Vermeer Corp. history; agricultural & construction equipment including round baler, trenchers, boring & environmental equipment.
Hours & Admission Prices: Call for hours. &

Perry

CARNEGIE LIBRARY MUSEUM, 1123 Willis Ave., Perry, IA 50220. Tel.: 515-465-7713 & 9941.
Founded: 1903.
Congressional District: 4
Key Personnel: Chm. (V), Philip L. Stone.
Personnel Profile: Part-Time Volunteers 15.
Governing Authority: city of Perry.
Institution Type/Description: Library Museum: built in 1904. Listed on the National Register of Historic Places.
Collections: books including midwest literature, women's fiction, children's books, & books on literacy and libraries; early life of area women; town's former courthouse.
Hours & Admission Prices: Tours: Thurs.-Sat. by appointment. No charge; donations accepted. &
Attendance: 750 (estimated)

FOREST PARK MUSEUM AND ARBORETUM, 14581 K Ave., Perry, IA 50220-6379. Tel.: 515-465-3577. Fax: 515-465-3579.
E-mail: conservation@co.dallas.ia.us
Web Site: www.conservation.co.dallas.ia.us
Founded: 1953.
Congressional District: 4
Key Personnel: C.E.O., Mike Wallace; Cur., Archivist & Education, Pete Malmberg.
Personnel Profile: Full-Time Paid 1; Interns 1.
Governing Authority: county. Parent Institution: Dallas County Conservation Department, Perry, IA 50220. Subsidiary Institution: The Voas Nature Area/Voas Museum, Minburn, IA. Tax-exempt.
Institution Type/Description: History Museum.
Collections: early transportation; farm machinery; small hand tools; railroading; blacksmith shop.
Research Fields: archaeological excavations; early farmstead; cultural resource surveys & mapping.
Facilities: 100-vol. library Perry Newspapers; 6,500 sq. ft. exhibit space; arboretum; visitor center; picnic area; county conservation headquarters.
Activities: arts festivals; concerts; formal education programs; historical programs; guided tours; lectures; loan & temporary exhibitions. Museum Sponsors: Arboretum Accolades - concert; Scenic History Drive; antique roadshows (evaluations); family fun days; old fashioned Christmas.
Publications: quarterly newsletter.
Hours & Admission Prices: May-Oct. Mon.-Fri. 9-4:30, Sat.-Sun. 1-4:30; Nov.-April by appointment only. No charge.
Attendance: 3,700 (estimated)

Peterson

PRAIRIE HERITAGE CENTER, 4931 Yellow Ave., Peterson, IA 51047-7528. Tel.: 712-295-7200.
E-mail: occb@iowatelecom.net
Web Site: prairieheritagecenter.org
Institution Type/Description: History Museum.
Collections: exhibits & artifacts relating to the prairie.
Hours & Admission Prices: Wed.-Fri. 9-4, Sat.-Sun. 1-4.

Polk City

BIG CREEK HISTORICAL MUSEUM, 112 Third St., Polk City, IA 50226. Mailing Address: P.O. Box 201, Polk City, IA 50226. Tel.: 515-965-3828.
Governing Authority: Tax-exempt.
Institution Type/Description: History Museum: housed in the city hall building; built in 1863. Listed on the National Register of Historic Places.
Collections: local history & culture; photographs; period furnishings; personal artifacts.
Hours & Admission Prices: 2nd Sun. of month 1-3; other times by appointment. &

Pomeroy

THE KALEIDOSCOPE FACTORY, 104 S. Main St., Pomeroy, IA 50575-7736. Tel.: 712-468-2420.
E-mail: chelp@ncn.net
Web Site: kaleidoscopefactory.com
Key Personnel: Head Kaleidoscope Maker, Leonard Olson
Institution Type/Description: History Museum.
Collections: hand-crafted kaleidoscopes, dippers & spurtles.
Facilities: Museum-related items for sale.

Activities: woodturning demonstrations.
Hours & Admission Prices: Tours: Tues. & Thurs. 1-9, Sat. 10-5; other times by appointment. No charge.

POMEROY TORNADO MUSEUM, 114 S. Ontario St., Pomeroy, IA 50575. Tel.: 515-574-1615.
Institution Type/Description: Science Museum.
Collections: science & history of tornadoes; photographs; wind farms.
Activities: self-guided tours of local wind farms; group tours.
Hours & Admission Prices: Call for hours. &

Prairie City

NEIL SMITH NATIONAL WILDLIFE REFUGE PRAIRIE LEARNING CENTER, 9981 Pacific St., Prairie City, IA 50228-7820. Mailing Address: P.O. Box 399, Prairie City, IA 50228-3400. Tel.: 515-994-3400.
E-mail: buffalo@tallgrass.org
Web Site: www.tallgrass.org
Key Personnel: Pres. Friends of Prairie Learning Center, Mark Lyle; Park Ranger, Al Murray; Park Ranger, Hallie Runeussen
Institution Type/Description: Wildlife Refuge.
Collections: bison & elk; interactive exhibits; photographs; film.
Facilities: nature trails. Museum-related items for sale.
Activities: film.
Hours & Admission Prices: Refuge: dawn to dusk. Learning Center: Jan.-March Mon.-Sat. 9-4; April-Dec. Mon.-Sat. 9-4, Sun. 12-5. No charge. Closed New Year's Day; Thanksgiving; Christmas. &

Prescott

KLINE MUSEUM, 112 6th Ave., Prescott, IA 50859. Mailing Address: 605 5th Ave., Prescott, IA 50859.
Founded: 1988.
Key Personnel: Dir., Randy Cooper
Institution Type/Description: History Museum.
Collections: local history; period vehicles including 1911 Carter Car, 1929 Ford Model A fire truck; Ford Model T truck; farm machinery; local memorabilia.
Hours & Admission Prices: Memorial Day to Labor Day Sun. 1-4; other times by appointment. &
Attendance: 25 (estimated)

Princeton

BUFFALO BILL CODY HOMESTEAD, 28050 230th Ave., Princeton, IA 52768-9713. Mailing Address: 14910 110th Ave., Davenport, IA 52804-9020. Tel.: 563-225-2981. Fax: 563-381-2805.
E-mail: conservation@scottcountyiowa.com
Web Site: www.scottcountyiowa.com
Founded: 1970.
Congressional District: 1
Key Personnel: C.E.O., Roger Kean; Museum Shop Mgr., Marilyn McCool.
Personnel Profile: Full-Time Paid 1; Part-Time Paid 2.
Governing Authority: county. Parent Institution: Scott County Conservation Bd. Tax-exempt: 501(c)(3).
Institution Type/Description: Historic House: 1847 boyhood home of Buffalo Bill Cody.
Collections: middle 19th-century furnishings; photos; Indian artifacts; farm implements; buffalo.
Facilities: Gift items for sale.
Activities: guided tours.
Hours & Admission Prices: April-Oct. daily 9-5. Adults $2; children 16 & under no charge.
Attendance: 7,000 (estimated)

Quad Cities

ROCK ISLAND ARSENAL MUSEUM, 1 Rock Island Arsenal, Bldg. 60, Quad Cities, IA 61299. Tel.: 309-782-5021.
E-mail: kris.g.leinicke.civ@mail.mil
Institution Type/Description: History Museum.
Collections: Rock Island Arsenal & Arsenal Island history; military & civilian small arms; hands-on exhibitions.
Hours & Admission Prices: Call for hours. &

Quimby

GRAND MEADOW HERITAGE CENTER AND MUSEUM, 767 610th St., Quimby, IA 51049-7053. Tel.: 712-447-6201.
Institution Type/Description: History Museum.
Collections: local history & culture; period furnishings; Native American artifacts; photographs; blacksmith shop & tools.
Activities: Annual Event: Heritage Festival in September.
Hours & Admission Prices: June-Sept. Sun. 1-5; other times by appointment. No charge, donations accepted.

Red Oak

BURLINGTON NORTHERN DEPOT & WWII MEMORIAL MUSEUM, 305 S. 2nd St., Red Oak, IA 51566-2655. Tel.: 712-623-6340.
E-mail: jacky@depothill.net
Web Site: depothill.net
Founded: 2003.
Key Personnel: Dir., Chm. (V) & Pres. (V), Jacky Adams.
Personnel Profile: Full-Time Volunteers 3.
Governing Authority: Parent Institution: Depot Hill Historical District, LLC. Tax-exempt.
Institution Type/Description: History Museum: housed in the Burlington Northern Depot; built in 1903.
Collections: local history & culture; period furnishings; personal artifacts; photographs; military artifacts.
Facilities: theater; 35-seat auditorium.
Activities: Annual Event: Tractor Boys in May.
Hours & Admission Prices: Mon.-Fri. 10 to noon; other times by appointment. No charge; donations accepted. &
Attendance: 2,000

MONTGOMERY COUNTY HISTORY CENTER, 2700 N. 4th St., Red Oak, IA 51566-1369. Tel.: 712-623-2289.
Key Personnel: Dir., David McFarland
Institution Type/Description: History Museum.
Collections: local history & culture; period furnishings; personal artifacts; photographs.
Activities: special events.
Hours & Admission Prices: Tues.-Sun. 12-4; other times by appointment. Adults $3, children $2; members no charge.

Rock Rapids

LYON COUNTY HISTORICAL SOCIETY MUSEUM COMPLEX, 110 1/2 N. Story St., Rock Rapids, IA 51246. Mailing Address: P.O. Box 322, Rock Rapids, IA 51246. Tel.: 712-472-2962.
Founded: 1972.
Congressional District: 5
Key Personnel: Pres. (V), Albert Van Holland
Governing Authority: Tax-exempt.
Institution Type/Description: History Museum: housed in the former Rock Island Depot.
Collections: local history & culture; caboose; livery stable; windmill; Victorian house; photographs; personal artifacts.
Hours & Admission Prices: Sun. & holidays 2-5; other times by appointment. Adults $1.
Attendance: 125 (estimated)
Membership: Individual $15.

Rockwell City

CALHOUN COUNTY MUSEUM, 150 Hwy. 20 E., Rockwell City, IA 50579. Mailing Address: P.O. Box 368, Jolley, IA 50551-0368. Tel.: 712-297-5081 & 8139 (summer). Fax: 712-297-5216.
E-mail: jmjolley@iowatelecom.net
Founded: 1956.
Congressional District: 6
Key Personnel: Pres., Marlene Johnson; 1st Vice Pres., Iola Zimbeck; Sec., Marjorie Hepp; Treas., Toni Kerns; Museum Shop Mgr., JoAnn Maguire.
Personnel Profile: Part-Time Paid 1; Part-Time Volunteers 25.
Governing Authority: bd. of directors. Tax-exempt: 501(c)(3).
Institution Type/Description: General Museum.
Collections: farming & period farm machinery; medical & dental equipment; clothing; irons; bottles; typewriters; books; dishes; pioneer equipment; photographic equipment; period furnishings; 1,000 salt & pepper shakers; 100 birds & small animals; local genealogical records; manuscripts;

scrapbooks; musical instruments; period carpenter's tools; Mickey Mouse; quilts; military uniforms & artifacts; books.
Research Fields: county history; preservation of artifacts; genealogy.
Facilities: 1,450-vol. library of books from early homes & schools available for research; 600 books of tax records.
Activities: guided tours; permanent & temporary exhibitions; pioneer crafts for sixth grade classes; tour & pioneer history of county for second, third & fourth grade classes.
Publications: cemetery booklets, County Tours; book, Calhoun County History, Sesquicentennial.
Hours & Admission Prices: May-Oct. Tues. & Sat.-Sun. 1-4. No charge; donations accepted. &
Attendance: 2,500 (estimated)
Membership: Annual $5; Life $50.

Sac City

SAC CITY MUSEUM, 1301 W. Main St., Sac City, IA 50583. Mailing Address: 111 N. 7th, Sac City, IA 50583. Tel.: 712-662-4349.
E-mail: kaychris@mchsi.com
Web Site: saccountyiowa.com
Founded: 1984.
Congressional District: 5
Governing Authority: Sac County Historical Society. Tax-exempt.
Institution Type/Description: History Museum.
Collections: costumes; period artifacts; popcorn ball; farm tools & equipment. Historic Village: country store, chapel, drug store, telephone office, alumni hall, country school, volunteer garage, doctor's office, post office; hardware store.
Publications: annual newsletter.
Hours & Admission Prices: Memorial Day-Labor Day Sat.-Sun. 2-4:30. No charge; donations accepted.
Attendance: 2,500 (accurate)
Membership: Annual $5.

Saint Ansgar

ST. ANSGAR HERITAGE MUSEUM, 126 W. Fourth St., Saint Ansgar, IA 50472. Mailing Address: P.O. Box 214, Saint Ansgar, IA 50472. Tel.: 641-713-2776.
Institution Type/Description: History Museum.
Collections: local history & culture; period furnishings; genealogy; personal artifacts.
Hours & Admission Prices: April-Sept. Wed.-Fri. 10-4. No charge.
Attendance: 500 (estimated)

Scotch Grove

EDINBURGH "GHOST TOWN" AND MUSEUM, 13838 Edinburgh Rd., Scotch Grove, IA 52310. Tel.: 563-487-3711.
E-mail: iajchs@aol.com
Institution Type/Description: History Museum.
Collections: local history & culture; period furnishings; photographs; documents. Historic Buildings: log cabin; schoolhouse; church; depot; blacksmith shop.
Hours & Admission Prices: May-Sept. call for hours.

Shelby

CARSTENS 1880 MEMORIAL FARMSTEAD, 32409 - 380th St., Shelby, IA 51570. Mailing Address: P.O. Box 302, Shelby, IA 51570-0302. Tel.: 712-554-2638 & 2341.
E-mail: info@carstensfarm.com
Web Site: www.carstensfarm.com
Institution Type/Description: Farming History Museum.
Collections: farming history & heritage; early farm tools & equipment; tractors; period artifacts.
Activities: special events; demonstrations. Annual Event: Farm Days in September.
Hours & Admission Prices: By appointment. Adults $5, students & seniors $2.

Sheldon

PRAIRIE QUEEN MUSEUM, 319 10th St., Sheldon, IA 51201. Tel.: 712-324-5108 & 3235.
Institution Type/Description: History Museum: housed in the former Carnegie Library building; built in 1908.
Collections: local history & culture; early industry; Native American artifacts; military equipment & uniforms; school memorabilia; personal artifacts; photographs; Sheldon Hall of Fame.
Hours & Admission Prices: Mon. 6pm-8pm, Tues. & Thurs. 1-4, Sat. 1-3; other times by appointment. No charge.

WANSINK ART GALLERY, PRAIRIE SCHOOLHOUSE & PIONEER HOME, 423 Park St., Sheldon, IA 51201. Tel.: 712-324-3371.
E-mail: hptuttle@nethtc.net
Institution Type/Description: Historic Buildings & Art Gallery.
Collections: fine art; paintings; period furnishings; local history; historic buildings.
Activities: art classes; special events; educational programs.
Hours & Admission Prices: Call for hours. &

Shell Rock

SHELL ROCK COMMUNITY HISTORICAL MUSEUM, 127 E. Adair St., Shell Rock, IA 50670-9713. Mailing Address: P.O. Box 57, Shell Rock, IA 50670-0057. Tel.: 319-885-4478 & 6687.
Founded: 2007.
Congressional District: 3
Key Personnel: Chm. (V), Sherri Willey; Museum Shop Mgr., Linda McCann.
Personnel Profile: Part-Time Volunteers 40.
Governing Authority: Parent Institution: Shell Rock Community Historical Society. Tax-exempt.
Institution Type/Description: Historic House Museum: housed in a 1920s craftsman-style home.
Collections: period furnishings; maps; personal artifacts; Shell Rock High School artifacts.
Publications: quarterly newsletter, Shell Rock Historical Society.
Hours & Admission Prices: May-Oct. Sat.10-2; other times by appointment. No charge; donations accepted. &
Attendance: 200 (accurate)
Membership: Single $15; Family $25.

Shenandoah

GREATER SHENANDOAH HISTORICAL SOCIETY, 800 W. Sheridan Ave., Shenandoah, IA 51601-1645. Mailing Address: P.O. Box 182, Shenandoah, IA 51601. Tel.: 712-246-1669.
E-mail: gshmuseum@hotmail.com
Web Site: www.greatershenandoahhs.org
Founded: 1971.
Congressional District: 5
Key Personnel: Dir., Sallie Brownlee; Pres. (V), Ron Qestmann.
Personnel Profile: Part-Time Paid 1; Part-Time Volunteers 5.
Governing Authority: Tax-exempt.
Institution Type/Description: History Museum.
Collections: local history & culture; period furnishings; personal artifacts; photographs.
Hours & Admission Prices: March-Dec. Tues.-Fri. 1-4; other times by appointment. No charge; donations accepted. &
Attendance: 650 (estimated)
Membership: Individual $10; Family $25. Business: Bronze $75; Silver $150; Gold $200; Lifetime $1,000.

Sibley

MCCALLUM MUSEUM & BRUNSON HERITAGE HOME, 5th St. & 8th Ave., Sibley, IA 51249. Mailing Address: 724 3rd Ave, Sibley, IA 51249-1606. Tel.: 712-754-3882.
E-mail: jstoff1@hotmail.com
Web Site: www.osceolacountyia.com/info/museums.htm
Founded: 1956.
Congressional District: 6
Key Personnel: Dir., Jan Stofferan; Pres. (V), Shirley Swenson.
Personnel Profile: Part-Time Paid 1; Part-Time Volunteers 10.
Governing Authority: municipal. Parent Institution: Osceola County Historical Association. Subsidiary Institution: Tracy House, Ocheyedan, IA; DeBoer Museum, Ashton, IA. Tax-exempt.
Institution Type/Description: History Museum.
Collections: Civil War guns, swords, bayonets, ammunition cases & uniforms; World War I & II items; china; glass; farm & household equipment; furniture; agriculture; Indian artifacts; paintings; dolls; photos; books; 1908 auto buggy; sleigh, surrey; quilts. Historic Buildings: Brunson Heritage House; Rogers House; De Boer Grocery Museum.
Research Fields: Civil War; Osceola County History.
Facilities: 75-vol. library.

Activities: guided tours; permanent exhibitions.
Hours & Admission Prices: May-Sept. Sun. 1:30-4:30; other times by appointment. No charge; donations accepted. Closed holidays. ♿
Attendance: 1,000 (estimated)
Membership: Individual $15; Family $25; Patron $50; Contributor $100; Corporate $200.

Sigourney

DUMONT MUSEUM, 20545 255th St., Sigourney, IA 52591-8352. Tel.: 641-622-2592 & 9937.
E-mail: oliver@lisco.com
Web Site: www.dumontmuseum.com
Founded: 1994.
Governing Authority: Tax-exempt.
Institution Type/Description: History Museum.
Collections: local history & culture; over 100 tractors; horse drawn equipment; buggies; toys; dolls; Roy Rogers memorabilia; household artifacts; Lionel train layout.
Hours & Admission Prices: May-Oct. Sat.-Sun. 10-5; other times by appointment. Adults $8. ♿
Membership: Individual $45.

Sioux City

DOROTHY PECAUT NATURE CENTER, 4500 Sioux River Rd., Sioux City, IA 51109-1657. Tel.: 712-258-0838. Fax: 712-258-1261. Facebook: Dorothy Pecaut Nature Center.
E-mail: dsnyder@sioux-city.org
Web Site: www.woodburyparks.com
Founded: 1995.
Institution Type/Description: Nature Center.
Collections: Loess Hills natural history; live native reptile & fish; hands-on exhibits.
Hours & Admission Prices: Tues.-Sat. 9-5, Sun. 1-5. No charge; donations accepted. Closed New Year's Day; Thanksgiving; Christmas Eve & Day. ♿
Attendance: 47,276 (estimated)
Membership: Individual $20; Family $35; Wildlife $50; Woodland $100; Prairie $250; Wetland $500; Distinguished Conservationist $1,000.

✳ LOREN D. CALLENDAR GALLERY, (M), City Hall, 405 6th St., Sioux City, IA 51101-1255. Tel.: 712-279-6174. Fax: 712-252-5615.
Web Site: www.siouxcitymuseum.org
Governing Authority: Parent Institution: Sioux City Public Museum.
Institution Type/Description: Art Gallery.
Collections: photographs; Sioux City history.
Hours & Admission Prices: Mon.-Fri. 8-4:30. No charge. Closed holidays.

MID AMERICA MUSEUM OF AVIATION AND TRANSPORTATION, 2600 Expedition Ct., Sioux City, IA 51111. Mailing Address: P.O. Box 3525, Sioux City, IA 51102-3525. Tel.: 712-252-5300. Fax: 712-222-1688.
E-mail: airmuseum@longlines.com
Web Site: midamericaairmuseum.org
Founded: 1991.
Congressional District: 4
Key Personnel: Dir., Larry L. Finley; Chm. (V), Ray Edgington.
Personnel Profile: Part-Time Paid 1; Interns 2.
Governing Authority: Parent Institution: MAMAT Volunteer Board of Directors. Tax-exempt.
Institution Type/Description: Aviation & Transportation Museum.
Collections: aviation & transportation history; aircraft; bicycles; early automobiles; motorcycles; fire engine; Boeing 727-200; military artifacts & uniforms; early toys.
Facilities: 30,000 sq. ft. exhibition space.
Activities: temporary classic aircraft visits. Museum Sponsors: Fly In Breakfast.
Publications: email newsletters.
Hours & Admission Prices: April-Sept. Thurs.-Tues. 9-5; Oct.-March Sun. 12-4, Mon. & Thurs.-Sat. 10-4. Adults $6, seniors 55 & over and active military $5, children 5-14 $3; discounts to groups of 8 or more; children 4 & under no charge. Closed New Year's Day; Easter; Memorial Day; Independence Day; Thanksgiving; Christmas. ♿
Attendance: 4,000 (accurate)
Membership: Individual $24; Family $36; Lifetime Individual $500; Lifetime Family $700.

✳ SERGEANT FLOYD MUSEUM & WELCOME CENTER, (M), 1000 Larsen Park Rd., Sioux City, IA 51103-4914. Tel.: 712-279-0198. Fax: 712-279-6934.
E-mail: scpm@sioux-city.org
Web Site: www.siouxcitymuseum.org
Governing Authority: Parent Institution: Sioux City Public Museum. Tax-exempt.
Institution Type/Description: Maritime History Museum.
Collections: Siouxland maritime history; 1932 Army Corps of Engineers work boat; photographs.
Facilities: Museum-related items for sale.
Hours & Admission Prices: Temporarily closed. ♿
Attendance: 26,583 (accurate)

✳ SIOUX CITY ART CENTER, (M), 225 Nebraska St., Sioux City, IA 51101-1712. Tel.: 712-279-6272, ext. 208. Fax: 712-255-2921. Facebook: Sioux City Art Center.
E-mail: kwelch@sioux-city.org
Web Site: www.siouxcityartcenter.org
Founded: 1914.
Congressional District: 4
Key Personnel: Dir., Al Harris-Fernandez; Pres. (V), John Wagner; Chm. (V), Jeff Baldus; Exhibitions & Collections Coord., Shannon Sargent; Sec., Kjersten Welch.
Personnel Profile: Full-Time Paid 8; Part-Time Paid 7; Part-Time Volunteers 769.
Volunteer Hours: 4,435
Operating Expenses: 794,664
Operating Income: 802,703
Governing Authority: municipal. Parent Institution: City of Sioux City. Subsidiary Institution: Art Center Assoc. of Sioux City. Tax-exempt: 501(c)(3).
Institution Type/Description: Art Museum.
Collections: work by contemporary regional, national & international artists.
Major Exhibits: Art for the Seasons, 10/19/13-2/2/14; James Goff, 10/19/13-2/2/14; An Impressive Collection: A Half-Century of Prints, 10/26/13-2/2/14; Quilt Art: International Expressions (T), 11/23/13-2/26/14; Youth Art Month Exhibition, 2/15/14-4/6/14; James Goff, 3/1/14-6/1/14; The Briar Cliff Review, 4/17/14-6/29/14.
Research Fields: Upper Midwest artists.
Facilities: permanent collection gallery; 4 temporary exhibition halls; children's hands-on gallery; 6 studios; non-lending library. Museum-related items for sale.
Activities: inter-museum loan; permanent & temporary exhibitions; lectures; gallery talks; formally organized education programs; docent program; tours. Museum Sponsors: ARTSPLASH Festival of the Arts, Labor Day weekend arts festival.
Publications: quarterly bulletin; catalogs for exhibitions & special programs; announcements.
Hours & Admission Prices: Tues.-Wed. & Fri.-Sat. 10-4, Thurs. 10-9, Sun. 1-4. No charge; donations accepted. Closed holidays. ♿
Attendance: 60,633 (accurate)
Membership: Senior Citizen $15; Individual $35; Household $50; Donor $100; Charter $250; Leader $500; Renaissance Society $1,000 & up.

THE SIOUX CITY LEWIS & CLARK INTERPRETIVE CENTER, 900 Larsen Park Rd., Sioux City, IA 51103-4916. Tel.: 712-224-5242. Fax: 712-224-5244.
E-mail: mpoole@siouxcitylcic.com
Web Site: www.siouxcitylcic.com
Key Personnel: Exec. Dir., Marcia Poole; Business Mgr., Russell Movall
Institution Type/Description: History Museum.
Collections: artifacts & memorabilia pertaining to Lewis & Clark.
Hours & Admission Prices: Tues.-Wed. & Fri. 9-5, Thurs. 9-8, Sat.-Sun. 12-5. No charge. Closed New Year's Day; Easter; Thanksgiving; Christmas.

✳ SIOUX CITY PUBLIC MUSEUM, (M), 607 4th St., Sioux City, IA 51101-1634. Tel.: 712-279-6174. Fax: 712-252-5615.
E-mail: scpm@sioux-city.org
Web Site: www.siouxcitymuseum.org
Founded: 1886.
Congressional District: 4
Key Personnel: Dir., Steven D. Hansen; Chm. (V), Miles Patton, Jr.; Cur. History, Grace Linden; Cur. Education, Theresa Weaver-Basye; Exhibits Designer, Matt Anderson; Devel. Coord., Mary Green-Warnstadt; Welcome Center Supvr., Kathy Meisner; Administrative Asst., Deanna Mayo.
Personnel Profile: Full-Time Paid 5; Part-Time Paid 11; Part-Time Volunteers 150; Interns 4.

Governing Authority: municipal. Parent Institution: City of Sioux City, IA. Subsidiary Institution: Sioux City Museum & Historical Association. Tax-exempt.
Institution Type/Description: General Museum: Sergeant Floyd River Museum & Welcome Center; Loren D. Callendar Gallery.
Collections: Indian artifacts; national, state & local history; archives; archaeology; mineralogy; paleontology; military; costumes.
Research Fields: local history.
Facilities: 2,000-vol. library on state & local history available for use on premises; archives; classroom. Cards, pottery, jewelry, wood carvings & books for sale.
Activities: oral history program; lectures; films; hobby workshops; formally organized education programs; inter-museum loan, permanent, temporary & traveling exhibitions; school loan service.
Publications: book, Sioux City, A Pictorial History; Sioux City History, 1980-2002.
Hours & Admission Prices: Museum: Tues.-Sat. 9-5, Sun. 1-5. Sergeant Floyd River Museum & Welcome Center daily 10-4. No charge; donations accepted. Closed holidays. &
Attendance: 50,741 (accurate)
Membership: Senior Citizen $15; Individual $20; Family $30; Supporting $50; Patron & Business $100.

South Amana

COMMUNAL AGRICULTURE MUSEUM, 505 P St., South Amana, IA 52334. Mailing Address: Amana Heritage Society, P.O. Box 81, Amana, IA 52203-0081. Tel.: 319-622-3567.
Web Site: www.cr.nps.gov/nr/travel/amana/agr.htm
Institution Type/Description: Agriculture Museum.
Collections: period agricultural implements; photographs.
Hours & Admission Prices: May-Sept. Mon.-Sat. 10-5, Sun. 12-5.

Spencer

PARKER HISTORICAL SOCIETY OF CLAY COUNTY, 7 Grand Ave., Spencer, IA 51301. Mailing Address: P.O. Box 91, Spencer, IA 51301-0091. Tel.: 712-262-3304. Fax: 712-262-3304.
E-mail: parkermuseum@smunet.net
Founded: 1960.
Key Personnel: Dir., Cindy McGranahan.
Personnel Profile: Part-Time Paid 3.
Governing Authority: Subsidiary Institution: Clay County Heritage Center. Tax-exempt.
Institution Type/Description: Local History Museum: housed in an historic house, built in 1916.
Collections: local history & culture; photographs; personal artifacts.
Hours & Admission Prices: Mon.-Wed. & Fri. 10-4, Thurs. 10-6, Sat. 10-3, Sun. 12-4; other times by appointment. Adults $5, children 5-12 $3; members & preschool no charge. &

PARKER MUSEUM, 300 E. Third St., Spencer, IA 51301. Mailing Address: 220 E. Third St., Spencer, IA 51301. Tel.: 712-262-3304.
Institution Type/Description: History Museum.
Collections: local history & culture; period furnishings; personal artifacts; photographs.
Hours & Admission Prices: Tues.-Fri. 11:30-3:30; other times by appointment. No charge; donations accepted.

Spillville

BILY CLOCK MUSEUM/ANTONIN DVORAK EXHIBIT, 323 S. Main, Spillville, IA 52168. Mailing Address: P.O. Box 258, Spillville, IA 52168-0258. Tel.: 563-562-3569.
E-mail: bilyclocks@mchsi.com
Web Site: www.bilyclocks.org
Founded: 1923.
Congressional District: 2
Key Personnel: Dir. & Museum Shop Mgr., Georgiann Eckheart.
Personnel Profile: Full-Time Paid 1; Part-Time Paid 5; Part-Time Volunteers 1.
Governing Authority: municipal. Tax-exempt.
Institution Type/Description: Clock Museum.
Collections: hand carved clocks. Historic Building: c.1859 former home of composer Antonin Dvorak; summer of 1893.
Activities: guided tours; permanent exhibitions.
Hours & Admission Prices: March-April & Nov. Sat.-Sun. 10-4; May-Oct. daily 9-5; call for other times. Adults $6, children 7-12 $4; special school & group rates; discount to AAA members; children under 7 & members no charge. Closed New Year's Day; Easter; Thanksgiving; Christmas.

Attendance: 14,000 (estimated)
Membership: Individual $30; Seniors $40; Dual $50; Lifetime $500; Business $1,000.

Stanton

SWEDISH HERITAGE & CULTURAL CENTER, 410 Hilltop Ave., Stanton, IA 51573. Mailing Address: P.O. Box 231, Stanton, IA 51573-0231. Tel.: 712-829-2840. Fax: 712-829-2393.
E-mail: shcc@myfmtc.com
Web Site: www.stantoniowa.com
Founded: 1993.
Key Personnel: Dir. & Chm. (V), Marlene Kennon.
Personnel Profile: Part-Time Volunteers 20.
Governing Authority: Parent Institution: Stanton Historical Society. Tax-exempt.
Institution Type/Description: Swedish Heritage Center.
Collections: Swedish heritage, culture & history; personal artifacts; photographs; documents; Swedish immigration & settlements; one-room restored country school.
Publications: quarterly newsletter.
Hours & Admission Prices: April-Nov. Tues.-Sat. 1-4; Dec.-March Fri.-Sun. 1-4; other times by appointment. Adults $2. &
Attendance: 500
Membership: Individual $10; Family $25; Sustaining $50; Corporate $100.

State Center

WATSON'S GROCERY STORE MUSEUM, 106 W. Main St., State Center, IA 50247. Mailing Address: P.O. Box 156, State Center, IA 50247. Tel.: 641-483-3002 & 485-3959.
Key Personnel: Dir., Everett Halsted; Pres. (V), Mike Riemenschneider
Governing Authority: Parent Institution: State Center Development Association. Subsidiary Institution: State Center Historical Society. Tax-exempt.
Institution Type/Description: Historic Building: housed in a former grocery store; built in 1895. An historic building.
Collections: period furnishings & artifacts; photographs.
Activities: Annual Event: Watson's Fall Festival in October.
Hours & Admission Prices: Memorial Day to Labor Day Sat.-Sun. 1-4; other times by appointment. No charge; donations accepted.
Attendance: 2,000 (estimated)

Storm Lake

BUENA VISTA HISTORICAL SOCIETY, 214 W. 5th St., Storm Lake, IA 50588-2346. Tel.: 712-732-4955.
Key Personnel: Pres. (V), Marjorie Nuelieb; Office Mgr., Gwen Henrich.
Personnel Profile: Part-Time Paid 1.
Operating Expenses: 18,500
Operating Income: 19,000
Governing Authority: Tax-exempt.
Institution Type/Description: Historical Society Museum.
Collections: local history & culture; period furnishings; paintings; photographs; personal artifacts. Historic Buildings: school house; log cabin.
Activities: fundraiser.
Publications: Historical Society Musings.
Hours & Admission Prices: Mon.-Fri. 12-4; other times by appointment. Adults $2, children 12 & under $1. &
Attendance: 1,999 (accurate)
Membership: Single $10; Family $15; Business $25.

HARKER HOUSE, 328 Lake Ave., Storm Lake, IA 50588-2435. Mailing Address: P.O. Box 368, Aurelia, IA 51005. Tel.: 712-732-3267.
E-mail: info@harkerhouse.com
Web Site: www.harkerhouse.com
Key Personnel: Pres. (V), Roger Redig
Institution Type/Description: Historic House: built in 1875.
Collections: local history & culture; Harker family history, furnishings, & personal artifacts; photographs; period clothing.
Hours & Admission Prices: June-Aug. Sat.-Sun. 2-4; special tours available upon request for groups over 10.
Attendance: 263 (accurate)

LIVING HERITAGE TREE MUSEUM, W. Lakeshore Dr., Sunset Park, Storm Lake, IA 50588. Tel.: 712-732-3780; 888-752-4692.
E-mail: info@stormlakeunited.com
Web Site: visitstormlake.com
Institution Type/Description: Tree Museum.

Collections: tree history & lineage.
Activities: audio tours.
Hours & Admission Prices: Daily dawn to dusk.

WITTER GALLERY, 609 Cayuga St., Storm Lake, IA 50588-2239. Tel.: 712-732-3400.
E-mail: wittergallery@yahoo.com
Web Site: thewittergallery.org
Founded: 1972.
Congressional District: 6
Key Personnel: Gallery Dir., Ron Stevenson.
Personnel Profile: Part-Time Paid 2; Part-Time Volunteers 100; Interns 2.
Governing Authority: nonprofit organization. Tax-exempt.
Institution Type/Description: Art Gallery.
Collections: oil paintings by Ella Witter; woodcuts; lithographs by Dorothy Skewis.
Activities: lectures; workshops; instructional art classes; programs for children; performances; artists receptions; temporary & traveling exhibitions.
Publications: monthly newsletter; exhibition catalogs.
Hours & Admission Prices: Summer: Tues.-Wed. & Fri. 1-5, Thurs. 1-6, Sat. 10-2; Winter: Tues.-Wed. & Fri. 1-5, Thurs. 1-7, Sat. 10-2. No charge, donations accepted. Closed national holidays. &
Attendance: 11,000 (accurate)
Membership: Active $25; Sustaining $50; Supporting $100; Sponsor $250; Benefactor $500; Patron $1,000 & up.

Story City

MUSEUMS OF STORY CITY, 619 Grand Ave., Story City, IA 50248-1412. Mailing Address: P.O. Box 104, Story City, IA 50248-0104. Tel.: 515-460-1749. Facebook: Story City Historical Society.
E-mail: storycityhistory@gmail.com
Web Site: www.storycity.net
Formerly: Bartlett Museum and Carriage House
Founded: 1982.
Key Personnel: Dir., Kate Feil; Pres. (V), Cindy Spurlock.
Personnel Profile: Part-Time Volunteers 10.
Institution Type/Description: Historic House Museum: built in 1903. Listed on the National Register of Historic Places.
Collections: local history; period furnishings from 1903 1920; personal artifacts; photographs; pioneer artifacts; jail & business artifacts; papers. Historic Building: 1860 one-room country school.
Activities: guided tours; special events.
Hours & Admission Prices: Wed.-Fri. 12-5. No charge; donations accepted.
Membership: Student $10; Contributing $25; Sustaining $50; History Enthusiast $250; Heritage Preservationist $1,000; Life Time $2,500.

Strawberry Point

WILDER MEMORIAL MUSEUM, 123 W. Mission, Strawberry Point, IA 52076. Mailing Address: Box 206, Strawberry Point, IA 52076-0206. Tel.: 563-933-4615.
E-mail: director@wildermuseum.org
Web Site: www.wildermuseum.org
Founded: 1970.
Congressional District: 1
Key Personnel: Pres., Dean Knight; Dir., Angela Beenken.
Personnel Profile: Part-Time Paid 5; Part-Time Volunteers 4.
Governing Authority: municipal; nonprofit. Tax-exempt.
Institution Type/Description: Historical Museum.
Collections: Victorian furniture & art glass; hanging lamps; 800 period dolls; period furnishings; military artifacts from Revolutionary War through Vietnam War; prairie farming; impressionist art.
Activities: guided tours; programs.
Publications: brochures; newsletter.
Hours & Admission Prices: May & Sept.-Oct. Sat. 10-4, Sun. 1-4; Memorial Day-Labor Day Sun.-Thurs. 1-4, Fri.-Sat. 10-4; other times by appointment. Family $10, adults $4, students $2; discounts for senior citizens on Wed., AAA members & adult groups of 10 or more; pre-school & members no charge. Participating Blue Star Museum. &
Attendance: 675 (estimated)

Swedesburg

SWEDISH AMERICAN MUSEUM, 107 James Ave., Swedesburg, IA 52652. Tel.: 319-254-2317.
E-mail: swedish@iowatelecom.net
Founded: 1991.

Congressional District: 2
Key Personnel: Pres. (V), Jean Anderson; Dir., Louise Unkrich; Museum Shop Mgr., Norma Lindeen
Governing Authority: Parent Institution: Swedish Heritage Society.
Institution Type/Description: Swedish American Museum.
Collections: Swedish heritage, culture, & history; personal artifacts; photographs.
Facilities: library; stuga (Swedish cottage) & huckster wagon building. Museum-related items for sale.
Publications: quarterly newsletter, Swedish Heritage Society.
Hours & Admission Prices: Mon.-Tues. & Thurs.-Sat. 9-4. No charge; donations accepted. Closed New Year's Day; Thanksgiving; Christmas. &
Attendance: 4,933 (accurate)
Membership: Individual $5; Family $20; Friend $50; Supporting $100; Benefactor $500 & up.

Tipton

CEDAR COUNTY HISTORICAL SOCIETY & MUSEUM, 1094 Hwy. 38, Tipton, IA 52772. Mailing Address: P.O. Box 254, Tipton, IA 52772. Tel.: 563-886-2899.
E-mail: cchsmus@iowatelecom.net
Institution Type/Description: Historical Society Museum.
Collections: local history & culture; period furnishings; personal artifacts; photographs; Charles Gabriel memorabilia; early clothing; Native American tools; military artifacts.
Hours & Admission Prices: Tues. & Thurs. 10-4, Sat. 10-2.

OLD CEDAR COUNTY JAIL, 118 W. 4th St., Tipton, IA 52772. Mailing Address: 707 King Ave., Stanwood, IA 52337-9619. Tel.: 563-886-2131.
E-mail: kwhitll601@gmail.com
Web Site: www.oldcedarcountyjail.com
Governing Authority: Parent Institution: Cedar County Friends of Historic Preservation, Inc.
Institution Type/Description: Historic Building: housed in the former county jail & sheriff's residence. Listed on the National Register of Historic Places.
Collections: local history & culture; period furnishings; photographs.
Hours & Admission Prices: 1st Sat. each month 10-2.

Toledo

TAMA COUNTY HISTORICAL MUSEUM, 200 N. Broadway, Toledo, IA 52342-1308. Tel.: 641-484-6767.
E-mail: tracers@pcpartner.net
Founded: 1942.
Congressional District: 1
Key Personnel: Dir. & Pres. (V), Joyce Wiese; Vice Pres., Christine Draisey; Treas., Gail Kilstofte; Exec. Sec., Bonnie Grimmius.
Personnel Profile: Full-Time Volunteers 12; Part-Time Volunteers 11.
Volunteer Hours: 1,665
Governing Authority: society. Tax-exempt.
Institution Type/Description: Local History Museum: housed in 1869 former Tama County Jail.
Collections: pioneer & Indian artifacts; Mesquakie Indian clothing & tools; county newspapers on micro-film, county records & census rolls; genealogies; extensive genealogical collection of Tama County, state of Iowa and beyond; one room schoolhouse; music room; patriotic room; military room; professional room; 1870 one-room schoolhouse. Historic Building: 1860s log house.
Research Fields: genealogy; county history; Indians.
Facilities: 2,000-vol. genealogical library with microfilms of newspapers, county records & census of all counties of Iowa; reading room.
Activities: guided tours; lectures; preservation of Tama County pioneer artifacts; permanent exhibitions.
Publications: quarterly, Tama County Museum News.
Hours & Admission Prices: Tues.-Sat. 1-4:30; other times by appointment. Bus tours available. No charge; donations accepted. Closed holidays. &
Attendance: 1,750 (estimated)
Membership: Individual $15; Family & Business $25.

Traer

TRAER HISTORICAL MUSEUM, 514 2nd St., Traer, IA 50675-1139. Tel.: 319-478-2346.
E-mail: traermuseum@hotmail.com
Founded: 1989.
Institution Type/Description: History Museum: highlights the life of "Tama Jim" Wilson, U.S. Secretary of Agriculture.

Collections: Wilson's career & family history; northern Tama County & Traer history; Endicott carved folk art collection.
Major Exhibits: Answering the Call to Colors: North Tama County in the Civil War, 1/14-4/15.
Publications: Soldier Life: Many Must Fall.
Hours & Admission Prices: By appointment. No charge; donations accepted.
&

TRAER SALT & PEPPER SHAKER GALLERY, 411 Second St., Traer, IA 50675. Tel.: 319-231-7654.
E-mail: saltandpepper@traer.net
Web Site: www.traer.com
Institution Type/Description: Shaker Gallery.
Collections: over 16,000 sets of salt & pepper shakers.
Hours & Admission Prices: March-Nov. Tues.-Sat. 1-5; groups & other times by appointment. Adults $3, children 5-12 $1; children under 5 no charge. Closed major holidays.

Urbandale

LIVING HISTORY FARMS, 2600 111th St., Urbandale, IA 50322-3792. Tel.: 515-278-5286. Fax: 515-278-9808.
E-mail: info@lhf.org
Web Site: www.livinghistoryfarms.org
Founded: 1967.
Congressional District: 3
Key Personnel: C.E.O. & Pres., Ruth Haus; Chm., Terry Garner; Vice Pres. Devel., Jim Dietz-Kilen; Treas., Don Brush; Dir. Interpretation, Janet Clair Dennis; Vice Pres. Mktg. & Communications, Jennie Deerr; Finance & Museum Shop Mgr., Elaine Raleigh.
Personnel Profile: Full-Time Paid 23; Part-Time Paid 70; Part-Time Volunteers 1,000; Interns 20.
Governing Authority: nonprofit.
Institution Type/Description: Living History Museum.
Collections: pioneer artifacts; farm machines; quilt collection; 1875 businesses & stores; Victorian Flynn House, barn & Tangen home; farm implement dealer.
Research Fields: midwestern agriculture & rural life.
Facilities: visitor center; 1700 Iowa Indian farm; 1850 & 1900 working farms; 1875 town of Walnut Hill including working trades; Henry Wallace Exhibit Center; William Murray Conference Center; picnic shelter. Museum-related items for sale.
Activities: organized education programs for children; hands-on activities; farming at 1700, 1850 & 1900 farms; special events; organized education programs for undergraduate or graduate college students.
Publications: annual brochures; newsletter, Salmanack.
Hours & Admission Prices: May-Oct. call for hours. Adults $12, senior citizens & military $11, children $7. &
Attendance: 130,000 (accurate)
Membership: Individual $50; One Plus One $60; Grandparent & Family $75; Family Plus $115.

Van Meter

BOB FELLER MUSEUM, 310 Mill St., Van Meter, IA 50261. Mailing Address: P.O. Box 95, Van Meter, IA 50261-0095. Tel.: 515-996-2806; 866-996-2806. Fax: 515-996-2952.
E-mail: info@bobfellermuseum.org
Web Site: www.bobfellermuseum.org/
Founded: 1995.
Key Personnel: Mgr., Scott Havick
Institution Type/Description: Sports Museum: housed in the hometown of Cleveland Indians pitcher, Bob Feller, 1936-1956; member of the Baseball Hall of Fame.
Collections: personal artifacts; photographs; baseball memorabilia.
Facilities: Museum-related items for sale.
Activities: autograph signings; special events.
Publications: newsletter.
Hours & Admission Prices: Oct.-March Tues.-Sat. 10-3, Sun. 12-4; April-Sept. Tues.-Sat. 10-5, Sun. 12-4. Adults $5, seniors & school-aged children $3; discounts to groups with appointment.

Vinton

FRANK G. RAY HOUSE, 912 First Ave., Vinton, IA 52349-1712. Mailing Address: Benton County Historical Society, P.O. Box 22, Vinton, IA 52349.
Institution Type/Description: Historic House Museum: built in 1893. Listed in the National Register of Historic Places.

Collections: Victorian architecture with Queen Anne features; carriage house.
Hours & Admission Prices: June-Aug. Sat.-Sun. 1-4.

HORRIDGE HOUSE, 612 First Ave., Vinton, IA 52349-1705. Mailing Address: PO Box 22, Vinton, IA 52349-0022. Tel.: 319-472-4574. Fax: 319-472-4574.
Institution Type/Description: Historic House Museum: c.1860.
Collections: period furnishings & artifacts.
Hours & Admission Prices: Library: June-Aug. Wed. & Sat.-Sun. 1-4; Sept.-May Wed. 1-4.

Walcott

IOWA 80 TRUCKING MUSEUM, I-80 Exit 284, Walcott, IA 52773. Mailing Address: 505 Sterling Dr., Walcott, IA 52773. Tel.: 563-468-5500. Fax: 563-468-5506.
E-mail: joni.waller@iowa80group.com
Web Site: www.iowa80truckingmuseum.com
Institution Type/Description: Trucking Museum.
Collections: trucking history; period trucks; toy trucks; trucking memorabilia.
Facilities: Museum-related items for sale.
Hours & Admission Prices: Memorial Day to Labor Day Mon.-Sat. 9-5, Sun. 12-5; Sept.-May Wed.-Sat. 9-5, Sun. 12-5. No charge; donations requested.
&
Attendance: 12,400 (estimated)

Wall Lake

ANDY WILLIAMS BIRTHPLACE, 102 E. First St., Wall Lake, IA 51466-7707. Mailing Address: P.O. Box 566, Wall Lake, IA 51466-0566. Tel.: 712-664-2119.
Institution Type/Description: Historic House Museum: housed in the birthplace of Andy Williams.
Collections: Williams family history; period furnishings; personal artifacts; photographs.
Hours & Admission Prices: Memorial Day to Labor Day Sat.-Sun. 2-4; other times by appointment.

Walnut

MONROE #8 ONE-ROOM COUNTRY SCHOOLHOUSE, 610 Highland St., Walnut, IA 51577. Mailing Address: c/o Walnut Historical Society, 210 Redwood Rd., Walnut, IA 51577. Tel.: 712-784-2100 & 2244.
Institution Type/Description: Historic Building: c.1920 schoolhouse.
Collections: local history & culture; period furnishings.
Hours & Admission Prices: Memorial Day to Labor Day Sat.-Sun. 1-4; other times by appointment.

WALNUT CREEK HISTORICAL MUSEUM, 304 Antique City Dr., Walnut, IA 51577. Tel.: 712-784-2100 & 2244.
Institution Type/Description: History Museum.
Collections: local history & culture; period furnishings; personal artifacts; clothing; photographs.
Hours & Admission Prices: Memorial Day to Labor Day Sat.-Sun. 1-4; other times by appointment.

Wapello

LOUISA COUNTY HERITAGE CENTER, 609 N. James L. Hodges Ave., Wapello, IA 52653. Tel.: 319-527-5247.
E-mail: lchs@louisacomm.net
Web Site: louisacountyhistory.com
Institution Type/Description: History Museum.
Collections: local history & culture; period furnishings; Native American artifacts; photographs; military artifacts; early tools & clothing; genealogy.
Hours & Admission Prices: Call for hours. &

Waterloo

✱ GROUT MUSEUM DISTRICT, (M), 503 South St., Waterloo, IA 50701-1517. Tel.: 319-234-6357. Fax: 319-236-0500.
Web Site: www.groutmuseumdistrict.org
Founded: 1933.
Congressional District: 1
Key Personnel: Exec. Dir., Billie K. Bailey; Chm. & Pres. (V), Barbara Corson; Cur. Exhibits, Robin Venter; Dir. Devel. & Mktg., Cyd McHone; Mktg. Coord. & Graphic Designer, Shelby Sitzmann; Mgr. Collections,

Lorraine Ihnen; Devel. & Visitor Service Asst., Nancy Kinter; Mgr. Russell House, Annette Freeseman; Imaginarium Mgr., Alan Sweeney; Mgr. Operations, Diane Popelka; Archivist, Catreva Manning; Coord. Veterans Project, Bob Neymeyer; Education Asst., Jane Ryan; Exhibit Technician, William Bisbee; Outreach Coord., Jason Dornbush; Museum Shop Mgr., Judith Slaikeu.

Personnel Profile: Full-Time Paid 21; Part-Time Paid 8; Part-Time Volunteers 20; Interns 8.

Governing Authority: nonprofit organization. Subsidiary Institutions: Grout Museum of History and Science; Bluedorn Science Imaginarium; Rensselaer Russell House Museum; Snowden House; Sullivan Brothers Iowa Veterans Museum. Tax-exempt: 501(c)(3).

Institution Type/Description: General Museum.

Collections: early Native American artifacts; industrial history; textiles, costumes; Pioneer Hall historical dioramas including log cabin, tool shed, carpenter shop, blacksmith shop, country store, apothecary shop; genealogy; anthropology; paleontology; geology; astronomy; manuscript collections. Physical Sciences: Bluedorn Science Imaginarium. Historic House: Rensselaer Russell House.

Facilities: 950-vol. library; planetarium; 200-seat conference room; 65-seat theatre; 250-seat mess hall; rental facilities; 3 classrooms. Museum-related items for sale.

Activities: tours; formally organized education programs; permanent & temporary exhibitions; planetarium programs; history & science traveling trunk program; science demonstrations.

Publications: Museum Calendar; newsletter; membership brochure; general brochure; Teacher's Guide to Programs; summer activities guide; group tour manual; annual report; visitors guide.

Hours & Admission Prices: Tues.-Sat. 9-5. Sullivan Brothers Iowa Veterans Museum & Grout Museum of History & Science: adults 14 & over $10, children 4-13, active military & veterans $5; children 3 & under no charge. Bluedorn Science Imaginarium & Rensselaer Russell House Museum: adults & children 4 & over $5 (per site); children 3 & under no charge. &

Attendance: 33,630 (accurate)

Membership: Individual $50; Individual Plus $60; Family & Grandparent $75; Family & Friends $125; Family & Friends Plus $250; Advocate $500; Historian $1,000.

NATIONAL WRESTLING HALL OF FAME DAN GABLE MUSEUM, 303 Jefferson St., Waterloo, IA 50701. Tel.: 319-233-0745. Fax: 319-233-3477.
E-mail: dgmstaff@nwhof.org
Institution Type/Description: Sports Museum.
Collections: wrestling history & artifacts; photographs; personal artifacts; Hall of Fame Inductees.
Activities: Museum Sponsors: Induction Ceremony.
Hours & Admission Prices: Mon.-Fri. 9-5; other times by appointment. Adults $6, students 17 & under $3; children 6 & under no charge.

PHELPS YOUTH PAVILION, Waterloo Center for the Arts, 225 Commercial St., Waterloo, IA 50701-1313. Tel.: 319-291-4490. Fax: 319-291-4270.
E-mail: museum@waterloo-ia.org
Web Site: www.phelpsyouthpavilion.org
Founded: 2008.
Key Personnel: Dir., Carolyn Carpenter; Museum Shop Mgr., Maureen Newbill.
Governing Authority: Parent institution: Waterloo Center for the Arts. Tax-exempt.
Institution Type/Description: Children's Museum.
Collections: hands-on exhibitions.
Activities: special events; classes; workshops; educational programs.
Hours & Admission Prices: Tues.-Sat. 10-5, Sun. 1-5. Admissions $5 per person; active military, members, children under one & Association of Children's Museums Reciprocal Program no charge. &
Attendance: 25,000 (estimated)

WATERLOO CENTER FOR THE ARTS, 225 Commercial St., Waterloo, IA 50701-1313. Tel.: 319-291-4490. Fax: 319-291-4270.
E-mail: museum@waterloo-ia.org
Web Site: www.waterloocenterforthearts.org
Formerly: Waterloo Museum of Art
Founded: 1947.
Congressional District: 3
Key Personnel: Dir., Cammie V. Scully; Chm. Cultural & Arts Commission (V), Tom Langlas; Pres. Friends of the Art Center (V), Barb Krizek; Visitor Svcs. Mgr., Maureen Newbill; Education Dir., Bonnie Winninger; Cur., Kent Shankle; Registrar, Marlene Ackerman; Digital Arts Mgr., Chawne Paige; Building Mgr., Mike Guild; Coord. Public Programs, Chad Allen;

Finance Mgr., Paulette Hawkenson; Develop & Marketing Dir., Shannon Farlow; Maintenance Supvr., Lonzo Coleman; Sec., Maureen Hastings; Sec., Nita Hodapp.

Personnel Profile: Full-Time Paid 8; Part-Time Paid 45; Part-Time Volunteers 100; Interns 2.

Governing Authority: municipal; nonprofit. Parent Institution: City of Waterloo Cultural & Arts Commission. Tax-exempt.

Institution Type/Description: Art Museum.

Collections: word by midwest regional artists; American decorative arts; Caribbean Art; Haitian Art; Grant Wood collection.

Research Fields: midwest regional artists; Haitian & Caribbean art; international folk art; American decorative art.

Facilities: 2,000-vol. library of American art & art of the Midwest material available to the public by appointment; 385-seat theater; 60-seat theater; 250 to 300-seat auditorium; ceramics classroom; art classroom; youth pavilion. Museum-related items for sale.

Activities: guided tours; lectures; films; concerts; arts festival; theater; study clubs; hobby workshops; education programs; docent program; loan, traveling, & temporary exhibitions; school programs K-8 grades; Northeast Iowa Print Club; youth hands-on learning experiences. Museum Sponsors: Annual Holiday Arts Festival; Rooftop Jazz & Blues; Performing Arts series; Arti Gras.

Publications: Exhibition & program announcements; exhibition catalogues; quarterly newsletter; class brochures; monthly calendars.

Hours & Admission Prices: Tues.-Sat. 10-5, Sun. 1-5. No charge; donations accepted. Closed New Year's Day; Memorial Day; Independence Day; Labor Day; Veterans Day; Thanksgiving & day after; Christmas Day & day after. &

Attendance: 100,000 (estimated)

Membership: Student & Senior $30; Individual $40; Family & Dual $50; Supporting $75; Contributing $100; Corporate $200-$1,000; Sustaining $250; Patron $500; Benefactor $1,000.

Wayland

WAYLAND & MIDWEST MEMORIES MUSEUMS, 217 W. Main St., Wayland, IA 52654. Tel.: 319-331-5986.
E-mail: midwestmemories@farmtel.net
Web Site: midwestmuseum.blogspot.com
Institution Type/Description: History Museum.
Collections: local history & culture; agricultural & farm artifacts; one-room school memorabilia; period furnishings; photographs; milking machines; cream separators; butter churns; replica tractor.
Hours & Admission Prices: Tues. & Thurs. 9-4 by appointment.

West Bend

GROTTO OF THE REDEMPTION, 300 N. Broadway, West Bend, IA 50597. Mailing Address: P.O. Box 376, West Bend, IA 50597. Tel.: 515-887-2371. Fax: 515-887-2372. Facebook: Grotto of the Redemption.
E-mail: info@westbendgrotto.com
Web Site: westbendgrotto.com
Founded: 1912.
Governing Authority: Parent Institution: Diocese of Sioux City, IA. Tax-exempt.
Institution Type/Description: Religious History Museum.
Collections: religious history; life of Christ; sculptures; video; precious & semiprecious stones from around the world; early photographs & newspaper articles; tools.
Facilities: campground. Gift items for sale.
Activities: guided tours; video.
Hours & Admission Prices: Grotto: daily until 10pm. Museum & Gift Shop: Mon.-Sat. 10-4, Sun. 12-4; other times by appointment. No charge; donations accepted. Closed New Year's Eve & Day; Easter; Thanksgiving; Christmas Eve, Day & week. &
Attendance: 25,000 (estimated)

WEST BEND HISTORICAL SOCIETY & MUSEUM, 7 3rd St., S.E., West Bend, IA 50597. Mailing Address: P.O. Box 23, West Bend, IA 50597. Tel.: 515-200-9234.
E-mail: dbanwart@ncn.net
Founded: 1972.
Congressional District: 4
Key Personnel: Chm. (V) & Pres. (V), Pat Lauck.
Personnel Profile: Part-Time Volunteers 20.
Governing Authority: Tax-exempt.
Institution Type/Description: Historical Society Museum.
Collections: local history, heritage & culture; early agriculture & farm

equipment; period furnishings & clothing. Historic Buildings: school house; post office; sod house.
Activities: school tours; holiday collections; tree show; tour of homes. Museum Sponsors: Quilt Raffle; May Day Tea; Apple Pie Days; Pillow Cleaning Day; Parade Float.
Hours & Admission Prices: By appointment. Adults $4, students $2. &
Attendance: 1,000 (estimated)
Membership: Individual $5; Business $10.

West Branch

HERBERT HOOVER NATIONAL HISTORIC SITE, 110 Parkside Dr., West Branch, IA 52358. Mailing Address: P.O. Box 607, West Branch, IA 52358-0607. Tel.: 319-643-2541. Fax: 319-643-7864.
Web Site: www.nps.gov/heho
Founded: 1965.
Congressional District: 2
Key Personnel: Museum Shop Mgr., Kristin Gibbs.
Governing Authority: federal. Parent Institution: National Park Service, Dept. of the Interior. Tax-exempt.
Institution Type/Description: Historic Site: birthplace of Herbert Hoover and graves of President & Mrs. Hoover.
Collections: Herbert Hoover's childhood; late 19th century Americana; blacksmithing & related trades; prairie environment & management; National Park system resources & management; cultural & natural history. Historic House & Buildings: 1871 Hoover Birthplace; 1856 Friend's Meetinghouse; 1853 School House; late 19th & early 20th-century homes.
Research Fields: Herbert Hoover's childhood.
Facilities: 430-vol. library 1874-1885; reading room; picnic area. Publications, national park themes & products of demonstration from the blacksmith shop for sale.
Activities: guided tours; lectures; films; concerts; formally organized education programs for children, adults & undergraduate college students; loan, permanent & temporary exhibitions.
Publications: brochure; guidebook.
Hours & Admission Prices: Daily 9-5. No charge; donations accepted. Closed New Year's Day; Thanksgiving; Christmas. &
Attendance: 153,000 (accurate)

HERBERT HOOVER PRESIDENTIAL LIBRARY-MUSEUM, (M), 210 Parkside Dr., West Branch, IA 52358-9685. Mailing Address: P.O. Box 488, West Branch, IA 52358-0488. Tel.: 319-643-5301. Fax: 319-643-6045.
E-mail: hoover.library@nara.gov
Web Site: www.hoover.archives.gov
Founded: 1962.
Congressional District: 2
Key Personnel: Dir., Thomas Schwartz; Cur., Marcus Eckhardt; Registrar, Jennifer Pedersen; Photo Archivist, Lynn Smith; Outreach Archivist, Matthew Schaefer; Internet Archivist, Craig Wright; Museum Shop Mgr., Pamela Hinkhouse.
Personnel Profile: Full-Time Paid 15; Part-Time Paid 4; Part-Time Volunteers 35; Interns 3.
Governing Authority: federal. Parent Institution: National Archives & Records Admin., Washington, DC. Subsidiary Institution: Hoover Presidential Library Assoc. Tax-exempt: 170(b)(1)(a).
Institution Type/Description: Presidential Library.
Collections: personal papers; government records; photographs; motion picture films; audio & video tapes; sound recordings; head of state gifts; gifts from private citizens; political campaign items; personal & family memorabilia; journalism; atomic energy; civil aviation; public administration.
Research Fields: life, times, career & presidential administration of President Hoover & associates.
Facilities: 25,000-vol. library; 180-seat auditorium. Documents, slides, prints, posters, books, exhibit catalogs & other museum-related items for sale.
Activities: guided tours; lectures; films; permanent, temporary & traveling exhibitions; organized educational programs for children, adults, undergraduate & graduate students; scholarly conferences.
Publications: brochure; exhibition catalogs; books.
Hours & Admission Prices: Daily 9-5. Adults $6, senior citizens $3; members & children under 16 no charge. Closed New Year's Day; Thanksgiving; Christmas. &
Attendance: 70,000 (estimated)
Membership: Hoover Association $40 & up.

West Des Moines

THE BENNETT SCHOOL, 4001 Fuller Rd., West Des Moines, IA 50061. Mailing Address: West Des Moines Historical Society, 2001 Fuller Rd., West Des Moines, IA 50265-5528. Tel.: 515-277-6652.
Web Site: thejordanhouse.org
Governing Authority: Parent Institution: West Des Moines Historical Society.
Institution Type/Description: Historical Society Museum: housed in c.1926 one room school.
Collections: period furnishings; photographs.
Activities: tours.
Hours & Admission Prices: By appointment. Tours: $3 per person.

THE JORDAN HOUSE, 2001 Fuller Rd., West Des Moines, IA 50265-5528. Mailing Address: c/o West Des Moines Historical Society, P.O. Box 65563, West Des Moines, IA 50265. Tel.: 515-227-6652.
E-mail: wdmhs@valleyjunction.com
Web Site: thejordanhouse.org
Governing Authority: Parent Institution: West Des Moines Historical Society. Subsidiary Institution: Bennett School, Fuller Rd. & 50th St., West Des Moines, IA.
Institution Type/Description: Historic House Museum: housed in the former home of James Jordan, the founder of Valley Junction (later renamed West Des Moines); the home served as a station on the Underground Railroad. National Register of Historic Places.
Collections: period furnishings; personal artifacts; Underground Railroad.
Activities: rental facilities; special events.
Hours & Admission Prices: May-Sept. Wed. & Sat. 1-4, Sun. 2-5; groups of 10 or more by appointment. Adults $3, children $1; members no charge.

West Union

FAYETTE COUNTY HISTORICAL & GENEALOGICAL SOCIETY, 100 N. Walnut St., West Union, IA 52175-1347. Tel.: 563-422-5797.
E-mail: info@wuhistorical.com
Formerly: Fayette County Helpers Club & Historical Society
Founded: 1975.
Congressional District: 2
Key Personnel: Pres., Frances Bowden; Admin., Phyllis Holmstrom.
Personnel Profile: Full-Time Volunteers 1; Part-Time Volunteers 9.
Volunteer Hours: 8,500
Operating Expenses: 10,000
Operating Income: 10,000
Governing Authority: nonprofit organization. Tax-exempt.
Institution Type/Description: Historical & Preservation Society.
Collections: materials & artifacts relating to the history & heritage of Fayette County, Iowa; cemetery records; genealogical materials & records; manuscript collections. Historic Building: Pleasant Ridge #5 Schoolhouse.
Research Fields: Fayette County history & genealogy.
Facilities: 200-vol. library, 1850-present, Fayette County historical material & typed census records by townships available for use by prior arrangement; reading room; meeting facilities.
Activities: lectures; temporary exhibitions; museum tours.
Publications: quarterly, letter to members; historical.
Hours & Admission Prices: Jan. to mid-Dec. Mon.-Fri. 10-3; other times by appointment. No charge: donations accepted. Closed national holidays. &
Attendance: 1,500 (estimated)
Membership: Individual $10; Life $100.

Williams

THE HEMKEN COLLECTION, 202 Main St., Williams, IA 50271. Tel.: 515-854-2749.
E-mail: ddhemken@ncn.net
Web Site: www.the-hemken-collection.org
Institution Type/Description: Automobile Museum.
Collections: automobiles from a 1914 to cars from the 1970s including a Model T, Chevys, Lincolns, Hudsons & Packards; auto memorabilia.
Hours & Admission Prices: May-Oct. Wed. & Fri. 1-5; other times by appointment.

Winfield

WINFIELD HISTORICAL SOCIETY AND MUSEUM, 114 S. Locust St., Winfield, IA 52659-9586. Mailing Address: P.O. Box 184, Winfield, IA 52659-0184. Tel.: 319-257-6974.
Web Site: www.winfieldhistoricalsociety.com

Founded: 1998.
Key Personnel: Dir., Judy Rawson.
Volunteer Hours: 500
Institution Type/Description: Historical Society Museum.
Collections: local history & culture; photographs; personal artifacts; newspaper press; Senator William Carden memorabilia.
Hours & Admission Prices: Mon. 10-12; other times by appointment. No charge; donations accepted.
Attendance: 400 (estimated)

Winterset

JOHN WAYNE BIRTHPLACE, 216 S. 2nd St., Winterset, IA 50273-1910. Tel.: 515-462-1044. Fax: 515-462-3289.
E-mail: director@johnwaynebirthplace.museum
Web Site: www.johnwaynebirthplace.museum
Formerly: Birthplace of John Wayne
Founded: 1982.
Congressional District: 5
Key Personnel: Pres., Joe Zuckschwerdt; Exec. Dir., Brian Downes; Dir. & Museum Shop Mgr., Carolyn Farr; Treas., Rebecca Kile; Sec., Wayne Davis.
Personnel Profile: Part-Time Paid 9.
Governing Authority: nonprofit. Tax-exempt: 501(c)(3).
Institution Type/Description: Historic House & Preservation Project: c.1907 frame house, birthplace of film star John Wayne.
Collections: John Wayne memorabilia; family photographs.
Activities: guided tours.
Publications: brochures.
Hours & Admission Prices: Daily 10-4:30. Adults $7, seniors $6, children $3; babies in arms no charge. &
Attendance: 40,000 (estimated)
Membership: $45 per year.

MADISON COUNTY HISTORICAL SOCIETY, 815 S. 2nd Ave., Winterset, IA 50273-2108. Mailing Address: P.O. Box 15, Winterset, IA 50273-0015. Tel.: 515-462-2134.
E-mail: mchistory@historyonthehill.com
Web Site: www.historyonthehill.com
Founded: 1904.
Congressional District: 5
Key Personnel: Pres. (V), Robert Young; Cur., Jared McDonald; Vice Pres., Sally Oldham; Treas., Tim Waddingham.
Personnel Profile: Full-Time Paid 1; Part-Time Paid 5; Part-Time Volunteers 139.
Governing Authority: nonprofit. Tax-exempt: 701(b)(1)(A).
Institution Type/Description: Historic Society Museum Complex.
Collections: articles belonging to Madison County settlers; Indian artifacts; uniforms; tools; clothing; furniture; large mineral & rock collection; new barn structure housing farm equipment. Historic Structures: 1850 log school; 1850 log post office; 1880 church; 1856 brick house; 1856 stone barn; depot; 1920s country schoolhouse; neighborhood store; attorney's office.
Activities: guided tours.
Publications: newsletter; The Delicious Apple; Three River Country; Scenic Madison County, Iowa; George Washington Carver.
Hours & Admission Prices: May-Oct. Mon.-Sat. 11-4, Sun. 1-5. Museum or House: adults $3; members no charge. Combination Ticket: $5, group rate $4; members & children under 12 no charge. &
Attendance: 10,000 (estimated)
Membership: Annual $10; Life $150.

WINTERSET ART CENTER, 224 S. John Wayne Dr., Winterset, IA 50273. Mailing Address: P.O. Box 325, Winterset, IA 50273-0325. Tel.: 515-210-3286.
E-mail: wacjnarland@aol.com
Web Site: www.wintersetartcenter.org
Founded: 1958.
Congressional District: 5
Key Personnel: Chm., Jerrold Narland; Sec., Ethel Lee Osborne; Public Rels., Joe Held.
Governing Authority: nonprofit. Tax-exempt: 170(b)(1)(A).
Institution Type/Description: Art Museum: housed in c.1854 building used as an Underground Railway stop during the Civil War.
Collections: mixed media art.
Research Fields: history relative to George Washington Carver.
Activities: arts festivals; hobby workshops; organized education programs; temporary exhibitions.

Hours & Admission Prices: Mon.-Sat. 10-4 & by appointment. No charge; donations accepted.
Attendance: 3,000 (estimated)
Membership: Junior & Student $10; Regular $20; Family $45.

KANSAS

(348 listings)

Abilene

DICKINSON COUNTY HERITAGE CENTER, (M), 412 S. Campbell St., Abilene, KS 67410-2905. Tel.: 785-263-2681. Fax: 785-263-0380.
E-mail: heritagecenterdk@sbcglobal.net
Web Site: heritagecenterdk.com
Founded: 1928.
Congressional District: 2
Key Personnel: Dir. & C.E.O., Jeff Sheets; Pres. (V), Thelma Lexow; Museum Shop Mgr., Twila Jackson.
Personnel Profile: Full-Time Paid 1; Part-Time Paid 5; Part-Time Volunteers 50; Interns 1.
Governing Authority: nonprofit organization. Parent Institution: Dickinson County Historical Society. Tax-exempt: 501(c)(3).
Institution Type/Description: History Museum.
Collections: agriculture; cattle drive memorabilia; toys; carnivals; household items; musical instruments; county archives; newspapers & photographs; cemetery surveys; census records. Historic Structures: 1901 C.W. Parker Carousel; pioneer log cabin; 1900 telephone exchange building; blacksmith shop.
Research Fields: early settling of the West; county & state history; genealogy.
Facilities: Gift items for sale.
Activities: guided tours; summer lecture series; special events; pioneer camp; school groups; traveling trunks. Museum Sponsors: Annual Heritage Day; Christmas in the Cabin; Annual Quilt Show.
Publications: quarterly magazine, Gazette.
Hours & Admission Prices: Winter: Mon.-Fri. 9-3, Sat. 10-5, Sun. 1-5; Summer: Mon.-Fri. 9-4, Sat. 10-8, Sun. 1-5. Adults $4, seniors 62 & over $3, children 2-14 $2; discounts to groups & AAM members & telephone pioneers. Carousel Rides $2. Closed New Year's Day; Thanksgiving; Christmas. &
Attendance: 14,000 (accurate)
Membership: Individual $15; Family $25; Institutional $50; Contributing $50-$999; Life $1,000.

EISENHOWER PRESIDENTIAL LIBRARY, MUSEUM AND BOYHOOD HOME, 200 S.E. 4th St., Abilene, KS 67410-2900. Mailing Address: P.O. Box 339, Abilene, KS 67410-0339. Tel.: 785-263-6700. Fax: 785-263-6718. Facebook: Ike Library.
E-mail: eisenhower.library@nara.gov
Web Site: www.eisenhower.archives.gov
Founded: 1945.
Congressional District: 1
Key Personnel: Dir., Karl Weissenbach; Cur., William D. Snyder; Museum Shop Mgr., Carol Needham.
Personnel Profile: Full-Time Paid 40; Part-Time Paid 20; Part-Time Volunteers 50; Interns 10.
Governing Authority: federal. Parent Institution: National Archives and Records Administration, Washington, DC 20408. Tax-exempt: 170(b)(1)(A).
Institution Type/Description: Presidential Library & Museum.
Collections: personal papers; government records; photographs; motion picture films; audio & video tapes; sound recordings; head of state gifts; gifts from private citizens; political campaign items; personal & family memorabilia; 35,000 serials. Historic House: 1887 Eisenhower family home.
Research Fields: life, times, career & presidential administration of President Eisenhower.
Facilities: 25,000-vol. library; 2 auditoriums. Document facsimiles, slides; prints; posters; books, exhibit catalogs & museum object reproductions for sale.
Activities: guided tours; lectures; films; permanent, temporary & traveling exhibitions; organized educational programs for children, adults and undergraduate & graduate students.
Publications: general information brochure; newsletter, Overview.
Hours & Admission Prices: June-July daily 8-5:45; Aug.-May daily 9-4:45. Museum: adults $10, seniors $9, children 6-15 $2; children 5 & under no charge. Closed New Year's Day; Thanksgiving; Christmas. &
Attendance: 163,554 (accurate)
Membership: Friends of the Eisenhower Library Foundation: One Star up to $50; Two Star $51-$100; Three Star $101-$500; Four Star $501-$1,000; Five Star $1,001-$10,000; Presidential $10,000 & up.

GREYHOUND HALL OF FAME, 407 S. Buckeye, Abilene, KS 67410-2925. Tel.: 785-263-3000. Fax: 785-263-2604.
E-mail: info@greyhoundhalloffame.com
Web Site: greyhoundhalloffame.com
Founded: 1963.
Congressional District: 1
Key Personnel: Mgr., Kathryn Lounsbury; Bd. Pres., Tom Taplin.
Personnel Profile: Full-Time Paid 1; Part-Time Paid 2.
Governing Authority: nonprofit organization. Tax-exempt: 501(c)(3).
Institution Type/Description: Sports Museum.
Collections: historic items relating to the sport of greyhound racing & the greyhound animal.
Research Fields: history.
Facilities: 800-vol. library of statistics & history available for use on premises; theater. Books, gifts & museum-related items for sale.
Activities: guided tours; films; permanent & temporary exhibitions.
Hours & Admission Prices: Daily 9-5. No charge; donations accepted. Closed New Year's Day; Thanksgiving; Christmas. &
Attendance: 50,000 (estimated)
Membership: Hall of Fame Club: Assoc. $100; Full $250.

LEBOLD MANSION, 106 N. Vine St., Abilene, KS 67410. Tel.: 785-263-4356.
E-mail: lebold@sbcglobal.net
Institution Type/Description: History Museum: housed in the former home of banker, politician, & entrepreneur C. H. Lebold; built in 1880. Listed on the National Register of Historic Places.
Collections: local history; period furnishings; personal artifacts; paintings; books.
Hours & Admission Prices: Tues.-Sat. 10-4. Adults $10, children 6-16 $5.

MUSEUM OF INDEPENDENT TELEPHONY, 412 S. Campbell, Abilene, KS 67410-2905. Tel.: 785-263-2681. Fax: 785-263-0380.
Web Site: www.heritagecenterdk.com
Founded: 1973.
Congressional District: 2
Key Personnel: C.E.O. & Dir., Jeff Sheets.
Personnel Profile: Full-Time Paid 2; Part-Time Volunteers 50.
Governing Authority: Parent Institution: Dickinson County Historical Society. Tax-exempt: 501(c)(3).
Institution Type/Description: Telephonic History Museum.
Collections: early artifacts of communications; telephones from the primitive to the modern; old telephone business office; old phone booths; switchboards; sheet music; photographs; interactive exhibits.
Research Fields: telephone company history & the history of independent telephone companies.
Facilities: 1,000-vol. library of books on telephone history; outdoor museum, 1927 Mack Line Truck.
Activities: guided tours; lectures; films; gallery talks; permanent & traveling exhibitions.
Publications: Tales of Telephony.
Hours & Admission Prices: Memorial Day-Labor Day Mon.-Fri. 9-4, Sat. 9-8; Winter: Mon.-Fri. 9-3, Sat. 10-5, Sun. 1-5. Adults 16 & over $4, seniors 62 & over $3, children 2-15 $2; discounts to groups, AAA & AAM members. Closed New Year's Day; Thanksgiving; Christmas. &
Attendance: 10,904 (estimated)
Membership: Individual $10; Family $15; Corporate $100.

THE SEELYE MANSION & PATENT MEDICINE MUSEUM, 1105 N. Buckeye Ave., Abilene, KS 67410-1942. Mailing Address: P.O. Box 337, Abilene, KS 67410-0337. Tel.: 785-263-1084.
E-mail: terryt@access-one.com
Web Site: www.seeleymansion.org
Institution Type/Description: Historic House: former home of Dr. and Mrs. A.B. Seelye; built in 1905.
Collections: Mansion: period artifacts & furnishings. Medicine Museum: A.B. Seelye Medical Company artifacts.
Hours & Admission Prices: Mon.-Sat. 10-6, Sun. 1-5. Adults $10, children 6-12 $5. Closed Christmas.

Alden

AT & SF DEPOT, Alden, KS 67512-0158. Mailing Address: P.O. Box 158, Alden, KS 67512-0158. Tel.: 620-534-2425.
E-mail: prflrcraft@aol.com
Founded: 1970.
Congressional District: 1
Key Personnel: Dir., Sara Fair Sleeper.

Governing Authority: county. Parent Institution: Rice County Historical Museum. Subsidiary Institution: Coronado Quivira Museum. Tax-exempt.
Institution Type/Description: Transportation Museum: housed in c.1872 Atchison, Topeka & Santa Fe railroad building & contents.
Collections: railroad items.
Activities: tours.
Hours & Admission Prices: By appointment only. No charge.
Attendance: 4 (estimated)

Alma

WABAUNSEE COUNTY HISTORICAL MUSEUM, 227 Missouri, Alma, KS 66401. Mailing Address: P.O. Box 387, Alma, KS 66401-0387. Tel.: 785-765-2200.
E-mail: wabcomuseum@embarqmail.com
Web Site: www.wabaunsee.org
Founded: 1968.
Congressional District: 5
Key Personnel: Pres. (V), John Hund; Cur., Alan Winkler; Asst. Cur., Linda Maas.
Personnel Profile: Part-Time Paid 2; Part-Time Volunteers 6.
Volunteer Hours: 420
Operating Expenses: 51,000
Operating Income: 70,500
Governing Authority: society. Tax-exempt: 501(c)(3).
Institution Type/Description: Historical Society Museum: housed in 100-year old native stone building.
Collections: Gen. Lewis Walt's display; Native American artifacts; old time school room display; old time doctor's office; photos; guns; buggies; clothes, costumes; Main Street U.S.A.; general store; barber shop; shoe & harness shop; 1923 REO fire truck; post office; blacksmith shop; ladies mercantile shop; railroad depot; Cane collection; farm machinery; genealogy; Civil War artifacts; William M. Meyers Civil War artifacts.
Research Fields: oral history project; school records.
Facilities: library.
Activities: guided tour of museum; lectures; permanent & temporary exhibitions; fall historical tour of Wabaunsee County; annual meeting historical speaker.
Publications: quarterly newsletter; pamphlets.
Hours & Admission Prices: March-Nov. Tues.-Sat. 10-12 & 1-4; Dec.-Feb. Tues.-Wed. 10-12 & 1-4. Suggested Donations: adults 22-64 $2; seniors & students no charge. Closed major holidays. &
Attendance: 1,920 (accurate)
Membership: Individual $25; Family $50; Donor $75.

Anthony

HISTORICAL MUSEUM OF ANTHONY INC., 526 W. Main St., Anthony, KS 67003. Mailing Address: P.O. Box 185, Anthony, KS 67003. Tel.: 620-842-3852.
Founded: 1984.
Key Personnel: Pres. (V), Kathy Francis.
Personnel Profile: Part-Time Paid 2.
Governing Authority: Tax-exempt.
Institution Type/Description: History Museum: housed in the former Santa Fe Railroad Depot; built in 1928.
Collections: local history & culture; period furnishings; personal artifacts; photographs; clothing; farm equipment & machinery; railroad memorabilia; 1950 fire truck; fire memorabilia; hospital memorabilia.
Publications: yearly membership letter.
Hours & Admission Prices: Thurs.-Sat. 10-12 & 1-4; other times by appointment. &
Attendance: 500 (estimated)
Membership: Active Individual $10; Family or Buusiness $25; Life $100.

Argonia

SALTER MUSEUM, 220 W. Garfield, Argonia, KS 67004. Mailing Address: P.O. Box 126, Argonia, KS 67004-0126. Tel.: 620-435-6376.
Founded: 1961.
Congressional District: 4
Key Personnel: Pres. Argonia & Western Sumner Historical Society, Troy Bookless; Docent, Carol Pearce; Chief Cur., Mary Beth Bookless.
Governing Authority: nonprofit organization. Affiliated with Argonia and Western Sumner County Historical Society. Tax-exempt.
Institution Type/Description: Historic House Museum: 1884 home of America's first woman mayor, Mrs. Susanna M. Salter.
Collections: 19th-century home furnishings. Historic Building: 1924 church building.

Research Fields: Sumner County history.
Activities: guided tours.
Publications: brochure.
Hours & Admission Prices: By appointment. No charge; donations accepted.
&

Membership: Annual $5; Life $100.

Arkansas City

CHEROKEE STRIP LAND RUSH MUSEUM, 31639 US 77, Arkansas City, KS 67005. Mailing Address: P.O. Box 778, Arkansas City, KS 67005-0778. Tel.: 620-442-6750. Fax: 620-441-4332.
E-mail: hferguson@arkansascityks.gov
Web Site: www.arkansascityks.gov
Founded: 1966.
Congressional District: 5
Key Personnel: City Mgr., Steve Archer; Chm., Jerry Hooley; Dir., Heather Ferguson.
Personnel Profile: Full-Time Paid 1; Part-Time Paid 1; Part-Time Volunteers 10.
Governing Authority: nonprofit organization. Tax-exempt.
Institution Type/Description: History Museum.
Collections: Cherokee Strip Land Rush of Sept. 16, 1893; historical artifacts from 1800s-1920s; Indian & military artifacts; Bryson Paddock Wichita Indian archaeological site replica (1700-1753). Historic Buildings: Hardy Jail; Bland Schoolhouse.
Research Fields: Cherokee Strip Run; Cherokee Outlet; Chilocco Indian School.
Facilities: 100-vol. library of books on the Cherokee Strip available for research; Cowley County Genealogical Society Collection. Western & Indian items for sale.
Activities: guided tours; lectures; traveling exhibitions. Annual Events: Mountain Man Living History Encampment; Pioneer Festival and Western Heritage Days.
Publications: Between the Rivers, Vols. 1-3; Images of America Cherokee Strip Land Rush; Arkansas City - Images of America Series.
Hours & Admission Prices: Tues.-Sat. 10-5. Adults $4.50, senior citizens $3.50, children 6-12 $2; discount to AAM & AAA members & senior citizens; children under 6 & members no charge. Closed major holidays. &
Attendance: 10,428 (accurate)
Membership: Individual $15; Family $25; Extended Family $30; Benefactor $50; Business $100; Corporate $200.

Arma

SCOTTY'S CLASSIC CAR MUSEUM, 302 N. 9th St., Arma, KS 66712-9520. Tel.: 620-347-8387. Fax: 620-249-5555.
E-mail: memrylane@yahoo.com
Web Site: www.scottysclassiccars.com
Key Personnel: Owner, Scotty Bitner
Institution Type/Description: History Museum.
Collections: classic & period automobiles.
Hours & Admission Prices: Tues.-Sat. 11-5; other times by appointment.

Ashland

PIONEER-KRIER MUSEUM, 430 W. 4th, Ashland, KS 67831. Mailing Address: P.O. Box 862, Ashland, KS 67831-0862. Tel.: 620-635-2227. Fax: 620-635-2227 (call first).
E-mail: pioneer@ucom.net
Web Site: www.pioneer-krier.com
Founded: 1966.
Congressional District: 1
Key Personnel: Pres., George Krier; Vice Pres., Jim Baker; Dir., Tony Maphet; Chm. (V), Milton Hughes.
Personnel Profile: Part-Time Paid 1; Part-Time Volunteers 25.
Governing Authority: society. Parent Institution: Clark County Historical Society. Tax-exempt: 501(c)(3).
Institution Type/Description: Pioneer & Aerobatic Museum.
Collections: dishes, glassware & other kitchen equipment; dolls & toys; bits & spurs; guns; pre-1900 musical instruments; blacksmith shop; furniture; handcrafts; military memorabilia; pioneer pictures; church furniture; farm machinery; pioneer era saddles & tack; fossils from local area; Indian artifacts; display of aerobatic airplanes & trophies; musical instruments; pioneer doctor & hospital equipment; business office of pioneer undertaker & abstractor; early day bank; general store; barbed wire; manuscripts.
Research Fields: old cattle trails; prehistoric fossils; pioneer graves; pioneer life; cattle industry.
Facilities: 750-vol. library of books on genealogy, Kansas history, family records & collection of 1884-1991; reading room. Books & museum-related items for sale.

Activities: school study tours; permanent & temporary exhibitions.
Publications: books, 6-vol., Notes on Early Clark County, Kansas; booklets, Souvenir of Ashland & Clark County, 1884-1909; Kings & Queens of the Range; Pictures of Trail Drives & Ranches, 1894-1904; Cattle Ranching South of Dodge City - the Early Years 1870-1920; Sitka.
Hours & Admission Prices: Tues.-Fri. 10-12 & 1-5. No charge; donations accepted. Closed New Year's Day; Thanksgiving; Christmas. &
Attendance: 2,000 (accurate)
Membership: Life $25.

Atchison

AMELIA EARHART BIRTHPLACE MUSEUM, 223 North Ter., Atchison, KS 66002-2525. Mailing Address: 609 Meridian Rd., Chester, NE 68327-7004. Tel.: 913-367-4217.
E-mail: aemuseum@att.com
Web Site: ameliaearhartmuseum.org
Founded: 1984.
Congressional District: 2
Key Personnel: Chm. (V), Carole Sutton; Museum Shop Mgr., Jan Coyle.
Personnel Profile: Full-Time Paid 1; Part-Time Paid 1; Part-Time Volunteers 40.
Governing Authority: private; nonprofit organization. Parent Institution: The Ninety-Nines, Inc., 4300 Amelia Earhart Rd., Will Rogers Airport, Oklahoma City, OK 73159. Tax-exempt: 501(c)(3).
Institution Type/Description: Historic House: 1861 Gothic Revival cottage, birthplace of Amelia Earhart.
Collections: period furnishings & memorabilia; life of Amelia Earhart & other women pilots.
Research Fields: Amelia Earhart & other women pilots, past & present.
Facilities: 500-vol. library of books on aviation & women aviators. Museum-related items for sale.
Activities: docent program; guided tours; lectures; participatory exhibits. Museum Sponsors: Women in History Month; Amelia Earhart Festival & Birthday Celebration; seasonal open houses.
Publications: annual Amelia Earhart picture calendar; Amelia Earhart cookbook.
Hours & Admission Prices: Feb. 16-Dec. 14 Mon.-Fri. 9-4, Sat. 10-4, Sun. 1-4; Dec. 15-Feb. 15 Wed.-Sat. 10-4, Sun. 1-4. No charge; donations accepted. Closed New Year's Day; Christmas. &
Attendance: 12,000 (estimated)

ATCHISON COUNTY HISTORICAL SOCIETY, (M), 200 S. 10th St., Atchison, KS 66002-2772. Mailing Address: P.O. Box 201, Atchison, KS 66002-0201. Tel.: 913-367-6238.
E-mail: gowest@atchisonhistory.org
Web Site: www.atchisonhistory.org
Founded: 1967.
Congressional District: 2
Personnel Profile: Full-Time Paid 1; Part-Time Volunteers 2; Interns 1.
Governing Authority: society; nonprofit organization. Branch Museum: Independence Creek Lewis and Clark Historic Site, 19917 314th Rd., P.O. Box 201, Atchison, KS 66002. Tax-exempt: 501(c)(3).
Institution Type/Description: Historical Society Museum: housed in c.1880 Santa Fe Depot.
Collections: county history & culture; firearms from Revolutionary War & Civil War-modern era; military items; childhood memorabilia, clothing from Amelia Earhart; Lewis & Clark artifacts.
Research Fields: Atchison County, Kansas.
Facilities: small group meeting facilities.
Activities: guided tours; lectures; study clubs.
Publications: bimonthly newsletter, Go West.
Hours & Admission Prices: Mon.-Fri. 8-5, Sat. 9-5, Sun. 12-5. No charge; donations accepted. &
Attendance: 24,000 (estimated)
Membership: Scout $5; Pony Express $10; Overland Stage & Pioneer $15; Homesteader $25; Gov. Glick $50; AT & SF $100; Lewis & Clark $1,000; Senator J.J. Ingalls $5,000.

EVAH C. CRAY HISTORICAL HOME MUSEUM, 805 N. 5th St., Atchison, KS 66002-1807. Tel.: 913-367-3046.
Institution Type/Description: Historic House: housed in a 25-room Victorian era mansion; built in 1882.
Collections: local history & culture; Victorian furnishings; personal artifacts; photographs; carriage house.
Facilities: theater. Museum-related items for sale.
Hours & Admission Prices: April 10-4; May-Oct. Mon.-Sat. 10-4, Sun. 1-4; Nov. 27-Dec. 17 Fri.-Mon. call for hours. Adults $3.

THE MUCHNIC GALLERY, 704 N. 4th St., Atchison, KS 66002-1924. Tel.: 913-367-4278. Fax: 913-367-2939.
E-mail: atchisonart@gmail.com
Web Site: www.atchison-art.org/muchnic
Founded: 1970.
Congressional District: 2
Key Personnel: Dir. & Cur., Deborah Geiger; Pres. (V), Patty Boldridge.
Personnel Profile: Part-Time Paid 4; Interns 1.
Governing Authority: nonprofit organization. Parent Institution: Atchison Art Association. Tax-exempt.
Institution Type/Description: Art Gallery: housed in 1885 Victorian brick residence.
Collections: paintings & lithographs of John Falter, John S. Curry, Jim Hamil, Thomas Hart Benton, Robert Sudlow & Jack O'Hara; furniture & paintings of the Muchnic family.
Activities: guided tours; permanent & traveling exhibitions.
Publications: brochure, Tour Guide of Home; newsletter.
Hours & Admission Prices: Gallery: Wed. 1-5, Sat.-Sun. 1-5, Special Exhibits: Sat.-Sun. 1-5; special tours available. No charge, donations accepted.
Attendance: 1,500 (accurate)
Membership: $25; $50; $100; $500.

Atwood

RAWLINS COUNTY MUSEUM, 308 State, Atwood, KS 67730. Tel.: 785-626-3885.
Governing Authority: Parent Institution: Rawlins County Historical Society.
Institution Type/Description: History Museum.
Collections: local history & culture; period furnishings; personal artifacts; photographs; Gov. Mike Hayden life & career; cameras; tools; Hopi kachina dolls; clocks; mural; paintings.
Hours & Admission Prices: Mon.-Fri. 9-1 & 1-4. No charge; donations accepted.

Augusta

AUGUSTA HISTORICAL MUSEUM & GENEALOGY, 303 State, Augusta, KS 67010-1103. Tel.: 316-775-5655.
E-mail: augustahm@aol.com
Web Site: www.augustahistoricalsociety.net
Founded: 1938.
Congressional District: 5
Key Personnel: Pres., Eldon Foreman; Dir., Rachelle Meinecke.
Personnel Profile: Full-Time Paid 1; Part-Time Paid 1; Part-Time Volunteers 7.
Governing Authority: county. Tax-exempt.
Institution Type/Description: General Museum.
Collections: pioneer relics; items from late 1800s settlements. Historic Building: 1868 log house.
Research Fields: genealogy.
Activities: permanent exhibitions.
Publications: Cabin Chronicle.
Hours & Admission Prices: Mon.-Fri. 11-3, Sat.-Sun. 1-4. No charge; donations accepted. Closed holidays.
Attendance: 8,500 (accurate)
Membership: Pioneer $25; Settler $50; Patron $100.

KANSAS MUSEUM OF MILITARY HISTORY, 135 S. Hwy. 77, Augusta, KS 67010-7681. Tel.: 316-775-1425.
E-mail: info@kmmh.org
Web Site: www.kmmh.org
Key Personnel: Museum Dir. & Pres., John Lara
Institution Type/Description: Military Museum.
Collections: aviation & military artifacts & memorabilia.
Hours & Admission Prices: April-Sept. daily 1-5; Oct.-March Sat.-Sun. 1-5.

Baldwin City

OLD CASTLE MUSEUM, 511 Fifth St., Baldwin City, KS 66006. Mailing Address: Baker University, P.O. Box 65, Baldwin City, KS 66006-0065. Tel.: 785-594-8380. Fax: 785-594-2522.
E-mail: jen.mccollough@bakerU.edu
Founded: 1953.
Congressional District: 2
Key Personnel: Dir. Archives & Museum, Jen McCollough.
Governing Authority: Parent Institution: Baker University, 515 5th, Baldwin City, KS 66006. Tax-exempt: 170(b)(1)(A).
Institution Type/Description: Historic Buildings Complex.
Collections: Midwest Native American display; Southwest Native American pottery; pioneer artifacts; 19th century print shop; early dentist & doctor exhibit; ironstone china pieces; silver & pewter items; replica of Kibbee Cabin, 1857. Historic Buildings: 1857 Santa Fe Trail Post Office; 1858 3-story stone structure.
Research Fields: Santa Fe Trail.
Facilities: library.
Activities: individual & group tours by appointment.
Publications: pamphlets, Old Castle Museum; Kibbee Cabin; Palmyra Post Office.
Hours & Admission Prices: Temporarily closed. No charge; donations accepted.
Attendance: 5,436 (accurate)

QUAYLE BIBLE COLLECTION, Collins Library-Spencer Quayle Wing, 518 8th St., Baldwin City, KS 66006-0065. Mailing Address: Collins Library, P.O. Box 65, Baldwin City, KS 66006-0065. Tel.: 785-594-8393. Fax: 785-594-6721.
E-mail: quayle@bakeru.edu
Web Site: www.bakeru.edu/quayle
Founded: 1925.
Congressional District: 3
Key Personnel: Dir., Kay Bradt.
Personnel Profile: Part-Time Paid 2; Interns 1.
Operating Expenses: 2,622
Operating Income: 3,402
Governing Authority: Baker University. Tax-exempt.
Institution Type/Description: Rare Bible Museum.
Collections: 600 rare Bibles or portions thereof, including illuminated manuscripts & incunabula.
Major Exhibits: Illuminating the Bible: Woodcuts & Engravings, 9/13-7/14.
Facilities: 100-vol. library of books by William A. Quayle & reference materials on items in the collection available for use by special permission of the director; reading room.
Activities: guided tours with advance notice of 2 weeks; lectures; gallery talks; formally organized education programs for undergraduate college students affiliated with Baker University; permanent & temporary exhibitions.
Publications: book, The Catalog of The William Alfred Quayle Bible Collection; flyers; exhibit booklet.
Hours & Admission Prices: Sat.-Sun. 1-4; other times by appointment. No charge; donations accepted. Closed university holidays. &
Attendance: 400 (estimated)

Barnes

BARNES STATE BANK MUSEUM, 107 W. Railroad Ave., Barnes, KS 66933. Mailing Address: P.O. Box 94, Barnes, KS 66933. Tel.: 785-763-4569.
E-mail: historictrust@barnesks.net
Web Site: www.barnesks.net
Founded: 2007.
Congressional District: 4
Key Personnel: Dir., Cindy Hiesterman; Chm. (V) & Museum Shop Mgr., Gloria Moore; Pres. (V), Todd Frye.
Personnel Profile: Part-Time Volunteers 16.
Governing Authority: Parent Institution: Barnes Historic Trust. Tax-exempt.
Institution Type/Description: Historic Building: built in 1911. Listed on the National Register of Historic Places.
Collections: local history; period furnishings & artifacts; photographs.
Hours & Admission Prices: Mon.-Sat. 12-4. No charge; donations accepted.
Attendance: 540 (accurate)

Baxter Springs

BAXTER SPRINGS HERITAGE CENTER AND MUSEUM, 740 East Ave., Baxter Springs, KS 66713. Mailing Address: P.O. Box 514, Baxter Springs, KS 66713-0514. Tel.: 620-856-2385.
E-mail: heritagectr@embarqmail.com
Web Site: www.baxterspringsmuseum.org
Founded: 1962.
Congressional District: 14
Key Personnel: Pres. (V), Jim Hall; Treas., Earleene Spaulding.
Personnel Profile: Full-Time Paid 1; Full-Time Volunteers 2; Part-Time Volunteers 100.
Governing Authority: private; nonprofit organization. Tax-exempt: 501(c)(3).
Institution Type/Description: History Museum.
Collections: Civil War; 1863 Baxter Springs Massacre; school & education; Native American; mining; industry & business; domestic life; WWI & WWII.
Research Fields: historical Civil War battle site.
Facilities: Museum-related items for sale.
Activities: docent program; guided tours; lectures.

Publications: quarterly newsletter, Cassion Tracks.
Hours & Admission Prices: April-Oct. Mon.-Sat. 10-4:30, Sun. 1-4:30; Nov.-March Thurs.-Sat. 10-4:30, Sun. 1-4:30; other times by appointment. No charge. &
Attendance: 4,000 (estimated)
Membership: Yearly $5; Life Single $40; Life Couple $70.

Belleville

BOYER MUSEUM OF ANIMATED CARVINGS, 1205 M St., Belleville, KS 66935-3069. Tel.: 785-527-5884.
E-mail: boyermuseum@sbcglobal.net
Web Site: www.kansastravel.org/boyergallery.htm
Formerly: Boyer Gallery
Founded: 1997.
Key Personnel: Owner, Paul Boyer; Museum Shop Mgr., Ann Lewellyn.
Personnel Profile: Part-Time Volunteers 2.
Institution Type/Description: Mechanical Sculpture Museum.
Collections: animated wood carvings; animated metal & wire sculptures; photographs; paintings.
Hours & Admission Prices: May-Sept. Wed.-Sat. 1-5; other times by appointment (weather permitting); call to confirm hours. Adults $5, children 6-12 $2.
Attendance: 2,000 (estimated)

HIGHBANKS HALL OF FAME & NATIONAL MIDGET AUTO RACING MUSEUM, 1204 H St., Belleville, KS 66935. Mailing Address: P.O. Box 264, Belleville, KS 66935-0264. Tel.: 785-527-2526.
E-mail: highbanksmuseum@nckcn.com
Web Site: www.highbanks-museum.org
Founded: 1997.
Key Personnel: Cur., Bob Blazek; Pres. (V), Justin Sly.
Governing Authority: Tax-exempt.
Institution Type/Description: Hall of Fame & Racing Museum.
Collections: midget auto racing history; photographs; personal artifacts; Hall of Fame Inductees.
Activities: Annual Event: Induction Ceremony.
Hours & Admission Prices: May-Sept. Tues.-Sun. 10-5; Oct.-April Wed.-Sun. & holidays 11-4; other times by appointment. No charge; donations accepted. &
Attendance: 2,400 (accurate)
Membership: Individual $25; Family $40; Business $50.

REPUBLIC COUNTY HISTORICAL SOCIETY MUSEUM, 615 28th St., Belleville, KS 66935-2469. Tel.: 785-527-5971.
E-mail: repcomuse@nckcn.com
Web Site: www.nckcn.com/homepage/repcomuse/home.htm
Founded: 1985.
Key Personnel: C.E.O., Cur. & Archivist, Sherrie Larson; Pres. (V), Nancy Holt; Treas., David Bowersox.
Personnel Profile: Full-Time Paid 1; Part-Time Volunteers 6.
Governing Authority: historical society. Tax-exempt: 501(c)(3).
Institution Type/Description: Historical Society Museum.
Collections: pioneer historical artifacts; tool collection; railroad caboose. Historic Buildings: rural schoolhouse; log cabin; country church; round limestone smokehouse.
Facilities: 300-vol. library of history & genealogy.
Activities: Museum Sponsors: Soup Supper in January; Ice Cream Social in July.
Publications: newsletter 3 times per year, Illumination.
Hours & Admission Prices: April-Nov. Mon.-Fri. 1-5, Sun. 1:30-4:30; Dec.-March Mon.-Fri. 1-5. Adults $3; members & children under 10 no charge. Closed major holidays. &
Attendance: 2,420 (estimated)
Membership: Individual $10; Family $15; Business $25.

Beloit

LITTLE RED SCHOOLHOUSE-LIVING LIBRARY, Roadside Park, N. Walnut & Hwy. 24, Beloit, KS 67420. Mailing Address: P.O. Box 582, 123 N. Mill, Beloit, KS 67420-0582. Tel.: 785-738-2717.
Founded: 1976.
Congressional District: 106
Key Personnel: Chm. (V), Mildred Peterson.
Governing Authority: nonprofit. Parent Institution: City of Beloit. Subsidiary Institution: Alpha Pi Chapter of Delta Kappa Gamma. Tax-exempt: 501(c)(3)(6).

Institution Type/Description: Education Museum: housed in 1874 one-room schoolhouse.
Collections: early school books & rhythm band instruments; manuscripts; recitation bench; country school desks; pot belly stove; teaching materials; bells; slate boards; globes; maps; water pails; dipper; dinner buckets; syrup pails; dunce stool; pictures; piano; library shelves; teacher's desk.
Research Fields: early day schools of Mitchell Co.
Facilities: 323-vol. library of old classroom texts & early day children's books. Pamphlets, postcards & other museum-related items for sale.
Activities: guided tours; retired teachers (many who taught in one-room schools) hold sessions for field trips; activities as they were done in the late 1800s & early 1900s; inter-museum loan & temporary exhibitions; meetings; parties; clubs.
Hours & Admission Prices: May to Oct. 1 Fri.-Mon. 1:30-4:30; other times by appointment. No charge; donations accepted.
Attendance: 379 (accurate)

MITCHELL COUNTY HISTORICAL SOCIETY MUSEUM, 402 W. 8th, Beloit, KS 67420. Tel.: 785-738-5355.
E-mail: mchs@yahoo.com
Institution Type/Description: Historical Society Museum.
Collections: local history & culture; period furnishings; military artifacts; photographs; personal artifacts; early tools; dentistry equipment; scrapbooks; business records.
Hours & Admission Prices: Sun. 1-5, Mon.-Tues. 10-4; other times by appointment. No charge; donations accepted.

Blue Rapids

BLUE RAPIDS HISTORICAL SOCIETY AND MUSEUM, #36 Public Square, Round Town Square, Blue Rapids, KS 66411. Tel.: 785-363-7949.
E-mail: brhissoc@yahoo.com
Founded: 2007.
Key Personnel: Cur., Patricia Osborne; Pres. (V), Nolan Sump.
Personnel Profile: Part-Time Volunteers 6.
Volunteer Hours: 2,100
Governing Authority: Tax-exempt: 501(c)(3).
Institution Type/Description: Historical Society Museum.
Collections: local history, business & culture; period furnishings; personal artifacts; photographs; farm tools & equipment; gypsum mining; Ice Age monument & information.
Major Exhibits: Czech Heritage, 1/14-4/30/14; Marshall County Traits, 5/14-8/14; War, 9/14-12/14.
Research Fields: local genealogy.
Activities: cemetery walk; quarterly speakers; potluck dinners. Annual Events: Czech Festival; Veterans Day Parade; Voices of the Past Cemetery Tour.
Publications: quarterly newsletter; e-newsletter.
Hours & Admission Prices: Sat. 9-12; other times by appointment. No charge; donations accepted. &
Attendance: 1,000 (estimated)
Membership: Individual $10; Lifetime $100.

Bonner Springs

THE NATIONAL AGRICULTURAL CENTER & HALL OF FAME, 630 N. 126th St., Bonner Springs, KS 66012-9045. Tel.: 913-721-1075. Fax: 913-721-1202.
E-mail: info@aghalloffame.com
Web Site: www.aghalloffame.com
Founded: 1958.
Congressional District: 3
Key Personnel: Exec. Dir., Tim Daugherty; Chm. (V), Robert Carlson; Pres. (V), Joel Ebbertt; Cur., Kate Alexander; Dir. Education, Lee Sigley; Museum Shop Mgr., Angela Cannizzano.
Personnel Profile: Full-Time Paid 4; Part-Time Paid 15; Part-Time Volunteers 15; Interns 1.
Governing Authority: nonprofit organization. Tax-exempt: 501(c)(3).
Institution Type/Description: Agriculture Museum.
Collections: harvesting & planting equipment; steam traction engines; rural living items; farm equipment & tools; buggies, wagons & trucks; National Farmers Memorial; Farm Town including one-room school, train depot, museum & blacksmith shop.
Facilities: 200-seat auditorium; nature trail; early 1900s rural village replica. Museum-related items for sale.
Activities: arts festivals; guided tours; formal education programs for children; permanent & temporary exhibitions. Annual Events: threshing exhibition; Antique Tractor Pull; Artist-of-the-Month; Santa's Express.
Publications: quarterly newsletter, Update.
Hours & Admission Prices: mid-March to Nov. Tues.-Sat. 9-5, Sun. 1-5. Adults

$7, senior citizens over 62 $6, children 5-16 $3; discounts for individual programs, groups, AAA, AAM & ICOM members and Newcomer's Packets; members & children under 5 no charge. &
Attendance: 25,000 (estimated)
Membership: Booster Club $25; Honorary Farmer $100; Honor Acre $200; Sustaining $1,000; Sponsoring $5,000; Foundation $50,000.

WYANDOTTE COUNTY MUSEUM, (M), 631 N. 126th St., Bonner Springs, KS 66012-9046. Tel.: 913-573-5002.
E-mail: pschurkamp@wycokck.org
Web Site: www.wycomuseum.org
Formerly: Wyandotte County Historical Society and Museum
Founded: 1889.
Congressional District: 6
Key Personnel: C.E.O. & Dir., Trish Schurkamp; Cur., Jennifer Laughlin.
Personnel Profile: Full-Time Paid 3; Part-Time Volunteers 3.
Governing Authority: Unified government of Wyandotte County and Kansas City, Kansas. Tax-exempt: 501(c)(3).
Institution Type/Description: History Museum.
Collections: Emigrant Tribal material; textiles and crafts; photographs; costumes pertaining to Wyandotte County; decorative arts; manuscript collections; archives.
Research Fields: Wyandotte County & adjacent areas; genealogy.
Facilities: library; 100-seat auditorium. Museum-related items for sale.
Activities: guided tours; lectures; permanent exhibitions; fourth grade education program.
Hours & Admission Prices: Mon.-Fri. 9-4, Sat. 9-12. No charge; donations accepted. Closed holidays. &
Attendance: 300 (accurate)

Burlingame

BURLINGAME SCHUYLER MUSEUM, 117 S. Dacotah, Burlingame, KS 66413-1225. Mailing Address: P.O. Box 74, Burlingame, KS 66413-0074. Tel.: 785-654-3170.
E-mail: museum@burlingamemuseum.org
Web Site: www.burlingamemuseum.org
Founded: 2001.
Key Personnel: Chm. (V), Pres. (V) & Dir., Carolyn Strohm.
Personnel Profile: Part-Time Volunteers 75.
Volunteer Hours: 240
Operating Expenses: 17,897
Operating Income: 18
Governing Authority: Parent Institution: Burlingame Historical Society. Tax-exempt.
Institution Type/Description: History Museum: former home of The Schuyler Grade School, built in 1902.
Collections: local history & culture; arrowheads; Native American artifacts; military artifacts; coal mining.
Research Fields: genealogy.
Facilities: Museum-related items for sale.
Activities: programs for school children.
Publications: quarterly newsletter.
Hours & Admission Prices: Wed., Fri. & Sun. 1-4, Sat. 10-4. No charge; donations accepted. Closed occasional holidays. &
Attendance: 1,500 (estimated)
Membership: Individual $20; Family & Business $25.

Burlington

THE COFFEY COUNTY HISTORICAL MUSEUM, (M), 1101 Neosho St., Burlington, KS 66839-1656. Tel.: 620-364-2653. Fax: 620-364-8933. Facebook: Coffey Museum.
E-mail: visit.us@coffeymuseum.org
Web Site: www.coffeymuseum.org
Founded: 1965.
Key Personnel: Dir., Shirley Gorge-Logan; Pres. Bd. (V), Carol Sunseri; Vice Pres. Bd., Retha Sleezer; Treas. (V), Brenda Grace Klubeck; Bd. Member, Eileen Coker; Bd. Member, Victor Edelman; Bd. Member, Maureen Eggleston; Bd. Member, Tom Allen; Administrative Asst., Erin Burdick.
Personnel Profile: Full-Time Paid 2; Part-Time Volunteers 30.
Volunteer Hours: 1,655
Governing Authority: private; nonprofit organization. Tax-exempt: 501(c)(3).
Institution Type/Description: Historical Society Museum.
Collections: Kansas history; Coffey County history & genealogy; pioneer history.
Research Fields: genealogy; personal histories.
Facilities: genealogy library.
Activities: guided tours; meeting room rental; building rental; traveling exhibitions; fundraisers. Annual Event: Kansas Days in January.

Publications: quarterly newsletter, Timelines.
Hours & Admission Prices: Summer: Mon.-Fri. 10-5, Sat.-Sun. 1-4; Winter: Mon.-Fri. 10-5. No charge; donations accepted. Closed national & state holidays.
Attendance: 3,500 (estimated)
Membership: Annual $5; Business $15; Lifetime $100.

Bushton

BUSHTON MUSEUM, 218 S. Main St., Bushton, KS 67427. Tel.: 620-562-3411.
Institution Type/Description: History Museum.
Collections: local history; period furnishings; personal artifacts; photographs; books.
Hours & Admission Prices: Call for hours.

Caney

CANEY VALLEY HISTORICAL SOCIETY, 310 W. 4th St., Caney, KS 67333. Mailing Address: P.O. Box 354, Caney, KS 67333-0354. Tel.: 620-879-5131. Fax: 620-879-5131. Facebook; Caney Valley Historical Society.
E-mail: cvhistsoc@terraworld.net
Founded: 1984.
Congressional District: 4
Key Personnel: Pres., Nancy Roe.
Governing Authority: Tax-exempt: 501(c)(3).
Institution Type/Description: General Museum.
Collections: education; medical; post office; military; municipal; religious; industry; organizations; domestic; early businesses; meeting room with kitchen & genealogy room; photographs; Osage Indian artifacts; 28 ft. wall mural; farming tools.
Research Fields: early community events history; genealogy.
Facilities: Museum-related items for sale.
Activities: temporary exhibitions; children's programs; banquet. Museum Sponsors: Christmas Program.
Publications: quarterly newsletter.
Hours & Admission Prices: Mon.-Fri. 9-12 & 12:30-3. No charge; donations accepted. Closed New Year's Eve & Day; Memorial Day; Independence Day; Christmas Eve & Day. &
Attendance: 550 (estimated)
Membership: Individual $10.

SAFARI ZOOLOGICAL PARK, CR 1425, Caney, KS 67333. Mailing Address: 1751a CR 1425, Caney, KS 67333. Tel.: 620-879-2885.
E-mail: safaripark@terraworld.net
Web Site: www.safaripark.org
Key Personnel: Dir., Tom Harvey; Chm. (V), Randy Smith
Institution Type/Description: Zoo.
Collections: over 100 species of animals including tigers, lions, wolves, foxes, tropical birds, bears, & monkeys.
Activities: guided tours.
Hours & Admission Prices: Summer: daily call for hours. Adults $10, children 2-12 and seniors 60 & over $8.

Chanute

MARTIN AND OSA JOHNSON SAFARI MUSEUM, 111 N. Lincoln Ave., Chanute, KS 66720-1819. Tel.: 620-431-2730. Fax: 620-431-2730.
E-mail: osajohns@safarimuseum.com
Web Site: www.martinandosa.com
Founded: 1961.
Congressional District: 5
Key Personnel: Pres. Bd. Trustees, Tim Cunningham; Dir., Conrad G. Froehlich; Cur., Jacqueline L. Borgeson; Store & Office Mgr., Shirley Rogers-Naff.
Personnel Profile: Full-Time Paid 3; Part-Time Volunteers 40; Interns 1.
Governing Authority: private. Tax-exempt: 501(c)(3).
Institution Type/Description: Biographical Museum: located in historic Santa Fe Train Depot.
Collections: Johnson Collection: films, photographs, manuscripts & artifacts of Martin & Osa Johnson; ethnological collections from West Africa, East Africa, Borneo & South Pacific; natural history library; natural history art collection, originals & hand colored lithographs.
Research Fields: ethnology; anthropology; natural history; exploration; African art; photographic & film resources of E. Africa, South Pacific, Borneo 1917-1936.

Facilities: natural history & exploration library available on premises; archives. Museum-related items for sale.
Activities: guided tours; film showings; education programs. Annual Events: Film Festival in April.
Publications: I Married Adventure; Four Years in Paradise; quarterly newsletter, Martin & Osa Johnson Safari Museum(R) Wait-A-Bit News.
Hours & Admission Prices: Mon.-Sat. 10-5, Sun. 1-5. Adults $6, senior citizens & students over 12 $4, children 6-12 $3; discounts to prearranged bus tours, KMA, ICOM, AASLH, AAM, MPMA and AAA members; children under 6 accompanied by adult no charge. Closed New Year's Day; Easter; Independence Day; Thanksgiving; Christmas. &
Attendance: 6,000 (accurate)
Membership: Single $35; Family $55; Wanderer $70; Adventurer & Business $100; Explorer $250 & up.

Chapman

KANSAS AUTO RACING MUSEUM, 1205 Manor Rd., Chapman, KS 67431. Mailing Address: P.O. Box 549, Chapman, KS 67431. Tel.: 785-922-6644. Fax: 785-922-6684.
E-mail: karm@eaglecom.net
Web Site: kansasautoracingmuseum.com
Founded: 1998.
Congressional District: 2
Key Personnel: Dir., Doug Thompson.
Governing Authority: Tax-exempt.
Institution Type/Description: Auto Racing Museum.
Collections: auto racing history & cars; first NASCAR trophy; photographs; personal artifacts; first NHRA trophy.
Activities: Grassroots Racing TV Show.
Publications: annual, Grassroots Racing Show.
Hours & Admission Prices: Mon.-Sat. 9-5, Sun. by appointment. Adults $5. Closed New Year's Day; Easter; Thanksgiving; Christmas. &
Attendance: 4,000 (accurate)

Cheney

EAGLE VALLEY RAPTOR CENTER, 927 N. 343rd St., W., Cheney, KS 67025. Tel.: 316-393-0710.
E-mail: raptorcare@aol.com
Web Site: www.eaglevalleyraptorcenter.org
Key Personnel: Dir. Programs, Ken Lockwood
Institution Type/Description: Wildlife Center.
Collections: Harris Hawk; Bald & Golden Eagles; Great Horned Owl; Peregrine Falcon; Barn Owl; Barred Owl; Red & Gray Screech Owls; American Kestrel; Eurasian Eagle Owl; Turkey Vulture.
Activities: educational programs.
Hours & Admission Prices: Call for hours. Donation: $5 per person.

Cherryvale

CHERRYVALE MUSEUM, 215 E. 4th St., Cherryvale, KS 67335-2102. Mailing Address: 420 N. Depot St., Cherryvale, KS 67335. Tel.: 620-336-3350.
Institution Type/Description: History Museum.
Collections: local history; period furnishings; personal artifacts; early glass & china; Bender memorabilia; spinning wheel; clothing; photographs.
Hours & Admission Prices: April-Oct. Sun. 2-4; other times by appointment. No charge.

Chetopa

CHETOPA HISTORICAL MUSEUM, 406 Locust, Chetopa, KS 67336. Mailing Address: P.O. Box 648, Chetopa, KS 67336-0648. Tel.: 620-236-7121.
Founded: 1881.
Congressional District: 5
Key Personnel: Dir., Cur. & Museum Shop Mgr., Fannie Bassett.
Governing Authority: society. Affiliated with the Labette County Historical Society, Parsons, 67357. Tax-exempt: 501(c)(3).
Institution Type/Description: History Museum.
Collections: local history; local papers on microfilm from 1869-2007.
Hours & Admission Prices: April-Oct. Tues.-Wed. & Fri. 1-4. No charge; donations accepted. &
Attendance: 450 (estimated)
Membership: Individual $2.

Clay Center

CLAY COUNTY HISTORICAL MUSEUM, 2121 7th St., Clay Center, KS 67432-1509. Tel.: 785-632-3786.
E-mail: ccmuseum@twinvalley.net
Institution Type/Description: History Museum.
Collections: Clay County history; paintings; period furnishings; military artifacts; musical instruments; minerals; gems; clothing; toys.
Facilities: library.
Hours & Admission Prices: Tues.-Sat. 10-5; other times by appointment.

Clearwater

CLEARWATER HISTORICAL SOCIETY, 149 N. 4th, Clearwater, KS 67026-9764. Tel.: 620-584-2444.
E-mail: museum@sktc.net
Web Site: www.clearwaterhistoricalsociety.com
Institution Type/Description: History Museum.
Collections: local history & culture; period furnishings; personal artifacts; photographs.
Hours & Admission Prices: Sun. 1-4. &

Clifton

CLIFTON COMMUNITY HISTORICAL SOCIETY, 105 Clifton St., Clifton, KS 66937-9780. Mailing Address: P.O. Box 5, Clifton, KS 66937-0005.
Founded: 1976.
Key Personnel: Pres. (V), Mary Veesart; Treas., Erma Bouley.
Governing Authority: private; nonprofit organization. Tax-exempt.
Institution Type/Description: General Museum: housed in Old Missouri Pacific Depot & Caboose.
Collections: period artifacts from Clifton & surrounding areas including military & scout clothing; 19th century photos; country school records; china; tools; newspapers; film of Clifton, Kansas newspapers from start to June 24, 1993.
Research Fields: genealogy from old newspapers.
Facilities: 10-vol. family genealogy library available to the public; school corner. Museum-related items for sale.
Activities: guided tours. Museum Sponsors: Christmas Party-meal.
Publications: quarterly newsletter, Courier.
Hours & Admission Prices: By appointment only. No charge; donations accepted. Closed New Year's Day; Easter; Independence Day; Thanksgiving; Christmas. &
Attendance: 62 (estimated)
Membership: Family & Business $5; Life $100.

Coffeyville

BROWN MANSION, 2019 S. Walnut, Coffeyville, KS 67337-6819. Mailing Address: P.O. Box 843, Coffeyville, KS 67337-0843. Tel.: 620-251-0431. Fax: 620-251-5448.
Web Site: www.brownmansion.com
Founded: 1904.
Congressional District: 4
Key Personnel: Pres., Kris Crane; Vice Pres., Darla Thornburg; Mgr. & Cur., Rob Burrows.
Personnel Profile: Full-Time Volunteers 10.
Governing Authority: nonprofit society. Parent Institution: Coffeyville Historical Society. Tax-exempt.
Institution Type/Description: Historic House.
Collections: period furniture, carpets, china & silver; leaded Tiffany glass accents; Tiffany chandelier.
Facilities: Museum-related items for sale.
Activities: guided tours.
Hours & Admission Prices: Tours: Mon., Thurs.-Fri. 10, 12, 2, Sat.-Sun. 11, 1, 3, 5. Adults $6.50, children 7-17 $4.50; children 6 & under no charge. Closed Easter; Thanksgiving; Christmas; during special events.
Attendance: 7,500 (estimated)

COFFEYVILLE AVIATION HERITAGE MUSEUM, 2002 N. Buckeye St., Coffeyville, KS 67337. Mailing Address: P.O. Box 774, Coffeyville, KS 67337-0774. Tel.: 620-515-0232.
Institution Type/Description: Aviation History Museum.
Collections: aviation history; aircraft; photographs; personal artifacts.
Hours & Admission Prices: Sat. 10-4, Sun. 1-4; other times by appointment. No charge; donations accepted.
Membership: Annual $25.

DALTON DEFENDERS MUSEUM, 113 E. 8th, Coffeyville, KS 67337-5803. Mailing Address: P.O. Box 843, Coffeyville, KS 67337-0843. Tel.: 620-251-5944.
E-mail: chamber@coffeyville.com
Web Site: www.daltondefendersmuseum.com
Founded: 1954.
Congressional District: 4
Key Personnel: Pres., Kris Crane; Vice Pres., Darla Thornburg; Museum Mgr., John Alvey; Museum Mgr., Wendy Alvey.
Personnel Profile: Full-Time Volunteers 15.
Governing Authority: nonprofit. Parent Institution: Coffeyville Historical Society. Tax-exempt.
Institution Type/Description: History Museum.
Collections: mementos from the Dalton Raid and from Dalton Gang members; Wendell Willkie mementos; Walter Johnson mementos; Native American artifacts; artifacts & photographs from the early history of Coffeyville.
Facilities: Museum-related items for sale.
Activities: guided tours.
Hours & Admission Prices: Daily 10-4; other times by appointment. Adults $6.50, children $4.50; discounts to AAA members; children 7 & under no charge. Closed Easter; Thanksgiving; Christmas.
Attendance: 9,000 (estimated)

Colby

THE PRAIRIE MUSEUM OF ART & HISTORY, (M), 1905 S. Franklin, Colby, KS 67701-3710. Tel.: 785-460-4590. Fax: 785-460-4592.
E-mail: prairiem@st.tel.net
Web Site: www.prairiemuseum.org
Founded: 1959.
Congressional District: 1
Key Personnel: Dir., Sue Ellen Taylor.
Personnel Profile: Full-Time Paid 4; Part-Time Paid 4; Part-Time Volunteers 20.
Governing Authority: society; nonprofit organization. Thomas County Historical Society. Tax-exempt: 501(c)(3).
Institution Type/Description: Historical Society Museum & Archives.
Collections: articles relating to the area's history from the period of the homesteaders who lived in sod homes to World War II; the Kuska Collection, containing 2,000 dolls; signed Tiffany & Sevres, Capo De Monte, Royal Vienna, Satsuma, Ridgway, Wedgwood, Limoge & Meissen; glass including cut, Redford, Steigel, Stuben, Galle & Cameo; furniture; textiles; silver; books; memorabilia.
Research Fields: pertaining to the collection.
Facilities: library of books, manuscripts, documents & photography pertaining to Thomas County; reading room. Books, arts & crafts of the area & museum-related items for sale.
Activities: guided tours; lectures; films; gallery talks; formally organized education programs; docent program; loan, permanent & temporary exhibitions; hobby workshops. Museum Sponsors: Sam White Day; folk artists; Horse Powered Harvest.
Publications: quarterly newsletter, Prairie Winds; books, Land of the Windmills; Golden Jubilee; Golden Heritage of Thomas County Kansas; A History of Thomas County, Kansas, 1885-1964.
Hours & Admission Prices: April-Oct. Mon.-Fri. 9-5, Sat.-Sun. 1-5; Nov.-March Tues.-Fri. 9-5, Sat.-Sun. 1-5. Adults $8, senior citizens $6, children 6-16 $2; members & Sun. no charge. Closed New Year's Day; Easter; Thanksgiving; Christmas. ♿
Attendance: 15,000 (accurate)
Membership: Individual $15; Family $25; Lifetime $500.

Coldwater

COMANCHE COUNTY HISTORICAL MUSEUM, 105 W. Main St., Coldwater, KS 67029. Tel.: 620-582-2108.
Institution Type/Description: History Museum.
Collections: local history & culture; period furnishings; personal artifacts; military artifacts; Native American artifacts; Western memorabilia; photographs.
Hours & Admission Prices: Call for hours.

Concordia

BROWN GRAND THEATRE, 310 W. 6th St., Concordia, KS 66901. Mailing Address: P.O. Box 347, Concordia, KS 66901. Tel.: 785-243-2553.
Institution Type/Description: Historic Building: built in 1907. Listed on the National Register of Historic Places.
Collections: theatre history; period furnishings; paintings; photographs; posters.
Hours & Admission Prices: Memorial Day to Labor Day Tues.-Fri. 10-12 & 1-4, Sat. 10-2; Sept.-May Tues.-Fri. 10-12 & 1-4. Self-Guided Tour: $2. Guided Tours: adults $5, children under 12 $3; discounts to groups of 10 or more.

CLOUD COUNTY HISTORICAL SOCIETY MUSEUM, 635 Broadway, Concordia, KS 66901-2914. Tel.: 785-243-2866.
E-mail: museum@cloudcountyks.org
Web Site: www.cloudcountyks.org
Founded: 1959.
Congressional District: 1
Key Personnel: Dir., Cynthia Reimann; Pres., Dana Brewer; Sec., Aline Luecke; Treas., Betty Losh.
Personnel Profile: Part-Time Paid 1; Part-Time Volunteers 14; Interns 1.
Governing Authority: county. Parent Institution: Cloud County Historical Society. Museum Annex & Wall Mural, 6th St. & Hwy. 81, Concordia, KS. Tax-exempt.
Institution Type/Description: Historical Society Museum: housed in 1908 Andrew Carnegie Library.
Collections: prisoner of war camp, military casket flags; 1793 letter from Martha Washington to niece, Fanny Bassett; clothing; dolls; medical, architectural, royal families, dental & school textbooks; Indian artifacts; household goods; farm tools; books; pictures & photographs; early undertaking equipment; long distance telephone switchboard; PBX cordboard; telephone memorabilia; barbed wire collection; absorption type wavemeter; 1936 amateur radio WGWXY; 1928 Lincoln-Page biplane with 90 HP Curtis engine; horse drawn farm machinery including 1898 horseless carriage; 1903 Holsman car; printing presses; linotype; railroad nails; candy making equipment; period toys; toothpick holders; gem & mineral collection; Concordia Daily Kansas 1900-1918; 1870-2005 microfilm of Concordia Blade-Empire; newspaper microfilm: Jamestown 1881-1985, Miltonvale 1882-2001, Clyde 1878-2003, Glasco 1883-2001; Concordia Kansas 1918-1983. Museum Annex: wall mural including 6,400 carved bricks.
Research Fields: Cloud County history; genealogy.
Activities: guided tours of county interest areas; quarterly meetings. Annual Events: Open House; dinner meeting with elected officers in October.
Publications: quarterly newsletter, Cloud Comments.
Hours & Admission Prices: Tues.-Fri. 1-5, Sat. 10-5; call to confirm; other times by appointment. No charge; donations requested. ♿
Attendance: 6,000 (accurate)
Membership: Annual $5; Life $15-$100.

NATIONAL ORPHAN TRAIN MUSEUM, (M), 300 Washington St., Concordia, KS 66901. Mailing Address: P.O. Box 322, Concordia, KS 66901. Tel.: 785-243-4471. Facebook: Orphan Train Depot.
E-mail: orphantraindepot@gmail.com
Web Site: www.orphantraindepot.com
Formerly: Orphan Train Heritage Society of America
Founded: 1987.
Congressional District: 1
Key Personnel: Cur., Amanda Wahlmeier.
Personnel Profile: Full-Time Paid 1; Part-Time Volunteers 2.
Governing Authority: Tax-exempt.
Institution Type/Description: History Museum.
Collections: orphan train history; photographs; personal artifacts.
Research Fields: orphan train riders.
Facilities: research center.
Activities: educational programs.
Publications: quarterly newsletter.
Hours & Admission Prices: Tues.-Fri. 10-12 & 1-4, Sat. 10-4. Adults $5, children $3; discounts to groups of 10 or more; members no charge. ♿
Attendance: 2,200 (estimated)
Membership: Annual $25; Lifetime $250.

Cottonwood Falls

CHASE COUNTY HISTORICAL SOCIETY, INC., 301 Broadway, Cottonwood Falls, KS 66845. Mailing Address: Box 375, Cottonwood Falls, KS 66845-0375. Tel.: 620-273-8500. Fax: 620-273-8500 (call first).
E-mail: cscohist@sbcglobal.net
Web Site: chasecountyhistoricalmuseum.com
Founded: 1934.
Key Personnel: Pres. (V), Dan Riggs; Dir. & Cur., Geneva Lawrence; Financial Dir. & Museum Shop Mgr., Jacqualine Couch.
Personnel Profile: Part-Time Paid 3; Part-Time Volunteers 3.

Governing Authority: county government; nonprofit. Tax-exempt.
Institution Type/Description: Historical Museum.
Collections: old tools; household items; obituaries; marriage licenses, cemetery records; Knute Rockne crash artifacts; pioneer school room; early dentist office; miniature mule team sculpture; outdoor covered cooks wagon; original county jail.
Major Exhibits: Victorian Women's Dress, 1/14-2/14.
Research Fields: genealogy; local history.
Facilities: 16,000 sq. ft. exhibit space.
Activities: guided & self-guided tours; special receptions.
Publications: Chase County Historical Sketches, Vols. 1, 2, 3, 4; Roots & Patchwork.
Hours & Admission Prices: Tues.-Sat. 12-4. Suggested Donations: $2. Closed Memorial Day; Independence Day; day before Thanksgiving & Thanksgiving; Christmas Eve through New Year's Day.
Attendance: 2,000 (estimated)
Membership: Student $10; Chase County $25; Non-Chase County & Organization $35; Life 500.

FLINT HILLS GALLERY, 321 Broadway St., Cottonwood Falls, KS 66845-2884. Tel.: 620-273-6454. Fax: 620-273-8235.
Institution Type/Description: Art Gallery.
Collections: paintings; sculpture; cowboy art; western furnishings.
Hours & Admission Prices: Mon.-Sat. 10-3; other times by appointment.

RONIGER MEMORIAL MUSEUM, 315 Union St., Cottonwood Falls, KS 66845. Mailing Address: P.O. Box 70, Cottonwood Falls, KS 66845-0070. Tel.: 620-273-6310 & 6412. Fax: 620-273-6335.
E-mail: deroy10@sbcglobal.net
Founded: 1959.
Key Personnel: Pres. & Cur., David E. Croy; Sec. & Treas., John Roniger.
Personnel Profile: Full-Time Paid 1.
Governing Authority: county. Tax-exempt.
Institution Type/Description: History Museum: located on lawn of 100-year old courthouse.
Collections: Indian artifacts; stuffed native animals; local historical mementoes.
Hours & Admission Prices: Tues.-Wed. & Fri.-Sun. 1-5; other times by appointment. No charge; donations accepted. &
Attendance: 2,000

Council Grove

KAW MISSION STATE HISTORIC SITE, 500 N. Mission, Council Grove, KS 66846-1433. Tel.: 620-767-5410. Fax: 620-767-5816.
E-mail: kawmission@kshs.org
Web Site: www.kawmission.org
Founded: 1951.
Congressional District: 5
Key Personnel: Historic Site Admin., Mary Honeyman; Pres. (V), Jeremiah Hershberger.
Personnel Profile: Full-Time Paid 1; Part-Time Paid 1; Part-Time Volunteers 35.
Governing Authority: state. Parent Institution: Kansas State Historical Society, 6425 S.W. Sixth St., Topeka, KS 66615. Tel. 913-296-3251. Tax-exempt: 501(c)(3).
Institution Type/Description: Historical Society Museum: 1850-1851, 2-story building.
Collections: 19th century America; Plains Indians.
Research Fields: History of Kaw or Kansas Native American Indian Tribe, Santa Fe Trail & Council Grove.
Facilities: reconstructed Indian hut; meeting room; education center.
Activities: guided tours; permanent exhibitions. Museum Sponsors: Annual Wah-Shun-Gah days activities in June; Kaw Nation Inter-tribal Pow-wow; Kaw Councils, education program series.
Publications: quarterly newsletter, Toh-Po-Ska.
Hours & Admission Prices: Wed.-Sat. 9-5. Adults $2, students $1; children under 5 & members no charge. Closed state holidays. &
Attendance: 7,500 (accurate)
Membership: Individual $10; Business & Family $30; Benefactor $50; Sustaining $100.

Cunningham

CUNNINGHAM HISTORICAL MUSEUM, 100 N. Main, Cunningham, KS 67035.
Key Personnel: Chm. (V), Bill Parker; Pres. (V), Mike McGovney; Museum Shop Mgr., Donna Glenn.

Governing Authority: Tax-exempt.
Institution Type/Description: History Museum.
Collections: local history & culture; period furnishings; personal artifacts; photographs; railroad memorabilia. Historic Building: early jail.
Hours & Admission Prices: Fri. 9:30-4:30. No charge; donations accepted.
Attendance: 500 (estimated)
Membership: Individual $5; Lifetime $50.

Delphos

DELPHOS MUSEUM, 101 N. Washington St., Delphos, KS 67436. Mailing Address: P.O. Box 338, Delphos, KS 67436. Tel.: 785-523-4540.
Congressional District: 1
Key Personnel: Dir., Mary Ballou; Pres. (V), C.J. Ballou.
Governing Authority: Tax-exempt.
Institution Type/Description: History Museum: housed in the former Brock Auto repair building.
Collections: local history & culture; period furnishings; personal artifacts; photographs; barbed wire.
Hours & Admission Prices: Mon. & Fri. 9-12, Tues. 12-4, Sat. 9-12 & 1-4; other times by appointment. No charge; donations accepted.

Derby

DERBY HISTORICAL SOCIETY, 208 N. Westview, Derby, KS 67037. Mailing Address: P.O. Box 1054, Derby, KS 67037-1054. Tel.: 316-788-7740.
Founded: 1995.
Key Personnel: Pres. (V), Jason Crippen; Treas., Darrell Butterfield.
Personnel Profile: Part-Time Volunteers 30; Interns 3.
Governing Authority: Tax-exempt.
Institution Type/Description: History Museum.
Collections: local history & culture; period furnishings; personal artifacts; photographs.
Activities: special events.
Hours & Admission Prices: mid-April to Oct. Sat. 10-2. No charge; donations accepted.
Attendance: 1,500 (estimated)

Dighton

LANE COUNTY HISTORICAL MUSEUM, 333 N. Main St., Dighton, KS 67839. Mailing Address: P.O. Box 821, Dighton, KS 67839-0821. Tel.: 620-397-5652. Fax: 620-397-5652. Facebook: Lane County Historical Society.
E-mail: lchm@st-tel.net
Web Site: www.lanecountymuseum.com
Founded: 1976.
Congressional District: 1
Key Personnel: Pres., Joel Herndon; Chm., Donna Morgan; Dir., Curator, Museum Shop Mgr., Sonya Reed.
Personnel Profile: Full-Time Volunteers 12; Part-Time Paid 2; Part-Time Volunteers 8.
Governing Authority: society; nonprofit. Parent Institution: Lane County Historical Society. Tax-exempt.
Institution Type/Description: History Museum.
Collections: early Lane County photos; articles used in Lane County; manuscript collections. Historic House: sod house.
Research Fields: local history.
Facilities: 50-vol. library of books pertinent to Lane County available for use within the library; reading room; meeting room. Handcrafted items & books for sale.
Activities: guided tours; lectures; films; gallery talks; arts festivals; hobby workshops; permanent, temporary & traveling exhibitions; school loan service.
Publications: quarterly newsletter, Friends of Museum.
Hours & Admission Prices: Memorial Day to Labor Day Tues.-Sat. 1-5, Sun. 2-5; Sept.-May Tues.-Sat. 1-5. No charge; donations accepted. Closed legal holidays. &
Attendance: 1,200 (estimated)
Membership: Individual & Contributing $10; Business $15; Lifetime $100.

Dodge City

BOOT HILL MUSEUM, INC., (M), Front St., Dodge City, KS 67801. Tel.: 620-227-8188. Fax: 620-227-7673.
E-mail: info@boothill.org
Web Site: www.boothill.org

Founded: 1947.
Congressional District: 1
Key Personnel: Gen. Mgr., Lara Brehm; Chm. Bd., Kerri Baker; Cur. Exhibits & Interpretation, Karen Pankratz; Cur. Collections & Education, Kathie Bell.
Personnel Profile: Full-Time Paid 11; Part-Time Paid 105; Part-Time Volunteers 100.
Governing Authority: nonprofit. Subsidiary Institution: Boot Hill Museum Store. Tax-exempt: 501(c)(3).
Institution Type/Description: Western History Museum: located on Boot Hill.
Collections: 19th-century business & home furnishings, equipment; tools; agricultural equipment; weapons; photographs. Historic Buildings: 1878 cattleman's home; 1880 carriage shop; 1865 Fort Dodge jail; 1917 school; 1931 depot.
Research Fields: 19th-century business & social life in southwest Kansas; decorative arts; historic preservation; history of Dodge City.
Facilities: 1,200-vol. library of historical books & documents on Dodge City available for use on premises; theater. Museum-related items for sale.
Activities: permanent & temporary exhibits; school loan service; craft demonstrations; inter-museum loans; formally organized education programs for children; docent program; Long Branch Saloon Revue & medicine shows.
Publications: brochure; exhibit catalog; gallery guides; book, Dodge City: Up Through A Century in Story & Pictures.
Hours & Admission Prices: Winter: Mon.-Sat. 9-5, Sun. 1-5; Memorial Day to Labor Day daily 8-8. Summer: adults $10; children 4 & under no charge. Off-Season: adults $9; children 4 & under and members no charge. Closed New Year's Day; Thanksgiving; Christmas. &

Attendance: 81,311 (accurate)
Membership: Senior 62 & over $40; Single $45; Senior Family $60; Family $65; Marshal (Business) $250.

HOME OF STONE (THE MUELLER-SCHMIDT HOUSE 1881), A FORD COUNTY MUSEUM, CURATED BY FORD COUNTY HISTORIC SOCIETY, (M), Ave. A & Vine St., Dodge City, KS 67801. Mailing Address: P.O. Box 131, Dodge City, KS 67801-0131. Tel.: 620-227-6791.
E-mail: glaughead@sbcglobal.net
Web Site: www.skyways.org/orgs/fordco
Founded: 1965.
Congressional District: 116
Key Personnel: Pres. (V), George Laughead, Jr.; Financial Dir., Sonja Hughes; Archivist, Ann Warner; Museum Shop Mgr., Janice Klein.
Personnel Profile: Full-Time Volunteers 2; Part-Time Paid 5; Part-Time Volunteers 10.
Governing Authority: private; nonprofit organization. Parent Institution: Ford County Historical Society. Tax-exempt: 501(c)(3).
Institution Type/Description: Historic House.
Collections: original kerosene lamps, clocks; Pioneer Mother Room; photographs.
Facilities: 500-vol. library; 1,200 sq. ft. exhibit space. Museum-related items for sale.
Activities: guided tours; temporary exhibitions. Annual Events: Victorian Christmas Tea; Spring Open House.
Publications: monthly newsletter, Ford County Historical Society, Inc.
Hours & Admission Prices: Memorial Day to Labor Day Mon.-Sat. 9-5, Sun. 2-4. Adults $3.
Attendance: 4,000 (accurate)
Membership: Individual $10; Lifetime $50.

THE KANSAS TEACHERS' HALL OF FAME, 603 5th St., Dodge City, KS 67801-1674. Mailing Address: P.O. Box 1674, Dodge City, KS 67801-1674. Tel.: 620-225-7311.
E-mail: ksteachers@thof.kscoxmail.com
Web Site: www.teachershallfamedodgecityks.org
Founded: 1977.
Key Personnel: Dir., J. Dennis Doris.
Personnel Profile: Part-Time Volunteers 13.
Governing Authority: private; nonprofit organization. Tax-exempt.
Institution Type/Description: Outstanding Teachers Museum.
Collections: pictures of outstanding teachers inducted into the Hall of Fame; textbooks & other items pertinent to early day classrooms.
Facilities: library of textbooks available for use on premises. Museum-related items for sale.
Activities: Annual Event: Induction Ceremony & Banquet.
Hours & Admission Prices: April-May & Sept.-Oct. Mon.-Sat. 11-4, Sun. 1-5; Memorial Day-Labor Day Mon.-Sat. 8:30-6, Sun. 1-5. No charge; donations accepted. &
Attendance: 7,500 (estimated)

Douglass

DOUGLASS HISTORICAL MUSEUM, 318 S. Forest, Douglass, KS 67039. Mailing Address: P.O. Box 95, Douglass, KS 67039-0095. Tel.: 316-746-2319 & 747-2166.
Founded: 1950.
Congressional District: 5
Key Personnel: Dir. & Cur., Frances Renfro.
Personnel Profile: Part-Time Volunteers 6.
Governing Authority: society. Parent Institution: Historical Society. Tax-exempt.
Institution Type/Description: Pioneer Museum.
Collections: Indian artifacts; tools; costumes; kitchen display; school-room; church display; Victorian sitting room; millinery & dress shop; doctors office; photographs; cookbooks; farming tools; medical items; toys; family history files.
Research Fields: town, business & people of Douglass; local genealogy.
Facilities: library of local & pioneer history books, photographs, cemetery maps & 1884-1984 microfilm of Douglass Tribune available for use on premises; Osage hunting grounds.
Activities: guided tours; temporary exhibitions.
Publications: The Douglass Story; cookbook, Family Favorites.
Hours & Admission Prices: Mon., Wed. & Fri. 9-11 & 1-3; other times by appointment. No charge; donations accepted. Closed New Year's Eve & Day; Christmas Eve, Day & week. &
Attendance: 750 (estimated)
Membership: Patron $25; Life Time $500.

Edgerton

JOHNSON COUNTY MUSEUM, 18745 Dillie Rd., Edgerton, KS 66021. Tel.: 913-893-6645. Fax: 913-882-9730.
Institution Type/Description: History Museum: housed in a former one-room school house; built in 1869.
Collections: local history & culture; period furnishings; personal artifacts; photographs.
Hours & Admission Prices: Tues.-Sun. 1-5. No charge.

Edna

EDNA HISTORICAL MUSEUM, 100 S. Delaware, Edna, KS 67342. Mailing Address: P.O. Box 368, Edna, KS 67342-0368.
Founded: 1978.
Congressional District: 2
Key Personnel: Pres. (V), Kenneth E. Cary; Treas., Hazel Stone; Cur., Ronald Neidigh.
Personnel Profile: Part-Time Volunteers 6.
Governing Authority: private; nonprofit organization. Tax-exempt: 501(c)(3).
Institution Type/Description: History Museum: housed in the old First National Bank building.
Collections: early Edna history; newspapers on microfilm; area family histories.
Activities: guided tours.
Hours & Admission Prices: May-Oct. Fri. 1-4, Sat. 9-12; other times by appointment. No charge; donations accepted. Closed all holidays.
Attendance: 100 (estimated)
Membership: Annual $1.

El Dorado

BUTLER COUNTY HISTORY CENTER & KANSAS OIL MUSEUM, 383 E. Central Ave., El Dorado, KS 67042-2133. Tel.: 316-321-9333. Fax: 316-321-3619.
E-mail: history@kansasoilmuseum.org
Web Site: www.kansasoilmuseum.org
Founded: 1956.
Congressional District: 5
Key Personnel: Chm. (V), Diana Edmiston; Exec. Dir., Mindy Tallent; Museum Educator, Renea Albert; Museum Shop Mgr., Marsha Dekker.
Personnel Profile: Full-Time Paid 2; Part-Time Paid 1; Part-Time Volunteers 50.
Governing Authority: nonprofit organization. Parent Institution: Butler County Historical Society. Subsidiary Institution: Kansas Oil Museum. Tax-exempt: 501(c)(3).
Institution Type/Description: History Museum.
Collections: Butler County & Kansas oil history; objects & documents related to the history of the Kansas oil industry; oil field equipment & restored buildings; objects related to county development by ranching & farming; Flint Hills flora & fauna. Historic Structures: 1858 Conner Log Cabin; 1930s oil field lease house; oil tank car; 1930s cable tool drilling rig; 1950

rotary drilling rig; late 1920s reconstructed grocery store, doctor's office & print shop; 1917 lease house; 1890s one-room schoolhouse.
Research Fields: petroleum & local history.
Facilities: 4,500-vol. library pertaining to petroleum, technology & local history available for use on premises only; nature conservation center; reading room. Museum-related items for sale.
Activities: guided tours; lectures; films; permanent, temporary & traveling exhibitions; educational programs; special events.
Publications: newsletter, The Crownblock; books related to county history; Illustrated Directory of Kansas Oilmen; Stone Arch Bridges of Southern Butler County; video, Oil in Kansas.
Hours & Admission Prices: May-Sept. Mon.-Sat. 10-4; Oct.-April Tues.-Fri. 10-4, Sat. 12-5. Adults $4; members no charge. Closed major holidays. &
Attendance: 9,500 (accurate)
Membership: Individual $25; Family $35; Benefactor $75; Business $120; Corporate $300.

COUTTS MUSEUM OF ART, (M), 110 N. Main St., El Dorado, KS 67042-2016. Mailing Address: P.O. Box 1, El Dorado, KS 67042. Tel.: 316-321-1212. Fax: 316-321-1215. Facebook: Coutts Museum.
E-mail: rseel@couttsmuseum.org
Web Site: couttsmuseum.org
Founded: 1970.
Congressional District: 4
Key Personnel: Pres. Bd., Jeremy Sundgren; Exec. Dir, Rod Seel; Vice Pres., Jackie Clark.
Personnel Profile: Full-Time Paid 2; Part-Time Paid 1; Part-Time Volunteers 35.
Governing Authority: nonprofit organization. Tax-exempt: 501(c)(3).
Institution Type/Description: Fine Art Museum: housed in 1917 bank building.
Collections: paintings; sculptures; items of historical interest.
Research Fields: historical sites.
Activities: talks; tours; organizational meetings; art shows, concerts & promotions.
Publications: quarterly newsletter, The Exhibitionist.
Hours & Admission Prices: Tues.-Fri. 9-5, Sat. 12-4; groups of 12 or more by appointment. No charge; donations accepted. Closed federal & state holidays. &
Attendance: 7,800 (accurate)
Membership: Single $30; Couple & Patron $55; Corporate & Supporter $100; Underwriter $250; Benefactor $500 & up.

THE ERMAN B. WHITE GALLERY, 901 S. Haverhill Rd., El Dorado, KS 67042. Tel.: 316-321-2222; 800-794-0188.
Institution Type/Description: Art Gallery.
Collections: works by regional & national artists; paintings; prints; drawings; sculptures; ceramics; photographs.
Activities: special events.
Hours & Admission Prices: Sept.-May call for hours.

Elkhart

MORTON COUNTY HISTORICAL SOCIETY MUSEUM, (M), 370 E. Hwy. U.S. 56, Elkhart, KS 67950. Mailing Address: P.O. Box 1248, Elkhart, KS 67950-1248. Tel.: 620-697-2833. Fax: 620-697-4390.
E-mail: mtcomuseum@elkhart.com
Web Site: www.mtcoks.com/museum
Founded: 1987.
Institution Type/Description: Historical Society Museum.
Collections: local history & culture; period furnishings; personal artifacts; photographs; paintings; pioneer & Native American artifacts.
Hours & Admission Prices: Tues.-Fri. 1-5; other times by appointment. No charge; donations accepted.

Ellis

ELLIS RAILROAD MUSEUM, 911 Washington, Ellis, KS 67637. Mailing Address: P.O. Box 82, Ellis, KS 67637. Tel.: 785-726-4493. Fax: 785-726-3294.
E-mail: allaboard@ellisrailroadmuseum.com
Web Site: www.ellisrailroadmuseum.com
Institution Type/Description: Railroad Museum.
Collections: railroad artifacts; over 5,000 sq. ft. working model train layout; Union Pacific caboose; over 1,200 dolls.
Activities: miniature train rides.
Hours & Admission Prices: Museum: April-Oct. Tues.-Sat. 10-4; Nov.-March Tues.-Sat. 11-4. Adults 12 & over $2, children 5-12 $1; discounts to groups of 8 or more; children under 5 no charge. Miniature Train Rides: Memorial Day to Labor Day Mon.-Sat. 11am, 1pm, 3pm, & 5pm, Sun. 1pm, 3pm, 5pm. Adults 12 & over $2, children 5-12 $1; children under 5 no charge.

WALTER P. CHRYSLER BOYHOOD HOME AND MUSEUM, 102 W. 10th, Ellis, KS 67637. Tel.: 785-726-3636. Fax: 785-726-3653.
E-mail: chrysler55@eaglecom.net
Web Site: www.chryslerboyhoodhome.com
Founded: 1954.
Congressional District: 1
Key Personnel: Dir., Dena Patee; Pres. (V), Michael Downing; Treas., David McDaniel.
Personnel Profile: Part-Time Paid 3.
Governing Authority: private; nonprofit organization. Tax-exempt: 501(c)(3).
Institution Type/Description: History Museum.
Collections: personal artifacts; jewelry; books; photographs.
Major Exhibits: Chrysler 700c 3/8 Scale Model, 10/12-12/14.
Facilities: Museum-related items for sale.
Activities: guided tours.
Hours & Admission Prices: Memorial Day to Labor Day Tues.-Sat. 9-4, Sun. 1-4; Labor Day to Memorial Day Tues.-Sat. 11-3, Sun. 1-4. Adults $4, seniors $3, children 6-11 $2; children under 5 no charge. Closed New Year's Day; Easter; Thanksgiving; Christmas.
Attendance: 1,200 (estimated)
Membership: Individual $25; Family $35.

Ellsworth

HODGDEN HOUSE MUSEUM COMPLEX, 104 W. South Main, Ellsworth, KS 67439-3232. Mailing Address: P.O. Box 144, Ellsworth, KS 67439-0144. Tel.: 785-472-3059.
E-mail: echs@eaglecom.net
Formerly: Ellsworth County Museum Complex
Founded: 1961.
Congressional District: 1
Key Personnel: C.E.O., Phyllis Dolenzal.
Personnel Profile: Full-Time Paid 1; Part-Time Paid 2; Part-Time Volunteers 6.
Governing Authority: society. Parent Institution: Ellsworth County Historical Society. Tax-exempt.
Institution Type/Description: Museum Complex.
Collections: folklore; Indian; agriculture; archaeology; costumes. Historic Buildings: 1876 Hodgden house; livery stable; country schoolhouse; church; Union Pacific caboose; log cabin; Ft. Harker Guardhouse and officer's house; 2 train depots; jail.
Facilities: 200-vol. library of local newspapers, file books & books used by pioneers available for use on microfilm. Books & pamphlets for sale.
Activities: Annual Event: Cowtown Days in September.
Publications: Sharing History; brochure
Hours & Admission Prices: Tues.-Sat. 9-5, Sun. 1-5. Adults $3 (includes entrance to Fort Harker Museum Complex). Closed New Year's Day; Easter; Thanksgiving; Christmas. &
Attendance: 3,000 (estimated)
Membership: Individual $10; Family $20; Business $25; Corporate $100.

ROGERS HOUSE MUSEUM GALLERY, 102 E. Main St., Ellsworth, KS 67439. Tel.: 785-472-5674.
Founded: 1968.
Congressional District: 1
Key Personnel: Dir., Robert Rogers.
Governing Authority: individual operation.
Institution Type/Description: Art Museum: housed in 1870, American House, cowboy hotel.
Collections: original paintings; prints; art.
Facilities: original paintings & fine art prints for sale.
Activities: permanent & temporary exhibitions.
Publications: books, The Great West; Quill of the Kansan; Country Neighbor; Art Observations.
Hours & Admission Prices: Open by appointment; call for information. No charge; donations accepted. &

Emporia

DAVID TRAYLOR ZOO OF EMPORIA, 75 Soden Rd., Emporia, KS 66801-8702. Mailing Address: P.O. Box 928, Emporia, KS 66801-0928. Tel.: 620-341-4365. Fax: 620-341-4367.
E-mail: emporiazoo@emporia-kansas.gov
Web Site: www.emporiazoo.org
Founded: 1934.

Congressional District: 5
Key Personnel: Dir., Lisa Keith; Pres., Mike Alpers; Museum Shop Mgr., Lori Heavener.
Personnel Profile: Full-Time Paid 4; Part-Time Paid 4; Interns 2.
Governing Authority: municipal. Parent Institution: City of Emporia. Tax-exempt.
Institution Type/Description: Zoo: located in Soden's Grove Park, one of Emporia's earliest parks.
Collections: native & exotic animals; Eurasian Black Vultures; mountain lion; eagles; bobcat; lemurs; tamarins; waterfowl; birds of prey; birds; badger; bison; prairie dogs; llamas; mule deer; cranes; reptiles; amphibians & arthropods.
Research Fields: waterfowl propagation; species survival plans; turtles.
Facilities: zoological park.
Activities: guided tours & animal presentations for groups & organizations by appointment.
Publications: newsletter, Keeping In Touch.
Hours & Admission Prices: Winter: daily 10-4:30; Summer: Mon.-Tues. & Thurs.-Sat. 10-4:30, Wed. & Sun. 10-8. No charge; donations accepted. Closed Thanksgiving, Christmas, New Year's Day. &
Attendance: 81,552 (accurate)
Membership: Emporia Friends of the Zoo: Individual $20; Family $30; Sponsor $50; Patron $100; Benefactor $500; Endowment $1,000.

EMPORIA STATE UNIVERSITY - JOHNSTON GEOLOGY MUSEUM, 1200 Commercial St., Emporia, KS 66801-5087. Mailing Address: ESU Cram Science Hall, 14th and Merchant St., Emporia, KS 66801. Tel.: 620-341-5330. Fax: 620-341-6055.
Web Site: www.emporia.edu/earthsci/museum/museum.htm
Founded: 1983.
Key Personnel: Dir., Michael Morales.
Governing Authority: public university; nonprofit.
Institution Type/Description: Geology Museum.
Collections: Kansas vertebrate & invertebrate fossils; materials collected from the Hamilton Quarry, which includes fossil flora, insects, fish & amphibians from one layer of Pennsylvanian strata.
Facilities: 1,800 sq. ft. exhibit space.
Hours & Admission Prices: Academic Year: Mon.-Fri. 8am-10pm, Sat. 8am to noon. No charge; donations accepted. Closed holidays.
Attendance: 750 (estimated)

LYON COUNTY HISTORICAL SOCIETY AND MUSEUM, (M), 118 E. 6th Ave., Emporia, KS 66801-3922. Tel.: 620-340-6310.
E-mail: lycomu@osprey.net
Founded: 1937.
Congressional District: 5
Key Personnel: Exec. Dir., J. Greg Jordon; Pres. (V), Lisa Goldstein; Chm. (V), Annette Rice; Education Coord., Laura Dodge; Registrar, Jake Dalton; Asst. Registrar, Clerk & Museum Store Mgr., Carolyn Eckstrom.
Personnel Profile: Full-Time Paid 1; Part-Time Paid 4; Part-Time Volunteers 30; Interns 1.
Governing Authority: society. Parent Institution: Lyon County Historical Society. Subsidiary Institution: Lyon County Historical Museum. Tax-exempt: 501(c)(3).
Institution Type/Description: County General History Museum: housed in 1904 Carnegie Library.
Collections: costumes; local, county archives & artifacts; State archives & documents; Gilson scrapbook collection; manuscripts; family histories; photographs.
Research Fields: Lyon County history; genealogical research.
Facilities: library & archives available for use on premises; research room. Kansas artists' work, Kansas made & museum-related articles for sale.
Activities: lectures; tours; multi-media & slide & VCR presentations; rotating exhibits; participatory education programs; research service.
Publications: Lyon County Lines; Lyon County Historical Society Happenings.
Hours & Admission Prices: June-Aug. Tues.-Fri. 10-5, Sat. 1-5; Sept.-May Tues.-Sat. 1-5; other times by appointment. No charge; donations accepted. Closed major holidays. &
Attendance: 10,000 (estimated)
Membership: General $15; Contributing $25; Sustaining $50; Benefactor $100.

THE NATIONAL TEACHERS HALL OF FAME, (M), 1200 Commercial, #4017, Emporia, KS 66801. Tel.: 620-341-5660; 800-96-TEACH. Fax: 620-341-5912.
E-mail: hallfame@emporia.edu
Web Site: www.nthf.org

Founded: 1989.
Congressional District: 2
Key Personnel: Dir., Dr. J. Phillip Bennett; Chm. (V), Lindy Whetzel.
Personnel Profile: Full-Time Paid 1; Part-Time Volunteers 1.
Governing Authority: Tax-exempt.
Institution Type/Description: Hall of Fame & Teaching History Museum.
Collections: Hall of Fame Inductees; early days of teaching; school desks; class attendance records; teacher contracts; period textbooks; one-room school.
Facilities: Museum-related items for sale.
Activities: Annual Event: Induction Ceremony.
Hours & Admission Prices: Call for hours. No charge; donations accepted. &
Attendance: 300 (estimated)

NORMAN R. EPPINK ART GALLERY, EMPORIA STATE UNIVERSITY, 1200 Commercial, Emporia, KS 66801-5057. Tel.: 620-341-5246. Fax: 316-341-6246.
E-mail: reichenb@emporia.edu
Web Site: www.emporia.edu/m/www/art/eppink.htm
Founded: 1939.
Congressional District: 5
Key Personnel: C.E.O., Michael Shomrock; Dir., Roberta Eichenberg.
Personnel Profile: Full-Time Paid 1; Part-Time Paid 4.
Governing Authority: state. Parent Institution: Emporia State University. Tax-exempt: 170(b)(1)(A).
Institution Type/Description: University Art Gallery.
Collections: contemporary drawings & prints; artifacts; paintings; sculpture.
Facilities: library; nature center; field research station; reading room; 400-seat auditorium; theater; classrooms; cafeteria.
Activities: guided tours; lectures; gallery talks; TV & radio programs; formally organized education programs for undergraduate & graduate students affiliated with Emporia State University; loan, temporary & traveling exhibitions.
Publications: annual, National Invitational Drawing Exhibition Catalog; annual, Art Faculty Exhibition catalog.
Hours & Admission Prices: Mon.-Fri. 9-4. No charge; donations accepted. Closed university holidays. &
Attendance: 11,000 (estimated)

RED ROCKS HISTORIC SITE, 927 Exchange St., Emporia, KS 66801-3040. Tel.: 620-342-2800. Fax: 620-342-2800.
E-mail: redrocks@kshs.org
Web Site: kshs.org/p/william-allen-whitehouse/11953
Formerly: William Allen White State Historic Site
Founded: 2003.
Congressional District: 1
Key Personnel: Site Admin., Jennifer Baldwin; Chm. (V), Dr. Steve Haught.
Personnel Profile: Part-Time Paid 1; Part-Time Volunteers 15; Interns 1.
Governing Authority: Parent Institution: Kansas Historical Society.
Institution Type/Description: Historic House: housed in the former home of William Allen White, nationally known newspaperman & author.
Collections: family history; personal artifacts; period furnishings; photographs.
Research Fields: White family history; Emporia history.
Hours & Admission Prices: April-Oct. Thurs.-Sat. 10-5, Sun. 1-5. Adults $5, college students & children 6-18 $3; discounts to groups; members, active military & children 5 and under no charge.
Attendance: 2,000 (estimated)

RICHARD H. SCHMIDT MUSEUM OF NATURAL HISTORY, 1200 Commercial St., Emporia State Univ., Emporia, KS 66801-5057. Mailing Address: Dept. of Biological Sciences, Box 4050, Emporia State Univ., Emporia, KS 66801. Tel.: 620-341-5311. Fax: 620-341-5607.
E-mail: wjensen1@emporia.edu
Web Site: www.emporia.edu/smnh/
Founded: 1959.
Congressional District: 5
Key Personnel: Dir., Dr. William Jensen.
Governing Authority: state. Affiliated with Emporia State University. Tax-exempt: 501(a).
Institution Type/Description: Natural History Museum.
Collections: skins & mounted specimens of birds, mammals; ornithology; mammalogy; ichthyology; herpetology.
Research Fields: ornithology; mammalogy.
Facilities: nature center; field research station.
Activities: guided tours; lectures; formally organized education programs for children & undergraduate college students; permanent, temporary & traveling exhibitions.

Publications: Kansas School Naturalist.
Hours & Admission Prices: Mon.-Fri. 8-5. No charge. Closed school holidays.
&

TOAD HOLLOW DAYLILY FARM, 1534 Rd. 170, Emporia, KS 66801-8125. Tel.: 620-343-8655.
Founded: 1996.
Institution Type/Description: Botanical Garden.
Collections: plants; flowers; paintings; statues; photography.
Activities: Annual Events: Iris Tour in May; Daylily Tour in June.
Hours & Admission Prices: May-July Tues.-Sun. 10-7; other times by appointment. No charge; donations accepted.
Attendance: 550 (estimated)

Erie

MEM-ERIE HISTORICAL MUSEUM, 225 S. Main St., Erie, KS 66733-1334. Mailing Address: 18505 140th Rd., Erie, KS 66733-4206. Tel.: 620-244-5452.
E-mail: 20farmcats@sbcglobal.net
Founded: 1994.
Key Personnel: Pres. (V), Claudia Obenhaus; Sec., Jeanette Harris; Treas., Pat Richey.
Personnel Profile: Part-Time Volunteers 30.
Governing Authority: private; not-for-profit organization. Tax-exempt.
Institution Type/Description: Historical Society Museum: site of first Masonic Hall in 1860s.
Collections: furnishings, photographs, personal artifacts & documents dating from founding of Erie in 1860s; recreational artifacts; books on Erie family history; men's & women's military suits; World War I & II GAR roster; Santa Fe & Kay railroad artifacts.
Research Fields: houses; jail; businesses; genealogy.
Facilities: library; educational facilities for grade school children. Pencils & books for sale.
Activities: guided tours; educational programs for grade school children; study clubs; temporary exhibits. Annual Events: Old Soldiers & Sailors Reunion; East Hill Cemetery Tour in October.
Publications: newspaper; newsletter.
Hours & Admission Prices: May-Sept. Wed. & Fri. 103; Oct.-April by appointment. No charge; donations accepted. &
Attendance: 684 (accurate)
Membership: Single $10; Family $25.

Eureka

GREENWOOD COUNTY HISTORICAL SOCIETY & MU-SEUM, 120 W. 4th St., Eureka, KS 67045-1445. Mailing Address: P.O. Box 86, Eureka, KS 67045-0086. Tel.: 620-583-6682.
E-mail: gwhistory@sbcglobal.net
Founded: 1973.
Congressional District: 4
Key Personnel: Pres., Mike Pitko; Vice Pres., Mike French; Sec., Hazel Russell; Treas., Sue Williams; Historian, Dru Land.
Personnel Profile: Full-Time Paid 1; Part-Time Volunteers 6.
Governing Authority: nonprofit organization. Tax-exempt.
Institution Type/Description: Historical Society Museum.
Collections: dolls; tools; pictures; 1868-2007, newspapers; furniture; dishes; bottles; genealogy.
Research Fields: genealogy; town & county history.
Facilities: 450-vol. library; 3,100 sq. ft. exhibit space.
Activities: guided tours. Museum Sponsors: Open House.
Publications: semi-annual newsletter.
Hours & Admission Prices: Mon.-Fri. 10-12 & 1-4, Sat. by appointment only. No charge; donations accepted. Closed national holidays. &
Attendance: 1,100 (estimated)
Membership: Single $10; Couple & Corporate $15; Sustaining $25; Life $200.

Florence

HARVEY HOUSE MUSEUM, 221 Marion, Florence, KS 66851 1263. Mailing Address: P.O. Box 143, Florence, KS 66851-0143. Tel.: 620-878-4296.
Web Site: www.florenceks.com
Founded: 1971.
Congressional District: 5
Key Personnel: Pres., Judy Mills; Vice Pres., Edna Robinson; Treas., Phoebe Janzen; Sec., Marjorie Jackson; Trustee, Bob Harris.
Personnel Profile: Part-Time Volunteers 10.
Governing Authority: nonprofit organization. Parent Institution: Florence

Historical Society. Subsidiary Institution: City of Florence, KS. Tax-exempt: 501(c)(3).
Institution Type/Description: Historic Building: 1878 first Fred Harvey Restaurant-Hotel.
Collections: period furniture; pictures; dishes; tools; Santa Fe Way Car (caboose); baggage wagon.
Research Fields: Santa Fe railroad; Harvey Houses; local history; Fred Harvey.
Activities: guided tours; permanent exhibitions; limited number of dinners served annually.
Publications: booklets, Harvey House; Century of Pride; newsletter, Florence Historical Society News; book, Florence Historical Society Cookbook.
Hours & Admission Prices: By appointment. No charge; donations accepted. Closed New Year's Day; Easter; Thanksgiving; Christmas. &
Attendance: 700 (estimated)
Membership: Member $10; Life $100.

Fort Dodge

FORT DODGE MUSEUM - KANSAS SOLDIERS HOME, 714 Sheridan - Unit 128, Fort Dodge, KS 67843. Tel.: 316-227-2121.
Institution Type/Description: History Museum.
Collections: local history; photographs; personal artifacts.
Hours & Admission Prices: Museum: daily 1-4. Grounds: daily dawn to dusk.

Fort Leavenworth

FORT LEAVENWORTH HISTORICAL SOCIETY, Gift Shop-Post Museum, 100 Reynolds Ave., Fort Leavenworth, KS 66027. Mailing Address: P.O. Box 3356, Fort Leavenworth, KS 66027-0356. Tel.: 913-651-7440. Facebook: Fort Leavenworth Historical Society.
E-mail: flhsgs@kc.rr.com
Web Site: www.ftlvnhistsoc.org
Founded: 1950.
Congressional District: 2
Personnel Profile: Full-Time Paid 3; Part-Time Paid 2.
Governing Authority: society; nonprofit organization. Parent Institution: United States Army, Ft. Leavenworth. Tax-exempt: 501(c)(3).
Institution Type/Description: Historical Society.
Collections: local history; period artifacts; photographs.
Research Fields: local history.
Facilities: Museum-related items for sale.
Activities: guided tours; programs.
Publications: monographs on the history of Fort Leavenworth.
Hours & Admission Prices: Mon.-Fri. 9-4, Sat. 11-4. No charge; donations accepted. &
Attendance: 50,000 (accurate)
Membership: Family & Individual $5 annually.

* **FRONTIER ARMY MUSEUM,** 100 Reynolds Ave., Fort Leavenworth, KS 66027-2334. Tel.: 913-684-3767. Fax: 913-684-3192.
E-mail: leav-fam@conus.army.mil
Web Site: usacac.army.mil/CAC2/CSI/FrontierArmyMuseum.asp
Founded: 1938.
Key Personnel: Dir., Stephen J. Allie; Exhibit Mgr., George Moore; Collection Mgr., Russ Ronspies; Museum Shop Mgr., Lois Kaftner.
Personnel Profile: Full-Time Paid 4; Part-Time Volunteers 200.
Governing Authority: federal. Administered by United States Army. Parent Institution: Center of Military History. Tax-exempt: 170(b)(1)(A).
Institution Type/Description: Military Museum: located at Fort Leavenworth.
Collections: evolution of military technology; 19th-20th century Fort Leavenworth; military uniforms & equipment 1804-present; military horse-drawn vehicles; evolution of military education.
Research Fields: history of Fort Leavenworth & the U.S. Army in the development of the trans-Mississippi West.
Facilities: 1500-vol. library of books & other research material related to Museum mission available for use on premises only. Museum-related items for sale.
Activities: guided tours; traveling & temporary exhibits; living history programs; multi-media history presentations.
Publications: pamphlets: Frontier Army Museum; Self-Guided Tours of Fort Leavenworth.
Hours & Admission Prices: Mon.-Fri. 9-4, Sat. 10-4. No charge; donations accepted. Closed federal holidays. &
Attendance: 25,000 (estimated)

Fort Riley

FIRST TERRITORIAL CAPITOL OF KANSAS, Bldg. 693, Huebner Rd., K-18, Fort Riley, KS 66442. Mailing Address: Box 2122, Fort Riley, KS 66442-0122. Tel.: 785-784-5535 & 238-1666.
Web Site: www.kshs.org
Founded: 1928.
Congressional District: 2
Key Personnel: Cur., Ron Tedder.
Personnel Profile: Part-Time Paid 1; Part-Time Volunteers 20.
Governing Authority: state. Parent Institution: Kansas History Center, 6425 S.W. 6th Ave., Topeka, KS 66615-1099. Tax-exempt.
Institution Type/Description: Historic House Museum: 1855 two-story stone structure that served as the first territorial capitol of Kansas.
Collections: local history items; Civil War cavalry accoutrements; weapons; Indian artifacts; photographs; early Pawnee city & pre-civil war artifacts; 1858 Bogus Law Book.
Research Fields: local & political history.
Facilities: nature trails. Museum-related items for sale.
Activities: lectures; films; slide programs; nature trail activities; special school student history days. Special Events: living history festival weekend; Buffalo Soldiers event.
Publications: The Five Day Capitol.
Hours & Admission Prices: March-Oct. Fri.-Sun. 1-5; Winter: call for hours. No charge; donations accepted. Closed major holidays. &
Membership: Student & Senior $10; Individual $12; Family $20; Business & Organization $50; Sustainer $100; Benefactor $500.

U.S. CAVALRY MUSEUM, Bldg. 205, Fort Riley, KS 66442. Mailing Address: Bldg. 500 Huebner Rd., Fort Riley, KS 66442. Tel.: 785-239-2737. Fax: 785-239-6243.
Founded: 1957.
Congressional District: 2
Key Personnel: C.E.O., Robert J. Smith, Ph.D.; Exhibit Specialist, Ron Doyle.
Personnel Profile: Full-Time Paid 4; Part-Time Volunteers 33.
Governing Authority: federal. Parent Institution: U.S. Army. Tax-exempt.
Institution Type/Description: Military Museum: housed in 1855 building used as a hospital, 1855-1890 & as post headquarters 1890-1948.
Collections: 1776-1950 historical artifacts of U.S. Cavalry.
Research Fields: 1776-1950 U.S. Cavalry.
Facilities: 4,500-vol. library of books on the Cavalry & the U.S. Army available for use on premises; reading room.
Activities: guided tours; lectures.
Publications: quarterly newsletter, Bugle Notes.
Hours & Admission Prices: Mon.-Sat. 9-4:30, Sun. 12-4:30. No charge; donations accepted. Closed New Year's Day; Easter; Thanksgiving; Christmas. &
Attendance: 49,176 (accurate)

Fort Scott

FORT SCOTT NATIONAL HISTORIC SITE, Old Fort Blvd., Fort Scott, KS 66701. Mailing Address: P.O. Box 918, Fort Scott, KS 66701-0918. Tel.: 620-223-0310. Fax: 620-223-0188.
E-mail: fosc_superintendent@nps.gov
Web Site: www.nps.gov/fosc
Founded: 1978.
Congressional District: 2
Key Personnel: Supt., Betty Boyko; Chief Interpretation & Resource Management, Kelley Collins; Program Coord., Galen Ewing; Cooperating Assoc. Coord., Barak Geertsen.
Personnel Profile: Full-Time Paid 13; Part-Time Paid 22; Part-Time Volunteers 360; Interns 4.
Governing Authority: federal. National Park Svc. Tax-exempt.
Institution Type/Description: Historic Site: 1842 restored & reconstructed Fort Scott.
Collections: Historic Buildings: 1842-53 Officers Quarters; Post Hospital; Barracks; Post Headquarters; Stable; Guardhouse; Magazine; Bakery; Quartermaster Storehouse; Well Canopy.
Research Fields: Fort Scott as a frontier military post; Bleeding Kansas; Civil War; Kansas railroad years 1865-1873.
Facilities: library. Books, pamphlets & postcards for sale.
Activities: self-guided tours; living history demonstrations; special events; weekend conducted activities during the summer.
Publications: booklet, Fort Scott on the Indian Frontier.
Hours & Admission Prices: April-Nov. daily 8-5; Dec.-March daily 9-5. No charge. Closed New Year's Day; Thanksgiving; Christmas. &
Attendance: 26,650 (accurate)

Fredonia

STONE HOUSE GALLERY, 320 N. 7th St., Fredonia, KS 66736-1337. Mailing Address: P.O. Box 355, Fredonia, KS 66736-0355. Tel.: 620-378-2052.
E-mail: stonehouse320@embarqmail.com
Founded: 1967.
Congressional District: 2
Key Personnel: Pres. (V), Linda Sewell; Chm. (V), Denise Guthrie; Dir., Admin. & Museum Shop Mgr., Victoria Black Starr.
Personnel Profile: Part-Time Paid 1; Part-Time Volunteers 18.
Governing Authority: society; nonprofit organization. Parent Institution: Fredonia Arts Council, Inc.
Institution Type/Description: Art Museum: housed in 1872 Stone House.
Collections: visual art collection.
Research Fields: quarterly art council newsletter.
Facilities: Gift items for sale.
Activities: monthly art exhibits; guided tours; gallery talks; hobby workshops; formally organized education programs; docent program or council; loan, temporary & traveling exhibitions; summer art programs. Museum Sponsors: Children's Art Exhibit in April; Area High School Art Exhibit in May; Area Artist Exhibits in October; Sounds of Christmas in December.
Publications: monthly art council newsletter.
Hours & Admission Prices: Mon.-Fri. 10-2; other times by appointment. No charge; donations accepted. Closed Federal holidays. &
Attendance: 1,000 (estimated)
Membership: Friend $25; Member $50; Contributor $100; Supporter $250; Patron $500; Benefactor $1,000 & up.

WILSON COUNTY HISTORICAL SOCIETY MUSEUM, 420 N. 7th, Fredonia, KS 66736-1315. Tel.: 620-378-3965.
E-mail: wilcohisoc@twinmounds.com
Founded: 1961.
Congressional District: 2
Key Personnel: C.E.O. & Pres. (V), Emma Crites; Vice Pres., Joan Richardson; Sec., Mary Jean Browne; Finance Officer, Joe Bambick; Museum Shop Mgr., E. Nadine Dishman.
Personnel Profile: Part-Time Paid 1; Part-Time Volunteers 8.
Governing Authority: society. Tax-exempt.
Institution Type/Description: General Museum: located in old county jail.
Collections: pioneer relics related to farming, industry, schools, churches, household; quilts; American Indian artifacts; costumes; children's toys, books & clothing; photographs; archives; taped interviews; programs; glass; special war commemorative displays; manuscript collections: obituary notebooks & card file index, county cemetery indexes, birth & marriage records, family genealogies; 1930 Federal Census index.
Research Fields: archives; local history; American Indian artifacts; genealogy.
Facilities: 425-vol. library on history; Civil War records; genealogy; maps available for use on premises.
Activities: guided tours; lectures; films; permanent & temporary exhibitions.
Publications: newsletter; book, Fredonia Cemetery Index; Wilson County 1881/1890 Atlas; 1930 Federal Census Index.
Hours & Admission Prices: Mon.-Fri. 12-4:30. No charge; donations accepted. Closed national holidays.
Attendance: 550 (accurate)
Membership: Historical Society $15.

Galena

GALENA MINING & HISTORICAL MUSEUM, 319 W. Seventh St., Galena, KS 66739-1211. Mailing Address: P.O. Box 372, Galena, KS 66739-0372. Tel.: 620-783-2192. Fax: 620-783-1974.
Founded: 1984.
Congressional District: 2
Key Personnel: Pres. (V), Gene Russell; Treas., Don Noe.
Personnel Profile: Part-Time Volunteers 4.
Governing Authority: private; nonprofit organization.
Institution Type/Description: Mining Museum: housed in the old Katy Train Depot.
Collections: lead & zinc mine artifacts; tools; lamps; pictures; paintings; mineral specimens; locomotive; caboose; helicopter; tank.
Facilities: Museum-related items for sale.
Activities: guided tours.
Hours & Admission Prices: Mon.-Sat. 9-11:30 & 1-3:30; Winter: reduced hours in the afternoon. No charge; donations accepted. Closed New Year's Day; Memorial Day; Independence Day; Labor Day; Thanksgiving; Christmas.
Attendance: 1,200 (estimated)
Membership: Yearly Adult $10; Lifetime $100.

Galva

GALVA HISTORICAL MUSEUM, 204 S. Main, Galva, KS 67443. Mailing Address: P.O. Box 505, Galva, KS 67443. Tel.: 620-654-3343.
Founded: 1992.
Congressional District: 1
Key Personnel: Chm. (V), Naomi Ford.
Governing Authority: Tax-exempt.
Institution Type/Description: History Museum.
Collections: local history & culture; photographs; period furnishings; personal artifacts; school memorabilia; newspapers.
Hours & Admission Prices: 1st Sun. each month 2-4, 3rd Thurs. each month 7 pm-9 pm, 2nd Sat. each month 10 am to noon; other times by appointment. No charge; donations accepted. &

Garden City

FINNEY COUNTY KANSAS HISTORICAL MUSEUM, 403 S. 4th, Garden City, KS 67846. Mailing Address: P.O. Box 796, Garden City, KS 67846-0796. Tel.: 620-272-3664. Fax: 620-272-3662. Facebook: Finney County Museum.
E-mail: fico.historical@gcnet.com
Web Site: www.finneycounty.org/history.asp
Founded: 1948.
Congressional District: 1
Key Personnel: Dir. & C.E.O., Steve Quakenbush; Pres., Barbara Goss; Asst. Dir., Laurie Oshel; Mgr. Collections, Yadira Hernandez; Mgr. Collections, Todd Roberts; Education & Museum Shop Mgr., Johnetta Hebrlee.
Personnel Profile: Full-Time Paid 4; Part-Time Paid 8; Part-Time Volunteers 4.
Governing Authority: society. Parent Institution: Finney County Historical Society, Inc. (Kansas). Tax-exempt: 501(c)(3).
Institution Type/Description: History Museum.
Collections: newspaper clippings; farm & agricultural equipment; pioneer pictures; photographs of Finney County; 19th- & 20th-century Western Kansas artifacts; textile & clothing; Finney County cattle industry history; repository for information on local participation in Ford Foundation study on new immigration; Western Kansas fossils & stone implements; social & cultural history; pioneer artifacts. Historic Building: 1884-era folk Victorian house; late 19th century one-room schoolhouse.
Major Exhibits: Garden City Then & Now, 11/13-6/14.
Research Fields: pioneer history; sugar beet & cattle industry of Finney County; immigration.
Facilities: 1,000-vol. library of the history of Southwest Kansas.
Activities: guided tours; permanent & temporary exhibitions; children's history workshops; oral history projects; media presentations; film series; exhibit programs.
Publications: quarterly bulletin, The Sequoyan; Finney Co. History Vol. 1 & 2. Conquest of Southwest Kansas; book, Constant Frontier, The Continuing History of Finney County, Kansas; Those Who Served...Finney County Veterans; Buffalo Jones: Citizen of the Kansas Frontier.
Hours & Admission Prices: Winter: daily 1-5; Summer: Mon.-Sat. 10-5, Sun. 1-5. No charge; donations accepted. &
Attendance: 24,000 (accurate)
Membership: Individual $15; Family $25; Business $50.

LEE RICHARDSON ZOO IN FINNUP PARK, 312 E. Finnup Dr., Garden City, KS 67846-6561. Tel.: 620-276-1250. Fax: 316-276-1259.
E-mail: zoo.department@gardencityks.us
Web Site: www.leerichardsonzoo.org
Founded: 1927.
Congressional District: 1
Key Personnel: Dir., Kathy Sexson; Society Pres., Matt Fields; Deputy Dir., Kristi Newland.
Personnel Profile: Full-Time Paid 25; Part-Time Paid 8; Part-Time Volunteers 33; Interns 1.
Governing Authority: municipal. Parent Institution: City of Garden City. Subsidiary Institution: Friends of the Lee Richardson Zoo. Tax-exempt.
Institution Type/Description: Zoo, Arboretum & Nature Center.
Collections: 300 animals; birds; reptiles; aviary.
Facilities: 1,200-vol. library of zoology, conservation & horticulture books; 125-seat auditorium; botanical garden; snack bar; classrooms; conservation center. Museum-related items for sale.
Activities: guided tours; lectures; loan, permanent, traveling & participatory exhibitions; docent program; formal education programs; mobile vans; school loan service; broadcast programs; summer zoo camp; weekly children's story hour. Annual Events: Earth Day Fair; Blues at the Zoo (Blues concert); Boo at the Zoo; A Wild Affair; Tumbleweed Festival.

Publications: quarterly newsletter, Zoo Gnus.
Hours & Admission Prices: April-Sept. 2 daily 8-6:30; Sept. 3-March daily 8-4:30. Vehicles: March-Nov. $10; members, daily 8am-10am & Dec.-Feb. no charge. Pedestrians: no charge. AZA reciprocal. Closed New Year's Day; Thanksgiving; Christmas. &
Attendance: 183,000 (estimated)
Membership: Individual $30; Family & Grandparent $50; Patron $130; Sustaining $240.

Gardner

GARDNER HISTORICAL MUSEUM, 204 W. Main St., Gardner, KS 66030. Mailing Address: P.O. Box 442, Gardner, KS 66030. Tel.: 913-856-4447.
E-mail: gardnerhistoricalmuseum@centurylink.net
Web Site: www.gardnerhistoricalmuseum.com
Governing Authority: nonprofit organization. Subsidiary Institution: Bray House, 207 W. Shawnee, Gardner, KS 66030. Tax-exempt: 501(c)(3).
Institution Type/Description: History Museum: housed in the Herman B. Foster house; built in 1893. Listed on the National Register of Historic Places.
Collections: local history, heritage & culture; period furnishings; personal artifacts; photographs.
Hours & Admission Prices: Gardner: Fri. 4-7, Sat.-Tues. 1-4; tours by appointment. Bray House: Tues.-Wed. & Sat. 1-4. No charge; donations accepted. &
Membership: Single 65 & over $30; Single under 65 $35; Family $40.

Garnett

ANDERSON COUNTY HISTORICAL MUSEUM, 406 W. 4th Ave., Garnett, KS 66032. Mailing Address: P.O. Box 183, Garnett, KS 66032. Tel.: 785-867-2966 & 448-5740.
Founded: 1968.
Congressional District: 5
Key Personnel: Chm. & Pres. (V), Dorothy L. Lickteig.
Personnel Profile: Full-Time Volunteers 7; Part-Time Volunteers 10.
Governing Authority: society; county. Tax-exempt.
Institution Type/Description: Local History Museum: 1886 home & carriage house of Dr. Harris; Longfellow school building.
Collections: guns; history of early settlers; Indian artifacts; furniture; church & school items; farm tools; genealogy records; early diggings; local histories; family stories & books; clothing; military artifacts; country store.
Research Fields: pertaining to collections; genealogy; church, marriage, birth & census records; town & it's citizens.
Facilities: 500-vol. library.
Activities: guided tours; formally organized education programs for adults; permanent & temporary exhibitions.
Publications: quarterly newsletter, Regional Museum News.
Hours & Admission Prices: Oct.-May Tues.-Sat. 1-4. No charge; donations accepted. &
Attendance: 1,800 (accurate)
Membership: Individual $5; Lifetime $50.

THE WALKER ART COLLECTION OF THE GARNETT PUBLIC LIBRARY, 125 W. 4th Ave., Garnett, KS 66032-1313. Mailing Address: Library, 125 W. 5th Ave., Garnett, KS 66032. Tel.: 785-448-5496. Fax: 913-448-3936 & 5555.
E-mail: joyce@garnettks.net
Web Site: www.garnettks.net
Founded: 1965.
Congressional District: 5
Key Personnel: Chm. (V), Barbara Foltz; Dir., Robert Cugno.
Personnel Profile: Full-Time Paid 1; Part-Time Paid 6; Part-Time Volunteers 5.
Governing Authority: municipal. Parent Institution: City of Garnett. Tax-exempt: 501(c)(3).
Institution Type/Description: Art Gallery.
Collections: 19th-20th century oil paintings & works on paper; regional artists.
Facilities: library.
Activities: guided tours; lectures; docent program; participatory & loan exhibitions; school loan service; slide & audio tour of the collection.
Hours & Admission Prices: Mon.-Tues. & Thurs. 10-8, Wed. & Fri. 10-5:30, Sat. 10-4. No charge. Closed New Year's Day; Presidents' Day; Memorial Day; Independence Day; Labor Day; Veterans Day; Thanksgiving; Christmas. &
Attendance: 10,000 (estimated)

Girard

HISTORICAL MUSEUM OF CRAWFORD COUNTY, Summit St. and Buffalo St., Girard, KS 66743-1543. Mailing Address: P.O. Box 132, Girard, KS 66743. Tel.: 620-724-4570.
Founded: 1975.
Institution Type/Description: History Museum: housed in the former St. John's Episcopal Church. Listed on the National Historic Register.
Collections: local history & culture; period clothing; personal artifacts; photographs.
Hours & Admission Prices: Temporarily closed.

Glasco

OSBORNE COUNTY HISTORICAL MUSEUM, 929 N. 2nd St., Glasco, KS 67445. Mailing Address: P.O. Box 572, Glasco, KS 67445-0572. Tel.: 785-346-2881 & 2798.
Institution Type/Description: History Museum.
Collections: county history & culture; photographs.
Hours & Admission Prices: Memorial Day to Labor Day Mon.-Thurs. 2-4; other times by appointment.

Goddard

LAKE AFTON PUBLIC OBSERVATORY, 25,000 W. 39th St., S., (Mac Arthur Rd.), Goddard, KS 67052. Mailing Address: 1845 Fairmont, Wichita, KS 67260-0032. Tel.: 316-978-3191. Fax: 316-978-3350.
E-mail: observatory@wichita.edu
Web Site: webs.wichita.edu/lapo
Founded: 1980.
Congressional District: 4
Key Personnel: Dir., Greg Novacek; Program Mgr., Robert Henry.
Personnel Profile: Part-Time Paid 4; Part-Time Volunteers 10.
Governing Authority: nonprofit organization. Tax-exempt: 501(c)(3).
Institution Type/Description: Astronomy Museum.
Collections: telescopes; astronomy exhibits of celestial objects & concepts.
Research Fields: photometry of variable stars.
Facilities: observation room; outdoor observing pad; classrooms.
Activities: guided tours; computer games; lectures; organized educational programs for children, undergraduate & graduate college students affiliated with Wichita State University; participatory exhibits; school loan service.
Publications: annual brochures; public & school programs.
Hours & Admission Prices: March-Sept. Fri.-Sat. call for hours; Oct.-Feb. Fri.-Sat. 7:30 pm-10 pm. Adults 13 & over $5, children 6-12 $3; children under 6 & members no charge. Closed New Year's Eve & Day; Christmas Eve, Day & week. &
Attendance: 4,000 (estimated)
Membership: Student & Senior $15; Individual $20; Family $30.

TANGANYIKA WILDLIFE PARK, 1000 S. Hawkins Lane, Goddard, KS 67052. Tel.: 316-794-8954. Fax: 316-794-2153.
E-mail: twp@twpark.com
Web Site: www.twpark.com
Key Personnel: Dir., Jim Fouts; Asst. Dir., Matt Fouts
Institution Type/Description: Zoo.
Collections: wildlife & their habitats; lions; camels; lemurs; penguins; zebra; rhinoceros; kangaroos; pandas; giraffe; leopards; tigers; monkeys; hands-on exhibits.
Hours & Admission Prices: April & Oct. Fri.-Sun. 10:30-5; May-Sept. daily 9-5. Adults $13.95, seniors 60-89 $10.95, children 3-12 $8.95; seniors 90 & over and children under 3 no charge. &

Goessel

MENNONITE HERITAGE & AGRICULTURAL MUSEUM, 200 N. Poplar St., Goessel, KS 67053. Mailing Address: P.O. Box 231, Goessel, KS 67053-0231. Tel.: 620-367-8200.
E-mail: mhmuseum@mtelco.net
Web Site: skyways.lib.ks.us/museums/goessel
Formerly: Mennonite Heritage Museum
Founded: 1974.
Congressional District: 5
Key Personnel: Pres. & Chm. (V), Steven Banman; Dir., Cur. & Museum Shop Mgr., Marjorie J. Shoemaker; Treas., Aileen Esau.
Personnel Profile: Part-Time Paid 3; Part-Time Volunteers 25.
Governing Authority: nonprofit; board of directors. Affiliated with the Mennonite Immigrant Historical Foundation. Tax-exempt.
Institution Type/Description: History Museum.

Collections: late 1800s & early 1900s Kansas agriculture; Mennonite immigrants; manuscripts; clothing; household goods; books; farm machinery including threshing machines of various periods; machinery related to wheat industry; birds in taxidermy display. Historic Buildings: 1906 The Preparatory School; 1910-1935 The Goessel State Bank; 1911 Friesen House; 1875 South Bloomfield School (one-room); 1875 Krause House; 1902 Schroeder Barn; 1874 Immigrant House replica.
Major Exhibits: Folk Art Collection, Spring 2014.
Research Fields: history of the German Russian Mennonites.
Facilities: Books & gift items for sale.
Activities: guided tours; lectures; country threshing days; permanent & temporary exhibitions. Annual Event: Annual Antique and Classic Car Show in June.
Publications: newsletter; Church Book of the Alexanderwohl Mennonite Church; Church Records of Old Flemish or Groningen Mennonisten Societaet in West Prussia; In Earlier Days: A History of Goessel, Kansas; From Pluma Moos to Pie Cookbook; They Sought a New Land.
Hours & Admission Prices: March-April & Oct.-Nov. Tues.-Sat. 12-4; May-Sept. Tues.-Sat. 10-5; groups of 10 or more by appointment. Adults $4, children 7-12 $2; discounts to AAM members; members no charge. Closed major holidays. &
Attendance: 1,100 (estimated)
Membership: Individual $15; Couple $30; Family $40; Life $500.

Goodland

HIGH PLAINS MUSEUM, 1717 Cherry, Goodland, KS 67735-3200. Tel.: 785-890-4595. Fax: 785-890-4532.
E-mail: museumsir@goodlandks.us
Web Site: www.highplainsmuseum.org
Founded: 1959.
Congressional District: 1
Key Personnel: Dir., Karen Anderson.
Personnel Profile: Full-Time Paid 1; Part-Time Paid 3.
Governing Authority: municipal. Parent Institution: City of Goodland. Tax-exempt: 501(c)(3).
Institution Type/Description: History Museum.
Collections: Plains Indians artifacts; pioneer life; Rock Island History in Sherman County; Sherman County History; farming tools; 1902 rope-driven automobile; first patented helicopter in America; six local history dioramas.
Research Fields: aviation history in Northwest Kansas; rainmaking in the Great Plains; Rock Island railroad.
Facilities: 3,000 sq. ft. exhibit space. Local handcrafts & books for sale.
Activities: guided tours; lectures; participatory & temporary exhibitions.
Publications: brochures.
Hours & Admission Prices: June-Aug. Sun. 1-5, Mon. & Wed.-Sat. 9-5, Sept.-May Mon. & Wed.-Sat. 9-5. No charge; donations accepted. Closed major holidays. &
Attendance: 3,000 (accurate)

SHERMAN COUNTY HISTORICAL SOCIETY - THE ENNIS-HANDY HOUSE, 202 W. 13th St., Goodland, KS 67735-2806. Tel.: 785-899-6773.
Founded: 1975.
Congressional District: 3
Personnel Profile: Part-Time Volunteers 5.
Governing Authority: Tax-exempt.
Institution Type/Description: Historic House Museum: built in 1907.
Collections: local history & culture; period furnishings; personal artifacts; photographs. Historic Buildings: 1907 Victorian home; one room rural school house.
Facilities: memorial gazebo.
Activities: tours; ice cream socials; garage sales.
Publications: books, They Came to Stay Vol. I-IV; quarterly newsletter; miscellaneous publications.
Hours & Admission Prices: Wed.-Mon. 1-5. Suggested Donations: adults $5, senior citizens $4, children 3-11 $3; discounts to school groups. Closed Thanksgiving; Christmas.
Attendance: 350 (accurate)
Membership: Individual $15; Business $25; Lifetime $250.

Great Bend

BARTON COUNTY HISTORICAL MUSEUM & VILLAGE, 85 S. Hwy. 281, Great Bend, KS 67530. Mailing Address: P.O. Box 1091, Great Bend, KS 67530-1091. Tel.: 620-793-5125. Fax: 620-793-5125 (call first).
E-mail: bchsdirector@gmail.com
Web Site: www.bartoncountymuseum.org

Formerly: Barton County Historical Society Village & Museum
Founded: 1963.
Congressional District: 1
Key Personnel: C.E.O. & Chm., Beverly Komarek; Pres. (V), Dotty Keenan; Registrar, Frances Wasson.
Personnel Profile: Part-Time Paid 3; Part-Time Volunteers 35.
Governing Authority: society; nonprofit organization. Tax-exempt.
Institution Type/Description: Historic Village.
Collections: dolls; wedding dresses; photographs; Civil War & other military items; farm implements; period artifacts; general store; barber & beauty shop; telephone switchboard; quilt shop; blacksmith shop; Santa Fe National Historic Trail. Historic Buildings: 1875 stone house; 1898 church; c.1900 schoolhouse; c.1900 post office; 1910 railway depot; furnished 1950 Lustron House.
Research Fields: county history.
Facilities: 400-vol. library available to public; 1,600-vol. research library; classroom; 15,000 sq. ft. exhibit space; research area; conservation center; interpretive facility. Museum-related items for sale.
Activities: guided tours; lectures; organized education programs; docent program; participatory & temporary exhibitions.
Publications: quarterly newsletter, The Village Crier.
Hours & Admission Prices: Tues.-Fri. 10-5, Sat.-Sun. 1-5. Adults $4; members no charge. Closed most major holidays. &
Attendance: 5,000 (estimated)
Membership: Annual $20; Family $30; Contributor $100-$499; Benefactor $500; Life $1,000.

GREAT BEND-BRIT SPAUGH ZOO & GREAT BEND RAP-TOR CENTER, 2123 N. Main St., Great Bend, KS 67530. Mailing Address: P.O. Box 215, Great Bend, KS 67530. Tel.: 620-793-4226. Fax: 620-791-5001.
E-mail: scott@greatbendzoo.com
Web Site: greatbendzoo.com
Founded: 1952.
Key Personnel: Dir., Scott Gregory.
Personnel Profile: Full-Time Paid 5; Part-Time Paid 3.
Governing Authority: city. Tax-exempt.
Institution Type/Description: Zoo.
Collections: wildlife including jaguar, tigers, monkeys, porcupines, lions, leopards, bison, bears, lemurs, bobcats, Bald Eagles; owls; birds.
Hours & Admission Prices: Mon.-Thurs. 9-4:30, Fri.-Sun. 10-7. No charge; donations accepted. &
Attendance: 55,000 (estimated)
Membership: Zoological Society $25.

SHAFER GALLERY - BARTON COMMUNITY COLLEGE, (M), 245 N.E. 30 Rd., Great Bend, KS 67530-9107. Tel.: 800-722-6842; 620-792-9342.
E-mail: barnesd@bartonccc.edu
Web Site: www.bartonccc.edu/community/artsentertainment/shafergallery
Key Personnel: Dir., David E. Barnes.
Personnel Profile: Full-Time Paid 1.
Governing Authority: Parent Institution: Barton Community College. Tax-exempt.
Institution Type/Description: Art Gallery.
Collections: sculpture; paintings; bronzes.
Facilities: 7,709 sq. ft. exhibit space.
Activities: permanent & temporary exhibitions; lectures; musical performances.
Hours & Admission Prices: Mon.-Fri. 10-5, Sun. 1-4; groups by appointment. No charge; donations accepted. Closed college-related holidays. &
Attendance: 7,000 (estimated)

Greensburg

BIG WELL, 315 S. Sycamore, Greensburg, KS 67054-1758.
Founded: 1937.
Key Personnel: Museum Shop Mgr., Rich Stephenson.
Personnel Profile: Full-Time Paid 1; Part-Time Paid 5.
Governing Authority: Parent Institution: Greensbury Chamber of Commerce.
Institution Type/Description: General Museum.
Collections: local history; Pallasite meteorite; hand dug well c.1887.
Facilities: Museum-related items for sale.
Activities: guided tours upon request.
Hours & Admission Prices: Memorial Day to Labor Day 8-8; Winter: Mon.-Sat. 9-5, Sun. 1-5. Adults $2, children $1.50. Closed Thanksgiving; Christmas.
Attendance: 42,672 (accurate)

Grenola

GRENOLA HISTORICAL SOCIETY - GRENOLA ELEVA-TOR MUSEUM, 313 N. Main St., Grenola, KS 67346. Mailing Address: P.O. Box 111, Grenola, KS 67346. Tel.: 620-358-3241.
E-mail: dorothykeplinger@yahoo.com
Founded: 1989.
Personnel Profile: Part-Time Paid 1; Part-Time Volunteers 4.
Governing Authority: Tax-exempt.
Institution Type/Description: Historical Society Museum: housed in the former Grenola Mill and Elevator.
Collections: local history & culture; genealogical materials: obituary collection & cemetery index; farm equipment; buggies; blacksmith tools; household artifacts; wagons; photographs; yearbooks.
Research Fields: genealogy.
Hours & Admission Prices: Sat.-Sun. 1-5. No charge; donations accepted. &
Attendance: 200 (estimated)
Membership: Lifetime $10.

Halstead

HALSTEAD HERITAGE MUSEUM & DEPOT, 116 E. First, Halstead, KS 67056-1713. Mailing Address: P.O. Box 88, Halstead, KS 67056-0088. Tel.: 316-835-2267.
Founded: 1985.
Congressional District: 4
Key Personnel: Pres. (V), Philip Adams; Sec., Carolyn Williams; Museum Shop Mgr., Marjory Hensley.
Personnel Profile: Part-Time Volunteers 20.
Governing Authority: Parent Institution: Halstead Historical Society. Tax-exempt.
Institution Type/Description: History Museum. housed in the former Halstead Railway Station; built in 1917.
Collections: depot & railroading history; local history; photographs; personal & military artifacts; biographies of local Civil War veterans; Native American arrowheads.
Major Exhibits: Buller Photography Collection, Summer 2014.
Research Fields: local genealogy; community development.
Facilities: Museum-related items for sale.
Activities: special events.
Publications: books; newsletter, The Halstead Times.
Hours & Admission Prices: Sat.-Sun. 2-5; other times by appointment. No charge; donations accepted. &
Attendance: 550 (accurate)
Membership: Single $10; Family $20; Business $50; Corporate $100; Friend $500; Associate $1,000; Benefactor $5,000.

KANSAS LEARNING CENTER FOR HEALTH, 505 Main St., Halstead, KS 67056-2233. Mailing Address: P.O. Box 288, Halstead, KS 67056-0288. Tel.: 316-835-2662. Fax: 316-835-2755. Facebook: Kansas Learning Center for Health.
E-mail: brenda@learningcenter.org
Web Site: www.learningcenter.org
Founded: 1965.
Congressional District: 4
Key Personnel: C.E.O. & Dir., Brenda S. Sooter; Pres. (V), Lois Loflin.
Personnel Profile: Full-Time Paid 3; Part-Time Paid 6; Part-Time Volunteers 12.
Volunteer Hours: 360
Operating Expenses: 300,000
Operating Income: 538,000
Governing Authority: nonprofit organization. Tax-exempt: 501(c)(3).
Institution Type/Description: Health Museum.
Collections: displays depicting the human body & the way it functions.
Facilities: 61-seat auditorium. Health education materials for sale.
Activities: guided tours; lectures; films; gallery talks; formally organized education programs; permanent & temporary exhibitions; school assemblies; professional development workshops.
Publications: quarterly newsletter; annual report.
Hours & Admission Prices: Mon.-Fri. 9-4. Full Day Program: $8. Half Day Program $5. Self-Guided Tour: $3. Closed New Year's Day; Easter; Memorial Day; Independence Day; Labor Day; Thanksgiving; Christmas. &
Attendance: 28,000 (accurate)

Hanover

HOLLENBERG PONY EXPRESS STATION STATE HISTORIC SITE, 2889 23rd Rd., Hanover, KS 66945-8901. Fax: 785-337-2635. Facebook: Hollenberg Pony Express Station State Historic Site.
E-mail: hollenberg@kshs.org
Web Site: www.kshs.org/hollenberg
Formerly: Hollenberg Station State Historic Site
Founded: 1857.
Congressional District: 1 & 2
Key Personnel: Cur., Duane R. Durst; Site Admin., Katherine McCartney; Friends Group Pres., Gary Minge.
Personnel Profile: Part-Time Paid 3; Part-Time Volunteers 18.
Governing Authority: state. Parent Institution: Kansas History Center, 6425 S.W. 6th Ave., Topeka, KS 66615-1099. Tax-exempt.
Institution Type/Description: Historic Building: 1857 Pony Express station, comprised of general store, tavern, stage station, post office & home.
Collections: period items; Oregon trail period; Pony Express items; tools; cart maps; biographical data on Gerat Hollenberg 1823-74; weaponry; pioneer items; furnished 1860 kitchen; store room.
Research Fields: Pony Express; Oregon & California trails.
Facilities: visitors center.
Activities: tours. Museum Sponsors: Victorian Tea in April; Pony Express Re-ride in June; Pony Express Festival in August; Museum Day in Sept.; Christmas Open House on Dec. 6, 2014.
Publications: pamphlet.
Hours & Admission Prices: April-Sept. Wed. & Sat. 1-4, Thurs.-Sat. 10-5. Adults $3, students $1; children under 5 & KSHS members no charge, school groups $2, adult groups $1. Closed holidays.
Attendance: 847 (accurate)
Membership: Student 1 yr. $30; Student 2 yr. $55; Individual 1 yr. $40; Individual 2 yr. $75; Household 1 yr. $60; Household $2 yr. $115; Organization 1 yr. $50; Organization 2 yr. $95.

Harper

HARPER CITY HISTORICAL SOCIETY, 804 E. 12th St., Harper, KS 67058-1804. Mailing Address: 708 W. 14th, Harper, KS 67058-1528. Tel.: 620-896-7877.
Founded: 1959.
Congressional District: 5
Key Personnel: Pres., Mary Helen Baker; Sec., Gail Bellar; Treas., Karen Walker.
Personnel Profile: Part-Time Volunteers 5.
Governing Authority: society. Tax-exempt: 170(b)(1)(A).
Institution Type/Description: Historical Society Museum: housed in a former German Apostolic Church; built in 1887.
Collections: furniture; period artifacts; early kitchen; tools; Bibles; square grand piano; clothing. Historic Building: c.1889 Runnymede Church.
Facilities: library of Harper Advocate newspapers & quarterlies of Kansas Historical Society of Topeka available by request.
Activities: guided tours; permanent exhibitions.
Publications: annual newsletter.
Hours & Admission Prices: By appointment. No charge; donations accepted.
Attendance: 300 (estimated)
Membership: Individual & Business $10.

Hays

BERENS' ANTIQUE FARM MACHINERY, 1915 Holmes Rd., Hays, KS 67601-2520. Tel.: 785-735-9364.
Institution Type/Description: Farm Machinery Museum.
Collections: horse drawn grain wagons; early farm tractors; John Deere two cylinder tractors; horse drawn buggies; barn lanterns; gas cans; buckets; cattle yokes; farm equipment.
Hours & Admission Prices: By appointment. No charge.

ELLIS COUNTY HISTORICAL SOCIETY, 100 W. 7th St., Hays, KS 67601-4429. Tel.: 785-628-2624. Fax: 785-628-0386.
E-mail: office@elliscountyhistoricalsociety.org
Web Site: www.elliscountyhistoricalmuseum.org
Founded: 1971.
Congressional District: 1
Key Personnel: Pres., Tom Drees; Treas., Brad Boyer; Cur., Elisha Beck; Archivist & Dir., Janet Johannes; Dir. & Museum Shop Mgr., Sharon Behrman.
Personnel Profile: Full-Time Paid 4; Part-Time Paid 3; Part-Time Volunteers 10.
Governing Authority: private; nonprofit organization. Tax-exempt: 501(c)(3).

Institution Type/Description: History Museum.
Collections: archives; photographs; area history from 1867 to present; history of Wild West; Volga-German & other immigrant groups; agriculture & business; personal articles. Historic Building: chapel.
Research Fields: early settlement; wild west period; Volga-German history; local history.
Facilities: library; 4,000 sq. ft. exhibit space; archives. Museum-related items for sale.
Activities: Annual Events: Independence Day Celebration; Oktoberfest; Christmas Open House; annual meeting; Pioneer Day.
Publications: quarterly newsletter, Homesteader.
Hours & Admission Prices: June-Aug. Tues.-Fri. 10-5, Sat. 1-5; Sept.-May Tues.-Fri. 10-5. Adults $4; members no charge.
Attendance: 3,000 (accurate)
Membership: Student under 18 $3; College Student $10; Single $15; Family $25; Friend $75; Homesteader $200; Sponsorship $250; Settler: Lifetime $500, Couple $750.

FORT HAYS STATE HISTORIC SITE, 1472 Hwy. 183 Alt., Hays, KS 67601-9212. Tel.: 785-625-6812. Fax: 785-625-6812. Facebook: Fort Hays State Historic Site.
E-mail: thefort@kshs.org
Web Site: www.kshs.org/fort_hays
Founded: 1965.
Congressional District: 1
Key Personnel: Exec. Dir. KSHS, Jennie Chinn; Pres. (V), Mike Smith; Supt., Robert Wilhelm; Museum Shop Mgr., Connie Schmeidler.
Personnel Profile: Full-Time Paid 1; Part-Time Paid 2; Part-Time Volunteers 45.
Governing Authority: state. Parent Institution: Kansas Historical Society, 6425 S.W. 6th, Topeka, KS 66615. Tel. 785-272-8681. Tax-exempt.
Institution Type/Description: Military Museum; Visitors & Tourist Information Center.
Collections: uniforms; accoutrements; utensils; Indian artifacts; weapons; excavated bottles & tools; insignias. Historic Structures: 1872 furnished Guardhouse; 1867 Blockhouse; 1867 two officers' quarters.
Research Fields: fort history.
Facilities: library of Fort Hays records & documents on microfilm available for viewing on premise only; 40 acres of native grassland. Museum-related items for sale.
Activities: guided tours; lectures; films; docent programs; formally organized education programs for undergraduate college students; folk art workshops; permanent exhibitions. Museum Sponsors: Fort Hays Anniversary Celebration; Christmas Program.
Publications: quarterly newsletter, Post Returns; brochures, Bugle Calls at Fort Hays, Buffalo Soldiers at Fort Hays, Fort Hays, Conflict on the Plains; Educational Programs at Fort Hays.
Hours & Admission Prices: Tues.-Sat. 9-5. Adults $5, students $1; KHS, Friends and children 5 & under no charge. Closed legal holidays.
Attendance: 13,000 (estimated)
Membership: Society of Friends of Historic Fort Hays: Student $10; Individual $15; Family $20; Organization & Business $30; Life $250 & up.

HAYS ARTS CENTER GALLERY, 112 E. 11th St., Hays, KS 67601. Tel.: 785-625-7522.
Key Personnel: Exec. Dir., Brenda K. Meder
Institution Type/Description: Art Gallery.
Collections: paintings; sculpture; photographs.
Activities: permanent & temporary exhibitions.
Hours & Admission Prices: Mon.-Fri. 10-4, Sat. 10-1.

STERNBERG MUSEUM OF NATURAL HISTORY, (M), Fort Hays State University, 3000 Sternberg Dr., Hays, KS 67601-2006. Tel.: 785-628-5516. Fax: 785-628-4518.
E-mail: rebarrick@fhsu.edu
Web Site: sternberg.fhsu.edu
Founded: 1926.
Congressional District: 1
Key Personnel: C.E.O. & Cur. Paleontology, Reese Barrick; Cur. Mammals & Adjunct Cur. Birds & Mammals, E. Finck; Cur. Plants, J.R. Thomasson; Asst. Cur. Birds, G. Farley; Asst. Cur. Insects, R. Packauskas; Museum Educator, Brian Bartels; Chief Cur. & Cur. Vertebrate Paleontology, R.J. Zakrzewski; Collection Mgr., M. Eberle; Exhibits Dir., G. Walters; Office Mgr., A. Klein; Public Rels., M. Kellerman; Operations Mgr., James Helget; Education Asst., Thea Haugen; Exhibit Technician, Beatrice Bauer; Mgr. Visitor Svcs., Brad Penka; Cur. Emeritus, Gene Fleharty; Cur. Emeritus, H. Reynolds; Assoc. Cur. Herptiles, T. Taggart; Assoc. Cur. Herptiles, C. Schmidt; Adjunct Cur. Herptiles, Joe Collins; Adjunct Cur. Vertebrate Paleontology, Mike Everhart; Adjunct Cur. Birds & Mammals, E. Finck;

Adjunct Cur. Vert Paleontology, K. Shimada; Adjunct Cur. Vert Paleontology, B. Schumacher; Asst. Cur. Fishes, W. Stark; Building Maintenance, G. Beilman.
Personnel Profile: Full-Time Paid 8; Part-Time Paid 6; Part-Time Volunteers 175.
Governing Authority: Parent Institution: Fort Hays State University. Subsidiary Institution: Sternberg Museum Foundation. Tax-exempt: 501(c)(3).
Institution Type/Description: Natural History Museum.
Collections: natural history of the Great Plains; herbarium; insects; birds; amphibians & reptiles; mammals; fossils; fish.
Research Fields: ornithology; mammalogy; herpetology; paleobotany; plant taxonomy; entomology; ichthyology; paleontology.
Facilities: 15,000-vol. research library; separate laboratory operation; classrooms.
Activities: guided tours; lectures; training programs; docent program; discovery room; field trips; education programs for children & teachers.
Publications: infrequent, processed material, Occasional Papers of the Sternberg Museum of Natural History; periodical, Fort Hays Studies; annual report.
Hours & Admission Prices: Memorial Day to Labor Day Mon.-Sat. 9-6, Sun. 1-6; Sept.-May Tues.-Sat. 9-6, Sun. 1-6. Adults $8, seniors $6, children $5; museum & ASTC members no charge. &

Attendance: 38,279 (accurate)
Membership: Student & Senior Citizens $20; Individual $30; Senior Family $45; Family $55; Sponsor $100; Curators $250; Directors $500; Lifetime $1,000. Corporate: Mammoth $500; Xiphactinus $1,000; Plesiosaur $5,000; Mosasaur $10,000; Tyrannosaurus $25,000.

Herington

HERINGTON HISTORICAL SOCIETY & MUSEUM, INC. - SE DICKINSON COUNTY, 800 S. Broadway, Herington, KS 67449-3060. Tel.: 785-258-2842.
E-mail: heringtonmuseum@att.net
Formerly: Tri-County Historical Society & Museum, Inc.
Founded: 1975.
Congressional District: 74
Key Personnel: Dir., Museum Shop Mgr. & Membership Chm., Jolene Bradford; Pres., Phyllis Smith; Sec., Joan Mattan; Treas., Helen Mitchell.
Personnel Profile: Part-Time Paid 1; Part-Time Volunteers 13.
Governing Authority: society; nonprofit corporation. Parent Institution: Kansas State Historical Society. Subsidiary Institution: Dickinson County Historical Society. Tax-exempt: 501(c)(3).
Institution Type/Description: General & Rock Island Railroad History Museum.
Collections: agricultural & historical artifacts; archives; genealogy books; Herington collection; military uniform collections; Rock Island railroad collection; slides; photographs; Herington Times newspapers back to 1884 on micro-film; Herrington Sun newspapers May 13, 1920-Dec. 31, 1931 on microfilm; Hope, KS Dispatch 1885-1979.
Research Fields: genealogy; historical data for the public & for writers; city history; World War II Herington Army air field; Rock Island depot history; Herington Times newspaper, May 1920 to Dec. 1931; Herington Cemetery record; Dickinson, Morris & Marion counties Federal census, 1865-1920; Herington High school history, graduates' addresses; Herington's Sunset Hill Cemetery records; Clark's Creek Cemetery records; St. Paul Lutheran Church cemetery records, Ramona, Kansas; St. Paul Lutheran Church records, Shadybrook, KS; St. John's Catholic Church cemetery records, Herington; St. Johns Lyon Creek church records 1861-present.
Facilities: 25-vol. family genealogy books, 9-vol. community history books; 8-vol. area church history booklets, 10-vol. area church history, 42-vol. club scrapbooks & minutes all available to the public. Gift items for sale.
Activities: guided tours; lectures; docent program. Museum Sponsors: annual banquet; special displays; 4-H Fair & Carnival; tours of the Rock Island railroad baggage car annex & caboose.
Publications: newsletter, Herington Historical Society & Museum Inc.
Hours & Admission Prices: Tues.-Fri. 1-5. No charge; donations accepted. Closed New Years Eve & Day; Thanksgiving; Christmas Eve, Day & week. &
Attendance: 500 (estimated)
Membership: Student $5; Individual $12.50; Family $20; Life $200; Commercial $250.

Hiawatha

AG MUSEUM & WINDMILL LANE, 301 E. Iowa St., Hiawatha, KS 66434-9826. Tel.: 785-742-3702. Fax: 785-742-3330.
E-mail: bchsdirector@yahoo.com
Web Site: www.bckshistory.com
Key Personnel: Dir. & Cur., Eric Oldham; Pres. (V), Jere Bruning.

Personnel Profile: Full-Time Paid 1; Part-Time Paid 4.
Institution Type/Description: Historical Society Museum.
Collections: horse-drawn combine; buggy; tractors; period cars; the first Brown County post office; sleigh; Brown County artifacts.
Hours & Admission Prices: Tues.-Fri. 10-4, Sat. 10-2. Adults $5.
Attendance: 500 (estimated)
Membership: Individual $15; Family & Business $30; Life $200.

MEMORIAL AUDITORIUM, 611 Utah St., Hiawatha, KS 66434-2319. Tel.: 785-742-3330. Fax: 785-742-3330.
E-mail: bchsdirector@yahoo.com
Web Site: www.bckshistory.com
Founded: 1978.
Congressional District: 2
Key Personnel: Dir., Eric Oldham; Pres. (V), Jere Bruning.
Personnel Profile: Full-Time Paid 1; Part-Time Paid 4.
Governing Authority: Tax-exempt.
Institution Type/Description: Historical Society Museum.
Collections: period artifacts & clothing.
Major Exhibits: Our Eves, Our Stories: America's Greatest Generation (T), 11/13-1/14; The Bison: American Icon (T), 4/6/14-5/25/14.
Hours & Admission Prices: May 2-Oct. Mon.-Fri. 10-12 & 1-3, Sat. 10-2. Adults $5. &
Attendance: 500
Membership: Individual $15; Family & Business $30; Lifetime $200.

Highland

YOST ART GALLERY, Highland Community College, 101 N. Elmira, Highland, KS 66035. Tel.: 785-442-6000.
E-mail: jtyler@highlandcc.edu
Key Personnel: Dir., Janet Tyler
Institution Type/Description: Art Gallery.
Collections: works by regional photographers.
Hours & Admission Prices: Mon.-Fri. 8:30-4.

Hill City

GRAHAM COUNTY HISTORICAL SOCIETY, 103 E. Cherry, Hill City, KS 67642. Tel.: 785-421-2543.
Institution Type/Description: Historical Society Museum.
Collections: local history & culture; period artifacts; photographs.
Facilities: archives.
Activities: research.
Hours & Admission Prices: Fri.-Sat. 1-4.

Hillsboro

HILLSBORO MUSEUMS, 501 S. Ash St., Hillsboro, KS 67063-1531. Mailing Address: P.O. Box 125, Hillsboro, KS 67063-0125. Tel.: 620-947-3775.
E-mail: hillsboro_museums@yahoo.com
Web Site: www.hillsboro-museums.com
Formerly: Hillsboro Historical Society & Museum
Founded: 1958.
Congressional District: 1
Key Personnel: C.E.O., Stan R. Harder.
Personnel Profile: Full-Time Paid 3; Part-Time Paid 2; Part-Time Volunteers 20.
Governing Authority: municipal; nonprofit. Parent Institution: City of Hillsboro. Subsidiary Institution: The Mennonite Settlement Museum; The William F. Schaeffler House Museum; Hillsboro Museums Visitor Center. Tax-exempt.
Institution Type/Description: History Museum.
Collections: agriculture; folklore; preservation project. Mennonite Settlement Museum: Russian & Polish Mennonite immigrant village life in Kansas. Historic Structures: Peter Paul Loewen House; Kreutziger Country Schoolhouse; William Schaeffler House; Jacob Friesen Dutch Flouring Windmill replica; Mennonite Settlement Museum.
Research Fields: Russian Mennonite immigration to North America.
Activities: guided tours to the Mennonite Settlement Museum and the William F. Schaeffler House Museum; special events. Museum Sponsors: Independence Day Celebration; Schaeffler House concerts; Weihrachtsfest.
Publications: periodic brochures & handbills; local histories; annual newsletter; centennial history publication, Hillsboro: City on the Prairie 1884-1984.
Hours & Admission Prices: March-Dec. Tues.-Fri. 10-12 & 1:30-4, Sat.-Sun. 2-4. Adults $3, students $1; discounts to AAM members. Closed holidays.
Attendance: 3,000 (estimated)

　　　　　　　　　　　　　　　　　THE OFFICIAL MUSEUM DIRECTORY

Membership: Friends $5.

Hoisington

HOISINGTON HISTORICAL SOCIETY MUSEUM, 120 E. 2nd St., Hoisington, KS 67544. Mailing Address: Hoisington Historical Society, 358 West 8th St., Hoisington, KS 67544. Tel.: 620-653-4320, 2857 & 4683.
E-mail: customerservice@hoisingtonhistoricalsociety.org
Web Site: www.hoisingtonhistoricalsociety.org
Founded: 1995.
Key Personnel: Pres. (V), Lon Palmer
Governing Authority: Tax-exempt.
Institution Type/Description: Historical Society Museum: building built in 1905.
Collections: local history & culture; period furnishings; photographs; personal artifacts.
Publications: quarterly newsletter.
Hours & Admission Prices: 1st & 3rd Sat. each month 1-3. No charge; donations accepted.
Attendance: 300 (estimated)
Membership: Single $5; Family $10.

Holton

JACKSON COUNTY HISTORICAL SOCIETY, 216 New York Ave., Holton, KS 66436-1738. Tel.: 785-364-4991.
Founded: 1979.
Congressional District: 2
Key Personnel: Chm. (V), Anna Wilhelm; Pres. (V), Steve Banaka.
Personnel Profile: Part-Time Volunteers 25.
Governing Authority: Tax-exempt.
Institution Type/Description: Historical Society Museum.
Collections: local history & culture; personal artifacts; photographs; Victorian era clothing; lace.
Publications: quarterly newsletter, The Jacksonian.
Hours & Admission Prices: May-Oct. Fri. 10-4; other times by appointment. No charge; donations accepted. &
Attendance: 300 (accurate)
Membership: Individual $25; Family $40; Supporting & Business $75; Individual Life $400; Business Life $600.

Howard

BENSON HISTORICAL MUSEUM, 145 S. Wabash, Howard, KS 67349.
Key Personnel: Chm. (V), Gleneva Winn.
Governing Authority: Parent Institution: Elk County Historical Society. Tax-exempt.
Institution Type/Description: History Museum.
Collections: local history & culture; period furnishings; personal artifacts; photographs; historic buildings; 1916 Tumley oil pull tractor; dolls.
Hours & Admission Prices: By appointment. No charge; donations accepted.

Hugoton

STEVENS COUNTY GAS AND HISTORICAL MUSEUM, 905 S. Adams, Hugoton, KS 67951-2817. Mailing Address: P.O. Box 87, Hugoton, KS 67951-0087. Tel.: 620-544-8751.
E-mail: svcomus@pld.com
Founded: 1961.
Congressional District: 1
Key Personnel: Pres., Stanley McGill; Cur. & Museum Shop Mgr., Gladys Renfro.
Personnel Profile: Part-Time Paid 3.
Governing Authority: county. Tax-exempt: 501(c)(3).
Institution Type/Description: History Museum Complex: 1913 original A.T.S.F. depot country store, including barber shop & grocery store.
Collections: period furnishings; American Indian artifacts; early furnishings; art; gas industry equipment; model drilling rigs; horse drawn machinery vehicles; active gas well; agricultural & automotive artifacts. Historic Buildings: 1888 schoolhouse; 1886-1887 house; 1905-1906 Second Methodist Church; first county jail; 1887 South Harmony School.
Research Fields: American Indian artifacts.
Activities: guided tours; school groups; meeting places for local clubs.
Publications: brochures, Sixty Years of Development of the Hugoton Fields; The Hugoton Stony Meteorite, Stevens County, Kansas; Buddy Heaton; single sheets of museum complex; postcards.
Hours & Admission Prices: June-Aug. Mon.-Fri. 9:30-11:30 & 1-5, Sat. 2-4,

Sun. by appointment; Sept.-May Mon.-Fri. 1-5, Sat. 2-4, Sun. by appointment. No charge; donations accepted. Closed Easter; Memorial Day; Labor Day; Columbus Day; Veterans Day; Thanksgiving; Christmas. &
Attendance: 1,217 (accurate)
Membership: Annual $1; Life $10.

Humboldt

HUMBOLDT HISTORICAL MUSEUM, 416 N. Second, Humboldt, KS 66748-1402. Mailing Address: P.O. Box 63, Humboldt, KS 66748-0063. Tel.: 620-473-5055.
E-mail: rrthompson504@yahoo.com
Web Site: www.humboldtksmuseum.com
Founded: 1966.
Key Personnel: Dir., Roland E. Thompson; Sec., Michelle McDown; Treas., Ellen Lee.
Personnel Profile: Full-Time Volunteers 15; Part-Time Volunteers 5.
Governing Authority: Tax-exempt.
Institution Type/Description: History Museum.
Collections: historic farm equipment; furnished kitchen, dining & bedroom displays; Civil War cannon; Humboldt's original jail cell; horse-drawn adult & infant hearse; collection of scale model horse-drawn wagons; clothing; quilts; toys; medical instruments; books; photographs; paintings.
Hours & Admission Prices: Open year round by appointment. No charge. &
Attendance: 1,525 (accurate)
Membership: Lifetime $10.

Hutchinson

HUTCHINSON ART CENTER, (M), 405 N. Washington, Hutchinson, KS 67501-4852. Tel.: 620-663-1081. Fax: 620-663-6367.
E-mail: hutchart2@hac.kscoxmail.com
Web Site: hutchinsonartcenter.org
Founded: 1949.
Congressional District: 1
Key Personnel: Dir., Mark L. Rassette; Pres. (V), Jane Dronberger; Museum Shop Mgr., Beth Kammerer.
Personnel Profile: Part-Time Paid 2; Part-Time Volunteers 2.
Operating Expenses: 132,000
Operating Income: 132,000
Governing Authority: private; nonprofit organization. Subsidiary Institution: Vignettes. Tax-exempt: 501(c)(3).
Institution Type/Description: Art Museum.
Collections: 19th-20th century American & European art.
Facilities: 6,611 sq. ft. exhibit space; classroom; auditorium. Museum-related items for sale.
Activities: guided tours; lectures; loan exhibitions; arts and humanities events.
Publications: quarterly newsletter, Gallery Notes.
Hours & Admission Prices: Tues.-Fri. 9-5, Sat.-Sun. 1-5. No charge. &
Attendance: 5,000 (accurate)
Membership: Senior Citizens $20; Individual $30; Artist $35; Family $40; Contributor $100; Business $200; Friend $250; Donor $500; Patron $1,000; Benefactor $5,000.

HUTCHINSON ZOO, 6 Emerson Loop E., Hutchinson, KS 67501-7500. Mailing Address: P.O. Box 1567, Hutchinson, KS 67504. Tel.: 620-694-2693. Fax: 620-694-2654. Facebook: Hutchinson Friends of the Zoo.
E-mail: janad@hutchgov.com
Web Site: www.hutchgov.com/zoo
Founded: 1986.
Key Personnel: Dir., Jana Durham; Cur., Kiley Buggeln; Gift Shop Mgr., Nelda Petering.
Personnel Profile: Full-Time Paid 5; Part-Time Paid 1; Part-Time Volunteers 94.
Governing Authority: Parent Institution: City of Hutchinson. Tax-exempt.
Institution Type/Description: Zoo.
Collections: 300 birds, mammals, reptiles & amphibians.
Facilities: Zoo-related items for sale; rental facilities.
Activities: animal encounters; train rides; educational programs.
Hours & Admission Prices: Daily call for hours. No charge; donations accepted. Closed New Year's Day; Thanksgiving; Christmas.
Attendance: 48,851
Membership: Single $25; Family $40.

KANSAS COSMOSPHERE AND SPACE CENTER, (M), 1100 N. Plum, Hutchinson, KS 67501-1418. Tel.: 620-662-2305. Fax: 620-662-3693.
E-mail: info@cosmo.org

Web Site: www.cosmo.org
Founded: 1962.
Congressional District: 4
Key Personnel: C.E.O., Richard Hollowell; Pres. & C.O.O., Jim Remar; Controller, Steven Birdsall; Museum Retail Mgr., Phyllis Cole.
Personnel Profile: Full-Time Paid 30; Part-Time Paid 35; Part-Time Volunteers 87.
Governing Authority: public; nonprofit foundation. Tax-exempt: 501(c)(3).
Institution Type/Description: Space Museum.
Collections: U.S. & Soviet space artifacts including Mercury, Gemini & Apollo spacecrafts; Vostok, Voskhod & Soyuz spacecraft; spacesuits; a lunar module; rocket engines; SR-71 Blackbird; German V-1 and V-2 rockets; actual Apollo 13 command module Odyssey; planetarium; Hall of Space Museum.
Research Fields: space sciences & related astronomy.
Facilities: digital dome theater; planetarium; classrooms; NASA teacher resource center.
Activities: astronomy & scientific programs; teacher in-service programs; lectures; films; space science discovery workshops; tours; classes; Kansas Adventures in Outer Space-KAOS (space science education with NASA training activities; Road Scholar (international camp experience); adult astronaut adventure; Boy and Girl Scout programs.
Hours & Admission Prices: Summer & Christmas Breaks: Mon.-Sat. 9-7, Sun. 12-5; Fall, Spring & Winter: Mon.-Thurs. 9-6, Fri.-Sat. 9-7, Sun. 12-6. All Day Mission Pass: adults $18, senior citizens 60 & over and children 3-12 $16; discounts to AAM members; children 2 & under no charge. Single Venue: adults $8.50, senior citizens 60 & over and children 3-12 $7.50; discounts to AAM members. Closed Easter; Thanksgiving; Christmas. &
Attendance: 120,000 (accurate)
Membership: Senior 60 & over and Student $40; Individual $45; Senior Family 60 & over $70; Family $80; Mercury $150; Gemini $250; Apollo $500; Shuttle $1,000.

KANSAS UNDERGROUND SALT MUSEUM, 3504 E. Ave. G, Hutchinson, KS 67501-8284. Mailing Address: P.O. Box 1864, Hutchinson, KS 67504-1864. Tel.: 620-662-1425; 866-755-3450. Fax: 620-259-6134.
Web Site: www.undergroundmuseum.org
Key Personnel: Dir. Operations, Gayle Ferrell; Volunteer Coord., Tonya Gehring; Maintenance, Dave Unruh; Visitor Svcs. Coord., Colleen McCallister
Institution Type/Description: Mining Museum.
Collections: mining equipment, artifacts & memorabilia.
Hours & Admission Prices: Tues.-Sat. 9-6, Sun. 1-6. Adults $14.35, seniors 60 & over, active military & AAA members $12.75; Reno County residents & children 4-12 $9.05.

RENO COUNTY MUSEUM, (M), 100 S. Walnut, Hutchinson, KS 67501-7406. Mailing Address: P.O. Box 664, Hutchinson, KS 67504-0664. Tel.: 620-662-1184. Fax: 620-662-0236.
Web Site: renocomuseum.org
Founded: 1961.
Congressional District: 1
Key Personnel: Exec. Dir., Linda Schmitt; Chief Cur., Jamin Landavazo.
Governing Authority: nonprofit organization. Parent Institution: Reno County Historical Society. Tax-exempt: 501(c)(3).
Institution Type/Description: County History Museum.
Collections: artifacts relating to the history of Reno County, Kansas.
Research Fields: Reno County history.
Facilities: research library; meeting room. Museum-related items for sale.
Activities: permanent & temporary exhibits; special events; education programs.
Publications: quarterly journal, Legacy: The Journal of the Reno County Historical Society.
Hours & Admission Prices: Museum: Tues.-Fri. 9-5, Sat. 11-5. Office: Mon.-Fri. 8-5. No charge; donations accepted. &
Attendance: 22,129 (accurate)
Membership: Information available upon request.

Independence

INDEPENDENCE HISTORICAL MUSEUM & ART CENTER, 123 N. 8th, Independence, KS 67301-3501. Mailing Address: P.O. Box 294, Independence, KS 67301-0294. Tel.: 620-331-3515.
E-mail: museum123@cableone.net
Web Site: independencehistoricalmuseum.org
Formerly: Independence Museum
Founded: 1882.
Congressional District: 5

Key Personnel: Pres., Ray Rothgeb; Dir., Sylvia Augustine; Museum Shop Mgr., Ellie Culp.
Personnel Profile: Full-Time Paid 1; Part-Time Paid 1; Part-Time Volunteers 80.
Governing Authority: nonprofit. Parent Institution: Ladies Library & Art Association. Tax-exempt.
Institution Type/Description: History & Art Museum.
Collections: coin glass; cigar store Indian; 12' statue of Miss Justice; 1850's barber shop; oil history room; military room; fishing room; school room; doctor's office; paintings; period clothing; general store; bedroom; kitchen; black pottery; blacksmith shop; early fire equipment; western room; 1869 log cabin.
Activities: concerts; arts festivals; study clubs; permanent, temporary & traveling exhibitions; private artists exhibits; quilt exhibits; artists' workshops & meetings.
Publications: monthly newsletter.
Hours & Admission Prices: Tues.-Sat. 10-4; call for special tours. Adults $3; discounts to NARM members; members no charge. Closed national holidays. &
Attendance: 2,500 (accurate)
Membership: Single $25; Couple $40; Reciprocal $150; Supporter $250; Sponsor $500; Contributor $1,000; Benefactor $2,500; Sustaining $5,000.

INDEPENDENCE SCIENCE AND TECHNOLOGY CENTER, 125 S. Pennsylvania, Independence, KS 67301-3525. Tel.: 620-331-1999; 800-882-3606. Facebook: Independence Science & Technology Center.
E-mail: indyscitech@valnet.net
Founded: 1992.
Key Personnel: Dir., Amy Finney; Chm. (V), Ned Stichman; Pres. (V), Lloyd Harding.
Personnel Profile: Part-Time Paid 1.
Governing Authority: Tax-exempt.
Institution Type/Description: Science Center.
Collections: science & physics exhibits; telescopes; microscopes.
Facilities: meeting rooms.
Activities: educational activities.
Publications: newsletter, SCI-LINES.
Hours & Admission Prices: Mon.-Sat. 1-5. Admission $3; members & children under 3 no charge.
Membership: Senior Citizen $12; Single & Student $15; Senior Family $24; Family $30; Corporate $500.

LITTLE HOUSE ON THE PRAIRIE MUSEUM, 2507 CR 3000, Independence, KS 67301-7265. Tel.: 620-289-4238. Facebook; Little House on the Prairie Museum.
E-mail: lhopmuseumks@gmail.com
Web Site: www.littlehouseontheprairiemuseum.com
Formerly: Little House on the Prairie Historic Site
Founded: 1974.
Key Personnel: Dir., Michelle Martin; Pres. (V), Jean Kurtis Schodorf.
Governing Authority: Tax-exempt.
Institution Type/Description: Historic House: official site of Little House on the Prairie from Laura Ingalls Wilder's books.
Collections: period artifacts & memorabilia.
Hours & Admission Prices: April -Oct. Mon.-Sat. 10-5, Sun. 1-5. No charge; donations accepted.
Attendance: 20,000 (estimated)

RALPH MITCHELL ZOO, Riverside Park, Oak & Park St., Independence, KS 67301. Mailing Address: P.O. Box 9, Independence, KS 67301-0009. Tel.: 620-332-2513.
Institution Type/Description: Zoo.
Collections: 200 animals & birds.
Activities: train rides.
Hours & Admission Prices: Call for hours.

Ingalls

SANTA FE TRAIL MUSEUM, 204 S. Main St., Ingalls, KS 67853. Mailing Address: P.O. Box 74, Ingalls, KS 67853-0074. Tel.: 620-335-5220.
E-mail: dlmkwend@ucom.net
Founded: 1973.
Congressional District: 1
Key Personnel: Pres. (V), Audrey Maxwell; Sec. & Treas., Linda Hirschler; Museum Shop Mgr., Debbie Milne.
Personnel Profile: Part-Time Paid 2; Part-Time Volunteers 5.
Governing Authority: nonprofit organization. Tax-exempt: 501(c)(3).

Institution Type/Description: Historical Site & Local History Museum: housed in two Santa Fe railroad depot buildings.

Collections: exhibits pertaining to the area including: china; glass; silver; antiques; furniture; military & agricultural items; religious articles; old school textbooks; original pump from Soule Canal Irrigation project.

Activities: guided tours; loan exhibitions.

Hours & Admission Prices: May-Oct. Mon.-Sat. 9-11 & 1-4; other times by appointment; Nov.-April by appointment only. No charge; donations accepted. ♿

Attendance: 256 (accurate)

Iola

ALLEN COUNTY HISTORICAL SOCIETY, 20 S. Washington Ave., Iola, KS 66749-3204. Tel.: 620-365-3051.

E-mail: info@allencountyhistory.org

Web Site: www.allencountyhistory.org

Founded: 1956.

Congressional District: 2

Key Personnel: Exec. Dir. & Cur., Elyssa Jackson; Pres. (V), Leon Smith.

Personnel Profile: Full-Time Paid 1; Part-Time Volunteers 30.

Governing Authority: society, with county contract. Tax-exempt: 501(c)(3).

Institution Type/Description: Local History Museum.

Collections: Allen County memorabilia; A.E. Gibson collection of negatives & photos.

Research Fields: local history & gas boom in Kansas.

Activities: tours & special exhibits.

Publications: quarterly newsletter, Gaslight.

Hours & Admission Prices: May-Oct. Tues.-Sat. 12:30-4; Nov.-April Tues.-Sat. 2-4. No charge; donations accepted. ♿

Attendance: 1,888 (accurate)

Membership: Annual $10; Life $100.

ALLEN COUNTY JAIL MUSEUM, 203 N. Jefferson Ave., Iola, KS 66749. Mailing Address: 20 S. Washington, Iola, KS 66749. Tel.: 620-365-3051. Facebook; Allen County Jail Museum.

E-mail: info@.allencountyhistory.org

Web Site: www.allencountyhistory.org

Formerly: Old Jail Museum

Founded: 1956.

Congressional District: 2

Key Personnel: Dir., Elyssa Jackson; Pres. (V), Leon Smith.

Personnel Profile: Full-Time Paid 1; Part-Time Volunteers 31.

Governing Authority: Tax-exempt.

Institution Type/Description: Historic Building: housed in the former Allen County Jail; built in 1869. Listed on the National Register of Historic Places.

Collections: local history; period furnishings; solitary confinement cell; steel cage cell block; jailer's quarters.

Publications: quarterly newsletter, Gaslight.

Hours & Admission Prices: Tours: May-Sept. Tues.-Sat. 1:30; other times by appointment. No charge; donations accepted.

Attendance: 325 (accurate)

Membership: Individual $10; Life $100.

THE MAJOR GENERAL FREDERICK FUNSTON BOYHOOD HOME AND MUSEUM, 14 S. Washington Ave., Iola, KS 66749-3204. Tel.: 620-365-3051. Facebook; The Major General Frederick.

E-mail: info@allencountyhistory.org

Web Site: www.allencountyhistory.org

Founded: 1995.

Congressional District: 2

Key Personnel: Exec. Dir. & Cur., Elyssa Jackson; Pres. (V), Leon Smith.

Personnel Profile: Full-Time Paid 1; Part-Time Volunteers 30.

Governing Authority: private; nonprofit. Subsidiary Institution: Allen County Historical Society, Iola. Tax-exempt: 501(c)(3).

Institution Type/Description: History Museum: housed in the c.1860 Frederick Funston childhood residence, originally located on a homestead approximately five miles north of Iola.

Collections: the home is restored in Victorian decor typical of the 1880s & 1890s; artifacts & furniture that were used during Frederick Funston's boyhood as well as items pertaining to his botanical explorations and military career.

Research Fields: life & career of Frederick Funston, his family & relations; the era during which he lived; Spanish American War; Philippines' Insurrection; 1906 San Francisco earthquake & fire.

Facilities: library; theater. Toys, postcards, clothing & local items for sale.

Activities: docent program; films; guided tours; lectures.

Publications: quarterly newsletter, Gaslight.

Hours & Admission Prices: May-Oct. Tues.-Sat. 12:30-4; Nov.-April Tues.-Sat. 2-4; other times by appointment. No charge; donations accepted. ♿

Attendance: 460 (accurate)

Membership: Annual $10; Lifetime $100.

Jennings

CZECH MEMORIAL MUSEUM, 114 S. Kansas Ave., Jennings, KS 67643. Tel.: 785-678-2470.

Founded: 1970.

Key Personnel: Chm. (V), Neoma Tacha

Governing Authority: Tax-exempt.

Institution Type/Description: History Museum: housed in a former United Methodist Church Royal Neighbor Lodge Bldg.

Collections: local history, heritage & culture; period artifacts; photographs.

Hours & Admission Prices: By appointment. No charge; donations accepted. ♿

Attendance: 1,400 (estimated)

Jetmore

HAUN MUSEUM, Rte. 2, Jetmore, KS 67854. Tel.: 620-357-8794.

Institution Type/Description: History Museum.

Collections: local history & culture; period furnishings; photographs; books; personal artifacts; early farm & fire equipment.

Hours & Admission Prices: Memorial Day to Labor Day Sat. 9-12 & 1-5, Sun. 1-5; other times by appointment.

Jewell

PALMER MUSEUM, 108 S. Washington, Jewell, KS 66949. Mailing Address: P.O. Box 282, Jewell, KS 66949. Tel.: 785-428-3466 & 3335.

Founded: 1991.

Key Personnel: Chm. (V), Roberta Holbren.

Personnel Profile: Part-Time Volunteers 8.

Governing Authority: Chamber of Commerce. Parent Institution: Jewell Chamber of Commerce.

Institution Type/Description: General Museum.

Collections: bound copies of the Jewell County Republican; printing machines; city history artifacts; high school trophies before 1964; 1872 wedding dress made in Scotland.

Activities: temporary exhibitions.

Hours & Admission Prices: By appointment. No charge; donations accepted.

Attendance: 480 (estimated)

Johnson

STANTON COUNTY MUSEUM, (M), 104 E. Highland, Johnson, KS 67855. Mailing Address: P.O. Box 806, Johnson, KS 67855. Tel.: 620-492-1526. Fax: 620-492-1785.

Institution Type/Description: History Museum.

Collections: county history & heritage; personal artifacts; photographs; period artifacts.

Activities: educational programs; research.

Hours & Admission Prices: Memorial Day to Labor Day & Dec. Mon.-Fri. 10:30-12 & 1-5, Sun. 1-4; other times by appointment.

Junction City

GEARY COUNTY HISTORICAL SOCIETY & MUSEUM, 530 N. Adams St., Junction City, KS 66441. Mailing Address: P.O. Box 1161, Junction City, KS 66441. Tel.: 785-238-1666. Fax: 785-238-1666 (call first). Facebook: Geary History.

E-mail: gearyhistory@gmail.com

Web Site: gchsweb.org

Founded: 1983.

Key Personnel: Exec. Dir., Sarah Moppin; Pres. (V), Florence Whitebread; Museum Shop Mgr., Sue Steinfort.

Personnel Profile: Full-Time Paid 1; Part-Time Paid 3.

Governing Authority: Parent Institution: Geary County Historical Society. Subsidiary Institutions: St. Joseph Historic Church and Cemetery, lower McDowell Creek Rd., Geary County, KS; Starcke House, 306 W. 5th, Junction City, KS 66441; Spring Valley Historic Site, K-18 & Spring Valley Rd., Junction City, KS 66441. Tax-exempt.

Institution Type/Description: Historical Society Museum.

Collections: local history, industry & culture; period furnishings; personal artifacts; photographs.

Major Exhibits: Playtime, 9/13-10/14.

Research Fields: genealogy; family and local history.
Facilities: micro film & reader/printer-computers; archives containing manuscripts & documents.
Activities: research.
Publications: bimonthly newsletter, Geary Glimmers.
Hours & Admission Prices: Tues.-Sun. 1-4. No charge; donations accepted. &
Attendance: 10,000 (estimated)
Membership: Senior $10; Individual $15; Family $25; Business $35, $50, $75; Sustaining $50; Lifetime $500.

JUNCTION CITY ARTS COUNCIL GALLERY, 107 W. Seventh St., Junction City, KS 66441. Mailing Address: P.O. Box 403, Junction City, KS 66441-0403. Tel.: 785-762-2581.
Institution Type/Description: Art Gallery.
Collections: works by local, state & national artists.
Activities: special events; permanent & temporary exhibits.
Hours & Admission Prices: Tues.-Fri. 9-5, Sat. 12-4.

MILFORD NATURE CENTER, 3415 Hatchery Dr., Junction City, KS 66441-8651. Tel.: 785-238-5323. Fax: 785-238-5775.
E-mail: pat.silovsky@ksoutdoors.com
Web Site: www.kdwp.state.ks.us
Founded: 1989.
Key Personnel: Dir., Pat Silovsky.
Personnel Profile: Full-Time Paid 2; Part-Time Paid 4; Part-Time Volunteers 4.
Governing Authority: state. Subsidiary Institution: Kansas Department of Wildlife & Parks.
Institution Type/Description: Nature Center.
Collections: wildlife & aquatic life of Kansas including snakes, amphibians, turtles, lizards, prairie dogs, birds of prey, bobcats, fish, turtles, & insects; butterflies; native animal furs.
Facilities: nature trails; educational facilities.
Activities: birdwatching; demonstrations; educational programs; wildlife rehabilitation; seasonal butterfly house; fish hatchery.
Publications: newsletter, On Tracks.
Hours & Admission Prices: April-Sept. Mon.-Fri. 9-4:30, Sat.-Sun. 1-5; Oct.-March Mon.-Fri. 9-4:30. No charge; donations accepted. &
Attendance: 19,347 (estimated)
Membership: Friends: Individual $10; Family $15; Sponsor $100.

SPRING VALLEY HISTORIC SITE, K-18 & Spring Valley Rd., Junction City, KS 66441. Mailing Address: P.O. Box 1161, Junction City, KS 66441. Tel.: 785-238-1666.
Institution Type/Description: Historic Site.
Collections: local history; period furnishings; personal artifacts; photographs. Historic Buildings: barn; cabin; 1857 Wetzel's log cabin church; 19th century one-room schoolhouse.
Activities: group tours.
Hours & Admission Prices: Call for hours.

STARCKE HOUSE, 306 W. 5th, Junction City, KS 66441. Mailing Address: P.O. Box 1161, Junction City, KS 66441. Tel.: 785-238-1666.
Institution Type/Description: Historic House Museum: housed in the former home of pioneer jeweler & watch-maker Andrew Vogler; built in 1880s.
Collections: period furnishings; personal artifacts; photographs; household items.
Hours & Admission Prices: Call for hours.

Kanopolis

FORT HARKER MUSEUM COMPLEX, 309 W. Ohio St., Kanopolis, KS 67454. Mailing Address: P.O. Box 144, Ellsworth, KS 67439-0144. Tel.: 785-472-3059.
E-mail: echs@eaglecom.net
Founded: 1961.
Congressional District: 1
Key Personnel: Chm. (V) & Pres. (V), Phyllis Dolezal.
Personnel Profile: Full-Time Paid 1; Part-Time Paid 2; Part-Time Volunteers 6.
Governing Authority: society. Parent Institution: Ellsworth County Historical Society, Ellsworth, KS 67439. Tax-exempt.
Institution Type/Description: Military Museum.
Collections: documents; files; 1870-present, guns & material; uniforms; Indian artifacts; train depot. Historic House: 1867 Fort Harker Guardhouse; Jr. & Commanding Officer's quarters.

Facilities: Pamphlets, papers & museum-related items for sale.
Activities: Museum Sponsors: Frontier Military Living History in July.
Publications: brochure, Sharing History.
Hours & Admission Prices: April & Oct. Tues.-Fri. & Sun. 1-5, Sat. 10-5; May-Sept. Tues.-Sat. 10-5, Sun. 1-5; Nov.-March Sat. 10-5, Sun. 1-5. Adults $3 (includes entrance to Hogden House Museum complex); discounts to groups; members no charge. Closed New Year's Day; Easter; Memorial Day; Independence Day; Labor Day; Thanksgiving; Christmas. &
Attendance: 3,000 (estimated)
Membership: Individual $10; Family $20; Business $25; Corporate $100.

Kansas City

GRINTER PLACE STATE HISTORIC SITE, 1420 S. 78th St., Kansas City, KS 66111-3208. Tel.: 913-299-0373. Fax: 913-788-8046. Facebook: Grinter Place.
E-mail: grinter@kshs.org
Web Site: www.kshs.org/grinter_place
Formerly: Grinter Place Museum
Founded: 1971.
Congressional District: 2
Key Personnel: Site Admin., Joe Brentano.
Governing Authority: state. Parent Institution: Kansas State Historical Society. Tax-exempt.
Institution Type/Description: Historic House: built by Moses Grinter.
Collections: 1860s-1890s household furnishings.
Research Fields: Delaware Indians in Kansas; interaction of cultures in 19th century Kansas including Europeans, Africans & Delaware Indians.
Activities: guided tours; crafts festivals; quilt show; school programs; outreach. Museum Sponsors: Apple Fest in Fall.
Publications: GPF News.
Hours & Admission Prices: Thurs.-Sat. 9:30-5. Adults $3, students $1; members, and children 5 & under no charge. Closed state holidays. &
Attendance: 1,000 (estimated)
Membership: Individual $10; Family $15.

STRAWBERRY HILL ETHNIC MUSEUM & CULTURAL CENTER, 720 N. 4th St., Kansas City, KS 66101-2908. Tel.: 913-371-3264.
Web Site: strawberryhillmuseum.org
Founded: 1988.
Personnel Profile: Part-Time Volunteers 30.
Governing Authority: Tax-exempt.
Institution Type/Description: History Museum.
Collections: period furnishings; culturally specific artifacts.
Publications: quarterly, Strawberry Hill Vine.
Hours & Admission Prices: Sat.-Sun. 12-5; other times by appointment. Adults $7, children 6-12 $3; children under 6 no charge. Closed New Year's Day; Easter; Mother's Day; Father's Day; Christmas. &
Attendance: 4,000
Membership: Associate $25-$99; Individual $100-$249; Patron $250-$499; Cultural $500-$999; Display $1,000-$4,999; Life $5,000 & up.

UNIVERSITY OF KANSAS MEDICAL CENTER, CLENDENING HISTORY OF MEDICINE LIBRARY AND MUSEUM, 3901 Rainbow Blvd., Kansas City, KS 66160. Tel.: 913-588-7098. Fax: 913-588-7060.
E-mail: ccrenner@kumc.edu
Web Site: clendening.kumc.edu
Founded: 1945.
Key Personnel: Dir., Christopher Crenner; Rare Book Librarian, Dawn McInnis.
Governing Authority: state. Parent Institution: University of Kansas.
Institution Type/Description: Medical Museum.
Collections: microscopes; Egyptian amulets; surgical instruments; obstetrical forceps; brass acupuncture manikins; diagnostic doll; British touch pieces; 19th century American medical artifacts.
Facilities: 29,000-vol. library of medicine history available for inter-library loan except rare & fragile items which are to be used on premises only; reading room; research by appointment only.
Hours & Admission Prices: Museum: daily 8-4:30. Library: Mon. & Wed. 9-1, Tues. & Thurs. 12-4; other times by appointment. No charge. Closed New Year's Day; Martin Luther King Jr. Day; Easter; Memorial Day; Independence Day; Labor Day; Thanksgiving & day after; Christmas. &
Attendance: 1,000

Kingman

KINGMAN COUNTY HISTORICAL MUSEUM, 400 N. Main, Kingman, KS 67068-1304. Mailing Address: P.O. Box 281, Kingman, KS 67068-0281. Tel.: 620-532-5274.
E-mail: kcomuseum@gmail.com
Founded: 1969.
Congressional District: 4
Key Personnel: Pres., Gayle Dye.
Personnel Profile: Full-Time Volunteers 3; Part-Time Volunteers 12.
Governing Authority: nonprofit. Tax-exempt: 501(c)(3).
Institution Type/Description: History Museum: housed in 1888 City Hall.
Collections: World War I books & uniforms; World War II uniforms; Indian artifacts; salt mine; farm tools; family histories; fire engines; dresses; one-room school; records of county; early days band music; musical instruments; hospital surgical instruments; early furniture; dental equipment; old Cessna airplane parts; mural of Cessna plane & stage coach; church organ; death & birth records.
Research Fields: anything pertaining to Kansas; Kingman County.
Facilities: approx. 150-vol. library of Kansas history, Kingman County history & records of early school teachers available for use on premises; reading room. Postcards, personalized notes, replicas of Kingman KS buildings & museum publications for sale.
Activities: guided tours; lectures; permanent & temporary exhibitions. Museum Sponsors: program talks to Rotary, school classes, Boy & Girl Scouts.
Publications: book, History of Kingman County; Centennial History of Kingman KS; Centennial Cookbook of Kingman KS.
Hours & Admission Prices: Fri. 9-4; other times by appointment. No charge; donations accepted. Closed Christmas. &
Attendance: 2,500 (estimated)
Membership: Annual $25.

Kinsley

EDWARDS COUNTY HISTORICAL MUSEUM, Hwy. 50 & 56, #183, Kinsley, KS 67547-0064. Mailing Address: P.O. Box 64, Kinsley, KS 67547-0064. Tel.: 620-659-2420.
E-mail: librarian281942@yahoo.com
Web Site: edwardscountymuseum.info
Founded: 1967.
Congressional District: 1
Key Personnel: Pres. (V), Robert Cross; Cur., Mike Williams.
Personnel Profile: Full-Time Paid 1; Part-Time Paid 1.
Governing Authority: nonprofit organization. Parent Institution: Edwards County Historical Society. Tax-exempt: 501(c)(3).
Institution Type/Description: History Museum.
Collections: agriculture; costumes; Indian artifacts; farm implements; late 1880s period furnishings; sod house replicas.
Facilities: wood-framed church available for meetings & weddings. Museum-related items for sale.
Activities: guided tours; permanent & temporary exhibitions.
Hours & Admission Prices: May-Sept. Mon.-Sat. 9-5, Sun. 1-5; other times by appointment. No charge; donations accepted. &
Attendance: 2,695 (estimated)
Membership: Individual $5 & up.

La Crosse

BARBED WIRE MUSEUM, W. 1st St., La Crosse, KS 67548. Mailing Address: P.O. Box 578, La Crosse, KS 67548-0578. Tel.: 785-222-9900.
E-mail: barbedwire@rushcounty.org
Web Site: www.rushcounty.org/barbedwiremuseum
Founded: 1971.
Congressional District: 110
Key Personnel: C.E.O. & Pres. (V), Bradley R. Penka.
Personnel Profile: Part-Time Paid 2; Part-Time Volunteers 10.
Volunteer Hours: 250
Operating Expenses: 10,000
Operating Income: 10,000
Governing Authority: Parent Institution: Kansas Barbed Wire Collectors Association. Tax-exempt.
Institution Type/Description: Barbed Wire Museum.
Collections: over 2,000 varieties of barbed wire; tools & items related to barbed wire; fencing tools.
Facilities: reference library; research center.
Hours & Admission Prices: May-Sept. Mon.-Sat. 10-4:30, Sun. 1-4:30. No charge; donations accepted. &
Attendance: 1,000 (estimated)
Membership: Individual $25; Family $50; Sustaining $100; Lifetime $1,000.

RUSH COUNTY HISTORICAL SOCIETY, INC., 202 W. 1st St., La Crosse, KS 67548. Mailing Address: P.O. Box 473, La Crosse, KS 67548-0473. Tel.: 785-222-2719 & 3403.
Web Site: rushcounty.org
Formerly: Post Rock Museum
Founded: 1963.
Congressional District: 1
Key Personnel: Pres. (V), Lawrence Erbes; Treas., Ron Sandstrom.
Governing Authority: society. Parent Institution: Rush County Historical Society Inc., La Crosse, KS 67548. Branch Museums: Rush County Historical Museum; Nekoma State Bank Museum; Post Rock Museum. Tax-exempt.
Institution Type/Description: History Museum.
Collections: post rock products; quarrying tools; history of post rock; rolling pins; area historical memorabilia. Bank Museum: bank artifacts; teller stations; photographs. Historic Buildings: Post Rock House; Train Depot.
Research Fields: stories & uses of the post rock.
Activities: permanent exhibitions.
Hours & Admission Prices: May-Sept. Mon.-Sat. 10-4:30, Sun. 1-4:30; other times by appointment. No charge; donations accepted. &
Attendance: 1,500 (estimated)
Membership: Rush County Historical Society Inc.: Annual $5; Life $25.

La Cygne

LA CYGNE HISTORICAL SOCIETY MUSEUM, 300 N. Broadway, La Cygne, KS 66040. Mailing Address: RR 1, P.O. Box 9, La Cygne, KS 66040. Tel.: 913-757-2101.
Institution Type/Description: Historical Society Museum.
Collections: local history & culture; Native American artifacts; early machinery; photographs; personal artifacts.
Hours & Admission Prices: Sat.-Sun. 1-4.

Lakin

KEARNY COUNTY HISTORICAL MUSEUM, 111 S. Buffalo, Lakin, KS 67860. Mailing Address: P.O. Box 329, Lakin, KS 67860-0329. Tel.: 620-335-7448.
Founded: 1974.
Institution Type/Description: History Museum.
Collections: county history; Santa Fe Trail; period artifacts. Historic Buildings: depot; schoolhouse; 1882 house.
Hours & Admission Prices: Tues.-Fri. 9-4. No charge; donations accepted. &
Membership: Life $20.

Lansing

LANSING HISTORICAL MUSEUM, 115 E. Kansas Ave., Lansing, KS 66043-1667. Tel.: 913-250-0203. Facebook: Lansing Historical Museum.
E-mail: lphillippi@sbcglobal.net
Web Site: www.lansing.ks.us
Founded: 1992.
Congressional District: 2
Key Personnel: Site Supvr., Laura Phillippi.
Personnel Profile: Full-Time Paid 1; Part-Time Volunteers 20.
Volunteer Hours: 326
Operating Expenses: 59,748
Governing Authority: city of Lansing. Tax-exempt.
Institution Type/Description: Historic Building: housed in the restored Atchison, Topeka & Santa Fe depot built in 1887.
Collections: area history; railroad artifacts; photographs; period farming equipment.
Publications: The Whistle Stop.
Hours & Admission Prices: Tues.-Fri. 10-5, Sat. 10-3, Sun. 1-4. No charge; donations accepted. Closed federal holidays. &
Attendance: 746 (estimated)

Larned

CENTRAL STATES SCOUT MUSEUM, 815 Broadway, Larned, KS 67550-2525. Mailing Address: P.O. Box 392, Larned, KS 67550-0392. Tel.: 620-285-6427; 6431.
Founded: 1982.
Congressional District: 1
Key Personnel: Dir., C.E.O. & Chm. (V), Charles Sherman; Pres. (V), Jack Dipman
Governing Authority: Parent Institution: Pawnee Valley Scouts Inc. Tax-exempt.

Institution Type/Description: Scouting Museum.
Collections: scouting memorabilia; personal artifacts.
Activities: camping area; cabins; halls; showers for youth groups; bunkhouse camping.
Hours & Admission Prices: Adults $3, youth $2. &
Attendance: 484 (accurate)
Membership: $10-$500.

FORT LARNED NATIONAL HISTORIC SITE, 1767 KS Hwy. 156, Larned, KS 67550. Tel.: 620-285-6911. Fax: 620-285-3571.
E-mail: george_elmore@nps.gov
Web Site: www.nps.gov/fols
Founded: 1964.
Congressional District: 1
Key Personnel: Supt., Kevin McMurry; Cur., George Elmore; Museum Shop Mgr., Mike Seymour.
Personnel Profile: Full-Time Paid 14; Part-Time Paid 10; Part-Time Volunteers 250.
Governing Authority: Parent Institution: U.S. Dept. of Interior. Subsidiary Institution: National Park Service, Washington, DC 20240. Tax-exempt.
Institution Type/Description: Historic Site: historic fort on Santa Fe Trail.
Collections: military items of the period 1859-1878; 80,000 recovered artifacts. Historic Buildings: officers' quarters; quartermaster storehouse; commanding officer's quarters; two commissary storehouses; shops building containing the post bakery, carpenter, paint, saddlers, wheelwright and blacksmith shops; infantry barracks; combination barracks/post hospital; reconstructed blockhouse/guardhouse.
Research Fields: 1859-1878 U.S. Military; military uniforms, weapons, accoutrements; 1859-1878 fort furnishings; 1859-1878 history of Ft. Larned.
Facilities: 500-vol. library including photographs, slides, archival material, microfilm records for research on premises; visitor center with museum exhibits; picnic area; hiking trails. Books for sale.
Activities: guided & self-guided tours; lectures; exhibitions; history talks and living history demonstrations throughout the summer on weekends; self-guided history/nature trail; audio-visual program. Site Sponsors: Santa Fe Trail Days; Old Time Independence Day Celebration; Fall Candlelight Tour; Christmas Open House.
Publications: handbook, Fort Larned
Hours & Admission Prices: Daily 8:30-4:30. No charge; donations accepted. Closed New Year's Day; Thanksgiving; Christmas. &
Attendance: 40,000 (accurate)

*** SANTA FE TRAIL CENTER, (M),** 1349 K-156 Hwy., Larned, KS 67550-5347. Tel.: 620-285-2054. Fax: 620-285-7491.
E-mail: museum@santafetrailcenter.org
Web Site: www.santafetrailcenter.org
Founded: 1974.
Congressional District: 1
Key Personnel: Pres. (V), Tom Seltmann; Dir. & Cur., Anna Bassford; Education Dir. & Archivist, Katie Keckeisen; Office Mgr., Linda Revello; Supt. Bldgs. & Grounds, Farley Goertzen.
Personnel Profile: Full-Time Paid 3; Part-Time Paid 1; Part-Time Volunteers 379.
Governing Authority: society. Parent Institution: Fort Larned Historical Society, Inc. Tax-exempt: 501(c)(3).
Institution Type/Description: History Museum.
Collections: prehistoric & historic artifacts. Historic Buildings: 1908 L'Dora School from Frizell, KS; 1929 Santa Fe Railroad Depot from Frizell, KS; c.1870 Stone Cooling House; 1906 Escue Chapel CME Church.
Research Fields: economics & history of the Santa Fe Trail; Pawnee County history & genealogy.
Facilities: 2,000-vol. library of books dealing with the Santa Fe Trail & Pawnee County, Kansas, available for use in library reading room; auditorium. Museum-related items for sale.
Activities: guided tours; lectures; films; gallery talks; formally organized education programs; docent program; permanent & temporary exhibitions.
Publications: newsletter, Trail Ruts; pamphlet.
Hours & Admission Prices: Tues.-Sat. 9-5, Mon. by appointment. Adults $4, student 12-18 $2.50, children 6-11 $1.50; discounts to organized school groups; members no charge. Closed New Year's Day; Thanksgiving; Christmas. &
Attendance: 7,325 (accurate)
Membership: Student $10; Individual $30; Friend $50; Nonprofit & Civic Groups $100; Patron $100-$350; Business $250 & up; $1 A Day $365; Sustaining $370-$999; Benefactor $1,000 & up.

Lawrence

KU BIODIVERSITY INSTITUTE - KU NATURAL HISTORY MUSEUM, The University of Kansas, 1345 Jayhawk Blvd., Dyche Hall, Lawrence, KS 66045-7505. Tel.: 785-864-4540. Fax: 785-864-5335.
E-mail: naturalhistory@ku.edu
Web Site: naturalhistory.ku.edu
Founded: 1866.
Congressional District: 2
Key Personnel: Dir., Dr. Leonard Krishtalka; Chm. (V), Jann Rudkin; Assoc. Dir. Administration, Jordan Yochim; Asst. Dir. Informatics, James Beach; Cur. Mammalogy, Robert Timm; Cur. Ornithology, Townsend Peterson; Cur. Ornithology, Robert Moyle; Collection Mgr. Ornithology, Mark Robbins; Cur. Entomology, Caroline Chaboo; Cur. Entomology, Michael Engel; Cur. Entomology, Andrew Short; Museum Specialist Entomology, Jennifer Thomas; Collection Mgr. Entomology, Zack Falin; Cur. Invertebrate Paleontology, Bruce Lieberman; Collection Mgr. Invertebrate Paleontology, Una Farrell; Cur. Invertebrate Zoology, Daphne Fautin; Cur. Botany, Craig C. Freeman; Cur. Botany, Mark Mort; Collection Mgr. Botany, Caleb Morse; Cur. Herpetology, Rafe Brown; Collection Mgr. Herpetology, Andrew Campbell; Cur. Ichthyology, E.O. Wiley; Collection Mgr. Ichthyology, Andy Bentley; Cur. Vertebrate Paleontology, Larry Martin; Collection Mgr. Vertebrate Paleontology, Desui Miao; Preparator, Vertebrate Paleontology, David Burnham; Cur. Paleobotany, Tom Taylor; Cur. Paleobotany, Edith Taylor; Collection Mgr. Paleobotany, Rudolph Serbet; Cur. Parasitology, Kirsten Jensen; Cur. Archaeology, Mary Adair; Research Scientist, Global Biodiversity Science, Jorge Soberon; Accountant, Don Shobe; Dir. Exhibits, Bruce Scherting; Exhibits Asst., Gregory Ornay; Dir. Education, Teresa MacDonald; Program Specialist Education, Rebecca Lampe; Dir. Communications, Jen Humphrey; Asst. to Dir., Lori Schlenker.
Personnel Profile: Full-Time Paid 70; Part-Time Paid 20; Part-Time Volunteers 12.
Governing Authority: university. Parent Institution: The University of Kansas. Subsidiary Institution: R.L. McGregor Herbarium, 2045 Constant Ave., Univ. of Kansas, Lawrence, 785-864-4493; Entomology Div., 1501 Crestline Dr., Suite 140, Univ. of Kansas, Lawrence, 785-864-2234; Invertebrate Paleontology Div., 120 Lindley Hall, Univ. of Kansas, Lawrence, 785-864-2747; The Archaeology Research Center, 1340 Jayhawk Blvd., Spooner Hall, Lawrence, KS, 785-864-2675. Tax-exempt: 501(c)(3).
Institution Type/Description: Natural History Museum: housed in c.1901 Romanesque Revival building, listed on National Register of Historic Places.
Collections: eight and a half million specimens pertaining to botany, paleobotany, invertebrate paleontology, entomology, mammalogy, ornithology, herpetology, ichthyology, vertebrate paleontology; Great Plains animals & habitats including panorama of North American plants & animals; 7th Cavalry horse Comanche; live snakes, & bees; endangered species; 1.5 million archaeological artifacts.
Research Fields: systematic & evolutionary biology; ecology; environmental informatics; ecological niche modeling.
Facilities: 50,000 sq. ft. exhibit space; 75,000 sq. ft. of research collections & associated office space.
Activities: lectures; films; organized education programs for adults, children & undergraduate or graduate college students affiliated with Univ. of Kansas; training programs for professional museum workers and elementary & secondary science teachers; summer workshops for young people.
Publications: papers; monographs; publications; Nature in Kansas series; newsletters.
Hours & Admission Prices: Tues.-Sat. 9-5, Sun. 12-5. Suggested Donations: adults $5, children 6-18 & seniors $3; museum members, KU students, staff & faculty and children under 6 no charge. Closed most federal holidays. &
Attendance: 46,000 (estimated)
Membership: Senior & Youth $20; Individual $30; Household $40; Patron $75; Contributor $150; Curator's Circle $300 & up.

PRAIRIE PARK NATURE CENTER, 2730 S.W. Harper, Lawrence, KS 66046. Tel.: 785-832-7980.
Institution Type/Description: Nature Center.
Collections: plants; birds; insects; reptiles; amphibians; fish; mammals; photographs; fossils.
Facilities: nature trails; 71 acre park.
Activities: educational programs; summer camps; special events.
Hours & Admission Prices: Tues.-Sat. 9-5, Sun. 1-4. No charge.

THE ROBERT J. DOLE INSTITUTE OF POLITICS, (M), 2350 Petefish Dr., Lawrence, KS 66045-7555. Tel.: 785-684-4900. Fax: 785-684-1414.
E-mail: doleinstitute@ku.edu
Web Site: www.doleinstitute.org/vistors.html
Key Personnel: Dir., William B. "Bill" Lacy; Assoc. Dir. Programming, Jonathan Earle; Assoc. Dir. Outreach, Barbara Ballard; Senior Archivist, Dole Archive, Morgan Davis; Dir. Facilities & Events, Lawrence D. Bush; Media & Exhibits Archivist, Judy Sweets; Communications & Events Coord., Cori Ast; Asst. to Dir., Maggie Mahoney; Friends of Dole Institute, Lori Hutfles; Asst. Archivist, Catherine "Cat" C. Riggs; Asst. Archivist, Robert Lay; Mktg., Alison Heath Carther; Dir. Devel., Shawn McDaniel
Institution Type/Description: History Museum.
Collections: artifacts & memorabilia pertaining to Senator Robert Dole's life.
Hours & Admission Prices: Mon.-Sat. 9-5, Sun. 12-5. No charge; donations accepted. Closed New Year's Day; Thanksgiving; Christmas.

✱ SPENCER MUSEUM OF ART, THE UNIVERSITY OF KANSAS, (M), 1301 Mississippi St., Lawrence, KS 66045-7500. Tel.: 785-864-4710. Fax: 785-864-3112. TDD: 800-776-3777 (Kansas Relay).
E-mail: spencerart@ku.edu
Web Site: www.spencerart.ku.edu
Founded: 1928.
Congressional District: 2
Key Personnel: Dir., Saralyn Reece Hardy; Assoc. Dir. & Sr. Cur. Prints & Drawings, Stephen Goddard; Cur. European & American Arts, Susan Earle; Cur. Global Contemporary & Asian Art, Kris Ercums; Cur. Global Indigenous Art, Casey Mesick; Dir. Academic Prog., Celka Straughn; Asst. Dir. Collections, Janet Dreiling; Exhibition Designer, Richard Klocke; Information & Image Mgr., Robert Hickerson; Public Rels. Liaison, Gina Kaufmann-Smith; Dir. External Affairs, Margaret Perkins-McGuinness; Dir. Education, Kristina Walker; Dir. Internal Opers., Jennifer Talbott; Graphic Designer, Tristan Telander; Asst. to Dir, Traci Furan.
Personnel Profile: Full-Time Paid 24; Part-Time Paid 14; Part-Time Volunteers 160; Interns 7.
Governing Authority: university. Parent Institution: The University of Kansas. Tax-exempt: 170(b).
Institution Type/Description: Art Museum.
Collections: European & American painting, sculpture, prints, drawings, photographs, decorative arts; textiles; Japanese paintings & prints; Chinese paintings & sculpture; American & African art; ethnographic artifacts including Native American, African, Latin American & Australian cultures.
Major Exhibits: James Turrell, 9/13-5/14.
Research Fields: history of art in relation to teaching; use of permanent collections & exhibits.
Facilities: over 170,000-vol. library of art & architecture. Art books & museum-related items for sale.
Activities: guided tours; lectures; films; classes for children & adults; concerts; inter-museum loans, permanent, temporary & traveling exhibitions; international artist in residency program; interdisciplinary programs across the university.
Publications: annual journal, The Register of the Spencer Museum of Art; annual, Murphy Lecture in Art Series; exhibitions catalogues; collection catalogues; biannual newsletter.
Hours & Admission Prices: Tues. & Fri.-Sat. 10-4, Wed.-Thurs. 10-8, Sun. 12-4. No charge; donations accepted. Discount to AAM members in bookstore. Closed holidays. ♿
Attendance: 142,000 (estimated)
Membership: Student $15; Senior Citizen $35; Friend $50-$199; Donor $200-$499; Patron $500-$999; Benefactor $1,000-$2,499; Fellow $2,500-$4,999; Cornerstone $5,000 & up. Corporate: Friend $150-$299; Donor $300-$499; Patron $500-$999; Benefactor $1,000-$1,499; Associate $1,500-$2,499; Fellow $2,500-$4,999; Cornerstone $5,000 & up.

WATKINS MUSEUM OF HISTORY, 1047 Massachusetts St., Lawrence, KS 66044-2961. Tel.: 785-841-4109. Fax: 785-841-9547.
E-mail: info@watkinsmuseum.org
Web Site: www.watkinsmuseum.org
Formerly: Watkins Community Museum of History
Founded: 1972.
Congressional District: 3
Key Personnel: Dir., Steven J. Nowak; Cur. Collections Mgr., Brittany Keegan; Coord. Education Programs, Abby Magariel; Business Mgr., John Jewell.
Personnel Profile: Full-Time Paid 3; Part-Time Paid 2; Part-Time Volunteers 21; Interns 3.
Governing Authority: nonprofit society. Parent Institution: Douglas County Historical Society. Tax-exempt: 170(b)(1)(A).

Institution Type/Description: Local History Museum: housed in 1888 Richardsonian Romanesque building built by J.B. Watkins.
Collections: photographs & illustrations; documents including scrapbooks, maps & pamphlets; textile items including clothing, lace & quilts; toys, dolls & other personal items; tools & agricultural equipment; 1916 Miburn electric car; artifacts surviving William Quantrill's 1863 raid on Lawrence; 1879 reconstructed children's playhouse; 1900 surrey; Mexican war cannon Old Sacramento.
Research Fields: county & local history and genealogy; Bleeding Kansas, Civil War & Quantrill's Raid; 19th-20th century history; social history, Native Americans, local culture.
Facilities: archives of local history for use on premises; temporary & permanent exhibits.
Activities: lecture series; youth programming; workshops; historical programs; walking tours; concerts.
Publications: quarterly newsletter.
Hours & Admission Prices: Museum: Tues.-Wed. & Fri.-Sat. 10-4, Thurs. 10-8. Research Room: by appointment. No charge; donations accepted. Closed holidays. ♿
Attendance: 10,518 (accurate)
Membership: Student $15; Senior 60 & over $30; Individual $40; Household $60; Conservator $100-$249; Preservationist $250-$499; Patron $500-$999; Benefactor $1,000-$2,499; Fellow $2,500-$4,999; JB Watkins Director's Circle $5,000 & up.

Leavenworth

C.W. PARKER CAROUSEL MUSEUM, 320 S. Esplanade, Leavenworth, KS 66048-1585. Tel.: 913-682-1331.
E-mail: j-m-reinhardt@sbcglobal.net
Web Site: www.firstcitymuseums.org/carousel_main.html
Founded: 2005.
Congressional District: 2
Key Personnel: Dir., Jerry Reinhardt
Governing Authority: Parent Institution: Leavenworth Historical Museum Assoc. Tax-exempt.
Institution Type/Description: History Museum.
Collections: carousel artifacts & memorabilia; 3 operating carousels.
Publications: quarterly newsletter.
Hours & Admission Prices: Feb.-Dec. Thurs.-Sat. 11-5, Sun. 1-5. Tours: adults $6, children $3; discounts to groups of 20 or more; members no charge. Carousel Ride: $1.25. Closed New Year's Eve & Day; Easter; Independence Day; Thanksgiving; Christmas Eve & Day. ♿
Attendance: 23,000 (estimated)
Membership: Individual $20; Family $30; Friend $50; Benefactor $100; Corporate $200; Sponsor $500; Patron $1,000.

CARNEGIE ARTS CENTER, 601 S. 5th St., Leavenworth, KS 66048. Tel.: 913-651-0765.
Web Site: www.carnegieartscenter.org
Key Personnel: Dir., Michael Winer
Institution Type/Description: Art Gallery.
Collections: paintings; sculpture; photographs.
Activities: special events; gallery openings; educational classes.
Hours & Admission Prices: Mon.-Thurs. 1-7, Fri. 10-1. No charge; donations accepted.
Membership: Individual $40.

CARROLL MANSION MUSEUM HOME OF LEAVENWORTH COUNTY HISTORICAL SOCIETY, 1128 5th Ave., Leavenworth, KS 66048-3213. Tel.: 913-682-7759. Fax: 913-682-7759.
E-mail: leavenworthhistory@sbcglobal.net
Web Site: www.leavenworthhistory.org
Founded: 1964.
Congressional District: 2
Key Personnel: Pres., Dick Gervasini; Interim Dir., Beverly Lynch; Museum Shop Mgr., Hazel May Fackler.
Personnel Profile: Part-Time Paid 1; Part-Time Volunteers 50.
Volunteer Hours: 4,000
Governing Authority: Parent Institution: Leavenworth County Historical Society. Tax-exempt.
Institution Type/Description: History Museum: housed in 1867 Carroll Mansion.
Collections: furniture; silver; china; costumes; baths; Victorian furnishings & local artifacts; local history.
Research Fields: Leavenworth County & Kansas history.
Activities: guided tours; city tours; lectures; school programs & tours.
Publications: bimonthly, Historical Society Gazette.
Hours & Admission Prices: Year round: Tues.-Sat. 10:30-4:30. Adults $5,

seniors $4, children 5-12 $3; discounts to AAM & AAFLH members; children under 5 & members no charge. Closed holidays & inclement weather.
Attendance: 3,000 (estimated)
Membership: Student & Military $15; Individual & Senior 60 & over $20; Benefactor $500.

FIRST CITY MUSEUM, 743 Delaware St., Leavenworth, KS 66048-2472. Tel.: 913-682-1866. Fax: 913-682-1866.
E-mail: first_city_museum@yahoo.com
Web Site: www.firstcitymuseums.org
Founded: 1988.
Key Personnel: Pres., John Sanders; Dir., Jerry Reinhardt; Treas., Audrey Sanders; Museum Shop Mgr., Nancy Klemp.
Governing Authority: Parent Institution: Leavenworth Historical Museum Assoc. Tax-exempt.
Institution Type/Description: History Museum.
Collections: local history; photographs.
Facilities: Books for sale.
Activities: Museum Sponsors: Open Houses.
Hours & Admission Prices: Thurs. 1-5 & by appointment. No charge; donations accepted. &
Attendance: 335 (estimated)
Membership: Leavenworth Historical Museum Association: Individual $20; Family $30; Friend $50; Benefactor $100; Corporate $200; Sponsor $500; Patron $1,000.

NATIONAL FRED HARVEY MUSEUM, 624 Olive St., Leavenworth, KS 66048-2653. Tel.: 913-682-1866.
E-mail: fredharveymuseum@lvnworth.com
Web Site: www.firstcitymuseums.org/fredharvey_main.html
Institution Type/Description: Historic House Museum.
Collections: Harvey family history; personal artifacts; period furnishings.
Hours & Admission Prices: Tours by appointment.

Lecompton

CONSTITUTION HALL, 319 Elmore, Lecompton, KS 66050. Mailing Address: P.O. Box 198, Lecompton, KS 66050-0198. Tel.: 785-887-6520. Fax: 785-887-6520.
E-mail: consthall@kshs.org
Web Site: www.lecomptonkansas.com
Founded: 1986.
Personnel Profile: Full-Time Paid 1.
Governing Authority: Parent Institution: Kansas State Historical Society.
Institution Type/Description: History Museum.
Collections: local history & culture; photographs; period furnishings; personal artifacts.
Facilities: Museum-related items for sale.
Activities: special events.
Hours & Admission Prices: Wed.-Sat. 9-5, Sun. 1-5; other times by appointment. Adults $3, students $1; members, active military and children 5 & under no charge. Closed state holidays. &

TERRITORIAL CAPITAL-LANE MUSEUM, 393 N. 1900 Rd., Lecompton, KS 66050-4119. Mailing Address: P.O. Box 68, Lecompton, KS 66050. Tel.: 785-887-6148 & 6285. Fax: 785-887-6148.
E-mail: lanemuseum@aol.com
Web Site: www.lecomptonkansas.com
Founded: 1969.
Congressional District: 2
Key Personnel: C.E.O. & Pres., Paul M. Bahnmaier; Chm. (V) & Education, Charlene Winter; Dir., Rich McConnell; Treas. & Gift Shop Mgr., Betty Leslie; Cur., Arlene Simmons; Archivist, Mae Holderman; Membership, Iona Spencer; Public Rels., Opal Goodrick; Registrar, Helen Norwood; Security, Bob Weeks.
Personnel Profile: Part-Time Paid 2; Part-Time Volunteers 60.
Governing Authority: Lecompton Historical Society, not-for-profit organization. Tax-exempt: 501(c)(3).
Institution Type/Description: History Museum: located on foundation of proposed capitol building of Kansas.
Collections: Kansas territorial history through the Lane University era; quilts; early pioneer artifacts.
Research Fields: Lecompton history.
Facilities: 250-vol. library of territorial books & territorial maps of Kansas available to the public; Lane University chapel. Museum-related items for sale.
Activities: guided tours. Annual Event: Territorial Day in June.

Publications: quarterly newsletter, Bald Eagle.
Hours & Admission Prices: Wed.-Sat. 11-4, Sun. 1-5; tours by appointment; call 913-887-6285 for more information. No charge; donations accepted. &
Attendance: 5,300 (estimated)
Membership: Adult $10; Family $14; Life $100.

Lenexa

LEGLER BARN MUSEUM, 14907 W. 87th St., Lenexa, KS 66215-4135. Tel.: 913-492-0038.
E-mail: info@leglerbarn.org
Web Site: www.leglerbarn.org
Founded: 1983.
Key Personnel: Pres. (V), Todd Crow.
Personnel Profile: Part-Time Paid 1; Part-Time Volunteers 25.
Governing Authority: city. Parent Institution: Lenexa Historical Society. Tax-exempt.
Institution Type/Description: Historic Building: housed in stone barn built by Adam Legler in 1864.
Collections: period furnishings; personal artifacts.
Facilities: research library.
Hours & Admission Prices: Tues.-Sat. 10-4. Suggested Donation: adults $2, children $1. Closed holidays.
Attendance: 3,000 (estimated)
Membership: Lenexa Historical Society: Individual $20; Family $30; Lifetime $200; Corporate $250; Family Lifetime $300; Corporate Lifetime $2,500.

Leoti

WICHITA COUNTY HISTORICAL SOCIETY AND THE MUSEUM OF THE GREAT PLAINS, 201 N. 4th St., Leoti, KS 67861. Mailing Address: P.O. Box 1561, Leoti, KS 67861-1561. Tel.: 620-375-2316.
E-mail: museum@wichitacountymuseum.org
Web Site: www.wichitacountymuseum.org
Founded: 1982.
Congressional District: 1
Key Personnel: Dir., Curtis Walk.
Personnel Profile: Part-Time Volunteers 2.
Governing Authority: Tax-exempt.
Institution Type/Description: History Museum.
Collections: local history & culture; period furnishings; personal artifacts; photographs; 1892 Victorian home.
Hours & Admission Prices: Tues.-Fri. 1-5, Sat.-Sun. 2-5; other times by appointment. No charge; donations accepted. &
Attendance: 1,200 (estimated)
Membership: Adults $5; Lifetime: Individual $50; Family $100.

Liberal

BAKER ARTS CENTER, 624 N. Pershing Ave., Liberal, KS 67901-3115. Tel.: 620-624-2810. Fax: 620-624-7726.
E-mail: bakerartscenter@sbcglobal.net
Web Site: www.bakerartscenter.org
Institution Type/Description: Art Gallery.
Collections: paintings; sculpture; photography; quilts; children's hands-on exhibits.
Facilities: 2,000-vol. library.
Activities: classes; workshops.
Hours & Admission Prices: Tues.-Fri. 9-12 & 1-5, Sat. 2-5. No charge.

MID-AMERICA AIR MUSEUM, 2000 W. 2nd St., Liberal, KS 67901. Mailing Address: P.O. Box 2199, Liberal, KS 67905-2199. Tel.: 620-624-5263. Fax: 316-624-5454.
E-mail: jim.bert@cityofliberal.org
Web Site: www.liberalairmuseum.com
Formerly: Liberal Air Museum
Founded: 1987.
Congressional District: 1
Key Personnel: Exec. Dir., Jim Bert; Chm. (V), Steven Graham; Museum Shop Mgr., Cindy Chance.
Personnel Profile: Full-Time Paid 4; Part-Time Volunteers 50.
Governing Authority: Parent Institution: City of Liberal. Tax-exempt: 501(c)(3).
Institution Type/Description: Aviation Museum.
Collections: aircraft & artifacts relating to military & civilian aviation; photographs; personal collections; Hall of Aviation Science featuring hands-on displays.
Facilities: 20-vol. library of aviation books; 1,000-vols. of aviation magazines,

available for research; 250-seat auditorium; 80,000 sq. ft. exhibit space. Museum-related items for sale.

Activities: films; guided tours; traveling exhibitions.

Publications: monthly newsletter, Mid-America Air Museum News; quarterly newsletter, Flying Colors.

Hours & Admission Prices: Mon.-Fri. 8-5, Sat. 10-5, Sun. 1-5. Adults $7, senior citizens $5, children 6-18 $3; discounts for groups of 10 or more; children 5 & under and members no charge. Closed New Year's Day; Thanksgiving; Christmas. &

Attendance: 12,000 (accurate)

Membership: Individual $25; Family $50; Sustaining $100; Patron $250; Donor $500.

SEWARD COUNTY HISTORICAL MUSEUM, (M), 567 E. Cedar, Liberal, KS 67901-3865. Tel.: 620-624-7624.

Key Personnel: Exec. Dir., JoAnne Mansell

Institution Type/Description: History Museum.

Collections: county history; personal artifacts; period furnishings.

Hours & Admission Prices: Memorial Day to Labor Day Mon.-Sat. 9-6, Sun. 1-6; Sept.-May Tues.-Sat. 9-5, Sun. 1-5. Dorothy's House & Land of Oz: adults $5, senior citizens & children 6-18 $3.50; children 5 & under no charge.

Lincoln

CRISPIN'S DRUG STORE MUSEUM, 161 E. Lincoln Ave., Lincoln, KS 67455. Tel.: 785-524-5383.

E-mail: rxmuseumist@yahoo.com

Web Site: crispinsdrugstoremuseum.com

Founded: 2007.

Key Personnel: Dir., Jack D. Crispin, Jr.

Governing Authority: Parent Institution: Crispin Antiquarian Foundation. Tax-exempt.

Institution Type/Description: History Museum.

Collections: pharmaceutical history; period furnishings; personal artifacts; photographs.

Research Fields: pharmacy.

Hours & Admission Prices: By appointment. No charge; donations accepted.

Attendance: 300 (estimated)

Membership: Crispin Antiquarian Foundation: Basic $10; Patron $25; Grand Patron $50; Benefactor $100; Lifetime $500; Business Sponsor $1,000.

KYNE HOUSE MUSEUM, 216 W. Lincoln Ave., Lincoln, KS 67455. Mailing Address: P.O. Box 85, Lincoln, KS 67455-0085. Tel.: 785-524-9997.

E-mail: lchs10@att.net

Web Site: www.lincolncohistmuseum.com

Founded: 1978.

Congressional District: 1

Key Personnel: Pres. (V), Kathy Lupfer-Nielsen; Dir., Andrew Anderson.

Personnel Profile: Part-Time Paid 1; Part-Time Volunteers 14.

Governing Authority: Tax-exempt.

Institution Type/Description: Historic House Museum: built in 1885.

Collections: Lincoln County history, artifacts, culture & events; farm equipment; dolls; Indian artifacts; military artifacts; blacksmith; a hotel lobby; period store; doctor's office; tack & saddle artifacts; undertaker's equipment; 1920s furniture; F.A. Cooper's paintings, life, drawings, & metal etchings; genealogical records.

Facilities: library; research room. Museum-related items for sale.

Activities: guided tours; loan & participatory exhibits; educational programs for Lincoln County 2nd & 5th graders. Museum Sponsors: Lincoln Day & Lincoln Look-A-Like Contest.

Hours & Admission Prices: Kyne House: Tues. & Thurs. 1-4, Sat. 10-1. Suggested Donation $2. Marshall-Yohe House: by appointment. Adults $5, groups $4. &

Attendance: 1,400 (estimated)

Membership: Individual $10; Family $15; Business & Institution $25.

LINCOLN COUNTY HISTORICAL SOCIETY, 216 W. Lincoln Ave., Lincoln, KS 67455-1920. Mailing Address: P.O. Box 85, Lincoln, KS 67455-0085. Tel.: 785-524-9997.

E-mail: lchs10@att.net

Web Site: www.lincolncohistmuseum.com

Founded: 1978.

Congressional District: 1

Key Personnel: Pres., Kathy Lupfer-Nielsen; Treas., Brenda Peterson; Dir., Andrew Anderson.

Personnel Profile: Part-Time Paid 1; Part-Time Volunteers 14.

Volunteer Hours: 652

Governing Authority: private; nonprofit organization. Branch Institution: Marshall-Yohe House, Lincoln, KS 67455. Tax-exempt: 501(c)(3).

Institution Type/Description: General Museum.

Collections: Lincoln County history, artifacts, culture & events; farm equipment; dolls; Indian artifacts; military artifacts; blacksmith display; a hotel lobby; old time store; doctors office; tack & saddle room; an undertaker's display; 1920s furniture; F.A. Cooper's paintings, life story, drawings & metal etchings; Dolcette.

Research Fields: brochures.

Facilities: library; research room. Museum-related items for sale.

Activities: guided tours; loan & participatory exhibits; educational programs for Lincoln County 5th graders at Marshall-Yohe House. Annual Events: Lincoln Day Brunch for Abe Lincoln look-alike contestants & judges in February; Secretary Week luncheon at Marshall-Yohe House in June; Topsy School Program.

Publications: annual newsletter.

Hours & Admission Prices: Kyne House Museum: Tues. & Thurs. 1-4, Sat. 10-1. Suggested Donation: $2. Marshall-Yohe House: by appointment. Adults $5, groups $4. &

Attendance: 1,400 (estimated)

Membership: Individual $10; Family $15; Business & Institution $25.

POST ROCK SCOUT MUSEUM, 161 E. Lincoln Ave., Lincoln, KS 67455-2050. Tel.: 785-524-5383. Facebook: Post Rock Scout Museum.

Web Site: www.postrockscoutmuseum.com

Key Personnel: Owner, Kathie Crispin

Institution Type/Description: Scout Museum.

Collections: scouting history; Girl Scout uniforms from 1918 to present; badges; photographs; personal artifacts.

Activities: youth project area.

Publications: semiannual newsletter.

Hours & Admission Prices: By appointment. No charge; donations accepted.

Attendance: 350 (estimated)

Membership: $10; $25; $50; $100; $500; $1,000.

Lindsborg

BIRGER SANDZEN MEMORIAL GALLERY, 401 N. 1st St., Lindsborg, KS 67456-1813. Mailing Address: P.O. Box 348, Lindsborg, KS 67456-0348. Tel.: 785-227-2220. Fax: 785-227-4170.

E-mail: fineart@sandzen.org

Web Site: www.sandzen.org

Founded: 1957.

Congressional District: 4

Key Personnel: Sandzen Foundation Pres. & C.E.O., Tremenda Dillon; Dir., Larry L. Griffis; Cur., Ron Michael; Sec., Muriel Gentine.

Personnel Profile: Full-Time Paid 3; Full-Time Volunteers 14; Part-Time Paid 1; Part-Time Volunteers 11.

Governing Authority: nonprofit. Tax-exempt: 501(c)(3).

Institution Type/Description: Art Museum. Built in memory of Swedish-American painter and printmaker Birger Sandzen.

Collections: oils, watercolors, prints, archives of letters, papers & photographs by Birger Sandzen; family & contemporaries for research purposes; Henry Varnum Poor; Doel Reed; Lester Raymer; Swedish-American artists; paintings, ceramics, sculpture & graphics by well-known artists; Carl Milles fountain; oriental art.

Facilities: library containing art books; archives; recital area.

Activities: guided tours by appointment; lectures; gallery talks; concerts; loan exhibitions; college receptions.

Publications: book, Birger Sandzen: An Illustrated Biography; The Graphic Work of Birger Sandzen, by Charles Pelham Greenough, 3rd; a catalogue of the graphic work of Birger Sandzen; Sandzen and the New Land, an exhibition catalogue.

Hours & Admission Prices: Tues.-Sun. 1-5. No charge. Closed New Year's Eve & Day; Memorial Day; Independence Day; Thanksgiving; Christmas Eve & Day. &

Attendance: 11,227 (accurate)

Membership: Strom $30; Wallin $35, Jaderborg $50; Lofgren $100; Brase $250; Thorsen $500; Milles $1,000; Greenough $2,000 & up.

MCPHERSON COUNTY OLD MILL MUSEUM AND PARK, 120 Mill St., Lindsborg, KS 67456-2815. Mailing Address: P.O. Box 94, Lindsborg, KS 67456-0094. Tel.: 785-227-3595. Fax: 785-227-2810.

E-mail: oldmillmuseum@hotmail.com

Web Site: www.oldmillmuseum.org

Founded: 1959.

Congressional District: 4
Key Personnel: C.E.O. & Dir., Lorna Nelson; Museum Shop Mgr., Lenora Lynam.
Personnel Profile: Full-Time Paid 3; Full-Time Volunteers 2; Part-Time Paid 7; Part-Time Volunteers 75; Interns 1.
Governing Authority: county. Parent Institution: McPherson County. Tax-exempt.
Institution Type/Description: History Museum & Historic Site.
Collections: history; agriculture; Native American artifacts; industrial; folklore; natural history; geology; archives. Historic Buildings: 1898 Smoky Valley Roller Mill; 1904 Swedish Pavilion.
Research Fields: local & regional history with emphasis on Swedish & Mennonite heritage of area.
Facilities: library of county & local history information available for use on premises; reading room.
Activities: Special Events: Millfest, guided tours of 1898 flour mill in May; Heritage Christmas in December.
Publications: brochures & Mill history booklet; McPherson County Genealogical resources Guide.
Hours & Admission Prices: Mon.-Sat. 9-5. Adults $2, children 6-12 $1; discounts to school & group tours; children under 6 no charge. Closed New Year's Day; Thanksgiving; Christmas. &
Attendance: 10,000 (estimated)

RED BARN STUDIO MUSEUM, 212 S. Main, Lindsborg, KS 67456-2614. Tel.: 785-227-2217.
Web Site: www.redbarnstudio.org/
Founded: 1988.
Congressional District: 1
Key Personnel: Dir., Marsha Howe; Pres. (V), Todd Ray.
Personnel Profile: Part-Time Paid 4; Part-Time Volunteers 10.
Governing Authority: Parent Institution: Raymer Society for the Arts. Tax-exempt.
Institution Type/Description: Art Museum.
Collections: works by Lester Raymer; paintings; prints; ceramics; metalwork; woodcarving; furniture; jewelry.
Publications: quarterly, Raymer Society Newsletter.
Hours & Admission Prices: Tues.-Sun. 1-4; other times by appointment. No charge; donations accepted.
Attendance: 2,000 (estimated)
Membership: Student $15; Individual $25; Family $35; Friend $50; Donor $100; Sponsor $250; Patron $500; Benefactor $1,000.

Logan

DANE G. HANSEN MEMORIAL MUSEUM, (M), 110 W. Main St., Logan, KS 67646. Mailing Address: P.O. Box 187, Logan, KS 67646-0187. Tel.: 785-689-4846. Fax: 785-689-4892.
E-mail: hansenmuseum@ruraltel.net
Web Site: www.hansenmuseum.org
Founded: 1973.
Congressional District: 1
Key Personnel: Dir., Shirley A. Henrickson; Pres. Museum Bd., Carol Bales.
Personnel Profile: Full-Time Paid 2; Part-Time Paid 4.
Governing Authority: nonprofit organization. Tax-exempt: 501(c)(3).
Institution Type/Description: Art Museum.
Collections: Oriental art; gun, coin & spoon collection; work by Kansas artists; Hansen Family memorabilia.
Major Exhibits: Association of American Artists: Art by Subscription (T), 12/6/13-2/2/14.
Facilities: meeting space.
Activities: guided tours; arts festivals; permanent & traveling exhibitions.
Publications: quarterly newsletter; brochure, Hansen Family; brochure, Museum Plaza.
Hours & Admission Prices: Mon.-Fri. 9-12 & 1-4, Sat. 9-12 & 1-5, Sun. & holidays 1-5. No charge. Closed New Year's Day; Thanksgiving; Christmas. &
Attendance: 8,000 (estimated)
Membership: Benefactor $10; Patron $25; Sustaining $50.

Louisburg

CEDAR COVE FELINE SANCTUARY & EDUCATION CENTER, 3783 Hwy. K68, Louisburg, KS 66053. Tel.: 913-837-5515; 816-739-0363.
Web Site: www.saveoursiberians.org
Founded: 2000.
Key Personnel: Acting Dir. & Pres., Steve Klein
Institution Type/Description: Wildlife Refuge & Education Center.

Collections: tigers; cougars; bobcats; caracals; lions; leopards; wolves; white-nosed coati.
Activities: guided tours.
Hours & Admission Prices: April-Oct. Sat.-Sun. 10-3; Nov.-March Sat.-Sun. 11-3. Admission 3 & over $5. &

Lucas

GARDEN OF EDEN AND CABIN HOME, 305 E Second St., Lucas, KS 67648. Mailing Address: P.O. Box 57, Lucas, KS 67648. Tel.: 785-525-6395.
E-mail: info@garden-of-eden-lucas-kansas.com
Web Site: www.garden-of-eden-lucas-kansas.com
Founded: 1907.
Key Personnel: Pres., John Hachmeister; Vice Pres., Doug Hickman; Mgr., Lynn Schneider.
Personnel Profile: Part-Time Paid 6; Part-Time Volunteers 20.
Governing Authority: Parent Institution: Friend's of S.P. Dinosaur's Garden of Eden, Inc. Tax-exempt.
Institution Type/Description: Historic House Museum: housed in the former home of S.P. Dinsmoor. Listed on the National Register of Historic Places.
Collections: concrete sculptures; limestone log cabin.
Facilities: picnic area.
Hours & Admission Prices: March-April daily 1-4; May-Oct. daily 10-5; Nov.-Feb. Sat.-Sun. 1-4. Adults $7, children 6-12 $2; discounts to groups; children 5 & under no charge. Closed major holidays.
Attendance: 10,000 (estimated)

GRASSROOTS ART CENTER, 213 S. Main St., Lucas, KS 67648. Mailing Address: P.O. Box 304, Lucas, KS 67648-0304. Tel.: 785-525-6118. Facebook: Grassroots Art Center.
E-mail: grassroots@wtciweb.com
Web Site: www.grassrootart.net
Founded: 1991.
Congressional District: 1
Key Personnel: Dir., Rosslyn Schultz; Pres. (V), Janice Zamecnik; Museum Shop Mgr., Peg Gilbert.
Personnel Profile: Part-Time Paid 3; Part-Time Volunteers 12.
Volunteer Hours: 4,118
Operating Expenses: 102,476
Operating Income: 132,656
Governing Authority: Parent Institution: Lucas Arts & Humanities Council, Inc. Tax-exempt: 501(c)(3).
Institution Type/Description: Art Museum.
Collections: works by Midwest outsider artists.
Research Fields: outsider, intuitive, self-taught, & recycled art.
Activities: spring workshop. Museum Sponsors: Paper Artistry in April.
Publications: spring & fall newsletters.
Hours & Admission Prices: April & Oct. Thurs.-Mon. 1-4; May-Sept. Mon.-Sat. 10-5, Sun. 1-5; Nov.-March Thurs.-Sat. 1-4. Adults $6, children 6-12 $2; discounts to groups of 10 or more; members no charge. Closed winter holidays. &
Attendance: 6,800 (accurate)
Membership: Single $15; Family $25; Education $50; Visionary $100; Preservation $250; Exhibit $500; Program $1,000; Special Projects $5,000.

Lyndon

OSAGE COUNTY HISTORICAL SOCIETY RESEARCH CENTER, 631 Topeka Ave., Lyndon, KS 66451. Mailing Address: P.O. Box 361, Lyndon, KS 66451-0361. Tel.: 785-828-3477.
E-mail: researchosagechs@embarqmail.com
Web Site: www.osagechs.org
Founded: 1965.
Congressional District: 5
Key Personnel: C.E.O. & Pres. (V), Eileen Matzek Davis; Vice Pres., Marilyn Sanders; Treas., Ann Rogers.
Personnel Profile: Part-Time Volunteers 18.
Volunteer Hours: 4,000
Operating Expenses: 12,000
Operating Income: 12,000
Governing Authority: society. Parent Institution: Osage County Kansas. Tax-exempt.
Institution Type/Description: History Museum & Genealogy Library.
Collections: Osage County historical items; mining & railroad artifacts; civil & criminal court cases; 1864-1930 newspapers; 1865-1925 census.
Research Fields: local history; genealogy.
Facilities: museum at Lyndon.
Activities: Museum Sponsors: genealogical workshop.
Publications: quarterly, Hedge Post; book, Stories of Osage County & Its

Families; reprints, Annals of Lyndon: Early Days in Kansas; Along the Santa Fe & Lawrence Trails: Old Ridgeway; Council City 1854-55; Superior 1856; Burlingame 1856-64; County Atlas 1879; index, Probate County record; index, Osage County marriages & divorces; Osage County cemetery records; Osage County Murders; Sac & Fox Chief Keokuk in his Time; Mokohoko Chief of Sac & Fox Indians.

Hours & Admission Prices: Lyndon Museum & Research Center: April-Oct. Wed.-Sat. 1-5; other times by appointment. No charge; donations accepted. Closed holidays. ☒

Attendance: 850 (estimated)

Membership: Individual & Family $10; Business $15.

Lyons

RICE COUNTY HISTORICAL SOCIETY CORONADO QUIVIRA MUSEUM, (M), 105 W. Lyon, Lyons, KS 67554-2703. Tel.: 620-257-3941.

E-mail: director@cqmuseum.org

Web Site: www.cqmuseum.org

Formerly: Coronado Quivira Museum

Founded: 1927.

Congressional District: 1

Key Personnel: Dir., Charlene Akers; Pres. (V), Shirley Fair.

Personnel Profile: Full-Time Paid 1; Part-Time Paid 3; Part-Time Volunteers 23.

Governing Authority: society; nonprofit. Parent Institution: Rice County Historical Society. Branch Museum: Alden Santa Fe Depot Museum, Alden, KS 67512. Tax-exempt.

Institution Type/Description: General Museum: housed in 1910-1911 Carnegie library building.

Collections: Coronado & Quiviran Indian artifacts; Wichita Indians; Santa Fe Trail; 1541 Coronado chain mail; pre-1934 pioneer artifacts; anthropology; archaeology; pioneer; 20th century history; archives; genealogy; research library collections.

Major Exhibits: Apron Strings - Tics to the Past, 1/28/14-3/16/14; Kansas Women in Politics, 10/14-12/14.

Research Fields: Coronado & Quiviran Indian artifacts; pioneers; Santa Fe Trail; Wichita Indians; Rice County genealogy.

Activities: self-guided tours; lectures; films; permanent & temporary exhibitions; children's hands-on exhibits.

Publications: Chase (KS.) America; My One-Half Mile of the Santa Fe Trail; From the Little Arkansas to the Big Arkansas.

Hours & Admission Prices: Tues.-Sat. 9-5. Out of county visitors: ages 13 & over $3, children 6-12 $2; discounts to AAM members; children under 5 no charge. Closed major holidays. ☒

Attendance: 3,500 (estimated)

Membership: Individual $15; Family $25; Business $40; Donor $75 & up.

Manhattan

FLINT HILLS DISCOVERY CENTER, (M), 315 S. Third St., Manhattan, KS 66505. Mailing Address: P.O. Box 706, Manhattan, KS 66505. Tel.: 785-587-2726.

Web Site: www.flinthillsdiscovery.org

Founded: 2012.

Key Personnel: Dir., Fred Goss; Museum Shop Mgr., Penny Cullers.

Governing Authority: Parent Institution: City of Manhattan, Kansas Parks & Recreation Dept.

Institution Type/Description: History Museum.

Collections: Flint Hills history, ecology, & culture of the region; hands-on exhibitions.

Major Exhibits: Football (T), 1/17/14-5/31/14; Ice Age Imperials (T), 6/14/14-9/14/14; Forces II, 9/27/14-1/25/15.

Activities: hands-on exhibitions; rental facilities; special events; educational programs.

Hours & Admission Prices: Labor Day to Memorial Day Mon.-Wed. & Fri.-Sat. 10-5, Thurs. 10-8, Sun. 12-5. Adults $9, military, students & seniors 65 & over $7, children $4; children under 2 no charge. ☒

Membership: Military Youth $9.50; Youth 2-17 $10; Military/Veteran Adult, Educator, College Student & Senior 65 & up $20.25; Adult 18-24 $22.50.

GOODNOW MUSEUM STATE HISTORIC SITE, 2301 Claflin Rd., Manhattan, KS 66502. Mailing Address: 2309 Claflin Rd., Manhattan, KS 66502-3421. Tel.: 785-565-6490. Fax: 785-565-6491.

E-mail: ccollins@rileycountyks.gov

Web Site: www.kshs.org

Founded: 1969.

Congressional District: 2

Key Personnel: Contact, D. Cheryl Collins.

Personnel Profile: Part-Time Volunteers 10.

Governing Authority: state. Parent Institution: Kansas State Historical Society, 6425 S.W. 6th Ave., Topeka, KS 66615. Tel.: 785-272-8681. Subsidiary Institution: Riley County Historical Society & Museum. Tax-exempt.

Institution Type/Description: Historic House: home of pioneer Kansas educator, Isaac Tichenor Goodnow and his wife, Ellen Denison Goodnow.

Collections: artifacts relating to Goodnow family & early Kansas education programs, systems & schools.

Activities: guided tours.

Publications: brochure.

Hours & Admission Prices: Sat.-Sun. 2-5 & by appointment. No charge; donations accepted. Closed Easter; national holidays.

Attendance: 1,359 (accurate)

HAROLD M. FREUND AMERICAN MUSEUM OF BAKING, 1213 Bakers Way, Manhattan, KS 66502-4555. Mailing Address: P.O. Box 3999, Manhattan, KS 66505-3999. Tel.: 785-537-4750. Fax: 785-537-1493.

E-mail: information@aibonline.org

Web Site: www.aibonline.org

Founded: 1982.

Key Personnel: Librarian, Tammy L. Popejoy.

Governing Authority: nonprofit. Parent Institution: American Institute of Baking, Manhattan, KS. Tax-exempt: 501(c)(3).

Institution Type/Description: Baking Museum.

Collections: history of baking & milling.

Research Fields: bakery products; contributions by specific individuals & companies to the baking industry.

Facilities: library; 900 sq. ft. exhibit space.

Activities: group tours by appointment.

Hours & Admission Prices: Mon.-Fri. 8-5. No charge. Closed major holidays.

Attendance: 900 (estimated)

HARTFORD HOUSE MUSEUM, 2309 Claflin Rd., Manhattan, KS 66502-3421. Tel.: 785-565-6490.

Web Site: www.rileycountyks.gov/museum

Founded: 1974.

Congressional District: 2

Key Personnel: Dir., D. Cheryl Collins; Pres. (V), Lynn Berry.

Personnel Profile: Part-Time Volunteers 100.

Governing Authority: society. Parent Institution: Riley County Historical Society & Museum, 2309 Claflin Rd., Manhattan, KS 66502. Tax-exempt.

Institution Type/Description: Historic House: restored 1855 pre-fabricated house shipped on the Hartford Steamboat to Manhattan, KS.

Collections: Victorian period furnishings.

Publications: booklet; brochure.

Hours & Admission Prices: Tues.-Fri. 8:30-5 subject to availability of staff, Sat.-Sun. 2-5. No charge; donations accepted. Closed national holidays.

Attendance: 1,500 (estimated)

KANSAS STATE UNIVERSITY HERBARIUM, Div. of Biology, Ackert Hall, Manhattan, KS 66506-4900. Tel.: 785-532-6619. Fax: 785-532-6653.

E-mail: herbarium@ksu.edu

Web Site: www.ksu.edu/herbarium

Founded: 1875.

Congressional District: 2

Key Personnel: Dir., Carolyn Ferguson, Ph.D.

Personnel Profile: Full-Time Paid 1; Part-Time Paid 4.

Governing Authority: university. Tax-exempt.

Institution Type/Description: Herbarium.

Collections: botany; 180,000 plant specimens.

Research Fields: systematic botany; plant evolution.

Facilities: 3,000-vol. library of books on systematic botany available on premises.

Activities: formally organized education programs for undergraduate & graduate college students affiliated with Kansas State University; specimen identification available to the public; research in systematic botany.

Hours & Admission Prices: Mon.-Fri. 8-5. No charge. Closed national holidays. ☒

THE KANSAS STATE UNIVERSITY INSECT ZOO, 1500 Denison Ave., Manhattan, KS 66506. Tel.: 785-532-2847.

E-mail: insect@ksu.edu

Web Site: www.k-state.edu/butterfly/index.htm

Institution Type/Description: Insect Museum.

Collections: many species of tropical insects including tarantulas & spiders, scorpions & other arthropods.

Hours & Admission Prices: Tues.-Sat. 12-6; other times by appointment. Adults $2, senior citizens $1.50.

MANHATTAN ARTS CENTER, 1520 Poyntz Ave., Manhattan, KS 66502-4147. Tel.: 785-537-4420. Fax: 785-539-3356.
E-mail: director@manhattanarts.org
Web Site: www.manhattanarts.org
Founded: 1996.
Key Personnel: Exec. Dir., Penny Senften; Devel. Dir., Amy Gross; Education & Mktg. Dir., Kim Belanger.
Governing Authority: private; nonprofit organization. Tax-exempt.
Institution Type/Description: Art Museum.
Collections: Gordon Parks photograph collection, From the Huge Silence: A Century of Life in a Small Kansas Town; Grandma Layton drawing; F. Remington sculpture, Bronco Buster; Remington cowboy.
Facilities: 125-seat auditorium; classroom; theatre.
Activities: concerts; lectures; rental gallery; temporary exhibitions; theater.
Publications: magazine, City Arts.
Hours & Admission Prices: Mon.-Fri. 12-5, Sat. 1-4. No charge; donations accepted. Closed Christmas. &
Attendance: 5,000 (estimated)
Membership: Fan $25-$99; Aficionado $100-$249; Director $250-$499; Master Artist $500-$999; Maestro $1,000 & up.

* **MARIANNA KISTLER BEACH MUSEUM OF ART AT KANSAS STATE UNIVERSITY, (M),** 701 Beach Lane, Manhattan, KS 66506-0601. Tel.: 785-532-7718. Fax: 785-532-7498.
E-mail: beachart@ksu.edu
Web Site: beach.k-state.edu
Founded: 1996.
Congressional District: 2
Key Personnel: Dir., Linda Duke; Pres. (V), Jackie Hartman; Asst. Dir. Operations, Cindi Morris; Asst. Cur., Liz Seaton; Registrar & Collections Mgr., Sarah Price; Business Mgr., Martha Scott; Coord. Pub. Programs & Events, Adrianne Russell; Exhibitions Designer, Lindsay Smith; Sr. Educator, Kathrine Walker Schlageck.
Personnel Profile: Full-Time Paid 10; Part-Time Paid 15; Part-Time Volunteers 25; Interns 5.
Governing Authority: nonprofit. Parent Institution: Kansas State University, Manhattan, KS. Tax-exempt: 501(c)(3).
Institution Type/Description: Art Museum.
Collections: art of Kansas & the Mountain Plains region with emphasis on printmaking & painting; works by John Steuart Curry; 20th- & 21st-century American printmaking.
Research Fields: 20th-21st century art of Kansas & the Mountain Plains region; 20th-century Kansas printmaking; 20th-21st century American printmaking.
Facilities: 4,000-vol. library; 10,600 sq. ft. exhibit space; 100-seat auditorium; classroom. Museum-related items for sale.
Activities: docent program; formal education programs for children & undergraduate or graduate college students; guided tours; lectures; loan & traveling exhibitions; school loan service; pre-professional museum training; teachers workshops.
Publications: newsletter, In Sight; exhibition catalogues.
Hours & Admission Prices: Tues.-Sat. 10-5, Sun. 12-5; reduced hours during academic breaks. No charge; donations accepted. Closed major holidays. &
Attendance: 32,273 (accurate)
Membership: Student $10; Individual $40-$99; Family $60-$99; Supporter $100-$249; Sponsor $250-$499; Benefactor $500-$999; Director $1,000 & up.

PIONEER LOG CABIN, City Park, 11th & Poyntz, Manhattan, KS 66502. Mailing Address: 2309 Claflin, Manhattan, KS 66502-3421. Tel.: 785-565-6490. Fax: 785-565-6491.
Web Site: www.rileycountyks.gov/museum
Founded: 1914.
Congressional District: 2
Key Personnel: Pres. (V), Lynn Berry; Dir., D. Cheryl Collins.
Personnel Profile: Part-Time Paid 1.
Governing Authority: Society. Parent Institution: Riley County Historical Society. Tax-exempt: 501(c)(3).
Institution Type/Description: History Museum: housed in log cabin.
Collections: farm & shop tools; log cabin pioneer home.
Activities: tours.
Publications: newsletter.
Hours & Admission Prices: April-Oct. Sun. 2-5; other times by appointment. No charge; donations accepted. &

Attendance: 500 (estimated)
Membership: Riley County Historical Society: Individual: Friend $10-$49; Sponsor $50-$74; Sustainer $75-$99; Patron $100 & up. Family: Friend $15-$54; Sponsor $55-$79; Sustainer $80-$110; Patron $110 & up.

RILEY COUNTY HISTORICAL MUSEUM, (M), 2309 Claflin Rd., Manhattan, KS 66502-3421. Tel.: 785-565-6490. Fax: 785-565-6491.
Web Site: www.rileycountyks.gov/museum
Founded: 1914.
Congressional District: 2
Key Personnel: Dir., D. Cheryl Collins.
Personnel Profile: Full-Time Paid 3; Part-Time Paid 5; Part-Time Volunteers 131; Interns 1.
Governing Authority: Parent Institution: Riley County and Riley County Historical Society. Branch Museums: Hartford House Museum; Goodnow House State Historic Site; Pioneer Log Cabin; Wolf House; Rocky Ford School; Randolph Jail. Tax-exempt: 501(c)(3).
Institution Type/Description: History Museum.
Collections: household equipment, furniture, tools, weapons, musical instruments, toys, glass, china, school equipment, church equipment, doctor's & dentist's equipment; local Indian artifacts; dolls; transportation; hardware; pictures, photos, newspapers, books relating to local history; manuscript collections.
Research Fields: history of Riley County.
Facilities: 4,000-vol. library of county history. Prints of historic buildings, books, stationery & cards for sale.
Activities: guided tours; lectures; films; drama; formally organized education programs; docent program; permanent & temporary exhibitions; intern program.
Publications: newsletter; brochure; books on Riley County.
Hours & Admission Prices: Museum: Tues.-Fri. 8:30-5, Sat.-Sun. 2-5. Library: by appointment. No charge; donations accepted. Closed national holidays. &
Attendance: 10,000 (estimated)
Membership: Riley County Historical Society: Individual: Friend $10-$49; Sponsor $50-$74; Sustainer $75-$99; Patron $100 & up; Life $300. Family: Friend $15-$54; Sponsor $55-$79; Sustainer $80-$110; Patron $110 & up.

STRECKER-NELSON GALLERY, 406 1/2 Poyntz Ave., Manhattan, KS 66502. Tel.: 785-537-2099. Fax: 785-539-2139.
E-mail: gallery@kansas.net
Web Site: www.strecker-nelsongallery.com
Institution Type/Description: Art Gallery.
Collections: paintings; sculpture.
Activities: temporary exhibits.
Hours & Admission Prices: Mon.-Sat. 10-6.

SUNSET ZOOLOGICAL PARK, (M), 2333 Oak St., Manhattan, KS 66502-3824. Tel.: 785-587-2737. Fax: 785-587-2730.
E-mail: shoemaker@cityofmhk.com
Web Site: www.sunsetzoo.com
Founded: 1933.
Congressional District: 2
Key Personnel: C.E.O., Scott Shoemaker.
Personnel Profile: Full-Time Paid 15; Part-Time Paid 15; Part-Time Volunteers 160; Interns 10.
Governing Authority: municipal government. Parent Institution: City of Manhattan. Tax-exempt.
Institution Type/Description: Zoo.
Collections: home to 220 animals representing 104 species from around the world featuring: red panda, snow leopard, chimpanzee, maned wolves, Caribbean flamingos & cheetahs.
Research Fields: conservation; Conservation Action Partnership (CAP) Paraguay.
Facilities: 48 acres; concession. Museum-related items for sale.
Activities: 14 special events each year.
Publications: quarterly magazine; annual report.
Hours & Admission Prices: April-Oct. daily 9:30-5; Nov.-March daily 12-5. Adults, senior citizens & students $4, children 3-12 $2; discounts to groups; members no charge. &
Attendance: 64,500 (accurate)
Membership: Senior Citizen $20; Individual $40; Individual Plus $45; Family $50; Zoo Buff $75; Director's Club Associate $250; Director's Club Patron $500; Director's Club Benefactor $1,000.

WOLF HOUSE MUSEUM, 630 Fremont, Manhattan, KS 66502-5820. Mailing Address: Riley County Historical Society, 2309 Claflin Rd., Manhattan, KS 66502. Tel.: 785-565-6490. Fax: 785-565-6491.
Web Site: www.rileycountyks.gov/museum
Founded: 1983.
Congressional District: 2
Key Personnel: Pres. (V), Lynn Berry; Dir., D. Cheryl Collins; Cur., Edna Williams.
Personnel Profile: Part-Time Paid 3; Part-Time Volunteers 50.
Governing Authority: society. Parent Institution: Riley County Historical Society. Tax-exempt.
Institution Type/Description: Historic House: 1868 boarding house.
Collections: c.1830-1900 Wolf collection; 1880s artifacts.
Research Fields: 19th-century furniture; boarding houses.
Activities: Museum Sponsors: special events in late spring & Christmas.
Publications: brochure.
Hours & Admission Prices: Sat.-Sun. 2-5. No charge; donations accepted.
Attendance: 1,500 (estimated)
Membership: Riley County Historical Society: Individual: Friend $10-$49; Sponsor $50-$74; Sustainer $75-$99; Patron $100 & up; Life $300. Family: Friend $15-$54; Sponsor $55-$79; Sustainer $80-$110; Patron $110 & up.

WONDER WORKSHOP CHILDREN'S MUSEUM, 506 S. 4th St., Manhattan, KS 66502. Tel.: 785-776-1234.
E-mail: wonder@kansas.net
Web Site: www.wonderworkshop.org
Key Personnel: Exec. Dir., Richard Pitts; Pres., David Griffin
Institution Type/Description: Children's Museum.
Collections: hands-on exhibits.
Activities: workshop; special events; after-school programs.
Hours & Admission Prices: Mon.-Sat. 10-2. Admission $3; children one & under no charge. &

Mankato

JEWELL COUNTY HISTORICAL MUSEUM, 118 N. Commercial St., Mankato, KS 66956-2207. Tel.: 785-786-3337.
E-mail: jchs1870@hotmail.com
Web Site: www.jewellcountyhistory.com
Founded: 1959.
Congressional District: 6
Key Personnel: C.E.O., Roger Fedde; Pres. (V), Leon Boden; Cur. & Museum Shop Mgr., Janelle Greene.
Personnel Profile: Part-Time Paid 1; Part-Time Volunteers 3.
Governing Authority: nonprofit. Tax-exempt.
Institution Type/Description: Agriculture Museum.
Collections: farm implements; county history; Federal & State census records, 1870-1920; political & military memorabilia; turn of the century period rooms; general store; post office; pharmacy; doctor's office; old limestone jail; blacksmith shop; geology; oral history recordings of county residents.
Facilities: library of county newspapers on microfilm.
Activities: quarterly education programs; guided tours; temporary exhibitions. Museum Sponsors: annual Antique Farm Machinery Show and Threshing Bee; Arts & Crafts Festival.
Publications: quarterly newsletter; family histories of people of the county.
Hours & Admission Prices: Wed.-Fri. 1-5; other times by appointment. No charge; donations accepted. Closed holidays.
Attendance: 500 (estimated)
Membership: Individual $5; Family $10; Life $100.

Marion

MARION CITY MUSEUM, 623 E. Main St., Marion, KS 66861-1800. Tel.: 620-382-9134 & 3703.
Institution Type/Description: History Museum: housed in the former Baptist church building; built in 1887.
Collections: local history; 19th century dresses, needlework, dolls, toys & wooden washing machines; photographs; personal artifacts; period furnishings.
Hours & Admission Prices: May to mid-Oct. Tues.-Sat. 10-2, Sun. 12-2; other times by appointment.

Marquette

KANSAS MOTORCYCLE MUSEUM, 120 N. Washington, Marquette, KS 67464. Tel.: 785-546-2449.
Key Personnel: Cur., LaVona Engdahl
Institution Type/Description: Motorcycle Museum.

Collections: over 100 early motorcycles; racing memorabilia; photographs; posters; trophies.
Hours & Admission Prices: Mon.-Sat. 10-5, Sun. 11-5; groups by appointment. No charge; donations accepted.

MARQUETTE MUSEUM, 202 N. Washington, Marquette, KS 67464.
Governing Authority: Parent Institution: Marquette Historical Society. Tax-exempt.
Institution Type/Description: Historical Society Museum: building built in 1910.
Collections: local history & culture; period furnishings; photographs; personal artifacts.
Hours & Admission Prices: May-Sept. Sun. 2-5. &
Attendance: 1,200

RANGE SCHOOL MUSEUM, 204 N. Washington, Marquette, KS 67464.
Institution Type/Description: History Museum: housed in a former country school; built in 1906.
Collections: local history; period furnishings; photographs.
Hours & Admission Prices: May-Sept. Sun. 2-5.
Attendance: 1,000 (estimated)

Marysville

DOLL HOUSE MUSEUM, 912 Broadway, Marysville, KS 66508-1805. Mailing Address: 1107 Pony Express Hwy., Marysville, KS 66508. Tel.: 785-562-3029. Fax: 785-562-2990.
E-mail: candcdoll@yahoo.com
Founded: 1997.
Key Personnel: Dir., Lois Cohorst; Pres. (V), Deb Krashaar.
Personnel Profile: Part-Time Volunteers 2.
Governing Authority: Tax-exempt: 501(c)(3).
Institution Type/Description: Doll, Toy, & Indian Artifacts Museum.
Collections: dolls; toys; Native American artifacts; rocking horses; pedal toys; doll houses.
Research Fields: dolls.
Activities: storytelling; educational touring; doll meetings; convention.
Publications: books, The Mystery of Mary McEivin; The Legend of Edwina Farris Dolls.
Hours & Admission Prices: By appointment. Suggested Donation: $5 per person. &
Attendance: 650 (estimated)

KOESTER HOUSE MUSEUM, 209 N. 8th, Marysville, KS 66508-1637. Tel.: 785-562-2417.
Institution Type/Description: Historic House Museum: housed in the former home of banker, Charles F. Koester; c.1876.
Collections: period furnishings; personal artifacts; portraits; clothing; toys; books; white bronze sculptures.
Facilities: gardens.
Hours & Admission Prices: May-Nov. Tues.-Sun. 10-12 & 1-4:30; other times by appointment.

MARSHALL COUNTY HISTORICAL SOCIETY, 1207 Broadway, Marysville, KS 66508-1845. Tel.: 785-562-5012.
E-mail: mchs@bluevalley.net
Web Site: skyways.lib.ks.us/museums/mchc
Founded: 1980.
Congressional District: 1
Governing Authority: Tax-exempt.
Institution Type/Description: Historic Building: housed in the county courthouse; built in 1891.
Collections: county history; books; photography.
Facilities: library.
Hours & Admission Prices: Museum: Memorial Day to Sept. 15 daily 1-4; Winter: Mon.-Fri. 1-4. Library: Mon.-Fri. 1-4.
Attendance: 700 (estimated)

PONY EXPRESS ORIGINAL HOME - STATION #1, 106 S. 8th St., Marysville, KS 66508-1832. Tel.: 785-562-3825.
Institution Type/Description: Historic Building: built in 1859 by Joseph Cottrell; original home station along the Pony Express route.
Collections: Pony Express mail service history.
Hours & Admission Prices: April-Oct. Mon.-Sat. 9-4, Sun. 12-4.
Attendance: 2,132 (accurate)

McCracken

MCCRACKEN HISTORICAL MUSEUM, 200 Main St., McCracken, KS 67556. Mailing Address: P.O. Box 342, McCracken, KS 67556-0342. Tel.: 785-394-2540 & 2446.
E-mail: yawgerr@ruraltel.net
Institution Type/Description: History Museum.
Collections: local history & culture; period furnishings; personal artifacts; photographs.
Hours & Admission Prices: By appointment.

McPherson

MCPHERSON MUSEUM & ARTS FOUNDATION, (M), 1111 E. Kansas Ave., McPherson, KS 67460. Tel.: 620-241-8464. Fax: 620-241-2676.
E-mail: carla@mcphersonmuseum.com
Web Site: mcphersonmuseum.com
Founded: 1984.
Congressional District: 1
Key Personnel: Exec. Dir., Carla Barber; Pres., Marck Cobb.
Personnel Profile: Full-Time Paid 2; Part-Time Paid 1; Part-Time Volunteers 2.
Governing Authority: McPherson Museum & Arts Foundation. Tax-exempt: 501(c)(3).
Institution Type/Description: General Museum.
Collections: Oriental porcelains; Native American artifacts; Pleistocene fossils; rocks & minerals; meteorites; period furniture & clocks; pioneer farm tools & household goods; art.
Research Fields: local history; paleontology.
Activities: lectures; research; open microphone event for poetry, music & short stories; writers' conference; student writers' competition.
Publications: newsletter, The Diamond; brochures.
Hours & Admission Prices: Tues.-Sat. 9-5, Sun. 1-5. Adults $7.50, seniors 65 & up and students 4 & up w/ID $5.50; children 3 & under and members no charge. Closed major holidays.
Attendance: 4,000 (estimated)
Membership: Railroader & Clubs $35; Bagpiper $50; Refiner $75; Light Capital Club $100; World Travelers' Club $250; Diamond Club $500; Director's Circle $501 & up.

Meade

DALTON GANG HIDEOUT, 502 S. Pearlette St., Meade, KS 67864. Mailing Address: P.O. Box 515, Meade, KS 67864-0515. Tel.: 620-873-2731; 800-354-2743.
E-mail: daltonhideout@yahoo.com
Web Site: oldmeadecounty.com
Founded: 1941.
Congressional District: 1
Key Personnel: Mgr., Marc S. Ferguson.
Personnel Profile: Full-Time Paid 1.
Governing Authority: Parent Institution: Meade County Historical Society.
Institution Type/Description: Historic House Museum: housed in the former home of Eva Dalton, sister of the Dalton Gang.
Collections: House: period furnishings. Barn: 95 ft. long tunnel to the house; period artifacts.
Facilities: Barn serves as entrance to the house. Museum-related items for sale.
Publications: Historical Society Newsletter.
Hours & Admission Prices: Mon.-Sat. 9-5, Sun. 1-5. Admission $5.
Attendance: 9,000 (accurate)

MEADE COUNTY HISTORICAL SOCIETY MUSEUM, 200 E. Carthage, Meade, KS 67864-0893. Mailing Address: P.O. Box 893, Meade, KS 67864-0893. Tel.: 620-873-2359. Fax: 620-873-2359. Facebook: Meade County Historical Museum.
E-mail: meademuseums@yahoo.com
Web Site: www.meadecountymuseum.com
Founded: 1969.
Congressional District: 1
Key Personnel: C.E.O., Norman Dye; Vice Chm., Glen Lauppe; Administration, Nancy Ohnick.
Personnel Profile: Part-Time Paid 4; Part-Time Volunteers 2.
Governing Authority: society; nonprofit. Parent Institution: Meade County Historical Society, Inc. Tax-exempt.
Institution Type/Description: General Museum.
Collections: American Indian artifacts; 400-600 year old artifacts from archaeological dig; general store; barber shop; harness shop; blacksmith shop; sod house; parlor, kitchen, bedroom of turn-of-the-century home; covered wagon; agricultural exhibits; photographer's studio; bank; doctor's office; school & chapel; Rock Island & Pacific Railroad; Livery Barn; county schoolhouse; farm machinery.
Research Fields: Meade County history.
Facilities: 500-vol. library, to be used on premises; microfilms of all 1878-2005 county newspapers; reading room. County history books & other western history books for sale.
Activities: guided tours; formally organized education programs for children; permanent exhibitions.
Publications: annual newsletter; book, Meade County History.
Hours & Admission Prices: Tues.-Sat. 9-5, Sun. 1-5. No charge; donations accepted. Closed major holidays.
Attendance: 5,000 (estimated)
Membership: Ranch Hands $25; Pioneer $100; Homesteaders $250; Trail Boss $500.

Medicine Lodge

CARRY A. NATION HOME AND STOCKADE MUSEUM, 209-211 W. Fowler, Hwy. 160, Medicine Lodge, KS 67104. Tel.: 620-886-3553.
Founded: 1950.
Congressional District: 7
Key Personnel: C.E.O. & Treas., Dorothy Reed; Museum Shop Mgr., Ann Bryan; Museum Shop Mgr., Janet Hindman.
Personnel Profile: Part-Time Paid 2; Part-Time Volunteers 4.
Governing Authority: society. Parent Institution: City of Medicine Lodge. Tax-exempt.
Institution Type/Description: Historic House Museum: 1880-1903 home of Carry A. Nation.
Collections: furniture & original articles used by the Nations; Carry A. Nations' bed, organ, hat, dresser, valise and purse; stockade; period artifacts; old jail. Historic Building: 1877 log cabin.
Activities: guided tours.
Publications: pamphlet, A Short Biography of Carry A. Nation.
Hours & Admission Prices: Summer: 10:30-5; Winter: 1-4. Adults $5, senior citizens $4, children 6-17 $3; children under 5 no charge.
Attendance: 500 (estimated)

Minneapolis

OTTAWA COUNTY HISTORICAL MUSEUM, 110 S. Concord St., Minneapolis, KS 67467-2322. Tel.: 785-392-3621.
E-mail: otcomu@networksplus.net
Founded: 1959.
Key Personnel: Dir., Mr. Jettie Condray
Governing Authority: county. Tax-exempt.
Institution Type/Description: County History Museum.
Collections: local history & culture; George Washington Carver artifacts; military; fossils; period furnishings; Eisenhauer dolls; personal artifacts; photographs; Lewis & Clark; beaver trapping; Silvisaurus condrayi dinosaur artifacts.
Research Fields: George Washington Carver in Minneapolis, KS 1880-1884, at the ages of 16-20.
Hours & Admission Prices: Tues.-Sat. 10-12 & 1-5. No charge; donations accepted.
Attendance: 3,000 (estimated)

Montezuma

STAUTH MEMORIAL MUSEUM, (M), 111 N. Aztec St., Montezuma, KS 67867-0396. Mailing Address: P.O. Box 396, Montezuma, KS 67867-0396. Tel.: 620-846-2527. Fax: 620-846-2810.
E-mail: stauthm@ucom.net
Web Site: stauthmemorialmuseum.org
Founded: 1996.
Congressional District: 1
Key Personnel: Dir., Financial Dir. & Public Rels., Kim Legleiter.
Personnel Profile: Full-Time Paid 1; Part-Time Paid 2.
Governing Authority: private; nonprofit organization. Parent Institution: Claude & Donald Stauth Foundation. Tax-exempt: 501(c)(3).
Institution Type/Description: Decorative Arts Museum.
Collections: decorative art, woodcarvings, sculpture, ivory carvings, musical instruments, real & ceremonial weapons, foreign coins & currency, fur wall hangings, animal hides, cloisonne & jewelry collected from around the world; over 10,000 slides; big game animal trophies from Canada, Alaska & Norway, as well as other parts of North America; western bronze sculptures including Frederic Remington bronze reproductions.
Facilities: 850-vol. library of National Geographic magazines; 9,000 sq. ft. exhibit space; 100-seat community room serves as an auditorium, classroom, theatre & meeting room.

Activities: films; guided tours; lectures; traveling & temporary exhibitions.
Publications: traveling exhibition schedule.
Hours & Admission Prices: Tues.-Sat. 9-12 & 1-4:30, Sun. 1:30-4:30. No charge; donations accepted. Closed New Year's Day; Easter; Independence Day; Thanksgiving; Christmas. &
Attendance: 3,248 (accurate)

Mulvane

MULVANE HISTORICAL MUSEUM, 300 W. Main, Mulvane, KS 67110-1779. Mailing Address: P.O. Box 17, Mulvane, KS 67110-1779. Tel.: 316-777-0506.
Web Site: mulvanedepot.com
Founded: 1983.
Institution Type/Description: History Museum.
Collections: local history; photographs; period artifacts.
Hours & Admission Prices: Tues.-Sat. 10-4. No charge; donations accepted. Closed holidays.
Attendance: 1,300 (estimated)
Membership: Annual $5; Lifetime $50.

Neodesha

NORMAN #1 OIL WELL MUSEUM, 106 S. First St., Neodesha, KS 66757-1802. Tel.: 620-325-5316. Fax: 316-325-5316.
E-mail: norman1@terraworld.net
Founded: 1967.
Key Personnel: Pres., Dan Railsback; Dir., Jackie Clark.
Personnel Profile: Full-Time Paid 1; Part-Time Paid 1; Part-Time Volunteers 3.
Governing Authority: municipal; nonprofit organization. Tax-exempt.
Institution Type/Description: Historical & Oil Museum: first commercial oil well in the mid-continent oil fields.
Collections: full sized derrick; oil well equipment; farming tools & equipment; clown clothing & props used by two former Neodesha High School graduates that have since retired from Ringling Bros. Circus; Indian tools & artifacts; WPA dolls, standard oil refinery Osage Indian artifacts.
Facilities: small chapel; RV park. Museum-related items for sale.
Activities: guided tours; formal education programs for children; temporary exhibitions.
Publications: Little Bear Tracks; Cho-O-Nee to High Iron.
Hours & Admission Prices: Tues.-Sat. 10-5. No charge; donations accepted. Closed Independence Day; holidays. &
Attendance: 1,000 (estimated)
Membership: Roughneck $15; Wildcats $25; Roustabout $50; Jughound $100; Swamper $250; Oil Tycoon over $250.

Ness City

NESS COUNTY HISTORICAL MUSEUM, 123 S. Pennsylvania, Ness City, KS 67560-1907. Mailing Address: P.O. Box 512, Ness City, KS 67560-0512. Tel.: 785-798-3298.
Founded: 1930.
Congressional District: 1
Personnel Profile: Part-Time Paid 1; Part-Time Volunteers 20.
Governing Authority: Tax-exempt.
Institution Type/Description: History Museum.
Collections: county history; photographs; period furnishings; obituaries; tableware; quilts; tools.
Hours & Admission Prices: Tues.-Fri. 1-5; other times by appointment. Closed holidays.
Membership: Life $25.

New Century

CAF HEART OF AMERICA WING EDUCATION CENTER, 6 Aero Plaza, New Century, KS 66031. Tel.: 913-907-7902.
E-mail: hoacafinfo@yahoo.com
Web Site: www.kcghostsquadron.org
Institution Type/Description: Military Aviation History Museum.
Collections: military aviation history; early aircraft.
Activities: special events.
Hours & Admission Prices: By appointment. &

Newton

HARVEY COUNTY HISTORICAL SOCIETY, 203 N. Main, Newton, KS 67114-3442. Mailing Address: P.O. Box 4, Newton, KS 67114-0004. Tel.: 316-283-2221. Facebook: Harvey County Historical Society.
E-mail: info@hchm.org
Web Site: www.hchm.org
Founded: 1962.
Congressional District: 4
Key Personnel: Dir., Debra Hiebert; Chm. (V), Nancy Krehbiel; Archivist, Jane Jones; Cur., Kris Schmucker; Photo Registrar, Linda Koppes; Office Mgr., Lana Myers.
Personnel Profile: Part-Time Paid 4; Part-Time Volunteers 30.
Governing Authority: board of directors. Tax-exempt.
Institution Type/Description: Historical Society Museum: housed in 1903 Carnegie Library Building.
Collections: railroad; photo & paper archives; textiles; immigrants.
Facilities: 1,000-vol. library of books & documents of city & county history; reading room. Museum-related items for sale.
Activities: guided & self-guided tours; permanent & temporary exhibitions; educational programs; Speaker's Bureau programs; reunions.
Publications: quarterly newsletter.
Hours & Admission Prices: Tues.-Fri. and 1st & 3rd Sat. 10-4. General Admission: no charge. Programs: fees vary. Closed most major holidays.
Attendance: 1,200 (estimated)
Membership: Individual $20; One Plus One, Couple & Family $25; Sustaining Member $50; Business $100.

Nickerson

HEDRICK'S EXOTIC ANIMAL FARM, 7910 N. Roy L. Smith Rd., Nickerson, KS 67561-9049. Tel.: 620-422-3296; 800-618-9577.
Institution Type/Description: Animal Farm.
Collections: cattle; camels; ostriches; rheas; zebras; antelope; kangaroos; giant tortoises; giraffes.
Activities: petting zoo; pony rides.
Hours & Admission Prices: By appointment.

North Newton

KAUFFMAN MUSEUM, Bethel College, 2801 N. Main, North Newton, KS 67117-1700. Mailing Address: Bethel College, 300 E. 27th St., North Newton, KS 67117-8061. Tel.: 316-283-1612.
E-mail: kauffman@bethelks.edu
Web Site: www.bethelks.edu/kauffman
Founded: 1941.
Congressional District: 4
Key Personnel: Dir., Rachel K. Pannabecker; Pres. (V), David Stucky; Cur. Education, Andrea Schmidt Andres; Cur. Exhibits, Charles Regier; Museum Technician, David Kreider.
Personnel Profile: Full-Time Paid 3; Part-Time Paid 3; Part-Time Volunteers 85; Interns 1.
Governing Authority: Parent Institution: Bethel College & Kauffman Museum Assoc. Tax-exempt.
Institution Type/Description: Cultural & Natural History Museum.
Collections: 40,000 items with emphasis on the natural and cultural history of Central Plains. Historic Building: historic farmstead.
Research Fields: Mennonite history.
Activities: lectures, guided tours, concerts; special, permanent & traveling exhibitions.
Publications: newsletter.
Hours & Admission Prices: Tues.-Fri. 9:30-4:30, Sat.-Sun. 1:30-4:30. Adults $4, children 6-16 $2; members no charge. &
Attendance: 7,000 (estimated)
Membership: Individual $30; Family $50; Patron $100; Fellow $500; Benefactor $1,000; Founder $5,000.

MENNONITE LIBRARY AND ARCHIVES, Bethel College, 300 E. 27th St., North Newton, KS 67117-0531. Tel.: 316-284-5304; 800-522-1887 ext. 304. Fax: 316-284-5843.
E-mail: mla@bethelks.edu
Web Site: www.bethelks.edu/mla
Founded: 1938.
Congressional District: 5
Key Personnel: Archivist, John D. Thiesen; Librarian, Barbara A. Thiesen.

Personnel Profile: Part-Time Paid 3; Part-Time Volunteers 7.
Governing Authority: college; church. Parent Institution: Bethel College. Affiliated with Mennonite Church USA. Tax-exempt.
Institution Type/Description: Religious Museum.
Collections: Anabaptist & Mennonite manuscripts, prints, paintings, lithographs; archives.
Research Fields: Anabaptist & Mennonite studies; peace studies; regional history.
Facilities: 35,000-vol. library of books on Mennonite & Anabaptist history, life & principles available for inter-library loan; reading room.
Activities: guided tours; lectures.
Publications: Mennonite Life.
Hours & Admission Prices: Mon.-Thurs. 10-12 & 1-5, first & third Thurs. evenings of the month 6:30-9:30. No charge; donations accepted. Closed national holidays. &
Attendance: 927 (estimated)

Norton

NORTON COUNTY HISTORICAL SOCIETY & MUSEUM, 105 E. Lincoln, Norton, KS 67654. Mailing Address: P.O. Box 303, Norton, KS 67654-0303. Tel.: 785-877-5107.
Institution Type/Description: Historical Society Museum.
Collections: county history; photographs; prehistoric animal bones; 1948 meteor strike history.
Hours & Admission Prices: May-Sept. Wed. & Sat. 2-4; other times by appointment.

STATION 15, Water Tower Park, W. Hwy. 36, Norton, KS 67654-0097. Mailing Address: Norton Area Chamber of Commerce, 104 S. State, 205 S. State, Norton, KS 67654-0097. Tel.: 785-877-2501. Fax: 785-877-3300.
E-mail: nortoncc@ruraltel.net
Web Site: discovernorton.com
Founded: 1961.
Congressional District: 1
Key Personnel: Exec. Dir., Tara Vance; Chm. (V), Jim Ray.
Governing Authority: Norton Travel & Tourism.
Institution Type/Description: Historic Site.
Collections: papier mache figures; replica of stagecoach & wagon train depot of 1859; costumed figures; Indian artifacts.
Publications: handouts, excerpts from Kansas Historical Quarterlies & writings of Horace Greeley and Albert Dean Richardson.
Hours & Admission Prices: Daily 24 hours. No charge. &
Attendance: 1,000 (estimated)

Oakley

FICK FOSSIL & HISTORY MUSEUM, 700 W. 3rd St., Oakley, KS 67748-1256. Tel.: 785-671-4839. Fax: 785-672-3497.
E-mail: fickmuseum@ruraltel.net
Web Site: www.DiscoverOakley.com
Founded: 1972.
Congressional District: 1
Key Personnel: Pres., Alice Lindeman; Dir., Janet Bean; Registrar, Ruth Clark; Tour Guide, Nadine Kuasnicka.
Personnel Profile: Full-Time Paid 1; Part-Time Paid 2; Part-Time Volunteers 3.
Governing Authority: municipal. Parent Institution: City of Oakley. Tax-exempt.
Institution Type/Description: Geology, Paleontology & History Museum.
Collections: locally found fossils; fossil art; minerals; rocks; wood carvings; general store with period items; local history; historical photos; sod house; replica of Oakley 1886 Union Pacific Depot; military; depression glass.
Research Fields: Monument Rocks in Logan & Gove Counties of Kansas.
Facilities: 2,000 photographs showing 100 years of history.
Activities: guided tours; temporary exhibits.
Publications: brochure.
Hours & Admission Prices: Summer: Mon.-Sat. 9-5, Sun. 1-5; Winter: Mon.-Sat. 9-12 & 1-5. No charge; donations accepted. Closed holidays. &
Attendance: 10,000

Oberlin

DECATUR COUNTY LAST INDIAN RAID MUSEUM, 258 S. Penn Ave., Oberlin, KS 67749-2245. Tel.: 785-475-2712.
E-mail: lirm@sbcglobal.net
Web Site: skyways.lib.ks.us/museums/lirm
Founded: 1958.
Congressional District: 1

Key Personnel: Pres., Dana Marintzer; Vice Pres., Chris Koerperich; Cur., Sharleen Wurm; Sec. & Treas., Barbara Dehlinger.
Personnel Profile: Full-Time Paid 2; Part-Time Paid 1; Part-Time Volunteers 4.
Governing Authority: society; nonprofit organization. Tax-exempt.
Institution Type/Description: History Museum: located near the site of 1878 last Indian raid on Kansas soil.
Collections: local history; manuscripts; Indian artifacts; Western art; dolls; quilts.
Research Fields: local history; Indian raid of 1878.
Facilities: Museum-related items for sale.
Activities: guided tours; lectures; hobby workshops; formally organized education programs; permanent collections.
Publications: weekly newspaper column; brochures on Indian Raid; quarterly newsletter.
Hours & Admission Prices: Museum: April-Nov. Tues.-Sat. 10 to noon & 1-4. Office: Dec.-March Tues.-Thurs. 9:30-12 & 1-4:30. Adults $5, children 6-12 $3; children 6 & under and members no charge. Closed holidays. &
Membership: Individual $10; Family $20.

Olathe

DEAF CULTURAL CENTER AND WILLIAM J. MARRA MUSEUM, (M), 455 E. Park St., Olathe, KS 66061-5436. Tel.: 913-782-5808. Facebook: Deaf Cultural Center.
E-mail: deafcc@att.net
Web Site: www.deafculturalcenter.org
Founded: 2005.
Congressional District: 3
Key Personnel: Exec. Dir., Sandra Kelly; Pres., Terry Hostin; Treas., David Wilcox; Volunteer Coord., Laurel Sack; Museum Shop Mgr., Tia Rivard.
Personnel Profile: Part-Time Paid 10; Part-Time Volunteers 60.
Governing Authority: Parent Institution: Deaf Cultural Center Foundation. Tax-exempt.
Institution Type/Description: History Museum.
Collections: history & culture of the deaf and hard of hearing; photographs.
Major Exhibits: Deaflympics, 1/14-12/14.
Research Fields: history of Kansas School for the Deaf.
Facilities: Museum-related items for sale.
Activities: educational programs; sign language classes.
Publications: Hand Signs: A Broader Vision.
Hours & Admission Prices: Tues.-Fri. 10-4, Sat. 10-3. No charge; donations accepted. Closed New Year's Day; Independence Day; Thanksgiving; Christmas. &
Attendance: 2,500 (estimated)
Membership: Nonprofit & Individual: $50, $100, $250, $500; Small Business: $100, $250, $350; Corporate: $1,000, $2,500, $5,000, $10,000.

ENSOR PARK AND MUSEUM, 18995 W. 183rd St., Olathe, KS 66062-9278. Tel.: 913-592-4141.
E-mail: larryw0hxs@yahoo.com
Web Site: www.ensorparkandmuseum.org
Formerly: Ensor Farmsite and Museum
Founded: 1975.
Personnel Profile: Full-Time Paid 1; Part-Time Paid 2.
Governing Authority: Parent Institution: City of Olathe, KS. Tax-exempt.
Institution Type/Description: Historic Buildings.
Collections: Ensor family history; period furnishing; handmade textiles; fabric art; amateur radio artifacts 1922-1972; photography equipment; metal lathe; farm equipment; industrial arts; metal sculpture by Alexander Calder; period furnishings; history of high school industrial arts instruction. Historic Buildings: 1890 barn; 1892 farmhouse.
Research Fields: Marshall H. Ensor 1899-1970; Loretta Ensor 1904-1991.
Facilities: dark room.
Activities: school tours; special events.
Hours & Admission Prices: May-June & Sept.-Oct. Sat.-Sun. 1-5; other times by appointment. No charge; $2 donation requested.
Attendance: 1,000 (estimated)

ERNIE MILLER NATURE CENTER, 909 N. Hwy. 7, Olathe, KS 66061-4040. Tel.: 913-764-7759. Fax: 913-764-0109.
Web Site: www.erniemiller.com
Key Personnel: Dir., Bill McGowan.
Personnel Profile: Full-Time Paid 6; Part-Time Paid 10; Part-Time Volunteers 15.
Governing Authority: Parent Institution: Johnson County Park & Recreation District.
Institution Type/Description: Nature Center.
Collections: live animals; natural, cultural, & environmental history.
Facilities: nature trails; amphitheater.
Activities: educational programs; hiking; summer camp.

Hours & Admission Prices: Center: April-May & Sept.-Oct. Mon.-Sat. 9-12 & 1-5, Sun. 1-5; June-Aug. Mon.-Sat. 9-12 & 1-5; Nov.-Feb. Mon.-Sat. 9-12 & 1-4:30, Sun. 12:30-4:30. Trails: daily dawn to dusk. No charge; donations accepted. &
Attendance: 35,000 (accurate)

MAHAFFIE STAGECOACH STOP AND FARM HISTORIC SITE, 1200 E. Kansas City Rd., Olathe, KS 66061-3002. Tel.: 913-971-5111. Fax: 913-971-5114.
E-mail: mahaffie@olatheks.org
Web Site: www.mahaffie.org
Founded: 1977.
Congressional District: 3
Key Personnel: Site Mgr., Tim Talbott; Volunteer Coord., Wendy Burkett; Museum Shop Mgr., Alexis Radil.
Personnel Profile: Full-Time Paid 7; Part-Time Paid 13; Part-Time Volunteers 136.
Governing Authority: city government, Olathe, KS.
Institution Type/Description: Historic Site & House: 1865 Mahaffie house and stagecoach stop on Santa Fe Trail.
Collections: furnishings; agricultural equipment & machinery.
Research Fields: stagecoaches; Oregon Trail; Santa Fe Trail; local & state history; Mahaffie family.
Facilities: 600-vol. library pertaining to history available for research; reading room.
Activities: guided tours; lectures; docent programs; educational programs; off-site programs & lectures; living history programs.
Publications: newsletter.
Hours & Admission Prices: April-May & Nov.-Dec. Wed.-Sat. 10-4, Sun. 12-4; June-Oct. Thurs.-Sat. 10-4; other times by appointment. Call for admission prices. &
Attendance: 38,000 (estimated)
Membership: Individual $35; Family $75.

TRADITIONS FIRE COMPANY & MUSEUM, The Great Mall of the Great Plains, 20175 W. 151st St. #175, Olathe, KS 66061. Tel.: 913-829-4319.
Founded: 2009.
Governing Authority: Tax exempt: 501(c)(3).
Institution Type/Description: Firefighting Museum.
Collections: firefighting history, equipment & rescue apparatus; photographs; personal artifacts; 1955 American Lafrance ladder truck.
Hours & Admission Prices: Thurs.-Fri. 10-2 & 5-8, Sat. 10-7, Sun. 12-6.

Osawatomie

JOHN BROWN MUSEUM STATE HISTORIC SITE, 10th & Main Sts., Osawatomie, KS 66064. Mailing Address: P.O. Box 37, Osawatomie, KS 66064-0037. Tel.: 913-755-4384. Fax: 913-755-4164.
E-mail: adaircabin@kshs.org
Web Site: www.kshs.org/john_brown
Formerly: John Brown State Historic Site
Founded: 1854.
Congressional District: 2
Key Personnel: Site Admin., Grady Atwater.
Governing Authority: state. Parent Institution: Kansas State Historical Society & City of Osawatomie. Tax-exempt.
Institution Type/Description: Historic House & Site: 1854 Adair Cabin, used as headquarters by John Brown 1855-58; 1856 Battle of Osawatomie; Underground Railroad stop.
Collections: weapons; pictures; antiques; fireplace cooking equipment.
Research Fields: abolitionists; slavery; Civil War; Bleeding Kansas.
Facilities: city park.
Activities: tours. Museum Sponsors: Annual reenactment of the Battle of Osawatomie; John Brown Jamboree in June; Christmas Program; Railroad Day in fall.
Publications: yearly program series, Border War Brochures; Safe Passages in Perilous Times; Territorial Kansas Heritage Alliance-John Brown Brochure.
Hours & Admission Prices: Tues.-Sat. 10-5, Sun. 1-5. Adults $3; students $1. Closed national holidays. &
Attendance: 5,000 (estimated)

Oskaloosa

OLD JEFFERSON TOWN - JEFFERSON COUNTY HISTORICAL SOCIETY, 703 Walnut, Hwy. 59, Oskaloosa, KS 66066. Mailing Address: P.O. Box 146, Oskaloosa, KS 66066-0146. Tel.: 785-863-2070.
Founded: 1966.
Congressional District: 2
Key Personnel: Pres. (V), Leann Chapman; Cur., Marilyn Sharkey.
Personnel Profile: Part-Time Paid 1; Part-Time Volunteers 15.
Governing Authority: nonprofit. Jefferson County Historical Society. Tax-exempt.
Institution Type/Description: Preservation Society: housed in seven c.1880 buildings, relocated to Old Jefferson Town.
Collections: country store, chapel & school with original furnishings; 1906 replica Jefferson County Courthouse bandstand; city jail; blacksmith shop. Historic House: 1880 Nincehelser House with original furnishings; John Stewart Curry, famous artist, boyhood home with art gallery.
Research Fields: local & family history.
Facilities: library containing records of families, census, cemeteries, churches, schools & files of newspapers pertinent to early history of Jefferson County.
Activities: guided tours; kids' workshop; craft demonstrations.
Publications: quarterly newsletter.
Hours & Admission Prices: Museum: May-Sept. Sat. 1-5, Sun. 1:30-5. Library: mid-May to Sept. Sat. 1-5, Sun. 1:30-5; Oct. to mid-May Sat. 1-5. Tours by appointment. Suggested Donation: adults $2. &
Attendance: 400 (estimated)
Membership: Annual $15; Life $225.

Oswego

OSWEGO HISTORICAL SOCIETY, INC., 410 Commercial St., Oswego, KS 67356-2018. Tel.: 620-795-4500.
E-mail: historyatoswego@embarqmail.com
Web Site: www.oswegohistory.org
Formerly: Oswego Historical Museum, Inc.
Founded: 1967.
Congressional District: 2
Key Personnel: Pres. (V) & Dir. (V), Richard W. Farris; Vice Pres. (V) & Dir. (V), Donna Strickland; Treas. & Dir. (V), Jolene Gaier; Dir. (V), Jean Snyder; Dir (V), Eleanor Monroe; Dir. (V), John Davis; Dir. (V), Phil Blair; Dir. (V), Cheryl Lewis; Dir. (V), Janet King.
Personnel Profile: Part-Time Paid 2; Part-Time Volunteers 4.
Governing Authority: county; nonprofit. Tax-exempt.
Institution Type/Description: History Museum.
Collections: paintings; photographs; furnishings; personal artifacts; genealogy.
Facilities: 100-vol. library. Museum-related items for sale.
Activities: docent program; formal education programs for children; guided tours; temporary exhibitions.
Hours & Admission Prices: June-Oct. Mon.-Fri. 1-5. No charge; donations accepted. &
Attendance: 300 (estimated)
Membership: Individual $10.

Ottawa

OLD DEPOT MUSEUM, 135 W. Tecumseh, Ottawa, KS 66067. Mailing Address: P.O. Box 145, Ottawa, KS 66067-0145. Tel.: 785-242-1250. Fax: 785-242-1267.
E-mail: history@old.depot.museum
Web Site: www.olddepotmuseum.org
Founded: 1963.
Congressional District: 2
Key Personnel: C.E.O. & Dir., Deborah Barker; Chm. (V), Geoff Hanson; Treas., Greg Gilroy; Archivist, Ann Shepherd; Registrar, Laura Miller; Museum Asst., Phyllis Foster; Museum Shop Mgr., Kim Hanson.
Personnel Profile: Full-Time Paid 1; Part-Time Paid 4; Part-Time Volunteers 27; Interns 1.
Governing Authority: private; nonprofit organization. Subsidiary Institution: Dietrich Cabin, 5th & Main, Ottawa, KS 66067. Tax-exempt: 501(c)(3).
Institution Type/Description: Historical Society Museum.
Collections: period rooms display local manufacturing items & furniture, clothing & accessories. Historic Sites: c.1888 limestone passenger depot; c.1859 walnut cabin.
Research Fields: Underground Railroad through the area; emigrant tribes of Native Americans; 19th century school discrimination case.
Facilities: 22,000-vol. library; 7,000 sq. ft. exhibit space. Museum-related items for sale.

Activities: guided tours; loan & temporary exhibitions. Museum Sponsors: Kansas Day; Christmas Event; Esoteric and Terrible Order of Pie Eaters Conclave.
Publications: bimonthly newsletter, The Headlight.
Hours & Admission Prices: Tues.-Sat. 10-4, Sun. 1-4. Adults $3, students $1; discounts to groups; members & preschoolers no charge. Closed New Year's Day; Easter; Thanksgiving; Christmas. ♿
Attendance: 2,879 (accurate)
Membership: Individual $15; Family $35; 1937 $50; A.P. Elder $100; Ben Park $250; Jacob Dietrich $500; Train Master $1,000.

Overland Park

DEANNA ROSE CHILDREN'S FARMSTEAD, 13800 Switzer Rd., Overland Park, KS 66221-7803. Tel.: 913-897-2360.
E-mail: farmsteadquestservices@opkansasorg
Web Site: www.opkansas.org
Founded: 1978.
Key Personnel: Dir., Virgil Miles; Chm. (V), Heather Gilpin; Museum Shop Mgr., Stephanie Jones.
Personnel Profile: Full-Time Paid 7; Part-Time Paid 46; Part-Time Volunteers 90.
Governing Authority: Parent Institution: City of Overland Park. Subsidiary Institution: Friends of the Farmstead. Tax-exempt.
Institution Type/Description: Farmstead.
Collections: farm animals; birds of prey; butterfly garden; Kanza Indian artifacts; one-room schoolhouse; mining; general store; bank; ice cream parlor; gardens.
Facilities: nature trail.
Activities: rental facilities; pony rides; animal feedings; petting zoo; horse drawn hay rides; mining; bottle fed baby goats.
Hours & Admission Prices: April-May & Sept.-Oct. daily 9-5; June-Aug. Tues. & Thurs. 9-8, Wed. & Fri.-Mon. daily 9-5. Fri.-Sun. 2 & over $2; members & Mon.-Thurs. no charge.
Attendance: 400,000 (accurate)
Membership: Overland Park Resident $35; Non-Resident $50; Patron $100

KANSAS CITY JEWISH MUSEUM OF CONTEMPORARY ART - EPSTEN GALLERY, 5500 W. 123rd St., Overland Park, KS 66209. Tel.: 913-266-8414.
Key Personnel: Exec. Dir., Eileen Garry; Cur., Marcus Cain; Prog. & Devel. Asst., Abby Rufkahr.
Personnel Profile: Full-Time Paid 7.
Institution Type/Description: Art Gallery.
Collections: works by contemporary artists.
Hours & Admission Prices: Tues.-Fri. 11-4, Sat.-Sun. 1-4. No charge; donations accepted.

NERMAN MUSEUM OF CONTEMPORARY ART, (M), 12345 College Blvd., Overland Park, KS 66210-1283. Tel.: 913-469-3000. Fax: 913-469-2348.
E-mail: sgreco@jccc.edu
Web Site: www.nermanmuseum.org
Formerly: Johnson County Community College, Gallery of Art
Founded: 1969.
Congressional District: 3
Key Personnel: Pres. (V), Dr. Terry Calaway; Dir., Bruce Hartman; Dir. Devel., Katherine Allen; Dir. College Information, Julie Haas; Supvr. Security, Gus Ramirez.
Personnel Profile: Full-Time Paid 6; Part-Time Paid 16; Part-Time Volunteers 30.
Governing Authority: college. Parent Institution: Johnson County Community College. Tax-exempt.
Institution Type/Description: Contemporary Art Museum.
Collections: contemporary paintings; ceramics; works on paper; photography; Oppenheimer collection.
Research Fields: contemporary art.
Facilities: 12,000 sq. ft. exhibition space; 110-seat cafe; 200-seat auditorium; 2 classrooms. Museum-related items for sale.
Activities: docent guided tours; lectures; visiting artists; film series; children's studio classes; theatre; broadcast programs; organized education programs for children, adults & undergraduate students affiliated with Johnson County Community College; temporary, permanent, loan & traveling exhibitions.
Publications: exhibition & collection catalogues.
Hours & Admission Prices: Tues.-Thurs. & Sat. 10-5, Fri. 10-9, Sun. 12-5. No charge. Closed major holidays. ♿
Attendance: 100,000 (accurate)
Membership: Student $15; Individual $50; Friend $75; Contributor $125; Patron $250; Sponsor $500; Underwriter $1,000; Benefactor $5,000.

Paola

MIAMI COUNTY HISTORICAL MUSEUM, 12 E. Peoria, Paola, KS 66071-1707. Tel.: 913-294-4940.
E-mail: museum@mchgm.org
Web Site: www.thinkmiamicountyhistory.org
Formerly: Swan River Museum
Founded: 1979.
Institution Type/Description: History Museum.
Collections: Miami County, Civil War & Border Wars history; military artifacts; Native American & pioneer artifacts; period clothing; furniture; toys; sports equipment.
Hours & Admission Prices: Mon.-Sat. 10-4. No charge; donations accepted. Closed holidays.

Parker

PARKER HISTORICAL MUSEUM, 207 W. Main, Parker, KS 66072. Mailing Address: Parker Community Historical Society, P.O. Box 173, Parker, KS 66072-0173. Tel.: 913-898-4781.
E-mail: woodcarv@everestkc.net
Web Site: www.rootsweb.ancestry.com/~kspchs/
Key Personnel: Pres., John Riggs; Treas., Marilyn Rhoads.
Governing Authority: Parent Institution: Parker Community Historical Society.
Institution Type/Description: Historic Buildings.
Collections: Parker history; newspapers.
Facilities: Museum-related items for sale.
Activities: special events. Museum Sponsors: PRHS Alumni Banquet in April.
Hours & Admission Prices: By appointment.

Parsons

PARSONS HISTORICAL COMMISSION MUSEUM, 401 S. 18th St., Parsons, KS 67357-4220. Tel.: 620-421-7000 & 3694.
Formerly: Parsons Historical Society Museum
Founded: 1969.
Congressional District: 2
Key Personnel: Chm. (V), Lewis Hevel; Financial Dir., Betty Olmsted.
Personnel Profile: Part-Time Paid 9; Part-Time Volunteers 15.
Governing Authority: private; nonprofit organization. Owned by City of Parsons. Operated by bd. directors. Tax exempt.
Institution Type/Description: Historical Society Museum.
Collections: Parsons history from 1870 to present with an emphasis on MKT Railroad; local historic memorabilia.
Facilities: 300-vol. library; 7,000 sq. ft. exhibit space.
Activities: guided tours.
Hours & Admission Prices: May-Oct. Fri.-Sun. 1-4. No charge; donations accepted. ♿
Attendance: 1,200 (estimated)
Membership: Individual $1; Life $20.

Peabody

PEABODY HISTORICAL MUSEUM, 106 E. Division, Peabody, KS 66866. Mailing Address: 1556 E. 59, Peabody, KS 66866-9485. Tel.: 620-983-2174.
Founded: 1961.
Key Personnel: Hostess, Gwen Gaines; Pres. (V), Marilyn Jones.
Personnel Profile: Part-Time Volunteers 12.
Governing Authority: bd. of directors. Tax-exempt.
Institution Type/Description: History Museum.
Collections: period history; photographs. Historic Buildings: 1874 library; 1881 newspaper editor's home; 1920 printing building; 1904 barn.
Publications: biannual newsletter.
Hours & Admission Prices: Call for hours. No charge; donations accepted. ♿
Attendance: 498 (accurate)
Membership: Individual $10; Family $15; Lifetime $100.

Phillipsburg

FORT BISSELL MUSEUM, 501 Fort Bissell Ave., Phillipsburg, KS 67661-7116. Mailing Address: P.O. Box 53, Phillipsburg, KS 67661-0053. Tel.: 785-543-6212.
Web Site: www.phillipsburgks.us
Formerly: Old Fort Bissell
Founded: 1961.
Congressional District: 1
Key Personnel: Chm. (V) & Sec., Shelly Lare; Vice Pres., Connie Cox; Treas., Kathy Beard.

Personnel Profile: Part-Time Paid 1.
Governing Authority: society. Parent Institution: Phillips County Historical Society, Inc. Tax-exempt.
Institution Type/Description: Historical Society Museum.
Collections: artifacts from 1740 to present, guns; McDowell saddle & chaps. Historic Buildings: 1872 two log cabins; 1885 store; sod house; 1879 railroad depot; 1882 schoolhouse.
Research Fields: historical artifacts.
Activities: guided tours; permanent exhibitions.
Publications: brochures.
Hours & Admission Prices: Memorial Day to Labor Day Tues-Fri. 9-4, Sat. 9-2; tours by appointment. No charge; donations accepted. Closed Independence Day. &
Attendance: 1,500 (estimated)
Membership: Individual $5; Family $20; Business $25; Lifetime $100.

Pittsburg

CRAWFORD COUNTY HISTORICAL MUSEUM, 651 S. Hwy. 69, Pittsburg, KS 66762-8600. Tel.: 620-231-1440 & 3794.
Web Site: www.kansastravel.org/crawfordcountymuseum.htm
Founded: 1968.
Congressional District: 5
Key Personnel: C.E.O., Alan Ross.
Governing Authority: county; society. Operated by the Crawford County Historical Society, Inc. Tax-exempt: 501(c)(3).
Institution Type/Description: General Museum.
Collections: art; decorative arts; clothing; farming; government; household; industry; transportation; primitive; education; biology; taped local history collections; 1922 Marion steam shovel. Historic Buildings: 1902 grocery store; one-room schoolhouse.
Facilities: library of historical documents.
Activities: Museum Sponsors: annual Ice Cream Social; Open House in June.
Hours & Admission Prices: Thurs.-Sun. 12-4 or by appointment. No charge; donations accepted. &
Attendance: 2,895 (accurate)

Pleasanton

LINN COUNTY MUSEUM/GENEALOGY LIBRARY, 307 E. Park (Dunlap Park), Pleasanton, KS 66075. Mailing Address: P.O. Box 137, Pleasanton, KS 66075-0137. Tel.: 913-352-8739. Fax: 913-352-8739.
Formerly: Linn County Museum
Founded: 1973.
Congressional District: 5
Key Personnel: Pres. (V) & Museum Shop Mgr., Ola May Earnest.
Personnel Profile: Full-Time Volunteers 2; Part-Time Volunteers 5.
Governing Authority: nonprofit. Linn County Historical Society. Tax-exempt.
Institution Type/Description: History Museum.
Collections: panels containing pictures, maps, copies of documents, and explanations of events of Linn County history; artifacts from 1850-early 1900s; four period rooms; paintings by Linn County artists; authentic country store; genealogy library.
Research Fields: genealogy; Kansas & Missouri history.
Facilities: library of books on Linn County history; microfilm copies of all existing Linn County newspapers; display area emphasizing Linn County history; genealogy center with microfilm reader; 1920's era car museum & R.R. depot with related items. Books & other related items for sale.
Activities: guided tours; bus tours of historic sites in Pleasanton area; narrated school bus trips to historic sites. Annual Events: Kansas Day in January; Black History Month in February; Fashions of Yesterday seasonally.
Publications: newsletter; books, From Pioneering to the Present, Vols. I, II, III; reprint of 1928 Linn County History; booklets, Historic Linn County; Border Warfare in Southeastern Kansas; Linn County Cemeteries.
Hours & Admission Prices: Tues. & Thurs. 9-4, Sat.-Sun. 1-5 & by appointment. No charge; donations accepted. &
Attendance: 3,000 (estimated)
Membership: Annual $5; Life $25.

MINE CREEK BATTLEFIELD STATE HISTORIC PARK, 20485 Kansas Hwy. 52, Pleasanton, KS 66075-9549. Tel.: 913-352-8890.
E-mail: minecreek@kshs.org
Web Site: www.kshs.org/places/minecreek/index.htm
Founded: 1998.
Congressional District: 2
Governing Authority: Parent Institution: Kansas State Historical Society.
Institution Type/Description: Historic Site: location of the Civil War battle fought on Oct. 25, 1864.

Collections: Civil War history; personal artifacts; period furnishings & clothing; weapons; photographs.
Facilities: Museum-related items for sale.
Hours & Admission Prices: March-Oct. Wed.-Sat. 10-5. Adults $5, seniors & students $1; members and children 5 & under no charge. Closed state holidays. &

Pratt

KANSAS DEPARTMENT OF WILDLIFE, PARKS & TOURISM, 512 S.E. 25th Ave., Pratt, KS 67124-8174. Tel.: 620-672-5911, ext. 108 & 0708. Fax: 620-672-6020.
E-mail: mike.rader@ksoutdoors.com
Web Site: www.kdwpt.state.ks.us
Founded: 1903.
Congressional District: 4
Key Personnel: C.E.O., Mike Rader; Caretaker, Chris Shrack.
Personnel Profile: Full-Time Paid 1; Part-Time Paid 2.
Governing Authority: state. Tax-exempt.
Institution Type/Description: Nature Center.
Collections: live display of fish & reptiles species of Kansas; exhibits of birds & animals of Kansas; hands-on displays.
Facilities: aquarium.
Activities: school & group tours available upon request for museum & fish hatchery.
Publications: Kansas Wildlife & Parks Magazine; ON T.R.A.C.K.S.; self-guided museum tour brochures.
Hours & Admission Prices: Mon.-Fri. 8-5. No charge; donations accepted. &
Attendance: 4,300 (estimated)

PRATT COUNTY HISTORICAL SOCIETY MUSEUM, 208 S. Ninnescah St., Pratt, KS 67124-2715. Tel.: 620-672-7874.
E-mail: pchsmuseum@sctelcom.net
Web Site: prattcountymuseum.org/
Founded: 1967.
Congressional District: 1
Key Personnel: Pres. (V), Marvin Proctor; Vice Pres., Ross Hoener; Treas., Linda Brehm; Bd. Sec., Ulanda Stiebben; Office Mgr., Sandra Hettrick; Cur., Marsha Brown.
Personnel Profile: Part-Time Paid 1; Part-Time Volunteers 40.
Governing Authority: private; nonprofit organization. Tax-exempt: 501(c)(3).
Institution Type/Description: General Museums.
Collections: fossils; Plains Indian artifacts; Pratt County pioneer period rooms; Miss Kansas display; military artifacts; turn-of-the-century general store; jail, carpenter shop, post office; 1890s Main St; railroad history.
Research Fields: family histories; history of business buildings & homestead land owners.
Facilities: 2,218-vol. library; 2,200 sq. ft. exhibit space. Museum-related items for sale.
Activities: formal education programs for children; guided tours; temporary & loan exhibitions; school loan service. Museum Sponsors: Open House in Fall; Miss KS Pageant week open house; Christmas Open House; historical programs 4 times a year.
Publications: newsletter, Pratt County Historical Society.
Hours & Admission Prices: Mon.-Fri. 1-4, Sat.-Sun. 1-3; other times by appointment. No charge; donations accepted. Closed New Year's Day; Easter; Memorial Day; Independence Day; Labor Day; Thanksgiving; Christmas. &
Attendance: 3,000 (estimated)
Membership: Annual $10; Life $75.

Republic

PAWNEE INDIAN VILLAGE STATE HISTORIC SITE, 480 Pawnee Trail, Republic, KS 66964-8057. Tel.: 785-361-2255. Fax: 785-361-2255.
E-mail: piv@kshs.org
Web Site: www.kshs.org
Founded: 1967.
Congressional District: 1
Key Personnel: Historic Property Cur., Richard Gould; Chm. (V), Narveen Brzon; Pres. (V), Beth Carlgren.
Personnel Profile: Full-Time Paid 1; Part-Time Paid 3; Part-Time Volunteers 2.
Governing Authority: state. Admin. by Kansas State Historical Society, 6425 S.W. 6th Ave., Topeka, KS 66615-1099. Tel.: 785-272-8681. Tax-exempt: 501(c)(3).
Institution Type/Description: Archaeology Museum: located on the preserved Pawnee Site.
Collections: original lodge floor; hearth; stone; metal tools & implements;

bone hoe blades made from the scapula of a bison; trade gun; artifacts to illustrate Pawnee life & customs; dioramas.
Research Fields: Pawnee Indians.
Facilities: nature walk; picnic grounds.
Activities: guided tours.
Hours & Admission Prices: Wed.-Sat. 9-5; other times by appointment. Adults $3, seniors & students $2; members, military, and children 5 & under no charge. Closed major holidays. &
Attendance: 15,000 (accurate)
Membership: Kansas State Historical Society: Student $15; Individual $25; Family $35.

Russell

DEINES CULTURAL CENTER, 820 N. Main St., Russell, KS 67665-1932. Tel.: 785-483-3742. Fax: 785-483-4397.
E-mail: deinescenter@russellcity.org
Web Site: deinesculturalcenter.org
Founded: 1990.
Congressional District: 1
Key Personnel: Dir., Shannon Trevethan Marvel Castor.
Personnel Profile: Part-Time Paid 3.
Governing Authority: municipal. Parent Institution: City of Russell. Tax-exempt.
Institution Type/Description: Art Center.
Collections: E. Hubert Deines wood engravings; 20th-century art; ceramics.
Activities: temporary & traveling exhibitions.
Hours & Admission Prices: Tues.-Fri. 12-5, Sat.-Sun. 1-5. No charge; donations accepted. &
Attendance: 2,500 (estimated)
Membership: Supporting $25; Patron $50; Sustaining $100; Advocate $500; Benefactor $1,000.

GERNON HOUSE & BLACKSMITH SITE, 818 N. Kansas St., Russell, KS 67665. Mailing Address: P. O. Box 245, Russell, KS 67665. Tel.: 785-483-3637.
E-mail: rchs@ruraltel.net
Web Site: www.russellkshistory.com
Congressional District: 1
Institution Type/Description: Historic House Museum: built in 1872.
Collections: local history & culture; period furnishings; personal artifacts; early clothing.
Hours & Admission Prices: Memorial Day to Labor Day Sat. 11-4, Sun. 1-4; other times by appointment. Donations accepted.

HEYM-OLIVER HOUSE - RUSSELL COUNTY HISTORICAL SOCIETY, 503 N. Kansas St., Russell, KS 67665. Tel.: 785-483-3637.
Institution Type/Description: Historic House Museum: housed in the former home of Nicholas Heym; built in 1878.
Collections: local history & culture; period furnishings; photographs.
Hours & Admission Prices: Summer: Sat.-Sun. 1-4. Adults $2.

OIL PATCH MUSEUM, 100 Edwards Ave., Russell, KS 67665. Tel.: 785-483-6640.
E-mail: rchs@rural.net
Web Site: www.russellkshistory.com
Key Personnel: Dir., Kay Homewood.
Governing Authority: Parent Institution: Russell County Historical Society. Subsidiary Institution: Fossil Station Museum, 331 Kansas, Russell, KD 67665.
Institution Type/Description: Oil History Museum.
Collections: western Kansas oil history; oil storage tank; drilling & pumping equipment.
Hours & Admission Prices: June-Aug. daily 4-8; Winter: by appointment. Donation Requested.
Attendance: 250 (estimated)

RUSSELL COUNTY HISTORICAL SOCIETY/FOSSIL STATION MUSEUM, 331 Kansas St., Russell, KS 67665-2019. Mailing Address: P.O. Box 245, Russell, KS 67665-0245. Tel.: 785-483-3637.
E-mail: rchs@russellks.net
Web Site: www.russellkshistory.com
Founded: 1969.
Congressional District: 1
Key Personnel: Pres., Kay Homewood; Sec. & Treas., Aldean Banker.
Personnel Profile: Part-Time Paid 1; Part-Time Volunteers 4.

Governing Authority: county government; nonprofit. Parent Institution: Fossil Station Museum. Subsidiary Institutions: Gernon House, 818 Kansas St., Russell, KS 67665; Oil Patch Museum. Jct. I-70 & U.S. 281, Russell, KS 67665; Heym Oliver House, 503 Kansas St., Russell, KS 67665. Tax-exempt: 501(c)(3).
Institution Type/Description: County History Museum & Preservation.
Collections: Native American artifacts; historic furniture & household items from 1850-1930; cable tool & rotary oilfield equipment from 1900-1965.
Research Fields: arts & entertainment history of Russell County history.
Facilities: 50-vol. genealogical library; 1,500 sq. ft. exhibit space.
Activities: guided tours; loan, travel & temporary exhibits; school loan service.
Hours & Admission Prices: Memorial Day to Labor Day Sun. 1-4, Mon.-Sat. 11-4; other times by appointment. No charge; donations accepted.
Attendance: 503 (accurate)
Membership: Life $25.

Russell Springs

BUTTERFIELD TRAIL HISTORICAL MUSEUM, 515 Hilts, Russell Springs, KS 67764. Tel.: 785-751-4247.
Founded: 1964.
Congressional District: 1
Key Personnel: Pres., Jarett Haremza; Museum Shop Mgr., Andrea Plummer.
Governing Authority: nonprofit. Sponsored by The Butterfield Trail Association & Historical Society of Logan County, Kansas, Inc. Tax-exempt.
Institution Type/Description: History Museum: housed in 1887 Logan County Courthouse & Jail.
Collections: period furniture; artifacts relating to the passage through & settlement of the High Plains; fossils; Indian relics; dishware; agricultural implements; family heirlooms.
Research Fields: Logan County history; Butterfield Overland Dispatch & the Smoky Hill Trail.
Facilities: 250-vol. library of Kansas law books, periodicals, old local newspapers & county records available for use on the premises; reading room. Curios, books by local authors & newspaper reprints on local history for sale.
Activities: guided tours; permanent exhibitions. Museum Sponsors: annual Butterfield Trail Ride; annual Old Settlers Picnic & Butterfield Day.
Publications: semiannual newsletter, Butterfield Happenings.
Hours & Admission Prices: May to Labor Day Mon.-Sat. 9-12 & 1-5, Sun. 1-5. No charge; donations accepted.
Attendance: 1,000 (accurate)
Membership: Annual $10; Life $100.

Sabetha

ALBANY HISTORICAL SOCIETY, INC., 415 Grant, Sabetha, KS 66534-2317. Tel.: 785-284-3446 & 3529.
Web Site: www.albanydays.org
Founded: 1965.
Congressional District: 2
Key Personnel: Pres., Alex Dawdy; Vice Pres., Alan Meyer; Sec., Gary Yess.
Personnel Profile: Part-Time Volunteers 20.
Governing Authority: society. Tax-exempt.
Institution Type/Description: General Museum: housed in 1867 two-story stone schoolhouse.
Collections: period vehicles; agricultural machinery; train equipment; military; two airplanes. Historic Buildings: 1866-1867 Albany School; Berwick School wood structure; Rock Island Depot; Union Pacific Caboose; 1866 frame dwelling; c.1870 Most House; machinery building; car building; shelter house; two-story log cabin; post office; operative blacksmith shop; sawmill; print shop; shoe & leather shop.
Research Fields: local history.
Facilities: 500-vol. library of school books, histories & magazines available for use on premises under supervision of director.
Activities: guided tours; permanent exhibitions. Society Sponsors: Threshing Bee; Arts & Crafts.
Publications: book, History of Albany, Recipes & Remembrances.
Hours & Admission Prices: Memorial Day-Labor Day Sat.-Sun. 2-5; other times by appointment. No charge; donations accepted. &
Attendance: 3,000 (estimated)
Membership: Active $2.50; Contributing $5; Sustaining & Business $15; Life $100.

Saint Francis

CHEYENNE COUNTY MUSEUM, U.S. Hwy. 36 W., Saint Francis, KS 67756. Mailing Address: P.O. Box 611, Saint Francis, KS 67756-0611. Tel.: 785-332-2504.
Founded: 1985.

Personnel Profile: Part-Time Volunteers 2.
Institution Type/Description: History Museum.
Collections: county history; personal artifacts; furnishings.
Hours & Admission Prices: May-Sept. Wed.-Fri. 1-4; other times by appointment. No charge; donations accepted. &

Membership: Individual $10.

Saint John

ST. JOHN SCIENCE MUSEUM, 312 N. Main St., Saint John, KS 67576-1733. Tel.: 620-549-3818.
Institution Type/Description: Science Museum.
Collections: hands-on exhibits.
Activities: demonstrations; school groups.
Hours & Admission Prices: Call for hours; groups by appointment. No charge; donations accepted.

Saint Marys

INDIAN PAY STATION MUSEUM, 111 E. Mission, Saint Marys, KS 66536-1526. Mailing Address: c/o City of Saint Marys, P.O. Box 130, Saint Marys, KS 66536. Tel.: 785-437-6600.
Institution Type/Description: Historic Building Museum: built in 1857 as an Indian Agency for the Pottawatomie; later used to receive payments for lands taken from them. Listed on the National Register of Historic Places.
Collections: early 1800 artifacts; recreated general store.
Hours & Admission Prices: Daily 1-4; other times by appointment.

Saint Paul

OSAGE MISSION - NEOSHO COUNTY MUSEUM, 203 Washington St., Saint Paul, KS 66771. Tel.: 620-449-2320.
E-mail: museum@osagemission.org
Web Site: osagemission.org
Institution Type/Description: History Museum.
Collections: local history & culture; period furnishings; personal artifacts; photographs.
Activities: research.
Hours & Admission Prices: Tues.-Sat. 9-2; other times by appointment.

Salina

ROLLING HILLS WILDLIFE ADVENTURE, (M), 625 N. Hedville Rd., Salina, KS 67401-9764. Tel.: 785-827-9488.
E-mail: vickee@rollinghillswildlife.com
Web Site: www.rollinghillswildlife.com
Founded: 1999.
Congressional District: 1
Key Personnel: Dir., Kathy Tolbert; Cur., Sandy Walker; Dir. Operations, Jeff Parker; Dir. Education, Anita Butler; Dir. Devel. & Mktg., Vickee Spicer; Exec. Asst., Debra Preston; Group Sales Mgr. & Special Events, Debbie Tasker; Conference Center Mgr., Gail Vance; Museum Shop Mgr., Erin Swelter.
Personnel Profile: Full-Time Paid 35.
Governing Authority: Tax-exempt.
Institution Type/Description: Zoo, Natural History Museum & Art Gallery.
Collections: 105 species of animals.
Publications: quarterly newsletter, Animal Tales.
Hours & Admission Prices: Summer: daily 8-5; winter: daily 9-5. Zoo: adults $10.95, senior citizens 65 & over $9.95, children 3-12 $5.95; members & children under 3 no charge. Museum: adults $9.95, senior citizens 65 & over $8.95, children 3-12 $4.95; members & children under 3 no charge. Combo: adults $13.95, senior citizens 65 & over $12.95, children 3-12 $7.95; members & children under 3 no charge. Closed New Year's Day; Christmas Eve & Day. &
Attendance: 119,016 (accurate)

✳ **SALINA ART CENTER, (M),** 242 S. Santa Fe, Salina, KS 67402-0743. Mailing Address: P.O. Box 743, Salina, KS 67402-0743. Tel.: 785-827-1431. Fax: 785-827-0686. Facebook: Salina Art Center.
E-mail: info@salinaartcenter.org
Web Site: www.salinaartcenter.org
Founded: 1978.
Congressional District: 1
Key Personnel: Pres., Brandon Ebert; Exec. Dir. & Cur., Bill North; Cinema, Heather Greene.
Personnel Profile: Full-Time Paid 4; Part-Time Paid 5; Interns 2.
Governing Authority: nonprofit organization. Tax-exempt: 501(c)(3).

Institution Type/Description: Art Center.
Collections: temporary exhibitions.
Facilities: film theatre; classroom.
Activities: films; artist residencies; tours; lectures; concerts; outreach education; interdisciplinary adult & children's workshops & classes; hands-on Discovery Area; visual art exhibition; interactive exhibits.
Publications: exhibition catalogues; Exhibits & Programs guide; annual report.
Hours & Admission Prices: Wed.-Sat. 12-5, Sun. 1-5. No charge; donations accepted. Closed major holidays. &
Attendance: 18,600 (accurate)
Membership: Individual $50; Contributor $100; Friend $250; Activist $500; Patron $750; Sustainer $1,000; Advocate $2,500; Partner $5,000; Visionary $10,000.

✳ **SMOKY HILL MUSEUM, (M),** 211 W. Iron Ave., Salina, KS 67401-2613. Mailing Address: P.O. Box 101, Salina, KS 67402-0101. Tel.: 785-309-5776. Fax: 785-826-7414.
E-mail: museum@salina.org
Web Site: www.smokyhillmuseum.org
Founded: 1983.
Congressional District: 1
Key Personnel: C.E.O., Brad Anderson; Dir., Susan Hawksworth; Registrar, Jennifer Toelle; Cur. Collections, Lisa Upshaw; Administrative Asst., Pauline Fallis; Cur. Exhibits, Joshua Morris; Education Coord., Nona Miller; Museum Shop Mgr., Judy Kuasnicka.
Personnel Profile: Full-Time Paid 6; Part-Time Paid 2; Part-Time Volunteers 160; Interns 2.
Governing Authority: city. Parent Institution: Salina Arts & Humanities Commission; City of Salina. Subsidiary Institution: Friends of the Smoky Hill Museum. Tax-exempt.
Institution Type/Description: History Museum: housed in 1937 federal building.
Collections: history; archives.
Research Fields: history; archives.
Activities: tours; changing exhibitions; historical presentations; school programs.
Publications: newsletter, Heritage Express.
Hours & Admission Prices: Tues.-Fri. 12-5, Sat. 10-5, Sun. 1-5. Closed all major holidays. No charge. &
Attendance: 20,129 (accurate)
Membership: Friend $35; Participant $50; Patron $75-$199; Advocate $200-$499; Supporter $500-$999; Benefactor $1,000-$2,499; Partner $2,500-$4,999; Corporate $5,000 & up.

YESTERYEAR MUSEUM, 1100 W. Diamond Dr., Salina, KS 67401-9542. Tel.: 785-825-8473.
E-mail: ckf@yesteryearmuseum.org
Governing Authority: nonprofit organization. Parent Institution: Central Kansas Flywheels.
Institution Type/Description: History Museum.
Collections: American history & heritage; 1880s print shop; 1869 Swedish Bible; Boy Scout memorabilia; toys.
Activities: Museum Sponsors: By Gone Days in Spring; Antique Engine Show and Antique Tractor Pull in Fall.
Hours & Admission Prices: Call for hours.

Scandia

SCANDIA MUSEUM, 409 4th St., Scandia, KS 66966. Tel.: 785-335-2620.
Founded: 1947.
Volunteer Hours: 120
Governing Authority: Tax-exempt.
Institution Type/Description: History Museum.
Collections: Scandinavian history & heritage; farm machinery; tools; buggies; war memorabilia; photographs; scrapbooks; genealogy.
Hours & Admission Prices: Memorial Day to Labor Day Mon.-Sat. 1-4; other times by appointment. No charge; donations accepted. &
Attendance: 500 (estimated)

Scott City

EL QUARTELEJO MUSEUM, 902 W. 5th St., Scott City, KS 67871. Tel.: 602-872-5912.
Governing Authority: Parent Institution: Scott County Historical Society.
Institution Type/Description: History Museum.
Collections: local history & culture; photographs; paintings; fossils; personal artifacts.
Hours & Admission Prices: Mon.-Fri. 1-5; other times by appointment. No charge.

KEYSTONE GALLERY, 401 U.S. Hwy. 83, Scott City, KS 67871-8013. Tel.: 620-872-2762. Facebook: Keystone Gallery.
E-mail: keystone@keystonegallery.com
Web Site: www.keystonegallery.com
Founded: 1991.
Congressional District: 1
Key Personnel: Dir., Barbara Shelton
Institution Type/Description: Fossils & Art Gallery.
Collections: Kansas fossils; local history; paintings; monument rocks.
Research Fields: paleontological.
Facilities: Museum-related items for sale.
Hours & Admission Prices: Call for hours. No charge; donations accepted. Closed Thanksgiving; Christmas.
Attendance: 5,000 (estimated)

Sedan

EMMETT KELLY HISTORICAL MUSEUM, 202 E. Main, Sedan, KS 67361-1629. Tel.: 620-725-3470.
Web Site: www.emmettkellymuseum.com
Founded: 1967.
Congressional District: 5
Key Personnel: Chm. (V), Roger Floyd; Museum Shop Mgr., Darla Loyd.
Personnel Profile: Part-Time Paid 1; Part-Time Volunteers 10.
Governing Authority: municipal; museum board. Tax-exempt.
Institution Type/Description: Clown & Historical Museum.
Collections: clowns; circus; decanters; county & historical items; quilts; period print shop; paintings of Chautauqua Co. artists; Civil War, World War I & II show cases; Native American artifacts; shell collection from Guam; 50 period radios & consoles.
Research Fields: circus; clowns local history.
Publications: brochures.
Hours & Admission Prices: June-Aug. 15: Tues.-Fri. 10-12 & 1-5, Sat. 10-5. No charge; donations accepted. &
Attendance: 1,500 (estimated)

Seneca

NEMAHA COUNTY HISTORICAL SOCIETY, 113 N. 6th St., Seneca, KS 66538-1748 Tel.: 785-336-6366.
E-mail: nchs@rainbowtel.net
Web Site: nemahacountyhistorical.com
Institution Type/Description: Historical Society Museum: housed in the former Nemaha County Jail and Sheriff's home; built in 1879.
Collections: local history, industry & culture; personal artifacts; period furnishings; photographs.
Hours & Admission Prices: Memorial Day to Labor Day Wed.-Sat.

Shawnee

JOHNSON COUNTY MUSEUM, (M), 6305 Lackman Rd., Shawnee, KS 66217-9740. Tel.: 913-715-2550. Fax: 913-715-2565. TDD: 913-782-7188.
E-mail: jcmuseum@jocogov.org
Web Site: www.jocomuseum.org
Founded: 1967.
Congressional District: 3
Key Personnel: Dir., Mindi C. Love; Cur. Collections, Anne Jones; Cur. Education, Jennifer Crane; Cur. Interpretation, Matt Gilligan; Collections Mgr., Russ Czaplewski.
Personnel Profile: Full-Time Paid 7; Part-Time Paid 3; Part-Time Volunteers 50; Interns 3.
Governing Authority: county. Subsidiary Institution: Friends of Johnson County Museum; Museum of History; Lanesfield School Historic Site; 1950's All-Electric House. Tax-exempt.
Institution Type/Description: History Museum.
Collections: Johnson County history 1820s-present; suburbia.
Research Fields: local history; suburban history.
Activities: tours; speakers; films; workshops; permanent & changing exhibits.
Publications: newsletter, The Album; book, Johnson County Kansas: A Pictorial History, 1825-2005.
Hours & Admission Prices: Museum: Mon.-Sat. 10-4:30. No charge. 1950s All-Electric House: Mon.-Sat. 1-4. Adults $2, children 12 & under $1. Lanesfield Historic Site: Fri.-Sat. 1-5 & by appointment. No charge. Closed legal holidays. &
Attendance: 54,904 (accurate)
Membership: Individual $25; Family $50; Supporting $100; Patron $250; Benefactor $500.

SHAWNEE TOWN 1929, (M), 11501 W. 57th St., Shawnee, KS 66203-2225. Tel.: 913-248-2360. Fax: 913-248-2363. Facebook: Shawnee Town 1929.
E-mail: cpautler@cityofshawnee.org
Web Site: www.shawneetown.org
Founded: 1966.
Congressional District: 3
Key Personnel: Dir., Charles Pautler; Museum Shop Mgr., Royal Krueger.
Personnel Profile: Full-Time Paid 4; Part-Time Paid 5; Part-Time Volunteers 80.
Governing Authority: municipal. Parent Institution: City of Shawnee, Shawnee, KS. Tax-exempt.
Institution Type/Description: History Museum.
Collections: history of Shawnee from 1840s to present.
Research Fields: local history.
Facilities: rental facilities for parties, meetings & weddings.
Activities: concerts; docent program; guided tours; permanent exhibit on the history of Shawnee, KS; youth & adult educational programs. Annual Events: Old Shawnee Days; Craft Fair; Barbecue Contest; Living History Weekends; Garden Party; Tomato Roll; Historical Hauntings; Christmas Around Town.
Hours & Admission Prices: March-Oct. Tues.-Sat. 10-4:30. Self-Guided Tours: adults $3, students 5-17 $1; discounts to AAM members & Time Travelers; members & children under 6 no charge. Closed Memorial Day; Independence Day; Labor Day. &
Attendance: 120,000 (estimated)
Membership: Friends $35; Business $100.

WONDERSCOPE CHILDREN'S MUSEUM OF KANSAS CITY, 5700 King, Shawnee, KS 66203-2708. Tel.: 913-287-8888. Fax: 913-268-4608.
E-mail: info@wonderscope.org
Web Site: www.wonderscope.org
Founded: 1989.
Congressional District: 3
Key Personnel: C.E.O & Dir., Fred G. Andrews; Chm. (V), Dan Zmijewski; C.O.O., John Lowe.
Personnel Profile: Full-Time Paid 4; Part-Time Paid 15; Part-Time Volunteers 25.
Governing Authority: Tax-exempt.
Institution Type/Description: Children's Museum.
Collections: hands-on exhibits.
Activities: workshops; special events; educational programs; temporary & traveling exhibits.
Hours & Admission Prices: June-Aug. Mon.-Sat. 10-5, Sun. 12-5; Sept.-May Tues.-Fri. 10-4, Sat. 10-5, Sun. 12-5. Admission 3-63 $7, senior citizens 64 & over $6, children 1-2 $4; discounts to ACM members; children under 1 & members no charge. Closed major holidays. &
Attendance: 63,780 (accurate)
Membership: Playtime $80; Wonderful $125.

Stafford

QUIVIRA NATIONAL WILDLIFE REFUGE AND VISITOR CENTER, 1434 N.E. 80th St., Stafford, KS 67578. Tel.: 620-486-2393.
E-mail: quivira@fws.gov
Founded: 1955.
Institution Type/Description: Wildlife Refuge.
Collections: wildlife & their habitats; local history; photographs; mounted wildlife.
Activities: hands-on activities.
Hours & Admission Prices: Visitors Center: Mon.-Fri. 7:30-4. Refuge: daily sunrise to sunset. No charge. &

STAFFORD COUNTY HISTORICAL SOCIETY MUSEUM, 100 N. Main, Stafford, KS 67578-1343. Tel.: 620-234-5664.
E-mail: staffordcountymuseum@earthlink.net
Web Site: home.earthlink.net/~mjhathaway61
Founded: 1976.
Congressional District: 1
Key Personnel: Pres. (V), Marion Hearn.
Personnel Profile: Full-Time Paid 1; Part-Time Volunteers 20.
Volunteer Hours: 1,174
Operating Expenses: 56,581
Operating Income: 39,730
Governing Authority: Tax-exempt.
Institution Type/Description: Historical Society Museum & Library.
Collections: local history; personal artifacts; photographs; genealogy; 29,000 glass plate negatives.

Facilities: library.
Activities: conservation workshops; book discussion groups. Museum Sponsors: Open House.
Publications: biannual newsletter, Reflections.
Hours & Admission Prices: Mon.-Fri. 9-3:30; weekends & special tours by appointment. No charge; donations accepted.
Attendance: 1,000 (estimated)
Membership: Family $15; Sponsor $25; Donor $100; Contributor $250; Sustaining $500; Patron $1,000; Royal Patron $2,000.

Stockton

FRANK WALKER MUSEUM, 921 S. Cedar, Stockton, KS 67669. Tel.: 785-425-7217.
Governing Authority: Parent Institution: Rooks County Historical Society.
Institution Type/Description: History Museum.
Collections: local history & culture; dolls; Lorenzo Fuller's instruments; medical equipment; farm tools; clothing; machinery.
Hours & Admission Prices: Tues., Thurs. & Sat. 10-4; other times by appointment. No charge; donations accepted.

WALLER-COOLBAUGH 20TH-CENTURY HOUSE, 421 N. Walnut, Stockton, KS 67669. Mailing Address: 319 Gracie St., Stockton, KS 67669. Tel.: 785-425-7227.
Institution Type/Description: Historic House Museum: built in 1905.
Collections: local history & culture; period furnishings; personal artifacts; photographs.
Hours & Admission Prices: Call for hours.

Strong City

TALLGRASS PRAIRIE NATIONAL PRESERVE, 2480 KS Hwy. 177, Strong City, KS 66869. Tel.: 620-273-8494 & 6034. Fax: 620-273-8950.
Founded: 1996.
Governing Authority: Parent Institution: Department of Interior. Subsidiary Institution: National Park Service. Tax-exempt.
Institution Type/Description: Nature Preserve.
Collections: local cultural & natural history; early farm & ranch machinery. Historic Buildings: 1881 ranch house; limestone barn; outbuildings; one-room schoolhouse.
Facilities: nature trails.
Activities: orientation film; educational programs; special events; demonstrations.
Hours & Admission Prices: Preserve & Trails: daily 24 hours. Buildings: daily 8-4:30. No charge; donations accepted. Closed most holidays.
Attendance: 18,500 (estimated)

Studley

COTTONWOOD RANCH, 14432 E. US Hwy. 24, Studley, KS 67740-4135. Mailing Address: 746 S. Rd. 148 E., Hoxie, KS 67740-4130. Tel.: 785-627-5866.
Governing Authority: Parent Institution: Kansas State Historical Society.
Institution Type/Description: Historic Site: built by John Fenton Pratt from 1885-1896.
Collections: Pratt family history; photographs; personal artifacts; period furnishings.
Hours & Admission Prices: Jan. 5-Feb. Fri.-Sat. 10-4, Sun. 1-4; March-Nov. Wed.-Sat. 9-5, Sun. 1-5. Adults $2, seniors & students $1; active military & children under 5 no charge. Closed state holidays. &

Sublette

HASKELL COUNTY HISTORICAL SOCIETY, North Fairground Rd., Sublette, KS 67877. Mailing Address: P.O. Box 101, Sublette, KS 67877-0101. Tel.: 620-675-8344.
Formerly: The Haskell County Historical Museum
Founded: 1983.
Congressional District: 1
Key Personnel: Pres. (V), James Groth; Dir., Darlene Groth.
Personnel Profile: Full-Time Paid 1.
Governing Authority: private; nonprofit organization. Tax-exempt: 501(c)(3).
Institution Type/Description: General Museum.
Collections: artifacts; rocks & minerals; American Indian artifacts.
Facilities: library; 10,000 sq. ft. exhibit space.
Activities: workshops; meetings.
Hours & Admission Prices: Tues.-Sat. 1-5. No charge; donations accepted. Closed major holidays &

Attendance: 650 (accurate)
Membership: Annual $5; Sustaining $15; Life $100.

Sylvan Grove

YESTERDAY HOUSE MUSEUM, 118 S. Main St., Sylvan Grove, KS 67481. Mailing Address: P.O. Box 68, Sylvan Grove, KS 67481. Tel.: 785-526-7270.
Governing Authority: Parent Institution: Sylvan Grove Historical Society.
Institution Type/Description: History Museum.
Collections: local history & culture; photographs; period furnishings; barbed wire.
Hours & Admission Prices: May-Oct. Sat.-Sun. 1-5; other times by appointment.

Syracuse

HAMILTON COUNTY MUSEUM, Hwy. 50 & Gates St., Syracuse, KS 67878. Mailing Address: 102 N. Gates, P.O. Box 923, Syracuse, KS 67878-0923. Tel.: 620-384-7496. Facebook: Hamilton County Historical Society Museum.
E-mail: historic@pld.com
Web Site: www.hamiltoncountymuseum.org
Founded: 1966.
Congressional District: 122
Key Personnel: Mgr., Joanice Jantz; Pres. (V) Bd., Jason Ochs; Vice Pres., Eddie George; Sec., Marcia Ashmore; Treas., JoDean Hawkins; Member, Marcus Ashlock; Member, Rusty Wharton; Member, Charles Whitaker.
Personnel Profile: Full-Time Paid 1.
Operating Expenses: 27,000
Operating Income: 30,000
Governing Authority: bd. Parent Institution: Hamilton County Historical Society.
Institution Type/Description: History Museum.
Collections: local history & culture; photographs; period furnishings; personal artifacts; Native American; barber shop; general store; guns; military artifacts.
Hours & Admission Prices: Tues.-Sat. 9-12 & 1-4. No charge; donations accepted. Closed New Year's Day; Labor Day; Thanksgiving; Christmas.
Attendance: 900 (estimated)
Membership: Senior Citizen 65 & over $10; Single $15; Family $20; Business $50; Family Lifetime $200.

Tonganoxie

TONGANOXIE COMMUNITY HISTORICAL SOCIETY, 201 W. Washington St., Tonganoxie, KS 66086. Mailing Address: P.O. Box 785, Tonganoxie, KS 66086-0785. Tel.: 913-369-3835.
E-mail: tchs@att.net
Web Site: www.tonganoxiehistoricalsociety.org
Founded: 1981.
Congressional District: 2
Personnel Profile: Part-Time Volunteers 15.
Governing Authority: Tax-exempt.
Institution Type/Description: Local History Museum.
Collections: local history & culture; photographs; period furnishings & clothing; personal artifacts.
Activities: educational programs.
Hours & Admission Prices: April-Oct. Wed. 9-12, Sun. 1-4. No charge; donations accepted. &
Attendance: 376 (estimated)
Membership: Single $10; Family $15; Life $100.

Topeka

ALICE C. SABATINI GALLERY-TOPEKA AND SHAWNEE COUNTY PUBLIC LIBRARY, (M), 1515 W. 10th, Topeka, KS 66604-1304. Tel.: 785-580-4515 & 4400. Fax: 785-580-4496.
E-mail: sbest@tscpl.org
Web Site: www.tscpl.org/gallery
Founded: 1870.
Congressional District: 2
Key Personnel: C.E.O., Gina Millsap; Coord., Kari Zimmerman; Gallery Dir., Sherry L. Best; Deputy Dir., Robert Banks; Museum Shop Mgr., Laura Schmidt.
Personnel Profile: Full-Time Paid 4; Part-Time Volunteers 11.
Governing Authority: municipal. Parent Institution: Topeka & Shawnee County Public Library. Tax-exempt. 170(b)(1)(A).
Institution Type/Description: Fine Arts Gallery.

Collections: period & contemporary glass paperweights; art Nouveau glass & ceramics; contemporary ceramics; American prints; Kansas paintings, drawings, photography, fiber, & sculpture; Topeka collectors: West African sculpture, East African beadwork, southwestern reliquary figures, Chinese pewters, & books as art; archives; historic Shawnee County.
Major Exhibits: Topeka Competition 31, 12/13-1/14; Kansas Burns, 2/14-3/14; The Printed Images, 4/14-6/14; Ramp It Up!, 7/14-8/14; Woodworking, 11/14-1/15.
Research Fields: glass & ceramics.
Facilities: 30,000-vol. library on the arts; reading room.
Activities: lectures; 1/2-inch videotapes; gallery talks; concerts; inter-museum loan & temporary exhibitions.
Hours & Admission Prices: Mon.-Fri. 9-9, Sat. 9-6, Sun. 12-9. No charge. Closed New Year's Day; Martin Luther King Jr. Day; Washington's Birthday; Memorial Day; Independence Day; Labor Day; Veterans Day; Thanksgiving; Christmas. &
Attendance: 28,000 (estimated)
Membership: Friend of Art $30; Illustrator $40; Emerging Artist $60; Master Artist $200; Lifetime Art Benefactor $3,000 & up.

BROWN V. BOARD OF EDUCATION NATIONAL HISTORIC SITE, 1515 S.E. Monroe St., Topeka, KS 66612-1143. Tel.: 785-354-4273.
Web Site: www.nps.gov/brvb
Founded: 1992.
Institution Type/Description: Historic Site.
Collections: artifacts & memorabilia from the Supreme Court decision to end segregation.
Hours & Admission Prices: Daily 9-5. No charge; donations accepted. Closed New Year's Day; Thanksgiving; Christmas.

CHARLES CURTIS HOUSE MUSEUM, 1101 SW Topeka Blvd., Topeka, KS 66612-1602. Tel.: 785-597-5380 & 357-1371.
Web Site: www.charlescurtismuseum.com
Founded: 1993.
Institution Type/Description: Historic House: former home of Senator Charles Curtis; built in 1879.
Collections: Curtis family history; personal artifacts; period furnishings.
Hours & Admission Prices: Sat. 11-3; other times by appointment. Adults $5.

THE COLLECTIVE GALLERY, 3121 S.W. Huntoon, Topeka, KS 66604. Tel.: 785-234-4254.
E-mail: barbara.peters@att.net
Web Site: thecollectiveartgallery.com
Institution Type/Description: Art Gallery.
Collections: works by regional artists; paintings; sculpture.
Hours & Admission Prices: Wed.-Fri. 12-4, Sat. 10-2.

COMBAT AIR MUSEUM, INC., Forbes Field, J St., Hangar 602, Topeka, KS 66619. Mailing Address: P.O. Box 19142, Topeka, KS 66619-0142. Tel.: 785-862-3303. Fax: 785-862-3304.
E-mail: combatairmuseum@aol.com
Web Site: www.combatairmuseum.org
Founded: 1976.
Congressional District: 2
Key Personnel: Chm. & Pres., Gene Howerter.
Personnel Profile: Full-Time Paid 1; Part-Time Paid 1; Part-Time Volunteers 35.
Governing Authority: nonprofit organization. Tax-exempt: 501(c)(3).
Institution Type/Description: Military Aviation Museum: 1942 Topeka Army Air Field-later Forbes AFB, now Forbes Field.
Collections: military aircraft, weapons, uniforms & vehicles; insignia; photographs; medals; period rooms. Historic Aircraft: JN-4D Jenny (replica); MIG-15; MIG-17; 1939 O-47B; C-47 Skytrain; Harvard MK IV; T-33A; F-86H Sabre; US-2A Tracker; EC-121T Super Constellation; F-84F Thunderstreak; F11F-1 Tiger, Blue Angel #5; F-101B Voodoo; RU-8D Seminole; F-4D Phantom, F3D Skyknight; UH-1H Iroquois Huey; UH-1M Iroquois Huey; CH-54 Tarhe Sky Crane; NCH-53 Sea Stallion; F-105D Thunderchief; TA-4J Skyhawk; Meyers OTW #1; SNB-5; UC-61K Forwarder; F9F-5 Panther; BF-109 Movie Mockup; 1914 Taube (1/2 scale replica); Hiller UH-12A; BT-13A Valiant; MiG-21.
Facilities: Gift items for sale.
Activities: guided tours.
Publications: bimonthly newsletter, Plane Talk.
Hours & Admission Prices: Jan.-Feb. daily 12-4:30; March-Dec. Mon.-Sat. 9-4:30, Sun. 12-4:30; last admission 3:30. Adults $6, active military & children 4-17 $4; discounts to AAA members; children under 5 & members no charge. Closed New Year's Day; Easter; Thanksgiving; Christmas. &
Attendance: 10,000 (accurate)

Membership: Individual $30; Family $40; Lifetime $1,000.

THE GREAT OVERLAND STATION, 701 N. Kansas Ave., Topeka, KS 66608-1260. Mailing Address: P.O. Box 8792, Topeka, KS 66608-0792. Tel.: 785-232-5533. Fax: 785-232-6259.
E-mail: info@greatoverlandstation.com
Web Site: greatoverlandstation.com
Founded: 2000.
Congressional District: 2
Key Personnel: Pres. & C.O.O., Bette Allen; Chm., Robert St. John; Dir. Capital Campaign, Beth Fager; Program Mgr., Genevieve Nichols; Business Mgr., Jeannie Rose; Events Mgr., Amanda Beach.
Personnel Profile: Full-Time Paid 2; Full-Time Volunteers 1; Part-Time Paid 6; Part-Time Volunteers 30.
Governing Authority: private; nonprofit organization. Parent Institution: Railroad Heritage, Inc. Tax-exempt: 501(c)(3).
Institution Type/Description: Railroad Heritage Museum: housed in the historic Union Pacific Depot.
Collections: people of the railroad; Oregon Trail & Santa Fe Railway history; local history; underground railroad; Vice President Charles Curtis; Santa Fe railcars. Historic Building: Welda Depot.
Activities: rental facilities; tours; Harvey house luncheons & tours; member events. Museum Sponsors: Kansas Hall of Fame Gala; Topeka Railroad Festival; Santa Arrives on Union Pacific.
Publications: newsletter.
Hours & Admission Prices: Tues.-Sat. call for hours. Adults $5, seniors & military $4, children $2; members no charge. Closed national holidays. &
Attendance: 23,000 (estimated)
Membership: Family $35; Bronze $100; Silver $250; Gold $500; Platinum $1,000; Diamond $2,500.

HOLLEY MUSEUM OF MILITARY HISTORY, Ramada Hotel & Convention Center, 420 S.E. 6th St., Topeka, KS 66606. Tel.: 785-272-6204. Fax: 785-224-5034.
Key Personnel: Cur., Vernon Fisher
Institution Type/Description: Military History Museum.
Collections: military artifacts including WWI & WWII; U.S. Navy; Air Force One; Kansas war heroes.
Hours & Admission Prices: Daily 10-8. No charge; donations accepted.

KANSAS CHILDREN'S DISCOVERY CENTER, (M), Gage Park, 4400 S.W. 10th Ave., Topeka, KS 66604. Tel.: 785-783-8300. Fax: 785-783-7662.
Web Site: www.kansasdiscovery.org
Founded: 2011.
Key Personnel: Exec. Dir., Joanne Morrell; Dir. Operations, Carolyn Chinn Lewis; Dir. Education & Programs, Margaret Hennessey Springe; Visitor Svcs. & Membership, Maureen Washatka; Mktg., Andrea Etzel.
Personnel Profile: Full-Time Paid 5; Part-Time Paid 10; Part-Time Volunteers 175; Interns 1.
Governing Authority: Tax-exempt.
Institution Type/Description: Children's Museum.
Collections: hands-on exhibitions.
Activities: summer camps; educational programs; field trips; birthday parties.
Hours & Admission Prices: Tues.-Sat. 10-5, Sun. 1-5. Admission $7.25, seniors 65 & over $6.25; discounts to ACM members; children under one and members no charge.
Membership: Grandparent $75; Family $85; Grandparent Plus One $100; Family Plus One $110; Reciprocal $125; Reciprocal Plus One $150.

*** KANSAS MUSEUM OF HISTORY, (M),** 6425 S.W. Sixth Ave., Topeka, KS 66615-1099. Tel.: 785-272-8681, ext. 401. Fax: 785-272-8682.
E-mail: KansasMuseum@kshs.org
Web Site: www.kshs.org
Founded: 1875.
Congressional District: 2
Key Personnel: Exec. Dir. Kansas State Historical Society, Jennie Chinn; Chm. & Pres. (V), Dean Carlson; Museum Dir., Robert J. Keckeisen; Asst. Dir., Rebecca J. Martin; Cur., Blair D. Tarr; Cur., Laurel Fritzsch; Exhibits Dir., Chris Prouty; Education & Outreach Dir., Mary Madden; Registrar, Nikaela Zimmerman; Museum Shop Mgr., Jeannie Lloyd.
Personnel Profile: Full-Time Paid 11; Part-Time Paid 2; Part-Time Volunteers 130; Interns 2.
Governing Authority: state. Parent Institution: Kansas State Historical Society. Branch Museums: 1851 Kaw Mission, Council Grove; 1855 First Territorial Capitol of Kansas, Fort Riley; 1857 Hollenberg Pony Express Station,

Hanover; 1846 Native American Heritage Museum, Highland; 1838 Shawnee Mission, Kansas City; Pawnee Indian Village, Republic; 1865 Fort Hays & Frontier Historical Park, Hays; 1854 John Brown Memorial Park, Osawatomie; 1870 Marais des Cygnes Massacre Memorial Park, Pleasanton; 1857 Grinter Place, Kansas City; 1860 Goodnow House, Manhattan; Cottonwood Ranch, Studley; Constitution Hall, Lecompton. Tax-exempt: 501(c)(3).
Institution Type/Description: History Museum.
Collections: ethnological & archaeological materials; costumes; decorative arts; furniture; textiles; transportation vehicles; military equipage; agricultural implements; tools representing 19th-century trades, crafts, paintings, drawings & prints.
Research Fields: social, cultural, political, railroad & military history, material culture of Kansas & the West.
Facilities: Museum-related items for sale.
Activities: guided tours; permanent, temporary & traveling exhibitions; adult & youth workshops; field trips; lectures; school loan service; slide & tape programs; craft demonstrations; films; internship program with Kansas colleges & universities; traveling resource trunk program for schools.
Publications: occasional exhibit catalogs; culture essays.
Hours & Admission Prices: Museum: Tues.-Sat. 9-5, Sun. 1-5. Historic Sites: Tues.-Sat. 10-5, Sun. 1-5. Adults $6, students $4; members no charge. Closed New Year's Day; Thanksgiving; Christmas; state holidays. &
Attendance: 52,202 (accurate)
Membership: Individual $40; Household $50; Sustainer $100; Sponsor $250; Benefactor $1,000 & up.

KANSAS STATE ARCHIVES, Center for Historical Research, Reference Dept., 6425 S.W. Sixth Ave., Topeka, KS 66615-1099. Tel.: 785-272-8681, ext. 117.
Key Personnel: Dir. State Archives Div., Patricia Michaelis; Exec. Dir. Kansas Historical Society, Jennie Chinn.
Personnel Profile: Full-Time Paid 22; Part-Time Paid 5; Part-Time Volunteers 21; Interns 2.
Governing Authority: Parent Institution: Kansas Historical Society. Support Organization: Kansas Historical Foundation. Tax-exempt.
Institution Type/Description: Archives.
Collections: local history & culture; genealogical records; photographs; newspapers; government & military records; maps; books; pamphlets; unpublished non-government records including letters, diaries, business & original records.
Research Fields: Kansas history; Kansas state government; genealogy.
Hours & Admission Prices: Call for hours. &
Attendance: 5,215

✱   **MULVANE ART MUSEUM, (M),** 17th & Jewell Sts., Topeka, KS 66621-1150. Mailing Address: 1700 S.W. College Ave., Topeka, KS 66621. Tel.: 785-670-1124. Fax: 785-670-1329.
E-mail: mulvane.info@washburn.edu
Web Site: www.washburn.edu/mulvane
Founded: 1922.
Congressional District: 2
Key Personnel: Dir., Cindi Morrison; Cur. Collections & Exhibitions, Carol Emert; Cur. Education, Kandis Barker; Asst. Cur. Education, Jane Hanni; Preparator, Michael Allen; Artlab Supvr., Josh Davis; Office Asst., Delene VanSickel.
Personnel Profile: Full-Time Paid 8; Part-Time Paid 1; Part-Time Volunteers 150; Interns 1.
Governing Authority: university. Parent Institution: Washburn University of Topeka. Tax-exempt.
Institution Type/Description: Art Museum.
Collections: 19th- to 21st-century American art; 16th- to 20th-century European prints; contemporary representational, figurative & narrative work; young & emerging eastern Kansas artists.
Research Fields: contemporary representational, figurative and narrative work in all media.
Facilities: William I. Koch Fine Arts Library located in university's Mabee Library; ArtLab. Museum-related items for sale.
Activities: lectures, gallery talks & guided tours; music events; art after school & art in-school outreach programs; fall, spring & summer art classes for K-adult. Annual Events: Mountain Plains Fine Art Fair.
Publications: Exhibition catalogues; brochures; newsletter; annual report.
Hours & Admission Prices: Tues. 10-7, Wed.-Fri. 10-5, Sat.-Sun. 1-4. No charge; donations accepted. Closed major holidays. &
Attendance: 60,000 (accurate)
Membership: Student & Senior Citizen $25; Individual $35; Young Contemporaries $55-$75; Dual & Family $60; Director's Circle $125; Silver Circle $250; Gold Circle $500; Platinum Circle $1,000.

MUSEUM OF THE KANSAS NATIONAL GUARD, 6700 S.W. Topeka Blvd., Bldg. 301, Topeka, KS 66619-0285. Mailing Address: P. O. Box 19285, Topeka, KS 66619-0285. Tel.: 785-862-7203.
E-mail: kngmuseum@aol.com
Web Site: www.kansasguardmuseum.org
Founded: 1997.
Congressional District: 2
Key Personnel: Pres. (V), BG Ed Gerhardt, (Ret.)
Personnel Profile: Part-Time Volunteers 41.
Governing Authority: Tax-exempt.
Institution Type/Description: Army and Air National Guard Museum.
Collections: Kansas National Guard history & heritage; personal artifacts; period furnishings; military uniforms, weapons & equipment; medals; photographs; military weapons & equipment.
Publications: semi-annual newsletter, Kansas National Guard Museum Newsletter.
Hours & Admission Prices: Tues.-Sat. 10-4. No charge, but donations accepted. &
Attendance: 12,000 (accurate)
Membership: Individual $20; Major Donor $100 & up.

OLD PRAIRIE TOWN AT WARD-MEADE HISTORIC SITE, 124 N.W. Fillmore, Topeka, KS 66606-1171. Tel.: 785-368-3888. Fax: 785-368-3890.
E-mail: john.bell@snco.us
Web Site: parks.snco.us
Formerly: Historic Ward-Meade Park
Key Personnel: Dir., John Bell.
Personnel Profile: Full-Time Paid 2; Part-Time Paid 7; Part-Time Volunteers 100.
Governing Authority: municipal; nonprofit. Parent Institution: City of Topeka.
Institution Type/Description: History Museum.
Collections: clothing; toys; furniture; art work; mid-1800s to 1900s physician's & dentist's equipment; 1800s church; replica log cabin; drugstore; dentist & physician's office; general store. Historic Building: 1870s Victorian home.
Facilities: botanical garden; 35-seat soda fountain; visitor center. Museum-related items for sale.
Activities: docent program; guided tours. Annual Events: Apple Festival; A Night Under the Stars Campout; summer concert; Scary on the Prairie in October; Holiday Happenings.
Publications: bimonthly newsletter, Prairie Sun.
Hours & Admission Prices: Park: daily 8am to dusk. General Store & Drugstore: Mon.-Sat. 10-4, Sun. 12-4. Guided Tours: Mon.-Fri. 10, 12 & 2, Sat.-Sun. 12 & 2. Tours: adults $4.50, senior citizens $4, children 6-12 $2; discounts to school groups; children under 5 & botanical garden no charge. Stores closed federal holidays. &
Attendance: 70,000 (estimated)

TOPEKA ZOOLOGICAL PARK, 635 S.W. Gage Blvd., Topeka, KS 66606-2066. Tel.: 785-368-9180. Fax: 785-368-9152.
E-mail: zoo@topeka.org
Web Site: www.topeka.org/zoo
Founded: 1933.
Congressional District: 2
Key Personnel: Dir. Zoo, Brendan Wiley; Mgr. Zoo Operations, Fawn Moser; Education Cur., Dennis Dinwiddie; Zoo Veterinarian, Dr. Shirley Llizo; Animal Care Supvr., Shanna Simpson.
Governing Authority: municipal. Parent Institution: City of Topeka. Tax-exempt: 170(b)(1)(A).
Institution Type/Description: Zoo.
Collections: mammals; birds; reptiles; amphibians; fish; invertebrates; natural history collection of skulls; egg shells.
Research Fields: zoo animal nutrition, behavior & reproduction.
Facilities: 1,000-vol. reference library; tropical rain forest; great ape facility; small animal exhibit; classroom; concessions. Zoo-related items for sale.
Activities: guided tours; Discovery Carts; films; TV & radio programs; organized classes for children & adults; docent program; Explorer Post; audio-visual loans to teachers.
Publications: magazine published 3 times annually, Zooreka.
Hours & Admission Prices: Daily 9-5. Adults $5.75, senior citizens 65 & up $4.75, children 3-12 $4.25; discounts to AZA members; members and children 2 & under no charge. Closed New Year's Day; Christmas. &
Attendance: 180,000 (accurate)
Membership: Senior $25; Individual $30; Zoo for Two $40; Single Family $45; Family $50; Special Friend $70; Super Zoo Buff $145.

Towanda

THE MUSEUM IN TOWANDA, 110 S. Third St., Towanda, KS 67144-8924. Tel.: 316-536-2243.
E-mail: lbrunson@cox.net
Key Personnel: Pres., Yvonne Brunson
Institution Type/Description: History Museum.
Collections: local history & culture; Towanda High School memorabilia; early photographs; Whitewater Valley Falls Farm memorabilia; personal artifacts.
Hours & Admission Prices: Thurs.-Sat. 1-5; group tours by appointment. No charge; donations accepted.

PARADISE DOLL MUSEUM, 119 S. Sixth St., Towanda, KS 67144-9040. Tel.: 316-536-2710. Fax: 316-536-2780.
E-mail: fbrush@cox.net
Founded: 1989.
Key Personnel: Pres. & Museum Shop Mgr., Barbara Brush.
Personnel Profile: Full-Time Volunteers 1.
Institution Type/Description: Doll Museum.
Collections: over 5,000 dolls.
Hours & Admission Prices: Tues.-Sat. 1-5. No charge; donations accepted. &
Attendance: 2,000 (estimated)

Tribune

HORACE GREELEY MUSEUM, 214 E. Harper, Tribune, KS 67879. Tel.: 620-376-4996.
E-mail: museum1@fairpoint.com
Founded: 1976.
Congressional District: 122
Personnel Profile: Part-Time Volunteers 6.
Governing Authority: Parent Institution: Greeley County Historical Society. Tax-exempt.
Institution Type/Description: History Museum: housed in the former county courthouse.
Collections: local history & culture; period furnishings; personal artifacts; photographs.
Research Fields: early Greeley County Republican newspapers.
Hours & Admission Prices: Thurs. 9-4; other times by appointment. Donations requested.
Attendance: 900 (estimated)
Membership: Annual $20; Life $75.

Ulysses

GRANT COUNTY MUSEUM AKA HISTORIC ADOBE MUSEUM, (M), 300 E. Oklahoma, Ulysses, KS 67880-2542. Mailing Address: P.O. Box 906, Ulysses, KS 67880-0906. Tel.: 620-356-3009. Fax: 620-356-5082.
E-mail: ulyksmus@pld.com
Web Site: www.historicadobemuseum.org
Founded: 1978.
Congressional District: 1
Key Personnel: Dir. & Chm. (V), Ginger Anthony; Pres. (V), Pam Meile; Treas., Jim Hickok.
Personnel Profile: Full-Time Paid 3; Part-Time Paid 2; Part-Time Volunteers 1.
Governing Authority: county. private; nonprofit organization. Tax-exempt: 501(c)(3).
Institution Type/Description: General Museum.
Collections: Indian artifacts; turn of the century furnishings; toys; clothing; quilts; local & family history.
Activities: guided tours; formal education programs for children.
Hours & Admission Prices: Mon.-Fri. 10-5, Sat.-Sun. 1-5. No charge; donations accepted. Closed holidays. &
Attendance: 9,000 (estimated)
Membership: Life $25.

Valley Center

VALLEY CENTER HISTORICAL MUSEUM, 112 N. Meridian, Valley Center, KS 67147. Tel.: 316-755-0783.
Institution Type/Description: History Museum.
Collections: local history & culture; period furnishings; household artifacts; photographs; scrapbooks; personal artifacts.
Facilities: research center.
Hours & Admission Prices: Call for hours. No charge; donations accepted.
Membership: Student $5; Individual $10; Family $20.

WaKeeney

TREGO COUNTY HISTORICAL SOCIETY MUSEUM, Fairgrounds, 128 N. 13th/Hwy. 283, WaKeeney, KS 67672. Mailing Address: P.O. Box 132, WaKeeney, KS 67672-0132. Tel.: 785-743-2964.
E-mail: tregohistorical@ruraltel.net
Founded: 1975.
Key Personnel: Dir., Marjean Deines; Pres. (V), Stan Deines.
Personnel Profile: Full-Time Paid 1; Part-Time Paid 1; Part-Time Volunteers 9.
Institution Type/Description: Historical Society Museum.
Collections: local history & culture; period furnishings; personal artifacts; photographs; medical, dental & optometrist equipment; newspapers; county records; military artifacts; farm equipment & tools.
Hours & Admission Prices: Tues. & Fri. 10-12 & 1-5, Wed.-Thurs. 1-5, Sat. 10-12 & 1-4, Sun. 1-4. No charge; donations accepted. &
Attendance: 600 (estimated)
Membership: Annual $10; Life $150.

Wakefield

WAKEFIELD MUSEUM ASSOCIATION, 604 6th St., Wakefield, KS 67487. Mailing Address: 604 6th St., P.O. Box 193, Wakefield, KS 67487. Tel.: 785-461-5516.
E-mail: wakefieldmuseum@eaglecom.net
Web Site: www.wakefieldmuseum.com
Founded: 1973.
Key Personnel: Pres., Joy Shandy; Cur., James Beck; Treas., Sharon Babst.
Personnel Profile: Part-Time Paid 1; Part-Time Volunteers 58.
Governing Authority: private; nonprofit organization. Subsidiary Museums: Sunnyslope School, Wakefield, KS; Saints John & George Episcopal Church, Wakefield, KS; Republican Valley Farm Museum. Tax-exempt.
Institution Type/Description: General Museum.
Collections: history of Wakefield; genealogy files; music; textiles; hats; toys; furniture; governor's desk & clothing; books; pictures; newspapers.
Research Fields: genealogy.
Facilities: library. Museum-related items for sale.
Activities: guided tours for children; temporary exhibitions. Annual Event: Christmas Bazaar.
Publications: winter newsletter, Christmas Bazaar; spring newsletter, Annual Newsletter.
Hours & Admission Prices: Jan.-March Sat.-Sun. 1-4; April-Dec. Wed.-Sun. 1-4. No charge; donations accepted. Closed New Year's Day; Thanksgiving; Christmas; Easter. &
Attendance: 841 (accurate)
Membership: Individual $7.50; Family $15.

Wallace

FORT WALLACE MUSEUM, 2655 Hwy. 40, Wallace, KS 67761. Mailing Address: P.O. Box 53, Wallace, KS 67761. Tel.: 785-891-3564.
Web Site: www.ftwallace.org
Founded: 1955.
Congressional District: 1
Key Personnel: Pres. (V), Jayne Humphrey Pearce
Governing Authority: Parent Institution: Fort Wallace Memorial Association. Tax-exempt.
Institution Type/Description: History Museum.
Collections: local history & culture; period furnishings; personal artifacts; photographs; barbed wire folk art; Indian Wars military artifacts.
Hours & Admission Prices: Summer: Mon.-Sat. 9-5, Sun. 1-5. No charge; donations accepted. &
Attendance: 3,000 (estimated)

Wamego

THE COLUMBIAN THEATRE FOUNDATION, INC., 521 Lincoln Ave., Wamego, KS 66547-1633. Tel.: 785-456-2029. Fax: 785-456-9498.
E-mail: boxoffice@columbiantheatre.com
Web Site: columbiantheatre.com; ozmuseum.com
Formerly: The Columbian Theatre, Museum & Art Center
Founded: 1990.
Congressional District: 2
Key Personnel: Exec. Dir., Clint Stueve; Pres. (V), Jaimee Hoobles; Museum Shop Mgr., Brooke Rinolt.
Personnel Profile: Full-Time Paid 4; Part-Time Paid 10; Part-Time Volunteers 200.

Governing Authority: nonprofit organization. Museums: The Columbian Theatre & Museum, Wamengo, KS; The OZ Museum, Wamengo, KS. Tax-exempt: 501(c)(3).
Institution Type/Description: Decorative Arts Museum: housed in 1895 building to display the 1893 Chicago World's Fair paintings bought by J.C. Rogers; a museum that celebrates the cultural legacy of The Wizard of Oz.
Collections: six historic 16' x 11' murals that hung in the rotunda of the government building at the 1893 Chicago World's Fair, representing the promise & prosperity of America at the end of the 19th century; collection of 20,000 pieces of The Wizard of Oz artifacts & memorabilia.
Research Fields:
Activities: workshops; live performance.
Hours & Admission Prices: The Columbian Theatre & Museum: Tues.-Fri. 10-5, Sat. 10-3. Suggested Donation: $5. The OZ Museum: March-Sept. Mon.-Sat. 10-5, Sun. 12-5; Oct.-Feb. Mon.-Sat. 10-3, Sun. 12-3. Adults $7, children $4. Closed New Year's Day; Easter; Thanksgiving; Christmas. &
Attendance: 60,000 (estimated)
Membership: Individual $50; Family $100; Columbian Assets $150; Corporate Eagles $250; Roger's Circle $500; Roger's Circle Patron $1,500.

OZ MUSEUM, 511 Lincoln, Wamego, KS 66547-1633. Tel.: 866-458-TOTO (toll free). Fax: 785-456-9498.
E-mail: ozmuseum@wamego.net
Web Site: www.ozmuseum.com
Founded: 2003.
Personnel Profile: Full-Time Paid 1; Part-Time Paid 4; Part-Time Volunteers 13.
Governing Authority: Parent Institution: Columbian Theater Foundation. Tax-exempt: 501(c)(3).
Institution Type/Description: Movie Museum.
Collections: over 2,000 artifacts from 1900 to present; Wizard of Oz memorabilia; early silent films; Baum books; Oz Parker Brothers board games.
Facilities: Museum-related items for sale.
Hours & Admission Prices: Mon.-Sat. 10-5, Sun. 12-5. Adults 13 & over $7, children 4-12 $4; discounts to groups, military & AAM members; children 3 & under no charge. &
Attendance: 30,000 (accurate)
Membership: Lion $50; Tin Man $100; Scarecrow $150; Dorothy $250; Good Witch $500; Wizard $1,000.

WAMEGO HISTORICAL SOCIETY & MUSEUM, E. 4th St., City Park, Wamego, KS 66547. Mailing Address: P.O. Box 84, Wamego, KS 66547-0084. Tel.: 785-456-2040.
E-mail: wamegomuseum@wamego.net
Web Site: wamegohistoricalmuseum.org
Founded: 1974.
Personnel Profile: Part-Time Paid 3; Part-Time Volunteers 30.
Governing Authority: Tax-exempt.
Institution Type/Description: Historical Society Museum: housed in a KS Pacific Railroad mechanics rock building; built in 1862.
Collections: local history & culture; period furnishings; personal artifacts. Historic Buildings: one room school house; a rock jail; log cabin; Dutch mill.
Hours & Admission Prices: April-Oct. Mon.-Sat. 10-12 & 1-4, Sun. 1-4; Nov.-March Tues.-Sun. 1-4; groups by appointment. Adults $4; discounts to seniors, military, groups, students & AAA members; members no charge. &
Attendance: 3,243 (accurate)
Membership: Single $15; Family $25.

Washington

WASHINGTON COUNTY HISTORICAL SOCIETY, 216 Ballard, Washington, KS 66968-1901. Tel.: 785-325-2198.
Founded: 1982.
Key Personnel: Pres. (V), Jack Barley; Editor of Newsletter, Librarian & Genealogy, Jo Rippe; Treas., Arlene Dague; Archivist & Genealogy, Mary Alice Pacey.
Personnel Profile: Part-Time Volunteers 27.
Governing Authority: county. Tax-exempt: 501(c)(3).
Institution Type/Description: Historical Society Museum.
Collections: artifacts from the 1860s to 2000s.
Research Fields: genealogy.
Publications: quarterly newsletter.
Hours & Admission Prices: Mon.-Fri. 9-4, Sun. 1-4. No charge; donations accepted. Closed New Year's Day; Thanksgiving; Christmas. &
Attendance: 2,500 (accurate)
Membership: Individual & Family $10; Life $100; Life Couple $150.

Wellington

CHISHOLM TRAIL MUSEUM, 502 N. Washington, Wellington, KS 67152-4061. Tel.: 620-326-3820 & 7466.
Founded: 1964.
Congressional District: 4
Key Personnel: Pres. (V), Richard M. Gilfillan; Sec., Nancy McNett; Treas., Cur. Archives & Librarian, Doris Dwyer.
Personnel Profile: Part-Time Volunteers 8.
Governing Authority: nonprofit. Tax-exempt: 170(b)(1)(A).
Institution Type/Description: History Museum: housed in first hospital in Wellington.
Collections: china; glass; silver; folk art; costumes; toys & dolls; parlor; school room; old kitchen; barber shop; doctor's office; military; grandmother's room; woolly mammoth skull & bones; cattle trail artifacts & history.
Research Fields: genealogy.
Facilities: 600-vol. library of old & rare books; 8,000 genealogical & historical printed items about Wellington & Sumner County available on premises under supervision. Handwork, Kansas books & pioneer curios for sale.
Activities: guided tours; gallery talks; permanent & temporary exhibitions.
Hours & Admission Prices: mid-April to Memorial Day Sat.-Sun. 1-4; June-Oct. daily 1-4, Nov. Sat.-Sun. 1-4. No charge; donations accepted.
Attendance: 2,000 (estimated)
Membership: Annual $10; Patron $100.

PANHANDLE RAILROAD MUSEUM, 425 E. Harvey, Wellington, KS 67152-3065. Tel.: 620-399-8611.
Founded: 2005.
Key Personnel: Dir. & Museum Shop Mgr., Perry H. Wiley; Chm. (V), Sherry Wiley.
Personnel Profile: Full-Time Volunteers 2.
Institution Type/Description: Railroad Museum.
Collections: railroad history & memorabilia; early furnishings; uniforms; safety equipment; clocks & watches.
Hours & Admission Prices: By appointment. No charge; donations accepted. &
Attendance: 325 (estimated)

Westmoreland

ROCK CREEK VALLEY HISTORICAL SOCIETY, 507 Burkman St., Westmoreland, KS 66549. Mailing Address: P.O. Box 13, Westmoreland, KS 66549-0013. Tel.: 785-457-0100.
E-mail: museum@westmorelandkshistory.org
Web Site: www.westmorelandkshistory.org
Founded: 1978.
Governing Authority: Parent Institution: Rock Creek Valley Historical Society. Tax-exempt.
Institution Type/Description: Historical Society Museum.
Collections: local history & culture; period furnishings; personal artifacts; photographs; historic buildings.
Hours & Admission Prices: April-Nov. Tues.-Sun. 1-4; other times by appointment. No charge; donations accepted.
Membership: Individual $5; Lifetime $100.

Wichita

BOTANICA, THE WICHITA GARDENS, 701 Amidon, Wichita, KS 67203-3199. Tel.: 316-264-0448. Fax: 316-264-0587. Facebook: Botanica Wichita.
E-mail: mmiller@botanica.org
Web Site: www.botanica.org
Founded: 1985.
Congressional District: 4
Key Personnel: C.E.O., Marty Miller; Pres. (V), Justus Fugate; Cur., Pat McKernan; Public Rels., Kristin Marlett; Volunteer Coord., Jodi McArthur; Facilities Coord., Linda Keller; Memberships, Kathy Osler; Dir. Finance, Paula Englert.
Personnel Profile: Full-Time Paid 13; Part-Time Paid 4; Part-Time Volunteers 375; Interns 3.
Governing Authority: municipal; nonprofit organization. Tax-exempt: 501(c)(3).
Institution Type/Description: Botanical Gardens.
Collections: annual & perennial plants; sculptures.
Facilities: 2,700-vol. reference library available to public; 300-seat auditorium; butterfly house. Cards, china, tools, yard ornaments & jewelry for sale.
Activities: guided tours; lectures; organized education programs; summer concerts; docent program. Museum Sponsors: Poster Contest; Photo Competition; Downing Children's Garden Dinner Gala; Christmas Display.

Publications: newsletter; magazine.
Hours & Admission Prices: April-Oct. Mon.-Sat. 9-5, Sun. 1-5; Nov.-March Mon.-Sat. 9-5. Adults $7, seniors $6, children 3-12 $5; discounts to active military & American Public Garden Assoc. members; children under 2 no charge. American Horticultural Society reciprocal program. Closed Martin Luther King Jr. Day; Presidents' Day; Thanksgiving; Christmas Eve & Day. ♿
Attendance: 166,390 (accurate)
Membership: Individual Plus One $45; Family $55; Family Plus One $70.

COLEMAN FACTORY OUTLET AND MUSEUM, 235 N. St. Francis, Wichita, KS 67202. Tel.: 316-264-0836.
Institution Type/Description: History Museum.
Collections: Coleman company history & products including irons, lanterns & camping equipment.
Facilities: 200 sq. ft. exhibit space.
Hours & Admission Prices: Mon.-Fri. 9-6, Sat. 9-1. No charge.

EXPLORATION PLACE, INC., (M), 300 N. McLean Blvd., Wichita, KS 67203-5901. Tel.: 316-660-0600. Fax: 316-660-0670. Facebook: Exploration Place, Inc.
E-mail: cbluml@exploration.org
Web Site: www.exploration.org
Formerly: Exploration Place, Inc. dba the Children's Museum of Wichita
Founded: 1984.
Congressional District: 4
Key Personnel: Pres., Jan. B. Luth; Chairperson, Greg Sevier; Dir. Mktg., Christina Bluml; Dir. Devel., Sharon Miles; Dir. Exhibits, Lynn Corona; Museum Shop Mgr., Christina Sim.
Personnel Profile: Full-Time Paid 26; Part-Time Paid 10; Part-Time Volunteers 925.
Volunteer Hours: 9,127
Operating Expenses: 3,904,687
Operating Income: 4,290,487
Governing Authority: private; not-for-profit organization. Tax-exempt: 501(c)(3).
Institution Type/Description: Science Museum.
Collections: interactive exhibits.
Major Exhibits: MathAlive! (T), 2/14-5/4/14; Tony Hawk Rad Science (T), 5/24/14-9/14; EWW! What's Eating You? (T), 9/13/14-1/14/15.
Facilities: interactive exhibits and Presentation Theatre, Boeing Dome Theater & Planetarium; MiniGolf course; Exploration Park. Explore store; snack bar.
Activities: interactive exhibits.
Hours & Admission Prices: Jan.-mid-March Tues.-Sat. 10-5, Sun. 12-5; mid-March-Dec. Mon.-Sat. 10-5, Sun. 12-5. Adults $9.50, senior citizens 65 & over $8, youth 4-11 $6, children 2-4 $3; discounts to Reciprocal Program, ACM & ASTC members; members no charge. Minigolf or CyberDome Theater: adults $5, seniors $4, youth $3; discounts to groups of 15 or more. Closed Thanksgiving; Christmas. ♿
Attendance: 210,678 (accurate)
Membership: Student $20; Single Plus $50; Family $75; Premium $145; Traveler $250; Voyager $500; Explorer Society $1,000.

FISCH HAUS GALLERY, 524 S. Commerce, Wichita, KS 67202-4610. Tel.: 316-263-6770.
E-mail: info@fischhaus.com
Web Site: www.fischhaus.com
Institution Type/Description: Art Gallery.
Collections: works by regional & national artists; paintings; sculpture.
Activities: temporary exhibitions; special events.
Hours & Admission Prices: Sept.-June last Fri. each month 7pm-10pm; other times by appointment.

GREAT PLAINS NATURE CENTER, 6232 E. 29th St. N., Wichita, KS 67220-2200. Tel.: 316-683-5499. Fax: 316-688-9555.
E-mail: lorrie_beck@fws.gov
Web Site: www.gpnc.org
Founded: 2000.
Key Personnel: Dir., Lorrie Beck; Volunteer Coord., Heidi Bowen; Museum Shop Mgr., Charlene Van Walleghen.
Personnel Profile: Full-Time Paid 5; Part-Time Paid 11; Part-Time Volunteers 200.
Governing Authority: Parent Institutions: US Fish & Wildlife Service, Kansas Dept. of Wildlife, Parks & Tourism, City of Wichita Park & Recreation Department. Subsidiary Institution: Friends of GPNC.
Institution Type/Description: Nature Center.
Collections: wildlife & their habitats; plants of the Great Plains region.

Facilities: Museum-related items for sale.
Activities: educational programs.
Hours & Admission Prices: Mon.-Sat. 9-5. No charge; donations accepted. Closed holidays. ♿
Attendance: 160,702 (accurate)
Membership: Educator, Senior & Student $15; Individual $25; Family $40; Pronghorn $100; Bald Eagle $250; Bison $500.

GREAT PLAINS TRANSPORTATION MUSEUM, 700 E. Douglas, Wichita, KS 67202-3506. Tel.: 316-263-0944.
Web Site: www.gptm.us
Founded: 1983.
Congressional District: 4
Key Personnel: Vice Pres., Steve Corp; Pres. (V), John Gries; Financial Dir., Gale Meek; Public Rels., Affairs & Devel., J. Harvey Koehn; Membership, Fred Tefft; Security, Norman Walters; Museum Shop Mgr., David Meek.
Personnel Profile: Part-Time Volunteers 20.
Governing Authority: private; nonprofit organization. Parent Institution: Wichita Chapter-National Railway Historical Society. Tax-exempt: 501(c)(3).
Institution Type/Description: Transportation Museum.
Collections: locomotives; rolling stock; railroad prints, signs, lanterns, & tools.
Facilities: 2,000-vol. library of railroad history; 50,800 sq. ft. exhibit space. Museum-related items for sale.
Activities: guided tours; loan & temporary exhibitions. Museum Sponsors: Train Show with Model R.R. Hobbyists.
Publications: monthly newsletter, Great Plains Dispatcher.
Hours & Admission Prices: April-Oct. Sat. 9-4, Sun. 1-4; Nov.-March Sat. 9-4. Adults $5, children 3-12 $3; children under 3 no charge. Closed Christmas.
Attendance: 2,979 (accurate)
Membership: Regular $20; Family $30.

THE KANSAS AFRICAN AMERICAN MUSEUM, 601 N. Water St., Wichita, KS 67203-3833. Tel.: 316-262-7651. Fax: 316-265-6953.
E-mail: info@tkaamueum.org
Web Site: tkaam.org
Founded: 1972.
Congressional District: 4
Key Personnel: Interim Exec. Dir., Lisa Dodson; Pres. (V), Carol Cole.
Personnel Profile: Full-Time Paid 2; Part-Time Paid 2.
Governing Authority: private; nonprofit organization. Tax-exempt.
Institution Type/Description: Art & Culture Museum.
Collections: documents, programs & visual art forms of African American life & culture.
Facilities: 500-vol. library.
Hours & Admission Prices: Tues.-Fri. 10-5, Sat. 12-4. Adults $5.50, students & seniors $4.50, children 5-17 $2.50; members and children 5 & under no charge. Closed major holidays.
Attendance: 12,000 (estimated)
Membership: The Cradle Roll Club $20; The Live Wire Class $35; Mission Club $50; Friendship Club $100; Missionary Society $350; Calvary Club $500; Harmony Motion Club $1,000; Trustee Club $5,000.

KANSAS AVIATION MUSEUM, (M), 3350 George Washington Blvd., Wichita, KS 67210-2194. Tel.: 316-683-9242; 877-683-9242. Fax: 316-683-0573.
E-mail: info@kansasaviationmuseum.org
Web Site: www.kansasaviationmuseum.org
Founded: 1990.
Congressional District: 4
Key Personnel: Exec. Dir., Lon Smith; Pres. (V), Ron Williams.
Personnel Profile: Full-Time Paid 5; Part-Time Paid 2; Part-Time Volunteers 125.
Governing Authority: nonprofit organization. Tax-exempt: 501(c)(3).
Institution Type/Description: Aviation Museum: housed in c.1935 art deco municipal air terminal administration building.
Collections: vintage airplanes manufactured in Kansas; photographic collection of early development of aviation industry.
Facilities: library of books, manuals & videos; 20,000 sq. ft. exhibit space. Museum-related items for sale.
Activities: guided tours. Museum Sponsors: Kansas Aviation Museum Gala in November.
Publications: quarterly newsletter; booklet, Prairie Runways; various brochures.
Hours & Admission Prices: Mon.-Sat. 10-5, Sun. 12-5. Adults $8, seniors 65 & over $7, children 4-12 $6; discounts to military & AAA members; children under 3 no charge. Closed New Year's Day; Easter; Thanksgiving; Christmas.

Attendance: 20,000 (accurate)
Membership: Student $20; Individual $40; Family & Grandparent $65; Crew $100-$299, Co-Pilot $300-$999, Pilot $1,000-$2,999; Wing Leader $3,000-$4,999; Ace $5,000-$9,999; Commander $10,000 & up.

KANSAS FIREFIGHTERS MUSEUM, 1300 S. Broadway, Wichita, KS 67211. Tel.: 316-264-3616.
Founded: 2000.
Key Personnel: Pres. (V), B.K. Owens.
Personnel Profile: Part-Time Volunteers 20.
Governing Authority: Tax-exempt.
Institution Type/Description: Firefighting History Museum: housed in Engine House No. 6; built in 1910. Listed on the National Register of Historic Places.
Collections: firefighting history & equipment; 1921 American LaFrance fire truck; 1926 Reo Speed Wagon; 1909 horse drawn fire wagon; early firehouse furnishings; photographs.
Hours & Admission Prices: Call for hours. Adults $3; members no charge.
Attendance: 450 (estimated)
Membership: Individual $15; Family $25.

KANSAS SPORTS HALL OF FAME AT THE WICHITA BOATHOUSE, 515 S. Wichita St., Wichita, KS 67202-3633. Tel.: 316-262-2038. Fax: 316-263-2539.
Web Site: www.kshof.org
Founded: 1961.
Key Personnel: Pres. & C.E.O., Ted Hayes; Events Mgr., Laura Hartley; Mktg. Mgr., Jordan Poland.
Personnel Profile: Full-Time Paid 3; Part-Time Paid 6; Interns 1.
Governing Authority: Tax-exempt.
Institution Type/Description: Sports Museum.
Collections: Kansas sports heroes; sports history; photographs.
Activities: induction ceremony; rental facilities; virtual sports game.
Hours & Admission Prices: Mon.-Fri. 10-4. No charge; donations accepted. Closed New Year's Day; Easter; Thanksgiving; Christmas. &
Attendance: 10,000 (accurate)
Membership: Individual $25; Family $75.

MID-AMERICA ALL INDIAN CENTER, 650 N. Seneca, Wichita, KS 67203-3204. Tel.: 316-350-3340. Fax: 316-262-4216.
Web Site: www.theindiancenter.org
Founded: 1975.
Congressional District: 4
Key Personnel: Exec. Dir., April Scott; Museum Dir., Deborah Roseke; Education Dir., Crystal Flannery-Bachicha.
Governing Authority: nonprofit organization. Parent Institution: City of Wichita. Tax-exempt: 170(b)(1)(A), 501(c)(3), 509(c)(2).
Institution Type/Description: Native American Museum: located on the site of old Indian council grounds.
Collections: Native American art & artifacts from North America, specializing in traditional Plains artifacts & art, contemporary works & paintings.
Research Fields: Native American life, art & religion; life & works of Blackbear Bosin; life & works of other plains artists.
Facilities: 1,250-vol. library of books pertaining to Native Americans available for use on premises only; Blackbear Resource Center. Native American arts & crafts for sale.
Activities: guided tours; lectures; gallery talks; rental gallery; docent program.
Publications: The Keeper of the Plains.
Hours & Admission Prices: Tues.-Sat. 10-4. Adults $7, military, student 13 & up w/ID and seniors 55 & over $5, children 6-12 $3, children under 6 no charge. Closed New Year's Eve & Day; Easter; Thanksgiving; Christmas. &
Attendance: 60,000 (estimated)
Membership: Willow Family $5; Sage Individual $25; Sage Individual Plus One $40; Sage Family $50; Cedar Individual $60; Cedar Family $100; Sweet Grass $250; Bear Root $500; Sweet Corn $1,000.

MIDWEST HISTORICAL AND GENEALOGICAL LIBRARY, 1203 N. Main, Wichita, KS 67203-3614. Mailing Address: P.O. Box 1121, Wichita, KS 67201-1121. Tel.: 316-264-3611.
Institution Type/Description: Library.
Collections: local history & culture; genealogical records; books; magazines; manuscripts; photographs.
Hours & Admission Prices: Tues. & Sat. 9-4, call to confirm.

MUSEUM OF WORLD TREASURES, (M), 835 E. First St. N., Wichita, KS 67202-2700. Tel.: 316-263-1311.
Web Site: www.worldtreasures.org/
Founded: 2001.

Congressional District: 4
Key Personnel: Founder, Dr. Jon Kardatzke; Pres. & C.E.O., Mike Noller; Chm. (V), Bill Williams, IV; Operations Mgr., Penny Moore; Dir. Education, Jillian Overstake-Forsberg; Cur. Collections, LaWanda Smith; Cur. Exhibits., Timothy Howard.
Personnel Profile: Full-Time Paid 5; Part-Time Paid 4; Part-Time Volunteers 30; Interns 2.
Governing Authority: nonprofit organization. Tax-exempt.
Institution Type/Description: General Museum.
Collections: prehistoric animals & cultures; early civilizations; meteorites & crystals; early American history; American Presidents; European Royalty; Civil War to Vietnam War artifacts; American frontier; American pop culture; sports; historic composer; historic authors.
Facilities: banquet facilities; youth activity center. Museum-related items for sale.
Activities: guided tours; birthday parties; club & business meetings; special events; camp-ins; summer camps; youth education programs.
Publications: quarterly, Treasure Times.
Hours & Admission Prices: Mon.-Sat. 10-5, Sun. 12-5. Adults 13-59 $8.95, seniors 60 & over $7.95, children 4-12 $6.95; members & children under 4 no charge. Closed Easter; Thanksgiving; Christmas. &
Attendance: 34,958 (accurate)
Membership: Senior $36; Senior Couple $42; Single $43; Single Plus $49; Household & Grandparent (Grandparents & Grandchildren) $79; Grandparent Plus & Household Plus $99.

✱ OLD COWTOWN MUSEUM, (M), 1865 W. Museum Blvd., Wichita, KS 67203-3295. Tel.: 316-219-1871. Fax: 316-858-7968.
Web Site: www.oldcowtown.org
Founded: 1950.
Congressional District: 4
Key Personnel: Exec. Dir., David Flask; Cur., Teddy Barlow; Business Affairs, Yvonne Kirker; Mktg. Dir., Angela Cato; Education & Interpretation Coord., J. Anthony Horsch; Volunteer Coord., Jacky Goerzen; Gift Store Mgr. & Rentals Coord., David Abbott.
Governing Authority: nonprofit corporation. Parent Institution: Historic Wichita Sedgwick County, Inc. Tax-exempt: 501(c)(3).
Institution Type/Description: History Museum: 1865-1880 era of Wichita & Sedgwick County Kansas.
Collections: 1865-1880 history of Wichita & Sedgwick County; textiles; clothing; tools; agricultural & documentary artifacts; transportation & recreational artifacts; societal & ceremonial artifacts; packages & containers; natural specimens; architectural fragments; a study collection; archives.
Research Fields: commercial & social activities of groups in early Wichita & Sedgwick County; regional history.
Facilities: archives; microfilms of 1860-1880 Wichita & area newspapers; snack bar. Museum-related items for sale.
Activities: living history programs; special events; Teaching in a One-Room School program; 19th century blacksmithing demonstrations & workshops; re-enactments of 1870s County Fair & 1876 Independence Day; 1880s Living History Farm; interpretive tours; lecture series; Girl Scout living history summer program; formally organized education programs; membership & volunteer programs.
Publications: The Chronicle Newsletter; volunteer newsletter.
Hours & Admission Prices: Tues.-Sat. 10-5. Adults $7.75, senior citizens $6.50, youth 12-17 $6, children 4-11 $5.50; discounts to AAM & ICOM members & groups of 15 or more; children under 4 & member adults no charge. Closed New Year's Day; Thanksgiving; Christmas. &
Attendance: 51,890 (accurate)
Membership: Single $30; Companion $45; Family & Grandparent $55. Pioneer Society: Member $100-$999; JR Mead $1,000-$4,999; Marshall Murdock $5,000 & up.

SEDGWICK COUNTY ZOO, 5555 Zoo Blvd., Wichita, KS 67212-1643. Tel.: 316-660-9453. Fax: 316-942-3781.
E-mail: mreed@scz.org
Web Site: www.scz.org
Founded: 1971.
Congressional District: 4
Key Personnel: Dir., Mark C. Reed; C.F.O., Vickie Moore; Pres. (V), Stanley G. Andeel; Deputy Dir., Jim Marlett; Visitor Svcs. Mgr., Steve Fairchild; Volunteer Coord., Bridget Landers.
Personnel Profile: Full-Time Paid 108; Part-Time Paid 41; Part-Time Volunteers 41; Interns 5.
Governing Authority: county; nonprofit. Parent Institution: County of Sedgwick & Sedgwick County Zoological Society, Inc. Tax-exempt: 501(c)(3).
Institution Type/Description: Zoo.
Collections: animals; plants.
Research Fields: embryo transfer; reproductive studies.

Facilities: over 1,000-vol. library of reference & animal material; botanical garden; zoological park; educational facilities. Gift items for sale.

Activities: guided tours; lectures; films; organized education programs for children, adults & undergraduate or graduate students affiliated with Friends University or Wichita State University; docent program; participatory exhibits. Museum Sponsors: Zoobilee; Halloween Party; Kid's Zoobilee; other events all year long.

Publications: bimonthly newsletter, Zoo Tracks; annual report.

Hours & Admission Prices: Winter: daily 10-5; Summer: daily 8:30-5. Adults and children 12 & over $12, children 4-11 $7.50, school groups $4 each; discounts to AZA members; children under 3 & members no charge. &

Attendance: 548,919 (accurate)

Membership: Family $95; Household Plus $115; Sponsor $165; Associate & Contributing $355 & up.

✳ **ULRICH MUSEUM OF ART, (M),** Wichita State University, 1845 Fairmount St., Wichita, KS 67260-0046. Tel.: 316-978-3664. Fax: 316-978-3898. Facebook: Ulrich Museum of Art.

E-mail: ulrich@wichita.edu

Web Site: www.ulrich.wichita.edu

Formerly: Edwin A. Ulrich Museum of Art

Founded: 1974.

Congressional District: 4

Key Personnel: Interim Dir., Teresa Veazey; Chm. (V), Kelly Callen; Cur. Education, Aimee Geist; Asst. Dir. Finance & Mgmt., Linda Doll; Mgr. Public Rels., Madeline McCullough; Designer & Preparator, James Porter; Coord. Special Projects, Carolyn Copple.

Personnel Profile: Full-Time Paid 8; Part-Time Paid 7; Interns 2.

Governing Authority: nonprofit. Parent Institution: Wichita State University. Tax-exempt: 501(c)(3).

Institution Type/Description: University Art Museum.

Collections: outdoor sculpture collection; 20th-21st century art, paintings, works on paper & electronic art; contemporary artists; new media works; video art.

Research Fields: modern & contemporary art.

Facilities: 10,000 sq. ft. exhibit space.

Activities: lectures; gallery talks; guided tours; organized education programs for undergraduate & graduate college students and adults 55 & over; artists-in-residence programs; hands-on workshops; films; conferences; docent program.

Publications: books, Art of Our Time: Selections from the Ulrich Museum of Art, Wichita State University; Not So Cute & Cuddly: Dolls and Stuffed Toys in Contemporary Art; The Sculptor's Clay, Sculpture by Duane Hanson, The John Philip Kassebaum Collection (ceramics); Tobi Kahn Correspondence; American Visions: The Paintings of Sandy Walker; Beyond the Museum Walls: The Martin H. Bush Outdoor Sculpture Collection; catalogues: Subversive Domesticity, Portrait Prints (1599-1641), Force I sculptures & paintings, The Persistence of Abstraction; David Reed - Leave Yourself Behind: Paintings and Special Projects, 1967-2005; Poets on Painters, 2006.

Hours & Admission Prices: Tues.-Fri. 11-5, Sat.-Sun. 1-5. No charge; donations accepted. Closed major & university holidays. &

Attendance: 21,005 (accurate)

Membership: WSU Students no charge; WSU Faculty, Staff, & Seniors 55 & over $30; Individual $50; Dual & Family $60-$99; Friend $100-$249; Supporter $250-$499; Sustainer $500-$749. Salon Circle: Curator $750-$2,499; Director $2,500-$4,999; Trustee $5,000-$9,999; Benefactor $10,000-$19,999; Grand Gallery $20,000 & up.

✳ **WICHITA ART MUSEUM, (M),** 1400 W. Museum Blvd., Wichita, KS 67203-3296. Tel.: 316-268-4921 & 4976. Fax: 316-268-4980.

E-mail: info@wichitaartmuseum.org

Web Site: www.wichitaartmuseum.org

Founded: 1935.

Congressional District: 4

Key Personnel: Chm., Paula Downing; Dir., Patricia McDonnell; Dir. Education, Courtney Spousta; Registrar, Leslie Servantez; Museum Shop Mgr., Kevin Bishop.

Personnel Profile: Full-Time Paid 21; Part-Time Paid 21.

Governing Authority: municipal. Subsidiary Institution: Friends of the Wichita Art Museum Inc. Tax-exempt.

Institution Type/Description: Art Museum.

Collections: Roland P. Murdock Collection of American Art; John W. & Mildred L. Graves Collection; Paul Ross Charitable Foundation Collection of American Painting; L.S. & Ida L. Naftzger Collection of American & European Prints & Drawings; M. C. Naftzger collection of Charles M. Russell paintings, drawings & sculpture; Gwendolyn Houston Naftzger collection of Porcelain birds; Kurdian & Beren collection of Pre- Colum-

bian Mexican artifacts; European porcelain & faience; Ablah collection of British watercolors; Misco collection of Art Works; F. Price Cossman collection of Steuben Glass.

Major Exhibits: The Woodblocks of Herschel C. Logan, 11/4/13-4/27/14; Downton Abbey, 12/14/13-5/4/14; George Catlin's American Buffalo (T), 2/14-5/11/14; Kansas Artists in the Max and Icee Moxley Collection, 5/10/14-10/5/14; American Moderns, 1910-1960: From O'Keeffe to Rockwell (T), 9/27/14-1/4/15.

Research Fields: American Art.

Facilities: 10,000-vol. library of books, art journals & archives available on premises; restaurant. Museum-related items for sale.

Activities: guided tours by appointment; lectures; gallery talks; films; educational programs.

Publications: bimonthly magazine; exhibition catalogues & educational brochures.

Hours & Admission Prices: Sun. 12-5, Tues.-Sat. 10-5; guided tours by appointment. Adults $7, seniors 60 & over $5, students & youth 5-17 $3; discount to AAM members; children under 5, Sat. & members no charge. Closed national holidays. &

Attendance: 50,402 (accurate)

Membership: Student with ID $20; Individual $40; Household Plus $95-$149; Donor $150-$349; Sponsor $350-$599; Patron $600-$999; Benefactor $1,000-$4,999; Director's Circle $5,000.

WICHITA CENTER FOR THE ARTS, 9112 E. Central, Wichita, KS 67206-2506. Tel.: 316-634-2787. Fax: 316-634-0593.

E-mail: arts@wcfta.com

Web Site: www.wcfta.com

Founded: 1920.

Congressional District: 4

Key Personnel: Exec. Dir., Howard W. Ellington; Gallery Admin. & Scholastic Art Coord., Amy Reep; Chm., Carol Wilson; Dir. Education, Kathy Sweeney; Dir. Theatre, John Boldenow; Business Mgr., Shawna Thompson.

Personnel Profile: Full-Time Paid 12; Part-Time Paid 31; Interns 2.

Governing Authority: nonprofit organization. Tax-exempt: 501(c)(3).

Institution Type/Description: Art Center.

Collections: paintings; drawings; fine prints; enamels; sculpture; pottery; porcelains; textiles; 20th-century decorative arts.

Research Fields: 20th-century American decorative art; works of Bruce Moore.

Facilities: 3,000-vol. library of art reference & magazines; school of theatre & performing arts; school of visual arts; professional community theater.

Activities: lectures; films; consultation service; concerts; professional community theatre for adults & children; formally organized education programs in performing & visual arts for children & adults; rotating schedule of permanent & temporary exhibitions; national competitive exhibitions.

Publications: bimonthly newsletter; exhibition catalogues; school of theatre & performing arts & school of visual arts catalogues; annual roster & report; activities; bulletins.

Hours & Admission Prices: Tues.-Fri. 10-5, Sat.-Sun. 1-5. No charge; donations accepted. Closed national holidays. &

Attendance: 53,484 (accurate)

Membership: Student $20; Individual $35; Family $60; Contemporaries $75; Patron $150; Sponsor $250; Associate $500; Donor $1,000; Benefactor $5,000 & up.

✳ **WICHITA-SEDGWICK COUNTY HISTORICAL MUSEUM ASSOCIATION, (M),** 204 S. Main, Wichita, KS 67202-3796. Tel.: 316-265-9314. Fax: 316-265-9319.

E-mail: wschm@wichitahistory.org

Founded: 1939.

Congressional District: 4

Key Personnel: Dir., Eric M. Cale; Pres., Kelly Callen.

Personnel Profile: Full-Time Paid 5; Part-Time Paid 2; Part-Time Volunteers 3.

Governing Authority: society. Tax-exempt: 501(c)(3).

Institution Type/Description: History Museum: housed in 1892 old City Hall.

Collections: local history; costumes; Indian artifacts; industrial; toys; period rooms.

Research Fields: local business; transportation; cultural history.

Facilities: 1,000-vol. library of books on Wichita & Kansas history; reading room. Books & note paper for sale.

Activities: guided tours; lectures; films; gallery talks; docent program; permanent, temporary & traveling exhibitions.

Publications: biannual newsletter, Heritage.

Hours & Admission Prices: Tues.-Fri. 11-4, Sat.-Sun. 1-5. Adults $4, children $2; discounts to AAM & ICOM members; members no charge. Closed national holidays. &

Attendance: 8,900 (accurate)

Membership: Individual $25; Family $35; Donor $50; Contributing $75; Sustaining $150; Patron $250 & up.

Wilson

WILSON CZECH OPERA HOUSE CORPORATION, FOUN-DATION, INC. & HOUSE OF MEMORIES MUSEUM, 415 27th St., Old Hwy. #40, Wilson, KS 67490-0271. Mailing Address: P.O. Box 271, Wilson, KS 67490-0271. Tel.: 785-658-3505 & 3343.
Founded: 1986.
Congressional District: 1
Key Personnel: C.E.O., Pres. & Chm., Libbie Sebesta; Vice. Pres., Laverne Libal; Dir. House of Memories Museum, Pres. (V), & Museum Shop Mgr., Jean T. Kingston; Treas., Ida Mae Goodman; Sec. & City Delegate, Joe E. Vocasek; Security, Bill Seifers; Security, Tim Heard; Asst. & Tour Guide Museum Shop Mgr., Una Joyce Podlena.
Personnel Profile: Full-Time Paid 1; Full-Time Volunteers 3; Part-Time Paid 1; Part-Time Volunteers 3.
Governing Authority: private; nonprofit organization. Parent Institution: Wilson Czech Opera House Corp. Subsidiary Institution: House of Memories Museum (Basement). Tax-exempt: 501(c)(3).
Institution Type/Description: History Museum.
Collections: items from Czech heritage; trophies; memorabilia; hair wreath; 1st phone book; hand tools; glassware for Czechoslovakia; costumes.
Research Fields: House of Memories inventory.
Facilities: library; 450-seat auditorium; 200-seat restaurant; 1,500 sq. ft. exhibit space. Gift items for sale.
Activities: art festivals; films; formal education programs for college students; guided tours; hobby workshops; loan & temporary exhibitions; study clubs; theater; broadcast programs. Museum Sponsors: Czech Festival in July; Alumni Dinners; Foreign Exchange week.
Publications: brochure; History of Wilson; History of J.T. Hastings.
Hours & Admission Prices: Mon.-Sat. 10-12 & 1-4, Sun. 1-4. No charge; donations accepted. &
Attendance: 575 (estimated)
Membership: Individual $10.

Winfield

THE COWLEY COUNTY HISTORICAL SOCIETY, 1011 Mansfield St., Winfield, KS 67156-3557. Tel.: 620-221-4811.
E-mail: cchsm@kans.com
Web Site: www.cchsm.com
Founded: 1931.
Congressional District: 4
Key Personnel: Pres. (V), Jerry Aistrup; Museum Shop Mgr., Jane Reeves.
Personnel Profile: Full-Time Paid 1; Part-Time Paid 1; Part-Time Volunteers 10.
Governing Authority: society; nonprofit organization. Tax-exempt: 501(c)(3) & 170(b)(1)(A).
Institution Type/Description: General Museum.
Collections: manuscripts; glassware; costumes; household equipment from pioneer days.
Research Fields: early history of the community.
Facilities: 5,000-vol. library of history books; reading room; period rooms; archives. Museum-related items for sale.
Activities: guided tours; lectures; slides; formally organized education programs for children; permanent & temporary exhibitions; oral history program.
Publications: newsletter.
Hours & Admission Prices: Tues.-Sun. 1-4. No charge; donations accepted. Closed New Year's Day; Easter; Memorial Day; Labor Day; Thanksgiving; Christmas. &
Attendance: 8,000 (accurate)
Membership: Active $15-$25; Sustaining $35; Organization $50; Life $200; Patron $500; Benefactor $1,000.

Yates Center

WOODSON COUNTY HISTORICAL SOCIETY, 208 W. Mary, Yates Center, KS 66783-1728. Mailing Address: 602 S. Kalida, Yates Center, KS 66783-1508. Tel.: 620-625-2626.
E-mail: rcall1@cox.net
Founded: 1965.
Congressional District: 5
Key Personnel: Pres. (V), Pete Watts; Vice Pres., Ron Call; Dir. & Treas., Linda Call; Cur., Geri Town.
Personnel Profile: Part-Time Volunteers 3.
Governing Authority: society. Tax-exempt.
Institution Type/Description: Historical Society Museum.
Collections: agriculture; Indian artifacts; farm machinery. Historic Buildings: 1877 church; country school; 1866 log cabin.

Research Fields: family histories.
Activities: bimonthly dinner meeting.
Hours & Admission Prices: June-Sept. Mon.-Wed. & Fri.-Sat. 10-3. No charge; donations accepted. &
Attendance: 360 (accurate)
Membership: Annual $2; Life $25; Memorial Gifts (for name on plaque) $100.

KENTUCKY

(247 listings)

Alexandria

CAMPBELL COUNTY LOG CABIN MUSEUM, 890 Clayridge Rd., Alexandria, KY 41001. Tel.: 859-466-0638.
E-mail: kennethareis@yahoo.com
Founded: 1983.
Key Personnel: Owner, Dir. & Museum Shop Mgr., Kenneth A. Reis.
Personnel Profile: Part-Time Volunteers 7.
Governing Authority: private.
Institution Type/Description: Local History & Agriculture Museum.
Collections: local history & culture; period furnishings; personal artifacts; photographs; early farm tractors & equipment; steam engine; tools; household artifacts; maps; school books; blacksmith shop; broom makers shop.
Activities: Museum Sponsors: Back Road's Farm Tour in July; Open House in July.
Hours & Admission Prices: Daily 10am to dusk. No charge; donations accepted. &
Attendance: 1,000 (estimated)

Ashland

HIGHLANDS MUSEUM & DISCOVERY CENTER, 1620 Winchester Ave., Ashland, KY 41101-7639. Tel.: 606-329-8888. Fax: 606-324-3218.
E-mail: info@highlandsmuseum.com
Web Site: www.highlandsmuseum.com
Founded: 1984.
Congressional District: 4
Key Personnel: C.E.O., Cur. & Registrar, Carolyn P. Warnock; Exec. Dir., Carol Rice Allen; Pres., Chris Pullem; Treas., David Griffith; Museum Shop Mgr., Sheila Rice.
Personnel Profile: Full-Time Paid 1; Full-Time Volunteers 1; Part-Time Paid 4; Part-Time Volunteers 10; Interns 2.
Governing Authority: private; nonprofit organization. Tax-exempt: 501(c)(3).
Institution Type/Description: Children's & History Museum.
Collections: period clothing from 1850s to modern era; transportation; communication; industry; country music; medical & military from WWII to present; cultural & industrial heritage of the Highlands region of Appalachia.
Facilities: educational facilities; 200-seat community hall; 15,000 sq. ft. exhibit space; portable planetarium; reception area. Museum-related items for sale.
Activities: concerts; dance recitals; docent program; formal education programs for children; guided tours; lectures; loan, participatory, temporary, traveling exhibitions; rental gallery. Annual Events: dining with past; auction; book & attic sale; Christmas House Tours.
Publications: monthly newsletter, The Highlander.
Hours & Admission Prices: Mon. by appointment, Tues.-Sat. 10-5, 1st Fri. of month 5-8. Adults $6.50, senior citizens & children $5; discounts to AAM members, ASTC reciprocal & groups; children under 2 & members no charge. Closed New Year's Day; Memorial Day; Independence Day; Labor Day; Thanksgiving; Christmas. &
Attendance: 26,000 (accurate)
Membership: Campers $25; Excavators $55; Basic Family $75; Explorers $100-$249; Discoverers $250-$499; Adventurers $500-$999; Founders $1,000 & up.

Auburn

THE AUBURN MUSEUM, 433 W. Main St., Auburn, KY 42206. Tel.: 270-542-4677.
Web Site: www.auburnhistoricalsociety.com
Institution Type/Description: Historical Society Museum.
Collections: local history & culture; period furnishings; personal artifacts; clothing; photographs.
Hours & Admission Prices: Mon.-Fri. 1-4; other times by appointment. Adults $5, children $2; children 6 & under and members no charge.
Membership: Individual $10.

Augusta

THE ROSEMARY CLOONEY HOUSE, 106 E. Riverside Dr., Augusta, KY 41002. Mailing Address: P.O. Box 197, Augusta, KY 41002. Tel.: 866-898-8091.
E-mail: info@rosemaryclooney.org
Founded: 2005.
Congressional District: 4
Institution Type/Description: History Museum: housed in the former home of singer & actress, Rosemary Clooney.
Collections: Rosemary Clooney's life & career; photographs; personal artifacts; period furnishings.
Facilities: Museum-related items for sale.
Hours & Admission Prices: Tues.-Fri. 11-3, Sat. 11-5, sun. 1-5; other times by appointment. Adults $5, members $4; discounts to AAM members.
Attendance: 97,000 (estimated)
Membership: $25.

Barbourville

APPALACHIAN FOOTHILLS FIRE HISTORICAL SOCIETY, N. Allison Ave., Barbourville, KY 40906. Mailing Address: 208 Sycamore Dr., Barbourville, KY 40906. Tel.: 606-627-8385.
Governing Authority: private; nonprofit organization. Tax-exempt: 501(c)(3).
Institution Type/Description: Historical Society Museum.
Collections: local fire service history; photographs; personal artifacts; fire trucks & equipment.
Hours & Admission Prices: Call for hours.

DR. THOMAS WALKER STATE HISTORIC SITE, 4929 KY 459, Barbourville, KY 40906-7232. Tel.: 606-546-4400. Fax: 606-546-4400.
Founded: 1931.
Key Personnel: Park Mgr., Andy Teasley.
Personnel Profile: Full-Time Paid 2; Part-Time Paid 2.
Governing Authority: state. Tax-exempt.
Institution Type/Description: Historic Site.
Collections: replica of the first cabin built in Kentucky by Dr. Thomas Walker.
Hours & Admission Prices: Wed.-Sun. Gift shop closed Nov.-Mar. 1. No charge.
Attendance: 15,000 (estimated)

KNOX HISTORICAL MUSEUM, 196 Daniel Boone Dr., Barbourville, KY 40906-1164. Mailing Address: P.O. Box 1446, Barbourville, KY 40906-5446. Tel.: 606-627-6856.
E-mail: khm1446@hotmail.com
Founded: 1988.
Key Personnel: Pres. (V), Michael C. Mills.
Personnel Profile: Part-Time Volunteers 6.
Governing Authority: Tax-exempt.
Institution Type/Description: History Museum.
Collections: local history & culture.
Publications: quarterly, The Knox Countian.
Hours & Admission Prices: Wed. 10-4. No charge; donations accepted.
Membership: $15; $20; $30; $50; $100; $500.

Bardstown

HEAVEN HILL DISTILLERIES BOURBON HERITAGE CENTER, 1311 Gilkey Run Rd., Bardstown, KY 40004. Tel.: 502-337-1000.
Web Site: www.bourbonheritagecenter.com
Institution Type/Description: History Museum.
Collections: company history; bourbon-making; distilleries equipment.
Activities: guided tours.
Hours & Admission Prices: March-Dec. Mon.-Sat. 10-5, Sun. 12-4. Tours: call for pricing. Closed New Year's Eve & Day; Easter; Election Day; Thanksgiving; Christmas.

MY OLD KENTUCKY HOME STATE PARK, 501 E. Stephen Foster Ave., Bardstown, KY 40004-2205. Mailing Address: P.O. Box 323, Bardstown, KY 40004-0323. Tel.: 502-348-3502; 800-323-7803. Fax: 502-349-0054. TDD: 502-348-3502.
Web Site: www.kystateparks.com
Founded: 1922.
Congressional District: 2
Key Personnel: Park Supt., Alice Heaton; Pres. (V), Dr. Harry Spalding; Museum Shop Mgr., Gail Downs.

Governing Authority: state. Parent Institution: Kentucky Dept. of Parks, Capital Plaza Tower, Frankfort, KY 40601. Tel. 502-564-2172. Subsidiary Institution: My Old Kentucky Home Foundation. Tax-exempt.
Institution Type/Description: Historic House: 1818 home of Judge John Rowan, where Stephen Foster wrote My Old Kentucky Home.
Collections: original Rowan furniture; china.
Facilities: 200-vol. library of books available for research by special permission from Parks Commissioner; theater; visitor center. Crafts, glass & china for sale.
Activities: guided tours.
Hours & Admission Prices: Daily 9-5. Adults $7, children 6-12 $3.50. Closed New Year's Day; Thanksgiving; Christmas week.
Attendance: 100,000 (estimated)

OLD BARDSTOWN VILLAGE - CIVIL WAR MUSEUM OF THE WESTERN THEATER, 310 E. Broadway, Bardstown, KY 40004. Tel.: 502-349-0291.
Web Site: www.civil-war-museum.org
Institution Type/Description: History Museum.
Collections: local history & culture; women of the Civil War; period clothing & furnishings; photographs; personal artifacts; historic buildings; paintings.
Hours & Admission Prices: March-Oct. daily 10-5; Nov. to mid-Dec. Fri.-Sun. 10-5. Adults $10, children 6-15 $5; children 5 & under no charge. Closed Thanksgiving; Christmas.

OSCAR GETZ MUSEUM OF WHISKEY HISTORY AND THE BARDSTOWN HISTORICAL MUSEUM, 114 N. Fifth St., Bardstown, KY 40004-1449. Tel.: 502-348-2999.
E-mail: whiskeymuseum@bardstowncable.net
Web Site: www.whiskeymuseum.com
Founded: 1984.
Congressional District: 2
Key Personnel: Chm. (V), Thomas C. Dawson; Cur., Mary Ellyn Hamilton.
Personnel Profile: Full-Time Paid 1; Part-Time Paid 3; Part-Time Volunteers 1.
Governing Authority: nonprofit organization. Tax-exempt.
Institution Type/Description: Local History Museum: housed in the former St. Joseph College and Seminary; built c.1819.
Collections: documentaries, memorabilia, & artifacts pertaining to the history of the American whiskey industry since its beginnings in the mid-18th century, through the Prohibition Era; Native American artifacts.
Research Fields: old whiskey memorabilia.
Facilities: 50-vol. library of books on whiskey.
Activities: permanent exhibitions. Annual Events: On the Lawn - Wine & Cheese Tasting in June; Antique Bourbon Auction in September; Louisville Wheelman Finish.
Publications: books, Pictorial History of Whiskey; Whiskey, An American Pictorial History.
Hours & Admission Prices: May-Oct. Mon.-Fri. 10-5, Sat. 10-4, Sun. 12-4; Nov.-April Tues.-Sat. 10-4, Sun. 12-4. No charge; donations accepted. Guided Group Tours: $3 per person. Call for holiday closings.
Attendance: 10,040 (accurate)

Barlow

BARLOW HOUSE MUSEUM, 509 Broadway St., Barlow, KY 42024. Mailing Address: P.O. Box 400-509 Broadway, Barlow, KY 42024. Tel.: 270-334-3691 & 3010.
Institution Type/Description: Historic House Museum.
Collections: Barlow family history; period furnishings; personal artifacts; photographs.
Hours & Admission Prices: Mon., Fri., and 2nd & 4th Sun. 1-4; other times by appointment. Adults 13 & over $3; children under 12 no charge.

Barstown

NATIVE AMERICAN MUSEUM, 310 E. Broadway, Barstown, KY 40004-1566. Tel.: 502-349-0291.
E-mail: museumrow@barstowncable.net
Formerly: Neal Spalding Native American Museum
Founded: 1981.
Congressional District: 2
Key Personnel: Pres. (V), Michael A. Thomas.
Personnel Profile: Full-Time Paid 1; Full-Time Volunteers 3; Part-Time Volunteers 3.
Governing Authority: Tax-exempt.
Institution Type/Description: History Museum.
Collections: Native American history & culture; replica longhouse, teepee, reed & straw house, & adobe house; period furnishings; personal artifacts.
Hours & Admission Prices: March-Oct. daily 10-5, Nov.- Dec. 15 Fri.-Sun. 10-5. Tour groups year-round by appointment only.

Benham

KENTUCKY COAL MUSEUM, (M), 231 Main St., Benham, KY
40807. Mailing Address: P.O. Box A, Benham, KY 40807. Tel.:
606-848-1530. Fax: 606-848-1546. Facebook: KY Coal
Museum/Portal 31.
E-mail: kycoalmuseumportal31@kctcs.edu
Web Site: kycoalmuseum.com
Formerly: Kentucky Coal Mining Museum
Founded: 1994.
Congressional District: 5
Key Personnel: Dir., Phyllis Sizemore; Pres. (V), Dr. W. Bruce Ayers.
Personnel Profile: Full-Time Paid 3; Part-Time Volunteers 20.
Governing Authority: Tax-exempt.
Institution Type/Description: Coal Mining Museum.
Collections: history of Appalachian coal mining & company towns; historic
photography; tools & equipment; Loretta Lynn exhibit.
Facilities: Museum-related items for sale.
Activities: video; Portal 31 underground mine tour.
Publications: Coal Town Curriculum.
Hours & Admission Prices: Adults $8, senior citizens 62 & over $6, high
school & college students $5, elementary school children $4; children 2 &
under no charge. Portal 31: additional fee charged. Open some holidays;
call for availability. &
Attendance: 20,000 (estimated)

Berea

BEREA COLLEGE BURROUGHS GEOLOGICAL MUSEUM,
Main St., Berea, KY 40404. Mailing Address: CPO 2191, Berea,
KY 40404. Tel.: 859-985-3351 & 3893. Fax: 859-985-3303.
E-mail: larry_lipchinsky@berea.edu
Web Site: www.berea.edu
Founded: 1920.
Congressional District: 6
Key Personnel: Dir., Chm. (V) & Cur., Prof. Zelek L. Lipchinsky.
Personnel Profile: Full-Time Volunteers 1.
Governing Authority: college. Parent Institution: Berea College. Subsidiary
Institution: Berea College Geology Dept. Tax-exempt: 501(c)(3).
Institution Type/Description: Science Museum.
Collections: geological & archaeological specimens.
Activities: permanent exhibits; guided tours. Educational material for sale.
Hours & Admission Prices: Mon.-Fri. 9-5. No charge; donations accepted. &
Attendance: 2,000 (estimated)

BEREA COLLEGE, DORIS ULMANN GALLERIES, (M), Cor-
ner of Chestnut & Elipse St., Berea, KY 40403. Mailing Address:
CPO 2162, Berea, KY 40404. Tel.: 859-985-3530.
E-mail: meghan_doherty@berea.edu
Web Site: www.berea.edu/art/doris-ulmann-galleries
Founded: 1975.
Congressional District: 6
Key Personnel: Chm. Art Dept., Lisa Kriner, M.F.A.; Dir., Meghan C. Doherty,
Ph.D.
Personnel Profile: Full-Time Paid 1; Part-Time Paid 2.
Governing Authority: college. Parent Institution: Berea College. Subsidiary
Institution: Berea College Art Dept. Tax-exempt: 501(c)(3).
Institution Type/Description: College Art Galleries.
Collections: crafts; paintings; Doris Ulmann photographs of Appalachian
people & craftsmen; African art; European, Asian & American prints; 13
Kress paintings; C.C. Coyle paintings; Frank Long paintings; Chinese
robes; contemporary prints.
Facilities: print study room.
Activities: intermuseum loan; temporary & traveling exhibitions.
Hours & Admission Prices: Fall-Spring Mon.-Wed. 8-6, Thurs. 8-8, Fri. 8-5,
Sat.-Sun. 1-5; Summer: Mon.-Sat. 1-5. No charge. Closed college holidays.
&
Attendance: 3,000 (estimated)

BEREA COLLEGE WEATHERFORD PLANETARIUM, Sci-
ence Bldg., Berea, KY 40404. Mailing Address: C.P.O. Box 2191,
Berea, KY 40404. Tel.: 859-985-3277. Fax: 859-985-3303.
E-mail: amer_lahamer@berea.edu
Web Site: www.physics.berea.edu
Founded: 1985.
Congressional District: 6
Key Personnel: Chm., Amer S. Lahamer.
Governing Authority: college. Parent Institution: Berea College. Subsidiary
Institution: Berea College Physics Dept. Tax-exempt: 501(c)(3).

Institution Type/Description: Planetarium.
Collections: astronomy.
Activities: planetarium shows.
Hours & Admission Prices: Sept.-May Sun. 4 pm; other times by appointment.
Adults $1, $10 per group (seats 50). &
Attendance: 1,000 (accurate)

Blackey

C.B. CAUDILL STORE & HISTORY CENTER, 7822 Hwy. 7,
Blackey, KY 41804. Tel.: 606-633-3281.
Institution Type/Description: History Museum.
Collections: local history, culture & life; early farm machinery; coal mining
equipment; kitchen utensils; glassware; patent medicines; period furnish-
ings; personal artifacts.
Hours & Admission Prices: Call for hours.

Bowling Green

**BRIMS - BARREN RIVER IMAGINATIVE MUSEUM OF
SCIENCE,** 1229 Center St., Bowling Green, KY 42101-3426.
Mailing Address: P.O. Box 71, Bowling Green, KY 42102-0071.
Tel.: 270-843-9779.
E-mail: b.r.i.m.s@insightbb.com
Founded: 1994.
Key Personnel: Dir., Charles Phillips; Chm. (V) & Pres. (V), Jeff Moore.
Personnel Profile: Part-Time Paid 1; Part-Time Volunteers 2; Interns 2.
Governing Authority: bd. of directors; nonprofit organization. Tax-exempt.
Institution Type/Description: Science Museum.
Collections: sciences & technology.
Activities: special events; Bridge Building Sponsor.
Publications: quarterly newsletter.
Hours & Admission Prices: Thurs.-Sat. 10-3, Sun. 1-4; groups by appointment.
Adults $5, students $4.
Attendance: 7,000.
Membership: Annual $50.

HARDIN PLANETARIUM, Western Kentucky University, Dept. of
Physics & Astronomy, 1906 College Heights Blvd., #11077,
Bowling Green, KY 42101-1077. Tel.: 270-745-4044. Fax: 270-
745-2014.
E-mail: ronn.kistler@wku.edu
Web Site: physics.wku.edu/planetarium.html
Founded: 1967.
Congressional District: 2
Key Personnel: Dir., Dr. Richard Gelderman; Coord., Ronn Kistler.
Personnel Profile: Full-Time Paid 2.
Governing Authority: state; university. Parent Institution: Western Kentucky
University, Bowling Green, KY 42101. Tax-exempt.
Institution Type/Description: Planetarium, Observatory & Astronomy Mu-
seum.
Collections: meteorite; astronomy photographs.
Facilities: 200-vol. library of astronomy books; 160-seat auditorium.
Activities: guided tours; lectures; films; formally organized programs for
children, adults & undergraduate college students affiliated with Western
Kentucky University; temporary, traveling & permanent exhibitions; tour-
ing school presentations.
Hours & Admission Prices: Sept.-May Mon.-Fri. 8-4:30 during academic
session. No charge. Public Shows: open all year Tues. & Thurs. 7pm, Sun.
2pm; closed major holidays. No charge. &
Attendance: 15,000

HISTORIC RAILPARK TRAIN MUSEUM - L&N DEPOT, (M),
401 Kentucky St., Bowling Green, KY 42101-1260. Tel.: 270-745-
7317. Fax: 270-782-3398.
E-mail: info@historirailpark.com
Web Site: www.historicalrailpark.com
Founded: 2007.
Congressional District: 2
Key Personnel: Pres. (V), Richard Webber; Dir., Bethany Sutton.
Personnel Profile: Full-Time Paid 1; Part-Time Paid 6; Part-Time Volunteers
35.
Governing Authority: Parent Institution: Friends of L&N Depot, Inc. Tax-
exempt.
Institution Type/Description: Railroad Museum.
Collections: vintage railroad equipment & artifacts.
Major Exhibits: Civil War Railroad, 9/11-4/14.
Publications: newsletter, Main Line.
Hours & Admission Prices: April-Oct. Mon.-Sat. 9-5, Sun. 1-4, Nov.-March

Tues.-Sat. 9-5, Sun. 1-4. Adults $12, seniors 55 & over $10, children 5-12 $6; children under 4 no charge. &

Attendance: 11,000 (accurate)
Membership: Single $25; Family $50.

THE KENTUCKY MUSEUM, (M), 1906 College Heights Blvd. #11092, Bowling Green, KY 42101-1092. Tel.: 270-745-2592. Fax: 270-745-6264.
E-mail: timothy.mullin@wku.edu
Web Site: www.wku.edu/kentuckymuseum
Formerly: The Kentucky Museum
Founded: 1929.
Congressional District: 2
Key Personnel: Dir., Timothy J. Mullin; Chm. (V), Phillip Bale; Cur. Exhibits, Donna Parker; Registrar & Cur. Collections, Sandra Staebell; Education Cur., Christy Spurlock; Exhibits Technician, Tony Thurman; Artist-in-Residence, Lynne Ferguson; Devel. Officer, John Perkins; Museum Asst., Lynn Claycomb; Museum Asst., Deborah Cole.
Personnel Profile: Full-Time Paid 9; Part-Time Paid 16; Part-Time Volunteers 8; Interns 2.
Volunteer Hours: 530
Operating Expenses: 585,000
Operating Income: 590,000
Governing Authority: bd. of Regents. Parent Institution: Western Kentucky University. Subsidiary Institution: Kentucky Museum Board of Directors. Tax-exempt: 501(c)(3).
Institution Type/Description: General Museum & Historic House.
Collections: period furniture; Shaker decorative arts; toys; traditional tools; musical instruments; anthropology; textiles; Civil War; fine & decorative arts. Historic Building: 1815 log house.
Research Fields: Kentucky history; Shaker history; Civil War history, upper south; folk life; Chinese culture; decorative arts; historic landscape; history of costume.
Facilities: 80,000 sq. ft. exhibition space.
Activities: guided tours; lectures; student research projects; permanent & temporary exhibits; workshops; children's activities; volunteer organization.
Publications: brochure; catalogues; newsletter.
Hours & Admission Prices: Mon.-Sat. 9-4, Sun. 1-4; groups by appointment. Adults $10, seniors $5; discounts to AAM, AASLH & KAM members; children 5 & under and members no charge. Closed major holidays. &
Attendance: 84,000 (accurate)
Membership: Individual $35.

NATIONAL CORVETTE MUSEUM, 350 Corvette Dr., Bowling Green, KY 42101-9134. Tel.: 270-781-7973. Fax: 270-781-5286.
E-mail: strode@corvettemuseum.com
Web Site: www.corvettemuseum.org
Founded: 1994.
Congressional District: 2
Key Personnel: Dir., Wendell Strode.
Governing Authority: nonprofit. Tax-exempt: 501(c)(3).
Institution Type/Description: Automobile Museum.
Collections: over 70 Corvettes including classics, race cars, & design cars; photographs; movies & videos; advertisements; scale models; memorabilia.
Facilities: 165-seat theater. Museum-related items for sale.
Activities: school programs.
Publications: magazine, America's Sports Car.
Hours & Admission Prices: Daily 8-5. Museum: adults $10, seniors $8; children 6-16 $5; discounts AAA members & GM employees; members and children under 6 no charge. Closed New Year's Day; Easter; Thanksgiving; Christmas Eve & Day. &
Attendance: 130,000
Membership: Online $25; Individual $50; Family $100; Business $250; Senior Individual $500; Lifetime $1,500; Business Lifetime $2,500.

RIVERVIEW AT HOBSON GROVE, 1100 W. Main Ave., Bowling Green, KY 42101-4894. Tel.: 270-843-5565. Fax: 270-843-5557.
E-mail: riverview.at.hg@att.net
Web Site: www.bgky.org/riverview/
Founded: 1972.
Congressional District: 2
Key Personnel: Dir., Laura Southard; Chm., Lynda Neale; Pres., Sharlene Mitchell.
Personnel Profile: Full-Time Paid 1; Part-Time Paid 4; Part-Time Volunteers 50; Interns 1.
Governing Authority: municipal government; nonprofit. Tax-exempt: 501(c)(3).
Institution Type/Description: Historic House Museum: housed in a c.1872 Italianate architecture home which was used to store Confederate munitions during the Civil War.
Collections: concentration on the Victorian family living in southern Kentucky during the 1860s-1890s; personal artifacts & decorative arts pieces.
Research Fields: 19th-century recreation, servant life and death & mourning customs; the life of Victorian women; Civil War.
Facilities: tea & luncheon committee caters by reservation. Museum-related items for sale.
Activities: docent program; formal education programs for children; guided tours; lectures; temporary exhibitions; children's Manners classes. The Other Victorians: 19th-Century Servant Life Tour; Riverview Chautauqua Series features monthly programs, activities, & exhibits related to the time period.
Publications: quarterly newsletter, The Riverview Observer.
Hours & Admission Prices: Feb.-Dec. Tues.-Sat. 10-4, Sun. 1-4. Families $14, adults $7, students $2.50; discount to groups, veterans & members; children under 6 no charge. Closed holidays.
Attendance: 3,600 (accurate)
Membership: Individual $15; Family $25; Contributor $50; Sustaining $100; Patron $101 & up.

WESTERN KENTUCKY UNIVERSITY GALLERY, Rm. 441 Ivan Wilson Center for Fine Arts, Bowling Green, KY 42101-1000. Mailing Address: 1906 College Heights Blvd., Art Dept., Bowling Green, KY 42101-1000. Tel.: 270-745-3944. Fax: 270-745-5932.
E-mail: brent.oglesbee@wku.edu
Web Site: www.wku.edu/art/
Congressional District: 13
Key Personnel: Dept. Head, Brent Oglesbee.
Governing Authority: university. Parent Institution: Western Kentucky University. Subsidiary Institution: Art Dept. Tax-exempt.
Institution Type/Description: Art Gallery.
Collections: contemporary works.
Activities: gallery talks; temporary & traveling exhibitions.
Hours & Admission Prices: Mon.-Fri. 8-4:30. No charge. Closed between exhibitions. &
Attendance: 3,500 (estimated)

Burlington

DINSMORE HOMESTEAD FOUNDATION, (M), 5656 Burlington Pike, Burlington, KY 41005-8668. Mailing Address: P.O. Box 453, Burlington, KY 41005-0453. Tel.: 859-586-6117. Fax: 859-334-3690.
E-mail: mcdonaldmj@fuse.net
Web Site: www.dinsmorefarm.org
Founded: 1986.
Congressional District: 4
Key Personnel: Exec. Dir., Marty McDonald; Chm., Howard Tankersley; Education Coord., Cathy Collopy; Program & Events Asst., Elizabeth Tankersley; Homestead Asst., Sue Clare.
Governing Authority: private; nonprofit organization. Tax-exempt: 501(c)(3).
Institution Type/Description: History Museum: housed in c.1842 Federal farmhouse.
Collections: furniture; books; photographs; paintings; textiles; family correspondence journals on microfilm; early Boone County history.
Research Fields: African-American slavery; local plant life; paint & wallpaper.
Facilities: library of microfilmed family papers; nature center; 1,500 sq. ft. exhibit space. Museum-related items for sale.
Activities: docent program; formal education programs; guided tours; hobby workshops.
Publications: quarterly newsletter, The Dinsmore Dispatch.
Hours & Admission Prices: April-Dec. 15 Wed. & Sat.-Sun. 1-5; other times by appointment. Adults $5, senior citizens $3, students 5-17 $2; members no charge.
Attendance: 3,500
Membership: Student & Senior Citizen $20; Individual $35; Family $60; Friend $100; Julia Dinsmore Society $250-$2,500.

Cadiz

JANICE MASON ART MUSEUM, 71 Main St., Cadiz, KY 42211-9101. Mailing Address: P.O. Box 303, Cadiz, KY 42211-0303. Tel.: 270-522-9056.
E-mail: jmam@bellsouth.net
Web Site: www.jmam.org
Founded: 1998.
Congressional District: 1
Personnel Profile: Part-Time Paid 2; Part-Time Volunteers 30.

Governing Authority: Tax-exempt.
Institution Type/Description: Art Museum.
Collections: works by regional, national & international artists.
Activities: classes; educational programs.
Hours & Admission Prices: Tues.-Sat. 10-4, Sun. 1-4. No charge; donations accepted. Closed holidays. ♿
Attendance: 5,000 (accurate)
Membership: Individual $35.

Campbellsville

THE FRIENDSHIP SCHOOL HOUSE, 300 Ingram Ave., Campbellsville, KY 42718-1625. Tel.: 270-465-5410, 5106, 2055.
Web Site: campbellsvilleky.com/what-to-see/historical/
Institution Type/Description: History Museum.
Collections: historic school memorabilia from 1918-1955.
Hours & Admission Prices: 1st Sun. of the month 1-4 or by appointment.

GREEN RIVER LAKE VISITOR'S CENTER, 544 Lake Rd., Campbellsville, KY 42718. Tel.: 270-465-4463.
Governing Authority: Parent Institution: US Army Corps of Engineers. Tax-exempt.
Institution Type/Description: Visitor's Center.
Collections: local history; Native American artifacts; aquarium; water safety; video; live turtles. Historic Building: Atkinson Griffon log house.
Facilities: theater; nature trails.
Activities: hiking; video.
Hours & Admission Prices: Daily 7:30-5, call to verify. No charge. ♿
Attendance: 12,800 (estimated)

JACOB HIESTAND HOUSE-TAYLOR COUNTY MUSEUM, 1075 Campbellsville Bypass, Campbellsville, KY 42718-8835. Tel.: 270-789-4343. Facebook: Jacob Hiestand House-Taylor County Museum.
E-mail: smithgorin@windstream.net
Web Site: taylorcounty.us/history
Founded: 1992.
Congressional District: 1
Key Personnel: Chm. (V) & Pres. (V), Betty Gorin; Volunteer Sec., Debbie Gilpin
Governing Authority: Tax-exempt.
Institution Type/Description: Historic House Museum: built in 1823.
Collections: period furnishings; Taylor county history; spinning wheels; quilt patterns; photo exhibit; looms; tools.
Research Fields: local history.
Facilities: plantation house; restrooms; cemetery; servant's quarters; spring house; detached kitchen; black history room; country store.
Activities: programs; musical affairs.
Publications: annual newsletter.
Hours & Admission Prices: Tues.-Sat. 10-3. Tours: $3; discounts to groups; children under 12 no charge. Closed New Year's Day; Memorial Day; Independence Day; Labor Day; Christmas. ♿
Attendance: 1,000 (estimated)
Membership: Annual $10.

Carlisle

BLUE LICKS PIONEER MUSEUM, 10299 Maysville Rd., Carlisle, KY 40311. Mailing Address: Blue Licks Battlefield State Resort Park, P.O. Box 66, Mount Olivet, KY 41064. Tel.: 800-443-7008. Fax: 859-289-5409.
E-mail: bluelicks@ky.gov
Web Site: www.parks.ky.gov
Founded: 1928.
Congressional District: 7
Key Personnel: Resort Park Mgr., Michael Schwendau; Business Mgr., Erik Unthank; Park Naturalist, Cur. & Museum Shop Mgr., Paul Tierney; Museum Shop Mgr., Jean Dillon.
Personnel Profile: Full-Time Paid 1; Part-Time Paid 3.
Governing Authority: state. Affiliated with Kentucky Dept. of Parks, Capital Plaza Tower, Frankfort, KY 40601. Tax-exempt.
Institution Type/Description: History Museum.
Collections: Fort Ancient artifacts; fossils; mastodon & musk ox bones; pioneer tools & artifacts; folklore; Native American; Revolutionary War; natural history; geological; archaeology; archives; area history.
Activities: guided tours; gallery talks; audiovisuals; special events & programs. Museum Sponsors: Guided Battlefield Walks in summer; Revolutionary War Reenactment & 18th Century Programs & Demonstrations in August.
Hours & Admission Prices: March 15-Nov. 15 Mon.-Sat. 9-5, Sun. 1-5. Admission $2; children under 6 no charge. ♿
Attendance: 10,000 (estimated)

KENTUCKY DOLL & TOY MUSEUM, 106 W. Main St., Carlisle, KY 40311. Tel.: 859-289-3344.
Key Personnel: Cur., Jan Taylor
Institution Type/Description: Toy & Doll Museum.
Collections: late 19th & early 20th century dolls & toys; dollhouses.
Activities: research.
Hours & Admission Prices: Mon.-Wed. 10-2; other times by appointment. Adults $2; children under 12 no charge.

NICHOLAS COUNTY HISTORY MUSEUM & L&N PASSENGER DEPOT, 101 E. Market St., Carlisle, KY 40311. Mailing Address: P.O. Box 222, Carlisle, KY 40311. Tel.: 859-289-4200.
Institution Type/Description: Historic Building: housed in a former depot built in 1910.
Collections: local history & culture; period furnishings; photographs; personal artifacts.
Hours & Admission Prices: Call for hours.

Carrollton

BUTLER-TURPIN STATE HISTORIC HOUSE, General Butler State Resort Park, 1608 Hwy. 227, Carrollton, KY 41008-8051. Mailing Address: P.O. Box 325, Carrollton, KY 41008. Tel.: 502-732-4384; 866-462-8853.
E-mail: generalbutler@ky.gov
Web Site: www.generalbutler.com
Congressional District: 4
Key Personnel: Park Mgr., Eddie Moore.
Personnel Profile: Part-Time Paid 2.
Governing Authority: Parent Institution: Kentucky Department of Parks. Tax-exempt.
Institution Type/Description: Historic House: former home of the Butler family, built in 1859.
Collections: military documents; furniture; objects; paintings.
Hours & Admission Prices: May-Sept. daily by appointment. Adults $10.
Attendance: 2,000 (estimated)

Cave City

DINOSAUR WORLD, 711 Mammoth Cave Rd., Cave City, KY 42127-8437. Tel.: 270-773-4345. Fax: 270-773-5303.
E-mail: dinosaurworlds@gmail.com
Web Site: www.dinosaurworld.com
Founded: 2004.
Institution Type/Description: Natural History Museum.
Collections: over 150 life-size dinosaurs.
Activities: fossil dig; boneyard; movie cave; monmouth garden; prehistoric museum.
Hours & Admission Prices: Daily 8:30am. Adults $12.75, seniors over 60 $10.75, children 3-12 $9.75. Closed Thanksgiving; Christmas.
Attendance: 80,000 (estimated)

FLOYD COLLINS MUSEUM, 1240 Old Mammoth Cave Rd., Cave City, KY 42127. Tel.: 270-773-3366.
Institution Type/Description: History Museum.
Collections: life & death of cave explorer Floyd Collins in 1925; replica cave; photographs; period artifacts.
Hours & Admission Prices: Call for hours.

MAMMOTH CAVE WAX MUSEUM, Hwy. 70 W., 901 Mammoth Cave Rd., Cave City, KY 42127. Mailing Address: P.O. Box 678, Cave City, KY 42127-0678. Tel.: 270-773-3010.
Key Personnel: Dir., Wesley Odle
Institution Type/Description: Wax Museum.
Collections: wax figures.
Hours & Admission Prices: Call for hours.

MAMMOTH CAVE WILDLIFE MUSEUM, 409 E. Happy Valley St., Hwy. 90, Cave City, KY 42127. Mailing Address: P.O. Box 236, Cave City, KY 42127. Tel.: 270-773-2255.
Web Site: www.mammothcave.com/guntown/wildlife.htm
Institution Type/Description: Wildlife Museum.
Collections: tigers; lions; bears; leopards; Snow Leopard; deer; sheep; ox; marine life; birds.
Facilities: 14,000 sq. ft. exhibit space.

Activities: education programs.
Hours & Admission Prices: Call for hours.

Clay City

RED RIVER HISTORICAL SOCIETY MUSEUM, 4541 Main Street, Clay City, KY 40312. Mailing Address: Box 195, Clay City, KY 40312-0195. Tel.: 606-663-9930.
Founded: 1966.
Congressional District: 7
Key Personnel: Dir., Larry G. Meadows; Cur. Archaeology, John Faulkner; Cur. Photography, Steve Abner; Cur. History, Jim Spencer.
Governing Authority: nonprofit.
Institution Type/Description: General Museum: located in 1889 National Bank.
Collections: general display of local articles.
Research Fields: local Indian archaeology.
Activities: guided tours; lectures; films.
Publications: Natural Bridge in the Kentucky Mountains.
Hours & Admission Prices: Sat.-Sun. & holidays 10-6 or by appointment (call 606-663-4000). No charge; donations accepted.
Attendance: 5,500
Membership: Annual $10.

Clermont

BERNHEIM ARBORETUM AND RESEARCH FOREST, Hwy. 245, Clermont, KY 40110. Mailing Address: P.O. Box 130, Clermont, KY 40110-0130. Tel.: 502-955-8512. Fax: 502-955-4039.
E-mail: roger@bernheim.org
Web Site: www.bernheim.org
Founded: 1929.
Congressional District: 2
Key Personnel: Exec. Dir., Mark Wourms, Ph.D.; Pres. Bd. Trustees, Frank B. Hower, III; Dir. Education, Claude Stephens; Dir. Horticulture, Dena Rae Garvue; Dir. Mktg., Margaret Zurkahlen; Dir. Operations, Roger Fauver; Museum Shop Mgr., Deborah P'Pool Midget.
Personnel Profile: Full-Time Paid 39; Part-Time Paid 10; Part-Time Volunteers 300; Interns 5.
Governing Authority: nonprofit organization. Parent Institution: Isaac W. Bernheim Foundation. Tax-exempt: 501(c)(3).
Institution Type/Description: Nature Museum & Arboretum.
Collections: 4,000 varieties of plants in arboretum; exhibits on native flora & fauna; research forest.
Research Fields: study aids; university projects.
Facilities: 2,300-vol. library of horticulture & nature books; 14,000-acre wildlife refuge; arboretum; nature center; research forest.
Activities: planned walking lectures, workshops, seminars; planned school activities including fieldtrips, teacher workshops & a self-guided tour; art & cultural events.
Publications: Geology of Bernheim Forest; History of Bernheim Forest; Nature Trail Guide Sheet, Road & Trail Guide, cards; Wildflower Check List of Bernheim; Bird Check List; Plant Materials List.
Hours & Admission Prices: Daily. Admission: Sat.-Sun. & holidays $5 per vehicle; discounts to AAM, AABGA & AHS members under reciprocal visitation agreement; Mon.-Fri. & members no charge. Closed New Year's Day; Christmas. &
Attendance: 300,000 (accurate)
Membership: Individual $25; Family $35; Orchid Sponsor $50; Columbine Club $100; Holly Patron $500; Tuliptree Council $1,000; Oak Society $5,000; Director's Circle $10,000.

Clinton

HICKMAN COUNTY MUSEUM, 221 E. Clay St., Clinton, KY 42031-1224. Mailing Address: P.O. Box 284, Clinton, KY 42031. Tel.: 270-994-5530.
Institution Type/Description: Historic House Museum: housed in the former home of Captain Henry C. Watson of the Confederate 7th Kentucky Infantry Regiment; c.1870.
Collections: local history; period furnishings; medical & military artifacts; sports memorabilia; government; business; Native American artifacts.
Hours & Admission Prices: Wed.-Sat. 1-4; other times by appointment. No charge; donations accepted.

Cloverport

CLOVERPORT DEPOT MUSEUM, 415 E. Houston St., Cloverport, KY 40111.
Formerly: Cloverport Community Museum
Institution Type/Description: History Museum.
Collections: local history & culture; photographs; period furnishings; personal artifacts; books.
Hours & Admission Prices: By appointment. No charge.

Columbia

TRABUE RUSSELL HOUSE, 201 Jamestown St., Columbia, KY 42728. Tel.: 270-384-2501 & 6183.
Institution Type/Description: Historic House Museum: built in 1821 by Daniel Trabue.
Collections: local history; period furnishings; photographs.
Activities: guided tours.
Hours & Admission Prices: By appointment.

Columbus

COLUMBUS-BELMONT CIVIL WAR MUSEUM, Columbus-Belmont State Park, 350 Park Rd., Columbus, KY 42032. Mailing Address: P.O. Box 9, Columbus, KY 42032-0009. Tel.: 270-677-2327. Fax: 270-677-4013. TDD: 270-677-2327.
E-mail: cindy.lynch@.ky.gov
Web Site: www.kystateparks.com/agencies/parks/columbus.htm
Founded: 1934.
Congressional District: 1
Key Personnel: Park Mgr., Cindy Lynch.
Personnel Profile: Part-Time Paid 1.
Governing Authority: state. Parent Institution: the Kentucky Dept. of Parks, Capital Plaza Tower, Frankfort, KY 40601. Tax-exempt.
Institution Type/Description: History Museum.
Collections: Civil War artifacts; pioneer relics of Old Columbus; Indian artifacts; audio-visuals; hospital artifacts.
Activities: school groups & organization visits. Museum Sponsors: Special Event in October.
Hours & Admission Prices: May-Labor Day daily; Labor Day-Oct. weekends; other times by appointment. Adults $4, children $3; discount to senior citizens & groups. &
Attendance: 6,000 (accurate)

Corbin

HARLAND SANDERS MUSEUM & CAFE, 688 US Hwy. 25W., Corbin, KY 40701. Tel.: 606-528-2163.
E-mail: frankie.bostick@jrninc.com
Governing Authority: Parent Institution: JRN Inc.
Institution Type/Description: General Museum: housed in the original restaurant of Kentucky Fried Chicken; built in 1937. Listed on the National Register of Historic Places.
Collections: artifacts & memorabilia from the early days of Kentucky Fried Chicken.
Hours & Admission Prices: Daily 10-10. No charge.

KENTUCKY NATIVE AMERICAN HERITAGE MUSEUM, 4116 Cumberland Falls Hwy., Corbin, KY 40701. Tel.: 606-526-5635.
E-mail: sioux80@msn.com
Web Site: www.knahm.org
Institution Type/Description: History Museum.
Collections: Native American history & culture; personal artifacts; photographs.
Hours & Admission Prices: Call for hours.

Covington

BEHRINGER-CRAWFORD MUSEUM, (M), 1600 Montague Rd., Devou Park, Covington, KY 41011-5648. Tel.: 859-491-4003. Fax: 859-491-4006. Facebook: Behringer-Crawford Museum.
E-mail: info@bcmuseum.org
Web Site: www.bcmuseum.org
Founded: 1950.
Congressional District: 4
Key Personnel: Exec. Dir., Laurie Risch; Cur. Exhibits & Collections, Tiffany Hoppenjans; Pres., John Boh; Dir. Education, Regina Siegrist; Museum Svcs. Coord. & Museum Shop Mgr., Linda Schneider.

Personnel Profile: Full-Time Paid 4; Part-Time Paid 3; Part-Time Volunteers 60; Interns 3.
Governing Authority: nonprofit organization. Tax-exempt: 501(c)(3).
Institution Type/Description: History & Culture Museum.
Collections: northern Kentucky history & culture; transportation; paleontology; prehistoric & historic archaeology; mineralogy; textiles; costumes; glass; military artifacts; Ohio River heritage; taxidermy; fine, decorative, & folk art.
Research Fields: historic archaeology; transportation history; northern Kentucky history; Civil War; Ohio Valley.
Facilities: 860-vol. library; 12,500 sq. ft. exhibit space. Books & children's items for sale.
Activities: guided tours; lectures; radio programs; organized education programs; participatory, loan & temporary exhibitions.
Publications: weekly newsletter; weekly postings on social media.
Hours & Admission Prices: Tues.-Sat. 10-5, Sun. 1-5. Adults $7; discount to AAM, SEMC & Kentucky Heritage Alliance members; members no charge. Closed national holidays. &
Attendance: 25,000 (estimated)
Membership: Student, Senior & Teacher $25; Individual $40; Household $55; Contributing $100; Supporting $250; Patron $500.

JAMES A. RAMAGE CIVIL WAR MUSEUM, 1402 Highland Ave., Covington, KY 41011-3743. Tel.: 859-344-1145.
E-mail: ramagecivilwarmuseum@gmail.com
Institution Type/Description: Civil War Museum.
Collections: Civil War history, artifacts, & memorabilia; Fort Wright history.
Activities: special events; group tours. Annual Events: Northern Kentucky Veterans' Day Event in November; Civil War Christmas in December.
Hours & Admission Prices: Fri.-Sat. 10-5, Sun. 12-5. Closed holidays.

THE RAILWAY EXPOSITION CO., INC., 315 W. Southern Ave., Covington, KY 41015-1180. Mailing Address: 212 Wyoming Ave., Cincinnati, OH 45215-4308. Tel.: 513-761-3500.
Founded: 1975.
Congressional District: 4
Key Personnel: Vice Pres., William F. Sprague; Public Rels., Roberta Sprague; Museum Shop Mgr., Corrie Reade-Hale.
Personnel Profile: Part-Time Volunteers 50.
Governing Authority: nonprofit organization. Tax-exempt: 501(c)(3).
Institution Type/Description: Railway Museum.
Collections: railroad passenger cars & locomotives; railroad memorabilia; paper items of railroad history.
Research Fields: passenger car restoration.
Facilities: library of early railroad records; 30-seat lecture area. Museum-related items for sale.
Activities: guided tours; lectures; films; participatory exhibits.
Publications: monthly newsletter, Trainsheet.
Hours & Admission Prices: May-Oct. Sun. 12:30-4:30; special tours by appointment. Adults $4, children $2; discounts to groups. Closed holiday weekends.
Membership: Regular $20; Family $25; Business $150; Life $200.

Crestwood

YEW DELL BOTANICAL GARDENS, 6220 Old LaGrange Rd., Crestwood, KY 40014. Tel.: 502-241-4788.
Web Site: www.yewdellgardens.org
Key Personnel: Exec. Dir., Paul E. Cappiello, Ph.D.
Institution Type/Description: Botanical Gardens: listed on the National Register of Historic Places.
Collections: plants; flowers; trees.
Activities: educational programs; special events.
Hours & Admission Prices: May to mid-Dec. Tues.-Wed. & Fri.-Sat. 10-4, Thurs. 10-8, Sun. 12-4; mid-Dec. to March Mon.-Fri. 10-4. Adults $7, seniors over 55 $5; members, active military & children under 12 no charge.

Cynthiana

CYNTHIANA - HARRISON COUNTY MUSEUM, 124 S. Walnut St., Cynthiana, KY 41031-1592. Mailing Address: P.O. Box 411, Cynthiana, KY 41031-0411. Tel.: 859-234-7179.
Founded: 1994.
Institution Type/Description: History Museum.
Collections: local history & culture; military; agriculture; industry; religion; medical artifacts.
Facilities: Museum-related items for sale.
Publications: Chronicles of Cynthiana; Cromwell's Comments; This Old House.

Hours & Admission Prices: Fri.-Sat. 10-5; other times by appointment. No charge.

Danville

CONSTITUTION SQUARE STATE HISTORIC SITE, 134 S. Second St., Danville, KY 40422-1802. Mailing Address: c/o Kentucky Department of Parks, 500 Metro St., 10th Fl. CPT, Frankfort, KY 40601. Tel.: 859-239-7089. Fax: 859-239-7894.
Web Site: www.ky.parks.ky.gov
Founded: 1937.
Congressional District: 6
Key Personnel: Park Mgr., Jack Bailey.
Personnel Profile: Full-Time Paid 3; Part-Time Paid 2; Part-Time Volunteers 5.
Governing Authority: state. Branch of Kentucky Dept. of Parks, Capital Plaza Tower, Frankfort, KY 40601. Tel.: 502-564-8110. Tax-exempt.
Institution Type/Description: Historic Site.
Collections: personal items of Isaac Shelby, first Governor of Kentucky. Historic Buildings: c.1792 Post Office; 1785 Grayson's Tavern; c.1816-1817 Watts-Bell House; 1817 Fisher's Row houses with gallery & visitor center; replicas of original log courthouse, site of first 10 constitutional conventions; jail; Presbyterian meetinghouse; restored c.1820 brick schoolhouse; c.1820 Alban Goldsmith House.
Facilities: picnic area.
Activities: guided tours. Annual Events: Brass Band Festival in June; Historic Constitution Square Festival in Sept.
Hours & Admission Prices: March-Dec. Thurs.-Sat. 10-4. Call for admission prices & tours. &
Attendance: 60,000 (estimated)

THE GREAT AMERICAN DOLLHOUSE MUSEUM, 344 Swope Ave., Danville, KY 40422. Tel.: 859-236-1883.
E-mail: lori@thedollhousemuseum.com
Web Site: www.thedollhousemuseum.com
Founded: 2008.
Key Personnel: Cur., Lori Kagan-Moore; Sculptor & Dollmaker, Nicola Cooper; Diversity Advisor, J.H. Atkins; Web Designer, Jon Sachs; Consultant Asian Design & History, Akiki Otake; Set Designer, Ruth Neeman
Institution Type/Description: Dollhouse & Miniatures Museum.
Collections: dollhouses displayed in a 1900s village setting.
Hours & Admission Prices: Tues.-Sat. 11-5. Adults $7.50, seniors $6.60, children $5. &

McDOWELL HOUSE MUSEUM, (M), 125 S. 2nd St., Danville, KY 40422-1801. Tel.: 859-236-2804. Fax: 859-236-2804 (press star twice).
E-mail: mcdhse@kih.net
Web Site: www.mcdowellhouse.org
Formerly: McDowell House and Apothecary Shop
Founded: 1939.
Congressional District: 5
Key Personnel: C.E.O., Dr. Charles Martin; Dir., Carol J. Senn; Asst. Dir., Alberta Moynahan; Museum Shop Mgr., Anna Ingram.
Personnel Profile: Part-Time Paid 4; Part-Time Volunteers 4; Interns 1.
Governing Authority: Parent Institution: McDowell House Museum, Inc. Tax-exempt.
Institution Type/Description: Historic Building: housed in the former home & shop of pioneer surgeon, Ephraim McDowell; built from 1792-1820.
Collections: medical; herbarium; paintings; books.
Research Fields: herbarium; medical & social history.
Facilities: 200-vol. library concerning Ephraim McDowell & Jane Todd Crawford, early medical books & books printed before 1830 available for use on the premises. Literature, slides & postcards for sale.
Activities: guided tours; lectures; permanent exhibitions; special events.
Publications: annual newsletter; The Life & Times of Ephraim McDowell; historical data catalog; video tour of McDowell House for sale.
Hours & Admission Prices: March-Oct. Mon.-Sat. 10-12 & 1-4, Sun. 2-4 (last tour at 3:30); Nov.-Feb. Tues.-Sat. 10-12 & 1-4, Sun. 2-4. Adults $7, senior citizens 62 & over $5, students over 12 $3; discounts to prearranged groups of 10 or more & AAA members; children 5 & under and members no charge. Closed New Year's Day; Easter; Thanksgiving; Christmas.
Attendance: 1,595 (estimated)
Membership: Individual $25; Family $50; Sponsor $100; Jane Todd Crawford Circle $250-$500; Life $500; Ephraim McDowell Society $1,000; Corporate $1,000.

Dawson Springs

DAWSON SPRINGS MUSEUM AND ART CENTER, INC., 127 S. Main St., Dawson Springs, KY 42408-1713. Mailing Address: P.O. Box 107, Dawson Springs, KY 42408-0107. Tel.: 270-797-3503.
E-mail: thomas1958@bellsouth.net
Founded: 1986.
Congressional District: 1
Key Personnel: C.E.O., Kathy Beshears; Exec. Dir., Sylvia Lynn Thomas; Pres., Kathy Lyon; Chm. (V), Shirley Menser.
Personnel Profile: Full-Time Volunteers 1; Part-Time Volunteers 40.
Governing Authority: private; nonprofit organization. Tax-exempt: 501(c)(3).
Institution Type/Description: Art & History Museum: housed in the 1907 Romanesque style Commercial Bank.
Collections: history of Dawson Springs, which was a leading spa of the south from late 19th-century to the Depression era; cultural heritage: The Spa Days, The Outwood Days, History of the Dawson Springs Independent School System & The Coal Mining Days; Japanese art; woodblock prints; Kimono; 150 period cassette tapes; 1,200 photographs.
Facilities: 70-vol. history library; 535 sq. ft. exhibit space.
Activities: loan & traveling & temporary exhibitions.
Publications: brochure, The Dawson Springs Museum & Art Center.
Hours & Admission Prices: Feb.-Dec. Tues.-Fri. 1-4. No charge; donations accepted. Closed major holidays. &
Attendance: 2,200 (estimated)
Membership: Single $10; Family $15; Supporting $50.

Eddyville

ROSE HILL - LYON COUNTY MUSEUM, Water St., Eddyville, KY 42038. Mailing Address: P.O. Box 811, Eddyville, KY 42038. Tel.: 270-388-2924.
Institution Type/Description: Historic House Museum: built c.1834.
Collections: local & regional history; period furnishing; personal artifacts; photographs.
Facilities: Gift items for sale.
Activities: guided tours.
Hours & Admission Prices: mid-May to mid-Oct. Wed.-Sun. 1-4. Adults $5; children no charge.

Elizabethtown

BLACK HISTORY GALLERY, 602 Gallery Place, Elizabethtown, KY 42701. Mailing Address: Elizabethtown Tourism & Convention Bureau, 1030 N. Mulberry St., Elizabethtown, KY 42701. Tel.: 270-765-2175.
Institution Type/Description: History Museum.
Collections: African American history & culture; photographs; biographies; newspaper & magazine articles; personal artifacts.
Hours & Admission Prices: Temporarily closed.

BROWN-PUSEY HOUSE, 128 N. Main St., Elizabethtown, KY 42701-1415. Tel.: 270-765-2515.
E-mail: brownpuseyhouse@windstream.net
Web Site: brownpuseyhouse.org
Founded: 1923.
Key Personnel: Dir., Twylane Van Lahr; Pres. (V), Morris Miller.
Personnel Profile: Full-Time Paid 1; Part-Time Paid 5; Part-Time Volunteers 3.
Governing Authority: Tax-exempt.
Institution Type/Description: Historic House Museum: built in 1825.
Collections: local history & culture; personal artifacts; period furnishings; medical equipment & literature; photographs.
Major Exhibits: Community House: Local Artists' Renderings, 2/14-5/14; Here Come the Bride: 80 Years of Weddings at the Brown-Pusey House, 5/14-10/14; Christmas Greetings: Holiday Cards from the Brown-Pusey Collections, 11/14 1/15.
Research Fields: history of dermatology; history of Elizabethtown, Hardin County & surrounding areas.
Facilities: genealogical & local history library; museum.
Activities: rental facilities; community events.
Hours & Admission Prices: Tues.-Sat. 10-4; groups by appointment. No charge; donations accepted. Closed most major holidays.
Attendance: 16,000 (estimated)
Membership: Friend of the Genealogical Library $25.

HARDIN COUNTY HISTORY MUSEUM, 201 W. Dixie Ave., Elizabethtown, KY 42701-1533. Tel.: 270-763-8339.
E-mail: info@hardinkyhistory.org
Web Site: www.hardinkyhistory.org
Institution Type/Description: History Museum.
Collections: artifacts, documents & other memorabilia relating to Hardin County.
Hours & Admission Prices: By appointment.

LINCOLN HERITAGE HOUSE, 212 Freeman Lake Park Rd., Elizabethtown, KY 42701-2702. Tel.: 270-769-3916; 800-437-0092.
Institution Type/Description: Historic House Museum: housed in the home of Hardin Thomas; the first section was built in 1789 & the second part in 1805 with the help of President Abraham Lincoln's father, Thomas Lincoln.
Collections: Lincoln family history; period furnishings; personal artifacts; photographs.
Hours & Admission Prices: June-Oct. 1 Tues.-Sun. 10-5.

ONE ROOM SCHOOL HOUSE MUSEUM, Freeman Lake Park, Blue Heron Way, Elizabethtown, KY 42701. Mailing Address: c/o Elizabethtown Tourism & Convention Bureau, 1030 N. Mulberry St., Elizabethtown, KY 42701. Tel.: 800-437-0092.
Institution Type/Description: Historic House Museum: built in 1892.
Collections: local history & culture; period furnishings; personal artifacts; photographs.
Hours & Admission Prices: Call for hours.

SWOPE'S CARS OF YESTERYEAR, 1100 N. Dixie Ave., Elizabethtown, KY 42701-2534. Tel.: 270-765-2181. Fax: 270-763-6187.
E-mail: fwswope@swope.com
Web Site: www.swopemuseum.com
Founded: 1999.
Congressional District: 2
Key Personnel: Dir., Bill Swope; Cur., Sue Marski; Hostess, Judy Asbury; Hostess, Linda Snyder.
Personnel Profile: Part-Time Paid 3.
Governing Authority: Parent Institution: Swope Auto Center. Subsidiary Institution: Swope Auto Center.
Institution Type/Description: Auto Museum.
Collections: 56 period cars from 1910-1969.
Facilities: Cars for sale.
Hours & Admission Prices: Mon.-Sat. 10-5. No charge. Closed holidays. &
Attendance: 9,000 (estimated)

Elkhorn City

ELKHORN CITY RAILROAD MUSEUM, 100 Pine St., Elkhorn City, KY 41522. Tel.: 606-754-8300.
E-mail: elkhorncityrailroadmuseum@yahoo.com
Web Site: elkhorncityrrm.tripod.com
Institution Type/Description: Railroad Museum.
Collections: railroad history; railroad artifacts & memorabilia.
Hours & Admission Prices: March-Nov. Tues.-Sat. 10-4, Sun. 12-4.

Erlanger

ERLANGER HISTORICAL SOCIETY, 3319 Crescent Ave., Erlanger, KY 41018. Mailing Address: P.O. Box 18062, Erlanger, KY 41018-0062. Tel.: 859-727-2630.
Web Site: erlangerhistoricalsociety.org
Founded: 1990.
Congressional District: 4
Key Personnel: Chm. (V), Patricia A. Hahn; Pres. (V), John Scheben
Institution Type/Description: Historic Building: built in 1877 by the Southern Railway Company.
Collections: railroad history; period artifacts.
Facilities: picnic area; playground.
Activities: rental facilities; arts & crafts vendors; music. Museum Sponsors: Heritage Day in September.
Publications: newsletter; history book.
Hours & Admission Prices: March-Nov. Sat. 12-4. No charge; donations accepted. &
Attendance: 5,000 (estimated)
Membership: Individual $10; Corporate $100.

Evarts

CLOVERFORK MUSEUM, 203 Philpot Ln., Evarts, KY 40828.
E-mail: evelynphilpot@hillbillymail.com

Institution Type/Description: Historic House Museum: housed in the former home of coal miner, Jack Taylor and his family; built c.1930.
Collections: local history & culture; period furnishings; mining artifacts; memorabilia; copper moonshine still; photographs; Memory Wall.
Activities: special events.
Hours & Admission Prices: By appointment.

Fairview

JEFFERSON DAVIS STATE HISTORIC SITE, 258 Pembroke-Fairview Rd., Fairview, KY 42221. Mailing Address: Box 157, Fairview, KY 42221-0157. Tel.: 270-889-6100. Fax: 270-889-6102.
E-mail: ron.sydnor@ky.gov
Founded: 1924.
Congressional District: 1
Key Personnel: Site Supt., Ron Sydnor.
Personnel Profile: Full-Time Paid 1; Part-Time Paid 5.
Governing Authority: state. Affiliated with the Kentucky Dept. of Parks, Capital Plaza Tower, Frankfort, KY 40601.
Institution Type/Description: State Park Monument.
Collections: life of Confederate President, Jefferson Davis.
Facilities: Museum related items for sale.
Activities: guided tours; gallery talks.
Hours & Admission Prices: April to Nov. 15 Tours: 9-5. Adults $5, senior citizens & military $4, children 12 & under $3. &
Attendance: 25,000 (estimated)

Flemingsburg

FLEMING COUNTY COVERED BRIDGE MUSEUM, 119 E. Water St., Flemingsburg, KY 41041. Mailing Address: P.O. Box 12, Flemingsburg, KY 41041. Tel.: 606-845-1223.
Institution Type/Description: History Museum.
Collections: local history & culture; period furnishings; personal artifacts; photographs.
Hours & Admission Prices: March-Dec. Wed. 10-4, Sat. 12-4; other times by appointment. No charge; donations accepted.

Fordsville

FORDSVILLE L&N DEPOT COMMUNITY MUSEUM, 32 Ridge Rd., Fordsville, KY 42343. Mailing Address: P.O. Box 18, Fordsville, KY 42343. Tel.: 270-276-5400.
E-mail: fordsvilledepot@gmail.com
Institution Type/Description: Railroad Depot Museum.
Collections: local history; railroad artifacts; period furnishings; photographs; personal artifacts.
Hours & Admission Prices: Sun. 2-4; call for additional hours.

Fort Campbell

DON F. PRATT MEMORIAL MUSEUM, 5702 Tennessee Ave., Fort Campbell, KY 42223-5919. Mailing Address: P.O. Box 2133, Fort Campbell, KY 42223-2133. Tel.: 270-798-3215.
Web Site: www.fortcampbell.com/pratt.php
Founded: 1956.
Congressional District: 7
Key Personnel: Museum Dir., Daniel Peterson; Historian, John O'Brien, III; Museum Technician, John Foley; Exhibits Specialist, Jim Spencer.
Personnel Profile: Full-Time Paid 4; Full-Time Volunteers 2; Part-Time Volunteers 9; Interns 3.
Governing Authority: federal. Affiliated with U.S. Army Museums System, Office of the Center of Military History, Department of Army, Washington, DC 20315: Tax-exempt.
Institution Type/Description: Airborne Military Museum.
Collections: Fort Campbell, KY from 1942 to present; history of 101st ABN Division; 11th Airborne Division; 12th & 14th Armored Divisions; 173rd Airborne Brigade; U.S., German, Japanese, Iraqi & Vietnamese uniforms; equipment & documents relating to U.S. military heritage; WWII 20th Armored Division, U.S. Army.
Research Fields: U.S. Military; U.S. Airborne history; 101st Airborne Div. history.
Facilities: library of books & manuscripts pertaining to Fort Campbell Airborn, 101st & 11th Airborne Divisions history, 12th & 14th Armored Divisions & 173rd Airborne Brigade.
Activities: guided tours; lectures; films; student programs; loan, permanent & temporary exhibitions.
Publications: pamphlets, 101st Airborne Div. (Air Assault), Fort Campbell.

Hours & Admission Prices: Mon.-Sat. 9:30-4:30; guided tours of 10 or more by appointment. No charge; donations accepted. Closed New Year's Day; Christmas. &
Attendance: 55,000 (accurate)

Fort Knox

GENERAL GEORGE PATTON MUSEUM OF LEADERSHIP, 356 Fayette Ave., Bldg. 4554, Fort Knox, KY 40121. Mailing Address: P.O. Box 1304, Fort Knox, KY 40121. Tel.: 502-624-3729. Fax: 502-624-4333.
Web Site: www.generalpatton.org
Formerly: Patton Museum of Cavalry and Armor
Founded: 1948.
Congressional District: 2
Key Personnel: C.E.O., Robert Keats; Cur., Nathan C. Jones; Mgr. Collections, Amber Hills; Mgr. Collections, O.B. Edens.
Personnel Profile: Full-Time Paid 4; Part-Time Volunteers 25; Interns 2.
Governing Authority: federal. Affiliated with U.S. Army ROTC Cadet Command Center of Military History, Training & Doctrine Command. Tax-exempt.
Institution Type/Description: Military Museum.
Collections: tanks; armored vehicles; firearms; medals & decorations; Gen. Patton memorabilia; paintings; military equipage; photographs.
Research Fields: Gen. George S. Patton, Jr.; Army leadership; ROTC; Fort Knox.
Facilities: library of material on U.S. & leadership development reference books; archives & photograph collection.
Activities: films; temporary exhibitions; tours; educational programming.
Hours & Admission Prices: Mon.-Fri. 9-4:30, Sat.-Sun. & holidays l0-4:30. No charge; donations accepted. Closed New Year's Eve & Day; Easter; Thanksgiving; Christmas Eve & Day. &
Attendance: 150,000 (accurate)

Fort Mitchell

VENT HAVEN MUSEUM, 33 W. Maple Ave., Fort Mitchell, KY 41011-2616. Tel.: 859-341-0461.
E-mail: venthaven@insightbb.com
Web Site: www.venthavenmuseum.net
Founded: 1973.
Congressional District: 4
Key Personnel: Cur., Jen Dawson.
Governing Authority: nonprofit organization. Tax-exempt.
Institution Type/Description: Theater Museum.
Collections: 790 ventriloquial figures; library of rare books; manuscripts; films; records; play bills.
Activities: guided tours; permanent exhibitions; videotape presentation.
Hours & Admission Prices: May-Sept. by appointment only. Admission Donation: $5.
Attendance: 1,000

Fort Thomas

FORT THOMAS MILITARY AND COMMUNITY MUSEUM, 940 Cochran Ave., Fort Thomas, KY 41075. Tel.: 859-572-1225 & 815-8481.
Web Site: www.ftthomas.org
Institution Type/Description: Military History Museum.
Collections: local history; military artifacts & equipment; personal artifacts; clothing; photographs.
Hours & Admission Prices: Wed.-Sun. 12-4; other times by appointment.

Frankfort

CAPITAL CITY MUSEUM, 325 Ann St., Frankfort, KY 40601-2803. Tel.: 502-696-0607.
E-mail: frankforthistory@gmail.com
Web Site: www.capitalcitymuseum.com
Founded: 2002.
Congressional District: 6
Key Personnel: Collection Mgr., John Patrick Downs; Asst. Cur., Russ Hatter.
Personnel Profile: Full-Time Paid 2; Part-Time Paid 12.
Governing Authority: Parent Institution: City of Frankfort Dept. Parks, Recreation & Historic Sites. Tax-exempt.
Institution Type/Description: History Museum: housed in the former Gayle Drug Store.
Collections: local history; period furnishings; personal artifacts; political memorabilia; whiskey distilling industry.

Research Fields: history of Frankfort and Franklin County, Kentucky.
Facilities: research library; special & themed tours.
Hours & Admission Prices: Mon.-Sat. 10-4. No charge. ♿
Attendance: 16,000 (accurate)

CLYDE E. BUCKLEY WILDLIFE SANCTUARY, 1305 Germany Rd., Frankfort, KY 40601-8257. Tel.: 859-873-5711. Fax: 859-873-5711.
E-mail: twilliams@audubon.org
Web Site: www.audubon.org
Founded: 1967.
Key Personnel: Sanctuary Mgr., Tim Williams.
Personnel Profile: Full-Time Paid 1; Interns 3.
Governing Authority: private; nonprofit society. Parent Institution: The National Audubon Society. Tax-exempt: 501(c)(3).
Institution Type/Description: Wildlife Sanctuary & Nature Center.
Collections: various plants & animals native to the area, including birds, reptiles & amphibians; terrariums & aquariums; insects; collection of the prints of Ray Harm; hands-on exhibits & displays.
Research Fields: flora & fauna of the Inner Bluegrass area.
Facilities: 200-vol. library of books & magazines available for use on premises; nature & conservation center; Marion E. Lindsey Bird Blind, which has one-way windows to permit visitors to observe the feeding of the birds; nature trails; information area. Nature books, stationery & bird related items for sale.
Activities: guided tours; lectures; slide presentations; environmental workshops; organized education programs & activities for children & adults; permanent & temporary exhibitions; intern program.
Publications: sanctuary information & history brochure; internship brochure.
Hours & Admission Prices: Sanctuary: Wed.-Fri. 9-5, Sat.-Sun. 9-6. Museum: Sat.-Sun. 1-6; other times by appointment. Adults $4, children $3; members no charge. Closed holidays. ♿
Attendance: 10,000
Membership: Individual $35; Family $50; Patron $150 & up.

THE GOVERNOR'S MANSION, 704 Capitol Ave., Frankfort, KY 40601-3448. Tel.: 502-564-8004. Fax: 502-564-5022.
E-mail: ann.evans@ky.gov
Web Site: www.governorsmansion.ky.gov
Formerly: The Executive Mansion
Founded: 1914.
Congressional District: 6
Key Personnel: Exec. Dir., Ann Evans.
Governing Authority: state. Parent Institution: Finance & Administration Cabinet. Subsidiary Institution: Historic Properties. Tax-exempt.
Institution Type/Description: Historic House: c.1914 Beaux-Arts style 25-room residence of 23 of Kentucky's governors.
Collections: furnishings; Kentucky paintings; period silver; Chinese porcelains.
Activities: guided tours.
Publications: brochure.
Hours & Admission Prices: Tues. & Thurs. 9-11. No charge; donations accepted. Closed legal holidays. ♿

KENTUCKY DEPARTMENT OF FISH & WILDLIFE RESOURCES-SALATO WILDLIFE EDUCATION CENTER, 1 Sportsman's Lane, Frankfort, KY 40601-3951. Tel.: 502-564-7863, ext. 4407; 800-858-1549. Fax: 502-564-6508.
E-mail: salato@ky.gov
Web Site: fw.ky.gov
Founded: 1995.
Congressional District: 4
Key Personnel: Commissioner, Dr. Jonathan W. Gossett.
Personnel Profile: Full-Time Paid 6; Part-Time Paid 8; Part-Time Volunteers 6.
Governing Authority: state. Parent Institution: Kentucky State Government. Subsidiary Institution: Kentucky Dept. Fish & Wildlife Resources, Frankfort, KY 40601. Tax-exempt.
Institution Type/Description: Wildlife Museum.
Collections: live animals & birds; animal artifacts & native plants.
Research Fields: preserved specimens fish, amphibians.
Facilities: wildlife education center; zoological park; picnic area; hiking trails; fishing lakes.
Activities: special programming; self-guided tours; scavenger hunts; hiking; fishing.
Publications: magazine, Kentucky Afield-The Magazine; technical papers on fish & wildlife; newsletter.
Hours & Admission Prices: mid-Feb. to mid-Dec. Tues.-Fri. 9-5, Sat. 10-5. No charge; donations accepted. Closed state holidays. ♿
Attendance: 100,000 (accurate)

✳ **KENTUCKY HISTORICAL SOCIETY, (M),** 100 W. Broadway, Frankfort, KY 40601-1931. Tel.: 502-564-1792. Fax: 502-564-4701.
E-mail: KHS@ky.gov
Web Site: www.history.ky.gov
Founded: 1836.
Congressional District: 6
Key Personnel: Dir., Kent Whitworth; Asst. Dir., Scott Alvey.
Personnel Profile: Full-Time Paid 54; Part-Time Paid 5; Part-Time Volunteers 52; Interns 7.
Governing Authority: state. Parent Institution: Tourism, Arts & Heritage Cabinet. Branch Museums: Thomas D. Clark Center for Kentucky History; Kentucky Military History Museum (Old State Arsenal); Old State Capitol. Tax-exempt: 501(c)(3).
Institution Type/Description: History Museum.
Collections: decorative arts; textiles; social & political history material related to Kentucky history; military artifacts.
Research Fields: genealogy; Kentucky history; military history.
Facilities: 80,000-vol. library of books, 3,000-vol. of genealogies, 10,000-reels of microfilm, a complete set of Kentucky legislative journals, manuscript collection & Kentucky tax records, available for use on premises only. Gift items for sale.
Activities: guided tours; interpretive programs & school activities by schedule; permanent, temporary & traveling exhibitions; local history; Historymobile.
Publications: quarterly, The Register; quarterly, Kentucky Ancestors; quarterly newsletter, The Chronicle.
Hours & Admission Prices: See website for hours. Adults $4; members no charge. ♿
Attendance: 70,000 (accurate)
Membership: See website for information.

✳ **KENTUCKY MILITARY HISTORY MUSEUM, (M),** 128 E. Main St., Frankfort, KY 40601. Mailing Address: 100 W. Broadway, Frankfort, KY 40601-1931. Tel.: 502-564-3265. Fax: 502-564-4054.
Web Site: history.ky.gov
Founded: 1974.
Congressional District: 6
Key Personnel: Cur., Bill Bright.
Personnel Profile: Full-Time Paid 3; Part-Time Paid 1; Part-Time Volunteers 2; Interns 3.
Governing Authority: state. Parent Institution: Kentucky Historical Society. Subsidiary Institution: Kentucky Dept. of Military Affairs (Kentucky National Guard). Tax-exempt.
Institution Type/Description: Military Museum: housed in 1850 State Arsenal, built for Kentucky Militia.
Collections: Kentucky-related ordnance; flags; uniforms; personal items; archival materials.
Research Fields: general Kentucky military history; history of Kentucky militia & national guard; Civil War; 20th century.
Activities: guided tours; hands-on programs for school groups; living history programs in schools; permanent, temporary & traveling exhibitions.
Publications: brochures; exhibit guides; calendar.
Hours & Admission Prices: Tues.-Sat. 10-5; call to confirm. Adults $4; members no charge. Closed New Year's Eve & Day; Easter; Thanksgiving; Christmas Eve & Day. ♿
Attendance: 15,000 (accurate)
Membership: Kentucky Historical Society: Student $20; Senior $35; Individual $40; Senior Family $45; Family & Institutional $50; Friend $100; Benefactor $250.

KENTUCKY STATE CAPITOL, Capital Ave., Frankfort, KY 40601. Mailing Address: Historic Properties, Berry Hill Mansion, 700 Louisville Rd., Frankfort, KY 40601-3304. Tel.: 502-564-3000 & (Tour Information) 564-3449. Fax: 502-564-6505.
E-mail: david.buchta@ky.gov
Web Site: www.historicproperties.ky.gov
Founded: 1910.
Congressional District: 6
Personnel Profile: Full-Time Paid 2; Part-Time Paid 1; Interns 2.
Governing Authority: state. Parent Institution: Finance & Administration Cabinet. Subsidiary Institution: Historic Properties.
Institution Type/Description: Historic Building: c.1910 Kentucky's fourth Statehouse.
Collections: period furnishings; murals; early 20th-century decorative arts; First Ladies miniature collection.
Activities: guided tours; rotating exhibits; self-guided walking tour.
Publications: brochure.

Hours & Admission Prices: Mon.-Fri. 8-4:30. No charge. Closed New Year's Day; Easter; Thanksgiving; Christmas Eve & Day. &

Attendance: 100,000 (estimated)

KENTUCKY STATE POLICE MUSEUM, 633 Chamberlin Ave., Frankfort, KY 40601. Tel.: 502-875-1625.

E-mail: quartermaster@ksppa.com
Web Site: www.ksppa.com
Institution Type/Description: Police Museum.
Collections: Kentucky's State Police history & artifacts; photographs; personal artifacts.
Hours & Admission Prices: Mon.-Thurs. 8-4, Fri. 8-3.

LIBERTY HALL HISTORIC SITE, (M), 202 Wilkinson St., Frankfort, KY 40601-1826. Tel.: 502-227-2560; 888-516-5101. Fax: 502-227-3348.

E-mail: libhall@dcr.net
Web Site: www.libertyhall.org
Founded: 1937.
Congressional District: 6
Key Personnel: Coord. Education, Jennifer Koach; Cur., Kate Hesseldenz; Museum Shop Mgr., Becky Shipp.
Personnel Profile: Full-Time Paid 3; Part-Time Paid 4; Part-Time Volunteers 35; Interns 1.
Governing Authority: private; nonprofit organization. Parent Institution: Liberty Hall Inc. Tax-exempt: 501(c)(3).
Institution Type/Description: Historic Site: housed on the site of Senator John Brown's home, Kentucky's first US senator.
Collections: late 18th- & 19th-century furniture; family portraits; china; silver; books; Brown family furnishings; art; decorative art; kitchen & household artifacts; family documents. Historic Buildings: Liberty Hall, a 1796 Georgian house; Orlando Brown house, an 1835 Greek Revival home.
Research Fields: life in late 18th-century to early 19th-century Kentucky & slavery in this time period.
Facilities: 1,000-vol. library of history & genealogy books available for use by prior arrangement; garden wedding facilities; botanical garden.
Activities: guided tours; formal education programs; hobby workshops; lectures; loan, permanent & temporary exhibitions; 19th-century Christmas; docent program.
Publications: books, Liberty Hall & Orlando Brown; The Browns of Liberty Hall; Kentucky Courthouses; Kentucky Ante-Bellum Portraits; seasonal newsletter, Liberty Hall Historic Site Gazette.
Hours & Admission Prices: Tours: March to mid-Dec. Tues.-Sat. 10:30, 12, 1:30, 3; Sun. 1:30, 3. Adults $4, seniors 60 & over $3, children 5-18 $1; members and children 4 & under no charge. Closed New Year's Day; Thanksgiving; Christmas.
Attendance: 4,276 (accurate)

❋ **THE OLD GOVERNOR'S MANSION, (M),** 420 High St., Frankfort, KY 40601-2175. Mailing Address: 700 Capitol Ave., Frankfort, KY 40601-3410. Tel.: 502-564-3449. Fax: 502-564-4099.
Founded: 1798.
Congressional District: 6
Key Personnel: Dir., David Buchta.
Personnel Profile: Full-Time Paid 4; Part-Time Paid 2; Part-Time Volunteers 4.
Governing Authority: state. Parent Institution: Finance & Administration cabinet, Historic Properties. Tax-exempt.
Institution Type/Description: Historic Building: 1798 transitional Federal-Georgian style Old Governor's Mansion.
Collections: historic furniture & artifacts that belonged to former Governors; furniture; paintings; portraits; silver items.
Activities: guided tours.
Publications: brochure.
Hours & Admission Prices: Mon.-Tues. & Thurs. 1:30-3:30. No charge. &
Attendance: 6,000 (estimated)

VEST-LINDSEY HOUSE, 401 Wapping St., Frankfort, KY 40601-2607. Tel.: 502-564-3000 & 6980. Fax: 502-564-6505.
Founded: 1978.
Congressional District: 6
Key Personnel: Dir. & Cur., David Buchta.
Personnel Profile: Full-Time Paid 2.
Governing Authority: state of Kentucky.
Institution Type/Description: Historic House: c.1820 12-room home located in a four-block area of 28 historic homes & churches; official state meeting house.
Collections: early 1800s furnishings.

Activities: guided tours.
Publications: brochure.
Hours & Admission Prices: By appointment. No charge. Closed state & national holidays. &

Franklin

AFRICAN AMERICAN HERITAGE CENTER, 500 Jefferson St., Franklin, KY 42134-1728. Mailing Address: PO Box 369, Ansted, WV 25812. Tel.: 270-598-9986. Fax: 270-586-5719.
E-mail: africanamericanh@bellsouth.net
Web Site: www.aahconline.org
Institution Type/Description: History Museum: listed on the National Register of Historic Places.
Collections: local history & culture pertaining to African Americans in Simpson County.
Hours & Admission Prices: Mon.-Fri. 9-12 & 1-4:30, Sat. 9-11. Closed holidays.

OCTAGON HALL MUSEUM, 6040 Bowling Green Rd., Franklin, KY 42134. Tel.: 270-791-0071.
Web Site: www.octagonhall.com
Institution Type/Description: Historic House Museum: housed in an eight-sided home used during the Civil War by Confederate and Federal troops; it also served as a Civil War hospital.
Collections: Civil War history; personal artifacts; period furnishings; photographs.
Hours & Admission Prices: Wed.-Sat. 9-11 & 1-3:30. Adults $5, children 6-12 $1. Closed major holidays.

SIMPSON COUNTY ARCHIVES & MUSEUM, 206 N. College St., Franklin, KY 42134-1826. Tel.: 270-586-4228. Fax: 270-586-4429. Facebook: Simpson County Archives & Museum.
E-mail: oldjail@comcast.net
Web Site: www.simpsoncountykyarchives.com
Personnel Profile: Part-Time Paid 12; Part-Time Volunteers 10.
Governing Authority: Tax-exempt.
Institution Type/Description: History Museum: housed in the old jail and jailer's residence.
Collections: old scrapbooks; family Bibles; assorted manuscripts; census records; maps; county records 1819 to current; arrowheads; 1800s furnishings; clothing; military uniforms; agricultural artifacts.
Publications: quarterly newsletter, Jailhouse Journal.
Hours & Admission Prices: Mon.-Fri. 9-4, Sat. 10-2; other times by appointment. No charge, but donations accepted. Closed federal holidays.
Attendance: 1,500 (estimated)
Membership: Individual $10; Family $15.

Georgetown

CARDOME CENTER - MUSEUM OF THE WRITTEN WORD, 800 Cincinnati Rd., Ste. 3, Georgetown, KY 40324. Tel.: 502-863-1575.
Web Site: www.cardomecenter.com
Governing Authority: Parent Institution: Cardome Academy Association, Inc.
Institution Type/Description: Communications Museum.
Collections: history of language; poems; essays; newspapers; magazines; novels.
Hours & Admission Prices: Call for hours.

GEORGETOWN COLLEGE ART GALLERIES, (M), 400 E. College St., Georgetown, KY 40324-1628. Tel.: 502-863-8399.
E-mail: laura_stewart@georgetowncollege.edu
Web Site: www.georgetowncollege.edu/galleries
Key Personnel: Dir. & Cur. Collections, Laura Stewart.
Personnel Profile: Full-Time Paid 1; Part-Time Paid 10.
Governing Authority: Parent Institution: Georgetown College. Tax-exempt.
Institution Type/Description: Art Gallery.
Collections: art exhibitions.
Hours & Admission Prices: Mon.-Fri. 12-4:30; other times by appointment. No charge. &
Attendance: 2,500 (estimated)

GEORGETOWN-SCOTT COUNTY MUSEUM, 229 E. Main St., Georgetown, KY 40324-1759. Tel.: 502-863-6201.
E-mail: museum.scottco@yahoo.com
Founded: 1992.
Key Personnel: Dir., David E. Rice.

Governing Authority: Tax-exempt.
Institution Type/Description: History Museum.
Collections: local history & culture; photographs; personal artifacts; video.
Activities: video.
Publications: newsletter.
Hours & Admission Prices: Wed.-Sat. 10-4; other times by appointment. No charge; donations accepted. &

SCOTT COUNTY ARTS & CULTURAL CENTER & VISITOR'S WELCOME CENTER, 117 N. Water St., Georgetown, KY 40324. Mailing Address: Scott County Arts Consortium, Inc., P.O. Box 1126, Georgetown, KY 40324. Tel.: 502-570-8366.
E-mail: artscenter@bellsouth.net
Web Site: www.scottcountyartworks.org
Founded: 2006.
Institution Type/Description: Art Museum: housed in an historic house built in 1870s.
Collections: works by local & regional artists.
Activities: classes; educational programs.
Publications: e-newsletter.
Hours & Admission Prices: Tues.-Sat. 12-4; other times by appointment.

WARD HALL, 1782 Frankfort Pike, Georgetown, KY 40324. Mailing Address: P.O. Box 1957, Georgetown, KY 40324-6957. Tel.: 859-879-9393.
Web Site: www.wardhall.net
Founded: 1979.
Congressional District: 6
Key Personnel: Chm., David Stuart.
Governing Authority: individual operation.
Institution Type/Description: Historic House: 1853 Classical Greek Revival House.
Collections: period furnishings.
Research Fields: Bluegrass houses & traditions; antebellum houses.
Activities: guided tours.
Publications: booklet, Ward-Johnson Families of Central Kentucky; Scott County History Book; The Bluegrass of Kentucky; information sheets on architectural features of the house.
Hours & Admission Prices: By appointment only. No Charge.

Glasgow

SOUTH CENTRAL KENTUCKY CULTURAL CENTER, 200 W. Water St., Glasgow, KY 42141-1738. Mailing Address: PO Box 1714, Glasgow, KY 42142. Tel.: 270-651-9792. Fax: 270-651-2806.
E-mail: sckculturalcenter@glasgow-ky.com
Web Site: www.kyculturalcenter.org
Founded: 1988.
Key Personnel: Dir., Gayle B. Berry; Chm. (V), Kristen Bale.
Personnel Profile: Full-Time Paid 1; Part-Time Paid 3.
Governing Authority: Parent Institution: Barren County Historical Foundation, Inc. Tax-exempt.
Institution Type/Description: History Museum housed in the old Kentucky Pants factory.
Collections: local history & cultural pertaining to Barren County.
Hours & Admission Prices: Mon.-Fri. 9-4, Sat. 9-2. No charge; donations accepted. &
Attendance: 9,000 (accurate)
Membership: Friend $25; Cultural Friend $100; Corporate Friend $250; Life Friend $1,000; Beula C. Nunn Friend $5,000.

Golden Pond

USDA FOREST SERVICE - LAND BETWEEN THE LAKES, 100 Van Morgan Dr., Golden Pond, KY 42211-9001. Tel.: 270-924-2000. Fax: 270-924-2060.
E-mail: lblinfo@fs.fed.us
Web Site: www.lbl.org
Founded: 1963.
Congressional District: 1
Key Personnel: Gen. Mgr., Bill Lisowsky.
Governing Authority: federal. USDA Forest Service
Institution Type/Description: Historic Building & Site, Park & Nature Center, Planetarium & Natural History Museum.
Collections: period furnishings; farm tools. Historic Structures: 1850 The Homeplace, 16 structures comprising a living history farm.
Research Fields: regional & local history; recreation, cultural & natural resource management.

Facilities: Golden Pond Visitor Center; theatre; observatory; planetarium; The Nature Station; Homeplace-1850; The Elk & Bison Prairie.
Activities: guided group tours, scheduled on request; AV presentations; educational exhibits; information services; variety of programmed learning & skill-developing activities; trails for handicapped; special weekends; arts & crafts festival; group camps; horseback riding; student intern & apprentice programs; organized education programs for undergraduate & graduate college students; living history demonstrations.
Publications: Spring Wildflowers; Summer and Fall Wildflowers; Ancient Man; Amphibians and Reptiles of Land Between The Lakes; Lichens and Ferns of Land Between The Lakes; Trees and Shrubs of Land Between The Lakes; Mushrooms of Land Between the Lakes; 1850s Cookbook; Tennessee's Iron Industry Revisited: The Stewart County Story; leader guide.
Hours & Admission Prices: The Homeplace-1850: March & Nov. Wed.-Sat. 9-5, Sun. 10-5; April-Oct. Mon.-Sat. 9-5, Sun. 10-5. Adults $4, children 5-12 $2; discounts to groups of 100 or more by appointment; children 4 & under no charge. &
Attendance: 1,300,000 (estimated)
Membership: Land Between the Lakes Association: Sponsor $30; Sustaining $50; Patron $100; Business $250; Corporate $1,000.

Gravel Switch

FORKLAND COMMUNITY CENTER - LINCOLN MUSEUM, 16479 Forkland Rd., Gravel Switch, KY 40328. Tel.: 859-936-7489 & 2061.
E-mail: info@forklandcomctr.org
Web Site: www.forklandcomctr.org
Key Personnel: Chm. (V), Wayne Thurman.
Governing Authority: Parent Institution: Forkland Community Center. Tax-exempt.
Institution Type/Description: History Museum.
Collections: Lincoln family history; local history & culture; period furnishings; photographs; personal artifacts.
Activities: genealogy research; special events. Museum Sponsors: Forkland Festival in October.
Publications: local history & genealogy books.
Hours & Admission Prices: May-Oct. Sat. 12-4; other times by appointment. No charge; donations accepted.

Greenville

DUNCAN CENTER MUSEUM & ART GALLERY, 122 S. Cherry St., Greenville, KY 42345-1234. Mailing Address: P.O. Box 289, Greenville, KY 42345-0289. Tel.: 270-338-2605. Facebook: Duncan Center Museum & Art Gallery.
E-mail: duncan.museum@gmail.com
Web Site: www.duncancentermuseum.com
Formerly: Duncan Cultural Center Museum and Art Gallery
Founded: 1989.
Congressional District: 15
Key Personnel: Dir., Shara Sumner; Chm. (V), Wesley Harris
Governing Authority: Tax-exempt.
Institution Type/Description: History Museum & Art Gallery.
Collections: Duncan family history; period furnishings; personal artifacts; coal mining; Victorian artifacts.
Activities: rental facilities; special events. Annual Events: Mother's Day Victorian Tea in May; Pictures With Mr. & Mrs. Claus in December.
Publications: annual yearbook in Dec.
Hours & Admission Prices: Mon.-Fri. 11-4. No charge; donations encouraged. &
Attendance: 3,000 (estimated)
Membership: Annual $25; Life $500; Patron $500; Benefactor $1,000.

Guthrie

ROBERT PENN WARREN BIRTHPLACE, Third & Cherry Sts., Guthrie, KY 42234. Mailing Address: P.O. Box 296, Guthrie, KY 42234. Tel.: 270-483-2683.
Founded: 1989.
Key Personnel: Dir., Jeane Moore, Museum Shop Mgr., Melba W. Smith.
Personnel Profile: Part-Time Paid 1.
Governing Authority: Tax-exempt.
Institution Type/Description: Historic House Museum: housed in the birthplace of Robert Penn Warren.
Collections: Warren's life & career; personal artifacts; period furnishings; photographs.
Hours & Admission Prices: Tues.-Sat. 11:30-3:30; other times by appointment. No charge. &
Attendance: 3,250 (estimated)

Hardinsburg

BRECKINRIDGE COUNTY HISTORICAL SOCIETY MU-SEUM, 204 E. 3rd St., Hardinsburg, KY 40143. Mailing Address: P.O. Box 498, Hardinsburg, KY 40143. Tel.: 270-756-5216.
Founded: 1984.
Key Personnel: Pres., Melissa Kampars
Institution Type/Description: Historical Society Museum: housed in a 1920s home.
Collections: local history & culture; photographs; period artifacts.
Publications: quarterly newsletter, Breckinridge County Historical Society.
Hours & Admission Prices: Wed. 10-2, Sat. 1-4. No charge. &

Membership: Individual $15; Lifetime $150.

Harrodsburg

MORGAN ROW MUSEUM AND HARRODSBURG-MERCER COUNTY RESEARCH LIBRARY, 220 S. Chiles St., Harrodsburg, KY 40330-1631. Mailing Address: P.O. Box 316, Harrodsburg, KY 40330-0316. Tel.: 859-734-5985.
Formerly: Morgan Row Museum and Research Center
Founded: 1907.
Key Personnel: Pres. (V), Jerry Sampson.
Governing Authority: society. Affiliated with Harrodsburg Historical Society. Tax-exempt.
Institution Type/Description: Historic House: 1800s Row House built by Joseph Morgan.
Collections: an original Chester Harding portrait of Daniel Boone; antique furniture; displays of rifles; miniature furniture; examples of early glass, china & silver; original documents & personal items related to the early history of Harrodsburg & Mercer County; Genealogy Library: family files, cemetery records, tax & census, newspapers, books of history, original documents of local families. Historical Building: 1800 Old Mud Meeting House, built for the first Dutch Reformed Church west of the Alleghenies.
Research Fields: genealogy; early Kentucky history.
Facilities: genealogical & research library of rare books, documents, maps, family & subject files, census records & material dealing with the early history of this area available for research on premises with a member of the society present; microfilm reader; meeting room.
Activities: permanent displays of portraits by Kentucky artists.
Publications: 4-vols. Mercer County Cemetery Records; 1831-1850 Mercer County Marriage Bonds & Consents; Old Mud Meeting House.
Hours & Admission Prices: Tues. 10-4, Wed.-Sat. 1-4. Museum: no charge; donations accepted. Library: $5.
Attendance: 900 (estimated)

OLD FORT HARROD STATE PARK MANSION MUSEUM, 100 S. College St., Harrodsburg, KY 40330-1508. Mailing Address: P.O. Box 156, Harrodsburg, KY 40330-0156. Tel.: 859-734-3314. Fax: 859-734-0794.
E-mail: fortharrod@ky.gov
Web Site: www.parks.ky.gov/findparks/recparks/fh
Founded: 1934.
Congressional District: 6
Key Personnel: Park Mgr., David Coleman.
Personnel Profile: Full-Time Paid 3; Part-Time Paid 16; Part-Time Volunteers 1.
Governing Authority: state. Parent Institution: Kentucky Department of Parks, Capital Plaza Tower, Frankfort, KY 40601.
Institution Type/Description: Museum & Historic House: 1813 Greek revival mansion; replica of original fort built by James Harrod.
Collections: archaeology; archives; costumes; decorative arts; glass; history; Indian artifacts; music; President Abraham Lincoln memorabilia.
Activities: guided tours; permanent exhibitions; manuscript collections.
Hours & Admission Prices: Museum: April-Oct. Wed.-Sat. 10-5, Sun. 12-5. Fort: March-Nov. Mon.-Sat. 9-5, Sun. 12-5; Dec.-Feb. Mon.-Fri. 8-4:30. April-Oct.: adults $5, children $3; Nov.-March adults $3, children $2. Park: closed New Year's Eve & Day; Thanksgiving; Christmas Eve, Day & week. &
Attendance: 30,000 (accurate)

SHAKER VILLAGE OF PLEASANT HILL, 3501 Lexington Rd., Harrodsburg, KY 40330-8846. Tel.: 859-734-1549. Fax: 859-734-7278.
E-mail: lcurry@shakervillageky.org
Web Site: www.shakervillageky.org
Founded: 1961.
Congressional District: 6
Key Personnel: C.E.O. & Pres., Madge B. Adams; Vice Pres. & Cur., Larrie

Spier Curry; Chm. Bd. (V), James G. Kenan; Dir. Human Resources, Candace Parker; Mgr. Interpretation & Education, Susan Lyons Hughes; Mgr. Historic Farm, Ralph E. Ward; Mgr. Craft Store, Lorrin Ingerson; Mgr. Preservation, Mike McGinnis.
Personnel Profile: Full-Time Paid 90; Part-Time Paid 70; Part-Time Volunteers 45.
Governing Authority: nonprofit organization. Tax-exempt: 501(c)(3).
Institution Type/Description: Historic Village Museum: over 3,000 acres former Shaker farmland includes 34 original 19th century buildings.
Collections: 34 historic early 19th-century buildings; furniture; artifacts of Shakers; tools & equipment for agriculture, woodworking & other 19th-century industries; textiles; manuscripts; photographs.
Research Fields: Shaker & Utopian social history of the 19th-century; 19th-century agriculture, trades & religious history.
Facilities: 1,500-vol. library of material relating to the history of the Shakers & 19th-century Utopian societies, 3,000 acres with 40 miles of multi-use trails; meeting & conference rooms; overnight accommodations & dining room. Shaker furniture reproductions & publications; handmade Kentucky crafts for sale.
Activities: self-guided & guided tours; lectures; films; music & dance interpretation; historic agriculture; interpreted horse-drawn carriage tours; crafts events; temporary & permanent exhibitions; organized educational programs; workshops; hiking & horseback trails; nature hikes; riverboat excursions. Museum Sponsors: seasonal special events. Annual Events: craft fair; antique show & sale; Chamber Music Festival; dry stone wall building competition.
Publications: books, Pleasant Hill & Its Shakers; Welcome Back to Pleasant Hill; We Make You Kindly Welcome; Pleasant Hill in the Civil War; annual calendar, Keepsake Art Calendar.
Hours & Admission Prices: April-Oct. daily 10-5. Village Tour: adults 13 & over $15, youth 6-12 $5. Riverboat Excursions: children 13 & over $10, youth 6-12 $5; discounts to groups; children under 6 accompanied by a parent & members no charge. Nov.-March tour hours & prices reduced.
Attendance: 51,700 (accurate)
Membership: member $50; Partner $100; Supporter $250; Contributor $500; Benefactor $1,000.

Hartford

THE BLUEGRASS MOTORCYCLE MUSEUM, 5608 US Hwy. 231 N., Hartford, KY 42347-9583. Tel.: 270-298-7764.
Key Personnel: Owner, Jack Embry; Owner, Nancy Embry
Institution Type/Description: Motorcycle Museum.
Collections: period motorcycles.
Hours & Admission Prices: Tues.-Fri. 10-5, Sat. 10-3; other times by appointment.

OHIO COUNTY MUSEUM & OHIO COUNTY VETERANS MUSEUM, 415 Mulberry St., Hartford, KY 42347. Tel.: 270-298-3444.
Institution Type/Description: History Museum.
Collections: local history & culture; period furnishings; personal artifacts; photographs; military artifacts, uniforms & memorabilia. Historic Buildings: log cabin; one-room school house; home; Elfie Autry's Little Country store.
Hours & Admission Prices: County Museum: May-Oct. Thurs.-Fri. 1-4. Veterans Museum: Sat. 9-4; other times by appointment.

Hawesville

HANCOCK COUNTY MUSEUM, 110 River St., Hawesville, KY 42348. Mailing Address: P.O. Box 605, Hawesville, KY 42348. Tel.: 270-927-8721.
Founded: 1988.
Congressional District: 2
Key Personnel: Dir., James H. Fallin.
Personnel Profile: Part-Time Volunteers 50.
Governing Authority: Tax-exempt.
Institution Type/Description: History Museum: housed in the former railroad station; built in 1902.
Collections: local history & culture; railroad artifacts; photographs; period artifacts; farming; river traffic; schools; family life; paintings.
Publications: newsletter.
Hours & Admission Prices: April-Oct. Sun. 2-4; other times by appointment. No charge; donations accepted. Closed holidays. &
Attendance: 200 (estimated)
Membership: Student $3; Individual $10; Family $15; Supporting $25; Donor $100-$249; Contributor $250-$499; Life $500-$999.

Hazard

BOBBY DAVIS MUSEUM AND PARK, 234 Walnut St., Hazard, KY 41701-1852. Tel.: 606-439-4325. Bobby Davis Museum and Park.
E-mail: bdwalnutforest@msn.com
Formerly: Perry County Museum
Founded: 1984.
Key Personnel: Dir., Martha Quigley; Pres. (V), Anne Gilbert; Museum Shop Mgr., Sydney Francis.
Personnel Profile: Full-Time Paid 2; Part-Time Volunteers 5.
Governing Authority: Parent institution: City of Hazard. Tax-exempt.
Institution Type/Description: History Museum.
Collections: local history; photographs; historical documents; oral history tapes; period artifacts; herb & medicinal gardens.
Research Fields: Main Street.
Facilities: gardens.
Activities: Art on main.
Hours & Admission Prices: Mon.-Fri. 8-4; no charge, donations accepted. &
Attendance: 1,500 (estimated)

Henderson

JOHN JAMES AUDUBON MUSEUM, 3100 U.S. Hwy. 41 N., Henderson, KY 42420-2055. Mailing Address: P.O. Box 576, Henderson, KY 42419-0576. Tel.: 270-826-2247. Fax: 270-826-2286. TDD: 270-826-2247.
E-mail: audubon@ky.gov
Web Site: parks.ky.gov/findparks/recparks/au/
Founded: 1938.
Congressional District: 1
Key Personnel: Cur., Alan Gehret.
Personnel Profile: Full-Time Paid 6; Part-Time Paid 2; Part-Time Volunteers 8.
Governing Authority: state. Affiliated with Kentucky Dept. of Parks, Capital Plaza Tower, Frankfort, KY 40601.
Institution Type/Description: State Park Museum: located on migratory bird route.
Collections: works, prints & personal memorabilia of John James Audubon; exhibits about Audubon's life and work; archives; paintings; botany; costumes; history; natural history; manuscript collections.
Facilities: nature observatory; discovery room devoted to study of birds; nature trails. Wildlife prints, printed matter & works of wildlife artists for sale.
Activities: guided tours; lectures; gallery talks; arts festivals; nature trail tours; over-night lodging available; formally organized education programs for children; wildlife observation area.
Publications: The Warbler.
Hours & Admission Prices: Jan.-Feb. Thurs.-Sun. 10-5; March-Nov. daily 10-5. Family $10, adults $4, children 6-12 $2.50; discounts to groups; members, Friends of Audubon members & children under 6 no charge. Closed New Year's Eve & Day; Martin Luther King Jr. Day; Thanksgiving & day after; Christmas week. &
Attendance: 12,000 (estimated)
Membership: Student $5; Individual $20; Family $35; Donor $100.

Hickman

CARNEGIE LIBRARY MUSEUM AND VISITOR CENTER, 312 Main St., Hickman, KY 42050.
Institution Type/Description: Historic Building Museum: built in 1906. Listed on the National Register of Historic Places.
Collections: local history & culture; river & rail transportation history; photographs; period furnishings; Civil War artifacts.
Hours & Admission Prices: Daily 9-4.

Highland Heights

MUSEUM OF ANTHROPOLOGY, (M), University Drive, Northern Kentucky Univ., 216 Landrum Academic Center, Highland Heights, KY 41099. Tel.: 859-572-1569. Fax: 859-572-5566.
E-mail: voelkerjl@nku.edu
Web Site: anthropologymuseum.nku.edu
Founded: 1976.
Congressional District: 6
Key Personnel: Dir., Dr. Judy Voelker.
Personnel Profile: Part-Time Paid 2.
Governing Authority: Northern Kentucky University. Tax-exempt.
Institution Type/Description: Anthropology Museum.
Collections: Ohio Valley archaeological materials; contemporary West Africa; contemporary Southwestern Indians; contemporary Southeastern Indians; contemporary Northwest Coast; Huichol Indians; New Guinea; Mexico; Latin America; hominid fossil casts; comparative vertebrate collection.
Research Fields: Ohio Valley archeology.
Facilities: 200-vol. library of archaeological reports, research manuals, journals available for research by special arrangement; field research station; separate laboratory operation; classrooms.
Activities: lectures; films; slide shows; demonstrations; guided tours; gallery talks; formally organized education programs for school & citizen groups; museum methods course for college students; loan, permanent & temporary exhibitions.
Publications: reports; contract archaeology.
Hours & Admission Prices: By appointment only. No charge. Closed university holidays; Christmas. &
Attendance: 800 (estimated)

NORTHERN KENTUCKY UNIVERSITY ART GALLERIES, Nunn Dr., Highland Heights, KY 41099. Tel.: 859-572-5148 & 5421. Fax: 859-572-6501.
E-mail: knight@nku.edu
Web Site: www.nku.edu/~art/galleries.html
Founded: 1968.
Congressional District: 4
Key Personnel: Dir., David J. Knight.
Personnel Profile: Full-Time Paid 1; Interns 3.
Governing Authority: public university; nonprofit. Parent Institution: State of Kentucky. Tax-exempt.
Institution Type/Description: Art Museum.
Collections: concentration on regional & national artists and varied media.
Facilities: library of various art exhibition catalogues; 2,700 sq. ft. exhibit space.
Activities: lectures; participatory & traveling exhibits.
Hours & Admission Prices: mid-Jan. to mid-Dec. Mon.-Fri. 9-9, other hours by appointment. No charge; donations accepted. Closed legal holidays; spring break. &
Attendance: 18,000 (estimated)

Hodgenville

ABRAHAM LINCOLN BIRTHPLACE NATIONAL HISTORI-CAL PARK, 2995 Lincoln Farm Rd., Hodgenville, KY 42748-9707. Tel.: 270-358-3137. Fax: 270-358-3874. Facebook: Lincoln Birthplace NPS.
E-mail: abli_superintendent@nps.gov
Web Site: www.nps.gov/abli
Founded: 1916.
Congressional District: 2
Key Personnel: Supt., William Justice.
Personnel Profile: Full-Time Paid 7; Full-Time Volunteers 2; Part-Time Paid 3; Part-Time Volunteers 3; Interns 1.
Governing Authority: federal. Parent Institution: National Park Service, Department of the Interior.
Institution Type/Description: Historic Site: birthplace of Abraham Lincoln.
Collections: Lincoln family Bible; tools & household objects relating to pioneer living of the early 19th century; graphics. Historic Buildings: c.1911 memorial building; c.1840 cabin.
Research Fields: early life of Lincoln & the environment of his birth.
Facilities: 300-vol. library of books on Abraham Lincoln, Kentucky history & the Civil War available for use on application to the Superintendent; 80-seat auditorium.
Activities: self-guided tours; programs; films; permanent exhibitions; ranger guided tours.
Publications: folder covering basic park history.
Hours & Admission Prices: Memorial Day-Labor Day daily 8-6:45; Sept.-May daily 8-4:45. Visitor Center: call for hours. No charge; donations accepted. Closed New Year's Day; Thanksgiving; Christmas. &
Attendance: 200,000 (estimated)

THE LINCOLN MUSEUM, 66 Lincoln Sq., Hodgenville, KY 42748-1551. Tel.: 270-358-3163.
E-mail: abc@lincolnmuseum-ky.org
Web Site: www.lincolnmuseum-ky.org
Founded: 1988.
Congressional District: 22
Key Personnel: Dir., Iris LaRue; Museum Shop Mgr., Charlotte Blair.
Personnel Profile: Full-Time Paid 2; Part-Time Paid 3; Part-Time Volunteers 25.
Governing Authority: Tax-exempt.
Institution Type/Description: History Museum.
Collections: 12 dioramas depicting Lincoln's life; paintings; photographs.

Hours & Admission Prices: Mon.-Sat. 8:30-4:30, Sun. 12:30-4:30. Adults $3; discounts to AAA members; members no charge. &

Attendance: 30,000 (accurate)

Membership: Individual $100.

Hopkinsville

CHEROKEE TRAIL OF TEARS COMMEMORATIVE PARK & HERITAGE CENTER, 100 Trail of Tears Dr., Hopkinsville, KY 42240. Tel.: 270-886-7503.

Institution Type/Description: Cherokee Indian History Museum.

Collections: Cherokee Indian culture & history; personal artifacts; period furnishings; photographs.

Activities: special events; walking trails.

Hours & Admission Prices: Center: Tues.-Sat. 10-3.

PENNYROYAL AREA MUSEUM, (M), 217 E. Ninth St., Hopkinsville, KY 42240-3448. Mailing Address: P.O. Box 1093, Hopkinsville, KY 42241-1093. Tel.: 270-887-4270. Fax: 270-887-4271.

E-mail: pennyroyal.museum@gmail.com

Founded: 1975.

Congressional District: 10

Key Personnel: Dir., Donna K. Stone; Education & Program Dir., Janet Bravard.

Personnel Profile: Full-Time Paid 2; Part-Time Paid 2; Part-Time Volunteers 25.

Governing Authority: municipal. Parent Institution: City of Hopkinsville, KY. Tax-exempt: 501(c)(3).

Institution Type/Description: History Museum.

Collections: pioneer bedroom; household & farm tools; Black heritage objects; Mogul farm wagon; L & N railroad items; Railway Express Wagon; 1909 Model 10 Buick; 1926 Chevrolet Pumper Truck (fire); historic clothing; asst. room & business settings; Edgar Cayce artifacts; tobacco farming exhibit.

Research Fields: Kentuckian & Kentucky New Era Newspaper files.

Facilities: Locally-crafted items & Kentucky gift items for sale.

Activities: guided tours; lectures; formally organized education program for adults & undergraduate college students affiliated with Murray State University & Hopkinsville Community College, permanent & temporary exhibitions; school loan service.

Publications: booklet, Building of a Monument; Pennyroyal Rambler; Pictorial History of Hopkinsville & Christian County, Gateway II; Edgar Cayce's Hometown; Hopkinsville Postcard History; assortment of local genealogical research books, as available.

Hours & Admission Prices: Tues.-Fri. 9-5, Sat. 10-3. Adults $2, senior citizens & children $1; discounts for prearranged group tours & AAM members; members & prearranged museum professionals no charge. Closed New Year's Day; Memorial Day; Independence Day; Labor Day; Thanksgiving; Christmas. &

Attendance: 17,000 (accurate)

Membership: Individual $15; Senior Individual $12; Senior Couple $20; Family $25; Contributing & Sustaining $100; Lifetime $1,000.

Horse Cave

AMERICAN CAVE MUSEUM & HIDDEN RIVER CAVE, 119 E. Main St., Horse Cave, KY 42749-1112. Mailing Address: P.O. Box 409, Horse Cave, KY 42749-0409. Tel.: 270-786-1466. Fax: 270-786-1467.

E-mail: acca@cavern.org

Web Site: cavern.org

Founded: 1977.

Key Personnel: Exec. Dir., David G. Foster; Administrative Asst., Shannon L. Johnson; Mktg. & Community Outreach, Peggy A. Nims

Institution Type/Description: Natural Science Museum.

Collections: prehistoric cave explorers; modern cave exploration; cave lighting; ground water in America; Cave County living; history of Mammoth & Horse Cave; cave wars; story of Floyd Collins.

Hours & Admission Prices: Memorial Day-Labor Day daily 9-7; Sept.-May daily 9-5. Adults 16 & over $15, youth 12-15 $10, children 3-11 $7; active military no charge. Closed New Year's Eve & Day; Thanksgiving; Christmas Eve & Day. &

Membership: Student $15; Regular $25; Family $35; Supporter $50; Sustainer $100; Guarantor $200; Benefactor $500; Patron $1,000.

KENTUCKY DOWN UNDER, 3700 L & N. Tpke. Rd., Horse Cave, KY 42749. Mailing Address: P.O. Box 189, Horse Cave, KY 42749-0189. Tel.: 800-762-2869; 270-786-2634.

E-mail: info@kdu.com

Web Site: www.kdu.com

Key Personnel: Dir., Judy Austin; Dir. Mktg., Melissa McGuire; Animal Crew Mgr., April Hatcher; Bookkeeping, Vicki Fancher; Gift Shop Mgr., Courtney Eaton

Institution Type/Description: Animal Park.

Collections: animals from Australia & the U.S. including Red Kangaroo, Emu, Kookaburra, Blue Tongue Skink, Papuan Frogmouth, Rainbow Lorikeet.

Hours & Admission Prices: Kentucky Caverns: April-May 26 daily 8-5; May 27-Aug. 11 daily 8-6; Aug. 12-March daily 9-4. Animal Areas: March 12-March 31 & Aug. 12-Oct. daily 9-4; April-May 26 daily 8-5; May 27-Aug. 11 daily 8-6. Adults $22, seniors 62 & over $19.30, children 5-14 $13; discounts to AAA members; children 4 & under and active military with ID no charge.

Irvington

IRVINGTON DEPOT AND MUSEUM, 243 N. First St., Irvington, KY 40146. Mailing Address: P.O. Box 22, Irvington, KY 40146. Tel.: 270-547-3835.

Institution Type/Description: Historic Building: built c.1920.

Collections: local history & culture; period furnishings; personal artifacts; photographs.

Hours & Admission Prices: Call for hours.

Jackson

BREATHITT COUNTY MUSEUM, INC., 329 Broadway St., Jackson, KY 41339-1040. Tel.: 606-666-4159.

E-mail: breathittmuseum@bellsouth.net

Web Site: breathittmuseum.com

Founded: 1980.

Congressional District: 5

Key Personnel: Dir., Janie Griffith; Chm. & Pres. (V), Grace Warrix.

Personnel Profile: Full-Time Paid 1.

Governing Authority: Tax-exempt.

Institution Type/Description: Appalachian History Museum.

Collections: county history & culture; photographs; personal artifacts; farming; mining; food preparation; textiles; logging; education; military.

Activities: classes.

Hours & Admission Prices: Mon., Wed. & Fri. 9-3:30. &

Jeffersontown

JEFFERSONTOWN HISTORICAL MUSEUM, 10635 Watterson Tr., Jeffersontown, KY 40299-3850. Tel.: 502-261-8290.

E-mail: bethw@jeffersontownky.com

Web Site: www.jeffersontownky.gov

Founded: 1997.

Key Personnel: Dir., Beth Wilder; Asst. Cur. Education, Ester Harlow.

Personnel Profile: Full-Time Paid 2.

Governing Authority: city.

Institution Type/Description: History Museum.

Collections: artifacts & photographs from the early times in Jeffersontown; over 1,500 folk dolls.

Major Exhibits: That's My J-Town, 1/21/14-6/20/14; 1920s Jeffersontown, 7/14-12/19/14.

Hours & Admission Prices: Mon.-Fri. 10-5. No charge. Closed holidays. &

Attendance: 5,000 (accurate)

Knifley

THE GILES SOCIETY, 6112 Elkhorn Rd., Knifley, KY 42753. Mailing Address: 380 Spout Springs Rd., Knifley, KY 42753. Tel.: 270-849-8803. Fax: 270-849-0547.

E-mail: keysha.tucker@yahoo.com

Web Site: www.gilessociety.org

Founded: 1996.

Congressional District: 1

Key Personnel: Pres. (V), Keysha Tucker; Treas., Gayla Baker.

Personnel Profile: Full-Time Volunteers 1; Part-Time Volunteers 12.

Governing Authority: private; nonprofit organization.

Institution Type/Description: Historical Society Museum: housed in the former home of author Janice Holt Giles and her husband Henry Giles. Listed on the National Register of Historic Homes.

Collections: Giles' furnishings including Janice's desk & typewriter; books written by Janice Holt Giles; personal artifacts.

Facilities: library.

Activities: arts festivals; guided tours; readings by budding Kentucky authors; pottery workshops for children. Museum Sponsors: plant exchange.

Publications: newsletter 3 times annually.
Hours & Admission Prices: June-Sept. Sat.-Sun. 1-5. No charge; donations accepted.
Attendance: 700 (estimated)
Membership: Basic $30; Friends $31-$50; 40 Acres and No Mule $51-$100; Shady Grove $101-$250; Around Our House $251-$500; A Little Better Than Plumb $501-$1,000; The Enduring Hills $1,001-$5,000; The Believers $5,001 & up.

La Grange

OLDHAM COUNTY HISTORY CENTER, (M), 106 N. Second Ave., La Grange, KY 40031-1102. Tel.: 502-222-0826. Fax: 502-222-7115.
Web Site: www.oldhamcountyhistoricalsociety.org
Founded: 1959.
Congressional District: 4
Key Personnel: Exec. Dir., Nancy Theiss; Pres., Robert Martin.
Personnel Profile: Full-Time Paid 1; Part-Time Paid 4; Part-Time Volunteers 87.
Governing Authority: Tax-exempt.
Institution Type/Description: History Museum & Center.
Collections: documents, books & maps pertaining to Oldham County; videos; interactive exhibits.
Facilities: archives.
Activities: lectures; school & family programs; archaeology programs.
Publications: Oldham County Kentucky: The First Century 1824-1924; Life at the River's Edge; History by Ford: Recipes and Stories From Oldham County Families; Graves and Cemeteries: Oldham County.
Hours & Admission Prices: Tues.-Sat. 10-4. No charge; donations accepted. Closed legal holidays. &
Attendance: 22,000 (accurate)
Membership: Individual $25; Family $50; Donor $100; Friend $250; Patron $500; Director's Circle $1,000.

Lancaster

GARRARD COUNTY JAIL MUSEUM, 103 Stanford St., Lancaster, KY 40444. Tel.: 859-792-9452.
E-mail: garrardjail@yahoo.com
Web Site: www.garrardcounty.ky.gov
Institution Type/Description: Historic Building: county jail built in 1873. Listed on the National Register of Historic Places.
Collections: local history; jail cell; jailers living quarters; solitary confinement chamber; inmate history; military artifacts.
Hours & Admission Prices: Call for hours.

GOVERNOR WILLIAM OWSLEY HOUSE - "PLEASANT RETREAT", 656 Stanford Rd., Lancaster, KY 40444-9543. Mailing Address: P.O. Box 571, Lancaster, KY 40444-0571. Tel.: 859-792-2500.
E-mail: director@owsleyhouse.org
Web Site: owsleyhouse.org/public/Welcome.html
Key Personnel: Exec. Dir., Travis Rose
Institution Type/Description: Historic House Museum: housed in the home of Kentucky's 16th Governor.
Collections: Owsley family history; period furnishings; portraits; quilts; personal artifacts.
Activities: educational programs.
Hours & Admission Prices: Tues.-Sun. 10-5.

Leitchfield

GRAYSON COUNTY HISTORICAL SOCIETY, 122 E. Main St., Leitchfield, KY 42754. Mailing Address: P.O. Box 84, Leitchfield, KY 42755. Tel.: 270-230-8989.
Founded: 1976.
Key Personnel: Pres. (V), Ken Robinson.
Personnel Profile: Full-Time Paid 1; Part-Time Volunteers 10.
Governing Authority: Subsidiary Institutions: Walnut Grove School; Edwards School. Tax-exempt.
Institution Type/Description: Historical Society Museum: housed in the Jack Thomas House.
Collections: local history & culture; period furnishings; photographs; personal artifacts. Historic Building: Buchanan log cabin.
Facilities: library.
Activities: research.
Publications: newsletter, 3 times a year.
Hours & Admission Prices: Tues.-Fri. 10-4. No charge; donations accepted. &

Attendance: 160 (estimated)
Membership: Annual $15.

Lexington

AMERICAN SADDLEBRED MUSEUM, 4083 Iron Works Pkwy., Lexington, KY 40511-8401. Tel.: 859-259-2746. Fax: 859-255-4909.
E-mail: museum@asbmuseum.org
Web Site: www.asbmuseum.org
Founded: 1962.
Key Personnel: Exec. Dir., Tolley Graves; C.E.O. & Pres. (V), Thomas E. Erffmeyer; Museum Shop Mgr., Lynn Morris.
Personnel Profile: Full-Time Paid 3; Part-Time Paid 7; Part-Time Volunteers 25; Interns 2.
Governing Authority: nonprofit organization. Tax-exempt: 501(c)(3).
Institution Type/Description: Natural History Museum.
Collections: art, artifacts & development of the American Saddlebred horse; paintings; trophies; saddles; photographs; memorabilia; tack.
Major Exhibits: Oak Hill Farm - A Jewel of the Bluegrass, 2/14/14-1/15.
Research Fields: American Saddlebred Horse; registered horse breed.
Facilities: 3,000-vol. library. Museum-related items for sale.
Activities: permanent special & interactive exhibitions; multi-image theater presentation; videos.
Publications: Commitment.
Hours & Admission Prices: Memorial Day-Labor Day daily 9-6; Sept.-Oct. & mid-March to Memorial Day daily 9-5; Nov. to mid-March Wed.-Sun. 9-5. Adults $16 (includes Kentucky Horse Park); discounts to groups, Kroger Plus, military, AAM, AAA & AARP members; children under 6 & members no charge. &
Attendance: 30,000 (accurate)
Membership: Youth $10; Young Adults $20; Individual $35; Family $50; Contributing $125; Business/Farm/Association $150; Dual Contributing $200; Life $1,000.

* **THE ART MUSEUM AT THE UNIVERSITY OF KENTUCKY, (M),** 405 Rose St., Lexington, KY 40506-0241. Tel.: 859-257-5716. Fax: 859-323-1994. Facebook: Art Museum UK.
E-mail: artmuseum@uky.edu
Web Site: finearts.uky.edu/art-museum
Formerly: University of Kentucky Art Museum
Founded: 1976.
Congressional District: 6
Key Personnel: Interim Dir. & Dir. Grants & Assets, Amy Nelson; Dir. Education, Deborah Borrowdale-Cox; Registrar, Barbara Lovejoy; Preparator, Alan Rideout; Cur., Janie Welker; Mktg. & Membership, Lyndi Van Deursen; Publications & Public Rels., Dorothy Freeman; Head Gallery Attendant, Judith Brin.
Personnel Profile: Full-Time Paid 8; Part-Time Paid 9; Part-Time Volunteers 35; Interns 6.
Governing Authority: state. Parent Institution: University of Kentucky, Rose St., Lexington 40506. Tax-exempt.
Institution Type/Description: Art Center.
Collections: 14th- to 20th-century painting, sculpture, works of art on paper; decorative arts; pre-Columbian, African & Asian art; photography; regional art.
Major Exhibits: Wide Angle: American Photography from the Collection, 1/6/14-4/27/14; Cross Fire Contemporary: Wood Fired Ceramics, 1/25/14-4/12/15; Wells Fargo Exhibition (Landscapes), 5/18/14-8/17/14; Conversations (Pairs from the Permanent Collection), 9/7/14-12/23/14.
Research Fields: related to collections.
Activities: special loan exhibitions; educational activities related to collections & exhibitions.
Publications: newsletter, Educational Materials; exhibition catalogs.
Hours & Admission Prices: Tues.-Thurs. & Sat.-Sun. 12-5, Fri. 12-8. Adults $8; discounts to AAM, ICOM & NARM members; members, students and UK alumni, faculty & staff no charge. Closed university holidays. &
Attendance: 27,300 (estimated)
Membership: Faculty, Teacher, & Staff $30; Individual $50; Family $60; Supporting $100; Patron $250; Sustaining $500; Benefactor $1,000.

ASHLAND, THE HENRY CLAY ESTATE, 120 Sycamore Rd., Lexington, KY 40502-1842. Tel.: 859-266-8581. Fax: 859-268-7266. Facebook: Ashland, The Henry Clay Estate.
E-mail: tjordan@henryclay.org
Web Site: www.henryclay.org
Founded: 1926.
Congressional District: 6
Key Personnel: Exec. Dir., Timothy W. Jordan; Pres., Nancy I. Bishop; Cur., Eric Brooks; Museum Shop Mgr., Judy Ogger.

Personnel Profile: Full-Time Paid 4; Part-Time Paid 7; Part-Time Volunteers 100.
Volunteer Hours: 4,078
Governing Authority: nonprofit organization. Parent Institution: Henry Clay Memorial Foundation. Tax-exempt: 501(c)(3).
Institution Type/Description: Historic House: c.1806 estate of Henry Clay; c.1856 house of James Clay.
Collections: family furnishings.
Facilities: 17 acres of greenspace; 1/2 acre formal garden.
Activities: guided tours; lectures; videos; special events.
Publications: newsletter & Web site.
Hours & Admission Prices: Feb. by appointment; March-Dec. Tues.-Sat. 10-4, Sun. 1-4. Adults $10, children $4; discounts to AAA members & groups; members no charge. Closed major holidays. &
Attendance: 12,584 (accurate)
Membership: Single $50; Family $75; Donor $250; Contributing $100; Supporting $500; Sustaining $1,000.

AVIATION MUSEUM OF KENTUCKY INC., 4029 Airport Rd., Blue Grass Airport, Lexington, KY 40510-9682. Tel.: 859-231-1219. Fax: 859-381-8739.
E-mail: info@aviationky.org
Web Site: aviationky.org
Formerly: Kentucky Aviation History Round Table
Founded: 1995.
Congressional District: 6
Key Personnel: Chm. Bd., Jim McCormick; Pres., Jerry Van Der Meer, Jr., M.D.; Archivist, Dennis Sparks; Museum Shop Mgr., Paula Fey.
Personnel Profile: Full-Time Paid 1; Full-Time Volunteers 50; Part-Time Volunteers 60.
Governing Authority: private; nonprofit. Subsidiary Institution: The Kentucky Aviation Hall of Fame. Tax-exempt: 501(c)(3).
Institution Type/Description: Aviation Museum.
Collections: early Kentucky aviation history; space & exploration; general aviation history; helicopter; Cessna 150; Navy A-4; Marine F-4; Air Force T-38; WACO RNF Biplane; Air Force RF-101; Navy F-14.
Research Fields: Navy F-14 Tomcat; Army Cobra helicopter.
Facilities: 1,000-vol. library; 75-seat auditorium; field research station. Museum-related items for sale.
Activities: docent program; formal education programs for children; guided tours; hobby workshops; lectures; loan, temporary & traveling exhibitions; participatory aircraft exhibits; flight simulator; TV & radio programs. Museum Sponsors: Aviation Camps for children 10-15 in June & July.
Publications: monthly newsletter, AMK News.
Hours & Admission Prices: Tues.-Sat. 10-5, Sun. 1-5. Adults $7, senior citizens $6, students $5; discounts to AAM members & groups. &
Attendance: 45,000 (estimated)
Membership: Student $15; Individual $30; Family $40.

BODLEY-BULLOCK HOUSE, 200 Market St., Lexington, KY 40507-1030. Tel.: 859-259-1266.
Institution Type/Description: History Museum: built c.1814 for Lexington Mayor Thomas Pindell, shortly after sold to General Thomas Bodley, a veteran of the War of 1812. Served as headquarters for both Union and Confederate forces during the Civil War.
Collections: personal artifacts; furnishings.
Facilities: garden.
Activities: tours.
Hours & Admission Prices: By appointment. Closed holidays.

EXPLORIUM OF LEXINGTON, 440 W. Short St., Lexington, KY 40507-1206. Tel.: 859-258-3253. Fax: 859-258-3255.
E-mail: explore@explorium.com
Web Site: www.explorium.com
Formerly: Lexington Children's Museum
Founded: 1990.
Congressional District: 6
Key Personnel: Exec. Dir., Lee Ellen Martin; Office Mgr., Brittany Plassmann; Programming Coord. & Cur., Catie Richwine; Community Outreach Coord., Morgan Brotherton; Education Coord., Melissa Brown; Visitor Service Coord., Jon Ricker.
Personnel Profile: Full-Time Paid 5; Part-Time Paid 8; Part-Time Volunteers 5; Interns 3.
Governing Authority: nonprofit organization. Tax-exempt: 501(c)(3).
Institution Type/Description: Children's Museum: located in Victorian Square.
Collections: hands-on exhibits.
Facilities: 20,000 sq. ft. exhibit space; multi-use room: changing exhibits; fabrication shop. Museum-related items for sale.

Activities: participatory exhibits; programs; demonstrations; workshops; camps; Museum Sponsors: Parent's Night Out.
Publications: e-newsletter; social media; brochures; education pamphlets.
Hours & Admission Prices: Tues., Thurs.-Sat. 10-5, Wed. 10-8, Sun. 1-5; Mon. Memorial Day-Labor Day 10-5. Admission $8; discounts to senior citizens and active duty military; members & children under one no charge. Closed Easter; Tues.-Fri. the week after Labor Day; Thanksgiving; Christmas Eve & Day. &
Attendance: 51,000 (accurate)
Membership: Teacher $30; Grandparent $70; Family $90; Family Plus $125.

HEADLEY-WHITNEY MUSEUM, 4435 Old Frankfort Pike, Lexington, KY 40510-9657. Tel.: 859-255-6653. Fax: 859-255-8375.
E-mail: hwmuseum@headley-whitney.org
Web Site: www.headley-whitney.org
Founded: 1968.
Congressional District: 6
Key Personnel: Dir. & Cur., Amy Gundrum; Cur. Education, Lauren Hunter-Smith.
Personnel Profile: Full-Time Paid 2; Part-Time Paid 4; Part-Time Volunteers 8; Interns 2.
Governing Authority: nonprofit organization. Tax-exempt: 501(c)(3).
Institution Type/Description: Decorative Arts Museum.
Collections: glass & metalwork; Kentucky-made silver; jewelry & bibelots of founder/designer George W. Headley.
Facilities: 1,500-vol. library of Decorative art books, periodicals & catalogues; reading room; picnic area; 8 acres. Museum-related items for sale.
Activities: changing temporary exhibitions; lectures; concerts; education, travel, tour & volunteer programs; special cultural events; school outreach programs; Symposia; outreach programming.
Publications: book, The Headley Treasure of Bibelots & Boxes; newsletter, The Polished Jewel; exhibition catalogues.
Hours & Admission Prices: March-Dec. Wed.-Fri. 10-5, Sat.-Sun. 12-5. Adults $10, seniors & students $7; discounts for AAM, ICOM, AAA, KMHA & SEMC members; members no charge. Blue Star Program. Closed federal holidays. &
Attendance: 15,000 (estimated)
Membership: Individual $55; Family $85; Patron $150; Benefactor $400; Directors Circle $2,500.

HUNT-MORGAN HOUSE, 201 N. Mill St., Lexington, KY 40507-1034. Mailing Address: 253 Market St., Lexington, KY 40507-1031. Tel.: 859-253-0362. Fax: 859-259-9210.
E-mail: info@bluegrasstrust.org
Web Site: bluegrasstrust.org
Founded: 1955.
Key Personnel: Dir., Sheila Ferrell; Museum Shop Mgr., Jason Sloan.
Governing Authority: nonprofit organization. Parent Institution: Blue Grass Trust. Tax-exempt: 501(c)(3).
Institution Type/Description: Historic House: housed in a c.1814 Federal-style home located in Gratz Park historic district.
Collections: period furniture; decorative objects; portraits.
Activities: guided tours; docent program.
Hours & Admission Prices: March to mid-Dec. Wed.-Fri. & Sun. 1-5, Sat. 10-4. Adults $7, students & children 3 & up $4; members no charge. Closed major holidays.
Attendance: 5,000 (estimated)
Membership: Individual $50; Family $100; Hunt-Morgan Society $250; Clay Lancaster Society $500; Carolyn Reading Hammer Society $1,000.

INTERNATIONAL MUSEUM OF THE HORSE, (M), 4089 Iron Works Pkwy., Lexington, KY 40511-8483. Tel.: 859-259-4232. Fax: 859-225-4613. TDD: 859-233-4303.
E-mail: info@kyhorsepark.com
Web Site: www.imh.org
Founded: 1978.
Congressional District: 6
Key Personnel: Dir., Bill Cooke; Art Dir., Gina Gibson; Cur. Collections, Amy Beisel; Cur. Asst., Melissa Williams; Asst. Dir., Travis Robinson.
Personnel Profile: Full-Time Paid 6; Part-Time Paid 1; Part-Time Volunteers 25; Interns 1.
Governing Authority: Commonwealth of Kentucky. Parent Institution: Kentucky Horse Park, 4089 Iron Works Pkwy., Lexington 40511. Tax-exempt.
Institution Type/Description: Equine Museum.
Collections: equine history collection covering all breeds worldwide.
Research Fields: equine history; Kentucky history.
Facilities: library of equine history available for in-house research during museum operating hours; reading room.

Activities: lectures; films; gallery tours; education program; docent programs; inter-museum loans; education packets for primary & secondary levels.
Hours & Admission Prices: March 15-Oct. daily 9-5; Nov.-March 14 Wed.-Sun. 9-5. Adults $16, children 7-12 $8; discounts to AAM, SEMC, AASLH & KMHA members; children 6 & under no charge. &

Attendance: 200,000 (estimated)

KENTUCKY PRO FOOTBALL HALL OF FAME, 3364 Leestown Rd., Lexington, KY 40511. Tel.: 859-276-3488.
E-mail: info@kyprofootballhof.com
Web Site: www. kyprofootballhof.org
Institution Type/Description: Sports Museum.
Collections: sports history & memorabilia; photographs; personal artifacts.
Hours & Admission Prices: Mon.-Fri. 9-4. No charge.

THE LEXINGTON CEMETERY, 833 W. Main St., Lexington, KY 40508-2094. Tel.: 859-255-5522. Fax: 859-258-2774.
Web Site: www.lexcem.org
Founded: 1849.
Key Personnel: Pres. & Gen. Mgr., Daniel R. Scalf; Asst. Gen. Mgr., Mark C. Durbin.
Personnel Profile: Full-Time Paid 25; Part-Time Paid 6.
Governing Authority: private, nonprofit, non-sectarian corporation. Tax-exempt: 501(c)(13).
Institution Type/Description: Union & Confederate Civil War Cemetery.
Collections: cemetery monuments from mid-1800's to present; Henry Clay Memorial.
Research Fields: genealogy; horticulture.
Facilities: botanical garden; arboretum; files open to public for genealogy research.
Activities: self-guided tours.
Publications: brochure, Self-Guided Historical Walking Tour; children's brochure, Self Guided Tree Tour.
Hours & Admission Prices: Cemetery: daily 8-5. Office: Mon.-Fri. 8-4, Sat. 8am-12pm. No charge. &

LEXINGTON HISTORY MUSEUM, 215 W. Main St., Lexington, KY 40507. Tel.: 859-254-0530.
E-mail: info@lexingtonhistorymuseum.org
Web Site: www.lexingtonhistorymuseum.org
Founded: 1999.
Congressional District: 6
Key Personnel: Pres. & C.E.O., J.K. Millard; Chm. (V), Foster Ockerman, Jr.; Vice Chm., Stephen G. Amato; Treas., William Ambrose; Museum Shop Mgr., Paige Prewitt.
Personnel Profile: Full-Time Paid 2; Part-Time Volunteers 42; Interns 2.
Governing Authority: private; nonprofit organization. Parent Institution: Lexington History Museum, Inc. Tax-exempt.
Institution Type/Description: History Museum.
Collections: historical artifacts related to Lexington & the Bluegrass area from 1792 to present.
Facilities: 100-vol. library; rental facilities; 160-seat auditorium; 8,000 sq. ft. exhibit space. Museum-related items for sale.
Activities: docent program; films; formal education programs for adults; guided tours; hobby workshops; lectures; temporary exhibitions; rental facilities; family activities.
Publications: quarterly newspaper, The Bluegrass Historian.
Hours & Admission Prices: Temporarily closed. &
Attendance: 10,299 (accurate)
Membership: Student $10; Individual $38; Family $50; Weathervane Directors $100; Lafayette Circle $200; Phoenix Friends $500; Thomas D. Clark Society $1,000.

THE LIVING ARTS AND SCIENCE CENTER, INC., 362 N. Martin Luther King Blvd., Lexington, KY 40508-1889. Tel.: 859-252-5222 & 255-2284. Fax: 859-255-7448.
E-mail: info@lasclex.org
Web Site: www.lasclex.org
Founded: 1968.
Congressional District: 8
Key Personnel: Pres. (V), Yajaira Aich; Dir., Heather Lyons; Coord. Educational Outreach, Katherine Bullock; Cur., Elaine Quave; Art Education, Molly Wilson; Bookkeeper, LeAnn Jenkins.
Personnel Profile: Full-Time Paid 6; Part-Time Paid 33; Part-Time Volunteers 10.
Governing Authority: nonprofit organization. Affiliated with Lexington Arts & Cultural Council. Tax-exempt: 501(c)(3) & 170(b)(1)(A).
Institution Type/Description: Children's Art and Science Center.

Collections: art & science exhibits; native Kentucky plants; butterfly garden.
Facilities: 600-vol. library of books on art, environmental & natural science & education; classrooms.
Activities: art & science classes for children; outreach programs; loan, temporary & traveling exhibitions; teachers workshops; lectures; festivals; special education classes for underserved populations, at risk youth & special needs populations; traveling science kits; traveling Starlab Planetarium; field trip programs for K-12 in the arts & sciences.
Publications: exhibition brochures; teacher reference materials; quarterly schedule of classes; quarterly newsletter & program schedule, Imagine That!
Hours & Admission Prices: Academic Year: Mon.-Fri. 8:30-5, Sat. 10-2; Summer: Mon.-Fri. 8-5. No charge; donations accepted. Closed New Year's Day; Independence Day; Thanksgiving; Christmas. &
Attendance: 25,000 (estimated)
Membership: Individual $30; Family $50; Friend $100; Supporter $250; Patron $500.

MARY TODD LINCOLN HOUSE, 578 W. Main, Lexington, KY 40507-1642. Mailing Address: P.O. Box 132, Lexington, KY 40588-0132. Tel.: 859-233-9999. Fax: 859-252-2269.
E-mail: mtlhouse@windstream.net
Web Site: www.mtlhouse.org
Founded: 1968.
Congressional District: 6
Key Personnel: Dir., Gwen Thompson; Museum Shop Mgr., Glenna Holloway.
Personnel Profile: Full-Time Paid 1; Part-Time Paid 8; Part-Time Volunteers 5.
Governing Authority: nonprofit organization. Parent Institution: Kentucky Mansions Preservation Foundation, Inc., 578 W. Main, Lexington, KY 40588. Tax-exempt.
Institution Type/Description: Historical & Preservation Society Museum: housed in 1803-1806 Inn occupied by Todd family from 1832-1849.
Collections: Lincoln book collection; rare old leather bound books; personal articles of Mary Todd Lincoln & the Lincoln family; julep cup collections; furnishings & furniture of the period; family portraits.
Activities: guided tours.
Publications: The Courier Newsletter.
Hours & Admission Prices: March 15-Nov. Mon.-Sat. 10-4; last tour at 3pm. Adults $10, children 6-12 $4; tours of 15 or more $8 per person; discounts to AAA & KMPF members. Closed holidays. &
Attendance: 8,500 (estimated)
Membership: Individual $25; Family $40; Voting Membership $125; Corporate $1,000.

PHOTOGRAPHIC ARCHIVES, University of Kentucky, Special Collections & Archives, M.I. King Library, 104 A King Bldg., Lexington, KY 40506. Tel.: 859-257-2654. Fax: 859-257-6311.
E-mail: jasonf@.uky.edu
Web Site: www.uky.edu/libraries/special/av
Founded: 1978.
Congressional District: 6
Key Personnel: Photographic Archivist, Jason Flahardy.
Governing Authority: university. Parent Institution: University of Kentucky. Tax-exempt: 170(b)(1)(A).
Institution Type/Description: Library of Photographic Material.
Collections: 100,000 photographs documenting the history of photography and the history of Kentucky, Appalachia and surrounding areas; manuscript collections.
Research Fields: history of photography, photographers; general history of city, state, region.
Facilities: 80,000-vol. library; manuscript materials available for qualified researchers upon request; reading room; 80-seat auditorium.
Activities: lectures; concerts; loan & temporary exhibitions.
Publications: exhibition catalogs.
Hours & Admission Prices: Mon.-Fri. 8-5. No charge. Closed legal & academic holidays. &
Attendance: 2,000

TRANSYLVANIA MUSEUM, 300 N. Broadway, Lexington, KY 40508-1797. Tel.: 859-233-8229. Fax: 859-233-8171.
E-mail: jday@transy.edu
Web Site: www.transy.edu/homepages/museum
Founded: 1802.
Key Personnel: Dir., Dr. James Day.
Governing Authority: university. Affiliated with the Transylvania University. Tax-exempt: 101.
Institution Type/Description: Science Museum: housed in 1833 Greek Revival University Bldg.

Collections: early 19th-century scientific apparatus; equipment used in teaching 19th-century medicine; mounted bird skins. Historic Buildings: 1780 Patterson Cabin; 1833 Old Morrison.
Research Fields: early scientific instruments; natural history.
Activities: guided tours; formally organized education programs for students.
Hours & Admission Prices: By appointment. No charge. Closed national holidays.
Attendance: 150

WAVELAND STATE HISTORIC SITE, 225 Waveland Museum Lane, Lexington, KY 40514-1618. Tel.: 859-272-3611. Fax: 859-272-4834.
E-mail: ron.bryant@ky.gov
Web Site: www.parks.ky.gov
Founded: 1957.
Congressional District: 6
Key Personnel: Park Mgr., Ron D. Bryant.
Personnel Profile: Full-Time Paid 2; Part-Time Paid 3.
Governing Authority: state. Parent Institution: Kentucky Department of Parks, Capital Plaza Tower, Frankfort, KY 40601. Tax-exempt.
Institution Type/Description: Historic Site & House: 1847 Greek Revival Mansion.
Collections: period furniture; portraits; china. Historic Buildings: servants' quarters with fully equipped fireplace kitchen; smokehouse; ice house.
Facilities: playground; picnic area.
Activities: guided tours; lectures; formally organized education programs for children; permanent & temporary exhibitions.
Hours & Admission Prices: Mon.-Sat. 10-5, Sun. 1-5. Adults $7, seniors $6, students $4; discount to groups; children under 6 no charge.
Attendance: 44,366 (accurate)

WHEELER MUSEUM, Kentucky Horse Park, 3870 Cigar Lane, Lexington, KY 40511. Tel.: 859-225-6704. Fax: 859-258-9033.
Web Site: www.ushja.org
Founded: 2009.
Governing Authority: Parent Institution: USHJA. Tax-exempt.
Institution Type/Description: Equestrian History Museum.
Collections: hunter jumper industry history; trophies; ribbons; equestrian equipment; photographs.
Hours & Admission Prices: Call for hours. No charge; donations accepted.
Attendance: 550 (estimated)

London

MOUNTAIN LIFE MUSEUM, Levi Jackson Park, 998 Levi Jackson Mill Rd., London, KY 40744-8325. Tel.: 606-330-2130. Fax: 606-330-2123. TDD: 606-330-2130.
Web Site: www.kystateparks.com/agencies/parks/levijack.htm
Founded: 1929.
Congressional District: 5
Key Personnel: Dir., Recreational Parks, Museums and Shrines & Park Supt., Ben Sizemore; Museum Supt., Ella Goodin.
Personnel Profile: Full-Time Paid 1; Part-Time Paid 3.
Governing Authority: state. Parent Institution: Kentucky Department of Parks, Capital Plaza Tower, Frankfort, KY 40601. Tax-exempt.
Institution Type/Description: History Museum.
Collections: archives; manuscripts; costumes; decorative arts; folklore; glass; history; Indian artifacts; preservation project; textiles; transportation; primarily 1860s era farming tools & household implements.
Facilities: Mountain-made crafts for sale.
Activities: guided tours; permanent exhibitions.
Hours & Admission Prices: April-Oct. call for hours. Adults $3.50, children $2.50; discounts to groups; children 2 & under no charge.
Attendance: 10,967 (accurate)

Louisa

FRED M. VINSON MUSEUM AND WELCOME CENTER, 315 E. Madison St., S.E., Corner Court House Sq., Louisa, KY 41230. Tel.: 606-638-0078.
Institution Type/Description: Historic House Museum: housed in the former home of Frederick Moore Vinson, 13th Chief Justice of the U.S. Supreme Court.
Collections: Fred Vinson's life & career; personal artifacts; photographs; period furnishings.
Hours & Admission Prices: Call for hours.

Louisville

ALLEN R. HITE ART INSTITUTE, (M), Univ. of Louisville, Belknap Campus, Schneider Hall, Louisville, KY 40292. Tel.: 502-852-4483. Fax: 502-852-6791.
E-mail: john.begley@louisville.edu
Web Site: www.art.louisville.edu
Founded: 1946.
Congressional District: 3
Key Personnel: Professor, Dir. & Chm., James Grubola; Professor Emeritus, William Morgan; Professor Emeritus, Donald R. Anderson; Professor Emeritus, Henry J. Chodkowski; Professor Emeritus, Suzanne Mitchell; Professor Emeritus, Stephanie J. Maloney; Professor, John Whitesell; Asst. Professor, Che Rhodes; Assoc. Professor, Jay M. Kloner; Assoc. Professor, Linda Gigante; Professor, Lida Gordon; Assoc. Professor, Barbara Hanger; Professor, Steven Skaggs; Professor, Ying Kit Chan; Dir. Speed Art Museum, Peter Morrin; Assoc. Professor, Mark Priest; Asst. Professor, Mary Carothers; Asst. Professor, Mitch Eckert; Assoc. Professor, Stow Chapman; Assoc. Professor, Moon-He Baik; Gallery Dir. & Adjunct Professor, John Begley; Asst. Professor, Benjamin Hufbauer; Asst. Professor, Christopher Fulton; Art Librarian, Gail Gilbert; Professor Emeritus, Dario A. Covi; Assoc. Professor Emerita, Nancy L. Pearcy; Asst. Professor, Todd Burns; Asst. Professor, Scott Massey; Unit Business Mgr., Linda Rowley; Facilities Coord., Wesley Kent; Designer-in-Residence, Leslie Friesen; Asst. Professor, Susan Jarosi; Asst. Professor, Delin Lai.
Personnel Profile: Full-Time Paid 27; Part-Time Paid 16; Part-Time Volunteers 10.
Governing Authority: Parent Institution: University of Louisville. Subsidiary Institution: Hite Art Institute; Cressman Center for Visual Art, 1st St. at Main, Louisville. Tax-exempt: 101(6).
Institution Type/Description: Art Institute.
Collections: University Art Collection; 15th- to 20th-century European & American prints, drawings & paintings; objects focused on Louisville area.
Facilities: 80,000-vol. library of art books available for inter-library loan; non-circulating reference library; 350,000 slide visual resources collection.
Activities: lectures; formally organized education programs for undergraduate & graduate college students; temporary exhibitions.
Publications: exhibition catalogues; graduate art history student journal, Parnassus.
Hours & Admission Prices: Mon.-Fri. 9-4:30, Sat. 10-2, Sun. 1-6. No charge; donations accepted. Closed national holidays.
Attendance: 48,000 (estimated)

THE BRENNAN HOUSE HISTORIC HOME AND GARDENS, 631 S. Fifth St., Louisville, KY 40202. Tel.: 502-540-5145.
E-mail: director@thebrennanhouse.org
Web Site: thebrennanhouse.org
Governing Authority: nonprofit organization. Tax-exempt: 501(c)(3).
Institution Type/Description: Historic House Museum: housed in the former home of inventor Thomas Brennan and his wife, Anna Bruce; built in 1868. Listed on the National Register of Historic Places.
Collections: Brennan family history; personal artifacts; period furnishings; photographs; family portraits; books; medical office of son, Dr. J.A.O. Brennan including furnishings.
Facilities: banquet facilities.
Activities: slide show; educational programs; rental facilities.
Hours & Admission Prices: Call for hours.

CONRAD-CALDWELL HOUSE MUSEUM, 1402 St. James Ct., Louisville, KY 40208-2127. Tel.: 502-636-5023.
E-mail: conradcaldwellhouse@gmail.com
Web Site: conrad-caldwell.org
Founded: 1987.
Congressional District: 3
Key Personnel: Exec. Dir., Allison Wroblewski; Pres., John Crum.
Personnel Profile: Full-Time Paid 1; Part-Time Paid 2; Part-Time Volunteers 10.
Governing Authority: Parent Institution: St. James Court Historical Foundation, Inc. Tax-exempt.
Institution Type/Description: Historic House Museum.
Collections: period furnishings; stained glass; personal artifacts.
Research Fields: late Victorian art, architecture & culture.
Facilities: two rental halls.
Activities: tours.
Hours & Admission Prices: Sun. & Wed.-Fri. 12-4, Sat. 10-4; other times by appointment. Adults $10, senior citizens $6, students $4.
Attendance: 5,000 (accurate)
Membership: Docent $15; Student $20; Individual $35; Family $50; Donor $100; Benefactor $500; Corporate $1,500.

FARMINGTON HISTORIC PLANTATION, 3033 Bardstown Rd., Louisville, KY 40205-3019. Tel.: 502-452-9920. Fax: 502-456-1976.
E-mail: farmington@historichomes.org
Web Site: www.historichomes.org/farmington
Founded: 1958.
Congressional District: 3
Key Personnel: Exec. Dir., Andrea Pridham; Regent, Davis Boland; Treas., Amanda McWane.
Personnel Profile: Full-Time Paid 3; Part-Time Volunteers 75.
Governing Authority: nonprofit organization. Parent Institution: Historic Homes Foundation. Tax-exempt: 501(c)(3).
Institution Type/Description: Historic Site: house built in 1816.
Collections: 1830 period furnishings; Lincoln artifacts.
Research Fields: Speed family history; Jeffersonian architecture; African-American slave community & oral history; 19th-century social history.
Facilities: visitor's center. Crafts & reproductions of decorative pieces for sale.
Activities: guided tours; education programs. Annual Events: Harvest Festival; Derby Breakfast; An Evening in the Garden; Home for the Holidays.
Publications: membership newsletter; Calendar of Special Events.
Hours & Admission Prices: Tues.-Sat. 10-4:30. Adults $9, senior citizens 60 & up and military $8, students 6-18 $4; discounts to groups with reservation & AAA members; children under 6 & members no charge. Closed New Year's Day; Easter; Mother's Day; Father's Day; Kentucky Derby Day; Independence Day; Thanksgiving; Christmas Eve & Day.
Attendance: 14,000 (estimated)
Membership: Individual $35; Family $50; Friend $100; Donor $250; Patron $500; Benefactor $1,000.

THE FILSON HISTORICAL SOCIETY, INC., (M), 1310 S. Third St., Louisville, KY 40208-5506. Tel.: 502-635-5083. Fax: 502-635-5086.
E-mail: markweth@filsonhistorical.org
Web Site: www.filsonhistorical.org
Formerly: Filson Club Historical Society, Inc.
Founded: 1884.
Congressional District: 3
Key Personnel: C.E.O. & Dir., Mark V. Wetherington; Pres. (V), J. Mcauley Brown.
Personnel Profile: Full-Time Paid 15; Part-Time Paid 5; Part-Time Volunteers 5; Interns 2.
Governing Authority: nonprofit. Parent Institution: The Filson Historical Society, Inc. Tax-exempt: 501(c)(3).
Institution Type/Description: History Museum.
Collections: Kentucky, Ohio Valley & Upper South history; original manuscripts; landscapes; 1.8 million manuscripts; 50,000 photographs & prints; genealogical materials; printed family histories; local business records; 400 portraits including artists Matthew Harris Jouett, Chester Harding, John Wesley Jarvis, Joseph H. Bush, G.P.A. Healy, John J. Audubon, George Caleb Bingham & Nicola Marschall; frontier, Lewis and Clark expedition; antebellum; Civil War; sculpture.
Major Exhibits: United We Stand Divided We Fall: The Civil War in the Ohio Valley Region, 1/12-12/14.
Research Fields: history of Kentucky, Ohio River Valley region & the Upper South; genealogy.
Facilities: 50,000-vol. library of books, pamphlets, maps, 1.8 million manuscripts, 50,000 photos & prints, 400 portraits, newspapers & microfilm available for use on premises.
Activities: lectures; exhibit openings; permanent & temporary exhibitions; self-guided tours; educational programs for school groups;
Publications: journal, Ohio Valley History; members' magazine, The Filson; book, Dear Brother, Letters of William Clark to Jonathan Clark; museum guides.
Hours & Admission Prices: House & Museum: Mon.-Fri. 9-5. No charge. Library: Mon.-Fri. 9-5, 1st Sat. of month 9-4. Adults $10; discount to members. Special Collections: Mon.-Fri. 9-5, 1st Sat. of month 9-4. Adults $10; discount to members. Closed national holidays. ♿
Attendance: 28,000 (estimated)
Membership: Student $25; Individual $50; Boone $100; Audubon $250; Clay $500; Shelby $1,000; Clark $2,500; Filson $5,000.

FRAZIER HISTORY MUSEUM, (M), 829 W. Main St., Louisville, KY 40202-2619. Tel.: 502-753-5663; 866-886-7103. Fax: 502-412-8148.
E-mail: info@fraziermuseum.org
Web Site: www.fraziermuseum.org
Formerly: Frazier International History Museum
Founded: 2004.
Key Personnel: Exec. Dir., Madeleine H. Burnside, Ph.D.; Chm. (V), Owsley Brown Frazier; Vice Pres. Operations & Finance, Craig S. Mooney; Dir. Public Rels. & Mktg., Krista McHone
Governing Authority: Tax-exempt.
Institution Type/Description: History Museum.
Collections: arms; armor; personal artifacts; history of Armouries.
Facilities: 120-seat auditorium; 48-seat theater; cafe; rooftop garden; education center. Museum-related items for sale.
Activities: audio tour; group tours; educational field trips.
Publications: newsletter, Compass.
Hours & Admission Prices: Mon.-Sat. 9-5, Sun. 12-5. Adults $9.50, seniors 60 & over $7.50, children under 14 & students $6; discounts to AAM members; children under 5 no charge. Audio Tour: additional $3. Closed Thanksgiving; Christmas.
Attendance: 93,955 (accurate)
Membership: Individual $50; Dual $65; Family & Grandparents $75; Contributing $100; Sustaining $250; Patron $500; The Legend's Society $1,000.

GHEENS SCIENCE HALL AND RAUCH PLANETARIUM, Univ. of Louisville, Belknap Campus, Louisville, KY 40292. Mailing Address: Rauch Planetarium, University of Louisville, Louisville, KY 40292. Tel.: 502-852-6664 & 7597. Fax: 502-852-0831.
E-mail: planet@louisville.edu
Web Site: www.louisville.edu/planetarium
Founded: 1962.
Congressional District: 3
Key Personnel: Dir., Rachel Connolly; Technical Coord., Drew Foster; Mktg., Dorothy J. Vittitow; Devel., Paula Campbell; Operations Mgr., Paula McGuffey.
Personnel Profile: Full-Time Paid 3; Part-Time Paid 2.
Governing Authority: state. Affiliated with University of Louisville. Tax-exempt: 501(c)(3).
Institution Type/Description: Planetarium.
Collections: astronomy; space science.
Facilities: Books & gadgets pertaining to astronomy for sale.
Activities: lectures; films; formally organized education programs for children, adults & undergraduate college students.
Publications: quarterly newsletter.
Hours & Admission Prices: Tues.-Sat. call for hours. Adults $7, children, seniors & students $5; members no charge. ♿
Attendance: 91,000 (accurate)
Membership: Student & Senior $35; Individual $45; Family $75.

HISTORIC HOMES FOUNDATION, INC., 3110 Lexington Rd., Louisville, KY 40206-3002. Tel.: 502-899-5079. Fax: 502-899-5016.
Web Site: www.historichomes.org
Founded: 1957.
Congressional District: 3
Key Personnel: Pres. HHF, Butch Shaw; Dir. Farmington Historic Plantation, Diane Carman-Young; Dir. Whitehall House & Gardens, Merrill Simmons; Bookkeeper, Jodi Skees.
Personnel Profile: Full-Time Paid 4; Part-Time Paid 5; Part-Time Volunteers 120; Interns 2.
Governing Authority: nonprofit organization. Subsidiary Institutions: Farmington, 3033 Bardstown Rd., Louisville, KY 40205; Thomas Edison House, 729 E. Washington St., Louisville 40202; Whitehall, 3110 Lexington Rd., Louisville. Tax-exempt: 501(c)(3).
Institution Type/Description: Historic Houses.
Collections: Kentucky & Southern made furnishings. Historic Houses: 1810 Farmington; 1850s Thomas Edison house; 1860s Whitehall.
Research Fields: archaeology; local history; archaeology; historic landscapes; Abraham Lincoln; Thomas Edison.
Facilities: classrooms; meeting rooms; facilities for private parties. Gift items for sale.
Activities: guided tours; lectures; arts festivals; antique shows; formally organized education programs for adults & children; docent program or council; temporary exhibitions; garden tours. Foundation Sponsors: Derby Breakfast.
Publications: quarterly, newsletter; calendar.
Hours & Admission Prices: Farmington: Tues.-Sat. 10-4:30, Sun. 1:30-4:30; last tour 3:45. Adults $9, senior citizens $8, students $4; children 5 & under no charge. Whitehall Mon.-Fri. 10-2. Adults $5, senior citizens $4; children 5 & under no charge. Thomas Edison House: Tues.-Sat. 10-2 or by appointment. Adults $5, senior citizens $4, students $3; children under 5 no charge. Group Tours: minimum 25. $3 per person. Closed New Year's Day; Easter; Derby Day; Thanksgiving; Christmas Eve & Day. ♿
Attendance: 29,000 (estimated)

Membership: Individual $35; Family $50; Patron $60; Friend $100; Benefactor $200; Donor $250.

KENTUCKY DERBY MUSEUM, (M), 704 Central Ave., Gate 1, Churchill Downs, Louisville, KY 40208-1212. Tel.: 502-637-1111. Fax: 502-636-5855.
E-mail: info@derbymuseum.org
Web Site: www.derbymuseum.org
Founded: 1985.
Congressional District: 3
Key Personnel: Exec. Dir., Lynn Ashton; Pres. & Chm. (V), Brooks Bower; Dir. Finance, Jim Tougher; C.O.O., Sherry Crose; Dir. Retail, Katie Stephenson; Dir. Facilities, Dan Shomer; Museum Educator, Ronnie Dreistadt; Dir. Public Rels., Wendy Trienan; Membership Coord., Carla Grego; Dir. Klein Family Learning Center, Sherry Stanley.
Personnel Profile: Full-Time Paid 34; Part-Time Paid 34; Part-Time Volunteers 65.
Governing Authority: nonprofit organization. Tax-exempt: 501(c)(3).
Institution Type/Description: Thoroughbred Racing & Equine Art, History & Science: located at Gate 1 of Churchill Downs, a National Historic Landmark.
Collections: Thoroughbred racing & equine artifacts; memorabilia; photographs; archives; artwork; trophies.
Major Exhibits: Student Art Show, 1/14-3/14; Horseplay, 4/14-12/14.
Research Fields: Thoroughbred racing with emphasis on the Kentucky Derby.
Facilities: research library of books on horses; multimedia presentation facility; restaurant. Gift items for sale.
Activities: guided tours; formally organized education programs; permanent & changing exhibitions.
Publications: newsletter, The Inside Track; Churchill Downs, America's Most Legendary Racetrack.
Hours & Admission Prices: March 15- Nov. Mon.-Sat. 8-5, Sun. 11-5; Dec.-March 14 Mon.-Sat. 9-5, Sun. 11-5. Adults $14, senior citizens $13, children 3-11 $5; discounts to AAM members; members & children under 5 no charge. Insiders Tour: $10. Backside Track Tours: $10. Closed Oaks Day; Derby Day; Thanksgiving; Christmas. &
Attendance: 215,000 (accurate)
Membership: Call to Post $45; First Turn $125; Backstretch $250; Finish Line $500-$750; Farm $500-$1,000; Aristides $1,000-$2,000; Corporate: Oaks $750; Derby $1,250; Winners Circle $2,000.

KENTUCKY MUSEUM OF ART AND CRAFT, (M), 715 West Main, Louisville, KY 40202-2633. Tel.: 502-589-0102. Fax: 502-589-0154.
E-mail: admin@kentuckyarts.org
Web Site: www.kentuckyarts.org
Formerly: Kentucky Museum of Arts & Design
Founded: 1981.
Congressional District: 3
Key Personnel: Chm. (V), Marlene M. Grissom; Pres. (V), Mary Stone; Dir., Kevin O'Brien; Deputy Dir. & Cur., Brion Clinkingbeard; Museum Shop Mgr., David McGuire.
Personnel Profile: Full-Time Paid 17; Part-Time Paid 3; Part-Time Volunteers 12; Interns 3.
Governing Authority: nonprofit organization. Tax-exempt: 501(c)(3).
Institution Type/Description: Art & Craft Museum.
Collections: decorative arts.
Research Fields: contemporary crafts & folk art.
Facilities: educational workshop space. Crafts made by Kentucky artists for sale.
Activities: guided tours; lectures; arts festivals; loan, traveling & temporary exhibitions; annual craft conference; educational workshops.
Publications: annual newsletter; annual exhibition catalog.
Hours & Admission Prices: Mon.-Fri. 10-5, Sat. 11-5. Adults $5, seniors & military $4; members, students with ID & children under 12 no charge. Closed most public holidays; Derby Day. &
Attendance: 63,000 (accurate)
Membership: Individual $40; Family $75; Patron $150; Fellow $250; Benefactor $500.

∗ **LOCUST GROVE, (M),** 561 Blankenbaker Lane, Louisville, KY 40207-7100. Tel.: 502-897-9845. Fax: 502-897-0103.
E-mail: lghh@locustgrove.org
Web Site: www.locustgrove.org
Founded: 1964.
Congressional District: 3
Key Personnel: Dir., Carol Ely; Pres., Susan Reigler; Treas., Christopher Green; Programs, Mary Beth Williams; Museum Shop Mgr., Jennifer Jansen.

Personnel Profile: Full-Time Paid 4; Part-Time Paid 10; Part-Time Volunteers 140; Interns 5.
Operating Expenses: 405,134
Operating Income: 433,427
Governing Authority: Tax-exempt: 501(c)(3).
Institution Type/Description: Historic Site: housed in the former home of the Croghan family & General George Rogers Clark; built 1790s.
Collections: early Kentucky furniture; tools; maps; surveying equipment. Historic Buildings: mansion; outbuildings.
Research Fields: 1778-1849 Clark & Croghan family, furniture & accessories; early Kentucky & regional history and lifestyle; American Revolution in the West; slave life in the upper south.
Facilities: visitors center; gardens. Museum-related items for sale.
Activities: guided tours; lectures; concerts; pioneer skills workshops; organized education programs for adults & children; docent program; permanent exhibitions; weekend events; video presentations. Annual Event: 18th Century Market Fair.
Publications: newsletter; brochure; book, George Rogers Clark & Locust Grove.
Hours & Admission Prices: Mon.-Sat. 10-4:30, Sun. 1:30-4:30. Adults $8, seniors $7, children $4; discounts to National Trust, AAA, AASLH, ICOM & AAM members; members & children under 6 no charge. Closed New Year's Eve & Day; Easter; Derby Day; Thanksgiving; Christmas Eve & Day. &
Attendance: 26,619 (accurate)
Membership: Individual $35; Family $50.

LOUISVILLE FIRE DEPARTMENT LEARNING CENTER & MUSEUM, 3228 River Park Dr., Louisville, KY 40210. Tel.: 502-574-3731.
Institution Type/Description: Fire Fighting History Museum.
Collections: fire fighting history & equipment; hands-on exhibitions; uniforms; helmets; badges; tools.
Activities: educational programs; group tours.
Hours & Admission Prices: By appointment.

LOUISVILLE SCIENCE CENTER, 727 W. Main St., Louisville, KY 40202-2681. Tel.: 502-561-6100. Fax: 502-561-6145.
E-mail: lscinfo@louisvilleky.gov
Web Site: www.louisvillescience.org
Founded: 1872.
Congressional District: 3
Key Personnel: C.E.O. & Dir., JoAnna Haas; Chm. (V), Laurie Schalow; Mng. Dir. Visitor Experiences, Theresa Mattei; Museum Shop Mgr., Toph Bryant.
Personnel Profile: Full-Time Paid 38; Part-Time Paid 44; Part-Time Volunteers 886.
Governing Authority: private; nonprofit. Tax-exempt: 501(c)(3).
Institution Type/Description: Science & Technology Center: housed in 1878 five-building structure.
Collections: archaeology findings; geology items; prehistory items; history artifacts; mineralogy; natural history exhibits; science technology; space science artifacts; clothing, costume & seashell collections; electricity, sound, light & color displays; dental health displays; Egyptian mummy & related artifacts.
Research Fields: education; archaeology; geology; paleontology; zoology; history.
Facilities: 1,000-vol. reference library available for use in building only; 40,000 sq. ft. exhibit space; IMAX theatre; video conferencing room; classrooms; special function room; lunchroom; restaurant. Science & math kits, books & science equipment & other museum-related items for sale.
Activities: guided tours; lectures; IMAX films; gallery talks; workshops; formally organized education programs for school children; family, group & adult training programs; volunteer programs; teacher institutes; permanent, temporary & traveling exhibitions; outreach kits; assembly programs; camp-ins; educational videoconferences.
Publications: quarterly newsletter; membership & guide brochures; educator's guides; special program announcements; tourism & promotional brochures; annual report.
Hours & Admission Prices: Mon.-Thurs. & Sun. 9:30-5, Fri.-Sat. 9:30-9. Museum/IMAX: adults $15, senior citizens & children 2-12 $12. IMAX only: adults $8, senior & children 2-12 $7, members $6. Museum only: adults $12, senior citizens & children 2-12 $10; discounts to ASTC members, sponsored groups, school groups; tourists & visitors with qualifying coupons; members & children under 2 no charge. Closed Thanksgiving; Christmas Eve & Day. &
Attendance: 500,000 (estimated)
Membership: Individual $69; Household $75; Super Scientist $99; Dual $140. Underwriter Membership: Benefactor $150; Patron $250; Supporter $500; Advocate $1,000.

LOUISVILLE SLUGGER MUSEUM & FACTORY, 800 W. Main St., Louisville, KY 40202-2637. Tel.: 502-585-5226. Fax: 502-585-1179.
E-mail: museum@slugger.com
Web Site: www.sluggermuseum.com
Founded: 1996.
Key Personnel: Exec. Dir., Anne Jewell; Museum Shop Mgr. & Retail Dir., Whitney Pfister; Dir. Operations, Deana Lockman.
Personnel Profile: Full-Time Paid 12; Part-Time Paid 20.
Governing Authority: private corporation. Parent Institution: Hillerich & Bradsby Co., 800 W. Main St., Louisville, KY.
Institution Type/Description: Baseball Museum and Bat Factory.
Collections: valuable & rare baseball bats; baseball related artifacts.
Research Fields: baseball.
Facilities: 15,000 sq. ft. exhibit space; 87-seat large screen theatre. Museum-related items for sale.
Activities: films; education programs for children; guided tours; lectures; loan, participatory & traveling exhibitions; theatre; broadcast programs; batting cages; interactive exhibits.
Hours & Admission Prices: Mon.-Sat. 9-5, Sun. 12-5. Adults $10, senior citizens $9, children $5; discounts to groups & AAM members. Closed New Year's Day; Thanksgiving; Christmas Day. &
Attendance: 220,000 (accurate)

LOUISVILLE STONEWARE COMPANY MUSEUM, 731 Brent St., Louisville, KY 40204. Tel.: 800-626-1800.
Web Site: www.louisvillestoneware.com
Institution Type/Description: Company History Museum.
Collections: company history; hand painted & handmade pottery.
Activities: factory tour; paint your own pottery.
Hours & Admission Prices: Factory Tours: Mon.-Fri. 10:30-1:30; groups of 8 or more by appointment. Tours: adults $7, seniors 65 & over $6, students & children 6-12 $5; children 5 & under no charge. Tour & Paint Your Own Pottery: adults $30, seniors 65 & over $24, students & children 6-12 $23; children 5 & under no charge.

LOUISVILLE VISUAL ART ASSOCIATION, 609 W. Main St., 2nd Floor, Louisville, KY 40202. Tel.: 502-584-8166. Fax: 502-896-2148.
E-mail: shannon@louisvillevisualart.org
Web Site: www.louisvillevisualart.org
Founded: 1909.
Congressional District: 3
Key Personnel: Exec. Dir., Shannon Westerman; Pres. (V), Robert Hallenberg; Vice Pres., Anne O'Daniel; Treas., John Lewis; Education Coord., Jackie Pallesen; Sec., Rick Sneed.
Personnel Profile: Full-Time Paid 6; Part-Time Paid 4; Interns 4.
Governing Authority: nonprofit organization.
Institution Type/Description: Art Gallery: 1860 neo-classical building once housed public water systems first pumping station.
Collections: structures; contemporary art.
Research Fields: regional & national contemporary art.
Facilities: library of art publications, catalogues & magazines available to the public; classrooms.
Activities: tours; lectures; workshops; organized education programs for adults & children; internships with Louisville metro area institutions; loan & traveling exhibitions; slide registry & consultation service; temporary exhibitions.
Publications: newsletter; art auction calendar.
Hours & Admission Prices: Mon.-Fri. 9-5, Sun. 12-4. Adults $3, seniors & students $2; members & children under 12 no charge. Closed New Year's Day; Easter; Independence Day; Thanksgiving; Christmas Eve & Day. &
Attendance: 150,000 (accurate)
Membership: Student & Senior $25; Individual $40; Family $60; Friend $100; Patron $300; Benefactor $500.

* **LOUISVILLE ZOOLOGICAL GARDEN,** 1100 Trevilian Way, Louisville, KY 40213-1559 Mailing Address: P.O. Box 37250, Louisville, KY 40233-7250. Tel.: 502-459-2181. Fax: 502-459-2196.
Web Site: www.louisvillezoo.org
Founded: 1963.
Congressional District: 3
Key Personnel: Exec. Dir., John Walczak; Senior Veterinarian, Dr. Roy Burns, D.V.M.; Society Pres., Thomas P. O'Brien, III; Foundation Pres., Mark F. Wheeler; Asst. Dir., Mark Zoeller; General Cur., Steven Wing; Dir. Communications, Deborah R. Sebree; Business. Mgr., Carol Miller; Mktg.

Dir., Maureen Horrigan; Dir. Devel., Jill Gorsky; Education Cur., Marcelle Gianelloni; Museum Shop Mgr., Kathy Kline.
Personnel Profile: Full-Time Paid 135; Part-Time Paid 150; Part-Time Volunteers 1,427; Interns 3.
Governing Authority: municipal; city. Tax-exempt: 501(c)(3).
Institution Type/Description: Zoo & Botanical Gardens.
Collections: 1,300 animals; birds; mammals; reptiles; amphibians; invertebrates; fish; exotic environmental habitats; plants; gardens; ornamental displays.
Research Fields: raptor rehabilitation program; exotic bird breeding; behavior studies with South American woolly monkeys; pioneered first successful transfer of an exotic equine, Grant's Zebra; embryo to a surrogate mother (quarter horse); artificial insemination research; Cuban crocodile; Timber rattlesnake; Black-footed ferret breeding program 1991-present; breeding & husbandry of other exotic species; animal enrichment & behavioral training programs.
Facilities: 135 acres; HerpAquarium; zoo education center; restaurant. Museum-related items for sale.
Activities: interactive exhibits; lectures; festivals; organized education programs; docent programs; community presentations; special promotional events; seasonal scenic train & tram rides; permanent exhibitions; indoor & outdoor viewing; discovery trail; researcher's station; carousel; miniature train; formal education programs: in-house & outreach throughout the state; Backyard Action Hero & youth incentive.
Publications: quarterly newsletter, Trunkline; member e-mails; map inserts; Backyard Action Hero Guidebooks; education guides.
Hours & Admission Prices: March-Labor Day daily 10-5; Sept.-Feb. daily 10-4. Adults 12-59 $11.95, senior citizens 60 & over and children 3-11 $8.95; discounts to AZA, reciprocal Zoos & Aquaria members; children under 3 & members no charge. Closed New Year's Day; Thanksgiving; Christmas. &
Attendance: 810,546 (accurate)
Membership: Individual $45; Family $75; Keeper $95; Family Dual $140; Curator $150; Keeper Dual $180; Conservation Partner $350; Nature's Guardian $500; Wildlife Champion $1,000.

MUHAMMAD ALI CENTER, 144 N. Sixth St., Louisville, KY 40202-2939. Tel.: 502-584-9254. Fax: 502-589-4905.
E-mail: info@alicenter.org
Web Site: www.alicenter.org
Founded: 1997.
Congressional District: 3
Key Personnel: Pres. & C.E.O., Donald Lassere; Sr. Dir. Public Rels & External Affairs, Jeanie Kahnke.
Governing Authority: Tax-exempt.
Institution Type/Description: History Museum.
Collections: Muhammad Ali's life & history; photographs; personal artifacts; poetry; drawings.
Facilities: library & archives; auditorium; classrooms; rental facility; theater. Museum-related items for sale.
Activities: education outreach; public programs & global initiatives.
Publications: membership newsletter, In Your Corner.
Hours & Admission Prices: Tues.-Sat. 9:30-5, Sun. 12-5. Adults $9, seniors 65 & over $8, military & students $5, children 6-12 $4; discounts to groups; members & children 5 & under no charge. Closed New Year's Eve & Day; Easter; Memorial Day; Independence Day; Columbus Day; Thanksgiving; Christmas Eve & Day. &
Attendance: 135,000 (accurate)
Membership: Student & Senior 65 & over $25; Adult $45; Family $75.

MUSEUM OF THE AMERICAN PRINTING HOUSE FOR THE BLIND, (M), 1839 Frankfort Ave., Louisville, KY 40206-3148. Tel.: 502-895-2405, ext. 365; 800-223-1839, ext. 365. Fax: 502-899-2363.
E-mail: museum@aph.org
Web Site: www.aph.org
Formerly: Callahan Museum
Founded: 1994.
Congressional District: 3
Key Personnel: Dir., Micheal A. Hudson; Mgr. Collections, Anne Rich.
Personnel Profile: Full-Time Paid 2; Part-Time Paid 2; Part-Time Volunteers 1.
Governing Authority: Parent Institution: American Printing House for the Blind. Tax-exempt.
Institution Type/Description: History Museum.
Collections: artifacts & interpretive materials related to the history of the education of blind & visually impaired people; history of the American Printing House for the Blind (founded 1858); historical tactile books, tactile writing & printing equipment, education aids for visually impaired.
Major Exhibits: Child in a Strange Country: Helen Keller & The History of Education for People who are Blind or Visually Impaired (T), 1/13-1/18.

Research Fields: blindness education; printing history.
Facilities: 4,600 sq. ft. exhibition space; reception area; orientation room.
Activities: guided tours; interactive & hands-on exhibits; audio; print, and Braille descriptions; traveling exhibits; education programs.
Publications: brochures: Museum; a history of the American printing house for the blind.
Hours & Admission Prices: Mon.-Fri. 8:30-4:30, Sat. 10-3. Plant Tours: Mon.-Thurs. 10 & 2. No charge; donations accepted. Closed New Year's Day; Derby Day; Memorial Day; Independence Day; Labor Day; Thanksgiving; Christmas. &
Attendance: 5,000 (accurate)

NATIONAL SOCIETY OF THE SONS OF THE AMERICAN REVOLUTION, (M), 1000 S. Fourth St., Louisville, KY 40203-3292. Tel.: 502-589-1776. Fax: 502-589-1671.
Web Site: www.sar.org
Founded: 1889.
Key Personnel: Exec. Dir., Joe Harris.
Personnel Profile: Full-Time Paid 15; Part-Time Paid 2; Part-Time Volunteers 50.
Governing Authority: nonprofit organization. Parent Institution: National Society SAR. Subsidiary Institution: State Societies & chapters; SAR Genealogical Research Library, 809 W. Main St., Louisville, KY 40202. Tax-exempt: 501(c)(3).
Institution Type/Description: National Historical Society.
Collections: 18th-century memorabilia relating to the American Revolutionary War; 18th-century fine arts, prints, engravings, paintings, statuary; manuscripts.
Research Fields: American Revolutionary War; genealogy; fine arts relating to the 18th century.
Facilities: 58,000-vol. library of genealogy & history pertaining to the Revolutionary War available for researcher or student on premises only; reading room. Items relating to history & S.A.R. memorabilia for sale.
Activities: guided tours; lectures; films; gallery talks; formally organized education programs for children & adults; docent programs; permanent, temporary & traveling exhibitions.
Publications: quarterly, The S.A.R. Magazine.
Hours & Admission Prices: Museum & Library: Mon.-Fri. 9:30-4:30. Museum: No charge. Library Research: $5 per day; SAR & DAR members no charge. Closed New Year's Eve & Day; Independence Day; Thanksgiving; Christmas Eve & Day; national holidays. &
Attendance: 5,000 (estimated)
Membership: Friends of the Library $25.

THE NICOL & EISENBERG ARCHAEOLOGICAL COLLECTION, Southern Baptist Theological Seminary, 2825 Lexington Rd., Louisville, KY 40280. Tel.: 502-897-4039. Fax: 502-897-4036.
E-mail: campinfo@sbts.edu
Web Site: www.sbts.edu
Formerly: The Joseph A. Callaway Archaeological Museum
Founded: 1963.
Key Personnel: Librarian, Bruce Keisling.
Governing Authority: college. Parent Institution: Southern Baptist Seminary. Tax-exempt: 501(c)(3) & 170(b)(1)(A).
Institution Type/Description: Biblical Archaeology Museum.
Collections: archaeology; glass; textiles; sculpture; excavation materials from Jericho; Ai; Raddana; Machaerus; numismatics; pottery; papyri; Egyptian mummy.
Research Fields: archaeology; glass; textiles; sculpture; numismatics.
Facilities: 200,000-vol. library available for inter-library loan; reading room.
Activities: guided tours; films.
Hours & Admission Prices: Closed indefinitely. &
Attendance: 7,000 (estimated)

PORTLAND MUSEUM, 2308 Portland Ave., Louisville, KY 40212-1036. Tel.: 502-776-7678. Fax: 502-776-9874.
E-mail: pmuse@iglou.com
Web Site: www.goportland.org
Founded: 1978.
Congressional District: 3
Key Personnel: Exec. Dir., Nathalie Taft Andrews; Pres., Christian Trabue; Chm. (V), Sally Craven; Asst. to Dir & Museum Shop Mgr., Jessica Dawkins.
Personnel Profile: Full-Time Paid 3; Part-Time Paid 1; Part-Time Volunteers 25; Interns 1.
Governing Authority: nonprofit organization. Tax-exempt: 501(c)(3).
Institution Type/Description: General Museum: housed in 1850 Beach Grove residence of William Skene.

Collections: local, historic & contemporary artifacts; children's photography; family home movie archive; historic photographs & documents related to Portland neighborhood.
Research Fields: local history; maritime history of the Western Waters; women's & family history.
Facilities: 10,000 sq. ft. exhibit space; 20-seat theater; educational facilities. Readers, workbooks, handmade crafts, hand-printed books & other gift items for sale.
Activities: guided tours; lectures; films; concerts; theater; broadcast programs; organized education programs for children & adults; school loan service; temporary exhibitions.
Hours & Admission Prices: Tues.-Fri. 10-4:30. Adults $7, senior citizens $6, students 6 & over $5; discounts to AAA & AAM members; children 5 & under no charge. &
Attendance: 5,000 (estimated)

RIVERSIDE, THE FARNSLEY-MOREMEN LANDING, 7410 Moorman Rd., Louisville, KY 40272-4572. Tel.: 502-935-6809. Fax: 502-935-6821.
E-mail: info@riverside-landing.org
Web Site: www.riverside-landing.org
Founded: 1993.
Congressional District: 3
Key Personnel: Dir. & Mgr., Patti Linn; Chm. (V), Reba Doutrick; Museum Shop Mgr., Heather French.
Personnel Profile: Full-Time Paid 4; Part-Time Paid 3; Part-Time Volunteers 60.
Governing Authority: private; nonprofit organization. Tax-exempt: 501(c)(3).
Institution Type/Description: Historic House Museum.
Collections: Farnsley & Moremen families personal artifacts; 19th century Kentucky farm life; historic farm life on the Ohio River. Historic Building: 1837 Farnsley Moremen House.
Research Fields: Louisville, KY historical interiors & decorative arts, 1840s & 1880s.
Facilities: 250-vol. library; 200-seat auditorium; educational facilities; 3,000 sq. ft. exhibit space; 300 acre historic farm. Museum-related items for sale.
Activities: concerts; docent program; formal education programs; guided tours; lectures; temporary exhibitions. Annual Events: Ice Cream Social; Riverside Heritage Festival; Plant & Herb Sale; A Riverside Christmas.
Publications: quarterly newsletter, Riverside Review.
Hours & Admission Prices: Tues.-Sat. 10-4:30, Sun. 1-4:30. Family $15, adults $6, senior citizens $5, students & children $3; members no charge. Closed major holidays. &
Attendance: 25,000 (estimated)
Membership: Individual $20; Family $35.

SPECIAL COLLECTIONS, UNIVERSITY OF LOUISVILLE LIBRARIES, Ekstrom Library, University of Louisville, 2301 S. 3rd St., Louisville, KY 40292. Tel.: 502-852-6752. Fax: 502-852-8734.
E-mail: special.collections@louisville.edu
Web Site: special.library.louisville.edu
Founded: 1967.
Congressional District: 3
Key Personnel: Dir. University Archives & Records Center, William J. Carner; Assoc. Cur., Amy Purcell; Cur. Rare Books, Delinda Buie.
Personnel Profile: Full-Time Paid 4; Part-Time Paid 2; Part-Time Volunteers 4.
Governing Authority: university. Parent Institution: University of Louisville. Affiliated with University of Louisville Libraries, Louisville, KY 40292. Tax-exempt.
Institution Type/Description: Historic Research Institute.
Collections: 1,500,000 photographs; small equipment collection; fine art print collection; photographically illustrated books; manuscripts; fine prints.
Research Fields: history of photography; documentary photography; photography as a fine art; history of Louisville, KY.
Facilities: 75,000-vol. library; reading room; reference service.
Activities: lectures; temporary & traveling exhibitions.
Publications: book, For Love of Learning.
Hours & Admission Prices: Mon.-Fri. 9-5. No charge; donations accepted. &

＊　THE SPEED ART MUSEUM, (M), 2035 S. Third St., Louisville, KY 40208-1812. Tel.: 502-634-2700. Fax: 502-636-2899. TDD: 502-634-2706.
E-mail: info@speedmuseum.org
Web Site: www.speedmuseum.org
Founded: 1925.
Congressional District: 3
Key Personnel: Dir. & C.E.O., Charles L. Venable; C.O.O., Lisa Betson Resnik; Chm., Todd P. Lowe; Chief Cur., Ruth Cloudman; Registrar,

Charles Pittenger; Dir. Education, Cynthia Moreno; C.F.O. & Business Mgr., David C. Knopf; Dir. Visitor Experience, Mindy Johnson; Museum Shop Mgr., Kristina Griesshaber.
Personnel Profile: Full-Time Paid 53; Part-Time Paid 32; Part-Time Volunteers 229; Interns 12.
Governing Authority: nonprofit organization. Tax-exempt: 501(c)(3).
Institution Type/Description: Art Museum.
Collections: over 14,000 works from antiquity to the present of European & American decorative arts; paintings; sculpture; graphic arts; antiquities; African arts; Native American arts.
Research Fields: paintings; sculpture; graphics.
Facilities: 13,000-vol. art library; art learning center; workshop; lecture rooms; restaurant; 350-seat auditorium. Books, reproductions & museum-related items for sale.
Activities: guided tours; lectures; films; gallery talks; concerts; family fests; school programs; teacher in-services; education programs for children & adults; docent program; inter-museum loan, permanent, temporary & traveling exhibitions. Museum Sponsors: The Alliance, Collectors groups.
Publications: Exhibition Catalogs; quarterly members' program guide; handbook.
Hours & Admission Prices: 74452 &
Attendance: 74,452 (accurate)
Membership: Individual $50; Family & Dual $70; Reciprocal $130; Supporter $250. Patron Circle: Silver $500-$999; Gold $1,000-$2,499; Platinum $2,500-$4,999; Connoisseurs: Donor Circle $5,000-$9,999; Curator Circle $10,000-$24,999; Director Circle $25,000-$49,999; Artist Circle $50,000 & up. Corporate: Friends $1,000-$2,499; Trustees $2,500-$4,999; Benefactors $5,000-$7,499; Guarantors $7,500-$9,999; Leaders $10,000-$24,999; Season Sponsors $25,000 & up.

THOMAS EDISON HOUSE, 729-731 E. Washington St., Louisville, KY 40202-1050. Tel.: 502-585-5247. Fax: 502-585-5231.
E-mail: edisonhouse@historichomes.org
Web Site: www.historichomes.org
Founded: 1978.
Congressional District: 3
Key Personnel: Exec. Dir., Kristen Lutes.
Personnel Profile: Full-Time Paid 1; Part-Time Paid 1; Part-Time Volunteers 27; Interns 1.
Governing Authority: nonprofit organization. Parent Institution: Historic Homes Foundation, Inc. Tax-exempt.
Institution Type/Description: Historic House.
Collections: Edison's inventions, including phonographs, kinetoscope & bulbs.
Research Fields: history, science & communication.
Facilities: Gift items for sale.
Activities: guided tours; films, interactive experiences.
Publications: foundation newsletter, The Foundation.
Hours & Admission Prices: Tues.-Sat. 10-2, or by appointment. Adults $5, senior citizens $4, students $3; discounts to AAA members; children under 5, Historic Homes Foundation & members no charge (not including special events). Closed New Year's Day; Thanksgiving; Christmas Eve & Day. &
Attendance: 5,500 (estimated)
Membership: Individual $75; Family $100; Donor $250; Supporter $500; Benefactor $1,000.

21C MUSEUM, (M), 700 W. Main St., Louisville, KY 40202-2634. Tel.: 502-217-6300. Fax: 502-217-6347.
E-mail: agreystites@21cmuseum.org
Web Site: www.21cmuseum.org
Founded: 2005.
Key Personnel: Dir., Alice Gray Stites.
Personnel Profile: Full-Time Paid 5; Interns 1.
Institution Type/Description: Art Gallery.
Collections: works by regional, national & international contemporary artists.
Hours & Admission Prices: Call for hours. No charge. &

Madisonville

HISTORICAL SOCIETY OF HOPKINS COUNTY, 107 Union St., Madisonville, KY 42431-2529. Tel.: 270-821-3986.
Founded: 1974.
Congressional District: 1
Institution Type/Description: Historical Society Museum.
Collections: local history & culture; period furnishings; personal artifacts; photographs; books; newspapers. Historic Building: cabin where Gov. Ruby Laffoon was born.
Facilities: library.
Hours & Admission Prices: Mon.-Fri. 1-5.

Mammoth Cave

MAMMOTH CAVE NATIONAL PARK, U.S. Dept. Interior, 10 miles I65 exit 53, Mammoth Cave, KY 42259. Mailing Address: P.O. Box 7, Mammoth Cave, KY 42259-0007. Tel.: 270-758-2180 & 2181. Fax: 270-758-2447.
E-mail: maca_information@nps.gov
Web Site: www.nps.gov/maca/home.htm
Founded: 1941.
Congressional District: 2
Key Personnel: Supt., Sarah Craighead; Chm. (V), Eddie Wells; Education Coord., Cheryl Messenger.
Governing Authority: federal. National Park Service, Dept. of the Interior, Washington, DC 20240. Tel.: 202-343-4621. Tax-exempt.
Institution Type/Description: National Park: 52,000+ acre park with over 390 miles of underground passageways, some exposed for public viewing.
Collections: natural & cultural history collections with objects related to the prehistory, history & biological diversity of the park.
Research Fields: hydrology; geology; biology; archaeology; paleontology; ethnology; history; speleology; ecology; geomorphology; meteorology.
Facilities: 550-vol. library pertaining to natural history available for research on premises; visitor center; field research station; 200-seat cafeteria & restaurant. Books & other museum related items for sale.
Activities: guided tours; formally organized education programs for children; hiking; underground cave tours.
Publications: brochures.
Hours & Admission Prices: March 12 to mid-June & Sept.-Oct. daily 8-6; mid-June to mid-Aug. daily 8-7; Nov.-March 11 8:45-5. Tours 1 to 6 hours long depending on time of year, call for information on reservations & fees at 877-444-6777 or www.recreation.gov. Discounts to children 6-12, groups of 12 or more, school groups, Golden Age/Access card holders. Closed Christmas.
Attendance: 1,800,000 (estimated)
Membership: Friends: Student $15; Troglobite $25; Caver $50; Lantern $100; Big Woods $250; Echo River $500; Mammoth $1,000.

Marion

BEN E. CLEMENT MINERAL MUSEUM, (M), 205 N. Walker St., Marion, KY 42064. Mailing Address: P.O. Box 391, Marion, KY 42064-0391. Tel.: 270-965-4263; 877-965-4263 (Toll Free). Facebook: Ben E. Clement Mineral Museum.
E-mail: beclement@att.net
Web Site: www.clementmineralmuseum.org
Founded: 1996.
Congressional District: 1
Key Personnel: Dir. & Museum Shop Mgr., Tina Walker; Chm. (V), Bill Frazer.
Personnel Profile: Part-Time Paid 2; Part-Time Volunteers 20.
Operating Expenses: 44,790
Operating Income: 53,000
Governing Authority: Tax-exempt.
Institution Type/Description: Mineral Museum.
Collections: local history; minerals; geology; mining equipment; historical pictures.
Activities: educational classes; mineral digs; scouting activities; group tours. Museum Sponsors: Annual Gem, Mineral, Fossil & Jewelry Show.
Hours & Admission Prices: June-Sept. Mon.-Sat. 10-3, Oct.-May Wed.-Sat. 10-3; other times by appointment. Adults $5. Closed New Year's Day; Thanksgiving; Christmas. &
Attendance: 4,000 (estimated)
Membership: $20 per person.

CRITTENDEN COUNTY HISTORICAL MUSEUM, 124 E. Bellville St., Marion, KY 42064-1410. Mailing Address: 124 E. Bellville St., P.O. Box 25, Marion, KY 42064. Tel.: 270-965-9257.
Founded: 1967.
Congressional District: 1
Governing Authority: Tax-exempt.
Institution Type/Description: Historic Building: housed in the first church in Marion, built in 1881.
Collections: local history & culture; military uniforms; 200 year old loom; spinning wheels; farm equipment; telephone switchboard; photographs; period clothing.
Hours & Admission Prices: April-Oct. Wed.-Sat. 10-3; other times by appointment. No charge.
Attendance: 800 (accurate)
Membership: Individual $10.

Mayfield

ICE HOUSE GALLERY, 120 N. 8th St., Mayfield, KY 42066-1804. Tel.: 270-247-6971.
E-mail: icehousearts@att.net
Web Site: www.icehousearts.org
Institution Type/Description: Art Gallery.
Collections: works by local & regional artists.
Facilities: classroom. Gallery-related items for sale.
Activities: educational programs; special events; workshops; temporary exhibitions.
Hours & Admission Prices: Tues.-Fri. 10-4:30, Sat. 10-1. No charge. &

Maysville

ALBERT SIDNEY JOHNSTON HOUSE, 503 S. Court St., Maysville, KY 41056. Tel.: 606-759-7411.
Institution Type/Description: Historic House Museum: housed in the birthplace of Confederate General Johnston.
Collections: General Johnston's life & military career; period furnishings; personal artifacts; photographs.
Hours & Admission Prices: Guided Tours: April-Nov. by appointment.

HARRIET BEECHER STOWE SLAVERY TO FREEDOM MUSEUM, 2124 Old Main St., Maysville, KY 41096. Mailing Address: P.O. Box 184, Old Washington, KY 41096. Tel.: 606-564-0250 & 9419.
E-mail: orloffgmiller@mac.com
Institution Type/Description: History Museum.
Collections: local history, heritage, & culture; slavery; Civil War artifacts; photographs; period furnishings; personal artifacts.
Activities: guided tours.
Hours & Admission Prices: Sat. 10-4, Sun. 12-4.

KENTUCKY GATEWAY MUSEUM CENTER, (M), 215 Sutton St., Maysville, KY 41056-1109. Tel.: 606-564-5865. Fax: 606-564-4372.
E-mail: museum@kygmc.org
Web Site: www.kentuckygatewaymuseumcenter.org
Formerly: Mason County Museum & Museum Center
Founded: 1878.
Congressional District: 4
Key Personnel: Registrar, Pat King; Dir., Dewey Applegate; Cur. Books, Art & Artifacts, Sue Ellen Grannis; Business Mgr., Gayle McKay; Mktg. & Sales Coord. & Museum Shop Mgr., Paula Ruble; Researcher, Caye Chamness; Researcher, Myra Hardy; Accounting, Joyce Weigott; Cur. Education, James Shires, Ph.D.; Reference Registrar, Anne Pollitt.
Personnel Profile: Full-Time Paid 2; Part-Time Paid 7; Part-Time Volunteers 60.
Governing Authority: nonprofit. Tax-exempt: 170(b)(1)(A).
Institution Type/Description: Art Gallery, Genealogical & Historical Library & Museum: housed in 1881, restored library building.
Collections: paintings; geology & early Indian items; papers & documents; slides, photos & artifacts; dioramas of the earliest settlement; miniatures.
Research Fields: genealogy; Kentucky history; Civil War.
Facilities: 7,500-vol. library of genealogical & historical material for use on premises; reading room; theater; KSB Miniatures Collection. Books & museum-related items for sale.
Activities: guided tours; educational activities; loan, permanent, temporary & traveling exhibitions; pioneer life movies.
Publications: quarterly newsletter; books: Maysville, KY Its Past & Present; Towns of Mason County, KY; Bicentennial Minutes Maysville, KY 1787-1987.
Hours & Admission Prices: Tues.-Fri. 10-4, Sat. 10-4. Museum: adults $10, students $2; discounts to AAA members; AASLH members no charge. Library: adults $5. Closed New Year's Day; Easter; Mother's Day; Father's Day; Independence Day; Thanksgiving; Christmas. &
Attendance: 5,000 (estimated)
Membership: Individual $30; Family $50; Wormald Society Individual $130; Wormald Society Family $140; Corporate Business $200-$500.

NATIONAL UNDERGROUND RAILROAD MUSEUM, 38 W. 4th St., Maysville, KY 41056. Mailing Address: P.O. Box 421, Maysville, KY 41056. Tel.: 606-564-3200 & 4413.
Web Site: llbierbower.org
Founded: 1995.
Institution Type/Description: History Museum.
Collections: Underground Railroad history; African American history, slavery & ancestry; photographs; period furnishings; personal artifacts.
Hours & Admission Prices: Wed. & Fri.-Sat. 10-3; other times by appointment.

OLD CHURCH MUSEUM, 2028 Old Main St., Maysville, KY 41056. Mailing Address: Maysville-Mason Co CVB, The Cox Building, 2 E. Third St., Maysville, KY 41056. Tel.: 606-759-7411 & 564-9419. Fax: 606-564-9416.
E-mail: info@maysvilleky.net
Web Site: www.cityofmaysville.com
Institution Type/Description: History Museum: housed in a former Methodist Episcopal Church - South; built in 1848.
Collections: nondenominational religious artifacts; spinning wheels; period furnishings; personal artifacts; photographs.
Hours & Admission Prices: Call for hours.

Middlesboro

ALEXANDER ARTHUR MUSEUM, 2215 Cumberland Ave., Middlesboro, KY 40965. Tel.: 606-248-2482.
Institution Type/Description: History Museum.
Collections: history of Alexander Arthur & his family; personal artifacts; period furnishings; photographs.
Hours & Admission Prices: Call for hours.

BELL COUNTY HISTORICAL SOCIETY, 207 N. 20th St., Middlesboro, KY 40965. Mailing Address: P.O. Box 1344, Middlesboro, KY 40965-3144. Tel.: 606-242-0005.
Institution Type/Description: Historical Society Museum.
Collections: local history & cultural heritage; period furnishings; personal artifacts; photographs; documents.
Activities: special events; lectures; seminars.
Hours & Admission Prices: Tues.-Sat. 10-3.

CUMBERLAND GAP NATIONAL HISTORICAL PARK, 91 Bartlett Park Rd., Middlesboro, KY 40965. Tel.: 606-248-2817. Fax: 606-248-7276.
E-mail: cuga_superintendent@nps.gov
Web Site: www.nps.gov/cuga
Founded: 1955.
Congressional District: 5
Governing Authority: federal. Affiliated with the U.S. Department of Interior, National Park Service. Tax-exempt.
Institution Type/Description: History Museum.
Collections: Cumberland Gap history; botany; archaeology. Historic Houses: 1903-1951 Hensley Settlement.
Research Fields: westward expansion; Civil War; Appalachian culture.
Facilities: research archives by appointment.
Activities: Park wide lectures; films; permanent exhibitions; campfire programs; nature walks; on-site tours; sound/slide programs; loan films & sound/slide programs.
Hours & Admission Prices: Visitor Center and Pinnacle: daily 8-5. Park Gates: April-May & Sept.-Oct. 8-7; June-Aug. 8 am-9 pm; Nov.-March 8-5. Park: no charge. Wilderness Road Campground: user fee charged. Visitor Center closed: New Year's Day; Christmas. &
Attendance: 85,000 (estimated)

Monticello

WILLIAM CRENSHAW KENNEDY, JR. MEMORIAL MUSEUM, 75 N. Main St., Monticello, KY 42633-1439. Mailing Address: P.O. Box 67, Monticello, KY 42633-0067. Tel.: 606-340-2300.
E-mail: museum123@windstream.net
Web Site: waynecountykymuseum.com
Key Personnel: Dir. & Cur., Harlan Ogle.
Governing Authority: Parent Institution: The Wayne County Historical Society. Tax-exempt.
Institution Type/Description: History Museum.
Collections: local history & culture relating to the heritage of Monticello & Wayne County; cave exhibit.
Hours & Admission Prices: Tues.-Sat. 10-4; other times by appointment. No charge; donations accepted. &
Attendance: 4,000 (accurate)

Morehead

CLAYPOOL-YOUNG ART GALLERY - MOREHEAD STATE UNIVERSITY, Claypool Young Bldg., Morehead, KY 40351. Tel.: 606-783-5446.
E-mail: j.reis@moreheadstate.edu
Key Personnel: Dir., Jennifer Reis
Institution Type/Description: Art Gallery.
Collections: works of contemporary art.
Activities: special events.
Hours & Admission Prices: Mon.-Fri. 8-4.

THE KENTUCKY FOLK ART CENTER, (M), 102 W. First St., Morehead, KY 40351-1723. Tel.: 606-783-2204. Fax: 606-783-5034.
E-mail: g.barker@morehead-st.edu
Web Site: www.kyfolkart.org
Founded: 1994.
Congressional District: 6
Key Personnel: C.E.O., Garry G. Barker; Cur., Adrian Swain; Chm. (V), Jean M. Dorton; Museum Store Mgr., Tammy F. Stone.
Personnel Profile: Full-Time Paid 5; Part-Time Volunteers 50; Interns 2.
Governing Authority: private; nonprofit organization. Parent Institution: Kentucky Folk Art Center, Inc. Subsidiary Institution: Morehead State University. Tax-exempt.
Institution Type/Description: Folk Art Museum.
Collections: contemporary folk art.
Facilities: 5,000 sq. ft. exhibit space.
Activities: lectures; films; workshops; outreach programs; organized education programs for children; traveling exhibitions.
Publications: annual report; newsletter.
Hours & Admission Prices: Mon.-Sat. 9-5. Adults 12 & over $3, seniors & children under 12 $2; members no charge. &
Attendance: 8,000 (estimated)
Membership: Senior Citizen & Student $15-$24; Individual $25-$34; Family $35-$99; Patron $100-$499; Sustaining $500-$999; Benefactor $1,000-$5,000; Live $5,000 & up.

Morganfield

CAMP BRECKINRIDGE MUSEUM & ARTS CENTER, 1116 N. Village Rd., Morganfield, KY 42437. Mailing Address: P.O. Box 60, Morganfield, KY 42437-0060. Tel.: 270-389-4420. Fax: 270-389-3546.
E-mail: campbreckinridge@bellsouth.net
Web Site: www.breckinridge-arts.org
Founded: 2000.
Congressional District: 1
Key Personnel: Dir., Vicki Ricketts.
Personnel Profile: Full-Time Paid 1; Part-Time Paid 3.
Governing Authority: Parent Institution: Union County Fiscal Court.
Institution Type/Description: History Museum.
Collections: military history; murals; photographs; paintings.
Hours & Admission Prices: Tues.-Fri. 10-3, Sat. 10-4, Sun. 1-4. &

Mount Sterling

MONTGOMERY COUNTY HISTORY MUSEUM, 36 Broadway St., Mount Sterling, KY 40353. Mailing Address: Montgomery County Historical Society, P.O. Box 861, Sterling, KY 40353. Tel.: 859-498-4669.
Institution Type/Description: History Museum.
Collections: local history & culture; period furnishing; personal artifacts; photographs.
Hours & Admission Prices: Fri.-Sat. 10-2.

Munfordville

HART COUNTY HISTORICAL MUSEUM, 109 Main St., Munfordville, KY 42765. Mailing Address: P.O. Box 606, Munfordville, KY 42765-0606. Tel.: 270-524-0101.
E-mail: hartmuseum@scrtc.com
Web Site: www.hartcountymuseum.org
Founded: 1968.
Congressional District: 2
Key Personnel: Pres. (V), Nathaniel Crenshaw; Museum Shop Mgr., Carolyn Short.
Personnel Profile: Part-Time Paid 2.
Volunteer Hours: 300

Operating Expenses: 25,000
Operating Income: 25,000
Governing Authority: Parent Institution: Hart County Historical Society board. Tax-exempt.
Institution Type/Description: History Museum: housed in the historic 1893 Chapline Building.
Collections: historic artifacts, photographs & documents on the history of Hart County.
Research Fields: local history; Civil War; genealogy.
Activities: quarterly meetings with speaker. Museum Sponsors: Civil War Days.
Publications: quarterly, Hart County Historical.
Hours & Admission Prices: Mon.-Fri. 9-4, Sat. 8-4. No charge; donations accepted. Closed major holidays. &
Attendance: 3,000 (estimated)
Membership: Student $10; Individual $25; Family $40.

Murray

UNIVERSITY ART GALLERIES, MURRAY STATE UNIVERSITY, Price Doyle Fine Arts Center, 15th & Olive Sts., 6th Fl., Murray, KY 42071. Mailing Address: 604 Fine Arts Center, Murray, KY 42071-3342. Tel.: 270-809-3052 & 6734.
E-mail: tmccalment@murraystate.edu
Web Site: www.murraystate.edu/artgallery
Founded: 1971.
Congressional District: 1
Key Personnel: Dir., Tina McCalment.
Personnel Profile: Full-Time Paid 1; Part-Time Paid 1; Interns 5.
Governing Authority: university. Parent Institution: Art Department, Murray State University. Tax-exempt.
Institution Type/Description: University Art Gallery.
Collections: 1,200-work permanent collection including Harry L. Jackson print Collection; Asian Cultural Exchange Foundation of Asian Art & artifacts; WPA print collection; MSU student art collection; magic silver photography collection.
Facilities: two lecture halls.
Activities: guided tours; lectures; gallery talks; formally organized education programs for children, adults & undergraduate & graduate students of Murray State University; loan, temporary & traveling exhibitions.
Publications: exhibition catalogues.
Hours & Admission Prices: Sept.-May Mon.-Fri. 8-5, Sat.-Sun. 1-4; Summer Mon.-Fri. 9-4. No charge. Closed university holidays. &
Attendance: 12,000 (accurate)

WRATHER WEST KENTUCKY MUSEUM, Murray State University, 100 Wrather Museum, Murray, KY 42071-3315. Tel.: 270-809-4771. Fax: 270-809-4485.
Web Site: www.murraystate.edu/info/wrather/wrather.htm
Founded: 1982.
Congressional District: 1
Key Personnel: C.E.O., Kate A. Reeves.
Personnel Profile: Full-Time Paid 1; Part-Time Volunteers 1; Interns 1.
Governing Authority: university. Parent Institution: Murray State University. Tax-exempt.
Institution Type/Description: History Museum.
Collections: guns; period furniture & tools; Murray State University memorabilia.
Facilities: 260-seat auditorium.
Activities: permanent & traveling exhibitions.
Hours & Admission Prices: Mon.-Fri. 8:30-4, Sat. 10-1. No charge; donations accepted. Closed university holidays. &
Attendance: 20,200 (accurate)

Nancy

MILL SPRINGS BATTLEFIELD VISITOR CENTER AND MUSEUM, 9020 W. Hwy. 80, Nancy, KY 42544. Mailing Address: Mill Springs Battlefield Association, P.O. Box 282, Nancy, KY 42544 0282. Tel.: 606 636-4045. Fax: 606-636-4050.
E-mail: info@millsprings.net
Web Site: www.millsprings.net
Founded: 2006.
Congressional District: 5
Key Personnel: Dir., Stephen McKinney; Pres. (V), Bruce Burkett, D.V.M.
Personnel Profile: Full-Time Paid 3; Part-Time Paid 3; Part-Time Volunteers 3.
Governing Authority: Parent Institution: Mill Springs Battlefield Association, Inc. Tax-exempt.
Institution Type/Description: History Museum.

Collections: Civil War history; Battle of Mill Springs; artifacts & exhibits pertaining to the Battle of Mill Springs. Historic Houses: Brown-Lanier House; West Metcalfe House.

Research Fields: American Civil War.

Facilities: library; visitor center; nature trails; Civil War driving & walking trails.

Activities: special events. Annual Events: Anniversary Observance of Battle of Mill Springs; Memorial Day Services; Living History Weekend; Ghostwalk; Christmas Open House.

Publications: membership newsletter, The Zollie Tree.

Hours & Admission Prices: Daily 10-4. Adults $5, seniors & military $3, students $2; members no charge. &

Attendance: 10,000 (estimated)

Membership: Student $5; Individual $25; Sustaining $140; Corporate $480.

New Castle

HENRY COUNTY HISTORY CENTER & MUSEUM - THE CAPLINGER HOUSE, 219 S. Main, New Castle, KY 40050. Mailing Address: P.O. Box 570, New Castle, KY 40050. Tel.: 502-845-0999.

E-mail: henrycountyhisto@bellsouth.net

Founded: 1978.

Congressional District: 4

Key Personnel: Dir., E.T. Smith; Pres. (V), Debbie Feemster.

Personnel Profile: Full-Time Volunteers 1.

Governing Authority: Parent Institution: Henry County Historical Society. Tax-exempt.

Institution Type/Description: History Museum: house built in 1863.

Collections: local history & culture; period furnishings; photographs; personal artifacts.

Hours & Admission Prices: Mon.-Fri. 9-4; groups by appointment. No charge; donations accepted.

Attendance: 1,100 (estimated)

Membership: Individual $35; Family $50; Life $1,000.

New Haven

KENTUCKY RAILWAY MUSEUM, INC., 136 S. Main St., New Haven, KY 40051-6355. Mailing Address: P.O. Box 240, New Haven, KY 40051-0240. Tel.: 800-272-0152; 502-549-5470. Fax: 502-549-5472.

E-mail: kyrail@bardstown.com

Web Site: kyrail.org

Founded: 1954.

Congressional District: 4

Key Personnel: Exec. Dir. & Pres., Greg Mathews; Chm. Bd. (V), Charlie Buccola; Office Mgr., Kim Maupin; Marketing & Public Rels., Lynn Dawson; Maintenance, William Ward; Museum Store Mgr., Christopher Cecil.

Personnel Profile: Full-Time Paid 5; Part-Time Paid 4; Part-Time Volunteers 60.

Governing Authority: county; nonprofit organization. Tax-exempt: 501(c)(3).

Institution Type/Description: Railroad Transportation Museum.

Collections: railroad engines; steam & diesel; passenger & freight car; blueprints; memorabilia; special rail equip.; John B. Hundley model train collection; model trains; photographs; L & N steam wrecker; railroad related articles. Historic Railway Structures: c.1912, wooden L & N combine; c.1920, Pullman Solarium Car; Monon #32, BL-2 diesel locomotive; c.1905, Pacific, 4-6-2 L & N #152 steam locomotive; CF-7 Santa Fe locomotive; GP7 locomotive.

Research Fields: railroad history, equipment, restoration, operation, maintenance.

Facilities: library of Trains magazine, Railfan magazine, industrial publications pertaining to railroad history, Poor's manual & archives of L & N law records available for research on premises; technical & historical books. Railway & other museum-related items for sale.

Activities: 22-mile roundtrip rail excursions; permanent & temporary exhibitions. Museum Sponsors: Railroad Museum Fundraising Banquet in November.

Publications: newsletter, The Station Lamp; annual report; brochures.

Hours & Admission Prices: Train Rides: March-May & Aug.-Dec. Sat. 11-2, Sun. 2; June-July Tues. & Fri. 1. Museum: adults $5, children 2-12 $2. Train Tickets: adults $17.50, children $12.50; discounts to AAA members; museum members no charge. Package rates available. &

Attendance: 40,000 (accurate)

Membership: Member $45; Family $60; Patron & Life $500.

Newport

NEWPORT AQUARIUM, One Aquarium Way, Newport, KY 41071-1679. Tel.: 859-261-7444. Fax: 859-261-5888.

Web Site: www.newportaquarium.com

Institution Type/Description: Aquarium.

Collections: aquatic life from around the world.

Activities: educational programs; shows.

Hours & Admission Prices: Daily 10-6. Adults $20, children 2-12 $13; children under 2 no charge.

Nicholasville

NATIONAL SOFTBALL MUSEUM, 101 NSA Way, Nicholasville, KY 40340. Mailing Address: P.O. Box 7, Nicholasville, KY 40340. Tel.: 859-887-4114. Fax: 859-887-4874.

E-mail: nsahdqtrs@aol.com

Web Site: www.playnsa.com

Key Personnel: Pres. & C.E.O., Hugh Cantrell

Institution Type/Description: History Museum.

Collections: sports memorabilia; photographs; personal artifacts.

Activities: special events.

Hours & Admission Prices: Call for hours.

OLD JESSAMINE COUNTY JAIL, 200 S. Main St., Nicholasville, KY 40356. Tel.: 859-885-4500.

Web Site: www.jessamineco.com

Institution Type/Description: Historic Building: 1870 jail.

Collections: local history; jail cell. Historic Building: 1873 jailers residence.

Hours & Admission Prices: Call for hours.

Olive Hill

NORTHEASTERN KENTUCKY MUSEUM, 1385 Carter Caves Rd., Olive Hill, KY 41164-8295. Tel.: 606-286-6012.

E-mail: nekymuseum@atcc.net

Web Site: www.kymuseum.org

Founded: 1972.

Institution Type/Description: History Museum.

Collections: rocks & fossils; early native art & artifacts; pioneer tools & weapons; Civil War, World War I & II artifacts.

Hours & Admission Prices: Spring to Summer daily 9-5, Winter by appointment. No charge.

Owensboro

INTERNATIONAL BLUEGRASS MUSIC MUSEUM, 117 Daviess St., Owensboro, KY 42303-4201. Mailing Address: 207 E. 2nd St., Owensboro, KY 42303-4201. Tel.: 270-926-7891. Fax: 270-689-9440. Facebook: Rompfest.

E-mail: gabrielle@bluegrassmuseum.org

Web Site: www.bluegrassmuseum.org

Founded: 1990.

Congressional District: 2

Key Personnel: C.E.O. & Exec. Dir., Gabrielle Gray; Chm., Terry Woodward.

Personnel Profile: Full-Time Paid 5; Full-Time Volunteers 2; Part-Time Paid 6; Part-Time Volunteers 300; Interns 1.

Governing Authority: private; nonprofit. Tax-exempt: 501(c)(3).

Institution Type/Description: Music Museum: housed in renovated c.1895 three-story brick Victorian storefront building attached to River Park Performing Arts Center.

Collections: artifacts; library & archival holdings; all tangible & recorded aspects of the history & development of bluegrass music; musical instruments; clothing; accessories; posters; flyers; recorded sound in all formats; fine art related to bluegrass music; radio & recording technology; bluegrass CD collection.

Research Fields: bluegrass professional musicians' oral histories; history of bluegrass music, c.1700-1985; interpretive presentation on bluegrass festivals, 1965-1985.

Facilities: 3,000-vol. library of books & periodicals relating to bluegrass & country music history under development; 20,000 sq. ft. exhibit space; classroom; auditorium. Recorded music & other museum-related items for sale.

Activities: open jam sessions; workshops; lectures; temporary & participatory exhibits; films; docent program; concerts; festival; video oral history program. Museum Sponsors: Bluegrass in the Schools.

Publications: quarterly newsletter, Bluegrass Legacy.

Hours & Admission Prices: Tues.-Sat. 10-5, Sun. 1-4; other times by appointment. Adult $5, children 16 & under $2; discounts to groups; members and children 6 & under no charge. &

Attendance: 48,000 (estimated)
Membership: Student $25; Base $45; Couple $80; Band & Organization $150; Silver Lifetime $1,000; Gold Lifetime $5,000.

OWENSBORO MUSEUM OF FINE ART, INC., (M), 901 Frederica St., Owensboro, KY 42301-3052. Tel.: 270-685-3181. Fax: 270-685-3181.

E-mail: mail@omfa.us
Web Site: www.omfa.us
Founded: 1977.
Congressional District: 2
Key Personnel: C.E.O. & Dir., Mary Bryan Hood; Chm., B. Dean Stanley; Registrar, Anthony Hardesty; Dir. Operations, Jason Hayden; Business Mgr., Jamie Scheffer.
Personnel Profile: Full-Time Paid 6; Part-Time Paid 10; Part-Time Volunteers 400.
Governing Authority: nonprofit organization. Subsidiary Institution: Owensboro Museum of Fine Art Foundation, Inc. Tax-exempt: 501(c)(3).
Institution Type/Description: Art Museum.
Collections: 19th- & 20th-century American, English & French paintings, drawings, graphics & sculpture; 14th -through 18th-century American & Asian decorative arts; contemporary American art; 19th-century German stained glass; 20th-century Appalachian Folk Art; 20th-century studio glass; Atrium Sculpture Court. Historic Buildings: 1905 Carnegie Library; pre-Civil War era mansion 1859.
Research Fields: 19th-20th century American artists & craftsmen with emphasis on Kentucky and the region; Appalachian Folk Art.
Facilities: interior sculpture court; two outdoor sculpture parks; classrooms. Artist's market.
Activities: guided tours; lectures; films; gallery talks; formally organized education program for children & adults; docent program; permanent, temporary & traveling exhibitions; interactive children's art studio; marionette theatre.
Publications: newsletter; exhibition catalogues & brochures.
Hours & Admission Prices: Tues.-Thurs. 10-4, Fri. 10-7, Sat.-Sun. 1-4. Suggested Donations: adult $2, children $1; discounts to AAM members. Closed New Year's Day; Memorial Day; Independence Day; Labor Day; Christmas. &
Attendance: 72,000 (estimated)
Membership: Friends of the OMFA Foundation: Junior $5; Adult $25; Family $45; Contributor $75; Donor $150; Patron $300; Supporter $600; Sponsor $1,000; Sustainer $2,500; Benefactor $5,000; Fellow $10,000.

OWENSBORO MUSEUM OF SCIENCE AND HISTORY, 122 E. 2nd St., Owensboro, KY 42303-4108. Tel.: 270-687-2732. Fax: 270-687-2738.

E-mail: information@owensboromuseum.com
Web Site: owensboromuseum.com
Founded: 1966.
Congressional District: 2
Key Personnel: Exec. Dir., Kathy Olson; Cur. Exhibits, Chris Norton; Government, Wendell H. Ford; Coord. Education Center, Ron Mayhew.
Personnel Profile: Full-Time Paid 5; Part-Time Paid 2; Part-Time Volunteers 50.
Governing Authority: public school district; nonprofit organization. Tax-exempt: 501(c)(3).
Institution Type/Description: General Museum.
Collections: agricultural & industrial technology; social & natural history; paleontology; geology; physical science.
Research Fields: archaeology; local & state history; paleontology.
Facilities: 250-vol. library of natural science & history research material available to students on the premises; archive of photographs & manuscripts; hands-on science discovery center; nature center.
Activities: guided tours; lectures; films; gallery talks; arts festivals; temporary exhibitions; school loan service.
Publications: quarterly newsletter: Mammoth Happenings.
Hours & Admission Prices: Sun. 1-5, Mon. 10-8, Tues.-Sat. 10-5. Admission $3; discounts to ASTC, AAM, SEMC & KAM members; members and children 2 & under no charge. Closed New Year's Eve & Day; Thanksgiving; Christmas. &
Attendance: 70,000 (estimated)
Membership: Student $5; Single $20; Family $40; Contributing $50; Donor $100; Supporting $250; Patron $500; Sustaining $1,000.

Paducah

LLOYD TILGHMAN HOUSE & CIVIL WAR MUSEUM, 631 Kentucky Ave., Paducah, KY 42001. Tel.: 270-575-5477.

Institution Type/Description: Historic House Museum: housed in the former home of General Lloyd Tilghman; built in 1852.
Collections: local history; Civil War artifacts; photographs; personal artifacts; military weapons.
Hours & Admission Prices: April-Nov. call for hours; other times by appointment. Adults $5. &

NATIONAL QUILT MUSEUM, 215 Jefferson St., Paducah, KY 42001-0714. Mailing Address: P.O. Box 1540, Paducah, KY 42002-1540. Tel.: 270-442-8856. Fax: 270-442-5448.

E-mail: info@quiltmuseum.org
Web Site: www.quiltmuseum.org
Formerly: Museum of the American Quilter's Society
Founded: 1991.
Congressional District: 1
Key Personnel: C.E.O., Frank Bennett; Pres. (V), Ann Hazelwood; Vice Pres., Donna Wilder; Registrar & Cur. Collections, Judith Schwender; Museum Shop Mgr., Pam Hill.
Personnel Profile: Full-Time Paid 9; Part-Time Paid 7; Part-Time Volunteers 207; Interns 1.
Governing Authority: private; nonprofit organization. Tax-exempt: 501(c)(3).
Institution Type/Description: Arts & Quilt Museum.
Collections: contemporary quilting from 1980 to present.
Research Fields: contemporary quilt making.
Facilities: 20,000 sq. ft. exhibit space; classrooms. Quilt-related books & handcrafted gift items for sale.
Activities: docent program; formal education programs for adults & children with course credit from West Kentucky Community & Technical College; guided tours; lectures; loan, temporary & traveling exhibitions; hands-on activities; special events.
Publications: quarterly, Friends of Museum Newsletter; exhibit catalogues, Museum Quilts: The Founder's Collection; book, New Quilts from an Old Favorite.
Hours & Admission Prices: March-Nov. Mon.-Sat. 10-5, Sun. 1-5; Dec.-Feb. Mon.-Sat. 10-5. Adults $11, seniors $9, youth 13 & over $5; discounts to groups; members no charge. Closed New Year's Day; Easter; Thanksgiving; Christmas Eve & Day. &
Attendance: 39,000 (accurate)
Membership: Friends Program: Individual Basic $30; Dual $60; Benefactor $100; Donor $250; Patron $500; President's Club $1,000; Program Sponsor $3,500-$10,000 & up.

PADUCAH RAILROAD MUSEUM, 3rd & Washington Sts., 2nd Fl., N.C. & St. L Railway Freight Office, Paducah, KY 42001. Mailing Address: 3409 Central Ave., Paducah, KY 42001. Tel.: 270-442-4032 & 443-7084.

Institution Type/Description: Railroad Museum.
Collections: railroad history & equipment; photographs; model train layout; No. 1518 steam locomotive.
Hours & Admission Prices: Sat. 10-4; other times by appointment. No charge.

RIVER DISCOVERY CENTER, (M), 117 S. Water St., Paducah, KY 42001-0787. Tel.: 270-575-9958. Fax: 270-444-9944.

E-mail: jharris@riverdiscoverycenter.org
Web Site: www.riverdiscoverycenter.org
Formerly: River Heritage Museum
Founded: 1990.
Congressional District: 1
Key Personnel: Exec. Dir., Julie Harris; Chm., Alex Edwards; Education, E.J. Abell.
Personnel Profile: Full-Time Paid 2; Part-Time Paid 3; Part-Time Volunteers 5.
Governing Authority: private; nonprofit organization. Tax-exempt: 501(c)(3).
Institution Type/Description: General Museum.
Collections: river & nautical memorabilia; Civil War artifacts; riverboat, steamboat, towboat & paddlewheel models.
Facilities: 148-seat auditorium; 2,300 sq. ft. exhibit space. Museum-related items for sale.
Activities: films; guided tours; participatory exhibits; rental gallery; school loan service. Annual Event: Marine Industry Day.
Publications: quarterly newsletter, The Anchor.
Hours & Admission Prices: April-Nov. Mon.-Sat. 9:30-5, Sun. 1-5. Adults $7,

senior citizens $6.50, children $5; discounts to groups; members no charge. Closed Thanksgiving; Christmas Eve & Day. &
Attendance: 15,000 (accurate)
Membership: First Mate $50; Crew $100; Engineer $250; Pilot $500; Captain $1,000.

WHITEHAVEN TOURIST WELCOME CENTER, 1845 Lone Oak Rd., Paducah, KY 42001-7903. Tel.: 270-554-2077.
E-mail: whitehaven.wc@ky.gov
Founded: 1983.
Congressional District: 1
Key Personnel: Supvr., Regina Topp.
Personnel Profile: Full-Time Paid 5.
Governing Authority: nonprofit. Parent Institution: Commonwealth of Kentucky. Subsidiary Institution: Dept. of Travel Devel. Tax-exempt: 501(c)(3).
Institution Type/Description: Historic House: c.1860 Classical Revival Mansion.
Collections: memorabilia from home of Alben Barkley, Vice President of the United States, 1949-1953.
Activities: guided tours; organized education programs for children.
Publications: quarterly newsletter, Whitehaven.
Hours & Admission Prices: Daily 8-4:30; tours every half hour 1-4. No charge. Closed New Year's Eve & Day; Thanksgiving & day after; Christmas Day & day after. &
Attendance: 23,000 (accurate)

WILLIAM CLARK MARKET HOUSE MUSEUM, 121 S. 2nd St. in Market House Sq., Paducah, KY 42001-0789. Mailing Address: P.O. Box 12, Paducah, KY 42002-0012. Tel.: 270-443-7759.
E-mail: info@markethousemuseum.com
Web Site: markethousemuseum.com
Founded: 1968.
Congressional District: 1
Key Personnel: Exec. Dir., Penny Baucum Fields.
Personnel Profile: Full-Time Paid 1; Part-Time Paid 1; Part-Time Volunteers 22.
Governing Authority: nonprofit organization. Tax-exempt: 501(c)(3).
Institution Type/Description: History Museum: built in 1905 the Market House was used as a farmers market.
Collections: complete two story Victorian Gingerbread woodwork interior of 1877 List Drug Store; 1913 American LaFrance Fire Truck; fire-related items; Irvin S. Cobb & Vice President Alben W. Barkley exhibits; silver service; wheel & fog bell from U.S.S. gunboat, Paducah; exhibits & artifacts of the War between the states; KY Orphan's Brigade exhibit; 1850 parlor set used by U.S. Grant; Native American artifacts.
Research Fields: Paducah & Kentucky history 1820-present.
Facilities: 4,800 sq. ft. exhibit space.
Activities: self or guided tours; temporary & rotating exhibitions.
Hours & Admission Prices: March to mid-Dec. Mon.-Sat. 12-4. Adults $4, children 6-11 $1; children under 6 no charge. Closed major holidays. &
Attendance: 10,000 (estimated)

YEISER ART CENTER, 200 Broadway, Paducah, KY 42001-0732. Tel.: 270-442-2453.
E-mail: jewhite@theyeiser.org
Web Site: www.theyeiser.org
Founded: 1957.
Congressional District: 1
Key Personnel: Exec. Dir., Joshua E. White; Pres. Bd. (V), Jane Gamble; Administrative Asst., John Paul Henry.
Personnel Profile: Part-Time Paid 2; Part-Time Volunteers 40.
Governing Authority: nonprofit organization. Tax-exempt: 501(c)(3).
Institution Type/Description: Art Museum: housed in 1905 Market House.
Collections: European, American, Asian & African 19th & 20th century works of art.
Facilities: Local & regional fine art & craft for sale.
Activities: lectures; participatory & temporary exhibitions. Annual Events: International Photographs Exhibit; International Fibers Exhibit.
Publications: monthly newsletter.
Hours & Admission Prices: Tues.-Sat. 10-4. No charge; donations accepted. Closed major holidays. &
Attendance: 10,000 (estimated)
Membership: Student $25; Friend $50; Contributor $100; Partner $250; Patron $500; Sponsor $1,000. Business: Friend $100; Donor $250; Patron $500; Leader $1,000.

Paris

DUNCAN TAVERN HISTORIC CENTER, 323 High St., Paris, KY 40361-2002. Tel.: 859-987-1788.
Institution Type/Description: History Museum: built in 1788.
Collections: local history & culture; period furnishings; personal artifacts; genealogy.
Facilities: genealogy library.
Hours & Admission Prices: Tours: April to mid-Dec. Wed.-Sat. 1:30; other times by appointment. Adults $8, senior citizens $6, children 6-12 $2; children under 6 no charge.

HOPEWELL MUSEUM, (M), 800 Pleasant St., Paris, KY 40361-1734. Tel.: 859-987-7274. Fax: 859-987-7274.
E-mail: hopewellmuseum@yahoo.com
Web Site: www.hopewellmuseum.org
Founded: 1995.
Congressional District: 6
Key Personnel: Dir., Nancy Smith.
Personnel Profile: Full-Time Paid 1; Part-Time Volunteers 45.
Governing Authority: private; nonprofit. Parent Institution: Historic Paris-Bourbon County, Inc. Tax-exempt: 501(c)(3).
Institution Type/Description: History & Art Museum: housed in a 1909 Beaux-Arts style post office building.
Collections: local, regional & state art, artists, art history and general history; art, artifacts & papers from Paris and Bourbon County, Kentucky within a state & national context.
Research Fields: local history; noted local artists; distilling; hemp; equine racing industries for history exhibits.
Activities: docent program; films; lectures; loan & temporary exhibitions.
Publications: bimonthly newsletter, Hopewell Museum Post.
Hours & Admission Prices: Wed.-Sat. 12-5, Sun. 2-4. Adults $3; students, children & members no charge. &
Attendance: 3,978 (accurate)
Membership: Senior 65 & over $30; Senior & Spouse or Individual $35; Family $45; Supporter & Corporation $100; Patron $250; Sponsor $500; Benefactor 1,000.

Perryville

PERRYVILLE BATTLEFIELD STATE HISTORIC SITE, 1825 Battlefield Rd., Perryville, KY 40468-0296. Mailing Address: P.O. Box 296, Perryville, KY 40468. Tel.: 859-332-8631. Fax: 859-332-2440. TDD: 859-332-8631.
E-mail: joan.house@ky.gov
Web Site: www.perryvillebattlefield.org
Formerly: Perryville Battlefield Museum
Founded: 1965.
Congressional District: 6
Key Personnel: Park Supt., Kurt Holman; Program Coord., Joan House.
Personnel Profile: Full-Time Paid 4; Full-Time Volunteers 12; Part-Time Paid 4; Interns 1.
Governing Authority: state. Parent Institution: Kentucky Dept. of Parks, Capital Plaza Tower, Frankfort, KY 40601. Tax-exempt.
Institution Type/Description: Civil War Museum.
Collections: artifacts from the Battle of Perryville.
Major Exhibits: The Hard Hand of War, 6/12-1/15.
Facilities: interpretive exhibits.
Activities: tours; self-guided walking tour; living history programs. Museum Sponsors: battle reenactment in October.
Hours & Admission Prices: Grounds: March-Dec. daily. Museum: daily 9-5. Adults $3.50, children under 12 $2.50; discount to groups of 10 or more.
Attendance: 50,000 (estimated)

Petersburg

CREATION MUSEUM, 2800 Bullittsburg Church Rd., Petersburg, KY 41080-9364. Mailing Address: P.O. Box 510, Hebron, KY 41048-0510. Tel.: 888-582-4253.
Web Site: creationmuseum.org
Founded: 2007.
Key Personnel: Dir., Dan Mangus.
Governing Authority: Parent Institution: Answers in Genesis.
Institution Type/Description: Natural History Museum.
Collections: life-sized dinosaur models; fossils; minerals; waterfalls; frogs; fish; turtles; bugs.
Facilities: 180-seat theater; 1,000-seat auditorium; observatory; petting zoo; nature trails; planetarium. Books for sale.
Activities: workshops; videos; petting zoo; dinosaur digs in Montana.

Hours & Admission Prices: Mon.-Fri. 10-6, Sat. 9-6, Sun. 12-6. Adults $29.95, senior citizens 60 & over $23.95, children 5-12 $15.95; children under 5 no charge.
Attendance: 300,000 (accurate)

Pikeville

BIG SANDY HERITAGE CENTER, 773 Hambley Blvd., Pikeville, KY 41501-9078. Mailing Address: P.O. Box 1041, Pikeville, KY 41502-1041. Tel.: 606-218-6050.
E-mail: everett.johnson@bigsandyheritage.org
Web Site: www.bigsandyheritage.org
Key Personnel: Cur., Everett Johnson
Institution Type/Description: History Museum.
Collections: local history & culture relating to Pike County & the region.
Hours & Admission Prices: Mon.-Fri. 10-5, Sat.-Sun. evenings by appointment. Adults $3, seniors & children $2.

Prestonburg

EAST KENTUCKY SCIENCE CENTER, 7 Bert Combs Dr., Prestonburg, KY 471653. Tel.: 606-889-8260.
Institution Type/Description: Science Center.
Collections: hands-on exhibitions.
Hours & Admission Prices: Center: Tues.-Fri. 1-4. Planetarium Show: Tues.-Fri. 2pm, Sat. 12:30 & 2pm. Laser Show: Tues.-Fri. 3:15, Sat. 12-4. Adults $6, students & seniors $4; children under 4 no charge.

Princeton

ADSMORE MUSEUM, 304 N. Jefferson St., Princeton, KY 42445-1551. Tel.: 270-365-3114. Fax: 270-365-3310. Facebook: Adsmore House & Gardens.
E-mail: adsmoremuseum@gmail.com
Web Site: www.adsmore.org
Founded: 1986.
Key Personnel: Cur., Rebecca Pool.
Personnel Profile: Full-Time Paid 1; Part-Time Paid 7.
Governing Authority: nonprofit. PNC Bank Trust. Subsidiary Institution: Caldwell Co. Library Dist. Tax-exempt.
Institution Type/Description: Living History Museum.
Collections: period furnishings; silver; china; linens; clothing; decorative accessories; photographs; letters; toys.
Major Exhibits: Spring/Katherine's Birthday (1907), 2/25/14-4/5/14; Easter Luncheon (1905), 4/8/14-5/17/14; Selina's Engagement (1907), 5/20/14-6/28/14; Selina's Wedding to Gov. Osborne (1907), 7/1/14-8/9/14; Black Patch War (1906), 8/12/14-9/20/14; Home from Washington, D.C. (1914), 9/23/14-11/8/14; Victorian Christmas (1901), 11/11/14-12/31/14.
Research Fields: family history.
Facilities: 3,500 sq. ft. exhibit space. Museum-related items for sale.
Activities: concerts; guided tours; candlelight tours; teas.
Publications: brochure.
Hours & Admission Prices: Feb. 25-Dec. Tues.-Sat. 11-4. Adults $7, senior citizens 65 & over $6, children 6-12 $2; discounts to groups, AAA members & active duty military and their families. Closed Jan. 1-Feb. 23; Easter; Independence Day; Thanksgiving; Christmas Eve & Day. &
Attendance: 7,000 (estimated)

CALDWELL COUNTY RAILROAD MUSEUM, 116 Edwards St., Princeton, KY 42445-2217. Tel.: 270-365-0580.
Founded: 1995.
Congressional District: 1
Institution Type/Description: Railroad Museum.
Collections: railroad history; railroad artifacts & memorabilia.
Facilities: Railroad-related items for sale.
Hours & Admission Prices: May-Dec. Wed.-Sun. 1-4; groups by appointment. No charge; donations accepted. &
Attendance: 500 (estimated)
Membership: Annual $10.

Providence

PARKER WARNER HISTORIC MUSEUM, 500 S. Broadway, Providence, KY 42450-1638. Mailing Address: 48 Park St., Clay, KY 42404.
Founded: 2000.
Key Personnel: Dir., Paul Cowan; Dir., David Fraser; Dir., Lowell Childress.
Personnel Profile: Part-Time Volunteers 3.
Governing Authority: Tax-exempt.

Institution Type/Description: History Museum: built in 1885.
Collections: local history & culture; period furnishings; personal artifacts; photographs.
Publications: cemetery books; Wagon Wheel
Hours & Admission Prices: April-Nov. Thurs.-Sat. 1-4. No charge; donations accepted. Closed holidays.
Attendance: 55 (estimated)
Membership: Webster County Historical & Genealogy Society: Individual $15.

Renfro Valley

KENTUCKY MUSIC HALL OF FAME AND MUSEUM, 2590 Richmond Rd., Renfro Valley, KY 40473. Mailing Address: P.O. Box 85, Renfro Valley, KY 40473-0085. Tel.: 606-256-1000. Fax: 606-256-2989.
E-mail: info@kentuckymusicmuseum.com
Web Site: kentuckymusicmuseum.com
Founded: 2002.
Key Personnel: Chm. Bd., Roy Martin; Exec. Dir., Robert Lawson
Institution Type/Description: Music Museum.
Collections: Kentucky music history & heritage; instruments; photographs; costumes; Kentucky stars & entertainers.
Facilities: Museum-related items for sale.
Activities: educational programs; special events; induction ceremony.
Hours & Admission Prices: Mon.-Sat. 10-6, Sun. 9-3. Adults $7.50, seniors $7, children 6-12 $4.50; discounts to groups; children under 6 & teachers no charge. &

Richmond

FORT BOONESBOROUGH MUSEUM, 4375 Boonesboro Rd., Richmond, KY 40475-9333. Tel.: 859-527-3131. Fax: 859-527-3328. TDD: 859-527-3131.
E-mail: phil.gray@ky.gov
Web Site: parks.ky.gov/findparks/recparks/fb
Founded: 1974.
Congressional District: 5
Key Personnel: Dir. Living History & Museum Shop Dir., Bill Farmer; Parks Mgr., Phil Gray; Cur., Jerry Raisor.
Personnel Profile: Full-Time Paid 12; Part-Time Paid 60.
Governing Authority: state. Affiliated with KY Dept. of Parks, Capital Plaza Tower, Frankfort, KY 40601. Tax-exempt.
Institution Type/Description: History Museum.
Collections: objects & archives relating to the early Euro-American settlement of Kentucky; local Native American artifacts; Kentucky River history and navigation; indigenous plant species; history of Daniel Boone & the Transylvania Land Company; 1906 River Lock & Dam.
Research Fields: Kentucky River Valley history.
Facilities: river walk trails; children's theatre; interpretive center; living fort surrounding museum; campgrounds; picnic area; snack bar. Fort-made crafts & museum-related items for sale.
Activities: living history demonstrations & instruction; native botanical restorations; special events for school programs & educational seminars; guided tours.
Hours & Admission Prices: April-Oct. daily 9-5. Adults $7, children $5; discount to groups. &
Attendance: 60,000

IRVINTON HOUSE MUSEUM, 345 Lancaster Ave., Richmond, KY 40475. Tel.: 859-626-1422.
Institution Type/Description: History Museum.
Collections: local history; photographs; personal artifacts; period furnishings.
Hours & Admission Prices: Call for hours.

WHITE HALL STATE HISTORIC SITE, 500 White Hall Shrine Rd., Richmond, KY 40475-9159. Tel.: 859-623-9178. Fax: 859-626-8489.
E-mail: whitehall@ky.gov
Web Site: parks.ky.gov/findparks/histparks/wh
Founded: 1971.
Congressional District: 6
Key Personnel: Dir., Kathleen White; Cur., Lashe Mullins.
Personnel Profile: Full-Time Paid 3; Part-Time Paid 2.
Governing Authority: state; Kentucky State Park System. Tax-exempt.
Institution Type/Description: Historic House: 1798 Georgian style building, added to in 1861 in the Italianate style, with 44 rooms & eight levels. Home of Cassius M. Clay, Ambassador to Russia during the 1860's under Abraham Lincoln.

Collections: period furnishings; Kentucky history from late 1700s to early 1900s; Cassius M. Clay & the Clay family history.
Research Fields: Cassius M. Clay & the Clay family.
Facilities: picnic area. Museum-related items for sale.
Activities: guided tours; concerts; internships; temporary exhibitions. Annual Events: Halloween Ghost Walk; theater event in October; self guided holiday tour in December; A Victorian Christmas at White Hall.
Hours & Admission Prices: April-Oct. Wed.-Sun. 9-4; Nov.-March by appointment. Adults $7, children 6-12 $4; discounts to Kentucky Junior Historical Society, groups of 10 or more, AAA members & military with ID.
Attendance: 5,000 (estimated)

Rosine

BILL MONROE BIRTHPLACE, 6210 Hwy. 62 E., Rosine, KY 42370. Mailing Address: P.O. Box 22, Hartford, KY 42347. Tel.: 270-274-9181.
Institution Type/Description: Historic House Museum: housed in the birthplace & childhood home of the father of Bluegrass music, Bill Monroe.
Collections: Bill Monroe's family & career; period furnishings; personal artifacts; photographs; instruments.
Hours & Admission Prices: Mon.-Sat. 9-4, Sun. 1-4. No charge; donations accepted.

Russellville

1817 SADDLE FACTORY MUSEUM, 280 E. 4th St., Russellville, KY 42276-1822. Tel.: 270-726-9559.
Institution Type/Description: History Museum: housed in a former saddle factory built in 1817 by the Caldwell brothers.
Collections: saddle industry history; drawings; personal artifacts; period furnishings; Civil War artifacts; photographs.
Hours & Admission Prices: Call for hours.

Salyersville

MAGOFFIN COUNTY PIONEER VILLAGE AND MUSEUM, 239 S. Church St., Salyersville, KY 41465. Mailing Address: P.O. Box 222, Salyersville, KY 41465. Tel.: 606-349-1607.
E-mail: magoffin@foothills.net
Institution Type/Description: History Museum.
Collections: pioneer history & culture; period furnishings; period artifacts; photographs; 15 log buildings.
Activities: demonstrations. Annual Event: Magoffin County Founders Day.
Hours & Admission Prices: Call for hours.

Sandy Hook

LAUREL GORGE CULTURAL HERITAGE CENTER, Old Rte. 7 & 32, Old Laurel Curves Rd., Sandy Hook, KY 41171. Mailing Address: P.O. Box 653, Sandy Hook, KY 41171. Tel.: 606-738-5543.
E-mail: info@laurelgorge.com
Web Site: www.laurelgorge.com
Institution Type/Description: History Museum.
Collections: local history & culture; period furnishings; photographs; personal artifacts; Paleo Indian artifacts; interactive exhibits.
Facilities: nature trails.
Activities: special events; hiking. Annual Events: Ancestor Photo Month in February; Wild Flower Walk in April; Wildflower Walk in April; Birding Hike in May; Rhododendron Festival; Fall Crafts Show & Bazaar in September; Quilt Show in September; Keith Whitley Exhibit in September; An Autumn Walk through Time in September; Haunted Trail in October.
Hours & Admission Prices: Call for hours.

Scottsville

ALLEN COUNTY HISTORICAL SOCIETY MUSEUM, 301 N. Fourth St., Scottsville, KY 42164. Tel.: 270-237-3026.
Institution Type/Description: Historical Society Museum.
Collections: local history & culture; period furnishings; personal artifacts; photographs.
Activities: research.
Hours & Admission Prices: Mon.-Thurs. 11-4; other times by appointment.

Shepherdsville

BULLITT COUNTY HISTORY MUSEUM, Courthouse, Shepherdsville, KY 40165. Mailing Address: P.O. Box 206, Shepherdsville, KY 40165. Tel.: 502-921-0161.
Governing Authority: Tax-exempt: 501(c)(3).
Institution Type/Description: History Museum.
Collections: local history, heritage & culture; photographs; period artifacts.
Hours & Admission Prices: Mon.-Fri. 8-4. No charge.

Slade

KENTUCKY REPTILE ZOO, 200 L&E Railroad, Slade, KY 40376. Tel.: 606-663-9160. Fax: 606-663-6917.
E-mail: reptilezoo@bellsouth.net
Web Site: www.kyreptilezoo.org
Founded: 1990.
Institution Type/Description: Reptile Museum.
Collections: 85 species & over 100 individuals on display.
Facilities: Museum-related items for sale.
Activities: venom extraction; outreach programs; internships; school field trips.
Hours & Admission Prices: March to mid-May & Labor Day-Oct. Fri.-Sun. 11-6; Memorial Day-Sept. daily 11-6. Adults $6, children 3-15 $4; discounts to AAA members, seniors & groups of 10 or more; children under 3 no charge.

South Union

SOUTH UNION SHAKER VILLAGE, (M), 850 Shaker Museum Rd., South Union, KY 42283. Mailing Address: P.O. Box 177, Auburn, KY 42206-0177. Tel.: 502-542-4167 & 7734. Fax: 502-542-7558.
E-mail: shakmus@logantele.com
Web Site: www.shakermuseum.com
Formerly: Shaker Museum at South Union
Founded: 1960.
Congressional District: 3
Key Personnel: Pres., Arthur Cleavinger; Dir. & Cur., Tommy Hines; Museum Shop Mgr., Bonnie Eilers.
Personnel Profile: Full-Time Paid 3; Part-Time Paid 6; Part-Time Volunteers 25; Interns 1.
Governing Authority: nonprofit; board of directors. Parent Institution: Shakertown Revisited, Inc. Tax-exempt.
Institution Type/Description: Historic Site: located on the site of 1807 South Union Shaker Village.
Collections: Shaker furniture; tools; textiles; boxes; baskets; straw bonnets; wooden ware; photographs. Historic House & Buildings: 1824 Centre House, used by the church family; c.1834 preservatory; 1847 Steam House; 1869 tavern; 1846 Ministry Shop; 1875 Grain Barn; 1917 store.
Research Fields: records of South Union Colony; history of Shakers.
Facilities: Museum & Shaker-related gifts for sale.
Activities: guided tours; lectures for groups; festivals; permanent & temporary exhibitions.
Publications: newsletter, South Union Messenger.
Hours & Admission Prices: March-Nov. Sun. 1-5, Tues.-Sat. 9-5; Dec.-Feb. Tues.-Sat. 10-4. Adults $8, children 6-12 $4; discounts to AAA members; members & children under 6 no charge. Closed New Year's Eve & Day; Thanksgiving; Christmas Eve & Day.
Attendance: 10,000 (accurate)
Membership: Individual $40; Family $50; North Family $100; East Family $300; Centre Family $500; Jasper Springs Society $1,000.

Springfield

LINCOLN HOMESTEAD STATE PARK, 5079 Lincoln Park Rd., Springfield, KY 40069-9504. Tel.: 859-336-7461. Fax: 859-336-0659. TDD: 606-336-7461.
Web Site: www.state.ky.us/agencies/parks/linchome.htm
Founded: 1936.
Congressional District: 2
Key Personnel: Museum Shop Mgr. & Park Mgr., Robert Bartholomai.
Personnel Profile: Part-Time Paid 1; Part-Time Volunteers 2.
Governing Authority: state. Affiliated with Kentucky Department of Parks, Division of Museums and Shrines, Capital Plaza Tower, Frankfort, KY 40601. Tax-exempt.
Institution Type/Description: Park Museum.
Collections: archives; natural history. Historic Buildings: Francis Berry House; Lincoln Home; blacksmith shop.
Facilities: outdoor museum.
Activities: guided tours; lectures; permanent exhibitions.

Hours & Admission Prices: May-Sept. daily 10-5:30. Adults $2, children $1.50; discounts to groups of 10 or more. &

Attendance: 4,000 (estimated)

Staffordsville

MOUNTAIN HOMEPLACE, 745 Ky. Route 2275, Staffordsville, KY 41256. Mailing Address: P.O. Box 809, Paintsville, KY 41240. Tel.: 606-297-1850.

Web Site: www.visitpaintsvilleky.com/homeplace

Institution Type/Description: History Museum.

Collections: recreated mid-nineteenth century farm including a blacksmith shop, one room schoolhouse, church, cabin, barn, & farm grounds; Appalachian history.

Facilities: amphitheater. Museum-related items for sale.

Activities: demonstrations; guided tours; special events.

Hours & Admission Prices: April-Dec. Tues.-Sat. 9-5. Adults $6, seniors $5, children $4.

U.S. 23 COUNTRY MUSIC HIGHWAY MUSEUM, 100 Stave Branch, Staffordsville, KY 41256-9001. Mailing Address: P.O. Box 809, Paintsville, KY 41240. Tel.: 606-297-1469.

E-mail: info@us23countrymusichwymuseum.com

Web Site: www.us23countrymusichwymuseum.com

Founded: 2005.

Institution Type/Description: Country Music History Museum.

Collections: life & career of country music entertainers; photographs; personal artifacts; recordings.

Activities: Annual Event: Kentucky Apple Festival.

Hours & Admission Prices: Call for hours.

Stanford

WILLIAM WHITLEY HOUSE STATE HISTORIC SITE, 625 William Whitley Rd., Stanford, KY 40484-9770. Tel.: 606-355-2881. Fax: 606-355-2778.

Web Site: parks.ky.gov/statehistoricsites/ww/index.htm

Founded: 1938.

Congressional District: 5

Key Personnel: Park Mgr., Jack C. Bailey.

Personnel Profile: Full-Time Paid 3; Part-Time Paid 2; Part-Time Volunteers 2.

Governing Authority: state. Parent Institution: Kentucky Dept. of Parks, Capitol Plaza Tower, Frankfort, KY 40601. Tax-exempt.

Institution Type/Description: Historic House: c.1792 William Whitley House, one of the first brick homes west of the Alleghenies.

Collections: period furnishings; gun powder horn & shoulder belt.

Facilities: picnic area; playground. Museum-related items for sale.

Activities: guided tours.

Publications: book, The William Whitley House.

Hours & Admission Prices: May-Dec. Wed.-Sat. 9-5, Sun. 12-6; other times by appointment. Closed Thanksgiving; Christmas. &

Attendance: 5,000 (estimated)

Stanton

GLADIE CULTURAL ENVIRONMENTAL LEARNING CENTER, 3451 Sky Bridge Rd., Stanton, KY 40380. Tel.: 606-663-8100.

Institution Type/Description: History Museum.

Collections: local history & cultural heritage; geology; photographs.

Activities: special events.

Hours & Admission Prices: March to mid-Nov. daily 9-5:30; mid-Nov. to March call for reduced hours.

Stearns

MCCREARY COUNTY MUSEUM, 100 Henderson St., Stearns, KY 42647. Mailing Address: P.O. Box 452, 1 Henderson St., Stearns, KY 42647-0452. Tel.: 606-376-5730. Fax: 606-376-5332. Facebook: McCreary County Museum.

E-mail: director@highland.net

Web Site: www.mccrearycountymuseum.com

Formerly: Stearns Museum

Founded: 1988.

Congressional District: 5

Key Personnel: Volunteer Coord., Shane Gilreath; C.E.O., William Coffey; Chm. (V) & Coord., Freida Staley; Museum Shop Mgr., Peggy Strunk.

Personnel Profile: Full-Time Paid 1; Part-Time Paid 1; Part-Time Volunteers 5.

Governing Authority: Parent Institution: McCreary County Heritage Foundation. Tax-exempt.

Institution Type/Description: History Museum: housed in the old Stearns Coal and Lumber Company Corporate Headquarters, built in 1907.

Collections: Stearns Coal and Lumber Company employee & company archives; photographs; family files; McCreary County artifacts & history.

Research Fields: business, county & family history.

Publications: newsletter, Museum News.

Hours & Admission Prices: April & Nov. Thurs.-Sat. 11-4; May-Oct. Tues.-Sat. 9-4, Sun. 11-4. Adults 12-59 $5, seniors 60 & over $4, children 6-12 $3; discounts to KY Museum & Heritage Alliance members, local residents & school groups; members and children 5 & under no charge. Closed Thanksgiving. &

Attendance: 12,000 (estimated)

Membership: Individual $25; Dual $35; Family $50; Friend $100; Patron $200.

Tompkinsville

OLD MULKEY MEETINGHOUSE STATE HISTORIC SITE, 38 Old Mulkey Park Rd., Tompkinsville, KY 42167-6781. Tel.: 270-487-8481. Fax: 270-487-8481. Facebook: Old Mulkey Meetinghouse State Historic Site.

E-mail: sheila.rush@ky.gov

Web Site: www.parks.ky.gov

Founded: 1804.

Key Personnel: Park Mgr., Sheila Rush.

Personnel Profile: Full-Time Paid 1; Part-Time Paid 1; Part-Time Volunteers 15.

Governing Authority: Parent Institution: Kentucky State Parks.

Institution Type/Description: Historic Site: built in 1804.

Collections: local history & culture; period furnishings.

Activities: Various. See website.

Hours & Admission Prices: April-Nov. daily 9-5. No charge. &

Attendance: 20,000 (estimated)

Union

BIG BONE LICK STATE PARK MUSEUM, 3380 Beaver Rd., Union, KY 41091. Tel.: 859-384-3522. Fax: 859-384-4775.

Web Site: www.parks.ky.gov

Founded: 1971.

Congressional District: 4

Key Personnel: Park Mgr., Bertie Lucas.

Personnel Profile: Full-Time Paid 7; Part-Time Paid 2.

Governing Authority: state. Branch of Kentucky Dept. of Parks, Capital Plaza Tower, Frankfort, KY 40601. Tel. 502-564-3811. Tax-exempt.

Institution Type/Description: Historic Site.

Collections: mastodon & bison bones.

Research Fields: archaeology.

Facilities: theater. Crafts for sale.

Activities: gallery talks; formally organized education programs for children & adults.

Hours & Admission Prices: Grounds: dawn-dusk. Museum: Jan.-March Thurs.-Sun. 8-4:30; April-Dec. Mon.-Thurs. 8-4:30, Fri.-Sun. 9-5. No charge.

Attendance: 5,000 (estimated)

Van Lear

COAL MINERS' MUSEUM/VAN LEAR HISTORICAL SOCIETY, INC., 78 Miller's Creek Rd., Van Lear, KY 41265. Mailing Address: P.O. Box 369, Van Lear, KY 41265-0369. Tel.: 606-789-8540.

E-mail: vanleartourism@yahoo.com

Web Site: www.vanlearkentucky.com

Founded: 1984.

Congressional District: 5

Key Personnel: Pres., Debra B. Music; Dir. & Vice Pres., Tina S. Webb.

Personnel Profile: Full-Time Volunteers 1; Part-Time Volunteers 12.

Governing Authority: private; nonprofit organization. Parent Institution: Van Lear Historical Society, Inc. Tax-exempt: 501(c)(3).

Institution Type/Description: Historical Society Museum: housed in the former office building of The Consolidation Coal Company.

Collections: artifacts from 1908 to present; mining implements; recreation of doctor's office with original equipment; Van Lear country music room; Van Lear veterans room; Van Lear post office; restored town jail; Van Lear school-related memorabilia; 1950s store; early town diorama.

Facilities: library; resource center. Museum-related items for sale.

Activities: guided tours; mobile vans; teacher in-service program; history classes for Prestonsburg Community College. Annual Event: Van Lear

Town Celebration; 10th Annual Haunted House; 10th Annual Breakfast with Santa; Chili Cook-Off.
Publications: quarterly newsletter, The Bankmule.
Hours & Admission Prices: Nov.-Sept. by appointment. Adults $5, senior citizens $4; members and children 5 & under no charge. &
Attendance: 500 (estimated)
Membership: Annual $15.

Vanceburg

VANCEBURG RAILROAD DEPOT MUSEUM, 218 Main St., Vanceburg, KY 41179. Tel.: 606-796-0238.
Institution Type/Description: Historic Building: housed in a former railroad depot; built in 1910.
Collections: local, railroad, & military history and artifacts; letters; scrapbooks; clothing; tools; photographs; books; newspapers; caboose.
Hours & Admission Prices: Call for hours.

Versailles

BLUEGRASS RAILROAD MUSEUM, 175 Beasley Rd., Versailles, KY 40383-8992. Mailing Address: P.O. Box 27, Versailles, KY 40383-0027. Tel.: 859-873-2476; 800-755-2476 (outside KY). Fax: 859-873-0408.
E-mail: bluegrassrailroad@yahoo.com
Web Site: www.bgrm.org
Founded: 1976.
Key Personnel: Exec. Dir. & Pres., John Penfield.
Personnel Profile: Full-Time Volunteers 4; Part-Time Volunteers 5.
Governing Authority: nonprofit organization.
Institution Type/Description: Railroad Museum.
Collections: railroad history; railroad artifacts.
Activities: 11 mile 1.5 hr. round-trip train ride in period railroad coaches to Young's High Bridge over the Kentucky River.
Publications: newsletter, The Connecting Rod.
Hours & Admission Prices: Call for hours. Train rides: adults $11.50, senior citizens $10.50, children 2-12 $9.50; members & children under 2 no charge. &
Attendance: 7,500 (estimated)
Membership: Regular $30; Family $45.

JACK JOUETT HOUSE, 255 Craig's Creek Rd., Versailles, KY 40383-9649. Tel.: 859-873-7902.
E-mail: info@jouetthouse.org
Web Site: www.jouetthouse.org
Key Personnel: Exec. Dir., Michael Lynch.
Personnel Profile: Full-Time Paid 1; Part-Time Paid 2.
Governing Authority: Tax-exempt.
Institution Type/Description: Historic House Museum: housed in a Federal-style house, built in 1797.
Collections: period furnishings; personal artifacts.
Hours & Admission Prices: April-Oct. Wed. & Sat.-Sun. 1-5; other times by appointment. No charge; donations accepted.

NOSTALGIA STATION TOY AND TRAIN MUSEUM, 279 Depot St., Versailles, KY 40383. Tel.: 859-873-2497.
Institution Type/Description: Toy Museum: housed in a restored 1911 railroad station.
Collections: toys; railroad memorabilia; late 1950s Lionel store display; reproduction of a 1926 standard gauge Lionel store display.
Hours & Admission Prices: Wed.-Sat. 10-5, Sun. 1-5. Adults $3.50, seniors 62 & over $3, children $1; children 3 & under no charge. Closed major holidays.

WOODFORD COUNTY HISTORICAL SOCIETY MUSEUM, 121 Rose Hill, Versailles, KY 40383-1221. Tel.: 859-873-6786.
E-mail: woodford@qx.net
Web Site: www.woodfordkyhistory.org
Founded: 1966.
Key Personnel: Library Asst., Lorraine Bradenburg.
Personnel Profile: Part-Time Paid 2; Part-Time Volunteers 7.
Governing Authority: Tax-exempt.
Institution Type/Description: History Museum.
Collections: Woodford County history; genealogy; Civil War memorabilia; quilts; clothing; furniture; spinning wheel; photographs; paintings.
Facilities: library.
Activities: research.
Hours & Admission Prices: Tues.-Sat. 10-4. No charge.

West Liberty

MEMORY HILL FOUNDATION MUSEUM, 89 Memory Hill Lane, West Liberty, KY 41001. Tel.: 859-743-4482.
Institution Type/Description: History Museum.
Collections: local history & culture; period furnishings; photographs. Historic Buildings: 8 log cabins; colonial house.
Hours & Admission Prices: Call for hours.

Wickliffe

WICKLIFFE MOUNDS STATE HISTORIC SITE, 94 Green St., Wickliffe, KY 42087. Mailing Address: P.O. Box 155, Wickliffe, KY 42087-0155. Tel.: 270-335-3681. Facebook: Wickliffe Mounds State Historic Site.
E-mail: wickliffemounds@ky.gov
Web Site: www.parks.ky.gov
Founded: 1932.
Congressional District: 1
Key Personnel: Park Mgr., Carla Hildebrand.
Personnel Profile: Full-Time Paid 2; Part-Time Paid 1; Part-Time Volunteers 2.
Volunteer Hours: 371
Operating Expenses: 121,488
Operating Income: 20,098
Governing Authority: state government. Parent Institution: Commonwealth of Kentucky, Tourism, Arts and Heritage Cabinet, Department of Parks. Tax-exempt.
Institution Type/Description: Archaeology Museum.
Collections: block excavations & artifact collections of Mississippian culture; mounds.
Research Fields: archaeology; anthropology.
Facilities: welcome center; trails; picnic area. Gift items for sale.
Activities: self-guided tours; special events.
Hours & Admission Prices: March 16-Nov. 15 Wed.-Sun. 9-4:30. Adults $5, children 6-12 $4, children 3-5 $1; discount to groups, active duty military & AAM, AAA members. &
Attendance: 5,678 (accurate)
Membership: Individual $8; Family $25.

Williamsburg

CUMBERLAND INN & MUSEUM, 649 S. 10th St., Williamsburg, KY 40769-1647. Tel.: 800-315-0286; 606-539-3100.
E-mail: museum@cumberlandinn.com
Web Site: www.cumberlandinn.com/mus.htm
Key Personnel: Gen. Mgr., David Maggard
Institution Type/Description: History Museum.
Collections: Cumberland College archives; stamps; arrowheads; coins; nutcrackers; Henkelmann Life Science; The Carl Williams Cross collection.
Hours & Admission Prices: Call for hours. Adults $4, seniors over 65 $3, children 6-12 $2, children 5-1 $1.

WHITLEY COUNTY HISTORICAL & GENEALOGICAL SOCIETY, 529 Main St., Williamsburg, KY 40769. Mailing Address: P.O. Box 536, Williamsburg, KY 40769-9634. Tel.: 606-549-7089.
E-mail: whitleycountyhis@bellsouth.net
Web Site: www.wchgs.org
Founded: 1996.
Key Personnel: Pres (V), Pat Jones; Museum Shop Mgr., Mary Alice Siler.
Personnel Profile: Part-Time Volunteers 8.
Governing Authority: nonprofit organization. Tax-exempt.
Institution Type/Description: Historical Society Museum: housed in the Old Williamsburg Depot.
Collections: local history; photographs; books.
Facilities: library.
Activities: research.
Publications: newsletter, Whitley Branches
Hours & Admission Prices: Wed. 10-1, 1st & last Sat. each month 10-12; other times by appointment. No charge; donations accepted.
Attendance: 509 (estimated)
Membership: Annual $20.

Winchester

BLUEGRASS HERITAGE MUSEUM, 217 S. Main St., Winchester, KY 40391-2455. Mailing Address: P.O. Box 147, Winchester, KY 40392-0147. Tel.: 859-745-1358.
E-mail: bgheritage@bellsouth.net
Web Site: www.bgheritage.com

Key Personnel: Dir., Sandy Stults; Pres., Gardner Wagers
Institution Type/Description: History Museum.
Collections: area history; Bluegrass culture & history; Native American artifacts.
Facilities: Museum-related items for sale.
Hours & Admission Prices: Mon.-Sat. 12-4.

LOUISIANA

(240 listings)

Abbeville

ALLIANCE CENTER MUSEUM AND ART GALLERY, 200 N. Magdalen Square, Abbeville, LA 70510-4645. Tel.: 337-898-4114.
Institution Type/Description: General Museum.
Collections: local history & culture; genealogy; photographs; documents; period artifacts; paintings.
Hours & Admission Prices: Tues. & Sat. 10-3, Wed.-Fri. 10-5.

DEPOT AT MAGDALEN PLACE, 201 W. Lafayette St., Abbeville, LA 70510. Tel.: 337-740-2112. Fax: 337-893-5983.
Web Site: www.magdalenplace.com
Institution Type/Description: Historic Building: housed in a former train depot; built in 1894.
Collections: local history & culture; period furnishings; photographs.
Facilities: Gift items for sale.
Hours & Admission Prices: Mon.-Sat. 10-5.

SAM GUARINO AND SON BLACKSMITH SHOP MUSEUM, 304 S. State St., Abbeville, LA 70510. Tel.: 337-893-8550.
Institution Type/Description: History Museum: housed in a former blacksmith shop; built in 1913.
Collections: local history; blacksmithing & farming tools and equipment; wagon wheels; tractors.
Hours & Admission Prices: By appointment.

Alexandria

* **ALEXANDRIA MUSEUM OF ART, (M),** 933 Second St., Alexandria, LA 71301-8322. Mailing Address: P.O. Box 1028, Alexandria, LA 71309-1028. Tel.: 318-443-3458. Fax: 318-443-0545. Facebook: AMOA Downtown.
E-mail: catherine@themuseum.org
Web Site: www.themuseum.org
Founded: 1977.
Congressional District: 6 & 8
Key Personnel: Exec. Dir., Catherine M. Pears; Pres. (V), Robert Radcliffe, Sr.; Devel. & Community Rels. Officer, Sarah Cortell Vandersypen; Education & Outreach, Anne Reid; Education & Outreach, Cindy Blair; Facilities & Communications, Natalie Walker; Visitor Svcs., Jenny Gallent.
Personnel Profile: Full-Time Paid 5; Part-Time Paid 3; Part-Time Volunteers 120.
Governing Authority: nonprofit organization. Parent Institution: Louisiana State University at Alexandria. Tax-exempt: 501(c)(3).
Institution Type/Description: Art Museum: housed in c.1900 Bank Building.
Collections: contemporary works in sculpture, on paper & painting; North Louisiana Folk crafts; contemporary Louisiana artists.
Major Exhibits: Reflections: African American Life from the Myrna Colley-Lee Collections, 12/13-2/14.
Research Fields: modern & contemporary American art.
Facilities: 3,500-vol. research library & archive; classroom; garden; multimedia auditorium. Gift items for sale.
Activities: guided tours; lectures; films; gallery talks; study clubs; formal education programs for children & adults; education programs for students affiliated with local & area colleges; temporary, loan & traveling exhibitions; school loan service; summer arts program; education program for visitors with Alzheimer's or dementia & their caregivers.
Publications: exhibit catalogs; quarterly newsletters; annual report.
Hours & Admission Prices: Tues.-Fri. 10-5, Sat. 10-4. Adults $4, seniors, students & military $3, children under 12 $2; discounts to groups, NARM, SEMC & LAM members; members no charge. Closed legal holidays. &
Attendance: 8,803 (accurate)
Membership: Senior & Student $25; Individual $30; Household $50; Patron $100; Sponsor $250; Sustainer $500; Bronze Circle $1,000; Silver Circle $2,500; Gold Circle $5,000; Platinum Circle $10,000.

ALEXANDRIA ZOOLOGICAL PARK, 3016 Masonic Dr., Alexandria, LA 71301-4240. Mailing Address: P.O. Box 6015, Alexandria, LA 71307-6015. Tel.: 318-473-1143, ext. 0. Fax: 318-473-1149.
E-mail: info@thealexandriazoo.com
Web Site: www.thealexandriazoo.com
Founded: 1926.
Key Personnel: Gen. Cur., Carla Oncay
Institution Type/Description: Zoo.
Collections: over 600 animals.
Facilities: 33 acres.
Activities: train rides.
Hours & Admission Prices: Daily 9-5. Adults 13 & over $7.50, children 4-12 $5.50, seniors 65 & over $4.50; discounts to groups of 15 or more; children 3 & under and FOTAZ members no charge. Train Rides: $1.50; children under 1 no charge. Closed New Year's Day; Thanksgiving; Christmas.

ARNA BONTEMPS AFRICAN AMERICAN MUSEUM, 1327 Third St., Alexandria, LA 71301-8248. Tel.: 318-473-4692. Fax: 318-473-4675.
E-mail: admin@arnabontempsmuseum.com
Web Site: www.arnabontempsmuseum.com
Governing Authority: nonprofit organization.
Institution Type/Description: Historic House Museum: housed in the boyhood home of poet, author, anthologist & librarian, Arna Bontemps. Listed on the National Register of Historic Places.
Collections: Bontemps family history; African American history & culture.
Activities: educational programs.
Hours & Admission Prices: Tues.-Fri. 10-4, Sat. 10-2. No charge; donations accepted.

KENT PLANTATION HOUSE, 3601 Bayou Rapides Rd., Alexandria, LA 71303-3629. Tel.: 318-487-5998. Fax: 318-442-4154.
E-mail: admin@kenthouse.org
Web Site: kenthouse.org
Key Personnel: Dir., Alice V. Scarborough; Pres. (V), Carolyn Pate.
Personnel Profile: Full-Time Paid 4; Part-Time Paid 1; Part-Time Volunteers 100.
Institution Type/Description: Historic House Museum: built c.1796. Listed on the National Register of Historic Places.
Collections: period furnishings; photographs; personal artifacts.
Hours & Admission Prices: Mon.-Sat. 9-5. Adults $6, military, AAA members and seniors 65 & over $5, children 6-12 $2; discounts to groups; children 5 & under no charge.
Attendance: 20,000 (estimated)

LOUISIANA STATE UNIVERSITY AT ALEXANDRIA ART GALLERY, 8100 Hwy. 71 S., Alexandria, LA 71302. Tel.: 318-473-6449.
E-mail: rdeville@lsua.edu
Key Personnel: Dir., Roy V. de Ville
Institution Type/Description: Art Gallery.
Collections: works by local, state, & regional artists.
Activities: temporary exhibitions.
Hours & Admission Prices: Mon.-Fri. 8-12. No charge. &

T.R.E.E. HOUSE - THE RAPIDES EXPLORATORY EDUCATION HOUSE, 1403 Third St., Alexandria, LA 71301-8250. Tel.: 318-619-9394. Fax: 318-619-9395.
E-mail: questions@kidstreehouse.org
Web Site: www.kidstreehouse.org
Governing Authority: Tax-exempt: 501(c)(3).
Institution Type/Description: Children's Museum.
Collections: hands-on exhibits.
Activities: educational programs.
Hours & Admission Prices: Tues.-Fri. 9-3, Sat. 9-4. Admission $3.50; members & children under 2 no charge. Closed major holidays.

Angola

LOUISIANA STATE PENITENTIARY MUSEUM, (M), Hwy. 66, Angola, LA 70712. Mailing Address: General Delivery, Angola, LA 70712-9999. Tel.: 225-655-2592. Fax: 225-655-2842.
E-mail: lspmuseu@bellsouth.net
Web Site: angolamuseum.org
Founded: 1997.
Key Personnel: Dir., Marsha Lindsey.
Personnel Profile: Full-Time Paid 4.

Governing Authority: nonprofit organization. Parent Institution: Louisiana State Penitentiary Museum Foundation.
Institution Type/Description: Penitentiary Museum.
Collections: prison artifacts; clothing; newspapers; photographs; a jail cell; inmate weapons; farm tools & equipment; movie memorabilia.
Hours & Admission Prices: Mon.-Fri. 8-4:30, Sat. 9-5, Sun. 10-5. No charge; donations accepted. Closed New Year's Day; Easter; Independence Day; Thanksgiving; Christmas; major holidays. &
Attendance: 14,400 (estimated)

Avery Island

JUNGLE GARDENS, Hwy. 329, Avery Island, LA 70513. Mailing Address: P.O. Box 126, Avery Island, LA 70513. Tel.: 337-369-6243. Fax: 337-369-6245.
E-mail: junglegardens@bellsouth.net
Web Site: www.junglegardens.org
Founded: 1935.
Institution Type/Description: Garden & Bird Sanctuary.
Collections: plants & flowers; birds; deer; alligators; raccoons.
Facilities: 170-acre site; nature trails.
Activities: bike riding, walking trails, bird watching.
Hours & Admission Prices: Daily 9-5. Adults $8, children $5; discounts to groups; children under 6 no charge. Island Toll: $1.

MCILHENNY COMPANY & VISITOR CENTER, Hwy. 329, Avery Island, LA 70513. Tel.: 800-634-9599; 337-365-8173.
E-mail: linda.clause@tabasco.com
Web Site: www.tabasco.com
Institution Type/Description: Company Museum.
Collections: company history; Tabasco brand products; period advertisements & artifacts; photographs; bottling & packaging machines & equipment.
Facilities: visitor center. Museum-related items for sale.
Activities: pepper sauce factory tours.
Hours & Admission Prices: Daily 9-4. Island Toll: $1. Visitor Center: no charge. Closed major holidays. &
Attendance: 100,000 (estimated)

Baker

HERITAGE MUSEUM & CULTURAL CENTER, 1606 Main St., Hwy. 19, Baker, LA 70714. Mailing Address: P.O. Box 707, 1606 Main St., Baker, LA 70704-0707. Tel.: 225-774-1776. Fax: 225-775-5635.
E-mail: bakermuseum@bellsouth.net
Web Site: www.bakerheritagemuseum.org
Governing Authority: municipal government.
Institution Type/Description: History Museum.
Collections: history & heritage of Baker; documents; railroad car. Historic Buildings: restored 1906 Victorian Cottage; one room school; rural store; train depot; chapel; barn.
Activities: exhibit box program; cultural activities; speakers.
Publications: monthly newsletter, Musings.
Hours & Admission Prices: Mon.-Fri. 9-4. No charge; donations accepted.

Barksdale AFB

8TH AIR FORCE MUSEUM, 88 Shreveport Rd., Barksdale AFB, LA 71110. Mailing Address: P.O. Box 75, Barksdale AFB, LA 71110-0075. Tel.: 318-456-5553 & 752-0055.
E-mail: info@8afmuseum.com
Web Site: 8afmuseum.com
Institution Type/Description: Military Museum.
Collections: Air Force history; photographs; military aircraft; personal artifacts.
Facilities: Museum-related items for sale.
Hours & Admission Prices: Daily 9:30-4. Closed New Year's Day; Thanksgiving; Christmas.

Bastrop

SNYDER MUSEUM & CREATIVE ARTS CENTER, 1620 E. Madison Ave., Bastrop, LA 71220-4062. Tel.: 318-281-8760.
E-mail: snydermuseum@gmail.com
Web Site: snydermuseum.com
Founded: 1974.
Congressional District: 5
Key Personnel: Exec. Dir., Emily Graves; Pres. (V), Melissa Smith.
Personnel Profile: Part-Time Paid 1; Part-Time Volunteers 1.

Governing Authority: nonprofit. Tax-exempt: 501(c)(3).
Institution Type/Description: Local History Museum & Art Center.
Collections: Indian artifacts; Civil War items; archival materials; photograph collection; vintage clothing collection; farm implement collection; early home & store items.
Research Fields: local history.
Facilities: library.
Activities: guided tours; lectures; traveling, loan & temporary exhibitions. Museum Sponsors: fundraisers; grants.
Hours & Admission Prices: Jan. to mid-Dec. Tues.-Fri. 9-4. No charge; donations accepted. Closed New Year's Day; Independence Day; Thanksgiving week. &
Attendance: 1,000 (estimated)
Membership: Individual $10; Family $25; Patron $50; Sustaining $100; Benefactor $250; Corporate $500.

Baton Rouge

ALFRED C. GLASSELL JR. EXHIBITION GALLERY, LSU School of Art, Shaw Center for the Arts, 100 Lafayette St., Baton Rouge, LA 70801. Tel.: 225-389-7180.
E-mail: artgallery@lsu.edu
Web Site: www.glassellgallery.org
Founded: 2005.
Key Personnel: Dir., K. Malia Krolak.
Personnel Profile: Full-Time Paid 1; Part-Time Paid 1.
Governing Authority: Parent Institution: Louisiana State University. Tax-exempt.
Institution Type/Description: Art Gallery.
Collections: works by local, national & international contemporary artists.
Hours & Admission Prices: Tues.-Fri. 10-5, Sat.-Sun. 12-5. No charge; donations accepted. &

BATON ROUGE GALLERY CENTER FOR CONTEMPORARY ART, City Park Pavilion, 1515 Dalrymple Dr., Baton Rouge, LA 70808-1037. Tel.: 225-383-1470. Fax: 225-336-0943. Facebook: Baton Rouge Gallery.
E-mail: info@batonrougegallery.org
Web Site: www.batonrougegallery.org
Founded: 1966.
Key Personnel: Dir., Jason Andreasen
Institution Type/Description: Art Gallery.
Collections: works by contemporary artists; photographs; paintings; sculptures.
Publications: The Art & Artists of Baton Rouge Gallery.
Hours & Admission Prices: Tues.-Sun. 12-6. No charge. &
Membership: Senior & Student $20; Friend $40; Dual $75; Patron $150; Collector $250; Connoisseur $500; Visionary $1,000.

BATON ROUGE ZOO, 3601 Thomas Rd., Baton Rouge, LA 70807. Mailing Address: P.O. Box 60, Baker, LA 70704-0060. Tel.: 225-775-3877. Fax: 225-775-3931. Facebook: BRECSBRzoo.
E-mail: info@brzoo.org
Web Site: www.brzoo.org
Formerly: BREC's Baton Rouge Zoo
Founded: 1970.
Congressional District: 6
Key Personnel: Dir., Phillip L. Frost; Dir. Mktg. & Devel., Kaki Heiligenthal; Gen. Cur., Sam Winslow; Cur. Education, Jennifer Shields; Cur. Birds, Sam Moran; Cur. Hoofstock, John Marshall; Cur. Carnivores & Primates, Erin Dauenhauer-Decota; Administrative Svcs. Coord., Lois Cook; Vet. Tech., Holly Taylor; Commissarian, Melissa Prisk; Zoo Veterinarian, Dr. Gordon Pirie; Souvenir Shop Mgr., Carroll Shirey; Guest Svcs. Mgr., Vicki Jones; Concessions Mgr., Gilda Conrad; Membership & Events Coord., Tara Brown; Receptionist, Kim Lodrigue.
Personnel Profile: Full-Time Paid 67; Part-Time Paid 37; Part-Time Volunteers 57.
Governing Authority: municipal. Parent Institution: Baton Rouge Recreation & Parks Commission. Tax-exempt: 501(c)(3).
Institution Type/Description: Zoological Park.
Collections: mammals; birds; reptiles; amphibians.
Research Fields: animal behavior work with LSU Vet. School; embryo transfer - Bongo, Asian elephant artificial insemination.
Facilities: 900-vol. library of encyclopedias on animals & books on vertebrate groups, specific orders & habitat, diet, medication & reproduction of animals & groups of animals represented at the zoo, available for research on premises only; auditorium. Zoo-related items for sale.
Activities: guided tours; lectures; concerts; formally organized education programs for children, adults & graduate students affiliated with LSU; mobile vans.

Publications: newsletter.
Hours & Admission Prices: Daily 9:30-5. Adults & teens $8.25, senior citizen 65 & over $7.25, children 2-12 $5.25; children one & under, AZA members & Friends of the Zoo no charge. Closed New Year's Day; Thanksgiving; Christmas Eve & Day. &
Attendance: 266,260 (accurate)
Membership: Individual $35; Household $55; Individual Plus 3 $65; Honorary Keeper $79; Safari Club $160; Director's Circle $260.

BRECS BLUEBONNET SWAMP NATURE CENTER, 10503 N. Oak Hills Pkwy., Baton Rouge, LA 70810. Tel.: 225-757-8905.
Institution Type/Description: Nature Center.
Collections: wildlife & their habitats.
Hours & Admission Prices: Tues.-Sat. 9-5, Sun. 12-5; group tours by appointment. Adults $3, seniors 65 & over and college students $2.50; children 2 & under no charge.

*** BREC'S MAGNOLIA MOUND PLANTATION,** 2161 Nicholson Dr., Baton Rouge, LA 70802-8105. Tel.: 225-343-4955. Fax: 225-343-6739.
E-mail: information@magnoliamound.org
Web Site: www.magnoliamound.org
Formerly: Magnolia Mound Plantation
Founded: 1968.
Congressional District: 4
Key Personnel: Dir., John Sykes; Bd. Chm., Nancy Dougherty; Museum Shop Mgr., Pauline Poole.
Personnel Profile: Full-Time Paid 4; Part-Time Paid 9; Part-Time Volunteers 50; Interns 4.
Governing Authority: municipal. Parent Institution: Baton Rouge Park & Recreation Commission. Tax-exempt: 501(c)(3).
Institution Type/Description: Historic Houses: 1791-1830 plantation house & outbuildings.
Collections: furnishings dating from 1780-1830.
Research Fields: restoration; architecture; archaeology; family & social history; period cooking; Louisiana history.
Facilities: visitors center; interpretive exhibits. Gift items for sale.
Activities: guided tours; docent program; formally organized education programs for adults & children; open hearth & bake oven cooking demonstrations.
Publications: quarterly newsletter; docent manual; walking tour booklet; books, Magnolia Mound Kitchen Book; Magnolia Mound, a Louisiana River Plantation; A Year in the Garden; children's workbook.
Hours & Admission Prices: Mon.-Sat. 10-4, Sun. 1-4; tours on the hour; last tour at 3pm. Adults $10, senior citizens 65 & over $8, children 3-17 $4; discounts to groups, LAM, BREC, military, teachers, AAA & AAM members; children under 3 no charge. Closed New Year's Day; Mardi Gras Day; Easter; Independence Day; Thanksgiving; Christmas.
Attendance: 21,656 (accurate)
Membership: Friend $35; Overseer $100; Settler $250; Planter $500; Duplantier Society $1,000; Constance Joyce Duplantier $2,500; Armand Allard Duplantier Circle $5,000.

THE ENCHANTED MANSION, A DOLL MUSEUM, 190 Lee Dr., Baton Rouge, LA 70808-4953. Mailing Address: 172 Lee Dr., Ste. 3, Baton Rouge, LA 70808-5088. Tel.: 225-769-0005. Fax: 225-766-6822.
E-mail: temansion@tem.brcoxmail.com
Web Site: www.enchantedmansion.org
Founded: 1995.
Congressional District: 6
Key Personnel: C.E.O., Rosemary Sedberry; Office Mgr. & Museum Shop Mgr., LuLu Ragusa.
Personnel Profile: Full-Time Paid 2; Part-Time Paid 4; Part-Time Volunteers 1.
Governing Authority: private; nonprofit organization. Tax-exempt.
Institution Type/Description: Toy & Doll Museum.
Collections: a life-size Victorian doll house; doll collection.
Activities: birthday parties; workshops for children & adults; meeting room.
Hours & Admission Prices: Thurs.-Sat. 10-5. Adults $4.50, senior citizens $3.50, children under 15 $2; discount to groups of 15 or more; handicapped, children under 2 & members no charge. Closed Thanksgiving; Christmas. &
Attendance: 5,000 (estimated)
Membership: Baby Doll $20; Rag Doll $35; Teddy Bear $100; China Head $250; Queen Anne Doll $500; Angel $1,000.

LAURENS HENRY COHN, SR. MEMORIAL PLANT ARBORETUM, 12206 Foster Rd., Baton Rouge, LA 70811-1231. Mailing Address: 12206 Foster Rd., Baton Rouge, LA 70811. Tel.: 225-775-1006. Fax: 225-775-1006.
Web Site: www.brec.org/cohn
Founded: 1965.
Congressional District: 6
Key Personnel: Mgr. Horticulture, K. Ed Norred.
Personnel Profile: Full-Time Paid 8.
Governing Authority: society. East Baton Rouge Parish Recreation & Park Commission, 6201 Florida Blvd., Baton Rouge, LA 70811. Tel. 504-272-9200. Tax-exempt.
Institution Type/Description: Arboretum & Botanical Garden.
Collections: native plant collection; rare & unusual plants of the region; plants from other countries; herbs; herbs of fragrance; Japanese maple collection.
Research Fields: wildflowers; tropicals; adaptable species.
Facilities: classrooms; greenhouses.
Activities: guided tours; lectures; plant recording system; seed exchange program.
Publications: newsletter, Arbornotes; Plantnotes Flyer; Review & Prospectus; Guidenotes.
Hours & Admission Prices: Daily 8-5. No charge. &
Attendance: 2,500 (accurate)

*** LOUISIANA ART & SCIENCE MUSEUM - IRENE W. PENNINGTON PLANETARIUM, (M),** 100 River Rd. S., Baton Rouge, LA 70802-5730. Tel.: 225-344-5272 & 9478. Fax: 225-344-9477.
E-mail: lasm@lasm.org
Web Site: www.lasm.org
Founded: 1960.
Congressional District: 4
Key Personnel: Chm., Gary Phillips; Pres. & Exec. Dir., Carol S. Gikas; Asst. Dir., Sam Losavio; Comm. Coord., Tara Kistler; Dir. Planetarium, Jon Elvert; Dir. Devel., Pamela Sills; Museum Cur., Elizabeth Chubbuck Weinstein; Publications Mng. & Editor, Sheri Gibson; Cur. Art Education, Tammy Johnston; Cur. Science Education, Nita Mitchell; Collections Mgr., Jennifer Mayer; Chm. (V) & Special Events Coord., Leslie Charleville; Museum Shop Mgr., Paula Taylor; Membership Sec., Barbara Miller; Planetarium Mgr. & Technical Dir., David Kors; Operations Mgr., Claire Sassic; Planetarium Educator, Chandra Weathers; Planetarium Educator, Sheree Westerhaus; Bldg. Supvr., Robert Gourgues; Asst. Dir. Devel., Katie Allen.
Personnel Profile: Full-Time Paid 25; Part-Time Paid 35; Part-Time Volunteers 75; Interns 2.
Governing Authority: nonprofit organization. Parent Institution: Louisiana Art & Science Museum. Tax-exempt: 501(c)(3).
Institution Type/Description: Art & Science Museum: housed in renovated & expanded former Illinois Central Railroad Station; space theater.
Collections: Louisiana modern & contemporary art from 1900 to present; American & European art from 18th-21st centuries including Ivan Mestrovic sculptures & drawings, Charles Burchfield, John Marin, Dale Chihuly; photographs; ethnographic art including Inuit, Tibetan, Native American, African, & pre-Colombian; antiquities including a Ptolemaic-era mummy; scientific artifacts including meteorites & star charts; hands-on children's exhibits.
Major Exhibits: Fritz Bullman: An American Abstractionist, 10/13-1/14.
Research Fields: art history, particularly Louisiana modernists; American contemporary art; astronomy.
Facilities: planetarium; theatre; auditorium; classrooms. Museum-related items for sale.
Activities: permanent & traveling exhibitions; large format films; planetarium sky shows; art & astronomy lectures; school classes & tours; public education programs for adults & children.
Publications: monthly publication, Connect; special art exhibition catalogues.
Hours & Admission Prices: Tues.-Fri. 10-3, Sat. 10-5, Sun. 1-4. Planetarium: Tues. Fri. 10 3, Sat. 10 8, Sun. 1 4. Galleries: adults $7.25; discounts to seniors, children, AAA & ASTC members; members no charge. Galleries Plus Theater: adults $9; discounts to members, seniors, children, AAA & ASTC members. Closed major holidays. &
Attendance: 198,807 (accurate)
Membership: Student & Teacher $25; Friend $50; Family I $75; Family II $100; Contributor $150; Patron $300; Supporting $500; Sustaining $1,000; Premier $2,500.

LOUISIANA MUD PAINTINGS, 16950 Strain Rd., Baton Rouge, LA 70816-1823. Tel.: 225-275-5126.
Institution Type/Description: Art Gallery.
Collections: paintings by Henry Neubig using mud & clay found in Louisiana.

Hours & Admission Prices: Tues.-Sat. 10-5.

LOUISIANA STATE MUSEUM - BATON ROUGE, 660 N. 4th St., Baton Rouge, LA 70802. Tel.: 225-342-5428.
Institution Type/Description: History Museum.
Collections: local history, culture & industry; period furnishings; personal artifacts; photographs.
Activities: group tours; rental facilities.
Hours & Admission Prices: Tues.-Fri. 10-5, Sat. 9-5, Sun. 1-5. No charge.

LOUISIANA STATE UNIVERSITY HERBARIUM, A257 Life Sciences Annex, LSU, Baton Rouge, LA 70803-1715. Mailing Address: 202 Life Sciences Building, Dept. of Biological Sciences, LSU, Baton Rouge, LA 70803-1715. Tel.: 225-578-8564. Fax: 225-578-2597.
E-mail: leu@lsu.edu
Web Site: www.herbarium.lsu.edu
Founded: 1869.
Congressional District: 6
Key Personnel: Dir., Lowell E. Urbatsch; Co-Dir., Meredith Blackwell; Collections Mgr., Jennifer Kluse.
Personnel Profile: Full-Time Paid 1; Part-Time Paid 4.
Governing Authority: university. Parent Institution: Louisiana State University. Subsidiary Institution: Dept. of Biological Sciences. Tax-exempt.
Institution Type/Description: Herbarium.
Collections: vascular plants of Louisiana & U.S.; Asteraceae; Fabaceae; Poaceae; tropical America.
Research Fields: morphological & biosystematic studies of plants; monographic botany; Louisiana area flora; biochemical systematics; compositae; DNA systematics.
Facilities: library.
Activities: scientific studies; research; identification of vascular plants; classes in taxonomy & ecology; local field trips.
Hours & Admission Prices: Mon.-Fri. 8:30-4:30. No charge; donations accepted.
Attendance: 150 (estimated)

*** LOUISIANA STATE UNIVERSITY MUSEUM OF ART, (M),** Shaw Center for the Arts, 100 Lafayette St., Baton Rouge, LA 70801-1201. Tel.: 225-389-7200. Fax: 225-389-7219.
E-mail: radam14@lsu.edu
Web Site: www.lsu.edu/lsumoa
Founded: 1959.
Congressional District: 6
Key Personnel: Exec. Dir., Jordana Pomeroy, Ph.D.; Chm. (V), Fran Harvey; Asst. Dir., Fran Huber; Assoc. Dir. Devel., Fairleigh Jackson; Cur., Natalie Mault; Dir. Mktg., Renee Payton; Administrative Asst., Becky Abadie; Museum Shop Mgr., LeAnn Russo.
Personnel Profile: Full-Time Paid 9; Part-Time Paid 16; Part-Time Volunteers 15; Interns 2.
Governing Authority: state. Parent Institution: Louisiana State University. Tax-exempt: 501(c)(3).
Institution Type/Description: Art Museum.
Collections: American contemporary art; decorative arts; fine arts; photography; prints; Asian, European & American art; 18th-20th century American & European ceramics, drawings, paintings; furniture; sculpture; Newcomb College ceramics; regional silver; Chinese jade; American photographers.
Major Exhibits: Tiffany Glass: Painting With Color and Light, 11/9/13-2/6/14; Rooted Communities: The Art of Nari Ward, 2/14-8/14; The Visual Blues: The Harlam Renaissance (T), 3/14-7/14.
Research Fields: decorative arts; paintings; graphics; sculpture; works on paper; photography.
Facilities: 500-vol. library of material pertaining to paintings, prints, drawings & decorative arts.
Activities: guided tours; lectures; films; gallery talks; inter-museum loan, permanent, temporary & traveling exhibitions.
Publications: exhibit catalogs; Whispers From the Stone; The James R. & Ann A. Peltier Collection of Chinese Jade; ARTALK.
Hours & Admission Prices: Tues.-Wed. & Fri.-Sat. 10-5, Thurs. 10-8, Sun. 1-5. Adults $5; discounts to AAM & NARM members; members no charge. Closed New Year's Day; Mardi Gras; Easter; Thanksgiving; Christmas Eve & Day.
Attendance: 32,500 (estimated)
Membership: Student $15; Individual $40; Dual & Family $60; Patron $100; Sustaining $250; Benefactor $500; Endowment Society $1,000; Corporate: $2,500; $5,000; $10,000.

*** LOUISIANA'S OLD STATE CAPITOL, (M),** 100 North Blvd., Baton Rouge, LA 70801-1502. Mailing Address: State of Louisiana, Secretary of State, P.O. Box 94125, Baton Rouge, LA 70804-9125. Tel.: 225-342-0500; 800-488-2968. Fax: 225-342-0316.
E-mail: osc@sos.louisiana.gov
Web Site: www.sos.louisiana.gov/osc
Founded: 1994.
Congressional District: 6
Key Personnel: Dir., Suzette Crocker; Public Rels., Nancy Chesson; Museum Shop Mgr., Charlotte Wall.
Personnel Profile: Full-Time Paid 21; Part-Time Paid 6.
Governing Authority: state. Parent Institution: Office of the Secretary of State, Baton Rouge, LA. Tax-exempt.
Institution Type/Description: Political History Museum.
Collections: political & governmental history of Louisiana; memorabilia; audio & video tapes.
Research Fields: history, politics & government, 1780's-present.
Facilities: theater. Museum-related items for sale.
Activities: arts festivals; docent program; films; guided tours; lectures; loan, temporary, traveling, interactive & participatory exhibitions; formal education programs; rental gallery; theater; broadcast programs. Annual Events: Red Stick Animation Festival; Veterans Day; Santa in the Senate.
Hours & Admission Prices: Tues.-Sat. 9-4. No charge. Closed most state holidays.
Attendance: 26,500 (accurate)
Membership: Call for information.

LSU STUDENT UNION ART GALLERY, (M), 210 LSU Student Union, Louisiana State University, Baton Rouge, LA 70803. Mailing Address: LSU Student Union Art Gallery, P.O. Box 25123, Baton Rouge, LA 70803. Tel.: 225-578-5162. Fax: 225-578-4329.
E-mail: unionartgallery@lsu.edu
Key Personnel: Dir., Judy Stahl
Institution Type/Description: Art Gallery.
Collections: paintings; drawings; lithographs; silkscreens; photographs; sculptures.
Activities: special events; educational programs.
Hours & Admission Prices: Mon.-Fri. 10-6, Sun. 1-5. No charge. Closed major holidays.

MUSEUM OF AFRICAN AMERICAN HISTORY, 538 S. Boulevard, Baton Rouge, LA 70802-6442. Tel.: 225-343-4431.
E-mail: oswafricanamericanmuseum@gmail.com
Founded: 2001.
Congressional District: 6
Key Personnel: Founder & Cur., Sadie Roberts-Joseph.
Governing Authority: Tax-exempt.
Institution Type/Description: History Museum.
Collections: minority inventors & inventions; African art; period artifacts; wood carvings; stationary trolley.
Activities: Annual Events: Dr. Martin Luther King Jr. Holiday in January; State Juneteenth Celebration in June; Veterans Day Celebration in November.
Hours & Admission Prices: Wed.-Sat. 10-5; other times by appointment. Adults $4; discounts to AAM & ICOM members. Closed major holidays.
Attendance: 1,500
Membership: Student $8; Individual $35; Friend $100; Life $200.

MUSEUM OF NATURAL SCIENCE, 119 Foster Hall, LSU, Baton Rouge, LA 70803. Tel.: 225-578-2855. Fax: 225-578-3075.
E-mail: museum@lsu.edu
Web Site: appl003.lsu.edu/natsci/lmnh.nsf/index
Founded: 1936.
Congressional District: 6
Key Personnel: Dir., Dr. Frederick H. Sheldon; Cur. Fishes, Dr. Prosanta Chakrabarty; Cur. Birds, Dr. J.V. Remsen, Jr.; Cur. Paleontology, Dr. Judith A. Schiebout; Cur. Anthropology, Dr. Rebecca A. Saunders; Cur. Herpetology, Dr. Christopher C. Austin; Dir. Education, Dr. Sophie Warny; Cur. Animals, Dr. Mark Hafner; Cur. Genetic Resource, Dr Robb Brumfield.
Personnel Profile: Full-Time Paid 17; Part-Time Paid 6; Part-Time Volunteers 8.
Governing Authority: university. Parent Institution: Louisiana State University. Tax-exempt: 501(c)(3).
Institution Type/Description: Natural History Museum.
Collections: ornithology; mammalogy; ichthyology; herpetology; frozen tissues; paleontology; archaeology; anthropology.

Research Fields: ornithology; mammalogy; ichthyology; herpetology; systematics; zoogeography; evolutionary biology; ecology; behavior; paleontology; archaeology; anthropology.
Facilities: vertebrate research collections; biochemical systematics laboratories; morphology analysis laboratory; karyology laboratory; sound analysis laboratory; computer laboratory; reprint library; classrooms; dermestid colony; paleontological collections; archaeological & anthropological collections.
Activities: research programs by staff & graduate students; permanent & temporary exhibits.
Publications: Occasional papers, Museum of Natural Science.
Hours & Admission Prices: Mon.-Fri. 8-4. No charge. Closed university holidays. &
Attendance: 32,000
Membership: Friends of the LSU Museum of Natural Science (Support group) $25.

OLD ARSENAL MUSEUM, 900 Capitol Lake Dr., Baton Rouge, LA 70802. Mailing Address: P.O. Box 94125, Baton Rouge, LA 70804-9125. Tel.: 225-342-0401. Fax: 225-342-5577.
E-mail: arsenal@sos.louisiana.gov
Web Site: www.sos.louisiana.gov/oam
Formerly: Old Arsenal Powder Magazine Museum
Key Personnel: Dir., Gregory Leggio.
Personnel Profile: Full-Time Paid 1.
Governing Authority: state. Parent Institution: Louisiana Secretary of State, Baton Rouge, LA. Tax-exempt.
Institution Type/Description: History Museum: built in 1838; used by the US military during the Mexican & Civil Wars. Listed on the National Register of Historic Places.
Collections: Louisiana heritage; military history; photographs.
Activities: guided group tours; participatory exhibits.
Hours & Admission Prices: Tues.-Sat. 9-4. No charge. Closed most state holidays. &
Attendance: 3,107 (accurate)

OLD GOVERNOR'S MANSION, 502 North Blvd., Baton Rouge, LA 70802. Mailing Address: P.O. Box 908, Baton Rouge, LA 70821. Tel.: 225-387-2464. Fax: 225-343-3989. Facebook: Old Governor's Mansion.
E-mail: info@fhl.org
Web Site: www.oldgovernorsmansion.org
Formerly: Old Louisiana Governor's Mansion
Founded: 1930.
Congressional District: 6
Key Personnel: Dir., Carolyn Bennett; Chm. (V), Doug Cochran; Museum Shop Mgr., Selena Grant; Dir. Education, Amelia Gilmore.
Personnel Profile: Full-Time Paid 5; Part-Time Paid 6; Part-Time Volunteers 20; Interns 3.
Volunteer Hours: 1,800
Operating Expenses: 469,262
Operating Income: 469,981
Governing Authority: Parent Institution: Foundation for Historical Louisiana with the State of Louisiana. Tax-exempt.
Institution Type/Description: Historic Mansion: built in 1930. Listed on the National Register of Historic Places.
Collections: local history & culture; period furnishings; photographs; personal artifacts; political history & artifacts; Magnolia Cemetery.
Research Fields: preservation; history of LA government; conservation; architectural; cultural.
Activities: school tours; heritage lectures; lunch & learn; group tours; children's camps.
Hours & Admission Prices: Tours: Tues.-Fri. 10-4. Adults $7; discounts to AAM & ICOM members; members no charge. &
Attendance: 5,000 (estimated)
Membership: Young Preservationist & Inherit Baton Rouge $35; Family $50-$99; Preservation Associate $100-$249; Preservation Principal $250-$1,000.

ROBERT A. BOGAN FIRE MUSEUM, 427 Laurel St., Baton Rouge, LA 70801-1810. Tel.: 225-344-8558. Fax: 225-344-7777.
E-mail: katherine@acgbr.com
Web Site: www.artsbr.org
Formerly: Old Bogan Firefighters Museum
Founded: 1924.
Congressional District: 6
Key Personnel: Deputy Dir., Katherine Scherer.
Personnel Profile: Part-Time Volunteers 5.

Governing Authority: nonprofit. Tax-exempt: 501(c)(3).
Institution Type/Description: Fire-Fighting Museum.
Collections: fire helmets, extinguishers, alarms boxes & horns; trophies; 1919 Type 17 Motor Aerial Truck with 75 ft. ladder; original brass poles; photographs.
Activities: guided tours; organized education programs for children.
Hours & Admission Prices: By appointment. No charge; donations accepted. &
Attendance: 3,000 (estimated)

RURAL LIFE MUSEUM & WINDRUSH GARDENS, 4560 Essen Lane, Baton Rouge, LA 70809-3424. Mailing Address: P.O. Box 80498, Baton Rouge, LA 70898-0498. Tel.: 225-765-2437. Fax: 225-765-2639.
E-mail: rurallife@lsu.edu
Web Site: rurallife.lsu.edu
Founded: 1970.
Congressional District: 6
Key Personnel: Exec. Dir., David Floyd; Interpretive Specialist, Catherine White; Registrar & Conservator, David Nicolosi; Dir. Mktg., Elizabeth McInnis; Devel. Dir., Tonja Normand.
Personnel Profile: Full-Time Paid 5; Part-Time Paid 5; Part-Time Volunteers 150; Interns 1.
Governing Authority: public university; nonprofit. Parent Institution: Louisiana State University. Tax-exempt: 501(c)(3).
Institution Type/Description: History Museum.
Collections: local history exhibits; glass; ceramics; metalwares; farm equipment & tools; tools for various trades; textiles; furniture; medical items; lighting devices; logging, hunting, trapping, & fishing items; transportation; Civil War items; toys. Historic Buildings: c.1835 general store; c.1835 overseer's house; c.1855 kitchen; c.1830-1840 sick house; c.1853 schoolhouse; c.1835 blacksmith's shop; c.1835 slave cabins; c.1870 country church; c.1840 pioneer's cabin & corncrib; c.1870 dogtrot house; cane grinder & sugarhouse; grist mill; acadian house; shotgun house.
Facilities: 1,200-vol. library of books, periodicals & related material pertaining to Louisiana & the South available for use on premises only; botanical garden.
Activities: guided tours; lectures; organized education programs for children, undergraduate, & graduate students; docent program.
Publications: museum guidebook.
Hours & Admission Prices: Daily 8:30-5. Adults & children 12-61 $7, seniors 62 & over $6, children 5-11 $5; children under 5 no charge. Closed New Year's Day; Easter; Thanksgiving; Christmas Eve & Day. &
Attendance: 15,000 (accurate)

U.S.S. KIDD VETERANS MEMORIAL, 305 S. River Rd., Baton Rouge, LA 70802-6220. Tel.: 225-342-1942. Fax: 225-342-2039.
E-mail: info@usskidd.com
Web Site: www.usskidd.com
Founded: 1981.
Congressional District: 6
Governing Authority: state; nonprofit. Affiliated with Louisiana State Department of Culture & Louisiana Naval War Memorial Commission & Foundation. Tax-exempt: 501(c)(3).
Institution Type/Description: Maritime Museum & Historic Ship.
Collections: U.S. Navy shipboard destroyer operations including weapons, electronics, lifestyle, uniforms, books & documents; maritime history represented by ship models, paintings, photographs & various artifacts; CBI display featuring restored P-40 fighter plane; documents, photographs, & memorabilia associated with Flying Tigers. Historic Ship: 1942 USS KIDD-Fletcher Class; a replica of USS Constitution's gun deck, Old Iron Sides; Corsair A-7 aircraft in tribute to Vietnam Veterans.
Research Fields: destroyer operations; design & development.
Facilities: 500-vol. library of blueprints, documents, technical manuals, training manuals, publications & nautical charts relating to United States Navy destroyer; 100-seat theater; 30-seat snack bar. Museum-related items for sale.
Activities: guided tours; lectures; docent program; temporary, traveling & loan exhibitions; school loan service; shipboard overnight camping for youth groups.
Publications: quarterly newsletter, KIDD's Compass; book, U.S.S. KIDD (DD-661) Technical History; documentary, Fletcher Class destroyers.
Hours & Admission Prices: Daily 9-5. Ship & Museum: adults $8, senior citizens 60 & over $7, active military with ID $6, children 5-12 $5; discounts for groups of 20 or more; members & children 4 and under no charge. Museum only: adults $5, children 5-12 $4; children 4 & under no charge. Closed Thanksgiving; Christmas Eve & Day; New Year's Eve & Day. &
Attendance: 53,303

Membership: Individual $30; Couple $40; Family $50; Extended Family $75; Contributor $100; Supporter $250; Patron-Bronze $500; Patron-Silver $750; Patron-Gold $1,000; Platinum Lifetime $1,500.

USS KIDD VETERANS MEMORIAL & MUSEUM, 305 S. River Rd., Baton Rouge, LA 70802-6220. Tel.: 225-342-1942. Fax: 225-342-2039.
Institution Type/Description: Military Museum: housed on a U.S. Navy destroyer.
Collections: destroyer history; military equipment; personal artifacts; photographs.
Facilities: Museum-related items for sale.
Hours & Admission Prices: Call for hours. Ship & Museum: adults 13 & over $7, seniors 60 & over $6, children 5-12 $4; children 4 & under no charge. Museum: adults 13 & over $4, children 5-12 $3; children 4 & under no charge.

Belle Chasse

THE TULANE UNIVERSITY MUSEUM OF NATURAL HISTORY, 3705 Main St., Bldg. A-3, Belle Chasse, LA 70037-3001. Tel.: 504-394-1711. Fax: 504-394-5045.
E-mail: hank@museum.tulane.edu
Web Site: www.museum.tulane.edu
Key Personnel: Cur. Fish, Dr. Henry L. Bart; Collections Mgr., Nelson Rios; Adjunct Cur. Mammals, Dr. Craig Hood
Institution Type/Description: Natural History Museum.
Collections: invertebrates; fish; amphibians; reptiles; birds; mammals; vertebrate fossils.
Activities: school groups; visiting scientists.
Hours & Admission Prices: Mon.-Fri. 8:30-5 by appointment.

Bernice

BERNICE DEPOT MUSEUM, Fourth and Louisiana St., Bernice, LA 71222. Mailing Address: P.O. Box 633, Bernice, LA 71222-0633. Tel.: 318-285-2433.
Key Personnel: Museum Shop Mgr., Gladys Harkins
Institution Type/Description: Historic Building: built in 1899.
Collections: railroad history & memorabilia; period artifacts; tools; wooden ox yoke.
Activities: special events.
Hours & Admission Prices: Mon.-Fri. 10-12 & 1-3. No charge; donations accepted.

Bossier City

ARK-LA-TEX MARDI GRAS MUSEUM, 2101 E. Texas St., Bossier City, LA 71111. Mailing Address: P.O. Box 6432, Bossier City, LA 71171-6432. Tel.: 318-218-2865.
Institution Type/Description: Mardi Gras History Museum.
Collections: Mardi Gras history & costumes; photographs.
Activities: special events.
Hours & Admission Prices: By appointment.

Broussard

ZOO OF ACADIANA, 5601 Hwy. 90 E., Broussard, LA 70518. Mailing Address: 116 Lakeview Dr., Broussard, LA 70518-8004. Tel.: 337-837-4325. Fax: 337-837-4253.
E-mail: wild@zooofacadiana.org
Key Personnel: Co Owner, George Oldenburg; Co Owner, Marleen Oldenburg
Institution Type/Description: Zoo.
Collections: over 500 animals representing more than 125 species.
Facilities: Museum-related items for sale.
Activities: train rides; special events; school groups. Annual Events: Zoolebrate; Eggstravaganzoo; Boo at the Zoo; Safari of Lights.
Hours & Admission Prices: Jan.-Nov. daily 9-5; Dec. daily 9-4. Safari of Lights: Dec. 5pm-9pm. Adults 13-54 $9.25, seniors 55 & over $8.25, children 3-12 $5.50; children 2 & under no charge. Closed New Year's Day; Easter; Thanksgiving; Christmas.

Carville

NATIONAL HANSEN'S DISEASE MUSEUM, Bldg. 12, Carville Historic District, 5445 Point Clair Rd., Carville, LA 70721-2119. Mailing Address: 1770 Physicians Dr., Baton Rouge, LA 70816. Tel.: 225-642-1950. Fax: 225-642-1949.
E-mail: NHDPmuseum@hrsa.gov
Web Site: www.hrsa.gov/hansens/museum/default.htm
Founded: 1996.
Key Personnel: Dir., Elizabeth Schexnyder; C.E.O., Dr. James Krahenbuhl; Interim Cur., Vicki Joseph.
Personnel Profile: Full-Time Paid 1; Part-Time Paid 1; Part-Time Volunteers 1.
Governing Authority: Parent Institution: National Hansen's Disease Programs. Tax-exempt.
Institution Type/Description: History & Medical Museum.
Collections: medical & social history; photographs; PHS Hospital artifacts.
Publications: With Love in Their Hearts: 1896-1996 Daughters of Charity at Carville; 100 Years - Carville Centennial.
Hours & Admission Prices: Tues.-Sat. 10-4. No charge; donations accepted. Closed federal holidays. &
Attendance: 2,000 (accurate)

Charenton

CHITIMACHA MUSEUM, 3289 Chitimacha Trail, Charenton, LA 70523. Mailing Address: P.O. Box 661, Charenton, LA 70523-0661. Tel.: 337-923-4830.
Web Site: www.chitimacha.org
Institution Type/Description: Native American Museum.
Collections: Chitimacha history & culture; Native American artifacts; baskets.
Activities: videos.
Hours & Admission Prices: Tues.-Sat. 9-4:30. No charge. &

Columbia

THE SCHEPIS, LOUISIANA ARTISTS MUSEUM, 106 Main St., Columbia, LA 71418. Mailing Address: P.O. Box 743, Columbia, LA 71418-0743. Tel.: 318-649-9931.
Key Personnel: Cur., Jane Meredith
Institution Type/Description: Art Museum: built by John Schepis c.1916. Listed on the National Register of Historic Places.
Collections: works by Louisiana artists.
Hours & Admission Prices: Call for hours. No charge; donations accepted.

Cottonport

COTTONPORT MUSEUM, 220 Cottonport Ave., Cottonport, LA 71327. Tel.: 318-876-3517.
Institution Type/Description: History Museum.
Collections: local history & culture; photographs; personal artifacts; period furnishings; murals; tools; documents.
Activities: special events.
Hours & Admission Prices: Call for hours.

Covington

INSTA-GATOR RANCH & HATCHERY, 23440 Lowe Davis Rd., Covington, LA 70435-6512. Tel.: 985-892-3669; 888-448-1560.
Institution Type/Description: Alligator Farm.
Collections: over 2,000 alligators of all sizes & ages; Louisiana alligator industry; Cajun heritage & culture.
Facilities: Museum-related items for sale.
Activities: educational programs; guided tours; video
Hours & Admission Prices: Daily call for hours. Adults $16, military $14, children 12 & under $10; discounts to groups.

Crowley

CROWLEY ART ASSOCIATION & GALLERY, 220 N. Parkerson, Crowley, LA 70526-5003. Tel.: 337-783-3747. Fax: 337-783-3747.
E-mail: gallerythe@bellsouth.net
Web Site: www.crowleyartgallery.com
Founded: 1980.
Congressional District: 7
Key Personnel: C.E.O., Hurley Gautreaux; Vice Pres., Virgie LeBlue; Museum Shop Mgr., Becky Faulk.

Personnel Profile: Full-Time Paid 1; Full-Time Volunteers 50; Part-Time Paid 1; Part-Time Volunteers 20.
Governing Authority: nonprofit organization. Parent Institution: Crowley Art Association. Tax-exempt.
Institution Type/Description: Art Association & Gallery.
Collections: pottery & crafts.
Facilities: library of art history & educational books, available for use by members only. Paintings in all media, crafts, jewelry, porcelain & needlework for sale.
Activities: guided tours; arts festivals; hobby workshops; formal education programs for children & adults; docent program; participatory exhibits. Association Sponsors: two outdoor arts & crafts shows; children's & adult workshops; poster contests; special exhibits & receptions; one outdoor show & one indoor show.
Publications: monthly, CAA Newsletter; annually, CAA Yearbook.
Hours & Admission Prices: Mon.-Fri. 10-4. No charge. Closed New Year's Day; Independence Day; Thanksgiving; Christmas week. &

Membership: Adult $20; Gallery $24; Family $25.

CRYSTAL RICE HERITAGE FARM, 6428 Airport Rd, Crowley, LA 70526-1604. Mailing Address: P.O. Box 1425, Crowley, LA 70527-1425. Tel.: 337-783-6417. Fax: 337-788-0123.
E-mail: dwrighth@cs.com
Web Site: www.crystalrice.com
Formerly: Crystal Rice Plantation
Founded: 1970.
Congressional District: 7
Key Personnel: C.E.O., Diane Hoffpauer; Treas., Elaine Wright; Dir. & Museum Shop Mgr., Redell Miller.
Governing Authority: nonprofit.
Institution Type/Description: Classic Car Museum & Historical Society: c.1848 plantation home.
Collections: period furniture & implements; rice samples; pictures; dolls; classic cars.
Research Fields: rice industry; crawfish production.
Facilities: rice research station.
Activities: tours.
Hours & Admission Prices: Mon.-Fri. 9-3 by appointment. Adults $10, senior citizens $6, students $4.50; discounts to AAA members; members no charge. Closed all holidays.
Attendance: 1,000 (accurate)

RICE INTERPRETIVE CENTER, J.D. MILLER MUSIC RE-CORDING STUDIO MUSEUM AND FORD AUTOMOTIVE MUSEUM, 425 N. Parkerson Ave., Crowley, LA 70526. Mailing Address: P.O. Box 1463, Crowley, LA 70527-1463. Tel.: 337-783-0824. Fax: 337-783-4331.
E-mail: charlotte.jeffers@crowley-la.com
Web Site: www.crowley-la.com
Founded: 2006.
Key Personnel: Tourism Coord., Charlotte R. Jeffers; Chm., Mayor Greg Jones.
Governing Authority: city. Tax-exempt.
Institution Type/Description: History Museum.
Collections: local history & culture; photographs; videos; J.D. Miller's music & life; personal artifacts; early machinery; growing & production; Ford automotive; Rice industry & history; memorabilia; two 1923 Model T's; 1928 Model A; music recording studio equipment & instruments.
Activities: video; hands-on exhibits; guided tours.
Publications: Crowley, Louisiana Visitors Travel Guide.
Hours & Admission Prices: Mon.-Fri. 8-4; other times by appointment. No charge. &
Attendance: 4,845 (accurate)

Darrow

HOUMAS HOUSE PLANTATION AND GARDENS, 40136 Hwy. 942, Darrow, LA 70725-2302. Tel.: 225-473-9380. Fax: 225-473-7891.
Institution Type/Description: Historic House.
Collections: local history & culture; period furnishings; personal artifacts. photographs; paintings.
Facilities: restaurant.
Activities: guided tours.
Hours & Admission Prices: Mon.-Tues. 9-5, Wed.-Sun. 9-7. Tours: Mansion & Gardens $20. Gardens & Grounds: $10.

DeQuincy

DEQUINCY RAILROAD MUSEUM, 400 Lake Charles Ave., DeQuincy, LA 70633. Tel.: 337-786-2823.
E-mail: dequincyrailroadmuseum@centurylink.net
Web Site: www.dequincyrailroadmuseum.com
Founded: 1976.
Congressional District: 7
Key Personnel: Pres. (V), Gary Cooper; Museum Shop Mgr., Evalin Hester.
Personnel Profile: Part-Time Paid 1.
Governing Authority: board of directors. Tax-exempt.
Institution Type/Description: Railroad Museum.
Collections: local history & culture; railroad artifacts & records; period furnishings; photographs; steam engine; passenger car; caboose; mow speeders; books.
Research Fields: Kansas City Southern and Missouri Pacific miscellaneous roads.
Facilities: library; archives.
Activities: Museum Sponsors: Louisiana Railroad Days Festival in April.
Hours & Admission Prices: Tues.-Sat. 10-5. No charge; donations accepted. Closed city holidays.
Attendance: 8,332 (accurate)
Membership: Individual $5; Family $10; Lifetime $100.

DeRidder

THE LOIS LOFTIN DOLL MUSEUM, 120 S. Washington Ave., DeRidder, LA 70634-4062.
Institution Type/Description: Doll Museum.
Collections: over 3,000 dolls from around the world.
Activities: Annual Event: Louisiana Doll Festival.
Hours & Admission Prices: Tues.-Sat. 10-4.

Destrehan

DESTREHAN PLANTATION, 13034 River Rd., Destrehan, LA 70047-5202. Tel.: 985-764-9315; 877-453-2095. Fax: 985-725-1929.
E-mail: info@destrehanplantation.org
Web Site: www.destrehanplantation.org
Key Personnel: Site Mgr., Nancy Robert
Institution Type/Description: Plantation: a National Historic Landmark, established in 1787.
Collections: period furnishings; personal artifacts; photographs.
Hours & Admission Prices: Daily 9-4. Adults $18, children 6-16 $7; discounts to groups. Closed major holidays.

Donaldsonville

RIVER ROAD AFRICAN AMERICAN MUSEUM, (M), 406 Charles St., Donaldsonville, LA 70346-3312. Mailing Address: P.O. Box 266, Donaldsonville, LA 70346-0266. Tel.: 225-474-5553.
E-mail: kathe@africanamericanmuseum.org
Web Site: www.africanamericanmuseum.org
Key Personnel: Dir., Kathe Hambrick
Institution Type/Description: History Museum.
Collections: local African American history & culture; books; photographs; period artifacts.
Hours & Admission Prices: Wed.-Sat. 10-5, Sun. 1-5; groups of 10 or more by appointment. Museum or School & Church Tour: adults $4. Heritage Tour: adults $25.

Edgard

EVERGREEN PLANTATION, 4677 Hwy. 18, Edgard, LA 70049. Tel.: 985-497-3837.
E-mail: evergreenplantation@gmail.com
Web Site: evergreenplantation.org
Institution Type/Description: Historic Plantation: listed on the National Register of Historic Places.
Collections: plantation history & culture. Historic Buildings: plantation house; 22 slave cabins.
Activities: guided tours.
Hours & Admission Prices: Guided Tours: Mon.-Sat. 9:30, 11:30, & 2. Adults $20, children 5-18 $6; discounts to students; military, seniors & groups of 10 or more; children under 5 no charge. Closed New Year's Eve & Day; Thanksgiving & day after; Christmas Eve & Day.

Erath

THE ACADIAN MUSEUM, 203 S. Broadway, Erath, LA 70533-4003. Tel.: 337-233-5832. Fax: 337-235-4382.
E-mail: info@acadianmuseum.com
Web Site: www.acadianmuseum.com
Founded: 1990.
Congressional District: 7
Key Personnel: Dir., Warren A. Perrin.
Governing Authority: Parent Institution: Acadian Heritage & Culture Foundation, Inc. Tax-exempt.
Institution Type/Description: History Museum.
Collections: local history & culture; paintings; photographs; books; maps; artifacts.
Hours & Admission Prices: Mon.-Fri. 1-4; other times by appointment. No charge.

Eunice

EUNICE DEPOT MUSEUM, 220 S. C.C. Duson Dr., Eunice, LA 70535-7808. Tel.: 337-457-6540 & 2565.
Institution Type/Description: Historic Site: housed in the building where C.C. Duson sold the first land sites & named the town after his wife Eunice; 1893-1894. Listed on the National Register of Historic Places.
Collections: town history; Cajun Mardi Gras & music; period toys; railroad artifacts; pioneer farming; Native American culture.
Facilities: Museum-related items for sale.
Activities: craft workshops.
Hours & Admission Prices: Tues.-Sat. 8-12 & 1-5. No charge.

PRAIRIE ACADIAN CULTURAL CENTER, 250 W. Park Ave., Eunice, LA 70535-4628. Tel.: 337-457-8499.
Institution Type/Description: History Museum.
Collections: Cajun life, music & heritage; photographs.
Facilities: auditorium.
Activities: educational programs; special events; demonstrations.
Hours & Admission Prices: Tues.-Fri. 8-5, Sat. 8-6. Closed Christmas.

Ferriday

DELTA MUSIC MUSEUM, 218 Louisiana Ave., Ferriday, LA 71334-2828. Mailing Address: State of Louisiana, Secretary of State, P.O. Box 94125, Baton Rouge, LA 70804-9125. Tel.: 318-757-9999. Fax: 318-757-1973.
E-mail: deltamusic@sos.louisiana.gov
Web Site: www.sos.louisiana.gov/dmm
Founded: 2001.
Congressional District: 5
Key Personnel: Dir., Judith Bingham.
Personnel Profile: Full-Time Paid 1; Part-Time Paid 6.
Governing Authority: state. Parent Institution: Louisiana Secretary of State, Baton Rouge, LA. Tax-exempt.
Institution Type/Description: Music Museum.
Collections: local & music history; Louisiana-Mississippi Delta region culture & music; mannequins; interactive music kiosks; Jerry Lee Lewis; Mickey Gilley; Jimmy Swaggart; Mississippi River Delta musicians including Leon (Pee Wee) Whittaker, Jimmie Davis, Conway Twitty, & Aaron Neville.
Facilities: theater. Museum-related items for sale.
Activities: loan, traveling & interactive exhibits; live performances; concerts; films; guided tours; theater; arts festivals; dance recitals; docent program; lectures. Annual Event: Delta Music Festival in April.
Hours & Admission Prices: Mon.-Fri. 9-4; student groups by appointment. No charge. Closed most state holidays. &
Attendance: 4,067 (accurate)
Membership: Call for information.

Folsom

GLOBAL WILDLIFE CENTER, 26389 Hwy. 40, Folsom, LA 70437. Tel.: 985-796-3585.
Founded: 1991.
Key Personnel: Museum Shop Mgr., Beth Kuhnau.
Governing Authority: Tax-exempt.
Institution Type/Description: Wildlife Center.
Collections: over 4,000 animals from around the world.
Activities: birthday parties.
Hours & Admission Prices: Daily call for hours. Suggested Donations: adults $17, seniors $13, children $10; discounts to groups. &
Attendance: 350,000

Membership: Family $100; Family Plus $150; Lifetime $1,000.

Fort Polk

FORT POLK MILITARY MUSEUM, 7881 Mississippi Ave., Fort Polk, LA 71459. Tel.: 337-531-7905 & 4840.
Key Personnel: Dir., Fred Adolphus
Institution Type/Description: Military History Museum.
Collections: local history; military equipment; personal artifacts; photographs.
Hours & Admission Prices: Tues.-Sat. 9-4. No charge. Closed federal holidays.

FORT POLK MUSEUM ACTIVITY, Bldg. 917 S. Carolina Ave., Fort Polk, LA 71459. Mailing Address: P.O. Box 3916, Fort Polk, LA 71459-0916. Tel.: 337-531-7905 & 4840. Fax: 337-531-4202.
E-mail: binghamd@polk.army.mil
Formerly: Fort Polk Historical Holding Area
Founded: 1972.
Congressional District: 4
Key Personnel: Historian & Cur., David S. Bingham.
Personnel Profile: Full-Time Paid 1.
Governing Authority: federal. Tax-exempt.
Institution Type/Description: Military Museum.
Collections: artifacts & memorabilia pertaining to the history of Camp/Fort Polk & Divisions stationed there since 1941; items from WWII, Korean War & Vietnam; 1940-1944 Louisiana Maneuvers; 7,000 military photographs; 2-acre outdoor park displaying: armor; helicopters; artillery; vehicles; missiles.
Research Fields: U.S. Army military history from WWII to present.
Facilities: 6,000-vol. library of books & publications on U.S. army history, military equipment data & museum professional publications available for research on premises; outdoor display park.
Activities: lectures upon request; permanent exhibitions.
Hours & Admission Prices: Wed.-Fri. 10-2, Sat.-Sun. 9-4. No charge. Closed major holidays. &
Attendance: 15,000 (estimated)

Franklin

GREVEMBERG HOUSE MUSEUM, 407 Sterling Rd., Hwy. 322, Franklin, LA 70538-0400. Mailing Address: P.O. Box 400, Franklin, LA 70538-0400. Tel.: 337-828-2092. Fax: 337-828-2028.
E-mail: info@grevemberghouse.com
Web Site: www.grevemberghouse.com
Founded: 1972.
Congressional District: 3
Key Personnel: Pres., Pam Heffner; Treas. & Public Rels., Didi Battle; Archivist, Margie Luke; Lead Interpreter, Craig Landry.
Personnel Profile: Part-Time Paid 2; Part-Time Volunteers 17.
Governing Authority: private; nonprofit organization. Tax-exempt: 501(c)(3).
Institution Type/Description: Historical Society Museum: housed in c.1851 Greek-revival townhouse. Listed on National Register of Historic Places.
Collections: 1820-1870 furniture; 19th century life in south Louisiana; Civil War artifacts.
Research Fields: translation of Grevemberg family papers; period furnishings.
Facilities: 5,000 sq. ft. exhibit space.
Activities: guided tours. Annual Event: Victorian Christmas Celebration.
Publications: semiannual newsletter, Landmark Lagniappe.
Hours & Admission Prices: Daily 10-4. Adults $10, senior citizens & students 12-18 $8, children under 12 $5; discounts to groups of 20 or more; members no charge. Closed New Year's Day; Good Friday; Easter; Thanksgiving; Christmas Eve & Day. &
Attendance: 262 (accurate)
Membership: Foundation $35; Pillar (couple) $60; Conservator $75; Gothic $100; Victorian $250; Queen Anne $500; Corinthian $1,000-$1,499; Greek Revival $1,500-$1,999; Italianate $2,000-$3,999; Landmark $4,000 & up.

OAKLAWN MANOR, 3296 E. Oaklawn Dr., Franklin, LA 70538-3218. Tel.: 337-828-0434. Fax: 337-828-1937.
E-mail: oaklawnmanor@yahoo.com
Web Site: oaklawnmanor.com
Founded: 1837.
Congressional District: 3
Key Personnel: C.E.O., Murphy J. Foster, Jr.
Institution Type/Description: Historic House Museum.
Collections: local history & culture; period furnishings; personal artifacts.
Hours & Admission Prices: Tues.-Sun. 10-4. Adults $15, students $10; discounts to groups of 10 or more.
Attendance: 500 (estimated)

YOUNG SANDERS CENTER, 104 Commercial St., Franklin, LA 70538-5427. Mailing Address: P.O. Box 545, Franklin, LA 70538-0545.
E-mail: ysc1861@aol.com
Web Site: www.youngsanders.org
Institution Type/Description: History Museum.
Collections: artifacts; documents; journals; period maps; photographs.
Hours & Admission Prices: Mon.-Fri. 9-5. No charge. &

Franklinton

WASHINGTON AREA MUSEUM FOUNDATION/VARNADO STORE MUSEUM, (M), 936 Pearl St., Franklinton, LA 70438-1736. Mailing Address: P.O. Box 184, Franklinton, LA 70438-0184. Tel.: 985-795-0680. Fax: 985-795-0680.
E-mail: varnadostoremuseum@franklinton.net
Web Site: www.varanadostoremuseum.org
Founded: 1996.
Key Personnel: Pres., Mary Jo Poole; Dir., Terry Seal.
Governing Authority: Tax-exempt.
Institution Type/Description: History Museum.
Collections: local history & heritage.
Facilities: Museum-related items for sale.
Activities: lectures; demonstrations; group tours; school field trips. Museum Sponsors: Street Fair in Spring; Christmas Festival in December.
Publications: members newsletter.
Hours & Admission Prices: Fri. by appointment, Sat. 10-4, Sun. 1-4. No charge; donations accepted. &
Attendance: 1,500 (estimated)
Membership: Student $5; Teacher's Circle & Individual $25; Family $35; Corporate, Institution & Organization $100; Bronze $300; Silver $500; Gold $1,000.

Frierson

YOGIE AND FRIENDS EXOTIC CAT SANCTUARY, 128 Fob Lane, Frierson, LA 71027-2065. Tel.: 318-795-0455.
Key Personnel: Exec. Dir., Jenny Senier
Institution Type/Description: Wildlife Sanctuary.
Collections: rescued tigers, lions, leopards, cougars, & servals.
Facilities: Museum-related items for sale.
Activities: special events; birthday parties; guided tours.
Hours & Admission Prices: Sat. 12-5.

Frogmore

FROGMORE COTTON PLANTATION & GINS - COTTON THEN & NOW, 11054 Hwy. 84, Frogmore, LA 71334-4655. Tel.: 318-757-2453 & 3333. Fax: 318-757-6535.
Web Site: www.frogmoreplantation.com
Personnel Profile: Full-Time Paid 2.
Institution Type/Description: History Museum.
Collections: plantation history; agriculture; industry; slave culture.
Hours & Admission Prices: March 8-May & Sept.-Nov. 15 Mon.-Fri. 9-3, Sat. 10-2; June-Aug. Mon.-Fri. 9-1; Winter: call for hours; other times by appointment. Historical Tour: adults 19 & over $12, students $5; children 5 & under no charge. Modern Tour: adults 19 & over and students $5; children 5 & under no charge. Complete Tour: adults 19 & over $15, students $5; children 5 & under no charge. &
Attendance: 15,000 (estimated)

Garyville

SAN FRANCISCO PLANTATION, 2646 Hwy. 44 (River Rd.), Garyville, LA 70051. Mailing Address: P.O. Box 950, Garyville, LA 70051-0950. Tel.: 888 322 1756; 985-535-2341. Fax: 985-535-5450.
E-mail: sanfran@rtconline.com
Web Site: www.sanfranciscoplantation.org
Founded: 1977.
Congressional District: 3
Institution Type/Description: History Museum: plantation built in 1856.
Collections: period furnishings; personal artifacts; photographs.
Activities: rental facilities; special events.
Hours & Admission Prices: April-Oct. daily 9:30-5; Nov.-March daily 9:40-4:40. Adults $15, students 17 & under $10; discounts to military & AAA members; children 5 & under no charge. Closed New Year's Day; Mardi Gras Day; Thanksgiving; Christmas.

Gibsland

BONNIE AND CLYDE AMBUSH MUSEUM, 2419 Main St., Gibsland, LA 71028. Mailing Address: P.O. Box 39, Gibsland, LA 71028. Tel.: 318-843-1934.
Web Site: www.bonnieandclydemuseum.com
Key Personnel: Owner, Col. Charles Heard; Owner, Ken M. Holmes, Jr.
Institution Type/Description: History Museum: housed in the former Ma Canfield's Cafe, the last place Bonnie Parker & Clyde Barrow visited before they were killed.
Collections: photographs; Clyde's Remington shotgun; Bonnie's red tam; ambush mural; bullet-ridden V-8 Ford replica used in the 1967 movie.
Hours & Admission Prices: Daily 10-6. Adults $7, seniors, active duty military & children 12 and under $5.

Gonzales

TEE JOE GONZALES MUSEUM, 217 W. Main St., Gonzales, LA 70737-2811. Mailing Address: 120 S. Irma Blvd., Gonzales, LA 70737-3604. Tel.: 225-647-9549. Fax: 225-647-9557.
E-mail: lisa@gonzalesla.com
Institution Type/Description: History Museum.
Collections: early settlers; period furnishings; photographs; clothing.
Hours & Admission Prices: By appointment only. No charge; donations accepted. Closed major holidays.

Greenwood

GATORS & FRIENDS - ALLIGATOR PARK AND EXOTIC ZOO, 11441 US Hwy. 80, Greenwood, LA 71033-2106. Tel.: 318-938-1199.
Institution Type/Description: Zoo.
Collections: alligators; miniature horses; kangaroos; lemurs; deer; goats; llamas; a camel; Scottish cow.
Activities: petting zoo; animal feedings; birthday parties.
Hours & Admission Prices: May-Aug. daily 10-6; Sept.-April Wed.-Sun. 10-6. Adults 13 & over $7.95, children 3-12 $5.95; children 2 & under no charge. Animal Feed: $2.

Gretna

GRETNA HISTORICAL SOCIETY MUSEUM, 209 Lafayette St., Gretna, LA 70053. Mailing Address: P.O. Box, Gretna, LA 70054. Tel.: 504-362-3854. Fax: 504-368-8236.
Key Personnel: Pres. (V), Paul Coles; Museum Shop Mgr., Ken Hunter.
Governing Authority: Tax-exempt.
Institution Type/Description: Historical Society Museum: housed in the former home of Claudius Strehle and Catherine Nousz; built in 1840.
Collections: local history & culture; furnishings from the late 1800s to 1939; personal artifacts; photographs; 1876 steam pumper; blacksmith shop.
Hours & Admission Prices: Call for hours.
Membership: $10 per year.

Hammond

LOUISIANA CHILDREN'S DISCOVERY CENTER, 113 N. Cypress, Hammond, LA 70401. Mailing Address: P.O. Box 1765, Hammond, LA 70404-1765. Tel.: 985-340-9150. Fax: 985-340-9156.
E-mail: anette@lcdcofhammond.org
Web Site: www.lcdcofhammond.org
Founded: 2010.
Key Personnel: Exec. Dir., Anette A. Kirylo; Pres. (V), Erica Williams; Museum Shop Mgr., Brandon Cutrer.
Personnel Profile: Full-Time Paid 2; Part-Time Paid 17; Part-Time Volunteers 2; Interns 3.
Governing Authority: Tax-exempt.
Institution Type/Description: Children's Museum.
Collections: hands-on exhibitions.
Activities: special events; birthday parties; rental facilities.
Hours & Admission Prices: Tues.-Sat. 10-6, Sun. 1-5. Admission $5.50 per person, seniors 60 & over $4.50; members and children under one no charge. &
Attendance: 33,000 (accurate)

SOUTHEASTERN LOUISIANA UNIVERSITY CONTEMPORARY ART GALLERY, 100 E. Stadium, Hammond, LA 70402. Tel.: 985-549-5080.
Institution Type/Description: Art Gallery.
Collections: works by national & regional contemporary artists.
Activities: lectures; workshops.
Hours & Admission Prices: Mon.-Tues. & Thurs.-Fri. 8-4:30, Wed. 8-8, Sat. 10-2.

Haughton

TOUCHSTONE WILDLIFE AND ART MUSEUM, 3386 Hwy. 80 E., Haughton, LA 71037. Tel.: 318-949-2323.
Institution Type/Description: Wildlife & Art Museum.
Collections: over 1,000 mounted animals from around the world; Native American artifacts; Civil War; World War I & II.
Facilities: Museum-related items for sale.
Hours & Admission Prices: Feb.-Sept. Tues.-Sat. 10-4:30; Oct.-Jan. Thurs.-Sat. 10-4:30. Adults $5; children 5 & under no charge. &

Homer

THE HERBERT S. FORD MEMORIAL MUSEUM, 519 S. Main St., Homer, LA 71040-3955. Mailing Address: P.O. Box 157, Homer, LA 71040-0157. Tel.: 318-927-9190.
E-mail: FordMuseum@bellsouth.net
Web Site: ford.claiborneone.org
Founded: 1982.
Congressional District: 4
Key Personnel: Pres. (V), David Watson; Project Dir., Linda Volentine.
Personnel Profile: Part-Time Paid 1.
Institution Type/Description: History Museum.
Collections: history & culture of north central Louisiana; Herbert Ford's personal artifacts; period Native American artifacts; military artifacts; 1920s oil boom.
Hours & Admission Prices: Mon., Wed. & Fri. 9-12 & 1-4; other times by appointment. Families $5, adults $3, children $1.
Attendance: 1,350 (estimated)
Membership: Basic $20; Friend $50; Supporting $100; Sustaining $250; Bronze Patron $500; Silver Patron $1,000; Gold Patron $2,500; Diamond Patron $5,000 & up.

Houma

BAYOU TERREBONNE WATERLIFE MUSEUM, 7910 Park Ave., Houma, LA 70364-3285. Mailing Address: P.O. Box 3678, Houma, LA 70361. Tel.: 985-580-7200.
Web Site: www.houmaterrebonne.org/waterlife.asp
Key Personnel: Dir., Ann Picon
Institution Type/Description: History Museum.
Collections: south Louisiana's cultural, industrial & ecological waterlife.
Hours & Admission Prices: Tues.-Fri. 10-5. Adults 12 & over $3, seniors $2.50, children 2-11 $2.

REGIONAL MILITARY MUSEUM, (M), 1154 Barrow St., Houma, LA 70360-5608. Mailing Address: P.O. Box 10247, Station 1, Houma, LA 70363-0247. Tel.: 985-873-8200.
E-mail: rmmuseum@triparish.net
Web Site: regionalmilitarymuseum.com
Founded: 2005.
Congressional District: 3
Key Personnel: Pres. & Chm. Bd., C.J. Christ; Vice Pres., Will Theriot; Sec. & Treas., Shirley Bourg Kenneth Royston.
Personnel Profile: Full-Time Volunteers 25; Part-Time Paid 1; Part-Time Volunteers 20; Interns 2.
Governing Authority: Tax-exempt.
Institution Type/Description: Military Museum.
Collections: military vehicles & armament; uniforms of the Vietnam, WWI, WWII, & Korean era; photographs; paintings; biographies; airplanes; landing craft LCVP.
Research Fields: veteran biographies.
Facilities: library of military-related books, magazines, newspapers, microfilm & videos.
Activities: monthly lecture.
Publications: bimonthly, Now Hear This; monthly, Roundtable Discussion.
Hours & Admission Prices: Mon.-Fri. 10-4. Adults $3, seniors $2; members & active vets no charge. Closed New Year's Day; Thanksgiving; Christmas. &
Attendance: 8,500 (estimated)
Membership: Bronze $25; Silver $50; Gold $100; Platinum & Life $1,000.

SOUTHDOWN PLANTATION HOUSE/THE TERREBONNE MUSEUM, 1208 Museum Dr., Houma, LA 70360-6072. Mailing Address: P.O. Box 2095, Houma, LA 70361-2095. Tel.: 985-851-0154. Fax: 985-868-1476.
E-mail: info@southdownmuseum.org
Web Site: www.southdownmuseum.org
Formerly: Terrebonne Historical and Cultural Society
Founded: 1972.
Key Personnel: Pres. (V), Doug Holloway; Vice Pres, Paul Labat; Sec., Dale Norred; Exec. Dir., Rachel Cherry; Museum Shop Mgr., Melva Fournier.
Personnel Profile: Full-Time Paid 2; Part-Time Paid 2; Part-Time Volunteers 100.
Governing Authority: private; nonprofit organization. Parent Institution: Terrebonne Historical & Cultural Society. Tax-exempt: 501(c)(3).
Institution Type/Description: General Museum: housed in the Southdown Plantation House.
Collections: history & culture of Terrebonne Parish, South Louisiana including Cajuns & Native Americans; sugar plantation history; 135 Boehm & Doughty porcelain birds; re-creation of U.S. Senator Allen J. Ellender's private office in Washington, D.C.; Minor family furniture; native people of Louisiana; art gallery; history & culture of South Louisiana residents including Cajuns & Native Americans.
Facilities: picnic area; outdoor pavilion. Gift items for sale.
Activities: docent program; guided tours; loan exhibitions. Annual Events: Marketplace Arts, Crafts & Food Festival in spring & fall; Louisiana Philharmonic Orchestra Concert & Jazz Brunch in spring; Taste of A Cajun Christmas in winter.
Publications: quarterly, THACS Newsletter; Southdown Plantation Rental Booklet.
Hours & Admission Prices: Tues.-Sat. 10-4, Sun. 12-4. Adults $10, senior citizens & students $8, children 6-18 $5; discounts to groups; children under 6 & members no charge. Closed New Year's Eve & Day; Mardi Gras; Good Friday; Independence Day; Thanksgiving; Christmas Eve & Day. &
Attendance: 31,179 (accurate)
Membership: Individual $20; Family $35; Contributing $50; Donor $100; Supporting $150.

Jackson

PORT HUDSON STATE HISTORIC SITE, 236 Hwy. 61, Jackson, LA 70748-4217. Tel.: 225-654-3775; 888-677-3400 (Toll Free). Fax: 225-654-4413.
E-mail: porthudson@crt.la.gov
Founded: 1982.
Personnel Profile: Full-Time Paid 8; Part-Time Paid 1; Part-Time Volunteers 6.
Governing Authority: Parent Institution: Louisiana State Parks. Tax-exempt.
Institution Type/Description: Historic Site: a National Historic Landmark.
Collections: Civil War history; period artifacts & furnishings; military weapons & uniforms; photographs.
Facilities: nature trails.
Activities: guided tours; hiking; special events; demonstrations.
Hours & Admission Prices: Tues.-Sat. 9-5; groups by appointment. Adults $4; seniors 62 & over and children 12 & under no charge. Closed New Year's Day; Thanksgiving; Christmas. &
Attendance: 25,000 (estimated)
Membership: Annual Pass $80.

Jeanerette

JEANERETTE BICENTENNIAL PARK AND MUSEUM, 500 E. Main St., Jeanerette, LA 70544-3712. Mailing Address: P.O. Box 1011, Jeanerette, LA 70544. Tel.: 337-276-4408. Fax: 337-276-9557.
E-mail: jbpmuseum@bellsouth.net
Web Site: www.jeanerettemuseum.com
Founded: 1976.
Key Personnel: Dir., Darlene Derise.
Governing Authority: Tax-exempt.
Institution Type/Description: History Museum.
Collections: local history; sugar can industry history; natural wildlife; period furniture; Mardi Gras; French embroidery & crochet; videos; early sewing artifacts; music room; cypress room.
Facilities: library. Museum-related items for sale.
Activities: videos.
Hours & Admission Prices: Call for hours. Adults $3, children under 12 $1. &
Attendance: 1,938 (accurate)

Jennings

W.H. TUPPER GENERAL MERCHANDISE MUSEUM, 311 N. Main St., Jennings, LA 70546-5341. Tel.: 337-821-5532.
E-mail: tuppermuseum@bellsouth.net
Web Site: www.tuppermuseum.com
Key Personnel: Dir., Polly Henry.
Personnel Profile: Part-Time Paid 3.
Governing Authority: Parent Institution: City of Jennings. Tax-exempt.
Institution Type/Description: History Museum.
Collections: period general store furnishings & merchandise including clothing, school supplies, tools, medicines, & toys; Native Indian basketry.
Hours & Admission Prices: Mon.-Fri. 9-5. Adults $3, students $1. Closed major holidays. &
Attendance: 2,000 (estimated)

ZAM - ZIGLER ART MUSEUM, 411 Clara St., Jennings, LA 70546-5235. Tel.: 337-824-0114. Fax: 337-824-0120.
E-mail: zigler-museum@charter.net
Web Site: www.ziglerartmuseum.com
Founded: 1963.
Congressional District: 7
Key Personnel: Dir. & Museum Shop Mgr., Dolores Spears; Chm. (V), Burt Tietje; Pres. (V), Gregory Marcantel; Museum Shop Mgr., Susan Bergeaux.
Personnel Profile: Full-Time Paid 1; Part-Time Paid 2; Part-Time Volunteers 10.
Governing Authority: nonprofit. Tax-exempt: 501(c)(3).
Institution Type/Description: European & American Art.
Collections: art collection including work by Jean-Louis Vergne, Charles Sprague Pearce, Louis Jambor, George Inness, Helen Turner, Robert Rucker, John Constable, James McNeil Whistler, Albrecht Durer, Camille Pissaro, Maurice deVlaminck, Robert Henri, John James Audubon, Albert Bierstadt, R.A. Blakelock, Sir John Everette Millias, Sir Anthony Wan Dyck & Japanese Woodblock Prints (18th-19th century); wildlife & natural history art.
Research Fields: Chenier Culture Jefferson Davis Parish History.
Activities: tours; art-related workshops; monthly guest artist shows.
Publications: catalog of Zigler Museum's collection of works by William Tolliver.
Hours & Admission Prices: Tues.-Sat. 10-4, Sun. 1-4. Adults $5, children $2; discounts to tour groups and ICOM & AAM members; members no charge. Closed major holidays. &
Attendance: 4,385 (estimated)
Membership: Individual $20; Family $30; Supportive $50; Patron $100-$500; Corporate $500-$1,000; Lifetime $5,000.

Kaplan

LE MUSEE DE KAPLAN, 405 N. Cushing Ave., Kaplan, LA 70548. Tel.: 337-643-1528.
Web Site: www.kaplanla.com
Institution Type/Description: History Museum.
Collections: local history; Cajun cultural history; works by local artists; personal artifacts.
Hours & Admission Prices: Wed.-Sat. 9-1.

Kenner

MARDI GRAS MUSEUM, 415 Williams Blvd., Rivertown, Kenner, LA 70062. Tel.: 504-468-7231.
Web Site: www.rivertownkenner.com
Institution Type/Description: General Museum.
Collections: 150 years of history from New Orleans to Acadiana; videos & memorabilia depicting King Cake traditions, balls, parades, French Quarter & Mardi Gras.
Activities: float & costume-making demonstrations.
Hours & Admission Prices: Tues.-Sat. 9-5. Adult $3, senior citizen 60 & over $2.50, children 2-12 $2.

Lafayette

ACADIAN CULTURAL CENTER - LAFAYETTE, 501 Fisher Rd., Lafayette, LA 70508-2033. Tel.: 337-232-0789.
Institution Type/Description: History Museum.
Collections: Acadian history & culture; period furnishings; personal artifacts; photographs.
Activities: special events; guided boat tours in spring & fall; educational programs.
Hours & Admission Prices: Daily 8-5. No charge. Closed Mardi Gras; Christmas.

ALEXANDRE MOUTON HOUSE/LAFAYETTE MUSEUM, 1122 Lafayette St., Lafayette, LA 70501-6838. Tel.: 337-234-2208. Fax: 337-234-2208.
Founded: 1954.
Congressional District: 3
Key Personnel: Hostess & Guide, Phyllis Goff; Hostess & Guide, Joyce Boutin.
Personnel Profile: Part-Time Paid 2.
Governing Authority: nonprofit organization. Affiliated with the Lafayette Museum Assn. Tax-exempt: 501(c)(3).
Institution Type/Description: Historic Building.
Collections: period furniture & textiles c.1850; historical documents; manuscripts; portraits; Civil War & Indian artifacts; Mardi Gras costumes.
Activities: guided tours; permanent & temporary exhibitions.
Publications: cookbooks, First You Make a Roux; Let Us Entertain You; illustrated brochure, The Lafayette Museum.
Hours & Admission Prices: Tues.-Sat. 10-4, Sun. 1-4. Adults $3, senior citizens $2, students $1; discounts to groups; school tours no charge. Closed New Year's Day; Mardi Gras; Easter; Independence Day; Christmas. &
Membership: $25-$99; $100-$249; $250-$499; $500 & up.

CATHEDRAL OF ST. JOHN THE EVANGELIST, 914 St. John St., Lafayette, LA 70501. Mailing Address: 515 Cathedral St., Lafayette, LA 70501-6701. Tel.: 337-232-1322 & 1325. Fax: 337-232-1379.
E-mail: info@saintjohncathedral.org
Web Site: www.saintjohncathedral.org
Founded: 1821.
Congressional District: 7
Key Personnel: Dir., Chm. (V) & Museum Shop Mgr., Janice McNeil; Asst., Cheryl Luke; Gift Shop Mgr., Brady LeBlanc.
Personnel Profile: Part-Time Paid 1; Part-Time Volunteers 24.
Volunteer Hours: 1,404
Operating Expenses: 6,358
Operating Income: 10,577
Governing Authority: Parent Institution: Diocese of Lafayette. Tax-exempt.
Institution Type/Description: Religious Museum: housed in an historic church; built in 1916. Listed on the National Register of Historic Properties.
Collections: parish history; religious artifacts & furnishings; documents; photographs; statues; paintings; Italian Nativity.
Major Exhibits: Religious Art by Faye Drobnic & Students, 11/13-2/14; Religious Art by Ms. Patty Ardoin, 3/14-6/14.
Facilities: Gift items for sale.
Publications: magazine, Acadiana Catholic; Travelhost of Louisiana; travel guide, BonTemps; calendar of events.
Hours & Admission Prices: By appointment. Tours: Mon.-Thurs. 9-12 & 1-4, Fri. 9-12. Adults $3, senior citizens $2, children 12 & under $1; discounts to groups of 20 or more. &
Attendance: 2,000 (accurate)

CHILDREN'S MUSEUM OF ACADIANA, 201 E. Congress St., Lafayette, LA 70501-6919. Tel.: 337-232-8500. Fax: 337-232-8167.
E-mail: info@cmalaf.org
Web Site: www.childrensmuseumofacadiana.org
Founded: 1990.
Congressional District: 7
Key Personnel: Exec. Dir., Marvita Hudson; Visitor Svcs., Rachel Dafford; Finance/Grant Dir., Suzanne Calais Oge.
Personnel Profile: Full-Time Paid 1; Part-Time Paid 14; Part-Time Volunteers 385; Interns 1.
Operating Expenses: 484,333
Operating Income: 438,561
Governing Authority: Tax-exempt.
Institution Type/Description: Children's Museum.
Collections: hands-on exhibits.
Activities: performances; workshops.
Publications: newsletter.
Hours & Admission Prices: Tues.-Sat. 10-5. Admission $5; children one & under no charge. &
Attendance: 44,500 (accurate)
Membership: Senior Citizen $75; Individual $100; Basic Local $100; Basic Family 125; Family Plus $175.

LAFAYETTE SCIENCE MUSEUM, (M), 433 Jefferson St., Lafayette, LA 70501-7013. Tel.: 337-291-5544. Fax: 337-291-5464.
Web Site: www.lafayettesciencemuseum.org
Formerly: Lafayette Natural History Museum and Planetarium

Founded: 1969.
Congressional District: 7
Key Personnel: Dir., Kevin Krantz; Cur. Planetarium, David Hostetter; Cur. Collections, Dr. Deborah J. Clifton; Cur. Education, Dawn Edelen; Museum & Planetarium Technician, Paul McCasland; Asst. Planetarium Cur., Charlotte Guillot; Sec., Karen Miller; Tour Coord., Edi Gilbert; Exhibit Guide, Keith Richard; Receptionist, Likassina Brown; Laborer II, Brian Knuckles; Cashier, C.C. Brown.
Personnel Profile: Full-Time Paid 11; Part-Time Paid 7; Part-Time Volunteers 23.
Governing Authority: municipal. Parent Institution: Lafayette Consolidated Government. Subsidiary Institution: Lafayette Science Museum Foundation. Tax-exempt: 501(c)(3).
Institution Type/Description: Science Museum & Planetarium.
Collections: living & preserved floral & faunal specimens of Louisiana area in excess of 55,000 items; material-culture of Louisiana's Native Americans and other groups including Creole, Spanish, French & African-American including contemporary photographic documentation & present-day crafts; Louisiana Acadian Textile Collection.
Research Fields: regional culture.
Facilities: 10,000 vol. library; 72-seat planetarium; 13,000 sq. ft. exhibit space; archives resources file; 150 seat auditorium; 2 classrooms.
Activities: self-guided tours; lectures; films; hands-on workshops; formally organized education programs for children & adults; permanent, temporary & traveling exhibitions; school loan service.
Publications: Checklist of the Vascular Flora of Louisiana; Checklist of the Birds of Louisiana; Checklist and Key to the Amphibians and Reptiles of Louisiana; Life in the Atchafalaya Swamp; Louisiana Ferns and Fern Allies; How Men Cook; exhibition catalogs; Travailler, C'est Trop Dur: The Tools of Cajun Music; Image of Lafayette Chitimacha Notebook; Craft Talk; Palmetto Braiding: The Folk Art of Elvina Kidder.
Hours & Admission Prices: Tues.-Fri. 9-5, Sat. 10-6, Sun. 1-6. Adults $5, seniors $3, children & chaperones $2, school groups outside of Lafayette Parish $1 per student; discounts to AAM members; members no charge. Closed Mardi Gras; Easter; Thanksgiving; Christmas. &
Attendance: 45,000 (estimated)
Membership: Student & Senior Citizen $30; Individual $35; Family & Grandparent $50; Patron $150; Group & Corporate $300; Benefactor $750; Lifetime $1,000.

LARC'S ACADIAN VILLAGE, 200 Greenleaf Dr., Lafayette, LA 70506-7400. Tel.: 337-981-2364; 800-962-9133. Fax: 337-988-4554. Facebook: LARC's Acadian Village.
E-mail: megan@acadianvillage.org
Web Site: www.acadianvillage.org
Founded: 1976.
Institution Type/Description: Folk Life and History Museum.
Collections: local history & culture; photographs; furnishings. Historic Homes: Bernard House c.1800; Billeaud House; Castille House c.1860; Leblac House, built between 1821 & 1856; St. John House c.1840; Thibodaux House c.1820. Doctor's Museum: office of the first resident dentist, Dr. Hypolite Salles.
Hours & Admission Prices: Jan.-Oct. Tues.-Sat. 10-4. Adults $8, students $6; discounts to groups; military no charge. Closed major holidays. &

PAUL AND LULU HILLIARD UNIVERSITY ART MUSEUM, UNIVERSITY OF LOUISIANA AT LAFAYETTE, (M), 710 E. St. Mary Blvd., Lafayette, LA 70503-2332. Mailing Address: P.O. Drawer 42571, Lafayette, LA 70504-2571. Tel.: 337-482-2278. Fax: 337-262-1268.
E-mail: artmuseum@louisiana.edu
Web Site: museum.louisiana.edu
Formerly: University Art Museum University of Southwestern Louisiana Campus; University Art Museum, University of Louisiana at Lafayette
Founded: 1968.
Congressional District: 7
Key Personnel: Dir., Mark A. Tullos, Jr.; Registrar, Ramona East; Visitor Svcs. & Volunteer Coord., Cindy Hamilton; Cur. Exhibitions & Collections, Dr. Lee Gray; Security, Preparator & Bldg. Maintenance, Kerry Frey; Security, Hugo Boute; Administrative Asst., Debby Mayne.
Personnel Profile: Full-Time Paid 7; Part-Time Paid 4; Part-Time Volunteers 50; Interns 2.
Governing Authority: nonprofit organization. Parent Institution: University of Louisiana at Lafayette. Tax-exempt: 501(c)(3).
Institution Type/Description: Art Museum.
Collections: The Louisiana Collection: 19th to 20th-century drawings, paintings, photographs & sculpture; Henry Botkin collection: paintings, drawings, collages 1928-1981; 19th to 20th-century Japanese prints; southern outsider art; naive art.

Research Fields: Louisiana photography; outsider art; mid-century abstract expressionism.
Facilities: 11,000 sq. ft. exhibit space; one acre sculpture garden.
Activities: guided tours; lectures; films; gallery talks; concerts; traveling exhibitions.
Publications: catalogues; calendars; retrospectives; books; seasonal newsletter.
Hours & Admission Prices: Tues.-Thurs. 9-5, Fri. 9-12, Sat. 10-5. Adults $5, senior citizens $4, students 5-17 $3; discounts to NARM, AAM & ICOM members & groups of 20 or more; members no charge. Closed major holidays. &
Attendance: 30,000 (accurate)
Membership: Senior Citizen $25; Individual $40; Family $55; Donor $100; Patron $250; Contributing $500; Benefactor $1,000 & up.

VERMILIONVILLE, (M), 300 Fisher Rd., Lafayette, LA 70508-2028. Tel.: 337-233-4077; 866-992-2968. Fax: 337-233-1694. Facebook: Vermilionville.
E-mail: curator@bayouvermiliondistrict.org
Web Site: www.vermilionville.org
Founded: 1990.
Governing Authority: Parent Institution: Lafayette Parish Bayou Vermilion District. Tax-exempt.
Institution Type/Description: Folklife Park.
Collections: local history & culture; historic village containing seven restored original homes; watershed.
Facilities: restaurant; rain garden. Museum-related items for sale.
Activities: live music; films; free event days.
Publications: newsletter, La Recolt.
Hours & Admission Prices: Tues.-Sun. 10-4. Adults $10, seniors 65 & over $8, students 6-18 $6; discounts to AAA members; members & children under 6 no charge. Closed New Year's Eve & Day; Martin Luther King Day; Mardi Gras Day; Memorial Day; Labor Day; Thanksgiving; Christmas Eve & Day. &
Membership: Senior Citizen $20; Individual $25; Couple $45; Family & Grandparents $55; Vermilionaire $120.

Lafitte

LOUISIANA MARINE FISHERIES MUSEUM, 580 Jean Lafitte Blvd., Lafitte, LA 70067-5108. Tel.: 504-689-3497.
Institution Type/Description: Cultural Heritage Museum.
Collections: 3,000 yr. old dugout canoe; Lafitte Skiff; trenasse digger; period boats; crawfish farming; hunting; cypress lumbering; moss picking; ranching & farming; period fishing & trapping implements; photographs.
Hours & Admission Prices: Wed.-Sun. 10-4. Admission $1.

Lake Charles

ABERCROMBIE GALLERY & GRAND GALLERY, McNeese State University, Ryan & Sale Sts., Lake Charles, LA 70609. Mailing Address: MSU Dept. of Visual Arts, P. O. Box 92295, Lake Charles, LA 70609. Tel.: 337-475-5060. Fax: 337-475-5927.
E-mail: hkelley@mcneese.edu
Web Site: www.mcneeseartonline.org
Formerly: Abercrombie Gallery
Founded: 1984.
Congressional District: 7
Key Personnel: Dir., Heather Ryan Kelley; Chm., Lynn Reynolds.
Governing Authority: Parent Institution: State of Louisiana, McNeese State University. Tax-exempt.
Institution Type/Description: Art Gallery.
Collections: works by local, regional, & national contemporary artists.
Major Exhibits: Deborah Luster & Pinky Bass, 1/14; 27th Annual McNeese National Works on Paper Exhibition, 3/14-4/14; Senior Exhibition, 5/14; Annual Faculty Show, 8/14-10/14.
Activities: temporary exhibitions.
Publications: catalog: Works on Paper.
Hours & Admission Prices: Mon.-Fri. 9-4:30. No charge. Closed school holidays. &

CHILDREN'S MUSEUM OF LAKE CHARLES, INC., (M), 327 Broad St., Lake Charles, LA 70601-4223. Tel.: 318-433-9420. Fax: 318-433-0144.
E-mail: dan@child-museum.org
Web Site: www.child-museum.org
Founded: 1988.
Congressional District: 7
Key Personnel: Exec. Dir., Dan Ellender; Asst. Dir., Allyson Blackwell; Education & Program Dir., Erin Bentley.

Personnel Profile: Full-Time Paid 1; Part-Time Paid 6; Part-Time Volunteers 20.
Governing Authority: nonprofit organization. Tax-exempt: 501(c)(3).
Institution Type/Description: Children's Museum.
Collections: hands on learning center including Kid's Town, USA.
Facilities: 9,000 sq. ft. exhibit area; nature center; computer lab; shadow room; art space. Museum-related items for sale.
Activities: guided tours; organized education programs for students; docent program; participatory, traveling & temporary exhibitions; birthday parties.
Publications: quarterly museum newsletter.
Hours & Admission Prices: Mon.-Sat. 10-5. Admission 2 & over $7.50, active military $6.75, senior citizens 55 & over $5.75; discounts to groups of 10 or more with a reservation; children under 2 & members no charge. Closed major holidays. &
Attendance: 24,000 (accurate)
Membership: Family $40; Family Plus & Grandparents $50.

IMPERIAL CALCASIEU MUSEUM, INC., 204 W. Sallier St., Lake Charles, LA 70601-5844. Tel.: 337-439-3797. Fax: 337-439-6040.
E-mail: impmuseum@bellsouth.net
Web Site: www.imperialcalcasieumuseum.org
Founded: 1963.
Congressional District: 7
Key Personnel: Exec. Dir., Susan H. Reed.
Personnel Profile: Full-Time Paid 1; Part-Time Paid 9.
Governing Authority: nonprofit. Tax-exempt: 501(a).
Institution Type/Description: Local History Museum.
Collections: Victorian period furnishings: parlor, kitchen, bedroom, country store; barber shop; pharmacy; dolls & toys; Audubon prints; Louisiana game bird exhibit; Civil War & World War I collection; historical photographs; paddleboat steering wheels; period musical instruments; Attakapa Indian artifacts; glass bottles; 16th-18th century apothecary jars; period shaving mugs.
Research Fields: historical, 5 parish area, comprising the original Imperial Calcasieu Land Grant.
Facilities: 300-vol. library of old school books & bibles; Gibson-Barham fine arts gallery; 100-seat auditorium. Cookbooks & sculpture for sale.
Activities: guided tours; permanent & temporary exhibitions. Museum Sponsors: special events for children June to August.
Hours & Admission Prices: Tues.-Sat. 10-5. Adults $2, children $1; members no charge. Closed major holidays. &
Attendance: 6,800 (estimated)
Membership: Family $50; Associate $100; Patron $300; Corporate $500 & up.

USS RADFORD NATIONAL NAVAL MUSEUM, 604 N. Enterprise Blvd., Lake Charles, LA 70606. Mailing Address: 482 Windyville Rd., Spencer, WV 25276.
Web Site: www.ussradford446.org
Founded: 2001.
Key Personnel: Pres. (V), Chuck Parsons.
Governing Authority: Parent Institution: USS Radford 446, Inc. Tax-exempt.
Institution Type/Description: Naval Military Museum.
Collections: WWII, Korea, Vietnam & Cold War memorabilia; Admirals original uniforms; QH 50 D.A.S.H.; pre WWII dress whites; personal artifacts; weapons; Japanese War memorabilia; part of USS Helena CL 50; DesRon 21 ship displays.
Activities: overnight stays.
Publications: triannual Radford newsletter.
Hours & Admission Prices: Mon.-Fri. 10-3, Sat.-Sun. 10-4; bus & family tours by appointment. &
Attendance: 18,000 (estimated)

Lake Providence

LOUISIANA STATE COTTON MUSEUM, 7162 Hwy. 65 N., Lake Providence, LA 71254-5226. Mailing Address: State of Louisiana, Secretary of State, P.O. Box 94125, Baton Rouge, LA 70804-9125. Tel.: 318 559 2041. Fax: 318 559 2217.
E-mail: cotton@sos.louisiana.gov
Web Site: www.sos.louisiana.gov/lscm
Founded: 1992.
Congressional District: 5
Key Personnel: Dir., Harriet Bridges.
Personnel Profile: Full-Time Paid 2; Part-Time Paid 2.
Governing Authority: state. Parent Institution: Louisiana Secretary of State, Baton Rouge, LA. Tax-exempt.
Institution Type/Description: Cotton Museum.

Collections: history & heritage of cotton cultivation; cottons influence on life in Louisiana; photographs.
Facilities: Museum-related items for sale.
Activities: guided tours.
Hours & Admission Prices: Tues.-Sat. 10-4. No charge. Closed most state holidays. &
Attendance: 4,573 (accurate)
Membership: Call for information.

Leesville

MUSEUM OF WEST LOUISIANA, 803 S. Third St., Leesville, LA 71446-4703. Tel.: 337-239-0927.
Founded: 1987.
Key Personnel: Museum Shop Mgr., Mary Cleveland.
Personnel Profile: Full-Time Volunteers 1; Part-Time Paid 1.
Governing Authority: Parent Institution: Vernon Parish Police Jury. Subsidiary Institution: Museum Association of West Louisiana. Tax-exempt.
Institution Type/Description: History Museum.
Collections: archaeological artifacts; logging implements; railroad memorabilia; P.O.W. paintings; WWII artifacts; toys; photographs; quilts; clothing; cooking; household items; furniture; history, culture, folk art, & resources of Vernon Parish and the West Central area of the Louisiana Territory; scale models of various buildings from 1800s.
Facilities: Museum-related items for sale.
Activities: picnics; special events; tours.
Hours & Admission Prices: Tues.-Sun. 1-5; other times by appointment. No charge; donations accepted. Closed major holidays. &
Attendance: 8,000 (accurate)
Membership: Student & Senior Citizen $10; Individual $15; Family $25; Patron $50; Benefactor $100; Corporate Donor $250; Angel $1,000.

Lockport

BAYOU LAFOURCHE FOLKLIFE & HERITAGE MUSEUM, 110 Main St., Lockport, LA 70374. Mailing Address: P.O. Box 416, Lockport, LA 70374. Tel.: 985-532-5909. Fax: 985-532-1108.
E-mail: bayoulafourchefo@bellsouth.net
Web Site: www.bayoumuseum.org
Institution Type/Description: History Museum.
Collections: local history & cultural heritage; natural history; photographs; personal artifacts; period furnishings.
Hours & Admission Prices: Tues. & Thurs. 10-4; other times by appointment.

Long Leaf

SOUTHERN FOREST HERITAGE MUSEUM, 77 Longleaf Rd., Long Leaf, LA 71448. Mailing Address: P.O. Box 101, Long Leaf, LA 71448-0101. Tel.: 318-748-8404. Fax: 318-748-8410.
E-mail: sfhm@centurytel.net
Web Site: www.forestheritagemuseum.org
Founded: 1992.
Congressional District: 5
Key Personnel: Dir., Claudia Troll, M.Ed.; Pres., Bobby Sebastian; Public Rels., Buck Vandersteen; Treas., Harold Elliott; Security, Charles Hudson.
Personnel Profile: Full-Time Paid 1; Part-Time Paid 5; Part-Time Volunteers 20; Interns 3.
Governing Authority: private; nonprofit organization. Tax-exempt: 501(c)(3).
Institution Type/Description: History Museum.
Collections: early 20th-century sawmill town history; forestry & lumber history; tools; 1910 planer mill & sawmill; c.1915 machine shop; 3 steam locomotives; steam logging equipment including 1919 Clyde skidder & two 1919 McGiffert loaders; archives.
Facilities: 500-vol. library; 57-acre site; 7,500 sq. ft. exhibit space; 50-seat theater; arboretum; nature trail. Museum-related items for sale.
Activities: machine shop demonstrations; railroad ride; docent program; guided tours; participatory exhibits; theater. Annual Events: Heritage Day; Legends & Lore.
Publications: quarterly newsletter, Edgings & Trimmings.
Hours & Admission Prices: Tues.-Sat. 9-4, Sun. 1-4, Mon. by appointment. Adults $8, students $4; discounts to members & groups. See website for holiday closings. &
Attendance: 6,100 (accurate)
Membership: Senior & Teacher $20; Individual $25; Family $50; Contributing $100; Supporting $250; Sustaining $500; Wall of Honor $1,000 & up.

Madisonville

LAKE PONTCHARTRAIN BASIN MARITIME MUSEUM, **(M),** 133 Mabel Dr., Madisonville, LA 70447-9301. Tel.: 985-845-9200. Fax: 985-845-9201.
E-mail: info@lpbmm.org
Web Site: www.lpbmm.org
Founded: 2001.
Key Personnel: Dir., Don Lynch; Educator, Stephanie Imel; Administrative Asst., Melanie Waddell; Museum Shop Mgr., Sharon Street.
Personnel Profile: Full-Time Paid 1; Full-Time Volunteers 1; Part-Time Paid 2; Part-Time Volunteers 12; Interns 1.
Governing Authority: private; nonprofit organization. Tax-exempt: 501(c)(3).
Institution Type/Description: Maritime Museum.
Collections: nautical & cultural heritage of Lake Pontchartrain Basin. Historic Buildings: Tchefuncte River lighthouse, c.1837; Tchefuncte River light-keepers cottage, c.1887.
Research Fields: shoreline erosion; maritime Louisiana history & culture.
Facilities: rental facilities.
Activities: boatbuilding classes; lightkeepers overnight program; aquatic robotics; ROV summer camps; birthday parties; field trips. Annual Event: Madisonville Wooden Boat Festival in October.
Publications: quarterly newsletter, Shipways.
Hours & Admission Prices: Tues.-Sat. 10-4, Sun. 12-4. Adults $5, seniors & children 6-12 $3; discount to groups; members and children under 6 & uniformed military no charge. Closed New Year's Day; Mardi Gras; Thanksgiving; Christmas. &
Attendance: 30,000 (estimated)
Membership: Student & Veteran $20; Individual $30; Family $50; Seaman $100-$499; Steward $500-$999; Captain $1,000-$4,999; Commodore $5,000-$10,000.

MADISONVILLE MUSEUM, 201 Cedar St., Madisonville, LA 70477. Mailing Address: P.O. Box 160, Madisonville, LA 70447-0160. Tel.: 985-845-2100.
Founded: 1991.
Key Personnel: Dir. & Museum Shop Mgr., Ginger Stanga.
Personnel Profile: Part-Time Paid 2; Part-Time Volunteers 4.
Governing Authority: town. Tax-exempt.
Institution Type/Description: History Museum.
Collections: local wildlife; Civil War; Native American artifacts.
Hours & Admission Prices: Sat.-Sun. 12-4. No charge; donations accepted.
Attendance: 600 (estimated)

OTIS HOUSE - FAIRVIEW-RIVERSIDE SATE PARK, 119 Fairview Dr., Madisonville, LA 70447. Tel.: 985-845-3318; 888-677-3247.
Institution Type/Description: Park & Historic House: built in the 1880s. Listed on the National Register of Historic Places.
Collections: local history & culture; photographs; period furnishings; period artifacts.
Hours & Admission Prices: Wed.-Sun. 9-5. Adults $2; children 3 & under and seniors 62 & over no charge. Closed New Year's Day; Thanksgiving; Christmas.

Mansfield

MANSFIELD FEMALE COLLEGE MUSEUM, 101 Monroe St., Mansfield, LA 71052. Mailing Address: State of Louisiana, Secretary of State, P.O. Box 94125, Baton Rouge, LA 70804-9125. Tel.: 318-871-9978. Fax: 318-871-9978.
E-mail: barbara.valentine@sos.la.gov
Web Site: www.sos.louisiana.gov/mfcm
Founded: 2003.
Congressional District: 4
Key Personnel: Dir., Barbara Valentine.
Personnel Profile: Part-Time Paid 1.
Governing Authority: state. Tax-exempt.
Institution Type/Description: History Museum.
Collections: history of women's education in northwest Louisiana & De Soto Parish; photographs; furnishings; personal artifacts.
Facilities: library.
Activities: docent program; guided tours; traveling exhibitions.
Hours & Admission Prices: Tues.-Fri. 8-4:30. No charge. Closed most state holidays. &
Attendance: 1,414 (accurate)

MANSFIELD STATE HISTORIC SITE - CIVIL WAR BATTLEFIELD, 15149 Hwy. 175, Mansfield, LA 71052-4774. Tel.: 318-872-1474; 888-677-6267. Fax: 318-871-4345.
E-mail: mansfield@crt.state.la.us
Web Site: www.crt.state.la.us
Formerly: Mansfield State Commemorative Area
Founded: 1957.
Congressional District: 4
Key Personnel: Park Mgr., Scott Dearman.
Personnel Profile: Full-Time Paid 4.
Governing Authority: state. Parent Institution: Louisiana Office of State Parks. Tax-exempt.
Institution Type/Description: Civil War Battlefield
Collections: documents; letters; pictures; diaries of soldiers; Civil War artifacts; uniforms; weapons.
Research Fields: Battle of Mansfield; Red River Campaign of 1864; Civil War in Louisiana; Civil War.
Facilities: books, letters, diaries & official records of Union & Confederate Armies available for use on the premises; visitor center; picnic area; theater; battlefield trail; restrooms.
Activities: school tours; lectures; programs for students on school days by appointment; living history events & programs; self-guided battlefield tours.
Publications: park brochure.
Hours & Admission Prices: Daily 9-5. Adults 13-61 $4, children 12 & under and senior citizens 62 & over no charge. Closed New Year's Day; Thanksgiving; Christmas. &
Attendance: 11,850 (accurate)

Mansura

LA COMMISSION DES AVOYELLES, INC. - DR. JULES CHARLES DESFOSSE HOUSE, 1832 L'Eglise St., Mansura, LA 71350. Mailing Address: P.O. Box 26, Hamburg, LA 71339. Tel.: 318-964-2152.
Founded: 1974.
Congressional District: 5
Key Personnel: Pres. (V), Carlos Mayeux.
Personnel Profile: Part-Time Paid 1; Part-Time Volunteers 1.
Governing Authority: town. Parent Institution: La Commission des Avoyelles, Inc. Tax-exempt: 501(c)(3).
Institution Type/Description: Historic House: housed in a French Colonial home containing mud walls; built c.1790. Listed on the National Register of Historic Places.
Collections: local history & culture; period artifacts & furnishings.
Hours & Admission Prices: Wed.-Thurs. 8-3, Fri. 8-2. Adults $2, children $1. Closed major holidays.
Attendance: 2,500 (estimated)

LOUISIANA 4-H MUSEUM, 8592 Hwy. 1, Ste. 2, Mansura, LA 71350. Tel.: 318-964-2245. Fax: 318-964-2259.
Institution Type/Description: History Museum.
Collections: 4-H history; photographs; personal artifacts; hands-on exhibitions.
Hours & Admission Prices: Mon.-Fri. 8:30-4; other times by appointment. Admission $3; discounts to groups; children under 3 no charge.

Many

FORT JESUP STATE HISTORIC SITE, 32 Geoghagan Rd., Many, LA 71449. Tel.: 318-256-4117; 888-677-5378.
E-mail: fortjesup@crt.state.la.us
Institution Type/Description: Historic Site: housed on the grounds of Fort Jesup; built in 1822. A National Historic Landmark.
Collections: local military history; personal artifacts; period furnishings.
Facilities: Museum-related items for sale.
Activities: guided tours.
Hours & Admission Prices: Daily 9-5. Adults $2; seniors 62 & over and children 12 & under no charge. Closed New Year's Day; Thanksgiving; Christmas.

Marksville

HYPOLITE BORDELON HOME, 242 Tunica Dr. W., Marksville, LA 71351. Mailing Address: P.O. Box 585, Marksville, LA 71351-0767. Tel.: 318-253-0284.
Founded: 1976.
Congressional District: 8
Key Personnel: Chm. (V), Clyde M. Neck

Governing Authority: Tax-exempt.
Institution Type/Description: Historic House Museum: built in 1820. Listed on the National Register of Historic Places.
Collections: local history & culture; period artifacts & furnishings; photographs.
Hours & Admission Prices: Tues.-Sat. 9-5. Adults $2, children $1.
Attendance: 900 (estimated)
Membership: Annual $15.

MARKSVILLE STATE HISTORIC SITE, 837 Martin Luther King Dr., Marksville, LA 71351-2478. Tel.: 318-253-8954; 888-253-8954.
E-mail: marksville@crt.state.la.us
Web Site: www.crt.state.la.us/parks/iMarksvle.aspx
Institution Type/Description: Historic Site. Designated as a National Historic Landmark.
Collections: 3,300 ft. semi-circular earthwork.
Hours & Admission Prices: Tues.-Sat. 9-5; groups by appointment. Adults $4; seniors 62 & over and children 12 & under no charge. Closed New Year's Day; Thanksgiving; Christmas.

TUNICA-BILOXI CULTURAL & EDUCATIONAL RESOURCES CENTER, 150 Melancon Rd., Marksville, LA 71351. Mailing Address: P.O. Box 1589, Marksville, LA 71351-1589. Tel.: 318-240-6451. Fax: 318-253-7711.
E-mail: earlii@tunica.org
Web Site: www.tunica.org
Formerly: Tunica-Biloxi Native American Museum
Founded: 1989.
Congressional District: 6
Key Personnel: CERC Dir., Earl J. Barbry, Jr.; C.F.O., Doug Burke; Security, Police Chief Harold Pierite, Sr.; Museum Shop Mgr., Melissa Barbin.
Personnel Profile: Full-Time Paid 5.
Governing Authority: Federally recognized Native American Tribal Council.
Institution Type/Description: History Museum.
Collections: French & Indian artifacts.
Facilities: library; 1,500 sq. ft. exhibit space; 40-seat large screen theater. Museum-related items for sale.
Hours & Admission Prices: Call for hours. Adults $5, children $3; discounts to groups, seniors & veterans.
Attendance: 2,400 (estimated)

Marrero

BARATARIA PRESERVE, JEAN LAFITTE NATIONAL HISTORICAL PARK AND PRESERVE, 6588 Barataria Blvd., Marrero, LA 70072-7526. Tel.: 504-689-3690. Fax: 504-689-7897. Facebook: Barataria Preserve, Jean Lafitte National Historical Park and Preserve.
Web Site: www.nps.gov/jela/
Founded: 1978.
Congressional District: 3
Key Personnel: Supervisory Park Ranger, Alentia Scott; Education Coord., Stacy Lafayette; Volunteer Coord., Stephanie Click; Publications Coord., Kristy Wallisch; Bookstore Mgr., Julie Castille.
Governing Authority: federal. Administered by the National Park Service, Washington, DC 20240. Tax-exempt.
Institution Type/Description: Park Visitor Center, Folkway & Natural History Museum.
Collections: artifacts; traps; fishing equipment; moss gathering equipment; pirogues; cypress logging industry; national history.
Research Fields: folkways of S. Louisiana; natural history & environment of park; prehistoric Indians Resource management.
Facilities: 50-seat auditorium; boardwalk trails; observation decks in swamp & marsh; interpretive waysides; environmental education center.
Activities: guided tours; film; nature trails; picnic area; permanent exhibits; pre-registration for school group interpretation; canoe trails; fishing & hunting in season; curriculum based programs for K-3, 4-5, 6-8 & 9-12 grades.
Publications: trail brochure; site bulletins.
Hours & Admission Prices: Daily 9-5. No charge; donations accepted. Closed Christmas; Mardi Gras.
Attendance: 35,000

Marthaville

REBEL STATE HISTORIC SITE & LOUISIANA COUNTRY MUSIC MUSEUM, 1260 Hwy. 1221, Marthaville, LA 71450-3459. Tel.: 318-472-6255; 888-677-3600. Fax: 318-472-9315.
E-mail: rebel@crt.state.la.us
Web Site: www.crt.state.la.us
Formerly: Rebel State Commemorative Area Louisiana Country Music Museum
Founded: 1981.
Congressional District: 5
Personnel Profile: Full-Time Paid 3; Part-Time Paid 2.
Governing Authority: state. Louisiana Office of State Parks, P.O. 44426, Baton Rouge, LA 70804. Tel. 504-342-8105. Tax-exempt.
Institution Type/Description: Louisiana Country Music Museum: housed in a treble clef shaped building; Rebel State Historic Site; resting place of the Unknown Confederate Soldier.
Collections: Louisiana country music; musical instruments; costumes of performers; memorabilia & graphic representations of the history of music; juke boxes; victrolas.
Research Fields: folk traditions of popular & country music; gospel music.
Facilities: 1,500-capacity amphitheater.
Activities: guided tours; lectures; concerts; formally organized education programs for children; permanent & temporary exhibitions.
Hours & Admission Prices: Daily 9-5. Adults $4; seniors 62 & over, children 12 & under and school groups no charge. Closed New Year's Day; Thanksgiving; Christmas.
Attendance: 7,710 (accurate)

Martinville

AFRICAN AMERICAN MUSEUM, 125 S. New Market St., Martinville, LA 70582. Mailing Address: P.O. Box 379, Saint Martinville, LA 70582-0379. Tel.: 337-394-2230. Fax: 337-394-2244.
Institution Type/Description: History Museum.
Collections: African American history & culture; photographs; personal artifacts; period furnishings; mural.
Hours & Admission Prices: Daily 10-4:30. Adults $3; children 12 & under no charge.

Melrose

MELROSE PLANTATION, Melrose General Delivery, 3533 Hwy. 119, Melrose, LA 71452. Mailing Address: Box 2248, Natchitoches, LA 71457-2248. Tel.: 318-379-0055. Fax: 318-352-6848. Facebook: Melrose Plantation.
E-mail: info@melroseplantation.org
Web Site: www.melroseplantation.org
Founded: 1796.
Congressional District: 5
Key Personnel: Exec. Dir. & Facilities Mgr., Molly Dickerson; Pres. (V), Martha Maynard; Museum Shop Mgr., Betty Metoyer.
Personnel Profile: Full-Time Paid 1; Part-Time Paid 9; Part-Time Volunteers 30.
Governing Authority: society. Parent Institution: Association for the Preservation of Historic Natchitoches. Tax-exempt.
Institution Type/Description: Historic House: 1833 early Louisiana type plantation home of Marie Therese Coin; her children freed slaves.
Collections: furnishings dating from 1796-early 1900s; Alberta Kinsey paintings; library of authors who stayed at plantation; Hunter Family book collection; plantation gardens. Historic Houses: c.1796 Yucca; c.1800 The African House featuring Clementine Hunter murals; The Weaving House; The Bindery; The Writer's Cabin; c.1796 Ghana.
Facilities: library of books & crafts.
Activities: Museum Sponsors: annual Historical Tour of Natchitoches & the Cane River Country in Oct.; annual Melrose Arts & Crafts Show in April.
Publications: pamphlets.
Hours & Admission Prices: Tues.-Sun. 10-5; other times by appointment. Adults $10, children $5; discounts to AAM members & groups of 15 or more by appointment; members no charge. Closed Easter; Independence Day; Thanksgiving; Christmas.
Attendance: 9,500 (accurate)
Membership: Student $15; Individual $35; Family $60; Sustaining $125; Contributing $250; Supporting $500; Donor $1,000; Patron $2,500.

Merryville

THE MERRYVILLE MUSEUM, 628 N. Railroad St., Merryville, LA 70653. Tel.: 337-825-0101.
Institution Type/Description: History Museum.
Collections: local history & culture; period furnishings; personal artifacts; photographs.
Hours & Admission Prices: Sat.-Sun. 2-4; other times by appointment.

Minden

DORCHEAT HISTORICAL MUSEUM, 116 Pearl St., Minden, LA 71055. Tel.: 318-377-3002.
Key Personnel: Dir., Schelley Brown Francis
Institution Type/Description: History Museum.
Collections: local history & culture; period furnishings; photographs; personal artifacts.
Hours & Admission Prices: Tues.-Fri. 10-1 & 2-4; other times by appointment. No charge.

GERMANTOWN COLONY AND MUSEUM, 120 Museum Rd., Minden, LA 71055-7331. Mailing Address: P.O. Box 178, Minden, LA 71058. Tel.: 318-377-6061.
Institution Type/Description: History Museum.
Collections: local history & culture; period furnishings; personal artifacts; photographs.
Hours & Admission Prices: March-Dec. Wed.-Sat. 10-3; other times by appointment. Adults $3, children under 12 $.50; discounts to groups.

Monroe

BIEDENHARN MUSEUM & GARDENS, (M), 2000 Riverside Dr., Monroe, LA 71201-4268. Tel.: 318-387-5281; 800-362-0983. Fax: 318-387-8253. Facebook: Biedenharn Museum & Gardens.
E-mail: director@bmuseum.org
Web Site: www.bmuseum.org
Founded: 1971.
Congressional District: 5
Key Personnel: Pres., Henry Biedenharn, III; Exec. Dir., Ralph Calhoun; Museum Shop Mgr., Mona Orloski.
Personnel Profile: Full-Time Paid 5; Part-Time Paid 25.
Governing Authority: private foundation. Parent Institution: Emy-Lou Biedenharn Foundation. Tax-exempt: 501(c)(3).
Institution Type/Description: Historic House, Garden & Museum Complex: Bible Museum: c.1914 Southern house, home of Joseph A. Biedenharn, first bottler of Coca-Cola; 1946 formal Elsong Gardens; 1971 Biblical Museum; Coca-Cola Museum.
Collections: Biedenharn Home: marble statue of cherub musicians; fireplace with andirons & grate from Spain; 18th-century Meissen porcelain musicians; hand-woven rug from France; Steinway piano; portraits; miniatures; Waterford crystal chandeliers; high testered beds; 1700s silver collection; period artifacts; art objects; Coca-Cola memorabilia; Model T Coca-Cola delivery truck; reproduction soda fountain. Elsong Gardens: Italian Garden; Water Garden; cast iron statues; porcelain fountain of Catherine the Great. Bible Museum: Exhibition of texts & artifacts structured to demonstrate the contribution of biblical text to the enrichment of Western culture, with special emphasis upon the role of the Bible in American life. Elsong Garden: 1946, formal English gardens; Biblical garden; 1990 octagonal conservatory with its Victorian design.
Research Fields: Biblical topics.
Facilities: 2,000-vol. library of biblically related books & manuscripts; biblical garden; formal gardens & conservatory; 50-seat auditorium & 60-seat meeting room.
Activities: guided tours; lectures; annual garden symposium & plant sale; concerts; permanent, traveling & temporary exhibits.
Publications: booklets, 1993: Elsong Garden; 1994: The Story of Joseph Augustus Biedenharn & The Bottling of Coca-Cola.
Hours & Admission Prices: Tues.-Sat. 10-5; groups of 10 or more by appointment. Adults $6, children $4. Closed New Year's Day; Easter; Independence Day; Thanksgiving; Christmas Eve & Day. &
Attendance: 30,000 (accurate)
Membership: Annual $20.

CHENNAULT AVIATION AND MILITARY MUSEUM, 701 Kansas Lane, Monroe, LA 71203-4775. Mailing Address: State of Louisiana Secretary of State, P.O. Box 94125, Baton Rouge, LA 70804-9125. Tel.: 318-362-5540. Fax: 318-362-5545.
E-mail: nell.calloway@sos.la.gov
Web Site: www.sos.louisiana.gov/camm

Formerly: Aviation and Military Museum of Louisiana
Founded: 1995.
Congressional District: 5
Key Personnel: Dir., Nell Calloway.
Personnel Profile: Full-Time Paid 1; Part-Time Paid 3.
Governing Authority: state. Parent Institution: Louisiana Secretary of State, Baton Rouge, LA. Tax-exempt.
Institution Type/Description: Military Museum.
Collections: aviation & military history; Delta Airlines history; Selman Field memorabilia.
Facilities: theater. Museum-related items for sale.
Activities: films; guided tours; lectures; loan, temporary & traveling exhibitions; theater. Annual Events: Veterans Day Program; Memorial Day Program.
Hours & Admission Prices: Tues.-Sat. 9-4. No charge. Closed most state holidays. &
Attendance: 27,763 (accurate)
Membership: Call for information.

LOUISIANA PURCHASE GARDENS AND ZOO, 1405 Bernstein Park Rd., Monroe, LA 71202-5545. Mailing Address: P.O. Box 123, Monroe, LA 71210. Tel.: 318-329-2400.
E-mail: kim.dooley@ci.monroe.la.us
Web Site: www.monroezoo.org
Key Personnel: Cur. Education, Kimberly Dooley
Institution Type/Description: Gardens & Zoo.
Collections: mammals; birds; reptiles; amphibians; fish; invertebrates; gardens.
Activities: special events; boat rides; rental facilities; birthday parties; zoo outreaches Roar'n Snores Overnight.
Hours & Admission Prices: Daily 10-5. Adults 13-64 $4.50, seniors 65 & over and children 3-12 $3; discounts to groups of 10 or more; children 2 & under no charge. Boat Rides: March 2-Oct. 10-4:30. Admission $2. Closed New Year's Day; Thanksgiving; Christmas.

MASUR MUSEUM OF ART, 1400 S. Grand St., Monroe, LA 71202-2012. Tel.: 318-329-2237. Fax: 318-329-2847.
E-mail: info@masurmuseum.org
Web Site: www.masurmuseum.org
Founded: 1963.
Congressional District: 5
Key Personnel: Dir., Evelyn P. Stewart; Chief Cur., Benjamin Hickey; Cur. Education, Jenny Burnham; Office Asst., Kaitlin Sanson; Bookkeeper, Christal Winfield.
Personnel Profile: Full-Time Paid 3; Part-Time Paid 2; Part-Time Volunteers 30.
Governing Authority: municipal; nonprofit organization. Parent Institution: City of Monroe. Subsidiary Institution: Twin City Art Foundation. Tax-exempt.
Institution Type/Description: Art Museum.
Collections: paintings; prints; sculpture; photographs; media art.
Major Exhibits: Biological Regionalism: Monroe, Louisiana, USA, 10/13-2/14; Narratives Near & Far: Selections from the Wells Fargo Collection, 10/13-2/14; 51st Annual Juried Competition with Juror Kelly Shindler, 2/14-6/14; Greely Myatt: Some More (T), 6/14-10/14; Timelines: Jenny Ellerbe & Northeast Louisiana, 10/14-2/15.
Research Fields: 20th-century art;19th-20th-century Louisiana art.
Facilities: classrooms.
Activities: permanent, temporary & traveling exhibitions; lectures; gallery talks; guided tours; films; studio art classes; docent program; drop-in activities for children; visiting curator program; artist-in-residence program.
Publications: monthly newsletter, Museum News; temporary exhibition brochures; catalogue for annual juried competition; monthly e-newsletter; exhibition catalogues; children's activity book.
Hours & Admission Prices: Tues.-Fri. 9-5, Sat. 12-5. No charge. Closed national holidays.
Attendance: 20,000 (estimated)
Membership: Student Artist $20; Individual $40; Family $60; Single Patron $95; Patron $150; Grand Patron $250; Donor $500. Corporate: Corporate Member $300; Corporate Donor $500; Corporate Sponsor $1,000; Corporate Leader $2,500; Event Sponsor $5,000.

NSCDA IN LOUISIANA; ISAIAH GARRETT LAW OFFICE, 520 S. Grand St., Monroe, LA 71201-7314. Mailing Address: 2204 Island Dr., Monroe, LA 71201. Tel.: 318-387-5691.
Founded: 1840.
Congressional District: 5
Key Personnel: Chm. (V), Margaret Lauve; Chm. (V), Jody Johnston.

Governing Authority: city. Parent Institution: The National Society of the Colonial Dames of America. Tax-exempt.
Institution Type/Description: Historic House: housed in a former law office; built in 1840.
Collections: local history & culture; period furnishings; personal artifacts; photographs.
Hours & Admission Prices: By appointment. No charge; donations accepted.
Attendance: 40 (estimated)

NORTHEAST LOUISIANA CHILDREN'S MUSEUM, 323 Walnut St., Monroe, LA 71201-6711. Tel.: 318-361-9611. Fax: 318-361-9613.
E-mail: nelcm@nelcm.org
Web Site: www.nelcm.org
Key Personnel: Exec. Dir., Julia Bland
Institution Type/Description: Children's Museum.
Collections: hands-on exhibits.
Facilities: Museum-related items for sale.
Activities: traveling & permanent exhibitions; monthly events; volunteering; birthday parties.
Hours & Admission Prices: Tues.-Fri. 9-2, Sat. 10-5. Admission 1 & over $5; discounts to groups of 15 or more.

NORTHEAST LOUISIANA DELTA AFRICAN AMERICAN HERITAGE MUSEUM, 1051 Chennault Park Dr., Monroe, LA 71203. Tel.: 318-323-3745.
Web Site: www.nldaahm.com
Founded: 1994.
Congressional District: 34
Key Personnel: Exec. Dir., Lorraine Slacks; Sec., Patricia Hudson.
Governing Authority: Parent Institution: Ouachita African American Historical Society. Tax-exempt.
Institution Type/Description: History Museum.
Collections: historical artifacts; African art; visual art; furniture.
Research Fields: History.
Activities: student workshops; films; lectures.
Publications: newsletter, Tall Talk.
Hours & Admission Prices: Tues.-Sat. 10-4. Adults $2. &
Attendance: 10,000 (estimated)
Membership: Senior Citizen $10; Family $25; Organization $100.

Moreauville

ADAM PONTHIEU GROCERY STORE AND BIG BEND POST OFFICE MUSEUM, 8554 Louisiana Hwy. 451, Moreauville, LA 71355. Tel.: 318-997-2465.
Institution Type/Description: History Museum.
Collections: local history; period furnishings; early post office & grocery store artifacts.
Hours & Admission Prices: Tues. & Thurs.-Fri. 7:30-3:30, Sun. 12-4:30; other times by appointment. Adults $2, children K-12 $1.

Morgan City

INTERNATIONAL PETROLEUM MUSEUM AND EXPOSITION - RIG MUSEUM, 111 First St., Morgan City, LA 70380. Mailing Address: P.O Box 1988, Morgan City, LA 70381. Tel.: 985-384-3744. Fax: 985-384-3047.
E-mail: rigmuseum@petronet.net
Web Site: rigmuseum.com
Institution Type/Description: Petroleum Industry Museum.
Collections: offshore petroleum drilling rig & industry history.
Hours & Admission Prices: Mon.-Sat. 10-2; groups by appointment. Adults $5, seniors $4, children under 12 $3.50; discounts to groups; children under 5 no charge.

Mound

SOUTHERN HERITAGE AIR FOUNDATION MUSEUM, Vicksburg-Tallulah Rgnl. Airport, Mound, LA 71282. Mailing Address: 179 VTR Airport Rd., Tallulah, LA 71282. Tel.: 601-415-1902. Facebook: Southern Heritage Air Foundation.
E-mail: info@southernheritageair.org
Web Site: www.southernheritageair.org
Founded: 2003.
Governing Authority: Tax-exempt.
Institution Type/Description: Aviation History Museum.
Collections: aviation history; aircraft; photographs; aviation artifacts; WWII artifacts.

Facilities: rental facility.
Activities: special events; plane rides; rental facility. Annual Events: Air Shows.
Hours & Admission Prices: Tues.-Sat. Adults $8, seniors and children 18 & under $5. Closed major holidays. &
Attendance: 10,000 (estimated)

Natchitoches

FORT ST. JEAN BAPTISTE STATE HISTORIC SITE, 155 Jefferson, Natchitoches, LA 71457-4350. Tel.: 318-357-3101. Fax: 318-357-7055.
E-mail: fortstjean@crt.state.la.us
Web Site: www.crt.state.la.us
Formerly: Fort St. Jean Baptiste State Commemorative Area
Founded: 1982.
Congressional District: 5
Key Personnel: Cur., James Prud'Homme; Mgr., Justin French; Interpretive Ranger, Tommy Adkins; Interpretive Ranger, Rhonda Gauthier.
Personnel Profile: Full-Time Paid 6; Part-Time Paid 5; Part-Time Volunteers 10.
Governing Authority: state. The Dept. of Culture, Recreation & Tourism, Office of State Parks, Baton Rouge, LA 70821. Tel.: 504-342-8111.
Institution Type/Description: Historical Fort: reconstruction of 1732 fort & related buildings.
Collections: barracks; guardhouse; servant & kitchen huts; officers' quarters; church; powder magazine; warehouse; bastion; videos.
Facilities: visitor center.
Activities: guided tours; videos. Museum Sponsors: daily living history activities; annual programs March & December.
Publications: pamphlet.
Hours & Admission Prices: Daily 9-5. Adults $4; children under 12 & senior citizens 62 & over no charge. Closed New Year's Day; Thanksgiving; Christmas. &
Attendance: 14,283 (accurate)

THE LEMEE HOUSE, APHN Headquarters, 310 Jefferson St., Natchitoches, LA 71457-4355. Mailing Address: P.O. Box 2248, Natchitoches, LA 71457-2248. Tel.: 318-581-8042. Fax: 318-352-6846.
E-mail: info@melroseplantation.org
Web Site: www.melroseplantation.org
Founded: 1834.
Congressional District: 5
Key Personnel: Dir., Adam T. Foreman; Pres (V), Martha Maynard; Museum Shop Mgr., Betty Metoyer.
Personnel Profile: Full-Time Paid 1.
Governing Authority: society. Parent Institution: The Association for the Preservation of Historic Natchitoches. Tax-exempt.
Institution Type/Description: Historic House: c.1833 house bought by Alexis Lemee in 1849 to serve as his home & the Union Bank of New Orleans.
Collections: period furnishings & artifacts; paintings; lamps; 1792 map by French engineer J. F. Broutin; c.1805 Seth Thomas clock; Rena Phillips fountain.
Facilities: rental facilities.
Activities: meetings for APHN, Les Aimes & Leche organizations.
Publications: semi-annual, Calico Bells.
Hours & Admission Prices: Oct. call for hours. &
Attendance: 1,800 (estimated)
Membership: Student $15; Individual $35; Family $60; Sustaining $125; Contributing $250; Supporting $500; Donor $1,000; Patron $2,500.

LOUISIANA SPORTS HALL OF FAME AND NORTHWEST LOUISIANA HISTORY MUSEUM, 800 Front St., Natchitoches, LA 71457. Tel.: 318-357-2492.
Institution Type/Description: Sports History Museum.
Collections: local history; sports equipment; photographs; personal artifacts; Hall of Fame Inductees.
Activities: school & group tours.
Hours & Admission Prices: Tues.-Sat. 10-4:30, Sun. 1-5; groups by appointment. Adults $5, students, senior citizens & active military $4; children 12 & under no charge. Closed state holidays.

MINOR BASILICA OF THE IMMACULATE CONCEPTION, 145 Church St., Natchitoches, LA 71457-4624. Tel.: 318-352-3422. Fax: 318-352-3822.
E-mail: church4321@catholic.org
Formerly: Immaculate Conception Catholic Church
Founded: 1728.

Congressional District: 31
Key Personnel: Pastor, Rev. Irion St. Romain; Sec., Susan Chesal.
Personnel Profile: Part-Time Volunteers 5.
Governing Authority: church. Parent Institution: Immaculate Conception Church. Tax-exempt.
Institution Type/Description: Historic Building & Museum: c.1856 Immaculate Conception Church and 1885 Rectory.
Collections: 1880 church bell; Austrian stained glass windows; French fittings & furnishings; chandeliers; spiral staircase with no center support; 18th-20th century church artifacts. Historic Buildings: 1885 Rectory; 1855 Old Seminary, now named the Bishop Martin Museum containing records; church artifacts; bells.
Hours & Admission Prices: Church: daily 9-4. Bishop Martin Museum: call 318-352-3422 for appointment. No charge; donations accepted.
Attendance: 500 (estimated)

NATCHITOCHES NATIONAL FISH HATCHERY, 615 South Dr., Natchitoches, LA 71457-3056. Tel.: 318-352-5324. Fax: 318-352-8082.
E-mail: jan_dean@fws.gov
Founded: 1931.
Congressional District: 4
Key Personnel: Deputy Project Leader, Dr. Jan Dean; Biologist, Tony Brady; Maintenance Mechanic, Dennis W. Scarbrough, Jr.; Office Asst., Lana J. Litton.
Personnel Profile: Full-Time Paid 4.
Governing Authority: federal. Dept. of the Interior, U.S. Fish & Wildlife Service, Regional Office, 1875 Century Blvd., Atlanta, GA 30303, Tel. 404-679-4157. Tax-exempt.
Institution Type/Description: Aquarium & Fish Hatchery.
Collections: live displays of fishes of the southeastern United States & Louisiana; reptile tank containing alligators & turtles.
Facilities: aquarium.
Activities: guided group tours; lectures; films & slides.
Publications: brochure.
Hours & Admission Prices: Daily 8-3, group tours: call for appointment. No charge. Closed all federal holidays. &
Attendance: 10,000 (estimated)

New Iberia

BAYOU TECHE MUSEUM, 131 E. Main St., New Iberia, LA 70560. Mailing Address: P.O. Box 14151, New Iberia, LA 70562-4151. Tel.: 337-606-5977. Fax: 337-369-2346.
E-mail: bayoutechemuseum@gmail.com
Web Site: bayoutechemuseum.org
Founded: 1992.
Congressional District: 3
Key Personnel: Dir. & Public Rels., Marcia Patout; Pres. (V), Larry Hensgens; Treas., Art Mixon.
Personnel Profile: Full-Time Paid 1; Full-Time Volunteers 15; Part-Time Paid 2.
Governing Authority: private; nonprofit organization.
Institution Type/Description: History Museum.
Collections: Iberia Parish history & culture; industry; period artifacts; paintings; photographs.
Major Exhibits: James Lee Burke Extension, 12/13-1/14; Religion, 1/14-2/14; Business, 1/14-2/14.
Research Fields: Spanish history related to Iberia Parish; early settlement.
Facilities: 12-seat theater.
Activities: school programs.
Hours & Admission Prices: Thurs.-Sat. 10-4; other times by appointment. Adults $4, senior citizens $3, children $2. Closed major holidays. &
Membership: Individual $20; Family $30.

CONRAD RICE MILL, 307 Ann St., New Iberia, LA 70560. Mailing Address: P.O. Box 10640, New Iberia, LA 70562. Tel.: 800-551-3245.
E-mail: info@conradricemill.com
Web Site: www.conradricemill.com
Institution Type/Description: Historic Building.
Collections: mill history; period equipment.
Facilities: Gift items for sale.
Activities: group tours.
Hours & Admission Prices: Mon.-Sat. 9-5; groups by appointment. Adults $4, seniors $3.50, children 3-11 $2.25.

RIP VAN WINKLE GARDENS ON JEFFERSON ISLAND, 5505 Rip Van Winkle Rd., New Iberia, LA 70560-8167. Tel.: 337-359-8525. Fax: 337-359-8526.
E-mail: jislgdns@earthlink.net
Web Site: www.ripvanwinklegardens.com
Founded: 1978.
Key Personnel: Mgr., Michelle Richard.
Personnel Profile: Full-Time Paid 10.
Institution Type/Description: Botanical Garden & Historic Houses: c.1870 house built by actor Joseph Jefferson; semi-tropical 25 acre landscape garden.
Collections: furniture; paintings; decorative accessories relating to Victorian lifestyle of 19th & 20th century occupants of the house.
Facilities: 600-vol. library of books pertaining to horticulture and Joseph Jefferson; 25-acre botanical garden; 15,000 sq. ft. exhibit space. Books on horticulture & Victorian lifestyle for sale.
Activities: guided tours; lectures.
Hours & Admission Prices: Tours: daily 9-4. Adults $10, children & senior citizens $8; discounts to groups; children under 8 & members no charge. Closed New Year's Day; Thanksgiving, Christmas Eve & Day. &
Attendance: 40,000 (accurate)

THE SHADOWS-ON-THE-TECHE, 317 E. Main St., New Iberia, LA 70560-3728. Tel.: 337-369-6446. Fax: 337-365-5213.
E-mail: shadows@shadowsontheteche.org
Web Site: www.shadowsontheteche.org
Founded: 1961.
Congressional District: 3
Key Personnel: Chm. Property Council, Taylor Barras; Dir., Patricia Kahle; Cur. Education, Catherine T. Schramm; Curatorial Technician, Yvonne Leblanc.
Personnel Profile: Full-Time Paid 3; Part-Time Paid 7; Part-Time Volunteers 75.
Governing Authority: nonprofit organization. Property of the National Trust for Historic Preservation, 1785 Massachusetts Ave., N.W., Washington, DC 20036.
Institution Type/Description: Historic House Museum: 1834 plantation home & restored landscape.
Collections: decorative arts; textiles & costumes; furnishings.
Research Fields: social history of pre-Civil War South & south central Louisiana; historic landscape.
Facilities: meeting space; gardens. Books & other museum-related items for sale.
Activities: guided tours; school programs; lectures; special events; summer interpreter program for high school students; teacher workshops; thematic tours & exhibits; living history programs.
Publications: The Shadows-on-the-Teche Cookbook; Fine Things Are Without Value: Inside the Shadows.
Hours & Admission Prices: Mon.-Sat. 9-4:30. Adults $10, senior citizens 65 & over $8, students 6-17 $6.50; discounts to groups and AAM & ICOM members; National Trust for Historic Preservation members; Friends of the Shadows & children under 6 no charge. Closed major holidays. &
Attendance: 18,591 (accurate)
Membership: Friends of the Shadows: Individual $25; Family $50; Sustaining $100; Contributing $250; Supporting $500; Donor $1,000; Patron $2,500; Benefactor $5,000. Corporate: $500, $1,000 & $1,500.

New Orleans

AFRICAN AMERICAN MUSEUM, 1418 Gov. Nicholls St., New Orleans, LA 70116-2344. Tel.: 504-566-1136.
Institution Type/Description: Art and History Museum: housed in the Treme Villa, built in 1828-29.
Collections: African beadwork; costumes; masks; textiles; musical instruments; divination objects.
Hours & Admission Prices: Wed.-Sat. 11-4. Adults $5, students & seniors $3, children 6-12 $2. &

AMERICAN ITALIAN CULTURAL CENTER, 537 S. Peters St., New Orleans, LA 70130-1628. Tel.: 504-522-7294. Fax: 504-522-1657.
Web Site: www.americanitalianculturalcenter.com
Formerly: American Italian Renaissance Foundation
Founded: 1978.
Key Personnel: Dir., Ashley Rice; Chm (V), Frank Maselli.
Personnel Profile: Full-Time Paid 1; Part-Time Volunteers 4.
Governing Authority: Parent Institution: American Italian Renaissance Foundation. Tax-exempt.
Institution Type/Description: History Museum.

Collections: history & culture of Italian Americans in the Southeast; photographs; articles; family histories; memorabilia; genealogy records.
Activities: Italian language & culture classes. Annual Event: Louisiana American Italian Sports Hall of Fame Induction Banquet.
Publications: quarterly digest, Italian American Digest.
Hours & Admission Prices: Tues.-Fri. 10-4. Adults $8, seniors & students $5; members no charge. Closed holidays. &
Attendance: 1,000 (estimated)
Membership: Level I $50; Level 2 $75; Level 3 $250; Level 4 $500; Level 5 $1,000; Level 6 $5,000.

AMISTAD RESEARCH CENTER, INC., Tilton Hall-Tulane University, 6823 St. Charles Ave., New Orleans, LA 70118-5665. Tel.: 504-862-3222. Fax: 504-862-8961. Facebook: Amistad Research Center.
E-mail: lhampto3@tulane.edu
Web Site: www.amistadresearchcenter.org & amistadresearchcenter-.blogspot.com
Founded: 1966.
Congressional District: 2
Key Personnel: Exec. Dir., Lee Hampton; C.E.O., Dr. Andred Jefferson; Dir. Reference & Library Svcs., Christopher Harter; Dir. Processing, Laura Thomson; Senior Processing Asst., Shannon Burrell; Archives & Library Asst., Andrew Salinas.
Personnel Profile: Full-Time Paid 9; Part-Time Paid 2; Interns 1.
Governing Authority: Tax-exempt.
Institution Type/Description: History Museum.
Collections: art; archives & manuscripts; photographs; digital archives.
Major Exhibits: The Free Southern Theater and the Black Arts Movement in the South, 1/14-3/14; Behind Every Good Movement: Women of the Civil Rights Era, 4/14-6/14; Rising Up II: The Life and Work of Hale Woodruff, 7/14-10/14.
Facilities: 13,000 sq. ft. exhibit space.
Publications: quarterly newsletter, Amistad Reports.
Hours & Admission Prices: Mon.-Fri. 8:30-4:30. No charge; donations accepted. &
Attendance: 1,771 (accurate)
Membership: Retiree & Student $20; Phillis Wheatly Club $40; Family & Sojourner Truth Club $55; Frederick Douglass Club $100; Carter G. Woodson Club $250; Harriet Tubman Club $500; Cinque Club $1,000; Clifton H. Johnson Club $5,000

AUDUBON AQUARIUM OF THE AMERICAS, (M), 1 Canal St., New Orleans, LA 70130-1175. Mailing Address: P.O. Box 4327, New Orleans, LA 70178-4327. Tel.: 800-774-7394; 504-581-4629. Fax: 504-565-3010.
E-mail: mcalhoun@audoboninstitute.org
Web Site: www.audoboninstitute.org
Founded: 1990.
Congressional District: 91
Key Personnel: C.E.O. & Pres., L. Ronald Forman; Dir., Karyn Kearney; Cur., John Hewitt; Public Rels., Meghan Calhoun; Security, Rodney Daniels; Museum Shop Mgr., Debra McGuire.
Personnel Profile: Full-Time Paid 160; Part-Time Paid 118; Part-Time Volunteers 356.
Governing Authority: municipal; partnership; nonprofit organization. Parent Institution: Audubon Institute, P.O. Box 4327, New Orleans, LA 70178-4327. Tax-exempt: 501(c)(3).
Institution Type/Description: Aquarium.
Collections: fresh & salt water fish; invertebrates; birds; reptiles; amphibians; marine mammals.
Research Fields: exotic animal behavior; endangered species preservation.
Facilities: library books on aquatic life, research & aquarium husbandry; 122,000 sq. ft. exhibit space; 354-seat, 3D Energy IMAX theatre; classrooms; restaurant. Museum-related and educational items for sale.
Activities: arts festivals; formal education programs for adults & children; mobile vans; traveling & participatory exhibits; training programs for professional museum workers.
Publications: quarterly membership magazine, Audubon Up Close; annual report.
Hours & Admission Prices: Tues.-Sun. 10-5. Aquarium: adults $19.95, senior citizens $15.95, children 2-12 $12.95; members no charge. IMAX: Mon.-Sat. 10-7. Adults $9.95, senior citizens $8.95, children 2-12 $6.95; member no charge. Closed Mardi Gras; Christmas Eve & Day. &
Attendance: 990,000 (estimated)
Membership: Senior Individual $60; Senior Couple $75; Individual $90; Individual Plus One $115; Family $135; Family Plus One $160; Safari Krewe $225; Wildlife Partner $300; Golden Eagle $450;

AUDUBON INSECTARIUM, 423 Canal St., New Orleans, LA 70130. Mailing Address: P.O. Box 4327, New Orleans, LA 70178. Tel.: 800-774-7394; 504-581-4629.
Institution Type/Description: Nature Center.
Collections: insects; insect history; butterfly garden.
Facilities: theater.
Hours & Admission Prices: Tues.-Sun. 10-5. Adults 13-64 $15, seniors 65 & over $12, children 2-12 $10.

AUDUBON ZOO, (M), 6500 Magazine St., New Orleans, LA 70118-4848. Tel.: 504-861-4629. Fax: 504-865-7332. Facebook: Audubon Zoo.
E-mail: air@auduboninstitute.org
Web Site: www.auduboninstitute.org
Formerly: Audubon Park and Zoological Garden
Founded: 1914.
Congressional District: 89
Key Personnel: Pres. & C.E.O., L. Ronald Forman; Exec. Vice Pres. & Mng. Dir., Larry Rivarde; Exec. Vice Pres. Devel., Laurie Conkerton; Sr. Vice Pres. & Dir. Animal Husbandry, John Hewitt; Vice Pres. & Gen. Cur., Joel Hamilton; Vice Pres. Mktg., Chimene Grant; Vice Pres. Gift Shops, Debra McGuire.
Governing Authority: municipal; partnership; nonprofit organization. Parent Institution: Audubon Nature Institute, Inc., P.O. Box. 4327, New Orleans, LA 70178-4327. Tax-exempt: 501(c)(3).
Institution Type/Description: Zoo and Park.
Collections: mammals; birds; reptiles; amphibians; fish; invertebrates.
Research Fields: exotic animal behavior; endangered species preservation.
Facilities: 2,125-vol. library; cafeteria. Museum-related items for sale.
Activities: arts festivals; formal education programs for adults and children; hobby workshops; lectures; mobile vans; participatory exhibits; training programs for professional museum workers. Annual Events: Zoo-to-Do; Earth Fest; Swamp Fest; Boo at the Zoo; Member Night Zoobilation.
Publications: quarterly membership magazine; annual report; education calendar; conservation report.
Hours & Admission Prices: Tues.-Fri. 10-4, Sat.-Sun. 10-5. Adults $17.50, senior citizens 65 & up $13, children 2-12 $12; members no charge. Closed Mardi Gras; Thanksgiving; Christmas. &
Attendance: 717,000 (estimated)

BACKSTREET CULTURAL MUSEUM, 1116 St. Claude Ave., New Orleans, LA 70116-2330. Mailing Address: 1116 Henriette Delille St., New Orleans, LA 70116. Tel.: 504-522-4806.
E-mail: info@backstreetmuseum.org
Web Site: www.backstreetmuseum.org
Key Personnel: Exec. Dir., Sylvester Francis
Institution Type/Description: Folk-Life and History Museum.
Collections: local history & culture; artifacts; exhibits; memorabilia; films & videos depicting Mardi Gras Indians.
Activities: Museum Sponsors: Treme Call Out and Dance; Mardi Gras Open House; All Saints Day Parade in November.
Hours & Admission Prices: Tues.-Sat. 10-5. Adults $8.

BEAUREGARD-KEYES HOUSE, 1113 Chartres St., New Orleans, LA 70116-2504. Tel.: 504-523-7257. Fax: 504-523-7257.
Web Site: www.bkhouse.org
Key Personnel: Dir., Marion S. Chambers; Pres. Keyes Foundation, Gary R. Williams.
Personnel Profile: Full-Time Paid 2; Part-Time Volunteers 2; Interns 1.
Governing Authority: Parent Institution: Keyes Foundation.
Institution Type/Description: Historic House Museum: home of wealthy auctioneer Joseph LeCarpentier, built in 1826.
Collections: period furnishings; personal artifacts; garden; period dolls; tea pots; folk costumes.
Facilities: Museum-related items for sale.
Activities: guided tours.
Publications: Beauregard-Keyes House; Lunch With Mrs. Keyes.
Hours & Admission Prices: Mon.-Sat. 10-3. Adults $10, students & senior citizens $9, children 6-12 $4; discounts to AAM members; children under 6 no charge. Closed major holidays; Mardi Gras.

BLAINE KERN'S MARDI GRAS WORLD, 1380 Port of New Orleans Pl., New Orleans, LA 70130-1805. Tel.: 800-362-8213; 504-361-7821.
Institution Type/Description: General Museum.
Collections: Mardi Gras history, customs, costumes, & floats; float designing & building.
Activities: video.

Hours & Admission Prices: Daily 10-6. Adults $18, seniors $14, children $11; discounts to groups. &

CALLAN CONTEMPORARY, 518 Julia St., New Orleans, LA 70130. Tel.: 504-525-0518. Fax: 504-525-0516.
Formerly: Gallery Bienvenu
Key Personnel: Owner, Steven Callan; Dir., Borislava Callan
Institution Type/Description: Art Gallery.
Collections: works by contemporary artists; paintings; sculpture.
Hours & Admission Prices: Tues.-Sat. 10-5; other times by appointment.

CATHOLIC CULTURAL HERITAGE CENTER/OLD UR-SULINE CONVENT/ST. LOUIS CATHEDRAL, 1100 Chartres St., New Orleans, LA 70116-2505. Mailing Address: 615 Pere Antoine Alley, New Orleans, LA 70116. Tel.: 504-525-9585. Fax: 504-525-9583.
E-mail: saintlouiscathedral-no@arch-no.org
Web Site: stlouiscathedral.org
Founded: 2004.
Congressional District: 2
Key Personnel: Dir., Very Rev. Philip G. Landry; Chm. (V), Barbara Windhorst; Museum Shop Mgr., Jolie Sekinger.
Personnel Profile: Full-Time Paid 3; Part-Time Paid 1; Part-Time Volunteers 5.
Governing Authority: Parent Institution: Archdiocese of New Orleans. Tax-exempt.
Institution Type/Description: Religious Museum: built in 1752.
Collections: religious artifacts; oil paintings; religious statues; bronze busts; manuscripts; drawings.
Major Exhibits: The Archbishop Wore Combat Boots, 10/5/13-5/26/14.
Facilities: archives; gardens.
Hours & Admission Prices: Tours: Mon.-Sat. 10-4. Adults $10, seniors $9, students $5; discounts to groups of 20 or more; members no charge. &
Attendance: 56,000 (accurate)
Membership: Individual $60; Family $75; Special Friend $100.

COLLINS C. DIBOLL ART GALLERY, 4th Fl. Monroe Library, 6363 St. Charles Ave., New Orleans, LA 70118-6143. Tel.: 504-864-7248.
E-mail: gallery@loyno.edu
Web Site: www.loyno.edu/dibollgallery
Key Personnel: Gallery Dir., Karoline Schleh
Institution Type/Description: Art Gallery.
Collections: paintings.
Hours & Admission Prices: Mon.-Sat. 10-6, Sun. 12-6. No charge.

CONFEDERATE MUSEUM, 929 Camp St., New Orleans, LA 70130-3907. Tel.: 504-523-4522. Fax: 504-523-8595.
E-mail: memhall@aol.com
Web Site: www.confederatemuseum.com
Founded: 1891.
Congressional District: 1
Key Personnel: Cur., Pat Ricci.
Personnel Profile: Full-Time Paid 1; Full-Time Volunteers 5; Part-Time Paid 2; Part-Time Volunteers 1; Interns 1.
Governing Authority: nonprofit. Parent Institution: Memorial Hall, Inc. Tax-exempt.
Institution Type/Description: Military Museum: housed in 1890 Memorial Hall.
Collections: pictures; paintings; weapons; silver; uniforms; flags; medical instruments; memorabilia of early Louisiana and Civil War history; personal effects of Jefferson Davis, Beauregard, Bragg, Lee & other Civil War leaders.
Research Fields: Louisiana history; Civil War.
Facilities: Books & other gift items for sale.
Activities: tours; permanent exhibitions; educational programs.
Publications: Louisiana Historical Association Books on Louisiana History & Civil War.
Hours & Admission Prices: Tues.-Sat. 10-4. Adults $8, children 7-14 $5; discounts to groups; children under 7 no charge.
Attendance: 15,000

CONTEMPORARY ARTS CENTER, 900 Camp St., New Orleans, LA 70130-3908. Tel.: 504-528-3805. Fax: 504-528-3828.
Web Site: www.cacno.org
Founded: 1976.
Congressional District: 2
Key Personnel: Exec. & Artistic Dir., Jay Weigel; Pres., Bennett Davis; Cur. Visual Arts, Dan Cameron.

Personnel Profile: Full-Time Paid 13; Part-Time Paid 8.
Governing Authority: private; nonprofit organization. Tax-exempt.
Institution Type/Description: Arts Center.
Collections: contemporary art.
Facilities: theater; performance spaces.
Activities: music, dance & theatre performances; temporary exhibitions; educational programs for children & adults.
Hours & Admission Prices: Wed.-Mon. 11-5. Adults $5, senior citizens & students $3; discounts to groups; members and children 15 & under no charge. Performance & special events prices may vary. Closed most major holidays. &
Attendance: 125,000 (estimated)
Membership: Student & Artist $25; Individual $35; Family & Couple $55; Friend $80; Collector's Club $175; Center Stage $250; Patron Now $500; Silver Circle $1,000; President's Council $5,000.

DEGAS HOUSE, 2306 Esplanade Ave., New Orleans, LA 70119-2502. Tel.: 504-821-5009.
Web Site: www.degashouse.com
Institution Type/Description: Historic House: former home of the French Impressionist painter Edgar Degas.
Collections: paintings.
Hours & Admission Prices: Guided tours by appointment only. Suggested donations: adults $10, seniors $8, children & students $5.

FORT PIKE STATE HISTORIC SITE, 27100 Chef Menteur Hwy., New Orleans, LA 70129-3106. Mailing Address: P.O. Box 44426, Baton Rouge, LA 70804-4426. Tel.: 504-255-9171; 888-662-5703. Fax: 504-662-0147.
E-mail: fortpike_mgr@crt.state.la.us
Web Site: www.crt.state.la.us
Formerly: Fort Pike State Commemorative Area
Founded: 1934.
Congressional District: 1
Key Personnel: Historic Site Mgr., Michelle Lewis.
Personnel Profile: Full-Time Paid 4; Part-Time Paid 3; Part-Time Volunteers 10.
Governing Authority: state. Parent Institution: Office of State Parks, Louisiana. Tax-exempt: 501(c)(3).
Institution Type/Description: Military Museum: housed in 1818-1827 fort built by U.S. government.
Collections: government relics and displays of armaments, dress and battle orders from the period 1812-1865; Louisiana history.
Research Fields: military history of War of 1812 & Civil War.
Facilities: picnic facilities; day-use recreation area.
Activities: guided tours; lectures; permanent exhibitions.
Publications: brochures.
Hours & Admission Prices: Tues.-Sat. 9-5. Admission $4; seniors 62 & over and children 12 & under no charge. Closed New Year's Day; Thanksgiving; Christmas.
Attendance: 13,500 (estimated)

✳ **GALLIER HOUSE, (M),** 1132 Royal St., New Orleans, LA 70116. Mailing Address: P.O. Box 56836, New Orleans, LA 70156-6836. Tel.: 504-525-5661. Fax: 504-568-9735.
E-mail: hgrimagallier@aol.com
Web Site: www.hgghh.org
Founded: 1971.
Key Personnel: Exec. Dir., Mamie Sterkx Gasperecz; Deputy Dir., Carolyn Bercier; Chm. Bd. The Woman's Exchange, Julie Breitmeyer; C.F.O., Steven Smith; Pres. (V), Marilee Hovet; Dir. Devel., Lisa Samuels; Dir. Communications, Nadine Segari.
Personnel Profile: Full-Time Paid 5; Part-Time Paid 4; Part-Time Volunteers 3.
Governing Authority: Parent Institution: The Woman's Exchange. Branch Museum: Hermann-Grima, 820 St. Louis St., New Orleans, LA 70112. Tax-exempt: 501(c)(3).
Institution Type/Description: Historic House: housed in the former home of architect, James Gallier, Jr.; built in 1857.
Collections: Gallier's life & family; personal artifacts; period furnishings; garden; slave quarters.
Research Fields: buildings; furnishings; gardens; life style, cooking methods during the period 1830-1860 in New Orleans; Creole foods & Christmas celebrations; social customs; urban archaeology; funeral practices; courtship & marriage customs; childhood; free blacks; urban slavery.
Facilities: Museum-related items for sale.
Activities: guided tours; lectures; education programs; docent programs; children's workshops; summer camp programs; special events.

Publications: Women Who Cared: The 100 Years of the Christian Woman's Exchange; quarterly newsletter; Creole cookery.

Hours & Admission Prices: Tours: Mon. Thurs.-Sat. 10-2; other times by appointment. Adults $12, students, senior citizens & children 8-18 $10; discounts to AAM, ICOM, LAM, National Trust & SEMC members; members & children under 8 no charge. Closed major holidays.

Attendance: 3,000 (estimated)

Membership: Individual $25; Family $50; Patron $100, $500, $1,000; Corporate $100, $500, $1,000, $5,000.

✻   **HERMANN-GRIMA, (M),** 820 St. Louis St., New Orleans, LA 70112-3416. Mailing Address: P.O. Box 56836, New Orleans, LA 70156-6836. Tel.: 504-525-5661. Fax: 504-568-9735.

E-mail: hgrimagallier@aol.com

Web Site: www.hgghh.org

Founded: 1971.

Congressional District: 2

Key Personnel: Exec. Dir., Mamie Sterkx Gasperecz; Deputy Dir., Carolyn Bercier; Chm. Bd. The Woman's Exchange, Julie Breitmeyer; Pres. (V), Marilee Hovet; C.F.O., Steven Smith; Devel. Assoc., Jennifer Daly; Dir. Devel., Lisa Samuels; Dir. Communications, Nadine Segari.

Personnel Profile: Full-Time Paid 5; Part-Time Paid 7; Part-Time Volunteers 14; Interns 3.

Governing Authority: Parent Institution: The Woman's Exchange. Branch Museum: Gallier House, 1132 Royal St., New Orleans, LA 70116. Tax-exempt: 501(c)(3).

Institution Type/Description: Historic House: built in 1831

Collections: 1830-1860 furnishings; textiles; needlework; cooking implements; restored 1830 open-hearth kitchen.

Research Fields: buildings; furnishings; gardens; life style, cooking methods during the period 1830-1860 in New Orleans; Creole foods & Christmas celebrations; social customs; urban archaeology; funeral practices; courtship & marriage customs; childhood; free blacks; urban slavery.

Facilities: rental facilities. Gift items for sale.

Activities: guided tours; lectures; formally organized education programs for school groups & adults; internships; docent program; period cooking demonstrations; adult education programs; children's workshops; summer camp programs; rental facilities. Museum Sponsors: open hearth cooking October to May.

Publications: Women Who Cared: The 100 Years of the Christian Woman's Exchange, quarterly newsletter, Creole cookery.

Hours & Admission Prices: Tours: Mon.-Tues. & Thurs.-Fri. 10, 11, 12, 1 & 2, Sat. 12, 1, 2 & 3; other times by appointment. Adults $12, students, senior citizens & children 8-18 $10; discounts to AAM, ICOM, LAM, National Trust & SEMC members; members & children under 8 no charge. Closed major holidays. &

Attendance: 12,000 (accurate)

Membership: Individual $25; Family $50; Patron $100, $500, $1,000; Corporate $100, $500, $1,000, $5,000.

✻   **THE HISTORIC NEW ORLEANS COLLECTION, (M),** 533 Royal St., New Orleans, LA 70130-2113. Tel.: 504-523-4662. Fax: 504-598-7108. Facebook: The Historic New Orleans Collection.

E-mail: wrc@hnoc.org

Web Site: www.hnoc.org

Founded: 1966.

Congressional District: 2

Key Personnel: Exec. Dir., Priscilla O'Reilly-Lawrence; C.F.O., Michael Cohn; Dir. Museum Programs, John H. Lawrence; Manager Admin. Svcs., Kathy Slimp; Dir. Systems, Carol Bartels; Mgr. Mktg., Teresa Devlin; Mgr. Collections, Warren Woods; Dir. Research Center, Alfred Lemmon; Dir. Publications, Jessica Dorman; Museum Shop Mgr., Michelle Gaynor.

Personnel Profile: Full-Time Paid 82; Part-Time Paid 9; Part-Time Volunteers 78.

Volunteer Hours: 4,294

Operating Expenses: 9,221,278

Operating Income: 1,194,521

Governing Authority: nonprofit organization. Parent Institution: Kemper & Leila Williams Foundation. Branch Museum: The Williams Research Center, 410 Chartres St., New Orleans, LA. Tax-exempt: 501(c)(3).

Institution Type/Description: History Museum & Research Center.

Collections: New Orleans, Louisiana & the Gulf South history & culture including books, pamphlets, manuscript materials, paintings, prints, drawings, maps, photographs, & period artifacts; historic buildings.

Major Exhibits: Occupy New Orleans! Voices from the Civil War, 10/20/13-9/20/14; Daguerreotypes to Digital: A Presentation of Photographic Processes, 11/13-2/14; Civil War Battlefields & National Parks: Photographs by A.J. Meek, 11/13-4/14; New Orleans in Movies, 2/14-6/14; The Musical Legacy of the Boswell Sisters, 3/14-10/14; Creole World, 4/14-11/14; Andrew Jackson & the Battle of New Orleans, 11/14-3/15.

Research Fields: all aspects of Louisiana history & culture; architecture; river life; maps; French Quarter; Mardi Gras; Battle of New Orleans; Civil War; Louisiana artists; cemeteries; New Orleans imprints; land tenure; performing arts; plantation & family papers; jazz; literature.

Facilities: Research Center: library. Museum-related items for sale.

Activities: tours; gallery talks; docent program; temporary exhibitions; publishing; seminars; lectures; symposia; educational programming.

Publications: Historic New Orleans Collection Quarterly; Tennessee Williams Annual Review; books, Vaudechamp in New Orleans; Printmaking in New Orleans; Common Routes: St. Domingue, Louisiana; A British Eyewitness at the Battle of New Orleans: The Memoir of Royal Navy Admiral Robert Aitchison, 1808-1827; From Louis XIV to Louis Armstrong: A Cultural Tapestry; George L. Viavant: Artist of the Hunt; Charting Louisiana: Five hundred years of maps; Queen of the South: New Orleans, 1853-1862: The Journal of Thomas K. Wharton; Complementary Visions of LA Art: The Laura Simon Nelson at the Historic New Orleans Collection; Haunter of Ruins: The Photography of Clarence John Laughlin; Jazz Scrapbook: Bill Russell and Some Highly Musical Friends; Bibliography of New Orleans Imprints, 1764-1864; Vicksburg: Southern City Under Siege; Nelly Custis Lewis's Housekeeping Book; The Buildings of the Historic New Orleans Collection; Encyclopedia of New Orleans Artists, 1718-1918; Southern Travels: Journal of John H.B. Latrobe, 1834; Music in the Street: Photographs of New Orleans by Ralston Crawford; Josephine Crawford: An Artist's Vision; Furnishing Louisiana: Creole and Acadian Furniture, 1735-1835; Unfinished Blue: Memories of a New Orleans Music Man; In Search of Julien Hudson: Free Artist of Color in Pre-Civil War New Orleans; Drawn to Life: Al Hirschfield and the Theater of Tennessee Williams; Ernie K-Doe: The R&B Emperor of New Orleans; A Company Man: The Remarkable French-Atlantic Voyage of a Clerk for the Company of the Indies; Perique: Photographs by Charles Martin.

Hours & Admission Prices: Tues.-Sat. 9:30-4:30, Sun. 10:30-4:30. Tours: Tues.-Sat.: 10, 11, 2, 3, Sun. 11, 2, 3. Tour: $5; discounts to AAM members; members, changing exhibitions & research areas no charge. Closed major holidays. &

Attendance: 31,000 (estimated)

Membership: Founder Individual $35; Founder Family $65; Merieult Society $100; Mahalia Society $250; Jackson Society $500; Laussat Society $1,000; Bienville Circle $5,000.

HOUSE OF BROEL'S VICTORIAN MANSION AND DOLL-HOUSE MUSEUM, 2220 St. Charles Ave., New Orleans, LA 70130. Tel.: 504-522-2220 & 494-2220. Fax: 504-524-6775.

E-mail: info@houseofbroel.com

Web Site: www.houseofbroel.com

Institution Type/Description: Historic Mansion: built c.1850.

Collections: local history & culture; dollhouses; period furnishings; Mardi Gras memorabilia; Asian art; photographs.

Activities: rental facilities.

Hours & Admission Prices: Mon.-Sat. 10-5. Adults $10, children $5.

THE JACKSON BARRACKS MILITARY MUSEUM, 6400 St. Claude Ave., New Orleans, LA 70117-1456. Tel.: 504-278-8242.

E-mail: jbmuseum@la.ngb.army.mil

Web Site: www.la.ngb.army.mil

Key Personnel: Cur., Stan Amerski

Institution Type/Description: Military Museum.

Collections: uniforms; honors & decorations dating back to the American Revolution; war letters, diaries & personal artifacts; tanks; fighter planes; helicopters; antiaircraft batteries; cannons.

Facilities: theater.

Activities: films; award ceremonies; concerts; historic society meetings; official functions.

Hours & Admission Prices: Mon.-Fri. 8-4, Sat. by appointment; group tours by appointment. No charge. Closed holidays.

JEAN LAFITTE NATIONAL HISTORICAL PARK & PRE-SERVE, (M), 419 Decatur St., New Orleans, LA 70130-1035. Tel.: 504-589-3882. Fax: 504-589-3851.

E-mail: kathy_lang@nps.gov

Web Site: www.nps.gov/jela

Founded: 1978.

Congressional District: 1

Key Personnel: Supt., Carol A. Clark.

Personnel Profile: Full-Time Paid 1.

Governing Authority: federal. Administered by the National Park Service, Washington, DC 20240. Branch Units: Acadian, Chalmette, Barataria Preserve & New Orleans. Tax-exempt.

Institution Type/Description: History Museum.

Collections: natural history; archaeology; military; furnishings; household items; tools & equipment; cultural & natural resources of the Mississippi Delta region; cultural resources related to the Acadian people. Historic House: c.1832 Malus Beauregard home.

Research Fields: Battle of New Orleans; War of 1812; historic & prehistoric archaeology in area; American Indians; ethnic groups of Mississippi Delta Region; environment of park; National Cemetery; natural history of area; history of French Quarter & New Orleans.

Facilities: 500-vol. library on Battle of New Orleans, historical military books, films, photos, maps & information on surrounding area & National Parks; photo files; archives; three Acadian Unit visitor centers: Lafayette, 150-seat theater; Eunice, 125-seat video area; Thibodaux, 200-seat theater, 25-seat video area; Bataria Preserve Visitor Center exhibits; 65-seat theater; New Orleans Visitor Center exhibits; 40-seat multipurpose area. Chalmette Battlefield, 50-seat theater, visitor center. Publications & gift items for sale.

Activities: canoe treks; videos; self-guided tours; dramatic presentation.

Publications: NPS handbooks.

Hours & Admission Prices: Daily 9-5. No charge. Closed New Year's Day; Mardi Gras; Christmas. ₠

Attendance: 391,019 (estimated)

✳ **LONGUE VUE HOUSE & GARDENS, (M),** 7 Bamboo Rd., New Orleans, LA 70124-1007. Tel.: 504-488-5488. Fax: 504-486-7015.

E-mail: lcosta@longuevue.com

Web Site: www.longuevue.com

Founded: 1980.

Congressional District: 1

Key Personnel: Dir., Tony Chauveaux; Admin., Ribby Ferguson; Bd. Chair, Lynne Stern; Asst. Cur., Lenora Costa; Accountant, Patrick Nedd; Devel., Jen Gick; Operations & Sales, Anna Bell Jones; Dir. Programs, Hilairie Schackai.

Personnel Profile: Full-Time Paid 12; Part-Time Paid 19; Part-Time Volunteers 50; Interns 3.

Governing Authority: nonprofit organization. Parent Institution: Longue Vue Foundation. Tax-exempt.

Institution Type/Description: Historic House & Gardens: 1939-42, Palladian style, Longue Vue, home of cotton broker Edgar Bloom Stern & Edith Rosenwald Stern, daughter of Sears Roebuck financier & philanthropist, Julius Rosenwald.

Collections: 17th to 20th-century European decorative arts; archival materials related to house & gardens; gardens designed by Ellen Biddle Shipman.

Major Exhibits: Silver, 1/14-4/14; Margaret Hull, 4/14-7/14.

Research Fields: architecture; history; decorative arts; fine arts; horticulture & landscape gardening.

Facilities: 100-vol. library of horticulture, landscape architecture, architecture, decorative arts, archival design material, photographs & family papers available for research; 80-seat auditorium. Decorative arts, horticulture objects & publications related to the house & gardens for sale in museum shop.

Activities: guided tours; lectures; films; gallery talks; docent program or council; permanent & changing exhibitions.

Publications: exhibition catalogs; guide book; quarterly newsletter; rack cards.

Hours & Admission Prices: Tues.-Sat. 10-5, Sun. 1-5. Garden: adults $7. House & Gardens: adults $10, children & students $5; discounts to senior citizens, students, tour groups, AABGA, AAA & AAM members; members no charge. American Horticultural Society reciprocal admission program. Closed national holidays. ₠

Attendance: 45,000 (accurate)

Membership: Individual $35; Member & Guest $50; Family $75; Nonprofit & Small Business $100; Long Viewer $1,000.

LOUISIANA CHILDREN'S MUSEUM, 420 Julia St., New Orleans, LA 70130-3606. Tel.: 504-523-1357. Fax: 504-529-3666.

E-mail: srobinson@lcm.org

Web Site: www.lcm.org

Founded: 1981.

Congressional District: 2

Key Personnel: Exec. Dir., Julia W. Bland; Mng. Dir., Simonne Robinson; Dir. Education, Cat Bacelieri; Museum Shop Mgr., Edward Begnaud.

Personnel Profile: Full-Time Paid 17; Part-Time Paid 15; Part-Time Volunteers 150; Interns 1.

Governing Authority: nonprofit. Tax-exempt: 501(c)(3).

Institution Type/Description: Children's Museum.

Collections: hands-on exhibits.

Facilities: 30,000 sq. ft. exhibit space; educational facilities. Museum-related items for sale.

Activities: special weekend & summer programs; hobby workshops; organized education programs for children; participatory exhibits.

Publications: quarterly newsletter, Hands-On; membership & visitor brochures; annual report.

Hours & Admission Prices: Winter: Tues.-Sat. 9:30-4:30, Sun. 12-4:30; Summer: Mon.-Sat. 9:30-5, Sun. 12-5. Admission $8; children under one year no charge. Closed New Year's Day; Mardi Gras; Easter; Independence Day; Thanksgiving; Christmas. ₠

Attendance: 140,000 (estimated)

Membership: Family I $55; Magician $150; Magician Plus $200.

✳ **LOUISIANA STATE MUSEUM,** 751 Chartres St., New Orleans, LA 70116-3205. Mailing Address: P.O. Box 2448, New Orleans, LA 70176-2448. Tel.: 800-568-6968. Fax: 504-568-4995.

E-mail: asmith@crt.la.gov

Web Site: lsm.crt.state.la.us

Founded: 1906.

Congressional District: 2, 4 & 6

Key Personnel: Interim Dir., Robert Wheat; Chm. (V), Rosemary Ewing; Dir. Curatorial Svcs., Dawn Hammatt; Dir. Interpretive Svcs., Whitney Babineaux; Dir. Collections & Museum Div., Greg Lambousey; Dir. Mktg. & Public Rels., Arthur Smith; Museum Historian, Dr. Karen Leathem; Cur. Decorative Arts, Katie Hall Burlison; Cur. Louisiana Historical Center, Sarah-Elizabeth Gundlach; Cur. Costumes & Textiles, Wayne Phillips; Cur. Visual Arts, Tony Lewis; Cur. Science & Technology, Polly Rolman; Registrar, Jennae Biddiscombe; Asst. Registrar, Beth Sherwood; Dir. Museum Branch, Gloria La Coste; Dir. Museum Branch, William Stark.

Personnel Profile: Full-Time Paid 72; Part-Time Volunteers 140; Interns 8.

Governing Authority: state. Parent Institution: Dept. of Culture, Recreation & Tourism, State of Louisiana. Tax-exempt: 170(b)(1)(A).

Institution Type/Description: Historical Museum Complex: six National Historic Landmark buildings located in the New Orleans French Quarter, Baton Rouge, Patterson, Thibodaux & Natchitoches.

Collections: fine, decorative & folk art; costumes, textiles, jazz music; photographic & inventive arts of Louisiana; science & technology; military history; paintings & portraiture; 1704-1803 Louisiana Aviation Colonial Archives. Historic Houses: 1795 The Cabildo; 1791 The Presbytere; 1850 The Lower Pontalba Building; 1788 Madame John's Legacy; 1835 The New Orleans Branch of the U.S. Mint; 1842 Creole House; 1842 Jackson House; 1839 Old Arsenal; Louisiana Sports Hall of Fame & Sports History Museum; The Louisiana State Museum, Patterson; E.D. White Historic Site.

Research Fields: Louisiana history: social, cultural, science, economic, technology, religion, folklore, music, politics, ethnic, racial.

Facilities: 40,000-vol. historical research library & archives; three auditoriums; museum learning center; rental space. Museum-related items for sale.

Activities: tours; lectures; educational programs; consultation to Louisiana historical agencies; special & traveling exhibitions; conservation; restoration; research.

Publications: special exhibition & collection catalogs; section in monthly magazine, Cultural Vistas; Louisiana Life.

Hours & Admission Prices: The Louisiana State Museum-Baton Rouge: Tues.-Sat. 9-5, Sun. 12-5. The Cabildo: Tues.-Sun. 10-4. Presbytere: Fri.-Sun. 10-4. Old U.S. Mint, 1850 House: call for hours. The Cabildo & Presbytere: adults $6, students, seniors & active military $5; discounts to Louisiana Assoc. of Museums, Louisiana Museum Foundation, Friends of the Cabildo and AAM & ICOM members; children 12 & under, school groups & members no charge. Madame John's Legacy, Louisiana State Museum-Baton Rouge and Louisiana State Museum-Patterson: no charge. Combination tickets to all LSM properties available. Closed legal & state holidays. ₠

Attendance: 308,616 (accurate)

Membership: Friends of the Cabildo: Statewide $15; Individual $25; Family $35; Friend $50; Sustaining $75; Contributing $100; Louisiana Museum Foundation $2,000.

MUSEE CONTI WAX MUSEUM, 917 Rue Conti, New Orleans, LA 70112-3409. Tel.: 504-581-1993; 800-233-5405. Fax: 504-566-7636.

Web Site: www.get-waxed.com

Founded: 1963.

Institution Type/Description: Wax Museum.

Collections: 154 life-size figures depicting New Orleans' history, legend & scandal.

Facilities: banquet facilities.

Activities: themed parties; receptions; dinners; tours; special events.

Hours & Admission Prices: Mon. & Fri.-Sat. 10-4. Adults $6.75, senior citizens $6.25, children 4-17 $5.75; discounts to groups & AAA members. Closed Mardi Gras; Thanksgiving; Christmas Eve, Day & week. ₠

MUSEUM OF THE AMERICAN COCKTAIL, 1504 Oretha Castle Haley Blvd., New Orleans, LA 70113. Tel.: 504-569-0405.
Governing Authority: nonprofit organization.
Institution Type/Description: History Museum: housed in an 1823 French Quarter town house.
Collections: liquor history; period liquor bottles; cocktail shakers & memorabilia; early swizzle sticks & Tiki cups; bartending; drink recipes; books; prohibition-era literature; glassware.
Activities: seminars.
Hours & Admission Prices: Mon.-Sat. 10-7, Sun. 12-6. Adults $10.

NATIONAL SHRINE OF BLESSED FRANCIS XAVIER SEE-LOS, 919 Josephine St., New Orleans, LA 70130. Tel.: 504-525-2495. Fax: 504-581-9181.
E-mail: bmiller@seelos.org
Web Site: www.seelos.org
Founded: 1959.
Key Personnel: Dir., Rev. Byron Miller, C.SS.R.; Museum Shop Mgr. & Admin., Joyce Bourgeois.
Personnel Profile: Full-Time Paid 1; Part-Time Paid 4; Part-Time Volunteers 30.
Governing Authority: Parent Institution: Redemptorists - Denver Province. Tax-exempt.
Institution Type/Description: Religious Museum.
Collections: portrait of Father Seelos used in Rome for his beatification; religious paintings; tapestries; photographs; religious artifacts; Father Seelos' personal artifacts; life-sized bronze statue of Father Seelos; hand-carved wooden statues of saints from Germany; 1891 German organ.
Publications: Seelos Center News.
Hours & Admission Prices: Shrine: Mon.-Fri. 9-3, Sat. 10:30-3:30, Sun. between masses. Shop: Mon.-Fri. 9-3, Sat. 10:30-3:30. Shrine: no charge; donations accepted. &
Attendance: 36,000 (estimated)

THE NATIONAL WORLD WAR II MUSEUM, (M), 945 Magazine St., New Orleans, LA 70130-3813. Tel.: 504-528-1944. Fax: 504-527-6088.
E-mail: info@nationalww2museum.org
Web Site: www.nationalww2museum.org
Formerly: The National D-Day Museum
Founded: 1991.
Congressional District: 2
Key Personnel: C.E.O. & Pres., Dr. Gordon H. Mueller; Chm. (V), Richard Adkerson; Sr. Vice Pres. Institutional Advancement, Michael Carrol; Vice Pres. & C.F.O., Rebecca Mackie; Vice Pres. & C.O.O., Stephen Watson; Dir. Education, Kenneth Hoffman; Sr. Vice Pres. Capital Projects, Bob Farnsworth; Dir. Security, Dave Heidenthal; Sr. Dir. History & Research, Keith Huxen; Dir. Facilities, Matt Gardner; Visitor Svcs. Mgr., Kelly Bules; Dir. Collections & Exhibitions, Thomas Czekanski; Assoc. Vice Pres. Operations, Paul Parrie; Assoc. Vice Pres. Mktg., Jonah Langenbeck; Dir. Sales, Ruth Katz; Dir. Membership & Annual Fund, Terri Burton; Dir. Retail Svcs., Chris Michel.
Personnel Profile: Full-Time Paid 210; Part-Time Paid 102; Part-Time Volunteers 350; Interns 6.
Governing Authority: private; nonprofit organization. Tax-exempt: 501(c)(3).
Institution Type/Description: History Museum.
Collections: artifacts, archival and audio/visual materials relating to the American experience in WWII.
Major Exhibits: Propaganda Posters of World War II, 11/13-2/14; The Japanese American Experience in World War II, 3/14-10/14; The Arsenal of Democracy (T), 11/14-5/15.
Research Fields: World War II history.
Facilities: educational facilities; 40,000 sq. ft. exhibit space; 240-seat theater; cafe. Museum-related items for sale.
Activities: docent program; films; formal education programs for adults, college students & children; guided tours; lectures; loan, traveling & temporary exhibitions; rental gallery; theater; training programs for professional museum workers.
Publications: quarterly newsletter, V-mail.
Hours & Admission Prices: Daily 9-5. Adults $22, seniors 65 & over $19, children, college students & US military with ID $13, members no charge. Combination tickets available. Closed New Year's Day; Mardi Gras; Thanksgiving; Christmas. &
Attendance: 375,000 (accurate)
Membership: Friend $50; Friend Plus One $90; Family $160; Advocate $250.

NEW ORLEANS BOTANICAL GARDEN, 1 Palm Dr., New Orleans, LA 70124. Tel.: 504-483-9488. Fax: 504-483-9485.
E-mail: nobgmail@yahoo.com
Web Site: garden.neworleanscitypark.com
Key Personnel: Dir., Paul M. Soniat; Museum Shop Mgr., Jessica Mathews.
Personnel Profile: Full-Time Paid 11; Part-Time Paid 6; Part-Time Volunteers 60.
Governing Authority: Parent institution: New Orleans City Park
Institution Type/Description: Botanical Garden.
Collections: plants; trees; shrubs; perennials; annuals; herbs.
Activities: special events.
Hours & Admission Prices: Tues.-Sun. 10-4:30. Adults 12 & over $6, children 5-12 $3; children under 5 and members no charge. &

NEW ORLEANS FIRE DEPT. MUSEUM & EDUCATIONAL CENTER, 1135 Washington Ave., New Orleans, LA 70130-5632. Tel.: 504-658-4713. Fax: 504-896-4756.
Founded: 1992.
Key Personnel: C.E.O., Chief Warren E. McDaniels; Cur., Archivist & Museum Shop Mgr., Michael Williams; Public Rels., Capt. Norman Woodridge.
Personnel Profile: Full-Time Paid 1; Full-Time Volunteers 1.
Governing Authority: municipal. Tax-exempt: 501(c)(3).
Institution Type/Description: Fire-Fighting Museum: c.1852 firehouse.
Collections: history of fire fighting in New Orleans.
Facilities: 240-vol. library; 2,300 sq. ft. exhibit space; 30-seat theater; 30-seat auditorium. Museum-related items for sale.
Activities: guided tours; hobby workshops; loan & temporary exhibitions; safety programs; interactive exhibits.
Hours & Admission Prices: Mon.-Fri. 9-2 by appointment. No charge; donations accepted. Closed New Year's Eve & Day; Martin Luther King Jr. Day; Mardi Gras; Good Friday; Memorial Day; Independence Day; Thanksgiving; Christmas Eve & Day.
Attendance: 5,200 (accurate)
Membership: Institution $20.

✳   NEW ORLEANS MUSEUM OF ART, (M), One Collins Diboll Circle, New Orleans, LA 70124-4605. Mailing Address: P.O. Box 19123, New Orleans, LA 70179-0123. Tel.: 504-658-4100. Fax: 504-658-4199.
E-mail: gasprodites@noma.org
Web Site: www.noma.org
Founded: 1910.
Congressional District: 1
Key Personnel: C.E.O. & Dir., Susan M. Taylor; Chm., David Edwards; Chm. (V), Julie George; Asst. Dir. Administration, Gail Asprodites; Dir. Devel., Brooke Minto; Facilities Mgr., Karl Oelkers; Mgr. Sculpture Garden, Pamela Buckman; Asst. Dir. Art, Lisa R. McCord; Asst. Dir. Education, Allison Reid; Cur. African Art, William Fagaly; Controller, Leo Sayer; Cur. Photography, Russell Lord; Cur. Native American & Pre-Columbian Art & Registrar, Paul Tarver; Cur. Modern & Contemporary Art, Miranda Lash; Cur. Education, Tracy Kennan; Graphics Coord. & Web Master, Aisha Champagne; Dir. Communications & Mktg., Allison Govaux; Arts Quarterly Editor, Taylor Murrow; Museum Shop Mgr., Helen Redmann; Librarian, Shelia Cork; Registrar, Jennifer Ickes; Grants Officer, Christina Carr.
Personnel Profile: Full-Time Paid 53; Full-Time Volunteers 2; Part-Time Paid 8; Part-Time Volunteers 150; Interns 4.
Governing Authority: municipal. Tax-exempt: 501(c)(3).
Institution Type/Description: Art Museum.
Collections: Old Master paintings of various schools; Kress collection of Italian Renaissance & Baroque paintings; Chapman H. Hyams collection of Barbizon & Salon paintings; pre-Columbian masterworks from Mexico, Central & South America; Latin American & Spanish colonial paintings & sculptures; 20th-century English & European art, including Surrealism & School of Paris; Japanese Edo period paintings; African art; photography; graphics; Glass Collection; 19th- & 20th-century U.S. & Louisiana paintings & sculptures; Latter-Schlesinger collection of English & continental portrait miniatures; collections of African, oceanic northwest coast American Indian & 20th-century European & American paintings & sculptures; The Matilda Geddings Gray Foundation collection of works by Peter Carl Faberge; Rosemunde E. & Emile Kuntz Federal & Louisiana period rooms; 16th to 20th-century French art; Sydney and Walda Besthoff Sculpture Garden, contemporary sculpture.
Research Fields: pre-Columbian; Ancient, European & American glass; 19th-century Louisiana painting; Latin American Spanish Colonial art; European art; African art; portrait miniatures; Japanese art; photography; contemporary art.
Facilities: 7,500-vol. library of general art, slide library available by appointment; 171,500 sq. ft. exhibition space; 5 acre sculpture garden; 220-seat auditorium; studio classrooms. Museum-related items for sale.
Activities: guided tours; lectures; films; gallery talks; adult art classes; arts festivals; formally organized education programs for children; concerts;

docent program; inter-museum loan, permanent & temporary exhibitions; art therapy program in public schools.
Publications: Arts Quarterly; handbook of the permanent collection & special exhibition catalogs.
Hours & Admission Prices: Tues.-Sun. 10-5, Fri. 10-9. Adults $10, senior citizens 65 & over & students $8, children 7-17 $6, children under 6 no charge. &
Attendance: 162,291 (accurate)
Membership: Individual $60; Dual/Family $75; Sustaining $125; Benefactor & Young Fellows (single) $250; Young Fellows (couple) $400; Advocates $500; Fellows $1,500; Family Circle $2,500.

NEW ORLEANS PHARMACY MUSEUM, 514 Chartres St., New Orleans, LA 70130-2110. Tel.: 504-565-8027. Fax: 504-565-8028. Facebook: New Orleans Pharmacy Museum.
E-mail: nopharmsm@aol.com
Web Site: www.pharmacymuseum.org
Founded: 1950.
Congressional District: 2
Key Personnel: Pres. (V), Anthony D'Angelo; Dir., Liz Sherman; Asst. Cur., Eboni Evans; Public Rels., Keith Ribbeck.
Personnel Profile: Full-Time Paid 1; Part-Time Paid 1; Part-Time Volunteers 3.
Governing Authority: municipal government; nonprofit. Friends of Historical Pharmacy. Tax-exempt: 501(c)(3).
Institution Type/Description: Pharmacy & Medicine Museum: housed in 1823 building constructed for Louis Joseph Dufilho, Jr., first licensed pharmacist in the U.S.
Collections: history of pharmacy, medicine, & health care during the 19th century; apothecary bottles & jars; drugs & herbs; medical devices; trade journals; pharmacopoeias; Civil War surgical instruments; cosmetics; c.1855 soda fountain; archives; gris-a-gris potions used by voodoo practitioners.
Research Fields: pharmacology; history of voodoo, gris-gris & pharmacy; patent medicines; European, American, African & Asian approaches & contributions to the discipline of botanical medicine especially in the profession of pharmacy & medicine.
Facilities: garden with medicinal herbs; function areas for rental. Museum-related items for sale.
Activities: docent program; formal education programs for children, adults & undergraduate & graduate students; guided tours; lectures; loan exhibitions; rental gallery.
Publications: quarterly newsletter, RX News.
Hours & Admission Prices: Call for hours. Adults $5, senior citizens & students $4; discounts to AAM & AAA members; children under 6 & members no charge. Closed New Year's Day; Mardi Gras; Easter; Independence Day; Labor Day; Thanksgiving; Christmas.
Attendance: 20,000 (estimated)
Membership: Individual $50; Family $75; Lifetime & Corporate $1,000.

NEWCOMB ART GALLERY, (M), Woldenberg Art Center, Tulane University, Bldg. 8, Corner of Newcomb Pl. & Drill Rd., New Orleans, LA 70118-5698. Tel.: 504-865-5328. Fax: 504-865-5329.
E-mail: gallery@tulane.edu
Web Site: www.newcombartgallery.tulane.edu
Founded: 1996.
Congressional District: 1
Key Personnel: Interim Dir., Jeremy Jernigan; Accounting I, Melissa Russell; Sr. Cur., Sally Main; Coord. Mktg. & Membership, Teresa Parker Farris; Registrar, Thomas Strider; Visitor Svcs., Beau Box; Coord. Education Programs, Laura Ledet.
Personnel Profile: Full-Time Paid 5; Part-Time Paid 2; Part-Time Volunteers 5; Interns 3.
Governing Authority: college. Parent Institution: Tulane University. Subsidiary Institution: Administrators of the Tulane University Education Fund. Tax-exempt.
Institution Type/Description: Art Museum.
Collections: Newcomb pottery & related crafts; paintings; sculpture; works by Newcomb faculty & alumnae; Tiffany windows. Tulane University Art Collection: 19th-20th century European & American paintings; sculpture; works on paper with emphasis on Louisiana artists.
Major Exhibits: Women, Art, and Social Change: The Newcomb Pottery Enterprise (T), 10/13-3/14; The Newcomb Pottery Enterprise (T), 10/21/13-3/6/14.
Facilities: library of fine arts available to public; slide library open to scholars & researchers in the fine arts field; art gallery.
Activities: gallery talks; symposia; workshops.
Publications: Newcomb Pottery and Crafts; Ida Kohlmeyer: Systems of Color; Carrie Mae Weems: The Louisiana Project; From Society to Socialism: The

Art of Caroline Durieux; In Company with Angels: Seven Rediscovered Tiffany Windows; Patricia Cronin: All Is Not Lost.
Hours & Admission Prices: Tues.-Fri. 10-5, Sat.-Sun. 11-4. No charge. Closed New Year's Eve & Day; Mardi Gras; Thanksgiving; Christmas Eve & Day. &
Attendance: 10,000 (estimated)
Membership: General $50; Sustaining $100; Advocate $250; Patron $500; Sadie Irvine Circle $1,000; Gertrude Roberts Smith Circle $2,500; Joseph Meyer Circle $5,000; Mary Given Sheerer Circle $10,000.

THE OGDEN MUSEUM OF SOUTHERN ART, 925 Camp St., New Orleans, LA 70130-3907. Tel.: 504-539-9600. Fax: 504-539-9602. Facebook: Ogden Museum.
E-mail: info@ogdenmuseum.org
Web Site: www.ogdenmuseum.org
Founded: 1994.
Key Personnel: Dir., William Andrews; Chm. (V), Henry Shane; Treas., Lloyd "Sonny" Shields; Education, Ellen Balkin; Dir. Public Rels., Sue Strachan; Registrar, Archivist & Cur., Bradley Sumrall; Deputy Dir. & Cur. Music, Libra LaGrone; Museum Shop Mgr., Rachel Ford.
Personnel Profile: Full-Time Paid 14; Part-Time Paid 9; Part-Time Volunteers 1; Interns 3.
Governing Authority: public university. Parent Institution: University of New Orleans. Tax-exempt: 501(c)(3).
Institution Type/Description: Art Museum.
Collections: visual art of 15 southern states & District of Columbia from 1733 to present.
Major Exhibits: Lee Diegard: Trespass, 1/16/14-4/6/14; Steffen Thomas Rediscovered (T), 1/16/14-4/6/14; Juan Logan: Leisure Spaces, 2/13/14-7/20/14; Contemporary Photographs and 19th Century Processes, 4/17/14-7/13/14; Shadows of History: Photographs of the Civil War from the Collection of Julia Norrell (T), 4/17/14-7/13/14; Paul Kmlecki: One Place (T), 7/24/14-9/21/14; Rolland Golden: An Alternate Vision (T), 8/2/14-9/21/14; Louisiana Contemporary presented by Regions Bank, 8/2/14-10/5/14; Self-Processing: Shelby Lee Adams, Linda Burgess, Michael McGraw, Michael Meads, Blake Boyds, Pinky Bass, 10/4/14-1/4/15; Self-Taught, Outsider and Visionary Art from the Richard Gaspen Collection, 10/4/14-1/4/15; Prospect 3 Basquait and the Bayou, 10/24/14-1/15.
Facilities: library; theater. Museum-related items for sale.
Activities: docent program; guided tours; children & adult workshops; family programs; Ogden After Hours Music Series; panel discussions; Sippin in Seersucker; O What A Night Gala; permanent, traveling & juried exhibitions; Art of the Cup; The Art of Giving; Artist Spotlight Series; educational outreach programs.
Hours & Admission Prices: Wed. & Fri.-Mon. 10-5, Thurs. 10-5 & 6-8. Adults $10, senior citizens 65 & over and college students $8, children 17 & under $5; discounts to groups; children under 5, members & Louisiana residents Thurs. 10-5 no charge. Closed New Year's Day; Mardi Gras; Memorial Day; Independence Day; Thanksgiving; Christmas. &
Attendance: 75,000 (estimated)
Membership: University of New Orleans Students $15; Individual $50; Family & Dual $75; Supporting $125-$249; Partners $250-$499; Curator's Circle $500-$999.

PITOT HOUSE MUSEUM, 1440 Moss St., New Orleans, LA 70119-2904. Tel.: 504-482-0312.
E-mail: info@louisianalandmarks.org
Web Site: www.louisianalandmarks.org
Founded: 1964.
Key Personnel: Pres., Thomas M. Ryan; Dir., Elizabeth Burger.
Personnel Profile: Full-Time Paid 1; Part-Time Paid 1; Interns 7.
Governing Authority: municipal; nonprofit. Parent Institution: Louisiana Landmarks Society. Tax-exempt.
Institution Type/Description: Historic House: 1799 French West Indies Country house.
Collections: 1790-1840, Louisiana cultural history items.
Facilities: botanical garden.
Activities: guided tours; heritage education programs; service learning partnership with Tulane Univ. Preservation students; preservation technology workshops. Annual Events: Life on the Bayou; Vino on the Bayou with Cork & Bottle Wine.
Publications: quarterly newsletter, Preservation.
Hours & Admission Prices: Wed.-Sat. 10-3. Adults $7, senior citizens & children $5; discounts to groups & National Trust members; members no charge. Closed major holidays.
Attendance: 1,000 (estimated)
Membership: Loyalist $25; Advocate $40; Guardian $250-$499; Protector $500-$999; Sustainer $1,000-$4,999; Preserver $5,000-$9,999; Champion $10,000 & up.

PRESERVATION RESOURCE CENTER OF NEW ORLEANS, 923 Tchoupitoulas St., New Orleans, LA 70130-3819. Tel.: 504-581-7032. Fax: 504-636-3073.
E-mail: prc@prcno.org
Web Site: www.prcno.org
Founded: 1974.
Congressional District: 2
Key Personnel: Exec. Dir., Patricia H. Gay.
Personnel Profile: Full-Time Paid 30; Part-Time Paid 4; Part-Time Volunteers 250; Interns 10.
Governing Authority: Tax-exempt: 501(c)(3).
Institution Type/Description: Historical & Preservation Society: housed in 1853 Gothic Revival style building designed by James H. Dakin.
Collections: archive, photograph & slide collection on architecture, historic preservation in historic districts in New Orleans.
Facilities: 44,000-entry cataloguing articles on landscape architecture & renovation available to members; meeting room.
Activities: guided tours; lectures; historic home tours; architecture symposium; home ownership & renovation seminars.
Publications: newspaper magazine 9 times annually, Preservation in Print; books, New Orleans: Life in an Epic City; New Orleans's Favorite Shotguns.
Hours & Admission Prices: Mon.-Fri. 9-5. No charge; donations accepted. &
Attendance: 2,500 (estimated)
Membership: Friend $35; Individual $40; Dual $60; Household $100; Preserver $300; Conservator $600; President's Circle $1,200.

SAINTS HALL OF FAME MUSEUM, The Superdome, 1500 Poydras St., New Orleans, LA 70112-1216. Mailing Address: 415 Williams Blvd., Kenner, LA 70062. Tel.: 504-450-9893.
Key Personnel: Dir., Ken Trahan
Institution Type/Description: Sports Museum.
Collections: New Orleans team memorabilia, photographs & videos.
Hours & Admission Prices: NFL Game Days: 3 hours before the game & 45 minutes after the game; other times by appointment. Adult $7; discounts to groups of 20 or more; season ticket holders no charge. &

SOUTHERN FOOD AND BEVERAGE MUSEUM, 1504 Oretha C. Haley Blvd., New Orleans, LA 70113. Tel.: 504-569-0405. Fax: 504-587-7944.
E-mail: info@southernfood.org
Web Site: www.southernfood.org
Founded: 2008.
Congressional District: 2
Key Personnel: Dir., Liz Williams; Chm. (V), James Carter; Museum Shop Mgr., Joseph Sunseri.
Personnel Profile: Full-Time Paid 3; Part-Time Paid 2; Interns 10.
Governing Authority: nonprofit organization. Tax-exempt.
Institution Type/Description: Cultural History Museum.
Collections: southern culture & history of food & drink; cultural heritage; oral histories; videos; paintings; photographs.
Facilities: library.
Activities: special events; classes; tastings; videos; demonstrations; lectures; research.
Publications: weekly newsletter; online magazine, Okra.
Hours & Admission Prices: Mon.-Sat. 10-7, Sun. 12-6. Adults $10, students, seniors & military $5; discounts to AAA members; members no charge. &
Attendance: 35,000 (estimated)
Membership: Out of Town $45; Individual $50; Friends & Family $75; Kitchen Cabinet $125.

New Roads

THE JULIEN POYDRAS MUSEUM & ARTS CENTER, 500 W. Main St., New Roads, LA 70760. Mailing Address: P.O. Box 462, New Roads, LA 70760. Tel.: 225-638-6575. Fax: 225-638-6578. Facebook: Julien Poydras Center.
E-mail: e.angelique.bergeron@gmail.com
Web Site: pointecoupeehistoricalsociety.org
Key Personnel: Exec. Dir., Angelique Bergeron.
Governing Authority: Parent Institution: Pointe Coupee Historical Society.
Institution Type/Description: Art Museum: housed in the former Poydras High School building; built in 1924.
Collections: art & history exhibitions.
Facilities: rental facilities; 250-seat auditorium.
Activities: rental facilities; classes; temporary exhibitions.
Hours & Admission Prices: Wed.-Fri. 1-5, Sat. 10-3.

POINTE COUPEE MUSEUM, 8348 False River Rd., New Roads, LA 70760. Mailing Address: P.O. Box 462, New Roads, LA 70760. Tel.: 225-638-7788.
E-mail: e.angelique.bergeron@gmail.com
Founded: 1979.
Key Personnel: Pres. (V), Angelique Bergeron, Ph.D.; Museum Shop Mgr., Winona Sicard.
Personnel Profile: Part-Time Paid 2.
Governing Authority: Parent Institution: Pointe Coupee Parish Police Jur.
Institution Type/Description: Historic Building.
Collections: local history & culture; period furnishings; personal artifacts.
Activities: Museum Sponsors: Open House in April & October.
Hours & Admission Prices: Daily 10-3; other times by appointment. No charge; donations accepted.
Attendance: 1,000 (estimated)

Newellton

WINTER QUARTERS STATE HISTORIC SITE, 4929 Hwy. 608, Newellton, LA 71357-6314. Tel.: 888-677-2784; 318-766-3530.
Institution Type/Description: Historic Site: built in 1805.
Collections: local history & culture; period furnishings; photographs; personal artifacts.
Hours & Admission Prices: By appointment. Adults $2; seniors 62 & over and children 12 & under no charge. Closed New Year's Day; Thanksgiving; Christmas.

Oil City

LOUISIANA STATE OIL & GAS MUSEUM, 200 S. Land Ave., Oil City, LA 71061. Mailing Address: State of Louisiana, Secretary of State, P.O. Box 94125, Baton Rouge, LA 70804-9125. Tel.: 318-995-6845. Fax: 318-995-6848.
E-mail: oil@sos.louisiana.gov
Web Site: www.sos.louisiana.gov/lsoagm
Formerly: Caddo-Pine Island Oil & Historical Society Museum
Founded: 1965.
Congressional District: 4
Key Personnel: Dir., Coe McKenzie.
Personnel Profile: Full-Time Paid 2; Part-Time Paid 1.
Governing Authority: nonprofit organization. Parent Institution: Louisiana Secretary of State Office. Tax-exempt: 501(c)(3) & 170(b)(1)(A).
Institution Type/Description: Oil and Gas Museum.
Collections: local history & culture; Caddo Indians; oil field equipment.
Facilities: library. Museum-related items for sale.
Activities: guided tours; organized education programs for children.
Hours & Admission Prices: Wed.-Fri. 10-4. No charge. Closed major federal & state holidays.
Attendance: 1,415 (accurate)
Membership: Call for information.

Olla

CENTENNIAL CULTURAL CENTER, 2962 Front St., Olla, LA 71465. Mailing Address: P.O. Box 896, Olla, LA 71465-0896. Tel.: 318-495-7988. Fax: 318-495-7988.
E-mail: cultural@centurytel.net
Web Site: www.culturalcenter.us
Key Personnel: Dir., Shawna Cockerham.
Governing Authority: nonprofit organization. Tax-exempt: 501(c)(3).
Institution Type/Description: History Museum.
Collections: Olla, Tullos, & Urania history; photographs; period artifacts; oral histories.
Hours & Admission Prices: Mon.-Fri. 9-1.

Opelousas

LOUISIANA ORPHAN TRAIN SOCIETY, INC., 610 Garland Ave., Opelousas, LA 70570. Tel.: 337-945-4691.
E-mail: hdupre2433@aol.com
Web Site: laorphantrain.com
Founded: 2009.
Key Personnel: Dir., Pres. (V) & Devel., Harold Dupre; Treas. & Archivist, Florella Inhern.
Personnel Profile: Part-Time Volunteers 16.
Governing Authority: private; nonprofit organization. Tax-exempt: 501(c)(3).
Institution Type/Description: History Museum.
Collections: Orphan Train history; personal artifacts; photographs; clothing & shoes.

Facilities: 15-vol. library of books; 4,000 sq. ft. exhibit space.
Publications: quarterly newsletter.
Hours & Admission Prices: Tues.-Fri. 10-3, Sat. 10-2. Adults $5, senior citizens, students & children $3. Closed New Year's Eve, Day & week; Christmas Day & week. &
Attendance: 500
Membership: Individual $10; Husband & Wife $15; Lifetime $100.

OPELOUSAS FIRE MUSEUM, 109 N. Union St., Opelousas, LA 70570. Tel.: 337-948-2543.
Institution Type/Description: Firefighting History Museum: housed in the former headquarters of the Hope Hook & Ladder Company.
Collections: local firefighting history & equipment; early fire truck & hose cart; photographs; personal artifacts.
Hours & Admission Prices: Call for hours.

OPELOUSAS MUSEUM AND INTERPRETIVE CENTER, (M), 315 N. Main St., Opelousas, LA 70570-6201. Tel.: 337-948-2589. Fax: 337-948-2592.
E-mail: museum@cityofopelousas.com
Web Site: www.cityofopelousas.com
Founded: 1992.
Congressional District: 7
Personnel Profile: Full-Time Paid 2; Part-Time Paid 1; Part-Time Volunteers 4.
Governing Authority: municipal government; nonprofit. Parent Institution: City of Opelousas. Tax-exempt: 501(c)(3).
Institution Type/Description: History Museum.
Collections: history & culture of the Opelousas area from prehistoric times to present.
Research Fields: zydeco music; medical & local history.
Facilities: 3,500 sq. ft. exhibit space. Books on the Opelousas area & musical & Cajun humor tapes for sale.
Activities: films; formal education programs for adults; guided tours; lectures; temporary exhibitions; broadcast programs. Annual Events: Birthday; Cultural A-Fair; Christmas Flower Show.
Publications: bimonthly, Dust & Cobwebs.
Hours & Admission Prices: Mon.-Fri. 8-4:30, Sat. 10-3. Tour Buses: $3 per person; discounts to AAM & ICOM members. Closed Easter; Thanksgiving; Christmas. &
Attendance: 6,000 (estimated)
Membership: Friends of the Museum: Student 5 volunteer hours; Individual $10-$50; Business $75-$150; Corporate $200-$500.

Patterson

LOUISIANA STATE MUSEUM-PATTERSON, 118 Cotten Rd., Patterson, LA 70392. Mailing Address: P.O. Box 38, Patterson, LA 70392-0038. Tel.: 985-399-1268. Fax: 985-399-9910.
E-mail: glacoste@crt.state.la.us
Web Site: lsm.crt.state.la.us/
Formerly: Wedell Williams Memorial Aviation Museum
Founded: 1975.
Congressional District: 50
Key Personnel: Museum Div. Dir., William Stark; Administrative Program Specialist, Gloria LaCoste.
Personnel Profile: Full-Time Paid 3.
Governing Authority: state; nonprofit. Subsidiary Institution: Wedell-Williams & Cypress Sawmill Foundation. Tax-exempt: 501(c)(3).
Institution Type/Description: Aviation Museum.
Collections: 067 aircraft & aviation memorabilia, emphasizing Louisiana natives & industry. Cypress Sawmill exhibit includes a variety of artifacts & a film of this early regional industry (1872-1930).
Major Exhibits: Cleo Scott: The Wildlife Carvings of a Louisiana Artist, 10/13-8/14.
Research Fields: aviation; cypress lumber industry.
Facilities: 24-seat theater.
Activities: guided tours; films; temporary & permanent exhibitions; adult & family programs.
Publications: quarterly newsletter.
Hours & Admission Prices: Tues.-Sat. 9:30-4. No charge; donations accepted. Closed state holidays. &
Attendance: 43,000 (accurate)
Membership: Individual $25; Family $35; Friend $100; Patron $300; Benefactor $500; Gold Benefactor $1,000.

Pioneer

POVERTY POINT STATE HISTORIC SITE, 6859 Hwy. 577, Pioneer, LA 71266. Tel.: 318-926-5492; 888-926-5492.
E-mail: povertypoint@crt.la.gov
Institution Type/Description: Historic Site: archaeological site dated from 1700 & 1100 B.C. A National Historic Landmark.
Collections: local history; archaeology; photographs.
Facilities: 400 acres.
Activities: special events; educational programs; guided tours.
Hours & Admission Prices: Daily 9-5; groups by appointment. Adults $4; seniors 62 & over and children 12 & over no charge. Closed New Year's Day; Thanksgiving; Christmas.

Plaquemine

IBERVILLE MUSEUM, (M), 57735 Main St., Plaquemine, LA 70764-2564. Mailing Address: P.O. Box 701, Plaquemine, LA 70765-0701. Tel.: 225-687-7197. Fax: 225-687-3060.
E-mail: ibervillemuseum@yahoo.com
Key Personnel: Dir., Bethany Cardinal
Institution Type/Description: History Museum: built in 1848. Listed in the National Register of Historic Places.
Collections: Iberville history & culture; photographs; America's wars; personal artifacts.
Hours & Admission Prices: Tues.-Sat. 10-4, Sun. by appointment. Adults 13 & over $2, children 6-12 $1; discounts to groups; teachers no charge.

PLAQUEMINE LOCK STATE HISTORIC SITE, 57730 Main St., Plaquemine, LA 70764-2530. Tel.: 225-687-7158; 877-987-7158.
E-mail: plaqlock@crt.state.la.gov
Web Site: www.lastateparks.com
Founded: 1982.
Governing Authority: Parent Institution: State of Louisiana, Dept. of Culture, Recreation & Tourism, Office of State Park.
Institution Type/Description: Historic Site. Listed on the National Register of Historic Places.
Collections: local history & culture; photographs.
Hours & Admission Prices: Tues.-Sat. 9-5. Adults $4; seniors 62 & over and children 12 & under no charge. Closed New Year's Day; Thanksgiving; Christmas.
Attendance: 5,000 (estimated)

Port Allen

* **WEST BATON ROUGE MUSEUM, (M),** 845 N. Jefferson Ave., Port Allen, LA 70767-2417. Tel.: 225-336-2422. Fax: 225-336-2448. Facebook: West Baton Rouge Museum.
E-mail: contact_us@wbrmuseum.org
Web Site: westbatonrougemuseum.com
Founded: 1968.
Congressional District: 6
Key Personnel: Dir., Julia Rose; Chm. (V), Ellis Gauthier; Vice Chm., Sue Blanchard; Cur., Lauren Davis; Education Cur., Jeannie Luckett; Museum Shop Mgr., Tommy McMorris.
Personnel Profile: Full-Time Paid 6; Part-Time Paid 9; Part-Time Volunteers 146; Interns 1.
Governing Authority: Parent Institution: West Baton Rouge Museum Board. Tax-exempt: 501(c)(1). Subsidiary Institution: West Baton Rouge Historical Association. Tax-exempt: 501(c)(3).
Institution Type/Description: Regional History Museum.
Collections: 19th- & 20th-century lifestyles in a Louisiana sugar parish; West Baton Rouge artifacts and memorabilia; raw sugar production. Historic Buildings: c.1904 model sugar mill; c.1850 slave cabin; c.1830 French-Creole cottage; 1880 share cropper cabin; Civil Rights era field workers' dwelling, c.1950-60; 20th century shotgun house; 20th century plantation store.
Research Fields: sugar industry; south Louisiana history.
Facilities: 6 acre grounds. Museum-related items for sale.
Activities: guided tours; lectures; education programs for children & adults; folklife and Louisiana artists exhibitions; festivals.
Publications: newsletter, Ecoutez.
Hours & Admission Prices: Tues.-Sat. 10-4:30, Sun. 2-5. Adults $4, students, military & seniors $2; discounts offered to AAM, AAA, LAM members & other tourism coupon holders; WBR Historical Assoc. members, citizens of WBR parish & members no charge. Closed major holidays. Business Office: Mon.-Fri. 9-4:30. &
Attendance: 20,000 (accurate)
Membership: Student $5; Member $10; Patron $35; Friends $100.

Rivertown

RIVERTOWN MUSEUMS, 2020 Fourth St., Rivertown, LA 70062. Tel.: 504-468-7231.
Institution Type/Description: Science Museum & Planetarium.
Collections: space science; hands-on exhibits.
Hours & Admission Prices: Groups: Tues.-Fri. by appointment. General Public: Sat. 11-4. Space Station & Science Complex: adults $5, seniors $4, children $3. Planetarium & Megadome: adults $6, seniors & children $5. Combination tickets available.

Robeline

ADAI INDIAN NATION CULTURAL CENTER, 4460 Hwy. 485, Robeline, LA 71469-4946. Tel.: 877-472-1007; 318-472-1007.
Institution Type/Description: Historic Site.
Collections: Native American history, culture, & personal artifacts; early dwellings.
Facilities: theater. Museum-related items for sale.
Hours & Admission Prices: Daily 9-5. Adults $6.50, children 3-12 $4.75; children under 3 no charge. Closed New Year's Day; Easter; Thanksgiving; Christmas.

LOS ADAES STATE HISTORIC SITE, 6354 Hwy. 485, Robeline, LA 71469. Tel.: 318-472-9449; 888-677-5378.
E-mail: fortjesup@crt.la.gov
Institution Type/Description: Historic Site: a National Historic Landmark.
Collections: local history; period artifacts; photographs; Native American artifacts; Spanish & French artifacts.
Activities: children's programs; wildlife viewing.
Hours & Admission Prices: Temporarily closed.

Ruston

LINCOLN PARISH MUSEUM & HISTORICAL SOCIETY, (M), 609 N. Vienna St., Ruston, LA 71270-3842. Tel.: 318-251-0018. Fax: 318-251-0018.
E-mail: lpmuseum@bellsouth.net
Web Site: www.lincolnparishmuseum.org
Founded: 1975.
Congressional District: 5
Key Personnel: C.E.O., Chm. (V) & Pres. (V), William Davis Green; Vice Pres., Dr. Barry Johnson; Dir., Margaret Anne Emory; Treas., Travis DeFreese; Sec., Linda Graham.
Personnel Profile: Full-Time Paid 1.
Volunteer Hours: 755
Operating Expenses: 52,541
Operating Income: 43,165
Governing Authority: nonprofit organization. Subsidiary Institution: Absalom Autrey House Museum. Tax-exempt: 501(c)(3).
Institution Type/Description: Historical Society Museum: housed in 1886 Victorian home.
Collections: history & culture of North Central Louisiana from prehistoric to present; early dolls & toys. Historic House: 1849 Autrey House log cabin.
Research Fields: restoration of log houses; north Louisiana quilting; WPA textiles.
Facilities: 100-seat conference room; 4,000 sq. ft. exhibit space.
Activities: guided tours; arts festivals; temporary & traveling exhibitions; rental facilities.
Publications: biannual newsletter; monthly article in local newspaper; material on the Kidd-Davis House & The Absalom Autrey House.
Hours & Admission Prices: Tues.-Fri. 10-4. No charge; donations accepted. Closed New Year's Eve & Day; Independence Day; Thanksgiving; Christmas Eve, Day & week. &
Attendance: 3,000 (accurate)
Membership: Individual $15; Friend $25; Donor $50; Sponsor $100; Sustaining $250; Patron $500.

LOUISIANA MILITARY MUSEUM, 201 Memorial Dr., Ruston, LA 71270-3955. Mailing Address: State of Louisiana, Secretary of State, P.O. Box 94125, Baton Rouge, LA 70804-9125. Tel.: 318-251-5099. Fax: 318-251-5088.
E-mail: military@sos.louisiana.gov
Web Site: www.sos.louisiana.gov/lmm
Congressional District: 5
Key Personnel: Dir., Ernest Stevens.
Personnel Profile: Full-Time Paid 1; Part-Time Paid 2.
Governing Authority: state. Parent Institution: Louisiana Secretary of State, Baton Rouge, LA. Tax-exempt.
Institution Type/Description: Military Museum.
Collections: military history from Civil War to present; period artifacts; military artifacts including weapons, uniforms, flags, banners, medals & badges.
Research Fields: area veterans.
Activities: guided tours; temporary exhibitions; special events.
Hours & Admission Prices: Tues.-Sat. 10-4. No charge. Closed most state holidays.
Attendance: 6,318 (accurate)

LOUISIANA TECH MUSEUM, Louisiana Tech University, Ruston, LA 71272. Tel.: 318-257-2935 & 3660. Fax: 318-257-2579.
E-mail: pcarter@latech.edu
Founded: 1982.
Congressional District: 5
Key Personnel: Devel., Jonathan Donehoo; Education, Joan Marie Edinger; Public Rels., Sally R. Hollis; Archivist, Peggy Carter.
Personnel Profile: Part-Time Paid 2; Part-Time Volunteers 5.
Governing Authority: public university; nonprofit. Parent Institution: Louisiana Tech University, Ruston. Tax-exempt.
Institution Type/Description: University Museum.
Collections: Indian artifacts; paleontologic & geologic collections; pre-World War II weapons; Roman & Egyptian archaeology.
Research Fields: Indian archaeology.
Facilities: 500 sq. ft. exhibit space.
Activities: formal education programs for children, undergraduate & graduate students; guided tours; lectures; loan exhibitions.
Hours & Admission Prices: Mon.-Fri. 9-4. No charge; donations accepted. Closed New Year's Day; Thanksgiving; Christmas; university holidays. &
Attendance: 400 (estimated)

Saint Bernard

ISLENO MUSEUM COMPLEX - LOS ISLENOS HERITAGE & CULTURAL SOCIETY, 1357 Bayou Rd., Saint Bernard, LA 70085. Tel.: 504-676-3098. Fax: 504-676-3491.
Web Site: www.losislenos.org
Founded: 1980.
Key Personnel: Dir., William Hyland; Pres. (V), Dorothy L. Benge; Museum Shop Mgr., Celie Robin.
Personnel Profile: Full-Time Paid 1; Part-Time Paid 1.
Governing Authority: Tax-exempt.
Institution Type/Description: History Museum.
Collections: Spanish heritage, culture, music & language; period furnishings; personal artifacts; photographs.
Activities: special events.
Hours & Admission Prices: Call for hours. &
Attendance: 35,000 (estimated)
Membership: $10.

Saint Francisville

AUDUBON STATE HISTORIC SITE, 11788 Hwy. 965, Saint Francisville, LA 70775. Mailing Address: P.O. Box 546, Saint Francisville, LA 70775-0546. Tel.: 225-635-3739; 888-677-2838. Fax: 225-784-0578.
E-mail: audubon@crt.state.la.us
Web Site: www.crt.state.la.us
Formerly: Audubon State Commemorative Area
Founded: 1947.
Congressional District: 6
Key Personnel: Historic Site Mgr., John House.
Personnel Profile: Full-Time Paid 6; Part-Time Paid 6.
Governing Authority: state; nonprofit organization. Affiliated with Louisiana Office of State Parks, P.O. Drawer 1111, Baton Rouge, LA 70821. Tax-exempt.
Institution Type/Description: State Park Museum: housed in 1806 Oakley House.
Collections: Federal Period furnishings; 1st edition Audubon prints; early American lighting devices.
Research Fields: archaeological program on the plantation complex; new archival work on house & grounds.
Facilities: library of books on life of Audubon available for use on the premises.
Activities: guided tours; lectures; permanent exhibitions.
Publications: brochure, Oakley.
Hours & Admission Prices: Oakley House & Grounds: Tues.-Sat. 9-5. Adults $8; senior citizens 62 & over $6, children 6-17 $4; children 5 & under no charge. Closed New Year's Day; Thanksgiving; Christmas.
Attendance: 30,000 (accurate)

ROSEDOWN PLANTATION, 12501 Hwy. 10, Saint Francisville, LA 70775. Tel.: 888-376-1867; 225-635-3332.
Institution Type/Description: Historic House.
Collections: local history; agriculture; period furnishings; photographs; gardens.
Facilities: gardens.
Hours & Admission Prices: Daily 9-5. Adults $10; senior citizens 62 & over $8; students 6-17 $4; children 5 & under no charge. Closed New Year's Day; Thanksgiving; Christmas.

Saint Martinville

LONGFELLOW EVANGELINE STATE HISTORIC SITE, 1200 N. Main St., Saint Martinville, LA 70582-3516. Tel.: 888-677-2900; 337-394-3754. Fax: 337-394-3553. Facebook: Longfellow Evangeline State Historic Site.
E-mail: longfellow@crt.state.la.us
Web Site: www.crt.state.la.us/Parks/ilongfell.aspx
Formerly: Oliver House & Interpretive Center
Founded: 1931.
Congressional District: 3
Key Personnel: Historic Site Mgr., Reinaldo Barnes.
Personnel Profile: Full-Time Paid 8; Part-Time Paid 4.
Governing Authority: state; nonprofit. Parent Institution: Louisiana Culture Recreation & Tourism Office of State Parks P.O. Drawer 1111, Baton Rouge, LA. Tax-exempt.
Institution Type/Description: History Museum.
Collections: 18th & 19th-century furniture; Acadian cypress looms; 19th-century textiles; woodworking tools; agricultural implements pertaining to cotton & sugar plantation; Spanish moss artifacts; 19th & 20th-century ironstone & porcelain.
Research Fields: 19th-century plantation life; sugar cane economy; Acadian & Creole culture; Acadian & French migration to southwest Louisiana.
Facilities: 150-acre state commemorative area along the Bayou Teche; visitor center; picnic area; vegetable gardens; formal historic garden; education center.
Activities: tours of plantation house & Acadian cabin; interpretive center; education center; demonstration.
Publications: brochure, Welcome to Longfellow-Evangeline State Commemorative Area.
Hours & Admission Prices: Tues.-Sat. 9-5. Adult $4; senior citizens over 62, children 12 & under no charge. Closed New Year's Day; Thanksgiving; Christmas. &
Attendance: 27,000 (accurate)

MUSEUM OF THE ACADIAN MEMORIAL, 121 S. New Market St., Saint Martinville, LA 70582. Mailing Address: P.O. Box 379, Saint Martinville, LA 70582-0379. Tel.: 337-394-2258. Fax: 337-394-2260.
E-mail: info@acadianmemorial.org
Web Site: www.acadianmemorial.org
Founded: 2001.
Key Personnel: Bd. Mem., Michelle Verret Johnson; Museum Shop Mgr., Jane G. Bulliard
Institution Type/Description: History Museum.
Collections: local history & culture; personal artifacts; photographs; period furnishings; mural.
Activities: audio tour.
Hours & Admission Prices: Daily 10-4:30. Adults $3; members no charge.
Attendance: 12,000 (estimated)

Scott

FLOYD SONNIER'S BEAU CAJUN ART GALLERY, 1010 St. Mary St., Scott, LA 70583. Mailing Address: P.O. Box 397, Scott, LA 70583. Tel.: 337-237-7104.
Institution Type/Description: Art Gallery: built in 1918.
Collections: drawings; prints; period furnishings.
Publications: book, From Small Bits of Charcoal - The Life and Works of a Cajun Artist.
Hours & Admission Prices: Wed.-Fri. 10-5, Sat. 10-4; other times by appointment.

Shreveport

ARK-LA-TEX SPORTS MUSEUM OF CHAMPIONS, 400 Caddo St., Shreveport, LA 71101. Tel.: 318-221-8445. Fax: 318-227-2442.
Institution Type/Description: Sports Museum.
Collections: history of 110 area athletes; photographs; personal artifacts.

Hours & Admission Prices: Sat.-Sun. during events at the convention center.

GARDENS OF THE AMERICAN ROSE CENTER, 8877 Jefferson Paige Rd., Shreveport, LA 71119-5402. Tel.: 318-938-5534.
E-mail: carol@ars-hg.org
Web Site: www.ars.org
Institution Type/Description: Garden.
Collections: over 65 rose gardens; 20,000 rose bushes; plants; sculptures; fountains.
Activities: Annual Events: Official Rose Bloom Season April to October; Christmas in Roseland November to December.
Hours & Admission Prices: Mon.-Sat. 9-5, Sun. 1-5. Adults $5.50, seniors 65 & over $4.50, children 5-12 $3.50; discounts to groups of 10 or more; children under 5 no charge.

LOUISIANA STATE EXHIBIT MUSEUM, (M), 3015 Greenwood Rd., Shreveport, LA 71109-4640. Mailing Address: State of Louisiana, Secretary of State, P.O. Box 94125, Baton Rouge, LA 70804-9125. Tel.: 318-632-2020. Fax: 318-632-2056.
E-mail: lsem@sos.louisiana.gov
Web Site: www.sos.louisiana.gov/lsem
Founded: 1937.
Congressional District: 4
Key Personnel: Dir., Wayne Waddell; Education, Cynthia Grogan; Cur., Nita Cole.
Personnel Profile: Full-Time Paid 10; Part-Time Paid 7.
Governing Authority: state. Parent Institution: Louisiana Secretary of State, Baton Rouge, LA. Tax-exempt.
Institution Type/Description: History & Art Museum.
Collections: murals; dioramas; period artifacts; agriculture; industrial; archaeology; glass; china; Native American artifacts; folk culture.
Research Fields: archaeology, prehistory, history; Louisiana Native American culture.
Facilities: auditorium.
Activities: permanent, temporary & traveling exhibitions; art workshops; films; concerts; formal education programs for children; rental gallery; lectures; guided tours.
Hours & Admission Prices: Mon.-Fri. 9-4. No charge. Closed most state holidays. &
Attendance: 13,723 (accurate)
Membership: Call for information.

*** MEADOWS MUSEUM OF ART OF CENTENARY COLLEGE, (M),** 2911 Centenary Blvd., Shreveport, LA 71104-3335. Tel.: 318-869-5169. Fax: 318-869-5730.
E-mail: ddufilho@centenary.edu
Web Site: www.centenary.edu
Founded: 1976.
Congressional District: 4
Key Personnel: C.E.O., Dr. Kenneth Schwab; Pres. (V), Grace Bareikis; Dir., Diane Dufilho; Cur., Bruce Allen; Museum Shop Mgr., Lorraine Soffer; Museum Education, Connie Blake; Collections Mgr., Kathy Brodnax; Student Registrar, Amanda Schiffner; Officer Mgr. & Internal Accountant, Neeta Kaji.
Personnel Profile: Full-Time Paid 3; Part-Time Paid 3; Part-Time Volunteers 90.
Governing Authority: college. Parent Institution: Centenary College of Louisiana. Tax-exempt.
Institution Type/Description: Art Museum.
Collections: Indochina collection by French academic artist Jean Despujols features the people & lands of French Sado-China between 1936-37; paintings by George Grosz, Alfred Maurer, William Glackens, & Ernest Lawson; Latin American works on paper featuring Diego Rivera, David Sigueiros, Emilio Amero & Jose Clemente Orozco; American works on paper include Reginald Marsh, Mary Cassatt & Isabel Bishop; African American, Haitian, Inuit & tribal works.
Research Fields: Southeast Asia 1900-1940; French artists 1915-1940.
Facilities: 8,500 sq. ft. exhibition space; art resource center. Museum-related items for sale.
Activities: guided tours; lectures; films; documentary films; gallery talks; concerts; docent program; permanent & temporary exhibitions.
Publications: booklet.
Hours & Admission Prices: mid-Aug. to July Tues.-Wed. & Fri. 12-4, Thurs. 12-5, Sat.-Sun. 1-4. No charge; donations accepted. Fee for some special exhibits. Closed New Year's Day; Easter; Memorial Day; Labor Day; Thanksgiving; Christmas. &
Attendance: 15,000 (estimated)
Membership: Individual $25; Family $40; Sustaining $100; Patron $250; Donor & Corporate $500.

THE MULTICULTURAL CENTER OF THE SOUTH, 520 Spring St., Shreveport, LA 71101. Mailing Address: P.O. Box 305, Shreveport, LA 71101-0305. Tel.: 318-424-1380. Fax: 318-424-1384.
E-mail: jgatlin-mccs@comcast.net
Web Site: mecsouth.org
Founded: 1999.
Congressional District: 4
Key Personnel: Dir. Programs, Janice Gatlin
Institution Type/Description: Multicultural Center.
Collections: African American; Asian; Chinese; Cambodian; Creole; Cajun; East Indian; Filipino; French; Greek; German; Hispanic/Latino; Italian; Iranian/Persian; Irish; Jewish; Japanese; Korean; Native American; Pacific Islanders; Pakistani; Scottish; Slavic; Thai; Vietnamese; photographs; paintings; drawings; costumes.
Activities: educational programs; special events; cultural celebrations; community programs.
Hours & Admission Prices: Gallery: Tues.-Fri. 10-4. Guided Tours: Sat. by appointment. Adults $3, seniors & students $2. Closed holidays.

THE MUSEUM OF AMERICAN FENCING, 1413 Fairfield Ave., Shreveport, LA 71101. Tel.: 318-227-7575.
E-mail: andy@museumofamericanfencing.com
Web Site: museumofamericanfencing.com
Institution Type/Description: Fencing Museum.
Collections: fencing history & artifacts; photographs; paintings; uniforms & equipment; personal artifacts; books.
Hours & Admission Prices: Mon.-Thurs. 3-8, Sat. 9am to noon.

PIONEER HERITAGE CENTER, LSU-SHREVEPORT, One University Place, Shreveport, LA 71115-2301. Tel.: 318-797-5339. Fax: 318-797-5110.
E-mail: marty.young@lsus.edu
Web Site: www.lsus.edu/pioneer
Founded: 1977.
Congressional District: 4
Key Personnel: Dir., Marvin R. Young, II; Pres. (V), Dr. Vincent J. Marsala; Treas., Michael Ferrell.
Personnel Profile: Full-Time Paid 1; Part-Time Volunteers 5.
Governing Authority: state. Parent Institution: Louisiana State University in Shreveport. Tax-exempt: 501(c)(3).
Institution Type/Description: History Museum Complex.
Collections: Northwest Louisiana & 19th-Century Red River regional history; medical implements; blacksmith & carpenter tools; period furnishings; textiles; clothing; books; archival documents & photographs; 300-pcs. 19th-century agricultural implements. Historic Structures: log single pen blacksmith shop; Thrasher House; log dogtrot; Caspiana House; plantation cottage; detached kitchen; doctor's office (shotgun house); Webb & Webb Plantation Commissary; blacksmith shop; riverfront mission.
Research Fields: folklife & history of Northwest Louisiana; history of medicine; humanities.
Facilities: library containing early medical books & plantation journals; 750-seat auditorium; educational facilities; classroom; conservation lab area in University Technology Center building; snack shop. University adjunct facilities available.
Activities: guided tours; lectures; films; broadcast programs; organized education programs for children, adults, undergraduate & graduate students; docent program; training programs for professional museum workers; participatory exhibits. Museum Sponsors: Suitcase museums: workshops for students & teachers; teaching packets; slide & tape presentations. Annual Events: Authors in April fundraiser; History Fair co-sponsored by LSU-Shreveport History Club.
Publications: annual brochure; teachers' packets; docent's manual; pamphlets; local history publications.
Hours & Admission Prices: Feb. to mid-Dec. Tues.-Fri. by appointment; other times for special events. Admission $2; discounts to AAM members; children under 5 no charge. Closed major holidays.
Attendance: 5,000 (accurate)
Membership: Pioneer Heritage Society $125; Sponsor $500 & up.

R.S. BARNWELL MEMORIAL GARDEN AND ART CENTER, 601 Clyde Fant Pkwy., Shreveport, LA 71101-3207. Tel.: 318-673-7703. Fax: 318-673-7707.
E-mail: director@barnwellcenter.com
Web Site: www.barnwellcenter.com
Founded: 1970.
Congressional District: 4
Key Personnel: Pres., Dr. Joe White; Museum Shop Mgr., Barbara White; Sec., Wendy Liles.

Personnel Profile: Full-Time Paid 3; Part-Time Volunteers 10.
Governing Authority: municipal. Parent Institute: City of Shreveport. Tax-exempt.
Institution Type/Description: Garden & Art Center.
Collections: paintings & tropical environment for plants.
Activities: lectures; films; arts festivals; study clubs; hobby workshops; formally organized education programs; traveling exhibitions.
Publications: quarterly report, Barnwell Beacon.
Hours & Admission Prices: Tues.-Fri. 10-4, Sat. 10-5, Sun. 1-5. No charge; donation accepted. Closed New Year's Day; Martin Luther King, Jr. Day; Presidents' Day; Good Friday; Easter; Memorial Day; Independence Day; Labor Day; Thanksgiving; Christmas. &
Attendance: 30,000 (estimated)
Membership: Individual $15; Family $25; Organization $50; Supporting $100; Patron $200; Gold $500.

THE R.W. NORTON ART GALLERY, (M), 4747 Creswell Ave., Shreveport, LA 71106-1899. Tel.: 318-865-4201. Fax: 318-869-0435.
E-mail: gallery@rwnaf.org
Web Site: www.rwnaf.org
Founded: 1946.
Congressional District: 4
Key Personnel: C.E.O. & Pres. Bd. Control, M. Lewis Norton; Dir. Public Rels., Sec. & Treas., Jerry M. Bloomer; Bldg. & Grounds Supt., Gerry Ward.
Personnel Profile: Full-Time Paid 29; Part-Time Volunteers 1.
Governing Authority: nonprofit organization. Parent Institution: R.W. Norton Art Foundation. Tax-exempt: 501(c)(3).
Institution Type/Description: Art Museum & Gallery.
Collections: American painting, sculpture, glass, silver, miniatures, including works by western artists Charles M. Russell & Frederic Remington; European paintings, sculpture, miniatures, tapestries; Wedgwood collection; dolls dressed in authentic fashions of Louisiana from 1720-1920; rare books; atlases including elephant folio edition of Audubon Birds of America; complete set of John Gould ornithology works; firearms collection.
Research Fields: 19th- & early 20th-century American & European art.
Facilities: 15,000-vol. library of fine arts, history, literature, ornithology, genealogy; James M. Owens Memorial Collection of Early Americana, available for use by public on premises. Exhibition catalogs for sale.
Activities: guided tours; lectures; films; gallery talks; formally organized education programs for children & adults; permanent, temporary & traveling exhibitions.
Publications: collection brochures; exhibition catalogs; catalogs of the Charles Russell & Wedgwood collections.
Hours & Admission Prices: Tues.-Fri. 10-5, Sat.-Sun. 1-5. Library: by appointment. No charge; donations accepted. Closed national holidays. &
Attendance: 26,609 (accurate)

SCI-PORT: LOUISIANA'S SCIENCE CENTER, 820 Clyde Fant Pkwy., Shreveport, LA 71101-3667. Tel.: 318-424-3466. Fax: 318-222-5592. Facebook: Sci-Port.
E-mail: kwissing@sciport.org
Web Site: www.sciport.org
Formerly: Sci-Port Discovery Center
Founded: 1994.
Congressional District: 4
Key Personnel: Pres. & C.E.O, Ann Fumarolo; Dir. Devel. & Mktg., Jennifer McMenamin; Museum Shop Mgr., Jacqui Brumley.
Personnel Profile: Full-Time Paid 35; Part-Time Paid 35; Part-Time Volunteers 70; Interns 2.
Governing Authority: nonprofit organization. Tax-exempt: 501(c)(3).
Institution Type/Description: Science Museum.
Collections: science, technology, engineering & math.
Major Exhibits: Treasure! (T), 1/15/14-5/11/14.
Facilities: 92,000 sq. ft. exhibit space; classroom; 173-seat IMAX Dome theatre.
Activities: teachers workshop, camps; scout events; daily changing programs; educational programs; birthday parties; rental facilities.
Publications: e-newsletter.
Hours & Admission Prices: Memorial Day-Labor Day Mon.-Sat. 9-5, Sun. 12-5; After Labor Day-Memorial Day Tues.-Sat. 9-5, Sun. 12-5. Sci-Port: adults $13, children 3-12, seniors & military $10; discounts to groups or more; members no charge. IMAX: adults $9.50, children 3-12, seniors & military $8.50, members $6; discounts to groups of 15 or more. Center & IMAX: adults $23, seniors & military $19, children 3-12 $18; discounts to groups of 15 or more. Closed Easter; Thanksgiving; Christmas. &
Attendance: 170,000 (estimated)

Membership: Discovery $125; Family Plus $150; Friend $175; Donor Friend Supporting $250; Patron Friend Supporting $500.

SHREVEPORT WATER WORKS MUSEUM, 142 N. Common St., Shreveport, LA 71101-2614. Mailing Address: State of Louisiana, Secretary of State, P.O. Box 94125, Baton Rouge, LA 70804-9125. Tel.: 318-221-3388.
Web Site: www.sos.louisiana.gov/swwm
Formerly: McNeill Street Pumping Station Museum
Congressional District: 4
Key Personnel: Chm. (V), Dale Ward.
Personnel Profile: Part-Time Paid 2.
Governing Authority: state. Parent Institution: Louisiana Secretary of State, Baton Rouge, LA. Tax-exempt.
Institution Type/Description: Water Works Museum.
Collections: period steam water pumping engines from 1898; community water development; water works history.
Activities: guided tours.
Hours & Admission Prices: Tues. 10-2, Wed.-Sat. 10-4. No charge. Closed most state holidays.
Attendance: 835 (accurate)
Membership: Call for information.

SOUTHERN UNIVERSITY MUSEUM OF ART IN SHREVE-PORT, 3050 Martin Luther King, Jr. Dr., Shreveport, LA 71107. Tel.: 318-670-6000.
Institution Type/Description: Art Museum.
Collections: works of art by Africans, African Americans & their descendants; personal artifacts; sculptures; jewelry; prints; paintings.
Activities: educational programs.
Hours & Admission Prices: Tues.-Fri. 10-5, Sat. 10-4; groups by appointment. No charge.

SPRING STREET HISTORICAL MUSEUM, 525 Spring St., Shreveport, LA 71101-3231. Mailing Address: State of Louisiana, Secretary of State, P.O. Box 94125, Baton Rouge, LA 70804-9125. Tel.: 318-424-0964. Fax: 318-424-0964.
E-mail: mloschen@sos.la.gov
Web Site: www.sos.louisiana.gov/sshm
Founded: 1977.
Congressional District: 4
Key Personnel: Dir., Marty Loschen.
Personnel Profile: Full-Time Paid 2.
Governing Authority: state. Parent Institution: Louisiana Secretary of State, Baton Rouge, LA. Tax-exempt.
Institution Type/Description: History Museum.
Collections: history of Shreveport & surrounding area; portraits of early settlers; maps; written documents; artifacts; newspaper illustrations; 19th century architecture; costumes & textiles.
Research Fields: women's suffrage; prohibition; the Depression; northwest Louisiana history.
Facilities: Museum-related items for sale.
Activities: docent program; guided tours; lectures; participatory, loan, temporary & traveling exhibitions.
Hours & Admission Prices: Wed.-Fri. 10-4; group tours by appointment. No charge. Closed most state holidays.
Attendance: 1,356 (accurate)
Membership: Call for information.

STAGE OF STARS MUSEUM, 705 Elvis Presley Ave., Shreveport, LA 71101. Mailing Address: 9068 Melody Ln., Shreveport, LA 71118. Tel.: 318-222-9391; 800-551-8682.
Institution Type/Description: History Museum.
Collections: local history and culture; photographs; personal artifacts.
Hours & Admission Prices: Call for hours.

STEPHENS AFRICAN-AMERICAN MUSEUM, 2810 Lindholm, Shreveport, LA 71108-2610. Tel.: 318-635-2147. Fax: 318-635-2147.
Founded: 1994.
Key Personnel: C.E.O., Chm. (V) & Pres. (V), Spencer Stephens; Museum Shop Mgr., Gwendolyn Frazier.
Personnel Profile: Part-Time Paid 2.
Governing Authority: private; nonprofit organization. Parent Institution: Spencer R. Stephens Foundation Inc. Tax-exempt: 501(c)(3).
Institution Type/Description: Art Museum.
Collections: African-American art including paintings, sculptures, prints & drawings.

Facilities: library; 1,200 sq. ft. exhibit space; classrooms.
Activities: arts festival; formal education programs; guided tours; lectures; temporary exhibitions. Annual Event: Back to School Art Festival.
Publications: newsletter, African-American Museum.
Hours & Admission Prices: Tues.-Sat. 12-4. Adults $2, senior citizens $1.75, students & children $1; discounts to AAM & ICOM members; members no charge. Closed Independence Day; Easter; Christmas. &
Attendance: 15,000 (estimated)

WALTER B. JACOBS MEMORIAL NATURE PARK, 8012 Blanchard Furrh Rd., Shreveport, LA 71107-8310. Tel.: 318-929-2806. Fax: 318-929-3718.
E-mail: rscarborough@caddo.org
Web Site: www.caddoparks.com/memorial.cfm
Founded: 1976.
Congressional District: 4
Key Personnel: Dir., Larry R. Raymond; Sr. Park Naturalist, Rusty Scarborough; Park Naturalist, Rachel Demascal; Park Naturalist, Kimberly Warren; Park Naturalist, Stacy Gray; Park Naturalist, Halya Onel.
Personnel Profile: Full-Time Paid 3; Part-Time Paid 3.
Governing Authority: county; nonprofit. Operated by the Caddo Parish Commission Parks & Recreation Dept. Tax-exempt.
Institution Type/Description: Nature Park & Interpretive Center.
Collections: natural history of northwest Louisiana; leaves of woody plants; 52 taxidermy mounts of North American waterfowl; tanned skins & skulls of north Louisiana mammals; 1,000 pieces of large & small petrified wood; Pleistocene gravel; 2,000 35mm slides.
Research Fields: plants & animals of the park; reproductive biology of local amphibians.
Facilities: library of natural history material including books; 800 sq. ft. exhibit space; classroom; pavilion workshop.
Activities: guided tours; lectures; films; slide shows; organized education programs; temporary & participatory exhibits; summer environmental day camp; discovery boxes available for loan to teachers: Earth Science Treasure Chest; Fantastic Fossils; Insects; The Bones Box; The Tree Trunk; Reptiles & Amphibians; Butterflies & Hummingbirds; Wetlands; The Wonder of Birds; Wildflowers; Dinosaurs; Mammal Tracks & Trails.
Hours & Admission Prices: Wed.-Sat. 8-5, Sun. 1-5. No charge. Closed New Year's Day; Easter; Thanksgiving; Christmas. &
Attendance: 13,063 (accurate)

Slidell

SLIDELL CULTURAL CENTER, 2055 Second St., Slidell, LA 70458-3430. Mailing Address: P.O. Box 828, Slidell, LA 70459-0828. Tel.: 985-646-4375. Fax: 985-646-4231.
E-mail: kbergeron@cityofslidell.org
Web Site: slidell.la.us
Founded: 1989.
Congressional District: 1
Personnel Profile: Full-Time Paid 3; Part-Time Paid 2.
Governing Authority: municipal government; nonprofit. Parent Institution: City of Slidell. Subsidiary Institution: Dept. of Cultural & Public Affairs. Tax-exempt.
Institution Type/Description: Art Center.
Collections: works by local & regional artists.
Facilities: 600-vol. library; 1,328 sq. ft. exhibit space; meeting room.
Activities: concerts; guided tours; traveling & temporary exhibitions; performing artists residency training. Annual Event: City Arts Awards.
Publications: quarterly newsletter, Bravo!
Hours & Admission Prices: Call for hours. No charge; donations accepted. Closed New Year's Day; Martin Luther King Jr. Day; Mardi Gras; Good Friday; Memorial Day; Independence Day; Labor Day; Thanksgiving; Christmas Eve & Day. &
Attendance: 7,000 (estimated)

SLIDELL MUSEUM, 2020 First St., Slidell, LA 70458-3402. Mailing Address: P.O. Box 828, Slidell, LA 70459-0828. Tel.: 985-646-4380. Fax: 985-646-6107.
E-mail: cityslidell@charter.net
Web Site: www.slidell.la.us
Founded: 1987.
Congressional District: 1
Key Personnel: Financial Dir., Sharon Howes; Cur., Reinhard Dearing.
Personnel Profile: Part-Time Paid 2.
Governing Authority: municipal.
Institution Type/Description: History Museum: housed in the city's original town hall & jail built in 1907.

Collections: history of Slidell from its founding to present; war between the states; Louisiana's military leaders.
Facilities: 300-vol. library.
Activities: guided tours; traveling exhibitions.
Hours & Admission Prices: Tues.-Sat. 9-4. No charge; donations accepted. Closed city holidays.
Attendance: 1,500 (accurate)

Sorrento

LOUISIANA POTTERY MUSEUM, 6470 Hwy. 22, Cajun Village, Sorrento, LA 70778. Tel.: 225-675-5572.
E-mail: lapottery@cox.net
Web Site: www.louisianapottery.com
Founded: 1999.
Key Personnel: Dir., Judy L. Starrett.
Personnel Profile: Full-Time Volunteers 1.
Institution Type/Description: General Museum: housed in an Acadian style home; c.1830.
Collections: pottery; etchings; hand-blown glass; pine needle baskets; hand-carved wooden ducks & boats; books.
Activities: temporary exhibits; classes.
Hours & Admission Prices: Tues.-Sun. 10-5. No charge. &

Sulphur

BRIMSTONE MUSEUM, 900 S. Huntington St., Sulphur, LA 70663-4420. Tel.: 337-527-0357. Fax: 337-527-0359.
E-mail: trahan@brimstonemuseum.org
Web Site: www.brimstonemuseum.org
Founded: 1975.
Congressional District: 7
Key Personnel: Pres., Randall Broussard; Dir., Thomas Trahan.
Personnel Profile: Full-Time Paid 1; Part-Time Paid 1; Part-Time Volunteers 5.
Governing Authority: society; nonprofit. Parent Institution: Brimstone Historical Society. Tax-exempt: 501(c)(3).
Institution Type/Description: History Museum: housed in 1915 railroad station.
Collections: sulphur mining by Frasch process; mid-1800 medical instruments; 1876 bell clock; 1800s clothing; sulphur mining photographs; renovated railway caboose.
Facilities: 1,950 sq. ft. exhibit space.
Activities: guided tours; permanent, temporary & traveling exhibitions, summer programming.
Hours & Admission Prices: By appointment. No charge; donations accepted. Closed New Year's Eve & Day; Good Friday; Memorial Day; Independence Day; Labor Day; Thanksgiving weekend; Christmas Eve & Day. &
Attendance: 7,600 (accurate)
Membership: Student & Senior $10; Individual $20; Family $45; Sponsor $250; Friend $350; Patron $500; Historian $750; Curator $1,000; Founder $2,500 & up.

HENNING CULTURAL CENTER, 923 S. Ruth St., Sulphur, LA 70663. Tel.: 337-527-0537.
Institution Type/Description: Art Center.
Collections: works by local & national artists.
Activities: traveling exhibitions.
Hours & Admission Prices: Mon.-Fri. 10-12 & 1-5, Sat. 10-2.

Tallulah

HERMIONE MUSEUM, 315 N. Mulberry, Tallulah, LA 71282-3828. Mailing Address: P.O. Box 268, Tallulah, LA 71284-0268. Tel.: 318-574-0082. Fax: 318-574-0082.
E-mail: hermionemuseum@bayou.com
Founded: 1994.
Congressional District: 5
Key Personnel: Dir. & Pres. (V), John Earl Martin; Cur. & Museum Shop Mgr., Geneva Williams; Pres. (V), Charles M. Finlayson.
Personnel Profile: Full-Time Volunteers 2; Part-Time Volunteers 1; Interns 1.
Volunteer Hours: 1,000
Operating Expenses: 12,500
Operating Income: 12,500
Governing Authority: nonprofit. Parent Institution: Madison Historical Society, Inc. Tax-exempt.
Institution Type/Description: Historic House Museum.
Collections: local history; aviation artifacts; Civil War; Black & White history.
Research Fields: local history archives.
Activities: special events.
Publications: 25 books & booklets.

Hours & Admission Prices: Tues.-Fri. 10-4. No charge; donations accepted. &
Attendance: 1,000 (estimated)
Membership: Individual $25; Associate $50; Patron $100; Super Patron $500; Lifetime $750.

Tangipahoa

CAMP MOORE CONFEDERATE CEMETERY AND MUSEUM, 70640 Camp Moore Rd., Tangipahoa, LA 70465. Mailing Address: P.O. Box 25, Tangipahoa, LA 70465-0025. Tel.: 985-229-2438.
E-mail: manager@campmoorela.com
Web Site: www.campmoorela.com
Founded: 1993.
Key Personnel: Dir., Kevin J. Miller.
Personnel Profile: Full-Time Volunteers 1; Part-Time Paid 1; Part-Time Volunteers 6.
Governing Authority: Tax-exempt.
Institution Type/Description: History Museum.
Collections: Camp Moore artifacts; soldiers' letters & journals; newspapers; photographs.
Facilities: library.
Activities: research. Annual Event: Camp Moore Reenactment in November.
Publications: quarterly newsletter.
Hours & Admission Prices: Wed.-Sat. 10-3. Adults $3, students $2; discounts to F.O.C.M. members; children under 6 no charge. Closed New Year's Day; Thanksgiving; Christmas.
Attendance: 1,500 (estimated)
Membership: Friends of Camp Moore: Individual $10; Family $30.

Thibodaux

∗ E. D. WHITE HISTORIC SITE, 2295 LA Hwy. 1, Thibodaux, LA 70301. Tel.: 985-447-0915. Fax: 985-447-4249.
Institution Type/Description: Historic House: housed in the former home of Gov. Edward Douglas White, and his son, Chief Justice Edward Douglass White. A National Historic Landmark.
Collections: White family history; Bayou Lafourche area history; Chitimacha Indians; Acadian settlers; sugarcane plantations; slavery; period furnishings; photographs.
Activities: guided tours.
Hours & Admission Prices: Tues.-Sat. 10-4:30. No charge; donations accepted. Closed major holidays.
Attendance: 3,513 (accurate)
Membership: Individual $15; Family $25; Business $50.

Vacherie

LAURA: A CREOLE PLANTATION, 2247 Hwy. 18, River Rd., Vacherie, LA 70090. Tel.: 888-799-7690; 225-265-7690.
Institution Type/Description: Historic House: built in 1840.
Collections: Creole history & culture; period furnishings; personal artifacts; photographs; historic buildings.
Activities: guided tours.
Hours & Admission Prices: Guided Tours: daily. Adults $15, students 6-17 $5; children 5 & under no charge. Closed New Year's Day; Mardi Gras Day; Easter; Thanksgiving; Christmas.

OAK ALLEY PLANTATION, 3645 Hwy. 18, Vacherie, LA 70090. Tel.: 225-265-2151. Fax: 225-265-7035.
E-mail: contactus@oakalleyplantation.com
Web Site: www.oakalleyplantation.org
Founded: 1839.
Congressional District: 3
Key Personnel: Dir., Zeb Mayhew, Jr.; Chm. (V), Shelby Saer
Institution Type/Description: Historic Mansion.
Collections: local history & culture; period furnishings; personal artifacts; photographs.
Hours & Admission Prices: Mon.-Fri. 10-4, Sat.-Sun. 10-5. Adults 19 & over $18, students 13-18 $7.50, children 6-12 $4.50. Closed New Year's Day; Mardi Gras Day; Thanksgiving; Christmas.
Attendance: 166,000 (accurate)

ST. JOSEPH PLANTATION, 3535 Hwy. 18, Vacherie, LA 70090. Tel.: 225-265-4078. Fax: 225-265-4843.
Institution Type/Description: Historic House.
Collections: local history & culture; personal artifacts; period furnishings; photographs.
Facilities: Museum-related items for sale.

Hours & Admission Prices: Mon.-Sat. 9:30-5. Adults $15, youth 13-18 $7, children 6-12 $5; children under 6 no charge. Closed New Year's Day; Easter; Independence Day; Labor Day; Thanksgiving; Christmas Eve & Day.

Ville Platte

LOUISIANA ARBORETUM, STATE PRESERVATION AREA, 1300 Sudie Lawton Lane, Ville Platte, LA 70586-7527. Tel.: 337-363-6289 & 888-677-6100. Fax: 337-363-5616.
E-mail: arboretum@crt.la.us
Web Site: www.louisianaarboretum.wordpress.com
Founded: 1961.
Congressional District: 8
Key Personnel: Interpretive Naturalist, Eric Bush; Cur., Kim Hollier; Horticulturist, Emma Debenport.
Personnel Profile: Full-Time Paid 5; Part-Time Paid 2.
Governing Authority: state. Affiliated with the Dept. of Culture, Recreation & Tourism, Office of State Parks.
Institution Type/Description: Arboretum.
Collections: native Louisiana flora; herbarium.
Facilities: 2 outdoor classrooms; 6 miles of interpretive nature trails.
Activities: tours by appointment; 150 planned public interpretive programs.
Publications: tree checklist; bird checklist.
Hours & Admission Prices: Daily 9-5. No charge. Closed New Year's Day; Thanksgiving; Christmas. &
Attendance: 7,500 (accurate)
Membership: Friends of the Arboretum $10.

LOUISIANA SWAMP POP MUSEUM, 205 N.W. Railroad Ave., Ville Platte, LA 70586. Tel.: 337-363-0900.
Key Personnel: Dir., Janie Knighten
Institution Type/Description: History Museum.
Collections: local history & culture; stage costumes; photographs; records; framed autographs; memorabilia.
Hours & Admission Prices: Fri.-Sat. 10-3.

Wallace

WHITNEY PLANTATION MUSEUM, 5099 Hwy. 18, Wallace, LA 70049. Tel.: 504-586-0003; 617-406-9225.
E-mail: jim@ornana.com
Web Site: whitneyplantation.com
Key Personnel: Owner & Cur., John Cummings
Institution Type/Description: Historic Plantation: 1790 home.
Collections: plantation history & culture.
Activities: guided tours.
Hours & Admission Prices: By appointment.

Washington

WASHINGTON MUSEUM & TOURIST CENTER, 404 N. Main St., Washington, LA 70589. Mailing Address: P.O. Box 597, Washington, LA 70589-0597. Tel.: 337-826-3627.
E-mail: towtourism@bellsouth.net
Web Site: washington-la.org
Founded: 1972.
Congressional District: 7
Key Personnel: Dir. Tourism, Raynold Soileau; Cur., Deborah Joubert; Cur., Lienola Chelette.
Personnel Profile: Part-Time Paid 3.
Governing Authority: nonprofit organization. Parent Institution: Town of Washington. Tax-exempt.
Institution Type/Description: History Museum.
Collections: artifacts; tools; 19th-century documents; ledgers; Indian baskets; fans; 19th-century stethoscope; photographs; firemen's hats.
Facilities: reading room.
Activities: guided tours.
Publications: brochures & maps of historic sites.
Hours & Admission Prices: Daily 8-12 & 1-4. No charge. &
Attendance: 2,000 (estimated)

West Monroe

OUACHITA RIVER ART GALLERY, 308 Trenton St., West Monroe, LA 71291. Tel.: 318-322-2380.
Founded: 1971.
Institution Type/Description: Art Gallery.
Collections: paintings; drawings; ceramics; photographs.

Activities: workshops; special events.
Hours & Admission Prices: Tues.-Sat. 10-5.

Westwego

WESTWEGO HISTORICAL MUSEUM, 275 Sala Ave., Westwego, LA 70094-3650. Tel.: 504-341-3161. Fax: 504-341-2570.
Key Personnel: Museum Coord., Lori Guin
Institution Type/Description: History Museum.
Collections: period furnishings; hardware store.
Hours & Admission Prices: Mon.-Tues. & Thurs.-Fri. 8-4, Wed. & Sat. 10-4; groups by appointment. Adults $3, seniors & children under 12 $2; discounts to groups of 10 or more; society members no charge.

White Castle

NOTTOWAY, 31025 Louisiana Hwy. 1, White Castle, LA 70788. Tel.: 866-527-6884; 225-545-2730. Fax: 225-545-8632.
Institution Type/Description: Historic House: built in 1859. Listed on the National Register of Historic Places.
Collections: local history & culture; period furnishings; personal artifacts; paintings; photographs.
Facilities: rental facilities; restaurant.
Activities: guided tours.
Hours & Admission Prices: Daily 9-4. Adults $15, children under 12 $6.

Winnfield

LOUISIANA POLITICAL MUSEUM AND HALL OF FAME, 499 E. Main St., Winnfield, LA 71483-3224. Tel.: 318-628-5928. Fax: 318-628-2551. Facebook: Friends of the Louisiana Political Museum.
E-mail: lapolmus@bellsouth.net
Web Site: www.lapoliticalmuseum.com
Founded: 1993.
Congressional District: 5
Key Personnel: Dir., Carolyn Phillips.
Personnel Profile: Full-Time Paid 3.
Governing Authority: state. Tax-exempt.
Institution Type/Description: Political History Museum.
Collections: Louisiana politicians & politics; Hall of Fame inductees includes senators, congressmen & representatives; political families; media personnel.
Hours & Admission Prices: Mon.-Fri. 9-5, Sat. by appointment. No charge; donations accepted. &
Attendance: 7,500 (estimated)

Zachary

ZACHARY HISTORIC VILLAGE, 4524 Virginia St., Zachary, LA 70791. Mailing Address: P.O. Box 1144, Zachary, LA 70791. Tel.: 225-654-1912.
Web Site: www.zachry.la.nww.net/historical
Key Personnel: Dir., Lois Hastings
Institution Type/Description: Historic Village.
Collections: local history; period furnishings; Native American artifacts; period tools & equipment. Historic Buildings: Victorian house; magic house; farm house; barn.
Activities: school groups.
Hours & Admission Prices: Mon.-Fri. 9-4; groups by appointment. Suggested Donation: $2 per person.

MAINE

(240 listings)

Allagash

ALLAGASH HISTORICAL SOCIETY MUSEUM, 456 Dickey Rd., Allagash, ME 04774-4113. Tel.: 207-398-3148 & 3278. Fax: 207-398-3148.
E-mail: marilynm@fairpoint.net; hmcb@fairpoint.net
Founded: 1975.
Congressional District: 2
Key Personnel: Dir., Linda McBreairty; Pres. (V), Marilyn McBreairty
Governing Authority: Parent Institution: Allagash Historical Society. Tax-exempt.

Institution Type/Description: History Museum.
Collections: local history & culture; period furnishings; personal artifacts; photographs.
Activities: fundraisers; book sales.
Hours & Admission Prices: Memorial Day to Labor Day Fri.-Sun. 1-5; other times by appointment. No charge; donations accepted. &
Attendance: 25 (estimated)
Membership: Individual $5; Business $10.

Alna

WISCASSET, WATERVILLE & FARMINGTON RAILWAY MUSEUM, 97 Cross Rd., Alna, ME 04535. Mailing Address: P.O. Box 242, Alna, ME 04535-0242. Tel.: 207-563-2516.
E-mail: info@wwfry.org
Web Site: www.wwfry.org
Founded: 1989.
Congressional District: 1
Key Personnel: Pres., Stephen T. Zuppa; Membership, Frances Hernandez; Treas., James Patten; Archivist, Bruce Wilson; Museum Shop Mgr., Linda Zollers.
Personnel Profile: Part-Time Paid 1; Part-Time Volunteers 200.
Governing Authority: private; nonprofit organization. Tax-exempt: 501(c)(3).
Institution Type/Description: Operating Narrow Gauge Railroad Museum.
Collections: two foot gauge railroad equipment; books; photographs; documents; period artifacts; steam locomotives.
Facilities: 300-vol. library; operating restored two foot gauge railroad; 5,000 sq. ft. exhibit space. Museum-related items for sale.
Activities: docent program; films; guided tours; training programs for professional museum workers. Annual Events: Picnic; Halloween Trains; Christmas Trains.
Publications: bimonthly, WW&F Newsletter; book, WW&F Musings; children's book, Harry's Train; The Twenty-Four Inch Gauge Railroad at Bridgton, Maine.
Hours & Admission Prices: Memorial Day to Columbus Day Sat.-Sun. 9-5; Oct.-May Sat. 9-5. Museum: no charge. Steam Train Rides: adults $7, seniors $6, children $4; children 3 & under no charge. &
Attendance: 5,500 (accurate)
Membership: Individual $30; Life $300.

Ashland

ASHLAND LOGGING MUSEUM, INC., 267 Garfield Rd., Ashland, ME 04732-5105. Mailing Address: P.O. Box 631, Ashland, ME 04732-0866. Tel.: 207-435-6679. Fax: 207-435-6579.
Web Site: www.townofashland.com/Ashland_Logging_Museum.htm
Founded: 1964.
Congressional District: 14
Key Personnel: Pres. (V), Robert Sawyer, V; Vice Pres., Robert Sawyer, IV; Cur., Ed Chase.
Governing Authority: nonprofit. Tax-exempt: 501(c)(3).
Institution Type/Description: Logging Museum: housed in a reproduction of an early logging camp.
Collections: 1903 Lombard steam & gas log haulers; log hauler sleds; bateau; tote wagon; pine log with king's arrow mark; tamarack ship's knee; machinery & hand tools from past years of lumbering activity.
Facilities: two machine sheds; logging camp.
Hours & Admission Prices: Fri. 1-4; other times by appointment. No charge; donations accepted.
Attendance: 500

Athens

ATHENS HISTORICAL SOCIETY MUSEUM, Academy St., Athens, ME 04912. Mailing Address: P.O. Box 295, Athens, ME 04912. Tel.: 207-654-2393.
Institution Type/Description: Historical Society Museum: housed in the former Somerset Academy building. Listed on the National Register of Historic Places.
Collections: local history & culture; Somerset Academy memorabilia from 1846-1967; local town reports; maps; photographs; scrapbooks; period artifacts.
Hours & Admission Prices: Call for hours.

Auburn

ANDROSCOGGIN HISTORICAL SOCIETY, 2 Turner St. Unit 8, Auburn, ME 04210-5978. Tel.: 207-784-0586.
E-mail: androhs@myfairpoint.net

Web Site: www.rootsweb.ancestry.com/~meandrhs
Founded: 1923.
Congressional District: 2
Key Personnel: Pres., David C. Young; Membership Sec., Bruce A. Hall; Treas., Susan F. Sturgis.
Personnel Profile: Part-Time Paid 1; Part-Time Volunteers 4.
Governing Authority: society. Tax-exempt.
Institution Type/Description: Historical Society Museum.
Collections: pioneer items; early settlers; Indian artifacts; dishes; furniture; clothing; small household items; photographs; maps; manuscripts; diaries; early documents; local history & genealogy.
Research Fields: local family & town history.
Facilities: 3,900-vol. library of local, county & Maine history books.
Activities: guided & self guided tours; lectures; permanent exhibitions. Museum Sponsors: annual dinner.
Publications: newsletter of the Androscoggin Historical Society Inc.; Androscoggin History.
Hours & Admission Prices: Call for hours. No charge; donations accepted. Closed national holidays. &
Attendance: 200 (accurate)
Membership: Individual $20; Family $30; Life $200. Corporate Patron I $50; Corporate Patron II $100; Corporate Sponsor $150; Corporate Sustaining $250.

Augusta

BLAINE HOUSE, 192 State St., Augusta, ME 04330-6406. Tel.: 207-624-7500.
Congressional District: 1
Key Personnel: Dir., Paula Benoit.
Personnel Profile: Full-Time Paid 4; Part-Time Paid 3.
Governing Authority: state.
Institution Type/Description: Historic House: 1830-1833 Blaine House, governor's mansion of Maine.
Collections: furnishings; silver from the battleships USS Maine 1895-98, 1905-22.
Facilities: 300-vol. private library of James G. Blaine.
Activities: guided tours.
Publications: The Blaine House Brochure.
Hours & Admission Prices: Tues.-Thurs. 2-4; tours on the half-hour. No charge. Closed national holidays. &
Attendance: 10,500 (accurate)

CHILDREN'S DISCOVERY MUSEUM, 171 Capitol St., Ste. 2, Augusta, ME 04330-4615. Tel.: 207-622-2209.
E-mail: info@childrensdiscoverymuseum.org
Web Site: www.childrensdiscoverymuseum.org
Founded: 1992.
Congressional District: 1
Key Personnel: Dir. & Museum Shop Mgr., Carne Arsenault; Chm. (V) & Pres. (V), Melissa Merfeld.
Personnel Profile: Full-Time Paid 1; Part-Time Paid 5; Part-Time Volunteers 4.
Governing Authority: Tax-exempt.
Institution Type/Description: Children's Museum.
Collections: hands-on exhibits.
Hours & Admission Prices: Tues.-Thurs. 10-4, Fri.-Sat. 10-5, Sun. 11-4. Children $5.50, adults $4.50; infants under 12 months no charge. &
Attendance: 25,000 (estimated)
Membership: Adult & Child $55; Family $75; Reciprocal to ACM & ASTC $125.

KENNEBEC HISTORICAL SOCIETY, 107 Winthrop St., Augusta, ME 04332. Mailing Address: P.O. Box 5582, Augusta, ME 04332-5582. Tel.: 207-622-7718. Fax: 207-622-7718.
E-mail: mail@kennebechistorical.org
Web Site: www.kennebechistorical.org
Institution Type/Description: Historical Society Museum.
Collections: local history & culture; photographs; manuscripts; period furnishings.
Activities: special events.
Hours & Admission Prices: Wed.-Fri. 10-2.

✱　**MAINE STATE MUSEUM, (M),** State House Complex, 230 State St., Augusta, ME 04333-0083. Mailing Address: 83 State House Station, Augusta, ME 04333-0083. Tel.: 207-287-2301. Fax: 207-287-6633. TTY: 711.
E-mail: maine.museum@maine.gov
Web Site: www.mainestatemuseum.org
Founded: 1837.

Congressional District: 1
Key Personnel: Dir., Bernard Fishman; Deputy Dir., Sheila McDonald; Chm. (V), Charles J. Micoleau; Chief Scientist, David Work; Cur. Photography, Fine Arts & Archives, Deanna Bonner Ganter; Art Dir., Don Bassett; Chief Cur. History & Decorative Arts, Laurie LaBar; Cur. Historical Collections, Kate McBrien; Registrar, Paula Work; Chief Educator, Joanna Torow; Chief Archaeologist, Bruce Bourque; Museum Store Mgr., Michelle Lagueux.
Personnel Profile: Full-Time Paid 15; Part-Time Paid 11; Part-Time Volunteers 90; Interns 5.
Governing Authority: state; independent state agency. Subsidiary Institution: Friends of the Maine State Museum. Tax-exempt.
Institution Type/Description: General Museum.
Collections: history; natural history; anthropology; ethnology; geology; marine; mineralogy; science; technology; art.
Major Exhibits: The Benjamin House Painted Murals, 2/11-2/16; Treasures of Mt. Mica, 6/11-6/16; Maine Voices from the Civil War, 6/13-6/15.
Research Fields: Maine history; regional natural history; regional archaeology; paleontology.
Facilities: 2,600-vol. library of research materials available for use on premises; conservation center. Items and books relating to Maine and the region for sale.
Activities: guided tours; lectures; formally organized education programs for children, families & undergraduate college students; inter-museum loan; permanent & temporary exhibitions; school loan service.
Publications: quarterly newsletter, Broadside; books, pamphlets & reports relating to collections.
Hours & Admission Prices: Tues.-Fri. 9-5, Sat. 10-4. Adults $2; discounts to AAM members; members no charge. Closed holidays. &
Attendance: 49,023 (accurate)
Membership: Senior Citizen & Student $25; Individual $30; Family & Household $40; Supporting $75; Sustaining $125; Contributing & Corporate $250; Curator's Circle $500; Director's Circle $1,000.

OLD FORT WESTERN, 16 Cony St., Augusta, ME 04330-5200. Tel.: 207-626-2385. Fax: 207-626-2304.
E-mail: oldfort@oldfortwestern.org
Web Site: www.oldfortwestern.org
Founded: 1922.
Congressional District: 1
Key Personnel: Dir. & Cur. Collections, Linda J. Novak; Chm., Richard Freeman.
Personnel Profile: Full-Time Paid 1; Part-Time Paid 15; Part-Time Volunteers 25.
Governing Authority: board of trustees. Parent Institution: City of Augusta. Tax-exempt.
Institution Type/Description: Historic House.
Collections: recreation of fort complex with 1922 blockhouses; 1754 main house with reproduction blockhouses & picket-work; watchboxes & palisade; 1754-1810 military, store & residential artifacts.
Research Fields: local history as it relates to the Kennebec Valley & New England, with emphasis on the 18th century.
Facilities: Gift items for sale.
Activities: tours; special events; demonstrations; lectures; pre-scheduled school & group programs.
Publications: occasional publications on research topics; booklet series on local history subjects; newsletter.
Hours & Admission Prices: Memorial Day to Labor Day daily 1-4; Sept.-Columbus Day Sat.-Sun. 1-4; other times by appointment. Adults $6, children 6-16 $4; discounts to AAM members, active military & veterans; children under 6 & members no charge. &
Attendance: 13,000 (accurate)
Membership: The Beverly Hewins Society $25-$99; The Robert Hotelling Society $100-$299; Mary Maher McCarthy Society $300-$499; The Gannett Society $500 & up.

Bangor

BANGOR MUSEUM AND HISTORY CENTER, 159 Union St., Bangor, ME 04401-6147. Tel.: 207-942-1900. Fax: 207-942-1910.
E-mail: info@bangormuseum.org
Web Site: www.bangormuseum.org
Formerly: Thomas A. Hill Historical House and Civil War Museum
Founded: 1864.
Congressional District: 2
Key Personnel: Exec. Dir., Jennifer Pictou; Pres. (V), Michael Aube; Dir. Museum Operations, Dana Lippitt.
Personnel Profile: Full-Time Paid 2; Part-Time Paid 1; Part-Time Volunteers 10; Interns 1.
Governing Authority: society. Parent Institution: Bangor Historical Society. Tax-exempt: 501(c)(3).

Institution Type/Description: History Museum.
Collections: restored 19th-century rooms; local history; portraits; manuscripts; photographs; Civil War artifacts. Historic House: Thomas A. Hill House & Civil War Museum.
Research Fields: history of Penobscot River area.
Facilities: 1,200-vol. library of Maine history.
Activities: lectures; permanent & temporary exhibitions; Mt. Hope Cemetery tour; candlelight ghost walks.
Hours & Admission Prices: June-Sept. Tues.-Fri. 10-4; other times by appointment. Adults $5, senior citizens $4; discounts to students, AAA, AAM & NEMA members; children & members no charge. Closed national holidays.
Attendance: 2,800 (accurate)
Membership: Friend $25; Household $50; River Driver $100; Master Mariner $250; Lumber Baron $500; Queen City Club $1,000 & up.

BANGOR POLICE MUSEUM, 240 Main, Bangor, ME 04401. Mailing Address: 15 Hudson Rd., Alton, ME 04468. Tel.: 207-947-7384.
Founded: 1978.
Governing Authority: Parent Institution: Bangor Police Dept.
Institution Type/Description: Police History Museum.
Collections: police department history from 1700s to present; personal artifacts; uniforms; photographs; newspaper clippings; mobile one-person jail.
Hours & Admission Prices: Mon.-Fri. 8-5.
Attendance: 2,000 (estimated)

COLE LAND TRANSPORTATION MUSEUM, 405 Perry Rd., Bangor, ME 04401-6725. Mailing Address: 359 Perry Rd., Bangor, ME 04401-6723. Tel.: 207-990-3600. Fax: 207-990-2653. Facebook: Cole Land Transportation Museum.
E-mail: mail@colemuseum.com
Web Site: www.colemuseum.org
Founded: 1990.
Congressional District: 2
Key Personnel: Chm. (V) & Pres. (V), Garret E. Cole; Projects Mgr., Christopher Thorne; Museum Shop Mgr., Pat Trice.
Personnel Profile: Full-Time Paid 2; Part-Time Paid 6; Part-Time Volunteers 80.
Governing Authority: nonprofit organization. Tax-exempt.
Institution Type/Description: Transportation Museum & the WWII, Korean & Vietnam Veterans Memorial.
Collections: period cars, trains, buses; military artifacts, equipment & vehicles.
Facilities: Books & pamphlets related to transportation & museum-related items for sale.
Activities: docent program; films; guided tours; participatory exhibits; TV programs.
Publications: books, As A Practical Matter: A Biography of Galen Cole, Allie Cole: A Maine Pioneer, The Cole Company: Started at Enfield Station in Maine, Allie Cole: The Man That Maine Made, Lest We Forget: A Pictorial History of Maine's Veterans, 1861-1995, The Collection - picture book of the major pieces in the museum; Quiet Courage - Stories of Maine Veterans.
Hours & Admission Prices: May-Nov. 11 daily 9-5. Adults $7, senior citizens $5; discounts to AAA members; students & children under 19 no charge. &
Attendance: 17,410 (accurate)

HOSE 5 FIRE MUSEUM, 247 State St., Bangor, ME 04401-5418. Mailing Address: P.O. Box 25, Bangor, ME 04401. Tel.: 207-945-3229.
Web Site: www.bangormaine.gov/cs_ps_hose5museum.php
Founded: 1994.
Institution Type/Description: Fire Museum.
Collections: fire fighting artifacts; restored fire trucks including 1930 McCann Pumper, a 1917 Garford Pumper, a 1952 Mack LS Pumper, a 1947 Jeep, a 1939 Seagraves Pumper, Bangor's old Engine #2; fire gear; hand & breathing apparatus; alarm boxes; wooden water mains; photographs.
Activities: tours.
Hours & Admission Prices: Call for hours; tours by appointment. No charge; donations accepted.

MAINE AVIATION HISTORICAL SOCIETY - MAINE AIR MUSEUM, 98 Maine Ave., Bangor, ME 04401. Mailing Address: P.O. Box 2641, Bangor, ME 04401. Tel.: 207-941-6757.
Key Personnel: Dir., Charles W. Byrum.
Personnel Profile: Part-Time Volunteers 6.
Volunteer Hours: 210
Operating Expenses: 2,000

Operating Income: 2,500
Governing Authority: Tax-exempt.
Institution Type/Description: History Museum.
Collections: military history; local aviation history; civil, commercial, military, & recreational flying.
Major Exhibits: MAM Air Show (T), 6/14.
Publications: quarterly newsletter.
Hours & Admission Prices: Memorial Day to Labor Day Sat. 10-4, Sun. 12-4; call for additional hours. Adults $3, children $1; discounts for members & active duty military. &
Attendance: 300 (estimated)
Membership: Individual $25; Family $35; Life $500.

MAINE DISCOVERY MUSEUM, 74 Main St., Bangor, ME 04401-6304. Tel.: 207-262-7200.
E-mail: nparker@mainediscoverymuseum.org
Web Site: www.mainediscoverymuseum.org
Founded: 2001.
Congressional District: 2
Key Personnel: Exec. Dir., Niles Parker.
Governing Authority: private; nonprofit organization. Tax-exempt: 501(c)(3).
Institution Type/Description: Children's Museum.
Collections: hands-on exhibits.
Facilities: Museum-related items for sale.
Activities: educational programs.
Hours & Admission Prices: Winter: Tues.-Sat. 10-5:30, Sun. 12-5; Summer: call for extended hours; groups by appointment. Adults $7.50; discounts to groups; members and children one & under no charge. ACM reciprocal program & ASTC Passport program. &
Attendance: 65,000 (estimated)
Membership: You & Me $60; Grandparent $70; Household $90; Reciprocal $125.

UNIVERSITY OF MAINE MUSEUM OF ART, (M), 40 Harlow St., Bangor, ME 04401-5102. Tel.: 207-561-3350. Fax: 207-561-3351. Facebook: University of Maine Museum of Art.
Web Site: www.umma.umaine.edu
Founded: 1946.
Congressional District: 2
Key Personnel: Asst. Museum Coord. & Membership Mgr., Kathryn Jovanelli; Exhibits Preparator, Sean Flannigan, Museum Dir., George Kinghorn; Education Coord., Eva Wagner; Museum Technician, Aaron Pyle.
Personnel Profile: Full-Time Paid 5; Part-Time Paid 4; Part-Time Volunteers 3; Interns 8.
Operating Expenses: 380,000
Operating Income: 380,000
Governing Authority: state. Parent Institution: University of Maine, Orono. Tax-exempt: 170(b)(1)(A).
Institution Type/Description: Art Museum: housed in downtown Bangor's historic Norumbega Hall.
Collections: 18th to 20th-century American & European graphics & paintings; modern & historic Maine art; contemporary art.
Major Exhibits: Hannah Cole, 1/17/14-3/22/14; Piranesi to Picasso: Prints from the Permanent Collection, 1/17/14-3/22/14; Kenny Cole, 1/17/14-3/22/14; Amy Becler, 4/4/14-6/7/14; Jay Kelly, 4/4/14-6/17/14; Maya Brodsky, 6/20/14-9/20/14; Jillian Mayer, 6/20/14-9/20/14.
Facilities: classroom.
Activities: modern & contemporary art exhibitions; guided tours; gallery talks; films; education programs for community; museums by mail; traveling exhibition service.
Publications: exhibition catalogues.
Hours & Admission Prices: Mon.-Sat. 10-5. No charge; donations accepted. Closed major holidays. &
Attendance: 13,000 (accurate)
Membership: Educator $35; Individual $50; Dual & Household $60; Patron $100; Sponsor $250; Benefactor $500; Corporate $1,000.

Bar Harbor

ABBE MUSEUM, (M), 26 Mount Desert St. & Sieur de Monts Spring, Acadia National Park, Bar Harbor, ME 04609. Mailing Address: P.O. Box 286, Bar Harbor, ME 04609-0286. Tel.: 207-288-3519. Fax: 207-288-8979.
E-mail: info@abbemuseum.org
Web Site: www.abbemuseum.org
Founded: 1928.
Congressional District: 2
Key Personnel: C.E.O., Cinnamon Catlin-Legutko; Pres. Bd., Sandy Wilcox, Ph.D.; Vice Pres., Ann Cox-Halkett; Museum Shop Mgr., Allison Shank; Business Mgr., John Brown; Program Coord., Raney Bench; Collections Mgr., Julia Clark; Public Affairs, Hannah Whalen.
Personnel Profile: Full-Time Paid 7; Part-Time Paid 8; Part-Time Volunteers 6; Interns 3.
Governing Authority: board of trustees; nonprofit organization. Branch Museum: 26 Mount Desert St., Bar Harbor, ME 04609. Tel.: 207-288-3519. Tax-exempt.
Institution Type/Description: Archaeology, Anthropology & Ethnology Museum: located on site at Sieur de Monts Spring, within Acadia National Park and downtown Bar Harbor at 26 Mount Desert Street.
Collections: Maine Native American culture, history & archaeology; past & present Indian artifacts & handicrafts; basketry.
Major Exhibits: Twisted Path III, 2/14-12/14; Wabamaki Student Art Show, 4/14-10/14.
Research Fields: archaeology.
Facilities: 500-vol. library available for use within the museum by appointment. Bulletins & books related to artifacts of Mount Desert Island & the Frenchman Bay area for sale.
Activities: gallery talks; school programs; crafts classes & demonstrations; library; inter-museum loan, permanent & temporary exhibitions; archaeology field schools. Museum Sponsors: Native American Festival.
Publications: The Handycraft of The Modern Indians of Maine; Uses of Birchbark in the Northeast; Brief Description of Birchbark Canoe Building; Indian Games, Toys & Pastimes of Maine & The Maritime Indians; Island in Time; Dogs of the Northeastern Woodland Indians; The Indian Shell Heap: Archaeology of the Ruth Moore Site.
Hours & Admission Prices: Sieur de Monts: May-Oct. daily 10-5. Mount Desert St.: May-Oct. daily 10-5; Nov.-April Thurs.-Sat. 10-4. Adults $6, children $3; AAM & museum members no charge. &
Attendance: 23,741 (accurate)
Membership: Student $20; Individual $40; Household $65; Sweetgrass $125; Birch Bark $300; Brown Ash $500; Quillwork $1,500.

BAR HARBOR HISTORICAL SOCIETY, 33 Ledgelawn Ave., Bar Harbor, ME 04609-1303. Tel.: 207-288-0000 & 3807. Facebook: Bar Harbor Historical Society.
E-mail: bhhistorical@gwi.net
Web Site: barharborhistorical.org
Founded: 1946.
Congressional District: 2
Key Personnel: Pres. (V), Sherwood Carr; Cur., Debbie Dyer.
Personnel Profile: Part-Time Paid 1; Part-Time Volunteers 6.
Governing Authority: historical society. Tax-exempt.
Institution Type/Description: Local History Museum.
Collections: photographs; artifacts; maps; records; deeds; manuscripts; hotel registers; newspapers, 1881 to date; early business account books & records of '47 fire; scrapbooks; video cassette of early moving pictures & stereopticon views; microfilm of local newspapers 1850-present.
Major Exhibits: Civil War, 12/11-1/15.
Research Fields: early history of town & area.
Facilities: research material available.
Activities: meetings; programs.
Publications: quarterly newsletter.
Hours & Admission Prices: mid-June to mid-Oct. Mon.-Sat. 1-4. No charge; donations accepted. Closed holidays.
Attendance: 1,500 (accurate)
Membership: Regular $20; Patron $50.

BAR HARBOR WHALE MUSEUM, 52 West St., Bar Harbor, ME 04609-1858. Mailing Address: c/o Allied Whale, College of the Atlantic, 105 Eden St., Bar Harbor, ME 04609. Tel.: 207-288-0288. Fax: 207-288-3218.
E-mail: whalemuseum@coa.edu
Web Site: barharborwhalemuseum.org
Founded: 1991.
Key Personnel: Dir. & Cur., Toby Stephenson; Museum Shop Mgr., Mindy Viechnicki
Governing Authority: Parent Institution: College of the Atlantic. Tax-exempt.
Institution Type/Description: Natural History Museum.
Collections: Gulf of Maine sea life and history.
Research Fields: marine studies.
Activities: educational programs.
Hours & Admission Prices: June & Sept.-Oct. daily 10-8; July-Aug. daily 9-9. No charge; donations accepted. &
Attendance: 70,000 (accurate)

GEORGE B. DORR MUSEUM OF NATURAL HISTORY, 105 Eden St., College of the Atlantic, Bar Harbor, ME 04609-1136. Tel.: 207-288-5395. Fax: 207-288-2917.
E-mail: museum@coa.edu
Web Site: www.coa.edu/dorr-museum-microsite.htm
Formerly: The Natural History Museum
Founded: 1982.
Congressional District: 2
Key Personnel: College Pres., Darron Collins; Dir., Dr. Stephen Ressel; Program Dir., Dianne Clendaniel.
Personnel Profile: Part-Time Paid 3.
Governing Authority: nonprofit. Parent Institution: College of the Atlantic. Tax-exempt.
Institution Type/Description: Natural History Museum.
Collections: whale skeletons; taxidermy specimens; exhibits on relationships between people & the environment, mammals, birds & plants of Maine.
Research Fields: local natural history; exhibit preparation; design of interpretative programs.
Facilities: teaching area. Books & museum-related items for sale.
Activities: participatory programs for children & adults; lectures; slide shows; inter-museum loan, temporary & traveling exhibitions; summer field studies for children.
Publications: brochure; newsletter.
Hours & Admission Prices: Tues.-Sat. 10-5. No charge; donations accepted. &
Attendance: 10,000 (estimated)

MOUNT DESERT OCEANARIUM, 1351 State Rte. 3, Bar Harbor, ME 04609. Mailing Address: P.O. Box 696, Southwest Harbor, ME 04679. Tel.: 207-288-5005.
E-mail: theoceanarium@earthlink.com
Web Site: theoceanarium.com
Founded: 1972.
Congressional District: 2
Key Personnel: C.E.O. & Co Dir., David K. Mills; Co Dir. & Museum Shop Mgr., Audrey S. Mills.
Personnel Profile: Full-Time Paid 3; Part-Time Paid 6; Interns 4.
Institution Type/Description: Oceanarium.
Collections: touch tank; lobster room; over 20 tanks of sea life from the coast of Maine; scallop tank; whale exhibit; sea exhibit including tides, waves, shells, sea salts, weather, survival, seaweeds, seagulls; fishing gallery; replica of wheelhouse; fishing gear diorama; lobster hatching process; harbor seals program; live salt marsh tours.
Research Fields: lobsters.
Facilities: aquarium. Books & sea related items for sale.
Activities: guided tours; lectures; organized education programs for children & college students affiliated with Kalamazoo College, Penn State University, Southampton College, Gordon College or Eastern Nazarene College; participatory exhibits; lobster & fishing programs. Museum Sponsors: unusual seafood festivals; Safety at Sea programs.
Publications: annual newsletter.
Hours & Admission Prices: Call for hours & admission prices. &

SIEUR DE MONTS SPRINGS NATURE CENTER, Acadia National Park, Rte. 233, Eagle Lake Rd., Bar Harbor, ME 04609. Mailing Address: P.O. Box 177, Bar Harbor, ME 04609-0177. Tel.: 207-288-3338. Fax: 207-288-8813.
E-mail: acadia_information@nps.gov
Founded: 1916.
Congressional District: 2
Key Personnel: Museum Cur., John McDade; Supt., Sheridan Steele; Museum Shop Mgr., Ann Cummings.
Personnel Profile: Full-Time Paid 1; Part-Time Paid 2; Part-Time Volunteers 1.
Governing Authority: federal. Parent Institution: National Park Service. Subsidiary Institution: Acadia National Park. Tax-exempt.
Institution Type/Description: Nature Center & Botanical Gardens.
Collections: Acadia National Park & Mount Desert Island natural history specimens; William H. Proctor invertebrate collection; Ralph H. Long ornithological slides; Harold White Odonata collection.
Research Fields: natural history of Acadia National Park.
Facilities: Sieur de Monts Springs Nature Center; Wild Gardens of Acadia.
Activities: permanent & temporary exhibitions.
Hours & Admission Prices: May call for hours; June-Aug. 9-5; Sept. to early Oct. 9-4. No charge. &
Attendance: 65,500 (estimated)

WILLIAM OTIS SAWTELLE COLLECTIONS AND RESEARCH CENTER, Acadia National Park, 20 McFarlund Hill Dr., Bar Harbor, ME 04609. Mailing Address: Acadia National Park, P.O. Box 177, Bar Harbor, ME 04609-0177. Tel.: 207-288-8729. Fax: 207-288-8709.
E-mail: rebecca_cole-will@nps.gov
Web Site: www.nps.gov/acad/historyculture/collections.htm
Founded: 1916.
Congressional District: 2
Key Personnel: Supt., Sheridan Steele.
Personnel Profile: Full-Time Paid 1; Part-Time Volunteers 5.
Governing Authority: federal government. Parent Institution: National Park Service. Subsidiary Institution: Acadia National Park. Tax-exempt.
Institution Type/Description: Research Center.
Collections: 1.5 million artifacts; archival documents, natural history specimens, & archeological materials pertaining to Acadia National Park; Saint Croix Island, International Historic Site; Town of Cranberry Isles; Naval Security Group Activity, Winter Harbor; Carroll family; Mount Desert Island; Maine; New France (Acadia).
Research Fields: genealogy; New France; History of Acadia National Park; natural history of Acadia National Park; Cranberry Isles; Saint Croix Island; archaeological; NSGA; Schoodic; Winter Harbor.
Facilities: collections & research center.
Activities: research.
Hours & Admission Prices: Tues.-Fri. 8:30-3:30 by appointment only. No charge. &
Attendance: 70 (accurate)

Bath

* **MAINE MARITIME MUSEUM, (M),** 243 Washington St., Bath, ME 04530-1638. Tel.: 207-443-1316. Fax: 207-443-1665.
E-mail: reservations@maritimeme.org
Web Site: mainemaritimemuseum.org
Founded: 1963.
Congressional District: 1
Key Personnel: Exec. Dir., Amy Lent; Dir. Finance, Jackie Berry; Cur., Dir. Library & Archivist, Nathan Lipfert; Cur. Exhibits, Chris Hall; Dir. Public Programs, Jason Morin; Volunteer Coord., Ann Harrison; Education Coord., James Nelson.
Personnel Profile: Full-Time Paid 12; Part-Time Paid 9; Part-Time Volunteers 200; Interns 2.
Governing Authority: nonprofit organization. Tax-exempt: 501(c)(3).
Institution Type/Description: Maritime Museum & Historic Shipyard.
Collections: Maine maritime history: paintings; decorative & folk arts; photographs; ship plans; charts; maps; shipping & shipbuilding records; models; half models; tools; instruments; trade goods; seamen's possessions; small boats; Percy & Small shipyard historic structures; small craft center. Historic House: c.1880 Donnell House.
Research Fields: Maine maritime history; Percy & Small 19th-century shipyard with 5 original buildings.
Facilities: 10,000-vol. library on Maine & maritime history; Percy & Small 19th-century shipyard; Maine Coast & Lobstering Exhibit building. Museum-related items for sale.
Activities: tours; permanent & temporary exhibitions; school programs; visiting vessels & tall ships; annual symposium on maritime history; evening lecture & film series; shipyard crafts demonstrations; waterfront activities & Kennebec River cruises; volunteer program; wooden boat building classes.
Publications: quarterly newsletter; books, Maritime History of Bath & the Kennebec River Region; Half-hull Modeling, Lobstering & the Maine Coast; The Skolfields & Their Ships; The Tancook Whalers; The Wessaweskeag Thorndikes; boat plans; The Pattens of Bath; A Shipyard in Maine: Percy & Small and the Great Schooners; A Doryman's Day; Snow Squall.
Hours & Admission Prices: Daily 9:30-5. Adults $12, seniors $11, children 4-17 $9; discounts to groups, AAM, ICOM & CAMM members; children under 4, museum members & staff no charge. Closed New Year's Day; Thanksgiving & Christmas. &
Attendance: 40,000 (accurate)
Membership: Individual $35; Family $65; Sustaining $125; Shipwright $500; Downeaster $1,000.

Belfast

BELFAST HISTORICAL SOCIETY AND MUSEUM, 10 Market St., Belfast, ME 04915-6555. Tel.: 207-338-9229.
E-mail: info@belfastmuseum.org
Web Site: belfastmuseum.org
Founded: 1955.

Institution Type/Description: Historical Society Museum.
Collections: local maritime history; Percy Sanborn paintings; photographs; postcards; scrapbooks; maps.
Hours & Admission Prices: late June to Labor Day Tues.-Sat. 11-4. No charge; donations accepted.

Belmont

GREENE PLANTATION HISTORICAL SOCIETY, 169 Howard Rd., Belmont, ME 04952. Tel.: 207-342-5208.
E-mail: mareshme@fairpoint.net
Founded: 1989.
Congressional District: 43
Key Personnel: Dir. & Archivist, Isabel Morse Maresh; Pres., R. Lenfest; Treas., Mary J. Smith
Governing Authority: private; nonprofit organization. Tax-exempt: 501(c)(3).
Institution Type/Description: Historical Society Museum: housed in a former one-room schoolhouse; built in 1908. Listed on the National Register of Historic Places.
Collections: local history & culture; period artifacts; photographs.
Activities: Annual Event: Remembrance Days.
Hours & Admission Prices: By appointment. No charge; donations accepted.
Attendance: 75 (estimated)
Membership: Adults $5.

Bethel

BETHEL HISTORICAL SOCIETY, (M), 10 Broad St., Bethel, ME 04217-0012. Mailing Address: P.O. Box 12, Bethel, ME 04217-0012. Tel.: 207-824-2908; 800-824-2910. Fax: 207-824-0882.
E-mail: info@bethelhistorical.org
Web Site: www.bethelhistorical.org
Formerly: Bethel Historical Society Regional History Center
Founded: 1966.
Congressional District: 2
Key Personnel: Pres. (V) & Chm. (V), William D. Andrews; Exec. Dir., Randall H. Bennett; Registrar, Jacalyn Bell; Museum Shop Mgr., Danna B. Nickerson.
Personnel Profile: Full-Time Paid 1, Part-Time Paid 2, Part-Time Volunteers 100; Interns 1.
Governing Authority: society; nonprofit organization. Parent Institution: Bethel Historical Society. Tax-exempt: 501(c)(3).
Institution Type/Description: Regional History Center: housed in Robinson House; built in 1821.
Collections: early & mid-19th century furnishings; archival material relating to Oxford County & the White Mountains, Northern New England. Historic Building: 1813 federal style house.
Research Fields: Oxford County life; the White Mountains, Northern New England.
Facilities: 3,000-vol. library of books, pamphlets & journals relating to Oxford Co., Maine history & the White Mountains; technical library & research room available only on premises; 100-seat auditorium. Museum-related items for sale.
Activities: guided tours; lectures; films; gallery talks; formally organized educational programs; permanent & temporary exhibitions; training programs for professional museum workers; docent program; craft demonstrations; conferences; special events.
Publications: The Courier; The Broad Street Herald.
Hours & Admission Prices: Jan.-April Tues.-Thurs. 10-4; May-June & Sept.-Dec. Tues.-Fri. 10-4; July-Aug. Tues.-Fri. 10-4, Sat. 1-4. Adults $3; discounts to AAM members; members no charge. &
Attendance: 5,000 (estimated)
Membership: Senior, Student & Teacher $10; Individual $20; Family $40; Benefactor $150; Life $300; Life Couple $400.

DR. MOSES MASON HOUSE, 14 Broad St., Bethel, ME 04217. Mailing Address: P.O. Box 12, Bethel, ME 04217. Tel.: 207-824-2908; 800-824-2910.
Institution Type/Description: Historical Society Museum: housed in the former home of Dr. Moses Mason; built in 1813.
Collections: Dr. Mason's life & career; personal artifacts; period furnishings; photographs.
Activities: Annual Event: Christmas with the Masons in December.
Hours & Admission Prices: July-Aug. Tues.-Sun. 1-4; Sept.-June by appointment. Adults $3, children 6-12 $1.50; member no charge.

MAINE MINERAL AND GEM MUSEUM, (M), 99 Main St., Bethel, ME 04217. Mailing Address: P.O. Box 500, Bethel, ME 04217. Tel.: 207-824-3036.
Web Site: mainemineralmuseum.com
Institution Type/Description: Mineral & Gem Museum.
Collections: minerals; gemstones; mining history; photographs; meteorites; fossils.
Hours & Admission Prices: Call for hours.

Biddeford

BIDDEFORD HISTORICAL SOCIETY, McArthur Library, 270 Main St., Biddeford, ME 04005. Mailing Address: P.O. Box 200, Biddeford, ME 04005-0200.
Institution Type/Description: Historical Society Museum.
Collections: local history & culture; city records from 1628 to 1932; personal diaries; scrapbooks.
Activities: society meetings.
Hours & Admission Prices: Call for hours.

Bingham

OLD CANADA ROAD HISTORICAL SOCIETY, 16 Sidney St., Bingham, ME 04920. Mailing Address: P.O. Box 742, Bingham, ME 04920. Tel.: 207-672-3440.
E-mail: ocrhs@oldcanadaroad.org
Web Site: www.oldcanadaroad.org
Founded: 2001.
Institution Type/Description: Historical Society Museum.
Collections: local history & culture; photographs; period artifacts.
Research Fields: Upper Kennebec Valley, ME history & culture which includes Bingham, Moscow, Caratunk, The Folks, Concord & Pleasant Ridge.
Hours & Admission Prices: Fri. 1-5, Sat. 11-4. No charge; donations accepted.

Blue Hill

BLUE HILL HISTORICAL SOCIETY, Water St., Blue Hill, ME 04614. Mailing Address: P.O. Box 710, Blue Hill, ME 04614-0710.
Web Site: www.bluehillhistory.org
Founded: 1902.
Institution Type/Description: Historical Society Museum: housed in Holt House, built in 1815.
Collections: period furniture & clothing; photographs; records & memorabilia from the 18th-20th centuries.
Hours & Admission Prices: July-Sept. Tues.-Fri. 1-4, Sat. 11-2.

THE JONATHAN FISHER MEMORIAL, INC., 44 Mines Rd., Blue Hill, ME 04614. Mailing Address: P.O. Box 537, Blue Hill, ME 04614-0537. Tel.: 207-374-2459. Fax: 207-374-5082.
E-mail: info@jonathanfisherhouse.org
Web Site: www.jonathanfisherhouse.org
Formerly: Parson Fisher House
Founded: 1954.
Congressional District: 2
Key Personnel: Pres. (V), Brad Emerson
Governing Authority: nonprofit corporation; Jonathan Fisher Memorial, Inc. Tax-exempt.
Institution Type/Description: Historic House: 1814 Jonathan Fisher House.
Collections: paintings; woodcuts; art work & furniture made by Jonathan Fisher; diaries, journals & books; manuscripts.
Facilities: homestead of Jonathan Fisher.
Activities: guided tours; permanent exhibitions.
Publications: book, Scripture Animals by Jonathan Fisher; Jonathan Fisher-Maine Parson by Mary Ellen Chase.
Hours & Admission Prices: July 5-Sept. 7 Wed.-Sat. 1-4; Sept. 13-Oct. 19 Fri.-Sat. 1-4; other times by appointment. No charge; donations accepted. &
Attendance: 350 (estimated)
Membership: Annual $30; Sustaining $50; Contributing $100; Life $1,000.

Boothbay

BOOTHBAY RAILWAY VILLAGE, (M), 586 Wiscasset Rd., Boothbay, ME 04537. Mailing Address: P.O. Box 123, Boothbay, ME 04537-0123. Tel.: 207-633-4727. Fax: 207-633-4733 (call first).
E-mail: staff@railwayvillage.org
Web Site: www.railwayvillage.org

Founded: 1965.
Congressional District: 1
Key Personnel: Dir., Robert Ryan; Sec. & Museum Shop Mgr., Maureen H. Stormont.
Personnel Profile: Full-Time Paid 7; Part-Time Paid 5; Part-Time Volunteers 6.
Governing Authority: nonprofit organization. Tax-exempt.
Institution Type/Description: Transportation Museum.
Collections: antique auto display; railroad memorabilia; transportation; The Thorndike Railroad Station; Freeport Station; period engines; turn-of-the-century displays; vintage steam operated narrow gauge locomotives; period vehicles.
Facilities: Early New England gifts & books on period autos & railroads for sale.
Activities: temporary exhibitions. Museum Sponsors: Steam train rides; special events.
Publications: newsletter, The Village Dispatch.
Hours & Admission Prices: Memorial Day to mid-June Sat.-Sun. 9:30-5; mid-June to mid-Oct. daily 9:30-5. Adults $10, children $5; members no charge. &

Attendance: 25,000 (accurate)
Membership: Individual $50; Family & Grandparent $60; Patron $85; Supporting $125.

Boothbay Harbor

BOOTHBAY REGION ART FOUNDATION, INC., One Townsend Ave., Boothbay Harbor, ME 04538-1765. Mailing Address: P.O. Box 124, Boothbay Harbor, ME 04538-0124. Tel.: 207-633-2703.
E-mail: braf@boothbayartists.org
Web Site: www.boothbayartists.org
Founded: 1964.
Congressional District: 1
Key Personnel: Pres., Jennifer Litchfield.
Personnel Profile: Full-Time Paid 1; Part-Time Volunteers 30.
Governing Authority: nonprofit; board of trustees. Tax-exempt.
Institution Type/Description: Art Gallery.
Collections: local artists.
Activities: art exhibits; school art show.
Hours & Admission Prices: Mon.-Sat. 10-5, Sun. 1-5. No charge; donations suggested.
Attendance: 10,000 (estimated)
Membership: Individual $45; Contributing $50; Exhibiting Artist $65; Patron $100 & up.

BOOTHBAY REGION HISTORICAL SOCIETY, 72 Oak St., Boothbay Harbor, ME 04538. Mailing Address: P.O. Box 272, Boothbay Harbor, ME 04538-0272. Tel.: 207-633-0820.
E-mail: brhs@gwi.net
Web Site: www.boothbayhistorical.org
Founded: 1967.
Key Personnel: Dir., Barbara Rumsey.
Personnel Profile: Part-Time Paid 2; Part-Time Volunteers 50.
Governing Authority: Tax-exempt.
Institution Type/Description: Historical Society Museum: house built in 1874.
Collections: Reed family history & furnishings; personal artifacts; photographs; archives; genealogical records of local families.
Major Exhibits: Boothbay Goes to War 1861-1865, 12/13-5/14.
Research Fields: local history.
Facilities: archives.
Activities: talks; open houses; periodical articles.
Publications: biannual newsletter.
Hours & Admission Prices: Thurs.-Sat. 10-2. No charge; donations accepted.
Attendance: 1,500 (estimated)
Membership: $15; $25; $50; $150.

Bradley

MAINE FOREST AND LOGGING MUSEUM, 686 Government Rd., Bradley, ME 04411. Mailing Address: P.O. Box 104, Bradley, ME 04411-0104. Tel.: 207-974-6278. Facebook: Maine Forest and Logging Museum.
E-mail: info@leonardsmills.com
Web Site: www.leonardsmills.com
Founded: 1960.
Key Personnel: Pres., Anette Rodrigues.
Governing Authority: Tax-exempt.
Institution Type/Description: Forestry Museum.
Collections: late 18th century life & history; period furnishing & artifacts; photographs; wood-related equipment.

Facilities: water-powered sawmill; blacksmith shop; barn; cabin; hovel; trapper's cabin; machinery buildings.
Activities: tours; educational programs; mill & daily life demonstrations. Annual Event: Living History Days.
Hours & Admission Prices: April-Oct. by appointment. Museum: no charge; donations accepted; Seasonal Events: admission varies; members no charge to most events.
Membership: Student $10; Individual $25; Family $40; Corporate $100; Sustaining $250; Benefactor $500; Oliver Leonard Club $1,000.

Brewer

BREWER HISTORICAL SOCIETY'S CLEWLEY MUSEUM, 199 Wilson St., Brewer, ME 04412-2029. Tel.: 207-989-6165.
Web Site: brewerhistoricalsociety.org
Founded: 1977.
Personnel Profile: Part-Time Volunteers 10.
Governing Authority: Tax-exempt: 501(c)(3).
Institution Type/Description: History Museum.
Collections: local history & culture; Maine's Underground Railroad; Civil War; photographs; Gen. Joshua Lawrence Chamberlain artifacts.
Facilities: resource center.
Activities: groups tours; educational programs for community & schools.
Publications: e-newsletter.
Hours & Admission Prices: Museum: by appointment. No charge; donations accepted.
Membership: Individual $20; Family $30.

Bridgton

BRIDGTON HISTORICAL SOCIETY, (M), 5 Gibbs Ave., Bridgton, ME 04009. Mailing Address: P.O. Box 44, Bridgton, ME 04009-0044. Tel.: 207-647-3699 (Gibbs Ave. museum) & 9954 (Narramissic).
E-mail: info@bridgtonhistory.org
Web Site: www.bridgtonhistory.org
Founded: 1953.
Congressional District: 1
Key Personnel: Pres. (V), Ned Allen.
Personnel Profile: Part-Time Paid 2; Part-Time Volunteers 25; Interns 1.
Governing Authority: nonprofit corporation. Subsidiary Institutions: Gibbs Avenue Museum, Tel. 207-647-3699; Narramissic, the Historic Peabody Fitch Farm, Ingalls Rd., South Bridgton, Tel. 207-647-9954. Tax-exempt: 501(c)(3).
Institution Type/Description: Historical Society Museum & Historic House Museum: housed in 1902 former firehouse.
Collections: artifacts; decorative arts; lumber & textile industries' mechanical implements; photographs; 18th- & 19th-century rural New England documents; narrow-gauge railway documents. Historic House: 1797 Narramissic.
Research Fields: genealogical; local architecture & industries; social history.
Facilities: library of local & state history, school records, maps, 1750s-present newspaper files & local archival records available for use by arrangement with the curator or librarian.
Activities: guided tours; films; craft demos & lectures.
Publications: book, Bridgton History: 1768-1968; Rediscover Brighton's Main Street.
Hours & Admission Prices: Call for hours & admission prices. &
Membership: Student & Retiree $10; Individual $15; Family $20; Contributing $30; Sustaining $50; Patron $100; Life $500.

THE RUFUS PORTER MUSEUM AND CULTURAL HERITAGE CENTER, 67 N. High St., Bridgton, ME 04009-1111. Mailing Address: P.O. Box 544, Bridgton, ME 04009-0544. Tel.: 207-647-2828.
E-mail: rufusportermuseum@myfairpoint.net
Web Site: www.rufusportermuseum.org
Founded: 2005.
Key Personnel: Pres., Nelle Ely; Dir., Diane Hoppe; Cur., Julie Lindberg.
Personnel Profile: Interns 2.
Governing Authority: Tax-exempt.
Institution Type/Description: History Museum.
Collections: paintings by Rufus Porter.
Hours & Admission Prices: June-Oct. Wed.-Sat. 12-4; other times by appointment. Adults $8, seniors & students $7; discounts to groups and AAM & ICOM members; children 12 & under no charge. &
Attendance: 350 (estimated)
Membership: Individual $25; Family $40; Lombard Circle $100; Farnsworth Circle $250; Perley Circle $500; Kitson Circle $1,000; Rufus Porter Circle $1,000 & up.

Brunswick

*** BOWDOIN COLLEGE MUSEUM OF ART, (M),** Walker Art Bldg., Brunswick, ME 04011. Mailing Address: 9400 College Station, Brunswick, ME 04011-8494. Tel.: 207-725-3275. Fax: 207-725--3762.
E-mail: artmuseum@bowdoin.edu
Web Site: www.bowdoin.edu/art-museum
Founded: 1811.
Congressional District: 1
Key Personnel: Assoc. Dir., Martina Duncan; Asst. Dir. & Museum Communications, Suzanne Bergeron; Curatorial Asst., Andrea Rosen; Post-Doctoral Curatorial, Andrew W. Mellon; Fellow, Sarah Montross; Registrar, Laura J. Latman; Cur., Joachim Homann; Asst. to Dir., Victoria Baldwin-Wilson; Tech. & Prep., Jose L. Ribas; Asst. Preparator, Jo Hluska; Asst. to Registrar, Michelle Henning; Museum Shop Mgr., Liza Nelson.
Personnel Profile: Full-Time Paid 9; Part-Time Paid 2; Part-Time Volunteers 10; Interns 1.
Governing Authority: college. Parent Institution: Bowdoin College. Tax-exempt: 101(6).
Institution Type/Description: Art Museum: housed in 1894, Walker Art Building, designed by Charles Follen McKim, located on the campus of Bowdoin College.
Collections: Assyrian sculpture; Greek & Roman antiquities; European & American paintings, prints, drawings, sculpture & decorative arts; Kress Study collection; Molinari Medals & Plaquettes; Winslow Homer memorabilia; Far Eastern ceramics; African & pre-Columbian sculpture.
Research Fields: American & European art; Winslow Homer; medals & plaquettes; antiquities.
Facilities: collections available for use by scholars. Catalogues of collections & special exhibitions, art books, postcards & posters for sale.
Activities: guided tours; lectures; gallery talks; formally organized education programs for adults & undergraduate college students; inter-museum loan, permanent, temporary & traveling exhibitions.
Publications: catalogues of the permanent collections & temporary exhibitions.
Hours & Admission Prices: Tues.-Wed. & Fri.-Sat. 10-5, Thurs. 10-8:30, Sun. 1-5. No charge; donations accepted. Closed holidays. ♿
Attendance: 45,000 (accurate)
Membership: Individual $50; Dual $75; Sponsor $150; Patron $500; Benefactor $1,000; Director's Circle $2,500; James Bowdoin III Round Table $5,000.

ICON CONTEMPORARY ART, 19 Mason St., Brunswick, ME 04011. Tel.: 207-725-8157.
Institution Type/Description: Art Gallery.
Collections: works by contemporary artists.
Activities: temporary artists.
Hours & Admission Prices: Mon.-Fri. 1-5, Sat. 1-4. No charge.

*** THE PEARY-MACMILLAN ARCTIC MUSEUM, (M),** Bowdoin College, 9500 College Station, Brunswick, ME 04011-8495. Tel.: 207-725-3416 & 3062. Fax: 207-725-3499. Facebook: Arctic Museum.
Web Site: www.bowdoin.edu/arctic-museum
Founded: 1967.
Congressional District: 1
Key Personnel: Dir., Dr. Susan A. Kaplan; Technician & Designer, David R. Maschino; Museum Outreach & Svcs. Coord., Amy Hawkes; Cur., Dr. Genevieve LeMoine; Sec. to Dir., Kristi Clifford; Asst. Cur., Anne Witty; Exhibit Tech., Steve Bunn.
Personnel Profile: Full-Time Paid 5; Part-Time Paid 2; Part-Time Volunteers 17; Interns 2.
Governing Authority: college. Parent Institution: Bowdoin College. Tax-exempt: 501(c)(3).
Institution Type/Description: College Museum.
Collections: arctic exploration, ecology, anthropology, archeology; prehistoric & historic Inuit (Eskimo) artifacts; contemporary Inuit arts & crafts; arctic flora & fauna; exploration equipment; nautical equipment; archives; photographs.
Major Exhibits: Spirits of Land, Air, and Water: Antler Carvings from the Robert & Judith Toll Collection, 4/10/13-2/23/14.
Research Fields: archeology; cultural anthropology; ecology of Arctic North America & Scandinavia.
Facilities: 3,000-vol. library pertaining to arctic exploration & research. Inuit arts & crafts, books & other related items for sale.
Activities: guided tours; lectures; films; docent program; organized education programs for children, undergraduate & graduate students affiliated with Bowdoin College.

Hours & Admission Prices: Tues.-Sat. 10-5, Sun. 2-5. No charge; donations accepted. Closed national holidays.
Attendance: 14,277 (accurate)

PEJEPSCOT HISTORICAL SOCIETY, 159 Park Row, Brunswick, ME 04011-2005. Tel.: 207-729-6606. Fax: 207-729-6012.
E-mail: director@pejepscothistorical.org
Web Site: www.pejepscothistorical.org
Founded: 1888.
Congressional District: 1
Key Personnel: Pres. (V), Ms. Marion Small.
Personnel Profile: Full-Time Paid 1; Part-Time Paid 2; Part-Time Volunteers 60; Interns 1.
Governing Authority: society. Branch Museums: Skolfield-Whittier House Museum, 161 Park Row, Brunswick; Joshua Chamberlain House Museum, 226 Maine St., Brunswick. Tax-exempt: 501(c)(3).
Institution Type/Description: Historical Society Museums.
Collections: regional history items; domestic artifacts; garments; early craft & industry items; local history archives; Joshua L. Chamberlain research collection. Historic Houses: 1858 Captain Alfred Skolfield-Whitter House; Joshua L. Chamberlain House; area genealogy.
Research Fields: local history; Maine; Civil War & Joshua L. Chamberlain; Victorian period; industrial history; Franco-American history; women's history.
Facilities: archive room.
Activities: lectures; permanent & temporary exhibitions; tours; slide show talks; school programs; community outings.
Publications: newsletter.
Hours & Admission Prices: Skolfield-Whittier House: Memorial Day to Columbus Day Thurs.-Sat. Tours 11 & 2. Adults $7.50, children $2.50; discounts to AAM members & groups; members & military in uniform no charge. Chamberlain House: Memorial Day to Columbus Day Tues.-Sat. 10-4. Adults $7.50, children $2.50; discounts to AAM members & groups; members & military in uniform no charge. Combination tickets to both houses available. Pejepscot Museum: Wed.-Sat. 10-4, call to confirm. No charge; donations accepted. Closed holidays. ♿
Attendance: 12,000 (accurate)
Membership: Individual $35; Family $50; Contributing $75 & up.

Bryant Pond

WOODSTOCK IIISTORICAL SOCIETY, 70 S. Main St., Bryant Pond, ME 04219-6424. Tel.: 207-665-2450.
Founded: 1979.
Key Personnel: Pres. (V), Olive Risko; Treas., Paul Billings; Cur., Larry Billings.
Personnel Profile: Full-Time Volunteers 1; Part-Time Volunteers 20.
Governing Authority: private; nonprofit. Tax-exempt: 501(c)(3).
Institution Type/Description: Historical Society Museum.
Collections: paintings & artifacts from the town of Woodstock, Maine.
Facilities: library of books pertaining to items of local interest. Museum-related items for sale.
Activities: guided tours. Annual Event: History Day in August.
Hours & Admission Prices: June-Sept. Sat. 1-4. No charge; donations accepted.
Attendance: 250 (estimated)
Membership: Annual $2; Life (over 54) $15; Life (under 55) $25.

Bucksport

BUCKSPORT HISTORICAL SOCIETY, INC., 379 Main St., Bucksport, ME 04416. Mailing Address: P.O. Box 798, Bucksport, ME 04416-0798. Tel.: 207-469-0924.
Founded: 1964.
Congressional District: 2
Key Personnel: Vice Pres., Mrs. Arthur M. Joost.
Governing Authority: nonprofit. Tax-exempt: 501(c)(3).
Institution Type/Description: General Museum: housed in 1874 Old Maine Central Railroad Station.
Collections: marine; military; naval; transportation; tools; toys; clothing; furniture; quilts; pictures; town history; newspapers; Admiral Perry collection; manuscripts.
Research Fields: local history.
Facilities: 200-vol. library of local history books & ship logs available for research by special request; reading room.
Activities: permanent & temporary exhibitions.
Publications: booklets, Did You Know?; Jonathan Buck of Bucksport; pamphlet, Bucksport Historical Society.
Hours & Admission Prices: July-Aug. Wed.-Fri. 1-4; other times by appointment. Admission $1; members no charge. ♿

Attendance: 350 (accurate)
Membership: Annual $5; Sustaining $10; Life $100.

NORTHEAST HISTORIC FILM, 85 Main St., Bucksport, ME 04416. Mailing Address: P.O. Box 900, Bucksport, ME 04416-0900. Tel.: 207-469-0924.
E-mail: nhf@oldfilm.org
Web Site: www.oldfilm.org
Founded: 1986.
Key Personnel: Exec. Dir., David S. Weiss.
Personnel Profile: Full-Time Paid 2; Part-Time Paid 5.
Governing Authority: Tax-exempt.
Institution Type/Description: Historic Building: housed in a 1916 cinema, The Alamo Theatre.
Collections: motion picture history; cultural history of movie-going; videos; projectors; cameras; editing equipment; posters; photographs; theater seats; tickets; advertisements.
Facilities: library; study center. Museum-related items for sale.
Activities: group tours; shows current Hollywood features; research. Museum Sponsors: Film Symposium in summer.
Hours & Admission Prices: Mon.-Fri. 9-4; call for additional hours. No charge; donations accepted. &
Attendance: 15,000 (estimated)
Membership: $20; $35; $50; $60; $100; $150; $250; $500; $1,000.

Burlington

STEWART M. LORD MEMORIAL HISTORICAL SOCIETY, INC., Rte. 188, Burlington, ME 04417. Mailing Address: P.O. Box 367, Howland, ME 04448-0367. Tel.: 207-732-3129.
E-mail: info@smlmhs.org
Web Site: www.smlmhs.org
Founded: 1968.
Key Personnel: Cur., Fern P. Cummings.
Personnel Profile: Part-Time Volunteers 10.
Governing Authority: society.
Institution Type/Description: History Museum.
Collections: pictures; documents; carriages; lumbering & household equipment of 1800s; Maine residents from 1800 to present; farming. Historical Buildings: 1850 general store, 1844 tavern; Page Building: wagons, surry boat; W.W. Building.
Research Fields: local history.
Activities: permanent & temporary exhibitions.
Publications: biannual newsletter.
Hours & Admission Prices: July-Labor Day Sun. 2-4; other times by appointment. No charge; donations accepted. &
Attendance: 145 (estimated)
Membership: Annual $5.

Calais

DR. JOB HOLMES COTTAGE AND MUSEUM - ST. CROIX HISTORICAL SOCIETY, 527 Main St., Calais, ME 04619. Tel.: 207-454-3061.
E-mail: schs@stcroixhistorical.org
Web Site: stcroixhistorical.com/holmescottage
Institution Type/Description: Historic House Museum: housed in the former home & office of Dr. Job Holmes from 1846-1864; built c.1790s.
Collections: local history; Dr. Job Holmes' life & career; medical instruments & equipment; personal artifacts; period furnishings.
Hours & Admission Prices: July-Aug. call for hours. No charge; donations accepted.

Cape Elizabeth

PORTLAND HEAD LIGHT MUSEUM, 1000 Shore Rd., Cape Elizabeth, ME 04107. Mailing Address: P.O. Box 6260, Cape Elizabeth, ME 04107. Tel.: 207-799-2661. Fax: 207-799-2800.
E-mail: cephl@aol.com
Web Site: www.portlandheadlight.com
Institution Type/Description: Historic Building: c.1790 lighthouse.
Collections: local history; lighthouse lenses.
Facilities: Museum-related items for sale.
Hours & Admission Prices: mid-April to May & Nov.-Dec. Sat.-Sun. 10-4; Memorial Day to Oct. daily 10-4. Adults $2, children 6-18 $1; children under 6 no charge. &

Caribou

THE NYLANDER MUSEUM OF NATURAL HISTORY, 657 Main St., Caribou, ME 04736-4431. Tel.: 207-493-4209. Fax: 207-498-3954.
E-mail: nylander@mainerr.com
Web Site: www.nylandermuseum.org
Founded: 1938.
Congressional District: 2
Key Personnel: Exec. Dir., Jeanie L. McGowan; Pres. Bd. Trustee, Deborah Nichols; Aide & Gift Shop Mgr., Liz Maifield.
Personnel Profile: Part-Time Paid 2; Part-Time Volunteers 15.
Governing Authority: municipal. Parent Institution: City of Caribou. Tax-exempt.
Institution Type/Description: Natural History Museum.
Collections: fossils; shells; mineralogy; geology; early man artifacts; mounted birds; mounted mammals of Maine; wetlands; Native American medicinal herb garden
Research Fields: geology; conchology; paleontology; artifacts of early man.
Facilities: resource library with books pertaining to native wild flowers, fresh water mollusks, fossils in the area & Northeastern Native Americans available for use in the museum by reservation only. Museum-related items for sale.
Activities: guided tours; permanent, temporary & traveling exhibitions; pre-arranged school programs throughout the year; traveling kits for science education programs.
Hours & Admission Prices: May -Oct. Mon.-Sat. 12:30-4:30; Nov.-April Wed. 9-5, Tues. & Thurs. by appointment. No charge; donations accepted. &
Attendance: 701 (accurate)
Membership: Friend $20; Patron $50; Donor $100.

THOMAS HERITAGE HOUSE, 444 Main St., Caribou, ME 04736. Mailing Address: P.O. Box 446, Caribou, ME 04736. Tel.: 207-496-3011. Fax: 207-493-3188.
Institution Type/Description: Historic House Museum.
Collections: local history & culture; period furnishings; personal artifacts; photographs.
Hours & Admission Prices: By appointment.

Casco

RAYMOND-CASCO HISTORICAL SOCIETY MUSEUM, Shadow Lane, Rte. 302, Casco, ME 04015. Mailing Address: P.O. Box 1055, Raymond, ME 04071. Tel.: 207-655-7672.
E-mail: info@raymondcascohistory.org
Web Site: raymondcascohistory.org
Governing Authority: Tax-exempt.
Institution Type/Description: Historical Society Museum.
Collections: local history & culture; period artifacts; photographs.
Hours & Admission Prices: Call for hours. No charge; donations accepted. &
Attendance: 800 (estimated)
Membership: Individual $20.

Castine

CASTINE SCIENTIFIC SOCIETY AKA WILSON MUSEUM, (M), 120 Perkins St., Castine, ME 04421. Mailing Address: P.O. Box 196, Castine, ME 04421-0196. Tel.: 207-326-9247. Fax: 207-326-9237.
E-mail: info@wilsonmuseum.org
Web Site: www.wilsonmuseum.org
Founded: 1921.
Congressional District: 2
Key Personnel: Dir., Patricia Hutchins; Administrative Asst., Debra Morehouse.
Personnel Profile: Full-Time Paid 3; Part-Time Paid 8; Part-Time Volunteers 20.
Governing Authority: nonprofit organization. Tax-exempt: 501(c)(3).
Institution Type/Description: History Museum.
Collections: Historic Buildings: 1921 Wilson Museum; 1763-83 John Perkins House: furnished with heirlooms & period pieces; blacksmith shop: working smithy.
Research Fields: local history; geology.
Facilities: library pertaining to anthropology, archaeology & local history for use by appointment; 6,000 sq. ft. exhibit space.
Activities: guided tours; blacksmith & living history demonstrations; lectures; workshops; permanent exhibitions.
Publications: newsletter, Wilson Museum Bulletin.
Hours & Admission Prices: Perkins House & Blacksmith Shop: July-Aug.

Wed. & Sun. 2-5. Perkins House: $5 per person; discounts to National Trust, NEMA Institutional staff, AAM & AAA members. Blacksmith Shop: no charge. Wilson Museum: May 27 to Sept. Mon.-Fri. 10-5, Sat.-Sun. 2-5. No charge; donations accepted.
Attendance: 4,000 (estimated)
Membership: Active $25; Family $40; Life $500.

Chebeague Island

MUSEUM OF CHEBEAGUE HISTORY, 137 South Rd., Chebeague Island, ME 04017-3100. Tel.: 207-846-5237.
E-mail: history@chebeague.net
Congressional District: 1
Governing Authority: Parent Institution: Chebeague Island Historical Society. Tax-exempt.
Institution Type/Description: History Museum: built in 1871.
Collections: local history & culture; photographs; furnishings; personal artifacts.
Research Fields: Chebeague Island history & genealogy.
Activities: research; permanent exhibitions; house tours; lectures.
Publications: biannual, The Sloops Log; annual newsletter, Chebeague Island.
Hours & Admission Prices: Tues.-Sun. 1-6. No charge; donations accepted.
Attendance: 1,200 (accurate)
Membership: Individual $5 & up.

Cherryfield

CHERRYFIELD NARRAGUAGUS HISTORICAL SOCIETY, 88 River Rd., Cherryfield, ME 04622. Mailing Address: P.O. Box 122, Cherryfield, ME 04622-0122. Tel.: 207-546-2076.
E-mail: info@cherryfieldhistorical.com
Web Site: www.cherryfieldhistorical.com
Institution Type/Description: Historical Society Museum.
Collections: local history; period furnishings; personal artifacts.
Hours & Admission Prices: Call for hours.

Columbia Falls

RUGGLES HOUSE SOCIETY, 1/4 mile off U.S. Rte. 1, 146 Main St., Columbia Falls, ME 04623. Mailing Address: 298 Tenan Lane, Cherryfield, ME 04622-4334. Tel.: 207-483-4637 & 546 7903. Facebook: The Ruggles House.
E-mail: etenan@ruggleshouse.org
Web Site: www.ruggleshouse.org
Founded: 1949.
Congressional District: 2
Key Personnel: Pres., Larry D. Smith; Sec., Ellen Tenan.
Personnel Profile: Part-Time Paid 3; Part-Time Volunteers 12.
Governing Authority: society; bd. dirs. Tax-exempt: 501(c)(3).
Institution Type/Description: Historic House Museum: 1818 Ruggles House.
Collections: genealogical records; photographs; maps; news files; glass; hand-carved woodwork; period furniture; children's toys & furniture; tools; farm; lumber; ship building.
Activities: guided tours.
Publications: leaflet; newsletter.
Hours & Admission Prices: June to mid-Oct. Mon.-Sat. 9:30-4:30, Sun. 12-4:30. Adults $5, children $2. Closed Independence Day; Labor Day.
Attendance: 1,800 (estimated)
Membership: $10 & up.

Corinth

CORINTH HISTORICAL SOCIETY, Old Grange Hall, 306 Main St., Corinth, ME 04427. Mailing Address: P.O. Box 541, Corinth, ME 04427-0541. Tel.: 207-285-7885. Facebook: Corinth Historical Society.
E-mail: tristrambryan@yahoo.com
Web Site: www.angelfire.com/me2/corinthhistorical
Founded: 1985.
Congressional District: 2
Key Personnel: Pres. & Cur., Betty LaForge.
Personnel Profile: Part-Time Volunteers 10.
Governing Authority: private; nonprofit organization. Tax-exempt: 501(c)(3).
Institution Type/Description: Historical Society Museum: built in 1907.
Collections: early plows & farm implements; period horse drawn fire wagon & hearse; trolley seats; period toys; school desks; military artifacts; WWI, WWII & Korean war uniforms; Civil War guns; period sewing machines, washing machines & kitchen items; quilts; handmade children's clothing & wedding dress; ladies' hats, haircombs, & fans.

Research Fields: Corinth history; genealogy of Corinth families.
Facilities: library. Museum-related items for sale.
Activities: Annual Events: 4th Grade Children's History Day for Corinth Grade School; Children's History Tour.
Publications: quarterly newsletter, Corinth Historical Society.
Hours & Admission Prices: June 4 to Sept. 3 Wed. 2-7; other times by appointment. No charge; donations accepted.
Attendance: 200 (estimated)
Membership: Individual $10; Family $15; Benevolent Donor $25; Lifetime $200; Benefactor $500.

Deer Isle

DEER ISLE-STONINGTON HISTORICAL SOCIETY, 416 Sunset Rd., Rte. 15A, Deer Isle, ME 04627. Mailing Address: P.O. Box 652, Deer Isle, ME 04627-0652. Tel.: 207-348-6400.
Web Site: www.dis-historicalsociety.org
Founded: 1959.
Congressional District: 2
Key Personnel: Pres. (V), Claudette O. Kydd.
Personnel Profile: Part-Time Volunteers 35.
Governing Authority: bd. of trustees; nonprofit. Tax-exempt 501(c)(3).
Institution Type/Description: History Museum.
Collections: preservation project; marine; archives; agriculture; costumes; manuscripts; Indian artifacts; basketry; genealogical. Historic House: 1830 Salome Sellers Home; new post & beam exhibit barn.
Research Fields: genealogy; local history; Americana; Native American artifacts.
Facilities: 350-vol. library of marine & historic books; family genealogies; reading room.
Activities: permanent & temporary exhibitions.
Publications: semi-annual newsletter; booklet, Salome Sellers; Images of America: Deer Isle and Stonington, Arcadia, 2004; Postcard History Series; Deer Isle and Stonington, Arcadia 2008.
Hours & Admission Prices: mid-June to mid-Sept. Wed.-Fri. 1-4; Oct.-May Wed. & Fri. 1-4 for research only. No charge; donations accepted.
Attendance: 400 (estimated)
Membership: Individual $15; Family $35; Supporting $50.

Dennysville

ACADEMY/VESTRY MUSEUM - DENNYS RIVER HISTORICAL SOCIETY, 115 Main St., Dennysville, ME 04628. Mailing Address: P.O. Box 11, Dennysville, ME 04628-0011. Tel.: 207-726-3905 & 5258. Facebook: Dennys River Historical Society.
E-mail: drhs@pioneerwireless.net
Founded: 1987.
Congressional District: 2
Key Personnel: Pres., Ronald A. Windhorst; Treas., Richard H. Hobart; Dir. Programs & Devel., Colin J.C. Windhorst; Cur. & Sec., Melinda Jaques.
Personnel Profile: Part-Time Volunteers 22.
Governing Authority: private; nonprofit organization. Parent Institution: Dennys River Historical Society. Subsidiary Institution: Lincoln Memorial Library. Tax-exempt: 501(c)(3).
Institution Type/Description: History Museum.
Collections: history of Eastern Maine from prehistoric times to present; Native American artifacts; late 18th-19th century village history; photographs; furniture; costumes; rural industry & transportation; family papers; archives.
Major Exhibits: Take Time to be Holy: Books of Worship at Dennys River, 5/14-9/14.
Research Fields: 19th century costumes; ; photographs; family history; handmade map collection of eastern Maine; architectural & building history in Eastern Maine; documenting family archival collections on Dennys River; rural industry & commerce; WWI in rural Maine.
Facilities: 150-vol. library; reading room; 1,000 sq. ft. exhibit space.
Activities: formal education programs; guided tours; lectures & seminars; school loan service; study clubs; temporary exhibitions. Annual Event: Summer Historical Tours.
Publications: occasional monographs; quarterly newsletter, Dennys River Historical Society Newsletter.
Hours & Admission Prices: Lincoln Memorial Library: Tues.-Fri. 1-4. Academy/Vestry Museum: Memorial Day to Columbus Day Sat. 1-4; other times by appointment. No charge; donations accepted. Closed national & state holidays.
Attendance: 245 (accurate)
Membership: Individual $15; Family $25.

Dexter

DEXTER HISTORICAL SOCIETY CAMPUS, 3 Water St. & 12 Church St., Dexter, ME 04930. Mailing Address: P.O. Box 481, Dexter, ME 04930-0481. Tel.: 207-924-5721.
E-mail: info@dexterhistoricalsociety.com
Web Site: dexterhistoricalsociety.com
Formerly: Dexter Historical Society Museum
Founded: 1966.
Congressional District: 2
Key Personnel: Dir. & Cur., Richard Whitney.
Personnel Profile: Part-Time Paid 1; Part-Time Volunteers 6; Interns 1.
Governing Authority: nonprofit. Tax-exempt.
Institution Type/Description: General Museum: housed in 1854 Grist Mill, located on the site 1818 canal.
Collections: artifacts of local historical interest; pictures; manuscripts; antiques; genealogy. Historic Buildings: 1854 Grist Mill, 1825 Millers House, 1845 one room schoolhouse; 1836 former Town Hall (Woolen Mill Office) now Abbott Museum.
Research Fields: local history & genealogy.
Facilities: 1,000-vol. library. Museum-related gift items for sale.
Activities: arts festivals; permanent exhibitions; seasonal public displays; outreach programs.
Publications: Dexter: Spirit of An Age; Our Neighborly Neighbors, Rural Dexter 1800-2000; Bubbles in the Sun; Growing Up In Dexter, Maine.
Hours & Admission Prices: mid-June to Labor Day Mon.-Fri. 10-4, Sat. 1-4, Sun. & holidays by appointment. Abbott Museum: Memorial Day to Columbus Day Mon.-Sat. 10-4; Oct.-May Wed.-Fri. 1-4, Sat. 10-4. No charge; donations accepted. Closed Independence Day. &
Attendance: 2,000 (estimated)
Membership: Annual $10.

Dover-Foxcroft

BLACKSMITH SHOP MUSEUM, 103 Dawes Rd., Dover-Foxcroft, ME 04426-3732. Tel.: 207-564-8618.
E-mail: dlockwood3@myfairpoint.net
Founded: 1963.
Key Personnel: Pres. (V), Mary Annis; Cur., Dave Lockwood; Treas., James Annis; Sec., Susan Burleigh
Governing Authority: society. Affiliated with Dover-Foxcroft Historical Society. Tax-exempt.
Institution Type/Description: Historic Building: 1863 Blacksmith Shop.
Collections: early blacksmith tools & equipment.
Research Fields: smithing in late 19th-century.
Activities: permanent exhibitions.
Publications: brochure.
Hours & Admission Prices: May-Oct. daily 8-8. No charge; donations accepted.
Attendance: 250 (estimated)
Membership: Individual $10.

OBSERVER BUILDING MUSEUM - DOVER-FOXCROFT HISTORICAL SOCIETY, Union Square, Dover-Foxcroft, ME 04426. Mailing Address: 28 Orchard Rd., Dove-Foxcroft, ME 04426. Tel.: 207-564-0820.
E-mail: mannis@myfairpoint.net
Web Site: rootsweb.com/~medfhs
Founded: 1963.
Congressional District: 2
Key Personnel: C.E.O., Chm. (V) & Pres (V), Mary Annis; Cur., Dennis Lyford; Cur., Dave Lockwood.
Personnel Profile: Part-Time Volunteers 12.
Volunteer Hours: 4,000
Operating Income: 10,000
Governing Authority: Parent Institution: Dover-Foxcroft Historical Society. Subsidiary Institution: Blacksmith Shop Museum, 100 Dawes Rd., Dover-Foxcroft, ME. Tax-exempt.
Institution Type/Description: Historical Society Museum: housed in the former county newspaper company building; built in 1854. Listed on the National Register of Historic Places.
Collections: local history & culture; photographs; period artifacts & memorabilia; town records; historic buildings.
Major Exhibits: Art of Amry Elizabeth Greeley, 6/14-9/14; WWI, 6/14-9/14.
Research Fields: local history; genealogy.
Publications: quarterly newsletter, Shiretown Conserver.
Hours & Admission Prices: Summer: Thurs. 11-2; other times by appointment. No charge; donations accepted.
Attendance: 400 (estimated)
Membership: Individual $10; Senior over 65 $7.

Dresden

1761 POWNALBOROUGH COURTHOUSE, 23 Courthouse Rd., Rte. 128, Dresden, ME 04342. Mailing Address: P.O. Box 61, Wiscasset, ME 04578-0061. Tel.: 207-737-2504 & 882-6817.
E-mail: lcha@wiscasset.net
Web Site: www.lincolncountyhistory.org
Founded: 1954.
Congressional District: 1
Key Personnel: Court House Chm., Steve Eagles; Pres. (V), John Rienhardt.
Personnel Profile: Part-Time Volunteers 25.
Governing Authority: nonprofit. Parent Institution: Lincoln County Historical Association, Inc. Federal St., Wiscasset, ME 04578. Tax-exempt.
Institution Type/Description: Historic Building: 1761 Pownalborough Court House.
Collections: mid 1700s-1900, pre-Revolutionary War court room; original court room, family rooms, tavern & ice making.
Research Fields: archaeological; Fort Shirley; period cemetery; 1761 court house.
Facilities: nature trails; picnic area.
Activities: guided tours; special exhibits; ice making. Special Events: Memorial Day Observance; Fall & Winter lecture series.
Publications: newsletter, The Lincoln County Chronicle; brochure, The Pre-Revolutionary Pownalborough Court House; 19th Century Law and Order: The 1811 Old Jail and 1839 Jailer House Museum; Preserve Local History: The Lincoln County Historical Association.
Hours & Admission Prices: Memorial Day to June & Sept.-Columbus Day Sat. 10-4, Sun. 12-4; July-Aug. Tues.-Fri. 10-4. Adults $4; members no charge.
Attendance: 1,000 (estimated)
Membership: Single $25; Family $35; Supporting $60; Sustaining $100.

East Vassalboro

VASSALBORO HISTORICAL SOCIETY MUSEUM, Rte. 32, East Vassalboro, ME 04935. Mailing Address: P.O. Box 43, East Vassalboro, ME 04935. Tel.: 207-923-3533 & 3505.
Founded: 1962.
Key Personnel: Cur., Julie Lyon
Institution Type/Description: Historical Society Museum.
Collections: local history & culture; period furnishings; personal artifacts; photographs.
Facilities: library.
Hours & Admission Prices: May-Nov. 2nd & 4th Sun. of the month 1-4; other times by appointment. &

Easton

FRANCIS MALCOLM SCIENCE CENTER, 776 Houlton Rd., Easton, ME 04740. Mailing Address: P.O. Box 186, Easton, ME 04740. Tel.: 207-488-5451. Fax: 207-488-2951.
Institution Type/Description: Science Center.
Collections: life & environmental science; animal & plant life.
Facilities: planetarium.
Activities: planetarium & nature programs.
Hours & Admission Prices: late Aug. to mid-June Tues.-Fri. 9-3 by appointment.

Eastport

BORDER HISTORICAL SOCIETY, 14 Key St., Eastport, ME 04631. Mailing Address: P.O. Box 95, Eastport, ME 04631-0095. Tel.: 207-853-2963.
E-mail: borderhistoricalsociety@yahoo.com
Web Site: borderhistoricalsociety.com
Key Personnel: Pres., Phyllis Siebert; Museum Shop Mgr., Eleanor Norton; Museum Shop Mgr., Leasa Garvin.
Governing Authority: Tax-exempt.
Institution Type/Description: Historical Society Museum.
Collections: local history & culture; photographs.
Hours & Admission Prices: Call for hours. No charge; donations accepted.

RAYE'S MUSTARD MILL MUSEUM, 83 Washington St., Eastport, ME 04631. Mailing Address: P.O. Box 2, Eastport, ME 04631. Tel.: 207-853-4451; 800-853-1903.
Institution Type/Description: Historic Building: housed in a working stone ground mustard mill.
Collections: mill history & equipment; photographs; personal artifacts.
Facilities: Museum-related items for sale.
Hours & Admission Prices: Tours: Mon.-Fri. 10-3 by appointment.

TIDES INSTITUTE & MUSEUM OF ART, 43 Water St., Eastport, ME 04631. Mailing Address: P.O. Box 161, Eastport, ME 04631. Tel.: 207-853-4047.
Key Personnel: Dir., Hugh French; Dir. Programs, Jude Valentine; Exhibitions, Kristin McKinlay
Institution Type/Description: Historic Building: housed in a former bank building; built in 1887.
Collections: paintings; prints; photographs; sculpture; Passamaquoddy & Micmac basketry; ship models; maps; furniture; musical instruments; books; graphic arts.
Facilities: library. Museum-related items for sale.
Activities: special events.
Hours & Admission Prices: Tues.-Sun. 10-4.

Ellsworth

STANWOOD WILDLIFE SANCTUARY, Rte. 3 289 High St., Ellsworth, ME 04605. Mailing Address: P.O. Box 485, Ellsworth, ME 04605-0485. Tel.: 207-667-8460.
E-mail: birdsacre@hotmail.com
Web Site: www.birdsacre.com
Founded: 1959.
Key Personnel: Pres. (V) & Museum Shop Mgr., Grayson Richmond; Chm. (V), Donald Knowles.
Personnel Profile: Full-Time Volunteers 1; Part-Time Volunteers 6.
Governing Authority: nonprofit. Tax-exempt: 501(c)(3).
Institution Type/Description: Homestead Museum & Nature Center: home of Cordelia Stanwood, pioneer naturalist, photographer & conservationist.
Collections: Victorian furnishings; bird mounts; nests & eggs; pioneer photography; paintings.
Facilities: library; picnic areas; nature trails. Gift items for sale.
Activities: museum tours; on & off site educational programs for children; natural history programs.
Publications: annual newsletter.
Hours & Admission Prices: Sanctuary: daily dawn-dark. Homestead Museum & Nature Center: June to mid-Oct. daily 10-4. No charge; donations accepted. Closed Independence Day; Labor Day. &
Attendance: 10,000 (estimated)

THE TELEPHONE MUSEUM, 166 Winkumpaugh Rd., Ellsworth, ME 04605-3035. Mailing Address: P.O. Box 1377, Ellsworth, ME 04605-1377. Tel.: 207-667-9491. Fax: 207-667-9491 (call first).
E-mail: switchboard@downeast.net
Web Site: www.thetelephonemuseum.org
Founded: 1983.
Key Personnel: Dir. & Pres. (V), Sandra Galley; Treas., David Thompson.
Personnel Profile: Full-Time Volunteers 2; Part-Time Volunteers 20.
Governing Authority: private; nonprofit organization. Tax-exempt: 501(c)(3).
Institution Type/Description: Communications Museum.
Collections: history of communications; hands-on exhibits; electro-mechanical telephone switching systems; switchboards; telephone sets; central office equipment; outside plant equipment; technical documentation; archival material.
Research Fields: early telephone lines in Hancock County, Maine.
Facilities: library; 2,000 sq. ft. exhibit space.
Activities: guided tours; participatory & temporary exhibits. Annual Event: Family Day.
Publications: biannual newsletter, The Pole Line; annual membership directory, Subscriber Directory.
Hours & Admission Prices: June & Oct. by appointment only; July-Sept. Thurs.-Sun. 1-4. Adults $10, children $5. &
Attendance: 350 (estimated)
Membership: Individual $30; Family $50; Organization $75; Participating $100; Sustaining $250; Life $1,000.

WOODLAWN: MUSEUM GARDENS & PARK, 19 Black House Dr., Ellsworth, ME 04605-2320. Mailing Address: P.O. Box 1478, Ellsworth, ME 04605-1478. Tel.: 207-667-8671. Fax: 207-667-7950.
E-mail: director@woodlawnmuseum.org
Web Site: www.woodlawnmuseum.org
Formerly: Woodlawn Museum/The Black House
Founded: 1928.
Congressional District: 2
Key Personnel: Exec. Dir., Joshua C. Torrance; Chm. (V), Sandra Blake-Leonard.
Personnel Profile: Full-Time Paid 2; Part-Time Paid 3; Part-Time Volunteers 7; Interns 3.
Governing Authority: nonprofit organization. Parent Institution: Hancock County Trustees of Public Reservations. Tax-exempt.
Institution Type/Description: Historic House: 1827 Black Family House.
Collections: original furnishings & outbuilding; formal garden; history; archives; paintings; decorative art.
Research Fields: social history of Down East, ME; local history; decorative & fine art; personal artifacts; ME lumbering history.
Facilities: 1,000-vol. library of Col. Black's personal books; 180-acre estate; 2 miles of walking trails & gardens.
Activities: guided tours; special events; workshops; lectures. Museum Sponsors: Weekly Teas in July & August; Ellsworth Antiques show at Woodlawn; Holiday Tours in December.
Publications: newsletter; annual journal.
Hours & Admission Prices: May & Oct. Tues.-Sun. 1-4; June-Sept. Tues.-Sat. 10-5, Sun. 1-4. Tours every hour. Adults $10, children $3; discount to AAA & AAM members; members & public park no charge. Closed Independence Day.
Attendance: 10,000 (estimated)
Membership: Student $5; Individual $35; Household $40; Supporter $125; Sponsor $250; Patron $500; Benefactor $1,000.

Fairfield

FAIRFIELD HISTORY HOUSE - THE COTTON-SMITH HOUSE, 42 High St., Fairfield, ME 04937. Tel.: 207-453-2998.
E-mail: fhs2@myfairpoint.net
Web Site: WWW.fairfieldmehistoricalsociety.net
Founded: 1973.
Congressional District: 2
Key Personnel: Pres. (V), Douglas Cutchin.
Personnel Profile: Part-Time Volunteers 10.
Volunteer Hours: 1,100
Operating Expenses: 9,151
Operating Income: 11,057
Governing Authority: Parent Institution: Fairfield Historical Society. Tax-exempt.
Institution Type/Description: Historic House: built c.1890.
Collections: local history & culture; period furnishings; personal artifacts; photographs.
Activities: research; special events. Annual Events: Quilt Show in July; Christmas Open House in December.
Hours & Admission Prices: Tues. 9-12 & 1-4; 2nd Sat. 9-12 & 1-4; other times by appointment. No charge; donations accepted. Closed Dec. thru Feb.
Attendance: 400 (estimated)
Membership: Senior $15; Individual $25; Family $45; Life $250.

Falmouth

THE FALMOUTH HISTORICAL SOCIETY, 60 Woods Rd., Falmouth, ME 04102. Mailing Address: 2 Homestead Ln., 190 U.S. Rte. 1, PMB 367, Falmouth, ME 04105-1313. Tel.: 207-781-4727.
E-mail: falmouthhistorical@myfairpoint.net
Web Site: www.falmouthmehistory.org
Key Personnel: Pres., Scott McLeod; Vice Pres., Janice Delima
Institution Type/Description: Historical Society Museum.
Collections: local history & culture; photographs; personal artifacts.
Hours & Admission Prices: Call for hours.
Membership: Student $5; Individual $15; Family $25; Supporting $50 & up.

Farmington

NORDICA HOMESTEAD, 116 Nordica Lane, Farmington, ME 04938. Mailing Address: c/o Franklin County Savings Bank, P.O. Box 825, Farmington, ME 04938-0825. Tel.: 207-778-2042.
Web Site: www.lilliannordica.com
Founded: 1927.
Congressional District: 2
Key Personnel: C.E.O. & Pres. (V), Tom Sawyer; Vice Pres. & Publicity Dir., Marion Smith; Treas., Cindy Wright.
Personnel Profile: Full-Time Volunteers 2.
Governing Authority: society. Nordica Memorial Association, Inc. Tax-exempt.
Institution Type/Description: Historical Society Museum: housed in c.1840 Nordica Homestead.
Collections: items relating to the life of Lillian Nordica, opera singer; opera & concert gowns by Worth; record containing all 14 recordings she is known to have made.
Research Fields: material related to Lillian Nordica.
Facilities: 300-vol. library of books, magazines & programs on Maine available by application to publicity director. Photographs, postcards, note paper, books, records & biography of Nordica for sale.

Activities: guided tours; permanent exhibitions.
Publications: brochure.
Hours & Admission Prices: June-Sept. 15 Tues.-Sat. 10-12 & 1-5, Sun. 1-5; Sept.16-Oct. 15 by appointment only. Adults $2, children $1; discounts to AAA members & school children groups; children under 6 no charge. &
Attendance: 700 (estimated)
Membership: Annual $5; Life $100; Patron $500.

Franklin

FRANKLIN HISTORICAL SOCIETY, Rte. 200, Sullivan Rd., Franklin, ME 04634. Mailing Address: P.O. Box 317, Franklin, ME 04634-0317. Tel.: 207-565-3635. Fax: 207-565-3323.
Founded: 1960.
Congressional District: 2
Key Personnel: Pres., Dania Stager-Snow; Vice Pres., William Robertson; Cur. & Museum Shop Mgr., Helen Cantor.
Governing Authority: society; nonprofit organization. Tax-exempt.
Institution Type/Description: Historical Society Museum: housed in Old East Franklin Church.
Collections: clothing; uniforms; picture albums; farming tools; lumbering & shipbuilding items; granite-work tools; household artifacts.
Facilities: library; 120-seat auditorium.
Activities: guided tours; organized education programs for children. Annual Events: Antique Fair in July; scholarship drive; educational workshop, Franklin-Past, Present & Future in June & July.
Hours & Admission Prices: July-Labor Day Sat. 2-4. No charge; donations accepted. &
Attendance: 90
Membership: Individual $3.

Freeport

DESERT OF MAINE & BARN MUSEUM, 95 Desert Rd., Freeport, ME 04032. Tel.: 207-865-6962.
E-mail: info@desertofmaine.com
Web Site: www.desertofmaine.com
Institution Type/Description: History Museum.
Collections: local history; farm tools & equipment; sand paintings; photographs; mining. Historic Building: 1783 barn.
Facilities: nature trails; picnic area. Gift items for sale.
Activities: walking tours; mining.
Hours & Admission Prices: May to mid-Oct. daily 9-5:30. Adults $10.50, teens 13-16 $7.75, children 4-12 $6.75.

FREEPORT HISTORICAL SOCIETY, 45 Main St., Freeport, ME 04032-1212. Tel.: 207-865-3170.
E-mail: info@freeporthistoricalsociety.org
Web Site: www.freeporthistoricalsociety.org
Founded: 1969.
Congressional District: 1
Key Personnel: Pres. (V), Tori Baron; Exec. Dir., Christina White.
Personnel Profile: Full-Time Paid 1; Part-Time Paid 2; Part-Time Volunteers 40; Interns 1.
Governing Authority: nonprofit organization. Tax-exempt: 501(c)(3).
Institution Type/Description: Historical Society Museum.
Collections: manuscripts; photographs; artifacts & furnishings. Historic Buildings: 1830 Harrington House; 1810 Rodick/Pettengill Farm.
Research Fields: saltwater farming; Maine coastal life & natural history; Freeport history.
Facilities: local history archives.
Activities: interpretation of historic sites; exhibits; school programs; workshops & demonstrations; walking tours; history trail; lectures.
Publications: quarterly newsletter, Freeport Historical Society Newsletter; book, Tides of Change; maps & Freeport related books; local history publications; A Window Through Time: Pettengill Farm and the Soul of New England.
Hours & Admission Prices: Office: Mon.-Fri. 8:30-5. Archives: Winter: Wed.-Fri. 9-5; Summer: Mon.-Fri. 9-5; other times by appointment. Suggested Donation: $3 per person.
Attendance: 3,000 (estimated)
Membership: Senior Citizen & Student $10; Individual $25; Family $35; Contributing & Business $50; Supporting $75; Corporate & Sustaining $150.

Frenchboro

FRENCHBORO LIBRARY AND THE FRENCHBORO HISTORICAL SOCIETY, Schoolhouse Hill, Frenchboro, ME 04635. Tel.: 207-334-2924.
Founded: 1979.
Key Personnel: Pres. (V), Sandra Lunt; Museum Shop Mgr., Donna Hasal.
Personnel Profile: Part-Time Volunteers 10.
Institution Type/Description: History Museum & Library.
Collections: local history & culture; period artifacts; photographs; books; early documents & records.
Hours & Admission Prices: Library: daily. Society: mid-June to Sept. Mon.-Fri. 1-5. No charge; donations accepted.
Membership: Individual $10; Lifetime $25; Island Archivist $50; Island Historian $100.

Friendship

FRIENDSHIP MUSEUM INC., 1 Martin Point Rd., Friendship, ME 04547. Mailing Address: P.O. Box 226, Friendship, ME 04547.
E-mail: info@friendshipmuseum.org
Web Site: www.friendshipmuseum.org
Founded: 1965.
Personnel Profile: Part-Time Paid 6; Part-Time Volunteers 17.
Institution Type/Description: History Museum: housed in a former one-room schoolhouse, 1851-1923.
Collections: local history & culture; photographs; personal artifacts.
Hours & Admission Prices: late June to Labor Day Mon.-Sat. 1-4, Sun. 2-4; Sept. to Columbus Day Sat. 1-4, Sun. 2-4. No charge; donations accepted.
Membership: Senior & Student $5; Individual $10; Family $25; Friend $50; Sponsor $100; Patron $250.

Fryeburg

FRYEBURG FAIR FARM MUSEUM, 113 N. Fryeburg Rd., Fryeburg, ME 04037. Tel.: 207-935-3268. Fax: 207-935-3662.
E-mail: info@fryeburgfair.org
Web Site: www.fryeburgfair.com
Founded: 1970.
Congressional District: 2
Key Personnel: Cur., Edward W. Jones; Asst. Cur., Diane L. Jones.
Governing Authority: nonprofit organization. Parent Institution: Oxford Agricultural Society, Fryeburg, ME. Tax-exempt.
Institution Type/Description: Agricultural Museum: housed in 1832 barn & carriage house.
Collections: farming & related industries; period woodstove; farm tools & implements; carriages; barn; carriage shed; one room schoolhouse; blacksmith shop; smokehouse; tool shed.
Research Fields: identification & facts on old Fryeburg & surrounding area industries & crafts; identification of old tools & artifacts; safeguards.
Activities: guided tours by appointment; demonstrations; temporary exhibits; cider making.
Hours & Admission Prices: early Oct. during Fair Week daily 9-9; school tours by appointment. &
Attendance: 50,000

FRYEBURG HISTORICAL SOCIETY, 511 Main St., Fryeburg, ME 04037. Tel.: 207-935-4192.
Key Personnel: Cur., Edward W. Jones
Institution Type/Description: Historical Society Museum.
Collections: local history & culture; period furnishings; paintings; early newspapers; photographs; diaries; scrapbooks.
Activities: research.
Hours & Admission Prices: Museum: Thurs. 1-4; other times by appointment. Library: Wed. & Fri. 9-12, Thurs. 1-4.

THE PALMINA F. & STEPHEN S. PACE GALLERIES OF ART, (M), Leura Hill Eastman Performing Arts Center, 18 Bradley St., Fryeburg, ME 04037. Mailing Address: Fryeburg Academy, 745 Main St., Fryeburg, ME 04037. Tel.: 207-935-9232.
E-mail: jday@fryeburgacademy.org
Founded: 2009.
Congressional District: 1
Key Personnel: Dir., John M. Day.
Personnel Profile: Part-Time Volunteers 2.
Governing Authority: Parent Institution: Fryeburg Academy. Tax-exempt.
Institution Type/Description: Art Gallery.
Collections: paintings; sculptures.

Hours & Admission Prices: Mon.-Fri. 9-1; other times by appointment. No charge. &

Attendance: 5,000 (estimated)

Gorham

BAXTER HOUSE MUSEUM, 67 South St., Gorham, ME 04038. Tel.: 207-839-3878. Fax: 207-839-7749.

Founded: 1908.

Congressional District: 1

Key Personnel: Guide, Chm. (V) & Pres. Bd. Library Trustees (V), Linda M. Frinsko.

Personnel Profile: Part-Time Paid 1; Part-Time Volunteers 24.

Volunteer Hours: 72

Operating Expenses: 1,927

Operating Income: 6,600

Governing Authority: municipal. Parent Institution: Baxter Library Board & Town of Gorham. Tax-exempt: 501(c)(3).

Institution Type/Description: Historic House: c.1797 Baxter House.

Collections: local history & culture; period furniture; early records.

Publications: brochure.

Hours & Admission Prices: June-Aug. Tues. & Thurs. 10-1; other times by appointment. No charge; donations accepted. &

Attendance: 300 (estimated)

GORHAM HISTORICAL SOCIETY, 28 School St., Gorham, ME 04038. Mailing Address: P.O. Box 141, Gorham, ME 04038.

E-mail: society@gorhamhistorical.com

Web Site: gorhamhistorical.com

Institution Type/Description: Historical Society Museum.

Collections: local history; photographs.

Facilities: archives.

Activities: research.

Hours & Admission Prices: April-Oct. Thurs. 10-2; other times by appointment.

USM ART GALLERY, 37 College Ave., Gorham Campus, Gorham, ME 04038-1032. Tel.: 207-780-5460 & 5008. Fax: 207-780-5759. TDD: 207-780-5646.

E-mail: ceyler@maine.edu

Web Site: www.usm.maine.edu/~gallery

Founded: 1965.

Key Personnel: Dir. Exhibits & Programs, Carolyn Eyler.

Personnel Profile: Full-Time Paid 1; Part-Time Volunteers 1; Interns 3.

Governing Authority: university. Parent Institution: University of Southern Maine. Tax-exempt: 501(c)(3).

Institution Type/Description: Art Museum.

Collections: paintings; sculpture; graphics; photographs.

Research Fields: contemporary art.

Activities: lectures; films; gallery talks; concerts; arts festivals; inter-museum loan, temporary & traveling exhibitions.

Publications: exhibition catalogs.

Hours & Admission Prices: Academic Year: Tues.-Sun. 12-4. No charge; donations accepted. &

Gouldsboro

GOULDSBORO HISTORICAL SOCIETY, 452 U.S. Rte. 1, Gouldsboro, ME 04607. Mailing Address: P.O. Box 94, Gouldsboro, ME 04607. Tel.: 207-963-7155.

E-mail: c.hodge@myfairpoint.net

Founded: 1984.

Congressional District: 2

Key Personnel: Pres. (V), Beatrice C. Buckley

Governing Authority: Subsidiary Institutions: Old Townhouse Museum; West Bay Church & Friendship Hall. Tax-exempt.

Institution Type/Description: Historical Society Museum: housed in the former Gouldsboro town hall; built in 1884.

Collections: local history & culture; period furnishings; personal artifacts; photographs.

Activities: monthly society meetings May to October.

Publications: annual newsletter.

Hours & Admission Prices: July-Aug. Sat. 2-4. &

Membership: Individual $7.50; Family $15; Contributor $25; Sponsor $50; Patron $100; Life $150.

Gray

GRAY HISTORICAL SOCIETY, 1 Main St., Gray, ME 04039. Mailing Address: P.O. Box 544, Gray, ME 04039-0544. Tel.: 207-657-4783.

E-mail: pkttaylor@aol.com

Institution Type/Description: Historical Society Museum: housed in the Old Fire Barn.

Collections: photographs; books; letters; manuscripts; period artifacts; Portland-Lewiston Interurban Railroad, 1914-1933; Pennell Institute from 1878; Mayall Woolen Mills from the 1790s; 19th century tools, household artifacts & clothing.

Hours & Admission Prices: By appointment.

Greenville

MOOSEHEAD HISTORICAL SOCIETY AND MUSEUM, 444 Pritham Ave., Greenville, ME 04441-1116. Mailing Address: P.O. Box 1116, Greenville, ME 04441-1116. Tel.: 207-695-2909.

E-mail: mooseheadhistory@myfairpoint.net

Web Site: mooseheadhistory.org

Founded: 1961.

Key Personnel: Dir., Candy Canders Russell

Governing Authority: nonprofit organization. Tax-exempt.

Institution Type/Description: Historical Society Museum.

Collections: local & regional heritage & history; military artifacts; period clothing & furnishings. Historic Houses: 1899 Victorian mansion; Eveleth-Crafts-Sheridan House.

Hours & Admission Prices: June-Sept. Wed.-Fri. 1-4. Adults $4.

Attendance: 2,000 (estimated)

MOOSEHEAD MARINE MUSEUM, 12 Lily Bay Rd., Greenville, ME 04441. Mailing Address: P.O. Box 1151, Greenville, ME 04441-1151. Tel.: 207-695-2716. Fax: 207-695-2367. Facebook: Moosehead Marine Museum.

E-mail: katahdin2@myfairpoint.net

Web Site: katahdincruises.com

Founded: 1977.

Congressional District: 2

Key Personnel: Dir., Pres. (V) & Museum Shop Mgr, Maynard Russell.

Personnel Profile: Part-Time Paid 9; Part-Time Volunteers 6.

Governing Authority: nonprofit organization. Parent Institution: Moosehead Marine Museum, Inc. Tax-exempt.

Institution Type/Description: Marine Museum.

Collections: steamboat memorabilia; photographs; logging & lumber artifacts; Moosehead area history; Katahdin, a 115' steamboat converted to diesel, built in 1914.

Activities: Katahdin cruises; special events; weddings; charter cruises from late June to mid-Oct.

Hours & Admission Prices: Call for hours & cruise rates. Museum: no charge; donations accepted. &

Attendance: 7,500 (accurate)

Membership: Single $25; Family $40.

Hallowell

HARLOW GALLERY, KENNEBEC VALLEY ART ASSOCIATION, 160 Water St., Hallowell, ME 04347-1315. Tel.: 207-622-3813. Facebook: Harlow Gallery.

E-mail: kvaa@harlowgallery.org

Web Site: harlowgallery.org

Founded: 1963.

Congressional District: 1

Key Personnel: C.E.O., Deborah Fahy; Pres. (V), Patricia O'Brien; Vice Pres., Perry McCourtney; Treas., Karen Johnson; Program Dir., Nancy Barron; Sec., Marie Giguere.

Personnel Profile: Part-Time Paid 5; Part-Time Volunteers 60; Interns 1.

Governing Authority: nonprofit. Tax-exempt.

Institution Type/Description: Art Gallery.

Collections: works by Maine artists.

Activities: monthly exhibitions; gallery talks; poetry reading; art workshops.

Publications: bimonthly newsletter.

Hours & Admission Prices: Wed.-Sat. 12-6. No charge; donations accepted. Closed most holidays.

Attendance: 7,000

Membership: Student $15; Artist Sliding Scale $30-$80; Community $25, $50, $150, $250.

Hampden

HAMPDEN HISTORICAL SOCIETY - KINSLEY HOUSE, 83
Main Rd. S., Hampden, ME 04444. Mailing Address: P.O. Box
456, Hampden, ME 04444. Tel.: 207-862-2562.
E-mail: hampdenhistorical@gmail.com
Founded: 1970.
Key Personnel: Pres. (V), Ken Rowell.
Personnel Profile: Part-Time Volunteers 10.
Governing Authority: Tax-exempt.
Institution Type/Description: Historical Society Museum: housed in the former
home of Martin Kinsley; built in 1794.
Collections: local history & culture; period furnishings; personal artifacts;
photographs.
Facilities: archives.
Activities: research.
Publications: quarterly newsletter.
Hours & Admission Prices: April-Oct. Tues. 10-4; other times by appointment.
No charge; donations accepted.
Membership: Student 15-18 $3; Single $15; Family & Supporting $25;
Corporate $75; Life 65-79 $100; Life 55-64 $200; Patron $500.

Harpswell

HARPSWELL HISTORICAL SOCIETY MUSEUM, 929
Harpswell Neck Rd., Harpswell, ME 04079. Mailing Address: 852
Harpswell Neck Rd., Harpswell, ME 04079. Tel.: 207-833-6322.
Institution Type/Description: History Museum.
Collections: local history & culture; period furnishings; personal artifacts;
photographs.
Activities: special events.
Hours & Admission Prices: Memorial Day to Columbus Day Sun. 2-4.

Harrison

SCRIBNER'S MILL, 244 Scribner's Mill Rd., Harrison, ME 04040.
Mailing Address: P.O. Box 323, Harrison, ME 04040. Tel.: 207-
583-6455.
Web Site: www.scribnersmill.org
Key Personnel: Pres. (V), Rick Johnson
Governing Authority: Tax-exempt.
Institution Type/Description: Historic Buildings: mill & homestead; built in
1847.
Collections: local history; period machinery; historic mill & homestead.
Activities: sawmill demonstrations; homestead tours; educational programs.
Annual Event: Back to the Past in August.
Hours & Admission Prices: Summer: Sat.-Sun. call to confirm. No charge;
donations accepted. Back to the Past Event: $6 per person.
Attendance: 1,000 (estimated)
Membership: Student $5; Individual $15; Sustaining $25; Family $30; Corpo-
rate & Patron $50; Benefactor $100; Life $500; Jessie P. Scriber Club
$1,000.

Hinckley

L.C. BATES MUSEUM, (M), Good Will-Hinckley Home For Boys
& Girls, Rte. 201, 14 Easler Rd., Hinckley, ME 04944-0159.
Mailing Address: Good Will-Hinckley Home For Boys & Girls,
P.O. Box 159, Hinckley, ME 04944-0159. Tel.: 207-238-4250. Fax:
207-238-4007.
E-mail: lcbates@gwh.org
Web Site: www.gwh.org/html/lcbatesmuseum.htm
Founded: 1889.
Congressional District: 2
Key Personnel: Exec. Dir., Cur. & Museum Shop Mgr., Deborah Staber; Chm.
(V), John Willey; Museum Educator, Serena Sandborn; Museum Educator,
Leigh Rose.
Personnel Profile: Full-Time Paid 1; Part-Time Paid 3; Part-Time Volunteers
46; Interns 7.
Governing Authority: nonprofit organization. Parent Institution: Good Will-
Hinckley Home Association. Tax-exempt.
Institution Type/Description: Historic Building: built in 1903.
Collections: Native American; natural history; archaeological stone; contem-
porary art; mineralogy; military artifacts; local pioneer artifacts; agricultural
antiques; Good Will-Hinckley history & archives; wildlife dioramas.
Major Exhibits: Looking At The Past Through Photography, 1/14-4/15/14;
Maine Northern Skies: Clear Light, 5/15/14-10/15/14.
Research Fields: American social history; archives of Good Will-Hinckley.
Facilities: classrooms; nature trails; 700-acre wildlife area; arboretum.

Activities: guided tours; permanent exhibitions; nature trail; outreach programs; school groups. Museum Sponsors: Junior Naturalist programs, after
school & summers.
Publications: book; newsletter.
Hours & Admission Prices: April-Nov. Wed.-Sat. 10-4:30, Sun. 1-4:30;
Dec.-March Wed.-Sat. 10-4:30; other times by appointment. Adults $3,
youth under 18 $1; discounts to AAM & AAA members, MEMA & group
tours; members no charge.
Attendance: 17,500 (accurate)
Membership: Individual $15; Family $25-$50; Group $51-$100; Patron
$101-$500; Benefactor $501-$1,000; Corporate $1,000 & up.

Houlton

AROOSTOOK COUNTY HISTORICAL & ART MUSEUM, 109
Main St., Houlton, ME 04730-2123. Tel.: 207-532-2519.
Institution Type/Description: Art & History Museum: housed in a Colonial
Revival building; built in 1903. Listed on the National Register of Historic
Places.
Collections: county history & culture; period furnishings; personal artifacts;
photographs.
Hours & Admission Prices: Call for hours.

Indian Island

PENOBSCOT NATION MUSEUM, 12 Down St., Indian Island,
ME 04468. Tel.: 207-827-4153.
Institution Type/Description: Native American Museum.
Collections: Native American history, culture, & artifacts; personal artifacts;
photographs; baskets; woodcarvings; paintings.
Activities: tours.
Hours & Admission Prices: Mon.-Thurs. 9-2, Sat. 10-3.

Island Falls

ISLAND FALLS HISTORICAL SOCIETY, 16 Nina Sawyer Ln.,
Island Falls, ME 04747. Mailing Address: P.O. Box 204, Island
Falls, ME 04747. Tel.: 207-463-2264.
E-mail: rdrew@katahdin.lib.me.us
Founded: 1990.
Key Personnel: Pres., Gregory Ryan.
Personnel Profile: Part-Time Volunteers 25.
Institution Type/Description: Historical Society Museum.
Collections: local history; period furnishings; personal artifacts; photographs.
Historic Buildings: Tingley house, c.1900; potato house.
Publications: quarterly newsletter.
Hours & Admission Prices: June-Dec. call for hours; school groups by
appointment. No charge; donations accepted.
Attendance: 300 (estimated)
Membership: Individual $10; Family $25.

Islesboro

ISLESBORO HISTORICAL SOCIETY AND MUSEUM, 388
Main Rd., Islesboro, ME 04848. Mailing Address: P.O. Box 301,
Islesboro, ME 04848-0301. Tel.: 207-734-6733.
E-mail: info@islesborohistorical.org
Web Site: www.islesborohistorical.org
Founded: 1964.
Congressional District: 2
Key Personnel: Pres. (V), Judith Kaminski.
Personnel Profile: Part-Time Paid 1; Part-Time Volunteers 51.
Governing Authority: society. Tax-exempt.
Institution Type/Description: Historical & Preservation Society Museum:
housed in 1894 Town Hall.
Collections: photographs; textiles; boats; vehicles; paintings & drawings; farm
& household equipment; Islesboro artifacts including a phone exchange;
two post office letter boxes; c.1900 telephone booth; 100-year old weaving
loom; manuscripts; scrapbooks; arts; crafts.
Research Fields: 19th to early 20th century Maine.
Facilities: library; stage.
Activities: arts festivals; loan, permanent & temporary exhibitions; historical
presentations; concerts; research.
Publications: books, History of Islesboro; History of Islesborough, Maine,
1764-1893; History of Islesboro, Maine 1893-1983; The Summer Cottages
of Islesboro - 1890-1930.
Hours & Admission Prices: Society: by appointment. Museum: July-Aug.
Sat.-Wed. 12:30-4:30. No charge; donations accepted.
Attendance: 650 (accurate)

Membership: Regular $10; Contributing $25; Sustaining $50; Patron $100.

SAILOR'S MEMORIAL MUSEUM, Grindle Point, Islesboro, ME 04848. Mailing Address: P.O. Box 76, Islesboro, ME 04848-0076. Tel.: 207-734-2253. Fax: 207-734-8394.
Founded: 1936.
Congressional District: 2
Key Personnel: Town Mgr., Damaris A. Diffin.
Personnel Profile: Part-Time Paid 1; Part-Time Volunteers 6.
Governing Authority: municipal. Tax-exempt.
Institution Type/Description: Maritime Museum: housed in 1850 Lighthouse.
Collections: marine; history. Historic House: 1850, Grindle Point Keeper's House.
Activities: permanent exhibitions.
Hours & Admission Prices: mid-June to Labor Day Tues.-Sun. 10-4. No charge; donations accepted. &
Attendance: 1,500 (estimated)

Islesford

ISLESFORD HISTORICAL MUSEUM, Little Cranberry Island, Islesford, ME 04646. Mailing Address: Acadia National Park, P.O. Box 177, Bar Harbor, ME 04609-0177. Tel.: 207-288-8729. Fax: 207-288-8709.
E-mail: rebecca_cole-will@nps.gov
Web Site: www.nps.gov/acad/planyourvisit/islesfordhistoricalmuseum.htm
Founded: 1919.
Congressional District: 2
Key Personnel: Supt., Sheridan Steele.
Personnel Profile: Full-Time Paid 1; Part-Time Paid 2; Part-Time Volunteers 2.
Governing Authority: federal government. Parent Institution: National Park Service. Subsidiary Institution: Acadia National Park. Tax-exempt.
Institution Type/Description: History Museum.
Collections: artifacts & archival documents pertaining to the maritime history & settlement of the Cranberry Isles; ship models; genealogy of Cranberry Isles settlers.
Research Fields: genealogy; New France, maritime & town history.
Activities: permanent & temporary exhibitions.
Publications: One Man's Museum: History of the Islesford Collection.
Hours & Admission Prices: mid-June to Sept. Mon.-Sat. 9-12 & 12:30-3:30, Sun. 10:45-12 & 12:30-3:30. No charge; donations accepted.
Attendance: 14,935 (accurate)

Jackman

JACKMAN - MOOSE RIVER VALLEY HISTORICAL SOCIETY, 574 Main St., Jackman, ME 04945. Mailing Address: P.O. Box 875, Jackman, ME 04945.
Founded: 1990.
Congressional District: 2
Key Personnel: Dir. & Pres. (V), Howard Paradise; Sec., Deborah Theriault.
Personnel Profile: Part-Time Volunteers 1.
Governing Authority: Tax-exempt.
Institution Type/Description: Historical Society Museum.
Collections: local history & culture; period furnishings; personal artifacts; photographs.
Major Exhibits: POW Camp Artifacts (T), 2012-2013.
Publications: annual newsletter; book, History of the Moose River Valley; History of Long Pond (ME) by Howard Paradise; Prisoners, Pulpwood and Potatoes, the Story of German POWs in Maine (1944-1946) by W. Randall; society brochure.
Hours & Admission Prices: late May to Sept. call for hours. No charge; donations accepted. &
Attendance: 401 (estimated)
Membership: Individual Juniors $2; Individual Adults $5.

Jay

JAY HISTORICAL SOCIETY, 14 Main St., Jay, ME 04239. Mailing Address: 287 Main St., Jay, ME 04239, Tel.: 207-897-4876. Fax: 207-897-9420.
E-mail: joffice@jay-maine.org
Web Site: www.jay-maine.org
Founded: 1971.
Key Personnel: Pres. (V), Dorothy White.
Governing Authority: Tax-exempt.
Institution Type/Description: Historical Society Museum: housed in a Federal style home built in the 1820s.
Collections: local history; early tools; period furnishings.

Hours & Admission Prices: By appointment.
Attendance: 110
Membership: $2

Jonesport

MAINE COAST SARDINE HISTORY MUSEUM, 34 Mason Bay Rd., Jonesport, ME 04649. Tel.: 207-497-2961.
E-mail: ronniep6@myfairpoint.net
Web Site: www.mainesardinemuseum.org
Key Personnel: Dir. & Chm. (V), Ronnie Peabody; Dir., Mary Peabody; Dir., Gary Ray.
Personnel Profile: Part-Time Volunteers 2.
Governing Authority: nonprofit organization. Tax-exempt: 501(c)(3).
Institution Type/Description: History Museum.
Collections: sardine industry history; photographs; sardine cans & paper labels; tools; machinery; factory history & artifacts.
Hours & Admission Prices: Memorial Day to mid-Oct. Tues., Thurs. & Sun. 10-4; other times by appointment. Adults $4; discount to groups. &
Attendance: 264 (accurate)

Kennebunk

* **BRICK STORE MUSEUM, (M),** 117 Main St., Kennebunk, ME 04043-7088. Tel.: 207-985-4802. Facebook: Brick Store Museum.
E-mail: info@brickstoremuseum.org
Web Site: www.brickstoremuseum.org
Founded: 1936.
Congressional District: 1
Key Personnel: Exec. Dir., Christopher Farr; Pres. Bd. Trustees, Stephen P. Spofford; Archivist, Rosalind Magnuson; Registrar & Collections Mgr., Kathryn Hussey; Assoc. Dir. & Exhibitions Cur., Cynthia Walker.
Personnel Profile: Full-Time Paid 2; Full-Time Volunteers 2; Part-Time Volunteers 75; Interns 2.
Governing Authority: nonprofit organization. Tax-exempt: 501(c)(3).
Institution Type/Description: Local History Museum Complex: housed in 1825 William Lord's Brick Store building & three adjacent restored 19th-century buildings.
Collections: costumes; marine history; tools; early American utensils & decorative arts; manuscripts; documents; fine arts; photographs; Kenneth Roberts materials; Booth Tarkington materials; Abbott Graves, Thomas Badger & John Brewster paintings.
Research Fields: maritime; decorative arts; local history.
Facilities: 3,000-vol. library of books on history & genealogy. Publications for sale.
Activities: lectures; seminars; architectural tours; school programs; field trips; workshops; cell phone audio tours.
Publications: newsletter; local history journal; exhibit catalogs; books, Agreeable Situations: Society, Commerce & Art in South Maine 1780-1830; Sketch of An Old River: Shipbuilding On The Kennebunk; handbooks of architectural styles; An Anchor to Windward: The Maine Connection; Quiet, Well-Kept for Sensible People: The Development of Kennebunk Beach, 1860-1920; Trunks, Textiles and Transits: Manufacturing on the Mousam River; Windows on the Past: An Illustrated Guide to Kennebunk History through Architecture (2010); Learning is an Ornament: Education in Kennebunk from its Earliest Settlements to the Formation of School Administrative District #71 (2009).
Hours & Admission Prices: Tues.-Fri. 10-4:30, Sat. 10-1. Suggested Donation: $5; discounts to AAM, ICOM, NEMA & AAA members; members no charge. Closed holidays. &
Attendance: 5,036 (accurate)
Membership: Voyagers Society $5; Student & Senior $20; Individual $30; Senior Couple $35; Family $40; Business Partner Program $150-$1,000.

HEARTWOOD GALLERY - HEARTWOOD COLLEGE, 123 York St., Kennebunk, ME 04043-7105. Tel.: 207-985-0985.
E-mail: hca@heartwoodcollegeofart.org
Web Site: heartwoodcollegeofart.org/galleries.html
Institution Type/Description: Art Gallery.
Collections: paintings; sculpture; photographs.
Hours & Admission Prices: Mon.-Fri. 9-4.

Kennebunkport

KENNEBUNKPORT HISTORICAL SOCIETY, 125 North St., Kennebunkport, ME 04046. Mailing Address: P.O. Box 1173, Kennebunkport, ME 04046-1173. Tel.: 207-967-2751.
E-mail: kporths@roadrunner.com

Web Site: www.kporthistory.org
Founded: 1952.
Congressional District: 1
Key Personnel: Pres., Richard Litchfield; Treas., David Reid; Business Mgr., Kirsten Camp.
Personnel Profile: Full-Time Paid 1; Part-Time Volunteers 100.
Governing Authority: nonprofit organization. Tax-exempt: 501(c)(3).
Institution Type/Description: Historic House & History Museum: housed in 1853 Greek Revival Nott House; c.1900 District 5 Schoolhouse.
Collections: local crafts, art, shipbuilding & archives; genealogy; manuscript collections; over 2,000 photographs; old gowns; Clark shipyard office maritime collection; period furnishings; original jail cells of town; Nott House Victorian gardens. Historic Buildings: 1900 schoolhouse; Clark shipbuilding office; Louis D. Norton artist studio.
Research Fields: genealogy; land titles; local history; shipbuilding & maritime history.
Facilities: research library in schoolhouse.
Activities: lectures; temporary exhibitions; local research; house tours; historic village walking tours. Annual Events: re-creation of Nott House gardens in spring; Christmas Prelude in December.
Publications: newsletter, The Log.
Hours & Admission Prices: Town House School: Research Wed. & Fri. 10-1. Pasco Exhibit Center: July-Sept. Tues.-Fri. 10-4, Sat. 10-1; Oct.-June Tues.-Fri. 10-4. Nott House: July-Aug. Wed.-Fri. 10-4, Sat. 10-1; Sept. to Columbus Day Thurs.-Fri. 10-4, Sat. 10-1. No charge; donations accepted. Closed holidays.
Attendance: 2,000 (estimated)
Membership: Individual $40; Household $60; Business $100; Life $1,000.

SEASHORE TROLLEY MUSEUM, (M), 195 Log Cabin Rd., Kennebunkport, ME 04046-1690. Mailing Address: P.O. Box A, Kennebunkport, ME 04046-1690. Tel.: 207-967-2712. Fax: 207-967-0867.
E-mail: busi.ofc@neerhs.org
Web Site: www.trolleymuseum.org
Founded: 1939.
Congressional District: 1
Key Personnel: Chm. (V), James D. Schantz; Exec. Dir., Sally Bates; Treas. & Comptroller, Jeffrey N. Sisson; Museum Shop Mgr., Gayle Dion; Museum Shop Mgr., Birnie Bisnette.
Personnel Profile: Full-Time Paid 5; Full-Time Volunteers 4; Part-Time Paid 3; Part-Time Volunteers 100.
Governing Authority: nonprofit organization. Owned by New England Electric Railway Historical Society, Inc. Subsidiary Institution: National Streetcar Museum at Lowell. Tax-exempt: 501(c)(3).
Institution Type/Description: Electric Railway History & Technology Museum.
Collections: city trolley cars; interurban electric cars; early streetcars; rapid transit cars; trackless trolley coaches; buses; plows & sweepers; work cars; mail & express cars; locomotives; demonstration electric railway.
Research Fields: publications on equipment & corporate history.
Facilities: 10,000-vol. library of books, blueprints, photographs & documents; visitor center. Railroad & trolley related books & souvenirs for sale.
Activities: guided tours; lectures; films; workshops; training programs for professional museum workers; permanent & temporary exhibitions; special events; demonstration railway with 3 mile rides featuring different trolley cars.
Publications: bimonthly newsletter: Dispatch; annual report.
Hours & Admission Prices: May & Oct. Sat.-Sun. 10-5; Memorial Day to Columbus Day daily 10-5. Adults $10, seniors over 60 $8, children 6-16 $7.50; discounts to groups and AAM & ARM members; members and children 5 & under no charge. &
Attendance: 19,000 (estimated)
Membership: Students, Senior Citizens, Military & Handicapped $30; Regular $35; Family $50; Sustaining $60; Contributing $120; Life $900; Benefactor $1,200.

Kingfield

THE SKI MUSEUM OF MAINE, 256 Main St., Kingfield, ME 04947. Mailing Address: P.O. Box 359, Kingfield, ME 04947. Tel.: 207-265-2023.
E-mail: info@skimuseumofmaine.org
Web Site: www.skimuseumofmaine.org
Key Personnel: Exec. Dir., Bruce Miles
Institution Type/Description: Ski Museum.
Collections: ski history & heritage; wooden skiis; snowboards; early ads & posters; leather boots; photographs; journals; Hall of Fame.
Hours & Admission Prices: Call for hours.

STANLEY MUSEUM, INC., 40 School St., Kingfield, ME 04947. Mailing Address: P.O. Box 77, Kingfield, ME 04947-0077. Tel.: 207-265-2729. Fax: 207-265-4700.
E-mail: maine@stanleymuseum.org
Web Site: www.stanleymuseum.org
Founded: 1981.
Congressional District: 2
Key Personnel: Dir., Pres. & Chm. (V), D. Howard Randall, Jr.; Vice Chm., John Linderman; Archivist, Jim Merrick; Museum Shop Mgr., Kim Richmond White.
Personnel Profile: Full-Time Paid 2; Full-Time Volunteers 3; Part-Time Paid 3; Part-Time Volunteers 10; Interns 2.
Governing Authority: nonprofit organization. Branch Site: Stanley Museum in Lower Stanley Village, Estes Park, CO 80517. Tax-exempt: 501(c)(3).
Institution Type/Description: History & Transportation Museum: housed in 1903 school designed by F.E. Stanley, inventor of the steam car.
Collections: steam cars; violins; painting; airbrush art; photography; historical materials documenting inventions & activity.
Research Fields: steam cars & photography; Northeast art; violins; airbrush.
Facilities: archive material pertaining to cars, photography & family history; educational facilities. Museum-related items for sale.
Activities: guided tours; lectures; films; concerts; arts festivals; hobby workshops; docent program; participatory, loan & temporary exhibitions; art classes; programs at the Stanley Hotel, Estes Park, CO.
Publications: quarterly magazine, The Stanley Museum; books, Historic Touring, Early Tale of Steam Travel; Reflections of Transportation & Communication, An Evening with R. Buckminister Fuller; Stanley Family Reunion: A Family History Transcript; The Genealogy of the Locomobile Steam Carriage; The Stanleys: Renaissance Yankees; Innovations in Industry & the Arts; Mr. Stanley of Estes Park, 2000; The Stanley Steamer: America's Legendary Steam Car, 2004; Bravo Stanley, 2006.
Hours & Admission Prices: June-Oct. Tues.-Sun. 1-4; Nov.-May Tues.-Fri. 1-4; other times by appointment. Adults $4, seniors $3, children $2; discounts to AAM, AAA, & ICOM members; members no charge. &
Attendance: 45,000 (accurate)
Membership: Individual $45; Family $65; Contributing $100; Supporting & Business $375; Associate $750; Benefactor $1,500.

Kittery

KITTERY HISTORICAL & NAVAL MUSEUM, 200 Rogers Rd. Ext., Kittery, ME 03904-1458. Mailing Address: P.O. Box 453, Kittery, ME 03904. Tel.: 207-439-3080. Fax: 207-439-3080.
E-mail: kitterymuseum@netzero.net
Web Site: kitterymuseum.com
Founded: 1976.
Congressional District: 1
Key Personnel: Dir., Wayne Manson; Museum Shop Mgr. & Program Chair, Barbara Estes.
Personnel Profile: Full-Time Volunteers 1; Part-Time Volunteers 20.
Governing Authority: nonprofit organization. Tax-exempt: 501(c)(3).
Institution Type/Description: Naval & History Museum.
Collections: local history items; documents; ship models; decorative arts; crafts; regional archaeology finds; shipbuilding; maritime.
Research Fields: local maritime history; town history.
Facilities: library by appointment only; 3,000 sq. ft. exhibit space.
Activities: guided tours; lectures; organized educational programs.
Publications: quarterly newsletter, The Bosun's Pipe.
Hours & Admission Prices: June to Columbus Day Tues.-Sat. 10-4; Nov.-May Wed. & Sat. 10-4 or by appointment. Adults $3, senior citizens $2, children 7-15 $1.50; discounts to groups, AAM, ICOM & AAA members; members no charge. &
Attendance: 1,200 (accurate)
Membership: Associate $15; Family $25; Contributing $50; Sustaining $100; Donor $250; Patron $500; Benefactor $1,000.

Lewiston

BATES COLLEGE MUSEUM OF ART, (M), 75 Russell St., Lewiston, ME 04240-6044. Tel.: 207-786-6158. Fax: 207-786-8335.
E-mail: museum@bates.edu
Web Site: www.bates.edu/museum.xml
Founded: 1986.
Congressional District: 2
Key Personnel: Dir., Dan Mills; Cur., William Low; Cur. Education, Anthony Shostak.
Personnel Profile: Full-Time Paid 4; Part-Time Paid 2.
Governing Authority: college. Parent Institution: Bates College. Tax-exempt.
Institution Type/Description: Art Museum.

Collections: works of art from around the world; Marsden Hartley Memorial Collection of drawings & memorabilia; art in all media particularly American & European art; photography.
Research Fields: Marsden Hartley drawings & memorabilia.
Activities: permanent, temporary & traveling exhibitions; gallery talks; integration of visiting artists/scholars into liberal arts curriculum; tours; special events for campus & community; Thousand Words Project & educational programs primarily for secondary education students.
Publications: catalogs, Xiaoze Xie: Amplified Moments (1993-2008); Tale Spinning; Emerging Dis/Order; Ninety-Nine Drawings by Marsden Hartley; Eight Poems & One Essay by Marsden Hartley; Documenting China; Robert Indiana: The Hartley Elegies.
Hours & Admission Prices: Mon.-Sat. 10-5. No charge. Closed major holidays. &
Attendance: 16,000 (accurate)
Membership: Student $10; Member $30 & $50; Patron $100.

MUSEUM L-A, 35 Canal St., Lewiston, ME 04240-7775. Mailing Address: P.O. Box A7, Lewiston, ME 04240. Tel.: 207-333-3881. Fax: 207-376-3353.
E-mail: info@museumla.org
Web Site: museumla.org
Founded: 1996.
Key Personnel: Exec. Dir., Rachel Desgrosseilliers.
Personnel Profile: Full-Time Paid 3; Part-Time Paid 4.
Governing Authority: private; nonprofit organization. Tax-exempt.
Institution Type/Description: History Museum.
Collections: history of Lewiston & Auburn communities; local textile, shoe & brick making industries; personal artifacts; local music history.
Facilities: Museum-related items for sale.
Activities: educational programs; lectures; workshops; special events; group tours.
Publications: spring & fall newsletter.
Hours & Admission Prices: Tues.-Fri. 10-4, Sat. 12-4; groups by appointment. Adults $5, seniors 62 & over $4, students $3; discounts to groups of 10 or more. Closed New Year's Day; Thanksgiving; Christmas Eve & Day. &

UNIVERSITY OF SOUTHERN MAINE - FRANCO-AMERICAN, 51 Westminster St., Rm. 153, Lewiston, ME 04240. Tel.: 207-753-6545.
E-mail: franco@usm.maine.edu
Web Site: usm.maine.edu/franco
Institution Type/Description: History Museum.
Collections: Franco-American culture & history; photographs; government; religion; education; industry; sports; arts.
Hours & Admission Prices: Mon.-Thurs. 9-4.

Liberty

THE DAVISTOWN MUSEUM, 58 Main St., #4, Liberty, ME 04949. Mailing Address: Hulls Cove Office, P.O. Box 144, Hulls Cove, ME 04644-0144. Tel.: 207-288-5126. Fax: 207-288-2725.
E-mail: curator@davistownmuseum.org
Web Site: www.davistownmuseum.org
Founded: 2000.
Key Personnel: C.E.O., Harold G. Brack.
Personnel Profile: Full-Time Volunteers 2; Part-Time Paid 2; Interns 1.
Governing Authority: Tax-exempt.
Institution Type/Description: History & Art Museum.
Collections: history of tool making; New England's maritime culture; Native American history; sculpture; works of local & regional artists.
Facilities: library; garden.
Activities: special events; internships. Annual Event: Art Show.
Publications: series, Hand Tools in History.
Hours & Admission Prices: Summer: Wed.-Fri. & Sun. 11-5, Sat. 10-5; call for additional seasonal hours. No charge; donations accepted.
Attendance: 1,000 (estimated)
Membership: Student $10; Individual $25; Family $30; Associate $50; Contributing $100; Sustaining $500; Partner $1,000; Benefactor $5,000.

Limestone

AROOSTOOK NATIONAL WILDLIFE REFUGE, 97 Refuge Rd., Limestone, ME 04750. Tel.: 207-328-4634.
Founded: 1998.
Institution Type/Description: Wildlife Refuge.
Collections: wildlife & their habitat; black bear; moose; river otters; mink; fish; beavers.
Hours & Admission Prices: Call for hours.

Lincolnville

SCHOOL HOUSE MUSEUM LINCOLNVILLE HISTORICAL SOCIETY, 33 Beach Rd. (Rte. 173), Lincolnville, ME 04849. Mailing Address: P.O. Box 204, Lincolnville, ME 04849-0204. Tel.: 207-789-5445.
E-mail: history@sent.com
Web Site: www.lincolnvillehistory.org/
Institution Type/Description: Historical Society Museum: housed in a one room school built in 1892.
Collections: period furnishings; photographs; genealogy notebooks.
Hours & Admission Prices: June to early Oct. Mon., Wed. & Fri. 1-4. No charge.

Littleton

SOUTHERN AROOSTOOK AGRICULTURAL MUSEUM, 1664 U.S. Rte. 1, Littleton, ME 04730. Tel.: 207-538-9300.
E-mail: info@oldplow.org
Web Site: www.oldplow.org
Institution Type/Description: Agriculture Museum.
Collections: agricultural history & equipment; tractors; photographs.
Activities: special events.
Hours & Admission Prices: By appointment.

Livermore

WASHBURN NORLANDS LIVING HISTORY CENTER, 290 Norlands Rd., Livermore, ME 04253-3807. Tel.: 207-897-4366. Fax: 207-897-4963.
E-mail: norlands@norlands.org
Web Site: www.norlands.org
Founded: 1973.
Congressional District: 2
Key Personnel: Dir., Sheri Leahan.
Personnel Profile: Part-Time Paid 4; Part-Time Volunteers 15.
Governing Authority: Tax-exempt.
Institution Type/Description: Living History Museum.
Collections: Historic Buildings: the Mansion, Library, Meeting House & School House containing artifacts relating to the Washburn Family & 19th century rural Maine.
Research Fields: 19th century rural farm life; Washburn Family history; Livermore history.
Activities: guided tours; interactive living history programs for students & groups; wedding rentals. Seasonal Events: Maple Sugaring & Pancake Breakfast; Christmas at Norlands.
Hours & Admission Prices: Summer Drop-in Living History Tours: July-Aug. Tues. & Thurs.; Tours & Programs: March-Dec. by appointment. Adults $10.
Attendance: 3,500 (estimated)
Membership: Individual $25.

Lovell

LOVELL HISTORICAL SOCIETY - KIMBALL-STANFORD HOUSE, Rte. 5, Lovell, ME 04051. Mailing Address: P.O. Box 166, Lovell, ME 04051. Tel.: 207-925-3234.
E-mail: lovellhist@fairpoint.net
Web Site: lovellhistoricalsociety.org
Institution Type/Description: Historical Society Museum: house built in 1839.
Collections: local history & culture; period artifacts; genealogical records; early maps.
Activities: special events.
Hours & Admission Prices: Tues.-Wed. 9-4, Sat. 9-12; other times by appointment. No charge.

Lubec

LUBEC HISTORICAL SOCIETY, 135 Main St., Lubec, ME 04652. Mailing Address: 155 Main St., Lubec, ME 04652.
E-mail: lubechistoricalsociety@yahoo.com
Founded: 1985.
Congressional District: 2
Key Personnel: Pres. (V), Barbara Sellitto.
Volunteer Hours: 150
Governing Authority: Tax-exempt.
Institution Type/Description: Historical Society Museum.
Collections: local history; period artifacts & furnishings; photographs.
Publications: annual newsletter.

Hours & Admission Prices: June 20-Oct. 15 Mon.-Wed. & Fri. 9-3. No charge; donations accepted. &

Attendance: 1,100 (accurate)

Membership: Individual $10.

QUODDY HEAD STATE PARK, 973 S. Lubec Rd., Lubec, ME 04652. Tel.: 207-733-0911 (Park season) & 941-4014 (Off season).

Web Site: www.maine.gov/cgi-bin/online/doc/parksearch/index.pl

Institution Type/Description: State Park Museum: located on the easternmost point of land in the U.S.

Collections: lighthouse; geological; botanical.

Facilities: 481 acres; hiking trails; picnic sites.

Activities: hiking; whale watching.

Hours & Admission Prices: May 15 to Oct. 15.

Machias

BURNHAM TAVERN MUSEUM, 14 Colonial Way, Machias, ME 04654. Mailing Address: 1027 N. Lubec Rd., Lubec, ME 04652. Tel.: 207-255-6930.

E-mail: info@burnhamtavern.com

Web Site: www.burnhamtavern.com

Founded: 1910.

Congressional District: 2

Key Personnel: Sec., Ruth H. Ahrens, Ed.D.

Personnel Profile: Part-Time Volunteers 10.

Governing Authority: society. Affiliated with Hannah Weston Chapter, Daughters of the American Revolution. Tax-exempt.

Institution Type/Description: Historic Building: c.1770 Burnham Tavern.

Collections: pre-1830 period furnishings; local history.

Research Fields: local history.

Activities: guided tours; permanent exhibitions.

Hours & Admission Prices: mid-June to Sept. Mon.-Fri. 9:30-3:30. Adults $5; discounts to MPBN members. Closed Independence Day; Labor Day.

Attendance: 1,000 (estimated)

Membership: Annual $10.

Machiasport

GATES HOUSE, MACHIASPORT HISTORICAL SOCIETY, 344 Port Rd., Machiasport, ME 04655. Mailing Address: P.O. Box 301, Machiasport, ME 04655-0301. Tel.: 207-255-8461.

E-mail: folknoter@maineline.net

Web Site: www.gateshouse.org

Founded: 1964.

Congressional District: 2

Key Personnel: C.E.O. & Pres. (V), Al Sousa; Recording Sec., Celeste Sherman; Treas., Barbara Maloy.

Personnel Profile: Part-Time Volunteers 20.

Governing Authority: nonprofit organization. Tax-exempt.

Institution Type/Description: Historic House: c.1810 Gates House.

Collections: marine artifacts; period furniture & clothing; genealogy; early post cards of local area; photographs; carpenter's tools.

Research Fields: local history & genealogy; maritime history.

Facilities: library; reading room.

Activities: lectures; films; junior members history research classes.

Publications: booklet, The Whitneyville & Machiasport Railroad; descriptive brochures of Machiasport; biannual newspaper, The Tide; Sails to Rails & Beyond: A Brief Study of Transportation in the Machias Valley of Maine 1763-1983.

Hours & Admission Prices: July-Aug. Tues.-Fri. 12:30-4:30; Sept. Sat. 12:30-4:30; other times by appointment. No charge; donations accepted.

Attendance: 500 (estimated)

Membership: Family $20; Life $300.

Madawaska

MARTIN ACADIAN HOMESTEAD, 137 Saint Catherine St., Madawaska, ME 04756-1423. Tel.: 207-728-6412. Fax: 207-728-6412. Facebook: Martin Acadian Homestead.

E-mail: martinacadianhomestead@yahoo.com

Web Site: visitaroostook.com

Congressional District: 2

Key Personnel: Pres. & Chm. (V), Lois Muller; Treas. & Museum Shop Mgr., Lisa Schroeder; Dir., Paul Muller.

Personnel Profile: Full-Time Volunteers 2; Part-Time Volunteers 6.

Governing Authority: private. National Park Service.

Institution Type/Description: Historic House Museum: built 1823-1860. Listed on the National Register of Historic Places.

Collections: Martin family history; Acadian architecture; period furnishings.

Research Fields: genealogy; Acadian history, restoration & preservation.

Facilities: library.

Activities: guided tours; lectures; temporary exhibitions; school tours. Annual Event: Acadian Festival Open House in August; Martin Family Reunion in August.

Hours & Admission Prices: Daily 10-4 by appointment. No charge; donations accepted. Closed Easter; Thanksgiving; Christmas. &

Attendance: 300 (estimated)

TANTE BLANCHE MUSEUM, U.S. #1, Madawaska, ME 04756-1165. Mailing Address: Madawaska Public Library, 393 Main St., Madawaska, ME 04756-1165. Tel.: 207-728-6412. Fax: 207-728-6412.

E-mail: ljmuller@ymail.com

Web Site: madawaskahistorical.org

Founded: 1968.

Congressional District: 2

Key Personnel: Pres. (V), Lois Muller; Chm. (V), Paul Muller; Treas., Celine Lausier.

Governing Authority: society; nonprofit. Parent Institution: Madawaska Historical Society. Subsidiary Institution: Genealogical Research Center. Tax-exempt: 501(c)(3).

Institution Type/Description: History Museum.

Collections: Acadian & French Canadian artifacts & records. Historic Structure: Acadian Cross Historic Shrine.

Major Exhibits: Acadian Family Mass - St. David Church, 6/14; World Acadian Congress Acadian Reenactment, Aug. 14, 2014.

Research Fields: genealogy.

Facilities: library, for use on premises only.

Activities: guided tours; lectures; temporary exhibitions; educational scholarship; family reunions. Museum Sponsors: Acadian Festival in August.

Publications: annual newsletter.

Hours & Admission Prices: June-Sept. Wed.-Sun. 12-4. No charge; donations accepted.

Attendance: 3,000 (estimated)

Milbridge

MILBRIDGE HISTORICAL SOCIETY AND MUSEUM, Main St., Milbridge, ME 04658. Mailing Address: P.O. Box 194, Milbridge, ME 04658-0194. Tel.: 207-546-4471.

Web Site: milbridgehistoricalsociety.org

Founded: 1989.

Key Personnel: Museum Shop Mgr., Ellen Strout.

Governing Authority: Tax-exempt.

Institution Type/Description: Local History Museum.

Collections: town's shipbuilding history; local industries; genealogy; period artifacts.

Activities: special events; historical programs June-Oct., 2nd Tues. of month; art exhibitions July-August.

Publications: biannual newsletter.

Hours & Admission Prices: June to early Sept. Tues. & Sat.-Sun. 1-4. No charge; donations accepted. &

Attendance: 400 (accurate)

Membership: Individual $10; Life $100.

Millinocket

NORTH LIGHT GALLERY, 256 Penobscot St., Millinocket, ME 04462-1510. Tel.: 207-723-4414.

E-mail: artnorthlight@gmail.com

Web Site: www.artnorthlight.com

Founded: 2005.

Key Personnel: Founder, Marsha Donahue

Institution Type/Description: Art Gallery.

Collections: paintings; drawings; sculpture; crafts.

Activities: special events; temporary & permanent exhibits.

Hours & Admission Prices: Mon.-Sat. 10-6; other times by appointment. &

Milo

MILO HISTORICAL SOCIETY, 23 Park St., Milo, ME 04463-1315. Tel.: 207-943-2268. Facebook: Milo Historical Society.

E-mail: amonroeart@gmail.com

Web Site: www.milohistorical.org

Founded: 1972.

Key Personnel: Dir., Gwen Bradeen; Cur., Virgil Valente.

Personnel Profile: Part-Time Volunteers 6.

Governing Authority: Tax-exempt.
Institution Type/Description: Historical Society Museum.
Collections: town history & heritage; photographs.
Activities: educational programs; public programs; group tours.
Hours & Admission Prices: June-Aug. Tues.-Fri. 1-3; other times by appointment. No charge; donations accepted.
Attendance: 100 (estimated)
Membership: General $5; Patron $25.

Monhegan

THE MONHEGAN MUSEUM, (M), 1 Lighthouse Hill, Monhegan, ME 04852. Tel.: 207-596-7003.
E-mail: museum@monheganmuseum.org
Web Site: www.monheganmuseum.org
Founded: 1968.
Congressional District: 1
Key Personnel: Dir. & Pres., Edward Deci; Cur., Jennifer Pye; Cur., Emily Grey.
Personnel Profile: Part-Time Paid 3; Part-Time Volunteers 60.
Volunteer Hours: 2,000
Operating Expenses: 153,985
Operating Income: 144,641
Governing Authority: society. Parent Institution: Monhegan Historical & Cultural Museum Association, Monhegan, ME. Tax-exempt.
Institution Type/Description: Art & History Museum: housed in 19th-century lighthouse & outbuildings.
Collections: fishing & lobster trapping gear & equipment; wildlife displays; topographical & ecological maps of the island; Indian artifacts; local historical memorabilia; Monhegan Art; lighthouse exhibits.
Major Exhibits: The Famous & the Forgotten, 7/14-9/14.
Research Fields: local history.
Activities: permanent & temporary exhibitions.
Publications: A Spirit of Wonder: Monhegan Artists & the 1913 Armory Show
Hours & Admission Prices: June 24-June 30 & Sept. daily 1:30-3:30; July-Aug. daily 11:30-3:30. No charge; donations requested.
Attendance: 6,000 (estimated)
Membership: Student $2; Regular $15; Fellow $25; Sponsor $100; Patron $250; Benefactor $1,000.

Mount Desert

MOUNT DESERT ISLAND HISTORICAL SOCIETY, 373 Sound Dr., Mount Desert, ME 04660. Mailing Address: P.O. Box 653, Mount Desert, ME 04660-0653. Tel.: 207-276-9323. Fax: 207-276-4204.
Web Site: www.mdihistory.org
Founded: 1931.
Congressional District: 2
Key Personnel: Pres. (V), Kathleen W. Miller; Exec. Dir., Charlotte Singleton.
Personnel Profile: Full-Time Paid 1; Part-Time Paid 1; Part-Time Volunteers 86.
Governing Authority: society; board of directors. Tax-exempt.
Institution Type/Description: Local History Museum.
Collections: period artifacts; books & genealogical records of early Mount Desert Island; furniture; deeds; 2 historic buildings.
Research Fields: genealogy of Mount Desert Island.
Facilities: library of old history books, Bibles & textbooks available for use upon request; research room.
Activities: guided tours; monthly programs; permanent exhibitions.
Publications: biannual newsletter; annual history journal; annual magazine, The Mount Desert Island Historical Society.
Hours & Admission Prices: Somesville Museum: July to mid-Sept. Tues.-Sat. 1-4. Sound School House: June-Aug. Tues.-Sat. 10-4; Sept.-May Mon.-Fri. 10-4. No charge; donations accepted.
Attendance: 3,000 (estimated)
Membership: Individual $25; Family $40; Artifact $60; Archivist $100; Curator's Circle $500; Heirloom Society $1000; Honor Roll $5000.

Naples

NAPLES HISTORICAL SOCIETY, (M), 1757 Village Green Ln., Rte. 302, Naples, ME 04055. Tel.: 207-693-3103.
E-mail: nhs@fairpoint.net
Web Site: www.napleshs.org
Founded: 1972.
Key Personnel: Dir., Meryl J. Watson; Cur., Merry Watson; Cur. Dillingham Collection, Richard Doyle.
Personnel Profile: Part-Time Volunteers 3.
Institution Type/Description: Naples History Museum.

Collections: local history & culture; books; photographs; Native American artifacts.
Activities: Museum Sponsors: Wheels and Water Show in June; Wheels and Water Antique Transportation Show in June; Native American Pow Wow in September.
Publications: brochures, Cumberland and Oxford Canal; Bay of Naples Hotel, The Idol.
Hours & Admission Prices: Sept.-June. No charge; donations accepted.
Attendance: 100 (estimated)
Membership: Individual $5.

New Gloucester

SHAKER MUSEUM, 707 Shaker Rd., New Gloucester, ME 04260-2652. Tel.: 207-926-4597.
E-mail: usshakers@aol.com
Web Site: www.shaker.lib.me.us
Founded: 1931.
Congressional District: 1
Key Personnel: Dir., Leonard L. Brooks; C.E.O., Sr. Frances Carr; Cur., Michael S. Graham; Librarian, Charles E. Rand; Museum Shop Mgr., Wendy Thoren.
Personnel Profile: Full-Time Paid 4; Part-Time Paid 7; Part-Time Volunteers 50.
Governing Authority: nonprofit organization. Parent Institution: United Society of Shakers. Tax-exempt: 501(c)(3).
Institution Type/Description: Religious Museum & Library: located at an active Shaker religious community, comprising of 18 buildings on 1,900 acres dating from the 1760s to 1960s.
Collections: Shaker furniture; tinware; woodenware; folk art; decorative arts; tools; farm implements; Shaker books, manuscripts, ephemera books, manuscripts dealing with religious communities.
Research Fields: research into all aspects of Shakerism, information disseminated through The Shaker Quarterly & 17 publications; herbal medicine; The Koreshan Unity; Muggletonians; Free Will Baptists; Christian Israelitism.
Facilities: 3,000-vol. library available to the public of Shaker (8,000 MSS, 3,300 ephemera); 545-vol Koreshan Unity; 9,795-vol. religion, agriculture & herbs; botanical garden; 7,500 sq. ft. exhibit space. Gift items for sale.
Activities: guided tours; lectures; concerts; organized educational programs; craft workshops & demonstrations; loan & traveling exhibitions; nature walks. Annual Events: Friends of the Shakers Meeting; Apple Saturdays; Maine Festival of American Music; Honorable Harvest; Maine Farm Day; Christmas Fair.
Publications: The Shaker Quarterly; Shaker Your Plate; Gift Drawing & Gift Song; The Sabbathday Lake Shakers; Holy Land; Poems & Prayers; Life in the Christ Spirit; All Things Anew; In The Eye of Eternity; Shakerism For Today; Ingenious & Useful; Growing Up Shaker.
Hours & Admission Prices: Memorial Day to Columbus Day Mon.-Sat. 10-4:30. Tour: adults $7, children $2; discounts to groups & NEMA members; members no charge.
Attendance: 9,082 (accurate)
Membership: Individual $25; Family $35; Sponsor $100; Patron $200; Life $1,000.

New Harbor

COLONIAL PEMAQUID, Colonial Pemaquid State Historical Site, New Harbor, ME 04554. Mailing Address: P.O. Box 117, New Harbor, ME 04554-0117. Tel.: 207-677-2423 (April-Oct); 207-624-6075 (Off-Season).
Web Site: www.friendsofcolonialpemaquid.org
Founded: 1970.
Congressional District: 1
Key Personnel: Mgr., Kelsie Tardif; Pres. (V), Bob Howell; Museum Shop Mgr., Carol Ring.
Personnel Profile: Full-Time Paid 6; Part-Time Volunteers 35.
Governing Authority: state. Affiliated with State of Maine, Bureau of Parks & Lands, Augusta, ME 04333. Tax-exempt.
Institution Type/Description: Archaeological Dig Site.
Collections: Indian & Colonial artifacts; replica of 1692 Fort William Henry.
Research Fields: history; archeology.
Facilities: 70-vol. library of books pertaining to history of Pemaquid.
Activities: guided group tours; lectures; formally organized education programs for children; permanent exhibitions.
Hours & Admission Prices: Memorial Day to Labor Day daily 9-6. Adults $2; children under 12 & seniors over 65 no charge.
Attendance: 100,000 (estimated)
Membership: Single $10; Family $25.

THE FISHERMEN'S MUSEUM, Lighthouse Park, End of Rte. 130, New Harbor, ME 04554. Mailing Address: P.O. Box 263, New Harbor, ME 04554-0263. Tel.: 207-677-2726.
Founded: 1972.
Key Personnel: Dir., John Allan; Pres., Robert Cushing; Chm. (V), Barbara Marshall.
Personnel Profile: Part-Time Paid 4; Part-Time Volunteers 20.
Governing Authority: nonprofit organization. Tax-exempt: 501(c)(3).
Institution Type/Description: Historical & Preservation Society: housed in old 1827 lighthouse keeper's house.
Collections: artifacts; period pictures; charts; models of fishing boats.
Activities: permanent exhibitions.
Hours & Admission Prices: June to mid-Oct. Mon.-Sat. 10-5, Sun. 11-5. No charge; donations accepted. &
Attendance: 46,479 (accurate)

New Portland

NOWETAH'S AMERICAN INDIAN MUSEUM & STORE, 2 Colegrove Rd., New Portland, ME 04961-3821. Tel.: 207-628-4981. Facebook: Nowetah's Indian Store and Museum.
Web Site: www.nowetahs.webs.com
Founded: 1969.
Key Personnel: Owner & Cur., Mrs. Nowetah Cyr; Tour Guide & Visitor Speaker, Thomas Cyr.
Personnel Profile: Full-Time Volunteers 2.
Institution Type/Description: Native American Museum.
Collections: American Indian art from the United States, Canada, & South America with special focus on the Abenaki of Maine; over 600 Native American sweetgrass & brown ash splint baskets; quill embroidered moccasins; bark moose calls & cradleboards; moose hair & quill embroidered pipe bags; hunting decoys; musical instruments including drums & flutes; dolls; carvings; peace-pipes; war bonnets; birch bark hunter canoe; 12 ft. long wood dug-out racing canoe.
Facilities: Museum-related items for sale.
Publications: books, The Indian Massacre of 1724 at Norridgewock, Maine; The History of Indian Wampum; How to Smoke Deerhide - The Old Indian Way; How to Weave an Indian Rug; The Ancient Wisdoms & Knowledge of the Abenaki Indians.
Hours & Admission Prices: Daily 10-5. No charge; donations accepted. Closed Thanksgiving; Christmas. &
Attendance: 1,731 (accurate)

New Sweden

LARS NOAK BLACKSMITH SHOP, LARSSON/OSTLUND LOG HOME & ONE-ROOM CAPITOL SCHOOL, Station Rd., New Sweden, ME 04762. Mailing Address: P.O. Box 50, New Sweden, ME 04762-0050. Fax: 207-896-3199.
E-mail: info@maineswedishcolony.info
Web Site: www.maineswedishcolony.info
Founded: 1989.
Congressional District: 2
Key Personnel: Pres. (V), William Duncan; Vice Pres., Matt Grandy; Dir., Allen Kampe; Dir., Linnea Helstrom; Dir., Ralph Ostlund; Sec., Jean Duncan; Treas., Alwin Espling; Correspondence Sec., Sylvia Kamps; Asst. Treas. & Museum Gift Shop Mgr., Rena Hultgren.
Personnel Profile: Full-Time Volunteers 2; Part-Time Volunteers 25.
Governing Authority: society; nonprofit. Parent Institution: Maine's Swedish Colony, Inc. Tax-exempt.
Institution Type/Description: Historic Village.
Collections: blacksmith shop tools & equipment; woodworking & wheel-making equipment; period furnishings; c.1870 farm log home & buildings; one-room schoolhouse with furnishings & books; blacksmith shop.
Research Fields: history of Maine's Swedish Colony.
Facilities: Gift items for sale.
Activities: living museum activities.
Publications: newsletter to members; historical booklets; monographs; calendars; poetry.
Hours & Admission Prices: By appointment. &
Attendance: 4,500 (estimated)
Membership: Initiation Fee $50; Associate Membership $5; Life Membership $100.

NEW SWEDEN HISTORICAL SOCIETY, 116 Station Rd., New Sweden, ME 04762-3523. Mailing Address: P.O. Box 33, New Sweden, ME 04762. Tel.: 207-896-5200.
Web Site: www.maineswedishcolony.info
Founded: 1925.

Key Personnel: Pres., Gary Dickinson; Sec., Janice McDougal; Treas., Pat Williams; Museum Shop Mgr., Gloria Ringdahl.
Personnel Profile: Part-Time Volunteers 20.
Governing Authority: society; nonprofit corporation. Tax-exempt.
Institution Type/Description: Historical Society Museum: replica of original colony capital which burned in 1971.
Collections: historical collection, from life of early settlers of Maine's Swedish Colony in New Sweden, Maine founded 1870; household furnishings, textile equipment, farm equipment, photographs & portraits of immigrants; early furnishings; restored immigrant cottage, Lindsten Stuga; blacksmith shop; farm house.
Research Fields: family history & culture.
Facilities: picnic area.
Activities: Swedish language lessons. Museum Sponsors: Midsummer Celebration; Founder's Day Celebration; video series on Swedish-American topics in July.
Publications: newsletter.
Hours & Admission Prices: Memorial Day to Labor Day Mon.-Fri. 12-4, Sat.-Sun. 1-4; Sept.-May by appointment. No charge; donations accepted. &
Attendance: 500 (estimated)
Membership: Individual $10; Family $20.

Newcastle

NEWCASTLE HISTORICAL SOCIETY - TANISCOT ENGINE HOUSE, Main St., Newcastle, ME 04553. Mailing Address: P.O. Box 482, Newcastle, ME 04553. Tel.: 207-563-3347.
E-mail: newcastlehistoricalsociety@hotmail.com
Web Site: newcastlemainehistoricalsociety.org
Institution Type/Description: Historical Society Museum.
Collections: local history & culture; period furnishings; personal artifacts; photographs; town & tax records; books.
Hours & Admission Prices: By appointment.

Newfield

19TH CENTURY WILLOWBROOK VILLAGE, (M), 70 Elm St., Newfield, ME 04056. Mailing Address: P.O. Box 28, Elm St., Newfield, ME 04056-0028. Tel.: 207-793-2784.
E-mail: director@willowbrookmuseum.org
Web Site: www.willowbrookmuseum.org
Founded: 1970.
Congressional District: 1
Key Personnel: C.E.O., Amelia E. Chamberlain; Chm. & Pres. (V), Doug King; Museum Shop Mgr., Debbie Albert.
Personnel Profile: Full-Time Paid 1; Part-Time Paid 10; Part-Time Volunteers 210; Interns 1.
Governing Authority: individual operation. Tax-exempt.
Institution Type/Description: History Museum.
Collections: arts & crafts; washing machines; churns; looms; shoe & harness making; carriages and sleighs; bicycles; gasoline engines; wheelright; farm tools & equipment; costumes; musical instruments; toys; scales; barber shop; print shop; carpenter shop; horse stock building; cooperage. Historic Buildings: 1813 restored William Durgin Homestead; 1856 Dr. Isaac Trafton Homestead; 1810 schoolhouse; 1849 restored Concord stagecoach; 1894 restored Armitage-Herschell carousel powered by original steam engine.
Facilities: sandwich shop; ice cream parlor; country store; picnic area. Gift items for sale.
Activities: guided tours to organized educational groups; children's programs; demonstration & displays of heritage crafts & trades.
Publications: biannual newsletter, Willowbrook; annual calendar of events; annual report.
Hours & Admission Prices: Memorial Day to Columbus Day Thurs.-Mon. 10-5. Adults $10, seniors $7.50, students 6-18 $5; discounts to groups over 20, AAM members & local historical society members; children under 6 & members no charge. &
Attendance: 7,582 (accurate)
Membership: Individual $25; Dual $40; Family $60; Friend/Library $120; Patron/Business $250; Sponsor $1,000; Benefactor $2,000.

Nobleboro

NOBLEBORO HISTORICAL SOCIETY, 198 Center St., Nobleboro, ME 04555. Mailing Address: P.O. Box 122, Nobleboro, ME 04555-0122. Tel.: 207-563-5376.
E-mail: sheldon@tidewater.net
Web Site: www.nobleborohistoricalsociety.org
Founded: 1978.

Congressional District: 1
Key Personnel: Pres. & Cur., Mary K. Sheldon; Treas., Eleanor O'Donnell.
Personnel Profile: Part-Time Volunteers 20.
Governing Authority: society; nonprofit. Tax-exempt.
Institution Type/Description: Historical Society Museum: housed in an 1818 schoolhouse.
Collections: agricultural, forestry, Indians, sailing vessels, home utensils, schoolbooks, history books & industry artifacts; genealogical records; maps; cemeteries & town records.
Research Fields: local history; artifacts.
Facilities: 25-vol. library of local history books available to the public; 1,200 sq. ft. exhibit space; 40-seat room with small screen.
Activities: formal education for adults; temporary exhibitions. Annual Events: one day event for all school children; 5 society meetings.
Publications: biannual newsletter, Spring & Fall; brochure; news articles in weekly newspapers.
Hours & Admission Prices: July-Aug. Sat. 1:30-4:30 & for special exhibits; also by appointment. No charge; donations accepted.
Attendance: 450 (estimated)
Membership: Children $1; Single Adult $10; Family $15; Life $200.

Norridgewock

NORRIDGEWOCK HISTORICAL SOCIETY, 11 Mercer Rd., Norridgewock, ME 04957. Mailing Address: P.O. Box 903, Norridgewock, ME 04957. Tel.: 207-634-5032 & 5064.
Institution Type/Description: Historical Society Museum: housed in a former school for women; built in 1832.
Collections: local history & culture; personal artifacts; period furnishings; photographs.
Activities: special events in summer.
Hours & Admission Prices: Memorial Day to Labor Day Sat. 11-1.

North Amity

A.E. HOWELL WILDLIFE CONSERVATION CENTER & SPRUCE ACRES REFUGE, 101 Lycette Rd., North Amity, ME 04471-5114. Tel.: 207-532-6880. Fax: 207-532-6880 (call first).
E-mail: eagleman1008@earthlink.net
Web Site: www.spruceacresrefuge.org
Key Personnel: Vice Pres., Janet M. Easter; Dir., Kim Keehn
Institution Type/Description: Wildlife Refuge.
Collections: native wildlife; plants.
Hours & Admission Prices: mid-May to Oct. 15 Tues.-Sat. 10-4 by appointment. Adults $10; children under 16 no charge.

North Yarmouth

SKYLINE FARM CARRIAGE MUSEUM, 95 The Lane, North Yarmouth, ME 04097. Tel.: 207-829-9203.
E-mail: info@skylinefarm.org
Web Site: www.skylinefarm.org
Governing Authority: nonprofit organization. Tax-exempt: 501(c)(3).
Institution Type/Description: History Museum.
Collections: period carriages & sleighs; historic buildings; horse-drawn transportation history; New England's cultural, social & economical development.
Activities: educational & recreational programs.
Hours & Admission Prices: Call for hours.

Northeast Harbor

GREAT HARBOR MARITIME MUSEUM, (M), 124 Main St., Northeast Harbor, ME 04662. Mailing Address: P.O. Box 145, Northeast Harbor, ME 04662-0145. Tel.: 207-276-5262 & 5650 (Off Season).
E-mail: sydr@me.com
Web Site: www.greatharbormaritimemuseum.org
Formerly: Great Harbor Collection
Founded: 1982.
Congressional District: 5
Key Personnel: Co Chm. (V), Sydney Roberts Rockefeller; Co Chm. (V), Carl E. Kelley.
Personnel Profile: Part-Time Paid 1; Part-Time Volunteers 4.
Governing Authority: private; nonprofit organization. Tax-exempt: 501(c)(3).
Institution Type/Description: Maritime Museum.
Collections: regional boat models; boat builders; boat design; education; arts; artifacts.
Research Fields: oral history; lobstering; boat building; steamships; small boats; health of our oceans.

Facilities: 2,700 sq. ft. exhibit space. Museum-related items for sale.
Activities: educational art programs; island exploration; lectures; arts festivals; films; formal education programs for children; participatory exhibits. Annual Events: Exhibit Openings; Christmas Festival.
Hours & Admission Prices: Mon.-Sat. 10-5. Suggested Donation: $3. ♿
Attendance: 5,000 (estimated)

Norway

NORWAY HISTORICAL SOCIETY MUSEUM, 471 Main St., Norway, ME 04268. Tel.: 207-743-7377.
Key Personnel: Cur., Charles Longley
Institution Type/Description: Historical Society Museum.
Collections: local history & culture; works by local artists; period clothing; snowshoes; Norway's architectural history; china; photographs.
Activities: educational programs; special events; research.
Hours & Admission Prices: June-Aug. Tues. & Sat. 9 to noon; Sept.-May Sat. 9 to noon; other times by appointment.

Oakfield

OAKFIELD RAILROAD MUSEUM, Station St., Oakfield, ME 04763. Mailing Address: Oakfield Historical Society, P.O. Box 176, Oakfield, ME 04763-0176. Tel.: 207-757-8575.
E-mail: oakfieldmuseum@pwless.net
Web Site: www.oakfieldmuseum.org
Founded: 1986.
Key Personnel: Pres., Arthur Collier; Chm. (V), Pres. (V), & Museum Shop Mgr., Alberta McDonald.
Personnel Profile: Part-Time Volunteers 30.
Governing Authority: Tax-exempt.
Institution Type/Description: Railroad Museum.
Collections: early railroad transportation; photographs; period signs; signal lanterns; maps; telegraph equipment; newspapers; a Hand Car; Motor Car; C-66 caboose; small scale model railroad station.
Facilities: Museum-related items for sale.
Activities: tours; special events.
Publications: annual calendar.
Hours & Admission Prices: late May to Labor Day Sat.-Sun. 1-4. No charge; donations accepted.
Attendance: 250 (estimated)
Membership: Student $2; Individual $4; Family $10; Friend of Society $30.

Oakland

MACARTNEY HOUSE MUSEUM, 25 Main St., Oakland, ME 04963. Mailing Address: P.O. Box 59, Oakland, ME 04963-0059. Tel.: 207-465-7549.
Founded: 1979.
Key Personnel: Pres. (V), Alberta Porter; Treas., Richard Lord; Cur., Ruth W. Wood.
Personnel Profile: Part-Time Volunteers 60.
Governing Authority: private; nonprofit organization.
Institution Type/Description: History Museum: housed in the former home of mill owner, Leonard Cornforth, built 1815.
Collections: Oakland area history; period furniture; artifacts.
Facilities: research library.
Publications: biannual newsletter.
Hours & Admission Prices: Summer: Wed. 1:30-4:30; other times by appointment. No charge; donations accepted.
Membership: Individual $5; Family $7.50; Life $50.

Ogunquit

BARN GALLERY, Shore Rd. & Bourne Lane, Ogunquit, ME 03907. Mailing Address: P.O. Box 529, Ogunquit, ME 03907-0529. Tel.: 207-646-8400.
Governing Authority: Parent Institution: Ogunquit Arts Collaborative. Tax-exempt: 501(c)(3).
Institution Type/Description: Art Gallery.
Collections: works by regional artists; paintings; sculpture.
Activities: workshops; lectures; special events; educational programs.
Hours & Admission Prices: Call for hours.

OGUNQUIT HERITAGE MUSEUM, 86 Obeds Lane, Ogunquit, ME 03907. Mailing Address: P.O. Box 875, Ogunquit, ME 03907. Tel.: 207-646-0296.
Web Site: ogunquitheritagemuseum.org
Founded: 2001.

Institution Type/Description: History Museum: housed in the former home of Captain James Winn; c.1780.
Collections: local history; period furnishings; personal artifacts.
Research Fields: Ogunquit history.
Hours & Admission Prices: June-Sept. Tues.-Sat. 1-5. No charge. &

OGUNQUIT MUSEUM OF AMERICAN ART, (M), 543 Shore Rd., Ogunquit, ME 03907-0815. Mailing Address: P.O. Box 815, Ogunquit, ME 03907-0815. Tel.: 207-646-4909. Fax: 207-646-6903.
E-mail: rcrusan@ogunquitmuseum.org
Web Site: ogunquitmuseum.org
Founded: 1952.
Congressional District: 1
Key Personnel: Dir., Ron Crusan; Pres. (V), Michael Kenslea; Coord. Visitor Svcs. & Museum Shop Mgr., Susan Joy Sager; Financial Operations & Membership Coord., Marsha Sibley.
Personnel Profile: Full-Time Paid 1; Part-Time Paid 6; Part-Time Volunteers 20; Interns 2.
Governing Authority: nonprofit organization. Tax-exempt: 501(c)(3).
Institution Type/Description: Art Museum.
Collections: 19th-century to present American art.
Facilities: outside sculpture garden. Museum-related items for sale.
Activities: permanent exhibitions; individual & group shows.
Publications: annual illustrated catalog; museum bulletin.
Hours & Admission Prices: May-Oct. Mon.-Fri. 9:30-3:30. Adults $10, seniors & students $9; children under 12 & members no charge. Closed Labor Day. &
Attendance: 16,000 (accurate)
Membership: Individual $35; Dual $60; Family $85; Associate $150; Supporting $250; Donor $500; Partner's Circle $1,000; Benefactor $5,000; Corporate $10,000.

Old Orchard Beach

OLD ORCHARD BEACH HISTORICAL SOCIETY, 4 Portland Ave., Old Orchard Beach, ME 04064. Mailing Address: P.O. Box 464, Old Orchard Beach, ME 04064. Tel.: 207-934-9319.
E-mail: oobhistsoc@maine.rr.com
Web Site: www.harmonmuseum.org
Founded: 1954.
Congressional District: 1
Key Personnel: Museum Trustee, Daniel Blaney; Pres., Arthur Guerin; Vice Pres., Charles Davis; Sec., Arlene Hanson; Treas., Stanley Quinlan; Cur., Jeanne Guerin; Projects Mgr. & Researcher, Janet Hamilton
Institution Type/Description: Historical Society Museum.
Collections: local history & culture; archives; photographs.
Major Exhibits: Cemeteries of O.O.B. on Private Property, 6/14-9/14; Churches-Temples-Tabernacle of O.O.B. Past & Present, 6/14-9/14; Fire Room: Photo Display of Major Fires thru the Years 1907-2004 Police Dept. Photos, 6/14-9/14.
Activities: group tours.
Hours & Admission Prices: June 24 to Labor Day Tues.-Fri. 11-4, Sat. 10-2; other times by appointment. No charge; donations accepted.
Attendance: 450 (estimated)
Membership: Individual $10; Family $15; Life $25.

Old Town

OLD TOWN MUSEUM, 353 Main St., Old Town, ME 04468-1536. Mailing Address: P.O. Box 375, Old Town, ME 04468-0375. Tel.: 207-827-7256.
E-mail: eustis@infionline.net
Web Site: oldtownmuseum.com
Founded: 1976.
Congressional District: 2
Key Personnel: Exec. Bd. Member, Richard Eustis; Pres. (V), Bill Osborne; Museum Shop Mgr., Mary Giboleau.
Personnel Profile: Part-Time Volunteers 15.
Governing Authority: nonprofit. Tax-exempt: (501)(c)(3).
Institution Type/Description: History Museum: housed in c.1928 former Church.
Collections: Civil War items; period telephones; costumes; glass ceramics; photographs; logging artifacts.
Facilities: Museum-related items for sale.
Activities: guided tours; lectures; films; organized educational programs for children; loan exhibitions.
Hours & Admission Prices: May-Oct. Fri.-Sun. 1-5. No charge; donations accepted. &
Attendance: 750 (accurate)

Membership: Regular $20; Civic Group & Corporate $100.

Orland

ORLAND HISTORICAL SOCIETY, Castine Rd., Rte. 175, Orland, ME 04472. Mailing Address: P.O. Box 242, Orland, ME 04472. Tel.: 207-469-2476.
Founded: 1966.
Key Personnel: Pres. (V), Cindi Kimball; Treas., JoAnn Carlson.
Personnel Profile: Part-Time Volunteers 3.
Governing Authority: society. Tax-exempt.
Institution Type/Description: General Museum: housed in 1800s store.
Collections: local historical artifacts.
Facilities: library of historical documents pertaining to early settlement of Orland available by appointment.
Activities: guided tours.
Publications: Orland Historical Highlights
Hours & Admission Prices: July-Sept. Tues. & Sat. 2-4; other times by appointment. No charge.
Attendance: 100 (estimated)
Membership: One Year $5; Two Years $10; Three Years $15; Five Years $20; Life $125.

Orono

THE FAY HYLAND BOTANICAL PLANTATION, 5751 Murray Hall, Univ. of Maine, Orono, ME 04469. Tel.: 207-581-2540. Fax: 207-581-2537.
Founded: 1934.
Key Personnel: Chm., Dr. Christopher S. Campbell.
Governing Authority: university.
Institution Type/Description: Arboretum.
Collections: trees & shrubs.
Research Fields: reproductive biology & biosystematics of woody plants.
Facilities: botanical garden; arboretum.
Activities: formally organized education programs for undergraduate college students.
Hours & Admission Prices: Daily dawn-dusk. No charge.

HUDSON MUSEUM, THE UNIVERSITY OF MAINE, (M), 5746 Collins Center for the Arts, Orono, ME 04469. Tel.: 207-581-1901. Fax: 207-581-1950.
E-mail: hudsonmuseum@umit.maine.edu
Web Site: www.umaine.edu/hudsonmuseum/
Founded: 1986.
Congressional District: 2
Key Personnel: Dir., Gretchen Faulkner; Registrar, Susan M. Smith.
Personnel Profile: Full-Time Paid 1; Part-Time Paid 1; Part-Time Volunteers 12; Interns 2.
Governing Authority: state government. Parent Institution: University of Maine. Tax-exempt.
Institution Type/Description: Anthropology Museum.
Collections: ethnology collections from Oceania, Africa, Central, North & South America, Arctic, Asia; archaeological material from North, Central & South America.
Research Fields: ethnographic research in Northeastern United States.
Activities: guided tours; lectures; family program; collectors' workshop; intermuseum loans; permanent & temporary exhibitions. Annual Events: Maine Indian Basketmakers sale & demonstration.
Publications: gallery guides; Bibliographic Guide to the Native Peoples of Maine; exhibition catalogues.
Hours & Admission Prices: Mon.-Fri. 9-4, Sat. 11-4. No charge; donations accepted. Closed federal & state holidays. &
Attendance: 67,427 (estimated)
Membership: Student, Educator & Senior $25; Basic $40; Contributor $100-$249; Supporter $250-$499; Palmer Circle $500 & up.

PAGE FARM & HOME MUSEUM - THE UNIVERSITY OF MAINE, 5787 Museum Barn, Portage Rd., Orono, ME 04469-5787. Tel.: 207-581-4100.
Web Site: www.umaine.edu/pagefarm/
Key Personnel: Dir., Patricia Hemer
Institution Type/Description: History Museum.
Collections: Maine farming history & cultural heritage.
Activities: special events.
Hours & Admission Prices: Tues.-Sat. 9-4. Closed holidays.

Orrington

CURRAN HOMESTEAD LIVING HISTORY FARM & MUSEUM, 372 Fields Pond Rd., Orrington, ME 04474. Mailing Address: P.O. Box 107, Orrington, ME 04474-0107. Tel.: 207-945-9311. Fax: 207-942-9914.
E-mail: irv@bangorlettershop.com
Web Site: curranhomestead.org
Founded: 1991.
Congressional District: 2
Key Personnel: Dir., Bruce R. Bowden; Pres. (V), Richard A. Stockford.
Personnel Profile: Full-Time Volunteers 1; Part-Time Volunteers 25.
Governing Authority: Tax-exempt.
Institution Type/Description: Living History Farm & Museum.
Collections: farm history; period furnishings & artifacts; farm equipment & tools; vegetable & herb gardens; photographs. Historic Structures: early 20th century farmhouse & barn; 19th-century blacksmith shop.
Activities: guided tours; demonstrations; traditional-skills.
Publications: occasional newsletter.
Hours & Admission Prices: Call or see website for hours or to arrange tours.
Attendance: 1,500 (estimated)

Owls Head

OWLS HEAD TRANSPORTATION MUSEUM, 117 Museum St., Rte. 73, Owls Head, ME 04854. Mailing Address: P.O. Box 277, Owls Head, ME 04854-0277. Tel.: 207-594-4418. Fax: 207-594-4410.
E-mail: info@ohtm.org
Web Site: www.ohtm.org
Founded: 1974.
Congressional District: 1
Key Personnel: Dir., Charles Chiarchiaro; Cur. & Dir. Education, Ethan Yankura; Chm., James S. Rockefeller, Jr.
Personnel Profile: Full-Time Paid 12; Part-Time Paid 2; Part-Time Volunteers 200; Interns 2.
Governing Authority: nonprofit organization. Tax-exempt: 501(c)(3).
Institution Type/Description: Transportation Museum.
Collections: pioneer air & ground vehicles.
Research Fields: evolution of transportation.
Facilities: library; reading room; nature park. Gift items for sale.
Activities: guided tours; lectures; films; gallery talks; formally organized education programs; operational weekend displays.
Publications: quarterly newsletter, The Strut and Axle; collections catalog; History of the Museum.
Hours & Admission Prices: Daily 10-5. Adults $10, seniors 65 & over $8; children under 18 & members no charge. Closed New Year's Day; Thanksgiving; Christmas.
Attendance: 96,377 (accurate)
Membership: Individual $40; Family $60; Participating $100; Supporting $250; Sustaining $500; Benefactor $1,000; Life $5,000.

Patten

PATTEN LUMBERMEN'S MUSEUM, INC., 61 Shin Pond Rd., Patten, ME 04765. Mailing Address: P.O. Box 300, Patten, ME 04765-0300. Tel.: 207-528-2650. Fax: 207-528-2650.
E-mail: curator@lumbermensmuseum.org
Web Site: www.lumbermensmuseum.org
Founded: 1963.
Key Personnel: Dir. & Cur., Rhonda R. Brophy; Pres., Donald Shorey; Vice Pres., Bud Blumenstock.
Personnel Profile: Full-Time Paid 1; Full-Time Volunteers 40; Part-Time Paid 3; Part-Time Volunteers 30; Interns 1.
Governing Authority: nonprofit organization. Tax-exempt: 501(c)(3).
Institution Type/Description: Logging & Lumbering History Museum.
Collections: lumberman, carpenter, wheelwright, cooper & blacksmith tools; horse drawn logging equipment; bateaux; Lombard loghaulers; photographs; paintings; working models of saw mills; early trucks; 1820 logging camp; double camp; single camp; blacksmith shop.
Facilities: reception center. Books, postcards & memorabilia for sale.
Activities: guided tours; permanent exhibitions. Annual Event: Bean Hole Bean Dinner in August.
Hours & Admission Prices: Memorial Day to June Fri.-Sun. 10-4; July-Oct. Tues.-Sun.10-4. Adults $8, seniors $7, children 7-12 $3; discounts to groups, AAM & AAA members; members & children under 6 no charge.
Attendance: 3,800 (estimated)
Membership: Individual $10; Family $20; Friend $25; Steward $50; Supporter $75; Sustaining $100; Sponsor $150; Patron $200; Benefactor $500; Guardian $1,000.

Peaks Island

FIFTH MAINE REGIMENT MUSEUM, 45 Seashore Ave., Peaks Island, ME 04108-1311. Mailing Address: P.O. Box 41, Peaks Island, ME 04108-0041. Tel.: 207-766-3330. Fax: 207-766-5514.
E-mail: fifthmaine@portland.twcbc.com
Web Site: www.fifthmainemuseum.org
Founded: 1954.
Congressional District: 1
Key Personnel: Pres., Sharon McKenna; Treas., Nancy Cuthbertson.
Personnel Profile: Full-Time Paid 1; Part-Time Paid 1; Part-Time Volunteers 50; Interns 1.
Governing Authority: private; nonprofit. Tax-exempt: 501(c)(3).
Institution Type/Description: Military Museum: housed in the Fifth Maine Regiment Memorial Hall, built in 1888 as a Civil War Memorial & Reunion Hall; listed in the National Register of Historic Places.
Collections: Civil War with special emphasis on the Fifth Maine soldiers & Maine's role; Peaks Island history from c.1600 to present; military artifacts.
Research Fields: war time activities of the Fifth Maine Regiment; post-war lives of many of the Fifth Maine soldiers; Peaks Island history.
Facilities: library; resource room; 100-seat auditorium; 1,528 sq. ft. exhibit space; dining room. Civil War & local history books, maps, T-shirts and caps for sale.
Activities: concerts; docent program; formal education programs for children & college students; guided tours; lectures; temporary & permanent exhibitions. Annual Events: fair; Civil War reenactment.
Publications: semi-annual newsletter, Fifth Maine News; An Island at War: The Peaks Island Military Reservation 1942-1946.
Hours & Admission Prices: Mon.-Fri. 12-4, Sat.-Sun. 11-4. Suggested Donation: adults $5; discounts to AAM, MAM, & NEMA, AASLH and Civil War Preservation Trust members.
Attendance: 6,000 (accurate)
Membership: Individual $15; Family $30; Patron $50; Hundred Club $100; Corporate $250.

THE UMBRELLA COVER MUSEUM, (M), 62-B Island Ave., Peaks Island, ME 04108. Tel.: 207-939-0301. Facebook: Umbrella Cover Museum.
E-mail: info@umbrellacovermuseum.org
Web Site: www.umbrellacovermuseum.org
Founded: 1996.
Key Personnel: Dir. & Cur., Nancy 3. Hoffman
Institution Type/Description: Umbrella Cover Museum.
Collections: over 730 umbrella covers.
Activities: guided tours.
Publications: book, Uncovered And Exposed, A Guide to the Umbrella Cover Museum.
Hours & Admission Prices: Summer: Tues.-Sat. 10-1 & 2-5, Sun. 10-1. No charge; donations requested. Peaks Island is a 20 minute ferry ride from Portland.
Attendance: 2,700 (accurate)
Membership: $5-$100.

Pemaquid

HARRINGTON MEETING HOUSE MUSEUM, 278 Harrington Rd., Pemaquid, ME 04558. Mailing Address: 239 Harrington Rd., Pemaquid, ME 04558. Tel.: 207-677-2193.
Congressional District: 1
Governing Authority: Tax-exempt: 501(c)(3).
Institution Type/Description: Historic House Museum: built in 1772.
Collections: local history & culture; photographs; early tools; period clothing; instruments; maps; documents; personal artifacts.
Hours & Admission Prices: July-Aug. Mon., Wed. & Fri. 2-4:30. No charge; donations accepted.

Phillips

PHILLIPS HISTORICAL SOCIETY, 8 Pleasant St., Phillips, ME 04966. Mailing Address: P.O. Box 216, Phillips, ME 04966-0216. Tel.: 207-639-3111 & 2888 (Tours).
Founded: 1959.
Congressional District: 2
Key Personnel: Pres., Dennis Atkinson.
Personnel Profile: Part-Time Volunteers 12.
Governing Authority: society. Subsidiary Institution: Sandy River & Rangeley Lakes R.R. Tax-exempt: 501(c)(3).
Institution Type/Description: General Museum: housed in 1832 house owned by important local families connected with town's history.

Collections: S.R. & R.L. railroad relics & railroad park; Portland glass; antique furniture; maps; documents; photographs; antique tools; archives; history; art exhibit; rural schoolroom replica; household implements; farming implements; quilts; collection of bound local newspapers & town histories available for use at the society.

Research Fields: narrow gauge railroad history, Sandy River & Rangeley Lakes Railroad.

Facilities: Document Room of family histories. Postcards & assorted gifts for sale.

Activities: guided tours; permanent exhibitions.

Publications: Pease History of Phillips.

Hours & Admission Prices: June-Aug. first & third Sun. 1-3; other times by appointment. Railroad: June-Sept. 1st & 3rd Sun. No charge; donations accepted. &

Attendance: 1,000 (estimated)

Membership: Single $10; Family $20; Lifetime $100.

Phippsburg

PHIPPSBURG HISTORICAL SOCIETY, Parker Head Rd., Phippsburg, ME 04562. Mailing Address: P.O. Box 21, Phippsburg, ME 04562-0021.

Institution Type/Description: Historical Society Museum.

Collections: local history & culture; period furnishings; personal artifacts; photographs; clothing.

Hours & Admission Prices: Summer: Mon.-Fri. 2-4.

Pittsfield

THE DEPOT HOUSE MUSEUM, 114 Central St., Pittsfield, ME 04967. Mailing Address: Pittsfield Historical Society, P.O. Box 181, Pittsfield, ME 04967-0181. Tel.: 207-487-4926.

Institution Type/Description: Historic Building: built in 1886.

Collections: local history & culture; depot artifacts; personal artifacts; photographs.

Hours & Admission Prices: April-Oct. Tues.-Sat. 10-1; Nov.-March by appointment.

Pittston

MAJOR REUBEN COLBURN HOUSE, 33 Arnold Rd., Pittston, ME 04345-5145. Mailing Address: Bureau of Parks & Lands, 22 State House Station, Augusta, ME 04333.

Governing Authority: Parent Institution: Maine Bureau of Parks and Lands. Subsidiary Institution: Arnold Expedition Historical Society.

Institution Type/Description: Historic House: built in 1765. Listed on the National Register of Historic Places.

Collections: local history; personal artifacts; period furnishings; historic batteaux; canoes.

Hours & Admission Prices: Summer: Sat.-Sun. call for hours.

Poland

POLAND HISTORICAL SOCIETY MUSEUM, 1231 Maine St., Poland, ME 04274. Tel.: 207-998-4601. Fax: 207-998-2002.

Institution Type/Description: Historical Society Museum: housed in a former schoolhouse.

Collections: local history & culture; period artifacts; photographs.

Activities: special events.

Hours & Admission Prices: July-Aug. call for hours.

Poland Spring

MAINE STATE BUILDING, 37 Preservation Way, Poland Spring, ME 04274. Mailing Address: P.O. Box 444, Poland Spring, ME 04274. Tel.: 207-998-4142.

E-mail: polandspringpreservation@gmail.com

Web Site: www.polandspringps.org

Governing Authority: Parent Institution: Poland Spring Preservation Society.

Institution Type/Description: Historic Building: built in 1893 for the Chicago Worlds Fair with materials sent from Maine to Chicago. The structure was later dismantled and moved back to Maine by train in 1895.

Collections: building history; books; paintings; period furnishings.

Facilities: Museum-related items for sale.

Activities: special events.

Hours & Admission Prices: Memorial Day to Columbus Day Tues.-Sat. 9-4.

Port Clyde

MARSHALL POINT LIGHTHOUSE MUSEUM, 179 Marshall Point Rd., Port Clyde, ME 04855. Mailing Address: P.O. Box 247, Port Clyde, ME 04855-0247. Tel.: 207-372-6450.

Web Site: www.marshallpoint.org

Key Personnel: Chm., Bob Sierer; Dir., Jim Quinn

Institution Type/Description: History Museum.

Collections: local quarry history; tool; photographs; fishing in St. George; miniature lobster trap buoys; lighthouse memorabilia; area history.

Facilities: Museum-related items for sale.

Hours & Admission Prices: May Sat.-Sun. 1-5; Memorial Day to Columbus Day Sun.-Fri. 1-5, Sat. 10-5. No charge.

Porter

PARSONSFIELD-PORTER HISTORICAL SOCIETY, 92 Main St., Porter, ME 04068. Mailing Address: P.O. Box 250, Parsonsfield, ME 04047-0250. Tel.: 207-625-7019. localhistorymatters-.blogspot.com.

E-mail: pphs@parsonsfieldporterhistorical.org

Web Site: www.parsonsfieldporterhistorical.org

Founded: 1946.

Congressional District: 2

Key Personnel: Pres., Sylvia P. Wilson; Vice Pres., Patricia Turner; Sec., Janice M. Iler; Treas., Meredith Shea; Auditor, Cynthia Berube.

Personnel Profile: Part-Time Volunteers 10.

Governing Authority: society. Tax-exempt: 501(c)(3).

Institution Type/Description: Historic House Museum: Parsonsfield-Porter History House.

Collections: historical memorabilia. Historic Structures: Porter-Parsonsfield Covered Bridge; 1824 Porter Old Meeting house, built by Bullockites, a local Baptist group of the time.

Research Fields: local history; genealogy.

Facilities: library of historical, genealogical & town reports available for use by permission.

Activities: field trips; slide shows. Museum Sponsors: Open House 1st Sat. in June & 4th Sun. in Sept.; Meetings April-Oct. 4th Sat.

Publications: quarterly newsletter, Porter Meeting House History; books, Porter Maine - 200 Years; Cemeteries of the Town of Porter, Maine; Parsonfield, Maine A Town of Many Villages, 1785-2012.

Hours & Admission Prices: April-Nov. by appointment. Research Room: Mon. 10-1. No charge; donations accepted. Closed holidays. &

Attendance: 350 (estimated)

Membership: Student (under 18) $5; Individual $10; Couple $15.

Portland

CHILDREN'S MUSEUM & THEATRE OF MAINE, 142 Free St., Portland, ME 04101-3961. Mailing Address: P.O. Box 4041, Portland, ME 04101-0241. Tel.: 207-828-1234, ext. 221. Fax: 207-828-5726.

E-mail: suzanne@kitetails.com

Web Site: www.kitetails.org

Founded: 1977.

Congressional District: 1

Key Personnel: C.E.O. & Exec. Dir., Suzanne Olson; Pres. (V), Michael Bockrque; Dir. Exhibits, Christopher Sullivan; Dir. Finance, Sara Merrill; Devel., Cathy Prichard; Mktg. & Public Rels. Mgr., Lucy Bangor.

Personnel Profile: Full-Time Paid 10; Part-Time Paid 13; Part-Time Volunteers 50; Interns 2.

Governing Authority: nonprofit organization. Tax-exempt: 501(c)(3).

Institution Type/Description: Children's Museum: housed in modified four-story historic business building.

Collections: interactive & hands-on exhibits including fire safety, marine, supermarket, bank, space shuttle, star lab, giant globe; one of the largest camera obscura's in U.S.; multicultural & early childhood programming; climbing wall; environmental education; sailing ship; car repair shop.

Research Fields: learning methods.

Facilities: theater; cafe.

Activities: self-guided tours; demonstrations; science shows; planetarium shows; artist-in-residence workshops; musical performances; clubs; school tours; birthday parties; overnights; summer day camps; children's theatre classes & performances.

Publications: quarterly newsletter; membership & museum brochures; annual report.

Hours & Admission Prices: Memorial Day-Labor Day & school vacation weeks Mon.-Sat. 10-5, Sun. 12-5; Sept.-May Tues.-Sat. 10-5, Sun. 12-5. Adults $9; discounts to groups and AAA & ACM reciprocal members; 1st Fri. each month 5pm-8pm, children under 18 months & members no charge.

Closed New Year's Day; Easter; Independence Day; Thanksgiving; Christmas Eve & Day. &

Attendance: 100,000 (accurate)

Membership: Caregiver option $10; Plus option $30; Just the Two of Us $65; Grandparent $75; Family $95; Reciprocal $125; Sponsoring $175; Corporate $250-$5,000; Benefactor $500.

INSTITUTE OF CONTEMPORARY ART AT MAINE COLLEGE OF ART (ICA AT MECA), 522 Congress St., Portland, ME 04101-3378. Tel.: 207-879-5742, ext. 229. Fax: 207-780-0816.

E-mail: ica@meca.edu

Web Site: www.meca.edu/ica

Founded: 1983.

Congressional District: 1

Key Personnel: Dir., Lauren Fensterstock; Assoc. Cur., Linda Lambertson.

Personnel Profile: Full-Time Paid 2; Part-Time Paid 1; Interns 4.

Governing Authority: college. Parent Institution: Maine College of Art. Tax-exempt: 501(c)(3).

Institution Type/Description: Art Museum: housed in 1904 Beaux Art design Porteous Building.

Collections: contemporary art.

Research Fields: contemporary art.

Facilities: 3,300 sq. ft. exhibit space.

Activities: guided tours; lectures; films; organized education programs for undergraduate or graduate college students; traveling exhibitions.

Publications: exhibition catalogues; Translation/Seduction/Displacement Beyond Decorum: The Photography of Ike Ude; William Pope.L: The Friendliest Black Artist in America; Wenda Guifro: Art from Middle Kingdom to Biological Millennium; Revisioning Portland.

Hours & Admission Prices: Wed. & Fri.-Sun. 11-5, Thurs. 11-7. No charge; donations accepted. Closed legal holidays. &

Attendance: 20,000 (accurate)

INTERNATIONAL CRYPTOZOOLOGY MUSEUM, 11 Avon St., Portland, ME 04101. Mailing Address: P.O. Box 4311, Portland, ME 04101.

E-mail: lcoleman@maine.rr.com

Web Site: cryptozoologymuseum.com

Key Personnel: Dir. & Founder, Loren Coleman; Asst. Dir., Jeff Meuse; Coord. Special Events, Hannah DeLong; Treas., Caleb Cone-Coleman; Chief Tour Coord., Sarah McCann; Coord. Arts Exhibits, Andy Finkle.

Governing Authority: nonprofit organization.

Institution Type/Description: Cryptozoology Museum.

Collections: photographs; paintings; cryptids including Sasquatch & Loch Ness Monster; sculptures.

Facilities: Museum-related items for sale.

Activities: school field trips; group tours.

Publications: books.

Hours & Admission Prices: Mon. 12-4, Wed.-Sat. 11-4, Sun. 12-3:30; other times by appointment. Adults 13 & over $7, children 12 & under $5; children under one no charge. Closed New Year's Day; Thanksgiving & day before; Christmas.

MAINE HISTORICAL SOCIETY, (M), 489 Congress St., Portland, ME 04101-3414. Tel.: 207-774-1822. Fax: 207-775-4301.

E-mail: info@mainehistory.org

Web Site: www.mainehistory.com

Founded: 1822.

Congressional District: 1

Key Personnel: Dir., Stephen Bromage; Dir. Devel., Nan Cumming; Interim Dir. Finance, David Sullins; Head Library Svcs., Nicholas Noyes; Head Cur., John Mayer.

Personnel Profile: Full-Time Paid 20; Part-Time Paid 16; Part-Time Volunteers 125; Interns 10.

Governing Authority: society. Subsidiary Institution: Center for Maine History. Tax-exempt.

Institution Type/Description: History Museum.

Collections: manuscripts; architectural & engineering drawings; photographs; prints; broadsides; paintings; textiles; costumes; ceramics; glass; metalware; original furnishings & memorabilia of the Wadsworth & Longfellow families. Historic Building: Wadsworth-Longfellow House, 1785 boyhood home of poet Henry Wadsworth-Longfellow.

Major Exhibits: This Rebellion: Maine in the Civil War, 6/28/13-5/26/14.

Research Fields: Maine history & genealogy.

Facilities: over 60,000-vol. library of books & 2 million manuscripts; 11,000 artifacts of Maine history & genealogy available for use on the premises; reading room.

Activities: lectures; tour program; temporary exhibitions.

Publications: quarterly newsletter, Maine Historical Society Quarterly; monographs; guide book; postcards.

Hours & Admission Prices: Library: Tues.-Sat. 10-4. Maine Historical Society Museum: May-Oct. Mon.-Sat. 10-5, Sun. 12-5; Nov.-April Mon.-Sat. 10-5. Wadsworth-Longfellow House: May-Oct. Mon.-Sat. 10-5, Sun. 12-5. Museum: adults $8, seniors over 65, students & AAA members $7, children 6-17 $2; discounts to groups of 5 or more. House & Museum: adults $12, seniors over 65, students & AAA members $10, children 6-17 $3; discounts to groups of 5 or more, North American reciprocal & National Trust for Historic Preservation. Library $10; members no charge. Closed state & federal holidays. &

Attendance: 17,500 (accurate)

Membership: Student $20; Individual $40; Family $50; Friend $100; Supporter $250; 1822 Associate $500 & up.

MAINE NARROW GAUGE RAILROAD CO. & MUSEUM, 58 Fore St., Portland, ME 04101-4842. Tel.: 207-828-0814. Fax: 207-879-6132.

E-mail: info@mainenarrowgauge.org

Web Site: www.mainenarrowgauge.org

Founded: 1993.

Key Personnel: Exec. Dir., Allison Tevsh Zittel; Pres. (V), Jerry Angier; Mgr. Visitor Svcs., Christina Aliquo.

Personnel Profile: Full-Time Paid 2; Part-Time Paid 3; Part-Time Volunteers 84.

Governing Authority: bd. of trustees. Parent Institution: Maine Narrow Gauge Railroad and Industrial Heritage Trust.

Institution Type/Description: Railroad Museum.

Collections: two-foot gauge rolling stock; period rail cars; steam & diesel locomotives; railroad artifacts.

Facilities: Museum-related items for sale.

Activities: birthday parties; train rides; special events. Museum Sponsors: Polar Express in December.

Publications: Two Foot Flyer.

Hours & Admission Prices: Museum: March-April Sat.-Sun. 10-4; May-Oct. daily 10-4; call for additional open hours during school vacation weeks. Museum: adults 13 & over $3, seniors & children 3-12 $2, children 2 & under no charge. Train Rides: 10am, 11am, 12pm, 1pm, 2pm, & 3pm. Museum & Train Rides: adults $10, senior citizens $9, children 3-12 $6; children under 2 no charge. Closed New Year's Day; Thanksgiving; Christmas. &

Attendance: 40,000 (estimated)

Membership: Individual $35; Family $65; Individual Pass $70; Family Pass $100; Lifetime $350.

THE MUSEUM OF AFRICAN CULTURE, 13 Brown St., Portland, ME 04101-3934. Tel.: 207-871-7188. Fax: 207-773-1197. Facebook: The Museum of African Culture.

E-mail: africart@museumafricanculture.org

Web Site: www.museumafricanculture.org

Formerly: The Museum of African Tribal Art

Founded: 1998.

Congressional District: 1

Key Personnel: Pres., Gail S. Edgerly; Dir., Oscar O. Mokeme; Treas., Barbara Payson; Sec., Lila C. Hunt.

Personnel Profile: Full-Time Paid 1; Full-Time Volunteers 9; Part-Time Paid 1; Part-Time Volunteers 5; Interns 5.

Governing Authority: private; nonprofit organization. Tax-exempt.

Institution Type/Description: Art Museum.

Collections: African art from ancient to modern times; artifacts; sub-Saharan African tribal art.

Major Exhibits: The Spirits of Igbo Masks (T), 11/13-11/14; Earthen Vessels (T), 1/14-12/14; Traditional Medicare (T), 1/14-12/14; The Spirits of Cameroon, 2/14-3/14; Water Spirits, 4/14-5/14; The Role of Ancestors, 7/14-9/14; Spirits of Dryness, 10/14-12/14.

Research Fields: traditional medicine; traditional use of masks; ritual power of intentions.

Facilities: 600 sq. ft. exhibit space.

Activities: guided tours; loan & temporary exhibitions; fire ceremony.

Publications: monthly newsletters.

Hours & Admission Prices: Tues.-Fri. 10:30-4, Sat. 12-4. Adults $5; discounts to AAM & ICOM members; members no charge. Closed New Year's Day; Easter; Thanksgiving; Christmas.

Attendance: 10,000 (estimated)

Membership: Student $15; Individual $35; Family $50; Friend $100; Supporter $250; Patron $500; Benefactor $1,000; Director's Circle $2,000 and up.

PORTLAND FIRE MUSEUM, 157 Spring St., Portland, ME 04104. Mailing Address: P.O. Box 1743, Portland, ME 04104-1743. Tel.: 207-772-2040.
E-mail: history@portlandfiremuseum.com
Web Site: www.portlandfiremuseum.com
Institution Type/Description: Fire-Fighting Museum.
Collections: original horse stalls; 1938 McCann pumper engine; 1848 Crockett hand tub & teel; 1867 Button hand pumper; 1857 Jeffords hand pumper; photographs; paintings; flags & banners.
Hours & Admission Prices: 1st Fri. each month 6pm.

✻　**PORTLAND MUSEUM OF ART, (M),** Seven Congress Square, Portland, ME 04101-1119. Tel.: 207-775-6148. Fax: 207-773-7324. Facebook: Portland Museum of Art.
E-mail: info@portlandmuseum.org
Web Site: www.portlandmuseum.org
Founded: 1882.
Congressional District: 1
Key Personnel: Dir., Mark Bessire; Chm., John F. Isacka; Pres., Anna H. Wells; Dir. Public Rels., Kristen Levesque; Cur. Contemporary Art., Jessica May; Cur. American Art, Karen Sherry; Registrar, Lauren Silverson; Dir. Learning & Interpretation, Dana Baldwin; Deputy Dir. Finance & Operations, Elena Murdock; Museum Shop Mgr., Sally Struever.
Personnel Profile: Full-Time Paid 49; Part-Time Paid 54; Part-Time Volunteers 291; Interns 10.
Volunteer Hours: 9,500
Operating Expenses: 5,650,578
Operating Income: 5,652,350
Governing Authority: nonprofit organization. Tax-exempt: 501(c)(3).
Institution Type/Description: Art Museum.
Collections: European collection including Renoir, Degas, Monet, Picasso; Maine works by Homer, Wyeth, Hartley; 18th & 19th century European & American works. Historic Houses: 1801 McLellan House, 1911 L.D.M. Sweat Memorial Galleries, & 1983 building by Pei Cobb Freed & Partners; Winslow Homer's studio, 1884.
Major Exhibits: American Vision: Photographs from the Collection of Owen & Anna Wells, 12/21/13-2/23/14; Fine Lines: American Drawings from the Brooklyn Museum (T), 1/30/14-4/27/14; Andrea Sulzer, 4/26/14-8/24/14; Richard Estes' Realism (T), 5/22/14-9/7/14; Aaron Stephan, 9/6/14-1/4/15; Treasures from the Berger Collection: British Art, 1400-2000 (T), 10/2/14-1/4/15.
Research Fields: American decorative arts; 19th century glass; the McLellan, Clapp, Dearborn, Wingate & Sweat families; Maine artists; Winslow Homer.
Facilities: 200-seat auditorium; cafe; rental facility. Books, reproductions, catalogues, postcards, prints & selected examples of contemporary jewelry & glass for sale.
Activities: guided tours; lectures; films; gallery talks; concerts; cultural tours; formally organized education programs; docent program; outreach programs; permanent, temporary & traveling exhibitions; inter-museum loan.
Publications: brochures; monthly bulletins; exhibition catalogs.
Hours & Admission Prices: Memorial Day-Columbus Day Fri. 10-9, Sat.-Thurs. 10-5; Oct.-May Tues.-Thurs. & Sat.-Sun. 10-5, Fri. 10-9. Adults $12, senior citizens 65 & over and students with ID $10, youth 13-17 $6; discounts to AAM & ICOM members; children 12 & under, members & Fri. 5-9 no charge. Closed New Year's Day; Thanksgiving; Christmas. ♿
Attendance: 177,824 (accurate)
Membership: Individual $50; Family $75; Contributing $140; Donor $250; Patron $500; Director's Circle $1,500; 1882 Circle $10,000 & up.

PORTLAND OBSERVATORY MUSEUM, 138 Congress St., Portland, ME 04101. Mailing Address: Greater Portland Landmarks, 93 High St., Portland, ME 04101. Tel.: 207-774-5561.
E-mail: jpollick@portlandlandmarks.org
Web Site: www.portlandlandmarks.org
Founded: 1964.
Key Personnel: Dir., Hilary Bassett; Pres. (V), Marjorie Getz; Museum Shop Mgr., Jennifer Pollick.
Personnel Profile: Full-Time Paid 3; Part-Time Paid 6; Part-Time Volunteers 65; Interns 2.
Governing Authority: Parent Institution: Greater Portland Landmarks. Tax-exempt.
Institution Type/Description: Portland Observatory: built in 1807. A National Historic Landmark.
Collections: local maritime heritage & history; 86-foot high tower; photographs.
Activities: educational programs; rental facilities; public tours.
Hours & Admission Prices: Guided Tours: Memorial Day to Columbus Day daily 10-5. Adults $8, seniors & students $7, children 6-16 $5; discounts to Portland residents & AAA members; members & children under 6 no charge.
Attendance: 10,000 (accurate)

SOUTHWORTH PLANETARIUM, 96 Falmouth St., Portland, ME 04103-4864.
Formerly: Portland Planetarium
Founded: 1969.
Congressional District: 1
Key Personnel: Mgr., Edward Gleason; Dir., Jerry LaSala
Institution Type/Description: Planetarium.
Collections: astronomy.
Research Fields: Novas.
Facilities: Museum-related items for sale.
Hours & Admission Prices: Call for hours and admission prices.

TATE HOUSE MUSEUM, 1267 Westbrook St., Portland, ME 04102-1934. Tel.: 207-774-6177. Fax: 207-774-6198.
E-mail: info@tatehouse.org
Web Site: www.tatehouse.org
Founded: 1931.
Congressional District: 1
Key Personnel: Bd. Chm., Anita Jones; Museum Shop Mgr., Joan Hatch.
Personnel Profile: Part-Time Paid 1; Part-Time Volunteers 80.
Governing Authority: Tax-exempt: 501(c)(3).
Institution Type/Description: Historic House: 1755 George Tate House.
Collections: 18th-century furnishings; herb gardens; architecture; decorative arts.
Research Fields: family of George Tate; the mast trade; decorative arts; local history; Georgian architecture; historic gardening.
Facilities: herb gardens; meeting & rental facilities.
Activities: guided tours; school tours; Wednesday garden tours; rental facilities; architecture tours.
Publications: Books, Tate House, Crown of the Mast Trade; An Herbal of 18th Century Gardens at Tate House; This is Stroudwater; quarterly newsletter, Tate House Gazette.
Hours & Admission Prices: June-Oct. Wed.-Sat. 10-4, Sun. 1-4. Adults $8, seniors $6, children 6-12 $3; discount to members. Closed Independence Day; Labor Day.
Attendance: 2,000 (estimated)
Membership: Individual $35; Family $50; Patron $100.

UNIVERSITY OF NEW ENGLAND ART GALLERY, (M), 716 Stevens Ave., University of New England, Portland, ME 04103-2693. Tel.: 207-221-4499. Fax: 207-523-1901.
E-mail: azill@une.edu
Web Site: www.une.edu/artgallery
Formerly: Payson Gallery
Founded: 1977.
Congressional District: 1
Key Personnel: Dir., Anne B. Zill.
Personnel Profile: Full-Time Paid 1; Part-Time Paid 3; Part-Time Volunteers 2; Interns 1.
Governing Authority: private university.
Institution Type/Description: Fine Art Gallery.
Collections: 19th-20th century art; contemporary arts; photographs; paintings; prints; sculpture.
Activities: formal education programs for UNE, adults & children; lectures; loan & temporary exhibitions.
Hours & Admission Prices: Wed. & Fri.-Sun. 1-4, Thurs. 1-7. No charge; donations accepted. Closed Easter; Independence Day; Thanksgiving; Christmas.
Attendance: 10,000 (estimated)
Membership: $35; $50; $100; $500; $1,000.

VICTORIA MANSION, (M), 109 Danforth St., Portland, ME 04101-4504. Tel.: 207-772-4841. Fax: 207-772-6290.
E-mail: information@victoriamansion.org
Web Site: www.victoriamansion.org
Founded: 1941.
Congressional District: 1
Key Personnel: Pres. (V), Michael Stone; Dir., Robert Wolterstorff; Deputy Dir. Administration, Julia Kirby; Office Mgr., Timothy Brosnihan; Site Mgr. & Education Asst., Katie Worthing; Dir. Education, Tracy Quimby; Museum Shop Mgr., Alice Dwyer Ross.
Personnel Profile: Full-Time Paid 3; Part-Time Paid 10; Part-Time Volunteers 60.

Governing Authority: nonprofit organization. Tax-exempt: 501(c)(3).
Institution Type/Description: Historic House Museum: National Historic Landmark.
Collections: the first and only known surviving commission of Gustave Herter, 1858-1860; original house contents including furniture, gas lighting fixtures, wall paintings, artworks, carpets, stained glass, porcelain, textiles & architecture.
Research Fields: 19th & early 20th-century material culture, architecture & the decorative arts & social history.
Facilities: Carriage House: gift-items for sale such as Victorian-era style items, Victorian Mansion memorabilia & books.
Activities: guided tours; temporary exhibitions; educational programs: lectures & displays. Museum Sponsors: special installations & tours in December.
Publications: newsletter 3 times per year; brochures; A Guide to Victoria Mansion.
Hours & Admission Prices: May-Oct. Mon.-Sat. 10-4, Sun. 1-5, group tours by appointment. Adults $15, children 6-17 $4; discounts to senior citizens, groups, AAM, AAA members; children under 6 & members no charge. &
Attendance: 17,759 (accurate)
Membership: Individual $35; Family & Household $65; Supporting $125; Herter Circle $125 & up; Hester Circle $250 & up; Morse Assoc. $1,000 & up.

Presque Isle

THE NORTHERN MAINE MUSEUM OF SCIENCE, UNIVERSITY OF MAINE, Folsom Hall, 181 Main St., Presque Isle, ME 04769-2844. Mailing Address: P.O. Box 285, Presque Isle, ME 04769. Tel.: 207-768-9482. Fax: 207-768-9553.
E-mail: mcgowanj@polaris.umpi.maine.edu
Web Site: www.umpi.maine.edu/info/nmms/about.htm
Key Personnel: Dir., Dr. Kevin McCartney; Cur. Chemistry, Michael Knopp, Ph.D.; Cur. Herbarium, Robert J. Pinette, Ph.D.; Cur. Mathematics, Richard Kimball; Cur. Collections, Jeanie McGowan
Institution Type/Description: Science Museum.
Collections: fresh-water sea shells; local forestry specimens; biology; geology; chemistry; physics; agriculture.
Activities: tours.
Hours & Admission Prices: Daily 7am-10pm. No charge. Closed university holidays & breaks.

PRESQUE ISLE AIR MUSEUM, Northern Maine Regional Airport, 650 Airport Dr., Ste. 4, Presque Isle, ME 04769. Tel.: 207-764-2542. Fax: 207-764-2544.
E-mail: piairmuseum@fcmail.com
Founded: 1999.
Congressional District: 1
Key Personnel: Pres. (V), Nate Grass
Governing Authority: Tax-exempt.
Institution Type/Description: History & Pictorial Museum.
Collections: historical artifacts & photographs relating to the 1930s, WWII, the Cold War & the Missile era.
Hours & Admission Prices: Call for hours. No charge. &
Attendance: 10,000 (estimated)

Prospect

FORT KNOX STATE HISTORIC SITE AND PENOBSCOT NARROWS OBSERVATORY, 711 Fort Knox Rd., Prospect, ME 04981-3125. Mailing Address: P.O. Box 456, Bucksport, ME 04416-0456. Tel.: 207-469-7719 & 6553. Fax: 207-469-6906.
E-mail: fofk1@aol.com
Web Site: fortknox.maineguide.com
Founded: 1923.
Congressional District: 2
Key Personnel: Exec. Dir., Leon Seymour; Pres. (V), Chris Popper; Mgr., Mike Wilusz.
Personnel Profile: Full-Time Paid 1; Part-Time Paid 8; Part-Time Volunteers 13.
Governing Authority: state. Affiliated with Bureau of Parks & Lands Augusta, ME 04333. Subsidiary Institution: Friends of Fort Knox. Tax-exempt.
Institution Type/Description: Historic Site: 1844 Fort Knox.
Collections: 19th-century fort; Rodman cannons.
Facilities: picnic area; activity field.
Activities: guided tours; special tours by arrangement. Museum Sponsors: Civil War Reenactment; weekend special events.
Publications: folder, Maine Fort Histories.
Hours & Admission Prices: May-Oct. call for hours & admission prices. &
Attendance: 50,000 (estimated)

Membership: Individual $20; Vehicle (includes all occupants) $40.

Rangeley

RANGELEY LAKES HISTORICAL SOCIETY, Main St., Rangeley, ME 04970. Mailing Address: P.O. Box 521, Rangeley, ME 04970. Tel.: 207-864-5571.
E-mail: palmer@rangeley.org
Institution Type/Description: Historical Society Museum: housed in a Classical Revival style building; built c.1905. Listed on the National Register of Historic Places.
Collections: local history & culture; period furnishings; personal artifacts; photographs; early records; railroading; logging; fishing & hunting.
Activities: educational programs.
Hours & Admission Prices: July-Sept. Mon.-Sat. 10-2.

RANGELEY LAKES REGION LOGGING MUSEUM, 221 Stratton Rd., Rangeley, ME 04970. Mailing Address: P.O. Box 154, Rangeley, ME 04970-0154. Tel.: 207-864-3939 & 5551. Facebook: Rangeley Logging Museum.
E-mail: info@rlrlm.org
Web Site: www.rlrlm.org
Founded: 1979.
Congressional District: 2
Key Personnel: Pres. (V) & Dir., Ronald J. Haines; Vice Pres. & Festival Coord., Stephen A. Richard; Cur., Archivist, & Folklorist, Dr. Margaret Yocom; Treas., Carolyn Nobbs; Public Rels. & Museum Shop Mgr., Carol Sullivan; Sec., June Aleck.
Personnel Profile: Part-Time Paid 1; Part-Time Volunteers 60.
Governing Authority: private; nonprofit organization; governing body, volunteer bd. of dirs. Tax-exempt: 501(c)(3).
Institution Type/Description: Logging Museum.
Collections: logging in the western mountains of Maine; early photographs; woodcarving; paintings; textile arts; history; folklife; folk arts of logging & logging communities.
Research Fields: history of western Maine loggers, their families & communities.
Facilities: 30-vol. library; community meeting space. Museum-related items for sale.
Activities: lectures; group tours; nature walks & trails; children's area. Special Events: Logging Festival; Knit & Craft Show; Auction; Apple Festival.
Publications: books, Logging in the Maine Woods: The Paintings of Alden Grant; Working the Woods; Art Reproductions of the Paintings of Alden Grant; notecards featuring the paintings of Alden Grant.
Hours & Admission Prices: See website for hours. No charge; donations accepted. &
Attendance: 1,400 (estimated)
Membership: Individual $10; Family $20.

RANGELEY PUBLIC LIBRARY, 7 Lake St., Rangeley, ME 04970. Mailing Address: P.O. Box 1150, Rangeley, ME 04970-1150. Tel.: 207-864-5529. Fax: 207-864-2523.
E-mail: info@rangeleylibrary.com
Web Site: www.rangeleylibrary.com
Institution Type/Description: Library: housed in a Romanesque Revival style building; c.1909. Listed on the National Register of Historic Places.
Collections: books; photographs.
Activities: special events.
Hours & Admission Prices: Tues. 10-7, Wed.-Fri. 10-4:30, Sat. 10-2.

WILHELM REICH MUSEUM - ORGONON, Dodge Pond Rd., Rangeley, ME 04970. Mailing Address: P.O. Box 687, Rangeley, ME 04970-0687. Tel.: 207-864-3443. Fax: 207-864-5156.
E-mail: wreich@rangeley.org
Web Site: www.wilhelmreichtrust.org
Founded: 1960.
Institution Type/Description: Historic House Museum: housed in the home, laboratory & research center of physician & scientist, Wilhelm Reich, M.D.
Collections: Reich's life & work; inventions & scientific equipment; personal artifacts; paintings; sculptures.
Facilities: 175 acres; nature trails; Orgone Energy Observatory. Books for sale.
Activities: video. Annual Event: Summer Conferences.
Publications: Orgonomic Functionalism, 6 vol.
Hours & Admission Prices: Conference Center: Mon.-Fri. 9-2. Observatory: July-Aug. Wed.-Sun. 1-5; Sept. Sat. 1-5; other times by appointment. Adults $6; discounts to AAM & ICOM members; children 12 & under no charge.
Attendance: 3,000 (estimated)

Readfield

READFIELD HISTORICAL SOCIETY AND MUSEUM, 759 Main St., Readfield Depot, Readfield, ME 04355. Mailing Address: P.O. Box 354, Readfield, ME 04355. Tel.: 207-685-4662.
E-mail: readfieldhistorical@gmail.com
Web Site: www.readfieldhistorical.org
Founded: 1986.
Key Personnel: Dir. & Pres. (V), Florence Drake.
Personnel Profile: Part-Time Volunteers 20; Interns 4.
Governing Authority: Tax-exempt.
Institution Type/Description: History Museum.
Collections: local history & culture; period artifacts; photographs.
Research Fields: history; genealogy.
Facilities: archives.
Activities: research.
Publications: biannual newsletter.
Hours & Admission Prices: June to mid-Sept. Thurs. & Sat. 10-2. No charge; donations accepted. &
Attendance: 200 (estimated)
Membership: Senior $5; Individual $10; Life $100.

Rockland

COASTAL CHILDREN'S MUSEUM, 75 Mechanic St., Rockland, ME 04841. Tel.: 207-596-0300.
E-mail: info@coastalchildrensmuseum.org
Web Site: www.coastalchildrensmuseum.org
Key Personnel: Chm. (V) & Pres. (V), Elaine Wilson; Chm. (V), Felicity Bowelitch.
Personnel Profile: Part-Time Paid 1; Part-Time Volunteers 3.
Governing Authority: Subsidiary Institution: Association of Children's Museum. Tax-exempt.
Institution Type/Description: Children's Museum.
Collections: hands-on exhibitions.
Activities: birthday parties; educational programs; special events.
Publications: biannual newsletter.
Hours & Admission Prices: Wed.-Sat. 10-4, Sun. 1-4. Discounts to ACM members. Closed New Year's Day; Easter; Independence Day; Labor Day; Thanksgiving; Christmas Eve & Day. &

MAINE LIGHTHOUSE MUSEUM, One Park Dr., Rockland, ME 04841. Mailing Address: P.O. Box 1116, Rockland, ME 04841-1116. Tel.: 207-594-3301. Fax: 207-596-6549.
E-mail: dot@mainelighthousemuseum.org
Web Site: www.mainelighthousemuseum.org
Formerly: Shore Village Museum
Founded: 2005.
Congressional District: 1
Key Personnel: Dir., Dorothy Black; Chm. (V), Paul Dilger; Museum Shop Mgr., Deb McNeil.
Personnel Profile: Full-Time Volunteers 2; Part-Time Paid 1; Part-Time Volunteers 25.
Governing Authority: Parent Institution: Maine Lighthouse Museum, Inc. Tax-exempt.
Institution Type/Description: Maritime Museum.
Collections: lighthouse artifacts; related books; marine items from the U.S. Coast Guard including working foghorns, lights, lenses; scrimshaw lobstering tools; ship models; items of local historic interest; lighthouse lenses.
Research Fields: local history; documentation records dating from late 1860s.
Activities: guided tours; lectures; gallery talks; loan, permanent & temporary exhibitions; special programs for children.
Publications: newsletter.
Hours & Admission Prices: June-Oct. Mon.-Fri. 9-5, Sat.-Sun. 10-4; Nov.-May Thurs.-Fri. 9-5, Sat. 10-4. Adults $5, seniors $4; members, Coast Guard & children under 12 no charge. &
Attendance: 12,000 (estimated)
Membership: Basic $25; Family $50; Supporting $100; Sustaining $250; Sponsor $500; Lightkeeper $1,000.

∗ WILLIAM A. FARNSWORTH LIBRARY AND ART MUSEUM, INC. DBA FARNSWORTH ART MUSEUM, (M), 16 Museum St., Rockland, ME 04841-2867. Tel.: 207-596-6457. Fax: 207-596-0509.
E-mail: writeus@farnsworthmuseum.org
Web Site: www.farnsworthmuseum.org
Founded: 1948.
Congressional District: 1
Key Personnel: Dir., Christopher J. Brownawell; Pres. (V), Frederic R. Kellogg; Chief Cur, Michael Komanecky; Business Mgr., Cathy Knowles; Registrar, Angela Waldron; Museum Shop Mgr., Wendy Kirklian.
Personnel Profile: Full-Time Paid 25; Part-Time Paid 35; Part-Time Volunteers 250.
Governing Authority: board of directors. Tax-exempt: 501(c)(3).
Institution Type/Description: Art Museum.
Collections: 19th-20th century paintings including works by George Bellows, Thomas Cole; Thomas Eakins; Childe Hassam, Winslow Homer, Fitz Henry Lane, Willard Metcalf, Rockwell Kent, Alex Katz, Robert Indiana, Louise Nevelson, & 3 generations of the Wyeth family; photographs; prints. Historic Houses: c.1850 Farnsworth Homestead; c.1820 Olson House.
Research Fields: Maine & American art.
Facilities: 5,000-vol. library; archives; general art reference area; educational facilities. Museum-related items for sale.
Activities: guided tours; lectures; films; gallery talks; concerts; permanent, temporary & traveling exhibitions.
Publications: books; exhibition catalogs; annual reports.
Hours & Admission Prices: Jan.-March Wed.-Sun. 10-5; April-May Tues.-Sun. 10-5; June-Oct. Wed. 10-8, Thurs.-Tues. 10-5; Nov.-Dec. Tues.-Sun. 10-5. Adults $12, senior citizens & students $10; discounts to AAM & ICOM members; members & New England Consortium of Museums members no charge.
Attendance: 66,281 (accurate)
Membership: Student $30; Senior 65 & over $50; Individual $60; Senior Dual $75; Dual $85; Partner $150; Steward $250; Sustainer $500; Circle $1,000; Benefactor $2,500; Fellow $5,000; President's Council $10,000.

Rockport

CENTER FOR MAINE CONTEMPORARY ART, 162 Russell Ave., Rockport, ME 04856. Mailing Address: P.O. Box 147, Rockport, ME 04856-0147. Tel.: 207-236-2875, ext. 304. Fax: 207-236-2490.
E-mail: info@cmcanow.org
Web Site: www.cmcanow.org
Formerly: Maine Coast Artists
Founded: 1952.
Congressional District: 1
Key Personnel: Dir., Suzette McAvoy; Mgr. Operations & Communications, Paula Blanchard; Chm. (V) Bd. Trustees, Marilyn Moss Rockefeller.
Personnel Profile: Full-Time Paid 2; Part-Time Paid 1; Part-Time Volunteers 50; Interns 1.
Governing Authority: nonprofit corporation. Tax-exempt.
Institution Type/Description: Contemporary Art.
Research Fields: Maine artists & their work; current Maine artists.
Facilities: 4-story historic building; 3 galleries; Artlab for workshops.
Activities: workshops; lectures; temporary exhibitions; evening programs; art tours; works of Maine artists.
Publications: exhibition catalogs.
Hours & Admission Prices: Galleries: May-Dec. Tues.-Sat. 10-5, Sun. 1-5. Offices: Mon.-Fri. 9-5. Donations Requested: $5 per person; members no charge. &
Attendance: 10,000 (estimated)
Membership: Artist $30; Individual $40; Household $60; Sustaining $120 & up; Business Supporter $250.

CONWAY HOMESTEAD & CRAMER MUSEUM, 223 Union St. #747, Rockport, ME 04856. Tel.: 207-236-2257.
E-mail: crhs@midcoast.com
Web Site: www.conwayhouse.org
Founded: 1960.
Congressional District: 1
Key Personnel: Pres., Frank Carr.
Personnel Profile: Part-Time Volunteers 45.
Governing Authority: Parent Institution: Camden-Rockport Historical Society. Tax-exempt.
Institution Type/Description: History Museum; listed on the National Register of Historic Places.
Collections: local history & culture; period furnishings; personal artifacts; photographs. Historic Buildings: Conway House, c.1770; 18th century barn; blacksmith shop; 1820 Maple Sugar House.
Activities: workshops; seminars; summer camp.
Hours & Admission Prices: Call for hours. Adults $5; discounts to AAA members & other local historical societies; members no charge. &
Attendance: 1,254 (estimated)
Membership: Individual $20.

Rumford

GREATER RUMFORD HISTORICAL SOCIETY, 145 Congress St., Rumford, ME 04276. Tel.: 207-364-2540.
E-mail: rhs@gwi.net
Founded: 1976.
Key Personnel: Pres. (V), Jane W. Peterson; Cur., David Gawtry; Archives, Dru Breton.
Personnel Profile: Part-Time Volunteers 2.
Governing Authority: Tax-exempt: 401(c)(3).
Institution Type/Description: Historical Society Museum & Archives.
Collections: local history & culture; period artifacts; photographs; early tools; books.
Facilities: archives.
Activities: monthly meetings & programs. Annual Event: Lawn, Plant & Food Sale in summer.
Publications: books.
Hours & Admission Prices: June-Aug. Sat. 9-2. No charge; donations accepted.
Attendance: 258 (accurate)
Membership: Individual $10.

Saco

SACO MUSEUM, (M), 371 Main St., Saco, ME 04072-1520. Tel.: 207-283-3861. Fax: 207-283-0754.
E-mail: museum@sacomuseum.org
Web Site: www.sacomuseum.org
Formerly: York Institute Museum
Founded: 1867.
Congressional District: 1
Key Personnel: Dir., Jessica Routhier; Exec. Dir., Leslie Rounds; Collections Mgr., Marie O'Brien; Program & Education Mgr., Camille Smalley.
Personnel Profile: Full-Time Paid 3; Part-Time Paid 3; Part-Time Volunteers 13; Interns 2.
Governing Authority: nonprofit organization. Parent Institution: Dyer Library Association. Tax-exempt: 501(c)(3).
Institution Type/Description: History & Art Museum: housed in an historic building built in 1926 by John Calvin Stevens.
Collections: Colonial & Federal period fine & decorative arts; regional paintings & furniture; works by John Brewster, Jr., Gibeon Elden Bradbury & Charles Henry Granger; pilgrim's progress.
Research Fields: Maine history; 18th to 20th-century American Art; decorative arts research.
Activities: guided tours; inter-museum loan, permanent & temporary exhibitions; programs include Artist-in-Residence courses in fine arts, lecture series, teacher workshops & gallery talks.
Publications: occasional monographs.
Hours & Admission Prices: Tues.-Thurs. & Sun. 12-4, Fri. 12-8, Sat. 10-4. Adults $5, seniors $3, students & children $2; discounts to NEMA, AAA & AAM members; members & Fri. 4-8 no charge. Closed legal holidays. &
Attendance: 10,305 (accurate)
Membership: Individual $35.

Scarborough

SCARBOROUGH HISTORICAL MUSEUM, 647 U.S. Rte. 1 Dunstan, Scarborough, ME 04074. Mailing Address: P.O. Box 156, Scarborough, ME 04070-0156. Tel.: 207-885-9997.
E-mail: scarboroughhist@maine.rr.com
Founded: 1961.
Congressional District: 1
Key Personnel: Pres. (V), Rodney Laughton.
Personnel Profile: Part-Time Volunteers 9.
Governing Authority: society. Parent Institution: Scarborough Historical Society, Inc. Tax-exempt: 170(b)(1)(A).
Institution Type/Description: History Museum.
Collections: tools; colonial farmhouse memorabilia; old school records & pictures; old town valuation books & records.
Major Exhibits: Fans, 1/1/14-12/20/14; Schools, 1/1/14-2/20/14; Roger Dearing Murals, 1/1/14-2/20/14.
Facilities: over 500-vol. library of books pertaining to local & state history, Scarborough town record books, bibles & old school books.
Activities: guided tours; lectures; permanent & temporary exhibitions.
Publications: quarterly newsletter, Owascoag Notes.
Hours & Admission Prices: Tues. 9-12; other times by appointment. No charge; donations accepted. &
Attendance: 225 (estimated)
Membership: Student $1; Individual $10; Sponsor $25; Life $75; Benefactor $100.

Seal Cove

THE SEAL COVE AUTO MUSEUM, (M), 1414 Tremont Rd., Seal Cove, ME 04674. Mailing Address: P.O. Box 106, Seal Cove, ME 04674. Tel.: 207-244-9242. Fax: 207-244-9772.
E-mail: info@sealcoveautomuseum.org
Web Site: www.sealcoveautomuseum.org
Founded: 1963.
Key Personnel: Exec. Dir., Roberto M. Rodriguez; Pres. (V), Barbara Fox.
Personnel Profile: Full-Time Paid 2; Part-Time Paid 2; Part-Time Volunteers 20; Interns 2.
Governing Authority: Tax-exempt.
Institution Type/Description: Transportation Museum.
Collections: antique cars & motorcycles.
Activities: educational programs. Museum Sponsors: Electric Car Day; Steam Car Day; Mud Day; Spooky Car Days; License Plate Puzzle Day.
Publications: e-newsletter.
Hours & Admission Prices: May-Oct. daily 10-5. Adults $5, senior citizens $4, children 12 & under $2; discounts to AAM, ICOM & AAA members; members no charge. &
Attendance: 14,000 (accurate)
Membership: $30; $50; $100; $250; $500. Business: $100-$1,000.

Searsport

* **PENOBSCOT MARINE MUSEUM, (M),** 5 Church St., Searsport, ME 04974-3351. Mailing Address: P.O. Box 498, Searsport, ME 04974-0498. Tel.: 207-548-2529. Fax: 207-548-2520.
E-mail: museumoffices@pmm-maine.org
Web Site: www.penobscotmarinemuseum.org
Founded: 1936.
Congressional District: 1
Key Personnel: Exec. Dir., Liz Lodge; Cur., Benjamin A.G. Fuller; Dir. Education, Betty Schopmeyer; Business Mgr., Matthew Timney.
Personnel Profile: Full-Time Paid 11; Full-Time Volunteers 1; Part-Time Paid 19; Part-Time Volunteers 18; Interns 1.
Governing Authority: nonprofit corporation. Tax-exempt: 501(c)(3).
Institution Type/Description: Maritime Museum: housed in the homes of three former shipmasters and the original Town Hall.
Collections: paintings; prints; ship models; builders half models; small craft collection; shipbuilding tools; navigational instruments; charts; 19th-century American & Oriental furnishings; log books; manuscripts; genealogy; whaling memorabilia. Historic Houses: 1805/1837 Fowler House; 1845 Old Town Hall; 1860 Merithew House; 1880 Nichols House; 1843 Education Center.
Research Fields: Maine & regional maritime history.
Facilities: 12,000-vol. library on maritime history available for use on premises.
Activities: permanent & temporary exhibitions; guided tours; school loan service; inter-museum loans; special art exhibitions; degree programs with Univ. of Maine-Orono & Colby College. Annual Event: Fall Regional History Conference.
Publications: catalogs; annual report; newsletter, The Bay Chronicle.
Hours & Admission Prices: Mon.-Sat. 10-5, Sun. 12-5. Family $18, adults $8, children 7-15 $3; discounts to local guests of Inns, groups, AAM & CAMM members; children under 7 & members no charge. &
Attendance: 18,000 (accurate)
Membership: Friend $40; Supporter $65; Contributor $150; Sponsor $250; Patron $500; Benefactor $1,000; President's Circle $2,000.

Sedgwick

SEDGWICK-BROOKLIN HISTORICAL SOCIETY, 575 N. Sedgwick Rd., Sedgwick, ME 04676. Mailing Address: P.O. Box 171, Sedgwick, ME 04676-0063. Tel.: 207-359-8086.
Founded: 1963.
Congressional District: 2
Key Personnel: Pres. (V), Anne P. Dentino.
Personnel Profile: Part-Time Volunteers 10.
Governing Authority: board of trustees; nonprofit organization. Tax-exempt: 501(c)(3).
Institution Type/Description: History Museum: housed in 1795 Rev. Daniel Merrill house.
Collections: archives; local history; tools; toys; photographs; horse-drawn hearses; barn; clothing; farm equipment. Historical Buildings: c.1795 Reverend Daniel Merrill House; 1793-1794 Sedgwick Town House & Common; 1821 Town Cattle Pound; Hearse House; 1798 rural cemetery; renovated barn using timber & boards from original structure.
Research Fields: archives; local history; genealogy.

Facilities: library of old account books, maps, genealogical material & local histories; reading room.
Activities: lectures; programs. Museum Sponsors: annual Ax Throwing Contest.
Publications: booklets, History of Brooklin, Maine, 1876; Our Tradition of Optimism; History of Sedgwick; map, Plan of Sedgwick (Township #4); The Cemeteries of Brooklin, Maine a Genealogist's Guide; Life and Times in a Coastal Village/Sedgwick, Maine 1789-1989; Sedgwick Cemeteries.
Hours & Admission Prices: July-Aug. Sun. 2-4; other times by appointment. No charge; donations accepted.
Attendance: 100 (estimated)
Membership: Single $10; Family $20; Enthusiastic $35; Stalwart $60; Committed $100.

Skowhegan

MARGARET CHASE SMITH LIBRARY, 56 Norridgewock Ave., Skowhegan, ME 04976-1204. Tel.: 207-474-7133. Facebook: Margaret Chase Smith Library.
E-mail: mcsl@mcslibrary.org
Web Site: www.mcslibrary.org
Founded: 1982.
Congressional District: 2
Key Personnel: Dir., David Richards.
Personnel Profile: Full-Time Paid 4; Part-Time Paid 1; Part-Time Volunteers 1.
Volunteer Hours: 300
Operating Expenses: 343,344
Operating Income: 343,426
Governing Authority: Parent Institution: Margaret Chase Smith Foundation. Subsidiary Institution: University of Maine. Tax-exempt.
Institution Type/Description: Library & Archives.
Collections: life & career of Senator Margaret Chase Smith.
Major Exhibits: 1964 Presidential Campaign, 1/14-12/14.
Research Fields: 20th century American political history.
Facilities: library.
Activities: group tours; research; educational programs.
Publications: newsletter.
Hours & Admission Prices: Mon.-Fri. 10-4. No charge; donations accepted.
Attendance: 2,500 (estimated)

SKOWHEGAN HISTORY HOUSE MUSEUM & RESEARCH CENTER, 66 Elm St., Skowhegan, ME 04976-0832. Mailing Address: P.O. Box 832, Showhegan, ME 04976-0832. Tel.: 207-474-6632.
E-mail: melvinburnham@skowheganhistoryhouse.org
Web Site: skowheganhistoryhouse.org
Founded: 1937.
Congressional District: 93
Key Personnel: Dir. & Pres., Melvin Burnham; Sec., Bonnie Chamberlain; Treas., Patricia Horine; Cur., Lee Granville.
Personnel Profile: Part-Time Paid 2; Part-Time Volunteers 14.
Governing Authority: nonprofit organization. Parent Institution: Bloomfield Trust. Tax-exempt: 501(c)(3).
Institution Type/Description: Local History Museum: built in 1839.
Collections: furnishings; toys; Civil War; guns; early town records.
Major Exhibits: Connection With The Civil War, 12/11-1/14.
Research Fields: local history; historic documents; family histories.
Facilities: 500-vol. library of books on local history available for use on premises.
Activities: Museum Sponsors: Good Ole Days - How Horrible in April; Heritage Tea in June; Celebrating Coburn Legacy in September.
Publications: History House Society newsletter, By The River's Edge.
Hours & Admission Prices: June to mid-Oct. Tues.-Sat. 10-4; other times by appointment. No charge; donations accepted. Closed holidays.
Attendance: 511 (accurate)
Membership: Senior & Student $15; Individual $25; Family $50; Business $100; Benefactor $150; Lifetime $300.

South Berwick

COUNTING HOUSE MUSEUM, Main & Liberty Sts., South Berwick, ME 03908. Mailing Address: P.O. Box 296, South Berwick, ME 03908-0296. Tel.: 207-384-0000.
E-mail: info@oldberwick.org
Web Site: www.oldberwick.org
Institution Type/Description: Historic House Museum.
Collections: local history & culture; period furnishings & artifacts; photographs.
Facilities: Museum-related items for sale.

Hours & Admission Prices: June-Oct. Sat.-Sun. 1-4; other times by appointment.

* **HAMILTON HOUSE, (M),** 40 Vaughan's Lane, South Berwick, ME 03908-1711. Mailing Address: 141 Cambridge St., Boston, MA 02114-2702. Tel.: 207-384-2454; 617-227-3956 (Historic New England).
Web Site: www.historicnewengland.org
Founded: 1949.
Congressional District: 1
Key Personnel: Pres. & C.E.O., Carl Nold; Site Mgr., Brooke Steinhauser.
Governing Authority: society; nonprofit organization. Parent Institution: Historic New England, 141 Cambridge St., Boston, MA 02114. Tel. 617-227-3956. Tax-exempt: 501(c)(3).
Institution Type/Description: Historic House: c.1785 Hamilton House, Georgian estate overlooking the Salmon Falls River.
Collections: furniture & artifacts of the 18th & 19th centuries; colonial revival reproduction of wallpapers, personal & decorative objects belonging to Mrs. Tyson & her stepdaughter.
Facilities: garden.
Activities: guided tours; lectures & special events; concerts in the garden.
Publications: magazine, Historic New England.
Hours & Admission Prices: June-Oct. 15 Wed.-Sun. 11-5, last tour at 4pm. Adults $8; discounts to seniors, AAM, ICOM, AAA, WGBH members; members no charge.
Attendance: 6,310 (accurate)
Membership: National $35; Individual $45; Household $55; Garden & Landscape $75; Contributing and Library & School $100; Young Friends of Historic New England $100-$1,500; Friends of the Library and Archives $125; Historic Homeowner $200; Business & Ogden Codman Design Group $250; Appleton Circle $2,500 & up.

* **SARAH ORNE JEWETT HOUSE, (M),** 5 Portland St., South Berwick, ME 03908. Mailing Address: 141 Cambridge St., Boston, MA 02114-2702. Tel.: 207-384-2454; 617-227-3956 (Historic New England).
E-mail: jewetthouse@historicnewengland.org
Web Site: www.historicnewengland.org
Founded: 1931.
Key Personnel: Pres. & C.E.O., Carl Nold; Site. Mgr., Brooke Steinhauser.
Governing Authority: society; nonprofit organization. Parent Institution: Historic New England, 141 Cambridge St., Boston, MA 02114. Tel.: 617-227-3956. Tax-exempt: 501(c)(3).
Institution Type/Description: Historic House: 1774 Georgian residence of the celebrated regional author Sarah Orne Jewett.
Collections: furniture; artifacts; refurbished original wall paper & paneling.
Activities: guided tours; special events; lectures.
Publications: magazine, Historic New England.
Hours & Admission Prices: June-Oct. 15 Fri.-Sun. 11-5, (last tour at 4). Adults $5; discounts to seniors, AAM, ICOM, AAA, WGBH members; members no charge.
Attendance: 2,938 (accurate)
Membership: National $35; Individual $45; Household $55; Garden & Landscape $75; Contributing and Library & School $100; Young Friends of Historic New England $100-$1,500; Friends of the Library and Archives $125; Historic Homeowner $200; Business & Ogden Codman Design Group $250; Appleton Circle $2,500 & up.

South Bristol

SOUTH BRISTOL HISTORICAL SOCIETY, 2124 State Rte. 129, South Bristol, ME 04568. Mailing Address: P.O. Box 229, South Bristol, ME 04568. Tel.: 207-315-0558.
E-mail: sbhistorical@gmail.com
Web Site: southbristolhistoricalsociety.org
Founded: 1998.
Institution Type/Description: Historical Society Museum.
Collections: local history & culture; photographs; town records; genealogical files; period artifacts; postcards; manuscripts. Historic Building: one-room schoolhouse.
Facilities: genealogy research center.
Publications: biannual newsletter; books, Down On The Island, Up On The Main; A History of the Families and Their Houses: South Bristol, Maine.
Hours & Admission Prices: Jan.-May Thurs. 1-3; June-Aug. Fri. 1-4; other times by appointment. No charge; donations accepted.
Membership: Individual $10; Couple & Family $20; Sustaining $40; Life $250; Life Family & Couple $350.

South Paris

HAMLIN MEMORIAL LIBRARY & MUSEUM, 16 Hannibal Hamlin Dr., South Paris, ME 04281. Mailing Address: P.O. Box 43, Paris, ME 04271-0043. Tel.: 207-743-2980. Facebook: Hamlin Library.
E-mail: hamlinstaff@hamlin.lib.me.us
Web Site: www.hamlin.lib.me.us
Key Personnel: Museum Cur., Ann McDonald.
Personnel Profile: Part-Time Paid 1.
Governing Authority: Tax-exempt.
Institution Type/Description: Historic Building: built in 1822.
Collections: portraits of members of early Paris families; Lincoln-Hamlin campaign artifacts; local minerals & gems; grandfather clock; friendship quilt.
Hours & Admission Prices: April 1-Nov. 1 Tues.-Thurs. 11-5, Sat. 10-3. No charge; donations accepted.

Southport

HENDRICKS HILL MUSEUM, 417 Hendricks Hill Rd., Rte. 27, Southport, ME 04576. Mailing Address: P.O. Box 3, Southport, ME 04576-0003. Tel.: 207-633-1102.
Founded: 1988.
Congressional District: 1
Personnel Profile: Part-Time Volunteers 46; Interns 1.
Governing Authority: Tax-exempt: 501(c)(3).
Institution Type/Description: History Museum: housed in 1810 farmhouse.
Collections: period furnishings; 1850-1960 fishing equipment; photographs; genealogical. Boatshop: boats, tools, ice harvesting equipment.
Activities: tours.
Publications: annual newsletter; local history books.
Hours & Admission Prices: July to Labor Day Tues., Thurs. & Sat. 11-3; Sept. by appointment. No charge; donations accepted. &
Attendance: 382 (accurate)
Membership: $5; $10 & up.

Southwest Harbor

WENDELL GILLEY MUSEUM, (M), 4 Herrick Rd., Southwest Harbor, ME 04679-4431. Mailing Address: P.O. Box 254, Southwest Harbor, ME 04679-0254. Tel.: 207-244-7555. Fax: 207-244-5134.
E-mail: info@wendellgilleymuseum.org
Web Site: www.wendellgilleymuseum.org
Founded: 1979.
Congressional District: 2
Key Personnel: C.E.O. & Dir., Nina Z. Gormley; Pres., Tad Templeton; Exec. Vice Pres., Paul F. Haertel; Carver-in-Residence, Steven L. Valleau.
Personnel Profile: Full-Time Paid 2; Part-Time Paid 4; Part-Time Volunteers 6.
Governing Authority: nonprofit. Tax-exempt: 501(c)(3).
Institution Type/Description: Folk Art & Woodcarving Museum.
Collections: over 200 Wendell Gilley's woodcarvings; decorative birds & decoys; Audubon Birds of America facsimile; prints; wildlife art; Carroll S. Tyson's Birds of Mount Desert Island prints; miniature waterfowl carvings by A. Elmer Crowell; study & mounted bird specimens.
Facilities: library of material relating to wildlife & art topics; bird carving patterns & work of Wendell Gilley available for research by request; solar energy heating system; reading room. Carving tools, books & other bird-related items for sale.
Activities: guided tours; lectures; films; formally organized educational programs; permanent, loan & temporary exhibitions; movie rental to schools loan service; carving demonstrations & classes.
Publications: brochures; e-news.
Hours & Admission Prices: Jan.-April by appointment only; May & Nov.-Dec. Fri.-Sun. 10-4; June & Sept.-Oct. Tues.-Sun. 10-4; July-Aug. Tues.-Sun. 10-5. Adults $5, children 5-12 $2; discounts to groups, AAM & ICOM members; Maine Association of Museums & New England Museum Association members & members no charge. Closed national holidays. &
Attendance: 26,450 (accurate)
Membership: Individual $35; Family $50; Sustaining $100; Sponsor $250; Patron $500; Benefactor $1,000 and up.

Standish

* **MARRETT HOUSE, (M),** 40 Ossipee Trail E., Rte. 25, Standish, ME 04084-0003. Mailing Address: P.O. Box 3, Rte. 25, Standish, ME 04084-0003. Tel.: 207-882-7169. Fax: 207-882-7169.
Web Site: www.historicnewengland.org
Founded: 1944.
Key Personnel: Pres., Carl Nold; Regl. Mgr., Peggy Konitzky.
Governing Authority: society; nonprofit organization. Parent Institution: Historic New England, 141 Cambridge St., Boston, MA 02114. Tel.: 617-227-3956. Tax-exempt: 501(c)(3).
Institution Type/Description: Historic House: 1789 late Georgian house externally remodeled in the later Greek Revival & 19th century styles.
Collections: furnishings & family memorabilia spanning 150 years.
Facilities: perennial flower & herb garden.
Activities: guided tours; plant sale.
Publications: Historic New England Guide.
Hours & Admission Prices: June-Oct. 15 1st & 3rd Sat. of month 11-4. Adults $5; discounts to seniors, children, AAM, ICOM, AAA & WGBH members; Historic New England members no charge.
Attendance: 1,210 (accurate)
Membership: National $35; Individual $45; Household $55; Garden & Landscape $75; Contributing and Library & School $100; Historic New England Affiliate $100-$350; Young Friends of Historic New England $100-$1,500; Friends of the Library and Archives $125; Historic Homeowner $200; Business & Ogden Codman Design Group $250; Appleton Circle $1,750-$3,500.

Stockholm

STOCKHOLM HISTORICAL SOCIETY MUSEUM, 280 Main St., Stockholm, ME 04783. Tel.: 207-896-5812.
E-mail: jhede@mfx.net
Web Site: aroostook.me.us
Founded: 1976.
Congressional District: 2
Key Personnel: Pres. (V), Sandra Hara; Vice Pres., Albertine Dufour; Sec., Rosemary Hede; Treas., Membership and Librarian, Collection & Display, Linda Callison.
Personnel Profile: Part-Time Volunteers 7.
Governing Authority: society; nonprofit. Parent Institution: Stockholm Historical Society. Tax-exempt: 501(c)(3).
Institution Type/Description: Historical Society Museum: housed in 1900 store & post office.
Collections: photographs; tapes; town reports; oral & centennial (1881-1981) histories; town books; household items; farm & lumber implements.
Research Fields: family & house histories; town history.
Facilities: library of books available for use on premises; reading room.
Activities: permanent & temporary exhibitions; historical & cultural programs; video tapes of community events available for viewing at museum or taken on loan.
Publications: annual, Stockholm Historical Society Newsletter.
Hours & Admission Prices: July to early Sept. Wed.-Sun. 1:30-4:30; other times by appointment. No charge; donations accepted. &
Attendance: 200 (estimated)
Membership: Student $1; Contributing $5; Business $25; Life $100.

Thomaston

THE GENERAL HENRY KNOX MUSEUM, (M), 30 High St., Thomaston, ME 04861. Mailing Address: P.O. Box 326, Thomaston, ME 04861-0326. Tel.: 207-354-8062. Fax: 207-354-0886.
E-mail: info@knoxmuseum.org
Web Site: www.knoxmuseum.org
Founded: 1931.
Congressional District: 1
Key Personnel: Chm. (V), David Farmer; Museum Shop Mgr., Sandy Orluk.
Personnel Profile: Full-Time Paid 2; Full-Time Volunteers 20; Part-Time Volunteers 60; Interns 2.
Volunteer Hours: 500
Operating Expenses: 299,757
Operating Income: 235,588
Governing Authority: bd. of trustees. Tax-exempt.
Institution Type/Description: Historic House: 1795 Montpelier home of Major General Henry Knox.

Collections: furniture & artifacts of Colonial & Federal periods; Knox family memorabilia. Historic Building: 1824 Cole house.

Research Fields: Henry Knox; pre- & post-Revolutionary War American history; early settlements of Maine.

Facilities: library.

Activities: guided tours; lectures; encampments; family picnic. Special Events: concerts June to September; Knox birthday celebration in July; Revolutionary Encampment in August; Fall Harvest Weekend in October; Christmas Open House in December.

Publications: catalogue, The General Henry Knox Museum; semiannual newsletter, The Cannon; Montpelier: Through the Eyes of Tillman Crane and Friends; Montpelier: The Spot So Sacred to a Name So Great.

Hours & Admission Prices: June-Oct. Thurs.-Fri. 10-4; groups & other times by appointment. Family $20, adults $10, seniors $8, children 5-13 $4; discounts to AAA members; children under 5 & members no charge.

Attendance: 3,000 (accurate)

Membership: Individual $25; Basic Family $40; Revere Circle $100; Knox Circle $250; Washington Circle $500; Signature Society $1,000.

THOMASTON HISTORICAL SOCIETY, 80 Knox St., Thomaston, ME 04861-3714. Mailing Address: P.O. Box 384, Thomaston, ME 04861-0384. Tel.: 207-354-2295.

E-mail: katsmeow@roadrunner.com

Web Site: www.thomastonhistoricalsociety.com

Founded: 1971.

Congressional District: 1

Key Personnel: Historian, Margaret McCrea; Cur., Susan Devlin; Treas., Gerry Zwick; Sec., Aleta Kilborn; Finance, William Dashiell; Security, Galo Hernandez.

Personnel Profile: Part-Time Volunteers 8.

Governing Authority: society. Tax-exempt.

Institution Type/Description: Historic House Museum: housed in Henry Knox farmhouse; built in 1797.

Collections: paintings; furniture; maps of the town of Thomaston; photographs; journals; logbooks; historical books; ship models; Jonathan Cilley letters 1820-1867 & family letters thru mid 1860s; Revolutionary War, Civil War, World Wars I & II documents & memorabilia; 19th-century political artifacts; artifacts from 19th-century shipbuilding & seafarers lives; quarries & kilns; houses; notable residents of Thomaston.

Research Fields: local genealogy; maritime history; photographic journals; ship's logs.

Facilities: lecture hall.

Activities: guided tours; monthly lectures; cemetery tours. Annual Events: Knox Birthday Observance in July; 4th of July Parade Participation; Holiday House Tour in December.

Publications: annual newsletter; occasional, historical reprints: A Town That Went to Sea; History of Thomaston; Tall Ships, White Houses and Elms; A Thomaston Scrapbook; books, A Breach of Privilege: The Cilley Family Letters 1820-1867.

Hours & Admission Prices: June-Aug. Tues.-Thurs. 2-4, Sat. 1-3; Winter: Tues. 2-4 other tours by appointment. No charge; donations requested. &

Attendance: 380 (estimated)

Membership: Single $18; Family $25; Business $75.

Thorndike

BRYANT STOVE & MUSIC MUSEUM, 27 Stovepipe Alley, Thorndike, ME 04986. Tel.: 207-568-3665. Fax: 207-568-3666.

E-mail: sales@bryantstove.com

Web Site: bryantstove.com

Founded: 1982.

Congressional District: 2

Key Personnel: Owner, Joe Bryant.

Governing Authority: Parent Institution: Bryant Stove & Music, Inc.

Institution Type/Description: History Museum.

Collections: period stoves; early cars; player pianos; music boxes; dolls.

Hours & Admission Prices: Daily; groups by appointment. No charge; donations accepted. &

Attendance: 600 (estimated)

Union

MATTHEWS MUSEUM OF MAINE HERITAGE, Union Fairgrounds, Union, ME 04862. Mailing Address: P.O. Box 582, Union, ME 04862-0582. Tel.: 207-785-4330. Fax: 207-785-5145.

E-mail: mmomh@matthewsmuseum.org

Web Site: matthewsmuseum.org

Founded: 1965.

Congressional District: 1

Key Personnel: Chm. & Pres. (V), George R. Gross; Cur., Irene Hawes; Museum Shop Mgr., Clark Hooper.

Personnel Profile: Full-Time Volunteers 2; Part-Time Volunteers 30; Interns 1.

Governing Authority: society. Parent Institution: Knox Agricultural Society & Union Fair. Tax-exempt: 501(c)(3).

Institution Type/Description: Heritage Museum: one-room Hodge Schoolhouse (1864-1954).

Collections: artifacts of early Maine settlers; Maine life; agricultural; movie memorabilia; Moxie bottle stand; Moxie memorabilia.

Research Fields: tracing local families & farms; history of Maine; Moxie; 800-vol. library of old account, school & agriculture books & Civil War maps.

Facilities: picnic grounds by appointment only; restrooms available.

Activities: guided tours; hand-crafts exhibits; demonstrations; school field trips by appointment; Moxie bottle stand. Hodge School Sponsors: story readings for visiting children; Maine antiques festival; Union Fair on grounds Aug. 16-23.

Publications: books, revision of The Two Hundred Years of Union, history of Union, Maine; Edward Matthews Horse & Buggy Days; The Life & Efforts of the Evangelist Rev. Edward Smith Ufford; The First Century Union Fair, 1869-1969; Come Spring by Ben Ames Williams; Sibley's History of Union; various Moxie books.

Hours & Admission Prices: July-Aug. Wed.-Sat. 12-4. Adults $5, senior citizens and children 12 & over $3; discounts to groups; members & children under 12 no charge. Closed Independence Day & Aug. 9. &

Attendance: 4,000 (accurate)

Membership: Adult $5; Family $10; Business $25; Individual Life $100; Couple Life $150.

UNION HISTORICAL SOCIETY - ROBBINS HOUSE & OLD TOWN HOUSE, 343 Common Rd., Union, ME 04862. Mailing Address: P.O. Box 154, Union, ME 04862. Tel.: 207-785-5444.

E-mail: info@unionhistoricalsociety.org

Web Site: www.unionhistoricalsociety.org

Founded: 1972.

Congressional District: 1

Key Personnel: Pres. (V), Dan Day.

Personnel Profile: Part-Time Volunteers 13.

Governing Authority: independent. Tax-exempt.

Institution Type/Description: Historical Society Museum: housed in a Greek Revival home; built in 1847.

Collections: local history & culture; period furnishings; photographs; early telephone exchange switchboard; genealogy; Union Maine furniture.

Research Fields: genealogy of Union families; history of Union's historic houses; local business history.

Facilities: rental facilities.

Activities: monthly history programs. Annual Event: Founders Day Weekend in July.

Publications: newsletter; brochure; Come Spring; 200 Years in Union; Bridges to the Past; History of the Town of Union (reprint).

Hours & Admission Prices: Tues.-Wed. & Sat. 9-12; call to confirm; other times by appointment. No charge; donations accepted. &

Attendance: 700 (estimated)

Membership: Individual $5; Business $25; Life $100.

Van Buren

ACADIAN VILLAGE, 859 Main St., U.S. Rte. 1, Van Buren, ME 04785. Mailing Address: P.O. Box 165, Van Buren, ME 04785-0165. Tel.: 207-868-5042.

E-mail: mack53197@yahoo.com

Founded: 1976.

Key Personnel: Dir., Pres. (V) & Museum Shop Mgr., Anne L. Roy.

Personnel Profile: Full-Time Volunteers 1; Part-Time Paid 3.

Governing Authority: Tax-exempt.

Institution Type/Description: Historic Village: listed on the National Registry for the Preservation of Historical Landmarks.

Collections: Historic Houses: The Roy House & The Morneault House; Ouellette House; Sirols House.

Hours & Admission Prices: June 14-Sept. 15 daily 12-5. Adults $6, children $3; discount to AAA members.

Attendance: 700 (estimated)

Membership: Yearly $5; Lifetime $30.

Vinalhaven

THE VINALHAVEN HISTORICAL SOCIETY MUSEUM, 41 High St., Vinalhaven, ME 04863. Mailing Address: P.O. Box 339, Vinalhaven, ME 04863-0339. Tel.: 207-863-4410.

E-mail: vhhissoc@myfairpoint.net

Web Site: www.vinalhavenhistoricalsociety.org
Founded: 1963.
Congressional District: 1
Key Personnel: Pres., William Chilles; Dir., Susan Rodley; Vice Pres., Wyman Philbrook; Treas., Jacob Thompson.
Personnel Profile: Full-Time Paid 1; Part-Time Volunteers 7.
Governing Authority: nonprofit organization. Parent Institution: Vinalhaven Historical Society. Tax-exempt.
Institution Type/Description: History Museum: housed in 1838 church built in Rockland, moved to Vinalhaven Island in 1875.
Collections: granite industry; lobstering; fishing; sailboats; home appliances; farm & sea life; archives; costumes; folklore; manuscripts; Indian artifacts; genealogical records; 30 Microfilm rolls, 1872-1949 town property & poll tax books, 1884-1889 newspapers, 1850, 1860, 1870, 1880, 1900, 1910 & 1920 Vinalhaven census, 1787-1849 Penobscot-Castine district vessels; vital records 1789-1983; granite cutters union journal 1877-1939; Vinalhaven census 1930.
Facilities: library.
Activities: permanent & temporary exhibitions; school loan service.
Publications: annual spring newsletter, Vinalhaven Island (photographs) Arcadia, Images of America series.
Hours & Admission Prices: June & Sept. Mon.-Sat. 11-4; July-Aug. Mon.-Sat. 11-4, Sun. 12-3; other times by appointment. No charge; donations accepted.
Attendance: 1,600 (estimated)
Membership: Seniors 65 & over and Students $5; Individual $10; Family $15; Associate $30; Life $100.

Waldoboro

WALDOBOROUGH HISTORICAL SOCIETY MUSEUM, 1164 Main St., Waldoboro, ME 04572. Mailing Address: P.O. Box 110, Waldoboro, ME 04572.
E-mail: info@waldoborohistory.us
Web Site: waldoborohistory.us/
Founded: 1968.
Governing Authority: Tax-exempt.
Institution Type/Description: Historical Society Museum.
Collections: local history & culture; period clothing; personal artifacts; photographs; sewing machines; jewelry; quilts; rugs; dolls & toys; tools; early furnishings. Historic Building: one-room schoolhouse, c.1857.
Activities: educational programs.
Hours & Admission Prices: June-Sept. Wed.-Mon. 12-3. No charge; donations accepted.
Membership: Individual $10; Family $25; Patron $50; Sustaining Patron $150; Benefactor $500.

Warren

WARREN HISTORICAL SOCIETY, 225 Main St., Warren, ME 04864. Mailing Address: P.O. Box 11, Warren, ME 04864-0011. Tel.: 207-273-2726.
Founded: 1964.
Congressional District: 1
Key Personnel: Pres., Bruce Thornton; Cur., Barbara Larson; Genealogist, Diana Sewell.
Personnel Profile: Part-Time Volunteers 5.
Governing Authority: society. Tax-exempt.
Institution Type/Description: Regional History Museum.
Collections: history & growth of the Warren settlement & town; medical doctor's office; Warren archaeology artifacts; memorabilia & photos of wars; household furnishings; canal diagrams & history. Historic Buildings: c.1849 house & shed.
Research Fields: local history & traditions.
Facilities: 300-vol. library of historical material.
Activities: special events. Annual Event: Warren Day Celebration.
Publications: annual newsletter; Annals of Warren, 1605-1876; From Warren to the Sea, 1827-1852; Old Warren (photos), 1736-1936; Warren Cemeteries 1736-1985.
Hours & Admission Prices: Call for appointment. No charge; donations accepted.
Attendance: 300
Membership: Individual $5; Sustaining $10; Contributing $15 and over; Life $50.

Waterville

* **COLBY COLLEGE MUSEUM OF ART, (M),** 5600 Mayflower Hill, Waterville, ME 04901-8856. Tel.: 207-859-5600. Fax: 207-859-5606.
E-mail: museum@colby.edu
Web Site: www.colby.edu/museum
Founded: 1959.
Congressional District: 1
Key Personnel: Carolyn Muzzy Dir. & Chief Cur., Sharon Corwin; Chm. (V), Barbara Alfond; Asst. Dir. Operations, Gregory J. Williams; Assoc. Dir., Patricia King; Mirken Cur. Education, Lauren Lessing; Lunder Cur. American Art, Elizabeth Finch; Langlais Asst. Cur., Hannah Blunt; Administrative Sec., Karen Wickman; Curatorial Fellow, Elizabeth Speer; Mirken Coord. Education & Public Programs, Matthew Timme.
Personnel Profile: Full-Time Paid 8; Part-Time Paid 2; Part-Time Volunteers 27; Interns 5.
Governing Authority: college. Parent Institution: Colby College. Tax-exempt.
Institution Type/Description: Art Museum.
Collections: American Heritage Collection; Helen Warren and Willard Howe Cummings collection of American folk art; 18th-century American portraits; 19th-century weathervanes and landscapes; Jette collection of American painters of the Impressionist Period; John Marin collection; Alex Katz collection & archive; Lunder collection; James McNeill Whistler prints; 20th century & contemporary American painting, sculpture, photography; Bernat collection of Oriental ceramics; Colville collection of early Chinese art; Terry Winters prints.
Facilities: Museum-related items for sale.
Activities: guided tours; lectures; gallery talks; formally organized education programs for school children, adults & undergraduate college students; loan, permanent, temporary & traveling exhibitions.
Publications: Art at Colby Celebrating the Fiftieth Anniversary of the Colby College Museum of Art; With the Help of Friends The Colby College Museum of Art The First Fifty Years, 1959-2009; Colby Museum exhibition catalogs: Handbook of the Colby College Art Museum; Maine & Its Role in American Art; American Painters of the Impressionist Period; Maine Forms of American Architecture; Drawings from Maine Collections; Alex Katz at Colby College; 100 Works of the 20th Century at Colby; The John Marin Collection at the Colby College Museum of Art; Alex Katz: Collages; Currents1: Julianne Swartz; Currents2: Sam Van Aken; Currents3: Lihua Lei; The Skowhegan School of Painting and Sculpture: 60 Years.
Hours & Admission Prices: Tues.-Sat. 10-5, Sun. 12-5, tours by appointment. No charge. Closed holidays. &
Attendance: 20,137 (accurate)
Membership: Single $15; Family $25; Contributor $50; Sponsor $100; Subscriber $250; Benefactor $500; Patron $1,000.

REDINGTON MUSEUM, 62 Silver St., Unit B, Waterville, ME 04901-6524. Tel.: 207-872-9439. Facebook: The Redington Museum.
Web Site: www.redingtonmuseum.org
Founded: 1903.
Congressional District: 1
Key Personnel: Pres. Historical Society (V), Frederic P. Johnson; Cur., Sarah Sudgren.
Personnel Profile: Part-Time Paid 1; Part-Time Volunteers 25.
Governing Authority: society; nonprofit organization. Parent Institution: Waterville Historical Society. Tax-exempt.
Institution Type/Description: Historical Society Museum: housed in 1814 residence of Asa Redington.
Collections: house furnishings; early 19th-century tools & utensils; early American apothecary; Apothecary Museum; local portraits; local scenic photos 1855 to present; maps; charts; Indian artifacts; manuscripts; early weapons; complete file of the Waterville Mail newspaper; early toys; clothing; period furniture.
Research Fields: local & Civil War history.
Facilities: 300-vol. library of history books, chiefly on Maine & imported collection of early schoolbooks available for use on premises.
Activities: guided tours; lectures; formally organized educational programs; permanent exhibitions; cooperative program at Colby College for student research in local history.
Publications: picture book, Old Waterville.
Hours & Admission Prices: Memorial Day-Labor Day Tues.-Sat. Tours: 10, 11, 1 & 2. Adults $5; members & children under 18 when accompanied by an adult no charge. Closed holidays.
Attendance: 600 (estimated)
Membership: Single $20; Family $40; Friend $100; Sponsor $250; Patron $500; Benefactor $1,000.

Wells

HISTORICAL SOCIETY OF WELLS AND OGUNQUIT, Rte. 1, 938 Post Rd., Wells, ME 04090. Mailing Address: P.O. Box 801, Wells, ME 04090-0801. Tel.: 207-646-4775. Fax: 207-646-0832.
E-mail: wohistory@gwi.net
Web Site: www.historicalsocietyofwellsandogunquit.org
Founded: 1954.
Congressional District: 1
Key Personnel: Chm. (V), Lee Richheimer.
Personnel Profile: Part-Time Paid 1; Part-Time Volunteers 35.
Governing Authority: Tax-exempt: 501(c)(3).
Institution Type/Description: Historic Structure: 1862 Meeting House in Wells.
Collections: costumes; paintings; crafts; decorative arts; archives; photographs; genealogy.
Research Fields: genealogical & historical records.
Facilities: 1,100-vol. library; 200-seat auditorium; area for educational presentations. History books for sale.
Activities: guided tours; lectures; concerts; organized educational programs; docent program.
Publications: annual programs; guide to historic sites; quarterly newsletter.
Hours & Admission Prices: Memorial Day to Columbus Day Tues.-Thurs. 10-4; Oct.-May Wed.-Thurs. 10-4; other times by appointment. Requested Donation: $5; discounts to AAM members. &
Attendance: 600 (estimated)
Membership: Senior & Student $10; Individual $25; Family $35; Sustaining $50; Supporting Business $75; Patron $100; Benefactor $250-$500; Life $1,000.

Windham

WINDHAM HISTORICAL SOCIETY, 234 Windham Center Rd., Windham, ME 04062. Mailing Address: P.O. Box 1475, Windham, ME 04062. Tel.: 207-892-1433.
E-mail: info@windhamhistorical.org
Web Site: windhamhistorical.org
Founded: 1967.
Governing Authority: Tax-exempt: 501(c)(3).
Institution Type/Description: Historical Society Museum.
Collections: local history & culture; period artifacts; photographs.
Hours & Admission Prices: May-Nov. call for hours. No charge; donations accepted.
Attendance: 500 (estimated)
Membership: Individual $15; Family $25; Life $100.

Winterport

WINTERPORT HISTORICAL SOCIETY, 183 Main St., Winterport, ME 04496. Mailing Address: P.O. Box 342, Winterport, ME 04496. Tel.: 207-223-5556.
Key Personnel: Archivist, Theodora Weston
Institution Type/Description: Historical Society Museum.
Collections: local history & culture; photographs; personal artifacts.
Activities: educational programs.
Hours & Admission Prices: July-Aug. Tues. 2-4; other times by appointment.

Wiscasset

* **CASTLE TUCKER, (M),** 2 Lee St., Wiscasset, ME 04578-4121. Mailing Address: Historic New England, 141 Cambridge St., Boston, MA 02114-2702. Tel.: 207-882-7169; 617-227-3956 (Historic New England).
Web Site: www.historicnewengland.org
Founded: 1997.
Congressional District: 1
Key Personnel: Pres. & C.E.O., Carl Nold; Site Mgr., Peggy Konitzy.
Governing Authority: society. Parent Institution: Historic New England.
Institution Type/Description: Historic House: 1807 mansion overlooking Wiscasset Harbor.
Collections: mid-Victorian era furnishings & wallpaper typical of a wealthy sea-captain's home; free-standing elliptical staircase.
Research Fields: local history.
Activities: guided tours.
Publications: magazine, Historic New England.
Hours & Admission Prices: June-Oct. 15 Wed.-Sun. 11-5; last tour at 4pm. Adults $5; discounts for seniors, AAA, AAM, WBGH & ICOM members; members no charge.
Attendance: 3,934 (accurate)
Membership: National $35; Individual $45; Household $55; Garden &

Landscape $75; Contributing and Library & School $100; Young Friends of Historic New England $100-$1,500; Friends of the Library and Archives $125; Historic Homeowner $200; Business & Ogden Codman Design Group $250; Appleton Circle $2,500 & up.

THE 1811 LINCOLN COUNTY MUSEUM & OLD JAIL, 133 Federal St., Wiscasset, ME 04578. Mailing Address: P.O. Box 61, Wiscasset, ME 04578-0061. Tel.: 207-882-6817.
E-mail: lcha@wiscasset.net
Web Site: www.lincolncountyhistory.org
Founded: 1954.
Congressional District: 1
Key Personnel: Pres. (V), John Rienhardt.
Personnel Profile: Part-Time Paid 25.
Governing Authority: nonprofit. Parent Institution: Lincoln County Historical Association, Inc. Tax-exempt.
Institution Type/Description: Regional History Museum.
Collections: history; textiles; prints & photographic glass negatives of area. Historic Houses: 1809-1811 Old Lincoln County Jail; 1839 jailer's house.
Research Fields: jails; law; 19th century penal and criminal justice system; 19th century Wiscasset.
Facilities: reference library & regional histories; index of Maine design.
Activities: tours; lectures; workshops; school services; permanent & temporary exhibitions.
Publications: newsletter, The Lincoln County Chronicle; 18th Century Law & Order: 1761 Pownalborough Courthouse; 19th Century Law and Order: The 1811 Old Jail and 1839 Jailer House Museum; Preserve Local History; jail history brochure; The First Lincoln County Jail in Pownalborough; Histories of the Pownalborough Court House.
Hours & Admission Prices: Memorial Day to June & Sept. to Columbus Day Sat. 10-4, Sun. 12-4; July-Aug. Tues.-Fri. 10-4; other times by appointment. Adults $4; members no charge.
Attendance: 2,000 (estimated)
Membership: Student $2; Single $25; Family $35; Supporting $60; Sustaining $100.

MAINE ART GALLERY, 15 Warren St., Wiscasset, ME 04578-4032. Tel.: 207-882-7511.
E-mail: info@maineartgallery.org
Web Site: www.maineartgallery.org
Founded: 1958.
Congressional District: 1
Key Personnel: Pres. (V), Merlin Smith; Vice Pres., Barbara Vanderbilt; Gallery Mgr., Michele Roberge.
Personnel Profile: Full-Time Paid 1; Part-Time Paid 3; Part-Time Volunteers 10.
Governing Authority: nonprofit. Tax-exempt.
Institution Type/Description: Art Gallery.
Collections: paintings by Maine artists. Historic House: 1807 Old Academy.
Facilities: Art for sale.
Activities: guided tours; temporary exhibitions; monthly shows.
Publications: monthly catalog; newsletters; show announcements.
Hours & Admission Prices: Tues.-Sat. 10-4, Sun. 11-4. No charge; donations accepted.
Attendance: 1,600 (accurate)
Membership: Member $35; Couple $60; Sustaining $100; Patron $250.

MUSICAL WONDER HOUSE, 18 High St., Wiscasset, ME 04578-4118. Mailing Address: P.O. Box 604, Wiscasset, ME 04578-0604. Tel.: 207-882-7163. Fax: 207-882-6373.
Web Site: www.musicalwonderhouse.com/
Founded: 1963.
Key Personnel: Trustee & Founder, Danilo Konvalinka; Trustee, Joseph M. Villani.
Personnel Profile: Full-Time Paid 2; Part-Time Paid 1; Interns 1.
Governing Authority: individual operation.
Institution Type/Description: Mechanical Musical Instrument Museum: housed in 1852 sea captain's home.
Collections: musical boxes; gramophones; player & crank pianos; player & crank organs; talking machines; mechanical musical instruments.
Research Fields: mechanical musical instruments; preservation of antique music boxes, gramophones, player & crank pianos and organs.
Facilities: library of books in the field of mechanical music, player-piano & player-organ catalogues, talking machine literature & musical box literature available for use on premises. Museum-related items for sale.
Activities: guided tours; lectures; TV programs; permanent exhibitions. Museum Sponsors: Head Start programs; blindness programs; Voice of America.
Publications: Musical Wonder House Recordings; brochures.

Hours & Admission Prices: Memorial Day to Oct. Mon.-Sat. 10-5, Sun. 12-5. Tours: $10; $20; $40; discounts to seniors & AAA members. &
Attendance: 5,000 (estimated)

✱　**NICKELS-SORTWELL HOUSE, (M),** 121 Main St., Rte. 1, Wiscasset, ME 04578. Mailing Address: 141 Cambridge St., Boston, MA 02114-2702. Tel.: 207-882-7169; 617-227-3956 (Historic New England).
Web Site: www.historicnewengland.org
Founded: 1958.
Key Personnel: Pres. & C.E.O., Carl Nold; Site Mgr., Peggy Konitzky.
Governing Authority: society; nonprofit organization. Parent Institution: Historic New England, 141 Cambridge St., Boston, MA 02114. Tel.: 617-227-3956. Tax-exempt: 501(c)(3).
Institution Type/Description: Historic House: 1807 mansion designed with an elliptical stairway, lit by a skylight; Colonial Revival furnishings.
Collections: Sortwell family furnishings.
Activities: guided tours; special events.
Publications: magazine, Historic New England.
Hours & Admission Prices: June-Oct. 15 Fri.-Sun. 11-5, (last tour at 4). Adults $5; discounts to seniors, AAM, ICOM & AAA members; Historic New England members no charge.
Attendance: 1,691 (accurate)
Membership: National $35; Individual $45; Household $55; Garden & Landscape $75; Contributing and Library & School $100; Young Friends of Historic New England $100-$1,500; Friends of the Library and Archives $125; Historic Homeowner $200; Business & Ogden Codman Design Group $250; Appleton Circle $2,500 & up.

Woodland

LAGERSTROM HOUSE MUSEUM, Beckstrom Rd., Woodland, ME 04736. Mailing Address: Woodland Historical Society, 1149 New Sweden Rd., Woodland, ME 04736. Tel.: 207-493-4478 & 3081.
Institution Type/Description: Historic House: built in 1896.
Collections: family history; period furnishings & artifacts; photographs.
Hours & Admission Prices: July-Aug. Sun. 1-4; other times by appointment.

SNOWMAN SCHOOL HOUSE, Woodland Center Rd., Woodland, ME 04736. Mailing Address: Woodland Historical Society, 1149 New Sweden Rd., Woodland, ME 04736-5703. Tel.: 207-498-8430 & 3081.
Institution Type/Description: Historic Building: housed in a former school built in 1895 by carpenter & school teacher David Snowman.
Collections: local history & culture; period furnishings & artifacts; photographs.
Hours & Admission Prices: Memorial Day to June Sun. 1-4; other times by appointment.

Woolwich

WOOLWICH HISTORICAL SOCIETY MUSEUM, Nequasset Rd., Woolwich, ME 04579. Mailing Address: P.O. Box 98, Woolwich, ME 04579. Tel.: 207-443-4833 & 5684.
E-mail: whs@gwi.net
Web Site: woolwichhistoricalsociety.org
Key Personnel: Pres. (V), Debbie Locke
Institution Type/Description: Historical Society Museum.
Collections: local history & culture; personal artifacts; photographs.
Facilities: Museum-related items for sale.
Activities: educational programs; research.
Hours & Admission Prices: Call for hours.

Yarmouth

MUSEUM OF YARMOUTH HISTORY, Merrill Memorial Library, 215 Main St., Yarmouth, ME 04096. Mailing Address: P.O. Box 107, Yarmouth, ME 04096-0107. Tel.: 207-846-6259.
E-mail: mchaney@yarmouthmehistory.org
Web Site: www.yarmouthmehistory.org
Founded: 1958.
Congressional District: 1
Key Personnel: Exec. Dir., Michael Chaney; Chm. (V), Linda P. Grant.
Personnel Profile: Full-Time Paid 1; Part-Time Paid 3; Part-Time Volunteers 12; Interns 1.
Governing Authority: board of trustees. Parent Institution: Yarmouth Historical Society. Tax-exempt: 501(c)(3).

Institution Type/Description: Local History Museum.
Collections: local & maritime history; fine & decorative arts; business & domestic artifacts; organizational artifacts; photographs & documentary materials; historical recordings & literature. Historic Building: 1738 one-room Old Ledge Schoolhouse.
Research Fields: geology; genealogy; archaeology; sociology; industrial development; local history; social history; land use; architecture; maritime history.
Facilities: research room.
Activities: lectures & slide series; formally organized educational programs; oral history recording program; field trips; permanent & temporary exhibitions; special events; research.
Publications: newsletter, Yarmouth Historical Society; brochures; annual report.
Hours & Admission Prices: July-Aug. Mon.-Fri. 1-5; Sept.-June Wed.-Fri. 1-5, Sat. 10-5. No charge; donations accepted. Closed holiday weekends. &
Attendance: 3,000 (estimated)
Membership: Individual $25; Family $50.

York

MUSEUMS OF OLD YORK, 3 Lindsay Rd., York, ME 03909-1044. Mailing Address: P.O. Box 312, York, ME 03909-0312. Tel.: 207-363-4974. Fax: 207-363-4021.
E-mail: oyhs@oldyork.org
Web Site: www.oldyork.org
Formerly: Old York Historical Society
Founded: 1984.
Congressional District: 1
Key Personnel: Pres., Georgia Bennett; Dir., Joel Lefever; Dir. Devel., Laura Dehler; Dir. Education, Zoe Keefer-Norris; Registrar, Cynthia Young-Gomes; Program Specialist, Eileen Sewell; Office & Database Mgr., Katy Kreiger; Supvr. Bldgs. & Grounds, Jon Powers; Librarian, Virginia Spiller.
Personnel Profile: Full-Time Paid 6; Part-Time Paid 13; Part-Time Volunteers 300; Interns 3.
Governing Authority: nonprofit organization. Tax-exempt: 101(6); 501(c)(3).
Institution Type/Description: Historic Building Complex.
Collections: American furniture; decorative arts; manuscripts; ceramic & glass collection; tools & maritime artifacts; textiles. Historic Buildings: 1742 Emerson-Wilcox House; 1754 Jefferds' Tavern; 1740 John Hancock Warehouse; 1870 Marshall Store; 1745 schoolhouse; 1730 Elizabeth Perkins house; 1719 Old Gaol; 1820 Ramsdell house; 1834 Remick barn.
Research Fields: local history; colonial revival; decorative arts; archaeology; architecture; crime & punishment; genealogy.
Facilities: library/archives; visitors center. Museum-related items for sale.
Activities: guided tours; lectures; films; permanent & changing exhibitions; educational programs; summer children's camp; fellowship program; special events.
Publications: quarterly newsletters; annual educational materials; annual Proceedings of Elizabeth Perkins Fellowship Symposium.
Hours & Admission Prices: Exhibits: June-Oct. Mon.-Sat. 9:30-4. Office: Mon.-Fri. 8:30-4:30. Family $25, adults $12, senior citizens $10, children 6-16 $5; discounts to AAA & AAM members and groups; members & children under 6 no charge. Library: Thurs.-Fri. 9-12 & 1-5. Adults $5; members no charge. &
Attendance: 30,000 (estimated)
Membership: Individual $35; Family/Duel $60; Contributing & Business Friend $100; Sustaining & Business Associate $250; Business Patron $500; 1631 Partner & Business Partner $1,000.

York Harbor

✱　**SAYWARD-WHEELER HOUSE, (M),** Nine Barrell Lane Extension, York Harbor, ME 03911. Mailing Address: 141 Cambridge St., Boston, MA 02114-2702. Tel.: 207-384-2454; 617-227-3956.
Web Site: www.historicnewengland.org
Founded: 1977.
Key Personnel: Pres. & C.E.O., Carl Nold; Site Mgr., Brooke Steinhauser.
Governing Authority: society; nonprofit organization. Parent Institution: Historic New England, 141 Cambridge St., Boston, MA 02114. Tel.: 617-227-3956. Tax-exempt: 501(c)(3).
Institution Type/Description: Historic House: c.1718 house built on a slope overlooking the York River.
Collections: Queen Anne & Chippendale furniture; family portraits; china.
Activities: guided tours.
Publications: magazine, Historic New England.
Hours & Admission Prices: Tours: June-Oct. 15 2nd & 4th Sat. each month 11-5, (last tour at 4). Adults $5; discounts to groups & seniors, ICOM, AAM & AAA members; Historic New England members no charge.
Attendance: 1,308 (accurate)

Membership: National $35; Individual $45; Household $55; Garden & Landscape $75; Contributing and Library & School $100; Young Friends of Historic New England $100-$1,500; Friends of the Library and Archives $125; Historic Homeowner $200; Business & Ogden Codman Design Group $250; Appleton Circle $2,500 & up.

MARYLAND

(350 listings)

Aberdeen

ABERDEEN ROOM ARCHIVES & MUSEUM, 18 N. Howard St., Aberdeen, MD 21001. Mailing Address: P.O. Box 698, Aberdeen, MD 21001. Tel.: 410-273-6325.
E-mail: sayhello@aberdeenroom.org
Web Site: www.aberdeenroom
Founded: 1987.
Congressional District: 2
Key Personnel: Dir. & Cur., Charlotte Cronin; Museum Shop Mgr., James Lindsey.
Personnel Profile: Full-Time Volunteers 15.
Governing Authority: Tax-exempt.
Institution Type/Description: History Museum.
Collections: local history & culture; period furnishings; personal artifacts; photographs; yearbooks; newspapers.
Research Fields: newspaper, Harford Democrat & Aberdeen Enterprise, 1919-1986.
Activities: Museum Sponsors: Open Houses; Smithsonian, Journey Stories.
Publications: bimonthly, Museum Review; book, Sketches of Village to Town to City.
Hours & Admission Prices: 1st Sat. each month 11-3, Tues. & Thurs. 10-1. No charge; donations accepted. &
Attendance: 2,000 (estimated)

Accident

DRANE HOUSE, Old Cemetery Rd., Accident, MD 21520. Mailing Address: P.O. Box 190, Accident, MD 21520. Tel.: 301-746-6346. Fax: 301-746-7376.
E-mail: accidenttownhall@verizon.net
Congressional District: 1A
Institution Type/Description: Historic House: frontier plantation house; built 1798.
Collections: local history; period artifacts.
Hours & Admission Prices: By appointment. No charge; donations accepted.

Accokeek

THE ACCOKEEK FOUNDATION, INC., (M), 3400 Bryan Point Rd., Accokeek, MD 20607-9676. Tel.: 301-283-2113.
E-mail: info@accokeek.org
Web Site: www.accokeek.org
Founded: 1957.
Congressional District: 4
Key Personnel: Chm., Allen McCurry; Pres., Dr. Lisa Hayes; Museum Shop Mgr., Mary Alice Bonomo.
Personnel Profile: Full-Time Paid 20; Part-Time Paid 15; Part-Time Volunteers 210; Interns 5.
Volunteer Hours: 5,000
Governing Authority: nonprofit. Tax-exempt: 501(c)(3).
Institution Type/Description: Agriculture Museum.
Collections: outdoor living history museum of Colonial farming practices; sample crops, Indian corn & tobacco seed; preservation project; archaeology; botany; Native American artifacts; technology; entomology (nonliving)
Major Exhibits: Piscataway Connections to the Land (T), 10/14-12/14.
Research Fields: historical research into 18th-century crops & their culture; genetic research on crops & animals of the 18th century, the development of a blight-resistant American Chestnut; colonial life; land use.
Facilities: 500-vol. library of history & agriculture available for inter-library loan & for use on premises; field research station; boat dock; kayak launch.
Activities: guided tours; lectures; films; formally organized education programs for children & adults; permanent & temporary exhibitions; special events for Colonial crafts; daily activities of an 18th-century tobacco plantation.
Publications: newsletter.
Hours & Admission Prices: Park: open every day dawn to dusk. Colonial Farm: March thru Dec. Tues. - Sun. 10-4. Guided Tours: available by appointment only. No charge; donations accepted. Closed New Year's Day; Christmas. &

Attendance: 30,000 (estimated)
Membership: Individual $25; Family $45.

Annapolis

ANNAPOLIS MARITIME MUSEUM, 723 Second St., Annapolis, MD 21403-3323. Mailing Address: P.O. Box 3088, 723 Second St., Annapolis, MD 21403-0088. Tel.: 410-295-0104. Fax: 410-295-2962.
E-mail: office@maritime.org
Web Site: www.amaritime.org
Founded: 1997.
Congressional District: 40
Key Personnel: Dir., Jeff Holland.
Personnel Profile: Full-Time Paid 3; Part-Time Volunteers 250.
Governing Authority: bd. of directors. Tax-exempt.
Institution Type/Description: Maritime Museum.
Collections: maritime heritage of Annapolis & the Chesapeake Bay; maritime artifacts. Historic Structure: Thomas Point Shoal Lighthouse.
Activities: concerts; boat rides to lighthouse; lectures; educational programs; lighthouse tours; concerts.
Publications: quarterly newsletter.
Hours & Admission Prices: Museum: Thurs.-Sun. 12-4. Lighthouse Tours: call for hours. No charge; donations accepted. Boat Ride: adults 12 & over $70; children under 12 not admitted. &
Membership: Shipmate $35; Ensign $60; 1st Mate $100; Commander $500; Captain $1,000.

BANNEKER-DOUGLASS MUSEUM, (M), 84 Franklin St., Annapolis, MD 21401-2738. Tel.: 410-216-6180. Fax: 410-974-2553.
Web Site: www.bdmuseum.com
Founded: 1984.
Key Personnel: Exec. Dir., Joni Jones.
Personnel Profile: Full-Time Paid 5; Interns 2.
Governing Authority: Parent Institution: Governor's Office of Community Initiatives. Subsidiary Institution: MD Commission of African American History and Culture. Tax-exempt.
Institution Type/Description: African American Heritage Museum.
Collections: African American culture & history; photographs; African American art; books.
Facilities: library.
Activities: lectures; workshops; performances; educational programs.
Hours & Admission Prices: Tues.-Sat. 10-4. No charge. &

CHARLES CARROLL HOUSE, 107 Duke of Gloucester St., Annapolis, MD 21401-2526. Tel.: 410-269-1737. Fax: 410-269-1746.
E-mail: info@charlescarrollhouse.com
Web Site: www.charlescarrollhouse.com
Institution Type/Description: Historic House Museum: housed in the home of Charles Carroll, a signer of the Declaration of Independence in 1776.
Collections: personal artifacts; period furnishings; wine cellar.
Facilities: gardens; theater.
Activities: educational seminars; school programs; theater performances; wine events.
Hours & Admission Prices: June-Sept. Sat.-Sun. 12-4; groups by appointment. No charge; donations accepted. Closed Easter; Thanksgiving; Christmas Eve & Day.

CHASE-LLOYD HOUSE, 22 Maryland Ave., Annapolis, MD 21401-8006. Tel.: 410-263-2723.
E-mail: ck.05/03@yahoo.com
Founded: 1896.
Key Personnel: Dir., Carol Kelly; C.E.O. & Chm. (V), Molly Smith.
Personnel Profile: Full-Time Paid 3; Part-Time Paid 5.
Governing Authority: Parent Institution: Chase Home, Inc. Tax-exempt.
Institution Type/Description: Historic House Museum.
Collections: furnishings; personal artifacts.
Hours & Admission Prices: March-Dec. Mon.-Sat. 2-4. Adults $4; members & children under 6 no charge. Closed holidays.
Attendance: 600 (estimated)

THE CHESAPEAKE CHILDREN'S MUSEUM, (M), 25 Silopanna Rd., Annapolis, MD 21403-1117. Tel.: 410-990-1993. Fax: 410-990-1007.
E-mail: info@theccm.org
Web Site: www.theccm.org/
Founded: 1992.

Congressional District: 1
Key Personnel: C.E.O. & Dir., Deborah Wood.
Personnel Profile: Full-Time Paid 1; Full-Time Volunteers 1; Part-Time Paid 12; Part-Time Volunteers 20.
Governing Authority: Tax-exempt.
Institution Type/Description: Children's Museum.
Collections: hands-on exhibits.
Facilities: nature trail; herb garden.
Activities: please visit website at www.theccm.org or call for current newsletter.
Hours & Admission Prices: Summer: Thurs.-Tues. 10-5, Wed. groups by appointment; Winter: Thurs.-Tues. 10-4, Wed. groups by appointment. Admission 1 & over $4. ACM reciprocal membership. &
Attendance: 20,000 (estimated)
Membership: Basic $75; Reciprocal $150; Funded $225.

* **ELIZABETH MYERS MITCHELL ART GALLERY, ST. JOHN'S COLLEGE, (M),** 60 College Ave., Annapolis, MD 21401-1687. Mailing Address: P.O. Box 2800, 60 College Ave., Annapolis, MD 21404-2800. Tel.: 410-263-2371 & 626-2556. Fax: 410-626-2886.
Web Site: www.stjohnscollege.edu
Founded: 1989.
Congressional District: 4
Key Personnel: C.E.O. & Pres., Christopher Nelson; Chm. & Pres. (V), Cascy Pingle; Dir., Hydee Schaller; Treas., Bud Billups; Financial Dir., Barbara Goyette; Security, Timon K. Linn; Security, Mike Boston; Membership Coord., Kathy Dulisse; Membership Coord., Alexandra Fotos; Art Educator, Lucinda Dukes Edinberg.
Personnel Profile: Full-Time Paid 2; Part-Time Paid 7; Part-Time Volunteers 35; Interns 1.
Volunteer Hours: 450
Governing Authority: college; nonprofit. Parent Institution: St. John's College. Tax-exempt: 501(c)(3).
Institution Type/Description: College Art Gallery.
Major Exhibits: Dialogues: Words and Images in Art, 1500-1924, 1/31/14-4/6/14; St. John's College Community Art Exhibition 2014, 4/27/14-5/11/14; Less is More: Small Works in a Great Space (Mitchell Gallery National Juried Exhibition), 5/28/14-6/15/14; Annapolis Collections: Mitchell Gallery Celebrates 25 Years, 8/14-10/14; Along the Eastern Road: Hiroshige's Fifty-Three Stations of the Tokaido (T), 11/14-12/14.
Facilities: 1,825 sq. ft. exhibit space; lecture room; studios.
Activities: lectures; family programs; group tours; exhibit related workshops; loan, temporary & traveling exhibitions; studio courses in painting, life-drawing, sculpture, ceramics & photography; poetry writing workshops & readings; interpretive music programs; docent program; art express lunch-time tours; docent tours.
Publications: exhibition programs, Artline (for Mitchell Gallery membership); exhibit brochures & exhibition catalogues.
Hours & Admission Prices: Academic Year: Tues.-Thurs. & Sat.-Sun. 12-5, Fri. 12-5 & 7-8, call for details. No charge. &
Attendance: 11,000 (accurate)
Membership: Basic: Individual $50; Dual/Family $75; Membership Plus: Individual $100; Dual/Family $150; Supporting $250; Sustaining $500; Bennefactor $1,000.

HAMMOND-HARWOOD HOUSE ASSOCIATION, (M), 19 Maryland Ave., Annapolis, MD 21401-1626. Tel.: 410-263-4683. Fax: 410-267-6891.
E-mail: clively@hammondharwoodhouse.org
Web Site: www.hammondharwoodhouse.org
Founded: 1938.
Congressional District: 4
Key Personnel: Exec. Dir. & C.E.O., Mr. Carter C. Lively; Pres., Richardson A. Libby; Vice Pres., Louise Hayman.
Personnel Profile: Full-Time Paid 3; Full-Time Volunteers 3; Part-Time Paid 25; Part-Time Volunteers 270; Interns 2.
Governing Authority: nonprofit organization. Tax-exempt: 501(c)(3).
Institution Type/Description: Historic House: 1774 Hammond-Harwood House.
Collections: 18th-century furniture; decorative arts; paintings.
Research Fields: 18th-century architecture; 18th-century American & English decorative arts; Maryland & Annapolis history.
Facilities: Gift items for sale.
Activities: guided tours; special events.
Publications: books, Maryland's Way; The Hammond-Harwood House Cookbook; Hammond-Harwood House: An Illustrated History and Guide.
Hours & Admission Prices: April-Oct. Tues.-Sun. 12-5; Nov.-March by appointment. Adults $6, senior citizens $5.50, children 6-11 & Students $3;

discounts to groups, AAM & AAA members; children under 6 & members no charge. Closed New Year's Day; Thanksgiving; Christmas.
Attendance: 13,004 (estimated)
Membership: Individual $40; Contributor $50; Patron $100; Sponsor $250; Donor $500; Benefactor $1,000.

* **HISTORIC ANNAPOLIS FOUNDATION, (M),** 18 Pinkney St., Annapolis, MD 21401-1763. Tel.: 410-267-7619. Fax: 410-267-6189.
E-mail: info@annapolis.org
Web Site: www.annapolis.org
Founded: 1952.
Congressional District: 1
Key Personnel: Pres., John W. Guild; Vice Pres. Collections, Heather Ersts; Chm. (V), F. Joseph Rubino; Vice Pres. Advancement, Carrie Kiewitt; Dir. Garden, Mollie Ridout.
Personnel Profile: Full-Time Paid 11; Part-Time Paid 28; Part-Time Volunteers 450; Interns 5.
Governing Authority: private; nonprofit organization. Tax-exempt: 501(c)(3).
Institution Type/Description: Preservation & Education organization: housed in 5 historic sites.
Collections: architecture & decorative arts; Annapolis history. Historic Houses: 1765 William Paca House & Garden; c.1715 Shiplap House; barracks; waterfront warehouse; Historic Annapolis Museum at St. Clair Wright Center, c.1790.
Research Fields: Maryland & Annapolis history; 18th- & 19th-century architecture, household & business inventories.
Facilities: research center; history center. Museum-related items for sale.
Activities: guided tours; lectures; films; formally organized education program for children & adults; special events; architectural survey; audio walking tours of historic downtown Annapolis; research.
Publications: William Paca House & Garden; Archaeological Annapolis; Journal; tour guides.
Hours & Admission Prices: Adults $8; discounts to AAA & AAM members & senior citizens; members no charge. Closed New Year's Day; Thanksgiving; Christmas. &
Attendance: 428,000 (estimated)
Membership: Individual $50; Family & Patron $75; Sponsor & Small Business $125; Donor $250; Benefactor $500; Preservation Circle $1,000.

MARYLAND STATE ARCHIVES, (M), 350 Rowe Blvd., Annapolis, MD 21401-1686. Tel.: 410-260-6400. Fax: 410-974-3895. TDD: 800-735-2258.
E-mail: archives@mdsa.net
Web Site: www.msa.md.gov
Founded: 1935.
Congressional District: 4
Key Personnel: State Archivist, Edward C. Papenfuse; Dept. Archivist, Timothy Baker; Dir. Artistic Property, Elaine Rice Bachmann; Dir. Reference Svcs., Mike McCormick; Dir. Acquisition & Preservation, Kevin J. Swanson; Dir. Government Information Svc., Diane P. Evartt; Personnel, Richard Richardson; Dir. Special Collections, Rob Schoeberlein; Dir. Information Svcs., Wei Yang; Cur. Artistic Property, Alexander "Sasha" Lourie; Librarian, Christine Alvey; Registrar, Christopher Kintzel.
Personnel Profile: Full-Time Paid 100; Part-Time Volunteers 15; Interns 25.
Governing Authority: state. Tax-exempt.
Institution Type/Description: State Archival Institution.
Collections: colonial & state executive, legislative & judicial government records from 1634 to present; county probate, land & court records; private papers; church records; business records; state publications & reports; newspapers; maps; photographs; paintings; decorative arts; sculpture.
Research Fields: Maryland history; genealogy; American portraiture; decorative & fine arts.
Facilities: 14,000-vol. library of books on Maryland history, genealogy, U.S. history, archival & manuscript repositories available for use on premises; 100,000 cu. ft. of public government records; research room.
Activities: research, digitizing of land & other records.
Publications: Archives of Maryland (new series); Maryland Manual (biennial); finding aids; scholarly editions; historical essays.
Hours & Admission Prices: Tues.-Sat. 8:30-4:30. No charge; donations accepted. Closed state holidays; holiday weekends; first Sat. each month. &

MARYLAND STATE HOUSE, 100 State Circle, Annapolis, MD 21401. Tel.: 410-974-3400.
E-mail: elaineb@masa.net
Founded: 1772.
Institution Type/Description: Historic Building.
Collections: state history; paintings; John Shaw furniture; silver; statuary.

Hours & Admission Prices: Mon.-Fri. 9-5, Sat.-Sun. 10-9. Tours: 11 & 3. No charge. Closed New Year's Day; Thanksgiving; Christmas. &

UNITED STATES NAVAL ACADEMY MUSEUM, (M), 118 Maryland Ave., Annapolis, MD 21402-5034. Tel.: 410-293-2108. Fax: 410-293-5220.
Web Site: www.usna.edu/Museum
Founded: 1845.
Congressional District: 4
Key Personnel: Dir., Dr. J. Scott Harmon; Assoc. Dir. & Senior Cur., James W. Cheevers; Cur. Ship Models, Donald R. Preul; Cur. Beverly R. Robinson Collection, Laura Arrington.
Personnel Profile: Full-Time Paid 7; Part-Time Paid 1; Part-Time Volunteers 14.
Governing Authority: federal. Affiliated with the United States Naval Academy. Tax-exempt.
Institution Type/Description: Naval Museum: located at the U.S. Naval Academy.
Collections: paintings; prints; sculpture; historic ship models; naval uniforms; navigational instruments; naval weapons & gear; trophies of war; flags; china; silver; anthropological; numismatics; manuscripts & personal memorabilia of naval officers; over 6,000 naval battle & ship prints dating from 16th century to present; rare books; remains of John Paul Jones & artifacts.
Research Fields: naval history; naval & marine art; ship models; naval uniforms; instruments; equipment; weapons.
Facilities: 2,500-vol. library of books on U.S. Navy & Naval Academy history, marine architecture, naval flags, marine art & museology available for inter-library loan & on the premises.
Activities: hobby workshops; formally organized education programs for undergraduate students affiliated with USNA; permanent & temporary exhibitions.
Publications: various catalogs.
Hours & Admission Prices: Mon.-Sat. 9-5, Sun. 11-5. No charge; donations accepted. Closed New Year's Day; Thanksgiving; Christmas. &
Attendance: 170,000 (accurate)

Annapolis Junction

NATIONAL CRYPTOLOGIC MUSEUM, (M), 8290 Colony Seven Rd., Annapolis Junction, MD 20701. Mailing Address: 9800 Savage Rd. #6272, Fort Meade, MD 20755-6272. Tel.: 301-688-5849.
Web Site: www.nsa.gov/museum
Founded: 1993.
Congressional District: 3
Key Personnel: Cur., Patrick Weadon.
Personnel Profile: Full-Time Paid 4; Part-Time Volunteers 15.
Governing Authority: federal government; nonprofit. Tax-exempt.
Institution Type/Description: Cryptologic Museum.
Collections: cipher machines; cryptologic equipment; rare books; historical computers.
Research Fields: cryptologic history.
Facilities: library available to scholars & researchers. Gift items for sale.
Activities: guided tours; films; cell phone audio tours; youth programs.
Publications: brochures.
Hours & Admission Prices: Mon.-Fri. 9-4, 1st & 3rd Sat. 10-2. No charge. Closed federal holidays. &
Attendance: 50,000 (estimated)
Membership: National Cryptologic Museum Foundation: Sustaine $25; Contributor $100; Donor $500; Sponsor $1,000; Patron $5,000; Benefactor $10,000.

Arnold

HERBARIUM AT ANNE ARUNDEL COMMUNITY COLLEGE, 101 College Pkwy., Arnold, MD 21012-1895. Tel.: 410-541-2260.
Institution Type/Description: Herbarium.
Collections: plant specimens.
Facilities: library.
Activities: temporary exhibits.
Hours & Admission Prices: Mon.-Fri. by appointment.

Baltimore

THE ALBIN O. KUHN LIBRARY & GALLERY, UNIVERSITY OF MARYLAND-BALTIMORE COUNTY, 1000 Hilltop Cir., Baltimore, MD 21250. Tel.: 410-455-2232. Fax: 410-455-1567.
E-mail: beck@umbc.edu
Web Site: www.umbc.edu/library
Founded: 1975.
Congressional District: 7
Key Personnel: Chief Cur., Tom Beck; Head Info Technology, May Chang.
Personnel Profile: Full-Time Paid 3; Part-Time Paid 4; Part-Time Volunteers 4; Interns 2.
Governing Authority: public university; nonprofit. Parent Institution: University of Maryland-Baltimore County. Tax-exempt: 501(c)(3).
Institution Type/Description: University Museum.
Collections: over 1.5 million photographs, including 19th & 20th-century photographers Lewis Hine, Lotte Jacobi and photo-secessionists; Azriel Rozenfeld Science Fiction Research collection; Edward G. Howard collection of Marylandia; Merkle 18th & 19th-century English graphic satire; biological sciences archives including The American Society of Microbiology and The American Society of Cell Biology; rare books; manuscripts.
Research Fields: 19th and 20-century photographs; biological sciences; popular culture.
Facilities: 40,400-vol. library of books on photographs, microbiology and science fiction research; reading room; 4,420 sq. ft. exhibit space.
Activities: guided tours; formal education programs for the general public & campus community; loan, temporary & traveling exhibitions.
Publications: exhibition catalogues, Framing the Exhibition: Multiple Constructions; Eye of the Storm: Photographs by Mildred Grossman; Word & Image: Swiss Poster Design, 1955-1997; Visual Griots: Works by Four African-American Photographers; Fields of Vision: Women in Photography; Romantic Archaeologies: Some Images of the Age and Selected Women Writers; An American Vision: John G. Bullock and the Photo-Secession; A. Audrey Bodine; INTERMEDIA: The Dick Higgins Collection at UMBC; Typographically Speaking: The Art of Matthew Carter.
Hours & Admission Prices: Gallery: Sept.-Dec. 13 Mon.-Wed. & Fri. 12-4, Thurs. 12-8, Sat.-Sun. 1-5. Library: Mon.-Thurs. 8:30-10:30, Fri. 8:30-6, Sat. 12-6, Sun. 12-10:30. No charge; donations accepted. Closed major holidays. &
Attendance: 12,000 (estimated)
Membership: Student $15; Associate $25; Patron $50; Contributor $100; Sponsor $500; Benefactor $1,000 & up.

AMERICAN VISIONARY ART MUSEUM, 800 Key Hwy., Baltimore, MD 21230-3940. Tel.: 410-244-1900. Fax: 410-244-5858.
E-mail: info@avam.org
Web Site: www.avam.org
Founded: 1995.
Congressional District: 3
Key Personnel: Dir., Rebecca Alban Hoffberger; Dir. Education & Design, Theresa Segreti; Communications, Nick Prevas; C.F.O., Donna Katrinic; Exhibition Design, George Geary; Facility Rental, Michele Goldberg; Education, Amy Armiger; Membership & Devel. Assoc., Abby Baer; Registrar Assoc., Amanda Marsh Banks; Museum Shop Mgr., Ted Frankel.
Personnel Profile: Full-Time Paid 12; Part-Time Paid 30; Part-Time Volunteers 409; Interns 5.
Governing Authority: nonprofit corporation. Tax-exempt: 501(c)(3).
Institution Type/Description: Art Museum: located on 1.1 acre campus at Baltimore's Inner Harbor.
Collections: original works of art created by self-taught, intuitive artists; works range in scale from a 55 ft., 3 ton whirligig to mixed media assemblages, textiles, film & works on paper.
Research Fields: outsider art; visionary environments; Social Action and Art of Living.
Facilities: 1,000-vol. library of visionary, folk art & creativity reference books and visionary art catalogues; wildflower sculpture garden; theater; classrooms; 7,500 sq. ft. conference & party space; visionary center; 1,000-seat outdoor movie theater; restaurant. Museum-related items for sale.
Activities: workshops conducted by visionary artists; films; guided tours; lectures; temporary exhibitions; performances; conferences. Museum Sponsors: Annual Kinetic Sculpture Race.
Publications: annual magazine, Visions Magazine.
Hours & Admission Prices: Tues.-Sun. 10-6. Adults $15.95, senior citizens 60 & over $13.95, students $9.95; discounts to groups & museum employees; members, children 6 & under and Martin Luther King Jr. Day no charge. Closed Thanksgiving; Christmas. &
Attendance: 111,453 (accurate)
Membership: Student $25; Senior & Far-Out Fan $35; Single $50; Far-Out Couple & Senior Couple over 60 $55; Couple $75; Family $100.

THE BABE RUTH BIRTHPLACE AND SPORTS LEGENDS MUSEUM AT CAMDEN YARDS, (M), 216 Emory St., Baltimore, MD 21230-2235. Tel.: 410-727-1539. Fax: 410-727-1652.
E-mail: info@baberuthmuseum.com
Web Site: www.baberuthmuseum.com
Founded: 1974.

Congressional District: 3
Key Personnel: Exec. Dir., Michael L. Gibbons; Chm. (V), Thomas Winstead; Cur., Shawn Herne; Deputy Dir., John Ziemann; Dir. Sales & Mktg., John Hein.
Personnel Profile: Full-Time Paid 15; Full-Time Volunteers 100; Part-Time Paid 5; Part-Time Volunteers 50; Interns 2.
Governing Authority: nonprofit organization. Parent Institution: Babe Ruth Birthplace Foundation, Inc. Tax-exempt: 501(c)(3).
Institution Type/Description: Sports Museum.
Collections: 500 Home Run Club; life & career of Babe Ruth; autographed baseballs & bats; trophies; photographs; Baltimore Orioles & Baltimore Colts & Ravens memorabilia; Baltimore baseball history; Negro League exhibits; Maryland Baseball Hall of Fame; regional baseball archives for Maryland; video tapes of Baltimore Orioles Highlights; life of Babe Ruth film documentary; trophy collection; archives of Johnny Unitas.
Research Fields: local sports statistics & history.
Facilities: library of scrapbooks, record books, clippings, newspapers, magazines, photographs, film, audio tapes & phonograph records; theatre.
Activities: permanent, temporary & traveling exhibitions; films; lectures; educational programs; speakers bureau. Annual Events: Babe Ruth's Birthday; Fall gala; Game Day events.
Publications: newsletter; brochures, education; Visit The Museum.
Hours & Admission Prices: April-Oct. daily 10-5 (10-7 when Orioles play at home); Nov.-March Tues.-Sun. 10-5. Museum: Adults $6, senior citizens $4, children 3-12 $3; discounts to military, AAA & AAM members. Sports Legends at Camden Yards: adults $8, senior citizens $6, children 3-12 $5. Closed New Year's Day; Thanksgiving; Christmas. &
Attendance: 70,000 (accurate)
Membership: Children 5-12 $15; Individual $35; Family & Dual $50; Heavy Hitter $100; Sultan of Swat $150; Corporate $250-$2500.

BALTIMORE AMERICAN INDIAN CENTER, 113 S. Broadway, Baltimore, MD 21231-1727. Tel.: 410-675-3535. Fax: 410-675-6909.

E-mail: info@baic.org
Web Site: www.baic.org
Key Personnel: Exec. Dir., John Simermeyer
Institution Type/Description: History Museum.
Collections: Native American history & culture; sculpture; textiles; costumes; folk culture; archaeological artifacts; equipment; personal artifacts; tools.
Activities: Annual Events: Powwow in August.
Hours & Admission Prices: Call for hours.

BALTIMORE & OHIO RAILROAD MUSEUM, 901 W. Pratt St., Baltimore, MD 21223-2644. Tel.: 410-752-2490. Fax: 410-752-2499. Facebook: Baltimore & Ohio Railroad Museum.

E-mail: info@borail.org
Web Site: www.borail.org
Formerly: The B&O Railroad Museum
Founded: 1953.
Congressional District: 7
Key Personnel: Chm. Bd., Francis X. Smyth; Exec. Dir., Courtney B. Wilson; Vice Pres., Gino Gemignani, Jr.; Vice Pres., Gregory A. Farno; Vice Pres., John Magness; C.O.O., Stef Fay; Dir. Corporate Sponsorship & Events, Kathy Hargest; Grants Writer, James Smolinksi; Dir. Museum Svcs., Dana Kirn; Dir. Facilities, Steven Johnson; Mgr. Group Sales & Membership, Kelly Flanagan; Acting Supt. Railroad Operations & Chief Cur., David Shackelford; Dir. Operations, Travis Harry; Museum Shop Mgr., Eileen Blinzley.
Personnel Profile: Full-Time Paid 30; Part-Time Paid 20; Part-Time Volunteers 100.
Governing Authority: nonprofit organization. Parent Institution: B&O Railroad Museum, Inc. Tax-exempt: 501(c)(3).
Institution Type/Description: Transportation Museum: located on site of the historic Baltimore & Ohio Railroad's Mt. Clare Shops, site of the birthplace of American railroading.
Collections: over 120 locomotives & rolling stock; textiles; tools; lanterns; dining car china and silver; clocks & pocket watches; communication devices; signals; models; shop equipment; archival collections include corporate records; photographs; manuscripts; mechanical & engineering drawings; maps; artwork; rare books; audiovisual & film. Historic Buildings: 1884 Roundhouse; 1851 Mt. Clare Station; 1884 Annex; 1869 North Car Shop.
Major Exhibits: Civil War: The War Came By Train, 1/12-5/15; The War Came By Train, 11/13-5/15.
Facilities: research library & archives; theater car; snack area. Museum-related items for sale.
Activities: living history, docent, & educational programs for all ages; scheduled programs; short-line excursion train ride on 1 mile track; 40' HO model train layout; seasonal outdoor family activity area containing G-Scale train layout and seasonal miniature train ride; special events; rental facilities; carousel rides; indoor kids zone.
Publications: monthly e-newsletter, Train Mail.
Hours & Admission Prices: Mon.-Sat. 10-4, Sun. 11-4. Adults $16, senior citizens $14, children 2-12 $10; discount to groups of 20 or more with reservations and AAM & AAA members; members & children under 2 no charge. Closed New Year's Day; Easter; Memorial Day; Independence Day; Labor Day; Thanksgiving; Christmas Eve & Day. &
Attendance: 200,000 (estimated)
Membership: Senior $55; Individual $60; Dual Senior $65; Dual $75; Grand Family Senior $95; Family $100; Family & Friends $125; Lifetime: Society of the Iron Horse $5,000.

BALTIMORE CITY ARCHIVES, 2615 Mathews St., Baltimore, MD 21218-4705. Tel.: 410-396-0306.

Institution Type/Description: History Museum.
Collections: Baltimore history; pamphlets; photographs; posters; brochures.
Hours & Admission Prices: Call for hours.

BALTIMORE CITY FIRE MUSEUM, Old Town Mall, 414 N. Gay St., Baltimore, MD 21202-4134. Tel.: 410-727-2414.

Institution Type/Description: History Museum.
Collections: local firefighting history; furnishings; personal artifacts; fire equipment & tools; books; photographs; prints; drawings.
Facilities: Museum-related items for sale.
Hours & Admission Prices: Thurs. 9:30-12, Fri. 6:30pm-9:30pm, Sun. 1-4; other times by appointment.

BALTIMORE CIVIL WAR MUSEUM, 601 President St., Baltimore, MD 21202-4472. Tel.: 410-385-5188. Fax: 410-962-7058.

E-mail: museum_dept@mdhs.org
Web Site: www.mdhs.org
Founded: 1995.
Congressional District: 3
Key Personnel: Dir., Robert Rogers; Mgr. Museum, Frank Maurer; Chm., Henry Stansbury; Pres., Alex Fisher.
Personnel Profile: Full-Time Paid 1; Part-Time Volunteers 2; Interns 1.
Governing Authority: private; nonprofit organization. Parent Institution: Maryland Historical Society. Tax-exempt: 170(b)(1)(a).
Institution Type/Description: History Museum: located in the c.1850 President Street Station.
Collections: artifacts relating to Baltimore's role in the Civil War; MD railroad history; station's role in the underground railroad.
Facilities: 1,500 sq. ft. exhibit space. Museum-related items for sale.
Activities: guided tours; lectures; living history; walking tours.
Publications: Baltimore Civil War Museum Guide Book; newsletter.
Hours & Admission Prices: Library: Wed.-Sat. 10-4:30. Museum: Mon.-Sat. 10-5, Sun. 12-5. Adults $4, seniors, students with ID & children 13-18 $3; discounts to AAM members; children under 12 & members no charge. Closed Thanksgiving; Christmas. &
Attendance: 14,129 (accurate)
Membership: $50-$1,000.

BALTIMORE LITHUANIAN MUSEUM, 851-3 Hollins St., Baltimore, MD 21201-1003. Mailing Address: 3550 Benzinger Rd., Baltimore, MD 21229. Tel.: 301-774-3445; 410-646-0261.

E-mail: lituva@verizon.net
Formerly: Lithuanian Hall Museum
Founded: 1979.
Key Personnel: Dir. & Cur., Henry Gaidis; Assoc. Cur., Jon Burbulis; Recording Sec., Judy Baker; Treas., Maria Patlaba.
Governing Authority: nonprofit. Parent Institution: Lithuanian Hall Assoc., Inc. Tax-exempt.
Institution Type/Description: Lithuanian Heritage Museum.
Collections: Lithuanian heritage, culture, & history; photographs; personal artifacts; replica of c.1890 Lithuanian house.
Research Fields: historic military; ethnic & immigration.
Hours & Admission Prices: By appointment. No charge; donations accepted.
Attendance: 750 (estimated)

* THE BALTIMORE MUSEUM OF ART, (M), 10 Art Museum Dr., Baltimore, MD 21218-3827. Tel.: 443-573-1700. Fax: 443-573-1582. TDD: 410-396-4930.

E-mail: amannix@artbma.org
Web Site: www.artbma.org
Founded: 1914.

Congressional District: 3

Key Personnel: Dir., Doreen Bolger; Chm. (V), Frederick S. Koontz; Deputy Dir. Mktg. & Communications, Becca Seitz; Deputy Dir. Devel., Judith Gibbs; Deputy Dir. Curatorial Affairs & Sen. Cur Prints, Drawings & Photographs, Jay Fisher; Senior Cur. European Painting & Sculpture, Katherine Rothkopf; Cur. Textiles, Anita Jones; Librarian, Emily Rafferty; Dir. Communications, Anne Brown; Dir. Retail Operations, Deana Karras; Dir. Security, Fred Venhuizen; Cur. Decorative Arts and American Painting & Sculpture, David Park Curry; Deputy Dir. Education & Interpretation, Anne Manning; Deputy Dir. Finance & Planning, Christine Dietz.

Personnel Profile: Full-Time Paid 117; Part-Time Paid 19; Part-Time Volunteers 99; Interns 15.

Operating Expenses: 12,943,798

Operating Income: 11,218,019

Governing Authority: municipal. Tax-exempt: 501(c)(3).

Institution Type/Description: Art Museum: housed in 1929 building designed by John Russell Pope with later additions & adjoining sculpture gardens.

Collections: European paintings & sculpture from the Renaissance to present; 18th- to 20th-century American paintings; The Cone Collection, featuring works of Matisse & Picasso; modern & contemporary art featuring 20th-century American & European paintings & sculpture; prints; drawings, especially 19th-century French (The George A. Lucas Collection); photographs; American, European & Asian decorative arts & textiles; Maryland furniture & silver; pre-Columbian, Native American, African & Oceanic arts; 1st- to 3rd-century mosaics from Antioch, Syria.

Major Exhibits: Matisse's Marguerite: Model Daughter, 9/13-1/14; Morris Louis Unveiled, 9/13-2/14; Front Room: An-My Le, 10/13-2/14; Black Box: Gerard Byrne, 10/13-2/14; German Expressionism: A Revolutionary Spirit, 1/14-9/14; Baker Artist Awards 2013, 2/14-3/14; Black Box: Camille Henrot, 3/14-6/14; Front Room: Sterling Ruby, 3/14-6/14; On Paper: Figural Drawings From the Benesch Collection, 4/14-9/14; Grand Reopening of the American Wing, 11/14.

Research Fields: pertaining to collections.

Facilities: library; inter-library loan of art reference books; auditorium; sculpture gardens; cafe; education classroom. Art books & museum-related articles for sale.

Activities: gallery guide programs; lectures; concerts; dance recitals; education program for children & teachers; scholarly symposia; permanent & temporary exhibitions; docent-led & self-guided gallery tours; special programs for senior citizens, singles & adolescents at risk; special services for visually & hearing impaired upon advance request.

Publications: bimonthly calendar; newsletters; miscellaneous brochures; exhibition & collection catalogs; posters; postcards; bookmarks; quarterly members magazine.

Hours & Admission Prices: Wed.-Fri. 10-5, Sat.-Sun. 11-6. No charge. Special ticketed exhibitions. Closed New Year's Day; Independence Day; Thanksgiving; Christmas. &

Attendance: 205,000 (estimated)

Membership: Seniors $45; Donor $55, $80, $150, $250; Patron $500, $750, $1,000, $1,500; Benefactor $2,500, $3,500, $5,000, $7,500 & up.

BALTIMORE MUSEUM OF INDUSTRY, (M), 1415 Key Hwy., Inner Harbor South, Baltimore, MD 21230-5100. Tel.: 410-727-4808. Fax: 410-727-4869. Facebook: Baltimore Museum of Industry.

E-mail: tours@thebmi.org

Web Site: www.thebmi.org

Founded: 1977.

Congressional District: 3

Key Personnel: Exec. Dir., Roland H. Woodward; Bd. Chm., Stuart FitzGibbon; Deputy Dir., Carole Baker; Dir. Communications & Public Rels., Jessica Williams; Collections Mgr., Catherine Scott Dunkes; Visitor Svcs. & Museum Shop Mgr., Kelley Edelmann.

Personnel Profile: Full-Time Paid 15; Full-Time Volunteers 1; Part-Time Paid 30; Part-Time Volunteers 50; Interns 2.

Governing Authority: nonprofit organization. Tax-exempt: 501(c)(3).

Institution Type/Description: History & Industry Museum: housed in 1865 waterfront oyster cannery.

Collections: historical & modern industrial and business artifacts; tools; Maryland's popular culture; business records; manuscripts; photographs; films; tapes; books & serial publications; electric light bulb history from Edison's first installations in the 1880s; lamp from the Enola Gay; microscopic light bulb from a missile warhead; machine shop; blacksmith; print shop. Historic Vessel: 1906 S.T. Baltimore (steam tug).

Research Fields: industrial & business history; technology; labor & economic history; historical geography; Baltimore history.

Activities: self-guided & docent-guided tours; children's educational programs (Kids Cannery - role playing activity recreating 1883 oyster cannery workers; Children's Motor Works - children experience assembly line production while manufacturing toy model vehicle); temporary special exhibits & IPOD multi-media tours; lectures; concerts; internship opportunities.

Publications: newsletter, Nuts & Bolts; annual report; e-newsletter.

Hours & Admission Prices: Tues.-Sun. 10-4. Adults $12, seniors $9, children 7-18 & students with ID $7; discounts to AAM & ICOM members; children 6 & under and members no charge. Closed New Year's Day; Thanksgiving; Christmas Eve & Day. &

Attendance: 160,000 (estimated)

Membership: Senior Citizen $25; Individual $35; Household $55; Supporter $100-$249.

BALTIMORE PUBLIC WORKS MUSEUM, INC., 751 Eastern Ave., Baltimore, MD 21202-4369. Tel.: 410-396-5565. Fax: 410-545-6781.

E-mail: mari.ross@baltimorecity.gov

Web Site: www.baltimorepublicworksmuseum.org

Founded: 1982.

Congressional District: 1

Key Personnel: C.E.O., David E. Scott; Exec. Dir., Mari B. Ross; Dir. Education & Programs, Vince Pompa.

Personnel Profile: Full-Time Paid 4; Part-Time Paid 1; Interns 1.

Governing Authority: nonprofit organization. Parent Institution: Baltimore Dept. of Public Works. Tax-exempt: 501(c)(3).

Institution Type/Description: Urban Environmental History Museum: housed in restored c.1912 sewage pumping station.

Collections: approx. 2,000 c.1910 glass plate negatives documenting the construction of public works projects; c.1804 wooden water pipes; heavy equipment; gauges; meters; electrical material; tools; machinery.

Research Fields: Baltimore public works projects; urban environmental issues.

Facilities: library of books, journals & yearly reports on engineering & urban planning; 2,000 sq. ft. exhibit space. Gift items & books for sale.

Activities: guided tours; programs for high school & elementary school students; videos; internship programs; permanent exhibits.

Publications: Highlights in Public Works History.

Hours & Admission Prices: Tues.-Sun. 10-4. Adults $3, seniors, students & active military $2.50; discounts for groups of 10 or more & AAM members; Baltimore City public school students & members no charge. Closed holidays. &

Attendance: 20,000 (estimated)

Membership: Iron $150; Copper $450; Bronze $750; Silver $1,000; Gold $1,500.

BALTIMORE STREETCAR MUSEUM, INC., (M), 1901 Falls Rd., Baltimore, MD 21211. Mailing Address: P.O. Box 4881, Baltimore, MD 21211-0881. Tel.: 410-547-0264. Fax: 410-547-0264.

Web Site: www.baltimorestreetcar.org/

Founded: 1966.

Congressional District: 7

Key Personnel: Pres., John J. O'Neill; Exec. Vice Pres., Christopher M. McNally; Admin. Vice Pres., Edward M. Amrhein; Vice Pres. Engineering & Operations, John D. La Costa; Vice Pres. Cur., Mark E. Dawson; Treas., Christopher Howell; Comptroller, Paul W. Wirtz; Corp. Sec. & Recording Sec., Mark A. Hurley; Dir. Public Affairs, Andrew S. Blumberg; Museum Shop Mgr., Raymond L. Cannon.

Personnel Profile: Part-Time Volunteers 40.

Governing Authority: nonprofit organization. Tax-exempt: 501(c)(3).

Institution Type/Description: Transportation Museum: located on site of the Maryland & Pennsylvania RR Terminal in Baltimore.

Collections: 1880-1947 streetcars; 1859 & 1880 horsecars; 1913 crane; 1922 & 1940 trackless trolleys; 1945, 1947 & 1963 buses; 1952 line truck.

Research Fields: street railway & Baltimore history.

Facilities: library relating to Baltimore street railway history. Museum-related items for sale.

Activities: streetcar rides; guided tours; lectures; permanent exhibitions; video presentation.

Publications: newsletter, Live Wire; guide book; annual report; visitors' brochure.

Hours & Admission Prices: June-Oct. Sat.-Sun. 12-5; Nov.-May Sun. 12-5. Adults $7, children 4-11 & seniors 65 & over $5, maximum family $24; children under 4 no charge; valid passes issued by other Association of Railway Museums (ARM) honored for admission. Closed New Year's Day; Christmas. &

Attendance: 15,000 (estimated)

Membership: Senior Citizen $15; Student $17.50; Senior Family $20; Individual $30; Family $40.

BALTIMORE TATTOO MUSEUM, 1534 Eastern Ave., Baltimore, MD 21231-2330. Tel.: 410-522-5800. Fax: 410-522-4074.
Web Site: www.baltimoretattoomuseum.net/
Institution Type/Description: Tattoo Art Museum.
Collections: original artwork; tattooing equipment; lithographs; tattoo history.
Hours & Admission Prices: Mon.-Sat. 10-9, Sun. 11-7. No charge.

BALTIMORE'S BLACK AMERICAN MUSEUM, 1767 Carswell St., Baltimore, MD 21218-4908. Tel.: 410-243-9600.
Institution Type/Description: History Museum.
Collections: African American history & culture; photographs; personal artifacts.
Hours & Admission Prices: Sat. 8am-11pm, 1st Sun. of month 2-4; other times by appointment.

BENJAMIN BANNEKER HISTORICAL PARK & MUSEUM, 300 Oella Ave., Baltimore, MD 21228-5416. Tel.: 410-887-1081. Fax: 410-203-2747.
Web Site: www.benjaminbannecker.wordpress.com
Founded: 1998.
Congressional District: 7
Key Personnel: Dir., Justine Schaeffer; Chm. (V), Cynthia DeJesus; Museum Shop Mgr., Marilyn Cornish.
Personnel Profile: Full-Time Paid 2; Part-Time Paid 6; Part-Time Volunteers 50.
Governing Authority: private; nonprofit organization. Parent Institution: Baltimore County Government. Subsidiary Institution: The Friends of Banneker Historical Park. Tax-exempt: 501(c)(3).
Institution Type/Description: Historical Park & Museum.
Collections: historical & archaeological artifacts & prints; historical & cultural books; videos; wildlife habitats. Historic Building: farmhouse; log cabin.
Research Fields: early history of free African Americans & Native Americans; archaeology; the legacy of Benjamin Banneker; the natural environment native to our land; Colonial & early American history.
Activities: educational history tours & nature presentations; concerts; special events; permanent & temporary exhibitions.
Publications: newsletter, Banneker Journal.
Hours & Admission Prices: Tues.-Sat. 10-4. No charge; donations accepted. Charge for events & programs. Closed holidays. &
Attendance: 40,000 (estimated)
Membership: Full-Time Student & Senior Citizen $15; Individual $25; Family $40; Patron $100; Proprietor $250; Lifetime $400.

THE CARROLL MANSION, 800 E. Lombard St., Baltimore, MD 21202. Tel.: 410-605-2964. Fax: 410-528-1196.
E-mail: info@carrollmuseums.org
Web Site: www.carrollmuseums.org
Founded: 2002.
Key Personnel: Exec. Dir., Paula Hankins
Institution Type/Description: Historic House Museum: housed in the winter home of Charles Carroll, signer of the Declaration of Independence; built in 1828.
Collections: local history & culture; period furnishings; personal artifacts; photographs. Historic Building: 1828 Phoenix Shot Tower.
Activities: special events.
Hours & Admission Prices: Guided Tours: Sat.-Sun. 12-4. Adults $5, children 6-18, seniors 65 & over, students and military $4; children under 6 no charge.

CENTER FOR ART DESIGN AND VISUAL CULTURE, (M), 1000 Hilltop Circle, Fine Arts Bldg., 105, Baltimore, MD 21250. Tel.: 410-455-3188.
Governing Authority: nonprofit organization.
Institution Type/Description: Fine Arts Gallery.
Collections: painting; sculpture; drawing; printmaking; photography; graphic design; imaging & digital art; video; film; advertising; industrial design; architecture.
Activities: educational outreach programs; special events; guided tours.
Hours & Admission Prices: Tues.-Sat. 10-5. Closed UMBC holidays; New Year's Eve & Day, Christmas Eve, Day & week.

CYLBURN ARBORETUM, Cylburn Mansion, 4915 Greenspring Ave., Baltimore, MD 21209-4642. Tel.: 410-367-2217. Fax: 410-367-7112.
E-mail: info@cylburnassociation.org
Web Site: www.cylburnassociation.org
Founded: 1954.
Congressional District: 7
Key Personnel: Pres., Michael Smith; Exec. Dir., Nancy Hill; Dir. Nature Museum, Susan Patz; Chief Horticulturist, Melissa Grimm; Dir. Education, Lili Levy.
Personnel Profile: Full-Time Paid 1; Full-Time Volunteers 2; Part-Time Paid 2; Part-Time Volunteers 200.
Governing Authority: Parent Institution: City of Baltimore. Affiliated with Cylburn Arboretum, Inc. Tax-exempt: 501(c)(3).
Institution Type/Description: Natural History Museum: housed in a Victorian mansion of Renaissance revival style, built of stone from a nearby chromite mine & 207 acre arboretum.
Collections: birds; mammals; insects; plants; local material; rocks; minerals; arboretum; trees; magnolias; viburnums; maples; conifers; hollies.
Facilities: 207 acre arboretum; nature center; 250-seat auditorium; educational facilities.
Activities: guided tours; lectures; formally organized education programs; loan, permanent & temporary exhibitions; school loan service.
Publications: Association's newsletters, City Farms Newsletter; Trails Guide; bimonthly e-news, Cylburn Arboretum Association; quarterly e-news, Seasons.
Hours & Admission Prices: Call for hours. No charge. &
Attendance: 15,000 (estimated)
Membership: Individual $25; Household $35; Group $50; Sustaining $100; Life $500.

DR. SAMUEL D. HARRIS NATIONAL MUSEUM OF DENTISTRY, 31 S. Greene St., Baltimore, MD 21201-1504. Tel.: 410-706-7461. Fax: 410-706-8313.
E-mail: sswank@dentalmuseum.umaryland.edu
Web Site: www.dentalmuseum.org
Founded: 1996.
Congressional District: 7
Key Personnel: Dir. Operations, Cynthia Hollis; Cur., Dr. Scott D. Swank.
Personnel Profile: Part-Time Paid 2; Part-Time Volunteers 2; Interns 1.
Governing Authority: nonprofit organization. Parent Institution: University of Maryland, Baltimore. Tax-exempt.
Institution Type/Description: Dentistry Museum.
Collections: early dental instruments; furniture; equipment; early dentures; period artifacts; portraits.
Research Fields: dental history.
Facilities: library; archives; meeting & reception space.
Activities: guided tours; permanent, traveling & temporary exhibitions; educational programs.
Publications: book, The Baltimore College of Dental Surgery: The Pioneer of Dental Education.
Hours & Admission Prices: By appointment. Adults $7, seniors & students w/ID $6, children 3-12 $5; discounts to AAM members; members, active military & immediate family and children under 3 no charge. Closed most major holidays. &
Attendance: 10,000 (estimated)
Membership: Individual $50; Family $75; Dental Office $100.

EDGAR ALLAN POE HOUSE AND MUSEUM, 203 N. Amity St., Baltimore, MD 21223-2501. Mailing Address: 417 E. Fayette St., 8th Fl., Baltimore, MD 21202-3431. Tel.: 410-396-7932. Fax: 410-396-5662.
Web Site: www.eapoe.org/balt/poehse.htm
Founded: 1923.
Key Personnel: Dir., Kathleen Kotarba; Cur., Jeff Jerome.
Personnel Profile: Full-Time Paid 1; Part-Time Volunteers 5.
Governing Authority: municipal. Affiliated with the Baltimore City Preservation Commission.
Institution Type/Description: Historic House: 1830 home of Edgar Allan Poe.
Collections: period artifacts; Poe artifacts; lamps.
Activities: guided tours; lectures; slide & audiovisual presentations.
Hours & Admission Prices: April-Nov. Wed.-Sat. 12-3:30. Adults $3; night tours of the Poe grave & catacombs by special arrangement: $4 per person in groups of 25 & over. Closed national holidays.
Attendance: 6,000 (estimated)

EUBIE BLAKE NATIONAL JAZZ INSTITUTE AND CULTURAL CENTER, 847 N. Howard St., Baltimore, MD 21201-4605. Tel.: 410-225-3130. Fax: 410-225-3139.
E-mail: info@eubieblake.org
Web Site: www.eubieblake.org
Founded: 1983.
Congressional District: 7
Key Personnel: Exec. Dir., Troy Burton; Chm., John Clark Mayden.
Personnel Profile: Full-Time Paid 3; Part-Time Paid 16; Part-Time Volunteers 10.

Governing Authority: nonprofit. Tax-exempt: 501(c)(3).
Institution Type/Description: History Museum.
Collections: Eubie Blake memorabilia.
Activities: African-American heritage performance series.
Publications: quarterly newsletter, Ragtime.
Hours & Admission Prices: Wed.-Fri. 11-6, Sat. 11-3, Sun. by appointment. Admission $5. Closed federal holidays. &
Attendance: 10,000 (estimated)
Membership: Students $10; Individual $20; Family Members $50; Patrons $100

EVERGREEN MUSEUM & LIBRARY, JOHNS HOPKINS UNIVERSITY MUSEUMS, (M), 4545 N. Charles St., Baltimore, MD 21210-2693. Tel.: 410-516-0341. Fax: 410-516-0864.
E-mail: evergreenmuseum@jhu.edu
Web Site: www.museums.jhu.edu
Founded: 1990.
Congressional District: 3
Key Personnel: Dir., James Archer Abbott; Chm. (V), James R. Garrett; Coord. Membership, Rosalie Parker; Coord. Mktg., Heather Egan Stalfort; Tour Coord., Nancy Powers; Coord. Special Events, Anna Papierniak.
Personnel Profile: Full-Time Paid 7; Part-Time Paid 2; Part-Time Volunteers 75; Interns 3.
Governing Authority: university; board of trustees; nonprofit. Parent Institution: Johns Hopkins University. Tax-exempt: 501(c)(3).
Institution Type/Description: Historic House: housed in an 1850s Italianate mansion formerly owned by philanthropic Garrett family.
Collections: international art; French post-Impressionist paintings; furnishings; rare books; European & Oriental porcelain; Japanese Inro, Netsuke and Lacquer; Tiffany glass; carriage house.
Research Fields: rare books; Tiffany glass; oriental rugs & artifacts; American & English furniture; 20th century contemporary art.
Facilities: research library; theatre; formal gardens.
Activities: guided tours; Tea & Tour; concerts; symposia; lectures; meetings; temporary exhibitions; outdoor theatre productions; rental facility.
Publications: print & electronic newsletters, View From Evergreen; annual report.
Hours & Admission Prices: Tues.-Fri. 11-4, Sat.-Sun. 12-4; last tour 3pm. Adults $6, senior citizens $5, students with ID $3; discounts to groups, AAM & ICOM members, JHU faculty & staff; members & JHU students no charge. Closed holidays. &
Attendance: 11,600 (accurate)
Membership: Friend $50; Family $75; Supporter $125; Fellow $250; Patron $500; Benefactor $1,000.

FELLS POINT MARITIME MUSEUM, 1724 Thames St., Baltimore, MD 21231-3416. Tel.: 410-732-0278.
Web Site: www.mdhs.org
Institution Type/Description: Maritime Museum.
Collections: maritime history & artifacts; medical; law; religion; furnishings; tools; photographs; documents.
Facilities: library. Museum-related items for sale.
Activities: lectures; special events.
Hours & Admission Prices: Thurs.-Mon. 10-5. Adults $4, children, students & seniors $3; children 11 & under no charge.

FORT MCHENRY NATIONAL MONUMENT AND HISTORIC SHRINE, 2400 E. Fort Ave., Baltimore, MD 21230-5390. Tel.: 410-962-4290. Fax: 410-962-2500. TDD: 410-962-4290.
E-mail: fomc_superintendent@nps.gov
Web Site: www.nps.gov/fomc
Founded: 1933.
Congressional District: 3
Key Personnel: Supt., Tina Cappetta; Cur., Gregory Weidman; Historian, Scott Sheads; Chief Interpretation, Vince Vaise; Museum Shop Mgr., Karol Clark.
Personnel Profile: Full-Time Paid 25; Full-Time Volunteers 8; Part-Time Paid 15; Part-Time Volunteers 60; Interns 1.
Governing Authority: federal. Parent Institution: U.S. Dept. of Interior. Subsidiary Institution: National Park Service, Washington, DC 20240. Tax-exempt: 501(c)(3).
Institution Type/Description: Historic Site: Ware of 1812 fort, site of the bombardment that inspired Francis Scott Key to write "The Star-Spangled Banner", the National Anthem; fort also served during the American Civil War (1861-65) as a Union prison camp for Confederate soldiers and sympathizers, and as a hospital in WWI (1917-1925).
Collections: archives; military artifacts; U.S. Flags; Rodman Guns; Mrs. Reuben Ross Holloway manuscript collection relating to "The Star-Spangled Banner" becoming the National Anthem; Historic & Archaeological Research Project; c.1900 to the present, photographs. Historic Buildings: 1814 powder magazine & guardhouse; 1830s soldiers' barracks.

Research Fields: Battle of Baltimore; War of 1812; U.S. flag; history of Fort McHenry; "The Star-Spangled Banner."
Facilities: library of material relative to Fort's history, War of 1812 & National Park Service available on premises; visitor center. Publications & museum-related items for sale.
Activities: living history presentation; ranger programs; films; permanent exhibitions; research by appointment only; special events.
Publications: park folder, Fort McHenry National Monument & Historic Shrine.
Hours & Admission Prices: Memorial Day-Labor Day daily 9-6, rest of year daily 9-5. Adults 16 & over $7; senior citizens 62 & over with Golden Age Passport/Senior Pass and children 15 & under no charge. Research: by appointment. Closed New Year's Day; Thanksgiving; Christmas. &
Attendance: 600,000 (estimated)

FREDERICK DOUGLASS-ISAAC MYERS MARITIME PARK, 1417 Thames St., Baltimore, MD 21231. Tel.: 410-685-0295, ext. 487. Fax: 410-276-6347.
E-mail: mjews@douglassmyers.org
Web Site: www.douglassmyers.org
Institution Type/Description: History Museum & Park.
Collections: Douglass' life as an enslaved child & young man; African American heritage & maritime history; Chesapeake Marine Railway & Dry Dock Company; photographs; period furnishings; personal artifacts.
Activities: guided tours.
Hours & Admission Prices: Mon.-Fri. 10-4, Sat.-Sun. 12-4. Adults $5, seniors 60 & over $4, students 6-18 $2; discounts to groups; children 5 & under no charge.

GEPPI'S ENTERTAINMENT MUSEUM, 301 W. Camden St., Baltimore, MD 21201. Tel.: 410-625-7060. Fax: 410-625-7090. Facebook: Geppi's Museum.
E-mail: blaura@geppismuseum.com
Web Site: www.geppismuseum.com
Founded: 2006.
Key Personnel: Dir., Laura Bevens; C.E.O., Steve Geppi; Pres., Missy Geppi Bowersox.
Personnel Profile: Full-Time Paid 6; Part-Time Paid 5.
Governing Authority: Parent Institution: Diamond Comic Dist.
Institution Type/Description: History Museum.
Collections: American pop culture history; comic strips & books; early radio & television shows, films, & cartoons; toys; dolls; games.
Major Exhibits: Milestones, 2/13-1/14.
Activities: educational programs; special events.
Hours & Admission Prices: Tues.-Sun. 10-6. Adults $10, seniors 55 & over $9, students 5-18 $7; discounts on Thurs.; members & children under 4 no charge. Closed New Year's Day; Thanksgiving; Christmas. &
Attendance: 9,000 (estimated)
Membership: Friend $35; Friends $55; Generations $65; Family $75.

GLENN L. MARTIN MARYLAND AVIATION MUSEUM, 701 Wilson Point Rd., Hangar 5, Ste. 531, Baltimore, MD 21220-4238. Mailing Address: P.O. Box 5024, Baltimore, MD 21220-0024. Tel.: 410-682-6122.
E-mail: martinmuseum@gmail.com
Web Site: www.mdairmuseum.org
Founded: 1990.
Personnel Profile: Part-Time Volunteers 45.
Governing Authority: private; nonprofit organization. Tax-exempt: 501(c)(3).
Institution Type/Description: Aviation Museum.
Collections: Maryland's aviation history; aircraft & rocket models; aircraft; photographs.
Research Fields: Martin; Maryland aviation.
Facilities: archives by appointment.
Publications: quarterly newsletter.
Hours & Admission Prices: Wed.-Sat. 11-3. Adults $3; members no charge. &
Membership: Senior $25; Individual $35; Senior Couple $45; Family $50; Supporting $100; Patron $250; Lifetime $1,000. Corporate: Friend $1,000; Associate $2,500; Contributor $5,000; Benefactor $10,000; Leader $15,000.

GOLDSMITH MUSEUM OF CHIZUK AMUNO CONGREGATION, (M), 8100 Stevenson Rd., Baltimore, MD 21208-1899. Tel.: 410-486-6400.
E-mail: svick@chizukamuno.org
Founded: 2000.
Key Personnel: Exec. Dir., Ronald N. Millen
Institution Type/Description: Jewish Heritage Museum.
Collections: Jewish life, culture & history; religious artifacts.

Facilities: Museum-related items for sale.
Hours & Admission Prices: Call for hours. No charge; donations accepted.

THE HERITAGE MUSEUM, (I), Hamlet Court, 4509 Prospect Circle, Baltimore, MD 21216-1615. Tel.: 410-664-6711. Fax: 410-664-6711.
E-mail: heritagemuseum@usa.com
Founded: 1991.
Key Personnel: C.E.O., Steven Lee.
Personnel Profile: Full-Time Volunteers 1; Part-Time Paid 1; Part-Time Volunteers 3.
Governing Authority: private; nonprofit organization. Tax-exempt: 501(c)(3).
Institution Type/Description: Cultural, Historical & Environmental Museum.
Collections: traditional & contemporary art from Africa & the Americas; arboretum; rare & endangered, native & exotic plants.
Research Fields: early history of free African Americans & Native Americans; inventory & assessment of the natural resource of the Gwynns Falls wilderness park.
Facilities: 15,000-vol. library of history & culture of people of color; botanical garden; studio; field research station. Museum-related items for sale.
Activities: concerts; formal education programs for children; guided tours; lectures; temporary & traveling exhibitions; community programs.
Hours & Admission Prices: Mon.-Fri. 10-4:30. No charge; donations accepted. Charge for events & programs. Closed holidays.
Attendance: 900 (estimated)

HISTORIC SHIPS IN BALTIMORE, (M), Pier 1, 301 E. Pratt St., Baltimore, MD 21202-3134. Tel.: 410-539-1797, ext 422. Fax: 410-539-6238.
E-mail: crowsom@historicships.org
Web Site: historicships.org
Formerly: USS Constellation Museum & Baltimore Maritime Museum
Founded: 1999.
Congressional District: 7
Key Personnel: Exec. Dir., Christopher Rowsom; Chm. (V), Herbert Frerichs, Jr.; Dir. Finance, Robert Sandler; Membership, Dayna Aldridge; Education, Stan Berry; Public Rels., Laura Givens; Restoration Mgr., Paul Powichroski; Museum Shop Mgr., Audrey Monsberger.
Personnel Profile: Full-Time Paid 15; Part-Time Paid 40; Part-Time Volunteers 180
Governing Authority: private; nonprofit organization. Parent Institution: Living Classrooms Foundation. Tax-exempt: 501(c)(3).
Institution Type/Description: Maritime Museum: homeport to USS Constellation, USS Torsk, USCGC Taney, Lightship Chesapeake & the seven foot Knoll Lighthouse.
Collections: artifacts, photographs, documents & history related to the three ships that carry the name Constellation; personal artifacts; furnishings; U.S. Lighthouse Service, U.S. Navy & the U.S. Coast Guard artifacts. Historic Ships: U.S.C.G.C. Taney; U.S.S. Torsk; Light Ship Chesapeake; USS Constellation.
Research Fields: Civil War; US Naval History; sailor life; naval ordnance; wooden shipbuilding; maritime navigation; naval medicine; US Coast Guard history.
Facilities: 150-vol. library; classroom; 525 sq. ft. exhibit space. Museum-related items for sale.
Activities: tours; demonstrations; special programming; formal education programs; guided tours; lectures; participatory & temporary exhibitions; overnight adventures.
Publications: quarterly newsletter, The Deck Log.
Hours & Admission Prices: May & Sept.-Oct. 14 daily 10-5:30; June-Aug. call for hours; Oct. 15-April 30 daily 10-4:30. Adults $11-$18, senior citizens $8, children 6-14 $5; discount to groups, AAM members; children 5 & under and members no charge. Closed New Year's Day; Thanksgiving; Christmas. &
Attendance: 100,000 (estimated)
Membership: Individual $40; Regular Family $75; Large Family $125; Captain $250; Commodore $500; Admiral $1,000.

HOMEWOOD MUSEUM, (M), The Johns Hopkins Univ., 3400 N. Charles St., Baltimore, MD 21218-2608. Tel.: 410-516-5589. Fax: 410-516-7859.
E-mail: homewoodmuseum@jhu.edu
Web Site: www.museums.jhu.edu
Founded: 1987.
Congressional District: 7
Key Personnel: Dir., Winston Tabb; Pres. (V), Ross Jones; Cur., Catherine Rogers Arthur; Program Coord., Judith L. Proffitt; Coord. Membership, Rosalie Parker; Coord. Mktg. & Communications, Heather Egan Stalfort.

Personnel Profile: Full-Time Paid 5; Part-Time Paid 3; Part-Time Volunteers 50; Interns 4.
Governing Authority: university; advisory council. Parent Institution: Johns Hopkins University. Tax-exempt: 501(c)(3).
Institution Type/Description: Historic House Museum: 1801 Federal period home.
Collections: Federal period furnishings; porcelain; ceramics; tableware; textiles; furniture; prints; paintings; kitchen equipment; household items.
Research Fields: late 18th-to early 19th century domestic architecture; Carroll family history; late 18th- to early 19th century lifeways; 19th century games; Day in 1808, middle school program; slave/servant life in Baltimore; gardens & plants.
Facilities: 2,000-vol. research library, available to scholars. Museum-related items for sale.
Activities: guided tours; lectures; decorative arts workshops; volunteer program; membership program; holiday & seasonal events.
Hours & Admission Prices: Tues.-Fri. 11-4, Sat.-Sun. 12-4; last tour at 3:30. Adults $6, groups & senior citizens $5, students with ID $3; discounts to AAM & ICOM members and JHU faculty & staff; members & JHU students no charge. Closed holidays.
Attendance: 5,000 (accurate)
Membership: Friend $50; Supporter $125; Fellow $250; Patron $500; Benefactor $1,000.

IRISH SHRINE AND RAILROAD WORKERS MUSEUM, 920 Lemmon St., Baltimore, MD 21202. Tel.: 410-669-8154.
Web Site: www.irishshrine.org
Institution Type/Description: Historic Building Museum: built in 1848.
Collections: Irish-immigrant family life in the 1860s; Irish-American history, heritage & culture; period furnishings; personal artifacts; photographs.
Hours & Admission Prices: Sat. 11-2.

JAMES E. LEWIS MUSEUM OF ART, MORGAN STATE UNIVERSITY, (M), 2201 Argonne Dr., Baltimore, MD 21251. Mailing Address: The Carl J. Murphy Fine Arts Center, Rm. 242, 1700 E. Coldspring Lane, Baltimore, MD 21251. Tel.: 443-885-3030 & 3333. Fax: 443-885-8258.
E-mail: gabriel.tenabe@morgan.edu
Web Site: www.jelmamuseum.org
Founded: 1951.
Congressional District: 3
Key Personnel: Dir., Gabriel S. Tenabe; Cur., Dr. Diala Toure; Asst. Dir. Museum Studies & Historical Preservation, Robin Howard; Registrar, Nicole Paterson; Dir. Education, Dr. Tina Stevenson; Digitization Specialist, Richard Green.
Personnel Profile: Full-Time Paid 6; Part-Time Paid 1; Part-Time Volunteers 10; Interns 4.
Governing Authority: university. Affiliated with Morgan State University. Tax-exempt: 170(b)(1)(A).
Institution Type/Description: University Museum.
Collections: 19th- & 20th-century American and European sculpture; graphics; paintings; decorative arts; archaeology; African & Oceanian sculpture; Asian prints.
Research Fields: European, American, African & Afro-American art.
Activities: lectures; gallery talks; temporary, traveling & permanent exhibitions; visiting artists.
Publications: catalogs of collections & temporary exhibitions.
Hours & Admission Prices: Tues.-Fri. 10-4, Sat. 11-4, Sun. 12-4. No charge; donations accepted. Closed New Year's Eve & Day; Easter; Thanksgiving; Christmas break. &
Attendance: 40,000 (accurate)

✱ JEWISH MUSEUM OF MARYLAND, (M), 15 Lloyd St., Baltimore, MD 21202-4606. Tel.: 410-732-6400. Fax: 410-732-6451. Facebook: Jewish Museum of Maryland.
E-mail: info@jewishmuseummd.org
Web Site: www.jewishmuseummd.org
Formerly: Jewish Historical Society of Maryland
Founded: 1960.
Congressional District: 7
Key Personnel: Exec. Dir., Marvin Pinkert; Pres., Ira Papel; Asst. Dir., Deborah Cardin; C.F.O., Susan Press; Cur., Karen Falk; Collections Mgr., Jobi Zink; Dir. Education, Ilene Dackman-Alon; Museum Shop Mgr., Esther Weiner.
Personnel Profile: Full-Time Paid 11; Part-Time Volunteers 45; Interns 6.
Governing Authority: nonprofit organization. Tax-exempt: 501(c)(3).
Institution Type/Description: History Museum: housed in a 3-building complex, including history museum & the Lloyd St. Synagogue, built in 1845 and the B'nai Israel Synagogue built in 1876.

Collections: manuscripts; documents; photographs; ceremonial art; business records; ephemera; costumes; textiles; paintings; fine art; folk art; furniture; 6,500 objects related to regional Jewish history; 100,000 photographs and negatives; rare books, microfilms & materials on Jewish genealogy. Historic Buildings: 1876 Moorish Revival style synagogue; 1845 Greek Revival Synagogue.

Major Exhibits: Passages Through the Fire: Jews and the Civil War, 10/13-2/14; Project Mah Jongg (T), 3/14-6/14; The A-Maze-Ing Mendes Cohen, 9/14-8/15.

Research Fields: Jewish life related to Maryland & surrounding region pertaining to European and other antecedents; immigration; settlement; religious; business; social and urban history; Jewish religious architecture; material culture; decorative art; general Jewish history, culture & art related to or ancestral to the Jewish experience in Maryland from the 18th century to the present.

Facilities: 3,000-vol. library of local and American Jewish history; 4,500 sq. ft. exhibit space; educational facilities. Jewish art, ceremonial items, books & publications for sale.

Activities: gallery & historic building tours; walking tours; bus trips; lectures; films; concerts; study clubs; organized educational programs for children & adults; special public school programs; speakers' bureau; docent program; permanent, temporary & traveling exhibits.

Publications: exhibit catalogues; monographs; annual magazine, Generations; books; monthly e-newsletter, Museum Matters.

Hours & Admission Prices: Sun.-Thurs. 10-5, Fri. by appointment. Library & Archives by appointment. Adults $8, students $4, children 12 & under $3; discounts to groups, AAM, & GBHA members; members no charge. Closed major Jewish holidays. &

Attendance: 8,977 (accurate)

Membership: Student, Teacher, & Service Member $25; Senior $35; Individual $50; Family $75; Lombard Street Club $150; Living History Circle $250; Lloyd Street League $500; 1845 Society $1,000.

THE JOHNS HOPKINS UNIVERSITY ARCHAEOLOGICAL COLLECTION, 129/130 Gilman Hall, 3400 N. Charles St., Baltimore, MD 21218-2608. Tel.: 410-516-7561 & 6717. Fax: 410-516-4848.

E-mail: emaguire@jhu.edu

Web Site: neareast.jhu.edu/archaeo/

Founded: 1884.

Key Personnel: Dir. Dept. Near Eastern Studies, Dr. Betsy Bryan; Cur., Dr. Eunice Daughterman-Maguire.

Governing Authority: university. Parent Institution: The Johns Hopkins University. Tax-exempt.

Institution Type/Description: Art & Archaeology Museum.

Collections: Egyptian, Mesopotamian, Greek & Roman art objects & artifacts from c.4000 B.C.-500 A.D.

Activities: gallery talks on special request; loan, permanent & temporary exhibitions.

Publications: Ellen Reeder Williams, The Archaeological Collection of the Johns Hopkins University.

Hours & Admission Prices: Temporarily closed. &

Attendance: 700

LES HARRIS' AMARANTHINE MUSEUM, 2010 Clipper Park Rd., Baltimore, MD 21211. Tel.: 410-456-1343.

Web Site: amaranthinemuseum.org

Institution Type/Description: Art Museum.

Collections: Les Harris' paintings & sculptures.

Hours & Admission Prices: Sun. 12-3; other times by appointment. Suggested Donation: $5 per person.

MARSHY POINT NATURE CENTER, 7130 Marshy Point Rd., Baltimore, MD 21220. Tel.: 410-887-2817. Fax: 410-335-8995. TDD: 410-887-5319.

Institution Type/Description: Nature Center.

Collections: local wildlife & their habitats; photographs; natural heritage; butterfly garden.

Facilities: auditorium; classrooms; nature trails. Center-related items for sale.

Activities: educational programs; special events; guided canoe trips; summer camps; lectures. Museum Sponsors: Earth Day; Volunteer Appreciation Day; Popsicle Plunge; Holiday Council Party; Spring Festival; Fall Festival.

Hours & Admission Prices: Daily 9-5. Closed holidays.

MARYLAND ART PLACE, 8 Market Pl., Ste. 100, Baltimore, MD 21202. Tel.: 410-962-8565. Fax: 410-244-8017. TDD: 1-800-735-2258.

E-mail: map@mdartpeace.org

Web Site: www.mdartplace.org

Founded: 1981.

Congressional District: 29

Key Personnel: Exec. Dir., Amy Cavanaugh Royce; Program Mgr., Sofia Rutka; Devel. Mgr., Emily Sollenberger; Administrative Asst., Erin West.

Personnel Profile: Full-Time Paid 3; Part-Time Paid 1; Part-Time Volunteers 80; Interns 5.

Governing Authority: nonprofit organization. Tax-exempt: 501(c)(3).

Institution Type/Description: Art Museum.

Collections: temporary exhibitions of contemporary visual art & annual series of contemporary performing arts.

Facilities: 2,400 sq. ft. exhibit space.

Activities: participatory & traveling exhibitions; concerts; performances; films; guided tours; lectures; workshops. Annual Programs: Critics-in-Residence; Curator's Incubator; Young Blood.

Publications: quarterly newsletter; annual, Critics' Residency Catalog.

Hours & Admission Prices: Office: Tues.-Sat. 9-5. Gallery: Tues.-Sat. 11-5. No charge; donations accepted. Closed federal holidays. &

Attendance: 50,000 (estimated)

Membership: Artist $30; Individual $40; Dual $65; Contributor $100; Supporter $250.

MARYLAND HISTORICAL SOCIETY, (M), 201 W. Monument St., Baltimore, MD 21201-4674. Tel.: 410-685-3750. Fax: 410-962-7058.

E-mail: info@mdhs.org

Web Site: www.mdhs.org

Founded: 1844.

Congressional District: 7

Key Personnel: Pres. & C.E.O., Burton Kummerow; C.F.O., Dennis Elder; Chief Devel. Officer, Mark Letzer; Chief Cur., Alexandra Deutsch; Dir. Publications & Library Svcs., Patricia Anderson, Ph.D.; Dir. Education, Kristin Schenning.

Personnel Profile: Full-Time Paid 24; Part-Time Paid 12; Part-Time Volunteers 50.

Governing Authority: nonprofit organization.

Institution Type/Description: History Museum.

Collections: Maryland portraits & landscapes; furniture; silver; china; glass; costumes; clocks; uniforms; textiles; Chesapeake Bay maritime collection; maps; Civil War gallery; photographs, prints; architectural drawings; manuscripts including original draft of Star Spangled Banner; books; collections from the Baltimore City Life Museums.

Research Fields: Maryland history; material culture; maritime, architectural, fine & decorative arts.

Facilities: 60,000-vol. library of books; event spaces & meeting rooms. Pamphlets & books on Maryland history & gift items for sale.

Activities: guided tours; lectures; films; formally organized education programs; permanent & temporary exhibitions.

Publications: quarterly magazine, Maryland Historical Magazine; news bulletin, MdHS News, three times per year.

Hours & Admission Prices: Museum: Wed.-Sun. 10-5. Library: Wed.-Sat. 10-4:30. Museum: adults $6, senior citizens $5, students & children $4; members no charge. Library: $6. Closed New Year's Day; Thanksgiving; Christmas. &

Attendance: 18,000 (accurate)

Membership: Student, Teacher & Scholar $40; Individual $50; Household $65.

THE MARYLAND INSTITUTE, COLLEGE OF ART: DECKER, MEYERHOFF AND PINKARD GALLERIES, 1300 Mt. Royal Ave., Baltimore, MD 21217-4191. Tel.: 410-669-9200 & 225-2280. Fax: 410-225-2396.

Web Site: www.mica.edu

Founded: 1826.

Congressional District: 7

Key Personnel: Pres., Fred Lazarus, IV; Exhibitions Dir., Gerald Ross.

Governing Authority: board of trustees. Tax-exempt: 501(c)(3).

Institution Type/Description: Art Galleries: housed in 1896 Mt. Royal Railroad Station & Fox Building, former Cannon Shoe Factory.

Collections: George A. Lucas collection of 19th-century paintings.

Research Fields: American and European paintings, drawings, prints.

Facilities: over 39,000-vol. library of books on art, art history, design, photography, crafts & fine arts; graduate studio; auditorium; classrooms.

Activities: lectures; concerts; tours of the station building & Fox Building; changing exhibitions; performances.

Publications: exhibition catalogs; bulletins; posters.

Hours & Admission Prices: Call for hours. No charge. &

MARYLAND SCIENCE CENTER, 601 Light St., Baltimore, MD 21230-3803. Tel.: 410-685-5225. Fax: 410-545-5973. TDD: 410-962-0223.
E-mail: kszondy@marylandsciencecenter.org
Web Site: www.marylandsciencecenter.org
Formerly: Maryland Academy of Sciences
Founded: 1797.
Congressional District: 39
Key Personnel: C.E.O., Van R. Reiner; Chm. (V), Edward A. St. John; Dir. Education, Pete Yancone; Senior Scientist, Jim O'Leary; Senior Dir. Mktg., Christopher Cropper; C.O.O., Richard Hesse; Senior Dir. Guest Svcs., Lori Blau; Dir. Facilities, Shawn Chne; Museum Shop Mgr., Donna Plitt.
Personnel Profile: Full-Time Paid 100; Part-Time Paid 86; Part-Time Volunteers 154; Interns 10.
Governing Authority: nonprofit. Parent Institution: Maryland Academy of Sciences. Tax-exempt: 501(c)(3).
Institution Type/Description: Science & Technology Museum.
Collections: participatory exhibits on space; earth science; human body; early childhood education; science arcade; Hubble space telescope.
Facilities: 400-seat IMAX theater; 140-seat planetarium; 135-seat live theatre; classrooms; observatory; meeting rooms. Museum-related items for sale.
Activities: lectures; films; formally organized education programs for children & adults; informal workshops; scientific excursions; science demonstrations; traveling exhibitions; live demonstrations daily. Museum Sponsors: Student Science Seminars; scientific interest groups; career symposia.
Publications: education program catalogs & course descriptions; annual report; brochures; visitor map guide; monthly e-newsletter.
Hours & Admission Prices: Call or visit website for hours. Adults $16.95, senior citizens 62 & over $15.95, children 3-12 $13.95; discounts to AAM, ICOM & ASTC members; members & children under 3 no charge. Closed Thanksgiving; Christmas. &
Attendance: 700,000 (estimated)
Membership: Explorer $75; Voyager $100; Adventurer $125; Discoverer $150; Pioneer $200.

THE MARYLAND ZOO IN BALTIMORE, 1876 Mansion House Dr., Druid Hill Park, Baltimore, MD 21217. Tel.: 410-396-7102.
E-mail: info@marylandzoo.org
Web Site: www.marylandzoo.org
Formerly: The Baltimore Zoo
Founded: 1876.
Congressional District: 2
Key Personnel: C.E.O. & Pres., Donald Hutchinson; Veterinarian, Dr. Ellen Bronson; C.O.O., Karl Kranz; Gen. Cur., Mike McClure; Vice Pres. Human Resources, Eve Devine; Dir. Volunteers, Kerrie Kovaleski; Exec. Vice Pres. Institutional Advancement, Nancy Hinds; Dir. Public Rels., Jane Ballentine; Vice Pres. Education, Lori Finkelstein.
Personnel Profile: Full-Time Paid 214; Part-Time Volunteers 600.
Governing Authority: society. Maryland Zoological Society Inc. Tax-exempt: 501(c)(3).
Institution Type/Description: Zoo.
Collections: over 1,500 animals.
Research Fields: animal research, health, reproduction & behavior.
Facilities: 500-vol. library of zoo-oriented books available for research to qualified researchers; classrooms. Gift items for sale.
Activities: guided tours; lectures; films; formally organized education programs for children, adults & undergraduate college students; docent program or council.
Publications: triannual, Zoogram; School Services catalog; annual report.
Hours & Admission Prices: Daily 10-4. Adults $16.50, seniors $13.50; children 2-12 $11.50; discounts to AZA members; members & children under 2 no charge. Reciprocal discounts with other participating zoos & organizations. Closed Thanksgiving; Christmas. &
Attendance: 409,843 (accurate)
Membership: College $35; Individual & Senior $65; Grandparents $105; Family $109; Family Plus $125.

MOTHER SETON HOUSE, 600 N. Paca St., Baltimore, MD 21201-1995. Tel.: 410-523-3443.
Institution Type/Description: Religious Museum: housed in the home of Saint Elizabeth Ann Seton, the first American native-born canonized Saint of the Roman Catholic Church. Founder of Sisters of Charity of St. Joseph which later became the Daughters and Sisters of Charity in the US & Canada.
Collections: Mother Seton history; personal artifacts; period furnishings.
Hours & Admission Prices: Nov.-Feb. Sat.-Sun. 1-3; other times by appointment. No charge.

MOUNT CLARE MUSEUM HOUSE, 1500 Washington Blvd., Carroll Park, Baltimore, MD 21230. Tel.: 410-837-3262. Fax: 410-837-0251.
E-mail: info@mountclare.org
Web Site: www.mountclare.org
Founded: 1917.
Congressional District: 3
Key Personnel: Dir., Jane D. Woltereck; Pres. (V), Isabelle Obert; Museum Shop Mgr., Marguerite Ayers.
Personnel Profile: Full-Time Paid 1; Part-Time Paid 3; Part-Time Volunteers 15; Interns 2.
Governing Authority: municipal; nonprofit organization. Affiliated with The National Society of the Colonial Dames of America in the State of Maryland, 2715 Que St., N.W., Washington, DC 20007. Tax-exempt: 501(c)(3).
Institution Type/Description: Historic House: built in 1760.
Collections: 18th-century and early 19th-century silver; furniture; china; glassware; family portraits by Charles Willson Peale, Robert Feke, John Hesselius; many items original to mansion.
Research Fields: archaeological historical; 18th century topics & industries.
Facilities: 1,000-vol. library of genealogical books & records available on premises; rental facilities.
Activities: guided tours; permanent & temporary exhibits; lectures; educational programs; workshops; colonial camps; special exhibits; rental facilities. Museum Sponsors: Colonial Christmas Event.
Publications: quarterly newsletter, Notes From the Mount; booklet, Mount Clare; book, Mount Clare, Being an Account of the Seat Built by Charles Carroll, Barrister, Upon His Lands at Patapsco; Adventures, Cavaliers, Patriots; Mrs. Carroll's Favorite Receipts; Maryland's First Ladies of the White House.
Hours & Admission Prices: Tours: Thurs.-Sun. 11-3 on the hour. Adults $6, seniors $5, students $4; discounts to groups, tour operators, AAA & AAM members; members & children under 2 no charge. Closed major holidays.
Attendance: 5,700 (estimated)
Membership: Individual Friends $25-$49; Family $50-$99; Heritage $100-$249; Patrons $250-$749; Barrister $750 & up.

MUSEUM OF BALTIMORE LEGAL HISTORY, Clarence M. Mitchell Jr. Courthouse, 100 N. Calvert St., Rm. 243, Baltimore, MD 21202. Mailing Address: 101 W. Lombard St., Rm. 9442, Baltimore, MD 21201-2605. Tel.: 410-962-2820.
Founded: 1984.
Congressional District: 7
Key Personnel: Dir., Judge James F. Schneider
Institution Type/Description: Legal History Museum.
Collections: artwork; books; photographs; furnishings; tools; 18th century memorabilia of judges & lawyers.
Hours & Admission Prices: Mon.-Fri. 12-1 by appointment. Closed holidays. &
Attendance: 5,000

NATIONAL AQUARIUM, (M), 501 E. Pratt St., Pier 3, Baltimore, MD 21202-3103. Tel.: 410-576-3800. Fax: 410-576-8238. TDD: 410-727-3022.
E-mail: dpittenger@aqua.org
Web Site: www.aqua.org
Founded: 1981.
Congressional District: 3
Key Personnel: Exec. Dir., David M. Pittenger; Dept. Exec. Dir. Programs & Operations, and C.O.O., Paula Schaedlich; Deputy Exec. Dir. External Affairs, Kathy Sher; Chm., William R. Roberts; Deputy Exec. Dir. Finance, C.F.O. & Administration, Bruce Hoffberger; Senior Dir. Mktg., Denise Aranoff-Brown; Deputy Exec. Dir. Biological Programs, Dr. Brent Whitaker.
Personnel Profile: Full-Time Paid 250; Part-Time Paid 29; Part-Time Volunteers 797; Interns 55.
Governing Authority: nonprofit organization. Parent Institution: National Aquarium Institute. Tax-exempt: 501(c)(3).
Institution Type/Description: Aquarium.
Collections: marine life: over 14,000 fish, birds, amphibian, invertebrates & mammals.
Research Fields: husbandry of aquatic animals; water quality & treatment; veterinary medicine.
Facilities: aquarium; auditorium; classrooms; restaurant. Gift items for sale.
Activities: members' programs & trips; lecture series; events; school tours & programs; teacher training; outreach programs; conservation activities.
Publications: quarterly, Watermarks; annual report.
Hours & Admission Prices: Call for hours & admission prices. &
Attendance: 1,431,077 (accurate)

Membership: Individual $74; Senior $80; Couples $104; Family & Grandparents $149; Family Plus $199; Stingray Club $250; Dolphin Club $500.

THE NATIONAL GREAT BLACKS IN WAX MUSEUM, INC.,
1601-03 E. North Ave., Baltimore, MD 21213-1409. Tel.: 410-563-3404, ext. 17 & ext. 16. Fax: 410-675-5040 & 563-7806 (Exec. Office). Facebook: Great Blacks in Wax.
E-mail: jmartin@greatblacksinwax.org
Web Site: www.greatblacksinwax.org
Founded: 1983.
Congressional District: 7
Key Personnel: C.E.O., Dr. Joanne M. Martin; Deputy Dir., Jon Wilson; Devel. Dir., Karaleigh Henson; Museum Educator, Reba Bullock; Public & Media Rels., Ginger Williams; Gift Shop Mgr., Eric Cherry.
Personnel Profile: Full-Time Paid 10; Full-Time Volunteers 2; Part-Time Paid 7; Part-Time Volunteers 30; Interns 2.
Governing Authority: not-for-profit organization. Tax-exempt: 501(c)(3). Executive Office: 1649 E. North Ave., Baltimore, MD 21213.
Institution Type/Description: History & Wax Museum.
Collections: African-American history from ancient Africa to the present; African artifacts; over 100 wax figures of African ancestored Americans.
Major Exhibits: African American Contributions to the War of 1812, 1/13-12/14.
Research Fields: African-American history.
Facilities: 250-seat auditorium. Gift items for sale.
Activities: films; formal education programs for children; guided tours; lectures; rental gallery; traveling exhibitions.
Publications: quarterly, GBIW Newsletter.
Hours & Admission Prices: Tues.-Sat. 9-5, Sun. 12-6. Adults $12, senior citizens 55 & up, college students & children 12-17 $11, children 3-11 $10; discount to military, teachers, educators, government employees, AARP, AAM, ICOM, AAA members & groups of 10 or more; children under 3 & members no charge. Closed New Year's Day; Thanksgiving; Christmas.
Attendance: 200,000 (estimated)
Membership: Teacher $15; Senior & College Students $20; Individual $30; Mates $50; Family $75; Family Recognition $100; Contributor $150; Great Walk $250.

NATIONAL LACROSSE HALL OF FAME/US LACROSSE,
113 W. University Pkwy., Baltimore, MD 21210-3301. Tel.: 410-235-6882, ext. 122 & 133. Fax: 410-366-6735.
E-mail: info@uslacrosse.org
Web Site: www.laxmagazine.com
Formerly: The Lacrosse Museum & National Hall of Fame/US Lacrosse
Founded: 1959.
Key Personnel: Retail Operations & Museum Mgr., Michael Clark.
Personnel Profile: Full-Time Paid 65; Part-Time Paid 6; Part-Time Volunteers 12; Interns 2.
Governing Authority: nonprofit organization. Parent Institution: US Lacrosse. Tax-exempt: 501(c)(3).
Institution Type/Description: National Lacrosse Hall of Fame.
Collections: rare photographs & art; vintage equipment & uniforms; sculpture; trophies; memorabilia; artifacts; medals; prints; books; Hall of Fame plaques; periodicals; traces & documents oldest team sport native to the North American continent 500 years old from its roots in Native American religion to present day.
Research Fields: history of lacrosse.
Facilities: resource center. Books & museum-related items for sale.
Activities: Hall of Fame elections; magazine publishing; information & resource center. U.S. Lacrosse Sponsors: funding for U.S. National Team; National Development Center.
Publications: magazine published twelve times a year, Lacrosse Magazine.
Hours & Admission Prices: Call or see website for information.
Attendance: 6,000 (estimated)
Membership: Youth $25; High School $35; Adult $50.

NATIONAL MUSEUM OF CERAMIC ART AND GLASS,
2406 Shelleydale Dr., Baltimore, MD 21209-3242. Tel.: 410-764-1042. Fax: 410-764-1042.
Founded: 1989.
Congressional District: 3
Key Personnel: Dir., Chm. (V) & Education, Shirley B. Brown; Pres. (V), Richard Taylor; Vice Pres. & Membership, Paulyn Hyman; Financial Dir., Robert B. Brown; Devel., Bruce T. Taylor, M.D.
Personnel Profile: Full-Time Paid 1; Part-Time Paid 4; Part-Time Volunteers 1.
Governing Authority: nonprofit organization. Tax-exempt: 501(c)(3).
Institution Type/Description: Ceramic Art Museum.
Collections: ceramics; glass.

Activities: lectures; films; concerts; hobby workshops; education programs for middle school students; seminars; temporary exhibitions; special events.
Publications: quarterly newsletter.
Hours & Admission Prices: Please call for hours. No charge; donations accepted.
Attendance: 11,500 (estimated)
Membership: Students $15; Donor $25; Contribution $50; Associate $100; Sustainer $250; Supporter $500; Benefactor $1,000.

NATIONAL PINBALL MUSEUM,
608 Water St., Baltimore, MD 21202. Mailing Address: 614 Cannon Rd., Silver Spring, MD 20904. Tel.: 301-384-3802.
E-mail: inquiries@nationalpinballmuseum.org
Web Site: www.nationalpinballmuseum.org
Institution Type/Description: Pinball Museum.
Collections: pinball machine history, technology & art; hands-on exhibits.
Facilities: library. Museum-related items for sale.
Activities: demonstrations; lectures; workshops; outreach programs.
Hours & Admission Prices: Call for hours.

NORMAN AND SARAH BROWN ART GALLERY,
Jewish Community Center, 5700 Park Heights Ave., Baltimore, MD 21215-3930. Tel.: 410-542-4900, ext. 239.
Institution Type/Description: Art Gallery.
Collections: paintings; prints; sculpture; photographs; documents; drawings; books.
Facilities: community center; classrooms; garden. Museum-related items for sale.
Activities: temporary exhibitions.
Hours & Admission Prices: Mon.-Tues. 11-5, Wed.-Thurs. 3-5, Fri. 12-2:30, Sun. 12-5.

OLD ST. PAUL'S CEMETERY,
737 W. Redwood St., Baltimore, MD 21201-1011. Mailing Address: Old St. Paul's Parish Office, 309 Cathedral St., Baltimore, MD 21201-4410. Tel.: 410-685-3404. Fax: 410-385-0186.
Institution Type/Description: Cemetery.
Collections: cemetery monuments from early 1800's to early 1900s.
Hours & Admission Prices: Cemetery: by appointment. Office: Mon.-Fri. 11:30-1:30.

PORT DISCOVER CHILDREN'S MUSEUM, (M),
35 Market Place, Baltimore, MD 21202-4002. Tel.: 410-727-8120. Fax: 410-864-2729; 410-727-3042. TDD: 410-823-2551.
E-mail: info@portdiscovery.org
Web Site: www.portdiscovery.org
Formerly: Port Discovery, The Children's Museum in Baltimore
Founded: 1977.
Congressional District: 3
Key Personnel: C.E.O. & Pres., Bryn Parchman.
Personnel Profile: Full-Time Paid 27; Part-Time Paid 49; Part-Time Volunteers 5; Interns 2.
Governing Authority: private; nonprofit organization. Tax-exempt.
Institution Type/Description: Children's Museum.
Collections: hands-on exhibits.
Publications: calendar; newsletter, DreamScene.
Hours & Admission Prices: Memorial Day-Labor Day Mon.-Sat. 10-5, Sun. 12-5; Sept. Fri. 9:30-4:30, Sat. 10-5, Sun. 12-5; Oct.-May Tues.-Fri. 9:30-4:30, Sat. 10-5, Sun. 12-5. Admission $12.95; discounts for AAM members; passholders & children under 2 no charge. Closed Thanksgiving; Christmas.
Attendance: 250,000 (estimated)
Membership: Family & Grandparent $99; Family Plus $129.

REGINALD F. LEWIS MUSEUM OF MARYLAND AFRICAN AMERICAN HISTORY AND CULTURE,
830 E. Pratt St., Baltimore, MD 21202-4403. Tel.: 443-263-1800. Fax: 410-333-1138.
E-mail: emailus@maamc.org
Web Site: www.africanamericanculture.org
Founded: 1998.
Congressional District: 46
Key Personnel: Exec. Dir., A. Skipp Sanders; Exec. Asst., Sharon Harper; Chm. (V), Leslie King-Hammond, Ph.D.; Membership Coord., Zandra Carson; Dir. Collections, Michelle J. Wilkinson, Ph.D.; Exhibits Mgr., Dave Ferraro; Registrar, Deborah Nobles-McDaniel; Senior Accountant & Museum Shop Mgr., Bridget Lyday.
Personnel Profile: Full-Time Paid 18.

Governing Authority: private; nonprofit organization. Tax-exempt: 501(c)(3).
Institution Type/Description: History Museum.
Collections: African American heritage; personal artifacts; folk culture; paintings; audiovisual & film; sculpture; photographs; decorative arts.
Major Exhibits: Kinsey Collection of African American Art & Artifacts, 10/13-2/14.
Facilities: 200-seat auditorium; restaurant; theater; resource center; distance learning center; recording & listening studio. Museum & gift-related items for sale.
Activities: concerts; dance recitals; docent program; films; formal educational programs; guided tours; lectures; loan, participatory & traveling exhibitions; theater training programs for professional museum workers. Museum Sponsors: fundraising gala.
Publications: quarterly members newsletter, Journeys.
Hours & Admission Prices: Wed.-Sat. 10-5, Sun. 12-5. Cafe 10-4. Adults $8, senior citizens & students with valid I.D. $6, discount to groups; children under 6 & members no charge. Closed New Year's Day; Easter; Thanksgiving; Christmas. &
Membership: Student $20; Senior Charter $30; Individual Charter $35; Family $55; Contributor $100; Supporter $250; Partner $500; Friend $1,000.

RIPLEY'S BELIEVE IT OR NOT, 301 Light St., Light Street Pavilion, Baltimore, MD 21202. Tel.: 443-263-1800.
Web Site: www.ripleys.com/baltimore
Institution Type/Description: General Museum.
Collections: artifacts from around the world; personal artifacts; furnishings; paintings; photographs; sculpture; Mini Cooper covered in more than one million Swarovski crystals; Harry Potter's Hogwarts Castle made from 600,000 wooden matchsticks.
Hours & Admission Prices: Sun. 10-8, Mon.-Sat. 10-9.

SCHOOL 33 ART CENTER, 1427 Light St., Baltimore, MD 21230. Tel.: 410-263-4350. Fax: 410-837-6947.
E-mail: school33@promotionandarts.org
Web Site: www.school33.org
Founded: 1979.
Key Personnel: Dir., Randi Vega.
Governing Authority: Parent Institution: Baltimore Office of Promotion & the Arts.
Institution Type/Description: Art Gallery.
Collections: works by contemporary artists.
Facilities: classroom rental space.
Activities: classes; workshops; educational programs; special events.
Hours & Admission Prices: Gallery: Wed.-Sat. 9-5; call to confirm. No charge; donations accepted. &
Attendance: 6,500 (estimated)
Membership: Artist/Individual $35; Contributing $60; Patron $125; Sustaining $250; Benefactor $1,000.

SILBER ART GALLERY, GOUCHER COLLEGE, 1021 Dulaney Valley Rd., Baltimore, MD 21204-2780. Tel.: 410-337-6477. Fax: 410-337-6405. TDD: Maryland Relay System.
E-mail: laura.amussen@goucher.edu
Web Site: www.goucher.edu/rosenberg
Formerly: Rosenberg Gallery, Goucher College
Founded: 1885.
Congressional District: 2
Key Personnel: Dir. Exhibitions & Collection Coord., Laura Amussen.
Personnel Profile: Part-Time Paid 1; Interns 3.
Governing Authority: nonprofit. Parent Institution: Goucher College. Tax-exempt: 501(c)(3).
Institution Type/Description: Art Gallery.
Collections: contemporary & modern art; photography; prints; Asian art; Mexican ceramics.
Facilities: 1,000-seat theater; 225-seat auditorium.
Activities: lectures; temporary exhibitions of contemporary art. Annual Event: multi-disciplinary panel discussion.
Publications: quarterly, exhibition catalog.
Hours & Admission Prices: Tues.-Sun. 11-4. No charge. &
Attendance: 175,000 (estimated)

THE STAR-SPANGLED BANNER FLAG HOUSE, 844 E. Pratt St., Baltimore, MD 21202-4495. Tel.: 410-837-1793. Fax: 410-837-1812.
E-mail: info@flaghouse.org
Web Site: www.flaghouse.org
Founded: 1927.
Congressional District: 3

Key Personnel: Exec. Dir., Annelise Montone.
Governing Authority: board of directors; nonprofit organization. Parent Institution: The Star-Spangled Banner Flag House Association, Inc. Tax-exempt: 501(c)(3).
Institution Type/Description: History Museum: The Flag House, a National Historic Landmark, home of Mary Pickersgill, who sewed the Star-Spangled Banner which inspired Francis Scott Key to write the poem that became the National Anthem.
Collections: collection of 19th-century American documents, paintings, prints, watercolors; early 20th-century photographs; 19th-century American decorative arts; textiles. The Star-Spangled Banner Museum: flags; Mary Pickersgill artifacts; War of 1812 artifacts.
Research Fields: history & art history.
Facilities: 200-vol. library of books on the War of 1812 & the History of the Flag available on the premises by appointment. Flags, maps, American crafts & other museum-related items for sale.
Activities: guided tours; slide lectures; permanent & temporary exhibitions; living history program.
Publications: newsletter, The Star; annual report.
Hours & Admission Prices: Tues.-Sat. 10-4. Adults $8, senior citizens 55 & up & military $7, students K-12 $5; discount to AAM members & groups; members & children under 6 no charge. Closed major holidays. &
Attendance: 13,127 (accurate)
Membership: Individual $30; Family $40.

UNITED METHODIST HISTORICAL SOCIETY OF BALTIMORE WASHINGTON CONFERENCE, 2200 St. Paul St., Baltimore, MD 21218-5805. Tel.: 410-889-4458.
E-mail: rshindle@bwcumc.org
Web Site: www.lovelylanemuseum.com
Founded: 1855.
Congressional District: 3
Key Personnel: Dir., Robert W. Shindle; Pres., Rev. Emora T. Brannan.
Personnel Profile: Full-Time Paid 1; Part-Time Paid 1; Part-Time Volunteers 16; Interns 2.
Governing Authority: Parent Institution: Baltimore-Washington Conference of the United Methodist Church. Subsidiary Institution: Strawbridge Shrine Assoc., 2650 Strawbridge Ln., New Windsor, MD.
Institution Type/Description: Historical & Preservation Societies.
Collections: church history & heritage; religious artifacts; portraits.
Research Fields: Methodist history; local churches.
Facilities: library. Museum-related items for sale.
Activities: research; tours.
Publications: newsletter four times per year, 3rd Century Methodism.
Hours & Admission Prices: Thurs.-Fri. 10-4; other times by appointment. No charge; donations accepted. &
Attendance: 3,500 (estimated)

UNIVERSITY OF MARYLAND SCHOOL OF NURSING MUSEUM, 655 W. Lombard St., Rm. 727, Baltimore, MD 21201-1512. Tel.: 410-706-2822.
Web Site: nursing.umaryland.edu/museum
Formerly: University of Maryland School of Nursing Living History Museum
Founded: 1999.
Congressional District: 7
Key Personnel: Dir. & Cur., Daniel Caughey.
Personnel Profile: Part-Time Paid 1; Part-Time Volunteers 20.
Governing Authority: Parent Institution: University of Maryland School of Nursing.
Institution Type/Description: Nursing History Museum.
Collections: original nursing artifacts, historical photographs, letters & documents; audio & video presentations.
Hours & Admission Prices: Academic Year: Mon.-Tues. 10-2; other times by appointment. No charge. &
Attendance: 1,000 (estimated)

＊ **WALTERS ART MUSEUM, (M),** 600 N. Charles St., Baltimore, MD 21201-5185. Tel.: 410-547-9000 & 685-7823. Fax: 410-783-7969 & 752-4797 (curatorial) & 727-7591 (marketing).
E-mail: info@thewalters.org
Web Site: www.thewalters.org
Founded: 1931.
Congressional District: 7
Key Personnel: Exec. Dir., Dr. Julia Margari-Alexander; Chm., Andrea B. Laporte; Pres., Douglas Hamilton; C.O.O., Kathleen Basham; Chief Cur., Rob Mintz, Ph.D.; Cur. Renaissance & Baroque Art, Dr. Joaneath Spicer; Cur.-in-Charge Medieval Art & Manuscripts, Martina Bagnoli, Ph.D.; Dir. Conservation & Technical Res., Terry Drayman-Weisser; Deputy Dir. Audience Engagement, Jacqueline Tibbs Copeland; Deputy Dir. Devel., Joy

Heyrman; Dir. Mktg., Matt Fry; Deputy Dir. Curatorial Exhibition & Conservation, Dr. Nancy E. Zinn; Registrar, Joan-Elisabeth Reid; Museum Shop Mgr., Alice McAuliffe.
Personnel Profile: Full-Time Paid 148; Part-Time Paid 10; Part-Time Volunteers 242; Interns 6.
Operating Expenses: 13,621,543
Operating Income: 13,627,249
Governing Authority: municipal. Tax-exempt: 170(b)(1)(A).
Institution Type/Description: Art Museum.
Collections: arts from antiquity through 19th-century; decorative arts; paintings; sculpture; arms & armor; jewelry; manuscripts.
Major Exhibits: Designed for Flowers: Contemporary Japanese Ceramics, 2/23/14-5/11/14; The Janet & Walter Sondheim Artscape Prize 2014 Finalists, 6/20/14-8/17/14.
Research Fields: art history; archaeology; conservation & technical research of art objects.
Facilities: 80,000-vol. library of art history reference material available for inter-library loan & on premises by appointment; reading room; 400-seat auditorium; classrooms. Gift items for sale.
Activities: special exhibitions; guided tours; lectures; films; gallery talks; concerts; formally organized education programs; inter-museum loan, permanent, temporary & traveling exhibitions.
Publications: annual journal, The Journal of the Walters Art Museum; collection & exhibition catalogs; quarterly magazine for members; annual report.
Hours & Admission Prices: Wed. & Fri.-Sun. 10-5, Thurs. 10-9. Fees for some special exhibitions. Closed Independence Day; Thanksgiving; Christmas Eve & Day. &
Attendance: 170,133 (accurate)
Membership: Student $35; Educator and Senior 65 & over $45; Individual $50; Dual Educator $70; Dual Senior $70; Dual Family $75; Supporter $125; Sustainer $250-$499; Advocate $500-$999; Patron $1,000-$1,499; Curator's Circle $1,500-$2,499; Director's Circle $2,500-$4,999; Henry & William Walker's Circle $5,000-$9,999; Founder's Circle $10,000-$24,999; Benefactor $25,000-$49,999; President's Club $50,000-$200,000.

Bel Air

HAYS HOUSE MUSEUM, 324 Kenmore Ave., Bel Air, MD 21014. Tel.: 410-838-7691.
Governing Authority: Parent Institution: Historical Society of Harford County, Inc.
Institution Type/Description: Historic House Museum: built in 1788.
Collections: local history & culture; period furnishings; personal artifacts; photographs.
Activities: special events; rental facilities.
Publications: newsletter.
Hours & Admission Prices: March-Dec. Sun. 1-4; other times by appointment. Adults $3, students & seniors $2; children under 4 & members no charge. Closed holidays.

Beltsville

ABRAHAM HALL, 7612 Old Muirkirk Rd., Beltsville, MD 20705-1341. Tel.: 240-264-3415. TTY: 301-699-2544.
Institution Type/Description: Historic House Museum: built in 1889.
Collections: Black history & culture; period furnishings; personal artifacts; photographs.
Activities: educational programs.
Hours & Admission Prices: By appointment.

UNITED STATES NATIONAL AGRICULTURAL LIBRARY, 10301 Baltimore Ave., Beltsville, MD 20705-2326. Tel.: 301-504-5755.
Web Site: www.nal.usda.gov
Institution Type/Description: Library.
Collections: 2.3 million volumes; manuscripts; nursery & seed trade catalogs; photographs; posters from the 1500s to present.
Hours & Admission Prices: Mon.-Fri. 8:30-4:30. Special Collections: Mon.-Fri. 8:30-12 & 1-4 by appointment.

Berlin

CALVIN B. TAYLOR HOUSE MUSEUM, 208 N. Main St., Berlin, MD 21811. Mailing Address: Berlin Heritage Foundation, Inc., P.O Box 351, Berlin, MD 21811-0351. Tel.: 410-641-1019.
E-mail: taylorhousemuseum@verizon.net
Web Site: www.taylorhousemuseum.org
Founded: 1981.

Congressional District: 1
Key Personnel: Pres. (V), Jan Quick.
Personnel Profile: Part-Time Paid 1; Part-Time Volunteers 32.
Governing Authority: Parent Institution: Berlin Heritage Foundation, Inc. Tax-exempt.
Institution Type/Description: Historic House Museum.
Collections: period furnishings; personal artifacts; local memorabilia.
Facilities: 20-seat meeting room.
Activities: concerts on the lawn; fundraising dinners.
Publications: quarterly newsletter.
Hours & Admission Prices: Memorial Day to Oct. Mon., Wed. & Fri.-Sat. 1-4; other times by appointment. No Charge; donations accepted.
Attendance: 3,500 (estimated)
Membership: Individual $10; Family $15; Corporate $25.

Bethesda

DENNIS & PHILLIP RATNER MUSEUM, 10001 Old Georgetown Rd., Bethesda, MD 20814. Tel.: 301-897-1518.
Institution Type/Description: Art Museum.
Collections: paintings; sculpture; drawings; photographs.
Activities: traveling exhibitions; lectures.
Hours & Admission Prices: Sept.-July Sun. 10-4:30, Mon.-Thurs. 12-4; groups of 12 or more by appointment. No charge. Closed holidays.

DEWITT STETTEN, JR, MUSEUM OF MEDICAL RESEARCH, 1 Cloister Ct., Bldg. 60, Rm. 262, Office of Intramural Research, Bethesda, MD 20814. Tel.: 301-496-6610 & 8856. Fax: 301-402-1434.
E-mail: history@nih.gov
Web Site: www.pinterest.com/nihhistory; www.historyatnih.tumblr.com
Founded: 1986.
Congressional District: 8
Key Personnel: Acting Dir., Chris Wanjek, Ph.D.; Archivist, Barbara Harkins, M.L.S.; Cur., Michele Lyons, M.A.; Program Asst., Dee Andrews; Exhibits & Education, Henry Grasso, M.A.
Personnel Profile: Full-Time Paid 1; Part-Time Paid 3; Part-Time Volunteers 4.
Governing Authority: federal government; nonprofit organization. Foundation for Advanced Education in the Sciences, Inc. Parent Institution: National Institutes of Health. Tax-exempt.
Institution Type/Description: Medical Museum.
Collections: 20th-21st century biomedical research instruments & technologies; history of NIH; National Institutes of Health memorabilia; documents; photographs.
Research Fields: biomedical research.
Facilities: library of reference books, archives & museum collections available to public by appt.; educational facilities; 250-seat auditorium; 1,500-seat cafeteria. Must pass through NIH security gateway to enter campus.
Activities: guided tours; lectures.
Publications: brochures; books; articles.
Hours & Admission Prices: Clinical Center: daily 24 hours. Exhibits: daily 9-9. Campus access through security entrance. No charge. &
Attendance: 20,000 (estimated)

UNITED STATES NATIONAL LIBRARY OF MEDICINE, 8600 Rockville Pike, Bethesda, MD 20894. Tel.: 301-496-6308.
E-mail: custserv@nlm.nih.gov
Web Site: www.nlm.nih.gov
Institution Type/Description: Biomedical Library.
Collections: 12 million books, journals & manuscripts; audiovisuals.
Activities: research.
Hours & Admission Prices: Main Reading Room: Mon.-Fri. 8:30-5, Sat. 8:30-2. History of Medicine Reading Room: Mon.-Fri. 8:30-5. Tours: Mon.-Fri. 1:30.

Big Pool

FORT FREDERICK STATE PARK, 11100 Fort Frederick Rd., Big Pool, MD 21711-1313. Tel.: 301-842-2155. Fax: 301-842-0028.
E-mail: rstudy@dnr.state.md.us
Web Site: www.dnr.state.md.us
Founded: 1922.
Congressional District: 6
Key Personnel: Park Mgr., Angie Hummer; Asst. Park Mgr., Ben Sanderson; Administrative Specialist, Sherian Hose; Park Sec., Betsy Mellott; Ranger, Steve Robertson; Maintenance Chief, Kevin Zeigler; Ranger, Andy Simmons; Maintenance Tech, Dean Smook.
Personnel Profile: Full-Time Paid 8; Part-Time Paid 25; Part-Time Volunteers 150.

Governing Authority: state. A facility of the Maryland Dept. of Natural Resources, Tawes Office Bldg., Annapolis, MD 21401. Tax-exempt.
Institution Type/Description: Historic Building & Site.
Collections: Indian relics; artifacts; history of Fort Frederick; early firearms; Civil War rifle; Confederate bronze Napoleon cannon; household & farm equipment.
Research Fields: French & Indian War; Revolutionary War; Civilian Conservation Corps.
Facilities: visitor center; French & Indian war fort; Civilian Conservation Corps Museum.
Activities: guided tours; permanent exhibitions; living history programs; camping. Museum Sponsors: 18th Century Market Fair; French & Indian War Muster.
Publications: annual calendar of events.
Hours & Admission Prices: April-May & Sept.-Oct. Sat.-Sun. 10-5; Memorial Day-Labor Day daily 10-5. Adults $3 per car MD residents, $5 per car non-residents. &
Attendance: 197,000 (accurate)
Membership: Friends of Fort Frederick State Park, Inc. Annual: Student $5; Individual $10; Couple $15; Family $20; Contributing $100; Life: Regular $500; Sustaining $1,000; Endowing $2,000.

Boonsboro

BOONSBOROUGH MUSEUM OF HISTORY, (M), 113 N. Main St., Boonsboro, MD 21713-1007. Tel.: 301-432-6969. Facebook: Boonsborough Museum of History, Boonsboro, MD.
Founded: 1975.
Congressional District: 6
Key Personnel: Dir. & Owner, Douglas G. Bast
Governing Authority: private; nonprofit organization.
Institution Type/Description: History Museum.
Collections: local & state history; early lighting; Russian icons; ceramics & glassware; dinosaur bones; Egyptian animal mummies; Civil War artifacts. Historic Buildings: cabinetmaker's shop; 19th century general store.
Activities: guided tours; temporary exhibitions.
Hours & Admission Prices: May-Sept. Sun. 1-5; tours by appointment. Donations Requested: adults $4, tours & groups $3, children $1.50; discounts to AAM & ICOM members; local school groups no charge.
Attendance: 600 (estimated)

WASHINGTON COUNTY RURAL HERITAGE MUSEUM, 7313 Sharpsburg Pike (Rte. 65), Boonsboro, MD 21713-2431. Tel.: 240-420-1714.
E-mail: eoverdorff@washco-md.net
Web Site: www.ruralheritagemuseum.org
Founded: 2000.
Key Personnel: Pres., Phil Muritz.
Governing Authority: Parent Institution: Washington County Buildings, Grounds & Parks. Tax-exempt.
Institution Type/Description: History & Rural Heritage Museum and Village.
Collections: farm kitchen; parlor; country church; modes of travel including a Conestoga wagon, sleighs & sleds; farming equipment & artifacts; butchering; dairying; log cabins; homestead; church.
Facilities: Museum-related items for sale.
Activities: special events: Spud Fest in August; Track Show and Pull in May.
Hours & Admission Prices: Sat.-Sun. 1-4; other times by appointment. No charge; donations accepted. &
Attendance: 2,500 (estimated)
Membership: Single $15; Family $20; Sustain $30; Contributor $75; Benefactor $100; Corporate $300.

Bowie

BELAIR MANSION, 12207 Tulip Grove Dr., Bowie, MD 20715-2340. Tel.: 301-809-3089. Fax: 301-809-2308.
E-mail: museums@cityofbowie.org
Web Site: www.cityofbowie.org/museum
Founded: 1968.
Congressional District: 5
Key Personnel: Dir., Pamela Williams; Chm. (V), Jean Lancaster.
Personnel Profile: Full-Time Paid 5; Part-Time Paid 7; Part Time Volunteers 35; Interns 2.
Governing Authority: municipal; nonprofit. Parent Institution: City of Bowie Museums. Tax-exempt: 170(b)(1)(A).
Institution Type/Description: Historic House Museum: housed in c.1745 Georgian plantation of Governor Samuel Ogle, and later country estate of William Woodward.
Collections: furniture & decorative arts c.1730-1957.
Research Fields: decorative & social history of Maryland plantation houses, 1745-1955.

Facilities: 100-vol. library of local history & decorative arts. Postcards, books, notecards & museum-related items for sale.
Activities: docent program; guided tours; lectures; loan & temporary exhibitions. Annual Events: Bowie Heritage Day; Candlelight Tour.
Publications: newsletter, Musings.
Hours & Admission Prices: Tues.-Sun. 12-4. No charge; donations accepted. Closed major holidays. &
Attendance: 5,400 (accurate)
Membership: Individual $25; Family $35; Club & Group $50; Patron $125; Life $500.

BELAIR STABLE MUSEUM, 2835 Belair Dr., Bowie, MD 20715. Mailing Address: 12207 Tulip Grove Dr., Bowie, MD 20715-2340. Tel.: 301-809-3089. Fax: 301-809-2308.
E-mail: museums@cityofbowie.org
Web Site: www.cityofbowie.org/museum
Founded: 1968.
Congressional District: 5
Key Personnel: Dir., Pamela Williams; Museum Facility Mgr., Russell Davies; Chm. (V), Kelly L. Pierce.
Personnel Profile: Full-Time Paid 4; Part-Time Paid 3; Part-Time Volunteers 8.
Governing Authority: municipal; nonprofit. Parent Institution: City of Bowie Museums. Tax-exempt: 170(b)(1)(A).
Institution Type/Description: History Museum.
Collections: Thoroughbred racing artifacts; farming implements; furnishings & carriage collection.
Research Fields: history of Thoroughbred racing in America.
Facilities: 500-vol. library of local history & decorative arts. Postcards, books, notecards & museum-related items for sale.
Activities: docent program; guided tours; lectures; loan & temporary exhibitions. Annual Event: Bowie Heritage Day.
Publications: newsletter, Musings.
Hours & Admission Prices: Tues.-Sun. 12-4. No charge; donations accepted. Closed major holidays. &
Attendance: 3,700 (accurate)
Membership: Student $15; Individual $25; Family $35; Club & Group $50; Patron $125; Life $500.

BOWIE HERITAGE WELCOME CENTER, 8606 Chestnut Ave., Bowie, MD 20715. Mailing Address: 12207 Tulip Grove Dr., Bowie, MD 20715-2340. Tel.: 301-575-2488.
Web Site: www.cityofbowie.org/museum
Founded: 2006.
Congressional District: 5
Key Personnel: Dir., Pamela Williams; Facility Mgr., Ruth A. Murphy.
Personnel Profile: Full-Time Paid 1; Part-Time Paid 1.
Governing Authority: municipal; nonprofit. Parent Institution: City of Bowie Museums, Bowie, MD. Tax-exempt: 170(b)(1)(A).
Institution Type/Description: Children's Museum.
Collections: Bowie social & business history.
Research Fields: African-American.
Facilities: Museum-related items for sale.
Activities: participatory exhibits. Annual Events: Spring Fling; Fall Fest; Kids' Kaboose.
Hours & Admission Prices: Tues.-Sun. 10-4. No charge; donations accepted. &
Attendance: 5,000 (estimated)

BOWIE RAILROAD STATION MUSEUM, 8614 Chestnut Ave., Bowie, MD 20715-3732. Mailing Address: 12207 Tulip Grove Dr., Bowie, MD 20715-2340. Tel.: 301-809-3089. Fax: 301-809-2308.
E-mail: museums@cityofbowie.org
Web Site: www.cityofbowie.org/museum
Founded: 1994.
Congressional District: 5
Key Personnel: Dir., Pamela Williams; Chm. (V), Edward Maenner.
Personnel Profile: Full-Time Paid 1; Part-Time Paid 2; Part-Time Volunteers 20; Interns 1.
Governing Authority: municipal; nonprofit. Parent Institution: City of Bowie. Tax-exempt: 170(b)(1)(A).
Institution Type/Description: History Museum.
Collections: railroad artifacts; history of Huntington section of Old Bowie. Historic Structures: c.1910-1933 Pennsylvania Railroad Station; tower; passenger shed; 1922 N&W R.R. caboose.
Research Fields: rail history & development of Huntington section of Old Bowie.
Facilities: Postcards, books, notecards & museum-related items for sale.

Activities: docent program; guided tours; lectures; loan & temporary exhibitions. Annual Events: Bowie Heritage Day; Spring Fling; Fall Antiques Street Fest; Kids Kaboose.
Publications: newsletter, Musings.
Hours & Admission Prices: Tues.-Sun. 10-4. No charge; donations accepted. Closed major holidays. ♿
Attendance: 3,000 (accurate)
Membership: Student $2; Individual $5; Family, Club & Group $10; Contributor $20; Supporting $50; Patron $100; Life $500.

CITY OF BOWIE MUSEUMS, 12207 Tulip Grove Dr., Bowie, MD 20715-2340. Tel.: 301-809-3089. Fax: 301-809-2308.
E-mail: museums@cityofbowie.org
Web Site: www.cityofbowie.org/museum
Founded: 1968.
Congressional District: 5
Key Personnel: Dir., Pamela L. Williams; Cur., Samantha Dorsey.
Personnel Profile: Full-Time Paid 5; Part-Time Paid 7; Part-Time Volunteers 35.
Governing Authority: municipal; nonprofit. Parent Institution: City of Bowie. Subsidiary Institutions: Belair Mansion; Belair Stable Museum; Bowie Railroad Station; Bowie Heritage Children's Museum and Welcome Center; Prince George's Co. Genealogical Library; Radio-Television Museum. Tax-exempt: 170(b)(1)(A).
Institution Type/Description: History Museums.
Collections: focus on Belair Estate, railroad & development of the City of Bowie.
Research Fields: decorative & social history of Maryland plantation houses, 1745-1955; development of Maryland railroad towns; history of Thoroughbred racing.
Facilities: 1,000-vol. library of local history, decorative arts, & equine history.
Activities: Bowie Heritage Day.
Publications: quarterly newsletter, Musings; calendar of special events.
Hours & Admission Prices: Mansion & Stable: Tues.-Sun. 12-4. Railroad & Welcome Center: Tues.-Sun. 10-4. Office: Mon.-Fri. 9-5. No charge; donations accepted. ♿
Attendance: 11,000 (accurate)
Membership: Individual $20; Family, Club & Group $35; Corporate $50; Patron $125; Lifetime $500.

NATIONAL CAPITAL RADIO & TELEVISION MUSEUM, (M), 2608 Mitchellville Rd., Bowie, MD 20716-1392. Mailing Address: P.O. Box 1809, Bowie, MD 20717. Tel.: 301-390-1020. Fax: 301-809-2308.
E-mail: radiobelanger@comcast.net
Web Site: www.ncrtv.org
Founded: 1999.
Congressional District: 5
Key Personnel: Exec. Dir. & Cur., Brian Belanger; Pres., Christopher Sterling; Deputy Dir., Laurie A. Baty.
Personnel Profile: Full-Time Paid 1; Part-Time Volunteers 21.
Governing Authority: municipal; nonprofit organization. Tax-exempt.
Institution Type/Description: History Museum.
Collections: radio & television equipment; broadcast history 1900-present.
Research Fields: history of radio & television development.
Activities: lectures; screenings.
Publications: newsletter, Dials and Channels.
Hours & Admission Prices: Fri. 10-5, Sat.-Sun. 1-5. No charge; donations accepted. Closed major holidays.
Attendance: 1,800 (estimated)
Membership: Basic $25; Corporate $1,000.

PRINCE GEORGE'S COUNTY GENEALOGICAL SOCIETY LIBRARY, 12219 Tulip Grove Dr., Bowie, MD 20715. Mailing Address: P.O. Box 819, Bowie, MD 20718-0819. Tel.: 301-262-2063.
E-mail: pgcgs@juno.com
Web Site: www.pgcgs.org
Founded: 1969.
Key Personnel: Pres. (V), David Frederick, Ph.D.
Personnel Profile: Part-Time Volunteers 6.
Volunteer Hours: 432
Operating Expenses: 1,000
Operating Income: 1,000
Governing Authority: Parent Institution: Prince George's County Genealogical Society. Tax-exempt.
Institution Type/Description: History Museum.
Collections: periodicals; surname files; family records; Bible records.
Research Fields: family; Maryland history.

Facilities: 5,800-vol. library.
Activities: meetings featuring speakers; one day seminars.
Publications: quarterly bulletin, Prince George's County Genealogical Society; records, of land, tax, probate, cemetery, church & funeral home records.
Hours & Admission Prices: Wed. call for hours; other times by appointment. No charge. ♿
Attendance: 500 (estimated)
Membership: Individual $15; Family $22.

Boyds

BOYDS NEGRO SCHOOL MUSEUM, 19510 White Ground Rd., Boyds, MD 20841. Mailing Address: P.O. Box 161, Boyds, MD 20841.
E-mail: info@boydshistory.org
Web Site: www.boydshistory.org
Founded: 1979.
Governing Authority: Parent Institution: Boyds Historical Society.
Institution Type/Description: Historic Building Museum: housed in a former public school for African American students from 1895-1936.
Collections: local history & culture; period furnishings.
Hours & Admission Prices: By appointment. No charge.

KING BARN DAIRY MOOSEUM, S. Germantown Recreational Park, 18028 Central Park, Boyds, MD 20841. Mailing Address: P.O. Box 76, Boyds, MD 20841-0076. Tel.: 301-528-6530.
E-mail: info@mooseum.com
Web Site: www.mooseum.com
Institution Type/Description: Dairy Heritage Museum.
Collections: dairy heritage & history; hands-on exhibitions; local organizations & businesses.
Activities: educational programs; special events; groups tours.
Hours & Admission Prices: May-Oct. Sat. 10-4, 4th Sun. each month 1-4.

Brentwood

PRINCE GEORGE'S AFRICAN AMERICAN MUSEUM & CULTURAL CENTER (PGAAMCC), Gallery 110, 3901 Rhode Island Ave., Brentwood, MD 20722. Mailing Address: 4519 Rhode Island Ave., North Brentwood, MD 20722-1225. Tel.: 301-809-0440. Fax: 301-209-0594. Facebook: Prince Georges African American Museum.
E-mail: info@pgaamcc.org
Web Site: www.pgaamcc.org
Key Personnel: Dir., Dr. Jacqueline F. Brown; Dir. Media & Public Programs, Tracey Tolbert Jones
Institution Type/Description: African American History Museum.
Collections: local African American history & culture; period furnishings; personal artifacts; photographs.
Activities: educational programs; public programs.
Hours & Admission Prices: Tues.-Sat. 10-5; other times by appointment. Museum: no charge; donations accepted. Public Programs: $10 per person.
Attendance: 4,000 (estimated)
Membership: Students & Seniors $20; Individual $35; Family $50; Corporate $250.

Brunswick

BRUNSWICK RAILROAD MUSEUM AND C&O CANAL VISITORS CENTER, 40 W. Potomac St., Brunswick, MD 21716-1111. Tel.: 301-834-7100.
E-mail: contact@brrm.net
Web Site: brrm.net
Founded: 1974.
Congressional District: 6
Key Personnel: Cur., Rebecca O'Leary.
Personnel Profile: Part-Time Paid 1; Part-Time Volunteers 40.
Governing Authority: private; nonprofit organization. Parent Institution: Brunswick-Potomac Foundation Inc. Tax-exempt: 501(c)(3).
Institution Type/Description: Transportation & Social History Museum.
Collections: Ca 1880-1920 Baltimore & Ohio Railroad communications technology; small equipment; tools; vintage photographs; costume; ephemera; social history of town life in room dioramas; costume; baseball; medical & organizational exhibitions; African American railroad workers & wives of railroaders; interactive HO scale model of Baltimore & Ohio metropolitan branch line between DC & Brunswick; interactive telegraphy; Chesapeake & Ohio canal history, 1828-1924.
Research Fields: Baltimore & Ohio freight classification, roundhouse operations, communications & equipment; railroad social history & town life c.1880-1920.

Facilities: meeting rooms; children's activity room. Museum-related items for sale.
Activities: docent-led tours; canal hikes; lectures on railroad history; birthday parties. Special Events: Victorian Tea and Valentine workshop in February; Bell & History Days in April; Railroad Days in October; A Night at the Brunswick Railroad Museum Legends and Storytelling Tour in November; Victorian Christmas weekend following Thanksgiving.
Publications: newsletter, Rail Letter
Hours & Admission Prices: Fri. 10-2, Sat. 10-4, Sun. 1-4. Adults $7, senior citizens $5.40, children $4; discounts to groups, MD Passport Program, National Railway Historical Society & AAA members; children 3 & under, members and county docents no charge. Closed major holidays. &
Attendance: 8,000 (estimated)
Membership: Senior $20; Individual $25; Senior Partner $30; Partner $40; Family $60; Benefactor $75; Patron $125; Lifetime $1,500.

Burretsville

GATHLAND STATE PARK, 1 Mile West off Rte. 17, Burretsville, MD 20866. Mailing Address: c/o South Mountain Recreation Area, 21843 National Pike, Boonsboro, MD 21713-9535. Tel.: 301-791-4767. Fax: 301-791-0962. TDD: 301-974-3683.
Web Site: www.dnr.state.md.us/publiclands/western/gathland.html
Founded: 1958.
Key Personnel: Greenbrier Park Mgr., Dan Spedden; Shelter Mgr., Marge Magruder.
Personnel Profile: Full-Time Paid 2; Part-Time Paid 6; Part-Time Volunteers 270.
Governing Authority: state; nonprofit. Parent Institution: State Forest & Park Service. Subsidiary Institution: Friends of Gathland. Tax-exempt.
Institution Type/Description: Park Museum: located on the estate of George Alfred Townsend.
Collections: photographs; architectural pieces; Civil War guns; Civil War Battle of South Mountain.
Activities: Civil War living history weekends.
Publications: The Arch: News & Happenings of the Friends of Gathland State Park.
Hours & Admission Prices: Park: daily 8-sunset. Museums: April & Oct. Sat.-Sun.; May-Sept. daily. No charge. Closed Christmas. &
Attendance: 78,000 (estimated)
Membership: Individual $10; Family $25.

Cambridge

BRANNOCK EDUCATION & RESEARCH CENTER, 100 Maryland Ave., Cambridge, MD 21613. Tel.: 410-228-6938.
Institution Type/Description: Science Museum.
Collections: local history; early navigational instruments; ship models; Chesapeake Bay & local maritime memorabilia; maritime history.
Facilities: library.
Hours & Admission Prices: Call for hours.

HARRIET TUBMAN MUSEUM, 424 Race St., Cambridge, MD 21613. Tel.: 410-228-0401. Fax: 410-228-5641.
Institution Type/Description: History Museum.
Collections: Harriet Tubman's life & accomplishments; Underground Railroad history; photographs; period furnishings.
Hours & Admission Prices: Tues.-Fri. 10-3, Sat. 12-4.

HERITAGE MUSEUMS & GARDENS OF DORCHESTER, 1003 Greenway Dr., Cambridge, MD 21613-2009. Tel.: 410-228-7953. Facebook: Dorchester County Historical Society.
E-mail: dchs@verizon.net
Web Site: dorchesterhistory.org
Formerly: Dorchester County Historical Society
Founded: 1953.
Congressional District: 1
Key Personnel: Dir., Ann W. Phillips; Pres., Jeanne Bernard; Treas., Judi Leaming.
Personnel Profile: Full-Time Paid 1; Part-Time Volunteers 30.
Governing Authority: private; nonprofit organization. Tax-exempt: 501(c)(3).
Institution Type/Description: Historical Society Museum.
Collections: Dorchester County history from prehistoric to modern times with emphasis on the agricultural & development of the county; historic house includes period furnishings, dolls, portraits & personal artifacts of notable persons from the county's past; local farming & Indian artifacts. Historic Buildings: c.1760 two and a half story brick Georgian building with Greek Revival alterations; c.1790 brick federal Goldsborough Stable; c.1750 stronghouse, log outbuilding.

Facilities: 500-vol. library on history & genealogy available to the public; 1,000 document files including wills & deeds; more than 90 rolls of microfilm; 5,280 sq. ft. exhibit space.
Activities: guided tours; lectures; temporary exhibitions; bus tours; social events; afternoon teas; special area tours; civil war; heritage tour; quest for freedom.
Publications: newsletter published four times annually; tombstone records of Dorchester County; Bible records of Dorchester County.
Hours & Admission Prices: Tues.-Fri. 10-4, Sat. 10-2; other times by appointment. Admission $5; discounts to bus tours; members no charge. Closed major holidays. &
Attendance: 3,000 (estimated)
Membership: Individual $25; Family $45; Benefactor & Corporation $100.

RICHARDSON MARITIME MUSEUM, 401 High St., Cambridge, MD 21613-1804. Mailing Address: P.O. Box 1198, Cambridge, MD 21613-5198. Tel.: 410-221-1871. Fax: 410-228-5471.
E-mail: info@richardsonmuseum.org
Web Site: www.richardsonmuseum.org/
Key Personnel: Exec. Dir., Jane Devlin; Operating Mgr. Ruark Boatworks, Dan Cada; Cur., Melvin Hickman
Institution Type/Description: Maritime Museum.
Collections: life & work of James B. (Mr. Jim) Richardson; 50 Chesapeake Bay wooden boat models; watermen's equipment & gear; boatbuilding tools; photographs & artifacts from the George T. Johnson & Sons Shipyard.
Activities: special events.
Publications: newsletter.
Hours & Admission Prices: Wed. & Sun. 1-4, Sat. 10-4 or by appointment. Suggested Donation: $3. Closed New Year's Day; Easter; Independence Day; Thanksgiving; Christmas. &

Catonsville

CATONSVILLE HISTORICAL SOCIERTY, 1824 Frederick Rd., Catonsville, MD 21228. Mailing Address: P.O. Box 9311, Catonsville, MD 21228-0311. Tel.: 410-744-3034.
E-mail: info@catonsvillehistory.org
Web Site: www.catonsvillehistory.org
Institution Type/Description: History Museum.
Collections: local history & culture; period furnishings; photographs; personal artifacts; Native American artifacts; miniatures; early prints & paintings; model trains & layout.
Hours & Admission Prices: Call for hours.

SPRING GROVE HOSPITAL CENTER, ALUMNI MUSEUM, 55 Wade Ave., Garrett Bldg., Catonsville, MD 21228-4663. Tel.: 410-402-7786 & 6000. Fax: 410-402-7050.
Web Site: www.springgrove.com/history.html
Founded: 1995.
Congressional District: 7
Key Personnel: Chm. (V), Joseph Sanphilipo; Treas., Diane Johns; Sec., Ella Nora Hoerl.
Personnel Profile: Part-Time Volunteers 5.
Governing Authority: state; nonprofit. Parent Institution: Spring Grove Hospital Center. Tax-exempt: 501(c)(3).
Institution Type/Description: History Museum.
Collections: hand-made tools; furniture; artifacts; photographs dating back to 1897.
Facilities: library.
Activities: Annual Events: May Pole; Flea Market; Christmas party for patients.
Hours & Admission Prices: Thurs. 9am to noon; other times by appointment. No charge; donations accepted. Closed holidays.
Attendance: 132 (estimated)

Centerville

HISTORIC SITES CONSORTIUM OF QUEEN ANNE'S COUNTY, (M), 124 S. Commerce St., Centerville, MD 21617. Mailing Address: P.O. Box 62, Centerville, MD 21617-0062. Tel.: 410-758-3010.
E-mail: info@qachistory.org
Web Site: www.qachistory.org
Founded: 1995.
Key Personnel: Dir., Rebecca Marquardt.
Personnel Profile: Full-Time Paid 1; Part-Time Volunteers 55.
Governing Authority: Parent Institution: Queen Anne's County, MD. Tax-exempt.

Institution Type/Description: Historic Sites Preservation Consortium.
Collections: regional cultural history & heritage; historic structures.
Hours & Admission Prices: Office: Wed. 10-1. Historic Sites: call for hours. No charge; donations accepted.
Attendance: 3,239 (accurate)

Centreville

QUEEN ANNE'S MUSEUM OF EASTERN SHORE LIFE, 126 Dulin Clark Rd., Centreville, MD 21617. Mailing Address: P.O. Box 525, Centreville, MD 21617. Tel.: 410-758-8641. Facebook: Queen Anne's Museum of Eastern Shore Life.
E-mail: mesl@myshorelink.com
Web Site: www.historicqac.org/sites/mesl.htm
Institution Type/Description: History Museum.
Collections: local history & culture; period farm tools & equipment; Native American artifacts; household items; blacksmith shop; print shop.
Activities: special events.
Hours & Admission Prices: April-Oct. Sat.-Sun. 1-4; other times by appointment.

TUCKER HOUSE - QUEEN ANNE'S COUNTY HISTORICAL SOCIETY, 124 S. Commerce St., Centreville, MD 21617. Mailing Address: P.O. Box 62, Centreville, MD 21617. Tel.: 410-758-3010.
E-mail: info@qachistory.org
Web Site: qachistory.org
Founded: 1960.
Congressional District: 4
Key Personnel: Pres. (V), Rebecca Marquardt.
Personnel Profile: Part-Time Volunteers 5.
Governing Authority: Parent Institution: Queen Anne's County Historical Society. Tax-exempt.
Institution Type/Description: Historical Society Museum.
Collections: local history & culture; period furnishings; personal artifacts; genealogical records; photographs; paintings. Historic Buildings: c.1794 house; smoke house.
Facilities: research library.
Activities: special events.
Publications: triannual newsletter.
Hours & Admission Prices: May-Oct. 1st Sat. each month 10-2 by appointment. No charge; donations accepted.
Attendance: 200 (estimated)
Membership: Single $35; Family $45; Corporate $100.

WRIGHT'S CHANCE HOUSE MUSEUM - QUEEN ANNE'S COUNTY HISTORICAL SOCIETY & TUCKER HOUSE, 124 S. Commerce St., Centreville, MD 21617. Mailing Address: P.O. Box 62, Centreville, MD 21617. Tel.: 410-758-3010.
E-mail: info@qachistory.org
Web Site: qachistory.org
Formerly: Wrights Chance House Museum Queen Annex County Historical Society
Founded: 1960.
Congressional District: 4
Key Personnel: Pres. (V), Rebecca Marquardt.
Personnel Profile: Part-Time Volunteers 5.
Governing Authority: Parent Institution: Queen Anne's County Historical Society. Tax-exempt.
Institution Type/Description: Historic House Museum: built c.1744.
Collections: local history & culture; period furnishings; personal artifacts; photographs.
Activities: special events.
Hours & Admission Prices: May-Oct. 1st Sat. each month 10-2 by appointment. No charge; donations accepted.
Membership: Single $35; Family $45; Corporate $100.

Charlestown

TORY HOUSE/107 HOUSE, 343 Market St., Charlestown, MD 21914. Mailing Address: P.O. Box 52, Charlestown, MD 21914. Tel.: 410-287-8262.
Founded: 1974.
Congressional District: 5
Key Personnel: Pres. (V), Rebecca C. Phillips.
Personnel Profile: Part-Time Volunteers 7.
Governing Authority: Tax-exempt.
Institution Type/Description: Historic House Museum: built c.1810.
Collections: local history & culture; period furnishings; household artifacts; 19th century wooden water tower.

Hours & Admission Prices: May-Sept. 3rd Sun. each month 2-4; other times by appointment. No charge; donations accepted.
Attendance: 100 (estimated)

Chesapeake Beach

THE CHESAPEAKE BEACH RAILWAY MUSEUM, (M), 4155 Mears Ave., Chesapeake Beach, MD 20732. Mailing Address: P.O. Box 1227, Chesapeake Beach, MD 20732-1227. Tel.: 410-257-3892.
E-mail: cbrailway@co.cal.md.us
Web Site: www.cbrm.org
Founded: 1979.
Congressional District: 27
Key Personnel: Chief Cur., Harriet M. Stout; Administrative Asst., Correine E. Moore.
Personnel Profile: Full-Time Paid 2; Part-Time Paid 6; Part-Time Volunteers 35.
Governing Authority: society; county government; historical society. Calvert County Government; Calvert County Historical Society; Friends of the Chesapeake Beach Railway Museum. Tax-exempt: 501(c)(3).
Institution Type/Description: History Museum: housed in 1898-1899 Chesapeake Beach Railway Station.
Collections: archives; photographs; artifacts relating to the early railroad era of 1900-1935; artifacts relating to the resort, 1900-1971; local history; oral history collection.
Research Fields: history of the Chesapeake Beach Railway; local history; history of early Chesapeake resort town.
Facilities: 300-vol. library; archives; 1,000 sq. ft. exhibit space; study & research area; 20-seat AV area. Museum-related items & books for sale.
Activities: guided tours; lectures; films; concerts; organized education programs for children, adults & undergraduate or graduate college students; training programs for professional museum workers; temporary exhibitions. Museum Sponsors: Family Fun Day in Fall & Spring; Bay Breeze Summer Series June to September; Children's Summer Program Series June to August; Bayside Chats.
Publications: biannual, The Chesapeake Beach Railway Museum Newsletter, The Chesapeake Dispatcher.
Hours & Admission Prices: mid-March to March 31 & Nov. Sat.-Sun. 1-4; April-May & Sept.-Oct. daily 1-4; June-Aug. Mon.-Fri. 1-4, Sat.-Sun. 11-5; other times by appointment; call for groups & tours. No charge; donations accepted. Closed New Year's Day; Christmas. &
Attendance: 12,000 (accurate)
Membership: Individual $10; Family $15; Business $50; Life $100.

Chester

KIRWAN STORE AND HOUSE - KENT ISLAND HERITAGE SOCIETY, Rte. 552 - 641 Dominion Rd., Chester, MD 21619. Mailing Address: P.O. Box 321, Stevensville, MD 21666. Tel.: 410-758-2502.
Congressional District: 36
Personnel Profile: Part-Time Volunteers 4.
Governing Authority: Parent Institution: Kent Island Heritage Society.
Institution Type/Description: Historic House & Country Store: built in 1879.
Collections: local history & culture; period furnishings; photographs; country store contents from 1879-1920.
Activities: special events.
Hours & Admission Prices: May-Oct. 1st Sat. each month 12-4; other times by appointment. No charge; donations accepted.

Chestertown

CHESAPEAKE FARMS, 7319 Remington Dr., Chestertown, MD 21620. Tel.: 410-778-8400. Fax: 410-778-8405.
Institution Type/Description: Agricultural Research & Demonstration Area.
Collections: wildlife & their habitats; agricultural practices; wildlife management techniques.
Facilities: 3,000 acre farm.
Activities: demonstrations; wildlife observation.
Hours & Admission Prices: Call for hours.

CLIFFS SCHOOLHOUSE, Rte. 289, Quaker Neck Rd., Chestertown, MD 21620. Mailing Address: c/o Port of Chester Questers, 104 Birch Run Rd., Chestertown, MD 21620. Tel.: 410-778-9173 & 3098.
E-mail: cbcordes@verizon.net
Key Personnel: Pres. (V), Carol Cordes
Institution Type/Description: Historic Building: housed in a former one-room schoolhouse; built in 1878.

Collections: local history; period furnishings.
Hours & Admission Prices: May-Oct. 3rd Sat. each month 1-4. No charge.

HISTORICAL SOCIETY OF KENT COUNTY, INC., (M), 101 Church Alley, Chestertown, MD 21620-1505. Mailing Address: P.O. Box 665, Chestertown, MD 21620-0665. Tel.: 410-778-3499.
E-mail: director@kentcountyhistory.org
Web Site: www.kentcountyhistory.org
Founded: 1936.
Congressional District: 1
Key Personnel: Pres., Robert L. Bryan, Jr.; Museum Shop Mgr., Karen L. Emerson.
Personnel Profile: Part-Time Paid 3; Part-Time Volunteers 50; Interns 1.
Governing Authority: nonprofit organization. Tax-exempt: 501(c)(3).
Institution Type/Description: History Museum.
Collections: Indian artifacts; furniture; paintings; maps; special exhibits. Historic Structure: c.1780 house.
Research Fields: history & limited genealogical assistance.
Facilities: library.
Activities: permanent & temporary exhibitions; special tours; lectures; social functions. Museum Sponsors: First Friday Happy Hours 4-6pm; Country Driving Tour in June; Walking Tour of 18th, 19th, & 20th century homes in October.
Publications: Historic Houses of Kent County (1998, 2008); The Rolling Year on Maryland's Upper Eastern Shore (1985) Trumpington.
Hours & Admission Prices: May-Oct. Tues.-Fri. 10-4, Sat. 1-4; Nov.-April Tues.-Fri. 1-4; group tours by appointment. Walking Tour: no charge. Historic Group Tours: Historic District $15; House $4.
Attendance: 4,000 (estimated)
Membership: Individual $40; Family $75; Century $100; Lifetime $2,500.

KOHL GALLERY - WASHINGTON COLLEGE, Gibson Center for the Arts, 300 Washington Ave., Chestertown, MD 21620. Tel.: 410-778-2800; 800-422-1782.
Key Personnel: Cur., Donald McColl
Institution Type/Description: Art Gallery.
Collections: paintings; sculpture.
Hours & Admission Prices: Tues. 2-8, Wed.-Fri. 2-5, Sat. 11-4; groups of 10 or more by appointment.

Chevy Chase

AUDUBON NATURALIST SOCIETY, 8940 Jones Mill Rd., Chevy Chase, MD 20815-4799. Tel.: 301-652-9188. Fax: 301-951-7179.
E-mail: contact@audubonnaturalist.org
Web Site: www.audubonnaturalist.org
Founded: 1897.
Congressional District: 8
Key Personnel: C.E.O. & Exec. Dir., Neal Fitzpatrick; Pres., Anne Cottingham; Vice Pres., Kathy Rushing; Sec., Lois Schiffer; Financial Dir., Fred Bailey.
Personnel Profile: Full-Time Paid 18; Part-Time Paid 14; Part-Time Volunteers 400; Interns 9.
Governing Authority: society. Tax-exempt: 501(c)(3).
Institution Type/Description: Nature Center & Conservation Area: headquarters housed in c.1927 Georgian Revival brick and stone house, located on 40-acre Woodend Sanctuary.
Collections: marked tree specimens; Wilbur Fisk Banks Memorial collection of birds.
Research Fields: natural history; bird populations.
Facilities: 4,000-vol. non-circulating library containing natural history books & conservation materials; nature center; nature trails; educational facilities. Books, guides, bird feeders, optical equipment & related items for sale.
Activities: guided tours; lectures; organized education programs; self-guiding nature trails; teacher training programs in environmental education.
Publications: Audubon Naturalist News.
Hours & Admission Prices: Offices: Mon.-Fri. 9-5. Grounds: daily dawn-dusk. No charge; donations accepted. Closed federal holidays. &
Membership: Student $10; Individual & Senior Family $30; Family $40; Contributing $50; Sustaining $100; Supporting $150.

Clinton

POPLAR HILL HISTORIC HOUSE MUSEUM, 7606 Woodyard Rd., Clinton, MD 20735. Tel.: 301-856-0358.
E-mail: info@poplarhillonhlk.com
Web Site: www.poplarhillonhlk.com
Institution Type/Description: Historic House Museum.

Collections: local history & culture; period furnishings; photographs; personal artifacts.
Hours & Admission Prices: Thurs.-Fri. 10-4, Sun. 12-4; other times by appointment.

SURRATT HOUSE MUSEUM, (M), 9118 Brandywine Rd., Clinton, MD 20735-2501. Tel.: 301-868-1121. Fax: 301-868-8177. TDD: 301-699-2544.
E-mail: laurie.verge@pgparks.com
Web Site: www.surratt.org
Founded: 1976.
Congressional District: 4
Key Personnel: C.E.O., Dir. & Staff Historian, Laurie Verge; Pres. (V), Louise Oertly; Museum Shop Mgr., Joan Chaconas.
Personnel Profile: Full-Time Paid 2; Part-Time Paid 5; Part-Time Volunteers 40.
Governing Authority: bi-county agency chartered under the State of Maryland. Maryland-National Capital Park & Planning Commission, Dept. of Parks & Recreation for Prince George's County, Natural & Historical Resources Div. Tax-exempt: 501(c)(3).
Institution Type/Description: Historic House.
Collections: 1800-1865 furnishing & decorative arts; archival & photographic material relating to Lincoln assassination, Civil War era & mid-19th century life.
Major Exhibits: Between the Lines: Southern Maryland in the Civil War, 1/13-12/14.
Research Fields: Lincoln assassination.
Facilities: 2,000-vol. library of books on the Civil War & Lincoln assassination available for use by the public in-house, National archives files available on microfilm & extensive vertical file materials. Gift items & books for sale.
Activities: guided tours; lectures; docent program; temporary exhibitions; curriculum based school programs. Museum Sponsors: John Wilkes Booth Escape Route Tour in April & September.
Publications: monthly newsletter, Surratt Courier; occasional research materials.
Hours & Admission Prices: Wed.-Fri. 11-3, Sat.-Sun. 12-4. Adults $3, senior citizens $2, students 5-18 $1; discounts to groups, Civil War Trust, National Historical Trust, military, AAA & AAM members; children under 5 & members no charge. Closed Easter; Independence Day; Thanksgiving; Christmas. &
Attendance: 10,842 (accurate)
Membership: Individual $10 (each additional family member $5); Life $125.

Cockeysville

HISTORICAL SOCIETY OF BALTIMORE COUNTY, 9811 Van Buren Lane, Cockeysville, MD 21030-5022. Tel.: 410-666-1878.
E-mail: info@hsobc.org
Web Site: www.hsobc.org
Founded: 1959.
Congressional District: 2
Key Personnel: Bd. Pres. (V), Glenn Johnston; Admin., Adam Youssi; Mgr. Collections, Melissa Heaver.
Personnel Profile: Full-Time Paid 1; Full-Time Volunteers 2; Part-Time Paid 1; Part-Time Volunteers 30; Interns 3.
Governing Authority: society. Tax-exempt: 501 (c)(3).
Institution Type/Description: Historical Society Museum.
Collections: artifacts from early country homes & farms; manuscripts; agriculture; history; military; music; textiles; transportation; maps; photographs.
Research Fields: genealogy; mills; homes; cemetery transcriptions; manufacturing; land records; maps; 350 years of history.
Facilities: library of history books; archives.
Activities: guided tours; lectures; slides; permanent & temporary exhibitions; education trunks for schools.
Publications: quarterly pamphlet, History Trails; quarterly, newsletter.
Hours & Admission Prices: 2nd Wed. of month 6:30pm-8:30pm, Fri. 12-4, Sat. 10-2. Adults $5; members no charge. &
Attendance: 979 (accurate)
Membership: Senior $35; Senior Couple $45; Sustaining & Family $55; Contributor & Family Sustaining $65; Family Contributor $75; Benefactor $250; Corporate Bronze $500; Corporate Silver $1,000; Corporate Gold $1,500.

OREGON RIDGE NATURE CENTER AND PARK, 13555 Beaver Dam Rd., Cockeysville, MD 21030. Tel.: 410-887-1815. TDD: 410-887-5319.
E-mail: info@oregonridgenaturecenter.org
Web Site: www.oregonridge.org

Institution Type/Description: Nature Center & Park.
Collections: wildlife & their habitats; wildflowers; birds; wildlife management pond; archaeology; marble quarries; iron ore pits.
Facilities: 1,100 acre park; nature trails.
Activities: hiking trails.
Hours & Admission Prices: Tues.-Sun. 9-5.

College Park

THE ART GALLERY AT THE UNIVERSITY OF MARYLAND, COLLEGE PARK, (M), 1202 Art-Sociology Building, University of Maryland, College Park, MD 20742. Tel.: 301-405-2763. Fax: 301-314-7774.
E-mail: theartgallery@umd.edu
Web Site: www.artgallery.umd.edu
Founded: 1966.
Congressional District: 5
Key Personnel: Dir., John Shipman.
Personnel Profile: Full-Time Paid 2; Part-Time Paid 10; Interns 2.
Governing Authority: Regents. Parent Institution: University of Maryland, College Park. Tax-exempt: 501(c)(3).
Institution Type/Description: Art Gallery.
Collections: 20th-century American paintings & prints; African art; contemporary prints; Japanese prints; photography.
Research Fields: art history; art criticism; contemporary art.
Facilities: 40,000-vol. library; reference room.
Activities: lectures; gallery talks; inter-museum loan; temporary & traveling exhibitions.
Publications: Women Artists in Washington Collection, 1979; 350 Years of Art and Architecture in Maryland, 1984; Dreams, Lies, and Exaggerations: Photomontage in America, 1991; Sources: Multicultural Influences on Contemporary African American Sculptors, 1993; The Helen D. Ling Collection of Chinese Ceramics, 1995; Terra Firma, 1997; Willem de Looper: A Retrospective Exhibition 1966-1996, 1997; Russian Constructivist Roots: Present Concerns, 1997; Narratives of African American Art and Identity: The David C. Driskell Collection, 1998; Reframing Andy Warhol: Constructing American Myths, Heroes, and Cultural Icons, 1998; Close Enough: Photography by David Seymour (Chim), 1999; Handle with Care, Loose Threads in Fiber, 2000; Possible Futures: Science Fiction Art From The Frank Collection: Re-Reading Science Fiction Art, 2000; Prints by African American Artists from the Jean and Robert Steele Collection, 2002; Steven Cushner, Recent Paintings, 2002; Clarice Smith ReCollection 1978-2003, 2003; Andrew Dunnill, Extractions, 2005; Michael Platt, Just Above Water, 2006; Out of Place, 2007; Trajectories, Considering Time in Contemporary Art, 2008; Linn Meyers, Here Today, 2009.
Hours & Admission Prices: Aug.-May Mon.-Tues. & Thurs.-Sat. 11-4, Wed. 11-6. No charge; donations accepted. Closed national & university holidays. &
Attendance: 8,000 (accurate)

COLLEGE PARK AVIATION MUSEUM, 1985 Cpl. Frank Scott Dr., College Park, MD 20740-2000. Tel.: 301-864-6029. Fax: 301-927-6472.
E-mail: aviationmuseum@pgparks.com
Web Site: www.collegeparkaviationmuseum.com
Founded: 1982.
Congressional District: 5
Key Personnel: Museum Educator, Jane Welsh; Collection Mgr., Tiffany Davis.
Personnel Profile: Full-Time Paid 4; Full-Time Volunteers 12; Part-Time Paid 14; Part-Time Volunteers 84; Interns 3.
Governing Authority: county. Parent Institution: Maryland-National Capital Park & Planning Commission. Subsidiary Institution: Natural & Historical Resources Division. Tax-exempt.
Institution Type/Description: Aviation Museum: located on the grounds of College Park Airport.
Collections: photographs; College Park Airport history; models; art; interactive exhibits; World War I artifacts; uniforms; books; historic aircraft items; archaeology; footings & foundations of original hangars; original air mail hangar & compass rose; aircraft - repro. Wright B (1910); 1916 NJ4 Jenny; 1924 Berliner helicopter; 1932 Monocoupe; 1936 J2-Cub; Wright memorabilia; 1946 Ercoupe; 1941 Boeing Stearman; 1939 Taylorcraft; 1912 Bleriot; 1911 Curtiss pusher.
Research Fields: related to College Park Airport & its history; general aviation history; early WWI & pre WWI aviation history; early military aviation history.
Facilities: 2,000-vol. library; 15,000 sq. ft. exhibit space; auditorium. Museum-related items for sale.
Activities: films; intern program for college students; children programs; guided tours; lectures; loan & participatory exhibits; restoration & preser-

vation; research; archival preservation; school, camp & scout tours; Wright Bros. Wing Rib Making (1909); federally legislated Veterans History Project partner.
Publications: book, Maryland Aloft.
Hours & Admission Prices: Daily 10-5. Adults $4, seniors & groups $3, children under 2-18 $2; members & children under 2 no charge. Closed major holidays. &
Attendance: 50,000 (accurate)
Membership: Individual $25; Family (up to 4) $75.

THE DAVID C. DRISKELL CENTER, (M), 1214 Cole Student Activities Bldg., University of Maryland, College Park, MD 20742. Tel.: 301-314-2615. Fax: 301-314-0679.
E-mail: driskellcenter@umd.edu
Web Site: www.driskellcenter.umd.edu
Founded: 2001.
Key Personnel: Exec. Dir., Dr. Robert E. Steele; Deputy Dir., Dorit Yaron; Office Mgr., Veronica McDougal; Archivist, Jennifer Eidson
Institution Type/Description: Art Gallery.
Collections: works by African American artists including Romare Bearden, David C. Driskell, Sam Gillian, Jacob Lawrence, Keith Morrison, & Charles White.
Facilities: Museum-related items for sale.
Activities: special events; rental facilities; lectures; workshops; film festivals; speaker series; temporary & permanent exhibitions.
Hours & Admission Prices: Gallery: Mon.-Tues. & Thurs.-Fri. 11-4, Wed. 11-6. Office: Mon.-Fri. 8:30-4:30. Closed holidays.

NATIONAL ARCHIVES AT COLLEGE PARK, 8601 Adelphi Rd., College Park, MD 20740-6002. Tel.: 301-837-2000; 866-272-6272.
Web Site: www.archives.gov/index.html
Institution Type/Description: Archives.
Collections: US Federal Government documents & materials; films; military & naturalization records; photographs; interactive exhibits.
Activities: research.
Hours & Admission Prices: Research: Mon.-Tues. & Sat. 9-5, Wed.-Fri. 9-9. Closed Thanksgiving; Christmas.

THE NATIONAL MUSEUM OF LANGUAGE, 7100 Baltimore Ave., Ste. 202, College Park, MD 20740-3638. Tel.: 301-864-7071.
Web Site: www.languagemuseum.org
Founded: 1997.
Congressional District: 21
Key Personnel: Pres. (V), John Joseph Smith.
Personnel Profile: Part-Time Paid 1; Part-Time Volunteers 20; Interns 1.
Governing Authority: Tax-exempt.
Institution Type/Description: History Museum.
Collections: research materials relating to language.
Publications: NML Newsletter.
Hours & Admission Prices: Tues. & Sat. 12-4, 1st & 3rd Sun. of month 1-4. No charge; donations accepted.
Attendance: 800 (estimated)
Membership: Student & Retirees $25; Individual $50; Dual & Family $75; Supporting $120-$499; Sustainer $500-$999; Benefactor $1,000 & up.

NIXON PRESIDENTIAL MATERIALS STAFF, 8601 Adelphi Rd., Rm. 1360, College Park, MD 20740-6002. Tel.: 301-837-3290. Fax: 301-837-3202.
Web Site: nixon.archives.gov/index.php
Key Personnel: Dir., Tim Naftali; Deputy Dir., Marty McGann; Staff Cur., Edward R. Quick; Archivist, Shar Conway-Lanz.
Personnel Profile: Full-Time Paid 23; Part-Time Volunteers 3; Interns 3.
Governing Authority: federal.
Institution Type/Description: Presidential Museum.
Collections: Nixon administration materials created & received by the White House from 1969-1974.
Facilities: library.
Activities: research.
Hours & Admission Prices: Research Room: Mon.-Tues. & Sat. 9-5, Wed.-Fri. 9-9. Closed federal holidays.

UNION GALLERY, 1220 Stamp Student Union, Adele Stamp Memorial Union, The University of Maryland, College Park, MD 20742. Tel.: 301-314-8492.
E-mail: jmilad@umd.edu
Web Site: www.union.umd.edu/gallery

Key Personnel: Program Coord., Jackie Milad
Institution Type/Description: Art Gallery.
Collections: photographs; paintings.
Hours & Admission Prices: Fall & Spring Mon.-Thurs. 10-8, Fri. 10-6, Sat. 11-4; Summer Mon.-Thurs. 10-6, Fri.-Sat. 11-4.

Colton's Point

*** ST. CLEMENTS ISLAND AND PINEY POINT MUSEUMS,** (M), 38370 Point Breeze Rd., Colton's Point, MD 20626-2011. Tel.: 301-769-2222. Fax: 301-769-2225.
Web Site: www.stmarysmd.com/recreate/museums
Founded: 1975.
Congressional District: 1
Key Personnel: Dir., Debra L. Pence; Chm. (V), Barbara McWilliams; Site Supvr. SCI, Christina Barbour; Site Supvr. PPLM, April Havens; Mktg. & Programs, Kimberly Cullins; Exhibits Fabricator & Museum Tech, Tom Emery; Museum Shop Mgr., Carol Cribbs.
Personnel Profile: Full-Time Paid 5; Full-Time Volunteers 1; Part-Time Paid 15; Part-Time Volunteers 40.
Governing Authority: county government. Parent Institution: St. Mary's County Department of Recreation, Parks and Community Services. Subsidiary Institutions: Charlotte Hall Schoolhouse; Piney Point Lighthouse; Drayden African American Schoolhouse; U-1105 Underwater Submarine Shipwreck Preserve; Piney Point Lighthouse Keeper's Quarters; Piney Point Lighthouse Fuel Bldg.; 1943 U.S. Navy Workshop Bldg. Tax-exempt: 501(c)(3).
Institution Type/Description: Archaeology & History Museum: located on 1634 landing site of Maryland colonists; site of first Roman Catholic mass in English colonies.
Collections: archaic & woodland period Native Americans; 17th- to 20th-century material culture; maritime collections; documents.
Research Fields: archaeology pertaining to early Maryland; marine surveys of Potomac River; architectural contributions of 18th & 19th centuries.
Facilities: 300-vol. library of Maryland & general history books & vertical files, available for inter-library loan & for use by public; 60 acre site on St. Clements Island; 4,000 sq. ft. exhibit space; nature center. Gift items & books for sale.
Activities: guided tours; lectures; organized education programs for children & adults; docent program; participatory, temporary & traveling exhibitions; school loan service. Museum Sponsors: Children's Waterfront Festival; Blessing of the Fleet; Maryland Day; Christmas Exhibit.
Publications: quarterly newsletter, Finer Points.
Hours & Admission Prices: See website for hours. Adults $3, students 6-18 $1; discounts to AAM members; members & children under 6 no charge. Water Taxi Service: $5 per person. &
Attendance: 54,437 (accurate)
Membership: Individual $40; Family $60; Heritage $125 Patron $250; Benefactor $500. Corporate: Patron $250; Benefactor $500.

Columbia

HOWARD COUNTY CENTER OF AFRICAN AMERICAN CULTURE, 5434 Vantage Point Rd., Columbia, MD 21044-2644. Tel.: 410-715-1921. Fax: 410-715-8755.
E-mail: hccaacmd@juno.com
Web Site: www.nccaac.org
Founded: 1987.
Congressional District: 3
Key Personnel: Dir., Wylene Sims-Burch; Chm. (V), Everlene G. Cunningham; Museum Shop Mgr., Florence C. Smith.
Personnel Profile: Full-Time Volunteers 1; Part-Time Paid 6; Part-Time Volunteers 20.
Governing Authority: private; nonprofit organization. Tax-exempt.
Institution Type/Description: History Museum.
Collections: artifacts; murals.
Research Fields: County Underground Railroad.
Facilities: 3,000-vol. library.
Activities: book club; poetry hour; African American Artists' Alliance.
Publications: newsletter, Compass.
Hours & Admission Prices: Tues.-Fri. 10-5, Sat. 10-4, Sun. by appointment. Adults $4, children $2; members no charge.
Attendance: 5,000 (estimated)
Membership: Junior $10; Friends $25; Contributor $50; Patron $100; Life $500; Corporate $1,500.

THE ROUSE COMPANY FOUNDATION GALLERY - HOWARD COMMUNITY COLLEGE, Peter and Elizabeth Horowitz Visual and Performing Arts Center, 10901 Little Patuxent Pkwy., Columbia, MD 21044. Tel.: 443-518-1200.
Key Personnel: Art Dir., Rebecca Bafford; Asst. To Dir., Chaya Shapiro
Institution Type/Description: Art Gallery.
Collections: paintings; sculpture; photographs.
Hours & Admission Prices: Mon.-Fri. 10-8, Sat.-Sun. 12-5. Closed university holidays.

Crisfield

J. MILLARD TAWES MUSEUM, 3 Ninth St., Crisfield, MD 21817. Tel.: 410-968-2501.
Web Site: www.crisfieldheritagefoundation.org
Institution Type/Description: Maritime History Museum.
Collections: maritime history; hands-on crab shanty; period furnishings; photographs; personal artifacts.
Activities: guided tours.
Hours & Admission Prices: Call for hours. &

Crownsville

RISING SUN INN, 1090 Generals Hwy., Crownsville, MD 21032-1417. Tel.: 410-268-9249. Fax: 410-268-1994.
E-mail: ellanwt@aol.com
Web Site: marylanddar.org.annarundel/patriots.html
Founded: 1911.
Key Personnel: Dir., Ellan Thorson.
Personnel Profile: Part-Time Volunteers 20.
Governing Authority: Tax-exempt. Parent Institution: Maryland Historical Trust.
Institution Type/Description: Historic House Museum: housed in the chapter house of the Ann Arundel Chapter of the Daughters of the American Revolution; built c.1753. Listed on the National Register of Historic Places.
Collections: period artifacts; Colonial money & eyewear; pewter; cannonballs; early glassware from 1700-1900; doll house replica of Jonas Green House.
Hours & Admission Prices: 2nd Sun. of month 1-4; other times by appointment. Donation: adults $5; members no charge.
Membership: Friends of Rising Sun Inn: Senior $10; Single $15; Couple $25.

Cumberland

ALLEGANY COUNTY HISTORICAL SOCIETY, INC., (M), 218 Washington St., Cumberland, MD 21502-2827. Tel.: 301-777-8678. Fax: 301-777-8678. Facebook: Allegany County Historical Society.
E-mail: info@gordon-robertshouse.com
Web Site: www.gordon-robertshouse.com
Founded: 1937.
Congressional District: 1
Key Personnel: Exec. Dir., Evan Sloanker; Pres. (V), Nadeane Gordon; Educational Dir., Amber Butcher; Museum Shop Mgr., Mickey Miller; Asst., Lindsay Droll.
Personnel Profile: Full-Time Paid 1; Part-Time Paid 2; Part-Time Volunteers 15.
Governing Authority: society. Subsidiary Institution: Gordon-Roberts House. Tax-exempt: 501(c)(3).
Institution Type/Description: Historic House Museum: housed in a Second Empire-style home built by Josiah Hance Gordon; 1867.
Collections: American decorative arts from 1800-1930; dolls & toys.
Activities: guided tours; monthly exhibits; educational program; docent group; lifestyle tour with costumed tour docents; Museum Explorer Tour for children.
Publications: members' quarterly newsletter.
Hours & Admission Prices: Wed.-Sat. 10-4; bus tours available. Adults $7, seniors 60 & over $ 6, children 12 & under $5; members no charge. Closed major holidays.
Attendance: 8,000 (accurate)
Membership: Student $10; Individual $20; Joint $30; Family $35; Patron $100; Life Individual $300; Life Couple $450; Corporate $500.

ALLEGANY MUSEUM, 3 Pershing St., Cumberland, MD 21502-3042. Tel.: 301-777-7200 & 724-4339.
Web Site: alleganymuseum.org
Formerly: Allegany County Museum
Founded: 1985.
Congressional District: 6
Key Personnel: Pres. (V), Gary Bartik; Vice Pres. (V), Joseph H. Weaver.

Personnel Profile: Part-Time Volunteers 80.
Governing Authority: Tax-exempt.
Institution Type/Description: History Museum.
Collections: local history & culture; period furnishings; personal artifacts; photographs; industrial exhibits.
Hours & Admission Prices: March 17-Dec. Tues.-Sat. 10-4, Sun. 1-4. No charge; donations accepted. &
Attendance: 11,500 (estimated)
Membership: Sponsor $25-$99; Patron $100-$499; Benefactor $500 & up.

C. WILLIAM GILCHRIST MUSEUM OF THE ARTS, 104 Washington St., Cumberland, MD 21502.
E-mail: gilchristgallery@atlanticbb.net
Web Site: www.gilchristgallery.com
Institution Type/Description: Art Gallery.
Collections: local & state history and culture; paintings; sculpture.
Facilities: library.
Activities: rental facilities; permanent & temporary exhibitions.
Hours & Admission Prices: April-Dec. Fri.-Sun. 1-4.

F. BROOKE WHITING HOUSE MUSEUM, (M), 632 Washington St., Cumberland, MD 21502-2827. Mailing Address: 218 Washington St., Cumberland, MD 21502-2827. Tel.: 301-777-7782.
E-mail: info@whitinghouse.com
Web Site: www.thewhitinghouse.org
Congressional District: 1
Key Personnel: Exec. Dir., Evan Sconaker; Pres. (V), Nadeane Gordon.
Personnel Profile: Full-Time Paid 1; Part-Time Volunteers 12.
Governing Authority: Parent Institution: Allegany County Historical Society. Tax-exempt.
Institution Type/Description: Historic House: housed in the home of F. Brook Whiting I; built in 1911.
Collections: personal artifacts; oriental porcelains; artwork; period furnishings.
Hours & Admission Prices: Tours: May-Oct. 2nd & 4th Fri.-Sat. 2 & 3. Adults $5; members no charge.
Attendance: 25 (accurate)
Membership: Student $10; Individual $20; Joint Couple $30; Family $35; Patron/Business $100; Life Individual $300; Life Couple $450; Corporate $500.

GEORGE WASHINGTON'S HEADQUARTERS, Greene St., Cumberland, MD 21502. Mailing Address: Parks/Recreation City Hall, 57 N. Liberty St., Cumberland, MD 21502. Tel.: 301-759-6636. Fax: 301-759-3223.
E-mail: djohnson@ci.cumberland.md.us
Founded: 1925.
Congressional District: 6
Key Personnel: Dir. Parks & Recreation, Diane Johnson; Dir., Cathy McKenny.
Governing Authority: municipal. Subsidiary Institution: Daughters of the American Revolution (DAR) Cresap Chapter. Tax-exempt.
Institution Type/Description: Historic Building: 1755 log cabin built during the French & Indian War.
Collections: French & Indian War relics; early Allegany County history.
Activities: guided tours; lectures.
Hours & Admission Prices: by appointment only. No charge.

THE SAVILLE GALLERY, 9 N. Centre St., Cumberland, MD 21502. Tel.: 301-777-2787.
E-mail: arts@allconet.org
Web Site: www.alleganyartscouncil.org
Founded: 1975.
Congressional District: 6
Governing Authority: Tax-exempt.
Institution Type/Description: Art Gallery.
Collections: works by local, regional & national artists.
Hours & Admission Prices: mid-May to mid-Nov. Mon.-Fri. 9-5, Sat.-Sun. 11-4; mid-Nov. to mid-May Mon.-Fri. 9-5, Sat. 11-4. No charge. &

Denton

MUSEUM OF RURAL LIFE, 16 N. 2nd St., Denton, MD 21629-1004. Mailing Address: P.O. Box 514, Denton, MD 21629-0514. Tel.: 410-479-2055. Fax: 410-479-4513.
E-mail: jok@jokwalsh.com
Founded: 1952.
Key Personnel: Pres. Historical Society, J.O.K. Walsh; Dir., Cur. & Museum Shop Mgr., Carol D. Stockley; Treas., Carolyn D. Spicher.

Personnel Profile: Part-Time Paid 1.
Governing Authority: private; nonprofit organization. Tax-exempt: 501(c)(3)
Institution Type/Description: History Museum.
Collections: Caroline County, Maryland history; structures; furnishings; personal artifacts; photographs.
Research Fields: deeds; property; inventories; families of Caroline County, MD.
Facilities: 30-seat theater. Museum-related items for sale.
Activities: fund-raisers (approx. four per year); guided tours; lectures; rental gallery; temporary exhibitions of your own collection. Museum Sponsors: Student Discover Days for 3rd graders.
Publications: annual activities report, Museum Notes.
Hours & Admission Prices: Fri.-Sat. 10-3, Sun. 12-4. No charge; donations accepted. &
Attendance: 400 (accurate)
Membership: Individual $35.

Earleville

MOUNT HARMON PLANTATION, 600 Mount Harmon Rd., Earleville, MD 21919. Mailing Address: P.O. Box 65, Earleville, MD 21919. Tel.: 410-275-8819.
E-mail: info@mountharmon.org
Web Site: www.mountharmon.org
Key Personnel: Exec. Dir., Paige Howard
Institution Type/Description: Historic Plantation: housed in an 18th century manor home on a former tobacco plantation.
Collections: local history & heritage; nature preserve; boxwood garden; tobacco prize house; period furnishings.
Hours & Admission Prices: May-Oct. Thurs.-Sun. 10-3; other times by appointment.

Easton

* **ACADEMY ART MUSEUM, (M),** 106 South St., Easton, MD 21601-2949. Tel.: 410-822-2787. Fax: 410-822-5997.
E-mail: academy@academyartmuseum.org
Web Site: www.academyartmuseum.org
Founded: 1958.
Congressional District: 1
Key Personnel: Exec. Dir., Erik H. Neil; Chm., Richard Bodorff; Cur., Anke Van Wagenberg, Ph.D.; Dir. Programs, Janet Hendrick; Asst. to Dir., Marie Bradley; Dir. Devel., Beth Jones; Facilities Supervisor, Edward Robinson; Coord. Education, Constance Del Nero.
Personnel Profile: Full-Time Paid 7; Part-Time Paid 10; Part-Time Volunteers 225.
Governing Authority: nonprofit organization. Tax-exempt: 501(c)(3).
Institution Type/Description: Art Museum.
Collections: paintings; sculpture; prints; photographs.
Facilities: 3,000-vol. library of books on painting, sculpture, architecture & graphics available for use; classrooms; sculpture courtyard.
Activities: lectures; films; gallery talks; concerts; formally organized education programs for children & adults; temporary exhibitions. Academy Sponsors: art appreciation programs in public schools.
Publications: exhibition catalogues; quarterly magazine, Academy.
Hours & Admission Prices: Mon.-Sat. 10-4. Adults $3; AAM & museum members no charge. Closed New Year's Day; Easter; Memorial Day; Independence Day; Labor Day; Thanksgiving; Christmas. &
Attendance: 71,230 (accurate)
Membership: Individual $50; Family $65; Friend $100; Contributor $250; Sustaining $500; Patron $1,000; Benefactor $5,000.

PICKERING CREEK AUDUBON CENTER, 11450 Audubon Lane, Easton, MD 21601. Tel.: 410-822-4903. Fax: 410-822-5041.
Key Personnel: Dir., Mark Scallion
Institution Type/Description: Audubon Center.
Collections: wildlife & their habitats; trees; plants; flowers; agriculture; gardnes; farm tools & equipment; reptiles; mounted animals.
Facilities: library; 400-acre working farm; nature trails.
Activities: environmental education & outreach programs; walking trails; school groups.
Hours & Admission Prices: Trails: daily dawn to dusk. Office: Mon.-Fri. 9-5. No charge.

TALBOT HISTORICAL SOCIETY, 25 S. Washington St., Easton, MD 21601-3014. Tel.: 410-822-0773. Fax: 410-822-7911.
E-mail: director@hstc.org
Web Site: www.hstc.org
Formerly: Historical Society of Talbot County

Founded: 1954.

Congressional District: 3

Key Personnel: Pres. (V), Carla Howell; Cur., Beth Hansen; Office Mgr., Karen Clements.

Personnel Profile: Full-Time Paid 2; Part-Time Paid 3; Part-Time Volunteers 120.

Governing Authority: society. Tax-exempt.

Institution Type/Description: Historical Society Museum.

Collections: costumes; furniture; china; documents; decorative arts. Historic Buildings: 1795 frame dwelling; 1880 stove store; 1810 Federal Period townhouse.

Research Fields: local history.

Facilities: auditorium. Museum-related items for sale.

Activities: guided house tours; walking tours; family programs; vintage baseball team.

Publications: quarterly newsletter.

Hours & Admission Prices: Wed.-Sat. 10-4; call to confirm. Tours: adults $5, children 6-12 $2; discounts to groups of 20 or more; members and children 6 & under no charge. Closed major holidays. &

Attendance: 7,382 (accurate)

Membership: Individual $45; Family $65; 19th Century $100; 20th Century $250; 21st Century $500; James Neall Society $1.000.

Edgewater

HISTORIC LONDON TOWN AND GARDENS, (M), 839 Londontown Rd., Edgewater, MD 21037-2120. Tel.: 410-222-1919. Fax: 410-222-1918.

E-mail: londontown@historiclondontown.org

Web Site: historiclondontown.org

Founded: 1971.

Congressional District: 4

Key Personnel: Exec. Dir., Rod Cofield; Dir. Operations, Ken Schroeder; Chm. (V), Maureen T. Konschnik; Horticulturist, Nate Powers.

Personnel Profile: Full-Time Paid 6; Part-Time Paid 8; Part-Time Volunteers 100; Interns 3.

Governing Authority: Parent Institution: London Town Foundation. Tel. 410-222-1919. Tax-exempt: 501(c)(3).

Institution Type/Description: Historic Building & Botanical Gardens: c.1760 Georgian mansion & 8-acre botanical garden; significant archaeological site.

Collections: mid-18th century furnishings appropriate for a rural Maryland home & tavern; country Queen Anne & early Chippendale furniture, pewter, creamware & Hogarth prints; plant collection of native & exotic species.

Research Fields: 18th-century history of Anne Arundel County, Maryland; horticulture; archaeology.

Facilities: botanical garden. Museum-related items for sale.

Activities: museum reproduction program; guided house tours; lectures; concerts; plant sales; education programs for children; rentals for private functions; archaeology dig.

Publications: quarterly newsletter; London Town News.

Hours & Admission Prices: Tours: Jan.-Feb. Wed.-Fri. 10-4, Sun. 12-4; March-Dec. Wed.-Sat. 10-4, Sun. 12-4; other times by appointment. Gardens: Tues.-Sat. 10-3. Adults $10, senior citizens $9, children 7-18 $5; discounts to AAM members; members no charge. Additional fee for special events. Closed major holidays.

Attendance: 17,883 (accurate)

Membership: Individual $25; Family $50; Friend $100; Patron $250; Benefactor $500.

Elkton

HISTORICAL SOCIETY OF CECIL COUNTY, 135 E. Main St., Elkton, MD 21921-5955. Tel.: 410-398-1790.

E-mail: questions@cecilhistory.org

Web Site: www.cecilhistory.org

Founded: 1931.

Congressional District: 1

Key Personnel: Dir., Paula Newton.

Personnel Profile: Full-Time Volunteers 20; Part-Time Volunteers 20.

Governing Authority: society; board of officers & trustees. Tax-exempt.

Institution Type/Description: History Museum.

Collections: genealogy; manuscripts; over 1,000 Cecil County photographs; Cecil County newspapers, 1838-1980; governmental records including road books, commissioners' minutes, tax assessment books & slave register. Historic Buildings: c.1799 school; log house.

Research Fields: genealogical; Cecil County & Maryland history.

Facilities: 1,800-vol. library on Cecil County history.

Activities: tours by appointment; permanent & temporary exhibitions.

Publications: triannual bulletin, The Historical Society of Cecil County.

Hours & Admission Prices: Mon. & Thurs. 10-4, 1st & 4th Sat. each month 10-2. No charge. &

Attendance: 2,000 (accurate)

Membership: Student $10; Basic $20; Couple $25; Contributor $50; Benefactor $75; Curator's Circle $100; Director's Circle $200; Historian's Circle $300; Legacy Circle $500; Life $1,000.

Ellicott City

B&O RAILROAD MUSEUM: ELLICOTT CITY STATION, 2711 Maryland Ave., Ellicott City, MD 21043-4661. Tel.: 410-461-1945. Fax: 410-461-1944.

E-mail: sitemanager@borail.org

Web Site: www.ecborail.org

Founded: 1976.

Congressional District: 6

Key Personnel: Exec. Dir., Courtney B. Wilson; Site Mgr., Tom Hane.

Personnel Profile: Full-Time Paid 1; Part-Time Paid 4; Part-Time Volunteers 85.

Governing Authority: nonprofit. Owner: Howard County Rec. & Parks. Subsidiary Institution: B&O Railroad Museum, Inc. Tax-exempt: 501(c)(3).

Institution Type/Description: Transportation Museum: housed in 1831 B&O Railroad Station, first terminus in the United States; terminus is located at the end of the first 13 miles of track laid in the U.S.

Collections: 45 ft. long HO layout of the first 13 miles from Baltimore to Ellicott City; 1927 caboose; tools & transportation equipment.

Facilities: theater. Gift items for sale.

Activities: temporary & permanent exhibitions; presentations on history of railroading; living history program; special events; holiday train layouts. Museum Sponsors: Holiday Festival of Trains November to January.

Hours & Admission Prices: Wed.-Sun. 11-4. Adults $6, senior citizens 60 & over $5, children 2-12 $4; discounts to AAM members; members no charge. Closed major holidays. &

Attendance: 25,000 (accurate)

Membership: Senior $55; Individual $60; Dual Senior $65; Dual $75; Grand Family Senior $95; Family $100; Family & Friends $125.

ELLICOTT CITY COLORED SCHOOL, 8683 Frederick Rd., Ellicott City, MD 21043-4310. Tel.: 410-313-1427 & 5131.

Institution Type/Description: Historic Building: housed in a one-room schoolhouse, the first public school for black children in Howard County; built in 1880.

Collections: African American history; period furnishings; photographs; personal artifacts.

Activities: temporary exhibitions.

Hours & Admission Prices: Tours: April-Oct. Sat.-Sun. 3pm; other times by appointment.

ELLICOTT CITY HERITAGE ORIENTATION CENTER, 8334 Main St., Ellicott City, MD 21043-4604. Tel.: 410-313-5131.

Institution Type/Description: History Museum.

Collections: local history & culture; period furnishings; personal artifacts; photographs.

Hours & Admission Prices: Daily 11-4.

THE FIREHOUSE MUSEUM, 3829 Church Rd., Ellicott City, MD 21043. Mailing Address: Howard County Recreation & Park, 7120 Oakland Hills Rd., Columbia, MD 21046-1621. Tel.: 410-465-8500. Fax: 410-465-8817.

E-mail: jgalke@howardcountymd.gov

Web Site: www.howardcountymd.gov

Founded: 1991.

Key Personnel: Heritage Programs Supvr., Jacquelyn Galke.

Governing Authority: Parent Institution: Howard County Recreation & Park.

Institution Type/Description: Firefighting Museum: 1889 Ellicott City fire station.

Collections: original fire equipment including a two-wheeled hose cart put into service in 1893; fallen heroes dedication; firefighting gear.

Hours & Admission Prices: April-Dec. Sat.-Sun. 1-4. Private tours year-round by appointment. No charge; donations accepted. &

Attendance: 2,000 (estimated)

HOWARD COUNTY HISTORICAL SOCIETY, (M), 8328 Court Ave., Ellicott City, MD 21043. Mailing Address: P.O. Box 109, Ellicott, MD 21041-0109. Tel.: 410-750-0370. Fax: 410-750-0370.

E-mail: info@hchsmd.org

Web Site: www.hchsmd.org

Founded: 1957.
Congressional District: 3
Key Personnel: Pres. (V), Shelley D. Wygant; Dir., Lauren McCormack; Cur., Karen Griffith.
Personnel Profile: Full-Time Paid 1; Part-Time Paid 2; Part-Time Volunteers 15.
Governing Authority: private; nonprofit organization. Tax-exempt: 501(c)(3).
Institution Type/Description: Historical Society Museum, Archives & Library.
Collections: artifacts that reflect Howard County's history from 1700 to present; period furniture 1550-1890; weapons Revolution to Civil War.
Research Fields: Howard County history.
Facilities: library. Gift items for sale.
Activities: formal education programs for children & adults; guided tours on request; lectures. Annual Events: Dinner Dance, members & guests; Candlelight tour, members & guests; lectures, 4 times per year.
Publications: quarterly newsletter, The Legacy.
Hours & Admission Prices: Tues. & Sat. 1-4. No charge; donations accepted. Closed New Year's Day; Independence Day; Thanksgiving; Christmas.
Attendance: 1,400 (accurate)
Membership: Student $10; Single $30; Family $45; Sustaining & Corporate $100.

MT. IDA VISITOR CENTER, 3691 Sarah's Lane, Ellicott City, MD 21041. Mailing Address: P.O. Box 293, Ellicott City, MD 21041. Tel.: 410-465-8500.
Institution Type/Description: History Museum: house built in 1828.
Collections: local history; photographs; personal artifacts; period furnishings.
Hours & Admission Prices: Call for hours.

PATAPSCO VALLEY STATE PARK - THE AVALON VISITOR CENTER, 8020 Baltimore National Pike, Ellicott City, MD 21043. Tel.: 410-461-5005; 888-432-2267.
Institution Type/Description: Park & Visitor Center: housed in a 19th century stone dwelling.
Collections: Patapsco River history; period furnishings; photographs; personal artifacts; 1930s replica of the forest warden's office.
Hours & Admission Prices: Call for hours.

THOMAS ISAAC LOG CABIN, 8394 Main St., Ellicott City, MD 21043-4604. Tel.: 410-313-1413.
Institution Type/Description: Historic Building: built c.1780.
Collections: local history & culture; period furnishings; personal artifacts; photographs.
Hours & Admission Prices: Sat.-Sun. 1-4.

Emmitsburg

SETON HERITAGE MUSEUM & THEATER AT THE NATIONAL SHRINE OF SAINT ELIZABETH ANN SETON, 339 S. Seton Ave., Emmitsburg, MD 21727-9297. Tel.: 301-447-6606. Fax: 301-447-6061. Facebook: Elizabeth Ann Seton.
E-mail: office@setonheritage.org
Web Site: www.setonheritage.org
Founded: 1975.
Institution Type/Description: History Museum.
Collections: life & legacy of Saint Elizabeth Ann Seton; photographs; personal artifacts.
Facilities: theater. Museum-related items for sale.
Activities: religious & historically themed events.
Hours & Admission Prices: Tues.-Sun. 10-4:30; call to confirm.

Ewell

SMITH ISLAND CENTER, 20846 Caleb Jones Rd., Ewell, MD 21824. Tel.: 410-425-3351; 800-521-9189.
E-mail: smithisland@verizon.net
Web Site: www.smithisland.org
Institution Type/Description: History Museum.
Collections: local history, heritage, economic & social life; photographs; period furnishings.
Activities: special events.
Hours & Admission Prices: May-Oct. daily 12-4. Accessible by boat from Crisfield and Solomons.

Fort Meade

FORT GEORGE G. MEADE MUSEUM, 4674 Griffin Ave., Fort Meade, MD 20755-7047. Mailing Address: Attn: IMNE-MEA-M, Fort George G. Meade, MD 20755-5094. Tel.: 301-677-6966 & 7054. Fax: 301-677-2953.
E-mail: robert.s.johnson212.civ@mail.mil
Web Site: www.ftmeade.army.mil/museum/index.htm
Founded: 1963.
Congressional District: 6
Key Personnel: Cur., Robert S. Johnson; Pres. (V), (Ret) MAS David L. Burget; Exhibits Specialist, Barbara Taylor; Museum Technician, Richard Frank.
Personnel Profile: Full-Time Paid 3; Part-Time Volunteers 12.
Governing Authority: federal. Parent Institution: U.S. Army Center of Military History. Tax-exempt.
Institution Type/Description: Military Museum.
Collections: uniforms; weapons & equipment of the U.S. Army from the Revolution to the present; photographs & graphics; World War I & II research collections.
Research Fields: history of Fort Meade.
Facilities: library of military books, journals & unit histories particularly related to World Wars I & II; research area.
Activities: guided tours; lectures; permanent & temporary exhibitions.
Publications: Illustrated History of Fort George G. Meade.
Hours & Admission Prices: Wed.-Sat. 11-4, Sun. 1-4; other times by appointment. No charge. Closed national holidays. &
Attendance: 23,741 (accurate)

Fort Washington

FORT WASHINGTON PARK, 13551 Fort Washington Rd., Fort Washington, MD 20744-7044. Tel.: 301-763-4600. Fax: 301-763-1389. Facebook: Fort Washington Park.
E-mail: nace_fort_washington_park@nps.gov
Web Site: www.nps.gov/fowa
Founded: 1940.
Congressional District: 4
Key Personnel: Supt., Gopaul Noojibail; Park Mgr., Elizabeth Jackson.
Personnel Profile: Full-Time Paid 4; Part-Time Paid 2; Part-Time Volunteers 125; Interns 1.
Governing Authority: federal. Parent Institution: National Capital Parks-East, National Park Svc., U.S. Dept. of Interior, 1900 Anacostia Dr. S.E., Washington, DC 20020. Tax-exempt.
Institution Type/Description: Military Museum: housed in 1824 Fort Washington; 1817 Commandant's house.
Collections: military; history.
Facilities: research files on historic records available for use in office by request.
Activities: guided tours; lectures; living history demonstrations.
Publications: interpretive folder.
Hours & Admission Prices: Park: 8-sunset. Historic Fort & Visitor Center: 1st Sun. April to last Sat. Oct. daily 9-5; Nov.-March daily 9-4:30. Annual Pass $20; $5 per vehicle, buses & walk-in $3 per person. Closed New Year's Day; Thanksgiving; Christmas. &
Attendance: 297,000 (accurate)

Frederick

BARBARA FRITCHIE HOUSE, 154 W. Patrick St., Frederick, MD 21701. Tel.: 301-600-4047.
Institution Type/Description: Historic House: housed in the former home of Barbara Fritchie, heroine of John Greenleaf Whittier's poem from the Civil War.
Collections: Barbara Fritchie's life; local history; period furnishings; personal artifacts.
Hours & Admission Prices: Call for hours.

BJORLEE MUSEUM, 101 Clarke Pl., Frederick, MD 21701. Tel.: 301-360-2011. Fax: 301-360-2044.
Institution Type/Description: History Museum.
Collections: local history; school related artifacts; communication equipment; hearing aids from 1950s; teletypewriters & pagers.
Hours & Admission Prices: Tues. & Thurs. 12-5.

CATOCTIN CENTER FOR REGIONAL STUDIES, 7932 Opossumtown Pike, Frederick, MD 21702. Tel.: 301-624-2773.
Web Site: catoctincenter.frederick.edu
Governing Authority: Parent Institution: Frederick Community College.

Institution Type/Description: History Museum.
Collections: local history & cultural heritage; African American & Underground Railroad artifacts; freedom seeker J.W.C. Pennington.
Activities: special events; educational programs.
Hours & Admission Prices: By appointment.

THE CHILDREN'S MUSEUM OF ROSE HILL MANOR PARK AKA ROSE HILL MANOR PARK & MUSEUM, 1611 N. Market St., Frederick, MD 21701-4304. Tel.: 301-600-1646 (reservations) & 2743 (office). Fax: 301-600-2749. TDD: 301-600-2936.
E-mail: ksaavedra@frederickcountymd.gov
Web Site: rosehillmuseum.com
Founded: 1972.
Congressional District: 7A
Key Personnel: Pres. Museum Council (V), Joann Ramsburg; Museum Mgr., Kari Saavadra; Museum Shop Mgr., Shirley Swaim.
Personnel Profile: Full-Time Paid 1; Part-Time Paid 16; Part-Time Volunteers 48; Interns 1.
Volunteer Hours: 490
Operating Expenses: 12,200
Operating Income: 51,322
Governing Authority: county; nonprofit. Parent Institution: Frederick County Commissioners. Subsidiary Institution: Rose Hill Manor Park. Division of Parks & Recreation. Tax-exempt: 501(c)(3).
Institution Type/Description: Children's Museum.
Collections: 19th-century house & farm; gardens.
Major Exhibits: Storybooks & Art, 4/14-6/14; Emancipation: A Maryland Story 1864, 7/14-12/14; WWII Special Weekend Highlights, 8/14.
Research Fields: history of Manor House & owner.
Facilities: picnic area.
Activities: seasonal weekend festivals. Museum Sponsors: Easter Egg Roll; Living History Weekends; Civil War Encampment; holiday programs & activities.
Publications: e-newsletter, Rose Hill Record.
Hours & Admission Prices: April-Sept. Mon.-Sat. 11-4, Sun. 1-4; Oct.-Nov. Sat. 11-4, Sun. 1-4; other times by appointment. Adults $5, senior citizens 55 & over & children 2-18 $4; discounts to AAM & AAA members; museum employees & volunteers no charge.
Attendance: 13,449 (estimated)

DELAPLAINE VISUAL ARTS EDUCATION CENTER, 40 S. Carroll St., Frederick, MD 21701. Tel.: 301-698-0656, ext. 366.
E-mail: info@delaplaine.org
Web Site: delaplaine.org
Founded: 1986.
Key Personnel: Exec. Dir., Catherine Moreland.
Governing Authority: nonprofit organization.
Institution Type/Description: Art Gallery.
Collections: photographs; paintings; sculpture.
Activities: classes; workshops; lectures; special events; educational programs; rental facilities.
Hours & Admission Prices: Mon.-Sat. 9-5, Sun. 1-4. Closed New Year's Day; Easter; Memorial Day; Independence Day; Labor Day; Thanksgiving; Christmas.

FREDERICK COUNTY VOLUNTEER FIRE & RESCUE MUSEUM AND PRESERVATION SOCIETY, 527 N. Market St., Frederick, MD 21701-5242. Mailing Address: c/o Frederick County Volunteer Fire & Rescue Association, 5370 Public Safety Pl., Frederick, MD 21704-6728. Tel.: 301-631-3465. Fax: 301-694-6063.
E-mail: dsears@fredco-md.net
Web Site: www.fcvfra.com
Key Personnel: Pres., Mickey Fyock
Institution Type/Description: Fire-Fighting Museum.
Collections: fire-fighting equipment; personal artifacts.
Hours & Admission Prices: April 7-Nov. daily 10-2. No charge.

✶ THE HISTORICAL SOCIETY OF FREDERICK COUNTY, INC., (M), 24 E. Church St., Frederick, MD 21701-5402. Tel.: 301-663-1188. Fax: 301-663-0526.
E-mail: hshoaf@frederickhistory.org
Web Site: www.frederickhistory.org
Founded: 1892.
Congressional District: 6
Key Personnel: Pres., Ethel Lee "Pepper" Scotto; Acting Exec. Dir., Duane Doxzen; Treas., Harold Mohn; Cur., Carrie Blough; Program Coord.,

Kristen Butler; Coord. Museum Operations, Jennifer Winter; Asst. Dir., Duane K. Doxzen; Coord. Research Center, Rebecca Crago.
Personnel Profile: Full-Time Paid 5; Part-Time Paid 8; Part-Time Volunteers 120; Interns 8.
Governing Authority: Parent Institution: Historical Society of Frederick County. nonprofit. Tax-exempt: 501(c)(3).
Institution Type/Description: History Museum.
Collections: Frederick County history & culture; decorative & fine arts; historical artifacts; photographs; textiles; archives; genealogical artifacts. Historic Buildings: 1820s headquarters; 1790s house once owned by Chief Justice Roger B. Taney.
Major Exhibits: Huzza for Liberty! Frederick County in the Civil War, 1864, 6/13-12/14; Emancipation in Frederick County, 4/14-12/14.
Research Fields: Frederick County History.
Facilities: library; archives. Books for sale.
Activities: guided tours; lecture series; seminars; workshops; special exhibits; docent & intern programs; children's story hour; walking tours.
Publications: quarterly newsletter; semi-annual journal; e-newsletter; books on Frederick County history.
Hours & Admission Prices: Museum: mid-Jan. to March Tues.-Sat. 10-4; April-Dec. Tues.-Sat. 10-4, Sun. 1-4. Research Center: Wed. & Fri. 10-4, Thurs. 10-8. Taney House: April to mid-Dec. Sat. 10-4, Sun. 1-4. Adults $6, youth 12-18 $3; discounts to AAA & Maryland Passport members; members & children under 12 no charge. Closed New Year's Day; Easter; Memorial Day; Independence Day; Labor Day; Thanksgiving; Christmas.
Attendance: 8,000 (estimated)
Membership: Student $10; Educator & College Student $25; Individual $40; Family $75; Patron $100; Associate $250; Benefactor $500; Business $250-$1,000.

MONOCACY NATIONAL BATTLEFIELD, 5201 Urbana Pike, Frederick, MD 21704-7303. Mailing Address: 4632 Araby Church Rd., Frederick, MD 21704. Tel.: 301-662-3515. Fax: 301-668-7437. Facebook: Monocacy National Battlefield.
E-mail: tracy_evans@nps.gov
Web Site: www.nps.gov/mono
Founded: 1991.
Congressional District: 6
Key Personnel: Supt., Rick Slade; Chief Ranger, Jeremy Murphy; Facility Mgr., Al Kirkwood; Administrative Officer, Kathy Snider; Park Ranger & Collateral Curator, Tracy L. Evans; Museum Shop Mgr., Bob Casey.
Personnel Profile: Full-Time Paid 14; Part-Time Paid 6; Part-Time Volunteers 16.
Governing Authority: federal. Tax-exempt.
Institution Type/Description: Historic Site & Monument: dedicated to soldiers who fought in the Battle of Monocacy, July 9, 1864.
Collections: Civil War artifacts related to the Battle of Monocacy on July 9, 1864; militaria; uniforms; personal artifacts of soldiers.
Major Exhibits: 150th Anniversary of the Battle of Monocacy, 6/14-8/14; Maryland Emancipation, 7/14-11/14.
Research Fields: Civil War history, primarily Battle of Monocacy; African American history as related to pre-civil war slave village on battlefield property.
Facilities: 350-vol. library on Civil War history available for public use on premises only; walking trails. Museum-related items for sale.
Activities: ranger programs; living history programs; trails.
Publications: leaflet & map, Monocacy National Battlefield; park brochure; auto-tour folder; Junior Ranger activity folder; various trail brochures; Explorer Ranger; mobile application, Monocacy for iPhone.
Hours & Admission Prices: Daily 8:30-5. No charge; donations accepted. Closed New Year's Day; Thanksgiving; Christmas.
Attendance: 35,000 (estimated)

✶ NATIONAL MUSEUM OF CIVIL WAR MEDICINE, (M), 48 E. Patrick St., Frederick, MD 21701-5628. Mailing Address: P.O. Box 470, Frederick, MD 21705-0470. Tel.: 301-695-1864. Fax: 301-695-6823.
E-mail: museum@civilwarmed.org
Web Site: www.CivilWarMed.org
Founded: 1990.
Congressional District: 3
Key Personnel: Exec. Dir., George Wunderlich; Chm. (V), Gordon E. Dammann, D.D.S.; Pres. (V), Betsy Estilow; Treas., Meredith Harshman; Museum Shop Mgr., Judy Candela.
Personnel Profile: Full-Time Paid 10; Part-Time Paid 8; Part-Time Volunteers 50; Interns 2.
Governing Authority: private; nonprofit organization. Branch Museum: Pry

House Field Hospital Museum, 18906 Shepherds Town Pike, Antietam National Battlefield, Sharpsburg, MD 21782. Tax-exempt: 501(c)(3).
Institution Type/Description: History Museum: housed in Carty Building, former 1832 furniture store & funeral home.
Collections: over 3,000 Union & Confederate Civil War artifacts & documents related to medical history of the Civil War including surgical instruments, stretchers, uniforms & insignia, books, documents, manuscripts and personal effects from doctors, nurses & patients.
Research Fields: Civil War medicine.
Facilities: 500-vol. library on medical history of Civil War. Museum-related items for sale.
Activities: special events; living history; guided & educational tours; outreach programs. Annual Events: Civil War Medicine Conference in October; Jonathan Letterman Award in October.
Publications: quarterly journal.
Hours & Admission Prices: Mon.-Sat. 10-5. Adults $7.50, seniors 60 & over $7, children 10-16 $5.50; discounts to AAA members; members & children under 10 no charge. &
Attendance: 35,000 (accurate)
Membership: Medical Cadet (Student) $35; First Aid Station (Individual) $50; Ambulance (Family of 4) $75; Field Hospital $125; Brigade Hospital $250; General Hospital $500; Surgeon General $1,000; Corporate $5,000.

SCHIFFERSTADT ARCHITECTURAL MUSEUM, (M), 1110 Rosemont Ave., Frederick, MD 21701-4127. Tel.: 301-668-6088.
E-mail: fredcolandmarks@aol.com
Web Site: frederickcountylandmarksfoundation.org
Founded: 1974.
Congressional District: 6
Key Personnel: Pres., Alan Imhoff; Vice Pres., Joe Sweeney.
Personnel Profile: Part-Time Paid 1; Part-Time Volunteers 12; Interns 1.
Governing Authority: private; nonprofit organization. Parent Institution: Frederick County Landmarks Foundation. Tax-exempt: 501(c)(3).
Institution Type/Description: Architecture Museum: housed in c.1758 German sandstone farmhouse; wattle & daub construction; hand-hewn oak beams pinned with wooden pegs.
Collections: farming tools & household implements; original 5-plate jamb stove, wrought iron hardware; archaeology; vaulted cellar; squirrel tail bake oven; stone dry sink; wishbone (divided) central chimney; hearth; summer kitchen.
Facilities: library.
Activities: guided tours; demonstrations; public events; temporary exhibitions.
Publications: quarterly newsletter.
Hours & Admission Prices: April to mid-Dec. Sat.-Sun. 1-4; other times by appointment. Adults $5; discounts to members, AAA, seniors & students; children under 12 no charge.
Attendance: 20,000 (estimated)
Membership: Student & Senior $20; Individual $25; Family $35; Corporate $250; Life $500.

Freeland

MORRIS MEADOWS HISTORICAL MUSEUM, 1523 Freeland Rd., Freeland, MD 21053. Tel.: 800-643-7056.
Institution Type/Description: History Museum.
Collections: local history & culture; period furnishings; personal artifacts; photographs.
Facilities: rental facilities.
Activities: rental facilities; special events.
Hours & Admission Prices: By appointment. &

Frostburg

FROSTBURG MUSEUM, 69 Hill St., Frostburg, MD 21532. Mailing Address: P.O. Box 92, Frostburg, MD 21532-0092. Tel.: 301-689-1195. Facebook: Frostburg Museum.
E-mail: frostburgmuseum@verizon.net
Web Site: frostburgmuseum.org
Founded: 1976.
Congressional District: 6
Key Personnel: Cur., Elizabeth Eshleman.
Personnel Profile: Part-Time Volunteers 25; Interns 2.
Governing Authority: Tax-exempt.
Institution Type/Description: Historic Building: housed in the former Hill Street School; built in 1899.
Collections: local history; personal artifacts; photographs.
Activities: monthly artist exhibit with reception.
Hours & Admission Prices: Tues.-Sat. 12-5. No charge; donations accepted.
Attendance: 1,500 (estimated)
Membership: Senior & Student $10; Individual $20; Family $35; Sponsor $50;

Patron $100; Benefactor $250; Life $1,000. Business Level I $100; Business Level 2 $250.

FROSTBURG SCIENCE DISCOVERY CENTER, Compton Science Center, 1st Fl., 101 Braddock Rd., Frostburg, MD 21532. Tel.: 301-687-4723.
E-mail: FSDC@frostburg.edu
Web Site: www.frostburg.edu/FSDC
Formerly: Frostburg State University Planetarium
Institution Type/Description: Science Center.
Activities: educational programs.
Hours & Admission Prices: Call for hours. No charge; donations accepted. &

THRASHER CARRIAGE MUSEUM, 19 Depot St., Frostburg, MD 21532-1309. Mailing Address: c/o Allegany County Historical Society, 218 Washington St., Cumberland, MD 21502. Tel.: 301-689-3380. Fax: 301-689-3380.
E-mail: info@thethrashercarriagemuseum.com
Web Site: www.thethrashercarriagemuseum.com
Founded: 1992.
Congressional District: 1
Key Personnel: Exec. Dir., Jeff Nealis.
Personnel Profile: Full-Time Paid 1; Part-Time Paid 4; Part-Time Volunteers 12; Interns 2.
Governing Authority: county; nonprofit. Parent Institution: Allegany County Commissioners. Subsidiary Institution: Allegany County Historical Society. Branch Museum: Queen City Transportation Museum, 210 S. Centre St., Cumberland, MD 21502. Tax-exempt: 501(c)(3).
Institution Type/Description: Transportation Museum: housed in a renovated 1800s era warehouse.
Collections: 100 horse-drawn conveyances; phaetons; landaus; wagons; carts; surreys; sleighs; delivery wagons; hearse; buckboards; saddles; harnesses; bridles; carriage tools; early automobiles.
Research Fields: museum collection; socio-economic factors of the carriage era specific to Allegany County, Maryland.
Facilities: 7,000 sq. ft. exhibit space. Museum-related items for sale.
Activities: docent program; guided tours; lectures; temporary exhibitions.
Publications: Horse Drawn Vehicles: Carriages and the Community.
Hours & Admission Prices: Jan.-April by appointment; May 2-Oct. Thurs.-Sun. 12-2; Nov. to mid-Dec. Sat.-Sun. 12-2. Adults $4, students $2; discount to groups; children under 6 no charge. Closed New Year's Day; Martin Luther King Jr. Day; Presidents' Day; Thanksgiving; Christmas. &
Attendance: 18,000 (accurate)
Membership: Students $5; Individuals $10; Family $25; Patron $100; Lifetime $2,000; Corporate $10,000.

Fulton

AFRICAN ART MUSEUM OF MARYLAND, (M), 11711 E. Market Pl., Fulton, MD 20759-2596. Tel.: 301-490-6070.
E-mail: africanartmuseumofmd@verizon.net
Web Site: www.africanartmuseum.org
Formerly: Maryland Museum of African Art
Founded: 1980.
Congressional District: 3
Key Personnel: Founder & Dir., Doris H. Ligon; Chm. Bd. (V), Jean W. Toomer.
Personnel Profile: Full-Time Volunteers 1; Part-Time Volunteers 35.
Governing Authority: nonprofit organization. Subsidiary Institution: Baltimore/Washington Jazz Fest. Tax-exempt: 501(c)(3).
Institution Type/Description: Art Museum.
Collections: African art, including sculpture, textiles, masks, jewelry; musical instruments; household items.
Research Fields: traditional African art.
Activities: guided tours; lectures; films; formally organized education programs for children; school outreach programs; temporary & traveling exhibitions; special events.
Publications: newsletter, Museum Memos; program booklets for major exhibitions; quarterly jazz journal, The Quartet.
Hours & Admission Prices: Wed.-Sat 10-3; other times by appointment for groups. No charge, but donations accepted. &
Attendance: 45,000 (estimated)
Membership: Students & Seniors $20; Individual $25; Senior Family $35; Family $40; Community Organization & Nonprofit $175; Century Circle Life $1,000.

Gaithersburg

GAITHERSBURG COMMUNITY MUSEUM, 9 S. Summit Ave., Gaithersburg, MD 20877. Mailing Address: 506 S. Frederick Ave., Gaithersburg, MD 20877-2325. Tel.: 301-258-6160. Fax: 301-258-6329.
E-mail: museum@gaithersburgmd.gov
Web Site: www.gaithersburgmd.gov/museum
Founded: 1988.
Personnel Profile: Full-Time Paid 1; Full-Time Volunteers 7; Part-Time Paid 3; Part-Time Volunteers 20.
Governing Authority: city of Gaithersburg. Gaithersburg Historic Association.
Institution Type/Description: History Museum: housed in the former B&O train station freight house; built in 1884. Listed on the National Register of Historic Places.
Collections: local history & culture; period furnishings; personal artifacts; photographs; hands-on exhibitions; rolling stock train yard; rolling stock engine, caboose, RDC (Budd Car) passenger.
Facilities: Museum-related items for sale.
Activities: educational programs; hands-on exhibits; permanent & temporary exhibitions; genealogy research.
Hours & Admission Prices: Thurs.-Sat. 10-2; other times by appointment. No charge; donations accepted.
Attendance: 4,100 (accurate)

GAITHERSBURG - WASHINGTON GROVE V.F.D. FIRE MUSEUM, 13 E. Diamond Ave., Gaithersburg, MD 20877. Mailing Address: 801 Russell Ave, Gaithersburg, MD 20879. Tel.: 301-646-1222.
E-mail: firemansfund@gwgvfd.org
Web Site: www.gwgvfd.org
Founded: 1927.
Institution Type/Description: Fire Museum.
Collections: local history; Diamond-T 1941 pumper; helmets; uniforms; badges; department logs; radios; photographs; personal artifacts.
Activities: fire prevention programs; hands-on exhibits.
Hours & Admission Prices: Sat. 10-2. No charge; donations accepted. &

Galesville

CARRIE WEEDON SCIENCE CENTER, 911 Galesville Rd., Galesville, MD 20765-3101. Tel.: 410-222-1625. Fax: 410-867-0588.
E-mail: fieldtrip@carrieweedon.org
Web Site: www.carrieweedon.org
Founded: 1988.
Congressional District: 4
Key Personnel: Dir. & Lead Teacher, Sharon Hartge Solbert; Pres., Dorothy Chaney.
Personnel Profile: Full-Time Volunteers 1.
Governing Authority: college; nonprofit organization. Parent Institution: Anne Arundel County Public Schools. Tax-exempt.
Institution Type/Description: Natural History & Science Museum.
Collections: history of Anne Arundel County; freshwater aquaria; ornithology; horticultural gardens; insect collection; flora & fauna of the Chesapeake Bay.
Research Fields: estuarine biology.
Facilities: botanical garden; aquarium; classrooms.
Activities: formally organized education programs for children & undergraduate college students; permanent & traveling exhibitions; school loan service.
Hours & Admission Prices: Mon.-Fri. 8-4. Donations accepted. &
Attendance: 5,000 (estimated)
Membership: Family $25; Silver $50; Gold $100; Diamond $500; Lifetime $1,000; Benefactor $10,000 & up.

GALESVILLE HERITAGE MUSEUM - CARRIE WEEDON HOUSE, 988 Main St., Galesville, MD 20765-0373. Mailing Address: P.O. Box 373, Galesville, MD 20765-0373. Tel.: 410-867-9199.
E-mail: info@galesvilleheritagesociety.org
Web Site: www.galesvilleheritagesociety.com
Governing Authority: Parent Institution: Galesville Heritage Society, Inc.
Institution Type/Description: History Museum.
Collections: local history & culture; period furnishings; personal artifacts; photographs; books; video.
Facilities: library.
Hours & Admission Prices: April-Nov. Sun. 1-4; other times by appointment. No charge, donation accepted.

Germantown

BLACKROCK CENTER FOR THE ARTS, 12901 Town Commons Dr., Germantown, MD 20874. Tel.: 301-528-2260. Fax: 301-528-2266.
E-mail: info@blackrockcenter.org
Web Site: blackrockcenter.org
Institution Type/Description: Art Gallery.
Collections: paintings; sculpture; photographs.
Hours & Admission Prices: Mon.-Fri. 10-5; call for additional hours.

BUTTON FARM LIVING HISTORY CENTER, 16820 Black Rock Rd., Germantown, MD 20874. Mailing Address: P.O. Box 1366, Olney, MD 20830. Tel.: 301-916-7090.
Institution Type/Description: History Center.
Collections: 19th-century slave plantation life; Underground Railroad history; period furnishings; photographs.
Activities: educational programs; special events; school group tours. Annual Events: Earth Day; May Plant Sale; Juneteenth-Heritage Days Weekend; Montgomery County Farm Tour; Halloween Festival; Make A Difference Day; Fall Festival-Emancipation Weekend.
Hours & Admission Prices: Call for hours.

WATERS HOUSE HISTORY CENTER, 12535 Milestone Manor Ln., Germantown, MD 20876. Tel.: 301-515-2887.
Institution Type/Description: Historic House Museum: built c.1790.
Collections: Waters family history; county history & culture; period furnishings; photographs. Historic Buildings: bank barn; corn crib; carriage house.
Facilities: research library.
Activities: research.
Hours & Admission Prices: Wed. & Sat. 10-4; other times by appointment. No charge.

Girdletree

GIRDLETREE BARNES BANK, Snow Hill Rd., Girdletree, MD 21829.
Institution Type/Description: Historic Building: built in 1901.
Collections: local history & culture; period furnishings; walk-in vault; personal artifacts.
Hours & Admission Prices: May-Sept. Wed. & Sat. 1-4; groups by appointment.

Glen Echo

✻ **CLARA BARTON NATIONAL HISTORIC SITE, (M),** 5801 Oxford Rd., Glen Echo, MD 20812-1201. Tel.: 301-320-1410. Fax: 301-320-1415.
E-mail: gwmp_clara_barton_nhs@nps.gov
Web Site: www.nps.gov/clba
Founded: 1975.
Congressional District: 8
Key Personnel: Museum Cur., Kimberly Robinson.
Personnel Profile: Full-Time Paid 6; Part-Time Volunteers 12.
Governing Authority: federal. Parent Institution: U.S. Department of the Interior, National Park Service, George Washington Memorial Pkwy. Tax-exempt: 500.
Institution Type/Description: Historic House Museum: 1897-1912 home of Clara Barton, founder of the American Red Cross.
Collections: Victorian furnishings; Clara Barton artifacts; archives.
Major Exhibits: Clara Barton and the Civil War, 4/11-4/15.
Research Fields: Clara Barton; American Red Cross; Victorian America 1891-1912; women's history.
Facilities: 500-vol. research library; transcripts of Barton diaries & manuscripts available for use on premises. Books & gift items for sale.
Activities: Volunteer-in-Park program; guided tours of site; Park-as-Classrooms education program; interpretive special events; online Jr. Ranger program; online virtual tour.
Publications: park brochure.
Hours & Admission Prices: Daily 10-5. Tours: 10-4 on the hour; groups of 10 or more by appointment. No charge; donations accepted. Closed New Year's Day; Thanksgiving; Christmas.
Attendance: 17,000 (accurate)

Glenn Dale

DORSEY CHAPEL, 10704 Brookland Rd., Glenn Dale, MD 20769. Tel.: 240-264-3415. TTY: 301-699-2544.
Institution Type/Description: Historic Building: housed in a former African-American church; built in 1900.

Collections: chapel history; religious furnishings & artifacts.
Facilities: rental facilities.
Activities: guided tours; rental facilities.
Hours & Admission Prices: By appointment. Adults $3, seniors $2, children 5-18 $1; children 4 & under no charge.

MARIETTA HOUSE MUSEUM, 5626 Bell Station Rd., Glenn Dale, MD 20769-9120. Tel.: 301-464-5291. Fax: 301-464-5654. TDD: 301-699-2544.
E-mail: mary.amen@pgparks.com
Web Site: www.pgparks.com
Founded: 1978.
Congressional District: 5
Key Personnel: Natural & Historical Div. Chief, Gregory Kernan; Facility Mgr., Mary Amen; Chm. (V), Stacey Hawkins.
Personnel Profile: Full-Time Paid 3; Part-Time Paid 4; Part-Time Volunteers 15.
Governing Authority: county. Parent Institution: Maryland National Capital Park & Planning Commission, Natural & Historical Resources Div., Greenlanding Rd., Upper Marlboro, MD. Tax-exempt: 501(c)(3).
Institution Type/Description: Historic House: c.1815 Federal style brick home.
Collections: 19th-century decorative arts & furnishings; outbuildings; smokehouse; law office; root cellar.
Research Fields: county history; U.S. Supreme Court Associate Justice Gabriel Duvall; 19th-century history.
Facilities: educational facilities.
Activities: docent program; guided tours; lectures; living history encampment. Annual Events: Marching Through Time, A Multiperiod (from 1st century to 20th century) in April.
Hours & Admission Prices: Open by appointment only, please call ahead. Adults $3, senior citizens $2, students $1; discounts to groups; Metropolitan Washington Historic Houses Consortium & Prince George County History Consortium no charge. Closed major holidays. &
Attendance: 5,000 (accurate)

Grantsville

NEW GERMANY STATE PARK NATURE CENTER, 349 Headquarters Lane, Grantsville, MD 21536. Tel.: 301-895-5453.
Institution Type/Description: Nature Center.
Collections: wildlife & their habitats; environment; natural history; ecology; natural resources management.
Activities: educational programs; special events; guest speakers.
Hours & Admission Prices: Call for hours.

SPRUCE FOREST ARTISAN VILLAGE, 177 Casselman Rd., Grantsville, MD 21536. Tel.: 301-895-3332.
E-mail: artisans@spruceforest.org
Web Site: www.spruceforest.org
Key Personnel: Acting Dir., Lynn Lois; Chm. (V), Reita Marks.
Personnel Profile: Part-Time Paid 1; Part-Time Volunteers 12.
Volunteer Hours: 800
Operating Expenses: 33,603
Operating Income: 48,691
Governing Authority: Tax-exempt.
Institution Type/Description: History & Art Museum.
Collections: paintings; pottery; sculpture. Historic Buildings: log cabins; 1800s grist mill.
Activities: special events; workshops; group tours.
Publications: SFAV Newsletter.
Hours & Admission Prices: May-Dec. call for hours. No charge; donations accepted. &
Attendance: 15,000 (estimated)

Grasonville

CHESAPEAKE BAY ENVIRONMENTAL CENTER, 600 Discovery Lane, Grasonville, MD 21638. Tel.: 410-827-6694.
Key Personnel: Exec. Dir., Judy Wink.
Governing Authority: nonprofit organization.
Institution Type/Description: Environmental Education Center.
Collections: native woodlands; tidal marshes; meadows; wildlife & their habitats.
Facilities: 510 acre wildlife preserve.
Activities: school groups; educational & outreach programs.
Hours & Admission Prices: By appointment.

Greenbelt

GREENBELT MUSEUM, (M), 15 Crescent Rd., Rm. 110, Greenbelt, MD 20770-0805. Mailing Address: P.O. Box 1025, Greenbelt, MD 20768. Tel.: 301-474-1936 & 507-6582. Fax: 301-441-8248.
E-mail: museum@greenbeltmd.gov
Web Site: www.greenbeltmuseum.org
Founded: 1987.
Congressional District: 5
Key Personnel: Dir., Megan Searing Young.
Personnel Profile: Full-Time Paid 1; Part-Time Paid 1; Part-Time Volunteers 50; Interns 1.
Governing Authority: nonprofit organization. Parent Institution: City of Greenbelt. Subsidiary Institution: Friends of the Greenbelt Museum. Tax-exempt.
Institution Type/Description: History Museum.
Collections: 1930s & 1940s furnishings & furniture; toys; glass & china; kitchen equipment; textiles; drawings & print materials; Lenore Thomas' replica sculpture of Mother & Child statuette; Greenbelt history book; depression era, era of the New Deal; material culture of the 1930s & planned communities; World War II homefront experience.
Research Fields: decorative arts of the 1930s & 1940s; Great Depression; social history of the 1930s, education, recreation, political affairs; domestic technology, women & domestic work.
Facilities: Video-related, note cards, post cards & museum-related items for sale.
Activities: guided tours; lectures; docent program; temporary exhibitions; special exhibits in community center rotating every 18 months.
Publications: quarterly newsletter for members; Greenbelt: History of a New Town 1937-1997.
Hours & Admission Prices: Sun. 1-5; other times by appointment. Community Center Exhibit Room: Mon.-Sat. 9-9, Sun. 9-7. Tours: adults $3, senior & children $2; members & children under 12 no charge. &
Attendance: 2,500 (estimated)
Membership: Individual $25; Family $50; Roosevelt Club $500 & up.

NASA GODDARD SPACE FLIGHT CENTER, 8800 Greenbelt Rd., Greenbelt, MD 20771-2400. Mailing Address: Goddard Visitors Center, 8800 Greenbelt Road - Code 130, Greenbelt, MD 20771. Tel.: 301-286-2000 & 3978.
E-mail: gsfc-pao@listserv.gsfc.nasa.gov
Web Site: www.gsfc.nasa.gov/vc/index.html
Personnel Profile: Full-Time Paid 5; Part-Time Paid 4; Part-Time Volunteers 3.
Institution Type/Description: Space Science Museum.
Collections: space science; space-flight artifacts & photographs; hands-on exhibits; model-rocket launchings.
Facilities: Museum-related items for sale.
Activities: educational programs; workshops; public lectures. Museum Sponsors: Family Science Night.
Hours & Admission Prices: Flight Center: tours available to school, community & cultural groups by appointment. Visitor Center: by appointment. No charge.
Attendance: 39,000 (accurate)

Hagerstown

BEAVER CREEK SCHOOL MUSEUM, 9702 Beaver Creek Church Rd., Hagerstown, MD 21740. Mailing Address: c/o Washington County Historical Society, 135 W. Washington St., Hagerstown, MD 21740-4709. Tel.: 301-797-8782. Fax: 240-625-9498.
E-mail: info@washcomdhistoricalsociety.org
Web Site: www.washcomdhistoricalsociety.org
Founded: 1911.
Congressional District: 6
Key Personnel: Exec. Dir., Linda C. Irvin-Craig.
Personnel Profile: Full-Time Paid 1.
Governing Authority: Parent Institution: Washington County Historical Society. Tax-exempt.
Institution Type/Description: Historic Building: housed in a former two-room school; built in 1904.
Collections: period furnishings; books; slates; tools; hats; musical instruments; glassware; clothing.
Activities: Harvest Festival; Healing Waters for Veterans
Hours & Admission Prices: By appointment.

CHESAPEAKE AND OHIO CANAL NATIONAL HISTORICAL PARK, 1850 Dual Hwy., Ste. 100, Hagerstown, MD 21740-6622. Tel.: 301-739-4200. Fax: 301-739-6179.
Web Site: www.nps.gov/choh
Formerly: Chesapeake & Ohio Canal Tavern Museum

Founded: 1971.
Congressional District: 6 & 8
Key Personnel: Supt., Kevin Brandt; Acting Chief Interpretation, John Noel.
Personnel Profile: Full-Time Paid 135; Part-Time Volunteers 3,500; Interns 16.
Governing Authority: federal. Parent Institution: National Park Service, U. S. Department of the Interior. Branch Facilities: Georgetown Visitor Center, 1057 Thomas Jefferson St. N.W., Washington, DC 20007, Tel. 202-653-5190. Great Falls Tavern, 11710 MacArthur Blvd., Potomac, MD 20854, Tel. 301-767-3714. Brunswick Visitor Center, 40 W. Potomac St., Brunswick, MD 21716, Tel. 301-834-7100. Williamsport Visitor Center, 205 W. Potomac St., Williamsport, MD 21795, Tel. 301-582-0813. Hancock Visitor Center, 439 E. Main St., Hancock, MD 21750, Tel. 301-745-5877. Cumberland Visitor Center, Western Maryland Station, 13 Canal St., Cumberland, MD 21502, Tel. 301-722-8226.
Institution Type/Description: Historic Canal between Georgetown, District of Columbia to Cumberland, Maryland.
Collections: artifacts relating to construction and operation of canal, 1828-1924, including navigation, commerce and domestic items; archaeological specimens; historic images. Historic Structures: over 1,300 including Great Falls Tavern, 1828-1830, Monocacy Aqueduct, 1829-1833 and Paw Paw Tunnel, 1836-1850.
Research Fields: archaeology; Potomac, Chesapeake and Ohio Canal (1785-1924); transportation & regional history (Washington D.C., Central & Western Maryland).
Facilities: visitor centers located in Georgetown, Great Falls, Brunswick, Williamsport, Hancock, Cumberland & Sharpsburg; hiking & biking towpath; camping facilities.
Activities: interpretive walks and programs; living history demonstrations; canal boat rides; audiovisual programs.
Publications: Chesapeake and Ohio Canal, Official National Park Handbook.
Hours & Admission Prices: Park: daylight hours. Great Falls: cars $5, buses $25-$100. National Parks Passes honored. Visitor Center: closed major Federal holidays. &
Attendance: 3,928,697 (accurate)

CHRISTIAN HERITAGE MUSEUM, 14111 Pennsylvania Ave., Hagerstown, MD 21742. Tel.: 877-313-9002.
Institution Type/Description: Religious Museum.
Collections: English & American Bibles; religious art & artifacts.
Hours & Admission Prices: By appointment.

CONTEMPORARY SCHOOL OF THE ARTS AND GALLERY, 4 W. Franklin St., Hagerstown, MD 21740. Tel.: 301-791-6191.
Web Site: www.csagi.org
Institution Type/Description: Art Gallery.
Collections: paintings; sculpture.
Activities: after-school programs; classes.
Hours & Admission Prices: Mon.-Fri. 11-3, Sat.-Sun. 1-3.

DISCOVERY STATION AT HAGERSTOWN, 101 W. Washington St., Hagerstown, MD 21740-4709. Tel.: 301-790-0076; 877-790-0076 (Toll Free). Fax: 301-790-0045.
E-mail: info@discoverystation.com
Web Site: www.discoverystation.org
Founded: 1996.
Congressional District: 6
Key Personnel: Dir., C.E.O. & Museum Shop Mgr., B. Marie Byers; Chm. (V), Joseph Gerstner; Chm. (V), Patricia Beard; Pres. (V), Phil Kelly; Museum Shop Mgr., Ellen Gerke.
Personnel Profile: Part-Time Volunteers 42; Interns 1.
Volunteer Hours: 8,300
Operating Expenses: 292,350
Operating Income: 292,350
Governing Authority: Tax-exempt.
Institution Type/Description: History & Science Museum.
Collections: hands-on exhibits; science; technology; history; 15' Titanic & memorabilia, C&O Canal artifacts, early Muller organ; Vision; NASA; Moller.
Research Fields: space; Titanic.
Activities: educational programs; Saturday Plus Programs.
Hours & Admission Prices: Feb.-April Tues.-Sat. 10-4, Sun. 2-5; May-Jan. Tues.-Sat. 10-4. Adults $7, children 4-17 $6, seniors 55 & over and military $5; discounts to Travel Passport Program & ASTC members; members, children under 4 & Blue Star Museum members in the summer no charge. Closed New Year's Day; Easter; Mother's Day; Father's Day; Independence Day; Thanksgiving; Christmas Eve, Day & day after. &
Attendance: 12,000 (accurate)
Membership: Seniors & Military $5; Youth 4-17 $6; Adults $7; Family of 4 $80.

HAGERSTOWN AVIATION MUSEUM, 14235 Oak Springs Rd., Hagerstown, MD 21742. Tel.: 301-733-8717.
E-mail: info@hagerstownaviationmuseum.org
Web Site: www.hagerstownaviationmuseum.org
Founded: 2005.
Key Personnel: Pres. (V), John P. Seburn.
Personnel Profile: Part-Time Volunteers 20.
Governing Authority: Tax-exempt.
Institution Type/Description: Aviation Museum.
Collections: aviation history & heritage; photographs; models; engines; equipment; electronics; 19 aircraft.
Publications: annual magazine, The New Pegasus.
Hours & Admission Prices: Call for hours. No charge; donations accepted.
Membership: Individual $30; Contributor $50; Supporter $100; Patron $200; Leader $500; Benefactor $1,000 & up.

HAGERSTOWN ROUNDHOUSE MUSEUM, 300 S. Burhans Blvd., Hagerstown, MD 21740-5339. Mailing Address: P.O. Box 2858, Hagerstown, MD 21741-2858. Tel.: 301-739-4665. Fax: 301-739-5598.
E-mail: hrm@roundhouse.org
Web Site: www.roundhouse.org
Founded: 1990.
Key Personnel: Chm. (V), Blaine Snyder; Pres. (V), Rick Eyler; Museum Shop Mgr., Crystal Sprecher.
Personnel Profile: Part-Time Volunteers 150.
Governing Authority: Tax-exempt: 501(c)(3).
Institution Type/Description: Railroading History.
Collections: local railroad history & culture; photographs; model trains; period railway equipment.
Activities: Annual Event: Railroad Heritage Days 3rd weekend in May; Trains of Christmas, Dec.-Feb.
Publications: quarterly newsletter.
Hours & Admission Prices: Fri.-Sun. 1-5. Adults $5; discounts to groups of 20 or more; members no charge. Closed Easter; Memorial Day. &
Attendance: 10,000 (accurate)
Membership: Individual $25; Family $30.

JONATHAN HAGER HOUSE & MUSEUM, 351 N. Cleveland Ave., Hagerstown, MD 21740-4155. Tel.: 301-739-8393.
E-mail: hagerhouse@hagerstownmd.org
Web Site: www.hagerhouse.org
Founded: 1962.
Congressional District: 6
Key Personnel: Cur., John Bryan.
Personnel Profile: Full-Time Paid 1; Part-Time Paid 2; Part-Time Volunteers 2.
Governing Authority: municipal. Parent Institution: City of Hagerstown. Tax-exempt.
Institution Type/Description: Historic House: housed in 1739 Hager House constructed as a private frontier fort.
Collections: 18th-century furnishings; pottery; glass; china; 18th-century coins.
Research Fields: life of Jonathan Hager, founder of Hagerstown Maryland; private frontier forts in the mid-Atlantic.
Facilities: medical herbal gardens. Museum-related items for sale.
Activities: guided tours; formally organized education programs for children; craft demonstrations; concerts; Christmas celebration; permanent exhibitions.
Publications: booklet, Jonathan Hager, Founder; book, What God Does Is Well Done - The Jonathan Hager Files.
Hours & Admission Prices: April-Dec. Thurs.-Sat. 10-4; other times by appointment. Adults $3, senior citizens & groups of 20 or more $2, children 6-12 $1; children under 6 no charge. Closed New Year's Day, Easter, Thanksgiving week; Christmas.
Attendance: 13,000 (estimated)

MANSION HOUSE ART CENTER, 501 Highland Way, Hagerstown, MD 21740. Tel.: 301-797-6813.
Institution Type/Description: Art Gallery.
Collections: paintings; etchings; silk screen; lithographs; carvings;
Publications: monthly newsletter.
Hours & Admission Prices: Fri.-Sat. 11-4, Sun. 1-5. No charge; donations accepted.
Attendance: 1,500 (estimated)
Membership: Individual $25.

THE TRAIN ROOM AND MUSEUM, 360 S. Burhans Blvd., Hagerstown, MD 21740-5339. Tel.: 301-745-6681. Fax: 301-766-4697.
E-mail: trainroom@verizon.net
Web Site: www.the-train-room.com
Founded: 1988.
Key Personnel: Museum Shop Mgr., Charles Lee Mozingo
Institution Type/Description: Lionel Train & Toy Museum.
Collections: period Lionel & Marx trains; model railroading; train artifacts; toys; ives trains.
Facilities: Toy trains & train layout items for sale.
Hours & Admission Prices: Sun. 12pm-5pm, Mon. & Fri. 9-6, Tues., Thurs. & Sat. 9-5, Wed. 9-4.
Attendance: 5,000 (estimated)

WASHINGTON COUNTY ARTS COUNCIL, 34 S. Potomac St., Hagerstown, MD 21740-5513. Tel.: 301-791-3132.
E-mail: maryanneb@washingtoncountyarts.com
Web Site: www.washingtoncountyarts.com
Founded: 1968.
Congressional District: 6
Key Personnel: Dir., Mary Anne Burke; Pres., James G. Pierne; Gallery Shop Mgr., Ashley Durboraw.
Personnel Profile: Part-Time Paid 2; Part-Time Volunteers 6.
Institution Type/Description: Art Gallery.
Collections: works by regional artists; ceramics; metal; glass.
Facilities: Gallery-related items for sale.
Activities: temporary exhibitions.
Hours & Admission Prices: Tues.-Fri. 11-5, Sat. 10-4; other times by appointment. No charge; donations accepted. &
Membership: Student, Senior, Artist & Teacher $20; Individual $35; Family $50; Business $200; Sponsor $250; Patron $500.

WASHINGTON COUNTY HISTORICAL SOCIETY AND MILLER HOUSE MUSEUM, (M), 135 W. Washington St., Hagerstown, MD 21740-4709. Tel.: 301-797-8782. Fax: 240-625-9498.
E-mail: info@washcomdhistoricalsociety.org
Web Site: www.washcomdhistoricalsociety.org
Founded: 1911.
Congressional District: 6
Key Personnel: Exec. Dir., Linda C. Irvin-Craig; Pres. (V), Robert Savitt; Treas., William Maharay; Information Specialist & Registrar, Catherine Landsman.
Personnel Profile: Full-Time Paid 2; Part-Time Volunteers 21; Interns 1.
Governing Authority: private; nonprofit organization. Parent Institution: Washington County Historical Society. Branch Museums: Beaver Creek School Museum. Tax-exempt: 501(c)(3).
Institution Type/Description: Antique Museum: housed in Miller House, an 1825 Federal style townhouse.
Collections: clocks; Bell pottery; dolls; period furniture; costumes & textiles; toys; period automobiles; Civil War artifacts; silver; carriage house; garden. Historic Building: Beaver Creek School c.1904 two-room schoolhouse.
Research Fields: genealogy.
Facilities: 2,150-vol. library of material on genealogy, church records, local history, documents, deeds & photographs available for research on premises; 400 sq. ft. exhibit space.
Activities: lectures; guided tours; temporary exhibitions. Annual Events: Museum Ramble in April; Girls' Tea in May; Blues Fest Garden Party in June; Holiday Reception & Tour of Historic Houses of Worship in December.
Publications: quarterly newsletter, The Legacy.
Hours & Admission Prices: Library: Tues.-Fri. 9-4, Sat. by appointment. Beaver Creek: by appointment. No charge, donations accepted. Miller House: April-Dec. Wed.-Fri. 1-4, Sat. by appointment. Adults $5, senior citizens $3; members & children under 16 no charge. Closed major holidays.
Attendance: 2,500 (estimated)
Membership: Student & Senior $15; Single $20-$34; Family $35-$99; Supporters $100-$499; Patron $500 & up.

* **WASHINGTON COUNTY MUSEUM OF FINE ARTS, (M),** City Park, 401 Museum Dr., Hagerstown, MD 21740-6271. Mailing Address: P.O. Box 423, Hagerstown, MD 21741. Tel.: 301-739-5727. Fax: 301-745-3741.
E-mail: info@wcmfa.org
Web Site: www.wcmfa.org
Founded: 1929.
Congressional District: 6

Key Personnel: Dir., Rebecca Massie Lane; Pres., Mary Helen Strauch; Vice Pres., Howard Kaylor; Vice Pres., John Schnebly; Sec., John League; Treas. Bd., Alfred Martin; Collections & Exhibition Mgr., Jennifer Smith; Educator, Amy Hunt; Registrar, Linda Dodson; Museum Shop Mgr., Amy Hunt; Head Security, Ed Lewis.
Personnel Profile: Full-Time Paid 8; Part-Time Paid 9; Part-Time Volunteers 150; Interns 1.
Governing Authority: private. Tax-exempt: 501(c)(3).
Institution Type/Description: Art Museum.
Collections: American art with focus on 19th-20th century; collection of old masters; 19th century European paintings; Asian & African art.
Facilities: 4,500-vol. art reference library; music gallery; classrooms. Crafts, toys, museum replicas & jewelry for sale.
Activities: guided tours; lectures; concerts; education programs for children; art classes; inter-museum loan & permanent exhibitions.
Publications: quarterly, News Bulletin; special exhibition catalogs.
Hours & Admission Prices: Tues.-Fri. 9-5, Sat. 9-4, Sun. 1-5. No charge; donations accepted. Closed New Year's Eve & Day; Good Friday; Independence Day; Thanksgiving & day after; Christmas Eve & Day. &
Attendance: 50,000 (accurate)
Membership: Individual $35; Dual Senior $55; Dual & Family $60; Friend $150; Diana League or Business & Civic Association $500-$999; Decorative Arts Guild $1,000-$2,499; Federal Portrait Patron $2,500-$4,999; Singer's Sponsor $5,000-$9,999; Hudson River Society $10,000 & up.

WILLIAM M. BRISH PLANETARIUM, 820 Commonwealth Ave., Hagerstown, MD 21740-6836. Tel.: 301-766-2898.
E-mail: kopcochr@wcps.k12.md.us
Web Site: www.wbplanetarium.weekly.com
Key Personnel: Dir., Christopher Kopco
Institution Type/Description: Planetarium.
Collections: astronomy; space science.
Facilities: 70-seat planetarium.
Activities: educational programs.
Hours & Admission Prices: Sept.-May Tues. 7pm-8pm. Adults $3, children & students $2; senior citizens no charge.

Havre de Grace

HAVRE DE GRACE DECOY MUSEUM, 215 Giles St., Havre de Grace, MD 21078-3661. Tel.: 410-939-3739. Fax: 410-939-3775. TDD: 410-939-3739 5x.
E-mail: mlelledge@verizon.net
Web Site: www.decoymuseum.com
Founded: 1983.
Congressional District: 2
Key Personnel: Pres., Patrick Vincenti; Dir., C. John Sullivan; Vice Pres., Charles Packard; Sec., James Carroll; Treas., Ralph Hockman; Special Events Coord., Margaret Jones.
Personnel Profile: Full-Time Paid 1; Part-Time Paid 5; Part-Time Volunteers 80.
Governing Authority: nonprofit organization. Tax-exempt: 501(c)(3).
Institution Type/Description: Folk Art Museum.
Collections: hunting accessories; primarily hunting decoys; firearms; watercraft; clothing; tools & equipment associated with decoy carving; photographs; a decoy maker's shop equipment.
Facilities: carving classroom; research library; meeting room. Museum-related items for sale.
Activities: guided tours; public education programs; carving instruction. Annual Event: Decoy Festival.
Publications: quarterly news magazine, The Canvasback.
Hours & Admission Prices: Mon.-Sat. 10:30-4:30, Sun. 12-4. Adults $6, senior citizens $5, children 9-18 $2; discounts to groups & AAA members; members & children under 8 no charge. Closed New Year's Day; Easter; Thanksgiving; Christmas. &
Attendance: 10,182 (accurate)
Membership: Student $15; Individual $35; Family $50; Life $2,000.

HAVRE DE GRACE MARITIME MUSEUM, (M), 100 Lafayette St., Havre de Grace, MD 21078-3542. Tel.: 410-939-4800.
E-mail: museum@comcast.net
Web Site: www.hdgmaritimemuseum.org
Founded: 1988.
Congressional District: 34
Key Personnel: Dir. Operations, Elizabeth Ricci.
Personnel Profile: Full-Time Paid 1; Part-Time Paid 2; Part-Time Volunteers 30.
Governing Authority: bd. of directors. Tax-exempt.
Institution Type/Description: History Museum.

Collections: local maritime heritage; regional history; Native American artifacts; lifeways from the 1600s to present day life; maritime commerce.
Research Fields: regional history & archaeology.
Facilities: library; archives; environmental center; boat shop; classroom. Museum-related items for sale.
Activities: lectures; workshops; classes; guided tours; boat building; presentation series; special events; ecology; concerts; kid's corner storytime in summer.
Publications: quarterly newsletter.
Hours & Admission Prices: Wed.-Sat. 10-5, Sun. 1-5. No charge; donations accepted. Closed major holidays. &

Attendance: 11,000 (estimated)
Membership: Student $15; Individual $25; Family $50; Friend $1,000; Corporate $1,500.

STEPPINGSTONE FARM MUSEUM, (M), 461 Quaker Bottom Rd., Havre de Grace, MD 21078-1329. Tel.: 410-939-2299 & 2321.
E-mail: director@steppingstonemuseum.org
Web Site: www.steppingstonemuseum.org
Founded: 1970.
Congressional District: 35
Key Personnel: Exec. Dir. & Museum Shop Mgr., Angela Yau; Pres. (V), John Stanley.
Personnel Profile: Full-Time Paid 1; Part-Time Volunteers 200.
Governing Authority: private; nonprofit corporation. Tax-exempt: 501(c)(3).
Institution Type/Description: Arts & Crafts Museum: housed in late 18th-century two story stone dwelling.
Collections: 1880 to 1920 agricultural tools; 1880 to 1920 rural arts & crafts. Historic Buildings: 1771 Land of Promise historic house; c.1880 blacksmith shop.
Research Fields: agricultural history; rural arts & crafts; local history.
Facilities: outdoors rental facilities. Rural arts, crafts & other museum related items for sale.
Activities: guided tours; annual special programs; formally organized education programs for children; permanent & temporary exhibitions; school loan service; craft demonstrators to local colleges & churches; living history demonstrations.
Publications: newsletter, Steppingstones; booklet, The Land of Promise; The Rural Arts and Crafts of Harford County and Steppingstone Museum.
Hours & Admission Prices: May-Sept. Sat.-Sun. 1-4; other times by appointment. Adults $3; members & children under 12 no charge. Fees for special events vary. &
Attendance: 15,000 (estimated)
Membership: Single $15; Family $30; Sustaining $50; Corporate $150; Life $500.

SUSQUEHANNA MUSEUM OF HAVRE DE GRACE INC. AT THE LOCK HOUSE, 817 Conesteo St., Havre de Grace, MD 21078. Mailing Address: P.O. Box 253, Harve de Grace, MD 21078-0253. Tel.: 410-939-5780.
E-mail: lockhousemuseum@gmail.com
Web Site: www.thelockhousemuseum.org
Founded: 1970.
Congressional District: 34
Key Personnel: Exec. Dir., Ciera Fishes; Pres., Charlie Vasilakis; Vice Pres., Jim Sherring.
Personnel Profile: Part-Time Paid 1; Part-Time Volunteers 60.
Governing Authority: private; nonprofit organization. Tax-exempt: 501(c)(3).
Institution Type/Description: History Museum.
Collections: artifacts from mid-19th century to recent local history; Pivot Bridge over lock; Outlet Lock.
Facilities: 200-vol. library; educational facilities; 1,000 sq. ft. exhibit space. Gift items for sale.
Activities: concerts; docent program; formal education programs for children; guided tours; lectures; temporary exhibitions; training programs for volunteer professional museum workers. Annual Events: Pirate Gala Fundraiser; Candlelight Tour of historic homes in December; War of 1812 reenactment.
Publications: quarterly newsletter.
Hours & Admission Prices: mid-April to Oct. Fri.-Sun. 1-5; group tours by appointment. No charge; donations accepted. &
Attendance: 2,500 (accurate)
Membership: Student $15; Adult $30; Family $40; Sterling $75; Golden & Corporate $200.

Highland Beach

FREDERICK DOUGLASS MUSEUM AND CULTURAL CENTER, 3200 Wayman Ave., Highland Beach, MD 21403. Mailing Address: Highland Beach Historical Commission, 3202 Wayman Ave., Highland Beach, MD 21403. Tel.: 410-267-6960. Fax: 410-267-0091.
E-mail: jeanw57@aol.com
Web Site: highlandbeachmd.org
Founded: 1893.
Congressional District: 3
Key Personnel: Dir., Jean Langston.
Personnel Profile: Part-Time Volunteers 12.
Governing Authority: Parent Institution: Town of Highland Beach. Subsidiary Institution: Highland Beach Historical Commission. Tax-exempt.
Institution Type/Description: History Museum.
Collections: photographs; furniture; personal artifacts; documents newspaper & magazine articles; slides; films; videos.
Activities: temporary exhibits.
Hours & Admission Prices: By appointment. No charge; donations accepted.
Attendance: 2,500 (estimated)

Hollywood

SOTTERLEY PLANTATION, (M), 44300 Sotterley Lane, Hollywood, MD 20636. Mailing Address: Historic Sotterley, Inc., P.O. Box 67, Hollywood, MD 20636-0067. Tel.: 301-373-2280; 800-681-0850. Fax: 301-373-8474.
E-mail: officemanager@sotterley.org
Web Site: www.sotterley.org
Founded: 1961.
Congressional District: 1
Key Personnel: Exec. Dir., Nancy L. Easterling; Chm. Bd. Trustees & Pres. (V), Janice Briscoe; Education Dir., Jeanne Pirtle; Office Mgr., Kim Husick; Museum Shop Mgr., Ginger Newman-Askew.
Personnel Profile: Full-Time Paid 5; Part-Time Paid 5; Part-Time Volunteers 165.
Volunteer Hours: 14,000
Governing Authority: board of trustees. Tax-exempt.
Institution Type/Description: Historic House Museum: 1703 mansion & outbuildings illustrating Tidewater Plantation culture.
Collections: architecture, home furnishings; slave cabin; customs warehouse; smokehouse; formal gardens. Historic House: 1703 plantation house; botanical (Living).
Major Exhibits: The Enemy at Our Door! (T), 5/14.
Research Fields: architecture of buildings; archaeology of grounds; African-American experience; southern Maryland & national history from 1703-1961.
Facilities: gardens; farm; estuarine site; historic structures; trails.
Activities: guided & audio tours; colonial African-American history and environmental education for school groups; speaker series; specialty tours; summer camps. Annual Events: War of 1812 Living History Events; Gala in the Garden; Independence Celebration concert; Wine Festival; Ghost Tours; Holiday Candlelight Tours; Family Plantation Christmas; second Saturday series.
Publications: pamphlet; newsletters, Sotterley Times.
Hours & Admission Prices: Guided Plantation House Tours: May 1-Oct. 31 Tues.-Sat. 10-4, Sun. 12-4. Adults $10, seniors $8, children 6-12 $5; discount to National Trust for Historic Preservation members, military & AAA members; children 5 & under no charge. Self-guided grounds tour $3; Sotterley members with unlimited touring no charge. Please visit website for additional information. &
Attendance: 26,000 (accurate)
Membership: Individual $35; Family $65; Patron $150; Sponsor $300; Preserver $500; George Plater Society $1,200; Rousby Circle $2,500 & up.

Kennedyville

KENT COUNTY FARM MUSEUM, 13689 Turner's Creek Rd., Kennedyville, MD 21645. Mailing Address: P.O. Box 43, Kennedyville, MD 21645-0043. Tel.: 410-348-5543.
Web Site: www.kentcounty.com/farmmuseum
Institution Type/Description: Farm Museum.
Collections: farming history; agriculture machinery; period artifacts; household items.
Hours & Admission Prices: May-Oct. 1st & 3rd Sat. 10-4; other times by appointment. No charge; donations accepted.

Kensington

WASHINGTON D.C. TEMPLE VISITORS' CENTER, 9900 Stoneybrook Dr., Kensington, MD 20895. Tel.: 301-587-0144.
E-mail: vcwashington@ldschurch.org
Institution Type/Description: Religious Museum.
Collections: Christus statue; life of Christ; church beliefs & teachings; photographs.
Activities: guided tours.
Hours & Admission Prices: Daily 10-9. No charge.

Kingsville

JERUSALEM MILL VILLAGE MUSEUM AND VISITOR CENTER, 2813 Jerusalem Rd., Kingsville, MD 21087. Mailing Address: P.O. Box 237, Kingsville, MD 21087. Tel.: 410-877-3560.
E-mail: jerusalemmill@yahoo.com
Web Site: www.jerusalemmill.org
Institution Type/Description: History Museum.
Collections: local history & culture; period furnishings; early tools & equipment. Historic Buildings: 1772 grist mill; c.1770 gun shop; blacksmith shop; springhouse; c.1830 general store; 1865 Jericho covered bridge.
Activities: demonstrations; special events.
Hours & Admission Prices: Sat.-Sun. 1-4. No charge.

La Plata

AFRICAN AMERICAN HERITAGE SOCIETY OF CHARLES COUNTY, INC., 7485 Crain Hwy., La Plata, MD 20646. Mailing Address: P.O. Box 2250, La Plata, MD 20646. Tel.: 301-609-9909.
Governing Authority: nonprofit organization.
Institution Type/Description: History Museum.
Collections: African American history, culture & life; period furnishings; personal artifacts; photographs.
Activities: special events.
Hours & Admission Prices: By appointment.

LA PLATA TRAIN STATION MUSEUM, 101 Kent Ave., La Plata, MD 20646. Mailing Address: Historical Society of Charles County, Inc., P.O. Box 2806, La Plata, MD 20646. Tel.: 301-934-8836.
Founded: 1967.
Key Personnel: Pres. (V), Joyce Candland.
Personnel Profile: Part-Time Paid 1.
Institution Type/Description: History Museum: housed in the former Charles County railroad depot.
Collections: local history & culture; railroad artifacts; photographs; red caboose; period furnishings.
Publications: quarterly newsletter.
Hours & Admission Prices: Call for hours. No charge. &

LaVale

LAVALE TOLL GATE HOUSE, 14302 National Hwy., LaVale, MD 21502. Tel.: 301-777-5132.
Web Site: www.marylandnationalroad.org
Institution Type/Description: Historic Building: housed in a seven-sided Toll Gate House; built in 1811. Listed on the National Register of Historic Places.
Collections: historic building; early plaque indicating fees for wagons, animals & pedestrians.
Hours & Admission Prices: May-Oct. Sat.-Sun. 1:30-4:30; other times by appointment.

Lake Arbor

NORTHAMPTON PLANTATION AND SLAVE QUARTERS, Lake Overlook Dr., Lake Arbor, MD 20721. Tel.: 301-627-1286 & 454-1780. TDD: 301-699-2544 & 454-1472.
Institution Type/Description: History Museum.
Collections: replicas of two former slave quarters; African-American history & culture; archaeology.
Activities: tours; educational & interpretive programs.
Hours & Admission Prices: Call for hours. No charge.

Lanham

HOWARD B. OWENS SCIENCE CENTER AND CHALLENGER LEARNING CENTER, 9601 Greenbelt Rd., Lanham, MD 20706-3397. Tel.: 301-918-8750. Fax: 301-918-8753.
E-mail: howardb.owens@pgcps.org
Web Site: www1.pgcps.org/howardbowens
Founded: 1978.
Congressional District: 14
Key Personnel: Program Admin., Treesa Elam-Respass; Planetarium Dir., Patty Seaton.
Governing Authority: Prince George's County Public School District. Tax-exempt: 170(b)(1)(A).
Institution Type/Description: Science Museum & Planetarium.
Collections: interactive exhibits & instructional programs for the Prince George's County Public School System.
Research Fields: science education.
Facilities: 174-seat planetarium; classrooms; computer labs; demonstration area; learning center.
Activities: guided tours; lectures; formally organized education programs for children & adults; docent program; permanent & temporary exhibitions.
Publications: annual brochures; fliers.
Hours & Admission Prices: Center: call for hours. Planetarium Shows: Sept.-May 2nd Fri. each month 7:30pm; groups of 16 or more by appointment. Adults $5, students, teachers, senior citizens & military $3. &
Attendance: 30,000 (estimated)

SEABROOK SCHOOLHOUSE, 6116 Seabrook Rd., Lanham, MD 20706. Mailing Address: c/o Marietta House Museum, 5626 Bell Station Rd., Glenn Dale, MD 20769. Tel.: 301-464-5291. TTY: 301-446-3302.
Institution Type/Description: Historic Building Museum: housed in Seabrook's former one-room school, grades 1-7 until early 1950s.
Collections: local history & culture; period furnishings.
Activities: educational programs.
Hours & Admission Prices: By appointment.

Laurel

DINOSAUR PARK, 13201 Mid-Atlantic Blvd., Laurel, MD 20708. Tel.: 301-627-7755 & 1286. TTY: 301-699-2544.
Institution Type/Description: Park Museum.
Collections: fossils; trees & plants similar to those in prehistoric times; interpretive signs.
Activities: school group; fossil digs.
Hours & Admission Prices: 1st & 3rd Sat. 12-4.

THE LAUREL MUSEUM, (M), 817 Main St., Laurel, MD 20707-3429. Tel.: 301-725-7975. Fax: 301-725-2675.
E-mail: director@laurelhistoricalsociety.org
Web Site: www.laurelhistoricalsociety.org
Founded: 1996.
Congressional District: 5
Key Personnel: Pres. (V), Jhanna Levin; Vice Pres., Lisa Losito; Dir., Lindsey Baker; Museum Shop Mgr., Frieda Weise.
Personnel Profile: Full-Time Paid 1; Part-Time Paid 1; Part-Time Volunteers 30.
Governing Authority: private; nonprofit. Parent Institution: Laurel Historical Society. Tax-exempt: 509(c).
Institution Type/Description: History Museum: housed in an 1840 millworkers house.
Collections: concentration on Laurel history from local Native American times to present with emphasis on the social, financial, religious & cultural development of the city of Laurel and its people.
Facilities: library on local & Maryland history and genealogy; educational facilities. Books & video on Laurel history for sale.
Activities: guided tours; lectures; college internships for University graduate programs; temporary exhibitions.
Publications: bimonthly society newsletter, Laurel Light.
Hours & Admission Prices: Wed. & Fri. 10-2, Sun. 1-4; groups & other times by appointment. No charge. &
Attendance: 3,000 (estimated)
Membership: Seniors & Students $15; Individual $20; Family $25; Corporate $35; Lifetime $300.

MONTPELIER ARTS CENTER, 9652 Muirkirk Rd., Laurel, MD 20708. Tel.: 301-377-7800; 410-792-0664. Fax: 301-377-7801.
E-mail: montpelier.arts@pgparks.com
Web Site: arts.pgparks.com

Founded: 1979.
Institution Type/Description: Art Center.
Collections: works by contemporary artists; printmaking; sculpture; paintings; ceramics.
Facilities: classroom.
Activities: classes; workshops; educational programs; special events; rental facilities; guided tours.
Hours & Admission Prices: Daily 10-5. No charge. Closed major holidays.

MONTPELIER MANSION, (M), 9650 Muirkirk Rd., Laurel, MD 20708-2605. Tel.: 301-377-7817. Fax: 301-377-7818. TTY: 301-699-2544.
E-mail: mary.jurkiewicz@pgparks.com
Web Site: www.pgparks.com
Founded: 1976.
Congressional District: 5
Key Personnel: Pres. Volunteer Group, Friends of Montpelier, Helen Bailey; Dir. & Museum Shop Mgr., Mary Jurkiewicz; Chief, Natural & Historical Resources Div., Anthony Nolan.
Personnel Profile: Full-Time Paid 1; Part-Time Paid 6; Part-Time Volunteers 20.
Governing Authority: county. Parent Institution: Maryland-National Capital Park & Planning Commission, Natural & Historical Resources Div.
Institution Type/Description: Historic House: 18th-century Georgian mansion.
Collections: 18th- & early 19th-century decorative arts & furnishings; replica early 19th-century kitchen.
Research Fields: 18th-19th century gardens & landscapes; 18th century, early domestic life.
Facilities: 333-vol. library of agriculture, historic preservation & restoration, 18th- to 19th-century domestic life, local & state history. Gift items for sale.
Activities: concerts; docent program; guided tours; lectures; loan exhibitions; afternoon tea by reservation. Annual Events: Needleart Exhibit; Christmas Candlelight; Herb & Arts Festival.
Publications: quarterly newsletter, The Fireback; annual brochure, Montpelier Mansion.
Hours & Admission Prices: Self-Guided Tours: Mon.-Thurs. 11-3. Guided Tours: March-Nov. Sun. 12, 1, 2, & 3; Dec.-Feb. Sun. 1 & 2; group tours by appointment. Adults $3, senior citizens over 62 & groups of 10 or more $2, children 5-18 $1; discounts to members of Consortium of Historic House Museums of Metropolitan Washington, DC & members of Friends of Montpelier; active military & members no charge. Closed major holidays.
Attendance: 7,350 (accurate)
Membership: Annual $15; Life $200.

NATIONAL WILDLIFE VISITOR CENTER, PATUXENT RE-SEARCH REFUGE, 10901 Scarlet Tanager Loop, Laurel, MD 20708-4011. Tel.: 301-497-5763. Fax: 301-497-5765.
E-mail: patuxent@fws.gov
Web Site: www.fws.gov/northeast/patuxent
Founded: 1994.
Congressional District: 5
Key Personnel: Refuge Mgr., Brad Knudsen; Chm. Friends of Patuxent (V), Emmy Holdridge; Treas., Evelyn Adkins; Public Rels. & Education, Nell Baldacchino; Museum Shop Mgr. & Bookstore Mgr., Lisa Garrett.
Personnel Profile: Full-Time Paid 22; Part-Time Paid 2; Part-Time Volunteers 100; Interns 2.
Governing Authority: federal. Parent Institution: Department of the Interior, US Fish & Wildlife Service.
Institution Type/Description: Science & Environmental Education Center.
Collections: dioramas of gray wolves, sea otters, whooping cranes & canvas backs; endangered species; causes of decline; recovery efforts for extinct species; tools & equipment used by scientists in research studies & recovery techniques; impact of man on wildlife throughout the country.
Research Fields: environmental contaminants; endangered species recovery; wetlands & wildlife habitat management; migratory birds; amphibians.
Facilities: 520-vol. library of natural history, zoology, biographies & children's books; 230-seat auditorium; conference facilities; meeting rooms; 4,000 sq. ft. exhibit space; hiking trails. Museum-related items for sale.
Activities: films; guided tours; tram tours; participatory & traveling exhibitions; teacher workshops; environmental education programs for scouts, school classes & summer camp groups; wildlife observation. Annual Events: Wildlife Art Show & Sale in March; Refuge System Anniversary Celebration in March; Refuge Birthday Celebration in March; Earth Day; Fishing Day; National Fishing & Hunting Day; Patuxent Wildlife Festival in October.
Hours & Admission Prices: Daily 9-4:30. Tram Tours: mid-March to late June & late Aug. to mid-Nov. Sat.-Sun. 11:30, 1, 2, 3; late June to late Aug. Mon.-Fri. 11:30, 1, 2:30, Sat.-Sun. 11:30, 1, 2, 3. No charge; donations accepted. Tram Tour: fee charged. Closed federal holidays. &

Attendance: 100,000 (estimated)

SNOW HILL MANOR, 13301 Laurel-Bowie Rd., Laurel, MD 20708-1509. Tel.: 301-725-6037. Fax: 301-498-2053. TTY: 301-446-3302.
Institution Type/Description: Historic House Museum: housed in a home formerly owned by the Snowden family; built in 1764. Listed on the National Register of Historic Places.
Collections: local history & culture; period furnishings.
Facilities: rental facilities.
Activities: rental facilities; special events.
Hours & Admission Prices: By appointment.

Leonardtown

NORTH END GALLERY, 41652 Fenwick St., Leonardtown, MD 20650. Mailing Address: P.O. Box 1238, Leonardtown, MD 20650. Tel.: 301-475-3130.
E-mail: lindaepstein1@mac.com
Web Site: www.northendgallery.org
Institution Type/Description: Art Gallery.
Collections: paintings; photographs; drawings; sculpture; jewelry.
Activities: Annual Event: Community Show in summer.
Hours & Admission Prices: Tues.-Sat. 11-6, Sun. 12-4.

OLD JAIL MUSEUM - ST. MARY'S COUNTY HISTORICAL SOCIETY, (M), 4625 Courthouse Dr., Leonardtown, MD 20650. Mailing Address: P.O. Box 212, Leonardtown, MD 20650. Tel.: 301-475-2467.
E-mail: director@stmaryschs.org
Founded: 1951.
Congressional District: 5
Institution Type/Description: Historic Building: built in 1858.
Collections: county history; personal artifacts; photographs; period furnishings; cannon from the ship, Ark.
Hours & Admission Prices: By appointment. No charge.
Attendance: 4,800 (estimated)

TUDOR HALL - ST. MARY'S COUNTY HISTORICAL SOCI-ETY, (M), 41680 Tudor Pl., Leonardtown, MD 20650. Mailing Address: P.O. Box 212, Leonardtown, MD 20650. Tel.: 301-475-2467 & 9455.
E-mail: director@stmaryschs.org
Founded: 1951.
Key Personnel: Exec. Dir., Susan J. Wolf.
Governing Authority: board; nonprofit organization.
Institution Type/Description: Historical Society Museum.
Collections: local history & culture; period furnishings; personal artifacts; county newspapers; early church records.
Facilities: Museum-related items for sale.
Activities: research.
Publications: Chronicles of St. Mary's.
Hours & Admission Prices: By appointment. No charge; donations accepted. &
Attendance: 2,400 (estimated)

Lexington Park

PATUXENT RIVER NAVAL AIR MUSEUM, 22156 Three Notch Rd., Lexington Park, MD 20653-2008. Tel.: 301-863-1900. Fax: 301-863-5048.
E-mail: association@paxmuseum.com
Web Site: www.paxmuseum.com
Key Personnel: Pres., Adm. Ed Forsman, Capt. USN Ret.; Museum Shop Mgr., Don House.
Personnel Profile: Part-Time Paid 6; Part-Time Volunteers 6; Interns 3.
Institution Type/Description: Naval Air Museum.
Collections: aircraft; photographs; art; documents; Naval aviation history & artifacts; Naval aircraft development & technology.
Research Fields: Naval aviation history.
Activities: flight simulators.
Hours & Admission Prices: Tues.-Sun. 10-5. No charge; donations accepted. &
Membership: Individual $35; Gold Corporate $1,000.

Linthicum

BENSON HAMMOND HOUSE - ANN ARRUNDELL COUNTY HISTORICAL SOCIETY, INC., (M), 7101 Aviation Blvd., Linthicum, MD 21090. Mailing Address: P.O. Box 385, Linthicum, MD 21090-0385. Tel.: 410-768-9518. Facebook: Benson Hammond House.
Web Site: www.aachs.org
Founded: 1962.
Congressional District: 2
Key Personnel: Pres. (V), Pamela T. Duncan; Cur., Andrea Frazier; Museum Shop Mgr., Vicky Zephir.
Personnel Profile: Part-Time Volunteers 50.
Governing Authority: Parent Institution: Maryland Historical Trust; Subsidiary Institution: Maryland Aviation Administration. Tax-exempt.
Institution Type/Description: Historical Society Museum: listed on the National Register of Historic Places.
Collections: local history & culture; period furnishings; personal artifacts; photographs; dolls; picker checks.
Facilities: Museum-related items for sale.
Activities: Annual Events: Strawberry Festival; Day Lily Festival; Children's Halloween Party; Holiday Open House.
Publications: quarterly publication, Ann Arrundell County Historical Society, Inc. History Notes; books.
Hours & Admission Prices: Sat. 11-3, Sun. 12-3. Discount to Blue Star Museums program members. Closed Independence Day.
Attendance: 5,000 (estimated)
Membership: Student $10; Senior $25; Individual $40; Family $50; Patron $100; Life $500.

NATIONAL ELECTRONICS MUSEUM, (M), 1745 W. Nursery Rd., Linthicum, MD 21090-2906. Mailing Address: P.O. Box 1693, MS4015, Baltimore, MD 21203-1693. Tel.: 410-765-0230. Fax: 410-765-0240.
E-mail: nemuseum@gmail.com
Web Site: www.hem-usa.org
Formerly: Historical Electronics Museum
Founded: 1980.
Congressional District: 4
Key Personnel: Pres., Louis Butler; Dir., Michael Aurele Simons; Asst. Dir., Alice Donahue; Financial Dir., Larraine Clark.
Personnel Profile: Full-Time Paid 2; Part-Time Paid 3; Part-Time Volunteers 50; Interns 1.
Governing Authority: not-for-profit organization. Tax-exempt: 501(c)(3).
Institution Type/Description: History & Technology Museum.
Collections: breakthroughs in advanced electronics technology, with emphasis on radar, countermeasures & communications; radar systems; transmitters; receivers; antennas; satellites; test equipment; archives; hands-on/interactive displays teaching the fundamental principles of electricity & electronics.
Facilities: library; 20,000 sq. ft. exhibit space.
Activities: guided tours; participatory exhibits; hands-on exhibits for children; STEM programming.
Publications: quarterly newsletter.
Hours & Admission Prices: Mon.-Fri. 9-4, Sat. 10-2. Adults $3; discounts to members. Closed major holidays. &
Attendance: 28,000 (accurate)
Membership: Student $15; Individual $25; Family $30; Supporting $100; Life $1,000.

Lutherville

FIRE MUSEUM OF MARYLAND, INC., 1301 York Rd., Lutherville, MD 21093-6023. Tel.: 410-321-7500. Fax: 410-769-8433.
E-mail: info@firemuseummd.org
Web Site: www.firemuseummd.org
Founded: 1971.
Congressional District: 2
Key Personnel: Dir. & Cur., Stephen G. Heaver, Jr.; Pres., W. Lee Smith, III; Registrar & Museum Shop Mgr., Melissa M. Heaver.
Personnel Profile: Full-Time Paid 2; Full-Time Volunteers 2; Part-Time Paid 7; Part-Time Volunteers 15; Interns 5.
Governing Authority: private; nonprofit corporation. Tax-exempt: 501(c)(3).
Institution Type/Description: Fire Museum.
Collections: 40 pieces of period fire fighting apparatus, including hand-pulled, horse drawn & motorized from 1806-1957; period artifacts; photographs; operational fire alarm telegraph system; Great Baltimore fire of 1904; recreation of 1871 Balto City fire house with original facade.
Research Fields: fire-fighting apparatus & equipment; history of fire service; fire manufacturers.

Facilities: library of fire history, technical manuals, catalogs & annual reports; archives; children's discovery room; 25,000 sq. ft. exhibit area; movie theater; rental facilities; birthday parties. Museum-related items for sale.
Activities: guided tours by appointment; lectures; films; permanent & temporary exhibitions; active restoration of vehicles; audio tours.
Publications: apparatus catalog; color post cards; semi-annual newsletter; fire history books; No Reason to Burn; Baltimore Builders: 1823-1964.
Hours & Admission Prices: May-& Sept.-Dec. Sat. 10-4; June-Aug. Wed.-Sat. 10-4; tours by appointment. Adults $12, firefighters & senior citizens $10, children 2-18 $5; discounts for group tours & AAM members; members & children under 2 no charge. Closed Independence Day; Christmas Eve & Day. &
Attendance: 11,975 (accurate)
Membership: Jr. Firefighter $15; Firefighter & Senior Citizen $30; Individual $35; Family $55.

LUTHERVILLE HISTORICAL COLORED SCHOOL NUMBER 24 MUSEUM, 1426 School Lane, Lutherville, MD 21093. Tel.: 410-825-6114.
Web Site: www.luthervillecoloredschool.webs.com
Institution Type/Description: Historic Building Museum: housed in a former schoolhouse; built c.1900.
Collections: African-American history & culture; period furnishings; personal artifacts; photographs.
Hours & Admission Prices: By appointment.

Marbury

MATTAWOMAN CREEK ART CENTER, Smallwood State Park, Marbury, MD 20658. Mailing Address: P.O. Box 258, Marbury, MD 20658-0258. Tel.: 301-743-5159.
E-mail: mattawomanart@aol.com
Web Site: www.mattawomanart.org
Founded: 1989.
Governing Authority: Tax-exempt: 501(c)(3).
Institution Type/Description: Art Gallery.
Collections: works by regional, national & international artists; paintings; sculpture.
Activities: workshops; lectures; demonstrations; films; seminars.
Hours & Admission Prices: Gallery: Fri.-Sun. 11-4. Office: Mon. & Wed.-Thurs. 9-1. No charge; donations accepted. Closed Memorial Day; Thanksgiving. &
Attendance: 3,500 (estimated)

SMALLWOOD STATE PARK, 2750 Sweden Point Rd., Marbury, MD 20658-2102. Tel.: 301-743-7613; 888-432-2267. Fax: 301-743-9405. TTY: 866-804-7864.
E-mail: park-smallwood@dnr.state.md.us
Web Site: www.dnr.state.MD.us
Founded: 1954.
Congressional District: 1
Key Personnel: Asst. Park Mgr., William Moffat; Park Ranger, Nakia Johnson; Park Ranger, Stephen Youngkin.
Personnel Profile: Full-Time Paid 4; Part-Time Paid 12; Part-Time Volunteers 35.
Governing Authority: state. Parent Institution: Maryland Park Service, Dept. of Natural Resources, Tawes State Office Bldg., 580 Taylor Ave., Annapolis, MD 21401. Tax-exempt.
Institution Type/Description: State Park Museum & Historic House: Gen. Smallwood's retreat.
Collections: 18th-century furniture; historic 18th-century home & kitchen; restored 19th-century tobacco barn.
Facilities: picnic areas; boating ramps; marina; playground equipment; hiking trails; information center; kitchen & herb gardens.
Activities: guided tours; lectures; nature walks; cooking demonstrations; candlemaking.
Hours & Admission Prices: May-Sept. Sun. 1-5; groups by appointment. Park Service Charge: April-Oct. Sat.-Sun. & holidays $3 per person. &
Attendance: 10,000 (estimated)

Marion

ACCOHANNOCK INDIAN TRIBAL MUSEUM, 28325 Farm Market Rd., Marion, MD 21838. Mailing Address: Box 404, Marion, MD 21838-0404. Tel.: 410-623-2660.
E-mail: accohannock@verizon.net
Institution Type/Description: Native American Indian Cultural Museum.
Collections: Native American culture, traditions & history; personal artifacts.
Facilities: Museum-related items for sale.

Activities: special programs; demonstrations.
Hours & Admission Prices: Mon.-Fri. call for hours. Suggested Donation: adults $3; children under 6 no charge.

Massey

MASSEY AIR MUSEUM, INC., 33541 Maryland Line Rd., Massey, MD 21650. Tel.: 410-928-5270.
E-mail: masseyaero@dmv.com
Web Site: www.masseyaero.org
Founded: 2002.
Governing Authority: nonprofit organization.
Institution Type/Description: Air Museum.
Collections: early aircraft; local history.
Activities: special events. Museum Sponsors: Chili Fiesta Fly-In in April; Open Hangar Party in December.
Hours & Admission Prices: Tues.-Sat. 10-4; other times by appointment. Suggested Donation: $5 per person; members no charge. &

Middletown

MIDDLETOWN VALLEY HISTORICAL SOCIETY, 305 W. Main St., Middletown, MD 21769-7928. Mailing Address: P.O. Box 294, Middletown, MD 21769-0294. Tel.: 301-371-7582. Fax: 301-371-7582.
E-mail: j.dwighthut@juno.com
Founded: 1976.
Congressional District: 6
Key Personnel: Pres. (V), Mrs. Devra Boesch; Museum Shop Mgr., Edna Alice Hoffman.
Personnel Profile: Part-Time Volunteers 20.
Governing Authority: private; nonprofit association. Tax-exempt: 501(c)(3).
Institution Type/Description: Historical Society: housed in 1840 stone house.
Collections: local artifacts.
Facilities: genealogy research library for local families.
Hours & Admission Prices: May-Oct. Sun. 2-4; Nov.-April for special events only. No charge; donations accepted.
Attendance: 100 (estimated)
Membership: Single $15; Family $25; Life $200.

Mitchellville

NEWTON WHITE MANSION, 2708 Enterprise Rd., Mitchellville, MD 20721-2544. Tel.: 301-249-2004. Fax: 301-446-3302. TTY: 301-454-1472.
E-mail: nwmansion@pgparks.com
Web Site: www.pgparks.com
Institution Type/Description: Historic Mansion: housed in the former home of U.S. Navy Captain Newton H. White; built in 1938.
Collections: period furnishings; photographs; paintings.
Facilities: banquet facilities.
Activities: rental facilities.
Hours & Admission Prices: By appointment.

Monkton

LADEW TOPIARY GARDENS, 3535 Jarrettsville Pike, Monkton, MD 21111-1910. Tel.: 410-557-9570 (office) & 9466 (information). Fax: 410-557-7763.
E-mail: eemerick@ladewgardens.com
Web Site: www.ladewgardens.com
Founded: 1971.
Congressional District: 2
Key Personnel: C.E.O., Cur. House & Exec. Dir., Emily W. Emerick; Pres. (V), Dudley Mason; House Com., L.B. Boyce; Museum Shop Mgr., Betsey Barringer.
Personnel Profile: Full-Time Paid 13; Part-Time Paid 15; Part-Time Volunteers 400.
Governing Authority: nonprofit organization. Tax-exempt: 501(c)(3).
Institution Type/Description: Topiary Gardens & Manor House Museum.
Collections: art; silver; period artifacts; fox hunting objects & memorabilia; 3 centuries of decor. Historic Building: Manor House.
Research Fields: horticultural.
Facilities: 22-acre botanical garden; nature walk; 75-seat auditorium; picnic grounds. Gift items for sale.
Activities: horticultural lecture series; concert series. Annual Event: Christmas Open House.
Publications: newsletter three times annually.
Hours & Admission Prices: mid-April to Oct. Mon.-Fri. 10-4, Sat.-Sun.

10:30-5; group tours by appointment. Garden: adults $10, students & senior citizens $8, children $2; members no charge. House: additional charge. &
Attendance: 30,000 (accurate)
Membership: Individual $35; Garden Contributor $65; Garden Friend $100; Garden Patron $250; Garden Benefactor $500; Garden Leader $1,000.

Mount Airy

SWETCHARNIK ART STUDIO, 7044 Woodville Rd., Mount Airy, MD 21771-7934. Tel.: 301-829-0137.
E-mail: sara@swetcharnik.com
Web Site: www.swetcharnik.com
Founded: 1980.
Key Personnel: Dir. & Artist, William Swetcharnik; Project Coord. & Artist, Sara Morris Swetcharnik.
Governing Authority: private. Subsidiary Institutions: Latin American Art Resource Project; Art Resource Traditions.
Institution Type/Description: Art Museum.
Collections: works by William & Sara Swetcharnik; drawings; paintings; photographs; sculpture.
Facilities: educational facilities.
Activities: formal education programs; guided tours; lectures; loan, temporary & traveling exhibitions.
Publications: My Patriotic Primer Lesson I.
Hours & Admission Prices: By appointment only. No charge; donations accepted.

Mount Savage

EVERGREEN HERITAGE CENTER, 15603 Trimble Rd., N.W., Mount Savage, MD 21545. Tel.: 301-687-0664.
Web Site: www.evergreenheritagecenter.org
Institution Type/Description: Heritage Center.
Collections: local history & culture; period furnishings; photographs. Historic Buildings: log cabin; stone mansion; barn.
Activities: workshops; special events; educational programs.
Hours & Admission Prices: May-Oct. by appointment.

MOUNT SAVAGE MUSEUM BANK JAIL AND MINING BUILDING, Main St., Mount Savage, MD 21545. Mailing Address: Mount Savage Historical Society, P.O. Box 401, Mount Savage, MD 21545. Tel.: 301-876-7847.
Web Site: www.mountsavagehistoricalsociety.org
Institution Type/Description: History Museum.
Collections: local history & heritage; mining; iron works; brick & railroad industries; photographs.
Hours & Admission Prices: By appointment.

National Harbor

NATIONAL CHILDREN'S MUSEUM, 151 St. George Blvd., National Harbor, MD 20745. Tel.: 301-392-2400.
E-mail: info@ncm.museum
Web Site: www.ncm.museum
Institution Type/Description: Children's Museum.
Collections: hands-on exhibitions.
Activities: interactive exhibitions; educational programs.
Hours & Admission Prices: Call for hours.

New Windsor

STRAWBRIDGE SHRINE, 2650 Strawbridge Lane, New Windsor, MD 21776. Mailing Address: P.O. Box 388, New Windsor, MD 21776. Tel.: 410-635-2600.
E-mail: tours@strawbridgeshrine.org
Web Site: strawbridgeshrine.org
Institution Type/Description: History Museum: housed in a replica of a 1760 Methodist meeting house.
Collections: local history & culture; period furnishings; personal artifacts; photographs.
Hours & Admission Prices: By appointment. &

Newark

QUEPONCO RAILWAY STATION MUSEUM, 8378 Patey Woods Rd., Newark, MD 21841. Mailing Address: P.O. Box 146, Newark, MD 21841-0146. Tel.: 410-632-0950 & 641-0067.
Web Site: www.octhebeach.com/museum/Queponco.html

Founded: 1910.
Congressional District: 3
Key Personnel: Pres. (V), Ralph L. Mason, Jr.
Governing Authority: Tax-exempt.
Institution Type/Description: Railway Station Museum: housed in a 1910 depot.
Collections: railroad station history; period artifacts; railroad wrenches; hand car mover; scales; freight cart; locks; telegraph key; early telephone; hand carts; steamer trunk; strawberry crate; morse code.
Hours & Admission Prices: May-Oct. 1st & 3rd Sat. 1-4; other times by appointment. No charge; donations accepted. &

Attendance: 200 (estimated)
Membership: Single $25; Family $40; Corporate $100.

North Beach

BAYSIDE HISTORY MUSEUM, 9006 Dayton Ave., North Beach, MD 20714. Mailing Address: P.O. Box 348, North Beach, MD 20714. Tel.: 410-495-8386.
Founded: 2003.
Governing Authority: nonprofit organization.
Institution Type/Description: History Museum.
Collections: local history & culture; period furnishings; personal artifacts; photographs; paleontology; Boy Scout memorabilia.
Hours & Admission Prices: Sun. 1-4.

North Bethesda

THE MANSION AT STRATHMORE, 10701 Rockville Pike, North Bethesda, MD 20852-3224. Tel.: 301-581-5109.
Institution Type/Description: Art Museum.
Collections: works by local & national artists; sculpture.
Facilities: Gift items for sale.
Activities: educational programs; lectures; concerts; special events.
Hours & Admission Prices: Call for hours.

North East

UPPER BAY MUSEUM, 219 Walnut St., North East Town Park, North East, MD 21901. Mailing Address: P.O. Box 275, North East, MD 21901-0275. Tel.: 410-287-2675.
Institution Type/Description: History Museum.
Collections: local commercial & recreational fishing and hunting; decoys; period marine engines; photographs; personal artifacts.
Activities: Museum Sponsors: Decoy Show in spring.
Hours & Admission Prices: Memorial Day to Labor Day Sat. 10-4, Sun. 12-4; other times by appointment. No charge; donations accepted. &

Oakland

GARRETT COUNTY HISTORICAL MUSEUM, 107 S. 2nd St., Oakland, MD 21550-1519. Tel.: 301-334-3226.
E-mail: gchsmuseum@verizon.net
Web Site: www.garrettcountymuseum.com
Founded: 1969.
Congressional District: 6
Key Personnel: Pres., Robert Boal; Vice Pres., Jim Ashby; Sec., Alice Eary; Cur., Eleanor Callis; Asst. Cur., Brenda Gnegy.
Personnel Profile: Part-Time Paid 2; Part-Time Volunteers 20.
Governing Authority: society. Affiliated with Garrett County Historical Society. Tax-exempt.
Institution Type/Description: Historical Society Museum.
Collections: local Indian artifacts; period rooms; early cooking utensils; glassware; railroad exhibit; leather tools; weaving; bridal gowns; coal mining exhibit & tools; old dolls & carriages; U.S.S. Garrett County artifacts.
Research Fields: Garrett County.
Activities: guided tours by appointment; permanent & temporary exhibitions.
Publications: quarterly booklet, Glade Star.
Hours & Admission Prices: Mon.-Sat. 10-3. No charge; donations accepted. &
Attendance: 6,000 (estimated)
Membership: Single & Couple $20; Life $250.

OAKLAND B&O MUSEUM, 117 E. Liberty St., Oakland, MD 21550-1201. Mailing Address: 15 S. 3rd St., Oakland, MD 21550. Tel.: 301-334-3204. Fax: 301-334-4401.
E-mail: townofoak@gmail.com
Formerly: Oakland B&O Train Station
Founded: 1884.

Congressional District: 1A
Governing Authority: city. Subsidiary Institution: Oakland Heritage Community Foundation, Inc.
Institution Type/Description: Historic Train Station: built in 1884.
Collections: depot history; period furnishings; photographs; B&O Museum in Baltimore displays.
Activities: educational exhibitions.
Hours & Admission Prices: By appointment. No charge; donations accepted. &
Attendance: 41,465 (estimated)

Ocean City

OCEAN CITY LIFE-SAVING STATION MUSEUM, 813 S. Atlantic Ave., Ocean City, MD 21842. Mailing Address: P.O. Box 603, Ocean City, MD 21843-0603. Tel.: 410-289-4991.
E-mail: sandy@ocmuseum.org
Web Site: www.ocmuseum.org
Founded: 1978.
Congressional District: 10
Institution Type/Description: History Museum.
Collections: exhibits relating to the history of the U.S. Life-Saving Service & Ocean City history; shipwreck artifacts; sands of the world; Surf's Up.
Publications: triannual newsletter, Scuttlebutt.
Hours & Admission Prices: April & Nov.-Dec. Sat.-Sun. 10-4; May-Oct. daily 10-4; other times by appointment. Adults $3, children 6-17 $1; children under 6 & members no charge. &
Attendance: 18,454 (accurate)
Membership: Student $5; Individual $10; Family $30; Supporting $100; Sustaining $250; Benefactor $500.

WHEELS OF YESTERDAY, 12708 Ocean Gateway, Ocean City, MD 21842-9542. Tel.: 410-213-7329.
Founded: 1997.
Key Personnel: Cur., Jack Jarvis
Institution Type/Description: Transportation Museum.
Collections: period cars & fire engine; replica 1950s service station.
Facilities: Museum-related items for sale.
Hours & Admission Prices: Call for hours. Adults $5, children 12 & under $3; discounts to groups of 20 or more.

Oldtown

IRVIN ALLEN/MICHAEL CRESAP MUSEUM, 19015 Opessa St., S.E., Oldtown, MD 21555-9702. Tel.: 301-478-5848.
Institution Type/Description: Historic House Museum: built in 1764.
Collections: local history; period furnishings; personal artifacts.
Hours & Admission Prices: May-Oct. by appointment.

Owings Mills

SOLDIERS DELIGHT NATURAL ENVIRONMENT AREA, 5100 Deer Park Rd., Owings Mills, MD 21117. Tel.: 410-461-5005.
Institution Type/Description: Environment Area & Visitor Center.
Collections: plants; insects; rocks; minerals.
Facilities: nature trails.
Activities: educational programs; hiking trails.
Hours & Admission Prices: Visitor Center: Sat. 11-3; other times by appointment.

Oxford

OXFORD MUSEUM, INC., Morris and Market Sts., Oxford, MD 21654. Mailing Address: P.O. Box 131, Oxford, MD 21654-0131. Tel.: 410-226-0191.
E-mail: oxford_museum@verizon.net
Web Site: www.oxfordmuseum.org
Founded: 1964.
Congressional District: 1
Key Personnel: Pres. Bd. (V), Jan Mroczek; Dir., Ellen Anderson.
Personnel Profile: Part-Time Paid 1; Part-Time Volunteers 65.
Governing Authority: nonprofit organization. Tax-exempt.
Institution Type/Description: Local History Museum.
Collections: maritime.
Major Exhibits: Historic Newspapers, Spring 2014.
Research Fields: Oxford & Talbot County history.
Activities: lectures; slides; multimedia productions.
Publications: Port of Entry (Oxford History); Oxford Treasures - Then & Now; walking tour brochure.

Hours & Admission Prices: late April to May & Oct. to mid-Nov. Mon. & Fri.-Sat. 10-4, Sun. 1-4; June-Sept. Mon., Wed. & Fri.-Sat. 10-4, Sun. 1-4. No charge; donations suggested. &

Attendance: 4,000 (estimated)

Membership: Annual $25-$250.

Oxon Hill

OXON COVE PARK, 6411 Oxon Hill Rd., Oxon Hill, MD 20745-1100. Mailing Address: 1900 Anacostia Dr., S.E., Washington, DC 20020-6722. Tel.: 301-839-1176. Fax: 301-763-1066. TDD: 301-839-1783.

E-mail: vanessa-molineaux@nps.gov

Web Site: www.nps.gov/oxhi

Founded: 1967.

Congressional District: 4

Key Personnel: Acting Dir., Lisa Mendelson-Lelmini; Supt., Gale Hazelwood; Asst. Supt., Alex Romero; Park Mgr., Sharon Vanessa Molineaux.

Governing Authority: federal. Parent Institution: National Capital Parks-East, National Park Service, U.S. Dept. of the Interior, 1900 Anacostia Dr., S.E., Washington, DC 20020. Tax-exempt.

Institution Type/Description: Agriculture Museum: housed in c.1900 farm & farm outbuildings.

Collections: farming tools & implements; farm animals.

Research Fields: agriculture of 1900; equipment; techniques.

Facilities: picnic area.

Activities: self-guided tours; living history & farm chore demonstrations.

Publications: brochure; monthly news sheet, Farmer's Log & handbill.

Hours & Admission Prices: Daily 8-4:30. No charge; donations accepted. Closed New Year's Day; Thanksgiving; Christmas. &

Attendance: 100,000 (estimated)

OXON HILL MANOR, 6901 Oxon Hill Rd., Oxon Hill, MD 20745. Tel.: 301-839-7782. Fax: 301-839-4867. TTY: 301-446-6802.

E-mail: ohmanor@pgparks.com

Institution Type/Description: Historic House Museum: housed in the former home of nephews of George Washington, & the nephew of John Hanson, the first president elected by the Continental Congress under the Articles of Confederation; c.1928.

Collections: local history, heritage & culture; period furnishings.

Facilities: rental facilities.

Hours & Admission Prices: Mon. 1-4, Tues.-Fri. 9-4 by appointment.

Pasadena

HANCOCK'S RESOLUTION FARM, 2795 Bayside Beach Rd., Pasadena, MD 21056. Tel.: 410-255-4048.

Institution Type/Description: Historic Farm.

Collections: local history & culture; period furnishings; personal artifacts; photographs.

Activities: special events.

Hours & Admission Prices: April-Oct. Sun. 1-5. No charge; donations accepted. Closed Easter.

Perryville

RODGERS TAVERN, 259 Broad St., Perryville, MD 21903-0322. Mailing Address: P.O. Box 322, Perryville, MD 21903-0322. Tel.: 410-642-6066. Fax: 410-642-6391.

E-mail: townhall@perryvillemd.org

Web Site: www.Perryvillemd.org

Founded: 1956.

Congressional District: 1

Key Personnel: Chm. (V), Barbara Brown.

Personnel Profile: Full-Time Volunteers 1; Part-Time Volunteers 12.

Governing Authority: society. Town of Perryville. Tax-exempt.

Institution Type/Description: Historic House: pre-1743 building, operated by Rodgers family.

Collections: local history.

Activities: guided tours; fund-raising bazaars; annual meetings. Museum Sponsors: Spring Fling in May; Autumnfest in October; Colonial Christmas in December.

Publications: brochure, Historic Rodgers Tavern.

Hours & Admission Prices: Grounds: by appointment. Building: temporarily closed for renovation. No charge; donations accepted.

Attendance: 6,000 (estimated)

Pikesville

MARYLAND STATE POLICE MUSEUM, 1201 Reisterstown Rd., Pikesville, MD 21208-3898. Tel.: 410-653-4278. Fax: 410-653-4341.

E-mail: mringley@mdsp.org

Web Site: www.mdsp.org

Key Personnel: Coord., Margaret Ringley

Institution Type/Description: Police History Museum.

Collections: law enforcement history; weaponry; riot helmets; period uniforms; photographs; memorials.

Hours & Admission Prices: Mon.-Fri.

Piney Point

PINEY POINT LIGHTHOUSE MUSEUM AND HISTORIC PARK, 44720 Lighthouse Rd., Piney Point, MD 20674. Tel.: 301-994-1471.

Institution Type/Description: History Museum.

Collections: lighthouse history; U-1105 Black Panther shipwreck dive preserve; Potomac River maritime history; photographs; period artifacts.

Hours & Admission Prices: April-Sept. daily 10-5; Oct.-Jan. 1 Fri.-Mon. 12-4; call for special Christmas hours. Adults $3, seniors $2, children $1.50. Closed Easter; Thanksgiving; Christmas. &

Pocomoke City

COSTEN HOUSE & HALL-WALTON MEMORIAL GARDEN, 206 Market St., Pocomoke City, MD 21851. Mailing Address: P.O. Box 430, Pocomoke City, MD 21851. Tel.: 410-957-3110. Facebook: Costen House Museum.

E-mail: ritarae1@comcast.net

Founded: 1974.

Key Personnel: Dir., Rita Ullmann.

Personnel Profile: Part-Time Volunteers 6; Interns 2.

Volunteer Hours: 1,560

Governing Authority: Tax-exempt.

Institution Type/Description: History Museum: housed in the home of the first mayor of Pocomoke.

Collections: local history & culture; period furnishings; personal artifacts; photographs.

Hours & Admission Prices: Call for hours.

Attendance: 400 (estimated)

DELMARVA DISCOVERY CENTER, 2 Market St., Pocomoke City, MD 21851. Tel.: 410-957-9933.

E-mail: contact@delmarvadiscoverycenter.org

Web Site: delmarvadiscoverycenter.org

Founded: 2009.

Institution Type/Description: Cultural & Natural Heritage Center.

Collections: local history, natural heritage & culture; river ecology.

Facilities: rental facilities; classrooms. Museum-related items for sale.

Activities: educational programs; classes; group tours; presentations; rental facilities; river cruises; field trips; birthday parties; special events.

Hours & Admission Prices: Memorial Day to Labor Day daily 10-4; Sept.-May Tues.-Sat. 10-4, Sun. 1-4. Adults 18 & over $10, seniors 60 & over and students $8, children 4-17 $5; discounts to groups; members no charge. &

Attendance: 14,000 (accurate)

STURGIS ONE-ROOM SCHOOL MUSEUM, 209 Willow St., Pocomoke City, MD 21851. Mailing Address: P.O. Box 697, Pocomoke City, MD 21851-0697. Tel.: 410-957-1913.

E-mail: gatling144@verizon.net

Web Site: www.sturgismuseum.org

Key Personnel: Pres., James Gatling; Cur., Sudie Gatling

Institution Type/Description: History Museum: housed in a one-room African American school house used for first through seventh grades until 1937.

Collections: schoolhouse furnishings.

Hours & Admission Prices: May-Oct. Tues.-Sat. 1-4; other times by appointment. Adults $3, children $1.

Poolesville

JOHN POOLE HOUSE, 19923 Fisher Ave., Poolesville, MD 20837. Mailing Address: P.O. Box 232, Poolesville, MD 20837. Tel.: 301-972-8588.

E-mail: info@historicmedley.org

Web Site: www.historicmedley.org

Congressional District: 8
Institution Type/Description: Historic House Museum: housed in a former trading post built by John Poole, Jr.; built in 1793.
Collections: local history; period furnishings; linen; miniatures; notepaper; toys; books; local art.
Hours & Admission Prices: By appointment. No charge; donations accepted.

SENECA SCHOOLHOUSE MUSEUM, 16800 River Rd., Poolesville, MD 20837-0232. Mailing Address: Historic Medley District, Inc., P.O. Box 232, Poolesville, MD 20837-0232. Tel.: 301-972-8588.
E-mail: info@historicmedley.org
Web Site: www.historicmedley.org
Congressional District: 8
Key Personnel: Dir., Patty Cooper
Institution Type/Description: Historic Building Museum.
Collections: period furnishings & artifacts.
Facilities: Museum-related items for sale.
Activities: education programs; school groups.
Hours & Admission Prices: By appointment. No charge; donations accepted.

Port Deposit

PAW PAW MUSEUM, 98 N. Main St., Port Deposit, MD 21904-1210. Tel.: 410-378-4480.
E-mail: pawpawmuseum@gmail.com
Institution Type/Description: Historic Building: housed in a former Methodist Church; built in 1821.
Collections: local history & culture; personal artifacts; photographs; clothing; yearbooks; Civil War artifacts.
Facilities: Museum-related items for sale.
Hours & Admission Prices: May-Oct. 2nd & 4th Sun. of month 1-5.

Port Tobacco

PORT TOBACCO COURTHOUSE, Chapel Point Rd., Port Tobacco, MD 20677. Mailing Address: P.O. Box 302, Port Tobacco, MD 20677. Tel.: 301-934-4313.
Web Site: www.somd.com
Institution Type/Description: Historic Building Museum: housed in a former courthouse built in 1819. Listed on the National Register of Historic Places.
Collections: local history & heritage; tobacco industry; Civil War artifacts; personal artifacts; period furnishings; photographs.
Facilities: rental facilities.
Activities: special events; rental facilities.
Hours & Admission Prices: Call for hours.

PORT TOBACCO ONE ROOM SCHOOL, 7215 Chapel Point Rd., Port Tobacco, MD 20677. Mailing Address: P.O. Box 2770, La Plata, MD 20646-2770. Tel.: 301-934-9483.
E-mail: idcornette@gmail.com
Key Personnel: Dir. & Chm. (V), Dale Cornette
Institution Type/Description: History Museum: built in 1871.
Collections: schoolhouse furnishings; books; toys; period lunch pails.
Hours & Admission Prices: By appointment. No charge; donations accepted.
Attendance: 1,000 (estimated)

THOMAS STONE NATIONAL HISTORIC SITE, 6655 Rose Hill, Port Tobacco, MD 20677. Tel.: 301-392-1776. Fax: 301-934-8793.
Web Site: www.nps.gov
Institution Type/Description: Historic Site: housed in a 1770s Georgian mansion, former home of Thomas Stone, a Maryland signer of the Declaration of Independence.
Collections: local history; Thomas Stone's life; period furnishings; personal artifacts; photographs.
Facilities: 322 acre site.
Hours & Admission Prices: Call for hours.

Potomac

GREAT FALLS TAVERN MUSEUM AND VISITOR CENTER, 11710 MacArthur Blvd., Potomac, MD 20854. Tel.: 301-767-3714.
Institution Type/Description: Historic Building: built in 1831.
Collections: local history & culture; period furnishings; personal artifacts; photographs.
Activities: educational programs; boat rides.

Hours & Admission Prices: Daily 9-4:30. Closed New Year's Day; Thanksgiving; Christmas.

Preston

LINCHESTER MILL, 3390 Linchester Mill Rd., Preston, MD 21655. Tel.: 410-673-1910.
Web Site: www.tourcaroline.com
Institution Type/Description: Historic Mill: housed in the former mill that sold grain to George Washington's army during the Revolutionary War; c.1840. Listed on the National Register of Historic Places.
Collections: mill history; milling machinery.
Activities: Annual Event: Fall Festival in October.
Hours & Admission Prices: Call for hours.

Prince Frederick

BATTLE CREEK CYPRESS SWAMP SANCTUARY, 2880 Gray's Rd., Prince Frederick, MD 20678. Tel.: 410-535-5327.
Institution Type/Description: Nature Sanctuary.
Collections: wildlife & their habitat; seasonal wildflowers; 100 ft. canopy of cypress trees.
Facilities: 100-acre nature sanctuary.
Hours & Admission Prices: Tues.-Sat. 10-4:30, Sun. 1-4:30. Closed major holidays.

CALVERT COUNTY HISTORICAL SOCIETY, 70 Church St., Prince Frederick, MD 20678. Mailing Address: P.O. Box 358, Prince Frederick, MD 20678. Tel.: 410-535-2452. Fax: 410535-4660.
E-mail: cchsadmin@calverthistory.org
Web Site: www.calverthistory.org
Founded: 1953.
Key Personnel: Dir., Leila Boyer; Archivist, Karen Sykes.
Governing Authority: Parent Institution: Calvert County Historical Society. Tax-exempt.
Institution Type/Description: Historical Society.
Collections: local history & culture; period furnishings; personal artifacts; photographs; local history archives.
Research Fields: family & county history.
Publications: monthly newsletters, Postscripts, Letters from Linden; annual journal, The Calvert Historian.
Hours & Admission Prices: Tues.-Thurs. 10-3; other times by appointment.

Princess Anne

THE SOMERSET COUNTY HISTORICAL SOCIETY, INC. - TEACKLE MANSION, 11736 Mansion St., Princess Anne, MD 21853. Mailing Address: P.O. Box 181, Princess Anne, MD 21853-0181. Tel.: 410-651-2238.
E-mail: moreinfo@teacklemansion.org
Web Site: www.teacklemansion.org
Formerly: Olde Princess Anne Days, Inc.
Founded: 1958.
Congressional District: 1
Key Personnel: Pres., Gabe Stuckey; Vice Pres., Jill Hall; Museum Shop Mgr., Linda Alder.
Personnel Profile: Part-Time Volunteers 68.
Governing Authority: nonprofit organization. Tax-exempt: 170(b)(1)(A).
Institution Type/Description: General Museum: housed in 1801 Teackle Mansion, early 19th-century federal period building.
Collections: costumes; archives; decorative arts; furniture; paintings.
Facilities: 100-vol. library of historical books available for research by permission of directors; children's museum.
Activities: guided tours. Special Event: Olde Princess Anne Days in October, tour of Teackle Mansion & 20 local historic homes.
Hours & Admission Prices: April to mid-Dec. Thurs. & Sat.-Sun. 1-3; tours by appointment. Adults $5, students $2; discounts to members; children under 12 accompanied by adult no charge.
Attendance: 778 (estimated)
Membership: Student $15; Individual $25; Family $40; Business $50; Patron $100.

Queenstown

QUEENSTOWN COLONIAL COURTHOUSE, 100 Del Rhodes Ave., Queenstown, MD 21658. Mailing Address: P.O. Box 4, Queenstown, MD 21658. Tel.: 410-827-7646; 410-827-7661.
E-mail: qtowncom@crosslink.net

Governing Authority: Tax exempt.
Institution Type/Description: History Museum.
Collections: local history & culture; period furnishings; personal artifacts.
Activities: special events.
Hours & Admission Prices: May-Oct. Mon.-Fri. 8:30-4:30, 1st Sat. each month 10-2; Nov.-April Mon.-Fri. 8:30-4:30. No charge.
Attendance: 120 (estimated)

Ridgely

ADKINS ARBORETUM, (M), 12610 Eveland Rd., Ridgely, MD 21660. Mailing Address: P.O. Box 100, Ridgely, MD 21660-0100. Tel.: 410-634-2847. Fax: 410-634-2878.
E-mail: info@adkinsarboretum.org
Web Site: www.adkinsarboretum.org
Founded: 1980.
Congressional District: 1
Key Personnel: Dir., Ellie Altman; Pres., Sydney Doehler; Museum Shop Mgr., Robyn Affron.
Personnel Profile: Full-Time Paid 7; Part-Time Paid 4; Part-Time Volunteers 75; Interns 1.
Operating Expenses: 730,000
Operating Income: 730,000
Governing Authority: Tax-exempt.
Institution Type/Description: Arboretum.
Collections: native plants, trees, & wildflowers.
Research Fields: lady slipper orchids
Facilities: 400-acre garden; nature trails.
Activities: educational programs.
Publications: 3 times a year, Native Seed.
Hours & Admission Prices: Tues.-Sat. 10-4, Sun. 12-4. Adults $5, students 6-18 $2; members and children 5 & under no charge. Closed major holidays.
Attendance: 20,000 (accurate)
Membership: Individual $50; Household $75; Contributor $100; Supporter $250; Leon Andrus Society $1,000.

Riverdale Park

RIVERSDALE HOUSE MUSEUM, 4811 Riverdale Rd., Riverdale Park, MD 20737-1911. Mailing Address: 6005 48th Ave., Riverdale Park, MD 20737-2015. Tel.: 301-864-0420. Fax: 301-927-3498. TDD: 301-699-2544.
E-mail: riversdale@pgparks.com
Web Site: history.pgparks.com/sites_and_museums/riversdale_house_ museum.htm
Founded: 1949.
Congressional District: 22
Key Personnel: Dir., Edward Day; Pres. (V), Patrick Gossett; Div. Chief, Anthony Nolan; Museum Shop Mgr., Renee Kidd.
Personnel Profile: Full-Time Paid 2; Part-Time Paid 9; Part-Time Volunteers 61.
Governing Authority: nonprofit organization. Parent Institution: The Maryland National-Capital Park & Planning Commission. Subsidiary Institution: Natural and Historical Resources Division. Tax-exempt.
Institution Type/Description: National Historic Landmark. Historic House: 1803 five-part stucco covered brick manor, blending Belgian & American architectural styles built by Henri Stier & occupied by his daughter, Rosalie & her husband George Calvert and inherited by their son Charles Benedict. Later, residents included Senator Hiram Johnson (CA) & Senators Thaddeus & Hattie Caraway (AR).
Collections: personal & domestic artifacts; home furnishings; outbuildings; family correspondence; books.
Research Fields: home & life of Rosalie & George Calvert; early 1800s; African-American history; foodways; early 19th-century costume.
Facilities: visitors center.
Activities: guided tours; rental gallery; docent program; formal education programs; open-hearth cooking; garden & grounds tours; school groups; scout badge programs.
Publications: quarterly newsletter, The Riversdale Letter; quarterly, The Riversdale Docent; monthly volunteer newsletter, The Column.
Hours & Admission Prices: Sun. & Fri. 12:15-3:15; other times by appointment. Adults $3, senior citizens 60 & over $2, children $1; discounts to groups & members. Closed New Year's Day; Independence Day; Christmas.
Attendance: 11,000 (estimated)
Membership: Riversdale Historical Society: Basic $25; Supporting $50.

Rock Hall

ROCK HALL MUSEUM, Municipal Bldg., S. Main St., Rock Hall, MD 21661. Mailing Address: P.O. Box 367, Rock Hall, MD 21661. Tel.: 410-639-7611. Fax: 410-639-7298.
Web Site: www.rockhallmd.com/museum/index/php
Founded: 1976.
Key Personnel: Chm., Doug Francis.
Personnel Profile: Part-Time Volunteers 1.
Governing Authority: Parent Institution: Town of Rock Hall.
Institution Type/Description: History Museum.
Collections: local history & culture; period furnishings; photographs; model ships; decoy carvings; personal artifacts.
Hours & Admission Prices: Sat.-Sun. 11-3; other times by appointment. No charge; donations accepted.
Attendance: 500 (estimated)

TOCHESTER BEACH REVISITED MUSEUM, Main St., Rock Hall, MD 21661. Mailing Address: 21341 Virginia Ave., Chestertown, MD 21620-4326. Tel.: 410-778-5347.
E-mail: wbetts88@gmail.com
Founded: 1999.
Key Personnel: Cur., Bill Betts
Institution Type/Description: History Museum.
Collections: amusement park history & memorabilia; photographs; period artifacts.
Hours & Admission Prices: Sat.-Sun. 11-3; groups by appointment. No charge; donations accepted.
Attendance: 1,500 (accurate)

WATERMAN'S MUSEUM, 20880 Rock Hall Ave., Rock Hall, MD 21661-1407. Tel.: 410-778-6697. Fax: 410-639-2971.
E-mail: email@havenharbour.com
Web Site: www.havenharbour.com/hhwatmus.htm
Founded: 1993.
Congressional District: 7
Key Personnel: Mgr., Woodrow Loller
Governing Authority: Parent Institution: Haven Harbour Marina.
Institution Type/Description: History Museum.
Collections: oystering; crabbing; fishing; photographs; carvings; boats; shanty house replica.
Publications: brochure.
Hours & Admission Prices: Daily 8-5. No charge; donations accepted.
Attendance: 200 (estimated)

Rockville

JANE L. AND ROBERT H. WEINER JUDAIC MUSEUM, The Gildenhorn/Speisman Center for the Arts, 6125 Montrose Rd., Rockville, MD 20852-4860. Tel.: 301-881-0100 & 230-3711. Fax: 301-881-5512. TDD: 301-881-0012.
Web Site: www.jccgw.org
Founded: 1969.
Congressional District: 17
Personnel Profile: Full-Time Paid 2; Part-Time Volunteers 90.
Governing Authority: nonprofit organization. Parent Institution: Jewish Community Center of Greater Washington. Tax-exempt.
Institution Type/Description: Judaic Museum.
Collections: Judaica; ethnology; archaeology; art.
Facilities: library.
Activities: lectures; guided tours; films; concerts; dance recitals; workshops; organized education programs for children; inter-arts programs; inter-museum loans; permanent & traveling exhibitions.
Publications: exhibition notes & catalogue; Center Scene.
Hours & Admission Prices: Mon.-Thurs. 12-4 & 7:30 pm-9:30 pm, Sun. 2-5. No charge. Closed national & Jewish holidays.
Attendance: 9,000 (estimated)

LATVIAN MUSEUM, 400 Hurley Ave., Rockville, MD 20850-3121. Mailing Address: P.O. Box 67, Fabius, NY 13063. Tel.: 301-340-1914. Fax: 301-340-8732.
E-mail: lbergs@twcny.rr.com
Web Site: www.alausa.org
Founded: 1980.
Congressional District: 8
Key Personnel: Dir. & Cur., Lilita Bergs.
Personnel Profile: Part-Time Volunteers 12.
Governing Authority: nonprofit organization. Parent Institution: American

Latvian Association in the United States. Subsidiary Institution: Latvian Institute. Tax-exempt: 501(c)(3).
Institution Type/Description: Ethnic Museum: Latvian historic & cultural development from Ice Age to 20th Century.
Collections: textiles; costumes; farm implements; photographs; documents; folk art; military uniforms & medals; philately; numismatics.
Research Fields: Latvian history & culture; folk arts; political & military history.
Facilities: 1,200 sq. ft. exhibition space.
Activities: guided tours. Annual Event: Montgomery County Heritage Days in June.
Publications: brochure.
Hours & Admission Prices: By appointment. No charge; donations accepted. &
Attendance: 2,000 (estimated)

THE MONTGOMERY COUNTY HISTORICAL SOCIETY, INC., (M), 103 W. Montgomery Ave., Rockville, MD 20850-4212.

Mailing Address: 111 W. Montgomery Ave., Rockville, MD 20850-4212. Tel.: 301-340-2825 & 762-1492. Fax: 301-340-2871.
E-mail: info@montgomeryhistory.org
Web Site: www.montgomeryhistory.org
Founded: 1944.
Congressional District: 8
Key Personnel: Pres. (V), Jack Devine; Dir., Thomas A. Kuehhas; Librarian, Patricia Andersen; Dir. Collections, Joanna Church; Education & Volunteer Coord., Elizabeth Keaney; Office Mgr., Mary Martin.
Personnel Profile: Full-Time Paid 3; Part-Time Paid 5; Part-Time Volunteers 100; Interns 2.
Volunteer Hours: 3,500
Operating Expenses: 357,000
Operating Income: 362,000
Governing Authority: society. Subsidiary Institution: Research Library, 42 W. Middle Lane, Rockville, MD 20850. Tax-exempt: 501(c)(3).
Institution Type/Description: Historic House: 1815 Beall-Dawson House; 1852 Stonestreet Medical Museum.
Collections: 23,000 Montgomery County artifacts including archival records; photographs; maps; costumes; quilts; glass; furniture & medical instruments.
Major Exhibits: This Is The Way We Wash Our Clothes, 6/13-1/14; Montgomery Maternity: Pregnancy, Childbirth and Infant Care at the Turn of the Century, 9/13-6/14; The Colonial Revival in Montgomery County, 2/14-5/14; Civic Action in Montgomery County from the Colonial Era to the Present, 6/14-1/15.
Research Fields: Montgomery County history including genealogy, oral history, archaeology, historic preservation.
Facilities: 2,000-vol. library of books; 12,000 linear ft. of general history material on Maryland & Montgomery County available for use on premises; reading room. Museum-related items for sale.
Activities: guided & self-guided tours; genealogy club; lectures; walking tours; research assistance; preservation workshops.
Publications: biannual magazine, The Montgomery County Story; quarterly newsletter; brochures; county history books; monographs.
Hours & Admission Prices: Museum: Wed.-Sun. 12-4. Library: Wed.-Fri. 10-4, Sat. 12-4. Adults $5; members no charge.
Attendance: 6,000 (estimated)
Membership: Sligo $40; Seneca $60; Rock Creek $125; Patuxent $250; Potomac $500.

PEERLESS ROCKVILLE HISTORIC PRESERVATION, LTD., 29 Courthouse Sq., Rm. 110, Rockville, MD 20850.

Mailing Address: P.O. Box 4262, Rockville, MD 20849. Tel.: 301-762-0096.
E-mail: info@peerlessrockville.org
Web Site: www.peerlessrockville.org
Formerly: Peerless Rockville Collection & Research Library
Founded: 1974.
Congressional District: 17
Key Personnel: Dir., Mary A. van Balgooy; Pres., Erick Ledbetter.
Personnel Profile: Full-Time Paid 1; Part-Time Paid 2; Part-Time Volunteers 7; Interns 1.
Volunteer Hours: 1,157
Operating Expenses: 191,661
Operating Income: 244,412
Governing Authority: Tax-exempt.
Institution Type/Description: History Museum: housed in the Red Brick Courthouse.
Collections: local history, heritage & culture; photographs; period artifacts; business records; blueprints; oral histories; manuscripts; genealogy; publications.

Facilities: library.
Activities: research.
Hours & Admission Prices: Center & Library: Mon.-Fri. 10-1; other times by appointment. Office: Mon.-Fri. 9-3. No charge. &
Attendance: 12,000 (estimated)
Membership: Basic $35; Friend $50; Supporter $100; Sustainer $250; Sponsor $500; Benefactor $1,000; Patron $2,500.

Saint Leonard

JEFFERSON PATTERSON PARK & MUSEUM, (M), 10515 Mackall Rd., Saint Leonard, MD 20685-2433. Tel.: 410-586-8501.

Fax: 410-586-0080. TDD: 800-735-2258 (Maryland Relay).
E-mail: jppm@mdp.state.md.us
Web Site: www.jefpat.org
Founded: 1983.
Congressional District: 3
Key Personnel: Dir., Mark Thompson; Pres. (V), Pat Furey; Chief Conservator, Betty Seifert; Fiscal Officer, Denise America; Admin. Education, Kimberley Popetz; Admin. Research, Edward E. Chaney; Collections, Rebecca Morehouse; Museum Shop Mgr., Michele Parlett; Dir. Maryland Archaeological Conservation Laboratory, Patricia Samford; Education & Archaeology Specialist, Kate Dinnel; Head Conservation, Nichole Doub; Conservation Tech, Gareth McNair-Lewis; Sec., Sharon Raftery; Federal Cur., Sara Rivers-Cofield; Maintenance Supvr., Dimitrios Papadakis; Maintenance Mechanic, Stephen Embrey; Maintenance Asst., William Wyatt.
Personnel Profile: Full-Time Paid 31; Part-Time Paid 3; Part-Time Volunteers 30; Interns 3.
Governing Authority: state government. Parent Institution: Maryland Historical Trust. Subsidiary Institution: Friends of JPPM, Inc. Tax-exempt.
Institution Type/Description: Park & Museum.
Collections: life in the Chesapeake Bay area from prehistoric to historic times; prehistoric & historic artifacts; historic farm equipment & machinery; 75 archaeological sites on property; 560-acre waterfront park; Native American structures. Historic Structures: 1932 steer barn; 1936 tenant farmer complex.
Research Fields: prehistoric & historic archaeology; Southern Maryland history & agriculture; research & collections care; African American, European American & Native American rural life oral history research; War 1812.
Facilities: visitor center; archaeology & nature trails; laboratories; discovery room; gardens. Archaeology books, artifact replicas, farm toys & other related items for sale.
Activities: guided walking & wagon ride tours; lectures; films; concerts; organized education programs; docent program; public archaeology program. Museum Sponsors: Children's Day on the Farm; Celtic Festival; African-American Family Community Day; War of 1812 reenactment; American Indian Heritage Day; War of 1812 Tavern nights.
Publications: newsletter, Patterson Points; newsletter, MAC Lab; Popular Archaeology Series; annual report.
Hours & Admission Prices: April 15-Oct. 15 Wed.-Sun. 10-5. No charge; donations accepted. &
Attendance: 55,000 (accurate)
Membership: Individual $35; Family $50; Supporting $100; Associate $250; Sponsor $500; Sustaining $1,000; Patron $5,000; Founder $10,000.

Saint Mary's City

THE DWIGHT FREDERIC BOYDEN GALLERY, (M), St. Mary's College of Maryland, 18952 E. Fisher Rd., Saint Mary's City, MD 20686-3002. Tel.: 240-895-4246. Fax: 240-895-4958.

Web Site: www.smcm.edu/boydengallery
Founded: 1971.
Personnel Profile: Full-Time Paid 2; Part-Time Paid 4; Interns 1.
Governing Authority: Saint Mary's College of Maryland. Tax-exempt.
Institution Type/Description: Art Gallery.
Collections: 20th & 21st century paintings, sculpture, prints & drawings.
Facilities: 1,600 sq. ft. exhibit space.
Activities: lectures; loan & traveling exhibitions. Annual Event: Student Art Exhibit.
Hours & Admission Prices: Sept.-May Mon.-Fri. 11-5; Summer: call for hours. No charge; donations accepted. &
Attendance: 4,000 (estimated)

* HISTORIC ST. MARY'S CITY, (M), Rte. 5, Saint Mary's City, MD 20686. Mailing Address: P.O. Box 39, Saint Mary's City, MD 20686-0039. Tel.: 240-895-4990. Fax: 240-895-4968. TDD: 800-735-2258.

E-mail: hsmc@smcm.edu
Web Site: www.stmaryscity.org

Founded: 1966.
Congressional District: 1
Key Personnel: C.E.O., Regina Faden; Chm. (V), John McAllister, Jr.; Pres. (V), Bonnie Green; Dir. Research, Henry M. Miller; Dir. Communications & Mktg., Susan Wilkinson; Dir. Educational Programs, Elizabeth Nosek; Museum Shop Mgr., Cheryl Stevenson.
Personnel Profile: Full-Time Paid 47; Part-Time Volunteers 196; Interns 2.
Governing Authority: state. Parent Institution: State of Maryland. Tax-exempt: 170(b)(1)(A).
Institution Type/Description: Outdoor Living History Museum.
Collections: archaeology; architecture; anthropology. Re-created Historic Houses: 1676 State House; 17th-century plantation & outbuildings; 17th-century inn. Historic Ship: Maryland Dove.
Research Fields: Maryland & Colonial history & architecture; archaeology.
Facilities: library of books on Maryland history, architectural history & archaeology available for research purposes. Museum-related items for sale.
Activities: living history interpretations & historical vignettes; craft demonstrations; archaeological field school; internships; permanent & temporary exhibitions; craft workshops; lectures; educational & outreach programs; sail-training program.
Publications: St. Mary's Research series; St. Mary's City Archaeology series; brochures; quarterly newsletters; annual report; visitor's guide.
Hours & Admission Prices: mid-March to Nov. Adults $10, senior citizens $9, youth $6; members and children 5 & under no charge. &
Attendance: 45,000 (estimated)
Membership: Individual $30; Dual & Family $50; Patron $100; Contributor $250; Benefactor $500.

Saint Michaels

*** CHESAPEAKE BAY MARITIME MUSEUM, (M),** 213 N. Talbot St., Saint Michaels, MD 21663-0636. Mailing Address: P.O. Box 636, Saint Michaels, MD 21663-0636. Tel.: 410-745-2916. Fax: 410-745-6088. Facebook: Chesapeake Bay Maritime Museum.
E-mail: havefun@cbmm.org
Web Site: www.cbmm.org
Founded: 1965.
Congressional District: 1
Key Personnel: Pres., Langley R. Shook; Chm., C G Appleby; Vice Chm., Tom Seip; Treas., Jim Harris; Chief Cur., Ronald E. (Pete) Lesher; Vice Pres. Finances, Jean Brooks; Vice Pres. Operations, William Gilmore; Vice Pres. Communications, Tracey Munson; Vice Pres. Constituent Svcs., Rene Stevenson; Vice Pres. Devel., David Crosson.
Personnel Profile: Full-Time Paid 27; Part-Time Paid 1; Part-Time Volunteers 210; Interns 4.
Governing Authority: private; nonprofit organization. Tax-exempt: 501(c)(3) & 170(6)(1)(A).
Institution Type/Description: Regional Maritime History Museum.
Collections: art; small craft; tools of fisheries; waterfowling; boatbuilding; lighthouses; steamboats; shoreside trades; manuscripts; photographs; ship plans.
Research Fields: Chesapeake Bay culture.
Facilities: 10,000-vol. library of books on maritime history & Chesapeake Bay area; 18 acre campus.
Activities: guided tours; lectures; permanent & temporary exhibits; year-round formally organized education & apprentice programs. Museum Sponsors: festivals; boat auction; outdoor concerts.
Publications: quarterly magazine, CBMM; monographs, A Heritage in Wood; Maryland's Oyster Navy; Lambert Wickes: Pirate or Patriot; Chesapeake Bay Sloops; Chesapeake Bay Crabbing Skiffs; Notes on Chesapeake Bay Skipjacks; booklets, Bay Sailing Craft; It's How You Pick the Crab: An Oral Portrait of Eastern Shore Crab Picking; Beacons of Hooper Strait; books, John M. Barbers Chesapeake; From Pot Pie to Hell and Damnation, An Illustrated Gazetteer of Talbot County; My Life as an Oyster.
Hours & Admission Prices: April-May & Sept.-Oct. daily 10-5; June-Aug. daily 10-6; Nov.-March daily 10-4. Adults $13, seniors $10, children 6-17 $6; discounts to groups & military personnel; children under 6 & members no charge. Closed New Year's Day; Thanksgiving; Christmas. &
Attendance: 63,000 (accurate)
Membership: Introductory $55; Family $70; Contributor $100; Supporter $200; Benefactor $500; Sustaining $1,000; Life $2,500.

ST. MICHAELS MUSEUM, E. Chestnut St. & St. Mary's Sq., Saint Michaels, MD 21663. Tel.: 410-745-9561.
E-mail: stmichaelsmuseum@atlanticbb.net
Web Site: www.stmichaelsmuseum.org
Founded: 1964.
Key Personnel: Pres. (V), Marie Martin; Cur., Kate Fones.
Governing Authority: Tax-exempt.

Institution Type/Description: History Museum.
Collections: local history & culture; period furnishings; personal artifacts; photographs; family papers; documents.
Hours & Admission Prices: May-Oct. Mon. 10-1, Fri. & Sun. 1-4, Sat. 10-4; other times by appointment. Adults $3, youth 6-17 $1.
Membership: Individual $20; Family $30; Supporting $50.

Salisbury

ART INSTITUTE & GALLERY, 212 W. Main St., Ste. 101, Salisbury, MD 21801. Tel.: 410-546-4748. Fax: 410-546-2793.
E-mail: aiandg@comcast.net
Web Site: www.artinstituteandgallery.com
Institution Type/Description: Art Gallery.
Collections: works by regional & national artists; paintings; sculpture.
Activities: classes. Annual Event: juried exhibition.
Hours & Admission Prices: Mon.-Sat. 10-3.

CHARLES H. CHIPMAN CULTURAL CENTER, 325 Broad St., Salisbury, MD 21801. Mailing Address: P.O. Box 4374, Salisbury, MD 21803. Tel.: 410-860-9290.
Institution Type/Description: History Museum: housed in a former African American church; built in 1837.
Collections: local history, heritage & culture; period furnishings; photographs; personal artifacts.
Hours & Admission Prices: Call for hours.

EDWARD H. NABB RESEARCH CENTER FOR DELMARVA HISTORY & CULTURE AT SALISBURY UNIVERSITY, 1450 Wayne St., #190, Salisbury, MD 21804. Mailing Address: 1101 Camden Ave., Salisbury, MD 21801. Tel.: 410-543-6312. Fax: 410-677-5067.
E-mail: rcdhac@salisbury.edu
Web Site: nabbhistory.salisbury.edu
Founded: 1982.
Congressional District: 1
Key Personnel: Dir., Dr. Ray Thompson; Chm. (V), Michael Hitch.
Personnel Profile: Full-Time Paid 5; Part-Time Paid 2; Part-Time Volunteers 20; Interns 8.
Governing Authority: Parent Institution: Salisbury University Foundation. Tax-exempt.
Institution Type/Description: History Museum.
Collections: local historic artifacts; archives; books; newspapers; documents; microfilm; photographs; family history; maps; manuscripts.
Facilities: library; archives.
Activities: research.
Publications: newsletter, Shoreline.
Hours & Admission Prices: Galleries: Fall & Spring Mon., Wed. & Fri. 1-4. Research: adults $5; members, faculty, students & staff no charge. Closed holidays.
Attendance: 5,000 (estimated)
Membership: Individual $30; Life $1,000.

HERITAGE CENTRE AND PEMBERTON HALL, Pemberton Historical Park, Pemberton Dr., Salisbury, MD 21804. Mailing Address: P.O. Box 573, Salisbury, MD 21803-0573. Tel.: 410-860-0447. Fax: 410-860-1441.
Institution Type/Description: History Museum.
Collections: local history & culture; hands-on exhibits; period furnishings; replica 18th century Eastern Shore Virginia tobacco barn. Historic House: 1741 home.
Hours & Admission Prices: Thurs.-Sun. 12-4. Heritage Center: adults $2, children under 12 $1.

POPLAR HILL MANSION, 117 Elizabeth St., Salisbury, MD 21801. Tel.: 410-749-1776. Facebook: Poplar Hill Mansion.
E-mail: curator@poplarhillmansion.org
Web Site: www.poplarhillmansion.org
Founded: 1976.
Key Personnel: Dir., Sarah Meyers; Chm. (V), Aleta Davis.
Personnel Profile: Full-Time Paid 1.
Governing Authority: Parent Institution: Friends of Poplar Hill Mansion, Inc. Tax-exempt.
Institution Type/Description: Historic House Museum.
Collections: local history & culture; period furnishings; photographs; personal artifacts.

Activities: rental facilities. Annual Events: Valentine Tea; Spring Luncheon; Fall Tea; Winter Luncheon; Tea With Santa; Holiday Open House in December.
Hours & Admission Prices: Feb.-Dec. 1st & 3rd Sun. each month 1-4; other times by appointment. No charge. Private Tours at other times: adults $5; discounts to school groups. Closed city holidays.
Attendance: 1,500 (accurate)
Membership: Individual $25; Family $50; Patron & Business $100; Lifetime $500.

SALISBURY STATE UNIVERSITY GALLERIES, 1101 Camden Ave., Salisbury, MD 21801-6837. Tel.: 410-548-2547 & 6000. Fax: 410-548-3002.
Web Site: salisbury.edu
Founded: 1962.
Congressional District: 1
Key Personnel: Cur., Linda Shipp; Dir. Cultural Affairs, June Krell-Salgado.
Personnel Profile: Full-Time Paid 3; Part-Time Paid 6; Part-Time Volunteers 30; Interns 1.
Governing Authority: university; not-for-profit. Parent Institution: Salisbury State University. Affiliated with University of MD System. Tax-exempt: 170(b)(1)(A).
Institution Type/Description: University Art Gallery.
Collections: Maryland regional artists; post-Impressionist prints & drawings; landscape photographs; 19th & 20th-century American sculpture; Japanese block prints.
Facilities: Fulton Hall Gallery and Atrium Gallery.
Publications: quarterly newsletter.
Hours & Admission Prices: Fulton Hall Gallery: Sept.-May Tues.-Fri. 10-4. Atrium Gallery: Sept.-May Mon.-Wed. 10-4. No charge; donations accepted. Closed major holidays. &
Attendance: 20,000 (accurate)
Membership: Individual $30; Family $50; Benefactor $100; Supporting $500.

THE SALISBURY ZOOLOGICAL PARK, 755 S. Park Dr., Salisbury, MD 21804-5600. Mailing Address: P.O. Box 2979, Salisbury, MD 21802-2979. Tel.: 410-548-3188. Fax: 410-860-0919.
E-mail: salisburyzooed@gmail.com
Web Site: www.salisburyzoo.org
Founded: 1954.
Congressional District: 1
Key Personnel: Dir., Joel Hamilton; Chm. (V), Ronald G. Alessi, Sr.; Cur. Education, Leonora Dillon; Membership & Museum Shop Mgr., Mary Seemann.
Personnel Profile: Full-Time Paid 14; Part-Time Paid 14; Part-Time Volunteers 50; Interns 4.
Governing Authority: Parent Institution: City of Salisbury. Subsidiary Institution: Salisbury Zoo Commission. Affiliated with the Dept. of Public Works, City of Salisbury, Government Bldg., Salisbury, MD 21801. Tel. 410-548-3170. Tax-exempt: 501(c)(3).
Institution Type/Description: Zoo.
Collections: mammals; reptiles; waterfowl; several species of endangered animals.
Research Fields: ethology.
Facilities: 100-vol. library of books on animals available for educational reasons to people associated with the zoo.
Activities: guided tours; ZOO-TO-YOU program; lectures; films; formally organized education programs for children & undergraduate college students affiliated with Salisbury St. University; docent program; permanent exhibitions.
Publications: quarterly newsletter.
Hours & Admission Prices: Memorial Day-Labor Day daily 9-7:30; Sept.-May daily 9-4:30. No charge; donations accepted. Closed Thanksgiving; Christmas. &
Attendance: 192,000 (estimated)
Membership: Senior $20; Individual $30; Individual Plus $40; Family & Grandparent $45; Deluxe Family $70; Naturalist $100; Patron $500; Wildlife Benefactor $1,000.

*** THE WARD MUSEUM OF WILDFOWL ART, SALISBURY UNIVERSITY, (M),** 909 S. Schumaker Dr., Salisbury, MD 21804-8722. Tel.: 410-742-4988. Fax: 410-742-3107.
E-mail: wardinfo@salisbury.edu
Web Site: www.wardmuseum.org
Founded: 1975.
Congressional District: 1
Key Personnel: Exec. Dir., Lora Bottinelli; Chm. Bd., Sam Dyke; Dir. Education, Mark McMullen-Bushman; Events Coord., Eric Turner; Dir.

Outreach, Rose Taylor; Cur. & Folklorist, Dr. Cynthia Byrd; Museum Shop Mgr., Linda Davis.
Personnel Profile: Full-Time Paid 12; Part-Time Paid 5; Part-Time Volunteers 200; Interns 4.
Governing Authority: nonprofit. Parent Institution: The Ward Foundation, Inc. DBA The Ward Museum of Wildfowl Art an affiliated Foundation of Salisbury University. Tax-exempt: 501(c)(3).
Institution Type/Description: Art Museum.
Collections: decoys; decorative bird carvings; wildfowl paintings; fowling pieces & skiffs; hunting artifacts; documents & manuscripts; audiovisual archives.
Research Fields: folk art, history, maritime & art history; American historical economics; folklore & folklife.
Facilities: library; 31,000 sq. ft. exhibit space; theatre; education center; nature trails. Gift items for sale.
Activities: guided tours; lectures; films; gallery talks; arts festivals; workshops; formally organized education programs; loan, permanent & temporary exhibitions. Chesapeake Wildfowl Expo; Ward World Championship Wildfowl Carving Competition; Art in Nature Photography Festival.
Publications: magazine, Wildfowl Art; documentary films; exhibit catalogues.
Hours & Admission Prices: Mon.-Sat. 10-5, Sun. 12-5. Families $17 on Sun.; adults $7, seniors $5, children under 18 $3; discounts to AAM members; preschoolers, members, Salisbury Univ. staff, students & faculty no charge. Closed New Year's Day; Thanksgiving; Christmas. &
Attendance: 35,000 (estimated)
Membership: Personal $35; Family $60; Sponsor $150; Contributor $250; Heritage $500; Heritage Gold $1,000; Benefactor $5,000. Foreign Members: Canada add $10; Other Countries add $25.

Sandy Spring

SANDY SPRING MUSEUM, (M), 17901 Bentley Rd., Sandy Spring, MD 20860-1001. Tel.: 301-774-0022. Fax: 301-774-8149.
Web Site: www.sandyspringmuseum.org
Founded: 1980.
Congressional District: 8
Key Personnel: Exec. Dir., Allison Weiss; Pres., Marcia Ferranto.
Personnel Profile: Full-Time Paid 1; Part-Time Paid 5; Part-Time Volunteers 20; Interns 5.
Governing Authority: nonprofit organization. Tax-exempt: 501(c)(3).
Institution Type/Description: History Museum.
Collections: farm tools; china; costumes; domestic utensils; furniture; photographs; books; toys; documents; letters; barn; smithy.
Research Fields: local social & architectural history; 18th-century land patents; index of the Annals of Sandy Spring; genealogical research; historic homes; 18th & 19th century Quakers, education, slavery & emancipation.
Facilities: 1,300-vol. library containing 18th to 20th-century books & photographs for use on premises; rental facilities; performance space; garden.
Activities: guided tours; lectures; arts festivals; organized education programs for children & adults; docent program; children's summer craft program; summer art show; community service hours program; members' tours to area museums. Museum Sponsors: Strawberry Festival; Antique Show; Garden Club; Holiday Open House; Chamber music series; Greens Sale.
Publications: quarterly newsletter, The Sandy Spring Museum Legacy.
Hours & Admission Prices: Mon. & Wed.-Thurs. 9-4, Sat.-Sun. 12-4. Adults $5; members & children no charge. Closed New Year's Eve & Day; Labor Day; Christmas Eve, Day & week. &
Attendance: 10,000 (estimated)
Membership: Individual $40; Family $65; Educators $200; Collectors $500; Curators $1,000; Preservationists $2,500; Innovators $5,000; Historians $10,000; Stewards $25,000.

SANDY SPRING SLAVE MUSEUM & AFRICAN ART GALLERY, INC., 18524 Brooke Rd., Sandy Spring, MD 20860-1407. Tel.: 301-774-4066.
Institution Type/Description: Art & History Museum
Collections: African American history & art; personal artifacts; photographs; paintings; sculpture; period furnishings; instruments.
Hours & Admission Prices: Adults $5. No charge; donations accepted. Closed national holidays. &

Scotland

POINT LOOKOUT STATE PARK & MUSEUM, 11175 Point Lookout Rd., Scotland, MD 20687. Tel.: 301-872-5688.
Institution Type/Description: Park & History Museum.
Collections: local history; photographs; Civil War artifacts.
Hours & Admission Prices: Daily 6am to sunset.

Shady Side

CAPTAIN AVERY MUSEUM, INC., (M), 1418 E.W. Shadyside Rd., Shady Side, MD 20764-9713. Mailing Address: P.O. Box 89, Shady Side, MD 20764-0089. Tel.: 410-867-4486. Fax: 410-867-4486.
E-mail: captainavery@verizon.net
Web Site: captainaverymuseum.org
Formerly: Shady Side Rural Heritage Society, Inc.
Founded: 1988.
Congressional District: 5
Key Personnel: Exec. Dir., Jeff Holland; Chair, Prue Hoppin; Vice Chair, Melanie Turner; Treas., Susy Smith; Public Rels., Mavis Daly; Museum Shop Mgr., Lynn Bedard.
Personnel Profile: Full-Time Volunteers 2; Part-Time Paid 2; Part-Time Volunteers 150.
Governing Authority: nonprofit. Tax-exempt.
Institution Type/Description: Historic House: 1860 waterman's house & 1920s fishing club.
Collections: furniture; tools; 1860-1920 boats; artifacts; photographs; oral history.
Research Fields: 1860-1890 life of watermen of Chesapeake; 20th century history; local community history.
Facilities: library of history & genealogy books available to the public; 600 sq. ft. exhibit space. Gift items for sale.
Activities: festivals; guided tours; lectures; participatory & temporary exhibitions; bus trips. Annual Events: lecture series in winter.
Publications: Miss Ethel Remembers...; Capt. Salem Avery House Museum, Its History 1860-1990; Doc. The Life of Emily Hammond Wilson; Journey To Our Past; teachers activity guide, Seasons of a Chesapeake Bay Waterman; Passing Through Shady Side.
Hours & Admission Prices: Museum: April-Dec. Sun. 1-4. Library: Mon. 12-3. Grounds & Outdoor Exhibits: daily dawn to dusk. No charge; donations accepted. &
Attendance: 4,000 (accurate)
Membership: Senior $25; Individual $30; Family $65; Friend $125; Donor $200; Supporter $300; Patron $500; Benefactor $1,000; Director's Circle $1,000.

Sharpsburg

ANTIETAM NATIONAL BATTLEFIELD-VISITOR CENTER, 5831 Dunker Church Rd., Sharpsburg, MD 21782. Mailing Address: P.O. Box 158, Sharpsburg, MD 21782-0158. Tel.: 301-432-5124. Fax: 301-432-4590. TDD: 301-432-5124.
Web Site: www.nps.gov/anti
Founded: 1890.
Congressional District: 6
Key Personnel: Supt., John Howard.
Governing Authority: federal. Affiliated with Dept. of the Interior National Park Service. Tax-exempt.
Institution Type/Description: Historic Site: site of 1862 Civil War Maryland Campaign & battle of Antietam or Sharpsburg.
Collections: Civil War weapons; military uniforms & equipment; relics; documents; lithographs; photographs; military and personal papers of Henry Kyd Douglas; Captain James Hope paintings; Cope-Carman battlefield maps. Historic Houses: c.1840 Sherrick House; 1853 Dunker Church, rebuilt 1961; late 18th-century Piper House; Pry House, Gen. McClellan's headquarters; 19th-century Mumma Farm.
Research Fields: Civil War history.
Facilities: 1,200-vol. library of Civil War records & books relating to Maryland campaign & participants available for use on the premises; numerous monuments & markers. Publications for sale.
Activities: taped tours; films; firearm demonstrations; historical walks; living history programs; audiovisual programs; wayside exhibits.
Publications: leaflet & map, Antietam Battlefield.
Hours & Admission Prices: Visitor Center: daily 9-5. Day Pass: family $6, adults $4. Closed New Year's Day; Thanksgiving; Christmas. &
Attendance: 313,201 (accurate)

BARRON'S C&O CANAL MUSEUM, 5632 Mose Cir., Sharpsburg, MD 21782. Tel.: 410-583-5299.
Institution Type/Description: History Museum: housed in a country store.
Collections: local history & heritage; period artifacts; photographs.
Hours & Admission Prices: Sat.-Sun. 9-6. &

KENNEDY FARM HOUSE MUSEUM, 2406 Chestnut Grove Rd., Sharpsburg, MD 21782. Tel.: 202-537-8900.
Institution Type/Description: Historic House Museum: housed in the farmhouse that served as a staging area for John Brown and his army as they prepared for the Harpers Ferry raid in the summer of 1859. A National Historic Landmark.
Collections: local history; period furnishings; personal artifacts.
Hours & Admission Prices: May-Oct. by appointment.

Silver Spring

GEORGE MEANY MEMORIAL ARCHIVES, 10000 New Hampshire Ave., Silver Spring, MD 20903-1706. Tel.: 301-431-5451. Fax: 301-431-5455.
E-mail: ldeloach@nlc.edu
Web Site: www.nlc.edu/archives/home.html
Founded: 1980.
Congressional District: 5
Key Personnel: Dir., Pat Greenfield; Archivist, Lynda DeLoach; Archivist, Sarah M. Springer.
Governing Authority: nonprofit. Parent Institution: The George Meany Center for Labor Studies. Tax-exempt.
Institution Type/Description: History Museum: labor organizations.
Collections: George Meany permanent exhibit; records of AFL, CIO & AFL-CIO; labor arts.
Publications: journal, Labor's Heritage.
Hours & Admission Prices: Research by appointment: Mon.-Tues. & Fri. 8:30-5, Wed.-Thurs. 7:30-6. No charge. Closed federal holidays. &
Attendance: 3,000 (estimated)

NATIONAL CAPITAL TROLLEY MUSEUM, 1313 Bonifant Rd., Silver Spring, MD 20905-5955. Tel.: 301-384-6352. Fax: 301-384-2865.
Web Site: www.dctrolley.org
Founded: 1959.
Congressional District: 5
Key Personnel: C.E.O., Pres. (V) & Museum Shop Mgr., Ken Rucker; Treas., Charles Tirschman; Dir. Devel., Wesley Paulson.
Personnel Profile: Full-Time Volunteers 2; Part-Time Paid 3; Part-Time Volunteers 30.
Governing Authority: private; nonprofit organization.
Institution Type/Description: Transportation Museum.
Collections: electric street cars; postal cards; photographs; trolley era ephemera; demonstration railway.
Facilities: 600-vol. of railway history books; 65-seat auditorium; demonstration electric railway with two mile roundtrip. Museum-related items for sale.
Activities: docent program; films; formal education for children; participatory exhibits. Annual Events: Holly Trolley Fest; DC Transit Day.
Publications: bimonthly newsletter, The Headway Recorder.
Hours & Admission Prices: Call for hours. Adults $7.
Attendance: 15,155 (accurate)
Membership: Student $15; Individual $30.

NATIONAL MUSEUM OF HEALTH AND MEDICINE, (M), 2500 Linden Ln., Silver Spring, MD 20910. Mailing Address: 2460 Linden Ln., Bldg. 2500, Wilver Spring, MD 20910. Tel.: 301-319-3300 & 3349. Facebook: National Museum of Health and Medicine.
Web Site: www.medicalmuseum.mil
Formerly: Army Medical Museum
Founded: 1862.
Key Personnel: Dir., Adrianne Noe, Ph.D.; Admin., Kevin Monahan; Public Programs Mgr., Andrea Schierkolk; Public Affairs Officer, Tim Clarke, Jr.
Personnel Profile: Full-Time Paid 40; Part-Time Volunteers 16.
Governing Authority: federal. Dept. of Defense. Tax-exempt.
Institution Type/Description: Medical History Museum.
Collections: anatomy; anthropology; forensic sciences; history & sociology of medical science & technology; materia medica; medical research; military medicine; parasitology; pathology; paleopathology; public health.
Research Fields: anatomy; anthropology; ethnomedicine & ethnopharmacology; forensic sciences; history & sociology of medical science & technology; medical research; military medicine; parasitology; pathology; paleopathology; public health & health education.
Facilities: research collection of books & periodicals on medicine, medical research & public health; archives.
Activities: permanent & special exhibits; education programs for children, students, adults; volunteer docent program; guided tours.
Publications: pamphlet on exhibits; catalogues; Billings Microscope Collection, 2nd Ed.; monthly electronic newsletter.
Hours & Admission Prices: Daily 10-5:30. No charge. Closed Christmas. &
Attendance: 50,000 (estimated)

PYRAMID ATLANTIC ART CENTER, 8230 Georgia Ave., Silver Spring, MD 20910. Tel.: 301-608-9101. Fax: 301-608-9102.
E-mail: hello@pyramid-atlantic.org
Web Site: www.pyramidatlanticartcenter.org
Key Personnel: Dir., Jose Dominguez
Institution Type/Description: Art Gallery.
Collections: works by national & international artists.
Activities: temporary exhibitions; educational programs; special events.
Hours & Admission Prices: Tues.-Thurs. 12-6, Fri. 12-7, Sat. 10-5, Sun. 12-5. No charge.
Membership: Basic $40; Creative $100; Expressive $150; Innovative $500.

SEVENTH-DAY ADVENTIST CHURCH, 12501 Old Columbia Pike, Silver Spring, MD 20904. Tel.: 301-680-6310.
Institution Type/Description: Religious Museum.
Collections: church history; religious artifacts & furnishings. Ellen White Estate: period documents & memorabilia from one of the church founders.
Activities: guided tours.
Hours & Admission Prices: Church Tours: Mon.-Thurs. 9am; groups by appointment. No charge. Ellen White Estate: Mon.-Thurs.

Snow Hill

FURNACE TOWN LIVING HERITAGE MUSEUM, 3816 Old Furnace Rd., Snow Hill, MD 21863-3420. Mailing Address: P.O. Box 207, Snow Hill, MD 21863-0207. Tel.: 410-632-2032. Fax: 410-632-1735.
E-mail: furnacetown@gmail.com
Web Site: www.furnacetown.com/museum.htm
Founded: 1982.
Congressional District: 1
Key Personnel: Bd. Pres. (V), Kathy Fisher; Dir., C.E.O. & Museum Shop Mgr., Heather McAllen; Treas., Lee Chisholm.
Personnel Profile: Full-Time Paid 1; Part-Time Paid 14.
Governing Authority: private; nonprofit organization. Tax-exempt: 501(c)(3).
Institution Type/Description: Living Heritage Museum.
Collections: 1820-1850 village life; broommaking; blacksmithing; woodworking; gardening; weaving; spinning.
Research Fields: early 19th century iron manufacturing.
Facilities: 100-vol. library. Museum-related items for sale.
Activities: docent program; formal education programs for children; arts festivals; guided tours; lectures. Annual Event: Celtic Festival.
Publications: newsletter, Furnace Town Times.
Hours & Admission Prices: Buildings: April-Oct. daily 10-5. Grounds: daily 10-5. Adults $5, senior citizens over 60 & military with ID $4.50, children 2-18 $3; discount to AAA members; members no charge. &
Attendance: 10,077 (accurate)
Membership: Individual $25; Family $50; Collier $75; Ironmaster $100; Corporate & Supporter $150.

JULIA A. PURNELL MUSEUM, 208 W. Market St., Snow Hill, MD 21863-1059. Tel.: 410-632-0515. Fax: 410-632-0515. Facebook: Julia Purnell Museum.
E-mail: mail@purnellmuseum.com
Web Site: www.purnellmuseum.com
Founded: 1942.
Congressional District: 1
Key Personnel: Dir., Claire Otterbein; Bd. Pres. (V), Meme Suznavick.
Personnel Profile: Full-Time Paid 1; Part-Time Paid 2; Part-Time Volunteers 50.
Governing Authority: board of directors; nonprofit. Parent Institution: Town of Snow Hill. Tax-exempt: 501(c)(3).
Institution Type/Description: History Museum: housed in 1891 former Catholic Church.
Collections: area U.S. history; history of Worcester County; recreated general merchandise store; agricultural & domestic tools; Worcester County machinery; Victorian life; regional history of Pocomoke Indians of 16th century; fine & folk art.
Research Fields: material culture of Worcester County's place in U.S. & Maryland history; museum collection.
Facilities: 670-vol. library; 1,387 sq. ft. exhibit space; archives; conference area.
Activities: guided tours; docent program; formal education program for Salisbury State Univ. students; training programs for professional museum workers; concerts; participatory & temporary exhibitions; time travel trunk of cultural artifacts for children; heritage arts programs for adults & children; scavenger hunts; field trip programs; group tours; NYC excursions; summer concerts. Annual Events: Victorian Christmas; needlework show & contest; Del Marva Needle Art Show and Competition; Kids' Discovery Day; Julia Purnell's Birthday Party; Fiber Fest!
Publications: newsletter, The Sampler; children's artifacts coloring book; biography, Julia A. Purnell, A Life Embroidered with Love; People of the Pocomoke; Smoke on the Pocomoke; Historic Snow Hill Walking Tour; educators' packet; Snow Hill Calendar of Events.
Hours & Admission Prices: April-Oct. Tues.-Sat. 10-4, Sun. 1-4; Nov.-March by appointment. Adults $2, children $.50; discounts to AAM & ICOM members; members no charge. Package tours available for surrounding museums. Closed major holidays. &
Attendance: 4,000 (estimated)
Membership: Student $10; Individual $15; Family $25; Patron $50; Sponsor $100; Lifetime $500.

MT. ZION ONE-ROOM SCHOOL, 117 Ironshire St., Snow Hill, MD 21863. Mailing Address: 230 S. Washington St., Snow Hill, MD 21863. Tel.: 410-632-1265.
E-mail: kpfisher@intercom.net
Founded: 1964.
Key Personnel: Chm. (V), Robert A. Fisher.
Personnel Profile: Part-Time Volunteers 3.
Governing Authority: Parent Institution: Worcester County Board of Education. Tax-exempt.
Institution Type/Description: Historic Building: built in 1869.
Collections: local history & culture; period furnishings; photographs; personal artifacts; desks; quill pens.
Hours & Admission Prices: By appointment. Adults $2, children $.50. &
Attendance: 200 (estimated)

Solomons

ANNMARIE SCULPTURE GARDEN & ART CENTER, (M), 13480 Dowell Rd., Solomons, MD 20629. Mailing Address: P.O. Box 99, Dowell, MD 20629-0099. Tel.: 410-326-4640. Fax: 410-326-4887. Facebook: Annmarie Arts.
E-mail: info@annmariegarden.org
Web Site: www.annmariegarden.org
Founded: 1992.
Key Personnel: Dir., Stacey Hann-Ruff.
Governing Authority: Tax-exempt.
Institution Type/Description: Sculpture Park & Arts Center.
Collections: outdoor sculpture; garden.
Facilities: nature trails.
Activities: educational programs; art classes for all ages; painting; drawing; ceramics; glass; fiber; jewelry; workshops; special events; several rotating exhibits throughout the year. Museum Sponsors: Artsfest in Sept.; Garden in Lights from Dec.-Jan.
Hours & Admission Prices: Park: daily 9-5. artLAB: daily 1-4. Arts Building: Tues.-Sun. 10-5. Adults $5, senior citizens 65 & older $4, children 5-12 $3; discount to military; children under 5 & AMG members no charge. Closed Independence Day; Thanksgiving; Christmas Eve & Day. &
Attendance: 50,000 (estimated)
Membership: Individual $50; Family $75; Friend $100; Patron $250; Benefactor $500; Ambassador $1,000 & up.

✳ CALVERT MARINE MUSEUM, (M), 14200 Solomons Island Rd., Solomons, MD 20688. Mailing Address: P.O. Box 97, Solomons, MD 20688-0097. Tel.: 410-326-2042. Fax: 410-326-6691. TDD: 1-800-735-2258.
E-mail: information@calvertmarinemuseum.com
Web Site: www.calvertmarinemuseum.com
Founded: 1969.
Congressional District: 1
Key Personnel: Dir., C. Douglass Alves, Jr.; Chm. (V), Don McDougall; Business Mgr., Roxie Welch; Deputy Dir., Sherrod Sturrock; Cur. Estuarine Biology, David Moyer; Cur. Paleontology, Stephen J. Godfrey; Cur. Maritime History, Richard Dodds; Cur. Exhibits, James Langley; Dir. Devel., Vanessa Gill; Maintenance Supvr., Kenny Heard; Boat Capt., Don Prescott; Museum Shop Mgr., Maureen P. Baughman.
Personnel Profile: Full-Time Paid 28; Full-Time Volunteers 1; Part-Time Paid 34; Part-Time Volunteers 270; Interns 5.
Governing Authority: Calvert Marine Museum Society. Parent Institution: Calvert County Government. Tax-exempt: 501(c)(3); 170(b)(1)(A).
Institution Type/Description: Marine Museum.
Collections: maritime history; small craft; estuarine biological specimens; marine paintings, fossils. Historic Boat: 1899 log-built oyster boat. Historic Buildings: 1883 Drum Point Lighthouse on waterfront; 1934 seafood packing house; 1925 schoolhouse; 1828 Cove Point Lighthouse.

Research Fields: paleontology; maritime history; estuarine biology; southern Maryland; Chesapeake Bay.
Facilities: 11,000-vol. library; photographic & manuscript archives; commercial fisheries collections housed in seafood packing house. Local handicrafts, art, books, models & decorative items pertaining to museum's collection for sale.
Activities: guided tours; lectures; craft demonstrations; slide & film programs; field trips; formally organized educational programs; permanent & temporary exhibitions; boat rides.
Publications: quarterly newsletter, Bugeye Times.
Hours & Admission Prices: Daily 10-5. Adults $9, seniors & military $7, children $4; discount to AAM & CAMM members; members no charge. Closed New Year's Day; Thanksgiving; Christmas. &
Attendance: 73,025 (accurate)
Membership: Individual $40 & up; Family $60 & up; Sustaining $120 & up; Associate $275 & up; Patron $525 & up; Benefactor $1,000; Bugeye Society $1,500 & up. Corporate Dues: Sustaining $125 & up; Associate $250 & up; Patron $500 & up; Bugeye Society $1,000 & up.

CHESAPEAKE BIOLOGICAL LABORATORY VISITORS CENTER, 200 Farren Ave., Solomons, MD 20688. Mailing Address: P.O. Box 38, Solomons, MD 20688. Tel.: 410-326-7443.
Institution Type/Description: Environmental Heritage Museum.
Collections: local ecology & natural resources; photographs; weather.
Activities: educational programs; hands-on exhibitions.
Hours & Admission Prices: Visitor Center: May & Sept. to Dec. 8th Fri.-Sun. 12-4; Memorial Day to Labor Day Wed.-Sun. 12-4. Lab Tours: May to Dec. 8th Fri. 2pm. No charge.

Stevenson

STEVENSON UNIVERSITY ART GALLERY, 1525 Greenspring Valley Rd., Stevenson, MD 21153. Tel.: 443-352-4491. Facebook: Stevenson University Art Gallery.
E-mail: exhibitions@stevenson.edu
Founded: 1997.
Congressional District: 3
Governing Authority: .Parent Institution: Stevenson University. Subsidiary Instutions: St. Paul Companies, Greenspring, School of Design Gallery, Owings Mills
Institution Type/Description: Art Gallery.
Collections: works by regional artists; paintings; prints; sculpture; photographs.
Activities: lectures; temporary exhibitions.
Hours & Admission Prices: Mon.-Wed. & Fri. 11-5, Thurs. 11-8, Sat. 1-4. No charge; donations accepted. &

Stevensville

CRAY HOUSE, Cockey's Ln., Stevensville, MD 21666. Mailing Address: P.O. Box 321, Stevensville, MD 21666. Tel.: 410-758-2502.
Congressional District: 36
Personnel Profile: Part-Time Volunteers 4.
Institution Type/Description: Historic House Museum: housed in the former home of Nora Cray; built c.1809. Listed on the National Register of Historic Places.
Collections: local history & culture; period furnishings.
Activities: Museum Sponsors: Kent Island Day.
Hours & Admission Prices: May-Oct. 1st Sat. each month 12-4. No charge; donations accepted.

HISTORIC CHRIST CHURCH, 121 E. Main St., Stevensville, MD 21617. Mailing Address: Queen Anne's County Parks & Public Landing, 1945 4H Park Rd., Centreville, MD 21617-2172. Tel.: 410-604-2100.
Congressional District: 36
Personnel Profile: Part-Time Volunteers 3.
Governing Authority: Parent Institution: Queen Anne's County Parks. Tax-exempt.
Institution Type/Description: Historic Building: built in 1880.
Collections: local history, heritage & culture; religious furnishings & artifacts.
Activities: special events.
Hours & Admission Prices: By appointment. No charge; donations accepted.

HISTORIC STEVENSVILLE BANK - KENT ISLAND HERITAGE SOCIETY, 409 Love Point Rd., Stevensville, MD 21666. Mailing Address: P.O. Box 321, Stevensville, MD 21666. Tel.: 410-758-2502.
Formerly: Old Stevensville Post Office - Kent Island Heritage Society
Congressional District: 36
Personnel Profile: Part-Time Volunteers 4.
Governing Authority: Parent Institution: Kent Island Heritage Society.
Institution Type/Description: HistoricBuilding: built 1903.
Collections: local history & culture; period furnishings; photographs.
Activities: special events.
Hours & Admission Prices: May-Oct. 1st Sat. each month 12-4; other times by appointment. No charge; donations accepted.

STEVENSVILLE TRAIN DEPOT, Cockey's Ln., Stevensville, MD 21666. Mailing Address: P.O. Box 321, Stevensville, MD 21666. Tel.: 410-758-2502.
Congressional District: 36
Institution Type/Description: Historic Building: built in 1902.
Collections: local history & culture; period furnishings; photographs.
Activities: special events.
Hours & Admission Prices: May-Oct. 1st Sat. each month 12-4; other times by appointment. No charge; donations accepted.

Sudlersville

SUDLERSVILLE TRAIN STATION MUSEUM, 101 Linden St., Sudlersville, MD 21668. Mailing Address: P.O. Box 2, Sudlersville, MD 21668. Tel.: 410-438-3501.
E-mail: sudlersvillemuseum@gmail.com
Web Site: www.sudlersvillemuseum.org
Governing Authority: Parent Institution: Sudlersville Community Betterment Club, Inc. Tax-exempt.
Institution Type/Description: Historic Building: built in 1885.
Collections: local history & culture; period furnishings; Jimmy Foxx memorabilia.
Activities: Museum Sponsors: Old Fashion Christmas 1st & 2nd Sat in Dec.
Hours & Admission Prices: May-Oct. 1st & 3rd Sat. each month 10-2. No charge; donations accepted.

Suitland

AIRMEN MEMORIAL MUSEUM, 5211 Auth Rd., Suitland, MD 20746-4339. Mailing Address: P.O. Box 50, Temple Hills, MD 20757-0050. Tel.: 301-899-3500; 800-638-0594. Fax: 301-899-8136.
E-mail: staff@afsahq.org
Web Site: www.afsahq.org
Founded: 1986.
Congressional District: 4
Key Personnel: Exec. Dir. & C.E.O., John R. McCauslin.
Personnel Profile: Full-Time Paid 1.
Governing Authority: nonprofit. Parent Institution: Air Force Sergeants Association. Tax-exempt: 501(c)(3).
Institution Type/Description: Military Museum.
Collections: personal items & equipment used by enlisted personnel during their service with the Army Air Corps, Army Air Forces, & U.S. Air Force; archival holdings from 1907 particular emphasis on World War II color photography.
Research Fields: contributions by enlisted men & women in the Army Air Corps, Army Air Forces, & modern day Air Force.
Facilities: library of material pertaining to enlisted personnel; 6,000 sq. ft. exhibit space.
Activities: magazine, SERGEANTS.
Publications: The Airmen Heritage Series, video histories & educational monographs.
Hours & Admission Prices: Mon.-Fri. 8-5. No charge; donations accepted. Closed federal holidays. &

Sykesville

GATE HOUSE MUSEUM OF HISTORY, 7283 Cooper Dr., Sykesville, MD 21784. Tel.: 410-549-5150. Facebook: Sykesville and Eldersburg back in the day.
E-mail: mfraser@sykesville.net
Founded: 1904.
Congressional District: 8
Key Personnel: Dir., Curator, Dr. Mark Fraser; Chm. (V), Diana Stager.
Personnel Profile: Part-Time Paid 1; Part-Time Volunteers 17.

Volunteer Hours: 2,642
Operating Expenses: 18,712
Operating Income: 19,648
Governing Authority: Parent Institution: Town of Sykesville. Tax-exempt.
Institution Type/Description: History Museum.
Collections: local history & culture; period furnishings; personal artifacts; photographs.
Research Fields: local family genealogy & architecture.
Activities: special events; summer children's programs; genealogy training.
Publications: newsletter, Artifacts.
Hours & Admission Prices: Wed. & Fri. 10-5, Sun. 1-4. No charge; donations accepted.
Attendance: 2,346 (estimated)
Membership: Individual $25.

Thurmont

CATOCTIN IRON FURNACE & MANOR HOUSE RUINS, 14039 Catoctin Hollow Rd., Thurmont, MD 21788. Tel.: 301-271-7574; 800-830-3974.
Institution Type/Description: History Museum.
Collections: local history; iron-making process; slavery history.
Hours & Admission Prices: April-Oct. daily 8am to sunset; Nov.-March daily 10am to sunset.

CATOCTIN WILDLIFE PRESERVE & ZOO, 13019 Catoctin Furnace Rd., Thurmont, MD 21788-2134. Tel.: 301-271-4922 & 3180. Fax: 301-271-2673.
E-mail: administration@cwpzoo.com
Web Site: www.cwpzoo.com
Founded: 1933.
Key Personnel: Dir., Richard Hahn, C.A.P.; Pres. (V), Carole R. Brown; Vice Pres., Kelly Johnson; Asst. Dir. Zoological Affairs, June Bellizzi; Concessions Mgr., Brandi Owens.
Personnel Profile: Full-Time Paid 24; Full-Time Volunteers 1; Part-Time Paid 10; Part-Time Volunteers 15; Interns 6.
Governing Authority: corporation. Parent Institution: Global Wildlife Trust, Inc. Tax-exempt.
Institution Type/Description: Zoological Park.
Collections: animals from around the world; natural history; reptiles; gardens; Koi pond; safari.
Research Fields: reptilian propagation, primate propagation & in-situ conservation.
Facilities: outdoor stage; picnic pavilion; concession stand; classroom.
Activities: guided tours by appointment; lectures; sleepovers; birthday parties; 25-acre safari ride.
Publications: quarterly newsletter, Encounters.
Hours & Admission Prices: March & Nov. Sat.-Sun. 10-4 (weather permitting); April Mon.-Fri. 10-5, Sat.-Sun. 9-5; May daily 9-5; Memorial Day to Sept. Mon.-Fri. 9-5, Sat.-Sun. 9-6; Oct. daily 10-5. Adult $16.95, children $9.95; discounts for seniors, military & scheduled groups of 15 or more; members no charge. &
Attendance: 80,000 (estimated)
Membership: Child $27; Senior Citizen & Military $40; Adult $45; Family $134.

Towson

ASIAN ARTS & CULTURE CENTER, TOWSON UNIVERSITY, Asian Arts & Culture Center, Towson University, 8000 York Rd., Towson, MD 21252. Tel.: 410-704-2807. Fax: 410-704-4032.
E-mail: sshieh@towson.edu
Web Site: www.towson.edu/asianarts
Founded: 1971.
Congressional District: 2 & 3
Key Personnel: Dir., Mrs. Suewhei Shieh; Pres. (V), Anthony Montcalmo.
Personnel Profile: Full-Time Paid 1; Part-Time Paid 5; Part-Time Volunteers 20; Interns 1.
Governing Authority: university; nonprofit. Parent Institution: Towson University. Tax-exempt: 501(c)(3).
Institution Type/Description: University Arts Center.
Collections: concentration on Asian art from neolithic to modern times.
Facilities: 100-vol. library on Asian art; 2,000 sq. ft. exhibit space.
Activities: lectures; arts festivals; workshops; concerts; dance recitals; films; temporary, permanent, loan & traveling exhibitions. Annual Event: craft show & sale.
Publications: biannual newsletter.
Hours & Admission Prices: Academic Year: Mon.-Fri. 11-4, Sat. 1-4; call to confirm hours. Exhibits: no charge. Special Events: adults $10-$20; discounts to senior citizens, students, AAM members & museum members. Closed Easter; Christmas; national holidays. &

Attendance: 10,000 (estimated)
Membership: Individual $45; Family & Dual $75; Crane Club $100-$249; Tiger $250-$499; Phoenix Circle $500-$999; Dragon Circle $1,000 & up; Jade Circle $2,500.

HAMPTON NATIONAL HISTORIC SITE, 535 Hampton Lane, Towson, MD 21286-1397. Tel.: 410-823-1309. Fax: 410-823-8394.
Web Site: www.nps.gov/hamp
Founded: 1948.
Congressional District: 1
Key Personnel: Supt., Laurie Coughlan; Gen. Supt., Gay Vietzke; Volunteer Mgr., Kirby Shedlowski; Cur., Gregory Weidman; Museum Shop Mgr., Eileen Kalinoski.
Personnel Profile: Full-Time Paid 9; Part-Time Paid 1; Part-Time Volunteers 45; Interns 2.
Governing Authority: federal. Parent Institution: National Park Service. Affiliated with U.S. Dept. of the Interior, National Park Service, Washington, DC 20240. Tax-exempt.
Institution Type/Description: Historic House & Site: c.1783-1790 late Georgian Mansion including slave quarters located on agricultural-industrial complex.
Collections: furniture & decorative objects c.1760-1948; English landscape park & formal Italian gardens; specimen trees; 27 historic structures including dairy & slave quarters; archives; 5,000 historic photographs.
Research Fields: decorative arts; history; architecture; gardening; ethnography.
Facilities: formal gardens. Museum-related items for sale.
Activities: guided tours; self-guided walking tour; lectures. Museum Sponsors: Second Sunday Programs.
Publications: brochure; cookbook; guidebook; walking tour guide of gardens & grounds; site bulletins.
Hours & Admission Prices: Mansion: daily 9-4. Grounds: daily 9-5. Tours: daily 9-4; groups of 10 or more by appointment. No charge; donations accepted. Closed New Year's Day; Thanksgiving; Christmas. &
Attendance: 30,000 (estimated)
Membership: Historic Hampton, Inc.: Senior Citizen $25; Individual $30; Family $45; Sponsor $100; Patron $250.

Union Bridge

WESTERN MARYLAND RAILWAY HISTORICAL SOCIETY, 41 N. Main St., Union Bridge, MD 21791-9100. Mailing Address: P.O. Box 395, Union Bridge, MD 21791-0395. Tel.: 410-775-0150.
Web Site: www.westernmarylandrhs.com
Founded: 1967.
Key Personnel: Chm. (V), Stan Johnson; Pres. (V), Dennis Wertz; Museum Shop Mgr., Dick Liebno
Institution Type/Description: Historical Society Museum.
Collections: railroad history & artifacts; photographs.
Hours & Admission Prices: Sun. 1-4, Wed. 9-12 & 1-3; other times by appointment. No charge; donations accepted. Closed New Year's Day; Easter; Christmas.
Attendance: 1,000 (estimated)
Membership: U.S. $30; International $50.

Upper Fairmount

FAIRMOUNT ACADEMY HISTORICAL ASSOCIATION, Fairmount Rd., Upper Fairmount, MD 21867. Mailing Address: P.O. Box 133, Upper Fairmount, MD 21867. Tel.: 410-651-0781; 800-521-9189.
Institution Type/Description: History Museum.
Collections: local history & culture; period furnishings; photographs; personal artifacts; corn huskers; watermelon dredges; crab traps; tongs; medical equipment; Native American artifacts; farm implements; archeological remnants. Historic Buildings: c.1839 two-story schoolhouse; 1872 Knights of Pythias lodge; barbershop.
Activities: special events.
Hours & Admission Prices: By appointment.

Upper Marlboro

BILLINGSLEY HOUSE MUSEUM, 6900 Green Landing Rd., Upper Marlboro, MD 20772-7618. Tel.: 301-627-0730. Fax: 301-627-7085. TTY: 301-446-3302.
Institution Type/Description: Historic House Museum: built c.1740.
Collections: local history & culture; period furnishings; photographs.
Activities: rental facilities.
Hours & Admission Prices: Tours: Tues. & Fri. 9-3; call for additional hours.

DARNALL'S CHANCE HOUSE MUSEUM, M-NCPPC, 14800 Gov. Oden Bowie Dr., Upper Marlboro, MD 20772-3073. Tel.: 301-952-8010. Fax: 301-952-1773.
E-mail: susan.reidy@pgparks.com
Web Site: www.pgparks.com
Founded: 1988.
Congressional District: 5
Key Personnel: Natural & Historical Resources Div. Mgr., Anthony Nolan; Dir., Susan Reidy.
Personnel Profile: Full-Time Paid 2; Part-Time Paid 4; Part-Time Volunteers 10.
Governing Authority: state. Parent Institution: Maryland National Capital Park & Planning Commission. Subsidiary Institution: Natural & Historical Resources Div. Tax-exempt.
Institution Type/Description: Historic House: 1742, 18th century home of James and Lettice Wardrop.
Collections: archaeological artifacts; c. 18th C. underground brick burial vault; period furniture.
Research Fields: archaeology; burial vaults; county history.
Facilities: rental facilities.
Activities: guided tours; lectures; permanent, temporary & loan exhibitions; rental facilities; weddings; meetings. Annual Events: Colonial Tavern Dinners; Gingerbread House Contest & Show; Highland Tea; Pirate Fest.
Hours & Admission Prices: Tues.-Thurs. tours by appointment, Fri. & Sun. 12-4. Adults $3; discounts to AAM members. Closed major holidays. &
Attendance: 3,690 (accurate)

MERKLE WILDLIFE SANCTUARY & VISITOR CENTER, 11704 Fenno Rd., Upper Marlboro, MD 20772-8179. Tel.: 301-888-1377; 800-784-5380.
Web Site: www.dnr.state.md.us/publiclands/merkletrails.html
Institution Type/Description: Wildlife Sanctuary.
Collections: wildlife & their habitats; natural history; ecology.
Facilities: nature trails.
Activities: hiking; bird watching.
Hours & Admission Prices: Grounds: daily 7-sunset. Visitor Center: Sat.-Sun. 10-4.

MOUNT CALVERT HISTORICAL AND ARCHAEOLOGICAL PARK, 16801 Mt. Calvert Rd., Upper Marlboro, MD 20772. Tel.: 301-627-1286. TTY: 301-699-2544.
Institution Type/Description: History Museum.
Collections: local history & archaeology; Native American culture; African American history; War of 1812; Chesapeake Beach Railway; photographs; personal artifacts; pottery; early tools.
Activities: educational programs; research.
Hours & Admission Prices: April-Oct. Sat. 8:30-5, Sun. 12-4; other times by appointment.

THE PATUXENT RURAL LIFE MUSEUMS, Patuxent River Park, 16000 Croom Airport Rd., Upper Marlboro, MD 20772-8395. Mailing Address: 6706 Green Landing Rd., Upper Marlboro, MD 20772-7618. Tel.: 301-627-6074. Fax: 301-627-7085. TDD: 301-699-2544.
E-mail: mary.haley-amen@pgparks.com
Web Site: www.pgparks.com/places/eleganthistoric/patuxent_intro.html
Formerly: W. Henry Duvall Tool Museum
Founded: 1983.
Key Personnel: Dir., Mary Haley-Amen.
Personnel Profile: Full-Time Paid 1; Part-Time Paid 4; Part-Time Volunteers 10.
Governing Authority: county; nonprofit. Parent Institution: Patuxent River Park. Tax-exempt.
Institution Type/Description: Tool Museum.
Collections: farm tools; dental office; hand tools; storekeeper's supplies; domestic items; carpenter tools.
Facilities: 20-vol. library of tool guides.
Activities: guided tours.
Hours & Admission Prices: April-Oct. Sat.-Sun. 1-4; other times by appointment. Guided Tours: by appointment. Sat.-Sun. no charge. &
Attendance: 2,158 (accurate)

Waldorf

DR. SAMUEL A. MUDD HOUSE, 3725 Dr. Samuel Mudd Rd., Waldorf, MD 20601. Mailing Address: 14940 Hoffman Rd., Waldorf, MD 20601. Tel.: 301-274-4232 & 645-6870.
E-mail: dannyfluhart1@comcast.net

Key Personnel: Pres., Danny Fluhart; Museum Shop Mgr., Esther Sherwell.
Personnel Profile: Full-Time Paid 1; Part-Time Volunteers 15.
Governing Authority: Tax-exempt.
Institution Type/Description: Historic House Museum.
Collections: Dr. Mudd's life & career; period furnishings; personal artifacts; photographs.
Facilities: Museum-related items for sale.
Activities: special events. Museum Sponsors: Victorian Christmas 1st weekend in December.
Publications: newsletter 3 times per year.
Hours & Admission Prices: late March to late Nov. Wed. & Sat. 11-4, Sun. 12-4. Adults $7, children 6-16 $2; members no charge. Closed Easter.
Membership: Individual $15; Family $20; Contributing $25; Honorary $30; Life $200; Memorial $1,000.

PISCATAWAY INDIAN MUSEUM, 16816 Country Ln., Waldorf, MD 20601. Tel.: 240-432-7878.
E-mail: info@piscatawayindians.org
Web Site: www.piscatawayindians.org
Institution Type/Description: Native American History Museum.
Collections: Native American history, heritage & culture; period furnishings; replice longhouse; personal artifacts.
Hours & Admission Prices: Sun. 1-4; other times by appointment.

Walkersville

FOUNTAIN ROCK NATURE CENTER, 8511 Nature Center Place, Walkersville, MD 21793-8325. Tel.: 301-898-1460.
Web Site: www.recreater.com
Founded: 1990.
Key Personnel: Park Naturalist, Alice Nemitsas.
Personnel Profile: Full-Time Paid 1; Part-Time Paid 16; Part-Time Volunteers 40.
Governing Authority: county; nature council; nonprofit. Parent Institution: Frederick County Parks & Recreation. Tax-exempt: 501(c)(3).
Institution Type/Description: Nature Center.
Collections: wildlife mounts; working & observational honeybee hive; water spring; wetland, woodland & field habitats; 1872 battery of limestone kilns; hands-on exhibits; natural history; Native American artifacts; live animals including snakes, toads, turtles & insects.
Facilities: 22.5 acres in county park; nature trails.
Activities: nature programs; educational birthday parties.
Publications: The Recreater.
Hours & Admission Prices: 8am to sunset; groups by appointment. No charge. Nature Programs: adults $5, children $4. Closed holidays. &
Attendance: 10,000 (estimated)

WALKERSVILLE SOUTHERN RAILROAD MUSEUM, 34 W. Pennsylvania Ave., Walkersville, MD 21793-8505. Mailing Address: P.O. Box 651, Walkersville, MD 21793-0651. Tel.: 301-898-0899; 877-363-WSRR (toll free).
E-mail: musdir@wsrr.org
Web Site: www.wsrr.org/museum.htm
Founded: 1995.
Key Personnel: Dir., John Meise.
Personnel Profile: Full-Time Volunteers 1; Part-Time Volunteers 10; Interns 1.
Institution Type/Description: Railroad History Museum.
Collections: railroad history; model railroad; Pennsylvania Railroad magazine ads, timetables & documents. Historic Buildings: railroad station; freight house.
Activities: ride 1920s passenger car; locomotive facilities tours.
Publications: monthly online newsletter.
Hours & Admission Prices: Museum: call for hours. Train: May-June & Sept. Sat.-Sun. 11am & 2pm; Oct. Sat.-Sun. 11am, 1pm & 3pm; July-Aug. Sat. 11am & 2pm; additional special event hours. Adults $9, seniors over 55 $8, children $5; members and children under 3 no charge. &
Attendance: 12,000 (accurate)

Warwick

OLD BOHEMIA HISTORICAL SOCIETY, Bohemia Church Rd., Warwick, MD 21912. Mailing Address: P.O. Box 61, Warwick, MD 21912. Tel.: 302-328-4803.
Founded: 1953.
Congressional District: 1
Key Personnel: Pres. & Museum Shop Mgr., Margaret Matyniak; Pastor, Rev. Steven B. Giuliano.
Personnel Profile: Part-Time Volunteers 10.

Governing Authority: society. Parent Institution: Catholic Diocese Foundation. Tax-exempt: 501(c)(3).
Institution Type/Description: Historical Society Museum: 1704 Jesuit mission site, 1797 church, rectory with museum of religious artifacts (liturgical vessels, vestments, prayer books, devotional articles), barn with farm conveyances & tools, historic cemetery.
Collections: old furniture & church-related articles; patens; chalices; farm equipment; history & religious books; tools.
Activities: religious & educational activities.
Hours & Admission Prices: Call for hours. No charge; donations accepted.
Attendance: 700 (estimated)
Membership: Annual $15; 3 yr. $40; Benefactor $100.

Westernport

WESTERNPORT HERITAGE SOCIETY, 117 Maryland Ave., Westernport, MD 21562. Tel.: 301-359-0388.
Web Site: pages.prodigy.net/jimertz
Founded: 1989.
Congressional District: 1
Key Personnel: Pres. (V), Allan T. LaRue; Museum Shop Mgr., Mary Ann Imhoff.
Personnel Profile: Part-Time Volunteers 28.
Governing Authority: Parent Institution: Westernport Heritage Society, Inc. Tax-exempt.
Institution Type/Description: Heritage Society Museum.
Collections: local history & heritage; photographs; G-scale train; HO-scale railcars.
Publications: newsletter, Daybreak!
Hours & Admission Prices: 2nd & 4th Sat.-Sun. 1-4. No charge; donations accepted. &
Attendance: 315 (estimated)
Membership: Individual $5; Life $35.

Westminster

CARROLL COUNTY ARTS COUNCIL TEVIS AND COMMUNITY GALLERIES, 91 W. Main St., Westminster, MD 21157. Tel.: 410-848-7272. Fax: 410-848-8962.
E-mail: info@carrollcountyartscouncil.org
Web Site: www.carrollcountyartscouncil.org
Key Personnel: Exec. Dir., Sandy Oxx
Institution Type/Description: Art Gallery.
Collections: paintings; photographs; sculpture.
Facilities: 263-seat theater.
Activities: workshops; classes; educational programs.
Hours & Admission Prices: Mon.-Wed. & Fri.-Sat. 10-4, Thurs. 10-8. No charge.

CARROLL COUNTY FARM MUSEUM, (M), 500 S. Center St., Ste. 1, Westminster, MD 21157-5664. Tel.: 410-386-3880. Fax: 410-876-8544. MD Relay: 711/800-735-2258.
E-mail: ccfarm@ccg.carr.org
Web Site: www.carrollcountyfarmmuseum.org
Founded: 1965.
Congressional District: 6
Key Personnel: Park Supt., Dottie Freeman; Admin. Office Assoc., Susan Dell; Cur., Victoria Fowler; Events Coord., Bonnie Hood.
Personnel Profile: Full-Time Paid 7; Part-Time Paid 5; Part-Time Volunteers 100.
Governing Authority: county; nonprofit. Parent Institution: Carroll County Government. Tax-exempt.
Institution Type/Description: Agricultural & Historical Museum: housed in c.1852 historic building on 142 acres.
Collections: artifacts relating to the lifestyle of rural 19th-century Carroll County & rural America.
Facilities: 600-vol. library on agriculture, farm equipment, farm life & archives available for inter-library loan & to the public upon request; nature trail. Museum-related items for sale.
Activities: formal education programs for children; guided tours; artisan workshops; temporary exhibitions; weddings; receptions; reunions; picnics. Annual Events: Civil War Encampment; Fiddler's Convention; Old Fashioned July 4th Celebration; Spring Muster & Antique Fire Equipment Show; Living History Camp; Steam Show Days; The Maryland Wine Festival; Blacksmith Day; Fall Harvest Days; Holiday Visit; Specialty Teas; Surf & Turf Festival.
Publications: calendar of events; fliers; brochures; rack cards.
Hours & Admission Prices: May-June & Sept.-Oct. Sat.-Sun. 12-5; July-Aug. Tues.-Fri. 10-4, Sat.-Sun. 12-5; group tours by reservation only. Adults $5,

students 7-18 and senior citizens 60 & over $4; discounts to groups & AAM members; children 6 & under & members no charge. &
Attendance: 100,000 (accurate)
Membership: Individual $30; Family $60; Lifetime $300.

HISTORICAL SOCIETY OF CARROLL COUNTY, (M), 210 E. Main St., Westminster, MD 21157-5225. Tel.: 410-848-6494. Fax: 410-848-3596.
E-mail: hscc@carr.org
Web Site: hscc.carr.org
Founded: 1939.
Key Personnel: Exec. Dir., Timatha S. Pierce; Chm. (V), Tom Rasmussen; Treas., David Miller; Mgr. Operations, Linda Cunfer; Cur., Cathy Baty; Administrative Asst., Lin Conroy; Museum Shop Mgr., Debbie Leister.
Personnel Profile: Full-Time Paid 3; Part-Time Paid 2; Part-Time Volunteers 169; Interns 1.
Governing Authority: private; nonprofit organization. Tax-exempt: 501(c)(3).
Institution Type/Description: Historical Society Museum.
Collections: history & culture of Carroll County; photographs; clothing; quilts; ceramics & glass; furniture; clocks. Historic Houses: c.1807 Sherman-Fisher-Shellman House, contains personal artifacts depicting the life of a Pennsylvania German family; c.1820 Cockey's Tavern; Kimmey House; c.1800
Facilities: 1,000-vol. library; 1,200 sq. ft. exhibit space; learning center. Museum-related items for sale.
Activities: docent program; guided tours; lectures; broadcast programs; temporary exhibits; monthly box lunch talks. Annual Events: County Birthday Celebration; Maryland in the Civil War seminar; Antiques Appraisal Day; Shellman Birthday; Past Times for Children.
Publications: newsletter, Carroll Courier; various books available related to local history; Carroll History Journal.
Hours & Admission Prices: Office: Mon.-Fri. 8:30-5. Library: Tues.-Fri. 9:30-12:30 & 1-4, 2nd & 4th Sat. 9-12. Gallery: Tues.-Fri. 12:30-4:30, Sat.-Sun. by appointment. No charge; donations accepted. Sherman-Fisher Shellman House Museum: adults $3; members no charge. Closed New Year's Day; Martin Luther King Jr. Day; Presidents' Day; Good Friday; Memorial Day; Independence Day; Labor Day; Thanksgiving Day & day after; Christmas week. &
Attendance: 3,000 (estimated)
Membership: Senior Individual & Student $25; Individual $40; Senior Couple $45; Family $60; Business $250.

UNION MILLS HOMESTEAD & GRIST MILL, 3311 Littlestown Pike, Westminster, MD 21158-2137. Tel.: 410-848-2288.
E-mail: ejss61@aol.com
Web Site: www.unionmills.org
Founded: 1797.
Congressional District: 6
Key Personnel: Exec. Dir. & Museum Shop Mgr., Jane S. Sewell; Pres. Bd., Dr. Dawn Thomas.
Personnel Profile: Full-Time Paid 1; Part-Time Paid 2; Part-Time Volunteers 40.
Governing Authority: nonprofit foundation; Union Mills Homestead Foundation. Administered for Carroll County Commissioners. Tax-exempt: 501(c)(3).
Institution Type/Description: Historic House Museum: 1797 Shriver Homestead and Mill.
Collections: 18th-century mill & homestead containing original household & agricultural items.
Research Fields: Shriver family archives; social, industrial, political & architectural history surrounding the homestead.
Facilities: library of 19th-century books, periodicals & photographs available for research by personal request. Local craft items, cards & other museum-related items for sale.
Activities: classroom presentations; hands-on exhibits; guided tours; flower & plant market; grist mill grinding corn, wheat & buckwheat. Museum Sponsors: Flower Plant Market in May; Old Fashioned Ice Cream Sundae Social in July; Old-fashioned Corn Roast in August; Microbrewery Festival in September; fund-raiser events.
Publications: quarterly newsletter; pamphlet & booklet.
Hours & Admission Prices: May & Sept. Sat.-Sun. 12-4; June-Sept. 1 Tues.-Fri. 10-4, Sat.-Sun. 12-4; bus tours by appointment. Adults $5, children 6-12 $3; discounts to groups & senior citizens; members no charge. Combination ticket for house & mill: adults $5, children 6-12 $3. Closed Independence Day.
Attendance: 10,000 (estimated)
Membership: Individual $20; Husband & Wife $25; Family & Institutional $30; Life $200.

Wheaton

BROOKSIDE GARDENS, 1800 Glenallan Ave., Wheaton, MD 20902-1369. Tel.: 301-962-1400. Fax: 301-962-7878.
Web Site: www.brooksidegardens.org
Founded: 1969.
Key Personnel: Dir., Stephanie Oberle; Mgr. Plant Collection, Philip Normandy; Supvr., Joe Krout; Supvr. Enterprise, Ellen Hartranft.
Personnel Profile: Full-Time Paid 28; Part-Time Paid 25; Part-Time Volunteers 300.
Governing Authority: county. Affiliated with The Maryland National Capital Park & Planning Commission, 8787 Georgia Ave., Silver Spring, MD 20910. Tel.: 301-495-4500. Tax-exempt: 170(b)(1)(A).
Institution Type/Description: Botanical Garden.
Collections: flower gardens; hardy trees & shrubs; display conservatories.
Research Fields: testing & evaluation of plant collection from Japan.
Facilities: 3,000-vol. reference library of books, periodicals, handouts & catalogs available for research on premises; botanical gardens.
Activities: guided tours; lectures; formally organized education programs for children & adults; bus & van trips to regional horticultural institutions; special events.
Publications: horticulture information handouts, Adult Education Programs.
Hours & Admission Prices: Visitor Center: daily 9-5. North Conservatory: daily 10-5. Gardens: sunrise-sunset. No charge; donations accepted. Closed Christmas Day. &

Attendance: 320,000 (estimated)

Williamsport

MCMAHON'S MILL CIVIL WAR MILITARY & AMERICAN HERITAGE MUSEUM, 7900 Avis Mill Rd., Williamsport, MD 21795-2006. Tel.: 301-223-8778 & 9314.
Founded: 1965.
Key Personnel: Dir., William B. McMahon
Institution Type/Description: History Museum.
Collections: local history & heritage; military artifacts; period furnishings; personal artifacts; Civil War; World War I; World War II; weapons; ceramic art; early record players.
Hours & Admission Prices: By appointment. Adults $3; discounts to groups, AAM members, ICOM members.

Worton

AFRICAN-AMERICAN SCHOOLHOUSE MUSEUM, Rte. 297, St. James-Newtown Rd., Worton, MD 21678. Tel.: 410-810-1416, ext. 1.
Institution Type/Description: Historic Building: housed in a one-room schoolhouse; built in 1890.
Collections: local history, heritage & culture; photographs; oral histories; personal artifacts; period furnishings.
Publications: newsletter.
Hours & Admission Prices: Call for hours.

Wye Mills

WYE GRIST MILL, 900 Wye Mills Rd., Wye Mills, MD 21679. Mailing Address: P.O. Box 277, Wye Mills, MD 21679-0277. Tel.: 410-827-3850.
E-mail: oldwyemill@atlanticbbn.net
Web Site: oldwyemill.org
Institution Type/Description: History Museum.
Collections: Maryland's agricultural heritage; milling equipment; farm equipment. Historic Structure: working water-powered grist mill.
Research Fields: eastern shore agriculture; milling.
Hours & Admission Prices: mid-April to mid-Nov. Mon.-Sat. 10-4, Sun. 1-4. Tours: by appointment. Suggested Donation: $2.

MASSACHUSETTS

(448 listings)

Abington

DYER MEMORIAL LIBRARY, 28 Centre Ave., Abington, MA 02351-2228. Mailing Address: P.O. Box 2245, Abington, MA 02351-0745. Tel.: 781-878-8480.
E-mail: info@dyerlibrary.org
Web Site: www.dyerlibrary.org
Founded: 1932.
Congressional District: 11
Key Personnel: Dir., Joice Himawan; Librarian, Pamela Whiting.
Personnel Profile: Full-Time Paid 1; Part-Time Paid 3; Part-Time Volunteers 4.
Governing Authority: trust. Tax-exempt: 501(c)(3).
Institution Type/Description: General Museum.
Collections: local history & genealogy.
Research Fields: local history; genealogy.
Facilities: 15,000-vol. library of local history books & genealogy available for research on premises only; reading room.
Activities: guided tours; lectures; slides; formally organized educational programs; school loan service.
Publications: Old Abington in the American Revolution; The North Abington Riot; Heritage Trail.
Hours & Admission Prices: Tues.-Fri. 1-5, 2nd & 4th Sat. 12-4; other times by appointment. No charge; donations accepted. Closed holidays.
Attendance: 900 (estimated)

Acton

THE DISCOVERY MUSEUMS, (M), 177 Main St., Acton, MA 01720-3647. Tel.: 978-264-4200. Fax: 978-264-0210.
Web Site: www.discoverymuseums.org
Founded: 1982.
Congressional District: 5
Key Personnel: Pres. Bd. (V), Lees Stuntz; Exec. Dir., Neil H. Gordon; Dir. Education Science Discovery Museum, Denise LeBlanc; Assoc. Dir. Education Science Discovery Museum, Margaret Wimikates; Dir. Education, Amy Spencer; Dir. Finance & Business Admin., Kavita Katti; Dir. Devel. & Communications, Claudia Veitch; Dir. School Programs, Jill Foster; Dir. Mktg., Vicki Greene; Dir. Exhibits, Steve Roake.
Personnel Profile: Full-Time Paid 12; Part-Time Paid 55; Part-Time Volunteers 50; Interns 3.
Governing Authority: nonprofit organization. Parent Institution: The Discovery Museums, Inc. Subsidiary Institutions: The Children's Discovery Museum; The Science Discovery Museum. Tax-exempt: 501(c)(3).
Institution Type/Description: Children's Museums: consisting of The Children's Discovery Museum, housed in a 10-room Victorian house, and The Science Discovery Museum.
Collections: Children's Museum: hands-on interactive exhibit rooms including: Train Room, Discovery Ship, Bessie's Play Diner, & Sensations. Science Museum: interactive exhibits with scientific themes including Inventor's Workshop, Math Room, Light & Color Room, Water Room, Earth Science, Sound & Communication, Electricity, Magnets and Nature Balcony.
Facilities: 4 acres of wooded land; woodland path abutted by 200 acres of conservation land; discovery classroom; birthday parties; corporate events.
Activities: self-guided tours; organized education programs for children; teacher training; school outreach; participatory exhibits; collaborative partnerships with schools & universities; day camp in July & August; after-school discovery programs; workshops; Scout programs; weekly science drop-in workshops; science classes.
Publications: quarterly newsletter, Discovery Digest; flyer.
Hours & Admission Prices: Children's Discovery Museum: Tues.-Sun. 9-4:30. Science Discovery Museum: Tues.-Fri. 1-4:30, Sat.-Sun. 10-4:30. Adults & children $11.50, seniors 60 & over $10.50; members, teachers, & children under one no charge. Closed Independence Day; Labor Day; Thanksgiving Eve & Day; Christmas Eve & Day. &
Attendance: 140,000 (estimated)

Agawam

AGAWAM HISTORICAL & FIRE HOUSE MUSEUM, 35 Elm St., Agawam, MA 01001-2407. Mailing Address: P.O. Box 552, Agawam, MA 01001-0552. Tel.: 413-786-4631.
Institution Type/Description: Fire-Fighting Museum.
Collections: antique fire engines & apparatus; photographs.
Hours & Admission Prices: May-Dec. 2nd Sun. each month 1-4.

Amesbury

THE BARTLETT MUSEUM, 270 Main St., Amesbury, MA 01913. Mailing Address: P.O. Box 692, Amesbury, MA 01913-0016. Tel.: 978-388-4528.
E-mail: museum@bartlettmuseum.org
Web Site: bartlettmuseum.org
Founded: 1968.
Congressional District: 6
Key Personnel: Pres. (V), Richard Gale; Treas. (V), Steven Klomps; Dir., Hazel Kray.
Personnel Profile: Part-Time Paid 1.

Governing Authority: nonprofit organization. Tax-exempt.
Institution Type/Description: History Museum: housed in 1870 Old Victorian School House.
Collections: local artifacts; natural history; carriage trade; Victorian schoolroom; bird collection; Indian artifacts; old tools & implements; paintings of early Amesbury residents. Historic Building: c.1880 Salisbury Point Waiting Station.
Research Fields: genealogy; area sites.
Facilities: 100-vol. library of books written by local personages available by special permission only. Books, leaflets, brochures & maps for sale.
Activities: guided tours; lectures; films; concerts; arts festivals; hobby workshops; sound slide shows pertaining to carriages & churches of Amesbury; formally organized education programs; inter-museum loan, permanent & temporary exhibitions. Museum Sponsors: Genealogy Group.
Publications: quarterly newsletter.
Hours & Admission Prices: Memorial Day weekend to Labor Day Fri. & Sun. 1-4, Sat. 10-4; other times by appointment. Adults $3, children & senior citizens $1; AAM & museum members no charge.
Attendance: 1,000 (estimated)
Membership: Student $1; Adult $10; Family $20; Sustaining $25 & up.

JOHN GREENLEAF WHITTIER HOME, (M), 86 Friend St., Amesbury, MA 01913-2746. Mailing Address: P.O. Box 632, 86 Friend St., Amesbury, MA 01913-0014. Tel.: 978-388-1337. Fax: 978-388-1337.
E-mail: whittierhome@verizon.net
Web Site: www.whittierhome.org
Founded: 1898.
Congressional District: 6
Key Personnel: Pres. (V), Cynthia C. Costello; Museum Shop Mgr., Dianne Cole.
Personnel Profile: Full-Time Volunteers 1; Part-Time Paid 2; Part-Time Volunteers 50.
Governing Authority: nonprofit organization. Tax-exempt.
Institution Type/Description: Historic House: 1836 home of poet & abolitionist John Greenleaf Whittier.
Collections: manuscripts; books; pictures; original furniture & furnishings; Victorian gardens.
Research Fields: 19th century literature & poetry; life & work of J. G. Whittier; inventory of Whittier's works; local history.
Facilities: approx. 1,000-vol. library of literature & history books available on premises by appointment; garden. Postcards & books for sale.
Activities: guided tours; permanent exhibitions; public events; poetry readings; tea.
Publications: brochure.
Hours & Admission Prices: May-Oct. Sat. 10-4, last tour 3:30; other times by appointment. Adults $6, seniors & students $5, children 7-17 $3; discounts to AAM members & groups; members & children under 7 no charge. Closed Thanksgiving; Christmas.
Attendance: 600 (accurate)
Membership: Artists & Students $15; Active $25.

LOWELL'S BOAT SHOP, 459 Main St., Amesbury, MA 01913-4207. Tel.: 978-834-0050.
E-mail: info@lowellsboatshop.com
Web Site: www.lowellsboatshop.com
Founded: 1793.
Congressional District: 6
Key Personnel: Exec. Dir. & Museum Shop Mgr., Pam Bates; Chm., George O'Dell; Chm., Steven Batchelder; Pres., Sally McKay.
Personnel Profile: Full-Time Paid 1; Full-Time Volunteers 4; Part-Time Paid 1; Part-Time Volunteers 30; Interns 2.
Governing Authority: Parent Institution: Lowell's Maritime Foundation. Tax-exempt.
Institution Type/Description: Historic Buildings: housed in 19th century boat building shop.
Collections: fleet of dories; patterns; hand tools; business papers; working shop.
Research Fields: wooden boatbuilding of lower Merrimac Valley; relationship & impact of dories on Gulf of Maine fisheries.
Facilities: working 19th century shop & equipment; education center; river landing site.
Activities: boat building, on-the-water educational programs; dory model classes; seasonal rowing; tours.
Hours & Admission Prices: May-Oct. Tues.-Fri. 11-4, Sat.-Sun. 12-4; Nov.-April Tues.-Fri. 11-4. Office: Mon.-Fri. 10-5. Adults 15 & over $8, children 7-14 $6; discounts to members; children under 7 no charge. &
Attendance: 1,340 (estimated)

Membership: Single $35; Crew $60; Centennial $100; Quartermaster $250; Steward $500.

Amherst

AMHERST COLLEGE MUSEUM OF NATURAL HISTORY, Amherst College, 11 Barrett Hill Rd., Amherst, MA 01002-5000. Tel.: 413-542-2165. Fax: 413-542-2713.
E-mail: kwellspring@amherst.edu
Web Site: www.amherst.edu/museums/naturalhistory
Formerly: Pratt Museum of Natural History
Founded: 1848.
Key Personnel: Dir., Peter Crowley; Mgr. Collections, Kate Wellspring.
Personnel Profile: Full-Time Paid 1; Part-Time Paid 3; Interns 1.
Governing Authority: college. Parent Institution: Amherst College. Tax-exempt.
Institution Type/Description: Natural History Museum.
Collections: paleontology; geology; osteology; mineralogy; taxidermy; anthropology; ichnology.
Research Fields: vertebrate & invertebrate paleontology; ichnology; geology.
Facilities: library; laboratory.
Activities: prearranged guided group tours; formally organized education programs; permanent exhibitions.
Publications: self-guided tour sheets for Vertebrate Fossil & Mineral Exhibits; brochure; postcard.
Hours & Admission Prices: Tues.-Wed. & Fri.-Sun. 11-4, Thurs. 11-4 & 6-10. No charge. Closed holidays. &
Attendance: 25,000 (accurate)

AMHERST HISTORY MUSEUM AT THE SIMEON STRONG HOUSE - AMHERST HISTORICAL SOCIETY, (M), 67 Amity St., Amherst, MA 01002-2214. Tel.: 413-256-0678. Fax: 413-256-0672.
E-mail: info@amhersthistory.org
Web Site: www.amhersthistory.org
Founded: 1899.
Key Personnel: Pres. (V), Philip A. Shaver.
Personnel Profile: Part-Time Volunteers 25; Interns 4.
Governing Authority: society; nonprofit. Parent Institution: Amherst Historical Society. Tax-exempt: 501(c)(3).
Institution Type/Description: History Museum/History Site: housed in c.1750 Strong house. Listed on the National Register of Historic Places.
Collections: decorative arts; textiles; photographs; period clothing; household tools; folk arts; furniture; weaponry; tools; local history artifacts.
Major Exhibits: Amherst and the Civil War, 10/12-1/15.
Facilities: 18th-century garden. Museum-related items for sale.
Activities: guided tours; lectures; organized programs for children; temporary exhibitions. Museum Sponsors: Garden Tour; Victorian Christmas House; House Tour.
Publications: quarterly newsletter.
Hours & Admission Prices: May-Nov. Thurs.-Sat. 12-4. Suggested Donation: adults $5, seniors, students & children $3; children 6 & under and Amherst Historical Society members no charge.
Attendance: 2,400 (estimated)
Membership: Senior & Student $30; Individual $60; Family $100; Life $500.

EMILY DICKINSON MUSEUM: THE HOMESTEAD AND THE EVERGREENS, (M), 280 Main St., Amherst, MA 01002-2349. Tel.: 413-542-8161. Fax: 413-542-2152.
E-mail: info@emilydickinsonmuseum.org
Web Site: www.emilydickinsonmuseum.org
Formerly: Dickinson Homestead
Founded: 2003.
Congressional District: 1
Key Personnel: Exec. Dir., Jane H. Wald; Chm. (V), John Armstrong; Dir. Interpretation & Programming, Cindy Dickinson.
Personnel Profile: Full-Time Paid 3; Part-Time Paid 40; Part-Time Volunteers 10; Interns 2.
Governing Authority: private college; nonprofit. Parent Institution: Amherst College. Tax-exempt: 501(c)(3).
Institution Type/Description: Historic Houses: The Dickinson Homestead c.1813, birthplace & home of poet Emily Dickinson; The Evergreens 1856, home of the poet's brother Austin & sister-in-law Susan.
Collections: 19th-century furniture; decorative arts related to the Dickinson family.
Facilities: Gift items for sale.
Activities: docent program; guided tours; school programs; special programs; landscape tour; poetry discussion groups. Museum Sponsors: poetry walk in May; poetry garden series in July; birthday lecture in December.

Publications: Emily Dickinson: The Poet at Home c.2000; electronic newsletter; DVDs, The Poet at Home; Seeing New Englandly; member newsletter, Message From The Meadows.
Hours & Admission Prices: March-May & Sept.-Dec. Wed.-Sun. 11-4; June-Aug. Wed.-Sun. 10-5. Guided Tours: adults $10-$12; members no charge. Closed major holidays. &
Attendance: 14,000 (accurate)
Membership: Individual & Household $65.

THE ERIC CARLE MUSEUM OF PICTURE BOOK ART, (M),

125 W. Bay Rd., Amherst, MA 01002-3357. Tel.: 413-658-1100. Fax: 413-658-1139.
E-mail: info@carlemuseum.org
Web Site: www.carlemuseum.org
Founded: 2001.
Key Personnel: Exec. Dir., Alexandra Kennedy; Chief Cur., H. Nichols B. Clark; Chm. (V), Chris Milne; Dir. Devel., Rebecca Miller Goggins; Cur. Education, Courtney Waring; Dir. Finance & Administration, Andrea Powers; Registrar, Erica Jacobs; Mgr. Facilities, John Stark; Mgr. Mktg., Sandy Soderberg; Museum Shop Mgr., Eliza Brown.
Personnel Profile: Full-Time Paid 20; Part-Time Paid 17; Part-Time Volunteers 23; Interns 9.
Volunteer Hours: 1,963
Operating Expenses: 2,463,290
Operating Income: 2,463,579
Governing Authority: private; nonprofit organization. Tax-exempt: 501(c)(3).
Institution Type/Description: Art Museum.
Collections: original picture book art from around the world.
Major Exhibits: Seriously Serious: ADecade of Art & Whimsy by Mo Willems, 6/13-2/14; The Art of Eric Carle: Friends, 9/13-3/14; Illuminating Illustration: Picture Book Art Inspired by Illuminated Manuscripts, 10/13-5/14; Barbara McClintock & Natalie Merchant: Leave Your Sleep, 11/13-5/14; Lyle Crocodile & Friends: The Art of Bernard Waber, 3/14-6/14; Madeline at 75: The Art of ludwig Bemelmans, 11/14-2/15.
Research Fields: picture book art & artists; connection between visual & verbal literacy.
Facilities: 3,800-vol. library; 137-seat auditorium; 60-seat cafeteria; art studio; 6,000 sq. ft. exhibit space. Museum-related items for sale.
Activities: arts festival; concerts; live theatre & dance; docent program; films; formal education programs; guided tours; lectures; loan, participatory, traveling & temporary exhibitions; training programs; studio classes; teacher training; masters degree offered in children's literature in partnership with Simmons College. Annual Events: Summer Reading Festival With the Western Massachusetts Illustrators Guild; The Carle Honors Awards Event in the Field of Picture Book Art.
Publications: newsletter, Sowing the Seeds; exhibition catalogs.
Hours & Admission Prices: July-Aug. & MA School Vacation Weeks Mon.-Fri. 10-4, Sat. 10-5, Sun. 12-5; Sept.-June Tues.-Fri. 10-4, Sat. 10-5, Sun. 12-5. Adults $9, senior citizens & children $6; discount to groups & AAM members; members no charge. Closed New Year's Day; Independence Day; Thanksgiving; Christmas Eve & Day. &
Attendance: 49,820 (accurate)
Membership: Student $20; Senior $30; Teacher, Librarian & National Associate $35; Individual $45; Family $65; Art Associate & Organizational $125; Studio Sponsor $250; Picture Book Patron $500; Founder's Society $1,000; Director's Guild $2,500; Chairman's Circle $5,000.

HERTER ART GALLERY,

125a Herter Hall, University of Massachusetts Amherst, Amherst, MA 01003. Tel.: 413-545-0976.
Institution Type/Description: Art Gallery.
Collections: paintings; photographs; sculpture.
Hours & Admission Prices: Mon.-Fri. 11-4, Sun. 1-4. No charge.

✷ MEAD ART MUSEUM, (M),

Amherst College, 41 Quadrangle Dr., Amherst, MA 01002. Mailing Address: Amherst College, P.O. Box 5000, Amherst, MA 01002-5000. Tel.: 413-542-2335. Fax: 413-542-2117. Facebook; Mead Art Museum.
E-mail: mead@amherst.edu
Web Site: www.amherst.edu/mead
Founded: 1821.
Congressional District: 1
Key Personnel: Dir., Elizabeth E. Barker; Exec. Asst. to Dir. & Media & Mktg. Coord., Rachel Rogol; Sr. Cur. & Cur. Russian Art, Bettina Jungen; Head Education & Cur. Academic Programs, Pamela Russell; Postdoc Cur. Teaching Fellow in Japanese Prints, Bradley Bailey; Coord. Community Programs, Wendy Somes; Cur. & Education Asst., Sheila Flaherty-Jones; Study Room Supvr., Miloslava Waldman; Mgr. Collections, Stephen Fisher; Preparator, Timothy Gilfillan; Accounting Mgr., Karen Summers; Prepara-

tor, Tim Gilfillan; Head Security & Facility Mgr., Heath Cummings; Lead Security Officer, Nicholas Taupier.
Personnel Profile: Full-Time Paid 9; Part-Time Paid 42; Part-Time Volunteers 1; Interns 2.
Governing Authority: college. Parent Institution: Amherst College. Tax-exempt: 501(c)(3).
Institution Type/Description: Art Museum.
Collections: American paintings; sculpture; prints & photos; drawings & watercolors; decorative arts; additional works: Asian; Ancient Pre-Columbian; Mexican; African; Russian.
Research Fields: art history.
Facilities: auditorium.
Activities: lectures; films; formally organized education programs for undergraduate college students affiliated with Amherst, University of Massachusetts, Smith, Mt. Holyoke & Hampshire Colleges; inter-museum loan & temporary exhibitions.
Publications: occasional catalogs, brochures, calendars.
Hours & Admission Prices: Academic Season: Tues.-Thurs. & Sun. 9am to midnight, Fri. 9-8, Sat. 9-5. Academic Recess: Tues.-Thurs. & Sat.-Sun. 9-5, Fri. & 1st Thurs. each month 9-8. No charge; donations accepted. &
Attendance: 36,000 (estimated)

NATURAL HISTORY COLLECTIONS,

Univ. of Massachusetts, Rm. 146 Morrill 2, 622 N. Pleasant St., Amherst, MA 01002-1526. Mailing Address: 611 N. Pleasant St., Amherst, MA 01003. Tel.: 413-577-2303. Fax: 413-545-3243.
E-mail: bdumont@bio.umass.edu
Web Site: bcrc.bio.umass.edu/ummnh
Formerly: Museum of Zoology
Founded: 1863.
Congressional District: 1
Key Personnel: Dir., Elizabeth Dumont; Cur. Amphibians & Reptiles, Alan Richmond; Collection Mgr., Katherine Doyle.
Governing Authority: university. Affiliated with University of Massachusetts.
Institution Type/Description: Zoology Museum.
Collections: preserved animals of all groups.
Research Fields: anatomy; ecology; systematics.
Activities: formally organized education programs for undergraduate & graduate college students; loan exhibitions.
Hours & Admission Prices: Call for appointment. No charge. Closed state & national holidays. Exhibits not open to public. &

UNIVERSITY GALLERY, UNIVERSITY OF MASSACHUSETTS AT AMHERST, (M),

University Gallery, Fine Arts Center, University of Massachusetts, 151 Presidents Drive, Office 2, Amherst, MA 01003-9331. Tel.: 413-545-3670. Fax: 413-545-2018.
E-mail: ugallery@acad.umass.edu
Web Site: www.umass.edu/fac/universitygallery
Founded: 1975.
Congressional District: 1
Key Personnel: Dir., Loretta Yarlow; Communications, Thonsey Keopanya; Education Cur., Eva Fierst; Gallery Mgr., Craig Allaben; Collections Registrar & Preparator, Justin Griswold.
Personnel Profile: Full-Time Paid 3; Part-Time Paid 2; Part-Time Volunteers 2; Interns 3.
Governing Authority: university. Parent Institution: Univ. of Massachusetts. Tax-exempt.
Institution Type/Description: University Art Gallery.
Collections: 20th-century American photography, prints & drawings.
Research Fields: 20th-century prints, drawings, photographs, sculpture & painting.
Facilities: 6,533 sq. ft. exhibit space.
Activities: guided tours; lectures; films; gallery talks; formally organized education programs for undergraduate & graduate university students; temporary, traveling & loan exhibitions.
Publications: exhibition catalogues: Critical Perspectives in American Art, 1976; Selections From the Chase Manhattan Bank Art Collection, 1981; Bilge Friedlaender, 1976; Richard Fleischner, 1977; Late 19th Century American Drawings & Watercolors, 1977; Criticism of Photography, 1978; The Class of 1928 Photography Collection, 1978; Antonakos: Neons for the University of Massachusetts, 1978; Sam Gilliam: Indoor & Outdoor Paintings 1967-1978; John Walker, 1979; Al Souza: Photoworks 1974-1979; Sculpture on the Wall: Relief Sculpture of the Seventies, 1980; Jackie Ferrara, 1981; George Trakas, 1981; The Prints of Barnett Newman, 1983; Martin Puryear 1983; Mauro Staccioli, 1984; Domestic Tales, 1984; Ten, 1985; Mel Kendrick, 1985; Anish Kapoor 1987; Jeffrey Brosk: Prairie Dance, 1987; Beyond Light: Infrared Photography by Six New England Artists, 1987; Daniel Buren, 1988; Francesc Torres: Belchite/South Bronx,

1988; Allen Wexler: Dining Rooms & Furniture for the Typical House, 1989; Cristos Gianakos: Rampworks, 1989; In Site: Five Conceptual Artists from the Bay Area, 1991; brochure, Rita Myers: Phantom Cities, 1991; Ellen Phelan: From the Lives of Dolls, 1992; George Wardlaw: Exodus II; Alumni III, 1993; Traditional Artifacts from the South Pacific, 1993; Shirazeh Houshiary: Turning Around the Centre, 1994; Jin Soo Kim, 1994; Michele Blondel, 1994; In Vivo, 1995; Daisy Youngblood, 1996; Socks on my spoons: Ursula von Rydingsvard, 1996; The Thin Veneer: The Peoples of Bosnia & Their Disappearing Cultural Heritage, 1997; Something Else to See: Improvisational Bordering Styles in African-American Quilts, 1997; The Lois Beurman Torf Collection for the University of Massachusetts, 1997; Alumni IV, 1999; Primary Source: Roger Ackling, Dove Bradshaw & Sandy Gellis, 1999; Head to Toe: Impressing the Body, 2000; Soft White: Lighting Designs by Artists, 2000; Rolf Julius Black (Red), 2001; Natalie Alper, 2001; Brenda Zlamany, 2001; The Culture of Violence, 2002; Mandrake Tango, 2002.
Hours & Admission Prices: Academic Year Tues.-Fri. 11-4:30, Sat.-Sun. 2-5. No charge. &
Attendance: 12,600 (estimated)

YIDDISH BOOK CENTER, Harry & Jeanette Weinberg Bldg., 1021 West St., Amherst, MA 01002-3375. Tel.: 413-256-4900. Fax: 413-256-4700.
E-mail: yiddish@bikher.org
Web Site: www.yiddishbookcenter.org
Founded: 1980.
Congressional District: 1
Key Personnel: Pres., Aaron Lansky; Chm. (V), Lief D. Rosenblatt; Mgr. Visitors Svcs., Arielle Jackson; Museum Shop Mgr., Randi Silnutzer.
Personnel Profile: Full-Time Paid 21; Part-Time Paid 2; Part-Time Volunteers 7; Interns 18.
Governing Authority: Tax-exempt.
Institution Type/Description: Jewish Cultural Organization.
Collections: Eastern European Jewish history & culture, with the focus on Yiddish literature.
Facilities: 1,000,000-vol. library. Books for sale.
Activities: films; lectures; concerts; performances; exhibits; tours.
Publications: English-language magazine, Pakn Treger.
Hours & Admission Prices: Jan.-March Mon.-Fri. 10-4; April-Dec. Mon.-Fri. 10-4, Sun. 11-4. Suggested Donations: adults $8; members no charge. Closed Jewish & legal holidays. &
Attendance: 11,000 (accurate)
Membership: Member $36.

Andover

ADDISON GALLERY OF AMERICAN ART, (M), (I), Phillips Academy, Andover, MA 01810-4161. Mailing Address: 180 Main St., Andover, MA 01810-4166. Tel.: 978-749-4015. Fax: 978-749-4025.
E-mail: addison@andover.edu
Web Site: www.addisongallery.org
Founded: 1931.
Congressional District: 5
Key Personnel: Assoc. Dir. & Cur., Susan Faxon; Cur., Allison Kemmerer; Dir. Education, Rebecca Hayes; Dir. Museum Finance & Registration, Denise J.H. Johnson; Museum Shop Mgr., Anna Gesing.
Personnel Profile: Full-Time Paid 17; Part-Time Paid 15; Interns 2.
Operating Expenses: 3,090,000
Governing Authority: nonprofit. Parent Institution: Phillips Academy. Tax-exempt.
Institution Type/Description: Art Museum.
Collections: American art; paintings; sculpture; drawing; prints; 17th century-present photography; ship models.
Major Exhibits: An American in London: Whistler and the Thames (T), 2/14-4/14; Daze, 4/14-7/14; Loisaidas, 4/14-7/14; Lorna Simpson (T), 9/14-12/14; Dwish Tryon and American Tonalism, 9/14-1/15.
Research Fields: secondary school art education.
Activities: temporary exhibitions; lectures; films.
Publications: catalogs, Carroll Dunham Prints; Ipswich Days: Arthur Wesley Dow; William Wegman: Funney/Strange; Jennifer Bartlett: Early Plate Work; Sol Lewitt, 25 years of Wall Drawings 1968-1993; 1/4 in. scale models of American Sailing Ships; Addison Gallery of American Art: 65 years; Arthur Dove: A Retrospective; Joel Shapiro, 1971-1997; Terry Winters: Paintings, Drawings & Prints 1994-2004; Reinventing the West: The Photographs of Ansel Adams and Robert Adams; Trisha Brown: Art and Dance in Dialogue; John O'Reilly; Miracle in the Scrap Heap: The Sculpture of Richard Stankiewicz; The Treasures of the Addison Gallery of American Art; Sheila Hicks: 50 Years; American Vanguards; F. Holland Day.

Hours & Admission Prices: Tues.-Sat. 10-5, Sun. 1-5. No charge; donations accepted. &
Attendance: 30,000 (estimated)
Membership: Friends: $50; $100; $250; $500; $750; Director's Circle: $1,000; $2,500; $5,000; $10,000.

* **ANDOVER HISTORICAL SOCIETY, (M),** 97 Main St., Andover, MA 01810-3803. Tel.: 978-475-2236. Fax: 978-470-2741.
E-mail: info@andoverhistorical.org
Web Site: www.andoverhistorical.org
Founded: 1911.
Congressional District: 5
Key Personnel: Exec. Dir., Elaine Clements; Pres. (V), Douglas Mitchell; Museum Educator, Debra DeSmet; Public Rels. & Devel., Carrie Midura.
Personnel Profile: Full-Time Paid 4; Part-Time Paid 2; Part-Time Volunteers 120.
Governing Authority: board of directors; nonprofit organization. Tax-exempt: 501(c)(3).
Institution Type/Description: Local History Museum: c.1818-19 house & barn.
Collections: American & imported furnishings; costumes; textiles; household, agricultural & trade implements; photographs; genealogy & local history research materials. Historic House: c.1818 house & barn.
Research Fields: local history; architecture & decorative arts; genealogy.
Facilities: 3,500-vol. research library & archives, including Andover imprints, local & regional histories & documents, photographs, maps, architectural drawings, reference books & periodicals.
Activities: guided tours; lectures; educational programs; permanent & temporary exhibitions; workshops, outreach programs.
Publications: booklets, Andover A Century of Change: 1896-1996, The Townswoman's Andover, Historic Andover 325th Anniversary, Andover, Massachusetts 1946-1971; quarterly newsletter; walking tour brochures; catalog, Addison B. LeBoutillier: Andover Artist and Craftsman.
Hours & Admission Prices: Museum, Library & Archives: Tues.-Sat. 10-4. Office: Mon.-Fri. 9-5. No charge; donations accepted. Closed national holidays.
Attendance: 3,900 (accurate)
Membership: Student & Senior $20; Individual $25; Family & Dual $40; Contributor $75; Sustaining $100; Patron $250; Benefactor $500; Blanchard Circle $1,000.

ROBERT S. PEABODY MUSEUM OF ARCHAEOLOGY, 180 Main St., Andover, MA 01810. Tel.: 978-749-4490. Fax: 978-749-4495.
E-mail: rspeabody@andover.edu
Web Site: www.andover.edu/rspeabody/
Founded: 1901.
Congressional District: 5
Key Personnel: Dir., Ryan J. Wheeler; Chm. (V) & Peabody Advisory Committee, Marshall P. Cloyd; Registrar & Sr. Collection Mgr., Bonnie Sousa; Honorary Cur., Eugene C. Winter; Educator, Donald Slater; Educator, Lindsay Randall; Collection Mgr., Marla Taylor; Administrative Asst., Lesley Shahbazian.
Personnel Profile: Full-Time Paid 3; Part-Time Paid 2; Part-Time Volunteers 10; Interns 4.
Governing Authority: nonprofit organization. Parent Institution: Phillips Academy. Tax-exempt: 170(b)(1)(A).
Institution Type/Description: Archaeology Museum.
Collections: archaeology of North America, emphasis on Northeast, Southeast, Midwest, Southwest, Mexico & the Arctic; collections from Maine, Massachusetts, Etowah, GA., Yukon, Mexico & Pecos, NM; Highland Maya & North America ethnology; anthropology.
Research Fields: American archaeology; social organization; acculturation.
Facilities: 5,000-vol. library of books on archaeology, ethnology & physical anthropology.
Activities: speakers for both Phillips Academy & general public; permanent & temporary exhibits.
Publications: Papers of The R.S. Peabody Foundation (Inactive).
Hours & Admission Prices: by appointment only. No charge; donations accepted. Closed New Year's Day; Memorial Day; Independence Day; Labor Day; Thanksgiving; Christmas. &
Attendance: 6,800 (estimated)

Arlington

ARLINGTON CENTER FOR THE ARTS, Gibbs Center, 41 Foster St., Arlington, MA 02474-6813. Tel.: 781-648-6220.
E-mail: info@acarts.org
Institution Type/Description: Art Gallery.

Collections: paintings; sculpture.
Activities: classes; educational programs; special events.
Hours & Admission Prices: Mon.-Fri. 9-5; other times by appointment.

THE ARLINGTON HISTORICAL SOCIETY, 7 Jason St., Arlington, MA 02476-6410. Tel.: 781-648-4300.
E-mail: contact@arlingtonhistorical.org
Web Site: www.arlingtonhistorical.org
Founded: 1897.
Congressional District: 8
Key Personnel: Pres., Pamela Meister; Museum Admin., Doreen Stevens.
Personnel Profile: Part-Time Paid 2; Part-Time Volunteers 25.
Governing Authority: nonprofit corporation. Branch Museums: Jason Russell House & George Abbot Smith History Museum. Tax-exempt.
Institution Type/Description: Historic Building & Site.
Collections: photographic archives; archives; 17th to 19th-century furnishings; textiles; dolls; weaponry; regional & local history; historic house.
Research Fields: regional & local history.
Facilities: archives of books, photographs & documents; meeting space.
Activities: guided tours; changing exhibits; monthly programs; education program.
Publications: books; newsletter.
Hours & Admission Prices: April-Oct. Sat.-Sun. 1-4; other times by appointment. Adults $5, children under 12 $2; discounts to AAM, members of Lexington Historical Society & Massachusetts Teacher's Association; society members no charge.
Attendance: 700 (estimated)
Membership: Student $10; Adult $20; Family $40; Sustaining $75; Life $450; Donation $600.

CYRUS E. DALLIN ART MUSEUM, INC., 1 Whittemore Park, Arlington, MA 02474-1105. Tel.: 781-641-0747.
Web Site: www.dallin.org
Founded: 1995.
Congressional District: 7
Key Personnel: Co Chm., Roland Chaput; Co Chm., Heather Leavell
Governing Authority: Tax-exempt.
Institution Type/Description: Art Museum.
Collections: plaster models; plaster & bronze sculptures; C.E. Dallin archival material; silver & bronze medals.
Activities: group tours; elementary school outreach program. Museum Sponsors: Town Day, Art on the Green.
Publications: Cyrus E. Dallin and His Native American Work; Cyrus E. Dallin, Sculptor Frontier to Fame; Walking Tour of Dallin Sculptures.
Hours & Admission Prices: Wed.-Sun. 12-4. No charge; donations accepted. &

THE OLD SCHWAMB MILL, 17 Mill Lane, Arlington, MA 02476-4189. Tel.: 781-643-0554. Fax: 781-643-0640.
E-mail: info@oldschwambmill.org
Web Site: www.oldschwambmill.org
Founded: 1969.
Congressional District: 8
Key Personnel: Site Admin., Ed Gordon.
Personnel Profile: Part-Time Paid 2; Part-Time Volunteers 21.
Governing Authority: nonprofit. Parent Institution: The Schwamb Mill Preservation Trust, Inc. Tax-exempt: 501(c)(3).
Institution Type/Description: Industrial History Museum: housed in 1860 The Old Schwamb Mill, a water-powered mill.
Collections: working 19th-century assembly of belt-driven shaft & pulley operated woodworking machinery for production of oval & circular picture frames; woodworking hand tools; Hercules water turbine, sluice gates, stone bearings for water wheel & other parts of obsolete waterpower system; business & genealogical records of the five-generation German Schwamb Family & their Yankee & English Puritan predecessors over 350 years on the site. Historic Buildings: three 19th-century mill buildings.
Research Fields: architectural & business history recorded by Historic American Engineering Record of the U.S. Dept. of the Interior; local history; mill families' genealogy; history of picture frames; historic water-powered mill sites' markers along rails to trails bicycle path paralleling Mill Brook in Arlington and Lexington.
Facilities: Shaker workshops reproduction wooden furniture & made to order museum-quality carved picture frames for sale.
Activities: guided tours by appointment; PBS-TV nationally aired programs: Norm Abram's New Yankee Workshop, Boyhood of John Muir & This Old House; 19th century woodworking machinery demonstrations; MA Dept. of Education, Technology Summer Institute for public school teachers K-12.
Publications: brochure, Old Schwamb Mill; semiannual newsletter.
Hours & Admission Prices: Tues. & Sat. 11-3. No charge; donations accepted. Tour groups (6-10 people) $35; members no charge. Closed legal holidays.

Attendance: 1,000 (estimated)
Membership: Individual $30; Family $40; Sustaining $300 & up.

Ashfield

ASHFIELD HISTORICAL SOCIETY, 457 Main St., Ashfield, MA 01330. Mailing Address: P.O. Box 277, Ashfield, MA 01330-0277. Tel.: 413-428-4541.
E-mail: grace240@verizon.net
Web Site: www.ashfieldhistorical.org
Key Personnel: Pres. (V), Alden Gray; Cur., Grace Lesure; Museum Shop Mgr., Suzi Day
Institution Type/Description: Historic Building: built in 1830.
Collections: glass plate negatives; early Ashfield industries; 19th century period furnishings; military artifacts.
Activities: special events.
Hours & Admission Prices: June-1st weekend in Oct. Sat.-Sun. 11-1; other times by appointment.

Ashland

ASHLAND HISTORICAL SOCIETY, INC., 2 Myrtle St., Ashland, MA 01721-1106. Mailing Address: P.O. Box 145, Ashland, MA 01721-0145. Tel.: 508-881-8183.
Web Site: ashlandhistsociety.com
Founded: 1909.
Congressional District: 4
Key Personnel: Pres. (V), Clifford Wilson.
Governing Authority: society. Tax-exempt.
Institution Type/Description: History Museum.
Collections: genealogy; early books; newspapers; furniture; implements; portraits; glassware; china; household goods; manuscripts; school & town records. Historic House: 1748 house.
Research Fields: genealogy & vital record files.
Facilities: manuscripts on local history, Bibles, old school books, hymnals, microfilms of newspapers, 1869-1914 Ashland Advertiser & 1914-1915 Ashland Tribune available for use on premises; reading room.
Activities: guided tours; formally organized education programs for children; temporary exhibitions; public talks; presentations. Museum Sponsors: Open Houses.
Publications: books, History of Ashland, 1942; reprint maps; monthly newsletter.
Hours & Admission Prices: Wed. 7-9 by appointment. No charge. &
Attendance: 450 (estimated)
Membership: Annual $10.

Ashley Falls

THE ASHLEY HOUSE, Cooper Hill Rd., Ashley Falls, MA 01222. Mailing Address: P.O. Box 792, Stockbridge, MA 01262-0792. Tel.: 413-298-3239, ext. 3013. Fax: 413-298-5239.
E-mail: naumkeag@ttor.org
Web Site: www.thetrustees.org
Founded: 1972.
Congressional District: 1
Key Personnel: Cultural Site Mgr., Colleen Henry.
Personnel Profile: Part-Time Paid 3.
Governing Authority: privately-administered. Parent Institution: The Trustees of Reservations, 572 Essex St., Beverly, MA 01915. Tax-exempt.
Institution Type/Description: Historic House: 1735 oldest house in Berkshire County.
Collections: colonial furnishings; pottery; tools; African-American history.
Activities: guided tours.
Publications: illustrated booklet on Ashley family and house, 1982.
Hours & Admission Prices: July-Aug. Sat.-Sun. 1-3; other times by appointment. Adults $5; discounts to AAM, ICOM members; children 12 & under and members no charge.
Attendance: 800 (accurate)
Membership: Individual $47; Family $67.

Attleboro

ATTLEBORO AREA INDUSTRIAL MUSEUM, INC., 42 Union St., Ste. 2, Attleboro, MA 02703-2948. Tel.: 508-222-3918. Fax: 508-222-1498.
E-mail: info@industrialmuseum.com
Web Site: www.industrialmuseum.com
Founded: 1975.
Congressional District: 3

Key Personnel: Exec. Dir., George Shelton.
Personnel Profile: Part-Time Paid 2; Part-Time Volunteers 24; Interns 6.
Governing Authority: Tax-exempt.
Institution Type/Description: Industrial Museum.
Collections: local history.
Research Fields: jewelry industry; local industry.
Facilities: research library.
Publications: A Bit of Nostalgia.
Hours & Admission Prices: Thurs.-Fri. 10-4; groups by appointment. Adults $6, children $4. Closed Independence Day & week. &
Attendance: 3,000 (estimated)
Membership: Student & Senior $25; Individual $50; Family $100; Corporate $250; Benefactor $1,000.

ATTLEBORO ARTS MUSEUM, 86 Park St., Attleboro, MA 02703-2335. Tel.: 508-222-2644. Fax: 508-226-4401.
E-mail: office@attleboroartsmuseum.org
Web Site: www.attleboroartsmuseum.org
Formerly: Attleboro Museum, Center for the Arts
Founded: 1929.
Key Personnel: Pres. Bd. of Trustees, Nancy Aleo; Exec. Dir., Mim Fawcett; Museum Shop Mgr., Marion Volterra; Coord. Programs, Abby Roualdi; Office Mgr., Kerry St. Pierre; Tech Admin., Patrick Garriepy.
Personnel Profile: Full-Time Paid 3; Part-Time Paid 1; Part-Time Volunteers 50; Interns 1.
Governing Authority: nonprofit organization. Tax-exempt: 501(c)(3).
Institution Type/Description: Art Museum & Cultural Center.
Collections: American art & craft; prints; costumes; decorative arts; ceramics; photographs; paintings; Civil War flags, swords, medals.
Research Fields: regional art.
Activities: arts festivals; concerts; dance recital; classes for adults & children; visiting artist's workshops; school arts program; lectures; participatory & temporary exhibitions.
Publications: quarterly newsletter; announcements; quarterly museum school brochure.
Hours & Admission Prices: Summer: June-Aug. Tues.-Sat. 10-4; Sept.-May Tues.-Sat. 10-5. No charge; donations accepted. Closed national holidays & holiday weekends. &
Attendance: 10,000 (estimated)
Membership: Senior Citizen & Student $25; Artist & Individual $35; Family & Household $50; Associate $75; Supporting $125; Benefactor $250; Patron $500; Cornerstone $1,000.

CAPRON PARK ZOO, 201 County St., Attleboro, MA 02703-3510. Tel.: 774-203-1840. Fax: 508-223-2208.
E-mail: zoo@cityofattleboro.us
Web Site: www.capronparkzoo.com
Key Personnel: Dir., Jean Benchimol; Cur. Education, Mel Stoehrer
Institution Type/Description: Zoo.
Collections: animals from around the world.
Activities: educational programs; recreational activities.
Hours & Admission Prices: April to early Oct. daily 10-5; Oct. Mon.-Fri. 10-4, Sat.-Sun. 10-5; Nov.-March daily 10-4. Adults $5.50, children & seniors $3.75; discounts to military & Attleboro residents; children under 3 no charge. &

WOMEN AT WORK MUSEUM, Rte. 123, Attleboro, MA 02703. Mailing Address: P.O. Box 355, 35 County St., Attleboro, MA 02703-0006. Tel.: 508-222-4430.
E-mail: info@womenatworkmuseum.org
Web Site: www.womenatworkmuseum.org
Founded: 2003.
Key Personnel: Pres., Nancy Young; Treas., Kelly Fox.
Personnel Profile: Part-Time Paid 2; Interns 1.
Governing Authority: Tax-exempt.
Institution Type/Description: History Museum.
Collections: achievements of women around the world; paintings; quilts; photographs.
Activities: special events; educational programs.
Hours & Admission Prices: Sat. 11-4. No charge; donations accepted. &
Attendance: 1,000 (estimated)
Membership: Senior Citizen & Youth $20; Adult $35; Family $50; Supporting $100; Sustaining $250 & up.

Barnstable

OLDE COLONIAL COURTHOUSE, TALES OF CAPE COD, INC., Olde Colonial Courthouse, Rondezvous Lane & Rt. 6A, Barnstable, MA 02630. Mailing Address: P.O. Box 41, Barnstable, MA 02630-0041. Tel.: 508-362-8927. Fax: 508-362-9056.
Web Site: talesofcapecod.org
Founded: 1949.
Congressional District: 12
Key Personnel: Pres., Joe Berlandi; Vice Pres., Judith Arsenault; Treas., Ken Robinson.
Personnel Profile: Part-Time Volunteers 20.
Governing Authority: board of trustees. Tax-exempt: 170(b).
Institution Type/Description: Historic Site Museum & Olde Colonial Courthouse: Sachem Iyanough's gravesite dedicated to early Indians who befriended Pilgrims.
Collections: folklore; history; archaeology; Indian artifacts; films & slides of Cape Cod scenes; videotaped interviews.
Research Fields: oral & video folk tale recordings.
Facilities: collections of recordings of Cape Cod residents available for use in Cape Cod Room, Cape Cod Community College.
Activities: community cable TV show. Museum Sponsors: summer historic lecture series in July & August.
Publications: rare photographs; Cape Cod stories; legends; anecdotes; Cape Cod Trivia; Cape Cod Historical Almanac; Indian Rocks on Cape Cod; Henry's Cape Cod; Vikings to Vonnegut.
Hours & Admission Prices: By appointment only. Donations accepted. Lecture Series: July-Aug. Tues. 7:30pm.
Attendance: 1,250 (estimated)
Membership: Individual $15; Couple & Family $30; Sustaining $50; Life $200.

STURGIS LIBRARY, 3090 Main St., Barnstable, MA 02630. Mailing Address: Box 606, Barnstable, MA 02630-0606. Tel.: 508-362-6636. Fax: 508-362-5467.
Web Site: www.sturgislibrary.org
Founded: 1867.
Congressional District: 10
Key Personnel: Dir., Lucy Loomis; Pres. Bd. Trustees, Ellie Claus.
Personnel Profile: Full-Time Paid 3; Part-Time Paid 7; Part-Time Volunteers 30.
Governing Authority: nonprofit. Tax-exempt.
Institution Type/Description: Historic Building: housed in 1644 Rev. Lothrop House.
Collections: genealogy & local history; maritime history; Cape Cod History.
Research Fields: maritime history; Barnstable family genealogy; Cape Cod history.
Facilities: 65,000-vol. library includes Cape Cod, maritime history & genealogy; reading room. Postcards, books & gifts for sale.
Activities: monthly & temporary exhibitions.
Publications: quarterly newsletter; books, The History of the Sturgis Library; Nineteenth Century Literary Gentlemen.
Hours & Admission Prices: Mon. & Wed.-Fri. 10-5, Tues. 1-8, Sat. 10-4. No charge; donations accepted. Closed holidays. &
Attendance: 45,000 (accurate)

TRAYSER MUSEUM GROUP DBA COAST GUARD HERITAGE MUSEUM, 3353 Main St., Barnstable, MA 02630. Mailing Address: P.O. Box 161, Barnstable, MA 02630-0161. Tel.: 508-362-8521.
E-mail: cgheritage@comcast.net
Web Site: coastguardheritagemuseum.org
Formerly: Donald G. Trayser Memorial Museum/Barnstable County Customs House
Founded: 2004.
Congressional District: 10
Key Personnel: Pres., William Collette
Governing Authority: municipal. Town of Barnstable. Tax-exempt: 501(c)(3).
Institution Type/Description: History Museum: housed in 1856 Old Customs House.
Collections: history; marine articles; ship models; Indian artifacts; Coast Guard artifacts; blacksmith shop. Historic Building: 1800s jail.
Research Fields: early Barnstable.
Hours & Admission Prices: May-Oct. Tues.-Sat. 10-3. Adults $5; members, active Coast Guard and children 10 & under no charge. &
Attendance: 4,000 (estimated)
Membership: Individual $25; Family $40; Supporting $100; Sustaining $250.

Barre

BARRE HISTORICAL SOCIETY, INC., 18 Common St., Barre, MA 01005. Mailing Address: P.O. Box 755, Barre, MA 01005-0755. Tel.: 978-355-4978.
Founded: 1909.
Congressional District: 2
Key Personnel: Pres., Daniel E. Stevens; Cur., Bertyne Smith; Treas., Margaret Frost; Asst. Treas., Suzanne Fullam; Sec., Lester W. Paquin.
Personnel Profile: Part-Time Volunteers 11.
Volunteer Hours: 1,000
Governing Authority: society. Tax-exempt: 501(c)(3).
Institution Type/Description: Local History Museum: housed in 1839 home of Spencer Field; built by Elias Carter, architect.
Collections: 1763-1940 First Parish Church records; 1823-1969 United Methodist Church records; papers of Willard Broad; papers of Marshall D. Eaton; Houghton papers; manuscripts; 1859 12 passenger Abbott Downing stagecoach; Elm Street School for the Feeble Minded; 1848-1956 Broad St., Barre MA.
Major Exhibits: Dedication of Conserved Stagecoach, June 2014.
Research Fields: town history; genealogy.
Facilities: library of local history & genealogy available on premises with supervision.
Activities: guided tours; permanent exhibitions.
Hours & Admission Prices: Thurs. 10-12; other times by appointment. No charge; donations accepted. &
Attendance: 300 (estimated)
Membership: Individual $10; Family $20; Corporate $50; Life $250.

Becket

BECKET LAND TRUST HISTORIC QUARRY & FOREST, (M), Quarry Rd., Becket, MA 01223. Mailing Address: P.O. Box 44, Becket, MA 01223-0044. Tel.: 413-623-2100. Facebook: Becket Land Trust Historic Quarry and Forest.
E-mail: landtrust@becketlandtrust.org
Web Site: www.becketlandtrust.org
Founded: 1991.
Key Personnel: Dir., Dorothy Napp Schindel.
Governing Authority: nonprofit organization.
Institution Type/Description: History Museum.
Collections: Hudson-Chester Quarry history; mining; period truck; blacksmith's shop; wooden derrick; mining equipment; oral histories.
Facilities: 300-acre preserve; nature trails.
Activities: educational programs.
Hours & Admission Prices: Dawn-dusk. No charge.

Belchertown

THE STONE HOUSE MUSEUM, 20 Maple St., Belchertown, MA 01007-9416. Mailing Address: P.O. Box 1211, Belchertown, MA 01007-1211. Tel.: 413-323-6573.
E-mail: ttstockton@earthlink.net
Web Site: www.stonehousemuseum.org
Founded: 1903.
Congressional District: 1
Key Personnel: Pres. (V), Tom Stockton; Asst. Cur., Shirley Bock; Archivist, Cliff McCarthy.
Personnel Profile: Part-Time Volunteers 30; Interns 2.
Governing Authority: society. Parent Institution: The Belchertown Historical Assoc. Tax-exempt: 501(c)(3).
Institution Type/Description: Historic House Museum: 1827 stone house.
Collections: china; furniture; household goods; costumes; fabrics & hand work; jewelry; pewter; shaker; farm equipment; old vehicles; Rogers groups; toys; musical instruments; archives; manuscript collections; printing office; stone barn; portraits; carriages; sleighs; stagecoach; print shop.
Research Fields: ceramics; local families; accounting books for local business; genealogy.
Facilities: 500-vol. library of local history; genealogy. Local authors & religious works available for use on the premises by appointment.
Activities: guided tours; permanent exhibitions; living history programs; interactive school programs.
Publications: newsletter; brochures; cookbook.
Hours & Admission Prices: mid-May to mid-Oct. Sat. 2-5 & by appointment. Adults $5, seniors $4; discounts to AAM, ICOM & MTA members; members no charge.
Attendance: 2,000 (estimated)
Membership: Student under 18 $5; Senior 65 & over $8; Individual $15; Family $30; Friend $50; Patron $100; Benefactor $250 & up.

Bellingham

WHATCOM CHILDREN'S MUSEUM, 121 Prospect St., Bellingham, MA 98225-4401. Tel.: 360-733-8769.
Institution Type/Description: Children's Museum.
Collections: hands-on exhibits.
Hours & Admission Prices: Tues.-Wed. & Sun. 12-5, Thurs.-Sat. 10-5.

Berlin

BERLIN ART AND HISTORICAL SOCIETY, 4 Woodward Ave., Berlin, MA 01503. Mailing Address: P.O. Box 35, Berlin, MA 01503. Tel.: 978-838-2502.
Web Site: www.townofberlin.com/historical
Founded: 1950.
Key Personnel: Pres. (V), June Miller.
Personnel Profile: Part-Time Volunteers 7.
Governing Authority: municipal. Parent Institution: Berlin Art & Historical Society. Tax-exempt.
Institution Type/Description: General Museum.
Collections: archives; history; military; costumes; folklore; genealogy. Historic Houses: 1814 Powder House; 1870 Town Hall; 1805 Hearse House; 1790 Bullard House.
Research Fields: archives; history; military; costumes; folklore; genealogy.
Activities: lectures; temporary exhibitions; programs for historical society & schools.
Publications: newsletters, Heritage News; Berlin Historian.
Hours & Admission Prices: By appointment. No charge.
Membership: Junior $1; Adult $15; Family $25.

Beverly

BEVERLY HISTORICAL SOCIETY AND MUSEUM, (M), 117 Cabot St., Beverly, MA 01915-5196. Tel.: 978-922-1186. Fax: 978-922-7387.
E-mail: info@beverlyhistory.org
Web Site: www.beverlyhistory.org
Founded: 1891.
Congressional District: 6
Key Personnel: Dir., Susan Goganian; Pres. (V), Dan Lohnes; Mgr. Collections, Darren Brown.
Personnel Profile: Full-Time Paid 2; Part-Time Paid 2; Part-Time Volunteers 23; Interns 5.
Governing Authority: society. Branch Museums: John Balch House, 448 Cabot St., Beverly, MA; Rev. John Hale Farm, 39 Hale St., Beverly, MA. Tax-exempt: 501(c)(3).
Institution Type/Description: Historical Society Museum: housed in 1781 John Cabot Mansion.
Collections: decorative art; portraits; military; maritime; transportation; toys; dolls; genealogy; railroad photographs. Historic Houses: 1636 John Balch House; 1694 Rev. John Hale House.
Research Fields: Beverly history; maritime history; genealogy; New England transportation.
Facilities: 5,000-vol. library & manuscripts, including extensive shipping papers & log books.
Activities: guided tours; permanent & temporary exhibitions.
Publications: quarterly newsletter; Beverly Historical Review; books, Ryal Side from Early Days of Salem Colony; The Old Planters of Beverly; booklets, Men in the War of Independence; Made in Beverly.
Hours & Admission Prices: Museum: Tues. & Thurs.-Sat. 10-4, Wed. 1-9. Research-Galloupe Library: Tues. & Wed. Museum: adults $5, students & seniors $4; discounts to AAM & NEMA members; members & children under 16 no charge. Library: adults $5 per hour; members no charge.
Attendance: 2,487 (estimated)
Membership: Individual $30; Family $50; Patron $100; Sponsor $250; Benefactor $500; Lifetime $1,000.

Billerica

BILLERICA HISTORICAL SOCIETY - CLARA SEXTON HOUSE, 36 Concord Rd., Billerica, MA 01821. Mailing Address: P.O. Box 381, Billerica, MA 01821-0381. Tel.: 978-667-7020.
E-mail: billericahistorical@verizon.net
Web Site: www.billericahistorical.com
Institution Type/Description: Historical Society: built in c.1723.
Collections: artifacts; photographs.
Hours & Admission Prices: May-Oct. 1st Sun. each month 1-3.

Bolton

BOLTON HISTORICAL SOCIETY, INC., Sawyer House, 676
Main St., Bolton, MA 01740. Mailing Address: P.O. Box 211,
Bolton, MA 01740-0211. Tel.: 978-779-6392.
Web Site: www.boltonhistoricalsociety.org
Founded: 1962.
Congressional District: 5
Personnel Profile: Part-Time Volunteers 6.
Governing Authority: board of directors of society. Tax-exempt.
Institution Type/Description: General Museum: housed in c.1810 Sawyer
House & farm/barn blacksmith shop.
Collections: early industrial work; post-war articles; records & articles used in
early religious observances; family histories; early home articles; old farm
tools & machinery; old blacksmith shop with tools; letters; diaries;
documents; 18th & 19th century surveyors maps; 19th century medical
instruments; barn.
Research Fields: town history.
Facilities: archives containing local records, town & school department
reports, books & papers by local authors, genealogy, history & geographic
material of the area available for use on premises.
Activities: guided tours by appointment; lectures; formally organized education
programs for children; permanent & temporary exhibitions.
Publications: newsletters; program listings.
Hours & Admission Prices: Thurs. 1:30-3:30; tours by appointment. No
charge; donations accepted.
Attendance: 300 (estimated)
Membership: Single $20; Family $35; Supporting $50; Sustaining $100;
Benefactor $250.

Boston

**ANCIENT AND HONORABLE ARTILLERY COMPANY OF
MASSACHUSETTS,** 1 Faneuil Hall, Armory, 4th Fl., Boston,
MA 02109-1604. Tel.: 617-227-1638. Fax: 617-227-7221.
E-mail: ahac.curator@verizon.net
Web Site: www.ahac.us.com
Formerly: The Military Company of the Massachusetts
Founded: 1638.
Congressional District: 9
Key Personnel: Cur., Lt. Charles Fazio.
Personnel Profile: Full-Time Paid 1; Part-Time Paid 1.
Governing Authority: municipal. Tax-exempt: 501(c)(3).
Institution Type/Description: Military Museum.
Collections: archives; historical paintings; portraits; relics of wars; military
weapons.
Research Fields: 18th, 19th & 20th-century military history.
Facilities: 2,500-vol. library of military books & information available by
appointment.
Activities: lectures to groups of 25 or more.
Publications: company history brochure.
Hours & Admission Prices: Call for information. No charge, donations
accepted. &
Attendance: 40,000 (estimated)

ARNOLD ARBORETUM OF HARVARD UNIVERSITY, 125
Arborway, Boston, MA 02130-3500. Tel.: 617-524-1718. Fax:
617-524-1418.
E-mail: arbweb@arnarb.harvard.edu
Web Site: www.arboretum.harvard.edu
Founded: 1872.
Congressional District: 9
Key Personnel: Dir., Dr. Robert E. Cook; Finance & Admin., Andrea Nix.
Personnel Profile: Full-Time Paid 51; Part-Time Volunteers 100; Interns 14.
Governing Authority: university. Parent Institution: Harvard University. Tax-
exempt: 501(c)(3).
Institution Type/Description: Arboretum.
Collections: living collection of hardy trees & shrubs; herbarium; archives.
Research Fields: plant systematics; taxonomy; floristics.
Facilities: 40,000-vol. library of books on dendrology, botany, economic
botany, history, monographs & herbarium available for use to recognized
scholars or verified students.
Activities: guided tours; lectures; formally organized education programs for
adults, children, undergraduate & graduate college students; permanent &
temporary exhibitions. Museum Sponsors: Members Plant Sale in Septem-
ber; Lilac Sunday in May.
Publications: newsletter, Arnoldia.
Hours & Admission Prices: Visitor Center: Mon.-Fri. 9-4, Sat. 10-4, Sun. 12-4.
Library: Mon.-Sat. 10-4. Closed holidays. Grounds: daily sunrise-sunset.
No charge; donations accepted. &

Attendance: 250,000 (estimated)
Membership: Individual $35; Family $50; Sustaining $100; Sponsoring $200;
Benefactor $1,000.

THE ART INSTITUTE OF BOSTON MAIN GALLERY, 700
Beacon St., Boston, MA 02215-2598. Tel.: 617-585-6656. Fax:
617-437-1226.
Web Site: www.aiboston.edu
Founded: 1909.
Key Personnel: C.E.O., Bonnell Robinson.
Personnel Profile: Full-Time Paid 1; Part-Time Paid 1; Part-Time Volunteers 2;
Interns 2.
Governing Authority: university. Tax-exempt.
Institution Type/Description: Art Gallery.
Collections: paintings; sculpture; photographs.
Research Fields: design; illustration; fine arts; photography; sculpture; ceram-
ics.
Facilities: 1,600 sq. ft. exhibit space.
Activities: arts festivals; lectures; traveling & loan exhibitions.
Publications: exhibition posters; annual catalogues; postcards.
Hours & Admission Prices: Jan. 4-Dec. 23 Tues.-Fri. 12-6, Sat.-Sun. 12-5. No
charge. Closed major holidays. &
Attendance: 8,200 (estimated)

BARBARA AND STEVEN GROSSMAN GALLERY, School of
the Museum of Fine Arts, Boston, 230 The Fenway, Boston, MA
02115. Tel.: 617-369-3718.
Institution Type/Description: Art Gallery.
Collections: works by students, alumni & international artists.
Activities: special events; educational programs.
Hours & Admission Prices: Mon.-Wed. & Fri.-Sat. 10-5, Thurs. 10-8. Closed
holidays.

THE BOSTON ATHENAEUM, 10 1/2 Beacon St., Boston, MA
02108-3777. Tel.: 617-227-0270. Fax: 617-227-5266.
Web Site: www.bostonathenaeum.org
Formerly: Boston Athenaeum Library
Founded: 1807.
Congressional District: 9
Key Personnel: Pres., G. Marshall Moriarty; Acting Dir. & Librarian, Paula
Matthews; Assoc. Dir., John Lannon; Cur. Prints & Photographs, Catharina
Slautterback; Cur. Art, David Dearinger; Cur. Rare Books, Stanley E.
Cushing; Assoc. Cur. Paintings & Sculpture, Hina Hirayama; Conservator,
James Reid-Cunningham; Cur. Manuscripts & Coord. Community Affairs,
Stephen Nonack; Acquisition Librarian, Anthea Harrison Reilly; Head
Technical Svcs., Robert Kruse.
Personnel Profile: Full-Time Paid 34; Part-Time Paid 7; Part-Time Volunteers
48; Interns 5.
Governing Authority: nonprofit organization. Tax-exempt: 501(c)(3).
Institution Type/Description: Library with Art Collection: housed in 1847-49
library building.
Collections: paintings; sculpture; prints; drawings; photographs; daguerreo-
types; architectural drawings; archives; rare & illustrated books; manu-
scripts.
Research Fields: American & European history; fine & decorative arts; prints
& photographs.
Facilities: 700,000-vol. library of general books available for inter-library loan
& to the public upon request; reading room.
Activities: permanent & temporary exhibitions; lectures; concerts.
Publications: annual report; exhibition catalogues; monographs; library news-
letter.
Hours & Admission Prices: late May to early Sept. Mon. & Wed. 8:30-8, Tues.
& Thurs.-Fri. 8:30-5:30; Sept. 12-May 22 Mon. & Wed. 8:30-8, Tues. &
Thurs.-Fri. 8:30-5:30, Sat. 9-4. No charge. Closed major holidays. &
Membership: Associate Membership: Individual $110; Family $165. Family
Membership: Individual $220; Family $275. Life: Annual $275.

✱ **BOSTON CHILDREN'S MUSEUM, (M),** 308 Congress St.,
Boston, MA 02210-1034. Tel.: 617-426-6500. Fax: 617-426-1944.
TDD: 617-426-5466.
E-mail: info@BostonChildrensMuseum.org
Web Site: www.BostonChildrensMuseum.org
Formerly: The Children's Museum, Inc.
Founded: 1913.
Congressional District: 9
Key Personnel: Chm. Bd., Michael Yogman, M.D.; Pres. & C.E.O., Carole
Charnow; Sr. Vice Pres. & C.F.O., Amy Auerbach; Sr. Vice Pres. Research
& Program Planning, Leslie Swartz; Vice Pres. Corporate Devel. &

External Rels., Charlayne Murrell-Smith; Vice Pres. Family Learning & Early Childhood Programs, Jeri Robinson; Sr. Dir. Devel., Karin Blum; Dir. Public Rels. & Sponsorships, Jo-Anne Baxter.
Personnel Profile: Full-Time Paid 52; Part-Time Paid 38; Interns 5.
Governing Authority: nonprofit. Parent Institution: The Children's Museum. Tax-exempt: 501(c)(3).
Institution Type/Description: Children's Museum.
Collections: 50,000 artifacts; Native American archaeological & cultural artifacts with a focus on Northeast Woodland Indians & contemporary arts; 19th century Japanese merchant's house; global cultural artifacts; international folk dolls; toys; dolls; dollhouses; Boston & American social history; regional natural history.
Research Fields: pertaining to issues facing children & families in contemporary society; science; child development; multicultural education.
Facilities: 150-seat proscenium theater. Educational toys, games, books and museum-related items for sale.
Activities: educational programs; teacher, parent & student workshops and consulting services; neighborhood outreach; museum professional training; educational materials development; permanent, participatory, temporary & traveling exhibitions; Friday night performances; special events; KidStage, interactive theater.
Publications: newsletters; research reports; science books; multicultural books; audio tapes & teachers guides; books related to program areas; exhibit-related books, periodicals, & audiovisual materials.
Hours & Admission Prices: Fri. 10-9, Sat.-Thurs. 10-5. Admission $14; AAM members, other museum staff with ID, children under one, & members no charge. Closed Thanksgiving; Christmas. &
Attendance: 590,000 (estimated)
Membership: ACM Family (4 people) $150; Family Plus (6 people) $250; Library $350-$700; Family Donor Silver $500; Family Donor Gold $1,000; Corporate $600-$10,000 & up.

BOSTON FIRE MUSEUM, 344 Congress St., Boston, MA 02210-1204. Tel.: 617-338-9700.
E-mail: info@bostonfiremuseum.com
Web Site: bostonfiremuseum.com
Founded: 1983.
Congressional District: 9
Key Personnel: Chm. (V), Daniel O'Neill; Pres. (V), Paul Boudreau; Museum Shop Mgr., James Daly.
Personnel Profile: Part-Time Volunteers 20.
Governing Authority: not-for-profit organization. Parent Institution: Boston Sparks Association. Tax-exempt: 501(c)(3).
Institution Type/Description: Fire-Fighting History and Education Museum: housed in 1891 Congress Street Fire Station, a National Historic Landmark Building.
Collections: fire photographs, newspapers, printed material; helmets, fire alarm equipments; c.1905 hand-drawn American LaFrance Hook and Ladder Truck; 1826 Hunneman hand-pump; 1882 Christie/Amoskeag steam pumper; 1966 American LaFrance pumper; 1926 American LaFrance triple pump.
Research Fields: history of fire-fighting in the Boston area.
Facilities: 3,000 sq. ft. exhibit area. Museum-related items for sale.
Activities: tours for visitors to local fire stations; youth education.
Publications: monthly newsletter, The General Order.
Hours & Admission Prices: April-Oct. Fri. 10-4, Sat. 11-4; other times by appointment. No charge; donations accepted. &
Attendance: 3,500 (accurate)

BOSTON NATIONAL HISTORICAL PARK, (M), Charlestown Navy Yard, Boston, MA 02129. Tel.: 617-242-5648. Fax: 617-241-8650.
E-mail: david_vecchioli@nps.gov
Web Site: www.nps.gov/bost/
Founded: 1974.
Congressional District: 8
Key Personnel: Chief of Cultural Resources, Martin Blatt; Supt., Cassius Cash; Cur., David Vecchioli; Museum Specialist, Brandon Sexton.
Personnel Profile: Full-Time Paid 90; Part-Time Paid 80; Part-Time Volunteers 60; Interns 1.
Governing Authority: federal government. Parent Institution: National Park Service, U.S. Dept. of the Interior. Tax-exempt.
Institution Type/Description: Historic Houses & Sites.
Collections: Historic Sites & Structures include Old North church; Charlestown Navy Yard; Bunker Hill monument; Paul Revere house; Faneuil Hall; Old South meeting house; Old State House; Dorchester Heights; U.S.S. Constitution Museum; U.S.S. Cassin Young.
Research Fields: Boston history; Boston National Historical Park Sites; Boston Naval Shipyard.

Facilities: library; archives; visitor center; national park area.
Activities: visitor information; films; talks; guided tours; self-guided tours; museum sales; special events; educational programs; lecture series.
Publications: brochures; hardbooks.
Hours & Admission Prices: Daily 9-5. No charge. &
Attendance: 2,700,000 (accurate)

BOSTON PUBLIC LIBRARY, 700 Boylston St., Boston, MA 02116-2813. Tel.: 617-536-5400. Fax: 617-236-4306. TDD: 617-536-7055.
E-mail: ask@bpl.org
Web Site: www.bpl.org
Founded: 1852.
Congressional District: 9
Key Personnel: Cur. Arts Dept., Kimberly M. Tenney; Keeper Special Collections, Susan L. Glover.
Governing Authority: municipal; board of trustees. Parent Institution: City of Boston. Tax-exempt: 501(c)(3).
Institution Type/Description: Public Library with Art Collections.
Collections: mural decorations by Edwin A. Abbey, John Elliott, Pierre Puvis de Chavannes, John Singer Sargent; bronze doors by Daniel Chester French; sculptures by Frederick MacMonnies, Bela Pratt, Augustus & Louis Saint Gaudens, Francis Derwent Wood; paintings by Copley, Duplessis; dioramas depicting Dickens London, Alice in Wonderland, Arabian Nights & famous printmakers at work; Albert H. Wiggin Collection of 18th & 19th-century French and English prints; Old Masters prints & drawings; 18th & 19th-century American historical prints; 19th-century American photography by Bell, O'Sullivan, W.H. Jackson; British printmakers; American posters of the 90s; contemporary American printmakers; modern German prints; early French lithography; Holt Collection of photographs relating to Boston architecture; architectural drawings by Peabody & Stearns Cram & Ferguson, William G. Preston, Maginnis & Walsh, Charles Bulfinch, Charles Strickland; Charles J. Connick Collections of gouaches, photographs; bronzes, medals & graphics on Joan of Arc; Society of Arts & Crafts; Boston archives; 19th & 20th-century costumes & stage designs.
Research Fields: art & art history; architecture; history of photography; decorative arts & crafts of all countries & periods; New England artists; printing history; fine bookbinding.
Facilities: 138,000-vol. library representing all fields of art, 200,000 mounted pictures, 10,000 mounted photographs, 650,000 unmounted photographs, 150,000 postcards, 75,000 original prints, 170,000 prints by New England artists & books on Boston architecture available for research on premises; Index to Boston Architecture; 500,000 architectural plans; late 19th-century city building documents; microfilm sets on art, architecture, photography & the decorative arts; vertical & clipping files; Boston Pictorial Archive.
Activities: guided tours; lectures; films; concerts; demonstrations; gallery talks; inter-museum loan; temporary & traveling exhibitions.
Publications: books, A Handbook to the Art & Architecture of the Boston Public Library; A Survey of Boston Architectural Drawings & Photographs; Armstrong & Company: Artistic Lithographers; Etched in Sunlight: Fifty Years in the Graphic Arts (Samuel Chamberlain); The Lithographs of Stow Wengenroth, 1931-1972; Society of Arts & Crafts, Boston Exhibition Record 1897-1928; Afro-American Artists: A Bibliographical Directory; American Posters of the Nineties; Ralph Adams Cram, American Medievalist; The Work of Thomas W. Nason, N.A.; others in catalog of publications available upon request.
Hours & Admission Prices: June-Sept. Mon.-Thurs. 9-9, Fri.-Sat. 9-5; Oct.-May Sun. 1-5, Mon.-Thurs. 9-9, Fri.-Sat. 9-5. No charge. Closed national holidays. &
Attendance: 2,000,000

BOSTON SCULPTORS GALLERY, 486 Harrison Ave., Boston, MA 02118. Tel.: 617-482-7781.
E-mail: bostonsculptors@yahoo.com
Web Site: www.bostonsculptors.com
Founded: 1992.
Key Personnel: Admin., Jean Mineo.
Personnel Profile: Part-Time Paid 1.
Institution Type/Description: Art Gallery.
Collections: works by 36 contemporary northeast sculptors.
Hours & Admission Prices: Wed.-Sun. 12-6. No charge. Closed New Year's Day; Thanksgiving; Christmas. &

BOSTON UNIVERSITY ART GALLERY, 855 Commonwealth Ave., Boston, MA 02215-1303. Tel.: 617-353-3329. Fax: 617-353-4509.
E-mail: gallery@bu.edu
Web Site: www.bu.edu/ART
Founded: 1960.

Congressional District: 8
Key Personnel: Dir., Kate McNamara; Asst. Dir., Joshua Buckno.
Personnel Profile: Full-Time Paid 2; Interns 11.
Governing Authority: university. Parent Institution: Boston University, Boston, MA 02215. Tax-exempt: 170(b)(1)(a).
Institution Type/Description: Art Gallery.
Collections: paintings.
Research Fields: modern & contemporary art; New England art; history of photography.
Activities: lectures; gallery talks; concerts; formally organized education programs for undergraduate & graduate students affiliated with Boston University; training programs for professional museum workers; traveling exhibitions.
Publications: exhibition catalogues.
Hours & Admission Prices: mid-Jan. to mid-May & mid-Sept. to mid-Dec. Tues.-Fri. 10-5, Sat.-Sun. 1-5. No charge. Closed major holidays. &
Attendance: 7,450 (accurate)

BROMFIELD ART GALLERY, 450 Harrison Ave., Boston, MA 02118-2400. Tel.: 617-451-3605.
E-mail: gduehr@comcast.net
Web Site: www.bromfieldgallery.com
Founded: 1974.
Key Personnel: Pres. (V), Florence Montgomery; Gallery Mgr., Gary Duehr.
Personnel Profile: Part-Time Paid 1; Part-Time Volunteers 15; Interns 1.
Governing Authority: cooperative organization.
Institution Type/Description: Cooperative Art Gallery.
Collections: paintings; drawings; sculpture; photographs.
Facilities: 800 sq. ft. exhibit space.
Activities: guided tours; lectures; performances; rental gallery; poetry readings; members exhibits.
Hours & Admission Prices: Wed.-Sun. 12-5. No charge; donations accepted. Closed major holidays. &
Attendance: 3,000 (estimated)

CHASE YOUNG GALLERY, 450 Harrison Ave., No. 57, Boston, MA 02118. Tel.: 617-859-7222.
E-mail: mail@chaseyounggallery.com
Web Site: www.chaseyounggallery.com
Key Personnel: Dir. & Owner, Jane Young
Institution Type/Description: Art Gallery.
Collections: contemporary paintings, sculpture, & photographs.
Hours & Admission Prices: Tues.-Sat. 11-6, Sun. 11-4; other times by appointment.

THE COMMONWEALTH MUSEUM, 220 William T. Morrissey Blvd., Boston, MA 02125-3314. Tel.: 617-727-9268. Fax: 617-825-3613.
E-mail: commonwealthmuseum@sec.state.ma.us
Web Site: www.sec.state.ma.us/mus/museum/index.htm
Founded: 1986.
Congressional District: 9
Key Personnel: Sec. of Commonwealth, William Frances Galvin; Dir., Stephen Kenney, Ph.D.
Personnel Profile: Full-Time Paid 3; Part-Time Paid 2; Part-Time Volunteers 1; Interns 3.
Governing Authority: state. Parent Institution: Office of the Sec. of State. Tax-exempt: 170(c)(1).
Institution Type/Description: State History Museum.
Collections: local history & culture; period furnishings; photographs; personal artifacts.
Research Fields: Massachusetts state & local history; archives.
Facilities: lecture & conference room.
Activities: rotating, loan, participatory, temporary & traveling exhibitions; public tours; lectures; school loan service; broadcast programs; special events.
Publications: biannual newsletter; exhibition guides; educational materials; curriculum materials for teachers.
Hours & Admission Prices: Mon.-Fri. 9-4:45. No charge. Closed major holidays. &
Attendance: 10,000 (estimated)

COPLEY SOCIETY OF ART, 158 Newbury St., Boston, MA 02116. Tel.: 617-536-5049.
Institution Type/Description: Art Gallery.
Collections: paintings; sculpture; photographs.
Activities: temporary exhibitions.

Hours & Admission Prices: Tues.-Sat. 11-6, Sun. 12-5; other times by appointment.

THE GIBSON SOCIETY, INC. DBA GIBSON HOUSE MUSEUM, 137 Beacon St., Boston, MA 02116-1504. Tel.: 617-267-6338. Fax: 617-267-6338.
E-mail: info@thegibsonhouse.org
Web Site: thegibsonhouse.org
Founded: 1957.
Congressional District: 8
Key Personnel: Museum Administrator, Laura Gresh; Pres., Samuel H. Duncan; Treas., Robert S. Goodof; Museum Asst., Kyla Mackay-Smith; Consulting Cur., Wendy Swanton.
Personnel Profile: Full-Time Paid 1; Part-Time Paid 5; Part-Time Volunteers 5; Interns 1.
Governing Authority: society. Subsidiary Institution: Victorian Society in America, New England Chapter. Tax-exempt: 501(c)(3).
Institution Type/Description: Historic House Museum: 1859 Gibson House.
Collections: decorative arts; paintings; sculpture; photographs; clothing; Victorian period furniture.
Research Fields: Victorian decorative arts.
Facilities: meeting rooms.
Activities: guided tours; permanent exhibitions. Museum Sponsors: monthly lectures. Annual Events: Benefit Tea in March; Holiday Open House in December.
Publications: illustrated guide, Gibson House.
Hours & Admission Prices: Tours: Wed.-Sun. 1, 2 & 3. Adults $9, senior citizens & students $6, children $3; discount to Victorian Society (New England) & AAA members; members no charge. Closed national holidays.
Attendance: 3,500 (accurate)
Membership: Friend $50; Two Friends at the Same Address $75; Neighbor $100; Back Bay $150; Poet's Circle $250; Improper Bostonian $500; Proper Bostonian $1,000.

*** HISTORIC NEW ENGLAND, (M),** 141 Cambridge St., Boston, MA 02114-2799. Tel.: 617-227-3956. Facebook: Historic New England.
Web Site: www.historicnewengland.com
Founded: 1910.
Congressional District: 9
Key Personnel: Pres., Carl R. Nold; Site Mgr., Melinda Huff.
Governing Authority: nonprofit organization. Branch Museums: 1800 Barrett House, Main Street, New Ipswich, NH; 1907-34 Beauport Sleeper-McCann House, 75 Eastern Point Blvd., Gloucester, MA; 1690 Spencer-Peirce-Little Farm, Newbury, MA; 1846 Roseland Cottage, Rte. 169, Woodstock, CT; c.1750 Casey Farm, Rte. 1A, Saunderstown, RI; c.1740 Codman Estate, Codman Rd., Lincoln, MA; 1678 Coffin House, 14 High Rd., Rte. 1A, Newbury, MA; 1938 Gropius House, 68 Baker Bridge Rd., Lincoln, MA; c.1785 Hamilton House, 40 Vaughan's Ln., South Berwick, ME; 1774 Sarah Orne Jewett House, 5 Portland St., South Berwick, ME; 1807 Castle Tucker, Lee St., Wiscasset, ME; 1728 Cogswell's Grant. Spring St., Essex, MA; 1784 Governor John Langdon House, 143 Pleasant St., Portsmouth, NH; 1793 Lyman Estate, 185 Lyman St., Waltham, MA; 1789 Marrett House, Rte. 25, Standish, ME; 1807 Nickels-Sortwell House, 121 Main St., Wiscasset, ME; 1796 Otis House, 141 Cambridge St., Boston, MA; 1807 Rundlet-May House, 364 Middle St., Portsmouth, NH; 1718 Sayward-Wheeler House, 9 Barrell Ln., York Harbor, ME; 1796 Watson Farm, 455 North Rd., Jamestown, RI; c.1780 Winslow-Crocker House, 250 King's Hwy., Rte. 6A, Yarmouth Port, MA; c.1664 Jackson House, 76 Northwest St., Portsmouth, NH; 1821 Phillips House, 34 Chestnut St., Salem, MA; c.1770 Quincy House, 20 Muirhead St., Quincy, MA; c.1693 Arnold House, 487 Great Rd., Lincoln, RI; Historic New England also operates 11 properties that may be visited by appointment (617-227-3956) & are for architectural study. Tax-exempt: 501(c)(3).
Institution Type/Description: History Museum: housed in 1796 residence designed by Charles Bulfinch.
Collections: New England architectural artifacts; decorative arts; textiles; wallpapers; ceramics; glass; photographs; architectural drawings; ephemera; manuscripts; material culture; regional transportation;
Research Fields: architectural history; photographic history; decorative arts; preservation & conservation techniques for buildings & furniture.
Facilities: archives of 1,000,000 photographs; architectural drawings & other primary materials; auditorium.
Activities: guided tours; special events; lectures; school programs.
Publications: visitors guide; exhibition catalogues; Historic New England Magazine.
Hours & Admission Prices: Offices: Mon.-Fri. 9-5. Archives: by appointment. Otis House: Wed.-Sun. Tours 11-5, last tour 4:30. Admission $8; discounts to senior citizens, ICOM, AAA & AAM members; members no charge.
Attendance: 176,820 (accurate)

Membership: National $35; Individual $45; Household $55; Garden & Landscape $75; Contributing and Library & School $100; Young Friends of Historic New England $100-$1,500; Friends of the Library and Archives $125; Historic Homeowner $200; Business & Ogden Codman Design Group $250; Appleton Circle $2,500 & up.

* **THE INSTITUTE OF CONTEMPORARY ART/BOSTON, (M),** 100 Northern Ave., Boston, MA 02210-1870. Tel.: 617-478-3100. TDD: 617-927-6622.
E-mail: info@icaboston.org
Web Site: www.icaboston.org
Founded: 1936.
Congressional District: 9
Key Personnel: Pres. (V), Charles A. Brizius; Chm. (V), Paul Buttenwieser; Dir., Ellen Matilda Poss; Chief Cur., Barbara Lee; Sr. Curator, Jenelle Porter; Dir. Media, Branka Bogdanov; Dir. External Rels., Kelly Gifford; Chief Preparator, Tim Obetz; Dir. Retail Operations, Richard Gregg; Dir. Education, Monica Garza; Registrar, Janet Moore.
Personnel Profile: Full-Time Paid 48; Part-Time Paid 60; Part-Time Volunteers 45; Interns 15.
Governing Authority: nonprofit organization. Tax-exempt: 170(b)(1)(A).
Institution Type/Description: Art Museum.
Collections: contemporary works in various media.
Major Exhibits: Sandra and Gerald Fineberg Art Wall: Haegue Yans, 1/19/13-2/14.
Research Fields: 20th-21st century art.
Facilities: mediatheque; theater; art lab; digital studio. Museum-related items & exhibition videos for sale.
Activities: lecture series; poetry; literary readings; dance; theatre; performance; video & films; educational workshops.
Publications: books, Utopia Post Utopia: Configuration of Nature and Culture in Recent Sculpture and Photography; On the Passage of a Few People Through a Rather Brief Moment in Time: The Situationist International 1957-1972; Between Spring and Summer: Soviet Conceptual Art in the Era of Late Communism; Ulrike Rosenback: Video & Performance Art; Dissent: The Issue of Modern Art in Boston; The British Edge; Boston School; Inside the Visible; El Corazon Sangrante; Rachel Whiteread; Elvis and Marilyn: 2x Immortal; ICA Newsletter; Collectors Collect Contemporary 1990-99; Frieze; New Histories; Cornelia Parker Customized: Hot Rods, Low Riders & American Car; catalogues, Super Vision; Philip Lorca di Corcia; Anish Kapoor: Past, Present, Future; Tara Donovan; 20th Anniversary Limited Edition of Shepard Fairey: Supply & Demand; Damian Ortega: Do It Yourself; Dance/Draw; Charline von Hey 1.
Hours & Admission Prices: Tues.-Wed. & Sat.-Sun. 10-5, Thurs.-Fri. 10-9. Adults $15, senior citizens $13, students $10; discounts to AAM members; children under 17, families on last Sat. of month, Thurs. after 5 & members no charge. Closed New Year's Day; Thanksgiving; Christmas. &
Attendance: 200,000 (accurate)
Membership: Individual $65; Dual & Family $95; Associate $125; Friend $250; Patron $500; Advocate $1,000. Director's Circle: Fellow $2,000; Leader $5,000; Ars Longa $10,000.

* **ISABELLA STEWART GARDNER MUSEUM, (M),** 280 The Fenway, Boston, MA 02115-5809. Mailing Address: Two Palace Rd., Boston, MA 02115-5807. Tel.: 617-566-1401 & 278-5156. Fax: 617-264-6096.
E-mail: information@isgm.org
Web Site: www.gardnermuseum.org
Founded: 1903.
Congressional District: 9
Key Personnel: Chm. (V), John L. Gardner; Pres. (V), Barbara Hostetter; Dir., Anne Hawley; Treas., William Poorvu; Cur. Education, Margaret Burchenal; C.O.O., Peter Bryant; Dir. Public Rels., Katherine Armstrong; Dir. Mktg., Matt Montgomery; Dir. Devel., Helena Hartnett; Cur. Contemporary Art, Pieranna Cavalchini; Dir. Music, Scott Nickrenz; Consulting Cur. Landscape, Charles Waldheim.
Personnel Profile: Full-Time Paid 66; Part-Time Paid 106; Part-Time Volunteers 95; Interns 7.
Governing Authority: nonprofit organization. Tax-exempt: 501(c)(3).
Institution Type/Description: Art Museum: housed in 15th-century Venetian style palace.
Collections: paintings; sculpture; tapestries; stained glass; furniture; prints; textiles; decorative arts; music; horticulture installations; archives of letters to Founder.
Research Fields: pertaining to collections.
Facilities: library of books pertaining to the collections available by appointment only.
Activities: concerts; guided tours; lectures; scholarly symposia; educational programming; artist in residence program.
Publications: annual report; books; catalogues; e-newsletter.

Hours & Admission Prices: Tues.-Sun. 11-5; call for additional hours. Adults $12, senior citizens $10, college students with current ID $5; discounts to AAM & ICOM members; children under 18, members, anyone named Isabella, & on your birthday no charge. Reservation for guided tours 3 weeks in advance. Closed Independence Day; Thanksgiving; Christmas. &
Attendance: 170,000 (accurate)
Membership: Student $35; Nonresident $50; Individual $60; Family & Dual $85; Sustainer $120; Supporter $250; Contributor $500; Courtyard Circle $1,000.

JOHN F. KENNEDY PRESIDENTIAL LIBRARY & MUSEUM, (M), Columbia Point, Boston, MA 02125. Tel.: 617-514-1600. Fax: 617-514-1652. TDD: 617-514-1573.
E-mail: kennedy.library@nara.gov
Web Site: www.jfklibrary.org
Founded: 1979.
Congressional District: 9
Key Personnel: Chm. (V), Kenneth Feinberg; Pres. (V), Caroline Kennedy; Dir. Library & Museum, Tom Putnam; Treas., Marie Carbone; Cur., Stacey Bredhoff; Deputy Dir., James Roth; Devel., Ariadne Valsamis; Education, Nancy McCoy; Public Rels., Rachel Day Flor; Registrar, Kathryn Dodge; Archivist, Karen Adler Abramson; Museum Shop Mgr., Terri McGrath; Security, Norm Beland.
Personnel Profile: Full-Time Paid 52; Part-Time Paid 16; Part-Time Volunteers 20; Interns 20.
Governing Authority: federal. Parent Institution: National Archives and Records Administration, Washington, DC 20408. Tax-exempt: 170(b)(1)(A).
Institution Type/Description: History & Presidential Library.
Collections: the life & times of John F. Kennedy; films; memorabilia; tape recordings; personal papers; government records; political campaign items; presidential manuscripts; 32 million documents; 150,000 photographs; 11,000 serials; papers of Ernest Hemingway.
Major Exhibits: Freedom & Mercury Space Capsule, 10/12-12/15.
Research Fields: life, times, career & presidential administration of John F. Kennedy; middle 20th-century American politics & government; life & work of Ernest Hemingway.
Facilities: 70,000-vol. library pertaining to mid-century American politics & government available on premises for research only; 38,000 sq. ft. exhibit space; cafeteria; educational facilities; 700-seat auditorium; two 230-seat theaters. Museum-related items for sale.
Activities: lectures; formal education programs for children; temporary exhibitions; films; public forum series; teacher education institutes. Annual Events: PEN/Hemingway Awards Ceremony; Profile in Courage Essay Contest; Profile in Courage Awards Ceremony.
Publications: brochure; guide to collections; biannual newsletter, JFK Library Foundation Newsletter.
Hours & Admission Prices: Daily 9-5; groups by appointment. Adults $12, senior citizens & college students with ID $10, children 13-17 $9; discounts to groups, museums of Boston, NEMA & AAM members; members & children under 12 no charge. Closed New Year's Day; Thanksgiving; Christmas. &
Attendance: 220,000 (accurate)
Membership: Friends of the Kennedy Library: Individual $40; Family $60; Contributor $100; Benefactor $250; Leadership Circle $500; President's Circle $1,000.

KINGSTON GALLERY, 450 Harrison Ave., #43, Boston, MA 02118. Tel.: 617-423-4113.
Institution Type/Description: Art Gallery.
Collections: works by emerging artists.
Activities: temporary exhibitions.
Hours & Admission Prices: Wed.-Sun. 12-5; other times by appointment.

THE MARY BAKER EDDY LIBRARY, (M), 200 Massachusetts Ave., Boston, MA 02115-3017. Tel.: 617-450-7000. Fax: 617-450-7048. Facebook: The Mary Baker Eddy Library.
E-mail: librarymail@mbelibrary.org
Web Site: www.mbelibrary.org
Founded: 2002.
Congressional District: 8
Key Personnel: Pres. & Exec. Mgr., Lesley Pitts; Programs Producer, Jonathan Eder; Educational Programs Producer, Kelli Alvarez; Mgr. Mktg., Taryn McNichol; Senior Research Archivist, Judith A. Huenneke; Museum Shop Mgr., Katie Kimble.
Personnel Profile: Full-Time Paid 16; Part-Time Paid 8; Part-Time Volunteers 1; Interns 2.
Governing Authority: private; nonprofit organization. Tax-exempt: 501(c)(3).
Institution Type/Description: History Museum.

Collections: books; manuscripts; letters & correspondence; personal artifacts; photographs; films; paintings.
Major Exhibits: Transcending Boundaries, 1/14-12/14.
Research Fields: spirituality & health; American women in religious leadership; ideas & life of Mary Baker Eddy; quest for spirituality today.
Facilities: 12,000-vol. library; 75-seat restaurant; 13,500 sq. ft. exhibit space.
Activities: docent program; guided tours; lectures; participatory exhibits. Annual Events: Opening Our Doors; One World, arts educational program; First Night Boston; school vacation programs.
Publications: BLOGS; monthly e-newsletter.
Hours & Admission Prices: Tues.-Sun. 10-4. Adults $6, seniors, students & children 6-17 $4; discounts to AAM & NEMA members; children under 6 & donors no charge. Closed New Year's Day; Martin Luther King Jr. Day; Presidents' Day; Patriots' Day; Memorial Day; Independence Day; Labor Day; Thanksgiving; Christmas Eve & Day. &
Attendance: 100,000 (estimated)

MASSACHUSETTS COLLEGE OF ART + DESIGN - BAKALAR & PAINE GALLERIES, 621 Huntington Ave., Boston, MA 02115-5801. Tel.: 617-879-7333. Fax: 617-879-7340.
E-mail: galleryinfo@massart.edu
Web Site: www.massart.edu/galleries
Key Personnel: Dir., Lisa Tung
Institution Type/Description: Art Gallery.
Collections: works by contemporary artists.
Hours & Admission Prices: Mon.-Tues. & Thurs.-Fri. 12-6, Wed. 12-8, Sat. 12-5.

MASSACHUSETTS HISTORICAL SOCIETY, 1154 Boylston St., Boston, MA 02215-3695. Tel.: 617-536-1608. Fax: 617-859-0074.
E-mail: library@masshist.org
Web Site: www.masshist.org
Founded: 1791.
Congressional District: 9
Key Personnel: Pres., Dennis A. Fiori; Chm. Bd., Charles C. Ames; Librarian, Peter Drummey; Editor in Chief, C. James Taylor.
Personnel Profile: Full-Time Paid 42; Part-Time Paid 11; Part-Time Volunteers 8.
Governing Authority: board of trustees. Tax-exempt: 501(c)(3).
Institution Type/Description: Library.
Collections: archives; paintings; sculpture; American & New England history; manuscripts.
Facilities: library.
Activities: permanent & temporary exhibitions; lectures; tours; seminars; workshops.
Publications: Massachusetts Historical Society Review; newsletter, Miscellany; books.
Hours & Admission Prices: Library: Mon. & Wed.-Fri. 9-4:45, Tues. 9-7:45, Sat. 9-4. Exhibitions: Mon.-Sat. 10-4. No charge. Closed national holidays. &
Attendance: 8,750 (estimated)
Membership: Student $35; Educator (K-12) & Associate (Age 40 & Up) $75; Standard $150.

MILLS GALLERY - BOSTON CENTER FOR THE ARTS, 551 Tremont St., Boston, MA 02116-6338. Tel.: 617-426-8835.
Institution Type/Description: Art Gallery.
Collections: works by contemporary artists.
Activities: special events.
Hours & Admission Prices: Wed. & Sun. 12-5, Thurs.-Sat. 12-9. No charge.

MUSEUM OF AFRICAN AMERICAN HISTORY, 46 Joy St., Boston, MA 02114-4005. Mailing Address: 14 Beacon St., Ste. 719, Boston, MA 02108-3710. Tel.: 617-725-0022. Fax: 617-720-5225.
E-mail: history@maah.org
Web Site: www.maah.org
Founded: 1966.
Congressional District: 9
Key Personnel: Exec. Dir., Beverly Morgan-Welch; Chm. (V), James S. Hoyte.
Personnel Profile: Full-Time Paid 8; Part-Time Paid 7; Part-Time Volunteers 1; Interns 6.
Governing Authority: nonprofit organization. Tax-exempt: 501(c)(3).
Institution Type/Description: History Museum.
Collections: artifacts & archival material relating to the history of Afro-Americans in New England; photographs; papers on civil rights & civic organizations; Black family papers; sculptures; Civil War artifacts; artworks. Historic Buildings: 1806 African meeting house, the oldest extant Black church in the U.S.; 1835 Abiel Smith School, the first publicly funded

school for African Americans; the African Meeting House on Nantucket, 1820s; The Florence Higginbotham on Nantucket, 1845.
Research Fields: social history of African-American communities in New England; historic archaeology; oral history.
Facilities: 1,200-vol. library of literature on African American history available to students & faculty of Suffolk University & members; historic classroom.
Activities: gallery guided tours; lectures; films; guided walking tours of Black Heritage Trail (R); temporary exhibits; rental facilities.
Publications: electronic bulletin; occasional archaeology reports; brochure; Black Heritage Trail; guide, A Gathering Place for Freedom.
Hours & Admission Prices: Mon.-Sat. 10-4. Adults $5, youth 13-17 and seniors 62 & over $3; members no charge. Closed major holidays. &
Attendance: 230,000 (estimated)
Membership: Eunice Ross Senior & Student $15; Lewis Hayden Individual $25; Susan Paul Family $50; National Trust Fund $125; Maria Stewart Society $500; Frederick Douglas Society $1,000; Legacy Society $5,000.

✴　**MUSEUM OF FINE ARTS, (M),** 465 Huntington Ave., Boston, MA 02115-5597. Tel.: 617-267-9300. Fax: 617-369-3064. TDD: 617-267-9703.
E-mail: webmaster@mfa.org
Web Site: www.mfa.org
Founded: 1870.
Congressional District: 9
Key Personnel: Dir., Malcolm Rogers; Deputy Dir., Katherine Getchell; Chm. (V), Richard K. Lubin; Pres. (V), Sandra Moose; Deputy Dir., Maria Muller; Deputy Dir. & C.F.O., Mark Kerwin; Sr. Dir. Communications, Dawn Griffin; Chm. Prints, Drawings & Photographs, Clifford S. Ackley; Chm. Art of Asia, Oceania & Africa, Jane Portal; Chm. Art of the Americas, Elliot Davis; Chm. Art of the Ancient World, Rita Freed; Chm. Conservation & Collections Management, Matthew Siegal; Dir. Libraries & Archives and Museum Historian, Maureen Melton; Pres. SMFA & Deputy Dir. MFA, Christopher Bratton; Dir. Exhibitions & Design, Patrick McMahon; Chm. Contemporary Art & MFA Programs, Edward Saywell; Dir. Human Resources, Jane O'Reilly; Cur. Textiles & Fashion Arts, Pamela A. Parmal; Cur. Education, Barbara Martin; Cur. Musical Instruments, Darcy Kuronen; Dir. Facilities, David Geldart; Dir. MFA Publications, Emiko Usui; Dir. Intellectual Property, Debra La Kind; Museum Shop Mgr., Ellen Bragalone.
Personnel Profile: Full-Time Paid 570; Part-Time Paid 174; Part-Time Volunteers 1,080; Interns 120.
Governing Authority: nonprofit organization. Subsidiary Institution: School of the Museum of Fine Arts. Tax-exempt.
Institution Type/Description: Art Museum & School.
Collections: over 450,000 objects including Art of the Americas, Art of Europe, Contemporary Art, Art of Asia, Oceania & Africa, Art of the Ancient World, prints, drawings, photographs, textile & fashion arts, musical instruments.
Major Exhibits: Fashion and Jewelry from Hollywood's Golden Age, 9/10/13-5/26/14; Collecting Impressionism in Boston: The Story of a Love, 2/14-5/14; Permission to be Global: Latin American Art from the Ella Fontanals Cisneros Collection (T), 3/14-7/14; Quilts & Color: The Pilgrim/Roy Collection, 4/14-7/14; Jaime Wyeth (T), 7/14-12/14; Jasper Johns on Paper, 7/14-3/15; Where the Jinas Owell, 8/14-11/14; Isabel and Ruben Toledo, 8/14-1/15; Fashion & Jewelry from Hollywood's Golden Age, 9/14-3/15; Playing with Paper: Japanese Toy Prints, 9/14-5/15; Goya: Order and Disorder, 10/14-1/15.
Research Fields: art historical, archaeological & conservation science disciplines.
Facilities: 250,000-vol. library of art reference works available for inter-library loan & for use on the premises. Books, painting, sculpture, jewelry reproductions, postcards, note paper, miscellaneous gift items & reproductions of objects from the collections for sale.
Activities: guided tours; lectures; films; gallery talks; concerts; dance recitals; arts festivals; formally organized education programs; inter-museum loan, permanent, temporary & traveling exhibitions.
Publications: bimonthly, Preview; exhibition & permanent collection catalogues; annual report; art books.
Hours & Admission Prices: Museum: Wed.-Fri. 10-9:45, Sat.-Tues. 10-4:45. Adults $25, senior citizens and students 18 & over $23; youth 7-17 (Mon.-Fri. before 3pm) $10; discounts to AAM & ICOM members; youth 7-17 (Mon.-Fri. after 3, Sat.-Sun. & school holidays), members and children 6 & under no charge. Closed New Year's Day; Patriot's Day; Independence Day; Thanksgiving; Christmas. &
Attendance: 1,029,734 (accurate)
Membership: Supporter $75; Contributor $110; Ambassador $250; Sustainer $750; Leader $1,500; Patron $3,000; Patron Fellow $6,000; Patron Sponsor $12,000; Director's Circle Patron $30,000; President's Circle Patron $50,000; Chairman's Circle Patron $100,000.

✻ **MUSEUM OF SCIENCE, (M),** 1 Science Park, Boston, MA 02114-1099. Tel.: 617-589-0100 & 0222. Fax: 617-742-2246 & 589-0454. TTY: 617-589-0417.
E-mail: information@mos.org
Web Site: www.mos.org
Founded: 1830.
Congressional District: 7 & 8
Key Personnel: C.O.O., Wayne Bouchard; Chm. (V), Howard Messing; Dir. & Pres., Dr. Ioannis Miaoulis; Sr. Vice Pres. Strategic Initiatives, Larry Bell; Sr. Vice Pres. Advancement, Joan Hadly; Volunteer Service League Pres. (V), Damase Caouette; Vice Pres. Education, Paul Fontaine; Vice Pres. Visitor Svcs. & Operations, Jonathan Burke; Vice Pres. Mktg. & External Affairs, Cynthia Mackey; Vice Pres. Human Resources, Britton O'Brien; Vice Pres. Finance & System Svcs., John Slakey; Dir. Current Science & Technology, David Rabkin; Vice Pres. Research, Christine Cunningham.
Personnel Profile: Full-Time Paid 362; Part-Time Paid 262; Part-Time Volunteers 709; Interns 45.
Volunteer Hours: 56,556
Operating Expenses: 55,668
Operating Income: 55,701
Governing Authority: nonprofit organization. Tax-exempt: 501(c)(3).
Institution Type/Description: Science & Technology Museum.
Collections: mounted animal specimens; live animals; live plants; mineral, rock & gem specimens; fossils; paintings, prints & illustrations, scientific in context; sculptures; scientific instruments & inventions with a historical perspective; electrical devices & technological advancements; natural science & historical dioramas; models; transportation; botanical; human body organs & parts; photo & graphic panels detailing: bones, lungs, brain, heart, teeth, human reproductive process, living cells & DNA and hip repair; archeological artifacts; astronomical devices & photos.
Major Exhibits: Our Global Kitchen: Food, Nature, Culture (T), 12/22/13-4/13/14; Innovation in the Art of Food: Chef Ferran Adria (T), 2/14-5/14; 2Thextreme: Math Alive (T), 5/25/14-9/1/14; Maya: Hidden World Revealed (T), 10/12/14-5/3/15.
Facilities: 17,000-vol. library of science books available for inter-library loan & on premises; planetarium; electricity theatre; omni theater; reading room; cafeteria. Books, hobby material & other museum-related items for sale.
Activities: lectures; films; demonstrations; formally organized educational programs; standing exhibits.
Publications: members newsletter, Sparks; Annual Report; Engineering is Elementary; museum magazine.
Hours & Admission Prices: July 5 to Labor Day Fri. 9-9, Sat.-Thurs. 9-7; Sept.-July 4 Fri. 9-9, Sat.-Thurs. 9-5. Omni Theater, Planetarium, Laser Shows: call for hours. Exhibit Halls: adults $23, seniors $21, children $20; members no charge. Omni Theater, Planetarium & Laser: adults $10, seniors $9, children $8. Combo: adults $28, seniors $25.50, children $24; members no charge. Call 617-723-2500 or visit website for special holiday & vacation hours. Closed Thanksgiving; Christmas. &
Attendance: 1,419,558 (accurate)
Membership: Basic 2 $80; Basic 5 $120; Premier 2 $115; Basic 8 $150; Premier 5 $155; Premier 8 $185.

MUSEUM OF THE NATIONAL CENTER OF AFRO-AMERICAN ARTISTS, 300 Walnut Ave., Boston, MA 02119-1369. Tel.: 617-442-8614. Fax: 617-445-5525.
E-mail: bgaither@mfa.org
Web Site: www.ncaaa.org
Founded: 1969.
Congressional District: 9
Key Personnel: Co Chm., Margaret Burnham; Co Chm., Vivian Johnson, Ph.D.; Dir. & Cur., Edmund Barry Gaither; Asst. to Dir. & Registrar, Carol Murray.
Personnel Profile: Full-Time Paid 5; Part-Time Paid 5; Part-Time Volunteers 4.
Governing Authority: nonprofit. Parent Institution: National Center of Afro-American Artists, Boston, MA. Tax-exempt: 170(b)(1)(A).
Institution Type/Description: Art Museum.
Collections: paintings; prints & graphics by Afro-American artists; African art. Historic House: 19th-century house.
Research Fields: Afro-American, African & Afro Caribbean art.
Activities: guided tours; lectures; films; gallery talks; concerts; dance recitals; arts festivals; inter-museum loan, temporary & traveling exhibitions; educational school programs.
Publications: quarterly newsletter; annual report.
Hours & Admission Prices: Tues.-Sun. 1-5. Adults $4, students & senior citizens $3; members no charge.
Attendance: 10,000 (estimated)
Membership: Student $20; Individual $35; Family $55; Friend $85; Supporters $100; Donor $150; Contributor $250; Patron $500; Grand Patron $1,000.

✻ **NEW ENGLAND AQUARIUM CORPORATION, (M),** Central Wharf, Boston, MA 02110-3399. Tel.: 617-973-5200, ext. 0. Fax: 617-720-5098. TDD: 617-973-0223.
Web Site: www.neaq.org
Founded: 1957.
Congressional District: 9
Key Personnel: Pres., Howard "Bud" Ris; Chm., William Burgess; Exec. Vice Pres., Walter Flaherty; Vice Pres. Programs, Exhibits & Planning, William Spitzer; Dir. Visitor Experience, Deb Bobek; Vice Pres. Devel., Suellen Peluso; Vice Pres. Research, Scott Kraus; Vice Pres. Sales & Mktg., Jane Wolfson; Vice Pres. Finance & Operations, Joseph Zani; Dir. & Gen. Cur., John Dayton; Cur. Mammals, Kathy Streeter; Cur. Fishes, Steve Bailey; Dir. Design, Jim Duffey; Dir. Education, John Anderson; Assoc. Vice Pres. Conservation, Heather Tausig; Dir. Foundation & Government Support, Susan Thompson; Dir. MIS, Barbara Waller; Dir. Media Rels., Tony Lacasse; Dir. Gift Shop, Don Bors.
Personnel Profile: Full-Time Paid 240; Part-Time Paid 50; Part-Time Volunteers 500; Interns 100.
Governing Authority: nonprofit organization. Tax-exempt: 501(c)(3).
Institution Type/Description: Aquarium.
Collections: live aquatic displays; marine conservation; oceanography; fresh water conservation.
Research Fields: aquatic marine biodiversity: marine mammals, fisheries bycatch, oceanic fishes & turtles; animal husbandry & telemetry.
Facilities: 2,700-vol. library of marine sciences available for use on the premises; education center with classrooms & wet lab; cafe. Aquarium-related items for sale.
Activities: films; formally organized educational programs; lecture series on-line information via homepage; giant ocean tank dives; whale watch.
Publications: quarterly newsletter, Blue; teachers newsletter, Schooling.
Hours & Admission Prices: Summer: Mon.-Fri. 9-6, Sat.-Sun. 9-7; Winter: Mon.-Fri. 9-5, Sat.-Sun. 9-6. Adults $22.95, senior citizens $20.95, children 3-11 $15.95; discounts to groups; children under 3 & members no charge. Closed Thanksgiving; Christmas. &
Attendance: 1,300,000 (accurate)
Membership: Associate $85; Patron $135; Ambassador $185; Ocean Explorer $250; Conservation Society $500; Navigator Society $1,500.

NICHOLS HOUSE MUSEUM, (M), 55 Mount Vernon St., Boston, MA 02108-1330. Tel.: 617-227-6993. Fax: 617-723-8026
E-mail: info@nicholshousemuseum.org
Web Site: www.nicholshousemuseum.org
Founded: 1961.
Congressional District: 8
Key Personnel: Exec. Dir., Flavia Cigliano; Pres., David Beck.
Personnel Profile: Full-Time Paid 3; Part-Time Paid 4; Part-Time Volunteers 40; Interns 2.
Governing Authority: nonprofit organization. Tax-exempt: 501(c)(3).
Institution Type/Description: Historic House Museum: housed in former Beacon Hill home of Rose Standish Nichols.
Collections: early 1800s furnishings; sculpture; decorative arts.
Research Fields: textiles, ceramics, landscape gardening.
Facilities: library of books available on site by application only.
Activities: guided tours; lectures; permanent exhibitions.
Publications: books: Rose Nichols As We Knew Her; At Home on Beacon Hill: Rose Standish Nichols and Her Family; Lively Days - Margaret Homer Shurcliff.
Hours & Admission Prices: April-Oct. Tues.-Sat. 11-4; Nov.-March Thurs.-Sat. 11-4. Adults $8, students $5; discounts to MA Teachers Assoc., AAA, AAM, WGBH & ICOM members; children under 12 & members no charge. Closed holidays.
Attendance: 5,500 (accurate)
Membership: Single $45; Family $75; Sponsor $100; Donor $250; Corporate $500.

OLD SOUTH MEETING HOUSE, (M), 310 Washington St., Boston, MA 02108-4616. Tel.: 617-482-6439. Fax: 617-482-9621.
Web Site: www.oldsouthmeetinghouse.org
Founded: 1877.
Congressional District: 9
Key Personnel: Exec. Dir., Emily Curran; Pres., William G. Constable; Dir. Education, Aliza Saivetz; Asst. Dir., Janine Fabiano; Visitor Svcs. & Museum Shop Mgr., Karen Costello; Mktg. & Events Mgr., Robin DeBlosi.
Personnel Profile: Full-Time Paid 5; Part-Time Paid 9; Interns 2.
Governing Authority: nonprofit organization. Parent Institution: Old South Association in Boston. Tax-exempt: 501(c)(3).
Institution Type/Description: Historic Site: 1729 Old South Meeting House.
Collections: early American history; political & religious history related to Old South Meeting House.

Research Fields: historic preservation; issues involving freedom of speech; events leading to American Revolution, especially the Boston Tea Party.

Facilities: meeting hall; lecture hall; education room; 650-seat lecture & concert hall. Museum-related items for sale.

Activities: lectures; Music at the Meeting House Concert Series; forums; performances & walking tours; ongoing programs include Middays at the Meeting House weekly concert & lecture series; Summer Town Meeting reenactments; Partners in Public Dialogue public forum series; annual reenactment of the Boston Tea Party; wide array of educational programs for schools & community groups.

Publications: Old South Leaflets; newsletter; curriculum materials; An Architectural History of the Old South Meeting House.

Hours & Admission Prices: April-Oct. daily 9:30-5; Nov.-March daily 10-4. Adults $6, senior citizens 62 & over and students with ID $5, children 6-17 $1; members & children under 6 no charge. Closed New Year's Day; Thanksgiving; Christmas Eve & Day. &

Attendance: 78,000 (accurate)

Membership: Individual $35; Family $55; Old South Friend $100; Old South Donor $500; Phillis Wheatley Leadership Circle $1,000; Steeple Society $2,500; Boston Tea Party Patron $5,000.

OLD STATE HOUSE-THE BOSTONIAN SOCIETY, (M), 206 Washington St., Boston, MA 02109-1773. Tel.: 617-720-1713. Fax: 617-720-3289.

E-mail: adele@bostonhistory.org

Web Site: www.bostonhistory.org

Founded: 1881.

Congressional District: 9

Key Personnel: Exec. Dir., Brian Lemay; Bd. Chm. (V), Eric Hayes; Dir. The Old State House Museum, Rainey Tisdale; Dir. Finance & Admin., Linda Atlas; Collections Mgr., Marieke Van Damme; Devel., Gerrit Petersen; Office Mgr. & Collections and Library Asst., Adele Barbato; Mgr. Public Programs, Samantha Nelson; Visitor Svcs. Mgr., Erin Spencer; Facilities Mgr., Matthew Ottinger; Dir. Merchandising & Commercial Operations, Chuck Gordon; Distribution Mgr., S. Mark Edwards; Office Asst., Ashley Martin; Museum Shop Mgr., Nick Trainor.

Personnel Profile: Full-Time Paid 12; Part-Time Paid 28; Part-Time Volunteers 11; Interns 2.

Governing Authority: society. Tax-exempt: 501(c)(3).

Institution Type/Description: History Museum: 1713 Old State House.

Collections: Revolutionary & colonial artifacts; 19th century maritime & firefighting artifacts; domestic & commercial objects focusing on public life in Boston; paintings; manuscripts; photographs.

Research Fields: Boston history; architecture.

Facilities: 7,600-vol. library of Boston history books available to the public on a non-circulating basis; photo resource center; reading room. Museum-related items for sale.

Activities: lectures; gallery talks; walking tours.

Publications: pamphlets; newsletter.

Hours & Admission Prices: Daily 9-5. Adults $7, senior citizens & students $6, children 6-18 $3; discount to AAM members; MA school children, MA school teachers, NEMA members & active duty military no charge. Closed New Year's Day; Thanksgiving; Christmas.

Attendance: 105,538 (accurate)

Membership: Student $25; Senior Citizen $40; Individual $50; Family $80; Supporter $150; Benefactor $500; Life Benefactor $5,000.

✱ **PAUL REVERE HOUSE/PAUL REVERE MEMORIAL ASSOCIATION, (M),** 19 North Sq., Boston, MA 02113-2405. Tel.: 617-523-2338. Fax: 617-523-1775.

E-mail: staff@paulreverehouse.org

Web Site: www.paulreverehouse.org

Founded: 1907.

Congressional District: 8

Key Personnel: Exec. Dir., Nina Zannieri; Pres. (V), Paul Revere, Jr.; Cur., Edith Steblecki; Dir. Education, Emily Holmes; Dir. Research, Patrick Leehey; Interpretations & Visitor Svcs. Dir., Kristin Peszka.

Personnel Profile: Full-Time Paid 7; Part-Time Paid 15; Part-Time Volunteers 10; Interns 4.

Governing Authority: nonprofit organization. Parent Institution: Paul Revere Memorial Association. Branch Museum: The Pierce/Hichborn House, 29 N. Square, Boston, MA 02113. Tax-exempt: 501(c)(3).

Institution Type/Description: History Museum & Historic Houses: c.1680 Paul Revere House, Boston's oldest house; c.1711 Pierce-Hichborn House.

Collections: furniture & artifacts of the 17th-19th centuries; Revere engravings, documents, silver, family items & memorabilia.

Research Fields: social & political history of Boston, 1680 to present; material culture of the 17th- to 19th-centuries; Revere's life & work; immigrant history of the North End.

Facilities: education & visitor center. Gift items for sale.

Activities: guided tours; after-school & summer camp programs; neighborhood walking tours; special programs & events; permanent & temporary exhibitions; internship; research fellowship.

Publications: pamphlets; books; newsletter.

Hours & Admission Prices: Jan.-March Sun. & Tues.-Fri. 9:30-4:15; early April & Nov.-Dec. daily 9:30-4:15; mid-April to Oct. daily 9:30-5:15. Adults $3.50, college students & senior citizens $3, children $1; discounts to active military, museum professionals, AAM & ICOM members; members no charge. Closed New Year's Day; Thanksgiving; Christmas. &

Attendance: 248,781 (accurate)

Membership: Individual $20; Family $35; Supporting $50; Patron $100; Patriot $250.

PHOTOGRAPHIC RESOURCE CENTER AT BOSTON UNIVERSITY, 832 Commonwealth Ave., Boston, MA 02215-1205. Tel.: 617-975-0600. Fax: 617-975-0606.

E-mail: info@prcboston.org

Web Site: www.prcboston.org

Founded: 1976.

Key Personnel: Exec. Dir., Glenn Ruga; Pres. Bd. Dir., Vittorio Mezzaro.

Personnel Profile: Full-Time Paid 4; Part-Time Paid 1; Part-Time Volunteers 3; Interns 8.

Governing Authority: nonprofit organization. Tax-exempt: 501(c)(3).

Institution Type/Description: Photography Museum.

Collections: photographs.

Facilities: 4,700-vol. library of photo books & magazines.

Activities: lectures; temporary & traveling exhibitions; workshops; portfolio reviews. Annual Events: fundraising auctions.

Publications: 3 times a year, Loupe.

Hours & Admission Prices: Tues.-Fri. 10-5, Sat. 12-4. Suggested donation $3. Closed major holidays. &

Attendance: 10,000 (estimated)

Membership: Student & Faculty/Staff of Institutional Members $25; National $30; Seniors $40; Individual $50; Family $75; Supporting $150; Contributor $250; Benefactor $500; Institutional $600; Patron $1,000.

ROBERT KLEIN GALLERY, 38 Newbury St., 4th Fl., Boston, MA 02116. Tel.: 617-267-7997. Fax: 617-267-5567.

Key Personnel: Owner, Robert Klein; Dir., Eunice Hurd

Institution Type/Description: Art Gallery.

Collections: photographs.

Hours & Admission Prices: Tues.-Fri. 10-5:30, Sat. 11-5; other times by appointment.

RUBIN-FRANKEL GALLERY AT HILLEL HOUSE - BOSTON UNIVERSITY, 213 Bay State Rd., Boston, MA 02215-1499. Tel.: 617-353-7200. Fax: 617-353-7660.

E-mail: hillel@bu.edu

Web Site: www.bu.edu/hillel

Key Personnel: Dir., Holland Dieringer; Dir. Operations, Susan Sloane

Institution Type/Description: Art Gallery.

Collections: paintings; drawings; sculpture; photographs.

Activities: temporary exhibitions.

Hours & Admission Prices: Winter: Mon.-Fri. 9am-10pm, Sat. 8:30am-9pm, Sun. 3-9; Summer: Mon.-Fri. 10-4.

SHIRLEY-EUSTIS HOUSE, (M), 33 Shirley St., Boston, MA 02119-2725. Tel.: 617-442-2275. Fax: 617-442-2270.

E-mail: governorshirley@gmail.com

Web Site: www.shirleyeustishouse.org

Founded: 1913.

Congressional District: 9

Key Personnel: Exec. Dir., Patricia Violette; Pres., J. Archer O'Reilly; Treas., Marlowe Sigal.

Personnel Profile: Full-Time Paid 1; Part-Time Paid 4; Part-Time Volunteers 2.

Volunteer Hours: 180

Operating Expenses: 122,876

Operating Income: 183,210

Governing Authority: nonprofit organization. Parent Institution: Shirley-Eustis House Association. Tax-exempt: 501(c)(3).

Institution Type/Description: Historic House: 1747 Georgian country house designed by Peter Harrison, built by Royal Governor William Shirley; restored to Federal appearance, 1800; furnished according to Gov. William Eustis inventory, 1825; 1806 Gardner Carriage House.

Collections: decorative arts; French & Federal furnishings.

Research Fields: history of Shirley Place & its inhabitants, especially Gov. Shirley & Gov. Eustis; decorative arts of the Federal period; Georgian & Federal gardens.

Facilities: 3,500 sq. ft. exhibit space.

Activities: guided tours; lectures; loan exhibitions; educational programming.
Publications: newsletter, Shirley-Eustis Newsletter.
Hours & Admission Prices: Memorial Day to Labor Day Thurs.-Sun. 1-4; Sept.-Columbus Day Sat.-Sun. 1-4; Oct.-May by appointment. Adults $5, students & senior citizens $4; discounts to groups, NEMA, AAA, Blue Star & AAM members. Closed New Year's Day; Independence Day; Thanksgiving; Christmas.
Attendance: 2,000 (estimated)
Membership: Individual $25; Family $40; Business & Organization $100; Charter $250; Governors' Circle $500.

SKYWALK OBSERVATORY - DREAMS OF FREEDOM IMMIGRATION MUSEUM, The Prudential Center, 800 Boyleston St., Boston, MA 02199-8001. Tel.: 617-859-0648. Facebook: Skywalk Observatory.
E-mail: skywalk@topofthehub.net
Web Site: skywalkboston.com
Key Personnel: Skywalk Mgr., Laura Werneke.
Personnel Profile: Full-Time Paid 1.
Institution Type/Description: History Museum.
Collections: Boston's history & culture; photographs; personal artifacts.
Facilities: theater. Museum-related items for sale.
Activities: audio tour; 360-degree panoramic view of Boston.
Hours & Admission Prices: April-Nov. daily 10-10; Nov.-March daily 10-8. Adults $16, seniors 62 & over and university students with ID $13; children 12 & under $11; US military & Massachusetts Teachers Assoc. no charge. Closed Christmas.
Attendance: 250,000 (estimated)

THE SOCIETY OF ARTS & CRAFTS, 175 Newbury St., Boston, MA 02116. Tel.: 617-266-1810. Fax: 617-266-5654.
Key Personnel: Exec. Dir., Beth Ann Gerstein
Institution Type/Description: Art Gallery.
Collections: paintings; sculpture.
Activities: educational programs.
Hours & Admission Prices: Tues.-Sat. 10-6, Sun. 12-5; other times by appointment.

THE SPORTS MUSEUM, TD Garden, 100 Legends Way, Boston, MA 02114-1300. Tel.: 617-624-1235. Fax: 617-624-1238. Facebook: The Sports Museum.
E-mail: bcodagnone@dncboston.com
Web Site: www.sportsmuseum.org
Formerly: The Sports Museum of New England
Founded: 1977.
Congressional District: 8
Key Personnel: C.E.O. & Exec. Dir., Rusty Sullivan; Cur., Richard Johnson; Assoc. Cur., Brian Codagnone; Education & Public Program Mgr., Michelle Gormley.
Personnel Profile: Full-Time Paid 5; Part-Time Paid 4; Part-Time Volunteers 6; Interns 2.
Governing Authority: nonprofit organization. Tax-exempt: 501(c)(3).
Institution Type/Description: Sports Museum.
Collections: sports artifacts; films; documents archives; oral history; interactive exhibits.
Research Fields: sports history; sports media; sports equipment, uniforms, photos & video.
Facilities: 2,500-vol. library of general sport information; research archives.
Activities: special events. Annual Event: The Tradition fundraiser.
Hours & Admission Prices: Daily 10-4. Adults $10, members $5; discounts to AAM members & groups; children under 6 no charge. Closed during TD Garden events & major holidays. &
Attendance: 111,000 (estimated)

SUFFOLK RESOLVES HOUSE, 28 Glenmont Rd., Boston, MA 02135-3130. Tel.: 617-333-9700.
E-mail: askmhs@miltonhistoricalsociety.org
Web Site: www.miltonhistoricalsociety.org
Founded: 1904.
Governing Authority: nonprofit organization. Parent Institution: Milton Historical Society, Inc. Tax-exempt.
Institution Type/Description: Historic Society Museum: c.1774.
Collections: archives; furniture.
Activities: programs; workshops.
Publications: Hamilton, History of Milton; Morris-Webster, Story of the Suffolk Resolves; Bodsbinder, Judith, Margaret Sutermeister: Chronicling Seen and Unseen Worlds 1894-1909.
Hours & Admission Prices: By appointment, seasonal open houses. No charge; donations accepted. &

Membership: Individual $15; Family $25; Corporate $150; Life $300.

* **USS CONSTITUTION MUSEUM, (M),** Boston National Historical Park, Charlestown Navy Yard, Bldg. 22, Boston, MA 02129. Mailing Address: P.O. Box 291812, Boston, MA 02129-0215. Tel.: 617-426-1812. Fax: 617-242-0496.
E-mail: getinvolved@ussconstitutionmuseum.org
Web Site: www.asailorslifeforme.org; www.familylearningforum.org
Founded: 1972.
Congressional District: 8
Key Personnel: Pres., Anne Grimes Rand; Chm. (V), James Stokes; Dir. Finance & Admin., Jackie Hibbard; Dir. Retail Operations, Chris White; Dir. Exhibits, Robert Kiihne; Dir. Collections & Learning, Sarah Watkins; Dir. Devel., Laura O'Neill.
Personnel Profile: Full-Time Paid 25; Part-Time Paid 17; Part-Time Volunteers 25; Interns 10.
Governing Authority: private; nonprofit. Tax-exempt: 501(c)(3).
Institution Type/Description: History Museum.
Collections: naval & maritime history; works of art & memorabilia relating to USS Constitution; archives; manuscripts.
Research Fields: American naval history; Maritime history; social history; family learning.
Facilities: research library; microfilm of naval documents. Books, slides & nautical items for sale.
Activities: special programs for school groups by appointment; school outreach kits; audiovisual program; permanent & temporary collections.
Publications: biannual Constitution Chronicle; monthly member e-news; quarterly general e-news; Men of Iron, The Crew of USS Constitution in 1812; Old Ironsides activity book.
Hours & Admission Prices: April-Oct. daily 9-6; Nov.-March daily 10-5; groups by appointment. No charge; donations accepted. Closed New Year's Day; Thanksgiving; Christmas. &
Attendance: 404,000 (accurate)
Membership: Basic: Smithsonian Affiliate $20; Crew Member (Individual) $30; Quartermaster (Family) $50; Midshipman $75. Captain's Circle: Captain John Percival Society $100; Captain Silas Talbot Society $250; Commodore Edward Preble Society $500; Commodore William Bainbridge Society $1,000; Commodore Isaac Hull Society $2,500.

VILLA VICTORIA CENTER FOR THE ARTS, 85 W. Newton St., Boston, MA 02118-1523. Tel.: 617-927-1707. Fax: 617-236-7375.
Key Personnel: Dir., Javier Torres
Institution Type/Description: Art Gallery.
Collections: paintings; sculpture.
Activities: classes; educational programs; special events.
Hours & Admission Prices: Thurs.-Fri. 4-7, Sat. 1-4.

VILNA SHUL BOSTON'S CENTER FOR JEWISH CULTURE, 18 Phillips St., Boston, MA 02114-3711. Tel.: 617-523-2324. Fax: 781-459-2660.
E-mail: info@vilnashul.org
Web Site: www.vilnashul.org
Formerly: Boston Center for Jewish Heritage
Founded: 1995.
Key Personnel: Exec. Dir., Barnet Kessel; Pres. (V), Jack Swartz.
Personnel Profile: Full-Time Paid 1; Part-Time Paid 1; Part-Time Volunteers 12.
Governing Authority: private; nonprofit organization. Tax-exempt: 501(c)(3).
Institution Type/Description: Cultural Heritage Museum.
Collections: local Jewish culture, immigrants & diversity issues.
Research Fields: Jewish history of Boston; genealogy of Jewish Boston with a focus on the Vilna Shul.
Activities: concerts; docent program; films; guided tours; lectures; participatory exhibitions.
Publications: biannual newsletter, The Scribe.
Hours & Admission Prices: March-Nov. Wed.-Fri. 11-5, Sun. 1-5. Adults $5; members no charge. Closed Jewish holidays.
Attendance: 8,000 (accurate)
Membership: Student & Senior $18; Individual $36; Family $72; Contributor $180; Chei Society $360; Shamos Society $1,080.

ZOO NEW ENGLAND, One Franklin Park Rd., Boston, MA 02121-3255. Tel.: 617-541-5466. Fax: 617-989-2025.
E-mail: info@zoonewengland.com
Web Site: www.zoonewengland.com
Formerly: Franklin Park Zoo
Founded: 1905.
Congressional District: 7

Key Personnel: C.E.O. & Pres., John Linehan; Chm. Commonwealth Zoological Corp., Grace Fey; Museum Shop Mgr., Donna Roberts.

Personnel Profile: Full-Time Paid 121; Part-Time Paid 4; Part-Time Volunteers 80; Interns 5.

Governing Authority: state; nonprofit. Parent Institution: Commonwealth Zoological Corporation, Boston, MA. Tax-exempt.

Institution Type/Description: Zoo.

Collections: lions; western lowland gorillas; monkeys; baboons; giraffes; bali mynahs; Siberian cranes; zebras; camels; snow leopards; domestic animal farmyard; jaguars; African wild dogs; cougars; flamingos.

Research Fields: wildlife conservation.

Facilities: food service available. Gift items for sale.

Activities: education programs; guided tours.

Publications: newsletter, Wild Words.

Hours & Admission Prices: April-Sept. Mon.-Fri. 10-5, Sat.-Sun. 10-6; Oct.-March daily 10-4. Adults $17.95, seniors 62 & over $14.95, children 2-12 $11.95; members & children under 2 no charge. &

Attendance: 495,000 (accurate)

Membership: Individual $40; Family $60; Friend $100; Curator $250; Founder $1,000.

Bourne

APTUCXET TRADING POST MUSEUM, 24 Aptucxet Rd., Bourne, MA 02532-5434. Mailing Address: P.O. Box 3095, Bourne, MA 02532-0795. Tel.: 508-759-8167.

E-mail: bournehistoricalsociety@comcast.net

Web Site: www.bournehistoricalsociety.org

Founded: 1921.

Congressional District: 10

Key Personnel: C.E.O., Judith McAlister; Pres., Galon Barlow, Jr.; Site Mgr., Carol Wynne; Museum Shop Mgr., Mary Reid.

Personnel Profile: Part-Time Paid 4; Part-Time Volunteers 100; Interns 1.

Governing Authority: nonprofit organization. Parent Institution: Bourne Historical Society, Inc. Tax-exempt: 501(c)(3).

Institution Type/Description: Historic Building: trading post reconstructed on original 1627 site.

Collections: windmill; railroad station built for President Grover Cleveland; reconstruction of salt works; artifacts & written materials relating to the history of Bourne.

Research Fields: history; archaeology.

Facilities: picnic area. Books & museum-related items for sale.

Activities: guided tours; lectures; Maritime Week in April; Strawberry Festival in June; Colonial Day in July; Walk Through History in August; Massachusetts Archaeology Week in autumn; Wampanoag Day in September.

Publications: newsletter, Post Scripts.

Hours & Admission Prices: Memorial Day to Columbus Day Tues.-Sat. 11-4; other times by appointment. Adults $5, seniors $4, children 6-18 $2; discounts to groups, AAM and AAA & their guests; members no charge. &

Attendance: 6,000 (estimated)

Membership: Individual $25; Family $35; Institutional $50; Contributor $100; Business Supporter $200; Business Benefactor $400; Life $1,000.

Boylston

BOYLSTON HISTORICAL SOCIETY AND MUSEUM, 7 Central St., Boylston, MA 01505. Mailing Address: P.O. Box 459, Boylston, MA 01505. Tel.: 508-869-2720.

E-mail: info@boylstonhistory.org

Web Site: boylstonhistory.org

Founded: 1971.

Governing Authority: Tax-exempt.

Institution Type/Description: Historical Society Museum: housed in the former Old Town Hall; built in 1830.

Collections: local history & culture; period furnishings; personal artifacts; photographs.

Research Fields: local Boylston history.

Publications: quarterly newsletter, Potpourri.

Hours & Admission Prices: Tues. & Thurs. 9am to noon; other times by appointment. No charge; donations accepted.

Attendance: 450 (estimated)

Membership: Individual $15; Family $25; Dog Jack $50; Powder House $100; Old Pot $250.

* **WORCESTER COUNTY HORTICULTURAL SOCIETY/TOWER HILL BOTANIC GARDEN, (M),** 11 French Dr., Boylston, MA 01505-1008. Mailing Address: P.O. Box 598, Boylston, MA 01505-0598. Tel.: 508-869-6111 (hold or dial 10). Fax: 508-869-0314.

E-mail: thbg@towerhillbg.org

Web Site: www.towerhillbg.org

Founded: 1842.

Congressional District: 3

Key Personnel: Exec. Dir., Kathy Abbott; Pres., Christopher Reece; Dir. Horticulture, Joann Vieira; Librarian, Kathy Bell; Dir. Bldgs., Steve Fluet; Museum Shop Mgr., Judy Coughlin; Dir. Finance, Sharon Chauvin.

Personnel Profile: Full-Time Volunteers 22; Part-Time Paid 32; Part-Time Volunteers 150; Interns 2.

Governing Authority: Parent Institution: Worcester County Horticultural Society; nonprofit organization; society. Tax-exempt.

Institution Type/Description: Arboretum & Botanical Garden.

Collections: woody orchard 119 pre-20th century apple varieties; herbaceous & woody ornamentals suited to area, non-hardy plants in orangerie.; limina, Children's Garden.

Facilities: 8,000-vol. library of books on horticulture available for use by the public.

Activities: symposia; educational lectures; workshops; tours; seasonal flower shows; formally organized education programs; temporary exhibitions; New England School of Gardening accredited program.

Publications: quarterly newsletter; program guide.

Hours & Admission Prices: Mon. holidays & Tues.-Sun. 9-5. Adults $12, seniors $9, children 6-18 $7; discounts to members of AAM, AHS (American Horticultural Society) reciprocal admissions, AABGA, MA Horticultural Society, WICN & groups; members & children under 6 no charge. Closed New Year's Day; Thanksgiving; Christmas Eve & Day. &

Attendance: 60,000 (accurate)

Membership: Individual $55; Family & Dual $70; Friend $100; Library & Organization $150; Contributing $250; Supporting $500; Patron $1,000; Benefactor $3,000. Corporate: Friend $250; Contributing $500; Patron $1,000; Sponsor $2,500; Benefactor $5,000.

Braintree

BRAINTREE HISTORICAL SOCIETY, INC., 31 Tenney Rd., Braintree, MA 02184-6512. Tel.: 781-848-1640. Fax: 781-380-0731.

E-mail: bhsinc@braintreehistorical.org

Web Site: www.braintreehistorical.org

Founded: 1930.

Congressional District: 11

Key Personnel: Pres., Tom Fiorelli; Treas., Blaine Banker; Vice Pres., Military Archivist & Cur., James Fahey; Librarian & Archivist, Marjorie P. Maxham; Dir. Membership, Alan Weinberg; School Program, Gail Burns; Museum Shop Mgr., Ruth Powell.

Personnel Profile: Full-Time Paid 1; Part-Time Paid 1; Part-Time Volunteers 30.

Governing Authority: private; nonprofit organization. Subsidiary Museums: General Sylvanus Thayer Birthplace, 786 Washington St.; Gilbert Bean Museum; Watson Library and Research Center. Tax-exempt: 501(c)(3).

Institution Type/Description: Historical Society Museum.

Collections: decorative arts (American); 17-19th century American furniture; American folk art; 18-20th century decorative fans; American costumes & textiles; 2 early American homes (early 17-18th century). Historic Home: 1785 birthplace of Sylvanus Thayer.

Research Fields: manufacture of American hand fans; genealogy of local families; history of Braintree; life of Thomas A. Watson.

Facilities: genealogy library; herb gardens. Museum-related items for sale.

Activities: formal education programs for children; guided tours; lectures; temporary exhibitions.

Publications: quarterly members newsletter, The Lantern; gallery guides; exhibit catalogs; History of Braintree & General Sylvanus Thayer.

Hours & Admission Prices: Gilbert Bean Museum & Research Library: Thurs.-Sat. 10-4. Thayer Birthplace: April-Nov. Wed.-Fri. 1-3; Dec.-March by appointment only. Adults $3, children $2; discounts to AAA & AAM members; members no charge. Closed major holidays. &

Attendance: 7,000 (accurate)

Membership: Single $25; Family $50; Sustaining $100; Patron $500; Life $1,000.

Brewster

CAPE COD MUSEUM OF NATURAL HISTORY, INC., (M), 869 Rte. 6A, Brewster, MA 02631-1056. Tel.: 508-896-3867. Fax: 508-896-8844.
E-mail: rdwyer@ccmnh.org
Web Site: www.ccmnh.org
Founded: 1954.
Congressional District: 10
Key Personnel: Chm. Bd. Trustees, John McNair; Exec. Dir., Robert F. Dwyer; Museum Shop Mgr., Donna Durkee.
Personnel Profile: Full-Time Paid 4; Part-Time Paid 13; Part-Time Volunteers 250.
Governing Authority: nonprofit organization. Tax-exempt: 501(c)(3).
Institution Type/Description: Natural History Museum.
Collections: flora & fauna of Cape Cod; marine aquaria & displays; local archaeological collections; natural history exhibits; herbarium.
Research Fields: archaeology; flora & fauna of grounds; bird banding.
Facilities: 19,000-vol. library of natural sciences, conservation, natural history, ecology & reference books available to members of museum, teachers & children participating in educational programs; 3 nature trails; nature & conservation center; aquarium; 100-seat auditorium; classrooms. Prints, posters, cards, gift items & natural history books for sale.
Activities: guided tours; lectures; films; educational programs; permanent exhibitions; environmental science program; teacher workshops; field walks; consulting services; research; monthly art exhibits; off-site trips; PDP courses for teachers.
Publications: environmental education material.
Hours & Admission Prices: Jan. open for special programs only; Feb.-March Thurs.-Sun. 11-3; April-May & Oct.-Dec. Wed.-Sun. 11-3; June-Sept. daily 9:30-4. Additional hours during school vacation weeks. Adults $8, seniors 65 & over $7, children 3-12 $3.50; members & children under 3 no charge. Closed Federal holidays; Christmas Eve. &
Attendance: 31,000 (estimated)
Membership: Student $10; Individual $35; Family $60; Supporter $150; Sponsor $250; Patron $500; Benefactor $1,000; Visionary $5,000. Business: Good Neighbor $100; Corporate $250; Corporate Benefactor $500.

Brockton

BROCKTON FIRE MUSEUM, 216 N. Pearl St., Brockton, MA 02301-1712. Tel.: 508-583-1039.
Web Site: www.firemuseums.com
Founded: 1992.
Institution Type/Description: Fire-Fighting Museum.
Collections: area fire fighting history; fire-fighting artifacts; 1941 Strand Theater disaster memorial.
Hours & Admission Prices: 1st & 3rd Sun. of month 2-4; other times by appointment. Adults $2; children under 12 no charge.
Membership: Individual $20; Family $30; Business & Supporting $100; Life $200; Family Life & Benefactor $300.

✳ FULLER CRAFT MUSEUM, (M), 455 Oak St., Brockton, MA 02301-1340. Tel.: 508-588-6000. Fax: 508-587-6191.
E-mail: communications@fullercraft.org
Web Site: www.fullercraft.org
Formerly: Fuller Museum of Art
Founded: 1969.
Congressional District: 11
Key Personnel: Dir., Jonathan Fairbanks; Cur. Exhibitions & Collections, Jeffrey Brown; Business Mgr., Joyce Hochstrasser; Dir. Communications, Titilayo Ngwenya; Facilities Mgr., John Hastie; Mgr. Registrar & Collections, Cassandra Ortiz Dolzani; Asst. Cur., Beth McLaughlin; Devel. Assoc., Maria Francisco; Curatorial Asst., Michael McMillan; Museum Shop Mgr., Patty Dew.
Personnel Profile: Full-Time Paid 12; Part-Time Paid 30; Part-Time Volunteers 50; Interns 10.
Governing Authority: nonprofit organization. Tax-exempt: 501(c)(3).
Institution Type/Description: Craft Museum.
Collections: contemporary craft.
Research Fields: craft & art education.
Facilities: library; theater. Museum-related items for sale.
Activities: workshops; lectures; Meet the Maker weekends; tours; concerts; art classes; Family Days; outreach programs; docent program; inter-museum loan; temporary & traveling exhibitions.
Publications: newsletter; exhibition catalogues.
Hours & Admission Prices: Tues.-Wed. & Fri.-Sun. 10-5, Thurs. 10-9. Adults $8, senior citizens & students $5; discounts to AAM members; members, children under 12 & Thurs. 5-9 no charge. Closed New Year's Day; Christmas. &

Attendance: 18,000 (estimated)
Membership: Senior Basic 1 $30; Basic 1 $40; Senior Basic 2 $45; Basic 2 & Senior Friends 1 $50; Friends 1 $60; Basic 4 & Senior Friends 2 $65; Friends 2 $75; Friends 4 $100; Contributor $125; Supporter $250; Partner $500; Benefactor $1,000.

Brookline

BROOKLINE HISTORICAL SOCIETY, 347 Harvard St., Brookline, MA 02446-2907. Tel.: 617-566-5747.
E-mail: brooklinehistory@gmail.com
Web Site: www.brooklinehistoricalsociety.org
Founded: 1901.
Key Personnel: Pres., Ken Liss.
Governing Authority: society. Branch Museum: 1768 Putterham School, Newton & Grove Sts., Larz Anderson Park, off Newton St. Tax-exempt.
Institution Type/Description: Historical Society Museum: housed in c.1740 Edward Devotion House.
Collections: local furnishings; portraits; photos; Devotion family furniture; books; portraits; school furnishings.
Research Fields: local history.
Facilities: manuscripts pertaining to old Brookline, housed in the Brookline Public Library, available by consultation with accredited persons at the library.
Activities: guided tours; lectures; permanent exhibitions. Museum Sponsors: tours for schools.
Publications: pamphlet, Proceedings of the Brookline Historical Society.
Hours & Admission Prices: Call for hours. No charge; donations accepted.
Attendance: 500 (estimated)
Membership: Sponsor $30; Family $40; Friends of the Society $60; Benefactor $100.

FREDERICK LAW OLMSTED NATIONAL HISTORIC SITE, 99 Warren St., Brookline, MA 02445-5930. Tel.: 617-566-1689. Fax: 617-232-4073.
Web Site: www.nps.gov/frla
Founded: 1979.
Congressional District: 4
Key Personnel: Supt., Myra Harrison; Supervisory Park Ranger, Alan Banks; Site Mgr., Lee Farrow Cook.
Governing Authority: federal. Parent Institution: National Park Service, Dept. of the Interior, Interior Bldg., Washington, DC. Subsidiary Institution: Olmsted Center for Landscape Preservation. Tax-exempt.
Institution Type/Description: Historic Site: 1810 home & office of Frederick Law Olmsted.
Collections: 1857-1980 archival collection of landscape architectural design documents including drawings, plans, planting lists, lithographs, photographs & related records created by Frederick Law Olmsted, his sons, partners & successors.
Research Fields: landscape architecture; urban planning.
Facilities: archives available for research on premises by appointment only.
Activities: guided tours; permanent exhibitions.
Hours & Admission Prices: Grounds: daily dawn to dusk. &
Attendance: 4,500 (accurate)

GATEWAY ARTS GALLERIES, 62 Harvard St., Brookline, MA 02445. Tel.: 617-734-1577. Fax: 617-734-3199.
E-mail: gatewayarts@vinfen.org
Web Site: www.gatewayarts.org
Institution Type/Description: Art Gallery.
Collections: works by artists with disabilities; paintings.
Facilities: Gallery-related items for sale.
Activities: educational classes for artists with disabilities.
Hours & Admission Prices: Call for hours.

JOHN FITZGERALD KENNEDY NATIONAL HISTORIC SITE, 83 Beals St., Brookline, MA 02446-6010. Tel.: 617-566-7937. Fax: 617-730-9884. Facebook: JF Kennedy NHS.
E-mail: frla_kennedy_nhs@nps.gov
Web Site: www.nps.gov/joft
Founded: 1969.
Congressional District: 4
Key Personnel: Supt., Myra Harrison; Site Mgr., Lee Farrow Cook; Supervisory Park Ranger, Volunteer Coord. & Museum Shop Mgr., Jim Roberts; Lead Park Ranger, Sara Patton.
Personnel Profile: Full-Time Paid 2; Full-Time Volunteers 2; Part-Time Volunteers 16.
Governing Authority: federal. Parent Institution: National Park Service, U.S. Dept. of Interior, Washington DC 20240. Tax-exempt.

Institution Type/Description: Historic House Museum: housed in c.1917 home of Joe & Rose Kennedy, birthplace of our 35th president, John F. Kennedy.

Collections: furnishings as Mrs. Rose Kennedy remembers as they appeared May 29, 1917, the date of John F. Kennedy's birth.

Facilities: Museum-related items for sale.

Activities: ranger-guided tour; special school programs; permanent exhibitions.

Hours & Admission Prices: May 19-Oct. Wed.-Sun. 9:30-5, house tours every half hour; groups by appointment. No charge; donations accepted.

Attendance: 19,000 (accurate)

LARZ ANDERSON AUTO MUSEUM, 15 Newton St., Brookline, MA 02445-7406. Tel.: 617-522-6547. Fax: 617-524-0170.

Web Site: www.larzanderson.org

Formerly: Museum of Transportation

Founded: 1952.

Congressional District: 4

Key Personnel: Exec. Dir. & Cur., Sheldon Steele; Pres. (V), Michael Iandoli.

Personnel Profile: Full-Time Paid 5; Part-Time Volunteers 50.

Governing Authority: private; nonprofit. Tax-exempt: 501(c)(3), 509(a)(2).

Institution Type/Description: Transportation Museum.

Collections: period automobiles manufactured prior to 1942; gas, electric & steam cars; horse-drawn carriages & sleighs from the last quarter of the 19th-century; boneshakers, high wheel safeties, ordinaries, tricycles, bicycles, dating to the pre-war period; early period motorcycles; personal artifacts of Larz & Isabel Anderson; French lithograph collection from the Montaut school; automobilia; Packard Co. materials; Anderson vehicle & estate photographs; early fashion; original art.

Research Fields: historic & vintage autos (Edwardian, brass & nickel eras), 19th-century carriages & bicycles; Weld Garden, the estate of Larz & Isabel Anderson; Anderson collection of automobiles, carriages & sleighs; automobilia.

Facilities: 1,400-vol. library & photographic archives focusing on pre-war automobiles available for research.

Activities: rotating exhibits; guided tours; education programs; vintage, classic & antique auto shows; evening lecture series; car show lawn events.

Publications: quarterly newsletter, Carnotes; e-newsletter.

Hours & Admission Prices: Tues.-Sun. 10-4. Adults $10, children, seniors, students, & military $5; discounts to groups, AAA, AARP, WGBH, AAM, ICOM & MTA members; children under 6 & members no charge. Closed New Year's Day; Easter; Patriot's Day; Independence Day; Labor Day; Thanksgiving; Christmas.

Attendance: 34,000 (estimated)

Membership: Individual $30; Family $50; Contributing $100; Sponsoring $250; Patron $500. Car Club Membership: $100-$350.

Burlington

BURLINGTON HISTORICAL MUSEUM, 13 Bedford St., Burlington, MA 01803. Mailing Address: Town Hall, 29 Center St., Burlington, MA 01803-3058. Tel.: 781-272-1049 & 270-1600.

E-mail: archives@burlmass.org

Web Site: www.burlington.org

Founded: 1970.

Congressional District: 7

Key Personnel: C.E.O., Museum Guide & Museum Moderator, Joyce Fay; Co Chm. (V) Burlington Historical Commission, Mike Tredeau; Co Chm. & Museum Guide, Toni Faria; Archivist, Daniel McCormack.

Personnel Profile: Full-Time Volunteers 7; Part-Time Volunteers 10.

Governing Authority: municipal. Parent Institution: Town of Burlington. Tax-exempt.

Institution Type/Description: History Museum.

Collections: primitive agriculture; archives; costumes; slide collection; school furnishings; dolls; furniture; World War I memorabilia; toys; paintings; farm tools; Town of Burlington history; RCA memorabilia. Historic Building: restored one room schoolhouse, West School, archaeology.

Facilities: library of books of town history available for use for school projects. Museum-related items for sale.

Activities: temporary exhibitions; historical homes and sites tours; monthly meetings & programs; lectures; children's groups & school field trips.

Publications: historic house pamphlets; books, The History of Burlington, 1640-1950; Burlington, Part of a Greater Chronicle; Images of America-Burlington; Images of America-Burlington Firefighting; 1998-1999 Historic Preservation Survey of Burlington.

Hours & Admission Prices: Summer: Tues. & Sat. 10-2. Archives: 8-12 & 1-4:30. No charge; donations accepted.

Attendance: 600 (estimated)

Buzzards Bay

ABS INFORMATION COMMONS, Massachusetts Maritime Academy, 101 Academy Dr., Buzzards Bay, MA 02532-3405. Tel.: 508-830-5034. Fax: 508-830-5074.

E-mail: library@maritime.edu

Web Site: www.maritime.edu; library.maritime.edu

Formerly: Captain Charles H. Hurley Library

Founded: 1891.

Congressional District: 12

Key Personnel: Dir., Susan S. Berteaux; Pres., Richard G. Gurnon.

Personnel Profile: Full-Time Paid 4.

Governing Authority: state. Massachusetts Maritime Academy, Academy Dr., Buzzards Bay, MA 02532. Tax-exempt: 501(c)(3).

Institution Type/Description: Maritime Naval Museum & Library.

Collections: archives; manuscripts; over 100 scaled models set in base representing the sea & panoramic background.

Facilities: over 100,000-vol. library primarily in maritime transportation & engineering; reading rooms.

Activities: guided tours; lectures; formally organized education programs for undergraduate college students; permanent & temporary exhibitions.

Publications: magazine, Enterprise.

Hours & Admission Prices: Academic Year: Sun. 2-10, Mon.-Thurs. 7:30am-10pm, Fri. 7:30-4; extended hours during exams. No charge. Closed state & federal holidays.

Attendance: 83,900 (accurate)

NATIONAL MARINE LIFE CENTER, 120 Main St., Buzzards Bay, MA 02532-3221. Mailing Address: P.O. Box 269, Buzzards Bay, MA 02532-0269. Tel.: 508-743-9888. Fax: 508-759-5477. Facebook: National Marine Life Center.

E-mail: nmlc@nmlc.org

Web Site: www.nmlc.org

Formerly: The National Marine Life Center's Marine Animal Discovery Center

Founded: 1995.

Congressional District: 10

Key Personnel: Pres. & Exec. Dir., Kathy Zagzebski, M.E.M.; Chm. (V), Jeff Luce; Exec. Asst., Outreach Coord. & Museum Shop Mgr., Adele Raphael; Animal Care & Facilities Coord., Kate Shaffer; Science Dir. & Assoc. Veterinarian, Sea Rogers Williams, V.M.D.; Assoc. Veterinarian, Bridget Dunnigan.

Personnel Profile: Full-Time Paid 3; Part-Time Paid 4; Part-Time Volunteers 60; Interns 2.

Governing Authority: Tax-exempt.

Institution Type/Description: Marine Museum.

Collections: marine animals including sea turtles & seals; care & rehabilitation; marine science; ocean conservation.

Research Fields: marine wildlife health & conservation.

Facilities: Museum-related items for sale.

Activities: educational programs.

Publications: newsletter, NewsSplash; e-newsletter, E-Splash.

Hours & Admission Prices: Memorial Day-Labor Day daily 10-5; call for additional hours. No charge; donations accepted.

Attendance: 6,500 (estimated)

Membership: Youth $10; Individual $25; Family $50; Supporter $150; Contributor $250; Protector $500; Sponsor $1,000; Patron $1,500; Guardian $2,500.

Cambridge

CAMBRIDGE HISTORICAL SOCIETY, 159 Brattle St., Cambridge, MA 02138-3300. Tel.: 617-547-4252.

E-mail: info@cambridgehistory.org

Web Site: www.cambridgehistory.org

Founded: 1905.

Congressional District: 8

Key Personnel: Exec. Dir., Gavin Kleespies; Pres. (V), Richard Beatty; Treas., Andrew Leighton; Asst. Dir., Jasmine Laietmark.

Personnel Profile: Full-Time Paid 1; Part-Time Paid 2; Part-Time Volunteers 2; Interns 2.

Governing Authority: society; nonprofit. Tax-exempt: 501(c)(3).

Institution Type/Description: Historical Society Museum: 1685 building remodeled in early Georgian period.

Collections: artifacts; printed material & manuscripts; period furnishings.

Research Fields: local history; historic house interpretation.

Facilities: 300-vol. library pertaining to local history available on premises. Publications & postcards for sale.

Activities: guided tours; lectures; organized educational programs; internship programs.

Publications: periodical, Proceedings of the C-H-S; occasional monographs of Cambridge history; newsletter, Newetowne Chronicle.

Hours & Admission Prices: Mon.-Fri. by appointment. Adults $5, senior citizens & children $3; discounts to AAM & ICOM members; members no charge. Closed major holidays.

Attendance: 2,000 (estimated)

Membership: Individual $35; Family $75; Dana Fellow $100; Eliot Fellow $150; Sponsor $250; Patron $500; Benefactor $1,000.

✳ **HARVARD ART MUSEUMS, (M),** 32 Quincy St., Cambridge, MA 02138-3845. Tel.: 617-495-9400, ext. 0.

Web Site: www.harvardartmuseums.org

Founded: 1895.

Congressional District: 8

Key Personnel: Dir., Thomas W. Lentz; Deputy Dir., Maureen Donovan; Cur. Ancient Art and Head, Div. Asian & Mediterranean Art, Susanne Ebbinghaus; Cur. Chinese Art, Melissa A. Moy; Cur. Islamic & Later Indian Art, Mary McWilliams; Cur. Numismatics, Carmen Arnold-Biucchi; Cur. Paintings, Sculpture, & Decorative Arts and Head, Div. European & American Art, Stephan Wolohojian; Cur. American Art, Ethan Lasser; Cur. Drawings, William W. Robinson; Cur. Prints, Susan M. Dackerman; Chief Cur., Cur. Photographs and Acting Head, Div. Modern & Contemporary Art, Deborah Martin Kao; Cur. Busch-Reisinger Museum, Lynette Roth; Cur. Modern & Contemporary Art, Mary Schneider Enriquez; Admin. Archaeological Exploration of Sardis, Bahadir Yildirim; Dir. Center for the Technical Study of Modern Art, Carol Mancusi-Ungaro; Harvard Art Museums Archives, Megan Schwenke; Dir. Straus Center for Conservation & Technical Studies, Henry Lie; Dir. Collections Management, Jennifer Allen; Dir. Communications, Daron Manoogian; Head Academic & Pub. Programs, Jessica Levin-Martinez; Dir. Facilities Planning & Management, Peter Atkinson; Dir. Financial, Stephanie Schilling; Cons. Human Resources, Melissa Mooney; Dir. Digital Infrastructure & Emerging Technology, Jeff Steward; Dir. Institutional Advancement, Thomas Woodward; Dir. Security, Michael Kirchner; Dir. Visitor Svcs. & Facility Rentals, Sanja Cvjeticanin.

Personnel Profile: Full-Time Paid 126; Full-Time Volunteers 6; Part-Time Paid 44; Part-Time Volunteers 1; Interns 13.

Governing Authority: university. Parent Institution: Harvard University. Constituent Museums: Arthur M. Sackler Museum, Founded: 1985; Busch-Reisinger Museum, Founded: 1901. Fogg Museum, Founded: 1891. Tax-exempt: 501(c)(3).

Institution Type/Description: University Art Museums.

Collections: Fogg Museum: Western paintings, sculpture, decorative arts, photographs, prints, & drawings from the Middle Ages to present including Italian early Renaissance, British Pre-Raphaelite, & 19th century French art as well as 19th- & 20th century American paintings & drawings. Busch-Reisinger Museum: art from the German-speaking countries of Central & Northern Europe in all media and from all periods including German expressionism, Bauhaus materials, postwar & contemporary art. Arthur M. Sackler Museum: ancient, Asian, Islamic, & later Indian art including Greek vases & small bronzes and coins from the ancient Mediterranean world; Chinese jades, bronzes, & ceremonial weapons; Chinese & Korean ceramics; Japanese woodblock prints & calligraphy; works on paper from Islamic lands & India.

Research Fields: all media of fine arts. Arthur M. Sackler Museum: ancient, Asian, Islamic & Indian art. Busch-Reisinger Museum: Bauhaus, Walter Gropius & Lyonel Feininger. Fogg Museum: prints; drawings; conservation; photography; all areas of Western art.

Facilities: Opening fall 2014, uniting Fogg, Busch-Reisinger & Arthur M. Sackler museums, include 43,000 sq. ft. gallery space, 300 seat hall, art study center, classrooms, public education room, shop & cafe.

Activities: tours; lectures; gallery talks; seminars; concerts; permanent, loan & temporary exhibitions; formally organized education programs for Harvard undergraduate & graduate students; internships & fellowships for museum professionals; public education programs.

Publications: Index magazine print issue: 3 times per year; digital mazazine: daily; gallery guides; catalogues & other books.

Hours & Admission Prices: Galleries & museums closed for renovations until fall 2014. ♿

Attendance: 30,180 (accurate)

Membership: See website www.harvardartmuseums.org/members.

HARVARD MUSEUM OF NATURAL HISTORY, 26 Oxford St., Cambridge, MA 02138-2932. Tel.: 617-495-3045. Fax: 617-496-8206.

E-mail: hmnh@hmnh.harvard.edu

Web Site: www.hmnh.harvard.edu

Founded: 1995.

Congressional District: 8

Key Personnel: Exec. Dir. Harvard Museums of Science & Culture, Jane Pickering; Deputy Dir., Susan Shefte; Dir. Education, Wendy Derjue-Holzer; Assoc. Dir. Public Programs, Tom Scanlon; Dir. Exhibitions, Janis Sacco; Volunteer Coord., Carol Carlson; Dir. Communications, Blue Magruder; Dir. Travel Program, Lauren Bruck; Asst. Dir. Institutional Advancement, Brian Schubmehl; Museum Shop Mgr. & Assoc. Dir. Operations, Kevin Ebert.

Personnel Profile: Full-Time Paid 19; Part-Time Paid 21; Part-Time Volunteers 90; Interns 1.

Governing Authority: university. Parent Institution: Harvard University. Subsidiary Institutions: Harvard Museums of Science & Culture; Herbaria; Mineralogical Museum; Museum of Comparative Zoology. Tax-exempt: 501(c)(3).

Institution Type/Description: Natural History Museum.

Collections: glass flowers; zoological; mineral; collections of The Museum of Comparative Zoology, The Mineralogical Museum & The Harvard University Hebaria.

Major Exhibits: Mollusks: Shelled Masters of the Marine Realm, 3/12-2/14; Thoreau's Maine Woods: A Journey in Photographs with Scot Miller, 10/13-8/14; Island Biogeography, Fall 2014.

Facilities: educational facilities; 30,000 sq. ft. exhibit space. Museum-related items for sale.

Activities: docent program; formally organized educational programs; guided tours; lectures; rental gallery; temporary exhibitions; international travel program.

Publications: program & events calendar.

Hours & Admission Prices: Daily 9-5. Adults $12, senior citizens 65 & over and students $10, children 3-18 $8; discounts to NZMA, AAM & ICOM members; Harvard students with ID, members & children under 3 no charge. ASTC passport program. Closed New Year's Day; Thanksgiving; Christmas Eve & Day. ♿

Attendance: 220,000 (accurate)

Membership: Student & Senior $35; Individual $50; Household $85; Sustaining $200; Patron $500.

THE HARVARD UNIVERSITY HERBARIA, 22 Divinity Ave., Cambridge, MA 02138-2020. Tel.: 617-495-2365. Fax: 617-495-9484.

Web Site: www.huh.harvard.edu

Founded: 1864.

Key Personnel: Dir. Herbaria, Donald H. Pfister; Dir. Botany Libraries, Judith A. Warnement; Acting Mgr. Collections, Michaela Schmull.

Governing Authority: university. Affiliated with Harvard University.

Institution Type/Description: Herbarium.

Collections: botany; flowering plants; gymnosperms; ferns; cryptogams.

Research Fields: evolutionary & systematic botany; population biology; bibliography; botanical history.

Facilities: 291,000-vol. library of botany & horticulture books available for use on premises by researchers & students affiliated with Harvard University.

Activities: formally organized education program for undergraduate & graduate college students; inter-museum loans.

Publications: journal, Harvard Papers in Botany.

Hours & Admission Prices: Mon.-Fri. 9-5. Closed university holidays. ♿

LONGFELLOW HOUSE-WASHINGTON'S HEADQUARTERS NATIONAL HISTORIC SITE, 105 Brattle St., Cambridge, MA 02138-3499. Tel.: 617-876-4491. Fax: 617-497-8718.

Web Site: www.nps.gov/long

Founded: 1973.

Congressional District: 8

Key Personnel: Site Mgr., Beth Wear; Supt., Myra Harrison; Archives Specialist, Christine Wirth; Collections Mgr., David Daly.

Personnel Profile: Full-Time Paid 6; Part-Time Paid 4; Part-Time Volunteers 2; Interns 1.

Governing Authority: federal. Administered by the National Park Service, U.S. Dept. of the Interior, Washington, DC. Tax-exempt.

Institution Type/Description: Historic House & Site Museum: 1759 house, headquarters of George Washington, 1775-1776; later home of Henry Wadsworth Longfellow.

Collections: books; furnishings; paintings; works of art; manuscripts; letters; photographs; textiles; architectural fragments. Historic Building: 1844 Barn; 1844 carriage house.

Research Fields: Henry Wadsworth Longfellow & family; 19th-century American social & cultural history; 19th-century domestic life; decorative arts; 18th-century architecture; colonial revival movement; George Washington in Cambridge 1775-1776.

Facilities: Longfellow's personal library, manuscripts of various Longfellow family members available to researchers by appointment. Verse folders & various Longfellow related material for sale.

Activities: guided house tours; summer garden concert series; summer poetry readings.
Publications: Commemorative Conference; Friends of the Longfellow House Newsletter.
Hours & Admission Prices: June-Oct. Wed.-Sun. 10-4:30. No charge. National park passes honored. &

Attendance: 46,000 (accurate)

* **MIT MUSEUM, (M),** 265 Massachusetts Ave., Cambridge, MA 02139-4307. Mailing Address: Bldg. N51, 265 Massachusetts Ave., Cambridge, MA 02139-4307. Tel.: 617-253-5927. Fax: 617-253-8994. Facebook: MIT Museum.
E-mail: museum@mit.edu
Web Site: web.mit.edu/museum
Founded: 1973.
Congressional District: 8
Key Personnel: Dir., John Durant; Chm. (V), Prof. Phillip Sharp; Assoc. Dir., Mary Leen; Mgr. Exhibitions, Donald Stidsen; Dir. Public Rels., Josie Patterson; Dir. Technology, Allan Doyle; Dir. Cambridge Science Festival, P.A. d'Arboloff; Dir. Programs, Brindha Muniappan; Dir. Exhibitions, Alexander Goldowsky; Events Rentals Mgr., Karen Costello; Visitor Svcs. Mgr., Patricia Lane; Cur. Hart Nautical Collections, Kurt Hasselbalch; Cur. Science & Technology Collections, Deborah Douglas, Ph.D.; Collections Mgr. & Registrar, Joan Parks-Whitlow; Cur. Architecture & Design, Gary van Zante; Museum Shop Mgr., Claudia Majetich.
Personnel Profile: Full-Time Paid 29; Part-Time Paid 3; Part-Time Volunteers 25; Interns 5.
Volunteer Hours: 3,700
Operating Expenses: 4,000,000
Operating Income: 4,000,000
Governing Authority: university. Parent Institution: Massachusetts Institute of Technology, 77 Massachusetts Ave., Cambridge 02139. Branch Galleries: Francis Russell Hart Nautical Gallery, 55 Massachusetts Ave.; Margaret Hutchinson Compton Gallery, 77 Massachusetts Ave. Tax-exempt: 501(c)(3).
Institution Type/Description: Science & Technology Museum.
Collections: Architecture & Design collection: 1873-1968 student theses & projects; 1840-1920 rare drawings by European & American architects. Hart Nautical Gallery: late 19th to early 20th century New England technical history of ship & small craft design, construction & propulsion. Holography collection: 1940s inception to current artistic & technical evolution. MIT general collection: artifacts, visual & written materials documenting history of MIT & its role in the development of science, technology & engineering. Science & Technology Collection: record of 19th-20th century innovation.
Major Exhibits: 5000 Moving Parts, 11/20/13-11/20/14.
Activities: temporary & traveling exhibition programs; hands-on family programs.
Publications: exhibition brochures.
Hours & Admission Prices: Main facility at 265 Massachusetts Ave.: daily 10-5. Adults $10, youth under 18, students & seniors $5; discounts to AAM, NEMA, MTA members; MIT Community members, children under 5 & last Sun. of month no charge; evening programs with museum admission no charge. Hart Nautical Galleries: daily 10-5. No charge. Compton Gallery: daily 10-5. No charge. Closed holidays. &

Attendance: 150,000 (accurate)

MOUNT AUBURN CEMETERY, (M), 580 Mount Auburn St., Cambridge, MA 02138-5517. Tel.: 617-547-7105.
E-mail: info@mountauburn.org
Web Site: www.mountauburn.org
Governing Authority: private.
Institution Type/Description: Historic Site: founded in 1831. A National Historic Landmark.
Collections: cemetery monuments from 1831 to present; cemetery history; horticulture; art & architecture.
Research Fields: burial records.
Facilities: archives.
Activities: guided & self-guided tours; slide lectures; special events.
Publications: cemetery maps & informational materials; quarterly magazine, Sweet Auburn.
Hours & Admission Prices: Cemetery: May-Sept. daily 8-7; Oct.-April daily 8-5; groups by appointment. Office: Mon.-Fri. 8:30-4:30, Sat. 8:30-4.
Attendance: 200,000 (estimated)
Membership: Student & Senior $30; Individual $35; Supporter & Household $50; Contributor $100; Donor $250; Patron $500; 1831 Society $1,000.

MUSEUM OF COMPARATIVE ZOOLOGY, 26 Oxford St., Cambridge, MA 02138-2902. Tel.: 617-495-2460. Fax: 617-496-8308.
E-mail: cweisel@oeb.harvard.edu
Web Site: www.mcz.harvard.edu
Founded: 1859.
Congressional District: 8
Key Personnel: Dir., James Hanken.
Personnel Profile: Full-Time Paid 60; Part-Time Paid 80; Part-Time Volunteers 50.
Operating Expenses: 24,180,637
Operating Income: 24,180,637
Governing Authority: private university. Parent Institution: Harvard University. Subsidiary Institution: Harvard Museum of Natural History (HMNH), Cambridge, MA. Tax-exempt: 501(c)(3).
Institution Type/Description: Zoological Museum.
Collections: ornithology; herpetology; mammals; invertebrate zoology; invertebrate & vertebrate paleontology; entomology; marine biology; biological oceanography; population genetics; ichthyology; malacology.
Research Fields: entomology; invertebrate zoology; population genetics; invertebrate & vertebrate paleontology; herpetology; ichthyology; ornithology; marine biology; biological oceanography; malacology; mammalogy.
Facilities: library; laboratories; field station.
Activities: see Harvard Museum of Natural History listing for activities.
Publications: periodicals, Bulletin of the Museum of Comparative Zoology; Breviora.
Hours & Admission Prices: see Harvard Museum of Natural History for hours & admission. &
Membership: see Harvard Museum of Natural History listing for membership levels & prices.

* **PEABODY MUSEUM OF ARCHAEOLOGY & ETHNOLOGY, (M),** 11 Divinity Ave., Cambridge, MA 02138-2096. Tel.: 617-496-1027. Fax: 617-495-7535.
Web Site: www.peabody.harvard.edu
Founded: 1866.
Congressional District: 5
Key Personnel: Dir., Dr. Jeffrey Quilter; Deputy Dir. Administration, Catherine Cezeaux; Dir. Collections, Dr. Steven LeBlanc; Deputy Dir. Curatorial Admin. & Outreach, Dr. Pamela Gerardi; Registrar, Dr. Viva Fisher; Assoc. Cur., Dr. Castle McLaughlin; Assoc. Cur., Dr. Diana Loren; Assoc. Cur., Dr. Michele Morgan; Assoc. Cur., Dr. Patricia Capone; Collections Assoc., Susan Haskell; Head Conservator, T. Rose Holdcraft; Senior Collections Mgr., David De Bono Schafer; Conservator, Scott Fulton; Bldg. Mgr., Eugene Ayres; Exhibit Designer & Coord., Samuel Tager; Librarian, Janet Stein.
Personnel Profile: Full-Time Paid 45; Part-Time Paid 18; Part-Time Volunteers 20; Interns 15.
Governing Authority: university. Parent Institution: Harvard University. Tax-exempt: 501(c)(3).
Institution Type/Description: Anthropology Museum.
Collections: five million objects; archaeological & ethnographic collections from most areas of the world mainly North, South & Middle America; Old World (especially Paleolithic), European Iron Age; Near East archaeology; ethnography of Africa, North America & Oceania; photographic archives of over 500,000 images; paintings, prints & drawings; archival materials.
Major Exhibits: Storied Walls: Murals of the Americas, 10/13-10/14; Wiyohpiyata: Lakota Images of the the Contested West, 10/13-6/15; Digging Veritas: The Archaeology and History of Harvard's Indian College, 10/13-6/15.
Research Fields: archaeology; biological anthropology; social anthropology; ethnography; zooarchaeology.
Facilities: 225,260-vol. library of books on anthropology available for inter-library loan & for research on premises; classrooms; zooarchaeology lab; casting lab.
Activities: lectures; films; education programs for undergraduate & graduate students affiliated with Harvard University; inter-museum loan, permanent & temporary exhibitions; guided tours; preschool & youth education programs; workshops.
Publications: annual report; periodical monographs; scientific papers; memoirs; bulletins; exhibition catalogues.
Hours & Admission Prices: Daily 9-5. Adults $12, senior citizens & students with ID $10, children 3-18 $8; discounts to AAM & ICOM members; children under 3, Harvard students, members, Sun. 9-12, Sept.-May Wed. 3-5 no charge. Closed New Year's Day; Thanksgiving; Christmas Eve & Day. &
Attendance: 155,000 (estimated)
Membership: Senior & Student $25; Individual $40; Household & Dual $60; Sustaining $125; Supporting $250; Patron $500; Peabody Institute $1,000.

THE SEMITIC MUSEUM AT HARVARD UNIVERSITY, 6 Divinity Ave., Cambridge, MA 02138-2020. Tel.: 617-495-4631. Fax: 617-496-8904. Facebook: Semitic Museum.
E-mail: semiticm@fas.harvard.edu
Web Site: www.semiticmuseum.fas.harvard.edu
Founded: 1889.
Key Personnel: Dir. & Cur., Prof. Lawrence E. Stager; Cur. Cuneiform Collections, Prof. Piotr Steinkeller; Asst. Dir., Joseph A. Greene; Asst. Cur. Collections, Adam J. Aja; Museum Coord., Timothy R. Letteney; Dir. Publication, M.D. Coogan.
Personnel Profile: Full-Time Paid 2; Full-Time Volunteers 11; Part-Time Paid 15; Part-Time Volunteers 12; Interns 4.
Governing Authority: Faculty of Arts & Sciences, Harvard University, University Hall, Cambridge. Tax-exempt.
Institution Type/Description: Archaeological Museum: specializing in ancient Near East archaeology.
Collections: archaeology.
Research Fields: ancient Near Eastern languages & history; archaeology field work in Near East.
Activities: guided tours; lectures; special programs for groups by reservation; inter-museum loan, permanent, temporary & traveling exhibitions; White-Levy Program for Archaeological Publications: gives grants to publish terminated and unpublished archaeological field work from sites in the Levant, Mesopotamia, Aegean and Anatolia.
Publications: books, Harvard Semitic Series; Harvard Semitic Monographs, Studies in the Archaeology and History of the Levant; members' association newsletter; exhibition catalogs.
Hours & Admission Prices: Mon.-Fri. 10-4, Sun. 1-4. No charge; donations accepted. Closed holidays.
Attendance: 5,000 (estimated)
Membership: Individual & Family $35; Donor $50; Contributing $100; Supporting $250; Sustaining $500; Benefactor $1,000.

Canton

CANTON HISTORICAL SOCIETY, 1400 Washington St., Canton, MA 02021-2240. Tel.: 781-828-6957. Fax: 781-821-5780.
E-mail: historical@canton.org
Web Site: canton.org
Founded: 1893.
Key Personnel: Pres., Wallis Gibbs.
Governing Authority: nonprofit. Tax-exempt.
Institution Type/Description: History Museum.
Collections: Indian relics; manuscripts; James Bazin collection; genealogies; musical instruments; photographs.
Facilities: 500-vol. library of county & local history available by appointment with president of society; Indian burial ground; nature trails.
Activities: guided tours; lectures; formally organized educational programs; permanent & temporary exhibitions.
Hours & Admission Prices: Open select holidays & by request. No charge; donations accepted.
Attendance: 50
Membership: Annual $10; Life $100 or an outstanding contribution to the society.

MILTON ART MUSEUM, 900 Randolph St., Canton, MA 02021-1355. Tel.: 781-821-2222, ext. 2124. Fax: 781-575-9428.
E-mail: info@miltonartmuseum.org
Web Site: www.miltonartmuseum.org
Founded: 1986.
Congressional District: 9
Key Personnel: Chm. (V), Ellyn Moller.
Governing Authority: Tax-exempt.
Institution Type/Description: Art Museum.
Collections: Western & European prints; sculpture; lithographs; etchings; photography; Asian art & objects.
Hours & Admission Prices: Mon.-Thurs. 8-6:30, Fri. 8-4:30, Sat. 9-1. No charge; donations accepted. &
Attendance: 10,000 (estimated)
Membership: Annual $25.

MUSEUM OF AMERICAN BIRD ART AT MASS AUDUBON, 963 Washington St., Canton, MA 02021-2117. Tel.: 781-821-8853.
Formerly: Mass Audubon Visual Arts Center
Key Personnel: Dir., Amy Montague
Institution Type/Description: Art Museum & Nature Center
Collections: natural history art; photographs; wildflowers; hawks; meadow; forest; swamp.
Facilities: nature trails.
Activities: educational programs; permanent & temporary exhibitions.

Hours & Admission Prices: Gallery: Tues.-Sun. 1-5. Trail: Tues.-Sun. 9-5. Adults $4; members no charge.
Membership: Individual $48; Family $65; Supporter $70; Family Plus $80; Contributor $100; Protector $150; Sponsor $250; Patron $500; Guardian $750; President's Circle $1,250; Founder's Circle $2,500 & up.

Centerville

CENTERVILLE HISTORICAL MUSEUM, 513 Main St., Centerville, MA 02632-2913. Tel.: 508-775-0331. Fax: 508-862-9211.
E-mail: chsm@centervillehistoricalmuseum.org
Web Site: www.centervillehistoricalmuseum.org
Founded: 1952.
Key Personnel: C.E.O., Cur. & Devel., Randall Hoel; Pres., Diane Brooke; Treas., Robyn Simmons.
Personnel Profile: Full-Time Paid 1; Part-Time Volunteers 35; Interns 1.
Governing Authority: private; nonprofit organization. Tax-exempt: 501(c)(3).
Institution Type/Description: 19th century Cape Cod History Museum.
Collections: costumes; quilts; military artifacts & uniforms; marine artifacts; tools; toys; photographs; Dodge MacKnight watercolors; A.E. Crowell carved bird; decorative arts.
Research Fields: Civil War; Lt. Augustus D. Ayling.
Facilities: library. Museum-related items for sale.
Activities: concerts; docent program; films; guided tours; hobby workshops; lectures; temporary exhibitions; outreach programs. Annual Events: Open House Tours on holidays; Family Days; Old Home Week.
Publications: quarterly newsletter, Chequaquet Log.
Hours & Admission Prices: May-Dec. 15 Tues.-Sat. 12-4; other times by appointment. Adults $7, senior citizens & students $6; discounts to AAA & MTA members; members & children under 8 no charge. &
Attendance: 2,293 (accurate)
Membership: Individual $35; Dual $40; Family $50; Patron $80; Benefactor $125; Business $150.

Charlestown

BUNKER HILL MUSEUM, 43 Monument Sq., Charlestown, MA 02129-3430. Mailing Address: Boston National Historic Park, Charlestown Navy Yard, Charlestown, MA 02129-4543. Tel.: 617-242-7275.
Web Site: www.nps.gov/bost/historyculture/bhmuseum.htm
Founded: 1975.
Congressional District: 8
Key Personnel: Pres. & C.E.O. (V), Arthur L. Hurley; Chm. (V) & Museum Shop Mgr., Terry Savage.
Personnel Profile: Part-Time Volunteers 5.
Governing Authority: society; nonprofit organization. Charlestown Historical Society. Tax-exempt.
Institution Type/Description: History Museum: located across from the Bunker Hill Monument grounds.
Collections: local pottery; period gowns; history of Battle of Bunker Hill & community of Charlestown.
Research Fields: local history.
Facilities: 175-seat auditorium. Commemorative medals & crafts for sale.
Activities: guided tours; drama; inter-museum loan, permanent exhibitions.
Hours & Admission Prices: Daily 9-5. No charge; donations accepted. Closed New Year's Day; Thanksgiving; Christmas.
Membership: Individual $2; Family $15; Friend $50; Business $100.

Chatham

ATWOOD HOUSE MUSEUM, HOME OF THE CHATHAM HISTORICAL SOCIETY, 347 Stage Harbor Rd., Chatham, MA 02633-2229. Mailing Address: P.O. Box 709, Chatham, MA 02633-0709. Tel.: 508-945-2493. Fax: 508-945-1205.
E-mail: info@chathamhistorical.org
Web Site: www.chathamhistoricalsociety.org
Founded: 1923.
Congressional District: 12
Key Personnel: Chm., John King, II; Exec. Dir., Dennis McFadden; Museum Shop Mgr., Barbara Newberry.
Personnel Profile: Full-Time Paid 2; Part-Time Paid 1; Part-Time Volunteers 100.
Governing Authority: society. Parent Institution: Chatham Historical Society. Subsidiary Institution: Atwood House Museum. Tax-exempt: 501(c)(3).
Institution Type/Description: Historical Museum: housed in 1752 Atwood House and additions.
Collections: archives; period artifacts; Crowell birds; Sandwich glass; Stallknecht murals; Parian Ware; sea shells; manuscripts & publications of Joseph C. Lincoln; paintings by Harold Brett; portraits of local sea captains

& other local people by Frederick S. Wright; period tools; North Beach camp; herb garden; maritime paintings & artifacts; paintings & objects related to the China trade; commercial fishing history; Chatham's twin light lantern & Fresnel lens; maps & sea charts.
Major Exhibits: Constructing Wilderness, 6/14-10/14.
Research Fields: genealogy; local & maritime history.
Facilities: 300-vol. library of books & old reference books available for genealogical research available by appointment; research room; William Nickerson Wing.
Activities: guided tours; lectures; special & permanent exhibitions; summer children's programs; preserving your family papers classes. Museum Sponsors: Open House in June; Appraisal Day in June; Hearth Warming in December.
Publications: pamphlets, Old Chatham Houses; Chatham Since the Revolution; The Nickerson North Beach Camp; books, W. C. Smith History of Chatham, Mass., 1981; The History of Weir Fishing; Beyond The Bar: The Perilous Journey (history of fishing industry), 2007; Chatham Vital Records Vol. 1 to 1865; 1994 Vital Records, Vol. II, 1866-1910; Days to Remember; A Home on the Rolling Deep; Home Song; Weathering a Century of Change, Chatham, MA 1900-2000; Picturing Chatham-American Landscape Painting 1880-1977.
Hours & Admission Prices: June 21-June 28 & Sept. 2-Oct. 11 Tues.-Sat. 1-4; July 1-Aug. 3 Tues.-Fri. 10-4, Sat. 1-4; Adults $6, students $3; discount to AAM members; members & children 6 & under no charge. Closed Independence Day; Labor Day. &
Attendance: 3,700 (accurate)
Membership: Student Historian 18 & under $10; Individual $25; Family $50; Explorer $100; Discoverer $250; Heritage Society $500; Captain Atwood Circle $1,000.

Chester

CHESTER HISTORICAL SOCIETY, 15 Middlefield St., Chester, MA 01011.
Institution Type/Description: Historical Society Museum: housed in the former jail; built in 1840.
Collections: local history & culture; period furnishings; personal artifacts; photographs.
Activities: research.
Hours & Admission Prices: By appointment.

Chestnut Hill

LONGYEAR MUSEUM, (M), 1125 Boylston St., Chestnut Hill, MA 02467-1811. Tel.: 617-278-9000. Fax: 617-278-9003.
E-mail: letters@longyear.org
Web Site: www.longyear.org
Founded: 1923.
Congressional District: 4
Key Personnel: Exec. Dir., Sandra J. Houston; Pres. (V), Ellen Williams; Dir. Cur., Stephen R. Howard; Dir. Devel., Jody A. Wilkinson; Dir. Collections, Cheryl Moneyhun; Museum Shop Mgr., Holly Bates.
Personnel Profile: Full-Time Paid 23; Part-Time Paid 3; Interns 1.
Governing Authority: nonprofit organization; board of trustees. Parent Institution: Longyear Foundation. Branch Museums: Eight Mary Baker Eddy Historic Houses: 23 Paradise Rd., Swampscott, MA; Stinson Lake Rd., Rumney, NH; 133 Central St., Stoughton, MA; North Groton, NH; 277 Main St., Amesbury, MA; 62 N. State St., Concord, NH; 12 Broad St., Lynn, MA; 400 Beacon St., Chestnut Hill, MA. Tax-exempt: 501(c)(3).
Institution Type/Description: Historical Museum.
Collections: artifacts; manuscripts, letters; photographs; portraits; paintings; books, houses & furniture pertaining to the life & achievements of Mary Baker Eddy, founder of Christian Science.
Research Fields: Mary Baker Eddy & the history of Christian Science dating from 1821-1910; the Baker family; the Longyear family; papers and reminiscences of Christian Scientists active prior to 1911; the Longyear Foundation.
Facilities: 5,000-vol. library; 350 cu. ft. of archival material available for use by qualified researchers. Museum-related items for sale.
Activities: workshops; children's programs; school programs; lectures; speakers' bureau; walking tours; domestic & foreign travel; programs; concerts; documentary films.
Publications: newsletter, Report to Members; books, Pioneers in Christian Science; Christian Science in Germany; The Stoughton Years; A Precious Legacy: Christian Science Comes to Japan; The Human Life, 1906-1907 articles on Mary Baker Eddy; A Most Agreeable Man: Lyman Brackett; Genesis of a Poem; A Chronological Reference to Mary Baker Eddy's Books: Miscellaneous Writings, 1883-1896 and the First Church of Christ, Scientist and Miscellany; Homeward Part I: Lynn; Homeward Part II: Chestnut Hill; Violet Hay; films, The Onward and Upward Chain; Remem-

ber the Days of Old; "Who Shall Be Called?"; The Pleasant View Household: Working and Watching; The House on Broad Street: Finding a Faithful Few.
Hours & Admission Prices: Longyear Museum: Mon. & Wed.-Sat. 10-4, Sun. 1-4. Historic Houses: call for hours. No charge; donations accepted. Closed holidays. &
Attendance: 6,000 (accurate)
Membership: Student $10; Individual $35; Family $65.

MCMULLEN MUSEUM OF ART, BOSTON COLLEGE, Devlin Hall, 140 Commonwealth Ave., Chestnut Hill, MA 02467-3800. Tel.: 617-552-8587. Fax: 617-552-8577.
E-mail: artmusm@bc.edu
Web Site: bc.edu/artmuseum
Founded: 1976.
Congressional District: 4
Key Personnel: Dir., Dr. Nancy Netzer; Chm. (V), C. Michael Daley; Devel., Catherine Concanon; Asst. Dir., Mgr. Exhibitions & Collections, Diana Larsen; Asst. Dir., Admin. & Collections, John McCoy; Publications & Exhibitions Admin., Katherine Shugert; Museum Shop Mgr., Chris Bergin.
Personnel Profile: Full-Time Paid 6; Part-Time Paid 6; Part-Time Volunteers 12; Interns 6.
Governing Authority: private college; nonprofit. Parent Institution: Boston College. Tax-exempt.
Institution Type/Description: Art Museum.
Collections: 15th- to 17th-century European paintings, drawings, tapestries, prints, sculpture; Japanese prints; 19th- to 20th-century American paintings.
Facilities: 350-seat auditorium; educational facilities.
Activities: formal education programs for undergraduate & graduate students; guided tours; lectures; loan, temporary & traveling exhibitions.
Publications: triannual, Friends Newsletter; exhibition catalogs.
Hours & Admission Prices: Mon.-Fri. 11-4, Sat.-Sun. 12-5. No charge; donations accepted. Closed New Year's Day; Martin Luther King Jr. Day; Washington's Birthday; Good Friday; Easter; Memorial Day; Labor Day; Columbus Day; Christmas. &
Attendance: 75,000 (accurate)
Membership: Standard $50.

Clinton

MUSEUM OF RUSSIAN ICONS, (M), 203 Union St., Clinton, MA 01510-2903. Tel.: 978-598-5000. Fax: 978-598-5009.
E-mail: info@museumofrussianicons.org
Web Site: www.museumofrussianicons.org
Founded: 2006.
Key Personnel: Dir. & Chm. (V), Gordon B. Lankton; C.E.O., Kent Russell; Registrar, Laura Garrity-Arquitt; Museum Shop Mgr., Jocelyn Willis.
Personnel Profile: Full-Time Paid 5; Full-Time Volunteers 2; Part-Time Paid 4; Part-Time Volunteers 38; Interns 9.
Governing Authority: private; nonprofit organization. Parent Institution: Museum of Russian Icons, Inc. Subsidiary Institution: Lankton Charitable Corp. Tax-exempt: 501(c)(3).
Institution Type/Description: Art Museum.
Collections: works by Russian artists; paintings.
Research Fields: Russian history; Russian Icons.
Facilities: 2,500-vol. library; 20-seat restaurant; educational facility; 100-seat auditorium; 10,000 sq. ft. exhibit space. Museum-related items for sale.
Activities: arts festivals; concerts; dance recitals; films; docent program; educational programs; guided tours; workshops; lectures; loan, temporary, traveling & participatory exhibits; rental gallery; training programs for museum workers; broadcast programs.
Publications: quarterly newsletter, IcoNews; quarterly, Journal of Icon Studies.
Hours & Admission Prices: Tues.-Wed. & Fri. 11-3, Thurs. 11-7, Sat. 9-3; groups by appointment. Adults $7; discounts to groups and NEMA, AAM & ICOM members; members, students & children no charge. Closed New Year's Day; Independence Day; Thanksgiving; Christmas. &
Attendance: 15,000 (accurate)
Membership: Senior & Student $15; Individual $30; Family & Dual $45; Cossack $100; Boyar $250; Prince $500; Czar $1,000.

Cohasset

CAPTAIN JOHN WILSON HOUSE, 4 Elm St., Cohasset, MA 02025-1829. Mailing Address: 106 S. Main St., P.O. Box 627, Cohasset, MA 02025-0627. Tel.: 781-383-1434.
E-mail: cohassethistory@yahoo.com
Web Site: www.cohassethistoricalsociety.org
Founded: 1928.
Congressional District: 10

Key Personnel: Pres., Kathleen O'Malley; Exec. Dir., Lynne DeGiacomo; Historian, Rebecca Bates-McArthur.
Personnel Profile: Part-Time Paid 3; Part-Time Volunteers 50.
Governing Authority: society. Parent Institution: Cohasset Historical Society. Tax-exempt.
Institution Type/Description: Historic House Museum.
Collections: costumes; textiles; decorative arts; period furnishings.
Research Fields: decorative arts.
Activities: guided tours; permanent & temporary exhibits.
Publications: newsletter, Historical Highlights; Images of America/Cohasset.
Hours & Admission Prices: June-Labor Day Wed.-Fri. 1-4, Sat. 10-2. No charge; donations accepted.
Attendance: 800 (estimated)
Membership: Individual $25; Family $35; Sustaining $50; Patron $500; Benefactor $1,000.

MARITIME MUSEUM - COHASSET HISTORICAL SOCIETY,
6 Elm St., Cohasset, MA 02025. Mailing Address: P.O. Box 627, Cohasset, MA 02025-0627. Tel.: 781-383-1434.
E-mail: cohassethistory@yahoo.com
Founded: 1956.
Congressional District: 10
Key Personnel: Pres., Kathleen O'Malley; Exec. Dir., Lynne DeGiacomo; Historian, David Wadsworth.
Personnel Profile: Part-Time Paid 3; Part-Time Volunteers 10.
Governing Authority: society. Parent Institution: Cohasset Historical Society. Tax-exempt: 501(c)(3).
Institution Type/Description: Maritime Museum.
Collections: shipwreck; lifesaving equipment; 19th-century maritime artifacts; sailing ship models; paintings; early tools; Indian stone artifacts; local historic artifacts; mercantile; fishing; ship building; Cohasset seafaring history; Minot's Ledge Lighthouse.
Research Fields: maritime; military; local history.
Activities: guided tours; lectures; permanent exhibitions.
Publications: newsletter, Historical Highlights.
Hours & Admission Prices: Mid-June to Aug. Wed.-Fri. 1-4, Sat. 10-2. No charge; donations accepted.
Attendance: 800 (estimated)
Membership: Cohasset Historical Society: Individual $25; Family $35; Sustaining $50; Patron $500; Benefactor $1,000.

OUR WORLD CHILDREN'S GLOBAL DISCOVERY MUSEUM,
100 Sohier St., Cohasset, MA 02025. Tel.: 781-383-3198.
Founded: 2001.
Governing Authority: nonprofit organization.
Institution Type/Description: Children's Museum.
Collections: hands-on exhibitions.
Activities: educational programs; special events; birthday parties; classes.
Hours & Admission Prices: Tues.-Fri. 10-5, Sat. 10am-12pm. Admission $7. &

PRATT BUILDING - COHASSET HISTORICAL SOCIETY,
106 S. Main St., Cohasset, MA 02025-2097. Mailing Address: P.O. Box 627, Cohasset, MA 02025-0627. Tel.: 781-383-1434.
E-mail: cohassethistory@yahoo.com
Web Site: cohassethistoricalsociety.org
Founded: 1928.
Congressional District: 10
Key Personnel: Pres., Kathleen O'Malley; Exec. Dir., Lynne DeGiacomo; Historian, David Wadsworth.
Personnel Profile: Part-Time Paid 3; Part-Time Volunteers 50.
Governing Authority: Parent Institution: Cohasset Historical Society. Subsidiary Institution: Maritime Museum, Capt. John Wilson House. Tax-exempt.
Institution Type/Description: History Museum.
Collections: maritime; costumes; town history; Native American artifacts.
Research Fields: Cohasset history; local architectural surveys; theatre, professional & amateur archives; maritime.
Activities: lectures; permanent & temporary exhibits; research.
Publications: newsletter, Historical Highlights; Images of America: Cohasset.
Hours & Admission Prices: Mon.-Fri. 10-4. No charge; donations accepted. &
Attendance: 1,500 (estimated)
Membership: Cohasset Historical Society: Individual $25; Family $35; Sustaining $50; Patron $500; Benefactor $1,000.

SOUTH SHORE ART CENTER,
119 Ripley Rd., Cohasset, MA 02025-1744. Tel.: 781-383-2787. Fax: 781-383-2964.
E-mail: info@ssac.org
Web Site: www.ssac.org
Founded: 1954.
Key Personnel: Dir., Sarah Hannan

Institution Type/Description: Art Gallery.
Collections: works by regional & national contemporary artists; paintings; sculpture.
Major Exhibits: Made in America, 1/10/14-2/9/14; Love, 2/14/14-3/23/14; Organic Matters, 3/28/14-5/4/14; Faculty, 5/23/14-7/13/14; Blue Ribbon, 7/18/14-8/14; Print Show, 9/12/14-10/19/14.
Activities: classes; workshops.
Hours & Admission Prices: Mon.-Sat. 10-4, Sun. 12-4. No charge; donations accepted. &
Membership: Senior & Youth $25; Individual $35; Juried Artist $50; Family $55; Associate $100; Supporting $250; Sustaining $500; Patron $1,000; Benefactor $2,500.

Concord

CONCORD ART ASSOCIATION, (M),
37 Lexington Rd., Concord, MA 01742-2570. Tel.: 978-369-2578. Fax: 978-371-2496.
E-mail: gallery@concordart.org
Web Site: www.concordart.org
Founded: 1922.
Congressional District: 5
Key Personnel: Exec. Dir., Lili Ott; Chm. (V), Nancy Huggins; Dir. Education, Michele Kenna.
Personnel Profile: Full-Time Paid 1; Part-Time Paid 5; Part-Time Volunteers 80.
Governing Authority: nonprofit organization. Tax-exempt.
Institution Type/Description: Art Gallery & Sculpture Garden: building built in 1760.
Collections: early American portraits; paintings; sculpture; graphics; decorative arts.
Facilities: garden; facilities for wedding receptions, programs, meetings & workshops.
Activities: arts festivals; outdoor sculpture exhibits; permanent & temporary exhibitions.
Publications: newsletter; catalogs.
Hours & Admission Prices: Tues.-Sat. 10-4:30, Sun. 12-4. No charge; donations accepted. Closed holidays. &
Attendance: 8,500 (estimated)
Membership: Senior $35; Artist & Individual $50; Family $75; Patron $100; Life $500; Corporate $1,000.

❋ CONCORD MUSEUM, (M),
Cambridge Turnpike, at Lexington Rd., Concord, MA 01742-3711. Mailing Address: P.O. Box 146, Concord, MA 01742-0146. Tel.: 978-369-9763. Fax: 978-369-9660. Facebook: Concord Museum.
E-mail: cm1@concordmuseum.org
Web Site: www.concordmuseum.org
Founded: 1886.
Congressional District: 5
Key Personnel: Pres. (V), Churchill Franklin; Dir., Margaret Burke; Cur., David Wood; Dir. Devel., Sue Gladstone; Public Rels., Emer McCourt; Museum Shop Mgr., Judy Glam.
Personnel Profile: Full-Time Paid 10; Part-Time Paid 31; Part-Time Volunteers 250; Interns 2.
Governing Authority: society. Parent Institution: Concord Antiquarian Society. Tax-exempt: 501(c)(3).
Institution Type/Description: History & Decorative Arts Museum.
Collections: American decorative arts; paintings & graphics; Ralph Waldo Emerson's study; Henry David Thoreau; Revolution artifacts, Native American stone tools.
Major Exhibits: From the Minute Man to the Lincoln Memorial: The Timeless Sculpture of Daniel Chester French, 10/13-3/14; The Best Workman in the Shop: William Munroe and Concord Federal Furniture, 10/13-3/14; The Shot Heard 'Round the World: April 19, 1775, 4/14-9/14.
Research Fields: American decorative arts; Middlesex County history; Concord history.
Facilities: 1,400-vol. library of decorative arts & New England history books available on premises. Museum-related items for sale.
Activities: seminars/symposia; lectures; films; concerts; trips; formally organized education programs; permanent, temporary & traveling exhibitions; education program serving 7,000 school children from 55 Massachusetts communities & 19 states.
Publications: book, Concord Museum: Climate for Freedom; catalogs, Two Towns, Concord & Wethersfield, A Comparative Exhibition of Regional Culture 1635-1850; Decorative Arts from a New England Collection; bulletin, The Newsletter; catalogue, Forms to Sett on: A Social History of Concord Seating Furniture; From Musketaquid to Concord; sourcebook, Native American Sourcebook: A Teacher's Resource on New England Native Peoples; catalog, Harry Little's Concord: Public and Domestic

Architecture, 1914-1941; book, An Observant Eye: The Thoreau Collection at the Concord Museum.

Hours & Admission Prices: Jan.-March Mon.-Sat. 11-4, Sun. 1-4; April-May & Sept.-Dec. Mon.-Sat. 9-5, Sun. 12-5; June-Aug. daily 9-5. Adults $10, senior citizens & students $8, children $5; discounts to AAM, AAA & MTA members; members no charge. &

Attendance: 40,000 (estimated)

Membership: Individual $50; Family $65; Contributor $125; Patron $250; Benefactor $500.

LOUISA MAY ALCOTT'S ORCHARD HOUSE, (M), 399 Lexington Rd., Concord, MA 01742-3712. Mailing Address: P.O. Box 343, Concord, MA 01742-0343. Tel.: 978-369-4118. Fax: 978-369-1367.

E-mail: info@louisamayalcott.org
Web Site: www.louisamayalcott.org
Founded: 1911.
Congressional District: 5
Key Personnel: Exec. Dir., Jan Turnquist; Pres. (V), Beth Neeley Kubacki; Museum Shop Mgr., Sally Cody; Exec. Asst., Maria Powers; Education Dir., Lis Adams; Bldg. & Grounds Supt., Jay Powers.
Personnel Profile: Full-Time Paid 5; Part-Time Paid 35; Part-Time Volunteers 36.
Governing Authority: nonprofit organization. Parent Institution: Louisa May Alcott Memorial Association. Tax-exempt: 501(c) (3).
Institution Type/Description: History Museum: house where Louisa May Alcott wrote Little Women, and also the site of Bronson Alcott's School of Philosophy.
Collections: memorabilia of the Alcotts; manuscripts & publications, particularly by Louisa May Alcott & Bronson Alcott; paintings by May Alcott. Historic Buildings: 18th-century Orchard House; 1880 School of Philosophy.
Research Fields: Louisa Alcott; Bronson Alcott; Transcendental Movement in Concord.
Facilities: Museum-related items for sale.
Activities: guided tours; living history programs; children's programs; lecture series in School of Philosophy; school of philosophy can be rented for events.
Publications: Story of the Alcotts; The Poetry of Louisa May Alcott; American Transcendentalism; The Concord Summer School of Philosophy.
Hours & Admission Prices: April-Oct. Mon.-Sat. 10-4:30, Sun. 1-4:30; Nov.-March Mon.-Fri. 11-3, Sat. 10-4:30, Sun. 1-4:30. Adults $9, senior citizens 62 and over & college students with ID $8, youth 6-17 $5; discounts to families, military, teachers, AAM & MTA members; members & children under 6 no charge. Closed Easter; Thanksgiving; Christmas.
Attendance: 48,910 (accurate)
Membership: Junior 17 & under $10; Student & Senior $25; Individual $40; Household $60; Supporting $125; Philosophers Group $250; Transcendentalist Club $500.

MINUTE MAN NATIONAL HISTORICAL PARK, 174 Liberty St., Concord, MA 01742-1705. Tel.: 978-369-6993. Fax: 978-318-7800.

E-mail: mima_info@nps.gov
Web Site: www.nps.gov/mima
Founded: 1959.
Congressional District: 4 & 5
Key Personnel: Supt., Nancy Nelson; Cur., Teresa Wallace; Museum Shop Mgr., Linda Matte.
Personnel Profile: Full-Time Paid 25; Part-Time Volunteers 100.
Governing Authority: federal. National Park Service, U.S. Department of the Interior, Washington, DC 20240. Tax-exempt.
Institution Type/Description: National Park & Historic Houses: located along 1775 Battle Road Trail; 19th-century Wayside, Nathaniel Hawthorne & Alcott & Lothrop family home.
Collections: structures, furnishings, photographs & books; archaeology; Harriett M. Lothrop family papers. Historic Buildings: Hartwell Tavern; The Wayside home.
Research Fields: American Revolutionary War period & 19th-century New England literature.
Facilities: 1,200-vol. library available for use on premises upon request; 2 visitor centers; Colonial Living Interpretive Center. Pamphlets, booklets & other museum-related items for sale.
Activities: guided tours; lectures; permanent exhibitions; environmental study area; literary & patriotic 1883-1924; hiking on five-mile Battle Road Trail.
Publications: brochures.
Hours & Admission Prices: Park: daily sunrise to sunset. No charge; donations

accepted. Wayside Historic Houses: late-May to early Nov. Wed.-Sun, call for hours. Admission $5. Closed New Year's Day; Thanksgiving; Christmas. &

Attendance: 1,000,000 (estimated)

THE OLD MANSE, 269 Monument St., Concord, MA 01742-1837. Mailing Address: P.O. Box 572, Concord, MA 01742-0572. Tel.: 978-369-3909. Fax: 978-287-6154.

E-mail: oldmanse@ttor.org
Web Site: www.thetrustees.org/places-to-visit/greater-boston/old-manse.html
Founded: 1939.
Congressional District: 5
Key Personnel: Historic Site Mgr., Tom Beardsley; Dir. Historic Resources, Susan C.S Edwards.
Personnel Profile: Full-Time Paid 1; Part-Time Paid 15; Part-Time Volunteers 24.
Governing Authority: society. The Trustees of Reservations, Long Hill - 572 Essex St., Beverly, MA 01915. Tel. 978-921-1944. Tax-exempt: 501(c)(3).
Institution Type/Description: Historic House & Site: built in 1770 for Rev. William Emerson, town minister & patriot; located next to the North Bridge, site of the 1st major skirmish of the Revolutionary War, April 19, 1775. Later home to Emerson's grandson, Ralph Waldo, who drafted his first published work, Nature, here; home of Nathaniel & Sophia Hawthorne 1842-45.
Collections: original family furnishings.
Research Fields: American studies.
Facilities: Books for sale.
Activities: guided tours. Annual Events: Patriot's Day in April; Fall Festival in October; Halloween Event in October.
Hours & Admission Prices: mid-March to May & Nov.-Dec. Sat.-Sun. 12-5; Memorial Day to Oct. Tues.-Sun. 12-5; other times by appointment. Adults $8, senior citizens & students $7, children 6-12 $5; discounts for groups of 10 or more & for AAA, MTA & WGBH members; Trustees members & children under 6 no charge. &
Attendance: 11,400 (accurate)
Membership: Senior & Student $35; Individual $45; Friends & Family $65; Contributing $100; Supporting $150; Sustaining $300; Sponsor $600.

RALPH WALDO EMERSON HOUSE, 28 Cambridge Tpke., Concord, MA 01742-3700. Tel.: 978-369-2236.
Founded: 1930.
Key Personnel: Pres., Mrs. Nicholas Bancroft; Dir., Marie A. Gordinier.
Personnel Profile: Part-Time Paid 10.
Governing Authority: nonprofit organization. Parent Institution: Ralph Waldo Emerson Memorial Association. Tax-exempt: 501(c)(3).
Institution Type/Description: Memorial Museum: housed in 1835 Ralph Waldo Emerson House. Ralph Waldo Emerson's home for the greater part of his life 1835 to his death in 1882.
Collections: preservation project of Emerson's personal belongings; portraits; books; furniture.
Activities: guided tours; permanent exhibitions.
Hours & Admission Prices: mid-April to Oct. Thurs.-Sat. 10-4:30, Sun. 1-4:30; group tours by appointment. Adults $9, seniors & students $7; discounts to groups & AAM members; children under 6 no charge.
Attendance: 3,454 (accurate)

Cotuit

CAHOON MUSEUM OF AMERICAN ART, (M), 4676 Falmouth Rd., Cotuit, MA 02635. Mailing Address: P.O. Box 1853, Cotuit, MA 02635-1853. Tel.: 508-428-7581. Fax: 508-420-3709.
E-mail: rwaterhouse@cahoonmuseum.org
Web Site: www.cahoonmuseum.org
Founded: 1984.
Congressional District: 10
Key Personnel: Dir., Richard Waterhouse; Pres., Carol Wilgus; Museum Shop Coord., Stephanie Reeve.
Personnel Profile: Full-Time Paid 1; Part-Time Paid 5; Part-Time Volunteers 75.
Governing Authority: nonprofit organization. Tax-exempt: 501(c)(3).
Institution Type/Description: Art Museum: housed in 1775 Cape Cod Colonial tavern used for overnight stops on Hyannis-Sandwich Stagecoach line.
Collections: 19th-21st century American paintings; primitive artists Ralph & Martha Cahoon.
Research Fields: 19th to early 20th-century American & women artists; primitive painters, Ralph & Martha Cahoon.
Facilities: 1,500-vol. art library on American paintings specializing in artists of New England & American women artists. Notecards, prints, books & other art-related items for sale.

Activities: guided tours; lectures; artist demonstrations; art classes; special exhibitions; organized education programs; children's program.

Publications: quarterly newsletter, Spyglass; brochure; exhibition catalogue, American Paintings from Nature: Flower, Fruit & Leaf; Simple Pleasures: The Art of Martha Cahoon; Modern Primitives: Simple Art in a Complex Age; In the Beginning: The Decorated Furniture of Ralph and Martha Cahoon, 2005.

Hours & Admission Prices: Feb.-Dec. Tues.-Sat. 10-4, Sun. 1-4. Adults $5, seniors & students $4; discounts to teachers; members & children under 12 no charge. Closed major holidays.

Attendance: 2,100 (accurate)

Membership: Individual $40; Family $60; Contributor $100; Associate $250; Sponsor $500; Patron $1,000; Cahoon Society $1,500; Benefactor $5,000 & up.

COTUIT CENTER FOR THE ARTS, 4404 Falmouth Rd. (Rte. 28), Cotuit, MA 02635. Mailing Address: P.O. Box 2042, Cotuit, MA 02635. Tel.: 508-428-0669. Fax: 508-428-0633.

E-mail: info@cotuitcenterforthearts.org

Web Site: www.artsonthecape.org

Founded: 1995.

Congressional District: 10

Key Personnel: Dir., David Kuehn.

Governing Authority: Tax-exempt: 501(c)(3).

Institution Type/Description: Art Gallery.

Collections: works by contemporary artists.

Facilities: theater.

Activities: film; concerts; educational programs; musicals & plays; classes; special events; temporary exhibitions.

Hours & Admission Prices: Memorial Day to Columbus Day Mon.-Sun. 10-4; Oct.-May Mon.-Sat. 10-4. Gallery: no charge; donations accepted. Special Events: call for admission prices. Closed major holidays. &

Attendance: 30,000 (estimated)

Membership: Individual $55; Dual $100; Family $150; Contributing $250; Supporting $500 & up; Sustaining $1,000 & up; Patron $2,500; Benefactor $5,000 & up.

HISTORICAL SOCIETY OF SANTUIT AND COTUIT, 1148 Main St., Cotuit, MA 02635. Mailing Address: P.O. Box 1484, Cotuit, MA 02635-1484. Tel.: 508-428-0461.

E-mail: infohssc@verizon.net

Web Site: www.cotuithistoricalsociety.org

Founded: 1954.

Congressional District: 9

Key Personnel: Pres., Joyce Ginouves; Treas., Peggie Griffin Bretz; Museum Shop Mgr., Melanie Curtis.

Personnel Profile: Part-Time Volunteers 25.

Governing Authority: society. Tax-exempt: NZ 23-7177654.

Institution Type/Description: Historic House: 1800-1850 restored home of village carpenter.

Collections: historic furnishings. Historical Building: Dottridge Homestead.

Research Fields: old houses in village.

Facilities: herb garden. Museum-related items for sale.

Activities: temporary & loan exhibitions; educational tours; docent tours. Annual Events: Strawberry Festival; Taste of Cotuit Oktoberfest; Christmas in Cotuit in December.

Publications: annual historical paper; quarterly newsletter; book, monthly lecture series.

Hours & Admission Prices: Memorial Day to Labor Day Sat.-Sun. 1-5; Sept. to Columbus Day Sat. 1-5. No charge; donations accepted.

Attendance: 2,500 (estimated)

Membership: Student $15; Individual $40; Family $55; Contributor $100; Supporter $150; Benefactor $210; Lifetime $500.

Cummington

KINGMAN TAVERN HISTORICAL MUSEUM, 41 Main St., Cummington, MA 01026-9742. Mailing Address: P.O. Box 10, Cummington, MA 01026-0010. Tel.: 413-634-5527 & 8828 (administrative).

Web Site: hiddenhills.com/kingmantavern/

Founded: 1967.

Congressional District: 1

Key Personnel: Chm. (V), Carla Ness; Historic Commission, Stephanie Pasternak; Historic Commission, Stephen Howes; Historic Commission, Matthew Grallert; Historic Commission, Karen Westergaard.

Personnel Profile: Part-Time Paid 4; Part-Time Volunteers 8.

Governing Authority: municipal. Tax-exempt.

Institution Type/Description: Historical Museum: housed in 1800 frame building used as a post office, Masonic Lodge meeting hall & tavern.

Collections: furniture & furnishings of earlier inhabitants of the town; tavern room; examples of early American stenciling; tools; toys; costumes; a collection of twenty miniature rooms made by Alice C. Steele; 17 room tavern; barn; cider mill; carriage house; general store; farm machinery.

Research Fields: town histories & family genealogical material.

Facilities: 25-vol. library of reference material on early American tools, furniture & equipment available for use in museum library during regular hours or by appointment. Miniatures & museum-related items for sale.

Activities: films; slides; implement demonstrations.

Publications: books, Only One Cummington; Alice C. Steele's Miniature Rooms; Childhood Memories; Olive Thayer's Remembering Cummington; Aunt Teeks in Memoryland; vital records of Cummington.

Hours & Admission Prices: Aug. Sat. 2-5; other times by appointment. No charge; donations accepted. &

Attendance: 150 (estimated)

WILLIAM CULLEN BRYANT HOMESTEAD, 207 Bryant Rd., Cummington, MA 01026-9639. Tel.: 413-634-2244. Fax: 413-634-0376.

E-mail: bryanthomestead@ttor.org

Web Site: www.thetrustees.org

Founded: 1928.

Congressional District: 11

Key Personnel: C.E.O., Andy Kendall; Rgnl. Dir. The Trustees of Reservation, Jocelyn Forbush; Supt., Jim Caffrey; Site Interpreter, Susan Davidson.

Personnel Profile: Part-Time Paid 7; Part-Time Volunteers 15.

Governing Authority: nonprofit. Parent Institution: The Trustees of Reservations, Long Hill 572 Essex St., Beverly, MA 01915. Tel. 978-921-1944. Tax-exempt.

Institution Type/Description: Historic House: 1789 boyhood home & adult summer residence of famed poet William Cullen Bryant.

Collections: 3 generations of Bryant family furnishings; 19th century European & Middle East exotic travel memorabilia; rural life artifacts; barn.

Research Fields: literature; 19th-century travel; rural life.

Facilities: visitor's center. Museum-related items for sale.

Activities: guided tours; interpretive events; self guided nature walks; community meeting area.

Hours & Admission Prices: last week in June to Columbus Day Sat.-Sun. & Mon. holidays 1-5. Adults $5, children 6-12 $2.50; children under 6 & Trustees of Reservation members no charge. &

Attendance: 7,000 (estimated)

Membership: Membership in the Trustees of Reservations: Individual $45; Family $65.

Dalton

CRANE MUSEUM OF PAPERMAKING, Housatonic St. off Rte. 8 & 9, Dalton, MA 01226. Mailing Address: c/o Crane & Co., Inc., 30 South St., Dalton, MA 01226-1751. Tel.: 413-684-6481. Fax: 413-684-0817.

Web Site: www.crane.com

Governing Authority: Parent Institution: Crane & Co., Inc.

Institution Type/Description: Historic Building: housed in the former papermaking business of Crane and Co.'s Old Stone Mill, built in 1844. Crane produces the rag paper that US currency is printed on.

Collections: papermaking company history; papermaking process; financial instruments; stationery.

Hours & Admission Prices: May to mid-Oct. Mon.-Fri. 1-5; mid-Oct. to April Tues.-Thurs. 1-5. No charge.

Attendance: 1,000 (estimated)

Danvers

DANVERS ARCHIVAL CENTER, 15 Sylvan St., Danvers, MA 01923-2735. Tel.: 978-774-0554. Fax: 978-762-0251.

E-mail: trask@noblenet.org

Web Site: www.danverslibrary.org/archive

Founded: 1972.

Congressional District: 6

Key Personnel: Town Archivist, Richard B. Trask.

Personnel Profile: Full-Time Paid 1; Part-Time Paid 1; Part-Time Volunteers 2.

Volunteer Hours: 85

Operating Expenses: 13,000

Governing Authority: municipal government. Parent Institution: Peabody Institute Library of Danvers.

Institution Type/Description: Town Archives: housed in 1892, Peabody Institute Library.

Collections: books; pamphlets; manuscripts; maps; broadsides; photographs;

newspapers; printed or written media relating to the growth, development & history of Salem Village & Danvers.
Facilities: 7,000-vol. library on the history of witchcraft, Danvers, local history & biography available for use by the public; 130-seat auditorium. Gift items for sale.
Hours & Admission Prices: Mon. 1-7:30, Wed.-Thurs. & 1st. Sat. of month 9-12 & 1-5; 2nd & 4th Fri. of month 1-5. No charge. Closed state & national holidays. &
Attendance: 1,200 (estimated)

DANVERS HISTORICAL SOCIETY, 11 Page St., Danvers, MA 01923-2813. Mailing Address: Box 381, Danvers, MA 01923-0681. Tel.: 978-777-1666. Fax: 978-777-5028.
E-mail: dhs@danvershistory.org
Web Site: www.danvershistory.org
Founded: 1889.
Congressional District: 6
Key Personnel: Chm. (V), Richard Moody; Dir. Operations, Cathy Gareri; Office Mgr., Paula Ruta.
Personnel Profile: Full-Time Paid 1; Part-Time Paid 5; Part-Time Volunteers 30.
Governing Authority: society. Tax-exempt: 501(c)(3).
Institution Type/Description: Historical Society Museum.
Collections: furniture; tools; objects showing growth & development of town; archaeology; manuscripts; burial ground. Historic Houses: 1754 Jeremiah Page House, 11 Page St.; 1792 McIntyre Tea House; 1893 Glen Magna, 57 Forest St; 1850 Mrs. Day's Ideal Baby Shoe Shop; 17th century Israel Putnam birthplace, Maple Street.
Research Fields: local history.
Facilities: botanical garden; 130-seat auditorium. Books for sale.
Activities: guided tours; lectures; permanent & temporary exhibitions; school programs; community outreach & advocacy.
Publications: books, Danvers Historical Society Collections; Chronicles of Danvers; As The Century Turned: Photographic Glimpses of Danvers, Massachusetts, 1880-1910; On The Sands of Time: The Life of Charles Sutherland Tapley; The Devil Hath Been Raised: A Documentary History of the Salem Village Witchcraft Outbreak of March, 1692. Historical Glimpses of Danvers (2002).
Hours & Admission Prices: Society: Mon.-Fri. 9-1; other times by appointment. Donations accepted. Glen Magna & McIntyre Tea House: June-Labor Day Tues. & Thurs. 10-4; other times by appointment. Adults $5. &
Attendance: 11,500 (estimated)
Membership: Student (K-12) $10; Individual $20; Senior (65 & over) $18; Household $30; Sustaining $100; Benefactor $150.

REBECCA NURSE HOMESTEAD, 149 Pine St., Danvers, MA 01923-2693. Mailing Address: P.O. Box 456, Hathorne, MA 01937-0456. Tel.: 978-774-8799.
E-mail: president@rebeccanurse.org
Web Site: www.rebeccanurse.org
Founded: 1974.
Congressional District: 6
Key Personnel: Pres. & Bd. Chm., Marta Driscoll; Vice Pres., Jackson Tingle; Treas., William Quilnan; Clerk, Niamh Dolan; Bldg. Chm., Henry W. Rutkowski; Cur., Kathryn P. Rutkowski; Museum Shop Mgr., Candice Clemenzi.
Personnel Profile: Part-Time Paid 6; Part-Time Volunteers 20.
Governing Authority: society; nonprofit organization. Parent Institution: Danvers Alarm List Co., Inc. Tax-exempt: 501(c)(3).
Institution Type/Description: Historic House: 1678 home of Rebecca Nurse, hanged as a witch in 1692.
Collections: Massachusetts artifacts from 1650-1780; 17th & 18th century furnishings; Old Salem Village including early 18th century Saltbox house; reproduction 1672 Salem Village meeting house; barn; Nurse family cemetery.
Research Fields: 17th & 18th century Salem Village and Danvers; 1692 witch hysteria; 1770s militia; architecture.
Facilities: Museum-related items for sale.
Activities: guided tours; craft; militia activities and demonstrations by the Danvers Alarm List Company, a recreation of an 18th century militia; courses; lectures. Museum Sponsors: Annual 18th Century Field Day & Muster.
Publications: book, Rebecca Nurse-Saint but Witch Victim; quarterly newsletter, Rebecca Nurse Homestead Preservation Society.
Hours & Admission Prices: June 15 to Labor Day Fri.-Sun. 10-4; Sept.-Oct. Sat.-Sun. 10-4; other times by appointment. Adults $6.50, seniors $5, children under 16 $4.50; discount to school groups; members no charge.
Attendance: 4,000 (estimated)
Membership: Individual $25; Family $50; Sponsor $100; Patron $250.

Dartmouth

LLOYD CENTER FOR THE ENVIRONMENT, 430 Potomska Rd., Dartmouth, MA 02748. Tel.: 508-990-0505. Fax: 508-993-7868. Facebook: Lloyd Center for the Environment.
E-mail: fcallen@lloydcenter.org
Web Site: www.lloydcenter.org
Founded: 1978.
Congressional District: 4
Key Personnel: Exec. Dir., Rachel L. Stronach.
Governing Authority: nonprofit organization. Tax-exempt: 501(c)(3).
Institution Type/Description: Nature Center.
Collections: coastal, estuarine, & watershed environments; local sea life & their habitats; marine mammal skeletons; murals.
Research Fields: coastal ecosystems; endangered species.
Facilities: hiking trails. Museum-related items for sale.
Activities: observation area; touch tanks; educational programs; special events; seminars.
Hours & Admission Prices: June-Sept. Tues.-Sun. 10-4; Oct.-May Tues.-Sat. 10-4. No charge; donations accepted.
Attendance: 30,000 (estimated)

Dedham

DEDHAM HISTORICAL SOCIETY MUSEUM, (M), 612 High St., Dedham, MA 02026-1833. Mailing Address: P.O. Box 215, Dedham, MA 02027-0215. Tel.: 781-326-1385.
E-mail: society@dedhamhistorical.org
Web Site: dedhamhistorical.org
Founded: 1859.
Congressional District: 9
Key Personnel: Exec. Dir., Vicky L. Kruckeberg.
Personnel Profile: Full-Time Paid 1; Part-Time Paid 3; Part-Time Volunteers 65.
Governing Authority: society. Tax-exempt: 501(c)(3).
Institution Type/Description: History Museum & Genealogy Library.
Collections: 16th to 20th-century furniture; portraits & paintings; regional historical & industrial artifacts; Dedham & Chelsea pottery; needlework & costumes; manuscripts, newspapers & photo archive; genealogies & family histories; Katharine Pratt Silver; Dedham & Chelsea pottery & decorative arts; artists: Alvan Fisher, Lilian & Philip Hale, Henry Hitchings, Charles Mills, Harry Spiers, James Frothingham, Gilbert Stuart, & Jacob Wagner. Historic House: 1766 powder house.
Major Exhibits: Dedham and the Civil War, 10/13-10/14.
Research Fields: Norfolk County area history & New England genealogy.
Facilities: 20,000-vol. library of history & genealogical books & microform source materials, including partial IGI, available for use on premises; reading room.
Activities: museum tours; house tours; lecture series; educational programs for students; permanent & temporary exhibitions; courses on genealogy & antiques; special member events.
Publications: bimonthly newsletter; books, The Dedham Pottery & Earlier Robertson's Chelsea Potteries, Dedham reprinting; Dedham, 1635-1890; Building Dedham; Images of America, Dedham.
Hours & Admission Prices: Office: Tues.-Fri. 9-4. Museum: Tues.-Fri. 12-4, even dated Sat. 1-4. Museum: adults $2; DHS, NEMA, ICOM & AAM members no charge. Library: Tues. & Thurs. by appointment 9-4, even dated Sat. 1-4. Library: call for fees. Closed state & national holidays; Thanksgiving & day after; Christmas week.
Attendance: 2,000 (estimated)
Membership: Student $10; Individual $35; Family $50; Contributing $100; Life $750. Corporate: Contributor $150; Donor $250; Patron $500; Benefactor $1,200.

FAIRBANKS HOUSE, 511 East St., Dedham, MA 02026-3060. Tel.: 781-326-1170. Fax: 781-326-2147.
E-mail: homestead@fairbankshouse.org
Web Site: www.fairbankshouse.org
Founded: 1903.
Congressional District: 9
Key Personnel: Pres., Al Blood; Treas., Lynn Fairbank; Mgr. Business Operations, Lee Anne Hodson; Cur., Meaghan Siekman.
Personnel Profile: Full-Time Paid 1; Part-Time Paid 1; Part-Time Volunteers 12.
Governing Authority: nonprofit organization. Parent Institution: The Fairbanks Family in America. Tax-exempt: 501(c)(3).
Institution Type/Description: Historic House: c.1637-1641 home built for Jonathan & Grace Fairbanks and their family.
Collections: Fairbanks' family furnishings; personal artifacts.

Research Fields: 17th-century architectural research; Fairbanks family, allied families; life in America 1630-1905.
Facilities: Gift items for sale.
Activities: guided tours; lectures; docent program; formally organized education program for adults & school groups.
Publications: quarterly publication, Homestead Courier.
Hours & Admission Prices: May-Oct. Tues.-Sat. 10-5, Sun. 1-5. Tours given on the hour, last tour at 4pm. Family $35, adults $12, seniors $10, children 6-12 $6; discounts to AAM, AAA, MTA & NEMA members; members no charge.
Attendance: 3,000 (accurate)
Membership: Friend $20; Friend Family $25; Regular $35; Life $500.

MUSEUM OF BAD ART, 580 High St., Dedham, MA 02026-1845. Mailing Address: 73 Parker Rd., Needham, MA 02494-2038. Tel.: 781-444-6757.
E-mail: moba@museumofbadart.org
Web Site: www.museumofbadart.org
Founded: 1993.
Key Personnel: Interim Exec. Dir., Louise R. Sacco; Dir. Special Events, Garen Daley; Cur.-in-Chief, Mike Frank.
Personnel Profile: Part-Time Volunteers 5.
Governing Authority: nonprofit corporation. Branch Institution: MOBA Gallery, 55 Davis Sq., Somerville, MA; MOBA/Brookline Access TV, 55 Tappan St., Brookline, MA 02146.
Institution Type/Description: Art Museum.
Collections: bad works of art in all media.
Major Exhibits: Meet the Museum of Bad Art (T), 2/14-3/14.
Facilities: 1,500 sq. ft. exhibit space. Museum-related items for sale.
Activities: lectures; loan exhibitions; TV programs; mobile phone app.
Publications: email newsletter, Museum of Bad Art News; book, Museum of Bad Art: Art Too Bad To Be Ignored; Museum of Bad Art: Masterworks.
Hours & Admission Prices: Sun.-Thurs. 2-9, Fri.-Sat. 1-10. No charge; donations accepted.
Attendance: 7,500 (estimated)

Deerfield

* **HISTORIC DEERFIELD, INC., (M),** 84B Old Main St., Deerfield, MA 01342. Mailing Address: P.O. Box 321, Deerfield, MA 01342-0321. Tel.: 413-774-5581. Fax: 413-775-7220.
E-mail: tours@historic-deerfield.org
Web Site: www.historic-deerfield.org
Founded: 1952.
Congressional District: 1
Key Personnel: Pres., Philip Zea; Chm. (V), Anne K. Groves; Vice Pres. Museum Affairs, Anne Digan Lanning; Chm. Curatorial Dept., Amanda E. Lange; Public Historian, Barbara Mathews; Vice Pres. Business Affairs, Susan Martinelli; Dir. Museum Education & Interpretation, Amanda Rivera Lopez; Librarian, David C. Bosse; Museum Shop Mgr., Tina Harding; Supt. Properties Maintenance, George Holmes.
Personnel Profile: Full-Time Paid 50; Part-Time Paid 70; Part-Time Volunteers 115; Interns 8.
Governing Authority: nonprofit organization. Tax-exempt: 501(c)(3).
Institution Type/Description: Outdoor History Museum Village: consisting of 14 18th & 19th-century structures.
Collections: early New England Life; decorative arts; furniture; ceramics; silver; metalwork; 1998 Flynt Ctr of Early New England Life; textiles; costumes; paintings; 1814 Henry Needham Flynt Silver & Metalware. Historic Buildings: c.1760 Frary House; c.1733 Allen House; c.1733 Ashley House; c.1753 Sheldon House; c.1746 Wells-Thorn House; c.1799 Asa Stebbins House; c.1760-1800 Hall Tavern; c.1816 Hinsdale & Anna Williams House; c.1795 Barnard Tavern; c.1848 Rev. Johnson Farwell Moors House.
Research Fields: decorative arts; furniture; history; folklore; silver; ceramics; paintings; archaeology.
Facilities: 10,000-vol. library of history & decorative arts available on premises; reading room; restaurant. Publications, slides, needlework, postcards & museum reproductions for sale.
Activities: guided tours; lectures; films; elementary & secondary school tours; formally organized programs for college classes; workshops; forums; archaeological field school; family events; museum courses for adults.
Publications: books, Heritage Foundation Collection of Silver; Parson Ashley House; Early Settlement in the Connecticut Valley; booklets, Historic Deerfield: An Introduction; Five Colleges: Five Histories; In Debt to Shays: The Bicentennial of an Agrarian Rebellion; Historic Deerfield: A Portrait of Early America; Old Deerfield Massachusetts; Pursuing Refinement in Rural New England, 1750-1850, A Deerfield Sampler; record, Faded Memories: Songs of Deerfield; Delftware at Historic Deerfield, 1600-1800.
Hours & Admission Prices: April 17-Nov. 28 daily 9:30-4:30. Flynt Center of

Early New England Life: Jan.-March Sat.-Sun. Adults $12, children $5; discounts to groups of 20 or more with reservations & AAM members; members no charge. Closed Thanksgiving; Christmas Eve & Day. &
Attendance: 30,000 (estimated)
Membership: Individual $40; Family $60; Active $100; Contributing $150; Associate $250; Patron $500; Asher Benjamin Society $1,000, $2,000 & $5,000; Life $10,000.

MEMORIAL HALL MUSEUM, POCUMTUCK VALLEY MEMORIAL ASSOC., (M), 8 Memorial St., Deerfield, MA 01342. Mailing Address: Box 428, Deerfield, MA 01342-0428. Tel.: 413-774-7476, ext. 10. Fax: 413-774-5400. www.americancenturies.mass.edu.
E-mail: tneumann@deerfield.history.museum
Web Site: www.old-deerfield.org
Founded: 1870.
Congressional District: 2
Key Personnel: Dir. & C.E.O., Timothy C. Neumann; Dir. Youth Programs, Lynne Manring; Pres. (V), Carol Letson; Cur., Suzanne Flynt; Librarian, David Bosse; Museum Shop Mgr., Tom Mershon.
Personnel Profile: Full-Time Paid 8; Full-Time Volunteers 1; Part-Time Paid 25; Part-Time Volunteers 75.
Governing Authority: nonprofit organization. Affiliated with the Pocumtuck Valley Memorial Assn. Tax-exempt: 501(c)(3).
Institution Type/Description: History & Decorative Arts Museum: housed in 1798 Deerfield Academy Building.
Collections: early American furniture; American & English pewter; iron; tin & woodenware; glass; ceramics; farm tools; craftsmen's tools; tavern signs; textiles; architectural fragments; American paintings; musical instruments; toys; military equipment; manuscripts; Native American artifacts.
Facilities: 12,000-vol. library of books on local history, imprints, periodicals, manuscripts, 380 account & day books, school materials & diaries available for use on premises only. Books for sale.
Activities: lectures; gallery talks; concerts; formally organized education programs for children; permanent & temporary exhibitions; chamber theater group.
Publications: books, History of Deerfield; Short History of Deerfield; The Boy Captive of Old Deerfield; The Boy Captive in Canada; Deerfield Embroidery; Gathered & Preserved; Hadley Chests; The Allen Sisters: Pictorial Photographs 1885-1920; Poetry to the Earth: The Deerfield Arts and Crafts Movement.
Hours & Admission Prices: May Tues.-Fri. 10:30-4:30; May Sat.-Sun. & June-Oct. Tues.-Sun. 11-4:30. Adults $6, children & students 6-21 $3; discounts to museum employees, MA Teachers, NEMA & AAM members; members no charge. Library: call for hours. &
Attendance: 52,292 (accurate)
Membership: Annual $40; Family $60.

Dennis

CAPE COD MUSEUM OF ART, (M), 60 Hope Ln., off Rte. 6A, Dennis, MA 02638. Mailing Address: P.O. Box 2034, Dennis, MA 02638-5034. Tel.: 508-385-4477. Fax: 508-385-7933.
E-mail: info@ccmoa.org
Web Site: www.ccmoa.org
Formerly: Cape Museum of Fine Arts, Inc.
Founded: 1981.
Congressional District: 10
Key Personnel: Interim Exec. Dir., Cindy Nickerson; Pres. (V), Thomas N. George.
Personnel Profile: Full-Time Paid 8; Part-Time Paid 2; Part-Time Volunteers 200; Interns 3.
Governing Authority: nonprofit organization. Tax-exempt: 501(c)(3).
Institution Type/Description: Art Museum & Center.
Collections: 20th-century fine & contemporary art.
Research Fields: archives on artists; fine art of Cape Cod.
Facilities: auditorium; educational facilities. Museum-related items for sale.
Activities: exhibition & museum tours; lectures; studio arts for children; workshops; educational films; art & foreign film series; docent program; permanent & loan exhibitions; cultural travel program; art festival; Silent & Wet Art Auction; special events; volunteer & intern programs.
Publications: quarterly brochure; Art Matters; exhibition catalogs.
Hours & Admission Prices: Memorial Day to Columbus Day Mon.-Wed. & Fri.-Sat. 10-5, Thurs. 10-8, Sun. 12-5; Oct.-May Tues.-Wed. & Fri.-Sat. 10-5, Thurs. 10-8, Sun. 12-5. Adults $8; discounts to groups & AAM, ICOM, MASS, Teacher's Assoc., New England Consortium of Art Museums, & WGBH members; children & members no charge. Thurs. by donation. Closed New Year's Day; Thanksgiving; Christmas. &
Attendance: 29,961 (accurate)
Membership: Student, Artist & Teacher $30; Individual $50; Dual $75;

Sponsor $150; Patron $250; Sustaining $500; Associate $1,000; Fellow $2,500; Benefactor $5,000.

DENNIS HISTORICAL SOCIETY - 1736 JOSIAH DENNIS MANSE MUSEUM, 77 Nobscusset Rd., Dennis, MA 02638. Mailing Address: P.O. Box 705, Dennis, MA 02638. Tel.: 508-385-2232.
Founded: 1967.
Key Personnel: Chm. (V), Nancy Howes
Institution Type/Description: Historical Society Museum: housed in the former home of Rev. Josiah Dennis, for whom the town was named; built in 1736. Listed on the National Register of Historic Places.
Collections: local history & culture; period furnishings; personal artifacts; photographs; maritime artifacts.
Hours & Admission Prices: Call for hours. No charge; donations accepted. &
Attendance: 1,500 (accurate)

Dover

CARYL HOUSE AND FISHER BARN, 107 Dedham St., Dover, MA 02030-2223. Mailing Address: P.O. Box 534, Dover, MA 02030-0534. Tel.: 508-785-1832. Fax: 508-785-0789.
Web Site: www.doverhistoricalsociety.org
Founded: 1920.
Congressional District: 9
Key Personnel: Dir., Elisha F. Lee, Jr.; Vice Pres., Priscilla P. Jones.
Personnel Profile: Part-Time Volunteers 20.
Governing Authority: historical society. Parent Institution: Town of Dover. Tax-exempt: 501(c)(3).
Institution Type/Description: Historic House: built c.1777.
Collections: documents & artifacts of 1790s minister, doctor & families; restored barn, 1777; area agricultural artifacts.
Research Fields: history & culture of 1790s Dover & vicinity; early practice of medicine.
Hours & Admission Prices: April-June & Sept.-Nov. Sat. 1-4 & by appointment. No charge; donations accepted. Closed Sat. holidays.
Attendance: 500 (estimated)
Membership: Individual $35; Family $65.

DOVER HISTORICAL SOCIETY - SAWIN MUSEUM, 80 Dedham St., Dover, MA 02030. Mailing Address: P.O. Box 534, Dover, MA 02030-0534. Tel.: 508-785-1832.
Web Site: doverhistoricalsociety.org
Founded: 1895.
Congressional District: 4
Key Personnel: Pres., Elisha Lee; Vice Pres., Priscilla Jones; Cur. Sawin, Fay Bacher; Cur. Caryl, Barbara Palmer; Cur. Barn, Jack Hoehlen.
Personnel Profile: Part-Time Volunteers 15.
Governing Authority: nonprofit organization. Parent Institution: Dover Historical Society. Tax-exempt: 501(c)(3).
Institution Type/Description: Historical Society Museum.
Collections: documents & artifacts of Dover & vicinity; manuscripts.
Facilities: 700-vol. library of history & school books available on premises.
Activities: permanent & temporary exhibitions.
Hours & Admission Prices: April-June & Sept.-Nov. Sat. 1-4 & by appointment. No charge; donations accepted. Closed Sat. holidays.
Attendance: 200 (estimated)
Membership: Single $40; Family $65; Supporter $100; Patron $250; Benefactor $500.

Duxbury

ALDEN HOUSE HISTORIC SITE, 105 Alden St., Duxbury, MA 02332. Mailing Address: P.O. Box 2754, Duxbury, MA 02331-2754. Tel.: 781-934-9092. Fax: 781-934-9149. Facebook: Alden House Historic Site.
E-mail: aldenhouse@verizon.net
Web Site: www.alden.org
Formerly: Alden House Museum
Founded: 1906.
Congressional District: 10
Key Personnel: Dir., Matthew Vigneau; Pres. (V), Mary-Ruth Duquette; Sec., Bonnie Conant; Cur. & Museum Shop Mgr., James W. Baker.
Personnel Profile: Full-Time Paid 1; Part-Time Paid 1.
Governing Authority: nonprofit organization. Owned and operated by Alden Kindred of America, Inc., P.O. Box 2754, Duxbury, MA 02331. Tax-exempt: 501(c)(3).
Institution Type/Description: Historic House Museum: homestead of Pilgrims John Alden and Priscilla Mullins.

Collections: period furnishings; artifacts.
Research Fields: Plymouth Colony history.
Facilities: visitors center.
Activities: guided tours.
Publications: biannual newsletter.
Hours & Admission Prices: June-Sept. Wed.-Sat. 12-4. Adults $8, children 3-17 $5; members no charge. &
Attendance: 1,506 (accurate)
Membership: Individual $30; Family $55; Business $100.

THE ART COMPLEX MUSEUM, (M), 189 Alden St., Duxbury, MA 02332-3801. Mailing Address: P.O. Box 2814, Duxbury, MA 02331-2814. Tel.: 781-934-6634. Fax: 781-934-5117.
E-mail: info@artcomplex.org
Web Site: www.artcomplex.org
Founded: 1967.
Congressional District: 10
Key Personnel: Dir. & C.E.O., Charles A. Weyerhaeuser; Consulting Cur., Alice Hyland; Collections Mgr., Maureen Wengler; Education Coord., Sally Dean Mello; Contemporary Cur., Craig Bloodgood; Communications Coord., Laura Doherty; Asst. to Dir., Mary Curran; Librarian, Cheryl O'Neill; Community Coord., Doris Collins; Grounds & Maintenance, William Thomas; Preparator, Sue Aygarn-Kowalski.
Personnel Profile: Full-Time Paid 1; Part-Time Paid 11; Part-Time Volunteers 51.
Governing Authority: nonprofit organization. Parent Institution: Art Complex Inc. Tax-exempt: 501(c)(3).
Institution Type/Description: Art Museum.
Collections: American & European paintings; modern & Old Master prints; Shaker furniture & artifacts; Asian art; Japanese tea house; contemporary New England artists.
Facilities: 6,500-vol. library of books on art & artists.
Activities: lectures; gallery talks; concerts; permanent & traveling exhibitions; studio art classes & workshops; contemporary artists education program for all ages; concert series. Museum Sponsors: tea ceremony presentations 4 times per year.
Publications: book, The Lithographs of Ture Bengtz; catalogues, Master Prints 1850-1950, The Shakers, Pure of Spirit Pure of Mind; American Paintings; Rituals of the Land: Native American Art from the Colorado Plateau; Rufus Hathaway, Artist & Physician, Of Matter & Spirit: Dutch Prints of the 17th Century; New Horizons: 19th century American Marine painting; Tribute to Kojiro Tomita; Enduring Nature: Chinese Paintings from the Weyerhaeuser collection; Environmental Arts at the Art Complex Museum; Shaped with a Passion; Japanese Ceramics from the 1970s.
Hours & Admission Prices: Wed.-Sun. 1-4. No charge; donations accepted. Closed legal holidays. &
Attendance: 10,000 (estimated)

DUXBURY RURAL AND HISTORICAL SOCIETY, INC., 479 Washington St., Duxbury, MA 02331. Mailing Address: P.O. Box 2865, Duxbury, MA 02331-2865. Tel.: 781-934-6106. Fax: 781-934-5730.
E-mail: aarnold@duxburyhistory.org
Web Site: www.duxburyhistory.org
Founded: 1883.
Congressional District: 10
Key Personnel: Pres. (V), David Jenkins; Exec. Dir., Patrick Browne.
Personnel Profile: Full-Time Paid 1; Part-Time Paid 1; Part-Time Volunteers 75; Interns 1.
Governing Authority: society; executive committee. Tax-exempt.
Institution Type/Description: Historical Society Museum.
Collections: 19th-century decorative arts; textiles; relics of 1869 French-American cable. Historic Houses: 1808-1809, King Caesar Houses; 1808 Captain Gershom Bradford House; 1826 Drew House; Cedarfield on Clark's Island; 1807 Nathaniel Winsor House.
Research Fields: history from Pilgrim times to present; early 19th-century shipbuilding era.
Facilities: library of letters, ships logs & local history available by appointment to the Duxbury Rural & Historical Society.
Activities: guided tours in summer; lectures; permanent & temporary exhibitions.
Publications: books, The Duxbury Book, 1937-1987; Stopping Places Along Duxbury Roads; The Alden Family in the Alden House; Settlement and Growth of Duxbury 1828-1870; booklets, Roundabout Duxbury; A History of the Duxbury Rural & Historical Society; Tall Ships of Duxbury 1815-1850; The French Atlantic Cable 1869; Duxbury, A Guide.
Hours & Admission Prices: King Caesar House: July to Labor Day Wed.-Sun. 1-4. Adults $5; members no charge. Bradford House: July to Labor Day

Sun. 1-4. Adults $3; discounts to senior citizens; members no charge. Nathaniel Winsor House Office: Mon.-Fri. 9-4. Closed holidays.
Attendance: 1,852 (accurate)
Membership: Senior Citizen & Student $25; Single $30; Senior Couple $35; Family $45; Contributing $150; Sustaining $250; Life $1,000.

East Sandwich

THORNTON W. BURGESS SOCIETY GREEN BRIAR NATURE CENTER AND MUSEUM, 6 Discovery Hill Rd., East Sandwich, MA 02537. Tel.: 508-888-6870. Fax: 508-888-1919.
E-mail: info@thorntonburgess.org
Web Site: www.thorntonburgess.org
Formerly: Thornton W. Burgess Museum
Founded: 1976.
Congressional District: 12
Key Personnel: Exec. Dir., Gene A. Schott; Pres. (V), Wendy Maggio; Museum Shop Mgr., John Richmond.
Personnel Profile: Full-Time Paid 5; Part-Time Paid 21; Part-Time Volunteers 125.
Governing Authority: society; nonprofit. Parent Institution: Thornton W. Burgess Society, Inc. Branch: Green Briar Nature Center. Tax-exempt.
Institution Type/Description: History Museum & Nature Center: housed in 1780 Nye House.
Collections: published writings of Thornton W. Burgess including many first editions; autographs; original illustrations by Harrison Cady; herb & wild flower garden; taxidermy mounts, moths & butterflies; shells, rocks and minerals; fossils; photographs; historic buildings.
Major Exhibits: The Nature Photographs of Cheryl Calibano, 1/12/14-3/23/14; The Turtle Sisters - the Illustrations of Sisan Baur, 4/8/14-6/29/14; The Art of Chris Warrington, 7/13/14-9/14/14; Pastel Birds Nests and More by Eileen Casey, 9/21/14-11/16/14; The Amazing Thornton Burgess, 2/15; Fishing - The Richardson Collection, 8/15-11/16.
Research Fields: natural history of southeastern Massachusetts; related environmental issues.
Facilities: gardens; nature trails. Books & museum-related items for sale.
Activities: educational programs; summer story times; live animal programs; loan & temporary exhibitions.
Publications: biennial publication, Briar Patch Observer & program schedule.
Hours & Admission Prices: Jan.-March Tues.-Sat. 10-4; April-Dec. Mon.-Sat. 10-4, Sun. 1-4. No charge; donations accepted.
Attendance: 40,000 (estimated)
Membership: Individual $35; Family $50; Supporting $100.

Easthampton

MASSACHUSETTS AUDUBON AT CONNECTICUT RIVER VALLEY SANCTUARIES, 127 Combs Rd., Easthampton, MA 01027-9704. Tel.: 413-584-3009, ext. 12. Fax: 413-584-0250.
E-mail: mshanley@massaudubon.org
Web Site: www.massaudubon.org
Formerly: Massachusetts Audubon at Hampshire Sanctuaries
Founded: 1944.
Congressional District: 1
Key Personnel: Dir., Jonah Keane; Chm., Janet Bissel.
Personnel Profile: Full-Time Paid 7; Part-Time Paid 8; Part-Time Volunteers 150.
Governing Authority: society. Parent Institution: Massachusetts Audubon Society, S. Great Rd., Lincoln, MA 01773. Tax-exempt: 501(c)(3) & 170(b)(1)(A).
Institution Type/Description: Nature Center & Wildlife Sanctuary.
Collections: nature center & wildlife sanctuary.
Research Fields: botanical & geological in cooperation with Smith College & University of Massachusetts.
Facilities: 2,000-vol. library on natural science, conservation, ecology & ornithology available for use on premises; botanical garden; nature center; field research station; 80-seat auditorium. Books for sale.
Activities: guided tours, lectures, films, formally organized education programs for adults, children, families & students.
Publications: quarterly newsletter.
Hours & Admission Prices: Office: Mon.-Fri. 8:30-12:30. Grounds: Tues.-Sun. dawn-dusk. Adults $4, children 3-15 & senior citizens $3; members & children under 3 no charge.
Attendance: 20,000 (estimated)
Membership: Student $20; Individual $37; Family $47; Supporting $60; Defender $75; Donor $100; Sponsor $250; Patron $500; Leadership Friend $1,000.

Edgartown

FELIX NECK WILDLIFE SANCTUARY, 100 Felix Neck Dr., Edgartown, MA 02539. Mailing Address: P.O. Box 494, Vineyard Haven, MA 02568-0494. Tel.: 508-627-4850.
E-mail: felixneck@massaudubon.org
Web Site: www.massaudubon.org
Institution Type/Description: Wildlife Sanctuary.
Collections: wildlife & their habitats.
Facilities: nature trails; nature center.
Activities: summer camp; educational programs.
Hours & Admission Prices: Nature Center: Mon.-Sat. 9-4, Sun. 10-3; call for off-season hours. Trails: daily dawn to dusk. Adults $4, seniors & children 2-12 $3.

GUS BEN DAVID'S WORLD OF REPTILES AND BIRDS PARK, Batchedler Rd., Edgartown, MA 02539. Mailing Address: P.O. Box 1055, Oak Bluffs, MA 02557. Tel.: 508-627-5634.
Key Personnel: Dir., Gus Ben David
Institution Type/Description: Nature Park.
Collections: reptiles & birds including pythons, Rhinoceros Iguana; giant tortoise; crocodile lizard; fancy pigeons; pheasants; waterfowl; chickens.
Facilities: Museum-related items for sale.
Activities: educational programs & presentations; group tours; birthday parties.
Hours & Admission Prices: Tues.-Sun. 10-3. Admission: $5 per person.

MARTHA'S VINEYARD MUSEUM, (M), 59 School St., Edgartown, MA 02539. Mailing Address: P.O. Box 1310, Edgartown, MA 02539-1310. Tel.: 508-627-4441. Fax: 508-627-4436.
E-mail: frontdesk@mvmuseum.org
Web Site: www.mvmuseum.org
Formerly: Martha's Vineyard Historical Society
Founded: 1922.
Congressional District: 12
Key Personnel: C.E.O. & Dir., David Nathans; Chm. (V) & Pres. (V), Sheldon Hackney; Chief Cur., Bonnie Stacy; Asst. Cur., Anna Carringer; Genealogist, Catherine M. Mayhew; Dir. Devel., Noelle Colome; Education Dir., Nancy Cole; Asst. Mgr. Devel. & Membership Svcs., Jessica Barken; Dir. Finance & Museum Shop Mgr., Betsey Mayhew; Administrative Coord., Chris Bahara; Cur. Oral History, Linsey Lee.
Personnel Profile: Full-Time Paid 8; Part-Time Paid 5; Part-Time Volunteers 60; Interns 12.
Governing Authority: nonprofit organization. Tax-exempt: 501(c)(3).
Institution Type/Description: History Museum.
Collections: archaeology; archives; costumes; herbarium; history; Indian artifacts; industry; naval; outdoor museum; whaling industry; period artifacts; glass; china; wooden ware; domestic tools; ship models; Vineyard memorabilia; scrimshaw; Methodist Camp meeting ground; oral histories; 1856 Gay Head light. Historic Buildings: 1765 Thomas Cooke House; c.1845 Captain Francis Pease House.
Major Exhibits: A Taste for the Exotic: Tokens from Around the Globe, 5/13-4/14.
Research Fields: genealogy; whaling period; history of Martha's Vineyard; American Indians; glacial period geology; marine history; Island crafts; local oral history.
Facilities: 3,000-vol. library of books on Vineyard history, genealogical records, oral histories, archives including 500 linear feet of material, maritime & whaling history, photographic archives, maps & charts; reading room; garden. Books, postcards & guidebooks for sale.
Activities: guided tours; lectures; permanent & temporary exhibitions; educational programs.
Publications: quarterly journal, Dukes County Intelligencer.
Hours & Admission Prices: Fall & Spring: Mon.-Sat. 10-4; Summer: Mon.-Sat. 10-5. Winter: adults $6, seniors $5, children 6-15 $4; discounts to Blue Star Museum & AAM members; children under 6 & members no charge. Summer: adults $7, seniors $6, children 6-15 $4; children under 6 & members no charge. Closed major holidays.
Attendance: 5,000 (estimated)
Membership: Student $25; Individual $55; Family $75; Sustaining $125; Patron $250; Benefactor $500; President's Circle $1,000; Leadership $2,500; Steward $5,000.

Essex

COGSWELL'S GRANT, (M), 60 Spring St., Essex, MA 01929-1308. Tel.: 978-768-3632. Fax: 978-768-6274.
E-mail: cogswellsgrant@historicnewengland.org
Web Site: www.historicnewengland.org
Founded: 1998.

Key Personnel: Pres., Carl Nold; Site Mgr., Kristen Weiss.
Governing Authority: private; nonprofit organization. Parent Institution: Historic New England, Boston, MA. Tax-exempt: 501(c)(3).
Institution Type/Description: Historic House Museum: housed in an 18th century farmhouse used by American folk art collectors Bertram and Nina Fletcher Little as their summer home.
Collections: American folk art including decorative paintings, floor coverings, boxes, & New England pottery; Bertram & Nina Fletcher Little.
Research Fields: decorative arts.
Facilities: Museum-related items for sale.
Activities: guided tours; special events; lectures.
Publications: visitors guide; exhibition catalogues; magazine, Historic New England; monthly newsletter, What's Happening.
Hours & Admission Prices: June to Oct. 15 Wed.-Sun. 11-4. Adults $10; discounts to ICOM, AAA, AAM & WGBH members; members no charge.
Attendance: 6,605 (accurate)
Membership: National $35; Individual $45; Household $55; Garden & Landscape $75; Contributing and Library & School $100; Historic New England Affiliate $100-$350; Young Friends of Historic New England $100-$1,500; Friends of the Library and Archives $125; Historic Homeowner $200; Business & Ogden Codman Design Group $250; Appleton Circle $1,750-$3,500.

ESSEX SHIPBUILDING MUSEUM, 66 Main St., Essex, MA 01929-1343. Mailing Address: P.O. Box 277, Essex, MA 01929-0005. Tel.: 978-768-7541.
E-mail: info@essexshipbuildingmuseum.org
Web Site: www.essexshipbuildingmuseum.org
Founded: 1976.
Congressional District: 5
Key Personnel: Pres., Lee Spence; Treas., Sarah Willwerth-Dyer; Coord. Education, Nancy Dudley.
Personnel Profile: Part-Time Paid 4; Part-Time Volunteers 30.
Governing Authority: society. Parent Institution: Essex Historical Society & Shipbuilding Museum, Inc. Tax-exempt.
Institution Type/Description: Shipbuilding Museum: housed in 1835 schoolhouse, 1840's hearse house & Maritime History Museum; Historic Site: 1680-1860 burying ground; old Story Yard site where A.D. Story launched 397 fishing schooners.
Collections: old Story yard site, 1813-1985 the Story shipyard used for shipbuilding for 300 years; shipbuilding tools; photographs; shipbuilding memorabilia; archives; local history; half-models; the fishing schooner Evelina M. Goulart, built by A.D. Story in 1927; scale & builders' models from Smithsonian Institution, Washington, DC; Lewis H. Story, a Chebacco boat replica.
Research Fields: local shipbuilding history; owner-builder correspondence; plans, records, photos of individual vessels; photos of yards; drawings of construction details; builder's models for inboard carpentry.
Facilities: research library of shipbuilding books at 28 Main St.; archives; education center; meeting room. Books & other museum-related items for sale.
Activities: inter-museum loan; education programs; permanent & temporary exhibitions; lectures at museum & away for special interest groups; annual meeting; trips & tours.
Publications: booklets: Essex, the Shipbuilding Town; Essex Electrics; Dubbing, Hooping & Lofting: Shipbuilding Skills; book, A List of Vessels, Boats & Other Crafts Built in Essex During 1860-1980, 3rd ed., 1992.
Hours & Admission Prices: Call for hours. Adults $7, senior citizens $6, children $5; children under 6 & members no charge.
Attendance: 7,000 (estimated)
Membership: Senior Couple $35; Individual $35; Family $50.

Fall River

FALL RIVER HISTORICAL SOCIETY AND BORDEN MURDER MYSTERY MUSEUM, 451 Rock St., Fall River, MA 02720-3398. Tel.: 508-679-1071. Fax: 508-675-5754.
E-mail: curator@lizzieborden.org
Web Site: www.lizzieborden.org
Founded: 1921.
Congressional District: 10
Key Personnel: Cur., Michael Martins; Pres. (V), Jay Lambert; Asst. Cur., Dennis Binette.
Personnel Profile: Full-Time Paid 2; Part-Time Paid 2; Part-Time Volunteers 25; Interns 1.
Governing Authority: society. Tax-exempt: 501(c)(3).
Institution Type/Description: Historical Society Museum: housed in 1843 Granite House, used as underground railroad station c.1843-1860s.
Collections: glass; china; furniture; costumes; guns; marine; paintings; manuscripts; artifacts relating to the Lizzie Borden murder trial; 19th-century

decorative arts; Fall River newspapers 1850-1950 on microfilm; restored mill owners mansion.
Research Fields: local history; genealogy; Borden murder mystery; Lizzie Borden.
Facilities: 3,000-vol. library of general books available on premises; 1850-1950 newspapers on microfilm; reading room.
Activities: guided tours; lectures; permanent & temporary exhibitions.
Publications: The Fall River Society Quarterly Report-Newsletter.
Hours & Admission Prices: May & Oct. Tues.-Fri. 9-4; June-Sept. Tues.-Fri. 9-4, Sat.-Sun. 1-4:30. Adults $8, children 6-14 $6; discount to groups; children under 6 & members no charge. Closed holidays.
Attendance: 14,520 (accurate)
Membership: Full-Time student up to age 22 $10; Individual $25; Family (two adults & all children under 17 living at the same address) $40; Corporate $100; Life (single) $500, (couple) $800; Sponsor $500; Benefactor $1,000.

GRIMSHAW-GUDEWICZ ART GALLERY, Jackson Art Center, Bristol Community College, 777 Elsbree St., Fall River, MA 02720-7307. Tel.: 508-678-2811, ext. 2439. Fax: 508-730-3285.
E-mail: kathleen.hancock@bristolcc.doc
Web Site: www.bristol.mass.edu/gallery
Key Personnel: Dir., Kathleen Hancock
Institution Type/Description: Art Gallery.
Collections: sculpture; paintings.
Activities: educational programs.
Hours & Admission Prices: Mon., Wed. & Sat. 1-4, Tues. & Thurs.-Fri. 10-1. No charge.

THE MARINE MUSEUM AT FALL RIVER, INC., 70 Water St., Fall River, MA 02721-1598. Tel.: 508-674-3533. Fax: 508-674-3534.
Founded: 1968.
Congressional District: 4
Key Personnel: Chm. Bd., Dr. Robert Lawrence; Pres. Pro Tem, Margot Cottrell; Vice Pres., Sheila Salvo.
Personnel Profile: Part-Time Volunteers 2.
Governing Authority: nonprofit organization. Tax-exempt: 501(c)(3).
Institution Type/Description: Marine Museum: housed in restored textile mill machine shop.
Collections: Fall River Line Steamships (1847-1937) models: The Providence, The Puritan, The Plymouth, The Commonwealth, The Priscilla; artifacts & exhibits of The Titanic including a 28 ft. model & photos taken by Woods Hole Oceanographic Institution; The United Fruit Line Collection pertaining to the company's steam powered banana ships; William King Covell Collection including prints, negatives & glass slides relating to The Fall River Line.
Research Fields: steamships.
Facilities: 2,500-vol. library of maritime history available for use on premises by appointment; meeting room. Books & museum-related items for sale.
Activities: guided tours; lectures; slide presentations; formally organized educational programs; permanent & traveling exhibitions; social gatherings.
Publications: quarterly newsletter; brochures.
Hours & Admission Prices: Wed.-Sat. 10-3. Adults $6, senior citizens & children 5-12 $4; discounts to groups or 20 or more; members no charge. Closed New Year's Day; Thanksgiving; Christmas. &
Attendance: 28,000 (estimated)
Membership: Individual $30; Family $40; Sustaining $100; Admiral's Club $200-$1,000; Corporate $350; Life $2,500.

OLD COLONY & FALL RIVER RAILROAD MUSEUM, 2 Water St. at Battleship Cove, Fall River, MA 02720. Mailing Address: P.O. Box 3455, Fall River, MA 02722-3455. Tel.: 508-674-9340. Fax: 508-678-1220. Facebook: Old Colony and FR Railroad Museum.
E-mail: railroadjc@aol.com
Web Site: www.ocandfrrailroadmuseum.com
Founded: 1986.
Congressional District: 4
Key Personnel: Pres., Jay Chatterton.
Personnel Profile: Part-Time Volunteers 12.
Volunteer Hours: 2,890
Governing Authority: Tax-exempt.
Institution Type/Description: Railroad Museum.
Collections: railroad artifacts & memorabilia.
Publications: monthly newsletter.
Hours & Admission Prices: May-June & Sept. Sat.-Sun. 12-4; July-Aug. Fri.-Sun. 12-4. Adults $3, senior citizens 65 & over $2.50, children 5-12 $1.50; members & children under 5 no charge.

Attendance: 1,200 (accurate)
Membership: Individual $20; Family $30; Corporate $50.

USS MASSACHUSETTS MEMORIAL COMMITTEE, INC., (M), Battleship Cove, 5 Water St., Fall River, MA 02721-1540. Mailing Address: Battleship Cove, 5 Water St., P.O. Box 111, Fall River, MA 02722-0111. Tel.: 508-678-1100 & 1905; 800-533-3194. Fax: 508-674-5597.
E-mail: battleship@battleshipcove.org
Web Site: battleshipcove.org
Formerly: Battleship Massachusetts
Founded: 1965.
Congressional District: 10
Key Personnel: Exec. Dir., Bradley M. King; Pres. (V), Carl F. Sawejko; Dir. Finance, David W. Keyes; Cur., Christopher J. Nardi; Museum Shop Mgr., Michelle Cateon.
Personnel Profile: Full-Time Paid 24; Full-Time Volunteers 10; Part-Time Paid 24; Part-Time Volunteers 225.
Governing Authority: federal. Tax-exempt: 501(c)(3).
Institution Type/Description: Historic Ships Museum.
Collections: equipment & memorabilia connected with the operation of the USS Massachusetts, USS Lionfish & USS Joseph P. Kennedy, Jr. during World War II; PT Boat 796 & P.T. Boat 617; World War II Quonset Hut; LCM landing craft; T-28 aircraft; scale model World War II aircraft collection; UH-1E Viet Nam era helicopter; Russian Missile Corvette, Hiddensee; OSS convert operations semi-submersible boat.
Research Fields: naval history of World War II.
Facilities: 250-vol. library of the history of naval warfare especially during World War II & the history of the development of the battleship as a naval weapon; restaurant; banquet facilities. Flags, ship models, jewelry & sweatshirts for sale.
Activities: self-guided tours; lectures; films; permanent exhibitions.
Publications: quarterly newsletter.
Hours & Admission Prices: Spring daily 9-4:30. Summer daily 9-5. Adults $17, senior citizens 65 & up, military veterans & AAA members $15, children 6-12 $10.50, active duty military with ID $8.50; discounts to groups & Historic Naval Ship Association; members & children under 5 no charge. Closed New Year's Day; Thanksgiving; Christmas.
Attendance: 88,485 (accurate)
Membership: Individual $35; Family $65; Executive $250; Captain $500; Admiral $1,000.

Falmouth

FALMOUTH MUSEUMS ON THE GREEN, 55-65 Palmer Ave., Falmouth, MA 02540. Mailing Address: P.O. Box 174, Falmouth, MA 02541-0174. Tel.: 508-548-4857. Fax: 508-540-0968. Facebook: Falmouth Museums on the Green.
E-mail: fhs@cape.com
Web Site: www.FalmouthHistoricalSociety.org
Formerly: Falmouth Historical Society
Founded: 1900.
Congressional District: 12
Key Personnel: Exec. Dir., Mark A. Schmidt; Pres., Tamsen George; Cur., Amanda Wastrom; Museum Shop Mgr., Caroline Lloyd; Office Asst., Cathleen McDonnell.
Personnel Profile: Full-Time Paid 1; Full-Time Volunteers 35; Part-Time Paid 5; Part-Time Volunteers 200; Interns 2.
Governing Authority: society. Tax-exempt: 501(c)(3).
Institution Type/Description: Historical Society Museum.
Collections: two historic homes, barn, colonial garden; herb garden. Dr. Francis Wicks House: c.1790 contains 19th century French wallpaper, furnishings, china & paintings from 1790-1932; 1790-1840 doctor's office. Conant House: c.1730 exhibits on Falmouth history; video theater. Hallett Barn Visitor Center: whaling era memorabilia; life of Katharine Lee Bates; early farm implements. Education Center: lectures; programs; art exhibits.
Major Exhibits: 19th Century Dresses & Textiles, 5/14 10/14.
Research Fields: local history; genealogical; maritime.
Facilities: 500-vol. library of 18th- & 19th-century books on genealogy, history & religion available on premises; education center. Conant House: theater. Visitor Center: Museum-related items for sale.
Activities: guided tours; lectures; permanent & temporary exhibitions. Museum Sponsors: Guided Walking Tours June to October; Fridays for Families in July & August; Katharine Lee Bates Poetry Fest in August; Trolley Tours in September & October.
Publications: guides; books, Residential Falmouth; Hotels & Inns of Falmouth; Book of Holdings; quarterly journal, Spiritsail; newsletters & announcements; Falmouth (Images of America Series); Voice of the Tide, Biography of Katharine Lee Bates.
Hours & Admission Prices: June 9-Oct. 10 Tues.-Fri. 10-4, Sat. 10-1; other

times by appointment. Adults $5; discounts to AAM, MTA & NEMA members; children 13 & under, members & Falmouth residents on Fri. no charge. &
Attendance: 6,270 (estimated)
Membership: Individual $30; Family $50; Associate $75; Benefactor $150.

Fitchburg

❋ FITCHBURG ART MUSEUM, (M), 25 Merriam Pkwy., Fitchburg, MA 01420. Tel.: 978-345-4207. Fax: 978-345-2319.
E-mail: info@fitchburgartmuseum.org
Web Site: www.fitchburgartmuseum.org
Founded: 1925.
Congressional District: 2
Key Personnel: Dir., Nick Capasso; Pres. (V), Roderick Lewin; Dir. Corp. Member Svcs., Jane Keough; Dir. Membership & Public Rels., Janice Goodrow; Dir. Education, Laura Howick; Dir. Docents, Ann Descoteaux; Cur., Mary Tinti; Business Mgr., Sheryl Demers.
Personnel Profile: Full-Time Paid 7; Part-Time Paid 18; Part-Time Volunteers 20; Interns 6.
Governing Authority: nonprofit organization. Tax-exempt: 501(c)(3).
Institution Type/Description: Art Museum.
Collections: European & American 18th-20th century paintings, photographs, drawings & prints; decorative arts; 15th-20th century illustrated books; pre-Columbian, Asian, Ancient, African & Oceanic art.
Major Exhibits: Jeffu Warmouth: No More Funny Stuff, 2/9/14-6/1/14; 79th Regional Exhibition of Art & Craft, 6/22/14-8/31/14.
Facilities: studio; classroom; sculpture garden; docent & volunteer room.
Activities: guided tours; lectures; gallery talks; formally organized educational programs; inter-museum loan, permanent & temporary exhibitions; annual regional art & craft exhibition; bus excursions; banquet facilities; members' council; volunteer program; corporate art exchange program.
Publications: exhibition catalogs & posters; notices & invitations to members; museum calendar; newsletter.
Hours & Admission Prices: Wed.-Fri. 12-4, Sat.-Sun. 11-5. Adults $9, senior citizens & students $5; discount to AAM & ICOM members; members no charge. Closed major holidays. &
Attendance: 19,000 (accurate)
Membership: Individual $35; Family $50; Supporting $60; Contributor $100; Donor $250; Sponsor $500; Benefactor $1,000. Corporate $250-$1,000.

FITCHBURG HISTORICAL SOCIETY, 781 Main St., Fitchburg, MA 01420-3116. Mailing Address: P.O. Box 953, Fitchburg, MA 01420-0009. Tel.: 978-345-1157.
E-mail: fitchburghistoricalsociety@fitchburghistoricalsociety.com
Web Site: www.fitchburghistoricalsociety.org
Founded: 1892.
Congressional District: 4
Key Personnel: Dir., Shirley Wagner; Chm. & Pres. (V), Dan Mylott; Museum Shop Mgr., Kate Wells.
Personnel Profile: Part-Time Paid 2; Part-Time Volunteers 12.
Governing Authority: society. Tax-exempt: 501(c)(3).
Institution Type/Description: Local History Museum.
Collections: memorabilia of all U.S. wars; local portraits, inventions, maps, artifacts; glass & china; dolls, musical instruments; furniture; archival material of local industries.
Research Fields: genealogy; local history.
Facilities: library; exhibit hall.
Activities: guided tours; permanent & changing exhibits; monthly meetings; lectures.
Publications: quarterly newsletter.
Hours & Admission Prices: Mon.-Tues. 10-4, Wed. 10-6. Historical Society: no charge; donations accepted. Research Library: $10; students & members no charge. Closed holidays. &
Attendance: 2,000 (estimated)
Membership: Individual $30; Family $50; Sustaining & Business Member $100; Business Sponsor $250; Business Patron $500; Business Leader $750.

Framingham

DANFORTH MUSEUM OF ART, (M), 123 Union Ave., Framingham, MA 01702-8291. Tel.: 508-620-0050. Fax: 508-872-5542.
E-mail: dhagan@danforthmuseum.org
Web Site: www.danforthmuseum.org
Founded: 1975.
Congressional District: 5
Key Personnel: Pres. Bd. (V), Robert Martin; Dir., Katherine French; Dir. Finance & Operations, Mary Kiely; Museum School Mgr., Ashley Ocching;

Museum School Registrar, Catherine Sullivan; Dir. Education, Pat Walker; Membership & Mktg. Asst., Chelsea Long; Mktg. & Communications Dir., Debbie Hagan.
Personnel Profile: Full-Time Paid 10; Part-Time Paid 6; Part-Time Volunteers 30; Interns 5.
Governing Authority: nonprofit organization. Supported in part by Framingham State College & local municipal government. Tax-exempt: 501(c)(3).
Institution Type/Description: Art Museum & School.
Collections: paintings; drawings; graphics; sculpture; photographs.
Research Fields: 19th- & 20th-century American art.
Facilities: art library; lecture room. Museum-related items for sale.
Activities: guided tours; lectures; films; gallery talks; concerts; formally organized education programs; docent program; inter-museum loan, permanent, temporary & traveling exhibitions; art on the move activity kits for all grades; family days.
Publications: newsletter; exhibit announcements; exhibition catalogues; Danforth Museum of Art News; school newspaper; course listings and schedule of classes & workshops.
Hours & Admission Prices: Wed.-Thurs. & Sun. 12-5, Fri.-Sat. 10-5. Adults $11, seniors $9, students $8; discounts to AAM, ICOM & NEMA members; children under 17 & members no charge. &
Attendance: 30,000 (estimated)
Membership: Senior Citizens $40; Individual $50; Family $75; Friend $150; Supporter $250; Sponsor $500; Patron $1,000; Collector's Circle $2,500.

FRAMINGHAM HISTORY CENTER, 16 Vernon St., Framingham, MA 01701-4783. Mailing Address: P.O. Box 2032, Framingham, MA 01703-2032. Tel.: 508-872-3780. Fax: 508-872-3780.
E-mail: office@framinghamhistory.org
Web Site: www.framinghamhistory.org
Formerly: Framingham Historical Society & Museum
Founded: 1888.
Congressional District: 5
Key Personnel: C.E.O. & Dir., Anne Murphy; Pres., Sheryl Martin; Cur., Dana Dauterman Ricciardi, Ph.D.; Research, Frederic A. Wallace; Asst., Vanessa Prescott; Museum Shop Mgr., Grace Shwert; Museum Asst., Jane Whiting.
Personnel Profile: Part-Time Paid 3; Part-Time Volunteers 100.
Governing Authority: society. Tax-exempt: 501(c)(3).
Institution Type/Description: American History Museum.
Collections: 17th-20th centuries American objects of material culture, agricultural, domestic & industrial; archives; photographs; oral history; art & decorative art; books; diaries; manuscripts. Historic Buildings: 1837 Old Framingham Academy; 1873 Edgell Memorial library.
Research Fields: American history from colonial days to the present with focus on Framingham; genealogy.
Facilities: 2,500-vol. library of genealogical records & historical archives available for use by appointment.
Activities: guided tours; lectures; formally organized education programs for children; permanent & temporary exhibitions; special events.
Publications: quarterly, newsletter; book, Framingham: An American Town; exhibition catalogue, Zeal for Healing: A Framingham Trait.
Hours & Admission Prices: Academy & Edgell Memorial Library: Wed.-Sat. 12-4. No charge; donations accepted.
Attendance: 6,140 (accurate)
Membership: Senior & Student $15; Individual $25; Family $40; Friend $100; Supporter $250; Sustaining & Business $500.

* **GARDEN IN THE WOODS OF THE NEW ENGLAND WILD FLOWER SOCIETY, (M),** 180 Hemenway Rd., Framingham, MA 01701-2699. Tel.: 508-877-7630, ext. 0. Fax: 508-877-3658. TTY: 508-877-6553.
E-mail: information@newenglandwild.org
Web Site: www.newenglandwild.org
Founded: 1900.
Congressional District: 4
Key Personnel: Exec. Dir., Debbi Edelstein; Chm. (V), Deirdre C. Menoyo; Dir. Education, Bonnie Drexler; Dir. Horticulture, Mark Richardson; Dir. Conservation, William Brumback; Dir. Philanthropy, Tracey Willmott; Retail Mgr., Noni Macon.
Personnel Profile: Full-Time Paid 26; Part-Time Paid 28; Part-Time Volunteers 1,000; Interns 7.
Governing Authority: nonprofit organization. Parent Institution: New England Wild Flower Society. Tax-exempt: 501(c)(3).
Institution Type/Description: Botanical Garden.
Collections: botanical garden & woodland sanctuary with more than 1,000 species of native American plants; rare & endangered species; over 50,000 slides; digital images; books.
Research Fields: propagation; critical habitats; threatened & endangered native plant species; invasive plant management.
Facilities: 4,700-vol. library specializing in botany, gardening & natural history; 3,500 specimen herbarium; 45 acre botanic garden; education center; visitor center. Books, branded merchandise & plants for sale.
Activities: field trips; lectures; formally organized educational programs; Garden in the Woods tours for adults; nature walks for children; workshops & symposium on native plant conservation, horticulture, & plant identification; nature programs for children & school groups; classes; native plant education program.
Publications: magazine, New England Wild; annual, Conservation Notes; annual, New England Wild Flower Society Guide to Wild Flowers; Guide to Native Trees, Shrubs & Vines; biannual, Learn & Grow.
Hours & Admission Prices: April 15-July 3 Tues.-Wed., Sat.-Sun. & holiday Mon. 9-5, Thurs. 8-8; July 4-Oct. Tues.-Sun. & holiday Mon. 9-5. Guided Walking Tours: Tues.-Fri. 10 am, Sat.-Sun. 2 pm; tours by appointment. Adults $10, senior citizens 65 & over $7, youth 3-17 $5; members & children under 3 no charge. &
Attendance: 20,000 (estimated)
Membership: Associate $40; Individual $50; Individual Plus $60; Family & Friends $75; Contributor $115; Supporter $250; Sustainer $500.

Gardner

THE GARDNER MUSEUM, INC., 28 Pearl St., Gardner, MA 01440-2308. Tel.: 978-632-3277.
E-mail: info@thegardnermuseum.com
Web Site: www.gardnermuseuminc.com
Founded: 1978.
Congressional District: 4
Key Personnel: Pres., Scott Huntoon; Pres. Elect, Michael Gerry; Coord., Sally Sennott; Treas., Robert Venning; Asst. Treas., Thomas Mailloux; Historian, Robert Treptow; Recording Sec., Janet Stankaitis; Corresponding Sec., Jan Korhonen.
Personnel Profile: Part-Time Paid 1; Part-Time Volunteers 50.
Governing Authority: nonprofit organization. Tax-exempt: 501(c)(3).
Institution Type/Description: Local History Museum: housed in 1886 Richardson Romanesque brick building.
Collections: history & development of Gardner; silver workshop includes equipment, tools & finished pieces by Gardner craftsmen.
Research Fields: local history; local cultural interests.
Activities: guided tours; lectures; films; changing exhibits.
Publications: quarterly newsletter.
Hours & Admission Prices: March-Dec. Wed.-Sun. 1-4. Adults $3; AASLH members & members no charge. Closed occasional holidays. &
Attendance: 2,500 (estimated)
Membership: Student $10; Single $20; Family $30; Friend $35; Sponsor $50; Patron $100 & up; Corporate $200; Supporter $250; Contributor $500; Benefactor $1,000.

Georgetown

BROCKLEBANK MUSEUM, 108 E. Main St., Georgetown, MA 01833-2104. Mailing Address: Georgetown Historical Society, P.O. Box 376, Georgetown, MA 01833. Tel.: 978-352-8526.
E-mail: info@georgetownhistoricalsociety.com
Web Site: www.georgetownhistoricalsociety.com
Key Personnel: Pres. (V), Frederic Detwiller; Cur., Stephen Keene
Institution Type/Description: Historic House: built in the late 1600s.
Collections: local history; railroad memorabilia; Baldpate Inn china; shoe industry; period furnishings.
Hours & Admission Prices: late June to Columbus Day Sun. 2-5; other times by appointment. Adults $5, seniors 65 & over and students $3.

Gloucester

* **BEAUPORT, SLEEPER-MCCANN HOUSE, (M),** 75 Eastern Point Blvd., Gloucester, MA 01930-4433. Mailing Address: 141 Cambridge St., Boston, MA 02114-2702. Tel.: 978-283-0800; 617-227-3956 (Historic New England). Facebook: Beauport House.
Web Site: www.historicnewengland.org
Founded: 1942.
Congressional District: 6
Key Personnel: Pres. & C.E.O., Carl R. Nold; Site Mgr., Pilar Garro.
Governing Authority: society; nonprofit organization. Parent Institution: Historic New England, 141 Cambridge St., Boston, MA 02114. Tel.: 617-227-3956. Tax-exempt: 501(c)(3).
Institution Type/Description: Historic House: 1907-34 40-room designer showcase overlooking Gloucester Harbor, built by interior designer Henry Davis Sleeper.
Collections: American, European & Oriental decorative arts.
Activities: guided tours; lectures; special events; evening concerts; teas.

Publications: Beauport, magazine, Historic New England.
Hours & Admission Prices: Tours: June-Oct. 15 Tues.-Sat. 10-5, (last tour at 4). Adults $15; discounts to seniors, groups, AAM, ICOM, AAA, WGBH members.
Attendance: 5,914 (accurate)
Membership: National $35; Individual $45; Household $55; Garden & Landscape $75; Contributing and Library & School $100; Young Friends of Historic New England $100-$1,500; Friends of the Library and Archives $125; Historic Homeowner $200; Business & Ogden Codman Design Group $250; Appleton Circle $2,500 & up.

CAPE ANN HISTORICAL ASSOCIATION DBA CAPE ANN MUSEUM, (M), 27 Pleasant St., Gloucester, MA 01930-5909. Tel.: 978-283-0455. Fax: 978-283-4141.
E-mail: rondafaloon@capeannmuseum.org
Web Site: www.capeannmuseum.org
Founded: 1873.
Congressional District: 6
Key Personnel: Dir., Ronda Faloon; Pres. (V), John Cunningham; Museum Shop Mgr., Jeanette Smith.
Personnel Profile: Full-Time Paid 10; Part-Time Paid 7; Part-Time Volunteers 110; Interns 3.
Governing Authority: society. Tax-exempt: 501(c)(3).
Institution Type/Description: History, Art & Maritime Museum: c.1800 Federal House.
Collections: paintings & drawings by Fitz Henry (Hugh) Lane; paintings & sculpture by other Cape Ann artists; fine period furniture; silver; china; textiles; photographs; fisheries artifacts; 19th-century Gloucester waterfront diorama with schooner models; historic vessels; granite quarrying exhibit. Historic Houses: 1710 house; 1804 house.
Research Fields: 19th-20th century American art; maritime history; granite industry.
Facilities: 4,000-vol. library of local history, genealogy & the fishing industry available on premises; photo archives; artists' archives; 175-seat auditorium.
Activities: guided tours; lectures; concerts; permanent & temporary exhibitions.
Hours & Admission Prices: March-Jan. Tues.-Sat. 10-5, Sun. 1-4; group tours by appointment only. Adults $10, senior citizens, students, & Cape Ann residents $8; members & children under 6 no charge. Closed major holidays. &
Attendance: 20,000 (accurate)
Membership: Individual $45; Individual Plus $75; Contributor $100; Sponsor $250; Benefactor $500; Fitz Henry Lane Society $1,000, $2,500, $5,000.

HAMMOND CASTLE MUSEUM, 80 Hesperus Ave., Gloucester, MA 01930-5299. Tel.: 978-283-7673 & 2080. Fax: 978-283-1643.
Web Site: www.hammondcastle.org
Founded: 1930.
Congressional District: 6
Key Personnel: Acting Dir. & Cur. Education, John W. Pettibone; Pres. (V), Craig Lentz.
Governing Authority: nonprofit corporation. Owned by the Hammond Museum, Inc. Tax-exempt.
Institution Type/Description: Historic Building & Museum: castle built in 1928 by inventor John Hays Hammond, Jr., in the style of a combination of Roman, Medieval & Renaissance periods.
Collections: furniture; artifacts; tapestries; stained glass; paintings; icons; early American period furnishings; pipe organ containing over 8,600 pipes.
Facilities: Museum-related items for sale.
Activities: lecture series on Medieval, Gothic & Renaissance art & architecture; early music concert series; organ concert series & other classical concerts; special exhibitions; children workshops; guided tours.
Publications: guide book; gallery catalog; exhibition catalog; Hammond Biography.
Hours & Admission Prices: June-Sept. Tues.-Sun. 10-6; Labor Day to Oct. Sat.-Sun. 10-4; groups by appointment. Adults $10, senior citizens & college students w/ID $8, children 6-12 $6; children under 6 no charge.
Attendance: 69,000
Membership: Individual $20; Family & Dual $30; Contributing $50; Patron $100; Benefactor $250.

MARITIME GLOUCESTER, (M), 23 Harbor Loop, Gloucester, MA 01930-5004. Tel.: 978-281-0470. Fax: 978-281-0327.
E-mail: info@maritimegloucester.org
Web Site: www.maritimegloucester.org
Formerly: Gloucester Maritime Heritage Center
Key Personnel: Dir., Thomas Balf; Education Coord., Mary Kay Taylor.
Personnel Profile: Full-Time Paid 2; Part-Time Paid 8.

Governing Authority: Tax-exempt.
Institution Type/Description: Maritime Museum.
Collections: local maritime history & culture; boatbuilding; sealife.
Activities: hands-on activities.
Hours & Admission Prices: Memorial Day weekend to Oct. daily 10-5. Adults $8, seniors $6, children $4; discounts to families; members & children under 4 no charge.
Membership: Individual $35; Family $50; Corporate $250; Lifetime $1,000.

NORTH SHORE ARTS ASSOCIATION, 11 Pirate's Lane, Gloucester, MA 01930-3810. Tel.: 978-283-1857.
E-mail: arts@nsarts.org
Web Site: www.nsarts.org
Founded: 1922.
Key Personnel: Pres. (V), George Martin; Dir., Suzanne Gilbert.
Personnel Profile: Full-Time Paid 3; Part-Time Paid 3; Part-Time Volunteers 50; Interns 1.
Governing Authority: society; nonprofit. Tax-exempt.
Institution Type/Description: Art Gallery: housed in c.1870 barn.
Collections: contemporary regional & U.S. artists.
Research Fields: members since 1922.
Facilities: picnic area.
Activities: lectures; workshops; classes; art activities. Museum Sponsors: Annual Art Auction.
Publications: triannual newsletter, Horizon Line; annual newsletter, NSAA News.
Hours & Admission Prices: May-Oct. Mon.-Sat. 10-5, Sun. 12-5. No charge; donations accepted. &
Attendance: 12,000 (estimated)
Membership: Junior $10; Associate & Patron $40; Artist $75; Corporate Donor $250-$499; Corporate Supporter $500-$999; Corporate Sustainer $1,000-$2,499; Corporate Founder $2,500.

THE SARGENT HOUSE MUSEUM, 49 Middle St., Gloucester, MA 01930-5736. Tel.: 978-281-2432. Fax: 978-281-2432.
E-mail: sargenthouse@verizon.net
Web Site: sargenthouse.org
Founded: 1919.
Key Personnel: Dir., Martha Oaks; Pres. (V), Margaret Flavin; Treas., Amanda Hurd.
Personnel Profile: Part-Time Paid 3; Part-Time Volunteers 20.
Governing Authority: private; nonprofit organization. Tax-exempt: 501(c)(3).
Institution Type/Description: Historic House: 1782 Georgian style house built for early American philosopher, writer & activist Judith Sargent Murray (1751-1820).
Collections: early New England furniture; decorative & fine arts; special collection of works by 20th century artist John Singer Sargent (1856-1925).
Activities: concerts; formal education programs; guided tours; lectures; temporary exhibitions. Annual Events: Year-end holiday party & narrated walking tours.
Publications: biannual newsletter, The Dolphin.
Hours & Admission Prices: Memorial Day to Columbus Day Fri.-Mon. 12-4. Adults $7.50, senior citizens over 65 $5; students, children under 12 & members no charge.
Attendance: 2,500 (accurate)
Membership: Individual $25; Family $40; Donor $150; Life $1500; Corporate $5,000.

Grafton

WILLARD HOUSE AND CLOCK MUSEUM, INC., (M), 11 Willard St., Grafton, MA 01536-2011. Tel.: 508-839-3500. Fax: 508-839-3599.
E-mail: patrick@willardhouse.org
Web Site: www.willardhouse.org
Founded: 1971.
Congressional District: 3
Key Personnel: Dir., Patrick Keenan; Pres., Richard Currier.
Personnel Profile: Full-Time Paid 1; Part-Time Paid 6; Part-Time Volunteers 30.
Governing Authority: society. Tax-exempt: 501(c)(3).
Institution Type/Description: Horological Museum: housed in 1718 Willard Homestead and 1766 Clock Shop.
Collections: over 80 Willard clocks by Benjamin Willard, Simon Willard, Ephraim Willard, Aaron Willard, Benjamin Franklin Willard, Simon Willard Jr., Aaron Willard Jr. & Zabdiel Willard; family portraits, letters, patent rights; tools; 18th century furniture & furnishings; preservation project.
Research Fields: Willard clocks & Willard genealogy.
Activities: guided tours; lectures; formally organized education programs for

children; permanent & temporary exhibitions; annual antique show & annual Robinson lecture.
Publications: biannual member newsletter; exhibit catalogues.
Hours & Admission Prices: Jan.-March Fri.-Sat. 10-4, Sun. 1-4; April-Dec. Wed.-Sat. 10-4, Sun. 1-4. Adults $7, senior citizens $6, children $3; discounts to AAM & AAA members; Willard House members no charge. Closed all major holidays.
Attendance: 2,595 (accurate)
Membership: Individual $25; Family $40; Apprentice $55; Journeyman $100; Ephrain Willard Society $250; Benjamin Willard Society $500; Aaron Willard Society $1,000; Simon Willard Society $5,000; Robinson Society (Life) $10,000.

Granville

NOBLE & COOLEY CENTER FOR HISTORIC PRESERVA-TION, (M), 42 Water St., Granville, MA 01034. Mailing Address: P.O. Box 325, Granville, MA 01034-0325. Tel.: 413-357-8814. Fax: 413-357-6314.
E-mail: ncchp.org@gmail.com
Web Site: ncchp.org
Founded: 2005.
Congressional District: 1
Key Personnel: Dir., Liz Smith; Pres. (V), Matt Jones; Museum Shop Mgr., Carol Jones.
Personnel Profile: Part-Time Volunteers 12.
Governing Authority: Tax-exempt.
Institution Type/Description: Preservation Society: housed in the Noble & Cooley drum factory.
Collections: company history, machinery & equipment; photographs; toy & military drums.
Facilities: Museum-related items for sale.
Activities: special events.
Hours & Admission Prices: See website for hours. Adults $5; discounts to AAM members; members no charge.
Membership: Drummer Boy/Girl $35; Marching Band $55; Garage Band $100; Jazz Band $250; Rock Band $500; Concert Percussionist $1,000.

Greenfield

THE ASSOCIATION FOR GRAVESTONE STUDIES, Greenfield Corporate Center, 101 Munson St., Ste. 108, Greenfield, MA 01301-9675. Tel.: 413-772-0836. Fax: 413-772-0836 (call first).
E-mail: info@gravestonestudies.org
Web Site: www.gravestonestudies.org
Founded: 1977.
Key Personnel: Pres., Ian Brown; Admin., Andrea Carlin.
Personnel Profile: Part-Time Paid 2; Part-Time Volunteers 50.
Governing Authority: private; nonprofit. Tax-exempt.
Institution Type/Description: Preservation Society.
Collections: photographs; personal artifacts.
Research Fields: gravestone studies.
Facilities: library of books, pamphlets, photographs, & files pertaining to gravestone studies.
Activities: annual conference.
Publications: quarterly magazine, AGS Quarterly; annual journal, Markers; monthly e-newsletter.
Hours & Admission Prices: Office: Tues.-Thurs. 9-3. No charge. Closed holidays. &
Membership: Student $20; Senior $40; Individual $50; Institutional $60; Supporting $80; Sustaining $150; Contributing $250; Life $1,000.

HISTORICAL SOCIETY OF GREENFIELD, 43 Church St., Greenfield, MA 01302. Mailing Address: P.O. Box 415, Greenfield, MA 01302-0415. Tel.: 413-774-3663 & 5363.
E-mail: hsg1907@yahoo.com
Founded: 1907.
Congressional District: 1
Key Personnel: Pres. (V), Tim Blagg.
Governing Authority: society; board of directors. Tax-exempt.
Institution Type/Description: General Museum: housed in 1852 Museum Building.
Collections: photographs; books; local antiques & memorabilia; maps; paintings; artifacts.
Research Fields: local cultural history; genealogies.
Facilities: 600-vol. library of books & pamphlets pertaining to local & county history & genealogies available for research.
Activities: tours & local history lectures for students & other groups. Museum Sponsors: Open Houses & special events in July & August.

Publications: 1926, Historic Greenfield; 1953, Pictorial History of Greenfield; 1975, A New Pictorial History of Greenfield, Massachusetts.
Hours & Admission Prices: Open year-round by appointment; special events & open houses scheduled throughout the year; call for information. No charge.
Attendance: 600 (estimated)
Membership: Senior $12; Individual $15; Couple & Family $20; Sustaining & Business $50; Patron $300.

Groton

GROTON HISTORICAL SOCIETY, 172 Main St., Groton, MA 01450-1238. Mailing Address: P.O. Box 202, Groton, MA 01450-0202. Tel.: 978-448-0092.
E-mail: info@grotonhistoricalsociety.org
Web Site: www.grotonhistoricalsociety.org
Founded: 1894.
Congressional District: 5
Key Personnel: Pres. (V), John H. Ott; Cur., Bobbie Spigelman.
Personnel Profile: Part-Time Volunteers 14.
Governing Authority: society; nonprofit organization. Tax-exempt: 501(c)(3).
Institution Type/Description: Historic House: housed in the Governor Boutwell House; built in 1851.
Collections: household furnishings; Revolutionary & Civil War items; local history; textiles; Groton historical artifacts & papers; Middlesex County materials.
Activities: lectures; tours of historical sites.
Publications: quarterly newsletter, Then and Now; George S. Boutwell: Human Rights Advocate; Groton Houses.
Hours & Admission Prices: June-Sept. Sun. 2-4. No charge; donations accepted.
Attendance: 400 (estimated)
Membership: Individual & Family $35; GHS Sponsor $75; Corporate $250; Supporting $500; Sustaining $1,000.

Hadley

HADLEY FARM MUSEUM, 224 River Dr., Hadley, MA 01035-9641. Tel.: 413-584-7459.
Founded: 1930.
Key Personnel: Pres., Thomas West; Sec., Mrs. Glenn Clark.
Governing Authority: association. Tax-exempt: 501(c)(3).
Institution Type/Description: Farm Museum: housed in 1782 barn.
Collections: crafts; farm equipment; plows; winnowing machines; corn shellers; seeders; stage coach; sleighs.
Activities: guided tours.
Hours & Admission Prices: May-Oct. Sat.-Sun. 2-4; other times by appointment. Adults $5.
Attendance: 1,000 (estimated)
Membership: Individual $25; Family $40.

PORTER-PHELPS-HUNTINGTON FOUNDATION, INC., 130 River Dr., Hadley, MA 01035-9782. Tel.: 413-584-4699.
Web Site: www.pphmuseum.org
Founded: 1948.
Congressional District: 2
Key Personnel: Dir. & C.E.O., Susan J. Lisk; Pres., Tom Harris; Vice Pres. Bd., Dan Huntington Fenn, Jr.; Treas., Sidney Poritz; Clerk, Craig Malone.
Personnel Profile: Part-Time Paid 3; Part-Time Volunteers 15; Interns 1.
Volunteer Hours: 400
Operating Expenses: 64,574
Operating Income: 75,935
Governing Authority: not-for-profit organization. Tax-exempt: 501(c)(3).
Institution Type/Description: Historic House: 1752 Georgian home & grounds.
Collections: furniture; paintings; decorative arts; archives; photographs; clothing; household items belonging to the Porter, Phelps & Huntington families.
Research Fields: women's history; family history; clothing; genealogy.
Facilities: 500-vol. library of reference & history works available to the public; 100 linear ft. Porter-Phelps-Huntington family papers located at Amherst College archives.
Activities: concerts; guided tours; lectures.
Publications: newsletter, Forty Acres.
Hours & Admission Prices: mid-Oct. to mid-May Sat.-Wed. 1-4:30. Adults $5, children $1.
Attendance: 2,251 (accurate)
Membership: Levels: $25; $50; $100; $500; $1,000.

Hardwick

HARDWICK HISTORICAL SOCIETY, On Hardwick Common, Hardwick, MA 01037. Mailing Address: P.O. Box 492, Hardwick, MA 01037-0492. Tel.: 413-967-4002.
Web Site: townofhardwick.org
Founded: 1959.
Key Personnel: Pres., Randall Noble; Cur., Emily Bancroft; Sec., Anne Barnes.
Governing Authority: society. Tax-exempt.
Institution Type/Description: Local History Museum: housed in 1840 Old Brick School House.
Collections: period furniture and utensils; doll furniture; old books & papers; Indian artifacts; china; costumes; tools; local records; family pictures; memorabilia; historical portraits.
Research Fields: genealogy; history.
Facilities: 150-vol. library of local history books available on premises.
Activities: guided tours; lectures; films; gallery talks; special events; educational programs. Annual Events: Open House; Village Fair.
Publications: annual reports; annual newsletter; Village Tour brochures.
Hours & Admission Prices: By appointment. No charge; donations accepted.
Attendance: 320 (estimated)
Membership: Yearly $10; Life $100.

Harvard

FRUITLANDS MUSEUMS, (M), 102 Prospect Hill Rd., Harvard, MA 01451. Tel.: 978-456-3924. Fax: 978-456-8078.
Web Site: www.fruitlands.org
Founded: 1914.
Congressional District: 2
Key Personnel: Dir, Wyona Lynch-McWhite; Pres., Marie LeBlanc.
Personnel Profile: Full-Time Paid 5; Part-Time Paid 24; Part-Time Volunteers 10; Interns 4.
Governing Authority: nonprofit. Tax-exempt: 501(c)(3).
Institution Type/Description: History Museum.
Collections: American paintings (Hudson River School); decorative arts; ethnology; folklore; Indian artifacts; history; memorabilia of Thoreau, Emerson, Margaret Fuller & Lane. Historic Houses: c.1825 Fruitlands farmhouse used by Bronson Alcott & Transcendentalists; 1794 Shaker House; Shaker furnishings, products & material culture. Historic landscapes: Native American hunting ground, colonial garrison house & farm site.
Research Fields: paintings; history; shakers; transcendentalism; cultural landscape.
Facilities: 10,000-vol. library available by appointment only; luncheon facilities; picnic area; rental facilities; nature trails. Museum-related items for sale.
Activities: films; permanent & temporary exhibitions; concerts; lectures; field trips; nature walks; archaeological sites; natural history interpretation. Museum Sponsors: outdoor music series in summer.
Publications: booklets; quarterly newsletter, Under the Mulberry Tree; A New Eden; History's Daughter.
Hours & Admission Prices: Museums: May 14-Oct. Mon.-Fri. 11-4, Sat.-Sun. & holidays 11-5. Tours: by reservation. Grounds: daily 10-5. Library: by appointment only. Adults $10, seniors 60 and over & students with college ID $8, children 5-17 $4; discounts to AAM, NEMA & AAA members; members no charge. &
Attendance: 30,000 (accurate)
Membership: Senior & Student $30; Individual $50; Family $70; Patron $250.

Harwich

BROOKS ACADEMY MUSEUM, 80 Parallel St., Harwich, MA 02645-2716. Tel.: 508-432-8089.
E-mail: harwichhistoricalsociety@verizon.net
Web Site: www.harwichhistoricalsociety.org
Founded: 1954
Congressional District: 10
Key Personnel: Dir., Desiree Mobed; Pres., Jane Martin; Treas., Brian Michaelan; Recording Sec., Jane Michaels; Museum Shop Mgr., Jane Chase.
Personnel Profile: Part-Time Paid 1; Part-Time Volunteers 75.
Governing Authority: society; nonprofit. Parent Institution: Harwich Historical Society. Tax-exempt: 501(c)(3).
Institution Type/Description: Historical Society Museum: housed in c.1844 building that served as a private secondary school.
Collections: Harwich & Cape Cod history; photographs; marine material; glass collection; genealogical material; early cranberry industry equipment; deeds; maps; Indian arrow heads; hand tools; bound copies of 1872-1958, Harwich Independent.

Major Exhibits: Wychmere Harbor, 6/14-10/14; Harwick & World War I, 6/14-10/14; Charles D. Cahoon, 6/14-12/14.
Research Fields: Maritime research; preliminary work on Brooks family papers; Harwich history; Caleb Chase of Chase & Sanborn; Brooks Academy graduates; outerwear & underwear; white textiles & garments, 1860-1920.
Facilities: 100-vol. library; 50-seat auditorium; 5,000 sq. ft. exhibit space. Museum-related items for sale.
Activities: lectures; country auction; temporary exhibitions; docent program; formal education programs for children; guided tours.
Publications: quarterly newsletter, The Powderhouse Quarterly; book, Vital Records of Town of Harwich.
Hours & Admission Prices: late June to mid-Oct. Thurs.-Sat. 1-4. Adults $3; members no charge. Closed Independence Day. &
Attendance: 2,000 (accurate)
Membership: Junior $5; Regular $18; Household $25; Life $250.

BROOKS FREE LIBRARY, 739 Main St., Harwich, MA 02645-2752. Tel.: 508-430-7562. Fax: 508-430-7564.
E-mail: bfl_mail@clamsnet.org
Web Site: www.brooksfreelibrary.org
Founded: 1880.
Congressional District: 12
Key Personnel: Dir., Virginia A. Hewitt; Reference Librarian, Jennifer Pickett; Public Svcs. Librarian, Suzanne Martell; Youth Svcs. Librarian, Ann Carpenter; Principal Clerk, Mary Jo Metzger; Sr. Library Technician, Phil Inman; Sr. Library Technician, Pam Paine; Sr. Library Technician, Joanne Clingan.
Personnel Profile: Full-Time Paid 6; Part-Time Paid 5; Part-Time Volunteers 40.
Governing Authority: municipal. Tax-exempt.
Institution Type/Description: Library.
Collections: Rogers' statuary groups.
Facilities: 54,000-vol. library of books available for inter-library loan & for use on the premises; reference department; children's room; public access computers.
Hours & Admission Prices: Tues.-Thurs. 10-7, Fri.-Sat. 10-4. Local History Room: call for hours. No charge. Closed national & state holidays. &
Attendance: 82,569 (accurate)

Haverhill

HAVERHILL HISTORICAL SOCIETY, 240 Water St., Haverhill, MA 01830-6433. Tel.: 978-374-4626. Fax: 978-521-9176.
Web Site: haverhillhistory.org
Founded: 1897.
Congressional District: 6
Key Personnel: Pres. (V), Jay Cleary; Cur., Jan Williams; Program Coord., Stacey Fraser-deHaan; Weekend Supvr., Thomas Spitalere; Bookkeeper, Lisa O'Hearn.
Personnel Profile: Part-Time Paid 4; Part-Time Volunteers 8.
Governing Authority: nonprofit organization. Tax-exempt.
Institution Type/Description: History Museum.
Collections: 17th to 19th-century furniture, paintings & glass; Haverhill area history; lower Merrimack Valley archaeological artifacts. Historic Buildings: c.1850 Daniel A. Hunkins Shoe Shop; c.1710 John Ward House; 1814 Duncan House.
Research Fields: Lower Merrimack Valley; archeological collections.
Facilities: lecture hall. Museum-related items for sale.
Activities: guided tours; permanent & temporary exhibits; educational programs for children; history camp; lecture series; preschool program; after school programs.
Publications: quarterly newsletter.
Hours & Admission Prices: Summer: Tues.-Sun. 10-5; Winter: Tues.-Sat. 10-5. Adults $5, children $4; discounts to AAM members; members no charge. Closed major holidays. &
Attendance: 3,849 (accurate)
Membership: Individual $25; Family $40; Buttonwoods Benefactor $100.

Hingham

BARE COVE FIRE MUSEUM, 45 Bare Cove Park Dr., Hingham, MA 02043. Tel.: 781-749-0028.
E-mail: bcfm.hingham@gmail.com
Web Site: barecovefiremuseum.org
Founded: 1976.
Key Personnel: Pres. & Chief, David Clark; Vice Pres., Jack Crandall; Treas., Lisa Andre; Dir., Don Lincoln; Dir., Jaclyn Howard; Dir., Russell Clark; Clerk & Archivist, Geri Duff.
Personnel Profile: Full-Time Volunteers 2; Part-Time Volunteers 90.

Governing Authority: private; nonprofit organization. Tax-exempt.
Institution Type/Description: Fire-Fighting Museum.
Collections: hand & horse motored period fire engines; period artifacts; equipment; photos; records; 1920 era maxim fire engines; 1935 Ahrens-Fox pumping engine; restored 1850 era hand operated fire engine.
Facilities: library of records of past fire companies available for research; 300 sq. ft. exhibit space.
Activities: films; guided tours; lectures. Museum Sponsors: Open House.
Publications: quarterly seasonal newsletter, The Speaking Trumpet.
Hours & Admission Prices: Wed. 7:30pm-9:30pm; other times by appointment. No charge; donations accepted. Closed Christmas Eve & Day. &
Attendance: 300 (estimated)
Membership: Family $15 (w/Active); Active $35; Life $500.

HINGHAM HISTORICAL SOCIETY, 34 Main St., Hingham, MA 02043-2523. Mailing Address: P.O. Box 434, Hingham, MA 02043-0434. Tel.: 781-749-7721. Fax: 781-749-0091.
E-mail: director@hinghamhistorical.org
Web Site: www.hinghamhistorical.org
Formerly: Old Ordinary House Museum
Founded: 1914.
Congressional District: 10
Key Personnel: Dir., Suzanne Buchanan; Pres., Michael Studley; Museum Shop Mgr., Susan Achille.
Personnel Profile: Full-Time Paid 1; Part-Time Paid 1; Part-Time Volunteers 100; Interns 1.
Governing Authority: society.
Institution Type/Description: Historical Society.
Collections: architecture; 18th & 19th century furnishings; art; decorative arts; archival collections; photographs; toys.
Research Fields: local history & American Decorative Arts.
Facilities: 500-vol. library of history & genealogy books available for use by appointment.
Activities: guided tours; permanent & temporary exhibitions. Annual Event: House Tour in June.
Publications: Hingham Old & New; Two Hundred Years in South Hingham; By the Wayside; Hingham Colonial Industries; Profile of Hingham: News & Notes Out of the Ordinary; cookbook, Out of the Ordinary; Not All Is Changed.
Hours & Admission Prices: mid-June to early Sept. Tues.-Sat. 1:30-4:30; other times by appointment. Tours: 1:30, 2:30 & 3:30. Adults $5, children under 12 $3; members no charge.
Attendance: 3,000 (estimated)
Membership: Family $40; Sustaining Family $100; Patron Family $250; Sara Derby Society $500; Lincoln Society $1,000.

Holyoke

CHILDREN'S MUSEUM AT HOLYOKE, INC., 444 Dwight St., Holyoke, MA 01040-5842. Tel.: 413-536-7048. Fax: 413-533-2999.
E-mail: skelley@childrensmuseumholyoke.org
Web Site: www.childrensmuseumholyoke.org/
Founded: 1981.
Congressional District: 1
Key Personnel: C.E.O. & Pres. (V), Barry Waite; Exec. Dir., Susan Kelley; Visitor & Family Svcs. Coord., Diane West; Museum Shop Mgr., Margaret Boulais; Bookkeeper, Kathleen McCreary.
Personnel Profile: Full-Time Paid 1; Part-Time Paid 10; Part-Time Volunteers 2.
Governing Authority: nonprofit organization. Tax-exempt: 501(c)(3).
Institution Type/Description: Children's Museum.
Collections: hands-on exhibits including two-story curvy climber, waterworks area, exploration room, paperworks, tot lot, cityscape, TV studio; bubble works.
Research Fields: early childhood education.
Facilities: 16,500 sq. ft. exhibit & program space; 100-seat theater. Gift items for sale.
Activities: annual performance series; science & art workshops; preschool & after school programs; cultural celebrations; grade level appropriate & staff-facilitated elementary school group visits.
Hours & Admission Prices: Wed.-Sat. 10-4, Sun. 12-4. Adults & children $6, senior citizens $3; discounts to AYM members & groups; children under one & members no charge. Closed New Year's Day; Easter; Memorial Day; Independence Day; Labor Day; Thanksgiving; Christmas. &
Attendance: 40,000 (accurate)
Membership: Grandparent (including grandchildren) $30; You & Me $50; Family $75; Family Plus $100; Supporting $150; Sustaining $500; Lifetime $1,000.

HOLYOKE HERITAGE STATE PARK, 221 Appleton St., Holyoke, MA 01040-5714. Tel.: 413-534-1723.
E-mail: mass.parks@state.ma.us
Web Site: www.mass.gov/dcr/parks/central/hhsp.htm
Key Personnel: Supvr., Charlie Lotspeich
Institution Type/Description: Park Museum.
Collections: local history & culture; paper manufacturing; Holyoke's industrial history.
Hours & Admission Prices: Visitor's Center: Tues.-Sun. 10-4. No charge.

VOLLEYBALL HALL OF FAME, 444 Dwight St., Holyoke, MA 01040-5842. Tel.: 413-536-0926. Fax: 413-539-6673.
E-mail: info@volleyhall.org
Web Site: www.volleyhall.org
Key Personnel: Pres. (V), David C. Casey.
Governing Authority: Tax-exempt: 501(c)(3).
Institution Type/Description: Sports Museum.
Collections: volleyball memorabilia.
Hours & Admission Prices: Tues.-Sun. 12-4:30.

WISTARIAHURST MUSEUM, 238 Cabot St., Holyoke, MA 01040-3904. Tel.: 413-322-5660. Fax: 413-534-2344.
E-mail: boissellem@ci.holyoke.ma.us
Web Site: www.wistariahurst.org
Founded: 1959.
Congressional District: 1
Key Personnel: Dir., Melissa Boisselle; Cur., Penni Martorell; Events, Marjorie Latham.
Personnel Profile: Full-Time Paid 2; Part-Time Paid 5; Part-Time Volunteers 100; Interns 5.
Governing Authority: municipal. Parent Institution: city of Holyoke. Tax-exempt.
Institution Type/Description: Historic House: built in c.1868 26-room mansion & carriage house; former home of local silk manufacturer William Skinner with two additions made in 1914 & 1927.
Collections: furnishings; parquet flooring; leather wall coverings. Historic Structure: c.1874 carriage house; period furniture; costumes; archival materials; distinctive interior architectural detailing.
Major Exhibits: Echoes of Industry: The Death & Rebirth of Holyoke's Mills, 1/14-2/14; Deborah Baronas: Work & Culture, 3/14/4/14; A Genius for Place: American Landcapes of the Country Place era (T), 5/14-6/14.
Facilities: 100-seat music room.
Activities: guided tours; concerts; permanent, temporary & traveling exhibitions; school loan & outreach services; volunteer program; group & individual tours by appointment only; garden restoration.
Publications: Wistariahurst Museum Newsletter.
Hours & Admission Prices: Museum: Sat.-Mon. 12-4. Adults $7, seniors & students $5; discounts to MA Teachers Association & AAA members. Research Archives: Mon. 9-7, Thurs. 9-1; other times by appointment. Closed New Year's Day; Easter; Christmas. &
Attendance: 14,000 (estimated)
Membership: $20-$500.

Hull

HULL LIFESAVING MUSEUM INC., THE MUSEUM OF BOSTON HARBOR HERITAGE, 1117 Nantasket Ave., Hull, MA 02045-1310. Mailing Address: P.O. Box 221, Hull, MA 02045-0221. Tel.: 781-925-5433. Fax: 781-925-0992.
E-mail: info@hulllifesavingmuseum.org
Web Site: hulllifesavingmuseum.org
Founded: 1978.
Congressional District: 10
Key Personnel: C.E.O., Lory Newmyer; Dir. Maritime & New Program Devel., Edward P. McCabe; Dir. Advancement & Communications, Corinne Leung; Pres. (V), Robert MacIntyre; Treas., Diana Burns.
Personnel Profile: Full-Time Paid 10; Part-Time Paid 2; Part-Time Volunteers 150; Interns 1.
Governing Authority: private; nonprofit organization. Subsidiary Institutions: HLM Maritime Program, Hull, MA; HLM Maritime Program, Seaport District, Boston, MA. Tax-exempt: 501(c)(3).
Institution Type/Description: Maritime Museum: headquarters housed in restored c.1889 Point Allerton United States Life Saving Service Station.
Collections: artifacts; books; prints; clothing; lifesaving vessels & apparatus; photographs; logs; postcards; fleet of traditional small crafts.
Research Fields: history of Massachusetts Humane Society; U.S. lifesaving service; early United States Coast Guard; history & legend of Boston Harbor, it's islands, lighthouses & forts; maritime history of Hull.
Facilities: library. Museum-related items for sale.

Activities: year-round education, recreation & job-training programs; year-round open-water rowing programs in Boston; arts festivals; concerts; films; formal education programs; guided tours; hobby workshops; lectures; loan, participatory & temporary exhibitions; training programs for professional museum workers; broadcast programs. Museum Sponsors: Head of the Weir Rowing Race; Snow Row; Peddocks Island Row; Northeast Regional Youth Rowing Championship.
Publications: semiannual journal, The Station Log; bimonthly newsletter, The Messenger Line.
Hours & Admission Prices: Point Allerton Life-Saving Station: July-Oct. Mon., Wed. & Fri. 10-1, Sat.-Sun. 10-2; Nov.-June Mon., Wed. & Fri. 10-1; group tours & other times by appointment. Adults $5, senior citizens $3; AAM & ICOM members, children & members no charge. Closed National holidays. &
Attendance: 12,000 (estimated)
Membership: Individual $30; Family $50; Surfman $75; Keeper $100; Superintendent $250; Pt. Allerton Society $500.

Hyannis

THE CAPE COD MARITIME MUSEUM, (M), 135 South St., Hyannis, MA 02601-4014. Mailing Address: P.O. Box 443, Hyannis, MA 02601-0443. Tel.: 508-775-1723. Fax: 508-775-1706. Facebook: Cape Cod Maritime Museum.
E-mail: info@capecodmaritimemuseum.org
Web Site: www.capecodmaritimemuseum.org
Founded: 1998.
Key Personnel: Dir., Chris Galazzi; Pres. (V), David Scudder; Admin., Erin Trainor.
Personnel Profile: Full-Time Paid 3; Part-Time Paid 5; Part-Time Volunteers 15; Interns 1.
Operating Expenses: 330,373
Operating Income: 428,169
Governing Authority: private; nonprofit organization. Tax-exempt: 501(c)(3).
Institution Type/Description: Maritime Museum.
Collections: maritime history & art; books; drawings; models; photographs; 6 traditional small craft.
Major Exhibits: Risky Business, 3/14-4/14; Boatbuilders of Cape Cod, 3/14-12/14.
Research Fields: maritime culture of Cape Cod.
Facilities: 450-vol. library; educational facilities; 1,200 sq. ft. exhibit space. Museum-related items for sale.
Activities: formal education programs for adults & children; hobby workshops; lectures; loan & participatory exhibitions; boat shop with boat building classes. Annual Event: Maritime Festival.
Hours & Admission Prices: March-Dec. Tues.-Sat. 10-4, Sun. 12-4. Adults $5, senior citizens & students $4; discounts to AAM, MTA & NEMA members; members & children under 6 no charge. Closed most holidays. &
Attendance: 5,000 (estimated)
Membership: Individual $35; Family $50; Contributor $75; Supporter $100; Associate $250.

JOHN F. KENNEDY HYANNIS MUSEUM, 397 Main St., Hyannis, MA 02601-3914. Mailing Address: P.O. Box 100, Hyannis, MA 02601. Tel.: 508-790-3077. Fax: 508-775-7131.
Web Site: www.hyannis.com
Founded: 1992.
Key Personnel: C.E.O., Jessica Tinti
Institution Type/Description: History Museum.
Collections: over 80 photographs depicting the life of John F. Kennedy on Cape Code from 1934-1963.
Hours & Admission Prices: mid-Feb. to mid-April & Nov.-Dec. Thurs.-Sat. 10-4, Sun. 12-4; April-Memorial Day & day after Columbus Day to Oct. 31 Mon.-Sat. 10-4, Sun. 12-4; May to Columbus Day Mon.-Sat. 9-5, Sun. 12-5. Adults 18 & over $8, children 10-17 $4; discounts to groups 10 or more; children under 10 no charge. &

ZION UNION HERITAGE MUSEUM, INC., 276 North St., Hyannis, MA 02601-3826. Mailing Address: P.O. Box 2591, Hyannis, MA 02601-7591. Tel.: 508-790-9466.
Web Site: zionunionheritagemuseum.org
Institution Type/Description: History Museum.
Collections: artifacts & memorabilia pertaining to African-Americans & Cape Verdeans on Cape Cod.
Hours & Admission Prices: Feb. to mid-April & Nov.-Dec. Thurs.-Sat. 10-4, Sun. 12-4; mid-April to Oct. Tues.-Sat. 10-4, Sun. 12-4. Adults 18 & over $5, children 10-17 $2; children under 10 no charge.

Indian Orchard

THE TITANIC MUSEUM, 208 Main St., Indian Orchard, MA 01151-1132. Mailing Address: P.O. Box 51053, Indian Orchard, MA 01151-5053. Tel.: 413-543-4770. Fax: 413-583-3633.
E-mail: titanicinfo@titanichistoricalsociety.org
Web Site: www.titanic1.org/museum
Founded: 1963.
Governing Authority: Parent Institution: The Titanic Historical Society.
Institution Type/Description: History Museum.
Collections: artifacts & memorabilia from the Titanic.
Publications: quarterly, The Titanic Commutator.
Hours & Admission Prices: Mon.-Fri. 10-4, Sat. 10-3. Adults $4, children 6-11 $2; discounts to AAM & ICOM members; Titanic Historical Society members & children 5 and under no charge.
Membership: Bronze $50; Silver $60; Gold $100.

Ipswich

GREAT HOUSE AT CASTLE HILL ON THE CRANE ESTATE, 290 Argilla Rd., Ipswich, MA 01938. Mailing Address: 290 Argilla Rd., Ipswich, MA 01938. Tel.: 978-356-4351, ext. 4049.
E-mail: lmarshall@ttor.org
Founded: 1949.
Congressional District: 6
Key Personnel: Cultural Site Admin., Linda Marshall; Cultural Resources Mgr., Susan Hill-Dolan.
Governing Authority: private; nonprofit organization. Parent Institution: The Trustees of Reservations, Beverly, MA. Tax-exempt: 501(c)(3).
Institution Type/Description: Historic House Museum.
Collections: furnishings. Historic Buildings: 1928 estate.
Activities: guided tours; lectures; concerts; educational programs; guided house & landscape tours.
Publications: quarterly members magazine, Special Places.
Hours & Admission Prices: May-Oct. call for hours. Grounds: daily 8am to sunset. Adults $10; discounts to groups of 15 or more; members no charge. &
Attendance: 6,000 (estimated)
Membership: Individual $47; Family $67.

IPSWICH MUSEUM, 54 S. Main St., (Rte. 1A), Ipswich, MA 01938-2322. Tel.: 978-356-2811.
E-mail: admin@ipswichmuseum.org
Web Site: www.ipswichmuseum.org
Founded: 1890.
Congressional District: 6
Key Personnel: Dir., Wendy Evans; Pres., Terri Stephens; Treas., James Grimes.
Personnel Profile: Full-Time Paid 1; Part-Time Paid 3; Part-Time Volunteers 65; Interns 2.
Governing Authority: private; nonprofit organization. Tax-exempt.
Institution Type/Description: History Museum.
Collections: manuscripts; seafaring artifacts; furniture & arts from China; dolls & toys; costumes & textiles; military memorabilia; Arthur Wesley Dow paintings & prints; carriages; sleighs. Historic Houses: c.1677 John Whipple House with lace displays & artifacts of the 17th, 18th & early 19th century; 1800 John Heard House.
Research Fields: local history.
Facilities: library of local history, genealogy & early religious books. Museum-related items for sale.
Activities: walking & guided tours; docent training program; lectures; education programs for students & adults; temporary exhibitions. Annual Events: Harvest Supper; Halloween Party; Holiday Boutique & Party; Pub Night; Women's Luncheon.
Publications: pamphlets of Ipswich History; Ipswich Museum; quarterly newsletter.
Hours & Admission Prices: Feb.-April Sun. 2-4; May-Oct. Wed.-Sun. 12-4; Nov.-Dec. call for hours. One House: adults $7; children 6-12 $3; children under 6 & members no charge. Two Houses: adults $10, children 6-12 $3; children under 6 & members no charge.
Attendance: 2,000 (accurate)
Membership: Individual $40; Dual $60; Sustaining $150; Sponsor $250; Benefactor $500; Arthur Wesley Dow Circle $1,000-$2,500.

THE PAINE HOUSE AT GREENWOOD FARM, A PROPERTY OF THE TRUSTEES OF RESERVATIONS, 47 Jeffrey's Neck Rd., Ipswich, MA 01938. Mailing Address: 290 Argilla Rd., Ipswich, MA 01938. Tel.: 978-356-4351, ext. 4049.
E-mail: lmarshall@ttor.org

Web Site: www.ttor.org
Founded: 1993.
Congressional District: 6
Key Personnel: Cultural Site Admin., Linda Marshall; Cultural Resources Mgr., Susan Hill-Dolan.
Governing Authority: private; nonprofit organization. Parent Institution: The Trustees of Reservations, Beverly, MA. Tax-exempt: 501(c)(3).
Institution Type/Description: Historic House: housed in the former Paine family farmhouse; built in 1694.
Collections: local history & culture; indoor dairy; early Essex County furniture.
Activities: guided house & cultural landscape tours; self-guided walking tours.
Publications: quarterly membership magazine, Special Places.
Hours & Admission Prices: June-Oct. 1st Sat. each month; other times by appointment.
Attendance: 200 (estimated)
Membership: Individual $47; Family $67.

Kingston

MAJOR JOHN BRADFORD HOUSE, 50 Landing Rd., Kingston, MA 02364. Mailing Address: P.O. Box 22, Kingston, MA 02364-0022. Tel.: 781-585-6300.
E-mail: norman@jrvhs.org
Web Site: www.jrvhs.org
Founded: 1921.
Congressional District: 12
Key Personnel: Pres. (V), Norman P. Tucker.
Personnel Profile: Interns 8.
Volunteer Hours: 2,000
Operating Expenses: 104,529
Operating Income: 120,338
Governing Authority: nonprofit organization. Administered by Jones River Village Historical Society. P.O. Box 22, Kingston, MA. Tax-exempt: 501(c)(3).
Institution Type/Description: Historic House: 1714 Major John Bradford home.
Collections: child's room with doll collection; furnishings; crewel coverlets; 19th-century halfhulls & other maritime artifacts; period looms; pottery; 16th-19th century archaeological artifacts; institutional archives; historical documents; photographs; prints; books.
Research Fields: textiles; needlework.
Facilities: Museum-related items for sale.
Activities: guided tours; permanent exhibitions.
Publications: quarterly newsletter; exhibition catalogues.
Hours & Admission Prices: Summer: Sun. 9-11:30, tours by appointment. No charge; donations accepted.
Attendance: 2,500 (estimated)
Membership: Senior $15; Single $25; Family $35; Corporate $125.

Lancaster

FIRST CHURCH OF CHRIST UNITARIAN, 725 Main St., Lancaster, MA 01523. Mailing Address: P.O. Box 66, Lancaster, MA 01523-0066. Tel.: 978-365-2427. Fax: 978-368-0194.
E-mail: office@firstchurchlancasterma.org
Web Site: firstchurchlancasterma.org
Formerly: Fifth Meeting House
Founded: 1653.
Congressional District: 3
Key Personnel: Minister, Rev. Dr. Paul Hull; Church Admin., Karen Plaskon.
Governing Authority: church. Affiliated with First Church of Christ, Lancaster, MA. Tax-exempt.
Institution Type/Description: Religious Museum: housed in 1816 The First Church of Christ.
Collections: Historic Buildings: 1st Church of Christ; built 1816.
Publications: monthly bulletin; book, The Founding Mother; The Bulfinch Church in Lancaster; Precious Predecessors; A New England Village Church.
Hours & Admission Prices: Sun. 10:30; other times by appointment. No charge. &

Lawrence

ESSEX ART CENTER, 56 Island St., Lawrence, MA 01840-1889. Tel.: 978-685-2343. Fax: 978-688-0276.
E-mail: info@essexartcenter.com
Web Site: essexartcenter.com
Key Personnel: Exec. Dir., Leslie Costello
Institution Type/Description: Art Gallery.

Collections: paintings; sculpture.
Activities: special events; classes.
Hours & Admission Prices: Mon.-Thurs. 10-6.

LAWRENCE HERITAGE STATE PARK, 1 Jackson St., Lawrence, MA 01840-1613. Tel.: 978-794-1655. Fax: 978-794-9241. TDD: 978-794-1655.
E-mail: lawrence.heritage@state.ma.us
Web Site: www.mass.gov/dcr/parks/northeast/1whp.htm
Founded: 1986.
Congressional District: 5
Key Personnel: Park Supvr., Michael Mitchell; Chm. (V), Joe Bella; Visitor Svcs. Supvr. & Museum Shop Mgr., Jim Beauchesne; Maintenance, Tom Ceder; Maintenance, Will McDowell.
Personnel Profile: Full-Time Paid 4; Part-Time Paid 4; Part-Time Volunteers 3.
Volunteer Hours: 1,100
Operating Expenses: 319,369
Governing Authority: state. Parent Institution: Massachusetts Dept. of Conservation and Recreation.
Institution Type/Description: State Park Visitor's Center: housed in a former mill boardinghouse; built c.1847.
Collections: photographs; artifacts; models.
Research Fields: industrial & labor history.
Facilities: 8-vol. library of Lawrence History Curriculum Project material; picnic area.
Activities: guided tours; lectures; films; concerts; arts festivals; theater; temporary & permanent exhibitions; organized education programs for children; participatory & temporary exhibitions; school loan service; fishing. Museum Sponsors: Bread & Roses Labor Day Festival.
Hours & Admission Prices: No charge. Closed New Year's Day; Thanksgiving; Christmas. &
Attendance: 100,000 (estimated)
Membership: Friends of Lawrence Heritage State Park: Student & Senior Citizen $5; Adult $10.

LAWRENCE HISTORY CENTER: IMMIGRANT CITY ARCHIVES, 6 Essex St., Lawrence, MA 01840-1710. Tel.: 978-686-9230.
E-mail: director@lawrencehistory.org
Web Site: www.lawrencehistory.org
Formerly: Immigrant City Archives, Historical Society of Lawrence and Its People
Founded: 1978.
Congressional District: 5
Key Personnel: Exec. Dir., Barbara Brown; Pres. (V), Pamela Yameen; Asst. to Dir., Amita Kiley.
Personnel Profile: Full-Time Paid 1; Part-Time Paid 4; Part-Time Volunteers 15.
Governing Authority: nonprofit organization. Parent Institution: Immigrant City Archives, Inc. Tax-exempt: 501(c)(3).
Institution Type/Description: Historical Society Museum: housed in 1883 buildings.
Collections: Essex Company business & planning records; immigration; labor; urban planning; social organizations; churches; oral histories; photography; books; archives. Historic Buildings: 4 brick buildings in cobblestone courtyard.
Research Fields: local history; genealogy; ethnic; labor; women; urban planning; 19th-century technology; immigration.
Activities: lectures: ethnic, genealogy, labor, local history, workshops for teachers; school groups; tours; lectures: genealogy; cultural events.
Publications: booklet, Lawrence, Massachusetts: The Strike of 1912; Immigrant City Archives Cookbook; walking tour booklets; Lawrence, Massachusetts, Vols. I & II.; newsletter, Lawrence History News.
Hours & Admission Prices: Winter: Tues.-Fri. 9-4, Sat. 9-1 by appointment. Summer: Tues.-Fri. 9-4. Research: members no charge. Closed Independence Day & week; Christmas week. &
Attendance: 3,000 (estimated)
Membership: Senior $20; Individual $35; Family $50; Contributor $100; Sponsor $175; Patron $300.

Lenox

BERKSHIRE SCENIC RAILWAY MUSEUM, 10 Willow Creek Rd., Lenox, MA 01240. Mailing Address: P.O. Box 2195, Lenox, MA 01240-5195. Tel.: 413-637-2210. Fax: 413-637-4965.
E-mail: marketing@berkshirescenicrailroad.org
Web Site: www.berkshirescenicrailroad.org
Founded: 1984.

Key Personnel: Pres. (V), Richard Selva; Cur., John Trowill; Mktg., Pieter Lips; Public Rels., Pamela Green; Museum Shop Mgr., Cathleen Chittenden.
Personnel Profile: Part-Time Volunteers 30.
Governing Authority: private; nonprofit organization. Tax-exempt: 501(c)(3).
Institution Type/Description: Train Museum: housed in the restored 1903 Lenox station.
Collections: 1920s-1960s railroad equipment used by the New York, New Haven & Hartford Railroad; turn-of-the-century photographs of Berkshire Hills residences; vintage coaches.
Facilities: 3,000 sq. ft. exhibit space. Museum-related items for sale.
Activities: narrated train ride, Lenox Local. Museum Sponsors: Railroad Operation Lifesaver classes for elementary students; Fire Truck Displays; Antique Engines; Halloween & Santa Specials.
Publications: biannual newsletter, Along the River; annual report.
Hours & Admission Prices: Memorial Day-Oct. Sat.-Sun. & holidays 9-4. Train: Adults $15, seniors $14, children 4-14 $8; children under 4 & members no charge. Museum & Exhibits: no charge. &
Attendance: 8,500 (accurate)
Membership: Individual $20; Family $30; Patron $100; Benefactor $250; Lifetime $500.

THE MOUNT: EDITH WHARTON'S HOME, (M), 2 Plunkett St., Lenox, MA 01240-2704. Mailing Address: P.O. Box 974, Lenox, MA 01240-0974. Tel.: 413-551-5111. Fax: 413-637-0619.
E-mail: info@edithwharton.org
Web Site: edithwharton.org
Founded: 1979.
Key Personnel: Exec. Dir., Susan Wissler; Dir. Facilities, Ross Jolly; Public Programs Coord., Kelsey Mullen.
Personnel Profile: Full-Time Paid 10; Part-Time Paid 3; Part-Time Volunteers 30; Interns 1.
Volunteer Hours: 70
Institution Type/Description: Historic House: home designed by Pulitzer-Prize winning author Edith Wharton. A National Historic Landmark.
Collections: Edith Wharton's period artifacts, memorabilia & books.
Major Exhibits: Edith Wharten and the First World War, 5/14-10/14.
Facilities: library.
Activities: special events & programs.
Hours & Admission Prices: May-Oct. daily. Guided Tour: adults $18, seniors $17; children 18 & under no charge. &
Attendance: 30,000 (estimated)
Membership: Individual $50; Dual $75; Friend $125; Sustainer $250; Sponsor $500; Patron $1,000.

VENTFORT HALL MANSION AND GILDED AGE MUSEUM, 104 Walker St., Lenox, MA 01240. Mailing Address: P.O. Box 2424, Lenox, MA 01240. Tel.: 413-637-3206.
E-mail: info@gildedage.org
Web Site: gildedage.org
Governing Authority: Tax-exempt.
Institution Type/Description: Historic House: former home of Sarah Morgan, sister of J.P. Morgan, built in 1893. Listed on the National Register of Historic Places.
Collections: period artifacts & memorabilia from the 19th century; Gilded Age costumes.
Activities: summer play; summer tea; talk series; concerts; kids' summer camp.
Hours & Admission Prices: Summer: Mon.-Fri. 10-5, Sat.-Sun. 10-3; Winter: Sat.-Sun. 10-3. Adults $15, children 5-17 $7; members no charge. Closed New Year's Day; Easter; Thanksgiving; Christmas. &
Membership: Individual $40; Household $60.

Leominster

NATIONAL PLASTICS CENTER AND MUSEUM, 90 Main St., Leominster, MA 01453-5521. Tel.: 978-537-9529.
Web Site: www.plasticsmuseum.org
Founded: 1982.
Key Personnel: Cur., Marianne Zephir.
Personnel Profile: Full-Time Paid 5; Part-Time Paid 4; Part-Time Volunteers 4; Interns 2.
Governing Authority: Tax-exempt.
Institution Type/Description: Industrial Museum: housed in a school building built in 1888.
Collections: early molding machines; plastics industry history; recycling; hands-on exhibits; toys; plastics in medicine; period furniture; Plastics Hall of Fame.
Facilities: library; 10,000 sq. ft. exhibit space.
Activities: research; meetings; rental facility.

Hours & Admission Prices: Call for hours. Adults $5, children 4-11 & senior 65 & over $3; ASTC members & members no charge. &
Attendance: 12,000 (accurate)
Membership: $25; $50; $250; $500; $1,000.

Leverett

LEVERETT HISTORICAL SOCIETY, N. Leverett Rd., Leverett, MA 01054. Mailing Address: P.O. Box 57, Leverett, MA 01054-0057. Tel.: 413-548-9452.
Founded: 1963.
Congressional District: 1
Key Personnel: Dir., Edith Field.
Personnel Profile: Part-Time Volunteers 8.
Governing Authority: nonprofit.
Institution Type/Description: Historic House: 1820s schoolhouse.
Collections: local history items; oral history tapes; textiles; tools; photographs; schoolhouse memorabilia; charcoal kiln model.
Facilities: library of history, genealogy & town records available for research.
Publications: Leverett Massachusetts Historical Center; Historical Architectural Tour - 2004.
Hours & Admission Prices: Schoolhouse: Summer: Sun. 1-3; other times by appointment. Leverett Family Museum: Sat. 10-12; other times by appointment. No charge; donations accepted.
Attendance: 100 (estimated)
Membership: Individual or Family $10.

Lexington

LEXINGTON HISTORICAL SOCIETY, 13 Depot Sq., Lexington, MA 02420. Mailing Address: P.O. Box 514, Lexington, MA 02420-0005. Tel.: 781-862-1703.
E-mail: info@lexingtonhistory.org
Web Site: www.lexingtonhistory.org
Founded: 1886.
Congressional District: 7
Key Personnel: C.E.O., Susan Bennett; Pres. (V), William Mix; Museum Shop Mgr., Carla Fortmann.
Personnel Profile: Full-Time Paid 1; Full-Time Volunteers 1; Part-Time Paid 55; Part-Time Volunteers 100.
Governing Authority: society. Tax-exempt. 501(c)(3).
Institution Type/Description: Historical Society Museum.
Collections: American Revolution period furniture & artifacts; original documents & manuscripts; costumes; photographs; portraits; Lexington history; Battle of Lexington. Historic Houses: c.1700 Munroe Tavern; 1737 Hancock-Clarke House; 1710 Buckman Tavern.
Research Fields: local 18th-century history, Revolutionary War.
Facilities: research archives.
Activities: guided tours; lecture programs.
Publications: society newsletter; booklets; The Story of Buckman Tavern; The Munroe Tavern; The Hancock-Clarke House; The History of the Lexington Historical Society; Lexington Portraits 1734-1884 Lexington: A Century Of Photographs; Lexington: Birthplace of American Liberty.
Hours & Admission Prices: April-Oct. daily. One House: adults $7, children $5. Three Houses: adults $12, children 6-16 $8; members & children under 6 no charge.
Attendance: 30,000 (accurate)
Membership: Individual $35; Family $50; Sponsor $ 125; Corporate Patron $500; Life $1,000.

MUNROE CENTER FOR THE ARTS, 1403 Massachusetts Ave., Lexington, MA 02420-3828. Tel.: 781-862-6040. Fax: 781-674-2787.
Key Personnel: Admin., Nancy Sofen
Institution Type/Description: Art Gallery.
Collections: paintings; sculpture.
Activities: educational programs; classes; special events.
Hours & Admission Prices: Call for hours.

✳ **SCOTTISH RITE MASONIC MUSEUM & LIBRARY, (M),** 33 Marrett Rd., Lexington, MA 02421-5703. Tel.: 781-861-6559. Fax: 781-861-9846. TTY: 781-274-8539.
E-mail: info@monh.org
Web Site: www.nationalheritagemuseum.org
Formerly: National Heritage Museum
Founded: 1971.
Congressional District: 5
Key Personnel: Pres., John William McNaughton; Exec. Dir., Richard V. Travis; Dir. Exhibitions & Audience Devel., Hilary Anderson Stelling; Dir.

Collections, Aimee Newell; Museum Designer, Michael Rizzo; Archivist, Catherine Swanson; Public Rels., Linda Patch; Collections Mgr. & Registrar, Maureen Harper; Mgr. Library & Archives, Jeff Croteau.

Personnel Profile: Full-Time Paid 9; Part-Time Paid 6; Part-Time Volunteers 47.

Governing Authority: nonprofit. Parent Institution: Supreme Council Northern Masonic Jurisdiction of U.S.A. Tax-exempt: 501(c)(3).

Institution Type/Description: History Museum.

Collections: American prints; paintings; maps; decorative arts; clocks; furniture; costumes; photographs; memorabilia; Masonic & fraternal.

Research Fields: history of Freemasonry; symbolism of Masonic & fraternal organizations; patriotic iconography.

Facilities: 60,000-vol. library of books, manuscripts & maps; reading room; conference center; 400-seat auditorium; cafe.

Activities: lectures; concerts; workshops; educational & family programs; school tours; temporary & traveling exhibitions; group tours for adults.

Publications: quarterly calendar & newsletter; exhibit brochures; annual report; catalogs.

Hours & Admission Prices: Wed.-Sat. 10-4:30. No charge; donations accepted. Closed New Year's Day; Easter; Thanksgiving; Christmas. ⅋

Attendance: 39,174 (accurate)

Membership: Individual $40; Family $60; Contributor $100; Associate $250; Benefactor $500.

Lincoln

* **CODMAN ESTATE, (M),** 34 Codman Rd., Lincoln, MA 01773. Mailing Address: 141 Cambridge St., Boston, MA 02114-2702. Tel.: 617-994-6690.

Web Site: www.historicnewengland.org

Founded: 1968.

Congressional District: 5

Key Personnel: Pres. & C.E.O., Carl Nold; Site Mgr., Wendy Hubbard.

Governing Authority: society; nonprofit organization. Parent Institution: Historic New England, 141 Cambridge St. Boston, MA 02114. Tel.: 617-227-3956. Tax-exempt: 501(c)(3).

Institution Type/Description: Historic House: c.1740 Georgian mansion expanded in 1790s to resemble an English country seat. Home of Codman family through 1968.

Collections: architecture & furnishings of the Georgian, Federal, Victorian & Colonial Revival periods; landscaped grounds including trees & plantings; formal Italian garden; English cutting garden.

Facilities: rental facilities.

Activities: guided tours; lectures; special events & programs; rental facilities.

Publications: magazine, Historic New England.

Hours & Admission Prices: House: June-Oct. 15 2nd & 4th Sat. each month 11-5. Grounds: dawn to dusk. Adults $5; discounts to seniors, AAM, ICOM, AAA & WGBH members; Historic New England members no charge.

Attendance: 13,708 (accurate)

Membership: National $35; Individual $45; Household $55; Garden & Landscape $75; Contributing and Library & School $100; Young Friends of Historic New England $100-$1,500; Friends of the Library and Archives $125; Historic Homeowner $200; Business & Ogden Codman Design Group $250; Appleton Circle $2,500 & up.

* **DECORDOVA SCULPTURE PARK AND MUSEUM, (M),** 51 Sandy Pond Rd., Lincoln, MA 01773-2699. Tel.: 781-259-8355. Fax: 781-259-3650.

E-mail: info@decordova.org

Web Site: www.decordova.org

Founded: 1948.

Congressional District: 4

Key Personnel: Pres., Faith K. Parker; Exec. Dir., Dennis Kois; Dir. External Affairs, Nora Maroulis; Dir. Learning & Engagement, Julie Bernson; Dir. Institutional Support, Martha Winter; Head Individual Giving, Liz Herring; Sr. Cur., Dina Deitsch; Dir. Bldgs. & Grounds, Douglas Holston; Dir. Retail Operations, David Duddy.

Personnel Profile: Full-Time Paid 25; Part-Time Paid 30; Part-Time Volunteers 200; Interns 10.

Governing Authority: nonprofit organization. Tax-exempt: 501(c)(3).

Institution Type/Description: Contemporary Art Museum & Sculpture Park.

Collections: American art; 20th-century American painting, graphics, sculpture & photography; emphasis on artists associated with New England.

Research Fields: 20th-century American art with emphasis on contemporary New England artists; large-scale outdoor public sculpture.

Facilities: 38,000 sq. ft. exhibit space; 35-acre sculpture park; cafe. Art & Museum-related items for sale.

Activities: changing exhibitions; guided tours; lectures; individual & corporate membership program; outreach programs; children's programs; adult workshops.

Publications: catalogs; exhibition newsletter.

Hours & Admission Prices: Tues.-Sun. 10-5. Adults $14, seniors $12, students $10; discounts to Zipcar, WGBH, WBUR & AAM members; members, bicyclists, active duty military & their families, and children 12 & under no charge. ⅋

Attendance: 100,000 (estimated)

Membership: Individual $60; Dual $90; Family $95; Friend $125; Sponsor $250; Benefactor $500; Julian Club $1,000.

DRUMLIN FARM EDUCATION CENTER, 208 South Great Rd., Lincoln, MA 01773-4800. Tel.: 781-259-2200 & 2203. Fax: 781-259-7917.

E-mail: drumlinfarm@massaudubon.org

Web Site: www.massaudubon.org

Founded: 1955.

Congressional District: 5

Key Personnel: Pres., Laura Johnson; Dir., Christy Foote-Smith; Program Specialist, Kris Scopinich; Program Specialist, Karen Stein; Farmer Livestock, Caroline Malone; Farmer Crops, Matt Celona; Museum Shop Mgr., Ruth Smith.

Personnel Profile: Full-Time Paid 23; Part-Time Paid 60; Interns 7.

Governing Authority: nonprofit organization. Parent Institution: Massachusetts Audubon Society, Lincoln, MA. Tax-exempt: 501(c)(3).

Institution Type/Description: Living Farm Museum.

Collections: agriculture; aviary; botany; entomology; geology; herpetology.

Research Fields: bird banding; blue bird nesting monitoring; breeding; bird surveys.

Activities: guided tours; lectures; films; rental gallery; formally organized education programs; permanent exhibitions; summer natural history day camps.

Publications: program brochures; electronic newsletters.

Hours & Admission Prices: Tues.-Sun. 9-5; school tours by reservation only. Adults $6, senior citizens & children 2-12 $4; Lincoln residents & children under 2 no charge. Closed New Year's Eve & Day; Thanksgiving; Christmas. ⅋

Attendance: 150,000 (estimated)

Membership: Individual $40; Family $50; Supporting $60; Donor $100; Patron $500.

* **GROPIUS HOUSE, (M),** 68 Baker Bridge Rd., Lincoln, MA 01773-3105. Tel.: 781-259-8098. Fax: 781-259-9722.

E-mail: gropiushouse@historicnewengland.org

Web Site: www.historicnewengland.org

Founded: 1985.

Key Personnel: Pres., Carl Nold; Museum Site Mgr., Wendy L. Hubbard.

Personnel Profile: Full-Time Paid 1; Part-Time Paid 6.

Governing Authority: society; nonprofit organization. Parent Institution: Historic New England. Tax-exempt: 501(c)(3).

Institution Type/Description: Historic House: c.1937-38 Walter Gropius, one of the innovators of modern architecture, designed his family residence by combining Bauhaus principles with New England building materials.

Collections: furnishings made in Bauhaus workshops & brought from Germany; American fixtures & materials; industrial grade; decorative art.

Research Fields: decorative arts; modernism.

Facilities: visitor center. Museum-related items for sale.

Activities: guided tours; programs; lecture series; family events.

Hours & Admission Prices: Tours: June-Oct. 15 Wed.-Sun. 11-4 on the hour; Oct. 16-May Sat.-Sun. 11-5 on the hour. Adults $15, seniors $12, children & students $8; discounts to groups and AAM, AAA & WGBH members; members, Historic New England members & Lincoln residents no charge.

Attendance: 6,429 (accurate)

Membership: Individual $45; Household $55; Contributing $100; Historic Homeowner $200.

THE THOREAU INSTITUTE AT WALDEN WOODS, 44 Baker Farm, Lincoln, MA 01773-3004. Tel.: 781-259-4730. Fax: 781-259-4730.

E-mail: Jeff.Cramer@walden.org

Web Site: www.walden.org/institute

Founded: 1998.

Congressional District: 5

Key Personnel: Exec. Dir., Kathi Anderson; Cur. Collections, Jeffrey S. Cramer.

Personnel Profile: Full-Time Paid 7; Part-Time Paid 4; Interns 2.

Governing Authority: nonprofit. Parent Institution: Walden Woods Project. Tax-exempt: 501(c)(3).

Institution Type/Description: Historical Society Museum.

Collections: early editions of works by Henry David Thoreau; 19th-century

photographs; manuscripts; old maps, including original Thoreau survey maps; Ricketson bust of Thoreau; pencils made by Thoreau family business.
Research Fields: Thoreau and the other 19th-century transcendentalists.
Facilities: 50,000-vol. library of material by & about Thoreau & other New England Transcendentalists, Concord & New England history & 19th-century American literature available for use to serious readers & scholars; reading room; Helen and Scott Nearing papers; Paul Brooks collection.
Activities: lectures; formally organized educational programs.
Hours & Admission Prices: By appointment only. No charge; donations accepted. &

Longmeadow

RICHARD SALTER STORRS HOUSE, 697 Longmeadow St., Longmeadow, MA 01106-2215. Tel.: 413-567-3600.
E-mail: LCAsearch@aol.com
Founded: 1899.
Congressional District: 2
Key Personnel: C.E.O. & Pres. (V), Michael Gelinas; Cur. & Genealogist, Linda C. Abrams.
Personnel Profile: Part-Time Volunteers 12.
Governing Authority: society. Administered by the Longmeadow Historical Society. Tax-exempt: 501(c)(3).
Institution Type/Description: Historical Society Museum: housed in 1786 Storrs House.
Collections: early American furnishings; genealogy; local artifacts; decorative arts; history; military; costumes; toys; materials & genealogical memorabilia, 18th- & 19th-century manuscripts; theological & school; written memoirs.
Research Fields: town 18th- to 19th-century archives; genealogy.
Activities: tours & research by appointment. Annual Event: Long Meddowe Days Fair in May; Ghosts in the Graveyard in October.
Publications: assorted booklets of local historical facts, places & people; book, Reflections of Longmeadow; newsletter, Town Crier.
Hours & Admission Prices: By appointment. Adults $3; discounts to senior citizens, family reunion groups, children 12-18, AAM & ICOM members; members & children under 12 no charge.
Attendance: 800 (estimated)
Membership: Individual $15; Family $25; Life $500.

Lowell

* **AMERICAN TEXTILE HISTORY MUSEUM, (M),** 491 Dutton St., Lowell, MA 01854-4289. Tel.: 978-441-0400. Fax: 978-441-1412.
E-mail: sjackson@athm.org
Web Site: www.athm.org
Formerly: Merrimack Valley Textile Museum
Founded: 1960.
Congressional District: 5
Key Personnel: Chm. & Treas., Kenneth J. McAvoy; Pres & C.E.O., James S. Coleman; Vice Chm., Jan Russell; Vice Chm., Karl Spilhaus; Dir. Advancement, Linda Carpenter; Cur., Karen Herbaugh; Business Mgr., Steven Jackson; Dir. Interpretation, David Unger; Coord. Museum Education Svcs., Sue Bunker; Librarian, Clare Sheridan; Museum Shop Mgr., Ann Lochheao.
Personnel Profile: Full-Time Paid 15; Part-Time Paid 1; Part-Time Volunteers 45; Interns 2.
Governing Authority: nonprofit organization. Tax-exempt: 501(c)(3).
Institution Type/Description: Textile History Museum.
Collections: textiles; textile samples; costumes; tools; workplace artifacts; books; periodicals; manuscripts; business records; trade literature; ephemera; prints; photography; machinery; modern textile applications.
Research Fields: textiles & costume; design; business & labor; technology; community history.
Facilities: study areas; meeting rooms; restaurant. Museum-related items for sale.
Activities: gallery tours; special events; outreach; school programs; speakers bureau.
Publications: newsletter, Textile Times; leaflets; catalogues.
Hours & Admission Prices: Adults $8; discounts to AAM & NEMA members; members no charge. Closed New Year's Day; Martin Luther King, Jr. Day; Independence Day; Labor Day; Columbus Day; Thanksgiving; Christmas. &
Attendance: 52,000 (estimated)
Membership: Individual $50; Dual $65; Family $95; Contributing $125; Supporting $250; Patron $500; Leadership Circle $1,000-$10,000.

THE LOADING DOCK GALLERY, 122 Western Ave., Lowell, MA 01851.
E-mail: info@theloadingdockgallery.com
Web Site: www.theloadingdockgallery.com
Institution Type/Description: Art Gallery.
Collections: works by Western Avenue Studios Artist Association members.
Facilities: 1,450 sq. ft. exhibit space; rental facilities.
Activities: workshops; special events.
Hours & Admission Prices: Wed.-Sun. 11-4:30.

LOWELL NATIONAL HISTORICAL PARK, (M), 67 Kirk St., Lowell, MA 01852-1029. Tel.: 978-970-5000. TDD: 978-970-5002.
E-mail: lowe_reservations@nps.gov
Web Site: www.nps.gov/lowe
Founded: 1978.
Congressional District: 5
Key Personnel: Supt., Celeste Bernardo; Education & Dir. Tsongas Industrial History Center, Sheila Kirschbaum; Cur., David Blackburn; Public Information, Sue Andrews; Librarian, Jack Herlihy; Museum Shop Mgr., Joan Gagnon.
Personnel Profile: Full-Time Paid 109; Part-Time Volunteers 50.
Governing Authority: federal; nonprofit organization. Parent Institution: National Park Service. Subsidiary Institution: Boott Cotton Mills Museum, 115 John St., Lowell, MA. Tax-exempt.
Institution Type/Description: History Museum District.
Collections: ethnic & labor items; working textile equipment; mill girl letters & documents; historical engineering drawings & records. Historic Buildings: 37 preserved & restored buildings, construction dates ranging from 1824-1908.
Research Fields: labor, social & technology history.
Facilities: 3,500-vol. library of books, including informational folders, photographs & slides available to the public; visitor's center; educational facilities. Books & museum-related items for sale.
Activities: guided tours; special interest tours; lectures; organized programs for children; participatory exhibits; historic canal system & river tours; concerts; folk festivals.
Publications: park brochures; NPS Park Handbook.
Hours & Admission Prices: Visitor Center: Winter: daily 9-4:30; Summer: daily 9-5. Boott Cotton Mills Museum: Winter: daily 9:30-4:30; Summer: daily 9:30-5. Mill Girls & Immigrants Exhibit: call for hours. Adults $6, senior citizens & students $5, youth 6-16 $4; children under 5 no charge. Closed New Year's Day; Thanksgiving Day; Christmas. &
Attendance: 550,000 (estimated)

MIDDLESEX CANAL COLLECTION, Center for Lowell History, 40 French St., Lowell, MA 01852-1113. Tel.: 978-934-4998. Fax: 978-934-4995.
E-mail: martha_mayo@uml.edu
Web Site: library.uml.edu/clh
Founded: 1962.
Congressional District: 5
Key Personnel: Dir., Martha Mayo.
Governing Authority: University of Massachusetts Lowell. Affiliated with Middlesex Canal Association.
Institution Type/Description: History Museum/Center.
Collections: Middlesex Canal Assoc. archives; Middlesex Canal County records.
Activities: temporary exhibitions.
Publications: quarterly bulletin, Towpath Topics.
Hours & Admission Prices: Mon.-Fri. 9-6, Sat. 10-3. No charge. &
Attendance: 15,000 (estimated)
Membership: Associate $4; Proprietor $10.

NEW ENGLAND QUILT MUSEUM, (M), 18 Shattuck St., Lowell, MA 01852-1820. Tel.: 978-452-4207. Fax: 978-452-5405.
E-mail: director@nequiltmuseum.org
Web Site: www.nequiltmuseum.org
Founded: 1987.
Congressional District: 5
Key Personnel: Exec. Dir., Nora Burchfield.
Personnel Profile: Full-Time Paid 2; Part-Time Paid 6; Part-Time Volunteers 80; Interns 2.
Governing Authority: nonprofit organization. Tax-exempt.
Institution Type/Description: Quilt Museum: housed in landmark savings bank building.
Collections: period, contemporary & art quilts.
Research Fields: American quilts.

Facilities: 1,200-vol. library pertaining to quilting; educational facilities; 6,000 sq. ft. exhibit space; classroom. Gifts & quilts for sale.
Activities: guided tours; lectures; organized educational programs; participatory, loan, temporary & traveling exhibitions; volunteer program; community activities; special events.
Publications: NE Quilt Museum Quilts; CD catalogs of exhibitions; quarterly newsletter, Layers.
Hours & Admission Prices: May-Oct. Tues.-Sat. 10-4, Sun. 12-4; Nov.-April Tues.-Sat. 10-4. Adults $7, senior citizens & students $5; discounts to WGBH, AAM, MTA & AAA members; members no charge. Closed major holidays. &
Attendance: 8,000 (estimated)
Membership: Student & Senior $30; Individual $45; Household $55; Donor $100; Supporter: $250; Patron $500; Benefactor $1,000.

WHISTLER HOUSE MUSEUM OF ART, 243 Worthen St., Lowell, MA 01852-1874. Tel.: 978-452-7641. Fax: 978-454-2421.
E-mail: mlally@whistlerhouse.org
Web Site: www.whistlerhouse.org
Founded: 1878.
Congressional District: 5
Key Personnel: Chm. (V), Nancy Donahue; Chm. (V), Richard Donahue; Pres., Sara Bogosian; Exhibits & Gallery Mgr., James Dyment.
Personnel Profile: Full-Time Paid 1; Part-Time Paid 1; Part-Time Volunteers 4; Interns 3.
Governing Authority: society; nonprofit. Parent Institution: Lowell Art Association. Tax-exempt: 501(c)(3).
Institution Type/Description: Art Gallery: housed in 1823 Whistler House, birthplace of James A. M. Whistler.
Collections: American art including works by Whistler, John Singer Sargent & Arshile Gorky.
Research Fields: James A. M. Whistler; 19th-20th century American art; New England art; art museum education.
Facilities: 400-vol. library on James Whistler & New England artists; 100-séat auditorium.
Activities: walking & guided tours; lectures; films; gallery talks; concerts; arts festivals; formally organized education programs; permanent, temporary & traveling exhibitions.
Publications: calendars of events; newsletters; exhibition catalogues.
Hours & Admission Prices: Wed.-Sat. 11-4. Adults $5, senior citizens & students 6-21 $4; discounts to groups, AAA, AAM & NEMA members; children under 6, museum professionals & members no charge. Closed major holidays. &
Attendance: 7,000 (estimated)
Membership: Student $20; Senior $25; Individual $35; Family $45; Donor $100; Sponsor $250; Patron $500.

Lynn

GRAND ARMY OF THE REPUBLIC HALL MUSEUM, 58 Andrew St., Lynn, MA 01901-1102. Tel.: 781-477-7085.
Institution Type/Description: Historic Building: built in 1885 as a meeting place for General Frederick W. Lander Post No. 5. Listed on the National Register of Historic Places.
Collections: local history & culture; period furnishings; personal artifacts; photographs.
Hours & Admission Prices: By appointment.

LYNN MUSEUM & HISTORICAL SOCIETY, (M), 590 Washington St., Lynn, MA 01901-1406. Tel.: 781-581-6200. Fax: 781-581-6202.
E-mail: office@lynnmuseum.org
Web Site: www.lynnmuseum.org
Formerly: Lynn Historical Society
Founded: 1897.
Congressional District: 6
Key Personnel: Dir. & Cur., Kate Luchini; Pres., Joseph Scanlon; Asst. Dir., Abby Battis; Public Rels. & Administrative Support, JoAnn Baker.
Personnel Profile: Full-Time Paid 2; Full-Time Volunteers 2; Part-Time Paid 3; Part-Time Volunteers 87; Interns 4.
Governing Authority: society. Lynn Historical Society. Tax-exempt: 501(c)(3).
Institution Type/Description: Local History Museum: housed in old shoe factory building; Lynn Heritage State Park Visitor Center.
Collections: shoes & early shoe making tools; furniture; paintings; costumes; manuscripts; photographs; newspapers.
Research Fields: genealogy; local history; shoe industry.
Facilities: library & reading room for local history & genealogy; 75-seat meeting hall & function room; courtyard. Museum-related items for sale.

Activities: guided tours; lectures; school education program; permanent & temporary exhibitions.
Publications: Records of Ye Towne Meetings of Lynn, Vol. 1-7, 1691-1783; The Lynn Album: A Pictorial History; No Race of Imitators: Lynn & Her People, An Anthology; video, So Hear the Metropolis: The Story of Lynn Woods; A Lasting Impression, The Jan Metzelinger Story; Diners of the North Shore; Portrait of a City: Lynn on Film; The Civil War Diary of Lt. J.E. Hodgkins; J. Sanger Attill and His Craftsman; Lynn Beach Painters: Art Along the North Shore 1880-1920; Lynn Album II: A Pictorial History of the 20th Century; By Candlelight: Samplers in the Lynn Museum; The Brickyard: The Life, Death & Legend an Urban Neighborhood (2005); Arcadia Press: Lynn in the Victorian Era; The Great Fires of Lynn.
Hours & Admission Prices: Museum: Tues.-Wed. & Fri.-Sat. 12-4; Thurs. 12-8. Research & tours by appointment. Adults $5, children $2; discounts to AAA, AAM & ICOM members; members & Lynn School children 18 & under no charge. Closed state & national holidays. &
Attendance: 12,000 (estimated)
Membership: Student $10; Senior Individual $25; Individual $30; Senior Household $35; Household $45; Heritage Supporter $100; Benefactor $500; Lifetime $750; Dual Lifetime (Couple) $1,000.

LYNNARTS, INC., 25 Exchange St., Lynn, MA 01901-1423. Tel.: 781-598-5244. Fax: 781-599-8926. Facebook: LynnArts, Inc.
E-mail: jenashworth@lynnarts.org
Web Site: www.lynnarts.org
Key Personnel: Coord. Operations, Jennifer Ashworth
Institution Type/Description: Art Gallery.
Collections: paintings; sculpture.
Major Exhibits: Photography, 1/14-3/14; Paint, 3/14-5/14; Lynn Public Schools Art Show, 5/14-7/14; Members Exhibition, 7/14-8/14; Mixed Media, 8/14-10/14; Holiday Show & Sale, 10/14-12/14.
Activities: classes; special events; 21 workshops: adults & children; performances; open studios; private functions; gallery sales.
Hours & Admission Prices: Mon.-Mon. & Fri. 10-4, Thurs. 11-7, Sat. 11-4. No charge, donations accepted. Closed New Year's Eve & Day; Thanksgiving; Dec. 23-Jan. 1 for Christmas.
Membership: Student/Senior Citizens $15; Artist $25; Family $40; Sustaining $100; Patron $250; Benefactor $500.

Malden

MALDEN PUBLIC LIBRARY, 36 Salem St., Malden, MA 02148-5291. Tel.: 781-324-0218 & 388-0800. Fax: 781-324-4467.
E-mail: dmalgeri@maldenpubliclibrary.org
Web Site: www.maldenpubliclibrary.org
Founded: 1879.
Congressional District: 7
Key Personnel: Librarian, Dina G. Malgeri.
Personnel Profile: Full-Time Paid 23; Part-Time Paid 2.
Governing Authority: board of trustees; nonprofit organization. Tax-exempt.
Institution Type/Description: Public Library with Special Art Collection.
Collections: fine arts objects; oil paintings; sculpture.
Facilities: 201,189-vol. public library.
Activities: permanent & temporary exhibitions; concerts; workshops; lectures.
Publications: booklet, ArtFacts.
Hours & Admission Prices: June-Aug. Mon.-Wed. 9-9, Thurs. 9-1 & 2-6, Fri. 9-6; Sept.-May Mon.-Wed. 9-9, Thurs. 9-1 & 2-9, Fri.-Sat. 9-6. No charge; donations accepted. Closed national holidays. &
Attendance: 100,000 (estimated)

Manchester

MANCHESTER HISTORICAL MUSEUM, 10 Union St., Manchester, MA 01944-1553. Tel.: 978-526-7230. Fax: 978-526-0060.
E-mail: manchesterhistorical@verizon.net
Web Site: manchesterbytheseahistorical.org
Founded: 1886.
Congressional District: 6
Key Personnel: Dir., Vicki Cawley; Pres. (V), Donna Dussault; Vice Pres. (V), Fafa Diedrich; Archivist, Esther Proctor.
Governing Authority: nonprofit organization. Tax-exempt: 501(c)(3).
Institution Type/Description: Decorative Arts Museum.
Collections: 18th- & 19th-century Manchester made furniture; photographs.
Research Fields: local history.
Facilities: 200-vol. library of logbooks, diaries & unpublished papers on history of Manchester available for use on premises.
Activities: guided tours; lectures; informally organized education programs for children; temporary exhibitions.
Publications: newsletter.

Hours & Admission Prices: Tues.-Fri. 10-3; other times by appointment. Tours: adults $5; members no charge. &
Membership: Senior $10; Individual $25; Family $40.

Marblehead

MARBLEHEAD ARTS ASSOCIATION - KING HOOPER MANSION, 8 Hooper St., Marblehead, MA 01945-3213. Tel.: 781-631-2608. Fax: 781-639-7890.
E-mail: info@marbleheadarts.org
Web Site: marbleheadarts.org
Founded: 1922.
Congressional District: 6
Key Personnel: Dir., Deborah Greel; Pres. (V), James Regis.
Personnel Profile: Full-Time Paid 1; Part-Time Paid 2; Part-Time Volunteers 40; Interns 1.
Governing Authority: private; nonprofit organization. Tax-exempt: 501(c)(3).
Institution Type/Description: Art Association: housed in c.1728 colonial style mansion of Greenfield Hooper.
Collections: Hooper family furnishings; Marblehead pottery; 1940 prints; Steinway grand piano; wine cellar; historic mansion.
Facilities: educational facilities; rental facilities; garden. Museum-related items for sale.
Activities: arts festivals; concerts; dance recitals; education programs; guided tours by appointment; hobby workshops; lectures; participatory, loan & temporary exhibitions; rentals; theater; special events.
Publications: email newsletter; Marblehead Arts Association; brochure, Hooper Mansion; wedding rental brochure.
Hours & Admission Prices: Tues.-Sat. 10-5, Sun. 1-5. No charge; donations accepted. Large Tours: $3 per person. Closed New Year's Day; Easter; Thanksgiving; Christmas.
Attendance: 5,000 (accurate)
Membership: Student $15; Senior over 65 $30; Associate $35; Artist $50; Family $55.

MARBLEHEAD MUSEUM & HISTORICAL SOCIETY, 170 Washington St., Marblehead, MA 01945-3340. Tel.: 781-631-1768. Fax: 781-631-0917.
E-mail: mmhspeterson@yahoo.com
Web Site: www.marbleheadmuseum.org
Founded: 1898.
Congressional District: 6
Key Personnel: Dir., Pam Peterson; Pres., Richard Carlson; Treas., Richard Settlemeyer; Cur. Jeremiah Lee Mansion, Judy Anderson; Cur. Museum, Karen MacInnis; Registrar, Jean Fallon.
Personnel Profile: Full-Time Paid 2; Part-Time Paid 2; Part-Time Volunteers 100; Interns 2.
Governing Authority: society; nonprofit. Subsidiary Institutions: JOJ Frost Folk Art Gallery; Jeremiah Lee Mansion; G.A.R. and Civil War Museum. Tax-exempt: 170(b)(1)(A).
Institution Type/Description: History & Folk Art Museum.
Collections: 19th- & 20th-century decorative & folk arts; archives; manuscripts; maritime folk art paintings; historic Marblehead & maritime artifacts; American Colonial decorative arts. Historic Buildings: 1768 Jeremiah Lee Mansion; Grand Army of the Republic-Civil War Museum; J.O.J. Frost Folk Art Gallery.
Research Fields: genealogy; Marblehead history; maritime history; Jeremiah Lee Mansion history; Civil War; G.A.R. history.
Facilities: 500-vol. library of books on history & genealogy available on premises; archives; reading room; garden. Books, postcards, reproductions of J.O.J. Frost paintings for sale.
Activities: walking tours; guided tours; lectures; permanent & temporary exhibitions; docent program; children's enrichment program; study clubs; research. Annual Event: Garden Party; Gingerbread Festival; Wine Tasting Under the Tent.
Publications: brochure; annual report; calendar of events.
Hours & Admission Prices: Society: Tues.-Sat. 10-4. Lee Mansion: June-Oct. Tues.-Sat. 10-4. Mansion: adults $5. Frost Gallery & Civil War Museum: no charge; donations accepted. Closed major holidays.
Attendance: 5,500 (estimated)
Membership: Individual $30; Family & Dual $50; Friend $100; Sponsor $500; Patron $1,000; Benefactor $5,000.

1768 JEREMIAH LEE MANSION, 161 Washington St., Marblehead, MA 01945-3303. Mailing Address: Marblehead Museum & Historical Society, 170 Washington St., Marblehead, MA 01945. Tel.: 781-631-1768. Fax: 781-631-0917.
E-mail: mmhsanderson@yahoo.com
Web Site: www.marbleheadmuseum.org

Founded: 1909.
Congressional District: 6
Key Personnel: Pres. (V), Richard Carlson; Dir., Pam Peterson; Cur. Jeremiah Lee Mansion, Judy Anderson; Research & Collections Cur., Karen MacInnis.
Personnel Profile: Full-Time Paid 2; Part-Time Paid 1; Part-Time Volunteers 100.
Governing Authority: society; nonprofit. Parent Institution: Marblehead Museum & Historical Society. Tax-exempt: 501(c)(3).
Institution Type/Description: Decorative Arts & History Museum.
Collections: 18th- & 19th-century furniture; decorative arts; paintings; maritime & folk art; musical instruments; archives.
Facilities: garden.
Activities: guided tours; lectures; garden events; architecture tours & classes.
Publications: annual report; newsletter; calendar of events.
Hours & Admission Prices: Mansion: June-Oct. Tues.-Sat. 10-4. Adults $5, senior citizens $4.50; discounts to Massachusetts Teachers Assoc.; children, students, & members no charge.
Attendance: 3,000 (estimated)
Membership: Individual $30; Family & Dual $50; Friend $100; Sponsor $500; Patron $1,000; Benefactor $5,000.

Marion

THE MARION ART CENTER GALLERY, 80 Pleasant St., Marion, MA 02738. Tel.: 508-748-1266. Fax: 508-748-2759.
Institution Type/Description: Art Gallery.
Collections: paintings; sculpture.
Activities: special events; educational programs.
Hours & Admission Prices: Tues.-Fri. 1-5, Sat. 10-2. No charge.

MARION NATURAL HISTORY MUSEUM, 8 Spring St., Marion, MA 02738-1519. Mailing Address: P.O. Box 644, Marion, MA 02738-0011. Tel.: 508-748-2098.
E-mail: lizleid@msn.com
Web Site: www.marionmuseum.org
Key Personnel: Dir., Elizabeth Leidhold
Institution Type/Description: Natural History Museum.
Collections: wildlife including fish, turtles, waterfowl & birds of prey; Native American artifacts.
Hours & Admission Prices: Call for hours.

Marlboro

PETER RICE HOMESTEAD, HOME OF THE MARLBOROUGH HISTORICAL SOCIETY, INC., 377 Elm St., Marlboro, MA 01752-4518. Mailing Address: P.O. Box 513, Marlboro, MA 01752-0513. Tel.: 508-485-4763 & 251-1057.
E-mail: info@historicmarlborough.org
Web Site: www.historicmarlborough.org
Founded: 1962.
Congressional District: 3
Key Personnel: Pres., Janet Licht; Vice Pres., Peggy Ayres.
Governing Authority: society. Tax-exempt: 501(c)(3).
Institution Type/Description: Historic Museum: housed in 1688 Peter Rice homestead.
Collections: 19th-century furniture; industrial collection.
Research Fields: local history; cemeteries; local buildings; genealogy Marlborough families.
Facilities: library of books & manuscripts available for small donation.
Activities: monthly programs; special events; guided tours.
Publications: print & e-mail newsletter; brochure describing Homestead; brochure & map of Homestead indicating points of interest.
Hours & Admission Prices: By appointment and for special events & programs. No charge.
Attendance: 1,200 (estimated)
Membership: Annual $15; Life $150.

Marshfield

HISTORIC 1699 WINSLOW HOUSE, 634 Careswell St., Marshfield, MA 02050-5623. Mailing Address: P.O. Box 531, Marshfield, MA 02050-0531. Tel.: 781-837-5753.
E-mail: info@winslowhouse.org
Web Site: www.winslowhouse.org
Formerly: Isaac Winslow House
Founded: 1920.
Key Personnel: Exec. Dir., Mark A. Schmidt; Pres. (V), Cynthia Hagar Krusell; Cur., Karin J. Goldstein.

Personnel Profile: Full-Time Paid 1; Part-Time Paid 1.
Governing Authority: Parent Institution: Historic Winslow House Association. Tax-exempt.
Institution Type/Description: Historic House Museum: housed in the former home of Hon. Isaac Winslow, Marshfield's founding family; built in 1699.
Collections: period artifacts & furnishings.
Facilities: Museum-related items for sale.
Activities: special events.
Publications: newsletter, Careswell Chronicles.
Hours & Admission Prices: Wed.-Sun. 11-5. Adults $3, children $1; discounts for MA Teachers Association, WGBH, AAM & AAA members; members no charge. Closed Memorial Day; Independence Day; Labor Day.
Membership: Student & Senior $25; Individual (Basic) $40; Individual (Plus) & Family (Basic) $75; Family (Plus) $150; Sustaining $250; Winslow Society $1,000.

Mashpee

CAPE COD CHILDREN'S MUSEUM, 577 Great Neck Rd. S., Mashpee, MA 02649-3708. Tel.: 508-539-8788. Fax: 508-539-3285.
E-mail: info@capecodchildrensmuseum.org
Web Site: www.capecodchildrensmuseum.org
Founded: 1990.
Key Personnel: Exec. Dir., Barbara Cotton; Pres., John Cotton; Dir. Exhibits & Operations and Museum Shop Mgr., Holly Dayton; Controller, Carol Pardee; Dir. Mktg. & Administration, Lisa Sheehy.
Personnel Profile: Part-Time Paid 15; Part-Time Volunteers 10.
Governing Authority: nonprofit organization. Tax-exempt.
Institution Type/Description: Children's Museum.
Collections: hands-on exhibits.
Hours & Admission Prices: Summer: Mon.-Sat. 10-5, Sun. 12-5; Winter (after Labor Day):; Mon.-Fri. 10-3, Sat. 10-5, Sun. 12-5. Adults $7, senior citizens 60 & over $6; discounts to groups, military, ACM & PBS; members & children under one no charge. Closed New Year's Day; Easter; Independence Day; Labor Day; Thanksgiving; Christmas Eve & Day.
Attendance: 50,000 (estimated)
Membership: Family $99; ACM Reciprocal $125.

MASHPEE HISTORICAL COMMISSION, 13 Great Neck Rd., N., Mashpee, MA 02649-2521. Mailing Address: 16 Great Neck Rd., N., Mashpee, MA 02649-2528. Tel.: 508-539-1438.
E-mail: historic@ci.mashpee.ma.us
Web Site: www.ci.mashpee.ma.us
Key Personnel: Chm. (V), Lee Gurney; Historian, Rosemary Burns
Institution Type/Description: History Museum.
Collections: local history & culture; memorabilia & artifacts of Mashpee people, places & events.
Hours & Admission Prices: Mon. & Thurs. 10-2 & by appointment

MASHPEE TRIBAL COUNCIL MUSEUM, 483 Great Neck Rd., S., Mashpee, MA 02649-3707. Tel.: 508-477-0208. Fax: 508-477-1218.
Web Site: mashpeewampanoagtribe.com
Key Personnel: Chm., Cedric Cromwell; Sec., Marie Stone; Treas., Mark Harding
Institution Type/Description: History Museum. Listed on the National Register of Historic Places.
Collections: period artifacts; tools; baskets; hunting & fishing implements; weapons; domestic utensils.
Hours & Admission Prices: Call for hours.

Mattapoisett

MATTAPOISETT MUSEUM AND CARRIAGE HOUSE, (M), 5 Church St., Mattapoisett, MA 02739-2618. Mailing Address: P.O. Box 535, Mattapoisett, MA 02739-0535. Tel.: 508-758-2844.
E-mail: mattapoisett.museum@verizon.net
Web Site: www.mattapoisetthistoricalsociety.org
Founded: 1958.
Congressional District: 10
Key Personnel: Pres., Seth F. Mendell; Cur., Elizabeth Hutchison.
Personnel Profile: Part-Time Paid 1; Part-Time Volunteers 60.
Governing Authority: society; directors. Parent Institution: Mattapoisett Historical Society, Inc. Tax-exempt.
Institution Type/Description: History Museum: housed in the former Mattapoisett Christian Meeting House; built in 1821.
Collections: furniture; glass; guns; china; costumes; doll furniture; toys; canes; maritime & whaling items; period farm equipment; carpenters' tools;

cobblers' tools; ice-cutting equipment; baskets; kitchen equipment; blacksmith shop; model of saltworks; period horse-drawn vehicles; local, prehistoric stone artifacts; early spinning & weaving equipment; historic documents, maps & photographs. Historic Building: Carriage House.
Research Fields: local cemeteries; whaling paper & photographs; early houses.
Facilities: 100-vol. library of books on local history, genealogy & whaling available for use on premises.
Activities: guided tours; permanent & temporary exhibitions; children & adult programs; seasonal events; lectures; children's programs.
Publications: booklets, Historical Summary of Mattapoisett; Shipbuilders of Mattapoisett; Mattapoisett & Old Rochester; postcards, Memories of Mattapoisett.
Hours & Admission Prices: July-Aug. Wed.-Sat. 1-4; Sept.-June by appointment only. Adults $3, children 6 & over $1; discounts to groups; members no charge. &
Attendance: 800 (estimated)
Membership: Individual $20; Family $30; Corporate $100; Business $100; Life $150; Family Life $200.

Medford

ROYALL HOUSE ASSOCIATION, 15 George St., Medford, MA 02155-4513. Tel.: 781-396-9032.
E-mail: royallhouseevent@aol.com
Web Site: www.royallhouse.org
Founded: 1906.
Congressional District: 7
Key Personnel: Exec. Dir., Tom Lincoln; Vice Pres., John Woods.
Personnel Profile: Part-Time Paid 1; Part-Time Volunteers 20.
Governing Authority: nonprofit organization. Tax-exempt: 501(c)(3).
Institution Type/Description: Colonial History Museum.
Collections: period furniture & furnishings; oil paintings; extant slave quarters.
Activities: guided tours; lectures.
Publications: Royall House Reporter
Hours & Admission Prices: Tours: May 17-Oct. 26 Sat.-Sun. 1, 2, & 3. Families $16, adults $7, senior citizens & students $5; discounts to groups; children 17 & under & members no charge.
Attendance: 1,837 (accurate)
Membership: Student & Senior Citizen $20; Individual $25; Family $35; Single Life $300; Couple Life $400.

TUFTS UNIVERSITY ART GALLERY, (M), Aidekman Arts Center, 40 Talbot Ave., Medford, MA 02155. Tel.: 617-627-3518. Fax: 617-627-3121.
E-mail: artgallery@tufts.edu
Web Site: artgallery.tufts.edu
Founded: 1991.
Congressional District: 7 & 8
Key Personnel: Chm. (V), Laura Roberts; Chm. (V), Kenneth Aidekman; Dir. Galleries & Collections, Amy Ingrid Schlegel, Ph.D.; Preparator & Registrar, Doug Bell; Administrative Asst., Hannah Swartz.
Personnel Profile: Full-Time Paid 3; Part-Time Paid 22; Part-Time Volunteers 8.
Governing Authority: private university; nonprofit. Parent Institution: Tufts University. Tax-exempt: 501(c)(3).
Institution Type/Description: University Art Museum.
Collections: contemporary art; 19th- & 20th-century paintings, prints, sculpture, photographs; Greek & Roman antiquities; pre-Columbian art.
Research Fields: 21st-century art.
Facilities: 6,700 sq. ft. exhibit space.
Activities: lectures; loan, temporary & traveling exhibitions; educational outreach programs; guided tours; contemporary art circle.
Publications: exhibition catalogs, Imagenes e Historias/Images and Histories: Chicana Altar - Inspired Art; Friedel Dzubas: Critical Painting; Pattern Language: Clothing as Communicator; Barbara Zucker: Time Signatures; Ilya and Emilia Kabakov: The Center of Cosmic Energy; Empire and Its Discontents; Sacred Monsters: Everyday Animism in Contemporary Japanese Art and Anime; Questions Without Answers: A Photographic Prism of World Events, 1985-2010; Mexico Beyond Its Revolution/Mexico Mas Alla de Su Revolucion; Mildred's Lane: Renovating Walden; Lenore Malen: I Am The Animal.
Hours & Admission Prices: Tues.-Wed. & Fri.-Sun. 11-5, Thurs. 11-8. Suggested Donation: adults $3; discounts to AAM & ICOM members. Closed major & university holidays. &
Attendance: 9,000 (accurate)

Mendon

SOUTHWICK'S ZOO, 2 Southwick St., Mendon, MA 01756-1234. Tel.: 800-258-9182. Fax: 508-883-0242.
E-mail: betsey@southwickszoo.com
Web Site: southwickszoo.com
Founded: 1965.
Institution Type/Description: Zoo.
Collections: over 500 animals representing more than 100 species from around the world.
Facilities: 200 acre zoo; cafe; picnic area. Gift items for sale.
Activities: educational presentations; shows; field trips; pony & camel rides; kiddie rides; Parakeet feeding station; petting zoo; animal encounters.
Hours & Admission Prices: April 16-Oct. 16 daily 10-5. Adults 13 & over $20, senior citizens 62 & over and children 3-12 $15; discounts to AAM & AAA members; children 2 & under no charge. Combo Tickets: adults $28, children $25.
Attendance: 230,000 (estimated)
Membership: Senior & Child $40; Adult $60.

Middleborough

MIDDLEBOROUGH HISTORICAL ASSOCIATION, INC., 18 Jackson St., Middleborough, MA 02346-2469. Mailing Address: P.O. Box 304, Middleborough, MA 02346-0304. Tel.: 508-947-1969.
Founded: 1960.
Congressional District: 10
Key Personnel: Pres., Nancy Gedraitis; Sec., Cynthia McNair.
Personnel Profile: Part-Time Volunteers 8.
Governing Authority: society; nonprofit. Tax-exempt.
Institution Type/Description: History Museum.
Collections: G.A.R. collection; General & Mrs. Tom Thumb collection; Peirce Academy records; portraits attributed to Cephas Thompson; history; folklore; blacksmith shop; carriage shed including vehicles & early farming implements; manuscripts; ice harvesting; Maxim motor fire vehicle. Historic Houses: 1790 law office; 1700 Outhouse-Sproat Tavern; two 1820 houses.
Facilities: library of Lawrence B. Romaine's personal collection available for use under supervision; research room.
Activities: guided tours; permanent exhibitions.
Hours & Admission Prices: July-Sept. Wed. 12-3; other times by appointment. Adults $5, senior citizens $4, students $2; discounts to groups of 10 or more; children under 6 & Historical Association members no charge. Closed holidays.
Attendance: 400 (estimated)
Membership: Student $5; Single $10; Family $15; Participating $20; Sustaining $50; Benefactor $100; Life $200.

ROBBINS MUSEUM OF ARCHAEOLOGY, 17 Jackson St., Middleborough, MA 02346-2413. Mailing Address: P.O. Box 700, Middleborough, MA 02346-0700. Tel.: 508-947-9005. Fax: 508-947-9005.
E-mail: info@massarchaeology.org
Web Site: www.massarchaeology.org/museum
Founded: 1939.
Congressional District: 4
Key Personnel: Pres., Philip Graham; Vice Pres., Frederick Robinson; Museum Coord., David DeMello; Financial Officer, Daniel Lorraine.
Personnel Profile: Full-Time Volunteers 2; Part-Time Paid 1; Part-Time Volunteers 25; Interns 2.
Governing Authority: not-for-profit organization. Parent Institution: Massachusetts Archaeological Society, Inc. Tax-exempt.
Institution Type/Description: Archaeology Museum: housed in c.1920 factory building.
Collections: Native American archaeology, Paleo-Indian to present; Society archives; research library; ethnographic collection; Native American photo collection.
Research Fields: inventory of archaeological sites; collections in the town of Middleborough; native history; artifact from Southeastern Massachusetts.
Facilities: 3,000-vol. library of archaeological & historical books; journals of state society of archaeology, available for use by public; auditorium; educational facilities; 10,000 sq. ft. exhibit space; laboratory.
Activities: formal education programs for adults; educational programs & activities for children; loan & temporary exhibitions; semiannual & annual meetings of MAS.
Publications: biannual journal, Bulletin of the Massachusetts Archaeological Society; triannual, Newsletter of the Massachusetts Archaeological Society.
Hours & Admission Prices: Wed. 10-4, Thurs. 10-2. Adults $5, children $2. &

Attendance: 1,100 (estimated)
Membership: Massachusetts Archaeological Society, Inc.: Family $5 each; Senior Associate $10; Student Active $12; Senior Active (65 & over) & Associate $15; Individual Active $30; Institution & Foreign Active $40; Foreign Institution $45; Sustaining $50; Supporting $75; Patron $100; Life $500

Middleton

MIDDLETON HISTORICAL SOCIETY - LURA WOODSIDE WATKINS MUSEUM, 9 Pleasant St., Middleton, MA 01949. Mailing Address: P.O. Box 456, Middleton, MA 01949-0756. Tel.: 978-774-9301.
Founded: 1976.
Congressional District: 6
Key Personnel: Pres., Henry Tragert; Treas., Shirley Raynard; Education, Rita Kelley; Education, Gertrude Dearborn; Cur., Marge Watson.
Governing Authority: society; nonprofit. Parent Institution: Middleton Historical Society. Tax-exempt: 501(c)(3).
Institution Type/Description: Historical Society Museum.
Collections: 18th-century shoe making; early farm implements; 19th-century ladies' fashion; photographs; town government artifacts.
Research Fields: genealogy; historic houses.
Facilities: library available to the public for use on the premises; 50-seat auditorium; 2,000 sq. ft. exhibit space.
Activities: formal education programs for children; guided tours; temporary exhibitions; lectures.
Hours & Admission Prices: Sept.-May by appointment only, meetings 3rd Mon. each month. No charge; donations accepted.
Attendance: 500 (estimated)
Membership: Annual $10.

Milton

BLUE HILLS TRAILSIDE MUSEUM, 1904 Canton Ave., Milton, MA 02186-2335. Tel.: 617-333-0690, ext. 0. Fax: 617-333-0814.
E-mail: bluehills@massaudubon.org
Web Site: www.massaudubon.org
Founded: 1959.
Congressional District: 11
Key Personnel: Dir., Norman Smith; Museum Shop Mgr., Laura Liptak.
Personnel Profile: Full-Time Paid 8; Part-Time Volunteers 92.
Governing Authority: state; society. Operated by Massachusetts Audubon Society for Department of Conservation and Recreation. Tax-exempt.
Institution Type/Description: Natural History Museum & Environmental Education Center.
Collections: herpetology; zoology; geology; aviary; living collection of animals native to Blue Hills; natural history of Northeastern United States; archaeology; history of Blue Hills.
Research Fields: natural & human history of the Blue Hills Reservation; breeding study of Ambystomid salamanders; environmental education; snowy owl telemetry study.
Facilities: 700-vol. resource library; auditorium; hiking trails; workshop; cafeteria; interpretive pond walk. Museum-related items for sale.
Activities: lectures; special events; field trips; teacher training; internships; bird rehabilitation; formally organized education programs; permanent & temporary exhibitions; outreach school & community programs; summer & school vacation camps; weekend public programs; courses & workshops.
Publications: quarterly newsletter; monthly e-newsletter.
Hours & Admission Prices: Thurs.-Sun. 10-5. Adults $3, senior citizens 65 & over $2, children 2-12 $1.50; members no charge with presentation of MAS membership card. Closed New Year's Day; Thanksgiving; Christmas. &
Attendance: 125,000 (estimated)
Membership: Individual $44; Family $58.

FORBES HOUSE MUSEUM, (M), 215 Adams St., Milton, MA 02186-4215. Tel.: 617-696-1815. Fax: 617-696-1907.
E-mail: info@forbeshousemuseum.org
Web Site: forbeshousemuseum.org
Founded: 1964.
Congressional District: 11
Key Personnel: Exec. Dir., Robin M Tagliaferri; Chm. Bd. Trustees, Mary Wendell.
Personnel Profile: Full-Time Paid 2; Part-Time Paid 1; Part-Time Volunteers 60; Interns 2.
Governing Authority: nonprofit. Tax-exempt: 501(c)(3).
Institution Type/Description: Historic House: restored to 1870s-1880s.
Collections: 19th-century Forbes family furnishings & collections; Asian export porcelain, furniture & paintings; Empire & Victorian furniture;

American & European paintings, prints, & decorative arts; Abraham Lincoln & Civil War memorabilia; Lincoln log cabin replica.
Research Fields: Forbes family; U.S.-China trade history; U.S. economic development; Civil War & Abraham Lincoln; 19th-century decorative arts; garden of Asian export trees & shrubs; Victorian life.
Facilities: small library of books for use on premises; education center. Publications for sale.
Activities: special exhibitions; education programs; events; docent training; volunteer & internship opportunities; guided & group tours; lecture series.
Publications: quarterly newsletter.
Hours & Admission Prices: Tours: Wed. & Sat.-Sun. 1 & 3. Adults $8, senior citizens & students $5; discounts to AAM members; children under 5 no charge. Closed major holidays & during large events; call for details.
Attendance: 4,000 (estimated)
Membership: Senior & Students $25; Individual $30; Family $50; Donor $100; Sponsor $250; Steward $500; Benefactor $1,000.

Monterey

THE BIDWELL HOUSE MUSEUM, (M), 100 Art School Rd., Monterey, MA 01245. Mailing Address: P.O. Box 537, Monterey, MA 01245-0537. Tel.: 413-528-6888. Fax: 413-644-9997.
E-mail: bidwellhouse@gmail.com
Web Site: www.bidwellhousemuseum.org
Founded: 1990.
Congressional District: 1
Key Personnel: Pres., Robert Hoogs; Exec. Dir., Barbara Palmer.
Personnel Profile: Full-Time Paid 1; Part-Time Paid 1; Part-Time Volunteers 30; Interns 10.
Governing Authority: nonprofit organization. Tax-exempt.
Institution Type/Description: Historic House Museum: c.1750.
Collections: 18th-century decorative arts; redware; textiles; furniture; domestic & agricultural tools; lighting devices; quilts; treen.
Research Fields: 18th-century Berkshire life; mid-18th century religion.
Facilities: 1,000-vol. library of books available for use on premises; 200-acre grounds with trails & gardens; picnic area.
Activities: guided tours; lectures; concerts; workshops; trail guides; early winter gatherings; young intern programs.
Publications: newsletter, The Bidwell House.
Hours & Admission Prices: Memorial Day to Columbus Day Thurs.-Mon. 11-4. Adults $10, seniors $8, students $5; discounts to AAM & AAA members; children under 6 & members no charge.
Attendance: 1,185 (accurate)
Membership: Individual $35-$74; Family & Dual $75-$99; Benefactor $100-$149; Patron $150-$249; Old Manse Society $250-$499; Hargis & Brush Circle $500-$1,499; Bidwell Society $1,500 & $5,000.

Nantucket

ARTISTS ASSOCIATION OF NANTUCKET, 19 Washington St., Nantucket, MA 02554-3848. Mailing Address: P.O. Box 1104, Nantucket, MA 02554-1104. Tel.: 508-228-0722 (office) & 0294 (gallery). Fax: 508-228-9700.
Web Site: www.nantucketarts.org
Founded: 1945.
Congressional District: 13
Key Personnel: Mng. Dir., Cecil Barron Jensen; Gallery Dir., Robert Frazier; Gallery Asst., Susan Duane; Special Events & Membership Coord., Alison Cooley; Arts Program, Elizabeth Hunt O'Brien; Office Admin., Lucy Cobb.
Personnel Profile: Full-Time Paid 6; Part-Time Paid 1; Interns 6.
Governing Authority: nonprofit. Affiliated with Artists Association of Nantucket, Inc. Tax-exempt: 101(6).
Institution Type/Description: Art Gallery.
Collections: early 20th-century Nantucket art; contemporary art.
Facilities: Art for sale.
Activities: lectures; films; permanent & temporary exhibitions; arts & crafts workshops; adult & children's art classes. Association Sponsors: annual craft show; annual junior artists show; annual art auction; annual scholarship award.
Publications: brochure; newsletters; seasonal AAN calendar update.
Hours & Admission Prices: Mon.-Fri. 10-3; call for seasonal hours.
Attendance: 5,000
Membership: Patron $75 & up; Nantucket Artist $150.

EGAN MARITIME INSTITUTE, The Coffin School, 4 Winter St., Nantucket, MA 02554-3638. Tel.: 508-228-2505. Fax: 508-228-7069.
E-mail: egan@eganmaritime.org
Web Site: www.eganmaritime.org
Formerly: Egan Maritime Foundation

Founded: 1996.
Key Personnel: Dir., Jean H.M. Grimmer; Pres., Robert A. Egan.
Personnel Profile: Full-Time Paid 3; Full-Time Volunteers 8; Part-Time Paid 2; Part-Time Volunteers 6.
Governing Authority: private; nonprofit organization. Parent Institution: The Albert F. Eagan and Dorothy H. Egan Foundation, Inc. Tax-exempt: 501(c)(3).
Institution Type/Description: Maritime, Art & History Museum.
Collections: Nantucket history from the 18th-20th centuries.
Research Fields: Native Americans on Nantucket; history of the Coffin School; history of Maritime topics relating to Nantucket Island; American art; 19th-century women; Nantucket and Maritime literature; US. Coast Guard; Mass Human Society; Life-Saving & Shipwrecks.
Facilities: 500-vol. library; 150-seat auditorium; classroom. Museum-related items for sale.
Activities: concerts; films; formal educational programs; guided tours; lectures; temporary exhibitions; children's programs.
Publications: books relating to Nantucket; annual report.
Hours & Admission Prices: May-Oct. daily 10-4. Adults $5, student $3; discounts to AAM & ICOM members; members no charge.
Attendance: 5,000 (estimated)
Membership: Lifesaver $40-$99; Supporter $100-$249; Contributor $250-$999; Patron $1,000-$1,999; Benefactor $2,000 & up.

MARIA MITCHELL ASSOCIATION NATURAL HISTORY MUSEUM, 7 Milk St., Nantucket, MA 02554-2635. Mailing Address: 4 Vestal St., Nantucket, MA 02554-2609. Tel.: 508-228-0898.
E-mail: info@mmo.org
Formerly: Hinchman House Natural History Museum
Founded: 1903.
Key Personnel: Exec. Dir., Janet Shulte; Chm. (V), Malcolm MacNab, M.D.; Museum Shop Mgr., Cheryl Beaton.
Personnel Profile: Full-Time Paid 1; Part-Time Paid 2.
Governing Authority: Tax-exempt.
Institution Type/Description: Natural History Museum.
Collections: flowers; birds; natural history.
Facilities: Museum-related items for sale.
Activities: nature, bird-watching & marine ecology walks; environmental education.
Hours & Admission Prices: mid-June to early Sept. Mon.-Sat. 10-4; Oct. Fri.-Sat. 10-4. Adults $5, children $4; members no charge.
Attendance: 2,000 (estimated)
Membership: Individual $50; Family $100; $250; $500; $1,000; $2,500; $5,000.

* **NANTUCKET HISTORICAL ASSOCIATION, (M),** 15 Broad St., Nantucket, MA 02554-3502. Mailing Address: Box 1016, Nantucket, MA 02554-1016. Tel.: 508-228-1894, ext. 0. Fax: 508-228-5618.
Web Site: www.nha.org
Founded: 1894.
Congressional District: 10
Key Personnel: Exec. Dir., Dr. William Tramposch; Pres., Janet L. Sherlund; 1st Vice Pres., Kenneth L. Beaugrand; Treas., William J. Boardman; Curatorial & Library Assoc., Sarah Helm; Chief Cur., Ben Simons; Dir. Visitor Experience, Marjan Shirzad; Museum Shop Mgr., Georgina Winton.
Personnel Profile: Full-Time Paid 29; Part-Time Paid 50; Part-Time Volunteers 100; Interns 8.
Governing Authority: trustee; nonprofit. Branch Museums: 1847 Nantucket Whaling Museum; Peter Foulger Museum; 1686 Oldest House; 1805 Old Gaol; 1844 Hadwen House; 1746 Old Mill; 1838 Quaker Meetinghouse; Hose Cart House; Nantucket Historical Association Research Library; Whitney Gallery. Tax-exempt.
Institution Type/Description: History Museum.
Collections: items pertaining to whaling; Nantucket; maritime history; crafts; decorative arts; genealogy; fire fighting; paintings, prints & portraits; tools; furniture; manuscripts; photographs; local memorabilia.
Research Fields: whaling; local & maritime history; Quaker records; genealogy.
Facilities: 5,000-vol. research library of whaleship logbooks, maritime manuscripts, marine books; 45,000 photographs; genealogy & Nantucket history available for use on premises; reading room. Books on Nantucket history & whaling for sale.
Activities: guided tours; films; permanent, temporary & loan exhibitions; manuscript collections.
Publications: quarterly magazine, Historic Nantucket; books; exhibition catalogs; pamphlets.
Hours & Admission Prices: Museum: Feb. 2-April 14 & Nov. 2-Nov. 24

Sat.-Sun. 11-4; April 15-May 24 & Nov. 25-Dec. 1 daily 11-4; May 25-Nov. 1 & Dec.30-Dec. 31 daily 10-5; Dec. 6-Dec. 29 Fri.-Sun. 10-5. Museum & Historic Sites: adults $20, seniors & students with ID $18, youth 6-17 $5. Sites: adults $6, youth 6-17 $3. Closed New Year's Day; Thanksgiving. &

Attendance: 61,000 (accurate)

Membership: Individual $40; Supporting $60; Sustaining $100; Contributor $250; Hadwen Circle $500; Thomas Macy Associates $1,000; Mary Gardner Coffin Associates $5,000.

NANTUCKET LIGHTSHIP BASKET MUSEUM, 49 Union St., Nantucket, MA 02554-3869. Mailing Address: P.O. Box 2517, Nantucket, MA 02584-2517. Tel.: 508-228-1177. Fax: 508-228-7092.

E-mail: adminoffice@nantucketlightshipbasketmuseum.org

Web Site: www.nantucketlightshipbasketmuseum.org

Founded: 1997.

Key Personnel: Pres., Daryl Westbrook; Dir., Maryann Wasik; Vice Pres., Kathleen Myers; Vice Pres., Rafael Osona; Clerk, Peggy Kaufman; Treas., Eugene Collatz; Museum Shop Mgr., Beverly Barlow.

Personnel Profile: Full-Time Paid 2; Part-Time Paid 2; Part-Time Volunteers 65.

Governing Authority: private; nonprofit organization. Tax-exempt.

Institution Type/Description: History Museum: housed in c.1821 home.

Collections: lightship baskets from 1850s to present; photographs of early basketmakers; basketmakers workshop.

Research Fields: basketmaking history; the natural progression & evolution.

Facilities: 27-vol. library; 50-seat lecture area; 3,000 sq. ft. exhibit space; outside garden. Museum-related items for sale.

Activities: lectures; demonstrations; docent program; films; formal education programs; guided tours; loan, participatory & temporary exhibitions; broadcast programs. Annual Events: House & Garden Tour; Children's Event.

Publications: newsletter, The Newsletter.

Hours & Admission Prices: Memorial Day to Columbus Day Tues.-Sat. 10-4. Adults $4, senior citizens, students & children over 12 $2; discounts to AAM & ICOM members; members no charge. Closed Independence Day. &

Attendance: 3,000 (accurate)

Membership: Individual $25; Family $75; Cross Rip Friend $100; South Shoal Friend $500; Patron $1,000 & up.

NANTUCKET MARIA MITCHELL ASSOCIATION, 4 Vestal St., Nantucket, MA 02554-2609. Tel.: 508-228-9198. Fax: 508-228-1031.

E-mail: info@mmo.org

Web Site: www.mmo.org

Founded: 1902.

Congressional District: 9

Key Personnel: Pres. (V), Malcolm MacNab; Exec. Dir., Janet Schulte, Ph.D.; Dir. Education & Dir. Dept. Natural Science, Andrew McKenna-Foster; Dir. Astronomy, Vladimir Strelnitski, Ph.D.; Cur. Mitchell House, Jascin Leonardo-Finger; Museum Shop Mgr., Cheryl Comeau Beaton.

Personnel Profile: Full-Time Paid 6; Part-Time Paid 20; Part-Time Volunteers 20; Interns 21.

Governing Authority: nonprofit organization. Tax-exempt.

Institution Type/Description: Observatories, Natural Science & Historic House Museum.

Collections: herbarium; botany; zoology. Historic House: 1790 Maria Mitchell birthplace.

Research Fields: astronomy; natural science.

Facilities: 10,000-vol. library of natural science, astronomy & general science books available for loan.

Activities: guided tours; nature & bird walks; lectures; films; formally organized educational programs; aquarium; research training program for college undergraduates; permanent & temporary exhibitions.

Publications: annual report, The Nantucket Maria Mitchell Assn.; specialized publications of research in natural science field & research data in astronomical & astrophysics studies.

Hours & Admission Prices: Historic House & Natural Science Museum: mid-June to Labor Day Mon.-Sat. 10-4. Adults $10, children under 14 $8; library & members no charge. &

Attendance: 15,000 (estimated)

Membership: Individual & Senior Couple $50; Family $100; Contributing $250, $500, $1,000, $2,500, $5,000.

NANTUCKET SHIPWRECK & LIFESAVING MUSEUM, 158 Polpis Rd., Nantucket, MA 02554-2320. Mailing Address: 4 Winter St., Nantucket, MA 02554-3638. Tel.: 508-228-1885.

E-mail: info@nantucketshipwreck.org

Web Site: www.nantucketshipwreck.org

Founded: 1967.

Key Personnel: Exec. Dir. Egan Maritime Institute, Jean Grimmer; Pres. Bd., Allen Reinhard; Cur., Lisa McCandless.

Personnel Profile: Full-Time Paid 1; Part-Time Paid 16; Part-Time Volunteers 2.

Governing Authority: private; nonprofit organization. Parent Institution: Albert F. Egan Jr. & Dorothy H. Egan Foundation. Tax-exempt.

Institution Type/Description: History Museum.

Collections: period surfboats, beach carts & other life-saving apparatus used by the Massachusetts Humane Society, O.S.L.S.S., & U.S. Coast Guard; Fresnel lenses from Brant Point and Great Point Light; photographs.

Facilities: picnic area. Museum-related items for sale.

Activities: interpretive programs; guided tours; hands-on programs for children.

Hours & Admission Prices: Memorial Day to Columbus Day daily 10-4; other times by appointment. Adults $5, students & children $3; discounts to groups and AAM & ICOM members; members, active USCG, CAMM & museum professionals no charge. &

Attendance: 9,000 (accurate)

Membership: Lifesaver $40-$99; Supporter $100-$249; Contributor $250-$999; Patron $1,000-$1,999.

Natick

MUSEUM OF WORLD WAR II, (M), 8 Mercer Rd., Natick, MA 01760. Mailing Address: Administrative Office, 46 Eliot St., Natick, MA 01760-6042. Tel.: 508-651-1944. Fax: 508-655-1944.

E-mail: museumofworldwarii@yahoo.com

Web Site: www.museumofworldwarii.org

Founded: 1998.

Key Personnel: Dir. & Pres., Kenneth W. Rendell; Visitation & Veteran Affairs, Jeff Farrell.

Personnel Profile: Full-Time Paid 2; Full-Time Volunteers 2; Part-Time Volunteers 7.

Governing Authority: Tax-exempt.

Institution Type/Description: Military Museum.

Collections: over 100,000 artifacts relating to WWII including letters & documents, handguns & small spy weapons; photographs.

Major Exhibits: Battle of Britain, 9/15/13-3/15/14.

Research Fields: World War II.

Facilities: 10,000 sq. ft. exhibition space.

Activities: bi-weekly lectures.

Publications: World War II: Saving the Reality, 2009.

Hours & Admission Prices: Tues.-Sat. by appointment. No charge; donations accepted. &

Attendance: 10,500 (accurate)

NATICK HISTORICAL SOCIETY, 58 Eliot St., Natick, MA 01760-5542. Tel.: 508-647-4841.

E-mail: info@natickhistoricalsociety.org

Web Site: www.natickhistoricalsociety.org

Formerly: Historical Natural History & Library Society of Natick dba Natick Historical Society

Founded: 1870.

Congressional District: 10

Key Personnel: Exec. Dir., Roberto M. Rodriguez; Pres. (V), James W. Morley; Cur., Jessica Neuwirth.

Personnel Profile: Part-Time Paid 2; Part-Time Volunteers 18; Interns 2.

Governing Authority: society. Tax-exempt.

Institution Type/Description: Historical Society Museum.

Collections: John Eliot's Algonquin Bible; Native American artifacts; period artifacts; shoe factory; tools; textiles; maps; photographs; Vice Pres. Henry Wilson; Harriet Beecher Stowe; Horatio Alger, Jr.; ornithology.

Research Fields: John Eliot; Praying Indians; genealogy; Henry Wilson; local businesses & houses.

Facilities: archives.

Activities: tours; permanent exhibitions; programs; special events.

Publications: The Arrow; Images of America-Natick; Natick Air; From Many Backgrounds: The Heritage of the Eliot Church of South Natick; Home Town Natick: 1945-2000.

Hours & Admission Prices: Tues. 2-8:30, Wed. 1:30-4:30, Fri.-Sat. 10-12:30. No charge; donations accepted. &

Attendance: 1,300 (estimated)

Membership: Senior $10; Individual $30; Family $40; Contributing $75; Sponsoring $100; Sustaining $150.

New Bedford

ARTWORKS!, 384 Acushnet Ave., New Bedford, MA 02740-6238. Tel.: 508-984-1588.
E-mail: info@artworksforyou.org
Web Site: www.artworksforyou.org
Institution Type/Description: Art Gallery.
Collections: works by local, national & international artists.
Activities: educational classes & programs; artist talks; special events; workshops.
Hours & Admission Prices: Tues.-Sat. 12-5. No charge.

BUTTONWOOD PARK ZOO, 425 Hawthorn St., New Bedford, MA 02740-1418. Tel.: 508-991-6178. Fax: 508-979-1508.
E-mail: info@bpzoo.org
Web Site: bpzoo.org
Founded: 1894.
Key Personnel: Dir., Keith Lovett.
Personnel Profile: Full-Time Paid 22; Part-Time Paid 2; Interns 17.
Governing Authority: city.
Institution Type/Description: Zoo.
Collections: over 200 animals including elephants, bison, mountain lions, bears, eagles, seals, otters & farm animals.
Hours & Admission Prices: Daily 10-5. Adults New Bedford resident $6, non-resident $8; seniors & students New Bedford resident $4.50, non-resident $6; children 3-12 New Bedford resident $3, non-resident $4; children under 3 no charge. Closed New Year's; Thanksgiving; Christmas. &
Attendance: 210,000 (estimated)
Membership: Individual $45; Family $55.

MUSEUM OF MADEIRAN HERITAGE, 1 Funchal Pl., New Bedford, MA 02746. Mailing Address: 50 Madeira Ave., New Bedford, MA 02746-2343. Tel.: 508-994-2573. Fax: 508-992-5382.
E-mail: clubesss@comcast.net
Web Site: www.portuguesefeast.com
Founded: 1999.
Key Personnel: Chm. (V), Joseph V. Sousa.
Volunteer Hours: 1,300
Institution Type/Description: History Museum: dedicated to preserving the art and artifacts of the Portuguese island of Madeira.
Collections: Madeiran art & fine crafts including embroidery, lace, linens, pottery & weaving; traditional costumes.
Facilities: garden.
Hours & Admission Prices: May-Oct. Sun. 1-4. No charge; donations accepted. &
Attendance: 200 (estimated)

NEW BEDFORD ART MUSEUM, (M), 608 Pleasant St., New Bedford, MA 02740-6204. Tel.: 508-961-3072.
E-mail: info@newbedfordartmuseum.org
Web Site: www.newbedfordartmuseum.org
Key Personnel: Interim Dir., Kathryn Dinneen; Pres., Marilyn Whalley; Treas., Kevin Baptista; Museum Shop Mgr., Nooshin Navidi.
Personnel Profile: Full-Time Volunteers 1; Part-Time Paid 11; Part-Time Volunteers 1.
Governing Authority: Tax-exempt: 501(c)(3).
Institution Type/Description: Art Museum.
Collections: works by local, national & international artists; contemporary & historic art.
Major Exhibits: Color Code, 1/10/14-2/20/14; Surface Design, 3/8/14-3/18/14; University of Massachusetts at Dartmouth BFA Fine Arts Graduating Class Final Show, 4/5/14-4/28/14; New Works of Ron Rudnicki, 5/8/14-7/11/14; Collectors 3, 7/19/14-9/18/14; Synergy, Where Science & Art Meet, 10/3/14-12/29/14.
Activities: educational programs; lectures. Museum Sponsors: ArtMobile summer program.
Hours & Admission Prices: Wed.-Sun. 12-5. Adults $5, seniors & students $3; 2nd Tues. of month & members no charge. Closed holidays. &
Attendance: 5,000 (estimated)
Membership: Individual Senior/Student $15; Individual $25; Senior/Student Household $30; Household $50; Library/Business $100.

NEW BEDFORD FIRE MUSEUM, Old Station No. 4, 51 Bedford St., New Bedford, MA 02740-4815. Tel.: 508-992-2162.
Key Personnel: Dir., Larry Roy.
Governing Authority: nonprofit organization.
Institution Type/Description: Fire Museum: housed in 1867 fire station.
Collections: period fire trucks including 1840 hand-pump engine; uniforms; bells; period fire equipment, records & ledgers; fire fighting history.
Activities: interactive exhibits; historical & educational programs.
Hours & Admission Prices: July-Aug. Mon.-Sat. 12-4. Adults $3, seniors $2, children 6 & over $1; children under 6 no charge.

*** NEW BEDFORD WHALING MUSEUM, (M),** 18 Johnny Cake Hill, New Bedford, MA 02740-6398. Tel.: 508-997-0046. Fax: 508-997-0018.
E-mail: frontdesk@whalingmuseum.org
Web Site: www.whalingmuseum.org
Founded: 1903.
Congressional District: 10
Key Personnel: Chm. Bd., Janet Whitla; Pres. & CEO, James Russell; Vice Pres. Admin., Kristen Sniezek; Sr. Cur., Stuart W. Frank.
Personnel Profile: Full-Time Paid 20; Part-Time Paid 15; Part-Time Volunteers 137; Interns 4.
Governing Authority: nonprofit organization. Parent Institution: Old Dartmouth Historical Society. Subsidiary Institution: Kendall Institution. Tax-exempt: 501(c)(3) & 170(b)(1)(A).
Institution Type/Description: Whaling Museum.
Collections: history of International whaling; social, political, economic & cultural history of New Bedford area; whaling gear & tools; 89-1/2 foot half-scale model of whaling bark Lagoda which can be boarded; ship models; scrimshaw; paintings of whaling, local area & people; 66-foot complete skeleton of a blue whale; interactive exhibits on whales & conservation.
Research Fields: history of American & Arctic whaling; Arctic and local history.
Facilities: 70,000-vol. library of whaling & local history, including over 2,600 ships' logs of whaling voyages, available for research; reading room; 250-seat auditorium. Books, prints & reproductions for sale.
Activities: guided tours; lectures; films; gallery talks; formally organized education programs for children; docent program or council; inter-museum loan; permanent & temporary exhibitions.
Publications: books; newsletter, Bulletin from Johnny Cake Hill.
Hours & Admission Prices: April-Oct. daily 9-5; Nov.-March Tues.-Sat. 9-4, Sun.11-4. Adults $14, seniors 65 & over $12, students $9, children $6; children 5 & under no charge. Closed New Year's Day; Thanksgiving; Christmas. &
Attendance: 92,000 (accurate)
Membership: Student $20; Individual $40; Dual $60; Family $75.

THE ROTCH-JONES-DUFF HOUSE & GARDEN MUSEUM, INC., 396 County St., New Bedford, MA 02740-4934. Tel.: 508-997-1401. Fax: 508-997-6846.
E-mail: info@rjdmuseum.org
Web Site: www.rjdmuseum.org
Founded: 1984.
Congressional District: 13
Key Personnel: Exec. Dir., Kate Corkum; Pres. (V), William J. Hewitt; Treas., Nathanel R. Brayton.
Personnel Profile: Full-Time Paid 2; Part-Time Paid 6; Part-Time Volunteers 60.
Governing Authority: nonprofit organization. Tax-exempt: 501(c)(3).
Institution Type/Description: Historic House & Garden: c.1834, Greek Revival house & formal gardens.
Collections: 1800-1980, period furnishings.
Research Fields: historic preservation; horticulture; family genealogy; 19th-20th century decorative arts & furnishings.
Facilities: Museum-related items for sale.
Activities: guided tours; lectures; concerts; organized education programs for adults & children; docent program. Annual Events: Horticultural Symposia and Dinner Dance in June; Fall Holiday Celebration.
Publications: newsletter, The Record: Spring, Summer & Fall calendar of events.
Hours & Admission Prices: Mon.-Sat. 10-4, Sun. 12-4. Adults $6, senior citizens and AAM & AAA members $5, children under 12 $2; members no charge. Closed major holidays. &
Attendance: 10,603 (accurate)
Membership: Individual $40; Family $75; Sustainer $100; Sponsor $250; Patron $500; Benefactor $1,000.

UNIVERSITY ART GALLERY, UNIVERSITY OF MASSA-CHUSETTS DARTMOUTH, College of Visual and Performing Art, 715 Purchase St., New Bedford, MA 02740-6341. Tel.: 508-999-8555. Fax: 508-999-8912.
E-mail: lantonsen@umassd.edu
Web Site: www.umassd.edu/universityartgallery
Founded: 1987.
Congressional District: 3
Key Personnel: Dir., Viera Levitt.
Personnel Profile: Full-Time Paid 1; Part-Time Paid 7; Interns 2.
Governing Authority: public university; nonprofit. Tax-exempt: 501(c)(3).
Institution Type/Description: Art Gallery.
Collections: paintings; sculpture; photographs.
Facilities: 60-seat auditorium; 2,600 sq. ft. exhibit space; educational facilities.
Activities: arts festivals; concerts; films; formal education programs for undergraduate & graduate college students; guided tours; lectures; loan & traveling exhibitions; broadcast programs. Annual Event: art auction.
Hours & Admission Prices: June-Aug. Mon.-Sat. 9-6, Sun. 12-5; Sept.-May daily 9-6. No charge. &

Attendance: 10,000 (estimated)

New Braintree

NEW BRAINTREE HISTORICAL SOCIETY, 10 Utley Rd., New Braintree, MA 01531-9800. Mailing Address: P.O. Box 112, New Braintree, MA 01531-0112. Tel.: 508-867-8608.
E-mail: NBHS@newbraintreehistoricalsociety.org
Web Site: www.newbraintreehistoricalsociety.org
Key Personnel: Pres., Tom Fiorelli
Institution Type/Description: Historical Society.
Collections: exhibits of town events; artifacts; historical materials; photographs.
Hours & Admission Prices: By appointment only.
Membership: Individual $5; Family $10; Life $100.

Newbury

***   COFFIN HOUSE, (M),** 14 High Rd., Newbury, MA 01951. Mailing Address: 141 Cambridge St., Boston, MA 02114-2702. Tel.: 978-462-2634.
Web Site: www.historicnewengland.org
Founded: 1929.
Congressional District: 6
Key Personnel: Pres. & C.E.O., Carl Nold; Site Mgr., Bethany Groff.
Governing Authority: society; nonprofit organization. Parent Institution: Historic New England, 141 Cambridge St., Boston, MA 02114. Tel.: 617-227-3956. Tax-exempt: 501(c)(3).
Institution Type/Description: Historic House: c. 1678 Coffin House.
Collections: 17th, 18th & 19-century domestic rooms in rural New England.
Activities: school programs; guided tours & special events.
Publications: magazine, Historic New England.
Hours & Admission Prices: June-Oct. 15 1st & 3rd Sat. each month 11-5; last tour at 4pm. Adults $5; discounts to seniors, AAM, ICOM, AAA, WGBH members. Call Historic New England for further information.
Attendance: 799 (accurate)
Membership: National $35; Individual $45; Household $55; Garden & Landscape $75; Contributing and Library & School $100; Young Friends of Historic New England $100-$1,500; Friends of the Library and Archives $125; Historic Homeowner $200; Business & Ogden Codman Design Group $250; Appleton Circle $2,500 & up.

***   SPENCER-PEIRCE-LITTLE FARM, (M),** 5 Little's Lane, Newbury, MA 01951-1802. Mailing Address: 141 Cambridge St., Boston, MA 02114-2702. Tel.: 978-462-2634.
Web Site: www.historicnewengland.org
Founded: 1986.
Key Personnel: Pres. & C.E.O., Carl Nold; Site Mgr., Bethany Groff.
Governing Authority: society; nonprofit organization. Parent Institution: Historic New England, 141 Cambridge St., Boston, MA 02114. Tel. 617-227-3956. Tax-exempt.
Institution Type/Description: Historic House: c.1690 rare stone & brick cruciform-plan house.
Collections: 18th-century glass & ceramic vessels; decorative arts; farm animals.
Facilities: nature trails; farmhouse.
Activities: guided tours; lectures; school programs; special events; family activities; hiking; Foster Farm program.
Publications: magazine, Historic New England.
Hours & Admission Prices: June-Oct. 15 Thurs.-Sun. 11-5. Adults $5, children

$4; discounts to seniors, AAM, ICOM & AAA members; Historic New England members no charge. Call for more information.
Attendance: 28,600 (accurate)
Membership: National $35; Individual $45; Household $55; Garden & Landscape $75; Contributing and Library & School $100; Young Friends of Historic New England $100-$1,500; Friends of the Library and Archives $125; Historic Homeowner $200; Business & Ogden Codman Design Group $250; Appleton Circle $2,500 & up.

Newburyport

THE CUSTOM HOUSE MARITIME MUSEUM, 25 Water St., Newburyport, MA 01950-2754. Mailing Address: Newburyport Maritime Society, Inc., 25 Water St., Newburyport, MA 01950. Tel.: 978-462-8681. Fax: 978-462-8740. Facebook: The CHMM.
E-mail: info@thechmm.org
Web Site: www.thechmm.org
Founded: 1968.
Congressional District: 6
Key Personnel: Chm., Michael Mroz.
Personnel Profile: Part-Time Volunteers 40.
Governing Authority: nonprofit organization. Parent Institution: Newburyport Maritime Society, Inc. Tax-exempt: Chapter #155 Sec. 12A.
Institution Type/Description: Maritime Museum: housed in 1835 Greek Revival style Custom House.
Collections: artifacts relating to the maritime history of Merrimac Valley; objects brought back from the Orient, Europe & the South Seas in the 1800s; ship models; navigational tools & instruments; coast guard artifacts; uniforms.
Research Fields: maritime history of Newburyport & lower Merrimac Valley.
Facilities: 350-vol. library of books on maritime history & related fields, available for use on premises by appointment. Museum-related items for sale.
Activities: permanent & rotating exhibitions; lectures; workshops; education programs.
Publications: quarterly newsletter.
Hours & Admission Prices: Jan.-April Sat. 10-4, Sun.-Mon. 12-4; May-Dec. Tues.-Sat. 10-4, Sun.-Mon holidays 12-4. Adults $7, senior citizens $5, children 6-18 $3.50; active military, members & children under 6 no charge.
Attendance: 13,500 (accurate)
Membership: Individual $40; Family $75; Sustaining $100 & up.

HISTORICAL SOCIETY OF OLD NEWBURY, CUSHING HOUSE MUSEUM, (M), 98 High St., Newburyport, MA 01950-3053. Tel.: 978-462-2681. Facebook; Historical Society of Old Newbury, Cushing House Museum.
E-mail: info@newburyhist.org
Web Site: www.newburyhist.org
Founded: 1877.
Congressional District: 6
Key Personnel: Co-Pres., Christopher Armstrong; Co-Pres., David Mack; Cur., Jay Williamson.
Personnel Profile: Full-Time Paid 1; Part-Time Paid 1; Part-Time Volunteers 12.
Governing Authority: nonprofit organization. Parent Institution: Historical Society of Old Newbury. Subsidiary Institution: Cushing House Museum. Tax-exempt: 501(c)(3).
Institution Type/Description: History Museum: housed in 1808 Cushing House, a Federalist mansion, home of Caleb Cushing, first envoy to China.
Collections: paintings; silver; furniture; china; miniatures; dolls; carriages; paperweights; needlework, costumes; glass; toys; clocks; military. Historic Buildings: 19th-century barn; summer house; Jacob Perkins Mint.
Research Fields: history of Newbury, Newburyport & West Newbury; agrarian, maritime & mercantile communities; furniture & decorative arts.
Facilities: 5,000-vol. library of log books, manuscripts, deeds, documents, vital statistics & ledgers available by appointment on premises; carriage house.
Activities: guided tours; lectures; permanent & temporary exhibitions.
Publications: books & pamphlets on local history; catalogs.
Hours & Admission Prices: Museum: May-Oct. Tues.-Fri. 10-4, Sat. 12-4. Library by appointment only. Adults $8, senior citizens $7, students & children under 12 $2; discounts to groups; members no charge. Closed holidays.
Attendance: 2,000 (estimated)
Membership: Individual $30; Dual $50; Sustaining $100; Benefactor $250; Patron $500; Diplomat $1,000.

Newton

*** HISTORIC NEWTON, JACKSON HOMESTEAD AND MUSEUM, DURANT-KENRICK HOUSE & GROUNDS, (M),** 527 Washington St., Newton, MA 02458-1433. Tel.: 617-796-1450. Fax: 617-552-7228. Facebook: Historic Newton.
E-mail: cstone@newtonma.gov
Web Site: historicnewton.org
Formerly: The Newton History Museum at The Jackson Homestead
Founded: 1950.
Congressional District: 4
Key Personnel: Dir. & C.E.O., Cynthia Stone; Pres. Newton Historical Soc. (V), Carl M. Cohen; Cur., Sara Goldberg; Cur. Education, Melissa Westlake.
Personnel Profile: Full-Time Paid 6; Part-Time Paid 3; Part-Time Volunteers 3; Interns 3.
Governing Authority: municipal. Parent Institution: City of Newton. Subsidiary Institution: The Newton Historical Society, Inc. Tax-exempt: 501(c)(3).
Institution Type/Description: History Museum.
Collections: 18th & 19th-century furnishings; domestic & personal artifacts; costumes; toys; manuscripts, maps & photographs; underground railroad. Historic Buildings: 1809 Jackson Homestead; 1734 Durant-Kenrick House & Grounds.
Major Exhibits: Charles J. Connick: Adventurer in Light and Color, 10/13-5/14; Confronting Our Legacy: Slavery and Anti-Slavery in the North, 10/13-12/14; Annie Cobb - Early Woman Architect, 10/13-12/14; 3 Centuries, 3 Families, Stories of Durant-Kenrick, 10/13-12/14; Toys & Games, 10/13-12/14.
Research Fields: Newton architecture; churches; railroads; families; education; 19th to 20th-century suburban development; historic maps of Newton.
Facilities: 500-vol. library & vertical files of Newton historical material available for use by appointment; reading room.
Activities: permanent & changing exhibitions on Newton as an early railroad suburb & the Homestead as a stop on the Underground Railroad; lectures & videos; school programs; neighborhood walking; house tours; research.
Publications: Agudas Achim: Newton's Oldest Synagogue; A Century of Lace.
Hours & Admission Prices: Tues.-Fri. 11-5, Sat.-Sun. 12-5; research by appointment. Adults $5, senior citizens & children 6-17 $3; discounts to Newton residents and AAA, NARM, AAM, ICOM & NEMA members; members no charge. &
Attendance: 9,000 (accurate)
Membership: Student & Senior $20; Individual $30; Household $45; Supporting $70; Century Club $150; copper Beech Society $500.

MAYYIM HAYYIM ART GALLERY, 1838 Washington St., Newton, MA 02466. Tel.: 617-244-1836, ext. 214. Fax: 617-244-1830.
E-mail: info@mayyimhayyim.org
Web Site: www.mayyimhayyim.org
Founded: 2001.
Key Personnel: Exec. Dir., Carrie Bornstein; Coord. Devel. & Events, Jody Comins.
Personnel Profile: Full-Time Paid 2; Part-Time Paid 4; Part-Time Volunteers 50; Interns 1.
Institution Type/Description: Art Gallery.
Collections: paintings; sculpture; drawings; photographs; pottery.
Activities: temporary exhibitions; educational programs.
Hours & Admission Prices: Call for hours. No charge; donations accepted. &

Newtonville

NEW ART CENTER, 61 Washington Park, Newtonville, MA 02460-1915. Tel.: 617-964-3424. Fax: 617-630-0081.
Web Site: www.newartcenter.org
Formerly: Newton Art Center
Founded: 1976.
Institution Type/Description: Art Gallery.
Collections: paintings; sculpture.
Hours & Admission Prices: Mon.-Fri. 9-5, Sat.-Sun. 1-5. No charge; donations accepted.

North Adams

MASS MOCA, 87 Marshall St., North Adams, MA 01247-2402. Mailing Address: 1040 Mass MoCA Way, North Adams, MA 01247-2499. Tel.: 413-664-4481. Fax: 413-663-8548.
E-mail: info@massmoca.org
Web Site: www.massmoca.org
Founded: 1999.
Congressional District: 1
Key Personnel: C.E.O., Joseph Thompson; Chm. (V), Hans Morris; Financial

Dir., Andrea Hockridge; Cur., Susan Cross; Cur., Denise Markonish; Mktg. & Public Rels., Jodi Joseph.
Personnel Profile: Full-Time Paid 65; Part-Time Paid 10; Part-Time Volunteers 2; Interns 12.
Governing Authority: private; nonprofit organization. Tax-exempt: 501(c)(3).
Institution Type/Description: Art Museum.
Collections: Sol Lewitt wall drawing retrospective; Anselm Kiefer sculptures & paintings.
Major Exhibits: Life's Work: Tom Phillips and Johnny Carrera, 3/23/13-2/14; Mark Dion: The Octagon Room, 3/23/13-2/14; One Minute Film Festival: 10 Years, 3/23/13-2/14; Jason Middlebrook: My Landscape, 5/13-4/14; Guillaume Leblon, 5/13-4/14; Joseph Montgomery: Five Sets Five Reps, 5/13-4/14; Guillanme LeBlon, 5/25/13-4/1/14; Jason Middlebrook, 5/25/13-4/1/14; Freedom: Just Another Word For..., 6/13-5/14; Izhar Patkin: The Wandering Veil (T), 12/13-9/14; Darren Waterston: Uncertain Beauty, 3/14-1/15; Dying of the Light: Film as Medium and Metaphors, 3/14-2/15; Teresita Fernandez: As Above So Below, 5/14-4/15.
Research Fields: contemporary art.
Facilities: 650-seat auditorium; 100-seat cafeteria; 200,000 sq. ft. exhibit space. Museum-related items for sale.
Activities: music & dance recitals; films; formal education programs for children & college students; loan exhibitions; theater.
Hours & Admission Prices: July-Aug. daily 10-6; Sept.-June Wed.-Mon. 11-5. Adults $15, students $10, children $5; discount to groups; members no charge. Closed Thanksgiving; Christmas. &
Attendance: 120,000 (estimated)
Membership: Student & Senior $50; Individual $65; Dual & Family $95; Contributor $200; Contemporary Circle $500; Director's Forum $1,000; Tantalum Circle $2,500; Lewitt Society $5,000.

NORTH ADAMS MUSEUM OF HISTORY AND SCIENCE, Western Gateway Heritage State Park, Bldg. 5A, 9 Furnace St. Bypass, North Adams, MA 01247-3820. Tel.: 413-664-4700.
E-mail: nahs@bcn.net
Personnel Profile: Part-Time Volunteers 20.
Institution Type/Description: Historical Society Museum.
Collections: artifacts; photographs.
Hours & Admission Prices: May-Oct. Thurs.-Sat. 10-4, Sun. 1-4; Nov.-April Sat. 10-4, Sun. 1-4. No charge; donations accepted. Closed holidays.
Attendance: 2,127 (accurate)
Membership: Single $10; Family $20.

North Andover

MUSEUM OF PRINTING, 800 Massachusetts Ave., North Andover, MA 01845-4544. Mailing Address: P.O. Box 5580, Beverly, MA 01915. Tel.: 978-686-0450.
E-mail: info@museumofprinting.org
Web Site: www.museumofprinting.org
Founded: 1978.
Key Personnel: Exec. Dir., Ted Leigh; Pres. (V), Kim Pickard.
Personnel Profile: Part-Time Volunteers 11; Interns 2.
Governing Authority: nonprofit organization. Tax-exempt: 501(c)(3).
Institution Type/Description: Printing History Museum.
Collections: printing history, equipment, & technology.
Facilities: library.
Activities: Museum Sponsors: Father's Day Printing Arts Fair.
Publications: newsletter.
Hours & Admission Prices: Fri.-Sat. 10-4; other times by appointment. Adults $5, students & seniors $3; members & children under 6 no charge. Closed national holidays. &
Attendance: 2,211 (accurate)
Membership: Individual $40; Sustaining $250; Lifetime $1,000.

NORTH ANDOVER HISTORICAL SOCIETY, (M), 153 Academy Rd., North Andover, MA 01845-4037. Tel.: 978-686-4035. Fax: 978-686-6616. Facebook; North Andover Historical Society.
E-mail: director.nahistory@gmail.com
Web Site: www.northandoverhistoricalsociety.org
Founded: 1913.
Congressional District: 6
Key Personnel: Dir. & C.E.O., Carol Majahad; Pres. (V), Kathy C. Stevens.
Personnel Profile: Full-Time Paid 1; Part-Time Paid 3; Part-Time Volunteers 49.
Volunteer Hours: 1,105
Operating Expenses: 193,592
Operating Income: 202,254
Governing Authority: society. Branch Museums: Johnson Cottage; 1715 The Parson Barnard House, 179 Osgood St. Tax-exempt: 501(c)(3).

Institution Type/Description: Historical Society: housed in the Samuel Dale Stevens Memorial Building.
Collections: genealogical & local history archival collections; Johnson Cottage; period furnishings; cooking utensils; lighting devices; tools; pewter. Historic Buildings: 1789 Johnson Cottage; c.1715 Parson Barnard House: preserved & furnished to show living styles of four early owners.
Research Fields: local history including Andover's involvement in the Salem Witch Trials of 1692.
Facilities: 20,000-piece archival collection of local & regional history by appointment; reading room.
Activities: guided tours; lectures; formally organized education programs for children; permanent & temporary exhibitions; Adventures in Time enrichment programs.
Publications: pamphlets, Early Owners of the Parson Barnard House & Their Times; The Parson Barnard House; An Eighteenth Century New England Herbal Sampler; The Glacier Geologic History of North Andover & the Surrounding Area; book, Andover, Symbol of New England; republished book, 1880 Sketches of Andover; A Good Inland Town: 1649-1940; A Brief History of North Andover, Massachusetts.
Hours & Admission Prices: Johnson Cottage: Tues.-Fri. 10-3; other times by appointment. Adults $7, senior citizens 65 & over, students and children $5; discounts to AAM, ICOM, NEMA, & MTA members; members no charge. Parson Barnard House: June.-Oct. 1st Sat. each month 11-3. Admission by donation. Research Library: $15 1st hour, $10 each additional hour; discounts to students; NEMA, AAM & ICOM members no charge. Closed holidays except Veterans Day.
Attendance: 3,000 (estimated)
Membership: Senior $15; Individual $20; Family $30; Patron $50; Sponsor $100; Benefactor $250; Donor $500; Life Donor $2,500.

THE STEVENS-COOLIDGE PLACE, A PROPERTY OF THE TRUSTEES OF RESERVATIONS, 139 Andover St., North Andover, MA 01845. Tel.: 978-689-9105.

E-mail: kblock@ttor.org
Founded: 1962.
Congressional District: 6
Key Personnel: Historic Resources Mgr., Susan Hill Dolan; Supt., Kevin Block; Pres. & C.E.O., Barbara Grickson; Cultural Resources Program Dir., Cindy Brockway.
Personnel Profile: Full-Time Paid 3; Part-Time Paid 2; Part-Time Volunteers 20; Interns 2.
Governing Authority: The Trustees of Reservations, Long Hill 572 Essex St., Beverly, MA 01915. Tel.: 978-921-1944. Tax-exempt: 501(c)(3).
Institution Type/Description: Historic House: early 19th century restored Colonial Revival style house with period gardens.
Collections: Chinese porcelain; decorative arts; period furnishings.
Research Fields: papers; photographs relating to Stevens & Coolidge families.
Facilities: 91 acres of formal gardens, pastures & woodlands.
Activities: guided tours, programs, lectures.
Hours & Admission Prices: Gardens: daily 8am to sunset. House: by appointment only. No charge; donations accepted.
Attendance: 1,750
Membership: Trustees of Reservations $20; Individual $47; Family $67; Contributing $125; Supporting $165; Sustaining $350; Sponsor $600; 1891 Society $1,000-$2,500.

North Easton

THE CHILDREN'S MUSEUM IN EASTON, (M), 9 Sullivan Ave., North Easton, MA 02356-1419. Mailing Address: P.O. Box 417, North Easton, MA 02356-0417. Tel.: 508-230-3789. Fax: 508-230-7130.

E-mail: paula@childrensmuseumeaston.org
Web Site: childrensmuseumineaston.org
Founded: 1988.
Congressional District: 9
Key Personnel: C.E.O., Paula J. Peterson; Chm., Maria O'Connell Unda; Museum Shop Mgr., Karen Frick.
Personnel Profile: Full-Time Paid 4; Part-Time Paid 15; Part-Time Volunteers 200; Interns 3.
Governing Authority: nonprofit. Tax-exempt. 501(c)(3).
Institution Type/Description: Children's Museum: located in fire station in town's historical district.
Collections: hands-on exhibits.
Activities: participatory exhibits; after school workshops & programs; school vacation activities; family performances; parenting workshops.
Publications: seasonal newsletter.
Hours & Admission Prices: Tues.-Fri. 9-5, Sat.-Sun. 12-5; call for holiday & summer hours. Admission $7.50; discounts to AAM members; children under one & ACM reciprocal members no charge. Closed major holidays.

Attendance: 55,000 (estimated)
Membership: Grandparent $60; Basic Family $80; ACM Reciprocal $125.

North Oxford

CLARA BARTON BIRTHPLACE MUSEUM, 66 Clara Barton Rd., North Oxford, MA 01537-1301. Mailing Address: P.O. Box 356, North Oxford, MA 01537-0356. Tel.: 508-987-2056, ext. 2006. Fax: 508-987-2002.

E-mail: clarabartonbirthplace@bartoncenter.org
Web Site: www.clarabartonbirthplace.org
Founded: 1921.
Congressional District: 2
Key Personnel: Mgr., Donna Joly.
Personnel Profile: Part-Time Paid 1.
Governing Authority: Owned by the Barton Center for Diabetes Education, Inc. Tax-exempt.
Institution Type/Description: Historic House Museum: c.1818-20 Clara Barton Birthplace.
Collections: Clara Barton & Barton family memorabilia; Civil War relics; American Red Cross history; 19th & early 20th century correspondence.
Research Fields: 19th-century history; American Red Cross; Civil War.
Facilities: library of books & correspondence by Clara Barton available for use by appointment; residential camping facilities available year-round; conference center; meeting facilities available for rent.
Activities: guided tours; workshops; special events; Barton Center for Diabetes Education, Inc. offers year-round programs for children with diabetes & their families.
Hours & Admission Prices: June-Aug. Fri.-Sun. 10-4; Sept. Sat. 10-4; other times by appointment. Adults $6, children 6-12 $3; discounts to groups, Mass. Teachers Assn., Red Cross volunteers, AAM & ICOM members; members no charge.
Attendance: 1,200 (estimated)
Membership: Individual $20; Family $30; Founding $100; Lifetime $1,000.

North Woburn

RUMFORD HISTORICAL ASSOCIATION, 90 Elm St., North Woburn, MA 01801. Mailing Address: # 11 Lowell St., Woburn, MA 01801.

Founded: 1877.
Congressional District: 7
Key Personnel: Pres. (V), Leonard Harmon.
Personnel Profile: Part-Time Volunteers 4.
Governing Authority: nonprofit organization. Tax-exempt: 501(c)(3).
Institution Type/Description: Historic House: 1714 Count Rumford's birthplace.
Collections: history; science.
Facilities: collection of books pertaining to Count Rumford's scientific works available for inter-library loan & for use by trustees' permission.
Activities: permanent & temporary exhibitions.
Hours & Admission Prices: Sat.-Sun. 1-4:30. No charge; donations accepted. Closed New Year's Day; Christmas.

Northampton

THE BOTANIC GARDEN OF SMITH COLLEGE, (M), Lyman Plant House & Conservatory, 16 College Lane, Northampton, MA 01063-6352. Tel.: 413-585-2740 & 2742. Fax: 413-585-2744.

E-mail: garden@smith.edu
Web Site: www.smith.edu/garden
Founded: 1895.
Congressional District: 1
Key Personnel: Dir., Michael Marcotrigiano.
Personnel Profile: Full-Time Paid 14; Part-Time Paid 2; Part-Time Volunteers 35; Interns 6.
Governing Authority: college; nonprofit. Parent Institution: Smith College, Elm St., Northampton, MA 01063. Tax-exempt.
Institution Type/Description: Botanical Garden: housed in 1890s greenhouses built by Lord & Burnham, on site of Smith College campus, a 125 acre arboretum.
Collections: greenhouse, herbaceous gardens; arboretum collections representing the world's flora; herbarium; rock garden.
Major Exhibits: Woods of the World, 1/14-12/14; Orchid's of the Marquis of Lothian, 5/14-12/14.
Research Fields: plant systematics & genetics.
Facilities: botanical garden; classrooms; herbarium; conservatory; arboretum.
Activities: guided tours; formally organized education programs for undergraduate college students; permanent & temporary exhibitions; audio tours.
Publications: Index Seminum; biannual newsletter, Botanic Garden News.

Hours & Admission Prices: Daily 8:30-4. No charge; donations accepted. &
Attendance: 100,000
Membership: Student $20; Individual $50; Household $75; Contributor $150; Sustainer $500; Patron $1,000; Champion $2,000.

CALVIN COOLIDGE PRESIDENTIAL LIBRARY & MUSEUM, 20 West St., Northampton, MA 01060-3713. Tel.: 413-587-1014, 1012 or 1011. Fax: 413-587-1015.
E-mail: coolidge@forbeslibrary.org
Web Site: www.forbeslibrary.org
Formerly: Calvin Coolidge Memorial Room, Forbes Library
Founded: 1920.
Congressional District: 1
Key Personnel: Archivist, Julie Bartlett.
Personnel Profile: Part-Time Paid 1; Part-Time Volunteers 5; Interns 1.
Governing Authority: board of trustees, nonprofit. Parent Institution: Forbes Library. Tax-exempt: 501(c)(3).
Institution Type/Description: Presidential Museum.
Collections: manuscripts, books, documents, scrapbooks, personal papers; photographs; portraits; memorabilia associated with Calvin Coolidge, 30th President, Vice President, & Governor of Massachusetts.
Research Fields: manuscripts indexed on computer database.
Facilities: 4,000-vol. library of books associated with Calvin Coolidge available to adult researchers; reading room.
Activities: guided tours; lectures; permanent & temporary exhibitions; manuscript collections.
Hours & Admission Prices: Mon. & Wed. 3-9, Tues. & Thurs. 1-5, Sat. by appointment. No charge. Closed major holidays. &

HISTORIC NORTHAMPTON, 46 Bridge St., Northampton, MA 01060-2428. Tel.: 413-584-6011. Fax: 413-584-7956.
E-mail: mailbox@historic-northampton.org
Web Site: www.historic-northampton.org
Founded: 1905.
Congressional District: 1
Key Personnel: Dir., Kerry W. Buckley; Pres., Ronald Story; Museum Asst., Marie Panik.
Personnel Profile: Full-Time Paid 2; Part-Time Paid 2; Part-Time Volunteers 10.
Governing Authority: private; nonprofit organization. Tax-exempt: 501(c)(3).
Institution Type/Description: History Museum.
Collections: furniture; paintings; decorative arts; documents of the upper Connecticut River Valley; collection of regional textiles & costumes; archaeology.
Research Fields: local history; upper Connecticut River Valley; regional costume & textiles.
Facilities: education center. Museum-related items for sale.
Activities: changing & permanent exhibitions; adult public programs; school programs; guided tours; seasonal events; volunteer program.
Publications: brochure, Historical Society, Damon Education Center; Weathervane booklet series: A President in a Two-Family House: Calvin Coolidge of Northampton; books, The Parson's House of the Northampton Historical Society; booklet, Native Peoples & Museums in the Connecticut River Valley: A Guide for Learning; maps; cemetery.
Hours & Admission Prices: Museum: Tues.-Sat. 10-5, Sun. 12-5. Adults $3; members no charge. Research: by appointment. No charge. &
Attendance: 2,728 (accurate)
Membership: Student & Senior Citizens $25; Individual $40; Family $60; Supporting $100; Patron $200; Business Membership: Business Member $100; Business Sponsor $250; Business Patron $500.

NORTHAMPTON CENTER FOR THE ARTS, 17 New South St., Northampton, MA 01060-4075. Mailing Address: P.O. Box 366, Northampton, MA 01061-0366. Tel.: 413-584-7327. Fax: 413-582-9014.
E-mail: ncfa@nohoarts.org
Web Site: nohoarts.org
Institution Type/Description: Art Gallery.
Collections: works by local artists.
Activities: special events; educational programs.
Hours & Admission Prices: Tues.-Fri. 11-4.

✳ **SMITH COLLEGE MUSEUM OF ART, (M),** Elm St. at Bedford Ter., Northampton, MA 01063. Tel.: 413-585-2760. Fax: 413-585-2782.
E-mail: artmuseum@smith.edu
Web Site: www.smith.edu/artmuseum
Founded: 1879.

Congressional District: 1
Key Personnel: Dir., Jessica F. Nicoll; Assoc. Dir. Museum Svcs., David Dempsey; Assoc. Dir. Curatorial Affairs & Cur. Painting & Sculpture, Linda Muehlig; Cur. Prints, Drawings, & Photographs, Aprile Gallant; Collections Mgr. & Registrar, Louise Laplante; Cur. Education, Ann Musser; Dir. Membership & Mktg., Margi Caplan; Museum Shop Mgr., Nan Fleming.
Personnel Profile: Full-Time Paid 23; Part-Time Paid 77; Part-Time Volunteers 73; Interns 3.
Governing Authority: college. Parent Institution: Smith College. Tax-exempt: 501(c)(3).
Institution Type/Description: Art Museum.
Collections: art from most periods & cultures with special emphasis on 17th to 21st-century European & American paintings & sculpture; drawings & prints from the Renaissance to contemporary; photographs; decorative arts; new media.
Major Exhibits: River of Gold, 10/13-1/14; The Eye is a Door: Photographs by Anne Winston Spirn, 1/14-8/14; Tara Donovan: Moire, 10/14-1/15.
Research Fields: 18th- to 20th-century European & American paintings, sculpture, graphics, decorative arts; Asian & African arts.
Facilities: library; cafe. Museum-related items for sale.
Activities: audio guide; guided tours; gallery talks; education programs for undergraduate & graduate college students; inter-museum loan, traveling & temporary exhibitions. Museum Sponsors: Family Days; Second Fridays.
Publications: exhibitions catalogs; SCheMA; gallery guides; online teacher packets.
Hours & Admission Prices: Tues.-Sat. 10-4, Sun. 12-4, 2nd Fri. of month 10-8. Adults $5, seniors 65 & over $4, students 13 & over $3, youth 6-12 $2; 2nd Fri. of month 4-8 & members no charge. Closed major holidays. &
Attendance: 33,455 (accurate)
Membership: Student & Recent Alumnae $20; Educator $35; Individual $50; Household $75; Contributor $150; Sustainer $500; Patron $1,000; Champion $2,000 & up.

Northborough

NORTHBOROUGH HISTORICAL SOCIETY, INC., 50 Main St., Northborough, MA 01532. Mailing Address: 300 Ball St., Northborough, MA 01532. Tel.: 508-393-6298 & 2343.
E-mail: nhs1906@verizon.net
Web Site: www.northboroughhistsoc.org
Founded: 1906.
Congressional District: 3
Key Personnel: Pres., Kevin Carroll; Cur., Ellen Racine; Historian, Bob Ellis.
Governing Authority: society. Tax-exempt.
Institution Type/Description: General Museum: housed in 1860 Baptist Church.
Collections: lighting; apothecary supplies; kitchen utensils; clothing; tools; war souvenirs; china; button and comb; manuscripts; genealogy; portraits; school items; toys; books; town history.
Facilities: library; archives; 150-seat auditorium.
Activities: guided tours; permanent & temporary exhibits; guest speakers.
Publications: monthly newsletter.
Hours & Admission Prices: May-June & Sept.-Oct. Sun. 2-4; others times by appointment. No charge; donations accepted.
Attendance: 871 (accurate)
Membership: Student $10; Individual $15; Family $35; Life $250.

Norton

BEARD & WEIL GALLERIES, (M), 26 E. Main St., Norton, MA 02766-2311. Tel.: 508-286-3574. Fax: 508-286-3565.
E-mail: lheureux_michele@wheatoncollege.edu
Web Site: wheatoncollege.edu/gallery
Formerly: Watson Gallery, Wheaton College
Founded: 1960.
Congressional District: 4
Key Personnel: Cur. Collections & Registrar, Leah Niederstadt.
Personnel Profile: Part-Time Paid 2; Interns 2.
Governing Authority: private college; nonprofit. Parent Institution: Wheaton College. Tax-exempt.
Institution Type/Description: College Art Gallery: located within the Norton, MA historic district.
Collections: concentration on American & European paintings, prints & drawings from 15th-21st centuries; period bronzes, coins, jewelry & sculpture; Native American baskets; decorative arts with emphasis on Wedgewood & American glass; contemporary American ceramics.
Research Fields: Native American baskets; drawings; sculpture; artifacts; textiles; decorative arts; prints & paintings; antiquities; glass.
Facilities: 50-vol. in-house reference library; 1,568 sq. ft. exhibit space.
Publications: biannual exhibition catalogues; collection catalogues.

Hours & Admission Prices: Mon.-Sat. 12:30-4:30. No charge. Closed during college vacations. &

Attendance: 2,000 (estimated)

Membership: Wheaton College Friends of Art: Alumnus (within 5 yrs. of graduation) $10; Member $25; Double $40; Sponsor $60-$99; Patron $100-$500.

MASSACHUSETTS GOLF MUSEUM, 300 Arnold Palmer Blvd., Norton, MA 02766-1365. Tel.: 774-430-9100. Fax: 774-430-9101.

E-mail: info@mgalinks.org

Web Site: www.mgalinks.org/about-us/history_golfmuseum.html

Key Personnel: Exec. Dir., Jesse Menachem

Institution Type/Description: Golf Museum.

Collections: Massachusetts golf history & heritage; golfers; trophies.

Activities: video; interactive exhibits.

Hours & Admission Prices: Mon.-Fri. 9-4. No charge. &

Norwell

SOUTH SHORE NATURAL SCIENCE CENTER, INC., 48 Jacobs Lane, Norwell, MA 02061-1149. Mailing Address: P.O. Box 429, Jacobs Lane, Norwell, MA 02061-0429. Tel.: 781-659-2559. Fax: 781-659-5924.

E-mail: ssnsc@comcast.net

Web Site: www.ssnsc.org

Founded: 1962.

Congressional District: 5

Key Personnel: Exec. Dir. & Devel. Officer, Martha B. Twigg; Chm. Bd. (V), Richard Johnson; Treas., Jon Bond; Business Mgr., Tracey Cooke; Museum Shop Mgr., Wendy Blomberg; Registrar, Karen Kurkoski.

Personnel Profile: Full-Time Paid 6; Part-Time Paid 20; Part-Time Volunteers 30; Interns 1.

Governing Authority: nonprofit organization. Tax-exempt: 501(c)(3).

Institution Type/Description: Nature Center/Conservation Area.

Collections: wildlife & archaeological exhibits; flora; 16 living habitats; native animal skins & liquids.

Research Fields: bird & salamander counts; acid rain monitoring; mammal & bird road kill research; endangered species.

Facilities: research library; nature center; 100-seat auditorium; educational facilities; educational resource files; outdoor amphitheater; 3 classrooms in main site; separate cabin classroom; kitchen. Nature-related items for sale.

Activities: walking trails; environmental lecture series; study clubs; hobby workshops; organized education programs for children, adults, & families; participatory exhibits; school loan service; teacher workshops; teaching; training; adult field trips; craft programs. Museum Sponsors: Native Plant Sale; Fall Corn Festival; Nature-theme birthday parties; Travel Teas; Parents 'n Tots; 4's & 5's; Kindergarten; scout programs; nature preschool; summer day camp programs; Summer Garden Tour; after school enrichment.

Publications: quarterly program guide; members newsletter; trail guides; miscellaneous brochures; current elementary & middle school curriculum & materials.

Hours & Admission Prices: Mon.-Sat. 9:30-4:30. Adults $7, seniors $5; discounts to AAM & NEMA members and Massachusetts teachers; members no charge. Closed New Year's Day; Easter; Memorial Day; Independence Day; Labor Day; Thanksgiving; Christmas. &

Attendance: 48,000 (estimated)

Membership: Individual $25; Family $65; Contributing $100; Corporate & Sustaining $150; Patron $500; Benefactor $1,000.

Norwood

THE F. HOLLAND DAY HOUSE & NORWOOD HISTORY MUSEUM, 93 Day St., Norwood, MA 02062-2118. Tel.: 781-762-9197.

E-mail: info@norwoodhistoricalsociety.org

Web Site: www.norwoodhistoricalsociety.org

Institution Type/Description: Historic House: former home of Fred Holland Day. Listed on the National Register of Historic Places.

Collections: period artifacts; furnishings; paintings; photographs.

Hours & Admission Prices: Tours: June-Aug. Sun. 1-4; group tours & other times by appointment. Adults $5; Norwood Historical Society members no charge.

Oak Bluffs

COTTAGE MUSEUM - MARTHA'S VINEYARD CAMP MEETING ASSOCIATION, 1 Trinity Park, Oak Bluffs, MA 02557. Mailing Address: P.O. Box 1176, 80 Trinity Park, Oak Bluffs, MA 02557-1176. Tel.: 508-693-0525. Fax: 508-696-8661.

E-mail: office@mvcma.org

Web Site: www.mvcma.org/museum.htm

Key Personnel: Gen. Mgr. & Exec. Dir., Bob Clermont.

Governing Authority: nonprofit organization.

Institution Type/Description: Historic House: 1800s campground cottage. Listed on the National Register of Historic Places.

Collections: local history; period furnishings; photographs; historical documents.

Facilities: Museum-related items for sale.

Hours & Admission Prices: Summer: Mon.-Sat. 10-4, Sun. 1-4. Adults $2, children 3-12 $.50; children under 3 no charge.

Orleans

FRENCH CABLE STATION MUSEUM IN ORLEANS, 41 S. Orleans Rd., Orleans, MA 02653. Mailing Address: Box 85, Orleans, MA 02653-0085. Tel.: 508-240-1735.

E-mail: info@frenchcablestationmuseum.org

Web Site: www.frenchcablestationmuseum.org

Founded: 1971.

Congressional District: 12

Key Personnel: Pres. (V), Joseph Manas.

Personnel Profile: Full-Time Volunteers 6; Part-Time Volunteers 26.

Governing Authority: nonprofit organization. Tax-exempt: 501(c)(3).

Institution Type/Description: Atlantic Cable Terminal; Communications Museum: housed in 1891 building, built to house cable laid from France to Orleans.

Collections: all the equipment necessary to send and receive submarine cable messages from the United States to Brest, France; Heurtley magnifier; five generations of siphon recorders; artificial cables and circuitry to permit duplex operation.

Activities: guided tours; permanent exhibitions.

Publications: museum tour book.

Hours & Admission Prices: June & Sept. Fri.-Sun. 1-4; July-Aug. Thurs.-Sun. 1-4. No charge; donations accepted.

Attendance: 1,050 (accurate)

Membership: Junior $1; Individual $10; Contributing $25; Supporting $100; Life $500.

Osterville

OSTERVILLE HISTORICAL MUSEUM, (M), 155 W. Bay Rd., Osterville, MA 02655-2427. Mailing Address: P.O. Box 3, Osterville, MA 02655-0003. Tel.: 508-428-5861.

E-mail: ohs@ostervillemuseum.org

Web Site: www.ostervillemuseum.org

Founded: 1931.

Congressional District: 10

Key Personnel: Exec. Dir., Jennifer Williams; Pres. (V), Kathleen Capo.

Personnel Profile: Part-Time Paid 1; Part-Time Volunteers 50.

Governing Authority: society. Tax-exempt: 501(c)(3).

Institution Type/Description: Historical Society Museum: housed in late 18th-century Jonathan Parker House; 18th-century Cammett House; Wooden Boat Museum; Colonial Garden.

Collections: furniture; textiles; glass; china; paintings; folk art; dolls; toys; old documents & pictures pertaining to the village of Osterville; China trade exhibit; wooden boat building of Crosby catboats; wooden boats; boat building tools.

Research Fields: local history & boat shops.

Facilities: 40-seat conference room.

Activities: lectures; gallery talks; permanent exhibitions; walking tours; children's programs & activities.

Publications: brochures; e-newsletters.

Hours & Admission Prices: June-Sept. Thurs.-Sat. 10-2; other times by appointment. No charge; donations accepted.

Attendance: 6,370 (accurate)

Membership: Individual $35; Family $60; First Mate $150; Commodore $300.

Oxford

OXFORD LIBRARY MUSEUM, 339 Main St., Oxford, MA
01540-1729. Tel.: 508-987-6003. Fax: 508-987-3896.
Founded: 1903.
Congressional District: 2
Key Personnel: Head Librarian, Timothy Kelley.
Governing Authority: municipal. Affiliated with Oxford Free Public Library.
Tax-exempt.
Institution Type/Description: Local History Museum.
Collections: Nipmuc prehistoric Indian Exhibit; Colonial New England;
Revolutionary & Civil War artifacts; historic relics and displays of Oxford's
past & present.
Research Fields: genealogical searches done on request.
Facilities: library of Revolutionary & Civil War records, genealogy records,
collection of Elliot P. Joslin, M.D., works of Clara Barton, Huguenot
records & turn-of-the-century Oxford newspapers available for use on
premises; meeting room.
Activities: guided tours with one week advanced request; formally organized
education programs for children; permanent exhibitions; research assistance.
Publications: booklets, Historical Resources; Miss Clara Barton, Founder of
the Red Cross; Dr. Elliott P. Joslin, World Renowned Physician in Diabetic
Research.
Hours & Admission Prices: By appointment only. No charge; donations
accepted. &

Attendance: 150 (estimated)

Paxton

**FINE ARTS CENTER AT MIRIAM HALL, ANNA MARIA
COLLEGE,** 50 Sunset Lane, Paxton, MA 01612-1106. Tel.:
508-849-3300, ext. 442. Fax: 508-849-3408.
Web Site: www.annamaria.edu
Formerly: Moll Art Center, Anna Maria College
Founded: 1972.
Key Personnel: Dir. Art Programs, Alice Lambert.
Personnel Profile: Full-Time Paid 1; Part-Time Paid 10.
Governing Authority: private college; nonprofit. Parent Institution: Anna Maria
College.
Institution Type/Description: Art Gallery.
Collections: paintings; photographs; sculpture.
Facilities: art classrooms; photo lab.
Activities: formal education programs for students; rental gallery; senior art
exhibit; student, faculty, & alumni exhibits; professional exhibits.
Hours & Admission Prices: Call for hours. No charge; donations accepted.
Closed major holidays. &

Attendance: 600 (estimated)

Peabody

**FELTON-SMITH HISTORIC SITE - PEABODY HISTORICAL
FIRE MUSEUM,** 38 Rear Felton St., Peabody, MA 01960.
Mailing Address: Peabody Historical Society, 33 Washington St.,
Peabody, MA 01960. Tel.: 978-531-0805. Fax: 978-531-7292.
E-mail: info@peabodyhistorical.org
Web Site: www.peabodyhistorical.org
Key Personnel: Exec. Dir., William Power
Institution Type/Description: Historic Buildings.
Collections: local history; period furnishings; personal artifacts; gardens.
Historic Buildings: Nathaniel Felton Senior House; Nathaniel Felton Junior
House; Smith Barn.
Activities: rental facilities; special events.
Hours & Admission Prices: Fire Museum & Senior House: April-Nov. 1 Mon.
& Fri. 10-3. Junior House: April-Nov. 1 Mon. & Fri. 11-3.

**GEORGE PEABODY HOUSE MUSEUM & PEABODY
LEATHERWORKERS MUSEUM,** 205 Washington St., Pea-
body, MA 01960. Tel.: 978-531-0355.
Web Site: www.peabodymuseums.org
Institution Type/Description: Historic House Museum: housed in the birthplace
of George Peabody, an international merchant & financier.
Collections: George Peabody's life & career; local history & culture; period
furnishings; personal artifacts; photographs; Peabody's leather history;
tanning.
Hours & Admission Prices: House: Mon.-Wed. & Fri.-Sat. 10-3, Thurs. 10-7,
Sun. 12-4. Museum: Tues.-Thurs. & Sat. 11-3. No charge. Closed New
Year's Eve & Day; Independence Day; Labor Day; Columbus Day;
Veterans Day; Thanksgiving; Christmas.

**OSBORNE-SALATA HOUSE, RUTH HILL LIBRARY & AR-
CHIVES,** 33 Washington St., Peabody, MA 01960. Mailing
Address: 35 Washington St., Peabody, MA 01960. Tel.: 978-531-
0805. Fax: 978-531-7292.
E-mail: info@peabodyhistorical.org
Web Site: www.peabodyhistorical.org
Institution Type/Description: Historic House Museum: c.1860.
Collections: local history & culture; period furnishings; personal artifacts.
Activities: special events; rental facilities.
Hours & Admission Prices: Wed. 1-4; other times by appointment. No charge;
donations accepted.

PEABODY HISTORICAL SOCIETY, (M), 35 Washington St.,
Peabody, MA 01960-5520. Tel.: 978-531-0805. Fax: 978-531-
7292.
E-mail: phs_m@juno.com
Web Site: www.peabodyhistorical.org
Key Personnel: Pres., William Power; Vice Pres., Andrew Metropolis; Treas.,
Thomas Zellen; Cur., Heather Leavell.
Governing Authority: nonprofit organization.
Institution Type/Description: Historical Society Museum: housed in the former
home of General Gideon Foster from 1818-1831.
Collections: Peabody's history & heritage; period children's artifacts.
Facilities: Museum-related items for sale.
Activities: Annual Event: Dinner Meeting in May.
Hours & Admission Prices: Tues.-Fri. 10-3, 1st & 3rd Sun. each month; other
times by appointment. No charge.

Pembroke

PEMBROKE HISTORICAL SOCIETY, INC., 116 Center St.,
Pembroke, MA 02359. Mailing Address: P.O. Box 122, Pembroke,
MA 02359-0239. Tel.: 781-293-9083.
E-mail: pembrokehistoric@aol.com
Web Site: www.townofpembrokemass.org/historicalsociety
Founded: 1950.
Congressional District: 10
Key Personnel: Pres. (V), Mark DiGiovanni; Treas., Suzanne Scroggins; Sec.,
Lauren Richmond; Cur., Stacey Curtin; Research Dir., Karen Proctor.
Governing Authority: society. Tax-exempt.
Institution Type/Description: Historical Society Museums.
Collections: tools; memorabilia & articles of Pembroke. Historic Buildings:
1685 Adah F. Hall Memorial House; School House; Friends Meeting
House.
Research Fields: genealogy; local history.
Facilities: archives.
Activities: Annual Events: Fish Fry; 5th Grade Children's Day.
Publications: quarterly newsletter.
Hours & Admission Prices: By appointment only. No charge; donations
accepted.
Attendance: 200 (estimated)
Membership: Senior $10; Single $20; Family $30; Corporate $100.

Petersham

FISHER MUSEUM OF FORESTRY, 324 N. Main St., Petersham,
MA 01366-9504. Tel.: 978-724-3302. Fax: 978-724-3595.
E-mail: hf-edu@fas.harvard.edu
Web Site: harvardforest.fas.harvard.edu/fisher-museum
Founded: 1930.
Congressional District: 2
Key Personnel: Dir. Harvard Forest, David R. Foster.
Personnel Profile: Full-Time Paid 1; Part-Time Paid 1; Part-Time Volunteers
45.
Governing Authority: university. Affiliated with Harvard University. Tax-
exempt: 501(c)(3).
Institution Type/Description: Forestry Museum.
Collections: historical & silviculture dioramas & case exhibits; history of land;
forest ecology.
Research Fields: ecology; land-use history; effects of disturbances; ecophysi-
ology; soils; paleoecology.
Facilities: 22,000-vol. library of forestry books & journals available for use on
premises only; reading room; nature trails.
Activities: permanent & temporary exhibitions; audiovisual presentations;
formally organized education programs for K-12 students, teachers, gradu-
ate & undergraduate students.
Publications: bulletins; papers.
Hours & Admission Prices: May-Oct. Mon.-Fri. 9-4, Sat.-Sun. 12-4; Nov.-
April Mon.-Fri. 9-4. No charge; donations accepted. Closed holidays. &

Attendance: 5,000 (estimated)

PETERSHAM HISTORICAL SOCIETY, INC., 10 N. Main St., Petersham, MA 01366. Mailing Address: P.O. Box 364, Petersham, MA 01366-0364.
Founded: 1912.
Congressional District: 2
Key Personnel: Pres. (V), James Baird; Cur., Christine Mandel.
Personnel Profile: Part-Time Volunteers 32.
Governing Authority: nonprofit organization. Tax-exempt: 501(c)(3).
Institution Type/Description: History Society Museum.
Collections: local genealogies; town & church records; local historical art by Nathan Negus, George Fuller, Erastus Salisbury Field, C. Frederick Bosworth, Caroline Negus Hildreth; file on Daniel Shays & Shays' Rebellion; local products; manuscripts.
Research Fields: genealogy & local history.
Facilities: 850-vol. library of local history, genealogy, church & town papers available by appointment on premises.
Activities: guided tours; lectures; occasional informally organized educational programs; permanent & temporary exhibitions.
Publications: book, Song of the Swampland; map of Petersham; 225th anniversary address; Petersham Soldiers in the French & Indian War; semi-annual newsletter; The History of Petersham (1948).
Hours & Admission Prices: April-Nov. Sun. 1-5, Mon.-Sat. by appointment. No charge.
Attendance: 400 (estimated)
Membership: Single $10; Family $20; Patron $50; Life $150.

Pittsfield

BERKSHIRE COUNTY HISTORICAL SOCIETY, INC. - ARROWHEAD, (M), 780 Holmes Road, Pittsfield, MA 01201-7199. Tel.: 413-442-1793. Fax: 413-443-1449.
E-mail: melville@berkshire.net
Web Site: www.mobydick.org
Founded: 1962.
Congressional District: 1
Key Personnel: C.E.O. & Museum Shop Mgr., Elizabeth Sherma; Pres. (V), Arthur Stein; Office Mgr., Judith Garrett.
Personnel Profile: Full-Time Paid 1; Part-Time Paid 4; Part-Time Volunteers 40.
Governing Authority: bd. of directors. Tax-exempt.
Institution Type/Description: History House Museum: Arrowhead, home of Herman Melville, 1850-1863. A National Historic Landmark.
Collections: period furnishings; costumes; early industries; Berkshire County artifacts; decorative arts; archives; Herman Melville memorabilia; manuscripts.
Research Fields: Berkshire County political, economic, cultural, religious & social history; Herman Melville.
Facilities: 750-vol. library of books; 14,000 photographs; 500 linear feet of manuscripts, postcards, maps & atlases. Museum-related items for sale.
Activities: guided tours; lectures; field trips; gallery talks; formally organized educational programs; permanent & temporary exhibitions; special events.
Publications: quarterly newsletter.
Hours & Admission Prices: Society: Mon.-Fri. 9:30-4. Library & Archives: by appointment. Arrowhead: Memorial Day-Oct. Fri.-Wed. 10:30-4; other times by appointment. Adults $12, children 6-18 $8; discount to members; AASLH, NEMA, & BSHL members and children under 5 no charge. Closed New Year's Day; Easter; Thanksgiving; Christmas.
Attendance: 4,500 (accurate)
Membership: Individual $35; Family $50; Supporting $75; Sustaining $150; Donor $500; Patron $1,000.

BERKSHIRE MUSEUM, 39 South St., Pittsfield, MA 01201-6169. Tel.: 413-443-7171, ext.10. Fax: 413-443-2135.
E-mail: krawson@berkshiremuseum.org
Web Site: www.berkshiremuseum.org
Founded: 1903.
Congressional District: 1
Key Personnel: Exec. Dir., Stuart A. Chase; Dir. Finance, Jon C. Provost; Chm. (V), Mary Quith; Natural Science Program Mgr., Scott LaGrecha; Natural Science Program Mgr., Scott Gervis; Mgr. Collections, Leanne Hayden; Dir. Education & Programs, Maria Mingalone; Mgr. Advancement & Communications, Chris Hayden; Bldg. & Security Mgr., Brian Warner.
Personnel Profile: Full-Time Paid 21; Part-Time Paid 27; Part-Time Volunteers 191; Interns 5.
Governing Authority: nonprofit organization. Tax-exempt: 501(c)(3).
Institution Type/Description: General Museum.
Collections: American 19th- & 20th-century paintings, sculpture, works on paper; British & European paintings; Greek, Roman, Chinese, Etruscan,

Egyptian art; birds; shells; minerals; biological & paleontological specimens; scale model dinosaurs; tenth scale dioramas of animals of the world; fresh & salt water touch tank; vivaria; historical artifacts of Berkshire County; ethnographic collections including Native American & Pacific; global & local history.
Research Fields: American 19th- & 20th-century art; natural science.
Facilities: library; 250-seat auditorium with 35mm projector; classrooms. Museum-related gifts for sale.
Activities: docent tours; lectures; films; gallery talks; performing arts; formally organized education programs; permanent, temporary & traveling exhibitions; special events.
Publications: quarterly calendar; annual report.
Hours & Admission Prices: Mon.-Sat. 10-5, Sun. 12-5; docent tours by appointment. Adults $11, children 3-18 $6; discounts to AAM, ICOM & CNECM members; ASTC members, members & children under 3 no charge. Closed New Year's Day; Memorial Day; Independence Day; Labor Day; Thanksgiving; Christmas.
Attendance: 85,000 (accurate)
Membership: Single $45; Dual $60; Family $75; Special Members $100-$500; Corporate Members $200 & up; Crane Society $1,000 & up.

✳ HANCOCK SHAKER VILLAGE, INC., (M), 1843 W. Housatonic St., Pittsfield, MA 01201. Mailing Address: P.O. Box 927, Pittsfield, MA 01202-0927. Tel.: 413-443-0188. Fax: 413-447-9357.
E-mail: info@hancockshakervillage.org
Web Site: www.hancockshakervillage.org
Founded: 1960.
Congressional District: 1
Key Personnel: Chm. (V), Ron Walter; C.E.O., Linda Steigleder; Bd. Trustees Chm., Mary Rentz; Cur. Collections, Lesley Herzberg; Dir. Interpretation & Education, Todd Burdick; Museum Shop Mgr., Kathy Kester.
Personnel Profile: Full-Time Paid 10; Part-Time Paid 30; Part-Time Volunteers 110; Interns 3.
Governing Authority: nonprofit organization. Tax-exempt: 501(c)(3).
Institution Type/Description: Historic Village Museum: 21 buildings with constructions dating back to 1790.
Collections: furniture & artifacts of Shaker communities; agricultural tools & equipment; paintings; crafts; textiles; manuscripts.
Research Fields: Shakers; utopian & communal society studies; American folk art; architecture.
Facilities: 1,000-vol. library of books pertaining to the Shakers; restaurant. Shaker publications & gifts for sale.
Activities: guided tours; lectures; films; docent program; workshops; organized education programs; graduate program in historic preservation through UMASS Amherst.
Publications: newsletter; calendar of events; flyers; site brochure.
Hours & Admission Prices: Village & Galleries: April-Oct. Adults $18, youth 13-17 $8; discounts to AAM & AAA members; members and children 12 & under no charge. Closed New Year's Day; Thanksgiving; Christmas.
Attendance: 60,000 (accurate)
Membership: Individual $50; Family & Joint $75; Contributing $125.

LICHTENSTEIN CENTER FOR THE ARTS, 28 Renne Ave., Pittsfield, MA 01201-4720. Tel.: 413-499-9348. Fax: 413-448-9811.
E-mail: berkart@taconic.net
Web Site: www.pittfield-ma.org
Founded: 1975.
Congressional District: 1
Key Personnel: Dir. Cultural Devel., Megan Whilden.
Personnel Profile: Full-Time Paid 2; Full-Time Volunteers 5; Part-Time Paid 12; Part-Time Volunteers 50; Interns 12.
Governing Authority: municipal; nonprofit. Parent Institution: City of Pittsfield. Subsidiary Institution: Friends of Berkshire Artisans, Inc. Tax-exempt: 501(c)(3).
Institution Type/Description: Art Gallery: housed in the former factory building of William Stanley, inventor of alternating current and the birthplace of General Electric.
Collections: 9 photo-realistic public murals by Berkshire Artisans public mural team & Daniel Galvez; glass window by glass sculptor Thomas Patti; sculpture works by Lyn Horton & Brendan Stechenni.
Facilities: 52-vol. library for artists; workshops; darkroom; 37,000 sq. ft. exhibit space; 160-member writers room.
Activities: guided tours; lectures. Annual Event: First Night Pittsfield on New Year's Eve.
Publications: monthly newsletter, Back Street Notes; annual, Berkshire Review.

Hours & Admission Prices: Wed.-Sat. 12-5. No charge; donations accepted. Closed legal holidays. &

Attendance: 50,000

Membership: Individual $10; Family $25; Institution $50; Contributing $100; Supporting $200; Sustaining $500; Patron $1,000; Angel $5,000; Trustee $10,000.

Plymouth

JABEZ HOWLAND HOUSE, 33 Sandwich St., Plymouth, MA 02360-3353. Tel.: 508-746-9590.

E-mail: mcharry@verizon.net

Web Site: www.pilgrimjohnhowlandsociety.org

Founded: 1897.

Congressional District: 10

Key Personnel: Pres. (V), Bradford Gorham; Museum Shop Mgr., Gail Dobbins.

Personnel Profile: Part-Time Paid 11; Part-Time Volunteers 2.

Governing Authority: nonprofit organization. Affiliated with the Pilgrim John Howland Society, Inc., 121 School St., Chelmsford, MA 01824. Tax-exempt: 501(c)(3).

Institution Type/Description: Historic House Museum: 1667 Howland House.

Collections: period household furnishings; textiles; letters by Churchill & US Presidents.

Research Fields: artifacts from archaeological digs.

Facilities: Cards & guidebooks for sale.

Activities: guided tours.

Publications: booklet, The Howland Quarterly; The Pilgrim, John Howland Soc. Centennial Book.

Hours & Admission Prices: Daily 10-4:30. Tours 10-4. Adults $5, students & seniors $4, children $1; discounts to AAA, AAM & ICOM members; members & children under 6 no charge.

Attendance: 2,898 (accurate)

Membership: Lineal descendants of Pilgrim John Howland: Annual, Marriage & Parentage $15; Entrance Membership Fee $20; Junior $25; Life $200.

JENNEY MUSEUM, 48 Summer St., Ste. 2, Plymouth, MA 02360-3400. Tel.: 508-747-4544. Fax: 508-747-4544.

E-mail: info@jenneymuseum.org

Web Site: www.jenneymuseum.org

Formerly: The Jenney Grist Mill

Founded: 2004.

Congressional District: 10

Key Personnel: Mgr., Nancy Martin; Volunteer, Leo Martin.

Personnel Profile: Full-Time Paid 1; Full-Time Volunteers 1; Part-Time Volunteers 10.

Institution Type/Description: Historic Site: established in 1636 by Pilgrim, John Jenney.

Collections: local history & culture.

Hours & Admission Prices: April-Nov. Mon. 12-5, Tues.-Sat. 9:30-5. Adults 13 & over $10, children 5-12 $8; children 4 & under no charge. Closed Easter; Thanksgiving.

Attendance: 10,000 (estimated)

Membership: Individual $25; Resident of Plymouth $35; Family $75; Sponsor $150; Benefactor $500 & up.

MAYFLOWER SOCIETY MUSEUM, 4 Winslow St., Plymouth, MA 02360-3313. Mailing Address: 67 Bay Shore Dr., Plymouth, MA 02360-2085. Tel.: 508-746-2590.

Founded: 1897.

Key Personnel: Cur., Judith A. MacDonald.

Governing Authority: society. Affiliated with General Society of Mayflower Descendants. Tax-exempt: 501(c)(3).

Institution Type/Description: Historic House: 1754 Mayflower Museum.

Collections: archives; history; 18th-century furnishings; manuscripts; quilts; needlepoint; antique fans; Staffordshire plates.

Facilities: genealogical library. Publications & museum-related items for sale.

Activities: guided tours; inter-museum loan & permanent exhibitions.

Publications: booklets, The Mayflower Society Quarterly; school materials.

Hours & Admission Prices: June & Sept. to mid-Oct. Fri.-Sun. 11-4; July-Aug. daily 11-4. Library: Mon.-Fri. 10-3:30. Museum: adults $5, senior citizens $4.50, children $1; discounts for AAA members; Plymouth residents & members no charge. Library: non-members $7 per day; members no charge. Closed Columbus Day; Memorial Day; Independence Day; Labor Day.

Attendance: 2,000 (estimated)

∗ PILGRIM HALL MUSEUM, 75 Court St., Plymouth, MA 02360-3891. Tel.: 508-746-1620. Fax: 508-746-3396.

E-mail: director@pilgrimhallmuseum.org

Web Site: www.pilgrimhallmuseum.org

Founded: 1820.

Congressional District: 10

Key Personnel: C.E.O., Patrick Browne; Pres. (V), Barrie Young; Assoc. Dir. & Cur., Stephen C. O'Neill; Dir. Devel., Robin Nutter; Dir. Visitor Svcs., Ann Young.

Personnel Profile: Full-Time Paid 4; Part-Time Paid 7; Part-Time Volunteers 40.

Governing Authority: society. Parent Institution: The Pilgrim Society. Tax-exempt: 501(c)(3).

Institution Type/Description: Historical Society & Museum.

Collections: fine & decorative arts; furniture; manuscripts; books; paintings; archeological artifacts.

Research Fields: Pilgrim history; Plymouth Colony; Town of Plymouth.

Facilities: library.

Activities: tours; lectures; permanent & temporary exhibitions; internships; research.

Publications: newsletter; books; pamphlets.

Hours & Admission Prices: Feb.-Dec. 30 daily 9:30-4:30. Adults $8, senior citizens $7, children $5; discounts to AAA members; members no charge. Closed Christmas. &

Attendance: 25,000 (accurate)

Membership: Student & Senior Citizen $25; Individual $35; Family & Dual $45; Sustaining $100; Supporting $250; Patron $500; Life $1,000; Dual Life $1,500.

PLIMOTH PLANTATION INC., (M), 137 Warren Ave., Plymouth, MA 02360-2436. Mailing Address: Box 1620, Plymouth, MA 02362-1620. Tel.: 508-746-1622, ext. 8601. Fax: 508-746-3407.

E-mail: jmonac@plimoth.org

Web Site: www.plimoth.org

Founded: 1947.

Congressional District: 12

Key Personnel: Chm. (V), Stephen Brodeur, Jr.; C.E.O. & Exec. Dir., John McDonagh; C.O.O., Ivan Lipton; Dir. Museum Programs, Elizabeth Lodge; C.F.O., Richard Lamontagne; Museum Shop Mgr., Ellie Donovan.

Personnel Profile: Full-Time Paid 125; Part-Time Paid 45; Part-Time Volunteers 100; Interns 9.

Governing Authority: nonprofit organization. Subsidiary Institution: Mayflower II. Tax-exempt: 501(c)(3).

Institution Type/Description: Outdoor Living History Museum.

Collections: archaeology; 17th-century English & Native American artifacts; house furnishings; tools; arms & armor; replica of the Mayflower ship; replica of 1627 Pilgrim Village & Wampanoag Homesite, including various costumes & furnishings; domestic livestock & horticultural practices of the period.

Research Fields: history of Plymouth Colony from 1620-1692; 17th-century English & Native American history & culture.

Facilities: 5,000-vol. library of imprints & manuscripts available for use by appointment; Craft Center; theatre; picnic area; Mayflower II. Books & museum-related items for sale. Period reproductions made for use on site and for sale.

Activities: lectures; films; demonstrations; re-creation, through first & third person interpretation, of daily life in 17th-century Plymouth; formally organized education programs for children, adults & undergraduate college students; temporary exhibits; outdoor period exhibits.

Publications: members newsletter; monographs; magazine, Plimoth Life.

Hours & Admission Prices: Mayflower II: April-Nov. daily 9-5. Adults $8, children 6-12 $6. Other sites: April-Nov. daily 9-5. Combination admission (ship & other sites): adults $25, seniors 62 & over $22; children 6-12 $15. Other sites (HomeSite, Village, Crafts Center & Visitor Center Exhibits: adults $21, seniors $19, children 6-12 $12; discount to groups; members, AAM members, Massachusetts Teacher's Association & children under 5 no charge.

Attendance: 350,000 (estimated)

Membership: Individual $40; One Plus One $55; Family $95; Contributing $135; Supporting $175; Benefactor $500; Life $3,000.

PLYMOUTH ANTIQUARIAN SOCIETY, 126 Water St., Plymouth, MA 02360. Mailing Address: Box 3773, Plymouth, MA 02361-3773. Tel.: 508-746-0012. Fax: 508-746-7908.

E-mail: pasm@verizon.net

Web Site: www.plymouthantiquariansociety.org

Founded: 1919.

Congressional District: 9

Key Personnel: Exec. Dir., Donna D. Curtin; Pres., Ron Lindeman.

Personnel Profile: Full-Time Paid 1; Part-Time Paid 6; Part-Time Volunteers 50.

Governing Authority: nonprofit. Tax-exempt: 501(c)(3).

Institution Type/Description: Historical Society Museum.
Collections: 17th-, 18th- & 19th-century furnishings, dolls, quilts; Chinese export porcelain; fans; toys; costumes, textiles. Historic Houses: 1677 Harlow Fort House; 1749 Spooner House; 1809 Hedge House.
Major Exhibits: Plymouth Samplers & Their Makers, 6/14-8/14.
Research Fields: 17th-, 18th-, 19th-century family lifestyles; decorative arts, costumes, & textiles; local history after the Pilgrims; gravestone studies.
Facilities: 17th-century craft classrooms; garden. Gift items for sale.
Activities: guided tours; changing exhibits; heritage craft demonstrations; education programs; 17th-century household arts program for children; lecture series; special events.
Publications: book, The Plymouth Colony Cook Book; The Peregrinations of Plymouth Rock; brochures; newsletter.
Hours & Admission Prices: Hedge House: Wed.-Sun. 2-6. Spooner House: Thurs.-Sat. 2-6. Harlow House: Thurs. 2-6; tour groups by appointment. Adults $5, children $2; discount to AAA, AAM, & NEMA members; members & Plymouth residents no charge. Closed Independence Day.
Attendance: 3,000 (estimated)
Membership: Individual $25; Family $40; Life $500.

RICHARD SPARROW HOUSE INC., 42 Summer St., Plymouth, MA 02360-3456. Tel.: 508-747-1240. Fax: 508-746-9521.
E-mail: director@sparrowhouse.com
Web Site: www.sparrowhouse.com
Founded: 1961.
Congressional District: 10
Key Personnel: Dir. & Museum Shop Mgr., Lois Atherton; Pres. (V), Violet Berry.
Personnel Profile: Full-Time Paid 1; Full-Time Volunteers 1; Part-Time Paid 2; Part-Time Volunteers 2.
Governing Authority: nonprofit organization. Tax-exempt.
Institution Type/Description: Historic House Museum: 1640 Richard Sparrow House; a half house with cross-summer beam construction, the house is the oldest in Plymouth. The adjoining 1720 half house, houses a craft shop, supporting local crafts.
Collections: 17th-century decorative arts.
Activities: guided tours. Museum Sponsors: pottery classes.
Publications: brochure.
Hours & Admission Prices: Daily 10-5. Adults $2, children $1.
Attendance: 4,000 (estimated)
Membership: Annual & Individual $10 & $25.

Provincetown

HUDSON D. WALKER GALLERY AT THE FINE ARTS WORK CENTER, 24 Pearl St., Provincetown, MA 02657. Tel.: 508-487-9960. Fax: 508-487-8873.
E-mail: general@fawc.org
Web Site: www.fawc.org
Key Personnel: Exec. Dir., Margaret Murphy
Institution Type/Description: Art Gallery.
Collections: works by Cape artists & Work Center Fellows; paintings; sculpture.
Activities: special events; summer workshop; educational programs.
Publications: newsletters.
Hours & Admission Prices: Mon.-Fri. 9-5.

PILGRIM MONUMENT AND PROVINCETOWN MUSEUM, One High Pole Hill Rd., Provincetown, MA 02657. Mailing Address: P.O. Box 1125, Provincetown, MA 02657-1125. Tel.: 508-487-1310. Fax: 508-487-4702.
E-mail: info@pilgrim-monument.org
Web Site: www.pilgrim-monument.org
Formerly: Cape Cod Pilgrim Memorial Association
Founded: 1892.
Congressional District: 13
Key Personnel: Pres., William Dougal; Exec. Dir., John McDonagh; Museum Shop Mgr., Bobo Hino.
Personnel Profile: Full-Time Paid 4; Part-Time Paid 10; Part-Time Volunteers 25.
Volunteer Hours: 250
Operating Expenses: 795,000
Operating Income: 1,100,000
Governing Authority: nonprofit organization. Parent Institution: Cape Cod Pilgrim Memorial Association. Tax-exempt: 501(c)(3).
Institution Type/Description: History Museum: located on the site of the 1907-10, Pilgrim Monument.
Collections: Mayflower pilgrims; whaling; fishing; Arctic Colonial & Victorian home items; theater history; exploration; pirate treasures; marine archaeology & shipwrecks; archives; photographs.

Major Exhibits: Forgotten Port, 4/14-11/14.
Research Fields: Mayflower family genealogies; Provincetown artists; local history; shipwrecks; civic organizations; Admiral McMillan's papers.
Facilities: 450-vol. library; gardens; conservation lab; picnic area. Museum-related items for sale.
Activities: lectures; annual Lighting of the Monument; educational tours; walking tours.
Publications: Provincetown Classics in History, Literature & Art; newsletter; annual report.
Hours & Admission Prices: April-May & mid-Sept. to Nov. daily 9-5; June-Sept. 15 daily 9-7. Adults $12, senior citizens 65 & over $9, students with ID $7, children 4-12 $4; discounts to MTA members & groups; children 3 & under no charge. &
Attendance: 85,000 (accurate)
Membership: Individual $35; Family $55; Business $100.

* **PROVINCETOWN ART ASSOCIATION AND MUSEUM, (M),** 460 Commercial St., Provincetown, MA 02657-2415. Tel.: 508-487-1750. Fax: 508-487-4372.
E-mail: info@paam.org
Web Site: www.paam.org
Founded: 1914.
Congressional District: 10
Key Personnel: C.E.O., Christine McCarthy; Pres. (V), James Banner; Asst. Dir. & Registrar, Peter Macara; Archivist, James Zimmerman; Education Coord., Lynn Stanley; Bookkeeper, Steven Roderick; Bldg. & Grounds, William Rigby.
Personnel Profile: Full-Time Paid 5; Part-Time Paid 7; Part-Time Volunteers 220; Interns 8.
Volunteer Hours: 3,120
Operating Expenses: 1,200,000
Operating Income: 1,200,000
Governing Authority: bd. of trustees. Tax-exempt: 501(c)(3).
Institution Type/Description: Art Museum.
Collections: 3,000 works of art; early Provincetown painters; emphasis on regional, national & international artists related to Provincetown.
Major Exhibits: PAAM 100: A Century of Creativity, 1/14-12/14; Provincetown in Print, Painting & Photography, 8/1/14-11/1/14; Karl Knaths, 8/8/14-11/2/14.
Research Fields: history of the arts in Provincetown & area.
Facilities: 1,200-vol. library of art books & magazines available for use by membership; studio classrooms. Museum-related items for sale.
Activities: lectures; films; concerts; art classes for children & adults; art festivals; symposiums; formally organized educational programs; inter-museum loan, permanent, temporary & traveling exhibitions; consignment auction; rental facility.
Publications: exhibition catalogs; catalog of the permanent collection; members' newsletters.
Hours & Admission Prices: Memorial Day to Sept. Mon.-Thurs. 11-8, Fri. 11-10, Sat.-Sun. 11-5; Oct.-April Thurs.-Sun. 12-5. Adults $7; discounts to AAA & AAM members; members & children under 12 no charge. &
Attendance: 52,000 (estimated)
Membership: Individual $50; Dual $85; Family $100; Corporate $250; Benefactor $1,000 & up.

THE SCHOOLHOUSE GALLERY, 494 Commercial St., Provincetown, MA 02657-2414. Tel.: 508-487-4800.
E-mail: mike@schoolhouseprovincetown.com
Web Site: schoolhouseprovincetown.com
Key Personnel: Dir., Mike Carroll
Institution Type/Description: Art Gallery.
Collections: works by modern & contemporary artists; paintings; photographs; printmaking.
Hours & Admission Prices: Daily by appointment.

Quincy

ADAMS NATIONAL HISTORICAL PARK, 135 Adams St., Quincy, MA 02169-1749. Tel.: 617-773-1177. Fax: 617-847-3015.
E-mail: kelly_cobble@nps.gov
Web Site: www.nps.gov/adam
Founded: 1946.
Congressional District: 11
Key Personnel: Supt., Marianne Peak; Cur., Kelly Cobble.
Personnel Profile: Full-Time Paid 15; Part-Time Paid 16; Part-Time Volunteers 10.
Governing Authority: federal. Parent Institution: National Park Service. Branch Museums: John Adams & John Quincy Adams Birthplaces, 133-141 Franklin St., Quincy, MA 02169. Off-Site Visitor Center, 1250 Hancock St., Quincy, MA. Tax-exempt: 501(c)(3).

Institution Type/Description: History Museum: housed in 1731 Adams family home.
Collections: paintings; sculpture; graphics; decorative arts; glass; numismatic. Historic Houses: 1870 stone library; 1873 carriage house.
Research Fields: history.
Facilities: 14,000-vol. library of historic books.
Activities: guided tours; lectures; permanent exhibitions; concerts; symposiums.
Publications: pamphlets, Adams National Historic Site; Grounds & Gardens. booklet, A Family's Legacy to America.
Hours & Admission Prices: April 19-Nov. 10 daily 9-5. Adults $5; children 16 & under no charge when accompanied by an adult. Golden Eagle, Golden Access & Golden Age passports honored. Tours begin at Visitor Center. ♿
Attendance: 130,000 (accurate)
Membership: Park Pass Annual Fee $10.

ADAMS NATIONAL HISTORICAL PARK-JOHN ADAMS AND JOHN QUINCY ADAMS BIRTHPLACES, 133-141 Franklin St., Quincy, MA 02269. Mailing Address: 135 Adams St., Quincy, MA 02169-1749. Tel.: 617-773-1177 & 770-1175. Fax: 617-847-3015.
Web Site: www.nps.gov/adam
Founded: 1897.
Congressional District: 11
Key Personnel: Supt., Marianne Peak; Cur., Kelly Cobble.
Personnel Profile: Full-Time Paid 15; Part-Time Paid 16; Part-Time Volunteers 10.
Governing Authority: federal. National Park Service. Tax-exempt: 501(c)(3).
Institution Type/Description: History Museum: housed in early 18th-century New England saltbox houses.
Collections: reproduction furnishings representing the lifestyles of the Adams family from the periods 1764-1784 & 1805-1807.
Research Fields: history; architecture.
Activities: guided tours.
Publications: pamphlet, John Adams & John Quincy Adams Birthplaces.
Hours & Admission Prices: April 19-Nov. 10 daily 9-5. Adults $5; Golden Eagle, Golden Age, Golden Access & children under 16 no charge. ♿
Attendance: 150,000 (estimated)
Membership: Annual Park Pass $10.

JOSIAH QUINCY HOUSE, (M), 20 Muirhead St., Quincy, MA 02170. Tel.: 617-994-5930. Fax: 617-227-9204.
Web Site: www.historicnewengland.org
Founded: 1937.
Congressional District: 11
Key Personnel: Pres., Carl Nold; Site Mgr., Leah Walczak.
Governing Authority: society; nonprofit organization. Parent Institution: Historic New England, 141 Cambridge St. Boston 02114. Tel.: 617-227-3956. Tax-exempt: 501(c)(3).
Institution Type/Description: Historic House: built in 1770 by Revolutionary War Colonel Josiah Quincy. The family produced three mayors of Boston & president of Harvard University.
Collections: period wall paneling & fireplaces; Quincy family furniture & memorabilia.
Activities: private heritage tours available with reservations; school programs.
Hours & Admission Prices: June 5 & Sept. 25 1-4; tours on the hour. Adults $5; members no charge.
Attendance: 415 (accurate)
Membership: National $35; Individual $45; Household $55; Garden & Landscape $75; Institutional $85; Contributing $100; Historic Homeowner $200; Supporting $250.

QUINCY HISTORICAL SOCIETY, Adams Academy Bldg., 8 Adams St., Quincy, MA 02169-2002. Tel.: 617-773-1144. Fax: 617-773-1872.
E-mail: info@quincyhistory.org
Web Site: quincyhistory.org
Founded: 1893.
Congressional District: 10
Key Personnel: Pres. (V), James P. Edwards; Exec. Dir., Edward Fitzgerald, Ph.D.
Governing Authority: society. Tax-exempt: 501(c)(3).
Institution Type/Description: Historical Museum: housed in 1872 Adams Academy building.
Collections: local artifacts & memorabilia; Quincy history from Native Americans to Howard Johnson's ice cream.
Research Fields: regional history; genealogy; shipbuilding; granite industry; Adams family.

Facilities: 10,000-vol. library of regional history books, pamphlets, manuscripts, maps & photographs.
Activities: guided tours; permanent & changing exhibitions; educational & cultural programs; research & publishing; intern program for college students.
Publications: books, Descendants of Edmund Quincy; Calendar of the Papers of General Joseph Palmer 1716-1788; Index to 350 Years of Quincy; The Birthplaces of Presidents John & John Quincy Adams; Wollaston of Mount Wollaston; New Beginnings-Quincy & Norfolk County; The Braintree Iron Works; Quincy's Legacy: Topics from Four Centuries of Massachusetts History; biannual periodical, Quincy History.
Hours & Admission Prices: Summer: Mon.-Fri. 9-4, Sat. 10-2; Winter: call for hours. Adults $3, seniors $1.50; members & children under 14 no charge. ♿
Membership: Student $10; Single $25; Family $35; Friend $50; Patron $100.

UNITED STATES NAVAL SHIPBUILDING MUSEUM, 739 Washington St., Quincy, MA 02169-7330. Tel.: 617-479-7900. Fax: 617-479-8792.
Web Site: www.uss-salem.org
Institution Type/Description: Naval Museum: housed aboard the USS Salem which served a 10 year career as flagship of the US Sixth Fleet in the Mediterranean and the Second Fleet in the Atlantic.
Collections: US history of shipbuilding & naval duty; U.S. Navy Cruiser Sailors Memorial; USS Salem Memorial; USS Newport News Memorial, USS Saint Paul Memorial; U.S. Navy SEALs; photographs; uniforms.
Activities: group tours; birthday parties; special functions; overnight adventure program; reunions; retirements.
Hours & Admission Prices: Temporarily closed. Adults $8, seniors & children 4-12 $6; children 3 & under no charge.

Reading

PARKER TAVERN, 103 Washington St., Reading, MA 01867-3523. Mailing Address: Reading Antiquarian Soc., P.O. Box 842, Reading, MA 01867. Tel.: 781-944-5056.
Web Site: parkertavern.org
Founded: 1916.
Congressional District: 7
Key Personnel: Pres. (V), Alan Ulrich.
Personnel Profile: Part-Time Volunteers 6.
Governing Authority: society. Reading Antiquarian Society, P.O. Box 842, Reading, MA 01867. Tax-exempt: 501(c)(3).
Institution Type/Description: History Museum: housed in 1694 Parker Tavern, saltbox style house. Headquarters for Scotch Highlanders prisoners of war during the American Revolution.
Collections: Reading artifacts; Reading history; house furnishings; manuscripts;
Research Fields: local history.
Activities: guided tours; lectures; permanent & temporary exhibitions; school loan service; fairs.
Hours & Admission Prices: May-Oct. Sun. 2-5. No charge; donations accepted.
Attendance: 700 (estimated)
Membership: Child $2; Adult $8; Couple $10.

Rehoboth

CARPENTER MUSEUM, (M), 4 Locust Ave., Rehoboth, MA 02769-2321. Mailing Address: P.O. Box 2, Rehoboth, MA 02769-0002. Tel.: 508-252-3031. Facebook: Carpenter Museum.
E-mail: carpentermuseum@gmail.com
Web Site: www.carpentermuseum.org
Key Personnel: Dir., Barbara Spencer; Cur., Laura Napolitano.
Personnel Profile: Part-Time Paid 2; Interns 2.
Governing Authority: Parent Institution: Rehoboth Antiquarian Society. Tax-exempt.
Institution Type/Description: History Museum.
Collections: Rehoboth, MA history & culture; archives; photographs; textiles; costumes; period artifacts; archaeological artifacts; ceramics & glass; farming & industrial equipment; furniture; metalwork; paintings; weapons; works on paper; woodworking tools. Historic Buildings: 1760 gambrel-style house; 1746 barn.
Activities: ladies tea & lecture; educational program for Rehoboth 3rd graders; family day; lectures; oral history projects; wine tasting & silent auction; craft show.
Publications: newsletter.
Hours & Admission Prices: March-Nov. Sun. 2-4, Tues. & Thurs. 10-3; other times by appointment. Suggested Donation: $3. ♿
Membership: Student $5; Individual $15; Couple $25; Family $35.

Rockport

ROCKPORT ART ASSOCIATION, 12 Main St., Rockport, MA
01966-1594. Tel.: 978-546-6604. Fax: 978-546-9767.
E-mail: rockportart@verizon.net
Web Site: www.rockportartassn.org
Founded: 1921.
Congressional District: 12
Key Personnel: Exec. Dir., Carol Linsky; Pres. (V), Kathryn Tuck.
Personnel Profile: Full-Time Paid 3; Part-Time Paid 4; Part-Time Volunteers 4.
Governing Authority: nonprofit organization. Tax-exempt: 501(c)(3).
Institution Type/Description: Arts Center.
Collections: New England artists; paintings; graphics; sculpture; photographs.
Historic Building: 1787 The Old Tavern.
Research Fields: RAA members past & present; Cape Ann artists.
Facilities: 500-vol. library of art books available for use on premises by
members only; studio. Cards, reproductions, publications & postcards for
sale.
Activities: community events; lectures; gallery talks; concerts; member &
invitational exhibits; painting classes for children & adults; sketch groups
for adults; adult workshops. Association Sponsors: annual fine arts auction;
annual Nativity Pageant.
Publications: quarterly, Newsletter & Calendar of Activities; annual, Catalog
of Exhibitions; book, Artists of the Rockport Art Association.
Hours & Admission Prices: Feb.-April Wed.-Fri. 10-4, Sat. 10-5, Sun. 12-5;
May Tues.-Fri. 10-4, Sat. 10-5, Sun. 12-5; June to Columbus Day Mon.-Sat.
10-5, Sun. 12-5; Oct.-Dec. Tues.-Sat. 10-4, Sun. 12-5. No charge; donations
accepted. Closed New Year's Day; Veterans Day; Thanksgiving; Christmas.
Attendance: 50,000
Membership: Youth $10; Senior Citizens over 65 $30; Associate $40; Family
$75; Patron $150; Business $250; Donor $500.

SANDY BAY HISTORICAL SOCIETY & MUSEUMS, INC., 40
King St., Rockport, MA 01966-1460. Mailing Address: P.O. Box
63, Rockport, MA 01966-0063. Tel.: 978-546-9533.
E-mail: info@sandybayhistorical.org
Web Site: www.sandybayhistorical.org
Founded: 1925.
Congressional District: 6
Key Personnel: C.E.O. & Pres. (V), Ingrid Brown.
Governing Authority: society. Tax-exempt: 501(c)(3).
Institution Type/Description: History Museum.
Collections: local history; glass; Indian artifacts; the Hannah Jumper display;
Dogtown relics; dolls; toys; granite quarry artifacts; marine; industrial;
archaeology; furniture; glass photographic plates; arts; manuscripts. His-
toric Buildings: 1832 Sewall-Scripture House & 1711 First Period Salt Box
The Old Castle.
Research Fields: Rockport & Cape Ann history.
Facilities: 350-vol. library of books, pamphlets, newspapers & clippings on
Cape Ann history available for use by arrangement with the curator; reading
room.
Activities: guided tours; lectures; permanent & temporary exhibitions.
Publications: books, Town on Sandy Bay; Planters Plea; Rockport as it Was;
Pigeon Cove its early Settlers & Their Farms 1702-1840; A Book of
Pictures; booklet, Thacher's Woe & Avery's Fall; Fish, Timber, Granite &
Gold; O Rare Harrison Cady.
Hours & Admission Prices: mid-June to mid-Sept. Mon.-Sat. 2-5; other times
by appointment. Admission $5; members no charge.
Attendance: 800
Membership: Annual $10; Family $15; Sustaining $25; Life $150.

Rowe

**THE KEMP-MCCARTHY MEMORIAL MUSEUM OF THE
ROWE HISTORICAL SOCIETY, INC.,** 282 Zoar Rd., Rowe,
MA 01367-9774. Mailing Address: P.O. Box 455, Rowe, MA
01367. Tel.: 413-339-4238.
Web Site: www.rowehistoricalsociety.org
Founded: 1963.
Congressional District: 1
Key Personnel: Pres., Melissa Quinn; Vice Pres., Katherine Heililmann; Sec.,
Sharon Hudson; Treas., Ellen Miller.
Personnel Profile: Part-Time Volunteers 18.
Governing Authority: society; bd. of trustees. Tax-exempt.
Institution Type/Description: Historical Society Museum.
Collections: Revolutionary era furnishings; town history, its people & their
way of life; artifacts; research files; manuscripts. Historic Structures: 1785
one-room schoolhouse; 1835 one-room schoolhouse.
Research Fields: local history; local genealogy; Hoosac Tunnel.

Facilities: handicapped accessible carriage house with flat-screen TV and bar
for meetings and events.
Activities: guided tours; permanent & temporary exhibitions; special summer
& fall events.
Publications: newsletter, The Bulletin, Massachusetts Rowe Historical News-
letter; books, History of Davis Mine, Story of Myrifield, Geologic History
of Rowe.
Hours & Admission Prices: July to 3rd Sun. Oct. 2-5; other times by
appointment. No charge. &
Attendance: 600 (estimated)
Membership: Annual Junior (18 & under) & Annual Senior $10; Annual
Individual $15; Annual Family $25; Organization $50; Lifetime $100;
Benefactor $250; Corporate Partner $500.

Rowley

ROWLEY HISTORICAL SOCIETY, 233 Main St., Rowley, MA
01969-1503. Mailing Address: P.O. Box 41, Rowley, MA 01969-
0041. Tel.: 978-948-7483.
E-mail: info@rowleyhistory.org
Web Site: www.rowleyhistory.org
Founded: 1918.
Congressional District: 6
Key Personnel: Pres., Susan G. Hazen; Vice Pres., G. Robert Merry; Sec.,
Kathleen Cousins; Treas., Elizabeth Hicken; Museum Shop Mgr., Shirley
G. Todd.
Personnel Profile: Part-Time Volunteers 20.
Governing Authority: society. Tax-exempt.
Institution Type/Description: Historical Building/Site Society Museum.
Collections: old English garden; toys; furniture & furnishings of early
America. Historic Houses: 1677 Platt-Bradstreet House; 1830 shoe shop;
c.1775 barn contains saws, planes, ice cutting equipment & farm utensils.
Activities: guided tours; social functions.
Publications: member newsletters.
Hours & Admission Prices: By appointment only, call 978-948-2858. Donation
$4; children under 12 & members no charge.
Membership: Student & Senior $7.50; Annual $10; Family $20; Life $100.

Roxbury

ROXBURY HERITAGE STATE PARK, 183 Roxbury St., John
Eliot Sq., Roxbury, MA 02119-1525. Tel.: 617-445-3399. Fax:
617-445-5883.
E-mail: antonio.menefee@state.ma.us
Web Site: www.mass.gov/dcr/parks/metroboston/rxhp.htm
Founded: 1901.
Congressional District: 9
Key Personnel: Dir., Antonio Menefee.
Personnel Profile: Full-Time Paid 3.
Governing Authority: society; nonprofit organization. Parent Institution: Mas-
sachusetts Dept. of Environmental Management. Tax-exempt: 501(c)(3).
Institution Type/Description: History Museum: housed in 1750 Dillaway-
Thomas House.
Collections: archives; history; military; photographs of Roxbury. Historic Site:
Pulpit Rock.
Research Fields: history.
Facilities: 200-vol. library of books on Roxbury history available for use by
appointment.
Activities: lectures; temporary exhibitions; guided tours; consultations.
Publications: newsletter.
Hours & Admission Prices: Pulpit Rock by reservation only. No charge.
Dillaway-Thomas House: Tues. & Sat. 10-5; other times by appointment.
No charge. Guided walking tours of Roxbury by appointment only. &
Attendance: 20,000 (estimated)

Salem

40 WHACKS MUSEUM: THE LIZZIE BORDEN STORY, 203
Essex St., Salem, MA 01970-3727. Mailing Address: P.O. Box
3069, Fort Mill, SC 29708. Tel.: 978-666-4416. Fax: 978-666-
4378.
Web Site: www.40whacksmuseum.com
Formerly: The True Story of Lizzie Borden
Key Personnel: Owner, Leonard Pickel; Owner, Jeanne Escher-Pickel; Owner,
John Trigg; Owner, Shelia Trigg
Institution Type/Description: History Museum.
Collections: history of the murder of Abby & Andrew Borden on Aug. 4, 1892;
trial & acquittal of their daughter, Lizzie Borden; photographs.
Facilities: Museum-related items for sale.
Activities: special events.

Hours & Admission Prices: June-Sept. & Nov. daily 11-6:30; Oct. Mon.-Thurs. 11-6:30, Fri.-Sun. 10-8:30. Adults $10, seniors 65 & over $9, children 8-12 $6; discounts to groups of 10 or more; children under 8 no charge.

THE HOUSE OF THE SEVEN GABLES, (M), 115 Derby St., Salem, MA 01970-5640. Tel.: 978-744-0991. Fax: 978-741-4350. TTY: 978-745-5391.
E-mail: kmclaughlin@7gables.org
Web Site: www.7gables.org
Founded: 1910.
Congressional District: 6
Key Personnel: Interim Exec. Dir., Kara McLaughlin; Pres. (V), Andy Meyers; Museum Shop Mgr., Everett Phillbrook.
Personnel Profile: Full-Time Paid 15; Part-Time Paid 69; Part-Time Volunteers 20; Interns 1.
Governing Authority: nonprofit organization. Tax-exempt: 501(c)(3).
Institution Type/Description: Historic Museum Site: located on Salem Harbor, this 2.5 acre area includes five original historic structures representing 17th-, 18th- and 19th-century architecture, Nathaniel Hawthorne and the lives of the people who lived on site.
Collections: furnishings; period artifacts; manuscripts. Historic Houses: 1668 The House of the Seven Gables; 1655 Retire Beckett House; 1682 Hathaway House; c.1750 Nathaniel Hawthorne's birthplace; 1830 Sea Captain's Counting House; Colonial gardens.
Research Fields: 17th-century architecture; early American decorative arts; Nathaniel Hawthorne; social history.
Facilities: visitor center; colonial gardens. Museum-related items for sale.
Activities: guided tours; theatrical programs; lecture; educational programs.
Publications: The House of Seven Gables; Chronicle of Three Old Houses; newsletter, Souvenirs.
Hours & Admission Prices: Mid-Jan.-June & Nov.-Dec. daily 10-5; July-Oct. daily 10-7. Adults $12.50, children 5-12 $7.50; discounts to seniors, and AAA & AAM members; members no charge. Closed New Year's Day; Thanksgiving; Christmas.
Attendance: 100,000 (estimated)
Membership: Senior Citizen & Student $30; Individual $45; Senior Couple $50; Family $65; Patron $150; Benefactor $250; President's Circle $500; Caroline Osgood Emmerton Society $1,000.

NEW ENGLAND PIRATE MUSEUM, 274 Derby St., Salem, MA 01970-3635. Tel.: 978-741-2800. Fax: 978-741-2902.
E-mail: salemwitchpirate@aol.com
Web Site: www.piratemuseum.com
Founded: 1995.
Institution Type/Description: History Museum.
Collections: New England pirate history; pirate treasures; pirate captains including Kidd, Blackbeard, Bellamy & Quelch.
Activities: school programs; walking tours include recreated dockside village, board a pirate ship, & bat cave. Annual Event: Haunted Happenings in October.
Hours & Admission Prices: April & Nov. Sat.-Sun. 10-5; May-Oct. daily 10-5. Haunted Happenings: extended hours. Adults $8, senior citizens 65 & over $7, children 4-13 $6; discount to groups.

✳   **PEABODY ESSEX MUSEUM, (M),** E. India Sq., Salem, MA 01970-3783. Mailing Address: 161 Essex St., Salem, MA 01970-3783. Tel.: 978-745-1876 & 9500.
E-mail: pem@pem.org
Web Site: pem.org
Founded: 1799.
Congressional District: 6
Key Personnel: Exec. Dir. & C.E.O., Dan L. Monroe; Co Chm. (V), Samuel T. Byrne; Co Chm. (V), Sean M. Healey; Pres. (V), Robert N. Shapiro; C.F.O., Monica Mackey; Deputy Dir., Josh Basseches; Chief Cur., Lynda Roscoe Hartigan; Chief Mktg. Officer, Jay Finney; Chief Education & Interpretation, Juliette Fritsch; Dir. Facilities & Security, Robert Monk; Mgr. Public Rels., April Swieconek; H.A. Crosby Forbes Cur. Asian Export Art, Karina Corrigan; Sarah Frasier Robbins Dir. Art & Nature Center and Cur. Natural History, Jane Winchell; Carolyn & Peter Lynch Cur. American Decorative Art, Dean Lahikainen; Russell W. Knight Cur. Maritime Art & History, Dr. Daniel Finamore; Cur. Native American Art, Karen Kramer Russell; Dir. Merchandising, Lynne Francis-Lunn; Ann C. Pingree Dir. The Phillips Library, Sidney E. Berger; Dir. Museum Collection Svcs., William Phippen; Dir. Exhibition Planning, Priscilla Danforth; Dir. Exhibition Research & Publishing, Kathleen Fredrickson; Head Registrar, Claudine Scoville.
Personnel Profile: Full-Time Paid 153; Part-Time Paid 97; Part-Time Volunteers 150; Interns 30.
Governing Authority: nonprofit organization. Tax-exempt: 501(c)(3) & 170(b)(1)(A).

Institution Type/Description: General Museum.
Collections: American art; Asian export art; African, Chinese, Japanese, Korean, Oceanic, Native American, Indian, & contemporary art; photography; textiles, costumes; natural history; maritime history & art; historic houses.
Major Exhibits: Artist Animal Collaborations, 9/13-3/14; Future Beauty: 30 Years of Japanese Fashion, 10/13-1/14; Impressionists on the Water, 11/13-2/14.
Research Fields: maritime art & history; Asian export art; Pacific ethnology; ethnic & folk art; American decorative art, architecture & history 17th century-present; North American archaeology; Chinese art; Indian art; photography.
Facilities: 400,000-vol. research library; reading room; 250-seat auditorium; classrooms; 93,000 sq. ft. exhibit space; gardens.
Activities: guided tours; lectures; gallery talks; educational programs; permanent & temporary exhibitions; outreach programs; special school tours; concerts; family festivals.
Publications: exhibitions & collections catalogues; books; monographic books & pamphlets; monthly members magazine.
Hours & Admission Prices: Tues.-Sun. 10-5. Historic House Tours (Ward, Crowninshield-Bentley, Gardner-Pingree) daily. Adults $15, seniors $13, students $11; discounts to AAM, ICOM & NEMA members; members, youth 16 & under and Salem residents no charge. Chinese House: adults $5. Closed New Year's Day; Thanksgiving; Christmas.
Attendance: 181,641 (accurate)
Membership: Student $40; Senior & Outside New England $50; Individual $60; Dual Senior $85; Dual $90; Family $95; Sponsor $180; Patron $350; Benefactor $600.

PHILLIPS HOUSE, 34 Chestnut St., Salem, MA 01970-3129. Mailing Address: 141 Cambridge St., Boston, MA 02114-2702. Tel.: 978-744-0440. Facebook: Phillips House Museum.
E-mail: info@phillipsmuseum.org
Web Site: www.historicnewengland.org
Founded: 1971.
Key Personnel: Pres. & C.E.O., Carl Nold; Site Mgr., Julie Arrison.
Governing Authority: private; nonprofit organization. Parent Institution: Historic New England, 141 Cambridge St., Boston, MA. Tax-exempt: 501(c)(3).
Institution Type/Description: Historic House Museum: housed in a 1821 Federal style mansion.
Collections: Chinese export porcelain; ship portraits; film archives; artifacts from the Phillips' extensive travels including Fiji throwing clubs, early 20th-century kitchen & pantry; oriental carpets; period cars & carriages. Historic Building: brick carriage house.
Facilities: Museum-related items for sale.
Activities: films; guided tours; special events; lectures. Annual Event: Antique Vehicle Meet.
Publications: newsletter; magazine, Historic New England.
Hours & Admission Prices: June-Oct. Tues.-Sun. 11-5; Nov.-May Sat.-Sun. 11-5; last tour at 4pm. Adults $5; discounts to AAM & ICOM members; members no charge.
Attendance: 4,820 (accurate)
Membership: National $35; Individual $45; Household $55; Garden & Landscape $75; Contributing and Library & School $100; Young Friends of Historic New England $100-$1,500; Friends of the Library and Archives $125; Historic Homeowner $200; Business & Ogden Codman Design Group $250; Appleton Circle $2,500 & up.

SALEM MARITIME NATIONAL HISTORIC SITE, 160 Derby St., Salem, MA 01970-5643. Tel.: 978-740-1680. Fax: 978-740-1685.
E-mail: sama_orientation_center@nps.gov
Web Site: www.nps.gov/sama
Founded: 1937.
Congressional District: 6
Key Personnel: Supt., Michael Quijano-West; Museum Cur., David Kayser; Chief Interpretation, Jonathan Parker.
Personnel Profile: Full-Time Paid 25; Part-Time Paid 12; Part-Time Volunteers 70; Interns 1.
Governing Authority: federal government. Parent Institution: National Park Service. Tax-exempt.
Institution Type/Description: Maritime, Commerce & World Trade Site: located on 9 acres on Salem Harbor.
Collections: Custom House: mid to late 19th-century office furniture & supplies; 19th-century documents. Public Stores & Scale House: mid to late 19th-century Customs equip. for weighing, measuring & inspecting cargo; trade items. Derby house: late 18th-century furniture; Derby family portraits. Narbonne house: archaeological fragments from the 17th-20th

century. Historic Buildings: 1765 Derby wharf; 1790 Central wharf; 1819 Custom house; 1819 Bonded warehouse; 1829 Scale house; c.1800 West India goods store. Historic Houses: 1761 Derby house; 1675 Narbonne house; 1780 Hawkes house.

Research Fields: commercial maritime history, particularly from the Revolutionary War through the War of 1812.

Facilities: 1,200-vol. library on maritime history available on premises. Postcards, maritime history publications & environmental publications for sale.

Activities: guided tours; 17-min. video program; permanent & temporary exhibits; interpretive programs; scheduled programs for school & special interest groups.

Publications: handbook, Maritime Salem in the Age of Sail.

Hours & Admission Prices: Daily 9-5. No charge; donations accepted. Guided tours $5. Closed New Year's Day; Thanksgiving; Christmas. &

Attendance: 652,000 (estimated)

SALEM WITCH MUSEUM, 19 1/2 Washington Sq. N., Salem, MA 01970-4096. Tel.: 978-744-1692. Fax: 978-745-4414.

E-mail: tinaj@salemwitchmuseum.com

Web Site: www.salemwitchmuseum.com

Founded: 1972.

Congressional District: 6

Key Personnel: C.E.O. & Public Rels., Bruce P. Michaud; Dir., Tina Jordan; Education, Stacy Tilney; Dir. Sales., Merry Ward.

Personnel Profile: Full-Time Paid 15; Part-Time Paid 30.

Governing Authority: individual operation; organized for profit.

Institution Type/Description: History Museum.

Collections: history of witches, witchcraft, witch hunts & witch trials.

Facilities: 125-seat theater. Books & gift items for sale.

Activities: broadcast/cable programs; organized education programs for children; training programs for professional museum workers. Museum Sponsors: Haunted Happenings.

Hours & Admission Prices: July-Aug. daily 10-7; Sept.-June daily 10-5. Call for admission prices, discount to AAM members & groups. Closed New Year's Day; Thanksgiving; Christmas. &

Attendance: 400,000

WITCH DUNGEON MUSEUM, 16 Lynde St., Salem, MA 01970. Tel.: 978-741-3570.

E-mail: salemwitchpirate@aol.com

Web Site: witchdungeon.com

Founded: 1979.

Institution Type/Description: History Museum.

Collections: Salem witch history.

Activities: witch trial performance recreated from a 1692 historical transcript; dungeon tour. Annual Event: Haunted Happenings in October.

Hours & Admission Prices: April-Nov. daily 10-6. Adults $8, seniors 65 & over $7, children 4-13 $6; discounts to groups.

WITCH HISTORY MUSEUM, 197-201 Essex St., Salem, MA 01970. Tel.: 978-741-7770. Fax: 978-741-1139.

E-mail: salemwitchpirate@aol.com

Web Site: witchhistorymuseum.com

Founded: 1997.

Institution Type/Description: History Museum.

Collections: Salem witch history; period furnishings.

Activities: live presentations; tours. Annual Event: Haunted Happenings in October.

Hours & Admission Prices: April-Nov. daily 10-6. Adults $8, seniors 65 & over $7, children 4-13 $6; discounts to groups.

THE WITCH HOUSE AKA THE JONATHAN CORWIN HOUSE, 310 1/2 Essex St., Salem, MA 01970. Tel.: 978-744-8815. Fax: 978-741-0578.

E-mail: info@witchhouse.info

Web Site: www.salemweb.com/witchhouse

Institution Type/Description: Historic House Museum: housed in the home of Judge Jonathan Corwin who served on the court for the Witchcraft Trials in 1692.

Collections: local history; 17th-century lifestyles, furnishings, & architecture.

Facilities: Museum-related items for sale.

Activities: Annual Event: Eerie Evenings at the Witch House.

Hours & Admission Prices: May to early Nov. daily 10-5. Guided Tours: adults $10.25, seniors $8.25, children 7-14 $6.25. Self-Guided Tours: adults $8.25, seniors $6.25, children 6-14 $4.25; children under 6 no charge.

Sandwich

*   **HERITAGE MUSEUMS & GARDENS, (M),** 67 Grove St., Sandwich, MA 02563-2110. Tel.: 508-888-3300. Fax: 508-888-9535. Facebook: Heritage Museums and Gardens.

E-mail: info@heritagemuseums.org

Web Site: www.heritagemuseum.org

Formerly: Heritage Plantation of Sandwich

Founded: 1969.

Congressional District: 10

Key Personnel: Chm. (V) Bd. Trustees, Charles Robinson; Dir. & C.E.O., Ellen Spear; Dir. Programming, Heather Mead; Dir. Collections, Jennifer Madden; Dir. Horticulture, Les Lutz; Dir. Mktg., Andrea Early; Museum Shop Mgr., Tom Bourne.

Personnel Profile: Full-Time Paid 27; Part-Time Paid 26; Part-Time Volunteers 180; Interns 2.

Operating Expenses: 3,800,000

Operating Income: 3,800,000

Governing Authority: nonprofit. Tax-exempt: 501(c)(3).

Institution Type/Description: General Museum.

Collections: American period automobiles; American paintings; sculpture; American folk art; American toys; working carousel; Currier & Ives prints; Native American artifacts; military miniatures; firearms; scrimshaw; living collections.

Major Exhibits: Wicked Plants, 4/14-9/14; Wicked Bugs, 4/14-10/14.

Research Fields: American history, industry, art & horticulture.

Facilities: library available by appointment only; theater; cafe. Museum-related items for sale.

Activities: lectures; special events; films; concerts; formally organized education programs for children; temporary exhibitions; adult workshops & classes.

Publications: annual report.

Hours & Admission Prices: April-Oct. see website for hours. Adults $18, children $8; discounts to groups, school groups, active military, MA teachers, AAM, NEMA, MTA & AAA members; members no charge. &

Attendance: 103,000 (accurate)

Membership: Individual $50; Dual $65; Family $100; Family Plus $125; Supporting $250; Patron $500; Benefactor $1,000; Corporate Partner $250, $500, $1,000, $1,500, $2,000.

THE OLD HOXIE HOUSE, Rte. 130, Water St., Sandwich, MA 02563-2280. Mailing Address: Sandwich Recreation Dept., P.O. Box 1336, Forestdale, MA 02644. Tel.: 508-888-4361. Fax: 508-888-5884.

E-mail: recreation@townofsandwich.net

Web Site: sandwichrec.com

Founded: 1960.

Congressional District: 12

Key Personnel: Dir., Guy J. Boucher.

Personnel Profile: Full-Time Paid 2; Part-Time Paid 10.

Governing Authority: municipal. Parent Institution: Town of Sandwich Selectmen. Tax-exempt.

Institution Type/Description: Historic House: c.1675 Hoxie House.

Collections: 17th-century furniture & furnishings.

Activities: lectured tours.

Hours & Admission Prices: mid-June to Columbus Day Mon.-Sat. 11-4:30, Sun. 1-4:30. Call for admission prices; discounts to bus tours. &

Attendance: 5,000 (accurate)

SANDWICH GLASS MUSEUM, 129 Main St., Sandwich, MA 02563. Mailing Address: P.O. Box 103, Sandwich, MA 02563-0103. Tel.: 508-888-0251. Fax: 508-888-4941.

E-mail: katie.campbell@sandwichglassmuseum.org

Web Site: www.sandwichglassmuseum.org

Founded: 1907.

Congressional District: 12

Key Personnel: Exec. Dir., Katharine H. Campbell; Pres. (V), David Leary; Cur., Dorothy Schofield; Museum Shop Mgr., Robert Ward.

Personnel Profile: Full-Time Paid 3; Part-Time Paid 5; Part-Time Volunteers 30.

Governing Authority: bd. of trustees of Sandwich Historical Society. Parent Institution: The Sandwich Historical Society. Tax-exempt: 501(c)(3).

Institution Type/Description: Glass Museum.

Collections: glass collection primarily of the Sandwich area; local historical artifacts.

Major Exhibits: Wish You Were Here, Postcards from Sandwich, 2/1/14-6/8/14; Swinging the Lamp - Ship Models from the USS Constitution

Model Shipwright Guild, 7/1/14-11/2/14; 5th Annual Glass Blowers Christmas, 11/15/14-12/30/14; Special Exhibit: Antique Dolls & Glass Toys, 11/15/14-12/30/14.

Research Fields: history of Sandwich glass; local history.

Facilities: glass reference library; auditorium. Sandwich Glass reproductions & reference books for sale.

Activities: glass & history symposiums during fall; museum docent program; permanent & temporary exhibitions; walking tours; glass & related materials; glassblowing demonstrations. Annual Events: Cape Cod Glass Show & Sale; Annual Fall Symposium. Museum Sponsor: Members' Movie Festivals in February & March.

Publications: quarterly bulletin, The Cullet; annual, The Acorn.

Hours & Admission Prices: Feb.-March Wed.-Sun. 9:30-4; April-Dec. daily 9:30-5. Library by appointment only. Adults $7, children 6-14 $1.25; discounts to groups; members no charge. Closed New Year's Day; Easter; Thanksgiving; Christmas Eve & Day.

Attendance: 70,000 (estimated)

Membership: Individual $25; Family $35; Sustaining Individual $60; Family Sustaining $75; Sponsor $125; Patron $275.

Saugus

SAUGUS IRON WORKS NATIONAL HISTORIC SITE, 244 Central St., Saugus, MA 01906-2188. Tel.: 781-233-0050. Fax: 781-231-7345.

E-mail: sair_iron_works_house@nps.gov

Web Site: www.nps.gov/sair

Founded: 1954.

Congressional District: 6

Key Personnel: Supt., Patricia S. Trap; Museum Cur., Carl Salmons-Perez; Museum Specialist, Janet Regan.

Governing Authority: federal. Administered by the National Park Service, U.S. Dept. of Interior, Washington, DC 20240. Tax-exempt.

Institution Type/Description: Historic Site: 17th century Iron Works House; reconstructed blast furnace, forge & slitting mill.

Collections: 17th-century cast & wrought iron products; tools; hardware; domestic objects; architectural fragments; furnishings; archival holdings; Native American artifacts.

Research Fields: 17th-century iron making technology.

Facilities: 1,000-vol. library of material on iron technology, colonial life, nature & environmental awareness available on premises; nature trail. Books, postcards & museum-related reproductions of colonial items for sale.

Activities: guided & self-guided tours; lectures; films; slide show; formally organized education programs for children; permanent exhibitions.

Hours & Admission Prices: April-Oct. daily 9-5. No charge.

Attendance: 23,000 (estimated)

Scituate

SCITUATE HISTORICAL SOCIETY, Laidlaw Historical Center, 43 Cudworth Rd., Scituate, MA 02066. Mailing Address: P.O. Box 276, Scituate, MA 02066-0276. Tel.: 781-545-1083 & 0942. Fax: 781-544-1249.

E-mail: director@scituatehistoricalsociety.org

Web Site: scituatehistoricalsociety.org

Founded: 1916.

Congressional District: 10

Key Personnel: Chm. (V), Elizabeth Miessner; Pres. (V), David Ball; Sec., Alison Short; Treas., Denise Castro.

Personnel Profile: Full-Time Volunteers 60; Part-Time Volunteers 50.

Governing Authority: bd. of trustees. Subsidiary Institutions: Old Stockbridge Grist Mill; 1810 Old Scituate Lighthouse; Abigail & Rebecca Bates House; Mann Farmhouse & Historical Museum; 1902 Lawson Tower; 1675 Old Oaken Bucket Homestead; 1795 barn; 1636 cattle pound; 1893 The Little Red School House; James House: Maritime & Irish Mossing Museum. Tax-exempt: 501(c)(3).

Institution Type/Description: Historic Building.

Collections: early Colonial; textiles; Indian artifacts; military; Gen. Lafayette's coach; local history; costumes; books & documents; early farming tools; cobbler's tools; sail loft; sail making tools; ship building tools; coaches; historic buildings.

Research Fields: buildings 150 years & older in town; genealogy; pertaining to the collection.

Facilities: 300-vol. library. Museum-related items for sale.

Activities: guided tours; lectures; formally organized education programs for children; field trips to museums by school & adult groups & scout organizations by appointment.

Publications: books, Early Planters of Scituate; Deane's History of Scituate; Old Scituate; Irish Mossers & Scituate Harbor Village; Briggs Shipbuilding on North River; early maps; brochure, Historic Scituate; pamphlet, Scituate 1636-1961. Images of America: Scituate; Scituate Lighthouse Guidebook; bimonthly, The Scituate Historical Society Newsletter.

Hours & Admission Prices: Cudworth House, Mann Farmhouse: Grist Mill, Old Oaken Bucket Homestead, Lawson Tower & Lighthouse: Summer: call for hours; organizations by appointment. Adults $2; children & New England Museum Assoc. no charge. Little Red Schoolhouse: daily 10-4; organizations by appointment.

Attendance: 7,500 (estimated)

Membership: Basic $25; Silver & Corporate Basic $100; Gold & Corporate Blue $250; Platinum & Corporate Gold $500

SCITUATE MARITIME AND IRISH MOSSING MUSEUM, 301 The Driftway, Scituate, MA 02066-1904. Mailing Address: P.O. Box 276, Scituate, MA 02066-0276. Tel.: 781-545-1083. Fax: 781-544-1249.

E-mail: director@scituatehistoricalsociety.org

Web Site: www.scituatehistoricalsociety.org

Founded: 1997.

Congressional District: 10

Key Personnel: Chm. (V), Elizabeth Miessner; Pres. (V), David Ball.

Personnel Profile: Part-Time Volunteers 20.

Governing Authority: private; nonprofit organization. Parent Institution: Scituate Historical Society, Scituate, MA. Tax-exempt: 501(c)(3).

Institution Type/Description: General Museum.

Collections: maritime history from 17th-century to present; North River shipbuilding; Scituate coast disasters; early U.S. Coast Guard; lighthouse establishment; Irish mossing; lifesaving services. Historic House: 1739 home of Captain Benjamin James.

Research Fields: history of Mossing industry, early lifesaving & shipwrecks in Scituate & along the South Shore.

Facilities: educational facilities. Museum-related items for sale.

Activities: guided tours; lectures; loan & participatory exhibits. Annual Event: Scituate Heritage Days.

Publications: bimonthly newsletter, Scituate Historical Society; books, Images of America: Scituate; Scituate Lighthouse Guidebook.

Hours & Admission Prices: June by appointment; July-Aug. Sat.-Sun. 1-4; Sept.-May Sun. 1-4; call for updated schedule. Adults $4, senior citizens $3; NEMA members, members, children 12 and under no charge.

Attendance: 5,000 (estimated)

Membership: Basic $25; Silver & Corporate Basic $100; Gold & Corporate Blue $250; Platinum & Corporate Gold $500.

Sheffield

THE SHEFFIELD HISTORICAL SOCIETY, 159-161 Main St., Sheffield, MA 01257. Mailing Address: P.O. Box 747, Sheffield, MA 01257. Tel.: 413-229-2694. Facebook: Sheffield Historical Society.

E-mail: shs@sheffieldhistory.org

Web Site: sheffieldhistory.org

Founded: 1972.

Governing Authority: nonprofit organization.

Institution Type/Description: Historical Society Museum: housed in the Dan Raymond House; built in 1774.

Collections: local history & culture; period furnishings; personal artifacts; early tools & equipment; ceramics; toys. Historic Buildings: 1816 Mark Dewey Hatter's Shop; c.1870 carriage barn; 1820 Parker Hall law office; 1876 education center; c.1838 smokehouse; 1834 old stone store.

Research Fields: genealogy.

Facilities: Museum-related items for sale.

Activities: educational programs; special events; workshops; permanent & temporary exhibitions; tours; community programming.

Hours & Admission Prices: Memorial Day to Columbus Day Tues.-Thurs. 11-4 by appointment.

Shelburne Falls

SHELBURNE FALLS TROLLEY MUSEUM, 14 Depot St., Shelburne Falls, MA 01370. Mailing Address: P.O. Box 272, Shelburne Falls, MA 01370. Tel.: 413-625-9443.

E-mail: trolley@sftm.org

Web Site: www.sftm.org

Founded: 1991.

Congressional District: 1

Personnel Profile: Part-Time Volunteers 10.

Governing Authority: Tax-exempt.

Institution Type/Description: History Museum.

Collections: local history; photographs; railroad & trolley artifacts; CV caboose; 1896 Wason trolley car. Historic Buildings: 1867 freight house.

Activities: special events; rental facilities; trolley car rides; pump car rides.
Publications: triannual, Transfer.
Hours & Admission Prices: Call for hours. Adults $3. &
Attendance: 3,150 (accurate)

Somerville

THE SOMERVILLE MUSEUM, One Westwood Rd., Somerville, MA 02143-1517. Tel.: 617-666-9810.
E-mail: somemuseum@gmail.com
Web Site: somervillemuseum.org
Founded: 1897.
Congressional District: 8
Key Personnel: Pres. Bd., Barbara Mangum; Dir., Evelyn M. Battinelli.
Personnel Profile: Full-Time Volunteers 1; Part-Time Volunteers 4.
Governing Authority: society, nonprofit. Parent Institution: Somerville Historical Society. Tax-exempt: 501(c)(3).
Institution Type/Description: General Museum: housed in c.1925 three-story, red brick neo-Federal style building.
Collections: 18th- to 20th-century American history; architectural artifacts; furniture; paintings; prints; photographs; glassware; military & industrial items; scientific & mechanical objects; contemporary art; multi-cultural objects.
Research Fields: material culture; humanities; art.
Facilities: research library pertaining to Somerville history; educational facilities.
Activities: guided tours; lectures; films; concerts; dance recitals; arts festivals; theater; study clubs; hobby workshops; broadcast programs; festivals; participatory, loan, temporary & traveling exhibitions; organized education programs for children, adults & undergraduate or graduate college students; training programs for professional museum workers; special exhibitions & programs.
Publications: brochure; newsletter, Lifting the Veil: The Burning of the Ursuline Convent.
Hours & Admission Prices: Sept.-June Thurs. 2-7, Fri. 2-5, Sat. 12-5. No charge; donations accepted. Closed national holidays.
Membership: Student & Senior $15; Individual $25; Family $35; Contributing $50; Supporting $100; Patron $250; Sustaining $500; Benefactor $1,000.

South Chelmsford

THE CHELMSFORD HISTORICAL SOCIETY, INC., 40 Byam Rd., South Chelmsford, MA 01824-3827. Tel.: 978-256-2311.
E-mail: chelmhist@comcast.net
Web Site: www.chelmhist.org
Founded: 1930.
Congressional District: 5
Key Personnel: Pres. (V), Carol Merriam; Cur., Judy Fichtenbaum; Museum Dir., Donald Patterschal.
Governing Authority: society. Tax-exempt: 501(c)(3) & 170(b)(1)(A).
Institution Type/Description: History Museum: housed in c.1663 Barrett-Byam homestead.
Collections: archives; Indian & war relics; Chelmsford Glass; clothing; model of country store.
Research Fields: genealogy; local history & industry.
Facilities: library of books, newspapers & magazines pertaining to history of Chelmsford, available for use on premises; reading room; gardens.
Activities: guided tours; lectures; study clubs; formally organized education programs for children; permanent & temporary exhibitions.
Publications: newsletter.
Hours & Admission Prices: By appointment & during special events. No charge; donations accepted. Closed New Year's Day; Memorial Day; Labor Day; Thanksgiving; Christmas. &
Attendance: 250 (estimated)
Membership: Senior & Student $6; Individual $10; Family $18; Life $100.

South Deerfield

MAGIC WINGS BUTTERFLY CONSERVATORY & GARDENS, 281 Greenfield Rd., South Deerfield, MA 01373-9790. Tel.: 413-665-2805. Fax: 413-665-4062.
E-mail: info@magicwings.net
Web Site: www.magicwings.com
Institution Type/Description: Conservatory & Garden.
Collections: 3,000 species of butterflies from around the world.
Hours & Admission Prices: Memorial Day-Labor Day daily 9-6; Sept.-May daily 9-5. Adults $12, senior citizens 62 & over $10, children 3-17 $8; children under 3 no charge.

South Hadley

MOUNT HOLYOKE COLLEGE ART MUSEUM, (M), Lower Lake Rd., South Hadley, MA 01075-1499. Tel.: 413-538-2245. Fax: 413-538-2144.
E-mail: artmuseum@mtholyoke.edu
Web Site: www.mtholyoke.edu/artmuseum
Founded: 1876.
Congressional District: 1
Key Personnel: Dir. Florence French Abbott, John R. Stomberg; Chm. (V), Susan Vicinelli; Cur., Wendy M. Watson; Cur. Asst., Rachel Beaupre; Asst. Cur. Skinner Museum, Aaron Miller; Business & Events Mgr., Debbie Davis; Andrew W. Mellon Coord. Academic Affairs, Ellen Alvord; Mgr. Collections, Linda Delone Best; Asst. Mgr. Collections, Jenny Lind; Digitization Specialist, Laura Weston; Preparator, Brian Kiernan; Art Advisory Bd. Fellow, Yingxi Gong; Sr. Administrative Asst., Caitlin Grey; Sr. Administrative Asst., Gail Parker.
Personnel Profile: Full-Time Paid 10; Part-Time Paid 19; Part-Time Volunteers 22; Interns 3.
Governing Authority: college; nonprofit. Parent Institution: Mount Holyoke College. Tax-exempt.
Institution Type/Description: Art Museum.
Collections: art of all periods with concentrations in ancient, medieval, Renaissance, Asian & American art; prints; drawings; photographs.
Major Exhibits: El Anatsui, 2/14-5/30/14.
Facilities: 20,000-vol. library on art history. Publications & cards for sale.
Activities: guided tours; lectures; gallery talks; docent program; formally organized educational programs for students grades 2-9; inter-museum loan; permanent, temporary & traveling exhibitions.
Publications: exhibition catalogues; Newsletter; semester calendars.
Hours & Admission Prices: Tues.-Fri. 11-5, Sat.-Sun. 1-5. No charge; donations accepted. &
Attendance: 11,050 (estimated)
Membership: Individual $30; Family & Dual $50; Contributor $100; Sponsor $250; Patron $500; Director's Circle $1,000; Benefactor $2,500 & up.

THE OLD FIREHOUSE MUSEUM, 4 N. Main St., South Hadley, MA 01075. Mailing Address: P.O. Box 387, South Hadley, MA 01075-0387. Tel.: 413-536-4970.
Founded: 1976.
Institution Type/Description: Fire-Fighting Museum.
Collections: fire-fighting gear; two 19th century hand pumpers; 1926 Dodge Fire Engine; historical documents & artifacts.
Major Exhibits: World War I, 5/14-9/14.
Hours & Admission Prices: May-June & Sept. Sun. 1:30-4; July-Aug. Wed. & Sun. 1:30-4. No charge; donations accepted.
Attendance: 100 (estimated)

THE SKINNER MUSEUM OF MOUNT HOLYOKE COLLEGE, 33 Woodbridge St., South Hadley, MA 01075-1138. Mailing Address: c/o Mount Holyoke College Art Museum, Lower Lake Rd., South Hadley, MA 01075. Tel.: 413-538-2245. Fax: 413-538-2144.
E-mail: artmuseum@mtholyoke.edu
Web Site: www.mtholyoke.edu/artmuseum
Founded: 1933.
Congressional District: 1
Key Personnel: Dir., John Stomberg; Cur., Wendy Watson; Asst. Cur., Aaron Miller.
Personnel Profile: Full-Time Paid 1.
Governing Authority: college. Parent Institution: Mount Holyoke College. Tax-exempt.
Institution Type/Description: History Museum.
Collections: agriculture; archaeology; archives; geology; glass; history; New England home furnishings. Historic Buildings: 1846 Church & School House from Prescott, MA.
Major Exhibits: Experiencing the Civil War: From the Battlefield to the Homefront, 7/6/11-4/15.
Hours & Admission Prices: May-Oct. Wed. & Sun. 2-5. No charge; donations accepted.
Attendance: 350 (estimated)

South Weymouth

SHEA FIELD NAVAL AVIATION HISTORICAL MUSEUM, 495 Shea Memorial Dr., South Weymouth, MA 02190-4009. Tel.: 781-534-1043.
E-mail: inquires@anapatriotsquadron.org
Web Site: ww.anapatriotsquadron.org

Founded: 1992.
Personnel Profile: Part-Time Volunteers 5.
Governing Authority: Parent Institution: ANA. Tax-exempt.
Institution Type/Description: Naval Aviation Museum.
Collections: Naval aviation history; personal artifacts; photographs; documents.
Hours & Admission Prices: last Sat. each month 9am-11am. No charge; donations accepted.

WEYMOUTH HISTORICAL SOCIETY - JASON HOLBROOK HOMESTEAD, 238 Park Ave., South Weymouth, MA 02190-2512. Tel.: 781-340-1022.
Institution Type/Description: Historical Society Museum: housed in the former home of actor, Jason Holbrook; c.1763.
Collections: local history & culture; photographs; period furnishings; personal artifacts; shoe making tools & machines; military artifacts; farm tools.
Hours & Admission Prices: Wed. 9-1; other times by appointment. No charge.

Springfield

HATIKVAH HOLOCAUST EDUCATION CENTER, 1160 Dickinson St., Springfield, MA 01108-3122. Tel.: 413-734-7700. Fax: 413-734-0919.
E-mail: administrator@hatikvah-center.org
Web Site: www.hatikvah-center.org
Institution Type/Description: History Museum.
Collections: Holocaust history; photographs; personal artifacts; documents; audiovisuals.
Facilities: Museum-related items for sale.
Activities: special events; educational programs; workshops. Annual Events: Yom HaShoah; Kristallnacht.
Hours & Admission Prices: Mon.-Fri. 9-2, Sun. by appointment. &
Attendance: 2,500 (accurate)
Membership: Individual $50; Family $100; Hatikvah Benefactor $500; Hatikvah Patron $1,000.

NAISMITH MEMORIAL BASKETBALL HALL OF FAME, 1000 W. Columbus Ave., Springfield, MA 01105-2518. Tel.: 413-781-6500. Fax: 413-781-1939.
Web Site: www.hoophall.com
Founded: 1959.
Congressional District: 2
Key Personnel: Pres., John Doleva; Vice Pres. Finance & Operations, Don Senecal; Vice Pres. Experience & Programming, Paul Lambert; Dir. Devel., Scott Zuffelato; Dir. Sports Media, Howie Davis; Cur., Michael Brooslin; Archivist & Librarian, Robin Deutsch; Mgr. Events & Group Programs, David Elkins; Sr. Dir. Mktg., Dean O'Keefe; Supvr. Educational Programs, Tom Ashe; Controller, Doug Moss; Dir. Retail Operations, Paula Thompson; Mgr. Facility Operations, James Mullen; Group Sales, Joe Hevey; Dir. Mktg. Partnerships, Steve Sullivan.
Personnel Profile: Full-Time Paid 35; Part-Time Paid 75; Part-Time Volunteers 1; Interns 4.
Governing Authority: nonprofit organization. Tax-exempt: 501(c)(3).
Institution Type/Description: Sports Museum.
Collections: history of basketball, from its beginnings in Springfield, MA in December 1891, to the present; high tech interactive exhibits.
Research Fields: basketball history.
Facilities: 5,000-vol. library pertaining to the history of basketball for research on premises. Gift items for sale.
Activities: films; organized education programs for children; participatory exhibits.
Publications: newsletter; Souvenir Book; annual yearbook.
Hours & Admission Prices: Call for hours. Adults $20, seniors 65 & over $15, youth 5-15 $14; discounts to groups; children 4 & under no charge. Closed New Year's Day; Thanksgiving; Christmas. &
Attendance: 300,000 (estimated)
Membership: Call for information.

PAN AFRICAN HISTORICAL MUSEUM USA, Tower Square, Mezzanine Level, 1500 Main St., #2, Springfield, MA 01115-1004. Tel.: 413-733-4823. Facebook: The Pan African Historical Museum USA.
E-mail: pahmusa1@juno.com
Web Site: www.pahmusa.mysite.com
Key Personnel: Exec. Dir., LuJuana Hood; Art Dir., Carl Edward Yates
Institution Type/Description: History Museum.
Collections: African American artifacts, artwork & sculptures.
Hours & Admission Prices: By appointment.

SPRINGFIELD ARMORY NATIONAL HISTORIC SITE, One Armory Sq., Ste. 2, Springfield, MA 01105-1299. Tel.: 413-734-8551 & 6477. Fax: 413-747-8062.
E-mail: james_woolsey@nps.gov
Web Site: www.nps.gov/spar/
Founded: 1870.
Congressional District: 2
Key Personnel: Supt., James Woolsey; Cur., James D. Roberts; Visitor Svcs., Joanne M. Gangi; Historian, Richard Colton; Conservator, Alex Mackenzie.
Personnel Profile: Full-Time Paid 12; Part-Time Paid 2; Part-Time Volunteers 3; Interns 3.
Governing Authority: federal. Parent Institution: National Park Service, U.S. Dept. of Interior, Washington, DC. Tax-exempt.
Institution Type/Description: Historic Building & Site Museum: housed in 1850 Arsenal.
Collections: 8,730 military small arms of U.S. & foreign manufacture spanning 3 centuries, available to researchers by appointment under the supervision of a professional staff member; firearms manufactured at Springfield Armory from 1790s to 1968, including experimental & prototypes; Civil War rifle-muskets; Model 1903; M-1 Garand; hand guns; machine guns; edged weapons; manufacturing process; inventions; impact of armory upon the city; 3,000 archaeological artifacts; 7,500 toy soldiers; 21,000 photographs 1930-1968.
Research Fields: 1794-1968, U.S. military firearms development & production; photographs 1930-1968; technical reports; notes; memoranda; contracts; architectural drawings c.1790-1968; S.A. Armorer newsletter; annual reports; historical summaries; manuscripts; post orders; S.A. histories; microfilm correspondence & payroll 1813-1934; N.P.S. & compiled subject files; periodicals & 1,200 books, mainly on production, testing & collection of military arms, U.S. military history & ordinance; oral history interviews.
Facilities: 1,500-vol. library & archives of military firearms books & other research materials; education center; conference space. Armory-related publications for sale.
Activities: programs for school groups; friends group; friends of the Springfield Armory Museum.
Publications: books; fact sheets; site brochure; Springfield Firearms posters; book, Marco Paul at the Springfield Armory; Arms for the Nations; videos.
Hours & Admission Prices: Daily 9-5. No charge. Closed New Year's Day; Thanksgiving; Christmas. &
Attendance: 20,000 (estimated)
Membership: Friends of the Springfield Armory $35.

*** SPRINGFIELD MUSEUMS - GEORGE WALTER VINCENT SMITH ART MUSEUM, (M),** 21 Edwards St., Springfield, MA 01103-1548. Tel.: 413-263-6800, ext. 302. Fax: 413-263-6897.
E-mail: info@springfieldmuseums.org
Web Site: www.springfieldmuseums.org
Founded: 1896.
Congressional District: 2
Key Personnel: Pres., Holly Smith-Bove; Chm. (V), Sam Hanmer; Dir., Heather Haskell; Vice Pres., Kathleen Simpson; Dir. Public Rels. & Mktg., Matt Longhi; Registrar, Diane Barbarisi; Museum Shop Mgr., Lin Jaciow.
Personnel Profile: Full-Time Paid 59; Part-Time Paid 42; Part-Time Volunteers 552; Interns 6.
Governing Authority: nonprofit organization. Parent Institution: The Springfield Museums, 21 Edwards St., Springfield, MA 01103. Tax-exempt: 501(c)(3).
Institution Type/Description: Decorative Arts Museum: housed in 1895 building designed by Renwick, Aspenwall & Renwick.
Collections: Asian decorative arts of the 17th, 18th & 19th centuries; Middle Eastern rugs; arms & armor; 19th-century American paintings; plaster casts.
Research Fields: Japanese arms & armor; Asian decorative arts.
Facilities: classrooms; art studios; hands-on art discovery center.
Activities: age-oriented docent-guided tours; courses for young people & adults; lectures; gallery talks; inter-museum loans; permanent & temporary exhibitions.
Publications: catalogs & gallery guides; quarterly membership newsletter; annual report.
Hours & Admission Prices: Tues.-Sat. 10-5, Sun. 11-5. Adults $15, admission includes entry to 5 museums at same location; discounts to AAM members; members no charge. Closed New Year's Day; Easter; Independence Day; Thanksgiving; Christmas. &
Attendance: 76,000 (estimated)
Membership: Basic $45; Dual $55; Household $75; Patron $125-$249; Sustaining $250-$499; Contributing $500-$999; Society of William Rice $1,000 & up.

* **SPRINGFIELD MUSEUMS - LYMAN & MERRIE WOOD MUSEUM OF SPRINGFIELD HISTORY, (M),** 21 Edwards St., Springfield, MA 01103-1548. Tel.: 413-263-6800, ext. 304. Fax: 413-263-6898.
E-mail: info@springfieldmuseums.org
Web Site: www.springfieldmuseums.org
Founded: 2009.
Congressional District: 2
Key Personnel: Pres., Holly Smith-Bove; Dir., Guy McLain; Chm. (V), Sam Hanmer; Vice Pres., Kathleen Simpson; Dir. Public Rels. & Mktg., Matt Longhi; Registrar, Diane Barbarisi; Head Research Library, Margaret Humberston; Museum Shop Mgr., Lin Jaciow.
Personnel Profile: Full-Time Paid 59; Part-Time Paid 42; Part-Time Volunteers 552; Interns 3.
Governing Authority: nonprofit organization. Parent Institution: Springfield Museums. Tax-exempt: 501(c)(3).
Institution Type/Description: History Museum.
Collections: Springfield-built automobiles; Indian motorcycles; local industry & culture during the 19th & 20th centuries; photographs; maps; manuscripts; genealogy.
Research Fields: industrial & commercial history of Springfield and the region in the 19th & 20th centuries.
Facilities: library; archives.
Activities: docent-guided tours; family programs; lectures; children & adult classes; video presentations; permanent & temporary exhibitions; hands-on Hasbro GameLand for families.
Publications: books; DVDs; exhibit catalogs; gallery guides; quarterly membership publication.
Hours & Admission Prices: Tues.-Sat. 10-5, Sun. 11-5. Adults $15 (admission includes entry to 5 museums); discounts to AAM members; members no charge. Closed New Year's Day; Easter; Independence Day; Thanksgiving; Christmas. &
Attendance: 65,000 (estimated)
Membership: Basic $45; Dual $55; Household $75; Patron $125; Sustaining $250; Contributing $500; Society of William Rice $1,000 & up.

* **SPRINGFIELD MUSEUMS - MICHELE & DONALD D'AMOUR MUSEUM OF FINE ARTS, (M),** 21 Edwards St., Springfield, MA 01103-1548. Tel.: 413-263-6800, ext. 302. Fax: 413-263-6889.
E-mail: info@springfieldmuseums.org
Web Site: www.springfieldmuseums.org
Founded: 1933.
Congressional District: 2
Key Personnel: Pres., Holly Smith-Bove; Vice Pres., Kathleen Simpson; Chm. (V), Sam Hanmer; Dir., Heather Haskell; Dir. Public Rels. & Mktg., Matt Longhi; Registrar, Diane Barbarisi; Museum Shop Mgr., Lin Jaciow.
Personnel Profile: Full-Time Paid 59; Part-Time Paid 42; Part-Time Volunteers 552.
Governing Authority: nonprofit organization. Parent Institution: Springfield Museums, 21Edwards St. Tax-exempt: 501(c)(3).
Institution Type/Description: Art Museum.
Collections: European & American paintings; sculpture; graphics; decorative art; Japanese prints.
Facilities: 5,000-vol. library on Asian art & general history of art available for use by appointment; 280-seat auditorium.
Activities: age-oriented docent-guided tours; courses for young people & adults; lectures; gallery talks; inter-museum loans; permanent & temporary exhibitions.
Publications: gallery guides; periodic exhibition catalogues; quarterly membership newsletter; annual report.
Hours & Admission Prices: Tues.-Sat. 10-5. Sun. 11-5. Adults $15, includes entry to 5 museums; discounts to AAM members; members no charge. Closed New Year's Day; Easter; Independence Day; Thanksgiving; Christmas. &
Attendance: 86,000 (estimated)
Membership: Basic $45; Dual $55; Household $75; Patron $125; Sustaining $250; Contributing $500; Society of William Rice $1,000 & up.

* **SPRINGFIELD MUSEUMS - SPRINGFIELD SCIENCE MUSEUM, (M),** 21 Edwards St., Springfield, MA 01103-1548. Tel.: 413-263-6800, ext. 325. Fax: 413-263-6884.
E-mail: info@springfieldmuseums.org
Web Site: www.springfieldmuseums.org
Founded: 1859.
Congressional District: 2
Key Personnel: Pres., Holly Smith-Bove; Chm. (V), Sam Hamner; Dir., David Stier; Vice Pres., Kathleen Simpson; Registrar, Diane Barbarisi; Dir. Public Rels. & Mktg., Matt Longhi; Museum Shop Mgr., Lin Jaciow.
Personnel Profile: Full-Time Paid 59; Part-Time Paid 42; Part-Time Volunteers 552; Interns 6.
Governing Authority: nonprofit organization. Parent Institution: Springfield Museums. Tax-exempt: 501(c)(3).
Institution Type/Description: Science Museum.
Collections: anthropology; archaeology; astronomy; ecology; ethnology; herpetology; ichthyology; lithology; mammalogy; mineralogy; ornithology; paleontology; physical science; technology; live animals.
Research Fields: ornithology; regional archaeology; Native American; oceanic; ethnographic.
Facilities: 3,000-vol. library of books covering all phases of natural science available for research on premises; planetarium; observatory; aquarium. Museum-related items for sale.
Activities: age-oriented docent-guided tours; courses for young people & adults; lectures; gallery talks; inter-museum loans; permanent & temporary exhibitions; hands-on exhibits.
Publications: brochures; quarterly membership newsletter; annual report.
Hours & Admission Prices: Tues.-Sat. 10-5, Sun. 11-5. Planetarium: hours vary; guided tours available; special events. Museum (includes entry to 5 museums at Quadrangle location). Adults $15; discounts to AAM members; children under 3 & members no charge. Planetarium: adults $3, children $2. Closed New Year's Day; Easter; Independence Day; Thanksgiving; Christmas. &
Attendance: 228,000 (estimated)
Membership: Basic $45; Dual $55; Household $75; Patron $125-$249; Sustaining Patron $250-$999; Contributing Patron $500-$999; Society of William Rice $1,000 & up.

THE ZOO IN FOREST PARK, 302 Sumner Ave., Springfield, MA 01108. Mailing Address: P.O. Box 80295, Springfield, MA 01138-0295. Tel.: 413-733-2251. Fax: 413-733-2330.
E-mail: john@forestparkzoo.com
Web Site: forestparkzoo.org
Founded: 1962.
Key Personnel: Dir., John Lewis, Jr.; Dir. Education, Alison Summers.
Personnel Profile: Full-Time Paid 8; Part-Time Paid 14; Part-Time Volunteers 12; Interns 8.
Governing Authority: Tax-exempt.
Institution Type/Description: Zoo.
Collections: wildlife.
Hours & Admission Prices: April to Columbus Day daily 10-6, weather permitting; Oct. 14-Nov. Sat.-Sun. 10-3:30. Adults $6.75, senior citizens & children 5-12 $4.75, children 1-4 $2.50; children under one no charge.
Attendance: 88,000 (estimated)

Still River

HARVARD HISTORICAL SOCIETY, Still River Baptist Church, 215 Still River Rd., Still River, MA 01467. Mailing Address: P.O. Box 542, Harvard, MA 01451-0542. Tel.: 978-456-8285.
E-mail: curator@harvardhistory.org
Web Site: www.harvardhistory.org
Founded: 1897.
Congressional District: 5
Key Personnel: Pres., Carlene Phillips; Cur., Karen Zaikis.
Personnel Profile: Part-Time Paid 1; Part-Time Volunteers 30.
Governing Authority: bd. of directors. Tax-exempt.
Institution Type/Description: History Museum.
Collections: local history; costumes; furniture; decorative arts; printed materials; photographs; books; cottage. Historic Building: 1832 Still River Baptist Church.
Research Fields: Harvard, Shaker & Transcendentalist history.
Facilities: archives.
Activities: permanent & temporary exhibitions; house tours; musical programs; school education.
Publications: Spring & Fall newsletters. The Harvard Album; Directions of a Town.
Hours & Admission Prices: Mon.-Wed. 2-7; other times by appointment. No charge; donations accepted. &
Attendance: 1,000 (estimated)
Membership: Individual $20; Family $30; Friend $50; Patron $100; Benefactor $500.

Stockbridge

BERKSHIRE BOTANICAL GARDEN, Rte. 102 & Rte. 183, Stockbridge, MA 01262. Mailing Address: Box 826, Stockbridge, MA 01262-0826. Tel.: 413-298-3926. Fax: 413-298-4897.
E-mail: mboxer@berkshirebotanical.org
Web Site: www.berkshirebotanical.org

Founded: 1934.
Congressional District: 1
Key Personnel: Exec. Dir., Molly Boxer; Chm., David Carls; Education Coord., Elisabeth Cary; Gift Shop Mgr., Joan Carter.
Personnel Profile: Full-Time Paid 11; Part-Time Paid 1; Part-Time Volunteers 45; Interns 2.
Governing Authority: nonprofit organization. Tax-exempt: 501(c)(3).
Institution Type/Description: Botanic Garden.
Collections: botanic garden; herb & perennial gardens; arboretum; vegetable display garden; lily pond; daylily & rose gardens; rock gardens; conifer garden; woodland preserve; children's garden.
Research Fields: informal research only.
Facilities: meeting hall; education center. Gift shop with garden-related items including herb products produced by volunteers on the grounds.
Activities: guided tours; lectures; workshops; horticulture certificate program; seasonal exhibits. Garden Sponsors: Plant Sale in May; Flower Show in August; Harvest Festival in October; Holiday Marketplace in December.
Publications: magazine, Cuttings.
Hours & Admission Prices: May to mid-Oct. daily 9-5. Adults $12, senior citizens & students $10; discounts to AAA, AABGA, & MA Teachers Assoc.; American Horticultural Society Reciprocal Admission Program; children 12 & under and members no charge. &
Attendance: 24,000 (estimated)
Membership: Student $30; Individual $50; Family $75; Supporter $150; Friend $250; Patron $500; Fence Club $1,000.

CHESTERWOOD, 4 Williamsville Rd., off Rte. 183, Glendale Sect., Stockbridge, MA 01262-0827. Mailing Address: P.O. Box 827, Stockbridge, MA 01262-0827. Tel.: 413-298-3579. Fax: 413-298-1065. Facebook: Chesterwood.
E-mail: chesterwood@savingplaces.org
Web Site: www.chesterwood.org
Founded: 1955.
Congressional District: 1
Key Personnel: Dir., Donna Hassler; Chm. Advisory Council, Carol Bosco Baumann; Office Mgr., Lisa Reynolds; Bldgs. & Grounds Supt., Gerard Blache; Curatorial Asst., Anne Cathcart; Bldgs. & Grounds Coord., Brian McElhiney.
Personnel Profile: Full-Time Paid 5; Part-Time Paid 1; Part-Time Volunteers 25; Interns 1.
Governing Authority: nonprofit organization. Parent Institution: National Trust for Historic Preservation, 1785 Massachusetts Ave., N.W., Washington, DC 20036. Tax-exempt: 501(c)(3).
Institution Type/Description: Historic Site: housed in the summer estate of Daniel Chester French (1850-1931).
Collections: works of Daniel Chester French: sculpture; paintings; pastels; sculpture tools & equipment; blueprints; drawings; papers; books; 18th & 19th-century American & European furniture & decorative arts; art works of DCF contemporaries; period country place garden; period woodland walk. Historic Buildings: 1898 studio; 1901 house; early 1800s barn.
Major Exhibits: Historic Artists' Homes & Studios, 10/14-3/15; T.C. Steele, Nashville, TN, 12/13-3/14; Winslow-Homes Studio, 3/14-6/14; Olana SHS, 7/14-11/14; Roger Brown Study, Chicago, IL, 12/14-3/15.
Research Fields: Daniel Chester French works, life, career & times; Lincoln Memorial Statue; Chesterwood & its collections; museum methodology; Beaux Arts sculpture; The American Renaissance; Gilded Age; landscape architecture; horticulture; Berkshire estates.
Facilities: library of books on Daniel Chester French, art history, sculpture, historic preservation, landscape architecture, decorative arts & museum studies; 122 acre site; conference space. Books, reproductions & other museum-related items for sale.
Activities: lectures; education programs; docent program; sculpture demonstrations; gallery & landscape tours; summer & winter intern program for undergraduate & graduate college students; permanent & temporary exhibitions; special events. Museum Sponsors: Contemporary Sculpture at Chesterwood; Antique Car Show in May; Americana Music in July & August.
Publications: educational brochures; annual catalogue & brochures; children's guide to Chesterwood; landscape guide; property guidebook.
Hours & Admission Prices: May-Oct. daily. Adults $16; members no charge. &
Attendance: 12,500 (estimated)
Membership: Individual $50; Family $75; Contributing $150; Benefactor $250; Margaret French Cresson Fellow $500; Daniel Chester French Society $1,000; Abraham Lincoln Laureate Society $2,500.

* **MERWIN HOUSE, TRANQUILITY, (M),** 14 Main St., Stockbridge, MA 01262. Mailing Address: P.O. Box 1543, Stockbridge, MA 01262-9701. Tel.: 617-994-6662. Fax: 413-298-0121.
Web Site: www.historicnewengland.org
Founded: 1966.
Congressional District: 1
Key Personnel: Site Mgr., Lisa Centola; Pres., Carl Nold.
Governing Authority: society; nonprofit organization. Parent Institution: Historic New England, 141 Cambridge St., Boston, MA 02114. Tel.: 617-227-3956. Tax-exempt: 501(c)(3).
Institution Type/Description: Historic House: c.1825 late Federal style period house.
Collections: European & American furniture & decorative arts.
Hours & Admission Prices: Jan. 5, Oct. 2 & Dec. 4 11-4. Tours on the hours. Adults $5; members no charge.
Attendance: 569 (accurate)
Membership: National $35; Individual $45; Household $55; Garden & Landscape $75; Contributing and Library & School $100; Historic New England Affiliate $100-$350; Young Friends of Historic New England $100-$1,500; Friends of the Library and Archives $125; Historic Homeowner $200; Business & Ogden Codman Design Group $250; Appleton Circle $1,750-$3,500.

THE MISSION HOUSE, 19 Main St., Stockbridge, MA 01262-9800. Mailing Address: P.O. Box 792, Stockbridge, MA 01262-0792. Tel.: 413-298-3239, ext. 3013. Fax: 413-298-5239.
E-mail: naumkeag@ttor.org
Web Site: www.thetrustees.org
Founded: 1948.
Congressional District: 1
Key Personnel: Cultural Site Mgr., Colleen Henry.
Personnel Profile: Full-Time Paid 1; Part-Time Paid 7.
Governing Authority: nonprofit organization. The Trustees of Reservations, 572 Essex St., Beverly, MA 01915. Tel. 508-921-1944. Tax-exempt.
Institution Type/Description: Historic House: housed in the former home of Rev. John Sergeant, first missionary to the Stockbridge Mohicans; built in 1743.
Collections: period furnishings; Rev. Sergeant's cupboard & chairs; Mohican-Munsee artifacts.
Research Fields: early Stockbridge town history; Stockbridge Mohicans.
Activities: guided tours.
Hours & Admission Prices: July-Aug. Sat.-Sun. 10-3. House: guided tours. Gardens: self-guided tours. Adults $6; discounts to AAM, ICOM members; children 12 & under and members no charge.
Attendance: 3,000 (accurate)
Membership: Senior & Student $37; Individual $47; Family $67.

NAUMKEAG, 5 Prospect Hill Rd., Stockbridge, MA 01262. Mailing Address: P.O. Box 792, Stockbridge, MA 01262-0792. Tel.: 413-298-3239, ext. 3013. Fax: 413-298-5239.
E-mail: naumkeag@ttor.org
Web Site: www.thetrustees.org
Founded: 1959.
Congressional District: 1
Key Personnel: Cultural Site Mgr., Colleen Henry.
Personnel Profile: Full-Time Paid 5; Part-Time Paid 20; Part-Time Volunteers 1.
Governing Authority: nonprofit organization. The Trustees of Reservations, 572 Essex St., Beverly, MA 01915. Tel. 617-921-1944. Tax-exempt.
Institution Type/Description: Historic House: 1886 shingle style house, designed by architect McKim, Mead, & White, the summer home of Joseph Hodges Choate, U.S. Ambassador to the Court of St. James. Original gardens designed by Fletcher Steele.
Collections: Americana & European decorative arts; Chinese export porcelain.
Research Fields: the Choate family; landscape architecture.
Facilities: 8 acres of formal promenades; gardens.
Activities: guided tours; special events; garden audio tours.
Publications: Naumkeag-History of House & Family.
Hours & Admission Prices: May 15 to Oct. daily 10-5. House & Garden: adults $15; discounts to AAM & ICOM members; children 12 & under and members no charge.
Attendance: 10,000 (accurate)
Membership: Senior & Student $37; Individual $47; Family $67.

*** NORMAN ROCKWELL MUSEUM, (M),** 9 Rte. 183, Stockbridge, MA 01262. Mailing Address: P.O. Box 308, Stockbridge, MA 01262-0308. Tel.: 413-298-4100, ext. 221. Fax: 413-298-4142. Facebook: Norman Rockwell Museum.
E-mail: inforequest@nrm.org
Web Site: www.nrm.org
Founded: 1967.
Congressional District: 1
Key Personnel: C.E.O. & Dir., Laurie Norton Moffatt; Chm. Bd., Thomas L. Pulling; Pres. Bd., Anne H. Morgan; Deputy Dir. & Chief Cur., Stephanie Haboush Plunkett; C.O.O., Terry Smith; Chief Devel. Officer, John Nathaniel Arata; Museum Shop Mgr., Mike Duffy; Mgr. Facilities, Wesley Shufelt; Cur. Rockwell Center for American Visual Studies, Joyce Schiller; Collections & Registrar Asst., Thomas Mesquita; Information Technology Mgr., Frank Kennedy; Mgr. Collections & Registration, Martin Mahoney; Mgr. Visitors Svcs., Joseph Aubert; Dir. Mktg. & Communications, Margit Hotchkiss; Cur. Education, Thomas Daly; Mgr. Traveling Exhibitions, Mary Melius; Dir. Human Resources, Holly Coleman.
Personnel Profile: Full-Time Paid 34; Part-Time Paid 38; Part-Time Volunteers 29; Interns 4.
Governing Authority: private educational institution; nonprofit; board of trustees. Tax-exempt: 501(c)(3).
Institution Type/Description: Art Museum: dedicated to education & art appreciation inspired by Norman Rockwell.
Collections: Norman Rockwell original art including paintings, drawings & sketches; artist's studio with working materials & memorabilia; over 100,000 item archive, including photographs, publications, correspondence & papers; sculptures by Peter Rockwell, son of artist Norman Rockwell.
Major Exhibits: Wendell Minor's America, 11/9/13-5/26/14; Dancing Princesses: The Picture Book Art of Ruth Sanderson, 11/23/2013-2/23/14.
Research Fields: Norman Rockwell's life, career & art; 20th-century social history interpretation, field of illustration.
Facilities: library, available for use by appointment; archives; 36 acres of grounds; auditorium; classrooms. Museum-related items for sale.
Activities: guided tours; permanent & changing exhibitions; lecture series; image rental program; educational programs; facilities rental; audio tours; hands-on art activities; art classes.
Publications: quarterly newsletter, The Portfolio; books, The Illustrator's Moment, Norman Rockwell: A Definitive Catalogue, Norman Rockwell's Four Freedoms, Norman Rockwell: A Centennial Celebration, Willie Was Different, The Norman Rockwell Museum at Stockbridge - A Guidebook; Pictures for the American People (published with the High Museum of Art, Atlanta, GA, 1999); American Chronicles, The Art of Norman Rockwell; Portfolio: Norman Rockwell Museum, 40 Years of Illustration Art; Frederic Remington and the American Civil War: A Ghost Story (2006); Ephemeral Beauty: Al Parker and the American Women's Magazine (2007); The Fantastical Faces of Peter Rockwell: A Sculptor's Retrospective (2009); Witness: The Art of Jerry Pinkney (2010); Everett Raymond Kinstler: Pulps to Portraits (2012); Istvan Banyai: Stranger in a Strange Land (2013).
Hours & Admission Prices: May-Oct. daily 10-5; Nov.-April Mon.-Fri. 10-4, Sat.-Sun. 10-5. May-Oct. admission includes studio tour: adults $16, college students $10, children 6-18 $5; discounts to NEMA, AAM, & ICOM members & bus tour groups; children 5 & under no charge. Closed New Year's Day; Thanksgiving; Christmas. &
Attendance: 120,000 (estimated)
Membership: Individual $50; Family & Dual $75; Illustrator's Roundtable $125; Gallery Guild $250; Artist's Forum $500; Studio Society $1,000; Painter's Council $2,500; Norman Rockwell Circle $5,000 & up. Business Memberships: Business Friends $100-$249; Business Roundtable $250-$499; Business Four Freedoms $500-$999; Business Studio Society $1,000-$2,499; Business Linwood Society $2,500-$4,999; Business Norman Rockwell Circle $5,000 & up.

STOCKBRIDGE LIBRARY HISTORICAL COLLECTION, 46 Main St., Stockbridge, MA 01262. Mailing Address: P.O. Box 119, Stockbridge, MA 01262-0119. Tel.: 413-298-5501. Fax: 413-298-0218.
E-mail: ballen@cwmars.org
Web Site: www.stockbridgelibrary.org
Founded: 1938.
Congressional District: 1
Key Personnel: Cur., Barbara Allen; C.E.O., Katherine O'Neil; Pres. (V), John Hyson; Cur. Asst., Joshua Hall; Chm. (V), Karen Marshall.
Personnel Profile: Full-Time Paid 1; Part-Time Paid 2; Part-Time Volunteers 8; Interns 1.
Governing Authority: nonprofit organization. Parent Institution: Stockbridge Library Association, Main St. Tax-exempt: 501(c)(3).
Institution Type/Description: History Museum.
Collections: inventions by Anson Clark; memorabilia of Cyrus W. Field; 1751 Jonathan Edwards table; local history; archives; Indian artifacts & documents; Rachel Field's doll, Hitty.
Research Fields: local history; village improvement; religion, economic history of small towns; D.D. Field family; documents of Native Americans.
Facilities: 1,500-vol. library of local history books & books by local authors available on premises; reading room.
Activities: permanent & temporary exhibitions; research facility.
Publications: e-newsletter Now & Then, Jan. March, May, July, Sept. & Nov.
Hours & Admission Prices: Tues., Wed., Fri. 9-5, Thurs. 9-1, Sat. 9-2. No charge; donations accepted. Closed national holidays.
Attendance: 2,500 (estimated)

Stoneham

STONE ZOO, 149 Pond St., Stoneham, MA 02180. Tel.: 781-438-5100.
Institution Type/Description: Zoo.
Collections: cougars; black bear; jaguars; coyotes; otters; birds of prey; lynx; reindeer; bald eagle; snow leopards; goats; gibbons.
Facilities: Gift items for sale.
Activities: educational programs; special events,
Hours & Admission Prices: Daily 10-4. Adults $13, seniors $11, children 2-12 $9; children under 2 no charge. &

Stoughton

STOUGHTON HISTORICAL SOCIETY, Lucius Clapp Memorial, 6 Park St., Stoughton, MA 02072. Mailing Address: Box 542, Stoughton, MA 02072-0542. Tel.: 781-344-5456.
E-mail: stoughtonhistoricalsociety@verizon.net
Web Site: www.stoughtonhistory.com
Founded: 1895.
Congressional District: 9
Key Personnel: Chm. (V), Joseph M. Mokrisky; Pres., Dwight Mac Kerron; Historian, Howard Hansen; Cur., Henry J. Herbowy; Archivist, Jack Sidebottom.
Personnel Profile: Part-Time Volunteers 30.
Governing Authority: society; nonprofit. Tax-exempt.
Institution Type/Description: History Museum: housed in Lucius Clapp Memorial.
Collections: artifacts pertaining to civic and cultural aspects of the town; records of Old Stoughton Musical Society & Musical Society in Stoughton; old music books & manuscripts.
Research Fields: Revolutionary War soldiers from the area; old music; history of Stoughton; genealogy of Stoughton families.
Facilities: library of local history documents available for use by appointment; reading room; 60-seat auditorium.
Activities: lectures; musical or historical dramatic programs.
Publications: 100th Anniversary (1995) Historical Booklet; quarterly newsletter; Stoughton - A Pictorial History; History of Old Stoughton Musical Society.
Hours & Admission Prices: Tues. 10-3, Thurs. 6pm-8pm; other times by appointment. No charge; donations accepted.
Attendance: 500 (estimated)
Membership: Individual $10; Family $15.

Stow

RANDALL LIBRARY, 19 Crescent St., Stow, MA 01775-1188. Tel.: 978-897-8572. Fax: 978-897-7379.
E-mail: stow@minlib.net
Founded: 1892.
Congressional District: 3
Key Personnel: Dir., Melissa Fournier; Chm. (V), Tim Reed.
Governing Authority: municipal.
Institution Type/Description: Historical Society Museum.
Collections: period furnishings; artifacts.
Research Fields: local history.
Facilities: 40,000-vol. library of books, periodicals & A-V materials available for use on premises; reading room.
Activities: permanent exhibitions.
Hours & Admission Prices: July-Aug. Tues.-Thurs. 10-8, Fri. 10-2; Sept.-June Tues.-Thurs. 10-8, Fri. 10-2, Sat. 10-5. No charge; donations accepted. Closed legal holidays. &
Attendance: 7,200 (estimated)

STOW WEST SCHOOL MUSEUM, Harvard Rd., Stow, MA 01775. Mailing Address: 380 Great Rd., Stow, MA 01775-2127. Tel.: 978-562-6843.
E-mail: kcthreads@earthlink.net
Founded: 1974.
Congressional District: 3
Key Personnel: Chm. (V), Karen C. Gray.
Personnel Profile: Part-Time Volunteers 5.
Governing Authority: private; nonprofit. Parent Institution: Stow Historical Commission. Tax-exempt: 501(c)(3).
Institution Type/Description: Historic House Museum: housed in an 1825 one-room schoolhouse.
Collections: furnishings; maps; old photos; merit awards; school books.
Research Fields: life of early teachers & scholars.
Activities: Museum Sponsors: special events June, July, Sept. & Oct.
Publications: coloring book; brochure.
Hours & Admission Prices: July & Sept.-Oct. call for hours. No charge; donations accepted.
Attendance: 150 (estimated)

Sturbridge

OLD STURBRIDGE VILLAGE, 1 Old Sturbridge Village Rd., Sturbridge, MA 01566-1198. Tel.: 508-347-3362. Fax: 508-347-0377. TDD: 508-347-5383.
E-mail: osv@osv.org
Web Site: www.osv.org
Founded: 1946.
Congressional District: 2
Key Personnel: Pres. & C.E.O., Jim Donahue; Chief of Staff, Renee M. Chambers; Trustees Chm., Donna DeCorleto; Vice Pres. Museum Program, Debra Friedman; Vice Pres. Collections & Research, J. Edward Hood; Vice Pres. Operations, Brad King; Vice Pres. Mktg. & Public Rels., Ann Lindblad; Dir. Human Resources, Angela Cheng-Cimini.
Personnel Profile: Full-Time Paid 43; Part-Time Paid 109; Part-Time Volunteers 228; Interns 26.
Governing Authority: nonprofit organization. Parent Institution: Old Sturbridge, Inc. Tax-exempt: 170(b)(1)(A) & 501(c)(3).
Institution Type/Description: History Museum.
Collections: over 40 historical buildings with historical landscape & gardens; 7 houses; district school; 2 meetinghouses; bank; store; law office; gristmill; carding mill; cider mill; sawmill; tin shop; pottery; working kiln; working farm, livestock & crops; artifacts to 1840; costumes; ceramics; iron; leather goods; farm & craft tools; militia accoutrements; scientific & medical equipment; glass; guns; decorative arts; textiles; clocks; furniture; vehicles; lighting devices; plants & herbs; manuscripts; maps; atlases; photographs.
Major Exhibits: Delightfully Designed: The Furniture & life of Nathan Lombard, 10/13-5/14.
Research Fields: rural community life; agriculture & horticulture; crafts; technology & industry; historical archaeology; social history; women's history; domestic life; economic history; architectural history; material culture; audience research & evaluation.
Facilities: 33,000-vol. research library of books & bound periodicals; conference center; visitor center; theater. Reproductions of museum originals, books, decorative arts & crafts for sale.
Activities: theme days; workshops; educational programs; overnights; children's museum; teacher workshops; demonstrations of working crafts including pottery, coopering, basketmaking, tin work, blacksmithing, shoemaking, bonnetmaking, spinning & weaving; hearth cooking; membership & volunteer program; History Learning Laboratory online tour; hands-on History Gateway for children & families.
Publications: quarterly magazine, Old Sturbridge Visitor; thematic leaflets; cookbook; guidebook; booklets & monographs on early New England social & material life, artifacts & customs; educational publications; postcards; brochures; annual report; exhibition catalogs; pictorial calendar; email newsletter; online collections database.
Hours & Admission Prices: Call for hours. Adults $24, senior citizens 65 & over $22, youth 3-17 $8; children under 3, members & second visit within 10 days no charge. Closed Christmas; Mon. & Tues. in winter (special Dec. schedule). &
Attendance: 250,000 (estimated)
Membership: Individual $50; Individual Plus One $80; Family $90; Family Plus One $120.

Sudbury

LONGFELLOW'S WAYSIDE INN, 72 Wayside Inn Rd., Sudbury, MA 01776-3224. Tel.: 978-443-1776. Fax: 978-443-8041. Facebook: Longfellows Wayside Inn.
E-mail: history@wayside.org

Web Site: www.wayside.org
Formerly: Howe's Tavern
Founded: 1716.
Congressional District: 5
Key Personnel: Pres. (V), Lily Gordon; Innkeeper, Steve Pickford.
Personnel Profile: Full-Time Paid 2; Part-Time Paid 4; Part-Time Volunteers 2.
Governing Authority: board of trustees; nonprofit organization. Tax-exempt: 501(c)(3).
Institution Type/Description: History Museum.
Collections: period furnishings. Historic Buildings: operative 1716 inn; Redstone School; Martha-Mary Chapel; grist mill; coach house.
Research Fields: tavern keeping; colonial life.
Facilities: 150-vol. library of works by early American authors; archives.
Activities: guided tours; lectures; school groups; recreation historic militia; fife & drum corp; permanent & temporary exhibitions.
Publications: Tales of a Wayside Inn; Wayside Inn Historama.
Hours & Admission Prices: School & Grist Mill: April-Nov. Wed.-Sun. 9-5. No charge, donations accepted. Closed Independence Day; Christmas.
Attendance: 15,000 (estimated)

SUDBURY HISTORICAL SOCIETY, 322 Concord Rd., Sudbury, MA 01776. Tel.: 978-443-3747. Fax: 978-443-3747.
E-mail: sudburyhist01776@verizon.net
Web Site: www.sudbury01776.net
Governing Authority: nonprofit organization.
Institution Type/Description: Historical Society Museum.
Collections: local history & culture; period furnishings; personal artifacts; photographs.
Facilities: Museum-related items for sale.
Activities: special events.
Hours & Admission Prices: Mon.-Fri. 1-5 by appointment. No charge; donations accepted. Closed holidays.

Sutton

SUTTON HISTORICAL SOCIETY MUSEUM & RUFUS PUTNAM MUSEUM, 4 Uxbridge Rd., Sutton, MA 01590. Tel.: 508-865-8722.
Institution Type/Description: Historical Society Museum.
Collections: local history & culture; period furnishings; personal artifacts; photographs.
Activities: guest speakers.
Hours & Admission Prices: Tues. 7pm-9pm.

Swampscott

ATLANTIC NO. 1 VETERANS FIREMEN'S ASSOCIATION, INC., 76 Burrill St., Swampscott, MA 01907-1915. Mailing Address: 41 Lynn Shore Dr., Lynn, MA 01902-4927. Tel.: 781-595-4050.
Founded: 1966.
Congressional District: 6
Key Personnel: Pres., Rowe Austin; Dir., Richard Maitland; Sec., Edna Maitland; Treas., Douglas B. Maitland.
Personnel Profile: Part-Time Volunteers 5.
Governing Authority: municipal; nonprofit organization. Tax-exempt.
Institution Type/Description: Fire-Fighting Museum.
Collections: period fire engine; 1845 Hunneman hand tub.
Activities: Museum Sponsors: Muster for Antique Engines during summer.
Hours & Admission Prices: Call for appointment. No charge.
Attendance: 100 (estimated)
Membership: Individual $5.

SWAMPSCOTT HISTORICAL SOCIETY (SIR JOHN HUMPHREY HOUSE), 99 Paradise Rd., Swampscott, MA 01907-1955. Tel.: 781-599-1297.
Founded: 1921.
Congressional District: 8
Key Personnel: Pres., Mary Cassidy; Acting Treas., Douglas Maitland.
Personnel Profile: Part-Time Volunteers 10.
Governing Authority: nonprofit organization. Tax-exempt: 170(b)(1)(A).
Institution Type/Description: Historic House Museum: 1637 Sir John Humphrey House.
Collections: memorabilia of Swampscott & its residents.
Activities: guided tours; local school tours. Museum Sponsors: special open house events during year; Independence Day celebration.
Publications: newsletter.
Hours & Admission Prices: By appointment. No charge; donations accepted.
Attendance: 500 (estimated)

Membership: Students $10; Individual $20; Family $40; Life $250.

Taunton

OLD COLONY HISTORICAL SOCIETY, (M), 66 Church Green, Taunton, MA 02780-3445. Tel.: 508-822-1622. Fax: 508-880-6317.
E-mail: OldColony@oldcolonyhistoricalsociety.org
Web Site: www.oldcolonyhistoricalsociety.org
Founded: 1853.
Congressional District: 9
Key Personnel: Dir., Jane M. Hennedy; Pres. (V), Cynthia Booth Ricciardi, Ph.D.; Cur., Christie Jackson; Asst. to Dir., Elizabeth Bernier; Archivist, Andrew D. Boisvert.
Personnel Profile: Full-Time Paid 3; Part-Time Paid 2; Part-Time Volunteers 38; Interns 1.
Governing Authority: society; nonprofit. Tax-exempt: 501(c)(3).
Institution Type/Description: Historical Society Museum: housed in 1852 Bristol Academy Building designed by Richard Upjohn.
Collections: Taunton regional artifacts from pre-history to present; Native American artifacts; portraits; textiles; furniture; toys; household & industrial goods; stove; tools; firefighting equipment; silver and pewter; Rogers Groups statues.
Research Fields: genealogy; local history; local decorative arts.
Facilities: 7,000-vol. library; archives.
Activities: guided tours; illustrated talks; internships; children's programs.
Publications: quarterly newsletter; book, A History of Taunton, Massachusetts.
Hours & Admission Prices: Tues.-Sat. 10-4. Adults $4, seniors & children 12-18 $2; discounts to Southeastern Massachusetts Convention & Visitors Center Bureau, AASLH, NEMA, AAM & ICOM members; members no charge. Genealogical Research Admission: $7. Closed Saturdays preceding Monday holidays. &
Attendance: 5,025 (accurate)
Membership: Student $20; Individual $35; Dual $55; Family $75; Supporting $100; Sustaining & Corporate $250; Life $500. (Educator's 10% discount)

Templeton

NARRAGANSETT HISTORICAL SOCIETY, 1 Boynton Rd., Templeton, MA 01468. Mailing Address: P.O. Box 354, Templeton, MA 01468-0354. Tel.: 978-939-2303.
E-mail: narragansetthistoricalsociety@yahoo.com
Web Site: www.narragansetthistoricalsociety.org
Founded: 1928.
Key Personnel: Pres. (V), Beth Arsenault; Historian, Harry Aldrich; Treas., Debra Caisse.
Governing Authority: nonprofit organization. Tax-exempt.
Institution Type/Description: Historical Society: housed in 1810 two-story brick house.
Collections: local history.
Facilities: reference library; meeting rooms; garden.
Activities: guided tours; permanent exhibitions.
Publications: quarterly newsletter.
Hours & Admission Prices: July-Aug. Sat. 2-5. Adults $3; members no charge.
Membership: Student $5; Individual $15; Family $25; Lifetime $100.

Topsfield

IPSWICH RIVER WILDLIFE SANCTUARY (MASSACHU-SETTS AUDUBON SOCIETY), 87 Perkins Row, Topsfield, MA 01983-1922. Tel.: 978-887-9264. Fax: 978-887-0875.
E-mail: ipswichriver@massaudubon.org
Web Site: www.massaudubon.org
Founded: 1951.
Key Personnel: Dir., Carol J. Decker; Pres., Laura Johnson; Chm. (V), Jon Panek; Adult Program Coord. & Volunteer Coord., Susan Baeslack; Property Mgr., Richard Wolniewicz; Sec., Janet Barnes; Office Mgr., Christina MacDougall; Master Naturalist, Bob Speare; Caretaker & Property Worker, Fred Goodwin; Education Coord., Lois Bairstow; Teacher & Naturalist, Scott Santino; Education Coord., Sally Willard.
Personnel Profile: Full-Time Paid 7; Part-Time Paid 3; Part-Time Volunteers 125.
Governing Authority: society. Parent Institution: Massachusetts Audubon Society, S. Great Rd., Lincoln, MA 01773. Tax-exempt.
Institution Type/Description: Wildlife Refuge and Bird Sanctuary.
Collections: wildlife & their habitats.
Facilities: non-lending library of natural history books; education center; meeting barn; camping. Gift items for sale.
Activities: bird watching; maple sugaring; camping & canoeing for MA Audubon members only; cross-country skiing; river ecology canoe trips; 10 mile hiking trails; natural history programs for adults & children; summer day camp.

Publications: Map of Sanctuary; tri-annual program brochure; day camp brochure; school services brochure; bird list.
Hours & Admission Prices: Sanctuary: Mon. holidays & Tues.-Sun. dawn-dusk. Adults $4, senior citizens & children 2-12 $3; members of MA Audubon no charge. Office: May-Oct. Tues.-Fri. 9-4, Sat.-Sun. 9-5; Nov.-April Tues.-Sun. 9-4.
Attendance: 30,000
Membership: Student $20; Individual $40; Family $50; Supporting $60; Defender $75; Donor $100; Protector $150; Sponsor $250; Patron $500; Leadership Friend $1,250.

TOPSFIELD HISTORICAL SOCIETY, One Howlett St., Topsfield, MA 01983-1409. Mailing Address: P.O. Box 323, Topsfield, MA 01983-0523. Tel.: 978-887-9724.
E-mail: membership@topsfieldhistory.org
Web Site: www.topsfieldhistory.org
Formerly: Parson Capen House and Gould Barn
Founded: 1894.
Congressional District: 6
Key Personnel: Pres. (V) & Museum Shop Mgr., Norman J. Isler; Vice Pres., Anne Barrett; Treas., Richard Carlson.
Personnel Profile: Part-Time Volunteers 20.
Governing Authority: society. Tax-exempt.
Institution Type/Description: Historic Houses: 1683 Parson Capen House & 1710 Gould Barn.
Collections: early American artifacts.
Facilities: library pertaining to historical society's collections & maps of local area; limited to Topsfield, MA region by appointment.
Publications: catalogue, Houses & Buildings of Topsfield, MA, 1902-1950; brochure: Topsfield and the Witchcraft Tragedy; History of Topsfield 1940; Topsfield Historical Collection vol. XXiV, 2009.
Hours & Admission Prices: mid-June to early Sept. Wed., Fri. & Sun. 1-4:30. Suggested Donation: $3; discounts to Essex National Heritage volunteers.
Attendance: 3,500 (estimated)
Membership: Individual $15; Couple $25; Family $35; Contributing $50; Life $350.

Truro

TRURO CENTER FOR THE ARTS AT CASTLE HILL, 10 Meetinghouse Rd., Truro, MA 02666. Mailing Address: P.O. Box 756, Truro, MA 02666-0756. Tel.: 508-349-7511. Fax: 508-349-7513.
E-mail: info@castlehill.org
Web Site: www.castlehill.org
Key Personnel: Exec. Dir., Cherie Mittenthal
Institution Type/Description: Art Gallery.
Collections: paintings; sculpture; pottery.
Activities: educational programs; lectures; concerts; special events.
Hours & Admission Prices: Call for hours.

Turners Falls

THE GALLERY AT HALLMARK, 85 Avenue A, Turners Falls, MA 01376-1152. Tel.: 413-863-0009.
E-mail: info@gallery.hallmark.edu
Web Site: gallery.hallmark.edu
Formerly: The Hallmark Museum of Contemporary Photography
Key Personnel: Founder & Pres., George J. Rosa, III; Gallery Mgr., Christine Pineo; Dir. Exhibitions, Lisa Robinson; Asst. to Dir. & Exhibitions Installer, Stanley Wheeler; Business Operations Mgr., Ed Martin; Webmaster, Matt Jacobson-Carroll.
Governing Authority: Parent Institution: Hallmark Institute of Photography.
Institution Type/Description: Photography Museum.
Collections: contemporary photography from national & international artists.
Hours & Admission Prices: Fri.-Sun. 1-5. No charge. Closed New Year's Day & week; Easter; Thanksgiving; Christmas Day & week.

GREAT FALLS DISCOVERY CENTER, 2 Avenue A, Turners Falls, MA 01376-1101. Tel.: 413-863-3221.
Web Site: www.greatfallsdiscoverycenter.org
Founded: 2004.
Institution Type/Description: Natural History Museum.
Collections: exhibits pertaining to Connecticut River Watershed.
Hours & Admission Prices: May-Sept. daily 10-4; Oct.-April Fri.-Sat. 10-4; other times by appointment. No charge; donations accepted. Closed New Year's Day; Thanksgiving; Christmas. &

Wakefield

WAKEFIELD HISTORICAL SOCIETY, 39 Prospect St., Wakefield, MA 01880. Mailing Address: P.O. Box 1902, Wakefield, MA 01880-5902. Tel.: 781-246-3070.
E-mail: wakefieldhistory@gmail.com
Web Site: www.wakefieldma.org
Founded: 1890.
Key Personnel: Pres. (V), Nancy Bertrand; Treas., David Gooch.
Governing Authority: society. Tax-exempt: 501(c)(3).
Institution Type/Description: Local History Museum.
Collections: Wakefield history.
Activities: lectures.
Hours & Admission Prices: Sept.-May second Tues. of month; other times by appointment. No charge.
Membership: Students $5; Individual $15; Family $20; Lifetime $100.

Waltham

CHARLES RIVER MUSEUM OF INDUSTRY & INNOVATION, 154 Moody St., Waltham, MA 02453-5302. Tel.: 781-893-5410. Fax: 781-891-4536.
E-mail: elln@ormi.org
Web Site: www.crmi.org
Founded: 1980.
Congressional District: 8
Key Personnel: Dir. & Exec. Dir., Elln Hagney; Pres. (V), Arthur Nelson; Treas., Kim Washisuo; Asst. Dir., Kim Kalen.
Personnel Profile: Full-Time Paid 1; Part-Time Paid 4; Part-Time Volunteers 15; Interns 1.
Governing Authority: nonprofit organization. Tax-exempt: 501(c)(3).
Institution Type/Description: Textile and Industrial Museum: museum of innovation.
Collections: textiles; automobile & steam engine; power generation; machine tools.
Research Fields: pertaining to collections; social & cultural history of industrial communities.
Facilities: 4,000-vol. library including trade catalogues, technical manuals & technical drawings; business records; horology collection.
Activities: guided tours; lectures; organized education programs for adults, children & undergraduate or graduate college students; participatory exhibits.
Publications: quarterly newsletter, The Innovator; Canoeing the Charles-Images & Field Notes 1902-1912.
Hours & Admission Prices: Thurs.-Sat. 10-5. Adults $5, seniors & students $3; discounts to WGBH, AAM, AAA & NAWCC members; children under 6 & members no charge. &
Attendance: 13,000 (estimated)
Membership: Student & Senior $15; Individual $25; Family $40. Corporate: Member $100; Supporter $250; Patron $500; Guarantor $1,000; Sponsor $2,500; Benefactor $5,000.

* **GORE PLACE, (M),** 52 Gore St., Waltham, MA 02453-6866. Tel.: 781-894-2798. Fax: 781-894-5745.
E-mail: goreplace@goreplace.org
Web Site: goreplace.org
Founded: 1935.
Congressional District: 8
Key Personnel: Pres. (V), James F. Hunnewell, Jr.; Exec. Dir., Susan Robertson; Collections Mgr., Lana Lewis; Education Specialist, Susan Katz; Education Specialist, Tamar Agulian; Devel., Catie Camp; Supt. Grounds, Scott Clarke.
Personnel Profile: Full-Time Paid 4; Part-Time Paid 13; Part-Time Volunteers 75; Interns 5.
Governing Authority: nonprofit organization. Tax-exempt: 501(c)(3).
Institution Type/Description: Historic House & Farm: 1806 22-room Gore Mansion.
Collections: period furniture & Gore pieces; early 19th century American agriculture. Historic Buildings: 1793 carriage house; 1806 mansion built by Christopher & Rebecca Gore; 1835 farmhouse, on a 50 acre historic site.
Major Exhibits: Over the River & Through the Woods, 1/14-3/14.
Research Fields: genealogy; landscape design; architectural history; local history; historic breeds of animals.
Facilities: library; research room; 50 acre site.
Activities: guided tours; lectures; special events for members; scholarly research; formally organized education programs for children.
Publications: descriptive brochures; newsletter; books, House Servants Directory; Christopher Gore: Federalist of Massachusetts 1758-1827; educational film, The Gores of Massachusetts; Setting Out to Best Advantage - At Home with Christopher & Rebecca Gore.

Hours & Admission Prices: Estate & Farm: daily. Mansion Tours: Mon.-Thurs. 1-3, Fri. 1-3 & 7-9, Sat. 12-4. Adults $12, children under 12 $6; discounts to AAM members & New England Museum Association members; members no charge. Closed Federal holidays. &
Attendance: 26,512 (accurate)
Membership: Waltham-Watertown Resident $35; Individual $40; Family $50; Friends of the Farm $100; Friends of the Mansion $250; Rebecca Gore Society $500; Christopher Gore Society $1,000.

* **LYMAN ESTATE AND GREENHOUSES, (M),** 185 Lyman St., Waltham, MA 02452-5645. Mailing Address: 141 Cambridge St., Boston, MA 02114-2702. Tel.: 617-994-5912 (Estate); 781-891-1985 (Greenhouse).
Web Site: www.historicnewengland.org
Founded: 1951.
Congressional District: 8
Key Personnel: Pres. & C.E.O., Carl Nold; Museum Operations Mgr., Melinda Huff; Mgr. Greenhouses, Lynn Ackerman.
Governing Authority: society; nonprofit organization. Parent Institution: Historic New England, 141 Cambridge St., Boston 02114. Tel.: 617-227-3956. Tax-exempt: 501(c)(3).
Institution Type/Description: Historic House: housed in the country estate of Boston merchant Theodore Lyman; built in 1793; greenhouses among the oldest in the country.
Collections: landscape, grounds & gardens; operating greenhouses containing 100 year-old camellia trees; grapevines; orchids; rare exotic plants.
Facilities: botanical garden; banquet facilities.
Activities: guided tours; lectures & special events; rental facilities.
Publications: magazine, Historic New England.
Hours & Admission Prices: Mansion: 3rd Sat. each month 10-1, last tour 12pm. Greenhouse: Wed.-Sat. 9:30-4; Dec. 15-July 15 Wed.-Sun. 9:30-4. Grounds: sunrise to sunset. House tours $5; discount to AAM members; Historic New England members no charge.
Attendance: 19,882 (accurate)
Membership: National $35; Individual $45; Household $55; Garden & Landscape $75; Contributing and Library & School $100; Young Friends of Historic New England $100-$1,500; Friends of the Library and Archives $125; Historic Homeowner $200; Business & Ogden Codman Design Group $250; Appleton Circle $2,500 & up.

THE ROSE ART MUSEUM OF BRANDEIS UNIVERSITY, (M), 415 South St., MS069, Brandeis Univ., Waltham, MA 02453-2728. Mailing Address: P.O. Box 549110, Waltham, MA 02454-9110. Tel.: 781-736-3434. Fax: 781-736-3439.
E-mail: rosemail@courier.brandeis.edu
Web Site: www.brandeis.edu/rose
Formerly: The Rose Art Museum
Founded: 1961.
Congressional District: 7
Key Personnel: Dir. Museum Operations, Roy Dawes; Collections Mgr. & Registrar, Kristin Parker; Dir. Academic Programs, Dabney Hailey.
Personnel Profile: Full-Time Paid 4; Part-Time Volunteers 11; Interns 7.
Governing Authority: university. Parent Institution: Operated by Brandeis University, Waltham, MA. Tax-exempt.
Institution Type/Description: Art Museum.
Collections: international modern & contemporary art.
Research Fields: modern & contemporary art.
Activities: temporary exhibitions; lectures; tours; members' events; catalog exchange program.
Publications: exhibition catalogs; brochures.
Hours & Admission Prices: Tues.-Sun. 12-5. No charge; donations accepted. Closed major holidays. &
Attendance: 13,000 (accurate)
Membership: Individual $50; Couple $75; Friend $125; Associate $250; Patron $500; Benefactor $1,000; Angel $2,500; Director's Circle $5,000; Corporate $5,000 & up; Founder's Circle $10,000.

WALTHAM HISTORICAL SOCIETY, 190 Moody St., Waltham, MA 02453-5384. Tel.: 781-891-5815.
E-mail: waynemccarthy@comcast.net
Web Site: www.walthamhistoricalsociety.org
Founded: 1913.
Congressional District: 5
Key Personnel: Co-Pres. & Communications, Wayne McCarthy; Co-Pres., Sheila E. FitzPatrick; Treas., Mary Selig; Recording Sec., Leona Lindsay; Corresponding Sec., Joseph Vizard; Cur., Michelle Morello; Asst. Cur., Winifred W. Kneisel.
Personnel Profile: Part-Time Volunteers 10.

Governing Authority: society. Annex: 240 Grove St., Waltham, MA 02453. Tax-exempt: 501(c)3.
Institution Type/Description: Local History Museum.
Collections: archives; costumes; watch industry; historic objects; agricultural & household tools; photographs.
Facilities: 1,000-vol. library of books, pamphlets, photographs, maps & clippings available for use on premises by appointment.
Activities: lectures; temporary exhibitions.
Publications: Waltham As a Precinct of Watertown & As a Town, 1630-1884; Waltham Industries; The Golden Door: The Story of Waltham's Economic Progress; occasional pamphlets; quarterly newsletter; Driving Tour of North Waltham.
Hours & Admission Prices: By appointment only. Annex: Sat. 9-1. No charge; donations accepted. Closed holidays.
Attendance: 400
Membership: Individual $25; Family $30; Life $150.

THE WALTHAM MUSEUM, 25 Lexington St., Waltham, MA 02452-4415. Tel.: 781-893-9020.
E-mail: thewalthammuseum@verizon.net
Web Site: walthammuseum.com
Founded: 1971.
Congressional District: 5
Key Personnel: Dir., Albert A. Arena, Sr.; Museum Shop Mgr., Louise Butler.
Personnel Profile: Part-Time Volunteers 12; Interns 6.
Governing Authority: nonprofit organization. Tax-exempt: 501(c)(3).
Institution Type/Description: Local History Museum.
Collections: local history & culture; period artifacts; watches; clocks; automobiles; bottles; stoves; radios; tools; lanterns; cloth; bicycles; steam engines; period fire engines; cameras; military artifacts; Native American arrowheads & tools; boilers; lathes; local sports exhibits; locally made items.
Research Fields: local history.
Facilities: more than 2,000-vol. library of books on Waltham available by appointment only.
Activities: guided tours; lectures; films; permanent exhibitions; slide shows for organizations; school group tours; cable TV program, This Was Waltham.
Publications: bimonthly newsletter.
Hours & Admission Prices: Tues.-Sat. 1-4:30; other times by appointment. Adults $4, seniors & children $2; discount to local non profits; members no charge. Closed holidays. &
Attendance: 1,000 (estimated)
Membership: Annual $25; Corporations $100.

Watertown

ARMENIAN LIBRARY AND MUSEUM OF AMERICA (ALMA), 65 Main St., Watertown, MA 02472-4400. Tel.: 617-926-2562. Fax: 617-926-0175.
E-mail: info@almainc.org
Web Site: www.almainc.org
Founded: 1971.
Congressional District: 8
Key Personnel: Chm. (V), Haig Der Manuelian; Exec. Dir., Beri Chekijian; Asst. Cur., Howayda Abu Affan; Treas. (V), Jae Erdekian; Cur., Susan Lind-Sinanian; Cur. & Program Mgr., Gary Lind-Sinanian.
Personnel Profile: Full-Time Paid 4; Part-Time Paid 3; Part-Time Volunteers 50; Interns 1.
Governing Authority: nonprofit organization. Tax-exempt: 501(c)(3).
Institution Type/Description: Cultural Museum.
Collections: Ethnic Armenian historical items; folk art; fine art; textiles; coins; photography; ceramics; religious art.
Research Fields: Armenian history; textile arts.
Facilities: 20,000-vol. library; 220-seat auditorium; classroom; 16,000 sq. ft. exhibit space. Museum-related items for sale.
Activities: concerts; docent program; films; formal educational programs; guided tours; lectures; loan & temporary exhibitions; rental gallery; theater.
Publications: bimonthly calendar & in-house news, ALMA Matters; ALMA Newsletter, published 3 times annually.
Hours & Admission Prices: Thurs.-Fri. 12-8, Sat.-Sun. 12-6, Wed. by appointment. Adults $7, seniors & students $3; ALMA members & children under 12 no charge. Closed New Year's Day; Easter; Thanksgiving; Christmas. &
Attendance: 7,000 (estimated)
Membership: Student $15; Individual $35; Family $50; Sustaining $100; Supporting $250; Contributing $500 & up.

PERKINS HISTORY MUSEUM, PERKINS SCHOOL FOR THE BLIND, 175 North Beacon St., Watertown, MA 02472-2790. Tel.: 617-924-3434. Fax: 617-926-2027.
E-mail: historymuseum@perkins.org

Web Site: perkins.pvt.k12.ma.us/museum
Formerly: Museum on the History of Blindness, Perkins School for the Blind
Founded: 2004.
Congressional District: 8
Key Personnel: Cur., Betsy L. McGinnity.
Personnel Profile: Part-Time Paid 1.
Governing Authority: private school. Parent Institution: Perkins School for the Blind. Tax-exempt.
Institution Type/Description: History Museum.
Collections: maps; educational aids for blind or deaf students from 19th-century to present; embossed books; period writing & communication devices; Perkins students including Laura Bridgman, Anne Sullivan, & Helen Keller; tactile globe.
Activities: tours by appointment; permanent exhibitions.
Hours & Admission Prices: Tues. & Thurs. 2-4; other times by appointment. No charge. &
Attendance: 875 (estimated)

THE PLUMBING MUSEUM, 80 Rosedale Rd., Watertown, MA 02471. Tel.: 617-926-2111.
E-mail: info@theplumbingmuseum.org
Web Site: www.theplumbingmuseum.org
Institution Type/Description: History Museum: housed in a former ice house; c.1860.
Collections: plumbing history & equipment; modern fixtures & techniques; photographs; period artifacts.
Activities: rental facilities.
Hours & Admission Prices: Mon.-Thurs. by appointment. Closed holidays.

Wayland

WAYLAND HISTORICAL SOCIETY, 12 Cochituate Rd., Wayland, MA 01778. Mailing Address: P.O. Box 56, Wayland, MA 01778-0056. Tel.: 508-358-7959.
E-mail: wayhistsoc@comcast.net
Web Site: www.waylandhistoricalsociety.org
Founded: 1954.
Congressional District: 5
Key Personnel: Pres., Jane Dunn; Cur., Lois Davis.
Personnel Profile: Part-Time Volunteers 25.
Governing Authority: nonprofit organization. Tax-exempt: 501(c)(3).
Institution Type/Description: General Museum.
Collections: costumes; furniture; local history; local Indian artifacts. Historic House: c.1740 Grout-Heard House.
Research Fields: local genealogy & archaeology; history of old houses.
Facilities: library of history books available for use on premises; reading room.
Activities: guided tours; lectures; formally organized education programs for children.
Publications: bimonthly newsletter.
Hours & Admission Prices: Tues. & Fri. 9:30 to noon. No charge; donations accepted. &
Attendance: 2,000 (estimated)
Membership: Junior $2; Single $15; Family $25; Contributing $50; Patron $100; Life $350.

Wellesley

DAVIS MUSEUM AT WELLESLEY COLLEGE, (M), Wellesley College, 106 Central St., Wellesley, MA 02481-8203. Tel.: 781-283-2051. Fax: 781-283-2064.
Web Site: www.davismuseum.wellesley.edu
Founded: 1889.
Congressional District: 10
Key Personnel: Ruth Gordon Shapiro '37 Dir., Lisa Fischman; Cur. Academic Programs, Elaine Mehalakes; Security Mgr., David Accorsini.
Personnel Profile: Full-Time Paid 8; Part-Time Paid 4; Interns 5.
Governing Authority: Wellesley College. Tax-exempt: 501(c)(3).
Institution Type/Description: Art Museum.
Collections: European & American paintings, sculpture, drawings, prints & photographs; Asiatic, African & pre-conquest Meso-American art.
Facilities: 17,250 sq. ft. exhibit space; 167-seat cinema; print, drawing & photo study; study gallery; seminar room.
Activities: permanent & temporary exhibitions; guided tours; lectures; organized education programs for undergraduate students.
Publications: exhibition catalogues; collection handbook.
Hours & Admission Prices: Summer: Tues.-Fri. 12-4, Sept. 9-Dec. 13 Tues. & Thurs.-Sat. 11-5, Wed. 11-8, Sun. 12-4. No charge. Closed holidays; Christmas Eve, Day & week; campus holidays. &
Attendance: 30,000 (estimated)

Membership: Recent Alumnae $15; Contributor $50; Sponsor $100; Donor $250; Supporter $500; Benefactor $1,000; Patron $2,500.

WELLESLEY HISTORICAL SOCIETY, INC., 229 Washington St., Wellesley, MA 02481-3105. Tel.: 781-235-6690. Fax: 781-239-0660.
E-mail: info@wellesleyhistoricalsociety.org
Web Site: www.wellesleyhistoricalsociety.org
Founded: 1925.
Congressional District: 4
Key Personnel: Exec. Dir., Robert Damon; Pres., John Celi.
Personnel Profile: Full-Time Paid 1; Part-Time Volunteers 35.
Governing Authority: society. Tax-exempt: 501(c)(3).
Institution Type/Description: Local History Museum: housed in 1824 Dadmun-McNamara house.
Collections: Archives, artifacts, books, costumes & textiles; fine arts and photographs relating to Wellesley from Native American to present date; Denton butterfly collection; Oldham lace.
Research Fields: Wellesley-people, places, buildings, things.
Facilities: 700-vol. research library pertaining to local history & the Civil War available for use on the premises. Lace reference library; local authors collection.
Activities: changing exhibitions; lecture programs; bus trips; programs for children.
Publications: newsletter; books; exhibit catalogues.
Hours & Admission Prices: Tues.-Thurs. 10-3. No charge; donations accepted. Research by appointment. Research: non-members $10. Closed holidays. &
Attendance: 2,000 (estimated)
Membership: Senior 65 & over $20; Individual $35; Family $50; Corporate $150.

Wellfleet

CAPE COD NATIONAL SEASHORE, 99 Marconi Site Rd., Wellfleet, MA 02667. Tel.: 508-255-8925. Fax: 508-240-3291.
E-mail: hope_morrill@nps.gov
Web Site: www.nps.gov/caco
Founded: 1961.
Congressional District: 10
Key Personnel: Chief Interpretation & Cultural Resources, Sue Moynihan; Cur., Hope Morrill.
Governing Authority: federal. Administered by National Park Service, U.S. Dept. of the Interior. Branch Units: Salt Pond Visitor Center, Eastham, MA; Province Lands Visitor Center, near Provincetown. Tax-exempt.
Institution Type/Description: Park Museum & Visitor Center.
Collections: history; natural history; archaeology. Historic Buildings: 18th-century, Atwood-Higgins House; 19th-century Capt. Edward Penniman Home; Old Harbor Life Saving Station; 3 sister Lighthouses.
Research Fields: historic & prehistoric archaeology; natural history.
Facilities: 1,500-vol. library of history & natural history books available for research only on premises. Books for sale at visitor centers.
Activities: guided tours & evening programs; formally organized environmental education program for children during school year; permanent exhibitions; self-guided nature trails open year-round.
Publications: The Archeology of Cape Cod National Seashore; Seashells of Cape Cod National Seashore; Common Trailside Plants of Cape Cod National Seashore; Augusta Penniman Journal of a Whaling Voyage 1864-1868; brochures.
Hours & Admission Prices: Headquarters: Mon.-Fri. 8-4:30. Salt Pond Museum: mid-June to Labor Day daily 9-5; Sept.-May 9-4:30. Salt Pond Visitor Center: Summer 9-5. Province Lands Visitor Center: mid-April to Nov. No charge; donations accepted. Beaches: season pass $45, vehicles $15, pedestrians & bicycles $3. &
Attendance: 500,000
Membership: Friends of Cape Cod National Seashore: Individual $20; Family $35; Sustaining $100.

WELLFLEET HISTORICAL SOCIETY MUSEUM, 266 Main St., Wellfleet, MA 02667. Mailing Address: P.O. Box 58, Wellfleet, MA 02667. Tel.: 508-349-9157 & 2247 (Cur.).
Web Site: www.wellfleethistoricalsociety.com
Founded: 1951.
Key Personnel: Cur., Joan Hopkins Coughlin; Staff, David Wright; Staff, Barbara Kennedy.
Personnel Profile: Full-Time Volunteers 1; Part-Time Paid 2.
Governing Authority: nonprofit organization. Parent Institution: Wellfleet Historical Society. Tax-exempt.
Institution Type/Description: Historic Building: c.1800 two-story Victorian building.

Collections: local history; whaling & shell-fishing artifacts; original Thoreau manuscript page from book Cape Cod; scrimshaw; dolls; deeds & manuscripts; maps; genealogy; folk art paintings; needlework; clothing; children's toys & furniture; photographs.
Research Fields: preservation of charcoal painting gift from historian Shebvah Rich; preservation of deeds, documents & town records; preservation of photograph collection onto disc.
Facilities: library & files available for research by public on premises only.
Activities: guided tours; organized education programs for children; loan & temporary exhibitions; historic walking tours.
Publications: pamphlet, Beacon; books, Billingsgate; A Pictorial History of Wellfleet; Famous Beds of Wellfleet; Looking Back at Wellfleet.
Hours & Admission Prices: June 24-Sept. 6 Tues. & Fri. 10-4, Wed.-Thurs. & Sat. 1-4. No charge; donations accepted.
Attendance: 400 (estimated)
Membership: Annual $5; Family $10; Life $100.

Wenham

* **WENHAM MUSEUM, (M),** 132 Main St., Wenham, MA 01984-1520. Tel.: 978-468-2377. Fax: 978-468-1763.
E-mail: info@wenhammuseum.org
Web Site: www.wenhammuseum.org
Founded: 1921.
Congressional District: 6
Key Personnel: Exec. Dir., Lindsay Diehl; Chm. (V), Charles R. Richey, Jr.; Museum Shop Mgr., Cynthia Novotny.
Personnel Profile: Full-Time Paid 4; Part-Time Paid 9; Part-Time Volunteers 250.
Governing Authority: nonprofit organization. Tax-exempt: 501(c)(3).
Institution Type/Description: History Museum: housed in late 17th-century Claflin-Richards House & modern exhibition building.
Collections: dolls; doll houses; miniatures; toys; toy soldiers; model trains; costumes; textiles; Wenham Lake Ice Industry; fans; embroidery & needlework; quilts; 1890-1921 photographs; early domestic & kitchen utensils; manuscript collections. Historic Buildings: 1840 Merrill shoe shop; c.1690 Claflin-Richards House; 1840 Merrill Shoe Shop.
Research Fields: Wenham history; costumes; dolls; genealogy of local families; ice cutting history.
Facilities: 1,000-vol. library; children's interactive room; reading room; 100-seat hall; classroom. Publications, postcards & gift related items for sale.
Activities: guided tours; lectures; family programs; artisan workshops; formally organized education programs for children; permanent & temporary exhibitions; hands-on exhibits.
Publications: quarterly calendar & newsletter; illustrated annual report; Wenham in Pictures & Prose, The Claflin-Richards House.
Hours & Admission Prices: Tues.-Sun. 10-4. Adults $7, children 1-16 $5; discounts to groups and New England Museum Association, Massachusetts Teachers Assoc. & AAA members; members no charge. Closed Martin Luther King Jr. Day; Presidents' Day; Veterans Day. &
Attendance: 35,000 (accurate)
Membership: Senior Citizen $25; Individual $30; Senior Citizen Dual $40; Family $65; Family Plus $90; Contributing $120; Sponsoring $250.

West Barnstable

HIGGINS ART GALLERY - CAPE COD COMMUNITY COLLEGE, 2240 Iyanough Rd., West Barnstable, MA 02668-1599. Tel.: 508-362-2131, ext. 4844. Fax: 508-375-4020.
E-mail: higginsartgalley@capecod.edu
Web Site: www.capecod.mass.edu
Founded: 1989.
Congressional District: 10
Key Personnel: Dir., Betty Carroll Fuller; Pres., Kathleen Schatzberg.
Personnel Profile: Part-Time Paid 2; Interns 4.
Governing Authority: nonprofit; public college. Tax-exempt: 501(c)(3).
Institution Type/Description: Art Museum.
Collections: paintings; prints; drawings; graphics; sculpture; mixed media.
Facilities: 750-seat auditorium; 200-seat cafeteria; classrooms; 600 sq. ft. exhibit space.
Activities: arts festival; workshops; demonstrations; formal education programs for adults (non-credit classes); formal education programs for undergraduate or graduate college students affiliated with Cape Cod Community College; lectures; theater; broadcast programs; live musicians, performers & children's games. Annual Events: Annual Tilden Arts Festival in April & May; Annual Artists in Residence.
Hours & Admission Prices: Sept.-May Mon.-Thurs. 11-2; also open prior to events in adjacent college auditorium. No charge. Closed all official national & MA holidays. &

Attendance: 2,000 (estimated)

WILLIAM BREWSTER NICKERSON CAPE COD HISTORY ARCHIVES, Cape Cod Community College, 2240 Iyannough Rd., West Barnstable, MA 02668-1532. Tel.: 508-362-2131, ext. 4445. Fax: 508-375-4020.
E-mail: mlabombard@capecod.edu
Web Site: www.capecod.edu/web/nickerson/home
Founded: 1971.
Congressional District: 10
Key Personnel: Archivist, Mary LaBombard.
Personnel Profile: Part-Time Paid 1; Part-Time Volunteers 4.
Governing Authority: public college. Parent Institution: Cape Cod Community College. Tax-exempt: 501(c)(3).
Institution Type/Description: Archives.
Collections: Cape Cod history from Plymouth Colony records to present; Cape Cod Community College & Hyannis Normal School archives; Cape Cod Chamber of Commerce collection; ships' papers; scrimshaw.
Facilities: 5,000-vol. library of history books; educational facilities.
Activities: Museum Sponsors: Cape Heritage Week.
Hours & Admission Prices: Mon.-Wed. 9-3:30; other times by appointment. No charge. Closed MA state & federal holidays. &
Attendance: 500 (accurate)

West Springfield

RAMAPOGUE HISTORICAL SOCIETY, 70 Park St., West Springfield, MA 01089-3318. Mailing Address: P.O. Box 826, West Springfield, MA 01090-0826. Tel.: 413-734-8322 & 732-6187.
Web Site: www.west-springfield.ma.us/
Founded: 1903.
Congressional District: 1
Key Personnel: Pres. (V), Phyllis A. Bertera; Cur., Raymond Wellspeake.
Personnel Profile: Part-Time Volunteers 10.
Governing Authority: society. Tax-exempt: 501(c)(3).
Institution Type/Description: Historic House: 1754 Josiah Day House, original all brick salt box style house.
Collections: 18th-century furniture & accessories with many original Day family possessions.
Research Fields: genealogy of Day family.
Facilities: meeting room.
Activities: guided tours; special events.
Publications: pamphlets; semiannual newsletter.
Hours & Admission Prices: June-Oct. Sat.-Sun. 12-3; other times by appointment. Adults $3; children under 12 & members no charge.
Attendance: 534 (accurate)
Membership: Individual $10; Patron $25; Sponsor $50; Sustaining $100; Individual Life $100; Life Couple $150.

STORROWTON VILLAGE MUSEUM, 1305 Memorial Ave., Eastern States Exposition Grounds, West Springfield, MA 01089-3578. Tel.: 413-205-5051 & 737-2443. Fax: 413-205-5054.
E-mail: storrow@thebige.com
Web Site: thebige.com
Founded: 1929.
Congressional District: 1
Key Personnel: C.E.O., Eugene Cassidy; Dir., Dennis Picard; Admin. Asst., Jacki Sullivan.
Personnel Profile: Full-Time Paid 3; Part-Time Paid 3; Part-Time Volunteers 140.
Governing Authority: nonprofit organization. Parent Institution: Eastern States Exposition. Tax-exempt.
Institution Type/Description: Historic Village.
Collections: Historic Buildings: 1767 Phillips House; 1794 Gilbert House; 1810 Schoolhouse; 1810 Eddy law office; 1834 Union Meeting House; late 18th century, Atkinson Tavern; 1822 townhouse; 1776 Potter Mansion; c.1850 blacksmith shop; general store.
Research Fields: herbs & their uses; early New England education & lifestyle.
Facilities: restaurant. Gift items for sale.
Activities: guided tours by costumed interpreters; craft workshops; early American summer day camp; school vacation programs; seasonal special events; formally organized education programs for children; permanent exhibits; classes.
Publications: booklets, Doorways To The Past; Grandmother's Button Box; quarterly newspaper, The Crackerbarrel; cookbook, Aunt Helen's Cookie Recipes; book, Memories of a Door; cookbooks, Out of the Crackerbarrel; Out of the Farmstead Kitchen; Aunt Helen's Cookie Recipes; exhibit guide, Helen Osborn Storrow: The Lady and Her Legacy.
Hours & Admission Prices: Museum: Feb. to 3rd week in June & Sept.-Dec.

by appointment only; 3rd week in June to Labor Day Tues.-Sat. 11-3. Admission $5; discounts to NEMA members; AAM members & children under 6 no charge. Administration building, restaurant & gift shops open year round. Closed holidays. &
Attendance: 500,000 (estimated)

West Tisbury

POLLY HILL ARBORETUM, 795 State Rd., West Tisbury, MA 02575. Mailing Address: P.O. Box 561, West Tisbury, MA 02575. Tel.: 508-693-9426. Fax: 508-693-5772.
E-mail: info@pollyhillarboretum.org
Web Site: www.pollyhillarboretum.org
Founded: 1998.
Institution Type/Description: Botanical Garden.
Collections: over 1,700 different types of plants & trees from around the world.
Facilities: visitor center.
Activities: lectures; tours; workshops; classes.
Hours & Admission Prices: Grounds: daily sunrise to sunset. Visitor Center: Memorial Day to Columbus Day daily 9:30-4. Guided Tours: daily 2pm; groups by appointment. Suggested Donation: adults $5; children 12 & under no charge.

West Yarmouth

ZOOQUARIUM, 674 Rte. 28, West Yarmouth, MA 02673-5103. Tel.: 508-775-8883.
E-mail: info@zooquariumcapecod.net
Web Site: www.zooquariumcapecod.net
Institution Type/Description: Zoo.
Collections: wildlife.
Hours & Admission Prices: Call for hours. Adults 10 & over $9.75, children 2-9 $6.75; children under 2 no charge.

Westfield

AMELIA PARK CHILDREN'S MUSEUM, 29 S. Broad St., Westfield, MA 01086. Mailing Address: P.O. Box 931, Westfield, MA 01086-0931. Tel.: 413-572-4014. Fax: 413-572-1206.
E-mail: fun@ameliaparkmuseum.org
Web Site: www.ameliaparkmuseum.org
Institution Type/Description: Children's Museum.
Collections: hands-on exhibits.
Activities: special events; birthday parties; play groups.
Hours & Admission Prices: Sun.-Mon. & Wed.-Thurs. 10-4, Fri.-Sat. 10-7. Adults & children $7, seniors $3.50; children under one no charge.

JASPER RAND ART MUSEUM, AT WESTFIELD ATHENAEUM, 6 Elm St., Westfield, MA 01085-2904. Tel.: 413-568-7833. Fax: 413-568-0988.
Founded: 1927.
Congressional District: 1
Key Personnel: Dir., Christopher Lindquist; Pres., Bob Brown.
Governing Authority: nonprofit organization. Affiliated with Westfield Athenaeum. Tax-exempt: 101(6).
Institution Type/Description: Art Gallery.
Collections: archives; paintings.
Activities: temporary & traveling exhibitions; permanent collections; art classes for children.
Hours & Admission Prices: July-Aug. Mon.-Thurs. 8:30-8, Fri. 8:30-5; Sept.-June Mon.-Thurs. 8:30-8, Fri.-Sat. 8:30-5. No charge. Closed holidays. &

Westford

THE BUTTERFLY PLACE, 120 Tyngsboro Rd., Westford, MA 01886. Mailing Address: P.O. Box 1541, Westford, MA 01886. Tel.: 978-392-0955. Fax: 978-256-1080.
E-mail: bflybut@aol.com
Web Site: www.butterflyplace-ma.com
Founded: 1985.
Congressional District: 5
Key Personnel: Dir. & C.E.O., George C. Weslie, Jr.; Museum Shop Mgr., Sylvia L. Leslie.
Personnel Profile: Full-Time Paid 1; Full-Time Volunteers 1; Part-Time Paid 9.
Institution Type/Description: Butterfly Garden.
Collections: butterflies & tropical species from around the world.
Facilities: Books, videos, t-shirts, & jewelry for sale.
Activities: video; special events.

Hours & Admission Prices: Feb. 14-March & Sept.-Oct. daily 10-4; April-Aug. daily 10-5. Adults $12, seniors 65 & over $10, children 3-12 $8; discounts to groups; children 2 & under no charge. &

Attendance: 42,000 (estimated)

THE WESTFORD MUSEUM & HISTORICAL SOCIETY, 2
Boston Rd., Westford, MA 01886. Mailing Address: P.O. Box 411, Westford, MA 01886-0411. Tel.: 978-692-5550.
E-mail: museum@westford.com
Web Site: www.westford.com/museum
Founded: 1958.
Key Personnel: Dir., Penny Lacroix.
Personnel Profile: Part-Time Paid 1; Part-Time Volunteers 45.
Operating Expenses: 10,000
Operating Income: 10,000
Governing Authority: Parent Institution: Westford Historical Society, Inc. Tax-exempt.
Institution Type/Description: History Museum: housed in the Westford Academy school house, built in 1794.
Collections: local history & culture relating to Westford including yearbooks, photographs; local memorabilia.
Major Exhibits: Forge Village: Its People & Places, 11/13-6/14; Parkerville: Its People & Places, 9/14-6/15.
Research Fields: genealogy; community; genealogy resources; inventory of historic homes; maps; historic newspapers.
Activities: research; genealogy; lectures; school programs.
Hours & Admission Prices: Mon. & Wed.-Fri. 9-1; other times by appointment. No charge; donations accepted. Closed holidays. &
Attendance: 2,210 (accurate)
Membership: Students & Seniors $15; Individual $30; Family $50; Bronze $100; Silver $250; Gold $500; Benefactor $1,000.

Weston

GOLDEN BALL TAVERN MUSEUM, (M), 662 Boston Post Rd.,
Weston, MA 02493-1511. Mailing Address: P.O. Box 223, Weston, MA 02493. Tel.: 781-894-1751. Fax: 781-861-6218.
E-mail: joanb5@aol.com
Web Site: www.goldenballtavern.org
Founded: 1964.
Congressional District: 5
Key Personnel: Dir., Dr. Joan P. Bines; Chm. (V), William W. Gallagher, III; Pres. (V), William Wiseman.
Personnel Profile: Full-Time Paid 1; Part-Time Paid 1; Part-Time Volunteers 100.
Governing Authority: private; nonprofit organization. Tax-exempt: 501(c)(3).
Institution Type/Description: Historic House: 1768 Georgian house.
Collections: 1768 house; blue & white china; glass; silver; furnishings belonging to the Jones family over 200 years occupation of the house; artifacts found in digs on property; 18th-century tavern and tavern ware; clothing; barn.
Research Fields: archaeology; decorative arts; early taverns; early house construction.
Publications: books, The Grapevine; The Tavern and the Tory; newsletter, Tavern Tidings.
Hours & Admission Prices: Tours: by appointment. Adults $5, senior citizens & students $2; members no charge.
Attendance: 1,350 (estimated)
Membership: Single $15; Family $25; Sustaining $50; Fellow $100; Sponsor $250; Life $1,000; Benefactor $5,000.

SPELLMAN MUSEUM OF STAMPS AND POSTAL HISTORY,
235 Wellesley St., Regis College, Weston, MA 02493-1545. Tel.: 781-768-7331. Fax: 781-768-7332.
E-mail: info@spellman.org
Web Site: www.spellman.org
Founded: 1960.
Congressional District: 5
Key Personnel: Chm. (V) & Pres. (V), Mark W. Gallagher; Education & Community Outreach, Henry Lukas; Cur. Philatelic Collections, George Norton; Museum Shop Mgr., Anne O'Keefe.
Personnel Profile: Part-Time Paid 3; Part-Time Volunteers 18.
Governing Authority: nonprofit organization. Tax-exempt: 501(c)(3).
Institution Type/Description: Philatelic Museum.
Collections: over 2,000,000 stamps & covers; general reference philatelic library; particular strengths: United States, European, Vatican, Great Britain, and colonies, Hong Kong, Russia, Central & South America; air mail stamps and covers.
Research Fields: philatelic & postal history.

Facilities: library of philately & postal history books; meeting rooms; post office. Museum-related items for sale.
Activities: courses; lectures; workshops; guided tours; children's stamp education program; philatelic club meetings.
Publications: newsletter; catalogs; brochure; books, Philatelic Art In America; The Postal History of the Holocaust.
Hours & Admission Prices: July-Aug. Thurs.-Sat. 12-5; Sept.-June Thurs.-Sun. 12-5. Adults $8, seniors & students $5, children 5-16 $3; discounts to AAM, ICOM, AAA & NEMA members; children 4 & under and members no charge. Closed New Year's Day, Easter; Memorial Day weekend; Labor Day weekend; Thanksgiving & day after; Christmas. &
Attendance: 6,395 (accurate)
Membership: Student 16 & under $30; National Associate Living more than 150 miles of Weston, MA $35; Individual $40; Household $75; Participating $100; Sustaining $150; Contributing $250; Sponsor $500; Benefactor $1,000.

WESTON HISTORICAL SOCIETY, INC., 358 Boston Post Rd.,
Weston, MA 02493. Mailing Address: P.O. Box 343, Weston, MA 02493-0002.
Founded: 1963.
Congressional District: 4
Key Personnel: Pres. (V), Pamela W. Fox.
Governing Authority: society. Parent Institution: Weston Historical Society. Tax-exempt: 501(c)(3).
Institution Type/Description: History Museum.
Collections: photographs; archives; town reports. Historic Building: c.1757 Josiah Smith tavern.
Research Fields: history of Weston.
Facilities: library & archives by appointment; reading room.
Activities: lectures; temporary exhibitions.
Publications: biannual, Bulletin; books, Once Upon a Pung; Random Recollections; One Town in the American Revolution, Weston, Mass.
Hours & Admission Prices: By appointment. No charge.
Membership: Family $25; Life $500.

Weymouth Landing

WEYMOUTH HISTORICAL MUSEUM, Tufts Library, 46 Broad
St., Weymouth Landing, MA 02188-1713. Tel.: 781-340-1022.
Institution Type/Description: History Museum.
Collections: local history & culture; photographs; period furnishings; personal artifacts; Native American dugout canoe; bug & butterflies from around the world; Native American artifacts; fire & police memorabilia; ladies fans.
Activities: group tours.
Hours & Admission Prices: Mon. 7pm-9pm; groups by appointment. No charge. Closed holidays.

Williamstown

CHAPIN LIBRARY OF RARE BOOKS, 96 School St., Williamstown, MA 01267-2423. Mailing Address: P.O. Box 426, Williamstown, MA 01267-0426. Tel.: 413-597-2462. Fax: 413-597-2929.
E-mail: chapin.library@williams.edu
Web Site: chapin.williams.edu
Founded: 1923.
Congressional District: 1
Key Personnel: Asst. Librarian, Wayne G. Hammond; Administrative Asst., Elaine Yanow; Preparation Supvr., Nancy Birkrem; Preparation Asst., Ted Gilley; Custodian, Robert L. Volz.
Personnel Profile: Full-Time Paid 4; Part-Time Paid 1.
Governing Authority: college. Parent Institution: Williams College. Tax-exempt: 501(c)(3).
Institution Type/Description: Library.
Collections: 50,000-vol. collection of rare printed books; 9th to 20th-century manuscripts; bookplates; historical prints; photos; original art work for book illustrations; 15th to 20th-century prints, bibliographical & historical reference books; four founding documents of the United States.
Facilities: library of prints, bibliographical & historical reference books available for use to adults & students.
Activities: permanent & temporary exhibitions; occasional loan exhibitions from other libraries & collectors; lectures; tours by appointment.
Publications: Catalogue of the Collection of Samuel Butler in the Chapin Library; A Short-Title List of the Books in the Chapin Library; The Graphic Art of C.B. Falls; British Book Illustration, 1924-1936; British Ecclesiastical Architecture; Finished by Hand; London: High Life & Low Life.
Hours & Admission Prices: This is our temporary location during renovations until summer of 2014. Mon.-Fri. 10-12 & 1-5. No charge. Closed national holidays except Independence Day.
Attendance: 4,000 (estimated)

*** STERLING AND FRANCINE CLARK ART INSTITUTE, (M),** 225 South St., Williamstown, MA 01267-2891. Mailing Address: P.O. Box 8, Williamstown, MA 01267-0008. Tel.: 413-458-9545. Fax: 413-458-2324.
E-mail: info@clarkart.edu
Web Site: www.clarkart.edu
Founded: 1950.
Congressional District: 1
Key Personnel: Dir., Michael Conforti; Pres. (V), Peter S. Willmott; Deputy Dir., Anthony G. King; Starr Dir. Research & Academic Programs, Darby English; Sr. Cur., Richard Rand; Center for Education in the Visual Arts, Michael Cassin; Registrar, Mattie Kelley; Preparator, Paul Dion; Librarian, Susan Roeper; Dir. Communications, Vicki Saltzman; Museum Shop Mgr., Rachelle Jones.
Personnel Profile: Full-Time Paid 69; Part-Time Paid 34; Interns 10.
Governing Authority: nonprofit organization. Tax-exempt.
Institution Type/Description: Art Museum.
Collections: Italian, Flemish, Dutch & French Old Master paintings from the 14th-18th centuries; French 19th-century paintings, including the Impressionists; selected 19th-century American artists; 19th-century sculpture; Old Master French & American prints & drawings; porcelain; English silver; British art.
Research Fields: Visual arts and their role in culture; theory; history and interpretation of works from all periods and genres of art.
Facilities: 250,000-vol. library of art reference books available for inter-library loan & on premises; reading room; 320-seat auditorium. Art books in print; reproductions, postcards & catalogs published by Institute for sale.
Activities: gallery talks; organized educational programs for children; graduate program for MA in art history in collaboration with Williams College; Clark Fellowship; permanent, temporary & traveling exhibitions; social networking art events; film, lecture & concert series.
Publications: catalogs of special exhibitions; guide book; collections catalogues; Clark studies in the visual arts series.
Hours & Admission Prices: July-Aug. daily 10-5; Sept.-June Tues.-Sun. & Mon. holidays 10-5. June-Oct. Adults $15; discounts to AAM & ICOM members, children under 18 & students; members no charge. Closed New Year's Day; Thanksgiving; Christmas.
Attendance: 191,433 (accurate)
Membership: Individual $60; Family $90; Sustainer $150; Contributor $250; Sponsor $500; Benefactor $1,000; Scholar's Circle $2,500; Curator's Circle $5,000, Director's Circle $10,000; Sterling Circle $25,000.

*** WILLIAMS COLLEGE MUSEUM OF ART, (M),** 15 Lawrence Hall Dr., Ste. 2, Williamstown, MA 01267-3248. Tel.: 413-597-2429. Fax: 413-597-5000.
E-mail: WCMA@williams.edu
Web Site: www.wcma.org
Founded: 1926.
Congressional District: 1
Key Personnel: Interim Dir., Katy Kline; Dir. Asst., Amy Tatro; Mgr. Membership & Special Events, Raymond Torrenti; Cur. Special Projects, Kathryn Price; Coord. Public Rels., Kim Hugo; Museum Shop Mgr., Michele Migdal; Dir. Museum Registration, Diane Hart; Dir. Education, Cynthia Way; Coord. Education Programs, Joann Harnden; Chief Preparator & Exhibition Designer, Hideyo Okamura; Preparator, Gregory Jay Smith; Preparator, Richard Miller; Operations Supvr., Dorothy Lewis; Asst. Registrar, Rachel Tassone; Coord. Mellon Academic Programs, Elizabeth Gallerani; Museum Store Asst., Christine Maher; Museum Security Officer, Michele Alice; Museum Security Officer, Nancy Gwozdz; Museum Security Officer, Joe Congello; Museum Security Officer, Bethany Munsell; Museum Security Officer, Suzanne Stefanik.
Personnel Profile: Full-Time Paid 20; Part-Time Paid 8; Part-Time Volunteers 50; Interns 21.
Governing Authority: college. Parent Institution: Williams College. Tax-exempt: 501(c)(3).
Institution Type/Description: Art Museum: housed in 1846 Greek Revival rotunda with 1983 & 1986 additions designed by Charles Moore & Robert Harper of Centerbrook Architects & Planners.
Collections: 18th- to 19th-century American art; modern & contemporary American & European art; 9th- to 11th-century South Asian sculpture; 15th- to 19th-century Indian painting & sculpture; African art; Assyrian relief sculptures; 15th- to 18th-century Spanish painting & furniture; medieval art; Italian Renaissance painting & sculpture; European & American prints & drawings; 20th-century American photography; Baroque painting & furniture; holdings of Maurice & Charles Prendergast.
Research Fields: Maurice & Charles Prendergast; modern & contemporary art; American art.
Facilities: auditorium; print study room.
Activities: lectures; gallery talks & tours; concerts; formally organized educa-

tion programs for undergraduate college students; inter-museum loan, permanent & traveling exhibitions; interpretive programs for children; changing exhibitions of modern & contemporary American, European & non-western art; openings for special exhibitions; sign language interpretation available.
Publications: exhibition catalogues, 2-4 per year.
Hours & Admission Prices: Tues.-Sat. 10-5, Sun. 1-5. No charge; donations accepted.
Attendance: 53,452 (accurate)
Membership: Individual $30; Dual & Family $50; Patron $100; Benefactor $250; Donor $500; WCMA Contemporary $1,000; Prof. Whitney Stoddard 1935 $1,500; Prof. Karl Weston 1896 $2,500; Prof. S. Lane Faison, Jr. 1929 $5,000.

Wilmington

WILMINGTON TOWN MUSEUM, (M), 430 Salem St., Wilmington, MA 01887-1211. Tel.: 978-658-5475.
E-mail: htavern@town.wilmington.ma.us
Web Site: www.town.wilmington.ma.us/tavern1.html
Institution Type/Description: History Museum: housed in the Col. Joshua Harnden Tavern.
Collections: local history & culture relating to the town of Wilmington.
Hours & Admission Prices: July to early Aug. Thurs.-Fri. 10-2; Sept.-June. Tues. & Thurs. 10-2.

Winchendon

WINCHENDON HISTORICAL SOCIETY, INC., 151 Front St., Winchendon, MA 01475-1521. Mailing Address: P.O. Box 279, Winchendon, MA 01475-0279. Tel.: 978-297-2142.
E-mail: info@winchendonhistoricalsociety.org
Web Site: www.winchendonhistoricalsociety.org
Founded: 1930.
Congressional District: 1
Key Personnel: Pres. (V), Don O'Neil; Cur., Rita Saveall.
Personnel Profile: Part-Time Volunteers 10.
Governing Authority: nonprofit organization. Parent Institution: Winchendon Historical Society. Tax-exempt.
Institution Type/Description: History Museum.
Collections: archives; manuscripts; toys.
Activities: permanent & temporary exhibitions.
Publications: biannual newsletter.
Hours & Admission Prices: June 2-Oct. Tues. & Thurs. 11-4, Wed. 10-4, Sun.1-4; other times by appointment. Tours: 1 & 2:30. Adults $5; members no charge.
Attendance: 300 (estimated)
Membership: Student $5; Senior $20; Individual $25; Family $40; Sponsor $50.

Winchester

GRIFFIN MUSEUM OF PHOTOGRAPHY, 67 Shore Rd., Winchester, MA 01890-2821. Tel.: 781-729-1158. Fax: 781-721-2765.
E-mail: photos@griffinmuseum.org
Web Site: www.griffinmuseum.org
Formerly: Arthur Griffin Center For Photographic Art
Founded: 1992.
Congressional District: 7
Key Personnel: Pres. (V), Peter Griffin; Exec. Dir., Paula Tognarelli; Vice Pres., John McConnell; Treas., Doug Marmon; Assoc. Dir., Frances Jakubek; Gallery Monitor, Martha Stone.
Personnel Profile: Full-Time Paid 2; Part-Time Paid 1; Part-Time Volunteers 20; Interns 4.
Governing Authority: private; nonprofit organization. Tax-exempt: 501(c)(3).
Institution Type/Description: Photography Museum.
Collections: slides & transparencies taken by photographer Arthur Griffin, 1904-present.
Facilities: 1,500 sq. ft. exhibit space; 90-seat theater. Museum-related items for sale.
Activities: formal educational programs; guided tours; photographer workshops; lectures; rental gallery; temporary, loan & traveling exhibitions. Annual Event: Juried Photography Exhibition.
Publications: quarterly newsletter, Focus.
Hours & Admission Prices: Tues.-Thurs. 11-5, Fri. 11-4, Sat.-Sun. 12-4. Adults $5, seniors $2; members, children, students & Thurs. no charge. Closed major holidays.
Attendance: 7,000 (estimated)
Membership: Student $25; Senior $40; Individual $50; Family & Dual $75; Friend $150; Contributor $250; Patron & Corporate $500; Benefactor & Senior Corporate $1,000; Major Benefactor $5,000.

Woods Hole

WOODS HOLE HISTORICAL MUSEUM, (M), 579 Woods Hole Rd., Woods Hole, MA 02543. Mailing Address: P.O. Box 185, Woods Hole, MA 02543. Tel.: 508-548-7270.
E-mail: woods_hole_historical@hotmail.com
Web Site: www.woodsholemuseum.org
Founded: 1974.
Personnel Profile: Part-Time Paid 2.
Governing Authority: Parent Institution: Woods Hole Public Library.
Institution Type/Description: History Museum.
Collections: local history & culture; photographs; personal artifacts; paintings; boats.
Activities: educational programs; boat building; traditional craft workshops; ship-in-bottle building.
Publications: biannual journal, Spritsail.
Hours & Admission Prices: Call for hours. No charge; donations accepted.
Membership: Individual $30; Family $50; Donor $100.

WOODS HOLE OCEANOGRAPHIC INSTITUTION, OCEAN SCIENCE EXHIBIT CENTER, 15 School St., Woods Hole, MA 02543-1126. Mailing Address: Mail Stop 45, Woods Hole O.I., Woods Hole, MA 02543. Tel.: 508-289-2663. Fax: 508-457-2147.
E-mail: information@whoi.edu
Web Site: www.whoi.edu
Founded: 1930.
Key Personnel: Dir., Susan Avery; Museum Shop Mgr., Kathy Patterson.
Personnel Profile: Full-Time Paid 2; Part-Time Paid 2; Part-Time Volunteers 10.
Governing Authority: Tax-exempt.
Institution Type/Description: Oceanographic Research Institution.
Collections: oceanographic research photographs, data, tools, instruments, vehicles, vessels.
Research Fields: all aspects of oceanography.
Publications: magazines, Woods Hole Currents published quarterly; Oceanus published biannually.
Hours & Admission Prices: April groups by appointment; May-Oct. Mon.-Sat. 10-4:30; Nov.-Dec. Tues.-Fri. 10-4:30. Suggested Donation $2. Closed New Year's Day; Easter; Thanksgiving; Christmas. &
Attendance: 30,000

WOODS HOLE SCIENCE AQUARIUM, 166 Water St., Woods Hole, MA 02543-1097. Tel.: 508-495-2001. Fax: 508-495-2382.
Web Site: aquarium.nefsc.noaa.gov
Formerly: Aquarium of the National Marine Fisheries Service
Founded: 1885.
Key Personnel: Aquarium Dir., David Radosh; Cur., George Liles; Dir. Science & Research, Dr. Nancy Thompson.
Personnel Profile: Full-Time Paid 2; Part-Time Volunteers 1; Interns 15.
Governing Authority: federal. Affiliated with U.S. Department of Commerce, National Oceanic & Atmospheric Administration, Northeast Fisheries Center, Water St. & National Marine Fisheries Svc., Washington, DC 20235. Tax-exempt.
Institution Type/Description: Aquarium.
Collections: fish; invertebrate fauna; displays relating to marine environment; problems & methods of fishery research & management; dry exhibits; pictorial displays.
Research Fields: fishery research by Northeast Fisheries Science Center.
Activities: seal feedings: daily 11 & 4.
Hours & Admission Prices: Tues.-Sat. 11-4; groups by appointment during school year. No charge; donations accepted. Closed federal holidays. &
Attendance: 80,000 (estimated)

Worcester

AMERICAN ANTIQUARIAN SOCIETY, 185 Salisbury St., Worcester, MA 01609-1634. Tel.: 508-755-5221. Fax: 508-753-3311.
E-mail: library@mwa.org
Web Site: www.americanantiquarian.org
Founded: 1812.
Congressional District: 3
Key Personnel: Pres., Ellen S. Dunlap; Andrew W. Mellon Cur. Graphic Arts, Lauren Hewes; Marcus A. McLorison Librarian & Cur. Manuscripts, Thomas Knoles; Dir. Outreach, James David Moran.
Personnel Profile: Full-Time Paid 35; Part-Time Paid 18; Part-Time Volunteers 25; Interns 15.
Governing Authority: society. Tax-exempt.
Institution Type/Description: Research Library.

Collections: Canadian, colonial & early American history up to 1877; graphic arts; newspapers & periodicals; manuscripts & archives; books; pamphlets; music; maps; printed ephemera.
Research Fields: colonial & early American history to 1877; literature; culture.
Facilities: library available for research; reading room.
Activities: guided tours; public programs; lectures; concerts; fellowships offered; K-12 programs, curriculum units & teacher enrichment.
Publications: semi-annual journal; newsletter; occasional bibliographical & source document books; newsletter, The Book.
Hours & Admission Prices: Mon.-Tues. & Thurs.-Fri. 9-5, Wed. 10-8. No charge; donations accepted. Closed national holidays; New Year's Eve; Thanksgiving weekend; Christmas Eve. &
Attendance: 5,975 (accurate)

BROAD MEADOW BROOK CONSERVATION CENTER AND WILDLIFE SANCTUARY, 414 Massasoit Rd., Worcester, MA 01604-3546. Tel.: 508-753-6087.
E-mail: bmbrook@massaudon.org
Institution Type/Description: Wildlife Sanctuary & Nature Center.
Collections: wildlife & their habitats; over 78 species of butterflies; flowers; plants; trees; local natural heritage.
Facilities: over 400 acres; nature center; nature trails.
Activities: summer camp; educational programs; rental facilities. Annual Event: Butterfly Festival.
Hours & Admission Prices: Nature Center: Tues.-Sat. 9-4, Sun. 12:30-4. Trails: daily dawn to dusk. Adults $4, seniors & children 2-12 $3; members no charge.

ECOTARIUM, (M), 222 Harrington Way, Worcester, MA 01604-1899. Tel.: 508-929-2700. Fax: 508-929-2701.
E-mail: info@ecotarium.org
Web Site: www.ecotarium.org
Formerly: Worcester Natural History Society
Founded: 1825.
Congressional District: 3
Key Personnel: Pres. & C.E.O., Stephen M. Pitcher; Chm. Bd. Trustees, Patty Eppinger; Dir. Mktg. & Devel., Jennifer Kent; Dir. Exhibits & Education, Alexander Goldowsky; Volunteer Coord., Betsy Maloney; Deputy Dir. Administration & Operations, Patricia Crawford.
Personnel Profile: Full-Time Paid 21; Part-Time Paid 18; Part-Time Volunteers 147; Interns 3.
Governing Authority: nonprofit organization. Parent Institution: Worcester Natural History Society. Tax-exempt: 501(c)(3).
Institution Type/Description: Science & Nature Museum.
Collections: mounted birds & mammals; fossils & minerals; physical & natural science; live animal exhibits; shells; historical journals.
Research Fields: successful breeding of animals in captivity.
Facilities: 55 acres; 20,000 sq. ft. exhibit space; nature trails; 103-seat planetarium; snack bar; classrooms; 225-seat auditorium. Educational toys, books & museum related items for sale.
Activities: lectures; films; formally organized education programs for children; temporary, participatory & permanent exhibitions; narrow gauge passenger railway; tree canopy walkway with zip line; walking trails; summer jazz concerts.
Publications: Siegfried the Dinosaur.
Hours & Admission Prices: Tues.-Sat. 10-5, Sun. 12-5. Adults $14, children 2-18, senior citizens, college students with ID $8; discounts to groups, AAM, ASTC, AAA, MTA, & NEMA members, and staff members of other museums; members & children under 2 no charge. ASTC Passport Program. Planetarium: $4. Closed New Year's Day; Easter; Thanksgiving; Christmas Eve & Day. &
Attendance: 131,566 (accurate)
Membership: Student $35; Individual $55; Dual $75; Family & Grandparent $100; Family Plus & Grandparent Plus $120.

HIGGINS ARMORY MUSEUM, 100 Barber Ave., Worcester, MA 01606-2464. Tel.: 508-853-6015 & 6012. Fax: 508-852-7697.
E-mail: higgins@higgins.org
Web Site: www.higgins.org
Founded: 1928.
Congressional District: 3
Key Personnel: Exec. Dir., Nikki Andersen; Pres. (V) & Chm. (V), James C. Donnelly; Paul S. Morgan Cur., Dr. Jeffrey Forgeng; Registrar, Barbara Edsall; Program Dir., Devon Kurtz; Museum Store Mgr., Anne Burke.
Personnel Profile: Full-Time Paid 21; Part-Time Paid 5; Part-Time Volunteers 20; Interns 5.
Governing Authority: nonprofit organization. Tax-exempt: 501(c)(3).
Institution Type/Description: Arms, Armor, History & Art Museum.

Collections: ancient, medieval & Renaissance arms, armor & related material culture.
Research Fields: areas pertaining to collections.
Facilities: 2,500-vol. library pertaining to arms & armor available by appointment; 130-seat auditorium; 15,000 sq. ft. exhibit space; classroom. Museum-related items for sale.
Activities: inter-museum loans; changing exhibitions; lectures & demonstrations on arms, armor & related topics; educational programs for groups by reservation; craft activities for children; outreach programs; concerts; guided tours; films; story readings; curatorial evaluations; participatory gallery with try-on armor, costumes, & modern protective devices; games; videos;
Publications: quarterly newsletter for members; teacher materials; gallery study-guide series; annual report.
Hours & Admission Prices: Tues.-Sat. 10-4, Sun. 12-4. Adults $9, children 6-16 $7; discounts to area college students, groups, NEMA & AAM members; members, museum employees & children under 5 no charge. Closed legal holidays.
Attendance: 52,227 (accurate)
Membership: Student & Senior $20; Individual $45; Family & Dual $60; Crusader $100; Squire $250; Knight $500; Corporate $500-$1,000; Renaissance Society $1,000 & up.

IRIS & B. GERALD CANTOR ART GALLERY-COLLEGE OF THE HOLY CROSS, One College St., Worcester, MA 01610-2322. Tel.: 508-793-3356. Fax: 508-793-3030. Facebook: Cantor Art Gallery-College of the Holy Cross.
E-mail: rhankins@holycross.edu
Web Site: www.holycross.edu/cantorartgallery
Founded: 1983.
Congressional District: 3
Key Personnel: Dir., Roger Hankins.
Personnel Profile: Full-Time Paid 2; Interns 8.
Governing Authority: college. Parent Institution: College of the Holy Cross. Tax-exempt: 501(c)(3).
Institution Type/Description: Art Gallery.
Collections: Rodin & other sculptures; contemporary photographs; contemporary paintings; Indonesian textiles.
Major Exhibits: Early American Encounters with Asia, 2/20/14-4/12/14; Annual Senior Majors Seminar Exhibition, 4/24/14-5/23/14; Our Fragile Home (T), 6/6/14-7/31/14; Ignatius of Loyola: Impelled by God's Spirit (T), 8/29/14-1024/14; Lost Landscapes of the Italian Nativity, 10/31/14-12/3014.
Facilities: classrooms; 2,000 sq. ft. exhibit space.
Activities: organized education programs for college students affiliated with Holy Cross College; loan & traveling exhibitions; visiting artists exhibits.
Publications: exhibition catalogs.
Hours & Admission Prices: Mon.-Fri. 10-5, Sat. 2-5. No charge. Closed school holidays.
Attendance: 3,000 (estimated)

✱ **WORCESTER ART MUSEUM, (M),** 55 Salisbury St., Worcester, MA 01609-3123. Tel.: 508-799-4406. Fax: 508-798-5646. Facebook: Worcester Art Museum.
E-mail: information@worcesterart.org
Web Site: www.worcesterart.org
Founded: 1896.
Congressional District: 3
Key Personnel: Dir., Matthias Waschek; Deputy Dir. & C.O.O., Tracy Caforio; Pres., Clifford J. Schorer; Dir. Devel., Jere Shea; Cur. Contemporary Art, Susan Stoops; Cur. Asian Art, Louise Virgin; Chief Preparator & Exhibition Designer, Patrick Brown; Head Education, Marcia Lagerwey; Librarian, Deborah Aframe; Dir. Operations, Francis Pedone; Museum Shop Mgr., Susan Giordano; Cafe Mgr., Laurie Krohn-Andros.
Personnel Profile: Full-Time Paid 72; Part-Time Paid 66; Part-Time Volunteers 162.
Governing Authority: nonprofit organization. Tax-exempt: 501(c)(3).
Institution Type/Description: Art Museum.
Collections: 30,000 objects spanning 5,000 years of art & culture, ranging from Egyptian antiquities & Roman mosaics to Impressionist paintings and pop art.
Research Fields: European & American painting; pertaining to temporary exhibitions; American Portrait Miniatures Gallery.
Facilities: 40,000-vol. library of art reference materials available for inter-library loan & on the premises; reading room; education wing with studios, classrooms & resource center; cafe. Gift items for sale.
Activities: guided tours; gallery talks; formally organized education programs for children, adults, & undergraduate college students affiliated with 12 local colleges & universities; inter-museum loan, permanent & temporary exhibitions; films; lectures.

Publications: handbook of the Collection; special exhibition catalogues.
Hours & Admission Prices: Wed.-Fri. & Sun. 11-5, third Thurs. of month 11-8, Sat. 10-5. Adults $14, senior citizens & full-time college students with current ID $12; discounts to WGBH, AAA, MTA, AAM & ICOM members; members and children 17 & under no charge. Closed holidays.
Attendance: 121,267 (accurate)
Membership: Basic: Individual $60; Family & Household $80; Friend $200-$299; Sponsor $300-$599; Fellow $600-$1,249; Stephen Salisbury $1,250 & up.

WORCESTER CENTER FOR CRAFTS, 25 Sagamore Rd., Worcester, MA 01605-3914. Tel.: 508-753-8183. Fax: 508-797-5626.
E-mail: wcc@worcestercraftcenter.org
Web Site: www.worcestercraftcenter.org
Founded: 1856.
Congressional District: 3
Key Personnel: Exec. Dir., Carol Donnelly; Dir. Finance & Administration, Tammy Nigosian; Head Ceramics Dept., Tom O'Malley; Head Glass Dept., Jacob Vincent; Head Metals Dept., Lauren Beaudoin; Head Fiber Arts, Patti Sims; Head Wood Dept., Tony Gardner; Gallery Dir. & Gallery Shop Mgr., Candace Casey; Mgr. Organization Advancement, Caitlin Barkoskic; Registrar, Bettie Carlson.
Personnel Profile: Full-Time Paid 5; Part-Time Paid 10; Part-Time Volunteers 25; Interns 2.
Governing Authority: bd. of trustees. Tax-exempt.
Institution Type/Description: Arts & Crafts Center.
Collections: 10 exhibitions annually.
Research Fields: clay; glass; metals; enameling; photography; fibers; wood; refinishing.
Facilities: Museum-related items for sale.
Activities: studio craft classes for adults, children & college students; professional craft school for career-oriented students; artists-in-residence program; lectures; films; gallery exhibition program; workshops; summer school; tours; annual craft fairs; craft events & exhibitions. Museum Sponsors: Festival of Crafts in November.
Publications: newsletters; session brochures; event postcards.
Hours & Admission Prices: Mon., Wed. & Fri. 10-5:30, Tues. & Thurs. 10-7:30. No charge; donations accepted. Closed national holidays.
Attendance: 25,000 (estimated)
Membership: Senior, Youth & Student $35; Individual $45; Family $60; Benefactor $1,000; Life $5,000.

WORCESTER HISTORICAL MUSEUM, (M), 30 Elm St., Worcester, MA 01609-2570. Tel.: 508-753-8278. Fax: 508-753-9070.
E-mail: info@worcesterhistory.net
Web Site: www.worcesterhistory.org
Founded: 1875.
Congressional District: 3
Key Personnel: Pres., David Nicholson; Exec. Dir., William Wallace.
Personnel Profile: Full-Time Paid 6; Part-Time Paid 9; Part-Time Volunteers 19; Interns 2.
Governing Authority: society. Subsidiary Institution: Salisbury Mansion, 40 Highland St., Worcester, MA 01609. Tel: 508-753-8278. Tax-exempt: 170(b)(1)(A).
Institution Type/Description: History Museum.
Collections: costumes; archives; military; decorative arts; graphics; manuscripts. Historic House: 1772 Salisbury Mansion restored to 1830s.
Research Fields: local history.
Facilities: 6,000-vol. library of local history & general books available.
Activities: guided tours; lectures; permanent & temporary exhibitions.
Publications: books, Proceedings Worcester Society of Antiquity & Proceedings Worcester Historical Society.
Hours & Admission Prices: Tues.-Wed. & Fri.-Sat. 10-4, 4th Thurs. of month 10-8:30. Adults $5, seniors 62 & over and college students $4; discounts to AAM members; children under 18 & members no charge. Closed major holidays.
Attendance: 22,000 (accurate)
Membership: Individual $40; Family $50; Contributor $50-$99; Sustaining $100-$249; Salisbury Mansion Associate $250-$499; Staples Society $1,000.

Yarmouth Port

EDWARD GOREY HOUSE, 8 Strawberry Lane, Yarmouth Port, MA 02675. Tel.: 508-362-3909.
E-mail: info@edwardgoreyhouse.org
Web Site: www.edwardgoreyhouse.org

Founded: 2002.
Congressional District: 9
Key Personnel: Dir., Rick Jones
Institution Type/Description: Historic House Museum: housed in the former home of artist & writer Edward Gorey.
Collections: life & work of Edward Gorey including sketches, unpublished works, paintings, & ink drawings; works of Hillaire Belloc, Edward Lear, T.S. Eliot.
Hours & Admission Prices: April 18-June 30 Thurs.-Sat. 11-4, Sun. 12-4; July 4-Oct. 13 Wed.-Sat. 11-4, Sun. 12-4; Oct. 18-Dec. 29 Fri.-Sat. 11-4, Sun. 12-4. Adults $5, students & seniors $3, children 6-12 $2; children under 6 & members no charge. &
Attendance: 9,000 (accurate)
Membership: Senior Citizen & Student $20; Individual $30; Donor $100; Patron $300; Benefactor $500.

HISTORICAL SOCIETY OF OLD YARMOUTH, 11 Strawberry Lane, Yarmouth Port, MA 02675-1726. Mailing Address: Box 11, Yarmouth Port, MA 02675-0011. Tel.: 508-362-3021.
E-mail: hsoy@comcast.net
Web Site: www.hsoy.org
Founded: 1953.
Congressional District: 12
Key Personnel: Pres., Charles Adams.
Personnel Profile: Full-Time Volunteers 50; Part-Time Paid 4; Part-Time Volunteers 50.
Governing Authority: corporation. Tax-exempt.
Institution Type/Description: Historical & Preservation Society: housed in 1840, Capt. Bangs Hallet House.
Collections: period furniture; ships; models; ships paintings; kitchen equipment.
Research Fields: history; genealogy.
Facilities: library & archives of books on genealogy & ships logs, maps, monographs; reading room; walking trails. Museum-related items for sale.
Activities: guided tours; lectures; walking tours; education programs.
Publications: books: History of Old Yarmouth, All Around the Common; brochures: The Town of Yarmouth: A History 1639-1989; Stories of Yarmouth Shipmasters; Yarmouth's Proud Packets; When South Yarmouth was Quaker Village; The Breeds and the Caretakers of their Englewood Legacy; West Yarmouth, A Village Ignored; Port of the Bay; Images In Time.
Hours & Admission Prices: June-Oct. 15 Thurs.-Sun. 1-3:30; other times by appointment. Suggested Donation: adults $3, children under 12 $.50; members no charge. Closed holidays.
Attendance: 2,100 (estimated)
Membership: Individual $15; Family $25; Business $30; Patron $50 & up.

✱ **WINSLOW CROCKER HOUSE, (M),** (Old King's Hwy.) 250 Rte. 6A, Yarmouth Port, MA 02675. Mailing Address: c/o Historic New England, 141 Cambridge St., Boston, MA 02114. Tel.: 617-994-6661.
Web Site: www.historicnewengland.org
Founded: 1956.
Congressional District: 10
Key Personnel: Pres., Carl R. Nold; Site Mgr., Melinda Huff.
Governing Authority: society; nonprofit organization. Parent Institution: Historic New England, 141 Cambridge St., Boston, MA 02114. Tel.: 617-227-3956. Tax-exempt: 501(c)(3).
Institution Type/Description: Historic House: c.1780 Georgian house.
Collections: thorough survey of American furniture styles from late 17th century to colonial revival; hooked & Oriental rugs; ceramics; pewter.
Activities: guided tours; special events; lectures.
Publications: Historic New England Magazine.
Hours & Admission Prices: June-Oct. 15 2nd & 4th Sat. of month 11-4. Adults $4; discounts to seniors, AAM, ICOM & AAA members; Historic New England members no charge.
Attendance: 466 (accurate)
Membership: National $35; Individual $45; Household $55; Garden & Landscape $75; Contributing and Library & School $100; Young Friends of Historic New England $100-$1,500; Friends of the Library and Archives $125; Historic Homeowner $200; Ogden Codman Design Group $250; Appleton Circle $2,500 & up.

MICHIGAN

(344 listings)

Acme

MUSIC HOUSE MUSEUM, 7377 U.S. 31 N., Acme, MI 49610. Mailing Address: 7377 U.S. 31 N., Box 297, Acme, MI 49610-0297. Tel.: 231-938-9300. Fax: 231-938-3650.
E-mail: info@musichouse.org
Web Site: www.musichouse.org
Founded: 1983.
Congressional District: 11
Key Personnel: Pres. (V), Dorothy Clore; Museum Shop Mgr., Patte Richards.
Personnel Profile: Full-Time Paid 1; Full-Time Volunteers 1; Part-Time Paid 7; Part-Time Volunteers 40.
Governing Authority: nonprofit organization. Tax-exempt: 501(c)(3).
Institution Type/Description: Music/Musical Instrument Museum.
Collections: automatic musical instruments; music boxes; nickelodeons; reproducing pianos; orchestrions; violano; reed organs; band, dance & theater organs; early phonographs, radios & TVs; turn-of-the-century village settings.
Research Fields: music & musical instruments from c.1850-1950.
Facilities: library; 6,000 sq. ft. exhibit space. Music-related items for sale.
Activities: guided tours; lectures; concerts; docent program; silent movies.
Publications: books; brochures; musical recordings.
Hours & Admission Prices: May-Oct. Mon.-Sat. 10-4, Sun. 12-4; Nov.-Dec. Sat. Adults $11, children 6-15 $4; discounts to AAA members, groups; children under 6 no charge. Call for special Christmas schedule. &
Attendance: 12,000 (estimated)
Membership: Single $35; Couples $40; Grandparents $45; Family $50; Friend $100; Member $250; Patron $500; Angel $1,000.

Ada

AVERILL HISTORICAL MUSEUM OF ADA, 7144 Headley St., Ada, MI 49301. Mailing Address: P.O. Box 741, Ada, MI 49301. Tel.: 616-676-9346.
Institution Type/Description: History Museum.
Collections: local history & culture; period furnishings; personal artifacts; photographs.
Facilities: Museum-related items for sale.
Hours & Admission Prices: March-Dec. Fri.-Sat. 1-4; other times by appointment. No charge; donations accepted.

Adrian

KLEMM GALLERY, Siena Heights University, 1247 E. Siena Heights Dr., Adrian, MI 49221-1755. Tel.: 517-264-7860 & 7864. Fax: 517-264-7738.
E-mail: ddaniels@sienaheights.edu
Web Site: www.studioangelico.com
Founded: 1919.
Key Personnel: Chm. Art Dept., Peter Barr, Ph.D.; Dir. Gallery, Deborah Danielson.
Personnel Profile: Full-Time Paid 1; Part-Time Paid 1.
Governing Authority: college. Parent Institution: Siena Heights University. Tax-exempt.
Institution Type/Description: Arts Center and Institute.
Collections: changing monthly exhibits.
Hours & Admission Prices: Sept.-May Tues.-Fri. 9-4, Sun. 12-4. No charge. Closed Easter; Christmas; semester breaks. &
Attendance: 1,000 (estimated)

Albion

ALBION COLLEGE DEPARTMENT OF ART AND ART HISTORY, 805 E. Cass St., Albion, MI 49224-1831. Mailing Address: Bobbitt Visual Arts Center, 611 E. Porter St., Albion, MI 49224-1887. Tel.: 517-629-0246 & 0000. Fax: 517-629-0752.
E-mail: jmerrild@albion.edu
Web Site: www.albion.edu
Formerly: Albion College Department of Visual Arts
Founded: 1835.
Key Personnel: Prof., Lynne Chytilo; Prof., Bille Wickre; Prof. & Dept. Chair, Anne McCauley; Asst. Prof., Michael Dixon; Visiting Asst. Prof., Ashley Feagin.

Personnel Profile: Full-Time Paid 1; Part-Time Paid 1.
Governing Authority: college. Affiliated with Albion College. Tax-exempt: 501(c)(3).
Institution Type/Description: Art Museum.
Collections: contemporary art; 15th-century to present prints; folk arts of the world; decorative arts; glass; paintings; sculpture; archaeology.
Facilities: 4,200-vol. library of art books available for inter-library loan.
Activities: guided tours; lectures; gallery talks; arts festivals; formally organized education programs for undergraduate college students; permanent, temporary, traveling & print exhibitions.
Hours & Admission Prices: Sept.-May Mon.-Thurs. 9-9, Fri. 9-5, Sat. 10-2. No charge. Closed school holidays & vacations. &
Attendance: 2,000 (estimated)

BRUECKNER MUSEUM OF STARR COMMONWEALTH,
13725 Starr Commonwealth Rd., Albion, MI 49224-9525. Tel.: 517-629-5591; 800-837-5591. Fax: 517-629-2317.
E-mail: info@starr.org
Web Site: www.starr.org
Founded: 1956.
Congressional District: 7
Key Personnel: Pres. & C.E.O., Dr. Martin L. Mitchell.
Governing Authority: nonprofit organization. Parent Institution: Starr Commonwealth; A branch of Starr Commonwealth for Boys. Tax-exempt.
Institution Type/Description: General Museum.
Collections: paintings; sculptures; drawings; prints; period furnishings; ethnic collection; minerals.
Activities: guided tours; permanent, temporary & traveling exhibitions.
Publications: Starr News.
Hours & Admission Prices: Mon.-Fri. 8:30-4:30, Sat.-Sun. by appointment. No charge; donations accepted.
Attendance: 1,000 (estimated)

GARDNER HOUSE MUSEUM,
509 S. Superior St., Albion, MI 49224-2137. Tel.: 517-629-5100.
E-mail: info@albionhistoricalsociety.org
Web Site: www.albionhistoricalsociety.org
Founded: 1958.
Congressional District: 3
Key Personnel: Chm. & Pres. (V), Andrew Zblewski.
Personnel Profile: Part-Time Volunteers 30.
Volunteer Hours: 1,500
Operating Expenses: 14,300
Operating Income: 14,500
Governing Authority: society. Parent Institution: Albion Historical Society. Tax-exempt: 170(b)(1)(A).
Institution Type/Description: Local History Museum: housed in c.1875 A.P. Gardner House.
Collections: furniture; domestic & commercial artifacts; manuscripts.
Research Fields: local history & genealogy.
Activities: guided tours; permanent & temporary exhibition.
Publications: Quarterly Newsletter to members; Albion's Banks & Bankers; A Short History of Albion, Michigan; 1866 Bird's Eye View reprint; Twelfth Night Christmas Cookies; Patchwork Quilts; Gardners House Museum.
Hours & Admission Prices: May-Sept. Sat.-Sun. 2-4. No charge; donations accepted. &
Attendance: 1,000 (estimated)
Membership: Junior (15) $2; Senior Citizen $3; Regular $5; Supporting $10; Sustaining $15; Contributing $30; Benefactor $50; Century $100; Endowment $300 & up.

KIDS 'N' STUFF CHILDREN'S MUSEUM, (M),
301 S. Superior, Albion, MI 49224-1752. Mailing Address: P.O. Box 718, Albion, MI 49224-0718. Tel.: 517-629-8023. Fax: 517-629-8024.
E-mail: info@kidsnstuff.org
Web Site: www.kidsnstuff.org
Founded: 2002.
Key Personnel: Dir., Kathy Fischer.
Personnel Profile: Full-Time Paid 1; Part-Time Paid 8.
Governing Authority: Tax-exempt.
Institution Type/Description: Children's Museum.
Collections: hands-on exhibits.
Activities: birthday parties; field trips; girl scout events; Kindermusik; pre-school enrichment; after school classes; special events; summer programs.
Hours & Admission Prices: Tues.-Sat. 10-4, Thurs. 1-7. Admission one & over $5; discounts to ACM members; members no charge. &
Attendance: 18,000 (accurate)
Membership: Grandparent $55; Family $65; ACM $120.

WHITEHOUSE NATURE CENTER,
Albion College, Albion, MI 49224-1887. Tel.: 517-629-0582. Fax: 517-629-0509.
E-mail: tcrupi@albion.edu
Web Site: www.albion.edu/naturecenter
Founded: 1972.
Congressional District: 3
Key Personnel: Dir., Tamara Crupi.
Personnel Profile: Full-Time Paid 1; Part-Time Paid 5; Part-Time Volunteers 5; Interns 1.
Governing Authority: private college; nonprofit. Parent Institution: Albion College. Tax-exempt.
Institution Type/Description: Nature Center & Arboretum.
Collections: wildlife & their habitats; plants; trees; flowers.
Facilities: 500-vol. library of nature & natural science material available to the public; educational facilities; lab.
Activities: guided tours; lectures; films; hobby workshops; organized education programs for children, adults, undergraduate & graduate college students.
Publications: brochure, Whitehouse Nature Center.
Hours & Admission Prices: Mon.-Fri. 9-4:30, Sat.-Sun. 12-5. No charge; donations accepted. Closed national holidays & some Albion College holidays.
Attendance: 12,500 (estimated)

Algonac

ALGONAC/CLAY HISTORICAL SOCIETY,
1240 St. Clair River Dr., Algonac, MI 48001-1472. Mailing Address: P.O. Box 228, Algonac, MI 48001-0228. Tel.: 810-794-9015.
E-mail: achs@algonac-clay-history.com
Web Site: www.algonac-clay-history.com
Governing Authority: nonprofit organization.
Institution Type/Description: Historical Society Museum.
Collections: local history & culture; photographs; personal artifacts; boat building; military.
Activities: special events.
Hours & Admission Prices: June-Aug. Wed. 7pm-9pm, Sat.-Sun. 1-4; Sept.-May Sat.-Sun. 1-4. No charge; donations accepted.
Membership: Individual $15; Family $20; Life Individual $150; Life Family $200.

Allegan

ALLEGAN COUNTY HISTORICAL SOCIETY AND OLD JAIL MUSEUM,
113 Walnut St., Allegan, MI 49010-1249. Tel.: 269-673-8292.
E-mail: oldjailmuseum06@yahoo.com
Web Site: www.alleganoldjail.org
Founded: 1962.
Congressional District: 9
Key Personnel: Pres., Amanda Strickfaden; Treas., Helen Setter.
Personnel Profile: Part-Time Volunteers 20.
Governing Authority: society. Parent Institution: Allegan County Historical Society. Tax-exempt.
Institution Type/Description: Regional History Museum: housed in 1906 former county jail & sheriff's home.
Collections: photographs; furniture; books; documents; post cards; dolls; historical period rooms; dental, medical & barber equipment; pioneer artifacts; school room; jail cells; Gen. Pritchard Civil War room.
Facilities: library. Books, goat's milk soap & other gift items for sale.
Activities: guided tours; lectures; research.
Publications: books, Early Times in Allegan Township; Allegan & Barry County Biog. History Book-1880; River & Lake, a History of Allegan County; Allegan County Atlases, 1873, 1895, 1913; History of Casco Twp.; Six Months Among the Indians; The Index to River & Lake plus Ghost Towns & Ghost Stories; Ottawa County Atlas & Gazetteer; Allegan County Atlas & Gazetteer; Atlas & Index 1864; Sketches of Early Homes, Churches, Schools and Residents of Allegan, Michigan.
Hours & Admission Prices: May-Aug. Fri. & Sat. 10-4; Sept.-April Sat. 10-4. No charge; donations accepted. Closed New Year's Day; Christmas.
Attendance: 2,800 (estimated)
Membership: Individual $10; Family $15.

Allendale

ENGINE HOUSE NO. 5 MUSEUM,
6610 Lake Michigan Dr., P.O. Box 188, Allendale, MI 49401-0188. Tel.: 800-632-6184; 616-895-4347. Fax: 616-895-7158.
Web Site: www.enginehouse5.org
Institution Type/Description: Fire Fighting History Museum.
Collections: fire fighting history & equipment; period furnishings; personal artifacts; 1921 La France; 1928 Ahrens-Fox pumper.

Hours & Admission Prices: Call for hours.

GRAND VALLEY STATE UNIVERSITY ART GALLERY, (M),
1121 Performing Arts Center, Allendale, MI 49401-9403. Tel.:
616-331-2563. Fax: 616-331-8565.
E-mail: bazuinc@gvsu.edu
Web Site: www.gvsu.edu/artgallery
Key Personnel: Dir., Henry Matthews; Asst. Dir., Cathy Marashi
Institution Type/Description: Art Gallery.
Collections: works by local & regional artists.
Hours & Admission Prices: Winter: Mon.-Wed. & Fri. 10-5, Thurs. 10-7;
Summer: Mon.-Fri. 10-4. Closed holidays.

Alpena

* **BESSER MUSEUM FOR NORTHEAST MICHIGAN, (M),**
491 Johnson St., Alpena, MI 49707-1496. Tel.: 989-356-2202. Fax:
989-356-3133.
E-mail: adozier@bessermuseum.org
Web Site: www.bessermuseum.org
Formerly: Jesse Besser Museum
Founded: 1962.
Congressional District: 1
Key Personnel: Dir., Chris Witulski; Pres. (V), Ed Black; Cur., Richard Clute,
M.A.; Facilities & Exhibits Mgr., Randy Shultz; Museum Shop Mgr.,
Amanda Dozier.
Personnel Profile: Full-Time Paid 5; Part-Time Paid 1; Part-Time Volunteers
100.
Governing Authority: nonprofit organization. Tax-exempt: 501(c)(3).
Institution Type/Description: General Museum.
Collections: 19th and 20th-century furnishings & decorative arts; Northern
Michigan archaeology; Great Lakes copper culture artifacts; agricultural;
lumbering implements; Michigan wildlife; 19th-century shop interiors;
Foucault Pendulum; geology & fossils; art, primarily 20th-century graphic
art; late 19th & 20th-century historical documents & photographs; manu-
scripts relating to northern Michigan; 18th- to 20th-century maps of the
Great Lakes. Preserved Buildings: 1890 log building; 1872 Maltz Exchange
bank; 1860s Homesteader's Line cabin; 1896 Green Township one-room
school; 1912 Spratt Church; 1928 fishing tug.
Research Fields: regional history; archaeology; art; zoology.
Facilities: reference library; planetarium; classrooms.
Activities: docent tours; lectures; gallery talks; juried art shows; craft festivals;
workshops; education programs; inter-museum loan, permanent, temporary
& traveling exhibitions; school loan service; art, science & history classes
for adults & children; planetarium shows.
Publications: book, The Town That Wouldn't Die; exhibit brochures; The
Limestone Sinkholes of Northeastern Michigan; educational brochures;
teacher curriculum guides to the museum.
Hours & Admission Prices: Mon.-Sat. 10-5. Adults $5, children, students &
senior citizens $3; discounts to active military, & AAA members; members,
children under 5, persons with disabilities & Wed. 3-5 no charge. Planetarium Programs: Sat. 2pm. Adults $3, seniors & children $2. Closed major
holidays. &
Attendance: 23,000 (accurate)
Membership: Senior $20; Individual $25; Senior Couple $35; Family $40;
Friend $100; Supporter $250; Patron $500; Leader $1,000.

GREAT LAKES MARITIME HERITAGE CENTER, 500 W.
Fletcher St., Alpena, MI 49707. Tel.: 989-356-8805. Fax: 989-354-
0144.
E-mail: thunderbay@noaa.gov
Web Site: www.thunderbay.noaa.gov
Founded: 2000.
Congressional District: 1
Governing Authority: Parent Institution: NOAA/NOS/ONMS & State of
Michigan.
Institution Type/Description: Maritime History Museum.
Collections: local maritime history & artifacts; full-size replica wooden Great
Lakes schooner; photographs.
Facilities: 9,000 sq. ft. exhibition space; 93-seat theater. Center-related items
for sale.
Publications: e-newsletter, Thunderstruck.
Hours & Admission Prices: Winter: Mon.-Sat. 10-5; Summer: daily call for
hours. No charge; donations accepted. Closed New Year's Day; Thanks-
giving; Christmas. &
Attendance: 80,000 (estimated)

Alto

BOWNE TOWNSHIP HISTORICAL MUSEUM, 8240 Alden
Nash, S.E., Alto, MI 49302. Mailing Address: P.O. Box 35, Alto,
MI 49302.
Founded: 1992.
Key Personnel: Pres. (V), Sally C. Johnson.
Personnel Profile: Part-Time Volunteers 16.
Governing Authority: Tax-exempt.
Institution Type/Description: History Museum: Historic buildings; one room
school house; carriage shed.
Collections: local history & culture; period furnishings; personal artifacts;
photographs; schoolbooks.
Facilities: 2-story building (former ladies aid hall); restored carriage shed.
Hours & Admission Prices: June-Sept. 1st Sun. each month 2-4; other times by
appointment. No charge; donations accepted.
Membership: Student $10; Individual $12; Family $25; Business/Professional
$100; Life Individual $250; Life Couple $400.

Ann Arbor

ANN ARBOR ART CENTER, 117 W. Liberty St., Ann Arbor, MI
48104-1380. Tel.: 734-994-8004. Fax: 734-994-3610.
E-mail: mklopf@annarborartcenter.org
Web Site: annarborartcenter.org
Founded: 1909.
Key Personnel: C.E.O. & Pres. (V), Marie Klopf; Mktg. Dir., Nicholas Farrell;
Gallery Shop Dir., Amy Cameron; Dir. Programs, Erika Villarreal Bunce;
Exhibitions Mgr., Nathan Rice; Financial Controller, Eric Wolff.
Personnel Profile: Full-Time Paid 12; Part-Time Paid 15; Part-Time Volunteers
200; Interns 12.
Governing Authority: nonprofit organization. Tax-exempt: 501(c)(3).
Institution Type/Description: Art Center.
Collections: multicultural exhibits.
Facilities: 150-vol. library of art history & reference books; classrooms.
Original art including oil paintings, watercolors, graphics & ceramics for
sale.
Activities: gallery talks; rental gallery; formally organized education programs
for children & adults; loan exhibitions; drop-in, hands-on activity studio
(Artventures).
Publications: quarterly newsletter.
Hours & Admission Prices: Mon.-Fri. 10-7, Sat. 10-6, Sun. 12-5. No charge.
Closed legal holidays. &
Attendance: 100,000
Membership: Individuals $50; Seniors $40; Household $100 & up.

ANN ARBOR HANDS-ON MUSEUM, 220 E. Ann St., Ann Arbor,
MI 48104-1445. Tel.: 734-995-5439. Fax: 734-995-1188.
E-mail: info@aahom.org
Web Site: www.aahom.org
Founded: 1982.
Congressional District: 15
Key Personnel: Dir., Mel Drumm; Museum Shop Mgr., Ari Morris.
Personnel Profile: Full-Time Paid 17; Part-Time Paid 40.
Governing Authority: Tax-exempt.
Institution Type/Description: Science Museum.
Collections: over 250 hands-on exhibits.
Activities: educational programs; field trips; outreach; distance learning;
summer classes; demonstrations; professional development; birthday par-
ties; science labs.
Hours & Admission Prices: Mon.-Sat. 10-5, Sun. 12-5. Admission 2 & over
$10; discounts to ASTC members & groups of 20 or more; members &
children under 2 no charge. Closed Memorial Day; Labor Day; Thanksgiv-
ing; Christmas. &
Attendance: 197,500 (estimated)
Membership: Family $75; Family Plus $100.

ARGUS PLANETARIUM, 601 W. Stadium Blvd., Ann Arbor, MI
48103-5812. Tel.: 734-994-1771. Fax: 734-994-1724.
E-mail: schaffer@aaps.k12.mi.us
Founded: 1956.
Key Personnel: Dir., Stephen A. Schaffer.
Personnel Profile: Part-Time Paid 1.
Governing Authority: public school district. Parent Institution: Ann Arbor
Public Schools. Tax-exempt.
Institution Type/Description: Planetarium.
Collections: Digistar 3 SP projector.
Facilities: 59-seat auditorium.

Activities: lectures; formally organized education programs for children & adults.
Hours & Admission Prices: By appointment. School groups outside of district $3 per student, $45 min. &
Attendance: 6,000

COBBLESTONE FARM MUSEUM, 2781 Packard Rd., Ann Arbor, MI 48108. Tel.: 734-794-7120.
Institution Type/Description: Historic Farmstead: built in 1845.
Collections: farm history; agriculture; pioneer life; religious artifacts; period furnishings & equipment.
Activities: guided tours. Museum Sponsors: Living History Days; Harvest Evening; Michigan Log Cabin Day; Spring on the Farm; Pioneer Living Program; Celebration of Independence; Country Christmas.
Hours & Admission Prices: Thurs. 11-2; other times by appointment.

GERALD R. FORD PRESIDENTIAL LIBRARY, 1000 Beal Ave., Ann Arbor, MI 48109-2114. Tel.: 734-205-0555. Fax: 734-205-0571.
E-mail: ford.library@nara.gov
Web Site: www.fordlibrarymuseum.gov
Founded: 1977.
Key Personnel: Dir., Elaine K. Didier; Supervisory Archivist, Geir Gundersen.
Personnel Profile: Full-Time Paid 11; Part-Time Paid 2; Interns 3.
Governing Authority: federal. Parent Institution: National Archives and Records Administration, Washington, D.C. Tax-exempt.
Institution Type/Description: Presidential Library.
Collections: federal government policies & national politics during the Cold War-era; Gerald R. Ford's presidential papers; 300,000 photographs; film, audiotape & videotape.
Facilities: library; 200-seat auditorium; classroom; 1,000 sq. ft. exhibit space.
Activities: lectures; formal education programs for undergraduate or graduate college students; temporary exhibitions; special events.
Publications: semi-annual newsletter, Gerald R. Ford Foundation Newsletter.
Hours & Admission Prices: Mon.-Fri. 8:45-4:45. No charge. Closed New Year's Day; Martin Luther King Jr. Day; Presidents' Day; Memorial Day; Independence Day; Labor Day; Columbus Day; Veterans Day; Thanksgiving; Christmas. &
Attendance: 10,000 (estimated)

HERBARIUM OF THE UNIVERSITY OF MICHIGAN, 3600 Varsity Dr., Ann Arbor, MI 48108-2228. Tel.: 734-615-6200. Fax: 734-998-0038.
Web Site: herbarium.lsa.umich.edu
Founded: 1921.
Congressional District: 2
Key Personnel: Dir., Paul E. Berry; Cur. Emeritus, Robert L. Shaffer; Cur. Emeritus, Edward Voss; Cur. Emeritus, Rogers McVaugh; Cur. Emeritus, William R. Anderson; Cur. Emeritus, Michael Wynne; Cur., A.A. Reznicek; Cur., Christopher Dick; Cur. Emeritus, Robert Fogel; Cur., Timothy James; Research Scientist, Florence S. Wagner; Research Scientist, Christiane Anderson; Asst. Research Scientist, Richard K. Rabeler.
Governing Authority: university. Parent Institution: University of Michigan. Tax-exempt.
Institution Type/Description: Herbarium.
Collections: all groups of plants & fungi.
Research Fields: plant systematics; phytogeography; botanical nomenclature.
Facilities: library of plant systematics & related areas available for use by request.
Activities: research in plant systematics & phytogeography; education programs for college students.
Publications: brochures, Contributions from the University of Michigan Herbarium; Flora Novo-Galiciana; Michigan Flora Pts., I, II, III.
Hours & Admission Prices: Mon.-Fri. 8:30-4. No charge; donations accepted.

* **KELSEY MUSEUM OF ARCHAEOLOGY, (M),** 434 S. State St., Ann Arbor, MI 48109-1390. Tel.: 734-764-9304. Fax: 734-763-8976.
E-mail: sherbert@umich.edu
Web Site: www.lsa.umich.edu/kelsey/
Founded: 1928.
Congressional District: 2
Key Personnel: Pres., Mary Sue Coleman; Provost & Exec. Vice Pres. Academic Affairs, Teresa A. Sullivan; Dir. & Cur., Sharon Herbert; Assoc. Dir. & Cur. Academic Outreach, Lauren Talalay; Cur. Conservation, Suzanne Davis; Conservator, Claudia Chemello; Cur. Greece & Near East, Margaret Root; Cur. Hellenistic & Roman Empire, Elaine Gazda; Cur. Dynastic Egypt, Janet Richards; Cur. Graeco-Roman Egypt, Terry Wilfong;

Coord. Museums Collections, Michelle Fontenot; Coord. Museum Exhibitions, Scott Meier; Coord. Museum Collections, Sebastian Encina; Editor, Peg Lourie; Coord. Museum Visitor Programs& Museum Shop Mgr., Todd E. Gerring; Gift Mgmt. & Graphic Artist, Lorene Stervier.
Personnel Profile: Full-Time Paid 12; Part-Time Paid 15; Part-Time Volunteers 15; Interns 3.
Governing Authority: state; university. Parent Institution: University of Michigan. Tax-exempt.
Institution Type/Description: Archaeology Museum.
Collections: art & artifacts; classical & Near Eastern archaeology, especially of Roman & early Byzantine Egypt; results of excavations at Roman sites in Egypt & Seleucia on the Tigris, Iraq; Greek & Roman inscriptions; Roman building materials; Byzantine & Islamic textiles; Roman sculpture & glass.
Research Fields: excavation in Tunisia, Italy, Egypt, Israel, Turkey, Georgia & Armenia; conservation; Roman & Hellenistic Egypt & the Near East; Greek & Roman sculpture; glass; numismatics; textiles & inscriptions; excavation in Nubia/Sudan.
Facilities: 5,000-vol. library of materials on archaeology & related subjects available for use by appointment. Gift-related items for sale.
Activities: guided tours; lectures; gallery talks; permanent & temporary exhibitions; traveling educational kits for teachers & special interest groups; field research labs; object study areas; classical archaeology graduate program; children's programs.
Publications: Bulletin of the Museums of Art & Archaeology; gallery guide; exhibition catalogs & brochures; associates newsletter; Kelsey Museum Studies Series.
Hours & Admission Prices: Jan. 2-Dec. 24 Tues.-Fri. 9-4, Sat.-Sun. 1-4. No charge; donations accepted. Closed major holidays. &
Attendance: 21,490 (accurate)
Membership: Students $10; Individual $35; Family $50; Contributor $100; Sponsor $250; Patron $500; Benefactor $1,000 & up.

KEMPF HOUSE MUSEUM, 312 S. Division St., Ann Arbor, MI 48104. Tel.: 734-994-4898.
E-mail: kempfhousemuseum@gmail.com
Web Site: kempfhousemuseum.org
Institution Type/Description: Historic House Museum: housed in the former home of Reuben & Pauline Kempf; built 1853. Listed on the National Register of Historic Places.
Collections: Kempf family & Ann Arbor history; period furnishings; personal artifacts; music studio; 1877 Steinway Concert piano; photographs.
Activities: guided tours.
Hours & Admission Prices: mid-Feb. to June & Sept.-Dec. Sun. 1-4; other times by appointment. No charge; donations accepted.

MATTHAEI BOTANICAL GARDENS AND NICHOLS ARBORETUM, 1800 N. Dixboro Rd., Ann Arbor, MI 48105-9741. Tel.: 734-647-7600. Fax: 734-998-6205.
Web Site: www.umich.edu/mbgna
Founded: 1907.
Congressional District: 5
Key Personnel: Dir., Robert E. Grese; Assoc. Dir., Karen Sikkenga.
Personnel Profile: Full-Time Paid 24; Part-Time Paid 4; Part-Time Volunteers 500; Interns 35.
Governing Authority: Parent Institution: University of Michigan. Subsidiary Institution: Nichols Arboretum, 1610 Washington Heights, Ann Arbor, MI. Tax-exempt.
Institution Type/Description: Botanical Garden.
Collections: cacti & succulents; bromeliads; prairie; wildflower garden; tropical & warm temperate species; native flora; roses; perennial garden; orchids; herb garden; garden of new world plants; insectivores display, constructed wetland; rock garden; ornamental grass garden; ethnobotanical trail; exotic & native trees & shrubs of north temperate regions; Appalachian plants; lilacs; peonies; oak-hickory woodland.
Research Fields: prairie, wetland & forest species; landscape & ecology & classification; various disciplines of biology, natural resources, cultural & interpretive arts.
Facilities: horticultural library; conservatory; formal & experimental gardens; herbarium; laboratories; classrooms; nature trails; auditorium.
Activities: docent guided tours of conservatory & trails; adult, youth & family educational activities; university classes; plant sales & lectures; special events; meeting facilities for mission-related organizations & garden clubs; cultural performances.
Publications: quarterly newsletter, Friends.
Hours & Admission Prices: Grounds: daily 6am-10pm. Conservatory: Wed. 10-8, Thurs.-Tues. 10-4:30. No charge. Parking: $1.20 per hour; no charge. American Horticultural Society reciprocity to botanical garden/arboretum members. &
Attendance: 100,000 (estimated)
Membership: Friends of Matthaei Botanical Gardens & Nichols Arboretum:

Student $10; Individual $45; Family $55; Sustaining $100; Sponsor $250; Benefactor $500; Director's Circle $1,000.

SINDECUSE MUSEUM OF DENTISTRY, (M), University of Michigan, 1011 N. University - G565 Dental Bldg., Ann Arbor, MI 48109-1078. Tel.: 734-763-0767. Fax: 734-615-1429. Facebook: Sindecuse Museum of Dentistry.

E-mail: dentalmuseum@umich.edu
Web Site: www.dent.umich.edu/sindecuse
Founded: 1992.
Key Personnel: Cur., Shannon O'Dell; Coord. Collections, Kathy Daniels.
Personnel Profile: Full-Time Paid 2; Part-Time Paid 2.
Volunteer Hours: 305
Operating Expenses: 335,382
Operating Income: 335,382
Governing Authority: public university. Parent Institution: School of Dentistry, University of Michigan. Tax-exempt.
Institution Type/Description: Dental Museum.
Collections: dental tools & equipment from 18th century to present; dentistry & oral health history; dental advertising; games; toys; photographs.
Major Exhibits: Inside the Dental Practice: 1860-1940, 12/11-1/20; Women Dentists: Changing the Face of Dentistry, 10/13-1/5/15; Dental Hygiene: A Century of Progress, 10/13-/1/7/16.
Research Fields: history of dentistry; University of Michigan Dental School alumni.
Facilities: 500-vol. library; 2,700 sq. ft. exhibit space.
Activities: guided tours by appointment; permanent & temporary exhibitions; university student training programs; lectures; research by appointment. Annual Event: Hall of Honor induction ceremony.
Hours & Admission Prices: Mon.-Fri. 8-6. No charge; donations accepted. Closed New Year's Eve & Day; Memorial Day; Independence Day; Labor Day; Thanksgiving & day after; Christmas Eve, Day & week. &
Attendance: 4,000 (estimated)

STEARNS COLLECTION OF MUSICAL INSTRUMENTS, University of Michigan School of Music, 1100 Baits Dr., Ann Arbor, MI 48109-2085. Tel.: 734-936-2891. Fax: 734-647-1897.

E-mail: stearns@umich.edu
Web Site: www.music.umich.edu/research/stearns_collection/index.htm
Founded: 1899.
Congressional District: 2
Key Personnel: Dir., Joseph Lam.
Personnel Profile: Part-Time Paid 2; Part-Time Volunteers 1.
Governing Authority: university. Affiliated with the University of Michigan School of Music. Tax-exempt.
Institution Type/Description: Musical Instrument Museum.
Collections: over 2,000 modern Western, non-Western & period musical instruments.
Research Fields: organological computer cataloguing.
Facilities: research laboratory.
Activities: guided tours; lectures; concerts; permanent & temporary exhibitions; University Sponsors: Lecture series held each semester.
Publications: newsletter.
Hours & Admission Prices: Mon.-Fri. 10-5. No charge; donations accepted. Guided Tours: seniors $1, groups of 25 $30; call 662-7790 for more information. &
Attendance: 660
Membership: Student & Senior Citizen $15; Friends $35; Sustaining $100; Patron $500; Benefactor $1000.

UNIVERSITY OF MICHIGAN MUSEUM OF ANTHROPOL-OGY, 4013 Ruthven Museums Bldg., 1109 Geddes, Ann Arbor, MI 48109-1079. Tel.: 734-764-0485. Fax: 734-763-7783.

E-mail: anthro-museum@umich.edu
Web Site: www.lsa.umich.edu/umma/
Founded: 1922.
Congressional District: 2
Key Personnel: Dir., Cur. Latin American Archaeology & Ethnohistory, Joyce Marcus; Cur. Mediterranean Archaeology, Robert Whallon; Cur. Asian Archaeology, Carla Sinopoli; Cur. North American Archaeology, John D. Speth; Cur. Environmental Archaeology, Kent V. Flannery; Cur. Great Lakes Archaeology, John O'Shea; Cur. Near East Archaeology, Henry T. Wright; Cur. Eastern North American Archaeology, Robin Beck.
Personnel Profile: Full-Time Paid 5; Part-Time Paid 8; Part-Time Volunteers 22.
Governing Authority: state. Parent Institution: University of Michigan. Tax-exempt: 501(c)(3).
Institution Type/Description: Anthropology Museum.
Collections: archaeology; ethnology; ethnobotany; zoo archaeology; human osteology; geological.

Research Fields: archaeology; ethnology; human & cultural evolution.
Facilities: ethnobotanical, zoo archaeological, palynological & sedimentological laboratories.
Activities: formally organized education programs for undergraduate & graduate college students; research in anthropology.
Publications: monographs, Occasional Contributions; Anthropological Papers; Memoirs; Technical Reports; electronic publications (CD-Rom).
Hours & Admission Prices: By appointment only. No charge. &

* THE UNIVERSITY OF MICHIGAN MUSEUM OF ART, (M), 525 S. State St., Ann Arbor, MI 48109-1354. Tel.: 734-764-0395. Fax: 734-764-3731.

E-mail: umma.info@umich.edu
Web Site: www.umich.edu/
Founded: 1946.
Congressional District: 2
Key Personnel: Dir., Joseph Rosa; Dir. Education, Ruth Slavin; Chief Administrative Officer, Kathryn Huss; Sr. Cur. Western Art, Carole Mc-Namara; Research Cur. Asian Art, Natsu Oyobe; Mgr. Collections & Exhibitions & Chief Registrar, Orian Neumann; Museum Store Mgr., Cynthia Carson.
Personnel Profile: Full-Time Paid 40; Part-Time Paid 25; Part-Time Volunteers 187; Interns 4.
Governing Authority: state; university. Parent Institution: The University of Michigan. Affiliated with Regents of The University of Michigan. Tax-exempt: 501(c)(3).
Institution Type/Description: Art Museum. Western painting & sculpture from the 12th century to the present; Old Master & contemporary prints and drawings; photography; Asian & African art; Islamic ceramics.
Collections: Western art from 6th century to the present prints and drawings; photographs; Asian art; African & Oceanic art; objects from the Islamic World.
Research Fields: permanent art collections & special exhibitions.
Facilities: classrooms; rental facilities.
Activities: guided tours; lectures; gallery talks; formally organized education programs for children & adults; extensive public programs (K-12; adult; university); permanent, temporary & traveling exhibitions; concerts & performance programs.
Publications: exhibition catalogs; Bulletin, Museums of Art and Archaeology; scholarly articles.
Hours & Admission Prices: See website for current hours. Suggested Donation: $5. Closed New Year's Day; Memorial Day; Independence Day; Labor Day; Thanksgiving; Christmas. &
Attendance: 203,000 (estimated)
Membership: Student $15; Individual $50; Household $75; Donor $125; Sponsor $250; Curator's Circle $500; Director's Circle Associate $1,000; Director's Circle Patron $2,500; Director's Circle Benefactor $5,000.

UNIVERSITY OF MICHIGAN MUSEUM OF NATURAL HISTORY, (M), 1109 Geddes Ave., Ann Arbor, MI 48109-1079. Tel.: 734-764-0480. Fax: 734-647-2767.

E-mail: ummnh.feedback@umich.edu
Web Site: www.ummnh.org
Founded: 1881.
Congressional District: 14
Key Personnel: Dir., Amy S. Harris; Asst. Dir. Exhibits, Eugene Dillenberg; Exhibit Designer, John B. Klausmeyer; Asst. Dir. Devel., Nora Webber; Key Admin. & Publicity Coord., Daniel Madaj; Mgr. Planetarium, Matthew P. Linke; Asst. Dir. Education, Kira Berman; Museum Shop Mgr., Kelly Sullivan; Office Mgr., Marina Mayne; Program Mgr., Brittany Chunn; Mgr. Library Outreach Program, Amanda Paige.
Personnel Profile: Full-Time Paid 11; Part-Time Paid 55; Part-Time Volunteers 3; Interns 4.
Governing Authority: university. Parent Institution: University of Michigan. Subsidiary Institution: College of Literature, Science & Arts. Tax-exempt: 501(c)(3).
Institution Type/Description: University Natural History Museum.
Collections: Hall of Evolution including exhibits on vertebrate & invertebrate paleontology; paleobotany; Michigan wildlife & ecology; ethnology & anthropology; geology; astronomy.
Research Fields: anthropology; systematics; comparative anatomy; evolution; paleontology.
Facilities: 22,261 sq. ft. exhibit space; classrooms; planetarium. Museum-related items for sale.
Activities: guided tours; planetarium shows; formally organized educational programs for undergraduate & graduate students affiliated with the University of Michigan; museum methods classes; docent program; lectures; temporary & permanent exhibits.
Publications: bimonthly newsletter, The Display Case; brochures.
Hours & Admission Prices: Mon.-Sat. 9-5, Sun. 12-5. Museum: no charge;

donations accepted. $10 nonrefundable reservation fee for groups of 10 or more. Planetarium Shows: $6 per person. Closed major holidays. &

Attendance: 100,000 (estimated)

Membership: Individual & Couple $40; Grandparents $50; Family $60.

WASHTENAW COUNTY HISTORICAL SOCIETY, 500 N. Main St., Ann Arbor, MI 48104-1027. Mailing Address: P.O. Box 3336, Ann Arbor, MI 48106-3336. Tel.: 734-662-9092.

E-mail: wchs-500@ameritech.net

Web Site: www.washtenawhistory.org

Founded: 1857.

Institution Type/Description: Historical Society Museum.

Collections: county history & cultural heritage; personal artifacts; photographs.

Hours & Admission Prices: Wed. & Sat.-Sun. 12-4; other times by appointment. No charge; donations accepted. &

Membership: Youth, Student and Senior 60 & over $10; Individual $15; Senior Couple $19; Family, Library & Organization $25; Business $50; Patron $100; Sponsor $250; Sustaining $500.

Arcadia

ARCADIA AREA HISTORICAL MUSEUM, 3340 Lake St., Arcadia, MI 49613-5157. Mailing Address: P.O. Box 67, Arcadia, MI 49613-0067. Tel.: 231-889-3389.

Web Site: www.arcadiami.com

Formerly: Arcadia Township Historical Commission

Founded: 1992.

Key Personnel: Chm., Edward Howard; Dir. & Pres. (V), Joyce Howard; Chm. (V), Lyle Matteson; Museum Shop Mgr., Keith McArthur.

Personnel Profile: Part-Time Volunteers 30.

Governing Authority: Parent Institution: Arcadia Township. Tax-exempt.

Institution Type/Description: Historical Society & Furniture Museum.

Collections: Lumbertown artifacts; Sawmill furniture factory; mirror works; historical books; shipping; Harriet Quimby.

Activities: Cracker Barrel sessions.

Publications: newsletter 2 times a year.

Hours & Admission Prices: June 25 to Labor Day Thurs.-Sat. 1-4, Sun. 1-3; other times by appointment. No charge; donations accepted. &

Attendance: 800 (estimated)

Membership: Individual $10; Family $15; Business & Organization $25; Silver Patron $50; Gold Patron $100; Benefactor & Lifetime $1,000; Corporate Benefactor $2,500.

Auburn Hills

WALTER P. CHRYSLER MUSEUM, One Chrysler Dr., CIMS 488.00.00, Auburn Hills, MI 48326-2766. Tel.: 888-456-1924 (U.S.); 248-944-0001. Fax: 248-944-0460.

Web Site: www.wpchryslermuseum.org

Founded: 1998.

Key Personnel: Pres. & C.E.O., Lori Pinter; Exec. Dir. & C.O.O., Jim Worton; Cur., Brandt Rosenbusch; Dir. Museum Operations, Doreen Wright.

Personnel Profile: Full-Time Paid 6; Part-Time Paid 3; Part-Time Volunteers 125.

Governing Authority: Parent Institution: Walter P. Chrysler Museum Foundation. Tax-exempt.

Institution Type/Description: Automotive Museum.

Collections: history & contribution of the founder to the automotive industry; 65 historic vehicles.

Facilities: 55,000 sq. ft. exhibit space.

Activities: Annual Events: Summer Cruise Nights; Halloween Event; Christmas Holiday Exhibit.

Publications: triannual, Forward Newzine.

Hours & Admission Prices: Tues.-Sat. 10-5, Sun. 12-5. Adults $8, senior citizens 62 & over $7, juniors 6-12 & groups of 15 and over $4, children 5 & under no charge. Closed New Year's Eve & Day; Easter; Independence Day; Thanksgiving; Christmas Eve & Day. &

Attendance: 30,000 (accurate)

Membership: Senior Plymouth $35; Plymouth $45; Seniors Dodge $50; Dodge $60; Senior Hudson $100; Hudson $125; Senior DeSoto $200; DeSoto $250; Senior LeBaron $400; LeBaron $500; Senior Imperial $800; Imperial $1,000.

Augusta

W.K. KELLOGG BIRD SANCTUARY OF MICHIGAN STATE UNIVERSITY, 12685 East C Ave., Augusta, MI 49012-9707. Tel.: 269-671-2510. Fax: 269-671-2474.

E-mail: scarroll@kbs.msu.edu

Web Site: kbs.msu.edu/birdsanctuary

Founded: 1927.

Key Personnel: Facilities Mgr., Karen Charleston; Environmental Education Coord., Tracey Kast; Office Mgr., Sarah Carroll.

Personnel Profile: Full-Time Paid 3; Part-Time Volunteers 50; Interns 1.

Governing Authority: university. Parent Institution: Michigan State University. Subsidiary Institution: Kellogg Biological Station. Tax-exempt.

Institution Type/Description: Bird Sanctuary.

Collections: avian; mammal; fish; botanical.

Research Fields: Waterfowl ecology.

Facilities: auditorium; educational resource center; trail. Books for sale.

Activities: guided tours; lectures; films; formally organized education programs for children; permanent exhibitions; self-guided trail; spruce lodge available for school, group & small meetings; rental facility.

Hours & Admission Prices: Grounds: May-Oct. daily 9-7; Nov.-April daily 9-5. Adults $4, senior citizens $2, children 2-12 $1; children under 2 no charge. Gift Shop: Mon.-Fri. 9-5, Sat.-Sun. 11-4. Gift Shop: closed major holidays. &

Attendance: 12,000 (accurate)

Membership: Student & Senior $15; Individual $20; Household $30; Grandparent $45.

Bad Axe

ALLEN HOUSE MUSEUM, 303 N. Port Crescent Ave., Bad Axe, MI 48413. Mailing Address: P.O. Box 62, Bad Axe, MI 48413. Tel.: 989-712-0050.

E-mail: huroncountyhistoricalsociety@yahoo.com

Institution Type/Description: History Museum: housed in the former home of Wallace E. Allen; built in 1902.

Collections: local history & culture; period furnishings; personal artifacts; photographs.

Hours & Admission Prices: Temporarily closed. June-July Sun. 2-5. No charge.

BAD AXE HISTORICAL SOCIETY MUSEUM, Old City Hall, 110 S. Hanselman St., Bad Axe, MI 48413. Tel.: 989-712-0050.

Institution Type/Description: Historical Society Museum.

Collections: local history & culture; period furnishings; personal artifacts; photographs.

Hours & Admission Prices: Jan.-April 3rd Sun. each month 2-5.

Baraga

BARAGA COUNTY HISTORICAL MUSEUM, 863 US 41 South, Baraga, MI 49908. Mailing Address: P.O. Box 567, Baraga, MI 49908-0567. Tel.: 906-353-8444. Fax: 906-353-8444.

E-mail: baragacountyhistory@gmail.com

Web Site: baragacountyhistoricalmuseum.com

Founded: 1965.

Congressional District: 1

Key Personnel: Pres. (V), Lowell Harshaw.

Personnel Profile: Full-Time Volunteers 4; Part-Time Volunteers 5.

Governing Authority: Parent Institution: Baraga County Historical Society, Inc. Tax-exempt.

Institution Type/Description: History Museum.

Collections: county history & culture; photographs; personal artifacts; railroad artifacts; logging; household; military; religious artifacts.

Major Exhibits: Finding Your Roots: Genealogy, 5/14-9/14; The Joy of Collecting, 5/14-9/14.

Hours & Admission Prices: June-Sept. Tues.-Sat. 11-3; other times by appointment. Adults $2, teens $1; children under 12 no charge. &

Attendance: 400 (accurate)

Membership: Annual $18; Life $100.

Battle Creek

* **ART CENTER OF BATTLE CREEK,** 265 E. Emmett St., Battle Creek, MI 49017-4601. Tel.: 269-962-9511. Fax: 616-969-3838.

E-mail: artcenterofbc@yahoo.com

Web Site: www.artcenterofbattlecreek.com
Founded: 1948.
Congressional District: 3
Key Personnel: Exec. Dir., Linda Holderbaum; Pres. (V), Tim Elliott.
Personnel Profile: Full-Time Paid 1; Part-Time Paid 2; Part-Time Volunteers 50.
Governing Authority: bd. of directors. Tax-exempt.
Institution Type/Description: Art Gallery.
Collections: Michigan art collection.
Major Exhibits: Class Act, 3/14; The Scroll of Remembrance - Miriam Brysk, 5/14; Michigan Artist Competition, 6/14.
Research Fields: art appreciation programs for grades 1-12; Michigan artists.
Facilities: library; archives. Museum-related items for sale.
Activities: art classes for all ages; workshops; guided tours; kidspace hands-on gallery; outreach art programs for schools, local adult day-care facilities & VA hospital.
Publications: quarterly, newsletter & class schedules.
Hours & Admission Prices: Tues.-Fri. 10-5, Sat. 11-3. Adults $3, seniors & students $2; discounts to AAM & ICOM members; Thurs. & members no charge. Closed national holidays. &
Attendance: 35,241 (accurate)
Membership: Student $25; Individual $50; Family $75; Friend $150; Patron $500; Benefactor $1,500; Director's Circle $2,500.

BINDER PARK ZOO, 7400 Division Dr., Battle Creek, MI 49014-9500. Tel.: 269-979-1351. Fax: 269-979-8834.
E-mail: info@binderparkzoo.org
Web Site: www.binderparkzoo.org
Founded: 1975.
Congressional District: 3
Key Personnel: Pres. & C.E.O., Gregory B. Geise; Chm. (V), James Grohalski; C.O.O., Stacey Lawson; Dir. Business & Finance, Amy Riegel; Dir. Wildlife Management, Jenny Barnett; Dir. Animal Health & Research, Thomas W. deMaar, D.V.M.; Dir. Conservation Education, Thomas Funke; Dir. Physical Plant, Eric McNamara; Operations Mgr., Vicki Taft; Mgr. Mktg., Kari Parker; Cur. Collections, Andi Kornak; Cur. Conservation Education, Kathy Fischer.
Personnel Profile: Full-Time Paid 50; Part-Time Paid 150; Part-Time Volunteers 214; Interns 10.
Governing Authority: nonprofit organization. Parent Institution: Binder Park Zoological Society, Inc. Tax-exempt: 501(c)(3).
Institution Type/Description: Zoological Park.
Collections: over 650 animals & 150 species from around the world.
Research Fields: animal management & diseases; education.
Facilities: 1,200-vol. library of natural history & animal management books available for research on premises only; picnic area; 125-seat auditorium; restaurant; amphitheater. Zoo-related items for sale.
Activities: guided tours; lectures; formal organized education programs for children & adults; docent program; permanent & traveling exhibitions; mobile vans; Z.O. & O. railroad rides; Zoomobile Outreach, animal discovery series.
Publications: quarterly newsletter, ZooGoer; newsletter, Zooviews.
Hours & Admission Prices: April-Oct. Mon.-Fri. 9-5, Sat. & holidays 9-6, Sun. 11-6. Adults $11.95, senior citizens 65 & over $10.95, children 2-10 $9.95; discounts to groups by appointment & AAA members; children under 2, AZA, zoo members & reciprocal zoo members no charge. &
Attendance: 320,000 (estimated)
Membership: Individual $50; Family & Grandparents $65; Sustaining $125-$249; Contributing $250-$499; Donor $500-$4,999; Life $5,000 & up.

KIMBALL HOUSE MUSEUM, 196 Capital Ave., N.E., Battle Creek, MI 49017-3925. Mailing Address: 165 N. Washington, Battle Creek, MI 49037-2929. Tel.: 269-965-2613. Fax: 269-660-9072.
E-mail: info@heritagebattlecreek.org
Web Site: www.heritagebattlecreek.org
Founded: 1966.
Congressional District: 3
Key Personnel: Pres., Charles Rose; Dir. Research, Mary Butler.
Governing Authority: society. Parent Institution: Heritage Battle Creek. Tax-exempt: 501(c)(3).
Institution Type/Description: Historic House: 1886 Victorian home.
Collections: Victorian life 1890-1910; household items; furnishings; tools; 1880 medical instruments relating to Battle Creek; Battle Creek historical items; herb garden; Battle Creek Sanitarium; Sojourner Truth.
Research Fields: cereal; industry; health; genealogy; anti-slavery movements; underground railroad.
Facilities: Gifts for sale.

Activities: guided tours; permanent & temporary exhibitions; educational programs.
Publications: bimonthly, Living History; annual, Heritage Battle Creek.
Hours & Admission Prices: April-Dec. Sun. 1-4; tours by appointment. Adults $3, children 12 & under $2; members no charge. &
Attendance: 20,000 (estimated)
Membership: Senior $35; Individual $50; Family $60.

KINGMAN MUSEUM, 175 Limit St., Battle Creek, MI 49037-2176. Tel.: 269-965-5117. Fax: 269-965-3330.
E-mail: droberts@kingmanmuseum.org
Web Site: www.kingmanmuseum.org
Founded: 2000.
Key Personnel: Dir. Operations, Donna C. Roberts; Pres. (V), Corey LaGro; Collections, Museum Shop Mgr., Jamie Schiltz; Education, Katy Avery.
Personnel Profile: Part-Time Paid 2; Part-Time Volunteers 168; Interns 1.
Governing Authority: private; nonprofit organization. Tax-exempt: 501(c)(3).
Institution Type/Description: Natural History Museum.
Collections: fossils; rocks & minerals; Native American artifacts; mounted animals.
Facilities: planetarium. Museum-related items for sale.
Activities: education programs; seasonal day programs; participatory & temporary exhibits; after school programs; outreach programs; discovery kits.
Hours & Admission Prices: Sat. 1-5. Families $20, adults $7, senior citizens, veterans & active duty military $6, student $5; ASTC members and children under 3 no charge; groups of 10 or more by appointment only; discounts to groups 10 or more; additional charge for Planetarium shows, programs or tours. &
Attendance: 14,912 (accurate)
Membership: Students, Senior Citizens & Military $20; Individual $30; Military Family $40; Family & Grandparents $50.

WOLVERINE FIRE COMPANY MUSEUM, 13280 Verona Rd., Battle Creek, MI 49014. Tel.: 616-968-2998.
Institution Type/Description: Fire Fighting History Museum.
Collections: fire fighting history & equipment.
Hours & Admission Prices: By appointment.

Bay City

ANTIQUE TOY AND FIREHOUSE MUSEUM, 3456 Patterson Rd., Bay City, MI 48706. Mailing Address: P.O. Box 188, Kawkawlin, MI 48631-0188. Tel.: 888-888-1270. Fax: 989-667-4772.
E-mail: info@toyandfiretruckmuseum.org
Web Site: toyandfiretruckmuseum.org
Founded: 2005.
Key Personnel: Dir., James Dobson.
Governing Authority: Tax-exempt.
Institution Type/Description: Toy & Firefighting Museum.
Collections: over 60 fire trucks; period toys including fire, police & rescue vehicles, Tonka, Buddy L, Nylent, Hess, & NASCAR.
Activities: special events; rental facilities.
Hours & Admission Prices: May-Sept. Sat.-Sun. 12-3; other times by appointment. Family $15, adults $7, seniors $6, children 5-17 $5; discounts to groups.
Membership: Individual $30; Family $50.

HISTORICAL MUSEUM OF BAY COUNTY, 321 Washington Ave., Bay City, MI 48708-5837. Tel.: 989-893-5733. Fax: 989-893-5741. Facebook: The Bay County Historical Society / The Historical Museum of Bay County.
E-mail: info@bchsmuseum.org
Web Site: www.bchsmuseum.org
Founded: 1919.
Congressional District: 10
Key Personnel: Dir. Opers. & Chief Historian, Ron Bloomfield; Pres. Bd., Judy Jeffers; Vice Pres. Bd., Stephen Kent; Cur. Exhibits & Education, Corrine Bloomfield.
Governing Authority: society. Parent Institution: Bay County Historical Society. Tax-exempt: 501(c)(3).
Institution Type/Description: County History Museum: housed in 1910 National Guard Armory.
Collections: books; manuscripts; maps; photographs; documents; artifacts relating to the history of Bay County, Michigan & the Great Lakes area.
Research Fields: local, regional & archaeological history; genealogy; sugar beet industry; World War II patrol craft history.
Facilities: 1,000-vol. library of books on Great Lakes & Michigan history for use on premises; reading room. Books & museum-related items for sale.

Activities: guided tours; lectures; museum to the schools program; historic dinner; tour to city hall; craft classes. Museum Sponsors: Tour of Homes Event; Quilt Show; River of Time Living History Encampment.
Publications: newsletter; occasional local history booklets & books; Recipe for a Community: The Historical and Culinary Growth of 19th Century Linwood, Michigan; The Historic Architecture of Bay County, Michigan.
Hours & Admission Prices: Mon.-Fri. 10-5, Sat. 12-4. Research Library: Tues.-Thurs. 1-5 (librarian on duty). No charge; donations accepted. Closed major holidays. &
Attendance: 61,000 (estimated)
Membership: Student & Senior $20; Individual $25; Family $50; Patron/Business Class $120; Heritage Circle $750.

Bay View

BAY VIEW HISTORICAL MUSEUM, Bay View Association Encampment 1715, Bay View, MI 49770. Mailing Address: P.O. Box 583, Petoskey, MI 49770-0583. Tel.: 231-347-6225. Fax: 231-347-4330.
Web Site: bayviewassoc.org
Founded: 1970.
Key Personnel: Pres., Lawrence R. Ternan; Exec. Dir., Rodney J. Slocum; Cur. & Archivist, Anne Lewis; Cur. & Archivist, Sophia McGee.
Personnel Profile: Part-Time Volunteers 13.
Governing Authority: private; nonprofit organization.
Institution Type/Description: Historic Site: a National Historic Landmark consisting of 12 buildings & 430 summer homes.
Collections: historic buildings.
Activities: guided tours; lectures. Annual Event: Chautauqua program in summer.
Publications: book, History of Bay View; reprint from 1894 Bayview Summer, City of Michigan.
Hours & Admission Prices: July-Aug. Sun. after church until 1 pm, Wed. 2:30-4:30; Sept.-June after church until 1pm. No charge; donations accepted.
Attendance: 1,000 (estimated)

Beaver Island

BEAVER ISLAND HISTORICAL SOCIETY, 26275 Main St., Beaver Island, MI 49782-5101. Mailing Address: P.O. Box 254, Beaver Island, MI 49782. Tel.: 231-448-2254.
E-mail: history@beaverisland.net
Web Site: history.beaverisland.net
Founded: 1957.
Congressional District: 11
Key Personnel: Pres., Doug Hartle; Vice Pres., Barry Pischner; Treas., Sandy Birdsall; Sec., Linda Warren.
Governing Authority: private; nonprofit corporation. Subsidiary Institution: Mormon Print Shop & Marine Museum. Tax-exempt: 501(c)(3).
Institution Type/Description: Local History Museum.
Collections: artifacts from 1850-present; maps of island; reprint of early Mormon newspaper; Mormon documents; Irish fishing & lumbering artifacts; logging; boat building; Irish migration to Beaver Island; period furnishings; photographs; 50' gill-net fishing boat; outdoor displays; the Protar home & material. Historic Buildings: 1852 Mormon print shop; marine museum on St. James Harbor.
Research Fields: local history; materials relevant to James J. Strang; Irish Immigration; fishing community; Native Americans; genealogy.
Facilities: archives.
Activities: guided tours; lectures; permanent exhibitions; expeditions to historic sites. Museum Sponsors: annual Museum Week.
Publications: journal, The Journal of Beaver Island History Vols. I, II, III, IV; newsletters; maps; Child of the Sea (1905 Memoir).
Hours & Admission Prices: Print Shop Museum & Maritime Museum: Father's Day to Labor Day Mon.-Sat. 11-5, Sun. 12-3. Both Museums: adults $3, children $1; discounts to groups of 15 or more; members no charge. &
Attendance: 3,500 (accurate)
Membership: Student/Senior $10; Single $15; Couples $25; Good Will $50; Contributing $100; Sustaining $150; Life $250; Patron $500; Honorary Historian $1,000.

Belding

BELDING MUSEUM AND THE BEL (BELDING EXPLORA-TION LAB), (M), 108 Hanover St., Belding, MI 48809-1726. Mailing Address: P.O. Box 45, Belding, MI 48809-0045. Tel.: 616-794-1900 ext. 425.
E-mail: mehneyj7881@gmail.com

Founded: 1987.
Key Personnel: Co Dir., Barb Fagerlin; Co Dir., Jill Mason; Sec., Jan Mehney.
Personnel Profile: Part-Time Volunteers 5.
Governing Authority: municipal. Parent Institution: City of Belding, 120 S. Pleasant St., Belding, MI 48809. Tax-exempt: 501(c)(3).
Institution Type/Description: History Museum.
Collections: history of Belding & the silk industry.
Facilities: 80-seat community room. Museum-related items for sale.
Activities: formal education programs; guided tours; temporary exhibitions. Annual Event: Spring Into the Past in May; Quilt Show in November.
Publications: quarterly newsletter.
Hours & Admission Prices: 1st Sun. of month 1-4. No charge. Closed New Year's Day; Easter; Independence Day; Thanksgiving; Christmas. &
Attendance: 750 (estimated)

Belleville

BELLEVILLE AREA MUSEUM, (M), 405 Main St., Belleville, MI 48111-2617. Tel.: 734-697-1944.
E-mail: kdallos@provide.net
Web Site: www.vanburen-mi.org
Founded: 1989.
Key Personnel: Dir., Diane Wilson; Pres. (V), Fred Hudson
Institution Type/Description: Local History Museum.
Collections: local history & culture; personal artifacts; period furnishings.
Hours & Admission Prices: May to Labor Day Mon. & Wed.-Fri. 12-4; Sept.-April Tues. 3-7, Wed.-Sat. 12-4. Adults $1, children $.50; members no charge. Closed holiday weekends. &
Membership: Student $5; Individual $15; Dual & Family $25; Sustaining $50; Patron $150; Life $250.

OAKWOODS METROPARK NATURE CENTER, 17845 Savage Rd., Belleville, MI 48111-9668. Tel.: 800-477-3182; 734-697-9181. Fax: 734-782-3956.
E-mail: kevin.arnold@metroparks.com
Web Site: metroparks.com
Founded: 1975.
Congressional District: 15
Key Personnel: Dir., Dave Moilanen; Chief Interpretive Svcs., C. Michael George; Supervising Interpreter, Kevin J. Arnold; Interpreter, Roni Hutchinson.
Personnel Profile: Full-Time Paid 2; Part-Time Paid 2.
Governing Authority: county. Affiliated with Huron-Clinton Metropolitan Authority, 1300 High Ridge Rd., Brighton, MI 48116-8001. Branch Museums: Indian Springs Metropark Nature Center, Clarkston, MI; Kensington Metropark Nature Center, Milford, MI; Stony Creek Metropark Nature Center, Romeo, MI; Metrobeach Metropark Nature Center, Mt. Clemens, MI; Wolcott Mill Metropark Interpretive Center, Washington, MI; Lake Erie Marshlands Museum, Brownstown, MI. Tax-exempt: 501(c)(3).
Institution Type/Description: Nature Center: located on 1818-1842 Wyandot Indian Reservation.
Collections: botanical; zoological.
Research Fields: local Indian & early white history.
Facilities: nature conservation center; classroom; 100-seat auditorium.
Activities: guided tours; Voyageur canoe tours; lectures; hobby workshops; formally organized education programs for children, adults & graduate students; school visits by interpreters with illustrated programs.
Publications: quarterly events calendar; trail map.
Hours & Admission Prices: June-Aug. Tues.-Sun. 10-5; Sept.-May Tues.-Fri. 1-5, Sat.-Sun. 10-5. No charge. Motor vehicle permit required: annual $25, senior citizens $15, daily $5. Closed New Year's Day; Thanksgiving; Christmas. &
Attendance: 100,000 (estimated)

Belmont

HYSER RIVERS MUSEUM, 6440 W. River Rd., N.E., Belmont, MI 49306. Mailing Address: c/o Plainfield Charter Township, 6161 Belmont Ave., N.E., Belmont, MI 49306-9609. Tel.: 616-364-8466 & 1182 (museum). Fax: 616-364-6537.
E-mail: suzannec@iserv.net
Web Site: www.commoncorners.com
Founded: 1973.
Congressional District: 3
Key Personnel: Pres. (V), Sue Carpenter.
Personnel Profile: Part-Time Volunteers 1.
Governing Authority: Parent Institution: Plainfield Township. Tax-exempt.
Institution Type/Description: Historic House Museum: housed in former home of pioneer surgeon & Civil War Captain William Hyser, built in 1852.
Collections: period artifacts & furnishings.

Activities: genealogy research.
Hours & Admission Prices: April-Dec. first Sun. of month 2-4:30; other times by appointment. No charge; donations accepted.

Benton Harbor

MORTON HOUSE MUSEUM, 501 Territorial, Benton Harbor, MI 49022-3238. Mailing Address: P.O. Box 173, Benton Harbor, MI 49023-0173. Tel.: 269-925-7011.
E-mail: mortonhousemuseum@yahoo.com
Web Site: www.mortonhousemuseum.com
Founded: 1966.
Congressional District: 44
Key Personnel: House Chm. & Pres., Denise Reeves; Vice Pres. & Trustee, Miriam Pede; Sec. & Treas., Cherie Messinger; Trustee, Deb Geib; Trustee, Stuart Boekeloo; Trustee, Gineen Wiley; Trustee, Marcia Smith; Trustee, Denise Tackett; Trustee, Peg Williamson; Trustee, Brenda Layne.
Personnel Profile: Part-Time Volunteers 15.
Governing Authority: nonprofit. Tax-exempt.
Institution Type/Description: Historic House Museum.
Collections: costumes 1830-1930; foreign small box collection.
Research Fields: shipping lines between Chicago & cities along Lake Michigan.
Facilities: meeting room.
Activities: guided tours; lectures; docent program. Museum Sponsors: Holidays at the Morton House.
Publications: quarterly newsletter.
Hours & Admission Prices: mid-April to Oct. Sun. 1-4. Adults $5; members no charge.
Attendance: 500 (estimated)
Membership: Senior Citizens & Students $15; Individual $25; Friends $100.

SARETT NATURE CENTER, 2300 Benton Center Rd., Benton Harbor, MI 49022-9704. Tel.: 269-927-4832. Fax: 616-927-2742.
E-mail: sarett@sarett.com
Web Site: www.sarett.com
Founded: 1970.
Congressional District: 14
Key Personnel: Dir., Dianne Braybrook; Pres. (V), Gene Maddock; Museum Shop Mgr., Mindy Walker.
Personnel Profile: Full-Time Paid 6; Part-Time Paid 4; Part-Time Volunteers 10; Interns 2.
Operating Income: 425,000
Governing Authority: Tax-exempt: 501(c)(3).
Institution Type/Description: Nature Center & Wildlife Sanctuary.
Collections: wet/dry forest & meadow; bog; prairie; fen.
Facilities: nature trails; Butterfly House April-Oct.
Activities: guided tours; lectures; films; formally organized education programs; temporary exhibitions.
Publications: newsletter.
Hours & Admission Prices: Tues.-Fri. 9-5, Sat. 10-5, Sun. 1-5. Adults $3; members & children under 12 no charge. Closed New Year's Day; Easter; Independence Day; Thanksgiving; Christmas. &
Attendance: 35,000 (accurate)
Membership: Individual $35; Family $40.

Benzonia

BENZIE AREA HISTORICAL MUSEUM, 6941 Traverse Ave., Benzonia, MI 49616. Mailing Address: P.O. Box 185, Benzonia, MI 49616-0185. Tel.: 231-882-5539. Fax: 231-882-4435.
E-mail: bmuseum@att.net
Web Site: www.benziemuseum.org
Founded: 1969.
Congressional District: 9
Key Personnel: Pres. Bd. Dir. (V), Jerry Slater; Museum Mgr., Dr. Louis Yock.
Personnel Profile: Full-Time Paid 1; Part-Time Volunteers 100.
Governing Authority: society; nonprofit. Parent Institution: Benzie Area Historical Society. Tax-exempt: 501(c)(3).
Institution Type/Description: Historical Museum: housed in 1884-1887 church building.
Collections: Benzie County artifacts from 1850s to present; Bruce Catton & AA car ferries, resorts, 1890 period rooms; logging; agriculture, Gwen Frostic, Civil War.
Research Fields: Benzie area history.
Facilities: library; meeting room. Books for sale.
Activities: guided tours; docent program.
Publications: quarterly newsletter.
Hours & Admission Prices: May-Dec. Mon.-Sat. 11-5. No charge; donations accepted. Closed major holidays. &

Attendance: 2,000 (estimated)
Membership: Family $35; Business & Professional $75; Patron $150.

Berrien Springs

BERRIEN COUNTY HISTORICAL ASSOCIATION, History Center at Courthouse Sq., 313 N. Cass St., Berrien Springs, MI 49103. Mailing Address: P.O. Box 261, Berrien Springs, MI 49103-0261. Tel.: 269-471-1202. Fax: 269-471-7412.
E-mail: kcyr@berrienhistory.org
Web Site: www.berrienhistory.org
Founded: 1967.
Congressional District: 4
Key Personnel: Pres. (V), Robert D. Sykora; Dir., Kathy A. Cyr; Cur., Robert Myers; Museum Svcs. Coord., Kristen Patzer.
Personnel Profile: Full-Time Paid 3; Part-Time Paid 1; Part-Time Volunteers 96.
Governing Authority: nonprofit corporation. Tax-exempt: 501(c)(3).
Institution Type/Description: Historic Building Museum Complex.
Collections: Berrien County history; repository for county court records & Clark equipment company; genealogical records for Berrien County. Historic Buildings: 1839 Berrien County courthouse; 1830 two-story log house; 1870 sheriff's residence; 1860-1873 county records building; old county jail.
Research Fields: local & regional history; historic preservation.
Facilities: library of Southwest Michigan history; Berrien County court records; Clark equipment company archives; Marx Music Company archives. Books, prints & history-related gift items for sale.
Activities: guided tours; formally organized education programs for adults & children; theatrical performances; permanent & temporary exhibitions; advisory service to local historical organizations; self-guided tour of the 1839 courthouse square.
Publications: books, Historical Sketches of Berrien County; books, Adeline & Julia: Growing Up in Michigan and on the Kansas Frontier-Diaries from 19th Century America; Millennial Visions and Earthly Pursuits: The Israelite House of David 1903-Present; Greetings From Berrien Springs; Story of Buchanan, a history; Greetings from Buchanan; Greetings from St. Joseph; Locomotives along the Lakeshore: The Railroads of Berrien County.
Hours & Admission Prices: June-Aug. Mon.-Sat. 10-5; Sept.-May Mon.-Fri. 10-5. No charge, donations accepted. Closed holidays. &
Attendance: 21,000 (accurate)
Membership: Individual $20; Family $30; Institutional & Contributing $40; Sustaining $50; Patron $100; Benefactor $500.

SIEGFRIED H. HORN ARCHAEOLOGICAL MUSEUM, Andrews University, Institute of Archaeology, Berrien Springs, MI 49104-0990. Tel.: 269-471-3273. Fax: 269-471-3619.
E-mail: hornmuseum@andrews.edu
Web Site: www.andrews.edu/archaeology
Founded: 1970.
Key Personnel: Cur., Constance Gane; Asst. to Cur., L.S. Baker, Jr.
Personnel Profile: Full-Time Paid 3; Part-Time Paid 2.
Governing Authority: church; nonprofit. Parent Institution: Andrews University. Tax-exempt: 501(c)(3).
Institution Type/Description: Archaeological Museum.
Collections: Middle East artifacts: pottery; tools; coins; textiles; papyri; tablets; cuneiform tablets; replicas for teaching purposes.
Research Fields: on-site archaeological excavations in Jordan, Israel.
Facilities: library of books, professional journals, archival pamphlets & clippings of early archaeology of Ancient Near East & slide collection; 1,200 sq. ft. exhibit space; classroom; photo lab; student study areas. Books, pamphlets, replicas, posters & T-shirts for sale.
Activities: guided tours; lectures; temporary, participatory & loan exhibitions.
Publications: quarterly newsletter; occasional monographs; Near East Archaeological Society Bulletin.
Hours & Admission Prices: Museum: temporarily closed for renovations. Temporary exhibit area open to the public Sat. 3-5 by appointment. No charge; donations accepted. Closed holidays. &
Attendance: 1,100 (estimated)

Big Rapids

JIM CROW MUSEUM - FERRIS STATE UNIVERSITY, (M), 820 Campus Dr., ASC 2108, Big Rapids, MI 49307-2225. Tel.: 231-591-5873. Fax: 231-591-2541.
E-mail: jimcrowmuseum@ferris.edu
Web Site: www.ferris.edu/jimcrow
Founded: 1996.
Congressional District: 2

Key Personnel: Dir., J. Andy Karafa, Ph.D.; Cur., David Pilgrim, Ph.D.
Personnel Profile: Full-Time Paid 1; Part-Time Paid 1; Part-Time Volunteers 20.
Governing Authority: university. Parent Institution: Ferris State University. Tax-exempt.
Institution Type/Description: History Museum.
Collections: early to mid 20th century artifacts; segregation memorabilia including Ku Klux Klan artifacts & pro-segregation signs, tickets, brochures, posters, musical records & magazines; civil rights artifacts; over 3,000 anti-Black caricature artifacts; 200 items related to the leaders & organizations of the civil rights movement.
Research Fields: race relations; prejudice; discrimination.
Facilities: 504 sq. ft. exhibit space.
Activities: formal education programs; guided tours.
Hours & Admission Prices: Mon.-Fri. 8-5 by appointment. No charge; donations accepted. Closed national holidays. &

Attendance: 1,500 (estimated)

MECOSTA COUNTY HISTORICAL MUSEUM, 129 S. Stewart Ave., Big Rapids, MI 49307-1968. Tel.: 231-592-5091 & 796-0368.
Founded: 1957.
Congressional District: 10
Key Personnel: Pres. (V), Fredda Hankes; Museum Co-Dir., Agnes Tornblom; Museum Co-Dir., Barb Johnson.
Personnel Profile: Part-Time Volunteers 12.
Governing Authority: state. Parent Institution: Mecosta County Historical Society. Tax-exempt.
Institution Type/Description: General Museum: housed in two-story frame Victorian house.
Collections: agriculture; costumes; folklore; Indian artifacts; numismatic; logging & lumbering exhibits; household items; toys.
Research Fields: agriculture; folklore.
Activities: guided tours; formally organized education programs for children; temporary exhibitions; traveling museum.
Hours & Admission Prices: May-Oct. Sat. 2-4; other times by appointment. No charge; donations accepted.
Attendance: 1,200 (estimated)
Membership: Individual $10; Patron $50; Fitch Phelps Estate $100.

Birmingham

BIRMINGHAM BLOOMFIELD ART CENTER, 1516 S. Cranbrook Rd., Birmingham, MI 48009-1855. Tel.: 248-644-0866. Fax: 248-644-7904.
E-mail: annievangelderen@bbartcenter.org
Web Site: www.bbartcenter.org
Founded: 1957.
Congressional District: 18
Key Personnel: Pres. & C.E.O., Annie Van Gelderen; Chm. Bd., Margaret E. Greene; Vice Pres. Programs, Cynthia Mills; Vice Pres. Finance, Gwenn Rosseau.
Personnel Profile: Full-Time Paid 5; Part-Time Paid 9; Part-Time Volunteers 250; Interns 2.
Governing Authority: municipal; nonprofit organization. Tax-exempt: 501(c)(3).
Institution Type/Description: Community Art Center.
Collections: Sol Le Witt, Wall Drawing #975 Isometric Outline; Marshall Fredricks, Black Elk Speaks.
Major Exhibits: TDB, 1/10-2/28/14; Larry Cressman, 3/14-5/2/14; Michigan Fine Arts Competition, 5/16-7/11/14; Linda Soberman, 7/25-8/22/14.
Facilities: studios; classrooms. Two and three dimensional work for sale.
Activities: studio art & art history classes; lectures; workshops; holiday shop; auctions & fundraising events; national & international travel programs; competitions. Museum Sponsors: Art Festival in spring.
Publications: exhibition materials; course catalogue; weekly e-blast communications.
Hours & Admission Prices: Mon.-Thurs. 9-6, Fri.-Sat. 9-5. No charge. Closed major holidays. &
Attendance: 20,000 (estimated)
Membership: Individual $50; Household $75; Community $100.

BIRMINGHAM HISTORICAL MUSEUM & PARK, (M), 556 W. Maple, Birmingham, MI 48009-3360. Tel.: 248-530-1928. Facebook; Birmingham Historical Museum & Park.
E-mail: museum@bhamgov.org/museum
Web Site: bhamgov.org/museum
Founded: 2001.
Congressional District: 9

Key Personnel: Dir., Leslie Pielack; Pres. (V), Catherine Tuczek; Chm. (V), Russell Dixon.
Personnel Profile: Part-Time Paid 3; Part-Time Volunteers 40; Interns 1.
Governing Authority: municipal; nonprofit. Parent Institution: City of Birmingham. Tax-exempt: 501(c)(3).
Institution Type/Description: Historical Society Museum.
Collections: local history from prehistoric to present; archives; personal artifacts; furnishings. Historic Buildings: c.1822 John West Hunter House reflects rural life in the Michigan wilderness in 1840s; c.1928 Allen House is a colonial revival structure built by Birmingham's First Mayor with changing exhibits that reflect the community's historic past.
Research Fields: local history & biographies; early Michigan pioneering history.
Facilities: 500-vol. library of local history books. Museum-related items for sale.
Activities: docent program; formal education programs for children; guided tours; special behind the scenes tours & themed events; lectures; temporary exhibitions. Museum Sponsors: IMLS/First Lady Michelle Obama's Let's Move! Museums & Gardens Initiative.
Publications: newsletter, Birmingham Heritage
Hours & Admission Prices: Wed.-Sat. 1-4; other times by appointment. Adults $5, seniors & students $3; children 5 & under & members no charge. Closed city holidays. &
Attendance: 5,000 (estimated)
Membership: Student $10; Individual $25; Family $40.

Bloomfield Hills

*   **CRANBROOK ART MUSEUM, (M),** 39221 Woodward Ave., Bloomfield Hills, MI 48304-5162. Mailing Address: P.O. Box 801, Bloomfield Hills, MI 48303-0801. Tel.: 248-645-3319. Fax: 248-645-3323.
E-mail: artmuseum@cranbrook.edu
Web Site: www.cranbrook.edu
Founded: 1927.
Congressional District: 19
Key Personnel: Dir., Gregory M. Wittkopp; Administrative Asst., Kim Larsen; Registrar, Roberta Frey Gilboe; Preparator, Mark Baker; Collections Fellow, Chad Alligood.
Personnel Profile: Full-Time Paid 4; Part-Time Volunteers 20; Interns 15.
Governing Authority: nonprofit. Parent Institution: Cranbrook Educational Community. Affiliated with Cranbrook Academy of Art. Tax-exempt: 501(c)(3).
Institution Type/Description: Art Museum.
Collections: 20th-century architectural drawings, ceramics, furniture, metalwork, paintings, sculpture, textiles; study collection of textiles; 19th- & 20th-century prints & ceramics. Historic House: Saarinen House, the 1930 home & studio of Eliel and Loja Saarinen.
Research Fields: 20th & 21st century art, crafts, decorative arts, architecture & design.
Facilities: 28,000 sq. ft. exhibition space; 205-seat auditorium; seminar room.
Activities: guided tours of exhibitions & architecture; lectures; films; gallery talks; docent program or council; inter-museum loan; permanent, temporary & traveling exhibitions.
Publications: exhibition catalogs, Inigo Manglano-Ovalle; Beautiful Scenes: Selections from the Cranbrook Archives; Dream Sites: A Visual Essay; Weird Science: A Conflation of Art & Science; What's Next Newsletter; 100 Treasures of Cranbrook Art Museum; Three Decades of Contemporary Art: The Dr. John & Rose M. Shuey Collection; Richard De Vore; Mark Newport: Superheros In Action; No Object Is an Island: New Dialogues with the Cranbrook Collection; Saarinen House & Garden: A Total Work of Art.
Hours & Admission Prices: Museum: Academic Year: Tues.-Fri. 10-5, Sat.-Sun. 11-5; Summer: Wed.-Sun. 11-5. Adults $8, senior citizens 65 & over $6, students with ID $4; discounts to AAM members; children under 12 & members no charge. Saarinen House Tours: May-Oct. 2pm. Adults $10. Closed New Year's Eve & Day; Easter; Memorial Day; Independence Day; Labor Day; Thanksgiving, Christmas Eve & Day. &
Attendance: 37,000 (estimated)
Membership: Individual $50; Household $75; Friends $150 Collector's Circle $250; Curator's Circle $500; Director's Circle $1,000; Governors' Circle $2,500; Trustees' Circle $5,000.

CRANBROOK HOUSE AND GARDENS AUXILIARY, 380 Lone Pine Rd., Bloomfield Hills, MI 48303-0801. Mailing Address: P.O. Box 801, Bloomfield Hills, MI 48303-0801. Tel.: 248-645-3149. Fax: 248-645-3151. Facebook: Cranbrook House and Gardens.
E-mail: csmith@cranbrook.edu; mkrygier@cranbrook.edu
Web Site: www.cranbrook.edu
Founded: 1971.

Key Personnel: Chm. House & Garden, Richard Lilley.
Personnel Profile: Full-Time Paid 1; Part-Time Paid 2.
Governing Authority: nonprofit organization. Affiliated with Cranbrook. Tax-exempt.
Institution Type/Description: Historic House: 1908 home of George Gough & Ellen Scripps Booth.
Collections: art objects; furnishings.
Activities: guided tours; lectures; films; concerts; docent program.
Publications: quarterly newsletter to membership; History of Cranbrook House.
Hours & Admission Prices: Gardens: May to Labor Day Mon.-Sat. 10-5, Sun. 11-5; Sept. daily 11-3; Oct. Thurs.-Sun. 11-3. House Tours: June-Aug. Thurs. 11 & 1:15, Fri. 11 & 1, Sun. 1 & 3; Sept.-Oct. Thurs. 11 & 1:15, Fri. 11, Sun. 1; group tours at other times by prior arrangement. Gardens: $6 per person. House Tour & Gardens: $10 per person; discounts to senior citizens; members no charge.
Attendance: 8,000 (estimated)
Membership: Student $30; Senior $35; Friend $50; Group $50; Contributing $125; Patron $250; Benefactor $500; Life $1,000.

＊ **CRANBROOK INSTITUTE OF SCIENCE,** 39221 Woodward Ave., Bloomfield Hills, MI 48304-5162. Mailing Address: P.O. Box 801, Bloomfield Hills, MI 48303-0801. Tel.: 248-645-3200. Fax: 248-645-3050.
E-mail: mstafford@cranbrook.edu
Web Site: science.cranbrook.edu
Founded: 1930.
Congressional District: 19
Key Personnel: Chm. Bd. Governors, Richard E. Warren; Dir., Dr. Michael D. Stafford; Cur. Collections, Cameron Wood; Head Mktg., Stephen Pagnani.
Governing Authority: nonprofit organization. Parent Institution: Cranbrook Educational Community. Tax-exempt: 501(c)(3) & 170(b)(1)(A).
Institution Type/Description: Science Museum.
Collections: mineralogy; anthropology; zoology; botany; geology; bats; ethnic & natural history.
Research Fields: botany; anthropology; science education; paleontology; mineralogy.
Facilities: observatory & planetarium; 235-seat auditorium; classrooms; cafe; herbarium; participation physics hall; discovery room. Museum-related items for sale.
Activities: instructional programs for school groups; science education classes for children & adults; workshops; lecture series; permanent & temporary exhibitions; inter-museum loan; field trips; special events on science topics; laser light shows.
Publications: bulletin series focusing on the Institute's areas of specialization, including Michigan Flora, Michigan Lichens, Birds of Southeast Michigan.
Hours & Admission Prices: Tues.-Thurs. 10-5, Fri-Sat. 10-10, Sun. 12-4. Adults $13, senior citizens & children 2-12 $9.50; discounts to AAM members, groups and Fri. & Sat. evenings; children under 2, ASTC members & members no charge. Additional fee for Planetarium & Laser programs. Closed New Year's Eve & Day; Easter; Memorial Day; Independence Day; Labor Day; Thanksgiving; Christmas Eve & Day. &
Attendance: 176,000 (accurate)
Membership: Senior $55; Individual $60; Grandparent $65; Family $75; Friends $95; Galactic Pass $125; Cranbrook All-Access Pass $250; Back Stage Pass $350.

Byron Center

GAINEY GALLERY - VAN SINGEL FINE ARTS CENTER, 8500 Burlingame, S.W., Byron Center, MI 49315. Tel.: 616-878-6801.
Key Personnel: Cur., Cindi Ford
Institution Type/Description: Art Gallery.
Collections: works by local artists; paintings; sculpture; prints; photographs.
Activities: educational programs.
Hours & Admission Prices: June-Aug. Mon.-Thurs. 12-5; Sept.-May Mon.-Fri. 12-5.

Cadillac

WEXFORD COUNTY HISTORICAL SOCIETY & MUSEUM, 127 Beech St., Cadillac, MI 49601-1901. Mailing Address: P.O. Box 124, Cadillac, MI 49601-0124. Tel.: 231-775-1717.
E-mail: info@wexfordcountyhistory.org
Web Site: wexfordcountyhistory.org
Founded: 1978.
Congressional District: 10
Key Personnel: Pres. (V), Rob Grostick; Pres., Kristine Anderson; Museum Shop Mgr., Thelma Merritt.

Personnel Profile: Part-Time Paid 1; Part-Time Volunteers 25; Interns 1.
Governing Authority: society; nonprofit. Tax-exempt.
Institution Type/Description: County Historical Society Museum: housed in former Carnegie Library of Cadillac.
Collections: local photographs; paintings of local historical features; logging equipment; county rural one-room school; early fire equipment; railroad memorabilia; general store, post office & barber shop exhibits.
Facilities: 5,000 sq. ft. exhibit space.
Activities: temporary exhibitions; tours; special events; community gatherings.
Publications: newsletter; brochure, A Walking Tour of Cadillac; Rural Schools of Wexford County; Indian Trail of Wexford County; The Shay Locomotive; Cadillac Vintage Postcard History.
Hours & Admission Prices: First Sat. in May 12-4; Memorial Day to Labor Day Wed.-Fri. 12-4, Sat. 10-4; Labor Day to mid-Dec. Fri.-Sat. 12-4. Adults $3, family $5; children under 16 & members no charge.
Attendance: 3,200 (estimated)
Membership: Individual $10; Family $25; Friend $100; Lifetime $250; Corporate Patron $500.

Calumet

COPPER COUNTRY FIREFIGHTERS HISTORY MUSEUM, 327 Sixth St., Calumet, MI 49913. Mailing Address: P.O. Box, Calumet, MI 49913. Tel.: 906-337-4579. Facebook: Copper Country Firefighters History Museum.
Web Site: coppercountryfirefightershistorymuseum.web.com
Formerly: Upper Peninsula Firefighters Memorial Museum
Founded: 1991.
Congressional District: 1
Key Personnel: Chm. (V) & Pres. (V), Paul Bracco.
Personnel Profile: Part-Time Volunteers 7.
Governing Authority: Tax-exempt.
Institution Type/Description: Firefighting History Museum: housed in the former Red Jacket Fire Station; built in 1900.
Collections: firefighting history & equipment; photographs; personal artifacts.
Hours & Admission Prices: mid-June to Labor Day Mon.-Sat. 1-4:30. Adults $2, family $5.
Attendance: 1,080 (estimated)
Membership: Individual $7; Family $10; Life $100.

COPPERTOWN U.S.A., 25815 Red Jacket Rd., Calumet, MI 49913-2904. Mailing Address: 56638 Calumet Ave., Calumet, MI 49913-1965. Tel.: 906-337-4354.
Web Site: www.uppermichigan.com/coppertown
Founded: 1973.
Congressional District: 11
Key Personnel: Pres. (V), Richard Dana.
Personnel Profile: Part-Time Paid 1; Part-Time Volunteers 12.
Governing Authority: nonprofit public corporation.
Institution Type/Description: Restored Mining Co. Complex: situated on the site of the former Calumet & Hecla Mining Co. headquarters complex.
Collections: recreated mineshaft; old railroad equipment; industrial technology & artifacts reflecting the mining of copper & the life of the deep-shaft miner; ethnic displays, reflecting the earlier inhabitants of the area; restored C & H Pattern Shop.
Facilities: visitor orientation center. Gifts & museum-related items for sale.
Hours & Admission Prices: June to mid-Oct. Mon.-Sat. 11-5. Adults $4, Golden Age Pass $3, children 6-15 $2; children under 6 no charge. &
Attendance: 3,000 (estimated)
Membership: Trammer $10; Miner $25; Mine Captain $60; C&H Stockholder $100; Agassiz Society Member $500 & up.

Caspian

IRON COUNTY HISTORICAL & MUSEUM SOCIETY, (M), 100 Brady Ave., Caspian, MI 49915. Mailing Address: P.O. Box 272, Caspian, MI 49915. Tel.: 906-265-2617. TDD: 906-265-2617.
E-mail: info@ironcountyhistoricalmuseum.org
Web Site: www.ironcountyhistoricalmuseum.org
Founded: 1962.
Congressional District: 1
Key Personnel: Pres., Bill Leonoff; Vice Pres. & Programs, Betty Petroski; Gallery Dir. & Treas., Shirley Carlson; Asst. Treas., Bernadette Passamani; Founder, Harold Bernhardt; Founder, Marcia Bernhardt.
Personnel Profile: Full-Time Paid 1; Part-Time Paid 2; Part-Time Volunteers 174.
Governing Authority: nonprofit organization. Tax-exempt: 501(c)(3).
Institution Type/Description: History Museum.
Collections: 2,000 piece hand-carved miniature logging camp; glass mining dioramas; lumbering & mining tools; farm tools; manuscripts. Historic

Buildings: 1890 Stager railroad depot; 1890 Bates Township log barn; 1890 Beechwood log cabin; 1896 one-room schoolhouse; c.1903-1937, Caspian Mine engine house; 1911 Koski log cabin; 1900 Johnson homestead cabin; 1920 Sharrard log lumber cabin; 1890 home of composer Carrie Jacobs-Bond; 1920 Mining Head frame on National Register of Historic Places. Le Blanc Memorial Art Gallery: 140 prints, 50 originals; wildlife art; 12,000 photographs; 8,000 maps (underground & surface). Giovanelli Art Gallery & Studio: Italianate style.

Research Fields: local communities; mining; lumbering; folklore; agriculture; one room schools; genealogy

Facilities: 200-vol. library of materials on mining, minerals & local lore available for use on premises; archives. Local publications, books & crafts for sale.

Activities: concerts; workshops; education programs for children; arts shows; permanent exhibition; special events. Museum Sponsors: Annual Events: Ethnic Events; Labor Day Family Picnic; Christmas Tree Galleria.

Publications: annual newsletter, Past-Present Prints; books, Iron River History; folklore books; The History of Iron County; Black Rock & Roses; The Jewel of Iron Country; Half-Pint Pete; Caspian, A Caring City 1918-1993; Barns, Farms and Yarns, 1999; Men, Mines and Memories 2001; Rural Schools Recollections, 2003; Forty Years of Sports, 2004; Pine to Popple: People and Places, 2007; Iron River, A Mining - Logging Town 1885-1925; Frames for the Future (1980); Women of Iron County Michigan 1882-2013, 2013.

Hours & Admission Prices: Memorial Day to Labor Day Mon.-Sat. 10-4; Sept. Mon.-Sat. 12-4; other times by appointment. Adults $8. Annual Pass: $25. ♿

Attendance: 5,000 (estimated)

Membership: Individual $25.

Centreville

NOTTAWA STONE SCHOOL, 204 E. Burr Oak St., Centreville, MI 49032-9620. Tel.: 616-467-5400 & 6155.

Founded: 1968.

Key Personnel: Pres., Richard A. Cripe.

Governing Authority: nonprofit organization. Tax-exempt.

Institution Type/Description: Historic Building: 1870 Stone School.

Collections: school furniture & furnishings.

Publications: Nottawa Stone School Guidelines.

Hours & Admission Prices: Sept.-June Mon.-Fri. No charge.

Charlotte

COURTHOUSE SQUARE ASSOCIATION, (M), 100 W. Lawrence Ave., Charlotte, MI 48813-1494. Mailing Address: P.O. Box 411, Charlotte, MI 48813-0411. Tel.: 517-543-6999. Fax: 517-543-6999.

E-mail: preserve@ia4u.net

Web Site: www.visitcourthousesquare.org

Founded: 1993.

Congressional District: 7

Key Personnel: C.E.O. & Museum Shop Mgr., Jeralyn Bohms; Pres. (V), Carol Ranville.

Personnel Profile: Part-Time Paid 2; Part-Time Volunteers 10.

Governing Authority: private; nonprofit. Tax-exempt: 501(c)(3).

Institution Type/Description: Historic Area: listed on the National Register of Historic Sites.

Collections: emphasis on Eaton County, Michigan & the surrounding mid-Michigan area; military items (Civil War to Vietnam); textiles (quilts & costumes); government records; original documentation for the building including bills of sale, contracts & architectural drawings; political collections; domestic tools; agricultural items; furniture. Historic Buildings: 1873 second empire-style sheriff's residence; 1885 Renaissance revival-style Eaton County courthouse.

Facilities: 2,500-vol. library of Eaton County letters & diaries and government records & documents; genealogical research library; 10,000 sq. ft. exhibit space; 120-seat courtroom; rental facilities.

Activities: guided tours; rental facilities; temporary exhibitions; rental facilities for weddings.

Publications: quarterly newsletter, The CSA Ledger.

Hours & Admission Prices: Mon.-Fri. 9-4; other times by appointment. Adults $1; discounts to groups & families; children 12 & under and members no charge. Closed major holidays.

Attendance: 2,000 (accurate)

Membership: Individual: Docent $25-$99; Archivist $100-$499; Curator $500-$999; Director $1,000 & up. Business: Scholar $250-$499; Historian $500-$999; Conservator I $1,000-$2,499; Conservator II $2,500 & up.

Chassell

CHASSELL HERITAGE CENTER, 42373 N. Hancock St., Chassell, MI 49916. Mailing Address: P.O. Box 331, Chassell, MI 49916-0331. Tel.: 906-523-1155 & 4612.

Web Site: www.einerlei.com/community/CHO.html

Key Personnel: Pres., Corinne Hauring

Institution Type/Description: Heritage Center.

Collections: local history & culture; period artifacts.

Activities: special events. Museum Sponsors: Strawberry Festival in July; Strawberry Fashion Show in July.

Hours & Admission Prices: July-Aug. Tues. 1-4, Thurs. 4-9.

Cheboygan

THE CHEBOYGAN COUNTY HISTORY CENTER, (M), 427 Court Street, Cheboygan, MI 49721-1908. Mailing Address: P.O. Box 5005, Cheboygan, MI 49721-5005. Tel.: 231-627-9597.

E-mail: museum@cheboyganhistorycenter.org

Web Site: www.cheboyganhistorycenter.org

Formerly: Cheboygan County Historical Museum Complex

Founded: 1971.

Key Personnel: Dir. & Pres., Lewis D. Crusoe; Membership, Kay Forster; Registrar, Sharon Ecker.

Personnel Profile: Part-Time Paid 1; Part-Time Volunteers 40.

Governing Authority: private; nonprofit organization. Tax-exempt: 501(c)(3).

Institution Type/Description: History Museum.

Collections: county history; lumbering; maritime; general store; schoolroom; military artifacts; photographs. Historic Buildings: 1880s sheriff residence & jail; log cabin.

Research Fields: history of area culture, fishing, logging & railroads.

Activities: guided tours; programs; children's museum day camp. Annual Events: Log Cabin Day in June; Yard Sale in August; Autumn Fest in October; Festival of Trees in December; Winter History Lecture Series; History House Tour; Cemetery Tour; Storytelling Around A Summer Fire.

Publications: quarterly newsletter, The Chronicle.

Hours & Admission Prices: Memorial Day-Sept. Tues.-Sat. 1-4. Adults $5; discounts to groups and AAM & AAA members; children & members no charge.

Attendance: 580 (accurate)

Membership: Academic $10; Senior $15; Individual $25; Dual $35; Family $45; Bronze $50-$99; Silver $100-$499; Gold $500-$999; Platinum $1,000 & up.

Chelsea

GERALD E. EDDY DISCOVERY CENTER, 17030 Bush Rd., Chelsea, MI 48118-9747. Mailing Address: Waterloo Recreation Area, 16345 McClure Rd., Chelsea, MI 48118. Tel.: 734-475-3170. Fax: 734-475-6421.

Web Site: www.michigan.gov/bnr

Founded: 1976.

Key Personnel: Dir., Kathy Kavanagh.

Governing Authority: state; nonprofit. Tax-exempt: 501(c)(3).

Institution Type/Description: Natural history: located on glacial moraine overlooking a kettle lake.

Collections: Precambrian-Carboniferous rocks from Great Lakes region; over 1,000 locally found Indian arrowheads, spear points & stone tools.

Facilities: 77-seat auditorium; 4,000 sq. ft. exhibit space; nature & conservation area; geology interactive room for young people. Geology books, nature reference books, jewelry & other museum-related items for sale.

Activities: guided tours; lectures; videos; slide show; arts festivals; theater; hobby workshops; organized education programs for children & adults. Museum Sponsors: seasonal weekend festivals.

Hours & Admission Prices: Jan.-March Tues.-Sat. 10-5; Summer: daily 10-6; Dec. Tues.-Sun. 10-5. No charge. Parking: daily $6, annual $24. Closed state holidays. ♿

Attendance: 30,859 (accurate)

Membership: Waterloo Natural History Association: Individual $10; Family $20; Patron $30; Sustaining $50; Benefactor $100; Life $500.

Chesaning

CHESANING HISTORICAL SOCIETY & MUSEUM, 602 W. Broad St., Chesaning, MI 48616. Mailing Address: P.O. Box 83, Chesaning, MI 48616. Tel.: 989-845-3155.

Institution Type/Description: Historical Society Museum.

Collections: local history & culture; period furnishings; personal artifacts; photographs.

Hours & Admission Prices: Jan.-March Mon. & Wed. 10 to noon; April-Dec. Mon.-Wed. 10 to noon, Sat. 1-4; other times by appointment.

Clarkston

CLARKSTON HERITAGE MUSEUM, 6495 Clarkston Rd., Clarkston, MI 48346-1501. Tel.: 248-922-0270.
E-mail: info@clarkstonhistorical.org
Web Site: www.clarkstonhistorical.org/museum.htm
Founded: 1999.
Key Personnel: Dir., Toni Smith; Pres. (V), Jennifer Arkwright.
Personnel Profile: Part-Time Paid 1; Part-Time Volunteers 45.
Governing Authority: Parent Institution: Clarkston Community Historical Society. Tax-exempt.
Institution Type/Description: Historical Society Museum.
Collections: local history & culture; photographs; documents; photographs; period clothing.
Publications: Clarkston Columns.
Hours & Admission Prices: Mon.-Wed. 10-9, Thurs.-Sat. 10-6. &
Attendance: 5,000 (estimated)

Clawson

CLAWSON HISTORICAL MUSEUM, 41 Fisher Ct., Clawson, MI 48017. Mailing Address: 425 N. Main, Clawson, MI 48017-1500. Tel.: 248-588-9169.
E-mail: historicalmuseum@cityofclawson.com
Web Site: cityofclawson.com/museum
Founded: 1973.
Congressional District: 18
Key Personnel: Chm. (V), Joyce McIntyre; Cur., Melodie Nichols.
Personnel Profile: Part-Time Paid 1; Part-Time Volunteers 10.
Governing Authority: municipal. Tax-exempt.
Institution Type/Description: History Museum.
Collections: furniture & household articles from the 1920s; photographs, newspapers & other memorabilia of Clawson.
Research Fields: history of Clawson.
Facilities: 100-vol. library of material on Clawson history, surrounding communities, Oakland County & the State of Michigan available for use on premises with supervision.
Activities: guided tours; lectures; films; docent program or council.
Publications: brochure; newsletter.
Hours & Admission Prices: Wed. & Sun. 1-4. No charge; donations accepted. Closed holidays.
Membership: $5.

Clinton Township

CLINTON TOWNSHIP HISTORICAL VILLAGE MUSEUM, 40700 Romeo Plank Rd., Clinton Township, MI 48038-2942. Tel.: 586-286-9173.
E-mail: gcthsnewsletter@yahoo.com
Institution Type/Description: Historical Society Museum.
Collections: artifacts & memorabilia pertaining to Clinton Township.
Hours & Admission Prices: Call for hours.

Coldwater

WING HOUSE MUSEUM, 27 S. Jefferson St., Coldwater, MI 49036. Tel.: 517-278-2871.
Institution Type/Description: History Museum: housed in the former home of Jay and Frances Chandler; built in 1875.
Collections: local history & culture; period furnishings; personal artifacts; photographs.
Activities: special events.
Hours & Admission Prices: By appointment.

Coloma

NORTH BERRIEN HISTORICAL MUSEUM, 300 Coloma Ave., Coloma, MI 49038-9724. Mailing Address: P.O. Box 207, Coloma, MI 49038-0207. Tel.: 269-468-3330. Fax: 269-468-4083. Facebook: North Berrien History.
E-mail: info@northberrienhistory.org
Web Site: www.northberrienhistory.org
Founded: 1966.
Congressional District: 6
Key Personnel: Exec. Dir., Tracy Gierada; Pres. (V), Scott Young.
Personnel Profile: Full-Time Paid 2; Part-Time Paid 1; Interns 1.

Governing Authority: Tax-exempt.
Institution Type/Description: History Museum.
Collections: Paw Paw Lakes history; fruit farming; cultural heritage.
Publications: quarterly newsletter.
Hours & Admission Prices: May-Sept. Tues.-Sat. 10-4; Oct.-April Tues.-Fri. 10-4; other times by appointment. No charge; donations accepted. Closed holidays. &
Attendance: 5,300 (accurate)
Membership: Student $8; Senior $10; Adult $15; Family $50.

Concord

HISTORIC MANN HOUSE, (M), 205 Hanover St., Concord, MI 49237. Mailing Address: Michigan Historical Museum, 702 W. Kalamazoo, P.O. Box 30740, Lansing, MI 48909-8240. Tel.: 517-373-1359.
Web Site: www.michiganhistory.org
Founded: 1970.
Governing Authority: state. Affiliated with the Michigan Historical Museum, 702 W. Kalamazoo, P.O. Box 30740, Lansing, MI 48909-8240. Tax-exempt: 170(b)(A).
Institution Type/Description: Historic House: 1880 Mann House.
Collections: furniture, furnishings & decorative arts from the 1840s to 1920s.
Research Fields: family records.
Activities: guided tours.
Publications: brochure; booklet, The Mann House in Historic Concord.
Hours & Admission Prices: Memorial Day-Labor Day Tues.-Sat. 10-4. No charge.
Attendance: 3,500

Constantine

JOHN S. BARRY HISTORICAL SOCIETY, 300 N. Washington, Constantine, MI 49042. Mailing Address: 485 Centerville Rd., Constantine, MI 49042-0068. Tel.: 269-435-5825.
Founded: 1945.
Congressional District: 4
Key Personnel: Pres. (V), Dr. Vercher; Vice Pres. (V), Robert Ray; Sec. (V), Jane Bickle.
Governing Authority: society; nonprofit. Tax-exempt.
Institution Type/Description: Historical Society Museum: housed in 1835-1847 Governor Barry House, home of Michigan's third elected governor.
Collections: furniture; cookware; needlework; lamps; dolls; clothing; musical instruments; jail cell.
Research Fields: genealogical research into the pioneer families of the community.
Activities: guided tours; lectures; special events. Annual Event: barbeque in September.
Hours & Admission Prices: Open by appointment only. No charge; donations accepted.
Attendance: 550
Membership: Individual $10.

Coopersville

THE COOPERSVILLE AREA HISTORICAL SOCIETY MUSEUM, (M), 363 Main St., Coopersville, MI 49404-1234. Tel.: 616-997-6978.
E-mail: budphoto@j2k.com
Web Site: www.coopersville.com/museum
Key Personnel: Dir., Lillian Budzynski; Cur., Jim Budzynski.
Governing Authority: private; nonprofit organization.
Institution Type/Description: Historical Society Museum.
Collections: railroad memorabilia; gold records & memorabilia belonging to Del Shannon, an early rock-n-roll star; period drugstore re-creation.
Hours & Admission Prices: Aug. to mid-Dec. Tues. 3-8, Wed. 10-1, Sat. 10-4, Sun. 1-4; mid-Dec. to July Tues. 3-8, Wed. 10-1, Sat. 10-4; other times by appointment. Suggested Donation: adults $1. &
Attendance: 5,000 (estimated)

Copper Harbor

* **FORT WILKINS HISTORIC COMPLEX, (M),** 15223 US 41, Fort Wilkins State Park, Copper Harbor, MI 49918. Mailing Address: P.O. Box 71, Copper Harbor, MI 49918-0071. Tel.: 906-289-4215. Fax: 906-289-4939.
E-mail: strittmatter@michigan.gov
Web Site: www.michigan.gov/historicfortwilkins
Founded: 1923.

Congressional District: 11
Key Personnel: Museum Shop Mgr., Barb Wachowski.
Personnel Profile: Full-Time Paid 3; Part-Time Paid 25; Part-Time Volunteers 1.
Governing Authority: state. Parent Institution: Michigan Dept. of History, Arts & Libraries. Subsidiary Institution: Michigan Historical Center. Tax-exempt.
Institution Type/Description: General Museum & Outdoor Museum Complex: including restored Fort Wilkins, the Pittsburgh & Boston Company mine site, Copper Harbor lighthouse.
Collections: military; copper mining; decorative arts c. 1844-1920. Historic Buildings: 1844 soldiers' quarters; mess hall; hospital; bakery; Sutler's store; powder magazine; 1849 lightkeeper's dwelling; 1866 lighthouse; 1868 range lighthouse.
Research Fields: mining, marine & early military of the Keweenaw Peninsula.
Facilities: 200-seat auditorium; classrooms. Books for sale.
Activities: permanent exhibits; living history; lectures; audiovisual interpretation.
Publications: booklet, The Fort Wilkins Story; Shipwrecks off Keweenaw; A Keweenaw Guide to Yesterday; Peas Upon a Trencher: A Study of Diet at Fort Wilkins; Lighting the Way: A History of the Copper Harbor Lighthouse; Fort Wilkins Yesterday and Today.
Hours & Admission Prices: mid-May to mid-Oct. daily 8-dusk. Recreation Passport: $10 annual fee for Michigan residents. Nonresident Passport: $8 daily, $29 annual. ♿
Attendance: 150,000 (estimated)

Crystal Falls

HARBOUR HOUSE MUSEUM, 17 N. 4th St., Crystal Falls, MI 49920-1205. Mailing Address: P.O. Box 65, Crystal Falls, MI 49920-0065. Tel.: 906-875-4341.
Web Site: www.harbourhousemuseum.org
Key Personnel: Pres. (V), Sally Westfahl; Cur., Gloria Frederickson
Governing Authority: Parent Institution: City of Crystal Falls. Tax-exempt.
Institution Type/Description: History Museum.
Collections: photographs; period clothing & memorabilia; school trophies; personal artifacts; family heirlooms; children's period toys; a miniature carousel; posters; local newspapers dating back to 1887; Ojibwe artifacts; Civil War to Desert Storm.
Research Fields: Diamond Drill; newspapers from 1887.
Activities: guided tours; special events.
Hours & Admission Prices: June-Sept. 4 Tues.-Sat. 10-2. Family $5, adults $2.
Attendance: 350 (estimated)
Membership: Individual $10.

Davison

THE ART CAFE', 217 Shoppers Alley, Davison, MI 48423-1424. Tel.: 248-210-0862.
E-mail: staff@artcafeonline.org
Web Site: www.artcafeonline.org
Founded: 2003.
Key Personnel: C.E.O., Cora Smilkovich.
Governing Authority: nonprofit organization. Tax-exempt: 501(c)(3).
Institution Type/Description: Art Gallery.
Collections: works by local artists; paintings; sculpture; prints; ceramics; watercolors.
Activities: workshops; Word Theatre workshops; art classes; temporary exhibits; special events.
Hours & Admission Prices: Check website for hours. No charge; donations accepted. ♿
Membership: Individual $10.

Dearborn

✳ **ARAB AMERICAN NATIONAL MUSEUM,** 13624 Michigan Ave., Dearborn, MI 48126-3519. Tel.: 313-582-2266. Fax: 313-582-1086.
E-mail: aanm@accesscommunity.org
Web Site: www.arabamericanmuseum.org
Founded: 2005.
Key Personnel: C.E.O. & Dir., Dr. Anan Ameri; Museum Shop Mgr., Janet Elias.
Governing Authority: Parent Institution: ACCESS. Tax-exempt.
Institution Type/Description: History Museum.
Collections: Arab culture; Arab contributions to science, medicine, mathematics & astronomy; Arab architecture & decorative arts; Arab American immigration from 1500 to present; Arab Americans in the U.S.
Activities: school tours; performance series; film festival; cultural competency workshops.
Publications: invisible exhibition catalog; Arab American Encyclopedia.
Hours & Admission Prices: Wed.-Sat. 10-6, Sun. 12-5. Adults $6, students, seniors 62 & over and children 6-12 $3; children 5 & under no charge. Closed New Year's Day; Thanksgiving; Christmas. ♿
Attendance: 50,000
Membership: Student & Senior $25; Individual $35; Family $65.

AUTOMOTIVE HALL OF FAME, INC., 21400 Oakwood Blvd., Dearborn, MI 48124-4078. Tel.: 313-240-4000. Fax: 313-240-8641.
Web Site: www.automotivehalloffame.org
Founded: 1939.
Key Personnel: Pres., Jeffrey K. Leestma; Chm., Jason Vines; Sec., Neil DeKoker; Treas., Ronald J. Martoia.
Personnel Profile: Full-Time Paid 6; Part-Time Paid 6; Part-Time Volunteers 20.
Governing Authority: nonprofit. Tax-exempt: 501(c)(3).
Institution Type/Description: Automotive Museum & Library.
Collections: histories of people, company & corporations.
Research Fields: automotive history.
Facilities: 4,000-vol. library.
Activities: guided tours; films.
Publications: quarterly, The Driving Spirit.
Hours & Admission Prices: Wed.-Sun. 9-5. Adults $6, seniors $5, children 5-18 $3. Closed national holidays. ♿
Attendance: 30,000 (accurate)
Membership: Friend $50; Innovator $100; Driving Spirit $500; Visionary $1,000; Creative Spirit $2,500; Prime Mover $5,000.

DEARBORN HISTORICAL MUSEUM, (M), 915 S. Brady Rd., Dearborn, MI 48124-2322. Tel.: 313-565-3000. Fax: 313-565-4848.
E-mail: jtate@ci.dearborn.mi.us
Web Site: www.cityofdearborn.org
Founded: 1950.
Congressional District: 15
Key Personnel: Chief Cur., Kirt Gross; Archives Specialist, Helen Mamalakis; Registrar, Robin Anderko.
Personnel Profile: Full-Time Paid 2; Part-Time Paid 5.
Governing Authority: municipal. Affiliated with the Dearborn Historical Commission, Div. of the City of Dearborn. Tax-exempt: 501(c)(3).
Institution Type/Description: Local History Museum: housed in 1839 Powder Magazine, now McFadden-Ross House; 1833 Commandant's Quarters; 1831 Gardner House.
Collections: local history artifacts; agriculture; costumes; military; textiles; archives; decorative arts; transportation; industry; manuscripts.
Research Fields: local history; agriculture; costumes; military; archives; Henry Dearborn; Orville L. Hubbard; Alex Pilch; John B. O'Reilly; Lucille McCollough.
Facilities: 2,000-vol. library of local, county, regional & state history; 1942-1978 mayoral papers of Orville L. Hubbard; censuses; 1905-present local newspapers; reading room. Historic reproductions for sale.
Activities: guided tours; lectures; films; study clubs; workshops; formally organized education programs for children & adults; docent program; inter-museum loan, permanent, temporary & traveling exhibitions; school loan service; Museum Guild.
Publications: book, The Bark Covered House; quarterly magazine, The Dearborn Historian; brochures.
Hours & Admission Prices: Commandant's Quarters: Tues.-Fri. 11-4, Sat. 9-1. McFadden-Ross House: Tues.-Fri. 9-5. Archives: Tues.-Fri. 9-4, Sat. by appointment. No charge; donations accepted. Closed holidays.
Attendance: 10,000 (estimated)
Membership: Individual $25; Family $35; Curator's Circle $50; Commandant's Assembly $125; General Henry Dearborn Club $250; Mayor's Circle $500.

✳ **THE HENRY FORD, (M),** 20900 Oakwood Blvd., Dearborn, MI 48124-4088. Tel.: 313-982-6100. Fax: 313-982-6250. TDD: 313-271-2455.
E-mail: elainem@thehenryford.org
Web Site: www.thehenryford.org
Formerly: Henry Ford Museum & Greenfield Village
Founded: 1929.
Congressional District: 16
Key Personnel: Pres., Patricia E. Mooradian; Chm. (V), S. Evan Weiner; Sr. Vice Pres. Operations & C.F.O., Denise Thal; Chief Information Officer, Michael Butman; Sr. Dir. Facilities Mgmt., Robert Hanna; Dir. Human Resources & Security, James Rankine; Dir. Workforce Devel., James Van

Bochove; Vice Pres. Collections & Experience Design, Christian W. Overland; Dir., Media & Film Rels., Wendy Metros; Vice Pres. Institutional Advancement, J. Spencer Medford; Sr. Dir. External Rels., George Moroz; Dir. Food Service & Catering, Jesse Eisenhuth; Sr. Dir. Natl. Retail Sales, Terri Anderson; Dir. Visitor Svcs., AmyLouise Bartlett; Dir. Historical Resources & Education, Paula Gangopadhyay; Dir. Greenfield Village & Henry Ford Museum, John Neilson; Chief Mktg. Officer, Carol Kendra; Dir. Historical Resources, Marilyn Zoidis.

Personnel Profile: Full-Time Paid 299; Part-Time Paid 1,084; Part-Time Volunteers 662; Interns 17.

Governing Authority: nonprofit organization. Tax-exempt: 501(c)(3).

Institution Type/Description: History Museum.

Collections: 1,000,000 artifacts & collections of artifacts; over 26 million printed & visual materials; 9,000 linear ft. of archival holdings; 80 historic structures.

Research Fields: agriculture; leisure & entertainment; industry; domestic life; transportation; communication.

Facilities: research library; 80 acre village; classrooms; food service; IMAX theater; research center. Museum-related items for sale.

Activities: concerts; study clubs; craft workshops; formally organized education programs; inter-museum loans, permanent, temporary & traveling exhibitions; Ford Rouge factory tours; summer & holiday evening programs. Museum Sponsors: Smart Fun field trips; summer festival.

Publications: on subjects related to museum collections & fields of research interest.

Hours & Admission Prices: Henry Ford Museum: daily 9:30-5. Greenfield Village: April 15-Oct. daily 9:30-5; Nov. Fri.-Sun. 9:30-5. Henry Ford Museum: adults $17, senior citizens 62 & over $15, children 5-12 $12.50; members & children under 5 no charge. Greenfield Village: adults $24, senior citizens 62 & over $22; children 5-12 $17.50; members & children under 5 no charge. Closed Thanksgiving; Christmas. &

Attendance: 1,600,000 (accurate)

Membership: Student $40; Individual $45; Companion $85; Family $145; Companion Flex $145; Family Flex $205.

HENRY FORD ESTATE-FAIR LANE, 4901 Evergreen Rd., Dearborn, MI 48128-2406. Tel.: 313-593-5590. Fax: 313-593-5243.

E-mail: grodgers@umd.umich.edu

Web Site: www.henryfordestate.org

Founded: 1957.

Key Personnel: Dir., Gary Rodgers; Group Tour Coord., Bernadette Trisko; Visitor Svcs. & Gen. Tours, Frank Gasiorek.

Personnel Profile: Full-Time Paid 12; Part-Time Paid 5; Part-Time Volunteers 160; Interns 2.

Governing Authority: public university; nonprofit. Parent Institution: University of Michigan. Subsidiary Institution: University of Michigan, Dearborn. Tax-exempt.

Institution Type/Description: Nature Center & Historic House: former home of Henry Ford.

Collections: decorative arts; mechanical equipment; books; landscapes.

Facilities: 1,500-vol. library of books owned by Ford family; cafe. Museum-related items for sale.

Activities: guided tours; lectures; films; organized education program for children. Museum Sponsors: Christmas programs.

Publications: biannual newsletter, The Fair Lane Advisor.

Hours & Admission Prices: Tours: Jan.-March Tues.-Sun. 1:30; April-Dec. Tues.-Sun. 10:30, 11:30, 12:30, 1:30, & 2:30. Adults 13-61 $12, senior citizens 62 & over and students $11, children 6-12 $8; children 5 & under no charge. Closed New Year's Day; Easter; Thanksgiving; Christmas. &

Attendance: 150,000 (estimated)

PADZIESKI ART GALLERY, 15801 Michigan Ave., Dearborn, MI 48126-2904. Tel.: 313-943-3095.

Institution Type/Description: Art Gallery.

Collections: works by local & regional artists.

Hours & Admission Prices: Tues.-Fri. 12-6, Sat. 12-5:30. No charge.

Decatur

HISTORIC NEWTON HOME, 20689 Marcellus Hwy., Decatur, MI 49045-9455. Mailing Address: 14916 Dutch Settlement, Marcellus, MI 49067. Tel.: 269-782-2008.

Founded: 1974.

Congressional District: 4

Key Personnel: Chm. Cass County Historical Commission, Abigail Schten; Sec., Marjorie Federowski.

Governing Authority: Parent Institution: Cass County Historical Commission.

Institution Type/Description: Historic House Museum: 1867 Newton Home.

Collections: Victorian furniture; one room school.

Activities: guided tours.

Publications: various publications available through the Cass County, MI, Historical Commission.

Hours & Admission Prices: Home: mid-March to mid-Nov. Sun. 1-4:30; tours by appointment all year. School: by appointment. No charge; donations accepted.

Attendance: 110 (estimated)

Deckerville

DECKERVILLE HISTORICAL MUSEUM, 2485 Black River St., Deckerville, MI 48427-9355. Mailing Address: 4028 N. Ruth Rd., Deckerville, MI 48427-9355. Tel.: 810-376-6695.

Web Site: www.deckervillelibrary.com/deckerville_historical_museum.htm

Founded: 1987.

Key Personnel: Dir. & Museum Shop Mgr., Rev. Joyce Reid; Chm. (V), Carol Walton; Pres. (V), Elaine Phillips

Governing Authority: Tax-exempt.

Institution Type/Description: History Museum.

Collections: area history; period dishes & furnishings; portraits of local families & businesses; Native American artifacts; farm items; books; scrapbooks; Deckerville Recorder newspaper from 1900-1970.

Publications: book, Thumb Area Local Indian History.

Hours & Admission Prices: June-Sept. Sat. 10-2 by appointment. Adults $3, children 6-12 $1; members & children under 5 no charge.

Attendance: 300 (estimated)

Delton

BERNARD HISTORICAL SOCIETY AND MUSEUM, 7135 Delton Rd., Delton, MI 49046. Mailing Address: P.O. Box 307, Delton, MI 49046-0307. Tel.: 269-623-3565.

Web Site: www. bernardmuseum.org

Founded: 1962.

Congressional District: 3

Key Personnel: C.E.O. & Pres. (V), Margery Martin.

Governing Authority: nonprofit organization. Tax-exempt: 501(c)(3).

Institution Type/Description: Local History Museum.

Collections: pioneer family artifacts; farm equipment; manuscripts. Historic Structures: 1873 one-room schoolhouse; 1933 old brick hospital; country store; blacksmith shop; 1865 Seamstress cottage.

Research Fields: history of one-room schools, many local family histories.

Facilities: 800-vol. library of history books, manuscripts & local documents available on premises. Museum-related items & books for sale.

Activities: guided tours; lectures; permanent exhibitions.

Publications: book, 1968, Years Gone By; booklet, The First Twenty Years; quarterly newsletter.

Hours & Admission Prices: Sept. Sun. 1-5. No charge; donations accepted. &

Attendance: 2,000 (estimated)

Membership: Annual $10; Life $100.

Detroit

ANNA SCRIPPS WHITCOMB CONSERVATORY, Belle Isle Park, Detroit, MI 48207-4345. Tel.: 313-822-2867. Fax: 313-821-5793.

E-mail: thiedep@cadtwr.ci.detroit.mi.us

Web Site: www.bibsociety.org

Founded: 1904.

Congressional District: 13

Key Personnel: Park Mgr., Keith Flournoy; Chm. (V), Janice Ellison; Pres. (V), Joseph Stanton; Floriculture Supvr., Leticia Bertaud.

Personnel Profile: Full-Time Paid 4; Part-Time Volunteers 80.

Governing Authority: municipal. Parent Institution: Floriculture Unit of the Belle Isle Division of the Detroit Department of Recreation, 65 Cadillac Sq., Suite 4000, Detroit, MI 48226. Tax-exempt.

Institution Type/Description: Botanical Garden & Conservatory.

Collections: tropicals; fern; palms, cactus; orchid display.

Facilities: conservation center; gardens.

Activities: guided tours; lectures; permanent exhibitions. Annual Event: Plant Sale in May.

Hours & Admission Prices: Daily 10-5. No charge; donations accepted. &

Attendance: 68,700 (estimated)

Membership: Annual: Senior (over 60) & Junior (under 15) $10; Individual $25; Family $35. Club & Individual: Sustaining $50; Donor $100; Benefactor $500. Business & Corporate: Sustaining $250; Donor $1,000; Benefactor $5,000.

CENTER GALLERIES, COLLEGE FOR CREATIVE STUDIES, 301 Frederick Douglas Ave., Detroit, MI 48202-4024. Tel.: 313-664-7800. Fax: 313-664-7880.
E-mail: mperron@collegeforcreativestudies.edu
Web Site: www.collegeforcreativestudies.edu/center_galleries
Founded: 1989.
Congressional District: 11
Key Personnel: Dir., Michelle M. Perron.
Personnel Profile: Full-Time Paid 1; Part-Time Paid 1; Interns 3.
Governing Authority: college; nonprofit organization. Parent Institution: College for Creative Studies-College of Art & Design, 201 E. Kirby, Detroit 48202. Tax-exempt: 501(c)(3).
Institution Type/Description: Art Gallery.
Collections: works by CCS student, alumni, & faculty.
Major Exhibits: 101UPX11: The Circle Paintings of Mark Sengbusch, 3/14.
Facilities: 3,000 sq. ft. gallery space.
Activities: guided tours; lectures; loan & traveling exhibitions.
Publications: brochures on exhibiting artists.
Hours & Admission Prices: Sept.-July Tues.-Sat. 10-5. No charge. Closed Independence Day; Thanksgiving; Christmas. &
Attendance: 12,000 (accurate)
Membership: General $25; Premier $100; Sustaining $1,500.

CHARLES H. WRIGHT MUSEUM OF AFRICAN AMERICAN HISTORY, 315 E. Warren Ave., Detroit, MI 48201-1443. Tel.: 313-494-5800. Fax: 313-494-5855.
Web Site: www.chwmuseum.org
Founded: 1965.
Congressional District: 16
Key Personnel: Pres., Juanita Moore; Chm. (V), Elizabeth Brooks; Treas., Edward J. Hannan; Chief Operating Officer, Tyrone Davenport; Chief Financial Officer, Ollette Boyd; Dir. Research & Exhibitions, Robert Smith; Chief Cur., Patrina Chatman.
Personnel Profile: Full-Time Paid 40; Part-Time Paid 14; Part-Time Volunteers 100.
Governing Authority: nonprofit organization. Tax-exempt: 501(c)(3).
Institution Type/Description: History Museum.
Collections: African & African American art, artifacts, photographs and other items pertaining to African & African American culture.
Research Fields: the provenance of the objects & photographic collections; preservation & conservation of the collections.
Facilities: 1,500-vol. library of monographs, manuscripts, oral histories, periodicals on the African American experience, the Underground Railroad & Civil Rights, with limited public access; 3 classrooms; 2 multi-purpose rooms; orientation amphitheater; 317-seat theater; video room. Museum-related & imported African items for sale.
Activities: African World Festival; variety of performing arts and concerts; docent program; films; workshops and lecture series programs for children, adults affiliated with Wayne State Univ.; guided tours; hobby workshops; lectures; loan, temporary & traveling exhibitions; theatre; training programs for professional museum workers. Museum Sponsors: African World Festival; Juneteenth Celebration; Kwanzaa Celebration and other annual programs.
Publications: quarterly newsletter, African American News; quarterly calendar, MAAH Calendar of Events.
Hours & Admission Prices: Tues.-Sat. 9-5, Sun. 1-5. Adults $8, children $5; discounts to AAM & ICOM members; members no charge. Closed major holidays. &
Attendance: 150,000 (estimated)
Membership: Student $5; Senior Citizen $15; Individual $35; Family $65; Contributor $150.

CURTIS MUSEUM, 14034 W. McNichols, Detroit, MI 48235. Mailing Address: 14022 W. McNichols, Detroit, MI 48235. Tel.: 313-341-1512. Fax: 313-341-1571.
E-mail: curtismuseum@sbcglobal.net
Web Site: www.curtismuseum.com
Institution Type/Description: History Museum.
Collections: life & family history of Dr. Austin W. Curtis; historical documents; photographs; personal artifacts.
Hours & Admission Prices: Tues.-Sat. 10-4; tours by appointment. Adults $5, senior citizens & children $3.

DETROIT ARTISTS MARKET, 4719 Woodward Ave., Detroit, MI 48201-1307. Tel.: 313-832-8540. Facebook: Detroit Artists Market.
E-mail: info@detroitartistsmarket.org
Web Site: www.detroitartistsmarket.org
Founded: 1932.

Key Personnel: Chm. Bd. Directors, Darcel Deneau; Exec. Dir., Nancy Sizer; Exhibitions Mgr., Peter Gahan; Mgr. Programs, Dalia Reyes.
Personnel Profile: Full-Time Paid 3; Part-Time Volunteers 7; Interns 1.
Governing Authority: nonprofit organization. Tax-exempt: 501(c)(3).
Institution Type/Description: Contemporary Art Gallery.
Collections: paintings; photographs; prints; drawings; sculpture; ceramics; glass; wood.
Major Exhibits: Three, 1/14-2/14; Scholarship Awards & Exhibition: Cranbrook, 3/14-4/14; All Media Exhibition, 4/14-5/14; 313 Detroit Birthday Show, 6/14-7/14; Built It - They Will Come, 8/14; DAM Photography, 9/14-10-14; Art for the Holidays, 11/14-12/14.
Facilities: 2,500 sq. ft. exhibit space.
Activities: films; formal education programs for adults & children; lectures; permanent & temporary exhibitions; gallery talks. Museum Sponsors: Annual Garden Sale.
Hours & Admission Prices: Tues.-Sat. 11-6. No charge; donations accepted. &
Attendance: 21,000 (estimated)
Membership: Student $20; Artist $30; Individual $40; Household $65; Associate $100; Patron $250; Devotee $500; Collectors' Circle $1,000; Directors' Circle $2,500.

DETROIT CHILDREN'S MUSEUM, 6134 Second Ave., Detroit, MI 48202-3404. Tel.: 313-873-8100.
E-mail: julie.johnsondcm@gmail.com
Web Site: detroitk12.org/childrens_museum
Formerly: Children's Museum/Detroit Public Schools
Founded: 1917.
Congressional District: 13
Key Personnel: Dir., Julie Johnson; Mgr. Collections, Don Bogart; All Subjects, David Lehner; All Subjects, Steve Mackenzie; Preschool Mgr., Meredith Gregory; Administrative Asst., Kemela Callahan.
Personnel Profile: Full-Time Paid 4; Part-Time Paid 1; Part-Time Volunteers 1; Interns 4.
Governing Authority: Parent Institution: Detroit Public School, 3011 W. Grand Blvd., Detroit, MI 48202.
Institution Type/Description: Children's Museum.
Collections: ethnology; natural history; science; folk arts; costumes; textiles; dolls; toys; musical instruments; tools & equipment.
Research Fields: ethnic studies.
Facilities: 2,000-vol. library of reference books available on premises; planetarium; classrooms.
Activities: class lessons; demonstrations; formally organized education programs for children; permanent exhibitions; ethnic studies programs; Museum-to-the-Schools program; teacher workshops; loan collections to schools.
Publications: museum services handbook; exhibit brochures; monthly calendar of activities; catalogs of lending collections; workshop resource sheets.
Hours & Admission Prices: Oct.-June Mon.-Fri. 9-4. Closed national holidays. &
Attendance: 43,000 (accurate)

DETROIT HISTORICAL SOCIETY, (M), 5401 Woodward Ave., Detroit, MI 48202-4097. Tel.: 313-833-1805. Fax: 313-833-5342.
Web Site: detroithistorical.org
Founded: 1921.
Congressional District: 15
Key Personnel: Exec. Dir. & C.E.O., Robert Bury; Pres. (V), Thomas C. Buhl, II; Mng. Dir., Kate Baker; Chief Curtorial Officer, Tobi Voigt; Dir. Mktg. & Sales, Robert Sadler.
Governing Authority: Subsidiary Institutions: Dossin Great Lakes Museum; Detroit Historical Museum. Tax-exempt.
Institution Type/Description: History Museum.
Collections: artifacts pertaining to Detroit, Michigan & the Great Lakes; including military & maritime history; costumes; decorative arts; toys; furniture; Woodlands Indian; industrial & automotive history; communications; domestic life; fine arts.
Research Fields: urban, social, industrial, architectural, military & marine history
Facilities: 1,500-vol. library of fashion reference books available for use by special request; auditorium; classrooms. Museum-related items for sale.
Activities: guided tours of exhibits & local historical sites; lectures; community neighborhood tours; history related workshops; education programs for children; inter-museum loan, permanent, temporary & traveling exhibitions; school tour service. Museum Sponsors: conferences; historic preservation; church & local history; Black Historic Sites committee.
Publications: newsletter; annual report; calendar of events.
Hours & Admission Prices: Detroit Historical Museum: Tues.-Fri. 9:30-4,

Sat.-Sun. 10-5. No charge; donations accepted. Dossin Great Lakes Museum: Sat.-Sun. 11-4. No charge; donations accepted. Closed national holidays. &
Attendance: 194,388 (accurate)
Membership: Researcher $25; Collector $45; Ambassador $65; Patron $150; Benefactor $300; Historian $500; Cobblestone $1,000.

✷ DETROIT INSTITUTE OF ARTS, (M), 5200 Woodward Ave., Detroit, MI 48202-4094. Tel.: 313-833-7900. Fax: 313-833-2357 & 3756. TDD: 313-833-1454.
E-mail: operator@dia.org
Web Site: www.dia.org
Founded: 1885.
Congressional District: 13
Key Personnel: Dir., Graham W.J. Beal; Exec. Vice Pres., C.O.O., Annmarie Erickson; Cur. American Art, Kenneth Myers.
Governing Authority: municipal. Parent Institution: the City of Detroit. Subsidiary Institution: Founders Society-DIA. Tax-exempt: 501(c)(3).
Institution Type/Description: Art Museum.
Collections: European, 20th-century, Ancient, Middle Eastern, Islamic, American, Asian, African, Oceanic, & Native American art; textiles; period rooms; Paul McPharlin Puppetry collection; graphic arts & photography; Diego Rivera Detroit Industry Frescoes; films; film posters.
Research Fields: all objects of art in permanent collection & prospective acquisition.
Facilities: 150,000-vol. non-circulating library of art books & art material available for inter-library loan & on premises; 1,200-seat auditorium; restaurant; cafeteria; 382-seat recital & lecture hall; classrooms; conservation lab. Museum-related items for sale.
Activities: guided tours; student tours; lectures; films; gallery talks; concerts; art & art appreciation classes; professional museum training program; permanent & temporary exhibitions; education programs; outreach programs; workshops; artist demonstrations; drawing in galleries; music. Museum Sponsors: Detroit Film Theatre.
Publications: exhibition & permanent collection catalogues; annual report; periodic bulletin; monthly DIA magazine.
Hours & Admission Prices: Tues.-Thurs. 9-4, Fri. 9-10, Sat.-Sun. 10-5. Suggested Donation: adults $8, children $5; DIA, AAM & ICOM members no charge. Closed holidays. &
Attendance: 266,527 (accurate)
Membership: Seniors $60; Individual $65; Companion $80; Family Plus $110; Affiliate $180; Patron $300; Contributor $600; Conservator $1,000; Supporting Associate $2,000; Sustaining Associate $3,000; Director's Associate $5,000; Chairman's Associate $10,000.

DETROIT REPERTORY THEATRE GALLERY, 13103 Woodrow Wilson, Detroit, MI 48238. Tel.: 313-868-1347. Fax: 313-259-8242.
E-mail: gsnow19543@aol.com
Web Site: www.detroitreptheatre.com
Key Personnel: Dir., Gilda Snowden
Institution Type/Description: Art Gallery.
Collections: works by Detroit-area artists; 2-D works.
Activities: temporary exhibitions.
Hours & Admission Prices: Thurs.-Sat. 8:30pm-11:30pm, Sun. 2-4 & 7:30pm-10:30.

DETROIT SCIENCE CENTER, 5020 John R. St., Detroit, MI 48202-4045. Tel.: 313-577-8400. Fax: 313-832-1623.
E-mail: info@sciencedetroit.org
Web Site: www.detroitsciencecenter.org
Founded: 1970.
Congressional District: 13
Key Personnel: Chm. Bd., Francois Castaing; Pres. & C.E.O., Kevin F. Prihod; C.O.O., Ida Tomlin; Dir. Public Rels. & Mktg., Kelly Fulford; C.F.O., Bob Seestadt; Dir. Facilities, Tom Mott; Vice Pres. Science Programs, Todd Slisher; Reservations, Lynell Moore; Dir. Exhibits, Ed Summers; Dir. Exhibits Content, Elizabeth Chilton.
Personnel Profile: Full-Time Paid 35; Part-Time Paid 40; Part-Time Volunteers 90.
Governing Authority: bd. of trustees; nonprofit organization. Tax-exempt: 501(c)(3), 509(a)(1).
Institution Type/Description: Science Museum.
Collections: hands-on exhibits.
Facilities: IMAX dome theater; planetarium; science stage.
Activities: large format films; science demonstrations; formally organized educational programs for children; computer educational programs; permanent & temporary exhibitions.
Publications: quarterly newsletter.

Hours & Admission Prices: Mon.-Fri. 9-3, Sat.-Sun. 10-6. Adults $7.95, senior citizens & children 2-12 $6.95. IMAX & Dassault Systemes Planetarium $5; discounts to ASTC members; members no charge. Closed New Year's Day; Memorial Day; Independence Day; Labor Day; Thanksgiving; Christmas Eve & Day. &
Attendance: 250,000 (accurate)
Membership: Student & Senior Citizen $25; Individual Plus $55; Grandparents $65; Family $70; Family Plus $90; Premium $170.

DOSSIN GREAT LAKES MUSEUM, 100 Strand Dr., Belle Isle, Detroit, MI 48207-4372. Tel.: 313-833-5538; 313-821-2661. Fax: 313-833-5342.
Web Site: www.detroithistorical.org
Founded: 1948.
Key Personnel: Exec. Dir., Robert Bury; Dir. Mktg. & Sales, Robert Sadler.
Governing Authority: municipal. A branch of the Detroit Historical Museum. Tax-exempt.
Institution Type/Description: Great Lakes History Museum.
Collections: ship models; paintings; period & modern nauticalia; photographs; information files of Great Lakes ships; restored interior from the former lake steamer City of Detroit III; pilothouse from the freighter William Clayford.
Research Fields: Great Lakes commerce.
Facilities: 700-vol. library of government & private records, data files, news clips, photos, negatives of Great Lakes ships & history, available for use on premises. Books, postcards & other museum-related items for sale.
Activities: guided tours; lectures; films; permanent & temporary exhibitions.
Publications: bimonthly magazine, Telescope.
Hours & Admission Prices: Sat.-Sun. 11-4. No charge; donations accepted. Closed major holidays.
Attendance: 70,000
Membership: Libraries/Schools/Associations $20; Regular $40; Benefactor $100; Life $350.

FIRST UNDERGROUND RAILROAD MUSEUM, 33 E. Forest Ave., Detroit, MI 48201-1813. Tel.: 313-831-4080.
Institution Type/Description: Historic Building: housed in the First Congregational Church of Detroit, used to hide slaves en route to freedom.
Collections: Underground Railroad history; photographs; personal artifacts.
Facilities: restaurant; rental facilities.
Activities: summer camp; electronic arts program.
Hours & Admission Prices: Tues.-Sat. 11-3. Adults $12, children $10; discounts to groups.

INTERNATIONAL GOSPEL MUSIC HALL OF FAME & MUSEUM, 18301 W. McNichols, Detroit, MI 48219-4112. Mailing Address: P.O. Box 19009, Detroit, MI 48219. Tel.: 313-444-8352.
E-mail: igmhfm@cs.com
Web Site: www.igmhf.org
Formerly: Gospel Music Hall of Fame & Museum
Founded: 1995.
Key Personnel: C.E.O., Chm. & Pres., David Gough; Vice Chm. & Dir. Devel., Ida Tomlin; Treas., Johnny Stewart; Sec., Carolyn Gough; Archivist & Chief Cur., Sherry Dupree; Program Mgr., Jean Anderson.
Personnel Profile: Full-Time Volunteers 2; Part-Time Volunteers 9.
Governing Authority: private; nonprofit organization. Tax-exempt: 501(c)(3).
Institution Type/Description: Gospel Music Museum.
Collections: archives; photographs; tapes; sound recordings; sheet music; songbooks; magazine clippings; artist biographies; films; musical instruments; clothing & memorabilia associated with the development of gospel music; sight & sound exhibits.
Research Fields: history on each inductee; provenance of the objects & photographic collections; preservation & conservation of the collections.
Facilities: 2,000-vol. library of reference gospel music; 1,200-seat auditorium; multi-purpose room; cafeteria. Museum-related items for sale.
Activities: formal education programs; guided tours; lectures; participatory, temporary & traveling exhibitions; broadcast programs; conferences. Museum Sponsors: Annual Induction Dinner; Black History Program.
Publications: quarterly newsletter, GMHF News Update; brochures; pamphlets.
Hours & Admission Prices: Thurs.-Sat. 10-3. Call for admission fee. Closed all holidays.
Attendance: 250 (estimated)
Membership: Association $50; Gold $100.

MICHIGAN SPORTS HALL OF FAME, Cobo Center, One Washington Blvd., Detroit, MI 48226. Mailing Address: P.O. Box 1073, Farmington, MI 48332-1073. Tel.: 248-473-0656. Fax: 248-473-0674. Facebook: Michigan Sports HOF.
E-mail: info@michigansportshof.org
Web Site: www.michigansportshof.org
Founded: 1955.
Congressional District: 19
Key Personnel: Exec. Dir., Jim Stark.
Personnel Profile: Full-Time Paid 1.
Governing Authority: state. Tax-exempt: 501(c)(3).
Institution Type/Description: Sports Museum.
Collections: bronze wall plaques & photos of each honoree.
Research Fields: history of each honoree.
Activities: annual induction dinner.
Publications: magazine, Annual Michigan Sports Hall of Fame Yearbook.
Hours & Admission Prices: Temporarily closed. &
Membership: Annual $50.

MOTORSPORTS HALL OF FAME OF AMERICA, Detroit Science Center, 5020 John R. St., Detroit, MI 48202. Mailing Address: P.O. Box 194, Novi, MI 48376-0194. Tel.: 248-349-7223. Fax: 313-832-1623.
E-mail: info@mshf.com
Web Site: www.mshf.com
Formerly: Motorsports Museum & Hall of Fame
Institution Type/Description: History Museum.
Collections: over 40 racing & high performance vehicles; photographs; racing memorabilia; 188 Hall of Fame inductees' lives & sculptures.
Hours & Admission Prices: Tues.-Fri. 9-3, Sat. 10-6, Sun. 12-6. Adults $13.95, seniors 60 & over and children 2-12 $11.95; children under 2 no charge. &

MOTOWN HISTORICAL MUSEUM, (M), 2648 W. Grand Blvd., Detroit, MI 48208-1237. Tel.: 313-875-2264. Fax: 313-875-2267.
E-mail: arawls@motownmuseum.com
Web Site: motownmuseum.org
Founded: 1985.
Key Personnel: Founder & Chm., Esther Gordy Edwards; C.E.O. & Exec. Dir., Allen C. Rawls; Chm. (V), Robin Terry.
Governing Authority: private; nonprofit organization. Tax-exempt: 501(c)(3).
Institution Type/Description: History Museum.
Collections: history of Motown Record Corp., recording artists, founder Berry Gordy & his family; archives; sheet music; recordings; photographs; drawings; prints; musical & recording equipment; costumes; personal memorabilia. Historic Buildings: Hitsville, U.S.A., the first headquarters of Motown Records; recording Studio A.
Research Fields: Motown Record Corp.; Berry Gordy & family; development of Motown's sound & its influence on popular culture; the company as it relates to Detroit's history & Motown's artists.
Facilities: 4,000 sq. ft. exhibit space. Museum-related items for sale.
Activities: docent programs; guided tours; participatory exhibits. Annual Event: Gala Fundraiser.
Hours & Admission Prices: Tues.-Sat. 10-6. Adults $10, senior citizens $8, children under 12 $8; discounts to family & group tours of 20 people. Closed New Year's Day; Easter; Memorial Day; Independence Day; Labor Day; Thanksgiving; Christmas Eve & Day. &
Attendance: 50,000 (accurate)

MUSEUM OF CONTEMPORARY ART DETROIT, 4454 Woodward Ave., Detroit, MI 48201-1822. Tel.: 313-832-6622. Fax: 313-832-4665.
E-mail: info@mocadetroit.org
Web Site: mocadetroit.org
Founded: 2006.
Key Personnel: Exec. Dir., Elysia Borowy-Reeder; Deputy Dir. & Special Projects, Rebecca Mazzei; Chair (V), Julie Reyes Taubman; Pres. (V), Marsha Miro; Exhibition Coord., Zeb Smith; Cur. Public Programs, Greg Baise; Operations & Support Asst., Lizzy Courtuis; Exhibitions Asst., Jonathan Rajewski, Education Asst., Kottie Gaydos; Store Assoc. & Production Asst., Ivan Gamboa; Administrative Asst., Andrea Friedley; Facility Rental Coord., Leto Rankine; Cur. Education, Katie McGowan; Dir. Operations, Marie Patton; Facilities Assoc., Chris Riddell; Graphic Designer, Dan DeMaggio.
Personnel Profile: Full-Time Paid 7; Part-Time Paid 6; Part-Time Volunteers 21; Interns 23.
Governing Authority: Tax-exempt.
Institution Type/Description: Contemporary Art Museum.
Collections: works by contemporary artists.

Major Exhibits: James Lee Byars: I Cancel My Works at Death, 2/14-5/14; Solo Series: Jose Lerma - Beautiful Crisis; Dara Friedman - Films; Steve Locke - There Is No One Left To Blame, 5/14-7/14; The People's Biennial, 9/14-1/15.
Facilities: cafe. Museum-related items for sale.
Activities: artist talks; lectures; panel discussions; performance-based art; films; readings; music events; children's art workshops.
Publications: special exhibition catalogs; annual arts & literature journal, DETROIT.
Hours & Admission Prices: Wed. & Sat.-Sun. 11-5, Thurs.-Fri. 11-8. Suggested Donation: $5. Closed New Year's Eve & Day; Martin Luther King Jr. Day; Memorial Day; Independence Day; Labor Day; Thanksgiving; Christmas Eve & Day. &
Attendance: 835,000 (estimated)
Membership: Student & Senior $25; Artist & Instructor $35; Individual $50; Dual $90; Family $100; Infant $250; Supporter $500; Contributor $1,000; Patron $2,500 & up.

THE N'NAMDI CENTER FOR CONTEMPORARY ART, 52 E. Forest Ave., Detroit, MI 48201. Tel.: 313-831-8700.
E-mail: nnamdicenter@gmail.com
Web Site: nnamdicenter.org
Key Personnel: Owner, George N'Namdi.
Governing Authority: nonprofit organization.
Institution Type/Description: Art Gallery.
Collections: works by contemporary artists.
Facilities: restaurant; theater. Gift items for sale.
Activities: temporary exhibitions; lectures; art invitationals; family events; performance art; Artist in Residence program; educational programs.
Hours & Admission Prices: Tues.-Sat. 12-6.

NATIONAL MUSEUM OF THE TUSKEGEE AIRMEN, Historic Fort Wayne, 6325 W. Jefferson Ave., Detroit, MI 48209. Mailing Address: P.O. Box 32549, Detroit, MI 48232-0549. Tel.: 313-843-8849. Fax: 313-843-1540.
E-mail: tamuseum@tuskegeeairmenmuseum.org
Web Site: tuskegeeairmenmuseum.com
Institution Type/Description: Military Museum.
Collections: model aircraft; military equipment, supplies & uniforms; photographs; Blacks in aviation during WWII.
Activities: video presentation; guided tours.
Hours & Admission Prices: April-Sept. Tues.-Sun. 10-3; other times by appointment. No charge; donations accepted.

PEWABIC POTTERY, 10125 E. Jefferson, Detroit, MI 48214-3138. Tel.: 313-626-2000. Fax: 313-626-2100.
E-mail: info@pewabic.org
Web Site: www.pewabic.org
Founded: 1903.
Congressional District: 13
Key Personnel: Exec. Dir., Ed Barbsido; Pres. (V), Mary Wisgerhof; Dir. Design & Retail, Christina Devlin.
Personnel Profile: Full-Time Paid 35; Part-Time Paid 10; Part-Time Volunteers 6; Interns 7.
Governing Authority: nonprofit organization. Tax-exempt: 501(c)(3).
Institution Type/Description: Arts & Crafts Museum & Pottery Factory: housed in 1907 William Buck Stratton-designed English style structure; built 1906-1907. Pottery founded in 1903 by Mary Chase Perry & Horace J. Caulkins. National Historic Landmark.
Collections: archival material-papers; blueprints; drawings; photographs; ceramics & tile collections of Mary Chase Stratton; art pottery; original arts & crafts furniture; contemporary ceramic art.
Major Exhibits: Joe Zajac, 3/13-4/28.
Research Fields: historic & contemporary ceramics, architectural tile, arts & crafts movements.
Activities: guided tours; organized education programs for children and adults; participatory, loan, temporary & traveling exhibitions.
Publications: Pewabic Pottery Newsletter.
Hours & Admission Prices: Mon.-Sat. 10-6, Sun. 12-4. Museum: no charge; donations accepted. Tours: adults $5. Workshops: adults $17. Closed New Year's Day; Martin Luther King Jr. Day; Easter; Memorial Day; Independence Day; Labor Day; Thanksgiving; Christmas.
Attendance: 40,000 (estimated)
Membership: Pewabic Society Inc.: Individual $40; Family $60; Supporter $100; Patron $250; Benefactor $500; Perry $1,000; Founder's $2,500; Caulkins $5,000.

WAYNE STATE UNIVERSITY ART GALLERIES, 150 Art Bldg., Dept. of Art, Detroit, MI 48202-3917. Tel.: 313-577-2980. Fax: 313-577-3491.
E-mail: art@wayne.edu
Web Site: www.art.wayne.edu
Founded: 1958.
Congressional District: 13
Key Personnel: Cur. & Dir. Art Exhibitions, Lisa Gonzalez.
Personnel Profile: Full-Time Paid 1; Part-Time Paid 4.
Governing Authority: state; university. Parent Institution: Wayne State University. Subsidiary Institution: Dept. of Art & Art History. Tax-exempt.
Institution Type/Description: Art Gallery.
Collections: paintings; photographs; sculpture.
Facilities: auditorium; conference center.
Activities: monthly exhibitions; concerts; lectures; formally organized education programs for adults, students and the community.
Publications: annual calendar of events.
Hours & Admission Prices: Academic Year: Tues.-Thurs. 10-6, Fri. 10-7; Summer: Tues.-Fri. 12-5. No charge. &
Attendance: 15,000 (estimated)
Membership: Senior & Student $15; Gallery Friend $30; Family $40; Contemporary Collector $150; Patron's Circle $500.

WAYNE STATE UNIVERSITY MUSEUM OF ANTHROPOL-OGY, 4841 Cass Ave., Detroit, MI 48201-1203. Mailing Address: Dept. of Anthropology, Wayne State University, 3054 Faculty Administration Bldg., 656 W. Kirby, Detroit, MI 48202. Tel.: 313-577-2598. Fax: 313-577-9759.
E-mail: t.bray@wayne.edu
Web Site: www.clas.wayne.edu/anthromuseum
Founded: 1958.
Congressional District: 13
Key Personnel: Dir., Tamara Bray, Ph.D.
Personnel Profile: Full-Time Paid 1; Part-Time Paid 3; Part-Time Volunteers 5.
Governing Authority: university. Affiliated with Wayne State University. Tax-exempt.
Institution Type/Description: Anthropology Museum.
Collections: local archaeology (historic & prehistoric); ethnographic.
Research Fields: archaeology; local anthropology; history
Facilities: 800-vol. library of books available for use by students; archaeology lab; manuscript archives; teaching lab.
Activities: formally organized education programs for undergraduate & graduate college students.
Hours & Admission Prices: Sept.-May Mon.-Thurs. 10-4, Fri. 10-2. No charge; donations accepted. &
Attendance: 1,200 (accurate)
Membership: Student $15; Senior Citizen $20; Individual $25; Affiliate $50; Patron $100; Benefactor $250.

THE WAYNE STATE UNIVERSITY MUSEUM OF NATURAL HISTORY, Biological Sciences Bldg., Rm. 1155, 5047 Gullen Mall, Detroit, MI 48202-3917. Tel.: 313-577-2872. Fax: 313-577-6891.
E-mail: wmoore@biology.biosci.wayne.edu
Web Site: bio.wayne.edu/outreach/natural_history.html
Founded: 1972.
Congressional District: 13
Key Personnel: Dir., William S. Moore; Cur. Vertebrates, Jesheskel Shoshani; Cur. Insects, Stanley K. Gangwere; Cur. Herbarium, D. Carl Freeman.
Personnel Profile: Part-Time Volunteers 3; Interns 2.
Governing Authority: university. Wayne State University. Parent Institution: State of Michigan. Tax-exempt.
Institution Type/Description: Natural History Museum.
Collections: plants; birds; mammals; insects; vertebrates & invertebrates.
Activities: guided tours; formally organized education programs for elementary & high school students; permanent exhibitions.
Hours & Admission Prices: By appointment. No charge. Closed holidays. &

Dexter

DEXTER AREA HISTORICAL SOCIETY & MUSEUM, 3443 Inverness, Dexter, MI 48130-1409. Tel.: 734-426-2519.
E-mail: dexmuseum@aol.com
Web Site: www.dextermuseum.org
Founded: 1976.
Key Personnel: Pres. Historical Society, Bene Fusilier; Museum Dir., Carol Jones; Genealogist, Nancy Van Blaricum; Museum Shop Mgr., Alice Pastalan.
Personnel Profile: Part-Time Voluntccrs 28.

Volunteer Hours: 3,072
Governing Authority: society; nonprofit. Tax-exempt: 501(c)(3).
Institution Type/Description: Historical Museum: housed in 1883 German Lutheran Church.
Collections: local history items; Civil War items; dolls; costumes; photographs; genealogy records; local business, organization & family archives.
Facilities: Museum-related items & folk art for sale.
Activities: guided tours; lectures; loan exhibitions; school tour service.
Publications: bimonthly newsletter.
Hours & Admission Prices: May-Dec. Fri.-Sat. 1-3. No charge; donations accepted.
Attendance: 2,500 (estimated)
Membership: Student $1.50; Individual $10; Family $15; Business $25; Patron $50 & up; Life $150.

Dowagiac

HEDDON MUSEUM, 414 West St., Dowagiac, MI 49047-1045. Mailing Address: 204 W. Telegraph St., Dowagiac, MI 49047-1241. Tel.: 269-782-5698.
E-mail: heddonmuseum@lyonsindustries.com
Web Site: heddonmuseum.org
Formerly: National Heddon Museum
Founded: 1996.
Congressional District: 4
Key Personnel: Dir. & Museum Shop Mgr., Joan Lyons; Dir., Don Lyons.
Personnel Profile: Part-Time Volunteers 3.
Governing Authority: private; nonprofit organization.
Institution Type/Description: History Museum: housed in James Heddon's Sons Fishing Tackle Co.
Collections: more than 1,800 lures; more than 160 reels; more than 200 rods; golf clubs; violin bows; ski poles; box kites; WWII military box kites and antennas, information & models of The Heddon aviation airplane.
Research Fields: James Heddon's Sons Fishing Tackle Co.
Facilities: library; 3,000 sq. ft. exhibit space. Museum-related items for sale.
Activities: guided tours.
Hours & Admission Prices: Tues. 6:30 p.m.-8:30 p.m., last Sun. of month 1:30-4; other times by appointment. No charge; donations accepted. Closed New Year's Eve; Thanksgiving Day; Christmas Eve & Day. &
Attendance: 500 (estimated)

MUSEUM AT SOUTHWESTERN MICHIGAN COLLEGE, (M), 58900 Cherry Grove Rd., Dowagiac, MI 49047-9726. Tel.: 269-782-1374 & 1000. Fax: 269-782-1460.
E-mail: museum@swmich.edu
Web Site: www.swmich.edu/museum
Formerly: Southwestern Michigan College Museum
Founded: 1982.
Congressional District: 6
Key Personnel: Dir., Steve Arseneau; Chm. (V), Chuck Timmons; Exhibit Designer, Tom J. Caskey; Educator, Jennifer Quail; Volunteer Coord., Jo Silvia.
Personnel Profile: Full-Time Paid 1; Part-Time Paid 3; Part-Time Volunteers 40.
Governing Authority: nonprofit; public university. Parent Institution: Southwestern Michigan College. Tax-exempt.
Institution Type/Description: History Museum & Science Center.
Collections: local history items; Round Oak Stove Company; Heddon fishing lures; local family items; local industry; Dowagiac Manufacturing Co.; Rudy Furnace Co.
Research Fields: local industry.
Facilities: 250-vol. library of state & local history material available to the public; educational facilities.
Activities: guided tours; organized education programs for children; loan & temporary exhibitions.
Publications: newsletter; Identification & Dating of Round Oak Heating Stoves.
Hours & Admission Prices: Tues.-Fri. 10-5, Sat. 11-3. No charge. Closed college & major holidays. &
Attendance: 7,000 (accurate)
Membership: Senior Individual $10; Senior Couple $15; Individual $20; Family $35.

Drummond Island

DRUMMOND ISLAND HISTORICAL MUSEUM, 33492 S. Water St., Drummond Island, MI 49726. Mailing Address: Box 293, Drummond Island, MI 49726-0293. Tel.: 906-493-5245.
E-mail: drhistmuseum@alphacomm.net
Founded: 1961.

Congressional District: 107
Key Personnel: Pres., Robert Newell; Township Supvr., Frank Sasso; Treas., K. C. Lowe; Sec., Martha Carlin; Dir., Harry Ropp; Dir., Gerry Bailey; Dir., Betty Bailey; Dir., Shelby Gibbons; Dir., Joan Riordan; Dir., Ann Statler; Dir., Clayton Ledy; Cur., Audrey Moser.
Personnel Profile: Full-Time Paid 1.
Governing Authority: society. Parent Institution: Drummond Island Historical Society.
Institution Type/Description: Local History Museum.
Collections: archives; geology; Indian artifacts; manuscripts; articles, books & pictures depicting Island history; specimen glass cases of fossils from prehistoric digs; collection of pre-glacial and other rock formations, including the Pudding Stone; Indian artifacts: basketry, beadwork and prayer books written in Chippewa-Ojibway; restored chimney & fireplace built by British Garrison 1815-1829; swords; pottery & shards; cannon balls; memorabilia of Finnish Colony, 1907: Finnish bibles & other books, costumes, looms, spinning wheels, skis; lumbering tools; sawmill town pictures; mill wheels; farming tools; household items; papier-mache full-size horse & buggy; early medical books; medical & dental instruments; musical instruments; marine corner; ships wheel; old wooden boats; compass; sailors sail-mending kit; sportsmen's gear; guns; fishing rods; decoys; early snowmobile engine; marine engines; snowshoes; home-made utility sleds; Finnish kick sled; antique bicycle; family portrait; period quilts with family names; manuscripts; scrapbooks; family albums; tourist lodge register; town records; toys; oral history tapes; videos; log cabin.
Facilities: library.
Activities: permanent exhibitions. Museum Sponsors: special summer programs for local residents & summer visitors; annual home tours in August.
Hours & Admission Prices: Memorial Day to mid-Oct. daily 1-5. No charge; donations accepted. &
Attendance: 3,000
Membership: Single $10; Family $15; Life $100.

Dryden

DRYDEN HISTORICAL SOCIETY MUSEUM, 5488 Dryden Rd., Dryden, MI 48428. Mailing Address: P.O. Box 93, Dryden, MI 48428.
E-mail: drydenhistoricalsociety@gmail.com
Web Site: drydenhistoricalsociety.webs.com
Founded: 1979.
Key Personnel: Chm. (V), Jeanine Risch; Pres. (V), Karen Broecker
Governing Authority: Tax-exempt.
Institution Type/Description: Historical Society Museum.
Collections: local history & culture; period furnishings; personal artifacts; photographs; plat maps; area tax logs. Historic Building: train depot.
Activities: Museum Sponsors: Plant Sale; Quilt Raffle; Christmas Cookie Sale.
Publications: books, Dryden History; Local Area Barns.
Hours & Admission Prices: Mon. 5:30pm-7pm. No charge; donations accepted.
Attendance: 300 (estimated)
Membership: Individual $10; Family $20; Business $50; Lifetime $100.

Dundee

OLD MILL MUSEUM, 242 Toledo St., Dundee, MI 48131-1246. Tel.: 734-529-8596.
E-mail: museum@dundeeoldmill.com
Web Site: www.dundeeoldmill.com
Founded: 1981.
Congressional District: 15
Key Personnel: Pres., Sara Alexin; Vice Pres., Mary Schultz; Sec., Meg Heinlen; Treas., Shirley Massingill; Archivist, Randi Kominek; Gift Shop Mgr. & Newsletter Editor, Grace Hudson.
Personnel Profile: Part-Time Paid 1.
Governing Authority: Tax-exempt.
Institution Type/Description: History Museum.
Collections: local history & culture; period artifacts; photographs; early furnishings.
Facilities: banquet facilities.
Activities: rental facilities; paranormal investigations; traveling exhibitions. Museum Sponsors: German Festival in March; Easter Brunch & Hunt in March; May Fly Music Festival in June; Kid's Halloween Party in October; Lunch with Mrs. Claus in December.
Publications: pamphlets.
Hours & Admission Prices: Fri.-Mon. 12-4. No charge; donations accepted. Closed major holidays. &
Attendance: 2,000 (estimated)
Membership: Seniors 55 & over $5; Individual $10; Family $20.

East Jordan

EAST JORDAN PORTSIDE ART & HISTORICAL MUSEUM, 01656 S-M66 Hwy., East Jordan, MI 49727. Mailing Address: P.O. Box 1355, East Jordan, MI 49727-1355.
Web Site: www.portsideartsfair.org/histsoc.htm
Founded: 1976.
Congressional District: 11
Key Personnel: C.E.O. & Pres. (V), Paula Vollbach; C.E.O. & Chm. (V), Kim Prebble.
Personnel Profile: Part-Time Volunteers 36.
Governing Authority: nonprofit organization. Parent Institution: East Jordan Historical Museum Society. Tax-exempt: 501(c)(3).
Institution Type/Description: Historical Society Museum.
Collections: pioneer artifacts & photographs; early lumbering & agricultural tools; turn-of-the-century clothing; opera house memorabilia; early railroad items & off-site engine; Purchase Prize collection of contemporary art.
Research Fields: family genealogies; early local businesses.
Activities: guided tours; art fair; student docent program; scavenger hunt in museum.
Publications: annual newsletter; brochure.
Hours & Admission Prices: June & Sept. Sat.-Sun. 1:30-4:30; July-Aug. Tues., Thurs. & Sat.-Sun. 1:30-4:30; other times by appointment. No charge; donations accepted. &
Attendance: 3,000 (estimated)
Membership: Family $10; Donor $35; Life $100.

East Lansing

ABRAMS PLANETARIUM, MICHIGAN STATE UNIVERSITY, 755 Science Rd., East Lansing, MI 48824. Tel.: 517-355-4676. Fax: 517-432-3838.
Web Site: www.pa.msu.edu/abrams/
Founded: 1964.
Congressional District: 6
Key Personnel: Interim Dir., John French; Planetarium Education Coord., Shane Horvatin.
Personnel Profile: Full-Time Paid 3; Part-Time Paid 4; Part-Time Volunteers 2.
Governing Authority: university. Parent Institution: Michigan State University. Tax-exempt.
Institution Type/Description: Planetarium.
Collections: historical items pertaining to astronomy & meteorites; manuscripts.
Facilities: sky theater. Books & other items for sale.
Activities: observing sessions; lectures; planetarium programs; formally organized education programs for children & adults; permanent & temporary exhibitions.
Publications: monthly, Sky Calendars.
Hours & Admission Prices: Display Hall: Mon.-Fri. 9-12 & 1-4:30. No charge. Programs: Fri.-Sat. 8pm, Sun. 2:30 & 4pm. Adults $3, students & senior citizens $2.50, children 12 & under $2. For public show information call 517-355-4672. Closed national holidays. &
Attendance: 32,000 (estimated)
Membership: Student $15; Individual $25; Family $50.

ELI AND EDYTHE BROAD ART MUSEUM, (I), 556 E. Circle Dr., 344 Student Svcs., East Lansing, MI 48824. Mailing Address: 556 E. Circle Dr., Rm. 344, East Lansing, MI 48824. Tel.: 517-884-3900. Fax: 517-884-3901.
E-mail: eebam@msu.edu
Web Site: broadmuseum.msu.edu
Founded: 2012.
Key Personnel: Dir., Michael Rush; Museum Shop Mgr., Stephanie Kribs.
Personnel Profile: Full-Time Paid 20; Part-Time Paid 69; Part-Time Volunteers 60; Interns 2.
Governing Authority: Parent Institution: Michigan State University. Tax-exempt.
Institution Type/Description: Art Museum.
Collections: works by contemporary artists; new media; photographs; works on paper.
Major Exhibits: The Genres, 7/13-8/14.
Activities: permanent & temporary exhibitions.
Hours & Admission Prices: Tues.-Thurs. 11-6, Fri. 12-9, Sat.-Sun. 10-5. No charge; donations accepted.
Membership: Student $25; MSU Faculty & Staff $50; Individual $60; Household $100; Broad Contemporaries $200; Broad & Beyond $250; Curator's Club $1,000; Director's Club $5,000.

* **KRESGE ART MUSEUM,** Michigan State University, East Lansing, MI 48824-1119. Tel.: 517-353-9834. Fax: 517-355-6577.
E-mail: kamuseum@msu.edu
Web Site: www.artmuseum.msu.edu
Founded: 1959.
Congressional District: 8
Key Personnel: Cur., April Kingsley; Communications, Christine Nichols; Registrar, Rachel Vargas; Educator, Cari Wolfe; Devel., Bridget Paff; Preparator, Norbert J. Freese; Museum Shop Mgr., Angelica Santos.
Personnel Profile: Full-Time Paid 8; Part-Time Paid 12; Part-Time Volunteers 150; Interns 1.
Governing Authority: state; university. Parent Institution: Michigan State University. Tax-exempt: 501(c)(3).
Institution Type/Description: Art Museum.
Collections: European & American painting; sculpture from ancient times to contemporary; 15th to 21st century prints; photographs.
Facilities: library available for use on premises.
Activities: lectures; concerts; tours; films; trips; symposia.
Publications: Kresge Art Museum Bulletin; newsletters; occasional catalogues.
Hours & Admission Prices: June-July Tues.-Fri. 11-5, Sat.-Sun. 12-5; Sept.-May Mon.-Wed. & Fri. 10-5, Thurs. 10-8, Sat.-Sun. 12-5. No charge; donations accepted. Closed holiday weekends. &
Attendance: 25,000 (estimated)
Membership: Full Time Student $15; Senior Citizen $25; Friend $35; Sponsor $60; Benefactor $100; Patron $250; Grand Patron $500; Honorary Curator $1,000; Honorary Director $2,000.

MICHIGAN STATE UNIVERSITY HERBARIUM, 166 Plant Biology Bldg., East Lansing, MI 48824-1312. Tel.: 517-355-4696. Fax: 517-353-1926.
E-mail: alan@msu.edu
Web Site: herbarium.msu.edu
Founded: 1863.
Congressional District: 6
Key Personnel: Dir. & Cur., Alan Prather; Asst. Cur., Alan Fryday; Sec., John Richardson.
Personnel Profile: Full-Time Paid 3; Part-Time Paid 1; Part-Time Volunteers 1.
Governing Authority: university. Parent Institution: Michigan State University. Tax-exempt: 501(c)(3).
Institution Type/Description: Herbarium.
Collections: vascular plants of all areas with emphasis on Michigan, alpine areas of North America, Mexico, South America, West Indies & Northern Borneo. Lichens & bryophytes of all area with emphasis on Patagonia & southern hemisphere island groups, the West Indies & Canary Islands.
Research Fields: systematic botany & mycology.
Facilities: 1,400-vol. reference library available for use on premises & to visiting investigators.
Activities: formally organized education programs for undergraduate & graduate college students; loan & exchange of specimens.
Hours & Admission Prices: Mon.-Fri. 8-5. No charge. Closed national holidays.

* **THE MICHIGAN STATE UNIVERSITY MUSEUM, (M),** 409 W. Circle Dr., East Lansing, MI 48824-1045. Tel.: 517-355-2370. Fax: 517-432-2846.
E-mail: jilda@msu.edu
Web Site: museum.msu.edu
Founded: 1857.
Congressional District: 6
Key Personnel: Chm. (V), Bill Trevarthan; Chm. (V) & Pres. (V), Denis Maybank; Dir., Gary Morgan; Cur. Folk Art., Dr. C. Kurt Dewhurst; Cur. Mammalogy, Dr. Barbara Lundrigan; Asst. Cur. Ornithology, Dr. Pamela Rasmussen; Specimen Preparator, Paula Hildebrandt; Col. Mgr. Biological Collection, Laura Abraczinskas; Cur. Anthropology, Dr. William Lovis; Cur. Vertebrate Paleontology, Dr. Michael Gottfried; Cur. History, Val Berryman; Extension Specialist & Cur. Education, Dr. Julie Avery; Cur. Exhibits, Juan Alvarez; Collections Mgr., Cultural Collections, Lynne Swanson; Public Rels. Coord., Lora Helou; Education Program Coord., Judy Smyth; Computer Coord., Sunny Wang; Collections Coord., Pearl Wong; Quilt Collections Asst. & Traveling Exhibitions Mgr., Beth Donaldson; Cur. Folk Arts, Dr. Marsha MacDowell; Asst. Cur. Folk Arts, LuAnne Kozma; Asst. Cur. Great Lakes Archaeology, Dr. Jodie O'Gorman; Curatorial Asst., Mary Worrall; Natl. Resources Conservation Specialist & Natl. Science Conservation Info. Specialist, James Harding; Special Projects Coord., Julie Levy-Weston; Museum Shop Mgr., Catherine Huddy; Facilities Mgr. & Volunteer Program Coord., Mike Secord.
Personnel Profile: Full-Time Paid 18; Part-Time Paid 21; Part-Time Volunteers 405; Interns 3.

Governing Authority: university. Parent Institution: Michigan State University. Tax-exempt: 501(c)(3).
Institution Type/Description: Science & Cultural History Museum.
Collections: Great Lakes history; archaeology; anthropology; vertebrate paleontology; folklife; modern vertebrates; agriculture; popular cultures; decorative arts.
Research Fields: vertebrate zoology; vertebrate paleontology; anthropology; folklife; popular culture; history.
Facilities: 100-seat auditorium. Items relating to exhibit programs for sale.
Activities: guided tours; interactive computer stations; TV & radio programs; formally organized education programs for undergraduate college students & graduate students affiliated with Michigan State University; training programs for professional museum workers; inter-museum loan, temporary, permanent & traveling exhibition. Annual Events: Great Lakes Folk Festival; Dinosaur Dash; Chocolate Party; Benefit Wine Tasting Party; reception openings; Darwin Day: Natural History ID Day.
Publications: anthropological, biological, cultural, educational, folklife & paleontological bulletins, catalogues & books published occasionally.
Hours & Admission Prices: Mon.-Fri. 9-5, Sat. 10-5, Sun. 1-5. Suggested Donations: adults $4, children $2; discounts to AAM & ICOM members. Closed New Year's Day; Easter; Memorial Day; Independence Day; Labor Day; Thanksgiving; Christmas. &
Attendance: 410,000 (estimated)
Membership: Individual & Family $25; Honorary Assistant Curator $100; Honorary Curator $250; Honorary Director $500; Honorary Trustee $1,000.

W.J. BEAL BOTANICAL GARDEN, 408 W. Circle Dr., Rm. 412, Michigan State University, East Lansing, MI 48824-1047. Tel.: 517-355-9582. Fax: 517-432-1090.
E-mail: telewski@cpa.msu.edu
Web Site: www.cpa.msu.edu/beal/index.htm
Founded: 1873.
Congressional District: 6
Key Personnel: Cur., Dr. Frank W. Telewski; Botanical Garden Technician, Peter Murray; Botanical Garden Technician, Hope Rankin.
Personnel Profile: Full-Time Paid 2; Part-Time Paid 2; Interns 2.
Governing Authority: Parent Institution: Michigan State University, Office of Campus Planning and Administration. Tax-exempt.
Institution Type/Description: Botanical Garden.
Collections: 4,500 species & varieties of herbaceous & woody plants arranged in systematic, economic, ecological & landscape groupings; botany; arboretum.
Research Fields: temperate forest ecology; ecology of endangered & threatened plants; culture of woody ornamentals.
Activities: guided tours by appointment.
Publications: annual index seminum; descriptive brochure.
Hours & Admission Prices: Daily sunrise-sunset. No charge. &
Attendance: 20,000 (estimated)

East Tawas

IOSCO COUNTY HISTORICAL MUSEUM, 405 W. Bay St., East Tawas, MI 48730-1103. Tel.: 989-362-8911.
E-mail: iosco.history@gmail.com
Web Site: ioscomuseum.org
Founded: 1976.
Congressional District: 11
Key Personnel: Pres., Leonard Wilkuski; Vice Pres., Regina Kelley; Recording Sec., Judy Clark; Treas., Lisa Ernst.
Personnel Profile: Part-Time Paid 1; Part-Time Volunteers 20.
Governing Authority: nonprofit. Parent Institution: Iosco County Historical Society. Tax-exempt: 501(c)(3).
Institution Type/Description: Historical Building: built by J.D. Hawks, 1st president of the Detroit & Mackinaw Railway; built in 1903.
Collections: displays of local history; china; clothing; furniture; tools; reference books; funeral records; coins; newspapers from 1800; photographs; fire department equipment; horse drawn hearse; stage coach; Indian birch canoe; carriage house; log mark collection; Detroit & Mackinaw Railroad display; U.S.S. western states artifacts; military room; local school's restored class photos; one room schoolhouse.
Research Fields: genealogy.
Facilities: reference library pertaining to local history. Gift items for sale.
Activities: guided tours; docent program; loan exhibitions.
Publications: quarterly membership newsletter.
Hours & Admission Prices: Jan.-March Sat. 1-4; April-June 15 & Sept. 4-Dec. Thurs.-Sat. 1-4; June 16-Sept. 3 Tues.-Sat. 10-4. Suggested Donation: $2. &
Attendance: 1,894 (accurate)
Membership: Senior Citizens over 64 $8; Individual $10; Family $12; Sustaining & Business $25; Patron $50; Benefactor $100 & up.

Edwardsburg

LAW ENFORCEMENT MEMORIAL ASSOCIATION, INC., Edwardsburg, MI 49112-0293. Mailing Address: P.O. Box 293, Edwardsburg, MI 49112-0293. Tel.: 847-409-8691.
E-mail: forgottenheroes@aol.com
Web Site: forgottenheroes-lema.org
Founded: 1989.
Congressional District: 8
Personnel Profile: Full-Time Volunteers 2; Part-Time Volunteers 27.
Governing Authority: private; nonprofit association. Tax-exempt: 501(c)(3).
Institution Type/Description: Law Enforcement Museum.
Collections: documentation information on deaths of law enforcement officers from 1717-present; information, as applicable, on the killers & all information on those who were executed for the crimes; memorabilia relative to the criminal justice system; badges; patches; photographs; uniforms.
Research Fields: crime-law enforcement deaths.
Facilities: library.
Publications: LEMA News.
Hours & Admission Prices: Museum under construction; call for information.
Membership: Regular $25; Supporter $100.

Elk Rapids

ELK RAPIDS AREA HISTORICAL SOCIETY, 301 Traverse St., Elk Rapids, MI 49629. Mailing Address: P.O. Box 2, Elk Rapids, MI 49629. Tel.: 231-264-5692.
E-mail: president@elkrapidshistory.org
Web Site: www.elkrapidshistory.org
Founded: 1986.
Key Personnel: Pres. (V), Dan LeBlond
Governing Authority: Tax-exempt.
Institution Type/Description: Historical Society Museum.
Collections: local history & culture; period furnishings; personal artifacts; photographs; Elk Rapids iron works replica; period clothing; early tools & furnishings; works by local artists.
Hours & Admission Prices: Memorial Day to Labor Day Tues., Thurs. & Sat.-Sun. 1-4. No charge; donations accepted. Closed major holidays. &
Attendance: 1,100 (estimated)
Membership: Single $15; Family $20; Business $30; Patron $55; Life $300.

Empire

EMPIRE AREA HERITAGE GROUP, 11544 S. La Core, Empire, MI 49630-9401. Mailing Address: Box 192, Empire, MI 49630-0192. Tel.: 231-326-5568.
E-mail: empiremuseum@centurytel.net
Web Site: www.empiremimuseum.org
Founded: 1972.
Congressional District: 9
Key Personnel: Pres. (V), David Taghon; Vice Pres., Leigh Payment; Sec., Anne Krawczak
Governing Authority: society; nonprofit. Tax-exempt: 501(c)(3).
Institution Type/Description: History Museum.
Collections: displays depicting logging, lumbering, sailing & the railroad; large barn with horse-drawn vehicles; blacksmith shop; woodworking displays; one-room school with audiovisual center; 1900 saloon; a 1911 Hose House (fire station) with hand pulled fire fighting equipment and related items; D.H. Day's personal artifacts; red Big Wheels; audiovisuals.
Research Fields: genealogy.
Facilities: Books, CD's & DVD's for sale.
Activities: videos; museum tours. Museum Sponsors: Heritage Day in October; Cultural Fair Sponsored by the National Park Service of Sleeping Bear Dunes.
Publications: annual newsletter, Empire Leader; Remembering Empire Through Pictures; history book, Some Other Day; Empire Museum Cookbook; the Boizard Letters - Civil War letters from 1856 to post war 1900; written & audio memories by local pioneers of the Empire Area.
Hours & Admission Prices: Memorial Day to June & Labor Day to mid-Oct. Sat.-Sun. 1-4; July-Aug. Thurs.-Tues. 1-4. Suggested Donation: family $5, adults $2. &
Attendance: 5,000 (estimated)
Membership: Single $10; Family $25; Business $50; Century Club $100.

SLEEPING BEAR DUNES NATIONAL LAKESHORE, 9922 Front St., (Hwy. M-72), Empire, MI 49630-9797. Tel.: 231-326-5134. Fax: 231-326-5382.
E-mail: slbe_interpretation@nps.gov
Web Site: www.nps.gov/slbe/
Founded: 1972.
Congressional District: 9
Key Personnel: Chief Naturalist, Lisa Myers; Supt., Dusty Shultz.
Personnel Profile: Full-Time Paid 30; Part-Time Paid 30; Part-Time Volunteers 50; Interns 5.
Governing Authority: federal. Parent Institution: National Park Service Department of the Interior, Washington, DC. Tax-exempt.
Institution Type/Description: Park Museum.
Collections: Great Lakes maritime items.
Research Fields: surfboats & rescue equipment used by the U.S. Life Saving Service; U.S Light House Service; Great Lake Maritime History.
Facilities: 150-vol. library of books on general natural history, Great Lakes maritime & geology available on request; 90-seat auditorium. Adult & children's natural & cultural history publications for sale.
Activities: guided tours; lectures; films; formally organized education programs for children, adults, undergraduate & graduate college students; permanent, temporary & traveling exhibitions.
Publications: brochure, Pierce Stocking Scenic Drive; brochure, South Manitou Island.
Hours & Admission Prices: Philip A. Hart Visitor Center: Memorial Day-Labor Day daily 8-6; Sept.-May daily 8:30-4. Cannery Boat Museum: Memorial Day to Labor Day daily 11-5; call for additional hours. Sleeping Bear Point Maritime Museum: Memorial Day to Labor Day daily 11-5. Annual Pass $20; Golden Age Pass (senior citizens) $10; Weekly Pass $10. &
Attendance: 1,200,000

Escanaba

DELTA COUNTY HISTORICAL SOCIETY MUSEUM, 16 Waterplant Rd., Escanaba, MI 49829-4052. Tel.: 906-789-6790.
E-mail: deltacountyhistsoc@sbcglobal.net
Web Site: www.deltahistorical.org
Founded: 1947.
Congressional District: 109
Key Personnel: Chm., Charles Lindquist; Archives, Karen Lindquist.
Personnel Profile: Part-Time Paid 4; Part-Time Volunteers 20.
Governing Authority: society. Parent Institution: Delta County Historical Society. Tax-exempt.
Institution Type/Description: Historical Society Museum & Archives.
Collections: agriculture; archives; Native American artifacts; marine; local historic artifacts; logging; military; railroad; lighthouse & keeper's home; surf boat & boathouse.
Facilities: archives.
Activities: programs; tours.
Publications: newsletter, Delta Historian.
Hours & Admission Prices: Archives: Memorial Day to Labor Day Mon.-Fri. 1-4; Labor Day to Memorial Day Mon. 1-4. Museum: Memorial Day to Labor Day open daily 11-4. Labor Day to Memorial Day by appointment. Museum & Lighthouse: family $5, adults $3, children under 12 no charge. Bus $75. Archives: no charge. Members no charge. &
Attendance: 7,400 (estimated)
Membership: Basic $25; Family $35; Commercial $100; Individual Life $200.

SANDPOINT LIGHTHOUSE, Sandpoint, Ludington Park, Escanaba, MI 49829. Mailing Address: Delta County Historical Society, 16 Water Plant Rd., Escanaba, MI 49829-4052. Tel.: 906-789-6790.
Web Site: www.deltahistorical.org
Founded: 1990.
Congressional District: 109
Key Personnel: Pres., John Noreus; Bd. of Dir., Peter Strom.
Personnel Profile: Part-Time Paid 6.
Governing Authority: Delta County Historical Society. Tax-exempt.
Institution Type/Description: Historical Society Museum: housed in turn-of-the-century lighthouse.
Collections: maps; charts; pictures; lantern room with 4th order Fresnel lens.
Activities: school & group tours by appointment.
Publications: brochure; grades 1-8, curricular materials; Delta Historian.
Hours & Admission Prices: June-Aug. daily 9-5, Sept. daily 1-4. Adults $1, students under 18 $.50; members & school groups no charge.
Attendance: 8,000 (estimated)
Membership: Basic $20; Family $30; Industrial $50; Commercial $100; Individual Life $200

WILLIAM BONIFAS FINE ARTS CENTER, 700 1st Ave. S., Escanaba, MI 49829-3703. Tel.: 906-786-3833. Fax: 906-786-3840.
E-mail: beth@bonifasarts.org

Web Site: www.bonifasarts.org
Founded: 1974.
Congressional District: 1
Key Personnel: Exec. Dir., Mollie Larsen; Pres. (V), Mrs. Petey Semmens; Information Rels. Asst., Beth Mercier; Visual Arts & Education Dir., Pasqua Warstler; Bookkeeper, Kaymary Rettig; Custodian, Jesse Farkas; Special Events, Beth Cox; Dir. Public Rels., Kathryn Morski.
Personnel Profile: Full-Time Paid 3; Part-Time Paid 3; Part-Time Volunteers 100; Interns 1.
Governing Authority: bd. of directors; nonprofit. Tax-exempt: 501(c)(3).
Institution Type/Description: Arts Center.
Collections: Great Lakes regional art.
Facilities: 240-seat theater; studio; classrooms.
Activities: concerts; arts festivals; musicals; drama; poetry readings; formally organized education programs for children & adults.
Publications: quarterly newsletter, Arts News.
Hours & Admission Prices: Tues.-Fri. 10-5:30, Sat. 10-3. Center: no charge. Classes: discounts to members. Closed major holidays. &
Attendance: 4,568 (estimated)
Membership: Annual $30; Family $40; Patron $50; Day Sponsor $100; Art Partner $300; Benefactor $500; Tower Society $1,000; Corporate Friend $1,500.

Farmington

GOVERNOR WARNER MANSION, 33805 Grand River Ave., Farmington, MI 48335-3431. Mailing Address: City of Farmington, 23600 Liberty St., Farmington, MI 48335. Tel.: 248-474-5500 ext. 2225 & ext. 2218. Fax: 248-473-7278. Facebook: Governor Mansion.
E-mail: warner_mansion@tds.net
Web Site: ci.farmington.mi.us
Founded: 1982.
Congressional District: 11
Key Personnel: Volunteer Coord., Jean Schornick; Volunteer Coord., Sharon Bernath; City Clerk, Susan Halberstadt; Dep. Clerk, Susan Wendel.
Personnel Profile: Part-Time Volunteers 31.
Governing Authority: Parent Institution: City of Farmington, MI. Tax-exempt: 501(c)(3).
Institution Type/Description: Historic House Museum: built in 1867.
Collections: family history; period furnishings; photographs; personal artifacts.
Major Exhibits: Quilt Exhibit, 4/14; Victorian Mourning, 10/14; Victorian Christmas Tradition, 12/14.
Activities: special events. Museum Sponsors: Open House & Quilt Exhibit in April; Fashion Show Fundraiser in May; Mansion Gardeners Plant Exchange in June; Founder's Festival Children's Festival & Flea Market, Jewelry Sale and Gov. Fred Warner Birthday Porch Party in July; Beautification Committee Flower Show & Gov. Warner Mansion "Tea & Talk" in August; Jazz Concert in September; Ghost Walk & Heritage Festival in October; Gingerbread House Workshop in November; Jaycee Tree Lighting & Mansion Open House in December.
Publications: monthly newspaper article, Mansion Musing; SWOCC cable video presentations.
Hours & Admission Prices: April-Dec. Wed. & 1st Sun. of month 1-5. Tours: adults $3, children $1; military & military families no charge.
Attendance: 2,400 (estimated)
Membership: Friends of the Mansion: Annual $10; Lifetime $100.

Farmington Hills

HOLOCAUST MEMORIAL CENTER, (M), 28123 Orchard Lake Rd., Farmington Hills, MI 48334-3738. Tel.: 248-553-2400. Fax: 248-553-2433.
E-mail: info@holocaustcenter.org
Web Site: www.holocaustcenter.org
Founded: 1984.
Key Personnel: Exec. Dir., Stephen M. Goldman; Chm. (V), Dr. Steven Grant, M.D.; Pres. (V), Gary Karp.
Personnel Profile: Full-Time Paid 14; Part-Time Paid 5; Part-Time Volunteers 170.
Institution Type/Description: History Museum.
Collections: films; artifacts; portraits & biographies; culture of European Jews; internment of Jews on Cyprus; emigration via Spain & Shanghai; efforts to retrieve hidden children & other rescue efforts; postwar pogroms in Poland & the Yedwabne case; migration of survivors to Israel & the United States.
Major Exhibits: Nazi Persecution of Homosexuals (T), 1/5/14-5/4/14; Holocaust Portfolio (T), 5/25/14-8/31/14; Todd Weinstein Retrospective (T), 9/21/14-12/22/14.
Facilities: research library; reading room; 250-seat auditorium; 1,500 sq. ft.

temporary exhibition gallery; 55,000 sq. ft. permanent exhibit space; Museum of European Jewish Heritage.
Activities: docent training; reading room; audiovisual presentations; guided tours; annual scholarly symposium; teacher workshop; guest lectures & audiovisual presentations; Holocaust Remembrance Day program; temporary & permanent exhibit guided tours.
Publications: annual newsletter; exhibit catalogues.
Hours & Admission Prices: Sun.-Thurs. 9:30-5, Fri. 9:30-3. Tours: Sun.-Thurs. 1pm. Adults $8, senior citizens, college & university students $6, students $5; discounts to AAM members; uniform services no charge. Closed Jewish holidays; most legal holidays. &
Attendance: 90,000 (estimated)
Membership: Middle School/High School Student $5; University/College Student $18; Individual $36; Family $50; Supporter $100-$249; Patron $250-$499; Guardian $500-$999; President's Club $1,000-$4,999; Righteous $5,000 & up.

MARVIN'S MARVELOUS MECHANICAL MUSEUM, 31005 Orchard Lake Rd., Farmington Hills, MI 48334-1384. Tel.: 248-626-5020. Fax: 248-626-7945.
E-mail: marvin@marvin3m.com
Web Site: www.marvin3m.com/
Founded: 1980.
Key Personnel: Owner, Marvin Yagoda
Institution Type/Description: Mechanical Machine Museum.
Collections: arcade machines; fortune teller machines; posters; model airplanes; robots; coin operated games; period games; carousels; nostalgic advertising.
Facilities: 5,500 sq. ft. exhibit space.
Activities: birthday parties; children's rides.
Hours & Admission Prices: Mon.-Thurs. 10-9, Fri.-Sat. 10am-11pm, Sun. 12-9. No charge. &

Fenton

FENTON HISTORICAL SOCIETY, 310 S. Leroy St., Fenton, MI 48430. Mailing Address: 310 S. Leroy St., Fenton, MI 48430. Tel.: 810-629-2570.
Institution Type/Description: Historical Society Museum: housed in the former office of A.J. Phillips; built in 1900.
Collections: local history; period furnishings; personal artifacts; photographs.
Hours & Admission Prices: Sun. 1-4; other times by appointment.

THE PIONEER MEMORIAL ASSOCIATION OF FENTON AND MUNDY TOWNSHIPS, 2436 N. Long Lake Rd., Fenton, MI 48430. Mailing Address: P.O. Box 154, Fenton, MI 48430-0154. Tel.: 810-955-3336.
E-mail: podunkpioneer@aol.com
Web Site: www.addorio.com/podunk
Founded: 1967.
Congressional District: 82
Key Personnel: Pres., MaryJane Pinkston; Vice Pres., Bill Pinkston; Treas., Ramona Green; Sec., Phyllis Heusted.
Governing Authority: nonprofit organization. Tax-exempt: 501(c)(3).
Institution Type/Description: Historic Society: housed in 1836 Indian Trading Post, now called Pioneer House.
Collections: period furniture & furnishings; old zinc bathtub with water heater attached; hand tools; old pictures & records; fog horn, steam whistle, steering wheel & life saver from the old passenger boat, City of Fenton; family history recordings. Historic Buildings: mid-18th century farm house; 1876 Phelps & Bigelow; 35 ft. windmill.
Research Fields: pioneer life; some genealogies.
Facilities: 20-vol. library of early school books, atlases & books by local authors available for use on premises.
Activities: guided tours; special tours for elementary school history classes. Museum Sponsors: Annual Fall Festival Pioneer Day.
Publications: brochure describing the pioneer house; semiannual newsletter.
Hours & Admission Prices: June-Aug. by appointment only; call Phyllis Yancy, 810-629-8747. Pioneer Day: adults $1, children $.50.
Attendance: 250 (estimated)
Membership: Single $3; Family $5; Contributing $10; Sustaining $25; Life $100.

Ferndale

FERNDALE HISTORICAL SOCIETY, 1651 Livernois, Ferndale, MI 48220. Tel.: 248-545-7606.
Web Site: www.ferndalehistoricalsociety.org
Founded: 1984.

Institution Type/Description: Historical Society Museum.
Collections: local history & culture; period furnishings; personal artifacts; photographs.
Hours & Admission Prices: Mon.-Wed. 10-1, Sat. 1-4 by appointment. No charge; donations accepted.
Attendance: 400 (accurate)
Membership: Individual $10; Family $25; Business $75; Lifetime $100.

Flat Rock

FLAT ROCK HISTORICAL SOCIETY - MUNGER GENERAL STORE MUSEUM, 25200 Gibraltar Rd., Flat Rock, MI 48134. Mailing Address: P.O. Box 337, Flat Rock, MI 48134. Tel.: 734-782-5220.
Governing Authority: Branch Museums: 1896 Flat Rock Hotel; 1874 Langs-Wagar House; Stofflet Carriage House.
Institution Type/Description: History Museum: housed in a former general store, 1875-1937.
Collections: local history & culture; period furnishings; photographs; personal artifacts.
Facilities: archives.
Activities: special events; temporary exhibitions.
Hours & Admission Prices: 2nd Sun. each month 1-4.

Flint

FLINT CHILDREN'S MUSEUM, 1602 W. University Ave., Flint, MI 48504. Tel.: 810-767-5437.
E-mail: discovery@flintchildrensmuseum.org
Web Site: www.flintchildrensmuseum.org
Founded: 1980.
Personnel Profile: Full-Time Paid 5; Part-Time Paid 7.
Governing Authority: Tax-exempt.
Institution Type/Description: Children's Museum.
Collections: hands-on exhibitions.
Activities: educational programs; special events.
Hours & Admission Prices: Jan.-May Tues.-Fri. 9-5, Sat. 10-5, Sun. 12-5; June-Dec. Tues.-Fri. 9-5, Sat. 10-5; other times by appointment. Admission $6; ASTC members & children under one no charge. ♿
Attendance: 47,000 (estimated)
Membership: Annual $75.

✱ FLINT INSTITUTE OF ARTS, (M), 1120 E. Kearsley St., Flint, MI 48503-1915. Tel.: 810-234-1695. Fax: 810-234-1692.
E-mail: info@flintarts.org
Web Site: www.flintarts.org
Founded: 1928.
Congressional District: 7
Key Personnel: Dir., John B. Henry, III; Pres. (V), Samuel M. Harris; Asst. Dir. Finance & Administration, Michael Melenbrink; Asst. Dir. Art School, Jeff Garrett; Coord. Membership, Valarie Shook; Facilities Mgr., Bryan Christie; Cur. Exhibitions, Tracee Glab; Cur. Education, Monique Desormeau; Cur. Collections & Exhibitions, Michael Martin; Public Rels., Miles Lam; Registrar, Pete Ott; Museum Shop Mgr., Cory Potter; Asst. Dir. Devel., Kathryn Sharbaugh.
Personnel Profile: Full-Time Paid 31; Part-Time Paid 70; Part-Time Volunteers 471; Interns 3.
Governing Authority: nonprofit organization. Tax-exempt: 501(c)(3).
Institution Type/Description: Art Museum & School.
Collections: 8,000 works of art; 15th to 21st century European paintings, sculpture, graphics & decorative arts; 18th-21st century American paintings, sculpture, graphics & decorative arts; Native American, African & Asian art.
Major Exhibits: Point of View - African American Art from the Elliott Perry Collection, 1/14-4/14; Kathleen Gilje: Portraits & Paintings, 5/14-7/14; The Art of Video Games, 10/14-12/14.
Research Fields: European paintings, sculpture, graphics & decorative arts; American paintings, sculpture, graphics & decorative arts; Native American, African & Asian art.
Facilities: 8,000-vol. library; 330-seat theater; auditorium; 25,000 sq. ft. exhibition space; cafe; classrooms. Museum-related items for sale.
Activities: inter-museum loan, permanent, temporary & traveling exhibitions; guided tours; lectures; films; concerts; sales & rental gallery; non-credit education programs for children & adults; workshops for educators; docent program; Friends of Modern Art & Founders Society volunteer events. Annual Events: Summer Art Fair; Winter Art Fair.
Publications: exhibition catalogs; annual reports; bimonthly newsletter; collection catalogues, Vanitas; Great Lakes Muse: American Scene Painting in the Upper Midwest, 1910-1960; Thy Brother's Keeper; American Art at the Flint Institute of Arts; Edmund Lewandowski: Precisionism and Beyond;

Blown Away: International Glass of the 21st Century; Magnificence and Awe: Renaissance and Baroque Art in the Viola E. Bray Gallery at the Flint Institute of Arts; Pure Abstraction: Selections from The Flint Institute of Arts.
Hours & Admission Prices: Mon.-Wed. & Fri. 12-5, Thurs. 12-9, Sat. 10-5, Sun. 1-5. Collections: no charge. Temporary Exhibits: adult $7, students & seniors $5; discounts to AAM & ICOM members; children under 12 & members no charge. Closed national holidays. ♿
Attendance: 151,443 (accurate)
Membership: Youth 2 1/2-12 & Student 13 to College $20; Individual $30; Dual $40; Family $50; Sustainer $100; Sponsor $250; Donor $500; Rubens $1,000 & up.

MOTT COMMUNITY COLLEGE FINE ARTS GALLERY, Visual Arts & Design Bldg., 1401 E. Court St., Flint, MI 48503-2089. Tel.: 810-762-0443. Fax: 810-232-3452.
Web Site: www.mcc.edu
Institution Type/Description: Art Gallery.
Collections: works by Michigan artists.
Activities: temporary artists.
Hours & Admission Prices: Mon.-Fri. 8-5.

ROBERT T. LONGWAY PLANETARIUM, 1310 E. Kearsley St., Flint, MI 48503-1987. Tel.: 810-237-3400. Fax: 810-237-3417.
E-mail: tshickles@sloanlongway.org
Web Site: www.sloanlongway.org
Founded: 1958.
Congressional District: 7
Key Personnel: Dir., Tim Shickles; Office Mgr., Pam Atwell; Lecturer & Instructor, Richard A. Walker; Science Program Presenter, Loe'l Murphy; Cur. Programs, Laurie Bone; Museum Shop Mgr., Judith Latreille; Receptionist, Cindy Goodall.
Personnel Profile: Full-Time Paid 6; Part-Time Paid 10.
Governing Authority: nonprofit. Parent Institution: Flint Cultural Corp. Tax-exempt.
Institution Type/Description: Planetarium.
Collections: astronomical artifacts; manuscripts.
Facilities: 282-seat auditorium; theater. Astronomy-related items for sale.
Activities: formally organized education programs for adults, children & undergraduate students.
Hours & Admission Prices: Public Astronomy & Laser Shows: Sat.-Sun; Mon.-Fri. by appointment. Adults $4-$6, children $3-$5 depending on program; members no charge. Closed national holidays. ♿
Attendance: 62,000 (estimated)
Membership: Student Pass $20; Share-a-Pass $40; Family Pass $85.

✱ SLOAN MUSEUM & LONGWAY PLANETARIUM, (M), 1221 E. Kearsley St., Flint, MI 48503-1988. Tel.: 810-237-3450. Fax: 810-237-3451.
E-mail: tshickles@sloanlongway.org
Web Site: www.sloanlongway.org
Founded: 1966.
Congressional District: 9
Key Personnel: Dir., Tim Shickles; Cur. Collections, Jane McIntosh; Mktg. & Facilities Rental Coord., Gregory Ahejew; Museum Shop Mgr., Judi Latreille.
Personnel Profile: Full-Time Paid 16; Part-Time Paid 18; Part-Time Volunteers 60; Interns 3.
Governing Authority: private; nonprofit organization. Parent Institution: Flint Cultural Center Corp., 1310 E. Kearsley St., Flint. Tax-exempt: 501(c)(3).
Institution Type/Description: Regional History & Science Museum and Planetarium.
Collections: automotive & horse-drawn carriages; local historical artifacts; natural history artifacts; archaeology; 80 vintage & prototype General Motors automobiles & related artifacts.
Research Fields: local & automotive history; carriage industry.
Facilities: library; 40,000 sq. ft. exhibit space; 100-seat auditorium; classroom; cafeteria. Museum-related items for sale.
Activities: guided tours; lectures; docent program; formal education programs for children & University of Michigan students; hobby workshops; rental gallery; loan, participatory, temporary & traveling exhibitions.
Hours & Admission Prices: Mon.-Fri. 10-5, Sat.-Sun. 12-5. Adults $6, senior citizens $5, children 3-11 $4; discounts to groups & AAM members; members & children 2 & under no charge. Closed New Year's Day; Easter; Memorial Day; Independence Day; Labor Day; Thanksgiving; Christmas. ♿
Attendance: 117,646 (accurate)
Membership: Student $20; Share A Pass $40; Family $85.

WHALEY HISTORIC HOUSE MUSEUM, 624 E. Kearsley St., Flint, MI 48503-1909. Tel.: 810-471-4714. Facebook: Whaley Historic House Museum.
E-mail: 1885@whaleyhouse.com
Web Site: www.whaleyhouse.com
Formerly: Whaley House Museum
Founded: 1976.
Key Personnel: Dir., Samantha Engel.
Governing Authority: Parent Institution: Whaley Historical House Association. Tax-exempt.
Institution Type/Description: Historic House: former home of Robert and Mary McFarlan Whaley.
Collections: Guilded-age artifacts; Whaley family artifacts; Flint Historical pieces.
Major Exhibits: A House in Mourning, 10/14-11/14; Christmas at Whaley, 12/14-1/15.
Publications: The Whaley Gazette quarterly newsletter.
Hours & Admission Prices: 1st & 3rd Sat. of month 10-1, Mon.-Fri. by appointment. Adults $5, children 12 & under & students $3; members receive complimentary tours.
Membership: Individual $20; Family $30; Contributing $50; Century $100; Corporate $200; Benefactor $500.

Flushing

FLUSHING AREA HISTORICAL SOCIETY, 431 W. Main St., Flushing, MI 48433. Tel.: 810-487-0814.
E-mail: fahs@att.net
Web Site: flushinghistorical.org
Governing Authority: nonprofit organization. Tax-exempt: 501(c)(3).
Institution Type/Description: Historical Society Museum: housed in a former depot; built in 1888.
Collections: local history & culture; period furnishings; personal artifacts; photographs; early documents.
Hours & Admission Prices: Jan.-April Tues. 10-1; May-Dec. Tues. 10-1, Sun. 1-4; other times by appointment. No charge. Closed holidays. &
Membership: Senior $2; Individual $3; Family $5; Supporting $10; Contributing $25; Sustaining $100; Life $150.

Frankenmuth

* **FRANKENMUTH HISTORICAL MUSEUM,** 613 S. Main, Frankenmuth, MI 48734-1689. Tel.: 989-652-9701. Fax: 989-652-9390.
E-mail: fhaoffice@airadv.net
Web Site: www.frankenmuthmuseum.org
Founded: 1963.
Congressional District: 7
Key Personnel: Dir., Jon Webb; Pres., Jaime Furbush; Collection Mgr., Mary Nuechterlein; Museum Shop Mgr., Lorraine Eckert; Administrative Asst., Ammiee Kotch.
Personnel Profile: Full-Time Paid 2; Part-Time Paid 5; Part-Time Volunteers 11.
Governing Authority: nonprofit organization. Parent Institution: Frankenmuth Historical Association. Tax-exempt: 501(c)(3).
Institution Type/Description: Regional History Museum.
Collections: regional ethnic artifacts; early pioneer artifacts; primary archival material dating from 1815-1890 manuscripts. Historic Structure: 1890 Fischer Hall theatre.
Research Fields: local history, Bavarian-American immigration.
Facilities: 400-vol. library of German books & periodicals available on premises only. Local craft items, cards & gifts for sale.
Activities: guided tours, performing & cultural arts; educational classes for area schools; inter-museum loan, permanent & temporary exhibitions; historical markers.
Publications: quarterly newsletter, Frankenmuth Historical Assn.; books, Frankenmuth Historical Assn.; Cookbook; Local History.
Hours & Admission Prices: Jan.-March call for hours; April-Dec. Mon.-Thurs. 10:30-5, Fri. 10:30-8, Sat. 10-8, Sun. 11-7. Self Guided Tours: family $5, adults $2, children $1. Guided Tours: adults $2.50, children $1.50; members no charge. Closed New Year's Day; Easter; Thanksgiving; Christmas. &
Attendance: 17,000 (accurate)
Membership: Student & Senior Citizen $15; Family & Individual $40; Sustaining $100; Life $500; Founder's Club $1,500.

LAGER MILL HISTORY OF BREWING MUSEUM, 613 S. Main, Frankenmuth, MI 48734. Tel.: 989-652-9701. Fax: 989-652-9390. Facebook; Lager Mill.
E-mail: fhaoffice@airadv.net

Founded: 2011.
Congressional District: 7
Key Personnel: Dir., Jon Webb; Museum Shop Mgr., Tom Taylor; Administrative Asst., Ammiee Kotch.
Personnel Profile: Full-Time Paid 1; Part-Time Paid 4.
Governing Authority: Parent Institution: Frankenmuth Historical Association Sales and Services, 701 Mill St., Frankenmuth, MI 48734.
Institution Type/Description: History Museum.
Collections: brewing history & artifacts.
Hours & Admission Prices: Call for hours. Adults $2, students $1. &
Attendance: 500 (estimated)

Franklin

FRANKLIN HISTORICAL MUSEUM, 26165 13 Mile Rd., Franklin, MI 48025. Mailing Address: P.O. Box 7, Franklin, MI 48025. Tel.: 248-538-0273.
Institution Type/Description: History Museum.
Collections: local history & culture; period furnishings; personal artifacts; photographs.
Hours & Admission Prices: 1st Sat. each month 1-3 by appointment.

Garden

FAYETTE HISTORIC STATE PARK, (M), 4785 II Rd., Garden, MI 49835-9411. Tel.: 906-644-2603. Fax: 906-644-2666.
Web Site: www.michigan.gov/fayette
Key Personnel: Unit Supvr., Dept. Natural Resources, Randall W. Brown.
Personnel Profile: Full-Time Paid 3.
Governing Authority: Parent Institution: Dept. of History, Arts & Libraries-Historical Center in cooperation with the Michigan Department of Natural Resources.
Institution Type/Description: Historic Town Museum: site of iron smelting town 1867-1891, owned & operated by Jackson Iron Co. includes nineteen structures.
Collections: Museum: scale model of the town; hotel; company office; town hall; homes of the company employees; furnace complex & kilns; machine shop; old cemetery; interpretive displays & signs.
Research Fields: charcoal iron industry; economic & cultural development & decline.
Facilities: campgrounds; picnic area; bike & hiking trails.
Activities: self-guided tour; guided tours; permanent exhibitions; overnight boat camping.
Publications: brochures; Fayette Historic Townsite.
Hours & Admission Prices: mid-May to mid-June & Sept. to mid-Oct. daily 9-5; mid-June to Aug. daily 8-9. Motor vehicle passport required. Non-Michigan car: daily $8. Annual Pass: $29. Passport: $10. &
Attendance: 62,500 (accurate)

Gaylord

CALL OF THE WILD MUSEUM, 850 S. Wisconsin Ave., Gaylord, MI 49735-1747. Tel.: 989-732-4336. Fax: 989-732-3749.
E-mail: callofthewildgaylord@live.com
Web Site: www.callofthewildgaylord.com
Founded: 1957.
Key Personnel: Bd. Directors, William C. Johnson; Bd. Directors, Judy Fleet; Pres., Janis Vollmer.
Personnel Profile: Full-Time Paid 4; Part-Time Paid 15.
Governing Authority: corporate; bd. of directors. Parent Institution: Call of the Wild, Inc.
Institution Type/Description: Natural History Museum.
Collections: 50 displays of North American wildlife, featuring natural recorded sounds.
Facilities: Collectibles & other museum-related items for sale.
Hours & Admission Prices: mid-June to Labor Day daily 9-9; Sept. to mid-June Mon.-Sat. 9:30-6, Sun. 11-5. Adults $7, seniors 62 & up $6.50, children 5-13 $4.50; discount to AAM members. Closed New Year's Day; Thanksgiving; Christmas. &
Attendance: 20,000 (estimated)
Membership: Children $9; Adults $14; Family $40

Gibraltar

GIBRALTAR HISTORICAL MUSEUM, 29450 Munro St., Gibraltar, MI 48197-9720. Tel.: 734-676-3900.
Institution Type/Description: Historical Society Museum.
Collections: artifacts & memorabilia pertaining to the city of Gibraltar.
Hours & Admission Prices: 1st Sun. each month 2-4.

Grand Haven

TRI-CITIES HISTORICAL MUSEUM, (M), 200 Washington Ave., Grand Haven, MI 49417-1357. Tel.: 616-842-0700. Fax: 616-842-3698. Facebook: Tri Cities Historical Museum.
E-mail: sradtke@tri-citiesmuseum.org
Web Site: www.tri-citiesmuseum.org
Founded: 1959.
Congressional District: 96
Key Personnel: Dir., Steven Radtke; Bd. Pres., Nancy Solon; Museum Shop Mgr., Kathy Oppenhuizen.
Personnel Profile: Full-Time Paid 4; Part-Time Paid 10; Part-Time Volunteers 75; Interns 1.
Governing Authority: society; nonprofit. Parent Institution: Tri-Cities Historical Society. Tax-exempt: 501(c)(3).
Institution Type/Description: Historical Museum: housed in 1871 Akeley building & 1870 Grand Trunk Railroad Depot.
Collections: archives; medical; farming; photographs; costumes; Native American artifacts; Victorian period rooms; textiles; lumbering; shipping; Coast Guard; railroads; saw mill.
Major Exhibits: Dancing Waters, 1/14-8/14; Portrait of the Past, 3/14-8/14.
Research Fields: local history; Michigan history.
Facilities: library of newspapers, scrapbooks & memoirs available by special arrangement; audiovisual room.
Activities: guided tours; quarterly programs; permanent & temporary exhibitions.
Publications: quarterly newsletter, The Packet; booklets, History of Grand Haven and Ottawa County; photo history books; In the Path of Destiny.
Hours & Admission Prices: Memorial Day to Labor Day Tues.-Sat. 10-8, Sun. 12-5; Sept.-May Tues.-Fri. 10-5, Sat.-Sun. 12-5. No charge; donations accepted. &
Attendance: 57,114 (accurate)
Membership: Student $10; Single Senior $15; Individual $20; Senior Couple $25; Family $35.

Grand Ledge

GRAND LEDGE AREA HISTORICAL SOCIETY MUSEUM, 118 W. Lincoln, Grand Ledge, MI 48837. Mailing Address: P.O. Box 203, Grand Ledge, MI 48837. Tel.: 517-627-5170. Fax: 517-627-5170.
E-mail: marnor1@comcast.net
Web Site: gdledgehistsoc.org
Founded: 1984.
Congressional District: 7
Key Personnel: Chm. (V), Marilyn Smith; Pres. Historical Society (V), Ethelen Herbstreit; Archives, Cindy Langenberg.
Personnel Profile: Part-Time Volunteers 76.
Volunteer Hours: 2,000
Operating Expenses: 3,758
Operating Income: 8,068
Governing Authority: Parent Institution: Grand Ledge Area Historical Society. Tax-exempt.
Institution Type/Description: Historical Society Museum: housed in the former home of local minister, Byron S. Pratt; built in 1880.
Collections: local history & culture; period furnishings; photographs; personal artifacts; genealogy records; registry of historic district homes; school records & annuals; tapes; videos; DVD's.
Major Exhibits: Things of Beauty - Paintings (T), 1/14-12/14; Composite High School Graduation Pictures 1900-1959 (T), 1/14-12/14; Points of Pride - Eaton County Schools (T), 2/14-9/14; Public Servants: Who's Who, 3/14-12/20/14.
Research Fields: genealogy; businesses.
Activities: research; school, garden & bus tours; monthly programs. Museum Sponsors: Holiday Traditions Tour of Homes; Victorian Days.
Publications: 5 books; quarterly newsletter.
Hours & Admission Prices: March-Dec. Sun. 2-4; Festival Days: 12-4; other times by appointment. No charge; donations accepted.
Attendance: 1,000 (estimated)
Membership: Youth or Senior $25; Individual $30; Business $45; Patron $50; Family $75; Organization $175; Life $350; Business Life $450.

Grand Rapids

CALVIN COLLEGE CENTER ART GALLERY, (M), Covenant Fine Arts Center, 1795 Knollcrest Cir., S.E., Grand Rapids, MI 49546. Tel.: 616-526-6271. Fax: 616-526-8551.
E-mail: jhz2@calvin.edu
Web Site: www.calvin.edu/centerartgallery
Founded: 1974.
Congressional District: 5
Key Personnel: Dir., Joel Zwart.
Personnel Profile: Full-Time Paid 2; Part-Time Paid 10.
Governing Authority: college. Parent Institution: Calvin College. Subsidiary Institution: Center Art Gallery. Tax-exempt.
Institution Type/Description: Art Gallery.
Collections: 17th & 19th-century Dutch paintings & drawings; 20th-century American & regional prints, drawings, paintings & sculpture.
Research Fields: relating to exhibitions.
Facilities: library; 1,000-seat auditorium; 240-seat recital hall; theater; classrooms.
Activities: guided tours; lectures; films; temporary, permanent, loan & traveling exhibitions.
Publications: catalogs relating to exhibits.
Hours & Admission Prices: Winter: Mon.-Tues. 9-5, Wed.-Fri. 9-9, Sat. 10-4. No charge. Closed during school vacations except for special exhibitions. &
Attendance: 12,000 (accurate)

COMMUNITY ARCHIVES AND RESEARCH CENTER, 223 Washington St., S.E., Grand Rapids, MI 49503-4314. Tel.: 616-456-3977. Fax: 616-929-1700.
E-mail: info@grmuseum.org
Web Site: www.grmuseum.org
Institution Type/Description: Archives.
Collections: archives; historical documents; period artifacts; photographs; Kent County records.
Facilities: archives.
Activities: research; group tours.
Hours & Admission Prices: By appointment.

FREDERIK MEIJER GARDENS & SCULPTURE PARK, 1000 E. Beltline Ave., N.E., Grand Rapids, MI 49525-5804. Tel.: 888-957-1580 (Toll Free); 616-957-1580. Facebook: Meijer Gardens.
E-mail: info@meijergardens.org
Web Site: www.meijergardens.org
Founded: 1995.
Key Personnel: Pres. & C.E.O., David S. Hooker
Institution Type/Description: Botanical Garden & Sculpture Park.
Collections: over 200 sculptures; gardens.
Facilities: nature trails; gardens; cafe. Museum-related items for sale.
Activities: tram tours; educational programming.
Publications: quarterly membership newsletter, Seasons.
Hours & Admission Prices: See website for hours. Adults $8; members no charge. Closed New Year's Day; Thanksgiving; Christmas. &
Attendance: 500,000 (accurate)
Membership: Visit website or call for information.

GERALD R. FORD PRESIDENTIAL MUSEUM, (M), 303 Pearl St., N.W., Grand Rapids, MI 49504-5353. Tel.: 616-254-0400 & 0367. Fax: 616-254-0386.
E-mail: ford.museum@nara.gov
Web Site: www.fordlibrarymuseum.gov
Founded: 1981.
Congressional District: 3
Key Personnel: Dir., Dr. Elaine Didier; Deputy Dir., James R. Kratsas; Cur., Donald Holloway; Exhibits Specialist, Bettina Cousineau; Registrar, James Draper; Education, Barbara McGregor; Public Rels., Kristin Mooney; Museum Shop Mgr., Janice Berling.
Personnel Profile: Full-Time Paid 11; Part-Time Paid 4; Part-Time Volunteers 35; Interns 2.
Governing Authority: federal. Affiliated with the National Archives & Records Administration, Washington, DC 20408. Branch Museum: Gerald R. Ford Library, 1000 Beal Ave., Ann Arbor, MI. Tax-exempt: 170(b)(1)(A).
Institution Type/Description: Presidential History Museum.
Collections: personal papers; government records; photographs; motion picture films; audio & video tapes; sound recordings; head of state gifts; gifts from private citizens; political campaign items; personal & family memorabilia; bicentennial folk art.
Research Fields: life, times, career & presidential administration of President Ford.
Facilities: 252-seat auditorium; classrooms. Museum-related items for sale.
Activities: guided tours; lectures; films; permanent, temporary & traveling exhibitions; organized education programs for children, adults, undergraduate or graduate students; docent program. Annual Events: Holiday Open House; Birthday Celebration; Tree Lighting.
Publications: quarterly newsletter; brochure.
Hours & Admission Prices: Daily 9-5. Adults $7, senior citizens & military $6,

college students $5, children 6-18 $3; discounts to groups; members and children 5 & under no charge. Closed New Year's Day; Thanksgiving; Christmas. &
Attendance: 311,122 (accurate)
Membership: Individual $35; Family $50-$99; Associate $100-$249; Sustaining $250-$499; Patron $500-$999; President's Cabinet $1,000 & up.

＊ GRAND RAPIDS ART MUSEUM, (M), 101 Monroe Center, N.W., Grand Rapids, MI 49503-2801. Tel.: 616-831-1000. Fax: 616-831-1001.
E-mail: pr@artmuseumgr.org
Web Site: www.artmuseumgr.org
Founded: 1910.
Congressional District: 5
Key Personnel: C.E.O. & Dir., Dana Friis-Hansen; Deputy Dir. Administration & C.F.O., Randy VanAntwerp; Dir. Mktg. & Public Rels., Kerri Vanderhoff; Registrar, Kathleen Ferres; Asst. Cur., Cindy Buckner; Retail Mgr., Susan Coombes.
Personnel Profile: Full-Time Paid 22; Part-Time Paid 38; Part-Time Volunteers 414; Interns 6.
Governing Authority: nonprofit organization. Tax-exempt: 501(c)(3).
Institution Type/Description: Art Museum.
Collections: European 17th to 20th century paintings; American 19th & 20th-century paintings, prints, drawings, photography; 20th-century design.
Research Fields: curatorial material relevant to the museum's collection; 19th-20th century European & American art.
Facilities: 5,000-vol. library of general art reference; auditorium. Museum-related items for sale.
Activities: permanent & temporary exhibitions; lectures; gallery talks; concerts; guided tours; education programs for children; docent program.
Publications: members' newsletter; annual report; exhibition catalogues.
Hours & Admission Prices: Tues.-Thurs. & Sat. 10-5, Fri. 10-9, Sun. 12-5. Adults $8, senior citizens 62 & over and college students with ID $7, youth 6-17 $5; discounts to AAM members; members & children under 6 no charge. Closed major holidays. &
Attendance: 319,459 (accurate)
Membership: Individual & 1 Guest $50; Family $65; Friend of Art $150; Fine Arts $500; Masters $1,000.

GRAND RAPIDS CHILDREN'S MUSEUM, 11 Sheldon Ave., N.E., Grand Rapids, MI 49503-3218. Tel.: 616-235-4726, ext. 100. Fax: 616-235-4728.
E-mail: sandersen@grcm.org
Web Site: www.grcm.org
Founded: 1992.
Congressional District: 3
Key Personnel: Exec. Dir., Robert Dean; Dir. Devel., Nancy Brozek; Mktg. & Membership Mgr., Beth Szberowski; Exhibits & Community Rels. Mgr., Jan Stone.
Personnel Profile: Full-Time Paid 10; Part-Time Paid 16; Interns 4.
Governing Authority: private; nonprofit organization. Tax-exempt: 501(c)(3).
Institution Type/Description: Children's Museum.
Collections: hands-on exhibits.
Facilities: 18,000 sq. ft. exhibit space.
Activities: hands-on exhibits; special programs for toddlers; exhibit-based programming.
Publications: newsletter, Play Times; calendar, What's Happening; annual report; brochures; press releases & press kits; flyers; rack cards; guest guides; exhibit poster; exhibit banners; Team Fun cards; Team Fun contracts. Electronic Publications: quarterly teacher newsletter; monthly E-Play update for members.
Hours & Admission Prices: Tues.-Wed. & Fri.-Sat. 9:30-5, Thurs. 9:30-8, Sun. 12-5. Adults $6.50; reciprocal membership & members no charge. &
Attendance: 180,166 (accurate)
Membership: Members: $40; $60; $85. Contributing Members: $125.

＊ GRAND RAPIDS PUBLIC MUSEUM, (M), 272 Pearl St., N.W., Grand Rapids, MI 49504-5371. Mailing Address: Van Andel Museum Center, 272 Pearl St., N.W., Grand Rapids, MI 49504-5371. Tel.: 616-929-1700. Fax: 616-929-1780. TDD: 616-456-3724.
E-mail: info@grmuseum.org
Web Site: www.grmuseum.org
Formerly: Public Museum of West Michigan
Founded: 1854.
Congressional District: 5
Key Personnel: Pres. & C.E.O., Dale A. Robertson; Foundation Chm. (V), Danny R. Gaydou; Foundation Vice Chm., Carol Van Andel; C.F.O. & Dir.

Admin., Karen Wilburn; Dir. Education & Interpretation, Marilyn Merdzinski; Dir. Mktg. & Public Rels., Kate Moore; Dir. Exhibits & Facilities, Tom Bantle; Membership Svcs. Mgr., Nicole Forward; Museum Shop Mgr., Jane Clas; Asst. to Pres./C.E.O., Leslie Milstead.
Personnel Profile: Full-Time Paid 18; Part-Time Paid 165; Part-Time Volunteers 176; Interns 8.
Governing Authority: Parent Institution: Grand Rapids Public Museum Foundation. Subsidiary Institution: Grand Rapids Public Museum. Tax-exempt.
Institution Type/Description: General Museum.
Collections: Grand Rapids-made furniture; c.1830-present decorative arts; industrial & agricultural artifacts; costumes; household textiles; ethnographic material; toys; dolls; games; natural history; & paleontology; mammals; birds. traveling exhibits on a routine basis. Historic Buildings: 1836 Calkins Law Office; 1869 Stilwill Horseshoeing Shop; 1853 Star School; 1895 Voigt House; Norton Mounds (Hopewell period).
Major Exhibits: Dinosaurs Unearthed (T), 1/1/14-4/27/14; Lego Architecture (T), 3/1/14-8/31/14; Beal Pirates (T), 10/14.
Research Fields: ethnology; local history; 19th & 20th-century American furniture; domestic material culture; natural history; Native American culture, Great Lakes Woodland.
Facilities: 5,000-vol. library of history & natural history material available for use on premises; archives; research center; planetarium; 260-seat auditorium; classrooms; 87,000 sq. ft. exhibit space; cafe. Museum-related items for sale.
Activities: guided tours; lectures; films; gallery talks; hobby workshops; formally organized education programs; docent program; permanent & temporary exhibitions; facility use program; concerts; planetarium shows.
Publications: membership newsletter; exhibition & collection catalogs.
Hours & Admission Prices: Museum: Tues. 9-8, Wed.-Sat. 9-5, Sun. 12-5. Adults $8, senior citizens $7, children 3-17 $3; discount to AAA & AAM members; museum members no charge. Planetarium: daytime $3 with admission. Closed New Year's Day; Easter; Thanksgiving; Christmas. &
Attendance: 247,371 (accurate)
Membership: Seniors 62 & over $5 discount of some levels; Individual $35; Individual Plus One & Dual $50; Family, Grandparents, Individual Plus Two, & Dual Plus One $65; Family Plus One, Grandparent Plus One & Dual Plus Two $80; Grandparent Plus Two & Family Plus Two $95; Whalewatchers $150; Stargazers $250; 1890 Club $500; Carousel Society $1,000.

JOHN BALL ZOO, 1300 W. Fulton St., Grand Rapids, MI 49504-6100. Tel.: 616-336-4301. Fax: 616-336-3907.
E-mail: info@jbzoo.org
Web Site: www.jbzoo.org
Formerly: John Ball Zoological Garden
Founded: 1949.
Congressional District: 5
Key Personnel: Exec. Dir. Zoological Society, Brenda Stringer; Zoo Dir., Andy McIntyre; Museum Shop Mgr., Theresa Danneffel.
Personnel Profile: Full-Time Paid 32; Part-Time Paid 15; Part-Time Volunteers 130; Interns 4.
Governing Authority: municipal; society. Parent Institution: Kent County, MI. Subsidiary Institution: John Ball Zoo Society.Tax-exempt.
Institution Type/Description: Zoological Gardens.
Collections: exotic & native animals including endangered species.
Research Fields: captive animal management.
Facilities: botanical garden; aquarium; 60-seat auditorium; children's zoo; zoological & herpetological laboratories; food concessions; classrooms. Zoo-related gift items for sale.
Activities: guided tours; lectures; films; formally organized education programs for children; on-site school for 6th grade; permanent exhibitions; zipline; ropes course. Museum Sponsors: programs for special education groups; lectures and programs for clubs, churches & charities.
Publications: Zoo News.
Hours & Admission Prices: Call for hours. Adults $8.50, senior citizens 62 & over $7.50, children 3-13 $6.50; members no charge. &
Attendance: 455,000 (accurate)
Membership: Individual $35; Household $70; Household + 2 $100.

KENDALL GALLERIES - KENDALL COLLEGE OF ART AND DESIGN, (M), Ferris State University, 17 Fountain St., N.W., Grand Rapids, MI 49503. Tel.: 800-676-2787.
E-mail: kcadgallery@ferris.edu
Web Site: www.kcad.edu
Founded: 1982.
Key Personnel: Dir. Exhibitions, Sarah Joseph; Asst. Dir. Exhibitions, Michele Bosak.
Personnel Profile: Full-Time Paid 1; Part-Time Paid 1; Interns 8.
Governing Authority: Parent Institution: Ferris State University. Tax-exempt.
Institution Type/Description: Art Galleries.

Collections: Kendall Gallery: works by artists from around the world & Kendall faculty. Gallery 114: works by visiting artists & graduate thesis exhibitions. Gallery 104: works by undergraduate students. Gallery 602: competitions for painting undergraduates.
Activities: lectures; special events.
Hours & Admission Prices: Call for hours. No charge; donations accepted.
Attendance: 31,000 (accurate)

URBAN INSTITUTE FOR CONTEMPORARY ARTS, 2 W. Fulton, Grand Rapids, MI 49503. Tel.: 616-454-7000, ext. 20. Fax: 616-459-9395.
E-mail: info@uica.org
Web Site: www.uica.org
Founded: 1977.
Congressional District: 5
Key Personnel: C.E.O. & Exec. Dir., Jeff Meeuwsen; Film Program Mgr., Ryan Dittmer; Mng. Dir., Janet Teunis; Graphic Designer, Ryan Greaves; Business Mgr., Stacie Carrizzi; Dir. Clay Program, Israel Davis; Mgr. Expressive Arts Program, Elizabeth Goddard; Mgr. Youth Programs, Becca Schaub; Public Rels., Phillip Meade; Dir. Devel., Jill May.
Personnel Profile: Full-Time Paid 8; Part-Time Paid 5; Part-Time Volunteers 200.
Governing Authority: nonprofit organization. Tax-exempt: 501(c)(3).
Institution Type/Description: Civic Art, Cultural Center.
Collections: works by contemporary artists.
Facilities: 170-seat theater; educational center.
Activities: music concerts; dance concerts; performance art events; films; loan & participatory exhibitions; theater; literature readings; interarts.
Publications: monthly, events calendar; quarterly newsletter.
Hours & Admission Prices: Tues.-Sat. noon-10, Sun. noon-7. Adults $8; members no charge. &
Attendance: 120,000 (accurate)
Membership: Student & Senior $25; Individual $40; Family $75; Reciprocal $150; Off the Wall $500 & up.

Grass Lake

WATERLOO AREA HISTORICAL SOCIETY, 13493 Waterloo-Munith Rd., Grass Lake, MI 49240. Mailing Address: P.O. Box 37, Stockbridge, MI 49285-0037. Tel.: 517-596-2254.
E-mail: info@waterloofarmmuseum.org
Web Site: waterloofarmmuseum.org
Founded: 1962.
Congressional District: 7
Key Personnel: Pres. (V), April Gasbarre; Treas., Arlene R. Kaiser; Museum Shop Mgr., Nancy Wisman.
Personnel Profile: Part-Time Paid 3; Part-Time Volunteers 150.
Governing Authority: society. Parent Institution: Waterloo Area Historical Society. Subsidiary Institution: Waterloo Farm Museum; Dewey School Museum. Tax-exempt.
Institution Type/Description: History Museum.
Collections: Historic Structures: 1855 Greek Revival farmhouse; 1840s log cabin; 1840s-1880 one-room schoolhouse; 1880s outbuildings.
Research Fields: early Michigan pioneer history.
Facilities: Gift items for sale.
Activities: guided tours; summer & fall classes in pioneer crafts; special events. Museum Sponsors: Living History Program for children in May & September; Eastern Woodland Indian Living History in June; Log Cabin Day in June; Pioneer Day & Dewey School Day in October; Christmas on the Farm and Dewey School in December.
Publications: booklets, Primer for Guides; material for classrooms; newsletters.
Hours & Admission Prices: Fri.-Sun. 1-5. Adults $3, senior citizens over 62 $2.50, Dewey School & children 5-17 $1; discounts to AAA, AAM & Michigan Historical Society members; members & children under 5 no charge.
Attendance: 8,000 (estimated)
Membership: Individual Senior $10; Senior Couple & Individual $15; Couple & Family $20; Contributing $40; Business Sponsors $50; Senior Life $150; Individual Life $200.

Grayling

*** HARTWICK PINES LOGGING MUSEUM, (M),** 4216 Ranger Rd., Grayling, MI 49738. Tel.: 989-348-2537. Fax: 989-344-6803.
Key Personnel: Site Historian, Rob Burg
Institution Type/Description: Logging History Museum.
Collections: logging history, tools & equipment; photographs.
Facilities: nature trails.

Activities: special events; seasonal activities; group & school tours.
Hours & Admission Prices: May-Sept. call for hours. Museum: no charge. Park: Recreation Passport required for park entry.

Greenville

THE FIGHTING FALCON MILITARY MUSEUM, 516 W. Cass St., Greenville, MI 48838. Tel.: 616-225-1940.
Key Personnel: Pres. (V), Bill Garlick.
Governing Authority: Tax-exempt.
Institution Type/Description: Military History Museum.
Collections: military history & artifacts; Army Air Force aircraft; personal artifacts; photographs.
Activities: special events; educational programs.
Hours & Admission Prices: Call for hours. No charge; donations accepted. &

Grosse Ile

GROSSE ILE HISTORICAL SOCIETY, East River and Grosse Ile Pkwy., Grosse Ile, MI 48138. Mailing Address: P.O. Box 131, Grosse Ile, MI 48138-0131. Tel.: 734-675-1250.
E-mail: gihsociety@yahoo.com
Web Site: www.gihistory.org
Founded: 1959.
Congressional District: 14
Key Personnel: Pres., Tammy Taylor; Mgr., Clare Koester.
Personnel Profile: Part-Time Volunteers 50.
Governing Authority: society. Tax-exempt.
Institution Type/Description: Historical Society Museum.
Collections: documents related to local history; furniture; clothing; artifacts; photos. Historic Buildings: 1906 lighthouse; 1873 customs house; 1904 railroad station; Naval Air Station Grosse Ile.
Research Fields: local history.
Activities: guided tours; permanent & temporary exhibits; membership programs.
Publications: newsletter; local publications; Arcadia: Grosse Ile; Arcadia: U.S. Naval Air Station, Grosse Ile; Grosse Ile Then and Now.
Hours & Admission Prices: March-Dec. Thurs. 10-12, Sun. 1-4; other times by appointment. No charge; donations accepted. Closed holidays.
Attendance: 1,500 (estimated)
Membership: Individual $10; Family $20; Associate $50; Patron $100, Life $300; Family Life & Corporate $500.

Grosse Pointe Shores

EDSEL & ELEANOR FORD HOUSE, (M), 1100 Lake Shore Rd., Grosse Pointe Shores, MI 48236-4106. Tel.: 313-884-4222. Fax: 313-884-5977.
E-mail: info@fordhouse.org
Web Site: www.fordhouse.org
Founded: 1978.
Congressional District: 14
Key Personnel: Pres., Kathleen Stiso Mullins; Chm. (V), Edsel B. Ford, II; Vice Pres., Communications, Ann Fitzpatrick; Dir. Education, Christopher Shires; Dir. Devel., Bernadette Banke; Cur., Josephine Shea; Vice Pres. Finance & Admin., Robert Seestadt; Dir. Group Tours Sales, Donna Buchanan; Gallery Shop Mgr., Matthew Peplinski.
Personnel Profile: Full-Time Paid 39; Part-Time Paid 51.
Governing Authority: nonprofit organization. Tax-exempt: 501(c)(3).
Institution Type/Description: Historic Building & Site: Family home of Edsel & Eleanor Ford, designed by architect Albert Kahn & built in 1926-29 with interiors from historic English homes & four rooms designed in the modern style by Walter Dorwin Teague. Gardens & grounds designed by Jens Jensen.
Collections: paintings; graphic arts; French & English period furniture; silver; glass; ceramics; historic textiles. Historic Buildings: 60-room Main House; Power House; Play House; Gate Lodge.
Research Fields: history of collection, house & grounds.
Facilities: 87 acres of gardens & grounds with formal Rose garden; tea room; 2,500 sq. ft. exhibition & meeting space. Museum-related items for sale.
Activities: guided tours; lectures; children's programs; traveling exhibitions.
Publications: illustrated history & guide book, Edsel & Eleanor Ford House; Edsel B. Ford, 1893-1943.
Hours & Admission Prices: Jan.-March Tues.-Sun. 12-4; April-Dec. Tues.-Sat. 10-4, Sun. 12-4. Adults $12, senior citizens $11, children 6-12 $8; discounts to AAM members; children 5 & under no charge. Grounds $5. Closed New Year's Day; Thanksgiving; Christmas. &
Attendance: 40,000 (estimated)
Membership: Grounds $25; Individual $30; Individual Plus $35; Family & Grandparent $45; Fan $100; Friend $250.

Hamtramck

UKRAINIAN-AMERICAN ARCHIVES & MUSEUM, 11756 Charest St., Hamtramck, MI 48212-3059. Tel.: 313-366-9764.
E-mail: ukrainianmuseum@sbcglobal.net
Web Site: www.ukrainianmuseumdetroit.org
Founded: 1958.
Key Personnel: Exec. Dir., Chrystyna Nykorak; Pres., Swetlana Leheta.
Personnel Profile: Full-Time Paid 1; Full-Time Volunteers 1; Part-Time Volunteers 9.
Governing Authority: nonprofit. Tax-exempt: 501(c)(3).
Institution Type/Description: History Museum & Archives.
Collections: art; culture; costumes; headdresses; ceramics; woodwork; photographs; musical instruments.
Research Fields: immigration.
Facilities: meeting rooms.
Activities: special events.
Publications: quarterly newsletter.
Hours & Admission Prices: Tues.-Fri. 9-5; other times by appointment. Adults $3. Closed holidays.
Attendance: 500 (estimated)
Membership: $10; $25; $35; $100; $1,000 & up.

Hanover

CONKLIN REED ORGAN & HISTORY MUSEUM, 105 Fairview, Hanover, MI 49241-0256. Mailing Address: P.O. Box 256, Hanover, MI 49241-0256. Tel.: 517-563-8927. Fax: 517-563-8927.
E-mail: hhahs@frontier.com
Web Site: www.conklinreedorganmuseum.org
Formerly: Lee Conklin Antique Organ Museum
Founded: 1977.
Congressional District: 2
Key Personnel: Pres., Betty Jo DeForest; Treas., Richard Tallis; Museum Gift Mgr., Ann Tallis.
Personnel Profile: Part-Time Paid 1; Part-Time Volunteers 20.
Governing Authority: nonprofit organization. Parent Institution: Hanover-Horton Area Historical Society Inc. Tax-exempt.
Institution Type/Description: Historical Society & Antique Reed Organ Museum: housed in 1911 Hanover High School.
Collections: reed organs; parlor organs; chapel organs; Melodeons; farm & home equipment; restored early 20th-century classroom; old telephone switchboard; loom; player piano; job press; dog treadmill; arrow heads; local memorabilia; early fire fighting equipment.
Research Fields: local history; Reed organs.
Facilities: 82 acre Heritage Park; nature trails; 100-seat auditorium. Handcrafted items & organ music tapes for sale.
Activities: nature trails; hay wagon rides; guided tours; working sawmill demonstrations; concerts; organ repair & maintenance workshops. Society Sponsors: Special May Event; 4th of July Celebration; Rust N Dust Event in August; Harvest Home & Garden Luncheon in September; Corn Maze in October; Bluegrass Bonanza October to March; Christmas Open House in December.
Publications: quarterly newsletter, Bulletin of the Hanover-Horton Area Historical Society, Inc.
Hours & Admission Prices: May-Oct. Sun. 1-5; call for additional hours. No charge; donations accepted. Closed Easter. &
Attendance: 3,359 (accurate)
Membership: Full-time Student & Senior Citizen $10; Individual $20; Family $25; Business & Senior Citizen Life $100; Life $250.

Harbor Beach

FRANK MURPHY MEMORIAL MUSEUM, 142 S. Huron Ave., Harbor Beach, MI 48441. Mailing Address: P.O. Box 113, Harbor Beach, MI 48441-0113. Tel.: 989-479-6477.
Web Site: harborbeachmi.org/frank_murphy
Key Personnel: Cur., Barb McGowan
Institution Type/Description: Historic House Museum: housed in the home of statesman, Frank Murphy.
Collections: personal artifacts; furniture; photographs.
Activities: tours.
Hours & Admission Prices: Memorial Day-Labor Day Tues.-Fri. 12-4, Sat.-Sun. 10-4 & by appointment. Adults $2, children $1.

THE GRICE HOUSE MUSEUM, 865 N. Huron Ave., Harbor Beach, MI 48441. Mailing Address: Harbor Beach Chamber of Commerce, P.O. Box 113, Harbor Beach, MI 48441-0113. Tel.: 989-479-6477.
Web Site: harborbeachchamber.com/grice.html
Institution Type/Description: Historic House: former home of the Grice family, built in 1875.
Collections: mid-nineteenth century kitchen, parlor, sewing room & bedroom; military artifacts room; history of the Great Lakes room; one-room schoolhouse.
Hours & Admission Prices: Memorial Day-Labor Day Tues.-Fri. 12-4, Sat.-Sun. 10-4.

Harbor Springs

ANDREW J. BLACKBIRD MUSEUM, 368 E. Main St., Harbor Springs, MI 49740-1514. Mailing Address: P.O. Box 678, Harbor Springs, MI 49740-0678. Tel.: 231-526-0612 (museum); 2104 (City of Harbor Springs). Fax: 231-526-6865.
E-mail: cityhs@freeway.net
Founded: 1952.
Congressional District: 11
Key Personnel: Mgr., Joyce Shagonaby; Pres. (V), Robert Shagonaby.
Personnel Profile: Part-Time Paid 1; Part-Time Volunteers 6.
Governing Authority: municipal. Tax-exempt.
Institution Type/Description: Indian Museum: housed in c. 1855 home of Andrew J. Blackbird.
Collections: art; tools; utensils; Odawa Indian clothing.
Activities: permanent & traveling exhibitions.
Publications: History of the Ottawa & Chippewa Indians of Michigan; Ottawa Quillwork on Birchbark; Ottawa Quillwork on Birchbark Edition 2007; Beadwork & Textiles of the Ottawa.
Hours & Admission Prices: Mon.-Fri. 10-4. No charge; donations accepted. &
Attendance: 600 (estimated)
Membership: $10; $25; $50; $100.

Hartland

FLORENCE B. DEARING MUSEUM - HARTLAND HISTORICAL SOCIETY, 3503 Avon St., Hartland, MI 48353. Mailing Address: P.O. Box 49, Hartland, MI 48353. Tel.: 810-299-7621.
E-mail: hrfoley@earthlink.net
Web Site: www.hartlandareahistory.org
Founded: 1971.
Congressional District: 8
Key Personnel: Pres., Hildy R. Foley; Cur., Michael Forster.
Personnel Profile: Part-Time Volunteers 15.
Governing Authority: Parent Institution: Hartland Area Historical Society. Tax-exempt.
Institution Type/Description: History Museum: housed in the former Hartland's town hall; built in 1891.
Collections: local history & culture; period furnishings; personal artifacts; photographs; native birds & animals exhibit.
Facilities: c.1892 town hall building.
Activities: Museum Sponsors: Fund Raising Dinner & Silent Auction in Feb.; Memorial Day in May; Potluck Dinner in May & Nov.; Ice Cream Social in July; Heritage Day in Sept.
Hours & Admission Prices: Sun. 2-4; groups by appointment. Adults in group tours $1; individual adults no charge; donations accepted.
Attendance: 750 (estimated)
Membership: Individual $10; Family $15; Business $50.

Hastings

HISTORIC CHARLTON PARK VILLAGE, MUSEUM AND RECREATION AREA, 2545 S. Charlton Park Rd., Hastings, MI 49058-8102. Tel.: 269-945-3775. Fax: 269-945-0390.
E-mail: info@charltonpark.org
Web Site: www.charltonpark.org
Founded: 1936.
Congressional District: 3
Key Personnel: Dir., Dan Patton; Cur., Claire Johnston; Supvr. Operations, Tom Campbell; Education Coord., Shannon Ritzer.
Personnel Profile: Full-Time Paid 6; Part-Time Paid 4; Part-Time Volunteers 1.
Governing Authority: county. Tax-exempt.
Institution Type/Description: History Museum.
Collections: rural life in Southern Michigan; agriculture & industrial equipment; crafts; transportation; decorative arts; domestic life; weapons; costumes; gasoline & steam engines; prehistoric settlers & local historical

records. Historic Structures: 1852 Bristol Inn; 1885 Church; 1869 Lee School; 1886 Hastings Township Hall; c.1890s Hall House; 1908 Hastings Mutual Insurance Building; 1858 Sixberry House; Blacksmith Shop; Carpenter Shop; General Store; Hardware Store; Print Shop; Charlton Gas & Steam Building; Barber Shop; sawmill.
Research Fields: 1827-1940 midwestern rural life.
Facilities: 1,000-vol. library of local history books available for research by request; nature trails; boat launch; rental facilities.
Activities: guided tours; lectures; organized education programs; living history village; permanent & temporary exhibitions; loan service; annual special events.
Publications: The Villager.
Hours & Admission Prices: Park: Memorial Day to Labor Day daily 8am to dusk. Museum & Village: Memorial Day-Labor Day daily 9-4; other times by appointment. Office: Mon.-Fri. 8-5. Special Events & Activities: adults $5, children 5-12 $3; children under 4 no charge. &
Attendance: 41,000 (estimated)

Hickory Corners

GILMORE CAR MUSEUM, (M), 6865 Hickory Rd., Hickory Corners, MI 49060-9788. Tel.: 269-671-5089. Fax: 269-671-5843.
E-mail: info@gilmorecarmuseum.org
Web Site: www.gilmorecarmuseum.org
Founded: 1966.
Key Personnel: Dir., Michael J. Spezia.
Personnel Profile: Full-Time Paid 10; Part-Time Paid 25; Part-Time Volunteers 80.
Governing Authority: private; nonprofit organization. Parent Institution: Genevieve & Donald S. Gilmore Foundation. Tax-exempt: 501(c)(3).
Institution Type/Description: Car Museum.
Collections: over 220 automobiles including an 1899 Locomobile, Duesenberg, Tucker '48, the Model T, & muscle cars of the 1960s; replica of 1940s diner; Disney movie set. Historic Buildings: c1890 train depot; barns; 1930s gas station.
Facilities: 90-acrea site; picnic facilities; interpretive center.
Activities: car shows; concourse d'Elegance; swap meets; special exhibits; auto restoration.
Publications: brochures; newsletter; schedule of events; exhibit book.
Hours & Admission Prices: Automotive Heritage Center: Mon.-Fri. 9-5, Sat.-Sun. 9-6. Adults $10, senior citizens $9, youth 7-15 $8; discounts to AAA members & groups; members & children 6 and under no charge. Historic Campus: April-Nov. call for hours. Closed New Year's Day; Easter; Thanksgiving; Christmas. &
Attendance: 49,000 (accurate)
Membership: Individual $35; Family $50; Contributor $100; Supporter $250; Benefactor $500; Life $1,000.

Holland

CAPPON & SETTLERS HOUSE MUSEUMS, 228 W. 9th St., Holland, MI 49423-3116. Mailing Address: 31 W. 10th St., Holland, MI 49423-3101. Tel.: 616-392-9084. Fax: 616-394-4756.
E-mail: hollandmuseum@hollandmuseum.org
Web Site: www.hollandmuseum.org
Founded: 1986.
Congressional District: 9
Key Personnel: Chm., Geoffrey Reynolds; Exec. Dir., Christopher Shires.
Personnel Profile: Part-Time Paid 1; Part-Time Volunteers 20.
Governing Authority: not-for-profit corporation. Parent Institution: Holland Historical Trust. Tax-exempt: 501(c)(3).
Institution Type/Description: Historic Houses: 1874 Italianate Style home of Holland's first mayor; 1867 cottage of a Great Lakes Ship's Carpenter.
Collections: original furnishings; millwork.
Research Fields: the Cappon family; 19th-century history Holland MI; 19th century immigrant history; 19th century decorative arts.
Activities: guided tours; education programs. Special Events: Ice Cream Social; Tulip Time, lamplight tours, Christmas tours.
Hours & Admission Prices: May-Sept. Fri.-Sat. 12-4; special hours during Tulip Time. Adults $5; discounts to AAM members; members & children under 6 no charge. &
Attendance: 1,794 (accurate)
Membership: Senior Individual (over 64) $20; Individual $25; Senior Couple $30; Family $35; Builder $75; Sponsor $100; Patron $250; Benefactor $500; Wichers Circle $1,000.

DE GRAAF NATURE CENTER, 600 Graafschap Rd., Holland, MI 49423-4549. Tel.: 616-355-1057. Fax: 616-355-1069.
E-mail: degraafnaturecenter@cityofholland.com
Web Site: www.degraaf.org

Founded: 1962.
Congressional District: 2
Key Personnel: Dir., Robert Venner; Naturalist, Mike Graves; Naturalist, Lisa McKellips; Naturalist, Erin Wildt.
Personnel Profile: Full-Time Paid 2; Part-Time Paid 2; Part-Time Volunteers 20; Interns 1.
Governing Authority: municipal; nonprofit. Parent Institution: City of Holland Leisure & Cultural Services. Tax-exempt.
Institution Type/Description: Nature Center.
Collections: Michigan rocks; US rocks; taxidermy including state birds & animals; bird nests; insects; woodland habitat diorama. Historic Building: log cabin c.1847.
Facilities: library of nature related materials available to the public. Books, magnifier boxes, & other gift items for sale.
Activities: guided tours; lectures; films; study clubs; hobby workshops; organized education programs for children; docent program; loan & temporary exhibitions; summer classes for children, grades K-12; regular Sat. programs for children & adults: field trip to natural areas, videos, natural history, tours. Museum Sponsors: Animal Day; Warbler Festival; Pioneer Christmas; Fall Festival.
Publications: quarterly newsletter, Friend News.
Hours & Admission Prices: Brower Interpretive Center: Tues.-Fri. 9-5, Sat. 10-5. No charge. Trails: dawn-dusk. School Programs: students $2.25. Closed holidays. &
Attendance: 13,000
Membership: Individual $20; Family $25; Sponsor $50; Sustaining $100; Contributing $250; Benefactor $500.

DEPREE ART CENTER & GALLERY, 275 Columbia Ave., Holland, MI 49423. Mailing Address: P.O. Box 9000, Holland, MI 49422-9000. Tel.: 616-395-7500. Fax: 616-395-7499.
E-mail: art@hope.edu
Web Site: hope.edu/academic/art
Founded: 1982.
Congressional District: 4
Key Personnel: Dir., Dr. Heidi Kraus.
Personnel Profile: Part-Time Paid 4.
Governing Authority: private college; nonprofit organization. Tax-exempt: 501(c)(3).
Institution Type/Description: Academic Art Center & Gallery.
Collections: paintings; sculptures.
Facilities: 1,300 sq. ft. exhibit space; 100-seat auditorium; 2 classrooms.
Activities: guided tours; lectures; docent program; formal education program for Hope College students; loan, temporary and traveling exhibitions.
Hours & Admission Prices: May-Aug. Mon.-Fri. 10-5; Sept.-April Mon.-Sat. 10-5, Sun. 1-5; summer hours apply during college breaks. No charge. Closed New Year's Day; Memorial Day; Thanksgiving; Christmas. &
Attendance: 20,500 (estimated)
Membership: Patron $50.

HOLLAND AREA ARTS COUNCIL, 150 E. 8th St., Holland, MI 49423-3504. Tel.: 616-396-3278. Fax: 616-396-6298.
Web Site: www.hollandarts.org
Founded: 1967.
Congressional District: 4
Key Personnel: Exec. Dir., Lorma Williams Freestone; Pres. Bd., Jay Frankhouse; Program Dir., Mary Sundstrom; Artistic Dir., Derek Johnson; Asst. to Dir., Tennina Miozza.
Personnel Profile: Full-Time Paid 3; Part-Time Paid 3; Part-Time Volunteers 6; Interns 2.
Governing Authority: nonprofit organization. Tax-exempt: 501(c)(3).
Institution Type/Description: Arts Council.
Collections: paintings; sculpture; photographs.
Research Fields: West Michigan Arts Registry: registry of artists located in the West Michigan area.
Facilities: 300-vol. library of arts & crafts books; educational facilities; dance studio. Gift items & books for sale.
Activities: arts festivals; concerts; dance recitals; docent program; arts education classes & workshops for children; guided tours; lectures; participatory exhibits; theatre. Annual Events: fundraiser; art competition exhibits; local arts exhibit.
Publications: newsletter, Artupdate.
Hours & Admission Prices: Mon.-Thurs. 10-8, Sat. 10-3. No charge; donations accepted. Closed New Year's Day; Memorial Day; Independence Day; Labor Day; Thanksgiving; Christmas. &
Attendance: 50,000
Membership: Student $25; Individual $45; Family $60; Patron $100; Donor $250; Sponsor $500; Benefactor $1,000 & up.

*** THE HOLLAND MUSEUM, (M),** 31 W. 10th St., Holland, MI 49423-3101. Tel.: 616-394-1362. Fax: 616-394-4756.
E-mail: hollandmuseum@hollandmuseum.org
Web Site: www.hollandmuseum.org
Founded: 1937.
Congressional District: 9
Key Personnel: Chm., Geoffrey Reynolds; Exec. Dir., Christopher Shires; Operations Mgr., Paula Dunlap; Museum Shop Mgr., Taylor Wise-Hawthorn.
Personnel Profile: Full-Time Paid 6; Part-Time Paid 10.
Governing Authority: nonprofit corporation. Parent Institution: Holland Historical Trust. Tax-exempt: 501(c)(3); 170(b)(1)(A); 509(a)(2).
Institution Type/Description: History Museum: housed in 1914 restored Federal Post Office building.
Collections: Dutch art & culture; early Dutch settlers & local history; the Netherlands collection of Dutch paintings, 17th-20th centuries; decorative arts including delftware; furniture; glassware; traditional Dutch costumes.
Research Fields: local history.
Facilities: research library; archives.
Activities: permanent & temporary exhibits; education programs.
Publications: quarterly newsletter; Review.
Hours & Admission Prices: Mon. & Wed.-Sat. 10-5. Adults $7, seniors $6, students $4; discounts to AAM & AAA members; members & children under 6 no charge. Closed major holidays. &
Attendance: 20,983 (accurate)
Membership: Senior 65 & over $20; Individual $25; Senior Couple 65 & over $30; Family $35; Builder $75; Sponsor $100; Patron $250; Benefactor $500; Wichers Circle $1,000.

WINDMILL ISLAND GARDENS, 1 Lincoln Ave., Holland, MI 49423. Tel.: 616-355-1030; 888-535-5792. Fax: 616-355-1035.
E-mail: windmill@cityofholland.com
Web Site: www.windmillisland.org
Formerly: Windmill Island Municipal Park
Founded: 1965.
Key Personnel: Dir., Ad vanden Akker.
Personnel Profile: Full-Time Paid 2; Part-Time Paid 55.
Governing Authority: municipal. Tax-exempt.
Institution Type/Description: Park Museum.
Collections: Little Netherlands Museum: Dutch culture; panoramic exhibit of town & country in Old Holland. Historic Structure: c.1761 restored Dutch windmill.
Facilities: flower garden.
Activities: guided tours; films; permanent exhibitions; wooden shoe dancers; Amsterdam street organ; period carousel; farm animals.
Publications: Brochure.
Hours & Admission Prices: late April to early Oct. daily 9:30-6. Tulip Time: daily 9-6. Adults $7.50, children 5-15 $4.50; discounts to groups of 20 or more; children 4 & under no charge. Ticket booth open 9:15-5. &
Attendance: 50,000 (accurate)

Houghton

A. E. SEAMAN MINERAL MUSEUM, 1404 E. Sharon Ave., Michigan Technological University, Houghton, MI 49931-1295. Tel.: 906-487-2572. Fax: 906-487-3027.
E-mail: tjb@mtu.edu
Web Site: www.museum.mtu.edu
Founded: 1902.
Congressional District: 110
Key Personnel: Dir. & Prof., Dr. Theodore J. Bornhorst; Cur. & Prof., Dr. George W. Robinson; Assoc. Cur., Dr. Chris Stefano; Museum Asst., Monica Rovano.
Personnel Profile: Full-Time Paid 4; Interns 2.
Governing Authority: university. A division of Michigan Technological University, Houghton, MI. Tax-exempt.
Institution Type/Description: Mineralogy Museum.
Collections: Great Lakes region, Michigan & worldwide minerals; native minerals.
Research Fields: mineralogy & crystallography. L.S. Copper District geology, mineralogy & mining history.
Facilities: 500-vol. library of books & periodicals pertaining to mineralogy & geology available for use on the premises by mineralogists.
Activities: guided tours & gallery talks by request; traveling & permanent exhibitions.
Hours & Admission Prices: Jan. 14-April Tues.-Fri. 9-4; May-Dec. 23 Mon.-Sat. 9-5. Adults $5, seniors 65 & over $4, college students with ID $3, junior 8-17 $2; discounts to groups; members, children under 8 & Tues. no charge. Closed Jan. 1-13 & Dec. 24-31; Independence Day; Memorial Day; Labor Day; Thanksgiving & day after. &

Attendance: 2,500 (accurate)
Membership: Basic $25; Crystal $100; Copper $200; Silver $500; Gold $1,000 & up.

ISLE ROYALE NATIONAL PARK, 800 E. Lakeshore Dr., Houghton, MI 49931-1869. Tel.: 906-482-0984. Fax: 906-482-8753.
E-mail: liz_valencia@nps.gov
Web Site: www.nps.gov/isro
Founded: 1975.
Congressional District: 11
Personnel Profile: Full-Time Paid 1; Part-Time Volunteers 1.
Governing Authority: federal. Parent Institution: National Park Service, Dept. of the Interior, Washington DC. Tax-exempt.
Institution Type/Description: National Park & Historic Site: 1890 Edisen Commercial Fishery complex. Historic Structure: 1855 Rock Harbor Lighthouse. Park Museum: located in Houghton, MI.
Collections: North American Indian baskets; 1928-1931 Isle Royale photographs; archaeological survey collection; manuscripts; shipwreck artifacts; commercial fishery equipment; 19th- & 20th-century summer home resort furnishings; 1840-1895 copper mining tools; geologic artifacts; herbarium. Historic Building: 1855 Rock Harbor Lighthouse on Isle Royale including Great Lakes shipping & maritime history of the area.
Research Fields: copper mining; shipwrecks; Indian baskets; commercial fishing; Isle Royale resorts; lighthouses.
Facilities: 1,000-vol. library pertaining to regional history, shipping, mining & Lake Superior available for research on premises by request June-Sept.; wilderness camping.
Activities: guided tours.
Hours & Admission Prices: Rock Harbor Visitor Center: May-June & Sept. call for hours; July-Aug. daily 8-6. Windigo Visitor Center May-June & Sept. call for hours; July-Aug. daily 8-4:30. Houghton Visitor Center: call for hours. Museum: by appointment. No charge; donations accepted.
Attendance: 2,000 (estimated)

Imlay City

IMLAY CITY HISTORICAL COMMISSION, INC., 77 Main St., Imlay City, MI 48444-1313. Tel.: 810-724-1111.
E-mail: bswihart1904@charter.net
Web Site: www.imlaycityhistoricalmuseum.org
Founded: 1978.
Congressional District: 10
Key Personnel: Pres. (V), Carla Jepsen; Museum Shop Mgr., Marilyn Swichart.
Personnel Profile: Part-Time Volunteers 6.
Operating Expenses: 16,445
Operating Income: 16,366
Governing Authority: Tax-exempt.
Institution Type/Description: History Museum: housed in the Grand Truck Depot.
Collections: photographs; local artifacts; military artifacts; farm tools; Bob Burman memorabilia; Imlay City High School graduating composites.
Research Fields: family history.
Publications: quarterly newsletter; 3rd grade coloring book; historic calendars & atlas.
Hours & Admission Prices: April-Dec. Wed. 9 to noon, Sat. 1-4. No charge; donations accepted. &
Attendance: 1,000 (accurate)
Membership: Individual $15; Couple/Family $20; Corporate Sponsorship $40.

Ironwood

IRONWOOD AREA HISTORICAL MUSEUM, 150 N. Lowell St., Ironwood, MI 49938-2032. Mailing Address: P.O. Box 553, Ironwood, MI 49938-0553. Tel.: 906-932-0287.
Web Site: www.ironwood.org
Founded: 1970.
Congressional District: 11
Key Personnel: Dir., Gary Harrington.
Personnel Profile: Part-Time Volunteers 23.
Governing Authority: city; nonprofit organization. Tax-exempt.
Institution Type/Description: History Museum: housed in former Chicago & Northwestern Depot.
Collections: iron mining memorabilia; local historical photographs; railroad memorabilia; early sports artifacts; household artifacts; Ironwood Chamber of Commerce artifacts.
Research Fields: iron mining maps of Gogebic Range.
Facilities: 100-vol. library of mining material.

Activities: guided tours; lectures; concerts; arts festivals; broadcast programs; participatory & temporary exhibitions. Museum Sponsors: Heritage Day; historical plays.
Publications: periodic newsletter.
Hours & Admission Prices: Memorial Day to Labor Day Mon.-Sat. 12-4; Sept.-May daily 12-4; Sept.-May by appointment. No charge; donations accepted. Closed Independence Day. &
Attendance: 2,000 (estimated)
Membership: Individual $15; Lifetime $250.

Ishpeming

U.S. SKI & SNOWBOARD HALL OF FAME, (M), 610 Palms, Ishpeming, MI 49849-1035. Mailing Address: P.O. Box 191, Ishpeming, MI 49849-0191. Tel.: 906-485-6323. Fax: 906-486-4570.
E-mail: twest@skihall.com
Web Site: www.skihall.org
Formerly: U.S. National Ski and Snowboard Hall of Fame & Museum
Founded: 1954.
Congressional District: 11
Key Personnel: C.E.O., J. Thomas West; Chm. (V), Bernard Weichsel; Admin., Ann Schroeder.
Personnel Profile: Full-Time Paid 3; Full-Time Volunteers 2; Part-Time Paid 1; Part-Time Volunteers 6.
Governing Authority: nonprofit organization. Affiliated with the United States Ski Association, P.O. Box 100, Park City, UT 84060. Tax-exempt.
Institution Type/Description: Sports Museum.
Collections: ski-related exhibits.
Research Fields: ski history.
Facilities: 1,000-vol. library of ski-related material, available for research on premises only. Gift items for sale.
Activities: guided tours; formally organized education programs for children; permanent exhibitions.
Publications: monthly e-newsletter; biannual induction program; 75 Years of Skiing; Skiing: Then & Now; Nine Thousand Years of Skis; The Flying Norseman; Midwest Skiing-A Glance Back; bimonthly, Skiing Heritage.
Hours & Admission Prices: Mon.-Sat. 10-5. No charge; donations accepted. Closed New Year's Day; Independence Day; Thanksgiving; Christmas. &
Attendance: 10,000 (estimated)
Membership: Regular $50; Carl Tellefsen Society $1,000.

Ithaca

GRATIOT COUNTY GENEALOGICAL LIBRARY, 228 W. Center St., Ithaca, MI 48847. Mailing Address: Gratiot County Historical & Genealogical Society, P.O. Box 73, Ithaca, MI 48847. Tel.: 989-875-6232.
Institution Type/Description: Genealogical Library: housed in the Peet/Miller house.
Collections: photographs; personal artifacts; diaries; local history.
Facilities: library.
Activities: research.
Hours & Admission Prices: May-Sept. Tues. 1-5, last Tues. each month 1-8; Oct.-April Tues. 1-5. Closed New Year's Day; Christmas.

GRATIOT COUNTY HISTORICAL MUSEUM, 129 W. Center St., Ithaca, MI 48847. Mailing Address: Gratiot County Historical & Genealogical Society, P.O. Box 73, Ithaca, MI 48847. Tel.: 989-466-5730.
Institution Type/Description: Historical Society Museum: housed in a Victorian-style house; built in 1881. Listed on the National Register of Historic Places.
Collections: local history & culture; period furnishings; personal artifacts; photographs; household artifacts; farm implements & tools.
Hours & Admission Prices: mid-May to Oct. Tues. 12-3. No charge; donations accepted.

Jackson

✱ **ELLA SHARP MUSEUM, (M),** Jackson, MI 49203-5094. Tel.: 517-787-2320. Fax: 517-787-2933. Facebook: Ella Sharp Museum.
E-mail: info@ellasharp.org
Web Site: www.ellasharp.org
Founded: 1965.
Congressional District: 2
Key Personnel: Exec. Dir., Amy Reimann; Dir. Collections, Judy Horn; Museum Shop Mgr., Florence Csage.
Personnel Profile: Full-Time Paid 10; Part-Time Paid 11; Part-Time Volunteers 100; Interns 1.

Governing Authority: nonprofit organization. Tax-exempt: 501(c)(3).
Institution Type/Description: Art & History Museum.
Collections: Jackson County Michigan history; Merriman-Sharp family artifacts. Historic Buildings: farmhouse; woodworking shop; schoolhouse; barns; general store complex; log house.
Research Fields: Victorian restoration, furnishings & decorative arts; leisure & recreation; domestic life; transportation; Jackson County Michigan history.
Facilities: studio classrooms; restaurant. Gift items for sale.
Activities: guided tours; lectures & demonstrations; films; formally organized education programs for children & adults; workshops; permanent & temporary exhibits; gallery talks; docent program; seasonal special events.
Publications: bimonthly newsletter; class bulletins; brochures; illustrated monographs of collections.
Hours & Admission Prices: Tues.-Wed. & Fri.-Sat. 10-5, Thurs. 10-7. Galleries: Adults $5, children 5-12 $3; Galleries & tour Adults $7, children 5-12 $5; discounts to AAA, NARM & Public Broadcasting members; members & military no charge. Closed major holidays. &
Attendance: 20,000 (estimated)
Membership: Student/Teacher $20; Individual $40; Dual $50; Family $60; Sustainer $125; Investor $250; Benefactor $500; Steward $1,000.

K.I. Sawyer AFB

K.I. SAWYER HERITAGE AIR MUSEUM, 402 3rd St., K.I. Sawyer AFB, MI 49841. Mailing Address: 500 S. 3rd St., Marquette, MI 49855. Tel.: 906-362-3531. Facebook: KI Sawyer Heritage Air Museum.
E-mail: bvick37@gmail.com
Web Site: www.kishamuseum.org
Founded: 1995.
Key Personnel: Pres., C.E.O. & Museum Shop Mgr., Robert H. Vick.
Personnel Profile: Part-Time Volunteers 5.
Volunteer Hours: 2,200
Governing Authority: Subsidiary Institution: Silverwing Community Center. Tax-exempt.
Institution Type/Description: Military History Museum.
Collections: Air Force history; military artifacts & aircraft; photographs; personal artifacts; memorials.
Facilities: Museum-related items for sale.
Activities: operator of community center.
Hours & Admission Prices: May-Oct. 1 Wed.-Sun. 1-5. No charge.
Attendance: 500 (estimated)
Membership: Annual $30.

Kalamazoo

AIR ZOO, 6151 Portage Rd., Kalamazoo, MI 49002-3003. Tel.: 269-382-6555; 866-524-7966 (Toll Free). Fax: 269-382-1044.
E-mail: airzoo@airzoo.org
Web Site: www.airzoo.org
Formerly: Kalamazoo Aviation History Museum
Founded: 1977.
Congressional District: 6
Key Personnel: Pres. & C.E.O., Robert E. Ellis; Chm. Bd., Preston S. Parish; Cur. & Registrar, Bill Painter; Dir. Operations, Kim Robinson; Dir. Mktg. & Human Resources, Debra Mixis; Volunteer & Community Resources Mgr., Tamra Stafford; Librarian, Carol Smith; Bldg. & Grounds Supt., Jim Ross; Museum Store Mgr., Meredith Martin; Public Rels. & Mktg. Mgr., Danielle Nicholl; Sr. Cur. Aircraft, Greg Ward.
Personnel Profile: Full-Time Paid 25; Part-Time Paid 35; Part-Time Volunteers 175; Interns 3.
Governing Authority: private; nonprofit organization. Tax-exempt: 501(c)(3).
Institution Type/Description: Aviation Museum.
Collections: Guadalcanal Memorial Museum & Monument; Michigan Aviation Hall of Fame; 90 historic aircraft.
Facilities: 5,000-vol. library of aviation-related books; 1,000 videotapes & DVDs; 170,000 sq. ft. exhibit space; restaurant; theatre; banquet facilities. Museum-related items for sale.
Activities: guided tours; education programs; monthly special events; annual events.
Publications: quarterly, Air Zoo Newsletter; Airmail News, monthly volunteer letter.
Hours & Admission Prices: Mon.-Sat. 9-5, Sun. 12-5. Adults $8; additional fees for rides & attractions. Closed Thanksgiving; Christmas Eve & Day. &
Attendance: 125,000 (accurate)
Membership: Basic $50; Family $125; Grandparents $150; Supporting $250; Patron $500; P-40 Society $1,000.

ALAMO TOWNSHIP MUSEUM-JOHN E. GRAY MEMORIAL MUSEUM, 8119 N. 6th St., Kalamazoo, MI 49009-8808. Mailing Address: 7180 N. 2nd St., Kalamazoo, MI 49009-8814. Tel.: 269-344-2107.
E-mail: williamsmichigan@earthlink.net
Web Site: home.earthlink.net/~tommaas/Alamo_Township_museum.htm
Founded: 1969.
Congressional District: 47
Key Personnel: Cur., Brian Smith.
Governing Authority: township. Tax-exempt.
Institution Type/Description: Historic House: built in 1865 as a Presbyterian Church.
Collections: farm implements; church records; tapes of former teachers & pioneers; school records; township records.
Research Fields: genealogy; Alamo Township Pioneers.
Facilities: 50-vol. library of historic records & manuscripts available for use by written request; picnic area.
Activities: guided tours; lectures; films; arts festivals; drama; radio & TV programs; training program for professional museum workers; loan, temporary & permanent exhibitions.
Publications: programs of events, Anvil.
Hours & Admission Prices: May-Oct. Tues. 1-3, Sat.-Sun. 2-4; groups by appointment. No charge; donations accepted. &
Attendance: 5,000 (estimated)

KALAMAZOO INSTITUTE OF ARTS, (M), 314 S. Park St., Kalamazoo, MI 49007-5102. Tel.: 269-349-7775, ext. 3001. Fax: 269-349-9313.
E-mail: museum@kiarts.org
Web Site: www.kiarts.org
Founded: 1924.
Congressional District: 6
Key Personnel: Exec. Dir., James A. Bridenstine; Pres. (V), Nancy Springgate; Dir. Finance & Personnel, George Baltmanis; Registrar, Robin Goodman; Dir. Museum Education, Susan Eckhardt; Dir. Devel., Joe Bower; School Dir., Denise Lisiecki; Dir. Facilities, Ron Boothby; Mktg. & Public Rels. Coord., Phil Meade; Membership Coord., Jay Simon; Librarian, Malcolm McBryde; Special Events & Volunteer Coord., Sandy Linabury; Museum Shop Mgr., Karyn Juergens.
Personnel Profile: Full-Time Paid 27; Part-Time Paid 62.
Governing Authority: nonprofit. Tax-exempt: 501(c)(3).
Institution Type/Description: Art Museum/Center & School. Focus: American Art.
Collections: 19th & 20th-century American art; 20th-century European art; 15th to 20th-century graphics; ceramics; small sculpture; photography; works on paper.
Research Fields: fine arts.
Facilities: 10,000-vol. library of books & periodicals. Works by Michigan artists & museum-related items for sale.
Activities: guided tours; lectures; films; art fair; docent program; formally organized education program for children & adults; school outreach program; bus trips.
Publications: quarterly newsletter; biennial report; exhibition catalogues; program brochures.
Hours & Admission Prices: Tues.-Sat. 10-5, Sun. 12-5. Adults $5, students $2; members no charge. Closed major holidays. &
Attendance: 109,000 (accurate)
Membership: Student $35; Individual $55; Family $85; Sustaining $125; Donor $175; Patron $250; Benefactor $500; Director's Circle: Founder $1,000; Director's Circle Leader $2,500; Director's Circle Visionary $5,000.

KALAMAZOO NATURE CENTER, INC., 7000 N. Westnedge Ave., Kalamazoo, MI 49009-6309. Tel.: 269-381-1574. Fax: 269-381-2557.
E-mail: lpanich@naturecenter.org
Web Site: www.naturecenter.org
Founded: 1960.
Congressional District: 3
Key Personnel: C.E.O. & Pres., Willard M. Rose, Ph.D.; Dir. Finance, Sue Sobeck; Dir. Camps, Jenny Metz; Dir. Exhibits & Public Programs, Jason Byler; Dir. School Programs, Pete Stobie; Vice Pres. Devel., Michelle Karpinski; Vice Pres. Offsite Programs, Sarah Reding; Museum Shop Mgr., Rose Norwood.
Personnel Profile: Full-Time Paid 27; Part-Time Paid 71; Part-Time Volunteers 250; Interns 3.
Governing Authority: nonprofit organization. Tax-exempt: 501(c)(3).
Institution Type/Description: Nature Center.
Collections: mammal; avian; botanical; rock; insect; farm implements; Indian arrowheads. Historic Buildings: 1858 restored Greek Revival homestead; 1820 log cabin & farmstead; free-flying butterfly house.
Research Fields: ornithology; ecology; human environment.
Facilities: community garden; 160-seat auditorium; self-guiding nature trails. Books, educational materials & gifts for sale.
Activities: guided tours; lectures; films; gallery talks; formally organized education programs for children, adults, undergraduate & graduate students; preschool program; wild animal care & rehabilitation program; avian research; living history programs; simple farm technology program; docent program or council; permanent & temporary exhibitions; ecotour program; wildlife viewing room. Museum Sponsors: classes for handicapped; inner-city programs; summer camp; off-site educational programs that travel to schools and other community organizations, traveling art exhibits; special festivals and events; coordinates program to place interpreters in Michigan State Parks.
Publications: bimonthly periodical, Nature Center News; wild animal rehabilitation manual; The Atlas of Breeding Birds of Michigan; occasional publications, Research Publications, Special Publications, Limited Edition Prints, The Birds of Michigan, Glimpsing the Whole: The Kalamazoo Nature Center Story; Of Woods and Other Things, Essays; essays, Ramblings, Reflections on Nature.
Hours & Admission Prices: Mon.-Sat. 9-5, Sun. 1-5. Adults $6, senior citizens & students $5, children 4-13 $4; discounts to AAM & AAA members; members & children 3 & under no charge. Closed New Year's Day; Thanksgiving; Christmas Eve & Day. &
Attendance: 265,995 (accurate)
Membership: Family $60.

＊ **KALAMAZOO VALLEY MUSEUM, (M),** 230 N. Rose St., Kalamazoo, MI 49007-5803. Mailing Address: P.O. Box 4070, Kalamazoo, MI 49003-4070. Tel.: 269-373-7990. Fax: 269-373-7997.
E-mail: wgouldmcelh@kvcc.edu
Web Site: www.kalamazoomuseum.org
Founded: 1927.
Congressional District: 3
Key Personnel: Dir., Bill McElhone; Special Events Coord., Chris Falk; Cur. Research, Thomas A. Dietz; Asst. Dir. Program Svcs., Elspeth Inglis; Exhibits Coord., Ron Cleveland; Planetarium Coord., Eric Schreur; Asst. Dir. Collections Svcs., Paula Metzner; Programs Coord., Annette Hoppenworth; Group Reservations Coord., Elizabeth Barker; Flight Dir., Challenger Center, Kathy Godin; Planning Asst., Lindsay Baker; Design Asst., Megan Burtzloff; Coord. Interpretation, Megan O'Kon.
Personnel Profile: Full-Time Paid 12; Part-Time Paid 20; Part-Time Volunteers 150; Interns 4.
Governing Authority: college. Parent Institution: Kalamazoo Valley Community College. Tax-exempt.
Institution Type/Description: Participatory Museum: history, science & technology.
Collections: history; science; technology; decorative arts; ethnology; photography; manuscript collections.
Major Exhibits: Michigan's Heritage Barns (T), 1/14-6/14; How People Make Things (T), 2/14-5/14; Robot Zoo (T), 6/14-8/14; Speedbump (T), 6/14-9/14.
Research Fields: history, science & technology related to collections & south-west Michigan.
Facilities: 60,000 sq. ft. exhibit space; Digistar planetarium; high definition video theater; learning center.
Activities: permanent & temporary exhibitions; interpretive programs for families; Challenger Learning Center space missions; preschool classes; school programs; High Definition Video Programs; Classic Movies & Digistar Planetarium programs.
Publications: seasonal calendar; brochures; magazine 3 times a year, MuseOn.
Hours & Admission Prices: Mon.-Sat. 9-5, Sun. & holidays 1-5. No charge. Closed Easter; Thanksgiving; Christmas Eve & Day. &
Attendance: 122,000 (accurate)

WESTERN MICHIGAN UNIVERSITY RICHMOND CENTER FOR VISUAL ARTS - GWEN FROSTIC SCHOOL OF ART, 1903 W. Michigan Ave., Kalamazoo, MI 49008-5213. Tel.: 269-387-2455. Fax: 269-387-2477.
E-mail: donald.desmett@wmich.edu
Founded: 2007.
Congressional District: 3
Key Personnel: C.E.O. College of Fine Arts, Dean Dan Guyette; Dir. School of Art, Tricia Hennessy; Dir. Exhibitions, Don Desmett; Cur. University Art Collection, Milinda Bagnall.
Personnel Profile: Full-Time Paid 3; Part-Time Paid 6; Part-Time Volunteers 25; Interns 2.

Governing Authority: university. Parent Institution: Western Michigan University. Tax-exempt.
Institution Type/Description: Art Galleries.
Collections: prints including intaglio, relief, litho, screenprint, European & American contemporary artists, as well as by younger avant-garde artists; 19th & 20th-century American & European fine & decorative arts.
Major Exhibits: Kate Teal: The Housed (T), 9/4/14-10/12/14; NYPOP Emerging Curators Series, 1/15-2/15.
Research Fields: modern & contemporary art.
Facilities: classrooms.
Activities: guided tours; lectures; films; gallery talks; formally organized educational programs for graduate & undergraduate students; temporary, traveling & loan exhibitions; visiting artists program; Sculpture Tour.
Publications: essays, Sculptural Concepts; Charismatic Abstraction; David Henderson; Man O' War; Yinka Shonibare, MBE: Film, Photography & Sculpture; Heroes Like Us?: Dulce Pinzon, Mark Newport, Jerry Kraens; Complex Conversations: Willie Cole Sculptures & Wallworks.
Hours & Admission Prices: May-July Mon.-Fri. 10-5; Sept.-April Mon.-Thurs. 10-6, Fri. 10-9, Sat. 12-6. No charge. &

Attendance: 8,000 (accurate)
Membership: Friends $15-$1,000.

Kaleva

BOTTLE HOUSE MUSEUM, 14551 Wuoksi Ave., Kaleva, MI 49645-9341. Mailing Address: P.O. Box 252, Kaleva, MI 49645-0252. Tel.: 231-362-2080. allartsmanistee.com.
E-mail: caasiala@jackpine.com
Web Site: kalevami.com
Founded: 1983.
Congressional District: 101
Key Personnel: Chm. & Pres. (V), Cynthia Asiala.
Personnel Profile: Part-Time Volunteers 40.
Governing Authority: Tax-exempt.
Institution Type/Description: Historic House: housed in a home built in 1941 with over 60,000 soft drink bottles.
Collections: local history & culture; lumbering; farming; homemaking; office machines; period furnishings; photographs; Makinen tackle; co-operative movement of Finnish Immigrants display honoring Robert Rengo, 50 year president of village.
Hours & Admission Prices: Memorial Day to Labor Day Sat.-Sun. 12-4; Sept.-Oct. Sat. 12-4. Suggested Donation: adults $3.
Attendance: 1,500 (estimated)
Membership: Individual $15; Business $25; Patron $50; Friend $100; Benefactor $200.

KALEVA DEPOT RAILROAD MUSEUM, 14420 Walta St., Kaleva, MI 49645. Mailing Address: P.O. Box 252, Kaleva, MI 49645. Tel.: 231-362-3480, 2080 & 3481 (summer).
E-mail: caasiala@jackpine.com
Web Site: kalevami.com
Formerly: Kaleva Train Depot Museum
Founded: 1997.
Congressional District: 101
Key Personnel: Chm. (V), Cynthia Asiala
Governing Authority: Parent Institution: Kaleva Historical Society. Tax-exempt.
Institution Type/Description: Historic Building: housed in a former railroad depot; built in 1908.
Collections: local history & culture; railroad artifacts; scale model of depot with train; photographs; restored switch engine on outdoor track; restored M & NE engine.
Activities: group tours.
Hours & Admission Prices: Memorial Day to Labor Day Sat. 12-4. No charge; donations accepted. &
Attendance: 200 (estimated)
Membership: Kaleva Historical Society: Individual $15.

Lake Linden

HOUGHTON COUNTY HISTORICAL MUSEUM SOCIETY, 53102 Hwy. M-26, Lake Linden, MI 49945. Mailing Address: P.O. Box 127, Lake Linden, MI 49945-0127. Tel.: 906-296-4121. Fax: 906-296-8006.
E-mail: president@houghtonhistory.org
Web Site: www.houghtonhistory.org
Founded: 1961.
Congressional District: 1
Personnel Profile: Full-Time Volunteers 1; Part-Time Volunteers 20.

Governing Authority: nonprofit organization. Subsidiary Institution: Copper Country Railroad. Tax-exempt: 170(b)(1)(A).
Institution Type/Description: History Museum: located on Calumet & Hecla Millsite.
Collections: copper country lithographs; local history; steam engine; wagons. Historic Buildings: fire station; church; railroad depot; schoolhouse; mining company garage; log cabin.
Research Fields: local history.
Facilities: 1,000-vol. library of local history available for use under supervision on premises.
Activities: guided tours; lectures; operates John H. Forster press; education programs for adults.
Publications: newsletter; annual report; books on local history.
Hours & Admission Prices: Office: Mon.-Fri. 10-12. Museum: call for hours. Adults $5, seniors & students 6-16 $3; members no charge. Train Rides: adults $5, students 6-16 & seniors $3. &
Attendance: 10,000 (estimated)
Membership: Annual $25; Family $40; Life $300.

Lansing

CARL G. FENNER NATURE CENTER, 2020 E. Mt. Hope Rd., Lansing, MI 48910-1905. Tel.: 517-483-4224. Fax: 517-377-0012. TDD: 517-483-4479.
E-mail: info@mynaturecenter.org
Web Site: www.mynaturecenter.org
Founded: 1959.
Key Personnel: Exec. Dir., Jason Meyer; Program Mgr., Katie Woodhams; Naturalist, Andrea Lazzari; Volunteer Coord., Dani Torcolacci.
Governing Authority: municipal. Parent Institution: Lansing Parks & Recreation Dept. Subsidiary Institution: Friends of Fenner Nature Center. Tax-exempt.
Institution Type/Description: Natural History Museum.
Collections: natural history; hands-on exhibits.
Research Fields: birds; native plants; interpretive techniques.
Facilities: 2,400-vol. library of natural history books for adults & children; reading room; classrooms; nature center; picnic site. Museum-related items for sale.
Activities: guided tours; family programs. Museum Sponsors: Fall Apple Butter Festival; Spring Maple Syrup Festival; children's nature camp in summer.
Publications: Friends of Fenner Newsletter.
Hours & Admission Prices: Grounds: daily 8am-dark. Visitor Center: Wed.-Fri. 10-4, Sat.-Sun. 12-4. No charge. Closed holidays. &
Attendance: 40,000 (estimated)
Membership: Senior/Student $25; Individual $30; Family $45; Grand Family $60; Explorer $100-$249; Adventurer $250-$499; Naturalist $500-$999; President's Club $1,000 & up.

IMPRESSION 5 SCIENCE CENTER, 200 Museum Dr., Lansing, MI 48933-1914. Tel.: 517-485-8116, ext. 39. Fax: 517-485-8125.
E-mail: wittenauer@impression5.org
Web Site: www.impression5.org
Founded: 1972.
Congressional District: 8
Key Personnel: Exec. Dir., Erik D. Larson; Dir. Exhibits, Cyrus Miller; Finance Officer, Sandra Dunnebacke; Dir. Education, Micaela Blazer.
Personnel Profile: Full-Time Paid 8; Part-Time Paid 20; Part-Time Volunteers 105.
Governing Authority: nonprofit organization. Tax-exempt: 501(c)(3) & 170(b)(1)(A).
Institution Type/Description: Interactive Hands-On Learning Center.
Collections: science & technology.
Research Fields: education; exhibit design.
Activities: formally organized education programs for students and families; permanent & traveling science exhibits; workshops; camp-ins; summer camp.
Publications: museum newsletter.
Hours & Admission Prices: Mon.-Fri. 10-5, Sat. 10-7, Sun. 12-5. Adults & children 5 & over $5, grandparents & seniors $4.50; discounts to children under 5, groups with advance reservation & AAA members; ASTC members & members no charge. Closed New Year's Day; Easter; Memorial Day; Independence Day; Thanksgiving; Christmas Eve & Day. &
Attendance: 100,000 (estimated)
Membership: Family & Grandparent $60; Family Plus $100.

LANSING ART GALLERY, 119 N. Washington Square, Ste. 101, Lansing, MI 48933. Tel.: 517-374-6400. Fax: 517-374-6385. Facebook: Lansing Art Gallery.
E-mail: lansingartgallery@gmail.com
Web Site: lansingartgallery.org
Founded: 1965.
Congressional District: 6
Key Personnel: C.E.O. & Exec. Dir., Catherine Babcock; Program Mgr., Jane Kramer; Gallery Coord, Sara Pulver.
Personnel Profile: Full-Time Paid 1; Part-Time Paid 4; Part-Time Volunteers 30; Interns 4.
Governing Authority: nonprofit organization. Tax-exempt.
Institution Type/Description: Art Gallery.
Collections: works by Michigan artists.
Major Exhibits: Linda Beeman, 1/8/14-2/22/14; Mid-Michigan Art Guild, 2/28/14-3/28/14; Art Scholarship Alert, 3/31/14-4/26/14; Mark Chatterley, 5/2/14-6/28/14; Lesa Doke, 7/2/14-8/28/14; Norwood Viviano, 9/6/14-10/30/14; Holiday Exhibition, 11/12/14-12/23/14.
Facilities: Gift items for sale.
Activities: monthly changing exhibitions & competitions; education outreach; fundraisers & special events; lectures & workshops. Museum Sponsors: Art Scholarship Alert High School Art Competition; Holiday Art Market.
Publications: exhibition announcements.
Hours & Admission Prices: Tues.-Fri. 10-4, Sat. & 1st Sun. of month 1-4. No charge; donations accepted. &
Attendance: 25,000 (estimated)
Membership: Artist $35; Individual $50; Family $65; Supporting $100-$249; Picasso Gallery $250-$499; da Vinci Gallery $500-$999; Rembrant Gallery $1,000-$2,499; Masterpiece Gallery $2,500 & up.

* **MICHIGAN HISTORICAL MUSEUM, MICHIGAN HISTORICAL CENTER, (M),** 702 W. Kalamazoo, Lansing, MI 48909-8240. Mailing Address: P.O. Box 30740, Lansing, MI 48909-8240. Tel.: 517-373-3559. Fax: 517-241-4738. TDD: 1-800-827-7007.
Web Site: www.michigan.gov/museum
Founded: 1879.
Congressional District: 6
Key Personnel: C.E.O., Sandra S. Clark; Chief Cur., Maria Quinlan Leiby; Cur. Collections, Scott Peters; Visitor Svcs., Alexandra Raven.
Personnel Profile: Full-Time Paid 18; Part-Time Paid 4; Part-Time Volunteers 100; Interns 5.
Governing Authority: state. Parent Institution: Michigan Historical Center. Branch Museums: 1884 Mann House; Michigan Iron Industry Museum, Negaunee. Historic Sites: Father Marquette Memorial, Straits State Park, St. Ignace; Fayette Townsite & State Park, Fayette; Fort Wilkins Historic Complex & State Park, Copper Harbor; Hartwick Pines Lumbering Museum & State Park, Grayling; Sanilac Petroglyphs, Sanilac Petroglyphs State Park, Cass City; The Walker Tavern, Cambridge Historic Park, Cambridge Junction; Civilian Conservation Corps. Museum, Higgins Lake State Park; Tawas Lighthouse. Tax-exempt: 170(c)(1).
Institution Type/Description: History Museum.
Collections: history of Michigan & Northwest Territory; prehistoric & aboriginal history; Michigan settlement; industry; ethnic influences; technology; transportation; agriculture; domestic life; occupational skills; decorative arts; folklore; ceramics; textiles; 20th-century artifacts.
Activities: guided & self-guided tours; formally organized education programs for children; inter-museum loan, permanent, temporary & traveling exhibitions; seminars; workshops.
Publications: books & pamphlets related to the history of Michigan & the old Northwest Territory.
Hours & Admission Prices: Mon.-Fri. 9-4:30, Sat. 10-4, Sun. 1-5. Adults $6, seniors $4, youth 6-17 $2; children 5 & under no charge. Field Museums & Historic Sites: call for information. Closed New Year's Day; Christmas; state holidays. &
Attendance: 150,000 (estimated)
Membership: Friends of Michigan History: Individual $40; Household $50; Contributing $60; Sponsoring $100; Other $101 & up.

MICHIGAN MUSEUM OF SURVEYING, 220 Museum Dr., Lansing, MI 48933-1905. Tel.: 517-484-2413. Fax: 517-484-3711.
E-mail: mispsinstitute@gmail.com
Web Site: www.misps.org
Founded: 1989.
Congressional District: 8
Key Personnel: Exec. Dir., Mollee J. Neff.
Governing Authority: Parent Institution: Michigan Society of Professional Surveying. Tax-exempt: 501(c)(3).
Institution Type/Description: Surveying History Museum.

Collections: land surveying history & instruments.
Facilities: 340-vol. library available to the public; 2,000 sq. ft. exhibit space.
Activities: temporary exhibitions; reenactment group performs at area festivals & events. Annual Event: Golf Outing.
Publications: semi-annual newsletter; annual magazine.
Hours & Admission Prices: Call for hours.
Attendance: 1,781 (accurate)

MICHIGAN WOMEN'S HISTORICAL CENTER & HALL OF FAME, 213 W. Malcolm X St., Lansing, MI 48933-2315. Tel.: 517-484-1880. Fax: 517-372-0170. Facebook: Michigan Women's Hall of Fame.
E-mail: info@michiganwomen.org
Web Site: www.michiganwomenshalloffame.org
Founded: 1983.
Key Personnel: Exec. Dir., Sandy Soifer.
Personnel Profile: Full-Time Paid 3.
Governing Authority: Parent Institution: Michigan Women's Studies Association. Tax-exempt: 501(c)(3).
Institution Type/Description: Women's History Museum & Hall of Fame.
Collections: Michigan women's history; over 260 women inducted into the Hall of Fame; personal artifacts; paintings; portraits.
Facilities: research library.
Hours & Admission Prices: Wed.-Sat. 12-4, 1st Sun. each month 2-4; groups by appointment. Adults $2.50, seniors $2, students 5-18 $1; discounts to AAM & ICOM members. Closed major holidays. &
Attendance: 1,800 (accurate)
Membership: Student $15; Individual $30; Friends $40; Family $50; Organization $75; Sponsor $100.

POTTER PARK ZOO, 1301 S. Pennsylvania Ave., Lansing, MI 48912-1646. Tel.: 517-483-4222. Fax: 517-483-3894.
E-mail: zoocontact@ingham.org
Web Site: www.potterparkzoo.org
Founded: 1917.
Key Personnel: Cur., Cynthia Wagner; Veterinarian, Dr. Tara Harrison.
Governing Authority: municipal government. Branch Institution: Potter Park Zoological Society. Tax-exempt: 501(c)(3).
Institution Type/Description: Zoo.
Collections: over 400 animals from around the world.
Research Fields: dietary of certain primate species.
Facilities: 150-vol. library of educational resource materials; zoological park. Gift items for sale.
Activities: arts festivals; docent program; films; formal education programs; guided tours; lectures; participatory & traveling exhibitions; special events. Exploration & Discovery Center available to rent.
Publications: bimonthly newsletter, Potter Park Zoological Society Newsletter.
Hours & Admission Prices: April-Oct. daily 9-6. Residents: adults $4, senior citizens 60 & over $3, children 3-12 $2; Non-Resident: adults $10, senior citizens 60 & up $3, children 3-16 $2; members & children under 3 no charge. Nov.-March daily 10-4. Adults $2, children 3-12 $1; members & children under 3 no charge. &
Attendance: 380,000
Membership: Senior $30; Individual $40; Grandparent $50; Family $60; Zoo Friend $135; Lifetime $1,000.

R.E. OLDS TRANSPORTATION MUSEUM, 240 Museum Dr., Lansing, MI 48933-1905. Tel.: 517-372-0529. Fax: 517-372-2901.
E-mail: autos@reoldsmuseum.org
Web Site: reoldsmuseum.org
Founded: 1981.
Congressional District: 8
Key Personnel: Dir., William Adcock; Pres. (V), James Perkins; Mgr., Kristi Schwartzly.
Personnel Profile: Full-Time Paid 1; Part-Time Paid 2; Part-Time Volunteers 40.
Governing Authority: bd. of trustees. Parent Institution: R.E. Olds Museum Association, Inc. Tax-exempt: 501(c)(3).
Institution Type/Description: Transportation Museum.
Collections: REO Speedwagons; wheels; aviation display; photos; 50 vehicles; memorabilia in showcases; 1897 Oldsmobile; two rocket cars; 1901 Curved Dash Olds; license plates.
Research Fields: Lansing-built transportation; R.E. Olds.
Facilities: meeting rooms. Books & gift items for sale.
Activities: guided tours; docent program; films. Participates in: Be A Tourist in Your Own Town. Museum Sponsors: The Car Capital Auto Show in July.
Publications: brochures; pamphlets; quarterly newsletter.
Hours & Admission Prices: April-Oct. Tues.-Sat. 10-5, Sun. 12-5; Nov.-March

Tues.-Sat. 10-5. Family of 5 $12, adults $6, senior citizens & students $4; discounts for groups; members no charge. Closed major holidays. &
Attendance: 15,000 (estimated)
Membership: Student & Senior Citizen $15; Individual $25; Family $35; Corporate $100; Life $1,000.

WOLDUMAR NATURE CENTER, 5739 Old Lansing Rd., Lansing, MI 48917-8503. Tel.: 517-322-0030. Fax: 517-322-9394. Facebook: Woldumar.
E-mail: lori@woldumar.org
Web Site: www.woldumar.org
Founded: 1966.
Congressional District: 56
Key Personnel: Exec. Dir., Lori McSweeney.
Personnel Profile: Interns 1.
Governing Authority: nonprofit organization. Parent Institution: Woldumar Nature Association. Tax-exempt: 501(c)(3).
Institution Type/Description: Nature Center Conservation Area & Historic Building.
Collections: bird & mammal live mounts & study skins; early pioneer farm implements & tools. Historic Building: 1860 log house, boyhood home of Darius Moon, architect.
Facilities: 169-seat auditorium; educational facilities; visitors center. Rocks, minerals & other museum-related items for sale.
Activities: guided tours; films; organized education programs for children & adults. Museum Sponsors: Pioneer; American Heritage Festival; Wildflower Weekend; Chili Winter Evening.
Publications: bimonthly newsletter, Woldumar News.
Hours & Admission Prices: Visitors Center: Tues.-Sat. 10-5 & Sun. 12-4. Office: Mon.-Fri. 9-5. Trails: dawn-dark. Donation: $1 per person; discounts to ANCA members & groups; members no charge. Closed New Year's Day; Easter weekend; Memorial Day weekend; Labor Day weekend; Thanksgiving; Christmas. &
Attendance: 40,000 (estimated)
Membership: Household $50.

Leland

LEELANAU HISTORICAL MUSEUM, 203 E. Cedar St., Leland, MI 49654-5015. Mailing Address: P.O. Box 246, Leland, MI 49654-0246. Tel.. 231-256-7475.
E-mail: info@leelanauhistory.org
Web Site: www.leelanauhistory.org
Founded: 1957.
Congressional District: 1
Key Personnel: Dir., Francie Gits; Pres. (V), Molly Crimmins.
Personnel Profile: Part-Time Paid 2; Part-Time Volunteers 14; Interns 2.
Governing Authority: nonprofit organization. Parent Institution: Leelanau Historical Society, Inc. Tax-exempt: 501(c)(3).
Institution Type/Description: Local History Museum.
Collections: traditional handicrafts; photographs; manuscripts; 19th-century period artifacts; local traditional Odawa arts.
Research Fields: history; genealogy; traditional & folk arts; Native American history.
Facilities: 4,000 sq. ft. exhibit space; archives.
Activities: guided tours; school & adult education programs; permanent & temporary exhibits; special activities.
Publications: quarterly newsletter.
Hours & Admission Prices: Wed.-Fri. 10-4, Sat. 10-3. No charge; donations accepted. &
Attendance: 5,000 (estimated)
Membership: $50; $100; $250; $500; $1,000.

Lincoln Park

LINCOLN PARK HISTORICAL MUSEUM, 1335 Southfield Rd., Lincoln Park, MI 48146-2370. Tel.: 313-386-3137.
E-mail: curator@lphistorical.org
Web Site: www.lphistorical.org
Founded: 1972.
Congressional District: 12
Key Personnel: Dir., Muriel Lobb; Pres., Jim Nelson; Cur., Jeff Day; Museum Shop Mgr., Lucille Stroh.
Personnel Profile: Part-Time Paid 2; Part-Time Volunteers 25.
Governing Authority: municipal. Parent Institution: City of Lincoln Park. Subsidiary Institution: Lincoln Park Historical Commission. Tax-exempt.
Institution Type/Description: Local & Regional History Museum: housed in former U.S. Post Office building; built in 1938.
Collections: history of Ecorse Township & City of Lincoln Park.
Research Fields: local history from 1700-2010 & story of Lincoln Park People.

Facilities: 2,500-vol. library; 800 newspapers from the time the village was established available for inter-library loan.
Activities: arts & crafts; lecture series. Museum Sponsors: special commemorative events & annual member meeting and dinner.
Publications: quarterly newsletter.
Hours & Admission Prices: Wed.-Thurs. & Sat. 1-5; other times by appointment. No charge; donations accepted. Closed New Year's Day; Martin Luther King Jr. Day; Easter; Memorial Day; Independence Day; Thanksgiving; Christmas. &
Attendance: 6,500 (estimated)
Membership: Student $5; Individual $10; Family $15; Patron $25; Life $200.

Lowell

JAMES C. VEEN OBSERVATORY OF THE GRAND RAPIDS AMATEUR ASTRONOMICAL ASSOCIATION, 3308 Kissing Rock Ave., S.E., Lowell, MI 49331-8918. Tel.: 616-897-7065.
E-mail: graaa@graaa.org
Web Site: www.graaa.org
Founded: 1970.
Key Personnel: Dir., David L. DeBruyn.
Personnel Profile: Part-Time Volunteers 50.
Governing Authority: Subsidiary Institution: Grand Rapids Public Museum. Tax-exempt.
Institution Type/Description: Observatory.
Collections: astronomy; space science; 17-inch Dobsonian telescope; 16-inch Schmidt Cassegrain telescope; 14-inch Schmidt Cassegrain robotic telescope.
Research Fields: asteroid tracking; variable star monitoring.
Activities: group tours. Observatory Sponsors: Veen Observatory Public Nights April to October.
Publications: quarterly e-newsletter, Inside Orbit.
Hours & Admission Prices: April-Oct. call for hours. Adults $3; discounts to Grand Rapids Public Museum Friend members.
Attendance: 1,500 (estimated)
Membership: Student $25; Adult $40; Family $50.

LOWELL AREA HISTORICAL MUSEUM, 325 W. Main St., Lowell, MI 49331-1609. Mailing Address: P.O. Box 81, Lowell, MI 49331-0081. Tel.: 616-897-7688. Fax: 616-897-7688. Facebook: Lowell Area Historical Museum.
E-mail: history@lowellmuseum.org
Web Site: lowellmuseum.org
Founded: 1989.
Congressional District: 3
Key Personnel: C.E.O. & Dir., Pat Allchin; Pres., James M. Doyle; Education, Luanne Kaeb; Treas., Cathy Haefner.
Personnel Profile: Part-Time Paid 3; Part-Time Volunteers 40.
Governing Authority: private; nonprofit organization. Tax-exempt: 501(c)(3).
Institution Type/Description: History Museum: listed on the National Register of Historic Places.
Collections: area history from early 1800s to present.
Research Fields: military & oral histories; local residents & obituaries.
Facilities: 750-vol. library; 2,244 sq. ft. exhibit space. Museum-related items for sale.
Activities: docent program; education programs; children's activity workshops; guided tours; lectures; participatory, temporary & traveling exhibitions; military history panels; grade school group tours; oral history panels.
Publications: quarterly newsletter, Lowell Area Historical Museum Newsletter; Where the Rivers Meet-A Pictorial Journey Through Historic Lowell, Michigan; Images of Lowell.
Hours & Admission Prices: Tues. & Sat.-Sun. 1-4, Thurs. 1-8. Adults $3, students & children $1.50; discounts to groups; members no charge. Closed holidays. &
Attendance: 20,000 (estimated)
Membership: Individual $15; Family $25; Business & Donor $50.

Ludington

MASON COUNTY HISTORICAL SOCIETY/HISTORIC WHITE PINE VILLAGE, (M), 1687 S. Lakeshore Dr., Ludington, MI 49431-8316. Tel.: 231-843-4808. Fax: 231-843-7089.
E-mail: info@historicwhitepinevillage.org
Web Site: www.historicwhitepinevillage.org
Founded: 1937.
Congressional District: 9
Key Personnel: Exec. Dir., Kate J. Arbogast; Pres. Bd. (V), Dr. William Anderson; Asst. to Exec. Dir., Carmen Tiffany.
Personnel Profile: Full-Time Paid 2; Part-Time Paid 1; Part-Time Volunteers 400.

Governing Authority: nonprofit organization. Parent Institution: Mason County Historical Society, Inc., 1687 S. Lakeshore Dr., Ludington 49431. Tax-exempt.

Institution Type/Description: Local History Museum & Reconstructed Historical Village.

Collections: lumber tools; farm tools; Civil War artifacts; textiles; clothing & laces; bibles; hymn books; school textbooks; maps; pattern glass; china; crockery; bottles; silver; marine exhibit; photographs; original paintings in oils; lithographs; prints; model trainferries; sailing vessels; folk art; housewares; dolls; toys. Historic Buildings: over 25 buildings including Rose Hawley Museum, Mason County Lumber Museum; Museum of Music; 1890 one-room school; frame courthouse; post office; hardware store; trapper's cabin; blacksmith shop; chapel; general store; Sugar House; Sports Hall of Fame; Maritime Museum; sawmill. Farmstead: farmhouse; agricultural equipment sheds; barn; stable; milk house; transportation artifacts; learning center; windmill; garden; outhouse.

Research Fields: local history; genealogy; industrial development; Great Lakes shipping; lumbering; agriculture; local business; maritime.

Facilities: 4,000-vol. research library of books, archives & genealogy references for research.

Activities: self-guided tours; permanent & temporary exhibits; special events; farm tool skills learning center; lumbering; farmstead. Museum Sponsors: Old Time Baseball matches.

Publications: newsletter, History Happening; Centennial Farms of Mason County; Pictorial History of Mason County; The Story of Ludington - Born of Logs, Nurtured by Carferries, Forged by Resilience.

Hours & Admission Prices: mid-April to May & Sept. to mid-Oct. Tues.-Sat. 10-5; Memorial Day to Labor Day Tues.-Sat. 10-5, Sun. 1-5. Families $25, adults $9, children 6-17 $6; discounts to AAA members, seniors & Great Lakes Energy Co-op cardholders; children 5 & under and MCHS members no charge. Additional charge for special events & food events. &

Attendance: 18,000 (accurate)

Membership: Individual $40; Family $60; Supporter $100; Sponsor $250; Patron $500; Benefactor $1,000.

Mackinac Island

MACKINAC STATE HISTORIC PARKS-FORT MACKINAC & MACKINAC ISLAND STATE PARK, 7029 Huron Rd., Mackinac Island, MI 49757. Mailing Address: P.O. Box 873, Mackinaw City, MI 49701-0873. Tel.: 906-847-3328 (Summer); 231-436-4100 (Winter). Fax: 231-436-4210. Facebook: Mackinac State Historic Parks-Fort Mackinac & Mackinac Island State Park.

E-mail: mackinacparks@michigan.gov
Web Site: www.mackinacparks.com
Founded: 1895.
Congressional District: 1
Key Personnel: Dir., Phil Porter; Chm. (V), Chuck Yob; Park Mgr., Sue Topham; Deputy Dir., Steven C. Brisson; Cur. Archaeology, Lynn Evans; Cur. Natural History, Jeffrey A. Dykehouse; Cur. Education, Katherine Mallory; Public Rels. & Mktg. Officer, Kelsey Schnell; Registrar, Brian Jaeschke; Grant Writer & Membership, Diane A. Dombroski; Group Sales Coord., Scott Witsman; Chief Finance & Accounting, Lana L. Cotton; Exhibit Designer, Martin Douglas; Museum Shop Mgr., Suzette Schmalzried; Human Resources Coord., Kenneth Fegan.
Personnel Profile: Full-Time Paid 43; Part-Time Paid 101; Part-Time Volunteers 290; Interns 4.
Volunteer Hours: 39,554
Operating Expenses: 7,083,152
Operating Income: 7,083,152
Governing Authority: state; nonprofit. Parent Institution: Mackinac Island State Park Commission. Tax-exempt.
Institution Type/Description: Historic Site: 1780-1895 Fort Mackinac.
Collections: archaeology; 19th-century furnishings; manuscripts; photographs; military items; blacksmithing items; Indian items; carriages; 1814 battlefield; military cemetery. Historic Buildings: 1780 Biddle house; 1838 Indian dormitory; 1869 renovated Benjamin blacksmith shop; 1829 Mission church; 1780 McGulpin house; 1954 renovated Beaumont Memorial; 14 restored original structures.
Facilities: visitor center; theater; restaurant; heritage & research facility; 1,773 acre park includes 70.5 miles of signed roads & trails. Site-related items, publications & reproductions for sale.
Activities: guided tours; lectures; films; interpretive exhibitions; organized education programs; docent program; musket & cannon-firing demonstrations; craft demonstrations; dramatic reenactments of historic events; research.
Publications: 70 publications in print. Catalog available.
Hours & Admission Prices: Mackinac Island State Park: daily. No charge. Fort Mackinac: May-Oct. daily 9:30-6; Spring & Fall call for reduced hours.

Adults $12, children 6-17 $7; discounts to AAM & ICOM members; members & children 5 & under no charge. &
Attendance: 192,062 (accurate)
Membership: Friends $65; Mackinac Heritage $75; Voyageur $85; Sentinel $175; Explorer $400; Commandant's Circle $600; Steward $1,000; Guardian $2,500; Patron $5,000; Benefactor $10,000.

MACKINAC STATE HISTORIC PARKS - RICHARD & JANE MANOOGIAN MACKINAC ART MUSEUM, Huron Rd., Mackinac Island, MI 49757. Mailing Address: P.O. Box 873, Mackinaw City, MI 49701-0873. Tel.: 906-847-3328 (summer); 231-436-4100. Fax: 231-436-4210. Facebook: Mackinac State Historic Parks.

E-mail: mackinacparks@michigan.gov
Web Site: www.mackinacparks.com
Founded: 2010.
Congressional District: 1
Key Personnel: Dir., Phil Porter; Chm. (V), Chuck Yob; Deputy Dir., Steven C. Brisson; Park Mgr., Sue Topham; Registrar, Brian Jaeschke; Cur. Education, Katherine Mallory; Exhibit Designer, Martin Douglas; Museum Historian, Craig Wilson; Cur. Archaeology, Lynn Evans; Museum Shop Mgr., Suzette Schmalzried; Grant Writer & Membership, Diane A. Dombroski.
Personnel Profile: Full-Time Paid 43; Part-Time Paid 101; Part-Time Volunteers 290; Interns 4.
Volunteer Hours: 39,554
Operating Income: 7,083,152
Governing Authority: state; nonprofit. Parent Institution: Mackinac Island State Park Commission. Tax-exempt.
Institution Type/Description: Art Museum: housed in a former Federal Indian Dormitory; built in 1838.
Collections: archaeological & ethnographical; costumes & textiles; decorative arts; paintings; photographs; prints; drawings; graphic arts; sculpture.
Major Exhibits: Contemporary Artists Show: Places of Mackinac, 5/14-10/14.
Activities: films; interpretive exhibitions; organized education programs; collaborative events with Mackinac Island Arts Council.
Publications: 70 publications.
Hours & Admission Prices: early May to early Oct. daily 10-5:30; Spring & Fall: call for hours. Adults $5.50, children 6-17 $4; members and children 5 & under no charge. Fort Mackinac tickets accepted. &
Attendance: 9,614 (accurate)
Membership: Friends $65; Mackinac Heritage $75; Voyageur $85; Sentinel $175; Explorer $400; Commandant's Circle $600; Steward $1,000; Guardian $2,500; Patron $5,000; Benefactor $10,000.

STUART HOUSE CITY MUSEUM, Market St., Mackinac Island, MI 49757-0906. Mailing Address: P.O. Box 906, Mackinac Island, MI 49757-0906. Tel.: 906-847-3553.

Founded: 1930.
Congressional District: 1
Key Personnel: Dir., Armand Horn; Cur. Collections, Daniel Seeley.
Personnel Profile: Part-Time Paid 3; Part-Time Volunteers 3.
Governing Authority: municipal.
Institution Type/Description: Historic House Museum: c.1817 home of American Fur Company resident Mgr. Robert Stuart.
Collections: records of the American Fur Co.'s Fur Post; natural history; Indian artifacts; furs & pelts; furnished rooms typical of the 1800s; trade items; tools; War of 1812 items.
Activities: self-guided tours; permanent exhibitions.
Publications: brochure.
Hours & Admission Prices: Call for hours. No charge; donations accepted.

Mackinaw City

*** MACKINAC ISLAND STATE PARK COMMISSION-MACKINAC STATE HISTORIC PARKS, (M),** 207 W. Sinclair Ave., Mackinaw City, MI 49701-9635. Mailing Address: P.O. Box 873, Mackinaw City, MI 49701-0873. Tel.: 231-436-4100; 906-847-3328. Fax: 231-436-4210; 906-847-3815. Facebook: Mackinac Island State Park Commission-Mackinac State Historic Parks.

E-mail: mackinacparks@michigan.gov
Web Site: www.mackinacparks.com
Founded: 1895.
Congressional District: 1
Key Personnel: Dir., Phil Porter; Commission Chm. (V), Chuck Yob; Deputy Dir., Steven Brisson; Cur. Education, Katherine Mallory; Cur. Archaeology, Lynn Evans; Registrar, Brian Jaeschke; Museum Historian, Craig Wilson; Cur. Natural History, Jeffrey A. Dykehouse; Public Rels. & Mktg. Officer, Kelsey Schnell; Group Sales Coord., Scott Witsman; Chief Finance &

Accounting, Lana L. Cotton; Museum Shop Mgr., Suzette Schmalzried; Grant Writer & Membership, Diane A. Dombroski; Exhibit Designer, Martin Douglas; Human Resources Coord., Kenneth Fegan.

Personnel Profile: Full-Time Paid 43; Part-Time Paid 101; Part-Time Volunteers 290; Interns 4.

Volunteer Hours: 39,554

Operating Expenses: 7,083,152

Operating Income: 7,083,152

Governing Authority: state. Affiliated with State of Michigan, Dept. of Natural Resources. Subsidiary Institutions: Mackinac State Historic Parks: Fort Mackinac; Mackinac Island State Park; Colonial Michilimackinac; Historic Mill Creek; Beaumont Museum; Biddle House; Mission Church; Old Mackinac Point Lighthouse; Richard and Jane Manoogian Mackinac Art Museum. Tax-exempt.

Institution Type/Description: Regional History Museum.

Collections: military & social history of the Straits of Mackinac region; archaeological artifacts from site of Fort Mackinac, Colonial Michilimackinac & Mill Creek; Great Lakes maritime materials; Native American artifacts; 110 historic buildings.

Research Fields: 18th-century fur trade in the Great Lakes region; American Indian-White relations; 17th-19th centuries; historical archaeology; British & American military in Great Lakes region, 1760-1895; French Colonial history in Great Lakes region; tourism in Upper Great Lakes, 1890s-present.

Facilities: 3,000-vol. library of books & periodicals including 250 reels of microfilm available for use by appointment; theater; restaurant; nature trails; concessions.

Activities: permanent & temporary exhibits; guided & self-guided tours; school outreach program; lectures; films; TV programs; seminars; craft demonstrations; historic reenactments; interpretive programs for children & adults.

Publications: 70 publications in print. Catalog available.

Hours & Admission Prices: Mackinac Island State Park: daily. Historic Sites: early May to mid-Oct. daily. Park: no charge. Historic Sites: adults $6-$12. &

Attendance: 345,584 (accurate)

Membership: Mackinac Heritage & Friend $65; Voyageur $85; Sentinel $175; Explorer $400; Commandant's Circle $600; Steward $1,000; Guardian $2,500; Patron $5,000; Benefactor $10,000.

MACKINAC STATE HISTORIC PARKS-COLONIAL MICHILIMACKINAC & OLD MACKINAC POINT LIGHTHOUSE, 102 W. Straits Ave., Mackinaw City, MI 49701. Mailing Address: Box 873, Mackinaw City, MI 49701-0873. Tel.: 231-436-4100. Fax: 231-436-4210. Facebook: Mackinac State Historic Parks-Colonial Michilimackinac & Old Mackinac Point Lighthouse.

E-mail: mackinacparks@michigan.gov

Web Site: www.mackinacparks.com

Founded: 1909.

Congressional District: 1

Key Personnel: Dir., Phil Porter; Park Supvr., Michael Sutton; Commission Chm. (V), Chuck Yob; Deputy Dir., Steven C. Brisson; Cur. Education, Katherine Mallory; Registrar, Brian Jaeschke; Museum Historian, Craig Wilson; Public Rels. & Mktg. Officer, Kelsey Schnell; Cur. Archaeology, Lynn Evans; Cur. Natural History, Jeffrey A. Dykehouse; Groups Sales Coord., Scott Witsman; Chief Finance & Accounting, Lana L. Cotton; Grant Writer & Membership, Diane A. Dombroski; Exhibit Designer, Martin Douglas; Museum Shop Mgr., Suzette Schmalzried; Human Resources Coord., Kenneth Fegan.

Personnel Profile: Full-Time Paid 43; Part-Time Paid 101; Part-Time Volunteers 290; Interns 4.

Volunteer Hours: 39,554

Operating Expenses: 7,083,152

Operating Income: 7,083,152

Governing Authority: state; nonprofit. Parent Institution: Mackinac Island State Park Commission. Tax-exempt.

Institution Type/Description: History Museum: housed in reconstructed French & British military outpost & fur-trading village founded in 1715.

Collections: archaeological; military artifacts; 18th-century furnishings. Historic Structures: 1892 Old Mackinac Point Lighthouse; 1715-1780 Fort Michilimackinac & 16 reconstructed buildings.

Facilities: research library; visitor center; theater; food service; Peterson Archaeology & History Center. Museum-related items for sale.

Activities: guided tours; lectures; films; organized education programs; docent program; black-powder & craft demonstrations; continuing archaeological excavations.

Publications: 70 publications in print. Catalog available.

Hours & Admission Prices: May-Oct. daily 9-6; Fall & Spring: call for reduced hours. Colonial Michilimackinac: adult $11, children 6-17 $6.50; discount to AAM & ICOM members; members & children under 5 no charge. Lighthouse: adult $6, children $4. &

Attendance: 102,533 (accurate)

Membership: Friends $65; Mackinac Heritage $75; Voyageur $85; Sentinel $175; Explorer $400; Commandant's Circle $600; Steward $1,000; Guardian $2,500; Patron $5,000; Benefactor $10,000.

MACKINAC STATE HISTORIC PARKS-HISTORIC MILL CREEK DISCOVERY PARK, 9001 U.S. 23 S., Mackinaw City, MI 49701. Mailing Address: Box 873, Mackinaw City, MI 49701-0873. Tel.: 231-436-4100. Fax: 231-436-4210. Facebook: Mackinac State Historic Parks-Historic Mill Creek Discovery Park.

E-mail: mackinacparks@michigan.gov

Web Site: www.mackinacparks.com

Founded: 1975.

Congressional District: 1

Key Personnel: Commission Chm., Chuck Yob; Dir., Phil Porter; Deputy Dir., Steven C. Brisson; Cur. Education, Katherine Mallory; Registrar, Brian Jaeschke; Cur. Natural History, Jeffrey A. Dykehouse; Cur. Archaeology, Lynn Evans; Public Rels. & Mktg. Officer, Kelsey Schnell; Group Sales Coord., Scott Witsman; Chief Finance & Accounting, Lana L. Cotton; Museum Shop Mgr., Suzette Schmalzried; Park Supvr., Michael Sutton; Human Resources Coord., Kenneth Fegan; Exhibit Designer, Martin Douglas.

Personnel Profile: Full-Time Paid 43; Part-Time Paid 101; Part-Time Volunteers 290; Interns 4.

Volunteer Hours: 39,554

Operating Expenses: 7,083,152

Operating Income: 7,083,152

Governing Authority: state; nonprofit. Parent Institution: Mackinac Island State Park Commission. Tax-exempt: 501(c)(3).

Institution Type/Description: History Site Museum: reconstructed 18th-century water-powered sawmill and support buildings.

Collections: archaeology; natural science; 18th-century technology; mill dam; 3 reconstructed historic buildings including 1790-1830 sawmill.

Facilities: visitor center; theater; cafeteria; nature center; nature trails; picnic area. Museum & nature related items for sale.

Activities: guided tours; organized education programs; craft demonstrations; forest demonstration areas; multi-media orientation program; climbing wall & zipline

Publications: 70 publications in print. Catalog available.

Hours & Admission Prices: May-mid-Oct. daily 9-5; Spring & fall: call for hours. Adults $8, children 6-17 $4.75; members & children under 6 no charge. &

Attendance: 40,712 (accurate)

Membership: Mackinac Heritage $56; Friends $65; Voyageur $85; Sentinel $175; Explorer $400; Commandant's Circle $600; Steward $1,000; Guardian $2,500; Patron $5,000; Benefactor $10,000.

Manistee

MANISTEE COUNTY HISTORICAL MUSEUM, 425 River St., Manistee, MI 49660-1522. Tel.: 231-723-5531.

E-mail: manisteemuseum@yahoo.com

Web Site: www.manisteemuseum.org

Founded: 1953.

Key Personnel: Dir., Mark Fedder.

Personnel Profile: Full-Time Paid 1; Part-Time Paid 3; Part-Time Volunteers 10.

Governing Authority: society. Tax-exempt: 501(c)(3).

Institution Type/Description: History Museum: housed in 1883 A.H. Lyman Drug Company.

Collections: local history; logging & lumbering; railroads; Great Lakes shipping & passenger boats; costumes; dolls; Civil War artifacts; pioneer days; country store; Victorian period rooms; barber shop; early drug store; historical photos; Indian artifacts; glass; folk material. Historic Building: 1881 Holly Water Works, W. First St.

Research Fields: general local & state history; local genealogy; Lake Michigan Maritime History.

Facilities: 500-vol. library of history books & 10,000 1868-1920 historical photographs. Postcards & booklets for sale.

Activities: guided tours; lectures; education programs; permanent & temporary exhibitions; school visiting services; tours for the handicapped or other special groups.

Publications: Manistee museum log.

Hours & Admission Prices: Jan.-March Thurs.-Sat. 10-5; April-May & Oct.-Dec. Tues.-Sat. 10-5; June-Sept. Mon.-Sat. 10-5. Families $7, adults $3, students $1. Closed national holidays except Independence Day.

Attendance: 10,000 (estimated)

Membership: Single $10; Family $15; Contributing $25; Sustaining $50; Patron $75.

Manistique

SCHOOLCRAFT COUNTY HISTORICAL SOCIETY, Deer St., Pioneer Park, Manistique, MI 49854. Mailing Address: P.O. Box 284, Manistique, MI 49854-0284. Tel.: 906-341-5045.
Formerly: Imogene Herbert Historical Museum
Founded: 1963.
Congressional District: 11
Key Personnel: Pres. (V), M. Vonciel Le Duc; Vice Pres., Paul Walker; Treas., Darlene Furmanek.
Personnel Profile: Part-Time Volunteers 20.
Volunteer Hours: 500
Operating Expenses: 2,500
Operating Income: 3,000
Governing Authority: nonprofit. Parent Institution: The Schoolcraft County Board of Commissioners. Tax-exempt.
Institution Type/Description: Local History Museum.
Collections: archives; geology; history; history & fiction pertaining to Upper Peninsula of Michigan; plat display; Indian artifacts; period fire engine display. Historic Buildings: 1883 restored log cabin; water tower.
Major Exhibits: Cameras of the Past, 6/14-9/14; Native American Items, 6/14-9/14.
Research Fields: Schoolcraft County history.
Activities: art shows. Museum Sponsors: Pioneer Day, last Saturday in June.
Publications: newsletter.
Hours & Admission Prices: June-Sept. Wed.-Sat. 1-4. Adults $1, children $.50; members no charge.
Attendance: 500 (estimated)
Membership: Annual $10; Life $100.

Marine City

CAPTAIN DAVID LESTER RESIDENCE, 406 S. Main St., Marine City, MI 48039-1628. Tel.: 810-765-5912. Fax: 810-765-5916.
E-mail: sandy@historicallesterhome.com
Web Site: www.historicallesterhome.com
Institution Type/Description: Historic House: former home of Captain David Lester.
Collections: period artifacts & memorabilia pertaining to Captain David Lester & his family.
Hours & Admission Prices: Tours: May-Dec. Sat.-Sun. 1 & 3.

COMMUNITY PRIDE & HERITAGE MUSEUM, 405 S. Main St., Marine City, MI 48039-1634. Mailing Address: P.O. Box 184, Marine City, MI 48039-0184. Tel.: 810-765-5446.
E-mail: marinecitymuseum@hotmail.com
Web Site: www.marinecitymuseum.org
Formerly: Marine City Pride & Heritage Museum
Founded: 1983.
Congressional District: 10
Key Personnel: Dir. (V), John Foley; Pres. (V), Gary Beals
Governing Authority: Tax-exempt.
Institution Type/Description: History Museum.
Collections: local history, heritage & culture; personal artifacts; period furnishings; business & commercial; blacksmith shop; shipbuilding.
Hours & Admission Prices: June-Oct. Sat.-Sun. 1-4; tours by appointment. No charge; donations accepted. &

Attendance: 300 (estimated)
Membership: Individual $10.

Marquette

THE DEVOS ART MUSEUM AT NORTHERN MICHIGAN UNIVERSITY, (M), 1401 Presque Isle Ave., Marquette, MI 49855-5305. Tel.: 906-227-1481. Fax: 906-227-2276.
E-mail: mmatusca@nmu.edu
Web Site: art.nmu.edu/devosartmuseum
Key Personnel: Dir. & Cur., Melissa Matuscak.
Personnel Profile: Full-Time Paid 1; Part-Time Volunteers 8; Interns 1.
Governing Authority: Parent Institution: Northern Michigan University. Tax-exempt.
Institution Type/Description: University Art Museum.
Collections: Japanese art & artifacts; Native American art & artifacts; 20th century illustrations; contemporary regional artists.
Activities: workshops; artist lectures; docent tours; film screenings.
Hours & Admission Prices: Mon.-Wed. & Fri. 10-5, Thurs. 12-8, Sat.-Sun. 1-4. No charge; donations accepted. Closed New Year's Eve & Day; Independence Day; Memorial Day; Christmas Eve, Day & week. &

Attendance: 10,000 (accurate)
Membership: Student & Retired $15; Friend $30; Family $50; Good Friend $100; Close Friend & Group of Friends $250; Special Friend $500; Best Friend $1,000.

MARQUETTE REGIONAL HISTORY CENTER, 145 W. Spring St., Marquette, MI 49855-4220. Tel.: 906-226-3571. Fax: 906-226-0919.
E-mail: marquettehistory@gmail.com
Web Site: marquettehistory.org
Formerly: Marquette County History Museum
Founded: 1918.
Congressional District: 11
Key Personnel: Bd. Pres. (V), Susan Hornbogen; Exec. Dir., Kaye Hiebel; Dir. Devel., Cristine Osier; Business Mgr., Jennifer Lammi; Research Librarian, Rosemary Michelin; Cur., Jo DeYoung.
Personnel Profile: Full-Time Paid 3; Part-Time Paid 2; Part-Time Volunteers 4; Interns 2.
Governing Authority: bd. of trustees. Tax-exempt: 501(c)(3).
Institution Type/Description: History Museum.
Collections: archaeology; minerals; logging; shipping; archives; ethnology; folklore; china; glass; silver; toys; costumes; tools; geology; marine; technology; transportation; mining, railroad, lumbering history of Marquette County; Ojibwe culture; hands-on exhibits.
Research Fields: history of Central Upper Peninsula & Upper Great Lakes; subjects pertaining to collections.
Facilities: 15,000-vol. library on history of Great Lakes area available by membership or research fee. Historical books & other museum-related items for sale.
Activities: lectures; formally organized education programs for adults; permanent & temporary exhibitions; educational services to schools; interactive exhibits.
Publications: pamphlets; quarterly magazine, Harlow's Wooden Man; books, The Grand Island Story; North to Lake Superior; Landlooker in the Upper Peninsula; Saga of Iron Mining in Michigan's Upper Peninsula; Dandelion Cottage.
Hours & Admission Prices: Mon.-Tues. & Thurs.-Fri. 10-5, Wed. 10-8, Sat. 10-3. Adults $7, seniors $6, students 12 & over $3, children $2; members no charge. Closed legal holidays.
Attendance: 9,100 (accurate)
Membership: Basic Individual $25; Supporting Individual & Basic Family $40; Supporting Family $65; Business $50-$375.

UPPER PENINSULA CHILDREN'S MUSEUM, 123 W. Baraga Ave., Marquette, MI 49855-4744. Tel.: 906-226-3911. Fax: 906-226-7065.
E-mail: nittner@chartermi.net
Web Site: www.upcmkids.org
Founded: 1991.
Key Personnel: Dir., Nheena Weyer Ittner; Dir. 8-18 Media, Dennis Whitley; Explainers Dir. & Gen. Programming Mgr., Jim Edwards; Mgr. Facilities & Exhibits, Aaron Sault.
Personnel Profile: Full-Time Paid 4; Part-Time Paid 5; Part-Time Volunteers 2; Interns 2.
Institution Type/Description: Children's Museum.
Collections: hands-on exhibits.
Facilities: Museum-related items for sale.
Activities: special programs; preschool & elementary school programs; leadership programs for youth 8-18.
Hours & Admission Prices: Mon.-Wed. & Sat. 10-6, Thurs. 10-7:30, Fri. 10-8, Sun. 12-5. Adults & children $5; discounts to Art Serve Michigan, MEA & AAA members. &

Attendance: 51,000
Membership: Family $125; Grandparent $150.

Marshall

AMERICAN MUSEUM OF MAGIC & LUND MEMORIAL LIBRARY, INC., 107 E. Michigan, Marshall, MI 49068-1543. Mailing Address: P.O. Box 5, Marshall, MI 49068-0005. Tel.: 269-781-7570.
E-mail: magic@americanmuseumofmagic.org
Web Site: americanmuseumofmagic.org
Founded: 1978.
Key Personnel: Pres. Bd. (V), Bradley Taylor; Dir., Jeff Taylor.
Personnel Profile: Full-Time Paid 1; Part-Time Paid 2; Part-Time Volunteers 6; Interns 2.
Governing Authority: bd. of directors. Subsidiary Institution: Library, 111 E. Mansion St., Marshall, MI 49068. Tax-exempt.

Institution Type/Description: Magic Museum: housed in 1869 building.
Collections: posters; photographs; advertising folders; recordings; coins & tokens; magician apparatus; costumes; scrapbooks; toys; games; figures; magic apparatus; magic memorabilia; magic illusions owned by Houdini, Blackstone, Henning, Thurston, & Kellar.
Research Fields: magic; magicians.
Facilities: 20,000-vol. library; archives.
Activities: guided tours; temporary exhibitions; public lectures; performances.
Hours & Admission Prices: April-May Thurs.-Sat. 10-4; June-Aug. Tues.-Sat. 10-4; Sept.-Oct. Thurs.-Sat. 10-4; other times by appointment. Adults $5, children 5-12 $3.50. Closed major holidays.
Attendance: 2,000 (accurate)

GRAND ARMY OF THE REPUBLIC HALL MUSEUM, 402 E. Michigan Ave., Marshall, MI 49068. Mailing Address: 107 N. Kalamazoo Ave., Marshall, MI 49068. Tel.: 269-781-8544.
E-mail: info@marshallhistoricalsociety.org
Web Site: marshallhistoricalsociety.org
Founded: 1903.
Personnel Profile: Part-Time Volunteers 4.
Governing Authority: Parent Institution: Marshall Historical Society. Tax-exempt.
Institution Type/Description: Historic Building: housed in the meeting place for Marshall Civil War veterans; built in 1902.
Collections: local history & culture; period furnishings; personal artifacts; Civil War; Spanish American War; World War I & II; photographs.
Hours & Admission Prices: Call for hours.
Attendance: 100 (estimated)

HONOLULU HOUSE MUSEUM, 107 N. Kalamazoo Ave., Marshall, MI 49068-1526. Tel.: 269-781-8544.
E-mail: info@marshallhistoricalsociety.org
Web Site: www.marshallhistoricalsociety.org
Founded: 1962.
Congressional District: 7
Key Personnel: Pres. (V), Ann Rhodes.
Personnel Profile: Part-Time Paid 3; Part-Time Volunteers 25.
Volunteer Hours: 13,000
Operating Expenses: 75,488
Operating Income: 84,164
Governing Authority: nonprofit organization. Parent Institution: Marshall Historical Society. Branch Museum: Capitol Hill School Children's Museum, Gar Hall. Tax-exempt: 170(b)(1)(A).
Institution Type/Description: History Museum: c.1860 Honolulu House.
Collections: 19th-century Midwest artifacts; decorative arts; folklore; military; letters, diaries, maps & legal documents pertinent to early settlement of Marshall.
Research Fields: 19th-century history of Marshall.
Facilities: 2,000-vol. library of local history available for research by special permission from curator. Books & museum-related items for sale.
Activities: guided tours; lectures; permanent exhibitions.
Publications: book, 19th-Century Homes of Marshall, Michigan; History of Marshall, Michigan.
Hours & Admission Prices: May-Oct. daily 11-4:30. Admission $5; discounts may apply. &
Attendance: 3,600 (estimated)
Membership: .

Marysville

MARYSVILLE HISTORICAL MUSEUM, 887 E. Huron Blvd., Marysville, MI 48040-1573. Mailing Address: 111 Delaware Ave., Marysville, MI 48040. Tel.: 810-364-6613.
E-mail: ashaw@cityofmarysvillemi.com
Web Site: www.cityofmarysvillemi.com/museum/index.htm
Key Personnel: Pres. (V), Kim Coggins
Institution Type/Description: History Museum.
Collections: local history & culture; photographs.
Hours & Admission Prices: June-Aug. Sun. 1:30-4. Suggested Donation: $1.
Attendance: 200 (estimated)

THE WILLS STE. CLAIRE AUTO MUSEUM, 2408 Wills St., Marysville, MI 48040-1978. Tel.: 810-987-2854.
E-mail: willsmuseum@sbcglobal.net
Web Site: www.willsautomuseum.org
Key Personnel: Dir., Terry Ernest; Sec., Pete Canjemi; Treas., Laurie Baker
Institution Type/Description: Automobile Museum: former Dow Chemical munitions factory built during WWII.
Collections: period cars.

Hours & Admission Prices: June-Aug. 2nd & 4th Sun. 1-5; Sept.-May 2nd Sun. 1-5. Adults $5. &

Mayville

MAYVILLE AREA MUSEUM OF HISTORY AND GENEAL-OGY, 2124 Ohmer Rd., Mayville, MI 48744. Mailing Address: P.O. Box 242, Mayville, MI 48744-0242. Tel.: 989-843-7185. Facebook: Mayville Museum.
E-mail: mayvillemuseum@hotmail.com
Web Site: www.rootsweb.ancestry.com/~mimmhs
Founded: 1972.
Congressional District: 7
Key Personnel: Dir., Pres. & Cur., Frank E. Franzel, Sr.; Vice Pres., Ron Johnson; Sec. & Treas., Fran Campbell.
Personnel Profile: Part-Time Volunteers 15.
Governing Authority: nonprofit organization. Tax-exempt.
Institution Type/Description: History & Genealogy Museum.
Collections: artifacts pertaining to the history of Mayville & the surrounding area, dating back to the founding of the village; family records; genealogies; obituary file; local cemetery readings. Historic Buildings: log cabin; one-room rural schoolhouse.
Research Fields: the history of the community; genealogical search.
Facilities: 500-vol. library of books on the history of the area; microfilm of all back issues of Mayville Monitor, the local weekly newspaper; reading room.
Activities: guided tours; early crafts; quilting; flea markets.
Publications: annual progress report; annual newsletter.
Hours & Admission Prices: May to Labor Day Fri.-Sat. 10-4; other times by appointment. No charge; donations accepted. &
Attendance: 800 (estimated)
Membership: Annual $5; Family $10; Life $100.

Menominee

MENOMINEE COUNTY HERITAGE MUSEUM, 904 11 Ave., Menominee, MI 49858-3044. Mailing Address: P.O. Box 151, Menominee, MI 49858-0151. Tel.: 906-863-9000.
Founded: 1967.
Congressional District: 1
Key Personnel: Dir., Michael Kaufman; Pres., Lou Ann Borski; Librarian, Amber Allard; Cur., Rebecca Beck.
Personnel Profile: Full-Time Volunteers 1; Part-Time Paid 2; Part-Time Volunteers 52.
Governing Authority: board. Parent Institution: Menominee County Historical Society, Inc. Tax-exempt: 501(c)(3).
Institution Type/Description: General Museum.
Collections: Indian artifacts; oral history tapes; logging; 1900 photographic studio; 1820-1915 Victorian furniture; military room; 1868-1968 probate court records; models; circus display; local history items; commercial fishing. Historic Building: c.1921 brick church with leaded glass windows made in Munich, Germany.
Research Fields: local history.
Facilities: 1,000-vol. library of books, court records, & newspapers.
Activities: guided tours.
Publications: manuscripts; biannual newsletter.
Hours & Admission Prices: Memorial Day-Sept. Mon.-Sat. 10-4. No charge; donations accepted.
Attendance: 5,000 (estimated)
Membership: Friend $15; Supporter $25; Contributor $50; Sustaining $100; Patron $250; Benefactor $500; Heritage $1,000 & up.

Middleville

HISTORIC BOWENS MILLS & PIONEER PARK, 200 Old Mill Rd., Middleville, MI 49333-9194. Mailing Address: 240 Old Mill Rd., Middleville, MI 49333-9194. Tel.: 269-795-7530. Fax: 269-795-7530.
E-mail: carleen@bowensmills.com
Web Site: www.bowensmills.com
Founded: 1978.
Key Personnel: Dir. Operations, Carleen Sabin; Dir. Operations, Owen Sabin.
Governing Authority: nonprofit organization. Tax-exempt.
Institution Type/Description: State Historic Site: 1864 water-powered grist & cider mill.
Collections: Indian artifacts; period tools; farm animals; family histories; firearms; knives; cooper's shop; cobbler's shop; water powered machine shop; blacksmith shop; sewing room; spinning & weaving loft. Historic Buildings: 1830 Plank house & covered bridge; 1850 one room school; 1880 barn; 1860 Victorian house.

Research Fields: area history.

Facilities: library of historic material available to the public on premises only. Antiques, collectibles & reproductions for sale.

Activities: guided tours; lectures; arts festivals; organized education programs for children & adults; temporary exhibitions; horse-drawn hay rides; Civil War & Revolutionary War reenactments; cider pressing demonstrations. Museum Sponsors: 14 Festivals per year; Mountain Man encampments.

Publications: historic booklets; books, Apple Cook Book; Corn Meal Cook Book.

Hours & Admission Prices: May-Dec. by appointment. Mill or Festivals: adults $5, children 5-11 $2; discounts to groups, AAM & ICOM members & groups. Horse-drawn rides with admission to Festivals.

Attendance: 30,000 (estimated)

Midland

ALDEN B. DOW HOME & STUDIO, 315 Post St., Midland, MI 48640-4099. Tel.: 989-839-2744; 866-315-7678 (Toll Free). Fax: 989-839-2611.

E-mail: info@abdow.org

Web Site: www.abdow.org

Key Personnel: House Mgr., Mary Lou Timmons

Institution Type/Description: Historic House Museum: housed in the former home & architectural studio of 20th century architect Alden B. Dow. A National Historic Landmark.

Collections: Dow family history; contemporary architecture; personal artifacts; furnishings.

Activities: educational programs; group tours.

Hours & Admission Prices: Mon.-Fri. 8-5. Tours: Mon.-Thurs. 2pm, Fri.-Sat. 10am. Adults $10, students $5. Children under 8 not admitted. Closed major holidays.

*** ALDEN B. DOW MUSEUM OF SCIENCE AND ART OF THE MIDLAND CENTER FOR THE ARTS, (M),** 1801 W. St. Andrews Rd., Midland, MI 48640-2656. Tel.: 989-631-5930. Fax: 989-631-7890.

E-mail: info@mcfta.org

Web Site: www.mcfta.org

Formerly: Arts Midland: Galleries and School

Founded: 1971.

Congressional District: 4

Key Personnel: C.E.O. & Pres., Michael Hayes; Chm., Julie Close; Dir., Bruce Winslow; Exec. Asst. & Special Events Mgr., Eminor Mills; Museum School Education Mgr. & Artist-in-Residence, Armin Mersmann; Mgr. Educational Programming, Debbie Anderson; Education Asst., Sarah Brandt.

Personnel Profile: Full-Time Paid 5.

Governing Authority: nonprofit organization. Parent Institution: Midland Center for the Arts. Subsidiary Institution: Science & Technology Museum. Tax-exempt: 501(c)(3).

Institution Type/Description: Science & Art Museum and School.

Collections: science & art exhibits.

Activities: guided tours; lectures; gallery talks; art festivals; education & outreach programs; permanent & traveling exhibitions.

Hours & Admission Prices: Tues.-Sat. 10-5, Sun. 1-5. Adults $8, children 4-14 $3; members & children under 3 no charge. ASTC reciprocal. Closed major holidays. &

Attendance: 60,000

Membership: Student $25; Single $40; Family $50; Supporter $100.

CHIPPEWA NATURE CENTER, 400 S. Badour Rd., Midland, MI 48640-8661. Tel.: 989-631-0830. Fax: 989-631-7070.

E-mail: dtouvell@chippewanaturecenter.org

Web Site: www.chippewanaturecenter.org

Founded: 1966.

Congressional District: 10

Key Personnel: Exec. Dir., Dick Touvell; Pres., Dr. Marianne McKelvy; Naturalist, Phil Stephens; Naturalist, Janea Little; Business Mgr., Chris Anderson; Office Mgr., Deana Beckham; Naturalist & Facilities Supvr., Tom Lenon; Mgr. Historical Programs, Kyle Bagnall; Dir. Education, Rachel Larimore; Dir. Volunteers & Outreach, Cathy Devendorf; Dir. Interpretation, Dennis Pilaske; Naturalist, Karen Breternitz; Newsletter Editor, Public Rels., Shelley Koop.

Personnel Profile: Full-Time Paid 20; Part-Time Paid 45; Interns 1.

Governing Authority: nonprofit organization. Tax-exempt.

Institution Type/Description: Nature & Historical Center.

Collections: archaeology; tools used by 19th-century settlers; birds; mammals; insects; ecosystem; cultural history. Historic Buildings: 1880 log schoolhouse; 1870 Homestead Farm with maple sugar house & log barn.

Research Fields: archaeology; botany; ornithology; natural resources management; local history.

Facilities: 2,000-vol. library of natural science & archaeology books; 85-seat auditorium; 1200 acres with 14 miles of nature trails; classrooms. Museum-related items for sale.

Activities: guided tours; lectures; films; formally organized education programs for adults, children & undergraduate students affiliated with local colleges; permanent & temporary exhibitions.

Publications: monthly newsletter, CNC News; occasional publications.

Hours & Admission Prices: Mon.-Fri. 8-5, Sat. 9-5, Sun. 1-5. No charge; donations accepted. Closed Thanksgiving; Christmas. &

Attendance: 50,000 (accurate)

Membership: College Student $18; Senior $20; Individual $24; Senior Citizen Family $28; Family $42; Meadow Society $100-$249; Woodland Society $250-$499; River Society $500 & up; Life Membership $1,000.

DOW GARDENS, 1809 Eastman Ave., Midland, MI 48640. Mailing Address: 1018 W. Main St., Midland, MI 48640-4292. Tel.: 989-631-2677; 800-362-4874. Fax: 989-631-0675.

E-mail: helmreich@dowgardens.org

Web Site: www.dowgardens.org

Founded: 1899.

Congressional District: 10

Key Personnel: Mng. Dir., Marty McGuire; Vice Pres., Michael L. Dow; Taxonomic Horticulturist, Richard Gillis; Dir. Visitors Center, Michelle Holmes.

Personnel Profile: Full-Time Paid 14; Part-Time Paid 4; Part-Time Volunteers 8; Interns 25.

Governing Authority: nonprofit organization. Parent Institution: Herbert H. & Grace A. Dow Foundation, 1018 W. Main St., Midland, MI 48640-4292. Tax-exempt.

Institution Type/Description: Botanical Garden: located on the Herbert H. Dow Estate.

Collections: tulip; woody plants; crab apple; rhododendron; bedding plants; herb garden; All-American garden; water falls; jungle walk; bridges; mazes; pines; ornamental pool.

Research Fields: propagation of native woody material, including Malus, Acer & Quercus; aquatic plant management; applied forestry management; insect phenology; insect & host interactions.

Facilities: library pertaining to horticulture, pest control, aquatics & herbs available for research on premises; reading room; classroom. Books & other museum-related items for sale.

Activities: guided tours; lectures; concerts; arts festivals; TV programs; radio programs; formally organized education programs for children, adults & undergraduate college students.

Publications: pamphlets & brochures; annual booklets, Bedding Plants Evaluation; Crab Apples & Pruning Booklet.

Hours & Admission Prices: April 15 to Labor Day daily 9-8:30; Sept.-Oct. daily 9-6:30; Nov.-April 14 daily 9-4:15. Adult $5, youth 6-17 $1. Closed New Year's Eve & Day; Thanksgiving; Christmas Eve & Day. &

Attendance: 260,500 (accurate)

MIDLAND COUNTY HISTORICAL SOCIETY, 1801 W. St. Andrews Rd., Midland, MI 48640-2656. Tel.: 989-631-5930, ext. 1300. Fax: 989-835-9120.

E-mail: skory@mcfta.org

Founded: 1952.

Congressional District: 10

Key Personnel: Chm., Floyd Andrick; Dir., Gary F. Skory; Business Mgr., Tammie Swinson; Educational Coord., Kristen Lences.

Personnel Profile: Full-Time Paid 2; Part-Time Paid 1; Part-Time Volunteers 250; Interns 4.

Governing Authority: nonprofit organization. Parent Institution: Midland Center for the Arts, Inc. 1801 W. St. Andrews, Midland, MI. Branch Museums: Herbert H. Dow Historical Museum, 3100 Cook Rd., Midland, MI; Bradley Home & Carriage House, 3200 Cook Rd., Midland, MI; The Midland County History Museum, 3417 W. Main St., Midland, MI. Tax-exempt: 501(c)(3).

Institution Type/Description: History Museum.

Collections: Midland & Michigan historical artifacts; silver; glass; textiles; music; photography; lumbering; decorative arts; chemistry; carriages & sleighs; 1890s Dow Chemical Co. reconstruction. Historic Buildings: 1874 vintage house; carriage house with wagon maker's annex.

Research Fields: lumbering; Victorian life; 19th-20th century Dow Chemical Co. history; early transportation; blacksmithing; genealogy.

Facilities: 500-vol. library of rare books available on premises only; auditorium; classrooms. Museum-related items for sale.

Activities: docent program; guided tours; lectures; gallery talks; formally

organized education programs for children; permanent & temporary exhibitions; theater; traveling & local history exhibitions; training programs for professional museum workers.
Publications: book, Building Arts in Midland, Michigan 1800s-1980; Salt of the Earth: A History of Midland County; quarterly newsletter; biannual journal, The Midland Log.
Hours & Admission Prices: Exhibitions: April-Oct. Thurs.-Sat. 11-4. Dow Museum, Vintage House & Carriage House: Wed.-Sat. 10-4. Adults $5, children $3; discounts to AAA members; members no charge. Closed New Year's Day; Good Friday; Memorial Day; Thanksgiving; Christmas. &
Attendance: 75,000 (estimated)
Membership: Student & Senior $35; Individual $40; Family $50; Collector $100; Researcher $250; Curator $500; Historian $1,000; John Larkin $5,000 & up.

Milan

THE HACK HOUSE MUSEUM, 775 County St., Milan, MI 48160-9701. Mailing Address: Milan Area Historical Society, P.O. Box 245, Milan, MI 48160-0245. Tel.: 734-439-1664.
Web Site: www.historicmilan.com
Founded: 1972.
Institution Type/Description: Historic House.
Collections: period artifacts & memorabilia from the Victorian era.
Hours & Admission Prices: May-Nov. Fri. 1-4. No charge; donations accepted.
Attendance: 600 (estimated)

Milford

KENSINGTON METROPARK NATURE CENTER, 2240 W. Buno Rd., Milford, MI 48380-4410. Tel.: 248-685-0603. Fax: 248-684-5836; 810-227-8917.
E-mail: jennifer.hollenbeck@metroparks.com
Web Site: www.metroparks.com
Founded: 1957.
Congressional District: 19
Key Personnel: Supervising Interpreter, Jennifer Hollenbeck; Park Interpreter, Michael Tucker; Park Interpreter, Michael Broughton; Interpreter, Laura Rogers; Interpreter, Andy Swift; Interpreter, Lynette Score.
Personnel Profile: Full-Time Paid 3; Part-Time Paid 2; Part-Time Volunteers 20.
Governing Authority: regional park authority. Affiliated with the Huron-Clinton Metropolitan Authority, 13000 High Ridge Dr., P.O. Box 2001, Brighton, MI 48114-9058. Tax-exempt.
Institution Type/Description: Nature Center.
Collections: natural science specimens for study & exhibit.
Facilities: 750-acre nature study area; nature trails.
Activities: guided tours; lectures; temporary exhibitions; school lecture programs.
Publications: Biennial Report.
Hours & Admission Prices: Mon. 1-5, Tues.-Sun. 10-5. Park: $30 annual; $7 daily. Closed Thanksgiving; Christmas. &
Attendance: 337,497 (estimated)

MILFORD HISTORICAL MUSEUM, 124 E. Commerce St., Milford, MI 48381-5300. Tel.: 248-685-7308.
E-mail: milfordhistory@hotmail.com
Web Site: www.milfordhistory.org
Founded: 1976.
Congressional District: 18
Key Personnel: Dir., Mary Lou Gharrity; Asst. Dir., Marlene Gomez.
Governing Authority: society; nonprofit. Parent Institution: Milford Historical Society. Tax-exempt: 501(c)(3).
Institution Type/Description: Historical Society Museum: housed in 1853 Greek Revival House.
Collections: local history & memorabilia; photographs; local cemetery records & property tax records.
Facilities: microfilm collection of 1871-present The Milford Times available to the public.
Activities: Museum Sponsors: Granny's Attic Sale in July; Annual Homes Tour in September; teas - spring & fall at Mary Jackson House.
Publications: history book, Ten Minutes Ahead of the Rest of the World; bimonthly newsletter.
Hours & Admission Prices: May to mid-Dec. Wed. & Sat. 1-4. No charge; donations accepted.
Attendance: 450 (estimated)
Membership: Students $5; Seniors $10; Individual $15; Family $25; Small Business $50; Silver & Lifetime $250; Gold $500; Platinum $1,000.

Mohawk

KEWEENAW COUNTY HISTORICAL SOCIETY, 670 Lighthouse Rd., Mohawk, MI 49950. Tel.: 906-289-4990 & 337-2244.
Web Site: www.keweenawhistory.org
Founded: 1981.
Congressional District: 1
Key Personnel: Pres. (V), Virginia Petermann Jamison.
Personnel Profile: Part-Time Volunteers 250.
Governing Authority: private; society; nonprofit. Tax-exempt: 501(c)(3).
Institution Type/Description: History Museum.
Collections: marine exhibits; copper mining; local photographs; local kitchen artifacts; period furnishings; fire fighting & automotive artifacts; turn-of-the-century school & Knights of Pythias Society Museum at Eagle Harbor; life-saving boats & equipment. Historic Buildings: 1850 Roman Catholic Church; 1870 copper miner's village of homes; life-saving station; turn-of-the-century blacksmith shop.
Research Fields: Lake Superior; copper mining; Knights of Phythias; general history of Keweenaw County; Roman Catholic Church artifacts from the Keweenaw & Houghton Counties; blacksmithing from the 1890s; maritime shipping & lifesaving; lighthouse.
Activities: guided tours; traveling & temporary exhibitions; school loan service; summer adventure history programs.
Publications: members' quarterly magazine, The Superior Signal.
Hours & Admission Prices: mid-June to mid-Oct. daily 10-5. Adults $5; members and children 13 & under no charge. Group rates available.
Attendance: 10,119 (accurate)
Membership: Historian $20; Copper Miner $35; Lightkeeper $75; Life $500.

Monroe

MONROE COUNTY HISTORICAL MUSEUM, 126 S. Monroe St., Monroe, MI 48161-2275. Tel.: 734-240-7780. Fax: 734-240-7788.
E-mail: andy_clark@monroemi.org
Web Site: www.co.monroe.mi.us/museum
Founded: 1939.
Congressional District: 16
Key Personnel: Education Coord., Lynn Reaume; Archivist, Christine L. Kull; Dir., H. Andrew Clark.
Personnel Profile: Full-Time Paid 2; Part-Time Paid 3; Part-Time Volunteers 100.
Governing Authority: county. Parent Institution: Monroe County Historical Commission. Branch Museums: Navarre-Anderson Trading Post; Papermill School and Country Store Museum. Tax-exempt: 170(b)(1)(A).
Institution Type/Description: History Museum: housed on the site of Gen. George A. Custer home.
Collections: local history artifacts: General George A. Custer; Woodland and Western Indians; tools; costumes; musical instruments; ceramics; medical & dental equipment; domestic appurtenances; manuscripts.
Research Fields: Michigan history, 1785-1940.
Facilities: 2,500-vol. library of printed & manuscript material pertaining to Monroe County & Southeast Michigan available for use by arrangement with archivist. History oriented items for sale.
Activities: guided tours; lectures; gallery talks; permanent & temporary exhibitions.
Publications: books, Legacy of the River Raisin; Brief History of Monroe; Escape to Frenchtown; Women on the Raisin.
Hours & Admission Prices: Jan.-April Wed.-Sat. 10-5; May-Dec. Wed.-Sun. noon-5 (may be seasonal). Suggested donation: adult $4, child $2. Closed New Year's Day; Easter; Thanksgiving; Christmas; county holidays. &
Attendance: 20,000 (accurate)

MONROE COUNTY LABOR HISTORY MUSEUM, 41 W. Front St., Monroe, MI 48161. Tel.: 734-693-0446.
E-mail: lwconnerjr@mail.com
Web Site: www.monroelabor.org
Key Personnel: Pres., Bill Conner
Institution Type/Description: Labor History Museum.
Collections: labor history; photographs; personal artifacts; period artifacts.
Hours & Admission Prices: Mon.-Fri. 9-5; other times by appointment. No charge.

VIETNAM VETERANS HISTORICAL MUSEUM, 1095 N. Dixie Hwy., Norman Heck Park, Monroe, MI 48162. Mailing Address: c/o Monroe County Historical Museum, 126 S. Monroe St., Monroe, MI 48161. Tel.: 734-240-7780.
E-mail: gpodhola0282@comcast.net
Founded: 2004.

Key Personnel: Chm. (V) & Museum Shop Mgr., Glenn R. Podhola.
Personnel Profile: Part-Time Volunteers 5.
Governing Authority: Parent Institution: Monroe County Historical Museum. Tax-exempt.
Institution Type/Description: Military History Museum.
Collections: Vietnam War history & artifacts; photographs; personal artifacts; magazine articles; medals & ribbons; military uniforms, weapons & helicopters; memorials.
Activities: guided tours.
Hours & Admission Prices: May 15-Sept. Wed. & Sat. 12-4; other times by appointment. No charge; donations accepted. &
Attendance: 600 (accurate)
Membership: Lifetime $10.

Montague

MONTAGUE MUSEUM & HISTORICAL ASSOCIATION, 8778 Ferry St., Montague, MI 49437. Tel.: 231-894-8249. Fax: 231-894-9955.
E-mail: michiganoh@aol.com
Founded: 1964.
Congressional District: 9
Key Personnel: Pres. (V), Sally Mclouth; Museum Shop Manager, James Haley.
Personnel Profile: Part-Time Volunteers 21.
Governing Authority: nonprofit organization. Parent Institution: City of Montague. Tax-exempt.
Institution Type/Description: General Museum: housed in a former United Methodist Church.
Collections: lumbering; Indian; military; farming; musical; religious; rocks; dolls; toys; local art; clocks; George Washington family artifacts; bicentennial; telephone exchange; doctors office; barber shop; Peruvian artifacts.
Research Fields: local history; 1890-1970 local newspaper.
Facilities: library of books on Michigan history & short story writings by Frank Adams.
Activities: guided tours; lectures; permanent exhibitions.
Hours & Admission Prices: Summer: Sat.-Sun. 1-5; private tours for groups available. No charge; donations accepted.
Attendance: 2,600 (estimated)
Membership: Single $6; Family $10; Life $100.

NUVEEN COMMUNITY CENTER FOR THE ARTS, 8697 Ferry St., Montague, MI 49437-1395. Tel.: 231-894-2787.
E-mail: nuveen@artscouncilofwhitelake.org
Web Site: www.artscouncilofwhitelake.org
Institution Type/Description: Art Gallery.
Collections: works by local artists.
Activities: demonstrations; classes; workshops; special events.
Hours & Admission Prices: Tues.-Fri. 12-5, Sat. 10-2.

Montrose

MONTROSE HISTORICAL & TELEPHONE PIONEER MUSEUM, 144 E. Hickory St., Montrose, MI 48457-9464. Mailing Address: P.O. Box 577, Montrose, MI 48457-0577. Tel.: 810-639-6644.
E-mail: staff@montrosemuseum.com
Web Site: www.montrosemuseum.com
Founded: 1980.
Key Personnel: Pres. (V), Joe Follett.
Personnel Profile: Part-Time Paid 1; Part-Time Volunteers 48.
Governing Authority: Parent Institution: Montrose Area Historical Assoc. Tax-exempt.
Institution Type/Description: History (local) & Telephone Museum.
Collections: local history & culture; Native American artifacts; photographs; genealogy; period telephone equipment; farming implements; household artifacts; old barber shop; pharmacy, loom.
Research Fields: genealogy; town history; area veterans; telephones.
Activities: genealogy research; speaker programs; research; temporary exhibitions.
Publications: Memory Lane Gazette; newsletter.
Hours & Admission Prices: Museum: Sun. 1-5. Office: Mon.-Tues. 9:30-3. Genealogy Room: call for appointment. No charge; donations accepted. Closed major holidays. &
Attendance: 2,735 (accurate)
Membership: Student $1; Individual $5; Family $7; Business $10; Life $60; Couple Life $100.

Mount Clemens

ANTON ART CENTER, 125 Macomb Place, Mount Clemens, MI 48043-5650. Tel.: 586-469-8666. Fax: 586-469-4529.
E-mail: information@theartcenter.org
Web Site: www.theartcenter.org
Formerly: The Art Center
Founded: 1969.
Personnel Profile: Full-Time Paid 2; Part-Time Paid 2; Part-Time Volunteers 100; Interns 3.
Governing Authority: Tax-exempt.
Institution Type/Description: Art Gallery.
Collections: works by local, regional & national artists; sculpture.
Facilities: classrooms. Museum-related items for sale.
Activities: educational programs; community outreach; special events.
Hours & Admission Prices: Tues.-Thurs. & Sat. 10-5, Fri. 10-6, Sun. 12-4. No charge; donations accepted. &
Attendance: 20,000 (estimated)
Membership: Student $20; Individual $40; Family $50; Schools $50; Organization $50; Friend $100;Patron $200; Corporation $$250; Sustaining $500; Benefactor $1,000.

CROCKER HOUSE MUSEUM & MACOMB COUNTY HISTORICAL SOCIETY, 15 Union St., Mount Clemens, MI 48043-5502. Tel.: 586-465-2488. Fax: 586-465-2932.
E-mail: crockerdirector@sbcglobal.net
Web Site: www.crockerhousemuseum.com
Founded: 1964.
Congressional District: 14
Key Personnel: Pres., Ross Champion; Dir., Kimberly Parr; Office Mgr., Marcia Swiderski; Museum Shop Mgr., Gladys Stevenson.
Personnel Profile: Full-Time Paid 1; Part-Time Paid 1; Part-Time Volunteers 40; Interns 2.
Governing Authority: society; nonprofit. Parent Institution: Macomb County Historical Society. Tax-exempt: 501(c)(3).
Institution Type/Description: Historical Society Museum: residence built 1869 Crocker House, a Victorian Italianate style house.
Collections: material relating to Macomb County history; mineral bath industry; local industry.
Research Fields: Macomb County history.
Facilities: 500-vol. library of county history & Crocker family genealogy available for use by appointment; reading room. Gift items for sale.
Activities: special projects; temporary exhibits; lectures; teas; historical programs; special events. Annual Events: Cemetery Walk & Funeral Tea; Garden Walk & Tea; Christmas events.
Publications: quarterly newsletter; books, Along the Huron; Crocker House; Beacon Tree; Warner Diary; Christian Clemens; Depression Days in Mount Clemens; Crocker House Museum; Made in Mount Clemens.
Hours & Admission Prices: Tues.-Thurs. 10-4, 1st Sun. of month 1-4. Suggested Donations & Special Events: adults $3, children $1; discounts to members.
Attendance: 2,500 (estimated)
Membership: Individual $20; Family $30; Contributing $50; Century $100; Corporate $200; Benefactor $500; Conservator $1,000 & up.

MICHIGAN TRANSIT MUSEUM, INC., 200 Grand Ave., Mount Clemens, MI 48043-5412. Mailing Address: P.O. Box 12, Mount Clemens, MI 48046-0012. Tel.: 586-463-1863.
E-mail: information@michigantransitmuseum.org
Web Site: www.michigantransitmuseum.org
Founded: 1973.
Congressional District: 10
Key Personnel: Chm. (V) & Pres. (V), Billie H. Henning; Treas., Gary J. Michaels.
Personnel Profile: Part-Time Volunteers 40.
Governing Authority: nonprofit organization. Tax-exempt: 501(c)(3).
Institution Type/Description: Transportation Museum: equipment housed at Selfridge Air National Guard Base & 1859, Grand Trunk Depot Railroad Museum, national historic site.
Collections: 1930, #11 Chicago, South Shore & South Bend Interurban coach; 1924 Chicago Rapid Transit El cars; 1895 Grand Trunk Western Railroad caboose; 1942 Alco Diesel-Electric Locomotive; railroad & transportation items; #4040, ex-USAF diesel electric Switch Locomotive; #268 ex-Detroit PCC streetcar; #4601 ex-Toronto PCC streetcar.
Research Fields: railroads; electric trolleys & interurbans of special interest.
Facilities: Prints, buttons, railroad-related souvenirs & booklets for sale.
Activities: films; temporary exhibitions.
Publications: monthly newsletter, Michigan Transit Museum Gazette.
Hours & Admission Prices: Depot Museum: Sat.-Sun. 1-4. No charge; donations accepted. Train rides: adults $7, children 4-12 $4. &

Attendance: 3,500 (estimated)
Membership: Associate $20; Active $30; Life $1,000.

Mount Pleasant

MUSEUM OF CULTURAL & NATURAL HISTORY, (M), 103 Rowe Hall, Corner of Bellows & East Campus Dr., Mount Pleasant, MI 48859. Tel.: 989-774-3829. Fax: 989-774-2612. Facebook: CMU Museum.
E-mail: cmuseum@cmich.edu
Web Site: www.museum.cmich.edu/
Founded: 1970.
Congressional District: 10
Key Personnel: Dir., Dr. Jay C. Martin; Cur. Natural History, Dr. Kirsten Nicholson; Mgr. Collections, Angela Riedel; Educator, Sheree Hall.
Personnel Profile: Full-Time Paid 4; Part-Time Paid 12; Part-Time Volunteers 10; Interns 2.
Governing Authority: university. Affiliated with Central Michigan University. Tax-exempt.
Institution Type/Description: University Museum.
Collections: archaeology; anthropology materials; Michigan Indian artifacts; rocks & minerals; paleontological materials; history materials & zoological collections; contemporary Native American art. Historical Building: 1901 one-room school.
Research Fields: Michigan archaeology & history; herpetology; mammalogy; ornithology.
Facilities: classrooms; labs.
Activities: guided tours; organized education programs for children, adults, undergraduate & graduate students affiliated with Central Michigan University; permanent & temporary exhibitions; school loan service.
Hours & Admission Prices: Mon.-Fri. 8-5, Sat.-Sun. 1-5. No charge; donations accepted. Closed national & university holidays. &
Attendance: 25,000 (estimated)

UNIVERSITY ART GALLERY - CENTRAL MICHIGAN UNIVERSITY, 251 E. Preston St., Mount Pleasant, MI 48859. Mailing Address: Wightman 132, Mount Pleasant, MI 48859. Tel.: 989-774-3800.
E mail: gochelas@cmich.edu
Web Site: www.uag.cmich.edu
Founded: 1970.
Congressional District: 99
Key Personnel: Dir., Anne Gochenour.
Personnel Profile: Full-Time Paid 1; Part-Time Paid 6.
Governing Authority: Parent Institution: Department of Art and Design, Central Michigan University. Tax-exempt.
Institution Type/Description: Art Gallery.
Collections: works by regional & national artists.
Major Exhibits: Global Matrix III (T), 1/14/14-2/8/14; 2014 Annual Juried CMU Student Art Exhibition, 2/14-3/14; Graphic Design BFA, 4/14; BFA, BAA, BA, BS Exhibitions, 4/14-5/14; Heroes, 9/14; Faculty Exhibition, 10/14; Solo Exhibition, 11/14; BFA, BAA, BA, BS Exhibitions, 11/14-12/14.
Hours & Admission Prices: Sept.-May Tues.-Fri. 11-6, Sat. 11-3; Summer: call for hours. No charge; donations accepted. &
Attendance: 9,727 (accurate)

ZIIBIWING CENTER OF ANISHINABE CULTURE & LIFE-WAYS, 6650 E. Broadway, Mount Pleasant, MI 48858-8950. Tel.: 989-775-4750. Fax: 989-775-4770.
Web Site: www.sagchip.org/ziibiwing
Key Personnel: Dir., Shannon Martin; Asst. Dir., Judy Pamp; Cur., William Johnson.
Governing Authority· Tax-exempt.
Institution Type/Description: General Museum.
Collections: culture & history of the Saginaw Chippewa Indian Tribe of Michigan & other Great Lakes Anishinabek; paintings; government documents.
Facilities: cafe. Museum-related items for sale.
Activities: permanent & temporary exhibits; educational workshops; lectures; special events; rental facilities; films.
Publications: quarterly newsletter, Noodagon; Diba Jimooying, Telling Our Story: A History of the Saginaw Chippewa Indian Tribe of Michigan.
Hours & Admission Prices: Mon.-Sat. 10-6. Adults $6.50, college students $4.50, senior citizens 60 & over, children 5-17 and active military $3.75; children under 4, members & teachers no charge. Closed New Year's Eve & Day; Thanksgiving; Christmas Eve & Day. &

Munising

ALGER COUNTY HISTORICAL SOCIETY, 1496 Washington St., Munising, MI 49862-1492. Tel.: 906-387-4308. Fax: 906-387-4188.
E-mail: algerchs@jamadots.com
Founded: 1966.
Congressional District: 11
Key Personnel: Pres. (V), Mary Jo Cook.
Personnel Profile: Part-Time Volunteers 20.
Governing Authority: nonprofit organization. Branch Museum: Grand Island Fur Trader's Cabin. Tax-exempt: 501(c)(3).
Institution Type/Description: Historical Society Museum.
Collections: pioneer artifacts from Alger County homes; portraits; blacksmith shop; fur trader's cabin.
Research Fields: Alger County history.
Facilities: 200-vol. library of books & 1,000 ephemera material library; 100-seat auditorium; research area. Museum-related items for sale.
Activities: guided tours; living history; annual barbeque; workshops; speakers.
Publications: quarterly magazine, Alger Footprints; Alger County Centennial History; Who Were Those People.
Hours & Admission Prices: Tues.-Sat. 12-3. No charge; donations accepted. &
Attendance: 1,600 (estimated)
Membership: Individual $10; Family $15; Contributing $25.

Muskegon

GREAT LAKES NAVAL MEMORIAL AND MUSEUM - USS SILVERSIDES SUBMARINE MUSEUM, (M), 1346 Bluff St., Muskegon, MI 49441-1089. Mailing Address: P.O. Box 1692, Muskegon, MI 49443-1692. Tel.: 231-755-1230. Fax: 231-755-5883.
E-mail: dherzhaft@glnmm.org
Web Site: www.glnmm.org
Founded: 1987.
Congressional District: 9
Key Personnel: Dir., Denise Herzhaft; Chm., Mark Fazakerley; Vice Chm., Don Morell; Mgr. Collections, Paul Garzelloni.
Personnel Profile: Full-Time Paid 3; Part-Time Paid 10; Part-Time Volunteers 30.
Governing Authority: nonprofit organization. Parent Institution: Great Lakes Naval Memorial and Museum. Tax-exempt: 501(c)(3).
Institution Type/Description: Historic Ship: World War II navy sub U.S.S. Silversides, served with the Pacific Fleet in waters controlled by the Japanese Empire: East China Sea & through key enemy shipping routes; USCGC McLane (WMEC 146).
Collections: United States World War II submarine; USCGC McLane; 1927 Coast Guard Cutter, enforced prohibition; diving bell; veterans photos & personal artifacts; USS Silversides SS-236; static displays.
Research Fields: U.S. maritime & naval history.
Facilities: Publications & gift items for sale.
Activities: guided tours; overnight camp-ins for groups of 20 to 100.
Publications: UPSCOPE
Hours & Admission Prices: April & Oct. Sat.-Sun. 10-5:30; June-Aug. daily 10-5:30. Museum: $6 per person. Tours: adults $15, seniors 62 & over $12.50, children 5-18 $10.50; children under 5, military, & veterans no charge. Overnight Program Camping for groups of 20-100 daily 6pm-9:30am, call for reservations. &
Attendance: 35,000 (estimated)

HACKLEY & HUME HISTORIC SITE, W. Webster Ave. & Sixth St., Muskegon, MI 49440. Mailing Address: 430 W. Clay Ave., Muskegon, MI 49440-1002. Tel.: 231-722-7578. Fax: 231-728-4119.
E-mail: dawn@lakeshoremuseum.org
Web Site: www.lakeshoremuseum.org
Founded: 1971.
Congressional District: 96
Key Personnel: Exec. Dir., John H. McGarry, III; Site Mgr., Dawn Willi; Pres. (V), Eric Gielow; Museum Shop Mgr., Peggy Jobe.
Personnel Profile: Full-Time Paid 2; Part-Time Paid 9; Part-Time Volunteers 80.
Governing Authority: nonprofit organization. Parent Institution: The Lakeshore Museum Center. Tax-exempt: 501(c)(3).
Institution Type/Description: Historic Houses: homes of C.H. Hackley & Thomas Hume.
Collections: period furniture & furnishings.
Research Fields: lumbering; life of Hackley & Hume families; Victorian architecture; decorative arts.
Facilities: visitor information center. Museum-related items for sale.

Activities: guided house tours; school programs; special events. Museum Sponsors: Holiday Tours in November & December.
Publications: brochure; newsletter.
Hours & Admission Prices: May-Oct. Wed.-Sun. 12-4; group tours by appointment; special holiday hours between Thanksgiving & Christmas. Adults $3; children 12 & under no charge. Closed major holidays.
Attendance: 7,095 (accurate)
Membership: Single $20; Family $30; Lumberjack $50; Explorer $100; Timber Cruiser $250; Pathfinder $500; Lumber Baron/Baroness $1,000.

LAKESHORE MUSEUM CENTER, (M), 430 W. Clay, Muskegon, MI 49440-1002. Tel.: 231-722-0278. Fax: 231-728-4119. Facebook: Lakeshore Museum Center.
E-mail: info@lakeshoremuseum.org
Web Site: www.lakeshoremuseum.org
Formerly: Muskegon County Museum
Founded: 1937.
Congressional District: 96
Key Personnel: C.E.O., John H. McGarry, III; Pres. (V), Sherry White; C.F.O., Cheryl Graves; Historic Sites Cur., Dani LaFleur; Historic Sites Mgr. & 'volunteer Coord., Erin Schnutz; Dir. Exhibits & Collections, Melinda Conley; Cur. Temporary Exhibits, Krista Menacher; Dir. Visitor Services, Melissa Horton; Cur. Education, Jacquelyn Funk; Asst. Cur. Education, Patrick Horn; Archivist, Beryl Gabel; Maintenance Supvr., George J. Raap; Dir. Communications, Joni Dorsett; Cur. Collections, Sharon McCullar; Museum Shop Mgr., Peggy Jobe.
Personnel Profile: Full-Time Paid 17; Part-Time Paid 15; Part-Time Volunteers 300; Interns 1.
Governing Authority: nonprofit organization. Subsidiary Institution: Hackley & Hume Historic Site. Tax-exempt: 501(c)(3).
Institution Type/Description: Historic Site/Building; Historic Museum; Children's Science Museum.
Collections: Indian, pioneer & lumbering artifacts; archaeology; geology; natural history; maritime history.
Major Exhibits: Bittersweet Harvest, 11/16/13-1/26/14.
Research Fields: lumbering; maritime history; Victorian architecture.
Facilities: library; archives.
Activities: special exhibits; free family programs; behind the scenes tours; education programs for schools & home school groups; programs for children.
Publications: bimonthly newsletter; brochure.
Hours & Admission Prices: Mon.-Fri. 9:30-4:30, Sat.-Sun. 12-4. No charge; donations accepted. Closed major holidays. &
Attendance: 46,555 (accurate)
Membership: Individual $20; Family $30; Lumberjack $50; Explorer $100; Timber Cruiser $250 & up; Pathfinder $500 & up; Lumber Baron/Baroness $1,000 & up.

MUSKEGON HERITAGE MUSEUM, 561 W. Western Ave., Muskegon, MI 49440-1042. Tel.: 231-722-1363.
Web Site: www.muskegonheritage.org
Founded: 1983.
Key Personnel: Dir., Allan Dake; Pres. (V), Jim Funnell.
Personnel Profile: Part-Time Volunteers 46.
Volunteer Hours: 1,460
Operating Expenses: 10,000
Operating Income: 12,000
Governing Authority: Parent Institution: Muskegon Heritage Association. Tax-exempt.
Institution Type/Description: History Museum: housed in the former Hathaway Building; built in 1910.
Collections: local history & industries; Corliss Steam engine; printing presses; woodworking machinery; logging tools; period artifacts; historic homes & buildings; pattern makers; foundries; industry & products manufactured in Muskegon.
Research Fields: Muskegon industrial history; historic homes; Muskegon businesses.
Facilities: rental facilities.
Activities: school & group tours; member events; rental facilities.
Publications: triannual newsletter, The Boomer; book, Muskegon's Industries (2013).
Hours & Admission Prices: mid-May to mid-Oct. Fri.-Sat. 11-4; Dec. 1st weekend 11-4; other times by appointment. Adults $4, students $2; members & Muskegon County school groups no charge. &
Attendance: 1,900 (accurate)
Membership: Family $15; Patron $25; Historian $50; Benefactor $100; Heritage $250; Preservation $500.

* **MUSKEGON MUSEUM OF ART, (M),** 296 W. Webster, Muskegon, MI 49440-1282. Tel.: 231-720-2570 & 2571. Fax: 231-720-2585.
Web Site: www.muskegonartmuseum.org
Founded: 1912.
Congressional District: 9
Key Personnel: Exec. Dir., Judith Hayner; Senior Cur., Dir. Collections & Exhibitions, E. Jane Connell; Public Rels., Marguerite Curran-Gawron; Pres., Larry Hines; Cur. Education, Catherine Mott; Collections Mgr. & Asst. Cur., Art Martin; Preparator, Keith Downie; Museum Store Mgr., Shawnee Larabee.
Personnel Profile: Full-Time Paid 7; Part-Time Paid 9; Part-Time Volunteers 200; Interns 1.
Governing Authority: public school district. Parent Institution: Muskegon Public Schools. Subsidiary Institution: Muskegon Museum of Art Foundation. Tax-exempt: 509(a)(1).
Institution Type/Description: Art Museum.
Collections: American & European paintings & graphic arts; photography; sculpture; Tiffany glass; modern studio glass.
Research Fields: 19th- & early 20th-century American & European art.
Facilities: 200-seat auditorium; classroom. Museum-related items for sale.
Activities: guided tours; lectures; films; gallery talks; concerts; loan exhibitions; study group; curriculum oriented programs for elementary & secondary students.
Publications: Pictures of the Best Kind: The First Century of the Muskegon Museum of Art.
Hours & Admission Prices: Winter: Wed. & Fri.-Sat. 10-4:30, Thurs. 10-8, Sun. 12-4:30. Summer: Tues. & Thurs. 10-6, Wed. & Fri.-Sat. 10-4:30, Sun. 12-4:30. Adults $7, college students $5; members no charge. Closed major holidays. &
Attendance: 29,165 (accurate)
Membership: Student $30; Educator & Artist $50; Individual $60; Household $70; Friend $150-$299; Patron $300-$499; Benefactor $500-$999; Ambassador $1,000-$2,499; Hackley Guild $2,500-$4,999; Collector's Circle $5,000-$9,999; Chairman's Society $10,000-$24,999; MMA Art Guild $25,000 & up.

Muskegon Heights

MUSKEGON COUNTY MUSEUM OF AFRICAN AMERICAN HISTORY, 7 E. Center St., Muskegon Heights, MI 49444. Mailing Address: 610 W. Western Ave., Muskegon, MI 49440. Tel.: 231-739-9500.
Governing Authority: Tax-exempt: 501(c)(3).
Institution Type/Description: History Museum.
Collections: African American history; photographs; paintings.
Hours & Admission Prices: Mon.-Sat. 2-5:30. No charge; donations accepted.

Newaygo

NEWAYGO COUNTY MUSEUM AND HERITAGE CENTER, (M), 12 Quarterline St., Newaygo, MI 49337. Mailing Address: P.O. Box 68, White Cloud, MI 49349. Tel.: 231-652-5003. Facebook: Newaygo County Museum.
E-mail: rbassett@newaygocountyhistory.org
Web Site: newaygocountyhistory.org
Formerly: Newaygo County Society of History and Genealogy
Founded: 1966.
Congressional District: 2
Key Personnel: Dir., Roxanne Bassett; Pres. (V), Toni Rumsey.
Personnel Profile: Part-Time Paid 3; Part-Time Volunteers 8.
Governing Authority: society. Tax-exempt: 501(c)(3).
Institution Type/Description: Regional Museum.
Collections: western Michigan history from prehistory to modern times; military uniforms; logging; Civil War items; costumes.
Research Fields: western Michigan; genealogy.
Facilities: museum building; archives office; church.
Publications: bimonthly newsletter.
Hours & Admission Prices: Memorial Day-Oct. Wed.-Sat. 12-4. No charge; donations accepted.
Attendance: 2,500 (accurate)
Membership: Individual $20.

Niles

FERNWOOD BOTANICAL GARDEN & NATURE PRESERVE, 13988 Range Line Rd., Niles, MI 49120-9042. Tel.: 269-695-6491 & 683-8653. Fax: 269-695-6688.
E-mail: visitor@fernwoodbotanical.org

Web Site: www.fernwoodbotanical.org
Founded: 1964.
Congressional District: 4
Key Personnel: C.E.O., Peter J. van der Linden; Pres. (V), Johanna Money; Museum Shop Mgr., Kathy Lawrence.
Personnel Profile: Full-Time Paid 11; Part-Time Paid 8; Part-Time Volunteers 250.
Governing Authority: nonprofit. Tax-exempt.
Institution Type/Description: Garden & Nature Preserve.
Collections: arboretum; botany; geology; herbarium; natural history; paintings; outdoor museum; textiles; zoology.
Research Fields: plant hardiness.
Facilities: library of horticultural & nature books available for use on premises; nature trails; garden; visitor's center; nature center.
Activities: formally organized community education programs; temporary exhibitions; lectures & workshops, nature study, gardening; guided tours by appointment; holiday lights event.
Publications: quarterly newsletter; booklets; annual report.
Hours & Admission Prices: Summer: Tues.-Sat. 10-6, Sun. 12-6; Winter: Tues.-Sat. 10-5, Sun. 12-5. Adults $6, seniors $5, students $4, children 6-12 $3; reciprocal benefits to other gardens & arboretum; members & children 5 & under no charge. Closed New Year's Eve & Day; Easter; Thanksgiving weekend; Christmas Eve, Day & week. &
Attendance: 30,000
Membership: Individual $25; Basic $40; Preferred $50; Plym Society $250.

FORT ST. JOSEPH MUSEUM, 508 E. Main St., Niles, MI 49120-2618. Tel.: 269-683-4702. Fax: 269-684-3930.
Web Site: fortstjosephmuseum.org
Founded: 1932.
Congressional District: 6
Key Personnel: Dir., Carol Bainbridge.
Personnel Profile: Full-Time Paid 1; Part-Time Paid 3.
Governing Authority: municipal. Parent Institution: City of Niles, MI. Tax-exempt.
Institution Type/Description: History Museum: housed in 1882 Henry A. Chapin Carriage House.
Collections: Ft. St. Joseph artifacts; 1881-83 pictographs by Sitting Bull & Rain-in-the-Face; local history; Plym/Quimby collection of Sioux Indian artifacts; 19th-century furnishings & decorative arts; early American tools & textiles; archives; glass; photographs.
Research Fields: local history; Fort St. Joseph.
Activities: permanent & temporary exhibits; children's activities; history hunts; group programs.
Hours & Admission Prices: Wed.-Fri. 10-4, Sat. 10-3; group tours by appointment. No charge; donations accepted. Closed holidays.
Attendance: 2,305 (accurate)

Northville

MILL RACE HISTORICAL VILLAGE, 215 Griswold Ave., Northville, MI 48167-1664. Tel.: 248-348-1845. Fax: 248-348-0056.
E-mail: mrv1845@yahoo.com
Web Site: www.millracenorthville.org
Founded: 1964.
Congressional District: 13
Key Personnel: Pres., Ed Gabrys; Vice Pres., Larry Last; Office Mgr., Abbie Holden; Archivist, Heidi Nielsen.
Personnel Profile: Part-Time Paid 3; Part-Time Volunteers 75.
Governing Authority: society; nonprofit. Operated by Northville Historical Society, 215 Griswold Ave., Northville, MI. Tax-exempt: 501(c)(3).
Institution Type/Description: Restored Village Museum.
Collections: Victorian furniture & furnishings; clothing; archives; Northville history, 1824-present. Historic Buildings: 1845 New School Church; 1849 Hunter House; 1868 Yerkes House; 1873 Wash Oak School; c.1890 Cottage House; c.1890 Hirsch Blacksmith Shop; c.1831 Cady Inn; c.1895 Interurban Station; c.1875 Mead General Store.
Research Fields: curriculum of one-room school; Northville community & family histories; photographs & other archive materials.
Facilities: 200-vol. library of community & surrounding area history, archives; meeting facilities; 400 sq. ft. exhibit space; 100-seat auditorium. Postcards & handcrafts for sale.
Activities: guided tours; lectures; docent program; organized education programs for children & adults; children's Christmas workshop. Museum Sponsors: Independence Day celebration; Victorian Festival in September; Christmas Walk-Village; Cemetery Walk in fall.
Publications: newsletter, Mill Race Quarterly.
Hours & Admission Prices: Grounds: dawn-dusk; Buildings: June to Oct. Sun.

1-4; other times by appointment. Office: Mon.-Fri. 9-1. No charge; donations accepted. Closed New Year's Day; Thanksgiving; Christmas. &
Attendance: 2,500 (estimated)
Membership: Crow's Nesters $20; Cobblers & Coopers $35; Globe Trotters $50; Bell Ringers $75; Fish Hatchers $125; Rail Riders $250; The Circle "Ns" $500.

Ontonagon

ONTONAGON COUNTY HISTORICAL SOCIETY MUSEUM, 422 River St., Ontonagon, MI 49953-1614. Tel.: 906-884-6165. Fax: 906-884-6165.
E-mail: ochs@jamadots.com
Web Site: www.ontonagonmuseum.org
Founded: 1957.
Congressional District: 32
Key Personnel: Pres. (V), Bruce Johanson.
Personnel Profile: Part-Time Paid 1; Part-Time Volunteers 37.
Governing Authority: society; nonprofit organization. Tax-exempt: 501(c)(3).
Institution Type/Description: History Museum.
Collections: period household items & furniture; musical instruments; pictures of the early settlers; early copper mining artifacts; crafts; tools; local archives; railroad display; firemen display.
Research Fields: mining; lumbering; genealogy.
Facilities: library of books, records & correspondence available for research. Gift items for sale.
Activities: guided tours.
Publications: books, Ontonagon Township Schools, 100 Years of History; This Ontonagon Country; This Land, the Ontonagon; In Days of Yore; Victoria, The Gem of Forest Hill; Murder and Mayhem Ontonagon; The River and the Land; The Mining Ventures of this Ontonagon Country; booklets on local industries & villages.
Hours & Admission Prices: Mon.-Fri. 10-5, Sat. 10-4, call for additional hours & to confirm winter hours. Adults $3 (first visit of season, thereafter no charge); discount for large groups; children 15 & under no charge. &
Attendance: 6,759 (accurate)
Membership: Junior 15 & under $2; Adult $15; Commercial & Family $25; Life $150.

Orchard Lake

GALERIA, 3535 Indian Trail, Orchard Lake, MI 48324-1623. Tel.: 248-683-0345.
Web Site: orchardlakeschools.org
Founded: 1963.
Congressional District: 18
Key Personnel: Dir. & Artist-in-Residence, Marian Owczarski.
Governing Authority: denominational group; nonprofit.
Institution Type/Description: Art Gallery: housed in former Michigan Military Academy.
Collections: Polish, Polish American & American art; Polish history.
Research Fields: contemporary Polish art.
Facilities: concert hall; lecture hall; reception area.
Activities: guided tours; lectures; concerts; permanent & traveling exhibitions.
Hours & Admission Prices: Mon.-Fri. by appointment, Sat.-Sun. 1-5. No charge; donations accepted. Closed Easter; Thanksgiving; Christmas. &
Attendance: 1,250 (estimated)

Oscoda

AUSABLE-OSCODA HISTORICAL SOCIETY & MUSEUM, 114 E. River Rd., Oscoda, MI 48750. Mailing Address: P.O. Box 679, Oscoda, MI 48750. Tel.: 989-739-2782.
Institution Type/Description: Historical Society Museum.
Collections: local history & culture; period furnishings; personal artifacts; photographs; Native American artifacts; railroad memorabilia; commercial fishing.
Hours & Admission Prices: April-May & Sept.-Oct. Sat. 11-4, Sun. 12-4; Memorial Day to Labor Day Thurs.-Sat. 10-5, Sun. 11-4; other times by appointment.

WURTSMITH AIR MUSEUM, E. Van Ettan St., Oscoda, MI 48750. Mailing Address: P.O. Box 664, Oscoda, MI 48750. Tel.: 989-739-7555. Facebook: Wurtsmith Air Museum.
E-mail: visit@wurtsmithairmuseum.org
Web Site: www.wurtsmithairmuseum.org/index.html
Formerly: Yankee Air Museum, Wurtsmith Div.
Founded: 1993.
Congressional District: 05

Key Personnel: Chm. (V), Don Gauvreau; Museum Shop Mgr. (V), J. Shuler.
Governing Authority: nonprofit organization. Tax-exempt.
Institution Type/Description: Military History Museum.
Collections: military aircraft, artifacts, & equipment; Memorial Wall; personal artifacts; photographs.
Facilities: library. Museum-related items for sale.
Activities: special events.
Hours & Admission Prices: May-Sept. Fri.-Sun. 11-3. Adults $5; members no charge.
Attendance: 1,650 (estimated)
Membership: Individual $45.

Ossineke

DINOSAUR GARDENS, LLC, 11160 U.S. 23 S., Ossineke, MI 49766. Mailing Address: 3010 E. Nicholson Hill Rd., Ossineke, MI 49766-9601. Tel.: 989-471-5477. Fax: 989-732-6951.
Web Site: dinosaurgardensllc.com
Founded: 1934.
Congressional District: 11
Key Personnel: Owner, Gary Stephan.
Governing Authority: individual operation.
Institution Type/Description: Paleontology Museum.
Collections: 25 life-size reproductions of prehistoric animals & birds (hand-sculptured by Paul N. Domke, Sr.) situated in natural outdoor settings; 80 ft., 60,000 lbs. Brontosaurus; flora; fauna; velociraptor sculpture created by artist Joe Donna.
Facilities: 40 acres of nature trails; snack bar; picnic area. Museum-related items for sale.
Activities: tours.
Publications: brochures.
Hours & Admission Prices: May 15 to late May daily 10-4; Memorial Day-Labor Day daily 9-6; Sept. Sat.-Sun. 10-4. Adults $6, children 6-11 $5, children under 6 $4; military discount. ♿
Attendance: 1,500 (estimated)

Owosso

CURWOOD CASTLE, 224 Curwood Castle Dr., Owosso, MI 48867-2723. Mailing Address: 301 W. Main St., Owosso, MI 48867-2915. Tel.: 989-725-0597.
Founded: 1975.
Congressional District: 10
Key Personnel: Chm. (V), Michael Erfourth.
Governing Authority: society; municipal; nonprofit organization. Parent Institution: City of Owosso. Subsidiary Institution: Owosso Historical Commission, City Hall, Owosso 48867. Tax-exempt.
Institution Type/Description: Historic Building: 1922 Norman castle, studio of novelist James Oliver Curwood.
Collections: Curwood furniture; furnishings of the early 1900s; books; paintings; manuscripts; documents.
Research Fields: James Oliver Curwood, author & conservationist.
Facilities: Museum-related items for sale.
Activities: guided tours; docent program & council.
Publications: brochures; pamphlets.
Hours & Admission Prices: Feb.-Dec. Tues.-Sun. 1-5, tour groups on off hours by appointment only. Adults $2, children $1. Closed holidays.
Attendance: 4,000

THE MOVIE MUSEUM, 318 E. Oliver St., Owosso, MI 48867-2351. Tel.: 989-725-7621.
E-mail: moviemuseum@aol.com
Founded: 1937.
Congressional District: 10
Key Personnel: C.E.O., Don Schneider; Pres. (V), Jason Hale; Dir., Don Nidefski; Advisor, Mike Nidefski; Advisor, Rodney Keay; Advisor, Alan Cook.
Personnel Profile: Full-Time Volunteers 1; Part-Time Volunteers 14.
Governing Authority: nonprofit organization. Tax-exempt: 501(c)(3).
Institution Type/Description: Movie Industry /Conflict Resolution Museum.
Collections: machinery; conflict resolution; books; costumes; photos; records; music; advertising; history; theater.
Research Fields: history of stage, records, radio, movies & TV; humanities history of social conflicts; conflict resolution.
Facilities: library; 100-seat auditorium; archives.
Hours & Admission Prices: Temporarily closed.
Membership: Senior & Student $7; Individual $15; Family $22.

SHIAWASSEE ARTS CENTER, 206 Curwood Castle Dr., Owosso, MI 48867-2723. Tel.: 989-723-8354. Fax: 989-729-9134.
E-mail: sac@shiawasseearts.org
Web Site: www.shiawasseearts.org
Governing Authority: nonprofit organization. Tax-exempt: 501(c)(3).
Institution Type/Description: Art Gallery.
Collections: works by Michigan artists.
Hours & Admission Prices: Tues.-Sun. 1-5. No charge.

STEAM RAILROADING INSTITUTE, 405 S. Washington St., Owosso, MI 48867-3523. Mailing Address: P.O. Box 665, Owosso, MI 48867-0665. Tel.: 989-725-9464. Fax: 989-723-1225.
E-mail: dlshorter@mstrp.com
Web Site: www.michigansteamtrain.com
Formerly: Michigan State Trust for Railway Preservation, Inc./ MSU Railroad Club
Founded: 1980.
Congressional District: 4
Key Personnel: Exec. Dir. & C.E.O., David Shorter; Pres., Aarne Frobom; Chm. (V), Jay Fulkerson; Vice Pres., Rich Greter; Treas., Chris Kurzweil; Chief Mechanical Officer, Kevin Mayer; Museum Shop Mgr., Carole Stevens.
Personnel Profile: Full-Time Paid 3; Full-Time Volunteers 10; Part-Time Paid 1; Part-Time Volunteers 40; Interns 4.
Volunteer Hours: 15,000
Operating Expenses: 2,000,000
Operating Income: 2,000,000
Governing Authority: private; nonprofit organization. Parent Institution: Michigan State Trust for Railway Preservation. Subsidiary Institution: Steam Railroading Institute. Tax-exempt: 501(c)(3).
Institution Type/Description: Railroad History Museum: housed in the former Ann Arbor Railroad machine shop; built in 1887.
Collections: operational 1941 2-8-4 steam locomotive from the Pere Marquette Railway; steam locomotive shop tools & reference materials; rolling stock & equipment from the steam era.
Research Fields: railroad history.
Facilities: 150-vol. library of locomotive technical materials; railroad machine shop. Locomotive-related items & art reproductions for sale.
Activities: guided tours; formal education programs; lectures; traveling exhibitions; demonstrations; train & weekend excursions. Annual Events: steam locomotive repair technique seminars; Hand on the Throttle North Pole Express.
Publications: small magazine published three times annually, Project 1225; published three times annually, Members Bulletin.
Hours & Admission Prices: Memorial Day to Labor Day Wed.-Sun. 10-5; Winter: Sat.-Sun. 10-5. Adults $5; discounts to AAM & museum members. Closed major holidays. ♿
Attendance: 20,000 (estimated)
Membership: Subscribing $35; Individual $45; Individual Plus $55; Family $75; Family Plus $95; Donor $175.

Oxford

NORTHEAST OAKLAND HISTORICAL MUSEUM, One N. Washington St., Oxford, MI 48371-4673. Tel.: 248-628-8413 & 391-1367.
E-mail: neohs@charterinternet.com
Web Site: www.orion.lib.mi.us/nohm
Founded: 1971.
Congressional District: 7
Key Personnel: Pres., Gerald Griffin; Vice Pres., Ron Brock; Cur., Marie English; Treas., Darryl Lambertson.
Personnel Profile: Part-Time Volunteers 10.
Governing Authority: society; nonprofit. Parent Institution: Northeast Oakland Historical Society. Tax-exempt: 501(c)(3).
Institution Type/Description: History Museum.
Collections: clothing; furniture; primitive farm and home items; tin shop from early Oxford hardware store; musical items; Mary Gregory cobalt glass; 2 WPA murals of Oxford size 7'x11'.
Research Fields: local history; genealogy; local cemeteries.
Facilities: library of atlases, history books & family histories available for use on premises. Books on local history & cemeteries & other items for sale.
Activities: guided tours; temporary exhibitions.
Hours & Admission Prices: June-Aug. Wed. & Sat. 1-4; Sept.-May Sat. 1-4. No charge; donations accepted.
Attendance: 1,000 (estimated)
Membership: Student $5; Individual $12; Life $125; Business & Organizations $500.

Paradise

GREAT LAKES SHIPWRECK MUSEUM, 18335 N. Whitefish
Point Rd., Paradise, MI 49768-9618. Mailing Address: 400 W.
Portage Ave., Sault Ste. Marie, MI 49783-1993. Tel.: 906-492-
3747. Fax: 906-492-3383. Facebook: Great Lakes Shipwreck
Museum.
E-mail: glshs@shipwreckmuseum.com
Web Site: www.shipwreckmuseum.com
Founded: 1978.
Congressional District: 1
Key Personnel: Site Mgr., Terry Begnoche; Devel. Officer, Sean Ley; Opera-
tions Mgr., Bruce Lynn.
Personnel Profile: Full-Time Paid 5; Part-Time Paid 29; Part-Time Volunteers
5.
Governing Authority: Tax-exempt. Subsidiary Institution: Whitefish Point
Lighthouse; Light Station, Whitefish Point, Lake Superior; US Weather
Bureau Bldg., Soo Locks Park, Sault Ste. Marie, MI.
Institution Type/Description: Maritime Museum.
Collections: maritime history; diving equipment; photographs.
Research Fields: scholarly & field research of Great Lakes shipwrecks with
emphasis on Lake Superior.
Activities: special events.
Publications: Shipwreck Journal, quarterly; Ghosts of the Shipwreck Coast
(The Art and Science of Mapping Lake Superior's Shipwreck Secrets).
Hours & Admission Prices: May-Oct. daily 10-6; other times by appointment.
Adults $12, children 6-17 $8; discounts to AAA members & active military;
children 5 & under no charge.
Attendance: 65,000 (accurate)
Membership: Individual 1 year $25, 3 years $70; Family 1 year $35, 3 years
$100; Business 1 year $100.

Peshawbestown

EYAAWING MUSEUM AND CULTURAL CENTER, (M), 2304
N. West Bayshore Dr., Peshawbestown, MI 49682-9365. Tel.:
231-534-7764 & 7768.
E-mail: museum@gtbindians.com
Web Site: www.gtbindians.org
Institution Type/Description: Native American History Museum.
Collections: Grand Traverse Band history, language, & culture; personal
artifacts; clothing; baskets.
Facilities: Center-related items for sale.
Hours & Admission Prices: mid-June to Labor Day Wed.-Sat. 10-4; Sept. to
mid-June call for hours.

Petoskey

CROOKED TREE ARTS CENTER, 461 E. Mitchell St., Petoskey,
MI 49770-2623. Tel.: 616-347-4337. Fax: 616-347-5414.
E-mail: liz@crookedtree.org
Web Site: www.crookedtree.org
Founded: 1971.
Congressional District: 1
Key Personnel: Exec. Dir., Liz Gowans Ahrens; Pres. (V), Kurt Wietzke; Cur.,
Membership & Museum Shop Mgr., Gail Hosner; Business Mgr., Donna
McDougall; Cultural Coord., Mary Wiklanski.
Personnel Profile: Full-Time Paid 7; Part-Time Volunteers 200; Interns 4.
Governing Authority: nonprofit organization. Tax-exempt: 501(c)(3).
Institution Type/Description: Arts Center: housed in 1890 Methodist Church.
Collections: paintings; prints; sculpture.
Research Fields: art of Ojibway, Odawa, & Nishnawbe Native American
peoples; Great Lakes maritime & history.
Facilities: 3,000 sq. ft. exhibition space; 260-seat theater; dance studio;
classrooms. Artwork by Michigan artists for sale.
Activities: guided tours; lectures; films; concerts; dance recitals; arts festivals;
theater; rental gallery; organized education programs for children & adults;
docent program; participatory & temporary exhibitions. Museum Sponsors:
March of the Arts; Art Auction; House Tour.
Publications: bimonthly newsletter, Art News; annual brochure, Native Ameri-
can Nishnawbe Festival.
Hours & Admission Prices: Mon.-Fri. 9-5, Sat. 10-4. Fee charged for concerts
& special events. Closed legal holidays. &
Attendance: 55,000 (estimated)
Membership: Individual $20; Family & Organization $40; Patron $50; Friend
$100; Trustee $500 & up.

LITTLE TRAVERSE REGIONAL HISTORICAL SOCIETY,
(M), 100 Depot St., Petoskey, MI 49770-2476. Mailing Address:
P.O. Box 2418, Petoskey, MI 49770. Tel.: 231-347-2620.
E-mail: info@petoskeymuseum.org
Web Site: www.petoskeymuseum.org
Founded: 1971.
Congressional District: 11
Key Personnel: Dir., Douglas W. Marshall, Ph.D.; Pres., Richard Clark.
Personnel Profile: Full-Time Paid 1; Part-Time Volunteers 15; Interns 1.
Governing Authority: state; nonprofit. Tax-exempt: 501(c)(3).
Institution Type/Description: Historical Society Museum.
Collections: local history; archives; Hemingway artifacts.
Research Fields: local history.
Facilities: 4,000-vol. library of local history books & newspapers from 1881
available for use by appointment.
Activities: walking tours; educational programs; school tours & programs;
permanent & temporary exhibitions.
Publications: newsletter; calendar.
Hours & Admission Prices: Memorial Day to mid-Oct. Mon.-Fri. 10-4, Sat.
1-4. Adults $2, children 5-17 $1; members no charge.
Attendance: 5,000 (estimated)
Membership: Individual $25; Family $40; Business $60; Life $500.

Plymouth

PLYMOUTH HISTORICAL MUSEUM, (M), 155 S. Main St.,
Plymouth, MI 48170-1635. Tel.: 734-455-8940. Fax: 734-455-
7797. Facebook: Plymouth Historical Museum.
E-mail: director@plymouthhistory.org
Web Site: www.plymouthhistory.org
Founded: 1962.
Congressional District: 13
Key Personnel: Exec. Dir., Elizabeth K. Kerstens; Pres., Pam Yockey; Museum
Shop Mgr., Paula Holmes.
Personnel Profile: Part-Time Paid 8; Part-Time Volunteers 150; Interns 2.
Volunteer Hours: 10,000
Operating Expenses: 135,576
Operating Income: 153,649
Governing Authority: society. Parent Institution: Plymouth Historical Society
Tax-exempt
Institution Type/Description: History Museum.
Collections: 19th-century historical artifacts; Daisy & Markham air rifles;
Alter car; manuscripts; Ford village industries; Petz Abraham Lincoln
collection; prints, photographs, statuary, relics, books & research files
relating to the life & study of Lincoln; Plymouth, Michigan community
history.
Major Exhibits: Vintage Project Runway, 1/31/14-6/8/14.
Research Fields: local history; Civil War history; genealogy; archives;
Lincoln.
Facilities: 1,500-vol. library of books, pamphlets & records available under
staff supervision on premises only. Local history books, pictorial histories &
craft items for sale.
Activities: guided tours; permanent, temporary & traveling exhibitions; chil-
dren's hands-on exhibits; social history classes K-12.
Publications: The Story of Plymouth; Pictures of Plymouth Past & Present;
Plymouth's First Century; Postcards of Plymouth; Plymouth's Air Rifle
Industry.
Hours & Admission Prices: Wed. & Fri.-Sun. 1-4. Adults $5, students 5-17 $2;
discounts to AAA members; members no charge. Closed holidays. &
Attendance: 15,000 (estimated)
Membership: Lincoln's Kids $15; Kellogg Park $25; Main Street $30; Daisy
$50; Alter Car $75; Pere Marquette $150; Lincoln Club $300.

Pontiac

CREATIVE ARTS CENTER, 47 Williams St., Pontiac, MI 48341-
1759. Tel.: 248-333-7849. Fax: 248-333-7841.
E-mail: pontiaccacangela@aol.com
Web Site: www.pontiac.mi.us/cac
Founded: 1964.
Congressional District: 6
Key Personnel: Dir., Angela Petroff.
Personnel Profile: Part-Time Paid 4; Part-Time Volunteers 10.
Governing Authority: nonprofit. Tax-exempt: 501(c)(3).
Institution Type/Description: Civic Art & Cultural Center: housed in 1898
Public Library of Oakland County.
Collections: regional, national & international artists.
Research Fields: Michigan artists registry; community artists activities;
minority arts programs; arts for the disabled.
Facilities: classrooms.

Activities: formally organized education programs for adults & children; community activities; art classes; summer art camp.
Publications: bimonthly announcements.
Hours & Admission Prices: Tues.-Thurs. 10-4:30; Sat. 10-3; other times by appointment. No charge; donations accepted. Closed holidays. &
Attendance: 3,000 (estimated)
Membership: Artist & Seniors $20; Family $50; Organizational $75.

MUSEUM OF NEW ART, 7 N. Saginaw St., Pontiac, MI 48342-2172. Mailing Address: 327 W. Second St., Rochester, MI 48307. Tel.: 248-210-7560.
E-mail: detroitmona@aol.com
Web Site: www.detroitmona.com
Key Personnel: Dir., Jef Bourgeau
Institution Type/Description: Art Gallery.
Collections: paintings; photographs; sculpture.
Hours & Admission Prices: Thurs.-Sat. 1-6; other times by appointment.

OAKLAND COUNTY PIONEER AND HISTORICAL SOCI-ETY, 405 Cesar E. Chavez Ave., Pontiac, MI 48342-1068. Tel.: 248-338-6732. Fax: 248-338-6731. Facebook: Oakland County pioneer and Historical Society.
E-mail: office@ocphs.org
Web Site: www.ocphs.org
Formerly: Pine Grove Historical Museum
Founded: 1874.
Congressional District: 6
Key Personnel: Pres. (V), Brian Golden; Museum Shop Mgr., Judy Hudalla.
Personnel Profile: Full-Time Volunteers 7; Part-Time Paid 1; Part-Time Volunteers 70; Interns 2.
Governing Authority: Parent Institution: Oakland County Pioneer & Historical Society. Tax-exempt: 501(c)(3).
Institution Type/Description: Historic House Museum: c.1845 Pine Grove, the Governor Moses Wisner Historic House.
Collections: pioneer family furnishings & artifacts; smokehouse. Pine Grove Museum: Michigan's 12th Governor, Moses Wisner's 19th century homestead; one-room schoolhouse, c.1865; summer kitchen; root cellar. The Carriage House: library; pioneer farm museum.
Research Fields: Oakland County history; genealogy; archaeology; architecture; local geography; businesses; industries; commerce; Civil War.
Facilities: research library & archives of print material, manuscripts, photographs, oral histories & maps relating to Oakland County. Gift items for sale.
Activities: guided tours; school groups can rent schoolhouse for field trips. Annual Events: Ice Cream Social; Victorian Christmas open house; monthly lecture series.
Publications: quarterly newsletter, Oakland Gazette; periodical, Small Talks; lecture series on historical topics.
Hours & Admission Prices: Jan. to mid-Dec. office & library Tues.-Thurs. 11-4; other times by appointment. Museum: tours by appointment. Admission $5; discounts to military. Closed Thanksgiving.
Attendance: 1,500 (estimated)
Membership: Individual & Nonprofit Organization $20; Family $35; Patron $100; Benefactor $200; Corporate Sponsor & Friends of OCPHS $500.

Port Austin

HURON CITY MUSEUMS, 7995 Pioneer Rd., Port Austin, MI 48467-9400. Mailing Address: c/o William Lloyd Phelps Foundation, 3169 Robina, Berkley, MI 48072-3816. Tel.: 989-428-4123; 313-640-0123. Fax: 989-428-4473.
E-mail: info@huroncitymuseums.org
Web Site: www.huroncitymuseums.org
Founded: 1950.
Key Personnel: Pres. (V), Kathryn H. Parcells.
Personnel Profile: Full-Time Paid 1; Full-Time Volunteers 1; Part-Time Paid 5.
Governing Authority: nonprofit organization. Parent Institution: William Lyon Phelps Foundation. Tax-exempt.
Institution Type/Description: General Museum.
Collections: local history & culture; photographs; personal artifacts; historic buildings.
Facilities: 12,000-vol. library.
Activities: guided tours.
Hours & Admission Prices: July-Aug. Fri.-Sat. 10-4; groups by appointment. Seven Gables or Village Museum: adults $6, senior citizens $5, youths 10-15 $3. Both Tours: adults $10, senior citizens $8, youths 10-15 $5; discounts to AAM & AAA members; children accompanied by an adult no charge. &
Attendance: 1,500 (estimated)

Port Hope

POINTE AUX BARQUES LIGHTHOUSE AND MUSEUM, 7320 Lighthouse Rd., Port Hope, MI 48468-9759. Mailing Address: P.O. Box 97, Port Hope, MI 48468. Tel.: 989-428-3035. Fax: 989-856-4505.
E-mail: president@pointeauxbarqueslighthouse.org
Founded: 2002.
Key Personnel: Pres. (V), Upton D. Bonner; Museum Shop Mgr., Carolyn Curtis.
Governing Authority: Tax-exempt.
Institution Type/Description: History Museum.
Collections: historical artifacts & memorabilia; lightkeepers tools & equipment. Historic Building: 1857 lighthouse.
Hours & Admission Prices: Call for hours. No charge; donations accepted. &
Attendance: 10,000 (accurate)
Membership: Individual $25.

Port Huron

THE KNOWLTON'S ICE MUSEUM, 317 Grand River Ave., Port Huron, MI 48060-3814. Tel.: 810-987-5441.
E-mail: knowltonsicemuseum@yahoo.com
Founded: 1987.
Governing Authority: Tax-exempt.
Institution Type/Description: General Museum.
Collections: natural ice harvesting industry history; tools; tongs; ice boxes; period cars & ice wagons; milk bottles; license plates; dolls & baby buggies.
Activities: videos; guided tours; community event programs.
Hours & Admission Prices: June-Sept. Thurs.-Sat. 11-5; Oct.-May Sat. 11-5; groups of 12 or more by appointment. Adults $4, children 6-10 $2; children under 6 no charge. Closed major holidays. &
Attendance: 3,050

PORT HURON MUSEUM, 1115 Sixth St., Port Huron, MI 48060-5346. Tel.: 810-982-0891. Fax: 810-982-0053.
E-mail: info@phmuseum.org
Web Site: www.phmuseum.org
Founded: 1968.
Congressional District: 10
Key Personnel: Exec. Dir., Susan Bennett; Pres. (V), Juanita Gittings; Cur. & Education Coord., Katherine Bancroft; Site Mgr. Huron Lightship, Gerald Rome; Dir. Business Operations & Museum Shop Mgr., Sheila Lindsey.
Personnel Profile: Full-Time Paid 5; Part-Time Paid 12; Part-Time Volunteers 200.
Volunteer Hours: 1,500
Operating Expenses: 791,432
Operating Income: 893,988
Governing Authority: nonprofit organization. Subsidiary Institution: Huron Lightship Museum; Thomas Edison Depot Museum; Fort Gratiot Lighthouse. Tax-exempt: 501(c)(3).
Institution Type/Description: General Museum: housed in 1904 Carnegie Library Building. Huron Lightship: National History Landmark. Edison Depot: Historic Train Depot; Fort Gration Light Station.
Collections: marine; history; natural history; surgeon's instruments from Ft. Gratiot, 1814-1879; lumbering pictures & tools; prehistoric Indian artifacts; models of old Great Lakes ships; reconstructed ship's Pilot house; archaeological material from local Indian site, Fort Gratiot site & Thomas Edison boyhood home; local & Canadian art; lightship, Huron. Historic Building: 1854 pioneer log house.
Major Exhibits: Famous Failures & Silver, 2/7/14-6/15/14.
Research Fields: Michigan & local history; Great Lakes marine history; local archaeological research, prehistoric & historic.
Facilities: 200-vol. library of Michigan & local history available for use on premises.
Activities: guided tours; lectures; gallery talks; inter-museum loan, permanent, temporary & traveling exhibitions.
Publications: membership newsletter.
Hours & Admission Prices: See website for hours. &
Attendance: 40,660 (accurate)
Membership: Student $25; Senior Citizens 60 & over $25; Individual $35; Family $45.

Port Sanilac

SANILAC COUNTY HISTORICAL VILLAGE AND MUSEUM, 228 S. Ridge St., Port Sanilac, MI 48469-9704. Mailing Address: P.O. Box 158, Port Sanilac, MI 48469-0158. Tel.: 810-622-9946.
E-mail: sanilacmuseum@gmail.com

Web Site: www.sanilaccountymuseum.org
Founded: 1964.
Congressional District: 8
Key Personnel: Pres. (V), Thomas Fisher.
Personnel Profile: Full-Time Paid 1; Part-Time Paid 3; Part-Time Volunteers 273.
Governing Authority: society; nonprofit organization. Parent Institution: Sanilac County Historical Society. Tax-exempt.
Institution Type/Description: Historic Village: housed on the former estate of Dr. Joseph Loop; built in 1853.
Collections: medical; agriculture; marine; Native American; Civil War; Victorian furnishings. Historic Buildings: 1872 Victorian Loop-Harrison Mansion; 1882 Banner Log Cabin, Platt's General Store; Dairy Museum; Carriage Barn; Decker Log Barn; Huckins One Room School; Hunting & Fishing Cabin; Native American Center; Ward Lake Cottage; Forestville Church; Hearse Shed; Deckerville Train Depot & Privy.
Research Fields: recording all cemeteries in county; genealogy-family county histories.
Facilities: library of local history, archives; theater; nature trail.
Activities: guided, school & private tours; permanent & temporary exhibitions; theatre; rental facilities. Annual Events: Sanilac Residents Day in May; Native American, Voyageurs & Settlers Weekend in June; Victorian Brunch/Crafters Fair/Agricultural Weekend in July; Children's Day in July & August; Civil War Days/Hunting, Fishing & Maritime Days in August; Wine & Cheese Tasting; Classical and Mosaic Concerts in the Museum Church September to June; House of Mourning/BooFest/Haunted Village in October; Christmas Open House in November; Victorian Dinners in December.
Publications: books, Shingle Shavers & Berry Pickers/The Loop-Harrison Family; This is from Walter.
Hours & Admission Prices: Museum: May-Sept. Wed.-Sun. 11-4. Office: May-Sept. Wed.-Sun. 11-4; Oct.-April Mon., Wed. & Fri. 10-2; other times by appointment. Archives & Genealogy Rooms: Wed. 10-2. Admission varies by event; discount to military. ⚬
Attendance: 3,300 (accurate)
Membership: Individual $25; Family $35.

Portage

TRAIN BARN MUSEUM, 10234 E. Shore Dr., Portage, MI 49002-7166. Tel.: 269-327-4016. Fax. 269-327-4012.
E-mail: custserv@trainbarn.com
Web Site: www.trainbarn.com
Institution Type/Description: Model Railroad Museum.
Collections: O, S, HO, & N gauge model railroading artifacts; O gauge layout; model railroads.
Hours & Admission Prices: Jan. 2 to late Nov. Wed.-Fri. 1-6, Sat. 10-5; late Nov. to Jan. 1 daily. Adults 16 & over $3, children 15 & under $1.

Rochester

MEADOW BROOK HALL, (M), 480 S. Adams Rd., Rochester, MI 48309-4401. Tel.: 248-364-6200. Fax: 248-364-6201.
E-mail: stobersk@oakland.edu
Web Site: www.meadowbrookhall.org
Founded: 1971.
Congressional District: 9
Key Personnel: Exec. Dir., Geoff Upward; Mgr. Merchandising & Database, Kelly Lenda; Mgr. Mktg. & Communications, Shannon O'Berski; Business Mgr., Robin Gardner; Assoc. Dir., Kim Zelinski; Coord. Museum Svcs., Meredith Long; Coord. Facility Operations, Nicole Thomas; Accounting Asst., Grace Campbell.
Personnel Profile: Full-Time Paid 9; Part-Time Paid 5; Part-Time Volunteers 450; Interns 2.
Governing Authority: public university; nonprofit. Parent Institution: Oakland University, Rochester, MI. Tax-exempt.
Institution Type/Description: Historic House Museum: housed in the Tudor-revival style mansion of Matilda Dodge Wilson, widow of automobile pioneer John Dodge & her second husband, Alfred Wilson, a Wisconsin lumber broker.
Collections: paintings; sculpture; prints & drawings; furniture; glass; porcelain & pottery; silver; linen & lace; fashion & costumes; rugs; textiles; archival materials including architectural plans; photographs; films; documents; artifacts; outbuildings.
Research Fields: 20th-century architecture & interior design; 20th-century social history; 20th-century fashion & costume; 19th-20th century Michigan & local history; automotive history.
Facilities: 100-vol. library of art & architecture books available for research only; 55,000 sq. ft. exhibit space; gardens. Museum-related items for sale.
Activities: docent program; formal education program for children & Oakland University students; guided tours; lectures; temporary & traveling exhibitions; creative writing program for children; classical concert series in collaboration with Oakland University. Museum Sponsors: Jazz in the Garden; dinner & a movie; holiday walk.
Publications: Meadow Brook.
Hours & Admission Prices: Guided Tours: June-Aug. daily 11:30, 12:30, 1:30, 2:30; Sept.-May Mon.-Fri. 1:30, Sat.-Sun. 11:30, 12:30, 1:30, 2:30. Adults $15, senior citizens $10; children 12 & under no charge. Closed New Year's Eve & Day Easter; Memorial Day; Independence Day; Labor Day; Thanksgiving; Christmas Eve, Day & week. ⚬
Attendance: 111,000 (accurate)

OAKLAND UNIVERSITY ART GALLERY, 2200 N. Squirrel Rd., 208 Wilson Hall, Rochester, MI 48309-4401. Tel.: 248-370-3005. Fax: 248-370-3377.
E-mail: goody@oakland.edu
Web Site: www.oakland.edu/ouag
Formerly: Meadow Brook Art Gallery
Founded: 1962.
Congressional District: 9
Key Personnel: Dir., Dick Goody; C.E.O., Ronald Sudol; Asst. to Dir., Jacky Leow.
Personnel Profile: Full-Time Paid 2; Part-Time Paid 4; Part-Time Volunteers 2; Interns 2.
Governing Authority: university. Parent Institution: Oakland University. Subsidiary Institution: College of Art & Science, Dept. of Art & Art History. Tax-exempt.
Institution Type/Description: Art Gallery.
Collections: African, Oceania, Indonesian, pre-Columbian art; contemporary American & European paintings, prints & graphics; outdoor sculptures.
Research Fields: Oriental, primitive, pre-Columbian & contemporary art.
Facilities: theater; concert hall; outdoor music pavilion; outdoor sculpture park.
Activities: lectures; gallery talks; film & slide presentations; music, dance performances in relation to art exhibitions; arts festivals; formally organized education programs for undergraduate college students; inter-museum loan; temporary & traveling exhibitions; school loan service.
Publications: exhibition catalogs.
Hours & Admission Prices: Tues.-Sun. 12-5. Special Events & Meadow Brook Theatre performances: Wed.-Fri. 7pm through 1st intermission, Sat.-Sun. 5pm through 1st intermission. No charge; donations accepted. ⚬
Attendance: 16,000 (estimated)
Membership: Student $15; Individual $30; Family $50; Friends $100; Artist's Circle $250; Curator's Circle $500; Director's Circle $1,000; Patron's Circle $2,000.

PAINT CREEK CENTER FOR THE ARTS, 407 Pine St., Rochester, MI 48307-1933. Tel.: 248-651-4110. Fax: 248-651-4110.
E-mail: comments@pccart.org
Web Site: www.pccart.org
Key Personnel: Exec. Dir., Suzanne Wiggins
Institution Type/Description: Art Gallery.
Collections: paintings; sculpture; pottery.
Facilities: Gallery-related items for sale.
Activities: educational programs; classes; outreach programs; special events. Annual Event: Art & Apples Festival.
Hours & Admission Prices: Mon.-Thurs. 9-9, Fri. 9-7, Sat. 10-4.

Rochester Hills

ROCHESTER HILLS MUSEUM AT VAN HOOSEN FARM, (M), 1005 Van Hoosen Rd., Rochester Hills, MI 48306-4555. Tel.: 248-656-4663. Fax: 248-608-8198. Facebook: Rochester Hills Museum.
E-mail: rhmuseum@rochesterhills.org
Web Site: www.rochesterhills.org
Founded: 1979.
Congressional District: 18
Key Personnel: Mayor, Bryan Barnett; Supvr. Interpretive Services, Patrick J. McKay; Museum Shop Mgr., Sue Thomasson.
Personnel Profile: Full-Time Paid 2; Part-Time Paid 4; Part-Time Volunteers 35; Interns 1.
Governing Authority: municipal; nonprofit. Parent Institution: City of Rochester Hills, MI. Tax-exempt.
Institution Type/Description: History Museum.
Collections: agricultural items; household furnishings; local archaeological artifacts; 20th-century clothing. Historic Buildings: 1840 Van Hoosen Farmhouse; 1848 schoolhouse; 1850 Red House; 1927 Dairy Barn; 1927 Calf Barn.
Research Fields: Van Doren coverlets; quilts; local history.

Facilities: 800 sq. ft. exhibit space; 60-seat auditorium; children's garden; 16 acre farm grounds. Museum-related items for sale.
Activities: guided tours; lectures; concerts; organized education programs; temporary & traveling exhibitions; summer day camps for children-archeology, science, farm. Museum Sponsors: Stonewall Pumpkin Festival.
Publications: quarterly, Museum Visitor.
Hours & Admission Prices: Fri.-Sat. 1-4. Adults $5, senior citizen & students $3; discounts to AAM members; members no charge. Closed major holidays. ♿
Attendance: 50,000 (estimated)
Membership: Individual $30; Individual Plus $40; Family $50; Business $250.

Rogers City

PRESQUE ISLE COUNTY HISTORICAL MUSEUM, 176 W. Michigan Ave., Rogers City, MI 49779-1638. Mailing Address: P.O. Box 175, Rogers City, MI 49779-0175. Tel.: 989-734-4121. Fax: 989-734-4121.
E-mail: bradleymuseum@yahoo.com
Web Site: www.thebradleyhouse.org
Founded: 1973.
Congressional District: 11
Key Personnel: Pres., David Nadolsky; Cur. & Museum Shop Mgr., Rose Buck; Cur., Mark Thompson.
Personnel Profile: Part-Time Paid 1; Part-Time Volunteers 60.
Governing Authority: private; nonprofit organization. Tax-exempt.
Institution Type/Description: General Museum: housed in 1914 home of Carl D. Bradley, owner of Bradley Transportation Line.
Collections: area historical artifacts; maritime exhibit; historical information regarding the Bradley Fleet; original ship model of s/s Carl D. Bradley, which sank in Lake Michigan 11/18/58 taking 33 area residents' lives.
Facilities: 2,000 sq. ft. exhibit space. Museum-related items for sale.
Activities: Annual Event: Christmas at the Bradley House.
Publications: semiannual newsletter.
Hours & Admission Prices: May-Sept. Tues.-Sat. 12-4. No charge; donations accepted.
Attendance: 2,500 (accurate)
Membership: Individual $15; Family $25; Sustaining $100 & up.

Royal Oak

DETROIT ZOOLOGICAL SOCIETY, (M), 8450 W. 10 Mile Rd., Royal Oak, MI 48067-3001. Tel.: 248-541-5717. Fax: 248-398-0504.
Web Site: www.detroitzoo.org
Founded: 1928.
Congressional District: 9
Key Personnel: Dir. & C.E.O., Ron L. Kagan; Chm. (V), Gail L. Warden.
Personnel Profile: Full-Time Paid 189; Part-Time Paid 31; Part-Time Volunteers 1,100; Interns 25.
Governing Authority: Parent Institution: Detroit Zoological Society. Tax-exempt.
Institution Type/Description: Zoo.
Collections: 56 mammal species; 58 bird species; 17 Fish (marine & freshwater) species; 9 invertebrate species; 79 amphibian species; 83 reptile species.
Research Fields: behavior; behavioral endocrinology; welfare assessment methods.
Facilities: 1,500-vol. library pertaining to natural history, scientific methods, animal welfare science & policy; aquarium. Natural history books and gift items for sale.
Activities: docent program; guided tours; lectures; mobile vans; Wild Adventure Simulator. Museum Sponsors: summer day camps; special event days.
Publications: monthly newsletter, Habitat.
Hours & Admission Prices: April-June & Labor Day weekend daily 9-5; July-Aug. Wed. 9-8, Thurs.-Tues. 9-5; Sept. 3-Oct. daily 10-5; Nov.-March daily 10-4. Adults 15-61 $14, senior citizens 62 & over and active military with ID $12, children 2-14 $9; discounts to groups; members & children under 2 no charge. Closed New Year's Day; Thanksgiving; Christmas. ♿
Attendance: 1,146,241 (accurate)
Membership: Individual $53; Individual Plus $70; Single Parent $75; Family & Grandparents $89; Supporter $160.

Saginaw

CASTLE MUSEUM OF SAGINAW COUNTY HISTORY, (M), 500 Federal Ave., Saginaw, MI 48607-1253. Tel.: 989-752-2861, ext. 301. Fax: 517-752-1533.
E-mail: info@castlemuseum.org
Web Site: www.castlemuseum.org
Founded: 1948.
Congressional District: 8
Key Personnel: Dir., Ken Santa; Dir., Irene Hensinger; Museum Shop Mgr., Sherri D. Greene.
Personnel Profile: Full-Time Paid 8; Part-Time Paid 5; Part-Time Volunteers 75; Interns 2.
Governing Authority: nonprofit organization. Parent Institution: Historical Society of Saginaw County, Inc. Tax-exempt: 501(c)(3).
Institution Type/Description: History Museum: housed in an 1898 Federal post office building in the style of a French Chateau.
Collections: photographs; costumes; archaeological repository; household furnishings; logging tools; fire equipment; items manufactured by local industries.
Research Fields: local area history; regional archaeology; Great Lakes fur trade; Michigan lumbering; architectural history; history of postal service; regional Native American history & culture; urban archaeology.
Facilities: meeting rooms; 80-seat community room; garden. Museum-related items for sale.
Activities: guided tours; organized education programs; workshops; lectures; films & video; living history programs; participatory archaeology programs; loan & temporary exhibitions; oral history programs.
Publications: newsletter.
Hours & Admission Prices: Tues.-Sat. 10-4:30, Sun. 1-4:30. Adults $1, children $.50; discounts to AAM members; members no charge. Closed major holidays. ♿
Attendance: 16,000 (accurate)
Membership: Individual $25; Contributing $50-$99; Century $100-$499; Museum Patron $500-$999; Museum Benefactor $1,000 & up.

CHILDREN'S ZOO AT CELEBRATION SQUARE, 1730 S. Washington Ave., Saginaw, MI 48601-2876. Tel.: 989-759-1408. Fax: 989-759-1328.
E-mail: info@saginawzoo.com
Web Site: saginawzoo.com
Formerly: Saginaw Valley Zoological Society
Founded: 1929.
Key Personnel: Exec. Dir., Nancy Parker; Mgr. Animal Collection, Megan Olmstead.
Personnel Profile: Full-Time Paid 15; Part-Time Paid 20; Part-Time Volunteers 100; Interns 5.
Governing Authority: Parent Institution: Saginaw Valley Zoological Society. Tax-exempt.
Institution Type/Description: Children's Zoo.
Collections: various mammals; reptiles & birds; native & exotic animals; bio diversity & artifacts.
Activities: train rides; pony rides; carousel; awareness amphitheater presentations; hands-on activities; interactive exhibits.
Publications: quarterly newsletter.
Hours & Admission Prices: April-Sept. daily 10-5; Oct. 1st three weekends Sat.-Sun. 10-5; Nov.-Feb. call for hours. Zoo: admission $7, senior citizens 65 & over on 1st Wed. of month, members & children under one no charge. Carousel or Train Rides: $2. ♿
Attendance: 95,000 (accurate)
Membership: Individual $40; Dual $50; Family & Grandparents $75.

MID-MICHIGAN CHILDREN'S MUSEUM, 315 W. Genesee, Saginaw, MI 48603. Mailing Address: P.O. Box 2283, Saginaw, MI 48605-2283. Tel.: 989-399-6626.
E-mail: info@midmicm.org
Web Site: www.midmicm.org
Founded: 2000.
Congressional District: 4
Key Personnel: Pres., Angela Barris
Institution Type/Description: Children's Museum.
Collections: hands-on exhibits.
Hours & Admission Prices: Mon. & Wed.-Sat. 10-5, Sun. 12-5. Admission $7, seniors 60 & over $6; children under 2 no charge. ♿
Attendance: 84,000

✱ SAGINAW ART MUSEUM, (M), 1126 N. Michigan Ave., Saginaw, MI 48602-4795. Tel.: 989-754-2491, ext. 208. Fax: 989-754-9387. Facebook: Saginaw Art Museum.
E-mail: info@saginawartmuseum.org
Web Site: www.saginawartmuseum.org
Founded: 1947.
Congressional District: 5
Key Personnel: Dir., Stacey Gannon; Deputy Dir. & Collection Mgr., Ryan

Kaltenbach; Facilities Mgr., Raul Servantes; Mktg. & Outreach Coord., Shelby Riggle; Chm. (V), Paul Furlo; Pres. (V), Stacie Rose.
Personnel Profile: Full-Time Paid 3; Part-Time Paid 1; Part-Time Volunteers 192.
Volunteer Hours: 767
Governing Authority: private; nonprofit organization. Tax-exempt: 501(c)(3).
Institution Type/Description: Art Museum: housed in former residence of Clark L. Ring family, building designed in 1904 by Charles Adams Platt.
Collections: 2,500 BCE to present international fine & applied art; E.I. Couse; Corot; Cropsey; Inness; Blakelock; Minor; Huntington; Demuth; Hassom; African; Etruscan; Chinese; Japanese; European; American; 14th century to present works on paper; archives of Michigan art; interactive Vision Area.
Research Fields: Michigan art & artists.
Facilities: 2,000-vol. art history library; studio & classroom; formal garden. Museum-related items for sale.
Activities: guided tours; lectures; films; gallery talks; workshops; seminars; formally organized education programs; permanent & temporary exhibitions; inter-museum loan; art classes; hands-on children's exhibits; facilities rental.
Publications: quarterly newsletter; catalogs; show announcements.
Hours & Admission Prices: Tues., Wed., Fri. & Sat. 11-5, Thurs. 11-8, Sun. Labor Day-Memorial Day 12-4. Adults $5, senior citizens 65 & over and students $3; discounts to AAM & ICOM members; members & youth under 15 no charge. Michigan Art Museums, NARM & American Horticultural Society reciprocal memberships. Closed national holidays. &
Attendance: 5,156 (accurate)
Membership: Student & Senior Citizen $25; Individual $50; Family $75; Garden $100; Affiliate $125; Patron $250; Platt's Circle $500; Ring Society $1,000.

Saint Clair Shores

SELINSKY-GREEN FARMHOUSE MUSEUM, 22500 Eleven Mile Rd., Saint Clair Shores, MI 48081-1312. Tel.: 586-771-9020. Fax: 586-771-8935.
E-mail: stachowm@libcoop.net
Institution Type/Description: Historic House.
Collections: period artifacts & memorabilia about rural life in St. Clair Shores during the late 19th & early 20th centuries.
Hours & Admission Prices: June-Aug. Wed. 1-4; Sept.-May Wed. & Sat. 1-4.

Saint Ignace

* **FATHER MARQUETTE NATIONAL MEMORIAL, (M),** 720 Church St., Saint Ignace, MI 49781-1729. Tel.: 906-643-8620. Fax: 906-643-9329. TDD: 800-827-7007.
Founded: 1980.
Congressional District: 11
Key Personnel: Park Mgr., Wayne Burnett.
Governing Authority: state; nonprofit. Tax-exempt.
Institution Type/Description: National Memorial.
Collections: sculpture.
Hours & Admission Prices: Memorial Day to Labor Day daily 10-8. No charge. &
Attendance: 14,810 (accurate)

Saint Johns

CLINTON COUNTY HISTORICAL SOCIETY PAINE-GILLAM-SCOTT MUSEUM, 106 Maple Ave., Saint Johns, MI 48879-1838. Mailing Address: P.O. Box 174, Saint Johns, MI 48879-0174. Tel.: 989-224-2894. Facebook: Paine Gillam Scott Museum.
E-mail: pgsmuseum@hotmail.com
Web Site: pgsmuseum.com
Founded: 1978.
Congressional District: 8
Key Personnel: Dir., Diane Carlson.
Personnel Profile: Full-Time Volunteers 2; Part-Time Volunteers 10.
Governing Authority: society; nonprofit. Parent Institution: Clinton Co. Historical Society. Tax-exempt: 501(c)(3).
Institution Type/Description: Historic House: Oldest Brick House in St. Johns.
Collections: local history; period furnishings; agricultural; industrial; medical.
Research Fields: county history.
Facilities: 90-vol. library of Michigan & Clinton County history material.
Activities: guided tours; organized education programs for children. Museum Sponsors: Mint Festival; Victorian Christmas.
Publications: quarterly genealogist newsletter; Clinton County Trails.
Hours & Admission Prices: April-Dec. Wed. 2-7, Sun. 1-4; other times by appointment. Suggested Donations: family $5, adults $2, children $1, discounts to military members. Closed legal holidays.

Attendance: 1,200 (accurate)
Membership: Friend $25.

CLINTON NORTHERN RAILWAY CENTER, 107 E. Railroad St., Saint Johns, MI 48879-1525. Mailing Address: 411 S. Clinton Ave., Saint Johns, MI 48879. Tel.: 989-224-6134. Facebook: Clinton Northern Railway Center.
E-mail: mccampbell60@gmail.com
Web Site: clintonnorthernrailway.org
Formerly: Depot Center for the Arts
Founded: 2005.
Key Personnel: Co Mgr. & Chm. (V), Gary McCampbell; Co Mgr. & Museum Shop Mgr., Jenny McCampbell.
Personnel Profile: Part-Time Volunteers 16.
Governing Authority: Parent Institution: Clinton County Arts Council. Tax-exempt.
Institution Type/Description: Railroad Museum: housed in the historic Grand Trunk Railroad Depot.
Collections: 4 historic railcars; early baggage wagons; railroad artifacts; demonstration tracks.
Hours & Admission Prices: May-Oct. Sun. 1-3. No charge; donations accepted. &
Attendance: 500 (estimated)

Saint Joseph

CURIOUS KIDS' MUSEUM AND DISCOVERY ZONE, 415 Lake Blvd., Saint Joseph, MI 49085-1231. Tel.: 269-983-2543. Fax: 269-983-3317.
E-mail: ckm@curiouskidsmuseum.org
Web Site: www.curiouskidsmuseum.org
Founded: 1988.
Congressional District: 6
Key Personnel: Pres., Maria Kibler; Mgr. Museum & Discovery Zone, Lori Marciniak; Dir. Programs & Outreach, Gina Mason; Exhibit Mgr. Discovery Zone, Josh Mason; Facilities Mgr. Curious Kids, Phil Rood; Floor Mgr. & Volunteer Coord., Myrna Piehl; Museum Shop Mgr., Becky Spear.
Personnel Profile: Full-Time Paid 4; Part-Time Paid 23; Part-Time Volunteers 40.
Governing Authority: private; nonprofit. Branch Museum: Discovery Zone, 333 Broad St., St. Joseph, MI 49085. Tel. 269-982-8500. Tax-exempt: 501(c)(3).
Institution Type/Description: Children's Museum: located in the former Memorial Hall.
Collections: over 100 hands-on exhibits in science, history & culture; medical; global child; rain forest; volcano; earthquake; space; simple machines; music; water exhibits; bridge building.
Research Fields: science, history, & culture technology.
Activities: family programs; day camps; curiosity camps; Girl & Boy Scout badges; outreach traveling exhibits & programs; curious classroom curriculum units; birthday parties; climbing wall; lighthouse climbing tower.
Publications: newsletter; brochure.
Hours & Admission Prices: June to Labor Day Mon.-Sat. 10-5, Sun. 12-5; Sept.-May Wed.-Sat. 10-5, Sun. 12-5. June to Labor Day: admission to both locations $10; discounts to ACM members; children under one & members no charge; Sept.-May admission to both locations $7; discounts to ACM members; children under one & members no charge. Closed New Year's Day; Easter; Memorial Day; Independence Day; Labor Day; Thanksgiving; Christmas Eve & Day. &
Attendance: 130,000 (accurate)
Membership: One Facility: Individual $25; Family & Grandparent $65. Both Facilities: Benefactor $130; Patron $500; Sponsor $1,000.

THE HERITAGE MUSEUM AND CULTURAL CENTER, 601 Main St., Saint Joseph, MI 49085-3354. Tel.: 269-983-1191.
Web Site: www.theheritagemcc.org
Formerly: Fort Miami Heritage Society
Founded: 1965.
Key Personnel: Dir., Christina H. Arseneau; Dir., Amy Zapal.
Personnel Profile: Full-Time Paid 1; Part-Time Paid 3; Part-Time Volunteers 25; Interns 1.
Governing Authority: Tax-exempt.
Institution Type/Description: History Museum.
Collections: local history & culture; photographs; period artifacts.
Major Exhibits: Southwest Michigan in the Civil War, 12/12-1/14.
Research Fields: St. Joseph & Benton Harbor; Michigan history.
Facilities: library; archives.
Publications: The Heritage Journal.

Hours & Admission Prices: Tues.-Fri. 10-4. Adults $5, children 6-18 $3; discount to AAM members; members & children 5 and under no charge. &
Membership: Senior $30; Individual $40; Family $60; Sponsor $150; Benefactor $500.

∗   KRASL ART CENTER, (M), 707 Lake Blvd., Saint Joseph, MI 49085-1313. Tel.: 269-983-0271. Fax: 269-983-0275.
E-mail: info1@krasl.org
Web Site: krasl.org
Founded: 1963.
Congressional District: 4
Key Personnel: Exec. Dir., Julia Gourley; Pres., Rick Dyer; Vice Pres., Marjorie Zibbel; Dir. Administration, Patrice Rose; Dir. Exhibitions & Collections, Tami Miller; Dir. Art Affairs, Sara Shambarger; Dir. Community Rels, Colleen Villa; Volunteer Coord., Cathie Pflaumer; Office Asst., Caryl Meister; Museum Shop Mgr., Brittany Stecker.
Personnel Profile: Full-Time Paid 6; Part-Time Paid 6; Part-Time Volunteers 375; Interns 2.
Governing Authority: nonprofit organization. Tax-exempt: 501(c)(3).
Institution Type/Description: Art Center.
Collections: sculpture.
Major Exhibits: Works on Paper from the Harmon & Harriet Kelley Collection of African American Art, 1/31/14-4/20/14; Krasl Art Center Biennial Sculptures Invitational, 4/25/14-6/22/14; Waves, 6/27/14-9/14/14; Krasl Art Fair on the Bluff, 7/12/14-7/13/14; Krasl Artisan Market, 11/7/14-11/13/14; Arnold Newman: 20th Century Luminaries in Art, Politics & Culture, 11/21/14-1/11/15.
Facilities: classrooms. Paintings, sculpture, ceramics, textiles & jewelry for sale.
Activities: guided tours; lectures; films; gallery talks; workshops; formally organized education programs for children & adults.
Publications: quarterly newsletter; class schedules; exhibition brochures; catalogues.
Hours & Admission Prices: Mon.-Wed. & Fri.-Sat. 10-4, Thurs. 10-9, Sun. 1-4. No charge; donations accepted. Closed holidays. &
Attendance: 30,000 (estimated)
Membership: Senior Citizen $28; Youth $30; Individual $35; Family $50; Sponsor $75-$99; Sustainer $100-$249; Benefactor $250-$499; Patron $500-$999; Guarantor $1,000-$2,499; Master $2,500-$4,999; Director's Choice $5,000-$9,999; Director's Circle $10,000 & up.

Sault Sainte Marie

KEMP MINERAL RESOURCES MUSEUM, Lake Superior State University, 650 W. Easterday Ave., Sault Sainte Marie, MI 49783-1656. Tel.: 906-632-6841, ext. 2267.
E-mail: kmrm@lssu.edu
Web Site: www.lssu.edu
Founded: 2007.
Key Personnel: Cur., Dr. David M. Knowles.
Governing Authority: Parent Institution: Lake Superior State University.
Institution Type/Description: Mineral Resources Museum.
Collections: rock & mineral collections.
Hours & Admission Prices: Call for hours.

LE SAULT DE SAINTE MARIE HISTORICAL SITES, INC., 501 E. Water St., Sault Sainte Marie, MI 49783-2038. Tel.: 906-632-3658. Fax: 906-632-9344.
E-mail: admin@saulthistorics.com
Web Site: www.saulthistoricsites.com
Founded: 1967.
Congressional District: 11
Key Personnel: Dir., Rich Brawley; Pres. (V), John P. Wellington; Bookkeeper, Charlotte Hendrickson.
Personnel Profile: Full-Time Paid 1; Part-Time Paid 20.
Governing Authority: nonprofit organization. Tax-exempt: 501(c)(3).
Institution Type/Description: Maritime Museum.
Collections: 1917 550 ft. Great Lakes straight deck bulk carrier; Great Lakes maritime items; mid-19th to early 20th-century items; ethnology; archives.
Facilities: aquarium; 25,000 sq. ft. exhibit space; two 25-seat theaters. Books & gift items for sale.
Activities: guided tours; films.
Publications: Ship's Bell.
Hours & Admission Prices: mid-May to June & Sept. to mid-Oct. Mon.-Sat. 10-4, Sun. 11-4; July to Aug. Mon.-Sat. 10-5, Sun. 11-5 Call for admission prices. &
Attendance: 28,100 (accurate)
Membership: Individual $25; Family $50; History Buff $75; Sustaining $100; Benefactor $250; Corporate $500; Lifetime $1,500 & up.

RIVER HISTORY MUSEUM, 531 Ashmun St., Sault Sainte Marie, MI 49783. Mailing Address: P.O. Box 627, Sault Sainte Marie, MI 49783. Tel.: 906-632-1999.
Institution Type/Description: History Museum.
Collections: local history & culture; period furnishings; photographs; personal artifacts.
Hours & Admission Prices: mid-May to Oct. Mon.-Sat. 11-5. Adults $7, children $3.50.

Sebewaing

LUCKHARD MUSEUM - THE INDIAN MISSION, 612 E. Bay St., Sebewaing, MI 48759-1644. Tel.: 989-883-2539 & 3730.
Founded: 1957.
Key Personnel: Dir., Jim Bunke.
Personnel Profile: Part-Time Volunteers 2.
Governing Authority: church; nonprofit. Owned by Michigan District of Missouri Lutheran Church, Ann Arbor, MI. Tax-exempt.
Institution Type/Description: Historic House: 1849 mission home.
Collections: household furnishings; kitchenware; religious artifacts; period tools & clothing; Native American birch bark canoe, arrowheads, headdress.
Facilities: library of church mission books & diaries.
Activities: guided tours; lectures.
Publications: brochure.
Hours & Admission Prices: June-Sept. 1st Sun. of month 2-4. Sebewaing Sugar Festival: call for hours. No charge; donations accepted.
Attendance: 40 (estimated)

Selfridge ANG Base

SELFRIDGE MILITARY AIR MUSEUM, (M), 127 WG/MU, 27333 C St., Bldg. 1011, Selfridge ANG Base, MI 48045-4901. Tel.: 586-307-5035. Fax: 586-307-6646.
Institution Type/Description: History Museum.
Collections: military history & heritage; aircraft; war memorabilia; photographs.
Hours & Admission Prices: April-Oct. Sat.-Sun. 12-4:30; other times by appointment. Adults $4, children 4-12 $3.

Seney

SENEY NATIONAL WILDLIFE REFUGE VISITOR CENTER, 1674 Refuge Entrance Rd., Seney, MI 49883-9509. Tel.: 906-586-9851, ext. 15. Fax: 906-586-3800.
E-mail: jennifer_mcdonough@fws.gov
Web Site: www.fws.gov/midwest/seney
Founded: 1965.
Congressional District: 1
Key Personnel: Pres. Seney Natural History Assoc., Dee Phinney; Museum Shop Mgr., Claudia Slater.
Personnel Profile: Part-Time Paid 2; Part-Time Volunteers 50; Interns 3.
Governing Authority: federal. Parent Institution: U.S. Dept. of the Interior, Fish & Wildlife Service. Tax-exempt.
Institution Type/Description: Nature Center.
Collections: wildlife refuge; study skin & full mounted specimens; dioramas of the common loon, eagle & wolf.
Research Fields: yellow rail research project; loon monitoring; swan monitoring; various bird surveys; sharptail grouse research; small mammal research.
Facilities: 75-seat auditorium; nature trails. Nature study books, study aids, wildlife posters & related items for sale.
Activities: educational programs; self-guided auto tour; permanent & temporary exhibitions; summer programs; high definition orientation DVD.
Publications: refuge brochure; hunting brochure; fishing brochure; bird checklist; newsletter.
Hours & Admission Prices: May 15-Oct. 15 daily 9-5. No charge; donations accepted. &
Attendance: 60,000 (estimated)
Membership: Seney Natural History Association: Retired & senior $5; Individual $10; Family $15; Contributing $25; Supporting $40; Business $75; Corporate $200; Lifetime $500.

South Haven

LIBERTY HYDE BAILEY MUSEUM, 903 Bailey Ave., South Haven, MI 49090. Mailing Address: P.O. Box 626, South Haven, MI 49090. Tel.: 269-637-3251. Facebook: Liberty Hyde Bailey.
E-mail: lhbm@south-haven.com

Web Site: libertyhydebailey.org
Founded: 1938.
Congressional District: 6
Key Personnel: Dir., John Linstrom; Dir., Lauren Denny; Chm. (V), Anne Long.
Personnel Profile: Part-Time Paid 2.
Governing Authority: Tax-exempt.
Institution Type/Description: Historic House Museum: housed in the former home of Liberty Hyde Bailey; built in 1857.
Collections: Bailey family history; period furnishings.
Facilities: library.
Hours & Admission Prices: April Thurs.-Sun. 11-4; May-Aug. Thurs.-Sun. 9-4; other times by appointment. Exhibits: adults $5.
Membership: $30-$1,000.

MICHIGAN MARITIME MUSEUM, 260 Dyckman Ave., South Haven, MI 49090-1065. Tel.: 269-637-8078; 800-747-3810. Fax: 269-637-1594. Facebook: Michigan Maritime Museum.
E-mail: info@michiganmaritimemuseum.org
Web Site: michiganmaritimemuseum.org
Founded: 1976.
Congressional District: 4
Key Personnel: Exec. Dir., Patti Montgomery-Reinert; Pres. Bd. (V), Gary Horton; Dir. Education & Admin., Katie Biell; Cur., Cobie Ball; Dir. Boat Shed, David Ludwig; Dir. Padnos Boat Shed, Paul D. Ludwig; Registrar, Judy Schlaack; Captain, Adele Arlitt; Coord. Programs, Amy Locker; Programs Asst., Amanda Morris; Librarian & Archivist, Kristen Edson; Docent Coord., Mary Stephens; Museum Shop Mgr. & Office Mgr., Mechele Kempski; Asst. Cur., Jed Jaworski.
Personnel Profile: Full-Time Paid 3; Part-Time Paid 2; Part-Time Volunteers 50.
Governing Authority: nonprofit organization; bd. of directors. Subsidiary Institution: Great Lakes Center for Maritime Studies. Tax-exempt.
Institution Type/Description: Maritime History Museum.
Collections: historical vessels; archival materials; photos; ship & small craft models; marine art; tools & technological implements; personal artifacts pertaining to Great Lakes regional maritime history; historic replica tall ship; commercial fishing tug; replica boathouse. Historic Buildings: 1873 lightkeeper's dwelling; 1900 U.S. Life Saving Station crew's quarters including service exhibit.
Major Exhibits: Mysteries Beneath the Waves, 5/14-12/15.
Research Fields: Great Lakes Maritime History.
Facilities: research library; visitor center.
Activities: guided tours; lectures; films; gallery talks; boatbuilding workshops; formally organized education programs for children, adults & undergraduate college students; training programs for professional museum workers; temporary & permanent exhibitions; sail on a tall ship; on-water experiences.
Publications: newsletter, Ship's Lamp.
Hours & Admission Prices: May-Oct. daily 10-5; call for additional hours. Adults $6, seniors citizens $5, children under 12 $4.50; members no charge. Closed New Year's Day; Christmas. &
Attendance: 30,000 (estimated)
Membership: Senior Citizen $25; Individual $30; Senior Couple $40; Family & Grandparents $50; Ensign $100; Captain $250; Business $300; Commodore $500; Keel Club $1,000.

South Range

COPPER RANGE HISTORICAL MUSEUM, Champion & Trimountain, South Range, MI 49963. Mailing Address: P.O. Box 148, South Range, MI 49963-0148. Tel.: 906-482-6125.
Web Site: www.pasty.com/crhm
Founded: 1988.
Congressional District: 107
Key Personnel: Pres. (V), Jean Pemberton.
Personnel Profile: Part-Time Volunteers 20.
Volunteer Hours: 500
Operating Expenses: 7,743
Operating Income: 10,433
Governing Authority: Tax-exempt.
Institution Type/Description: History Museum.
Collections: area history; furnishings; personal artifacts.
Major Exhibits: Replica of Genovesi Moving House in Baltic, 12/14.
Research Fields: area business; copper mines.
Activities: group tours upon request. Museum Sponsors: Open House in June; Annual Member Dinner in July.
Publications: newsletter 3 times per year.
Hours & Admission Prices: June & Sept.-Oct. Tues.-Fri. 12-3; July-Aug. Mon.-Fri. 12-3, Sat. by appointment. Adults $1; members no charge.

Attendance: 810 (accurate)
Membership: Individual $10; Family $15; Supporting $50; Senior Life $100; Life $150; Corporate Sponsor $1,000; Corporate Patron $5,000.

Tecumseh

TECUMSEH AREA HISTORICAL MUSEUM, (M), 302 E. Chicago Blvd., Tecumseh, MI 49286-1551. Mailing Address: P.O. Box 26, Tecumseh, MI 49286-0026. Tel.: 517-423-2374. Facebook: Tecumseh Area Historical Society.
E-mail: historictecumseh@gmail.com
Web Site: www.historictecumseh.org
Institution Type/Description: Historical Society Museum: housed in a gothic stone church c.1913.
Collections: local history & culture; photographs.
Publications: quarterly newsletter.
Hours & Admission Prices: April-Dec. Sat. 10:30-3:30; other times by appointment. No charge.
Membership: Individual $10; Senior Lifetime $50; Lifetime $100.

Tipton

HIDDEN LAKE GARDENS, 6214 Monroe Rd., (M-50), Tipton, MI 49287-9766. Tel.: 517-431-2060. Fax: 517-431-9148.
E-mail: court33@msu.edu
Web Site: www.hiddenlakegardens.msu.edu
Founded: 1926.
Congressional District: 2
Key Personnel: Mgr., Steven Courtney, N.P.D.; Sec. & Gift Shop Mgr., Cheryl Roe.
Governing Authority: university. Affiliated with Michigan State University, East Lansing, MI 48823. Tax-exempt.
Institution Type/Description: Arboretum.
Collections: over 2,500 species of woody plants arranged in systematic & use groupings; woodlands; regenerating forest.
Research Fields: plant breeding.
Facilities: 1,800-vol. library on horticulture & allied fields available for use on premises; botanical garden; nature center; nature trails; plant conservatory; 80-seat auditorium; classrooms. Books & prints for sale.
Activities: guided tours; lectures; formally organized education programs for adults; permanent exhibitions.
Hours & Admission Prices: April-Oct. daily 9-7; Nov.-March daily 9-4. Admission $3.
Attendance: 48,000 (accurate)
Membership: Single $30; Dual $40; Family & Friends $60.

Traverse City

DENNOS MUSEUM CENTER OF NORTHWESTERN MICHIGAN COLLEGE, (M), 1410 College Dr., Traverse City, MI 49686. Mailing Address: 1701 E. Front St., Traverse City, MI 49686-1055. Tel.: 231-955-1055. Fax: 231-995-1597.
E-mail: dmc@nmc.edu
Web Site: www.dennosmuseum.org
Founded: 1991.
Congressional District: 9
Key Personnel: Dir., Eugene A. Jenneman; Registrar, Kim Hanninen; Museum Shop Mgr., Terry Tarnow; Cur. Education, Jason Dake; Ops. Mgr., Gale Cook.
Personnel Profile: Full-Time Paid 6; Part-Time Paid 5; Part-Time Volunteers 120.
Governing Authority: nonprofit public college. Parent Institution: Northwestern Michigan College. Tax-exempt: 501(c)(3).
Institution Type/Description: College Art Museum.
Collections: contemporary Inuit sculpture & prints; Canadian Indian & Great Lakes Indian art; Japanese prints; 19th- & 20th-century American & European prints; contemporary German prints; outdoor sculpture; photography.
Research Fields: Inuit art.
Facilities: 300-vol. library of Inuit art publications & periodicals; educational facilities; 367 seat theater, 8,000 sq. ft. exhibit space. Museum-related items for sale.
Activities: guided tours; lectures; loan, participatory, temporary & traveling exhibits; concerts; dance recitals; docent program; films; formal education programs; theater.
Publications: quarterly newsletter, Inside.
Hours & Admission Prices: Mon.-Wed. & Fri.-Sat. 10-5, Thurs. 10-8, Sun. 1-5. Adults & senior citizens $6, students & children $4; discounts to groups, AAM & ICOM members; members no charge. Closed New Year's Day; Easter; Memorial Day; Independence Day; Labor Day; Thanksgiving; Christmas. &

Attendance: 60,000 (accurate)
Membership: Individual $40; Family $60; Friend $100; Associate $250; Curator's Circle $500; Director's Circle $1,000.

GREAT LAKES CHILDREN'S MUSEUM, 13240 S. West Bay Shore Dr., Traverse City, MI 49684. Mailing Address: P.O. Box 2326, Traverse City, MI 49685-2326. Tel.: 231-932-4526.
Web Site: www.greatlakeskids.org
Founded: 1998.
Congressional District: 4
Key Personnel: Pres. (V), Nick Trahir; Museum Admin., Martha Belfour; Museum Educator, Anne Drake; Museum Shop Mgr., Diane Hubert.
Governing Authority: Tax-exempt.
Institution Type/Description: Children's Museum.
Collections: hands-on exhibits; water & the Great Lakes.
Activities: birthday parties; member nights; science exploration. Museum Sponsors: Kids Safety Day; Kids Free Fishing Day.
Publications: quarterly newsletter, Wave Review; monthly email bulletin, Info Flash.
Hours & Admission Prices: Tues.-Sat. 10-5, Sun. 1-5. Adults $6; discounts to active duty military & their families with ID; children under 2 & members no charge. Closed New Year's Day Christmas Eve & Day. &
Attendance: 25,000 (accurate)
Membership: Individual & Preschool Provider $55; Grandparents $70; Family $80; Family with Babysitter $90; Three Generations $100; Patron $250; Benefactor $500; Corporate $1,000.

HISTORY CENTER OF TRAVERSE CITY, 322 Sixth St., Traverse City, MI 49684-2414. Tel.: 231-995-0313. Fax: 231-946-6750.
Web Site: www.traversehistory.org
Formerly: Grand Traverse Heritage Center
Founded: 2000.
Congressional District: 1
Key Personnel: Exec. Dir., Bill Kennis; Pres. (V) & Cur., Steve Harold; Chm. (V), Bob Knechel.
Personnel Profile: Full-Time Paid 2; Part-Time Paid 3; Part-Time Volunteers 50; Interns 4.
Governing Authority: municipal government. Parent Institution: City of Traverse. Subsidiary Institution: Friends of Con Foster Museum. Tax-exempt.
Institution Type/Description: Local History Museum.
Collections: American Indian artifacts; agriculture; local history; transportation; folk art; photographs; railroads; maritime.
Research Fields: local & regional history.
Facilities: temporary & permanent exhibits; classroom & reading room; reference library; archives; meeting rooms.
Activities: guided tours; lectures; special events; educational programs.
Publications: quarterly newsletter to members; occasional special event publications.
Hours & Admission Prices: Mon.-Sat. 10-4. Adults $5; children 6 & under and members no charge. &
Attendance: 40,000 (accurate)
Membership: Individual $25; Family $40; Friend $100; Patron $250; Sponsor $500; Life $1,000.

Trenton

TRENTON HISTORICAL MUSEUM, 306 St. Joseph, Trenton, MI 48183-2823. Mailing Address: 2800 Third St., Trenton, MI 48183-2918. Tel.: 734-675-2130.
Web Site: www.trentonhistoricalcommission.org
Formerly: The Moore House
Founded: 1962.
Congressional District: 16
Key Personnel: Dir., Linda Murdock; Vice Chm. (V), Nada Frost.
Personnel Profile: Part-Time Volunteers 15.
Governing Authority: municipal. Tax-exempt.
Institution Type/Description: Historic House Museum: c.1881 Victorian house.
Collections: Victorian era furnishings; Indian arrowheads; horse drawn carriages; Trenton Room. Historic Building & Carriage House.
Research Fields: files, local history information.
Facilities: library of reference books available for use on premises.
Activities: guided tours.
Hours & Admission Prices: Closed months of Jan. & Feb. Open Sat. 1-4. No charge, donations welcome.
Attendance: 300 (estimated)

Troy

TROY HISTORIC VILLAGE, (M), 60 W. Wattles Rd., Troy, MI 48098-4699. Tel.: 248-524-3570. Fax: 248-524-3572.
E-mail: museum@troymi.gov
Web Site: www.troyhistoricvillage.org
Founded: 1968.
Congressional District: 12
Key Personnel: Dir., Loraine Campbell; Pres. Historical Society, Cheryl Barnard; Program Dir., Anne Nagrant; Village Store Mgr., Judy Davy; Village Store Mgr., Barb Chambers.
Personnel Profile: Full-Time Paid 1; Part-Time Paid 5; Part-Time Volunteers 75.
Governing Authority: not-for-profit. Owned by City of Troy. Managed by Troy Historical Society. Tax-exempt.
Institution Type/Description: History Museum.
Collections: local agricultural equipment; early pioneer living collection; local archives & manuscripts; permanent & temporary exhibits. Historic Buildings: 1832 Caswell House; 1877 Poppleton School; c.1918 county store; c.1910 print shop; 1875 Wagon shop; c.1864 Town Hall; 1927 City Hall; 1820 log cabin; 1837 church; 1900 farm house.
Research Fields: local biographical.
Facilities: 500-vol. library. Gift items for sale.
Activities: guided tours; permanent & temporary exhibitions; educational programs. Special Events: Hanging of the Greens; Trick or Treating on the Green; Fall Farm Festival.
Publications: quarterly newsletter, The Troy-Village Press; Troy Museum & Historical Village Activities Book for Children; When Our Country Called: Stories of WWII Veterans from Troy and Troy Township; Troy: A City From the Corners; Troy Twp. Section: 1817-1877 History of Oakland County; The Life of Solomon Caswell.
Hours & Admission Prices: Tues.-Thurs. 10-3. Adults $3, children $2. Closed New Year's Eve & Day; Memorial Day; weekend; Independence Day; Labor Day weekend; Thanksgiving weekend; Christmas Eve & Day. &
Attendance: 25,000 (estimated)
Membership: Individual $20; Couple & Family $30; Nonprofit Institution $50; Business $100; Life $200.

University Center

*** MARSHALL M. FREDERICKS SCULPTURE MUSEUM, (M),** Arbury Fine Arts Center, Saginaw Valley State University, 7400 Bay Rd., University Center, MI 48710. Tel.: 989-964-7125. Fax: 989-964-7221.
E-mail: mfsm@svsu.edu
Web Site: www.marshallfredericks.org
Founded: 1988.
Congressional District: 8 & 10
Key Personnel: Chm. (V), Sue Vititoe; Dir., Marilyn L. Wheaton; Cur. Education, Andrea Ondish; Sr. Sec., Laurie Allison.
Personnel Profile: Full-Time Paid 5; Part-Time Paid 7; Part-Time Volunteers 1.
Governing Authority: college. Parent Institution: Saginaw Valley State University. Tax-exempt.
Institution Type/Description: Sculpture Museum.
Collections: plaster models; cast & carved sculptures; medals; jewelry; miniatures; drawings; photos; site models; moulds; armatures; tools; machinery; archival materials by Marshall M. Fredericks (1908-1998)
Research Fields: life & work of Marshall M. Fredericks, 1908-1998; American sculpture of the 20th century.
Facilities: 18,000 sq. ft. exhibit space. Museum-related items for sale.
Activities: guided tours; organized education programs for youth; youth summer art camps; docent program; film series.
Publications: exhibition catalogs; Friends of MFSM Newsletter.
Hours & Admission Prices: Mon.-Sat. 12-5. No charge. Closed national & university holidays. &
Attendance: 10,000 (estimated)
Membership: Senior & Student $35; Individual $50; Friend $100; Donor $250; Supporter $500; Patron $1,000; Benefactor $2,500.

Vicksburg

VICKSBURG DEPOT MUSEUM, 300 N. Richardson St., Vicksburg, MI 49097. Mailing Address: P.O. Box 103, Vicksburg, MI 49097. Tel.: 269-649-1733.
E-mail: vixmus1@yahoo.com
Governing Authority: Parent Institution: Vicksburg Historical Society.
Institution Type/Description: History Museum: housed in a former railroad depot; built in 1908.
Collections: local history & culture; period furnishings; telegraph equipment; personal artifacts; photographs; ice harvesting tools & equipment. Historic Structures: farm house; school house; barn; print shop; caboose; boxcar.

Facilities: Museum-related items for sale.
Activities: educational programs; special events.
Hours & Admission Prices: May-Dec.

Vulcan

IRON MOUNTAIN IRON MINE, Hwy. U.S. 2, Vulcan, MI 49892. Mailing Address: P.O. Box 177, Iron Mountain, MI 49801-0177. Tel.: 906-774-7914.
E-mail: ironmine@uplogon.com
Web Site: www.ironmountainironmine.com
Founded: 1956.
Congressional District: 11
Key Personnel: Dir., Eugene R. Carollo; Dir. Public Rels., Dennis Carollo.
Personnel Profile: Full-Time Paid 3; Part-Time Paid 8.
Governing Authority: individual operation.
Institution Type/Description: Mining Museum: located on Menominee Iron Range.
Collections: working mining machinery; diamond drills; stoper drills; water liner drills; underground locomotives.
Facilities: underground lighted cavern. Gift items for sale.
Activities: guided tours; formally organized education programs; permanent exhibitions; underground train rides.
Hours & Admission Prices: Memorial Day to mid-Oct. daily 9-5. Adults $14, children 6-12 $8; discount to members, AAM, ICOM, AARP, AAA & school groups; children 5 & under no charge. ♿
Attendance: 16,000 (estimated)

Wayne

CITY OF WAYNE HISTORICAL MUSEUM, 1 Town Square, Wayne, MI 48184-1637. Tel.: 734-722-0113.
E-mail: museum@ci.wayne.mi.us
Web Site: www.ci.wayne.mi.us/museum/php
Formerly: Wayne Village Hall
Founded: 1964.
Congressional District: 15
Key Personnel: Dir., Richard L. Story.
Personnel Profile: Part-Time Paid 1.
Governing Authority: municipal. Parent Institution: City of Wayne. Tax-exempt.
Institution Type/Description: Local History Museum: housed in 1878 Village Hall.
Collections: historical items related to the City of Wayne & vicinity; local area maps; published & unpublished historical articles; reference materials; newspapers; clippings; scrap books; photographs.
Research Fields: local history.
Facilities: over 500-vol. library of material on Wayne, Wayne County & Michigan history available for use on premises.
Activities: guided tours; permanent & temporary exhibits.
Hours & Admission Prices: Fri.-Sat. 1-4; groups by appointment. No charge; donations accepted. ♿
Attendance: 1,200 (accurate)

West Bloomfield

CHALDEAN CULTURAL CENTER, (M), 5600 Walnut Lake Rd., West Bloomfield, MI 48323-2370. Tel.: 248-681-5050. Fax: 248-681-9191.
E-mail: info@chaldeanculturalcenter.org
Web Site: www.chaldeanculturalcenter.org
Formerly: Chaldean Community Cultural Center
Founded: 2003.
Congressional District: 9
Key Personnel: Dir., Mary Romaya.
Governing Authority: Tax-exempt.
Institution Type/Description: Cultural Art Center.
Collections: Chaldean history, arts, culture & contributions; photographs; personal artifacts; paintings.
Hours & Admission Prices: Call for hours.

ORCHARD LAKE MUSEUM, 3951 Orchard Lake Rd., West Bloomfield, MI 48325. Tel.: 248-682-2279 & 757-2451.
E-mail: contact@gwbhs.com
Governing Authority: Parent Institution: Greater West Bloomfield Historical Society. Tax-exempt: 501(c)(3).
Institution Type/Description: History Museum: built in 1854.
Collections: local history & culture; period furnishings; photographs; personal artifacts; Native American artifacts; pioneer farmers; 17th-century dugout canoe; Roy C. Gamble's 1928 painting; Keego Cinema sign.

Hours & Admission Prices: 2nd Sun. each month 1-4; other times by appointment.

SHALOM STREET, 6600 W. Maple Rd., West Bloomfield, MI 48322-3003. Tel.: 248-432-5451.
E-mail: rmorais@jccdet.org
Web Site: www.shalomstreet.org
Key Personnel: Dir., Rabbi S. Robert Morais; Asst. Dir., Andrea Liberman.
Governing Authority: Parent Institution: Jewish Community Center of Metro Detroit.
Institution Type/Description: Children's Museum.
Collections: hands-on exhibits.
Facilities: 4,500 sq. ft. exhibit space.
Activities: special events; educational programs; birthday parties.
Hours & Admission Prices: Sept.-June Mon. & Wed.-Thurs. 10-7, Tues. 10-5, Sat. 10-3; Summer: see website for hours. No charge. ♿

Williamsburg

ARTCENTER TRAVERSE CITY, 5152 US Hwy. 31 N., Williamsburg, MI 49690-9316. Tel.: 231-941-9488; 866-242-0120 (Toll Free). Fax: 231-941-0886.
E-mail: patt@artcentertraversecity.com
Web Site: artcentertraversecity.com
Governing Authority: nonprofit organization.
Institution Type/Description: Art Gallery.
Collections: paintings; sculpture.
Activities: classes; workshops; educational programs; lectures; special events; temporary exhibitions.
Publications: newsletter.
Hours & Admission Prices: Call for hours.

Wyandotte

WYANDOTTE MUSEUM, 2610 Biddle Ave., Wyandotte, MI 48192-5208. Tel.: 734-324-7284. Fax: 734-324-7283.
E-mail: museum@wyan.org
Web Site: www.wyandotte.net
Founded: 1958.
Congressional District: 16
Key Personnel: Dir. Museums, Jody Chansuolme; Pres. (V), Ken Navarre; Museum Shop Mgr., Sandra Noble.
Personnel Profile: Full-Time Paid 1; Part-Time Paid 2; Part-Time Volunteers 2.
Governing Authority: municipal. Parent Institution: City of Wyandotte. Tax-exempt.
Institution Type/Description: History Museum: housed in 1896 Ford-MacNichol period home.
Collections: Americana; Wyandotte history; Wayne County history; shipbuilding; chemical industry collection. Historic House: 1857 Marx Museum period home; 1896 Ford-MacNichol period home.
Research Fields: local history.
Facilities: 2,000-vol. library of Americana, American history, Michigan history & Native American history available for use on premises.
Activities: guided tours; lectures; temporary exhibitions; school loan service.
Publications: monthly newsletter; pamphlets; brochures; books, Proudly We Record; Our Fame & Fortune in Wyandotte.
Hours & Admission Prices: Tours: Mon.-Fri. 9-5. Adults $2, students $.50; children under 13 & members no charge. Closed Easter; Labor Day; Thanksgiving; Christmas. ♿
Attendance: 999
Membership: Student $1; Adult $15; Contributing $25; Patron $50; Honorary Patron $100 & up.

Ypsilanti

FORD GALLERY - EASTERN MICHIGAN UNIVERSITY, Ford Hall, Ypsilanti, MI 48197. Tel.: 734-487-0465.
E-mail: gtom@emich.edu
Web Site: emich.edu/fordgallery
Key Personnel: Dir., Greg Tom
Institution Type/Description: Art Gallery.
Collections: paintings; sculpture; photographs; lithographs.
Activities: special events; temporary exhibitions.
Hours & Admission Prices: Mon. & Thurs. 10-5, Tues.-Wed. 10-7, Fri.-Sat. 10-2.

MICHIGAN FIREHOUSE MUSEUM AND EDUCATION CENTER, 110 W. Cross St., Ypsilanti, MI 48197-2445. Tel.: 734-547-0663. Fax: 734-547-0669.
E-mail: firemuseum@msn.com
Web Site: www.michiganfirehousemuseum.org
Founded: 1998.
Personnel Profile: Full-Time Paid 1; Part-Time Paid 6; Part-Time Volunteers 1.
Institution Type/Description: Fire-Fighting Museum.
Collections: fire-fighting equipment & memorabilia.
Activities: Annual Event: Fire Truck Muster in August.
Publications: newsletter, Michigan Firehouse Museum.
Hours & Admission Prices: Tues.-Sat. 10-4, Sun. 12-4. Adults $5, children 2-16 $3; children under 2 no charge. &
Attendance: 6,400 (accurate)
Membership: Individual $25; Family $35; Supporter $100.

YANKEE AIR FORCE, INC. (YANKEE AIR MUSEUM), Hangar 2, Willow Run Airport, Ypsilanti, MI 48198. Mailing Address: P.O. Box 590, Belleville, MI 48112-0590. Tel.: 734-483-4030. Fax: 734-483-5076.
E-mail: dick.stewart@yankeeairmuseum.org
Web Site: www.yankeeairmuseum.org
Founded: 1981.
Congressional District: 15
Key Personnel: Pres., Dick Stewart; Vice Pres., Lou Farkas; Treas., Tom Matz; Sec., Speed Gant; Aircraft Appearance Coord., Norman Ellickson; Cur. & Dir. Museum Activities, Gayle Drews; Membership, Jim Race; Public Rels., Bob Hynes; Security, William Tonak; Museum Shop Mgr., Dale Worcester.
Personnel Profile: Full-Time Paid 3; Full-Time Volunteers 12; Part-Time Paid 7; Part-Time Volunteers 450; Interns 6.
Governing Authority: nonprofit organization. Parent Institution: Eastern Michigan University & Michigan State University. Subsidiary Institutions: YAF Northeast Div., East Caldwell, NJ; YAF Saginaw Valley Div., Saginaw, MI; YAF Willow Run Div., Belleville, MI; YAF Wurtsmith Div., Oscoda, MI. Tax-exempt: 501(c)(3).
Institution Type/Description: Aeronautics Museum.
Collections: historic WWII era aircraft including B-17G, B-25D & C-47; costumes; textiles.
Activities: docent program; guided tours; temporary exhibitions; speakers' bureau; flyable historic aircraft participates in air shows & civic events throughout the U.S. & Canada. Museum Sponsors: Open House; Air Show on & off site; weekend event; Historic Flight Experience Rides May to October.
Publications: monthly magazine, Hangar Happenings; newsletter; 4 subsidiary publications.
Hours & Admission Prices: Outdoor Aircraft Display: call for hours. Flight Experience Rides: May-Oct. Museum: Wed.-Sat. 10-4, Sun. 12-4. &
Attendance: 28,000 (estimated)
Membership: Junior $24; Student $35; General $60; Senior Life (over 59) $500; Life $1,000; Contributor $2,000; Patron $5,000 & up.

YPSILANTI HISTORICAL MUSEUM, (M), 220 N. Huron St., Ypsilanti, MI 48197-2516. Tel.: 734-482-4990. Facebook: Ypsilanti Historical Society.
E-mail: yhs.museum@gmail.com
Web Site: www.ypsilantihistoricalsociety.org
Founded: 1960.
Congressional District: 2
Key Personnel: Pres. (V), Alvin Rudisill.
Personnel Profile: Part-Time Volunteers 30; Interns 2.
Operating Expenses: 50,000
Operating Income: 50,000
Governing Authority: society; municipal. Ypsilanti Historical Society. Affiliated with the City of Ypsilanti. Tax-exempt.
Institution Type/Description: Local History Museum.
Collections: archives; newspapers; Civil War letters; 35,000 cards; genealogical resource from earliest area settlers; city history; township history; photographs; Black history project; manuscripts; obituaries.
Research Fields: local history.
Facilities: archives.
Activities: permanent & temporary exhibitions.
Publications: quarterly newsletter, Gleanings.
Hours & Admission Prices: Museum: Tues.-Sun. 2-5. No charge; donations accepted. &
Attendance: 3,000 (estimated)
Membership: Single $15; Family $20; Sustaining $30; Business $75; Patron $100; Life $200.

YPSILANTI'S AUTOMOTIVE HERITAGE MUSEUM, 100 E. Cross, Ypsilanti, MI 48198-2936. Tel.: 734-482-5200. Fax: 734-480-2784.
E-mail: info@ypsiautoheritage.org
Web Site: www.ypsiautoheritage.org
Formerly: Ypsilanti's Automotive Heritage Museum & Miller Motors Hudson
Founded: 1995.
Key Personnel: Cur. & Museum Shop Mgr., Jack Miller.
Governing Authority: Tax-exempt.
Institution Type/Description: Automobile Museum.
Collections: vintage cars & memorabilia.
Hours & Admission Prices: Tues.-Sun. 1-4; Adults $5; children 13 & under no charge. Closed New Year's Day; Easter; Mother's Day; Memorial Day; Independence Day; Labor Day; Thanksgiving; Christmas. &

Zeeland

THE DEKKER HUIS/ZEELAND HISTORICAL MUSEUM, 37 E. Main St., Zeeland, MI 49464. Mailing Address: P.O. Box 165, Zeeland, MI 49464-0165. Tel.: 616-772-4079.
E-mail: info@zeelandmuseum.org
Web Site: www.zeelandmuseum.org
Founded: 1974.
Congressional District: 2
Key Personnel: Pres., Dorothy Voss; Cur., Suzy Frederick.
Personnel Profile: Part-Time Paid 2; Part-Time Volunteers 40.
Governing Authority: Parent Institution: Zeeland Historical Society. Tax-exempt.
Institution Type/Description: Historic House Museum: housed in former home of Dirk Dekker and his wife, built in 1876.
Collections: local history; period artifacts & furnishings.
Publications: member quarterly newsletter, Timeline.
Hours & Admission Prices: Call for hours. No charge; donations accepted.
Attendance: 2,500 (accurate)

MINNESOTA

(251 listings)

Aitkin

AITKIN COUNTY HISTORICAL SOCIETY, 20 Pacific St., S.W., Aitkin, MN 56431-1628. Mailing Address: P.O. Box 215, Aitkin, MN 56431-0215. Tel.: 218-927-3348.
E-mail: achs3348@embarqmail.com
Web Site: www.aitkin.com/achs
Founded: 1948.
Key Personnel: Dir. & Museum Shop Mgr., Gregory M. Leach; Chm. (V), Mary Rea; Pres. (V), William Stimac.
Personnel Profile: Part-Time Paid 1; Part-Time Volunteers 42.
Governing Authority: Branch Museum: Log Museum. Tax-exempt.
Institution Type/Description: Historical Society Museum: housed in the Northern Pacific Depot, built 1916. Listed on the National Register of Historic Places.
Collections: county history; photographs; documents.
Facilities: Museum-related items for sale.
Activities: educational programs.
Hours & Admission Prices: Wed. & Fri.-Sat. 10-4. Adults $2; members no charge. &
Attendance: 3,000 (accurate)
Membership: Senior $7.50; Individual $15; Family $20; Business $30; Memorial For Deceased $200.

THE JAQUES ART CENTER, (M), 121 Second St., N.W., Aitkin, MN 56431. Tel.: 218-927-2363.
E-mail: info@jaquesart.com
Institution Type/Description: Art Center.
Collections: paintings; sculpture; photographs.
Activities: permanent & temporary exhibits; workshops.
Hours & Admission Prices: Tues.-Sat. 11-4. No charge.

Albert Lea

FREEBORN COUNTY HISTORICAL MUSEUM, LIBRARY & VILLAGE, (M), 1031 Bridge Ave., Albert Lea, MN 56007-2205. Tel.: 507-373-8003. Fax: 507-552-1269.
E-mail: executivedirector@fchm.us
Web Site: www.fchm.us

Formerly: Freeborn County Historical Society
Founded: 1948.
Congressional District: 1
Key Personnel: Exec. Dir., Pat Mulso; Chm. (V), Kathy Mandt; Pres. Bd., Phil Hintermeister; Vice Pres., Dean Johnson; Museum Shop Mgr., Kathy Freese.
Personnel Profile: Full-Time Paid 2; Part-Time Paid 5; Part-Time Volunteers 150.
Volunteer Hours: 9,427
Operating Expenses: 169,601
Operating Income: 167,411
Governing Authority: society. Tax-exempt.
Institution Type/Description: History Museum & Historical Village.
Collections: pioneer artifacts; dolls; documents; family history; maps; rock 'n roll era singer Eddie Cochran archives & collection; Marion Ross (Mrs. Cunningham); MN Rock & Country Music Hall of Fame. Historic Buildings: 1853 log cabin; 1860 log houses; 1878 Lutheran Church; 1882 county schoolhouse.
Research Fields: genealogy & history of Freeborn County.
Facilities: Museum-related items for sale.
Activities: slide programs & programs for schools, churches, civic groups on home remedies, Native Americans, immigration, genealogy, etc.; guided tours for students of county & adult groups by appointment; special craft demonstrations; special events for elementary & high school students; classes for genealogists; programs & tours for nursing homes.
Publications: Images of America, Freeborn County, Minnesota; quarterly newsletter; Glimpses of Freeborn County, 1930-1980.
Hours & Admission Prices: Museum & Library: Wed.-Sat. 10-4. Village: May-Sept. Wed.-Sat. 10-4. Adults $5, students 12-18 $1; discounts to AAA members; members and children 11 & under no charge. Closed New Year's Day; Independence Day; Thanksgiving Day & day after; Christmas Eve through New Year's Day. Hours subject to change. &
Attendance: 12,000 (accurate)
Membership: Annual $25; Patron $50; Sustaining $100; Col. Albert Lea $250; Leader $500; Heritage Club $1,000.

Alexandria

DOUGLAS COUNTY HISTORICAL SOCIETY, 1219 Nokomis St., Alexandria, MN 56308-3712. Tel.: 320-762-0382. Fax: 320-762-9062.
E-mail: historic@dchsmn.org
Web Site: dchsmn.org
Key Personnel: Exec. Dir., Kim Dillon; Pres. (V), Bruce Haugen
Institution Type/Description: Historical Society Museum: housed in the former home of Senator Knute Nelson. Listed on the National Register of Historic Places.
Collections: period furnishings; personal artifacts.
Activities: special events.
Hours & Admission Prices: Tours: Mon.-Fri. 9-3. Admission $2. Research: Mon.-Fri. 9-4. &
Membership: Single $30; Family $50; Corporate Sponsor $250.

MINNESOTA LAKES MARITIME MUSEUM, 205 3rd Ave., W., Alexandria, MN 56308-1364. Mailing Address: P.O. Box 1216, Alexandria, MN 56308-3216. Tel.: 320-759-1114. Fax: 320-759-1101.
E-mail: boat@mnlakesmaritime.org
Web Site: mnlakesmaritime.org
Founded: 1998.
Key Personnel: Dir., Bruce T. Olson; Pres. (V), Fred Bursch.
Personnel Profile: Full-Time Paid 1.
Institution Type/Description: Maritime Museum.
Collections: history of the Letson House, Blakes Hotel, Geneva Hotel & Dickinson Inn; local resort history including Bedman's Beach resort; Larson boats & memorabilia; classic wood boats, featuring the Naphtha Launch, Frieda; history of Alexandria Boat Works, builder of wood fishing boats; period tackle & fish mounts; guide stories; fishing memorabilia; Duck boat; decoys; photographs.
Hours & Admission Prices: May 15-Oct. 15 Mon.-Sat. 10-5, Sun. 12-4; other times by appointment. Family $15, adults $6, seniors $5, students 5-17 $3; members no charge. &
Attendance: 4,000 (estimated)
Membership: Supporting $100; Contributing $250; Patron $500; Benefactor $1,000; Commodore Club $2,500; Boat Erickson $5,000.

RUNESTONE MUSEUM, 206 N. Broadway, Alexandria, MN 56308-1417. Tel.: 320-763-3160. Fax: 320-763-9705. Facebook: Runestone Museum Foundation.
E-mail: bigole@rea-alp.com
Web Site: www.runestonemuseum.org
Founded: 1958.
Congressional District: 7
Key Personnel: Museum & Exhibit Mgr., Jim Bergquist; Gift Shop Mgr., LuWanna Hintermeister.
Personnel Profile: Full-Time Paid 1; Part-Time Paid 2; Part-Time Volunteers 12.
Governing Authority: nonprofit. Tax-exempt.
Institution Type/Description: History & Youth Museum.
Collections: Discovery Room; Kensington Runestone; wildlife; Fort Alexandria; 1885 schoolhouse, church, cabins & general store; Native Americans; 40 ft. 3/4 scale Viking Ship replica Snorri.
Facilities: theater. Museum-related items for sale.
Activities: video; outdoor functions; outdoor audio tours in English and some in Norwegian.
Publications: Kensington Runestone Brochure; newsletter; A Holy Mission To Minnesota 600 Years Ago.
Hours & Admission Prices: Summer: Mon.-Sat. 9-4, Sun. 11-4; Winter: Mon.-Sat. 9-4. Families $15, adults $6, senior citizens $5, students 5-17 $3; discounts to AAA members; members no charge. Call to confirm admission prices. &
Attendance: 13,000 (accurate)
Membership: Student $10; Individual $25; Family $50; Business $75; Sustaining $100; Patron $500; Corporate $1,000.

Annandale

MINNESOTA PIONEER PARK MUSEUM, 725 Pioneer Park Trail, Annandale, MN 55302-3128. Tel.: 320-274-8489. Fax: 320-274-9612.
E-mail: pioneerp@lakedalelink.net
Web Site: pioneerpark.org
Founded: 1972.
Congressional District: 6
Key Personnel: Pres., Pete Axford.
Personnel Profile: Part-Time Paid 1; Part-Time Volunteers 52.
Governing Authority: nonprofit organization. Tax-exempt.
Institution Type/Description: Historic Site.
Collections: historical buildings & memorabilia from 1850-1940.
Activities: guided tours; school tours; day camps; craft fairs; reunions; weddings; ethnic church services. Museum Sponsors: Pioneer Festival; Halloween Festival; Fiddlers Festival; 1890s Festival; 4th of July Festival.
Publications: newsletter.
Hours & Admission Prices: Memorial Day to mid-Oct. Mon.-Fri. 10-4, Sat.-Sun. 12-4. Adults $5, senior citizens $4, children 6-16 $3; members and children 5 & under no charge.
Attendance: 8,000 (estimated)
Membership: Single $20; Family $30; Sponsor $300; Life $1,000.

Anoka

ANOKA COUNTY HISTORICAL SOCIETY, (M), 2135 Third Ave. N., Anoka, MN 55303-2258. Tel.: 763-421-0600. Fax: 763-323-0218.
E-mail: achs@ac-hs.org
Web Site: www.ac-hs.org
Founded: 1934.
Congressional District: 5
Key Personnel: Pres. (V), Paul Pierce, III; Exec. Dir., Todd Mahon.
Personnel Profile: Full-Time Paid 1; Part-Time Paid 8; Part-Time Volunteers 140; Interns 2.
Governing Authority: Anoka County Historical Society Board of Directors. Tax-exempt.
Institution Type/Description: County History Museum.
Collections: 1900 period furniture; manuscripts; textiles; period clothing; wedding quilts; dolls, rugs.
Research Fields: Anoka County History; Anoka County genealogical research tools.
Facilities: genealogical reference library; microfilm materials; early settlers reference material.
Activities: guided tours; films; temporary exhibitions; lectures. Museum Sponsors: Home and Garden Tour in July; Barn Dance in September.
Publications: bimonthly newsletter, History Center News.
Hours & Admission Prices: Tues. 10-8, Wed.-Fri. 10-5, Sat. 10-4. Adults $3; discounts to groups of 15 or more; members no charge. Library: no charge. &

Attendance: 15,000 (accurate)
Membership: Student and Senior 62 & over $7; Individual $15; Family $25.

Apple Valley

MINNESOTA ZOO, (M), 13000 Zoo Blvd., Apple Valley, MN 55124-4621. Tel.: 952-431-9200. Fax: 952-431-9336.
Web Site: www.mnzoo.org
Founded: 1978.
Congressional District: 3
Key Personnel: Dir. & C.E.O., Lee C. Ehmke; Gift Shop Mgr., Laurel Wright.
Personnel Profile: Full-Time Paid 185; Part-Time Paid 74; Part-Time Volunteers 650; Interns 45.
Governing Authority: state. Subsidiary Institution: Minnesota Zoo Foundation. Tax-exempt.
Institution Type/Description: Zoo.
Collections: 2,449 accessioned individual animals representing 445 species; 30,000 tropical rain forest to northern coniferous forest plant specimens.
Research Fields: animal health; physiological norms, providing specimen tissue, sera for outside research projects; animal management; exhibit evaluation; behavior projects with outside researchers; behavioral engineering; curriculum development & evaluation.
Facilities: library of biological & horticultural material available for research on premises only; zoological garden; botanical garden; trails; picnic area; theater; classrooms; cafeteria. Zoo-related items for sale.
Activities: guided tours; lectures; films; concerts; arts festivals; hobby workshops; TV programs; bird show; formally organized education programs for children, adults, undergraduate & graduate college students affiliated with Metropolitan University; traveling exhibits; loan & permanent exhibitions; zoomobile; trails; teacher workshops; grade school curriculum materials; special events & activities throughout the year.
Publications: bimonthly member newsletter, Zoo Tracks.
Hours & Admission Prices: Memorial Day-Labor Day daily 9-6; Sept. & May Mon.-Fri. 9-4, Sat.-Sun. 9-6; Oct.-April daily 9-4. Adults 13-64 $18, senior citizens 65 & over and children 3-12 $12; discounts to AZA members; members & children under 2 no charge. Parking: cars $6, buses $10, motor coaches $15; members no charge. Closed Thanksgiving; Christmas. &
Attendance: 1,338,581 (accurate)
Membership: Individual $55; Individual Plus $100; Household & Grandparent $130; Household Plus & Grandparent Plus $155; Friends Voyager $250.

Austin

MOWER COUNTY HISTORICAL SOCIETY, (M), Mower County Fairgrounds, 1303 6th Ave., S.W., Austin, MN 55912-2472. Tel.: 507-437-6082. Fax: 507-437-6082. Facebook: Mower County Historical Society.
E-mail: info@mowercountyhistory.org
Web Site: www.mowercountyhistory.org
Founded: 1947.
Congressional District: 1
Key Personnel: Dir., Dustin Heckman; Pres. (V), Garry Ellingson.
Personnel Profile: Full-Time Paid 3; Part-Time Volunteers 153.
Governing Authority: Tax-exempt.
Institution Type/Description: Historical Society Museum.
Collections: local history & culture; period furnishings; industry; photographs; Native American artifacts; tools; clothing; early firefighting equipment; horse-drawn carriages; railroading; steam locomotive; caboose; passenger & baggage cars; M-4 Sherman tank; Six-Mile Grove Lutheran Church replica; miniature three ring circus. Fairgrounds Buildings: 1891 Hormel building; Herald J. Williams, Sr. Indian building; Fire building; Pioneer building; Episcopal Church; Rahilly building; Excelsior School; Communications building; Oakland Depot; Agricultural building.
Facilities: library.
Hours & Admission Prices: Tues.-Fri. 10-4. Adults $5; children under 12 no charge.
Attendance: 30,000 (estimated)
Membership: Individual $25; Family $35; Club/Organization $50; Business $60; Keep the Doors Open Sponsor $100 & up; Ag Building Sponsor $250.

THE SPAM MUSEUM, 1101 N. Main St., Austin, MN 55912-3690. Tel.: 507-434-6370. Fax: 507-437-6721.
E-mail: spammuseum@hormel.com
Web Site: www.spam.com
Founded: 2001.
Key Personnel: Dir. & Cur., Tom Rypka; Museum Shop Mgr., Ariana Finholdt.
Personnel Profile: Full-Time Paid 2; Part-Time Paid 23.
Governing Authority: Parent Institution: Hormel Foods Corporation.
Institution Type/Description: General Museum.
Collections: 4,752 cans of SPAM from around the world; photographs; replica signage; recreated vintage products; late 1800s meat counter; letter from President Dwight D. Eisenhower; simulated SPAM production line.
Facilities: 16,500 sq. ft. exhibit space; 42-seat theater. Museum-related items for sale.
Activities: interactive quiz show.
Hours & Admission Prices: Mon.-Sat. 10-5, Sun. 12-5. No charge. Closed New Year's Day; Easter; Thanksgiving; Christmas Eve & Day. &
Attendance: 100,000 (estimated)

Baudette

LAKE OF THE WOODS COUNTY MUSEUM, 119 8th Ave., S.E., Baudette, MN 56623. Mailing Address: 206 8th Ave., S.E., Ste. 150, Baudette, MN 56623-2867. Tel.: 218-634-1200.
Founded: 1978.
Congressional District: 7
Key Personnel: C.E.O. & Cur., Marlys Hirst; Chm. (V), Dan Crompton.
Personnel Profile: Full-Time Paid 1; Part-Time Paid 1; Part-Time Volunteers 10.
Governing Authority: county; nonprofit organization. Parent Institution: Lake of the Woods Historical Society. Tax-exempt: 501(c)(3).
Institution Type/Description: County History Museum.
Collections: artifacts; archives; local history.
Research Fields: local history.
Facilities: library & archives.
Activities: permanent exhibitions; oral history.
Publications: newsletter, The LOWdown; books, Lake of the Woods Heritage; Baudette The First 100 Years; Memories of a Rolling Stone; Lake of the Woods County: A History of People, Places and Events.
Hours & Admission Prices: mid-May to Sept. Tues.-Fri. 10-4, Sat. 10-2; other times by appointment. No charge; donations accepted. Closed Labor Day. &
Attendance: 3,000 (estimated)
Membership: Single $15; Family $25; Business $30.

Becker

SHERBURNE HISTORY CENTER, 10775 27th Ave. S.E., Becker, MN 55308-4656. Tel.: 763-261-4433. Fax: 763-261-4437.
E-mail: mbrubaker@sherburnehistorycenter.org
Web Site: www.sherburnehistorycenter.org
Formerly: Sherburne County Historical Society
Founded: 1972.
Congressional District: 6
Key Personnel: Exec. Dir., Mike Brubaker; Pres., Brenda Peterson; Pres. (V), Jean Johnson.
Personnel Profile: Full-Time Paid 2; Part-Time Paid 2; Part-Time Volunteers 70.
Governing Authority: nonprofit organization. Tax-exempt: 501(c)(3).
Institution Type/Description: Historical Society Museum.
Collections: late 19th to early 20th-century costumes, decorative arts, farm machinery; photographs; maps; archival material. Historic Buildings: Herbert M. Fox House, Bailey Gas Station.
Research Fields: Sherburne County, Minnesota, History; genealogy.
Facilities: research library; archives.
Activities: loan, permanent & temporary exhibitions; lectures & outreach programs; internships; traveling exhibit trunks; historic walking trail.
Publications: quarterly newsletter, Historically Speaking.
Hours & Admission Prices: Tues.-Fri. 10-5, Sat. 10-4. No charge; donations accepted. Closed major holidays. &
Attendance: 8,800 (estimated)
Membership: Senior Citizen $15; Individual $30; Family $40; Business Booster $50; Patron $150; Ox Cart $500; Heritage $1,000.

Bemidji

BELTRAMI COUNTY HISTORY CENTER, 130 Minnesota Ave., S.W., Bemidji, MN 56601-4009. Tel.: 218-444-3376. Fax: 218-444-3377.
E-mail: depot@beltramihistory.org
Web Site: www.beltramihistory.org
Formerly: Beltrami County Historical Society
Founded: 1952.
Congressional District: 7
Key Personnel: Dir., Dan Karalus; Pres., Linda L. Lemmer; Vice Pres., Leo Soukup.
Personnel Profile: Part-Time Paid 1; Part-Time Volunteers 7; Interns 2.
Governing Authority: society. Parent Institution: Beltrami County Historical Society. Tax-exempt: 501(c)(3).
Institution Type/Description: History Museum.

Collections: photographs; period artifacts; manuscripts. Historic Buildings: 1898 log cabin; 1903 log school; 1912 first consolidated school in northern Minnesota.
Major Exhibits: Inventors and Inventions (T), 2/14-3/14.
Facilities: research library; history center; meeting room. Museum-related items for sale.
Activities: temporary & permanent exhibitions; educational programs. Museum Sponsors: Depot Day in August; History on the Move bus tour in fall.
Publications: newsletter, Depot Express; Beltrami County history.
Hours & Admission Prices: History Center: Nov.-March Sat. 10-4 or by appointment; April-Oct. Wed.-Sat. 12-4 or by appointment. Adults $5, seniors & students $4, children 12 & under $1; discounts to Time Travelers network; members no charge. Research Fee: non-members $15 per hour; members no charge. Closed New Year's Day; Good Friday; Memorial Day; Independence Day; Labor Day; Thanksgiving; Christmas Eve & Day. ⅃
Attendance: 10,000 (estimated)
Membership: Individual $25; Family $45; Sponsor $75; Benefactor $100; Patron $200.

HEADWATERS SCIENCE CENTER, 413 Beltrami Ave., Bemidji, MN 56601-3106. Mailing Address: P.O. Box 1176, Bemidji, MN 56619-1176. Tel.: 218-444-4472. Fax: 218-444-4473.
E-mail: contact@hscbemidji.org
Web Site: www.hscbemidji.org/index.htm
Institution Type/Description: Science Center.
Collections: hands-on exhibits; biological & life history processes.
Activities: outreach programs; demonstrations.
Hours & Admission Prices: Mon. & Wed.-Sat. 9:30-5, Sun. 1-5. Adults 12 & over $7, seniors 65 & over and military $6, children 2-11 $5; children under 2 no charge. Closed New Year's Day; Easter; Thanksgiving; Christmas.

Benson

SWIFT COUNTY HISTORICAL SOCIETY, 2135 Minnesota Ave., Bldg. 2, Benson, MN 56215-2101. Tel.: 320-843-4467.
E-mail: swiftmuseum@embarqmail.com
Founded: 1929.
Key Personnel: Exec. Dir., Marlys Gallagher; Archivist, Laura Puckett.
Personnel Profile: Part-Time Paid 3.
Governing Authority: society. Tax-exempt.
Institution Type/Description: History Museum.
Collections: newspapers including Appleton, Benson, Kerkhoven from 1880s, 1876, 1890s to present; period artifacts.
Research Fields: genealogy.
Facilities: library of newspapers available for use on premises.
Activities: outreach programs. Annual Events: Tribute to Veterans & School Week in May; Annual Museum Banquet in July; Christmas Event in December.
Publications: Swift County History Book; pictorial; Swift County Minnesota family histories.
Hours & Admission Prices: April-Sept. Tues.-Fri. 10-4:30, Sat. 10-3; Oct.-March Tues.-Fri. 10-4:30. No charge; donations accepted. ⅃
Attendance: 1,500 (estimated)
Membership: Senior Citizen & Individual $10; Family $15; Business $30.

Blaine

AMERICAN WINGS AIR MUSEUM, Anoka County Airport, Colorado Ln., Blaine, MN 55449-0322. Mailing Address: P.O. Box 490322, Blaine, MN 55449-0322. Tel.: 763-786-4146.
E-mail: lburgers@pro-ns.net
Web Site: www.americanwings.org/museum.htm
Institution Type/Description: Military History Museum.
Collections: military history & aircraft.
Activities: school outreach program.
Hours & Admission Prices: Tues. 4-8, Sat. 10-5; other times by appointment.

Bloomington

BLOOMINGTON HISTORICAL SOCIETY - OLD TOWN HALL HISTORY MUSEUM, 10200 Penn Ave. S., Bloomington, MN 55431. Mailing Address: 1800 W. Old Shakopee Rd., Bloomington, MN 55431. Tel.: 952-881-4114 & 4327.
Web Site: www.bloomingtonhistoricalsociety.org
Founded: 1964.
Congressional District: 5
Key Personnel: Pres. (V), Dir. & Museum Shop Mgr., Vonda Kelly; Museum Shop Mgr., Don Stiles.
Personnel Profile: Part-Time Volunteers 20.

Governing Authority: board of directors. Tax-exempt.
Institution Type/Description: Regional History Museum: housed in 1892 Old Town Hall.
Collections: artifacts pertaining to Bloomington; manuscripts; local history material.
Research Fields: local history.
Facilities: Printed literature for sale.
Activities: guided tours; lectures; meetings; inter-museum loan, permanent, temporary & traveling exhibitions.
Publications: 6 times per year, Bloomington Historical Society Newsletter; book, Bloomington on The Minnesota.
Hours & Admission Prices: Tues.-Thurs. & Sat. 12-4, Sun. 1-5. No charge; donations accepted. ⅃
Attendance: 2,500 (estimated)
Membership: Junior Historian $1; Contributing $10; Sustaining Member & Individual $25; Couple $35; Life Member $50; Patron $300; Corporate $600.

BLOOMINGTON THEATRE AND ART CENTER, 1800 W. Old Shakopee Rd., Bloomington, MN 55431-3071. Tel.: 952-563-8575. Fax: 952-563-8576. TDD: 952-563-8740.
E-mail: info@btacmn.org
Web Site: www.btacmn.org
Formerly: Bloomington Art Center
Founded: 1976.
Congressional District: 3
Key Personnel: Pres. (V), Mark Eaton; Exec. Dir., Andrea Specht; Treas., Rob Lunz; Vice Pres., Linda Batterson; Sec., Beth Albrecht; Dir. Exhibitions, Rachel Daly Flentje; Dir. Publicity, Nancy Lamberger.
Personnel Profile: Full-Time Paid 7; Part-Time Paid 2.
Volunteer Hours: 11,000
Operating Expenses: 1,500,000
Operating Income: 1,500,000
Governing Authority: nonprofit. Tax-exempt: 501(c)(3).
Institution Type/Description: Art Center.
Collections: paintings.
Major Exhibits: Transitory Patterns - Artists: Eleanor McGough and Claudia Poser, 1/10-14-2/14/14; Biannual Art in the Home, 2/20/14-4/4/14; Fruitful and Multiplying Curated by John Schuerman, 4/11/14-5/30/14; Triannual Invitational Ceramic Exhibition, 6/6/14-7/11/14; Impermanence - Artist: Sally Brown, 7/18/14-8/22/14; Dynamic Convergence - Artists: Megan Bell Honigman, Rory King, Spencer Silver, 8/28/14-10/3/14.
Facilities: classrooms; theaters; pottery & glass studios. Art works for sale.
Activities: guided tours; lectures; gallery talks; arts festivals; formally organized education programs; theater; visual & performing arts; literature; workshops.
Publications: e-newsletter; class catalog; season brochure.
Hours & Admission Prices: Mon.-Fri. 8am-10pm, Sat. 9-5, Sun.1-10. No charge; donations accepted. Closed holidays. ⅃
Attendance: 69,000 (estimated)
Membership: Individual $35; Household $50.

NWA HISTORY CENTRE, INC., 8101-34th Ave. S., Ste. B-747, Bloomington, MN 55425-1642. Tel.: 952-698-4478.
E-mail: info@nwahistory.org
Web Site: www.nwahistory.org
Founded: 2002.
Congressional District: 5
Key Personnel: Pres. (V), Bruce Kitt.
Personnel Profile: Part-Time Volunteers 75.
Governing Authority: Tax-exempt.
Institution Type/Description: Commercial Aviation History Museum.
Collections: Northwest Airlines history; uniforms; personal artifacts; photographs; airplane models; early documents.
Activities: special events. Museum Sponsors: Coffee & Conversation Events.
Publications: quarterly, Reflections.
Hours & Admission Prices: Mon.-Fri. 11-5, Sat. 9-1. No charge; donations accepted. Closed major holidays. ⅃
Attendance: 1,200 (estimated)
Membership: Email Newsletter $25; Postal Newsletter $50; Silver $100; Lifetime $500.

UNDERWATER ADVENTURES AQUARIUM, Mall of America, 120 E. Broadway, Bloomington, MN 55425-5511. Tel.: 852-883-0202; 888-348-3824. Fax: 952-883-0303.
Web Site: sharky.tv
Institution Type/Description: Aquarium.
Collections: sharks; stingrays; sea turtles.
Hours & Admission Prices: March 13-April 13 Mon.-Thurs. 9:30-8:30, Fri.

9:30-9:30, Sat. 9-9:30, Sun. 10-7; late-April to early March Mon.-Thurs. 10-8, Fri.-Sat. 9:30-8:30, Sun. 10-6:30. Adults 13 & over $18.99, children 3-12 $11.99; children under 2 no charge.

Blue Earth

FARIBAULT COUNTY HISTORICAL SOCIETY, 405 E. Sixth St., Blue Earth, MN 56013-2020. Tel.: 507-526-5421.
E-mail: fchs@bevcomm.net
Founded: 1948.
Congressional District: 2
Personnel Profile: Part-Time Volunteers 12.
Governing Authority: society. Tax-exempt: 170(b)(1)(A).
Institution Type/Description: Historical Society Museum.
Collections: house furnishings; kitchen artifacts; clothing; farm machinery; tools; military; medical; religious; books; Civil War; Luther Burbank. Historic Buildings: Wakefield House 1868; Good Shepherd Episcopal Church 1871; W.D.L. Church 1886; Krosh log house 1862; woodland school 1870; Guckeen post office 1901; blacksmith shop; period machine shed; general store.
Research Fields: local history; Faribault County genealogy.
Facilities: 100-vol. library of books on local history, available for research on premises with supervision.
Activities: guided tours; slide presentations; permanent exhibitions.
Publications: triannual newsletter.
Hours & Admission Prices: Tues.-Fri. 10-4; other times by appointment and during County Fair. Donations Accepted. Closed holidays.
Attendance: 2,000 (estimated)
Membership: Annual $20; Life $100.

Brainerd

CROW WING COUNTY HISTORICAL SOCIETY, 320 Laurel, Brainerd, MN 56401-3523. Mailing Address: P.O. Box 722, Brainerd, MN 56401-0722. Tel.: 218-829-3268. Fax: 218-828-4434.
E-mail: history@co.crowwing.us
Web Site: www.crowwinghistory.org/
Founded: 1927.
Congressional District: 8
Key Personnel: Pres. (V), Don Samuelson; Admin., Pamela Nelson.
Personnel Profile: Part-Time Paid 3; Part-Time Volunteers 25.
Governing Authority: society. Tax-exempt.
Institution Type/Description: Local History Museum.
Collections: Indian artifacts; industrial; paintings; archaeology; costumes; archives.
Research Fields: Crow Wing County history.
Facilities: library of genealogy & history books available for local history research; archives specializing in county & regional history.
Activities: guided tours.
Publications: biannual newsletter.
Hours & Admission Prices: Tues.-Sat. 10-3. Donations: adults $3; children, students & members no charge. Closed holidays. &
Attendance: 5,000 (estimated)
Membership: Individual $20; Family $40; Sustaining $50; Booster $100; Benefactor $250; Patron $500 & up.

NORTHLAND ARBORETUM, 14250 Conservation Dr., Brainerd, MN 56401. Mailing Address: P.O. Box 375, Brainerd, MN 56401. Tel.: 218-829-8770.
E-mail: arboretum@brainerd.net
Web Site: www.northlandarb.org
Founded: 1972.
Key Personnel: Dir., Dale E. Braddy.
Personnel Profile: Full-Time Paid 2; Part-Time Paid 2; Part-Time Volunteers 2,000.
Volunteer Hours: 8,895
Institution Type/Description: Arboretum.
Collections: regional flora & fauna.
Facilities: 500-acres; trails.
Activities: workshops; seminars.
Publications: quarterly newsletter, ARBLIFE; monthly e-newsletter, AR-BLINK.
Hours & Admission Prices: Daily dawn to dusk. Adults $3; members no charge. &
Attendance: 50,000 (estimated)
Membership: Individual $25; Family & Household $40; Contributing $50; Business & Supporting $100 & up; Paul Bunyan $500 & up; Arborist $1,000 & up; Lifetime $2,500 & up.

Breckenridge

WILKIN COUNTY HISTORICAL SOCIETY, 704 Nebraska Ave., Breckenridge, MN 56520-1547. Tel.: 218-643-1303.
Founded: 1965.
Congressional District: 7
Key Personnel: C.E.O. & Pres., Gordon Martinson; Treas., Ruth Poppel; Sec., Sylvia Peterson.
Personnel Profile: Part-Time Paid 2; Part-Time Volunteers 1.
Governing Authority: society. Parent Institution: Minnesota Historical Society. Tax-exempt.
Institution Type/Description: Local History Museum.
Collections: period artifacts; books; maps; school room; manuscripts; local newspapers & census records on microfilm; obituary files.
Research Fields: biographies; old newspapers; obituaries.
Activities: tours; open house.
Hours & Admission Prices: Tues.-Thurs. 1:30-4. No charge, donations accepted.
Attendance: 200 (estimated)
Membership: Annual $7.50, Couple $15; Life $100.

Browns Valley

SAM BROWN LOG HOUSE, 796 W. Broadway, Browns Valley, MN 56219-0013. Mailing Address: City Hall, Box 334, Browns Valley, MN 56219-0013. Tel.: 320-695-2110. Fax: 320-695-2127.
E-mail: tom.schmitz@prtel.com
Web Site: www.brownsvalleymn.com
Founded: 1932.
Congressional District: 2
Key Personnel: Pres. (V), Shirley Ecker; Clerk & Treas., Thomas A. Schmitz.
Personnel Profile: Part-Time Paid 1; Part-Time Volunteers 15.
Governing Authority: Parent Institution: Browns Valley Historical Society & city council.
Institution Type/Description: Historic House: 1863 Joseph & Sam Brown log house.
Collections: guns; clothing; Indian artifacts; furniture; archaeology; arts & crafts.
Research Fields: historical.
Publications: Early History of Browns Valley.
Hours & Admission Prices: Memorial Day weekend to Labor Day Fri.-Sun. 1-6 & holidays; other times by appointment. No charge; donations accepted.
Attendance: 500 (accurate)

Buffalo

WRIGHT COUNTY HISTORICAL SOCIETY, 2001 Hwy. 25 N., Buffalo, MN 55313. Mailing Address: P.O. Box 304, Buffalo, MN 55313-0304. Tel.: 763-682-7323. Fax: 763-682-7324.
Web Site: www.wrighthistory.org
Founded: 1942.
Congressional District: 6
Key Personnel: Pres., Leander Wetter; Coord. & Cur., Erin Storc; Archivist, Betty Dircks; Business Mgr., Sally Macnab.
Personnel Profile: Full-Time Paid 3; Part-Time Paid 2; Part-Time Volunteers 10.
Governing Authority: society; nonprofit. Tax-exempt: 501(c)(3).
Institution Type/Description: General Museum.
Collections: cultural, social, ethnic & craft themes specific to local history; the Nelsonian, a 32-piece one-man band; Hubert H. Humphrey's 1926 Model T Ford; library; archives; photographs; textiles; agricultural equipment; art; ceramics & metals.
Research Fields: local genealogy; Wright County & Minnesota history.
Facilities: library. Local interest books & museum-related items for sale.
Activities: organized education programs for children; docent program. Museum Sponsors: Children's Days in May.
Publications: books, Women of Wright County; Compendium of Plagues, Politics, Disasters, People, Places & Events; 101 Best Stories of Wright County; D.R. Farnham's History of Wright County; 2-vol. History of Wright County, Minnesota, 1915; 1909 N.W. Magazine; Wright County history reprinted from History of the Upper Mississippi Valley.
Hours & Admission Prices: Mon.-Fri. 8-4:30. No charge; donations accepted. &
Attendance: 8,000 (estimated)
Membership: Student $5; Individual & Family $10; Business $25.

Cambridge

ISANTI COUNTY HISTORICAL SOCIETY, 33525 Flanders St., N.E., Cambridge, MN 55008-4157. Tel.: 763-689-4229. Fax: 763-552-0740.
E-mail: ichsdirector@izoom.net
Web Site: www.ichs.ws
Founded: 1965.
Congressional District: 8
Key Personnel: C.E.O. & Dir., Kathleen J. McCully; Chm. (V) & Pres. (V), Kay Rodrigue.
Personnel Profile: Part-Time Paid 2; Part-Time Volunteers 250.
Governing Authority: county; nonprofit organization. Tax-exempt: 501(c)(3).
Institution Type/Description: Historical Society.
Collections: documents; manuscripts; photographs; oral histories; pioneer artifacts; 1870 to 1930, household & farm equip. Historic Structures: West Riverside school; Spencer Brook school. St. John's German Lutheran Church of Bradford; pioneer log cabin; blacksmith shop.
Research Fields: local history; Swedish immigrants.
Facilities: resource center with newspaper microfilm & oral history on county available for research at the ICHS Heritage Center.
Activities: slide lectures & other programs in county; local history input in county schools; Old Time school session; language & culture day camps for children (Swedish); journaling workshops for all ages; outdoor concerts dealing with folk culture/lore.
Publications: annual newsletter, Isanti Cuttings & monthly updates; Isanti County: Yesterday...Today...Tomorrow.
Hours & Admission Prices: Heritage Center: Mon.-Tues. & Fri. 9-4:30; other times by appointment. Historic Structures: open for special events; other times by appointment. No charge; donations accepted. &
Attendance: 1,500 (estimated)
Membership: Student & Senior Citizens over 60 $20; Individual $25; Family $40; Sustaining $55; Supporting $75; Patron $125; Bronze Patron $250; Gold Patron $500.

Cannon Falls

CANNON FALLS AREA HISTORICAL MUSEUM, 206 W. Mill St., Cannon Falls, MN 55009-2029. Mailing Address: P.O. Box 111, Cannon Falls, MN 55009-0111. Tel.: 507-263-4503. Fax: 507-263-4080.
E-mail: cannonfallsmuseum@gmail.com
Web Site: www.sites.google.com/site/cannonfallsmuseum
Founded: 1979.
Key Personnel: Dir., Zachary Wareham; Pres. (V), Steve Dabelow; Treas., Tom Monroe; Sec., Ilene Fox.
Personnel Profile: Part-Time Paid 2.
Governing Authority: society. Tax-exempt.
Institution Type/Description: Historical Society Museum: housed in 1888 town fire hall.
Collections: historical items from the Cannon Falls area.
Research Fields: history of local firms.
Activities: guided tours.
Hours & Admission Prices: Fri. 1-5, Sat. 10-4; groups by appointment. No charge; donations accepted.
Attendance: 700 (estimated)
Membership: Student $10; Single $25; Family $35; Business $100.

Chaska

MINNESOTA LANDSCAPE ARBORETUM, UNIVERSITY OF MINNESOTA, (M), 3675 Arboretum Dr., Chaska, MN 55318-9613. Tel.: 952-443-1400. Fax: 952-443-2521.
E-mail: arbinfo@umn.edu
Web Site: www.arboretum.umn.edu
Founded: 1958.
Congressional District: 2
Key Personnel: Dir., Edward L. Schneider; Pres. (V), Megan Dayton.
Governing Authority: university. Parent Institution: University of Minnesota, St. Paul, MN. 55108. Tax-exempt.
Institution Type/Description: Arboretum.
Collections: woody ornamentals; herbaceous ornamentals; fruit; natural landscapes; landscaped gardens; 18,000 books & periodicals in Andersen Horticultural Library.
Research Fields: woody plant breeding; cold hardiness studies; fruit breeding programs; plant stress studies; wetland restoration; ornamental grass breeding.
Facilities: 18,000-vol. non-circulating horticultural library; 1,047 acres of display gardens; natural areas; research lab; Snyder Building with conservatory; Clotilde Irvine Sensory Garden & Therapeutic Horticulture Center. Oswald Visitor Center: restaurant; learning center. Museum-related items for sale.
Activities: guided walking, bus & tram tours; lectures; films; hobby workshops; formally organized education programs for children, adults, families and undergraduate college students; research; excursions; Minnesota Extension Service Outreach.
Publications: bimonthly newsletter; fact sheets; books and videos; The Source List; plant publications.
Hours & Admission Prices: April to Oct. Mon.-Sat. 8-6, Sun. 10-6; Nov. to March Mon.-Sat. 8-4:30, Sun. 10-4:30. Visitors age 13 & up $13; University of Minnesota students, children 12 & under and members no charge. &
Attendance: 271,000 (accurate)
Membership: Solo $49; Duo $79; Duo +2 $99.

Chisholm

MINNESOTA DISCOVERY CENTER, 1005 Discovery Dr., Chisholm, MN 55719. Tel.: 218-254-7959; 800-372-6437. Fax: 218-254-7971.
E-mail: mai.vang@mndiscoverycenter.com
Web Site: www.mndiscoverycenter.com
Formerly: Ironworld
Founded: 1977.
Congressional District: 8
Key Personnel: Exec. Dir., Lisa Vesel; Archivist, Chris Welter; Cur., Mai Vang; Events Mgr., Tammy Jensen.
Personnel Profile: Full-Time Paid 20; Part-Time Paid 25.
Institution Type/Description: General Museum.
Collections: iron ore mining company records & artifacts; mine equipment; photographs; government records; personal & business records; social & civic organization records; historic buildings.
Research Fields: history, genealogy, labor history & immigration to Minnesota's Mesabi, Vermilion & Cuyuna iron ranges.
Facilities: 7,500-vol. library pertaining to Minnesota & mining history; theater; restaurant.
Activities: guided tours; special events; mini golf; field trips; music concerts; educational tours.
Hours & Admission Prices: Research Center: Tues.-Sat. 10-5. No charge. Park & Museum: Summer: Tues.-Wed. & Fri.-Sun. 10-7, Thurs. 10-9; Winter: Tues.-Wed. & Fri.-Sat. 10-5, Thurs. 10-9. Adults $5, students $3; members & Thurs. after 5pm no charge. Closed New Year's Day; Easter; Memorial Day; Independence Day; Labor Day; Thanksgiving; Christmas Eve & Day. &
Attendance: 24,000 (estimated)
Membership: Individual $10; Family $25.

MINNESOTA MUSEUM OF MINING, 900 W. Lake St., Chisholm, MN 55719-1736. Mailing Address: P.O. Box 271, Chisholm, MN 55719-0271. Tel.: 218-254-5543. Facebook: Minnesota Museum of Mining.
Web Site: www.mnmuseumofmining.org
Key Personnel: Pres., John Nelson
Institution Type/Description: Mining Museum.
Collections: mining history & equipment; underground mine & mining town replicas; trucks; steam shovels; early Mesabi Range lifestyle; simulated underground mine drift; log cabin; steam-driven diamond drill.
Activities: ride mining equipment.
Publications: brochure.
Hours & Admission Prices: Memorial Day to Labor Day Mon.-Sat. 9-5, Sun. 1-5; groups by appointment. Adults $5, seniors $4.50, students $3; children 5 & under no charge.

Cloquet

CARLTON COUNTY HISTORICAL SOCIETY, (M), 406 Cloquet Ave., Cloquet, MN 55720-1750. Tel.: 218-879-1938. Fax: 218-879-1938.
E-mail: director@carltoncountyhistory.org
Web Site: www.carltoncountyhistory.org
Founded: 1949.
Congressional District: 8
Key Personnel: Dir., Rachael E. Martin; Pres., Milo Rasmussen; Treas., James Dennie; Sec., Cassandra Brissett.
Personnel Profile: Part-Time Paid 4; Part-Time Volunteers 54.
Governing Authority: society. Affiliated with Minnesota Historical Society. Subsidiary Institution: Moose Lake Area Historical Society. Tax-exempt: 501(c)(3).
Institution Type/Description: History Museum.

Collections: Native American artifacts; agriculture; pioneer; railroad; dairy; business; logging & lumbering; fires of 1918 memorabilia; log cabin; barn; farm machinery.

Research Fields: local history.

Facilities: research library.

Activities: tours; permanent & temporary exhibits; narrated slide history interpretive programs.

Publications: newsletter, Society News; Crossroads in Time: A History of Carlton County, Minnesota; Reflections of Our Past: A Pictorial History of Carlton County; History of the Thomson Farming Area; History of the Pioneers of the Cromwell, MN area; Fire Storm: The Great Fires of 1918; A Hometown Album: Cloquet's Centennial Story; Stories of a Century; Reuben B. Carlton: Frontier Blacksmith and Visionary; A History of Mahtowa; Carlton Chronicles.

Hours & Admission Prices: Tues.-Sat. 9-4. Adults $2; children under 12 & members no charge.

Attendance: 1,500 (estimated)

Membership: Senior 60 & over $15; Adult $20; Family $30; Friend $50; Sponsor $100; Patron $500. Business: Basic $25; Friend $50; Sponsor $100; Patron $500.

Cokato

COKATO MUSEUM & AKERLUND PHOTOGRAPHY STU-DIO, 175 W. 4th St., Cokato, MN 55321-4852. Mailing Address: P.O. Box 686, Cokato, MN 55321-0686. Tel.: 320-286-2427. Fax: 320-286-5876. Facebook: Cokato Museum & Akerlund Photography Studio.

E-mail: cokatomuseum@embarqmail.com

Web Site: www.cokato.mn.us

Founded: 1976.

Congressional District: 6

Key Personnel: Dir., Mike Worcester; Pres. (V), Robert Gasch; Financial Dir., Dorene Erickson.

Personnel Profile: Full-Time Paid 2; Part-Time Paid 1; Part-Time Volunteers 60.

Volunteer Hours: 300

Operating Expenses: 98,000

Operating Income: 105,000

Governing Authority: municipal. Tax-exempt: 501(c)(3).

Institution Type/Description: History Museum & Historic Site.

Collections: period furnishings; catalogued 2 & 3 dimensional items (thru 1996) numbers 15,235; women's clothing from 1890-1920; Cokato businesses; Finnish culture; agricultural tools & equipment; immigration artifacts; photographs include 4,000 catalogued, plus 14,000 Akerlund photos & negatives; restored 1905 photography studio.

Facilities: library; 3,900 sq. ft. exhibit space.

Activities: concerts; guided tours; lectures; author readings; temporary exhibitions. Annual Events: annual meeting in February; Christmas potluck dinner in December; quarterly programs on local history topics.

Publications: quarterly newsletter, In the Midst Of; books: A Can of Cream: A History of Dairying in Stockholm Township, Wright County, Minnesota; Fire Fighters to the Rescue: A Century of Fires in Cokato, 1896-1996; The Cokato Canneries, 1904-1978; Cokato's First 125 Years, 1878-2003; Why Did They Take The Game Away From Us: The Story of Cokato High School Girls Basketball, 1903 to 1931.

Hours & Admission Prices: Tues.-Fri. 8:30-4:30, Sat. 8:30-3. No charge; donations accepted. Tours $2. Closed major holidays.

Attendance: 3,912 (accurate)

Membership: Individual $10; Family $15; Business $25; Supporter $50; Benefactor $100.

Comfrey

JEFFERS PETROGLYPHS HISTORIC SITE, (M), 27160 County Rd. 2, Comfrey, MN 56019-4430. Tel.: 507-628-5591. Fax: 507-628-5593.

E-mail: jefferspetroglyphs@mnhs.org

Web Site: www.jefferspetroglyphs.com

Founded: 1966.

Key Personnel: Site Mgr., Tom Sanders.

Personnel Profile: Full-Time Paid 2; Part-Time Paid 10; Part-Time Volunteers 8.

Governing Authority: state; nonprofit. Parent Institution: Minnesota Historical Society, 345 Kellogg Blvd. W., St. Paul, MN 55102. Tax-exempt.

Institution Type/Description: History Museum.

Collections: American Indian history & spirituality, 5000 B.C. - 600 A.D.; quartzite stone carvings.

Research Fields: archaeology; geology; prairie grasses & flowers; petroglyphs (carvings) history.

Facilities: library; 1,300 sq. ft. exhibit space; 50-seat theater. Museum-related items for sale.

Activities: formal education programs; guided tours; theater.

Publications: book, The Jeffers Petroglyphs.

Hours & Admission Prices: June-Aug. Mon. & Thurs.-Sat. 10-5, Sun. 12-5; Sept.-May group tours of 10 or more, call for appointment. Adults $6, senior citizens $5, children $4; discounts to groups; children under 6 & members no charge.

Attendance: 8,000 (accurate)

Membership: Senior Individual $45; Individual $50; Individual Plus & Senior Household $65; Household $75; Household Plus $95; Associate $125; Contributing $250; Sustaining $500; North Star Circle $1,000.

Crookston

POLK COUNTY HISTORICAL SOCIETY, 719 E. Robert St., Crookston, MN 56716-2043. Mailing Address: P.O. Box 214, Crookston, MN 56716-0214. Tel.: 218-281-1038.

E-mail: polkcomuseum@rrv.net

Web Site: www.mnhistoricnw.org/Polkchs.htm

Founded: 1933.

Congressional District: 7

Key Personnel: Pres. (V) & Cur., Gerald J. Amiot; Vice Pres., Jerry Wentzel; Dir. & Museum Shop Mgr., Alyson Leas.

Personnel Profile: Part-Time Paid 2; Part-Time Volunteers 80.

Governing Authority: county; nonprofit organization. Affiliated with Research Station. Tax-exempt.

Institution Type/Description: History Museum.

Collections: archives; agriculture; anthropology; textiles; Indian artifacts; photographs; industrial; music; medical; military; transportation; archaeology; graphics; costumes; folklore; manuscripts; period automobiles; agriculture equipment; schoolhouse; log cabin & blacksmith shop.

Research Fields: textiles; Indian artifacts; industrial; music; medical; military; transportation; archaeology; local county cemetery.

Facilities: library of books & documents available for use by special permission; 90-seat auditorium; ecumenical chapel; children's museum. Museum-related items for sale.

Activities: guided tours; lectures; gallery talks; country school programs; study clubs; hobby workshops; formally organized education programs for children, adults & undergraduate college students; permanent exhibitions.

Publications: newsletter.

Hours & Admission Prices: 3rd week in May to mid-Sept. Tues.-Sun. 12-4. No charge; donations accepted.

Attendance: 2,000 (estimated)

Membership: Individual $10; Family $15; Patron $25.

Crosby

CUYUNA IRON RANGE HERITAGE NETWORK, 101 First St., N.E., Crosby, MN 56441. Mailing Address: P.O. Box 272, Crosby, MN 56441-0272. Tel.: 218-546-6178 & 545-1166. Fax: 248-545-1166.

E-mail: cchps@crosbyironton.net

Formerly: Cuyuna Range Historical Society & Museum

Founded: 1970.

Congressional District: 7

Key Personnel: Pres. (V), Myrna Nelson.

Personnel Profile: Part-Time Paid 2; Part-Time Volunteers 2.

Governing Authority: board of directors; nonprofit. Parent Institution: Minnesota Historical Society. Tax-exempt.

Institution Type/Description: History Museum: housed in 1910 Soo Line Depot.

Collections: mining & logging memorabilia; tools, furniture, clothing, pictures, store items of settlers & immigrant families.

Research Fields: local history.

Activities: guided tours.

Publications: annual membership letter.

Hours & Admission Prices: June-Aug. Mon.-Sat. 10-4. Adults $1, children $.50. Closed holidays.

Attendance: 1,250 (estimated)

Membership: Annual $10; Sustaining $20 or more; Life $100.

Currie

END-O-LINE RAILROAD PARK & MUSEUM, 440 N. Mill St., Currie, MN 56123-1133. Mailing Address: P.O. Box 57, Slayton, MN 56172. Tel.: 507-763-3708. Fax: 507-763-3996. Facebook: Endoline.

E-mail: endoline@co.murray.mn.us

Web Site: www.endoline.com

Founded: 1972.
Congressional District: 1
Key Personnel: Museum Coord., Janet Timmerman.
Personnel Profile: Part-Time Paid 5; Part-Time Volunteers 4.
Governing Authority: Parent Institution: Murray County Government, Slayton, MN 56172.
Institution Type/Description: Railroad Museum.
Collections: historical artifacts; manual-operated turntable; diesel switcher & caboose; engine house; railroad memorabilia; hobo display; restored depot; 1875 Baldwin locomotive; Georgia Northern #102 steam engine; Fairmont motorcars; railroad water tower; section foreman's house; replica coal bunker; replica Lake Shetek Mills; general store; rural country school; Murray County Courthouse; 1870s Currie Presbyterian Church; American bison; M & St. L caboose.
Research Fields: railroads; local history.
Facilities: picnic facilities; nature trails; 6 mile bike & pedestrian trail. Gift items for sale.
Activities: guided tours/self-guided.
Hours & Admission Prices: Memorial Day-Labor Day Wed.-Sat. 10-5, Sun. 1-5; bus tours & groups by appointment. Adults 18 & over $5, students 6-17 $3; children under 5 no charge. ⑆
Attendance: 5,878 (accurate)

Detroit Lakes

BECKER COUNTY HISTORICAL SOCIETY, 714 Summit Ave., Detroit Lakes, MN 56501-2941. Mailing Address: P.O. Box 622, Detroit Lakes, MN 56502-0622. Tel.: 218-847-2938. Fax: 218-847-5048.
E-mail: mail@beckercountyhistory.org
Web Site: www.beckercountyhistory.org
Founded: 1882.
Congressional District: 7
Key Personnel: Pres. (V), Mike Wommer; Cur. & Mgr., Carrie Johnston.
Personnel Profile: Full-Time Paid 1; Part-Time Paid 2; Part-Time Volunteers 71.
Governing Authority: society. Tax-exempt.
Institution Type/Description: Local History Museum.
Collections: artifacts; relics; articles; implements; tools; materials; pictures; specimens pertaining to area history.
Research Fields: local history; genealogy; White Earth Indian Reservation.
Facilities: 2,200-vol. library of county newspapers, manuscripts, letters, photographs & old books available for use on premises.
Activities: permanent & temporary exhibitions.
Publications: quarterly newsletter; Pioneer History of Becker County.
Hours & Admission Prices: Tues.-Sat. 10-4. No charge; donations accepted. Closed national holidays. ⑆
Attendance: 7,500 (estimated)
Membership: Individual $25; Business $35; Family $50.

Duluth

CAF LAKE SUPERIOR SQUADRON 101 MUSEUM, Airport Rd., Duluth International Airport, Duluth, MN 55811. Mailing Address: 4931 Airport Rd., Duluth, MN 55811. Tel.: 218-733-0639.
E-mail: info@cafduluth.com
Web Site: www.cafduluth.com
Founded: 1999.
Institution Type/Description: Military History Museum.
Collections: military history & aircraft; photographs; uniforms; personal equipment; models.
Activities: flight simulator; special events.
Hours & Admission Prices: Call for hours. No charge; donations accepted. ⑆

DULUTH ART INSTITUTE, 506 W. Michigan St., Ste. 2, Duluth, MN 55802-1519. Tel.: 218-733-7560. Fax: 218-733-7506.
E-mail: getart@duluthartinstitute.org
Web Site: www.duluthartinstitute.org
Founded: 1907.
Congressional District: 68
Key Personnel: Exec. Dir., Kristin Dockart; Pres., Christine King; Cur., Anne Dugan; Dir. Education, Shannon Consino; Operations & Devel. Mgr., Laura Daugherty; Studio Mgr., Dave Lynas.
Personnel Profile: Full-Time Paid 4; Part-Time Paid 3; Part-Time Volunteers 100.
Governing Authority: Tax-exempt.
Institution Type/Description: Art Gallery.
Collections: works by regional artists.
Activities: educational program; artist services.

Publications: news magazine 3 times a year.
Hours & Admission Prices: Adults $12; members no charge. ⑆
Attendance: 50,000 (estimated)
Membership: Individual $40; Household $60.

DULUTH CHILDREN'S MUSEUM, 115 S. 29th Ave. W., Duluth, MN 55806. Tel.: 218-733-7543 & 7546. Fax: 218-733-7547.
E-mail: explore@duluthchildrensmuseum.org
Web Site: www.duluthchildrensmuseum.org
Founded: 1930.
Congressional District: 8
Key Personnel: C.E.O., Michael P. Garcia; Vice Pres. Operations, Programs & Collections, Rich Jaworski.
Personnel Profile: Full-Time Paid 2; Part-Time Paid 2; Part-Time Volunteers 12.
Governing Authority: nonprofit organization. Tax-exempt: 501(c)(3).
Institution Type/Description: Children's & Youth Museum: housed in 1892 French Norman style former railway station, the St. Louis County Heritage & Arts Center including two additional museums, an art institute & five performing arts organizations.
Collections: world cultures; childhood; immigration; two-story walk-through tree.
Research Fields: childhood; immigration.
Activities: formally organized educational programs for children; traveling, permanent & temporary exhibits; family adventure programs; volunteer program; hands-on exhibits; special programs & events; two-story walk-through tree. Museum Sponsors: Whole World Festival.
Publications: Museum Newsletter; books, A Childhood in Minnesota: Exploring the Lives of Ojibwe & Immigrant Families, 1880s-1920s; Growing Up in My Family: A Guide for Recording Information on Family History.
Hours & Admission Prices: Memorial Day to Labor Day Mon.-Wed. & Fri.-Sat. 9-5, Thurs. 9-8, Sun. 12-5; Sept.-May Tues.-Wed. & Fri.-Sat. 9-5, Thurs. 9-8, Sun. 12-5. Admission one & over $6.50; members, Association of Children's Museums, Association of Science & Technology Center reciprocal membership programs; children under one no charge. Closed New Year's Day; Easter; Thanksgiving; Christmas Eve & Day. ⑆
Attendance: 54,321 (accurate)
Membership: Discovery $55; Discovery Plus $115; Passport Discovery $130.

GLENSHEEN HISTORIC CONGDON ESTATE, (M), 3300 London Rd., Duluth, MN 55804-2010. Tel.: 218-726-8910; 888-454-GLEN. Fax: 218-726-8911.
E-mail: info@glensheen.org
Web Site: www.glensheen.org
Founded: 1979.
Congressional District: 8
Key Personnel: Dir., Daniel Hartman.
Personnel Profile: Full-Time Paid 6; Part-Time Paid 70; Part-Time Volunteers 15.
Governing Authority: university. Affiliated with the University of Minnesota, Minneapolis, MN 55414. Tax-exempt.
Institution Type/Description: Historic House & Site: 1905-08 Glensheen, 6.7-acre historic estate featuring 39-room Jacobean Revival mansion, built for Chester A. Congdon, along the shore of Lake Superior. Listed on National Register of Historic Places.
Collections: original furnishings; arts & crafts furniture; oriental rugs; stained glass windows; art glass; carriages & sleighs; carriage house, gardener's cottage, boathouse, formal gardens.
Research Fields: arts & crafts movement in the midwest; early 20th-century American painters; regional business & social history.
Facilities: Museum-related items for sale.
Activities: guided tours; sponsored events; special events; facility rentals.
Hours & Admission Prices: mid-May to mid-Oct. daily 9-5:30; mid-Oct. to mid-May Sat.-Sun. 9:30-3:30. Tours: family $45, adults $15-$26, senior citizens $14-$24, children $9-$15; discounts to AAM, ICOM & NTHP members; members & children 5 and under no charge. Closed New Year's Day; Thanksgiving; Christmas. ⑆
Attendance: 60,000 (accurate)
Membership: Individual $40.

GREAT LAKES AQUARIUM, 353 Harbor Dr., Duluth, MN 55802-2639. Tel.: 218-740-3474. Fax: 218-740-2020.
E-mail: info@glaquarium.org
Web Site: www.glaquarium.org
Institution Type/Description: Aquarium.
Collections: marine animals; touch tank.
Activities: special events.
Hours & Admission Prices: May-Oct. daily 10-6; Nov.-April daily 10-5.

Adults $15.50, seniors 62 & over $12.50, children 3-12 $9.50; children under 3 no charge.

KARPELES MANUSCRIPT LIBRARY MUSEUM, 902 E. First St., Duluth, MN 55805-2142. Tel.: 218-728-0630.
E-mail: kmuseumdut@aol.com
Web Site: www.rain.org/~karpeles/dulfrm.html
Key Personnel: Dir., Robert Wickham; Cur., Luis E. Rego
Institution Type/Description: Manuscript Library Museum.
Collections: original documents & manuscripts pertaining to history, music, science, literature & art.
Major Exhibits: The Intolerable Acts Exhibit, 1/14-4/30/14.
Activities: musical concerts; recitals; school outreach program; lectures; group tours.
Hours & Admission Prices: Tues.-Sun. 10-4. No charge.

LAKE SUPERIOR MARITIME VISITORS CENTER, 600 Lake Ave., S., Duluth, MN 55802-2322. Mailing Address: P.O. Box 177, Duluth, MN 55801-0177. Tel.: 218-720-5260, ext. 1. Fax: 218-720-5270.
E-mail: info@LSMMA.com
Web Site: www.LSMMA.com
Founded: 1973.
Congressional District: 8
Key Personnel: Dir., Thomas Holden; Public Rels., Beth M. Duncan.
Personnel Profile: Full-Time Paid 4; Part-Time Paid 1; Part-Time Volunteers 8.
Governing Authority: federal. Parent Institution: U.S. Army Corps of Engineers. Tax-exempt.
Institution Type/Description: Marine Museum: located at Duluth Ship Canal.
Collections: Upper Great Lakes commercial navigation history focused on Duluth-Superior harbor, ships, cargoes, shipwrecks, navigation locks, harbor development.
Research Fields: Great Lakes ships; harbor development.
Facilities: 100-seat auditorium; 14,000 sq. ft. exhibit space.
Activities: guided tours; lectures; films; organized education programs for children; participatory, loan & temporary exhibitions; school loan service.
Publications: bimonthly newsletter, The Nor'Easter; factsheets; brochures.
Hours & Admission Prices: Spring & Fall Sun.-Thurs. 10-4:30, Fri.-Sat. 10-6; Summer: daily 10-9; Winter: Fri.-Sun. 10-4:30. No charge; donations accepted. &
Attendance: 408,000 (accurate)
Membership: Family $40; Sustaining $75; Patron $100; Donor $150; Sponsor $250; Individual Life $1,000; Corporate $500.

LAKE SUPERIOR RAILROAD MUSEUM, (M), 506 W. Michigan St., Ste. 19, Duluth, MN 55802-1533. Tel.: 218-733-7590. Fax: 218-733-7596.
E-mail: museum@LSRM.org
Web Site: www.lsrm.org
Founded: 1974.
Congressional District: 8
Key Personnel: Dir., Ken Buehler; Pres. (V), Neal Vanstrom; Opers. Mgr., Richard Bergsrud; Cur., Tim Schandel; Museum Shop Mgr., Josh Miller.
Personnel Profile: Full-Time Paid 5; Part-Time Paid 2; Part-Time Volunteers 35.
Governing Authority: nonprofit organization. Subsidiary Institution: North Shore Scenic Railway. Tax-exempt: 501(c)(3).
Institution Type/Description: Railroad Museum: housed in 1891-92 Duluth Union Depot Building.
Collections: historic railroad equipment; steam locomotives; diesel locomotives; electric locomotives; passenger & freight cars; railway post office car; snowplows; steam wrecking crane; cabooses; china & silver; manuscripts; 8,000 photographs; 300 timetables; advertising; maps; blueprints; company records & scrapbooks. Historic Trains: 1862 William Crooks, locomotive & train; 1941-43 DM&IR Mallet steam locomotive.
Research Fields: regional railroad history.
Facilities: 1,000-vol. library of material primarily related to railroads & railroad equipment serving Minnesota available for research by special arrangement. Railroad books, china & related memorabilia for sale.
Activities: steam & diesel railroad excursions; youth volunteer/mentorship program; outreach programs; guided tours; lectures; films; gallery talks; loan, permanent, temporary & traveling exhibitions.
Publications: quarterly publication, The Junction.
Hours & Admission Prices: Memorial Day-Labor Day daily 9:30-6; Winter Sun. 1-5, Mon.-Sat. 10-5. Adults $12, children $6; members no charge. Closed New Year's Day; Easter; Thanksgiving; Christmas. &
Attendance: 100,000 (estimated)

Membership: Retired Railroad & Student $20; Individual $30; Family $45; Contributing $125; Sustaining $250. Call for Corporate Membership Information.

LAKE SUPERIOR ZOOLOGICAL GARDENS, 7210 Fremont St., Duluth, MN 55807-1854. Tel.: 218-730-4500. Fax: 218-723-3750.
E-mail: info@lszoo.org
Web Site: www.lszoo.org
Formerly: Lake Superior ZOO
Founded: 1923.
Key Personnel: Pres. Lake Superior Zoological Society, Dennis Ramberg; C.E.O., Sam Maida; Assoc. Parks Dir., Jeff Anderson; Museum Shop Mgr., Susan Wolniakowski.
Personnel Profile: Full-Time Paid 19; Full-Time Volunteers 90; Part-Time Paid 15; Part-Time Volunteers 45.
Governing Authority: Tax-exempt: 501(c)(3).
Institution Type/Description: Zoo.
Collections: Polar Shores; Northern Territory; Australian Connection; Primate Conservation Center.
Publications: quarterly newsletter, Wild Times.
Hours & Admission Prices: Summer: 10-5; Winter: 10-4. Adults 13 & over $10, seniors 62 & over $9, children 3-12 $5; discount to reciprocal members; members & children under 3 no charge. &
Attendance: 100,000 (accurate)
Membership: Individual $40; Family & Grandparent $65; Conservationist $90.

THE ST. LOUIS COUNTY HISTORICAL SOCIETY, 506 W. Michigan St., Duluth, MN 55802-1519. Tel.: 218-733-7580. Fax: 218-733-7585. Facebook: Saint Louis County Historical Society.
E-mail: history@thehistorypeople.org
Web Site: www.thehistorypeople.org; www.vets-hall.org
Founded: 1922.
Congressional District: 8
Key Personnel: Exec. Dir., JoAnne Coombe; Chm. (V) Veterans' Memorial Hall, Dennis Hughes; Pres., Jill Dupont; Mgr. Collections, Milissa Brooks Ojibway; Veterans' Memorial Hall Program Story Coord., John Potti; Veterans' Memorial Hall Program Mgr., Dwight Nelson; Administrative Svcs. Mgr., Julie Bolos; Exec. Asst., Susan Schwanekamp.
Personnel Profile: Full-Time Paid 3; Part-Time Paid 4; Part-Time Volunteers 12; Interns 7.
Volunteer Hours: 1,817
Operating Expenses: 494,378
Operating Income: 483,061
Governing Authority: nonprofit organization. Tax-exempt: 501(c)(3).
Institution Type/Description: History Museum: housed in restored Duluth Union Depot.
Collections: St. Louis County history & residents; Eastman Johnson paintings of Ojibwe Indians; northern Minnesota history; books, photographs, manuscripts, historical research material related to mining, shipping, lumbering & settlement of northeastern Minnesota history; artifacts & personal stories of area military veterans; 1892 historic building.
Research Fields: northeastern Minnesota history.
Facilities: archives located at the Northeast MN Historical Center at University of Minnesota, Duluth, 218-726-8526.
Activities: lectures; programs; workshops; special exhibits; antique appraisals; fundraisers.
Publications: society newsletter; books, Historic Sites & Place Names of the North Shore of Minnesota; We're Standing on Iron; Eastman Johnson's Lake Superior Indians; A County Built on Iron and Invincible: History of the Duluth Boat Club.
Hours & Admission Prices: Summer: daily 9:30-6; Winter: Mon.-Sat. 10-5, Sun. 1-5. Adults $12, children 3-13 $6; discounts to senior citizens, AAA & AARP members; members & children 2 & under no charge. Call 218-727-8025 to confirm holiday closures. &
Attendance: 129,582 (estimated)
Membership: Individual $30; Family $50; Supporter $75; Cornerstone $125; Benefactor $250; Patron $500; Directors' Circle $1,000.

TWEED MUSEUM OF ART, (M), Univ. of Minnesota Duluth, 1201 Ordean Ct., Duluth, MN 55812-3041. Tel.: 218-726-8222 & 7823. Fax: 218-726-8503. Facebook: Tweed Museum of Art.
E-mail: tma@d.umn.edu
Web Site: www.d.umn.edu/tma
Formerly: Tweed Gallery
Founded: 1950.
Congressional District: 8
Key Personnel: Dir., Ken Bloom; Public Rels., Christine Strom; Registrar,

Camille Doran; Museum Education, Susan Hudec; Technician, Eric Dubnicka; Exec. Sec., Kathy Sandstedt; Museum Shop Mgr., Barbara Boo.
Personnel Profile: Full-Time Paid 7; Part-Time Paid 3; Part-Time Volunteers 60; Interns 8.
Governing Authority: state; university. Parent Institution: University of Minnesota, Duluth, Board of Regents. Tax-exempt: 501(c)(3).
Institution Type/Description: Art Museum.
Collections: modern & contemporary works; paintings; drawings; works on paper; photography; sculpture & ceramics; Glenn C. Nelson international ceramics; 14th to 19th-century European & 19th to 20th-century American paintings, prints & sculpture; Potlatch Royal Canadian Mounted Police illustrations; French Barbizon School; Richard and Dorothy Nelson collection of American Indian art.
Research Fields: 19th to 20th-century American & 17th to 19th-century European paintings, works on paper, ceramics; modern & contemporary paintings, sculpture, photos, works on paper; American Indian art; American modernism.
Facilities: library & archives; lecture room; sculpture conservatory & courtyard. Museum-related items for sale.
Activities: guided tours; lectures; permanent, temporary & traveling exhibitions; gallery talks; artists-in-residence; education programs for children & adults.
Publications: exhibition & collection catalogs.
Hours & Admission Prices: Tues. 9-8, Wed.-Fri. 9-4:30, Sat.-Sun. 1-5. No charge, donations accepted. Closed university holidays. &
Attendance: 35,000 (accurate)
Membership: Student & Senior $10; Individual & Senior $35; Household $75; Friend $100; Associate $250; Society of Fellows $500; Leadership Circle $1,000.

Eden Prairie

WINGS OF THE NORTH, 14801 Pioneer Tr., Ste. 200, Eden Prairie, MN 55347. Tel.: 952-746-6100.
Governing Authority: Tax-exempt: 501(c)(3).
Institution Type/Description: Aviation History Museum.
Collections: aviation history; aircraft; photographs; personal artifacts; maps.
Activities: special events; educational programs.
Hours & Admission Prices: Call for hours.

Edina

EDINA HISTORICAL SOCIETY, 4711 W. 70th St., Edina, MN 55435-4059. Tel.: 612-928-4577.
E-mail: edinahistory@yahoo.com
Web Site: www.edinahistoricalsociety.org
Key Personnel: Pres., Dan Latham; Exec. Dir., Marci Matson.
Governing Authority: Subsidiary Institution: Edina Historical Center, Frank Tupa Park, 4918 Eden Ave., Edina, MN.
Institution Type/Description: Historical Society Museum.
Collections: Edina history & culture; photographs; early pioneers. Historic Buildings: one-room Cahill School built in 1864; Minnehaha Grange Hall built in 1879.
Hours & Admission Prices: Thurs. 9am-2pm, Sat. 10am-12pm; other times by appointment. &
Membership: Individual $15; Household $25.

MARGARET FOSS GALLERY - EDINA ART CENTER, 4701 W. 64th St., Edina, MN 55435-1501. Tel.: 952-903-5780. Fax: 952-903-5781.
E-mail: artcenter@ci.edina.mn.us
Web Site: www.edinaartcenter.com
Founded: 1977.
Key Personnel: Interim Dir., Michael Frey
Institution Type/Description: Art Gallery.
Collections: works by local, national & international artists.
Activities: special events.
Hours & Admission Prices: Mon.-Thurs. 9-8, Fri. 9-3:30, Sat. 9-1.
Membership: Individual $35; Family $50.

THE WORKS, 5701 Normandale Rd., Ste. 300, Edina, MN 55424-2402. Tel.: 952-848-4848.
E-mail: info@theworks.org
Web Site: www.theworks.org
Founded: 1995.
Governing Authority: Tax-exempt: 501(c)(3).
Institution Type/Description: Technology Discovery Center.
Collections: hands-on engineering exhibits for children 5-12.
Activities: hands-on activities for families & school groups.

Hours & Admission Prices: Sat. 10-4. Admission 3 & over $5.
Attendance: 40,000
Membership: Family $60.

Elbow Lake

GRANT COUNTY HISTORICAL SOCIETY, 115 2nd St., N.E., Hwy. 79E, Elbow Lake, MN 56531. Mailing Address: P.O. Box 1002, Elbow Lake, MN 56531-1002. Tel.: 218-685-4864.
E-mail: gcmnhist@runestone.net
Web Site: www.rootsweb.com/~mngrant/hist.htm
Founded: 1944.
Congressional District: 7
Key Personnel: Dir. & Cur., Patricia Benson; Chm. (V), Stuart Anderson.
Personnel Profile: Part-Time Paid 1; Part-Time Volunteers 12.
Operating Expenses: 42,200
Operating Income: 46,200
Governing Authority: society; nonprofit organization. Tax-exempt.
Institution Type/Description: Historical Society Museum.
Collections: Indian artifacts; agricultural implements; pioneer household wares; prehistoric fossil remains; military artifacts; manuscripts; microfilm of Grant County newspapers; Grant County census & naturalization records; Norwegian Lutheran Church records; Veterans Memorial Hall.
Research Fields: oxcart trails; Indian mounds & camp grounds; sites of prehistoric fossils; genealogy; cemeteries.
Facilities: 2,000-vol. library of bound volumes of Grant County newspapers, local history, military, biographies, school textbooks, township & county records; reading room.
Activities: guided tours; lectures; organized education programs for children & adults; permanent exhibitions.
Publications: Heritage of Grant County, MN (limited edition); triannual newsletter, Oxcart Tales.
Hours & Admission Prices: Memorial Day to Sept. Mon.-Fri. 10-12 & 1-4, Sat. 10-3; Oct.-May Mon.-Fri. 10-12 & 1-4. No charge; donations accepted. Closed New Year's Day; Easter; Memorial Day; Independence Day; Labor Day; Thanksgiving; Christmas. &
Attendance: 800 (estimated)
Membership: Individual $25; Family $35; Life $200.

Elk River

OLIVER KELLEY FARM, (M), 15788 Kelley Farm Rd., Elk River, MN 55330-6234. Tel.: 763-441-6896. Fax: 763-441-6302.
E-mail: kelleyfarm@mnhs.org
Web Site: www.mnhs.org/kelleyfarm
Founded: 1849.
Congressional District: 8
Key Personnel: Site Mgr., Bob M. Quist; C.E.O. & Dir. MHS, Nina M. Archabal.
Personnel Profile: Full-Time Paid 2; Part-Time Paid 16.
Governing Authority: private; not-for-profit organization. Parent Institution: Minnesota Historical Society. Tax-exempt: 501(c)(3).
Institution Type/Description: Living History Farm: 1876 home of Oliver Kelley, founder of the Grange in 1867.
Collections: Minnesota agricultural history, 1850-1876.
Hours & Admission Prices: May & Oct. Sat. 10-5, Sun. 12-5; Memorial Day to Labor Day Tues.-Sat. 10-5, Sun. 12-5; groups of 10 or more by appointment. Adults $8, senior citizens $6, children 6-17 $5; members & children under 6 no charge. Closed New Year's Eve & Day; Christmas. &
Attendance: 30,000 (accurate)
Membership: Senior $45; Individual $50; Senior Household $65; Household $75; Household Plus $95; Associate $145; Contributing $250; Sustaining $500; North Star Circle $1,000.

Ely

DOROTHY MOLTER MUSEUM, (M), Hwy. 169, Ely, MN 55731. Mailing Address: P.O. Box 391, Ely, MN 55731-0391. Tel.: 218-365-4451.
E-mail: rootbeerlady@frontiernet.net
Web Site: www.rootbeerlady.com
Institution Type/Description: History Museum: former home of Dorothy Molter.
Collections: local history & culture pertaining to the life of Dorothy Molter.
Hours & Admission Prices: Memorial Day-Labor Day Mon.-Sat. 10-5:30, Sun. 12-5:30. &

ELY-WINTON HISTORICAL SOCIETY, 1900 E. Camp St., Ely, MN 55731-1918. Tel.: 218-365-3226. Fax: 218-365-7207.
E-mail: ewhs@vcc.edu
Web Site: www.vcc/edu/ewhs
Formerly: Vermilion Interpretive Center
Founded: 1967.
Key Personnel: Dir., Margaret Sweet; Pres. (V), Patricia Koski.
Personnel Profile: Full-Time Paid 1; Part-Time Paid 2; Part-Time Volunteers 1.
Governing Authority: Tax-exempt.
Institution Type/Description: Historical Society Museum.
Collections: local history & culture; geology; Native American artifacts; logging; mining.
Activities: Museum Sponsors: History Night Programs in July.
Hours & Admission Prices: Tues.-Fri. 12-4. Adults $3, children 6-16 $2; discounts to AAM & ICOM members; members & children under 6 no charge. Closed New Year's Day; Easter; Labor Day; Thanksgiving; Christmas. &
Attendance: 1,800 (accurate)
Membership: Individual $10; Family $20; Patron $75; Life $200.

INTERNATIONAL WOLF CENTER, 1396 Hwy. 169, Ely, MN 55731-8129. Tel.: 218-365-4695. Fax: 218-365-3318.
E-mail: jedberg@wolf.org
Web Site: www.wolf.org
Founded: 1985.
Key Personnel: Operations & Finance Mgr., Linda Frisell; Dir. Information Svcs., Jess Edberg; Program Specialist, Tara Johnson; Store & Guest Svcs., Mary Jane Caspers; Wolf Cur., Lori Schmidt; Asst. Wolf Cur., Donna Prichard; Administrative Asst., Kirstin Nephew; Exec. Dir. (MN), Rob Schultz; Dir. Devel. (MN), Darcy Berus; Dir. Finance & Operations (MN), Sharon Reed; Communications Coord. (MN), Tom Myrick; Web Specialist (MN), Carissa Winter; Devel. & Member Assoc. (MN), Sarah Jansen; Membership Asst. (MN), Rebecca Mesa
Governing Authority: Tax-exempt.
Institution Type/Description: Nature Center.
Collections: wolves & their habitats, lives, & survival.
Major Exhibits: Wolves, Ravens & Humans in Culture and Mythology, 11/13-5/14.
Facilities: theater; observation area; auditorium. Museum-related items for sale.
Activities: daily programs; films; workshops; seminars; travel vacations; group visits; wolf watch cameras; facility rental; special events; online curriculum; videoconferencing programs; community events; children's area.
Publications: quarterly magazine, International Wolf; email newsletter, Wolf Chronicles.
Hours & Admission Prices: mid-May to Oct. daily; Nov. to mid-May Fri.-Sun. Adults 13 & over $9.50, seniors 60 & over $8.50, children 4-12 $5.50; discounts to groups of 9 or more, veterans & active military; members & children under 3 no charge. Closed major holidays. &
Attendance: 37,000 (estimated)
Membership: Individual $35; Family $60; International Family $75; Wolf Associate $125; Wolf Sponsor $500.

NORTH AMERICAN BEAR CENTER, 1926 Hwy. 169, Ely, MN 55731-8130. Mailing Address: P.O. Box 161, Ely, MN 55731-0161. Tel.: 218-365-7879; 877-365-7879.
E-mail: info@bear.org
Web Site: www.bear.org
Key Personnel: Mng. Dir. & Bear Cur., Donna Andrews; Assoc. Dir., Sharon Johnson; Membership Dir., Nancy Krause; Program Coord., Sharon Herrell; Group Coord., Glenn Krause
Institution Type/Description: Nature Center.
Collections: live bears & their habitats; black & brown bear mounts; videos; photographs.
Facilities: theater; nature trail. Museum-related items for sale.
Activities: observation areas; educational programs; special events.
Hours & Admission Prices: Feb.-April Fri.-Sat. 10-4, May 1-May 24 & Sept. 13-Oct. 23 daily 10-5; May 25-Sept. 12 daily 9-7; Oct. 24-Nov. 26 Fri.-Sat. 10-5; Dec. 26-Jan. 1 daily 10-4; other times by appointment. Adults $8.50, seniors 60 & over $7, children 3-12 $4.50; children under 3 & members no charge.
Attendance: 30,492 (estimated)
Membership: Student $25; Individual $40; Family & Grandparents $75; Bear Associate $150; Bear Sponsor $250; Bear Sustainer $500; Lifetime Individual $1,000; Lifetime Family $2,000.

Elysian

LESUEUR COUNTY HISTORICAL MUSEUM-CHAPTER 1, 301 N. 2nd St., Elysian, MN 56028. Mailing Address: P.O. Box 240, Elysian, MN 56028-0240. Tel.: 507-267-4202.
E-mail: skrenik@frontiernet.net
Founded: 1966.
Congressional District: 25
Key Personnel: Pres. (V), Shirley Krenik; Financial Dir., Michael LaFrance.
Personnel Profile: Part-Time Paid 8.
Governing Authority: county; nonprofit. Branch Museums: Ottawa Methodist Church-Chapter 2, Ottawa, MN; Geldner Saw Mill, Cleveland, MN; Le Sueur Museum-Chapter 3, Le Sueur, MN. All correspondence: Box 240, Elysian. Tax-exempt: 501(c)(3).
Institution Type/Description: General Museum: housed in c.1895 former Elysian School.
Collections: works by local artists; micrometeorite detector of Explorer I; 1900s general store & living quarters; agriculture; original & reproductions of works by Adolf Dehn, Lloyd Herfindahl, Albert Christ-Janer, Roger Preuss & David Maass; American Indian artifacts; Le Sueur County history; Little Crow display; Red Wing pottery; Civil War, World War I & World War II artifacts; birth, death, & marriage records; census information; church & cemetery records; microfilmed newspapers; atlases. Historic House: c.1869 log cabin.
Research Fields: local history; genealogy.
Facilities: 100-vol. library of local history books available upon request; field research station; genealogy center. Museum-related items for sale.
Activities: guided tours; lectures. Museum Sponsors: arts festivals; Ice Cream Socials; annual dinner for all members of the society.
Hours & Admission Prices: Closed for restorations. &
Attendance: 500 (estimated)
Membership: Individual $15; Couple, Family & Business $25.

Eveleth

UNITED STATES HOCKEY HALL OF FAME MUSEUM, 801 Hat Trick Ave., Eveleth, MN 55734-8640. Mailing Address: P.O. Box 679, Eveleth, MN 55734-0679. Tel.: 218-744-5167. Fax: 218-744-2590.
E-mail: dougp@ushockeyhallmuseum.com
Web Site: www.ushockeyhallmuseum.com
Formerly: United States Hockey Hall of Fame
Founded: 1969.
Congressional District: 8
Key Personnel: Exec. Dir., Doug Palazzari; Administrative Asst., Michelle Putzel; Museum Shop Mgr., Kerry Rich; Bd. Pres., Mitch Brunfelt.
Personnel Profile: Full-Time Paid 2; Part-Time Paid 9.
Governing Authority: nonprofit organization. Tax-exempt: 501(c)(3).
Institution Type/Description: Hockey Museum.
Collections: Cleve Bennewitz skate collection 1850-1921; 1960 & 1980 Olympics; 1998 women's hockey display; hockey from all cultures; paintings; prints; radar gun; original zamboni.
Research Fields: American hockey history.
Facilities: library temporarily closed; theater. Hockey-related items for sale.
Activities: temporary exhibits; hands-on exhibits.
Publications: brochures.
Hours & Admission Prices: Memorial Day to Labor Day Mon.-Sat. 9-5, Sun. 10-3; Sept.-May Fri. 10-5, Sat. 9-5, Sun. 10-3. Adults $8, senior citizens & children 13-17 $7, children 6-12 $6; discounts to hockey groups and AAA & AAM members; members and children 5 & under no charge. Closed New Year's Day; Easter; Thanksgiving; Christmas. &
Attendance: 15,000 (accurate)

Excelsior

EXCELSIOR-LAKE MINNETONKA HISTORICAL SOCIETY, 305 Water St., Excelsior, MN 55331. Mailing Address: P.O. Box 305, Excelsior, MN 55331-0305. Tel.: 952-221-4766.
E-mail: info@elmhs.org
Web Site: www.elmhs.org
Founded: 1972.
Key Personnel: Museum Committee Chm., Randy Lee Julian.
Personnel Profile: Part-Time Volunteers 15.
Governing Authority: society. A branch of the Minnesota State Historical Society & the Hennepin County Historical Society. Tax-exempt.
Institution Type/Description: Local History Museum.
Collections: maps; manuscripts; diaries; scrapbooks; pictures.
Research Fields: local history.
Facilities: 175-vol. library of village records from 1858 state & county reference histories, local church histories & general collection of Minnesota state laws available on premises only.

Activities: guided tours; lectures; gallery talks; formally organized education programs for children & adults; training programs; temporary exhibitions; slides.
Publications: The Excelsior Amusement Park; Eureka, Summer 1939; Happenings Around Excelsior; Happenings Around Wayzata; Historic Excelsior; Lake Minnetonka Historic Hotels; Lydia Ferguson Diary; Movies Come to Excelsior; Picturesque Deephaven; Picturesque Minnetonka; A Record of Old Boats; Sweet Sixteen and Then Some; Tales from Tonka; Walking The Trails of History; Hezekiah Brake, Excelsior Pioneer; Plat Map-Excelsior 1898; Map-Lake Minnetonka, 1896.
Hours & Admission Prices: May-Sept. Thurs. 3-6, Sat. 10-3. No charge.
Membership: Senior $10; Family $20; Silver $50; Gold $100.

Fairmont

PIONEER MUSEUM - MARTIN COUNTY HISTORICAL SOCIETY, 304 E. Blue Earth Ave., Fairmont, MN 56031-2865. Tel.: 507-235-5178. Fax: 507-235-5179.
E-mail: mch@frontiernet.net
Web Site: www.fairmont.org/mchs/
Founded: 1929.
Congressional District: 2
Key Personnel: C.E.O., Steve Berkeland; Exec. Dir., Lenny Tvedten; Vice Pres., Wendy Emler; Cur., James Marushin; Conservator, Sandy Nuss.
Personnel Profile: Full-Time Paid 1; Part-Time Paid 3; Part-Time Volunteers 50.
Governing Authority: society. Affiliated with Martin County Historical Society. Tax-exempt: 501(c)(3).
Institution Type/Description: County Historical Society Museum.
Collections: early history of Martin County.
Facilities: 600-vol. library of newspaper files available for research on premises.
Activities: permanent & temporary exhibitions.
Publications: newsletter 3 times a year.
Hours & Admission Prices: Mon.-Fri. 8:30-12 & 1-4:30, Sat.-Sun. special tours & by appointment. No charge; donations accepted. &
Attendance: 2,500 (estimated)
Membership: Copper $25-$49.99; Bronze $50-$249.99; Gold $500-$999.99; Platinum $1,000 & up.

Falcon Heights

GIBBS MUSEUM OF PIONEER AND DAKOTAH LIFE, 2097 W. Larpenteur Ave., Falcon Heights, MN 55113-5313. Mailing Address: Ramsey County Historical Society, 323 Landmark Center, 75 W. 5th St., Saint Paul, MN 55102. Tel.: 651-646-8629 & 659-0345. Fax: 651-659-0345.
E-mail: gibbs@rchs.com
Web Site: www.rchs.com
Formerly: Gibbs Farm Museum
Founded: 1949.
Congressional District: 4
Key Personnel: Interim Dir., John Lindley; Cur. & Archivist, Mollie Spillman.
Personnel Profile: Full-Time Paid 8; Part-Time Paid 15; Part-Time Volunteers 140.
Governing Authority: nonprofit organization. Parent Institution: Ramsey County Historical Society, 75 W. Fifth St. No. 323, St. Paul, MN 55102. Tel.: 612-222-0701. Tax-exempt: 501(c)(3).
Institution Type/Description: household artifacts housed in 1854-1974 Gibbs farmhouse.
Collections: history of the Gibbs family 1835-1940 & the relationship with the Dakota Native Americans; personal & decorative items; woodworking; early-mid 1800s Pioneer & Dakota Indian interpretation. Historical Buildings: 1880s Milan, MN one room schoolhouse; 1887 barn; replica of original soddy; Dakotah tipi & bark lodge.
Research Fields: Dakota Indian; agricultural techniques & family patterns; interpret urban fringe farming.
Facilities: 125-vol. library of local & Minnesota history books available for research; classrooms. Museum-related & handcraft items for sale.
Activities: guided tours; formally organized education programs for children & adults; day camps; docent program or council; loan, permanent, temporary & traveling exhibitions.
Publications: magazine, Ramsey County History; quarterly newsletter, History News & Notes.
Hours & Admission Prices: Memorial Day weekend-Labor Day: Wed.-Sun. 12-4; Sept. & Oct. weekends: 12-4; other times by appointment. Discounts to AAA members & groups for 15 or more; members no charge. &
Attendance: 33,000 (accurate)
Membership: Individual $45; House $50.

Faribault

PARADISE CENTER FOR THE ARTS, 321 Central Ave., Faribault, MN 55021. Tel.: 507-332-7372.
Key Personnel: Exec. Dir., Ryan Heinritz.
Governing Authority: nonprofit organization. Tax-exempt: 501(c)(3).
Institution Type/Description: Art Gallery.
Collections: works by regional artists.
Activities: lectures.
Hours & Admission Prices: Call for hours.

RICE COUNTY MUSEUM OF HISTORY, 1814 N.W. 2nd Ave., Faribault, MN 55021-3033. Tel.: 507-332-2121. Fax: 507-332-2121.
E-mail: rchs@rchistory.org
Web Site: www.rchistory.org
Founded: 1926.
Congressional District: 1
Key Personnel: Exec. Dir., Susan Garwood; Pres. & Chm. (V), Chip DeMann.
Personnel Profile: Full-Time Paid 1; Part-Time Paid 1; Part-Time Volunteers 86; Interns 1.
Governing Authority: Parent Institution: Rice County Historical Society. Subsidiary Institution: Northfield Historical Society, NFLD; Morristown Grist Mill, Morristown; 3R Landmark School, Lonsdale; Dundas Historical Society. Tax-exempt.
Institution Type/Description: History & Agricultural Museum Complex.
Collections: pioneer & Indian artifacts; turn-of-the-century main street; photographs. Historic Buildings: 1853 Alexander Faribault House; Holy Innocents Church; log house; Pleasant Valley rural schoolhouse; Harvest & Heritage Halls.
Research Fields: genealogy; natural & cultural history of Rice County.
Facilities: library; research center.
Activities: guided tours; slide presentation of history of Rice County; video presentation. Museum Sponsors: semi-annual banquets; Heritage Days in June; Rice County Fair in July.
Publications: newsletter, Rice County Historian.
Hours & Admission Prices: Museum: Mon.-Fri. 9-4, Sat. Memorial Day-Labor Day 10-2, other times by appointment. Museum: adults $3, senior citizens 55 & over and students $2, children 12 & under $1; members no charge. Faribault House: admission $2. &
Attendance: 6,000 (estimated)
Membership: Senior Citizen $20; Individual $30; Family $35; Patron $50; Business $200.

Farmington

DAKOTA CITY HERITAGE VILLAGE, Dakota County Fairgrounds, 4008 220th St. W., Farmington, MN 55024. Mailing Address: P.O. Box 73, Farmington, MN 55024-0073. Tel.: 651-460-8050. Fax: 651-463-6908.
E-mail: info@dakotacity.org
Web Site: www.dakotacity.org
Founded: 1979.
Key Personnel: Vice Pres. (V), Pearl Shirley; Museum Shop Mgr., Judy Storlie.
Personnel Profile: Part-Time Paid 2; Part-Time Volunteers 450; Interns 1.
Governing Authority: Parent Institution: Dakota County Agricultural Society. Tax-exempt.
Institution Type/Description: Heritage Village.
Collections: 1900-era rural village includes 24 buildings; period furnishings; agricultural equipment.
Facilities: 5 acre site.
Activities: special events; demonstrations; educational programs; celebrations. Annual Events: Dakota County Fair; WWII Re-enactment Event; Harvest Moon Festival; Village Holidays.
Publications: triannual newsletter, Party Line.
Hours & Admission Prices: May-Sept. Mon.-Sat.; groups of 15 or more by appointment. Adults $5. &
Attendance: 30,000 (estimated)
Membership: Individual $10; Family $15.

Fergus Falls

OTTER TAIL COUNTY HISTORICAL SOCIETY, 1110 Lincoln Ave. W., Fergus Falls, MN 56537-1029. Tel.: 218-736-6038. Fax: 218-739-3075.
E-mail: otchs@prtel.com
Web Site: www.otchs.org/
Founded: 1927.

Congressional District: 7
Key Personnel: Exec. Dir., Chris Schuelke; Cur. Collections, Kathy Evavold; Education Coord., Missy Hermes; Office Mgr., LeAnn Neuleib.
Personnel Profile: Full-Time Paid 3; Part-Time Paid 2; Part-Time Volunteers 100; Interns 1.
Governing Authority: private. Tax-exempt: 501(c)(3).
Institution Type/Description: Historical Society Museum.
Collections: 15,000 artifacts including American Indian artifacts; agriculture; Norwegian; Scandinavian; newspapers; photographs; maps; clippings; oral histories.
Research Fields: Otter Tail County and local history; genealogy.
Facilities: library; archives.
Activities: demonstrations; tours; educational services; annual meeting; genealogy & oral history workshops; history conference; rotating & special exhibits; traveling exhibits; research workshops; cemetery records; historic sites and car-tape tour; volunteer program.
Publications: books, The New Deal at the Grass Roots: Programs for the People in Otter Tail County; Homefires & Battlegrounds-A Military History of Otter Tail County; Old Clitherall's Storybook; O.T. Co. History in Brief; History of Fergus Falls; quarterly newsletter, The Otter Tail County Record; A History of the Pelican Lakes; bimonthly newsletter.
Hours & Admission Prices: June-Aug. Mon.-Fri. 9-5, Sat. 10-3; Sept.-May Mon.-Fri. 9-5. Research Library: Mon.-Fri. 9-5. Adults $3, children 5-11 $1; members no charge. &
Attendance: 9,000 (estimated)
Membership: Basic Individual $20; Supporting Individual or Family $35; Booster $50; Business $50 & up; Sustaining $100; Sponsor $150; Patron $250.

Fountain

FILLMORE COUNTY HISTORY CENTER AND GENEALOGY LIBRARY, 202 County Rd. No. 8, Fountain, MN 55935-8805. Tel.: 507-268-4449.
E-mail: fchc@frontier.com
Web Site: fillmorecountyhistory.wordpress.com
Founded: 1934.
Congressional District: 1
Key Personnel: Dir., Debra J. Richardson; Vice Pres., Flora Grabau; Pres. (V), Donald Boyum; Sec., Rita Joerg; Treas., Kathy Tesmer.
Personnel Profile: Full-Time Paid 1; Part-Time Paid 1; Part-Time Volunteers 15.
Governing Authority: corporation. Tax-exempt.
Institution Type/Description: History Museum.
Collections: 45,000 photographic negatives; agriculture; costumes; rural schoolhouse; late 1800-early 1900 home furnishings; obituary files; Civil War artifacts; archives; 1932 SkyScout airplane; Pietenpol Hangar; log cabin.
Research Fields: local & county history; genealogy.
Facilities: library of historical books & newspapers on microfilm; reading room.
Activities: guided tours; special programs; demonstrations.
Publications: quarterly newsletter, Rural Roots.
Hours & Admission Prices: Tues.-Sat. 9-4. Center: no charge; donations accepted. Library: call for fees. Closed major holidays. &
Attendance: 8,500 (estimated)
Membership: Student $5; Individual $12; Family $20; Associate & Patron $50; Life $200.

Franconia

FRANCONIA SCULPTURE PARK, 29836 St. Croix Trail, Franconia, MN 55074. Tel.: 651-257-6668.
E-mail: info@franconia.org
Web Site: www.franconia.org
Key Personnel: Site Mgr., Bobby Zokaites.
Governing Authority: nonprofit organization. Tax-exempt: 501(c)(3).
Institution Type/Description: Sculpture Park.
Collections: works by local, national & international contemporary artists.
Activities: educational programs; special events.
Hours & Admission Prices: Daily dawn to dusk. No charge.

Freeport

HEMKER PARK & ZOO, County Rd. 39, Freeport, MN 56331. Mailing Address: Box 262, Freeport, MN 56331-0262. Tel.: 320-836-2426. Facebook: Hemker Zoo.
E-mail: info@hemkerzoo.com
Web Site: www.hemkerzoo.com
Key Personnel: Owner, Joan Hemker; Dir., Heidi Roering

Institution Type/Description: Zoo.
Collections: wildlife.
Hours & Admission Prices: May-Oct. daily 10-6. Adults $7.25, seniors 60 & over $6.25, children 2-12 $5.25; infants under 2 no charge.

Fridley

BANFILL-LOCKE CENTER FOR THE ARTS, 6666 E. River Rd., Fridley, MN 55432-4229. Tel.: 763-574-1850. Fax: 763-502-6946.
E-mail: info@banfill-locke.org
Web Site: banfill-locke.org
Institution Type/Description: Art Gallery.
Collections: paintings; sculpture.
Facilities: Gallery-related items for sale.
Activities: educational programs; classes.
Hours & Admission Prices: Tues.-Sat. 10-4. No charge.

SPRINGBROOK NATURE CENTER, 100 85th Ave., N.E., Fridley, MN 55432. Mailing Address: 6431 University Ave., N.E., Fridley, MN 55432-4303. Tel.: 763-572-3588.
Web Site: www.springbrooknaturecenter.org
Founded: 1974.
Key Personnel: Dir., Siah L. St. Clair.
Personnel Profile: Full-Time Paid 3; Part-Time Paid 3; Part-Time Volunteers 25.
Governing Authority: municipal government; nonprofit. Parent Institution: City of Fridley.
Institution Type/Description: Nature Center: a 127-acre natural urban park.
Collections: central Minnesota natural history with emphasis on the Anoka County area.
Hours & Admission Prices: Summer: Mon.-Fri. 9-9, Sat. 10-6, Sun. 12-4; Winter: Mon.-Fri. 10-5, Sat. 9-5, Sun. 12-4. No charge. Closed all holidays. &
Attendance: 50,000 (accurate)

Glenwood

POPE COUNTY HISTORICAL SOCIETY, 809 S. Lakeshore Dr., Glenwood, MN 56334-9406. Tel.: 320-634-3293.
E-mail: popecountymuseum@gmail.com
Web Site: popecountymuseum.com
Founded: 1931.
Congressional District: 7
Key Personnel: Pres. (V), Mary Smith; Vice Pres., Jack Demorett; Dir., Merlin Peterson; Collections Mgr., Ann Grandy.
Personnel Profile: Part-Time Paid 4; Part-Time Volunteers 20.
Governing Authority: society; board of directors. Tax-exempt.
Institution Type/Description: General Museum.
Collections: county artifacts; county history & genealogy; Helbing collection of Indian artifacts; wildlife; communications; period rooms; trapper's cabin; newspapers dating to 1871; church records; census records; photographs; family histories; 1866 Pope County treasurer & assessors record; 1968, count ledgers, Wolf bounty, country schools; record books of organizations in the county; 44,000 biographical files including business, educational, professional & personal.
Research Fields: county genealogy.
Facilities: library of bound volumes of county newspapers from 1877 to present; genealogy library; meeting room; 8,000 sq. ft. exhibit hall; 5 acre campus with historic structures.
Activities: guided tours; lecture series. Museum Sponsors: A Treasure Hunt for Information.
Publications: brochures; newspaper articles; biannual newsletters.
Hours & Admission Prices: Tues.-Sat. 10-5. Adults $3, students 13-18 $1.50, children 6-12 $.50; summer season & members no charge; donations accepted. &
Attendance: 3,560 (accurate)
Membership: Individual $15; Family $30; Lifetime $100.

Grand Marais

COOK COUNTY HISTORICAL MUSEUM, 8 S. Broadway, Grand Marais, MN 55604-1293. Mailing Address: Cook County Historical Society, Box 1293, Grand Marais, MN 55604-1293. Tel.: 218-387-2883 & 9131.
E-mail: history@boreal.org
Web Site: www.cookcountyhistory.org
Founded: 1966.
Congressional District: 8

Key Personnel: C.E.O., Gene Erickson; Dir. Museum, Carrie McHugh; Dir. Johnson Heritage Post, Don Darison.
Personnel Profile: Full-Time Paid 1; Part-Time Paid 2; Part-Time Volunteers 20.
Governing Authority: county; society. Administered by Cook County Historical Society. Tax-exempt.
Institution Type/Description: General Museum & Art Museum: housed in 1896 lighthouse keeper's residence.
Collections: photographs; CCC camps; trapping; logging; textiles; folklore; military; agriculture; sports; Indian artifacts; medical; geology; voyageurs; music; commercial fishing; microfilm newspapers; census; 150 manuscripts on pioneer life; history of schools, churches, organizations, communities & industries; video; Johnson Heritage Post: art & artists on the north shore of Lake Superior.
Research Fields: pioneer life history of communities; obituaries; photographs.
Facilities: 4,000 sq. ft. exhibit space. Pamphlets, local books made by local society for sale.
Activities: permanent exhibitions; school loan service; County Fair exhibits.
Publications: historical pamphlets, Grand Marais; History of Gunflint Trail; books, Pioneer Faces & Places, Pioneers in the Wilderness; Law & Order Cook County 100 Years; Faces & Places II; newsletter, Overlook.
Hours & Admission Prices: June-Oct. Tues.-Sat. 11-4; Nov.-May Fri. 1-4, Sat. 10-2. No charge; donations accepted. &
Attendance: 15,000 (estimated)
Membership: Annual Single $10; Double $15.

Grand Portage

GRAND PORTAGE NATIONAL MONUMENT, 170 Mile Creek Rd., Grand Portage, MN 55605. Mailing Address: P.O. Box 426, Grand Portage, MN 55605-0426. Tel.: 218-475-0123; 218-475-0174.
E-mail: grpo_interpretation@nps.gov
Web Site: www.nps.gov/grpo
Founded: 1958.
Congressional District: 8
Key Personnel: Supt., Tim Cochrane; Chief Interpretation, Pam Neil; Chief Resource Management, Bill Clayton; Park Ranger, Jessica Barr.
Personnel Profile: Full-Time Paid 10; Part-Time Paid 10; Part-Time Volunteers 11; Interns 1.
Governing Authority: federal. Parent Institution: U.S. Dept. of Interior. Subsidiary Institution: National Park Service, Washington, DC 20240. Tax-exempt.
Institution Type/Description: Historical Site & Park Museum: housed in 18th-century reconstructed North West Company fur trading depot on its original site.
Collections: archaeological materials from the site; period fur trade artifacts; Chippewa Indian cultural materials.
Research Fields: exploration & fur trading centered on Grand Portage.
Facilities: 900-vol. library dealing with the fur trade including the exploration of the Canadian Northwest & the Chippewa Indians of the region, periodical collection, maps, charts, pamphlets, articles, slide & photograph collection with contemporary & historical information; nature trails; information center.
Activities: guided tours, audiovisual programs; Chippewa craft demonstrations; historical cooking, baking & musket demonstrations. Annual Event: fur trade rendezvous in August.
Publications: brochures; pamphlets.
Hours & Admission Prices: mid-May to mid-Oct. daily 8:30-5; groups by appointment. No charge; donations accepted. &
Attendance: 70,000 (estimated)

Grand Rapids

CHILDREN'S DISCOVERY MUSEUM, 2727 U.S. Hwy. 169 S., Grand Rapids, MN 55744. Mailing Address: P.O. Box 724, Grand Rapids, MN 55744-0724. Tel.: 218-326-1900; 866-236-5437 (Toll Free). Fax: 218-326-1934.
E-mail: office@cdmkids.org
Web Site: cdmkids.org
Founded: 1994.
Key Personnel: Dir., John A. Kelsch; Pres. (V), Douglas P. Miner.
Personnel Profile: Full-Time Paid 1; Part-Time Paid 5; Part-Time Volunteers 56.
Governing Authority: Tax-exempt.
Institution Type/Description: Children's Museum.
Collections: hands-on exhibits.
Facilities: Museum-related items for sale.
Activities: school service program.
Hours & Admission Prices: April-May Mon.-Sat. 10-5; Memorial Day to Sept.

daily 10-5; Oct.-March Fri.-Sat. 10-5. Admission $8; members no charge. Closed New Year's Day; Easter; Thanksgiving; Christmas. &
Attendance: 18,951 (accurate)
Membership: Youth $20; Adult $30; Family $40; Grandparent $50; Daycare & Resort $100.

* **FOREST HISTORY CENTER, (M),** 2609 County Rd. 76, Grand Rapids, MN 55744-8646. Tel.: 218-327-4482. Fax: 218-327-4715.
E-mail: foresthistory@mnhs.org
Web Site: mnhs.org/foresthistory
Founded: 1978.
Congressional District: 3
Key Personnel: Site Mgr., Jeff Johns; Site Supvr. & Retail Mgr., Becky Jennings.
Personnel Profile: Full-Time Paid 2; Part-Time Paid 18; Part-Time Volunteers 20; Interns 2.
Governing Authority: nonprofit organization. Parent Institution: Minnesota Historical Society. Tax-exempt: 501(c)(3).
Institution Type/Description: Historic Site, Historic Museum, Natural History Museum.
Collections: forestry-related artifacts; logging camp; c.1900 wanigan & batteau used in log drives; cabin; garage; firetower; warehouse.
Research Fields: natural history; human history; agriculture in northern forested areas of the Great Lakes states; Great Lakes environmental history; forest history of Great Lakes states.
Facilities: 750-vol. library; classrooms; 18,000 sq. ft. exhibit space; 30-seat theater; 5 mile nature trail. Forestry & history-related books & gifts for sale.
Activities: guided tours; lectures; films; organized education programs for children, adults & college students affiliated with Bemidji State University & Itasca Community College; docent program; training programs for professional museum workers; participatory & temporary exhibitions; interpretive programs; living history programs.
Hours & Admission Prices: 1st Tues. in June to Labor Day Tues.-Sat. 11-4. Adults $9, senior citizens $7, children 6-17 $6; discounts to AAA & MHS members, active military and students; members and children 5 & under no charge. &
Attendance: 20,000 (accurate)
Membership: Senior Citizen $45; Individual $50; Senior Household & Individual Plus $65; Household $75; Household Plus $95.

ITASCA COUNTY HISTORICAL SOCIETY, (M), 201 N. Pokegama Ave., Grand Rapids, MN 55744. Tel.: 218-326-6431.
E-mail: ichs@paulbunyan.net
Web Site: itascahistorical.org
Founded: 1948.
Congressional District: 8
Key Personnel: C.E.O. & Dir., Lilah J. Crowe; Chm. (V), Esther Hietala; Pres. (V), Leona Litchke; Museum Shop Mgr., Jeremy Anderson.
Personnel Profile: Full-Time Paid 1; Part-Time Paid 1; Part-Time Volunteers 20.
Governing Authority: society; nonprofit. Tax-exempt.
Institution Type/Description: Historical Society Museum.
Collections: history of Itasca County; manuscripts; papermaking; early life & career of Judy Garland.
Research Fields: county history.
Facilities: 700-vol. library of books pertaining to area; reading room. Books of local history & museum-related items for sale.
Activities: films; permanent & temporary exhibitions.
Publications: quarterly newsletter.
Hours & Admission Prices: Mon.-Fri. 9:30-5, Sat. 10-4; Summer: Mon.-Fri. 9:30-5, Sat. 10-4, Sun. 11-4. No charge; donations accepted. Closed national holidays. &
Attendance: 4,000 (accurate)

JUDY GARLAND MUSEUM, 2727 U.S. Hwy. 169 S., Grand Rapids, MN 55744. Mailing Address: P.O. Box 724, Grand Rapids, MN 55744-0724. Tel.: 800-664-5839.
E-mail: jgarland@uslink.net
Web Site: www.judygarlandmuseum.com/
Founded: 1994.
Key Personnel: Exec. Dir., John Kelsch; Pres. (V), Douglas P. Miner.
Personnel Profile: Full-Time Paid 1; Part-Time Paid 5; Part-Time Volunteers 52.
Governing Authority: Tax-exempt.
Institution Type/Description: Historic House Museum.
Collections: personal artifacts; photographs; furnishings; Emerald City carriage used in the Wizard of Oz; Garland's Tony Award. Historic House: Judy Garland's birthplace.

Research Fields: life & times of Judy Garland.
Activities: weddings. Annual Event: Judy Garland Festival in June.
Hours & Admission Prices: April-May Mon.-Sat. 10-5; Memorial Day to Sept. daily 10-5; Oct.-March Fri.-Sat. 10-5. Admission $7; members no charge. Closed New Year's Day; Easter; Thanksgiving; Christmas. ⅃
Attendance: 18,951 (accurate)
Membership: Youth $20; Adult $30; Family $40; Grandparent $50; Get Happy $100.

Granite Falls

YELLOW MEDICINE COUNTY HISTORICAL SOCIETY AND MUSEUM, Junction of Hwy. 67 & 23, Granite Falls, MN 56241. Mailing Address: Box 145, Granite Falls, MN 56241-0145. Tel.: 320-564-4479. Fax: 320-564-4479.
E-mail: ymchs@mvtvwireless.com
Founded: 1937.
Congressional District: 7
Key Personnel: Chm. (V), Jane Remiger; Dir. & Museum Shop Mgr., Brian Schulz; Vice Pres., Ernie Streich.
Personnel Profile: Part-Time Paid 1; Part-Time Volunteers 10.
Governing Authority: society. Parent Institution: Yellow Medicine County Historical Society. Subsidiary Institution: Yellow Medicine Machinery Museum. Tax-exempt: 170(b)(1)(A).
Institution Type/Description: History Museum.
Collections: geology; archaeology; pioneer artifacts; 8,000 year old bison bones. Historic Buildings: 1870 pioneer log cabin used as a general store; 1877 pioneer home; outside chapel.
Research Fields: genealogy
Facilities: 75-vol. library of history books available for use on premises; reading room. State historical publications, local area history books & other museum-related items for sale.
Activities: lectures; permanent & temporary exhibits. Museum Sponsors: special exhibits & demonstrations Memorial Day to Labor Day.
Publications: quarterly newsletter, Yellow Medicine County Heritage; History of Yellow Medicine County, 1914.
Hours & Admission Prices: May-Oct. Wed.-Fri. 9-4, Sat. 10-4; other times by appointment. No charge; donations accepted. ⅃
Attendance: 1,500 (estimated)
Membership: Individual $15; Couple $20; Family $25; Business $100.

Hanley Falls

MINNESOTA'S MACHINERY MUSEUM, 100 N. 1st St., Hanley Falls, MN 56245. Mailing Address: P.O. Box 70, Hanley Falls, MN 56245-0070. Tel.: 507-768-3522. Fax: 507-768-3522. Facebook: Minnesota's Machinery Museum.
E-mail: agmuseum@frontiernet.net
Founded: 1980.
Congressional District: 20
Key Personnel: Dir., Laurie Johnson; Chm. (V), Tom Offedahl.
Personnel Profile: Part-Time Paid 4.
Volunteer Hours: 115
Operating Expenses: 31,863
Operating Income: 36,415
Governing Authority: Parent Institution: Yellow Medicine County. Tax-exempt.
Institution Type/Description: Agriculture & Transportation Museum.
Collections: farm machinery; farm life history; period automobiles; railroad memorabilia; miniature & toy farm machinery; rural art; farm house rooms; quilts; tractors.
Major Exhibits: Threshing Show, Aug. 2 & 3 2014.
Facilities: Museum-related items for sale.
Activities: guided tours; school programs; demonstrations.
Hours & Admission Prices: May 15-Sept. 30 Mon.-Sat. 10-4, Sun. 1-4:30; other times by appointment. No charge; donations accepted. Closed holidays. ⅃
Attendance: 1,200 (estimated)
Membership: Individual $10; Sustaining $100; Corporate $250-$500.

Hastings

CARPENTER ST. CROIX VALLEY NATURE CENTER, 12805 St. Croix Trail, Hastings, MN 55033-9499. Tel.: 651-437-4359. Fax: 651-438-2908.
Founded: 1969.
Key Personnel: Exec. Dir., Jennifer Vieth; Pres. (V), Vickie Batroot; Museum Shop Mgr., Alan Maloney.
Personnel Profile: Full-Time Paid 7; Part-Time Paid 2; Part-Time Volunteers 100.

Governing Authority: Parent Institution: Thomas & Edna Carpenter Foundation; Subsidiary Institution: Carpenter St. Croix Valley Nature Center. Tax-exempt.
Institution Type/Description: Nature Center
Collections: wildlife habitat; native wildlife & plants.
Research Fields: bird banding; water quality monitoring.
Facilities: hiking trails.
Activities: birthday parties; educational programs; scout programs.
Publications: newsletter, St. Croix Current.
Hours & Admission Prices: Daily 8-4:30. No charge; donations accepted. Closed New Year's Day; Thanksgiving; Christmas. ⅃
Attendance: 25,000 (estimated)
Membership: Individual $25 & up.

Henderson

SIBLEY COUNTY HISTORICAL MUSEUM, 700 Main St. W., Henderson, MN 56044-7711. Mailing Address: P.O. Box 407, Henderson, MN 56044-0407. Tel.: 507-248-3434 & 3818.
Web Site: history.sibley.mn.us
Founded: 1940.
Congressional District: 2
Key Personnel: Chm., Jerome Petersen.
Personnel Profile: Part-Time Volunteers 30.
Governing Authority: society. Tax-exempt.
Institution Type/Description: History Museum: housed in 1884 residence.
Collections: agriculture; period furnishings; industrial; Sibley County, MN family histories. Historic Building: 1860 log cabin.
Research Fields: Joseph R. Brown Cemetery; Sibley County cemeteries; Sibley County family histories; microfilm census; naturalization; Sibley County newspapers & microfilms.
Facilities: library of bibles, maps, county newspapers & photo albums.
Activities: guided tours; formally organized education programs for children.
Publications: quarterly, Sibley County Historical Society Newsletter; books, Henderson to Fort Ridgely Trail: Bits and Pieces, A History of Sibley County; Boys in Blue, 1898-1902; Sibley County Plat maps, 1855-1997; Tales of the 10th Regiment Minnesota Volunteers, 1862-1863; Henderson Then & Now, 1852-1994; Joseph R. Brown and His Times; A History of St. Brendans Parish, The Village of Green Isle and First Irish Settlement; Barns of Sibley County.
Hours & Admission Prices: June-Oct. Sun. 2-5; other times by appointment. Adults $2, children $1; members no charge. ⅃
Attendance: 1,200 (estimated)
Membership: Annual $20; Family $25; Business $50.

Hendricks

LINCOLN COUNTY PIONEER MUSEUM, 610 W. Elm, Hendricks, MN 56136. Mailing Address: 406 S. Brook St., Hendricks, MN 56136. Tel.: 507-275-3537.
Founded: 1969.
Congressional District: 3
Key Personnel: Pres., Allen S. Johnson; Vice Pres. & Treas., Dr. Rolland Digre; Sec., Mr. Trygue Trooien.
Governing Authority: county; society. Tax-exempt.
Institution Type/Description: Historical Society Museum: housed in c.1899 train depot & Icelandic church.
Collections: train depot; farm machinery & tools; furniture; period military uniforms; country schoolhouse; an old opera house room; Lincoln County Courthouse artifacts; millinery shop; dolls; fifteen mannequins dressed in clothes from early 1900s; medical instruments; toys; buggies. Historic Buildings: 1918-1919 restored Sears Roebuck house; 1879 Icelandic Church.
Activities: guided tours.
Hours & Admission Prices: Wed.-Fri. 2-5 & 7-9, Sun. 2-5; other times by appointment. No charge; donations accepted. ⅃
Attendance: 600 (estimated)

Hibbing

HIBBING HISTORICAL SOCIETY AND MUSEUM, 400 E. 23rd St., Hibbing, MN 55746-1923. Tel.: 218-263-8522.
E-mail: hibbinghistory@mchsi.com
Web Site: www.hibbinghistory.com
Founded: 1976.
Congressional District: 8
Key Personnel: Pres. (V), Pru Lolich; Vice Pres., Leonard Hirsch.
Personnel Profile: Part-Time Paid 1; Part-Time Volunteers 4.
Volunteer Hours: 850
Operating Expenses: 23,900

Operating Income: 19,200
Governing Authority: society; nonprofit organization. Operated by Hibbing Historical Society. Parent Institution: First Settlers' Association. Affiliated with St. Louis County Historical Society, Duluth, MN 55802. Tax-exempt: 501(c)(3).
Institution Type/Description: History Museum: housed in 1936 Memorial Building.
Collections: children's artifacts; mining & mining tools; logging; paintings; clothing; research materials on local history; photos; Greyhound Bus history; manuscripts; photographs.
Research Fields: local history; genealogy.
Facilities: library of history books available for research on premises. Books, items of the area & postcards for sale.
Activities: guided tours; films; permanent & temporary exhibitions; outreach exhibits.
Publications: quarterly newsletter.
Hours & Admission Prices: Tues.-Fri. 10-2; other times by appointment. No charge; donations accepted. &
Attendance: 1,000 (estimated)
Membership: Senior Citizen $10; Individual $15; Family $30; Business $50; Sponsorship $100.

Hinckley

HINCKLEY FIRE MUSEUM, 106 Old Hwy. 61 S., Hinckley, MN 55037. Mailing Address: P.O. Box 40, Hinckley, MN 55037-0040. Tel.: 320-384-7338.
E-mail: info@hinckleyfiremuseum.com
Web Site: www.hinckleyfiremuseum.com
Founded: 1976.
Congressional District: 8
Key Personnel: Tour Guide & Museum Shop Mgr., Sandy Hinds; Pres. (V), Steve Johnson.
Personnel Profile: Full-Time Paid 1; Part-Time Volunteers 12.
Volunteer Hours: 800
Governing Authority: Tax-exempt.
Institution Type/Description: History Museum: housed in 1894, St. Paul & Duluth Railroad Depot.
Collections: artifacts; photographs; maps; newspapers pertinent to fire of Sept. 1894; household furnishings; depot agent's apartment; late 19th-century clothing.
Research Fields: 1837-1894 planning survey of logging operations; Ojibwe Indians in area; state relief effort; fire ecology.
Facilities: 100-seat auditorium; theater. Museum-related items for sale.
Activities: guided tours; lectures; films; concerts; arts festivals; drama; permanent & temporary exhibitions; interactive children's exhibit.
Publications: newsletter.
Hours & Admission Prices: May to mid-Oct. Tues.-Sun. 10-5; July-Aug. daily 10-5. Adults $5, senior citizens $4, students 13-18 $2, children 6-12 $1; members & children under 6 no charge.
Attendance: 7,000 (estimated)
Membership: Senior Citizen $10; Individual $15; Family $25; Patron & Business $50; Contributor $100 and up; Patron Business $100; Contributing Business $250 and up.

Hopkins

HOPKINS HISTORICAL SOCIETY MUSEUM, 33 14th Ave. N., Hopkins, MN 55343. Mailing Address: 1010 First St. S., Hopkins, MN 55343. Tel.: 952-548-6480.
Institution Type/Description: Historical Society Museum.
Collections: local history & culture; photographs; period furnishings; personal artifacts.
Hours & Admission Prices: Tues. & Thurs. 9am to noon; other times by appointment.

Hutchinson

MCLEOD COUNTY HISTORICAL SOCIETY, (M), 380 School Rd., N.W., Hutchinson, MN 55350-1430. Tel.: 320-587-2109.
E-mail: asa@hutchtel.net
Web Site: www.mcleodhistory.org/index.htm
Founded: 1940.
Congressional District: 2
Key Personnel: Exec. Dir., Lori Pickell-Stangel; Chm. (V) & Pres. (V), Stan Ehrke; Museum Asst., Peggy Paulson.
Personnel Profile: Full-Time Paid 1; Full-Time Volunteers 2; Part-Time Paid 1; Part-Time Volunteers 25; Interns 1.
Institution Type/Description: Historical Society Museum.
Collections: county history & culture; photographs; documents; art.

Research Fields: newsletter, McLeod Historical Society; historical booklets.
Facilities: library.
Activities: workshops.
Hours & Admission Prices: Mon. & Thurs.-Fri. 10-4, Sat. 1-4. Adults $3, seniors $2, students $1; members no charge. Closed major holidays. &
Attendance: 4,000 (estimated)
Membership: Individual $20; Family $35; Business $200; Lifetime $500.

International Falls

KOOCHICHING COUNTY HISTORICAL MUSEUM, (M), 214 6th Ave., International Falls, MN 56649-2336. Tel.: 218-283-4316. Fax: 218-283-2843.
Founded: 1958.
Congressional District: 8
Key Personnel: C.E.O., Edgar S. Oerichbauer; Pres. (V), Mike Williams.
Personnel Profile: Full-Time Paid 1; Part-Time Paid 5; Part-Time Volunteers 10.
Governing Authority: society. Tax-exempt: 501(c)(3).
Institution Type/Description: History Museum.
Collections: period items; artifacts; paintings; photographs; manuscripts; industry & history of Koochiching County.
Research Fields: logging & lumber; local history; Minnesota & Ontario Pulp and Paper Co.; International Lumber Co.
Facilities: library; 6,600 sq. ft. exhibit space.
Activities: meetings of the society; special events.
Publications: 8 times a year newsletter, Koochiching Chronicles.
Hours & Admission Prices: Mon.-Fri. 9-5. Adults $4, student 5-17 $2; discounts to groups; children under 5 & members no charge. &
Attendance: 5,000 (estimated)
Membership: Individual $15; Family $20; Contributing $50, $100, $250, $500.

VOYAGEURS NATIONAL PARK, 360 Hwy. 11 E., International Falls, MN 56649-8802. Tel.: 218-283-6600 (Park Headquarters).
Web Site: www.nps.gov/voya
Key Personnel: Supt., Michael M. Ward
Institution Type/Description: National Park Visitor Centers.
Collections: local history & culture.
Facilities: nature trails.
Activities: hiking.
Hours & Admission Prices: Park: year-round dawn-dusk. Visitor Centers: call for hours. No charge.

Kellogg

LARK TOYS & CAROUSEL, 171 Lark Lane, Kellogg, MN 55945-9629. Tel.: 507-767-3387. Fax: 507-767-4565.
E-mail: lark@wabasha.net
Web Site: larktoys.com
Key Personnel: Owner, Ron Gray; Owner, Kathy Gray; Owner, Scott Gray-Burlingham; Owner, Miranda Gray-Burlingham
Institution Type/Description: Toy Museum.
Collections: hand-carved carousel; costumes; period toys.
Hours & Admission Prices: Daily 9:30-5:30. Closed New Year's Day; Thanksgiving; Christmas.

Kenyon

GUNDERSON HOUSE, 101 Gunderson Blvd., Kenyon, MN 55946. Mailing Address: 510 4th St., Kenyon, MN 55946-1117. Tel.: 507-789-5936.
Founded: 1895.
Congressional District: 25
Key Personnel: Pres., Lois Easton.
Governing Authority: municipal. Operated by the Kenyon Area Historical Society. Tax-exempt.
Institution Type/Description: Historic House: 1895 Gunderson House. Listed on the National Register of Historic Places.
Collections: Victorian architecture; furnishings.
Research Fields: state & county history.
Activities: guided tours.
Hours & Admission Prices: 3rd Sat. of month. Adults $3, students $1; preschool no charge.
Attendance: 350
Membership: Senior Citizen $3; Individual $5; Family $7.50; Business $25; Life $100.

Lake Bronson

KITTSON COUNTY HISTORY CENTER MUSEUM, 332 E. Main St., Lake Bronson, MN 56734-3448. Tel.: 218-754-4100.
E-mail: history@wiktel.com
Founded: 1973.
Congressional District: 7
Key Personnel: Dir., Cindy Adams; Pres. (V), Dan Nordine; Business Officer, Melroe Gunnarson; Treas., Cecil Fossell.
Personnel Profile: Full-Time Paid 1; Part-Time Paid 2; Part-Time Volunteers 14.
Governing Authority: society; nonprofit organization. Tax-exempt: 501(c)(3).
Institution Type/Description: Historical Society Museum.
Collections: settlement of county from 1880-present.
Research Fields: local history.
Facilities: 1,000-vol. library of local, county & state information available for research by special permission from director; reading room; microfilm reader-printer with 172 microfilms containing local newspapers.
Activities: films; permanent & traveling exhibits; fundraising programs.
Publications: annual newsletter; annual calendar.
Hours & Admission Prices: April-May & Sept.-Nov. Mon.-Fri. 9-5; Memorial Day to Labor Day Mon.-Fri. 9-5, Sat.-Sun. & holidays 1-5; Dec.-March Tues.-Fri. 9-5. No charge; donations accepted.
Attendance: 4,000 (estimated)
Membership: Individual $15; Family $25.

Lakefield

JACKSON COUNTY HISTORICAL MUSEUM, 307 N. Hwy. 86, Lakefield, MN 56150-1259. Mailing Address: P.O. Box 238, Lakefield, MN 56150-0238. Tel.: 507-662-5505. Fax: 507-662-5505.
E-mail: jchs@frontiernet.net
Founded: 1931.
Congressional District: 22
Key Personnel: Pres., Mark Titus; Treas., John Hay; Mgr., Mike Kirchmeier.
Personnel Profile: Full-Time Paid 1; Part-Time Paid 2; Part-Time Volunteers 12.
Governing Authority: society. Member of the Minnesota State Historical Society. Parent Institution: Jackson County Historical Society. Tax-exempt.
Institution Type/Description: Local History Museum.
Collections: county newspapers; organizational records; family biographical files; Jackson County history; obituary files; cemetery records; photographs.
Research Fields: genealogy.
Facilities: 150-vol. research library.
Activities: local history tours.
Publications: quarterly newsletter; Jackson County History Vol. I, (1850-1910); Jackson County History Vol. II, (1910-1978).
Hours & Admission Prices: Mon.-Fri. 9:30-4:30, Sat. 8am to noon; other times by appointment. No charge; donations accepted.
Attendance: 400 (estimated)
Membership: Single $20; Family $30; Business $50.

Lanesboro

LANESBORO ARTS CENTER, 103 Parkway Ave. N., Lanesboro, MN 55949-0152. Mailing Address: P.O. Box 152, Lanesboro, MN 55949-0152. Tel.: 507-467-2446.
E-mail: info@lanesboroarts.org
Web Site: lanesboroarts.org
Formerly: Cornucopia Art Center
Founded: 1980.
Key Personnel: Exec. Dir., John Davis.
Personnel Profile: Full-Time Paid 3; Part-Time Paid 4; Part-Time Volunteers 150; Interns 1.
Institution Type/Description: Art Gallery.
Collections: paintings; sculpture; ceramics; glass; wood; fiber; metal; jewelry.
Facilities: Gallery-related items for sale.
Activities: educational programs; classes; workshops; temporary exhibitions; special events.
Hours & Admission Prices: Mon.-Thurs. 10-5, Fri.-Sat. 10-7, Sun. 11-3.

Le Sueur

LE SUEUR MUSEUM, 709 N. 2nd St., Le Sueur, MN 56058-1411. Tel.: 507-665-2050.
Web Site: lesueurcountyhistory.org/
Founded: 1975.
Congressional District: 1
Key Personnel: Pres. (V), Jean Haas.

Personnel Profile: Part-Time Paid 1; Part-Time Volunteers 12.
Governing Authority: county. Parent Institution: Le Sueur County Historical Society, P.O. Box 123, Le Center, MN 56057. Tax-exempt: 170(b)(1)(A).
Institution Type/Description: History Museum: housed in first school in Le Sueur.
Collections: genealogy; etchings of George T. Plowman; Green Giant Canning Co. records; photographs; canning history; history of veterinary medicine; war memorabilia from 1862 to present; 75 radios; early pharmacy.
Research Fields: genealogy; canning industry; local history; military.
Facilities: library of books on local history & genealogy available for research on premises only; reading room.
Activities: guided tours; lectures; films; loan & permanent exhibitions.
Publications: book, Le Sueur, Town on the River; brochure.
Hours & Admission Prices: Memorial Day to Labor Day Fri.-Sat. call for hours; other times by appointment. No charge; donations accepted. Closed holidays.
Attendance: 500 (estimated)
Membership: Le Sueur County Historical Society: Individual $20; Family $30.

Lino Lakes

WARGO NATURE CENTER, 7701 Main St., Lino Lakes, MN 55038-8741. Tel.: 651-429-8007. Fax: 651-429-8167.
Web Site: www.anokacountyparks.com/qlinks/wargonc/wargonc.htm
Key Personnel: Mgr., Lisa Gilliland; Program Coord., Deb Gallop; Recreation Specialist, Todd Murawski; Scheduler & Receptionist, Rhonda Lynch
Institution Type/Description: Nature Center.
Collections: native plant & animal life.
Hours & Admission Prices: April-Oct. Tues.-Fri. 8-4:30, Sat. 9-5, Sun. 12-5.

Litchfield

MEEKER COUNTY HISTORICAL SOCIETY MUSEUM & G.A.R. HALL, 308 N. Marshall Ave., Litchfield, MN 55355-2112. Tel.: 320-693-8911.
E-mail: mchsgar@hutchtel.net
Web Site: www.garminnesota.org
Founded: 1885.
Congressional District: 6
Key Personnel: Dir., James Milan; Pres., August Anderson.
Personnel Profile: Part-Time Paid 3; Part-Time Volunteers 3.
Governing Authority: society; nonprofit organization. Parent Institution: Minnesota Historical Society. Tax-exempt: 501(c)(3).
Institution Type/Description: Historic Building: c.1885 G.A.R. Hall.
Collections: firearms; Civil War items; crafts; clothing; photographs; Native American artifacts; clocks; log cabin; newspapers; tools; blacksmith shop; genealogy archives; toys.
Research Fields: Dakota War of 1862; Sioux Uprising; Grand Army of the Republic.
Facilities: library of Civil War files & research books; genealogical research information.
Activities: guided tours; special speakers; social events.
Publications: quarterly newsletter, Meeker County Historical Society.
Hours & Admission Prices: Tues.-Sun. 12-4. Adults $2; children & members no charge. Closed holidays except Memorial Day.
Attendance: 4,000
Membership: General $10; Business & Sustaining $25; Patron $125; Benefactor $500.

Little Falls

✻ CHARLES A. LINDBERGH HISTORIC SITE, (M), 1620 Lindbergh Dr. S., Little Falls, MN 56345. Tel.: 320-616-5421. Fax: 320-616-5423. Facebook: Lindbergh house.
E-mail: lindbergh@mnhs.org
Web Site: www.mnhs.org/lindbergh
Congressional District: 8
Key Personnel: Mgr., Melissa Peterson.
Personnel Profile: Full-Time Paid 1; Part-Time Paid 8; Part-Time Volunteers 4.
Governing Authority: state government. Parent Institution: Minnesota Historical Society, St. Paul, MN. Tax-exempt: 501(c)(3).
Institution Type/Description: Historic House Museum: housed in boyhood home of Charles A. Lindbergh, 1902-1920.
Collections: personal artifacts; furnishings; films.
Research Fields: Lindbergh's aviation, 1924-1927; Lindbergh's involvement in the antiwar effort, 1939-1940; general aviation history 1903-1940; Lindbergh's Pacific war experience, 1944; Lindbergh's environmental work 1960-1974.
Facilities: library; classroom; 2,500 sq. ft. exhibit space; 50-seat theater; visitor's center. Museum-related items for sale.

Activities: guided tours; films; hobby workshops; lectures; participatory & traveling exhibitions; theater; broadcast programs. Special Events: Children's Day; Family Fun Day; Air Show; Film Festivals.
Hours & Admission Prices: Memorial Day to Labor Day Thurs.-Sat. 10-5, Sun. 12-5; other times by appointment. Adults $8, senior citizens $7, students & children $6; discounts to groups; members no charge. &
Attendance: 10,000 (accurate)

CHARLES A. WEYERHAEUSER MEMORIAL MUSEUM, (M), 2151 S. Lindbergh Dr., Little Falls, MN 56345-0239. Mailing Address: P.O. Box 239, Little Falls, MN 56345-0239. Tel.: 320-632-4007.
E-mail: contactstaff@morrisoncountyhistory.org
Web Site: www.morrisoncountyhistory.org
Founded: 1936.
Congressional District: 7
Key Personnel: Dir., Jan Warner; Pres. (V), A. Arthur Warner; Gen. Mgr., Mary Warner; Collections Mgr., Ann Marie Johnson; Treas., Duane Welle.
Personnel Profile: Full-Time Paid 1; Part-Time Paid 3.
Governing Authority: society. Parent Institution: Morrison County Historical Society. Tax-exempt: 501(c)(3).
Institution Type/Description: Local History Museum.
Collections: county history artifacts; manuscripts; books; photos; archives.
Research Fields: genealogy; local history.
Facilities: 1,500-vol. library of biographies & early history books of Morrison County, Little Falls, local schools & churches, early pioneers & Indians; 1,900 sq. ft. exhibit space; 50-seat gathering space; nature area. Books & museum-related items for sale.
Activities: guided tours; films; permanent & temporary exhibitions; monthly board of directors meeting; history events; workshops; demonstrations; educational services. Museum Sponsors: Annual Meeting in September or October.
Publications: quarterly newsletter, Morrison County Historical Society Newsletter, "Little Falls on the Big River"; book, A Big Hearted Paleface Man: Nathan Richardson & the History of Morrison County, Minnesota.
Hours & Admission Prices: Tues.-Sat. 10-5; groups by appointment. No charge; donations accepted. Closed holidays. &
Attendance: 3,500 (estimated)
Membership: Individual $20; Family $25; Donor $50; Supporting $100; Patron $500; Benefactor $1,000 and up.

MINNESOTA FISHING MUSEUM AND EDUCATION CENTER, 304 W. Broadway, Little Falls, MN 56345-1535. Tel.: 320-616-2011.
E-mail: mnfm@mnfishingmuseum.com
Web Site: www.mnfishingmuseum.com/
Founded: 1998.
Key Personnel: Dir., Mavis Buker; Pres. (V), Jeff Doty.
Personnel Profile: Full-Time Paid 1; Part-Time Paid 1.
Governing Authority: Tax-exempt.
Institution Type/Description: Fishing Museum.
Collections: over 10,000 fishing artifacts.
Hours & Admission Prices: April 15-Sept. Tues.-Sat. 10-5, Sun. 12-4; Winter: call for hours. Adults $5, seniors 60 & over and students 6-17 $4; discounts to groups; members & children under 6 no charge. &
Attendance: 5,000 (accurate)
Membership: Individual $15; Family $25. Business & Organization: $100-$1,000.

Long Lake

WESTERN HENNEPIN COUNTY PIONEERS ASSOCIATION, 1953 W. Wayzata Blvd., Long Lake, MN 55356-9362. Mailing Address: Box 332, Long Lake, MN 55356-0332. Tel.: 952-473-6557.
E-mail: pioneer_museum@hotmail.com
Web Site: www.whcpa-museum.org
Founded: 1907.
Congressional District: 3
Key Personnel: Pres., Steve Kelley; Sec., Marion Merz; Treas., Lucia Stubbs; Cur., Russ Ferrin.
Governing Authority: private; board of directors. Tax-exempt.
Institution Type/Description: History Museum.
Collections: area transportation, records, tapes, clothing, tools, & pictures; Native American artifacts; farming; local newspaper files; scrap books; family histories; pictures; maps.
Facilities: records of local business & organizations available for research.
Activities: guided tours; lectures; museum programs.
Publications: quarterly newsletter.

Hours & Admission Prices: Museum: Sat. 10-2. Research Center: Sat. 10-1. No charge; donations accepted.
Attendance: 1,400 (estimated)
Membership: Individual $15; Family $20; Business $55.

Long Prairie

THE CHRISTIE HOUSE MUSEUM, 15 1st St., S., Long Prairie, MN 56347-1348. Mailing Address: c/o Chamber of Commerce, 42 N. 3rd St., Long Prairie, MN 56347. Tel.: 320-732-2514.
E-mail: chamber@longprairie.org
Institution Type/Description: Historic House: former home of Dr. George R. Christie. Listed on the National Register of Historic Places.
Collections: photographs; personal artifacts; furnishings.
Hours & Admission Prices: Memorial Day-Labor Day Wed.-Sun. 1:30-4:30.

TODD COUNTY HISTORICAL MUSEUM, 333 Central Ave., Long Prairie, MN 56347-1304. Tel.: 320-732-4426.
E-mail: tchmuseum@gmail.com
Founded: 1928.
Key Personnel: Dir. & Pres. (V), Shirley Lunceford; Cur., De Eberle
Institution Type/Description: History Museum.
Collections: local history & culture relating to Todd County.
Publications: quarterly newsletter.
Hours & Admission Prices: Wed. 11:30-6, Thurs.-Fri. 11:30-4. Adults $3; members no charge. &
Membership: Senior $7.50; Individual $10; Life $100.

Madison

LAC QUI PARLE COUNTY HISTORICAL SOCIETY, 250 8th Ave., S., Madison, MN 56256-1146. Tel.: 612-598-7678.
Founded: 1948.
Congressional District: 2
Key Personnel: C.E.O. & Pres. (V), Scotty Kuehl; Cur., Janet E. Liebl.
Personnel Profile: Part-Time Paid 2; Part-Time Volunteers 1.
Governing Authority: nonprofit corporation. Parent Institution: County Historical Society. Tax-exempt.
Institution Type/Description: General Museum.
Collections: Pioneer & Native American artifacts; manuscripts; military; textiles; Civil War; vehicles; tools; 300 personality dolls; salt & pepper shakers; mounted big game, wildlife and flora & fauna; genealogy; obituary index. Historic Buildings: 1870 log cabin with furnishings; c.1887 rural schoolhouse with equipment; machine shed with early farm machinery; poet Robert Bly studio, used from 1950s-1970s.
Research Fields: county family history.
Facilities: 1,500-vol. library of county & pioneer history books available for use on premises.
Activities: guided tours; lectures; informally organized education programs for children.
Publications: newsletter, Lake Talks.
Hours & Admission Prices: May-Oct. Mon.-Fri. 11-4, Sat. 11-3. Adults $3. Family History Research: $10 per hour. &
Attendance: 2,500 (estimated)
Membership: Individual $8; Family $15; Life $100.

Mankato

BLUE EARTH COUNTY HISTORICAL SOCIETY, 415 Cherry St., Mankato, MN 56001-3741. Tel.: 507-345-5566. Facebook: bechshistory.
E-mail: bechs@hickorytech.net
Web Site: www.bechshistory.com
Founded: 1901.
Congressional District: 1
Key Personnel: C.E.O. & Exec. Dir., Jessica Potter; Pres. (V), Randy Zellmer.
Personnel Profile: Full-Time Paid 4; Part-Time Paid 2; Part-Time Volunteers 100; Interns 5.
Volunteer Hours: 7,000
Operating Expenses: 179,000
Operating Income: 193,000
Governing Authority: society. Tax-exempt.
Institution Type/Description: History Museum.
Collections: American Indian & settlement artifacts; household items; furniture; clothing & textiles; fine & decorative arts; prints; photographs; transportation & local history artifacts; family history & research paper archives. Historic Houses: 1871 restored mansion & carriage house.
Research Fields: Blue Earth County prehistory & history.

Facilities: archives of cataloged books, letters, pamphlets, documents, memoirs, photographs, atlases, manuscripts, indexed scrapbooks, bound newspapers & vertical file materials of local history available for use on premises; reading room; formal Victorian gardens & herb garden. Historical books for sale.

Activities: guided tours; lectures; formally organized education programs for children; permanent & temporary exhibitions. Museum Sponsors: Ghosts From the Past; Christmas Open House; genealogy workshops; historical social events.

Publications: quarterly research journal, Historian; newsletters; genealogical data & original sources; county history booklet.

Hours & Admission Prices: Tues.-Sat. 10-4. Adult $5; members no charge. Closed legal holidays. &

Attendance: 20,000 (estimated)

Membership: Student & Individual Senior $20; Individual $25; Household $35; Contributor $50; Supporter $100; Business $100-$1,000; Booster $250; Advocate $500; Benefactor $1,000.

CHILDREN'S MUSEUM OF SOUTHERN MINNESOTA, 121 E. Cherry St., Mankato, MN 56002. Mailing Address: P.O. Box 3103, Mankato, MN 56002. Tel.: 507-386-0279. Fax: 507-388-1836.

E-mail: info@cmsouthernmn.org

Web Site: www.cmsouthernmn.org

Key Personnel: Exec. Dir., Peter Olson

Institution Type/Description: Children's Museum.

Collections: hands-on exhibitions.

Activities: rental facilities.

Hours & Admission Prices: Wed. 9-1, Thurs. & Sat. 9-5, Sun. 11-5, 1st Fri. each month 9-8. Admission one & over $5.

Membership: Family $50.

Mantorville

DODGE COUNTY HISTORICAL SOCIETY, 615 N. Main St., Mantorville, MN 55955-6129. Mailing Address: P.O. Box 456, Mantorville, MN 55955-0456. Tel.: 507-635-5508.

E-mail: dchs@kmtel.com

Web Site: www.dodgecohistorical.addr.com

Founded: 1876.

Congressional District: 32

Key Personnel: Pres. (V), Sue Harwood; Vice Pres. & Acting Treas., Bob Peterson; Sec., Vicki Peterson.

Personnel Profile: Part-Time Paid 1.

Governing Authority: county. Tax-exempt: 170(b)(1)(A).

Institution Type/Description: Historic Buildings & Site: Museum housed in 1869 former Episcopal church.

Collections: Civil War recruiting station; agriculture; paintings; costumes; military; preservation project; Indian artifacts; horse drawn hearse. Historic Houses: c.1860 Civil War recruiting station; 1858 Wasioja village limestone schoolhouse; 1883 country one-room schoolhouse; 1856 frame house.

Facilities: 300-vol. reference library available to the public on-site for educational & historical research; genealogy archives & newspapers.

Activities: guided tours.

Hours & Admission Prices: May 1-Oct. 15 Tues.-Sat. 10-4; Oct. 16-April 26 Thurs.-Sat. 10-4; other times by appointment for tours. Adults 13 & over $4, children 6-12 $2, seniors $1.50; members & children 5 and no charge. Research: hourly fee.

Attendance: 2,500 (estimated)

Membership: One Year $10; 5 Year $35; Lifetime $250.

MANTORVILLE RESTORATION, 407 Main St., Mantorville, MN 55955-6127. Mailing Address: P.O. Box 202, Mantorville, MN 55955-0311. Tel.: 507-269-8704.

Founded: 1963.

Congressional District: 1

Governing Authority: nonprofit organization. Parent Institution: Mantorville Restoration Assoc. Tax-exempt: 170(b)(1)(A).

Institution Type/Description: Preservation Project: 1856 Temporary Court House & Home.

Collections: mid 19th-century furniture, dishes, utensils, furnishings; historic buildings.

Research Fields: local history & buildings.

Facilities: library of local history books.

Activities: self-guided & guided tours; theater productions; concerts; special events; festivals. Museum Sponsors: Old Tyme Days; Marigold Days; Mulligan Stew.

Publications: tourism & tour guides.

Hours & Admission Prices: May-Oct. Tues.-Sun. 1-5. Requested Donation $2.

Attendance: 25,000 (estimated)

Membership: Single $5; Sustaining $10; Associate $25; Contributing & Institutional $50; Benefactor $500.

Maplewood

MAPLEWOOD NATURE CENTER, 2659 E. 7th St., Maplewood, MN 55119-3815. Tel.: 651-249-2170. Fax: 651-249-2189.

Web Site: www.ci.maplewood.mn.us

Founded: 1979.

Congressional District: 4

Key Personnel: Lead Naturalist, Ann Hutchinson.

Personnel Profile: Full-Time Paid 1; Part-Time Paid 6.

Governing Authority: Parent Institution: City of Maplewood. Subsidiary Institution: Friends of Maplewood Nature.

Institution Type/Description: Nature Center.

Collections: wildlife taxidermy; skins & skulls.

Facilities: library; nature trails; visitor center.

Publications: quarterly newsletter, Maplewood Seasons; Friends & Volunteer email newsletter.

Hours & Admission Prices: Visitor Center: Tues.-Sat. 8:30-4:30. No charge.

Attendance: 9,000 (accurate)

Marshall

LYON COUNTY HISTORICAL SOCIETY MUSEUM, 301 W. Lyon St., Marshall, MN 56258-1307. Tel.: 507-537-6580. Fax: 507-537-7699.

E-mail: lyoncomuseum@iw.net

Web Site: www.lyoncomuseum.org

Founded: 1934.

Key Personnel: Pres., Neal Ingbrightson; Vice Pres., Kathy Lozinski; Dir., Jennifer Andries; Sec. & Treas., Mary Becker.

Governing Authority: county. Tax-exempt.

Institution Type/Description: General Museum.

Collections: pioneer relics; Indian artifacts; geology; natural history; clocks; 1950s ice-cream/soda fountain, kitchen & other displays; history of Lyon County.

Research Fields: preservation of township & church records.

Facilities: library.

Activities: guided tours.

Publications: annual, The Lyon Tale; quarterly, newsletter.

Hours & Admission Prices: Feb.-Dec. Tues.-Fri. 10-5, Sat. 12-4. No charge; donations accepted. Closed national holidays. &

Membership: Individual $25; Couple/Family $35; Business $100; Benefactor $300

MUSEUM OF NATURAL HISTORY, Southwest Minnesota State University, 1501 State St., Marshall, MN 56258-3306. Tel.: 507-537-6178. Fax: 507-537-6151.

E-mail: betsy.desy@smsu.edu

Web Site: www.smsu.edu/campuslife/attractions/naturalhistorymuseum/

Founded: 1972.

Key Personnel: Dir., Dr. Elizabeth A. Desy.

Personnel Profile: Part-Time Paid 2; Part-Time Volunteers 5.

Governing Authority: public university; nonprofit. Tax-exempt: state agency.

Institution Type/Description: Natural History Museum.

Collections: concentration on animals & flora of southwest Minnesota.

Facilities: nature & conservation center; planetarium.

Activities: guided tours; participatory, temporary & traveling exhibitions.

Hours & Admission Prices: Academic Year: Mon.-Fri. 8-5; guided tours by appointment. No charge. Closed New Year's Day; Easter; Thanksgiving; Christmas.

Attendance: 800 (estimated)

SOUTHWEST MINNESOTA STATE UNIVERSITY ART MUSEUM, 1501 State St., Marshall, MN 56258. Tel.: 507-537-6266. Fax: 507-537-6577.

Institution Type/Description: Art Museum.

Collections: works by alumni, regional & international artists.

Hours & Admission Prices: Mon.-Thurs. 8am-11pm, Fri. 8-5, Sun. 2-11. Closed major holidays.

Mendota

* **SIBLEY HOUSE HISTORIC SITE, (M),** 1357 Sibley Memorial Hwy., Mendota, MN 55150. Mailing Address: P.O. Box 50772, Mendota, MN 55150-0772. Tel.: 651-452-1596. Fax: 651-405-6033.

E-mail: sibleyhouse@mnhs.org

Web Site: www.mnhs.org/places/sites/shs
Founded: 1910.
Congressional District: 4
Key Personnel: Pres. (V), Mrs. Marveen Minish.
Personnel Profile: Full-Time Paid 2; Part-Time Paid 14; Part-Time Volunteers 9.
Governing Authority: nonprofit organization. Parent Institution: Minnesota Historical Society. Subsidiary Institution: Sibley House Association. Tax-exempt: 501(c)(3).
Institution Type/Description: Historic House Museum: 1836 Henry Hastings Sibley house, dwelling & trading house for the American Fur Company factor, later Minnesota's first state governor's house; 1839 residence of trader, Jean Baptiste Faribault, which was a boarding house for visitors to the Fort Snelling area.
Collections: furnishings & material culture of the early to mid-19th century, Euro-American & Dakota; Indian artifacts; textiles & glassware.
Research Fields: history.
Facilities: meeting room in visitor center. Booklets, postcards & other museum-related items for sale.
Activities: guided tours & special weekend events.
Publications: newsletter, A Piece of History.
Hours & Admission Prices: May & Sept.-Oct. Sat. 10-4, Sun. 12:30-4; Memorial Day to Labor Day Sat. & holidays 10-4, Sun. 12:30-4; other times by appointment. Adults $6, senior citizens $5, children 6-17 $4; discounts to AAA, friends of the site & Minnesota Historical Society members; members & children under 6 no charge. ♿
Attendance: 5,000 (accurate)
Membership: Friends of Sibley Historic Site: Senior Citizen & Student $7; Individual $12; Household $18. Minnesota Historical Society: Senior Household $45; Individual $55; Household $65.

Minneapolis

AMERICAN SWEDISH INSTITUTE, (M), 2600 Park Ave., Minneapolis, MN 55407-1090. Tel.: 612-871-4907. Fax: 612-871-8682.
E-mail: info@asimn.org
Web Site: www.asimn.org
Formerly: American Institute of Swedish Arts, Literature and Science
Founded: 1929.
Congressional District: 5
Key Personnel: C.E.O. & Pres., Bruce N. Karstadt; Chm. (V), Steve Carlson; Cur. Collections, Curt Pederson; Publication Editor, Jenn Stromberg; Museum Shop. Mgr., Danielle Langehaug.
Personnel Profile: Full-Time Paid 18; Part-Time Paid 9; Part-Time Volunteers 350.
Governing Authority: society. Tax-exempt: 501(c)(3).
Institution Type/Description: Swedish-American Ethnic Museum.
Collections: Swedish-American & Swedish material culture including household articles; tools; clothing; textiles; furniture; immigrant & pioneer items; Swedish 17th- & 18th-century period artifacts; fine art; folklore; archives. Historic House: 1900 Chateauesque Mansion.
Research Fields: Swedish-American history & material culture; genealogical; Swedish immigration history; conservation of artifacts & archival material.
Facilities: classrooms.
Activities: guided tours in English & Swedish; lectures; programs; language classes; permanent & temporary exhibitions; tours; affiliate organizations in Minnesota & out of state areas; youth & family programs; summer camps.
Publications: bimonthly newsletter, ASI Posten.
Hours & Admission Prices: Jan.-Oct. Tues., Thurs.-Fri. & Sun. 12-5, Wed. 12-8, Sat. 10-5; Nov.-Dec. Sun.-Tues. & Thurs.-Fri. 12-5, Wed. 12-8, Sat. 10-5. Adults $6, senior citizens $5, children 6-18 $4; discounts to Minnesota Public Radio members; children under 6 & members no charge. Closed national holidays. ♿
Attendance: 70,000 (estimated)
Membership: Student $20; Non-Resident & Individual $35; Household $50; Sustaining $100; Patron $150; Turnblad $250; Linnaeus $500; Three Crowns $1,000; Life $3,000.

THE BAKKEN LIBRARY AND MUSEUM, 3537 Zenith Ave. S., Minneapolis, MN 55416-4623. Tel.: 612-926-3878. Fax: 612-927-7265. Facebook: The Bakken Museum.
E-mail: info@thebakken.org
Web Site: www.thebakken.org
Founded: 1975.
Congressional District: 5
Key Personnel: Exec. Dir., David J. Rhees; Chief Cur. Collections & Exhibits, Juliet Burba; Dir. Education, Steven Walvig; Dir. External Affairs, Annette Hanley; Dir. Special Projects, Beth Murphy; Dep. Dir. Programs, Kelly Finnerty; Dep. Dir. Admin., Susan Taple.

Governing Authority: nonprofit. Tax-exempt.
Institution Type/Description: Science Museum.
Collections: apparatus dating back to 1740, illustrating the use of electricity in medicine and biology; electric fish & other exhibits illustrating electrophysiology; books documenting the history of electricity, magnetism & their applications to medicine, 15th to early 20th centuries.
Research Fields: history & application of electromagnetism in the life sciences; animal magnetism & mesmerism.
Facilities: reading room; education & workshop facilities.
Activities: visiting scholars & researchers; summer and extension courses; workshops for children; Science Sat. for families; development workshops for elementary & high school science teachers.
Publications: historical brochures; newsletter; annual report; program brochures; guides.
Hours & Admission Prices: Library: Mon.-Fri. 9-5. Museum: Tues.-Wed. & Fri.-Sat. 10-5, Thurs. 10-8. Adults $7, seniors & students $5; children under 3 no charge. Closed major holidays. ♿
Attendance: 37,000 (accurate)
Membership: Senior Individual $25; Individual $30; Senior Dual $35; Dual $40; Family $65.

BELL MUSEUM OF NATURAL HISTORY, Univ. of Minnesota, 10 Church St., S.E., Minneapolis, MN 55455-0145. Tel.: 612-624-4112 & 7083. Fax: 612-626-7704.
E-mail: info@bellmuseum.org
Web Site: www.bellmuseum.org
Founded: 1872.
Congressional District: 5
Key Personnel: Dir., Susan Weller; Exhibits Coord., Tom Amble; Cur. Vascular Plants, Anita Cholewa; Cur. Breckenridge Chm. Ornithology, Bob Zink; Cur. Fungi & Lichens, David McLaughlin; Cur. Fishes, Andrew Simons; Assoc. Dir. Adult Education Programs & Media, Barbara Coffin; Assoc. Dir. K-12 Education & Family Program, Shoghig Berberian; Cur. Education, Kevin Williams; Cur. Exhibits, Donald T. Luce; Assoc. Dir. Communs. & Opers., Martin Moen; Devel., Shana Zaiser; Visitor Svcs. Coord. & Museum Shop Mgr., Emma Allen.
Personnel Profile: Full-Time Paid 21; Part-Time Paid 40; Part-Time Volunteers 78; Interns 2.
Governing Authority: university. Parent Institution: University of Minnesota. Tax-exempt: 501(c)(3).
Institution Type/Description: Natural History Museum.
Collections: regional, international & historical collections of birds, fishes, mollusks, mammals, amphibians and reptiles; seeds; frozen tissues; invertebrate & vertebrate fossils; botanical specimens including vascular plants, lichens, bryophytes, fungi & algae; natural history art; dioramas; hands-on exhibits.
Research Fields: ornithology; mammalogy; ichthyology; paleontology; herpetology; ethology; ecology; invertebrate zoology; evolution; systematics.
Facilities: 385-seat auditorium; classrooms; herbarium; research labs. Field guides, books, nature records & other museum-related gifts for sale.
Activities: guided tours; informal public courses; temporary exhibits; natural history art shows; family programs; lectures; workshops; loan collection; traveling exhibitions program; field trips; academic University courses.
Publications: technical reports; newsletter.
Hours & Admission Prices: Tues.-Wed. & Fri. 9-5, Thurs. 9-9, Sat. 10-5, Sun. 12-5. Adults $6, youths 3-17 & senior citizens $4; discounts for AAM & ICOM members; children under 3, Univ. of Minnesota students & employees no charge. Closed national holidays. ♿
Attendance: 65,493 (accurate)
Membership: Individual $30; Household $45; Sponsor $100; Patron $300; Benefactor $750.

BILL AND BONNIE DANIELS FIREFIGHTERS HALL & MUSEUM, 664-22nd Ave., N.E., Minneapolis, MN 55418. Tel.: 612-623-3817.
E-mail: firehallmuseum@msn.com
Web Site: firehallmuseum.org
Founded: 1994.
Personnel Profile: Part-Time Paid 1; Part-Time Volunteers 15.
Governing Authority: Tax-exempt.
Institution Type/Description: Firefighting History Museum.
Collections: firefighting history & equipment; photographs; hands-on exhibits.
Facilities: rental facilities. Museum-related items for sale.
Activities: firefighter's sliding pole; rental facilities; birthday parties; lectures; classes. Museum Sponsors: fire truck rides April-Oct.
Hours & Admission Prices: Sat. 9-4; other times by appointment. Adults $6, seniors 65 & over $5, children 3-12 $3. ♿
Attendance: 14,000 (estimated)

ELOISE BUTLER WILDFLOWER GARDEN AND BIRD SANCTUARY, Theodore Wirth Pkwy., Minneapolis, MN 55422. Mailing Address: MPRB Operations Center, 3800 Bryant Ave. S., Minneapolis, MN 55409-1029. Tel.: 612-230-6400 (Nov.-March) & 370-4903 (April-Oct.). Fax: 612-370-4898. Facebook: Eloise Butler Wildflower Garden.
E-mail: swilkins@minneapolisparks.org
Web Site: www.minneapolisparks.org
Founded: 1907.
Congressional District: 3
Key Personnel: Supt. of Parks, Jayne Miller; Cur. Garden, Susan Wikins; Coord. Environmental Education, Jaime McBride.
Personnel Profile: Full-Time Paid 1; Part-Time Paid 12; Part-Time Volunteers 100; Interns 2.
Governing Authority: municipal. Parent Institution: Minneapolis Park & Recreation Board.
Institution Type/Description: Wildflower & Bird Sanctuary.
Collections: native wildflowers; trees; shrubs.
Major Exhibits: Eloise Butler Wildflower Garden Florilegium Exhibit at the Minneapolis Central Library's Cargill Hall Exhibit Space, 4/1/14-5/31/14.
Facilities: 300-vol. library available for use on premises; nature center.
Activities: permanent & temporary exhibitions; self-guided brochure & trail.
Publications: guide book.
Hours & Admission Prices: mid-April to mid-Oct. daily 7:30-dusk. No charge; donations accepted.
Attendance: 50,000 (estimated)

✳ **FREDERICK R. WEISMAN ART MUSEUM, (M),** 333 E. River Rd., Minneapolis, MN 55455-0367. Tel.: 612-625-9494. Fax: 612-625-9630.
E-mail: wamdir@umn.edu
Web Site: www.weisman.umn.edu
Formerly: University Art Museum
Founded: 1934.
Congressional District: 5
Key Personnel: Dir., Lyndel King; Dir. Devel., Matt Nielsen; Museum Educator, Judi Petkau; Assoc. Dir. Communications & Mktg., Erin Lauderman; Accounts & Human Resources Mgr., Carol Stafford; Cur., Diane Mullin; Registrar, Karen Duncan; Collections Registrar, Laura Muessig; Museum Shop Mgr., Marissa Onheiber.
Personnel Profile: Full-Time Paid 26; Part-Time Paid 24; Part-Time Volunteers 88; Interns 5.
Governing Authority: university. Parent Institution: University of Minnesota. Tax-exempt.
Institution Type/Description: University Art Museum: housed in a stainless steel & brick building; designed by architect Frank Gehry.
Collections: 20th-century American painting, sculpture & prints; ancient to modern ceramics; Korean furniture.
Research Fields: 20th-century art; 18th-20th century painting, prints & decorative arts; archaeology.
Activities: guided tours; gallery talks; lectures; formally organized education programs for graduate students; inter-museum loan, temporary & traveling exhibitions; art rental program.
Publications: catalog of specific exhibitions; triannual newsletter.
Hours & Admission Prices: Tues. & Thurs.-Fri. 10-5, Wed. 10-8, Sat.-Sun. 11-5. No charge; donations accepted. Closed university holidays. &
Attendance: 130,000 (estimated)
Membership: Student $20; Individual $40; Dual $60; Sustainer $150; Collector $300; Founder $500; Patron $1,000; Director's Circle $1,500 & up. (Discounts to seniors, UMN faculty & staff, and UMAA members).

HENNEPIN HISTORY MUSEUM, (M), 2303 3rd Ave. S., Minneapolis, MN 55404-3505. Tel.: 612-870-1329. Fax: 612-870-1320.
E-mail: museum.info@hennepinhistory.org
Web Site: www.hennepinhistory.org
Founded: 1938.
Congressional District: 5
Key Personnel: Exec. Dir. & C.E.O., Jada Hansen; Chm. (V), Jan Callison; Cur., Jack Kabrud; Museum Shop Mgr., Alice Marks.
Personnel Profile: Full-Time Paid 12; Part-Time Paid 6; Part-Time Volunteers 15; Interns 3.
Governing Authority: nonprofit organization. Parent Institution: Hennepin County Historical Society. Tax-exempt: 501(c)(3).
Institution Type/Description: History Museum.
Collections: history of Minneapolis & Hennepin County, Minnesota.
Major Exhibits: A Century of Quilts: Quilts from the Permanent Collection (T), 12/13-3/14.
Research Fields: history of Hennepin County.

Facilities: research library.
Activities: permanent & temporary exhibitions; public programs.
Publications: quarterly newsletter; triannual magazine, Hennepin History.
Hours & Admission Prices: Tues. 10-2, Wed. & Fri.-Sun. 1-5, Thurs. 1-8. Adults $5, students & seniors $1; discounts to AAM & MN Assoc. of Museums' members; members no charge. Closed most federal holidays.
Attendance: 2,700 (accurate)
Membership: Student & Senior (no magazine) $20; Individual $35; Household $50; Corporate $75; Contributor $100; Archivist $250; Curator's Circle $500; Museum Club $1,000.

HENNEPIN MEDICAL HISTORY CENTER, 701 Park Ave., Blue Bldg. Lower Level, Minneapolis, MN 55415-1623. Tel.: 612-873-2512.
Web Site: hennepinmedicalhistory.org
Founded: 1976.
Congressional District: 5
Key Personnel: Chm. (V), Rondine Mehling.
Personnel Profile: Full-Time Paid 1; Part-Time Volunteers 30.
Institution Type/Description: History Museum.
Collections: photographs & artifacts relating to the medical center.
Hours & Admission Prices: Thurs. 10-2; other times by appointment. No charge.
Attendance: 800 (estimated)

INTERACT STUDIO'S INSIDE OUT GALLERY, 212 3rd Ave. N. #140, Minneapolis, MN 55401. Tel.: 612-339-5145.
E-mail: sandy@interactcenter.com
Web Site: interactcenter.com
Key Personnel: Mng. Dir., Sandy Moore
Institution Type/Description: Art Gallery.
Collections: paintings; sculpture; drawings.
Activities: workshop; educational programs.
Hours & Admission Prices: Call for hours.

JUXTAPOSITION ARTS, 2007 Emerson Ave. N., Minneapolis, MN 55411-2507. Tel.: 612-588-1148.
Key Personnel: Exec. Dir., DeAnna Cummings
Institution Type/Description: Art Gallery.
Collections: paintings; sculpture.
Hours & Admission Prices: Call for hours.

KATHERINE NASH GALLERY, 405 21st Ave. S., Minneapolis, MN 55455-0420. Mailing Address: Regis Center for Art, 405 21st Ave. S., Minneapolis, MN 55455. Tel.: 612-624-6518. Fax: 612-625-0152.
E-mail: nash@tc.umn.edu
Web Site: nash.umn.edu
Founded: 1973.
Congressional District: 5
Key Personnel: C.E.O., Clarence Morgan; Dir., Nicholas B. Shank.
Personnel Profile: Full-Time Paid 1; Part-Time Paid 8; Interns 10.
Governing Authority: public university; nonprofit. Parent Institution: University of Minnesota Dept. of Art. Tax-exempt.
Institution Type/Description: Art Gallery.
Collections: emphasis on local & regional artists and occasionally national artists.
Hours & Admission Prices: Tues.-Sat. 11-7. No charge. Closed University of Minnesota holidays & semester breaks. &
Attendance: 18,000 (estimated)

MCAD GALLERY, 2501 Stevens Ave., Minneapolis, MN 55404-4347. Tel.: 612-874-3667. Fax: 612-874-3704. TDD: 612-874-3800.
E-mail: gallery@mcad.edu
Web Site: www.mcad.edu
Formerly: Minneapolis College of Art and Design Gallery
Founded: 1886.
Key Personnel: Dir., Kerry Morgan.
Personnel Profile: Full-Time Paid 1; Part-Time Paid 1; Interns 1.
Governing Authority: college. Parent Institution: Minneapolis College of Art and Design. Tax-exempt: 501(c)(3).
Institution Type/Description: College Art Gallery.
Collections: prints; paintings; sculpture of the 20th century with focus on contemporary American art.
Activities: lectures; films; gallery talks; concerts; dance recitals; TV programs; formally organized education programs for undergraduate college students

affiliated with Minneapolis College of Art & Design; loan, temporary & traveling exhibitions.
Publications: exhibition catalogues.
Hours & Admission Prices: Mon.-Fri. 9-8, Sat. 9-5, Sun. 12-5. No charge. Closed holidays. &
Attendance: 6,000 (estimated)

MIDWAY CONTEMPORARY ART, 527 Second Ave., S.E., Minneapolis, MN 55414-1103. Tel.: 612-605-4504. Fax: 612-605-4538.
Web Site: www.midwayart.org
Founded: 2001.
Key Personnel: Exec. Dir., John Rasmussen.
Governing Authority: nonprofit organization.
Institution Type/Description: Art Gallery.
Collections: works by emerging & underrepresented contemporary artists; sculpture.
Activities: temporary exhibitions; lecture & film series; special events.
Publications: exhibition catalogs.
Hours & Admission Prices: Wed.-Sat. 11-5; groups by appointment. No charge. &
Attendance: 8,000 (estimated)

* **MILL CITY MUSEUM, (M),** 704 S. Second St., Minneapolis, MN 55401-2163. Tel.: 612-341-7555.
E-mail: mcm@mnhs.org
Web Site: www.millcitymuseum.org
Governing Authority: Parent Institution: Minnesota Historical Society.
Institution Type/Description: History Museum: former site of the Washburn A Mill.
Collections: artifacts & memorabilia pertaining to the flour milling industry.
Hours & Admission Prices: July-Aug. Mon.-Sat. 10-5, Sun. 12-5; Sept.-June Tues.-Sat. 10-5, Sun. 12-5. Adults $10, seniors & students $8, children 6-17 $5; MHS members no charge.

* **MINNEAPOLIS INSTITUTE OF ARTS, (M),** 2400 Third Ave. S., Minneapolis, MN 55404-3506. Tel.: 612-870-3000 & 3046. Fax: 612-870-3004. TDD: 612-870-3132.
E-mail: miagen@artsMIA.org
Web Site: www.artsMIA.org/MIA
Founded: 1883.
Congressional District: 5
Key Personnel: Pres. & Dir., Kaywin Feldman; Chm. Bd. Trustees, John Himle; Dir. Administrative Affairs, Michele L. Callahan; Deputy Dir. & C.O.O., Patricia J. Grazzini; Dir. Learning & Innovation, Katherine Milton; Deputy Dir. & Chief Cur. Japanese & Korean Art, Matthew Welch; Dir. Center for Alternative Museum Practice and Cur. Contemporary Art, Elizabeth Armstrong; Cur. Paintings, Patrick Noon; Head Registration, Brian Kraft; Dir. Merchandising, Mary Hele.
Personnel Profile: Full-Time Paid 240; Part-Time Paid 67.
Governing Authority: board of trustees. Tax-exempt: 501(c)(3).
Institution Type/Description: Art Museum.
Collections: European & American paintings & sculpture; decorative arts & period rooms; photography; textiles; prints & drawings; Oriental, Native American, African, Oceanic, Islamic & ancient art. Historic Buildings: Purcel-Cutts House; Prairie Schoolhouse.
Research Fields: pertaining to the general collection.
Facilities: 50,000-vol. library of publications pertaining to specialized areas of art history available for inter-library loan & for use on premises; 162,702 sq. ft. exhibit space; reading room; 285-seat auditorium; 50-seat theater; classrooms; restaurant. Books, periodicals, jewelry, posters & replicas of museum pieces for sale.
Activities: guided tours; lectures; films; gallery talks; arts festivals; study clubs; interactive video programs; formally organized education programs for children & adults & the visually impaired; docent program; formally organized education programs for undergraduate & graduate college students affiliated with regional colleges & universities; training programs for professional museum workers; inter-museum loan, permanent, temporary & traveling exhibitions; municipal governmental & school loan service.
Publications: guide, A Guide to the Galleries; exhibition catalogs; monthly magazine, ARTS; education & instructional workbooks.
Hours & Admission Prices: Tues.-Wed. & Fri.-Sat. 10-5, Thurs. 10-9, Sun. 11-5. No charge; donations accepted. Fees for some special exhibitions. Closed Independence Day; Thanksgiving; Christmas. &
Attendance: 450,000 (estimated)
Membership: Student, Senior & Non-Resident $30; Individual $45; Dual $60; Family $75; Arts Circle $100-$175; Ambassador $150; Contributor's Circle & Business League $250-$2,000; Patrons' Circle & Corporate Council $2,000.

MINNESOTA STREETCAR MUSEUM, 2330 W. 42nd St., Minneapolis, MN 55410. Mailing Address: P.O. Box 14467, University Station, Minneapolis, MN 55414-0467. Tel.: 952-688-7255 & 922-1096.
E-mail: mnscmuseum@gmail.com
Web Site: www.trolleyride.org
Founded: 2004.
Institution Type/Description: Transportation Museum.
Collections: electric railway history; street cars; papers; photographs.
Activities: special events.
Hours & Admission Prices: Call for hours.

THE MUSEUM OF RUSSIAN ART, (M), 5500 Stevens Ave. S., Minneapolis, MN 55419-1933. Tel.: 612-821-9045. Fax: 612-821-4392.
E-mail: bshinkle@tmora.org
Web Site: www.tmora.org
Founded: 2002.
Congressional District: 3
Key Personnel: Chm. (V), William McLaughlin; Pres. & Dir., Bradford Shinkle, IV; Special Events Mgr., Lynda Holker; Museum Shop Mgr., Melanie Brooks.
Personnel Profile: Full-Time Paid 4; Part-Time Paid 3; Part-Time Volunteers 75.
Governing Authority: private; nonprofit organization. Tax-exempt: 501(c)(3).
Institution Type/Description: Art Museum.
Collections: late 19th-century to 1991 Russian art; 20th-century Russian Socialist Realist art from the Soviet era 1934-1975.
Facilities: 500-vol. library; 15,000 sq. ft. exhibit space. Museum-related items for sale.
Activities: lectures; seminars; children's educational events; facility available for corporate & private events.
Publications: semiannual newsletter, View; themed exhibition brochures; English language catalogs & reference books.
Hours & Admission Prices: Mon.-Wed. & Fri. 10-5, Thurs. 10-8, Sat. 10-4, Sun. 1-5. Adults $7, discount to AAM members; members no charge. Closed New Year's Day; Memorial Day; Independence Day; Labor Day; Thanksgiving; Christmas Eve & Day. &
Attendance: 52,000 (accurate)
Membership: Individual $30; Partner $50; Family $65; Benefactor $125; Patron $250; Leadership $1,000.

SOAP FACTORY, 514 2nd St., S.E., Minneapolis, MN 55414-2105. Mailing Address: P.O. Box 581696, Minneapolis, MN 55458-1696. Tel.: 612-623-9176.
E-mail: info@soapfactory.org
Web Site: www.soapfactory.org
Key Personnel: Exec. Dir., Ben Heywood.
Governing Authority: Tax-exempt: 501(c)(3).
Institution Type/Description: Art Gallery.
Collections: works by contemporary artists.
Hours & Admission Prices: Call for hours.

SOO VISUAL ARTS CENTER, 2638 Lyndale Ave. S., Minneapolis, MN 55408. Tel.: 612-871-2263.
E-mail: info@soovac.org
Web Site: www.soovac.org
Key Personnel: Exec. Dir., Suzy Greenberg.
Governing Authority: nonprofit organization. Tax-exempt: 501(c)(3).
Institution Type/Description: Art Gallery.
Collections: paintings; sculpture.
Hours & Admission Prices: Wed.-Fri. 11-7, Sat.-Sun. 10-6.

TRAFFIC ZONE CENTER FOR VISUAL ARTS, 250 3rd Ave. N., Minneapolis, MN 55401. Tel.: 612-204-0012.
E-mail: trafficzoneart@gmail.com
Web Site: www.trafficzoneart.com
Institution Type/Description: Art Gallery.
Collections: works by mid-career visual artists.
Activities: special events.
Hours & Admission Prices: Call for hours.

TYCHMAN SHAPIRO GALLERY, Sabes Jewish Community Center, 4330 S. Cedar Lake Rd., Minneapolis, MN 55416. Tel.: 952-381-3400. Fax: 952-381-3401.
E-mail: info@sabesjcc.org
Web Site: www.sabesjcc.org

Institution Type/Description: Art Gallery.
Collections: paintings; sculpture; photographs.
Activities: lectures; workshops.
Hours & Admission Prices: Mon.-Thurs. 7:30 am-9:30 pm, Fri. 7:30-6, Sun. 8-3. Closed holidays.

* **WALKER ART CENTER, (M),** 1750 Hennepin Ave., Minneapolis, MN 55403-1169. Tel.: 612-375-7675 & 7676. Fax: 612-375-7567. TDD: 612-375-7585.
E-mail: olga.viso@walkerart.org
Web Site: walkerart.org/
Founded: 1927.
Congressional District: 5
Key Personnel: Pres., Andrew Duff; Dir., Olga Viso; C.O.A., Phillip Bahar; Chief Cur., Darsie Alexander; Registrar, Joe King; Sr. Cur. Performing Arts, Philip Bither; Cur., Siri Engberg; Chief of Audience Engagement/Communications & Cur., Andrew Blauvelt; C.F.O., Mary Polta; Dir. Education & Community Programs, Sarah Schultz; Dir. Mktg. & Public Rels., Ryan French; Chief Finance & Devel., Christopher Stevens; Dir. Merchandising, Michele Tobin; Librarian, Rosemary Furtak.
Personnel Profile: Full-Time Paid 120; Part-Time Paid 40; Part-Time Volunteers 150; Interns 16.
Governing Authority: nonprofit organization. Tax-exempt: 501(c)(3).
Institution Type/Description: Art Museum.
Collections: contemporary art including paintings, sculpture, drawings, prints & films; Minneapolis Sculpture Garden, 11-acre urban garden featuring 40 sculptures & conservatory with horticultural displays.
Research Fields: contemporary arts; film & the performing arts.
Facilities: library of 38,000 exhibition catalogues, 1,000 artist books, monographs, clipping files & taped interviews available for graduate student use; 385-seat theater; 344-seat cinema; 150-seat cafe; 85-seat restaurant; lecture room. Books, prints, posters, postcards, note cards, slides, jewelry & designer items for sale.
Activities: guided tours; lectures; films; gallery talks; concerts; dance; theater; music & performance art; formally organized education programs for children, teens, adults & undergraduate students; docent program; film program; inter-museum loan, permanent, temporary & traveling exhibitions.
Publications: monthly Calendar of Events; exhibition catalogs.
Hours & Admission Prices: Tues.-Wed. & Fri.-Sun. 11-5, Thurs. 11-9. Adults $10, seniors 65 & over $8, students & teens $6; discounts to AAM members; members, children under 12, Thurs. 5-9 & 1st Sat. of month no charge. &
Attendance: 759,347 (accurate)
Membership: Individual $60; Dual $70; Family $75; Friend $100; Contemporaries $175; Contributing $250-$25,000.

WELLS FARGO HISTORY MUSEUM, 90 S. 7th St., Skyway Level, Minneapolis, MN 55479-1100. Mailing Address: Wells Fargo Historical Services, 420 Montgomery St., MAC-A0101-106, San Francisco, CA 94163. Tel.: 612-667-4210. Fax: 612-316-4361.
Web Site: www.wellsfargohistory.com
Founded: 2000.
Key Personnel: Cur., Megan Schaack; Asst. Cur., Phyllis Thorne.
Governing Authority: Affiliated with Wells Fargo Bank.
Institution Type/Description: Company History Museum.
Collections: Concord stagecoach; Wells Fargo banking & express history; California Gold Rush; gold specimens & coins; working telegraphs; local history.
Facilities: Museum-related items for sale.
Activities: guided tours; imaginary rides on replica stagecoach; audiovisual programs; participatory & temporary exhibitions.
Publications: scholarly pamphlets.
Hours & Admission Prices: Mon.-Fri. 9-5. No charge. Closed bank holidays. &
Attendance: 15,447 (accurate)

Montevideo

CHIPPEWA COUNTY HISTORICAL SOCIETY, 151 Arnie Anderson Dr., Montevideo, MN 56265-2127. Mailing Address: P.O. Box 303, Montevideo, MN 56265-0303. Tel.: 320-269-7636.
Founded: 1936.
Congressional District: 6
Key Personnel: Pres. (V), Lee Hagemeyer; Museum Shop Mgr., Carol Westberg.
Personnel Profile: Full-Time Paid 1; Full-Time Volunteers 1; Part-Time Paid 3; Part-Time Volunteers 25.
Operating Expenses: 96,826

Operating Income: 105,523
Governing Authority: county. Parent Institution: Minnesota Historical Society. Tax-exempt.
Institution Type/Description: Historic Village Setting Museum: 23 buildings including 1882 church; rural school; 1870 log cabin.
Collections: Frank Stay collection; Rev. Stephen R. Riggs letters; Sioux Uprising; restored pre-1850s dugout canoe with interpretive material; 1,200 slides on county farmstead, barns & outbuildings; Henry Gippe collection; Civil War photographs. Historic House: 1901-1903 farm house & outbuildings, Swensson farm.
Research Fields: education; farm buildings
Facilities: 400-vol. library of school, religious & local history books. Museum-related gifts for sale.
Activities: guided tours on appointment; lectures; permanent exhibitions.
Publications: monthly newsletter.
Hours & Admission Prices: Chippewa City: Memorial Day to Labor Day Mon.-Fri. 9-5, Sat.-Sun. 1-5; Sept. Mon.-Fri. 9-5. Swensson Farm: Memorial Day to Labor Day Sun. 1-5 & by appointment. Adults $4, youth $2, children 5 & under & members no charge. &
Attendance: 5,000 (estimated)
Membership: Individual $15; Family $20; Sustaining $25; Contributing $50; Benefactor $100. Business Memberships: Bronze $50; Silver $75; Gold $100; Platinum $200.

Moorhead

COMSTOCK HISTORIC HOUSE, (M), 506 8th St. S., Moorhead, MN 56560-3504. Tel.: 218-291-4211.
E-mail: comstockhouse@gomoorhead.com
Web Site: www.mnhs.org/places/sites/ch/
Founded: 1975.
Congressional District: 7
Key Personnel: Chm. (V), Kathy Faeth.
Personnel Profile: Part-Time Paid 2; Part-Time Volunteers 4.
Governing Authority: society. Owned & operated by the Minnesota Historical Society. Subsidiary Institution: Comstock Historic House Society. Tax-exempt.
Institution Type/Description: Historical Society: housed in c.1882 Comstock home.
Collections: furnishings; artifacts of Solomon G. Comstock's family.
Research Fields: Ada Louise Comstock.
Facilities: library available to the public; meeting space available.
Activities: guided tours; concerts; organized education programs for children; docent program.
Publications: Comstock Comments.
Hours & Admission Prices: Memorial Day to Labor Day Thurs. 5pm-8pm, Sat.-Sun. 1-4:30. Adults $6, senior citizens $5, students & children 6-17 $4; discounts to AAM members; children under 6 no charge. Closed holidays.
Attendance: 600 (estimated)
Membership: Annual $55.

HISTORICAL AND CULTURAL SOCIETY OF CLAY COUNTY, (M), 202 1st Ave. N., Moorhead, MN 56560-1985. Mailing Address: P.O. Box 157, Moorhead, MN 56561-0157. Tel.: 218-299-5511. Fax: 218-299-5510.
E-mail: maureen.jonason@ci.moorehead.mn.us
Web Site: www.hcscconline.org
Formerly: Heritage-Hjemkomst Interpretive Center
Founded: 1957.
Congressional District: 7
Key Personnel: Interim Exec. Dir., Maureen Kelly Jonason; Archivist, Mark Peihl; Cur., Lisa Vedaa; Coord. Events, Tim Jorgensen; Coord. Visitor Svcs., Markus Krueger; Mktg., Brianne Carlsrud.
Personnel Profile: Full-Time Paid 6; Part-Time Paid 2; Part-Time Volunteers 60.
Governing Authority: nonprofit organization. Tax-exempt: 501(c)(3).
Institution Type/Description: History and Cultural Heritage Museum.
Collections: Clay County history & cultural heritage; Viking ship replica; Stave Church replica; 15,000 glass plate negatives from 1872 to late 1930s; clothing; guns; furniture; household items; pioneer artifacts; period files; paintings; pioneer cemetery. Historic Buildings: 1895 District 3 rural school house; 1870 Bergquist pioneer cabin.
Major Exhibits: The Vikings, 9/30/13-2/28/14.
Research Fields: Clay County, Red River Valley; Hjemkomst's journey.
Facilities: 300-vol. library & manuscripts available for research on premises; meeting rooms; auditorium. Gift items for sale.
Activities: guided tours; films; permanent, temporary & traveling exhibits; docent program. Museum Sponsors: Celtic Festival in March; Scandinavian Heritage Hjemkomst Festival in June; Midwest Viking and Festival; Pangea: Cultivate Our Cultures Festival in November.

Publications: quarterly newsletter, The Hour Glass.
Hours & Admission Prices: Mon. & Wed.-Sat. 9-5, Tues. 9-8, Sun. 12-5. Adults $8, seniors/students $7, youth 5-17 $6; discounts to ASTC & AAA members; members & children under 5 no charge. Closed New Year's Eve & Day; Easter; Thanksgiving; Christmas Eve & Day. &
Attendance: 30,000 (estimated)
Membership: Individual $30; Household $50; Booster $75; Heritage $125; Patron $250; Benefactor $500; Vanguard $1,000.

Mora

KANABEC HISTORY CENTER, 805 W. Forest Ave., Mora, MN 55051. Mailing Address: P.O. Box 113, Mora, MN 55051-0113. Tel.: 320-679-1665. Fax: 320-679-1673.
E-mail: center@kanabechistory.org
Web Site: www.kanabechistory.org
Founded: 1978.
Congressional District: 14, 18 & 19
Personnel Profile: Part-Time Paid 4; Part-Time Volunteers 25.
Governing Authority: nonprofit organization. Parent Institution: Kanabec County Historical Society. Tax-exempt: 501(c)(3).
Institution Type/Description: Historical Society Museum.
Collections: Kanabec County artifacts; photographs; manuscripts; newspapers; genealogies; cemetery records.
Research Fields: county history; genealogy; cemetery records; birth, death & marriage records; family histories.
Facilities: library; 37-acre site; interpretive nature trail; picnic facilities. Historical & fiction books & gift items for sale.
Activities: lectures; films; special programming; docent program.
Publications: quarterly newsletter.
Hours & Admission Prices: Tues.-Sat. 10-4:30, holidays 12:30-4:30. Adults $3, students K-12 $1; children under 5, KCHS members & during winter no charge. Children under 14 must be accompanied by an adult. Closed New Year's Day; Easter; Independence Day; Thanksgiving; Christmas Eve & Day. &
Attendance: 15,000 (accurate)
Membership: Individual $25; Family $40; Family Plus $50; Century $100; Sponsor $150; Corporate $250; Corporate Gold $300.

Morris

STEVENS COUNTY HISTORICAL SOCIETY MUSEUM, 116 W. 6th St., Morris, MN 56267-1922. Tel.: 320-589-1719. Fax: 320-589-1719 call ahead.
Web Site: www.stevenshistorymuseum.com
Founded: 1920.
Congressional District: 6
Key Personnel: Dir. & Museum Shop Mgr., Randee L. Hokanson; Historian Researcher, Tina Didreakson; Pres. (V) & Chm. (V), Ward B. Voorhees; Collections Registrar, Tom Vail.
Personnel Profile: Part-Time Paid 4; Part-Time Volunteers 23; Interns 3.
Governing Authority: society; state; county. Tax-exempt.
Institution Type/Description: Local History Museum: housed in 1905 Carnegie Library building.
Collections: clothing; photographs; war material; household furnishings; manuscripts; country church furnishings; farm tools. Historic Buildings: 1900 country school; log cabin.
Major Exhibits: The Denim Generation, 7/13-7/16.
Research Fields: genealogy & local history.
Facilities: 200-vol. library of old records of the area available for research under supervision; 140 linear ft. archival material; local newspapers 1876-present; 2,000 books & 5,000 photographs in supervised research area.
Activities: guided tours; lectures; films; radio programs; workshops; permanent & temporary exhibitions; bus trips. Museum Sponsors: EGA Show; Annual Art Show; Annual Holiday Open House.
Publications: bimonthly newsletter; coloring book; books, Syrup Pails & Overshoes: Stories from the Country Schools of Stevens County; An Honest Day's Work; The 40s: A Time for War & A Time for Peace; Celebrating 125 Years of Morris Cookbook; Images of Stevens County; The Wadsworth Trail (reprint).
Hours & Admission Prices: Mon.-Fri. 9-5; other times by appointment; call for additional weekend hours. No charge; donations accepted. &
Attendance: 5,000 (estimated)
Membership: Individual $36; Family $42; Sustaining $60; Business $62; Benefactor $112.

Morton

BIRCH COULEE BATTLEFIELD, Lower Sioux Agency Historic Site, 32469 Redwood County Hwy. 2, Morton, MN 56270. Tel.: 507-697-6321.
E-mail: birchcoulee@mnhs.org
Web Site: www.mnhs.org/places.sites/bc
Institution Type/Description: Historic Site: former site of the Battle of Birch Coulee.
Collections: recreated prairie; guide posts; battle sketches.
Hours & Admission Prices: May-Oct. daily dawn-dusk. No charge.

Mountain Lake

HERITAGE VILLAGE, County Rd. One, Mountain Lake, MN 56159. Mailing Address: P.O. Box 152, Mountain Lake, MN 56159-0152. Tel.: 507-427-3743.
E-mail: abdick@frontiernet.net
Formerly: Heritage House
Founded: 1972.
Congressional District: 22A
Key Personnel: Chm. Bd, Alvin Dick; Vice Chm., Harvey Buller; Treas., Betty Lou Ritzloff; Sec. & Museum Shop Mgr., Geneva Stoesz.
Personnel Profile: Part-Time Volunteers 25.
Governing Authority: private; nonprofit organization. Tax-exempt: 501(c)(3).
Institution Type/Description: Historic House & Village: 20 historic structures representing a late 1800s to early 1900s village.
Collections: immigration of Mennonite people from Russia, as well as Russian-German Lutherans.
Research Fields: history of collection.
Facilities: 21,000 sq. ft. exhibit space; landscape grounds. Museum-related items for sale.
Activities: docent program; guided tours; school loan service. Annual Event: Fall Festival featuring Mennonite food & entertainment in September.
Publications: annual newspaper supplement for the Sept. Heritage Fair.
Hours & Admission Prices: Memorial Day to Labor Day daily 1-5. Adults $5, children over 7 $3; members no charge.
Attendance: 2,000 (estimated)
Membership: Single $25; Family $40.

New Brighton

UNITED THEOLOGICAL SEMINARY OF THE TWIN CITIES, 3000 5th St., N.W., New Brighton, MN 55112. Tel.: 651-633-4311. Fax: 651-633-4315.
E-mail: cbjohnson@unitedseminary.edu
Web Site: www.unitedseminary.edu
Institution Type/Description: Religious Art Museum.
Collections: historical & contemporary works of art by local & national artists.
Activities: educational programs; lectures; talks.
Hours & Admission Prices: Mon.-Thurs. 9-9, Fri. 9-5.

New Ulm

AUGUST SCHELL BEWERY, 1860 Schell Rd., New Ulm, MN 56073-3834. Mailing Address: P.O. Box 128, New Ulm, MN 56073. Tel.: 800-770-5020; 507-354-5528. Fax: 507-359-9119.
E-mail: schells@schellsbrewery.com
Web Site: schellsbrewery.com
Founded: 1860.
Key Personnel: Dir. & Pres., Ted Marti.
Governing Authority: Tax-exempt.
Institution Type/Description: Company History Museum.
Collections: brewery & founding family history.
Hours & Admission Prices: Museum: Memorial Day-Labor Day daily 12-5; Winter: call for hours. Tours: Memorial Day-Labor Day Mon.-Thurs. 1, 2:30 & 4, Fri. 1, 2:30 & 4, Sat. noon, 1, 2, 3 & 4, Sun. 1, 2, 3, 4; Sept.-May Fri. 3, Sat. 1, 2, 3 & 4, Sun. 4. Museum: no charge. Tours: adults $3; children 12 & under no charge. &
Attendance: 30,000 (estimated)

BROWN COUNTY HISTORICAL SOCIETY, 2 N. Broadway, New Ulm, MN 56073-1714. Tel.: 507-233-2616. Fax: 507-354-1068.
E-mail: bchs@browncountyhistorymnusa.org
Web Site: browncountyhistorymnusa.org
Founded: 1930.
Congressional District: 2
Key Personnel: Exec. Dir., Robert Burgess; Pres., Lisa Besemer; Cur., Marilyn

Hesse; Research Librarian, Darla Gebhard; Museum Shop Mgr. & Office Mgr., Pam Kitzberger.
Personnel Profile: Full-Time Paid 1; Part-Time Paid 5; Part-Time Volunteers 85.
Governing Authority: society. Supported by Brown County. Tax-exempt: 501(c)(3).
Institution Type/Description: History Museum.
Collections: area history; Dakota & Ojibwe Indians; Dakota War of 1862; early biographical & genealogical files; representative collection of regional artists including Wanda & Anton Gag, Chris Heller & Alexander Schwendinger.
Research Fields: Native American-Dakota; Wanda Gag; US-Dakota war 1862; Turner Society; Hermann Monument.
Facilities: 1,000-vol. non-circulating library of books pertaining to Minnesota & local history; research library.
Activities: guided tours; lectures; gallery talks; historical markers; permanent & temporary exhibitions.
Publications: leaflets; brochures; maps; books; news notes.
Hours & Admission Prices: May-Oct. Mon.-Fri. 10-4, Sat. 10-3; Nov.-April Tues.-Fri. 10-4, Sat. 10-3. Adults $3; members, students & children no charge. Closed national holidays. &
Attendance: 6,894 (accurate)
Membership: Individual $25; Joint $30; Golden & Business $50; Century $100.

HARKIN STORE, (M), 66250 County Rd. 21, New Ulm, MN 56073. Mailing Address: P.O. Box 112, New Ulm, MN 56073-0112. Tel.: 507-354-8666 & 934-2160.
E-mail: harkinstore@mnhs.org
Web Site: www.mnhs.org/places/sites/hs
Institution Type/Description: Historic Site: housed in a former general store & post-office; built in 1867.
Collections: local history & culture; original store inventory; period artifacts.
Hours & Admission Prices: May & Sept.-Oct. Sat.-Sun. 10-5; Memorial Day to Labor Day Tues.-Sun. 10-5. Adults 18-64 $5, seniors 65 & over, college students and children 6-17 $3; NSHS & MHS members and children 5 & under no charge.

North Saint Paul

NORTH STAR MUSEUM OF BOY SCOUTING AND GIRL SCOUTING, (M), 2640 Seventh Ave. E., North Saint Paul, MN 55109-3103. Tel.: 651-748-2880.
E-mail: cnicholson@nssm.org
Web Site: www.nssm.org
Formerly: North Star Scouting Memorabilia, Inc.
Founded: 1976.
Congressional District: 4
Key Personnel: Exec. Dir., Claudia J. Nicholson; Chm., Ron Phillippo; Dir. Institutional Advancement, Shirley De La Torre.
Personnel Profile: Full-Time Paid 2; Part-Time Paid 2; Part-Time Volunteers 50; Interns 2.
Volunteer Hours: 500
Governing Authority: private; nonprofit. Tax-exempt.
Institution Type/Description: Boy Scout & Girl Scout Museum.
Collections: from inception of the Scouting program to the present; 150,000 artifacts; film; photographs; Scouting publications; uniforms; equipment; 5-state upper Midwest region especially Minnesota & Western Wisconsin.
Research Fields: Region 10-Boy Scouts of America; boy & girl scouts; girl guiding; councils; camps; history.
Facilities: 3,500-vol. library of Boy's Life & Scouting magazines and merit badge handbooks, Boy Scouting & Girl Scouting books; 8,000 sq. ft. exhibit space; community room.
Activities: films; temporary & traveling exhibitions; special events; lectures; classes & films; service opportunities for Boy Scouts & Girl Scouts. Annual Events: Veteran Scouter Reunion; national conferences; Minnesota State Fair.
Publications: newsletter, North Star Museum News.
Hours & Admission Prices: Visit website for hours. No charge; donations accepted. Closed New Year's Day; Memorial Day; Independence Day; Labor Day; Thanksgiving; Christmas. &
Attendance: 2,643 (accurate)
Membership: Individual $35; Family/Unit $50.

Northfield

CARLETON COLLEGE ART GALLERY, (M), One N. College St., Northfield, MN 55057-4001. Tel.: 507-646-4342. Fax: 507-646-7042.
E-mail: lbradley@carleton.edu

Founded: 1971.
Key Personnel: Dir. & Cur., Laurel Bradley.
Personnel Profile: Full-Time Paid 2; Part-Time Paid 16.
Governing Authority: private; college. Tax-exempt: 501(c)(3).
Institution Type/Description: Art Museum.
Collections: 1,400 art related items; 19th & 20th-century printmaking; contemporary photography; American paintings; Asian & ancient artifacts.
Activities: lectures; temporary & traveling exhibitions.
Hours & Admission Prices: Jan. to mid-June & Sept.-Nov. Mon.-Wed. 12-6, Thurs.-Fri. 12-10, Sat.-Sun. 12-4. No charge. Closed New Year's Day; Thanksgiving; spring break.
Attendance: 5,000 (estimated)

FLATEN ART MUSEUM OF ST. OLAF COLLEGE, (M), 1520 St. Olaf Ave., Northfield, MN 55057-1099. Tel.: 507-786-3556 & 3703. Fax: 507-786-3776.
E-mail: beckerj@stolaf.edu
Web Site: www.stolaf.edu/collections/flaten
Founded: 1976.
Congressional District: 1
Key Personnel: Dir., Jane Becker Nelson; Registrar, Mona Weselmann.
Personnel Profile: Part-Time Paid 16.
Governing Authority: private college. Parent Institution: St. Olaf College. Tax-exempt.
Institution Type/Description: College Art Museum.
Collections: contemporary & historic prints, paintings, photographs, mixed, sculpture, ceramics; art glass; Yoshida family contemporary Japanese prints particularly Yoshida Hodaka; Andy Warhol photographs; traditional Japanese prints; study collections: African sculpture; post-Romanian paintings, prints, sculpture; Southwest US pottery; Albert Christ-Janer prints; works from around the globe; Chinese posters from the Cultural Revolution.
Facilities: exhibit gallery; print-study room.
Activities: traveling & temporary exhibitions; lectures; films; concerts; graduating senior & apprentice exhibits.
Publications: exhibit catalogs; full color mailer; posters; brochures.
Hours & Admission Prices: Sept. to mid-April Mon.-Wed. & Fri. 10-5, Thurs. 10-8, Sat.-Sun. 2-5. No charge. Closed during school breaks. &
Attendance: 12,000 (accurate)

NORTHFIELD HISTORICAL SOCIETY MUSEUM, (M), 408 Division St., Northfield, MN 55057-2018. Tel.: 507-645-9268; 507-663-6080.
E-mail: nhsmuseum@rconnect.com
Web Site: www.northfieldhistory.org
Founded: 1975.
Congressional District: 1
Key Personnel: Pres., Jodi Lawson; Exec. Dir., Hayes Scriven; Treas., Chuck Sandstrom; Museum Shop Mgr., Dick Waters.
Personnel Profile: Full-Time Paid 1; Part-Time Paid 1; Part-Time Volunteers 80; Interns 12.
Governing Authority: nonprofit organization. Tax-exempt.
Institution Type/Description: Historical Society Museum: housed at the site of the original 1st National Bank of Northfield.
Collections: Northfield area artifacts; items relating to the 1876 attempted bank raid by the James/Younger Gang.
Research Fields: Northfield history; James/Younger Gang; railroad collection.
Facilities: 250-vol. library of books on local history & Jesse James available for research on premises under supervision. Books, notepaper, printed material & other museum-related items for sale.
Activities: guided tours; videotape TV programs; permanent, temporary & traveling exhibitions; adult & children's programming. Annual Events: Defeat of Jesse James Days in September; tour of homes in October.
Publications: quarterly newsletter, Scriver Scribbler.
Hours & Admission Prices: Memorial Day-Labor Day Mon.-Sat. 10-5, Sun. 1-4; Sept.-May Tues.-Sat. 10-4, Sun. 1-4. Guided Tours: Sat.-Sun. Adults & students $4, senior citizens $3, children $1.50; discounts to AAM members; members no charge. Closed New Year's Day; Easter; Independence Day; Thanksgiving; Christmas. &
Attendance: 12,000 (accurate)
Membership: Senior $35; Family $50; Building $100; Patron & Business $250; Sustaining $500; Life $1,000.

NORWEGIAN-AMERICAN HISTORICAL ASSOCIATION, St. Olaf College, 1510 St. Olaf Ave., Northfield, MN 55057-1097. Tel.: 507-786-3221. Fax: 507-646-3734.
E-mail: naha@stolaf.edu
Web Site: www.naha.stolaf.edu
Founded: 1925.

Congressional District: 1
Key Personnel: Admin. Dir., Kim Holland; Cur., Jeff Sauve.
Personnel Profile: Part-Time Paid 3; Part-Time Volunteers 1.
Governing Authority: board of directors. Tax-exempt: 501(c)(3).
Institution Type/Description: Archives.
Collections: Norwegian-American history.
Research Fields: Norwegian migration; Norwegian-American history.
Facilities: 7,000-vol. library of books, magazines, pamphlets & newspapers dealing with Norwegian-American history; archives available for research. Books for sale.
Activities: document preservation; scholarly publication program; occasional conferences held in Norway and the U.S.
Publications: Travel & Description series: Norwegian-American Studies; newsletter; Special Publication Series, Authors Series; Topical Studies series; Biographical Series.
Hours & Admission Prices: By appointment only. Admission: $15. &
Attendance: 300 (estimated)
Membership: Student $25; Associate $40; Sustaining $75; Patron $125; Life $500 (one payment); Institutional $500 (25 years).

Onamia

* **MILLE LACS INDIAN MUSEUM, (M),** U.S. Hwy. 169, Onamia, MN 56359. Mailing Address: 43411 Oodena Dr., Onamia, MN 56359-2259. Tel.: 320-532-3632.
E-mail: millelacs@mnhs.org
Web Site: www.mnhs.org/places/sites/mlim/index.html
Institution Type/Description: History Museum.
Collections: Mille Lacs Indian culture, language, music & dance.
Activities: workshops; special events; traditional cooking; birch-bark basketry & beadwork.
Hours & Admission Prices: Museum: April-May & Sept.-Oct. Wed.-Sat. 11-4; Memorial Day to Labor Day Tues.-Sat. 11-4. Adults $8, senior citizens & college students $7, children 6-17 $6; MHS members and children 5 & under no charge.

Owatonna

OWATONNA ARTS CENTER, 435 Garden View Lane, West Hills Complex, Owatonna, MN 55060. Mailing Address: P.O. Box 134, Owatonna, MN 55060-0134. Tel.: 507-451-0533. Fax: 507-446-0198.
E-mail: megan.proft@oacarts.org
Web Site: www.oacarts.org
Founded: 1974.
Congressional District: 1
Key Personnel: Exec. Dir., Megan Proft; Artistic Dir., Silvan Durben; Bd. Pres., Megan Kruggel.
Personnel Profile: Full-Time Paid 2; Part-Time Volunteers 35.
Governing Authority: nonprofit organization. Tax-exempt: 501(c)(3).
Institution Type/Description: Art Center.
Collections: Marianne Young 100-piece costume collection from 25 countries; prints & paintings; sculpture by Minnesota artists; 3 sculptures in garden by artists Paul Grandlund, John Rood & Richard Hammel.
Major Exhibits: Ellie Kingsbury Photograph, 1/5/14-1/26/14; William Bukowski: Cabin Series Paintings, 2/2/14-2/23/14; Owatonna Public Schools Student Exhibit, 3/2/14-3/30/14; Peacemakers Quilt Show, 4/6/14-4/27/14; 62nd Annual Steeve Co. Art Exhibition, 5/4/14-5/25/14; Christina Spencer: Nesting Exhibition, 5/4/14-5/25/14; Kimber Olson & Carolyn Halliday: Texture Exhibition, 6/1/14-6/29/14; Chair Exhibition, 7/6/14-7/27/14; Debra Blowers: Glacier Photos on Silk, 8/3/14-8/31/14; Mary Rose Gondek: Pencil Drawings, 8/3/14-8/31/14.
Facilities: library of art history primarily for members of arts center; 200-seat performing arts hall; outdoor sculpture garden.
Activities: guided tours; lectures; concerts; vocal & piano recitals; formally organized education programs; loan, permanent, temporary & traveling exhibitions; classes for youth & adults. Annual Event: Festival of Arts.
Publications: bimonthly newsletter.
Hours & Admission Prices: Tues.-Sun. 1-5. No charge. Closed holidays. &
Attendance: 25,000 (estimated)
Membership: Full-Time Student $25; Basic $50; Contributing $100-$249; Sustaining $250-$499; Patron $500-$999; Benefactor $1,000-$2,499; Cultural Advocate $2,500-$4,999; Philanthropist $5,000-$9,999; Leadership $10,000.

Park Rapids

NORTH COUNTRY MUSEUM OF ARTS, 301 Court Ave., Park Rapids, MN 56470-1421. Mailing Address: P.O. Box 328, Park Rapids, MN 56470-0328. Tel.: 218-237-5900.
Founded: 1977.
Congressional District: 8
Key Personnel: Chm., Louie Falk.
Personnel Profile: Full-Time Paid 1; Part-Time Paid 1; Part-Time Volunteers 6.
Governing Authority: nonprofit organization. Tax-exempt: 501(c)(3).
Institution Type/Description: Art Museum: housed in 1900 Hubbard County Courthouse.
Collections: late 15th to 19th-century old-school European paintings; Nigerian arts, crafts & artifacts; works by contemporary artists.
Research Fields: Nigerian arts, crafts, & artifacts (Yoruba, Ibo, Fulani & Hausa); late 15th to 19th-century European paintings.
Facilities: Works by regional artists & craftsmen for sale.
Activities: guided tours; special events; readings; concerts; lectures; classes; workshop; temporary exhibits.
Publications: quarterly newsletter; schedule of events.
Hours & Admission Prices: May-Oct. Tues.-Sun. 11-5. No charge; donations accepted.
Attendance: 4,000 (estimated)
Membership: Individual $15; Family $25.

Perham

ITOW VETERANS MUSEUM, 805 W. Main, Perham, MN 56573-1131. Tel.: 218-346-7678.
E-mail: info@itowmuseum.org
Web Site: www.itowmuseum.org
Founded: 2006.
Key Personnel: Project Mgr., Lina Belar
Institution Type/Description: History Museum.
Collections: videos & kiosks relating to veterans of World War I & II, the Korean War & the Vietnam War; personal artifacts; photographs.
Hours & Admission Prices: Mon.-Sat. 10-5, Sun. 1-4. Adults $4; Memorial Day, Veterans Day, Independence Day & veterans no charge.

Pine City

MINNESOTA HISTORICAL SOCIETY'S NORTH WEST COMPANY FUR POST, (M), 12551 Voyager Lane, Pine City, MN 55063. Mailing Address: P.O. Box 51, Pine City, MN 55063-0051. Tel.: 320-629-6356. Fax: 320-629-4667.
Congressional District: 8
Key Personnel: Site Mgr., Patrick Schifferdecker.
Personnel Profile: Full-Time Paid 1; Part-Time Paid 8; Part-Time Volunteers 20.
Governing Authority: private; nonprofit organization. Parent Institution: Minnesota Historical Society, 345 Kellogg Blvd. W., St. Paul, MN 55102-1906.
Institution Type/Description: History Museum: housed in recreated fur trading post & Ojibwe encampment.
Collections: Ojibwe history; fur trade.
Facilities: 2,800 sq. ft. exhibit space; visitor's center. Museum-related items for sale.
Activities: formal education programs for children; guided tours; hobby workshops; hands-on activities.
Hours & Admission Prices: June-Aug. Mon. & Thurs.-Sat. 10-5, Sun. 12-5; Sept.-Oct. Fri.-Sat. 10-5, Sun. 12-5. Adults $8, senior citizens $6, children $5; discounts to groups; members no charge. &
Attendance: 13,778 (accurate)
Membership: See Minnesota Historical Society.

Pipestone

PIPESTONE COUNTY HISTORICAL MUSEUM, 113 S. Hiawatha Ave., Pipestone, MN 56164-1664. Tel.: 507-825-2563. Fax: 507-825-2563.
E-mail: pipctymu@iw.net
Web Site: www.pipestoneminnesota.com/museum
Founded: 1880.
Congressional District: 2
Key Personnel: Dir., Susan Hoskins; Pres. (V), Curt Hess.
Personnel Profile: Full-Time Paid 1; Part-Time Paid 2; Part-Time Volunteers 50.
Governing Authority: nonprofit organization. Parent Institution: Pipestone County Historical Society. Tax-exempt: 501(c)(3).

Institution Type/Description: History Museum: housed in 1896 Old Pipestone City Hall.
Collections: pipes; quilled & beaded clothing from the Dakota & Ojibwa tribes; plains Indian saddles; period artifacts; American culture including tools, photographs, quilts, glassware, toys, farming implements, trade items, clothing & furniture; manuscripts.
Research Fields: local genealogy; history.
Facilities: 200-vol. library of newspapers, 200-vol. history publications, biographies, land deeds, local history publications & atlas available for research on premises under staff supervision; reading room; microfilm reader-printer with local newspapers; doctors records; Pipestone Indian School records, additional archival resources. Books & local craftwork for sale.
Activities: guided tours; lectures; gallery talks; formally organized education programs; temporary & traveling exhibitions; biannual Civil War Days Festival.
Publications: quarterly newsletter.
Hours & Admission Prices: Mon.-Sat. 10-5. Adults $3; children under 12 & members no charge. Tour Bus: Museum Tour or Step-On Tour $50. Closed New Year's Day; Presidents' Day; Easter; Memorial Day; Independence Day; Labor Day; Veterans Day; Thanksgiving; Christmas Eve & Day.
Attendance: 7,051 (accurate)
Membership: Friend $15-$24; Household $25-$49; Supporting $50-$99; Century $100-$249; Patron $250-$499; Benefactor $500 & up.

PIPESTONE NATIONAL MONUMENT, 36 Reservation Ave., Pipestone, MN 56164-1269. Tel.: 507-825-5464. Fax: 507-825-5466.
E-mail: pipe_interpretation@nps.gov
Web Site: www.nps.gov/pipe
Founded: 1937.
Congressional District: 1
Key Personnel: Supt., Glen Livermont.
Governing Authority: federal. Parent Institution: National Park Service. Tax-exempt: 501(c)(3).
Institution Type/Description: History Museum & Upper Midwest Indian Cultural Center.
Collections: Indian ceremonial pipes & pipestone objects; pipestone quarries; herbarium; insects; archives.
Research Fields: American Indian ceremonial pipes; Indian culture.
Facilities: 500-vol. library of books on Indians, the West, natural history & Bureau of Ethnology reports available for research on premises. Indian-made catlinite pipes & beadwork for sale.
Activities: guided tours; lectures; cultural demonstrations; slide presentations; films; self-guiding trail.
Publications: booklets, Pipestone: A History; Pipes on the Plains.
Hours & Admission Prices: Daily 8-5. Car: $5, Individual $3; children 15 & under & enrolled members of federally recognized American Indian tribes no charge. Closed New Year's Day; Thanksgiving; Christmas.
Attendance: 80,000
Membership: Annual $15.

Reads Landing

WABASHA COUNTY MUSEUM, WABASHA COUNTY HISTORICAL SOCIETY AT READS LANDING, 70537 206th Ave., Reads Landing, MN 55968. Mailing Address: P.O. Box 255, Lake City, MN 55041-0255.
E-mail: info@wabashacountyhistory.org
Web Site: www.wabashacountyhistory.org
Founded: 1965.
Congressional District: 2
Key Personnel: Pres., Helen Myers.
Personnel Profile: Part-Time Volunteers 10.
Governing Authority: nonprofit organization. Parent Institution: Wabasha County Historical Society. Tax-exempt: 501(c)(3).
Institution Type/Description: General Museum.
Collections: farm implements; horse drawn farm & home machinery, vehicles; history items; period furniture. Historic Building: 1870 brick and stone school.
Activities: guided tours; permanent exhibitions.
Publications: quarterly newsletter.
Hours & Admission Prices: May-Oct. Sat.-Sun. 1-4; groups by appointment. Adults $5, children 6-12 $3; children 5 & under no charge.
Attendance: 500 (estimated)
Membership: Individual $25; Family $40; Patron $100; Small Business $150; Corporate $250.

Red Wing

GOODHUE COUNTY HISTORICAL SOCIETY, (M), 1166 Oak St., Red Wing, MN 55066-2447. Tel.: 651-388-6024. Fax: 651-388-3577.
E-mail: goodhuecountyhis@qwestoffice.net
Web Site: goodhuehistory.org
Founded: 1869.
Congressional District: 2
Key Personnel: C.E.O., Amy Nelson; Outreach Coord., Diane Buganski; Collections Mgr., Johanna Grothe.
Personnel Profile: Full-Time Paid 2; Part-Time Paid 1; Part-Time Volunteers 86.
Volunteer Hours: 2,000
Operating Expenses: 235,000
Operating Income: 240,000
Governing Authority: nonprofit organization. Tax-exempt.
Institution Type/Description: Historical Museum.
Collections: photographs; manuscripts; biographies; historical books; immigration & early settlement; local government; pioneer artifacts; Dakota Indian artifacts; military history; natural history; industrial history; geology specimens; archaeology; clothing; sports; music; theatrical history; Red Wing stoneware & pottery; Nuremburg trial papers of Judge William Christianson of Red Wing, MN.
Research Fields: folklore; agriculture; archaeology; industrial history; natural history; paleontology; geology; medical history; mineralogy; textiles.
Facilities: approx. 300-vol. library of histories of Goodhue County, atlases, diaries, books on Minnesota, biographies, church publications & Bibles available for use on premises; reading room; historical slide & narrative programs, over 800 oral history tapes & transcripts. Indian-made articles, postcards, publications on geology, archaeology & Minnesota history for sale.
Activities: guided tours; formally organized education programs for children; permanent & temporary exhibitions.
Publications: quarterly, Goodhue County Historical News; The Sea Wing Disaster; 1894 Red Wing Stoneware Company catalog; 1877 plat map of Goodhue County; history of Red Wing Architecture; Goodhue County, Minnesota: A Narrative History; The Ghost Towns & Discontinued Post Offices of Goodhue County; Sky Crashers: A History of the Aurora Ski Club; Uncertain Lives: African-Americans and Their First 150 Years in the Red Wing, Minnesota Area.
Hours & Admission Prices: Call for hours. Adults $5; discounts to AAM & AASLH members; members no charge.
Attendance: 2,300 (estimated)
Membership: Annual $35-$500; Life $1,000.

Redwood

REDWOOD COUNTY MUSEUM, 913 W. Bridge St., Redwood, MN 56283. Tel.: 507-641-3329.
Founded: 1949.
Congressional District: 2
Key Personnel: Pres. (V), Bill Schwandt; Cur. & Museum Shop Mgr., Patricia Lubeck.
Personnel Profile: Full-Time Paid 1.
Governing Authority: nonprofit organization. Tax-exempt.
Institution Type/Description: General Museum.
Collections: county history; textiles; hand fans; period farm tools; Indian artifacts; natural history; clothing & handwork; doctor's office; one-room rural school; bridal room; military room; music; birds; animals; one cell jail.
Facilities: picnic area.
Activities: guided tours; lectures; permanent exhibitions.
Publications: monthly newsletter.
Hours & Admission Prices: May-Sept. Thurs.-Sun. 11-4. Adults $5, children $2; children under 5 no charge.
Attendance: 1,072 (accurate)
Membership: Individual $25; Couple $30; Lifetime $100.

Renville

HISTORIC RENVILLE PRESERVATION COMMISSION, 202 N. Main, Renville, MN 56284. Mailing Address: Box 681, Renville, MN 56284-0681. Tel.: 612-329-3545.
Founded: 1976.
Congressional District: 2
Key Personnel: Pres. (V), Mildred Zaske.
Operating Expenses: 865
Operating Income: 480
Governing Authority: nonprofit organization. Tax-exempt.
Institution Type/Description: Historical Society Museum.

Collections: pioneer tools & implements; instruments, journals & books belonging to early day doctors; school artifacts & memorabilia; military uniforms & items; Minnesota River Valley wildlife; Renville's oil & gas station artifacts. Historic Houses: 1904 City Jail.

Research Fields: Indian history; prominent settlers & business people; former riverboat city of Vicksburg.

Facilities: Renville County Genealogy Society research material available.

Activities: lectures; arts festivals; hobby workshops; permanent & temporary exhibitions. Commission Sponsors: Pioneer demonstrations; Renville celebration.

Publications: annual newsletter; monthly article in local paper; paperback, Adrian Looks Back (in cooperation with Renville Star Farmer News).

Hours & Admission Prices: June-Sept. Sun. 1-4 & special events. No charge; donations accepted. &

Attendance: 200 (accurate)

Membership: Single $2; Family $3.

Richfield

RICHFIELD HISTORICAL SOCIETY, 6901 Lyndale Ave., S., Richfield, MN 55423-2302. Tel.: 612-798-6140.

E-mail: staff@richfieldhistory.org

Web Site: www.richfieldhistory.org

Founded: 1967.

Key Personnel: Dir., Jodi Larson; Pres. (V), Bill Walker.

Personnel Profile: Part-Time Paid 1; Part-Time Volunteers 28; Interns 12.

Governing Authority: Tax-exempt.

Institution Type/Description: Historic House Museum: listed on the National Register of Historic Places.

Collections: local history & culture; photographs; period artifacts.

Research Fields: oral histories; post WW II cultural history; community history.

Facilities: History Center; library.

Activities: adult & student education programs; music series.

Publications: RHS quarterly newsletter, Richfield History; book, Richfield: Minnesota's Oldest Suburb.

Hours & Admission Prices: Wed. 12-8, Sat. 12-4; other times by appointment. No charge; donations accepted.

Attendance: 2,224 (accurate)

Membership: $20.

Rochester

MAYO CLINIC HERITAGE HALL, 2001 1st St., S.W., Mathews Grand Lobby, Rochester, MN 55905-0002. Tel.: 507-284-8540.

E-mail: mayoclinic.heritagehall@mayo.edu

Web Site: www.mayoclinic.org/heritage-hall

Key Personnel: Dir., Matthew Dacy

Institution Type/Description: Medical History Museum.

Collections: clinic history; medical instruments; photographs; personal artifacts.

Hours & Admission Prices: Mon.-Fri. 8-5. No charge.

OLMSTED COUNTY HISTORICAL SOCIETY DBA HISTORY CENTER OF OLMSTED COUNTY, 1195 W. Circle Dr., S.W., Rochester, MN 55902-6619. Tel.: 507-282-9447. Fax: 507-289-5481.

E-mail: director@olmstedhistory.com

Web Site: www.olmstedhistory.com

Founded: 1926.

Congressional District: 1

Key Personnel: Pres. (V), David Bastyr; Mgr. Collections, Cara Clarey; Accountant, Roxanne Ziecina; Program Coord., Leah Brey-Fratzke.

Personnel Profile: Full-Time Paid 9; Part-Time Paid 16; Part-Time Volunteers 400; Interns 3.

Governing Authority: society. Tax-exempt: 501(c)(3).

Institution Type/Description: History Museum & Historic Site.

Collections: general local & regional history; farm implements & machinery; manuscripts. Historic Buildings: 1910-1911 Mayowood; 1860 Stoppel Farmstead; 1860 Dee Cabin; 1900 Hadley School.

Major Exhibits: Our Human Landscape, 4/12-1/15; The Story of Us, 10/12-12/15.

Research Fields: Olmsted County history.

Facilities: 3,000-vol. library of history, technology, genealogy & local microfilm available for use under supervision; reading room; classrooms. Museum reproductions & books for sale.

Activities: guided tours; lectures; study clubs; formally organized education programs; permanent, temporary & traveling exhibitions; school loan service; Time Traveler's Program.

Publications: quarterly newsletter, The Scribe; biannual magazine, The Olmsted Historian.

Hours & Admission Prices: Museum: Tues.-Sat. 9-5. Adults $5, children under 15 $2; members no charge. Mayowood Tours: May 7-May 28 & Aug. 13-Oct. 15 Sat. hourly 11-2; June-Aug. 11 Tues.-Thurs. & Sat. hourly 11-2. Adults $12, children $6. Research Center: Tues-Thurs. & Sat. 9-5. Closed national holidays. &

Attendance: 28,000 (estimated)

Membership: Seniors 65 & over $35; Individual $40; Individual Plus $45; Partner & Business Supporter $50 & up; Senior Couple $60; Household (includes children under 17) $65; Business Booster $100 & up; Sponsor $250 & up; Benefactor $500 & up.

ROCHESTER ART CENTER, (M), 40 Civic Center Dr., S.E., Rochester, MN 55904-3773. Tel.: 507-282-8629. Fax: 507-282-7737.

E-mail: info@rochesterartcenter.org

Web Site: www.rochesterartcenter.org

Founded: 1946.

Congressional District: 1

Key Personnel: Exec. Dir., Sarah Stauder; Administrative Operations Dir., Joan Lovelace; Chief Cur., Kristopher Douglas; Facility Dir. & Head Preparator, Phillip Ahnen; Dir. Mktg. & Public Programs, Naura Anderson; Coord. Education, Jason Pearson; Dir. Devel. & Mktg, Sandy Thompson; Event Mgr., Emily Tweten; Membership & Mktg. Asst., Kayla Benson.

Personnel Profile: Full-Time Paid 7; Part-Time Paid 18; Part-Time Volunteers 20; Interns 8.

Governing Authority: nonprofit organization. Tax-exempt: 501(c)(3).

Institution Type/Description: Art Center.

Collections: paintings.

Facilities: classrooms; sculpture garden; video space.

Activities: tours; lectures; films; concerts; workshops; demonstrations; classes for children & adults; outreach programs in the community.

Publications: exhibition catalogs; education department brochures.

Hours & Admission Prices: Wed. & Fri.-Sat. 10-5, Thurs. 10-9, Sun. 12-5. Adults $5, seniors $3; discounts to AAM members; Olmsted County residents, members & students no charge. Closed holidays. &

Attendance: 20,000 (estimated)

Membership: Senior Citizen, Student & Non-Resident $30; Individual $40; Family $60; Supporting $100-$249; Sustaining $250-$499; Patron $500-$749; Benefactor $750-$999; Founder $1,000-$2,499; Sponsor $2,500 & up.

Rockford

ROCKFORD AREA HISTORICAL SOCIETY, 8131 Bridge St., Rockford, MN 55373. Mailing Address: P.O. Box 186, Rockford, MN 55373-0186. Tel.: 763-477-5383.

E-mail: storkhouse@cityofrockford.org

Web Site: www.rockfordmnhistory.org

Founded: 1986.

Key Personnel: Dir., Rebecca Mavencamp; Pres., Naomi Binsfeld.

Personnel Profile: Part-Time Paid 1; Part-Time Volunteers 100.

Governing Authority: Branch Museum: The Ames-Florida-Stork House. Tax-exempt.

Institution Type/Description: Historical Society Museum.

Collections: clothing & textiles; furniture; housewares; library & archival holdings; photographs.

Research Fields: Rockford history & genealogy.

Facilities: Property usable for group events up to 200 people.

Activities: children's programs; monthly educational classes. Museum Sponsors: Memorial Day Pie and Ice Cream Social; Christmas Tea.

Publications: online newsletters weekly & emails containing diary & letter entries.

Hours & Admission Prices: April-Sept. Tues.-Wed. 10-4, Thurs. 10-7; Oct.-March Tues.-Thurs. 10-4. Adults $3; members & children 6 and under no charge.

Attendance: 1,500 (accurate)

Membership: Individual $10; Family $25; Business $45; Sponsor $100.

Rogers

ELLINGSON CAR MUSEUM, 20950 Rogers Dr., Rogers, MN 55374-9191. Tel.: 763-428-7337. Fax: 763-428-4370.

E-mail: ecmmuseum@mm.com

Web Site: www.ellingsoncarmuseum.com

Founded: 1994.

Key Personnel: Owner, Scott Ellingson.

Personnel Profile: Full-Time Paid 1; Part-Time Paid 6.

Governing Authority: for profit.

Institution Type/Description: Car Museum.

Collections: over 100 cars from early 1900s to 1970.

Facilities: Museum-related items for sale.
Activities: outdoor car shows in summer.
Hours & Admission Prices: Daily 10-5. Adults $5. Closed holidays. &
Attendance: 25,000 (estimated)

Rollingstone

ROLLINGSTONE-LUXEMBOURG HERITAGE MUSEUM, 98 Main St., Rollingstone, MN 55969. Mailing Address: P.O. Box 63, Rollingstone, MN 55969-0063. Tel.: 507-689-2307 & 2096.
Founded: 1987.
Congressional District: 28
Key Personnel: Dir. & Chm. (V), Jean Kalmes; Pres., Diane Bronk.
Personnel Profile: Part-Time Volunteers 8.
Governing Authority: Tax-exempt.
Institution Type/Description: Heritage Museum.
Collections: local history & culture; early immigrants; photographs; period clothing & uniforms; quilts; paintings; tools; fire equipment; dolls & toys; Pieta statue; angel statue.
Research Fields: local genealogy.
Facilities: library.
Activities: Museum Sponsors: Treipenfest Banquet in January; Fall Fest in September.
Publications: newsletter.
Hours & Admission Prices: Sun. 1-4; other times by appointment. No charge. Closed Easter; Christmas. &
Attendance: 1,000 (estimated)
Membership: Individual $5; Family $10.

Roseau

ROSEAU COUNTY HISTORICAL MUSEUM AND INTERPRETIVE CENTER, 121 Center St. E., Ste. 101, Roseau, MN 56751-1127. Tel.: 218-463-1918. Facebook: Roseau County Historical Museum.
E-mail: rchsroseau@mncable.net
Web Site: www.roseaucohistoricalsociety.org
Founded: 1927.
Congressional District: 1
Key Personnel: Pres. (V), Glenn Holm; Cur. & Dir., Britt Dahl.
Personnel Profile: Full-Time Paid 1; Part-Time Paid 2; Part-Time Volunteers 80.
Governing Authority: Parent Institution: Roseau County Historical Society. Tax-exempt: 501(c)(3).
Institution Type/Description: Historical Society Museum & Interpretive Center.
Collections: Historic Building: 1898, restored Pinecreek church.
Research Fields: county & natural history; genealogy.
Facilities: library.
Activities: guided tours; traveling exhibits; workshops; research; adult & youth programs. Museum Sponsors: One Woman, Honoring Women in Roseau County; Annual Minnesota History Contest.
Publications: Books, Pioneers O Pioneers, History of early settlers in county 1885-1910; Remembrances, An Anthology of Roseau County Minnesota; Pioneers O Pioneers Book II; Roseau County Heritage Book; Roseau County Centennial Book; 75th Anniversary Cookbook; P is for Pioneer: An ABC Book for Children & Adults.
Hours & Admission Prices: Mon.-Fri. 9:30-5. Museum: no charge; donations accepted. Research Center: adults $5; discounts to AAM members. Closed national holidays. &
Attendance: 8,997 (accurate)
Membership: Individual $10; Family $15; Business $25; Sponsor $50; Patron $100; Benefactor $250.

Roseville

HARRIET ALEXANDER NATURE CENTER, 2520 N. Dale St., Roseville, MN 55113-3502. Tel.: 651-765-4262. Fax: 651-792-7160.
E-mail: hanc@ci.roseville.mn.us
Web Site: www.cityofroseville.com/index.aspx?nid=183
Key Personnel: Lead Naturalist & Recreation Supvr., Debbie Cash
Institution Type/Description: Nature Center.
Collections: live animals; anatomy models; furs; botanical.
Facilities: 52 acres.
Hours & Admission Prices: Tues.-Sat. 10-4, Sun. 1-4. No charge.

Saint Cloud

MINNESOTA AMATEUR BASEBALL HALL OF FAME, St. Cloud Civic Center, 2nd Fl., 10 4th Ave. S., Saint Cloud, MN 56303. Mailing Address: 541 Brook Lane, Saint Cloud, MN 56301-9611. Tel.: 320-252-8227. Fax: 320-230-3277.
E-mail: mnbaseballhof@charter.net
Web Site: www.mnamateurbaseballhof.com/
Founded: 1963.
Institution Type/Description: Sports Museum.
Collections: amateur baseball history; uniforms; caps; photographs; bats; balls; gloves; programs.
Activities: Museum Sponsors: Annual Banquet.
Hours & Admission Prices: Call for hours.

MUNSINGER AND CLEMENS GARDENS, 1300 Kilian Blvd., S.E., Saint Cloud, MN 56304-1647. Mailing Address: Friends of the Gardens, 101 S. 7th Ave., Ste. 100, Saint Cloud, MN 56301-4275. Tel.: 320-257-5959. Fax: 320-257-0657.
E-mail: info@co.stcloud.mn.us
Web Site: www.ci.stcloud.mn.us
Institution Type/Description: Botanical Gardens.
Collections: plants; trees; flowers.
Facilities: Museum-related items for sale.
Activities: special events; concerts.
Hours & Admission Prices: Spring to Fall daily 7am-10pm. No charge; donations accepted.

PARAMOUNT GALLERY, 913 W. St. Germain St., Saint Cloud, MN 56301-3460. Tel.: 320-257-3120. Fax: 320-257-3111.
E-mail: info@paramountarts.org
Web Site: www.paramountarts.org
Key Personnel: Dir., Ellen Nelson
Institution Type/Description: Art Gallery.
Collections: works by local & regional artists; paintings; sculpture.
Hours & Admission Prices: Mon.-Fri. 11-5, Sat. 10-2.

*** STEARNS HISTORY MUSEUM, (M),** 235 33rd Ave., S., Saint Cloud, MN 56301-3752. Tel.: 320-253-8424; 866-253-8424 (toll free). Fax: 320-253-2172. TDD: 320-253-8424.
E-mail: info@stearns-museum.org
Web Site: www.stearns-museum.org
Founded: 1936.
Congressional District: 6
Key Personnel: Pres. (V), Barclay Carrier; Deputy Dir., Ann Meline; Dir. Education, Debi Pack; Dir. Archives, John Decker; Archivist, Sarah Warmka; Business Mgr. & Museum Shop Mgr., Cynthia O'Konek; Facility Mgr., Glenn Liesch; Administrative Asst., Diane Smith; Archivist, Robert Lommel.
Personnel Profile: Full-Time Paid 9; Part-Time Paid 2; Part-Time Volunteers 100; Interns 2.
Governing Authority: society; nonprofit organization. Tax-exempt: 501(c)(3).
Institution Type/Description: History Museum.
Collections: biographies; research materials; artifacts; large miniature circus & carnival; library of oral history tapes; photographs; local history; dairy farming; resorts & tourism; natural history of central Minnesota; immigration; granite industry; Pan Car & Motor Company; local amateur & professional baseball.
Research Fields: genealogy; history of county & state.
Facilities: research library; archives; museum cultural center.
Activities: educational program; workshops & seminars.
Publications: bimonthly newsletter; annual report; books, topical history series.
Hours & Admission Prices: Mon.-Sat. 10-5, Sun. 12-5. Adults $5, children $4; discounts to Sept. Days Club, groups, Time Travelers Museum Network, AAA, AAM & ICOM members; children under 5 & members no charge. Closed major holidays. &
Attendance: 16,200 (accurate)
Membership: Senior $30; Individual $35; Family Plus $50; Business, Clubs & Organizations $150; Benefactor $250; Corporate $500; Patron's Circle $1,000.

Saint Joseph

BENEDICTA ARTS CENTER OF THE COLLEGE OF SAINT BENEDICT, 37 S. College Ave., Saint Joseph, MN 56374-2001. Tel.: 320-363-5777. Fax: 320-363-6097.
E-mail: jdubbeldeekuhn@csbsju.edu
Web Site: www.csbsju.edu/finearts

Founded: 1963.
Congressional District: 6
Key Personnel: Gallery Mgr., Jill Dubbeldee Kuhn; Exec. Dir. of Fine Arts, Brian Jose.
Governing Authority: college. Tax-exempt.
Institution Type/Description: Art Gallery.
Collections: Asian & New Guinea artifacts; contemporary prints, drawings, paintings, sculpture & ceramics.
Facilities: auditorium; theater.
Activities: guided tours; lectures; music, theater & dance presentations; temporary exhibits.
Hours & Admission Prices: Summer: call for hours; Sept.-May 11 Mon.-Sat. 10-9, Sun. 12-9. No charge. Closed holidays; school vacations. &
Attendance: 5,000 (estimated)

Saint Louis Park

PAVEK MUSEUM OF BROADCASTING, 3517 Raleigh Ave., Saint Louis Park, MN 55416-2625. Tel.: 952-926-8198. Fax: 952-929-6105.
E-mail: snr@museumofbroadcasting.org
Web Site: www.museumofbroadcasting.org
Founded: 1986.
Congressional District: 5
Key Personnel: Dir. & Chm. (V), Jeffrey T. Bakken.
Personnel Profile: Full-Time Paid 4; Part-Time Paid 2; Part-Time Volunteers 5.
Governing Authority: Tax-exempt.
Institution Type/Description: Technology Museum.
Collections: period radio, television, & broadcast equipment; history of broadcasting.
Facilities: Museum-related items for sale.
Activities: special events.
Publications: newsletter, Pavek Museum of Broadcasting.
Hours & Admission Prices: Wed.-Sat. 10-5; other times by appointment. Adults $6, students & senior citizens $5. Closed holidays.
Attendance: 10,053 (accurate)
Membership: Senior 62 & over $25; Individual $35; Group $70; Sparkle Circle $100-$499; Key Circle $500-$999; Innovators Circle $1,000-$4,999; Collector's Circle $5,000-$9,999; Power Circle $10,000-$49,999; Sustainers Circle $50,000 & up.

Saint Paul

*** ALEXANDER RAMSEY HOUSE, MINNESOTA HISTORICAL SOCIETY, (M),** 265 S. Exchange St., Saint Paul, MN 55102-2416. Tel.: 651-296-8760. Fax: 651-296-0100.
E-mail: ramseyhouse@mnhs.org
Web Site: www.mnhs.org
Founded: 1964.
Congressional District: 4
Key Personnel: C.E.O., Nina M. Archabal; Dir. & Security, Kevin Maijala; Education, Dana Heimark; Historic Site Admin., Jim Matson.
Personnel Profile: Full-Time Paid 3; Part-Time Paid 15; Part-Time Volunteers 22; Interns 1.
Governing Authority: nonprofit organization; semi-state agency. Parent Institution: Minnesota Historical Society. Tax-exempt.
Institution Type/Description: Historic House: Late 19th-century upper class Victorian mansion.
Collections: original furnishings of the Ramsey family.
Research Fields: Ramsey family history.
Activities: guided tours; lectures; organized education programs for children. Museum Sponsors: Victorian Holiday Program.
Hours & Admission Prices: Jan.-May & Sept.-Nov. 18 Fri.-Sat. & holidays 10-3; June-Aug. Tues.-Sat. & holidays 10-3; Nov. 19-Dec. Wed.-Sat. 10-3. Adults $8, senior citizens $6, children 6-17 $5; discounts for AAM members; children under 5 & members no charge. Closed Thanksgiving; Christmas Eve & Day. &
Attendance: 13,500 (accurate)
Membership: Senior Individual $45; Individual & Senior Household $55; Household $65; Associate Member $125.

AMERICAN ASSOCIATION OF WOODTURNERS GALLERY OF WOOD ART, 222 Landmark Ctr., 75 5th St. W., Saint Paul, MN 55102. Tel.: 651-484-9094.
Key Personnel: Dir., Tib Shaw; C.E.O., Phil McDonald.
Governing Authority: Parent Institution: American Association of Woodturners. Tax-exempt.
Institution Type/Description: Art Gallery.
Collections: works by wood artists including carvings, sandblasted, segmented, reconstructed & burned pieces.

Facilities: Gallery-related items for sale.
Activities: school & group tours.
Publications: American Woodturner.
Hours & Admission Prices: Tues.-Fri. 11-4, Sun. 12-3; other times by appointment. No charge; donations accepted. Closed New Year's Eve & Day; Easter; Memorial Day; Labor Day; Thanksgiving; Christmas Eve & Day. &
Attendance: 8,500 (accurate)

AMERICAN MUSEUM OF ASMAT ART AT ST. THOMAS, 2115 Summit Ave., Mail #57P, Saint Paul, MN 55105-1048. Tel.: 651-962-5512. Fax: 651-962-5861.
E-mail: jarisser@stthomas.edu
Web Site: www.stthomas.edu/art/asmat/
Founded: 1995.
Congressional District: 4
Key Personnel: Dir., Julie Risser; Chm. Bd. of Advisors, Gerald D. Brennan.
Governing Authority: private; not-for-profit. Tax-exempt: 501(c)(3).
Institution Type/Description: Ethnological Art Museum.
Collections: art, culture & carvings from the Asmat people of Western Papua & Indonesia.
Research Fields: Asmat art & culture.
Facilities: 220-vol. library on Asmat culture; 600 sq. ft. exhibit space.
Activities: guided tours; lectures; education programs for adults; loan, participatory, temporary & traveling exhibitions; docent program.
Publications: quarterly newsletter, Embodied Spirits; books, Asmat Sketch Book vol. 1-8, Making the Invisible Visible, Asmat Images, The Asmat: Myth & Ritual; The Asmat: Perception of Life in Art.
Hours & Admission Prices: Mon.-Wed. 10-4, Thurs. 10-8, Fri. 10-2, Sat.-Sun. 12-4. No charge; donations accepted. &
Attendance: 3,720 (accurate)
Membership: Individual $40; Family $60; Institutional $100 & up; Discovery $100-$499; Animating $500-$999; Sustaining $1,000 & up.

ARCHIVES - ARCHDIOCESE OF ST. PAUL AND MINNE-APOLIS, 226 Summit Ave., Saint Paul, MN 55102-2121. Tel.: 651-291-4429. Fax: 651-290-1629.
E-mail: archives@archspm.org
Founded: 1988.
Key Personnel: Dir. Records & Archives, Heather Lawton; Archivist, Steven T. Granger, C.A.
Governing Authority: church; nonprofit organization. Tax-exempt: 501(c)(3).
Institution Type/Description: Religious Museum.
Collections: over 1,000 linear ft. of archival materials; archdiocese business records; historic records of Catholic parishes, religious groups & institutions; photographs; maps & plans; religious artifacts.
Facilities: 300-vol. library on Catholic history.
Activities: guided tours; lectures; temporary exhibitions.
Hours & Admission Prices: By appointment.
Attendance: 25 (estimated)

CATHERINE G. MURPHY GALLERY - ST. CATHERINE UNIVERSITY, Visual Arts Bldg., 2004 Randolph Ave., Saint Paul, MN 55105-1789. Tel.: 651-690-6644. Fax: 651-690-6050. Facebook: Catherine G. Murphy Gallery - St. Catherine University.
E-mail: kmdaniels@stkate.edu
Web Site: www.stkate.edu/gallery
Founded: 1905.
Key Personnel: Dir., Kathleen M. Daniels.
Personnel Profile: Full-Time Paid 1; Part-Time Paid 2; Interns 1.
Governing Authority: Parent Institution: St. Catherine University. Tax-exempt.
Institution Type/Description: Art Gallery.
Collections: Clara Mairs; Adolph Dehn; Corita Kent; Cecilia Lieder; installations.
Hours & Admission Prices: Feb. 3-May 25 & Sept. 8-Dec. 18 Mon.-Fri. 8-8, Sat.-Sun. 12-6. No charge; donations accepted. Closed Easter break; Thanksgiving break. &
Attendance: 4,000 (accurate)

COMO PARK ZOO AND CONSERVATORY, 1225 Estabrook. Dr., Saint Paul, MN 55103-1022. Tel.: 651-487-8201. Fax: 651-487-8245. Facebook: facebook.com/comozooconservatory.
E-mail: comomarketing@stpaul.gov
Web Site: www.comozooconservatory.org
Formerly: St. Paul's Como Zoo
Founded: 1897.
Congressional District: 4
Key Personnel: Pres. Como Friends, Dr. Jackie Sticha; Consulting Veterinarian Prof. & Clinician, Univ. of Minn. College of Veterinary Medicine, Dr. Micky Trent; Animal Cur., John Dee; Campus Mgr., Michelle Furrer.

Personnel Profile: Full-Time Paid 84; Part-Time Paid 36; Part-Time Volunteers 1,200; Interns 30.
Governing Authority: Parent Institution: St. Paul Parks & Recreation; affiliated with Univ. of MN. College of Veterinary Medicine for Research & Teaching.
Institution Type/Description: Zoo.
Collections: mammals; birds; reptiles; endangered species; marine mammals.
Research Fields: propagation of large cats & orangutans; animal husbandry with students of veterinary medicine.
Facilities: 800-vol. library of natural history & taxonomic books available for research on premises; zoological park. Postcards, film, booklets & guide books for sale.
Activities: guided tours; lectures; TV & radio programs; docent program; permanent, temporary & traveling exhibitions; school loan service.
Publications: monthly bulletin, newsletter, Como Combo.
Hours & Admission Prices: April-Sept. daily 10-6; Oct.-March daily 10-4. No charge; donations accepted. &
Attendance: 2,230,000 (estimated)
Membership: $35-$500

DENLER ART GALLERY AT UNIVERSITY OF NORTH-WESTERN AT ST. PAUL, Totino Fine Arts Center, 2nd Fl., 3003 Snelling Ave., N., Saint Paul, MN 55113-1501. Tel.: 651-286-7560. Facebook: Denler Gallery.
E-mail: lbaleckson@unwsp.edu
Web Site: www.unwsp.edu/web/art-dept/denler-gallery
Formerly: Denler Art Gallery at Northwestern College
Key Personnel: Dir., Luke Aleckson
Institution Type/Description: Art Gallery.
Collections: artwork by local and national artists.
Hours & Admission Prices: Call for hours.

GOLDSTEIN MUSEUM OF DESIGN, (M), 364 McNeal Hall, 1985 Buford Ave., Saint Paul, MN 55108-6134. Tel.: 612-624-7434. Fax: 612-625-5762.
E-mail: gmd@umn.edu
Web Site: goldstein.design.umn.edu
Founded: 1976.
Congressional District: 5
Key Personnel: Dir., Lin Nelson-Mayson; Chm. (V), Tim Quigley; Asst. Cur., Jean McElvain; Registrar, Eunice Haucen; Grants Writer, Kathleen Campbell.
Personnel Profile: Full-Time Paid 3; Part-Time Paid 15; Part-Time Volunteers 25; Interns 1.
Volunteer Hours: 2,100
Operating Expenses: 403,271
Operating Income: 406,207
Governing Authority: university. University of Minnesota. Tax-exempt.
Institution Type/Description: Design Museum.
Collections: costumes; decorative arts; graphic design; textiles; furniture.
Research Fields: costume; textiles; decorative arts; graphic design; architecture; landscape architecture.
Activities: guided tours; gallery talks; lectures,
Publications: collections catalogue; catalogues for specific exhibitions.
Hours & Admission Prices: McNeal Hall: Tues.-Fri. 10-4. Rapson Hall: Mon.-Thurs. 9-9, Fri. 9-6, Sat.-Sun. 1-5. No charge; donations accepted. Closed major holidays & university holidays. &
Attendance: 13,000 (estimated)
Membership: Student $10; Senior $20; Individual $35; Household $50; Sponsor $150; Patron $250; Benefactor $500; Directors Circle $1,000.

HAMLINE UNIVERSITY, SOEFFKER GALLERY, Hamline University, Drew Fine Arts Center, Saint Paul, MN 55104-1284. Mailing Address: 1536 Hewitt Ave., Saint Paul, MN 55104-1284. Tel.: 651-523-2457. Fax: 651-523-3066.
E-mail: aaudeh@gw.hamline.edu
Web Site: www.hamline.edu/art/
Formerly: Hamline University Galleries, Department of Studio Arts & Art History
Founded: 1854.
Key Personnel: Registrar, Nicole Flam; Art Handler, John-Mark Schlink.
Governing Authority: Parent Institution: Hamline University. Tax-exempt: 501(c)(3).
Institution Type/Description: Art Gallery.
Collections: sculpture; paintings; archaeological; decorative arts.
Activities: lectures; films; gallery talks; docent-led exhibition tours; arts festivals; formally organized education programs for undergraduate college students; education programs & hands-on activities for children; inter-museum loan, permanent & temporary exhibitions.

Publications: exhibition catalogs; exhibition posters; Icons of Perfection: Figurative Sculpture from Africa.
Hours & Admission Prices: Mon.-Fri. 10-4. No charge. Closed national holidays.

＊ HISTORIC FORT SNELLING, (M), Ft. Snelling History Center, 200 Tower Ave., Saint Paul, MN 55111-4037. Tel.: 612-726-1171. Fax: 612-725-2429.
E-mail: ftsnelling@mnhs.org
Web Site: www.historicfortsnelling.org
Founded: 1970.
Key Personnel: Site Mgr., Tom Pfannenstiel.
Personnel Profile: Full-Time Paid 7; Part-Time Paid 55; Part-Time Volunteers 25.
Governing Authority: nonprofit organization. Parent Institution: Minnesota Historical Society. Tax-exempt.
Institution Type/Description: Historic Site.
Collections: 4 original & 14 reconstructed buildings.
Research Fields: Frontier social history & military material culture 1820s-1860s.
Facilities: history center; 3,000 sq. ft. exhibit space; 300-seat theater. Historical reproductions & books for sale.
Activities: guided tours; films; organized education for children; participatory exhibits; living history programs.
Hours & Admission Prices: Memorial Day to Labor Day Tues.-Sat. 10-5, Sun. 12-5; Sept.-Oct. Sat. 10-5. Adults $11, seniors & students with ID $9, children 6-17 $6; discounts to school group; children 5 & under and Minnesota Historical Society members no charge. &
Attendance: 100,000 (accurate)
Membership: Minnesota Historical Society: Teacher $40-$65; Individual $45-$60; Household $65-$95; Supporting $145-$500; Northstar Circle $1,200 & up

IFP MN CENTER FOR MEDIA ARTS, 2446 University Ave. W., Ste. 100, Saint Paul, MN 55114. Tel.: 651-644-1912. Fax: 651-644-5708.
E-mail: word@ifpmn.org
Web Site: www.ifpmn.org
Key Personnel: Exec. Dir., Jane Minton.
Institution Type/Description: Media Arts Gallery.
Collections: photographs; films.
Activities: classes.
Hours & Admission Prices: Mon. & Wed.-Sat. 10-5:30, Tues. 10-9.

＊ JAMES J. HILL HOUSE, (M), 240 Summit Ave., Saint Paul, MN 55102-2194. Tel.: 651-297-2555. Fax: 651-297-5655, TDD: 651-282-6073.
E-mail: hillhouse@mnhs.org
Web Site: www.mnhs.org/hillhouse
Founded: 1978.
Key Personnel: Site Mgr., Craig Johnson; Program Supervisor, Sara Scrimshaw.
Personnel Profile: Full-Time Paid 4; Part-Time Paid 16; Part-Time Volunteers 2.
Volunteer Hours: 624
Governing Authority: state agency; nonprofit. Parent Institution: Minnesota Historical Society, 345 Kellogg Blvd. W., St. Paul. Tax-exempt.
Institution Type/Description: Historic House: 1891 family home of James J. Hill, builder of the Great Northern Railway.
Collections: original architecture; decorative elements; technical systems; some original furnishings.
Hours & Admission Prices: Wed.-Sat. 10-3:30, Sun. 1-3:30. Adults $9, senior citizens 65 & over $7, students & children 6-17 $6; discounts to AAM members & groups; MHS members and children 5 & under no charge. Closed New Year's Day; Easter; Thanksgiving; Christmas. &
Attendance: 47,113 (accurate)
Membership: Senior $45; Individual $50; Individual Plus & Senior Household $65; Household $75; Household Plus $95.

JULIAN H. SLEEPER HOUSE, (M), 66 St. Albans St., S., Saint Paul, MN 55105-3501. Tel.: 651-225-1505.
Web Site: julianhsleeperhouse.com
Founded: 1993.
Key Personnel: C.E.O., Dr. Seth C. Hawkins; Chm. Bd. (V), Janet C. Mahoney.
Personnel Profile: Full-Time Volunteers 1; Part-Time Paid 1; Part-Time Volunteers 12.
Governing Authority: individual operation; nonprofit.

Institution Type/Description: Historic Site: 1884 Eastlake-Vernacular House, moved to present site in 1911, furnished & decorated in period style.
Collections: Gilded Age furnishings; James A. Garfield memorabilia; Roseville pottery; albums of vintage U.S. & Canadian picture postcards; Slovenian cultural & historical artifacts.
Research Fields: 19th century professional baseball; history of U.S. public speaking.
Activities: guided tours; lectures.
Publications: illustrated catalog of the James A. Garfield collection; annual calendar.
Hours & Admission Prices: Private residence, open by appointment only. Adults $7; discounts to members & AAM members.
Attendance: 1,785 (accurate)
Membership: Annual $35.

MACALESTER COLLEGE ART GALLERY, JANET WALLACE FINE ARTS CENTER, 1600 Grand Ave., Saint Paul, MN 55105-1899. Tel.: 651-696-6416. Fax: 651-696-6266.
E-mail: fitz@macalester.edu
Web Site: www.macalester.edu/gallery/index.html
Founded: 1964.
Key Personnel: Cur., Gregory Fitz.
Personnel Profile: Full-Time Paid 1; Interns 10.
Governing Authority: private college; nonprofit.
Institution Type/Description: College Art Gallery.
Collections: British & Asian ceramics; contemporary & historical prints, drawings & paintings; international arts.
Facilities: educational facilities; 2,500 sq. ft. exhibit space.
Hours & Admission Prices: Mon.-Wed. & Fri. 10-4, Thurs. 10-8, Sat.-Sun. 12-4. No charge. Closed college holidays. &

MARJORIE MCNEELY CONSERVATORY, 1225 Estabrook Dr., Saint Paul, MN 55103-1022. Tel.: 651-487-8201. Fax: 651-487-8255.
E-mail: comomarketing@stpaul.gov
Web Site: www.comozooconservatory.org
Formerly: Como Park Conservatory
Founded: 1915.
Congressional District: 4
Key Personnel: Mgr., Tina Dombrowski; Campus Mgr., Michelle Furrer; Museum Shop Mgr., Terri Scheunemann.
Personnel Profile: Full-Time Paid 9; Part-Time Paid 12; Part-Time Volunteers 540; Interns 4.
Governing Authority: municipal. Parent Institution: City of St. Paul. Subsidiary Institution: Como Park Zoo and Conservatory. Tax-exempt.
Institution Type/Description: Botanical Garden.
Collections: orchids; bonsai; bromeliads; Japanese garden; palms; tropical plants; ferns.
Activities: limited lectures; permanent exhibitions; plant-related workshops; flower shows.
Publications: newsletter, Views.
Hours & Admission Prices: April-Sept. daily 10-6; Oct.-March daily 10-4. No charge; donations accepted. &
Attendance: 2,230,000 (estimated)
Membership: Como Friends: Senior $35; Individual $45; Family $75; Family Advantage $125.

MINNESOTA AIR NATIONAL GUARD HISTORICAL FOUNDATION, INC., (M), 670 General Miller Dr., Saint Paul, MN 55111-0598. Mailing Address: P.O. Box 11598, Saint Paul, MN 55111-0598. Tel.: 612-713-2523. Fax: 612-713-2524.
E-mail: msp04332@isd.net
Web Site: www.mnangmuseum.org
Founded: 1980.
Congressional District: 8
Key Personnel: Chm. (V), Mark Ness; Vice Chm., Richard Weissner; Museum Shop Mgr., Kathleen Sundby.
Personnel Profile: Full-Time Paid 1; Full-Time Volunteers 50; Part-Time Paid 1; Part-Time Volunteers 50.
Governing Authority: private; nonprofit organization. Tax-exempt. 501(c)(3).
Institution Type/Description: Aviation Museum: located on grounds of the 133rd Airlift Wing, Minnesota Air National Guard.
Collections: aircraft & artifacts flown by the 109th Squadron, Minnesota Air National Guard; history of the Minnesota National Guard; photographs.
Research Fields: construction of reproduction Curtiss Jenny Restoration of 0-2, L-4.
Facilities: 5,000-vol. library; 25,000 sq. ft. exhibit space; classroom. Museum-related items for sale.
Activities: guided tours; formal education programs for children; hobby

workshops; birthday parties; participatory & temporary exhibitions. Museum Sponsors: Open Cockpit Days; State Fair Exhibit.
Publications: quarterly newsletter, The Historian.
Hours & Admission Prices: Summer: call for hours. No charge; donations accepted.
Attendance: 3,000 (estimated)
Membership: Individual $25; Family $40; Life $100.

MINNESOTA CHILDREN'S MUSEUM, 10 W. 7th St., Saint Paul, MN 55102-2453. Tel.: 651-225-6000. Fax: 651-225-6006.
E-mail: mcm@mcm.org
Web Site: www.mcm.org
Formerly: Minnesota's Aware House
Founded: 1979.
Congressional District: 4
Key Personnel: Pres., Dianne Krizan; Chm. (V), Bill McKinney.
Personnel Profile: Full-Time Paid 51; Part-Time Paid 39; Part-Time Volunteers 309.
Governing Authority: nonprofit organization. Tax-exempt: 501(c)(3).
Institution Type/Description: Children's Participatory Museum.
Collections: hands-on exhibits of multi-disciplinary nature for children 6 months-10 years; Our World, exploring neighborhoods and appreciating their diversity; Earth World, celebrating stewardship of the Earth; World Works, tools, problem-solving, invention & creativity; Habitot, infant-toddler learning landscape; Rooftop Art Park: art & nature.
Research Fields: children's environments; children's behaviors in exhibits; safety in interactive exhibits; customer segments.
Facilities: classrooms; performance area.
Activities: interactive, multidisciplinary exhibits; programs for children, families, teachers & care providers; volunteer program.
Publications: bimonthly newsletter; annual report; case statement.
Hours & Admission Prices: Summer: Mon.-Thurs. 9-4, Fri.-Sat. 9-8; Winter: Tues.-Thurs. 9-4, Fri. & Sun. 9-5, Sat. 9-8. Adults $8.95; discounts to AARP & ACM reciprocal membership program; members & children under 1 no charge. &
Attendance: 404,000 (estimated)
Membership: Nanny $20; Foster Care $50; Passport $89; Passport Deluxe $129; Explorer $250.

✻ MINNESOTA HISTORICAL SOCIETY, (M), 345 Kellogg Blvd. W., Saint Paul, MN 55102-1903. Tel.: 651-259-3000; 800-657-3773. Fax: 651-297-3343. TDD: 651-282-6073.
E-mail: director@mnhs.org
Web Site: www.mnhs.org
Founded: 1849.
Congressional District: 4
Key Personnel: C.E.O. & Dir., D. Stephen Elliott; Deputy Dir. External Rels., Andrea Kajer; MHS Pres., Pamela J. McClanahan; Chief Mktg. Officer, Lory Sutton; Dept. Dir. Programs, Pat Gaarder; Head Exhibits, Dan Spock; C.F.O., Peggy Ingison; Devel. Officer, Cassie Cramer; Museum Shop Mgr., Meta DeVine.
Personnel Profile: Full-Time Paid 360; Part-Time Paid 260; Part-Time Volunteers 1,700; Interns 35.
Governing Authority: society. Branches: Comstock House, Moorhead; Forest History Center, Grand Rapids; Fort Ridgely History Center, Fairfax State Park; Historic Fort Snelling, St. Paul; James J. Hill House, St. Paul; Jeffers Petroglyphs, Comfrey; Oliver H. Kelley Farm, Elk River; Lindbergh House History Center, Little Falls; Lower Sioux Agency History Center, Redwood Falls; Mill City Museum, Minneapolis; Mille Lacs Indian Museum, Onamia; Minnesota History Center, St. Paul; Minnesota State Capitol, St. Paul; North West Company Fur Post, Pine City; Alexander Ramsey House, St. Paul; Split Rock Lighthouse History Center, Two Harbors; Historic Forestville, Preston. Historic Sites owned by MHS & operated by other organizations: W.H.C. Folsom House, Taylors Falls; Harkin Store, New Ulm; Lac qui Parle Mission, Montevideo; Minnehaha Depot, Minneapolis; W.W. Mayo House, Le Sueur. Tax-exempt: 501(c)(3).
Institution Type/Description: Historic Site & History Center.
Collections: three-dimensional objects, manuscripts, Minnesota state archives; collections of historic & prehistoric archaeological materials to 19th & 20th-century material culture; paintings; maps; photographs & moving pictures; material documenting & describing the history of Minnesota & the Upper Mississippi Valley & Great Lakes Area.
Major Exhibits: American Spirits: The Rise and Fall of Prohibition, 11/1/13-2/23/14.
Research Fields: Minnesota; Upper Mississippi River Valley & Great Lakes area; family history; historic preservation; railroads & transportation; business; organizations; politics; wild ricing; Minnesota communities.
Facilities: 550,000-vol. reference library, research center; 44,000 sq. ft. exhibit

space; auditorium; classrooms; conservation labs; restaurant. Museum-related items for sale.

Activities: permanent, temporary & traveling exhibits; tours; lectures; conferences; public & school programs.

Publications: quarterly journal; membership newsletter & calendar, monthly newsletter for historical organizations; scholarly & general interest books & pamphlets; Minnesota history textbook for grades 5-7.

Hours & Admission Prices: Museum: Tues. 10-8, Wed.-Sat. 10-5, Sun. 12-5. Adults $11; discounts to AAM & ICOM members; members no charge. Library: Tues. 12-8, Wed.-Sat. 9-4, Sun. 12-4. Historic Sites: $5-$11. Call for hours & admissions for individual museums & historic sites. &

Attendance: 1,000,000 (estimated)

Membership: Teacher Individual $40; Senior Individual $45; Individual $50; Senior Household & Teacher Household $65; Household $75; Associate $145; Contributing $250; Sustaining $500; North Star Circle $1,200.

MINNESOTA MUSEUM OF AMERICAN ART, (M), 408 St. Peter St., Ste. 419, Saint Paul, MN 55102. Tel.: 651-266-1030. Fax: 651-291-2947.

E-mail: kmakholm@mmaa.org
Web Site: www.mmaa.org
Founded: 1927.
Congressional District: 4
Key Personnel: Chm. Bd. (V), Dave Kelly; Exec. Dir., Kristin Makholm; Mgr. External Rels., Sarah Suemnig; Cur., Theresa Downing; Visitor Svcs. & Member Coord., Ben Gessner.
Personnel Profile: Full-Time Paid 4; Part-Time Paid 3; Part-Time Volunteers 25.
Governing Authority: nonprofit corporation. Museum: 505 Landmark Center, 75 W. 5th St., St. Paul, MN 55102. Tax-exempt: 501(c)(3), 170(b)(1)(A) & 507(b)(1)(A).
Institution Type/Description: Art Museum.
Collections: American, late 19th- to 20th-century fine art & studio craft with emphasis on art of upper Midwest.
Research Fields: contemporary American crafts; drawings; Paul Howard Manship; late 19th & early 20th-century American art; art of upper Midwest.
Facilities: 2,000-vol. research library of source material pertaining to the collections; reference room; classroom; two studios.
Activities: community-based partnership programs; lectures; films; gallery talks; education programs; internships for college students; docent program; inter-museum loan, temporary & traveling exhibitions.
Publications: newsletter; educator curriculum materials; brochures; exhibition catalogs; annual report.
Hours & Admission Prices: Temporarily closed. &
Attendance: 35,000 (estimated)
Membership: Student & Senior $25; Individual $35; Household $50; Patron $75 and up.

MINNESOTA STATE CAPITOL HISTORIC SITE, (M), 75 Rev. Dr. Martin Luther King Jr. Blvd., Saint Paul, MN 55155-1605. Tel.: 651-296-2881. Fax: 651-297-1502.

E-mail: statecapitol@mnhs.org
Web Site: www.mnhs.org/statecapitol
Founded: 1969.
Key Personnel: Mgr., Brian Pease; Site Supvr., Jaymie Korman; Prog. Mgr., Linda Cameron; Program Administrative Asst., Candice Christensen.
Personnel Profile: Full-Time Paid 4; Part-Time Paid 20.
Governing Authority: private; nonprofit organization. Parent Institution: Minnesota Historical Society, 345 Kellogg Blvd. W., St. Paul, MN 55101. Tax-exempt: 501(c)(3).
Institution Type/Description: Historic Site: designed by 19th-century architect Cass Gilbert.
Collections: 1905 furniture; canvas murals; paintings; plaques; statues & busts; governor's portraits; historic battle flags.
Research Fields: historical figures; capitol construction labor history; political history; Civil War; women's history.
Facilities: cafe. Museum-related items for sale.
Activities: view legislature in action; formal education programs; guided tours; temporary exhibitions. Annual Events: Art, Architecture, Minnesota History, & State Government programs.
Publications: quarterly magazine, Minnesota History.
Hours & Admission Prices: Hourly Tours: Mon.-Sat. 10-3, Sun. 1-4. Suggested Donation: $5 per person. Special Events: adults $11-$9, senior citizens $8, children 6-17 $6; discounts to members. Closed holidays. &
Attendance: 217,232 (accurate)
Membership: Senior $50; Individual Plus & Senior Household $65; Household $75; Household Plus $95; Associate $145; Contributing $250; Sustaining $500; North Star Circle $1,000.

MINNESOTA TRANSPORTATION MUSEUM, INC., (M), 193 Pennsylvania Ave. E., Saint Paul, MN 55130-4319. Tel.: 651-228-0263. Fax: 651-293-0857.

E-mail: director@mtmuseum.org
Web Site: www.trainride.org
Founded: 1962.
Congressional District: 2, 4, 5 & 6
Key Personnel: Dir., Thomas Klein; Chm., Richard Mullen; Vice Chm., Eric Hopp; Sec., Noel Peht; Treas., Richard Hoppe.
Personnel Profile: Full-Time Paid 1; Full-Time Volunteers 2; Part-Time Volunteers 460.
Governing Authority: nonprofit organization. Affiliated with the Minnesota Historical Society. Branch Museums: 114 Depot Rd. Osceola, WI 54020; 193 E. Pennsylvania Ave., St. Paul, MN 55103. Tax-exempt: 501(c)(3).
Institution Type/Description: Operating Transportation Museum.
Collections: 1913 MA & CR Locomotive 100; NP Steam Locomotive 328; 1907 NP Steam Engine 2156; various diesel locomotives, coaches built 1900-1960 for various Minnesota railroad companies.
Research Fields: buses; electric steam & diesel railways of Minnesota & adjacent states.
Facilities: material about railroads, electric lines & buses of Minnesota available on premises.
Activities: guided tours; 10- & 20-mile diesel train rides.
Publications: quarterly magazine, Minnegazette.
Hours & Admission Prices: Minnehaha Depot: Memorial Day-Labor Day Sun. & holidays 12:30-4:30. No charge. Osceola & St. Croix Valley Railway: May to late-Oct. Sat.-Sun. & holidays 11, 1, & 2:30. Adults $17, children $8. Jackson Street Roundhouse: Wed. & Sat. 9-5. Adults $10, children $8; discounts to AAM members. &
Attendance: 25,000 (estimated)
Membership: Individual $40; Contributing $100; Sustaining $200.

MINNESOTA VETERINARY HISTORICAL MUSEUM, College of Veterinary Medicine, 1365 Gortner Ave., Rm. 143, Animal Science, Saint Paul, MN 55108-1010. Tel.: 612-625-1225.

E-mail: mvhm@umn.edu
Web Site: hist.cvm.vmn.edu
Founded: 1985.
Key Personnel: Pres. (V), Dr. Peter E. Poss.
Personnel Profile: Part-Time Paid 1; Part-Time Volunteers 4.
Governing Authority: bd. directors. Tax-exempt.
Institution Type/Description: Veterinary Museum.
Collections: veterinary history; early veterinary books & artifacts.
Hours & Admission Prices: Wed. 1-3. No charge. Closed holidays.
Attendance: 850 (estimated)

THE RAPTOR CENTER, College of Veterinary Medicine, University of Minnesota, 1920 Fitch Ave., Saint Paul, MN 55108-6108. Tel.: 612-624-4745. Fax: 612-624-8740. Facebook: The Raptor Center.

E-mail: raptor@umn.edu
Web Site: www.theraptorcenter.org
Founded: 1974.
Key Personnel: Exec. Dir., Dr. Julia Ponder; Education Program Mgr., Gail Buhl; Veterinarian, Dr. Michelle Willette; Volunteer Coord., Nancie Klebba
Governing Authority: Parent Institution: University of Minnesota Foundation. Tax-exempt.
Institution Type/Description: Conservation Center.
Collections: eagles, hawks, owls & falcons.
Research Fields: Avian clinical medicine & ecosystem health.
Hours & Admission Prices: Tues.-Fri. 10-4, Sat.-Sun. 12-4. No charge; donations accepted. Closed university holidays.
Attendance: 12,000 (estimated)

SAINT PAUL'S WESTERN SCULPTURE PARK, c/o Public Art Saint Paul, 351 Kellogg Blvd. E., Saint Paul, MN 55101. Tel.: 651-290-0921. Fax: 651-292-0345.

E-mail: johnhock@franconia.org
Web Site: www.publicartstpaul.org
Key Personnel: Dir., John Hock
Institution Type/Description: Sculpture Park.
Collections: contemporary sculpture.
Facilities: 6 acre park.
Activities: temporary exhibitions; educational programs for at-risk urban youth.
Hours & Admission Prices: Daily dawn to dusk.

THE SCHUBERT CLUB MUSEUM, (M), 75 W. 5th St., 2nd Fl., Saint Paul, MN 55102-1406. Mailing Address: 302 Landmark Center, 75 W. 5th St., Saint Paul, MN 55102. Tel.: 651-292-3267. Fax: 651-292-4317.
E-mail: kcooper@schubert.org
Web Site: www.schubert.org
Formerly: The Schubert Club Museum of Musical Instruments
Founded: 1972.
Congressional District: 4
Key Personnel: Exec. Dir., Barry Kempton; Museum Mgr., Kate Cooper.
Personnel Profile: Full-Time Paid 2; Part-Time Paid 6; Interns 1.
Governing Authority: private; nonprofit organization. Parent Institution: The Schubert Club. Tax-exempt: 501(c)(3).
Institution Type/Description: Musical Instruments Museum.
Collections: early keyboards; composer letters; photographs; music boxes; gamelan; musical innovation.
Activities: concerts; guided tours; lectures.
Hours & Admission Prices: Sun.-Fri. 12-4. No charge. &
Attendance: 8,500 (accurate)

❋ **THE SCIENCE MUSEUM OF MINNESOTA, (M),** 120 W. Kellogg Blvd., Saint Paul, MN 55102-1202. Tel.: 651-221-9444. Fax: 651-221-4777.
E-mail: info@smm.org
Web Site: www.smm.org
Founded: 1907.
Congressional District: 4
Key Personnel: Chm. Bd. (V), Richard C. Kelly; Pres., Eric Jolly, Ph.D.; Vice Pres. External Rels., Kathleen A. Wilson; Vice Pres. Finance & Admin., Duane J. Kocik; Vice Pres. Education, David Chittenden; Vice Pres. Mktg., Communications & Sales, Jane Eastwood; Senior Vice Pres., Mike Day; Dir. Devel., Leslie Cook; Vice Pres. Exhibits, Paul Martin; Cur. Paleontology, Dr. Kristi Curry Rogers; Div. Head Research & Collections, Ron Lawrenz; Museum Shop Mgr., Steve Fegley.
Personnel Profile: Full-Time Paid 333; Full-Time Volunteers 750; Part-Time Volunteers 796; Interns 21.
Governing Authority: nonprofit organization. Tax-exempt: 501(c)(3).
Institution Type/Description: Science Museum.
Collections: over 1,500,000 catalogued specimens in the fields of biology, anthropology, paleontology, geology.
Research Fields: ethnology; paleontology; biology; zoology; archaeology; geology; geography.
Facilities: nature center; research station; Omni Theater. Museum-related items for sale.
Activities: lectures; films; hobby workshops; formally organized education programs for children, adults, undergraduate, & graduate college students; volunteer program; permanent, temporary & school loan exhibits; demonstration programs; astronomy programs; outdoor education programs; Omni Theater programs.
Publications: membership newsletter; online newsletter for teachers; magazine, Big Frame; scientific publications, New Series; Monograph Series.
Hours & Admission Prices: Sun.-Wed. 9:30-5, Thurs.-Sat. 9:30-9. Combination Pass: adults $17, senior citizens 60 and over & children 4-12 $14.50. &
Attendance: 740,663 (accurate)
Membership: Senior $64; Dual $69; Household $89.

TWIN CITY MODEL RAILROAD MUSEUM, (M), 1021 Bandana Blvd. E., Ste. 222, Saint Paul, MN 55108. Tel.: 651-647-9628. Facebook: TCMRM.
E-mail: tcmrm@tcmrm.org
Web Site: tcmrm.org
Founded: 1934.
Key Personnel: Pres. (V), Rick Moore; Museum Shop Mgr., Paul Gruetzman.
Personnel Profile: Part-Time Volunteers 100; Interns 2.
Governing Authority: nonprofit organization. Tax-exempt.
Institution Type/Description: Model Railroad Museum.
Collections: railroading history; O-scale train layout; toy train layouts & artifacts.
Facilities: Museum-related items for sale.
Activities: special events. Annual Event: Model Railroad Show & Sale in spring and fall; Night Trains during holiday season; Circus Trains in October.
Publications: book, The Twin City Model Railroad Club/Museum.
Hours & Admission Prices: Tues.-Fri. 10-3, Sat. 10-5, Sun. 12-5. Adults $6; members no charge. &
Attendance: 30,000 (estimated)
Membership: Individual $25; Annual $100.

UNIVERSITY OF ST. THOMAS ART HISTORY, Mail 57P, 2115 Summit Ave., Saint Paul, MN 55105-1089. Tel.: 651-962-5560. Fax: 651-962-5861.
Web Site: www.stthomas.edu/arthistory
Founded: 1978.
Key Personnel: Chm. Art Historian, Mark Stansbury-O'Donnell; Chief Cur. & Clinical Faculty, Shelly Nordtorp-Madson.
Personnel Profile: Full-Time Paid 2; Interns 4.
Governing Authority: Parent Institution: University of St. Thomas. Tax-exempt.
Institution Type/Description: History & College Graduate Teaching Museum.
Collections: 20th-century American art; videos; oil paintings.
Research Fields: European; African; Colonial; Latin America; Scandinavian textiles; Pacific; textiles.
Facilities: education center.
Activities: graduate program in art history.
Publications: art history newsletter.
Hours & Admission Prices: Mon.-Sat. 9am-10pm, Sun. 12-10. No charge. Closed major holidays. &
Attendance: 3,500 (estimated)

Saint Peter

E. ST. JULIEN COX HOUSE, 500 N. Washington Ave. at Skaro St., Saint Peter, MN 56082-1979. Mailing Address: 1851 N. Minnesota Ave., Saint Peter, MN 56082-1727. Tel.: 507-934-2160 & 4309. Fax: 507-934-0172.
E-mail: cox@nchsmn.org
Web Site: www.nchsmn.org
Formerly: Nicollet County Historical Society: E. St. Julien Cox House
Founded: 1928.
Congressional District: 2
Key Personnel: Exec. Dir., Ben Leonard; Site Mgr., Richard Tostenson.
Governing Authority: society; nonprofit. Parent Institution: Nicollet County Historical Society. Tax-exempt: 501(c)(3).
Institution Type/Description: Historic House: c.1871 Gothic/Italianate style architecture was built by E. St. Julien Cox, a prominent lawyer & judge.
Collections: household goods; 1880s Victorian furniture; Cox family items; glassware; clothing; linens; art.
Research Fields: wallpaper & wood-working of the house; Victorian gardens.
Facilities: 50-vol. library containing Judge Cox's law & family books.
Activities: guided tours. Museum Sponsors: Christmas Teas in December; Christmas at the Cox House features holiday decor & old time traditions in December.
Publications: newsletter, The Crossing.
Hours & Admission Prices: By appointment only. Admission $4; members no charge.
Attendance: 1,000 (accurate)
Membership: Nicollet County Historical Society: Individual $30; Family $45; Sustaining $75; Patron $100; Supporter $200; Benefactor $500; Life $1,000.

HILLSTROM MUSEUM OF ART, (M), 800 W. College Ave., Saint Peter, MN 56082-1485. Tel.: 507-933-7171. Fax: 507-933-7205.
E-mail: dmyers@gustavus.edu
Web Site: www.gustavus.edu/finearts/hillstrom
Founded: 2000.
Congressional District: 1
Key Personnel: Dir., Donald Myers.
Personnel Profile: Part-Time Paid 2.
Governing Authority: private college. Tax-exempt: 501(c)(3).
Institution Type/Description: College Art Museum.
Collections: works by regional, national & international artists.
Facilities: 3,900 sq. ft. exhibit space.
Activities: loan & temporary exhibitions.
Hours & Admission Prices: Mon.-Fri. 9-4, Sat.-Sun. 1-5. No charge. Closed New Year's Day; Easter; Thanksgiving; Christmas; academic breaks.
Attendance: 5,300 (accurate)

NICOLLET COUNTY HISTORICAL SOCIETY, 1851 N. Minnesota Ave., Saint Peter, MN 56082-1727. Tel.: 507-934-2160. Fax: 507-934-0172.
E-mail: museum@nchsmn.org
Web Site: www.nchsmn.org
Formerly: Treaty Site History Center
Founded: 1928.
Congressional District: 2

Key Personnel: C.E.O., Ben Leonard; Pres. (V), Gary Schmidt; Harkin Store Mgr., Ruth Grewe.
Personnel Profile: Full-Time Paid 2; Part-Time Paid 14; Part-Time Volunteers 20; Interns 3.
Governing Authority: society; nonprofit. Parent Institution: Nicollet County Historical Society, Inc. Branch Museum: E. St. Julien Cox House, St. Peter, MN 56082. Tax-exempt: 501(c)(3).
Institution Type/Description: History Museum.
Collections: Nicollet County history; photographs; clothing; tools; household accessories; Civil War & World War II military; Native American & pioneer memorabilia; Native American clothing; Dakota history.
Research Fields: Nicollet County, Minnesota with special focus on the Treaty of 1851 between the Dakota Nation and the US government and how it affected the boundaries and all the people of Minnesota.
Facilities: 1,115-vol. library; archives contain manuscripts, photos & family histories of Nicollet County residents.
Activities: guided tours; organized education programs for children.
Publications: books, Early History of Nicollet County; Nicollet County Bicentennial Historic Markers; Old Traverse des Sioux: A History; quarterly newsletter, The Crossing,
Hours & Admission Prices: Tues.-Sat. 10-4, Sun. 1-4; other times & tours by appointment. Adults $4, children 6-18 $2; discounts to NCHS & AAM members; children 5 and under & members no charge. Combination Pass: E. Julien Cox House and Treaty Site History Center $6. &
Attendance: 5,000 (accurate)
Membership: Individual $30; Family $45; Sustaining $75; Patron $100; Business $200; Benefactor $500; Life $1,000.

WILLIAM & JOAN SODERLUND PHARMACY MUSEUM, 201 S. Third St., Saint Peter, MN 56082-2044. Mailing Address: 1801 Old Hwy. 8 N. W., Ste. 121, New Brighton, MN 55112-2307. Tel.: 800-603-8196 (Toll Free); 507-931-4410. Fax: 507-931-5434.
E-mail: bsoderlund@hickorytech.net
Web Site: www.villagedrug.com/index.php
Institution Type/Description: Pharmacy Museum.
Collections: American pharmacy history; period drugs; pharmacy history; drug store furnishings; soda fountain memorabilia; show globes.
Hours & Admission Prices: Mon.-Fri. 8:30-7, Sat. 8:30-2.

Sauk Centre

SINCLAIR LEWIS BOYHOOD HOME & INTERPRETIVE CENTER/MUSEUM, Interpretive Center Museum, 1220 Main St. S., Sauk Centre, MN 56378. Mailing Address: P.O. Box 25, Sauk Centre, MN 56378-0222. Tel.: 320-352-5201. Fax: 320-352-5202.
E-mail: sinclairlewisfoundation@gmail.com
Web Site: www.sinclairlewisfoundation.com
Founded: 1960.
Congressional District: 7
Key Personnel: Pres. (V) & Dir., Colleen Steffes.
Personnel Profile: Part-Time Paid 2; Part-Time Volunteers 10.
Governing Authority: nonprofit organization. Affiliated with the Sinclair Lewis Foundation. Sinclair Lewis House Museum, 810 Sinclair Lewis Ave., Sauk Centre, MN 56378. Tax-exempt: 501(c)(3).
Institution Type/Description: Historic House Museum: 1920 boyhood home of Sinclair Lewis.
Collections: House: books written by Sinclair Lewis; family furnishings & memorabilia; manuscripts. Center: family history; family artifacts; video presentation.
Facilities: library of English & foreign editions of novels by Sinclair Lewis; interpretive center. Museum-related items for sale.
Activities: Interpretive Center: video presentation. Sinclair Lewis House: guided tours; permanent exhibitions.
Hours & Admission Prices: Interpretive Center: Memorial Day-Labor Day Mon.-Fri. 8:30-4:30, Sat.-Sun. 9-5. House: June to Sept. 1 Tues.-Sat. 10-4. Interpretive Center: no charge, donations accepted. House Tours: adults & students $5, children 6-12 $3; discounts for AAA members; children 5 & under no charge. Call to confirm. &
Attendance: 4,000 (estimated)
Membership: Senior Citizen $10; Single $15; Couple $20; Business & Organization $35.

Saum

FIRST NEW CONSOLIDATED SCHOOL IN MINNESOTA, Saum Community Club, 41982 Pioneer Rd., N.E., Saum, MN 56650. Mailing Address: 12956 Twin Oaks Rd., N.E., Kelliher, MN 56650-9404. Tel.: 218-647-8531.
Founded: 1962.

Congressional District: 7
Key Personnel: Sec., Eva Stengel; Treas., Ione Smischney.
Personnel Profile: Part-Time Volunteers 10.
Volunteer Hours: 400
Operating Expenses: 14,100
Operating Income: 15,123
Governing Authority: board of directors; society. Saum Community Club. Affiliated with Beltrami Historical Society, P.O. Box 683, 130 Minnesota Ave. S.W., Bemidji, MN 56619. Tax-exempt.
Institution Type/Description: History Museum.
Collections: desks; pictures; fixtures; books & other school furnishings. Historic Buildings: 1903 original log school; 1912 Saum School; 1912 first new Consolidated School in MN; book, DVD & photos from 2012 centennial celebration.
Facilities: school books available for research on premises.
Activities: guided tours; research. Museum Sponsors: 100th Anniversary Celebration.
Hours & Admission Prices: By appointment only. No charge; donations accepted. &
Attendance: 200 (estimated)
Membership: Individual $5.

Shakopee

THE LANDING - MINNESOTA RIVER HERITAGE PARK, 2187 E. Hwy. 101, Shakopee, MN 55379-1750. Tel.: 763-694-7784. Fax: 952-403-9489.
Web Site: threeriversparks.org
Formerly: Historic Murphy's Landing
Founded: 1969.
Key Personnel: Site Supvr., Jefferson Spilman.
Governing Authority: Parent Institution: Three Rivers Park District.
Institution Type/Description: Living History Museum.
Collections: local history & culture; period furnishings; historic buildings.
Facilities: 88-acre park.
Activities: special events; rental facilities; educational programs.
Hours & Admission Prices: May 23-Sept. 7 Mon.-Fri. 10-4, Sat. & holidays 10-5, Sun. 12-5. Adults 18-64 $5, seniors & children $3. Closed New Year's Day; Easter; Thanksgiving; Christmas Eve & Day.

SCOTT COUNTY HISTORICAL SOCIETY, (M), 235 S. Fuller St., Shakopee, MN 55379-1320. Tel.: 952-445-0378; 888-325-2575. Fax: 952-445-4154.
E-mail: info@scottcountyhistory.org
Web Site: www.scottcountyhistory.org
Founded: 1968.
Personnel Profile: Full-Time Paid 2; Part-Time Paid 1; Part-Time Volunteers 30; Interns 3.
Institution Type/Description: Historical Society Museum.
Collections: Scott County Minnesota history; books; photographs; period artifacts; newspapers; local history; historic house.
Major Exhibits: Storied Treasurers, 8/13-8/14.
Facilities: gardens.
Activities: educational programs; special events; guided tours; temporary & permanent exhibitions.
Publications: quarterly newsletter.
Hours & Admission Prices: Tues.-Wed. & Fri. 9-4, Thurs. 9-8, Sat. 10-3. Adults $4, students $2; members & children under 5 no charge. Closed major holidays. &
Attendance: 6,093 (accurate)
Membership: Senior & Student $20; Individual $30; Household $40; Sponsor $75; Business $150.

Shevlin

CLEARWATER COUNTY HISTORICAL SOCIETY, 264 1st St. W., Shevlin, MN 56676. Mailing Address: P.O. Box 241, Bagley, MN 56621-0241. Tel.: 218-785-2000. Facebook: Clearwater County Historical Society.
E-mail: cchshist@gvtel.com
Web Site: mnhistoricnw.org
Founded: 1968.
Congressional District: 7
Key Personnel: C.E.O., Ken Braaten; Dir. & Museum Shop Mgr., Tamara Edevold.
Personnel Profile: Full-Time Paid 1; Part-Time Paid 2; Part-Time Volunteers 8.
Volunteer Hours: 1,500
Operating Expenses: 75,000
Operating Income: 75,000
Governing Authority: board of directors; nonprofit organization. Tax-exempt.

Institution Type/Description: Local History Museum.
Collections: agriculture; timber; lumbering; newspapers; military; religious artifacts. Historic Buildings: replica 1895 Gran church; 1890 log school, log house, WPA-built school, Great Northern Depot.
Research Fields: early history; forestry.
Activities: tours; art shows.
Publications: bimonthly newsletter, Clearwater History News.
Hours & Admission Prices: May-Aug. Tues.-Fri. 10-4, Sat. 10-2; Sept.-April Tues.-Fri. 10-4. No charge; donations accepted. Closed Christmas week. &
Attendance: 1,300 (estimated)
Membership: Individual $15; Family $20; Patron $30; Sustaining $50 & up; Benefactor $100 & up.

Slayton

MURRAY COUNTY HISTORICAL MUSEUM, 2480 29th St., Slayton, MN 56172. Mailing Address: P.O. Box 61, Slayton, MN 56172-0061. Tel.: 507-836-6533. Facebook: Murray County Hisorical Museum.
E-mail: museum@co.murray.mn.us
Web Site: murraycountyhistoricalsociety.org
Congressional District: 1
Key Personnel: Dir., Janet Timmerman.
Personnel Profile: Part-Time Paid 2; Part-Time Volunteers 3.
Governing Authority: Parent Institution: Murray County, MN.
Institution Type/Description: History Museum.
Collections: early radios & phonographs; period farm wagon, tools & machinery; Native American artifacts; family histories; newspapers; census records; military artifacts from Civil War to WWII; naturalization records.
Hours & Admission Prices: Feb.-Dec. Mon.-Fri. 10-5. No charge; donations accepted. Closed holidays. &

Sleepy Eye

SLEEPY EYE DEPOT MUSEUM, Sleepy Eye, MN 56085. Mailing Address: P.O. Box 544, Sleepy Eye, MN 56085-0544. Tel.: 507-794-5053.
E-mail: semuseum@sleepyeyetel.net
Web Site: thedepotlady.blogspot.com
Key Personnel: Dir., Debbie Joramo.
Governing Authority: Tax-exempt.
Institution Type/Description: Railroad Depot Museum: listed on the National Register of Historic Places.
Collections: railroad history; Sleepy Eye artifacts; Sleepy Eye Drum & Bugle Corps. memorabilia; vintage clothing.
Hours & Admission Prices: May-Dec. 10 Tues.-Sat. 10-4; other times by appointment. No charge; donations accepted.
Membership: Lone Eagle (Single) $15; Band (Family) $25; Trading Post (Business) $30; Buffalo Blanket (Giver of Warmth) $100; Great Teepee Maker (Shelter Giver) $500; The Turtle Mother Earth (Strength of Earth) $1,000.

Soudan

SOUDAN UNDERGROUND MINE STATE PARK, 1379 Stuntz Bay Rd., Soudan, MN 55782. Mailing Address: P.O. Box 335, Soudan, MN 55782-0335. Tel.: 218-753-2245. Fax: 218-753-2246.
E-mail: soudanmine.statepark@state.mn.us
Web Site: mndnr.gov/soudan
Founded: 1965.
Institution Type/Description: Historic Site & Underground Mine.
Collections: dry house; drill shop; crusher house; engine house; mining equipment & artifacts.
Hours & Admission Prices: Underground Mine Tour & High Energy Physics Lab Tour: call for hours & admission prices. &
Attendance: 37,000 (estimated)

South Saint Paul

DAKOTA COUNTY HISTORICAL SOCIETY, (M), Lawshe Memorial Museum, 130 3rd Ave., N., South Saint Paul, MN 55075-2002. Tel.: 651-552-7548 Fax: 651-552-7265.
E-mail: dakotahistory@co.dakota.mn.us
Web Site: www.dakotahistory.org
Founded: 1939.
Congressional District: 1, 3 & 4
Key Personnel: Exec. Dir., Chad Roberts; Pres., Mark Kaciszewski; Museum Shop Mgr., Maggie Langenfeld.
Personnel Profile: Full-Time Paid 2; Part-Time Paid 12; Part-Time Volunteers 95; Interns 2.

Governing Authority: board of directors. Subsidiary Institution: LeDuc Historic Estate, 1629 Vermillion St., Hastings, MN 55033. Tax-exempt.
Institution Type/Description: Historical Society Museum.
Collections: Indian artifacts; agriculture; trades & industry items; railroading; Civil War items; World War I & II items; genealogical records; maps; newspapers.
Research Fields: Dakota County history.
Facilities: 1,000-vol. research center containing all county newspapers on microfilm.
Activities: guided tours; traveling exhibits; lectures; living history events; permanent exhibitions; art & craft shows.
Publications: Three-year magazine, Over the Years.
Hours & Admission Prices: Wed. & Fri. 9-5, Thurs. 9-8, Sat. 10-3. Lawshe: no charge; donations accepted. LeDuc: mid-June to Oct. Wed.-Sat. 10-5, Sun. 1-5. Tours at 10, 11:30, 1, 2:30 & 4. Adults $6, seniors & military $5, students $3; members no charge. Closed national holidays. &
Attendance: 13,000 (accurate)
Membership: Individual $30; Family $50; Sustaining $100; Silver $250; Gold $500; Tower $1,000.

Spring Grove

BLUFF COUNTRY ARTISTS GALLERY, 111 W. Main St., Spring Grove, MN 55974. Tel.: 507-498-2787.
E-mail: bcagallery@springgrove.coop
Web Site: bluffcountryartistsgallery.org
Governing Authority: nonprofit organization. Tax-exempt: 501(c)(3).
Institution Type/Description: Art Gallery.
Collections: works by local artists; paintings; sculpture.
Activities: educational programs; classes; special events.
Hours & Admission Prices: Call for hours.

Spring Valley

SPRING VALLEY COMMUNITY HISTORICAL SOCIETY, INC., 220 W. Courtland St., Spring Valley, MN 55975-1232. Tel.: 507-346-7659.
E-mail: wilderinspringvalley@hotmail.com
Web Site: springvalleymnmuseum.org
Founded: 1956.
Congressional District: 1
Key Personnel: Pres. (V), Joseph Bezdicek; Vice Pres., Clarence Klenke; Dir. & Museum Shop Mgr., Julie Mlinar.
Personnel Profile: Part-Time Paid 8; Part-Time Volunteers 3.
Governing Authority: society. Tax-exempt.
Institution Type/Description: Pioneer History Museum, A Laura Ingalls Wilder site.
Collections: History Museum: pioneer artifacts. Church Museum: Wilder family photos; religious artifacts, Conley Camera collection. Historic Buildings: 1875-76 Methodist Church Museum; 1860 era Washburn-Zittleman House.
Facilities: Museum-related items for sale.
Activities: guided tours; video tour of church & house museum; formally organized education program for children.
Publications: History of the Methodist Church; History of Village of Washington, Fillmore County; 125-year History of Spring Valley, Minnesota; Spring Valley; The Laura Ingalls Wilder Connection.
Hours & Admission Prices: June-Aug. daily 10-4; Sept.-Oct. Sat.-Sun.; May-Oct. by appointment for groups. Church: adults $4, students $1.50. All Buildings: adults $6, students $2.50; members no charge.
Attendance: 2,000 (accurate)
Membership: Individual $10; Couple & Family $20; Life $100.

Stillwater

WASHINGTON COUNTY HISTORICAL SOCIETY WARDEN'S HOUSE MUSEUM, 602 N. Main St., Stillwater, MN 55082-4010. Mailing Address: P.O. Box 167, Stillwater, MN 55082-0167. Tel.: 651-439-5956.
E-mail: brent.peterson@wchsmn.org
Web Site: www.wchsmn.org
Founded: 1941.
Key Personnel: Pres., David Lindsey; Historic Sites Supvr., Kirsta Benson; Exec. Dir., Brent Peterson.
Personnel Profile: Full-Time Paid 2; Part-Time Paid 1; Part-Time Volunteers 10; Interns 3.
Governing Authority: society; nonprofit organization. Tax-exempt: 501(c)(3).
Institution Type/Description: History Museum: located in 1853 former State Prison Warden's Home.
Collections: lumberjack tools; pictures; costumes; Indian artifacts; military;

quilts; furniture; glassware; pottery; children's toys & games; prison artifacts; manuscripts.
Research Fields: Washington County history, prison history, rural schools, military & lumber company history.
Facilities: 500-vol. library of Minnesota and St. Croix River Valley history books available for use on premises; research department; 15,000 photos of area & residents.
Activities: guided & cemetery tours; vintage baseball club; temporary exhibitions.
Publications: brochure, Washington County Historical Museum, History of the Walden's House Museum; quarterly, Historical Whisperings; book, Washington: A History of the County, Pioneers of the Big Lake Community, Minnesota Beginnings.
Hours & Admission Prices: May-Oct. Thurs.-Sun. 1-5; tours by appointment. Adults $5, children 6-17 $1; discounts to AAA members; members no charge. &
Attendance: 5,000 (accurate)
Membership: Seniors $10; Adult $15; Family $25; Patron $50; Sustaining $100; Life $500.

Taylors Falls

THE FOLSOM HOUSE MUSEUM, (M), 272 W. Government Rd., Taylors Falls, MN 55084. Mailing Address: Box 333, Taylors Falls, MN 55084-0333. Tel.: 651-465-3125.
Formerly: The Historic W.H.C. Folsom House
Founded: 1978.
Congressional District: 8
Key Personnel: Site Mgr., William W. Scott; Pres., Sandra Berg.
Personnel Profile: Full-Time Volunteers 1; Part-Time Volunteers 45.
Governing Authority: society; nonprofit organization. Parent Institution: Minnesota Historical Society. Subsidiary Institution: Taylor Falls Historical Society. Branch Museums: 1852 Town House School. Tax-exempt.
Institution Type/Description: Historic House: 1854-55 Historic Folsom House.
Collections: original Folsom family furniture; complete collection of ledgers & day books from Folsom General Store 1851-1876; passenger & cargo journals from Folsom Steamboats.
Research Fields: local history.
Activities: guided tours. Annual Event: Christmas Lighting Festival in November.
Publications: brochure; biannual, Life & Times in Taylor Falls; book, Fifty Years in the Northwest, a sesquicentennial edition of WHC Folsoms book first published in 1888.
Hours & Admission Prices: Memorial Day to mid-Oct. Wed.-Mon. 1-4:30. Adults $4, students 6-12 $1; discounts to groups over 15; members, Minnesota Historical Society & Taylor Falls Historical Society members no charge.
Attendance: 4,000 (estimated)
Membership: Taylors Falls Historical Society: Individual $15; Household $25; Patron $35; Sustaining $100 & up.

Tofte

NORTH SHORE COMMERCIAL FISHING MUSEUM, 7136 W. Hwy. 61, Junction of Sawbill Trail & Hwy. 61, Tofte, MN 55615. Mailing Address: P.O. Box 2312, Tofte, MN 55615-2312. Tel.: 218-663-7050.
E-mail: info@commercialfishingmuseum.org
Web Site: www.commercialfishingmuseum.org
Key Personnel: Museum Mgr., Don Hammer
Institution Type/Description: Fishing Museum.
Collections: maritime customs & history; commercial fishing industry.
Activities: special events; educational programs.
Hours & Admission Prices: Fri. 12-5, Sat. 9-5. Adult $3, children 6-16 $1; discounts to groups; children under 6 no charge. &
Membership: Individual $20; Family $30; Business $50; Life $300.

Tower

BOIS FORTE HERITAGE CENTER, 1500 Bois Fort Rd., Tower, MN 55790-7800. Tel.: 218-753-6017.
E-mail: rberens@boisforte-NSN.gov
Web Site: boisforte.com/divisions/heritage_center.htm
Key Personnel: Dir. & Historic Preservation Officer, Rose Berens; Cur., William Latady; Gift Shop Supvr., Bev Miller; Interpreter, Rhonda Zuponcic
Institution Type/Description: History Museum.
Collections: Bois Forte Ojibwe Indian history & culture.
Hours & Admission Prices: Call for hours. Adults $5, seniors & children 4-12 $3; children 3 & under and Bois Forte Band members no charge.

Two Harbors

LAKE COUNTY HISTORICAL SOCIETY, 520 South Ave., Depot Bldg., Two Harbors, MN 55616. Mailing Address: P.O. Box 128, Two Harbors, MN 55616-0128. Tel.: 218-834-4898. Fax: 218-834-7198.
E-mail: lakehist@lakenet.com
Web Site: lakecountyhistoricalsociety.org
Founded: 1925.
Congressional District: 8
Key Personnel: C.E.O., Mel Sando; Pres. (V), Sam Gangi.
Personnel Profile: Full-Time Paid 3; Part-Time Paid 25; Part-Time Volunteers 10.
Governing Authority: self. Tax-exempt: 501(c)(3).
Institution Type/Description: History Museum & Historic Site.
Collections: local manuscripts; photographs on railroading, iron ore, lumbering; mineral collections; shipping & shipwreck artifacts. Depot Museum: general local history museum highlighting 1st shipment of iron ore from Minnesota; veterans room commemorating veterans 1800 on; two steam engines on display outside; 3M/Owan Museum, highlighting creation of Minnesota Mining & Manufacturing and the development & uses of its first successful product, sandpaper; 1892 Lighthouse & Harbor Museum, at Two Harbors Light Station, highlighting historical use & relationship to Lake Superior, commercial fishing, shipping iron ore, development of harbor; Edna G. Tugboat built in 1896, tour one of the last 3 steam powered coal burning tugs left in the U.S.
Research Fields: local oral histories; commercial fishing.
Facilities: 20-vol. library of historical manuscripts available for research on premises.
Activities: guided tours; permanent exhibitions; community education programs.
Publications: quarterly newsletter.
Hours & Admission Prices: May-Sept. daily 10-5; other times by appointment. Lighthouse 9-5 & 6-9. Adults $3, youth 9-17 $1; discounts to groups, AAM & ICOM members; children under 8 & members no charge. &
Attendance: 18,000 (estimated)
Membership: Individual $15; Family $25.

SPLIT ROCK LIGHTHOUSE HISTORIC SITE, (M), 3713 Split Rock Lighthouse Rd., Two Harbors, MN 55616-2020. Tel.: 218-226-6372. Fax: 218-226-6373.
E-mail: splitrock@mnhs.org
Web Site: www.mnhs.org
Founded: 1976.
Congressional District: 8
Key Personnel: Dir., Lee Radzak; Museum Shop Mgr., Gloria Rosenau.
Personnel Profile: Full-Time Paid 4; Part-Time Paid 30.
Governing Authority: society; nonprofit. Parent Institution: Minnesota Historical Society. Tax-exempt.
Institution Type/Description: Historic Site: located on 25 acres, site of 1910 Split Rock Light Station.
Collections: marine navigation.
Research Fields: U.S. Lighthouse service & marine aids to navigation relating to the Split Rock Lighthouse.
Facilities: library including records of the U.S. Lighthouse; 90-seat theater. Gift items for sale.
Activities: guided tours; films. Annual Events: Open House; interpretation, 1925 light station.
Publications: booklet, Split Rock: Epoch of a Lighthouse.
Hours & Admission Prices: mid-May to mid-Oct. daily 10-6. Visitors Center: Winter Sat.-Sun. 11-4. Adults $8, senior citizens $6, children 6-17 $5; children 5 & under, Minnesota Historical Society members no charge. State Park vehicle permit $7. Closed winter holidays.
Attendance: 101,000 (accurate)
Membership: Minnesota Historical Society: Senior $45; Individual $55; Senior Household $55; Household $65; Associate $125.

Victoria

LOWRY NATURE CENTER, 7025 Victoria Dr., Victoria, MN 55386-9668. Mailing Address: c/o Three Rivers Park District, 3000 Xenium Lane, N., Plymouth, MN 55441. Tel.: 763-694-7650. Facebook: Three Rivers Park District - Lowry Nature Center.
E-mail: lowrync@threeriversparkdistrict.org
Web Site: www.threeriversparkdistrict.org/parks/carver-park/lowry-nature-center.aspx
Key Personnel: Supvr., Allison Neaton; Interpretive Naturalist, Judy Englund
Institution Type/Description: Nature Center.
Collections: native plants & wildlife.

Hours & Admission Prices: Mon.-Sat. 9-5, Sun. 12-5.

Wabasha

NATIONAL EAGLE CENTER, 50 Pembroke Ave., Wabasha, MN 55981-1241. Mailing Address: P.O. Box 242, Wabasha, MN 55981-0242. Tel.: 651-868-4989; 877-332-4537. Fax: 651-565-5357.
E-mail: marybeth@nationaleaglecenter.org
Web Site: nationaleaglecenter.org
Key Personnel: Dir. Programming & Public Rels., MaryBeth Garrigan; Administrative Mgr., Catherine Cushing
Institution Type/Description: Nature Center.
Collections: Bald Eagles & their habitats; Native American artifacts; preserved wildlife.
Facilities: classrooms. Museum-related items for sale.
Activities: feeding programs; observation decks.
Hours & Admission Prices: March-Nov. Mon.-Thurs. 10-5, Fri.-Sun. 9-6; Dec.-Feb. daily 10-5. Adults $6, seniors 65 & over $5, students 4-17 $4; children under 3 & members no charge.

Wabasso

COUNTY CENTER HISTORICAL SOCIETY, 564 South St., Wabasso, MN 56293. Mailing Address: 1177 Duey St., Wabasso, MN 56293. Tel.: 507-342-5367.
Formerly: Wabasso Center Historical Society
Founded: 1973.
Congressional District: 2
Key Personnel: Pres., Armin Dallman; Treas., Merlin J. Goudy.
Personnel Profile: Part-Time Paid 1; Part-Time Volunteers 1.
Governing Authority: society; nonprofit organization. Tax-exempt.
Institution Type/Description: Historic Society Museum: housed in 1903 Knox Presbyterian Church.
Collections: household items; farm implements; books; pictures; rocks; guns; old newspapers.
Research Fields: sod houses; old artifacts.
Activities: guided tours; arts festivals; formally organized education programs for children; permanent exhibitions.
Hours & Admission Prices: Summer Wed.-Sun. 1-5; other times & tours by appointment. No charge; donations accepted. &
Attendance: 75
Membership: Individual $3.

Waconia

CARVER COUNTY HISTORICAL SOCIETY, 555 W. 1st St., Ste. A, Waconia, MN 55387-1223. Tel.: 952-442-4234. Fax: 952-442-2435.
E-mail: historical@co.carver.mn.us
Web Site: www.carvercountyhistoricalsociety.org/index.htm
Founded: 1940.
Congressional District: 3
Key Personnel: Exec. Dir., Wendy Petersen Biorn.
Personnel Profile: Full-Time Paid 2; Part-Time Paid 3; Part-Time Volunteers 50; Interns 1.
Governing Authority: society. Tax-exempt.
Institution Type/Description: History Museum.
Collections: pioneer agricultural tools; household tools; county business items; photographs; documents.
Research Fields: Carver county history; genealogy.
Facilities: microfilm reader/printer; microfilmed county newspapers; newspaper indexes; census records; printed material.
Activities: guided tours; historical research; seminars.
Publications: quarterly newsletter.
Hours & Admission Prices: Mon.-Fri. 10-4:30, Sat. 10-3; groups by appointment. No charge; donations accepted. &
Attendance: 10,600 (accurate)
Membership: Senior Citizen $10; Individual $15; Family $25; Sustaining $50.

Walker

CASS COUNTY MUSEUM & RESEARCH CENTER, 205 Minnesota Ave. W., Walker, MN 56484-2189. Mailing Address: P.O. Box 505, Walker, MN 56484-0505. Tel.: 218-547-7251.
E-mail: casscountymuseum@gmail.com
Web Site: www.casscountymuseum.org
Formerly: Cass County Museum & Pioneer School
Founded: 1937.

Congressional District: 8
Key Personnel: Pres., Dan Eikenberry; Dir., Renee Geving; Vice Pres., Lois Orton.
Personnel Profile: Part-Time Paid 2; Part-Time Volunteers 8.
Governing Authority: Cass County Historical Society. Tax-exempt.
Institution Type/Description: Historical Society Museum.
Collections: Indian handicraft & artifacts, mainly Ojibway/Chippewa.
Research Fields: Ojibway/Chippewa & Sioux artifacts from late 1800s to early 1900s.
Activities: guided tours.
Publications: quarterly newsletter.
Hours & Admission Prices: Memorial Day-Labor Day Mon.-Fri. 10-5. Families $9, adults $4, children $1; discounts to groups with reservations. &
Attendance: 1,300 (estimated)
Membership: Single $10; Family $15; Benefactor $20; Business $50; Friend $100-$249; Sponsor $250-$999; Patron $1,000.

Walnut Grove

LAURA INGALLS WILDER MUSEUM AND TOURIST CENTER, 330 8th St., Walnut Grove, MN 56180-1114. Tel.: 507-859-2358. Fax: 507-859-2933.
E-mail: lauramuseum@walnutgrove.org
Web Site: www.walnutgrove.org
Founded: 1974.
Congressional District: 2
Key Personnel: Pres., James Kleven; Dir. & Museum Shop Mgr., Amy Ankrum; Treas., Lori Julinson; Collections Mgr., Nicole Elzenga.
Personnel Profile: Full-Time Paid 3; Part-Time Paid 6; Part-Time Volunteers 14.
Governing Authority: nonprofit organization. Tax-exempt.
Institution Type/Description: General Museum: housed in c.1894 railroad station.
Collections: Walnut Grove history; Laura Ingalls Wilder memorabilia; quilts; furniture; toys; complete printing press; 250 dolls from 1860-1980s; native Minnesota grass & flowers. Historic House: Dugout, unique little chapel.
Major Exhibits: Little House TV Cast Reunion, 4/14-10/14.
Research Fields: Walnut Grove history.
Facilities: Museum-related items for sale.
Activities: self guided tours.
Publications: brochures; fan club newsletter; book, Laura Ingalls Wilder's Walnut Grove by William Anderson.
Hours & Admission Prices: April & Oct. Mon.-Sat. 10-4, Sun. 12-4; May & Sept. Mon.-Sat. 10-5, Sun. 12-5; June-Aug. daily 10-6. Adults $6, children 6-12 $3; members & children 5 and under no charge. Closed Easter; Thanksgiving; Christmas Eve & Day. &
Attendance: 17,500 (accurate)
Membership: Individual $15; Couple $20; Family $25; Lifetime $100 & up.

Warren

MARSHALL COUNTY HISTORICAL SOCIETY MUSEUM, 808 E. Johnson Ave., Warren, MN 56762. Mailing Address: P.O. Box 103, Warren, MN 56762-0103. Tel.: 218-745-4803.
E-mail: mchs@wiktel.com
Founded: 1933.
Congressional District: 7
Key Personnel: C.E.O. & Pres. (V), Delvin Potucek; Dir., Michael E. Johnson.
Personnel Profile: Part-Time Paid 2; Part-Time Volunteers 9.
Governing Authority: county; nonprofit organization. Tax-exempt.
Institution Type/Description: Historical Society Museum.
Collections: period farm machinery; business & household artifacts. Historic Buildings: 1886 log cabin; c.1880s schoolhouse; 1893 church; 1905 Soo line depot.
Research Fields: local history; agriculture.
Activities: Museum Sponsors: annual County Fair in July; Old Time School Days tour for 5th graders in September; monthly programs on historical topics.
Publications: self Portrait of Marshall County. book: Do you believe in UFO's History of the Old Mill State Park.
Hours & Admission Prices: May-Sept. Mon.-Fri. 9-5, Sat.-Sun. by appointment. No charge; donations accepted. Closed major holidays. &
Attendance: 2,100 (estimated)
Membership: Singles $10; Family $15.

Waseca

FARMAMERICA, THE MINNESOTA AGRICULTURAL IN-TERPRETIVE CENTER, 7367 360th Ave., Waseca, MN 56093-4414. Tel.: 507-835-2052. Fax: 507-835-2053.
E-mail: info@farmamerica.org
Web Site: farmamerica.org
Founded: 1978.
Congressional District: 23
Key Personnel: Chm. Bd. (V), Ed Frederick; Exec. Dir., James L. Gibson; Office Mgr., Crystal Paulson.
Personnel Profile: Full-Time Paid 1; Part-Time Paid 8; Part-Time Volunteers 450.
Volunteer Hours: 10,000
Governing Authority: nonprofit organization. Tax-exempt: 501(c)(3).
Institution Type/Description: Farm Equipment Museum.
Collections: farm machines; tools; crops; photo archives; decorative arts; c.1850 settlement farm; 1870-1880 settlement farm town hall, one room schoolhouse; feed mill; period gardens; blacksmith shop; c.1920 dairy farm; 1920s & 1930s farm; country church; grain mill.
Research Fields: Minnesota agriculture; local history.
Facilities: interpretive center; mini-museum; conference center; 10-acre native prairie.
Activities: festivals; self-guided tours; lectures.
Publications: quarterly newsletter, Over the Fence Post.
Hours & Admission Prices: June -Aug. Tues.-Fri. 9-2. No charge, donations accepted. &
Attendance: 20,000 (estimated)
Membership: Individual $25; Family $50; Sustaining $100; Contributing $250; Patron $500; Visionary $1,000.

WASECA COUNTY HISTORICAL SOCIETY, MUSEUM AND RESEARCH LIBRARY, (M), 315 2nd Ave., N.E., Waseca, MN 56093-2936. Mailing Address: P.O. Box 314, Waseca, MN 56093-0314. Tel.: 507-835-7700. Fax: 507-835-7811.
E-mail: director@historical.waseca.mn.us
Web Site: www.historical.waseca.mn.us
Founded: 1938.
Congressional District: 1
Key Personnel: Co Dir. & Museum Shop Mgr., Joan Mooney; Co Dir., Sheila Morris; Pres. (V), Jim King.
Personnel Profile: Part-Time Paid 5; Part-Time Volunteers 10.
Governing Authority: nonprofit organization; board of directors. Branch Museums: Hodgson Hall. Tax-exempt.
Institution Type/Description: Local History Museum.
Collections: agriculture; archives; costumes; decorative arts; photographs; maps.
Research Fields: local history.
Facilities: library of history books & local newspapers; reading room; gardens.
Activities: permanent exhibitions; educational programs; annual & special events; research.
Publications: quarterly newsletter, History Notes.
Hours & Admission Prices: Tues.-Fri. 9-5; other times by appointment. No charge; donations accepted. Research: $5 per day; members no charge. &
Membership: Individual $25; Family $35; Business & Organization $200; Benefactor $1,000.

Wayzata

MINNETONKA CENTER FOR THE ARTS, (M), 2240 North Shore Dr., Wayzata, MN 55391-9347. Tel.: 952-473-7361, ext. 16. Fax: 952-473-7363. Facebook: Art Centered.
E-mail: information@minnetonkaarts.org
Web Site: www.minnetonkaarts.org
Founded: 1952.
Key Personnel: Dir., Roxanne Heaton; Chm. (V), Tom Hull; Museum Shop Mgr., Robert Bowman
Institution Type/Description: Art Gallery.
Collections: paintings; sculpture.
Major Exhibits: Student Show, 1/14; Member Show, 5/14; Faculty Show, 9/14; Arts of the Holidays, 11/14-12/14.
Activities: educational programs; classes; summer camp; special events.
Hours & Admission Prices: Summer: Mon. & Fri. 9-4, Tues.-Thurs. 9am-9:30pm, Sat. 9-1; Winter: Mon., Fri. & Sat. 9-4, Tues.-Thurs. 9-1. &
Attendance: 20,000 (estimated)
Membership: Basic Household $50.

West Saint Paul

DODGE NATURE CENTER, 365 Marie Ave., W., West Saint Paul, MN 55118-3848. Tel.: 651-455-4531. Fax: 651-455-2575.
Web Site: www.dodgenaturecenter.org
Founded: 1967.
Key Personnel: Exec. Dir., Jason Sanders.
Personnel Profile: Full-Time Paid 18; Part-Time Paid 18; Part-Time Volunteers 400; Interns 12.
Governing Authority: private; not-for-profit organization. Tax-exempt: 501(c)(3).
Institution Type/Description: Nature Center.
Collections: arboretum; collections of pressed, mounted & identified plants & stuffed animals; live reptiles & amphibians; film & photos of history of nature center; heirloom plant garden.
Activities: restoration ecology; environment education; interpretive farm; heirloom garden; preschool.
Hours & Admission Prices: Mon.-Fri. 8-4:30 pre-registered school & other groups only, Sat. 10-4; occasional evening. Center Entrance: no charge. Special Programs: adults & senior citizens $2-$5, students & children $2-$4. &
Attendance: 42,000 (estimated)
Membership: Individual $30; Family $40; Supporting $60; Sustaining $100; Patron $500; Benefactor $1,000.

Wheaton

TRAVERSE COUNTY HISTORICAL SOCIETY, 1201 Broadway, Wheaton, MN 56296. Mailing Address: 601 1st Ave. S., Wheaton, MN 56296-1712. Tel.: 320-563-8520.
E-mail: cjuelich@frontiernet.net
Web Site: www.cityofwheaton.com
Founded: 1977.
Congressional District: 2
Key Personnel: Pres., Clarence Juelich; Sec., Julieann Fromeke.
Governing Authority: nonprofit organization. Tax-exempt: 501(c)(3).
Institution Type/Description: Historical Society Museum: housed in c.1906 railroad depot.
Collections: refurbished caboose; one-man band; theme rooms; kitchen; dental office; meat market; produce buying station; one room schoolhouse; farm machinery; church exhibit; furniture; toys; mounted animals; c.1882 organ; bicycle with wood rims & handlebars; 1907 surrey; 1924 gas truck; 1912 hand drawn fire & hose cart; 1926 Graham Page fire truck; photographs; newspaper printing equipment; printing press; rural schoolhouse; physician's office equipment; bank; post office; jewelry; military artifacts.
Hours & Admission Prices: Memorial Day-Labor Day Thurs.-Sun. 1-5. No charge; donations accepted. &
Attendance: 486 (estimated)
Membership: Individual $3; Life $25.

White Bear Lake

WHITE BEAR CENTER FOR THE ARTS GALLERY, 1280 N. Brich Lake Blvd., White Bear Lake, MN 55110. Tel.: 651-407-0597. Fax: 651-429-1569.
E-mail: wbca@whitebeararts.org
Web Site: whitebeararts.org
Institution Type/Description: Art Gallery.
Collections: paintings; drawings; sculpture.
Activities: classes; workshops; special events.
Hours & Admission Prices: Mon.-Fri. 9-1.

White Bear Township

TAMARACK NATURE CENTER, 5287 Otter Lake Rd., White Bear Township, MN 55110-5851. Mailing Address: c/o Ramsey County Parks & Recreation, 2015 N. Van Dyke St, Maplewood, MN 55109-3796. Tel.: 651-407-5350. Fax: 651-407-5354. Facebook: Tamarack Nature Center.
E-mail: tamarack@co.ramsey.mn.us
Web Site: www.parks.co.ramsey.mn.us/tamarack/Pages/tamarack.aspx
Key Personnel: Dir., May Vidas
Institution Type/Description: Nature Center.
Collections: wildlife.
Facilities: nature trails; dock.
Hours & Admission Prices: Trails: daily 30 minutes before sunrise to 30 minutes after sunset; Visitor Center: Mon.-Fri. 8-4:30, Sat. 9-5, Sun. 12-5; Discovery Hollow & Garden Mon.-Fri. 10-4, Sat. 10-4:30, Sun. 12-4:30. No charge; donations accepted

Willmar

KANDIYOHI COUNTY HISTORICAL SOCIETY, (M), 610 Hwy. 71, N.E., Willmar, MN 56201-2650. Tel.: 320-235-1881. Fax: 320-235-1881.
E-mail: kandhist@msn.com
Web Site: kandimuseum.com
Founded: 1897.
Congressional District: 7
Key Personnel: Exec. Dir., Jill Wohnoutka.
Personnel Profile: Full-Time Paid 1; Part-Time Paid 2; Part-Time Volunteers 20; Interns 1.
Governing Authority: nonprofit organization. Tax-exempt: 501(c)(3).
Institution Type/Description: History Museum.
Collections: transportation; preservation project; history, archives; manuscripts; books; newspapers; photo; Indian artifacts.
Research Fields: local & family history.
Facilities: 1,500-vol. library of state and local history books; research center; 450 linear ft. archives.
Activities: guided tours; lectures; gallery talks; concerts; arts festivals; intermuseum loan, permanent, temporary & traveling exhibitions; historic sites tours; informal education programs for adults & children; program speakers bureau.
Publications: quarterly, Kandi Express.
Hours & Admission Prices: Memorial Day-Labor Day Mon.-Fri. 9-5, Sat.-Sun. 1-5; Sept.-May Mon.-Fri. 9-5. Suggested Donation $2. &
Attendance: 7,500 (accurate)
Membership: Individual $15; Family $25; Friend $50; Supporter $100; Patron $250; Benefactor $500.

Windom

COTTONWOOD COUNTY HISTORICAL SOCIETY, 812 Fourth Ave., Windom, MN 56101-1657. Tel.: 507-831-1134. Fax: 507-831-2665.
E-mail: cchs@windomnet.com
Founded: 1901.
Congressional District: 5
Key Personnel: Pres., Thomas Wickie; Dir., Linda Fransen; Treas., Margaret McDonald; Sec., Janelle Kaye.
Personnel Profile: Part-Time Paid 5; Part-Time Volunteers 25.
Volunteer Hours: 2,823
Operating Expenses: 84,729
Operating Income: 120,527
Governing Authority: society; nonprofit organization. Tax-exempt.
Institution Type/Description: History Museum.
Collections: artifacts pertaining to the history of Cottonwood County including late 19th century immigrant items; Victorian era artifacts; military memorabilia; medical artifacts; photographs of local events and places; newspapers from 1871-present; farm-related items; general store items; post office; REA kitchen items; church items; books; paintings & prints and Indian stone tools.
Major Exhibits: Cottonwood County Student Art Show, 4/14; Horses of Many Colors & Friends by Deborah Ney, 5/14-8/14; Rural Images by Nan Kaufenberg, 9/14-11/14; Festival of Trees, 12/1/14-12/23/14.
Research Fields: local history; genealogical research, including county, state & federal census obits, plat books & many family histories; cemetery records.
Facilities: library of material on local history including census and tax records, country school records, plat maps & complete set of Civil War books available for use on premises; meeting room; copy machine; reader printer for microfilm & computer lab.
Activities: guided tours; lectures; bus tours; art receptions and an educational program entitled Tour Through County History for all third grade students. Host the Festival of Trees each December.
Publications: quarterly newsletter; book, Cottonwood County Courthouse, 1904-2004.
Hours & Admission Prices: Mon.-Fri. 8-4, Sat. 10-4. No charge; donations accepted. Research library: $2; members no charge. &
Attendance: 8,000 (estimated)
Membership: Individual $15-$45; Family (Couple) $25-$80; Business $40-$100. Life: Individual $200; Family (Couple) $300; Business $400.

Winnebago

WINNEBAGO AREA MUSEUM, 16 Main St. S., Winnebago, MN 56098. Mailing Address: P.O. Box 595, 16 Main St. S., Winnebogo, MN 56098-0595. Tel.: 507-893-4660.
E-mail: wmuseum@bevcomm.net
Founded: 1977.
Congressional District: 2
Key Personnel: Chm. (V), Lola Baxter; Treas., Patti McCreary; Sec., Shirley Hartman.
Personnel Profile: Part-Time Paid 1; Part-Time Volunteers 15.
Governing Authority: nonprofit organization. Tax-exempt: 501(c)(3).
Institution Type/Description: Historical Society & Archaeology Museum.
Collections: 900-1500 AD Oneota artifacts; 8000 BC-1000 BC Woodland artifacts; 1000 BC-1700 AD Archaic artifacts; paintings; pictures; pottery exhibit; bead work of the Chippewa & Sioux; 1890 bedroom; 1842-1854 gun exhibit; 1858-1876 military display; cemetery records for area; early kitchen & farm tools; papers; ledgers; land grant papers; clothing; costumes; genealogies; U.S. Census records from 1872-1874, 1897-1898 & 1903; index of local papers 1864-present; 1865-1900 handcraft display; papers on microfilm 1870-1947; period clothing 1860-1950; photographs of barns & houses in the local area; military display, 1858 civil war & Spanish-American World War I & II; photo & story of former canning factory 1924-1967; all cemetery records from 4 townships since 1858; family histories.
Research Fields: genealogy; Oneota history & 100-year farms in the area; early industry in area: tile plants, carriage manufacturing company, cigar companies, shingle factory, well drilling, steam powered saw mill.
Facilities: library of brochures & Indian cultures available for inter-library loan; classrooms.
Activities: guided tours; films; organized education programs for children; exhibits arranged for special events; temporary exhibitions. Museum Sponsors: Ex. Quilt show for 125th anniversary; Early Toys for Moto-Fest Celebration.
Publications: book, Reflections from Winnebago, Minnesota; cookbook, Are We Dunn Yet?
Hours & Admission Prices: Call for hours. No charge; donations accepted. &
Attendance: 930 (accurate)
Membership: Pioneer $25-$49.99; Partner $50-$99.99; Sustainer $100-$249.99; Heritage $250-$499.99; Legacy $500 & up.

Winona

ARCHES MUSEUM OF PIONEER LIFE, Hwy. 14, 9 mi. west of Winona, Winona, MN 55987-3434. Mailing Address: Winona County Historical Society, 160 Johnson St., Winona, MN 55987-3434. Tel.: 507-454-2723 & 523-2111. Fax: 507-454-0006.
E-mail: info@winonahistory.org
Web Site: winonahistory.org
Founded: 1964.
Congressional District: 1
Key Personnel: Exec. Dir., Mark F. Peterson; Chm. (V), Laurie Lucas; Cur., Jodi Brom; Asst. Dir., Jennifer Weaver.
Personnel Profile: Full-Time Paid 2; Part-Time Paid 4; Part-Time Volunteers 150.
Governing Authority: society. Parent Institution: Winona County Historical Society, 160 Johnson St., Winona, MN 55987. Tax-exempt.
Institution Type/Description: History Museum.
Collections: windmill; one-room schoolhouse; period farm & household articles; natural history. Historic Buildings: 1845-50 Log House; c.1900 Log Barn.
Activities: guided tours; permanent & temporary exhibitions.
Publications: bimonthly newsletter, The Argus.
Hours & Admission Prices: June-Aug. Wed.-Sun. 1-5. Adults $5, students $3; members no charge. &
Attendance: 600 (estimated)
Membership: Senior Citizen $25; Individual $30; Senior Household $35; Family $45; Contributing $100 & up.

MINNESOTA MARINE ART MUSEUM, 800 Riverview Dr., Winona, MN 55987-2272. Tel.: 866-940-6626 (Toll Free); 507-474-6626. Fax: 507-474-6625.
E-mail: amaus@mmam.org
Web Site: www.mmam.org
Key Personnel: Exec. Dir., Andrew J. Maus.
Governing Authority: Tax-exempt.
Institution Type/Description: Art Museum.
Collections: works by regional & national artists; Hudson River School, Impressionism, historical & contemporary.
Major Exhibits: Andrea Rich: An Abundance of Riches (T), 1/17/14-4/13/14; River Sojourn: Recent Work by Sara Lubinski (T), 3/7/14-5/11/14; William Bradford: Exhibitions of the Arts, 3/22/14-10/14; The Floating World: Ukiyo-e Prints (T), 4/18/14-6/29/14; X-Ray Vision Fish Inside Out (T), 5/16/14-8/10/14; Caused by the Rider: Nick Wroblewski, 7/6/14-10/10/14; Revisiting Twain's Mississippi: Chris Faust, 8/15/14-10/12/14; Classic Images: Ansel Adams Photographs, 10/17/14/1/11/15.
Hours & Admission Prices: Tues.-Sun. 10-5. Adults $7, students $3; members and children 4 & under no charge. Closed holidays. &

MISSISSIPPI

(162 listings)

Attendance: 20,000 (estimated)
Membership: Annual $25-$70.

WINONA COUNTY HISTORICAL SOCIETY, (M), 160 Johnson St., Winona, MN 55987-3461. Tel.: 507-454-2723. Fax: 507-454-0006.
E-mail: info@winonahistory.org
Web Site: www.winonahistory.org
Founded: 1935.
Congressional District: 34B
Key Personnel: Exec. Dir., Mark F. Peterson; Asst. Dir., Jennifer Weaver; Pres. (V), Mark Metzler; Archivist, Walt Bennick; Archivist, Marianne Masterbrook; Cur. Collections, Jodi Brom.
Personnel Profile: Full-Time Paid 2; Part-Time Paid 5; Part-Time Volunteers 150.
Governing Authority: private; nonprofit society. Branch Museums: Historic Bunnell House; Arches Museum of Pioneer Living. Tax-exempt: 501(c)(3).
Institution Type/Description: Local History Museum.
Collections: artifacts from prehistoric to present times.
Research Fields: county and area history; Upper Mississippi River lore.
Facilities: Laird Lucas Memorial Library containing materials on state, local and Mississippi River history; rental facilities. Gift items for sale.
Activities: guided tours; lectures; formally organized youth & adult education programs; permanent & temporary exhibitions; special shows; early craft programs.
Publications: bimonthly newsletter, The Augus; Scenes & Sites: Views from Winona County; books, Uses of History; Free Enterprise; Pioneers Forever; A History of St. Charles; 1874 Andreas Atlas of Minn.; I Grew Up in West Burns Valley; Timber Roots; Rivertown Winona.
Hours & Admission Prices: Historical Society Museum: Jan.-Feb. Mon.-Fri. 9-5; March-Dec. Mon.-Fri. 9-5, Sat.-Sun. 12-4. Bunnell House: June-Aug. Wed.-Sat. 10-5, Sun. 1-5. Adults $5, students $3; members no charge. Arches Museum: June-Aug. Wed.-Sun. 1-5. Closed New Year's Day; Easter; Thanksgiving; Christmas. &
Attendance: 25,000 (estimated)
Membership: Senior Citizens $25; Individual $30; Senior Household $35; Family $45; Contributing $100 & up. Business: 20 Employees & under $150; over 20 Employees $250.

Worthington

NOBLES COUNTY HISTORICAL SOCIETY & PIONEER VILLAGE, 407 12th St., Worthington, MN 56187-2471. Mailing Address: 407 12th St., Ste. 2, Worthington, MN 56187-2471. Tel.: 507-376-4431. Fax: 507-376-3005.
E-mail: nchs@frontiernet.net
Web Site: www.noblespioneervillage.com
Founded: 1933.
Congressional District: 2
Key Personnel: Pres., Jacoba Nagel; Dir. Pioneer Village, Roy Reimer; Office Mgr., Carolyn Soper.
Personnel Profile: Full-Time Paid 1; Part-Time Paid 2; Part-Time Volunteers 150; Interns 1.
Governing Authority: county; nonprofit organization. Tax-exempt: 501(c)(3).
Institution Type/Description: Historical Society Museum.
Collections: Nobles County Pioneer Village, 45 buildings on 6 acres; horse-drawn machinery; furniture; toys; books; tools; household items; clothing; arrowheads; harness; early power machinery; general store; medical; mortuary; gardens.
Facilities: 100-vol. general library available for research on premises.
Activities: Village: self-guided tours. Museum: guided tours.
Publications: Nobles County Historical Society, Pioneer Village News; quarterly newsletter.
Hours & Admission Prices: Memorial Museum: Mon.-Fri. 12-4; other times by appointment. No charge; donations accepted. Nobles County Pioneer Village: May-Sept. Mon.-Sat. 10-5, Sun. 1-5. Adults $6; discounts to groups; senior citizens 90 & over, children under 6 and members no charge. &
Attendance: 9,000 (estimated)
Membership: Settler $15; Sod-Buster $30; Homesteader $50; Prairie Builder & Business $100.

Amory

AMORY REGIONAL MUSEUM, (M), 801 Third St., S., Amory, MS 38821-5233. Tel.: 662-256-2761.
E-mail: suebrown@midsouth.com
Web Site: www.amoryms.us
Founded: 1976.
Congressional District: 1
Key Personnel: Dir., Bo Miller; Computer Operator, Sue Brown; Tour Guides, Gertrude Sanders; Tour Guides, Bettye Benedict.
Personnel Profile: Full-Time Paid 1; Part-Time Paid 4; Part-Time Volunteers 24; Interns 1.
Governing Authority: municipal; nonprofit organization. Parent Institution: City of Armory. Tax-exempt.
Institution Type/Description: History Museum: housed in 2-story Greek Revival brick hospital building.
Collections: archaeological artifacts; Chickasaw Indian artifacts exhibit; Lawrence E. (Rabbit) Kennedy memorabilia; historical pictures; photographs; office furniture; medical instruments & equipment; railroad coach; hospital & sterilization rooms; Frisco Railroad memorabilia; 1838-1840 regulator log cabin.
Research Fields: genealogy; history of Monroe County, Amory, Chickasaw Indians, Cotton Gin Port, railroads.
Facilities: 60-vol. library of research material, available for use on premises only; meeting rooms.
Activities: guided tours; slides; tapes; TV programs; permanent, temporary & traveling exhibitions.
Publications: brochure.
Hours & Admission Prices: Tues.-Fri. 9-5, Sat. 10-4, Sun. 1-5. No charge. Closed New Year's Day; Thanksgiving; Christmas. &
Attendance: 5,000 (estimated)

Annapolis

* **THE BARRACKS, (M),** 43 Pinkney St., Annapolis, MS 21401-1717. Tel.: 410-263-8550 & 267-8146.
Institution Type/Description: Historic Building Museum: housed in the former home of soldiers during the American Revolution.
Collections: local history; period furnishings; early weapons.
Activities: educational programs.
Hours & Admission Prices: Sat.-Sun. by appointment.

Baldwyn

MISSISSIPPI'S FINAL STANDS VISITOR AND INTERPRETIVE CENTER, 607 Grisham St., Baldwyn, MS 38824-8541. Tel.: 662-365-3969. Fax: 662-365-3322.
E-mail: finalstands@att.net
Web Site: www.finalstands.com
Formerly: Brice's Crossroads
Founded: 1998.
Congressional District: 1
Key Personnel: Cur., Edwina Carpenter.
Personnel Profile: Full-Time Paid 1; Part-Time Paid 1; Part-Time Volunteers 9.
Governing Authority: Parent Institution: Brice's Crossroads National Battlefield Commission. Tax-exempt.
Institution Type/Description: History Museum.
Collections: local history & culture; Civil War artifacts from the Brice's Crossroads battlefield; photographs; exhibits interpreting the Battles of Harrisburg and Old Town Creek (Tupelo).
Hours & Admission Prices: Visitor Center: Tues.-Sat. 9-5. Adults $5, children $3; discounts to groups. Closed major holidays. &
Attendance: 3,250 (estimated)
Membership: Individual $25; Battlefield Advocate $100; Battlefield Guardian $500; Battlefield Champion $1,000 & up.

Bay Saint Louis

ALICE MOSELEY FOLK ART AND ANTIQUE MUSEUM, 1928 Depot Way, Bay Saint Louis, MS 39520. Tel.: 228-467-9223. Facebook: Alice Moseley Folk Art and Antique Museum.
E-mail: alicemoseley@gmail.com

Web Site: www.alicemoseley.com
Founded: 2005.
Key Personnel: Bd. Pres., Geralyn Bleau.
Personnel Profile: Full-Time Paid 1; Part-Time Paid 2; Part-Time Volunteers 5.
Governing Authority: nonprofit. Tax-exempt: 501(c)(3).
Institution Type/Description: Folk Art & Antique Museum.
Collections: Alice Moseley's paintings; period furniture; pottery; prints; collectible glass; video; tools; toys.
Activities: guided tour; lectures; workshops; movies; art camp; classroom.
Hours & Admission Prices: Mon.-Sat. 10-4, Sun. 12-4; other times by appointment. No charge; donations accepted. Closed Easter; Thanksgiving; Christmas. &
Attendance: 12,000 (estimated)

KATE LOBRANO HOUSE - HANCOCK COUNTY HISTORI-CAL SOCIETY, 108 Cue St., Bay Saint Louis, MS 39520. Mailing Address: P.O. Box 3356, Bay Saint Louis, MS 39521. Tel.: 228-467-4090. Fax: 228-467-4090.
E-mail: hancockcountyhis@bellsouth.net
Web Site: www.hancockcountyhistoricalsociety.com
Key Personnel: Dir., Charles H. Gray
Institution Type/Description: Historical Society Museum: housed in the former home of Katherine Maynard Lobrano; built in 1896.
Collections: local history & culture; photographs; city archives; county, church & school records; newspapers; magazines.
Activities: special events.
Publications: publication, The Historian.
Hours & Admission Prices: Mon.-Fri. 10-12 & 1-3. No charge; donations accepted.

Belzoni

CATFISH MUSEUM AND VISITOR CENTER, 111 Magnolia St., Belzoni, MS 39038. Mailing Address: P.O. Box 145, Belzoni, MS 39038-0385. Tel.: 800-408-4838; 662-247-4838. Fax: 662-247-4805.
E-mail: catfish@belzonicable.com
Web Site: www.belzonims.com
Formerly: Catfish Capital Visitors Center and Museum
Key Personnel: Dir., Steve Anderson
Personnel Profile: Full-Time Paid 1; Part-Time Paid 1.
Governing Authority: Parent Institution: Balzoni Humphreys Devel. Foundation. Tax-exempt.
Institution Type/Description: History Museum.
Collections: local history & culture; photographs; paintings; personal artifacts; sculpture.
Facilities: Museum-related items for sale.
Hours & Admission Prices: Mon.-Fri. 9-5. No charge; donations accepted. &
Attendance: 3,000 (estimated)

THE ETHEL WRIGHT MOHAMED STITCHERY MUSEUM, 307 Central, Belzoni, MS 39038-3603. Mailing Address: P.O. Box 254, Belzoni, MS 39038. Tel.: 662-247-3633. Fax: 662-247-1433.
E-mail: hwilson493@aol.com
Web Site: www.mamasdreamworld.com
Key Personnel: Cur., Carol Mohamed Ivy
Institution Type/Description: Stitchery Museum: housed in the former home of Ethel Wright Mohamed, often called Mississippi's Grandma Moses of stitchery.
Collections: embroidered pictures.
Activities: guided tours.
Hours & Admission Prices: By appointment only. Admission $2; children under 12 no charge.

JAKE TOWN MUSEUM, 116 W. Jackson St., Belzoni, MS 39038-3514. Mailing Address: P.O. Box 145, Belzoni, MS 39038-0145. Tel.: 662-247-4838. Fax: 662-247-4805.
E-mail: catfish@belzonicable.com
Web Site: www.belzonims.com
Founded: 2008.
Personnel Profile: Part-Time Volunteers 30.
Governing Authority: Parent Institution: Belzoni-Humphreys Devel. Foundation. Tax-exempt.
Institution Type/Description: History Museum.
Collections: local history & culture; period furnishings; personal artifacts; photographs.
Hours & Admission Prices: Daily 10-4. No charge; donations accepted. &
Attendance: 1,500 (estimated)

Biloxi

BEAUVOIR, THE JEFFERSON DAVIS HOME AND PRESI-DENTIAL LIBRARY, 2244 Beach Blvd., Biloxi, MS 39531-5002. Tel.: 228-388-4400. Fax: 228-388-7800.
E-mail: director@beauvoir.org
Web Site: www.beauvoir.org
Founded: 1902.
Congressional District: 5
Key Personnel: Dir., Bert Hayes-Davis; Chm. (V), Richard V. Forte, Sr.; Cur., Richard Flowers; Business Mgr., George G. (Rusty) Trowbridge; Facilities Mgr., Quentin Kersten; Museum Shop Mgr., Rosemary Potter; Security Chief, Jay Peterson.
Personnel Profile: Full-Time Paid 6; Full-Time Volunteers 1; Part-Time Paid 17; Part-Time Volunteers 2.
Governing Authority: nonprofit. Mississippi Division, United Sons of Confederate Veterans, 2244 Beach Blvd., Biloxi, MS 39531. Tax-exempt: 501(c)(3).
Institution Type/Description: Historic Site: housed in the post-war home of Confederate President Jefferson Davis; built in 1853.
Collections: 19th-century period furnishings & decorative; Jefferson Davis family effects & papers; Civil War militaria; Confederate Veterans Cemetery.
Research Fields: Jefferson Davis; Jefferson Davis Soldiers' Home; Confederate military history; 19th-century Southern History.
Facilities: 6,000-vol. library; 51 acre complex; gardens. Museum-related items for sale.
Activities: self-guided tours; guided tours; educational outreach program; special events.
Publications: books, The Uncivil War; Adapt or Perish: The Life of Roger A. Pryor; Historic Beauvoir; booklets, Beauvoir, A Walk Through History; The Conspiracy Against Jefferson Davis; Jefferson Davis: The Unforgiven; Jefferson Davis: A Judicial Estimate; Jefferson Davis: The Making of a President; The Life of Jefferson Davis; The Autobiography of Jefferson Davis; The Chronicles of Beauvoir.
Hours & Admission Prices: Daylight Savings Time: daily 9-5; Standard Time: daily 9-4. Adults $9, seniors 65 & over, military & AAA members $7.50, children 6-18 $5. Closed Thanksgiving; Christmas. &
Attendance: 31,000 (estimated)
Membership: Individual $25; Family $50; Patron $100-$249; Sustaining $250-$499; Corporate $500 & up; Benefactor $500-$999; Sponsor $1,000-$4,999; President's Cabinet $5,000 & up.

MARITIME & SEAFOOD INDUSTRY MUSEUM, 115 First St., Biloxi, MS 39530. Mailing Address: P.O. Box 1907, Biloxi, MS 39533-1907. Tel.: 228-435-6320. Fax: 228-435-6309.
E-mail: schooner@maritimemuseum.org
Web Site: www.maritimemuseum.org
Founded: 1986.
Congressional District: 5
Key Personnel: Exec. Dir., Robin Krohn-David; Schooner Captain, Ron Reiter; Office Mgr., Megan Seymour; Museum Shop Mgr., Robert Sweeting.
Personnel Profile: Full-Time Paid 5; Part-Time Paid 20; Part-Time Volunteers 40.
Governing Authority: board of directors. Tax-exempt.
Institution Type/Description: Seafood Industry Museum: housed in Coast Guard station.
Collections: maritime crafts, fishing & boat building implements, early photographs of industry & watercraft; local history of city & seafood industry; hurricane history.
Research Fields: local history.
Facilities: 18,000 sq. ft. exhibit space. Museum-related items for sale.
Activities: annual museum festival; live demonstration of maritime crafts; trips available on replicated Biloxi oyster schooner.
Publications: member newsletter, Seafood Industry Museum.
Hours & Admission Prices: Mon.-Sat. 10-6, Sun. 12-6. No charge; donations accepted. &
Attendance: 150,000 (estimated)
Membership: Senior $10; Individual $25; Family $50; Patron $250; Lifetime $1,000.

OHR-O'KEEFE MUSEUM OF ART, (M), 386 Beach Blvd., Biloxi, MS 39530. Mailing Address: P.O. Box 248, Biloxi, MS 39533-0248. Tel.: 228-374-5547. Fax: 228-436-3641.
E-mail: director@georgeohr.org
Web Site: www.georgeohr.org
Formerly: George E. Ohr Arts & Culture Center
Founded: 1994.

Congressional District: 5
Key Personnel: Dir., Denny Mecham.
Personnel Profile: Full-Time Paid 9; Full-Time Volunteers 20; Part-Time Paid 6; Part-Time Volunteers 75; Interns 1.
Governing Authority: nonprofit organization. Tax-exempt: 501(c)(3).
Institution Type/Description: Art Museum.
Collections: George Ohr pottery; ceramics & pottery from around the world
Activities: guided tours; lectures; films; concerts; dance recitals; study clubs; organized education programs for children, adults, undergraduate & graduate students; docent program.
Publications: exhibition catalogs; exhibition announcements; e-blasts.
Hours & Admission Prices: Mon.-Fri. 9-4:30. Adults $10; members no charge.
Attendance: 56,800 (accurate)
Membership: Student $35; Adult $50; Family $75; Patron $100; Benefactor $500; Sustaining $1,000. Corporate Membership: Bronze $100; Silver $500; Gold $1,000; Platinum $5,000.

WEST END HOSE CO. NO. 3 MUSEUM AND FIRE EDUCATIONAL CENTER, 1046 Howard Ave., Biloxi, MS 39530. Tel.: 228-435-6119.
E-mail: jboney@biloxi.ms.us
Web Site: biloxi.ms.us/museums/firemuseum
Founded: 1990.
Key Personnel: Pres. (V), Larry Smith; Museum Shop Mgr., Joe Boney
Institution Type/Description: Fire Museum.
Collections: period fire-fighting tools, apparatus & photographs; 1880's hose cart; 1908 American LaFrance steam fire engine; 1923 American LaFrance chain-driven fire engine; memorabilia.
Hours & Admission Prices: Sat. 9-3; other times by appointment. No charge; donations accepted. &
Attendance: 1,000 (estimated)

Booneville

RAILS & TRAILS MUSEUM, 100 W. Church St., Booneville, MS 38829-3406. Tel.: 662-728-4130; 800-300-9302.
Institution Type/Description: Historic Building: housed in the Gulf, Mobile & Ohio Depot; built in 1913.
Collections: depot & railroad history; photographs; fossils; Native American artifacts; Civil War; documents; agriculture.
Hours & Admission Prices: Thurs.-Sat. 10-4; other times by appointment. No charge.

Brandon

RANKIN COUNTY HISTORICAL MUSEUM, 1475 W. Government St., Brandon, MS 39042. Mailing Address: P.O. Box 841, Brandon, MS 39043-0841.
E-mail: rchsinc@aol.com
Web Site: rankinhistory.com
Founded: 1978.
Congressional District: 3
Institution Type/Description: History Museum.
Collections: county history & culture; photographs.
Hours & Admission Prices: 1st Sat. each month 9-4, 1st Sun. each month 2-4. No charge; donations accepted.

Camp Shelby

MISSISSIPPI ARMED FORCES MUSEUM, Bldg. 850, Camp Shelby Training Site, Camp Shelby, MS 39407. Tel.: 601-558-2757 & 2347.
E-mail: chad.e.daniels@us.army.mil
Web Site: www.armedforcesmuseum.us
Founded: 2001.
Congressional District: 4
Key Personnel: Dir., Chad E. Daniels, M.A., M.S.; Pres. (V), MG Richard S. Poole, MSARNG (Ret.); Arms & Military Vehicle Conservator, MSG (Ret.) Glenn L. Husted, III, MSARNG; Archivist, Christy A. Jones; Registrar, Lisa Foster; Administrative Asst., Brenda Crowley.
Personnel Profile: Full-Time Paid 8; Part-Time Paid 5; Part-Time Volunteers 2.
Governing Authority: state. Subsidiary Institution: Mississippi Military Dept. Tax-exempt 501(c)(3).
Institution Type/Description: Military Museum.
Collections: monuments; military vehicles; weapons; history of Mississippi's veterans & training facilities; War of 1812; Mexican War; American Civil War; Spanish-American War; WWI & WWII; Korean War; Vietnam War; Gulf War; Global War on Terrorism.

Research Fields: military history.
Facilities: 4,500-vol. library; 16,000 sq. ft. exhibit space; theater.
Activities: simulations of real battlefield sights, sounds & smells.
Publications: newsletter, Mississippi Armed Forces Museum.
Hours & Admission Prices: Tues.-Sat. 9-4:30; groups by appointment. Photo ID required to enter post. No charge. Closed holidays. &
Attendance: 49,734 (accurate)
Membership: Annual $25; Sponsor $50; Benefactor $100; Gold Star $250; Minutemen $500; Patriot $1,000; Governor Shelby $2,500 & up.

Canton

CANTON MOVIE MUSEUMS, 147 N. Union St., Canton, MS 39046-3740. Mailing Address: P.O. Box 53, Canton, MS 39046. Tel.: 800-844-3369; 601-859-1307. Fax: 601-859-0346.
E-mail: canton@cantontourism.com
Key Personnel: Exec. Dir., Jo Ann Gordon
Institution Type/Description: Movie History Museum.
Collections: movie history & memorabilia; photographs; movie props; films; equipment; posters.
Hours & Admission Prices: Welcome Center: Mon.-Fri. 10-5, Sat. 10-2. One Movie Museum: adults $4, senior citizens $3, children & students $2. Two Movie Museums: adults $6, senior citizens $5, children & students $3. Multi-Cultural Museum: adults $3, students 12 & over $2, children 5-12 $1. All Museums: adults $7, senior citizens $6, children 5-12 $4.

CANTON MULTICULTURAL CENTER & MUSEUM, 147 N. Union St., Canton, MS 39046-3740. Mailing Address: P.O. Box 53, Canton, MS 39046-0053. Tel.: 601-859-1307. Fax: 601-859-0346.
Institution Type/Description: History Museum.
Collections: slavery; civil rights; early African American businesses; education; family; music.
Hours & Admission Prices: Mon.-Fri. 9-5, Sat. 10-2. Adults $3, students 12 & over $2, children K-11 $1.

Carrollton

MISSISSIPPI JOHN HURT HOME MUSEUM, Rte. 2, Rd. 109, Carrollton, MS 38917. Mailing Address: P.O. Box 1973, Carrollton, MS 38917. Tel.: 312-810-1954.
E-mail: mfhurt_wright74@yahoo.com
Web Site: www.msjohnhurtmuseum.com
Founded: 1999.
Key Personnel: Dir., Mary Hurt Wright
Institution Type/Description: Historic House Museum: housed in the former home of Blues Musician, Mississippi John Hurt.
Collections: Hurt's life & career; personal artifacts; period furnishings; photographs; guitars.
Activities: Annual Event: Mississippi John Hurt Jamfest in August & September.
Hours & Admission Prices: By appointment. Adults $5; children under 12 no charge.
Attendance: 300 (estimated)

Centreville

CAMP VAN DORN WORLD WAR II MUSEUM, 138 E. Main, Centreville, MS 39631. Mailing Address: P.O. Box 1113, Centreville, MS 39631-1113. Tel.: 601-645-9000.
E-mail: info@vandorn.org
Founded: 2005.
Key Personnel: Dir. & Pres. (V), Michael Stewart; Museum Shop Mgr., Patricia Rogers.
Governing Authority: Tax-exempt.
Institution Type/Description: Military Museum.
Collections: local & World War II history; photographs; personal artifacts.
Hours & Admission Prices: Mon.-Fri. and 1st & 3rd Sat. of the month 10-4. No charge; donations accepted. &

Choctaw

CHOCTAW MUSEUM OF THE SOUTHERN INDIAN, Hwy. 16 W., Choctaw, MS 39350. Mailing Address: Mississippi Band of Choctaw Indians, P.O. Box 6010, Choctaw, MS 39350-6010. Tel.: 601-656-5251. Fax: 601-650-3684.
E-mail: info@choctaw.org
Web Site: www.choctaw.org
Institution Type/Description: Native American History Museum.

Collections: exhibits & archives on the culture of Southeastern Indian tribes; photographs; personal artifacts.
Hours & Admission Prices: Call for hours.

Clarksdale

DELTA BLUES MUSEUM, #1 Blues Alley, Clarksdale, MS 38614-4336. Mailing Address: P.O. Box 459, Clarksdale, MS 38614-0459. Tel.: 662-627-6820. Fax: 662-627-7263.
E-mail: shelley@deltabluesmuseum.org
Web Site: www.deltabluesmuseum.org
Founded: 1979.
Congressional District: 2
Key Personnel: C.E.O. & Dir., Shelley Ritter; Chm. (V), Bill Gresham; Museum Shop Mgr., Christopher Coleman.
Personnel Profile: Full-Time Paid 5; Part-Time Paid 4; Part-Time Volunteers 35; Interns 2.
Governing Authority: nonprofit organization. Parent Institution: City of Clarksdale. Tax-exempt.
Institution Type/Description: Music Museum: housed in renovated freight depot.
Collections: books & periodicals on Blues music, artists; recordings of Blues artists; videotape, slide & sound programs on the Blues; art work; musical instruments; musicians personal artifacts.
Research Fields: American music & Black history.
Facilities: 90,000-vol. library of books, monographs, art reports, recordings & videotapes, reading room; 5,000 sq. ft. gallery space; classroom; outdoor stage.
Activities: lectures; permanent & temporary exhibitions; performances. Museum Co-Sponsors: Sunflower River Blues & Gospel Festival in August.
Publications: brochure; electronic newsletter.
Hours & Admission Prices: March-Oct. Mon.-Sat. 9-5; Nov.-Feb. Mon.-Sat. 10-5. Adults $7, children $5; discounts to groups, AAM, ICOM & Blues Society members; members & children under 6 no charge. Closed major holidays. &
Attendance: 26,500 (estimated)
Membership: Student & Senior $20; Individual $30; Family $40; Supporter $50-$149; Patron $150-$499; Benefactor $500-$999.

ROCK & BLUES MUSEUM, 113 E. Second St., Clarksdale, MS 38614-4205. Tel.: 901-605-8662.
Formerly: Rock 'N Roll & Blues Heritage Museum
Founded: 2005.
Governing Authority: Parent Institution: Rock N Roll Museum, Inc. Tax-exempt.
Institution Type/Description: History Museum.
Collections: blues and rock & roll history; memorabilia from the 1920s to 1960s; period furnishings; movie posters.
Facilities: Museum-related items for sale.
Activities: special events. Museum Sponsors: 2nd Street Blues Parties/Mini-Fests.
Hours & Admission Prices: April-Oct. Thurs.-Sat. 11-5; Sun. by appointment. Admission $5.
Membership: Annual $25.

Cleveland

FIELDING L. WRIGHT ART CENTER, Delta State University, Cleveland, MS 38733. Mailing Address: Box D 2, Delta State University, Cleveland, MS 38733. Tel.: 662-846-4720. Fax: 662-846-4726.
E-mail: artinfo@deltastate.edu
Web Site: www.deltastate.edu/academics/artsci/artdept.com
Founded: 1924.
Congressional District: 2
Key Personnel: Chm., Ron Koelher; Art Gallery Dir., Patricia Brown.
Governing Authority: university. Affiliated with Delta State University. Tax-exempt.
Institution Type/Description: Art Center.
Collections: prints by Kathe Kollwitz, Salvadore Dali, G. B. Piranesi; works of Walter Anderson; Marie Hull; Andrew Bucci; Japanese prints, woodblocks; sculptured Greek heads from Crete c.500 B.C.; medals struck in Paris Mint; works of Southern artists; works of Mississippi artists.
Research Fields: Southern art; Southern Folk art.
Facilities: library; reading room; classrooms.
Activities: lectures; films; gallery talks; formally organized education programs for children, undergraduate & graduate college students; temporary & traveling exhibitions.
Publications: announcements & publications for special exhibitions.

Hours & Admission Prices: during University Sessions: Mon.-Thurs. 8-8:30, Fri. 8-3:30. No charge. &
Attendance: 10,000

Clinton

MISSISSIPPI BAPTIST HISTORICAL COMMISSION, Mississippi College Library-College St., Clinton, MS 39058. Mailing Address: P.O. Box 4024, Clinton, MS 39058. Tel.: 601-925-3434. Fax: 601-925-3435.
E-mail: mbhc@mc.edu
Web Site: www.mc.edu/resources/mississippi_baptist_historical_collection/
Founded: 1887.
Congressional District: 4
Key Personnel: Librarian, Heather Weeden.
Personnel Profile: Full-Time Paid 1; Part-Time Paid 1.
Governing Authority: church. Parent Institution: Mississippi Baptist Convention.
Institution Type/Description: Religious Museum.
Collections: Baptist history; old church minute books; documents; manuscripts.
Research Fields: Baptist history.
Facilities: 570-vol. library of books, periodicals, church & associational minutes & vertical files, on Baptist history available for use on premises; reading room; microfilm reader.
Publications: books, A History of Mississippi Baptists, 1780-1970; Mississippi Baptist Convention ministers: current biographies; Highlights of Mississippi Baptist history.
Hours & Admission Prices: Mon.-Fri. 8-12 & 1:4:30. No charge. Closed some college vacations. &
Attendance: 150 (estimated)

Columbus

AMERICAN-INDIAN ARTIFACTS MUSEUM, 179 State Line Rd., Columbus, MS 39702-7134. Tel.: 662-251-1125; 800-327-2686.
Institution Type/Description: Native American History Museum.
Collections: Native American artifacts; personal artifacts.
Hours & Admission Prices: Call for hours. No charge, donations accepted.

COLUMBUS WAR MUSEUM, Columbus Municipal Complex, 1501 Main St., Columbus, MS 39701. Mailing Address: 364 Country Ln., Millport, AL 35576. Tel.: 205-557-0584.
E-mail: columbuswarmuseum@live.com
Founded: 2004.
Key Personnel: Dir., Wayne White; Pres. (V), Virginia White.
Personnel Profile: Part-Time Volunteers 7.
Institution Type/Description: Military History Museum.
Collections: military history & artifacts; personal artifacts; photographs.
Activities: tours. Annual Event: Veteran Ceremonies.
Hours & Admission Prices: Mon.-Thurs. 8-5:30. No charge; donations accepted. Closed federal holidays.
Attendance: 500 (estimated)

THE FLORENCE MCLEOD HAZARD MUSEUM, 316 Seventh St. N., Columbus, MS 39701-4680. Tel.: 662-329-3533. Fax: 662-329-1027.
Web Site: historic-columbus.org
Founded: 1959.
Congressional District: 2
Key Personnel: Dir., Heather Roland; Pres. (V), Libba Johnson.
Personnel Profile: Full-Time Paid 1; Part-Time Paid 30.
Governing Authority: society; nonprofit. Parent Institution: The Columbus & Lowndes County Historical Society.
Institution Type/Description: Historical Society Museum: housed in c.1847 Blewett-Harrison-Lee Home.
Collections: c.1833-1908 clothing; portraits; silver; furniture; china; glass; guns; documents; diaries; pictures; books; medical instruments.
Facilities: library of old newspapers, books, letters, documents, wills, scrapbooks, available for use on premises. Books & other museum-related items for sale.
Activities: guided tours; docent program; formally organized education programs for undergraduate college students affiliated with the Mississippi University for Women; permanent exhibitions.
Publications: book, I Remember When.
Hours & Admission Prices: Fri. 10-4; other times by appointment. Adults $5; students & children no charge. Closed Thanksgiving; Christmas.
Membership: Individual $10.

MISSISSIPPI UNIVERSITY FOR WOMEN, ARCHIVES AND MUSEUM, 1100 College St., W-1625, Columbus, MS 39701-5831. Tel.: 662-329-7332. Fax: 662-329-7348.
Web Site: www.muw.edu
Founded: 1978.
Congressional District: 5
Governing Authority: university. Affiliated with the Mississippi University For Women, College St., Columbus, MS. 39701. Tax-exempt.
Institution Type/Description: College Museum: housed in 1885 Orr Building.
Collections: University documents; photographs; memorabilia donated by alumni; rare books; china; artifacts; history of dance; early science & domestic science implements; dolls; manuscripts & documents. Historic Structure: university chapel with restored stained glass windows.
Research Fields: University history; history of women; business & government research in Tennessee Tombigbee Waterways, barges & old ports.
Publications: Finding Aid, 1979; Introductory Guide to the Tennessee Tombigbee Waterway Development Authority; printed guide to early photographs of students, faculty & buildings.
Hours & Admission Prices: Sun.-Thurs. 2-10, Fri. 7:30-5, Sat. 9-5. No charge.

ROSEDALE PLANTATION, 1523 9th St. S., Columbus, MS 39703. Tel.: 662-329-3533; 800-920-3533. Fax: 662-329-8969.
Institution Type/Description: Historic House Museum: c.1856.
Collections: local history & culture; period furnishings; personal artifacts.
Hours & Admission Prices: By appointment. Adults $7.50.

ROSEWOOD MANOR, 719 Seventh St. N., Columbus, MS 39701. Tel.: 662-328-7313 & 364-0705. Fax: 662-327-6217.
E-mail: hicks@columbursrosewood.com
Institution Type/Description: Historic House Museum: built in 1835. Listed on the National Register of Historic Places.
Collections: local history; period furnishings; personal artifacts; plantation chapel.
Hours & Admission Prices: By appointment. Adults $10; discounts to AAM & ICOM members.

STEPHEN D. LEE HOME MUSEUM, 316 Seventh St., N., Columbus, MS 39701-4680. Tel.: 662-327-8888; 800-920-3533.
Web Site: www.columbus-ms.org
Institution Type/Description: Historic House Museum: housed in the former home of Confederate Gen. Stephen D. Lee; built in 1847. Listed on the National Register of Historic Places.
Collections: local history; Civil War artifacts; period furnishings; photographs.
Activities: rental facilities.
Hours & Admission Prices: Fri. 10-4; other times by appointment. Adults $7.50.

TENNESSEE WILLIAMS BIRTHPLACE MUSEUM, 300 Main St., Columbus, MS 39701-4532. Tel.: 662-328-0222.
Web Site: www.muw.edu/tennesseewilliams
Institution Type/Description: Historic House Museum: housed in the home of playwright Tennessee Williams; c.1870.
Collections: Tennessee Williams' life & career; period furnishings; personal artifacts.
Hours & Admission Prices: Mon.-Sat. 8:30-5, Sun. 12-5.

Corinth

BLACK HISTORY MUSEUM OF CORINTH, 1109 Meigg St., Corinth, MS 38835. Tel.: 662-665-8500; 866-539-8500.
Institution Type/Description: History Museum.
Collections: local black history, memorabilia & artifacts; religious & school furnishings; African art; Corinth's first black mayor, E.S. Bishop; sports figures.
Hours & Admission Prices: Thurs.-Fri. 11-4; other times by appointment. No charge; donations accepted.

CORINTH CIVIL WAR INTERPRETIVE CENTER, 501 W. Linden St., Corinth, MS 38834. Tel.: 662-287-9273.
Web Site: www.nps.gov/shil
Institution Type/Description: History Museum.
Collections: Civil War history; personal artifacts; African American heritage; military artifacts.
Activities: orientation film.
Hours & Admission Prices: Daily 8:30-4:30.

CORINTH COCA-COLA MUSEUM, 305 E. Waldron St., Corinth, MS 38834-4756. Tel.: 662-284-4848.
Institution Type/Description: Company Museum.
Collections: Coca-Cola company history; memorabilia; period drink machines; photographs.
Hours & Admission Prices: Tues.-Fri. 10-4, Sat. 10-2. Adults $3, children 6-16 $1; children 5 & under no charge.

CROSSROADS MUSEUM, 221 N. Fillmore St., Corinth, MS 38834-5635. Tel.: 662-287-3120. Fax: 662-287-3120.
E-mail: director@crossroadsmuseum.com
Web Site: www.crossroadsmuseum.com
Formerly: Northeast Mississippi Museum
Founded: 1980.
Congressional District: 1
Key Personnel: Dir., Brandy Steen; Pres. (V), Bryan Clausel; Treas., Lila Wade.
Personnel Profile: Part-Time Paid 1; Part-Time Volunteers 3; Interns 1.
Governing Authority: private; nonprofit organization. Tax-exempt: 501(c)(3).
Institution Type/Description: History Museum.
Collections: fossils; American Indian; Civil War; railroad; photographs; local arts & crafts gallery; history of Northeast Mississippi with emphasis on the history of the city of Corinth & Alcorn County.
Facilities: 475-vol. library; 2,500 sq. ft. exhibit space. Museum-related items for sale.
Activities: guided & school tours; temporary exhibitions; summer children's camps. Museum Sponsors: Annual meeting of Museum membership; bimonthly education luncheons.
Publications: quarterly newsletter, NE MS Museum News.
Hours & Admission Prices: Tues.-Sat. 10-4, Sun. 1-4. Adults $5, senior citizens, military, students & members $3; discounts to groups and AAM & ICOM members; children 16 & under no charge. Closed Thanksgiving; Christmas.
Attendance: 6,500 (estimated)
Membership: Senior, Student & Educator $25; Friend $50; Supporter $100; Contributor $250; Sponsor $500; Patron $1,000.

JACINTO FOUNDATION, INC., Jacinto Courthouse, County Rd. 364, Corinth, MS 38382. Mailing Address: P.O. Box 1174, Corinth, MS 38835-1174. Tel.: 662-286-8662. Fax: 662-286-6500.
Founded: 1966.
Congressional District: 1
Key Personnel: C.E.O., Beth Whitehurst; Pres. (V), John C. Ross.
Personnel Profile: Part-Time Paid 4.
Governing Authority: nonprofit organization. Tax-exempt.
Institution Type/Description: Historic Foundation: housed in 1854 Jacinto Courthouse. Listed on the National Register of Historic Places.
Collections: local history & culture; period artifacts.
Facilities: nature & conservation center. Museum-related items for sale.
Activities: guided tours; drama. Foundation Sponsors: July 4th Festival.
Hours & Admission Prices: Tues.-Sat. 10-5, Sun. 1-5. No charge; donations accepted.

LAKE HILL MOTORS MUSEUM, 2003 Hwy. 72 E. Annex, Corinth, MS 38834. Tel.: 662-287-4451.
Web Site: www.lakehillmotors.com
Institution Type/Description: Motorcycle Museum.
Collections: early motorcycles.
Hours & Admission Prices: Mon.-Fri. 8-5, Sat. 8-3. No charge.

VERANDAH HOUSE, (M), 705 Jackson St., Corinth, MS 38834. Mailing Address: Corinth Visitor's Bureau, 215 N. Fillmore, Corinth, MS 38834. Tel.: 662-287-8300.
E-mail: tourism@corinth.net
Institution Type/Description: Historic House Museum: housed in the headquarters for Confederate Generals Braxton Bragg, John Bell Hood, Earl Van Dorn & Union Generals Henry W. Halleck & Granville Dodge; built in 1857. A National Historic Landmark.
Collections: Civil War history; period furnishings; personal artifacts.
Hours & Admission Prices: Tours by appointment only.

Crystal Springs

ROBERT JOHNSON HERITAGE & BLUES MUSEUM, 218 E. Marion Dr., Crystal Springs, MS 39059. Tel.: 601-892-7883. Fax: 601-892-7884.
E-mail: robertjohnsonblu@bellsouth.net
Web Site: www.robertjohnsonbluesfoundation.org/museum.html

Institution Type/Description: History Museum.
Collections: Robert Johnson & Delta Blues history; murals; guitars.
Facilities: Museum-related items for sale.
Hours & Admission Prices: Mon.-Fri. 10-5, Sat. 10-2; other times by appointment. Suggested Donation: adults $2, children $1.

DeKalb

KEMPER COUNTY HISTORICAL MUSEUM, Jackson St., De-Kalb, MS 39328. Mailing Address: 5217 Kipling Rd., DeKalb, MS 39328. Tel.: 601-743-5560. Fax: 601-743-2760.
Founded: 1982.
Key Personnel: Pres. Kemper Historical Society, Grace Gibson; Chm. & Pres., Charles Robbins.
Governing Authority: Tax-exempt.
Institution Type/Description: History Museum.
Collections: county history & culture; photographs; personal artifacts of U.S. Sen. John C. Stennis; early tools; county book authors.
Hours & Admission Prices: By appointment. No charge; donations accepted.
&

Attendance: 200 (estimated)
Membership: Family $10; Sustaining $25; Life $200.

Fayette

HARRISON HOUSE, 414 River Rd., Fayette, MS 39069. Mailing Address: P.O. Box 310, Fayette, MS 39069. Tel.: 601-493-3420. Fax: 601-786-6448.
Founded: 1900.
Key Personnel: Dir., M. Louise Coleman.
Personnel Profile: Part-Time Volunteers 1.
Institution Type/Description: Historic House Museum: built in 1900.
Collections: local history & culture; memorabilia; documents; manuscripts; photographs.
Facilities: Museum-related items for sale.
Hours & Admission Prices: Daily by appointment. Adults $5, children under 12 $3.
Attendance: 15 (accurate)

Flora

MISSISSIPPI PETRIFIED FOREST, 124 Forest Park Rd., Flora, MS 39071. Mailing Address: P.O. Box 37, Flora, MS 39071-0037. Tel.: 601-879-8189. Fax: 601-879-8165.
E-mail: info@mspetrifiedforest.com
Web Site: www.mspetrifiedforest.com
Founded: 1963.
Congressional District: 3
Key Personnel: Dir. & Owner, C. J. McNamara; Museum Shop Mgr., Deborah Shoemaker; Park Mgr.-Outdoor Museum, Doug Shoemaker.
Personnel Profile: Part-Time Paid 3.
Governing Authority: individual operation.
Institution Type/Description: Geology Museum.
Collections: petrified wood; mineral collections; vertebrate & invertebrate fossil collections; lapidary art; native wildlife.
Research Fields: identification of petrified wood.
Facilities: nature & conservation center; black light display; picnic area; campground. Lapidary equipment & supplies, books & gift items for sale.
Activities: permanent & temporary exhibitions; nature trail through Natural Petrified Forest; gemstone flume.
Publications: Mississippi Petrified Forest: A Place of Fascination; All In A Nutshell, history of vegetable ivory-tagua; Out of the Past, a scrapbook of Time.
Hours & Admission Prices: April to Labor Day daily 9-6; Sept.-May daily 9-5. Adults $7, seniors & students grades 1-12 $6; discounts to AAA members & groups of 15 or more; pre-school children no charge. Closed Thanksgiving; Christmas. &
Attendance: 15,000 (estimated)

Friars Point

NORTH DELTA MUSEUM, 748 2nd St., Friars Point, MS 38631. Tel.: 662-645-5063. Fax: 662-383-0057.
E-mail: flolarson@bellsouth.net
Key Personnel: Dir., Flo Larson.
Institution Type/Description: History Museum.
Collections: Civil War; Native American artifacts; personal artifacts; photographs.
Hours & Admission Prices: Tues.-Sat. 9-3; other times by appointment. Adults $4, children under 7 $2.

Fulton

JAMIE L. WHITTEN HISTORICAL CENTER, 100 Campground Rd., Fulton, MS 38843. Tel.: 662-862-5414.
Institution Type/Description: History Museum.
Collections: Appalachian Regional Commission; Tennessee Valley Authority; USDA & Natural Resources Conservation Service; National Park Service; U.S. Forest Service; photographs.
Facilities: 120-seat auditorium.
Activities: guided tours.
Hours & Admission Prices: April-May 25 & Sept.-Nov. 1 daily 7:30-4; May 26-Aug. daily 8:30-5; Nov.-March Mon.-Fri. 7:30-4.

Greenville

GOLDSTEIN NELKEN SOLOMON CENTURY OF HISTORY MUSEUM, 504 Main St., Greenville, MS 38701. Tel.: 662-332-4153.
Institution Type/Description: History Museum: housed in a temple built in 1906.
Collections: Jewish history & culture; photographs; personal artifacts.
Hours & Admission Prices: Mon.-Fri. 9-12.

GREENVILLE AIR FORCE BASE MUSEUM, 166 Fifth Ave., Mezzanine Level - 2nd Flr., Greenville, MS 38701. Tel.: 662-334-3121. Fax: 662-335-8757.
E-mail: dgant@greenvillems.org
Congressional District: 2
Governing Authority: Parent Institution: City of Greenville. Subsidiary Institution: Greenville Mid Delta Airport. Tax-exempt.
Institution Type/Description: Military History Museum.
Collections: military history from WWII to Cold War; military artifacts; photographs; personal artifacts.
Hours & Admission Prices: Daily 7-7. No charge. &
Attendance: 6,300 (estimated)

GREENVILLE HISTORY MUSEUM, 409 Washington Ave., Greenville, MS 38701-3617. Tel.: 662-335-5802.
Founded: 1988.
Key Personnel: Dir., Ben Nelken.
Personnel Profile: Full-Time Volunteers 1.
Institution Type/Description: History Museum: housed in the restored Miller Building.
Collections: memorabilia, photographs & newspapers relating to Greenville.
Hours & Admission Prices: Mon.-Fri. 9-5; other times by appointment. Adults $5, children $3.
Attendance: 2,000 (estimated)

OLD FIREHOUSE MUSEUM, 230 Main St., Greenville, MS 38701. Mailing Address: P.O. Box 897, Greenville, MS 38702-0897. Tel.: 662-378-1554. Fax: 662-378-1612.
Web Site: www.greenville.ms.us
Founded: 1994.
Congressional District: 2
Key Personnel: C.E.O., Mayor Heather Hudson; Treas., Tommy Jefcoat; Security, Chief Lester Carter.
Personnel Profile: Full-Time Paid 1; Full-Time Volunteers 1.
Governing Authority: municipal; nonprofit. Tax-exempt.
Institution Type/Description: General Museum.
Collections: photographs; firefighting equipment.
Facilities: 50-vol. library of manuals & directories; 500 sq. ft. exhibit space; interactive play area.
Activities: formal educational programs for children; guided tours; participatory exhibits.
Publications: semiannual newsletter, Firehouse News.
Hours & Admission Prices: By appointment. No charge. Closed New Year's Day; Martin Luther King Jr. Day; Memorial Day; Independence Day; Labor Day; Thanksgiving; Christmas. &
Attendance: 2,340 (accurate)

WINTERVILLE MOUNDS PARK AND MUSEUM, 2415 Hwy. 1 N., Greenville, MS 38703-9476. Tel.: 662-334-4684. Fax: 662-378-5559.
E-mail: wmounds@mdah.state.ms.us
Web Site: mdah.state.ms.us/hprop/winterville.html
Founded: 1960.
Congressional District: 4
Key Personnel: Dir., Dr. Mark Howell.

Personnel Profile: Full-Time Paid 5; Part-Time Volunteers 10.
Governing Authority: Parent Institution: Mississippi Department of Archives & History.
Institution Type/Description: Historic Site.
Collections: Native American artifacts.
Hours & Admission Prices: Grounds: daily dawn to dusk. Museum: Mon.-Sat. 9-5, Sun. 1:30-5. No charge; donations accepted.
Attendance: 15,000 (accurate)

Greenwood

BACK IN THE DAY MUSEUM, 200 Young St., Greenwood, MS 38930-4637. Tel.: 662-453-2742.
E-mail: deltablueslegendtours@yahoo.com
Web Site: deltablueslegendtours.com
Founded: 2006.
Key Personnel: Dir. & Museum Shop Mgr., Mary Ann Hoover.
Personnel Profile: Full-Time Volunteers 1.
Institution Type/Description: History Museum.
Collections: local history, heritage & culture; personal artifacts; period furnishings; photographs.
Hours & Admission Prices: Call for hours. No charge; donations accepted. ♿

MUSEUM OF THE MISSISSIPPI DELTA, (M), 1608 Highway 82 W., Greenwood, MS 38930-2725. Tel.: 662-453-0925. Fax: 662-455-7556.
E-mail: info@museumofthemississippidelta.com
Web Site: museumofthemississippidelta.com
Formerly: Cottonlandia Museum
Founded: 1969.
Congressional District: 2
Key Personnel: Exec. Dir., Cheryl A. Taylor; Pres., John Pittman; Vice Pres., Jere Stansel; Museum Shop Mgr., Tom Vlasic; Education & Art Coord., Jennifer Whites; Business Mgr., David Freeman.
Personnel Profile: Full-Time Paid 2; Part-Time Paid 2; Part-Time Volunteers 70.
Governing Authority: Parent Institution: Cottonlandia Educational & Recreational Foundation, Inc. Tax-exempt: 501(c)(3).
Institution Type/Description: General Museum.
Collections: Avent & Jones collection of polychrome ceramics from Humber-McWilliams site; Mississippian period Indians; historical regional agricultural implements & household furnishings; hands-on natural science room; contemporary Mississippi art; military history; artifacts from Lower-Yazoo Basin region; bone, bead, lithic & ceramic analysis of pre- & proto-historic Indian relics; European trade bead collection; wildflower garden; diorama of Mississippi Wetlands.
Research Fields: Mississippi archaeology: history, authors; regional archaeology: both pre- & proto-historic.
Facilities: banquet & lecture hall; meeting room.
Activities: guided tours; gallery talks; hobby workshops; docent program; permanent & temporary exhibitions; premises available for after-hour events; children's educational summer programs; scout programs.
Publications: research papers, The Southeastern Ceremonial Complex: Artifacts & Analysis; Early 16th-Century Glass Beads in the Spanish Colonial Trade.
Hours & Admission Prices: Mon.-Fri. 9-5, Sat. 10-4; other times by appointment. Adults $5, seniors $3.50, students $2; discounts to AAM & AAA members, Mississippi Museums Association, Southeast Museums Conference, Civil War Trust; members no charge. ♿
Attendance: 7,000 (estimated)
Membership: Student, Senior & Educator $25; Individual $35; Family $50; Patron $100; Sponsor $250; Benefactor $500; Director's Circle $1,000.

Gulfport

CENTER FOR MARINE EDUCATION AND RESEARCH, 10801 Dolphin Ln., Gulfport, MS 39503. Mailing Address: P.O. Box 207, Gulfport, MS 39502. Tel.: 228-896-9182. Fax: 228-896-9183.
E-mail: contactus@imms.org
Web Site: www.imms.org
Key Personnel: Exec. Dir. & Pres., Moby Solangi, Ph.D.
Governing Authority: nonprofit organization. Tax-exempt: 501(c)(3).
Institution Type/Description: Marine Science Museum.
Collections: marine mammal science, conservation & research; horseshoe crabs; fish; blue crabs; sea stars; sea urchins; tropical birds; reptiles; dolphins.
Research Fields: marine mammals.
Activities: demonstrations; touch tank; hands-on exhibitions; presentations; educational programs; special events.

Hours & Admission Prices: By appointment. Adults $8, children under 12 $6.

LYNN MEADOWS DISCOVERY CENTER, 246 Dolan Ave., Gulfport, MS 39507-1310. Tel.: 228-897-6039. Fax: 228-248-0071.
E-mail: lmdc@lmdc.org
Founded: 1998.
Key Personnel: Exec. Dir., Cynthia Minton
Institution Type/Description: Children's Museum.
Collections: hands-on exhibits.
Activities: educational programs.
Hours & Admission Prices: Tues.-Sat. 10-5. Admission $7, active military & seniors $5; children under one no charge. Closed New Year's Day; Thanksgiving; Christmas.

Harriston

THE FROG FARM SCULPTURE GARDEN, 186 Old Hwy. 61, Harriston, MS 39081. Mailing Address: P.O. Box 310, Fayette, MS 39069-0310. Tel.: 601-493-3420.
Web Site: www.artmarketing.com/gallery/frogfarm
Founded: 1998.
Personnel Profile: Full-Time Volunteers 1; Part-Time Volunteers 2.
Institution Type/Description: Folk Art Museum.
Collections: frogs, other amphibians, birds, & reptile sculptures.
Hours & Admission Prices: By appointment. No charge; donations requested.
Attendance: 100 (accurate)

Hattiesburg

AFRICAN AMERICAN MILITARY HISTORY MUSEUM, 305 E. Sixth St., Hattiesburg, MS 39401-2029. Tel.: 601-268-3220.
E-mail: museums@hattiesburg.org
Institution Type/Description: Military History Museum.
Collections: African American military men & women history; photographs; personal artifacts; paintings; murals; military uniforms; military medals & equipment.
Hours & Admission Prices: Call for hours.

HATTIESBURG AREA HISTORICAL SOCIETY AND HAHS MUSEUM, 723 Main Street, Hattiesburg, MS 39401-3431. Mailing Address: P.O. Box 1573, Hattiesburg, MS 39403-1573. Tel.: 601-582-5460; 268-0234.
E-mail: hahsmuseum@megagate.com
Web Site: www.hahsmuseum.org
Founded: 1970.
Congressional District: 5
Key Personnel: Pres. (V), Paula Harvey; Dir., Ursula Jones.
Personnel Profile: Part-Time Volunteers 35.
Governing Authority: private; nonprofit organization. Tax-exempt: 501(c)(3).
Institution Type/Description: History Museum.
Collections: area history; education; patriotism.
Research Fields: local building preservation.
Facilities: 183-vol. library; 3,800 sq. ft. exhibit space.
Activities: formal education programs for children; guided tours; temporary exhibitions. Museum Sponsors: programs.
Publications: newsletters; book, The History of Forrest County, MS.
Hours & Admission Prices: Mon., Tues. & Thurs. 2-5 & by appointment. No charge; donations accepted. ♿
Attendance: 500 (estimated)
Membership: Single $10; Couple $15; Life $100.

HATTIESBURG ARTS COUNCIL GALLERY, 723 Main St., Hattiesburg, MS 39401-3431. Mailing Address: P.O. Box 693, Hattiesburg, MS 39403-0693. Tel.: 601-583-6005.
E-mail: pattyhac@megagate.com
Founded: 1970.
Congressional District: 4
Key Personnel: Dir., Patty Hall; Pres. (V), Anita Price
Institution Type/Description: Art Gallery.
Collections: works by Mississippi artists.
Hours & Admission Prices: Mon.-Fri. 10-3. No charge. ♿
Attendance: 5,000

MUSEUM OF ART - THE UNIVERSITY OF SOUTHERN MISSISSIPPI, Dept. of Art and Design, 118 College Dr., #5033, Hattiesburg, MS 39406. Tel.: 601-266-5200.
E-mail: artmuseum@usm.edu

Web Site: www.usm.edu/visualarts/museum.php
Founded: 1978.
Key Personnel: Dir., Jan L. Siesling; Asst. Dir., Mark Rigsby.
Personnel Profile: Full-Time Paid 2.
Governing Authority: Parent Institution: University of Southern Mississippi. Tax exempt.
Institution Type/Description: Art Museum.
Collections: regional, international, historical & contemporary artwork.
Hours & Admission Prices: Jan.-May & Sept.-Dec. Tues.-Fri. 10-5, Sat. 10-4; June-Aug. Tues.-Fri. 12-5. No charge.
Attendance: 4,313 (accurate)

Hazlehurst

HAZLEHURST DEPOT MUSEUM, 138 N. Ragsdale Ave., Hazlehurst, MS 39083-3019. Mailing Address: P.O. Box 446, Hazlehurst, MS 39083-0446. Tel.: 601-894-3752. Fax: 601-894-3752.
Institution Type/Description: History Museum.
Collections: local history; railroad artifacts; Robert Johnson's personal artifacts.
Facilities: Museum-related items for sale.
Hours & Admission Prices: Daily. No charge. &

ROBERT JOHNSON HERITAGE HOUSE, 201 Downing St., Hazlehurst, MS 39083-3003. Tel.: 601-894-5777.
E-mail: robertjohnsonblu@bellsouth.net
Web Site: www.robertjohnsonbluesfoundation.org
Institution Type/Description: Historic House Museum: housed in the home of blues guitarist & singer, Robert Johnson; c.1900.
Collections: personal artifacts; period furnishings; photographs.
Facilities: Museum-related items for sale.
Hours & Admission Prices: Call for hours. No charge. &

Hernando

DESOTO COUNTY MUSEUM, 111 E. Commerce St., Hernando, MS 38632-2376. Tel.: 662-429-8852. Fax: 662-429-8852.
E-mail: info@desotomuseum.org
Web Site: www.desotomuseum.org
Founded: 2003.
Congressional District: 1
Key Personnel: Dir., Brian Hicks; Chm., Dr. Roma Thorn.
Personnel Profile: Full-Time Paid 1; Part-Time Paid 1; Part-Time Volunteers 35.
Governing Authority: private; nonprofit organization. Tax-exempt: 501(c)(3).
Institution Type/Description: History Museum.
Collections: DeSoto County history from 1541 to present; Native American artifacts; Civil War; personal artifacts.
Research Fields: local archaeology.
Facilities: 7,000 sq. ft. exhibit space; classroom. Museum-related items for sale.
Activities: school tours. Annual Event: DeSoto County Museum Anniversary & BBQ.
Hours & Admission Prices: Tues.-Sat. 10-5. No charge; donations accepted. Closed federal holidays. &
Attendance: 9,750 (estimated)
Membership: Basic: Student $5; Senior Citizen $10; Individual $15; Family $25; Corporate $100. Golden Circle: Friend $25-$99; Contributor $100-$199; Donor $200-$399; Sponsor $400-$999; Patron $1,000-$2,499; Major Benefactor $5,000 & up.

Holly Springs

IDA B. WELLS-BARNETT MUSEUM, 220 N. Randolph St., Holly Springs, MS 38635-2412. Tel.: 662-252-3232. Fax: 662-252-3232.
E-mail: idabwellsmuseum@gmail.com
Web Site: www.idabwells.org
Formerly: Ida B. Wells Family Art Gallery
Founded: 1996.
Congressional District: 1
Key Personnel: Dir. & C.E.O., Leona Harris; Pres. (V), Rev. Cornelia Booker.
Personnel Profile: Full-Time Volunteers 1; Part-Time Volunteers 8; Interns 2.
Governing Authority: Tax-exempt.
Institution Type/Description: History Museum: housed in the birthplace of journalist & women's activist, Ida B. Wells.
Collections: works by African & African American artists; personal artifacts; documents; Dr. David Stratmon African Art Collection 1948-1978.
Facilities: Museum-related items for sale.
Hours & Admission Prices: Mon.-Fri. 10-5, Sat. 12-5. Adults $4, children 12

& under $3; discounts to AAM members & members of other museums with membership card; members no charge. Closed New Year's Day; Easter; Thanksgiving; Christmas. &
Attendance: 13,500 (estimated)
Membership: Student & Senior Citizen $15; Individual $25; Family $35; Donor $100; Benefactor $250; Patron $500; Annual Sponsor $1,000.

MARSHALL COUNTY HISTORICAL MUSEUM, 220 E. College Ave., Holly Springs, MS 38635-3122. Mailing Address: P.O. Box 806, Holly Springs, MS 38635-0806. Tel.: 662-252-3669. Fax: 662-252-3669.
E-mail: marshallcomuseum@bellsouth.net
Founded: 1970.
Congressional District: 1
Key Personnel: Pres. (V) & Chm. (V), Bobby Mitchell; Dir. & Cur., Lois Swaney Shipp; Chm. (V) & Treas., Nancy Hutchins; Museum Shop Mgr., Jennifer Bone.
Personnel Profile: Full-Time Paid 1; Full-Time Volunteers 1; Part-Time Paid 2; Part-Time Volunteers 1.
Governing Authority: society. Tax-exempt.
Institution Type/Description: Historical Society Museum: housed in 1903 former Mississippi Synodical College Building.
Collections: Marshall County period artifacts; farm tools; period costumes & textiles; relics from eleven wars; Chickasaw Indian artifacts; physician office examining room equipment; wildlife room; early toys; period school artifacts; library; genealogy files.
Research Fields: genealogy.
Facilities: 3,000-vol. library of books gathered from family libraries, dating from 1715 to the present. Museum-related items for sale.
Activities: guided tours; films; lectures & slide lectures on local architecture; permanent & temporary exhibitions. Annual Events: Christmas in Holly Springs; Holiday House Tour in December.
Publications: quarterly newsletter, Marshall County Historical Society Newsletter; cookbook, A Taste in Time; books, It Happened Here; Holly Springs, Mississippi to the Year 1878; History of Holly Springs; History of Southeast Marshall County; Architectural Treasures of Holly Springs; Growing Up in Holly Springs; History of Southwest Marshall County & Byhalia; History of North Marshall County & Red Banks.
Hours & Admission Prices: Mon.-Fri. 10-5. Adults $5; members no charge. Closed Independence Day; Labor Day; Thanksgiving; Christmas Eve & Day. &
Attendance: 7,000 (estimated)
Membership: Individual $30.

RUST COLLEGE CENTER, 155 Rust Ave., Holly Springs, MS 38635. Tel.: 662-252-4590.
Web Site: www.rustcollege.edu
Institution Type/Description: African Art Museum.
Collections: tribal arts & fabrics.
Hours & Admission Prices: Mon.-Fri. 8-5; other times by appointment. No charge; donations accepted.

WALTER PLACE ESTATE, COTTAGES, AND GARDENS, 330 W. Chulahoma Ave., Holly Springs, MS 38635. Tel.: 662-252-2515. Fax: 662-252-2696.
Web Site: walterplace.com
Institution Type/Description: Historic House Museum: housed in the former home of General and Mrs. Ulysses S. Grant in 1862, during the Civil War.
Collections: local history & culture; period furnishings; personal artifacts; photographs; gardens; cottages.
Activities: guided tours.
Hours & Admission Prices: Tours: Mon.-Sat. 1pm. Admission $5-$20; discounts to groups.

Indianola

B.B. KING MUSEUM AND DELTA INTERPRETIVE CENTER, (M), 400 Second St., Indianola, MS 38751-2851. Tel.: 662-887-9539. Fax: 662-887-9478.
E-mail: info@bbkingmuseum.org
Web Site: www.bbkingmuseum.org
Founded: 2008.
Congressional District: 2
Key Personnel: Exec. Dir., Christopher D. Brown; Chief Administrative Officer, Malika Polk-Lee; Dir. Education, Dr. Gloria McIntosh; Dir. Entertainment, Robert Terrell; Dir. Facilities & Earned Income Operations, James McWilliams; Dir. Allstars Choir, Dr. Cheryl Weis.
Personnel Profile: Full-Time Paid 7; Part-Time Paid 7; Part-Time Volunteers 15.

Institution Type/Description: History Museum.
Collections: artifacts relating to B.B. King.
Hours & Admission Prices: Sun.-Mon. 12-5, Tues.-Sat. 10-5. Adults $12, seniors & students $5, school groups $3; discounts to AARP & AAA members, military & tour groups of 20 or more adults; children under 5 no charge. Closed Mon. Nov.-March. &
Attendance: 12,001 (accurate)
Membership: Young Riley $50; "Bee" and Lucille $75; Big Red Roadies $125; The King's Band $250; Blues Legend $500.

Iuka

TISHOMINGO COUNTY ARCHIVES & HISTORY MUSEUM (TCAHM), 203 E. Quitman St., Iuka, MS 38852-1938. Mailing Address: PO Box 273, Iuka, MS 38852-0273. Tel.: 662-423-3500.
E-mail: tishomingohistory@yahoo.com
Web Site: www.tishomingohistory.com
Formerly: Tishomingo County Mississippi Archives & History Museum aka Old Tishomingo County Courthouse
Founded: 2004.
Congressional District: 1
Key Personnel: Pres. (V), Cindy W. Nelson; Admin. Asst., Christina L. Thornton.
Personnel Profile: Part-Time Paid 1; Part-Time Volunteers 8.
Governing Authority: Parent Institution: Tishomingo County Historical & Genealogical Society. Tax-exempt.
Institution Type/Description: History Museum: housed in historic courthouse; built in 1870.
Collections: Indian artifacts; Civil War & other military memorabilia from the Battle of Iuka; local industry; period furnishings; photographs; local history; genealogy; medical office artifacts.
Activities: Annual Events: Iuka Heritage Day Festival; Family History Fair in May; Christmas at the Courthouse in December.
Publications: quarterly, Chronicles & Epitaphs from the Courthouse.
Hours & Admission Prices: Tues.-Fri. 10-4. No charge; donations accepted. Closed federal holidays. &
Attendance: 1,619 (accurate)
Membership: Regular $30; I Care $50; Business $100; Lifetime $1,000.

Jackson

EUDORA WELTY HOUSE, (M), 1119 Pinehurst St., Jackson, MS 39202-1812. Mailing Address: 1109 Pinehurst St., Jackson, MS 39202. Tel.: 601-353-7762. Fax: 601-354-5227.
E-mail: tours@eudoraweltyhouse.com
Web Site: www.eudoraweltyhouse.com
Founded: 2006.
Key Personnel: Dir., Bridget Edwards; Museum Shop Mgr., Megan Bankston.
Personnel Profile: Full-Time Paid 4; Part-Time Paid 1; Part-Time Volunteers 25.
Volunteer Hours: 1,500
Governing Authority: Parent Institution: Mississippi Dept. of Archives & History. Tax-exempt.
Institution Type/Description: Historic House Museum: housed in the former home of Eudora Welty. A National Historic Landmark.
Collections: Welty's life; books; period furniture; art.
Major Exhibits: Life Into Fiction: The Murder of Medgar Evers, 5/15/13-2/14/14; Where The Voice Came From, 5/15/13-2/14/14; Understanding the Medusa: Myth and Fairytales in the Fiction of Eudora Welty, 2/17/14-8/15/14.
Activities: Museum Sponsors: Welty's Birthday in April; Old Jackson by Candlelight Open House in December.
Hours & Admission Prices: Tues.-Fri. 9, 11, 1, & 3. Adults $5, students $3; discounts to groups & National Historic Trust members; children under 6 no charge. Closed state holidays.
Attendance: 3,750 (accurate)

FIRE MUSEUM & PUBLIC FIRE SAFETY EDUCATION CENTER, 355 Woodrow Wilson, Jackson, MS 39213. Tel.: 601-960-2433. Fax: 601-960-2432.
Key Personnel: Cur., Sherri Gibson
Institution Type/Description: Fire Museum.
Collections: fire fighting history & equipment; photographs; helmets & uniforms; 1904 horse-drawn steamer; 1917 chain-driven American LaFrance; 1936 Seagrave.
Hours & Admission Prices: Mon.-Fri. 9-5. No charge.

INTERNATIONAL MUSEUM OF MUSLIM CULTURES, 201 E. Pascagoula St., Ste. 102, Jackson, MS 39201-4114. Tel.: 601-960-0440.
Institution Type/Description: Islamic History & Culture Museum.
Collections: Islamic history & culture; photographs; period artifacts.
Facilities: Museum-related items for sale.
Hours & Admission Prices: Tues.-Sat. 10-5, Sun. 12-5.

JACKSON ZOOLOGICAL SOCIETY, INC., 2918 W. Capitol St., Jackson, MS 39209-4293. Tel.: 601-352-2581. Fax: 601-352-2594. Facebook: Jackson Zoo.
E-mail: info@jacksonzoo.org
Web Site: www.jacksonzoo.org
Founded: 1919.
Congressional District: 2
Key Personnel: Dir., Beth Poff; Animal Cur., David Wetzel; Museum Shop Mgr., Sheba Moses.
Personnel Profile: Full-Time Paid 35; Part-Time Paid 6; Part-Time Volunteers 18.
Governing Authority: private; nonprofit corporation. Parent Institution: Friends of the Jackson Zoo. Tax-exempt.
Institution Type/Description: Zoo.
Collections: endangered species.
Facilities: children's zoo; snack shop. Museum-related items for sale.
Activities: temporary & permanent exhibitions.
Publications: brochures, Jackson Zoological Park Zoo; biweekly e-blasts.
Hours & Admission Prices: Fall & Winter: daily 9-5; Summer: daily 8-4. Adults $10, senior citizens 65 & over $9, military $7, children 12 & under $6.75; members & children under 2 no charge. AZA reciprocal zoo admission. &
Attendance: 127,500 (accurate)
Membership: Individual $25; Companion $40; Family $65; Sponsor $100; Benefactor $250; Patron $500.

* **MANSHIP HOUSE MUSEUM, (M),** 420 E. Fortification St., Jackson, MS 39202-2340. Tel.: 601-961-4724. Fax: 601-354-6043.
E-mail: manship@mdah.state.ms.us
Web Site: www.mdah.state.ms.us/museum/manship
Founded: 1980.
Congressional District: 4
Key Personnel: C.E.O., Hank T. Holmes; Pres. Bd. Trustees, Kane Ditto; Dir., Marilynn Jones.
Personnel Profile: Full-Time Paid 2; Part-Time Volunteers 2.
Governing Authority: state. Parent Institution: Mississippi Dept. of Archives & History. Tax-exempt.
Institution Type/Description: Historic House Museum: 1857 restored Gothic Revival cottage of C.H. Manship.
Collections: 19th-century Mississippi furnishings & decorative arts.
Research Fields: 19th-century Mississippi architecture & decorative arts; local history; lifestyle of Charles Henry Manship & family, c.1888.
Facilities: 200-vol. library of books pertaining to architecture & decorative arts in the 19th century available for research in adjacent visitor center.
Activities: guided tours; concerts; formally organized education programs for children; docent program; summer workshops for children; video loan program; lectures. Annual Events: Christmas at the Manship House; Summer Dress.
Publications: brochure; guide book.
Hours & Admission Prices: Temporarily closed for renovation. &
Attendance: 8,000 (accurate)

MEDGAR EVERS HOME MUSEUM, 2332 Margaret Walker Alexander Dr., Jackson, MS 39213-6411. Tel.: 601-977-7710.
Institution Type/Description: Historic House Museum: housed in the former home of civil rights leader, Medgar Evers & the site of his assassination in 1963.
Collections: Medgar Evers' life & career; period furnishings; personal artifacts; photographs.
Hours & Admission Prices: Call for hours.

MISSISSIPPI AGRICULTURE & FORESTRY/NATIONAL AGRICULTURAL AVIATION MUSEUM, (M), 1150 Lakeland Dr., Jackson, MS 39216-4728. Tel.: 601-432-4500. Fax: 601-982-4292. Facebook: Mississippi Agriculture & Forestry Museum.
E-mail: lise@mdac.ms.gov
Web Site: msagmuseum.org
Founded: 1983.
Congressional District: 4

Key Personnel: Dir., Lise Foy; Dir. Operations, Sandy Havard; Dir. Administration, Sales & Visitor Svcs., Theresa Love.
Personnel Profile: Full-Time Paid 11; Part-Time Paid 3.
Governing Authority: state; nonprofit organization. Tax-exempt: 501(c)(3).
Institution Type/Description: General Museum.
Collections: history of agriculture, forestry & agricultural aviation in Mississippi; period artifacts; furnishings.
Facilities: 35,000 sq. ft. exhibit space; nature trail; 150-seat large screen theater; rental facilities. Museum-related items for sale.
Activities: arts festivals; concerts; docent program; guided tours; formal education programs; theater. Annual Events: Harvest Festival; Celtic Festival; Pumpkin Adventure; Easter Egg Hunt.
Hours & Admission Prices: Tues.-Sat. 9-5. Adults $5, senior citizens & military $4, children 3-18 $4; discounts to groups; children under 3 no charge. Additional fee for special events. Closed New Year's Day; Martin Luther King Jr. Day; Presidents' Day; Memorial Day; Independence Day; Labor Day; Veterans Day; Thanksgiving; Christmas. &
Attendance: 140,000 (accurate)
Membership: Individual $15; Family $40.

MISSISSIPPI CHILDREN'S MUSEUM, 2145 Highland Dr., Jackson, MS 39202. Mailing Address: P.O. Box 55409, Jackson, MS 39296-5409. Tel.: 601-981-5469; 877-793-5437. Fax: 601-709-2603.
E-mail: info@mississippichildrensmuseum.com
Web Site: www.mississippichildrensmuseum.com
Key Personnel: Pres., Susan Garrard; Dir. Education, Alicen Blanchard; Coord. Membership, Patti Reiss; Dir. Museum Experience, Charley Frye; Dir. Devel. & Mktg., Elaina Jackson; Dir. Finance & Administration, Johnny Kroeze; Dir. Programs, Chavanne McDonald; Mgr. Cafe, Deloris Washington
Institution Type/Description: Children's Museum.
Collections: hands-on exhibitions.
Facilities: cafe.
Activities: educational programs; special events; birthday parties.
Hours & Admission Prices: Tues.-Sat. 9-5, Sun. 1-6. Admission $8; children under one no charge.
Attendance: 250,000 (estimated)

MISSISSIPPI GOVERNOR'S MANSION, 300 E. Capitol, Jackson, MS 39201-3403. Tel.: 601-359-3175. Fax: 601-359-6473.
Web Site: www.mdah.state.ms.us
Founded: 1842.
Congressional District: 4
Key Personnel: Dir. Dept. Archives & History, H.T. Holmes; Cur., Lauren Miller.
Personnel Profile: Full-Time Paid 1; Part-Time Volunteers 50.
Governing Authority: state; nonprofit. Parent Institution: Mississippi Department of Archives & History. Tax-exempt: 170(b)(1)(A).
Institution Type/Description: Historic Building: 1842 Governor's Mansion, constructed 1839-1841.
Collections: 19th-century furniture & furnishings.
Research Fields: 19th-century furniture & furnishings.
Activities: guided tours; docent program.
Publications: brochure, Illustrated Guide to the Mississippi Governor's Mansion.
Hours & Admission Prices: Tues.-Fri. 9:30-11; tours given on the half-hour. No charge; donations accepted. &
Attendance: 15,224 (accurate)

MISSISSIPPI MUSEUM OF ART, (M), 380 S. Lamar St., Jackson, MS 39201-4007. Tel.: 601-960-1515. Fax: 601-960-1505.
Web Site: www.msmuseumart.org
Founded: 1911.
Congressional District: 4
Key Personnel: Dir., Betsy Bradley; Chm., Laurie Hearin McRee; Chief Resource Management, Stephanie Palmertree; Dir. Membership Svcs., Jillien Fry; Dir. Communications, Nina Moss; Finance Asst., Allison England; Chief Preparator, L.C. Tucker, Jr.; Asst. Preparator, Melvin Johnson; Chief Cur., Roger Ward; Dir. Operations, Mindy Kunz; Museum Store Mgr., Elizabeth Tyler; Visitor Svcs., Annette French; Graphic Design, Robin Smith; Preparator, Tom Jones; Dir. New Media, Julian Rankin; Registrar, Amber Schneider; Cur. Collection, Beth Batton; Curatorial Assoc., Kathleen Funches Varnell; Chief Public Participation, Jenny Tate; Events Coord., Kathryn Kelso; Dir. Events & Facility Rental, Marika Cackett; Dir. Family Programs, Carol Cox Peaster; Dir School Programs, Dorian Pridgen; Chief Security, James A. Steverson; Dep. Chief Security, Julia Stewart.

Personnel Profile: Full-Time Paid 24; Part-Time Paid 10; Part-Time Volunteers 50; Interns 3.
Operating Expenses: 2,000,000
Operating Income: 2,000,000
Governing Authority: nonprofit organization. Tax-exempt: 501(c)(3).
Institution Type/Description: Art Museum.
Collections: 19th- & 20th-century American & Mississippi art; mid 18th- to early 19th-century English painting; Native American basket collection; 19th- & 20th-century European & American prints; Southeastern folk art & photography; Oriental & ethnographic art; pre-Columbian art.
Major Exhibits: Italian Art from the Permanent Collection, 8/13-1/14; An Italian Palate: Paintings by Wyatt Waters, 10/13-1/14; Mississippi Watercolor Society Grand National Watercolor Exhibition, 10/13-1014; Gwendolyn A. Magee: Slave Series Quilts, 03/14-5/14; This Light of Ours: Activist Photographers of the Civil Rights Movement (T), 03/14-08/14; Norman Rockwell: Murder in Mississippi (T), 06/14-08/14; Spanish Sojourns: Robert Henri and the Spirit of Spain (T), 9/14-1/15.
Research Fields: 19th- & 20th-century American, Southern & Mississippi art; 18th- & 19th-century British painting; photography.
Facilities: 4,800-vol. library of art books available for research on premises only; restaurant; art instruction studios; garden. Museum-related items for sale.
Activities: guided tours; visitor services; lecture series; inter-museum loan; formally organized education programs; docent program; outdoor film series; outdoor music series; monthly classical music series; annual fundraising gala; annual consigned art sale & auction; new collector's club.
Publications: newsletter; exhibition catalogs.
Hours & Admission Prices: Tues.-Sat. 10-5, Sun. 12-5; call to confirm. Adults $5, seniors 60 & over $4, students $3; discounts to AAM & ICOM members; members and children 5 & under no charge. Subject to change during special exhibitions. Closed New Year's Day; Easter; Thanksgiving; Christmas. &
Attendance: 75,000 (accurate)
Membership: Senior $35; Individual $45; Senior Family & Dual $50; Family Dual $60; Supporting $100; Partner $250; Curator $500; Young Rembrandt $600; Rembrandt $1,200; Director's Rembrandt $2,500; Chairman's Rembrandt $5,000.

*   **MISSISSIPPI MUSEUM OF NATURAL SCIENCE, (M),** 2148 Riverside Dr., Jackson, MS 39202-1353. Tel.: 601-576-6000. Fax: 601-354-7227.
E-mail: libby.hartfield@mmns.state.ms.us
Web Site: www.mdwfp.com/museum
Founded: 1932.
Key Personnel: C.E.O., Libby Hartfield; Dir., Charles Knight; Asst. Dir., Angel Rohnke; Education Coord. & Project WILD Coord., Megan Fedrick; Exhibits Supvr., Ray Terry; Exhibits Supvr., Sam Biebers; Research/Collections Coord., Verity Mathis; Natural Heritage Program Coord., Andy Sanderson; Volunteer Coord., Ann Taylor; Gift Shop Mgr., Rebecca Jones.
Personnel Profile: Full-Time Paid 45; Part-Time Paid 9; Part-Time Volunteers 160.
Operating Expenses: 3,945,398
Operating Income: 4,361,775
Governing Authority: state. Parent Institution: Mississippi Dept. of Wildlife, Fisheries & Parks. Tax-exempt.
Institution Type/Description: Natural Science Museum.
Collections: herpetology; ichthyology; mammalogy; ornithology; invertebrate; paleontology & botany with emphasis on Mississippi material; dioramas & aquaria depicting Mississippi habitats.
Major Exhibits: A Forest Journey, 2/8/14-4/27/14; Nature's Numbers, 2/8/14-4/27/14; Animal Grossology, 5/17/14-12/30/14.
Research Fields: herpetology; ichthyology; mammalogy; paleontology; ornithology; invertebrate biology, botany; National Heritage Data.
Facilities: 15,000-vol. reference library of natural history material available for use on the premises; aquariums; 200-seat theater; 300 acre natural area; nature trails; preschool room.
Activities: guided tours; lectures; films; formally organized education programs for children, adults, undergraduate & graduate college students; permanent & temporary educational exhibitions; hands-on exhibitions. Museum Sponsors: special exhibits and kits on loan to Mississippi schools; Mississippi Natural Area Registry.
Publications: leaflets; newsletter; publications on Mississippi.
Hours & Admission Prices: Mon.-Fri. 8-5, Sat. 9-5, Sun. 1-5. Adults $6, seniors 60 & over $5, children 3-18 $4; members no charge. ASTC Passport Program participant. Closed New Year's Day; Memorial Day; Easter; Independence Day; Labor Day; Thanksgiving; Christmas Eve & Day. &
Attendance: 130,000 (accurate)
Membership: Individual $40; Family $65; Friend $100; Donor $250; Patron $500; Benefactor $750; Sustaining $1,000.

MISSISSIPPI SPORTS HALL OF FAME & MUSEUM, 1152 Lakeland Dr., Jackson, MS 39216-4701. Tel.: 601-982-8264; 800-280-FAME. Fax: 601-982-4702.
E-mail: generalinfo@msfame.com
Web Site: www.msfame.com
Founded: 1996.
Congressional District: 3
Key Personnel: Dir., Rick Cleveland; Chm. & Pres., Oscar Miskelly.
Personnel Profile: Full-Time Paid 4; Part-Time Paid 6; Part-Time Volunteers 33.
Governing Authority: Tax-exempt.
Institution Type/Description: Sports Museum.
Collections: sports history; hands-on exhibits.
Facilities: Museum-related items for sale.
Activities: rental facility; conference center; group tours; birthday parties; special events.
Hours & Admission Prices: Mon.-Sat. 10-4. Adults $5, seniors 60 & over and students 6-17 $3.50; discounts to active military & groups of 12 or more; members and children 5 & under no charge. ♿
Attendance: 35,000 (estimated)
Membership: Individual $25; Family $100; Life $1,000; Corporate $5,000.

MUSEUM OF MISSISSIPPI HISTORY/MISSISSIPPI CIVIL RIGHTS MUSEUM/MUSEUM DIVISION, (M), 618 E. Pearl St., Jackson, MS 39201. Mailing Address: Box 571, Jackson, MS 39205-0571. Tel.: 601-576-6800. Fax: 601-576-6815.
E-mail: info@2mississippimuseums.com
Web Site: www.mdah.state.ms.us
Formerly: Old Capitol Museum of Mississippi History
Congressional District: 3
Key Personnel: C.E.O., H.T. Holmes; Dir., Lucy J. Allen; Pres. (V), Kane Ditto; Dir. Exhibits, John Gardner; Dir. Collections, Cindy Gardner; Asst. Dir. Collections, Nan Prince; Dir. Education & Programs, Stacey Everett.
Personnel Profile: Full-Time Paid 10; Part-Time Volunteers 15.
Governing Authority: state. Parent Institution: Mississippi Dept. of Archives & History; 200 North St., Jackson, MS 39201. Tax-exempt.
Institution Type/Description: History Museum.
Collections: Civil War flags; quilts.
Research Fields: Mississippi history including textiles, armament and personal artifacts.
Activities: temporary, loan & traveling exhibitions; outreach educational programs; traveling trunk program for schools; on-line lesson plans.
Hours & Admission Prices: Temporarily closed. Temporary exhibits are on display in the William F. Winter Archives and History building.

MUSEUM OF THE SOUTHERN JEWISH EXPERIENCE, 4915 I-55 N., Ste. 204B, Jackson, MS 39206. Mailing Address: P.O. Box 16528, Jackson, MS 39236-6528. Tel.: 601-362-6357. Fax: 601-366-6293.
E-mail: information@msje.org
Web Site: www.msje.org
Founded: 1989.
Congressional District: 3
Key Personnel: C.E.O. & Pres., Macy B. Hart; Chm. (V), Rayman L. Solomon; Dir. History Dept. & Museum Projects Dir., Dr. Stuart Rockoff; Administrator, Kate Lubarsky.
Personnel Profile: Full-Time Paid 28; Part-Time Paid 3; Part-Time Volunteers 10; Interns 4.
Governing Authority: private; nonprofit organization. Parent Institution: Institute of Southern Jewish Life. Tax-exempt: 501(c)(3).
Institution Type/Description: Religious History Museum.
Collections: silver ritual Judaica & ceremonial items; Alsatian artifacts; furniture & architectural elements from Southern synagogues; historic photos; family & congregational papers; Bill Aron contemporary photos of Jewish life in the deep South; Southern Jewish artifacts & memorabilia.
Research Fields: Southern Jewish history, including family, institutions, congregations, communities & architecture; oral history of Southern Jews; Southern Jewish folklore & foodways.
Facilities: library of Judaica & general resource materials on Jewish experiences, history, museum education & Judaic texts; theater; auditorium. Museum-related items for sale.
Activities: formal education programs; guided tours; lectures; loan, temporary & traveling exhibitions.
Publications: newsletter, Circa; videos, Natchez Jewish Experience, May the Light Shine Forever.
Hours & Admission Prices: Daily by appointment. Adults $5; discounts to ICOM members & groups; members no charge. ♿
Attendance: 4,950 (estimated)

Membership: Friend $100; Patron $250; Sponsor $500; Benefactor $1,000; Humanitarium $5,000 & up.

MYNELLE GARDENS, 4736 Clinton Blvd., Jackson, MS 39209-2400. Tel.: 601-960-1894.
Institution Type/Description: Garden.
Collections: plants ranging from amaryllises, daylilies, gardenias, pinks & camellias; roses; Asiatic magnolias. Historic Houses: Westbrook House c.1917 & Greenbrook House c.1920.
Facilities: Museum-related items for sale.
Hours & Admission Prices: March-Oct. Mon.-Sat. 12-5:15, Sun. 12-4:15; Nov.-Feb. daily 8-4:15. Family pass $30, student pass $5, adults $4, children 4-12 $1. Closed New Year's Day; Martin Luther King Jr. Day; Independence Day; Thanksgiving & day after; Christmas. ♿

THE OAKS HOUSE MUSEUM, 823 N. Jefferson St., Jackson, MS 39202-4140. Tel.: 601-353-9339.
E-mail: oakshousemuseum@comcast.net
Web Site: theoakshousemuseum.org
Founded: 1960.
Congressional District: 4
Key Personnel: Dir. & Chm. (V), Linda Robertson.
Personnel Profile: Full-Time Volunteers 26; Part-Time Paid 1; Part-Time Volunteers 24.
Governing Authority: society. Parent Institution: National Society of the Colonial Dames in the State of Mississippi. Subsidiary Institution: The Oaks House Museum Corporation. Tax-exempt.
Institution Type/Description: Local History Museum: housed in c.1850 historic building site, The Oaks.
Collections: family life in Jackson, Mississippi from 1853-1863; furniture & furnishings of the period; restored Victorian gardens; camellia garden.
Activities: tours; holiday events; dramatic presentations; archaeology camps; seminars.
Publications: brochure; newsletter.
Hours & Admission Prices: Tues.-Sat. 10-3; other times by appointment. Adults $4.50, senior citizens 65 & over $4, children $3.50; discount to groups of 10 or more & AAA members. Closed major holidays. ♿
Attendance: 1,200 (estimated)

OLD CAPITOL MUSEUM, 100 S. State St., Jackson, MS 39201-4400. Mailing Address: P.O. Box 571, Jackson, MS 39205-0571. Tel.: 601-576-6920. Fax: 601-576-6981. Facebook: Old Capitol Museum.
E-mail: info@oldcapitolmuseum.com
Web Site: www.oldcapitolmuseum.com
Founded: 2009.
Congressional District: 3
Key Personnel: Dir., Clay Williams; Education, Maura Johnson; Education, Mike Stoll.
Personnel Profile: Full-Time Paid 4; Part-Time Paid 2; Part-Time Volunteers 15.
Governing Authority: state; nonprofit. Parent Institution: Mississippi Dept. of Archives & History. Tax-exempt: 501(c)(3).
Institution Type/Description: Historic Building: building served as Mississippi's statehouse from 1839-1903.
Collections: building history; state government; period artifacts; photographs; furnishings; House Chamber auditorium.
Facilities: 5,000 sq. ft. exhibit space; classroom. Museum-related items for sale.
Activities: docent program; formal education programs; guided tours; lectures.
Hours & Admission Prices: Tues.-Sat. 9-5, Sun. 1-5. No charge; donations accepted. Closed New Year's Day; Easter; Memorial Day; Labor Day; Thanksgiving; Christmas. ♿
Attendance: 40,000 (accurate)

RUSSELL C. DAVIS PLANETARIUM, 201 E. Pascagoula St., Jackson, MS 39201-4101. Mailing Address: P.O. Box 17, Jackson, MS 39205. Tel.: 601-960-1552. Fax: 601-960-1555.
E-mail: jfwilliams@city.jackson.ms.us
Web Site: www.thedavisplanetarium.com
Founded: 1978.
Congressional District: 4
Key Personnel: Mgr., John F. Williams.
Personnel Profile: Full-Time Paid 6; Part-Time Paid 1; Interns 1.
Governing Authority: municipal. Parent Institution: City of Jackson. Tax-exempt: 170(b)(1)(A).
Institution Type/Description: Planetarium.
Collections: hemisphere projection technology; hemisphere cinematography.

Research Fields: projection technology.
Facilities: 190-seat planetarium and hemispheric film theatre. Gift items for sale.
Activities: planetarium programs; hemispheric large-format films; laser light shows; formally organized education programs for children and adults; special school programs for students preschool-college.
Publications: annual program guide for teachers & group leaders.
Hours & Admission Prices: Call for current program, show, and ticket information. Large-format Films: adults $6.50, seniors $5.50, children $4. Sky Shows: adults $5.50, seniors $4.50, children $3. Closed New Year's Day; Martin Luther King Jr. Day; Presidents' Day; Easter; Memorial Day; Labor Day; Independence Day; Thanksgiving; Christmas Eve & Day. &
Attendance: 45,000 (accurate)

SMITH ROBERTSON MUSEUM & CULTURAL CENTER, 528 Bloom St., Jackson, MS 39202-4005. Tel.: 601-960-1457. Fax: 601-960-2070.
Founded: 1984.
Congressional District: 4
Key Personnel: Mgr. & Museum Shop Mgr., Pamela D. Junior; Sec., Euerla Christian Bester; Gift Shop Attendant, Mary Funches.
Personnel Profile: Full-Time Paid 3; Part-Time Volunteers 25; Interns 2.
Governing Authority: municipal. Tax-exempt: 501(c)(3).
Institution Type/Description: History Museum: housed in first school built with public funds for Blacks in Jackson, MS.
Collections: life, history & culture of Black Mississippians from territorial to the present day.
Research Fields: Black church in Mississippi.
Facilities: Museum-related items for sale.
Activities: guided tours; lectures; films; concerts; dance recitals; arts festivals; theater; study clubs; rental gallery; organized education programs for children; docent program; loan exhibitions; school loan service.
Publications: quarterly newsletter, Smith Robertson Museum News.
Hours & Admission Prices: Mon.-Fri. 9-5, Sat. 10-1. Adults $4.50, seniors $3, children under 18 $1.50. Closed New Year's Day; Martin Luther King Jr. Day; Memorial Day; Independence Day; Labor Day; Thanksgiving; Christmas. &
Attendance: 15,000 (estimated)
Membership: Individual $10; Family $25; Contributing $50; Corporate $250; Sustaining Corporate $500; Supporting Corporate $1,000.

WAR MEMORIAL BUILDING, 120 N. State St., Jackson, MS 39201-2810. Tel.: 601-354-7207.
Institution Type/Description: Military History Museum.
Collections: military artifacts; personal artifacts; photographs; replica of the Tomb of the Unknown Soldier.
Hours & Admission Prices: Daily. No charge.

Kosciusko

KOSCIUSKO MUSEUM AND VISITORS CENTER, Natchez Trace Pkwy., Kosciusko, MS 39090. Mailing Address: 101 N. Natchez St., Kosciusko, MS 39090. Tel.: 662-289-2981. Fax: 662-289-2986.
Key Personnel: Vice Pres., Tonya Threet
Institution Type/Description: History Museum.
Collections: local history & culture; photographs; personal artifacts.
Hours & Admission Prices: Daily 9-4. No charge. &

Laurel

* **LAUREN ROGERS MUSEUM OF ART, (M),** 565 N. 5th Ave., Laurel, MS 39440. Mailing Address: P.O. Box 1108, Laurel, MS 39441-1108. Tel.: 601-649-6374. Fax: 601-649-6379.
Web Site: www.LRMA.org
Founded: 1923.
Congressional District: 4
Key Personnel: Dir., George Bassi; Chm. (V), Robert Hynson; Pres. (V), Rosemary Norton; Bldg. Supt., Todd Sullivan; Dir. Mktg., Holly Green; Registrar, Tommie Rodgers; Cur., Jill Chancey, Ph.D.; Cur. Education, Mandy Buchanan; Librarian, Donna Smith; Business Mgr., JoLynn Helton; Dir. Devel., Allyn C. Boone; Visitor Svcs. Coord., Lizabeth Brumley.
Personnel Profile: Full-Time Paid 11; Part-Time Paid 4; Part-Time Volunteers 107; Interns 4.
Governing Authority: nonprofit organization. Parent Institution: Eastman Memorial Foundation. Tax-exempt: 509(a).
Institution Type/Description: Art Museum.
Collections: 19th-20th-century American & European paintings; drawings; 18th-century Japanese Ukiyo-e woodblock prints; Native American Indian baskets; 18th-century English Georgian silver.
Research Fields: visual arts; history.
Facilities: 20,000-vol. library with emphasis on art, local history, Mississippiana available for use by public on premises; reading room.
Activities: guided tours; lectures; films; gallery talks; concerts; docent program; permanent, temporary & traveling exhibitions; classes; workshops. Annual Events: Blues Bash; Fall Festival.
Publications: brochure; quarterly members newsletter; LRMA Handbook of the Collections; exhibition catalogs, By Native Hands: Woven Treasures from the Lauren Rogers Museum of Art.
Hours & Admission Prices: Tues.-Sat. 10-4:45, Sun. 1-4. No charge; donations accepted. Closed major holidays. &
Attendance: 37,000 (estimated)
Membership: Student $15; Associate $25; Friend $50; Sponsor $100; Patron $150; Donor $250; Benefactor $500; Grand Benefactor $1,000; Sustaining Benefactor $1,500; Laureate $2,500; Eastman $5,000.

VETERANS MEMORIAL MUSEUM, 920 Hillcrest Dr., Laurel, MS 39440-4726. Tel.: 601-428-4008.
Institution Type/Description: Military Museum.
Collections: military history; movies & documentaries; Honor Wall etched with names of veterans from Southeast Mississippi.
Facilities: library.
Hours & Admission Prices: Tues.-Fri. 10-4. No charge. &

Leakesville

GREENE COUNTY MUSEUM & HISTORICAL SOCIETY, Greene County House, 400 Main St., 4th Fl., Leakesville, MS 39451. Mailing Address: P.O. Box 841, Leakesville, MS 39451-0841. Tel.: 601-394-4343.
E-mail: museum@tds.net
Web Site: greenmuseum.org
Formerly: Greene County Courthouse Jail
Founded: 2004.
Congressional District: 4
Personnel Profile: Full-Time Volunteers 6.
Volunteer Hours: 1,560
Governing Authority: Tax-exempt.
Institution Type/Description: Historical Society Museum.
Collections: local history & culture relating to Greene County; early newspapers.
Publications: calendars.
Hours & Admission Prices: Mon.-Tues. & Thurs. 9-4, Wed., Fri. and first Sat. of each month 9-12. No charge; donations accepted. &
Attendance: 1,000 (estimated)
Membership: Individual $12.

Leland

HIGHWAY 61 BLUES MUSEUM, 307 N. Broad St., Leland, MS 38756-2744. Mailing Address: P.O. Box 251, Leland, MS 38756. Tel.: 662-686-2063.
Institution Type/Description: Blues Museum.
Collections: local history; Delta bluesmen & their music; photographs; personal artifacts.
Hours & Admission Prices: Mon.-Sat. 10-5; other times by appointment.

Louisville

THE AMERICAN HERITAGE "BIG RED" FIRE MUSEUM, 332 N. Church Ave., Louisville, MS 39339-2302. Mailing Address: 650 N. Church Ave., Louisville, MS 39339-2033. Tel.: 662-773-3421. Fax: 662-773-9183.
Web Site: www.taylorbigred.com
Founded: 1989.
Congressional District: 3
Key Personnel: C.E.O., W.A. "Lex" Taylor, III; Registrar, Kay Reynolds.
Governing Authority: private.
Institution Type/Description: Fire-Fighting Museum.
Collections: early fire equipment; late 1700 hand pumpers; hose reels; horse drawn ladder trucks; Cole Brother and Silsby steam pumpers; hand pulled reel & wagons; ALF pumpers; 1921 ladder trucks; 1924 Ahrens Fox; 1957 Seagrave ladder truck; taxidermic wildlife; Indian artifacts; model planes & ships; art.
Hours & Admission Prices: Mon.-Fri. by appointment only. No charge. Closed holidays.

Macon

NOXUBEE COUNTY HISTORICAL SOCIETY MUSEUM, 411
S. Jefferson St., Macon, MS 39341. Mailing Address: Noxubee
County Historical Society, P.O. Box 892, Macon, MS 39341. Tel.:
662-726-4456. Fax: 662-726-1041.
Institution Type/Description: Historical Society Museum.
Collections: local history & culture; Choctaw Indian artifacts; period furnish-
ings; personal artifacts; photographs.
Hours & Admission Prices: Temporarily closed for restructuring. No charge;
donations accepted.

McComb

BLACK HISTORY GALLERY, 819 Wall St., McComb, MS 39648.
Tel.: 601-684-1130.
Institution Type/Description: History Museum.
Collections: African American history, heritage & culture; period furnishings;
photographs; personal artifacts; books; charts.
Hours & Admission Prices: Call for hours.

MCCOMB CITY RAILROAD DEPOT MUSEUM, 108 N. Rail-
road Blvd., McComb, MS 39649-7220. Mailing Address: P.O. Box
7220, McComb, MS 39649-7220. Tel.: 601-684-2291. Facebook:
McComb City Railroad Depot.
E-mail: trainmaster@mcrrmuseum.com
Web Site: www.mcrrmuseum.com
Founded: 2000.
Congressional District: 3
Key Personnel: C.E.O. & Dir., Winnie Len Howell; Pres. (V), Laila McEwen;
Museum Shop Mgr., Bob Bellipanni.
Personnel Profile: Part-Time Paid 1; Part-Time Volunteers 50.
Governing Authority: city of McComb. Tax-exempt: 501(c)(3).
Institution Type/Description: Railroad Museum.
Collections: railroad memorabilia.
Hours & Admission Prices: Mon.-Fri. 12-4. No charge; donations accepted.
Closed official holidays. &
Attendance: 4,000 (estimated)
Membership: Individual $20; Family $40; Business $50.

Meridian

JIMMIE RODGERS MUSEUM, 1725 Highland Park Dr., Merid-
ian, MS 39307. Mailing Address: P.O. Box 4555, Meridian, MS
39304-4555. Tel.: 601-485-1808.
E-mail: jimmie_rodgers@hotmail.com
Founded: 1976.
Congressional District: 3
Key Personnel: C.E.O., Betty Lou Jones; Chm. (V), Greg Hatcher.
Personnel Profile: Part-Time Paid 3.
Governing Authority: municipal; nonprofit organization. Parent Institution:
Jimmie Rodgers Foundation. Tax-exempt: 170(b)(1)(A).
Institution Type/Description: Folk Art Museum.
Collections: items belonging to the late Jimmie Rodgers, father of country
music.
Activities: guided tours.
Hours & Admission Prices: Feb. to mid-Dec. Tues.-Sat. 10-4; other times by
appointment. Adults $10; discounts for groups, school groups, military,
Smithsonian & AAA members, children 8-18 and seniors 65 & over;
children under 8 no charge. Closed legal holidays; Thanksgiving week. &
Attendance: 7,000 (estimated)

MERIDIAN MUSEUM OF ART, 628 25th Ave., Meridian, MS
39301. Mailing Address: P.O. Box 5773, Meridian, MS 39302-
5773. Tel.: 601-693-1501. Facebook: Meridian Museum of Art.
E-mail: meridianmuseum@bellsouth.net
Web Site: meridianmuseum.org
Founded: 1969.
Congressional District: 3
Key Personnel: Dir., Kate Cherry; Pres., Dawn Gray.
Personnel Profile: Full-Time Paid 1; Part-Time Paid 2; Part-Time Volunteers
40.
Governing Authority: municipal; nonprofit. Affiliated with Meridian Art
Association. Tax-exempt: 501(c)(3).
Institution Type/Description: Art Museum Gallery.
Collections: Caroline Durieux lithographs; works by Will Barnet, Alexander
Van Laer, Marie Hull, Nell Blaine, George Ault, Mary Frank, Alice Neel,
Walter Anderson, William Hollingsworth, Karl & Mildred Wolfe, Andrew
Bucci, Homer Casteel; 18th-century British portraits; continental land-

scapes; Boehm porcelain; 18th century including Thomas Bardwell, Jean-
Marc Nattier, Thomas Phillips, & Sir Henry Raeburn; 17th century
including Sir Godfrey Kneller.
Major Exhibits: Membership, 12/11/13-2/1/14; Art Around Alabama, 2/5/14-
3/29/14; Helene & W. Ray Fielder, 4/9/14-5/14; People's Choice Art
Competition, 6/4/14-7/26/14; 41st Annual Bi-State Art Competition,
8/20/14-10/18/14.
Research Fields: museum outreach.
Facilities: classrooms; lecture room.
Activities: art classes for children & adults; artists group; lecture series;
workshops; school outreach program; after school classes.
Publications: biannual brochure; monthly newsletter.
Hours & Admission Prices: Aug. 20-Aug. 3 Wed.-Sat. 11-5. No charge;
donations accepted. Closed holidays. &
Attendance: 12,000 (estimated)
Membership: Individual $45; Dual & Family $75; Patron $150; Sponsor $300;
Benefactor $500; Grand Benefactor $1,000; Silver Circle $2,500; Gold
Circle $5,000; Platinum Circle $10,000.

Mississippi State

CHARLES H. TEMPLETON, SR. MUSIC MUSEUM, Mitchell
Memorial Library, Mississippi State, MS 39762. Mailing Address:
Mitchell Memorial Library, P.O. Box 5408, Mississippi State, MS
39762-5408. Tel.: 662-325-6634. Fax: 662-325-4263.
E-mail: scunetto@library.msstate.edu
Web Site: library.msstate.edu/templeton
Founded: 2006.
Governing Authority: Parent Institution: Mississippi State University Libraries.
Institution Type/Description: Music Museum.
Collections: musical instruments; records; sheet music.
Activities: group tours.
Hours & Admission Prices: Mon.-Fri. 9-4; groups by appointment. No charge.

COBB INSTITUTE OF ARCHAEOLOGY, College of Arts &
Sciences, Mississippi State, MS 39762-5542. Mailing Address:
P.O. Box AR, Mississippi State, MS 39762-5542. Tel.: 662-325-
3826. Fax: 662-325-8690.
E-mail: jds1@ra.msstate.edu
Web Site: www.cobb.msstate.edu
Founded: 1972.
Congressional District: 2
Key Personnel: Dir., Joe Seger; Administrative Sec., Kathleen Elliott; Cur.
Artifacts, Keith Baca.
Personnel Profile: Part-Time Paid 4.
Governing Authority: university. Parent Institution: Mississippi State Univer-
sity. Tax-exempt.
Institution Type/Description: Archaeology Museum.
Collections: Middle Eastern archaeological remains; southeastern U.S. ar-
chaeological materials.
Research Fields: Middle East & southeastern U.S. archaeology.
Facilities: 10,000-vol. library; research laboratories.
Activities: museum tours; lectures.
Publications: reports; occasional papers; biannual newsletter, Cobb Institute of
Archaeology.
Hours & Admission Prices: Call for hours. No charge; donations accepted.
Closed school holidays. &
Attendance: 3,000 (estimated)

DUNN-SEILER MUSEUM, (M), Dept. of Geosciences, Mississippi
State, MS 39762. Mailing Address: P.O. Box 5448, Mississippi
State, MS 39762-5448. Tel.: 662-325-3915. Fax: 662-325-9423.
E-mail: rclary@geosci.msstate.edu
Web Site: geosciences.msstate.edu/museum.htm
Founded: 1947.
Congressional District: 3
Key Personnel: Dir., Dr. Renee M. Clary; Cur. & Mgr. Collections, Amy Moe
Hoffman.
Governing Authority: university. Parent Institution: Mississippi State Univer-
sity. Tax-exempt.
Institution Type/Description: Natural History Museum.
Collections: Mesozoic & Cenozoic paleontology; Upper Cretaceous lepado-
morph barnacles; mineralogy; geology.
Research Fields: geology.
Activities: guided tours; lectures; formally organized education programs for
children, adults & undergraduate college students; permanent & temporary
exhibitions.
Hours & Admission Prices: Mon.-Fri. 8-5. No charge; donations accepted.
Closed school holidays & vacations. &

Attendance: 1,000 (estimated)

JOHN GRISHAM ROOM, MITCHELL MEMORIAL LIBRARY, Mississippi State University, 395 Hardy Rd., Mississippi State, MS 39762. Mailing Address: P.O. Box 5408, Mississippi State, MS 39762. Tel.: 662-325-2559.
E-mail: fcoleman@library.msstate.edu
Web Site: library.msstate.edu/grishamroom
Governing Authority: Parent Institution: Mississippi State University Libraries.
Institution Type/Description: History Museum.
Collections: author, John Grisham's writings & achievements; photographs; A Time to Kill manuscript; fan mail; personal letters; newspaper clippings.
Hours & Admission Prices: Mon.-Fri. 9-4. No charge.

MISSISSIPPI ENTOMOLOGICAL MUSEUM, Dept. of Entomology and Plant Pathology, Mississippi State, MS 39762. Mailing Address: P.O. Box 9775, Mississippi State, MS 39762-9775. Tel.: 662-325-2990. Fax: 662-325-8837.
Web Site: www.mississippientomologicalmuseum.org.msstate.edu
Key Personnel: Professor Entomology & Dir. Museum, Dr. Richard Brown; Cur., Terence L. Schiefer; Asst. Cur. & Scientific Illustrator, Joe A. MacGown.
Governing Authority: university. Parent Institution: Mississippi State University.
Institution Type/Description: Entomology Museum.
Collections: over 1,000,000 pinned specimens; H. E. Weed; Henry Dietrich, J. M. Langston, R. W. Harned, Gladys Hoke-Lobdell, E. W. Stafford, M. R. Smith, William H. Cross, Leon W. Hepner, and Charles Bryson; taxa from Central and South America, the Seychelles, New Caledonia, and Fiji Islands; the MacDonald collection, emphasizing Lepidoptera of Panama; Ross E. Hutchins collection of photographs.
Research Fields: taxa in Araneae, Acarina, Plecoptera, Homoptera, Coleoptera, Lepidoptera, and Diptera.
Hours & Admission Prices: Call for hours.

Natchez

* **GRAND VILLAGE OF THE NATCHEZ INDIANS, (M),** 400 Jefferson Davis Blvd., Natchez, MS 39120-5110. Tel.: 601-446-6502. Fax: 601-446-6503.
E-mail: gvni@mdah.state.ms.us
Web Site: mdah.state.ms.us
Founded: 1976.
Congressional District: 3
Key Personnel: Dir., James F. Barnett, Jr.; C.E.O., H.T. Holmes; Pres. (V), Kane Ditto; Sales Shop Mgr., Janice Sago; Historian, Rebecca Anderson.
Governing Authority: state. Affiliated with the State of Mississippi Department of Archives and History, Div. of Historic Properties, Old Capitol Green, P.O. Box 571, Jackson, MS 38205. Tax-exempt: 170(b)(1)(A).
Institution Type/Description: Anthropology Museum: 1700-1730, Natchez Indian ceremonial mound center.
Collections: artifacts collected from the site; Indian ceramics; stone implements; early 18th-century French trade items.
Research Fields: culture of the Natchez Indians; archaeological survey of prehistoric & historic sites within Mississippi.
Facilities: 50-vol. library of ethnographic & archaeological reports available for research on premises only; field research station; separate laboratory operation; 60-seat auditorium. Southeastern Indian crafts & books for sale.
Activities: guided tours; lectures; films; hobby workshops; school & group tours; permanent exhibitions; craft demonstration program.
Publications: The Natchez Indians.
Hours & Admission Prices: Mon.-Sat. 9-5, Sun. 1:30-5. No charge; donations accepted. Discounts to AAM members. Closed New Year's Day; Labor Day; Thanksgiving; Christmas.
Attendance: 35,000 (accurate)

HISTORIC JEFFERSON COLLEGE, 16 Old North St., Natchez, MS 39120. Mailing Address: P.O. Box 700, Washington, MS 39190-0700. Tel.: 601-442-2901. Fax: 601-442-2902.
E-mail: hjc@mdah.state.ms.us
Web Site: mdah.state.ms.us
Founded: 1971.
Congressional District: 4
Key Personnel: C.E.O., Hank Holmes; Dir., Robin Seage Person; Sec. & Sales Shop Mgr., Maxine Clay; Historian, Kay McNeil; Historian, Toni Avance.
Personnel Profile: Full-Time Paid 5; Part-Time Paid 1; Part-Time Volunteers 30.
Governing Authority: state; nonprofit organization. Parent Institution: the Mississippi Dept. of Archives & History, P.O. Box 571, State St., Jackson

39205. Tel.: 601-354-6218. Branch Museums: Windsor Ruins Historic Site, Claiborne County. Tax-exempt: 501(c)(3).
Institution Type/Description: Historic Site: Campus chartered in 1802 by Mississippi Territorial Legislature.
Collections: college related artifacts such as uniforms; books; weapons; cemetery. Historical Houses: 1817 East Wing; 1839 West Wing; 1836 West Kitchen; 1839 East Kitchen; c.1828 President's House; 1915 Raymond Hall; 1931 Prospere Hall; 1937 Carpenter Hall.
Research Fields: historical; preservation; restoration.
Facilities: 5,000-vol. library; picnic area; 40 acre nature trail. Gift items for sale.
Activities: self-guided tours daily; special programs; summer camps; lunchtime lecture series; educational program for students; Civil War reenactments; pioneer living history encampment.
Publications: alumni newsletter, Jefferson Military College.
Hours & Admission Prices: Mon.-Sat. 9-5, Sun. 1-5. No charge; donations accepted. Closed New Year's Day; Independence Day; Labor Day; Thanksgiving; Christmas.
Attendance: 15,000 (estimated)

MISSISSIPPI SOCIETY DAUGHTERS OF THE AMERICAN REVOLUTION ROSALIE HOUSE MUSEUM, 100 Orleans St., Natchez, MS 39120-3452. Tel.: 601-445-4555; 601-870-8840. Fax: 601-304-1376.
E-mail: rosaliemansion@yahoo.com
Web Site: www.rosaliemansion.com
Founded: 1898.
Key Personnel: Chm. (V), Pamela White; Museum Shop Mgr., Theresa Hill.
Personnel Profile: Full-Time Paid 1; Part-Time Paid 20; Part-Time Volunteers 30.
Governing Authority: Parent Institution: Mississippi State Society D.A.R. Tax-exempt.
Institution Type/Description: Historic Site Museum: housed in c.1820 house & gardens located on Old Fort Rosalie grounds.
Collections: original period furnishings including Belter and C. Lee furniture; decorative arts; historic structures.
Research Fields: genealogy library.
Facilities: library of genealogy and history books relating to Mississippi & Natchez; gardens.
Activities: guided tours; formally organized education programs for children. Annual Events: USS Mississippi on Memorial Day, Independence Day & Veterans Day.
Publications: Miss DAR Magazine.
Hours & Admission Prices: Daily 9-5; group tours available upon request. Adults $10, children $8; Mississippi DAR members & children under 6 no charge. Closed Easter; Thanksgiving; Christmas. &
Attendance: 25,000 (accurate)

NATCHEZ COSTUME & DOLL MUSEUM, 215 S. Pearl St., Natchez, MS 39120. Tel.: 601-442-6672; 800-647-6742. Fax: 601-446-8687.
Institution Type/Description: General Museum.
Collections: period dolls; costume gowns worn by queens of past pageants during Spring Pilgrimage.
Hours & Admission Prices: Mon.-Fri. 9-4, Sat. 10-4. Admission $6.

NATCHEZ MUSEUM OF AFRICAN-AMERICAN HISTORY AND CULTURE, 301 Main St., Natchez, MS 39120-3461. Mailing Address: P.O. Box 1844, Natchez, MS 39121-1844. Tel.: 601-445-0728.
Institution Type/Description: History Museum.
Collections: local history & culture depicting African-American life in Mississippi; photographs; artifacts.
Hours & Admission Prices: Tues.-Sat. 1-4:30. Adults $5, children $1.

New Albany

UNION COUNTY HERITAGE MUSEUM, (M), 114 Cleveland St., New Albany, MS 38652-4050. Mailing Address: P.O. Box 657, New Albany, MS 38652-0657. Tel.: 662-538-0014. Fax: 662-538-6019.
E-mail: uchm@ucheritagemuseum.com
Web Site: www.ucheritagemuseum.com
Founded: 1991.
Congressional District: 1
Key Personnel: Dir., Jill Smith; Pres. (V) & Museum Shop Mgr., Betsey Hamilton.
Personnel Profile: Full-Time Paid 2; Full-Time Volunteers 1; Part-Time Paid 1; Part-Time Volunteers 25.

Governing Authority: nonprofit. Parent Institution: Union County Historical Society.
Institution Type/Description: History Museum.
Collections: scale model of William Faulkner's birthplace; Faulkner memorabilia & local history; Faulkner Literary Garden; mounted wildlife.
Major Exhibits: The Way We Worked, 1/14.
Research Fields: William C. Faulkner.
Facilities: archives.
Publications: cookbook, Worth Savoring.
Hours & Admission Prices: Tues.-Fri. 9-4, Sat. 10-3. No charge; donations accepted. &

Attendance: 10,000 (estimated)
Membership: Single $15; Associate $25; Supporter $50; Sponsor $100; Patron $250; Donor $500; Contributor $1,000.

Ocean Springs

THE DOLL HOUSE, 809 Washington Ave., Ocean Springs, MS 39564. Tel.: 228-872-3971.
Institution Type/Description: Doll Museum.
Collections: dolls; stuffed animals; dollhouses.
Hours & Admission Prices: Tues.-Sat. 1-5.

G.I. MUSEUM, 5796 Ritcher Rd., Ocean Springs, MS 39564-2291. Tel.: 228-872-1943.
Web Site: www.gimuseum.com
Founded: 2005.
Key Personnel: Dir., Doug Mansfield.
Personnel Profile: Full-Time Volunteers 3; Part-Time Volunteers 3.
Governing Authority: Tax-exempt.
Institution Type/Description: Military Museum.
Collections: military artifacts & memorabilia.
Activities: Museum Sponsors: Living History Events in March & November.
Publications: 30 min. weekly television show, "Local Heros".
Hours & Admission Prices: 1st & 3rd Sun. of month & Wed. 10-5; other times by appointment. No charge; donations accepted. &

J.L. SCOTT MARINE EDUCATION CENTER AND AQUARIUM, GULF COAST RESEARCH LABORATORY COLLEGE OF MARINE SCIENCES, THE UNIVERSITY OF SOUTHERN MISSISSIPPI, 703 E. Beach Dr., Ocean Springs, MS 39564-5326. Tel.: 228-818-8890. Fax: 228-818-8894.
E-mail: marine.education@usm.edu
Web Site: www.aquarium.usm.edu
Founded: 1972.
Congressional District: 5
Key Personnel: Admin., Sharon H. Walker, Ph.D.; Coord. Education Programs, Susan Culipher-Ross; Administrative Asst., Johnette Bosarge; Aquarium Supvr. & Facilities Coord., Alex (Buck) Schesny; Museum Shop Mgr., Sylvia Covacevich.
Personnel Profile: Full-Time Paid 21; Part-Time Paid 3; Part-Time Volunteers 80; Interns 25.
Governing Authority: state. Parent Institution: Gulf Coast Research Laboratory, University of Southern Mississippi. Tax-exempt: 501(c)(3).
Institution Type/Description: Aquarium, Marine Museum & Education Center.
Collections: 71 aquariums; snake enclosures; vivarium with reptiles & amphibians; touch tank; displays of marine artifacts, including a whale skeleton; mounted fish specimens; crabs of the Gulf of Mexico; various interactive exhibits.
Research Fields: breeding habits of North American native fishes; development & implementation of pre-college & informal educational programs.
Facilities: teacher resource programs library; nature & conservation center; aquarium; 313-seat auditorium.
Activities: aquarium tour & habitat interpretation; family membership program; formally organized education programs for children, adults, undergraduate college students & graduate students during the academic year; overnight scouting programs; advisory series; numerous static exhibits.
Publications: various leaflets; monthly newsletter, Marine Briefs.
Hours & Admission Prices: Mon.-Sat. 9-4. Adults $4, senior citizens $3.50, children 3-17 $2.50; discounts to groups & AAA members; children under 3 & members no charge. &
Attendance: 75,000 (estimated)
Membership: Student & Senior Citizen $15; Individual $25; Family $50; Family Plus $60; Sponsor $75; Patron $100; Corporate $1,000.

❋ **WALTER ANDERSON MUSEUM OF ART, (M),** 510 Washington Ave., Ocean Springs, MS 39564-4632. Tel.: 228-872-3164. Fax: 228-875-4494.
E-mail: wama@walterandersonmuseum.org

Web Site: www.walterandersonmuseum.org
Founded: 1991.
Congressional District: 4
Key Personnel: Exec. Dir., Rosemary Roosa; Pres. Bd., Morris Strickland; Cur., Douglas Myatt; Museum Shop Mgr., Linda Johnson; Chm. (V), Glenna Gerard.
Personnel Profile: Full-Time Paid 6; Part-Time Paid 3; Part-Time Volunteers 46.
Governing Authority: nonprofit. Tax-exempt: 501(c)(3).
Institution Type/Description: Art Museum: located in Historic District of Ocean Springs.
Collections: concentration on the works of Mississippi artist Walter Inglis Anderson; murals; paintings; sculpture; linoleum block prints; ceramics; decorative arts; changing exhibits feature other artists whose works are related to Anderson's.
Research Fields: art & life of Walter Anderson.
Facilities: 3,000 sq. ft. exhibit space. Posters & illustrated books featuring works of Walter Anderson, children's books, art books, art & science-related merchandise, jewelry, T-shirts, educational toys & museum-related items for sale.
Activities: guided tours; lectures; arts festivals; concerts; dance recitals; films; docent program; formal education programs for adults & children; temporary & loan exhibits. Annual Events: Art Blooms at WAMA in May; Monthly Art Activities for Children; summer art camp.
Publications: quarterly newsletter, Motif.
Hours & Admission Prices: Mon.-Sat. 9:30-4:30, Sun. 12:30-4:30. Adults $10, seniors, students 16-24 with ID, AAA members & military $8, children 5-15 $5; discounts to groups with advance reservation, AAM, AFA, SEMC & MMA members; children under 5 & members no charge. Closed New Year's Day; Easter; Thanksgiving; Christmas Eve & Day. &
Attendance: 30,000 (accurate)
Membership: Student $20; Seniors $30; Individual $40; Family Senior $50; Family $60; Sustainer $100; Patron $250; Benefactor $500; President's Club $1,000.

WILLIAM M. COLMER VISITOR CENTER, 3500 Park Rd., Davis Bayou Area, Ocean Springs, MS 39564-9709. Tel.: 228-875-9057 & 0074 ext. 100. Fax: 228-872-2954 & 875-2358.
Web Site: www.nps.gov/guis/
Founded: 1971.
Congressional District: 5
Key Personnel: District Interpreter, Susan Blair.
Personnel Profile: Full-Time Paid 3; Part-Time Paid 1; Part-Time Volunteers 20; Interns 3.
Governing Authority: federal. U.S. Dept. of the Interior, National Park Service, Washington, DC (See separate listings for Fort Pickens Area-Gulf Islands National Seashore, Santa Rosa Island, FL & Gulf Islands National Seashore, Gulf Breeze, FL for more complete information.)
Institution Type/Description: National Park & Natural & Cultural History Museum: located on & part of Gulf Islands National Seashore.
Collections: items & artifacts pertaining to gulf ecosystem coastal fortifications & orientation to Gulf Islands; history of Ship Island.
Research Fields: natural history; history; archaeology.
Activities: permanent exhibitions.
Publications: pamphlets; newspaper.
Hours & Admission Prices: Daily 8:30-4:30. No charge; donations accepted. &
Attendance: 50,000 (estimated)

Oxford

L.Q.C. LAMAR HOUSE MUSEUM, 616 N. 14th St., Oxford, MS 38655. Tel.: 662-513-6071 & 232-2477.
Institution Type/Description: Historic House Museum: housed in the former home of Lucius Lamar, a Congressman, Senator, Secretary of the Interior, and Supreme Court Justice; built in 1888. A National Historic Landmark.
Collections: Lamar's life & career; period furnishings; personal artifacts; photographs.
Hours & Admission Prices: By appointment. Adults $5, children $1.

ROWAN OAK, HOME OF WILLIAM FAULKNER, (M), 916 Old Taylor Ave., Oxford, MS 38655. Mailing Address: Univ. of Mississippi Museum, P.O. Box 1848, University, MS 38677-1848. Tel.: 662-234-3284. Fax: 662-915-7035.
E-mail: wgriffit@olemiss.edu
Web Site: www.olemiss.edu
Founded: 1972.
Congressional District: 1
Key Personnel: Dir., Albert Sperath; Cur., William D. Griffith.

Personnel Profile: Full-Time Paid 6; Full-Time Volunteers 9; Part-Time Paid 4; Part-Time Volunteers 4; Interns 1.

Governing Authority: public university; nonprofit. Parent Institution: University of Mississippi. Tax-exempt: 501(c)(3).

Institution Type/Description: Historic House Museum: home of William Faulkner, 1930-1962. A National Historic Landmark & National Literary Landmark.

Collections: furniture; personal artifacts.

Hours & Admission Prices: Tues.-Sat. 10-4, Sun. 1-4. Adults $5; children no charge. Closed New Year's Eve & Day; Independence Day; Thanksgiving; Christmas Eve & Day. &

Attendance: 27,000 (accurate)

UNIVERSITY MUSEUM & HISTORIC HOUSES, THE UNIVERSITY OF MISSISSIPPI, (M), 5th St. & University Ave., Oxford, MS 38655. Mailing Address: P.O. Box 1848, University, MS 38677-1848. Tel.: 662-915-7073. Fax: 662-915-7035.

E-mail: museums@olemiss.edu

Web Site: www.olemiss.edu/depts/u_museum

Founded: 1939.

Congressional District: 1

Key Personnel: Dir., William Andrews; Chancellor, Dan Jones; Collections Mgr., William Griffith; Consultant in Classics, Dr. Aileen Ajootian; Preparator, Bob Pekala.

Personnel Profile: Full-Time Paid 5; Part-Time Paid 10; Part-Time Volunteers 10; Interns 2.

Governing Authority: university; state; nonprofit organization. Parent Institution: University of Mississippi. Subsidiary Institutions: Walton-Young Historic House; Rowan Oak, William Faulkner's Home. Tax-exempt.

Institution Type/Description: General Museum.

Collections: David M. Robinson collection of ancient Greek & Roman pottery; coins; sculpture; surgical instruments; architectural fragments; inscriptions; Sumerian clay tablets; Roman glass; Egyptian antiquities; John Millington & F.A.P. Barnard collection of 19th-century scientific instruments; Theora Hamblett paintings; antique dolls; Victorian memorabilia; West African art; folk art of the American South; technology.

Research Fields: ancient Greece; 19th-century physics & astronomy instruments; anthropology, technology, folk art.

Facilities: 1,500-vol. library available for use on premises; classroom; meeting room.

Activities: guided tours; lectures; organized programs; permanent, temporary & traveling exhibitions; traveling trunk shows; picture files & videotapes for schools.

Publications: newsletters 3 times a year; brochures; catalogs.

Hours & Admission Prices: Museum: Tues.-Sat. 9:30-4:30, Sun. 1-4:30. Walton-Young House: Fri. 10-12, 2-4. Rowan Oak: Tues.-Sat. 10-4, Sun. 1-4. Museum & Walton-Young House: no charge. Rowan Oak: adults $5; members no charge. Closed national & university holidays. &

Attendance: 19,000 (accurate)

Membership: Student $10; Individual $25; Family $40; Patron $100; Supporting & Corporate Member $250; Director's Circle & Corporate Supporting $500; Sponsor & Corporate Sponsor $1,000; Sustaining & Corporate Sustaining $5,000.

Pascagoula

SCRANTON NATURE CENTER, 3928 Nathan Hale Dr., Pascagoula, MS 39581-4727. Mailing Address: P.O. Drawer 908, Pascagoula, MS 39568-0908. Tel.: 228-938-6612.

Institution Type/Description: Nature Center.

Collections: owls; fossils; sea life; plants; rocks; minerals.

Facilities: nature trails.

Activities: educational programs.

Hours & Admission Prices: Tues.-Sat. 10-5. Adults $2, children 2-12 $1.

SCRANTON SHRIMP BOAT MUSEUM, River Park, Pascagoula, MS 39568. Mailing Address: P.O. Drawer 908, Pascagoula, MS 39568-0908. Tel.: 228-938-6612.

Institution Type/Description: History Museum: housed on a 70 ft. commercial shrimp boat.

Collections: period furnishings; bunk room; wheelhouse; boat history; fishing industry.

Hours & Admission Prices: Tues.-Sat. 10-4, Sun. 1-4. No charge.

Philadelphia

NESHOBA COUNTY-PHILADELPHIA HISTORICAL MUSEUM, 303 Water Ave. S., Philadelphia, MS 39350-2621. Mailing Address: P.O. Box 1482, Philadelphia, MS 39350. Tel.: 601-656-1284.

Formerly: Philadelphia Neshoba County Museum

Founded: 1992.

Congressional District: 4

Key Personnel: Chm. (V), Mack Alford.

Personnel Profile: Full-Time Paid 2; Part-Time Paid 1.

Governing Authority: county bd. of suprvrs. Tax-exempt.

Institution Type/Description: History Museum: housed in the home of George Pegram Woodward, c.1880.

Collections: county history; period furnishings.

Hours & Admission Prices: Mon.-Fri. 10-3. No charge; donations accepted.

Picayune

THE CROSBY ARBORETUM, MISSISSIPPI STATE UNIVERSITY, 370 Ridge Rd., Picayune, MS 39466-8151. Mailing Address: P.O. Box 1639, Picayune, MS 39466-1639. Tel.: 601-799-2311, ext. 21. Fax: 601-799-2372.

E-mail: drackett@ext.msstate.edu

Web Site: www.crosbyarboretum.msstate.edu

Founded: 1980.

Congressional District: 5

Key Personnel: Dir., Dr. Janine Conklin; Sr. Cur., Pat Drackett.

Personnel Profile: Full-Time Paid 4; Part-Time Paid 1; Part-Time Volunteers 20.

Governing Authority: nonprofit organization. Parent Institution: Mississippi State University. Subsidiary Institution: Coastal Research & Extension Center. Tax-exempt: 501(c)(3).

Institution Type/Description: Arboretum.

Collections: woody & herbaceous plants of the Pearl River Drainage Basin of Mississippi and Louisiana; photo archives; herbarium.

Research Fields: botany; ecology; landscape architecture; horticulture.

Facilities: Pinecote: 104-acres native plant center; nature trails; visitors center.

Activities: guided tours; field trips to natural areas; lectures; films; formally organized education programs; special events for members; seminars; native plant sales.

Publications: quarterly, News Journal; books, Native Woody Plant Species of the Pearl River Basin, Guide to the Natural Areas of the Crosby Arboretum; Checklist of the Native Woody Plants of the Pearl River Basin.

Hours & Admission Prices: Wed.-Sun. 9-5. Adults $4, senior citizens $3, children $2; discounts to AAM, ICOM members, arboretum members & American Horticultural Society Reciprocal Garden members; members no charge. Closed New Years Eve & Day; Thanksgiving; Christmas Eve & Day. &

Attendance: 5,500 (estimated)

Membership: Student $10; Individual $25; Family $35; Donor & Business $100; Patron $250-$999; Benefactor $1,000.

Piney Woods

LAURENCE C. JONES MUSEUM, Piney Woods School, 5096 Hwy. 49 S., Piney Woods, MS 39148. Mailing Address: P.O. Box 99, Piney Woods, MS 39148-0037. Tel.: 601-845-2214, ext. 2700. Fax: 601-845-2604.

E-mail: vwilson@pineywoods.org

Web Site: www.pineywoods.org

Founded: 1909.

Congressional District: 3

Key Personnel: C.E.O., Willie L. Crossley; Chm., Gen. Wallace Arnold; Chief of Staff, Renee Tillman.

Personnel Profile: Part-Time Paid 1; Part-Time Volunteers 1.

Governing Authority: private school; nonprofit. Parent Institution: The Piney Woods School. Tax-exempt: 501(c)(3).

Institution Type/Description: History Museum: housed in c.1922 Community House, The Piney Woods Country Life School.

Collections: Piney Woods Country Life School history; the life of its founder, Laurence C. Jones; school archives; African American artifacts.

Research Fields: oral history interviews of former students & employees.

Facilities: 245-vol. library of Piney Woods history; archives; 4,139 sq. ft. exhibit space.

Activities: guided tours; temporary exhibitions; videos.

Publications: quarterly newsletter, The Pine Torch; weekly, e-Torch; video, When Cotton Blossoms.

Hours & Admission Prices: Mon.-Fri. 8-5. No charge; donations accepted. Closed major holidays. &

Attendance: 5,000 (estimated)

Pontotoc

TOWN SQUARE POST OFFICE & MUSEUM, 59 S. Main St., Pontotoc, MS 38863-2824. Mailing Address: P.O. Box 141, Pontotoc, MS 38863-0141. Tel.: 662-488-0388. Fax: 662-488-0388.

Founded: 1998.

Key Personnel: Dir., Martha Coleman; Pres. (V) Pontotoc County Historical Society, James L. Roberts, Jr.

Personnel Profile: Part-Time Paid 1; Part-Time Volunteers 10.

Governing Authority: Parent Institution: Pontotoc County Historical Society. Tax-exempt.

Institution Type/Description: History Museum.

Collections: paintings by local artists; Chickasaw Indian artifacts; tools; period artifacts.

Facilities: Museum-related items for sale.

Hours & Admission Prices: Mon.-Fri. 10-4:30; other times by appointment. No charge; donations accepted.

Poplarville

PEARL RIVER COMMUNITY COLLEGE MUSEUM, (M), 101 Hwy. 11 N., Poplarville, MS 39470-2201. Tel.: 601-403-1000; 877-772-2338 (Toll Free).

Web Site: www.prcc.edu/museum

Key Personnel: Dir., Ronn Hague

Institution Type/Description: History Museum.

Collections: college & Pearl River history; documents; photographs.

Hours & Admission Prices: Mon.-Fri. 8-11 & 1-4; other times by appointment.

Port Gibson

BERNHEIMER HOUSE, 212 Walnut St., Port Gibson, MS 39150. Tel.: 601-437-2843; 800-735-3407.

Institution Type/Description: Historic House Museum.

Collections: local history; period furnishings.

Hours & Admission Prices: By appointment. Adults $5.

GRAND GULF MILITARY PARK, 12006 Grand Gulf Rd., Port Gibson, MS 39150-4549. Tel.: 601-437-5911. Fax: 601-437-2929.

E-mail: grandgulfpark@aol.com

Web Site: www.grandgulfpark.state.ms.us

Founded: 1958.

Congressional District: 4

Key Personnel: Chm., Robert St. John; Administrative Asst., Cathi Dodgen.

Personnel Profile: Full-Time Paid 7; Part-Time Paid 2; Part-Time Volunteers 6.

Governing Authority: state. Parent Institution: State of Mississippi. Tax-exempt.

Institution Type/Description: Military Museum & Historic Site.

Collections: archaeology; Civil War relics; Indian arrowheads; period buggies & coaches; rifle pits; gun emplacements; waterwheel; 42 RV camper pads; cemetery. Historic Houses: 1790 Spanish House; 1815 Dog Trot House; 1820 Wheeless House; 1830 cottage in park; 1868 Sacred Heart Catholic Church; carriage house; jailhouse; firehouse; 2 Civil War forts.

Research Fields: Civil War battlefields.

Facilities: library; reading room; picnic pavilion; observation tower.

Activities: lectures; tours; Civil War reenactments; living histories; full hook up camper pads; tent camping.

Publications: archaeological reports, The Confederate Upper Battery Site, Grand Gulf MS., Excavations, 1982; The Confederate Magazine at Fort Wade Grand Gulf, MS. Excavations, 1980-81.

Hours & Admission Prices: Daily 8-5. Museum & Grounds: adults $3, senior citizens $2, students K-12 $1; discount to groups with reservations; pre-schoolers no charge. Closed major holidays. &

Attendance: 80,000 (estimated)

Raymond

DUPREE HOUSE AND MAMIE'S COTTAGE, 2809 Dupree Rd., Raymond, MS 39154. Tel.: 601-857-6051; 877-629-6051.

Institution Type/Description: Historic House Museum: housed in the former home of Dr. H. T. T. Dupree; c.1850. Listed on the National Register of Historic Places.

Collections: Dupree family history; period furnishings; personal artifacts; photographs. Historic House: Mamie's cottage, c.1840.

Hours & Admission Prices: By appointment.

Ridgeland

CRAFTSMEN'S GUILD OF MISSISSIPPI & MISSISSIPPI CRAFT CENTER, 950 Rice Rd., Ridgeland, MS 39157-3040. Tel.: 601-856-7546. Fax: 601-856-7531.

E-mail: info@mscrafts.org

Web Site: www.mscrafts.org

Founded: 1973.

Congressional District: 4

Key Personnel: Exec. Dir., Nancy Perkins; Bd. Pres., Renee Flynt; Museum Shop Mgr., Carmen Castilla.

Personnel Profile: Full-Time Paid 5; Part-Time Paid 8; Part-Time Volunteers 100.

Governing Authority: nonprofit. Parent Institution: The Craftsmen's Guild of Mississippi, Inc. Subsidiary Institution: The Mississippi Crafts Center. Tax-exempt: 501(c)(3).

Institution Type/Description: Fine Crafts Gallery & Guild.

Collections: pottery; hand-woven items; quilts; baskets; jewelry; hand blown glass; woodcarvings; willow furniture; Indian basketry; metal sculpture; hand crafted toys.

Research Fields: Choctaw Indian culture.

Facilities: Craft items for sale.

Activities: lectures; arts festivals; organized education programs for children & adults; craft demonstrations on weekends. Museum Sponsors: Indian and Pioneer Heritage Festival; Chimneyville Crafts Festival.

Publications: books, The Language of Mississippi Choctaw Crafts, The Guild at Twenty-Five, A Portrait of the Craftmen's Guild of Mississippi; I Am a Craftsman: 40 at 40.

Hours & Admission Prices: Daily 9-5. No charge; donations accepted. Closed New Year's Day; Easter; Thanksgiving; Christmas. &

Attendance: 150,000 (estimated)

Membership: Friend & Student $50; Apprentice $100-$199; Journeyman $200-$499; Journeyman II $500-$999; Master Craftsmen $1,000-$2,499; Patron $2,500-$4,999; Benefactor $5,000-$9,999; Partner $10,000 & up.

Ripley

TIPPAH COUNTY HISTORICAL MUSEUM, 106 N. Siddell St., Ripley, MS 38663-2036. Tel.: 662-512-0099.

E-mail: tippahmuseum@dixie.net.com

Founded: 1997.

Congressional District: 1

Institution Type/Description: History Museum.

Collections: local history & culture; weapons; clothing; agricultural tools; Native American artifacts.

Major Exhibits: Tim Needham Handmade Toys, 7/14-9/14.

Facilities: 1800s house with restroom.

Activities: family genealogies.

Publications: quarterly e-newsletter.

Hours & Admission Prices: Mon.-Sat. 10-2. No charge; donations accepted. Closed holidays. &

Attendance: 650 (estimated)

Membership: Fellowship $25; Patron $50; Grand Patron $100.

Sandy Hook

JOHN FORD HOUSE, Hwy. 35 S., Sandy Hook, MS 19478. Mailing Address: 412 Courthouse Sq., P.O. Box 272, Columbia, MS 39429. Tel.: 601-731-3999.

Founded: 1809.

Governing Authority: Parent Institution: Marion County Historical Society. Tax-exempt.

Institution Type/Description: Historic House: housed in the former home of Rev. John Ford, a legislator & member of the state's Constitution Writing Committee; c.1805.

Collections: local history; period furnishings; personal artifacts; photographs.

Hours & Admission Prices: Sat.-Sun. 2-5 by appointment. Adults $4, children under 12 $3.

Attendance: 2,000 (estimated)

Senatobia

HERITAGE MUSEUM FOUNDATION OF TATE COUNTY, INC., 135 N. Front St., Senatobia, MS 38668-0375. Mailing Address: P.O. Box 375, Senatobia, MS 38668-0375. Tel.: 662-562-8715.

Founded: 1977.

Congressional District: 2

Key Personnel: Pres., Deborah Perkins; Sec., Janie Mortimer; Treas., Sara S. Henley.

Personnel Profile: Part-Time Volunteers 2.
Governing Authority: nonprofit organization. Tax-exempt: 170(b)(1)(A) & 509(a)(1).
Institution Type/Description: Historic Foundation Museum: housed in historic Tate County Courthouse.
Collections: general artifacts of the history of Tate County, MS.
Research Fields: Tate County local history.
Activities: self-guided tours.
Publications: brochure.
Hours & Admission Prices: Mon.-Fri. 8-5, except on courthouse holidays. No charge; donations accepted. &
Attendance: 150 (estimated)
Membership: Student $3; Individual $5; Couple $10; Life $1,000.

Starkville

OKTIBBEHA COUNTY HERITAGE MUSEUM, 206 Fellowship St., Starkville, MS 39759-3378. Tel.: 662-323-0211.
Web Site: www.oktibbehaheritagemuseum.com
Founded: 1979.
Congressional District: 4
Key Personnel: Chm. Bd. Trustees, Joan Wilson; Chm. (V), Marion Honsinger.
Personnel Profile: Part-Time Volunteers 45.
Governing Authority: board of trustees. Parent Institution: City of Starkville/Oktibbeha County. Tax-exempt.
Institution Type/Description: History Museum: housed in 1800s The G M & O Depot.
Collections: local history; county artifacts including maps, books, pictures, farm equipment, early dairy equipment, wedding dresses, military service uniforms & artifacts, quilts, cradles, Indian arrowheads; train exhibit including dated nails found in old cross ties, railroad stock, early tickets, time tables, telegraph system & 1800 & 1875 models of passenger trains; early radios; doctor's instruments case; period dolls & doll furniture; local sports figure artifacts; miniature trains; Coca-Cola memorabilia; local business; children's exhibits; Cool Papa Bell bust.
Facilities: meeting room.
Activities: guided tours.
Publications: brochure.
Hours & Admission Prices: Tues.-Thurs. 1-4, Sat. call for hours. No charge; donations accepted. &
Attendance: 3,000 (accurate)

Stennis Space Center

INFINITY SCIENCE CENTER, One Discovery Pl., Stennis Space Center, MS 39529. Mailing Address: 111 Court St., Bay Saint Louis, MS 39520. Tel.: 228-533-9025; 800-237-1821. Fax: 228-533-9232.
E-mail: brianalexander@visitinfinity.com
Web Site: www.visitinfinity.com
Founded: 2012.
Key Personnel: Sales & Mktg. Coord., Brian Alexander
Institution Type/Description: Science Center.
Collections: science & space explorations; hands-on exhibitions.
Facilities: restaurant; theatre. Gift items for sale.
Activities: educational programs; themed camps.
Hours & Admission Prices: Mon.-Sat. 10-4. Adults $8, seniors 55 & over and children 6-17 $6; discounts to groups of 20 or more; children 5 & under no charge. Stennis Space Center tours included with admission require photo ID for visitors 18 & over. &

Sturgis

BENCH MARK WORKS MOTORCYCLE MUSEUM, 3400 Earles Fork Rd., Sturgis, MS 39769. Tel.: 662-465-6444.
Web Site: www.benchmarkworks.com
Institution Type/Description: Motorcycle Museum.
Collections: early BMW motorcycles; photographs.
Activities: special events.
Hours & Admission Prices: Mon.-Thurs. 8-5.

Taylorsville

WATKINS MUSEUM, Eureka St., Taylorsville, MS 39168. Mailing Address: P.O. Box 358, Taylorsville, MS 39168-0358. Tel.: 601-785-6531. Fax: 601-785-2200.
E-mail: rosalynglenn@taylorsvillems.com
Founded: 1968.
Congressional District: 3

Key Personnel: Cur., Rosalyn Glenn.
Governing Authority: municipal.
Institution Type/Description: General Museum: housed in c.1900 Taylorsville Signal & General Store.
Collections: c.1837 hand printing press; early medical & farm equipment; manuscripts; period bottles; clothing; country store equipment; typewriter; catalogs; early 1900s house plans.
Facilities: 150-vol. library of books; magazines; newspapers available for use on premises only.
Activities: guided tours; permanent & temporary exhibitions.
Hours & Admission Prices: By appointment. No charge; donations accepted. &
Attendance: 100

Tougaloo

TOUGALOO COLLEGE, Tougaloo College Art Collection, 500 W. County Line Rd., Tougaloo, MS 39174-0578. Mailing Address: P.O. Box 578, Tougaloo, MS 39174-9799. Tel.: 601-977-7743. Fax: 601-977-7714.
E-mail: art@tougaloo.edu
Web Site: www.tougaloo.edu/artcolony
Founded: 1869.
Congressional District: 3
Key Personnel: Pres., Beverly Wade Hogan; Assoc. Professor Art, Bruce O'Hara; Professor & Dept. Chair Collections, Johnnie M. Maberry-Gilbert; Head Librarian, Orthello Moman; Archives Mgr., Minnie Watson.
Personnel Profile: Part-Time Volunteers 1.
Governing Authority: board of trustees; college. Affiliated with United Church Board for Homeland Ministries & The American Missionary Association, 132 W. 31st St., New York, NY 10001. Tel.: 212-239-8700. Tax-exempt.
Institution Type/Description: Historic Site/Building: located on historic Tougaloo College Campus. Housed in the Bennie Thompson Building.
Collections: European & African American & African art from 17th-21st century.
Facilities: 3,000-vol. library of books available for research with patron's card; reading room; auditorium; theater; classrooms; cafeteria.
Activities: guided tours; lectures; films; gallery talks; concerts; drama; rental gallery; formally organized education programs for undergraduate college students; cooperative with Mississippi Museum of Art; loan, permanent, temporary & traveling exhibitions.
Publications: 1968-1978 The Tougaloo College Art Collections; Silver Jubilee, 1963-1988 (a history); catalog, The Turbulent Years.
Hours & Admission Prices: Call for hours. &
Attendance: 1,000 (estimated)

Tunica

TUNICA MUSEUM, (M), One Museum Blvd., Tunica, MS 38676. Mailing Address: P.O. Box 1914, Tunica, MS 38676-1914. Tel.: 662-363-6631. Fax: 662-363-6651.
E-mail: drdick@tunicamuseum.com
Web Site: www.tunicamuseum.com
Founded: 1997.
Congressional District: 2
Key Personnel: Dir., Richard Taylor; Pres. (V), Bob Gann; Museum Shop Mgr., Darlene Griffith.
Personnel Profile: Full-Time Paid 3; Part-Time Paid 1.
Governing Authority: Tax-exempt: 501(c)(3).
Institution Type/Description: History Museum.
Collections: area history; Native American artifacts; railroad memorabilia.
Facilities: library. Museum-related items for sale.
Publications: newsletter.
Hours & Admission Prices: Tues.-Sat. 10-5, Sun. 1-5. No charge; donations accepted. &
Attendance: 30,000 (estimated)

Tupelo

ELVIS PRESLEY BIRTHPLACE AND MUSEUM, 306 Elvis Presley Dr., Tupelo, MS 38804-2812. Tel.: 662-841-1245. Fax: 662-690-6623.
E-mail: info@elvispresleybirthplace.com
Web Site: www.elvispresleybirthplace.com
Key Personnel: Exec. Dir., Dick Guyton
Institution Type/Description: History Museum: housed in the birthplace of Elvis Presley, born in 1935.
Collections: furnishings.
Facilities: picnic area; Memorial Chapel. Museum-related items for sale.
Activities: Museum Sponsors: Annual Fan Appreciation Day in August.

Hours & Admission Prices: May-Sept. Mon.-Sat. 9-5:30, Sun. 1-5; Oct.-April Mon.-Sat. 9-5, Sun. 1-5. House: adult $4, child $2. Museum: adult $8, child $4. Combined: adult $12, child $6; discounts to groups of 15 or more. Closed Thanksgiving; Christmas.

GUMTREE MUSEUM OF ART, 211 W. Main St., Tupelo, MS 38804-3917. Mailing Address: P.O. Box 786, Tupelo, MS 38802-0786. Tel.: 662-844-2787. Fax: 662-269-3296.
E-mail: kstafford@gumtreemuseum.org
Web Site: www.gumtreemuseum.com
Formerly: Tupelo Artist Guild Gallery
Founded: 1985.
Congressional District: 5
Key Personnel: Exec. Dir., Liz McIntosh; Chm. (V), Robert Reed.
Personnel Profile: Full-Time Paid 1; Part-Time Volunteers 15.
Governing Authority: nonprofit organization; local board. Tax-exempt: 501(c)(3).
Institution Type/Description: Art Museum.
Collections: paintings; photographs; sculptor; jewelry.
Activities: guided tours; lectures; study clubs; education programs for youth & adults.
Hours & Admission Prices: Tues.-Sat. 10-4. No charge; donations accepted. Closed New Year's Day; Independence Day; Thanksgiving; Christmas. &
Attendance: 13,200 (estimated)
Membership: Friend $40; Family & Friend $65; Bronze $100; Silver $250; Gold $500 & up.

HEALTHWORKS!, (M), 219 S. Industrial Rd., Tupelo, MS 38801. Tel.: 662-377-KIDS.
E-mail: healthworks@nmhs.net
Founded: 2000.
Institution Type/Description: Children's Health Museum.
Collections: health education; hands-on exhibitions.
Activities: special events; educational programs; field trips; rental facilities.
Hours & Admission Prices: Mon.-Fri. 8:30-4, Sat. 10-2. Admission $5; children under 2 no charge.

NATCHEZ TRACE PARKWAY, 2680 Natchez Trace Pkwy., Tupelo, MS 38804-9715. Tel.: 662-680-4025. Fax: 662-680-4036.
E-mail: christina_smith@nps.gov
Web Site: www.nps.gov/natr
Founded: 1938.
Congressional District: 1
Key Personnel: Supt., Cameron H. Sholly; Interpretive Specialist, Ernie Price.
Governing Authority: federal. Parent Institution: National Park Service. Tax-exempt.
Institution Type/Description: Visitor Center & Archival Study Collection.
Collections: Indian trails; 18th-century trade route; 1800 post road; archaeology; ethnology; geology; parkway construction 1930s-present. Historic Site: 1785 Mount Locust.
Research Fields: parkway construction.
Facilities: 1,400-vol. library of historical reference material; herbarium; picnic areas; trails.
Activities: orientation film; unscheduled talks.
Publications: folders, Parkway, Tupelo/Brices Crossroads; assorted informational handouts.
Hours & Admission Prices: Daily 8-4:30. No charge. Closed Christmas. &
Attendance: 51,991 (accurate)

OREN DUNN CITY MUSEUM, 689 Rutherford Rd., Tupelo, MS 38803. Mailing Address: P.O. Box 2674, Tupelo, MS 38803-2674. Tel.: 662-841-6438. Fax: 662-841-6458. Facebook: Oren Dunn City Museum.
E-mail: rae.mathis@tupeloms.us
Web Site: www.orendunnmuseum.org
Formerly: Tupelo City Museum
Founded: 1984.
Key Personnel: Advisory Bd., Dr. Mike Currie; Chm. Friends of the Museum, Boyd Yarbrough; Cur., Rae Mathis; Asst. Cur., Della Lentz; Operations Mgr., Jerry Duckett; Educator, Janice Anthony.
Personnel Profile: Full-Time Paid 2; Part-Time Paid 2; Part-Time Volunteers 5; Interns 2.
Governing Authority: city. Tax-exempt: 501(c)(3).
Institution Type/Description: History Museum.
Collections: local history & culture from 1870s to present; period Tupelo fire trucks; 1948 Lee County Book Mobile; Memphis trolley car turned diner; one room chapel; one room school; veterans' artifacts. Historic Building: 1870s Dog-Trot Cabin.

Activities: rental facilities; summer camp. Museum Sponsors: Dudie Burger Festival in May; Dogtrot Rockabilly Festival in October; Lighting of the Museum in December.
Hours & Admission Prices: Tues.-Fri. 9-4, Sat. 10-3. Adults $3, seniors 60 & over $2; children 4-16 $1.50. Closed City of Tupelo holidays. &
Attendance: 10,000 (accurate)
Membership: Student $10; Individual $20; Family $50; Patron $100; Organization $200-$1,000; Sponsor $500; Corporate $1,000.

TUPELO AUTOMOBILE MUSEUM, 1 Otis Blvd., Tupelo, MS 38804-4015. Tel.: 662-842-4242. Fax: 662-842-3734. Facebook: Tupelo Automobile Museum.
E-mail: info@tupeloauto.com
Web Site: www.tupeloauto.com
Founded: 2002.
Congressional District: 1
Key Personnel: Exec. Dir. & Chm. (V), Jane Spain; Dir., Phil Sullivan; Visitor Coord. & Museum Shop Mgr., Susan Stembridge.
Personnel Profile: Full-Time Paid 2; Full-Time Volunteers 1; Part-Time Paid 3; Part-Time Volunteers 6.
Governing Authority: Tax-exempt.
Institution Type/Description: Automobile Museum.
Collections: over 100 cars from 1880s to 1990s including movie & celebrity-owned vehicles; neon signs; period artifacts; Elvis memorabilia.
Facilities: 120,000 sq. ft. exhibit space. Gift items for sale.
Activities: tours; rental facilities; special events. Annual Events: Blue Suede Cruise in May.
Publications: quarterly newsletter, The Rear View Mirror.
Hours & Admission Prices: Mon.-Sat. 9-4:30, Sun. 12-4:30. Adult $10, seniors, military & AAA members $9, children 5-12 $5; discounts to groups of 10 or more; children 4 & under no charge. Closed New Year's Day; Easter; Thanksgiving; Christmas. &
Attendance: 20,000 (estimated)
Membership: Individual $35; Family $50; Contributor $100; Sustainer $250; Patron $500; Benefactor $1,000.

TUPELO VETERAN'S MUSEUM, 689 Rutherford Rd., Tupelo, MS 38801. Mailing Address: P.O. Box 7212, Tupelo, MS 38801. Tel.: 662-844-1515.
Key Personnel: Cur., Tony Lute
Institution Type/Description: Military History Museum.
Collections: local military history including the Civil War, World War I, World War II, Korean War, Vietnam War, Desert Storm & Iraqi Wars; personal artifacts; photographs; uniforms; medals; weapons; letters; battlefield organ; military vehicles.
Hours & Admission Prices: Call for hours.

Union

BOLER STAGECOACH INN MUSEUM, 205 E. Jackson Rd., Hwy. 492, Union, MS 39365. Mailing Address: P.O. Box 116, Decatur, MS 39327. Tel.: 601-635-3160.
E-mail: jmmoore@decatur.net
Founded: 1996.
Congressional District: 3
Key Personnel: Pres. (V), Nancy Moore
Governing Authority: Parent Institution: Foundation for Restoration and Preservation of Boler's Inn. Tax-exempt: 501(c)(3).
Institution Type/Description: Historic Building: c. 1856.
Collections: period furnishings & artifacts from 1830-1920; photographs; personal artifacts.
Hours & Admission Prices: By appointment. Adults $5, children $1.
Attendance: 60 (estimated)

Vicksburg

BIEDENHARN COCA-COLA MUSEUM, 1107 Washington St., Vicksburg, MS 39183-2959. Tel.: 601-638-6514. Fax: 601-636-5010.
E-mail: vburgfoundation@aol.com
Web Site: biedenharncoca-colamuseum.com
Founded: 1979.
Key Personnel: Exec. Dir., Nancy H. Bell; Chm. (V), Charlie Gholson; Treas., Denny Allman.
Personnel Profile: Full-Time Paid 1; Part-Time Paid 3; Part-Time Volunteers 4.
Governing Authority: private; nonprofit organization. Parent Institution: Vicksburg Foundation for Historic Preservation. Tax-exempt: 501(c)(3).
Institution Type/Description: History Museum: housed in the first Coca-Cola bottling company in the world, 1894.

Collections: history of Coca-Cola; the Biedenharn family; bottling works; Coca-Cola memorabilia; soda fountain; period bottles; advertising; signs; ice chests.
Facilities: 20-vol. library; 5,000 sq. ft. exhibit space; soda fountain. Museum-related items for sale.
Hours & Admission Prices: Mon.-Sat. 9-5, Sun. 1:30-4:30. Adults $3.50, students & children $2.50; discounts to groups, families, AAM & ICOM members; members no charge. Closed New Year's Day; Easter; Thanksgiving; Christmas.
Attendance: 25,000 (accurate)
Membership: Individual $25; Family $35; Contributing $45; Sustaining $55; Patron $100-$499; Gold Patron $500 & up.

CEDAR GROVE MANSION INN, 2200 Oak St., Vicksburg, MS 39180-4008. Tel.: 601-636-1000. Fax: 601-634-6126.
E-mail: info@cedargroveinn.com
Web Site: cedargroveinn.com
Founded: 1959.
Congressional District: 2
Key Personnel: C.E.O., Owner & Innkeeper, Colleen Small.
Governing Authority: privately owned.
Institution Type/Description: Historic Site & Building: 1840-1858 southern mansion located at the site of the Siege of Vicksburg.
Collections: original furnishings; Civil War cannonball embedded in parlor wall; Regina music box; Bohemian glass; original works of art; 15 Italian marble mantelpieces; Aubusson rugs; terra cotta statues; urns; pre-Civil War magazines.
Facilities: library of early books pertaining to law, medicine, history and botany available on premises; formal gardens; restaurant.
Activities: guided tours.
Hours & Admission Prices: Tours daily 9:30-11 & 1-4. Adults $6, children 6-12 $4; discounts to groups; children under 6 no charge.

CORNERS MANSION INN-A BED & BREAKFAST, 601 Klein St., Vicksburg, MS 39180. Tel.: 601-636-7421; 800-444-7421.
E-mail: cornersmansioninn@yahoo.com
Formerly: The Corners Mansion Inn
Founded: 1873.
Institution Type/Description: Historic House Museum: housed in a home built by John Alexander Klein for his daughter and her husband; built in 1873.
Collections: local & Klein family history; Susan Klein Bonham family bible; period furnishings; photographs; paintings.
Facilities: 16 guest rooms with private baths.
Activities: dinners, luncheons, special events.
Hours & Admission Prices: Daily 9:30-5:30 by appointment. Home Tours: $10 per person; overnight guests no charge.

DUFF GREEN MANSION, 1114 First East St., Vicksburg, MS 39180. Tel.: 601-636-6968 & 638-6662; 800-992-0037. Fax: 601-661-0079.
Web Site: www.duffgreenmansion.com
Institution Type/Description: Historic House Museum: served as a hospital during the Civil War for Confederate and Union soldiers; c.1856. Listed on the National Register of Historic Places.
Collections: local history & culture; period furnishings; personal artifacts; photographs.
Hours & Admission Prices: Daily 12-5 by appointment. Adults $6, students & children 5-12 $4; children under 5 no charge.

THE JACQUELINE HOUSE AFRICAN AMERICAN MUSEUM, 1325 Main St., Vicksburg, MS 39183-2647. Tel.: 601-636-0941.
Institution Type/Description: History Museum.
Collections: African American history & culture; photographs; books; manuscripts; music; posters; newspapers.
Activities: lectures; special events; rental facilities; film festivals.
Hours & Admission Prices: By appointment.

NATIONAL PARK SERVICE VICKSBURG NATIONAL MILITARY PARK-CAIRO MUSEUM, 3201 Clay St., Vicksburg, MS 39183-3495. Tel.: 601-636-2199 (Museum) & 0583 (Park). Fax: 601-638-7329.
Web Site: www.nps.gov/vick
Founded: 1980.
Congressional District: 2
Key Personnel: Museum Cur., Elizabeth Joyner; Chief Operations, Rick Martin; Park Supt., R. Michael Madell.
Personnel Profile: Full-Time Paid 30.

Governing Authority: federal. Affiliated with National Park Service. Tax-exempt.
Institution Type/Description: History Museum & Visitors Center: located on the site of the 1863 Union siege of the city of Vicksburg, MS.
Collections: official records of the Union and Confederate Armies & Navies; Historic House: 1800s Shirley House. Historic Restored Gunboat: U.S.S. Cairo Museum, containing artifacts from ironclad gunboat.
Research Fields: pertaining to those who fought in the Vicksburg campaign.
Facilities: 150-vol. library on Civil War naval history & artifacts available for use on premises; U.S.S. Cairo Museum. Civil War Naval literature for sale.
Activities: self-guided tours; lectures; films; audiovisual program; formally organized education programs for children & adults on and off premises; living history demonstration in summer.
Publications: book, Hardluck Ironclad; park folder; site bulletins on various topics pertaining to Civil War naval history & Vicksburg campaign.
Hours & Admission Prices: Cairo Museum: April 6-Oct. 30 9:30-6; Oct. 31-April 5 8:30-5. Park: $8 per car. Museum located inside park. Closed New Year's Day; Thanksgiving; Christmas.
Attendance: 500,000 (estimated)

OLD COURT HOUSE MUSEUM-EVA WHITAKER DAVIS MEMORIAL, 1008 Cherry St., Court Square, Vicksburg, MS 39183-2540. Tel.: 601-636-0741. Fax: 601-636-1004.
E-mail: societyhistorica@bellsouth.net
Web Site: www.oldcourthouse.org
Founded: 1947.
Congressional District: 1
Key Personnel: Dir. & Cur., George C. Bolm; Pres., Marlene Brooks; Vice Pres., John McHan.
Personnel Profile: Full-Time Paid 3; Part-Time Paid 3; Part-Time Volunteers 2.
Governing Authority: Sponsored by the Vicksburg & Warren County Historical Society. Tax-exempt.
Institution Type/Description: Historic Building: housed in the 1858 Old Warren County Courthouse. A National Historic Landmark.
Collections: Civil War & Indian artifacts; antebellum South; photographic glass negatives & prints of steamboats by Mack Moore; clothing; art; furniture.
Research Fields: genealogy; history pertaining to our local area.
Facilities: 1,400-vol. library of general material available for use by individuals paying a research fee. Postcards & museum-related items for sale.
Activities: Museum Sponsors: Old Court House Flea Market; Confederate Christmas Ball; Sacred Harp Singing.
Publications: annual newsletter; weekly newspaper column on local history.
Hours & Admission Prices: Mon.-Sat. 8:30-5, Sun. 1:30-5. Adults $5, senior citizens $4.50, students grades 1-12 $3; discounts to groups of 10 or more. Closed New Year's Day; Thanksgiving; Christmas Eve & Day.
Attendance: 32,763 (accurate)
Membership: Individual $25; Friend $50; Contributing $100; Supporting $500; Patron of History $1,000.

VICKSBURG BATTLEFIELD MUSEUM, 1010 Levee St., #A, Vicksburg, MS 39183-2569. Tel.: 601-638-6500. Fax: 601-638-8746.
E-mail: thegunboat@bellsouth.net
Web Site: vicksburgbattlefieldmuseum.net
Founded: 1993.
Congressional District: 2
Key Personnel: C.E.O., Lamar Roberts; Museum Shop Mgr., Sue Roberts.
Personnel Profile: Full-Time Paid 2; Part-Time Paid 2; Part-Time Volunteers 2.
Institution Type/Description: History Museum.
Collections: Civil War diorama recreating the battles of the Siege of Vicksburg including 2500 miniature soldiers; Civil War gunboat models; film documentary; models of U.S. Navy vessels named for Mississippians; paintings by Herb Mott; sculptures by Dan Rickardson; maps; clothing; photos.
Facilities: theater.
Activities: film presentation.
Hours & Admission Prices: Mon.-Sat. 9-5. Adults $5.50. Closed Easter; Thanksgiving; Christmas.
Attendance: 19,000 (accurate)

YESTERDAY'S CHILDREN ANTIQUE DOLL & TOY MUSEUM, 1104 Washington St., Vicksburg, MS 39183-2960. Tel.: 601-638-0650.
E-mail: mbakarich@aol.com
Web Site: www.yesterdayschildrenmuseum.com
Founded: 1986.
Key Personnel: Dir., Michael N. Bakarich; Cur., Carolyn C. Bakarich.
Personnel Profile: Full-Time Volunteers 2; Part-Time Volunteers 3.
Institution Type/Description: Toy and Doll Museum.

Collections: over 1,000 dolls including mid-1800 China heads to late 1800 French & German bisques, 1900 compositions, Madame Alexanders, Shirley Temples, Barbies, barefoot children, ventriloquist dolls, contemporary dolls from 1940 to present, & artist dolls. Boys' toys include pedal cars, trucks, trains, guns, soldiers, cars, planes, construction & farm equipment, G.I. Joes.
Facilities: 2,688 sq. ft. exhibit space. Museum-related items for sale.
Publications: brochure, Yesterday's Children Antique Doll & Toy Museum.
Hours & Admission Prices: Mon.-Sat. 10-4; other times by appointment. Adults $3, children $2; discounts to groups of 10 or more. Closed Thanksgiving; Christmas. &
Attendance: 12,000 (estimated)

Water Valley

THE WALTER VALLEY CASEY JONES RAILROAD MUSEUM, 105 Railroad Ave., Water Valley, MS 38965-3312. Tel.: 601-473-2849.
E-mail: gurnerjack@bellsouth.net
Web Site: www.watervalley.net/users/caseyjones/home.htm
Founded: 1994.
Key Personnel: Chm. (V) & Pres. (V), Jack Gurner, Jr.; Cur. & Museum Shop Mgr., J.K. Gurner, Sr.
Personnel Profile: Full-Time Volunteers 1; Part-Time Volunteers 3.
Governing Authority: Subsidiary Institution: Lion Club.
Institution Type/Description: Railroad Museum.
Collections: railroad memorabilia, hardware, documents, photography.
Hours & Admission Prices: Thurs.-Sat. 2-4; other times by appointment. No charge; donations accepted. &
Attendance: 200 (estimated)

West Point

HOWLIN' WOLF BLUES MUSEUM, 307 Westbrook St., West Point, MS 39773. Mailing Address: P.O. Box 1334, West Point, MS 39773. Tel.: 662-605-0770. Fax: 662-495-2007.
E-mail: rramsey@wpms.net
Web Site: www.wpnet.org/howlin_wolf.htm
Founded: 2005.
Congressional District: 1
Key Personnel: Dir., Richard Ramsey.
Personnel Profile: Full-Time Volunteers 1.
Institution Type/Description: Blues Museum.
Collections: Blues history; personal artifacts; photographs; life & career of Chester Arthur Burnett, The Howlin' Wolf; musical instruments; Howlin' Wolf statue.
Activities: special events. Museum Sponsors: Howlin' Wolf Festival each Labor Day weekend.
Hours & Admission Prices: Thurs.-Sat. 10-5; other times by appointment. No charge; donations accepted.
Attendance: 250 (estimated)

SAM WILHITE TRANSPORTATION MUSEUM, 210 Depot Dr., West Point, MS 39773. Mailing Address: 331 Jefferson St., West Point, MS 39773-2826. Tel.: 662-494-6385; 8910 (cell).
E-mail: terry.craig@bellsouth.net
Founded: 1994.
Congressional District: 3
Institution Type/Description: Transportation Museum.
Collections: early West Point transportation; 1927 Model T Ford car; model railroad; replica railroad ticket office; railroad clocks; telegraph; period telephones.
Hours & Admission Prices: Thurs.-Sat. 10-4. No charge; donations accepted.

Woodville

ROSEMONT PLANTATION, Rosemont Plantation, Hwy. 24 E., Woodville, MS 39669. Mailing Address: P.O. Box 814, Woodville, MS 39669-0814. Tel.: 601-888-6809. Fax: 601-888-3327.
E-mail: pbeacroft@aol.com
Web Site: www.rosemontplantation1810.com
Founded: 1971.
Congressional District: 3
Key Personnel: Owner, Percival T. Beacroft; Publicity & Tours, Jenny Angeline.
Personnel Profile: Full-Time Paid 2.
Governing Authority: individual operation.
Institution Type/Description: Historic House Museum: 1810, originally named Poplar Grove, family home of Jefferson Davis; family cemetery located on grounds.

Collections: Davis furniture, artifacts & portraits; furnishings; original house & reconstructed outbuildings.
Research Fields: genealogy of all descendants of Davis family compiled by Ernesto Caldeira.
Activities: guided tours; permanent exhibitions; Davis family reunion every two years.
Publications: brochure; newsletter, Davis Family.
Hours & Admission Prices: March to mid-Dec. Tues.-Fri. 10-4; groups by appointment; additional hours during Natchez Spring & Fall Pilgrimages. Suggested Donations: adults $10, students $4.
Attendance: 500

WILKINSON COUNTY MUSEUM, 203 Boston Row, Woodville, MS 39669. Mailing Address: P.O. Box 1055, Woodville, MS 39669-1055. Tel.: 601-888-7151. Fax: 601-276-9776.
E-mail: info@historicwoodville.com
Web Site: www.historicwoodville.org
Founded: 1971.
Congressional District: 4
Key Personnel: C.E.O., Ernesto Caldeira; Pres. (V), David Wilkerson.
Personnel Profile: Full-Time Volunteers 2; Part-Time Volunteers 38.
Governing Authority: nonprofit organization. Parent Institution: Woodville Civic Club, Inc. Branch Museum: The African American Museum, Bank St., Woodville, MS. Tax-exempt.
Institution Type/Description: History Museum: housed in a Greek Revival Temple style office & banking house of the West Feliciana Railroad; built in 1834.
Collections: decorative arts from county plantations & homes; photographs; archives; genealogical files; general area history.
Research Fields: Wilkinson County history.
Activities: lectures; loan, temporary & traveling exhibitions.
Publications: annual, The Journal of Wilkinson County History; reprints of 1899 cookbook; book, The Plantation World of Wilkinson County 1792-2012.
Hours & Admission Prices: Mon.-Fri. 10-12 & 2-4, Sat. 10-12.
Attendance: 4,900 (accurate)
Membership: Individual Adult $1; Friend $15; Donor $100; Supporter $101-$499; Patron $500-$999; Benefactor & Founders Plaque $1,000 & up.

Yazoo City

OAKES AFRICAN AMERICAN CULTURAL CENTER, 312 S. Monroe St., Yazoo City, MS 39194. Mailing Address: Yazoo County Fair & Civic League, P.O. Box 1192, Yazoo City, MS 39194. Tel.: 662-746-7984 & 5038. Fax: 662-746-6020.
E-mail: lmteam@bellsouth.net
Founded: 1993.
Congressional District: 2
Personnel Profile: Part-Time Paid 1.
Governing Authority: Parent Institution: Yazoo County Fair & Civic League, Inc. Tax-exempt.
Institution Type/Description: Cultural Center: housed in the former home of businessman & educator, A.J. Oakes. Listed on the National Register of Historic Places.
Collections: prominent African Americans in education, medicine, business, church, agriculture & community life; photographs; paintings; manuscripts.
Hours & Admission Prices: Fri. 9-12 & 1-3; other times by appointment. No charge; donations accepted.
Attendance: 350 (estimated)

SAM OLDEN YAZOO HISTORICAL SOCIETY MUSEUM, 332 N. Main St., Yazoo City, MS 39194-3958. Mailing Address: P.O. Box 575, Yazoo City, MS 39194. Tel.: 662-746-1815; 800-381-0662.
E-mail: yazoo@yazoo.org
Formerly: Yazoo Historical Museum
Congressional District: 2
Governing Authority: private. Tax-exempt.
Institution Type/Description: History Museum.
Collections: exhibits relating to the pre-historic, pioneer, ante-bellum, Victorian & modern time periods.
Hours & Admission Prices: Mon.-Fri. 8:30-4:30, Sat. 12-4. No charge; donations accepted.
Membership: Individual $10; Family $15; Business $25; Patron $50; Sustaining $100.

MISSOURI

(368 listings)

Agency

AGENCY FORD MUSEUM, 11351 Rte. FF, Agency, MO 64401.
Tel.: 816-253-9301.
E-mail: sharonrumpf@yahoo.com
Founded: 2000.
Personnel Profile: Part-Time Volunteers 2.
Institution Type/Description: History Museum.
Collections: agency history from 1836 to present.
Hours & Admission Prices: By appointment only. No charge.
Attendance: 250 (estimated)

Altenburg

LUTHERAN HERITAGE CENTER & MUSEUM OF THE PERRY COUNTY LUTHERAN HISTORICAL SOCIETY, 75
Church St., Altenburg, MO 63732. Mailing Address: P.O. Box 53,
Altenburg, MO 63732-0053. Tel.: 573-824-6070.
Formerly: The Perry County Lutheran Historical Society of Altenburg,
Missouri, Inc.
Founded: 1910.
Congressional District: 8
Key Personnel: Pres. (V), Warren Schmidt; Dir., Carla L. Jordan; Museum
Shop Mgr., Dolores Schmidt.
Personnel Profile: Part-Time Paid 1; Part-Time Volunteers 50; Interns 3.
Governing Authority: denominational group. Parent Institution: Perry County
Lutheran Historical Society. Branch Museums: 1839 Concordia Log Cabin
College; 1845 Big School. Tax-exempt.
Institution Type/Description: Religious Museum: located on the site of the
principal settlement of Saxon immigrants in 1839; historic roots of Lutheran
Church - Missouri Synod. Site of the 1839 Log College Concordia
Seminary.
Collections: religious books; bibles; sermon books; pictures; records; early
history of Saxon immigration, 1839; First By-laws, Trinity Lutheran
Church; Lutheran history; regional cultural German-American history;
period artifacts from 1838-1839.
Research Fields: Mo-Synod Lutheran Church history; Saxon immigration to
East Perry County, MO; genealogical research of synodical history & the
founding families of the 1938-39 immigration; family research.
Facilities: research library of religious books available for use on premises by
appointment.
Activities: lectures; self-guided & guided tours; presentations.
Publications: History of Trinity Lutheran Church; By-Laws of the Trinity
Church, Altenburg, MO; History of Trinity Church, Altenburg, MO; Except
the Corn Die, Koestering; DVD, From Faith to Faith.
Hours & Admission Prices: Daily 10-4. No charge; donations accepted. Closed
New Year's Eve & Day; Easter; Christmas Eve & Day.
Attendance: 6,000 (accurate)

Arcadia

IRON COUNTY HISTORICAL SOCIETY MUSEUM, 630 Hwy.
21, Arcadia, MO 63621. Mailing Address: P.O. Box 183, Ironton,
MO 63650-0183. Tel.: 573-546-3513.
E-mail: ironcohissoc@hotmail.com
Web Site: www.rootsweb.ancestry.com/~moichs
Founded: 1974.
Key Personnel: Pres., John Abney.
Personnel Profile: Part-Time Volunteers 24.
Institution Type/Description: Historical Society Museum.
Collections: local history & culture; photographs.
Hours & Admission Prices: April-Nov. Mon.-Sat. 10-4, Sun. 1-4; Dec.-March
Fri.-Sat. 10-4, Sun. 1-4. No charge; donations accepted.
Membership: Individual $10; Life $200.

Arrow Rock

ARROW ROCK STATE HISTORIC SITE, 39521 Visitor Center
Dr., Arrow Rock, MO 65320. Mailing Address: P.O. Box 1, Arrow
Rock, MO 65320. Tel.: 660-837-3330.
Web Site: mostateparks.com/park/arrow-rock-state-historic-park
Founded: 1923.
Congressional District: 6
Key Personnel: Dir., Michael Dickey.
Governing Authority: state. Parent Institution: Div. of State Parks, Missouri
Dept. of Natural Resources, P.O. Box 176, Jefferson City, MO 65102.
Tax-exempt.

Institution Type/Description: Village Museum Complex: c.1830-1870, town
located near eastern end of Santa Fe Trail.
Collections: period furnishings; paintings; manuscripts; textiles; artifacts;
exhibits interpreting the Boone's Lick region in the early 19th century.
Historic Houses: c.1840 general store; 1846 Dr. Matthew Hall house; 1837
George Caleb Bingham house; 1835 court house; 1830 Academy boarding
house; 1834 Arrow Rock tavern; 1870 jail house.
Research Fields: Missouri history; Santa Fe Trail; westward movement;
19th-century agriculture; 19th-century architecture; Missouri River trade;
historic preservation; African-American, Osage & Missouri Indian culture
& history.
Facilities: restaurant; recreation & camping areas. Gift items & supplies for
sale.
Activities: guided tours; seminars; crafts festivals; camping.
Publications: brochures; George Caleb Bingham House, Arrow Rock Site.
Hours & Admission Prices: Historic Site Grounds: daily 7 a.m. to 10 p.m.
Visitor Center: March-May & Sept.-Nov. daily 10-4; June-Aug. daily 10-5;
Dec.-Feb. Fri.-Sun. 10-4. No charge.
Attendance: 156,000 (accurate)

FRIENDS OF ARROW ROCK, INC., (M), 310 Main St., Arrow
Rock, MO 65320. Mailing Address: Box 124, Arrow Rock, MO
65320-0124. Tel.: 660-837-3231. Fax: 660-837-3230.
E-mail: office@friendsofarrowrock.org
Web Site: www.friendsofarrowrock.org
Founded: 1959.
Congressional District: 6
Key Personnel: Pres. (V), Dr. Thomas B. Hall, III; Admin., Kathy Borgman.
Personnel Profile: Full-Time Paid 2; Part-Time Paid 2.
Governing Authority: nonprofit corporation. Tax-exempt: 501(c)(3).
Institution Type/Description: Historic Buildings.
Collections: period artifacts; crafts. Historic Houses & Buildings: 1844 John P.
Sites gun shop; 1875 John P. Sites house; 1868 I.O.O.F. Hall; Dr. John
Sappington Museum; 1872 Christian Chapel; 1880 African American
properties: Brown's Chapel, Baptist Church; Black Masonic Lodge.
Facilities: theater; state park camping; restaurants; information center; craft
shop. Museum-related items for sale.
Activities: guided tours; children's hands-on tours; lectures; arts festival;
permanent exhibitions; demonstrations. Museum Sponsors: Craft Festival
in October; summer professional theater.
Publications: books, Arrow Rock Cook Book; Arrow Rock: Crossroads of the
Missouri Frontier; Dr. John Sappington of Saline County; Along the Old
Trail; Medicine on the Santa Fe Trail; Over the Santa Fe Trail in 1857.
Hours & Admission Prices: Tours: April-May & Sept.-Oct. Sat.-Sun. & by
appointment; Memorial Day to Labor Day daily 10, 11:30, 1:30 & 3; groups
by appointment. Adults $5, children $1.50; discounts to groups.
Attendance: 4,000 (estimated)
Membership: Supporting $30; Sustaining $50; Sponsor $100; Patron $250;
Benefactor $500; Life $1,000.

Ash Grove

OZARKS AFRO-AMERICAN HERITAGE MUSEUM, 107 W.
Main St., Ash Grove, MO 65604. Mailing Address: P.O. Box 265,
Ash Grove, MO 65604. Tel.: 417-672-3104.
E-mail: magberry@me.com
Web Site: www.oaahm.orrg
Founded: 2003.
Key Personnel: Dir., Fr. Moses Berry.
Personnel Profile: Part-Time Volunteers 1.
Governing Authority: Tax-exempt.
Institution Type/Description: History Museum.
Collections: African American history & culture; photographs; period furnish-
ings; personal artifacts; paintings; clothing.
Hours & Admission Prices: Tues.-Thurs. 9-1, Sat. 10-2; other times by
appointment.
Attendance: 500 (estimated)

Augusta

AUGUSTA HISTORICAL MUSEUM, 275 Webster St., Augusta,
MO 63332. Mailing Address: 498 Shell Rd., Augusta, MO 63332.
Tel.: 636-482-4558; 228-4303 & 4821. Fax: 636-228-4821.
Founded: 1992.
Congressional District: 109
Institution Type/Description: Historic House Museum: built in 1861. Listed on
the National Register of Historic Places.
Collections: local history & culture; period furnishings; personal artifacts;
lanterns; woodworking tools; railroad artifacts; Civil War; WWI; WWII;
municipal records.

Research Fields: 1800s German immigration.
Activities: Museum Sponsors: Open House Days.
Publications: annual activity report.
Hours & Admission Prices: May-Oct. Sun. 1-4; other times by appointment. No charge; donations accepted.
Attendance: 300 (estimated)
Membership: Individual $7; Couple $10.

Ava

DOUGLAS COUNTY HISTORICAL & GENEALOGICAL SO-CIETY, 401 E. Washington Ave., Ava, MO 65608. Mailing Address: P.O. Box 986, Ava, MO 65608-0986. Tel.: 417-683-5799 & 2536.
E-mail: eddie037@centurytel.net
Web Site: www.rootsweb.ancestry.com/~modougla/HistSoc/HISTSOC.htm
Founded: 1984.
Key Personnel: Pres. (V) & Museum Shop Mgr., Sharon Sanders; Sec., Treas. & Research, Pat Carmichael.
Personnel Profile: Part-Time Volunteers 6.
Volunteer Hours: 625
Operating Expenses: 7,881
Operating Income: 6,092
Governing Authority: Tax-exempt.
Institution Type/Description: History Museum: housed in the former Henry Wilson family home.
Collections: local history & culture; period furnishings; personal artifacts; photographs; arrowheads.
Facilities: library.
Publications: Douglas County journals.
Hours & Admission Prices: Summer: Wed.-Fri. 10-12; Winter: Wed.-Fri. 10-12, Sat. 10-2. No charge; donations accepted. &
Attendance: 800 (estimated)
Membership: Annual $20; Lifetime $175.

Bellefontaine Neighbors

GENERAL DANIEL BISSELL HOUSE, 10225 Bellefontaine Road, Bellefontaine Neighbors, MO 63137-2307. Mailing Address: Parks and Recreation, 41 South Central, Clayton, MO 63105. Tel.: 636-532-7298; 314-554-6224. TTY: 314-615-7840.
E-mail: jfoley@stlouisco.com
Web Site: www.stlouisco.com/ParksandRecreation/ParkPages/BissellHouse
Founded: 1960.
Key Personnel: Dir. & Site Supvr., Tom Ott.
Governing Authority: county; nonprofit. Parent Institution: St. Louis County Dept. of Parks & Recreation, 41 S. Central Ave., St. Louis, MO 63105.
Institution Type/Description: Historic House: Federal Style home of Gen. Daniel Bissell, Commander of the Upper Louisiana Territory, Revolutionary War soldier, gen. in the War of 1812.
Collections: Bissell family artifacts; 1812-1850 decorative arts & furnishings.
Research Fields: slavery in St. Louis; history of Fort Belle Fontaine & Bissell family.
Activities: guided tours.
Hours & Admission Prices: Tours: by appointment only. Park: 8 to one half hour past sunset.

Belton

BELTON GRANDVIEW AND KANSAS CITY RAILROAD CO., 502 E. Walnut St., Belton, MO 64012-2516. Tel.: 816-331-0630. Facebook: BGKCRR.
E-mail: info@beltonrailroad.org
Web Site: www.beltonrailroad.org
Institution Type/Description: Railroad Museum.
Collections: display equipment; freight cars; Club Car; instruction car; two static display steam locomotives.
Hours & Admission Prices: Train departures: May-Dec. Sat.-Sun. at 2. Adults $9.50-$14; children 2 & under no charge.

BELTON MUSEUM OF HISTORY, 512 Main St., Belton, MO 64012-2583. Tel.: 816-331-1905 & 322-3977.
Web Site: www.beltonhistoricalsociety.org
Institution Type/Description: History Museum.
Collections: local history & culture; photographs; Dale Carnegie; Carry Nation; Harry Truman.
Hours & Admission Prices: March-Dec. Tues. & Thurs. 1-4, Sat. 10-1. No charge; donations accepted.
Membership: Annual $15; Business $25; Lifetime $150.

Bloomfield

STARS AND STRIPES MUSEUM/LIBRARY, 17377 Stars and Stripes Way, Bloomfield, MO 63825-8487. Mailing Address: P.O. Box 1861, Bloomfield, MO 63825-0463. Tel.: 573-568-2055.
E-mail: stripes@newwavecomm.net
Web Site: starsandstripesmuseumlibrary.org
Key Personnel: Pres., Dr. Joe Baker
Institution Type/Description: Military Museum.
Collections: U.S. Armed Forces military newspapers from the Civil War to present; personal artifacts.
Facilities: Museum-related items for sale.
Hours & Admission Prices: Mon. & Wed.-Fri. 10-4, Sat. 10-2, Sun. 1-4; other times by appointment. Closed New Year's Day; Easter; Thanksgiving; Christmas. &

STODDARD COUNTY MUSEUM, 501 Center St., Bloomfield, MO 63825. Mailing Address: 15248 Palo Verde Lane, Dexter, MO 63841. Tel.: 573-568-2163.
Key Personnel: Pres., Anita Peters
Institution Type/Description: History Museum.
Collections: early farm machinery; 1875 period parlor & kitchen display.
Hours & Admission Prices: May-Oct. fourth Sun. of each month 2-4. No charge; donations accepted.

Blue Springs

THE DILLINGHAM-LEWIS MUSEUM, 101 S.W. 15th St., Blue Springs, MO 64015-3511. Mailing Address: Blue Springs Historical Society, P.O. Box 762, Blue Springs, MO 64013. Tel.: 816-797-4870.
E-mail: mpottereducation@comcast.net
Web Site: www.bluespringshistory.org
Founded: 1976.
Congressional District: 6
Key Personnel: Pres. (V), Mary Potter.
Personnel Profile: Full-Time Volunteers 1; Part-Time Volunteers 35.
Governing Authority: Parent Institution: Blue Springs Historical Society. Tax-exempt.
Institution Type/Description: House Museum: house built in 1906.
Collections: local history & culture; furnishings; photographs. Historic Building: 1879 hotel.
Facilities: heritage garden.
Publications: monthly newsletter.
Hours & Admission Prices: Sun. 1-3 by appointment. No charge; donations accepted. Closed holidays.
Attendance: 890 (estimated)

Bolivar

THE ELLA CAROTHERS DUNNEGAN GALLERY OF ART, 511 N. Pike, Bolivar, MO 65613-1568. Mailing Address: P.O. Box 468, Bolival, MO 65613-0468. Tel.: 417-326-3438.
E-mail: dunnegan@windstream.net
Web Site: www.dunnegangallery.com
Key Personnel: Dir., Jo Roberts
Institution Type/Description: Fine Arts Museum.
Collections: paintings.
Hours & Admission Prices: Mon., Wed. & Fri. 1-4, Sun. call for hours. No charge; donations accepted. Closed bank holidays. &

POLK COUNTY MUSEUM, 201 W. Locust St., Bolivar, MO 65613. Mailing Address: P.O. Box 423, Bolivar, MO 65613. Tel.: 417-326-6850.
E-mail: polkcountymuseum@hotmail.com
Founded: 1982.
Congressional District: 7
Key Personnel: Dir., C.E.O. & Chm. (V), Margaret Vest; C.E.O., Harlene Esther; Chm. (V), Nadine Hendrickson.
Personnel Profile: Part-Time Paid 1; Part-Time Volunteers 10.
Governing Authority: Parent Institution: Historical Society of Polk County. Tax-exempt.
Institution Type/Description: History Museum: housed in the former North Ward Elementary School building; built in 1903.
Collections: local history & culture; period furnishings; personal artifacts; photographs.
Research Fields: probate files.
Publications: newsletter 5 times a year; local history books.

Hours & Admission Prices: mid-May to mid-Sept. Mon.-Sat. 1-4. Adults $2, children 6-12 $1; children 5 & under no charge.
Attendance: 622 (estimated)
Membership: Historical Society: Individual $6-$14; Lifetime $50.

Boonesboro

BOONE'S LICK STATE HISTORIC SITE, State Rd. 187, Boonesboro, MO 65233. Mailing Address: Arrow Rock State Historic Site, P.O. Box 1, Arrow Rock, MO 65320. Tel.: 660-837-3330. Fax: 660-837-3300.
Web Site: www.mostateparks.com/booneslick.htm
Founded: 1960.
Congressional District: 6
Key Personnel: Dir., Michael Dickey.
Personnel Profile: Full-Time Paid 7; Part-Time Paid 1.
Governing Authority: state. Owned & operated by Div. of State Parks, Missouri Dept. of Natural Resources, P.O. Box 176, Jefferson City, MO 65102. Tax-exempt.
Institution Type/Description: Historic Site: c.1805 early salt manufacturing site.
Collections: housed at Arrow Rock State Historic Site: salt kettles, tools, wooden buckets, water wheel drive shaft excavated on site; outdoor interpretive exhibits; salt springs.
Research Fields: pioneer salt industry.
Facilities: interpretive pavilion; 1/4 mile trail; picnic area.
Hours & Admission Prices: Sunrise-sunset. Closed during periods of snow or ice. &
Attendance: 10,000 (estimated)

Branson

AMERICA'S PRESIDENCY MUSEUM AND GALLERY OF AMERICAN HISTORY, Majestic Bldg., Ste. 300, 2849 Gretna Rd., Branson, MO 65616-3387.
MO Tel.: 417-334-8683. Fax: 417-334-4927.
E-mail: amerpresidency@aol.com
Web Site: www.americanpresidentialmuseum.com
Founded: 2005.
Key Personnel: Dir., Stormy Lynn Snow; Chm. (V), James A. Levander; Museum Shop Mgr., Michele Shanan.
Personnel Profile: Full-Time Volunteers 2; Part-Time Paid 2; Part-Time Volunteers 4.
Governing Authority: Parent Institution: National Center for Presidential Studies. Tax-exempt.
Institution Type/Description: American History Museum.
Collections: America's Presidential history; photographs; personal artifacts.
Research Fields: American presidents; American history.
Activities: public programs; outreach; special events.
Publications: quarterly newsletter.
Hours & Admission Prices: Tues.-Sat. 9-5. Adults $10, seniors 55 & over $9; discounts to AAM & ICOM members; children 16 & under no charge. Closed Thanksgiving; Christmas. &
Attendance: 50,000 (estimated)
Membership: $10; $35; $50; $100.

BRANSON'S DINOSAUR MUSEUM, 2020 W. Hwy. 76, Branson, MO 65616. Tel.: 417-335-8739.
E-mail: info@bransondinosaurs.com
Web Site: www.bransondinosaurs.com
Institution Type/Description: Dinosaur Museum.
Collections: over 50 life-size dinosaurs; fossils; hands-on exhibits.
Facilities: theatre.
Activities: educational films; hands-on exhibits; dinosaur fossil dig; coloring stations.
Hours & Admission Prices: Call for hours & admission prices.

COOTER'S PLACE BRANSON, 1819 W. Hwy. 76, Ste. B, Branson, MO 65616-2296. Tel.: 417-336-3494.
Institution Type/Description: History Museum.
Collections: Dukes of Hazzard memorabilia, pictures, & props.
Hours & Admission Prices: Call for hours.

HOLLYWOOD WAX MUSEUM, 3030 W. Hwy. 76, Branson, MO 65616-8312. Tel.: 417-337-8277.
E-mail: chuck@hollywoodwax.com
Web Site: www.hollywoodwaxmuseum.com/branson
Key Personnel: General Mgr., Chuck O'Day
Institution Type/Description: Wax Museum.
Collections: wax figures & scenes; television & movie history.
Activities: birthday parties; school tours.
Hours & Admission Prices: Early Jan. to early March daily 8-8; early March to mid-May Sun.-Thurs. 8am-11pm, Fri.-Sat. 8am-12am; mid-May to early Sept. daily 8am-12am; early Sept. to early Jan. Sun.-Thurs. 8am-11pm, Fri.-Sat. 8am-12am. Adults $17.95; seniors 55+ $15.95; children 4-11 $8.95; discounts to online ticket buyers; children 4 & under no charge. &
Attendance: 100,000

TITANIC MUSEUM ATTRACTION, 3235 76 Country Blvd., (& Hwy. 165), Branson, MO 65616-3551. Mailing Address: 714 State Hwy. 248, Ste. 520, Branson, MO 65616. Tel.: 417-334-9500; 800-381-7670 (Toll Free). Facebook; Titanic Museum Attraction.
E-mail: info@titanicbranson.com
Web Site: www.titanicbranson.com
Formerly: Titanic Museum
Key Personnel: Co-owner & Dir., Mary Kellogg-Joslyn; Owner & C.E.O., John Joslyn
Institution Type/Description: History Museum.
Collections: Titanic history; photographs; lifeboat; period artifacts; clothing.
Hours & Admission Prices: Year round daily 9-5. Closed Christmas. &

VETERANS MEMORIAL MUSEUM, 1250 W. 76 Country Music Blvd., Branson, MO 65616-2211. Tel.: 417-336-2300. Fax: 417-336-2301.
E-mail: info@veteransmemorialbranson.com
Web Site: www.veteransmemorialbranson.com
Founded: 2000.
Institution Type/Description: History Museum.
Collections: American war veterans; personal artifacts; sculptures; memorabilia; photographs.
Hours & Admission Prices: Summer: daily 9-6; Winter: daily 9-5. Adults $14.95, veterans $12.95, students 13-17 $10, children 5-12 $5; children 4 & under no charge. Closed Christmas.

WORLD'S LARGEST TOY MUSEUM AND HAROLD BELL WRIGHT MUSEUM, 3609 W. Hwy. 76, Branson, MO 65616. Tel.: 417-332-1499. Fax: 417-332-0017. Facebook: World's Largest Toy Museum.
E-mail: mail@worldslargesttoymuseum.com
Web Site: www.worldslargesttoymuseum.com
Founded: 2001.
Key Personnel: Dir., Tom Beck
Institution Type/Description: Toy & Harold Bell Wright Historical Museum.
Collections: Toy Museum: toys from 1800s to present including planes; trains; boats; carriages; bears; fishing; toy soldiers; bicycles; lunch boxes; dolls; cast iron & tin wind-ups; pedal cars; cap guns; trucks; motorcycles; automobiles; Charles Dickens' A Christmas Carol window display. Harold Bell Wright Museum: paintings & handwritten copies of Harold Bell Wright's books; Civil War memorabilia; period guns; biography video.
Facilities: Museum-related items for sale.
Hours & Admission Prices: Jan.-March Mon.-Sat. 9-6; April-Dec. Mon.-Sat. 9-8. Toy Museum: adults $9.95, children $7.95. Wright Museum: adults $7.95, children $5.95. Closed Christmas.

Brentwood

BRENTWOOD HISTORICAL SOCIETY - BARLOW HOUSE, 8754 Rosalie Ave., Brentwood, MO 63144-2028. Tel.: 314-961-7948.
E-mail: bwdhistsoc@yahoo.com
Founded: 1985.
Key Personnel: Pres. (V), Dan Fitzgerald.
Personnel Profile: Part-Time Volunteers 6.
Institution Type/Description: Historical Society Museum: house built in 1910.
Collections: local history; photographs; newspaper articles; yearbooks; personal & official records.
Activities: research.
Hours & Admission Prices: Fri. 10am-12pm. No charge; donations accepted. &
Membership: Individual $10; Couple $15.

Bridgeton

PAYNE-GENTRY HOUSE, 4211 Fee Fee Rd., Bridgeton, MO 63044-2217. Tel.: 314-739-5599. Fax: 314-739-2484. Facebook: City of Bridgeton.
E-mail: clay@bridgetonmo.com
Web Site: bridgetonmo.com
Institution Type/Description: Historic House: former home of Elbridge and Mary Elizabeth Payne, built in 1870.
Collections: photographs; furnishings; personal artifacts; doctor's office including tools & equipment; herb garden.
Hours & Admission Prices: April-Nov. 1st Sun. each month 1-4; other times by appointment. Adults $3, seniors $2, children $1; children under 4 no charge. Candlelight Tour: 1st Sat.-Sun. Dec. 4-8pm.

Burfordville

BOLLINGER MILL STATE HISTORIC SITE, 113 Bollinger Mill Rd., Burfordville, MO 63739-9051. Tel.: 573-243-4591. Fax: 573-243-5385.
E-mail: bollinger.mill.state.historic.site@dnr.mo.gov
Web Site: www.mostateparks.com
Founded: 1967.
Congressional District: 10
Key Personnel: Site Admin., Lesley McDaniel; Park Maintenance Worker, Lee Haines; Historic Site Interpreter, Holly Kilpatrick.
Personnel Profile: Full-Time Paid 3; Part-Time Paid 3.
Governing Authority: state. Parent Institution: Missouri State Parks, P.O. Box 176, Jefferson City, MO 65102. Tax-exempt.
Institution Type/Description: Historic Site.
Collections: mill machinery. Historic Buildings: 1867 Bollinger Mill; c.1868 Burfordville covered bridge.
Research Fields: 19th-century milling; George Frederick Bollinger; covered bridges; Southeast Missouri history; Solomon Richard Burford; Sarah Daugherty; Cape County Milling Company.
Activities: self-guided & guided tours.
Publications: site brochures.
Hours & Admission Prices: April-Nov. Mon.-Sat. 10-4, Sun. 12-4; Dec.-March Wed.-Sat. 10-4, Sun. 12-4. Guided Tours: adults $4, children $2.50; discount to groups of 15 or more with advanced reservations; children under 6 no charge. Closed New Year's Day; Easter; Thanksgiving; Christmas. &
Attendance: 41,000 (estimated)

Butler

BATES COUNTY MUSEUM, 802 Elks Dr., Butler, MO 64730. Mailing Address: P.O. Box 164, Butler, MO 64730-0164. Tel.: 660-679-0134. Fax: 660-679-0134.
E-mail: director@batescountymuseum.org
Web Site: www.batescountymuseum.org
Founded: 1961.
Institution Type/Description: History Museum.
Collections: county history & culture; personal artifacts; period furnishings; paintings; photographs. Historic Buildings: 1905 Nyhart School; Wilcox School.
Facilities: Museum-related items for sale.
Activities: special events.
Hours & Admission Prices: April-Oct. Tues.-Sat. 10-4. Closed legal holidays. &

POPLAR HEIGHTS FARM, 208 N. Delaware St., Butler, MO 64730-1504. Tel.: 660-679-0764.
E-mail: info@poplarheightsfarm.org
Web Site: www.poplarheightsfarm.org
Institution Type/Description: Living History Farm.
Collections: local history & culture; period furnishings; paintings; photographs.
Facilities: library.
Activities: educational programs; demonstrations; research.
Hours & Admission Prices: Call for hours.

Camdenton

ORION CREATION XPO CENTER, 1163 S. Business Rte. 5, Camdenton, MO 65020. Mailing Address: P.O. Box 1733, Camdenton, MO 65020-1733. Tel.: 573-346-5516.
E-mail: billm@orioncenter.org
Web Site: www.creationxpo.com
Formerly: Orion DinoSpace Adventure Science Center

Founded: 2000.
Key Personnel: Dir., Chm (V), Museum Shop Mgr., William Mundhausen.
Personnel Profile: Full-Time Volunteers 2; Part-Time Volunteers 10.
Governing Authority: Parent Institution: Orion Center, Inc. Tax-exempt.
Institution Type/Description: Science Center.
Collections: hands-on science & natural history exhibits; space science; fossils; dinosaurs.
Activities: summer science camp; educational programs; special events; Starlab Planetarium programs.
Publications: newsletter, Creation Ammunition.
Hours & Admission Prices: April-May & Sept.-Oct. Fri.-Sat. 10-5; June-Aug. Tues.-Sat. 10-5; other times by appointment. Adults $7.
Attendance: 3,000 (accurate)
Membership: $25 Individual; $40 Grandparents; $50 Family.

Cameron

CAMERON RAILROAD DEPOT MUSEUM, 210 N. Walnut, Cameron, MO 64429. Mailing Address: 819 E. Prospect, Cameron, MO 64429. Tel.: 816-632-7414.
Institution Type/Description: History Museum.
Collections: local history & culture; railroad memorabilia.
Hours & Admission Prices: Call for hours.

Canton

LEWIS COUNTY HISTORICAL SOCIETY MUSEUM, 102 N. 4th St., Canton, MO 63435-1317. Tel.: 573-288-5713.
Key Personnel: Pres. (V), Cynthia S. Barker.
Personnel Profile: Part-Time Volunteers 9.
Governing Authority: Tax-exempt.
Institution Type/Description: Historical Society Museum.
Collections: local history, culture & memorabilia; personal artifacts; period furnishings; photographs; local family genealogy.
Hours & Admission Prices: Tues. & Thurs. 8-3; other times by appointment. No charge; donations accepted. &
Membership: Annual $10.

REMEMBER WHEN TOY MUSEUM, 19481 Rte. B, Canton, MO 63435. Tel.: 573-288-3995.
E-mail: robertwyatt11@yahoo.com
Founded: 1987.
Governing Authority: Parent Institution: Wyatt Industries.
Institution Type/Description: Toy Museum.
Collections: vintage Marx toys.
Publications: newsletter, Marx Toy & Train Collectors Club.
Hours & Admission Prices: By appointment. Adults $5, children $3; discount to school & church groups.
Attendance: 200 (estimated)

Cape Girardeau

CAPE RIVER HERITAGE MUSEUM, 538 Independence St., Cape Girardeau, MO 63703-6227. Tel.: 573-334-0405.
E-mail: crhm538@hotmail.com
Web Site: www.caperiverheritagemuseum.com
Founded: 1980.
Key Personnel: Dir., Bonnie Stepenoff; Pres. (V), Dr. John Holcomb.
Personnel Profile: Part-Time Paid 1; Part-Time Volunteers 10; Interns 1.
Volunteer Hours: 8,500
Operating Income: 10,000
Governing Authority: city. Tax-exempt.
Institution Type/Description: History Museum.
Collections: river-related artifacts; steamboats; period fire engine; photographs; postcards; prints; Oliver family memorabilia & papers; vintage clothing; quilts; musical instruments; uniforms.
Research Fields: local history.
Activities: lectures; book fairs; tag sales; writers' workshops.
Publications: annual, Heartland Heritage.
Hours & Admission Prices: mid-March to mid-Dec. Thurs.-Fri. 12-4, Sat. 11-4; other times by appointment. Adults $3.
Attendance: 2,000 (accurate)
Membership: Individual $20; Family $30; Heritage $106; Exhibit Sponsor $250.

GLENN HOUSE/HISTORICAL ASSOCIATION OF GREATER CAPE GIRARDEAU, 325 S. Spanish, Cape Girardeau, MO 63703-7442. Mailing Address: P.O. Box 1982, Cape Girardeau, MO 63702-1982.
Founded: 1972.
Congressional District: 10
Key Personnel: Pres. (V), Tom Grantham.
Personnel Profile: Part-Time Volunteers 30.
Governing Authority: private; nonprofit organization. Tax-exempt.
Institution Type/Description: Historic House Museum: housed in c.1883 Victorian home.
Collections: structures; furnishings.
Activities: guided tours & special events.
Publications: newsletter, Heritage Review & rackcard.
Hours & Admission Prices: May-Oct. Sat.-Sun. 1-4; Dec. Sat.-Sun. 1-4. Adults $5, children 12 & under $2; discounts to groups; members no charge.
Attendance: 2,000 (estimated)
Membership: Annual $15; Family $25; Corporate $100; Lifetime $1,000.

THE RED HOUSE INTERPRETIVE CENTER, 128 Aquamsi St., Cape Girardeau, MO 63701. Mailing Address: 410 Kiwanis Dr., Cape Girardeau, MO 63701. Tel.: 573-334-0757 & 335-1631.
Key Personnel: Dir., Brenda Schloss
Institution Type/Description: History Museum.
Collections: local history & culture; life of community founder, Louis Lorimier; early 1800s artifacts; Lewis & Clark; period furnishings; trading post; Native American.
Activities: special events.
Hours & Admission Prices: April-Nov. Sat. 10-4; other times by appointment. Adults $3, children $1.

ROSEMARY BERKEL AND HARRY L. CRISP II MUSEUM, (M), 518 S. Fountain, Southeast Missouri State University, Cape Girardeau, MO 63701. Mailing Address: One University Plaza, MS 7875, Cape Girardeau, MO 63701. Tel.: 573-651-2260.
E-mail: museum@semo.edu
Web Site: www.semo.edu/museum
Formerly: Southeast Missouri Regional Museum
Founded: 1976.
Key Personnel: Dir., Peter Nguyen; Cur. Collections, James Phillips; Cur. Education, Ellen Hahs; Administrative Asst., Peggy Haney; Outreach Specialist, Gary Tyler.
Personnel Profile: Full-Time Paid 5; Part-Time Paid 7; Interns 1.
Governing Authority: public university; nonprofit. Parent Institution: Southeast Missouri State University.
Institution Type/Description: University Museum.
Collections: fine art; American military history; regional history; university history; archaeology.
Activities: lectures related to exhibitions.
Publications: exhibition brochures.
Hours & Admission Prices: Summer: Tues.-Fri. 10-4; Winter: Tues.-Fri. 10-5, Sat.-Sun. 1-4. No charge; donations accepted. Closed holidays. &
Attendance: 10,000 (accurate)
Membership: Student $10; Individual $15; Family $25-$99; Benefactor $100-$499; Houck Associate $500-$999; Beckwith Associate $1,000-$1,999; Director's Circle $2,000 & up.

Carrollton

CARROLL COUNTY HISTORICAL SOCIETY, 510 N. Mason, Carrollton, MO 64633-2200. Tel.: 660-542-1511.
Founded: 1971.
Key Personnel: Pres. (V), Lillie Audsley.
Personnel Profile: Part-Time Volunteers 20.
Operating Expenses: 7,200
Operating Income: 4,500
Governing Authority: Tax-exempt.
Institution Type/Description: Historical Society Museum.
Collections: local history & culture; photographs; personal artifacts; period furnishings; arrowheads.
Hours & Admission Prices: First weekend in May through Sept. Sat.-Sun. 1-4; other times by appointment. No charge; donations accepted. &
Membership: Lifetime $10.

Carthage

CARTHAGE CIVIL WAR MUSEUM, 205 Grant St., Carthage, MO 64836. Tel.: 417-237-7060.
Institution Type/Description: History Museum.

Collections: Civil War history; mural; personal artifacts; photographs; African American & Native American artifacts.
Hours & Admission Prices: Call for hours. No charge.

PHELPS HOUSE, 1146 Grand Ave., Carthage, MO 64836-2832. Mailing Address: P.O. Box 375, Carthage, MO 64836-0375. Tel.: 417-358-1776.
Web Site: www.phelpshouse.org
Governing Authority: Parent Institution: Carthage Historic Preservation, Inc.
Institution Type/Description: Historic House.
Collections: period furnishings; photographs.
Hours & Admission Prices: Tours: April-Nov. Wed. 10-4.

POWERS MUSEUM, 1617 Oak St., Carthage, MO 64836. Mailing Address: P.O. Box 593, Carthage, MO 64836-0593. Tel.: 417-237-0456. Facebook: Powers Museum of Carthage.
E-mail: powersmuseum@att.net
Web Site: www.powersmuseum.com
Founded: 1982.
Congressional District: 7
Key Personnel: Dir., Michele Hansford; Pres. (V), Diane Sharits.
Personnel Profile: Full-Time Paid 1; Part-Time Volunteers 20; Interns 3.
Operating Expenses: 85,000
Operating Income: 85,000
Governing Authority: municipal; nonprofit. Parent Institution: City of Carthage. Tax-exempt.
Institution Type/Description: History Museum.
Collections: Carthage history; late 19th- to 20th-century decorative arts; 1870-1960 textiles & costumes; late 19th- to early 20th-century music; photographs; archives; medical equipment & books; family letters, 1880-1981; tri-state transportation history (i.e. Ozark Trails, Jefferson Highway, U.S. 66 & 71).
Major Exhibits: Missouri Dept. of Conservation & Carthage Sportsmen's Protective League (T), 11/14-12/14.
Research Fields: role of women in Carthage & related women's history; early highways; local history; limestone & lead/zinc mining of tri-state district; WWI; WWII to contemporary military action oral histories.
Facilities: 1,500-vol. library; educational facilities; 2,000 sq. ft. exhibit space; archives. Museum-related items for sale.
Activities: lectures; films; organized educational programs for children & adults; loan, temporary & traveling exhibitions; digital exhibits; co-op community events.
Hours & Admission Prices: mid-March to mid-Nov. Tues.-Sat. 10-4:30; mid-Nov. to mid-Dec. Wed.-Sat. 11-4; call to confirm. No charge; donations accepted. &
Attendance: 6,000 (accurate)
Membership: Single $25; Family/Couple $35.

Cassville

BARRY COUNTY MUSEUM, Hwy. 112 S., Cassville, MO 65625. Mailing Address: P.O. Box 338, Cassville, MO 65625-0338. Tel.: 417-847-1640. Fax: 417-847-1641.
E-mail: info@barrycomuseum.org
Web Site: www.barrycomuseum.org
Institution Type/Description: History Museum.
Collections: local history & culture; period furnishings; personal & military artifacts; photographs.
Hours & Admission Prices: Mon.-Sat. 9-5, Sun. 1-5. No charge. Closed most major holidays.

Centralia

CENTRALIA HISTORICAL SOCIETY MUSEUM, 319 E. Sneed St., Centralia, MO 65240-1341. Tel.: 573-682-5711.
Institution Type/Description: Historical Society Museum: housed in the former home of pharmacist, Robert Linwood Hope and his wife Belle Downing Hope; built in 1904.
Collections: local history & culture; Civil War artifacts; period furnishings; personal artifacts; photographs; local industry.
Hours & Admission Prices: Wed. & Sun. 2-4.

Chaffee

CHAFFEE HISTORICAL SOCIETY, 109 S. Main, Chaffee, MO 63740. Mailing Address: P.O. Box 185, Chaffee, MO 63740. Tel.: 573-887-6225.
E-mail: chaffeehistorical@gmail.com
Web Site: www.chaffeehistory.com

Institution Type/Description: Historical Society Museum.
Collections: local history; railroad artifacts; early clothing; newspapers; photographs; scrapbooks; local industry.
Hours & Admission Prices: Call for hours.

Chamois

TOWNLEY HOUSE MUSEUM, Third & Market Sts., Chamois, MO 65024. Mailing Address: P.O. Box 402, Linn, MO 65051-0402. Tel.: 573-897-2932.
E-mail: historic@osageconnect.net
Founded: 1991.
Congressional District: 4
Governing Authority: Parent Institution: Osage County Historical Society. Tax-exempt.
Institution Type/Description: Historical House Museum: built in 1856. Listed on the National Register of Historic Places.
Collections: local history & culture; period furnishings; personal artifacts; photographs.
Hours & Admission Prices: By appointment.
Membership: Individual $20; Joint & Business $25; Supporting $50; Life $500.

Charleston

MISSISSIPPI COUNTY HISTORICAL SOCIETY, 403 N. Main, Charleston, MO 63834-1028. Mailing Address: P.O. Box 312, Charleston, MO 63834-0312. Tel.: 314-683-4348.
Founded: 1966.
Congressional District: 8
Key Personnel: Pres. (V), Tom Graham.
Personnel Profile: Part-Time Volunteers 20.
Governing Authority: society. Tax-exempt.
Institution Type/Description: General Museum.
Collections: archaeology; archives; costumes; military; Indian artifacts; agricultural techniques & tools; documents related to the Jadwin Plan; historic home.
Facilities: Gift items for sale.
Activities: temporary art exhibitions; rental of home for receptions & parties.
Hours & Admission Prices: Tues. 1:30-3:30; other times by appointment only. Adults $3; members no charge.
Attendance: 2,000 (estimated)
Membership: Individual $10; Family $20; Life $100.

Chesterfield

FAUST PARK FOUNDATION, 15189 Olive Blvd., Chesterfield, MO 63017-1805. Tel.: 314-615-8383 & 7373. Fax: 314-615-8326. TDD: 314-615-7840.
E-mail: dwhite@stlouisco.com
Web Site: www.stlouisco.com/parks/carousel/carousel.htm
Formerly: Faust Cultural Heritage Foundation-The St. Louis Carousel
Founded: 1985.
Key Personnel: Chm. (V), Peggy Forrest; Museum Shop Mgr., Phyllis Goldberg.
Personnel Profile: Part-Time Paid 14.
Governing Authority: county. Parent Institution: St. Louis County Dept. of Parks & Recreation. Tax-exempt.
Institution Type/Description: General Museum.
Collections: carousel building for operating 1920s Dentzel carousel.
Research Fields: carousels; Dentzel Co. of Philadelphia.
Facilities: 200-seat gallery; 3,500 sq. ft. exhibit space. Museum-related items for sale.
Activities: arts festivals; rental gallery; festivals; park programs.
Hours & Admission Prices: Exhibits: Tues.-Sun. 10-4. Carousel Ride: $2; children one & under and members no charge. &
Attendance: 87,168 (accurate)
Membership: Senior $60; Family $70; Contributing $125; Sponsor $275; Patron $500.

FAUST PARK-THORNHILL HISTORIC SITE & FAUST HISTORIC VILLAGE, 15185 Olive Blvd., Chesterfield, MO 63017-1805. Tel.: 314-615-8328. Fax: 314-615-8325.
E-mail: jfoley@stlouisco.com
Web Site: www.stlouisco.com/parksandrecreation/parkpages/faust
Founded: 1968.
Key Personnel: Dir. Cultural Site, Jim Foley; Dir. Parks, Lindsey Swanick; Cur. Park, Jesse Francis; Office Coord., Rhonda Swagman.
Personnel Profile: Full-Time Paid 10; Part-Time Paid 5; Part-Time Volunteers 40; Interns 1.

Governing Authority: county. Parent Institution: St. Louis County Parks & Recreation. Tax-exempt.
Institution Type/Description: Historic House & Preservation Project.
Collections: Frederick Bates' personal artifacts, 1777-1825; Bates family library; 19th century period furniture; agricultural & archaeology artifacts; 19th century heritage gardens. Historic Buildings: c.1820 estate of Frederick Bates, second Governor of Missouri; historic structures.
Research Fields: Frederick Bates; early St. Louis and Missouri history; cultural heritage of St. Louis area.
Activities: guided tours; lectures; historic craft festivals; history hayrides; education programs for children & adults; carousel rides; summer concert series. Museum Sponsors: Craft Fair; holiday events.
Publications: calendar of events; self guided tour pamphlets.
Hours & Admission Prices: Thronhill Tours: Thurs.-Fri. by appointment. Adults $4; discounts to groups. Historic Village: June-Aug. last two full weekends 1-5. No charge; donations accepted. Carousel: Tues.-Sun. Rides: $2 per person. &
Attendance: 425,000 (accurate)

＊ **SOPHIA M. SACHS BUTTERFLY HOUSE,** Faust Park, 15193 Olive Blvd., Chesterfield, MO 63017. Tel.: 636-530-0076. Fax: 636-530-1516.
Web Site: www.butterflyhouse.org
Founded: 1996.
Key Personnel: Dir., Victoria Campbell.
Governing Authority: Parent Institution: Missouri Botanical Gardens.
Institution Type/Description: Conservatory.
Collections: butterflies; plants; flowers.
Activities: educational programs; special events; classes; rental facilities.
Hours & Admission Prices: Memorial Day to Labor Day daily 9-5; Sept.-May Tues.-Sun. 9-4. Adults $6, seniors 65 & over $4.50, children 3-12 $4; members and children 2 & under no charge. Closed New Year's Day; Columbus Day; Martin Luther King Jr. Day; Presidents' Day; Thanksgiving; Christmas.

Chillicothe

THE GRAND RIVER HISTORICAL SOCIETY & MUSEUM, 1401 Forest Dr., Chillicothe, MO 64601. Mailing Address: P.O. Box 154, Chillicothe, MO 64601-0154. Tel.: 660-646-3430 & 4323.
E-mail: chillicothemuseum@gmail.com
Web Site: grandriverhistoricalsociety.com
Founded: 1959.
Congressional District: 6
Key Personnel: Dir. & Pres. (V), Marvin Holcer; Treas., Laura O'Donnell; Cur., Pamela Clingerman; Sec., Nancy Hoyt.
Personnel Profile: Full-Time Paid 1.
Governing Authority: society. Tax-exempt 501(c)(3).
Institution Type/Description: Local History Museum.
Collections: period artifacts; local artifacts.
Major Exhibits: Better Than Sliced Bread, 7/7/13-7/7/16.
Research Fields: history & development of rural schools in county & areas of national backgrounds.
Activities: guided tours; lectures; demonstrations; school program; rental facilities.
Publications: quarterly, Harold; books, 100 Years in Livingston County; Rural & Small Town Schools in Livingston County.
Hours & Admission Prices: April-Oct. Sat.-Sun. 1-4; other times by appointment. No charge; donations accepted. &
Attendance: 2,000 (accurate)
Membership: Single $10; Family $15; Lifetime $30.

Clayton

HISTORIC HANLEY HOUSE, 7600 Westmoreland Ave., Clayton, MO 63105-3807. Mailing Address: The Center of Clayton, Parks Dept., 50 Gay Ave., Clayton, MO 63105. Tel.: 314-226-9893. Fax: 314-226-9326.
E-mail: sumlauf@ci.clayton.mo.us
Web Site: www.hanleyhouse.org
Key Personnel: Community Resource Coord., Sarah Umlauf.
Personnel Profile: Full-Time Paid 1; Part-Time Volunteers 10.
Governing Authority: municipal; not-for-profit. Tax-exempt: 501(c)(3).
Institution Type/Description: Historic House.
Collections: furnishings; family photographs & letters.
Activities: docent programs.
Hours & Admission Prices: Sat.-Sun. 12-4; other times by appointment. Adults $3; children under 12 no charge. Closed New Year's Day; Memorial Day; Independence Day; Christmas.

Attendance: 1,000 (estimated)

ST. LOUIS ARTISTS' GUILD, 2 Oak Knoll Park, Clayton, MO 63105-3008. Tel.: 314-727-6266 (gallery); 9599 (office). Fax: 314-727-9190.
E-mail: askus@stlouisartguild.org
Web Site: www.stlouisartistsguild.org
Founded: 1886.
Congressional District: 3
Key Personnel: Gallery Dir., Davide Weaver; Pres. (V), Brad Wastler; Museum Shop Mgr., Karen Roodman.
Personnel Profile: Full-Time Paid 3; Part-Time Paid 3; Part-Time Volunteers 200; Interns 1.
Governing Authority: nonprofit organization. Tax-exempt.
Institution Type/Description: Arts Center.
Collections: paintings; sculpture; graphics; photography; fine crafts.
Activities: lectures; films; gallery talks; concerts; arts festivals; temporary & traveling exhibitions; open competitive & invitational exhibits; classes; workshops.
Publications: quarterly newsletter; annual membership directory.
Hours & Admission Prices: Tues.-Sun. 12-4. No charge; donations accepted. Closed holidays. &
Attendance: 30,000 (estimated)
Membership: Junior $25; Individual $55; Family $75; Supporting $100.

Clinton

DORMAN HOUSE, 302 W. Franklin St., Clinton, MO 64735-2011. Mailing Address: 203 W. Franklin, Clinton, MO 64735. Tel.: 660-885-8414.
E-mail: hcmus1@centurylink.net
Key Personnel: Dir., Brenda Dehn; Pres. (V), Adele Bernard.
Personnel Profile: Full-Time Paid 1; Part-Time Paid 1.
Governing Authority: Parent Institution: Henry County Historical Society.
Institution Type/Description: Historic House Museum: housed in the former home of Judge J.G. Dorman; built in 1852. Listed on the National Register of Historic Places.
Collections: Dorman family history; period furnishings; personal artifacts; photographs.
Activities: guided tours; rental facilities.
Hours & Admission Prices: Call for hours. Adults $3; members no charge.
Membership: Individual $15; Family $25; Life $300; Family Life $500.

HENRY COUNTY MUSEUM AND CULTURAL ARTS CEN-TER, 203 W. Franklin St., Clinton, MO 64735-2008. Tel.: 660-885-8414. Fax: 660-890-2228.
E-mail: hcmus1@centurylink.net
Web Site: henrycountymomuseum.org
Founded: 1974.
Congressional District: 4
Key Personnel: Dir., Brenda Dehn; Pres., Adele Bernard; Museum Shop Mgr., Sarah Campbell; Genealogy Library, Pat Waugh.
Personnel Profile: Full-Time Paid 1; Part-Time Paid 1; Part-Time Volunteers 100.
Governing Authority: society. Parent Institution: Henry County Historical Society. Tax-exempt: 501(c)(3).
Institution Type/Description: Historical Museum: housed in 1886 restored Anheuser-Busch building. Listed on the National Register of Historic Places.
Collections: historic artifacts & furnishings; Louis Freund paintings; genealogical research room; period rooms; Courtney Thomas Collection; late 1800s to early 1900s village including 7 stores. Historic Buildings: 1850s Dog Trot Log House; 1887 Henry County Bank building.
Research Fields: genealogical records.
Facilities: kitchen; 100-seat hospitality room.
Activities: permanent exhibitions.
Publications: brochures; quarterly newsletter.
Hours & Admission Prices: April-Dec. Mon.-Sat. & special occasions 10-4. Adults $5, children 12 & under no charge. Closed Memorial Day; Independence Day; Labor Day; Thanksgiving; Christmas. &
Attendance: 4,093 (accurate)
Membership: Individual $15; Family $25; Sustaining $100; Life $300 Family Life $500.

Cole Camp

COLE CAMP MUSEUM, 108 S. Maple St., Cole Camp, MO 65325-1120. Mailing Address: P.O. Box 22, Cole Camp, MO 65325-0022. Tel.: 660-668-3037.
E-mail: museum@colecampmo.com
Web Site: colecampmo.com
Institution Type/Description: History Museum.
Collections: local history, culture & heritage; photographs; personal artifacts; period furnishings; Native American artifacts; military equipment; minerals; rocks; fossils; trapping industry; early tools.
Hours & Admission Prices: Call for hours.

Columbia

DAVIS ART GALLERY, 1414 E. Walnut St., Columbia, MO 65201. Mailing Address: Campus Box 2012, Stephens College, Columbia, MO 65215. Tel.: 573-876-7255. Fax: 573-876-7248.
E-mail: dscott@stephens.edu
Web Site: www.stephens.edu
Founded: 1962.
Congressional District: 8
Key Personnel: Cur., Dan Scott.
Governing Authority: college. Parent Institution: Stephens College. Tax-exempt.
Institution Type/Description: Art Gallery.
Collections: modern paintings & graphics; Melanesian sculpture.
Activities: formally organized education programs for undergraduate college students; temporary & traveling exhibitions.
Hours & Admission Prices: Mon.-Fri. 10-3; other times by appointment. No charge; donations accepted. Closed school holidays. &
Attendance: 500 (estimated)

GEORGE CALEB BINGHAM GALLERY - UNIVERSITY OF MISSOURI, A126 Fine Arts Center, Dept. of Art, Columbia, MO 65211-6090. Tel.: 573-882-3555. Fax: 573-884-6807.
E-mail: hannahrreeves@gmail.com
Web Site: binghamgallery.missouri.edu
Key Personnel: Dir., Hannah Reeves
Institution Type/Description: Art Gallery.
Collections: works by contemporary artists.
Activities: lectures; special events.
Hours & Admission Prices: Academic Year: Mon.-Fri. 8-5; Summer: Mon.-Fri. 7:30-4.

MASONIC LIBRARY AND MUSEUM, 6033 Masonic Dr., Ste. A, Columbia, MO 65202-6568. Tel.: 800-434-9804; 573-814-4663. Fax: 573-814-4660.
E-mail: bramsey@mohome.org & mphillippe@mohome.org
Web Site: www.mohome.org
Formerly: Masonic Home of Missouri
Founded: 2007.
Congressional District: 9
Key Personnel: Exec. Dir., Barbara Ramsey.
Governing Authority: Parent Institution: Masonic Home of Missouri. Subsidiary Institution: Missouri Masonic Museum. Tax-exempt.
Institution Type/Description: History Museum.
Collections: Masonic history & artifacts; photographs; personal artifacts.
Research Fields: Masonic history.
Hours & Admission Prices: Mon.-Fri. 9-4:30. No charge; donations accepted. &
Attendance: 375 (accurate)

MSA/GPC CRAFT STUDIO, N. 12 Memorial Union, Columbia, MO 65211. Tel.: 573-882-2889.
E-mail: craftstudio@missouri.edu
Formerly: Brady Gallery University of Missouri
Institution Type/Description: Art Gallery.
Collections: works by MU students, local & national artists.
Activities: workshops; temporary exhibitions; educational programs.
Hours & Admission Prices: Call for hours.

MUSEUM OF ANTHROPOLOGY, UNIVERSITY OF MIS-SOURI, 104 Swallow Hall, University of Missouri, Columbia, MO 65211. Tel.: 573-882-3573 & 3764. Fax: 573-884-3627.
E-mail: anthromuseum@missouri.edu
Web Site: anthromuseum.missouri.edu

Founded: 1939.
Congressional District: 9
Key Personnel: Dir., Michael J. O'Brien; Assoc. Cur., Candace A. Sall; Asst. Cur., Jessica Boldt; Asst. Cur., Brandy Tunmire; Mgr. Collections, Audrey Gayou; Administrative Asst., Christine Hudson.
Personnel Profile: Full-Time Paid 5; Part-Time Paid 6; Part-Time Volunteers 4; Interns 2.
Governing Authority: university. Parent Institution: University of Missouri. Tax-exempt: 501(c)(3).
Institution Type/Description: Anthropology Museum.
Collections: prehistoric Missouri archaeological artifacts; ethnographic material from native North American cultures in the Arctic, the Northwest coast, the Southwest, the Basin & the Great Plains; the Grayson archery collection & archery-related material from around the world.
Research Fields: archaeology; ethnography; ethnohistory.
Facilities: library; 22,000 sq. ft. curation center.
Activities: guided tours; lectures; outreach programs; permanent & temporary exhibits.
Publications: Museum Briefs; Monograph Series.
Hours & Admission Prices: Mon.-Fri. 9-4. Scheduled tours available. No charge; donations accepted. Closed university holidays. &
Attendance: 6,000 (accurate)

✱   MUSEUM OF ART AND ARCHAEOLOGY, UNIVERSITY OF MISSOURI, (M), University Ave. & Ninth St., 1 Pickard Hall, Columbia, MO 65211. Tel.: 573-882-3591. Fax: 573-884-4039.
E-mail: museumuser@missouri.edu
Web Site: maa.missouri.edu
Founded: 1957.
Congressional District: 8
Key Personnel: Dir., Alex W. Barker, Ph.D.; Asst. Dir., Coord. Membership, Mktg. & Museum Shop Mgr., Bruce Cox; Pres. (V), Robin LaBrunerie; Assoc. Cur. European & American Art, Mary Pixley; Assoc. Cur. Ancient Art, J. Benton Kidd; Registrar, Jeff Wilcox; Chief Preparator, Barbara Smith; Preparator, Larry Stebbing; Dir. Missouri Folk Arts Program, Lisa Higgins; Specialist, Deborah Bailey; Assoc. Educator, Cathy Callaway.
Personnel Profile: Full-Time Paid 7; Part-Time Paid 7; Part-Time Volunteers 37; Interns 1.
Governing Authority: university. Parent Institution: University of Missouri-Columbia. Tax-exempt: 170(b)(1)(A).
Institution Type/Description: University Art Museum.
Collections: ancient art & archaeology including Egypt, Palestine, Iran, Cyprus, Greece, Etruria, Rome, early Christian & Byzantine art; paintings; Kress study collection; 15th- to 20th-century European & American sculpture, drawings, prints & paintings; African, pre-Columbian, Chinese, Japanese, South & Southeast Asian art & artifacts.
Research Fields: old world archaeology; European & American paintings; sculpture graphics; South & Southeast Asia; Pre-Columbian & African art.
Facilities: 6,000-vol. library of sales & exhibition catalogs, museum bulletins & reference books available for use on premises & upon special request; 116-seat lecture hall. Museum-related items for sale.
Activities: guided tours; gallery talks; formally organized education programs for undergraduate & graduate students affiliated with University of Missouri, Columbia; educational outreach programs; permanent & temporary exhibitions.
Publications: Muse, Annual of the Museum of Art & Archaeology; magazine; Handbook of The Collections; Catalogue of Gandharan Art; British Comic Art, 1730-1830; The Art of the July Monarchy: France 1830 to 1848; Pasture to Polis: Art in the Age of Homer; Commitment: Fatherhood in Black America; The Samuel H. Kress Study Collection at the University of Missouri; New Light on a Dark Age; Corpus Vasorum Antiquorum; The Art of the Book; Testament of Time; Feeling, Thought and Spirit: The Ceramic Work of Glen Lukens; Golden Treasures: The Voyage of a Contemporary Italian Goldsmith in the Classical World.
Hours & Admission Prices: Tues.-Wed. & Fri. 9-4, Thurs. 9-8, Sat.-Sun. 12-4. No charge; donations accepted. Closed national & university holidays; New Year's Eve & Day; Christmas Eve, Day & week. &
Attendance: 33,733 (accurate)
Membership: Student $20; Senior $35; Individual $40; Household $60; Friend $100; Founder $250; Sponsor $500; Fellow $1,000; Patron $5,000.

ROGERS GALLERY - UNIVERSITY OF MISSOURI, 137 Stanley Hall, Columbia, MO 65211-6090. Tel.: 573-882-7224.
Institution Type/Description: Art Gallery.
Collections: works by Architectural Studies faculty, students & related professionals.
Activities: temporary exhibitions.
Hours & Admission Prices: Call for hours.

STATE HISTORICAL SOCIETY OF MISSOURI, Administrative Office, 1020 Lowry St., Columbia, MO 65201-7298. Tel.: 573-882-7083. Fax: 573-884-4950. Facebook: State Historical Society of Missouri.
E-mail: shsofmo@umsystem.edu
Web Site: shs.umsystem.edu
Founded: 1898.
Congressional District: 4
Key Personnel: Exec. Dir., Gary R. Kremer; Pres. (V), Stephen N. Limbaugh, Jr.; Museum Shop Mgr., Jeneva Pace.
Personnel Profile: Full-Time Paid 33; Part-Time Paid 5; Part-Time Volunteers 20; Interns 5.
Governing Authority: state. Tax-exempt.
Institution Type/Description: Art Museum.
Collections: paintings; sculpture; drawings; prints; photographs; maps; cartoons; manuscripts; Missouri newspapers, 1808-present.
Research Fields: Missouri, Trans-Alleghenian & Far Western history; American Indians.
Facilities: library of books, pamphlets & periodicals available for use in reading room.
Activities: guided tours; rotating exhibitions; lectures; research.
Publications: quarterly magazine, Missouri Historical Review; books & booklets on Missouri history; But I Forget That I Am A Painter and Not A Politician: The Letters of George Caleb Bingham; Fred Geary, Missouri Master of the Wood Cut; Thomas Hart Benton: Artist, Writer, Intellectual; Marking Missouri History; quarterly newsletter, Missouri Times; The Civil War in Missouri: Essays from the Missouri Historical Review, 1906-2006; Kansas City, America's Crossroads: Essays from the Missouri Historical Review, 1906-2006; Filling Leisure Hours: Essays from the Missouri Historical Review, 1906-2006; St. Louis from Village to Metropolis: Essays from the Missouri Historical Review, 1906-2006; A Rough Business: Fighting the Civil War in Missouri; Longer Than a Man's Lifetime in Missouri.
Hours & Admission Prices: Society & Art Gallery: Tues.-Fri. 8-4:45, Sat. 8-3:30. No charge; donations accepted. Closed national holidays; holiday weekends. &
Attendance: 4,250 (accurate)
Membership: Student & Teacher $25; Individual $30; International $40; Household $50; Contributing $75; Supporting $125; Sustaining $250; Patron $500; Corporate & Institutional $500; George Caleb Bingham Society $1,000.

WALTERS-BOONE COUNTY HISTORICAL MUSEUM AND VISITORS CENTER, 3801 Ponderosa St., Columbia, MO 65201-5460. Tel.: 573-443-8936. Fax: 573-875-5268. Facebook: Boone County Museum Galleries.
E-mail: boonecountymuseum@yahoo.com
Web Site: boonehistory.org
Key Personnel: Office Mgr., Joyce Klein; Curatorial Specialist, Chelsea Bacon
Institution Type/Description: History Museum.
Collections: county heritage & history; early pioneers; photographs.
Research Fields: genealogy; local history.
Facilities: genealogy library. Museum-related items for sale.
Activities: speaker's series. Museum Sponsors: Boone Piano Concerts; Holiday House Tour; Heritage Festival.
Publications: County Lines.
Hours & Admission Prices: Thurs.-Fri. & Sun. 12:30-4:30, Sat. 9:30-4:30. Guided tours $5, general admission $2; members no charge. &
Attendance: 24,000 (estimated)
Membership: Individual $30; Family $35; Supporting $60; Establishing Patron $100; Premium $200; Corporate $250; Individual Lifetime $300; Silver & Family Lifetime $500; Gold $750; Platinum $1,000.

Concordia

CONCORDIA MUSEUM, 802 S. Gordon St., Concordia, MO 64020-9363. Tel.: 660-463-2105.
E-mail: concordiamuseum@yahoo.com
Founded: 2003.
Key Personnel: Chm. (V), Donald Dittmer; Pres. (V), Deanna Rehmsmeyer; Museum Shop Mgr., Virginia Schnakenberg.
Personnel Profile: Part-Time Volunteers 14.
Governing Authority: Tax-exempt.
Institution Type/Description: History Museum.
Collections: local history & culture; period artifacts.
Hours & Admission Prices: Feb.-Oct. by appointment. No charge; donations accepted. &
Attendance: 180 (estimated)

Crestwood

SAPPINGTON HOUSE MUSEUM, 1015 S. Sappington Rd., Crestwood, MO 63126-1004. Tel.: 314-822-8171 (museum) & 9469 (library). Fax: 314-729-4794.
Founded: 1967.
Congressional District: 3
Key Personnel: Library Chm., Carol Bell; Chm. (V), Enid Barnes; Pres. (V), Kathy Saur; Museum Shop Mgr., Marilyn Fleming.
Personnel Profile: Full-Time Volunteers 1; Part-Time Paid 6; Part-Time Volunteers 100.
Governing Authority: municipal; not-for-profit. Affiliated with City of Crestwood, 1 Detjen Dr., Crestwood, MO. Tax-exempt: 501(c)(3).
Institution Type/Description: Historic House: Thomas Sappington House, 1808 Federal architecture.
Collections: decorative arts; Federal period furnishings.
Research Fields: decorative arts; American history; Missouri & local history.
Facilities: 2,000-vol. library of books on Americana history & decorative arts available for research; restaurant. Museum-related items for sale.
Activities: guided tours; speakers' bureau. Museum Sponsors: spring & fall exhibits; Christmas in 1808.
Publications: quarterly newsletter, Sappington House Volunteer Newsletter.
Hours & Admission Prices: Feb.-Dec. Wed.-Fri. 11-2, Sat. by appointment; last tour at 1:45. Adults $3, children 6-12 $1; members, school groups & scouts with prior reservations no charge. Closed New Year's Eve & Day; Good Friday & day after; Memorial Day weekend; Independence Day; Labor Day weekend; Thanksgiving Eve & Day; Christmas Eve & Day; Sat. preceding holidays.
Attendance: 1,250 (estimated)
Membership: Basic $10; Sponsoring $25; Patron $50.

Cuba

CRAWFORD COUNTY HISTORICAL SOCIETY & MUSEUM, 308 N. Smith St., Cuba, MO 65453-1166. Tel.: 573-885-6099.
Web Site: www.crawfordmomuseum.org
Founded: 1968.
Congressional District: 8
Institution Type/Description: Historical Society Museum.
Collections: local history & culture; photographs; period furnishings & clothing; personal artifacts.
Activities: special events; speakers. Annual Event: CubaFest in October.
Hours & Admission Prices: Tues.-Sat. 10-3, Sun. 12-3; tours by appointment.

Davisville

DILLARD MILL STATE HISTORIC SITE, 142 Dillard Mill Rd., Davisville, MO 65456-4014. Tel.: 573-244-3120.
E-mail: dillard.mill.state.historic.site@dnr.mo.gov
Web Site: www.mostateparks.com/dillardmill.htm
Founded: 1975.
Personnel Profile: Full-Time Paid 1; Part-Time Paid 2; Part-Time Volunteers 1.
Governing Authority: Parent Institution: Missouri Dept. of Natural Resources, Div. of State Parks.
Institution Type/Description: Historic Site.
Collections: local history & culture; photographs; water-powered gristmill.
Facilities: picnic sites; nature trail.
Hours & Admission Prices: March-Nov. Mon.-Sat. 10-4, Sun. 12-5; Dec.-Feb. Thurs.-Sat. 10-4, Sun. 12-4.
Attendance: 29,000 (accurate)

Defiance

DANIEL BOONE HOME AND BOONESFIELD VILLAGE, LINDENWOOD UNIVERSITY, 1868 Highway F, Defiance, MO 63341-1908. Tel.: 636-798-2005 & 2903. Fax: 636-798-2914.
E-mail: boonehome@lindenwood.edu
Web Site: www.lindenwood.edu/boone
Founded: 1803.
Congressional District: 9
Key Personnel: Dir. Operations, Pam Jensen.
Personnel Profile: Full-Time Paid 7; Full-Time Volunteers 4; Part-Time Paid 4; Part-Time Volunteers 200; Interns 2.
Governing Authority: Parent Institution: Lindenwood University.
Institution Type/Description: Historic House: 1803-1810 designed & built by Daniel Boone and patterned after his father's home in Pennsylvania. Historic village era 1801-1850; guided tour.
Collections: authentic documents, furniture and artifacts of the Boone family.
Facilities: Museum-related items for sale.
Activities: guided tours; permanent exhibitions; chapel weddings & christenings.

Publications: brochure, True Brief History of Daniel Boone; newspaper, Boonesfield Gazette.
Hours & Admission Prices: mid-April to Oct. daily 9-6; Nov. to mid-April daily 9-5. Tours: 9:30-4. Museum Shop: 9-5. Boone Home: adults $7, senior citizens $6, children $4; discounts to AAM members. Boone Home & Boonesfield Village: adults $12, senior citizens $10, children $6; discounts to AAM members. Closed New Year's Day; Easter; Thanksgiving; Christmas Eve & Day. ♿
Attendance: 50,000 (accurate)
Membership: Friend $15; Good Friend $30; Best Friend $45; Special Friend $55.

Diamond

GEORGE WASHINGTON CARVER NATIONAL MONUMENT, 5646 Carver Rd., Diamond, MO 64840-8314. Tel.: 417-325-4151. Fax: 417-325-4231.
E-mail: gwca_interpretation@nps.gov
Web Site: www.nps.gov/gwca
Founded: 1943.
Congressional District: 7
Key Personnel: Supt., James Heaney.
Governing Authority: federal. Parent Institution: National Park Service. Subsidiary: George W. Carver Birthplace Assn. Tax-exempt: 501(c)(3).
Institution Type/Description: National Monument.
Collections: historic materials relating to George Washington Carver; manuscript collections; herbarium. Historic House: c.1881 Moses Carver house.
Research Fields: history.
Facilities: 300-vol. library of biographies of George Washington Carver & local history books available for use by special arrangement; 3/4-mile nature trail; science discovery center. Books & postcards for sale.
Activities: guided tours; permanent & temporary exhibitions; film on the life of George Washington Carver; summer programs; Carver Science Discovery Center: living history programs; guided tours.
Publications: bulletin, Carver of Tuskegee.
Hours & Admission Prices: Daily 9-5. No charge; donations accepted. Closed New Year's Day; Thanksgiving; Christmas. ♿
Attendance: 50,000
Membership: George Washington Carver Birthplace Association: Student & Senior Citizen $15; Individual $35; Family $65.

WORLD'S LARGEST SMALL ELECTRIC APPLIANCE MUSEUM, 51 Hwy. 59, Diamond, MO 64840. Tel.: 417-793-7936.
E-mail: toastghost1@yahoo.com
Web Site: smallelectricappliancemuseum.com
Founded: 2008.
Key Personnel: Dir., C.EO. & Museum Shop Mgr., Richard Larison.
Personnel Profile: Full-Time Volunteers 1.
Institution Type/Description: Appliance Museum.
Collections: over 6,000 appliances including 700 toasters, percolator coffee pots, waffle irons, hot plates, blenders, mixers, razors, hair dryers, popcorn poppers, fans.
Hours & Admission Prices: Mon.-Sat. 9-6, Sun. 11-6; other times by appointment. No charge; donations accepted. ♿
Attendance: 200 (estimated)

Doniphan

CURRENT RIVER HERITAGE MUSEUM, 101 Washington St., Doniphan, MO 63935. Tel.: 573-996-5298.
E-mail: lynnmaples@doniphanmissouri.org
Web Site: www.doniphanmissouri.org
Institution Type/Description: History Museum.
Collections: local history & heritage; period furnishings; personal artifacts; photographs; early map & cameras.
Facilities: Museum-related items for sale.
Hours & Admission Prices: Mon.-Fri. 9-4, Sat. 9-12. Closed holidays.

Eagle Rock

PROMISED LAND ZOO, 32297 Hwy. 86, Eagle Rock, MO 65641-7108. Tel.: 417-271-3324. Fax: 417-271-3012.
E-mail: plzoo1@yahoo.com
Web Site: www.plzoo.com
Founded: 1999.
Institution Type/Description: Zoo.
Collections: zebras; dromedary camels; red kangaroos; ring-tailed lemurs; aoudad; miniature donkeys; pot-bellied pigs; Reeves Muntjac; American bison; Scimitar-horned Oryx; Nilgai antelope; Eland antelope; Fallow deer; American elk; emu; caracals; servals; large reptiles; tropical birds.

Hours & Admission Prices: Daily 9-7. Adults $14.95, children 2-12 and seniors 65 & over $9.95; foster families with ID no charge. &
Attendance: 15,000 (estimated)

East Prairie

TOWOSAHGY STATE HISTORIC SITE, East Prairie, MO 63845. Mailing Address: c/o Hunter-Dawson State Historic Site, P.O. Box 308, New Madrid, MO 63869-0308. Tel.: 573-748-5340.
E-mail: moparks@dnr.mo.gov
Web Site: www.mostateparks.com/towosahgy.htm
Congressional District: 7
Governing Authority: Parent Institution: Missouri State Parks. Tax-exempt.
Institution Type/Description: Archaeological Site.
Collections: local history & culture; ceremonial mounds.
Facilities: kiosk; trails.
Hours & Admission Prices: Call for hours. No charge.

Edina

KNOX COUNTY HISTORICAL SOCIETY MUSEUM, Court House, 107 N. 4th St., Edina, MO 63537. Mailing Address: 309 E. Marion, P.O. Box 75, Edina, MO 63537-1248. Tel.: 660-397-2349. Fax: 660-397-3331.
Founded: 1967.
Congressional District: 9
Key Personnel: Pres., Brent Karhoff.
Governing Authority: society. Tax-exempt.
Institution Type/Description: Local History Museum.
Collections: local history & culture; period artifacts; photographs.
Research Fields: genealogy.
Facilities: 100-vol. library of newspapers and books available for use on premises.
Activities: guided tours.
Hours & Admission Prices: By appointment. No charge; donations accepted. Closed legal holidays. &
Membership: Individual $2.

Ellsinore

ELLSINORE PIONEER MUSEUM, 11 S. Herren Ave., Ellsinore, MO 63937. Mailing Address: P.O. Box 74, Ellsinore, MO 63937-0074. Tel.: 573-322-5297.
E-mail: pioneermuseum@mail.com
Institution Type/Description: History Museum.
Collections: local history & culture; photographs.
Facilities: Museum-related items for sale.
Hours & Admission Prices: May-Aug. Wed. & Sat. 10-5.
Attendance: 300

Excelsior Springs

EXCELSIOR SPRINGS MUSEUM & ARCHIVES, 101 E. Broadway, Excelsior Springs, MO 64024-2513. Mailing Address: P.O. Box 144, Excelsior Springs, MO 64024-0144. Tel.: 816-630-0101.
E-mail: emuseum101@hotmail.com
Web Site: www.exsmo.com/museum
Formerly: Excelsior Springs Historical Museum
Founded: 1964.
Congressional District: 38
Key Personnel: Pres. (V) & Cur., Eric Woods
Governing Authority: Tax-exempt.
Institution Type/Description: History Museum.
Collections: Excelsior Springs history; personal artifacts; period furnishings; genealogy; newspapers, 1890s to present.
Major Exhibits: The Greatest Generation: Our Lives, Our Stories (T), 9/14.
Research Fields: genealogy.
Facilities: archives.
Publications: 3 times a year newsletter, Phunn.
Hours & Admission Prices: Tues.-Sat. 11-5; other times by appointment. Adults $2; children under 5 & members no charge. Closed New Year's Day; Memorial Day; Independence Day; Thanksgiving; Christmas. &
Attendance: 6,500 (accurate)
Membership: Regular $15; Family $35; Business $50; Patron $125; Lifetime $500.

Fayette

THE ASHBY-HODGE GALLERY OF AMERICAN ART, CENTRAL METHODIST UNIVERSITY, 411 Central Methodist Square, Fayette, MO 65248-0009. Tel.: 660-248-6324 & 6304 (office). Fax: 660-248-2622.
E-mail: dgebhardt@centralmethodist.edu
Web Site: www.centralmethodist.edu
Founded: 1993.
Congressional District: 6
Key Personnel: Collection Supvr., Joseph E. Geist, Ph.D.; Chief Docent, Virginia Monroe.
Personnel Profile: Full-Time Paid 1; Part-Time Paid 2; Part-Time Volunteers 35.
Governing Authority: private college; nonprofit. Parent Institution: Central Methodist University. Tax-exempt: 501(c)(3).
Institution Type/Description: Art Museum.
Collections: Midwestern regionalist artists including Thomas Hart Benton, Fred Shane, Grant Wood, Charles Banks Wilson, Robert MacDonald Graham, Jr., Emile Gruppe, Aaron Bohrod & Birger Sandzen; lithographs; oils; acrylic; drawings; watercolors & bronzes.
Facilities: 500-vol. library of art books & periodicals available to the public; 5,000 sq. ft. exhibit space; 52 acre campus.
Activities: docent program; guided tours; loan, temporary & traveling exhibitions. Annual Events: GALA celebration of opening of Gallery in October.
Publications: catalogue, Bingham in the Boonslick: A Bicentennial Celebration 1811-2011.
Hours & Admission Prices: Tues.-Thurs. 1:30-4:30; Special Exhibits: Sun. 1:30-4:30; groups by special arrangement. No charge. &
Attendance: 5,000 (accurate)
Membership: Angels $50-$99; Archangels $100-$199; Principalities $200-$299; Dominations $300-$499; Cherubim $500-$999; Seraphim $1,000.

THE STEPHENS MUSEUM, Central Methodist University, T. Berry Smith Hall, Fayette, MO 65248. Tel.: 660-248-6678 (office) & 6334 (museum). Fax: 660-248-2622.
E-mail: dlmorris@centralmethodist.edu
Web Site: cmu.edu
Founded: 1879.
Congressional District: 6
Key Personnel: Dir., Dana L. Morris, Ph.D.
Personnel Profile: Full-Time Paid 1; Part-Time Paid 3; Part-Time Volunteers 1.
Volunteer Hours: 120
Operating Income: 1,000
Governing Authority: college. Parent Institution: Central Methodist University. Tax-exempt.
Institution Type/Description: College Museum: housed in 1896 Romanesque educational building, T. Berry Smith Hall.
Collections: natural history; history & art; bird collection of 274 mounted specimens, 423 skins & 163 eggs; shells; mammals; geological displays; history of Boonslick Region; articles of the Methodist Church in Missouri; folk art; original tombstones of Daniel & Rebecca Boone; five original paint brushes of George Caleb Bingham; Lithograph John Wesley; artifacts commemorating the coronation of Czar Nicholas II, Russia, 1896; bird specimen now in extinction-passenger pigeon, Carolina parakeet; American Indian artifacts (arrowheads etc.).
Major Exhibits: Project Passenger Pigeon, 9/14.
Facilities: classrooms.
Activities: guided tours; lectures; temporary exhibitions; outreach programs (themed).
Hours & Admission Prices: Academic Year: Tues.-Thurs. 1-4; other times by appointment. Closed college vacations. No charge; donations accepted.
Attendance: 1,000 (estimated)

Fenton

FENTON HISTORY MUSEUM, 1 Church St., Fenton, MO 63026. Tel.: 636-326-0808.
Web Site: www.fentonhistory.com
Founded: 1993.
Governing Authority: Parent Institution: Fenton Historical Society.
Institution Type/Description: Historical Society Museum: housed in the former home of Frank M. Swantner; built in 1906.
Collections: local history & culture; period furnishings; personal artifacts; photographs.
Hours & Admission Prices: 2nd & last Sat. each month 12-3; other times by appointment.

Florida

MARK TWAIN BIRTHPLACE STATE HISTORIC SITE, 37352 Shrine Rd., Florida, MO 65283-2127. Tel.: 573-565-3449. Fax: 573-565-3718.

E-mail: mark.twain.birthplace.state.historic.site@dnr.mo.gov
Web Site: www.mostateparks.com
Founded: 1960.
Congressional District: 9
Key Personnel: Facility Mgr., Dustin Webb.
Personnel Profile: Full-Time Paid 4; Part-Time Paid 3.
Governing Authority: state. Owned & operated by Div. of State Parks, Missouri Dept. of Natural Resources, P.O. Box 176, Jefferson City, MO 65102. Tax-exempt.
Institution Type/Description: Historic House: 1835 Samuel L. Clemens birthplace.
Collections: period furnishings; paintings; manuscripts; Samuel Clemens memorabilia.
Research Fields: local history; Mark Twain.
Facilities: 800-vol. library of books, pamphlets, letters & other material relating to life & writings of Mark Twain available on premises; reading room.
Activities: guided tours; lectures; films; gallery talks; formally organized education programs for children & adults.
Publications: brochure; quarterly newsletter, Mark Twain Research Foundation Twainian.
Hours & Admission Prices: April-Oct. daily 10-5; Nov.-March Fri.-Sun. 10-4. No charge; donations accepted. Closed New Year's Day; Thanksgiving; Christmas. &
Attendance: 17,277 (accurate)

Florissant

FLORISSANT VALLEY HISTORICAL SOCIETY, 1896 S. Florissant Rd., Florissant, MO 63031. Mailing Address: P.O. Box 298, Florissant, MO 63032-0298. Tel.: 314-524-1100, 839-3626 & 921-5563.

E-mail: fredmary2@aol.com
Founded: 1958.
Congressional District: 9
Key Personnel: Pres., Joseph McDavid; Sec., Mary Kay Gladbach.
Personnel Profile: Part-Time Volunteers 25.
Governing Authority: board of directors; society; nonprofit organization. Tax-exempt: 501(c)(3).
Institution Type/Description: Historic House: 1790 Taille de Noyer log cabin; house enlarged to include 3 floors.
Collections: furnishings; personal artifacts; recreational artifacts; local library.
Facilities: 100-vol. library books on local history. Gift items for sale.
Activities: guided tours.
Publications: magazine, Florissant Valley Quarterly.
Hours & Admission Prices: March.-Dec. Sun. 1-4; group tours by appointment. Adults $2, children $1. Closed New Year's Day; Thanksgiving; Christmas.
Attendance: 2,500 (estimated)
Membership: Individual $15; Life $100.

FRIENDS OF OLD ST. FERDINAND, Friends of Old St. Ferdinand, Inc., One Rue St. Francois, Florissant, MO 63031. Tel.: 314-837-2110.

E-mail: oldstferdinandshrine@gmail.com
Web Site: www.oldstferdinandshrine.com
Formerly: Old St. Ferdinand's Shrine
Founded: 1958.
Congressional District: 2
Key Personnel: Pres., Douglas Riser; Sec. & Membership Chm., Geri Debo.
Personnel Profile: Full-Time Volunteers 4; Part-Time Volunteers 20.
Governing Authority: nonprofit corporation. Parent Institution: Friends of Old St. Ferdinand, Inc. Tax-exempt: 501(c)(3).
Institution Type/Description: History Museum: housed in historic site of church complex & community center from 1789-1957.
Collections: nun dolls in pre-Vatican II habits; church paraphernalia; early vestments; statues; pews; altars; 1789 tabernacle of St. Ferdinand Church; funeral cape worn by Father DeSmet; paintings; foot pump organ; 1855-1860 nun & student letters; ice skates; rings for hanging lamps; 1825 newspaper clippings; 1600-1955 multicultural artifacts; survival instruments & tools from 1700s to 1955; 1690 paintings from the House of Rueben; 1500s sanctuary lamp; wax figure of St. Valentine from 270 AD; artifacts from an ongoing archaeological dig; Mastadon bone found in Coldwater Creek 25-35,000 BC

Facilities: archival items. Books, maps & other museum-related items for sale.
Activities: self-guided & group tours by appointment.
Publications: newsletter, brochures.
Hours & Admission Prices: April-Nov. Sat. 9am to noon, Sun. 1-4; Dec.-March Sat. 9am to noon; other times by appointment. Tours: $3. Closed New Year's Day; Easter; Independence Day; Thanksgiving; Christmas. &
Attendance: 15,000 (estimated)
Membership: Individual $25; Family $50; Business & Institution $100.

Fort Leonard Wood

U.S. ARMY CHEMICAL CORPS MUSEUM, 495 S. Dakota Ave., Bldg. 1607, Fort Leonard Wood, MO 65473-8851. Tel.: 573-596-4221.

E-mail: usarmy.leonardwood.chemical-schl.mbx.museum@mail.mil
Web Site: www.wood.army.mil/ccmuseum/ccmuseum/
Key Personnel: Museum Specialist, Cynthia L. Riley
Institution Type/Description: Military Museum.
Collections: military history; photographs; documents.
Activities: educational programs.
Hours & Admission Prices: Mon.-Fri. 8-4, Sat. 10-4. Closed Federal holidays.

U.S. ARMY ENGINEER MUSEUM, 495 S. Dakota Ave., Bldg. 1607, Fort Leonard Wood, MO 65473-5165. Tel.: 573-596-0780.

Web Site: www.wood.army.mil/museum
Founded: 1972.
Congressional District: 4
Key Personnel: Dir., Frank McGrane; Museum Shop Mgr., Glenn A. Stines.
Personnel Profile: Full-Time Paid 5; Full-Time Volunteers 2; Part-Time Volunteers 4; Interns 1.
Governing Authority: federal. Parent Institution: U.S. Army, Fort Leonard Wood. Tax-exempt.
Institution Type/Description: Military History & Engineer Museum: including a restored WWII company compound consisting of 14 buildings on 50-acre site.
Collections: life & career of Gen. Leonard Wood; Fort Leonard Wood history; Army engineer training; firearms; uniforms; insignia; engineer vehicles; library archives; photographs.
Research Fields: military engineering history & evaluation of military training from 1940-present; POW's in the United States; minority military experiences.
Facilities: library pertaining to military history reference works; auditorium; educational facilities.
Activities: guided tours; lectures; films; broadcast & cable programs; organized education programs for children, adults & college students; docent program; loan & temporary exhibitions.
Publications: brochure, History of Fort Leonard Wood.
Hours & Admission Prices: Mon.-Fri. 8-4, Sat. 10-4. No charge. Closed federal holidays. &
Attendance: 150,000 (estimated)

U.S. ARMY MILITARY POLICE CORPS MUSEUM, (M), 495 S. Dakota Ave., Bldg. 1607, Fort Leonard Wood, MO 65473-8851. Tel.: 573-596-0604. Fax: 573-596-0603.

E-mail: leon.usampsmuseum@conus.army.mil
Web Site: www.wood.army.mil/usamps/organizations/dpo/museum.html
Formerly: Military Police Corps Regimental Museum
Founded: 1960.
Congressional District: 3
Key Personnel: Dir. & Cur., Jim Rogers.
Personnel Profile: Full-Time Paid 3.
Governing Authority: nonprofit organization. Parent Institution: U.S. Army MP School. Tax-exempt.
Institution Type/Description: Military Museum.
Collections: items pertaining to the history of the Military Police Corps; uniforms; weapons; flags; insignia; badges; military artifacts from allied countries.
Research Fields: military police history.
Facilities: 10,000 sq. ft. exhibit space. Museum-related items for sale.
Activities: guided tours; lectures; permanent exhibitions.
Publications: Military Police Corps History; various historical vignettes.
Hours & Admission Prices: Mon.-Fri. 8-4, Sat. 10-4. No charge; donations accepted. Closed holidays. &
Attendance: 58,000 (accurate)

Fortescue

HOLT COUNTY HISTORICAL SOCIETY, 115 Ada St., Fortescue, MO 64437. Mailing Address: P.O. Box 55, Mount City, MO 64470-0055. Tel.: 660-442-5949.
Institution Type/Description: Historical Society Museum: housed in a former Methodist Church.
Collections: local history & culture; military & school records; family histories; period artifacts; newspapers.
Facilities: library.
Activities: research.
Hours & Admission Prices: Call for hours.

Fredericktown

HISTORIC MADISON COUNTY MUSEUM, 122 N. Main St., Fredericktown, MO 63645. Tel.: 573-783-2722 & 4085.
E-mail: raskaggs_2000@yahoo.com
Web Site: Facebook: Historic Madison County
Key Personnel: Dir., Terry Wernecker; Chm. (V), Gary Lee.
Personnel Profile: Part-Time Volunteers 15.
Governing Authority: Parent Institution: Madison County Commission; Subsidiary Institution: Battle Re-enactment Group. Tax-exempt.
Institution Type/Description: History Museum.
Collections: local history & culture; period furnishings; photographs; personal artifacts.
Research Fields: genealogy; mining; Civil War.
Publications: monthly newsletter; Madison County history, veterans history, pictorial history & cemetery books.
Hours & Admission Prices: Tues. 1-4. No charge; donations accepted.
Attendance: 1,500 (estimated)
Membership: Annual $12.

Fulton

AUTO WORLD MUSEUM, 200 Peacock Dr., Fulton, MO 65251. Mailing Address: P.O. Box 128, Fulton, MO 65251-0128. Tel.: 573-642-2080.
E-mail: staff@autoworldmuseum.com
Web Site: autoworldmuseum.com
Founded: 1996.
Institution Type/Description: Automobile Museum.
Collections: automobile history & memorabilia; over 80 vehicles including Studebakers, Ford Quadracycle, 1903 English Hummerette & solar-powered cars.
Facilities: Museum-related items for sale.
Activities: videos.
Hours & Admission Prices: April-Dec. daily 9-5; other times by appointment. Adults $8, seniors $7, children under 13 $5.

KINGDOM OF CALLAWAY HISTORICAL SOCIETY, 513 Court St., Fulton, MO 65251-1901. Mailing Address: P.O. Box 6073, Fulton, MO 65251-6073. Tel.: 573-642-0570.
E-mail: museum@kchsoc.org
Web Site: kchsoc.org
Founded: 1960.
Institution Type/Description: Historical Society Museum.
Collections: local history & culture; period furnishings; photographs; personal artifacts.
Hours & Admission Prices: Tues.-Fri. 10-4. No charge; donations accepted.

NATIONAL CHURCHILL MUSEUM, (M), Westminster College, 501 Westminster Ave., Fulton, MO 65251-1299. Tel.: 573-592-5396. Fax: 573-592-5222.
E-mail: kit.freudenberg@churchillmemorial.org
Web Site: www.churchillmemorial.org
Formerly: Winston Churchill Memorial and Library
Founded: 1962.
Congressional District: 9
Key Personnel: Exec. Dir., Dr. Rob Havers; Asst. Dir. & Devel. Dir., Kit Freudenberg; Education & Public Programs Coord., Mandy Plybon; Mktg. Specialist, Caroline Slavin; Admin. Asst., Meda Young; Museum Store & Front Desk Mgr., Becky McCue.
Governing Authority: college. Affiliated with Westminster College, Westminster Ave. Tax-exempt.
Institution Type/Description: Historic Building: relocated & reconstructed Church of St. Mary the Virgin, Aldermanbury designed by Christopher Wren (original site: London, England). Historic Site: site on which Sir Winston Churchill delivered the Sinews of Peace Speech March 5, 1946.
Collections: Churchill memorabilia; Wren architecture; rare map collection; paintings; graphics; sculpture; numismatic; manuscripts; philatelic; motion pictures; photographs; sound recordings.
Research Fields: life of Sir Winston Churchill; World War II; Anglo-American relations; life & works of Sir Christopher Wren.
Facilities: research & microfilm libraries; theater. Museum-related items for sale.
Activities: guided tours; concerts; lectures; permanent & temporary exhibitions; religious services; festivals; self-guided tours; interactive exhibits.
Publications: periodic newsletter; visitor guide, The Words and the Man; John Findley Green Lecture Series; Crosby Kemper Lectures.
Hours & Admission Prices: Daily 10-4:30. Adults $7.50, senior citizens 65 & up and active military $6.50, youth 12-18 & college students $5.50, children 6-11 $4.50; discounts to AAM & AAA members; members, Westminster College Students, Calloway County Schools and children 5 & under no charge. Closed New Year's Day; Thanksgiving; Christmas. &
Attendance: 25,000
Membership: Subaltern $50; Member of Parliament $100; Junior Minister $250; First Lord of the Admiralty $500; Chancellor of the Exchequer $1,000; Prime Minister $2,500; Knight of the Garter $5,000.

Glasgow

GLASGOW MUSEUM, 381 County Rd. 220, Glasgow, MO 65254-9755. Mailing Address: 100 Market St., City Hall, Glasgow, MO 65254. Tel.: 660-338-9949.
Web Site: www.glasgowmo.com/aboutus_2.html
Institution Type/Description: History Museum: housed in a Presbyterian Church.
Collections: church & local history; Civil War artifacts; photographs.
Activities: tours; church service.
Hours & Admission Prices: Mid-May to mid-Oct. call for hours. No charge.
Attendance: 300 (estimated)

Glencoe

WABASH FRISCO AND PACIFIC ASSOCIATION, INC. - THE UNCOMMON CARRIER/MERAMEC RIVER - PALISADES ROUTE, 199 Grand Ave., Glencoe, MO 63038. Mailing Address: 1569 Ville Angela Lane, Hazelwood, MO 63042-1630. Tel.: 636-587-3538; 314-977-3568. Facebook: WFPRR.
E-mail: it1569djn@charter.net
Web Site: www.wfprr.com
Founded: 1939.
Congressional District: 2
Key Personnel: Pres., Stephen F. Marx; Treas., Michael E. Lorance; Sec., Thomas W. Heil.
Personnel Profile: Part-Time Volunteers 170.
Governing Authority: nonprofit organization. Tax-exempt: 501(c)(3).
Institution Type/Description: Mini-Steam Tourist Railway Museum: located on the site of the original mainline right of way of the Pacific Railroad; building began westward from St. Louis, MO in 1852.
Collections: ten 12-inch gauge steam locomotives, 1907-2005; 31 riding cars including flat & gondola cars; both 1925 Canadian Pacific (Treasure Island Railway) locomotives used in the Sesquicentennial in Philadelphia, PA; 4 diesel outline locomotives, 1945-2001; 2 cabooses; track maintenance cars; two 40 ft. highway trailers to pull equipment out of Meramec River flooding; lathes; milling machines; Bridgeport Mill; round house with turntable & eight tracks for storage of locomotives; car barn to store cars with 11 tracks; photographs; 1946 open top aluminum Hopper car built at Sedalia, MO shops of Mopac RR; 1950 steel Gondola car built at Springfield, MO shops of the SL-SF (Frisco) RR.
Research Fields: maintaining steam locomotives & riding cars; utilizing scientific evaluations to keep locomotives operational.
Activities: two-mile round trip train ride.
Publications: annual newsletter, Whiffenpoof; e-newsletter, Sunday Coupler.
Hours & Admission Prices: May-Oct. Sun. 11-4:15. Trains depart every 20 minutes. Admission: $4 per person; discount to groups; members & children under 3 no charge.
Attendance: 14,000
Membership: Student $10; Individual $25; Family $35.

Goldman

SANDY CREEK COVERED BRIDGE STATE HISTORIC SITE, Old Lemay Ferry Rd., Goldman, MO 63050. Mailing Address: 1050 Charles J. Becker Dr., Imperial, MO 63052-3524. Tel.: 636-464-2976. Fax: 636-464-3768.
Web Site: www.mostateparks.com

Founded: 1967.
Congressional District: 8
Key Personnel: DNR Dir., Bill Bryan; District Supvr., Delecia Huitt.
Personnel Profile: Part-Time Paid 1.
Governing Authority: state. Parent Institution: Missouri Dept. of Natural Resources, P.O. Box 176, Jefferson City, MO 65102. Subsidiary Institution: Div. of State Parks. Tax-exempt.
Institution Type/Description: Historic Site: 1872 covered bridge.
Collections: bridge.
Research Fields: covered bridges.
Facilities: picnic area; interpretive shelter.
Publications: brochure.
Hours & Admission Prices: Daily 8 am to half hour past sunset. No charge. &

Attendance: 110,000 (estimated)

Grandview

HARRY S TRUMAN NATIONAL HISTORIC SITE - TRUMAN FARM HOME, 12301 Blue Ridge Blvd., Grandview, MO 64030-1159. Mailing Address: 223 N. Main St., Independence, MO 64050-2804. Tel.: 816-254-2720 & 9929. Fax: 816-254-4491.
E-mail: larry_villalva@nps.gov
Web Site: www.nps.gov/hstr/
Founded: 1994.
Congressional District: 5
Personnel Profile: Full-Time Paid 17; Part-Time Paid 6; Part-Time Volunteers 14.
Governing Authority: federal. Parent Institution: U.S. Dept. of the Interior. Subsidiary Institution: National Park Service. Tax-exempt.
Institution Type/Description: Historic Site Museum: 1906-1917 home of former President Harry S Truman; family farm which he operated with his father.
Collections: period furniture & furnishings; outbuildings.
Research Fields: life of Harry S Truman emphasizing pre-1917; local agriculture & rural life.
Facilities: over 5 acres.
Activities: tours of farm home in summer.
Publications: bulletin; tour brochure.
Hours & Admission Prices: Farm: daily. Home: Memorial Day to Sept. Fri.-Sun. 9:30-4. Adults 16 & over $4. Closed New Year's Day; Thanksgiving; Christmas. &
Attendance: 4,992 (accurate)

Greenfield

DADE COUNTY MUSEUM, 429 W. Water St., Greenfield, MO 65661. Tel.: 417-637-0258.
E-mail: brenda@dadecountymohistoricalsociety.com
Web Site: dadecountymohistoricalsociety.com
Institution Type/Description: History Museum.
Collections: local history & culture; period furnishings; personal artifacts; photographs; early clothing.
Hours & Admission Prices: By appointment.

Hamilton

J.C. PENNEY MUSEUM AND BOYHOOD HOME, 312 N. Davis St., Hamilton, MO 64644-1145. Tel.: 816-583-2168. Fax: 816-583-4929.
Founded: 1974.
Key Personnel: Chm. (V) & Pres. (V), Dean Hales
Governing Authority: Tax-exempt.
Institution Type/Description: Historic House and Museum: former boyhood home of the entrepreneur J.C. Penney.
Collections: artifacts & memorabilia pertaining to J.C. Penney's life.
Hours & Admission Prices: Call for hours. No charge; donations accepted.
Attendance: 1,000 (estimated)

Hannibal

BECKY THATCHER HOME, 211 Hill St., Hannibal, MO 63401-3315. Tel.: 573-221-0822.
Governing Authority: Parent Institution: Mark Twain Home Foundation.
Institution Type/Description: Historic House Museum: housed in the former home of Laura Hawkins, Mark Twain's childhood sweetheart and inspiration for her character in his writings.
Collections: local history; period furnishings; photographs.
Facilities: Museum-related items for sale.
Hours & Admission Prices: Call for hours.

MARK TWAIN BOYHOOD HOME & MUSEUM, 120 N. Main St., Hannibal, MO 63401-3537. Tel.: 573-221-9010. Fax: 573-221-7975.
E-mail: henry.sweets@marktwainmuseum.org
Web Site: marktwainmuseum.org
Founded: 1936.
Congressional District: 9
Key Personnel: Exec. Dir., Henry H. Sweets, III; Pres. (V), David Mobley; Office & Gift Shop Mgr., Dena Ellis.
Personnel Profile: Full-Time Paid 4; Part-Time Paid 17; Part-Time Volunteers 35; Interns 3.
Governing Authority: nonprofit. Mark Twain Home Foundation. Tax-exempt.
Institution Type/Description: Historic House: 1844 Mark Twain's boyhood home.
Collections: Mark Twain letters, photographs, foreign language editions, 1st editions; Mark Twain popular culture; historic restorations; American history & literature. Historic Buildings: 1839 Pilaster House; 1839 Grant's Drugstore; 1840 John Marshall Clemens' Justice of the Peace Office; pre-1853 Becky Thatcher House.
Research Fields: Mark Twain; Hannibal.
Facilities: 3,530-vol. library of books & numerous publications pertaining to Mark Twain.
Activities: art exhibitions; lectures; educational programs; performing arts; teachers workshop.
Publications: bimonthly, The Fence Painter.
Hours & Admission Prices: Daily 9-5. Adults $11, seniors $9, children 6-17 $6; children 5 & under, members and Hannibal residents no charge. Closed New Year's Day; Easter; Thanksgiving; Christmas Eve & Day. &
Attendance: 45,000 (accurate)
Membership: Newsletter $15; Individual $25; Family $50; $100; $250; $500; $1,000; $2,500.

MOLLY BROWN BIRTHPLACE & MUSEUM, Hannibal Convention & Visitors Center, 505 N. Third St., Hannibal, MO 63401-3303. Tel.: 573-221-2477.
E-mail: vguide@visithannibal.com
Web Site: visitMollyBrown.com
Institution Type/Description: History Museum.
Collections: Molly Brown's life & family history; period furnishings; Titanic memorabilia.
Hours & Admission Prices: May Sat.-Sun. 10-4; Memorial Day to Labor Day daily 10-4. Adults $4, children $3; discounts to groups of 10 or more.

ROCKCLIFFE MANSION, 1000 Bird St., Hannibal, MO 63401-3436. Tel.: 573-221-4140; 877-423-4140 (Toll Free).
Web Site: www.rockcliffemansion.com
Institution Type/Description: Historic House.
Collections: period furnishings; photographs; personal artifacts.
Hours & Admission Prices: Daily 10-4. Adults $15. Closed Thanksgiving; Christmas.

TOM SAWYER DIORAMAS MUSEUM, 323 N. Main St., Hannibal, MO 63401-3540. Tel.: 573-221-3525.
Institution Type/Description: History Museum.
Collections: 16 hand-carved miniature scenes highlighting The Adventures of Tom Sawyer.
Hours & Admission Prices: By appointment.

Harrisonville

BURNT DISTRICT MUSEUM & ARCHIVES, 400 E. Mechanic, Ste. 203, Harrisonville, MO 64701. Tel.: 816-380-4396.
E-mail: cchsmo@gmail.com
Personnel Profile: Part-Time Paid 5; Part-Time Volunteers 6.
Institution Type/Description: History Museum.
Collections: local history; Battle of Morristown diorama; photographs; period artifacts; marriage licenses; obituaries.
Publications: quarterly newsletter.
Hours & Admission Prices: Mon.-Fri. 10-3:30; other times by appointment. No charge; donations accepted.
Attendance: 1,000 (estimated)
Membership: Youth $5; Single $20; Family $25; Corporate $40.

Hazelwood

FIRST DUE FIRE MUSEUM, St. Louis Mills Mall, 5555 St. Louis Mills Blvd., Hazelwood, MO 63042. Tel.: 314-227-5911.
Web Site: www.firstduefiremuseum.com

Institution Type/Description: Fire Fighting History Museum.
Collections: fire fighting history & equipment; photographs; uniforms; 1820 hose cart; breathing apparatus; helmets; hose.
Activities: fire safety education; hands-on stations.
Hours & Admission Prices: Fri.-Sat. 12-8, Sun. 12-6.

Hermann

DEUTSCHHEIM STATE HISTORIC SITE, 109 W. Second St., Hermann, MO 65041-1045. Tel.: 573-486-2200. Fax: 573-486-2249.
E-mail: deutschheim.state.historic.site@dnr.mo.gov
Web Site: www.mostateparks.com/deutschheim.htm
Founded: 1979.
Congressional District: 8
Key Personnel: Dir., Site Admin. & Museum Shop Mgr., Cynthia Browne.
Personnel Profile: Full-Time Paid 2; Part-Time Paid 4; Part-Time Volunteers 15.
Governing Authority: state. Affiliated with Div. of State Parks, Missouri Dept. of Natural Resources, P.O. Box 176, Jefferson City, MO 65102. Tax-exempt.
Institution Type/Description: German Heritage Museum.
Collections: period furnishings & decorative arts; furniture; books; household items & agricultural tools of 19th-century German immigrants in Missouri; 19th century gardening. Historic Houses: 1840 Pommer Gentner House; 1842 Strehly House; 1857 Winery; 1883 Barn.
Research Fields: German American daily life & folk ways; German immigration & settlement; decorative arts; Missouri in focus; German architecture; German gardens & culture.
Facilities: gardens. Museum-related items for sale.
Activities: guided tours; historic preservation activities; changing exhibits; shows; special events. Annual Events: Garden Tour in June; Weehnachtsfest - Christmas Celebration in December.
Publications: Missouri Germans & Slavery; A Midwest German Christmas (Cookbook); journal, Der Maibaum; 1852 News & Voices.
Hours & Admission Prices: Visitor Center: April-Oct. daily 10-4; Nov.-March Thurs.-Sun. 10-4. Tours: 10, 12:30 & 2:30. Adults $4, students $2.50, children 6-12 $1.50. Closed New Year's Day; Easter; Thanksgiving; Christmas.
Attendance: 6,000 (estimated)
Membership: Journal, Student & Senior Citizen $15; Friend $25; Family $35; Curator's Club $50; Director's Club $100.

HISTORIC HERMANN INC. - MUSEUM AT THE GERMAN SCHOOL, 312 Schiller St., Hermann, MO 65041-1154. Mailing Address: P.O. Box 105, Hermann, MO 65041-0105. Tel.: 573-486-2017.
E-mail: joyandcarol@centurytel.net
Web Site: www.historichermann.com
Founded: 1956.
Congressional District: 110
Key Personnel: Pres. (V), Steve Mueller; Sec., Carol Kallmeyer; Treas., Elaine Lalk.
Personnel Profile: Part-Time Volunteers 30.
Governing Authority: society; nonprofit organization. Parent Institution: Historic Hermann, Inc. Tax-exempt: 501(c)(3).
Institution Type/Description: Historic Museum: housed in 1871 German school building.
Collections: wine making & history of wineries; textiles; marine; Indian artifacts; restored loom; photographs river room; children's room; farm tools; kitchen utensils, spinning wheels; guns; old Bibles; books; restored 1886 pump organ; 6-piece bedroom set early 1800s; immigrant's trunk; steamboat models & equipment; early dolls & doll houses; heritage room; river room.
Research Fields: German-American history.
Facilities: 200-vol. library of German history & religion books; children's museum.
Activities: guided tours; guest lectures.
Publications: The German Settlement; Guardian.
Hours & Admission Prices: April-Oct. Thurs.-Tues. 10-4, Sun. 12-4. Adults $5, children 6-18 $3; discount to groups; members & children under 6 no charge. &

Attendance: 5,000 (accurate)
Membership: Individual $10; Couple $15; Business or Organization $20.

WHITE HOUSE HOTEL 1868, 232 Wharf St., Hermann, MO 65041. Mailing Address: P.O. Box 1, Williamsburg, MO 63388-0001. Tel.: 573-486-3200. Facebook: White House Hotel 1868.
E-mail: vickie@whitehousehotel1868.com
Web Site: www.whitehousehotel1868.com
Personnel Profile: Full-Time Paid 2; Part-Time Paid 2; Part-Time Volunteers 34.
Institution Type/Description: Living History Museum: housed in a former hotel; built in 1868.
Collections: local history & culture; period furnishings; personal artifacts; photographs.
Activities: guided tours.
Hours & Admission Prices: Call for hours. Adults $10.

Hermitage

JOHN SIDDLES WILLIAMS HOUSE - HICKORY COUNTY HISTORICAL SOCIETY, Museum St., Hermitage, MO 65668. Mailing Address: P.O. Box 248, Hermitage, MO 65668. Tel.: 417-722-4403.
Founded: 1980.
Congressional District: 4
Key Personnel: Pres. (V), Mackey E. Snyder.
Personnel Profile: Part-Time Volunteers 8.
Governing Authority: Tax-exempt.
Institution Type/Description: Historical Society Museum: listed on the National Register of Historic Places.
Collections: local history & culture; period furnishings; personal artifacts; photographs.
Activities: open house; fundraisers; community events.
Hours & Admission Prices: By appointment. No charge; donations accepted. &

Attendance: 120 (estimated)
Membership: Individual $5; Life $100.

Higginsville

CONFEDERATE MEMORIAL STATE HISTORIC SITE, 211 W. First St., Higginsville, MO 64037-8158. Tel.: 660-584-2853. Fax: 660-584-5134.
E-mail: confederate.memorial.state.historic.site@dnr.mo.gov
Web Site: www.mostateparks.com
Formerly: Confederate Home of Missouri
Founded: 1925.
Congressional District: 5
Key Personnel: Site Admin., Janae Fuller.
Personnel Profile: Full-Time Paid 5; Part-Time Paid 1; Part-Time Volunteers 1.
Governing Authority: state. Parent Institution: Missouri Dept. of Natural Resources, P.O. Box 176, Jefferson City, MO 65102. Subsidiary Institution: Div. of State Parks. Tax-exempt.
Institution Type/Description: Historic Site: Chapel, cemetery, family cottage & park are remnants of the Confederate Home of Missouri; opened in 1891.
Collections: documents; photographs; confederate home archives; early cemetery; monuments. Historic Building: chapel.
Research Fields: confederate homes; confederate cause in Missouri; late 19th & early 20th century architecture; social & cultural history; Civil War; personal & unit histories.
Activities: special events. Annual Event: Confederate Memorial Day in June; Summer Concert; Winter Concert.
Publications: quarterly CMFA newsletter.
Hours & Admission Prices: Tours: Thurs.-Sat. 9-4, Sun. 12-5. No charge; donations accepted. &
Attendance: 90,000 (estimated)

HARVEY J. HIGGINS HISTORICAL SOCIETY, 2113 S. Main St., Higginsville, MO 64037. Mailing Address: 1704 N. Main St., Higginsville, MO 64037. Tel.: 660-584-6474 & 3232. Facebook: Harvey Higgins Historical Museum.
Web Site: www.visitmo.com
Founded: 1992.
Congressional District: 4
Key Personnel: Pres. (V), Arlene Long; Dir., C.E.O. & Museum Shop Mgr., Loberta Runge.
Personnel Profile: Part-Time Volunteers 2.
Volunteer Hours: 300
Governing Authority: Tax-exempt.
Institution Type/Description: Historical Society Museum: housed the former Higginsville Depot; built in 1889. Listed on the National Register of Historic Places.
Collections: local history, culture artifacts; railroad memorabilia; Illinois Central Railroad caboose; personal artifacts.
Hours & Admission Prices: May-Oct. Wed.-Sat. 9-12; other times by appointment. No charge; donations accepted. &

Attendance: 100 (estimated)
Membership: Individual $5.

Imperial

MASTODON STATE HISTORIC SITE, 1050 Charles J. Becker Dr., Imperial, MO 63052-3524. Tel.: 636-464-2976. Fax: 636-464-3768.
E-mail: mastodon.state.historic.site@dnr.mo.gov
Web Site: www.mostateparks.com
Founded: 1976.
Congressional District: 3
Personnel Profile: Full-Time Paid 4; Part-Time Paid 4; Part-Time Volunteers 10; Interns 3.
Governing Authority: Parent Institution: Missouri Department of Natural Resources. Subsidiary Institution: Missouri Division of State Parks. Tax-exempt.
Institution Type/Description: History Museum & Historic Site: listed on the National Register of Historic Places.
Collections: natural & cultural history; archaeology; paleontology; Mastodon bones.
Facilities: 431-acres; picnic area; trails; youth campground.
Hours & Admission Prices: Site: daily 8am to sunset. Museum: March 15-Nov. 14 Mon.-Sat. 9-4:30, Sun. 12-4:30; Nov. 15-March 14 Mon. & Fri.-Sat. 11-4, Sun. 12-4. Adults $4; discounts to groups of 15 or more. Closed New Year's Day; Easter; Thanksgiving; Christmas. &
Attendance: 22,898 (accurate)

Independence

THE BINGHAM-WAGGONER ESTATE, 313 W. Pacific, Independence, MO 64051. Mailing Address: P.O. Box 1163, Independence, MO 64051-0663. Tel.: 816-461-3491. Fax: 816-461-1540.
E-mail: bwestate@att.net
Web Site: www.bwestate.org
Institution Type/Description: Historic House: former home of Missouri Civil War artist George Caleb Bingham.
Collections: furnishings; paintings.
Hours & Admission Prices: April-Oct. Mon.-Sat. 10-4, Sun. 1-4. Adults $6, senior citizens $5, children & students $3; members no charge.
Membership: Individual $15; Couple $25; Family $30; Life $1,000.

CHICAGO & ALTON RAILROAD DEPOT, 318 W. Pacific Ave., Independence, MO 64050. Tel.: 816-325-7955.
E-mail: info@chicagoalton1879depot.org
Web Site: chicagoalton1879depot.org
Founded: 1996.
Congressional District: 5
Key Personnel: Pres. (V), John S. Thornton.
Personnel Profile: Part-Time Volunteers 26.
Governing Authority: city. Tax-exempt.
Institution Type/Description: Historic Building: housed in a former railroad depot; built in 1879.
Collections: depot history; period furnishings; personal artifacts; C & A railroad artifacts.
Hours & Admission Prices: April-Oct. Sun. 12:30-4:30, Mon. & Thurs.-Sat. 9:30-4:30. No charge; donations accepted.
Attendance: 2,366 (accurate)
Membership: Individual $15; Couple $25; Family $30; Patron $100; Life $1,000.

CHILDREN'S PEACE PAVILION, (M), 1001 W. Walnut, Independence, MO 64050-3562. Tel.: 816-521-3033. Facebook; Children's Peace Pavilion.
E-mail: rbelcher@kidpeace.org
Founded: 1995.
Key Personnel: Dir., Rebecca Newcom Belcher, Ph.D.
Personnel Profile: Full-Time Paid 1; Part-Time Volunteers 10; Interns 1.
Governing Authority: nonprofit. Parent Institution: Peace Pathways. Tax-exempt.
Institution Type/Description: Children's Museum.
Collections: interactive exhibits.
Facilities: Museum-related items for sale.
Activities: special events; Girl Scout programs.
Publications: newsletter; Facebook page.
Hours & Admission Prices: Tues., Thurs. & Sat. by appointment only, Wed. & Fri. noon-4:30 no appointment needed. Scout Tours: Sat. 10-3. No charge; donations accepted. Closed New Year's Eve & Day; Memorial Day; Independence Day; Thanksgiving; Christmas Eve, Day & week. &

Attendance: 3,608 (accurate)

COMMUNITY OF CHRIST, The Temple, 1001 W. Walnut, Independence, MO 64050-3562. Tel.: 816-833-1000, ext. 2030. Fax: 816-521-3089.
E-mail: cloving@cofchrist.org
Web Site: www.cofchrist.org
Formerly: Reorganized Church of Jesus Christ of Latter Day Saints
Founded: 1830.
Congressional District: 5
Key Personnel: Admin. Asst., Joy Goodwin.
Personnel Profile: Part-Time Paid 1; Part-Time Volunteers 17.
Governing Authority: church. Supported by historic sites at Independence, MO; Kirtland, OH; Nauvoo, IL; Lamoni, IA; Beloit, WI; Far West, MO. Tax-exempt: 501(c)(3).
Institution Type/Description: Religious History Museum.
Collections: artifacts; manuscript & pictorial collections; archival materials related to the history of the church.
Research Fields: Latter Day Saint (Mormon) history.
Facilities: library; archives; reading room.
Activities: guided tours; films; permanent & temporary exhibitions.
Publications: Herald; The Restoration Witness.
Hours & Admission Prices: April-Oct. Mon.-Sat. 9-12 & 1:30-4:30, Sun. 1:30-4:30; Nov.-March Mon.-Fri. 9-12 & 1:30-4:30, Sat. 10-12 & 1:30-4:30. Prayer for Peace: daily 1pm. No charge. &
Attendance: 12,000 (accurate)

1859 JAIL, MARSHAL'S HOME & MUSEUM, 217 N. Main St., Independence, MO 64050-2804. Mailing Address: Jackson County Historical Society, P.O. Box 4241, Independence, MO 64051-4241. Tel.: 816-252-1892.
E-mail: info@jchs.org
Web Site: www.jchs.org
Founded: 1959.
Congressional District: 5
Key Personnel: Exec. Dir., Steve Noll; C.E.O., Brad Pace; Pres. (V), Ben Mann.
Personnel Profile: Full-Time Paid 1; Part-Time Paid 1; Part-Time Volunteers 33; Interns 1.
Governing Authority: nonprofit. Parent Institution: Jackson County Historical Society. Tax-exempt: 501(c)(3).
Institution Type/Description: History Museum: adjoining 1859 restored Jackson County Jail and federal style Marshal's House & office.
Collections: period furnishings; local history items; 12-cell county jail and dwelling for jailor; weapons & law enforcement tools; furnishings & household accessories; Civil War & Reconstruction period artifacts; Border War & Jesse James Brothers memorabilia. Historic Building: c.1870 one-room schoolhouse.
Research Fields: local history.
Activities: self-guided tours; permanent & changing exhibitions. Museum Sponsors: week of living history in one-room schoolhouse in spring.
Publications: magazine.
Hours & Admission Prices: April-Oct. Mon.-Sat. 10-4, Sun. 1-4. Adults $5, senior citizens 62 & over $4.50, children 5-15 $2; children under 5 & members no charge. Closed New Year's Day; Thanksgiving; Christmas.
Attendance: 11,000 (accurate)
Membership: Individual $35; Family $50; Donor $50; Patron $75; Patron Family $100; Guild $250.

HARRY S. TRUMAN LIBRARY & MUSEUM, 500 West U.S. Hwy. 24, Independence, MO 64050-1798. Tel.: 816-268-8200; 800-833-1225. Fax: 816-268-8295.
E-mail: truman.library@nara.gov
Web Site: www.trumanlibrary.org
Founded: 1957.
Congressional District: 5
Key Personnel: Dir., Michael J. Devine; Deputy Dir., Amy Williams; Museum Shop Mgr., Donna Denslow.
Personnel Profile: Full-Time Paid 22; Part-Time Paid 5; Part-Time Volunteers 100; Interns 15.
Governing Authority: federal. Parent Institution: National Archives & Records Administration, Washington, DC 20408. Subsidiary Institution: Office of Presidential Libraries. Tax-exempt: 170(b)(1)(A).
Institution Type/Description: Presidential Library.
Collections: personal papers; government records; still photographs; motion picture films; audio & video tapes; sound recordings; head of state gifts; gifts from private citizens; political campaign items; personal & family memorabilia; Truman Oval Office replica & the White House gallery.

Major Exhibits: The American President: Photographs from the Archives of the Associated Press, 11/24/13-2/2/14; Spies, Lies, and Paranoia: Americans in Fear, 3/15/14-10/26/14; Read My Pins: The Madeline Albright Collection, 11/30/14-2/22/15.
Research Fields: life, times, career & presidential administration of President Truman.
Facilities: 240-seat auditorium; meeting rooms; research room; museum galleries. Museum-related items for sale.
Activities: guided tours; lectures; films; permanent, temporary & traveling exhibitions; organized education programs for children, adults, undergraduate & graduate students.
Publications: general information brochure; booklet.
Hours & Admission Prices: Mon.-Sat. 9-5, Sun. 12-5. Adults $8, senior citizens 65 & up $7, children 6-15 $3, bus groups of 15 or more with reservation $5.75; discounts to AAM members; Honorary Fellows and children 5 & under no charge. Closed New Year's Day; Thanksgiving; Christmas. &
Attendance: 90,000 (accurate)
Membership: Basic $35; Family $50; Associate $120; Diplomat $250; Ambassador $500; Benefactor $750; Presidential $1000.

HARRY S TRUMAN NATIONAL HISTORIC SITE - TRUMAN HOME, 223 N. Main St., Independence, MO 64050-2804. Tel.: 816-254-2720 & 9929. Fax: 816-254-4491.
E-mail: larry_villalva@nps.gov
Web Site: www.nps.gov/hstr/
Founded: 1983.
Congressional District: 5
Key Personnel: Supt., Larry Villalva; Cur. Museum, Carol J. Dage.
Personnel Profile: Full-Time Paid 17; Part-Time Paid 6; Part-Time Volunteers 14.
Governing Authority: federal; nonprofit. Parent Institution: Dept. of Interior. Subsidiary Institution: National Park Service. Tax-exempt.
Institution Type/Description: National Historic Site & Building: former home of President Harry S Truman.
Collections: furnishings; clothing; personal possessions of Pres. Harry S Truman & Bess Wallace Truman; furnishings also reflect occupation of Bess Wallace Truman's grandparents, the Gates family, and Bess's mother and siblings, the Wallaces; four generations of furnishings & personal possessions from 1867-1982. Visitor Center: audio visual program on Pres. & Mrs. Truman.
Research Fields: personal possessions & lives of Pres. Harry S Truman & Bess Wallace Truman.
Facilities: Truman home. Visitor Center: bookstore.
Activities: ranger guided tours; site orientation films; summer walking tours of Truman neighborhood.
Publications: folder; bulletin.
Hours & Admission Prices: Visitor Center: daily 8:30-5. Truman House Tours: Tues.-Sat. 8:30-5. Adults 16 & over $4; children 15 & under no charge. Closed New Year's Day; Thanksgiving; Christmas Day. &
Attendance: 36,953 (accurate)

HARRY S. TRUMAN OFFICE & COURTROOM, 112 W. Lexington Ave., Independence, MO 64050. Mailing Address: Jackson County Historical Society, P.O. Box 4241, Independence, MO 64051-4241. Tel.: 816-252-7454.
E-mail: info@jchs.org
Web Site: jchs.org
Founded: 1973.
Congressional District: 4
Key Personnel: Dir. Parks & Recreation, Gary Salva; Supt. Heritage Programs & Museums, Gordon Julich.
Personnel Profile: Full-Time Paid 1; Part-Time Paid 2; Part-Time Volunteers 50.
Governing Authority: county. Affiliated Jackson County Historical Society, Tel.: 816-252-7454. Tax-exempt: 170(c)(1).
Institution Type/Description: Historic Site Museum: Harry S. Truman's office, 1933-1934; courtroom used by Truman during 2nd term as Presiding Judge of County Court.
Collections: Judge Truman's office & courtroom; memorabilia of Truman era.
Research Fields: life of Harry S. Truman, with emphasis on pre-Senatorial years.
Facilities: Museum-related items for sale.
Activities: guided tours; permanent & temporary exhibitions; audiovisual program.
Hours & Admission Prices: Temporary closed. Adults $2, youth $1; discounts to AAA members. &
Attendance: 886 (accurate)
Membership: Individual $20; Family $30.

JACKSON COUNTY HISTORICAL SOCIETY, 112 W. Lexington Ave., Independence, MO 64051-4241. Mailing Address: P.O. Box 4241, Independence, MO 64051-4241. Tel.: 816-461-1897.
E-mail: info@jchs.org
Web Site: www.jchs.org
Institution Type/Description: Historical Society Museum.
Collections: local history, heritage & culture; photographs; period artifacts.
Publications: magazine, Jackson County Historical Society Journal.
Hours & Admission Prices: Call for hours.
Membership: Individual $35; Family & Donor $50; Patron $75; Patron Family $100; Guild $250.

LEILA'S HAIR MUSEUM, 1333 S. Noland Rd., Independence, MO 64055-1303. Tel.: 816-833-2955.
E-mail: hairmuseum@aol.com
Web Site: www.hairwork.com/leila
Founded: 1995.
Congressional District: 5
Key Personnel: Dir. & C.E.O., Leila Cohoon; Museum Shop Mgr., Linda Gouldsmith; Museum Shop Mgr., Nancy Williams.
Personnel Profile: Full-Time Paid 2.
Governing Authority: Parent Institution: Independence College of Cosmetology.
Institution Type/Description: Art Museum.
Collections: 600 wreaths & over 2,000 pieces of jewelry; religious artifacts; photographs; postcards.
Hours & Admission Prices: Tues.-Sat. 9-4. Adults $6, senior citizens over 65 & children under 12 $3. Closed major holidays. &
Attendance: 2,000 (estimated)

MORMON VISITORS CENTER, 937 W. Walnut St., Independence, MO 64050-3646. Tel.: 816-836-3466. Fax: 816-252-6256.
E-mail: vcindepend@ldschurch.org
Web Site: www.ldschurch.org
Formerly: Independence Visitors Center
Founded: 1971.
Key Personnel: Dir., Elder Richard L. Adams.
Personnel Profile: Full-Time Volunteers 20; Part-Time Volunteers 8.
Governing Authority: church. Parent Institution: The Church of Jesus Christ of Latter-Day Saints, 50 E. North Temple, Salt Lake City, UT 84150. Tax-exempt.
Institution Type/Description: Religious History Museum.
Collections: art & period artifacts representing Mormon life in Missouri from 1831 to 1839.
Facilities: visitors center.
Activities: guided tours; lectures; films.
Hours & Admission Prices: Daily 9-9. No charge. &
Attendance: 57,000 (accurate)

NATIONAL FRONTIER TRAILS MUSEUM, 318 W. Pacific, Independence, MO 64050-4372. Tel.: 816-325-7575. Fax: 816-325-7579.
E-mail: nftminfo@indepmo.org
Web Site: frontiertrailsmuseum.org
Founded: 1990.
Congressional District: 5
Key Personnel: Finance Dir. City of Independence, James Harlow; Admin., David Aamodt; Museum Oper. Mgr., Debbie Stewart; Education & Events Mgr., Richard Edwards.
Personnel Profile: Full-Time Paid 4; Part-Time Paid 5; Part-Time Volunteers 17; Interns 2.
Governing Authority: municipal government; nonprofit. Parent Institution: State of Missouri; City of Independence.
Institution Type/Description: History Museum: Site is the location of an 1840s grist mill and public spring. The wagon caravans leaving from Independence in the 1840s used the spring on the property for water before starting out for Oregon, California or Santa Fe. Designated a Category 1 Interpretive Site for the Lewis and Clark National Historic Trail, Oregon National Historic Trail, California National Historic Trail, Mormon Pioneer National Historic Trail & the Santa Fe National Historic Trail.
Collections: trails history 1800-1870; early Independence western trails, archives & library; trades & craft tools; decorative arts; costumes; photographs. Historic Building: 19th century mill warehouse.
Research Fields: inventory of trails diaries and journals from our archives; information is used by a national project, Census of Overland Emigrant Diaries sponsored by the Oregon & California Trail Association.
Facilities: 3,000 vol. library; 3,500 sq. ft. exhibit space; 120-seat auditorium.

Books, 19th-century craft kits, handmade baskets, games, wood carvings, maps, metalware, reproductions, videos, museum-related items for sale.
Activities: guided tours; formal education programs.
Publications: The Trail Scout.
Hours & Admission Prices: Mon.-Sat. 9-4:30, Sun. 12:30-4:30. Adults $6, senior citizens $5, youth 6-17 $3; discounts to groups & tours; Friends of the National Frontier Trails Museum, Oregon-California Trails Assoc., Santa Fe Trail Assoc. members, children under 5 & members no charge. Closed New Year's Day; Thanksgiving; Christmas. &
Attendance: 17,000 (accurate)
Membership: Individual $15; Family & Couple $25; Benefactor $75; Business $100; Sponsor $250; Lifetime $500.

VAILE MANSION-DEWITT MUSEUM, 1500 N. Liberty, Independence, MO 64050-1821.
Independence, MO 64050. Tel.: 816-325-7111. Fax: 816-325-7932.
Web Site: www.visitindependence.com
Founded: 1983.
Congressional District: 5
Key Personnel: C.E.O. & Pres. (V), Jean Kimball.
Personnel Profile: Part-Time Volunteers 50.
Governing Authority: city government. Subsidiary Institution: Vaile Victorian Society. Tax-exempt.
Institution Type/Description: Historic House: c.1881 30-room Victorian Mansion.
Collections: 1880s furnishings.
Facilities: library. Gift items for sale.
Activities: guided tours; arts festivals; participatory, loan & temporary exhibitions. Annual Events: Opening Tea in March; Strawberry Festival in June; Spirit of Christmas Past.
Publications: quarterly mailer, Vaile Victorian Society.
Hours & Admission Prices: April-Oct.& late Nov.-Dec. Mon.-Sat. 10-4, Sun. 1-4. April-Oct. adults $6, senior citizens 62 & over $5, children 6-16 $3; children under 6 no charge. Nov.-Dec. adults $6, youth 6-16 $3; children under 6 no charge. Closed Easter; Mother's Day; Thanksgiving; Christmas.
Attendance: 10,000 (accurate)
Membership: Individual $15; Vaile Squire & Damsel $50; Vaile Lady or Colonel $100; Vaile Colonel & Lady $150; Life $500.

Jackson

THE OLIVER HOUSE, 224 E. Adams St., Jackson, MO 63755. Tel.: 573-243-3753.
Institution Type/Description: Historic House Museum: housed in the former home of Marie Watkins Oliver, co-designer of the Missouri state flag; built in 1896. Listed on the National Register of Historic Places.
Collections: Oliver family history; personal artifacts; period furnishings; photographs; piano; early clothing; portrait; sewing kit used to sew state flag.
Hours & Admission Prices: May, July & Oct. 1st Sun. 1:30-4:30; Dec. Sun. 1:30-4:30. Adults $5, children $1. Closed Christmas.

Jefferson City

COLE COUNTY HISTORICAL MUSEUM, 109 Madison, Jefferson City, MO 65101-3015. Tel.: 573-635-1850.
E-mail: cchs@socket.net
Web Site: www.colecohistsoc.org
Founded: 1941.
Personnel Profile: Part-Time Paid 1; Part-Time Volunteers 45.
Governing Authority: society. Tax-exempt.
Institution Type/Description: Historic House Museum: housed in 1871 Federal style row house (3 units), built by Gov. B. Gratz Brown, 20th governor of Missouri.
Collections: inaugural ball gowns of governors' wives; archives; wine glasses; historical 19th-century furniture; war relics.
Research Fields: history; genealogy.
Facilities: 900-vol. library of books on history & genealogy available only on premises. Limited section of local history books & postcards for sale.
Activities: guided tours; school tours; permanent & temporary exhibitions. Museum Sponsors: annual dinner meeting; Historic Suitcase-to-Go lectures.
Publications: membership newsletter.
Hours & Admission Prices: Feb. to mid-Dec. Tues.-Sat. 1-3. Adults $4, senior citizen $3, youth K-12 $2; members & children under 5 no charge. Special tours at other times if arranged in advance. Office: Tues.-Fri. 10-3. Closed national holidays except Independence Day. &
Attendance: 2,898 (accurate)
Membership: Senior Citizen $10; Individual $20; Family $30; Business $50; Life & Endowment $500; Patron $1,000.

JEFFERSON LANDING STATE HISTORIC SITE, 100 Jefferson St., Jefferson City, MO 65101. Tel.: 573-751-2854. Fax: 573-526-2927.
E-mail: dsp.state.museum@dnr.mo.gov
Web Site: www.missouristatemuseum.com
Founded: 1976.
Congressional District: 9
Key Personnel: Site Admin., Linda Endersby; Cur. Exhibits, Michele Blackmore; Tours, Jocelyn Korsch; Cur. Collections, Kate Keil.
Personnel Profile: Full-Time Paid 10; Part-Time Paid 10; Interns 2.
Governing Authority: state. Parent Institution: Missouri State Museum, Capitol Bldg., Rm. B-2, Jefferson City, MO 65101. Subsidiary Institution: Division of State Parks. Tax-exempt.
Institution Type/Description: Historic Site: Two buildings that were part of original steamboat landing on the Missouri River.
Collections: artifacts and documents dealing with Missouri State Capitol and the settlement of Jefferson City. Historic Buildings: c.1839 Lohman Building; c.1855 Union Hotel.
Research Fields: travel & life on the Missouri River; Jefferson City; development of river landing.
Facilities: theatre.
Activities: videos; special events; traveling & temporary exhibits; guided tours.
Publications: book, Jefferson Landing: Commercial Center of the Steamboat Era.
Hours & Admission Prices: Tues.-Sat. 10-4. No charge; donations accepted. Closed New Year's Day; Easter; Thanksgiving; Christmas. &
Attendance: 14,048 (accurate)

MISSOURI GOVERNOR'S MANSION, 100 Madison St., Jefferson City, MO 65101-3061. Mailing Address: P.O. Box 1133, Jefferson City, MO 65102-1133. Tel.: 573-751-2854.
Institution Type/Description: Historic Building: housed in the home for 34 governors and their families; built in 1871.
Collections: state & local history; photographs; paintings; period furnishings; personal artifacts.
Activities: guided tours.
Hours & Admission Prices: By appointment.

MISSOURI STATE MUSEUM, (M), 201 W. Capitol Ave., Jefferson City, MO 65101-1556. Mailing Address: 100 Jefferson St., Jefferson City, MO 65101. Tel.: 573-751-2854. Fax: 573-526-2927.
E-mail: dsp.state.museum@dnr.mo.gov
Web Site: www.missouristatemuseum.com
Founded: 1919.
Congressional District: 9
Key Personnel: Dir., Linda Endersby; Cur. Exhibits, Michele Blackmore; Cur. Collections, Katherine Keil; Tour & Museum Shop Mgr., Jocelyn Korsch.
Personnel Profile: Full-Time Paid 10; Part-Time Paid 15; Interns 2.
Governing Authority: state. Parent Institution: Missouri Dept. of Natural Resources, P.O. Box 176, Jefferson City, MO 65102. Subsidiary Institution: Div. of State Parks. Tax-exempt.
Institution Type/Description: General Museum.
Collections: Missouriana; artifacts; documents; specimens dealing with history & resources of Missouri.
Research Fields: cultural & natural history of Missouri.
Activities: guided tours; special events; permanent, temporary & traveling exhibits; exhibits workshop.
Publications: pamphlets; Souvenir Guide to Missouri's Capitol; The Thomas Hart Benton Mural in the Missouri State Capitol; book, Jefferson Landing: Commercial Center of the Steamboat Era.
Hours & Admission Prices: Daily 8-5. No charge; donations accepted. Closed New Year's Day; Easter; Thanksgiving; Christmas. &
Attendance: 458,746 (estimated)
Membership: Individual $15.

MISSOURI VETERINARY MEDICAL FOUNDATION VETERINARY MUSEUM, 2500 Country Club Dr., Jefferson City, MO 65109-1190. Tel.: 573-636-8612. Fax: 573-659-7175.
Web Site: www.movma.org
Formerly: Missouri Veterinary Museum
Institution Type/Description: Science Museum.
Collections: veterinary medicine history from 16th century to present; manuscripts; medical instruments; 1870s wooden operating table; 1920s Shikles vaccinating kit.
Publications: quarterly, Missouri Veterinary Medical Assoc.
Hours & Admission Prices: Mon.-Fri. 9-4, Sat. by appointment. No charge; donations accepted. Closed holidays.

Jennings

JENNINGS HISTORICAL SOCIETY, 8741 Jennings Station Rd., Jennings, MO 63136. Mailing Address: Jennings City Hall, 2120 Hord Ave., Jennings, MO 63136. Tel.: 314-381-6650.
Institution Type/Description: Historical Society Museum.
Collections: local history; slides; photographs; videos; newspaper articles; maps; flags; Fairview and Jennings High School yearbooks.
Activities: research.
Hours & Admission Prices: By appointment.

Joplin

DOROTHEA B. HOOVER HISTORICAL MUSEUM, (M), 504 Schifferdecker, Joplin, MO 64801-3321. Mailing Address: P.O. Box 555, Joplin, MO 64802-0555. Tel.: 417-623-1180. Fax: 417-623-2341.
E-mail: jopmusm@ipa.net
Web Site: www.joplinmuseum.org
Founded: 1973.
Congressional District: 7
Key Personnel: Bd. Pres., John Joines; Exec. Dir., Brad Belk; Resident Geologist, Dr. John Knapp; Administrative Asst., Dina Taylor; Cur., Christopher Wiseman.
Personnel Profile: Part-Time Paid 1; Part-Time Volunteers 50; Interns 4.
Governing Authority: society; nonprofit organization. Parent Institution: Joplin Historical & Mineral Museums Inc. Tax-exempt: 501(c)(3).
Institution Type/Description: Local History Museum.
Collections: Joplin's early history; Victorian era room settings; textiles; dolls; miniature circus; 1902 child's playhouse; 1927 LeFrance fire engine; Joplin Sports Hall of Fame; cookie cutters.
Research Fields: area history; area architecture.
Facilities: 300-seat auditorium; classroom; 9,564 sq. ft. exhibit space. Books & gift items for sale.
Activities: guided tours; lectures; docent program; internship program; temporary & traveling exhibitions; special events. Museum Sponsors: biannual toy train shows in March & November; Easter Egg Hunt; Open House with Linda Lindquist Baldwin; Breakfast with Santa.
Publications: quarterly newsletter; brochure, Joplin Museum Complex.
Hours & Admission Prices: Tues. 10-7, Wed.-Sat. 10-5, Sun. 2-5; call for holiday hours & group tour information. Families $5, adults $2; discounts to groups of 3 or more, AAM & ICOM members; Tues. & members no charge. Closed New Year's Day; Thanksgiving; Christmas.
Attendance: 15,000 (estimated)
Membership: Student $20; Single $35; Family $50; Patron $100.

GEORGE A. SPIVA CENTER FOR THE ARTS, 222 West 3rd St., Joplin, MO 64801-2513. Tel.: 417-623-0183. Fax: 417-623-3805.
E-mail: jmueller@spivaarts.org
Web Site: www.spivaarts.org
Founded: 1947.
Congressional District: 7
Key Personnel: Dir., Jo Mueller; Pres. (V), Ann Leach.
Personnel Profile: Full-Time Paid 2; Part-Time Paid 7; Part-Time Volunteers 366; Interns 2.
Governing Authority: nonprofit. Tax-exempt: 501(c)(3).
Institution Type/Description: Art Museum & Center.
Collections: paintings; sculpture; drawings.
Facilities: 250-vol. research art library; classroom. Museum-related items for sale.
Activities: temporary & traveling exhibitions; lectures; gallery talks; formally organized education programs; docent program; children's classes; studio workshops for adults.
Publications: bimonthly newsletter; exhibition catalog, PhotoSpiva.
Hours & Admission Prices: Tues.-Sat. 10-5, Sun. 1-5. No charge; donations accepted. Closed major holidays. &
Attendance: 10,000 (estimated)
Membership: Student $20; Educator, Working Artist and Senior 60 & over $30; Individual $35; Senior Couple $40; Family & Dual $50; Friend of the Arts $100; Spiva Circle $300 & up; Business Partner $300-$10,000.

TRI-STATE MINERAL MUSEUM, (M), 400 Schifferdecker Ave., Joplin, MO 64801. Mailing Address: P.O. Box 555, Joplin, MO 64802-0555. Tel.: 417-623-1180. Fax: 417-623-2341.
E-mail: jopmusum@ipa.net
Web Site: www.joplinmuseum.org
Founded: 1931.
Congressional District: 7
Key Personnel: Bd. Pres., John Joines; Exec. Dir., Brad Belk; Cur., Christopher Wiseman; Resident Geologist, Dr. John Knapp; Administrative Asst., Dina Taylor.
Personnel Profile: Part-Time Paid 1; Part-Time Volunteers 50; Interns 4.
Governing Authority: society; nonprofit organization. Parent Institution: Joplin Historical & Mineral Museums Inc. Tax-exempt: 501(c)(3).
Institution Type/Description: Mineralogical Museum.
Collections: lead & zinc ore dating back to Joplin's mining days; mining tools & equipment; mining models.
Research Fields: area minerals.
Facilities: Museum-related items for sale.
Activities: guided group tours; lectures; docent program; internship program; temporary & traveling exhibitions. Museum Sponsors: Tri-State Gem & Mineral Show.
Publications: semiannual newsletter; quarterly newsletter; brochure, Joplin Museum Complex.
Hours & Admission Prices: Tues. 10-7, Wed.-Sat. 10-5, Sun. 2-5; call for holiday hours & group tour information. Adults $2; discounts to groups of 3 or more, AAM & ICOM members; members no charge. Closed major holidays.
Attendance: 15,000 (estimated)
Membership: Student $20; Single $35; Family $50; Patron $100.

Kansas City

ADVERTISING ICON MUSEUM, (M), 4600 Madison Ave., Ste. 1500, Kansas City, MO 64112. Tel.: 816-960-5254. Fax: 816-399-6254.
E-mail: howardboasberg@bradv.com
Web Site: www.advertisingiconmuseum.com
Founded: 2005.
Key Personnel: Exec. Dir., Howard Boasberg; Chm. (V), Robert Bernstein; Pres. (V), Fred Pryor.
Personnel Profile: Part-Time Volunteers 1.
Governing Authority: Tax-exempt.
Institution Type/Description: Advertising Icon Museum
Collections: over 4,000 icons from late 1800s to present.
Hours & Admission Prices: Call for hours. &

AIRLINE HISTORY MUSEUM, Kansas City Downtown Airport, Hangar 9, 201 N.W. Lou Holland Dr., Kansas City, MO 64116-4223. Tel.: 816-421-3401; 800-513-9484. Fax: 816-421-3421.
Founded: 1985.
Key Personnel: Dir. Devel. & Business Mgr., Michael Dann; Exec. Asst., Gwyneth Bowen.
Personnel Profile: Full-Time Paid 2; Part-Time Volunteers 50.
Governing Authority: Tax-exempt.
Institution Type/Description: Airline History Museum.
Collections: Lockheed L1049 Super G Constellation; Martin 404; Douglas DC-3; propeller airline aircraft & artifacts 1920-1965; photographs; audio-visual displays; printed materials; uniforms; personal artifacts.
Research Fields: Airline Histories.
Facilities: Museum-related items for sale.
Activities: tours; special events. Museum Sponsors: summer dance & dinners; Hangar dances.
Publications: bi-monthly newsletter.
Hours & Admission Prices: Mon.-Sat. 10-4, Sun. 12-4. Adults 18-64 $10, senior citizens 65 & over and military $9, children 6-17 $6; children under 6 no charge. &
Attendance: 15,000 (estimated)
Membership: Annual $110.

ALEXANDER MAJORS HISTORIC HOUSE & MUSEUM, 8201 State Line Rd., Kansas City, MO 64114-2002. Tel.: 816-333-5556 & 444-1858.
E-mail: director@wornalhouse.org
Web Site: www.alexandermajors.com
Founded: 1980.
Key Personnel: Dir., Kandice Walker; Chm. (V), Alex Rossa.
Personnel Profile: Full-Time Paid 1; Part-Time Paid 5; Part-Time Volunteers 10.
Governing Authority: Parent Institution: Wornall House. Tax-exempt.
Institution Type/Description: Historic House: former home of Alexander Majors, co-founder of the Pony Express.
Collections: local history & culture; photographs; personal artifacts.
Activities: educational programs.
Hours & Admission Prices: April to Dec. Sat.-Sun. 1-4. Adults $6, seniors & students $5; discounts to seniors, groups & children; children 4 & under no charge. &

Attendance: 3,000
Membership: Single $25; Family $50; Benefactor $100 & up.

AMERICAN JAZZ MUSEUM, 1616 E. 18th St., Kansas City, MO 64108-1610. Tel.: 816-474-8463, ext. 207. Fax: 816-474-0074.
E-mail: info@kcjazz.org
Web Site: www.americanjazzmuseum.org
Founded: 1997.
Congressional District: 5
Key Personnel: C.E.O., Greg Carroll; Chm. (V), Nikki Newton; Museum Shop Mgr., Barbara Thomas.
Personnel Profile: Full-Time Paid 17; Part-Time Paid 12.
Governing Authority: bd. of directors. Tax-exempt.
Institution Type/Description: Jazz Museum.
Collections: jazz masters including Louis Armstrong, Duke Ellington, Ella Fitzgerald & Charlie Parker; artifacts from legendary jazz performers; Charlie Parker's saxophone.
Hours & Admission Prices: Tues.-Sat. 9-6, Sun. 12-6. AJM: Adults $8, children $3; discounts to AAM members & groups. AJM & NLBM: adults $10, children $5; discounts to groups & AAM members. &
Attendance: 300,000 (estimated)
Membership: Student $15; Senior $35; Jazz Solo $50; Jazz Duo $100; Jazz Trio $250; Jazz Quartet $500; Bebop $1,000; Kansas City Jazz $2,500.

AMERICAN ROYAL MUSEUM & VISITORS CENTER, 1701 American Royal Court, Kansas City, MO 64102-1097. Tel.: 816-569-4019. Fax: 816-221-8189. Facebook: The American Royal.
E-mail: kristiel@americanroyal.com
Web Site: www.americanroyal.com
Founded: 1992.
Key Personnel: C.E.O. & Pres., Robert Petersen; Chm. (V), Cynthia Savage; Dir. Education, Kristie Larson.
Personnel Profile: Full-Time Paid 2; Part-Time Volunteers 10.
Governing Authority: nonprofit organization. Parent Institution: American Royal Association. Tax-exempt: 501(c)(3).
Institution Type/Description: Agriculture Museum and Visitors Center.
Collections: collections from American Royal Livestock, Rodeo & Horse Shows; Kansas City stockyards; history of Kansas City & the West; cowboys.
Activities: guided tours; educational programs; agricultural displays.
Hours & Admission Prices: Tues.-Fri. 10-4; call for weekend hours. No charge; donations accepted. Closed holidays. &
Attendance: 15,341 (estimated)

AMERICAN TRUCK HISTORICAL SOCIETY, 10380 N. Ambassador Dr., Kansas City, MO 64153-1378. Mailing Address: P.O. Box 901611, Kansas City, MO 64190-1611. Tel.: 816-891-9900. Fax: 816-891-9903.
E-mail: courtney@aths.org
Web Site: www.aths.org
Founded: 1971.
Congressional District: 6
Key Personnel: Mng. Dir., Shelley Ruhlman; Exec. Dir., Bill Johnson; Pres., John Vannatta; Treas., Michael C. Colton.
Personnel Profile: Full-Time Paid 7; Part-Time Volunteers 1.
Volunteer Hours: 400
Operating Expenses: 991,467
Operating Income: 1,105,581
Governing Authority: nonprofit society. Tax-exempt: 501(c)(3).
Institution Type/Description: Truck Library.
Collections: Ernie Sternberg (Sterling Truck) Collection; PIE (Pacific Intermountain Express) Collection; White Motor Company (of Cleveland) Archives including nine other truck names under the White umbrella at one time; over 28,000 truck transport magazine titles from around the world; sales literature on all truck makes from turn-of-the-century forward; 5,000 owner, repair, parts & sales manuals for trucks; 110,000 photos of trucks dating from the early 1900s.
Facilities: 1,200-vol. library of truck-related books; 28,000 magazines; 120 videos available for use on premises.
Activities: Annual Events: Convention; Antique Truck Show.
Publications: bimonthly magazine, Wheels of Time; ATHS Show Time Collector's Special Issue.
Hours & Admission Prices: Mon.-Fri. 8-4:30. No charge. Closed national holidays. &
Attendance: 300 (estimated)
Membership: Youth $20; General $45; Family $55; Endowing $200 per year for 5 years; Company $225; Associate $500; Life $1,000.

ARABIA STEAMBOAT MUSEUM, 400 Grand Blvd., Kansas City, MO 64106-1111. Tel.: 816-471-1856. Fax: 816-471-1616.
Web Site: www.1856.com
Founded: 1991.
Congressional District: 5
Key Personnel: Pres. & Treas., Bob Hawley; Cur., Public Rels. & Archivist, David Hawley; Museum Shop Mgr., Florence Hawley.
Personnel Profile: Full-Time Paid 10; Part-Time Paid 15.
Governing Authority: private organization. Parent Institution: River Salvage, Inc.
Institution Type/Description: General Museum.
Collections: hardware; personal artifacts; tools; textiles; leather; glassware; china; tin; international goods; parts & equipment from the steamboat itself.
Facilities: 85-seat large screen theater; 50-seat restaurant. Museum-related items for sale.
Activities: guided tours; theater.
Publications: books, Treasure in a Cornfield; Treasures of the Steamboat Arabia, A Big Adventure for a Small Seed; Don't Let the Bedbugs Bite.
Hours & Admission Prices: Feb. 16-March Mon.-Sat. 10-4:30, Sun. 12-4:30 (last tour at 3); April-Dec. Mon.-Sat. 10-5:30 (tours every half-hour, last one 4), Sun. 12-5 (tours every half-hour, last one 3:30). Adults $14.50, senior citizens over 60 $13.50, children 4-14 $5.50; discounts to groups of 25 or more; children 3 & under no charge. Closed New Year's Day; Easter; Thanksgiving; Christmas Eve & Day. &
Attendance: 62,649 (accurate)

BATTLE OF WESTPORT VISITOR CENTER AND MUSEUM, 6601 Swope Pkwy., Kansas City, MO 64132. Mailing Address: Monnett Battle of Westport Fund, Inc., 6900 College Blvd., Ste. 510, Overland Park, KS 66211. Tel.: 913-345-2000. Fax: 913-345-2081.
E-mail: battleofwestport1864@yahoo.com
Web Site: www.battleofwestport.org
Founded: 2008.
Congressional District: 5
Key Personnel: Chm. (V), Daniel L. Smith.
Personnel Profile: Part-Time Volunteers 12.
Governing Authority: Parent Institution: Monnett Battle of Westport Fund, Inc. Tax-exempt: 501(c)(3).
Institution Type/Description: Civil War History Museum & Visitor Center.
Collections: Civil War history, documents & artifacts including the Battle of Westport on Oct. 21-23, 1864.
Research Fields: Civil War with emphasis on the Kansas/Missouri border region.
Activities: living history events; battlefield tours & reclamation; group tours. Battle-related publications for sale.
Publications: battle field tour maps.
Hours & Admission Prices: April-Oct. Thurs.-Sat. 1-5. No charge; donations requested.
Attendance: 1,000 (estimated)
Membership: Individual $30.

BELGER ARTS CENTER, 2100 Walnut St., Kansas City, MO 64108. Tel.: 816-474-3250. Fax: 816-221-1621.
E-mail: mdickens@belgerartscenter.org
Web Site: www.belgerartscenter.org
Founded: 2000.
Key Personnel: Dir., Evelyn Craft.
Governing Authority: Parent Institution: John & Maxine Belger Family Foundation. Tax-exempt.
Institution Type/Description: Art Gallery.
Collections: works by contemporary artists.
Hours & Admission Prices: Wed.-Fri. 10-4, Sat. 12-4; other times by appointment. No charge.
Attendance: 7,000 (estimated)

BRUCE R. WATKINS CULTURAL HERITAGE CENTER AND STATE MUSEUM, 3700 Blue Pkwy., Kansas City, MO 64130-2800. Tel.: 816-513-0700. Facebook: Bruce R. Watkins Cultural Heritage Center and State Museum.
E-mail: jesse.barnes@kcmo.org
Web Site: www.kcmo.org/CKCMO/Depts/ParksandRecreation/BruceR.Watkins/
Key Personnel: Exec. Dir., Jesse Barnes
Institution Type/Description: History Museum.
Collections: local history & culture; photographs.
Facilities: resource library; auditorium; children's workspace.
Hours & Admission Prices: Tues.-Sat. 10-6. No charge.

GRAND ARTS, 1819 Grand Ave., Kansas City, MO 64108-1815. Tel.: 816-421-6887. Fax: 816-421-1561.
E-mail: gallery@grandarts.com
Web Site: www.grandarts.com
Institution Type/Description: Art Gallery.
Collections: works by emerging & established artists; sculpture.
Activities: temporary exhibitions.
Hours & Admission Prices: Call for hours.

GREENLEASE ART GALLERY - ROCKHURST UNIVERSITY, 1100 Rockhurst Rd., Kansas City, MO 64110. Tel.: 816-501-4000; 800-842-6776.
Key Personnel: Dir., Anne Pearce
Institution Type/Description: Art Gallery.
Collections: paintings; sculpture.
Hours & Admission Prices: Thurs.-Sat. 12-5; other times by appointment.

HALLMARK VISITORS CENTER, 2450 Grand Blvd., Kansas City, MO 64108. Tel.: 816-274-3613.
E-mail: visitorscenter@hallmark.com
Web Site: www.hallmarkvisitorscenter.com
Founded: 1985.
Key Personnel: Dir., Regi Ahrens
Institution Type/Description: Company History Museum.
Collections: company history from 1910 to present; Hallmark greeting cards; Hallmark Keepsake Ornaments; Hallmark Hall of Fame television series; photographs; Hallmark characters; hands-on exhibits; television commercials; Emmy awards.
Activities: demonstrations; videos; interactive exhibits.
Hours & Admission Prices: Mon.-Fri. 10-4:30, Sat. 9:30-4:30. No charge. &
Attendance: 124,425 (accurate)

THE JOHN WORNALL HOUSE MUSEUM, 6115 Wornall Rd., Kansas City, MO 64113-1417. Tel.: 816-444-1858. Fax: 816-361-8165.
E-mail: kandice@wornallhouse.org
Web Site: www.wornallhouse.org
Founded: 1972.
Congressional District: 5
Key Personnel: Dir., Anna Marie Tutera; Chm. & Pres. Bd. (V), Alex S. Rosser; Museum Shop Mgr., Margueritte Milliken.
Personnel Profile: Full-Time Paid 1; Part-Time Paid 2; Part-Time Volunteers 75; Interns 1.
Governing Authority: nonprofit organization. Parent Institution: John Wornall House Museum, Inc. Tax-exempt: 501(c)(3).
Institution Type/Description: Historic House: c.1858 Greek Revival home used as a hospital during the Civil War.
Collections: Civil War era; furniture; textiles; ceramics; glass; silver; ironware; period herb garden.
Research Fields: social history of western Missouri in mid-19th century.
Facilities: Books, kitchen items, toys & craft items for sale.
Activities: guided tours; lectures; organized education programs; docent program; school loan service.
Publications: membership newsletter.
Hours & Admission Prices: Tues.-Sat. 10-4, Sun. 1-4. Adult $6, children 5-12 $3; discounts to groups & military personnel; children under 4 no charge. Closed holidays. &
Attendance: 6,286 (accurate)
Membership: Friends: Student & Senior $25; Individual $35; Family $50. Society of 1858: Contributing $100; Sponsoring $250; Sustaining $500; Benefactor $1,000 & up.

KALEIDOSCOPE, 2500 Grand, Kansas City, MO 64108. Mailing Address: P.O. Box 419580, Kansas City, MO 64141-6580. Tel.: 816-274-8301 & 8934. Fax: 816-274-3148.
E-mail: kaleidoscope@hallmark.com
Web Site: www.hallmarkkaleidoscope.com
Founded: 1969.
Congressional District: 5
Key Personnel: Admin., Regi Ahrens; Exhibit Mgr., Linda Avery; Volunteer Coord., Jeanine Ellis.
Personnel Profile: Full-Time Paid 7; Part-Time Paid 13; Part-Time Volunteers 80.
Governing Authority: company. Parent Institution: Hallmark Cards, Inc.
Institution Type/Description: Children's Participatory Art Exhibit.
Collections: hands-on exhibits.
Facilities: 8,000 sq. ft. exhibit space; art studio.

Activities: guided tours; participatory art experience for children & their families.
Hours & Admission Prices: Call for hours. No charge. Closed New Year's Day; Memorial Day; Independence Day; Labor Day; Thanksgiving; Christmas Eve & Day. &
Attendance: 195,000 (accurate)

KANSAS CITY ART INSTITUTE H&R BLOCK ARTSPACE, (M), 16 E. 43 St., Kansas City, MO 64111-1755. Mailing Address: 4415 Warwick Blvd., Kansas City, MO 64111. Tel.: 816-561-5563. Fax: 816-561-5993.
E-mail: rmsmith@kcai.edu
Web Site: www.kcai.edu/artspace
Founded: 1999.
Congressional District: 5
Key Personnel: Dir., Raechell Smith; Pres., Jacqueline Chanda.
Personnel Profile: Full-Time Paid 3; Part-Time Paid 7; Interns 2.
Governing Authority: Parent Institution: Kansas City Art Institute. Tax-exempt.
Institution Type/Description: Art Gallery & University Museum.
Collections: works of contemporary visual art.
Activities: temporary exhibitions.
Hours & Admission Prices: Tues.-Fri. 12-5, Sat. 11-5. No charge.
Attendance: 9,500 (accurate)
Membership:

KANSAS CITY FIRE MUSEUM, 30 W. Pershing Rd, Kansas City, MO 64108. Tel.: 816-474-0200.
Institution Type/Description: Fire Fighting History Museum.
Collections: fire fighting history & equipment; photographs; personal artifacts.
Activities: fire safety programs.
Hours & Admission Prices: Thurs.-Sat. 9:30-5:30, Sun. 12-5:30.

KANSAS CITY MUSEUM, 3218 Gladstone Blvd., Kansas City, MO 64123-1199. Tel.: 816-483-8300. Fax: 816-483-6050.
Web Site: kansascitymuseum.org
Founded: 1940.
Key Personnel: Dir., George Guastello; Chm. (V), Robert Regnier.
Personnel Profile: Full-Time Paid 12; Part-Time Paid 1; Part-Time Volunteers 48; Interns 6.
Governing Authority: Parent Institution: Union Station Kansas City, Inc. Tax-exempt.
Institution Type/Description: History Museum: housed in an urban estate built in 1910.
Collections: local & regional history; early costumes.
Facilities: 40-seat auditorium; visitor center.
Activities: special events; StoryTarium Programs include monthly featured films.
Publications: community curator series, Natural History on the Head; past exhibitions information; Loula Long Combs: Fashion Paper Doll Collection; What We Did for Love: AIDS Walk T-shirt Collection; Restoration of a Window; Pocket Museum.
Hours & Admission Prices: Visitor Center: Wed.-Sat. 10-4, Sun. 12-4. StoryTarium Programs: daily 2 p.m. No charge; donations accepted. &
Attendance: 25,000
Membership: Individual $25; Family $35; Corporate $50; R.A. Long Society $500.

KANSAS CITY ZOO, 6800 Zoo Dr., Kansas City, MO 64132-1711. Tel.: 816-513-5800. Fax: 816-513-5850.
E-mail: randywisthoff@fotzkc.org
Web Site: www.kansascityzoo.org
Formerly: Kansas City Zoological Gardens
Founded: 1909.
Congressional District: 5
Key Personnel: Exec. Dir. & C.E.O., Randy Wisthoff; Chm., Bill Crandall; Dir. Finance Officer, Linda Falk; Dir. Devel., Laura Berger; Dir. Mktg. & Membership, Julie Neemeyer; Dir. Living Collections, Sean Putney; Dir. Education, Debra Ryder; Veterinarian, Wm. Kirk Suedmeyer, D.V.M.; Asst. Dir. & C.O.O., Geoff Hall.
Personnel Profile: Full-Time Paid 128; Part-Time Paid 153; Part-Time Volunteers 458; Interns 20.
Governing Authority: municipal. Parent Institution: Friends of the Zoo, Inc., Kansas City, MO. Tax-exempt.
Institution Type/Description: Zoological Park.
Collections: 211 species; 1,114 specimens including endangered or threatened species; domesticated animals; African elephants, polar bears; Australian exhibit; African Forest & African Plains.
Major Exhibits: Green Revolution (T), 1/12-1/14.

Research Fields: animal behavioral studies & accumulation of physiological data from routine medical treatment; transabdominal technique & pupillary light response.
Facilities: 300-vol. research library of books including 6,500 items on animal sciences, natural history, ethology, exotic animal management & medicine available for research by appointment through the education department; 202 acre outdoor facility includes refreshment stands. Guidebooks & other zoo-related items for sale.
Activities: lectures; films; formally organized education programs for children & adults; volunteer program; permanent exhibitions; zoomobile; outreach program.
Publications: quarterly newsletter, Friends of the Zoo Newsletter; weekly, e-news.
Hours & Admission Prices: Daily 9:30-4. Adults $11.50, seniors $10.50, children 3-11 $8.50; discounts to groups & reciprocal Zoos Friends members; members & children under 3 no charge. Closed New Year's Day; Thanksgiving; Christmas. &
Attendance: 713,813 (accurate)
Membership: Individual $47; Individual Plus $58; Family $79; Sustaining $130. Discounted memberships for zoological district residents.

KEMPER MUSEUM OF CONTEMPORARY ART, (M), 4420
Warwick Blvd., Kansas City, MO 64111-1821. Tel.: 816-753-5784. Fax: 816-753-5806.
E-mail: communications@kemperart.org
Web Site: www.kemperart.org
Founded: 1994.
Congressional District: 5
Key Personnel: Chm., Mary Kemper Wolf; Exec. Dir., Barbara O'Brien; Dir. Finance & Accounting, Kathy Surber; Dir. Devel., Don Schreiner; Cur. & Head Adult Programs, Erin Dziedzic; Museum Educator, School & Family, Brenda Brinkhaus-Hatch; Dir. Mktg. & Communications, Margaret A. Keough; Facility & Operations Mgr., Paul Watts; Collections Mgr. & Registrar, Andrea Phillips; Coord. Membership, Sara Hale; Mgr. Accounting, Karla Ruud; Museum Shop Mgr., Raina Heinrich; Chief Preparator, Russell Horton; Museum Protection Mgr., Steve Crays; Mgr. Human Resources, Jean Hanover; Special Events Coord., Betty Kindler; Exec. Chef, Jennifer Maloney; Gen. Mgr. Cafe, Keith Goldman.
Personnel Profile: Full-Time Paid 26; Part-Time Paid 68; Part-Time Volunteers 152; Interns 10.
Operating Expenses: 4,813,548
Operating Income: 5,735,218
Governing Authority: nonprofit organization. Parent Institution: Kemper Museum Operating Foundation. Branch Museums: Kemper at the Crossroads (downtown); Kemper East. Tax-exempt: 501(c)(3).
Institution Type/Description: Contemporary Art Museum.
Collections: Bebe & Crosby Kemper Collection; contemporary art in all media.
Research Fields: contemporary art, design & architecture in all media.
Facilities: library; meeting room; cafe; 20,039 sq. ft. exhibition space.
Activities: films; formal education programs; guided tours; lectures; loan, temporary & traveling exhibitions.
Publications: exhibition catalogues; exhibition-related materials.
Hours & Admission Prices: Kemper Museum: Tues.-Thurs. 10-4, Fri.-Sat. 10-9, Sun. 11-5. Cafe: Tues.-Sun. 11-2:30, Fri.-Sat. 5:30-9:30. No charge; donations accepted. Kemper at the Crossroads: 1st Fri. 12-10. Kemper East: Tues.-Fri. 10-4. Closed New Year's Day; Independence Day; Thanksgiving; Christmas. &
Attendance: 130,000 (estimated)
Membership: Student $30; Solo $55; Duo $75; Family $150; Associate $300; Connoisseur $500; Patron & Business Partner $1,000; Business $2,500; Special Contributor $3,000; Corporate $4,000; Sustainer $5,000; Benefactor & Executive $10,000; Chairman $25,000.

LEEDY-VOULKOS ART CENTER, 2010 Baltimore Ave., Kansas
City, MO 64108-1914. Tel.: 816-474-1919.
Web Site: www.leedy-voulkos.com
Founded: 1985.
Key Personnel: Gallery Mgr. & Dir., Erin Woodworth; Museum Shop Mgr., Courtney Wasson
Institution Type/Description: Art Gallery.
Collections: works by local, regional & national artists.
Hours & Admission Prices: Gallery & Shop: Thurs.-Sat. 11-5. No charge.

THE MONEY MUSEUM-FEDERAL RESERVE BANK OF
KANSAS CITY, 1 Memorial Dr., Kansas City, MO 64198-0002. Tel.: 816-881-2683. Fax: 816-881-2569.
E-mail: KCtours@kc.frb.org
Web Site: www.kc.frb.org/moneymuseum

Formerly: The Roger Guffey Gallery-Federal Reserve Bank of Kansas City
Founded: 2001.
Key Personnel: Economic Education Analyst, Gigi Wolf.
Governing Authority: federal reserve system. Tax-exempt.
Institution Type/Description: Money Gallery: housed in the lobby of the Federal Reserve Bank of Kansas City; focuses on how money has evolved throughout banking history to present day.
Collections: The Harry S. Truman coin collection (453 coins dating back to George Washington's presidency through Jimmy Carter's term in office).
Facilities: 50-seat meeting room.
Activities: guided tours; lectures; films; participatory exhibits.
Publications: brochures.
Hours & Admission Prices: Mon.-Fri. 8:30-4:30. No charge. Closed banking holidays. &

NATIONAL MUSEUM OF TOYS AND MINIATURES, (M),
5235 Oak St., Kansas City, MO 64112-2824. Tel.: 816-235-8000. Facebook: Toy and Miniature Museum.
E-mail: juloa@toyandminiaturemuseum.org
Web Site: www.toyandminiaturemuseum.org
Formerly: Toy & Miniature Museum of Kansas City
Founded: 1981.
Congressional District: 5
Key Personnel: Exec. Dir., Jamie A. Berry; Collections Coord., Calleen Carver; Cur., Kristi Dobbins; Visitor Svcs. Assoc., Meg Hauser; Admin. & Mktg. Asst., Tony Julo; Community Devel. Coord., Cassie Mundt; Museum Educator, Laura S. Taylor.
Governing Authority: nonprofit organization. Tax-exempt: 501(c)(3).
Institution Type/Description: Toy & Miniature Museum: housed in 1911 Truman Mansion.
Collections: antique dolls; dollhouses; toys; contemporary scale miniatures.
Research Fields: antique toys & doll houses of German origin.
Facilities: 900-vol. library pertaining to toys & doll houses; 21,500 sq. ft. exhibit space. Reproduction paper dolls, collectors reference book & scaled miniature items for sale.
Activities: guided tours; lectures & slides; temporary exhibitions; classes.
Publications: quarterly newsletter; museum catalog; children's book.
Hours & Admission Prices: Closed for renovations. &
Attendance: 14,665 (accurate)
Membership: Individual $35; Grandparent $50; Family $60; Patron $100; Donor $250; Benefactor $500; Founder's Circle $1,000.

NATIONAL WORLD WAR I MUSEUM AT LIBERTY MEMO-
RIAL, (M), 100 W. 26th St., Kansas City, MO 64108-4616. Tel.: 816-888-8100. Fax: 816-888-8123.
E-mail: info@theworldwar.org
Web Site: www.theworldwar.org
Founded: 1926.
Congressional District: 5
Key Personnel: Chm. Bd. Trustees, Dr. Mary Cohen; Dir., Matthew Naylor; C.F.O., Jeffrey Walker; Controller, Mark Gunter; Senior Cur., Doran Cart; Vice Pres. Devel., Angela Tangen; Archivist, Jonathan Casey; Cur. Education, Lora Vogt.
Personnel Profile: Full-Time Paid 34; Part-Time Paid 10; Part-Time Volunteers 170; Interns 7.
Governing Authority: bd. of trustees. Tax-exempt: 501(c)(3).
Institution Type/Description: Military History Museum.
Collections: World War I history, veterans, documents & artifacts, 1914-1918; Pantheon de la Guerre mural conceived by French artists, Pierre-Carrier-Belleuse & Auguste Gorguet, 1914-1918; murals by Daniel MacMorris; artillery; weapons; uniforms; equipment; posters; photographs; letters; maps.
Research Fields: weapons, uniforms & posters, insignia of World War I; American involvement in the war.
Facilities: archives & library of maps, photographs, military unit histories & over 1,200 posters of World War I available for research.
Activities: permanent, temporary & traveling exhibitions.
Publications: information brochure; newsletter.
Hours & Admission Prices: Memorial Day to Labor Day daily 10-5; Sept.-May Tues.-Sun. & Mon. holidays 10-5. Adults $14, seniors & students $12, children 6-17 $8; discounts to AAM members; members no charge. Closed New Year's Day; Thanksgiving; Christmas Eve & Day. &
Attendance: 141,327 (accurate)
Membership: Student $25; Individual $35; Family $75; Patron $250; Supporter $500; Benefactor $1,000.

NEGRO LEAGUES BASEBALL MUSEUM, 1616 E. 18th St.,
Kansas City, MO 64108-1610. Tel.: 816-221-1920.
Key Personnel: Pres., Bob Kendrick; Chm. Bd. (V), Betty Brown.
Personnel Profile: Full-Time Paid 8.

Governing Authority: Tax-exempt.
Institution Type/Description: African-American Baseball History Museum.
Collections: African-American baseball history; photographs; bronze sculptures; baseball artifacts.
Facilities: Museum-related items for sale.
Activities: education programs; traveling exhibits; rental facility.
Publications: quarterly new magazine, Silhouettes.
Hours & Admission Prices: Tues.-Sat. 9-6, Sun. 12-6. Adults $10, seniors $9, children under 12 $6; discounts to groups; children under 5 no charge. &
Membership: Major Leaguer $25-$49; All Star $50-$99; MVP $100-$499; Hall of Fame $500-$999; Legacy $1,000 & up.

* **THE NELSON-ATKINS MUSEUM OF ART, (M),** 4525 Oak St., Kansas City, MO 64111-1873. Tel.: 816-751-1278. Fax: 816-561-4011. Facebook: Nelson Atkins.
Web Site: www.nelson-atkins.org
Founded: 1933.
Congressional District: 5
Key Personnel: Dir. & C.E.O., Julian Zugazagoitia; C.O.O., Karen L. Christiansen; Chm. Bd. Trustees, Sarah F. Rowland; Cur. Art of Ancient World, Dr. Robert Cohon; Dir. Conservation & Collections Mgmt., Elisabeth Batchelor; Mgr. Mktg. & Communications, Toni Wood; Sr. Conservator Paintings, Scott Heffley; Sr. Conservator Objects, Kathleen Garland; Investment & Compliance, William W. Markey, CPA; Finance & Accounting, Harland Hunt; Chief Registrar, Ann Erbacher; Visual Resources Librarian, Noriko Ebersole; Helen Jane & R. Hugh "Pat" Uhlmann Cur. Decorative Arts, Dr. Catherine L. Futter; Louis C. & Adelaide C. Ward Cur. European Painting & Sculpture, Ian Kennedy; Fred & Virginia Merrill Sr. Cur. American Indian Art, Gaylord Torrence; Museum Store Mgr., Brian Day; Coord. Mail Svcs., Mike Welch; Sanders Sosland Cur. Modern & Contemporary Art, Dr. Jan Schall; Cur. Exhibitions Management, Cindy Cart; Dir. Presentation, Steve Waterman; Consultative Cur. Medieval Art, Dr. Marilyn Stokstad; Mgr. Donor & Information Svcs., Kerry Peak; Mgr. Graphic Design, Michele Boeckholt; Dir. Advancement, Matthew Naylor; Mgr. Grants & Foundations, Ann Friedman; Chief Preparator, Mark Milani; Head, Imaging Svcs., Louis Meluso; Mgr. Event Planning, DeSaix Adams; Mgr. Engineering, Robert Bolt; Mgr. Maintenance, Angela Graham; Mgr. Facilities, Security & Visitor Svcs., Michael Cross; Dir. Human Resources, Kelly Summers; Dir. Education & Interpretive, Christine Minkler.
Personnel Profile: Full-Time Paid 193; Part-Time Paid 34; Part-Time Volunteers 500; Interns 12.
Governing Authority: nonprofit organization. Parent Institutions: The Nelson Gallery Foundation and The William Rockhill Nelson Trust. Tax-exempt: 501(c)(3).
Institution Type/Description: Art Museum.
Collections: European, American & ancient painting, sculpture, drawing, prints & decorative arts; Kansas City Sculpture Park including the Henry Moore Sculpture Garden; Asiatic, Native American, Oceanic & pre-Columbian art.
Major Exhibits: Impressionist France: Visions of Nation LeGray to Monet (T), 10/19/13-1/14.
Research Fields: European & American art; Asiatic art, emphasis on China.
Facilities: 145,000-vol. library of art reference books, monographs & catalogs available on premises; museum archives; reading room; 66 galleries; 9 period rooms; conservation laboratory; 510-seat auditorium; restaurant; Ford Learning Center. Art books, reproductions & other museum related-items for sale.
Activities: guided tours; lectures; films; gallery talks; concerts; formally organized education programs for children & adults; docent programs; permanent & temporary exhibitions; art & family fun weekend activities; pop-up events.
Publications: biannual members magazine; bimonthly calendar, Explore Art; member events calendar.
Hours & Admission Prices: Wed. 10-4, Thurs.-Fri. 10-9, Sat. 10-5, Sun. 12-5. No charge; donations accepted. Closed New Year's Day; Memorial Day; Independence Day; Thanksgiving; Christmas Eve & Day. &
Attendance: 400,000 (estimated)
Membership: Print Society: Individual $25; Couple $35; Sustaining $100; Patron $250. Friends of Art: Friend $65; Supporter $100; Contributor $150; Associate $250; Sponsor $500. National 100 miles outside of Kansas City $45. Society of Fellows: Patron $1,000; Sustaining $2,000; Nebon Society $3,000; Benefactor $5,000; Ambassador $10,000. Annual Business Council: Corporate Associate $3,000-$4,999; Corporate Contributor $5,000-$9,999; Corporate Director $10,000-$24,999; Corporate Sponsor $25,000-$49,000; Corporate Bronze Circle Sponsor $50,000-$99,999; Corporate Silver Circle Sponsor $100,000-$249,999; Corporate Gold Circle Sponsor $250,000 & up.

PIPER MEMORIAL MEDICAL MUSEUM, (M), 1000 Carondelet Dr., Kansas City, MO 64114-4673. Tel.: 816-943-2183. Fax: 816-943-2786.
E-mail: jmullen@carondelet.com
Founded: 1971.
Key Personnel: Cur., Joan Hilger-Mullen.
Personnel Profile: Part-Time Paid 1; Part-Time Volunteers 3.
Governing Authority: Parent Institution: St. Joseph Medical Center. Tax-exempt.
Institution Type/Description: Medical Museum.
Collections: St. Joseph Medical Center & Kansas City, Missouri, medical history from 1874 to present.
Hours & Admission Prices: Daily 8-8. No charge; donations accepted. &

SHOAL CREEK LIVING HISTORY MUSEUM, 7000 N.E. Barry Rd., Kansas City, MO 64156-1278. Tel.: 816-792-2655.
E-mail: parks@kcmo.org
Web Site: www.kcmo.org
Founded: 1976.
Congressional District: 6
Key Personnel: C.E.O., Pres. (V) & Museum Shop Mgr., Martha Edmunds; Chm. (V), Pam Payne.
Personnel Profile: Full-Time Volunteers 30; Part-Time Volunteers 92.
Governing Authority: municipal; nonprofit. Parent Institution: City of Kansas City, Missouri Dept. of Parks, Recreation and Boulevards. Subsidiary Institution: Shoal Creek Association. Tax-exempt.
Institution Type/Description: History Museum: comprised of 20 buildings dating from 1807-1890.
Collections: concentration on 19th-century daily Missouri life.
Research Fields: 19th century women's lives; history of Freemasonry & Masonic architecture.
Facilities: Period reproductions including clothes, millinery & other items related to daily life for sale.
Activities: docent program; formal education programs; guided tours. Annual Events: Civil War Battles; A Visit with St. Nicholas in December; Harvest Festival in November.
Publications: quarterly newsletter, The Mouse in the Attic.
Hours & Admission Prices: March-Dec. Mon.-Sat. 9-3. Tours, school programs & reenactments by appointment. No charge; donations accepted. Fees for programs & events. Closed government holidays. &
Attendance: 10,000 (estimated)
Membership: Individual $15; Family $25; Corporate $500.

THOMAS HART BENTON HOME AND STUDIO STATE HISTORIC SITE, 3616 Belleview, Kansas City, MO 64111-3808. Tel.: 816-931-5722. Fax: 816-931-5722.
E-mail: benton.home.state.historic.site@dnr.mo.gov
Web Site: www.mostateparks.com/benton.htm
Founded: 1977.
Congressional District: 5
Personnel Profile: Full-Time Paid 1; Part-Time Paid 1.
Governing Authority: state. Parent Institution: Div. State Parks, Missouri Dept. of Natural Resources, P.O. Box 176, Jefferson City, MO 65102. Tax-exempt.
Institution Type/Description: Historic House: home & studio of Thomas Hart Benton.
Collections: original furnishings relating to the life of Thomas Hart Benton.
Activities: guided tours; off-site presentations.
Publications: brochure, Thomas Hart Benton Home.
Hours & Admission Prices: Winter: Sun. 11-4, Mon. & Thurs.-Sat. 10-4; Summer: Sun. 12-5, Mon. & Thurs.-Sat. 10-4; groups of 15 or more by appointment. Adults $4, children 6-12 $2.50; discount to groups; children under 6 no charge. Closed New Year's Day; Thanksgiving; Christmas. &
Attendance: 3,904 (accurate)
Membership: Individual $35; Founder $50.

THORNHILL GALLERY AT AVILA UNIVERSITY, Dallavis Center, 11901 Wornall Rd., Kansas City, MO 64145-1007. Tel.: 816-501-3659. Fax: 816-846-4726.
E-mail: aylwardme@avila.edu
Web Site: www.avila.edu
Founded: 1978.
Congressional District: 5
Key Personnel: C.E.O., Susan Lawlor; Cur., Marci Aylward.
Personnel Profile: Part-Time Paid 3.
Governing Authority: college; nonprofit organization. Parent Institution: Avila University.
Institution Type/Description: Art Gallery.
Collections: works by national & international artists.

Facilities: 850 sq. ft. exhibit space.
Activities: traveling exhibits. Annual Event: High School Invitational Visual Arts Day.
Publications: annual calendar.
Hours & Admission Prices: Tues.-Fri. 12-3; other times by appointment. No charge; donations accepted. Closed New Year's Day; Easter; Memorial Day; Independence Day; Labor Day; Christmas; spring & fall breaks. &
Attendance: 2,400 (estimated)

TRAILSIDE CENTER, 9901 Holmes Rd., Kansas City, MO 64131-4205. Mailing Address: 712 W. 121st St., Kansas City, MO 64145-1009. Tel.: 816-942-3581.
E-mail: info@trailsidecenter.org
Web Site: www.trailsidecenter.org, www.newsantafe.org
Founded: 2004.
Key Personnel: Dir., Charles Loomis; C.E.O., Ann O'Hare.
Personnel Profile: Part-Time Volunteers 25.
Governing Authority: Parent Institution: The Historical Society of New Santa Fe. Tax-exempt.
Institution Type/Description: History Museum.
Collections: local history & culture; photographs; books; paintings.
Facilities: museum; meeting room; tourist information center.
Hours & Admission Prices: Mon.-Sat. 10-4, Sun. 12-4. No charge; donations accepted. &
Membership: Mule Skinner $25; Bull Whacker $50; Trail Boss $100; Freighter $250.

UNION STATION KANSAS CITY, INC., 30 W. Pershing Rd., Ste. 400, Kansas City, MO 64108-2410. Tel.: 816-460-2000. Fax: 816-460-2260.
E-mail: info@unionstation.org
Web Site: www.unionstation.org
Formerly: Science City at Union Station/Kansas City Museum
Founded: 2001.
Congressional District: 5
Key Personnel: C.E.O., George Guastello; Sr. Vice Pres. & C.F.O., Jerry Baber; Chm. Bd., Bob Regnier; Dir. Science City, Jeff Rosenblatt; Dir. Historic House, Christopher Leitch; Dir. Collections, Denise Morrison; Museum Shop Mgr., Sonia Filipek.
Personnel Profile: Full-Time Paid 40; Part-Time Paid 10; Part-Time Volunteers 300; Interns 14.
Governing Authority: nonprofit organization. Subsidiary Institution: Kansas City Museum. Tax-exempt: 501(c)(3).
Institution Type/Description: Science & History Museum.
Collections: archaeology; costumes & textiles; general history; historic & prehistoric Indians; archives; paleontology; technology; carriages & tack; Native American artifacts; railroad artifacts; period rail cars.
Research Fields: Kansas City history, science & technology education.
Facilities: 2,500-vol. library & archives available for research; 440-seat 15/70-3D theater; 200-seat stage theater; 150-seat dome planetarium; discovery area. Educational items for sale.
Activities: temporary exhibitions; theater shows; educational programs; special programs & events.
Publications: e-newsletter; pamphlets; brochures; school group programs.
Hours & Admission Prices: Summer & Spring: Tues.-Sat. 9:30-5, Sun. 12-5; Winter: Wed.-Fri. 9:30-4, Sat. 9:30-5, Sun. 12-5; call for holiday Mon. hours. Science City $10; members no charge. Theater: $8.50; discounts to members. KC Rail Experience: $7. Planetarium: $6. City Stage: fees vary. Discounts to AAM & ASTC members and groups of 15 or more; children 3 & under no charge. Closed Thanksgiving; Christmas. &
Attendance: 2,000,000 (estimated)
Membership: Individual $45; 2 Individuals $60; 3-4 Individuals $85; 5-8 Individuals $125.

UNIVERSITY OF MISSOURI-KANSAS CITY GALLERY OF ART, 5015 Holmes, Fine Arts Bldg. 203, 5100 Rockhill Rd., Kansas City, MO 64110-2499. Tel.: 816-235-1502. Fax: 816-235-5507. Facebook: UMKC Gallery of Art.
E-mail: umkcgallery@umkc.edu
Web Site: info.umkc.edu/art/umkcgallery/
Founded: 1975.
Congressional District: 5
Key Personnel: Gallery Art Coord., Davin Watne.
Governing Authority: university; nonprofit organization. Tax-exempt: 501(c)(3).
Institution Type/Description: Art Gallery.
Collections: paintings; sculpture; photographs.
Research Fields: contemporary.
Facilities: educational facilities.

Activities: lectures; formal education programs for undergraduate & graduate college students; guided tours; loan & traveling exhibitions. Museum Sponsors: workshop symposium.
Publications: catalogues.
Hours & Admission Prices: Mon. & Wed. 9-4, Tues. 11-4, Thurs. 11-4 & 5-7, Fri. 11-2. No charge; donations accepted. Closed holidays. &
Attendance: 10,000 (estimated)
Membership: Director $50; Associate $100; Host $250; Sponsor $500; Patron $1,000.

THE WESTPORT HISTORICAL SOCIETY AND THE 1855 HARRIS-KEARNEY HOUSE MUSEUM, 4000 Baltimore, Kansas City, MO 64111-7403. Tel.: 816-561-1821. Fax: 913-345-2081. Facebook: 1855 Harris-Kearney House Museum.
E-mail: westporthistorical@gmail.com
Web Site: www.westporthistorical.com/index.htm
Founded: 1950.
Congressional District: 5
Key Personnel: Dir., Jennifer McKeithen; Pres. (V), Alana Smith.
Personnel Profile: Part-Time Volunteers 25; Interns 2.
Governing Authority: Parent Institution: The Westport Historical Society. Tax-exempt.
Institution Type/Description: Historic House: housed in a Greek Revival mansion and former home of Col. John Harris & his wife Henrietta.
Collections: local history & culture; photographs; 19th century furnishings; personal artifacts.
Major Exhibits: Gilder Lehrman's Created Equal, Film Series, June 2014 (T).
Activities: Museum Sponsors: Living History Events & Exhibits April to November; Library Speaker Series monthly; monthly traveling exhibits
Publications: quarterly newsletter, Westport Historical Society.
Hours & Admission Prices: March to mid-Dec. Wed.-Sat. 1-5. Adults $6, students $4; discounts to society members, military & groups; children under 6 no charge. Closed major holidays.
Attendance: 1,500 (estimated)
Membership: Student 10-22 $15; Individual $35; Family & Spouse $45; Supporting $80; Patron $100; Business $150; Sponsor $300; Corporate $350; Life $700.

Kearney

JESSE JAMES FARM & MUSEUM, (M), 21216 Jesse James Farm Rd., Kearney, MO 64060-9343. Mailing Address: c/o Friends of the James Farm, P.O. Box 404, Liberty, MO 64069. Tel.: 816-628-6065. Fax: 816-628-6676.
E-mail: bbeckett@claycogov.com
Web Site: www.jessejames.org
Founded: 1978.
Congressional District: 6
Key Personnel: Historic Sites Admin., Beth Beckett.
Governing Authority: county. Parent Institution: Clay County Parks, Recreation & Historic Site. Tax-exempt.
Institution Type/Description: Historic Buildings & Site.
Collections: 1845 James farm; restored birthplace of Jesse James; 1858 Claybrook Plantation & restored house; Mt. Gilead Church & School; Jesse James Bank Museum.
Research Fields: life & career of Jesse W. James; Claybrook Plantation history; history of Mt. Gilead Church and School.
Activities: guided tours; laser disc show; simulated 19th-century school classes.
Hours & Admission Prices: May-Sept. daily 9-4; Oct.-April Mon.-Sat. 9-4, Sun. 12-4. Adults $8, seniors $7, children 8-15 $4.50; discounts to groups; children under 8 no charge. &
Attendance: 12,000 (estimated)
Membership: Friends of the James Farm: Border Ruffian $20; Bushwhacker $30; Road Agent $100; Long Rider $250.

KEARNEY HISTORIC MUSEUM, 101 S. Jefferson, Kearney, MO 64060-8503. Tel.: 816-903-1856.
E-mail: gerriantiques@aol.com
Founded: 2006.
Key Personnel: Dir., Gerri Spencer.
Governing Authority: Parent Institution: City of Kearney.
Institution Type/Description: History Museum.
Collections: local history & culture; period furnishings; personal artifacts; photographs.
Activities: special events; research.
Hours & Admission Prices: Thurs.-Sat. 10-2; groups by appointment. No charge; donations accepted. &
Attendance: 750 (estimated)

Kennett

DUNKLIN COUNTY MUSEUM, INC., 122 College, Kennett, MO 63857-2007. Mailing Address: 604 College, Kennett, MO 63857-2015. Tel.: 573-888-6620.
E-mail: cbbrown@clgw.net
Founded: 1976.
Congressional District: 10
Key Personnel: Chm. Bd. Dirs., Mrs. Charles B. Brown; Chm. (V), Sandra Brown.
Personnel Profile: Part-Time Volunteers 10.
Governing Authority: nonprofit. Tax-exempt: 509(a)(1).
Institution Type/Description: History Museum: housed in 1904 City Hall & Masonic Temple.
Collections: historical items; general store items; World War I & II uniforms & memorabilia; 1879 General Electric generator; manuscripts; primitive tools; pictures; nature display; 40 model engines showing power from wind, water, gas, steam, electricity; photographs of 1900 Bootheel area; Miles family personal artifacts; Mississippian culture of Dunklin County; stone & pottery artifacts.
Activities: guided tours; docent program; permanent & temporary exhibitions.
Hours & Admission Prices: March-Dec. Wed. 12-5, Sat. by appointment; tours by appointment. No charge; donations accepted.
Attendance: 1,500 (estimated)
Membership: Contributing $100; Patron $100-$1,000; Founder $1,000 & up.

Keytesville

GENERAL STERLING PRICE MUSEUM, 412 Bridge St., Keytesville, MO 65261-1016. Mailing Address: P.O. Box 40, Keytesville, MO 65261-0040. Tel.: 660-288-3204.
Founded: 1964.
Key Personnel: Pres., Janet Weaver; Treas., Kessie Friesz; Sec., Jan Canyon.
Personnel Profile: Part-Time Paid 2.
Governing Authority: nonprofit. Parent Institution: Friends of Keytesville, Inc. Tax-exempt.
Institution Type/Description: General Museum.
Collections: china; glass; silver; geology; fossilized insects; furniture; arrowheads; items belonging to Gen. Price; manuscript collections.
Facilities: library pertaining to the era of Gen. Price available on premises.
Activities: guided tours; permanent & temporary exhibitions. Annual Event: Sterling Price Day Celebration in September.
Publications: brochures, Gen. Sterling Price; book, Keytesville-150 Years, John Price Emigrant (Jamestown Colony) 1620 with some of his descendents; 1996 Keytesville Cookbook; 1976 Keytesville Cookbook; Hill House Letters, letters between William & Bettie Hill, 1864-1865; Dalton, MO history books; Dalton, MO history books.
Hours & Admission Prices: May 15-Oct. 15 Mon. & Thurs.-Fri. 2-5. No charge; donations accepted. Closed holidays. &
Attendance: 300 (accurate)
Membership: Individual $10; Couple $15.

Kimmswick

ANHEUSER ESTATE, 6000 Windsor Harbor Lane, Kimmswick, MO 63053. Mailing Address: P.O. Box 27, Kimmswick, MO 63053-0094. Tel.: 636-464-7407 & 1698. Fax: 636-464-7477. Facebook: Anheuser Museum & Estate.
E-mail: tbenack@cityofkimmswick.org
Web Site: cityofkimmswick.org
Institution Type/Description: Historic House Museum: housed in the former home of the Anheuser family; built in 1867.
Collections: Anheuser family & company history; personal artifacts; period furnishings; photographs; dinnerware; clothing; books.
Facilities: 23-acre estate; banquet facilities.
Activities: rental facilities; guided tours.
Hours & Admission Prices: Thurs. 12-4. Adults $5.

KIMMSWICK HISTORICAL SOCIETY MUSEUM, 6000 3rd St., Kimmswick, MO 63053. Mailing Address: P.O. Box 41, Kimmswick, MO 63053. Tel.: 636-464-8687.
Founded: 1977.
Key Personnel: Dir., Loretta Boemler.
Personnel Profile: Part-Time Volunteers 10.
Governing Authority: Parent Institution: Kimmswick Historical Society. Tax-exempt.
Institution Type/Description: Historical Society Museum.
Collections: local history & culture; photographs; period artifacts; maps.
Hours & Admission Prices: March-Dec. Sat.-Sun. 1-4. No charge; donations accepted. &

Attendance: 500 (estimated)
Membership: Individual $10; Lifetime $100.

King City

TRI-COUNTY MUSEUM & HISTORICAL SOCIETY, 508 N. Grand Ave., King City, MO 64463. Tel.: 660-535-4391. Fax: 660-535-4537.
Institution Type/Description: History Museum.
Collections: local history & culture; period furnishings; personal artifacts; photographs.
Activities: Annual Event: Living History Day in September.
Hours & Admission Prices: By appointment. No charge.

Kingsville

POWELL GARDENS, KANSAS CITY'S BOTANICAL GARDEN, 1609 N.W. U.S. Hwy. 50, Kingsville, MO 64061. Tel.: 816-697-2600. Facebook: Powell Gardens.
E-mail: info@powellgardens.org
Web Site: www.powellgardens.org
Institution Type/Description: Botanical Garden.
Collections: over 200 varieties of water plants; native prairie grasses & flowers; native woodland wildflowers; perennials; edible landscape garden.
Activities: performances; temporary exhibitions; school & youth groups; educational programs; special events. Museum Sponsors: Annual Festival of Butterflies.
Hours & Admission Prices: May-Sept. daily 9-6; Oct.-April daily 9-5. Admission: April-Oct. adults $10, seniors $9, children 5-12 $4. Nov.-March adults $7, seniors $6, children 5-12 $3. Closed New Year's Day; Thanksgiving; Christmas.

Kirksville

ADAIR COUNTY HISTORICAL SOCIETY, INC., 211 S. Elson St., Kirksville, MO 63501-3466. Tel.: 660-665-6502.
E-mail: peeve@cableone.net
Web Site: www.adairchs.org
Founded: 1976.
Congressional District: 9
Key Personnel: Pres., Mr. Charles Frost; Treas., Mrs. Denise Treasure; Cur., Archivist & Security, Mr. Walter Davison.
Personnel Profile: Full-Time Volunteers 2; Part-Time Volunteers 4.
Operating Expenses: 10,200
Operating Income: 9,600
Governing Authority: private; nonprofit organization. Tax-exempt: 501(c)(3).
Institution Type/Description: Historical Society Museum.
Collections: history of Adair county from prehistoric to modern times; Indian artifacts; artifacts from early settlers to modern times.
Research Fields: genealogy.
Facilities: 200-vol. library available for use on premises; 2,000 sq. ft. exhibit space. Books for sale.
Activities: guided tours; traveling exhibitions. Museum Sponsors: dinners.
Publications: quarterly newsletter, The Adair Historian; booklet, A Pocket Book History of Adair County; booklets, Sgt. John Shaver's The Last Roll Call; And There Arose A Mighty Wind.
Hours & Admission Prices: Wed.-Fri. 1-4. No charge; donations accepted. Closed New Year's Eve & Day; Memorial Day; Independence Day; Labor Day; Thanksgiving; Christmas Eve, Day & week.
Attendance: 224 (accurate)
Membership: Individual $15-$60; Family $25-$100; Sustaining $50-$200; Life $200-$600.

MUSEUM OF OSTEOPATHIC MEDICINE, (M), 800 W. Jefferson St., Kirksville, MO 63501-1443. Tel.: 660-626-2359. Fax: 660-626-2984.
E-mail: museum@atsu.edu
Web Site: www.atsu.edu/museum
Founded: 1978.
Congressional District: 9
Key Personnel: Dir., Jason Haxton; Cur., Debra Loguda-Summers; Asst. Dir., Robert Clement; Registrar, Kevin Allred.
Personnel Profile: Full-Time Paid 5; Full-Time Volunteers 1; Part-Time Paid 2; Part-Time Volunteers 5; Interns 1.
Governing Authority: nonprofit organization. Parent Institution: A.T. Still University of Health Sciences. Tax-exempt: 501(c)(3).
Institution Type/Description: Medical Museum Complex: housed in the education building on the campus of Kirksville College of Osteopathic Medicine.

Collections: artifacts, literature, memorabilia & material illustrative of the development of the osteopathic profession. Historic Houses: log cabin birthplace of A.T. Still, founder of osteopathy; c.1892 first classroom building.
Research Fields: 19th-century medicine; local history; osteopathic/medical history.
Facilities: 2,500-vol. library pertaining to osteopathic & related medical subjects available to the public. Books, pictures & gift items for sale.
Activities: guided tours; school discovery programs; health & wellness programs; loan, temporary, & traveling exhibitions.
Publications: brochure; semiannual newsletter; exhibition catalogues.
Hours & Admission Prices: Mon.-Wed. & Fri. 10-4, Thurs. 10-7, Sat. 12-4; other times by appointment. No charge; donations accepted. &
Attendance: 8,910 (accurate)
Membership: Student $5; Intern $15; Associate $25; Organization & Friend of the Museum $50; Patron $100; Life $1,000; Director's Circle $5,000.

TRUMAN STATE UNIVERSITY OBSERVATORY, University Farm, Boundary St., Kirksville, MO 63501. Mailing Address: Macgruder Hall 3168, Physics Dept., Kirksville, MO 63501. Tel.: 660-785-4594.
Institution Type/Description: Observatory.
Collections: astronomy; space science.
Activities: research; outreach programs.
Hours & Admission Prices: Call for hours.

Kirkwood

KIRKWOOD HISTORICAL SOCIETY - MUDD'S GROVE, 302 W. Argonne, Kirkwood, MO 63122. Mailing Address: P.O. Box 220602, Kirkwood, MO 63122. Tel.: 314-965-5151.
E-mail: info@kirkwoodhistoricalsociety.com
Web Site: www.kirkwoodhistoricalsociety.com
Founded: 1962.
Personnel Profile: Part-Time Volunteers 23.
Governing Authority: Parent Institution: Kirkwood Historical Society. Tax-exempt.
Institution Type/Description: Historical Society Museum: house built in 1860.
Collections: local history; period furnishings; photographs.
Activities: rental facilities.
Publications: quarterly newsletter; quarterly, Historical Review.
Hours & Admission Prices: Thurs. & Sun. 1-4. Adults $3; members no charge.
Attendance: 200 (estimated)
Membership: Historian $35; Kirkwood $70; Business $100; Henry T. Mudd $250; Lifetime $1,000.

Laclede

GENERAL JOHN J. PERSHING BOYHOOD HOME STATE HISTORIC SITE, 1100 Pershing Dr., Laclede, MO 64651. Mailing Address: P.O. Box 141, Laclede, MO 64651-0141. Tel.: 660-963-2525. Fax: 660-963-2520. Facebook: Pershing Home SHS.
E-mail: moparks@dnr.mo.gov
Web Site: www.mostateparks.com/gen-john-j-pershing-boyhood-home-state-historic-site
Founded: 1960.
Congressional District: 11
Key Personnel: Site Admin., Denzil R. Heaney.
Governing Authority: state. Parent Institution: Dept. of Natural Resources. Subsidiary Institution: Div. of State Parks. Tax-exempt.
Institution Type/Description: State Historic Site.
Collections: life & military career of Gen. John J. Pershing from early childhood to his death in 1948; personal artifacts; period furnishings; bronze statue of Gen. Pershing; Wall of Honor. Historic Building: c.1870 one room schoolhouse.
Research Fields: First World War; Punitive Expedition; Philippine Insurrection; Spanish American War; Russo-Japanese War; Buffalo soldiers of 9th & 10th cavalry; Indian Wars; Civil War in northern Missouri; military leaders of northern Missouri relating to the life & career of General John J. Pershing.
Facilities: library of books, pamphlets, clippings & documents available for use on premises; war memorial.
Activities: guided tours. Museum Sponsors: Pershing Days; Victorian Christmas Open House.
Publications: brochure, Gen. John J. Pershing Boyhood Home.
Hours & Admission Prices: April-Sept. Mon.-Sat. 10-4, Sun. 12-6; Oct.-March Tues.-Sat. 10-4. Tours: adults $4, youth 6-17 $3; discount to groups; children 5 & under no charge. Closed New Year's Day; Easter; Thanksgiving; Christmas. &

Attendance: 8,500 (estimated)

LOCUST CREEK COVERED BRIDGE STATE HISTORIC SITE, 16957 Dart Rd., Laclede, MO 64651. Mailing Address: P.O. Box 141, Laclede, MO 64651-0141. Tel.: 660-963-2525.
E-mail: moparks@dnr.mo.gov
Web Site: mostateparks.com/locust-creek-covered-bridge-state-historic-site
Founded: 1968.
Congressional District: 11
Key Personnel: Site Admin., Denzil R. Heaney.
Personnel Profile: Full-Time Paid 3; Part-Time Paid 1.
Governing Authority: state. Parent Institution: Missouri Dept. of Natural Resources, Div. of State Parks, P.O. Box 176, Jefferson City, MO 65102. Tax-exempt.
Institution Type/Description: Historic Site: 1868 151-foot covered bridge.
Collections: covered bridges; Pike's Peak ocean to ocean highway across Northern Missouri; Pershing transportation route.
Research Fields: history of covered bridges in America; local history; Pike's Peak ocean to ocean highway; Pershing transportation route.
Facilities: quarter mile foot trail to bridge.
Activities: guided tours as requested.
Publications: pamphlet on 4 covered bridges in Missouri.
Hours & Admission Prices: Daily sunrise to sunset. No charge; donations accepted.
Attendance: 18,400 (accurate)

Lamar

HARRY S TRUMAN BIRTHPLACE STATE HISTORIC SITE, 1009 Truman Ave., Lamar, MO 64759. Tel.: 417-682-2279. Fax: 417-682-6304.
E-mail: moparks@dnr.mo.gov
Web Site: mostateparks.com/harry-s-truman-birthplace-state-historic-site
Founded: 1959.
Congressional District: 4
Key Personnel: Site Admin., Beth BAzal.
Governing Authority: state. Parent Institution: Div. of Parks, Recreation & Historic Preservation, Missouri Dept. of Natural Resources, P.O. Box 176, Jefferson City, MO 65102. Tax-exempt.
Institution Type/Description: Historic House & Museum: housed in c.1880 Harry S. Truman birthplace.
Collections: period furnishings.
Activities: guided tours.
Publications: brochure, Harry S Truman Birthplace.
Hours & Admission Prices: Grounds: daily sunrise to sunset; Office: March-Oct. Wed.-Sat. 10-4, Sun. 12-4; Nov.-Feb. Wed.-Sat. 10-4. No charge; donations accepted. &
Attendance: 26,649 (accurate)

Laurie

OLD ST. PATRICK'S CHURCH AND MUSEUM, 810 Rte. O, Laurie, MO 65037. Mailing Address: P.O. Box 1098, Laurie, MO 65038. Tel.: 573-374-7855. Fax: 573-374-0627.
E-mail: parishsecretary@shrineofstpatrick.com
Key Personnel: Dir., Rev. Pat Dolan.
Personnel Profile: Part-Time Volunteers 60.
Governing Authority: Parent Institution: St. Patrick Catholic Church.
Institution Type/Description: Church Museum: founded by Irish immigrants; built in 1868. Listed on the National Register of Historic Places.
Collections: local history; historic documents & vestments.
Hours & Admission Prices: June-Aug. Sun. 11-1. No charge.
Attendance: 500 (estimated)

Lawson

WATKINS WOOLEN MILL STATE HISTORIC SITE & PARK, 26600 Park Rd., N., Lawson, MO 64062-8939. Tel.: 816-580-3387 & 3782. Fax: 816-580-3784.
E-mail: watkins.woolen.mill.state.historic.site@dnr.mo.gov
Web Site: www.mostateparks.com
Founded: 1964.
Congressional District: 6
Key Personnel: Facility Mgr., Mike Beckett; Historic Site Admin., Matt Carletti; Part Supt., Ron Sutton; Coord. Education, Melissa Hall; Coord. Period Clothing, Terri Gardner; Volunteer Coord., Amanda Coonce; Park Maintenance II, Russell Teague; Park Maintenance II, Farriel O'Dell; Park Maintenance II, Joe Green; Park Maintenance III, Tim Clifford; Clerk Typist, Linda Zink.

Personnel Profile: Full-Time Paid 11; Part-Time Paid 9; Part-Time Volunteers 60.
Governing Authority: state. Parent Institution: Div. of State Parks, Missouri Dept. of Natural Resources, P.O. Box 176, Jefferson City, MO 65102. Tax-exempt.
Institution Type/Description: Industrial & Agricultural Museum: 1839 Bethany, 19th-century plantation.
Collections: 60 textile production machines including carding engines, spinning jacks, looms, pliers & finishing machines; complete textile factory; over 30 archaeological sites of farm & factory support buildings & workers houses. Historic Buildings: 1860 Watkins woolen mill; 1850 W.L. Watkins family home; 1856 Franklin school (Octagonal); 1871 Mt. Vernon church; 1850-1880 miscellaneous outbuildings.
Research Fields: industrialization; agricultural history, 1820-1920; mid-1800s, textile industry.
Facilities: 600-vol. library of books, pamphlets, catalogs, manuscripts & photographs available on premises. Books for sale.
Activities: guided tours; lectures; exhibitions; special events.
Publications: brochure, Watkins Woolen Mill; walking tour guide; guidebook; video tape.
Hours & Admission Prices: Summer: Mon.-Sat. 9:30-5, Sun. 10:30-5; Winter: Mon.-Sat. 10-4:30, Sun. 11-4:30. Adults $2.50, children 6-12 $1.50; discounts to pre-scheduled groups of 15 or more; children under 6 no charge. Closed New Year's Day; Thanksgiving; Christmas. &
Attendance: 15,500 (accurate)

Lebanon

NATURE INTERPRETIVE CENTER, Bennett Spring State Park, Lebanon, MO 65536-6797. Mailing Address: Bennett Spring State Park, 26250 Hwy. 64A, Lebanon, MO 65536-6797. Tel.: 417-532-3925 & 4338. Fax: 417-532-7006.
Web Site: www.mostateparks.com/bennett.htm
Founded: 1969.
Congressional District: 7
Key Personnel: Dir., Diane Tucker.
Personnel Profile: Full-Time Paid 1; Part-Time Paid 1; Part-Time Volunteers 4.
Governing Authority: state. Div. of Parks & Historic Preservation, Missouri Dept. of Natural Resources, P.O. Box 176, Jefferson City, MO 65102. Tax-exempt.
Institution Type/Description: Natural History Museum.
Collections: botany; zoology; geology.
Research Fields: botany; zoology; geology; ecology; area history.
Facilities: nature center with displays; indoor classroom; work area; amphitheater; trails.
Activities: nature hikes; lectures; film & slide programs; formally organized education programs; permanent exhibitions; adult programs; demonstrations; outdoor education programs.
Publications: booklet, Self Guiding Trail; bird checklist.
Hours & Admission Prices: Feb. 25-Oct. daily 10-4. No charge; donations accepted. Closed Nov.-Feb. 24. &
Attendance: 35,000 (estimated)

ROUTE 66 MUSEUM AND RESEARCH CENTER, Lebanon-Laclede County Library, 915 S. Jefferson Ave., Lebanon, MO 65536. Tel.: 417-532-2148.
E-mail: mspangler@lebanon-laclede.lib.mo.us
Founded: 2004.
Governing Authority: Parent Institution: Lebanon-Laclede County Library. Tax-exempt.
Institution Type/Description: History Museum.
Collections: local history; recreated 1950s gas station & diner; period cars; Route 66 books, magazines & videos.
Hours & Admission Prices: Mon.-Thurs. 8-8, Fri.-Sat. 8-5. No charge; donations accepted.
Attendance: 9,500 (accurate)

Lecoma

MISSOURI UNIVERSITY OF SCIENCE AND TECHNOLOGY MINERALS COLLECTION, 125 McNutt Hall, Lecoma, MO 65401. Mailing Address: Geological Sciences & Engineering, 129 McNutt Hall, 1400 N. Bishop, Rolla, MO 65409-0140. Tel.: 573-341-4616. Fax: 573-341-6935.
E-mail: rocks@mst.edu
Web Site: gse.mst.edu
Formerly: University of Missouri-Rolla Minerals Museum
Founded: 1870.
Congressional District: 8

Key Personnel: Prof. & Head Geology & Geophysics, Dr. Francisca Oboh-Ikuenobe, Ph.D.
Governing Authority: university.
Institution Type/Description: Mineralogy Museum.
Collections: mineralogy; paleontology; geology.
Research Fields: economic geology.
Activities: guided tours; permanent exhibitions.
Hours & Admission Prices: Mon.-Fri. 8-5. No charge. Closed holidays. &
Attendance: 200 (estimated)

Lee's Summit

LEE'S SUMMIT HISTORICAL SOCIETY MUSEUM, 222 S.E. Main, Lee's Summit, MO 64063-2332. Mailing Address: 1923 S.E. 3rd St., Lee's Summit, MO 64063. Tel.: 816-525-9440.
E-mail: derbyshirelane@prodigy.net
Web Site: www.leessummithistory.net
Founded: 1965.
Key Personnel: Dir., Kathy Smith
Institution Type/Description: Historical Society Museum: housed in the 1905 train depot.
Collections: maps, photographs & artifacts pertaining to Lee's Summit; Civil War artifacts.
Research Fields: abstracts; family history.
Publications: brochure; exhibit booklet; local history books.
Hours & Admission Prices: Sat. 10-4; other times by appointment. No charge; donations accepted. &
Membership: Individual $10; Life $100.

MISSOURI TOWN 1855, 8010 E. Park Rd., Lee's Summit, MO 64064. Mailing Address: 22807 Woods Chapel Rd., Blue Springs, MO 64015-9799. Tel.: 816-503-4860. Fax: 816-795-7938. TDD: 800-735-2966.
E-mail: gjulich@jacksongov.org
Web Site: www.jacksongov.org
Founded: 1963.
Congressional District: 5
Key Personnel: Dir. Parks & Rec., Michele Newman; Program Specialist, Cindy Henley; Supt. Heritage Museums & Programs, Gordon Julich; Museum Admin., Gary Sutton.
Personnel Profile: Full-Time Paid 4; Full-Time Volunteers 6; Part-Time Paid 2; Part-Time Volunteers 150; Interns 1.
Governing Authority: county. Parent Institution: Jackson County Parks & Rec, 22807 Woods Chapel Rd., Blue Springs, MO 64015. Tax-exempt: 170(c)(3).
Institution Type/Description: Village Museum: over 25 structures c.1821-1860 from western Missouri.
Collections: agriculture; costumes; decorative arts; folklore; general; history; technology; textiles; furnishings; period gardens. Historic Buildings: 1855 mercantile, blacksmith shop & surrey; church; law office; livery stable; school & stagecoach stop tavern.
Research Fields: 1850-1860 pioneer lifestyles.
Facilities: wedding chapel; nature trails; surrounding park offers boating, fishing, camping & picnic facilities.
Activities: living history demonstrations; workshops; seminars; special events; volunteer program; weddings; corporate picnics; educational programs; youth activities; outreach educational programs. Museum Sponsors: annual Arts & Crafts Festival; Living History Week; Independence Day; Children's Day; Tradesmen Day Spirits From the Past; Christmas Open House; Special Interpretive weekends.
Publications: Guidebook: Missouri Town 1855, A Program in Architectural Preservation; brochures; quarterly newsletter.
Hours & Admission Prices: March-Nov. 15 Tues.-Sun. 9-4:30; Nov. 16-Feb. Sat.-Sun. 9-4:30. Adults $5, senior citizens & youths $3; discounts to groups with advance reservation; members & children under 5 no charge. Closed New Year's Day; Veterans Day; Truman's Birthday; Thanksgiving; Christmas. &
Attendance: 13,900 (accurate)
Membership: Individual $35; Family $45.

Lexington

BATTLE OF LEXINGTON STATE HISTORIC SITE, 1101 Delaware St., Lexington, MO 64067. Mailing Address: P.O. Box 6, Lexington, MO 64067-0006. Tel.: 660-259-4654. Fax: 660-259-2378. Facebook: Battle of Lexington SHS.
E-mail: battle.of.lexington.state.historic.site@dnr.mo.gov
Web Site: www.mostateparks.com/lexington
Founded: 1959.

Congressional District: 4
Key Personnel: C.E.O. & Site Admin., Janae Fuller.
Personnel Profile: Full-Time Paid 3; Part-Time Paid 6; Part-Time Volunteers 6.
Governing Authority: state. Parent Institution: State of Missouri, Dept. of Natural Resources, P.O. Box 176, Jefferson City, MO 65102. Tax-exempt.
Institution Type/Description: Historic House & Battlefield: 1853 Col. Oliver Anderson & Tilton Davis Home, located at site of Battle of the Hemp Bales.
Collections: Civil War artifacts; manuscripts; period furnishings; original Union trenches.
Research Fields: Civil War history; hemp cultivation; Missouri River Valley lifestyle; Lexington; the Anderson & Davis families; soldiers & units that fought at the battle.
Facilities: Gift items for sale.
Activities: guided tours; Civil War encampments & reenactments; living history events.
Publications: brochure, Battle of Lexington.
Hours & Admission Prices: Call for hours. Adults $4; children under 13 $2.50. Closed New Year's Day; Easter; Thanksgiving; Christmas. &
Attendance: 25,000 (estimated)

LEXINGTON HISTORICAL MUSEUM, 112 S. 13th St., Lexington, MO 64067-1402. Mailing Address: P.O. Box 121, Lexington, MO 64067. Tel.: 660-259-6313 & 2900.
E-mail: rslusher@yahoo.com
Web Site: www.lexingtonmuseum.org
Founded: 1924.
Congressional District: 3
Key Personnel: Dir., Roger Slusher; Pres. (V), Mike Kramer.
Personnel Profile: Part-Time Volunteers 35.
Governing Authority: Parent Institution: Lexington Historical Association. Tax-exempt.
Institution Type/Description: History Museum: housed in the 1846 Cumberland Presbyterian Church.
Collections: Osage Indians; the Pony Express; steamboats; Civil War; coal mining.
Publications: quarterly newsletter.
Hours & Admission Prices: Call 660-259-2900 for hours. Adults $3, students $2; members no charge.
Attendance: 500 (estimated)
Membership: Individual $20; Family $25.

Liberty

CLAY COUNTY HISTORICAL MUSEUM, 14 N. Main St., Liberty, MO 64068-1638. Tel.: 816-792-1849.
E-mail: info@claycountymuseum.org
Web Site: www.claycountymuseum.org
Founded: 1965.
Congressional District: 6
Key Personnel: Pres. (V), Carolyn Hatcher; Cur., Jay Thorne; Sec., Ann Cole.
Personnel Profile: Part-Time Paid 1; Part-Time Volunteers 25.
Governing Authority: nonprofit organization. Parent Institution: Clay County Museum & Historical Society, Inc., 14 N. Main, Liberty, MO 64068. Tax exempt: 501(c)(3).
Institution Type/Description: Local History Museum: housed in early drug store building; built in 1865.
Collections: medical & druggist equipment; household furniture dating from 1800-1930s; clothing & personal belongings; items relating to Civil War, Mexican War and World War I & World War II; Indian artifacts; history of local colleges.
Research Fields: local & regional history.
Facilities: archives.
Activities: guided tours; monthly meetings; special projects.
Publications: quarterly newsletter, Our Clay County Heritage.
Hours & Admission Prices: Feb.-Dec. Mon.-Sat. 1-4. No charge; donations accepted. Closed national holidays.
Attendance: 4,200 (estimated)
Membership: Bronze $20; Silver $30; Gold $50; Platinum $100 & up.

JESSE JAMES BANK MUSEUM, (M), 103 N. Water, Liberty, MO 64068-1736. Mailing Address: c/o Friends of the James Farm, P.O. Box 404, Liberty, MO 64069. Tel.: 816-736-8510. Fax: 816-736-8501. Facebook: Jesse James Bank Museum.
E-mail: historicsites@claycountymo.gov
Web Site: www.claycountymo.gov/Historic_Sites/Jesse_James_Bank_Museum
Founded: 1966.
Congressional District: 5
Key Personnel: Site Admin., Beth Beckett.

Personnel Profile: Full-Time Paid 2; Part-Time Paid 20; Part-Time Volunteers 1; Interns 1.
Governing Authority: county. Parent Institution: Clay County Dept. of Parks, Recreation & Historic Sites. Tax-exempt.
Institution Type/Description: Banking Museum: housed in 1858 old Liberty Bank building, site of James Gang bank robbery, Feb. 13, 1866.
Collections: Civil War banking; pictures & documents pertaining to the James Gang.
Research Fields: James; Younger; Quantrell; banking history.
Facilities: Gifts & books for sale.
Activities: guided tours; permanent exhibitions.
Publications: books, Good Bye, Jesse James; The Ancestry of Jesse James.
Hours & Admission Prices: Mon.-Sat. 10-4. Adults $6, senior citizens 62 & up $5.50, children 8-15 $3.50; discounts to groups; children under 8 no charge. &
Attendance: 4,420 (accurate)
Membership: Friends of the James Farm: Border Ruffian $20; Bushwhacker $30; Road Agent $100; Long Rider $250.

LIBERTY JAIL HISTORIC SITE, 216 N. Main, Liberty, MO 64068-1629. Tel.: 816-781-3188. Fax: 816-781-7311.
Formerly: Historic Liberty Jail Visitors Center Historic Site
Founded: 1963.
Key Personnel: Dir., Elder Douglas Brenchley.
Personnel Profile: Full-Time Volunteers 12; Part-Time Volunteers 4.
Governing Authority: church group. Parent Institution: Church of Jesus Christ of Latter-Day Saints, 50 E. N. Temple., Salt Lake City, UT 84150. Tax-exempt.
Institution Type/Description: Religious History Museum.
Collections: religious history items. Historic Building: 1859 jail.
Research Fields: history.
Facilities: 2 film rooms; visitors center.
Activities: guided tours; lectures; films; interactive video.
Publications: brochure, Liberty Jail.
Hours & Admission Prices: Daily 9-9. No charge. &
Attendance: 48,000 (accurate)

Linn

OSAGE COUNTY CULTURAL HERITAGE CENTER - ZEWICKI HOUSE MUSEUM, 402 Main St., Linn, MO 65051. Mailing Address: P.O. Box 402, Linn, MO 65051. Tel.: 573-897-2932.
E-mail: historic@osageconnect.net
Web Site: osagecounty.org
Founded: 1986.
Congressional District: 4
Key Personnel: Chm. (V), Roberta Schwinke; Pres. (V), Rev. David Means.
Personnel Profile: Part-Time Volunteers 8.
Governing Authority: Parent Institution: Osage County Historical Society. Tax-exempt.
Institution Type/Description: Historic House Museum: former home of dentist, Dr. E. T. Zewicki and his wife Amy; built in 1895.
Collections: family history; period furnishings; personal artifacts; photographs; Osage County history; Native American artifacts.
Major Exhibits: The Civil War in Missouri - A State Divided, 6/14/14-8/10/14.
Research Fields: genealogy; county history.
Facilities: library.
Activities: quarterly dinner meetings with programs.
Publications: monthly newsletter, Osage County Historical Society.
Hours & Admission Prices: April 2-Oct. Wed.-Thurs. 9-5, Sat. 9 to noon, Sun. 2-4; Nov.-Dec. 14 Wed.-Thurs. 9-5, Sat. 9 to noon. No charge; donations accepted. Closed holidays.
Attendance: 350 (estimated)
Membership: Individual $20; Joint & Business $25; Sustaining $50; Life $500.

Linn Creek

CAMDEN COUNTY HISTORICAL SOCIETY & MUSEUM, 206 S. Locust St., Linn Creek, MO 65052. Mailing Address: P.O. Box 19, Linn Creek, MO 65052-0019. Tel.: 573-346-7191.
Web Site: camdencountymuseum.com
Founded: 1962.
Congressional District: 9
Key Personnel: Chm. (V), Patricia Kitterman; Pres. (V), Daphne Jeffries; Museum Shop Mgr., Shirley Childers.
Personnel Profile: Full-Time Volunteers 2; Part-Time Volunteers 17.
Governing Authority: Parent Institution: Camden County Historical Society.
Institution Type/Description: History Museum.
Collections: local history & culture; photographs.

Publications: Camden County Historian; books, Before the Dam Waters; Camden County History; Early History of Camden County; The Meteorite in Decaturville; 2000 Glimpses of History.
Hours & Admission Prices: March-Oct. 10-4; other times by appointment. No charge; donations accepted. &

Membership: Individual $12; Family $15

Lone Jack

LONE JACK CIVIL WAR BATTLEFIELD MUSEUM & SOL-DIER'S CEMETERY, 301 S. Bynum Rd., Lone Jack, MO 64070-8508. Mailing Address: c/o Lone Jack Historical Society, P.O. Box 34, Lone Jack, MO 64070. Tel.: 816-697-8833.
Web Site: www.historiclonejack.org
Founded: 1964.
Congressional District: 4
Key Personnel: Chm. & Pres. (V), Alinder Miller.
Governing Authority: county. Affiliated with Jackson County Parks & Recreation, County Courthouse, Independence, MO 64050. Tax-exempt: 170(c)(1).
Institution Type/Description: Civil War Museum: housed in modern museum building on site of the Battle of Lone Jack, Aug. 16, 1862.
Collections: military; local history; Civil War; artifacts; dioramas; electric map.
Research Fields: Jackson County Civil War activity; local history; biography of soldiers at Lone Jack.
Facilities: picnic area.
Activities: guided tours; permanent & temporary exhibitions; living history interpretation. Annual Event: Commemoration in August.
Publications: pamphlet.
Hours & Admission Prices: March-Oct. Wed.-Sat. 10-4, Sun. 1-4; Nov.-Feb. Sat. 10-4, Sun. 1-4. Adult $3, children 6-12 $1; children under 6 no charge. Closed New Year's Day; Easter; Thanksgiving; Christmas. &
Attendance: 3,357 (accurate)

Macon

MACON COUNTY HISTORICAL SOCIETY, 1402 S. Missouri, Macon, MO 63552-4452. Mailing Address: P.O. Box 304, Macon, MO 63552-4452. Tel.: 660-395-0266 & 5135; 660-346-1455.
E-mail: mchsmuse@yahoo.com
Web Site: maconcountyhistoricalsociety.com
Founded: 1937.
Congressional District: 8
Key Personnel: Pres. (V), Ronald Watts, Sr.; Vice Pres., Harold Burkhardt; Treas., Ruth Master; Sec., Beth Watts; Tour Guide & Historian, Merlyn Amedei
Governing Authority: nonprofit organization. Tax-exempt: 501(c)(3).
Institution Type/Description: Historical Society Museum.
Collections: local history & culture; period furnishings; personal artifacts; photographs; military; mining; hospital artifacts; agriculture.
Activities: monthly educational, historical programs.
Publications: quarterly flyers.
Hours & Admission Prices: Thurs.-Fri. 2-4, Sat. 10am-12pm; other times by appointment. No charge; donations accepted. &
Attendance: 350
Membership: Individual $10; Couple $15.

Malden

BOOTHEEL YOUTH MUSEUM, (M), 700A N. Douglas, Malden, MO 63863-1510. Mailing Address: P.O. Box 182, Malden, MO 63863-0182. Tel.: 573-276-3600.
E-mail: info@bootheelyouthmuseum.org
Web Site: www.bootheelyouthmuseum.org
Founded: 1994.
Congressional District: 8
Key Personnel: Exec. Dir., Patsy Reublin; Chm. (V), Liz Provance; Museum Shop Mgr., Tyra Parker.
Personnel Profile: Full-Time Paid 3; Full-Time Volunteers 3; Part-Time Paid 5; Part-Time Volunteers 15.
Institution Type/Description: Children's Science Center.
Collections: hands-on exhibits.
Facilities: 22,000 sq. ft. exhibit space; 180-seat theater. Museum-related items for sale.
Activities: educational programs; birthday parties; rental facilities.

Hours & Admission Prices: Tues.-Sat. 10-4, Sun. 1-4, Mon. by appointment. Children 3-17 $5, adults $3; discounts to Assoc. of Children's Museums & groups of 8 or more. &
Attendance: 28,796 (accurate)
Membership: $50-$100.

MAAPS MILITARY MUSEUM, Malden Industrial Park Office, 3077 Mitchell Dr., Malden, MO 63863. Mailing Address: P.O. Box 411, Malden, MO 63863. Tel.: 573-276-2279. Fax: 573-276-2296.
E-mail: info@maaps.net
Web Site: maaps.net/museum.html
Founded: 1999.
Congressional District: 8
Institution Type/Description: Military History Museum.
Collections: military history, uniforms & memorabilia; Malden Army Airfield; Malden Air Base; photographs; personal artifacts; medals.
Hours & Admission Prices: Mon.-Fri. 8-12 & 1-5. No charge; donations accepted. &

MALDEN HISTORICAL MUSEUM, 201 N. Beckwith, Malden, MO 63863. Mailing Address: P.O. Box 142, Malden, MO 63863-0142. Tel.: 573-276-5008.
Institution Type/Description: History Museum.
Collections: local history & culture; photographs; period furnishings & clothing; personal artifacts; dolls.
Hours & Admission Prices: Wed.-Thurs. & Sat. 1:30-4:30; groups by appointment.

Mansfield

LAURA INGALLS WILDER-ROSE WILDER LANE HIS-TORIC HOME & MUSEUM, 3068 Hwy. A, Mansfield, MO 65704-8104. Tel.: 417-924-3626. Fax: 417-924-8580.
E-mail: info@lauraingallswilderhome.com
Web Site: www.lauraingallswilderhome.com
Founded: 1957.
Congressional District: 146
Key Personnel: Dir., Jean C. Coday; Treas., Jane K. Coday; Museum Shop Mgr., Kim Miller.
Personnel Profile: Full-Time Paid 2; Full-Time Volunteers 4; Part-Time Paid 20.
Governing Authority: private; nonprofit organization. Parent Institution: Laura Ingalls Wilder Home Association. Tax-exempt: 501(c)(3).
Institution Type/Description: History Museum: next to the home of author Laura Ingalls Wilder.
Collections: five of the original nine Little House books handwritten manuscripts; family photographs; Pa's fiddle; historic artifacts; organ; translations of Little House books; handwritten manuscripts; clothing.
Research Fields: history of Laura and Almanzo Wilder & Rose Wilder Lane.
Activities: guided tours.
Publications: brochures; biannual newsletter for association members, Rocky Ridge Review; mail order catalog.
Hours & Admission Prices: March-Nov. 15 Mon.-Sat. 9-5, Sun. 12:30-5; Nov. 16 to mid-Dec. by appointment only. Adults $10, senior citizens 65 & over $8, children 6-18 $6; children under 6 no charge. Closed Easter. &
Attendance: 55,000 (estimated)
Membership: Student $25; Individual $50; Supporter $100; Life: Sponsor $500; Patron $1,000; Benefactor $5,000.

MANSFIELD HISTORICAL SOCIETY & MUSEUM, 111 W. Park Square, Mansfield, MO 65704. Mailing Address: P.O. Box 374, Mansfield, MO 65704. Tel.: 417-924-4041.
E-mail: info@mansfieldhistorical.org
Web Site: mansfieldhistorical.org
Founded: 2003.
Key Personnel: Pres. (V), Donna Climer; Vice Pres. (V), Kathy Short.
Personnel Profile: Part-Time Volunteers 10.
Governing Authority: Tax-exempt.
Institution Type/Description: Historical Society Museum.
Collections: local history & culture; period furnishings; personal artifacts; photographs.
Hours & Admission Prices: Mon.-Fri. 10-3, Sat. 10-12; closed Oct. 15-Mar. 1; call for appointment for special tours (417-924-3396). No charge; donations accepted. &
Attendance: 1,000 (estimated)
Membership: Individual $12; Lifetime $120.

Marble Hill

BOLLINGER COUNTY MUSEUM OF NATURAL HISTORY, 207 Mayfield Dr., Marble Hill, MO 63764. Mailing Address: P.O. Box 676, Marble Hill, MO 63764. Tel.: 573-238-1174.
Founded: 2002.
Congressional District: 8
Key Personnel: Dir., Eva Dunn.
Personnel Profile: Part-Time Paid 2; Part-Time Volunteers 12.
Governing Authority: Tax-exempt: 501(c)(3).
Institution Type/Description: Natural History Museum.
Collections: local history & culture; photographs; personal artifacts; fossils; Lewis & Clark; Native American.
Facilities: Museum-related items for sale.
Hours & Admission Prices: Thurs.-Sat. 12-4:30; other times by appointment. Adults $2, children 18 & under $1.
Attendance: 4,800 (accurate)

Marceline

MAIN STREET USA CHARITABLE FOUNDATION, INC. - WALT DISNEY HOMETOWN MUSEUM, 120 E. Santa Fe Ave., Marceline, MO 64658-1144. Tel.: 660-376-3343.
E-mail: waltdisneymuseum@att.net
Web Site: www.waltdisneymuseum.org
Founded: 2001.
Congressional District: 8
Key Personnel: Dir., Kaye Malins; Chm. (V) & Museum Shop Mgr., Inez Johnson; Pres. (V), Greg Walton.
Personnel Profile: Part-Time Paid 1; Part-Time Volunteers 25.
Volunteer Hours: 2,450
Governing Authority: Tax-exempt.
Institution Type/Description: Walt Disney History & Railroad Museum.
Collections: personal letters from the early 1900s to the late 1960s; photographs of Roy & Walt Disney; Mickey Mouse dolls; Midget Autopia car; railroad artifacts; Marceline history.
Hours & Admission Prices: April-Oct. Tues.-Sat. 10-4, Sun. 1-4. Adults $5, children 6-10 $2.50; discounts to groups of 25 or more; children under 6 & Disney Fan Club members no charge. &
Attendance: 10,000 (estimated)
Membership: E.P. Ripley Park $75; Main Street USA $100; Dreaming Tree $250; Walts Barn $500; Boyhood Home $1,000; Santa Fe Depot $2,500.

Marshall

MARTIN COMMUNITY CENTER AND NICHOLAS BEAZLEY AVIATION MUSEUM, 1985 S. Odell, Marshall, MO 65340. Tel.: 660-886-2630. Fax: 660-886-2689.
E-mail: mccnbam@mmuonline.net
Web Site: nicholasbeazley.org
Institution Type/Description: Aviation History Museum.
Collections: aviation history; early airplanes.
Hours & Admission Prices: Sat. 10-4, Sun. 1-4.

SALINE COUNTY HISTORICAL SOCIETY MUSEUM, 101 N. Lafayette St., Marshall, MO 65340-1747. Tel.: 660-886-7546.
Institution Type/Description: Historical Society Museum.
Collections: artifacts & memorabilia pertaining to Saline County's history.
Hours & Admission Prices: April-Dec. Tues.-Fri. 9:30-12 & 1-4, Sat.-Sun. by appointment.

Marshfield

WEBSTER COUNTY HISTORICAL SOCIETY, 219 S. Clay St., Marshfield, MO 65706. Mailing Address: P.O. Box 13, Marshfield, MO 65706. Tel.: 417-468-7407 & 859-2036.
Founded: 1952.
Personnel Profile: Part-Time Volunteers 50.
Governing Authority: society bd. Branch: Research Center, 215 S. Clay, Marshfield, MO 65706. Tax-exempt: 501(c)(3).
Institution Type/Description: Historical Society Museum: housed in the former Carnegie Library; built in 1911.
Collections: local history & culture; period furnishings; personal artifacts; photographs; Edwin P. Hubble artifacts.
Major Exhibits: Quilt Exhibit, 10/14.
Research Fields: genealogy; local history.
Activities: Annual Event: Quilt Exhibit in October.
Publications: journal; books; quarterly newsletter.
Hours & Admission Prices: Jan.-March Fri. 1-4; April-Dec. Mon.-Sat. 1-4; other times by appointment. No charge; donations accepted.

Attendance: 2,200 (accurate)
Membership: Individual $5; Lifetime $50.

Maryland Heights

HISTORIC AIRCRAFT RESTORATION MUSEUM, 3127 Creve Coeur Mill Rd., Maryland Heights, MO 63146. Tel.: 314-434-3368. Fax: 314-878-6453.
Key Personnel: Dir., Al Stix
Institution Type/Description: Aircraft Museum.
Collections: classic, sport, & period aircraft.
Facilities: Museum-related items for sale.
Activities: airplane rides.
Hours & Admission Prices: Guided Tours: Sat.-Sun. 10-4; other times by appointment. Adults $10, children 5-12 $5; discounts to groups; children under 5 no charge. Airplane Rides $80-$110 per person.

Maryville

WARREN STUCKI MUSEUM OF BROADCASTING, Northwest Missouri State University, 800 University Dr., Maryville, MO 64468-6015. Tel.: 660-562-1163.
E-mail: rharris@nwmissouri.edu
Web Site: www.nwmissouri.edu
Congressional District: 6
Institution Type/Description: Broadcasting Museum.
Collections: 30 vintage radio sets; Edison phonograph with cylinder records; vintage television gear & recording equipment.
Hours & Admission Prices: Mon.-Fri. 8-5; other times by appointment. No charge. &

Memphis

DOWNING HOUSE/BOYER HOUSE & MEMPHIS DEPOT MUSEUM COMPLEX, 311 S. Main St., Memphis, MO 63555. Mailing Address: 314 Main St., Memphis, MO 63555. Tel.: 660-465-2275.
Founded: 1978.
Key Personnel: Chm. (V), Cur. & Museum Shop Mgr., Wilma June Kapfer
Governing Authority: Parent Institution: Scotland County Historical Society. Tax-exempt.
Institution Type/Description: History Museum: 14 room brick mansion built in 1858.
Collections: period furnishings; Ella Ewing; Scotland County's history; Civil War; aviation history; personal artifacts; railroad artifacts; Tom Horn history; school room with books and clothes; Pheasant Airplane built in 1927.
Publications: annual newsletter.
Hours & Admission Prices: April to Labor Day. Adults $5; children under 10 no charge.
Attendance: 750 (estimated)
Membership: Individual $3; Lifetime $25.

Mexico

AMERICAN SADDLEBRED HORSE MUSEUM, 501 S. Muldrow, Mexico, MO 65265-2082. Mailing Address: P.O. Box 398, Mexico, MO 65265-0398. Tel.: 573-581-3910.
E-mail: dkeller@audrain.org
Web Site: www.audrain.org/museums-saddlehorse.aspx
Founded: 1970.
Congressional District: 9
Key Personnel: Exec. Dir., Audrain County Histroical Society, Dana Keller; Cur., Tom Usnick.
Governing Authority: society; nonprofit organization. The Audrain County Historical Society. Branch Museum: Audrain Historical Museum. Tax-exempt.
Institution Type/Description: General History Museum: located on the grounds of the Audrain Historical Museum (see separate listing).
Collections: ribbons; bits; bridles; saddles; paintings; photographs; miscellaneous items connected with trainer Tom Bass; memorabilia from the Lee Brothers Stables, Cunningham Stables & Art Simmons Stables; George Ford Morris drawings.
Research Fields: American saddle horse.
Facilities: library of books magazines & photographs pertaining to the history of saddle horses; American Saddle Horse Register available.
Hours & Admission Prices: Tues.-Sat. 10-4, Sun. 1-4. Adults $5, children under 12 $3; members no charge. Closed national holidays. &
Attendance: 1,600 (accurate)

Membership: Individual $15; Family $30; Individual Life $300; Couple Life $500.

AUDRAIN HISTORICAL MUSEUM, GRACELAND, 501 S. Muldrow, Mexico, MO 65265-2082. Mailing Address: P.O. Box 398, Mexico, MO 65265-0398. Tel.: 573-581-3910. Fax: 573-581-7155.
E-mail: info@audrain.org
Web Site: www.audrain.org
Founded: 1952.
Congressional District: 9
Key Personnel: Exec. Dir., Audrain Historical Society, Dana Keller.
Governing Authority: society. Parent Institution: Audrain County Historical Society; nonprofit organization. Branch Museum: American Saddle Horse Museum. Tax-exempt.
Institution Type/Description: History Museum: housed in 1857 John P. Clark home, later lived in by Judge James E. Ross & his descendants, restored 1958.
Collections: furniture; paintings; piano; dolls; toys; antique glass; agriculture; costumes; history; Indian artifacts; industrial; medical; military; naval; 4,000 photographs, county related from 1837. Historic Building: Audrain country school, an original rural one-room schoolhouse.
Research Fields: genealogy; local history.
Facilities: 400-vol. library of local history books, pamphlets, genealogical & cemetery records.
Activities: guided tours; permanent & temporary exhibitions.
Publications: quarterly newsletter.
Hours & Admission Prices: Tues.-Sat. 10-4, Sun. 1-4. Adults $5, children 12 & under $3; members no charge. Closed holidays. &
Attendance: 1,600 (accurate)
Membership: Individual $15; Family $30; Individual Life $300; Couple Life $500.

Miami

VAN METER STATE PARK, 32146 N. Hwy. 122, Miami, MO 65344. Mailing Address: P.O. Box 47, Miami, MO 65344. Tel.: 660-886-7537. Fax: 660-886-7512.
Web Site: mostateparks.com/park/van-meter-state-park
Founded: 1932.
Key Personnel: Interpretive Resource Technician, Eric Fuemmeler.
Personnel Profile: Full-Time Paid 3; Part-Time Paid 4; Part-Time Volunteers 2.
Governing Authority: state. Parent Institution: State of Missouri Dept. of Natural Resources & Div. of State Parks. Tax-exempt.
Institution Type/Description: Park Museum & Visitor Center: located on Missouri Indian village archaeological site.
Collections: Missouri, Oneota, Woodland & Archaic Indian archaeological discoveries; early pioneer artifacts of Van Meter family.
Research Fields: archaeology of Missouri Indians.
Facilities: 40-seat auditorium. Books & gift items for sale.
Activities: guided tours; lectures; films; study clubs; organized education programs for children & adults; docent program; participatory exhibits.
Hours & Admission Prices: Park: daily 8-sunset. American Indian Cultural Center: May-Sept. Wed.-Sat. 10-4, Sun. 1-5; Oct.-March Sat. 10-4, Sun. 1-5; April Thurs.-Sat. 10-4, Sun. 1-5. No charge; donations accepted. &
Attendance: 50,000 (accurate)

Moberly

RANDOLPH COUNTY HISTORICAL SOCIETY HISTORY MUSEUM, 223 N. Clark St., Moberly, MO 65270-1540. Tel.: 660-263-9396.
E-mail: rchs@sbcglobal.net
Web Site: www.randolphhistory.com
Founded: 1976.
Key Personnel: Chm. (V), J.W. Ballinger.
Personnel Profile: Part-Time Volunteers 30.
Volunteer Hours: 3,000
Operating Expenses: 30,000
Operating Income: 30,000
Governing Authority: Subsidiary Institution: RR History Museum, 100 N. Sturgeon St., Moberly, MO. Tax-exempt.
Institution Type/Description: History Museum.
Collections: local history & culture; period furnishings; personal artifacts; photographs; military & railroad artifacts.
Activities: annual meeting with speakers; tours; meetings; radio history show; speakers bureau for class reunions.
Hours & Admission Prices: Mon. 10-3, Thurs. 1-3, Sat. 9am to noon; other times by appointment. Railroad Museum: April-Sept. 1-3. No charge; donations accepted.

Attendance: 2,000 (estimated)
Membership: Family $20; Patron $50; Business $100.

Monett

MONETT HISTORICAL SOCIETY, 705 E. Broadway, Monett, MO 65708. Mailing Address: P.O. Box 128, Monett, MO 65708. Tel.: 417-235-9030.
E-mail: historical@mo-net.com
Web Site: monetthistory.com
Founded: 2001.
Key Personnel: Pres. (V), Douglas Hobson
Governing Authority: Tax-exempt.
Institution Type/Description: History Museum.
Collections: local history & culture; period furnishings; personal artifacts; photographs.
Hours & Admission Prices: Tues. 1-3, Thurs. 6:30 pm-8:30 pm, Sat. 10-3. No charge; donations accepted. &
Attendance: 200 (estimated)
Membership: Individual $10; Life $100.

Montgomery City

GRAHAM CAVE STATE PARK, 217 Hwy T.T., Montgomery City, MO 63361-5509. Tel.: 573-564-3476. Fax: 573-564-2534.
E-mail: moparks@dnr.mo.gov
Founded: 1964.
Congressional District: 9
Key Personnel: Supt., Debra Ray.
Governing Authority: state. Parent Institution: Div. of Parks & Historic Preservation, Missouri Dept. of Natural Resources, Box 176, Jefferson City, MO 65102. Tax-exempt.
Institution Type/Description: State Park.
Collections: archaeology; anthropology; Indian artifacts.
Research Fields: archaeology; anthropology; Indian artifacts.
Facilities: interpretive center; campsites; hiking trails.
Activities: tours in summer.
Hours & Admission Prices: Park: daily 7am to sunset. Visitor Center: April-Oct. daily 9-11 & 12:30-4; Nov.-March Mon.-Fri. 9-11 & 12:30-4. No charge. &
Attendance: 55,000

Nelson

SAPPINGTON CEMETERY STATE HISTORIC SITE, Route AA, Nelson, MO 65347. Mailing Address: Arrow Rock State Historic Site, P.O. Box 1, Arrow Rock, MO 65320. Tel.: 660-837-3330. Fax: 660-837-3300.
E-mail: moparks@dnr.mo.gov
Web Site: mostateparks.com/park/sappington-cemetery-state-historic-site
Founded: 1969.
Congressional District: 6
Key Personnel: Site Admin., Michael Dickey.
Governing Authority: state. Operated by Division of Parks & Historic Preservation. Parent Institution: Missouri Department of Natural Resources, P.O. Box 176, Jefferson City, MO 65102. Subsidiary Institution: Division of State Parks. Tax-exempt.
Institution Type/Description: Historic Site: resting places of prominent Missourians, including Claiborne Fox Jackson, Dr. John Sappington & Gov. M.M. Marmaduke.
Collections: 2 acre cemetery.
Facilities: 2 acre cemetery.
Hours & Admission Prices: Daily sunrise to sunset. No charge.
Attendance: 5,000

Neosho

CROWDER COLLEGE-LONGWELL MUSEUM & CAMP CROWDER COLLECTION, 601 Laclede, Neosho, MO 64850-9165. Tel.: 417-455-5470. Fax: 417-455-5539.
Web Site: www.crowder.edu/about-crowder/longwell-museum
Founded: 1970.
Congressional District: 39
Governing Authority: college. Tax-exempt.
Institution Type/Description: Art Museum/Center & History Museum.
Collections: oil paintings by Ozark artist Daisy Cook; original prints by Birger Sandzen, Grant Wood, John Steuart Curry, Peter Hurd & Thomas Hart Benton; Japanese wood block prints; Chinese paper cuts; Camp Crowder, Army Signal Corp. collection.

Research Fields: family papers from Warren Cook, c.1850; archive letters.
Activities: lectures; films; study clubs; hobby workshops; broadcast programs; loan & temporary exhibitions; school loan service.
Publications: catalogues.
Hours & Admission Prices: Mon.-Fri. 9-4. No charge; donations accepted. Closed college holidays. &
Attendance: 800 (estimated)

Nevada

BUSHWHACKER MUSEUM, 212 W. Walnut St., Nevada, MO 64772-2341. Tel.: 417-667-9602.
E-mail: bushwhackerjail@sbcglobal.net
Web Site: www.bushwhacker.org
Founded: 1964.
Congressional District: 4
Key Personnel: C.F.O., Ed Morris; Pres. (V.), Jean McQueen; Dir. & Museum Shop Mgr., Will Tollerton.
Personnel Profile: Full-Time Paid 1; Part-Time Volunteers 1.
Governing Authority: nonprofit organization. Parent Institution: Vernon County Historical Society. Branch Museum: 212 W. Walnut, Nevada, MO 64772. Tax-exempt: 501(c)(3).
Institution Type/Description: History Museum.
Collections: archaeology; Indian artifacts; medicine; local history; Civil War & outlaw period; Victorian clothing; primitive tools; military history. Historic Buildings: 1860s Vernon County jail; 1871 jailer's home; 1920 Ford Agency Bldg.; Dr. Hornback's 1930s home & medical offices.
Research Fields: genealogy; local & regional history.
Facilities: archives.
Activities: guided tours; temporary & permanent exhibitions.
Publications: quarterly newsletter, Bushwhacker Musings.
Hours & Admission Prices: May-Oct. Wed.-Sat. 10-4. Adults $5, children 12-17 $2, children under 12 $1; discounts to AAA members; AASLH & members no charge. Closed Memorial Day; Independence Day; Labor Day. &
Attendance: 2,000 (accurate)
Membership: Individual $15; Family $25; Business & Benefactor $50.

New Bloomfield

NEW BLOOMFIELD AREA HISTORICAL SOCIETY, 313 Oak St., New Bloomfield, MO 65063. Mailing Address: P.O. Box 132, New Bloomfield, MO 65063-0132. Tel.: 573-491-0180.
E-mail: museum@newbloomfieldhistorical.org
Web Site: newbloomfieldhistorical.org
Institution Type/Description: Historical Society Museum.
Collections: local history & culture; personal artifacts; period furnishings; photographs.
Hours & Admission Prices: Sat. 10-1; other times by appointment.

New Madrid

HUNTER-DAWSON STATE HISTORIC SITE, Dawson Rd., Hwy. U, New Madrid, MO 63869. Mailing Address: P.O. Box 308, New Madrid, MO 63869-0308. Tel.: 573-748-5340. Fax: 573-748-7228.
E-mail: hunter-dawson.state.historic.site@dnr.mo.gov
Web Site: www.mostateparks.com
Founded: 1967.
Congressional District: 8
Key Personnel: Historic Site Specialist III, Michael Comer; Interpretive Resource Specialist, Vicki Jackson; Park Maintenance Worker I, Chadd Thomas.
Personnel Profile: Full-Time Paid 5; Part-Time Paid 3; Part-Time Volunteers 1.
Governing Authority: state. Div. of Parks & Historic Preservation, Missouri Dept. of Natural Resources, P.O. Box 176, Jefferson City, MO 65102. Tax-exempt.
Institution Type/Description: Historic House: c.1859 Hunter-Dawson House, a yellow cyprus frame house in the Greek Revival & Italianate style.
Collections: Period furnishings; 1860 collection of Mitchell & Rammelsburg furniture; three state champion trees, one over 300 years old. Historic building; 80% of original furnishings displayed.
Research Fields: historic restoration; 1860s southern culture and lifestyles; William W. Hunter papers.
Activities: guided tours; Civil War boot camp for children; Civil War outreach program; Victorian mourning customs. Museum Sponsors: Christmas Tour; Easter Egg Hunt.
Hours & Admission Prices: March-Nov. Mon.-Sat. 10-4, Sun. 12-4; Dec.-Feb. Tues.-Sat. 10-4. Adults $2.50, children 6-12 $1.50; children under 6 no charge. Closed New Year's Day; Easter; Thanksgiving; Christmas Day. &

Attendance: 8,098 (accurate)

NEW MADRID HISTORICAL MUSEUM, 1 Main St., New Madrid, MO 63869. Tel.: 573-748-5944.
E-mail: nmhmnm@yahoo.com
Web Site: www.newmadridmuseum.com
Founded: 1974.
Key Personnel: Dir. & Museum Shop Mgr., Jeff Grunwald; Pres. (V), Jan Farrenburg.
Personnel Profile: Part-Time Paid 4; Part-Time Volunteers 6.
Governing Authority: Tax-exempt.
Institution Type/Description: History Museum.
Collections: local history & culture; photographs; period furnishings; personal artifacts; quilts; New Madrid fault system & earthquakes of 1811 & 1812; military artifacts including the US Civil War.
Facilities: Museum-related items for sale.
Hours & Admission Prices: Call for hours. Adults $3, seniors $2.50, children $1.50, group rates available. &
Attendance: 8,000 (estimated)

New Melle

BOONE-DUDEN ARCHIVES AND MUSEUM, 3565 Mill St., New Melle, MO 63365-0082. Mailing Address: P.O. Box 82, New Melle, MO 63365-0082. Tel.: 636-987-2136.
E-mail: ida.g@outlook.com
Founded: 1986.
Key Personnel: Dir. & Museum Shop Mgr., Ida Gerdiman; Chm. (V), Ruth Busdieker.
Personnel Profile: Part-Time Volunteers 12.
Volunteer Hours: 352
Operating Expenses: 3,756
Operating Income: 3,825
Governing Authority: Parent Institution: Boone-Duden Historical Society. Tax-exempt.
Institution Type/Description: History Museum; house built in 1871.
Collections: local history; books; newspaper articles.
Publications: bimonthly, Boone-Duden Historical Society.
Hours & Admission Prices: Sun. 12:30-4; other times by appointment. No charge. &
Attendance: 83 (accurate)
Membership: High School Student $5; Household $20; Society Business $25-$500; Life $200 & up.

O'Fallon

DARIUS HEALD HOUSE, Fort Zumwalt Park, O'Fallon, MO 63366. Mailing Address: City of O'Fallon, 100 N. Main, O'Fallon, MO 63366. Tel.: 636-379-5614.
E-mail: mseymour@ofallon.mo.us
Web Site: www.ofallon.mo.us/dept_tourism_healdhome.htm
Institution Type/Description: Historic House Museum: built in the late 1800s.
Collections: Heald family history; period furnishings; photographs.
Hours & Admission Prices: By appointment.

O'FALLON HISTORICAL SOCIETY'S LOG CABIN MUSEUM, Civic Park, O'Fallon, MO 63366. Mailing Address: P.O. Box 424, O'Fallon, MO 63366. Tel.: 636-980-8015.
E-mail: info@ofallonmohistory.org
Web Site: ofallonmohistory.org
Founded: 1974.
Key Personnel: Pres. (V), Lewis Swinger.
Personnel Profile: Part-Time Volunteers 12.
Institution Type/Description: Historic Building Museum: built c.1870.
Collections: local history; photographs; period furnishings; early Wabash Railroad depot artifacts.
Hours & Admission Prices: Call for hours. No charge; donations accepted.

Odessa

ODESSA AREA HISTORICAL MUSEUM, 101 W. Phillips St., Odessa, MO 64076-1543. Tel.: 573-639-0762.
Institution Type/Description: History Museum.
Collections: local history & culture; photographs; yearbooks; postcards; business & school memorabilia; personal artifacts; newspapers; period clothing; phonebooks.
Hours & Admission Prices: Wed. & Sun. 3-6; other times by appointment.

Osceola

ST. CLAIR COUNTY HISTORICAL SOCIETY, 660 Main, Osceola, MO 64776. Mailing Address: P.O. Box 376, Osceola, MO 64776-0376. Tel.: 417-876-3925 & 7913.
E-mail: howardjoanmcp54@gmail.com
Web Site: www.stclaircountyhistoricalsociety.org
Founded: 2004.
Key Personnel: Pres. (V), Wade Harris; Interim Cur. & Museum Shop Mgr., Joan McPeak.
Personnel Profile: Part-Time Volunteers 50.
Governing Authority: Tax-exempt.
Institution Type/Description: Historical Society Museum.
Collections: local history & culture; period furnishings; personal artifacts; books; Native American artifacts.
Research Fields: genealogy-family.
Activities: genealogy research. Museum Sponsors: Gospel Music Concert in April; Readers Theater in Fall.
Publications: quarterly member newsletter.
Hours & Admission Prices: May-Oct.: Thurs. 12-4, Fri. 8am to noon, Sat. 10-2:30; Nov.-April Fri. 8am to noon; other times by appointment. No charge, donations accepted. &
Attendance: 480 (estimated)
Membership: Individual $10; Husband & Wife $15; Business $25.

Overland

OVERLAND HISTORICAL SOCIETY, 2404 Gass Ave., Overland, MO 63114. Tel.: 314-426-7027. Facebook: Overland Historical Society.
Web Site: www.overlandhistoricalsociety.com
Founded: 1976.
Key Personnel: Pres. (V), Linda Schuetz; Museum Shop Mgr., Sandy Jackson.
Personnel Profile: Part-Time Volunteers 15.
Governing Authority: Tax-exempt.
Institution Type/Description: Historical Society Museum.
Collections: local history & culture; period furnishings; personal artifacts; photographs.
Publications: newsletter, Log House Gazette.
Hours & Admission Prices: Call for hours. No charge; donations accepted.
Attendance: 400 (estimated)
Membership: Junior under 18 $1; Individual $15; Family $20; Business $25; Life $100.

Owensville

GASCONADE COUNTY HISTORICAL SOCIETY MUSEUM, 105 W. McFadden, Owensville, MO 65066. Mailing Address: P.O. Box 131, Hermann, MO 65041-0131. Tel.: 573-437-5617 (museum) 486-4028 (archives & library).
Founded: 1970.
Congressional District: 112
Key Personnel: Dir., Herb Lindroth; Cur., Shirley Lindroth
Governing Authority: Parent Institution: Gasconade County Historical Society. Tax-exempt.
Institution Type/Description: Historical Society Museum: housed in a former hotel; built in 1910.
Collections: local history & culture; period furnishings; personal artifacts; photographs.
Publications: quarterly newsletter.
Hours & Admission Prices: Mon. 9-1, Fri.-Sat. 11-3. No charge; donations accepted. &
Attendance: 700 (estimated)
Membership: Individual $20; Family & Civic or Service $30; Corporate $50. Life: Individual $150; Family $250.

Ozark

CHRISTIAN COUNTY HISTORICAL SOCIETY AND MUSEUM, 202 E. Church St., Ozark, MO 65721. Mailing Address: P.O. Box 442, Ozark, MO 65721-0442. Tel.: 417-988-7191.
E-mail: christiancohistorical@gmail.com
Web Site: www.christiancohistory.com
Key Personnel: Pres. & Webmaster, Shirley Scott; Vice Pres. & Volunteer Coord., Linda Myers; Sec., Culah Nixon; Treas., Avaline Harris; Property Mgmt. & Cur., John Nixon
Institution Type/Description: Historical Society Museum.
Collections: local history & culture; personal artifacts.
Hours & Admission Prices: April-Oct. Sun. 2-4; other times by appointment.

Paris

UNION COVERED BRIDGE STATE HISTORIC SITE, County Rd. C, Paris, MO 65275. Mailing Address: Mark Twain Birthplace, 37352 Shrine Rd., Florida, MO 65283-2127. Tel.: 573-565-3449. Fax: 573-565-3718.
E-mail: mark.twain.birthplace.state.historic.site@dnr.mo.gov
Web Site: www.mostateparks.com/unionbridge.htm
Founded: 1968.
Congressional District: 9
Key Personnel: Site Admin., Connie Ritter.
Personnel Profile: Full-Time Paid 1; Part-Time Paid 2.
Governing Authority: state. Owned & operated by Div. of State Parks, Missouri Dept. of Natural Resources, P.O. Box 176, Jefferson City, MO 65102. Tax-exempt.
Institution Type/Description: Historic Site: 1871 Union covered bridge.
Collections: newspaper clippings.
Research Fields: Monroe County Missouri, covered bridges.
Publications: brochure.
Hours & Admission Prices: Daily 8-5. No charge.
Attendance: 24,000 (estimated)

Perry

RALLS COUNTY HISTORICAL MUSEUM AND LIBRARY, 120 E. Main St., Perry, MO 63462. Mailing Address: P.O. Box 463, Perry, MO 63462-0463. Tel.: 573-248-6147 & 565-2025.
Institution Type/Description: History Museum.
Collections: local history & culture; period furnishings; personal artifacts; photographs.
Hours & Admission Prices: Wed. & Fri.-Sun. 10-5.

Pilot Grove

COOPER COUNTY HISTORICAL SOCIETY, 111 Roe St., Pilot Grove, MO 65276. Tel.: 660-834-3582.
Institution Type/Description: Historical Society Museum.
Collections: local history; cemetery records; family histories; genealogy & census records; school & church information.
Activities: research.
Hours & Admission Prices: late May to late Oct. Fri.-Sun. 1-5; other times by appointment.

Pilot Knob

FORT DAVIDSON STATE HISTORIC SITE, 118 E. Maple St., Pilot Knob, MO 63663-0509. Mailing Address: P.O. Box 509, Pilot Knob, MO 63663-0509. Tel.: 573-546-3454. Fax: 573-546-2713.
E-mail: fort.davidson.state.historic.site@dnr.mo.gov
Web Site: www.battleofpilotknob.org
Founded: 1969.
Congressional District: 8
Key Personnel: Pres. (V), Terry Cadenbach; Rgnl. Supvr., Delecia Huitt; Site Admin., Walter Busch; Cur., Brick Autry.
Personnel Profile: Full-Time Paid 4; Part-Time Paid 6; Part-Time Volunteers 1.
Governing Authority: state. Div. of State Parks, Missouri Dept. of Natural Resources, P.O. Box 176, Jefferson City, MO 65102. Tax-exempt.
Institution Type/Description: Historic Site: 1864 Battle of Pilot Knob; Civil War battlefield.
Collections: Civil War Artifacts; S.E. Missouri Civil War history.
Research Fields: primary focus on eastern Missouri Civil War history & Arcadian Valley local Civil War lore.
Facilities: walking tour; interpretive center.
Activities: periodic battle reenactments; living history.
Publications: brochures.
Hours & Admission Prices: April 16-Nov. daily 10-4; Dec.-April 15 Tues.-Sun. 10-4. No charge; donations accepted. Closed New Year's Day; Thanksgiving; Christmas. &
Attendance: 63,147 (accurate)

Pineville

MCDONALD COUNTY HISTORICAL MUSEUM, 302 Harmon, Pineville, MO 64856. Mailing Address: P.O. Box 572, Pineville, MO 64856-0572.
E-mail: info@mcdonaldcohistory.org
Institution Type/Description: History Museum.
Collections: local history & culture; period furnishings; personal artifacts; photographs.

Hours & Admission Prices: Fri.-Sat. 9-1.

Platte City

BEN FERREL 1882 MINI MANSION MUSEUM, 220 Ferrel St., Platte City, MO 64079-9511. Mailing Address: P.O. Box 102, Platte City, MO 64079. Tel.: 816-431-5121.
Founded: 1978.
Personnel Profile: Part-Time Paid 2; Part-Time Volunteers 4.
Governing Authority: Parent Institution: Platte County Missouri Historical Society. Tax-exempt.
Institution Type/Description: Historic House: built in 1882.
Collections: period furnishings & memorabilia.
Research Fields: genealogy.
Activities: Museum Sponsors: Christmas Tour in December.
Hours & Admission Prices: April 15-Oct. 15 Thurs.-Sat. 1-4; Oct. 16-April 14 by appointment. Adults $5; members no charge.
Attendance: 1,200 (accurate)
Membership: Student $1; Individual $20; Family $30; Corporate $50; Life over 75 $100; Life 56-75 $300; Life 1-55 $400. Contributing Donors: Contributor $50-$99; Sustaining $100-$149; Supporter $150-$199; Ambassador $200-$499; Benefactor $500 & up.

Pleasant Hill

PLEASANT HILL HISTORICAL SOCIETY, 125 Wyoming St., Pleasant Hill, MO 64080-1643. Mailing Address: P.O. Box 31, Pleasant Hill, MO 64080-0031. Tel.: 816-540-4010 & 529-6088.
E-mail: bevkennedy@kcweb.net
Web Site: www.orgsites.com/mo/pleasanthillhistoricalsociety
Founded: 1971.
Congressional District: 4
Key Personnel: Pres. (V), Beverly Kennedy
Governing Authority: Tax-exempt.
Institution Type/Description: Historical Society Museum.
Collections: local history & culture; period furnishings; photographs; personal artifacts; newspaper clippings including obituaries, births, & marriages; family history.
Research Fields: regional & family history.
Activities: tours; community group programs. Museum Sponsors: programs & meetings in January, April, July & October.
Publications: quarterly newsletter.
Hours & Admission Prices: Mon. 1:30-3:30, Wed. & Fri. 3:30-5:30, 2nd Sat. of month 10-3; other times by appointment. No charge; donations accepted.
Attendance: 310 (estimated)
Membership: Student $5; Adult $10; Business $25; Life $100.

Point Lookout

THE RALPH FOSTER MUSEUM, (M), College of the Ozarks, One Cultural Ct., Point Lookout, MO 65726. Mailing Address: P.O. Box 17, Point Lookout, MO 65726. Tel.: 417-690-3407. Fax: 417-690-2606.
E-mail: museum@cofo.edu
Web Site: www.rfostermuseum.com
Founded: 1930.
Congressional District: 7
Key Personnel: Dir. & Museum Shop Mgr., Annette J. Sain; C.E.O. & Pres., Dr. Jerry Davis; Cur., Jeanelle Ash; Cur., Thomas A. Debo; Cur., Mike Combs.
Personnel Profile: Full-Time Paid 4; Part-Time Paid 28.
Governing Authority: college. Parent Institution: College of the Ozarks. Tax-exempt: 501(c)(3).
Institution Type/Description: General Museum.
Collections: late 19th- & early 20th-century Ozarkiana; prehistory of the Ozark region; history of the Ozark Plateau; firearms; Rose O'Neill memorabilia; natural history; Si Siman music room dedicated to country, western & Ozark music; history of College of the Ozarks; military display, Conflicts of the 20th century.
Research Fields: history of the Ozark Plateau 1200 A.D.-present; weapons, dolls; glassware; textiles; late 19th- & early 20th-century tools.
Facilities: Lois Brownell Research library available for research with Director's permission.
Activities: Outreach program to pre-school, elementary & high schools.
Hours & Admission Prices: Feb. to mid-Dec. Mon.-Sat. 9-4:30; groups by appointment. Adults $6, seniors 62 & over and groups 20 or more $5; discounts to AAM & ICOM members; students no charge. Closed Thanksgiving week &
Attendance: 60,000 (estimated)

Poplar Bluff

MARGARET HARWELL ART MUSEUM, 421 N. Main St., Poplar Bluff, MO 63901-5107. Tel.: 573-686-8002. Fax: 573-686-8017.
E-mail: ethel@mham.org
Web Site: www.mham.org
Founded: 1981.
Congressional District: 8
Key Personnel: Dir., Tina M. Magill; Pres. (V), Nancy Buttry; Museum Shop Mgr., Gerry Vandervort.
Personnel Profile: Full-Time Paid 1; Part-Time Paid 1; Part-Time Volunteers 50.
Governing Authority: municipal government. Parent Institution: City of Poplar Bluff. Tax-exempt: 170(b)(1)(A).
Institution Type/Description: Art Museum: housed in 1883 mansion.
Collections: various forms of contemporary art media; mid 19th-century clothing and textiles; temporary exhibits change monthly & feature regional and nationally known artists.
Facilities: classrooms.
Activities: guided tours; lectures; organized educational programs; docent program.
Publications: quarterly newsletters.
Hours & Admission Prices: Tues.-Fri. 12-4, Sat.-Sun. 1-4. No charge; donations accepted. Closed national holidays. &
Attendance: 18,000 (estimated)
Membership: Student $5; Patron $25; Business $50; Contributing $100; Supporting $250; Life $2,000.

MOARK REGIONAL RAILROAD MUSEUM AKA THE POPLAR BLUFF RAILROAD MUSEUM, 303 Moran St., Poplar Bluff, MO 63901. Tel.: 573-785-4539.
Founded: 1991.
Congressional District: 8
Key Personnel: Pres. (V), David Silverberg
Governing Authority: Tax-exempt.
Institution Type/Description: Historic Building: built in 1928. Listed on the National Register of Historic Places.
Collections: local history & culture; railroad artifacts; period furnishings; baggage wagons; cabooses; large IIO scale model trains; photographs.
Facilities: Museum-related items for sale.
Activities: Annual Events: Amtrak's National Train Day in May; Iron Horse Festival in October.
Hours & Admission Prices: Sat. 1-4; groups by appointment. No charge; donations accepted.
Attendance: 1,900 (accurate)
Membership: Individual $12; Family $20.

Portage Des Sioux

CAF MISSOURI WING MUSEUM, St. Charles County Airport - Smartt Field, 6390 Grafton Ferry Rd., Portage Des Sioux, MO 63373. Mailing Address: P.O. Box 637, St. Charles, MO 63302. Tel.: 636-250-4515. Fax: 636-250-4515.
E-mail: pkf4@earthlink.net
Web Site: www.cafmo.org
Founded: 1982.
Institution Type/Description: Military History Museum.
Collections: military history, aircraft, equipment, uniforms & artifacts; photographs; personal artifacts.
Activities: special events.
Hours & Admission Prices: Thurs. & Sat. 10-2:30; call to confirm.

Princeton

CASTEEL-LINN HOUSE AND MUSEUM, 902 E. Oak St., Princeton, MO 64673-1255. Mailing Address: P.O. Box 1583, Palm Springs, CA 92263-1583. Tel.: 660-748-3905.
Founded: 1982.
Congressional District: 6
Key Personnel: C.E.O. & Dir. (V), N. P. Linn; Pres., Cur. & Public Rels. (V), Nancy Paige Linn; Archivist, Pamela Elizabeth Kidd; Security, Cy Linn.
Personnel Profile: Full-Time Volunteers 3; Part-Time Volunteers 2.
Governing Authority: private; not-for-profit organization.
Institution Type/Description: Historic House: built on a historic site where Civil War reunions were held.
Collections: library with concentration on world history, Missouri & Mercer County history; art and art history; china & glass; railroad china & antiques; antique toys; portraits; bronzes; period furniture.

Research Fields: Western art history; Missouri & Mercer County history; sociology.
Facilities: 6,500-vol. library available to scholars & researchers.
Activities: guided tours.
Publications: monthly newsletter, Great American West: Art & Books; books, Linn's Mercer County History Series (8 vols.).
Hours & Admission Prices: April-Dec. by appointment only. No charge. &
Attendance: 200 (estimated)

MERCER COUNTY GENEALOGICAL AND HISTORICAL SOCIETY, 601 Grant St., Princeton, MO 64673-1023. Mailing Address: P.O. Box 97, Princeton, MO 64673-0097. Tel.: 660-748-3725 & 4755. Fax: 660-748-3723.
Web Site: www.rootsweb.ancestry.com/~momercer/
Founded: 1965.
Congressional District: 6
Key Personnel: Mgr., Randi Ferguson.
Governing Authority: nonprofit. Tax-exempt: 501(c)(3).
Institution Type/Description: Local History Museum: housed in Mercer County Library.
Collections: history; period artifacts; costumes; furniture; china; geology.
Publications: reprint, 1888 History of Harrison & Mercer Counties, Mo.; The Pioneer Press; Linn's History of Mercer County, Mo.; 1911 Roger's History of Mercer County; The Mercer County History of 1883.
Hours & Admission Prices: By appointment only. No charge. Closed national holidays.
Membership: Individual $3.

Raytown

RAYTOWN HISTORICAL SOCIETY & MUSEUM, 9705 E. 63rd St., Raytown, MO 64133. Mailing Address: P.O. Box 16652, Raytown, MO 64133. Tel.: 816-353-5033.
E-mail: raytownhistorical@sbcglobal.net
Web Site: www.raytownhistoricalsociety.org
Institution Type/Description: Historical Society Museum.
Collections: local history & culture; period artifacts; photographs.
Hours & Admission Prices: Wed.-Sat. 10-4; other times by appointment. Adults $2; children under 12 no charge.

Republic

WILSON'S CREEK NATIONAL BATTLEFIELD, 6424 W. Farm Rd. 182, Republic, MO 65738-9492. Tel.: 417-732-2662, ext. 352. Fax: 417-732-1167.
E-mail: deborah_s_wood@nps.gov
Web Site: www.nps.gov/wicr; www.civilwarvirtualmuseum.org.
Founded: 1960.
Congressional District: 7
Key Personnel: Supt., T. John Hillmer; Chief Resource Mgmt., Gary Sullivan; Museum Cur., Deborah s. Wood
Governing Authority: federal. Parent Institution: U.S. Dept. of the Interior, National Park Service, Washington, DC. Virtual Museum: Trans-Mississippi West Theater. Tax-exempt.
Institution Type/Description: Military Museum: located on the battlefield on which the August 10, 1861 battle of Wilson's Creek occurred.
Collections: Battle of Wilson's Creek, American Civil War Trans-Mississippi West artifacts; documents, diaries, letters, discharge papers; mid-19th century cultural artifacts; flags; weapons; medical instruments; photographs. Historic House: 1852 The Ray House.
Research Fields: Battle of Wilson's Creek; the Civil War in Missouri & the Trans-Mississippi West.
Facilities: visitor's center.
Activities: formally organized education programs for children & undergraduate college students.
Hours & Admission Prices: Park: seasonal hours. Museum: Nov.-March daily 9-12 & 1-4. Park & Museum: adults 16 & over $5, maximum $10 per vehicle; school groups no charge. Closed New Year's Day; Thanksgiving; Christmas. &
Attendance: 300,000 (estimated)

Richmond

RAY COUNTY MUSEUM, 901 W. Royle St., Richmond, MO 64085-1545. Tel.: 816-776-2305.
E-mail: raycohistory@aol.com
Founded: 1973.
Key Personnel: Pres. (V), David Blyth; Cur., Linda Emley.
Governing Authority: Parent Institution: Ray County Historical Society. Tax-exempt.

Institution Type/Description: History Museum.
Collections: local history & culture; period furnishings; personal artifacts; photographs; Mormon history.
Hours & Admission Prices: Wed.-Sat. 10-4.
Membership: Individual $15; Lifetime $200.

Rolla

MISSOURI DEPARTMENT OF NATURAL RESOURCES, DIVISION OF GEOLOGY AND LAND SURVEY, 111 Fairgrounds Rd., Rolla, MO 65401-2909. Mailing Address: P.O. Box 250, Rolla, MO 65402-0250. Tel.: 573-368-2100 & 2118. Fax: 573-368-2111.
Web Site: www.dnr.mo.gov/geology
Formerly: Ed Clark Museum of Missouri Geology
Founded: 1963.
Congressional District: 8
Key Personnel: Dir. & State Geologist, Joseph A. Gillman; Public Rels., Hylan Beydler.
Governing Authority: State of Missouri. A branch of Div. of Geology & Land Survey, Missouri Dept. of Natural Resources. Tax-exempt.
Institution Type/Description: Geology Museum.
Collections: geology; mineralogy; paleontology; fossils; mastodon tusk; geologic maps; map making tools.
Research Fields: geology.
Facilities: labs & library of the Geological Survey Div.
Activities: permanent, temporary & traveling exhibitions.
Publications: pamphlets; reports; information circulars; books; brochures; geologic maps; posters.
Hours & Admission Prices: Mon.-Fri. 8-5. No charge. Closed national & Missouri holidays. &
Attendance: 3,000

PHELPS COUNTY HISTORICAL SOCIETY, 302 Third St., Rolla, MO 65402. Mailing Address: P.O. Box 1861, Rolla, MO 65402-1861. Tel.: 573-364-5977. Facebook: Phelps County Historical Society.
E-mail: pchs@rollanet.org
Founded: 1939.
Congressional District: 8
Key Personnel: Pres. (V), Dennis Peterman.
Governing Authority: private; nonprofit organization. Parent Institution: Missouri Benevolent Corp. Tax-exempt: 501(c)(3).
Institution Type/Description: Historical Society Museum: housed in Dillon Log House.
Collections: local history & culture; period furnishings; personal artifacts; photographs. Historic Buildings: Old Phelps County Courthouse; Old Phelps County Jail.
Publications: biannual newsletter, Phelps County Historical Society.
Hours & Admission Prices: By appointment. No charge; donations accepted.
Attendance: 500 (estimated)
Membership: Annual $15.

Saint Charles

FIRST MISSOURI STATE CAPITOL-STATE HISTORIC SITE, 200 S. Main St., Saint Charles, MO 63301-2855. Tel.: 636-940-3322. Fax: 636-940-3324.
E-mail: first.state.capitol.state.historic.site@dnr.mo.gov
Web Site: www.dnr.state.mo.us
Founded: 1971.
Congressional District: 2
Key Personnel: Site Admin., Victoria Love; Asst. Site Admin., Robert Adams; Interpretive Resource Tech., Sue Love.
Personnel Profile: Full-Time Paid 4; Part-Time Paid 7; Part-Time Volunteers 12.
Governing Authority: state. Parent Institution: Division of State Parks, Missouri Dept. of Natural Resources, P.O. Box 176, Jefferson City, MO 65102. Tax-exempt.
Institution Type/Description: Historic Buildings: c.1818 first Missouri State Capitol; temporary seat of government 1821-1826.
Collections: period furnishings; 11 restored rooms.
Research Fields: pre-1830 Missouri history.
Activities: guided tours; special events; living history demonstrations.
Publications: brochures.
Hours & Admission Prices: Summer: Mon.-Sat. 10-4, Sun. 12-4; Winter: call for hours; groups by appointment. Family $15, adults $4, children 6-11 $2.50; discounts to groups; children under 6 no charge. Closed New Year's Day; Easter; Thanksgiving; Christmas. &

Attendance: 52,000 (accurate)

FOUNDRY ART CENTRE, 520 N. Main Center, Saint Charles, MO 63301-2182. Tel.: 636-255-0270. Fax: 636-925-0345.
Web Site: foundryartcentre.org
Institution Type/Description: Art Gallery.
Collections: paintings; sculpture.
Activities: educational programs.
Hours & Admission Prices: Tues.-Thurs. 10-8, Fri.-Sat. 10-5, Sun. 12-4.

LEWIS & CLARK BOAT HOUSE AND NATURE CENTER, 1050 Riverside Dr., Saint Charles, MO 63301-3481. Tel.: 636-947-3199. Fax: 636-916-0240.
E-mail: lewisandclarkmuseum@yahoo.com
Web Site: www.lewisandclarkcenter.org
Founded: 1985.
Congressional District: 2
Key Personnel: Dir., Bill Brecht; Pres., Larry Kluesner; Museum Shop Mgr., Bob Learned.
Personnel Profile: Part-Time Paid 5; Part-Time Volunteers 200.
Governing Authority: private; nonprofit organization. Tax-exempt: 501(c)(3).
Institution Type/Description: History Museum.
Collections: dioramas; Missouri River history; replica of expedition boats; Lewis & Clark campsite; St. Charles history.
Facilities: 1,200 sq. ft. exhibit space. Museum-related items for sale.
Activities: formal education programs; guided tours; lectures; participatory exhibits; scout programs.
Hours & Admission Prices: Mon.-Sat. 10-5, Sun. 12-5. Adults $4, children under 17 $2; discounts to groups. Closed New Year's Day; Easter; Thanksgiving; Christmas. &
Attendance: 40,675 (accurate)

ST. CHARLES COUNTY HISTORICAL SOCIETY, INC., 101 S. Main St., Saint Charles, MO 63301-2802. Tel.: 636-946-9828.
E-mail: info@scchs.org
Web Site: www.scchs.org
Founded: 1956.
Congressional District: 2
Key Personnel: Archivist, Dorris Keeven-Franke.
Personnel Profile: Part-Time Paid 2; Part-Time Volunteers 25; Interns 2.
Governing Authority: nonprofit organization. Tax-exempt: 501(c)(3).
Institution Type/Description: Local History Museum.
Collections: archives; American Indian artifacts; St. Charles County Circuit & Probate Court records; indexes to St. Charles County Church records; tax records; naturalization; city directories & census.
Research Fields: archive.
Facilities: archives.
Activities: Society quarterly lunch meetings with guest speaker.
Publications: quarterly, Heritage; monthly, SCCHS Archives Newsletter.
Hours & Admission Prices: Library & Archives: Mon., Wed. & Fri. 10-3; 2nd & 4th Sat. 10-3. No charge. &
Attendance: 550 (accurate)
Membership: Student $10; Individual $25; Family $35.

Saint James

MARAMEC MUSEUM, THE JAMES FOUNDATION, Maramec Spring Park, 21880 Maramec Spring Dr., Saint James, MO 65559. Mailing Address: 320 S. Bourbeuse St., Saint James, MO 65559-1498. Tel.: 573-265-7124. Fax: 573-265-8770.
E-mail: jamesfoundation@centurytel.net
Web Site: maramecspringpark.com
Founded: 1971.
Key Personnel: Rgnl. Mgr., Danny Marshall; Dir. Interpretive Svcs., Lloyd Callies.
Personnel Profile: Full-Time Paid 1; Part-Time Paid 6.
Governing Authority: foundation. Parent Institution: The New York Community Trust. The James Foundation. Tax-exempt.
Institution Type/Description: History Museum.
Collections: model of Maramec Village, 1860; scale models of Old Maramec Iron Works 1826-1876, dioramas of iron works, grist mill, iron ore bank, charcoal pits & early Indian village; geology of springs, sinkholes, caves in Ozarks; archaeology related to Maramec & Indian culture; Indian artifacts; modes of transportation, trails & roads 1820-1860; life at the Iron Works; two aquaria of fish of Maramec Spring Branch; working rain/recharge model; wildlife display.
Research Fields: local history.
Activities: guided tours; group tours for students; illustrated talks to groups; permanent & temporary exhibits.

Publications: Frontier Iron, The Maramec Iron Works 1826-1876; Lucy Wortham James 1880-1938; brochures, Maramec Museum; Maramec Iron Works; Maramec Spring Park; The Rise & Fall of Maramec Iron Works.
Hours & Admission Prices: April & Oct. Sat.-Sun. 12-4; May Mon.-Fri. 10-3, Sat.-Sun. 12-4; June-Aug. Mon.-Sun. 11-5; Sept. Wed.-Sun. 12-4. No charge. &
Attendance: 30,496 (accurate)

Saint Joseph

*** THE ALBRECHT-KEMPER MUSEUM OF ART, (M),** 2818 Frederick Ave., Saint Joseph, MO 64506-2903. Tel.: 816-233-7003; 888-254-2787. Fax: 816-233-3413.
E-mail: frontdesk@albrecht-kemper.org
Web Site: www.albrecht-kemper.org
Founded: 1914.
Congressional District: 6
Key Personnel: Dir., Mr. Terry L. Oldham; Pres. (V), John Wilson; Registrar, Megan Benitz; Catering Mgr., Robyn Enright; Coord. Membership & Museum Shop Mgr., Chelsea Howlett-Weideman.
Personnel Profile: Full-Time Paid 4; Part-Time Paid 7; Part-Time Volunteers 200; Interns 2.
Governing Authority: nonprofit organization. Tax-exempt: 501(c)(3).
Institution Type/Description: Art Museum.
Collections: 18th to 21st century paintings, drawings, photographs & prints.
Facilities: 1,500-vol. noncirculating Bradley Art Library; reading room; classrooms; meeting rooms; 147-seat theatre/lecture hall; restaurant; formal rose garden. Museum-related items for sale.
Activities: guided tours; lectures; gallery talks; formally organized education programs; docent program; permanent & temporary exhibitions; space available for rent.
Publications: Catlin's Indians: The Kemper Portfolio; quarterly newsletter, Art Matters; exhibition catalog, Under The Influence: The Students of Thomas Hart Benton; William Christenberry: Retrospective; Native American Basketry: The Hartman Collection; The Art of Thomas King Baker; The Albrecht-Kemper Museum of Art, A History and Guide to the Collection.
Hours & Admission Prices: Tues.-Fri. 10-4, Sat.-Sun. 1-4. Adults $5, seniors 60 & over $2, students $1; members, AAM & ICOM members & children under 6 no charge. Closed major holidays. &
Attendance: 22,504 (accurate)
Membership: Student $15; Teacher $20; Nonresident $25; Individual $35; Family & Dual $55; Associate $100; Sustaining $300; Supporting $500; Business Sponsor $600; Sponsoring $1,000; Patron & Business Patron $1,500; Corporate Sponsor $2,500; Corporate Patron $5,000.

HEATON-BOWMAN-SMITH FUNERAL HOME MUSEUM, 3609 Frederick Ave., Saint Joseph, MO 64506-3033. Tel.: 816-232-3355.
Institution Type/Description: General Museum.
Collections: artifacts & memorabilia including wicker body basket used for Jesse James.
Hours & Admission Prices: Call for hours.

JESSE JAMES HOME MUSEUM, 1202 Penn St., Saint Joseph, MO 64503. Mailing Address: P.O. Box 1022, Saint Joseph, MO 64502-1022. Tel.: 816-232-8206. Fax: 816-232-3717.
E-mail: patee@ponyexpress.net
Web Site: www.ponyexpressjessejames.com
Founded: 1977.
Congressional District: 6
Key Personnel: Museum Dir., Gary Chilcote; Pres. (V), Thomas Duty; Cur. Collections, Doug Chilcote; Business Mgr., Amy Neely; Library & Archives, Carolyn Chilcote.
Personnel Profile: Full-Time Paid 3; Full-Time Volunteers 1; Part-Time Paid 3; Part-Time Volunteers 10.
Governing Authority: non-profit organization. Parent Institution: Pony Express Historical Association. Subsidiary Institution: Jesse James Home. Tax-exempt: 501(c)(3).
Institution Type/Description: Historic Building: 1879 house where Jesse James was killed in 1882.
Collections: Jesse James furnishings; the 1995 exhumation of Jesse James for DNA tests, including artifacts from the grave.
Research Fields: Jesse James & the James-Younger Gang.
Facilities: Gifts & museum-related items for sale.
Activities: permanent exhibitions.
Publications: monthly newsletter, Pony Express Mail.
Hours & Admission Prices: April-Oct. daily; Nov.-Mar. weekends only. Adults $4, seniors $3, students $2. Closed New Year's Day; Easter; Thanksgiving; Christmas Eve & Day. &
Attendance: 20,000 (estimated)

Membership: Individual $15; Sustaining $25; Corporate $30; Associate $50; Sponsor $100; Life $500.

NATIONAL MILITARY HERITAGE MUSEUM, 701 Messanie St., Saint Joseph, MO 64501-2219. Tel.: 816-233-4321. Fax: 816-279-9667.
E-mail: nmhm90@yahoo.com
Web Site: www.nationalmilitaryheritagemuseum.com
Founded: 1989.
Congressional District: 6
Key Personnel: Exec. Dir., Franklin A. Flesher.
Personnel Profile: Full-Time Volunteers 1; Part-Time Volunteers 6.
Governing Authority: Parent Institution: National Military Heritage Society, Inc. Tax-exempt.
Institution Type/Description: Military History Museum: housed in a former police station; built in 1890.
Collections: military & homefront artifacts from 1800s to present; photographs; awards & decorations.
Research Fields: military heritage & homefront history.
Hours & Admission Prices: Mon.-Fri. 9-4, Sat. 9-1. Adults $3; discounts to AAM members; students & members no charge.
Attendance: 3,000 (accurate)
Membership: Individual $25.

PATEE HOUSE MUSEUM, 1202 Penn St., Saint Joseph, MO 64503-2560. Mailing Address: P.O. Box 1022, Saint Joseph, MO 64502-1022. Tel.: 816-232-8206. Fax: 816-232-3717.
E-mail: patee@ponyexpress.net
Web Site: www.ponyexpressjessejames.com
Founded: 1963.
Congressional District: 6
Key Personnel: Museum Dir., Gary Chilcote; Pres., Thomas Duty; Business Mgr., Amy Neely; Cur. Collections, Doug Chilcote; Library & Archives, Carolyn Chilcote.
Personnel Profile: Full-Time Paid 3; Full-Time Volunteers 1; Part-Time Paid 3; Part-Time Volunteers 25.
Governing Authority: nonprofit organization. Parent Institution: Pony Express Historical Association. Subsidiary Institution: Jesse James Home. Tax-exempt: 501(c)(3).
Institution Type/Description: Western Museum: housed in 1858, Patee hotel, used as headquarters for the Pony Express in 1860 and during the Civil War. National Historic Landmark.
Collections: 1860, Hannibal & St. Joseph steam locomotive; first railroad mail car; cars; trucks; buggies; wagons; fire trucks; restored Pony Express office; railroad office; general store; printing shop; hotel lobby; apothecary shop; dental office; area Civil War headquarters including ball room used as war trial courtroom; 1854 Buffalo saloon; The Toy Carousel Shop; 1882 jail contains exhibit of local crimes & murders; 1903-1950 St. Joseph News-Press & Gazette; bound volumes; H-O gauge model railroad surrounded by 24 large vaudeville & silent movie posters from the turn-of-the-century belonging to the Great Renos acrobats & magicians; 4ft. bronze Pony Express statue by sculptor Avard Fairbanks; 1941 Allan-Herschel Hand-Carved Carousel; George Warfel art; "Westerners on Wood" 43 paintings of Jesse James & other westerners.
Research Fields: local history; Pony Express; Jesse James; Civil War.
Facilities: 4,500-vol. library of old texts on history, religion, available by application to the board. Museum-related items for sale.
Activities: lectures; gallery talks; carousel rides. Annual Events: Pony Express rerun in June.
Publications: monthly newsletter, Pony Express Mail.
Hours & Admission Prices: April-Oct. daily; Nov.-March Sat.-Sun. Adults $6, seniors 60 & over $5, students 6-17 $4; discounts to KCPT & National Railway Society members; members & children under 6 no charge. Closed New Year's Day; Easter; Thanksgiving; Christmas Eve & Day.
Attendance: 20,000 (estimated)
Membership: Regular $15; Sustaining $25; Corporate $30; Associate $50; Sponsor $100; Life $500.

PONY EXPRESS MUSEUM, (M), 914 Penn St., Saint Joseph, MO 64503-2544. Tel.: 816-279-5059 & 800-530-5930. Fax: 816-233-9370.
E-mail: pcdirector@ponyexpress.net
Web Site: www.ponyexpress.org
Founded: 1959.
Congressional District: 6
Key Personnel: Exec. Dir., Cindy Daffron; Pres. (V), Richard N. DeShon; Museum Shop Mgr., Michelle Mooney.
Personnel Profile: Full-Time Paid 2; Part-Time Paid 3; Part-Time Volunteers 3.

Governing Authority: nonprofit corporation. Tax-exempt: 501(c)(3).
Institution Type/Description: Historic Building Museum: 1858 original stables of the Pony Express, formally known as the Pike's Peak Stables, located on the site from which the first Pony Express rider left St. Joseph heading west to Sacramento.
Collections: exhibits relating to the Pony Express; manuscripts.
Research Fields: Pony Express; Westward Expansion.
Facilities: Pony Express & 19th-century western items for sale.
Activities: guided tours; lectures; films; permanent, temporary & traveling exhibitions; special events.
Publications: quarterly newsletter, Pony Trails.
Hours & Admission Prices: March 2-Nov. Mon.-Sat. 9-5, Sun. 11-4; Dec.-March 1 Mon.-Sat. 9-4. Adults $6, seniors $5, students & children 7-18 $3; discounts to groups, veterans, AAA & AARP members; children 6 & under and members no charge. Pony School: by appointment & during special events. Closed New Year's Eve & Day; Easter; Thanksgiving; Christmas Eve & Day.
Attendance: 39,000 (accurate)
Membership: Individual $40; Family $65; Trader $100; Scout $250; Trailblazer $500; Rider $1,000; Station Keeper $2,500; Founder $5,000.

ROBIDOUX ROW MUSEUM, 3rd & Poulin, Saint Joseph, MO 64501. Mailing Address: 217 W. Poulin St., Saint Joseph, MO 64501-1037. Tel.: 816-232-5861. Fax: 816-232-5861.
Web Site: www.robidouxrowmuseum.org
Founded: 1981.
Key Personnel: Dir. & Museum Shop Mgr., Clyde Weeks; Pres. (V), Bill Leppert; Museum Shop Mgr., Clyde Weeks.
Personnel Profile: Full-Time Paid 1; Part-Time Paid 1; Part-Time Volunteers 3.
Governing Authority: Parent Institution: Saint Joseph Historical Society. Tax-exempt.
Institution Type/Description: History Museum: housed in the city founded by Joseph Robidoux in 1843. Listed on the National Register of Historic Places.
Collections: local history; period artifacts; structures.
Publications: Saint Joseph Historical Society Journal.
Hours & Admission Prices: Feb.-April & Oct.-Dec. Tues.-Sat. 1-4; May-Sept. Tues.-Fri. 10-4, Sat.-Sun. 1-4. Adults $2.50, seniors 62 & over $2, students 6-18 $1; discounts to active military; members and children 5 & under no charge.
Attendance: 2,794 (accurate)
Membership: Individual $15; Family $20; Friend $25; Sustaining $50; Patron $100; Benefactor $500; Life $750.

* **ST. JOSEPH MUSEUM INC., (M),** 3406 Frederick, Saint Joseph, MO 64506-2913. Mailing Address: P.O. Box 8096, Saint Joseph, MO 64508-8096. Tel.: 816-232-8471. Fax: 816-232-8482. Facebook: St. Joseph Museum.
E-mail: sjm@stjosephmuseum.org
Web Site: www.stjosephmuseum.org
Founded: 1927.
Congressional District: 6
Key Personnel: Exec. Dir., Jacqueline A. Lewin; C.E.O., Pres. & Chm. (V), Clark W. Hampton; Cur. Collections, Sarah M. Elder; Head Security & Maintenance, Andy Meyer; Asst. Dir. Admin. & Retail Operations Mgr., Susan Noland; Head of Public Rels., Kathy Reno; Vol. Coord., Gift Shop Mgr., Joy Sander.
Personnel Profile: Full-Time Paid 5; Part-Time Paid 4; Part-Time Volunteers 50; Interns 2.
Governing Authority: municipal; nonprofit corporation. Parent Institution: St. Joseph Museums. Subsidiary Institutions: The Wyeth-Tootle Mansion; The Black Archives of St. Joseph; Mount Mora Cemetery; Round the Town Tours; Platte Purchase Publishers; The Glore Psychiatric Museum. Tax-exempt: 501(c)(3).
Institution Type/Description: Ethnology, Local, Natural, Ethnic & Medical History.
Collections: ethnology of North American Indians, Mezo-Americans & Philippines; vertebrate animals; geology; invertebrates; St. Joseph history; western expansion; Pony Express; Civil War; Jesse James; post Civil War & Victorian; manuscripts; psychiatric history; State Hospital #2 history.
Research Fields: Pony Express; archaeology; Jesse James; St. Joseph history; western ethnography; psychiatry; local black history.
Facilities: 7,000-vol. library of anthropology, ethnology, St. Joseph history, Western expansion, Pony Express, Civil War, Jesse James & natural history available by staff approval for use on the premises; reading room. Books, postcards & gift items relating to North American Indians, natural history & the Pony Express for sale.
Activities: guided tours; lectures; films; special events; formally organized

education programs for children; inter-museum loan, permanent, temporary & traveling exhibitions; school loan service.

Publications: quarterly newsletter, The Happenings; books, On the Winds of Destiny: A Biographical Look at Pony Express Riders; Old St. Jo: Gateway to the West 1799-1932; A Darkness Ablaze; St. Joe Road; Fishing on Deep River; As the Mockingbird Sang; A Home for the Minds Diseased: A History of State Hospital No. 2; history book by local authors.

Hours & Admission Prices: Mon.-Sat. 10-5, Sun. 1-5. Adults $5, children 7-15 $1; discounts to Blue Star Museum members; members no charge. Closed New Year's Eve & Day; Martin Luther King Jr. Day; President's Day; Memorial Day; Independence Day; Labor Day; Thanksgiving; Christmas Eve & Day. &

Attendance: 23,780 (accurate)

Membership: Student $20; Individual $50; Family $75; Contributor $100; Sponsor $250; Benefactor $500; Patron $1,000; Director's Circle $2,500; President's Circle $5,000; Trustee's Circle $10,000.

Saint Louis

A.K.C. MUSEUM OF THE DOG, (M), 1721 S. Mason Rd., Saint Louis, MO 63131-1518. Tel.: 314-821-3647. Fax: 314-821-7381.

E-mail: dogarts@aol.com
Web Site: www.museumofthedog.org
Founded: 1982.
Key Personnel: Exec. Dir., Barbara Jedda McNab; Pres. (V), Dorothy Welsh.
Personnel Profile: Full-Time Paid 4; Part-Time Paid 6.
Governing Authority: nonprofit. Tax-exempt: 501(c)(3).
Institution Type/Description: Fine Arts Museum: housed in c.1853 Jarville Estate.
Collections: dog related artwork; paintings; sculpture; decorative arts; works on paper; books; historic dog collars.
Research Fields: pure bred dogs.
Facilities: 1,500-vol. library available to the public; 14,000 sq. ft. exhibit space. Museum-related items for sale.
Activities: guided tours; lectures; organized education programs for children & adults.
Publications: triannual newsletter, Sirius.
Hours & Admission Prices: Tues.-Sat. 10-4, Sun. 1-5. Adults $5, senior citizens $2.50, children 5-14 $1; members & children under 5 no charge. Closed holidays. &
Attendance: 12,000
Membership: Student & Senior $25; Annual $35; Family $65; Participating $100; Supporting $500; Patron $1,000.

ANHEUSER-BUSCH BREWERY, 12th & Lynch St., Saint Louis, MO 63101. Tel.: 314-577-2626.

Web Site: www.budweisertours.com
Institution Type/Description: Company History Museum.
Collections: company history; Clydesdale horses; brewing process.
Facilities: Museum-related items for sale.
Activities: special events. Annual Event: Clydesdale Camera Days.
Hours & Admission Prices: Tours: Sept.-May Mon.-Sat. 10-4, Sun. 11:30-4; June-Aug. Mon.-Sat. 9-5, Sun. 11:30-5. No charge. Closed New Year's Eve & Day; M.L. King Day; President's Day; Easter; Veteran's Day; Thanksgiving, Christmas Eve, Day & day after. &

ARCHIVES & MUSEUM OF OPTOMETRY, 243 N. Lindbergh Blvd., Saint Louis, MO 63141-7851. Tel.: 800-365-2219. Fax: 314-991-4101.

E-mail: khebert@aoa.org
Web Site: www.aoa.org/about-the-aoa/archives-and-museum
Founded: 1966.
Governing Authority: Parent Institution: American Optometric Association Foundation.
Institution Type/Description: Optometry History Museum.
Collections: eyeglasses; contact lenses; cases; instruments; eyewear history; optical industry; vision testing methods; photographs.
Hours & Admission Prices: Call for hours. No charge.

ART SAINT LOUIS, 555 Washington Ave., Ste. 150, Saint Louis, MO 63101-1249. Tel.: 314-241-4810. Fax: 314-241-6933.

E-mail: info@artstlouis.org
Web Site: www.artstlouis.org
Key Personnel: Assoc. Dir., Robin Hirsch; Dir. Mktg. & Devel., Christine Malinee.
Governing Authority: nonprofit organizations. Tax-exempt: 501(c)(3).
Institution Type/Description: Art Gallery.
Collections: paintings; sculpture.
Activities: educational programs; temporary exhibitions.

Hours & Admission Prices: Mon. & Sat. 10-4, Tues.-Fri. 10-5. No charge. Closed holidays.

ATRIUM GALLERY LTD., 4728 McPherson Ave., Saint Louis, MO 63108-1918. Tel.: 314-367-1076. Fax: 314-367-7676.

E-mail: atrium@earthlink.net
Web Site: www.atriumgallery.net
Founded: 1986.
Key Personnel: C.E.O. & Dir., Carolyn P. Miles.
Personnel Profile: Full-Time Paid 2; Part-Time Paid 2.
Governing Authority: company; profit.
Institution Type/Description: Art Gallery.
Collections: exhibitions of contemporary art: paintings, drawings & sculpture by living artists.
Activities: formal education programs for adults; lectures.
Hours & Admission Prices: Tues.-Sat. 10-6, Sun. 12-4; other times by appointment. No charge. Closed New Year's Day; Memorial Day; Independence Day; Labor Day; Thanksgiving; Christmas. &
Attendance: 2,300 (estimated)

BELAS ARTES MULTICULTURAL CENTER AND ART GALLERY, 1854 Russell Blvd., Saint Louis, MO 63104. Tel.: 314-772-2787.

Key Personnel: Exec. Dir., Cileia Miranda-Yuen
Institution Type/Description: Art Gallery.
Collections: paintings; sculpture.
Activities: cultural events; educational programs.
Hours & Admission Prices: By appointment.

CAMPBELL HOUSE MUSEUM, 1508 Locust St., Saint Louis, MO 63103-1816. Tel.: 314-421-0325. Fax: 314-421-0113. Facebook: Campbell House Museum.

E-mail: andy@campbellhousemuseum.org
Web Site: www.campbellhousemuseum.org
Founded: 1943.
Congressional District: 1
Key Personnel: C.E.O., Andrew Hahn; Pres. (V), Frederick Z. Clifford, Jr.; Museum Shop Mgr., Mrs. Earl C. Lindburg.
Personnel Profile: Full-Time Paid 1; Full-Time Volunteers 1; Part-Time Paid 1; Part-Time Volunteers 45; Interns 5.
Governing Authority: nonprofit organization. Tax-exempt.
Institution Type/Description: Historic House: 1851 Campbell House.
Collections: decorative arts; 1840-1900 home furnishings (original); family photo albums; letters; Lucas Place archives.
Major Exhibits: Lucas Place: Lost Neighborhood of St. Louis' Gilded Age, 3/1-8/14.
Research Fields: Robert Campbell & family; fur trade, westward expansion; St. Louis history.
Activities: guided tours; lectures; school programs.
Publications: quarterly newsletter, Campbell House Courier.
Hours & Admission Prices: March-Dec. Wed.-Sat. 10-4, Sun. 12-4; other times by appointment. Adults $7; discounts to groups; children & members no charge. Closed national holidays.
Attendance: 6,000 (estimated)
Membership: Active $50; Century $100; Campbell Associate $300; 1851 Society $750; Lucas Place Partner $1,000; Museum Benefactor $2,500.

CARONDELET HISTORIC CENTER, 6303 Michigan Ave., Saint Louis, MO 63111-2504. Tel.: 314-481-6303.

E-mail: carondlt@stlouis.missouri.org
Founded: 1967.
Key Personnel: Ron Bolte
Institution Type/Description: History Museum.
Collections: Carondelet history & culture; photographs; military artifacts; furniture; personal artifacts.
Activities: special events.
Hours & Admission Prices: Tues.-Wed. & Fri 9:30 to noon, Sat. 10:30-2. Adults $2, children 12 & under $1; school groups no charge. &

CHATILLON-DEMENIL HOUSE FOUNDATION, 3352 DeMenil Place, Saint Louis, MO 63118-3211. Tel.: 314-771-5828.

E-mail: demenil@sbcglobal.net
Web Site: www.demenil.org
Founded: 1965.
Congressional District: 3
Key Personnel: Pres. (V), Ted Atwood; Dir., Lynn Josse.
Personnel Profile: Part-Time Paid 2; Part-Time Volunteers 12.

Governing Authority: nonprofit corporation. Affiliated with Chatillon-Demenil House Foundation. Tax-exempt: 501(c)(3).
Institution Type/Description: History Museum: housed in 1848 brick farm house of Henri & Odile Chatillon, 1863 Greek Revival addition by Dr. Nicolas N. DeMenil.
Collections: mid-Victorian furnishings; furniture; paintings by Nicolas & Sophie DeMenil; decorative arts; 1904 World's Fair artifacts.
Facilities: restaurant; garden area. Museum-related items for sale.
Activities: guided tours; weddings; meeting space; special events; private parties.
Publications: quarterly newsletter; brochures.
Hours & Admission Prices: Hourly tours: Wed.-Fri. 11-2, Sat. 11-3. Adults $5, children $2; discounts to members, groups, AAA, AAM & ICOM members; history teachers with ID no charge. Closed Jan. & national holidays.
Attendance: 3,000 (estimated)
Membership: Senior $15; Associate $40; Family $75; Madame Chouteau Associate $100; Henri Chatillon Associate $500; Nicolas DeMenil Society $1,000.

CITY MUSEUM, 701 N. 15th St., Saint Louis, MO 63103-1925. Mailing Address: 701 N. 15th St., Box 29, St. Louis, MO 63103. Tel.: 314-231-2489. Fax: 314-231-1009. Facebook: City Museum.
E-mail: info@citymuseum.org
Web Site: www.citymuseum.org
Founded: 1997.
Key Personnel: Creative Dir., Bob Cassilly; Museum Dir., Rick Erwin; everydaycircus, Jessica Hentoff; Archivist, Bruce Gerrie; Mirth, Mystery & Mayhem, Bill Christman; Museum Shop Mgr., Stephanie Von Drasek; World Aquarium, Leonard Sonnenschein.
Personnel Profile: Full-Time Paid 27; Part-Time Paid 40; Interns 2.
Governing Authority: private.
Institution Type/Description: General Museum.
Collections: historical artifacts; salvaged architectural relics; miniature train; circus artifacts.
Facilities: aquarium; 100-seat restaurant; rental space available. Museum-related items for sale.
Activities: hobby workshops; artist demonstrations; 10-story slide; ferris wheel; miniature indoor train; 5-story outdoor climbing structure; maze of indoor caves; Roof Atop the City; everydaycircus; Shoelace Factory; Museum of Mystery; Mirth and Mayhem; Architecture Museum; Skateless Park.
Hours & Admission Prices: Memorial Day to Labor Day Mon.-Thurs. 9-5, Fri.-Sat. 9-1, Sun. 11-5; Sept. to mid-March Wed.-Thurs. 9-5, Fri.-Sat. 9-1, Sun. 11-5. Museum: admission 3 & over $12. Roof Atop the City: $5. World Aquarium: $6; discounts to groups & Fri.-Sat. after 5pm. Closed New Year's Day; Easter; Thanksgiving; Christmas.
Attendance: 680,714 (accurate)
Membership: Level 1 $200; Level 2 $375; Gold Card $500.

CONCORDIA HISTORICAL INSTITUTE, 804 Seminary Place, Saint Louis, MO 63105-3014. Tel.: 314-505-7900. Fax: 314-505-7901. Facebook: Concordia Historical Institute.
E-mail: chi@lutheranhistory.org
Web Site: www.lutheranhistory.org
Founded: 1927.
Congressional District: 1
Key Personnel: Exec. Dir., Larry Lumpe; Pres. (V), Scott Meyer; Assoc. Dir. Archives & Library, Rev. Marvin A. Huggins; Business Mgr., Debbie Lower.
Personnel Profile: Full-Time Paid 3; Part-Time Paid 6; Part-Time Volunteers 15.
Governing Authority: church. Affiliated with The Lutheran Church, Missouri Synod, 1333 S. Kirkwood Rd., St. Louis, MO 63122. Tax-exempt: 501(c)(3).
Institution Type/Description: Religious History Museum.
Collections: Reformation & Lutheran medals and coins; costumes & crafts; works by Lutheran artists; materials pertaining to the Lutheran Church; archives; manuscripts; historical library; audiovisual material; 43 original Albrecht Durer woodcuts. Historic Building: Saxon Lutheran Memorial at Frohna, MO; Hill of Peace, Friedenberg at Perryville, MO.
Research Fields: Lutheranism in America; German-Americana; immigration & family history.
Facilities: 58,000-vol. library of books, pamphlets & periodicals relating to Lutheranism in America available for inter-library loan & upon application; reading room. New & used books for sale.
Activities: guided tours; lectures; inter-museum loan, temporary & traveling exhibitions.
Publications: quarterly magazine, Concordia Historical Institute Quarterly; Historical Footnotes.

Hours & Admission Prices: Mon.-Fri. 8:30-4. No charge, $3 donation suggested. Closed national holidays.
Attendance: 5,000 (estimated)
Membership: Subscription $40 (US) $50 (foreign); Active $50; Patron $100; Friend $250; Sponsor $600; Pacesetter $1,200; Life $5,000.

CONTEMPORARY ART MUSEUM ST. LOUIS, (M), 3750 Washington Blvd., Saint Louis, MO 63108-3612. Tel.: 314-535-4660. Fax: 314-535-1226.
E-mail: info@camstl.org
Web Site: www.camstl.org
Formerly: Forum for Contemporary Art
Founded: 1980.
Congressional District: 1
Key Personnel: Dir., Lisa Melandri; Chm., Pat Whitaker; Chief Cur., Dominic Molon; Asst. Cur., Kelly Shindler; Mgr. Education, Tuan Nguyen; Mgr. Exhibitions & Registrar, David Smith; Dir. Finance & Administration, Mary Walters; Dir. Devel., Emily Klimek; Grants & Sponsorships Mgr., Jessica Miller; Devel. & Special Events Mgr., Emily DeCenso; Mgr. Membership, Brie Alley; Dir. Programs & Audience Devel., Unitey Kull; Mgr. Public Programs & Interpretive Technology, Alex Elmestad; Coord. Public Rels. & Mktg., Ida McCall; Mgr. Visitor Svcs. & Museum Shop. Mgr., David Hartwell; Graphic Desinger, Todd Owyong; Coord. Facilities, B.J. Nebrida; Exhibitions Asst., Louis Nahlik.
Personnel Profile: Full-Time Paid 16; Part-Time Paid 19; Part-Time Volunteers 65; Interns 20.
Governing Authority: nonprofit organization. Tax-exempt: 501(c)(3).
Institution Type/Description: Contemporary Art Museum.
Collections: works by contemporary artists.
Facilities: 27,000 sq. ft. space.
Activities: exhibition programs; guided tours; lectures; films; concerts; organized educational programs for youth; cultural events.
Publications: exhibition catalogs; brochures; gallery guides; annual magazine, MESH; artist books.
Hours & Admission Prices: Wed. 11-6, Thurs.-Fri. 11-9, Sat.-Sun. 10-5. Adults $5, seniors $3; members, students with valid ID & children no charge. Closed national holidays.
Attendance: 22,400 (accurate)
Membership: Individual $55; Young Friend $65; Family $100; Contributor $250; Benefactor $500; Patron $1,000; Sustainer $2,500; Curator Circle $5,000; Director Circle $10,000; Chairman Circle $25,000 & up.

CRAFT ALLIANCE, 6640 Delmar Blvd., Saint Louis, MO 63130-4503. Tel.: 314-725-1177. Fax: 314-725-2068.
E-mail: gallery@craftalliance.org
Web Site: www.craftalliance.org
Founded: 1964.
Congressional District: 1
Key Personnel: Exec. Dir., Boo McLoughlin; Sr. Dir. Education & Exhibition Prog., Luanne Rimel; Opers. Dir., Lexi Glynias; Finance Mgr., Bob Lewis; Community Outreach Mgr., Robert Longyear; Studios Mgr., Dan Barnett; Museum Shop Mgr., Kris Richards.
Personnel Profile: Full-Time Paid 8; Part-Time Paid 40; Part-Time Volunteers 5.
Governing Authority: nonprofit organization. Tax-exempt: 501(c)(3).
Institution Type/Description: Arts & Crafts Museum.
Collections: craft arts; study collection.
Facilities: 100-vol. library of books & catalogues; studio classes and workshops; 1,500 sq. ft. exhibit space. Art & craft items for sale.
Activities: guided tours; lectures; traveling & loan exhibitions; formally organized education programs for children & adults; docent program. Annual Events: fundraising projects; annual Holiday Show.
Publications: 3 times annually, Newsletter/Class Brochure.
Hours & Admission Prices: Tues.-Thurs. 10-5, Fri.-Sat. 10-6, Sun. 11-5. Classes: daily. Exhibits: no charge; donations accepted. Closed national holidays.
Attendance: 12,000 (estimated)
Membership: National Associate $45; Senior Citizen $50; Basic Household $55; Family Sponsor $100; Supporter $250; Benefactor $500; Craft Partners Council $1,000.

DENTAL HEALTH THEATRE, 727 N. First St., Ste. 103, Saint Louis, MO 63102-2501. Tel.: 314-241-7391. Fax: 314-241-7391.
E-mail: info@ddhtstl.org
Web Site: www.ddhtstl.org
Formerly: Delta Dental Health Theatre
Founded: 1977.
Key Personnel: Exec. Dir., Shannon Woodcock; Pres. (V), Tom Flavin, D.D.S.
Personnel Profile: Full-Time Paid 1; Part-Time Paid 1.

Institution Type/Description: Children's Dental & Overall Health Museum.
Collections: oral health; healthy life habits; nutrition, diet & exercise; hands-on exhibits.
Activities: educational program; video; interactive games, presentations & discussions.
Hours & Admission Prices: Tues.-Sat. 9-3. Adults $1; discounts to AAM members; members no charge. &
Attendance: 10,000 (estimated)
Membership: Individual $50; Family $75; Family Share $150; Healthy Donor $250; Dragon $500; Wizard $1,000.

DES LEE GALLERY OF ART, Washington University School of Art, 1627 Washington Ave., Saint Louis, MO 63103. Tel.: 314-621-8735.
Institution Type/Description: Art Gallery.
Collections: works by contemporary artists.
Hours & Admission Prices: Call for hours.

EUGENE FIELD HOUSE AND ST. LOUIS TOY MUSEUM, (M), 634 S. Broadway, Saint Louis, MO 63102-1613. Tel.: 314-421-4689. Fax: 314-588-9468.
E-mail: info@eugenefieldhouse.org
Web Site: eugenefieldhouse.org
Founded: 1936.
Congressional District: 3
Key Personnel: Chm. & Pres. (V), William Piper; Dir., Kimberly Ann Larson.
Personnel Profile: Full-Time Paid 2; Part-Time Paid 1; Part-Time Volunteers 10; Interns 2.
Governing Authority: Eugene Field House Foundation, Inc., 634 S. Broadway, St. Louis, MO. 63102. Tax-exempt: 501(c)(3).
Institution Type/Description: Historic House: built 1845, birthplace of poet & toy collector, Eugene Field; home of Roswell Field, lawyer for Dred Scott & family during 1857 U.S. Supreme Court Decision.
Collections: manuscripts; 19th century furnishings & housewares; period toys & dolls.
Facilities: 200-vol. library on the works of Eugene Field; reference books on toys & dolls. Reproductions of period toys & dolls for sale.
Activities: guided tours; permanent & temporary exhibitions.
Publications: quarterly newsletter for members, Field Notes.
Hours & Admission Prices: March-Dec. Wed.-Sat. 10-4, Sun. 12-4; other times by appointment. Adults $5, children under 12 $1, discounts to groups & AAA members; E.F. Foundation members no charge. Closed major holidays.
Attendance: 7,000 (estimated)
Membership: Field Friend $50; Columnist Corner $75; Roswell Partner $150; Poet's Patron $250; 1845 Benefactor $1,000.

GALLERY FAB - UNIVERSITY OF MISSOURI, 201 Fine Arts Bldg., Saint Louis, MO 63121. Tel.: 314-516-6967.
E-mail: mcpeakk@umsl.edu
Institution Type/Description: Art Gallery.
Collections: works by local artists & students.
Hours & Admission Prices: Mon.-Thurs. 9-9, Fri. 9-5, Sat. 10-5, Sun. 12-8.

GALLERY 210 - UNIVERSITY OF MISSOURI-ST. LOUIS, 44 E. Dr., TCC, One University Blvd., Saint Louis, MO 63121-4400. Tel.: 314-516-5976.
Key Personnel: Dir., Prof. Jeane Tucker
Institution Type/Description: Art Gallery.
Collections: paintings; sculpture.
Hours & Admission Prices: Tues.-Sat. 11-5; other times by appointment.

GRANT'S FARM, 10501 Gravois Rd., Saint Louis, MO 63123-1808. Tel.: 314-843-1700.
Institution Type/Description: Zoo.
Collections: over 900 animals representing 100 species.
Activities: special events.
Hours & Admission Prices: Spring: Sat. 9-3:30, Sun. 9:30-3:30; Summer: Tues.-Fri. 9-3:30, Sat. 9-4, Sun. 9:30-4; Fall: Fri. 9:30-2:30, Sat.-Sun. 9:30-3:30.

GRIOT MUSEUM OF BLACK HISTORY, 2505 St. Louis Ave., Saint Louis, MO 63106-2324. Tel.: 314-241-7057. Fax: 314-241-7058.
E-mail: blkwaxmusm1@aol.com
Web Site: www.thegriotmuseum.com
Formerly: Blackworld History Museum

Founded: 1992.
Congressional District: 1
Key Personnel: Dir., Lois D. Conley.
Personnel Profile: Full-Time Paid 2; Part-Time Volunteers 10.
Governing Authority: bd. Parent Institution: Diaspora Connections Unlimited. Tax-exempt.
Institution Type/Description: Black History Museum.
Collections: Missouri's African-American history from slavery to present; photographs; wax sculptures; documents.
Facilities: Museum-related items for sale.
Activities: heritage preservation youth training program.
Hours & Admission Prices: Jan. 16-Dec. 14 Wed.-Sat. 10-5; other times by appointment. Adults $7, youth 13-18 and seniors 60 & over $3.50, children 12 & under $2.50; discounts to AAM & ICOM members; members no charge. &
Attendance: 7,530 (estimated)
Membership: Youth $15; Elder $25; Basic $35.

HISTORIC SAMUEL CUPPLES HOUSE, 3673 W. Pine Mall, Saint Louis, MO 63108-3303. Mailing Address: 3545 Lindell Blvd., Saint Louis, MO 63103. Tel.: 314-977-2666. Fax: 314-977-3581. Facebook: Historic Samuel Cupples House.
E-mail: sluma@slu.edu
Web Site: cupples.slu.edu
Founded: 1977.
Congressional District: 1
Key Personnel: Dir., Petruta Lipan; Cur., Kathleen Wang; Public Rels. Affairs, Mary Marshall; Business Mgr., Benjamin Faser; Dir. Facilities, Barth Breneman; Museum Shop Mgr., Nathanael Mooningham; Security, Dennis Thompson.
Personnel Profile: Full-Time Paid 6; Part-Time Paid 9; Part-Time Volunteers 2; Interns 2.
Governing Authority: private university. Parent Institution: Saint Louis University, St. Louis, MO. Subsidiary Institution: Saint Louis University Museum of Art, 3663 Linden Blvd., St. Louis, MO 63108. Tel.: 314-977-3399. Tax-exempt: 501(c)(3).
Institution Type/Description: Art & Decorative Art Museum: housed in a Richardsonian-Romanesque style mansion; built in 1888.
Collections: American & European glass, furniture & costumes; 16th-19th century European & American art.
Research Fields: decorative arts; American & European glass 1880-1940; American furniture; northern Renaissance art.
Facilities: 500-vol. library; educational facilities; 2,000 sq. ft. exhibit space. Museum-related items for sale.
Activities: formal education programs; guided tours; lectures; internships.
Publications: brochure, House History.
Hours & Admission Prices: Tues.-Sat. 11-4. Adults $5; members, students & children no charge. Closed national holidays.
Attendance: 5,400 (accurate)
Membership: Regular $50; Family $75; Contributing $100; Sustaining $150; Supporting $250; Associate $500; Sponsor $1,000; Patron $1,500; Fellow $2,500; Benefactor $5,000; Director's Circle $10,000.

HOLOCAUST MUSEUM & LEARNING CENTER, 12 Millstone Campus Dr., Saint Louis, MO 63146-5776. Tel.: 314-432-0020, ext. 3711.
Web Site: www.hmlc.org
Founded: 1995.
Institution Type/Description: History Museum.
Collections: pre-Holocaust life; the Holocaust; post-Holocaust; photographs; Jewish history & life; audio presentations.
Facilities: Books for sale.
Activities: lectures; temporary exhibitions; monthly film series.
Hours & Admission Prices: Mon.-Thurs. 9:30-4:30, Fri. 9:30-4, Sun. 10-4; other times by appointment. School Tours: sixth grade & up; not recommended for young children. No charge; donations accepted. &
Attendance: 30,000 (estimated)

JEFFERSON BARRACKS HISTORIC SITE, 345 North Rd., Saint Louis, MO 63125-4121. Tel.: 314-544-5714. Fax: 314-638-5009. TDD: 314-615-7840.
E-mail: dgonzales@stlouisco.com
Web Site: www.stlouisco.com/parks/j-b.html
Founded: 1826.
Key Personnel: Dir., J.D. Maguruny; Dir. St. Louis County Parks Recreation, Tom Ott; Chm. (V), Sue Bell; Cur., Daniel Gonzales; Exhibition & Maintenance, Mike Kladky; Museum Shop Mgr., Pat Galanos.

Personnel Profile: Full-Time Paid 3; Full-Time Volunteers 1; Part-Time Paid 1; Part-Time Volunteers 75; Interns 1.
Governing Authority: county government. Parent Institution: St. Louis County Dept. of Parks & Recreation. Tax-exempt.
Institution Type/Description: Historic Preservation Project: military post/museum.
Collections: artifacts related to history of Jefferson Barracks Military Post 1826-1946.
Major Exhibits: Homefront: Vmail to email, 4/14-12/14.
Research Fields: Westward Expansion; military history; Missouri military; U.S. Troop movement/expansion, St. Louis history.
Facilities: classrooms; 2,000 sq. ft. exhibit space. Museum-related items for sale.
Activities: guided tours; lectures; films; organized education programs for children & adults; docent program; traveling exhibition. Museum Sponsors: Story-telling Festival; Concerts.
Publications: JB Newsletter; Powder Magazine & Laborers House.
Hours & Admission Prices: Museums & historic buildings open Wed.-Sun. noon-4. No charge; donations accepted. Closed national holidays. &
Attendance: 250,000 (accurate)
Membership: Friends of Jefferson Barracks: Individual $10; Two Adults $15; Family $25.

∗ JEFFERSON NATIONAL EXPANSION MEMORIAL, (M), 11 N. 4th St., Saint Louis, MO 63102-1810. Tel.: 314-655-1700 & 1600. Fax: 314-655-1639. TDD: 1-800-735-2466.
E-mail: jeff_superintendent@nps.gov
Web Site: www.nps.gov/jeff
Founded: 1935.
Congressional District: 1
Key Personnel: Supt., Tom Bradley; Facility Mgr., Edwards Dodds; Pres. & C.E.O. JNPA, David Grove; Chief Museum Svcs. & Interpretation, Ann Honious; Cur. Cultural Resources, Kathryn Thomas.
Governing Authority: federal. Parent Institution: National Park Service, U.S. Dept. of Interior, 18th St. & Virginia, N.W., Washington, DC 20240. Subsidiary Institution: Jefferson National Parks Association. Tax-exempt.
Institution Type/Description: History Museum: Museum of Westward Expansion located beneath the Gateway Arch. Historic Building: Old Courthouse.
Collections: 19th-century Western American artifacts; American-Indian artifacts; local history; horse-drawn transportation; architectural; archives. Historic Buildings: 1839-1862 Old Courthouse with exhibit galleries & furnished courtrooms.
Research Fields: 19th-century westward expansion; St. Louis history; park administrative history.
Facilities: 4,900-vol. library on westward expansion & St. Louis history; 70mm Odyssey Theater; two 16mm/video theaters. History books, postcards & other museum-related items for sale.
Activities: permanent & temporary exhibitions; interpretive & educational programs; group tours; films; concerts; tram to top of Arch.
Publications: books on 19th-century westward expansion history & other park-related themes; park newspaper; site bulletins.
Hours & Admission Prices: Old Courthouse: daily 8-4:30. Museum of Westward Expansion: Memorial Day to Labor Day daily 8am-10pm; Sept.-May daily 9-6. Family $6, adults $3; additional fees for films & tram ride. Closed New Year's Day; Thanksgiving; Christmas. &
Attendance: 2,500,000 (estimated)

LACLEDE'S LANDING WAX MUSEUM, 720 N. Second St., Saint Louis, MO 63102-2519. Tel.: 314-241-1155.
Web Site: www.stlwaxmuseum.com
Institution Type/Description: Wax Museum.
Collections: over 200 wax figures.
Activities: special events.
Hours & Admission Prices: May-Oct. Sun.-Thurs. 11-8, Fri.-Sat. 11-10; Nov.-April Thurs.-Sun. 11-6. Adults $10, juniors 12-17 $7, seniors 55 & over $5, children 3-11 $3; discounts to groups of 15 or more.

∗ LAUMEIER SCULPTURE PARK & MUSEUM, (M), 12580 Rott Rd., Saint Louis, MO 63127-1212. Tel.: 314-615-5278. Fax: 314-615-5283.
E-mail: info@laumeier.org
Web Site: www.laumeier.org
Founded: 1976.
Congressional District: 3
Key Personnel: Exec. Dir., Marilu Knode; Chm. Bd., David Schlafly; Devel. Asst., Marie Oberkirsch; Devel. Officer, Jackie Chambers; Dir. Accounting, Mary Ruskin; Cur. Education, Karen Mullen; Cur. Educational Interpretation, Clara Collins Coleman; Mgr. Membership & Museum Svcs., Jennie Swanson; Cur. Exhibitions, Dana Turkovic; Mgr. Operations, Don Gerling;

Administrative Asst., Julia Norton; Librarian & Archivist, Joy Wright; Mgr. Parks Program Service, Kyra Kattenbrown; Exhibitions Preparator, Nick Lang.
Personnel Profile: Full-Time Paid 11; Part-Time Paid 5; Part-Time Volunteers 566; Interns 2.
Governing Authority: nonprofit. St. Louis County Parks. Tax-exempt: 503 (c)(3).
Institution Type/Description: Art Museum & Sculpture Park.
Collections: contemporary art; monumental and site-specific sculpture and related works; representative examples of Vito Acconci, Mark di Suvero, Jackie Ferrara, Charles Ginnever, Michael Heizer, Alexander Liberman, Beverly Pepper, and Ursula Von Rydingsvard; Ernest Trova sculptures.
Facilities: library; 105-acre open-air museum and park; educational center; amphitheater; trails.
Activities: temporary indoor & outdoor art exhibitions; permanent installations; educational programs and workshops for children and adults; guided tours; concert series; lecture; docent program; art fair; special events.
Publications: site map; exhibition brochures & catalogues; special event guides; quarterly newsletter.
Hours & Admission Prices: Sculpture Park: 8am to half hour past sunset. Indoor Galleries: Tues.-Fri. 10-5, Sat.-Sun. 12-5. Fee charged for special events only. Closed legal holidays. &
Attendance: 300,000 (estimated)
Membership: Dog(s) $15; Friends $45; Family $65; Friends & Neighbors $105; Casting Circle $125; Sculptor's Forum $300; Director's Circle $600; Collector's Circle $1,200; Laumeier Society $3,000.

THE MAGIC HOUSE, ST. LOUIS CHILDREN'S MUSEUM, (M), 516 S. Kirkwood Rd., Saint Louis, MO 63122-5926. Tel.: 314-822-8900, ext. 524. Fax: 314-822-8930.
E-mail: info@magichouse.org
Web Site: www.magichouse.org
Founded: 1979.
Congressional District: 2
Key Personnel: Pres., Elizabeth Fitzgerald; Chm., Harvey Wallace; C.F.O., Cheryl Darr; Dir. Community Education, Julie Tubbs; Dir. Design, Beth Hasek; Grant Mgr., Hedy Ehrlich; Mgr. Devel., Vicki Peckron; Dir. Human Resources, Elizabeth Hartman; Dir. Museum Education, Carla Krakoviak; Business Mgr., Anne Spinner; Dir. Visitor Svcs., Paula Burdge; Coord. Special Events, Marissa Cribbs; Mgr. Membership, Bethany Altis; Museum Shop Mgr., Sara Modray.
Personnel Profile: Full-Time Paid 30; Part-Time Paid 120; Part-Time Volunteers 100; Interns 1.
Operating Expenses: 4,842,183
Operating Income: 5,375,844
Governing Authority: nonprofit organization. Tax-exempt: 501(c)(3).
Institution Type/Description: Children's Museum: housed in 1901 home of George Lane Edwards.
Collections: hands-on, participatory exhibits.
Major Exhibits: Sid The Science Kid (T), 1/24/14-12/14; Sandcastle Beach, 5/14-7/14; Magical Theater Co., 7/14.
Activities: organized education programs for children; docent program; participatory exhibits; teacher workshops; outreach programs; field trips.
Publications: brochure; books: A Guide to The Magic House; A Taste of Science; Wonder House; Wonder Boat.
Hours & Admission Prices: Memorial Day to Labor Day Mon.-Thurs. & Sat. 9:30-5:30, Fri. 9:30-9, Sun. 11-5:30; Sept.-May Tues.-Thurs. 12-5:30, Fri. 12-9, Sat. 9:30-5:30, Sun. 11-5:30. Admission one & over $9.50; discounts to AAM members & groups. Closed Easter; Thanksgiving; Christmas. &
Attendance: 555,550 (accurate)
Membership: Value $75; Grandparents $100; Family $150; Grandparents Club $250; Supporting $500.

MARYVILLE UNIVERSITY - MORTON J. MAY FOUNDATION GALLERY, 650 Maryville University Dr., Saint Louis, MO 63141-7299. Tel.: 314-529-9381. Fax: 314-529-9940.
E-mail: artgallery@maryville.edu
Web Site: www.maryville.edu/morton-j-may-foundation-gallery.htm
Congressional District: 2
Key Personnel: Exec. Dir., Dr. Mark Lombardi, Ph.D.; Dir., John Baltrushunas, M.F.A.; Treas., Larry Hays, Ph.D.
Personnel Profile: Full-Time Paid 1; Part-Time Paid 1.
Governing Authority: nonprofit organization. Parent Institution: Maryville University. Tax-exempt.
Institution Type/Description: Art and Design Gallery.
Collections: paintings; prints; sculpture; photographs.
Activities: formal education programs for university students; lectures; guided tours; participatory & student exhibits.
Hours & Admission Prices: Mon.-Thurs. 8am-10pm, Fri.-Sat. 8-6, Sun. 2-10. No charge. &

Attendance: 30,000 (estimated)

MERAMEC CONTEMPORARY ART GALLERY, 11333 Big Bend Blvd., Saint Louis, MO 63122-2810. Tel.: 314-984-7632. Fax: 314-984-7920. TDD: 314-984-7127.

E-mail: mkeller@stlcc.edu
Key Personnel: Dir., Margaret Keller
Institution Type/Description: Art Gallery.
Collections: sculpture; paintings; drawings.
Facilities: 1,700 sq. ft. exhibit space.
Hours & Admission Prices: Mon.-Thurs. 9-9, Fri. 9-4, Sat. 10-5.

* MILDRED LANE KEMPER ART MUSEUM, (M), One Brookings Dr., Saint Louis, MO 63130-4862. Mailing Address: Campus Box 1214, One Brookings Dr., Saint Louis, MO 63130-4862. Tel.: 314-935-4523. Fax: 314-935-7282. Facebook: Kemper Art Museum.

E-mail: kemperartmuseum@wustl.edu
Web Site: www.kemperartmuseum.wustl.edu
Formerly: Washington University Gallery of Art
Founded: 1881.
Congressional District: 1
Key Personnel: Dir. & Chief Cur., Dr. Sabine Eckmann; Exhibition Preparator, Ron Weaver; Chief Registrar, Rachel Keith; Administrative Coord., John Foughty; Mng. Editor Publications, Jane Neidhardt; Facilities Mgr. & Preparator, Janet Hessel; Security Supvr., Michael Hesse; Assoc. Cur., Meredith Malone; Visitor Services & Events Coord. & Museum Shop Mgr., Indra Russell; Asst. Cur., Karen K. Butler; Cur. Pub. Art, Leslie Markle; Mgr. Education Programs, Allison Taylor.
Personnel Profile: Full-Time Paid 17; Part-Time Paid 64; Part-Time Volunteers 21; Interns 25.
Volunteer Hours: 87
Operating Expenses: 3,800,000
Operating Income: 3,800,000
Governing Authority: university. Parent Institution: Washington University. Tax-exempt: 501(c)(3).
Institution Type/Description: Art Museum.
Collections: 16th- to 20th-century European painting & sculpture; old master & modern prints; 19th- to 20th-century American painting & sculpture; antiquities; drawings; numismatics; contemporary art.
Major Exhibits: In the Aftermath of Trauma: Contemporary Video Installations, 1/14-4/14; On the Thresholds of Space-Making: Shinohara Kazuo and His Legacy, 1/14-4/14; Moving Parts: Time and Motion in Contemporary Art, 5/14-8/14; Drawing Ambiance: Alvin Boyarsky and the Architectural Association (T), 9/14-1/15; The Eisendrath Years, 9/14-1/15.
Research Fields: 19th, 20th & 21st-century European and American art.
Facilities: 150,000-vol. library of art history, fine arts, fashion design & architecture periodicals and catalogs available for inter-library loan & use by public for research.
Activities: guided tours; gallery talks; lectures; films; workshops; concerts.
Publications: semiannual, Calendar of Events; bulletin; exhibition catalogues.
Hours & Admission Prices: Mon. & Wed.-Sun. 11-5, first Fri. of month 11-8. No charge; donations accepted. &
Attendance: 29,557 (accurate)
Membership: Student $20; Faculty & Staff $40; Friend $60; Century Club $100-$249; Century Club Fellows $250-$499; Director's Committee $500-$999; William Greenleaf Eliot Society $1,000 & up.

MILLSTONE GALLERY AT COCA, 524 Trinity Ave., Saint Louis, MO 63130. Tel.: 314-725-6555.

E-mail: emcclure@cocastl.org
Institution Type/Description: Art Gallery.
Collections: contemporary art by regional & national artists.
Activities: workshops; lectures; classes; demonstrations; special events.
Hours & Admission Prices: Mon.-Fri. 10-5, Sat.-Sun. 12-5; other times by appointment.

MINIATURE MUSEUM OF GREATER ST. LOUIS, 4746 Gravois, Saint Louis, MO 63116-2437. Tel.: 314-832-7790.

E-mail: FZerb@aol.com
Web Site: www.miniaturemuseum.org
Founded: 1989.
Key Personnel: Pres. (V) & Museum Shop Mgr., Joanne Martin; Museum Shop Mgr., Fay Zerbolio.
Personnel Profile: Part-Time Volunteers 30.
Volunteer Hours: 2,380
Operating Expenses: 32,139
Operating Income: 44,030
Governing Authority: Tax-exempt.
Institution Type/Description: Miniature Museum.
Collections: miniature dollhouses & room; dolls; furniture; silver.
Activities: workshops; miniature shows.
Publications: quarterly newsletter.
Hours & Admission Prices: Wed.-Sat. 11-4, Sun. 1-4. Adults $5, seniors & youth 13-18 $4, children 12 & under $2; discounts to members; children under 2 no charge. Closed New Year's Eve & Day; Easter; Independence Day; Thanksgiving; Christmas Eve & Day. &
Attendance: 3,750 (estimated)
Membership: Club $13; Individual $15; Family $25; Supporting $50; Sustaining $100; Benefactor $500.

* MISSOURI BOTANICAL GARDEN, 4344 Shaw Blvd., Saint Louis, MO 63110-2291. Mailing Address: P.O. Box 299, Saint Louis, MO 63166-0299. Tel.: 314-577-5110. Fax: 314-577-9595.

Web Site: www.mobot.org
Founded: 1859.
Congressional District: 1
Key Personnel: Pres., Dr. Peter Wyse Jackson; Exec. Vice Pres., Robert Herleth; Chm. Bd. Trustees, W. Stephen Maritz; Pres. Members Bd., Parker McMillan; Vice Pres. Institutional Advancement, Donna McGinnis; Sr. Vice Pres. Research, Robert Magill, Ph.D.; Vice Pres. Information Technology & Chief Information Officer, Chuck Miller; Vice Pres. Education, Sheila Voss; Vice Pres. Horticulture, Andrew Wyatt; Vice Pres. Human Resources Management, Rebecca Ingram; Sr. Vice Pres. Gen. Svcs., Paul Brockmann; Sr. Vice Pres. Communications, Peggy Lents; Vice Pres. Financial Svcs. & Controller, Rick Angevine; Vice Pres. Center for Conservation & Sustainable Devel., Olga Martha Montiel; Volunteer Coord., Jackie Juras; Sr. Vice Pres. Retail Operations & Museum Shop Mgr., Peggy Lents; Dir. Shaw Nature Reserve, John Behrer; Dir. Butterfly House, Victoria Campbell; Dir. Earthways Center, Glenda Abney.
Personnel Profile: Full-Time Paid 379; Part-Time Paid 66; Part-Time Volunteers 1,829.
Volunteer Hours: 141,537
Operating Expenses: 39,809,532
Operating Income: 42,068,866
Governing Authority: nonprofit organization. Tax-exempt.
Institution Type/Description: Botanical Garden; located on the site of the garden of Henry Shaw, developed in 1850s. A National Historic Landmark.
Collections: herbarium including over 6 million plant specimens; manuscripts; Boehm porcelain; butterfly house; children's garden; archives. Historic Houses: 1849 Tower Grove House; 1850 Town House; 1859 Museum Building; 1882 Linnean House; 1858 Flora Gate (now Spink Pavilion); 1883 Mausoleum; 1960 Climatron (R).
Major Exhibits: Nature Connects 2.0 (Lego Brick Exhibit) (T), 5/23/14-9/9/14.
Research Fields: botany; horticulture.
Facilities: 178,500 vol. library of systematic botany & gardening available for inter-library loan & by permission of librarian; 79 acres of gardens & conservatories; field research stations; reading room; 400-seat auditorium; classrooms; restaurant; nature trails. Items related to gardening & botany, including books & plants, for sale.
Activities: guided tours; lectures; study clubs; public events; flower shows; formally organized education programs for adults, children, & graduate students affiliated with Washington University, St. Louis University, Southern Illinois University in Edwardsville & University of Missouri in St. Louis; docent program or council; inter-museum loan; tram tours; plant society show & sales.
Publications: quarterly: Annuals of the Missouri Botanical Garden, Novon; quarterly: Missouri Botanical Garden Bulletin; irregular: Monographs in Systematic Botany from The Missouri Botanical Garden.
Hours & Admission Prices: Memorial Day to Labor Day Wed.-Thurs. 9-9, Fri.-Tues. 9-5; Sept.-May daily 9-5. Grounds: Sun.-Tues. & Thurs.-Fri. 9-5, Wed. & Sat. 7-5. Adults 13-64 $8; discounts to St. Louis County residents; members & children under 12 no charge. Shaw Nature Reserve: call for information, 636-451-3512. Butterfly House: call for information 636-530-0076. Children's Garden: children 3-12 $5 (members), $3 (non-members), adults no charge. Some special events may require additional charge. Closed Christmas. &
Attendance: 897,625 (accurate)
Membership: Senior $60; Regular $65; Garden Plus $95; Family $150; Sponsoring $300; President's Assoc. $600; Henry Shaw Assoc. $1,200; Garden Fellows $2,500; Garden Ambassador $5,000; Peter H. Raven Society $10,000.

MISSOURI CIVIL WAR MUSEUM, 222 Worth Rd., Saint Louis, MO 63125. Tel.: 314-845-1861. Facebook: Missouri Civil War Museum.

Web Site: www.mcwm.org/

Founded: 2002.
Congressional District: 1
Key Personnel: Dir. & C.E.O., Mark Trout.
Personnel Profile: Full-Time Paid 2; Part-Time Paid 1; Part-Time Volunteers 6; Interns 2.
Institution Type/Description: Military History Museum.
Collections: Civil War history; personal artifacts; documents; photographs; uniforms; military equipment.
Facilities: Museum-related items for sale.
Activities: educational programs; research.
Hours & Admission Prices: Daily 9-5. Adults $5-$7; members no charge. Closed New Year's Day; Thanksgiving; Christmas. &
Attendance: 50,000
Membership: Annual $30-$500.

✶ MISSOURI HISTORICAL SOCIETY, (M), Lindell & De Baliviere, 5700 Lindell Blvd., Saint Louis, MO 63112. Mailing Address: P.O. Box 11940, Saint Louis, MO 63112-0040. Tel.: 314-454-3150. Fax: 314-746-4548.
E-mail: info@mohistory.org
Web Site: www.mohistory.org
Founded: 1866.
Congressional District: 1
Key Personnel: C.E.O. & Pres., Dr. Robert R. Archibald; Mng. Dir. Operations, Karen M. Goering; Mng. Dir. Institutional Advancement, Donna McGinnis; C.F.O., Harry Rich; Dir. Exhibitions & Research, Margaret Koch; Conservator, Linda Landry; Museum Shop Mgr., Susan Ponciroli; Mng. Dir. Community Education & Events, Melanie Adams; Cur. Special Projects, Carolyn Gilman; Sr. Cur., Shannon Berry; Mng. Dir. Museum Svcs., Katie Van Allen.
Personnel Profile: Full-Time Paid 125; Part-Time Paid 52; Part-Time Volunteers 121; Interns 24.
Governing Authority: nonprofit organization. Parent Institution: St. Louis Zoo-Museum Tax District; Library & Research Center, 225 S. Skinker Blvd. Tax-exempt: 501(c)(3).
Institution Type/Description: History Museum.
Collections: St. Louis, Missouri & Urban history; exploration of the West, Lewis and Clark; artifacts; broadsides & maps; costumes; manuscripts; photographs; painting; sculpture; Charles A. Lindbergh collection; Richard Gephardt Congressional papers.
Research Fields: history of St. Louis, Missouri, Mississippi Valley & western expansion.
Facilities: 70,000-vol. library & research center. Museum-related items for sale.
Activities: guided tours; lectures; formally organized community programs for adults & children; permanent, temporary & traveling exhibitions; community programs.
Publications: online magazine, Voices; magazine, Gateway; newsletter, Members' News; monographs; books.
Hours & Admission Prices: Museum: Tues. 10-8, Wed.-Mon. 10-5. No charge. Closed Thanksgiving; Christmas. Library & Research Center: Tues.-Fri. 12-5, Sat. 10-5. No charge. Closed major holidays; Sat. holidays. &
Attendance: 516,016 (accurate)
Membership: Senior $55; Regular $60; Family Plus $85; Contributing $150; Sustaining $250; Sponsoring $500; Thomas Jefferson Society $1,000-$25,000.

MISSOURI HISTORY MUSEUM, 5700 Lindell Blvd., Saint Louis, MO 63112. Mailing Address: P.O. Box 11940, Saint Louis, MO 63112-0040. Tel.: 314-746-4599.
Institution Type/Description: History Museum.
Collections: local history & culture; period furnishings; personal artifacts; photographs.
Activities: educational programs; special events.
Hours & Admission Prices: mid-May to mid-Sept. Tues. 10-8, Wed.-Mon. 10-6; mid-Sept. to mid-May Tues. 10-8, Wed.-Mon. 10-5.

MUSEUM OF CONTEMPORARY RELIGIOUS ART (MOCRA)-ST. LOUIS UNIVERSITY, (M), 3700 W. Pine Mall Blvd., Saint Louis, MO 63108-3306. Mailing Address: 221 N. Grand Blvd., Saint Louis, MO 63103-2006. Tel.: 314-977-7170. Fax: 314-977-2999. Facebook: MOCRA.SLU.
E-mail: mocra@slu.edu
Web Site: mocra.slu.edu
Founded: 1993.
Congressional District: 1
Key Personnel: C.E.O. & Cur., Rev. Terrence E. Dempsey, S.J.; Chm. (V), Jane Daggett Dillenberger; Devel. & Education, David Brinker; Registrar, Michelle Hake.

Personnel Profile: Full-Time Paid 3; Part-Time Paid 7.
Governing Authority: private university. Parent Institution: Saint Louis University. Tax-exempt: 501(c)(3).
Institution Type/Description: Interfaith Religious Contemporary Art Museum: housed in former Jesuit seminarian chapel.
Collections: contemporary art from all religious traditions which engages in a serious manner the religious and spiritual dimensions.
Research Fields: the intersection of the spiritual & religious dimensions in contemporary art.
Activities: guided tours; films; lectures; loan & traveling exhibitions.
Publications: exhibition catalogue, Bernard Maisner: Entrance to the Scriptorium.
Hours & Admission Prices: Tues.-Sun. 11-4. No charge; donations accepted. Closed New Year's Day; Martin Luther King Jr. Day; Easter; Memorial Day; Independence Day; Labor Day; Thanksgiving; Christmas; between exhibitions. &
Attendance: 2,600 (accurate)

MUSEUM OF TRANSPORTATION, 3015 Barrett Station Rd., Saint Louis, MO 63122-3398. Tel.: 314-615-8668. Fax: 314-615-8210.
E-mail: mbutterworth@stlouisco.com
Web Site: www.museumoftransport.org
Founded: 1944.
Congressional District: 2
Key Personnel: Cultural Site Mgr., Molly Butterworth; Pres. (V), Darryl Ross; Association Dir., Terri McEachern; Museum Shop Mgr., Sandra Williams.
Personnel Profile: Full-Time Paid 9; Part-Time Paid 6; Part-Time Volunteers 125.
Volunteer Hours: 35,419
Governing Authority: nonprofit organization; bd. of dirs. Parent Institution: St. Louis County Dept. of Parks & Recreation. Subsidiary Institution: Transport Museum Association. Tax-exempt: 509(c)(3) & 501(c)(3).
Institution Type/Description: Transportation Museum.
Collections: locomotives; railway cars; automobiles; streetcars; buses; trucks; horse drawn vehicles; aircraft; pipeline & communication devices; history of the technology & design of transportation and communication. Historic Site: roadbed of first railroad west of Mississippi River.
Research Fields: transportation history.
Facilities: 10,000-vol. library on transportation & communication available for research on premises; meeting rooms. Educational & museum related items for sale.
Activities: guided tours; education programs for adults & children; docent program.
Publications: newsletter, News & Views; papers; news releases.
Hours & Admission Prices: Memorial Day to Labor Day Mon.-Sat. 9-4, Sun. 11-4; Winter: Tues.-Sat. 9-4, Sun. 11-4. Adults $8, children $5; members no charge. Closed New Year's Day; Thanksgiving; Christmas. &
Attendance: 112,678 (accurate)
Membership: Active $35; Associate $50; Sustainer $100; Contributor $250; Sponsor $500; Patron $1,000; Life $5,000.

THE NATIONAL BLUES MUSEUM, 601 Washington Ave., Saint Louis, MO 63101.
E-mail: info@nationalbluesmuseum.org
Governing Authority: Tax-exempt: 501(c)(3).
Institution Type/Description: Blues Museum.
Collections: Blues history, music & musicians; photographs; personal artifacts.
Facilities: 100-seat theater.
Activities: educational programs; special events; workshops; seminars.
Hours & Admission Prices: Call for hours.

OAKLAND HOUSE - AFFTON HISTORICAL SOCIETY, 7801 Genesta, Saint Louis, MO 63123. Mailing Address: 9911 Vasel Dr., Affton, MO 63123. Tel.: 314-544-1006.
E-mail: historicdzn@aol.com
Web Site: www.afftonoaklandhouse.com
Founded: 1973.
Key Personnel: Pres. (V), Chas Brown.
Personnel Profile: Part-Time Volunteers 52.
Governing Authority: Parent Institution: Affton Historical Society. Tax-exempt.
Institution Type/Description: Historical Society Museum: housed in the former home of banker, Louis A. Benoist; built in 1850s. Listed on the National Register of Historic Places.
Collections: local history & culture; period furnishings; photographs.
Research Fields: Affton community & school; Benoist family.
Facilities: rental facilities.

Activities: guided tours. Annual Events: Santa House; Bunny Hutch; July 4th Breakfast & Program.
Publications: quarterly, Oakleaf.
Hours & Admission Prices: April-Oct. 3rd Sun. of month; other times by appointment. Adults $5; members no charge. &
Attendance: 13,150 (accurate)
Membership: Individual $15; Couples, Contributor & Organization $25; Friend $26-$100; Sustaining $101-$250; Benefactor $251-$500; Professional $501-$1,000; Patron $1,001 & up.

OLD CATHEDRAL MUSEUM, 209 Walnut, Saint Louis, MO 63102-2499. Tel.: 314-231-3250. Fax: 314-231-4280.
E-mail: oldcathedral@att.net
Web Site: www.catholic-forum.com/stlouisking
Founded: 1970.
Key Personnel: C.E.O., Rev. Msgr. Jerome Billing; Museum Shop, Mrs. Mary Dieterman.
Personnel Profile: Full-Time Paid 3; Part-Time Volunteers 6.
Governing Authority: church. Parent Institution: The Archdiocese of St. Louis. Subsidiary Institution: The Basilica of St. Louis, The King. Tax-exempt: 501(c)(3).
Institution Type/Description: Religious Museum: housed in 1834 Cathedral.
Collections: religious history items.
Facilities: Religious articles & other museum-related items for sale.
Activities: permanent exhibitions.
Publications: History of the Old Cathedral.
Hours & Admission Prices: Mon.-Fri. 9:30-2:30. No charge; donations accepted. Closed New Year's Day; Easter; Independence Day; Christmas. &
Attendance: 22,000 (estimated)

PHILIP SLEIN GALLERY, 4735 McPherson Ave., Saint Louis, MO 63108. Tel.: 314-621-4634.
Key Personnel: Owner, Philip Slein; Owner, Tom Bussmann; Dir., Bridget Melloy
Institution Type/Description: Art Gallery.
Collections: works by regional & national artists; paintings; drawings; printmaking; photographs; sculpture.
Hours & Admission Prices: Tues.-Sat. 10-5; other times by appointment.

THE PRINCIPIA SCHOOL OF NATIONS MUSEUM, 13201 Clayton Rd., Saint Louis, MO 63131-1099. Tel.: 314-514-3073.
Founded: 1930.
Key Personnel: Dir., Nancy Boyer-Rechlin.
Personnel Profile: Full-Time Paid 1; Part-Time Paid 1; Part-Time Volunteers 2.
Governing Authority: college. Parent Institution: The Principia Corp. Affiliated with Principia College, Elsan, IL and The Principia School, St. Louis, MO. Tax-exempt: 501(c)(3).
Institution Type/Description: General Museum & Arts and Crafts Museum.
Collections: American Indian & Pre-Columbian artifacts; objects of Asian art; decorative arts; textiles; costumes; dolls; pottery.
Activities: guided tours; formally organized education programs for undergraduate college students affiliated with Principia College & pre-K through high school students affiliated with the Principia School; permanent & temporary exhibitions; children's hands-on exhibits; visual aids program.
Publications: e-newsletter, The Principia School of Nations News.
Hours & Admission Prices: Sept.-June by appointment only. No charge. Closed school vacations.

✳   **SAINT LOUIS ART MUSEUM, (M),** One Fine Arts Dr., Forest Park, Saint Louis, MO 63110-1380. Tel.: 314-721-0072. Fax: 314-721-6172. TDD: 314-721-4807.
E-mail: publicrelations@slam.org
Web Site: www.slam.org
Founded: 1879.
Congressional District: 1
Key Personnel: Dir., Brent Benjamin; Pres. Bd. Commissioners, Barbara B. Taylor; Chm. Bd. Trustees, Jeffrey T. Fort; Assoc. Cur. Decorative Arts & Design, David Conradsen; Cur. Prints, Drawings, & Photographs, Elizabeth Wyckoff; Cur. Modern & Contemporary Art, Simon Kelly; Asst. Cur. Modern & Contemporary Art, Tricia Paik; Assoc. Cur. Asian Art, Philip Hu; Cur. European Art to 1800, Judith Mann; Controller & Asst. Dir. Finance, Carolyn Schmidt; Asst. Dir. Exhibitions & Collections, Linda Thomas.
Personnel Profile: Full-Time Paid 184; Part-Time Paid 67; Part-Time Volunteers 169.
Governing Authority: municipal. Special Zoo Museum District. Managed by Museum Commission. Tax-exempt.

Institution Type/Description: Art Museum: housed in both a 1904 World's Fair building & a recently opened building.
Collections: more than 30,000 works of art; Oceanic art; pre-Columbian art; ancient Chinese bronzes; European & American art of the late 19th & 20th centuries; 20th-century German painting; Max Beckman prints; Henri Matisse's "Bathers with a Turtle"; George Caleb Bingham's "Election Series"; Hans Holbein the Younger's "Mary Lady Guilford"; Vincent van Gogh's "Stairway at Auvers"; Bartolomeo Manfredi's "Apollo and Maryas".
Major Exhibits: Impressionist France: Visions of Nation from LeGray to Monet (T), 3/16/14-7/6/14; Atna: Sacred Gods of Polynesia, 10/12/14-1/4/15.
Research Fields: pertaining to the collections; featured exhibitions material; archaeology; textiles.
Facilities: 110,000-vol. library on art history, decorative arts, auction & museum catalogs, available for limited inter-library loan & on premises; teachers' resource center; reading room; study room for prints, drawings & photographs; 480-seat auditorium; classrooms; restaurant; conservation center. Art books, prints & other museum-related items for sale in Museum Shop.
Activities: guided tours; lectures; films; gallery talks; performances; formally organized education programs for children & adults; docent program; inter-museum loan. Museum Sponsors: featured exhibitions.
Publications: biennial report; quarterly magazine; quarterly programs & events guide; exhibition catalogues; collection/exhibition brochures; visitor guides.
Hours & Admission Prices: Tues.-Thurs. & Sat.-Sun. 10-5, Fri. 10-9. No charge; featured exhibitions no charge on Fri. Closed Thanksgiving; Christmas. &
Attendance: 346,457 (accurate)
Membership: Regular $65; Family $85; Family & Friends $150; Supporting $250; Associate $500; Sponsor $1,000; Beaux Arts Council $1,500 & up.

ST. LOUIS CARDINALS HALL OF FAME MUSEUM, Saint Louis, MO 63102-1727. Mailing Address: 700 Clark St., Saint Louis, MO 63102-1727. Tel.: 314-345-9600.
E-mail: halloffame@cardinals.com
Web Site: stlouis.cardinals.mlb.com/stl/hof/about_us.jsp
Founded: 1968.
Personnel Profile: Full-Time Paid 3; Part-Time Paid 18.
Governing Authority: Parent Institution: St. Louis Cardinals, LLC.
Institution Type/Description: Sports Museum.
Collections: over 100 years of baseball history; photographs; memorabilia; uniforms; Hall of Fame inductees; sports equipment; trophies; awards.
Hours & Admission Prices: Online museum. &
Attendance: 30,000 (estimated)

ST. LOUIS FIRE DEPARTMENT MUSEUM, 1421 N. Jefferson Ave., Saint Louis, MO 63106-2136. Tel.: 314-289-1933. Fax: 314-533-1681.
Web Site: www.stlouis-mo.gov/government/departments/publicsafety/fire
Congressional District: 1
Key Personnel: Museum Shop Mgr., Bob Pauly
Institution Type/Description: Fire Fighting Museum.
Collections: firefighting history, equipment & memorabilia; photographs; toy fire trucks.
Hours & Admission Prices: Mon.-Fri. 9-2. No charge. &

✳   **SAINT LOUIS SCIENCE CENTER, (M),** 5050 Oakland Ave., Saint Louis, MO 63110-1460. Tel.: 314-289-4400. Fax: 314-289-4420.
Web Site: www.slsc.org
Founded: 1985.
Congressional District: 1
Key Personnel: C.E.O. & Pres., Bert Vescolani; Chm. Bd. of Commissioners, Douglas Yaeger; Chm. Bd. of Trustees, Donna Wilkinson; Chief Institutional Advancement Officer, Patrick Williams; C.O.O. & C.F.O., Barbara Boyle; Mng. Dir. Mktg. & Communications, Cynthia Skaggs; Mng. Dir. Facilities, John Wharton; Mng. Dir. Staff Relations, Deborah Washington; Sr. Dir. Design & Galleries, Joe Seidler; Sr. Dir. Theater, Retail & Exhibitions, Jackie Mollet; Dir. Membership, Russ Hitzemann.
Personnel Profile: Full-Time Paid 130; Part-Time Paid 46; Part-Time Volunteers 493; Interns 106.
Governing Authority: metropolitan district. Parent Institution: Saint Louis Science Center subdistrict of the Metropolitan Zoological Park and Museum District; Saint Louis Science Center Foundation. Tax-exempt: 501(c)(3).
Institution Type/Description: Science/Technology Museum.

Collections: pre-Columbian North American Indian artifacts emphasizing Missouri; technology including aeronautics communications, lighting equipment, period radios, cameras, medical equipment; space artifacts; ethnology; geology; minerals; natural history including mammalogy, ornithology, entomology, herpetology & malacology.

Facilities: 56,903 sq. ft. exhibit space; resource center; 315-seat OMNIMAX(R) theater; planetarium & space station; 8 meeting rooms; 2 public meeting rooms. Museum-related items for sale.

Activities: boy & girl scout badge programs; formally organized education programs for children & adults; permanent, temporary & traveling exhibitions; gallery demonstrations; lecture programs; field trips; speakers bureau; portable planetarium; astronomy & outreach programs.

Publications: New Science; Opening Minds to Science; annual report.

Hours & Admission Prices: Memorial Day-Labor Day daily 9:30-5:30; Sept.-May daily 9:30-4:30. Museum: no charge. OMNIMAX(R) Theater: adults $9, children $8; discounts to AAM & ICOM members; members no charge. Closed Thanksgiving; Christmas. &

Attendance: 841,868 (accurate)

Membership: Science Supporter $50; Individual & Guest $60; Family & Friends $85; Family & Friends Plus $120; Family & Friends Max $175; Galileo Society $250; Newton Society $500; Albert Einstein Society $1,000; Albert Einstein Society Patron $2,500; Albert Einstein Society Fellow $5,000; Albert Einstein Society Presidents Council $10,000. Corporate Partners Associates $1,000; Corporate Partners Patron $2,500; Corporate Partners Fellow $5,000; Corporate Partners Presidents Council $10,000; Corporate Partners Leadership Council $25,000. Senior (60 and over) $5 discount on all member levels.

SAINT LOUIS UNIVERSITY MUSEUM OF ART, 3663 Lindell Blvd., Saint Louis, MO 63108. Mailing Address: 3545 Lindell Blvd., Saint Louis, MO 63103. Tel.: 314-977-2666. Fax: 314-977-3581. Facebook: Saint Louis University Museum of Art.

E-mail: sluma@slu.edu
Web Site: www.slu.edu/sluma.xml
Founded: 2002.
Congressional District: 1

Key Personnel: Dir., Petruta Lipan; Cur., Kathleen Wang; Public Rels. Affairs, Mary Marshall; Business Mgr., Benjamin Faser; Dir. Facilities, Barth Breneman; Security, Dennis Thompson.

Personnel Profile: Full-Time Paid 6; Part-Time Paid 12; Part-Time Volunteers 2; Interns 6.

Governing Authority: private university. Tax-exempt: 501(c)(3).

Institution Type/Description: Art Museum.

Collections: 16th-21st century European & American art; Japanese art; midwest Jesuit Mission; southeast Asian art.

Research Fields: collections; special exhibitions material.

Facilities: 7,500-vol. library; 20,000 sq. ft. exhibit space; educational facilities.

Activities: formal education programs; guided tours; lectures; loan, traveling & temporary exhibitions; internships.

Publications: exhibition catalogs & brochures.

Hours & Admission Prices: Wed.-Sun. 11-4. No charge; donations accepted. Closed New Year's Day; Easter; Thanksgiving; Christmas. &

Attendance: 8,100 (accurate)

Membership: Regular $50; Family $75; Contributing $100; Sustaining $150; Supporting $250; Associate $500; Sponsor $1,000; Patron $1,500; Fellow $2,500; Benefactor $5,000; Director's Circle $10,000; President's Circle $25,000.

SAINT LOUIS ZOO, (M), Forest Park, Saint Louis, MO 63110-1395. Mailing Address: 1 Government Dr., St. Louis, MO 63110-1395. Tel.: 314-781-0900. Fax: 314-647-7969.

E-mail: bonner@stlzoo.org
Web Site: www.stlzoo.org
Founded: 1910.
Congressional District: 1

Key Personnel: Chm. Zoological Park Commission, James F. Conway; Pres. & C.E.O., Jeffrey P. Bonner, Ph.D.; Exec. Asst., Kelly Fesler; Sr. Vice Pres. & Dir. Zoological Operations and the WildCare Institute, R. Eric Miller, D.V.M.; Dir. Research, Cheryl Asa, Ph.D.; Dir. Education, Louise Bradshaw; Vice Pres. Animal Collection, Jack Grisham; Asst. Gen. Cur., Bill Houston; Cur. Aquatics & Herps, Jeff Ettling; Cur. Mammals & Carnivores, Steve Bircher; Cur. Mammals & Primates, Ingrid Porton; Cur. Birds, Michael Macek; Cur. Mammals & Ungulates, Martha Fischer; Cur. Invertebrates, Ed Spevak; Cur. Children's Zoo, Alice Seyfried; Supvr. Life Support Systems, David Jarvis; Animal Registrar, Rae Lynn Haliday; Vice Pres. Business Operations & C.F.O., Steve Barth; Dir. Finance, Cassandra Ray; Dir. Facilities Management, Dan Snyder; Dir. Food Svc., Ken Stover; Dir. Housekeeping & Grounds, Ernest Frost; Dir. Purchasing & Distribution, Patrick Williamson; Dir. Guest Svcs., Joan Sisco; Dir. Business

Operations, Jim Madison; Vice Pres. External Rels., Cynthia S. Holter, CFRE; Dir. Devel., Jeffrey C. Huntington, CFRE; Dir. Planned Gifts, Lori Sullivan; Dir. Mktg., Ginnie Westmoreland; Dir. Public Rels., Susan Gallagher; Dir. Sales, Jennifer Poindexter; Dir. Corp. & Foundation Gifts, Diane Bauhof; Vice Pres. Internal Rels., Wyndel Hill; Dir. Safety & Risk Management, Philip Sonderman; Dir. Comp & Benefits, Tim Rakers; Dir. Security, Michael Siemers; Dir. Volunteer Svcs., Elaine Gill; Mgr. Internal Communications, Debbie Lammering; Payroll Specialist, Martha Michael; Dir. Human Resources, Dustin Deschamp; Vice Pres. Architecture & Planning, David McGuire.

Personnel Profile: Full-Time Paid 325; Part-Time Paid 170; Part-Time Volunteers 1,451; Interns 96.

Operating Expenses: 56,877,062

Operating Income: 57,760,789

Governing Authority: municipal by special Zoo Museum District. Managed by Zoological Commission. Tax-exempt.

Institution Type/Description: Zoo.

Collections: 129,176 specimens, 606 species of mammals, birds, reptiles, amphibians, fish & invertebrates.

Research Fields: zoo animal reproduction & conservation.

Facilities: 5,000-vol. library on zoology, conservation & related topics available for use by professional staff, area teachers & others by appointment; classrooms; auditorium; Endangered Species Research Center & Veterinary Hospital. Zoo-related items for sale.

Activities: guided tours; lectures; films; formally organized education programs for children, adults & undergraduate college students; training programs; permanent & temporary exhibitions.

Publications: bimonthly magazine, stlzoo.

Hours & Admission Prices: Summer: Mon.-Thurs. 8-5, Fri.-Sun. & holidays 8-7; Sept.-May 9-5. No charge; donations accepted. Closed New Year's Day; Christmas. &

Attendance: 3,019,597 (accurate)

Membership: Conservation Advocate, Student & Senior $50; Young Zoo Friends & Zoo-Goer $65; Family, Grandparents $85; Naturalist $120; Keeper $175; Zoologist $250; Curator $500; Director $750; Marlin Perkins Society, Animal Kingdom Club $1,000; Safari Circle $1,500; Wildlife Circle $2,500; Steward's Circle $5,000; Conservationist Circle $10,000; President's Circle $25,000.

SCOTT JOPLIN HOUSE - STATE HISTORIC SITE, 2658 Delmar Blvd., Saint Louis, MO 63103-1404. Tel.: 314-340-5790. Fax: 314-340-5793.

Web Site: www.mostateparks§scottjoplin.html
Founded: 1982.
Congressional District: 1

Personnel Profile: Full-Time Paid 3; Part-Time Paid 3; Part-Time Volunteers 9; Interns 1.

Governing Authority: state. Tax-exempt.

Institution Type/Description: Historic House Museum: c.1860. Listed on the National Register of Historic Places.

Collections: Joplin family history; African American cultural history; Ragtime history.

Research Fields: ragtime.

Facilities: rental facilities.

Activities: rental facilities; tours; educational & private events. Museum Sponsors: Live Ragtime Events.

Hours & Admission Prices: March-Oct. Mon.-Sat. 10-4, Sun. 12-4; Nov.-Feb. Tues.-Sat. 10-4. Adults $2.50, children 6-12 $1.50; discounts to groups; children 5 & under no charge. Closed New Year's Day; Easter; Thanksgiving; Christmas. &

Attendance: 12,500

SHELDON ART GALLERIES, 3648 Washington Blvd., Saint Louis, MO 63108-3610. Tel.: 314-533-9900. Fax: 314-533-2958.

E-mail: olg@thesheldon.org
Web Site: www.thesheldon.org/galleries.asp
Founded: 1998.

Key Personnel: Dir. & Cur., Olivia Lahs-Gonzales; Museum Shop Mgr., Rebecca Gunter.

Personnel Profile: Full-Time Paid 2; Interns 1.

Governing Authority: Parent Institution: The Sheldon Arts Foundation. Subsidiary Institution: The Sheldon Concert Hall. Tax-exempt.

Institution Type/Description: Art Gallery.

Collections: works by local, national & international artists; paintings; photographs; sculpture; children's art; drawings; video art.

Facilities: 7,000 sq. ft. exhibit space.

Activities: art exhibits; lectures; children's art workshops; temporary exhibitions.

Publications: Larry Fink: Attraction and Desire - 50 Years in Photography,

2011; Ralson Crawford and Jazz, 2011; Josephine Baker: Image and Icon, 2006; City of Gabriels: The History of Jazz in St. Louis, 2006; John Gossage: Secrets of Real Estate, 2008; The Sheldon: 100 Moments A Celebration of the 100th Anniversary (2012).
Hours & Admission Prices: Tues. 12-8, Wed.-Fri. 12-5, Sat. 10-2. No charge; donations accepted. Closed New Year's Eve & Day; Independence Day; Thanksgiving; Christmas Eve & Day. &
Attendance: 70,000 (estimated)
Membership: Member $125 & up; Patron $500 & up; Benefactor $1,000 & up; Louis Spiering Society $2,500 & up.

SOLDIERS MEMORIAL MILITARY MUSEUM, 1315 Chestnut St., Saint Louis, MO 63103-2317. Tel.: 314-622-4550. Fax: 314-622-4237.
Founded: 1938.
Congressional District: 1
Key Personnel: Museum & Collections Exec., Lynnea Magnuson, Ph.D.
Governing Authority: municipal. Tax-exempt.
Institution Type/Description: Military Museum.
Collections: uniforms; weapons; war mementos; photographs of St. Louis soldiers; local military history.
Facilities: auditorium; rental rooms.
Activities: programs; special memorial services on national holidays; veterans programs & events.
Hours & Admission Prices: Mon.-Fri. 9-4:30, Sat.-Sun. 10-3. No charge; donations accepted. Closed New Year's Day; Thanksgiving; Christmas. &

WHITE FLAG PROJECTS, 4568 Manchester Ave., Saint Louis, MO 63110. Tel.: 314-531-3442. Fax: 341-531-3474.
E-mail: info@whiteflagprojects.org
Web Site: www.whiteflagprojects.org
Governing Authority: nonprofit organization.
Institution Type/Description: Art Gallery.
Collections: works by national & international artists.
Hours & Admission Prices: Thurs.-Sat. 12-5.

WORLD CHESS HALL OF FAME, (M), 4652 Maryland Ave., Saint Louis, MO 63108. Tel.: 314-367-9243. Fax: 314-367-7501.
E-mail: info@worldchessof.org
Web Site: www.worldchesshof.org
Key Personnel: Dir., Susan Barrett
Institution Type/Description: Art Gallery.
Collections: chess-themed art by modern & contemporary artists.
Hours & Admission Prices: Tues. 10-8, Wed.-Sat. 10-5, Sun. 12-5. No charge; donations accepted. &
Membership: Pawn $40; Knight $75; Bishop $150; Rook $150; Queen $500; King $1,000.

Sainte Genevieve

BOLDUC HOUSE MUSEUM, 125 S. Main St., Sainte Genevieve, MO 63670. Tel.: 573-883-3105. Fax: 573-883-3415.
E-mail: bolduchouse@sbcglobal.net
Web Site: www.bolduchouse.org
Founded: 1958.
Congressional District: 3
Key Personnel: Dir., Lesley Barker; Museum Shop Mgr., Roscella Naeger.
Personnel Profile: Full-Time Paid 1; Part-Time Paid 11; Part-Time Volunteers 5; Interns 1.
Governing Authority: society; nonprofit. Parent Institution: The National Society of the Colonial Dames of America in the State of Missouri. Tax-exempt: 501(c)(3).
Institution Type/Description: Historic Houses Museum.
Collections: Historic Buildings: Louis Bolduc House, c.1792: period French furnishings. Bolduc LeMeilleur House: French & American architecture & construction styles; Federal style furnishings. Beaurais - Linden House: 19th-century furnishings; society headquarters, offices, library & archives. Stone Cottage: Shawnee & Delaware Indian artifacts.
Research Fields: French Colonial history, period furnishings & gardens.
Facilities: library; archives; French colonial garden; rental facilities. Gifts, books, plants & seeds from gardens for sale.
Activities: guided tours; permanent exhibitions; rental facilities; child-sized eastern woodland Indian village; ceramic classes; homeschool classes & events; living history events. Gardens: special events; tours; educational events; book signings. Annual Events: Stone Cottage - Shawnee & Delaware Indian Experience in eastern Missouri.

Hours & Admission Prices: Mon.-Thurs. 10-4, Fri. 10-7, Sat. 10-5, Sun. 12-4. Adults $8, seniors $6, students $2. Closed New Year's Eve & Day; Easter; Thanksgiving; Christmas Eve & Day.
Attendance: 6,609
Membership: Student $10; Ste. Genevieve County Resident $25; Individual $35; Ste. Genevieve County Family $45; Family $65.

FELIX VALLE STATE HISTORIC SITE, 198 Merchant St., Sainte Genevieve, MO 63670-1682. Mailing Address: P.O. Box 89, Ste. Genevieve, MO 63670-0089. Tel.: 573-883-7102. Fax: 573-883-9630. TDD: 800-379-2419.
E-mail: felix.valle.state.historic.site@dnr.mo.gov
Web Site: www.mostateparks.com
Founded: 1970.
Congressional District: 10
Key Personnel: Site Admin., Donna J. Rausch.
Personnel Profile: Full-Time Paid 3; Part-Time Paid 3; Interns 1.
Governing Authority: state. Parent Institution: Div. of State Parks, Missouri Dept. of Natural Resources, P.O. Box 176, Jefferson City, MO 65102. Tax-exempt.
Institution Type/Description: History Museum.
Collections: American Empire furniture; French colonial vertical-log buildings. Amoureux House: 9' x 11' diorama depicting village of Ste. Genevieve in 1832. Historic House: 1792 St. Gemme/Amoureux House; 1793 Delassus-Kern structure; 1818 Felix Valle House; 1819 Dr. Benjamin Shaw house.
Research Fields: early Ste. Genevieve & French-American period in Mississippi River Valley.
Facilities: restored & furnished home.
Activities: special events during summer. Museum Sponsors: French Christmas Open House.
Publications: brochure.
Hours & Admission Prices: April-Oct. Mon.-Sat. 10-4, Sun. 12-5; Nov.-March Thurs.-Sat. 10-4, Sun. 12-5. Adults $4, students $2.50; discount to groups; children under 6 no charge. Closed New Year's Day; Easter; Thanksgiving; Christmas.
Attendance: 13,000 (estimated)

STE. GENEVIEVE MUSEUM, Merchant & DuBourg St., Sainte Genevieve, MO 63670. Tel.: 573-883-3461.
Founded: 1935.
Congressional District: 8
Key Personnel: Dir., Dr. Deborah Woelich; Pres., James Baker; Treas., Dolores Koetting.
Personnel Profile: Part-Time Paid 2.
Governing Authority: Parent Institution: Museum Bd. Tax-exempt.
Institution Type/Description: History Museum.
Collections: folk art; Indian artifacts; historical relics from Ste. Genevieve; historical books; sketches; weapons; documents.
Facilities: Museum-related items for sale.
Hours & Admission Prices: April-Oct. Mon.-Sat. 10-4, Sun. 12-4; Nov.-March daily 12-4. Adults $2, students K-12 $.50; members no charge. Closed New Year's Eve & Day; Easter; Thanksgiving; Christmas Eve & Day. &
Attendance: 5,750 (estimated)
Membership: Single $10; Family $20; Business $50.

Salem

OZARK HERITAGE MUSEUM, 701 S. Main St., Salem, MO 65560. Tel.: 573-729-5707.
E-mail: kfiebelman@embarqmail.com
Institution Type/Description: History Museum.
Collections: local history & culture; period furnishings; personal artifacts; photographs; paintings; regional genealogy.
Hours & Admission Prices: By appointment. No charge.

OZARK NATURAL & CULTURAL RESOURCE CENTER, 202 S. Main, Salem, MO 65560. Mailing Address: P.O. Box 732, Salem, MO 65560. Tel.: 573-729-0029.
E-mail: oncrc@salemmo.com
Web Site: www.oncrc.org
Key Personnel: Dir., Jerry Craig
Institution Type/Description: Natural History Center.
Collections: natural & cultural history; photographs; period furnishings; personal artifacts.
Hours & Admission Prices: Call for hours.

Salisbury

CHARITON COUNTY HISTORICAL SOCIETY & MUSEUM, 115 E. 2nd St., Salisbury, MO 65281. Mailing Address: P.O. Box 114, Salisbury, MO 65281-0114. Tel.: 660-388-5941.
E-mail: museum@cvalley.net
Institution Type/Description: Historical Society Museum.
Collections: local history & culture; photographs; personal artifacts; genealogy; military artifacts; tools.
Activities: research.
Hours & Admission Prices: mid-April to mid-Oct. Tues. & Thurs.-Sat. 1-4, Wed. 1-6; other times by appointment.

Savannah

ANDREW COUNTY MUSEUM & HISTORICAL SOCIETY, 202 E. Duncan Dr., Savannah, MO 64485-1264. Mailing Address: P.O. Box 12, Savannah, MO 64485-0012. Tel.: 816-324-4720. Fax: 816-324-5271.
E-mail: acmuseum@stjoelive.com
Web Site: www.andrewcountymuseum.org
Founded: 1972.
Congressional District: 5
Key Personnel: Pres. (V), Bob Niebuhr; Dir., Glenn Uminowicz; Cur., Melissa Swindell; Museum Admin., Shirley Brown; Office Asst., Christy Sipes; Chm. (V), Bob Niebuhr.
Personnel Profile: Full-Time Paid 3; Part-Time Paid 2; Part-Time Volunteers 112; Interns 4.
Volunteer Hours: 2,512
Operating Expenses: 220,575
Operating Income: 236,039
Governing Authority: bd. of dirs. Parent Institution: Andrew County Historical Society. Tax-exempt.
Institution Type/Description: History Museum.
Collections: local history & culture from 1841 to present; farming; agriculture; business & technology; small town history.
Major Exhibits: Northwest Missouri Collects, 5/14-9/14.
Research Fields: local & regional history; genealogy; rural history.
Facilities: 14,000 sq. ft. museum bldg.
Activities: guided tours; docent program; loan & temporary exhibitions; educational programs; public programs; holiday event.
Publications: quarterly newsletter, Diggin History; seven cemetery inscription volumes; census information; marriage records; birth & death records; Andrew County: A Rural Way of Life (2012).
Hours & Admission Prices: Tues.-Sat. 10-4. Adults $3, students & seniors $2; Blue Star Museum members & children under 13 no charge. &
Attendance: 5,012 (accurate)
Membership: Student & Senior Citizen $8; Individual $12; Joint $20; Family $25; Patron $50; Benefactor $100.

Sedalia

DAUM MUSEUM OF CONTEMPORARY ART, State Fair Community College, 3201 W. 16th St., Sedalia, MO 65301-2188. Tel.: 660-530-5888. Fax: 660-530-5890. Facebook: Daum Museum.
E-mail: info@daummuseum.org
Web Site: www.daummuseum.org
Formerly: Goddard Gallery
Founded: 2001.
Congressional District: 4
Key Personnel: Dir., Thomas Piche, Jr.; Pres., Dr. Joanna Anderson; Cur. Education, Victoria Weaver; Registrar, Matthew Clouse; Office Mgr., Marcie Teter.
Personnel Profile: Full-Time Paid 4; Part-Time Paid 6; Part-Time Volunteers 40.
Governing Authority: Parent Institution: State Fair Community College.
Institution Type/Description: Contemporary Art Museum.
Collections: works from the last third of the 20th century to present.
Major Exhibits: Chris Gustin: Masterworks in Clay, 2/14-5/14.
Research Fields: contemporary art.
Activities: art & lecture series; workshops; symposium; children's busing program.
Publications: catalogs, Betty Woodman; Sculptural Clay Invitational; Jun Kaneko; Bay Area Ceramic Sculptures: Second Generation; Michiko Itatani; Vera Klement; Judy Onofrio; Jim Shrosbree; Steven Montgomery; Ruth Borgenight; Jeffrey Mongrain; Peter Callas; Ron Ehrlich; Albert Pfarr; Richard Deon; Elana Herzog; Marc Leuthold; Deanna Dikeman.
Hours & Admission Prices: Tues.-Fri. 11-5, Sat.-Sun. 1-5. No charge; donations accepted. &
Attendance: 26,000 (accurate)

Membership: Student $35; Individual $50; Family $60; Associate $250; Benefactor $500; Sponsor $1,000; Underwriter $2,500; Season $5,000.

PETTIS COUNTY HISTORICAL SOCIETY, 228 Dundee, Sedalia, MO 65301-2339.
Founded: 1943.
Congressional District: 4
Key Personnel: Pres., Dr. Rhonda Chalfant; Vice Pres., Ken Bird; Sec. & Treas., Clell Furnell.
Personnel Profile: Part-Time Volunteers 20.
Governing Authority: nonprofit organization. Tax-exempt.
Institution Type/Description: Historical Society Museum.
Collections: military; war; agriculture; farm life; costumes; science nature; Indian artifacts; history. Historic House: 1883 McVey School #3, known as the Little Red Schoolhouse, State Hwy. 50 East, Sedalia.
Activities: guided tours & lectures at McVey School; by appointment spring & summer.
Publications: bimonthly newsletter.
Hours & Admission Prices: March-Oct. Wed.-Sun. 1-4; other times by appointment. No charge; donations accepted. &
Attendance: 100 (estimated)
Membership: Individual $8; Family $12; Business $15.

Sibley

FORT OSAGE NATIONAL HISTORIC LANDMARK, 107 Osage St., Sibley, MO 64088. Mailing Address: 22807 Woods Chapel Rd., Blue Springs, MO 64015-9799. Tel.: 816-503-4860. Fax: 816-795-7938. TDD: 800-735-2966.
E-mail: gjulich@jacksongov.org
Web Site: www.jacksongov.org
Founded: 1948.
Congressional District: 6
Key Personnel: Dir. Parks & Recreation, Michele Newman; Interpretive Specialist, Kate Warfield; Supt. Heritage Museums & Programs, Gordon Julich; Museum Shop Mgr., Laura King.
Personnel Profile: Full-Time Paid 5; Part-Time Paid 1; Part-Time Volunteers 30.
Governing Authority: county. Parent Institution: Jackson County Parks & Recreation, 22807 Woods Chapel Rd., Blue Springs, MO 64015. Tax-exempt: 170(c)(3).
Institution Type/Description: Historic Site: 1808-1827 Fort Osage.
Collections: history; military; Indians; archaeology; westward expansion fur trade; Osage Indians; fur trade. Historic Buildings: restored trade house; military compound including soldiers quarters; private civilian quarters; trade room; blacksmith's shop.
Research Fields: history of Fort Osage; U.S. factory system; early fur trade; Osage Indian life; 1812 military.
Facilities: education center; park complex featuring picnic grounds and boat ramp.
Activities: tours; permanent & temporary exhibitions; living history interpretation; youth program; craft and skills demonstrations; volunteer program; other special events.
Publications: Fort Osage, Mother of the West.
Hours & Admission Prices: Tues.-Sun. 9-4:30. Adults $7, youth $4, senior citizens $3. Closed Martin Luther King Jr. Day; Veterans Day; Truman's Birthday; Thanksgiving; Christmas. &
Attendance: 11,500 (accurate)
Membership: Individual $35; Family $45.

Sikeston

THE SIKESTON DEPOT MUSEUM, 116 W. Malone St., Sikeston, MO 63801. Mailing Address: P.O. Box 182, Sikeston, MO 63801. Tel.: 573-481-9967.
E-mail: depotmuseum@sbcglobal.net
Institution Type/Description: History Museum: housed in a former railroad depot; built in 1916. Listed on the National Register of Historic Places.
Collections: local history & culture; period furnishings; photographs; personal artifacts; Native American artifacts; local & regional artists.
Hours & Admission Prices: Tues.-Sat. 10-4, Sun. 1-4. No charge; donations accepted.

Springfield

AIR AND MILITARY MUSEUM OF THE OZARKS, (M), 2305 E. Kearney, Springfield, MO 65803-4970. Tel.: 417-864-7997. Fax: 417-866-2448. Facebook: Ammo Museum.
Web Site: ammomuseum.com

Founded: 1992.
Congressional District: 7
Key Personnel: Chmn. & Pres., Raymond Hopper; Museum Dir., Ron Cutter.
Personnel Profile: Full-Time Volunteers 14; Part-Time Volunteers 20; Interns 2.
Volunteer Hours: 4,500
Operating Expenses: 25,000
Operating Income: 25,000
Institution Type/Description: Military Museum.
Collections: military history & memorabilia; military equipment; photographs; military library; artifacts.
Activities: historical/educational displays & programs; historical parades.
Publications: monthly newsletter.
Hours & Admission Prices: Tues.-Sat. 12-4; other times by appointment. Adults $5, children 6-12 $3; discounts to AAM members & military veterans; children 5 & under no charge. &

Attendance: 6,000 (accurate)
Membership: Single $25; Couple $36.

DICKERSON PARK ZOO, 1401 W. Norton Rd., Springfield, MO 65803-1023. Mailing Address: 3043 N. Fort Ave., Springfield, MO 65803-1079. Tel.: 417-833-1570. Fax: 417-833-4459.
E-mail: info@dickersonparkzoo.org
Web Site: www.dickersonparkzoo.org
Founded: 1923.
Congressional District: 7
Key Personnel: Dir., Michael Crocker; Pres. (V), B.J. Joplin; Zoological & Gen. Cur., John Collette; Museum Shop Mgr., Joni Baurichter.
Personnel Profile: Full-Time Paid 36; Part-Time Paid 24; Part-Time Volunteers 115; Interns 20.
Governing Authority: municipal. Branch of Springfield/Greene County Park Bd., 1923 N. Weller, Springfield, MO 65803. Tel.: 417-864-1049. Tax-exempt: 501(c)(3).
Institution Type/Description: Zoological Park.
Collections: gen. zoological collection of indigenous & exotic species.
Research Fields: elephant breeding biology & mgmt.
Facilities: 600-vol. library of biological & zoological text and reference material, available for use on certified research projects or school papers; reading room; classrooms.
Activities: lectures; films; concerts; broadcast programs; formally organized education programs; docent program or council; training programs for professional museum workers; special events.
Publications: bimonthly magazine, Wildtimes.
Hours & Admission Prices: April-Sept. daily 9-5; Oct.-March daily 10-4. Adults $8, children 3-12 $5; discounts to groups of 15 or more paying individuals & members of reciprocal zoos, AZA; members & children under 3 no charge. Closed New Year's Day; Thanksgiving; Christmas. &
Attendance: 218,949 (accurate)
Membership: Family, Plus One & Grandparents $65; Wildlife Club $75; Safari Club $100; Keepers Club $250; Curator's Club $500; Director's Club $1,000.

DISCOVERY CENTER OF SPRINGFIELD, 438 E. St. Louis St., Springfield, MO 65806-2312. Tel.: 417-862-9910, ext. 0 or ext. 700. Fax: 417-862-6898.
E-mail: efox@discoverycenter.org
Web Site: www.discoverycenter.org/
Founded: 1991.
Key Personnel: C.E.O., Emily Fox; Bd. Pres. (V), Allen Kunkel.
Personnel Profile: Full-Time Paid 14; Part-Time Paid 3; Part-Time Volunteers 50; Interns 5.
Governing Authority: nonprofit organization. Tax-exempt: 501(c)(3).
Institution Type/Description: Science Center.
Collections: hands-on exhibits.
Activities: lectures; arts festivals; organized education programs; participatory & temporary exhibits. Annual Events: Festival of Trees; Moonlight Bike Ride.
Hours & Admission Prices: Tues.-Thurs. 9-5, Fri.-Sat. 9-6, Sun. 1-5. Adults $12, seniors over 60 $10, children 3-15 $8; discounts to ASTC museum member; member no charge. &
Attendance: 100,000 (estimated)
Membership: Basic (1 or 2 people) $50; Family (up to 4) $75; Family plus (up to 6) $95; Corporate (50 single entry passes) $1,000.

DOLING MUSEUM, 301 E. Talmage, Springfield, MO 65803-7825.
Institution Type/Description: History Museum.
Collections: Doling Park history & memorabilia; amusement park & skating rink artifacts; photographs; hands-on exhibits.

Hours & Admission Prices: April-Oct. Tues. 2-6, Fri. 11-4, Sun. 1-5. No charge.

FLOWER PENTECOSTAL HERITAGE CENTER, 1445 N. Boonville Ave., Springfield, MO 65802. Tel.: 877-840-5200 (Toll Free).
E-mail: archives@ag.org
Web Site: www.ifphc.org
Formerly: Assemblies of God Archives
Founded: 1977.
Key Personnel: Dir., Darrin Rodgers.
Personnel Profile: Full-Time Paid 5; Part-Time Paid 3.
Governing Authority: Parent Institution: General Council of the Assemblies of God. Tax-exempt.
Institution Type/Description: Religious History Museum.
Collections: church history & artifacts; photographs; hands-on exhibits.
Publications: Assemblies of God Heritage.
Hours & Admission Prices: Mon.-Fri. 9-4:30. No charge. &
Attendance: 5,000 (estimated)

HISTORY MUSEUM ON THE SQUARE, 155 Park Central Sq., Springfield, MO 65806. Mailing Address: P.O. Box 2963, Springfield, MO 65801. Tel.: 417-831-1976. Fax: 888-965-9342.
E-mail: info@historymuseumonthesquare.org
Web Site: www.historymuseumonthesquare.org
Formerly: History Museum for Springfield-Greene County
Founded: 1975.
Congressional District: 7
Key Personnel: Exec. Dir., John E. Sellars; Cur., Joan Hampton-Porter.
Personnel Profile: Full-Time Paid 2; Part-Time Paid 1; Part-Time Volunteers 12; Interns 2.
Governing Authority: nonprofit. Tax-exempt: 501(c)(3).
Institution Type/Description: History Museum.
Collections: furnishings; tools & equipment; costumes; textiles; decorative arts; archives; photographs; dolls; toys; Civil War, World War I & World War II items; railroad & business collections; office equipment; household items.
Research Fields: local history; costumes; needlework; historic photographs.
Facilities: 6,075-vol. library pertaining to local history available to the public; 2,500 sq. ft. exhibit space; 40,000 item archival collection available to scholars. Historic reproduction gift items for sale.
Activities: guided tours; lectures; films; docent programs; loan, temporary & traveling exhibitions; organized education programs for undergraduate or graduate college students.
Publications: triannual newsletter, History Museum Bulletin; book, Crossroads at the Spring: A Pictorial History of Springfield, Missouri.
Hours & Admission Prices: Museum: Tues.-Sat. 10:30-4:30. Office: Mon.-Fri. 9-5. Suggested Donation: adults $5, students $3; discounts to groups; members no charge. Closed major holidays. &
Attendance: 10,000 (estimated)
Membership: Pioneer $25; Archivist $50; Historian $75; Founder $100.

MISSOURI INSTITUTE OF NATURAL SCIENCE, 2327 W. Farm Rd. 190, Springfield, MO 65810. Tel.: 417-883-0594.
E-mail: info@monatsci.com
Web Site: www.monatsci.com
Formerly: Natural History Museum of the Ozarks
Founded: 2001.
Congressional District: 7
Key Personnel: Dir., Matt Forir; Museum Dir., Mary Rauzi.
Governing Authority: nonprofit organization.
Institution Type/Description: Science Museum.
Collections: local natural history; fossils & shells from around the world; rocks, minerals & crystals; corals; Ice-Age mammoth; dire wolf; short-faced bear fossils from Riverbluff Cave.
Research Fields: cave/karst formation; paleontology; geology.
Facilities: Museum-related items for sale.
Activities: workshops; educational programs; lectures; birthday parties; seminars; presentations; movies; special events; hands-on exhibits; school & youth group field trips.
Hours & Admission Prices: Mon.-Fri. 8:30-4:30; call for additional hours. No charge; donations accepted. &

MISSOURI SPORTS HALL OF FAME, 3861 E. Stan Musial Dr., Springfield, MO 65809. Tel.: 417-889-3100; 800-498-5678. Fax: 417-889-2761.
E-mail: kari@mosportshalloffame.com
Web Site: www.mosportshalloffame.com

Founded: 1994.
Key Personnel: C.E.O. & Exec. Dir./Pres., Jerald Andrews; Vice Pres. Operations, Marty Willadsen; Dir. Sponsorship Devel., Craig Winegar; Operations & Mktg., Kari Crawford; Accounting, Dale Witte; Special Events & Admin., Sharyn Wagoner
Institution Type/Description: Hall of Fame.
Collections: hands-on exhibitions; autographed memorabilia; photographs; personal artifacts; jerseys; helmets; off-road baja truck.
Activities: indoor batting cages; hands-on exhibits; basketball shootout; football quarterback challenge; indoor batting simulator; NASCAR simulator; coaches challenge; call a play booth.
Hours & Admission Prices: Mon.-Sat. 10-4, Sun. 12-4. Adults $5, senior citizens $4, children 6-15 $3; discounts to groups of 10 or more; children 5 & under no charge. ♿
Attendance: 100,000 (estimated)

OZARKS GENEALOGICAL SOCIETY LIBRARY, 534 W. Catalpa, Springfield, MO 65807-1404. Mailing Address: P.O. Box 3945, Springfield, MO 65808-3945.
Founded: 1967.
Governing Authority: Subsidiary Institution: Library Center, 4653 S. Campbell, Springfield, MO.
Institution Type/Description: Historical Society Museum.
Collections: local family history & culture; period artifacts; photographs.
Research Fields: local history & genealogy.
Publications: quarterly, Ozar'kin.
Hours & Admission Prices: mid-March to Oct. Tues. 6pm-8:30pm, Wed. 1-4, Sat. 10-4; Nov. to mid-March Wed. 1-4, Sat. 10-4. No charge. Closed holidays; holiday weekends. ♿
Membership: Individual $18; Family $25.

RAILROAD HISTORICAL MUSEUM, INC., Grant Beach Park, 1300 N. Grant St., Springfield, MO 65802. Mailing Address: 1651 S. Roanoke Ave, Springfield, MO 65807-2085. Tel.: 417-865-6829.
E-mail: rrhistoricalmuseum@zoomshare.com
Web Site: rrhistoricalmuseum.zoomshare.com
Formerly: Railroad Historical Museum
Founded: 1985.
Congressional District: 7
Personnel Profile: Part-Time Volunteers 21; Interns 4.
Volunteer Hours: 1,800
Operating Expenses: 6,524
Operating Income: 15,939
Governing Authority: nonprofit organization. Tax-exempt.
Institution Type/Description: Railroad History Museum.
Collections: railroad history & artifacts; St. Louis San Francisco Locomotive No. 4524; Burlington Baggage Car; Chicago and Northwestern Commuter Car; Burlington Northern Caboose; equipment; photographs; paintings.
Hours & Admission Prices: Sat. 2-4; groups by appointment. No charge; donations accepted.
Attendance: 4,252 (accurate)

THE SOFTBALL MUSEUM, 2141 E. Pythian St., Killian Sports Complex, Springfield, MO 65802. Tel.: 417-837-5817.
Key Personnel: Dir., Mark Nelson.
Personnel Profile: Full-Time Paid 2.
Governing Authority: Parent Institution: Springfield ASA. Subsidiary Institution: Springfield Greene County Park Board. Tax-exempt.
Institution Type/Description: Sports Museum.
Collections: state, regional & national tournaments from 1920s; uniforms; photographs; trophies; Springfield & Missouri ASA Hall of Fame.
Hours & Admission Prices: Mon.-Fri. 11-5. No charge; donations accepted.
Attendance: 500 (estimated)

SPRINGFIELD ART MUSEUM, (M), 1111 E. Brookside Dr., Springfield, MO 65807-1899. Tel.: 417-837-5700. Fax: 417-837-5704.
E-mail: ArtMuseum@ci.springfield.mo.us
Web Site: www.springfieldmo.gov/art
Founded: 1928.
Congressional District: 7
Key Personnel: Dir., Nick Nelson; Chm. (V), Ron Hawley; Cur. Art, Sarah Buhr; Museum Educator, Kate Baird; Exhibit Mgr., Cindy Quayle; Exec. Sec., Tyra Knox.
Personnel Profile: Full-Time Paid 13; Part-Time Paid 1; Part-Time Volunteers 40; Interns 2.
Governing Authority: municipal. Parent Institution: City of Springfield, MO. Tax-exempt: 501(c)(3).

Institution Type/Description: Art Museum.
Collections: 18th, 19th & 20th-century American and European painting, sculpture, drawing, photography, prints & decorative arts.
Major Exhibits: Hooves, Tails & Claws: The Viviparous Quadrupeds of North America, 2/14-6/22/14.
Research Fields: related to collections & special exhibitions.
Facilities: 6,600-vol. library on art history, architecture & contemporary art; auction & museum catalogues available for use on premises; reading room; 400-seat auditorium; classrooms.
Activities: guided tours; lectures; films; concerts; dance recitals; formally organized art school offering classes for children & adults; inter-museum loan, permanent, temporary & circulating exhibitions.
Publications: bimonthly bulletin; Studio School of Art bulletins; temporary exhibition catalogues.
Hours & Admission Prices: Tues.-Wed. & Fri.-Sat. 9-5, Thurs. 9-8, Sun. 1-5. No charge; donations accepted. Closed City of Springfield Day; national holidays. ♿
Attendance: 43,848 (accurate)
Membership: Student & Senior over 55 $15; Member $25; Contributor $50; Sponsor $100; Corporate Member $250; Patron & Corporate Contributor $500; Benefactor & Corporate Sponsor $1,000; Corporate Patron $2,500; Corporate Benefactor $5,000.

Tarkio

TARKIO COLLEGE ALUMNI MUSEUM, 314 N. Main, Tarkio, MO 64491-1543. Mailing Address: P.O. Box 111, Tarkio, MO 64491-0111. Tel.: 660-736-4208.
E-mail: tcaa@asde.net
Web Site: www.tarkioalumni.org/museum.html
Key Personnel: Pres., Farleigh Joe Farley; Vice Pres., Marla Tollett-Ross; Sec., Sheridan Mires; Treas., Mary McAdams
Institution Type/Description: Alumni Museum.
Collections: photographs; memorabilia.
Hours & Admission Prices: Call for hours.

Trenton

GRUNDY COUNTY MUSEUM, 1100 Mabel St., Trenton, MO 64683. Tel.: 660-359-2411.
E-mail: schlarb@cebridge.net
Web Site: www.grundycountymuseum.org
Founded: 1976.
Key Personnel: Recording Sec., Phil Schlarb
Governing Authority: Tax-exempt: 501(c)(3).
Institution Type/Description: History Museum: built in 1895 by William McVay.
Collections: Grundy County history, culture & artifacts; 18th century dugout canoe; weaving loom; farm machinery; stationary vertical steam engine & boiler; lanterns, signals, uniforms, keys & photos; drawings of Trenton's past & classic buildings; letter written by Felix Grundy; old maps of Missouri; early wall cabinets; Civil War souvenirs; period costumes & furniture; Rock Island caboose; windmill; restored one-room schoolhouse & desks; military uniforms, weapons, & medals; Civil War, WWI, & WWII artifacts.
Hours & Admission Prices: Memorial Day to Oct. Sat.-Sun. & holidays 2-5; other times by appointment. Suggested Donation: adults $2; Life members no charge. ♿
Attendance: 1,269 (accurate)
Membership: Lifetime $100.

Tuscumbia

MILLER COUNTY HISTORICAL SOCIETY & MUSEUM, 2005 Highway 52, Tuscumbia, MO 65082. Mailing Address: P.O. Box 57, Tuscumbia, MO 65082-0057. Tel.: 573-369-3500.
E-mail: millercountymuseum@att.net
Web Site: www.millercountymuseum.org
Founded: 1980.
Key Personnel: Pres. (V), Joe Pryor; Dir., Karen Smith.
Personnel Profile: Full-Time Volunteers 1; Part-Time Volunteers 30.
Governing Authority: bd. of directors.
Institution Type/Description: Historical Society Museum.
Collections: local history & culture; photographs; period artifacts; memorabilia.
Research Fields: steamboat; genealogy.
Facilities: library.
Publications: quarterly newsletter.
Hours & Admission Prices: mid-May to mid-Oct. Mon., Wed. & Fri.-Sat. 10-4. No charge; donations accepted. Closed holidays. ♿

Attendance: 3,000
Membership: Individual $15; Lifetime $50.

Unionville

PUTNAM COUNTY MUSEUM, 201 S. 16th St., Unionville, MO 63565. Tel.: 660-947-2955.
Web Site: putnamcountyhistoricalsociety.com
Institution Type/Description: History Museum.
Collections: local history & culture; period furnishings; personal artifacts; photographs.
Hours & Admission Prices: Call for hours.

Van Buren

OZARK NATIONAL SCENIC RIVERWAYS, 404 Watercress Dr., Van Buren, MO 63965-9100. Mailing Address: P.O. Box 490, Van Buren, MO 63965-0490. Tel.: 573-323-8822. Fax: 573-323-4140 & 8823. TDD: 573-323-4270.
Web Site: www.nps.gov/ozar
Founded: 1972.
Congressional District: 8
Key Personnel: Supt., Reed Detring.
Personnel Profile: Full-Time Paid 1; Part-Time Paid 1.
Governing Authority: federal. Midwest Region, National Park Service, Washington DC. Subsidiary Institution: Alley Mill. Tax-exempt.
Institution Type/Description: Park Museum.
Collections: Ozark culture & milling equipment; herbarium; oral history tapes; photographs; insects; fish; civilian conservation Corps. Historic Building: mill.
Research Fields: CCC development of the area; Current River region history & uses; Civilian Conservation Corps. programs; folk life; material culture; Ozarks; regional natural history.
Facilities: 1,000-vol. library of books pertaining to natural history, regional histories & geology; visitor center; visitor contact station. Adult & children's cultural & natural history publications for sale.
Activities: evening programs; guided walks; cave tours in summer; cultural demonstrations of milling, quilting, johnboat making; folk music presentations.
Publications: brochures, Ozark Riverways; maps; pamphlets; site bulletins.
Hours & Admission Prices: Park: daily. Visitor Contact Station & Center: call for hours & information. Demonstrations April-Oct. Sat.-Sun.; other times by appointment. No charge. &
Attendance: 22,000 (estimated)

Versailles

MORGAN COUNTY HISTORICAL SOCIETY, 120 N. Monroe St., Versailles, MO 65084. Tel.: 573-378-5530.
Web Site: morgancohistory.org
Founded: 1965.
Key Personnel: Pres. (V), Barbara Barnard.
Personnel Profile: Part-Time Volunteers 40.
Governing Authority: board. Tax-exempt.
Institution Type/Description: County Historical Society: housed in the former Martin Hotel, built in 1877 & 1884.
Collections: local history & culture; photographs; period artifacts; military artifacts; tools; family & county histories; Martin family artifacts; hotel registers.
Hours & Admission Prices: May to mid-Oct. Mon.-Fri. 10-3; other times by appointment. Adults $3; handicapped no charge.
Attendance: 1,200 (estimated)
Membership: Individual $10.

Walker

OSAGE VILLAGE STATE HISTORIC SITE, Hwy. C, Walker, MO 64790. Mailing Address: 1009 Truman Ave, Lamar, MO 64759. Tel.: 417-682-2279. Fax: 417-682-6304.
E-mail: moparks@dnr.mo.gov
Web Site: mostateparks.com/osage-village-state-historic-site
Founded: 1984.
Congressional District: 4
Key Personnel: Site Admin., Beth Bazal.
Governing Authority: state. Parent Institution: Div. of State Parks, Recreation & Historic Preservation, Missouri Dept. of Natural Resources, P.O. Box 176, Jefferson City, MO 65102. Maintained by: Harry S. Truman Birthplace State Historic Site, 417-682-2279. Tax-exempt.
Institution Type/Description: Archaeological Site.

Collections: local history & culture; Native American artifacts.
Facilities: interpretive kiosk; interpretive trail.
Activities: self-guided tours; text & illustrations in a kiosk display.
Publications: brochure, Walking Trail Guide.
Hours & Admission Prices: Daily sunrise to sunset. No charge; donations accepted.
Attendance: 6,000 (estimated)

Walnut Shade

BONNIEBROOK MUSEUM, GALLERY & HOME OF ARTIST ROSE O'NEILL, 485 Rose O'Neill Rd., Walnut Shade, MO 65771. Tel.: 800-539-7437; 417-561-1509.
E-mail: oneillmuseum@aol.com
Web Site: www.roseoneill.org
Formerly: Bonniebrook Historical Society
Key Personnel: Pres. (V), Susan Scott; Museum Shop Mgr., Louise Williams.
Personnel Profile: Full-Time Volunteers 2; Part-Time Volunteers 3; Interns 2.
Governing Authority: Parent Institution: Bonniebrook Historical Society. Tax-exempt.
Institution Type/Description: Art Museum.
Collections: life, art & work of artist Rose O'Neill; personal artifacts; photographs; art; Kewpie ephemera.
Research Fields: art & life of artist Rose O'Neill.
Activities: Museum Sponsors: Annual Open House & Grand Opening of Newly Renovated Gallery in April; Thomas Hart Benton at Bonniebrook in July; Rose O'Neill's Faerie Gathering in September.
Publications: quarterly, The Bonniebrook News.
Hours & Admission Prices: April-Nov. Tues.-Sat. 9-4. Call for admission prices.
Attendance: 5,000 (estimated)
Membership: Individual $25.

Warrensburg

ARTHUR F. MCCLURE II ARCHIVES AND UNIVERSITY MUSEUM, (M), JCK Library 1470, 601 S. Missouri, Warrensburg, MO 64093-5040. Tel.: 660-543-4649 & 4404.
E-mail: vrichardson@ucmo.edu
Web Site: www.ucmo.edu/hist-anth/archives
Formerly: Central Missouri State University Archives and Museum
Founded: 1968.
Congressional District: 114
Key Personnel: Dir., John Sheets; Asst. Dir., Vivian Richardson.
Personnel Profile: Full-Time Paid 1; Part-Time Paid 1.
Governing Authority: university. Parent Institution: Central Missouri State University. Tax-exempt: 170(b)(1)(A).
Institution Type/Description: General Museum.
Collections: archaeology; ethnology; history; biology; Rohmiller shell collection; Haymaker Hispanic-American collection; Schmidt Meso-American; Nance Middle East collection; university history.
Research Fields: archaeology; cultural resource mgmt.; university history.
Facilities: 250-vol. gen. library available for use by permission of dir.
Activities: guided tours; temporary & traveling exhibitions; school loan service; lectures.
Publications: Museum Research Series, Archaeology.
Hours & Admission Prices: During University Sessions: Mon.-Fri. 9-12 & 1-4; other times by appointment. No charge. Closed holidays. &

JOHNSON COUNTY HISTORICAL SOCIETY MUSEUM, 302 N. Main St., Warrensburg, MO 64093-1554. Tel.: 660-747-6480.
E-mail: curator@jocomohistory.org
Web Site: jocomohistory.org
Founded: 1920.
Congressional District: 4
Key Personnel: Cur., Lisa Irle.
Personnel Profile: Full-Time Paid 1; Part-Time Paid 1; Part-Time Volunteers 30.
Governing Authority: nonprofit. Parent Institution: Johnson County Historical Society. Tax-exempt: 501(c)(3).
Institution Type/Description: Regional History Museum: 1838 federal style courthouse; Mary Miller Smiser Heritage Library; one room schoolhouse.
Collections: archives; folklore; lawn mowers; general historical & cultural items; Civil War; Whiteman AFB.
Major Exhibits: Civil War in Johnson County, Missouri, 1/12-5/14.
Research Fields: local history; 19th-century cultural artifacts.
Facilities: 500-vol. library of local and regional history books; photographs; clippings, available for use in library only; reading room. Museum-related items for sale.
Activities: guided tours; lectures; permanent & temporary exhibitions; workday activities for college students.

Publications: The Bulletin, Johnson County Historical Society.
Hours & Admission Prices: Mon.-Sat. 1-4; other times by appointment. Tours: adults $3; members no charge. Closed New Year's Eve & Day; Independence Day; Thanksgiving; Christmas Eve, Day & week. &
Attendance: 4,000 (estimated)
Membership: Individual $20; Family $30; Corporate $50; Lifetime $150.

Warrenton

SCHOWENGERDT HOUSE, 308 E. Booneslick Rd., Warrenton, MO 63383-2008. Tel.: 636-456-3820.
E-mail: schowhouse@history-warrencountymo.com
Governing Authority: Parent Institution: Warren County Historical Society.
Institution Type/Description: Historic House Museum: built in 1866. A National Historic Place.
Collections: Schowengerdt family history; period furnishings; personal artifacts; photographs.
Hours & Admission Prices: By appointment.

WARREN COUNTY HISTORICAL SOCIETY MUSEUM & LIBRARY, 102 W. Walton St., Warrenton, MO 63383. Tel.: 636-456-3820.
E-mail: museum@history-warrencountymo.com
Web Site: www.warrencountymohistory.com
Founded: 1970.
Congressional District: 9
Governing Authority: Tax-exempt.
Institution Type/Description: Historical Society Museum.
Collections: local history & culture; period furnishings; photographs; personal artifacts.
Research Fields: local history; genealogy; military; slaves.
Facilities: library.
Activities: educational programs.
Publications: quarterly newsletter.
Hours & Admission Prices: Thurs.-Fri. 10-4, Sun. 1-4. No charge; donations accepted. &
Membership: Individual $20; Business & Organization $50-$500; Life $30.

Washington

FIREHOUSE MUSEUM, 520 W. 5th St., Washington, MO 63090. Mailing Address: Washington Historical Society, P.O. Box 146, Washington, MO 63090-0146. Tel.: 636-390-0469.
Institution Type/Description: History Museum.
Collections: period fire engines; early automobiles.
Hours & Admission Prices: By appointment.

WASHINGTON HISTORICAL SOCIETY MUSEUM, Fourth & Market Sts., Washington, MO 63090. Mailing Address: P.O. Box 146, Washington, MO 63090-0146. Tel.: 636-239-0280.
Institution Type/Description: History Museum.
Collections: local history & culture; period artifacts; photographs; personal artifacts.
Hours & Admission Prices: March-Dec. Tues.-Sat. 10-4, Sun. 12-4. No charge.

Waynesville

THE OLD STAGECOACH STOP, 105 Lynn St., Waynesville, MO 65583. Mailing Address: P.O. Box 585, Waynesville, MO 65583-0585. Tel.: 573-762-9683.
Web Site: www.oldstagecoachstop.org
Founded: 1983.
Key Personnel: Pres. (V), Stephanie Nutt; Museum Shop Mgr., Jan Primas.
Personnel Profile: Part-Time Volunteers 15.
Governing Authority: Parent institution: Old Stagecoach Stop Foundation. Tax exempt.
Institution Type/Description: House Museum: listed on the National Register of Historic Places.
Collections: local history & culture; period rooms; 19th - 20th century furnishings collection; archeological collection; photographs.
Research Fields: Pulaski County history.
Publications: annual, Old Settlers Gazette; seasonal newsletter, OSS News.
Hours & Admission Prices: April-Sept. Sat. 10-4. No charge.
Attendance: 1,652 (accurate)
Membership: Individual $10; Family $15; Patron $25; Life $250.

PULASKI COUNTY MUSEUM & HISTORICAL SOCIETY, 301 A Historic Rte. 66, Waynesville, MO 65583. Mailing Address: P.O. Box 144, Waynesville, MO 65583. Tel.: 573-774-5368.
E-mail: mscott1108@cablemo.net
Web Site: www.oldpulaskicountycourthousemuseum.webs.com
Formerly: Pulaski County Courthouse Museum
Founded: 1969.
Congressional District: 122
Key Personnel: Society Pres. (V), Cur., & Museum Shop Mgr., Marge Scott.
Personnel Profile: Full-Time Volunteers 1; Part-Time Volunteers 1.
Governing Authority: state. Parent Institution: Pulaski County Historical Society. Tax-exempt: 501(c)(3).
Institution Type/Description: History Museum.
Collections: historical information on WWI, WWII & desert storm, Korean War, Civil War, Trail of Tears; toys; antique irons; arrowhead collection; farm equipment; art gallery; domestic room; medical room; original courtroom; military uniforms/quilts 1895; early history of Waynesville, Mo. in photos; DVDs.
Activities: Museum Sponsors: Open House in May.
Publications: Pulaski County Heritage quarterly historical society newsletter.
Hours & Admission Prices: April-Sept. Sat. 10-4; tours by appointment. No charge; donations appreciated.
Attendance: 1,538 (estimated)
Membership: Annual $15; Lifetime $50.

Webb City

THE CLUBHOUSE MUSEUM, 115 N. Madison St., Webb City, MO 64870. Mailing Address: Webb City Historical Society, P.O. Box 1, Webb City, MO 64870. Tel.: 417-673-5866.
Institution Type/Description: History Museum: housed in the former clubhouse for employees of the Southwest Missouri Electric Railway; built in 1910.
Collections: local history & culture; period furnishings; photographs; personal artifacts.
Hours & Admission Prices: By appointment.

Webster Groves

MAY GALLERY AT WEBSTER UNIVERSITY, (M), 8300 Big Bend Blvd., Webster Groves, MO 63119-3114. Mailing Address: 470 E. Lockwood Ave., St. Louis, MO 63119-3194. Tel.: 314-246-7673. Fax: 314-963-6924.
E-mail: mgallery@webster.edu
Web Site: www.webster.edu/maygallery
Founded: 1988.
Congressional District: 3
Key Personnel: Dir., Bill Barrett.
Personnel Profile: Part-Time Paid 2; Part-Time Volunteers 20.
Governing Authority: Parent Institution: Webster University. Tax-exempt.
Institution Type/Description: Art Gallery.
Collections: photographs.
Activities: special events.
Publications: mini show catalogs.
Hours & Admission Prices: Mon.-Fri. 9-9, Sat.-Sun. 12-5. No charge. &
Attendance: 4,500 (estimated)

West Plains

HARLIN MUSEUM, 505 Worcester Ave., West Plains, MO 65775. Mailing Address: P.O. Box 444, West Plains, MO 65775-0444. Tel.: 417-256-7801.
E-mail: staff@harlinmuseum.com
Web Site: harlinmuseum.com
Institution Type/Description: History Museum: housed in the home of former West Plains Mayor, James P. Harlin & his wife; built in 1889.
Collections: drawings; early farm tools; Native American artifacts; paintings; Lennis Broadfoot Painting Collection.
Hours & Admission Prices: Tues.-Sat. 12-4.

Weston

HERBERT BONNELL MUSEUM, 20755 Lamar Rd., Weston, MO 64098-9173. Mailing Address: P.O. Box 238, Weston, MO 64098-0238. Tel.: 816-386-5587 & 992-0102.
E-mail: belindak@embarqmail.com
Web Site: www.westonmo.com/history/museums_hist.html
Founded: 1985.
Key Personnel: Trust, Frank Green; Co Trust, Belinda Farris
Institution Type/Description: History Museum.

Collections: personal artifacts & memorabilia; farm equipment; 1929 Model A Coupe; horse-drawn equipment.
Hours & Admission Prices: May-Oct. Sat.-Sun. 1-5; other times by appointment. No charge; donations accepted.

NATIONAL SILK ART MUSEUM, The Saint George Hotel, 423 Main St., Weston, MO 64098-1249. Tel.: 816-640-9902. Fax: 816-640-9906.
E-mail: adrienne@thesaintgeorgehotel.com
Web Site: www.nationalsilkartmuseum.com
Founded: 2003.
Congressional District: 6
Key Personnel: Dir. & Cur., John Pottie; Pres. (V), Asst. Cur. & Museum Shop Mgr., Adrienne Haake; Librarian, Venessa Pottie.
Personnel Profile: Full-Time Volunteers 2.
Governing Authority: Tax-exempt.
Institution Type/Description: Art Museum.
Collections: over 200 French silk tapestries by artists from the 19th & 20th century.
Research Fields: silk weaving.
Facilities: art research library.
Publications: biannual newsletter.
Hours & Admission Prices: Sat. 11-5; groups & other times by appointment. &
Attendance: 30,000 (accurate)
Membership: Individual $25; Family $40; Corporate $100; Benefactor $250; Benefactor Gold $1,000.

WESTON HISTORICAL MUSEUM, 601 Main St., Weston, MO 64098-1207. Mailing Address: P.O. Box 266, Weston, MO 64098-0266. Tel.: 816-386-2977.
E-mail: director@westernhistoricalmuseum.org; curator@westernhistoricalmuseum.org
Web Site: www.westonhistoricalmuseum.org
Founded: 1960.
Congressional District: 6
Key Personnel: Dir., Barbara Fulk; Pres., Carl Felling; Museum Shop Mgr., Jeanne Braneff.
Personnel Profile: Part-Time Paid 3; Part-Time Volunteers 10.
Governing Authority: private. Tax-exempt.
Institution Type/Description: History Museum.
Collections: Native American moccasins; early physician instruments; period clothing; Civil War artifacts; newspaper office; household items; tools; glassware; china; furniture; historical documents; tobacco culture; Black history; toys.
Publications: quarterly members' newspaper.
Hours & Admission Prices: late March to early Dec. Tues.-Sat. 1-4, Sun. 1:30-4:30. No charge; donations accepted. Closed holidays.
Attendance: 3,000 (estimated)
Membership: Senior $10; Single $15; Family $25; Contributor $50; Patron $75; Benefactor $100.

Westphalia

WESTPHALIA HISTORICAL SOCIETY, 119 E. Main St., Westphalia, MO 65085. Mailing Address: P.O. Box 244, Westphalia, MO 68085-0244.
E-mail: whs65085@yahoo.com
Web Site: westphaliahistoricalsociety
Institution Type/Description: Historical Society Museum.
Collections: local history & culture; photographs; personal artifacts; period furnishings.
Hours & Admission Prices: April-Oct. Sun. 1-3.

Wildwood

HENCKEN HOUSE MUSEUM - WILDWOOD HISTORICAL SOCIETY, 18750 Hwy. 100, Wildwood, MO 63069-3000. Mailing Address: P.O. Box 125, Grover, MO 63040. Tel.: 636-458-3306.
E-mail: info@wildwoodhistoricalsociety.org
Web Site: wildwoodhistoricalsociety.org
Institution Type/Description: Historical Society Museum.
Collections: local history & culture; photographs; newspaper articles; period furnishings.
Hours & Admission Prices: Call for hours.

RIVER HILLS VISITOR CENTER, Dr. Edmund A. Babler Memorial State Park, 800 Guy Park Dr., Wildwood, MO 63005. Mailing Address: c/o Missouri State Parks, P.O. Box 176, Jefferson City, MO 65102-0176. Tel.: 636-458-3813. Fax: 636-458-9105.
Web Site: mostateparks.com/location/55677/river-hills-visitor-center
Founded: 1972.
Congressional District: 91
Key Personnel: Park Supt., Jeff Robinson.
Personnel Profile: Full-Time Paid 13; Part-Time Paid 15; Part-Time Volunteers 12.
Governing Authority: state. Parent Institution: Div. of State Parks, MO Dept. of Natural Resources, P.O. Box 176, Jefferson City, MO 65102. Tax-exempt.
Institution Type/Description: Park Museum & Visitor Center.
Collections: geology; habitats, fauna, flora; exhibits on the Missouri River Hills; slide show about the park.
Research Fields: natural history; ecology; environmental education.
Facilities: visitor orientation & observation area; auditorium; classroom; picnic facilities; hiking trails; primitive scout camping area, available by advance reservation by calling 314-458-3813; park headquarters.
Activities: nature walks; environmental education program for schools, grades K-12 & adults; swimming; camping; Jacob L. Babler Outdoor Education Center for campers with special handicaps, call 314-458-3048 for reservations; daily naturalist programs Memorial Day through Labor Day.
Hours & Admission Prices: April-Oct. Wed. -Sun. and Nov. & March Sat.-Sun. 8:30-4. No charge. Closed Dec.-Feb. &
Attendance: 393,000

ROCKWOODS RESERVATION, 2751 Glencoe Rd., Wildwood, MO 63038-1919. Tel.: 636-458-2236. Fax: 636-458-6726.
E-mail: anna-lisa.tucker@mdc.mo.gov
Founded: 1938.
Key Personnel: Naturalist, Anna Lisa Tucker; Naturalist, Shanna Raeker.
Personnel Profile: Full-Time Paid 2; Part-Time Paid 2; Part-Time Volunteers 25.
Governing Authority: state. Affiliated with the Missouri Dept. of Conservation, North Ten Mile Drive, Jefferson City, MO 65101. Tax-exempt.
Institution Type/Description: Conservation Education Center.
Collections: native animals exhibit; old limestone mine; conservation exhibits.
Facilities: interpretive hiking trails; nature & conservation center; 65-seat classroom.
Activities: lectures; nature programs; films; permanent, temporary & traveling exhibitions.
Publications: monthly newsletter, Conservation Connections.
Hours & Admission Prices: Visitor Center daily 8-5; Area open daily sunrise to 1/2 hour after sunset. No charge. Closed all state holidays. &
Attendance: 150,000 (estimated)

Willow Springs

FIRE MUSEUM OF MISSOURI, 908 E. Business Rte. 60-63, Willow Springs, MO 65793. Tel.: 417-469-4589.
E-mail: jlows@socket.net
Web Site: www.usfirehouse.com
Key Personnel: Owner, John Mathieu
Institution Type/Description: Fire-Fighting Museums.
Collections: 23 period fire engines; hose cart; photographs; fire extinguishers; Fire, Police, & Emergency Medical Service uniform patches; period cars from 1929 to 1972; over 3,000 soda pop bottles; soda pop & beer cans; period signs; vending machines from late 30s to early 60s.
Activities: patch trading.
Hours & Admission Prices: Adults $5, children 10-17 $3; children under 10 no charge.

Windsor

WINDOR HISTORICAL SOCIETY, 104 S. Franklin St., Windsor, MO 65360. Mailing Address: P.O. Box 111, Windsor, MO 65360. Tel.: 660-647-2345.
E-mail: winhissoc@socket.net
Web Site: windsormo.org
Institution Type/Description: Historical Society Museum.
Collections: local history & culture; period furnishings; personal artifacts; photographs.
Hours & Admission Prices: Fri.-Sun. 1-4.

MONTANA

(153 listings)

Anaconda

COPPER VILLAGE MUSEUM - ARTS CENTER OF DEER LODGE COUNTY - MARCUS DALY HISTORICAL SOCIETY, 401 E. Commercial, Anaconda, MT 59711-2360. Tel.: 406-563-2422 & 2220. Fax: 406-563-2422.
E-mail: copper_village@hotmail.com
Web Site: www.coppervillageartcenter.com
Founded: 1971.
Congressional District: 1
Key Personnel: Chief Exec. Dir. & Dir. Community Affairs, Susan Lanes; Pres. (V) & Chm. (V), Brian Tesson; Co-Pres., Barbara Beardslee; Dir. Education, Mary Johnson.
Personnel Profile: Part-Time Paid 4.
Governing Authority: Tax-exempt: 501(c)(3).
Institution Type/Description: Contemporary & Historic Art Gallery and Museum.
Collections: local history artifacts; house furnishings of 1900; ghost town mining artifacts; chronological development of area Indians, trappers, ranchers, miners; industry; bottles & artifacts from Anaconda's first drug store; antique doctor's furniture & books; barbed wire collection; renovated antique Victorian dollhouse; Marcus Daley Copper Empire Display: copper smelter, smelting process, workers.
Facilities: Museum-related items for sale.
Activities: guided tours; lectures; films; gallery talks; drama; monthly art exhibits; workshops; art classes; special festivals; summer recreational arts program for students ages 7-13. Museum Sponsors: annual champagne dinner & auction at Anaconda Elks Club; Art in Washoe Park in July.
Publications: quarterly newsletter.
Hours & Admission Prices: Tues.-Sat. 10-5. No charge; donations accepted. Closed legal holidays. &
Attendance: 8,000 (estimated)
Membership: Senior Citizen & Student $10; Individual $15; Family $25; Sponsor $35; Patron & Associate $50; Benefactor $100-$2,500.

Bainville

PIONEER'S PRIDE MUSEUM, 6013 Rd. 1011, Bainville, MT 59212-9625. Mailing Address: HC58 P.O. Box 63, Bainville, MT 59212. Tel.: 406-769-2064.
Institution Type/Description: History Museum.
Collections: pioneer bedroom, kitchen & music room; old jail from Mon-Dak; 1929 fire truck.
Hours & Admission Prices: Memorial Day-Labor Day Tues.-Sun. 1:30-4:30. No charge.

Baker

O'FALLON HISTORICAL MUSEUM, 723 S. Main St., Baker, MT 59313. Mailing Address: P.O. Box 285, Baker, MT 59313-0285. Tel.: 406-778-3265. Fax: 406-778-3967.
E-mail: ofmuseum@midrivers.com
Web Site: www.falloncounty.net
Founded: 1968.
Key Personnel: C.E.O., Reinhard Barth; Dir., Jody Strand.
Personnel Profile: Full-Time Paid 2.
Governing Authority: nonprofit organization. Tax-exempt.
Institution Type/Description: History Museum.
Collections: local homestead articles from 1908; Indian artifacts.
Activities: permanent exhibitions.
Publications: newsletter, O'Fallon Flashbacks; books, O'Fallon Flashbacks I & II - Homestead Recipe Cookbook; book, Yesterdays News.
Hours & Admission Prices: Daily. No charge; donations accepted. Closed legal holidays. &
Attendance: 1,300 (estimated)
Membership: Individual $15; Couples $20; Life $100.

Belt

BELT MUSEUM, 37 Castner St., Belt, MT 59412-8029. Mailing Address: P.O. Box 442, Belt, MT 59412-0442. Tel.: 406-277-3574.
E-mail: mkoontz@3rivers.net
Institution Type/Description: History Museum: housed in the old jail building, built in 1895.
Collections: historic jail cell; coal mine; records & photographs.

Hours & Admission Prices: Memorial Day-Labor Day Sat.-Sun. 12-4. No charge; donations accepted.
Attendance: 400 (estimated)

Big Timber

CRAZY MOUNTAIN MUSEUM, Exit 367 Cemetery Rd., Hwy. I-90, Big Timber, MT 59011. Mailing Address: P.O. Box 83, Big Timber, MT 59011-0083. Tel.: 406-932-5126.
E-mail: cmmuseum@mtintouch.net
Founded: 1990.
Key Personnel: Chm. & Pres. (V), Steve Harvey; Treas., Joan Van Daveer; Dir. & Cur., Rita Esp; Cur., Elli Hawks; Museum Shop Mgr., Christy Mosness.
Personnel Profile: Full-Time Paid 2; Part-Time Paid 2; Part-Time Volunteers 70.
Governing Authority: private; nonprofit organization. Tax-exempt: 501(c)(3).
Institution Type/Description: History Museum.
Collections: history of Sweet Grass County & surrounding area.
Research Fields: people & subjects of county.
Facilities: library; 4,000 sq. ft. exhibit space. Museum-related items for sale.
Activities: docent program; guided tours; hobby workshops; temporary exhibitions. Annual Events: Memorial Day Festival. Permanent Exhibits: 1 Room School; 1907 Model Big Timber; Norwegian Stabbur.
Publications: newsletter, The New Breeze.
Hours & Admission Prices: Memorial Day to Sept. Mon.-Sat. 10-4:30, Sun. 1-4:30. No charge; donations accepted. &
Attendance: 5,000 (estimated)
Membership: General $10; Donor $25; Founder $100.

Billings

BILLINGS CURATION CENTER, (M), 5001 Southgate Dr., Billings, MT 59101-4669. Tel.: 406-896-5213. Fax: 406-896-5317.
Web Site: www.mt.blm.gov/BCC
Founded: 1994.
Key Personnel: BLM State Archaeologist, Gary Smith; Cur., David K. Wade.
Personnel Profile: Full-Time Paid 2; Part-Time Volunteers 4; Interns 4.
Governing Authority: Parent Institution: Bureau of Land Mgmt. Subsidiary Institution: Montana State Office.
Institution Type/Description: Federal Agency Repository & Archaeology Site.
Collections: federal land artifacts; maps; fossils & artifacts; homesteading; mining; ghost towns; early northwest exploration.
Research Fields: history and pre-history of the Northern Plains.
Facilities: library; lab; archives.
Hours & Admission Prices: Mon.-Fri. 9-4:30. No charge. &

MOSS MANSION HISTORIC HOUSE MUSEUM, (M), 914 Division St., Billings, MT 59101-1921. Tel.: 406-256-5100.
E-mail: tours@mossmansion.com
Web Site: www.mossmansion.com
Founded: 1986.
Congressional District: 1
Key Personnel: Dir., Jenna Richter; House & Collections, Kendra Bruget; Pres. (V), Mike Nelson; 1st Vice Pres., Nate Allie; Sec. & Museum Shop Mgr., Lue Ponich; Treas., Amanda Nielson; Event Coord., Kelsey Palmer; Registrar, Karen Wegner.
Personnel Profile: Full-Time Paid 2; Part-Time Paid 1; Part-Time Volunteers 200; Interns 4.
Governing Authority: nonprofit organization. Parent Institution: Billings Preservation Society, Inc. Tax-exempt: 501(c)(3).
Institution Type/Description: Historic House Museum: housed in the Moss Mansion built in 1901-1903.
Collections: period furnishings; clothing & personal effects; city directories 1900-1949.
Research Fields: family history.
Facilities: classroom. Museum-related items for sale.
Activities: guided tours; hobby workshops. Annual Events: Spring Fest; Holiday at the Moss; Heritage Home Tour; Haunted House at the Moss; Party for Preservation; Murder Mysteries; Christmas at The Moss.
Publications: quarterly newsletter, Billings Preservation Society Newsletter; quarterly volunteer newsletter, MMVIP Newsletter.
Hours & Admission Prices: Summer: Mon.-Sat. 9-5, Sun. 1-4; Winter, Fall & Spring: daily 1-4. Adults $7, senior citizens & students $5, children $3; members no charge. Closed New Year's Day; Thanksgiving; Christmas. &
Attendance: 18,535 (accurate)
Membership: Student, Teacher & Senior 62 and over $30; Individual $40; Family $60; Preservationist $125; Patron $250; Benefactor $500; Founder $1,000.

MUSEUM OF WOMEN'S HISTORY, 2824 3rd Ave. N., Billings, MT 59101-1932. Mailing Address: 2822 3rd Ave. N., B-3, Billings, MT 59101-1934. Tel.: 406-248-2015. Facebook: Museum of Women's History.
E-mail: mowh@imt.net
Web Site: museumofwomenshistory.org
Founded: 1995.
Personnel Profile: Part-Time Volunteers 12; Interns 2.
Governing Authority: public; nonprofit organization. Tax-exempt: 501(c)(3).
Institution Type/Description: Women's History Museum.
Collections: women's history.
Research Fields: women in WWII.
Facilities: library; 1,500 sq. ft. exhibit space.
Activities: guided tours; lectures; broadcast programs.
Publications: quarterly newsletter, The Mirror.
Hours & Admission Prices: By appointment only. No charge; donations accepted. Closed legal holidays. &
Attendance: 350 (estimated)
Membership: Seniors & Students $20; Individual $30; Business $100.

NORTHCUTT STEELE GALLERY - MONTANA STATE UNIVERSITY, (M), 1500 University Dr., Billings, MT 59101. Tel.: 406-657-2903; 406-657-2324. Fax: 406-657-2187. Facebook: Northcutt Steele Gallery.
E-mail: leanne.gilbertson@msubillings.edu
Web Site: msubillings.edu/gallery
Founded: 1985.
Key Personnel: Dir., Dr. Leanne Gilbertson.
Personnel Profile: Full-Time Paid 1; Part-Time Paid 1; Interns 2.
Governing Authority: Parent Institution: Montana State University-Billings. Tax-exempt.
Institution Type/Description: Art Gallery.
Collections: paintings; sculpture.
Hours & Admission Prices: Sept.-May Mon.-Fri. 8-4. No charge.
Membership: AAM $125; AAMG $150.

RYNIKER-MORRISON GALLERY, Rocky Mountain College, 1511 Poly Dr., Billings, MT 59102-1796. Tel.: 800-877-6259.
E-mail: mcintoshart@imt.net
Institution Type/Description: Art Gallery.
Collections: works by RMC students, faculty & visiting artists.
Hours & Admission Prices: Mon.-Fri. 9-4.

✳ WESTERN HERITAGE CENTER, (M), 2822 Montana Ave., Billings, MT 59101-2305. Tel.: 406-256-6809. Fax: 406-256-6850.
E-mail: julie@ywhc.org
Web Site: www.ywhc.org
Founded: 1971.
Congressional District: 2
Key Personnel: Exec. Dir., Julie Dial; Community Historian, Kevin Kooistra-Manning; Business Mgr., Lisa Olinsted; Dir. Devel., Sherlynn Stewart; Bldg. Supt., Albert Gehring.
Personnel Profile: Full-Time Paid 4; Part-Time Paid 3; Part-Time Volunteers 20.
Governing Authority: county. Tax-exempt: 501(c)(3).
Institution Type/Description: Regional History Museum: housed in 1901 Parmly Billings Memorial Library Building.
Collections: material culture of the Yellowstone Valley; photographs & manuscripts.
Research Fields: history & contemporary characteristics of settlers in Eastern Montana & the Yellowstone River Valley.
Facilities: theater. Museum-related items for sale.
Activities: lectures; films; workshops; organized education programs for children & adults; docent program; changing exhibitions; oral history program; rental facilities; teachers professional development.
Publications: quarterly newsletter, Reflections; books, Images of Billings: A Photographic History, Pieces & Places of Billings: Local Monuments & Markers, Stories from An Open Country: Essays about The Yellowstone River Valley.
Hours & Admission Prices: Tues.-Sat. 10-5. Adults $5, seniors $3, children $1; members no charge. Closed major holidays. &
Attendance: 17,800 (estimated)
Membership: Senior $35; Individual $40; Couple $60; Family $100; Bronze $250; Silver $500; Gold $1,000; Platinum $2,000.

YELLOWSTONE ART MUSEUM, (M), 401 N. 27th St., Billings, MT 59101-1241. Tel.: 406-256-6804. Fax: 406-256-6817. Facebook; Yellowstone Art Museum.
E-mail: artinfo@artmuseum.org
Web Site: www.artmuseum.org
Founded: 1964.
Congressional District: 2
Key Personnel: Exec. Dir., Robyn G. Peterson; Pres., Kris Carpenter; Senior Cur., Bob Durden; Dir. Devel., Ryan Cremer; Dir. Education, Linda Ewert; Admin. & Finance Dir., Lisa Berke.
Personnel Profile: Full-Time Paid 13; Part-Time Paid 6; Part-Time Volunteers 305.
Volunteer Hours: 6,900
Operating Expenses: 1,448,788
Operating Income: 1,243,154
Governing Authority: private; nonprofit organization. Tax-exempt: 501(c)(3).
Institution Type/Description: Regional Art Museum
Collections: contemporary regional collection: paintings; sculpture; ceramics; photography; L.A. Huffman photographs; Poindexter Collection of New York abstract expressionists; minimalist & conceptual art; Snook collections of Will James; western historic art.
Major Exhibits: Hallowed Absurdities: Work by Ted Waddell, 9/13-1/14; Face to Face, 3/14-8/14; Un/conscious Beat: A Survey of Regional Surrealism, 3/14-8/14; Art of the Brick (T), 9/14-12/14; T.L. Solien: Toward the Setting Sun (T), 9/14-12/14.
Research Fields: contemporary art of the northern Rockies.
Facilities: 500-vol. library of visual arts available by appointment; reading room; 100-seat lecture gallery; education studio; art-in-residence studio; open storage facility. Museum-related items for sale.
Activities: guided tours; lectures; films; gallery talks; concerts; arts festivals; formally organized education programs for children & adults; docent program or council; inter-museum loan, temporary & traveling exhibitions; outreach education; artist residency consignment gallery.
Publications: newsletter; exhibition catalogs & brochures.
Hours & Admission Prices: Tues.-Wed. & Sat. 10-5, Thurs.-Fri. 10-8, Sun. 11-4. Adults $6; discounts to military, AAM, AARP, ICOM & AAA members; members no charge. North American Reciprocal group. Closed major holidays. &
Attendance: 23,537 (accurate)
Membership: Student & Senior Citizen $30; Individual $40; Family $65; Aficionado $125; Education Circle $300; Collection Circle $500; Exhibition Circle $1,000; Patron Circle $2,500.

YELLOWSTONE COUNTY MUSEUM, 1950 Airport Terminal Cir., Billings, MT 59105-1988. Tel.: 406-256-6811. Fax: 406-254-6031. Facebook: Yellowstone County Museum.
E-mail: ycm@tctwest.net
Web Site: yellowstonecountymuseum.org
Formerly: Peter Yegen Jr. Yellowstone County Museum
Founded: 1956.
Congressional District: 2
Key Personnel: Chm. (V), Charlie Yegen; Interim Dir., Rebecca Baken.
Personnel Profile: Full-Time Paid 1; Part-Time Volunteers 6.
Governing Authority: Yellowstone County. Tax-exempt.
Institution Type/Description: Montana & Western History Museum: portion of which is housed in an 1890s log structure belonging to Montana pioneer Paul McCormick.
Collections: Native American collection representing 27 tribes-includes beadwork, quillwork, clothing, toys, games, utilitarian items; sheep & cattle ranching industries; cowboy memorabilia; natural history; works by western artists, J.K. Ralston & LeRoy Greene; transportation artifacts, railroad, wagons; military artifacts, weaponry, uniforms, personal artifacts from the War of 1812 to the Korean War; textiles including afghans, quilts, handiwork, clothing 1840s-1950s; women's personal items; everyday household goods; occupational artifacts, medical & dental tools; prehistoric tools; photographs.
Major Exhibits: Charting the New Frontier: Mapping & Surveying in Yellowstone County, 1/2014-12/2014.
Research Fields: photo archives; internships; mentoring programs.
Facilities: 6,000 sq. ft. exhibit space; research center.
Activities: fundraising events; one feature exhibit per year; cattle drive & parade.
Publications: quarterly newsletter.
Hours & Admission Prices: Tues.-Sat. 10:30-5:30. No charge. Closed Easter; Thanksgiving; Christmas.
Attendance: 20,119 (accurate)
Membership: Senior $15; Individual $20; Family $25; Corporate $100; Lifetime $1,000.

ZOOMONTANA, 2100 S. Shiloh Rd., Billings, MT 59106-3908. Tel.: 406-652-8100. Fax: 406-652-9281. Facebook: ZooMontana.
E-mail: director@zoomontana.org
Web Site: www.zoomontana.org
Founded: 1993.
Congressional District: 111
Key Personnel: Dir., Jeff Ewelt; Education, Troy Paisley; Cur., Travis Goebel; Museum Shop Mgr., Katie Leger.
Personnel Profile: Full-Time Paid 11; Part-Time Paid 4; Part-Time Volunteers 150; Interns 5.
Volunteer Hours: 7,500
Operating Expenses: 835,878
Operating Income: 888,000
Governing Authority: private; nonprofit organization. Tax-exempt: 501(c)(3).
Institution Type/Description: Zoo.
Collections: native & exotic animals from the northern regions of the world; medicinal, exotic & native Montana plants.
Research Fields: restoration of wetlands and PZP vaccine.
Facilities: classroom; nature center; Botanical Garden; zoological park; nature trails; Science and Conservation Center. Museum-related items for sale.
Activities: concerts; docent program; formal education program; children's barn & garden; internships; guided tours; lectures; weddings; company picnics. Annual Events: Zoophoria in February; Roof, A Wild Affair in April; Highland Games in June; Zoograss in July; Boo At The Zoo in October; Chowder Challenge in October; Zoo Lights in December.
Publications: quarterly newsletter, Zoo Montana.
Hours & Admission Prices: May-Sept. daily 10-4; Oct.-April daily 10-2. Adults $7, senior citizens & military $5, children $4; discounts to groups of 12 or more & AZA members; members no charge. Closed Thanksgiving; Christmas. &
Attendance: 77,000 (estimated)
Membership: Individual $50; Family $75; Household $150.

Bozeman

AMERICAN COMPUTER MUSEUM, 2023 Stadium Drive - Unit 1-A, Bozeman, MT 59715-. Tel.: 406-582-1288. Fax: 406-587-9620.
E-mail: director@compustory.com
Web Site: www.compustory.com
Personnel Profile: Full-Time Volunteers 1; Part-Time Paid 2; Part-Time Volunteers 2.
Governing Authority: Tax-exempt.
Institution Type/Description: Computer Museum.
Collections: history of the information age; period office equipment including typewriters, Arithmometers, adding machines, furniture, electric fans; mainframe computers; printing press; telegraph; Apple computer history; programming; Si-Fi.
Facilities: theater.
Activities: tours; video presentation.
Hours & Admission Prices: June-Aug. daily 10-4; Sept.-May Tues.-Sun. 12-4. No charge; donations accepted. Closed New Year's Day; Easter; Thanksgiving; Christmas. &

CHILDREN'S MUSEUM OF BOZEMAN, 202 S. Willson Ave., Bozeman, MT 59715-4631. Tel.: 406-522-9087.
E-mail: info@cmbozeman.org
Web Site: www.cmbozeman.org
Founded: 2001.
Key Personnel: Exec. Dir., Susan Denson-Guy; Pres. (V), Marisa Bueno.
Personnel Profile: Full-Time Paid 4; Part-Time Paid 6; Part-Time Volunteers 100; Interns 1.
Institution Type/Description: Children's Museum.
Collections: hands-on exhibits.
Activities: special events; educational programs; parties.
Hours & Admission Prices: Mon.-Thurs. & Sat. 10-5, Fri. 10-8; groups of 12 or more by appointment. Admission $5; discounts to groups & ACM members; Fri. 5-8pm no charge.
Attendance: 21,187 (accurate)
Membership: Family $60; Family Plus $80; ACM $110.

GALLATIN HISTORICAL SOCIETY, 317 W. Main St., Bozeman, MT 59715-4576. Tel.: 406-522-8122. Fax: 406-522-0367.
E-mail: director@pioneermuseum.org
Web Site: www.pioneermuseum.org
Key Personnel: Exec. Dir., John C. Russell; Research Coord., Rachel Phillips.
Personnel Profile: Full-Time Paid 2; Part-Time Paid 1; Part-Time Volunteers 50.
Institution Type/Description: Historical Society Museum.

Collections: Gallatin Valley history & culture; Native American artifacts; farm & ranch tools; period automobiles; household items.
Activities: educational programs.
Publications: quarterly, Pioneer Museum.
Hours & Admission Prices: Memorial Day to Labor Day Mon.-Sat. 10-5; Winter: Tues.-Sat. 11-4. Adults $5; school groups, children 12 & under and members no charge. &
Attendance: 8,000
Membership: $40 & up.

HELEN E. COPELAND GALLERY, MSU School of Art, 242 Haynes Hall, Bozeman, MT 59717-3700. Mailing Address: School of Art, P.O. Box 173680, Bozeman, MT 59717-3680. Tel.: 406-994-2562. Fax: 406-994-3680. Facebook: Helen E. Copeland Gallery.
E-mail: hecgallery@gmail.com
Web Site: hecg.wordpress.com
Founded: 1974.
Key Personnel: Dir., Vaughan Judge.
Governing Authority: university. Parent Institution: Montana State University. Tax-exempt.
Institution Type/Description: Art Gallery.
Collections: print collections; international craft items; ceramics.
Activities: lectures; gallery talks; arts festivals; symposiums visiting artists; loan, temporary & traveling exhibitions.
Publications: brochures; catalogues.
Hours & Admission Prices: Mon.-Fri. 9-5. No charge. Closed state & federal holidays. &
Attendance: 5,000

* **MUSEUM OF THE ROCKIES, (M),** 600 W. Kagy Blvd., Bozeman, MT 59717-2730. Tel.: 406-994-2251. Fax: 406-994-2682.
E-mail: museum@montana.edu
Web Site: museumoftherockies.org
Founded: 1956.
Congressional District: 1
Key Personnel: Exec. Dir., Sheldon McKamey; Pres. Bd., Pat Downey; Cur. Paleontology, John R. Horner; Asst. Dir. & Cur. Art & Photography, Steven B. Jackson; Dir. Devel., Candace Strauss; Cur. History, Michael Fox; Dir. Exhibits, Patrick Leiggi; Mgr. Planetarium, Eric Loberg; Registrar, Pat Roath; Dir. Finance, Jeff Krauss; Museum Shop Mgr., Christine Stoppa.
Personnel Profile: Full-Time Paid 30; Part-Time Paid 6; Part-Time Volunteers 260.
Governing Authority: university. Parent Institution: Montana State University-Bozeman. Tax-exempt: 501(c)(3).
Institution Type/Description: General Regional Museum.
Collections: geology; astronomy; paleontology; history; Western art; living history.
Research Fields: paleontology.
Facilities: photo archives; planetarium; auditorium; classrooms; laboratories; nature area; 22-acre site owned by university. Museum-related items for sale.
Activities: guided tours; classes; field schools; field trips; lectures. Museum Sponsors: Gallery Openings.
Publications: Maia-A Dinosaur Grows Up; How to Build a Dinosaur; Dinosaurs Under the Big Sky; Digging up Dinosaurs; MOR News; monthly e-newsletter; triannual members calendar of events.
Hours & Admission Prices: Memorial Day-Labor Day daily 8-8; Sept.-May Mon.-Sat. 9-5, Sun. 12:30-5. Museum: adults $13, senior citizens 65 & over and military $10, MSU students & children 5-18 $9; discounts to AAA members; children 4 & under, members no charge. Planetarium $3. Closed New Year's Day; Thanksgiving; Christmas. &
Attendance: 163,037
Membership: Individual $30; Dual $50; Family & Grand Family $75; Sustaining $250; Director's Circle $500; Director's Guild $1,000.

Broadus

POWDER RIVER HISTORICAL MUSEUM AND MAC'S MUSEUM, 102 W. Wilson, Broadus, MT 59317. Mailing Address: P.O. Box 573, Broadus, MT 59317-0573. Tel.: 406-436-2977.
Founded: 1988.
Key Personnel: Dir., E. Lee Hubbard; Pres. (V), Ron Talcott.
Personnel Profile: Part-Time Paid 1.
Volunteer Hours: 1,529
Governing Authority: Tax-exempt.
Institution Type/Description: Local History Museum.
Collections: Powder River: photographs; books; Native American artifacts;

guns & ammunition; period automobiles; tractors & farm implements; military. Mac's Museum: arrowheads; birds' eggs; butterflies; geologic specimens; sea shells; vintage clothing.
Research Fields: local family files.
Facilities: school house; pioneer cabin & jail house.
Hours & Admission Prices: June-Sept. Mon.-Fri. 9-5; other times by appointment. No charge; donations accepted. &
Membership: Annual $10; Lifetime $50.

Browning

BLACKFEET HERITAGE CENTER & ART GALLERY, 333 Central Ave. W., Browning, MT 59417. Mailing Address: P.O. Box 1629, Browning, MT 59417. Tel.: 406-338-5661.
Web Site: www.blackfeetnationstore.com
Institution Type/Description: Native American History & Art Center.
Collections: Blackfeet history & culture; personal artifacts; paintings; sculpture.
Activities: art demonstrations.
Hours & Admission Prices: Summer: daily. No charge.

MUSEUM OF THE PLAINS INDIAN, 19 Museum Loop, Browning, MT 59417. Mailing Address: P.O. Box 410, Browning, MT 59417-0410. Tel.: 406-338-2230. Fax: 406-338-7404.
E-mail: mpi@3rivers.net
Web Site: www.iacb.doi.gov
Founded: 1941.
Congressional District: 14
Key Personnel: Cur., David Dragonfly.
Personnel Profile: Full-Time Paid 2.
Governing Authority: federal. Administered & operated by Indian Arts & Crafts Board, U.S. Dept. of Interior, Room 4004, Washington, DC 20240. Tax-exempt.
Institution Type/Description: Indian Art Museum.
Collections: historic, contemporary, social & ceremonial arts of the Northern Plains Indians; traditional costumes; painted tepees; murals; monument dedicated to a 1930 sign language conference.
Research Fields: historic & contemporary Native American arts of the United States.
Activities: guided tours by appointment; lectures; gallery talks; permanent, temporary & changing exhibitions.
Publications: exhibition brochures.
Hours & Admission Prices: June-Sept. daily 9-4:45; Oct.-May Mon.-Fri. 10-4:30. Adults $4, seniors $3, children $1; discounts to groups of 10 or more; children under 6 no charge. Closed New Year's Day; Thanksgiving; Christmas. &
Attendance: 25,000 (estimated)

Butte

BUTTE-SILVER BOW ARTS CHATEAU, 321 W. Broadway, Butte, MT 59701-9126. Tel.: 406-723-7600.
Web Site: www.artschateau.org
Founded: 1977.
Key Personnel: C.E.O., Glenn Bodish; Pres., Gretchen Miller.
Personnel Profile: Full-Time Paid 1; Part-Time Paid 5; Part-Time Volunteers 5.
Governing Authority: nonprofit organization. Parent Institution: Butte-Silver Bow Arts Foundation. Subsidiary Institution: Arts Chateau. Tax-exempt: 501(c)(3).
Institution Type/Description: Period Museum & Arts Center: housed in 1898 The Charles Clark Mansion.
Collections: contemporary regional art; Western art by Elizabeth Lochrie; dolls; vintage clothing; over 600 bells; period furniture collection.
Research Fields: local folklore.
Facilities: period museum; multi-purpose room. Museum-related items & original works by Montana artists for sale.
Activities: gallery talks; guided tours; lectures; slide program; workshops; art events; permanent & temporary art exhibits; history course; performing arts events. Museum Sponsors: Textile medium, Contemporary Quilt Exhibition & Competition; Satellite for Very Special Arts, Montana.
Publications: cookbook, Butte's Heritage Cookbook (ninth printing); newsletter; schedule of events.
Hours & Admission Prices: June-Aug. Tues.-Sun. 12-5; Sept.-May Tues.-Sat. 11-4, Sun. 12-5. Family $10, adults $3, AAA, & CAA members $2.50, senior citizens 50 & up $2, children 16 & under $1; discounts to AAM members & groups of 10 or more; members no charge. Closed major holidays.
Attendance: 4,900 (accurate)
Membership: Seniors & Students $10; Individual $15; Family $25; Contrib-

uting $50; Sustaining $100; Patron $500; Benefactor $1,000. Business: Contributing $100; Sustaining $500; Patron $750; Benefactor $1,000.

COPPER KING MANSION, 219 W. Granite St., Butte, MT 59701-9235. Tel.: 406-782-7580.
E-mail: esigl@in-tch.com
Web Site: www.thecopperkingmansion.com
Founded: 1966.
Congressional District: 2
Key Personnel: Gen. Mgr., John Thompson.
Personnel Profile: Full-Time Paid 1; Full-Time Volunteers 1; Part-Time Paid 10; Part-Time Volunteers 1.
Governing Authority: partnership.
Institution Type/Description: Historic House: 1884-1888, mansion built by William Andrews Clark including period furnishings.
Collections: Victorian furniture; crystal; cut glass; silver; porcelains; art.
Research Fields: local history.
Facilities: library of old literature, novels, encyclopedias available on premises only; catered parties, dinners & weddings. Arts & craft items for sale. Bed & Breakfast year round.
Activities: guided tours; lectures; gallery talks.
Hours & Admission Prices: May-Sept. daily 9-4; Oct.-April by appointment. Adults $7, students 6-18 $3.50; children under 6 no charge.
Attendance: 10,000 (estimated)

MAI WAH SOCIETY, 17 W. Mercury St., Butte, MT 59701-2019. Mailing Address: P.O. Box 404, Butte, MT 59703-0404. Tel.: 406-723-3231.
E-mail: info@maiwah.org
Web Site: www.maiwah.org
Personnel Profile: Part-Time Paid 1; Part-Time Volunteers 20.
Governing Authority: nonprofit organization. Tax-exempt.
Institution Type/Description: Asian Heritage Museum.
Collections: history & culture of Asian people in the Rocky Mountain West; photographs; general mercantile wares.
Facilities: Museum-related items for sale.
Activities: lectures; workshops; movies. Museum Sponsors: Chinese New Year Parade in February.
Hours & Admission Prices: late May to Sept. Tues.-Sat. 11-5. Adults $5; members no charge.
Attendance: 500 (estimated)
Membership: $25; $50; $100.

MINERAL MUSEUM, Montana Tech. of the University of Montana, 1300 W. Park St., Butte, MT 59701-8932. Tel.: 406-496-4414 & 4743. Fax: 406-496-4451.
E-mail: kscarberry@mtech.edu
Web Site: www.mbmg.mtech.edu/museum.htm
Founded: 1900.
Congressional District: 1
Key Personnel: Cur., Dr. Kaleb Scarberry; Asst. Cur., John Foley.
Personnel Profile: Full-Time Paid 2; Part-Time Paid 1.
Governing Authority: state. Parent Institution: Montana Tech. of the University of Montana. Subsidiary Institution: Montana Bureau of Mines & Geology. Tax-exempt: 501(c)(3).
Institution Type/Description: Mineralogy Museum.
Collections: 1,300 mineral specimens from around the world on display; fossils; relief map of Montana.
Research Fields: mineralogy.
Facilities: Museum-related items for sale.
Activities: guided tours; workshops.
Hours & Admission Prices: June 15-Sept. 15 daily 9-5; Winter: Mon.-Fri. 9-4. No charge; donations accepted. Closed state holidays. &
Attendance: 11,000 (estimated)

PICCADILLY TRANSPORTATION MEMORABILIA MUSEUM, 20 W. Broadway, Butte, MT 59701-9222. Tel.: 406-723-3034. Fax: 406-723-7425.
E-mail: info@piccadillymuseum.com
Web Site: www.piccadillymuseum.com
Institution Type/Description: Transportation & Advertising Art Museum.
Collections: transportation memorabilia from over 100 countries including highway & subway markers, license plates, vintage cars, advertising art.
Activities: special events.
Hours & Admission Prices: Memorial Day to Oct. 1 Mon.-Sat. 10-5; Fall & Winter: by appointment only. Suggested Donation: $3.

WORLD MUSEUM OF MINING, 155 Museum Way, Butte, MT 59703-0033. Mailing Address: P.O. Box 33, Butte, MT 59703-0033. Tel.: 406-723-7211. Fax: 406-723-2398. Facebook: World Museum of Mining.
E-mail: info@miningmuseum.org
Web Site: www.miningmuseum.org
Founded: 1963.
Congressional District: 1
Key Personnel: Dir., Tina Davis; Pres. (V), George Everett; Sec., Courtney McKee.
Personnel Profile: Full-Time Paid 2; Part-Time Paid 9; Part-Time Volunteers 25.
Governing Authority: nonprofit corporation. Tax-exempt: 501(c)(3).
Institution Type/Description: Mining Museum: located on the site of the Orphan Girl Mine, an early day zinc & silver mine, once owned by Butte Copper King Marcus Daly.
Collections: mining & ethic history of Butte and the surrounding region.
Research Fields: mining history, social & cultural history of Butte.
Facilities: 33 acres of grounds; 50 buildings with exhibits; picnic area; extensive photo archives available for research & purchase. Museum-related items for sale.
Activities: gold panning; scavenger hunt; underground mine tour.
Publications: book, The Orphan Girl Mine; Remembering Butte, Montana's Richest City (photobook about Butte's history).
Hours & Admission Prices: April-Oct. daily 9-5:30, last ticket sold at 4:30, grounds close at 5:30. Adults 19 & over $8.50, seniors $7.50, children 13-18 $6, children 5-12 $3; discounts to AAM, AAA & ICOM members, military and school groups; members and children 4 & under no charge. Underground Tours: adults $12, seniors $10, students $8, children $5.
Attendance: 25,000 (accurate)
Membership: Golden Age $15; Individual $25; Family $40; Nipper $60. Corporate: Copper $150-$499; Silver $500-$999; Gold $1,000-$4,999; Platinum $5,000-$9,999; Palladium $10,000.

Bynum

TWO MEDICINE DINOSAUR CENTER, 120 2nd Ave. S., Bynum, MT 59419. Mailing Address: P.O. Box 786, Bynum, MT 59419-0786. Tel.: 406-469-2211; 800-238-6873.
E-mail: dinoinfo@tmdinosaur.org
Web Site: www.tmdinosaur.org
Institution Type/Description: Research and Education Institution.
Collections: dinosaur exhibits; fossils.
Hours & Admission Prices: May & mid-Sept. to Sept. 30 Wed.-Sun. 10-5; June to mid-Sept. daily 9-6.

Charlo

NINEPIPES MUSEUM OF EARLY MONTANA, 69316 Hwy. 93, Charlo, MT 59824-9789. Tel.: 406-644-3435.
E-mail: ninepipesmuseum@yahoo.com
Web Site: ninepipesmuseum.org
Founded: 1997.
Key Personnel: Dir., Vern Cheff, Jr.; Chm. & Pres. (V), Rod Wamsley; Admin. & Museum Shop Mgr., Laurel Cheff.
Personnel Profile: Full-Time Volunteers 4; Part-Time Paid 1; Part-Time Volunteers 16.
Governing Authority: bd. of directors. Tax-exempt.
Institution Type/Description: History Museum: named after Chief Joseph Ninepipes, a Bitterroot Salish Chief.
Collections: Montana history; Native American artifacts; photographs; paintings; documents; mounted wildlife; Indian camp replica.
Publications: quarterly newsletter.
Hours & Admission Prices: March 1- May 1 & Sept. 2-Nov. 30 Thurs.-Sat. 11-3; May 2-Sept. 1 Mon.-Sat. 10-5. Adults $5, students $3, children 6-12 $2; discounts to seniors & groups of 20 or more; members & children under 6 no charge. Closed Easter; Thanksgiving; Christmas. &
Attendance: 5,500 (estimated)
Membership: Student $15; Individual $20; Family $35; Sustaining $100; Donor $500; Lifetime $1,000.

Chester

LIBERTY COUNTY MUSEUM, 230 Second St. E., Chester, MT 59522. Mailing Address: P.O. Box 417, Chester, MT 59522-0417. Tel.: 406-759-5256.
Founded: 1969.
Key Personnel: Chm. (V), Betty L. Marshall.
Personnel Profile: Part-Time Paid 4; Part-Time Volunteers 7.

Governing Authority: nonprofit. Tax-exempt.
Institution Type/Description: General Museum.
Collections: agriculture; archaeology; costumes; Indian artifacts; outdoor museum; one-room schoolhouse; old machinery; military & medical display; mock-trading company store; Great Northern R.R. display; working Blacksmith Shop.
Publications: book, Our Heritage.
Hours & Admission Prices: May to Labor Day daily 1-5 & 7-9. No charge; donations accepted.
Attendance: 500 (estimated)

LIBERTY VILLAGE ARTS CENTER & GALLERY, 410 W. Main St., Chester, MT 59522-0269. Mailing Address: P.O. Box 269, Chester, MT 59522-0269. Tel.: 406-759-5652. Fax: 406-759-5652.
E-mail: lvac@mtintouch.net
Web Site: libertyvillagearts.org
Founded: 1976.
Congressional District: 2
Key Personnel: Treas. (V), Laurie Lyders; Museum Shop Mgr., Diana Twedt.
Personnel Profile: Part-Time Paid 3; Part-Time Volunteers 11; Interns 1.
Governing Authority: state; nonprofit. Tax-exempt: 501(c)(3); 170(b)(1)(A).
Institution Type/Description: Folk Arts Center: housed in 1910 church building.
Collections: multi-media; sculpture & paintings.
Facilities: classrooms. Paintings, prints, pottery & jewelry for sale.
Activities: lectures; films; gallery talks; arts festivals; rental gallery; temporary & traveling exhibitions.
Hours & Admission Prices: mid-Jan. to Dec. Tues.-Wed. & Fri. 12:30-4:30, Thurs. 12:30-8. No charge; donations accepted. Closed New Year's Day; Easter; Independence Day; Thanksgiving; Christmas. &
Attendance: 3,500 (estimated)
Membership: Individual $25; Family $30; Friend of the Arts $50; Patron of the Arts $100; Benefactor $250 & up.

Chinook

BLAINE COUNTY MUSEUM, 501 Indiana, Chinook, MT 59523. Mailing Address: P.O. Box 927, Chinook, MT 59523-0927. Tel.: 406-357-2590. Fax: 406-357-2199.
E-mail: blmuseum@mtintouch.net
Web Site: www.chinookmontana.com
Founded: 1977.
Congressional District: 2
Key Personnel: Dir. & Cur., Jude Sheppard; Chm. (V), Stuart C. MacKenzie.
Personnel Profile: Full-Time Paid 1; Part-Time Paid 1; Part-Time Volunteers 3.
Governing Authority: Blaine County. Tax-exempt.
Institution Type/Description: History Museum: housed in 1915 former recreation center.
Collections: local history; multi-media presentation on the Nez Perce Indians & the Battle of the Bear's Paw; paleontology; old west artifacts; Indian culture; one room school; doctor's & dentist's office; Tar paper Homestead Shack; photographs.
Activities: guided tours.
Hours & Admission Prices: May & Sept. Mon.-Fri. 8-5; June-Aug. Mon.-Sat. 8-5, Sun. 12-5; Oct.-April Mon.-Fri. 1-5. No charge; donations accepted. &
Attendance: 4,350 (accurate)

Choteau

OLD TRAIL MUSEUM, 823 N. Main St., Choteau, MT 59422-9272. Tel.: 406-466-5332.
E-mail: otm@3rivers.net; oldtrail2@gmail.com
Web Site: oldtrailmuseum.org
Founded: 1986.
Congressional District: 11
Key Personnel: Chm. & Pres. (V), Polly Cunningham; Dir. & Museum Shop Mgr., Julie Ameline; Operations, Dave Wedum.
Personnel Profile: Part-Time Paid 2; Part-Time Volunteers 2.
Volunteer Hours: 208
Governing Authority: individual operation. Parent Institution: Chicago Field Museum. Subsidiary Institution: Project Exploration. Tax-exempt.
Institution Type/Description: Village Museum: housed in 1920 six building cleaning plant.
Collections: period artifacts; Indian artifacts; fossils; dinosaur exhibit; Metis cabin exhibit; schoolhouse; grizzly cabin; paleontology antechamber; early art studio; ice cream parlor. Historic Structures: two 1890 log cabins; stagecoach stop.
Major Exhibits: Blackfeet Tipi, 5/12-9/14.
Research Fields: fossils; Native American heritage.

Facilities: Victorian ice-cream parlor; art studio; bear cabin; school house; Metis cabin.
Activities: guided & self-guided tours; study clubs for children; loan, permanent & temporary exhibitions.
Publications: brochures; pamphlets; newsletter.
Hours & Admission Prices: Memorial Day to Labor Day daily 9-5. Admission $2; members no charge. &
Attendance: 3,000 (estimated)
Membership: Individual $20; Family $30; Sponsor $50; Sustaining $100; Life $1,000.

Circle

MCCONE COUNTY MUSEUM, 1507 Ave. B, Circle, MT 59215-0127. Mailing Address: P.O. Box 127, Circle, MT 59215-0127. Tel.: 406-485-2414.
Founded: 1953.
Congressional District: 2
Key Personnel: Pres. & Cur., Wendell Pawlowski.
Personnel Profile: Full-Time Paid 1; Part-Time Volunteers 4.
Governing Authority: nonprofit organization. Tax-exempt: 501(c)(3).
Institution Type/Description: History Museum.
Collections: history; county newspapers, 1914-present; animal husbandry; geology; Indians; wildlife exhibit; dinosaurs.
Research Fields: local history.
Facilities: reading room.
Activities: permanent exhibitions. Museum Sponsors: Open House.
Publications: books, Pioneers & Progress on the Prairie; The Depression Years: 1930-1939; Circle-Then & Now.
Hours & Admission Prices: May-Oct. 1 Mon.-Fri. 10-5, Sat.-Sun. call for hours. Adults $2. &
Membership: Adults $5.

Clancy

JEFFERSON COUNTY MUSEUM, 9 N. Main, Clancy, MT 59634-9547. Mailing Address: P.O. Box 50, Clancy, MT 59634-0050. Tel.: 406-933-5463. Fax: 406-933-5439.
Institution Type/Description: History Museum: housed in the 1898 Clancy Schoolhouse.
Collections: items pertaining to Jefferson County; exhibits on mining, ranching & railroading.
Hours & Admission Prices: Fri.-Sat. 1-5; other times by appointment. No charge; donations accepted.

Colstrip

SCHOOLHOUSE HISTORY & ART CENTER, 400 Woodrose St., Colstrip, MT 59323. Mailing Address: P.O. Box 430, Colstrip, MT 59323-0430. Tel.: 406-748-4822. Facebook: Schoolhouse History and Art Center.
E-mail: shac400@gmail.com
Web Site: www.colstripshac.org
Founded: 1996.
Key Personnel: Exec. Dir. & Museum Shop Mgr., Lu Shomate; Pres. (V), Tom Crippen.
Personnel Profile: Full-Time Paid 1; Part-Time Paid 2.
Governing Authority: Tax-exempt.
Institution Type/Description: History & Art Center: housed in former schoolhouse built in 1924.
Collections: period artifacts; furnishings; photographs.
Facilities: Museum-related items for sale.
Activities: children's activities; educational programs; art classes.
Hours & Admission Prices: Mon.-Fri. 11-5. No charge; donations accepted. &
Attendance: 2,500 (estimated)

Columbus

MUSEUM OF THE BEARTOOTHS, 440 E. 5th Ave. N., Columbus, MT 59019. Mailing Address: P.O. Box 1, Columbus, MT 59019. Tel.: 406-322-4588. Fax: 406-322-4588.
E-mail: predli@museumofthebeartooths.com
Web Site: www.museumofthebeartooths.com
Founded: 1993.
Key Personnel: Dir., Penny Redli; Pres. (V), Chuck Egan.
Personnel Profile: Full-Time Paid 1; Part-Time Volunteers 26.
Governing Authority: Parent Institution: Stillwater Historical Society. Tax-exempt.
Institution Type/Description: History Museum.

Collections: local history & culture; personal artifacts; photographs; period furnishings.
Facilities: Museum-related items for sale.
Hours & Admission Prices: April-May & Sept.-Dec. Mon.-Fri. 10-5; Memorial Day to Labor Day Mon.-Fri. 10-5, Sat.-Sun. 11-3. No charge; donations accepted. &
Attendance: 3,752 (accurate)

Conrad

CONRAD TRANSPORTATION AND HISTORICAL MUSEUM, 402 S. Virginia St., Conrad, MT 59425-2318. Mailing Address: P.O. Box 675, Conrad, MT 59425-0675. Tel.: 406-278-0178. Facebook: Conrad Transportation and Historical Museum.
E-mail: hbolson@3rivers.net
Web Site: www.conradmuseum.org
Founded: 1963.
Key Personnel: Dir. & Pres. (V), Harold Olson; Museum Shop Mgr., Charlotte Hovde.
Personnel Profile: Full-Time Volunteers 1; Part-Time Paid 1; Part-Time Volunteers 30; Interns 1.
Volunteer Hours: 4,250
Operating Expenses: 14,838
Operating Income: 17,084
Governing Authority: Parent Institution: Pondera History Assoc. Tax-exempt.
Institution Type/Description: Transportation & Historical Museum.
Collections: local history & culture; early cash register; period autos & steering wheels; license plates; petroleum products; gas pumps; oil dispensers; oil cans; auto accessories; petro memorabilia; drug store; blacksmith shop; homestead; business artifacts.
Major Exhibits: Local Notorious Outlaws, 7/14-8/14; Women's Suffrage, 7/14-8/14.
Activities: Museum Sponsors: October-Fest; Christmas Stroll.
Publications: biannual, Transportation Times.
Hours & Admission Prices: May-Oct. Mon.-Fri. 10-4, Sat. 1-4; Nov.-April by appointment. Adults $3 senior citizens $2; discounts to AAM members, MT Association of Museums & school groups; members no charge. Closed Independence Day. &
Attendance: 619 (accurate)
Membership: Individual $20; Family $25.

Crow Agency

LITTLE BIGHORN BATTLEFIELD NATIONAL MONUMENT, Interstate 90 & Hwy. 212, Crow Agency, MT 59022. Mailing Address: P.O. Box 39, Crow Agency, MT 59022-0039. Tel.: 406-638-3204. Fax: 406-638-2623.
E-mail: sharon_small@nps.gov
Web Site: www.nps.gov/libi
Founded: 1940.
Congressional District: 2
Key Personnel: Museum Cur., Sharon A. Small; Supt., Denice Swanke; Business Mgr., Charles Carroll.
Personnel Profile: Full-Time Paid 1; Part-Time Paid 2; Part-Time Volunteers 2; Interns 2.
Governing Authority: federal. Administered by National Park Service, Washington, DC. Tax-exempt.
Institution Type/Description: History Museum.
Collections: Indian & military items, manuscripts, photographs & documents associated with Gen. George A. Custer, the Battle of Little Big Horn & the Sioux War, 1876-90; Northern Plains Indians, Crow, Arikara, Lakota Sioux, Cheyenne, & Arapaho; Seventh U.S. Cavalry.
Research Fields: Battle of Little Big Horn; George A. Custer; Plains Indian Wars.
Facilities: library & archives; battlefield grounds; visitor center. Publications related to the battle of the Little Big Horn & the Indian Wars for sale.
Activities: permanent & temporary exhibitions; interpretive services, including historical talks, walks, demonstrations, self-guiding trail, park tour road & tape recorded audio stations; guided bus tours.
Publications: books.
Hours & Admission Prices: April-May & Sept.-Oct. daily 8-6; Memorial Day to July daily 8am-9pm; Aug. to Labor Day daily 8-8; Nov.-March daily 8-4:30. Vehicle $10, Individual $5; discounts to Golden Age & Annual pass holders. Closed New Year's Day; Thanksgiving; Christmas. &
Attendance: 350,000 (estimated)

Culbertson

CULBERTSON MUSEUM INC., 5860 Rd. 1021, Culbertson, MT 59218. Mailing Address: P.O. Box 95, Culbertson, MT 59218-0095. Tel.: 406-787-6320 & 5283. Fax: 406-787-6320.
Formerly: Culbertson Museum
Founded: 1990.
Key Personnel: Dir., Suzette Houle; Pres., Bruce Waldhausen; Treas., Jill Herness; Dir. Volunteers, Bernadette Raaum.
Personnel Profile: Part-Time Volunteers 30.
Governing Authority: private; nonprofit organization. Tax-exempt: 501(c)(3).
Institution Type/Description: History Museum.
Collections: pioneer & homesteader history in northeastern Montana & the Culbertson area, from the 1880s to the depression era & WWII; artifacts; photographs; cowboy, pioneer & homesteader artifacts; chuckwagon; one horse sleigh; buggy charcoal; gas domestic irons; period dishes; medical equipment; wildflower garden.
Research Fields: family histories.
Facilities: 50-vol. library of pictorial Montana area books, county & community history books; 6,000 sq. ft. exhibit space; garden.
Activities: guided tours; special events. Museum Sponsors: school tours in May & September; Birthday Bash; Chuckwagon Meal; County Fair Educational booth; Frontier Days parade float; Pie Social for Annual Threshing Bee.
Publications: annual newsletter; brochure.
Hours & Admission Prices: May-Sept. daily 8-4. No charge; donation requested. &
Attendance: 4,800 (estimated)
Membership: Individual $5; Family $15; Business $35.

Cut Bank

GLACIER COUNTY HISTORICAL SOCIETY & MUSEUM, 107 Old Kevin Hwy., Cut Bank, MT 59427. Mailing Address: P.O. Box 576, Cut Bank, MT 59427-0576. Tel.: 406-873-4904.
E-mail: gcmuseum@sofast.net
Web Site: www.glaciercountymt.org/museum
Founded: 1980.
Key Personnel: Pres. (V), Ken Finstad; Dir., Dennis Seglem
Governing Authority: Tax-exempt.
Institution Type/Description: Historical Society Museum.
Collections: Glacier County history & cultural heritage.
Hours & Admission Prices: Memorial Day to Labor Day Tues.-Sun. 10-5; other times by appointment. No charge; donations accepted. &
Membership: $25 & up.

Darby

DARBY PIONEER MEMORIAL MUSEUM, 101 E. Tanner, Darby, MT 59829. Mailing Address: P.O. Box 37, Darby, MT 59829-0037. Tel.: 406-821-3753 & 3748. Fax: 406-821-3244.
E-mail: darbymontana@usa.net
Web Site: darbymt.net
Founded: 1917.
Personnel Profile: Part-Time Volunteers 7.
Volunteer Hours: 117
Operating Expenses: 1,076
Operating Income: 1,880
Governing Authority: Parent Institution: Town of Darby. Tax-exempt.
Institution Type/Description: History Museum.
Collections: local history & culture; photographs & memorabilia of Darby and it's people; home & business artifacts.
Hours & Admission Prices: June-Sept. daily 12-4. No charge; donations requested. &
Attendance: 556 (accurate)

Deer Lodge

GRANT-KOHRS RANCH NATIONAL HISTORIC SITE, 266 Warren Lane, Deer Lodge, MT 59722-1002. Tel.: 406-846-2070, ext. 242. Fax: 406-846-3962.
E-mail: chris_ford@nps.gov
Web Site: www.nps.gov/grko/home.htm
Founded: 1972.
Congressional District: 1
Key Personnel: Chm. (V), Sales Shop Mgr. & Chief Interpretation, Julie Croglio; Facility Mgr., Alan Stewart; Admin. Officer, Anita Dore; Cur., Chris Ford; Museum Tech., Peggy Gow.
Personnel Profile: Full-Time Paid 11; Full-Time Volunteers 2; Part-Time Paid 8; Part-Time Volunteers 114; Interns 6.
Governing Authority: federal. U.S. Dept. of the Interior, National Park Service, Washington, DC 20240. Tax-exempt.

Institution Type/Description: National Historic Site: 1862-1890 23-room ranch house & 1861-1960 bunkhouse, barns & outbuildings.
Collections: 19th & 20th-century ranch equipment; horse drawn vehicles; blacksmith equipment; tack; bunkhouse furnishings; Victorian household furnishings; textiles; archives.
Research Fields: frontier cattle ranching.
Facilities: library on cattle ranching & household arts, primarily of the 19th century, available for use on premises. Postcards & books for sale.
Activities: guided tours; cattle ranching; draft horses; formally organized education programs for children.
Hours & Admission Prices: Memorial Day-Labor Day daily 9-5:30; Sept.-May daily 9-4:30. No charge; donations accepted. &
Attendance: 37,000 (accurate)

OLD MONTANA PRISON MUSEUMS, 1106 Main St., Deer Lodge, MT 59722-1426. Tel.: 406-846-3111. Fax: 406-846-3156.
E-mail: info@pcmaf.org
Web Site: pcmaf.org
Formerly: Old Montana Prison
Founded: 1980.
Congressional District: 1
Key Personnel: Dir., John O'Donnell; Pres., Dick Bauman.
Governing Authority: nonprofit organization. Parent Institution: Powell County Museum & Arts Foundation. Tax-exempt: 501(c)(3).
Institution Type/Description: Historic Monument & Complex: 1867 site, comprised of 12 structures surrounded by 1893 sandstone wall, serving as a territorial & state prison from 1871-1979.
Collections: contraband exhibits & display areas; photographs; archives; 1900-1980s vintage cars. Historic Building: 1912 cell house; original Montana territorial prison.
Research Fields: prison history.
Facilities: administrative building; prison yard; prison-related structures.
Activities: guided & self-guided tours; permanent & temporary exhibitions.
Publications: brochures; pamphlets; books.
Hours & Admission Prices: Wed.-Sun. 10-4. Adults $10, children 10-15 $6; discounts to groups, senior citizens, AAA & Good Sam members; members & children 9 & under no charge. &
Attendance: 52,600 (accurate)
Membership: Senior Citizen $15; Individual $20; Family $30; Commercial $75; Museum Patron $130.

POWELL COUNTY MUSEUM, 1106 Main St., Deer Lodge, MT 59722-1489. Tel.: 406-846-3111. Fax: 406-846-3156.
E-mail: info@pcmaf.org
Web Site: www.pcmaf.org
Founded: 1964.
Congressional District: 1
Key Personnel: Pres., Dick Bauman; Dir., John O'Donnell; Business Mgr., Julia Smith; Sec., Ed Gill.
Personnel Profile: Full-Time Volunteers 1; Part-Time Volunteers 5.
Governing Authority: nonprofit organization. Parent Institution: Powell County Museum & Arts Foundation. Branch Museums. Old Montana Territorial Prison; Yesterday's Playthings; Montana Auto Museum. Tax-exempt: 501(c)(3).
Institution Type/Description: Local History Museum.
Collections: area artifacts; mastodon fossils; trapping & mining tools; ranching artifacts; 26,000 photographs; railroad exhibit.
Research Fields: local history.
Facilities: visitor center. Museum-related items for sale.
Activities: guided tours available for groups. Museum Sponsors: local arts presentations; local theater group; professional summer theater.
Publications: newsletter, Museum Post; local history books; brochures.
Hours & Admission Prices: Memorial Day-Labor Day Wed.-Sun. 10-4. No charge; donations accepted. &
Attendance: 5,000 (estimated)
Membership: Senior Citizen $15; Individual $20; Family $30.

Dillon

BANNACK STATE PARK & TOWN SITE, 721 Bannack Rd., Dillon, MT 59725-9685. Tel.: 406-834-3413. Fax: 406-834-3548.
E-mail: bannack@smtel.com
Web Site: bannack.org
Founded: 1954.
Governing Authority: Parent Institution: MT Fish, Wildlife and Parks.
Institution Type/Description: Park & History Museum.
Collections: Bannack history; mining; cultural heritage; geology.
Activities: school programs; special events; camping; gold panning; tours; hiking; fishing; photography; bird watching.
Hours & Admission Prices: Park: May-Oct. daily 8am-9pm; Nov.-April daily

8-5. Town Site: May & Sept.-Oct. daily 8am to dusk; Memorial Day to Labor Day daily 8am-9pm; Nov.-April daily 8-5. Visitor Center: May-Aug. daily 10-6; Sept.-Oct. Sat.-Sun. 11-5. Out-of-State Residents: vehicle $5, bus & walk-ins $3; Montana residents no charge. Closed Christmas Eve & Day. &
Attendance: 36,000 (estimated)

BEAVERHEAD COUNTY MUSEUM, 15 S. Montana, Dillon, MT 59725-2433. Tel.: 406-683-5027.
E-mail: bvhdmuseum@bmt.net
Founded: 1947.
Congressional District: 1
Key Personnel: Dir., Lynn Giles; Sec., Ruth Little; Museum Shop Mgr., Joan McDougall.
Personnel Profile: Full-Time Paid 1; Part-Time Volunteers 30.
Governing Authority: nonprofit organization. Subsidiary Institution: County Mill. Tax-exempt: 501(c)(3).
Institution Type/Description: History Museum.
Collections: Indian artifacts, relics of mining & displays of business and household articles brought to the area by early settlers; branded boardwalk; rare photos of local pioneers; livestock & agriculture; natural history; Lewis & Clark exhibit. Historic Buildings: 1885 Homesteader Cabin; one room schoolhouse; 1908 Train Depot.
Research Fields: Montana history; Beaverhead County History.
Facilities: 100-vol. library of county history books available for research on premises; 4,500 photographic reprints indexed & identified; outdoor interpretive area; theater.
Activities: tours. Museum Sponsors: Traveling Trunk Program.
Publications: quarterly newsletter.
Hours & Admission Prices: May 30-Sept. 5 Mon.-Fri. 8-5, Sat. 9-5; Sept. 6-May 29 Mon.-Fri. 8-5. Adults $2, senior citizens & children 13-18 $1; discounts to AAM & ICOM members; members, school tours & children 12 & under no charge. &
Attendance: 14,000 (accurate)
Membership: Individual $17.50; Family $25; Business $45.

THE UNIVERSITY OF MONTANA WESTERN ART GALLERY & MUSEUM, Main Hall, 1st Fl., S. Entrance, Dillon, MT 59725-3511. Mailing Address: 710 S. Atlantic, Dillon, MT 59725. Tel.: 406-683-7011.
Web Site: www.umwestern.edu
Formerly: Western Montana College Gallery Museum
Founded: 1986.
Key Personnel: Dir., Randy Horst; Campus Security, Bob Campbell.
Personnel Profile: Part-Time Paid 3; Interns 1.
Governing Authority: university; nonprofit. Parent Institution: The University of Montana Western.
Institution Type/Description: Art Gallery & Museum: housed in 1893 main building.
Collections: wildlife trophies from Africa, Asia & North America (Seidensticker Collection); northwest regional artists.
Activities: temporary art exhibits.
Hours & Admission Prices: Mon.-Fri. 9-4. No charge. Closed holidays. &
Attendance: 2,000 (estimated)

Drummond

OHRMANN MUSEUM AND GALLERY, 6155 Hwy. 1, Drummond, MT 59832. Tel.: 406-288-3319.
E-mail: ohrmann@blackfoot.net
Web Site: www.ohrmannmuseum.com
Founded: 2002.
Institution Type/Description: Art Museum.
Collections: paintings, woodcarvings, bronzes & sculptures of Bill Ohrmann.
Hours & Admission Prices: Daily 10-5. No charge, donations accepted. &
Attendance: 1,500 (estimated)

East Glacier Park

JOHN L. CLARKE WESTERN ART GALLERY & MEMORIAL MUSEUM, 900 Montana Hwy. 49, East Glacier Park, MT 59434. Mailing Address: P.O. Box 141, East Glacier Park, MT 59434-0141. Tel.: 406-226-9238.
Founded: 1977.
Institution Type/Description: Art Gallery and Museum.
Collections: wood carvings & paintings by John L. Clarke.
Hours & Admission Prices: May-Sept. Mon.-Sat. 10-8, Sun. 10-5. No charge. &

East Helena

KLEFFNER RANCH, 305 Hwy. 518, East Helena, MT 59635-9602. Tel.: 406-495-9090 & 227-3521.
E-mail: kleffner@mt.net
Web Site: www.kleffnerranch.com
Institution Type/Description: Historic Site: former estate of William Child, an early Montana entrepreneur. Listed on the National Historic Register.
Collections: local history & culture; photographs.
Hours & Admission Prices: By appointment.

Ekalaka

CARTER COUNTY MUSEUM, 306 N. Main St., Ekalaka, MT 59324. Mailing Address: P.O. Box 445, Ekalaka, MT 59324-0445. Tel.: 406-775-6886.
E-mail: ccmuseum@midrivers.com
Web Site: cartercountymuseum.org
Founded: 1936.
Congressional District: 2
Key Personnel: Dir., Chioko Hammel; Asst. Dir., Marilyn Schultz; Cur., Nathan Carroll; Receptionist, Gwen Schultz.
Personnel Profile: Full-Time Paid 1; Part-Time Paid 3.
Governing Authority: county. Tax-exempt: 170(b)(1)(A).
Institution Type/Description: Natural History Museum.
Collections: mounted dinosaur skeletons; anatosaurus, triceratops & pachycephalosaurus skulls; nanotyrannus skull; ichthyosaur skeleton; local historical material; pictures; Indian artifacts; fluorescent mineral exhibit; Fossil Bison skull, Bison crassicornis; Civil War to Desert Storm artifacts.
Research Fields: vertebrate paleontology; archeology.
Facilities: 800-vol. library available on request.
Activities: permanent & temporary exhibitions; guided tours.
Publications: books, American Indian Collections, Shifting Scenes Vol. I, II, III, & IV.
Hours & Admission Prices: April-Nov. Mon.-Fri. 9-5, Sat.-Sun. 1-5; Dec.-March Tues.-Fri. 9-5, Sat.-Sun. 1-5. No charge; donations accepted. Closed legal holidays. &
Attendance: 5,500 (accurate)
Membership: Carter County Geological Society: Individual $5.

Forsyth

ROSEBUD COUNTY PIONEER MUSEUM, 1335 Main St., Forsyth, MT 59327. Mailing Address: P.O. Box 88, Forsyth, MT 59327-0088. Tel.: 406-346-7457.
Founded: 1966.
Key Personnel: Pres. (V), Cal MacConnel.
Governing Authority: county. Tax-exempt.
Institution Type/Description: Pioneer History Museum.
Collections: pioneer memorabilia.
Facilities: 100-vol. library of books & newspapers.
Hours & Admission Prices: May to mid-Sept. Mon.-Sat. 9-6, Sun. 1-6. No charge; donations accepted. &
Attendance: 2,000 (accurate)

Fort Benton

FORT BENTON MUSEUM OF THE UPPER MISSOURI, 1810 Front St., Fort Benton, MT 59442. Mailing Address: P.O. Box 262, Fort Benton, MT 59442-0262. Tel.: 406-622-5316. Fax: 406-622-3725.
E-mail: fbmuseums@mtintouch.net
Web Site: www.fortbenton.com
Founded: 1957.
Key Personnel: Chm. (V) & Pres. (V), John G. Lepley; Dir., Randal E. Morger; Museum Shop Mgr., Diane Vielleux; Museum Shop Mgr., Pam Schoonover.
Personnel Profile: Full-Time Volunteers 6; Part-Time Paid 10; Part-Time Volunteers 10.
Governing Authority: nonprofit organization. Parent Institution: River & Plains Society. Tax-exempt.
Institution Type/Description: History Museum: located in park near ruins of Fort Benton, fur trading post of 1846 American Fur Co.
Collections: artifacts; maps; models; pictures & dioramas of early trade routes; Lewis & Clark expedition; founding of Fort Benton; Indian artifacts. Historic House: 1866-67 I.G. Baker House.
Major Exhibits: "Karl Bodmer's America" Lithographs, 5/14-9/14; "No More Buffalo" Bob Scriver Sculptures, 5/14-9/14.
Research Fields: history of upper Missouri.
Facilities: library; archives. Museum-related items for sale.
Activities: guided tours; permanent exhibitions.

Publications: High Plains Chronicles.
Hours & Admission Prices: May-Sept. Mon.-Sat. 10:30-4:30, Sun. 12-4. Adults $10, children $1 (admission includes admittance to Museum of the Northern Great Plains, Museum of the Upper Missouri River, Upper Missouri River Breaks Interpretive Center & Old Fort Benton); members no charge. &
Attendance: 5,000 (estimated)
Membership: Individuals $30; Family $50; Lifetime $150.

MUSEUM OF THE NORTHERN GREAT PLAINS, 1205 20th St., Fort Benton, MT 59442. Mailing Address: P.O. Box 262, Fort Benton, MT 59442-0262. Tel.: 406-622-5316. Fax: 406-622-3725.
E-mail: fbmuseums@mtintouch.net
Web Site: www.fortbenton.com
Founded: 1989.
Key Personnel: Dir. & Pres. (V), John G. Lepley; Dir., Randal E. Morger; Museum Shop Mgr., Pam Schoonover; Museum Shop Mgr., Diane Vielleux.
Personnel Profile: Part-Time Paid 6; Part-Time Volunteers 10.
Governing Authority: nonprofit organization. Parent Institution: River & Plains Society, Inc. Subsidiary Institution: Museum Northern County Plains. Tax-exempt.
Institution Type/Description: History Museum.
Collections: concentration on the homesteading era on the Northern Great Plains, 1910-present.
Research Fields: Overholser Historical Research Center-history on people & places; research lab.
Facilities: library; archives. Museum-related items for sale.
Activities: guided tours; permanent exhibitions.
Publications: High Plains Chronicles; histories of Fort Benton.
Hours & Admission Prices: May-Sept. Mon.-Sat. 10:30-4:30, Sun. 12-4. Adults $10, children $1 (admission includes admittance to Museum of the Upper Missouri, Old Fort Benton, & the Upper Missouri River Breaks Interpretive Center); members no charge. &
Attendance: 5,000 (estimated)
Membership: Individual $30; Family $50; Life $150.

Fort Peck

FORT PECK INTERPRETIVE CENTER, Yellowstone Rd., Fort Peck, MT 59223. Mailing Address: P.O. Box 208, Fort Peck, MT 59223-0208. Tel.: 406-526-3493. Fax: 406-526-3593.
E-mail: michele.l.fromdahl@usace.army.mil
Web Site: www.nwo.usace.army.mil/html/Lake_Proj/fortpeck/museum.html
Key Personnel: Park Ranger, Michele Fromdahl.
Governing Authority: federal. Affiliated with the Corps of Engineers, Omaha District, 6014 USPO & Courthouse, Omaha, NE 68102. Tel.: 402-221-1221. Tax-exempt.
Institution Type/Description: Paleontology Museum.
Collections: paleontology from Cretaceous hellcreek & bearpaw formations; Fort Peck Dam construction history; area wildlife; aquariums.
Activities: guided tours; permanent exhibitions; guided tours; self-guided tours; film.
Hours & Admission Prices: May-Sept. daily 9-5; Oct.-April Tues.-Fri. 10-4; groups of 10 or more by appointment. No charge; donations accepted. &
Attendance: 30,000 (accurate)

Fromberg

CLARK'S FORK VALLEY MUSEUM, 101 East River St., Fromberg, MT 59029. Mailing Address: P.O. Box 103, Fromberg, MT 59029. Tel.: 406-668-7650.
Personnel Profile: Part-Time Volunteers 12.
Governing Authority: Tax-exempt.
Institution Type/Description: History Museum.
Collections: area artists & artisans; watercolors; drawings; photographs.
Facilities: Museum-related items for sale.
Activities: Museum Sponsors: Open House Reception.
Publications: A Tribute to Dr. T.J. Benson.
Hours & Admission Prices: Late June to Sept. Wed.-Sun. 11-3; groups by appointment. No charge; donations accepted. &
Attendance: 400 (estimated)

THE LITTLE COWBOY BAR & MUSEUM, 105 W. River, Fromberg, MT 59029. Mailing Address: P.O. Box 183, Fromberg, MT 59029-0183. Tel.: 406-668-9502.
Founded: 1990.
Key Personnel: Dir., Shirley Smith.
Governing Authority: nonprofit.

Institution Type/Description: General Museum.
Collections: local history, region & culture; western history.
Facilities: library of historical books.
Hours & Admission Prices: No charge; donations accepted.
Attendance: 5,000 (estimated)

Gardiner

YELLOWSTONE NATIONAL PARK, 200 Old Yellowstone Trail, Gardiner, MT 59030. Mailing Address: P.O. Box 168, Yellowstone Park, WY 82190-0168. Tel.: 307-344-2664. Fax: 406-848-9958. TDD: 307-344-2386.
Web Site: www.nps.gov/yell/
Founded: 1872.
Key Personnel: Supt., Daniel N. Wenk; Cur., Colleen E. Curry; Botanist, Heidi Anderson; Registrar, Brandon Sexton; Historian, Lee Whittlesey; Archivist, Anne L. Foster.
Personnel Profile: Full-Time Paid 7; Interns 6.
Governing Authority: federal. Parent Institution: National Park Service. Tax-exempt: 101(6).
Institution Type/Description: National Park Museums & Visitor Centers.
Collections: anthropology; archives; paintings; archaeology; botany; entomology; ethnology; geology; herbarium; history; Indian artifacts; mineralogy; biology; zoology; Thomas Moran watercolors & oil painting. Grant Village Visitor Center: wilderness exhibits; slide program. Fishing Bridge Visitor Center: lake ecology; wildlife & geology. Canyon Visitor Center: bison; geology; slide program. Norris Museum: geothermal exhibits and Albright Visitor Center. Mammoth Hot Springs: park history; paintings; photographs; natural & human history; wildlife exhibits; film on Yellowstone. Old Faithful Visitor Center: movie. Museum of the National Park Ranger: history of the ranger profession. Historic Sites: 1891-1913, 23 structures of Fort Yellowstone. Old Faithful Inn: Shaw & Powell Camping Company exhibit. Historic Structures: 1891-1923, Lake Hotel; 1936-1937, Mammoth Hot Springs Hotel; 1920 Roosevelt Lodge; 1917-1930, Old Faithful Lodge; 1919-1926, Lake Lodge; 1,000 historic structures.
Research Fields: management-oriented studies of wildlife, vegetation, geology; human history of park.
Facilities: 20,000-vol. library of history and natural history books available for inter-library loan and for use on premises; ranger stations; general stores; 90,000 item photo archive. Publications for sale.
Activities: permanent exhibitions; slide programs; guided walks & talks throughout the park in winter & summer; films.
Publications: publications by Yellowstone Association; interpretive leaflets.
Hours & Admission Prices: Horace M. Albright Visitor Center in Mammoth Hot Springs: early Oct. to late May 9-5, late May to Labor Day 8-7. Museum of the National Park Ranger (Norris): late May to Labor Day 9-6. Grant Village, Fishing Bridge, Canyon Visitor Centers: late May to Labor Day 8-7. Norris Geyser Basin Museum: late May to early Oct. 10-5. Old Faithful: late May to Labor Day 8-7, winter while road is open 9-5. Heritage & Research Center: Mon.-Fri. 8-5. Yellowstone Research Library Tues.-Fri. 9-4. Park's Archives & Museum Collections: by appointment to researchers. Call to confirm hours & dates, 307-344-2251. Park: $25 per car; Museums no charge. Closed Christmas. &
Attendance: 4,123,667 (estimated)
Membership: Golden Access Passport (for permanently disabled) no charge; U.S. Senior Citizen Golden Age Passport $10; Annual Area Pass (Yellowstone & Grand Teton National Parks) $40; National Parks Pass (free entrance to all national parks) $50; Golden Eagle Passport (free entrance to all national parks and other selected federal areas) $65.

Garryowen

CUSTER BATTLEFIELD MUSEUM, (M), Town Hall, 4185 Garryowen Rd., Garryowen, MT 59031-0200. Mailing Address: P.O. Box 200, Garryowen, MT 59031-0200. Tel.: 406-638-1876. Fax: 406-638-2019.
E-mail: info@custermuseum.org
Web Site: www.custermuseum.org
Founded: 1994.
Congressional District: 1
Key Personnel: C.E.O. & Founding Dir., Chris Kortlander.
Personnel Profile: Full-Time Paid 2; Full-Time Volunteers 2; Interns 2.
Governing Authority: private; nonprofit organization. Tax-exempt: 501(c)(3).
Institution Type/Description: Historic Site: the site of Sitting Bull's camp, where Major Reno's division of Lt. Col. George A. Custer's Seventh Cavalry attacked on June 25, 1876.
Collections: original artifacts; photographs; paintings; rare books; manuscripts related to Custer, the Battle of the Little Bighorn and other frontier subjects.
Research Fields: onsite research.

Facilities: 400-vol. library of books on western America; 2,500 sq. ft. exhibit space.

Activities: films; guided tours; lectures; temporary exhibitions; broadcast programs. Annual Events: Anniversary ceremonies in June.

Publications: book, Custer in Photographs.

Hours & Admission Prices: Memorial Day-Labor Day daily 8-7; Sept.-May daily 9-5. Adults $7.50; children under 12 no charge. &

Attendance: 50,000 (estimated)

Membership: Individual Plus Guest $30; Family & Dual $40; Contributing $100-$249; Sustaining $250-$499; Patron $500-$999; Eagle Society $1,000-$2,499; Golden Eagle Society $2,500-$4,999; President's Society $5,000 & up.

Glasgow

VALLEY COUNTY PIONEER MUSEUM, Hwy. #2 W., Glasgow, MT 59230. Mailing Address: P.O. Box 44, Glasgow, MT 59230-0044. Tel.: 406-228-8692.

E-mail: vcmuseum@nemontel.net

Web Site: valleycountymuseum.com

Founded: 1964.

Congressional District: 1

Key Personnel: Pres. (V) & Museum Shop Mgr., Mary Helland; Chm. (V), Norm Girard.

Personnel Profile: Part-Time Paid 5; Part-Time Volunteers 15.

Governing Authority: county; appointed board of directors. Parent Institution: Valley County Historical Society. Subsidiary Institution: Friends of the Pioneer Museum. Tax-exempt: 501(c)(3).

Institution Type/Description: History Museum.

Collections: pioneer memorabilia from northeast Montana; Indian artifacts & dioramas; wildlife exhibit; farm machinery; From Dinosaur Bones to Moonwalk; Progressive time history to the right: Fossils, Indians & Buffalo, Historic Dinosaurs, The Cattle & Sheep; The Railroad Irrigation, Wildlife, Veterans, Fort Peck Dam, Woodcarving, Flight & Technology with artifacts to fill in; Collection of Assiniboine Indian artifacts, including hand-crafted buckskin & beaded clothing, sacred medicine, warbonnet, hide tepee & dried foods.

Research Fields: history of local area & families.

Facilities: library of tapes, archives; tourist information center; county newspaper dating back over 100 years through 1936 acquired from State Historical Society after microfilmed; audio & transcribed family history to match family pictures.

Activities: guided tours; art shows.

Publications: brochures; summer newsletters; Valley County collectively published a three volume history in 1991 which includes 1200 family histories.

Hours & Admission Prices: May-Sept. Mon.-Sat. 8:30-5:30. Adults $3, children $2; discounts to Life members. &

Attendance: 4,000 (estimated)

Membership: Student $15; Family $25; Sustaining $30; Life $250.

Glendive

FRONTIER GATEWAY MUSEUM, (M), 201 State St., Glendive, MT 59330. Mailing Address: P.O. Box 1181, Glendive, MT 59330-1181. Tel.: 406-377-8168 & 365-2769.

E-mail: frontiermuseum@yahoo.com

Web Site: www.frontiergatewaymuseum.org

Founded: 1963.

Congressional District: 2

Key Personnel: Treas., Patty Atwell; Cur., Fayette Miller; Pres. (V), Mark Geiger.

Personnel Profile: Full-Time Volunteers 1; Part-Time Paid 2.

Governing Authority: county; nonprofit; board of trustees. Tax-exempt: 501(c)(3).

Institution Type/Description: History Museum.

Collections: fossils; mammoth; mastodon; buffalo, Indian; cattlemen; homesteaders; fashions; photographs.

Facilities: library; 10,080 sq. ft. exhibit space. Museum-related items for sale.

Activities: tours; demonstrations; workshops; lectures. Annual Events: Open House.

Publications: annual members newsletter.

Hours & Admission Prices: mid-May to mid-Sept. Mon.-Sat. 9-12 & 1-5, Sun. & holidays 1-5. No charge; donations accepted. &

Attendance: 1,828 (accurate)

Membership: Individual $5; Life $50.

GLENDIVE DINOSAUR & FOSSIL MUSEUM, 139 State St., Glendive, MT 59330. Mailing Address: P.O. Box 684, Glendive, MT 59330-0684. Tel.: 406-377-3228.

Web Site: www.creationtruth.org/museum

Institution Type/Description: History Museum.

Collections: full-size dinosaurs; fossils.

Hours & Admission Prices: Tues.-Sat. 10-5. Adults $7, seniors 60 & over and students 13 & over $6, children 2-12 $5; children under 2 no charge.

MAKOSHIKA DINOSAUR MUSEUM, 101 N. Merrill Ave., Glendive, MT 59330-1632. Tel.: 406-377-1637.

Founded: 2004.

Personnel Profile: Part-Time Volunteers 12.

Governing Authority: Tax-exempt.

Institution Type/Description: Paleontology Museum.

Collections: dinosaur history; fossil sculptures & casts.

Activities: educational programs; dinosaur exploration.

Hours & Admission Prices: Tues.-Sat. 10-5; other times by appointment.

MAKOSHIKA STATE PARK VISITOR CENTER, 1301 Snyder Ave., Glendive, MT 59330. Tel.: 406-377-6256.

Institution Type/Description: Park Visitor Center.

Collections: local history & geology; fossils; photographs; video.

Facilities: Museum-related items for sale.

Activities: video; hands-on exhibits.

Hours & Admission Prices: Memorial Day to Labor Day daily 10-6; Sept.-May daily 9-5. Park: $5 per vehicle, $1 per person.

Great Falls

***   C.M. RUSSELL MUSEUM, (M),** 400 13th St. N., Great Falls, MT 59401-1498. Tel.: 406-727-8787. Fax: 406-727-2402.

E-mail: sburt@cmrussell.org

Web Site: www.cmrussell.org

Founded: 1953.

Congressional District: 1

Key Personnel: Pres. Bd Dirs., Joe Masterson; C.O.O., Susan Johnson; Chief Cur., Sarah Burt; Collections Mgr., Brenda Kornick; Public Rels. & Mktg. Coord., Courtney Peterson; Museum Shop Admin., Donna Camp; Education & Public Programs, Kim Kapalka; Mgr. Museum Facility, Chuck Keen.

Personnel Profile: Full-Time Paid 12; Part-Time Paid 17; Part-Time Volunteers 334.

Governing Authority: nonprofit organization. Parent Institution: Trigg-C. M. Russell Foundation, Inc. Tax-exempt: 501(c)(3).

Institution Type/Description: Art Museum: located adjacent to the Russell home & log studio.

Collections: art of the American West with emphasis on C.M. Russell art: paintings & bronzes; contemporary western art; Plains Indian art & artifacts.

Research Fields: C.M. Russell; Western American art & history; Northern Plains Indians.

Facilities: 5,000-vol. library of publications on Western art, artists, history & Native Americans. Museum-related items for sale.

Activities: guided tours; docent program; lectures; library research; inter-museum loan, permanent, temporary & traveling exhibitions; hands-on gallery; junior visitor program; western history tour.

Publications: quarterly members' magazine.

Hours & Admission Prices: May-Sept. Tues.-Sun. 10-5; Labor Day to Memorial Day Wed.-Sat. 10-5; tours by appointment. Adults $9, senior citizens $7, students & children $4; discounts to AAM members; children under 5 & members no charge. Closed New Year's Day; Easter; Thanksgiving; Christmas. &

Attendance: 32,000 (accurate)

Membership: Student, Educator & Senior $25; Individual $40; Family $60; Sustaining $150; Patron $300; Director's Circle $500; Missouri Society $1,000; Bitterroot Society $1,500; Russell Society $2,500; Brush and Palette Society $5,000; Western Masters Society $10,000.

CHILDREN'S MUSEUM OF MONTANA, 22 Railroad Sq., Great Falls, MT 59401-4003. Tel.: 406-452-6661. Fax: 406-452-4462. Facebook: Children's Museum of Montana.

E-mail: info@childrensmuseumofmontana.org

Web Site: www.childrensmuseumofmontana.org

Key Personnel: Exec. Dir., Sandie Wright

Institution Type/Description: Children's Museum.

Collections: hands-on exhibits.

Activities: special events; summer camps.

Hours & Admission Prices: Mon.-Sat. 9:30-5. Per person $4; discounts to

groups; members no charge. Closed New Year's Day; Independence Day; Thanksgiving; Christmas.
Membership: Sustainer $50; Grandparent $55; Military $60; Family $75; Family Plus $90; Contributor $150; Benefactor $250.

GALERIE TRINITAS, 1301 20th St., S., Great Falls, MT 59405-4934. Tel.: 406-791-5367.
E-mail: mdriskell@ugf.edu
Web Site: www.ugf.edu/aboutus/galarie.htm
Founded: 1994.
Key Personnel: Chm., Marcia Driskell; Chm. (V), Virginia Wieck.
Personnel Profile: Part-Time Paid 1; Part-Time Volunteers 20.
Governing Authority: Parent Institution: University of Great Falls. Tax-exempt.
Institution Type/Description: Art Museum.
Collections: religious art; paintings; weaving; ceramics; religious artifacts.
Hours & Admission Prices: Tues. & Thurs. 12-3. No charge; donations accepted.

THE HISTORY MUSEUM, 422 2nd St. S., Great Falls, MT 59405-1816. Tel.: 406-452-3462. Fax: 406-761-3805.
E-mail: info@thehistorymuseum.org
Formerly: Cascade County Historical Society Museum
Founded: 1976.
Congressional District: 2
Key Personnel: Exec. Dir., Jim Meinert; Chm. (V), Jeanne Pugh; Archivist, Judy Ellinghausen; Bookkeeper, Sarah Schumacher; Coord. Membership, Kristen Bokovoy; Collections, Jane Boxengard.
Personnel Profile: Full-Time Paid 2; Part-Time Paid 4; Part-Time Volunteers 70.
Governing Authority: nonprofit organization. Tax-exempt.
Institution Type/Description: Historical Museum: housed in c.1929 International Harvester building.
Collections: north central Montana history.
Research Fields: MT-local & state history.
Facilities: archives; banquet facilities. Gift-related items for sale.
Activities: museum tours; exhibit programs; archival research; rental facilities.
Publications: newsletter; occasional local histories.
Hours & Admission Prices: Tues.-Fri. 10-5, Sat.-Sun. tours by appointment. No charge; donations accepted. Closed New Year's Day; Easter; Memorial Day; Independence Day; Labor Day; Thanksgiving; Christmas. &
Attendance: 7,500 (accurate)
Membership: Individual $35; Sustainer $75; Sponsor $125; Patron $250; Benefactor $500; Corporate $1,000.

LEWIS AND CLARK NATIONAL HISTORIC TRAIL INTERPRETIVE CENTER, 4201 Giant Springs Rd., Great Falls, MT 59405-0913. Tel.: 406-727-8733. Fax: 406-453-6157.
Web Site: www.fs.usda.gov/main/lcnf/learning
Founded: 1998.
Key Personnel: Dir., Elizabeth Casselli; Museum Shop Mgr., Sally Murphy.
Personnel Profile: Full-Time Paid 6; Part-Time Paid 3; Part-Time Volunteers 97; Interns 3.
Governing Authority: Parent Institution: USDA, Forest Service. Tax-exempt.
Institution Type/Description: History Museum.
Collections: Lewis & Clark history; photographs; period artifacts & reproductions.
Facilities: 158-seat theater; education room; nature trails. Museum-related items for sale.
Activities: school group programmings; summer day camps. Museum Sponsors: Winter Film Festival January to March; Anniversary Celebration in May; Riverside Voices Programs in summer; Lewis and Clark Festival in June; Voices in the Shadows in October; Discovery Lecture Series December to February.
Hours & Admission Prices: Memorial Day to Sept. daily 9-6; Oct.-May Tues.-Sat. 9-5, Sun. 12-5; groups by appointment. Adults 16 & over $8; children 15 & under no charge. Accept America The Beautiful, Annual, Senior & Access passes. Closed New Year's Day; Thanksgiving; Christmas. &
Attendance: 60,000 (accurate)

PARIS GIBSON SQUARE MUSEUM OF ART, 1400 1st Ave. N., Great Falls, MT 59401-3299. Tel.: 406-727-8255. Fax: 406-727-8256.
E-mail: info@the-square.org
Web Site: www.the-square.org
Founded: 1976.
Congressional District: 1
Key Personnel: Exec. Dir., Kathy Lear; Cur. Art, Bob Durden.

Personnel Profile: Full-Time Paid 6; Part-Time Paid 6; Part-Time Volunteers 250.
Governing Authority: nonprofit organization. Tax-exempt: 501(c)(3).
Institution Type/Description: Art Museum: housed in c.1895 stone building built as the first High School in Great Falls.
Collections: modern & contemporary art of the American Northwest including native American contemporary art; 20th-century American outsider art; special collections.
Research Fields: contemporary art; native & folk art; modernism.
Facilities: classrooms; meeting rooms. Works by Montana artists for sale.
Activities: guided tours; lectures; formally organized education programs for children & adults; docent program; temporary & traveling exhibitions; catered events.
Publications: educational publication & calendar; quarterly newsletter; posters; catalogues.
Hours & Admission Prices: Mon. & Wed.-Fri. 10-5, Tues. 10-5 & 7-9, Sat. 12-5. No charge; donations accepted. Closed national holidays. &
Attendance: 24,579 (accurate)
Membership: Student $20; Individual $30; Dual $35; Family $45; Sustainer $60; Sponsor $100; Benefactor $250; Patron $500; Director's Circle $1,000 & up.

UNIVERSITY OF GREAT FALLS LIBRARY EXHIBIT SPACE, 1301 20th St. S., Great Falls, MT 59405-4996. Tel.: 406-791-5375. Fax: 406-791-5395.
Institution Type/Description: Art Gallery.
Collections: works by contemporary regional artists, student & faculty.
Activities: visiting artist program; temporary exhibitions.
Hours & Admission Prices: Mon.-Fri. 8-5, Sat.-Sun. 2-5.

Hamilton

DALY MUSEUM - DALY MANSION PRESERVATION TRUST, 251 Eastside Hwy., Hamilton, MT 59840. Mailing Address: P.O. Box 223, Hamilton, MT 59840-0223. Tel.: 406-363-6004. Fax: 406-375-0048.
E-mail: april.johnson@dalymansion.org
Web Site: www.dalymansion.org
Founded: 1987.
Congressional District: 88
Key Personnel: Dir., April Johnson; Pres., Gina Wilson.
Personnel Profile: Full-Time Paid 2; Part-Time Paid 1; Interns 1.
Governing Authority: Parent Institution: Daly Mansion Preservation Trust. Tax-exempt.
Institution Type/Description: History Museum.
Collections: local history & culture.
Activities: Museum Sponsors: Community Band Concert; Kids in the Garden; Cross Country Meet; Shakespeare in the Park; Scottish/Irish Festival; Daly Days; Vintage Wine Auction.
Publications: quarterly members' newsletter.
Hours & Admission Prices: April-Oct. call for hours. Adults $9, seniors $8, youth 6-17 $6; discounts to AAA members; children under 6 no charge. &
Attendance: 10,000 (estimated)
Membership: Annual $35-$5,000.

RAVALLI COUNTY MUSEUM/BITTER ROOT VALLEY HISTORICAL SOCIETY, (M), 205 Bedford, Hamilton, MT 59840-2853. Tel.: 406-363-3338. Fax: 406-363-6588.
E-mail: rcmuseum@qwestoffice.net
Web Site: www.brvhsmuseum.org
Founded: 1955.
Key Personnel: Exec. Dir., Tamar Stanley; Pres. (V), Dan Rothlisbeqer; Program Coord., Sarah Monson; Archivist, Beverly Adams; Program Coord., Elizabeth Ettenger.
Personnel Profile: Full-Time Paid 1; Part-Time Paid 4; Part-Time Volunteers 60.
Governing Authority: private; nonprofit organization. Parent Institution: Bitter Root Valley Historical Society. Tax-exempt: 501(c)(3).
Institution Type/Description: History Museum.
Collections: Discovery Room-Native American and Lewis & Clark collection; Bitter Root Valley history; pioneer rooms; a trapper miner cabin; Bertie Lord & Ernst Peterson photographic collections; Civil War exhibit; children's exhibit; archives of newspapers from the valley, catalogued, historic books, historic photo collection, vertical file & periodicals from the 1800s to present day.
Major Exhibits: Natural History of the Bitter Root Valley, 9/13-2/14.
Facilities: 900-vol. library on Montana & Native Americans; educational facilities; 150-seat theater. Museum-related items for sale.
Activities: concerts; films; guided tours; weekly lectures; temporary & rotating

exhibitions; court room theater. Annual Events: Bitter Root Day; Veterans Day Observance; Art & Treasures Appraisal Event; An Afternoon of Cowboy Music & Poetry; poetry, music & art event; Thurs. Speakers Bureau; McIntosh Apple Day; Farmer's Market May to October.
Publications: newsletter, The Newsletter; The Bitter Root Trails I, II, III & IV; The Bitterrooter; Rocky Mountain Spotted Fever.
Hours & Admission Prices: Tues.-Wed. & Fri. 10-4, Thurs. 10-8, Sat. 9-1. Adults $3; discounts to seniors & NARM members; Thurs., Sat. & members no charge. &
Attendance: 15,000 (estimated)
Membership: Individual $40; Family $50; Bitterroot $100; Business $100 & up; Ponderosa $250; Sapphire $500; Trapper Peak $1,000 & up.

Hardin

BIG HORN COUNTY HISTORICAL MUSEUM, 1163 3rd St. E., Hardin, MT 59034-9720. Mailing Address: RRI Box 1206A, Hardin, MT 59034-9720. Tel.: 406-665-1671. Fax: 406-665-3068.
E-mail: diana@bighorncountyuseum.org
Web Site: www.bighorncountymuseum.org
Founded: 1979.
Congressional District: 2
Key Personnel: Dir., Diana Scheidt; Pres. (V), Beth Mehling; Treas., Merna Kincaid; Museum Asst., Bonnie Stark; Museum Shop Mgr., Joan Miller.
Personnel Profile: Full-Time Paid 3; Part-Time Paid 7; Part-Time Volunteers 35.
Governing Authority: private; nonprofit organization. Tax-exempt: 501(c)(3).
Institution Type/Description: History Museum.
Collections: horse drawn equipment; restored tractor & farm equipment; cultures that settled in this area including Crow Indians, the Northern Cheyenne Indians, Japanese, German, Russian, Korean & Norwegian. Historic Buildings: farmhouse; barn; 24 historic structures moved to site from around the county.
Facilities: library; 23 acre farm. Museum-related items for sale.
Activities: guided tours. Annual Events: Tractor Show; Kid's History Day in May; Auction in Sept.
Publications: quarterly newsletter, On the Big Horn.
Hours & Admission Prices: Memorial Day-Labor Day daily 8-6; Sept.-May Mon.-Fri. 9-5. Adults $5, seniors $4, students $3; school groups, members & children under 12 no charge. Closed New Year's Day; Thanksgiving; Christmas. &
Attendance: 25,000 (accurate)
Membership: Individual $15; Family $25; Business & Supporter $50; Contributor $100; Lifetime $500.

Harlowton

UPPER MUSSELSHELL HISTORICAL SOCIETY, 11 S. Central, Harlowton, MT 59036. Mailing Address: P.O. Box 364, Harlowton, MT 59036-0364. Tel.: 406-632-5519.
E-mail: museum@mtintouch.net
Web Site: harlowtonmuseum.com
Founded: 1985.
Congressional District: 14
Key Personnel: Chm. (V), R.C. Brown; Treas., Don Amundson.
Personnel Profile: Part-Time Paid 2; Part-Time Volunteers 6.
Governing Authority: private; nonprofit organization. Tax-exempt: 501(c)(3).
Institution Type/Description: General Museum.
Collections: local area items dating back to the dinosaur age.
Facilities: library; 10,000 sq. ft. exhibit space. Museum-related items for sale.
Activities: guided tours; loan exhibitions. Museum Sponsors: pot luck dinner.
Publications: annual newsletter.
Hours & Admission Prices: May-Aug. Tues.-Sat. 10-5, Sun. 1-5; Sept. call for hours. Adults $2. &
Attendance: 600 (estimated)
Membership: Annual $12; Life $100.

Havre

H. EARL CLACK MUSEUM, (M), Holiday Village Mall, 1753 US Hwy. 2 N.W. #1, Havre, MT 59501-3464. Tel.: 406-265-4000. Fax: 406-265-4000.
E-mail: clackmuseum@co.hill.mt.us
Founded: 1964.
Key Personnel: Dir. & Museum Shop Mgr., John Bruington; Chm. (V), Judi Dritshulas.
Personnel Profile: Part-Time Paid 2; Part-Time Volunteers 5.
Governing Authority: county. Parent Institution: Hill County. Tax-exempt.
Institution Type/Description: Local History Museum.
Collections: dinosaur fossils; history of Hill Country; Fort Assinniboine;

homestead era; development of communities in Hill Co.; Indian artifacts & history of Chippewa-Cree, Wahkpa Chu'gn buffalo jump site; excavations & artifacts; military.
Research Fields: local history; Indians; archaeology studies; paleontology.
Facilities: Western items & other museum-related items for sale.
Activities: guided tours; museum & buffalo jump.
Hours & Admission Prices: Memorial Day to Labor Day Mon.-Sat. 10-6, Sun. 12-5; Sept.-May daily 1-5. No charge; donations accepted. Closed major holidays. &
Attendance: 7,193 (accurate)
Membership: Individual $25.

Helena

EXPLORATION WORKS, 995 Carousel Way, Helena, MT 59601. Tel.: 406-457-1800. Fax: 406-457-5377.
E-mail: info@explorationworks.org
Web Site: www.explorationworks.org
Founded: 2007.
Key Personnel: Dir., Nikki Andersen; Chm. (V), John Cummings.
Personnel Profile: Full-Time Paid 3; Part-Time Paid 12.
Governing Authority: Parent Institution: Community Works, Inc. Tax-exempt.
Institution Type/Description: Children's Museum.
Collections: hands-on exhibits.
Facilities: 3,500 sq. ft. exhibition space; 3 classrooms; computer lab.
Activities: field trips; STEM & girl tech programs; school out & summer classes; adult education; member events.
Hours & Admission Prices: Tues. 10-8, Wed.-Sat. 10-5; other times by appointment. Adults $8, seniors 65 & over and students $6.50, youth under 18 $5.50; discounts to ASTC members; members & children under 2 no charge. Closed New Year's Day; Easter; Memorial Day; Independence Day; Labor Day; Thanksgiving; Christmas. &
Attendance: 36,523 (accurate)
Membership: Individual $35; Family $70; Grandparent $80; Sustaining $250$499; Investing $500-$999; Link $1,000 & up.

HOLTER MUSEUM OF ART, 12 E. Lawrence St., Helena, MT 59601-4019. Tel.: 406-442-6400. Fax: 406-442-2404.
E-mail: info@holtermuseum.org
Web Site: www.holtermuseum.org
Founded: 1987.
Congressional District: 1
Key Personnel: Exec. Dir., Caleb O. Fey; Pres., Madalyn Quinlan; Cur. Education, Sondra Hines; Education Asst., Hannah Gilbert; Exhib. Asst., Ben Pepka; Cur., Yvonne Seng; Visitor Svcs., David Spencer; Museum Shop Mgr., Jenny Gehl.
Personnel Profile: Full-Time Paid 5; Part-Time Paid 4; Part-Time Volunteers 14; Interns 2.
Governing Authority: nonprofit. Tax-exempt.
Institution Type/Description: Art Museum.
Collections: regional contemporary works of art.
Research Fields: contemporary art & crafts of Northwest region.
Facilities: library; 17,000 sq. ft. exhibit space; 2 classrooms
Activities: gallery talks; lectures; docent program; tours; classes; permanent, temporary & traveling exhibitions.
Publications: James Todd: Jazz Icons.
Hours & Admission Prices: Tues.-Sat. 10-5:30, Sun. 12-4. No charge; donations accepted. Closed major holidays. &
Attendance: 32,000 (accurate)
Membership: Individual $50-$500; General Business & Contributor $100, $250, $500; Patron $250; Benefactor $500; Guardian $1,000; Guardian Fellow $2,500.

*** MONTANA HISTORICAL SOCIETY, (M),** 225 N. Roberts, Helena, MT 59601-4514. Mailing Address: P.O. Box 201201, Helena, MT 59620-1201. Tel.: 406-444-2694. Fax: 406-444-2696.
E-mail: mhslibrary@mt.gov
Web Site: mhs.mt.gov
Founded: 1865.
Congressional District: 1
Key Personnel: Research Center Dir., Molly Kruckenberg; Program Mgr, Kirby Lambert; Sr. Cur., Jennifer Bottomly-O'Looney; Archivist, Jodie Foley; Museum Store Mgr., Rodric Coslet.
Personnel Profile: Full-Time Paid 50; Part-Time Paid 14; Part-Time Volunteers 110; Interns 2.
Governing Authority: state. Parent Institution: Montana Historical Society. Subsidiary Institution: Montana's Museum Original Governor's Mansion. Tax-exempt: 170(c)(1).
Institution Type/Description: Historical Society Museum.

Collections: human history of Montana & the region; archaeology; Native American material culture; costumes & textiles; decorative arts & furnishings; ranching, mining, transportation, agricultural & veterinary artifacts; weapons; American Western art including Mackay collection of C.M. Russell art; over 200,000 photographs of Montana & Western subjects; diaries, manuscripts, oral histories & documents; Montana newspapers from 1865; historical library of Montana & the West.
Research Fields: history, art & culture of Montana.
Facilities: 112,000-vol. library of Western & Montana history with public reading room; 20,000 sq. ft. exhibit space; photograph archives reference room; meeting room. Books & museum-related items for sale.
Activities: guided tours; lectures; gallery talks; docent program; formal education programs for adults & children; inter-library loan; curriculum resources for schools; web-based activities; permanent, temporary & traveling exhibits; tours of State Capitol & Original Governor's Mansion; museum services. Annual Events: Montana History Conference; Western Rendezvous of Art; Original Governor's Mansion Holiday Tours; Original Governor's Mansion Secret Garden Tour; Summer Under the Tent Activities Days.
Publications: quarterly scholarly journal, Montana The Magazine of Western History; quarterly newsletter, The Montana Post; books by the Montana Historical Society Press.
Hours & Admission Prices: Mon.-Wed. & Fri.-Sat. 9-5, Thurs. 9-8. Family $12, adults $5, children $1; members no charge. Closed holidays. &
Attendance: 40,256 (accurate)
Membership: Individual $45; Family $60; Explorer $100; Prospector $200; Homesteader $400; Patron $800; Benefactor $1,000.

MONTANA MASONIC MUSEUM, 425 N. Park Ave., Helena, MT 59601-5020. Mailing Address: P.O. Box 1158, Helena, MT 59624-1158. Tel.: 406-442-7774. Fax: 406-442-1321.
E-mail: mtglsec@grandlodgemontana.org
Web Site: www.grandlodgemontana.org
Institution Type/Description: History Museum.
Collections: Masonry in Montana; early Montana history; manuscripts; books.
Hours & Admission Prices: Mon.-Fri. 9-4.

Huntley

HUNTLEY PROJECT MUSEUM OF IRRIGATED AGRICULTURE, 770 Railroad Hwy., Huntley, MT 59037. Mailing Address: P.O. Box 353, Huntley, MT 59037-0353. Tel.: 406-348-2533.
E-mail: curator@huntleyprojectmuseum.org
Web Site: www.huntleyprojectmuseum.org
Key Personnel: Dir., Melissa Koch; Chm. (V), James Knapp.
Personnel Profile: Full-Time Paid 1; Part-Time Paid 2; Part-Time Volunteers 15.
Institution Type/Description: Agricultural & Homesteading History Museum.
Collections: everyday items of homesteader families; records & photographs of the communities; quilts; dresses; dishes & cookbooks; hand tools; saddles; farm machinery; handmade toys.
Facilities: 10 acre homestead with antique buildings; museum; picnic area.
Hours & Admission Prices: Tues.-Sat. 10-4. No charge. Suggested Donations: Families $5, adults $2, children & seniors $1. &
Attendance: 1,000 (accurate)
Membership: Individual $15; Family $25.

Hysham

TREASURE COUNTY 89'ERS MUSEUM, 325 Elliott Ave., Hysham, MT 59038. Mailing Address: P.O. Box 489, Hysham, MT 59038-0489. Tel.: 406-342-5252.
Web Site: treasurecountymuseum.org
Founded: 1989.
Key Personnel: Dir., Pat Miller; Chm.(V), Gary Perkins; Pres. (V), Linda Smith.
Personnel Profile: Part-Time Volunteers 12.
Governing Authority: Tax-exempt.
Institution Type/Description: History Museum.
Collections: local history; American heritage; period artifacts; fossils; Lewis and Clark.
Hours & Admission Prices: Memorial Day to Labor Day Mon.-Sat. 1-5; other times by appointment. No charge; donations accepted. &
Attendance: 400 (estimated)
Membership: Individual $10.

Jefferson City

TIZER BOTANIC GARDENS AND ARBORETUM, 38 Tizer Rd., Jefferson City, MT 59638. Mailing Address: P.O. Box 129, Jefferson City, MT 59638. Tel.: 406-933-8789.
E-mail: info@tizergardens.com
Web Site: www.tizergardens.com
Institution Type/Description: Arboretum.
Collections: annuals; roses; wildflowers; herbs; vegetables; perennials; bulbs.
Activities: self-guided tours.
Hours & Admission Prices: Gardens: May-Sept. daily 10-6. Nursery: April-Sept. daily 10-6. Adults $6; children 5 & under no charge.

Jordan

GARFIELD COUNTY MUSEUM, Montana Hwy. 200, Jordan, MT 59337. Mailing Address: 2251 Brusett Rd., Brusett, MT 59318. Tel.: 406-557-2517.
Founded: 1984.
Congressional District: 15
Key Personnel: Chm. (V), Pres. (V) & Museum Shop Mgr., Ruth Coulter.
Personnel Profile: Part-Time Volunteers 8.
Volunteer Hours: 370
Operating Expenses: 16,247
Operating Income: 18,766
Governing Authority: Tax-exempt.
Institution Type/Description: Dinosaur Museum.
Collections: Cretaceous fossils; T-rex skull; Triceratops replica; Pachycephalosaur domed skull.
Publications: annual, On the Banks of the Big Dig.
Hours & Admission Prices: June-Sept. daily 1-5. No charge; donations accepted. &
Attendance: 1,000 (estimated)
Membership: $5-$50.

Kalispell

CONRAD MANSION NATIONAL HISTORIC SITE MUSEUM, (M), Btwn. Third & Fourth Sts. on Woodland Ave., Kalispell, MT 59901. Mailing Address: P.O. Box 1041, Kalispell, MT 59903-1041. Tel.: 406-755-2166. Fax: 406-755-2176.
E-mail: conradmansion@centurytel.net
Web Site: www.conradmansion.com
Founded: 1975.
Congressional District: 1
Key Personnel: Exec. Dir., Gennifer Sauter; Pres. (V), Mark Norley; Vice Pres., Sue Corrigan; Museum Shop Mgr., Cindy Conner.
Personnel Profile: Full-Time Paid 1; Part-Time Paid 11; Part-Time Volunteers 12.
Governing Authority: city; nonprofit organization. Tax-exempt: 170(b)(1)(A).
Institution Type/Description: Historic House: Conrad Mansion is a Victorian style home completed in 1895 where 90% of furnishings are original to the Conrad family.
Collections: 1895 & earlier period furniture, clothing, toys, pictures & documents.
Facilities: 12,000 sq. ft. exhibit space. Gift items for sale.
Activities: guided tours; special events. Annual Event: Holiday Tours Thanksgiving to New Year's Eve by appointment.
Publications: brochures; book of family history; guide book.
Hours & Admission Prices: May 14-June 15 Wed.-Sun. 10-5; June 17-Oct. 12 Tues.-Sun. 10-5. Adults $10, senior citizens $9, children 12-17 $6, children 11 & under $4; discounts to groups, military, Montana PBS, AAM & AAA members; members no charge.
Attendance: 10,000 (estimated)
Membership: Student $25; Senior $35; Individual $40; Family $75; Business $200; Associate $200-$400; Corporate $300 & up; Mansion Patron $500-$999; Bronze $1,000-$4,999; Silver $5,000-$9,999; Gold $10,000 & up.

HOCKADAY MUSEUM OF ART, (M), 302 Second Ave. E., Kalispell, MT 59901-4942. Tel.: 406-755-5268. Fax: 406-755-2023.
E-mail: director@hockadaymuseum.org
Web Site: www.hockadaymuseum.org
Founded: 1968.
Congressional District: 1
Key Personnel: Pres. (V), Jim Kuhlman; Exec. Dir., Elizabeth Moss; Museum Shop Mgr., Sharon Staso.
Personnel Profile: Full-Time Paid 4; Part-Time Paid 2; Part-Time Volunteers 75.

Governing Authority: private; nonprofit organization. Tax-exempt.
Institution Type/Description: Art Museum.
Collections: contemporary & historic art and culture of Montana; portraits, paintings, prints, sculpture & pottery of Glacier National Park.
Facilities: art library available for use on premises; classroom. Museum-related items for sale.
Activities: guided tours; lectures; films; gallery talks; formally organized education programs for children & adults; permanent, temporary & traveling exhibitions. Museum Sponsors: Art Auction; Summer Arts Festival.
Publications: newsletter; flyers; exhibition catalog.
Hours & Admission Prices: Tues.-Sat. 10-5. Adults $5, seniors $4, college students $2; children K-12 & members no charge. Closed major holidays. &

Attendance: 20,000 (estimated)
Membership: Senior $35; Individual $40; Family $60; Associate $75; Friend of the Hockaday $125; Cultural Member $250; Heritage Member $500; Patron of the Arts $1,000 and up.

THE MUSEUM AT CENTRAL SCHOOL, (M), Northwest Montana Historical Society, 124 Second Ave., E., Kalispell, MT 59901. Tel.: 406-756-8381. Fax: 406-257-5719. Facebook: The Museum at Central School.
E-mail: history@yourmuseum.org
Web Site: www.yourmuseum.org
Formerly: Central School Museum
Founded: 1999.
Key Personnel: Exec. Dir., Gil Jordan; Accounting, Jan Woods; Office Mgr. & Membership, Kimberly Pinter; Museum Shop Mgr., Doreen Harper.
Personnel Profile: Full-Time Paid 1; Part-Time Paid 3; Part-Time Volunteers 60.
Volunteer Hours: 10,000
Governing Authority: Tax-exempt.
Institution Type/Description: History Museum: housed in the historic Central School building, which opened in 1894.
Collections: Native American artifacts; Northwest Montana records & documents; paintings; sculptures; photographs.
Publications: quarterly newsletter.
Hours & Admission Prices: Mon.-Fri. 10-5. Adults $5, seniors $4; children & members no charge. Closed New Year's Day; Independence Day; Thanksgiving; Christmas. &
Attendance: 12,000 (accurate)
Membership: Senior $20; Regular $30; Lifetime $750.

Lewistown

CENTRAL MONTANA HISTORICAL ASSOCIATION, INC. MUSEUM, 408 N.E. Main St., Lewistown, MT 59457-2019. Tel.: 406-535-3642.
E-mail: cmha@midrivers.com
Founded: 1955.
Congressional District: 2
Key Personnel: Pres. (V), Shirley Barrick; Vice Pres., Janet Lewellen.
Personnel Profile: Part-Time Volunteers 30.
Volunteer Hours: 1,020
Operating Expenses: 16,870
Operating Income: 14,768
Governing Authority: society. Affiliated with Central Montana Historical Association. Tax-exempt.
Institution Type/Description: History Museum.
Collections: geology; Native American artifacts; history.
Activities: permanent & temporary exhibitions.
Publications: book, Guarding the Carroll Trail.
Hours & Admission Prices: Memorial Day to Oct. daily 10-4; Nov.-May Mon.-Fri. 8-5; other times by appointment. No charge; donations accepted. &
Attendance: 4,000 (estimated)
Membership: Student $2; Senior Citizen $7.50; Single $10; Family $25; Organization $25; Contributing $50-$149; Sustaining $150-$499; Patron $500; Benefactor $1,000 & up.

LEWISTOWN ART CENTER, 323 W. Main St., Lewistown, MT 59457-2450. Mailing Address: P.O. Box 1018, Lewistown, MT 59457-1018. Tel.: 406-535-8278. Fax: 406-535-6024.
E-mail: lewistownartcenter@gmail.com
Web Site: www.lewistownartcenter.org
Founded: 1971.
Key Personnel: Exec. Dir., Nadine Robertson; Chm. (V), Paul Huff; Museum Shop Mgr., Mary Haight.
Personnel Profile: Part-Time Paid 4; Part-Time Volunteers 5.

Institution Type/Description: Art Center.
Collections: works by local & state artists.
Facilities: Museum-related items for sale.
Hours & Admission Prices: Tues.-Fri. 11:30-5:30, Sat. 10-4. No charge; donations accepted.
Membership: Senior/Student $15; Individual $30; Family $35; Advocate $50; Business $75; Youth Art Programs & Mt. Repertory Theater $100; Shakespeare - Parks $125; Gallery Show Sponsor $150; Agate $250; Silver $500; Gold $1,000; Sapphire $1,500; Diamond $2,500.

Libby

THE HERITAGE MUSEUM, 34067 US Hwy. 2 S., Libby, MT 59923. Mailing Address: P.O. Box 628, Libby, MT 59923-0628. Tel.: 406-293-7521.
E-mail: heritagemuseum@frontier.com
Web Site: www.libbyheritagemuseum.org
Institution Type/Description: History Museum.
Collections: local history & culture; exhibits on the Shay Locomotive, historic Jennings, Kootenai Indians & early mining; photographs.
Hours & Admission Prices: June-Aug. Mon.-Sat. 10-5, Sun. 1-5; other times by appointment. No charge; donations accepted.

Livingston

LIVINGSTON CENTER FOR ART AND CULTURE, 119 S. Main St., Livingston, MT 59047-2668. Tel.: 406-222-5222.
E-mail: admin@livingstoncenter.org
Web Site: www.livingstoncenter.org
Institution Type/Description: Art Gallery.
Collections: paintings; sculpture; photographs.
Activities: classes; educational programs.
Hours & Admission Prices: Tues.-Fri. 11-5, Sat. 11-4.

LIVINGSTON DEPOT CENTER, (M), 200 W. Park St., Livingston, MT 59047-2629. Mailing Address: P.O. Box 1319, Livingston, MT 59047-1319. Tel.: 406-222-2300. Fax: 406-222-2401.
Web Site: www.livingstondepot.org
Founded: 1985.
Congressional District: 1
Key Personnel: C.E.O. & Pres. (V), John Sullivan; Dir., Diana L. Seider.
Personnel Profile: Full-Time Paid 1; Part-Time Paid 2; Part-Time Volunteers 60.
Governing Authority: nonprofit organization. Tax-exempt: 501(c)(3).
Institution Type/Description: Historic Building: 1902 Northern Pacific railroad station.
Collections: objects related to the history of railroads in the Pacific Northwest.
Research Fields: railroad history; local history.
Facilities: 4,000 sq. ft. exhibit space. Western culture books & other gift items for sale.
Activities: railroad exhibition; temporary & traveling exhibitions; museum tours; lectures; special events & programs; annual railroad memorabilia show; outdoor Festival of the Arts & Holiday activities.
Publications: annual newsletter, Depot Center.
Hours & Admission Prices: May-Sept. Mon.-Sat. 9-5, Sun. 1-5; school & group tours by appointment; winter by event. Adults $3, senior citizens, students & children $2; members no charge. Closed New Year's Day; Thanksgiving; Christmas. &
Attendance: 25,000 (accurate)
Membership: Individual $30; Family $50; Centennial $100; Sponsor $250; Benefactor $500; Patron $1,000.

YELLOWSTONE GATEWAY MUSEUM OF PARK COUNTY, 118 W. Chinook, Livingston, MT 59047-2011. Tel.: 406-222-4184. Fax: 406-222-4146.
E-mail: museum@parkcounty.org
Web Site: yellowstonegatewaymuseum.org
Formerly: Park County Museum, House of Memories
Founded: 1976.
Congressional District: 1
Key Personnel: Dir., Paul Shea; Registrar, Karen Reinhart.
Governing Authority: county. Tax-exempt.
Institution Type/Description: Historical Society Museum: housed in 1906 North Side School.
Collections: tools used by early ranchers, miners, professionals; 1871-1950 Northern Pacific Railroad artifacts & photographs; 1869-1950 Yellowstone Park memorabilia; archaeological artifacts; clothing; household furniture; Wild West displays including Lewis & Clark, Calamity Jane & Buffalo Bill Cody; pioneer living artifacts & buildings; military artifacts from 1864 to WWII; 1868 Ft. Parker; mining & 10 stamp gold mill.

Research Fields: archaeology; local history; railroad history; genealogy.
Facilities: 350-vol. library of bound & unbound material, including history books & newspapers available for use on premises; theater; educational facilities. Local handcrafts & historical books for sale.
Activities: guided tours; radio programs; formally organized education programs; lectures; loan exhibitions.
Publications: biannual newsletter, News From the Red Caboose.
Hours & Admission Prices: May-Sept. daily 10-5; Oct.-May Thurs.-Sat. 10-5. Adults $5, senior citizens $4; discounts to AAM members; members & children under 18 no charge.
Attendance: 2,800 (accurate)
Membership: Individual $15; Family $25; Business $50; Gold $500; Platinum $1,000.

Lolo

HOLT HERITAGE MUSEUM, 6800 Lewis & Clark Tr., Lolo, MT 59847. Mailing Address: P.O. Box 129, Lolo, MT 59847-0129. Tel.: 406-273-6743. Fax: 406-273-6378.
E-mail: info@holtheritagemuseum.com
Web Site: www.holtheritagemuseum.com
Key Personnel: Owner, Bill Holt; Owner, Ramona Holt
Institution Type/Description: History Museum.
Collections: celebrity & collector western boots; saddles & tack; original artwork; wagon; Indian ceremonial display.
Hours & Admission Prices: By appointment.

Loma

EARTH SCIENCE MUSEUM, (M), 208 Broadway Ave., Loma, MT 59460. Mailing Address: P.O. Box 207, Loma, MT 59460-0207. Tel.: 406-739-4282.
Institution Type/Description: Science Museum.
Collections: gems; minerals; fossils; Native American artifacts; train memorabilia.
Facilities: Museum-related items for sale.
Hours & Admission Prices: Memorial Day to Labor Day daily 10-5. &

HOUSE OF A THOUSAND DOLLS, 106 First St., Loma, MT 59460. Mailing Address: P.O. Box 136, Loma, MT 59460-0136. Tel.: 406-739-4338.
Founded: 1979.
Key Personnel: C.E.O., Marion Britton.
Personnel Profile: Full-Time Volunteers 1.
Governing Authority: individual operation.
Institution Type/Description: Toy & Doll Museum.
Collections: dolls & toys from 1830 to present.
Hours & Admission Prices: By appointment only. Adults $1, children $.50.
Attendance: 600 (estimated)

Malmstrom AFB

MALMSTROM AFB MUSEUM, 341 Missile Wing/MU, 21 77th St. N., Ste. 144, Malmstrom AFB, MT 59402. Tel.: 406-731-2705. Fax: 406-731-2769.
E-mail: museum@malmstrom.af.mil
Web Site: www.malmstrom.af.mil/library/malmstrommuseum
Key Personnel: Dir., Curt Shannon
Governing Authority: Parent Institution: USAF.
Institution Type/Description: Military History Museum.
Collections: military history & aircraft; personal artifacts; photographs.
Hours & Admission Prices: Mon.-Fri. 10-4. Non-military members visiting the museum must go to the Visitor Center located at the 2nd Ave. North gate. No charge. &
Attendance: 9,500 (accurate)

Malta

BOWDOIN NATIONAL WILDLIFE REFUGE, 194 Bowdoin Auto Tour Rd., Malta, MT 59538. Tel.: 406-654-2863. Fax: 406-654-2866.
E-mail: bowdoin@fws.gov
Web Site: www.fws.gov/bowdoin/wildlife.html
Institution Type/Description: Wildlife Refuge.
Collections: wildlife & their habitats.
Activities: guided tours; video presentations; educational programs; wildlife presentations. Annual Events: International Migratory Bird Day; National Wildlife Refuge System Week; Enrichment Days in May.
Hours & Admission Prices: Mon.-Fri. 7:30-4. Closed Federal holidays.

GREAT PLAINS DINOSAUR MUSEUM AND FIELD STATION, (M), 405 N. 1st St. E., Malta, MT 59538. Mailing Address: P.O. Box 170, Malta, MT 59538. Tel.: 406-654-5300.
E-mail: dinosaur@itstriangle.com
Founded: 2002.
Key Personnel: Pres. (V), Carolyn Schmoeckel; Museum Shop Mgr., Sue Frary.
Personnel Profile: Part-Time Paid 1; Part-Time Volunteers 8.
Volunteer Hours: 5,327
Operating Expenses: 48,276
Operating Income: 46,333
Governing Authority: Parent Institution: Judith River Foundation, Inc. Tax-exempt.
Institution Type/Description: Geology, Mineralogy, Paleontology Museum.
Collections: fossil invertebrates; dinosaurs; fish; plants.
Research Fields: invertebrate; fish; plant; dinosaur fossils.
Facilities: Museum-related items for sale.
Activities: educational programs; lectures.
Hours & Admission Prices: May-Oct: Mon.-Sat. 10-5, Sun. 12:30-5. Adults $5; discounts to school groups; members no charge. &
Attendance: 2,500 (estimated)
Membership: $25; $100; $250; $500.

PHILLIPS COUNTY MUSEUM, 431 U.S. Hwy. 2 E., Malta, MT 59538. Mailing Address: P.O. Box 518, Malta, MT 59538. Tel.: 406-654-1037. Fax: 406-654-1037.
E-mail: pcm@itstriangle.com
Web Site: www.phillipscountymuseum.org
Personnel Profile: Full-Time Paid 1; Part-Time Paid 2.
Institution Type/Description: History Museum.
Collections: local history & culture; period furnishings; personal artifacts; photographs; Native American art; period clothing.
Hours & Admission Prices: Jan.-March Tues.-Sat. 11-5; April 15-Dec. Mon.-Sat. 10-5, Sun. 12:30-5. Adults $5, children over 5 $3; discounts to groups.

Martinsdale

BAIR FAMILY MUSEUM, 2751 MT Hwy. 294, Martinsdale, MT 59053. Tel.: 406-572-3314.
E-mail: info@bairfamilymuseum.org
Web Site: www.bairfamilymuseum.org
Key Personnel: Cur., Elizabeth Guheen
Institution Type/Description: History Museum.
Collections: Bair family history; personal artifacts; paintings.
Hours & Admission Prices: Memorial Day to Labor Day daily 10-5; Sept.-Oct. Wed.-Sun. 10-5. Adults $5, seniors 62 & over $3, children 6-16 $2; children 5 & under no charge.

Miles City

CUSTER COUNTY ART & HERITAGE CENTER, 85 Waterplant Rd., Miles City, MT 59301. Mailing Address: P.O. Box 1284, Miles City, MT 59301-1284. Tel.: 406-234-0635. Fax: 406-234-0637.
E-mail: ccartc@midrivers.com
Web Site: www.ccac.milescity.org
Founded: 1975.
Congressional District: 1
Key Personnel: Exec. Dir., Mark Browning; Pres. (V), Kathy Schlepp; Dir. Education, Jordan Pehler; Exec. Asst., Sandra Brunetti.
Personnel Profile: Full-Time Paid 3; Part-Time Paid 1; Part-Time Volunteers 35.
Governing Authority: nonprofit organization. Subsidiary Institution: CCAC Foundation. Tax-exempt: 501(c)(3).
Institution Type/Description: Visual Arts Center & Museum: housed in 1910 waterworks building.
Collections: 20th-century Western art with emphasis on contemporary work by Montana artists; photographs of L.A. Huffman, Edward S. Curtis, Evelyn Cameron; underground museum; former water holding tanks of Old Miles City water works.
Facilities: Museum-related items for sale.
Activities: temporary & traveling exhibitions; gallery talks; lectures; performing arts programs; workshops. Museum Sponsors: Quick Draw in park in May; Annual Art Auction in September; Annual Holiday Exhibit Nov. to Dec.
Publications: quarterly newsletter; exhibit brochures.
Hours & Admission Prices: May-Sept. Tues.-Sun. 9-5; Oct.-March Tues.-Sun. 1-5. No charge; donations accepted. Closed New Year's Day; Easter; Thanksgiving; Christmas. &
Attendance: 10,000 (accurate)

Membership: Student & Senior Citizen $15; Individual $25; Family $40; Business $60; Contributing $50; Sustaining $75; Sponsor $100; Patron $300; Benefactor $500.

RANGE RIDERS MUSEUM, 435 LP Anderson Rd., Miles City, MT 59301-4753. Tel.: 406-232-6146 & 234-2817.
E-mail: rangeriders@gmail.com
Founded: 1939.
Key Personnel: C.E.O. & Dir., Bunny Miller.
Personnel Profile: Full-Time Paid 1; Part-Time Paid 2; Part-Time Volunteers 20.
Governing Authority: Tax-exempt.
Institution Type/Description: History Museum: housed on the site of the 1876 Fort Keogh cantonment.
Collections: pioneer history; period furnishings & artifacts; photographs.
Activities: Museum Sponsors: Open House; Induction Ceremony.
Hours & Admission Prices: April-Oct. daily 8-5. Adults $7.50, seniors $5; members no charge. &
Attendance: 5,000 (estimated)
Membership: Individual $25; Sustaining $40; Contributing $75; Lifetime $300.

Missoula

AERIAL FIRE DEPOT AND SMOKEJUMPER VISITOR CENTER, 5765 W. Broadway St., Missoula, MT 59808. Tel.: 406-329-4934. Fax: 406-329-4955.
Web Site: www.fs.fed.us/fire/people/smokejumpers/missoula
Institution Type/Description: Firefighting Museum.
Collections: firefighting history, equipment, & artifacts; National Smokejumper Memorial; 1930s lookout tower replica; photographs; personal artifacts.
Activities: guided tours.
Hours & Admission Prices: Memorial Day to Labor Day daily 8:30-5; other times by appointment. No charge; donations accepted.

ELK COUNTRY VISITOR CENTER, 5705 Grant Creek Rd., Missoula, MT 59808-9394. Tel.: 406-523-4545.
Institution Type/Description: Visitor Center.
Collections: North American elk history; ecology; biology; game management; hunting; antlers.
Hours & Admission Prices: Jan.-May Mon.-Fri. 8-5, Sat. 10-5; June-Dec. Mon.-Fri. 8-6, Sat.-Sun. 9-6. No charge.

FAMILIES FIRST CHILDREN'S MUSEUM, 225 W. Front St., Missoula, MT 59802-4301. Tel.: 406-541-7529.
E-mail: info@familiesfirstmontana.org
Web Site: familiesfirstmontana.org
Formerly: Children's Museum Missoula
Founded: 2000.
Key Personnel: Dir., Coco Ballew.
Personnel Profile: Full-Time Paid 2; Part-Time Paid 10; Part-Time Volunteers 10; Interns 1.
Governing Authority: Parent Institution: Families First. Tax-exempt.
Institution Type/Description: Children's Museum.
Collections: hands-on exhibits.
Facilities: Museum-related items for sale.
Activities: special events; educational programs; interactive exhibits.
Hours & Admission Prices: Tues.-Sat. 10-5, Sun. 12-5. Admission $4.25; children under one no charge. ACM members reciprocal program. &
Attendance: 29,000 (estimated)
Membership: Grandparentship $50; Familyship $75; Friendship $125; Starship $250; Leadership $500.

✻ HISTORICAL MUSEUM AT FORT MISSOULA, (M), 3400 Captain Rawn Way, Missoula, MT 59804. Tel.: 406-728-3476. Fax: 406-543-6277.
E-mail: ftmslamuseum@montana.com
Web Site: www.fortmissoulamuseum.org
Founded: 1975.
Congressional District: 1
Key Personnel: Chm. (V), Michael Fussell; Pres. (V), Victor Machart; Exec. Dir., Robert M. Brown; Dir. Devel., Kristina Swanson; Cur., Nicole Webb; Dir. Education, Kristjana Eyjolfsson; Museum Aide, Sharon Garner; Asst. Dir., Carolyn Thompson.
Personnel Profile: Full-Time Paid 4; Part-Time Paid 3; Part-Time Volunteers 40; Interns 4.
Volunteer Hours: 12,381
Operating Expenses: 751,503

Operating Income: 620,660
Governing Authority: county. Parent Institution: Missoula County. Tax-exempt.
Institution Type/Description: Historical Museum Complex & Site: housed in 1911 brick quartermaster's warehouse, located on 32-acres (Historic District) at the core of what was Fort Missoula (1877-1947).
Collections: 19th & 20th-century furnishings, textiles, clothing, tools; USFS artifacts; logging tools; U.S. army uniforms, weapons & vehicles; alien detention records. Historical Buildings: 1878 Log NCO quarters; 1910 railroad depot; 1863 log church; 1933 USFS Lookout; 1907 Grant Creek school; 1910 USFS guard cabin; 1908 quartermaster's root house (cellar); 1940 U.S. Army warehouse; 1911 quartermaster's warehouse; 1941 internment camp barracks building; 1941 Alien Detention Camp headquarters building.
Major Exhibits: United We Will Win: World War II Posters That Mobilized A Nation, 4/12-2/14; Signs of the Times: A Trip Down Memory Lane, 2/13-1/14; Growing the Garden City: Missoula's First 150 Years, 4/14-1/16.
Research Fields: Missoula County & Fort Missoula history; history of forest management & timber production in western Montana; World War II Alien Detention Center.
Facilities: 500-vol. library of museum administration, western history & historical school books available for research by bona fide researchers; meeting room; Iris test gardens. Books related to Montana & western history for sale.
Activities: guided tours; gallery talks; formally organized education programs for children & adults; training programs for professional museum workers; loan, long term, changing & traveling exhibitions; lecture series; rental facilities. Museum Sponsors: Forest Day in April; Independence Day celebration.
Publications: booklet, The Military History of Fort Missoula; postcard series; Missoula, The Way It Was; Purple & Gold, A 60 Year History of Missoula County High School; An Alien Place: The Fort Missoula, Montana Detention Camp, 1941-1944.
Hours & Admission Prices: Memorial Day-Labor Day Mon.-Sat. 10-5, Sun. 12-5; Sept.-May Tues.-Sun. 12-5. Adults $3, seniors $2, students $1; discounts to AAA, AAM & ICOM members and museum professionals & associations; children under 6 & members no charge. &
Attendance: 45,000 (estimated)
Membership: Individual $30; Supporter $60; Patron $100; Historical $250; Contributor $500; Patron $1,000.

✻ MISSOULA ART MUSEUM, (M), 335 N. Pattee St., Missoula, MT 59802-4520. Tel.: 406-728-0447.
E-mail: museum@missoulaartmuseum.org
Web Site: www.missoulaartmuseum.org
Founded: 1975.
Congressional District: 1
Key Personnel: Exec. Dir., Laura J. Millin; Pres. (V), Liz Dybdal; Cur. Education, Renee Taaffe; Membership, Volunteer & Event Mgr., Anna Buxton; Cur. Exhibitions, Stephen Glueckert; Registrar, Theodore Hughes; Office Mgr., Pam Adams; Asst. Cur. & Preparator, John Calsbeek; Dir. Devel., Kay Grissom-Kiely; Dir. Mktg. & Communications, Katie Stanton.
Personnel Profile: Full-Time Paid 9; Part-Time Paid 1; Part-Time Volunteers 75; Interns 4.
Governing Authority: nonprofit organization. Tax-exempt: 501 (c)(3).
Institution Type/Description: Contemporary Art Museum: housed in 1903 Carnegie Library.
Collections: Art by artists who lived or worked in the Western US with an emphasis on contemporary Montana artists; contemporary American Indian art collection.
Research Fields: contemporary art of the western United States; Montana artists; contemporary American Indian art; contemporary artists of the region.
Facilities: library; classrooms. Museum-related items for sale.
Activities: summer art school; interactive tours; lectures; performances; gallery talks; concerts; formally organized education programs for children & adults; loan & traveling exhibitions; resource library.
Publications: books, Hmong Voices in Montana; Ernie Pepion: Dreams on Wheels; catalogs, Gennie Deweese Retrospective; Lynne Hull, Dreaming Missoula; Hamish Fulton; Cathy Weber's Grief Series; Corwin Clairmont; Halfway Between Here and There; Anne Appleby & Wes Mills; James Todd Retrospective 1941-2002; catalogue, Lela Autio; Stan Healty: Artist's Eye; Nancy Erickson: Recent Works; Native Perspectives on the Trail: A Contemporary American Indian Portfolio; John Armstrong: Engaged Abstractions; M.A. Papanek-Miller: A Snowman Cares for Our Memory of Water; Missoula Art Museum's 2009 Montana Triennial.
Hours & Admission Prices: Tues.-Thurs. 10-5, Fri.-Sun. 10-3. No charge; donations accepted. &
Attendance: 60,000 (accurate)
Membership: Artist, Student, Senior & Non-Resident of Missoula County $30;

Individual $40; Family & Dual $60; Friend $100; Patron $250; Benefactor $500 & up. Small Business $250; Business Friend $500; Business Patron $1,000; Business Benefactor $5,000 & up.

MONTANA MUSEUM OF ART & CULTURE, (M), Main Hall 006, University of Montana, 32 Campus Dr., Missoula, MT 59812. Tel.: 406-243-2019. Fax: 406-243-2797.

E-mail: museum@umontana.edu
Web Site: www.umt.edu/montanamuseum
Founded: 1956.
Congressional District: 1
Key Personnel: Dir., Barbara Koostra; Cur., Brandon Reintjes; Registrar, Lucy Capehart; Asst. to Cur., Bill Queen; Coord. Programs & Publications, Shawn Whitworth; Administrative Assoc., Erin Wilson.
Personnel Profile: Full-Time Paid 3; Part-Time Paid 7; Part-Time Volunteers 2; Interns 2.
Governing Authority: Parent Institution: The University of Montana. Tax-exempt.
Institution Type/Description: Art Museum.
Collections: Western & European Painters: Sharp, Paxson, Mauer, Chase, Fra Dana; American Painters: Krasner & Motherwell contemporary prints; Autio, Voulkos ceramic sculpture; Montana area artifacts; Chinese & Japanese art; Native American art with Montana emphasis.
Research Fields: art history; museology; historical & contemporary Montana artists; Native American artists; Asian art.
Activities: guided tours; lectures; films; arts festivals; organized educational programs for children & University of Montana students; participatory, temporary, traveling & loan exhibitions; school loan service.
Publications: exhibition catalogs.
Hours & Admission Prices: June-Aug. Wed.-Thurs. & Sat. 12-3, Fri. 12-6; Sept.-May Tues.-Wed. & Sat. 12-3, Thurs.-Fri. 12-6. Suggested Donations: $5. Closed Montana state holidays. ᕣ
Attendance: 9,000 (accurate)
Membership: UM Student $10; Friend $25-$49; Enthusiast $50-$99; Advocate $100-$249; Patron $250-$499; Visionary $500-$999; Champion $1,000 & up.

MONTANA NATURAL HISTORY CENTER, 120 Hickory St., Missoula, MT 59801-1820. Tel.: 406-327-0405. Fax: 406-327-0421.

E-mail: office@montananaturalist.org
Web Site: www.montananaturalist.org
Founded: 1991.
Key Personnel: Exec. Dir., Arnold Olsen, Ph.D.; Educational Dir., Lisa Bickell; Pres. (V), Hank Fischer; Coord. Community Programs, Christine Morris; Field Notes Coord. & Montana Naturalist Editor, Allison DeJong; Lead Naturalist, Brian William; Naturalist, Alyssa McLean; Volunteer Coord., Allison DeJong; Administrative Asst., Deb Jones; Administrative Asst., Caroline Romero.
Personnel Profile: Full-Time Paid 4; Part-Time Paid 18; Part-Time Volunteers 100; Interns 2.
Volunteer Hours: 2,600
Operating Expenses: 450,000
Operating Income: 450,000
Governing Authority: Tax-exempt: 501(c)(3).
Institution Type/Description: Natural History Center.
Collections: birds; reptiles; plants; mammals; insects.
Facilities: library.
Activities: Natural history programs for children & adults.
Publications: magazine 3 times a yr., Montana Naturalist.
Hours & Admission Prices: Tues.-Fri. 12-5, Sat. 12-4. Adults $4, children 13 & over $2, children 3-12 $.50; children under 3 & MNHC members no charge. ᕣ
Attendance: 2,500 (estimated)
Membership: Regular $50; Supporting $10; Student $25; Explorers Club $1,000.

MUSEUM OF MOUNTAIN FLYING, Missoula International Airport, Missoula, MT 59808. Mailing Address: 713 S. Third St., Missoula, MT 59801-2513. Tel.: 406-721-3644. Fax: 406-728-9280.

E-mail: phpc@montana.com
Web Site: museummountainflying.org
Founded: 1993.
Key Personnel: C.E.O. & Pres. (V), Stan Cohen.
Personnel Profile: Part-Time Volunteers 12.
Governing Authority: board. Tax-exempt.
Institution Type/Description: Aviation History Museum.

Collections: area mountain flying history; aircraft; hands-on exhibits.
Publications: quarterly newsletter.
Hours & Admission Prices: May-Sept. 10-4. Family $10, adult $3, children, seniors & military $2; members no charge. ᕣ
Attendance: 3,000 (estimated)
Membership: Single $25; Couple $35; Family $50; Life $1,000.

PHILIP L. WRIGHT ZOOLOGICAL MUSEUM AND UNIVERSITY OF MONTANA HERBARIUM, Division of Biological Sciences, University of Montana, 32 Campus Dr. #4824, Missoula, MT 59812. Tel.: 406-243-5222. Fax: 406-243-4184.

E-mail: dave.dyer@mso.umt.edu
Web Site: www.zoologicalmuseum.dbs.umt.edu
Founded: 1909.
Key Personnel: Assoc. Dean College Arts & Sciences, Dr. Charles Janson; Cur. Ornithology, Dr. Richard Hutto; Cur. Mammalogy, Dr. Kerry Foresman; Cur. Collections, David Dyer.
Personnel Profile: Full-Time Paid 1; Part-Time Paid 4; Part-Time Volunteers 3; Interns 2.
Governing Authority: public university; nonprofit. Parent Institution: University of Montana. Tax-exempt: 501(c)(3).
Institution Type/Description: Zoological Museum & Herbarium: the herbarium is housed in the Natural Sciences Building, which is a contributive structure within the University of Montana Historical District.
Collections: Museum: over 24,000 specimens of fish, birds & mammals from the Northern Rockies, particularly Montana; Chinese & Russian specimens. Herbarium: over 140,000 specimens of plants from the Northern Rocky Mountains, particularly Montana & to a lesser extent North America & the world.
Research Fields: small mammal population & systematic studies, particularly species of bats & shrews; neotropical bird studies; zooarchaeology; wildlife forensics.
Facilities: university library facilities; classrooms; laboratories; field research station.
Activities: formal education programs for undergraduate & graduate students affiliated with the University of Montana; programs are conducted in conjunction with the Montana Natural History Center & includes field trips, seminars, workshops, traveling natural history trunks & a weekly radio program.
Publications: annual newsletter.
Hours & Admission Prices: research & teaching collections not open to general public. ᕣ

ROCKY MOUNTAIN MUSEUM OF MILITARY HISTORY, Bldgs. T-310 & T-316 at Fort Missoula, Missoula, MT 59807. Mailing Address: P.O. Box 7263, Missoula, MT 59807-7263. Tel.: 406-549-5346. Facebook: Rocky Mountain Museum of Military History.

E-mail: info@fortmissoula.org
Web Site: www.fortmissoula.org
Key Personnel: Exec. Dir., Tate Jones.
Governing Authority: Tax-exempt: 501(c)(3).
Institution Type/Description: Military History Museum.
Collections: documents & artifacts from Civil War artillery to Vietnam-Era anti-tank missiles.
Hours & Admission Prices: June-Labor Day daily 12-5; Labor Day-June Sat.-Sun. 12-5 & by appointment.
Membership: Youth $10; Adult $25; Family $50; Business $100; Lifetime $500.

Moccasin

UTICA MUSEUM, 5443 Indiana Rd., Moccasin, MT 59462-9533. Tel.: 406-423-5531 & 5538.

Congressional District: 2
Key Personnel: Pres., George Keating; Treas., Katherine Hodge; Sec., Judy Ennis.
Personnel Profile: Full-Time Volunteers 15; Part-Time Paid 1; Part-Time Volunteers 10.
Governing Authority: nonprofit. Affiliated with the Utica Historical Society. Tax-exempt: 170(b)(1)(A).
Institution Type/Description: Historical Society Museum.
Collections: artifacts pertaining to early homesteaders & ranchers.
Research Fields: local history.
Facilities: reading room. Charles Russel prints, agate & yogo sapphire jewelry for sale.
Activities: permanent exhibitions.
Publications: book, Utica I & II.
Hours & Admission Prices: Memorial Day-Labor Day Sat.-Sun. 10-5; other times by appointment. No charge; donations accepted. ᕣ

Attendance: 300 (estimated)
Membership: Individual $5; Life $100.

Pablo

THE PEOPLE'S CENTER, 53253 Hwy. 93 W., Pablo, MT 59855. Mailing Address: P.O. Box 278, Pablo, MT 59855-0278. Tel.: 406-675-0160. Fax: 406-675-0160.
E-mail: peoplescenter@cskt.org
Web Site: www.peoplescenter.net
Institution Type/Description: Native American History Museum.
Collections: photographs & negatives, artifacts, beaded bags, stone tools & dance outfits pertaining to the Salish, Kootenai & Pend d'Oreille tribes.
Hours & Admission Prices: June-Sept. Mon.-Sat. 9-5; Oct.-May Mon.-Fri. 9-5. Adults $5, seniors & students $3; discounts to groups.

Philipsburg

GRANITE COUNTY MUSEUM & CULTURAL CENTER, 135 S. Sansome, Philipsburg, MT 59858. Mailing Address: P.O. Box 502, Philipsburg, MT 59858-0502. Tel.: 406-859-3020.
E-mail: stephens@blackfoot.net
Web Site: www.granitecountymuseum.com
Founded: 1994.
Key Personnel: Chm. (V) & Pres. (V), T.J. Vietor; Museum Shop Mgr., Esther McDonald
Governing Authority: Tax-exempt.
Institution Type/Description: History Museum.
Collections: American heritage; period artifacts; local history.
Facilities: rental facilities. Museum-related items for sale.
Publications: annual newsletter.
Hours & Admission Prices: May-Oct. 15 daily 10-4. Adults $4; members no charge. &
Attendance: 2,500 (accurate)
Membership: Individual $15; Life $100.

Plentywood

SHERIDAN COUNTY MUSEUM, 4262 Hwy. 16 S., Plentywood, MT 59254. Mailing Address: P.O. Box 191, Plentywood, MT 59254. Tel.: 406-765-2145.
Founded: 1968.
Key Personnel: Dir., Pat Tange; Chm. (V), Gordon Aus.
Personnel Profile: Part-Time Paid 3.
Governing Authority: Parent Institution: Sheridan County. Tax-exempt.
Institution Type/Description: History Museum.
Collections: local history & culture; paintings; photographs; furnishings.
Hours & Admission Prices: Memorial Day-Labor Day daily 10-5; other times by appointment. No charge; donations accepted. &

Polson

MIRACLE OF AMERICA MUSEUM INC., 36094 Memory Lane, Polson, MT 59860-8446. Tel.: 406-883-6804.
E-mail: info@miracleofamericamuseum.org
Web Site: miracleofamericamuseum.org
Founded: 1985.
Key Personnel: Dir., Cathleen Wilde; C.E.O., W. Gilbert Mangels; Treas. & Museum Shop Mgr., Ned Wilde.
Personnel Profile: Full-Time Volunteers 2; Part-Time Paid 0; Part-Time Volunteers 12.
Governing Authority: nonprofit organization. Tax-exempt: 501(c)(3).
Institution Type/Description: General Museum.
Collections: Native Americans; woodworking; cowboys; musical instruments; blacksmithing; agriculture; engines; tractors; toys & dolls; Civil War-present, guns & militaria; motorcycles; snow vehicles; pioneer village; timber industry; wildlife; schools; clothing; wooden boats; vintage autos & trucks; Flathead cherry industry; photographs.
Facilities: 2,000-vol. library; sheet music; 41,000 sq ft. exhibit space; 4 acre pioneer village. Gift items & books for sale.
Activities: docent program; formal education programs for children; guided tours; lectures; loan & temporary & participatory exhibitions. Annual Event: Living History Days in July.
Publications: quarterly newsletter; souvenir booklet; brochure.
Hours & Admission Prices: Mon.-Sat. 9-5, Sun. 1:30-6; other times by appointment; call to confirm holiday hours. Adults 13 & over $5, children 2-12 $2; discounts to AAA & media; members & children under 2 no charge. Closed Christmas. &
Attendance: 14,000 (estimated)

Membership: Family $25; Sustaining $100; Life $1,000.

POLSON-FLATHEAD HISTORICAL MUSEUM, 708 Main St., Polson, MT 59860-3225. Mailing Address: P.O. Box 206, Polson, MT 59860-0206. Tel.: 406-883-3049.
Web Site: www.polsonflatheadmuseum.org
Founded: 1964.
Key Personnel: Pres. (V), Lois Hart.
Governing Authority: board. Tax-exempt.
Institution Type/Description: History Museum.
Collections: Salish Kootenia Tribal history; homesteading; stagecoach; chuckwagon; buggies; 1884 trading post; guns; boat; wild animal display; farm equipment.
Publications: biannual newsletter.
Hours & Admission Prices: Memorial Day to Sept. 15 Mon.-Sat. 10-5. Adults $3, seniors $2.50; members no charge. &
Attendance: 2,500 (estimated)
Membership: Individual $15; Family $25; Lifetime $200.

Poplar

POPLAR MUSEUM, 210 US Hwy. 2 E., Poplar, MT 59255. Mailing Address: P.O. Box 157, Poplar, MT 59255-0157. Tel.: 406-768-5223.
Founded: 1967.
Key Personnel: Chm. (V), Pat Beck; Pres. (V), Ida Norgaard.
Personnel Profile: Part-Time Paid 1; Part-Time Volunteers 2.
Governing Authority: bd. of dirs. Tax-exempt.
Institution Type/Description: History Museum: housed in old Tribal Jail, built around 1920.
Collections: Frontier & Indian collections; photographs; personal artifacts.
Hours & Admission Prices: June to early Sept. Mon.-Fri. 11-4. No charge; donations accepted.
Attendance: 500 (estimated)

Pryor

CHIEF PLENTY COUPS MUSEUM, 1 Egdar Rd., Pryor, MT 59066. Mailing Address: P.O. Box 100, Pryor, MT 59066-0100. Tel.: 406-252-1289. Fax: 406-252-6668.
E-mail: plentycoups@plentycoups.org
Web Site: www.plentycoups.org
Founded: 1972.
Congressional District: 2
Key Personnel: State Parks Mgr. Region 5, Doug Habermann; Chief Plenty Coups Museum Park Mgr., Susan Stewart.
Personnel Profile: Full-Time Paid 1; Part-Time Paid 2; Part-Time Volunteers 2; Interns 1.
Governing Authority: state. Parent Institution: Montana Fish, Wildlife & Parks.
Institution Type/Description: Crow Indian Museum.
Collections: ethnographic materials of the Crow people; paintings; drawings; prehistoric artifacts. Historic Structure: c.1909 building.
Research Fields: prehistory & history of the Crow people.
Facilities: Books, prints, Crow beadwork & handicrafts for sale.
Activities: guided tours. Annual Event: Chief Plenty Coups Day of Honor.
Publications: annual newsletter, Friends of Chief Plenty Coups Association.
Hours & Admission Prices: Park: May-Sept. daily 8-8; Oct.-April Tues.-Sat. 8-5. Museum: May-Sept. daily 10-5. Admission: $3 per person; state residents no charge. &
Attendance: 12,000 (estimated)
Membership: Friends of Chief Plenty Coups Association $5; $25; $50; $100; $200 & up.

Red Lodge

CARBON COUNTY HISTORICAL SOCIETY AND MUSEUM, 224 N. Broadway, Red Lodge, MT 59068. Tel.: 406-446-3667. Fax: 406-446-1920.
Web Site: www.carboncountyhistory.com
Key Personnel: Dir., Donna Madson; Pres. (V), Paul Henry; Preservation Officer, Deb Hronek; Collections, Dana Wahlquist; Museum Store Mgr., Cathy Wiltgen.
Personnel Profile: Full-Time Paid 2; Part-Time Paid 2; Part-Time Volunteers 10.
Governing Authority: Tax-exempt.
Institution Type/Description: History Museum.
Collections: local history & culture; photographs; personal artifacts.
Hours & Admission Prices: Summer: Mon.-Sat. 10-6, Sun. 11-3; Winter:

Tues.-Fri. 10-5, Sat. 11-3; groups by appointment. Adults $3, students $2; discounts to AAA members; school groups and children 5 & under no charge.
Membership: Annual $15-$10,000.

Richey

RICHEY HISTORICAL SOCIETY, (M), 122 S. Main St., Richey, MT 59259. Mailing Address: Box 264, Richey, MT 59259-0264. Tel.: 406-773-5234.
Founded: 1973.
Key Personnel: Pres., Tristan Veverka; Sec., Agnes Sullivan; Cur. & Treas, Wanda Zuroff.
Personnel Profile: Part-Time Paid 2; Part-Time Volunteers 2.
Governing Authority: nonprofit organization. Tax-exempt: 501(c)(3).
Institution Type/Description: Historical Society Museum.
Collections: farm tools; ranch items; books & photographs; post office items; homestead shack; old time store; clocks; piano; reed organs; typewriters; musical instruments; kitchen & living room furniture; kitchen utensils; clothing; machinery; newspaper files; church & school items; local histories. Historic Buildings: 1913 Lisk Creek School #92; 1916 Old Richey Jail; 1916-1928 Richey National Bank; 1916 Homestead Shack; 1925 outhouse.
Facilities: library of newspapers, school textbooks, U.S. Post Office record books from Richey, Axtell, Sullivan & Paxton and other books, available for use on premises.
Activities: guided tours; permanent exhibitions.
Publications: Honyocker's Heritage; Richey School Memories: A History of Richey Schools 1916-1991; Richey 1916-1966; History of Community United Methodist Church of Richey 1914-1989.
Hours & Admission Prices: Memorial Day to Labor Day Tues. & Thurs.; other times by appointment. No charge; donations accepted.
Attendance: 200 (estimated)
Membership: Individual $10; Couple $15; Family $25.

Ronan

GARDEN OF THE ROCKIES MUSEUM, 45356 Hwy. 93 S., Ronan, MT 59864-9649. Tel.: 406-676-3390 & 0977.
Founded: 1985.
Key Personnel: Pres. (V), Andrew Holmlund.
Governing Authority: Tax-exempt.
Institution Type/Description: History Museum: housed in a church, built in the early 1900s.
Collections: one-room schoolhouse; tool shed; farm machinery building; doctor's instruments.
Hours & Admission Prices: Memorial Day-Labor Day Mon.-Fri. 11-4. No charge; donations accepted.
Membership: Individual $5; Patron $6-$49; Pioneer $50-$99; Heritage $100-$499; Lifetime $500 & up.

Roundup

MUSSELSHELL VALLEY HISTORICAL MUSEUM, 524 First W., Roundup, MT 59072-2437. Tel.: 406-323-1403 & 1662.
E-mail: dparrott@midrivers.com
Web Site: www.mvhm.us
Founded: 1972.
Congressional District: 2
Key Personnel: Pres., Bonnie DeMaio; Vice Pres., Tim Stevens; Asst. Dir. & Treas., Shirley Parrott; Sec., Phyllis Adolph.
Personnel Profile: Part-Time Volunteers 15.
Governing Authority: nonprofit organization. Tax-exempt: 501(c)(3).
Institution Type/Description: History Museum.
Collections: schoolhouse, post office and country store displays; Indian artifacts; fossils; crystals; coal mine; blacksmith shop; local wildlife; cowboys; newspapers from 1908 to present. Historic Structures: 1884 cabin; 1932 Peitenpol airplane; 1920 brick school building; Clovis to Coal-11,000 year community Identity project.
Research Fields: fossils; artifacts.
Facilities: 7,000 sq. ft. exhibit space. Russell prints, postcards & books of local origin for sale.
Activities: guided tours; arts festivals; formally organized education programs; permanent & temporary exhibitions.
Publications: book, Roundup on Musselshell; settlers' ethnic cookbook, Old Time Chuck; Horizons or the Musselshell.
Hours & Admission Prices: May-Sept. daily 1-5. No charge; donations accepted.
Attendance: 2,000 (accurate)
Membership: Annual $5; Initial Membership $10; Life $100.

Scobey

DANIELS COUNTY MUSEUM AND PIONEER TOWN, 7 W. County Rd., Scobey, MT 59263. Mailing Address: P.O. Box 133, Scobey, MT 59263-0133. Tel.: 406-487-5965 & 2061.
E-mail: dcmuseum@nemont.net
Web Site: scobey.org
Founded: 1965.
Congressional District: 1
Key Personnel: Pres., Mike Thierin; Vice Pres., Edgar Richardson; Exec. Dir. & Museum Shop Mgr., Mary Richardson; Dir., Jacki Oic; Dir., Gordy Blomquist; Dir., Kristi Shipstead; Dir., Justin Hanson; Dir., Frank Edwards; Dir. & Sec., Gorden Crandell.
Personnel Profile: Full-Time Volunteers 2; Part-Time Paid 1; Part-Time Volunteers 80.
Governing Authority: individual operation. Tax-exempt: 501(c)(3).
Institution Type/Description: Village Museum & Pioneer Town.
Collections: Daniels County history; cultural & heritage artifacts; house; bank; blacksmith shop; service station; general store; saloon; ladies' shop; shack; fire hall; doctor's & dentist's office; post office; implement building; funeral home; school; watch repair shop; tractors & cars; 3 churches; hotel; visitors center; theatre; period cars; City Hall.
Research Fields: Daniel County.
Facilities: community rooms; rental facilities.
Activities: guided tours; rental facilities. Museum Sponsors: threshing bee & antique show.
Hours & Admission Prices: Memorial Day to Labor Day daily 12:30-4:30; Sept.-May Tues.10-2, Fri. 1-4; other times by appointment. Tours: adults $7, children 6-11 $4; discounts to AAM members; members & children under 5 no charge. &
Attendance: 1,500 (estimated)
Membership: Individual $15; Family $40.

Seeley Lake

SEELEY LAKE HISTORICAL MUSEUM AND VISITOR CENTER, 2920 Mt. Hwy. 83 N., Seeley Lake, MT 59868-8605. Tel.: 406-677-2990. Facebook: Seeley Lake Historical Society.
E-mail: slhistory@blackfoot.net
Web Site: www.seeleyhistory.org
Founded: 2004.
Personnel Profile: Part-Time Volunteers 7.
Governing Authority: Parent Institution: Seeley Lake Historical Society. Tax-exempt.
Institution Type/Description: History Museum.
Collections: local history & culture; school artifacts; logging; Native American; photographs; documents.
Facilities: storage building.
Publications: Cabin Fever.
Hours & Admission Prices: Memorial Day to Labor Day daily 10-6; Winter: Tues.-Sat. 10-2. No charge; donations accepted. &
Attendance: 2,000 (estimated)
Membership: $30; $50; $100; $200.

Shelby

MARIAS MUSEUM OF HISTORY AND ART, 1129 1st St. N., Shelby, MT 59474. Tel.: 406-424-2551. Fax: 406-424-5422.
Founded: 1963.
Congressional District: 2
Key Personnel: C.E.O., Chm. (V) & Pres. (V), Larry Munson; Bd. Member & Vice Pres., Dean Hellinger; Bd. Member, Meredith Beckedahl; Bd. Member & Treas., Yvonne McAlpine; Bd. Member, Mary Mielke; Bd. Member & Sec., Carol Mundt; Bd. Member, Charlotte Hanson; Bd. Member, Ray Tomsheck; Bd. Member, Barbara Mercer; Bd. Member, Marian Hinds; Bd. Member, Merle Raph.
Personnel Profile: Part-Time Paid 3; Part-Time Volunteers 5.
Governing Authority: county; nonprofit. Tax-exempt.
Institution Type/Description: History Museum.
Collections: rocks; fossils; American Indian artifacts; military guns; bottles; toys; clothing; glassware; silver; musical instruments; cameras; phones; office furniture; rooms depicting early-day scenes; 1923 world champion boxing match memorabilia.
Research Fields: local history of school, churches, communities.
Facilities: library of early newspapers, books, magazines & maps.
Activities: temporary exhibitions; hands-on exhibits for children.
Publications: Toole County History Calendar: Toole County School Treasures.
Hours & Admission Prices: June-Aug. Mon.-Fri. 1-5 & 7-9, Sat. 1-5; Sept.-May Tues. 1-4; other times by appointment. No charge; donations accepted. &

Attendance: 1,327 (accurate)
Membership: Annual $1; Life $25.

Sidney

MONDAK HERITAGE CENTER, (M), 120 Third Ave., S.E., Sidney, MT 59270-4324. Tel.: 406-433-3500. Fax: 406-433-3503. Facebook: MonDak Heritage Center.
E-mail: mdhc@richland.org
Web Site: www.mondakheritagecenter.org
Founded: 1967.
Congressional District: 2
Key Personnel: Dir., Benjamin L. Clark; Museum Shop Mgr., Leann K. Pelvit.
Personnel Profile: Full-Time Paid 2; Part-Time Paid 1; Part-Time Volunteers 35.
Volunteer Hours: 3,200
Governing Authority: individual operation. Parent Institution: MonDak Historical & Art Society. Tax-exempt.
Institution Type/Description: Historical Museum & Art Gallery.
Collections: historical artifacts; original art of the western U.S.; street scene composed of homesteader town.
Research Fields: local history; genealogy.
Facilities: 250-vol. art library & 1,200-vol. history library.
Activities: gallery talks; arts festivals; hobby workshops; formally organized education programs; permanent & temporary exhibitions.
Publications: book of family histories, Courage Enough, Courage Enough II, Focus on Our Roots; quarterly newsletter.
Hours & Admission Prices: Tues.-Fri. 10-4, Sat. 1-4. No charge; donations accepted. &
Attendance: 8,000 (accurate)
Membership: Senior Citizen $15; Family $35; Business $100; Other $50-$1,000.

Stanford

JUDITH BASIN COUNTY MUSEUM, 93 3rd St., S., Stanford, MT 59479. Mailing Address: P.O. Box 315, Stanford, MT 59479-0038. Tel.: 406-566-2777 ext. 130.
Founded: 1966.
Congressional District: 2
Key Personnel: Pres. (V) & Dir., Florence E. Harris, Dir., Oliver Olson; Cur., Virginia Hayes; Dir., Gene Ernst; Dir., Tess Brady; Dir., Lorraine Boeck.
Personnel Profile: Full-Time Paid 1; Part-Time Paid 1.
Governing Authority: Judith Basin County. Tax-exempt.
Institution Type/Description: General Museum.
Collections: American Indian artifacts; period furnishings; 2,500-piece salt & pepper shaker collection; 50,000 piece button collection; Russell prints; period toys; clothes; canes; war items.
Hours & Admission Prices: June-Aug. Mon.-Fri. 10 to noon & 1-5. No charge; donations accepted. Closed major holidays. &
Attendance: 150 (estimated)

Stevensville

HISTORIC ST. MARY'S MISSION, West End of 4th St., Stevensville, MT 59870. Mailing Address: P.O. Box 211, Stevensville, MT 59870-0211. Tel.: 406-777-5734. Facebook: Historic St. Mary's Mission.
E-mail: stmary@cybernet1.com
Web Site: www.saintmarysmission.org
Key Personnel: Dir., Colleen Meyer; Pres. (V), Johni Steinke.
Personnel Profile: Part-Time Paid 1; Part-Time Volunteers 30.
Governing Authority: Parent Institution: Historic St. Mary's Mission, Inc. Tax-exempt.
Institution Type/Description: History Museum: founded in 1841 by Jesuit priests.
Collections: original hand-carved furnishings; Salish artifacts; historic documents; photographs.
Hours & Admission Prices: April 17-Oct. 15 Tues.-Sat. 10-4. Adults $7.
Attendance: 5,200 (estimated)

STEVENSVILLE MUSEUM, 517 Main St., Stevensville, MT 59870-2838. Mailing Address: P.O. Box 750, Stevensville, MT 59870-0750. Tel.: 406-777-1007 & 3546.
Institution Type/Description: History Museum.
Collections: local history & culture; photographs; Native American; period furnishings.
Hours & Admission Prices: Memorial Day to Labor Day Thurs.-Sat. 11-4, Sun. 1-4; groups by appointment. No charge, donations accepted. &

Superior

MINERAL COUNTY MUSEUM & HISTORICAL SOCIETY, 2nd Ave. E., Superior, MT 59872-0301. Mailing Address: P.O. Box 533, Superior, MT 59872-0533. Tel.: 406-822-3543.
E-mail: mchs1976@blackfoot.net
Web Site: mineralmtmuseum.com
Founded: 1975.
Congressional District: 1
Key Personnel: Pres. Historical Society, Peggy Temple; Vice Pres., Sue McLees; Cur. & Sec., Cathryn Strombo.
Personnel Profile: Part-Time Volunteers 2.
Governing Authority: county. Tax-exempt.
Institution Type/Description: Mineralogy Museum: housed in old hospital building; John Mullan & Mullan Trail Information Center.
Collections: history of Gold Rush; minerals; history of Milwaukee & Northern Pacific Railroad; Native American artifacts; tree nursery; mineral collection; manuscript collection; Captain John Mullan Papers, materials related to Mullan Military Road.
Research Fields: Western Montana Pioneers; Lewis & Clark Trail; collection of folklore; Captain John Mullan, U.S. Army engineer.
Facilities: library.
Activities: guided tours; films; formally organized education programs for children; loan, permanent & temporary exhibitions; costume loan.
Publications: quarterly, Mullan Chronicles.
Hours & Admission Prices: By appointment. No charge; donations accepted. &
Attendance: 1,000
Membership: Individual $5; Family $10.

Terry

PRAIRIE COUNTY MUSEUM AND EVELYN CAMERON GALLERY, 101 S. Logan, Terry, MT 59349. Mailing Address: P.O. Box 368, Terry, MT 59349-0426. Tel.: 406-635-4040.
Founded: 1974.
Key Personnel: Pres., Vadnae Greenfield; Pres. (V), Ruth Franks
Governing Authority: Parent Institution: Prairie County Board of Commissioners. Tax-exempt.
Institution Type/Description: Historic Building: housed in the 1916 State Bank of Terry.
Collections: local history & culture; early pioneer life; photographs; Terry Tribune Newspapers 1907 to present; obituary file 1907 to present; manuscripts 1880-2007.
Research Fields: genealogy; local history.
Facilities: library.
Hours & Admission Prices: Memorial Day to Labor Day Mon. & Wed.-Fri. 9-3, Sat.-Sun. 1-4; other times by appointment. No charge; donations accepted.
Attendance: 1,000 (estimated)

Thompson Falls

OLD JAIL MUSEUM, 109 S. Madison St., Thompson Falls, MT 59783. Mailing Address: P.O. Box 774, Thompson Falls, MT 59873-0774. Tel.: 406-827-4002.
E-mail: schs.thompsonfall@gmail.com
Web Site: saleeshhouse.org
Founded: 1980.
Congressional District: 13
Institution Type/Description: History Museum: former jail, sheriff's office & residence.
Collections: historical artifacts; photographs; maps.
Hours & Admission Prices: Mother's Day-Labor Day daily 12-4. Adults $2.
Attendance: 2,000 (estimated)

Three Forks

HEADWATERS HERITAGE MUSEUM, 202 S. Main, Three Forks, MT 59752. Mailing Address: P.O. Box 116, Three Forks, MT 59752-0116. Tel.: 406-285-4778. Facebook: Headqaters Heritage Museum.
E-mail: museumthreeforks@aol.com
Web Site: www.tfhistory.org
Founded: 1979.
Key Personnel: Dir., Cur. & Museum Shop Mgr., Robin Cadby-Sorensen; Pres. (V), Pat O'Brien Townsend; Treas., Patrick Finnegan.
Personnel Profile: Full-Time Volunteers 1; Part-Time Volunteers 49.
Governing Authority: Parent Institution: Three Forks Area Historical Society. private; nonprofit organization. Tax-exempt: 501(c)(3).

Institution Type/Description: History Museum.
Collections: local history artifacts & memorabilia; anvil from a fur trappers trading post, 1810; barbed wire; fossils; rocks; arrowheads; maps; agricultural tools; photographs; papers; canoe; 29 1/2 lb brown trout; quilts; railroad memorabilia.
Facilities: Museum-related items for sale.
Activities: guided tours; lectures; loan, temporary & traveling exhibitions; docent program. Annual Events: Christmas Stroll, Ice Cream Social.
Publications: quarterly newsletter; Book: Selected Papers of the 2010 Fur Trade Symposium at the Three Forks.
Hours & Admission Prices: June-Sept. Mon.-Sat. 9-5, Sun. 11-3; other times by appointment. No charge; donations accepted.
Attendance: 2,800 (estimated)
Membership: General Membership $25; Gallatin Patron $125; Madison Patron $250;Jefferson Patron $500; Missouri Patron $1,000; Headwaters Patron $5,000.

Townsend

BROADWATER COUNTY MUSEUM, 133 N. Walnut, Townsend, MT 59644-2324. Mailing Address: P.O. Box 614, Townsend, MT 59644. Tel.: 406-266-5252. Fax: 406-266-5252.
E-mail: bwcomuseum@mt.net
Web Site: www.broadwatercountymuseum.com
Founded: 1976.
Key Personnel: Pres. (V), Paul Putz; Cur., Linda Huth.
Personnel Profile: Full-Time Paid 1; Part-Time Paid 1.
Governing Authority: Tax-exempt.
Institution Type/Description: History Museum.
Collections: local history; mining; dental; homestead cabin; general store; blacksmith shop; early photographic equipment; fossils; Lewis and Clark; military; artifacts; genealogy.
Hours & Admission Prices: May 15-Sept. 15 daily 1-5. No charge; donations accepted. &
Attendance: 600 (estimated)
Membership: Individual $3; Family $5; Life $100.

Victor

VICTOR HERITAGE MUSEUM, Blake & Main St., Victor, MT 59875. Mailing Address: P.O. Box 610, Victor, MT 59875-0610. Tel.: 406-642-3997.
E-mail: victormuseum@cybernet1.com
Founded: 1989.
Congressional District: 1
Personnel Profile: Part-Time Volunteers 10.
Governing Authority: private; nonprofit organization. Tax-exempt: 501(c)(3).
Institution Type/Description: History & Heritage Museum: housed in the former NP Railroad Depot Building.
Collections: Victor's history from early 1900s; personal artifacts; folk culture.
Research Fields: Victor area residents before the 1930s.
Activities: Annual Events: Chocolate Tasting Party & Auction in December.
Publications: Bitterroot Trails IV; The Victor Community.
Hours & Admission Prices: Memorial Day to Labor Day Tues.-Sat. 1-4. No charge; donations accepted. &
Attendance: 500 (estimated)
Membership: Individual $10.

Virginia City

VIRGINIA CITY MADISON COUNTY HISTORICAL MUSEUM DBA VIRGINIA CITY J. SPENCER WATKINS MEMORIAL MUSEUM, 219 W. Wallace St., Virginia City, MT 59755. Mailing Address: P.O. Box 215, Virginia City, MT 59755-0215. Tel.: 406-843-5484. Fax: 406-843-5484.
E-mail: madsnian@3rivers.net
Founded: 1958.
Congressional District: 1
Key Personnel: C.E.O. & Dir., Daryl L. Tichenor.
Personnel Profile: Full-Time Volunteers 1.
Governing Authority: municipal. Tax-exempt: 501(c)(3).
Institution Type/Description: History Museum.
Collections: mounted animals; historical western items; furniture, mining & other things dating back to the 1800's.
Facilities: Books, gifts & Western paintings for sale.
Publications: book, Foot-Steps thru History
Hours & Admission Prices: May 15-Sept. daily 10-5. No charge; donations accepted.
Attendance: 20,000 (estimated)

Membership: Individual, Family & Business $25.

West Glacier

GLACIER NATIONAL PARK MUSEUM, Glacier National Park, West Glacier, MT 59936. Mailing Address: P.O. Box 128, West Glacier, MT 59936. Tel.: 406-888-7936. Fax: 406-888-7937.
E-mail: deirdre_shaw@nps.gov
Founded: 1930.
Congressional District: 2
Key Personnel: Cur., Deirdre Shaw.
Personnel Profile: Full-Time Paid 1; Part-Time Paid 2; Part-Time Volunteers 1.
Governing Authority: federal government agency.
Institution Type/Description: Park Museum.
Collections: geologic, bird, insect & animal specimens collected within the park; photographs, negatives, lantern slides, prints, film & oral histories of the park, Blackfeet Indians, visitors, residents & park personnel; clothing & furnishings documenting park history; herbarium.
Facilities: 14,000-vol. library of history & scientific material, available for use by the public.
Activities: tours & research by appointment only.
Hours & Admission Prices: Visitor Centers: call for hours. Research: Mon.-Fri. 8-4:30. No charge. Closed federal holidays.

West Yellowstone

EARTHQUAKE LAKE VISITOR CENTER, U.S. Hwy. 287, West Yellowstone, MT 59758. Mailing Address: P.O. Box 520, West Yellowstone, MT 59758-0520. Tel.: 406-682-7620. Fax: 406-823-6990.
Web Site: www.fs.fed.us/r1/gallatin
Formerly: Madison Canyon Earthquake Lake Visitor Center
Personnel Profile: Full-Time Paid 1; Part-Time Paid 3; Interns 1.
Institution Type/Description: History Museum.
Collections: 1959 earthquake history; geology; seismograph.
Activities: educational programs.
Hours & Admission Prices: Call for hours.
Attendance: 30,000 (estimated)

GRIZZLY & WOLF DISCOVERY CENTER, 201 S. Canyon, West Yellowstone, MT 59758. Mailing Address: P.O. Box 996, West Yellowstone, MT 59758-0996. Tel.: 800-257-2570; 406-646-7001. Fax: 406-646-7004.
E-mail: info@grizzlydiscoveryctr.com
Web Site: grizzlydiscoveryctr.org
Founded: 1993.
Personnel Profile: Full-Time Paid 15; Part-Time Paid 4; Part-Time Volunteers 4; Interns 1.
Governing Authority: nonprofit organization.
Institution Type/Description: Wildlife Preserve.
Collections: bears; wolves; taxidermic bear mounts; raptors; photographs.
Facilities: Museum-related items for sale.
Hours & Admission Prices: Daily 8am to dusk; call to confirm. Adults 13 & over $10.50, seniors 62 & over $9, children 5-12 $5; discounts to groups; children under 5 no charge. &
Attendance: 130,000 (estimated)
Membership: Senior $30; Family $65.

YELLOWSTONE HISTORIC CENTER MUSEUM, (M), 104 Yellowstone Ave., West Yellowstone, MT 59758. Mailing Address: P.O. Box 1299, West Yellowstone, MT 59758. Tel.: 406-646-1100. Fax: 406-646-1100.
E-mail: info@yellowstonehistoriccenter.org
Web Site: www.yellowstonehistoriccenter.org
Founded: 1999.
Key Personnel: Cur., Paul Shea.
Institution Type/Description: History Museum: housed in the historic Union Pacific Depot.
Collections: memorabilia collected by early visitors; historic transportation pieces; photographs; personal diaries & journals.
Hours & Admission Prices: Call for hours. Adults $5.
Membership: Individual $25; Family $50; Associate $100; Sponsor $250; Wonderland $1,000; Yellowstone $2,500; Heritage $5,000.

White Sulphur Springs

MEAGHER COUNTY HISTORICAL ASSOCIATION CASTLE MUSEUM, 310 1/2 2nd Ave., N.E., White Sulphur Springs, MT 59645. Mailing Address: P.O. Box 716, White Sulphur Springs, MT 59645-0389. Tel.: 406-547-2324.
Founded: 1967.
Congressional District: 1
Key Personnel: Pres., James Reed; Vice Pres., Vonnie Pederson; Sec., Marga Johnson; Treas., Polly Hanson.
Personnel Profile: Full-Time Paid 1; Part-Time Paid 1; Part-Time Volunteers 6.
Governing Authority: nonprofit organization. Tax-exempt.
Institution Type/Description: Historical Society Museum: housed in 1890 restored Victorian Mansion.
Collections: furniture & furnishings; Indian artifacts; carriages in new buildings.
Activities: guided tours.
Publications: The Man Who Built The Castle; Born To Be.
Hours & Admission Prices: Mid-May to mid-Sept. daily 10-6. Adults $5, children & senior citizens $2; discount to groups of 10 or more; members no charge. &
Attendance: 4,012 (estimated)
Membership: Individual $10.

Whitefish

STUMPTOWN HISTORICAL SOCIETY MUSEUM, 500 Depot St., Ste. 101, Whitefish, MT 59937-2567. Tel.: 406-862-0067.
Web Site: www.stumptownhistoricalsociety.org
Key Personnel: Exec. Dir., Jill Evans
Institution Type/Description: Historical Society Museum.
Collections: railroad & community artifacts; photographs.
Hours & Admission Prices: Mon.-Sat. 10-4.

WHITEFISH GALLERY - STUMPTOWN ART STUDIO, 145 Central Ave., Whitefish, MT 59937. Tel.: 406-862-5929. Fax: 406-862-5029.
E-mail: info@stumptownartstudio.org
Web Site: www.whitefishgallerynights.org
Institution Type/Description: Art Gallery.
Collections: paintings; sculpture; photographs.
Activities: special events; classes; educational programs; workshops.
Hours & Admission Prices: Call for hours.

Whitehall

JEFFERSON VALLEY MUSEUM, 303 S. Division, Whitehall, MT 59759. Mailing Address: P.O. Box 902, Whitehall, MT 59759-0902. Tel.: 406-287-7813. Facebook: Jefferson Valley Museum.
E-mail: jvmuseum@hotmail.com
Founded: 1990.
Key Personnel: Pres. (V), Janice Carmody; Sec., Roy Millegan.
Personnel Profile: Part-Time Volunteers 12.
Governing Authority: bd. of directors. Tax-exempt.
Institution Type/Description: History Museum: housed in a restored 1914 barn.
Collections: area history & heritage; genealogy; railroad memorabilia; personal artifacts; photographs; brand & spurs.
Research Fields: genealogy.
Activities: research.
Publications: annual newsletter, Tumbleweed.
Hours & Admission Prices: Memorial Day to Sept. 15 Tues.-Sun. 12-4; other times by appointment. No charge; donations accepted. &
Attendance: 750 (accurate)
Membership: Active $10; Family $20; Sustaining $25; Commercial $100; Life $250; Patron $500; Benefactor $1,000.

Wibaux

PIERRE WIBAUX MUSEUM, 112 E. Orgain Ave., Wibaux, MT 59353. Mailing Address: P.O. Box 72, Wibaux, MT 59353-0072. Tel.: 406-796-9969. Fax: 406-796-2625.
E-mail: wpmuseum@midrivers.com
Founded: 1969.
Congressional District: 38
Key Personnel: Chm. (V) & Sec., Lyda Schneidez; Pres. (V), Donna O'Conner.
Personnel Profile: Full-Time Paid 1; Part-Time Paid 2; Part-Time Volunteers 5.
Volunteer Hours: 50
Operating Expenses: 30,000
Operating Income: 34,679
Governing Authority: Parent Institution: Wibaux County. Tax-exempt.
Institution Type/Description: History Museum.
Collections: local history; personal artifacts; early settlers; period barber shop; livery stable; Montana Centennial Train Car; gardens.
Facilities: gardens; rental facilities. Museum-related items for sale.
Activities: children's activities; tours; rental facilities; weddings. Museum Sponsors: Teddy Bears Picnic.
Hours & Admission Prices: May-Sept. Mon.-Sat. 9-5, Sun. 1-5. No charge; donations accepted. &
Attendance: 200 (estimated)

Winifred

WINIFRED MUSEUM, 210 Main St., Winifred, MT 59489. Mailing Address: P.O. Box 181, Winifred, MT 59489-0181. Tel.: 406-462-5425. Fax: 406-462-5425.
E-mail: winimuse@mtintouch.net
Founded: 2005.
Institution Type/Description: History Museum.
Collections: over 3,000 Tonka toys; archaeological, homestead & military displays.
Hours & Admission Prices: Memorial Day weekend-Labor Day weekend Wed.-Sun. 12:30-5:30. No charge; donations accepted.

Wisdom

BIG HOLE NATIONAL BATTLEFIELD, 16425 Hwy. 43 W., Wisdom, MT 59761. Mailing Address: P.O. Box 237, Wisdom, MT 59761-0237. Tel.: 406-689-3155. Fax: 406-689-3151.
E-mail: biho_visitor_information@nps.gov
Web Site: www.nps.gov/biho
Founded: 1910.
Congressional District: 1
Key Personnel: Supt. & Unit Mgr., Steve Black.
Governing Authority: federal. Parent Institution: National Park Service, U.S. Dept. of Interior. Tax-exempt.
Institution Type/Description: Visitor Center Exhibit Hall: located on land which includes all major sites relating to the 1877 Battle of the Big Hole between the Nez Perce Indians & the 7th U.S. Infantry. Plus a small number of citizen volunteers.
Collections: historical & ethnological specimens of Nez Perce Indian culture; archaeology; 1870s army accoutrements, equipment, uniforms & firearms; manuscript collection.
Research Fields: Indian wars; Nez Perce War.
Facilities: 300-vol. library of books & microfilms available on premises; 35-seat auditorium. Publications for sale.
Activities: self-guided tours; lectures; films; gallery talks; permanent exhibitions; interpretive programs.
Publications: pamphlet, Big Hole National Battlefield, Montana; annual visitor guide.
Hours & Admission Prices: Winter & Spring: 10-5; Summer & Fall: 9-5. No charge; donations accepted. Closed New Year's Day; Martin Luther King Day; Presidents Day; Columbus Day; Veterans Day; Thanksgiving; Christmas. &
Attendance: 50,000 (accurate)

Wolf Point

WOLF POINT AREA MUSEUM INC., 203 U.S. Hwy. 2, Wolf Point, MT 59201-1507. Tel.: 406-653-1912 & 1958.
Formerly: Wolf Point Area Historical Society
Founded: 1972.
Congressional District: 2
Key Personnel: Dir., Keith Bryan; Pres. (V) & Museum Shop Mgr., Herman E. Shumway.
Governing Authority: nonprofit organization. Tax-exempt: 170(b)(1)(I).
Institution Type/Description: Historical Museum.
Collections: Western artifacts & pictures; Native American artifacts; ranching; homesteading; bead work; star quilt making; wood carving; free-hand paintings.
Activities: narrated video tape of museum exhibits at senior citizens center.
Hours & Admission Prices: Memorial Day to Labor Day Mon.-Fri. 10-5. No charge; donations accepted. &
Attendance: 1,100 (estimated)

NEBRASKA

(244 listings)

Ainsworth

THE COLEMAN HOUSE MUSEUM AND BROWN COUNTY HISTORICAL SOCIETY, 456 Old Highway #7, Ainsworth, NE 69210. Mailing Address: 339 North Ash St., Ainsworth, NE 69210-1612.
Founded: 1973.
Congressional District: 3
Key Personnel: Pres., Daniel W. Steele; Financial Dir., Marilyn A. Calver.
Personnel Profile: Part-Time Volunteers 22.
Governing Authority: private; nonprofit organization. Parent Institution: The Brown County Historical Society, Ainsworth, NE. Tax-exempt: 170(b)(1)(A).
Institution Type/Description: History Museum: housed in 1918 home.
Collections: history of Brown County, NE & Nebraska; books on local history.
Research Fields: local history & genealogy.
Activities: guided tours; quilting.
Publications: newsletter; Pioneer Doctors of Brown County; Tales of Brown County, Nebraska; Early History of Brown County, Nebraska; Postal History of Brown County, NE; reprinted pioneer stories of Brown, Keya Paha & Rock Counties, NE.
Hours & Admission Prices: Summer: Fri.-Sat. 11-4; other times by appointment. No charge; donations accepted. Closed Memorial Day; Independence Day; Thanksgiving; Christmas. &
Attendance: 259 (accurate)
Membership: Individual $10; Sustaining $12; Business $25; Lifetime $100.

SELLORS BARTON MUSEUM AKA LOG CABIN MUSEUM, 4th & Main St., City Park, Ainsworth, NE 69210-1213. Mailing Address: 43026 Finch Rd., Ainsworth, NE 69210. Tel.: 402-387-2740 (City Office).
Founded: 1936.
Congressional District: 43
Key Personnel: Dir., Carol Larson
Governing Authority: city of Ainsworth. Tax-exempt.
Institution Type/Description: History Museum: housed in log cabin built in 1936.
Collections: local history; furnishings; household artifacts; clothing; photographs.
Research Fields: local history & genealogy.
Facilities: picnic area.
Activities: Annual Event: Arts in the Park in July.
Hours & Admission Prices: late May to early Sept. Mon.-Fri. 10:30-4:30, Sat.-Sun. 1-5. No charge; donations accepted. &
Attendance: 500 (estimated)

Allen

DIXON COUNTY MUSEUM & HISTORICAL SOCIETY, 225 S. Clark St., Allen, NE 68710. Mailing Address: Box 95, Allen, NE 68710-0095. Tel.: 402-635-2582.
Web Site: www.visitdixoncounty.org
Founded: 1964.
Key Personnel: Pres. (V), Gloria Oberg
Governing Authority: Tax-exempt.
Institution Type/Description: History Museum.
Collections: county history; household & farming artifacts.
Research Fields: genealogy; Dixon County newspapers.
Hours & Admission Prices: June-Aug. Sun. 2-4; other times by appointment. No charge; donations accepted. &
Attendance: 350 (accurate)
Membership: Individual $5.

Alliance

CARNEGIE ARTS CENTER, 204 W. 4th St., Alliance, NE 69301-3332. Tel.: 308-762-4571. Fax: 308-762-4571.
E-mail: carnegieartscenter@bbc.net
Web Site: www.carnegieartscenter.com
Key Personnel: Dir., Lynne C. Messersmith.
Personnel Profile: Full-Time Paid 1; Part-Time Paid 2; Part-Time Volunteers 8.
Governing Authority: nonprofit organization. Tax-exempt.
Institution Type/Description: Art Gallery.
Collections: art exhibits.
Publications: quarterly activities newsletter.

Hours & Admission Prices: Tues.-Sat. 10-4, Sun. 1-4. No charge; donations accepted. Closed major holidays. &
Attendance: 3,500 (estimated)
Membership: Individual $25; Family $45.

KNIGHT MUSEUM & SANDHILLS CENTER, (M), 908 Yellowstone Ave., Alliance, NE 69301. Mailing Address: P.O. Box D, Alliance, NE 69301. Tel.: 308-762-2384. Fax: 308-763-1168.
E-mail: museum@cityofalliance.net
Web Site: knightmuseum.com
Formerly: Knight Museum of High Plains Heritage
Founded: 1965.
Congressional District: 49
Key Personnel: Dir., Becci Thomas.
Personnel Profile: Full-Time Paid 1; Part-Time Paid 2; Part-Time Volunteers 10.
Governing Authority: municipal. Parent Institution: City of Alliance. Tax-exempt: 170(b)(1)(A).
Institution Type/Description: General Museum.
Collections: local history & railroad items; Indian artifacts; sod house; one room school; genealogy.
Research Fields: Burlington Railroad; western Nebraska history; genealogy; World War II Alliance Air Base.
Facilities: library of historical & genealogical materials.
Activities: guided tours; lectures; programs. Museum Sponsors: monthly ethnic display June-August.
Publications: brochure.
Hours & Admission Prices: May-Sept. Mon.-Fri. 8-7, Sat. 10-6, Sun. 1-5; Oct.-April Mon.-Fri. 8-5, Sat. 10-5. No charge; donations accepted. &
Attendance: 15,250 (accurate)

SALLOWS MILITARY MUSEUM, 1101 Niobrara St., Alliance, NE 69301. Mailing Address: P.O. Box D, Alliance, NE 69301-0770. Tel.: 308-762-2385.
Institution Type/Description: Military Museum.
Collections: military history from WWI to present; Alliance Airbase artifacts; personal artifacts; photographs.
Hours & Admission Prices: Summer: Mon.-Fri. 1-4, Sun. 1:30-4:30; call for additional hours.

Arapahoe

FURNAS-GOSPER HISTORICAL SOCIETY & MUSEUM, 401 Nebraska Ave., Arapahoe, NE 68922. Mailing Address: P.O. Box 202, Arapahoe, NE 68922-0202. Tel.: 308-962-5236. Facebook: Historical Society of Furnas-Gosper Counties, Nebraska.
Founded: 1966.
Congressional District: 3
Key Personnel: Pres., Robert Trosper; Vice Pres., Keith Willets; Treas. & Sec., Lori Moore; Dir., Cathy Weber.
Personnel Profile: Part-Time Volunteers 11.
Governing Authority: county. Parent Institution: Furnas & Gosper Counties Historical Society. Tax-exempt.
Institution Type/Description: General Museum.
Collections: clothing; books; glassware; piano; sewing machines; crockery; roll top desk; flags; pictures.
Research Fields: 1874 Justice of Peace court records of Furnas Co.; 1900s records of Furnas Co.
Facilities: 100-vol. set of encyclopedias & medical books of pioneer doctors available for research on premises.
Activities: guided tours; permanent & temporary exhibitions.
Hours & Admission Prices: May-Sept. Sat.-Sun. 1-4; other times by appointment. No charge; donations accepted. &
Attendance: 1,200 (estimated)
Membership: Individual $10; Family $20; Corporate $75; Life $150.

Ashland

LEE G. SIMMONS CONSERVATION PARK AND WILDLIFE SAFARI, 16406 292 St., Ashland, NE 68003. Tel.: 402-944-9453.
Web Site: www.wildlifesafaripark.com
Institution Type/Description: Conservation Park.
Collections: North American animals including bison, elk, cranes, wolves & bears.
Activities: hands-on exhibitions.
Hours & Admission Prices: April-Oct. daily 9-5. Adults 12 & over $6.50, seniors 65 & over $5.50, children 3-11 $4.50; discounts to military; children 2 & under no charge.

STRATEGIC AIR & SPACE MUSEUM, 28210 West Park Hwy., Ashland, NE 68003-3525. Tel.: 800-358-5029; 402-944-3100, ext. 220. Fax: 402-944-3160. Facebook: Strategic Air & Space Museum.
E-mail: aroeber@strategicairandspace.com
Web Site: www.sasmuseum.com
Formerly: Strategic Air Command Museum
Founded: 1959.
Congressional District: 2
Key Personnel: Exec. Dir., Scott Tarry; Cur., Brian York; Deputy Dir., Ken Schroeder; Chief Structural Specialist, Mark Hamilton; Dir. Operations, Krystal Boose; Dir. Mktg., Angela Roeber; Museum Shop Mgr., J.C. Colson.
Personnel Profile: Full-Time Paid 18; Part-Time Paid 26; Part-Time Volunteers 55; Interns 5.
Volunteer Hours: 18,141
Governing Authority: private; nonprofit organization. Parent Institution: Strategic Air Command & Museum Memorial Society, Inc. Tax-exempt.
Institution Type/Description: Aviation & Space History Museum.
Collections: aircraft; missiles; aviation artifacts.
Major Exhibits: Math Alive (T), 9/13-1/14.
Research Fields: aerospace history; SAC history; formation of the USAF; WWII; Cold War; (STEM) science, technology, engineering, and mathematics.
Facilities: research library; cafeteria; gallery; aircraft restoration area; theater; children's interactive educational programs. Museum-related items for sale.
Activities: symposia; films; permanent & temporary exhibitions; education programs; group tours; field trips; workshops.
Publications: newsletters; quarterly educator newsletter, Wings; e-newsletter.
Hours & Admission Prices: Daily 10-5. Adults $12, children 4-12 $6; discounts to groups, AAA members, senior citizens & military; children 3 & under and members no charge. Closed New Year's Day; Easter; Thanksgiving; Christmas. &
Attendance: 100,000 (estimated)
Membership: Student $25; Individual $40; Military Family $50; Family $75; Community Group $250. Patron Levels: Bronze $250-$499; Silver $500-$999; Giving Society $1,000.

WILLOW POINT GALLERY & MUSEUM, 1431 Silver St., Ashland, NE 68003. Tel.: 402-944-3613; 800-861-4260. Fax: 402-944-3613.
E-mail: gr35419@windstream.net
Web Site: www.generoncka.com/
Key Personnel: Dir. & C.E.O., Gene Roncka; Museum Shop Mgr., Mary Roncka.
Personnel Profile: Part-Time Paid 4.
Institution Type/Description: Art Gallery.
Collections: wildlife & historic paintings by Gene Roncka.
Hours & Admission Prices: Mon.-Sat. 10-6, Sun. 1-4. No charge. &
Attendance: 8,234 (accurate)

Ashton

POLISH HERITAGE CENTER, INC., 226 Carlton Ave., Ashton, NE 68817. Mailing Address: P.O. Box 3, Ashton, NE 68817-0003. Tel.: 308-738-2249 & 2260.
E-mail: pp1335@nctc.net
Web Site: www.polishheritagecenter.com
Founded: 1997.
Congressional District: 4
Key Personnel: Mgr., Phyllis Piechota; Pres. (V), Larry Molczyk; Sec., Therese Hulinsky; Treas., Virginia Pokorski.
Personnel Profile: Part-Time Volunteers 15.
Governing Authority: bd. of directors. Tax-exempt.
Institution Type/Description: Polish Heritage Center.
Collections: Polish culture & traditions; books; music; local genealogical records; photographs; art.
Facilities: library. Museum-related items for sale.
Activities: dance performances; special events; educational events. Annual Event: Polish Fest in September; Wigilia Christmas Dinner in December.
Publications: newsletter, Polish Heritage Center, Inc.
Hours & Admission Prices: Sun. 2-4; other times by appointment. Admission $3; members no charge. &
Attendance: 750 (estimated)
Membership: Yearly $10; Lifetime $100.

Atkinson

STURDEVANT-MCKEE MUSEUM & FOUNDATION, 308 S. Main St., Atkinson, NE 68713-4982. Mailing Address: P.O. Box 225, Atkinson, NE 68713-0225. Tel.: 402-304-6628.
E-mail: sturdevantmckeemuseum@hotmail.com
Key Personnel: Dir., Diane Alden
Institution Type/Description: History Museum.
Collections: local history; architecture; furnishings; household artifacts.
Hours & Admission Prices: By appointment. No charge; donations accepted.

Auburn

NEMAHA VALLEY MUSEUM, 1423 19th St., Auburn, NE 68305-2350. Tel.: 402-274-3203.
Institution Type/Description: History Museum.
Collections: Nemaha Valley history & industry; quilts; artwork; family histories; school records; historic buildings.
Activities: permanent & temporary exhibits; research.
Hours & Admission Prices: April-Dec. Tues.-Sun. 1-4:30; other times by appointment. No charge; donations accepted. Closed holidays.

Aurora

EDGERTON EXPLORIT CENTER, 208 16th St., Aurora, NE 68818-3009. Tel.: 402-694-4032. Fax: 402-694-4035. Facebook: Edgerton Explorit Center.
E-mail: mary@edgerton.org
Web Site: www.edgerton.org
Founded: 1995.
Congressional District: 3
Key Personnel: Exec. Dir., Mary Molliconi; Sr. Educator, Daniel Glomski; Educator, Jessica Brock.
Governing Authority: private; nonprofit organization. Tax-exempt: 501(c)(3).
Institution Type/Description: Science Museum.
Collections: hands-on science exhibits for all ages; personal & work related materials of Dr. Harold E. Edgerton, an inventor, educator & explorer.
Facilities: educational facilities; 10,000 sq. ft. exhibit space; 75-seat large screen theater. Museum-related items for sale.
Activities: guided tours; hobby workshops; lectures; mobile vans; school loan service, participatory & traveling exhibitions.
Publications: quarterly newsletter, Explorit News; quarterly teachers program guide, The Science Edge.
Hours & Admission Prices: Mon.-Sat. 9-4, Sun. 1-5. &
Attendance: 16,722 (accurate)
Membership: You & Me $55; Grandparent $65; Family $70; Early Childhood Care Provider $125; Adventurer $150; Explorer $500; Inventor $1,000.

PLAINSMAN MUSEUM, 210 16th St., Aurora, NE 68818-3009. Tel.: 402-694-6531. Fax: 402-694-6531.
E-mail: plainsman@hamilton.net
Web Site: www.plainsmanmuseum.org
Founded: 1935.
Congressional District: 3
Key Personnel: Dir. & Cur., Megan Sharp; Asst. Dir. & Museum Shop Mgr., Gary Gustafson.
Personnel Profile: Full-Time Paid 2; Part-Time Volunteers 60.
Governing Authority: society. Parent Institution: Hamilton County Historical Society. Tax-exempt: 501(c)(3).
Institution Type/Description: Historical Society Museum.
Collections: murals & mosaics tracing the local & regional history from prehistoric times to the 1860s; photographs of Hamilton County; E.A. Carlson Collection; 1,200 dolls including the Doris Lord collection; iron toys; period rooms including prairie chapel, Main Street, Pioneer Hamilton county; farmstead including cow barn, windmill, blacksmith shop surrounded with vintage implements & vehicles; schoolhouse. Historic Buildings. 1859 log cabin; sod house; 1881 homestead house; 1876 home of General Delevan Bates, Civil War Medal of Honor winner.
Research Fields: genealogies; ethnic groups in Hamilton County, Nebraska.
Facilities: 600-vol. library of books & publications available by appointment for research on premises; microfilm of 1890-1947 Hamilton Co. newspapers; microfilm reader/printer available for general use; reading room.
Activities: guided tours; lectures; local oral histories; permanent & temporary exhibitions; hands-on activities; musicals.
Publications: brochures; quarterly newsletter.
Hours & Admission Prices: April-Oct. Tues.-Sat. 9-5, Sun. 1-5; Nov.-March Tues.-Sat. 9-5. Adults $6, senior citizens over 62 $4, youth 5-16 $2; discounts to groups of 30 or more; children under 5 & members no charge. Closed New Year's Day; Easter; Thanksgiving; Christmas. &
Attendance: 4,000 (estimated)

Membership: Individual $15; Family $25; Life $1,000.

Bancroft

JOHN G. NEIHARDT STATE HISTORIC SITE, (M), 306 W.
Elm, Bancroft, NE 68004-0344. Mailing Address: P.O. Box 344,
Bancroft, NE 68004-0344. Tel.: 402-648-3388; 888-777-4667. Fax:
402-648-3388. Facebook: John G. Neihardt State Historic Site.
E-mail: neihardt@gpcom.net
Web Site: www.neihardtcenter.org
Formerly: John G. Neihardt Center
Founded: 1976.
Congressional District: 1
Key Personnel: Dir. & Museum Cur., Nancy S. Gillis; Pres. (V), Dr. Jon Cerny.
Personnel Profile: Part-Time Paid 3; Part-Time Volunteers 10.
Volunteer Hours: 32
Governing Authority: state; nonprofit organization. Parent Institution: Ne-
braska State Historical Society. Tax-exempt: 501(c)(3).
Institution Type/Description: Preservation Project.
Collections: memorabilia of John G. Neihardt, poet laureate of Nebraska,
author of Black Elk Speaks & A Cycle of the West.
Research Fields: literature; history on Native Americans; mountain men;
western history.
Facilities: Neihardt Study building; Neihardt Center Museum; Sacred Hoop
Prayer Garden.
Activities: school tours; research, literary programs; lectures.
Publications: brochures, Nebraska Historical Society; semi-annual newsletter.
Hours & Admission Prices: March-Nov. Mon.-Sat. 9-5, Sun. 1:30-5; Dec.-Feb.
Mon.-Fri. 9-5. No charge; donations accepted. Closed New Year's Day;
Memorial Day; Independence Day; Labor Day; Thanksgiving; Christmas.
&
Attendance: 2,600 (estimated)
Membership: Individual $25; Family & Sustaining $30.

Bartlett

**WHEELER COUNTY HISTORICAL SOCIETY AND COURT-
HOUSE MUSEUM,** Maine St. between 2nd & 3rd Sts., Bartlett,
NE 68622. Mailing Address: 83713 Derner Ranch Ave., Bartlett,
NE 68622-3046. Tel.: 308-654-3424.
Institution Type/Description: History Museum.
Collections: local history; pioneer artifacts.
Hours & Admission Prices: By appointment. No charge.

Bassett

**ROCK COUNTY HISTORICAL SOCIETY & HISTORICAL
MUSEUM,** W. Hwy. 20, Bassett, NE 68714. Mailing Address: P.O.
Box 32, Bassett, NE 68714. Tel.: 402-684-3327 & 3908.
Institution Type/Description: History Museum.
Collections: local history & culture; photographs; personal artifacts; period
furnishings.
Hours & Admission Prices: May to Oct. 1 daily 9-5; other times by
appointment. No charge. Closed major holidays.

Bayard

BAYARD DEPOT MUSEUM, 103 S. Main St., Bayard, NE 69334.
Mailing Address: P.O Box 345, Bayard, NE 69334. Tel.: 308-586-
1496.
E-mail: barker@actcom.net
Web Site: bayardmuseum.tripod.com/home
Founded: 1993.
Congressional District: 3
Key Personnel: Chm. (V) & Pres. (V), Richard Clarke.
Personnel Profile: Part-Time Volunteers 10.
Governing Authority: Tax-exempt.
Institution Type/Description: History Museum.
Collections: Bayard history; pioneer machinery; tools; quilts; schoolroom;
kitchen & bedroom artifacts.
Hours & Admission Prices: Call for hours. No charge; donations accepted. &
Attendance: 900 (estimated)
Membership: Annual $10; Lifetime $100.

**CHIMNEY ROCK NATIONAL HISTORIC SITE AND VISI-
TORS CENTER, (M),** 9822 Rd. 75, Bayard, NE 69334. Mailing
Address: P.O. Box F, Bayard, NE 69334. Tel.: 308-586-2581. Fax:
308-586-2589.
E-mail: loren.pospisil@nebraska.gov
Web Site: www.nebraskahistory.org
Founded: 1994.
Congressional District: 3
Key Personnel: C.E.O. & Dir., Michael J. Smith; Site Supvr., Loren Pospisil;
Assoc. Dir. Education & Interpretation, Ann Billesbach; Historic Sites Mgr.,
Deb McWilliams.
Personnel Profile: Full-Time Paid 3; Part-Time Paid 1.
Governing Authority: Parent Institution: Nebraska State Historical Society.
Tax-exempt.
Institution Type/Description: Historic Site & Visitors Center.
Collections: local history & culture; early 19th century westward migration;
photographs; period artifacts.
Research Fields: overland trails
Hours & Admission Prices: Daily 9-5. Adults $3. Closed major holidays. &
Attendance: 25,000 (accurate)
Membership: Individual $40; Household $55; Contributing $150; Supporting
$250; Sustaining $500; Founder $1,000.

Beatrice

GAGE COUNTY HISTORICAL SOCIETY AND MUSEUM, 101
N. 2nd, Beatrice, NE 68310. Mailing Address: P.O. Box 793,
Beatrice, NE 68310-0793. Tel.: 402-228-1679. Facebook: Gage
County Museum.
E-mail: gagecountymuseum@beatricene.com
Web Site: gagecountymuseum.info
Founded: 1971.
Congressional District: 1
Key Personnel: Pres. (V), Scott Knispel; Vice Pres., Arnold Baeher; Dir., Lesa
Arterburn; Treas., Bob Jones; Cur. & Museum Shop Mgr., Rita Clawson.
Personnel Profile: Full-Time Paid 2; Part-Time Paid 1; Part-Time Volunteers
150.
Governing Authority: board of directors. Parent Institution: Gage County
Historical Society. Branch Museum: Filley Stone Barn, 13282 E. Scott Rd.,
Filley, NE 68357. Tax-exempt: 501(c)(3).
Institution Type/Description: County History Museum: housed in 1906 Burl-
ington Passenger Station.
Collections: railroad artifacts; agricultural implements; materials on the
industrial development; medical equipment; artifacts of the communities &
the people of Gage County. Historic Building: 1874 Filley Stone Barn.
Major Exhibits: County Seats from a Place Called Home, 4/22/14-9/26/14.
Research Fields: local history.
Facilities: local history research materials available for use on premises only.
Activities: guided tours; annual meeting; programs of historical nature;
lectures; special events. Annual Events at the Elijah Filley Stone Barn:
Living History Program; Harvest Festival in October.
Publications: quarterly newsletter; books, Queen City of the Blue, Beatrice,
NE; Gage County History; 1888 Beatrice Illustrated; Kansas City, Wyan-
dotte & Northwestern Railroad Co. - Beatrice, NE.
Hours & Admission Prices: June-Sept. Tues.-Sat. 9-12 & 1-5, Sun. 1:30-5;
Oct.-May Tues.-Fri. 9-12 & 1-5, Sun. 1:30-5. No charge; donations
accepted. Closed major holidays. &
Attendance: 6,000 (accurate)
Membership: Individual $15; Family $25; Trailblazer $50; Sodbuster $100;
Homesteader $250.

HOMESTEAD NATIONAL MONUMENT OF AMERICA, 8523
W. State Hwy. 4, Beatrice, NE 68310-6743. Tel.: 402-223-3514.
Fax: 402-228-4231.
E-mail: home_interpretation@nps.gov
Web Site: www.nps.gov
Founded: 1936.
Congressional District: 3
Personnel Profile: Full-Time Paid 14; Full-Time Volunteers 8; Part-Time Paid
2; Part-Time Volunteers 400; Interns 3.
Governing Authority: federal. Parent Institution: National Park Service.
Tax-exempt.
Institution Type/Description: Agriculture Museum: homestead era history.
Collections: artifacts pertaining to homesteading during the late 19th century.
Historic Buildings: 1867 Palmer-Epard Cabin; 1872 Brick School House.
Research Fields: U.S. land laws history; agricultural history; settlement &
emigration patterns; prairie restoration.
Facilities: 400-vol. library of general history on Westward expansion & land

reform, available for use on the premises; auditorium; 60,000 objects research collection. Books, postcards and slides for sale.
Activities: interpretive exhibits & programs; orientation film; education programs (by reservation) & teacher workshops; 3-mile self-guided trail through prairie & woodlands; audio description system available for visually impaired.
Publications: Homestead National Monument of America Handbook.
Hours & Admission Prices: Mon.-Fri. 8:30-5, Sat.-Sun. 9-5. No charge. Closed New Year's Day; Thanksgiving; Christmas. &
Attendance: 70,000 (accurate)

YESTERDAY'S LADY, 113 N. 5th St., Beatrice, NE 68310-3902. Tel.: 402-223-5121.
E-mail: info@yesterdayslady.com
Web Site: www.yesterdayslady.com
Key Personnel: Owner, Sue McLain
Institution Type/Description: Victorian Clothing Museum.
Collections: women's Victorian fashions including clothing, hats & artifacts.
Activities: educational programs; group tours.
Hours & Admission Prices: By appointment.

Bellevue

FONTENELLE FOREST, 1111 N. Bellevue Blvd., Bellevue, NE 68005-4008. Tel.: 402-731-3140. Fax: 402-731-2403. Facebook: Fontenelle Forest.
E-mail: info@fontenelleforest.org
Web Site: www.fontenelleforest.org
Formerly: Fontenelle Nature Association
Founded: 1913.
Congressional District: 2
Key Personnel: Exec. Dir., Laura Lenarz Shiffermiller; Dir. Facilities, Gene Ericson; Dir. Research & Stewardship, Jeanine Lackey; Dir. Education, Rick Schmid; Dir. Devel., Christi Churchill; Dir. Communications, Brad A. Watkins.
Governing Authority: nonprofit organization. Parent Institution: Fontenelle Nature Association Board of Directors. Subsidiary Institutions: Fontenelle Forest Nature Center & Neale Woods Nature Center. Tax-exempt: 501(c)(3).
Institution Type/Description: Nature Center: Fontenelle Forest Nature Center located on 1,400-acre natural forest; Neale Woods Nature Center located on 550 acre wooded & prairie reserve.
Collections: indigenous flora & fauna of the Missouri River Valley forest and adjacent Loess Hills; deciduous forest; wetlands; prairie; live animals.
Research Fields: urban deer population, movements & control.
Facilities: Fontenelle Forest Nature Center: 1,400 acres of woods & wetlands; 17 miles of trails; boardwalks. Buffett Forest Learning Center: classrooms. Museum-related items for sale. Hitchcock Wetlands Learning Center: classrooms; offices; exhibits. Neale Woods Nature Center: 550 acres of woods & prairies; 9 miles of trails. Jonas Interpretive Center: exhibits; classroom; offices. Millard Observatory: telescopes; star watch deck.
Activities: environmental educational programs for children through adults; teacher training workshops; children & family campouts; summer day camp; children's nature clubs; astronomy programs; guided hikes; field trips; canoe trips; nature photography; permanent & temporary exhibitions. Annual Events: Earth Day, Native American Music Festival; Halloween & Christmas special events.
Publications: quarterly newsletter; Self-Guided Trail Guide; Mammals of Eastern Nebraska; Checklist of Nebraska Birds; Annual Report; School Programs Guide.
Hours & Admission Prices: Daily 8-5. Adults $7, senior 62 & up $6, children 2-17 $5; children under 2 no charge. Closed New Year's Day; Thanksgiving; Christmas. &
Attendance: 90,000 (estimated)
Membership: Individual $35; Two Individuals $45; Family $55; Patron $100-$249; Supporting Patron $250-$499; Sustaining Patron $500-$999; Distinguished Patron $1,000-$2,499; Benefactor $2,500 & up.

SARPY COUNTY HISTORICAL MUSEUM, 2402 Clay St., Bellevue, NE 68005-3932. Tel.: 402-292-1880.
E-mail: info@sarpycountymuseum.org
Web Site: www.sarpymuseum.org
Founded: 1970.
Congressional District: 2
Key Personnel: Dir., Ben Justman; Pres., Brenda Carlisle.
Personnel Profile: Full-Time Paid 1; Part-Time Volunteers 50.
Governing Authority: historical society; nonprofit. Tax-exempt: 501(c)(3).
Institution Type/Description: Historical Society Museum.
Collections: Sarpy County history & geographical area from 1800 to present;

prehistoric Indian artifacts. Historic House: 1840 The Log Cabin. Historic Site: 1835 Moses Merrill mission; 1869 first Nebraska depot.
Research Fields: local history; county architecture; oral history; genealogy.
Facilities: library & archives on Nebraska & Sarpy County history.
Activities: guided tours; lectures; outreach program; permanent, temporary & special exhibitions.
Publications: 10 monthly newsletters.
Hours & Admission Prices: Tues.-Sat. 10-4, 1st Sun. each month 1-4. No charge; donations accepted. &
Attendance: 3,400 (accurate)
Membership: Individual $15; Household $25; Sustaining $25; Patron $100; Booster $150.

Benkelman

DUNDY COUNTY HISTORICAL SOCIETY, 522 Araphahoe, Benkelman, NE 69021. Mailing Address: P.O. Box 634, Benkelman, NE 69021-0634. Tel.: 308-423-2750.
Founded: 1970.
Congressional District: 4
Key Personnel: Co-Pres., Betty Deyle; Co-Pres., Paul Erdman; Vice Pres., Lanita Anderson; Sec. & Treas., Shirley Mullanix.
Personnel Profile: Part-Time Volunteers 7.
Governing Authority: society; board of directors. Tax-exempt.
Institution Type/Description: General Museum.
Collections: rural school mementos; homestead days; Indian artifacts; relics of open range era; replicas of brands; railroad depot; doll collection; bank & post office artifacts; drug store.
Research Fields: tapes & information on pioneer life research library.
Facilities: collection of rural school textbooks & historical books available for research by appointment; reading room.
Activities: guided tours; lectures; hobby display specials, twice during summer.
Publications: biannual newsletter.
Hours & Admission Prices: June-Sept. Thurs. 2-5; other times by appointment. No charge; donations accepted.
Attendance: 380 (accurate)

Boys Town

BOYS TOWN HALL OF HISTORY & FATHER FLANAGAN HOUSE, 14057 Flanagan Blvd., Boys Town, NE 68010. Tel.: 402-498-1185. Fax: 402-498-1159.
E-mail: thomas.lynch@boystown.org
Web Site: www.boystown.org
Founded: 1986.
Congressional District: 2
Key Personnel: C.E.O. & Mgr., Thomas J. Lynch; Collections Assoc., Mark Daniels.
Personnel Profile: Full-Time Paid 2; Part-Time Paid 3; Part-Time Volunteers 40; Interns 2.
Governing Authority: nonprofit organization. Parent Institution: Father Flanagan's Boys' Town. Tax-exempt: 501(c)(3).
Institution Type/Description: History Museum: located in Boys Town.
Collections: public & administrative archives; film, tape & video archives; 500,000 photograph collection.
Research Fields: history of Father Edward J. Flanagan & Boys Town; The Boys Town Journal.
Facilities: 200-vol. library, including yearbooks & newsletters from 1918 to present; 8,500 sq. ft. exhibit space; 20-seat theater.
Activities: guided tours; docent program; organized educational programs for children & undergraduate and graduate college students affiliated with University of Nebraska at Omaha and Creighton University; temporary & traveling exhibitions; biannual alumni convention; elderhostel program.
Hours & Admission Prices: Daily 10-4:30. No charge; donations suggested. Closed New Year's Day; Easter morning; Thanksgiving; Christmas. &
Attendance: 60,000 (estimated)

Bridgeport

PIONEER TRAILS MUSEUM, Hwy. 26 & 385, Bridgeport, NE 69336. Mailing Address: P.O. Box 134, Bridgeport, NE 69336-0134. Tel.: 308-262-1117.
E-mail: pioneertrailsmuseum@hotmail.com
Key Personnel: Cur., Bern Miller
Institution Type/Description: History Museum.
Collections: local history & culture; irrigation; black powder shoots; Native American history, culture & artifacts.
Hours & Admission Prices: Memorial Day to Labor Day Mon.-Sat. 10-6, Sun. 1-6. No charge.

Broken Bow

CUSTER COUNTY HISTORICAL SOCIETY, INC., 445 S. 9th St., Broken Bow, NE 68822-2015. Mailing Address: P.O. Box 334, Broken Bow, NE 68822-0334. Tel.: 308-872-2203.
E-mail: cchs4@live.com
Web Site: www.rootsweb.com/~necuster
Founded: 1962.
Congressional District: 3
Key Personnel: Pres., Lance Bristol; C.E.O., Dir. & Museum Shop Mgr., Carol Christen; Sec., Dee Adams.
Personnel Profile: Part-Time Paid 1; Part-Time Volunteers 24.
Governing Authority: society. Affiliated with Custer County Historical Society. Tax-exempt: 501(c)(3).
Institution Type/Description: General Museum.
Collections: history of Custer County; items used in sod houses; store items from Wescott, Gibbins & Bragg store, Comstock; microfilm reader; 230 rolls of microfilm; S.D. Butcher photographs; Custer County artifacts.
Research Fields: County history.
Facilities: library of local & state history; genealogical books available on premises; reading room.
Activities: guided tours; lectures.
Publications: biannual newsletter, Custer County Times.
Hours & Admission Prices: Winter: Mon.-Fri. 1-4; Summer: Mon.-Fri. 10-4. No charge; donations accepted. Closed national holidays. &
Attendance: 3,000 (estimated)
Membership: Individual $20; Sustaining $25; Patron $50; Benefactor $100.

Brownville

BROWNVILLE HISTORICAL SOCIETY MUSEUM, 213 Main St., Brownville, NE 68521. Mailing Address: Box 1, Brownville, NE 68321. Tel.: 402-825-6001.
Web Site: brownville-ne.com
Founded: 1956.
Congressional District: 1
Key Personnel: Pres. Brownville Historical Society, Dr. Charles Anderson.
Personnel Profile: Part-Time Paid 1; Part-Time Volunteers 24.
Governing Authority: nonprofit organization. U.S. Army Corps of Engineer's Dredge Meriwether Lewis Museum. Subsidiary Institution: Brownville Fine Arts Association; Brownville Village Theatre. Tax-exempt: 501(c)(3).
Institution Type/Description: Railroad History Center.
Collections: history; costumes; glass; agriculture; Indian artifacts; children's furniture; tools. Wheel Museum: depot; land office; dental office. Historic Houses: 1860-1880 John Carson House; 1868 Gov. Furnas House; 1958 Captain Bailey House.
Facilities: library.
Activities: guided tours; art & music festivals; drama; workshops; craft seminars.
Publications: triannual, The Bulletin; book, The Brownville Story; annual newspaper, The Brownville Banner.
Hours & Admission Prices: mid-May to mid-Oct. Fri.-Sun. 1-5. Adults $2, children $1; members no charge.
Attendance: 3,200 (estimated)

MERIWETHER LEWIS DREDGE MUSEUM, Brownville State Recreation Area, Brownville, NE 68321. Mailing Address: 73088 646th Ave., Brownville, NE 68321-6025. Tel.: 402-825-3341.
Founded: 1977.
Congressional District: 1
Key Personnel: Pres. (V), Harold Davis; Museum Shop Mgr., Clay W. Kennedy.
Personnel Profile: Full-Time Volunteers 1; Part-Time Paid 1; Part-Time Volunteers 10.
Governing Authority: foundation. Meriwether Lewis Foundation in connection with Peru State College. Tax-exempt: 501(c)(3).
Institution Type/Description: History Museum: housed in The Captain Meriwether Lewis, a former Corps of Engineers steam powered dredge.
Collections: river-related material.
Research Fields: Missouri River Steam Boats 1817 to present.
Activities: Museum Sponsors: River Rats Reunion.
Publications: brochure.
Hours & Admission Prices: Spring & Fall Sat.-Sun. 1-5; June to mid-Sept. daily 1-5; groups by appointment. Adults $3, children 6-12 $1; discount to members; children under 5 no charge.
Attendance: 5,000 (estimated)
Membership: First Mate $15, Chief $25; Ensign $50; Captain $100; Commodore $500; Admiral $1,000.

Burwell

FORT HARTSUFF STATE HISTORICAL PARK, 82036 Fort Ave., Burwell, NE 68823-6122. Tel.: 308-346-4715. Fax: 308-346-4715.
Web Site: www.ngpc.state.ne.us
Founded: 1874.
Congressional District: 3
Key Personnel: Park Supt., Jim Domeier; Museum Shop Mgr., Mary Hughes.
Personnel Profile: Full-Time Paid 1; Part-Time Paid 5; Part-Time Volunteers 15.
Governing Authority: state. Parent Institution: Nebraska Game & Parks Commission. Nebraska State Historical Parks Bureau, 2200 N. 33rd, Lincoln, NE 68503. Tel. 402-464-0641. Tax-exempt.
Institution Type/Description: Military Museum: housed in 1874-1881 Fort Hartsuff.
Collections: period furniture; archaeological collections; military uniforms and equipment; tools; firearms collection. Historic Structures: nine restored 1870s military structures.
Research Fields: Great Plains military history.
Facilities: picnic area.
Activities: orientations by pre-arrangement; lectures; permanent exhibitions; firearms demonstration. Museum Sponsors: living history demonstrations May to September.
Publications: brochure.
Hours & Admission Prices: Memorial Day-Labor Day daily 9-8; other times by appointment. Fort: adults $2, children under 12 $1. State park vehicle entry permit required: annual $25, daily $5. &
Attendance: 20,000 (estimated)

GARFIELD COUNTY HISTORICAL MUSEUM, 737 H St., Burwell, NE 68823-0545. Tel.: 308-346-4445 & 4307.
Web Site: www.visitburwell.org
Institution Type/Description: History Museum.
Collections: local history & culture; period furnishings; photographs; personal artifacts; general store; log cabin; pioneer heritage.
Hours & Admission Prices: May-Oct. Mon.-Fri. 1-5; Nov.-April Mon.-Fri. 9-11am; other times by appointment. &

Butte

BUTTE COMMUNITY HISTORICAL CENTER AND MUSEUM, 721 First St., Butte, NE 68722. Mailing Address: 39355 W. Hwy. 46, Wagner, SD 57380-7128. Tel.: 605-384-3509. Fax: 605-384-5460.
E-mail: frenchine@neb.rr.com
Founded: 1990.
Key Personnel: Co Founder, Dir. & Chm. (V), Mardell E. Schroeder.
Governing Authority: Parent Institution: City of Butte. Tax-exempt.
Institution Type/Description: History Museum.
Collections: area history & culture; geological artifacts; early settlers; maps; medical instruments; farming & ranchers; barbed wire.
Activities: Museum Sponsors: Annual Open House.
Hours & Admission Prices: April-Oct. by appointment. No charge; donations accepted. &
Attendance: 7,500 (estimated)

Cambridge

CAMBRIDGE MUSEUM, 612 Penn St., Cambridge, NE 69022. Mailing Address: P.O. Box 129, Cambridge, NE 69022-0129. Tel.: 308-697-4385.
E-mail: bettyboop1-39@hotmail.com
Founded: 1938.
Key Personnel: Pres. (V), Marilyn Kester; Museum Shop Mgr., Betty Kruger; Sec., Mae Groshong.
Personnel Profile: Part-Time Paid 2.
Governing Authority: public college; nonprofit. Tax-exempt.
Institution Type/Description: Local History Museum.
Collections: mounted birds; mounted animals; firearms; fossils; Indian artifacts; art pieces; photographs; dental room.
Facilities: library of school books, newspapers & magazines available to the public.
Hours & Admission Prices: April-Sept. Tues.-Sun. 1-5; Oct.-March Sat.-Sun. 1-5. No charge; donations accepted. &
Attendance: 1,050 (estimated)

Central City

MERRICK COUNTY HISTORICAL MUSEUM, 211 E. St., Central City, NE 68826-1326. Tel.: 308-946-2867.
Institution Type/Description: History Museum.
Collections: county history & culture; local artifacts.
Hours & Admission Prices: May-Oct. Sun. 2-4; other times by appointment. No charge; donations accepted.

Chadron

DAWES COUNTY HISTORICAL SOCIETY MUSEUM, 341 Country Club Rd., Chadron, NE 69337-7329. Mailing Address: P.O. Box 1319, Chadron, NE 69337-1319. Tel.: 308-432-4999; 308-638-7402.
E-mail: dchs@bbc.net
Founded: 1964.
Congressional District: 3
Key Personnel: C.E.O. & Pres., Bernard Cripps; Chm. (V), Belvadine Lecher; Exec. Dir. & Museum Shop Mgr., Phyllis Carlson.
Personnel Profile: Full-Time Volunteers 1; Part-Time Volunteers 50.
Operating Expenses: 56,200
Operating Income: 48,000
Governing Authority: public; society; nonprofit. Parent Institution: Dawes County Historical Society, Inc. Tax-exempt: 501(c)(3).
Institution Type/Description: Local History Museum.
Collections: exhibits emphasizing Dawes County history from prehistoric to modern times; Dawes County Cemetery records; World War I & World War II items; Spanish-American War items; fossils; marine biology; quilts; farm machinery; tools; player organ; fence maker; toys; furniture; 1895 bicycle; harmonium; egg case maker; guns; hub caps; pens; archives; blacksmith shop; railroad room; general store; hospital room; log house; log barn; C & NW Railroad caboose; antique car collection; Crawford & Chadron local newspapers 1886 to present; barber shop; family & history research library. Historic Buildings: c.1890 schoolhouse; rural church c.1890.
Research Fields: Dawes County history; family history; surrounding territory history & artifacts; area cemeteries.
Facilities: 3,400-vol. library of family, local, county & Nebraska history for research; original Dawes County marriage records 1884 through 1961; delayed birth registrations; obituaries from c.1920; postal appointment books 1940-1970 for all Nebraska counties & site locations; Dawes County civil & criminal record books from 1885; manuscripts; letters & diaries.
Activities: guided tours; films; temporary exhibitions; monthly historical programs open to the public; Frontier School for 3rd & 4th grade. Annual Event: History In Action in September.
Publications: quarterly newsletter.
Hours & Admission Prices: Mon.-Sat. 10-4, Sun. & holidays 1-5. No charge; donations accepted. &
Attendance: 3,752 (estimated)
Membership: Annual Student $5; Annual Adult $10; Annual Family $15; Life $100; Life Homestead $500; Life Heritage $1,000 & up.

ELEANOR BARBOUR COOK MUSEUM OF GEOLOGY, (M), Math & Science Building, Chadron State College, 1000 Main St., Chadron, NE 69337-2667. Tel.: 308-432-6377. Fax: 308-432-6434.
E-mail: mleite@csc.edu
Web Site: www.csc.edu
Founded: 1939.
Congressional District: 3
Key Personnel: Cur. & Professor Geoscience, Michael Leite.
Governing Authority: college. Affiliated with Chadron State College.
Institution Type/Description: Science Museum.
Collections: paleontology; archaeology; geology.
Research Fields: paleontology; geology.
Facilities: area available for display; preparation lab.
Activities: guided tours; permanent & temporary exhibitions.
Hours & Admission Prices: Sept. to early May Mon.-Fri. 8-4:30; other times by appointment. No charge. Closed spring break; major holidays. &
Attendance: 400 (estimated)

HIGH PLAINS HERBARIUM, 1000 Main St., Chadron, NE 69337-2667. Tel.: 308-432-6385.
Institution Type/Description: Herbarium.
Collections: ethnobotanical, medicinal plants; historic pharmaceutical specimens.
Hours & Admission Prices: Daily 8-5; tours by appointment. &

MARI SANDOZ HIGH PLAINS HERITAGE CENTER, (M), 1000 Main, Chadron, NE 69337-2667. Tel.: 308-432-6401. Fax: 308-432-6464. Facebook: Mari Sandoz High Plains Heritage Center.
E-mail: sandozcenter@csc.edu
Web Site: www.sandozcenter.com
Formerly: Mari Sandoz Room Museum
Founded: 2002.
Congressional District: 3
Key Personnel: Dir., Sarah Polak.
Personnel Profile: Full-Time Paid 1; Part-Time Paid 6; Part-Time Volunteers 1; Interns 4.
Governing Authority: Parent Institution: Chadron State College. Tax-exempt.
Institution Type/Description: History Museum.
Collections: books by Sandoz & local authors; manuscripts; photographs; tour maps; manuscripts; oral histories; Northern High Plains cattle ranching; local & regional authors.
Facilities: library.
Activities: guided tours; permanent & temporary exhibitions; museum studies program; in-school outreach. Museum Sponsors: History Day District Contest; Ranch Day; College Pow Wow.
Hours & Admission Prices: Mon.-Fri. 8-12 & 1-4, Sat. 9-12 & 1-4. No charge; donations accepted. Closed college holidays. &
Attendance: 8,000 (estimated)

MUSEUM OF THE FUR TRADE, (M), 6321 E. Hwy. 20, Chadron, NE 69337-5325. Tel.: 308-432-3843. Fax: 308-432-5943.
E-mail: museum@furtrade.org
Web Site: www.furtrade.org
Founded: 1949.
Congressional District: 3
Key Personnel: C.E.O. & Dir., Gail DeBuse Potter; Pres., James W. Carlson; Museum Shop Mgr., Ann M. Hanson; Editor, Dr. James A. Hanson.
Personnel Profile: Full-Time Paid 1; Full-Time Volunteers 2; Part-Time Paid 1; Part-Time Volunteers 63.
Governing Authority: nonprofit organization. Parent Institution: Museum Association of the American Frontier. Tax-exempt: 501(c)(3).
Institution Type/Description: History Museum.
Collections: trade goods; weapons & equipment; American Indian & Alaskan objects; native cultural materials of the North American fur trade, 1500-1900; Plains Indian garden. Historic Building: c.1837 James Bourdeaux Trading Post.
Research Fields: North American fur trade from 1500-1900; commercial interchange between native peoples & Europeans, Russians, and the Orient.
Facilities: 10,000-vol. library; 6,000 sq. ft. exhibit space; Plains Indian garden. Museum-related items for sale.
Activities: lectures; films; permanent exhibitions; treasure hunt for children.
Publications: scholarly journal, Museum of the Fur Trade Quarterly; Firearms of the Fur Trade, Vol. I; When Skins Were Money: A History of the Fur Trade; Little Chief's Gathering - the Smithsonian's G. K. Collection; Original of Charles Larpenteur: My Travels to the Rocky Mts. 1833-1872; The Hawken Rifle: Its Place in History; Metal Weapons, Tools & Ornaments of the Teton Dakota; The Buckskinners Cookbook; The Journals of David Adams; Texas Ranchman: Memoirs of John A. Loomis.
Hours & Admission Prices: May-Oct. daily 8-5; other times by appointment. Adults $5; discount to groups of 20 or more, AAA & AAM members; members & children with parents no charge. Audio tours available. &
Attendance: 59,750 (accurate)
Membership: Annual $15; Benefactor $1,000.

Champion

CHAMPION MILL STATE HISTORICAL PARK, US 6 W. to Spur 15A, Champion, NE 69023-0117. Mailing Address: 73122 338 Ave., Enders, NE 69027-2718. Tel.: 308-882-5860 & 394-5118.
Web Site: outdoornebraska.org
Founded: 1969.
Congressional District: 3
Key Personnel: Supt., Nik Johanson.
Personnel Profile: Full-Time Paid 2; Part-Time Paid 3; Part-Time Volunteers 2.
Governing Authority: state. Parent Institution: Historical Parks Section, Nebraska Game & Parks Commission, Box 30370, Lincoln, NE 68503. Tax-exempt.
Institution Type/Description: Historical Park: Nebraska's last functioning water-powered mill.
Collections: milling machinery; tools.
Research Fields: flour & feed milling; water-powered milling.
Facilities: working mill.

Activities: guided tours; organized education programs for children.
Hours & Admission Prices: Grounds: daily. Mill: Memorial Day-Labor Day Sat. 8-5, Sun. 1-4. Mill: no charge; donations accepted. Park permit required. Primitive camping fee $7. &

Attendance: 10,500 (estimated)

CHASE COUNTY HISTORICAL MUSEUM, 1711 Broadway, Champion, NE 69033-3032. Mailing Address: 220 2nd St., Champion, NE 69023. Tel.: 308-882-4601 & 1699.
E-mail: aghust@chase3000.com
Founded: 1963.
Congressional District: 43
Key Personnel: Pres., Byron Hust.
Personnel Profile: Part-Time Volunteers 15.
Governing Authority: Tax-exempt.
Institution Type/Description: History Museum.
Collections: county artifacts; period clothing; WWI & II uniforms; farm machinery; Indian artifacts; cameras; native grasses; guns & ammunition; woodworking tools.
Hours & Admission Prices: Mother's Day to Labor Day Sun. 1:30-4:30; groups by appointment. No charge; donations accepted. &
Attendance: 985 (accurate)
Membership: Individual $5; Life $100.

Chappell

CHAPPELL MUSEUM ASSOCIATION AND SUDMAN-NEUMANN HERITAGE HOUSE, 701 5th St., Chappell, NE 69129. Mailing Address: 1540 5th St., Chappell, NE 69129. Tel.: 308-874-2642.
Key Personnel: Dir., Deena Peters
Institution Type/Description: History Museum.
Collections: period furniture; household artifacts.
Hours & Admission Prices: May to Labor Day Sun. 2-4; other times by appointment. No charge.

Clarkson

CLARKSON HISTORICAL SOCIETY, 221 Pine St., Clarkson, NE 68629. Mailing Address: P.O. Box 121, 221 Pine St., Clarkson, NE 68629. Tel.: 402-892-3100.
Founded: 1967.
Congressional District: 23
Key Personnel: Chm. (V), Nancy Doernemann; Pres. (V), Ruth D. Waters.
Personnel Profile: Part-Time Volunteers 8.
Volunteer Hours: 1,200
Operating Expenses: 4,500
Operating Income: 5,000
Governing Authority: nonprofit organization. Tax-exempt: 501(c)(3).
Institution Type/Description: General Museum.
Collections: period artifacts; musical instruments; horsedrawn farm implements; early household artifacts; photographs; clothing; Czech heritage artifacts; military; over 300 farm toys.
Research Fields: photographs; family histories; town & county histories.
Activities: guided tours; lectures; films; permanent & temporary exhibitions. Museum Sponsors: Czech Day Celebration in June.
Hours & Admission Prices: April & Aug.-Sept. 1st Sun. of month 11:30-3; Memorial Day weekend Sun. 11:30-3; June 21-22 12-6; other times by appointment. No charge; donations accepted. &
Attendance: 1,000 (estimated)
Membership: Annual $5; Lifetime $50.

Clay Center

CLAY COUNTY MUSEUM & HISTORICAL SOCIETY, 316 W. Glenvil St., Clay Center, NE 68933-1153. Mailing Address: P.O. Box 191, Clay Center, NE 68933-0191. Tel.: 402-762-3563.
Web Site: www.oldtrusty.org/history.html
Founded: 1972.
Governing Authority: Tax-exempt.
Institution Type/Description: History Museum.
Collections: county history; one-room schoolhouse; railroad caboose; period tractors; steam & stationary engines; buggies; cars; trucks; fire equipment; blacksmith shop equipment; agricultural equipment.
Hours & Admission Prices: Mon.-Thurs. 10-4; other times by appointment. No charge; donations accepted. &
Attendance: 10,000 (estimated)
Membership: Single $10; Family $15.

Columbus

PLATTE COUNTY MUSEUM, 2916 16th St., Columbus, NE 68601-4200. Mailing Address: P.O. Box 31, Columbus, NE 68602-0031. Tel.: 402-564-1856. Facebook: Platte County NE Museum.
E-mail: museum@megavision.com
Web Site: www.megavision.net/museum
Key Personnel: Dir., Cheri Schrader.
Personnel Profile: Part-Time Paid 1.
Institution Type/Description: History Museum.
Collections: county history & artifacts. Historic Structures: 1857 log cabin; 1912 Challenger windmill.
Publications: monthly member newsletter.
Hours & Admission Prices: mid-May to Labor Day Fri.-Sun. 1-4; Oct.-April 1st Sun. each month 1-4. Adults $3; members & children under 14 no charge. &
Attendance: 1,800 (accurate)
Membership: Individual $15; Family $20; Friend of Platte County Museum $50 & up; Life $150; Family Life $200; Corporate Life $400.

Cozad

THE 100TH MERIDIAN MUSEUM, 206 E. 8th, Cozad, NE 69130-1834. Mailing Address: P.O. Box 325, Cozad, NE 69130-0325. Tel.: 308-784-1100. Facebook: 100th Meridian Museum.
Founded: 1994.
Congressional District: 3
Key Personnel: Pres., Curtis Sargent; Vice Pres., Maurice Andres; Treas. and Recording & Corresponding Sec., Julie Linn; Chm. (V) & Museum Shop Mgr., Judy Andres.
Personnel Profile: Part-Time Volunteers 100.
Volunteer Hours: 5,000
Operating Expenses: 5,000
Operating Income: 7,000
Governing Authority: private; nonprofit organization. Parent Institution: Cozad Historical Society. Tax-exempt: 501(c)(3).
Institution Type/Description: Historical Museum.
Collections: local history & archives; antique touring coach used by President Taft in Yellowstone in 1907; china & silver; farm equipment; telephone equipment; historical photographs; piano; 1927 Chevrolet truck.
Major Exhibits: Early Cozad Photography, 5/30/14-9/10/14; Sheet Music, 5/30/14-9/10/14; Genealogy, 5/30/14-9/10/14.
Research Fields: genealogy; archaeology.
Facilities: music center. Museum-related items for sale.
Activities: workshops; loan & temporary exhibitions.
Publications: The 100th Meridian Museum; brochures; newsletters.
Hours & Admission Prices: Memorial Day to Labor Day Tues.-Fri. 10-5, Sat. 10-4; other times by appointment. No charge; donations accepted. &
Attendance: 1,500 (estimated)
Membership: Family $20; Life $250.

ROBERT HENRI MUSEUM & HISTORICAL WALKWAY, 218 E. 8th St., Cozad, NE 69130-1834. Mailing Address: P.O. Box 355, Cozad, NE 69130-0355. Tel.: 308-784-4154.
E-mail: rhenri@cozadtel.net
Web Site: roberthenrimuseum.org
Institution Type/Description: History Museum.
Collections: pioneer schoolhouse; Pony Express station; local arts & crafts; historic walkway.
Hours & Admission Prices: June-Sept. Tues.-Sat. 10-5; other times by appointment. Adults $3, children 14 & under $1.

Crawford

CRAWFORD HISTORICAL MUSEUM, 337 2nd St., Crawford, NE 69339. Mailing Address: P.O. Box 165, Crawford, NE 69339. Tel.: 308-665-1732.
Institution Type/Description: History Museum.
Collections: local history; Native American artifacts; rocks; business ledgers; beaded handbags; photographs; local newspapers; pamphlets; posters; postcards.
Hours & Admission Prices: Memorial Day to Labor Day Mon.-Sat. 10-12 & 1-5. No charge; donations accepted.

HUDSON-MENG MUSEUM & ARCHAEOLOGY RESEARCH CENTER, 1811 Meng Dr., Crawford, NE 69339. Mailing Address: Mammoth Site of Hot Springs, P.O. Box 692, Hot Springs, SD 57747-0692. Tel.: 308-665-3900; 3907; 3090. Fax: 308-665-3908.
E-mail: news@mammothsite.org

Institution Type/Description: Archaeology Museum.
Collections: local history & culture; photographs.
Hours & Admission Prices: May 17-Labor Day daily 9-5. Adults $5, seniors 60 & over $4.50, children 5-12 $3; children 4 & under no charge.

NEBRASKA STATE HISTORICAL SOCIETY'S FORT ROB-INSON MUSEUM, (M), 3200 U.S. Hwy. 20, Crawford, NE 69339. Mailing Address: P.O. Box 304, Crawford, NE 69339-0304. Tel.: 308-665-2919. Fax: 308-665-2917.
E-mail: fortrob@bbc.net
Web Site: www.nebraskahistory.org
Formerly: Fort Robinson Museum
Founded: 1956.
Congressional District: 3
Key Personnel: C.E.O. & Exec. Dir., Michael J. Smith; Assoc. Dir., Ann Billesbach; Cur., Thomas R. Buecker; Chief Education & Research Officer, Lynne Ireland.
Personnel Profile: Full-Time Paid 2; Part-Time Paid 2; Part-Time Volunteers 1.
Governing Authority: state; society. Parent Institution: Nebraska State Historical Society. Tax-exempt: 501(c)(3).
Institution Type/Description: Military Museum: Fort Robinson history.
Collections: Indian artifacts; military items; archaeology. Historic Buildings: 1875 reconstructed log guardhouse; 1875 post adjutant's office; wheelwright shop; 1904-1906 blacksmith shop, harness repair shop; 1887 adobe officer's quarters; 1905 Post Headquarters, items related to Fort Robinson 1874-1948, Red Cloud Agency 1873-1877; 1908 veterinary hospital.
Research Fields: archaeology; anthropology; military; horses; history of the West.
Facilities: 400-vol. library of books on history of Nebraska & Western Americana, microfilm records available for use on premises by appointment. Publications & postcards for sale.
Activities: guided tours; lectures; permanent exhibitions; fifteen minute video on the history of Fort Robinson.
Publications: Nebraska History Magazine; Fort Robinson and the American West: 1874-1899; Fort Robinson and the American Century: 1900-1948; 1887 Adobe Barracks, Fort Robinson; The Cheyenne Outbreak Barracks, Fort Robinson; Self-Guided Tour of Fort Robinson.
Hours & Admission Prices: Memorial Day-Labor Day Mon.-Sat. 8-5, Sun. 9-5; Sept.-May Mon.-Fri. 8-5. Adults $2; discount to Time Travelers; members & children with adult no charge. &
Attendance: 11,000 (accurate)
Membership: Individual $40; Household $55; Contributing $100; Supporting $250; Sustaining $500; Founder $1,000.

* **TRAILSIDE MUSEUM OF NATURAL HISTORY, (M),** 3200 W. U.S. Hwy. 20, Crawford, NE 69339-3112. Mailing Address: Box 462, Crawford, NE 69339. Tel.: 308-665-2929. Fax: 308-665-2928.
E-mail: pnorman2@unl.edu
Web Site: www.trailside.unl.edu
Founded: 1961.
Institution Type/Description: Natural History Museum.
Collections: geological history of Western Nebraska; fossils; geology; mural & prehistoric animal posters.
Hours & Admission Prices: April-May & Sept.-Oct. Thurs.-Sun. 10-5; Memorial Day to Labor Day daily 9-6; other times by appointment. Adults 19 & over $3, children 5-18 $1; children 4 & under no charge. &

Crete

THE MAPLES HERITAGE COMPLEX AND BENNE MEMO-RIAL MUSEUM, 800 W. 13th St., Crete, NE 68333-2006. Mailing Address: P.O. Box 304, Crete, NE 68333. Tel.: 402-826-5270.
Web Site: www.creteheritage.org
Founded: 1976.
Personnel Profile: Part-Time Volunteers 10; Interns 2.
Governing Authority: bd. of directors. Parent Institution: Crete Heritage Society. Tax-exempt.
Institution Type/Description: History Museum.
Collections: exhibits that interpret Crete area history.
Hours & Admission Prices: Sun. 1-4; other times by appointment.

Crofton

LEWIS AND CLARK VISITOR CENTER, 55245 N.E. Hwy. 121, Crofton, NE 68730. Mailing Address: P.O. Box 710, Yankton, SD 57078-0710. Tel.: 402-667-2546. Fax: 402-667-2547.
E-mail: gavinspoint.nwo@usace.army.mil

Founded: 1976.
Personnel Profile: Full-Time Paid 1; Part-Time Paid 5; Part-Time Volunteers 3.
Governing Authority: federal; nonprofit. Parent Institution: U.S. Army Corps of Engineers. Tax-exempt.
Institution Type/Description: Visitor Center.
Collections: birds & fish of the Missouri River region; scientific instruments used to build the dam. History of the Missouri River, including geological, paleontological, cultural & natural history.
Facilities: 150-vol. library; 2,700 sq. ft. exhibit space; 50-seat theater. Museum-related items for sale.
Activities: formal education programs for children; guided tours; participatory exhibit.
Publications: brochure, Rack Card.
Hours & Admission Prices: March-May & Sept.-Nov. Mon.-Fri. 8-4:30; Memorial Day to Labor Day Sun.-Thurs. 8-6, Fri.-Sat. 8-7. No charge. &
Attendance: 27,429 (accurate)

Curtis

HANSEN MEMORIAL MUSEUM, 502 Prentiss, Curtis, NE 69025. Mailing Address: c/o City of Curtis, 201 Garlick Ave., Curtis, NE 69025. Tel.: 308-367-4122.
Web Site: www.curtis-ne.com/hansen/
Key Personnel: Pres (V), Tom Rue
Governing Authority: Parent Institution: City of Curtis. Tax-exempt.
Institution Type/Description: History Museum: housed in the former home of Anna Marie Hansen; built in 1911.
Collections: Hansen family history; personal artifacts; period furnishings; photographs.
Hours & Admission Prices: Call for hours.
Attendance: 54 (accurate)

Dakota City

DAKOTA COUNTY HISTORICAL SOCIETY & O'CONNOR HOUSE, Rural Rte. 1, Dakota City, NE 68731. Mailing Address: P.O. Box 971, Dakota City, NE 68731-0971. Tel.: 402-698-2288 & 2538.
Web Site: www.dakotacountyhistoricalsociety.com
Institution Type/Description: History Museum.
Collections: local artifacts; household items; furnishings; tractors; American Indian clothing. Historic House: O'Connor House, Homer, NE.
Hours & Admission Prices: Memorial Day to Labor Day Sat. 10-5, Sun. 12-5; other times by appointment. No charge; donations accepted.

Dalton

DALTON PRAIRIE SCHOONER MUSEUM, 109 U.S. Hwy. 385, Dalton, NE 69131. Tel.: 308-377-2637.
Institution Type/Description: History Museum.
Collections: community history; model railroad engines.
Hours & Admission Prices: Memorial Day to Labor Day Sat.-Sun. 1-4; other times by appointment. No charge.

David City

BONE CREEK MUSEUM OF AGRARIAN ART, (M), 575 "E" St., David City, NE 68632-1638. Tel.: 402-367-4488.
E-mail: artinfo@bonecreek.org
Web Site: www.bonecreek.org
Founded: 2007.
Personnel Profile: Full-Time Paid 1; Part-Time Volunteers 15.
Institution Type/Description: Art Museum: located in the boyhood hometown of artist Dale Nichols.
Collections: agrarian art; Nichols notes, drawings, & letters; works by Robert Bateman, Gary Ernest Smith, John Stewart Curry, Robert Gwathmey, Birger Sandzen, Winslow Homer & Thomas Hart Benton.
Research Fields: Agrarian art; regional art; American art.
Activities: lectures; workshops by artists exhibiting at the museum; permanent & temporary exhibitions.
Publications: Dale Nichols: Transcending Regionalism, 2011.
Hours & Admission Prices: Wed.-Sat. 10-4, Sun. 1-4. No charge; donations accepted. Closed major holidays. &
Attendance: 3,500 (estimated)
Membership: Souier $50; Harvester $100; Sod Buster $250; Homesteader $500; Pioneer $1,000; Sustaining Benefactor $2,500; Grand Benefactor $5,000.

Dorchester

SALINE COUNTY HISTORICAL SOCIETY, INC., 1145 State
Hwy. 33, Dorchester, NE 68343. Mailing Address: 561 County Rd.
S., Tobias, NE 68453-2007. Tel.: 402-946-2129 & 243-2356.
E-mail: rjrada@diodecom.net
Founded: 1956.
Congressional District: 1
Key Personnel: C.E.O. & Pres. (V), Judith K. Rada; Sec., Mary Ann Placek.
Personnel Profile: Part-Time Volunteers 20.
Governing Authority: society. Tax-exempt: 501(c)(3).
Institution Type/Description: Agriculture Museum.
Collections: agriculture; machinery; household items; books; paintings by
local artists; glass; clothing; In Loving Memory, burial customs & items
c.1890-1910. Historic Houses and Buildings: 1873 rural school; 1888
Center Hall; 1896 rural post office; 1868 home; 1868 log cabin; 1871 train
depot.
Research Fields: special exhibit, Saline County settler's funeral customs.
Activities: guided tours.
Publications: articles in all 4 County newspapers.
Hours & Admission Prices: Sun. 2-5; other times by appointment. No charge;
donations accepted. &
Attendance: 515 (accurate)
Membership: Individual $5.

Elkhorn

ELKHORN HISTORICAL SOCIETY, 20601 Glenn St., Elkhorn,
NE 68022. Mailing Address: P.O. Box 187, Elkhorn, NE 68022.
Tel.: 402-238-2571.
E-mail: smaltwn@yahoo.com
Web Site: www.elkhornhistory.org
Institution Type/Description: Historical Society Museum.
Collections: local history & culture; period furnishings; personal artifacts;
photographs.
Activities: Museum Sponsors: Elkhorn Days in June.
Hours & Admission Prices: Thurs. 1-3.

Ellsworth

CRESCENT LAKE NATIONAL WILDLIFE REFUGE, 10630
Rd. 181, Ellsworth, NE 69340. Tel.: 308-635-7851.
Institution Type/Description: Wildlife Refuge.
Collections: nesting & migratory birds.
Facilities: nature trails.
Activities: hiking.
Hours & Admission Prices: Call for hours.

Elm Creek

CHEVYLAND U.S.A. AUTO MUSEUM, 7245 Buffalo Creek Rd.,
Elm Creek, NE 68836-9802. Tel.: 308-856-4208.
E-mail: chevylandusa@rcom-ne.com
Founded: 1976.
Key Personnel: C.E.O., Monte Hollertz
Institution Type/Description: Automobile Museum.
Collections: over 110 cars including Chevys, Fords, Cadillacs & a 1928
Whippet from 1914 to 1975; motorcycles.
Hours & Admission Prices: May-Sept. daily 8-5; other times by appointment.
Adults $6, children 10-15 $2; discounts to AAM members; children under
10 no charge. &

Elmwood

BESS STREETER ALDRICH HOUSE & MUSEUM, 204 East F
St., Elmwood, NE 68349. Mailing Address: P.O. Box 167, Elm-
wood, NE 68349-0167. Tel.: 402-994-3855.
Web Site: www.bessstreeteraldrich.org
Founded: 1978.
Key Personnel: Dir., Teresa Lorensen; Pres. (V), Anna Jamrog.
Personnel Profile: Part-Time Paid 1; Part-Time Volunteers 20.
Governing Authority: Parent Institution: Bess Streeter Aldrich Foundation,
P.O. Box 167, Elmwood, NE.
Institution Type/Description: Historic House Museum: housed in the former
home of author, Bess Streeter Aldrich, 1881-1954.
Collections: House: life of Bess Aldrich; period furnishings; personal artifacts;
writings; history of the home; memorabilia; Elmwood's history. Museum:
personal letters; movie memorabilia; family history; manuscript; childhood
doll; video.

Facilities: House: books & museum-related items for sale.
Activities: Museum: video.
Hours & Admission Prices: House: May-Oct. Wed.-Thurs. & Sat.-Sun. 2-5;
Nov.-April Sat.-Sun. 2-5; groups by appointment. Museum: by appointment
only. House: adults $5, children 6-12 $3; children under 6 no charge.
Museum: no charge with paid house admission.
Attendance: 500 (estimated)
Membership: Spring Came on Forever $20; The Rim of the Prairie $50; A
Lantern in Her Hand $100.

Fairbury

FAIRBURY CITY MUSEUM, 1128 Elm St., Fairbury, NE 68352-
1427. Tel.: 402-671-6879.
E-mail: deltadawn63@windstream.net
Founded: 1955.
Key Personnel: Pres. (V), Ben McBride; Cur., Pa' Ren Sims.
Personnel Profile: Part-Time Paid 1; Part-Time Volunteers 7.
Governing Authority: Tax-exempt.
Institution Type/Description: History Museum.
Collections: period artifacts & memorabilia relating Fairbury's history.
Hours & Admission Prices: Sat.-Sun. 1-4; other times by appointment. No
charge; donations accepted. &
Attendance: 850 (estimated)

ROCK CREEK STATION STATE HISTORIC PARK, 57426
710th Rd., Fairbury, NE 68352. Tel.: 402-729-5777.
E-mail: ngpc.rock.creek.station@nebraska.gov
Web Site: www.outdoornebraska.org
Founded: 1980.
Congressional District: 3
Key Personnel: Supt., Wayne Brandt; Asst. Supt., Jeff Bargar.
Personnel Profile: Full-Time Paid 2; Part-Time Paid 6.
Governing Authority: state. Affiliated with the Nebraska Game & Parks
Commission, 2200 N. 33rd St., Lincoln, NE 68503. Tel. 402-464-0641.
Tax-exempt.
Institution Type/Description: Historic Site: Pony Express Station, stage &
freight station, road ranches along trail & toll bridge along the Oregon Trail.
Collections: blacksmith; wagons & stock; historical & interpretive displays.
Facilities: park; picnic area; nature trail; visitor's center; information theater;
campgrounds.
Activities: park tours; interpretive programs; slide programs; blacksmith
demonstrations.
Hours & Admission Prices: May 1 to mid-Sept. daily 10-5; mid-Sept. to Oct.
24 Sat.-Sun. 1-5. State park entry permit required; annual $20, daily $4 per
vehicle. &
Attendance: 48,000 (estimated)

ROCK ISLAND DEPOT RAILROAD MUSEUM, 910 2nd St.,
Fairbury, NE 68352. Tel.: 402-729-5131.
E-mail: fairburyridepot@alltel.net
Institution Type/Description: Railroad Museum.
Collections: railroad history; period furnishings; photographs; personal arti-
facts; historic building.
Facilities: Museum-related items for sale.
Activities: educational programs; special events.
Hours & Admission Prices: Wed.-Thurs. & Sun. 1-5; tours by appointment.

Fairmont

**FILLMORE COUNTY MUSEUM AND MCCLELLAN DRUG
STORE,** 601 Sixth Ave., Fairmont, NE 68354. Mailing Address:
P.O. Box 333, Fairmont, NE 68354. Tel.: 402-759-0597.
E-mail: dr.br@galaxycable.net
Governing Authority: Tax-exempt.
Institution Type/Description: History Museum: housed in the former office of
the Fairmont Creamery; built 1886; later sold to Dr. S. F. Ashby for his
medical practice.
Collections: Fillmore County history & artifacts; Fairmont Creamery &
Fairmont Foods; Ashby family medical practice; WWI & WWII; Fairmont
Army Airfield; books; period kitchen items; wedding dress; Spanish-
American War.
Hours & Admission Prices: By appointment. No charge; donations accepted.
Attendance: 350 (estimated)
Membership: Individual $10; Family $20; Life $75.

Falls City

RICHARDSON COUNTY HISTORICAL SOCIETY, 1401 Chase St., Falls City, NE 68355-2645. Mailing Address: P.O. Box 45, Falls City, NE 68355-0045. Tel.: 402-245-3481 & 4407.
E-mail: auxie@sentco.net
Founded: 1986.
Key Personnel: Pres. (V), Aaron Wing; Cur. & Museum Shop Mgr., JoAnn Auxier.
Personnel Profile: Part-Time Volunteers 8.
Governing Authority: Tax-exempt: 501(c)(3).
Institution Type/Description: Historical Society Museum.
Collections: local history & culture; period furnishings; personal artifacts; photographs.
Publications: annual newsletter.
Hours & Admission Prices: Mon.-Fri. 2-4; other times by appointment. Suggested Donations: adults $5, students $.50. Closed major holidays. ♿
Attendance: 2,396 (accurate)
Membership: Students 18 & under $5; Adult $10; Friend $25; Promoter $50; Century $100; Benefactor $500; Patron $1,000.

RICHARDSON COUNTY MILITARY HISTORY MUSEUM, 1700 Stone St., Falls City, NE 68355. Tel.: 402-245-4288.
Institution Type/Description: Military History Museum.
Collections: military history & artifacts; personal artifacts; photographs; war memorabilia; uniforms; medals; swords; WWI, WWII & Vietnam War artifacts; Veterans' Memorial Monument.
Hours & Admission Prices: Mon.-Fri. 8-5.

Fort Calhoun

FORT ATKINSON STATE HISTORICAL PARK, 7th & Madison, Fort Calhoun, NE 68023. Mailing Address: Box 240, Fort Calhoun, NE 68023-0240. Tel.: 402-468-5611. Fax: 402-468-5066.
E-mail: fort.atkinson@ngpc.ne.gov
Web Site: ngpc.state.ne.us
Founded: 1963.
Congressional District: 1
Key Personnel: Supt., John Slader; Asst. Supt., Jerry Farber.
Personnel Profile: Full-Time Paid 2; Part-Time Paid 6; Part-Time Volunteers 30.
Governing Authority: state. Div. of Nebraska Game & Parks Commission, Lincoln, NE 68503. Parent Institution: State of Nebraska. Tax-exempt.
Institution Type/Description: Historical Park: site of 1804 Lewis & Clark's Council on the Bluff & 1819-1827 U.S. Army's Ft. Atkinson.
Collections: artifacts from Fort period.
Research Fields: early log construction techniques; 1820's military history.
Facilities: interpretive center; records & reports of Fort period.
Activities: guided tours; reconstruction projects; changing exhibits. Museum Sponsors: events associated with early fur trade.
Publications: brochure.
Hours & Admission Prices: Park: Spring & Summer: daily 8-7; Fall & Winter: 8-sunset. Interpretative Center: May & Sept.-Oct. 21 Sat.-Sun. 10-5; Memorial Day to Labor Day Mon.-Fri. 10-5. Park Entry Permit: annual $14, daily $2.50. ♿
Attendance: 47,500 (estimated)

WASHINGTON COUNTY HISTORICAL ASSOCIATION, 102 N. 14th St., Fort Calhoun, NE 68023-3532. Mailing Address: P.O. Box 25, Fort Calhoun, NE 68023-0025. Tel.: 402-468-5740. Fax: 402-468-5741.
E-mail: info@wchamuseum.com
Web Site: www.wchamuseum.com
Founded: 1925.
Congressional District: 1
Key Personnel: Chm. (V), Michael O'Brien; Cur., Faith Norwood.
Personnel Profile: Full-Time Paid 1; Part-Time Paid 2; Part-Time Volunteers 20.
Governing Authority: society. Subsidiary Institutions: Frahm House, Fontanelle Town Hall, Fontanelle, NE. Tax-exempt: 501(c)(3).
Institution Type/Description: General Museum.
Collections: Fort Atkinson relics & records; pioneer farm implements; weapons associated with exploration & settlement of area; historic & pre-historic Indian artifacts; furniture; glass; toys; hand crafts; needle works; clothing; musical instruments; manuscripts. Historic Homes: 1905 Fred Frahm home, late 19th-century & turn-of-the-century furniture. 1855-1865 Fontanelle Town Hall: documents, letters; county records; genealogical records & material; Lewis & Clark diorama; photographs; maps.

Research Fields: Fort Atkinson history & pioneer history in Washington County; fur-trading era and Lewis & Clark Expedition on the Missouri and its tributaries.
Facilities: 1,300-vol. library of history books of the county; early textbooks; local biographical material; magazine files available for research on premises; three galleries; reading room; many records & information relative to Fort Atkinson 1819-1827. Books & booklets for sale.
Activities: guided tours; programs incorporating museum material; guided tours of Frahm House & Fontanelle Town Hall by appointment only.
Publications: booklet, Fort on the Prairie, Portal to the Plains; 1985 reprint with name index of History of Washington Co. Nebraska, 1875.
Hours & Admission Prices: Museum: Mon.-Wed. & Fri. 9-5, Thurs. 9-8, Sun. 12-4. Frahm House: Sat. 12-4; other times by appointment. Adults $3, children $2; discounts to Blue Star Museum members. Closed holidays. ♿
Attendance: 3,000 (estimated)
Membership: Friend $1-$299; Trailblazer Bronze $300; Silver $500; Gold $750; Platinum $1,000.

Franklin

FRANKLIN COUNTY MUSEUM, 1309 H Rd., Franklin, NE 68939-5168. Tel.: 308-425-3030. Fax: 530-425-3033.
E-mail: welovehistory@gtmc.net
Web Site: www.rootsweb.ancestry.com/~nefrankl/museum/fcmuseum
Founded: 1932.
Congressional District: 3
Key Personnel: Dir., Connie Osterbuhr; Chm. & Pres. (V), Jim Gorman.
Personnel Profile: Part-Time Volunteers 2.
Governing Authority: Parent Institution: Franklin County Historical Society, Inc. Tax-exempt.
Institution Type/Description: County History.
Collections: county history; pioneer & Native American artifacts; military; agriculture; one-room schoolhouse; Franklin Academy; Ol' Towne Main St.; Pierce Lyden; Frank Cyr; Clarence Mitchell.
Research Fields: Franklin County genealogy.
Activities: monthly board meetings; educational programs. Museum Sponsors: Memorial Day Weekend Events; Harvest Fest in October; Christmas Open House.
Publications: Franklin, Nebraska 1879-2002, The Best of the Good Life; These Good Old Golden Rural Days.
Hours & Admission Prices: April-May & Sept.-Dec. Sat.-Sun. 1-4; June-Aug. 1-5; other times by appointment. No charge; donations accepted. Closed major holidays.
Attendance: 1,400 (estimated)
Membership: Individual $5; Family $7.50.

Fremont

GALLERY 92 WEST, 92 W. Sixth St., Fremont, NE 68026-0335. Mailing Address: P.O. Box 335, Fremont, NE 68026-0335. Tel.: 402-721-7779.
Institution Type/Description: Art Gallery.
Collections: works by local & regional artists.
Activities: special events; permanent & temporary exhibits.
Hours & Admission Prices: Tues.-Sun. 1-4. No charge.

LOUIS E. MAY MUSEUM, 1643 N. Nye, Fremont, NE 68025-3327. Mailing Address: P.O. Box 766, Fremont, NE 68026-0766. Tel.: 402-721-4515. Fax: 402-721-8354.
E-mail: maymuseum@qwestoffice.net
Web Site: www.maymuseum.com
Founded: 1969.
Congressional District: 1
Key Personnel: Pres., Judy Ekeler; Cur. & Museum Shop Mgr., Jeff Kappeler.
Personnel Profile: Part-Time Paid 2; Part-Time Volunteers 75.
Governing Authority: county; society. Affiliated with Dodge County Historical Society. Tax-exempt.
Institution Type/Description: History Museum.
Collections: furniture; photographs; farm implements; rural school; general store; dolls. Historic House: 1874 Theron Nye residence.
Research Fields: historic sites; oral history; local history.
Facilities: library; 2 gardens.
Activities: guided tours; permanent & temporary exhibitions; art exhibits.
Publications: newsletter.
Hours & Admission Prices: April-Dec. Wed.-Sat. 1:30-4:30 (last tour 3:30). Adults $5, students 5-18 $1; children under 5 & members no charge. ♿
Attendance: 5,000 (accurate)
Membership: Student $5; Individual $10; Family $20; Contributing $50; Business $100; Life $250; Corporate Life $1,000.

LUENINGHOENER PLANETARIUM - MIDLAND UNIVERSITY, Swanson Hall of Science, 8th & Irving St., Fremont, NE 68025. Mailing Address: 900 N. Clarkson, Fremont, NE 68025. Tel.: 402-941-6353.
E-mail: clements@midlandu.edu
Founded: 1965.
Key Personnel: Dir., Dr. Greg Clements.
Personnel Profile: Part-Time Paid 1.
Governing Authority: Parent Institution: Midland University. Tax-exempt.
Institution Type/Description: Planetarium.
Collections: astronomy; space science.
Activities: special events; educational programs.
Hours & Admission Prices: By appointment. &
Attendance: 1,000 (accurate)

Genoa

GENOA HISTORICAL SOCIETY & MUSEUM, 402 Willard Ave., Genoa, NE 68640. Mailing Address: P.O. Box 279, Genoa, NE 68640-0279. Tel.: 402-993-2875 & 2330.
Institution Type/Description: Historical Society Museum.
Collections: Pawnee weapons, tools, & personal artifacts; Mormon Trail; U.S. Indian School; Pawnee Indian agency.
Hours & Admission Prices: Memorial Day to Labor Day Fri.-Sun. 1-5; other times by appointment. Requested Donations: adults $2, children $1.

GENOA U.S. INDIAN SCHOOL MUSEUM, 209 E. Webster Ave., Genoa, NE 68640. Mailing Address: P.O. Box 382, Genoa, NE 68640. Tel.: 402-993-6636 & 6055.
E-mail: nfcarls@hotmail.com
Web Site: www.megavision.net/genoamuseum
Congressional District: 3
Key Personnel: Pres. (V), Philip Swantek.
Personnel Profile: Part-Time Volunteers 40.
Governing Authority: Parent Institution: Genoa U.S. Indian School Foundation. Tax-exempt.
Institution Type/Description: Native American History Museum: housed in a former boarding school for Native American children from 1884-1934. Listed on the National Register of Historic Places.
Collections: Native American history & culture; wall murals; Tribal flags; campus model; personal artifacts; photographs.
Activities: groups tours. Museum Sponsors: Annual Celebration in August.
Publications: newsletter.
Hours & Admission Prices: Memorial Day to Labor Day Fri.-Sun. 1-5; other times by appointment. No charge; donations accepted. &
Attendance: 950 (accurate)
Membership: Individual $10; Family $25; Patron $35; Corporate $50; Lifetime $250.

Gering

LEGACY OF THE PLAINS MUSEUM, 2930 Old Oregon Trail, Gering, NE 69341. Tel.: 308-436-1989.
E-mail: legacyoftheplains@gmail.com
Web Site: www.legacyoftheplains.com
Formerly: Farm and Ranch Museum/North Platte Valley Museum
Founded: 1969.
Congressional District: 3
Key Personnel: Pres. (V), Jodi Ruzicka; Dir., Katie Bradshaw; Asst. Dir., Nancy Haney; Exec. Vice Pres. (V), Larry Hubbard; Vice Pres. (V), Jack Preston; Asst. Vice Pres. (V), George Schlothauer; Museum Shop Mgr., Becky Simpson; Museum Shop Mgr., Donna Considine.
Personnel Profile: Full-Time Paid 1; Full-Time Volunteers 3; Part-Time Paid 2; Part-Time Volunteers 100.
Governing Authority: Tax-exempt.
Institution Type/Description: History: western life migration; settlement; agriculture; ranching.
Collections: farm tractors & machinery; agricultural history; ephemera; books; photographs; Oregon Trail; Native American settlement; community history.
Research Fields: agriculture; Oregon Trail.
Facilities: library; archives.
Activities: 3rd weekend Sept.: Harvest Festival; 1st Sat. Dec.: High Plains Christmas.
Publications: quarterly newsletter.
Hours & Admission Prices: May-Sept. Mon.-Sat. 9-5, Sun. 1-5; Oct.-April Mon.-Fri. 9-5, also by appointment. Non-member adults $3; closed Thanksgiving, Christmas & New Years Day. &
Attendance: 10,000 (estimated)

Membership: Individual $30; Business $60.

OREGON TRAIL MUSEUM AND VISITOR CENTER, Scotts Bluff National Monument, 190276 Old Oregon Trail, Gering, NE 69341. Mailing Address: Scotts Bluff National Monument, P.O. Box 27, Gering, NE 69341-0027. Tel.: 308-436-9700. Fax: 308-436-7611.
Web Site: www.nps.gov/scbl
Founded: 1919.
Congressional District: 3
Key Personnel: Supt., Ken Maybery; Exec. Sec., Oregon Trail Museum Assn., Jolene Kaufman.
Governing Authority: federal. National Park Service, Dept. of Interior, Washington, DC 20240. Tax-exempt.
Institution Type/Description: National Monument: dominant natural feature of the North Platte Valley which has been a human migration corridor for centuries.
Collections: historical artifacts; paintings; vegetation; natural wildlife; geology; paleontology.
Research Fields: prairie restoration.
Facilities: 1,200-vol. library of historical material available for use on premises.
Activities: living history exhibits; conducted hikes; slide presentations.
Publications: book, Scotts Bluff Handbook.
Hours & Admission Prices: Summer: daily 9-5; Winter: daily 9-12 & 1-5. Non-commercial Vehicles $5; Hikers, Bicyclists & Motorcycles $3. &
Attendance: 150,000 (estimated)
Membership: See separate listing for Oregon Trail Museum Association.

SCOTTS BLUFF NATIONAL MONUMENT, 190276 Old Oregon Trail, Gering, NE 69341. Mailing Address: P.O. Box 27, Gering, NE 69341-0027. Tel.: 308-436-9700. Fax: 308-436-7611.
Web Site: www.nps.gov/scbl
Founded: 1919.
Congressional District: 3rd
Key Personnel: Supt., Ken Mabery; Gift Shop Mgr., Jolene Kaufmann.
Personnel Profile: Full-Time Paid 10; Part-Time Paid 7; Part-Time Volunteers 2; Interns 1.
Governing Authority: federal. Parent Institution: U.S. National Park Service.
Institution Type/Description: History Museum.
Collections: Oregon Trail history 1830-1880; prehistoric & historic objects; sketches & watercolors by pioneer photographer William Henry Jackson; historic & prehistoric Plains Indian pieces.
Research Fields: Oregon Trail history; William H. Jackson, pioneer photographer; fire management; prairie restoration; flora & fauna surveys; collections internship projects.
Facilities: reference library available to the public; 53-seat auditorium. Books, posters, postcards & stationery for sale.
Activities: lectures; films; hiking; Summit Road; temporary exhibitions.
Publications: informational brochures; handbook, Overland Migrations.
Hours & Admission Prices: Winter: daily 8-5; Summer: 8-7. Seven-day Pass: $5 per vehicle; Annual Pass: $15 per vehicle; senior pass & federal recreation lands pass no charge. &
Attendance: 121,000 (accurate)
Membership: Family & Tri-Park Pass $15.

WILDCAT HILLS NATURE CENTER, N.E. Hwy. 71, Gering, NE 69341. Mailing Address: P.O. Box 65, Gering, NE 69341. Tel.: 308-436-3777.
Institution Type/Description: Nature Center.
Collections: local history; local flora & fauna; fossils; paintings.
Activities: simulated fossil dig.
Hours & Admission Prices: Call for hours.

Gibbon

GIBBON HERITAGE CENTER, 2nd & Court St., Gibbon, NE 68840. Mailing Address: 411 Lawn Ave., Gibbon, NE 68840-6016. Tel.: 308-468-6109 & 5608.
Founded: 1975.
Congressional District: 36
Personnel Profile: Part-Time Volunteers 6.
Institution Type/Description: History Museum.
Collections: history of area schools & businesses; religious memorabilia; church.
Hours & Admission Prices: 1st Sun. of month 2-4; other times by appointment. No charge; donations accepted.
Attendance: 125 (estimated)

ROWE SANCTUARY AND THE IAIN NICOLSON AUDUBON CENTER, 44450 Elm Island Rd., Gibbon, NE 68840-4019. Tel.: 308-468-5282.
E-mail: kskaggs@audubon.org
Web Site: www.rowesanctuary.org
Governing Authority: Parent Institution: National Audubon Society.
Institution Type/Description: Wildlife Sanctuary & Audubon Center.
Collections: wildlife & their habitat; natural science.
Facilities: nature trails. Museum-related items for sale.
Activities: education programs.
Hours & Admission Prices: April 16-Feb. 14 Mon.-Fri. 9-5, Sun. 1-4; Feb. 15-April 15 daily 8-5:30. Closed holidays.

Goehner

SEWARD COUNTY HISTORICAL SOCIETY MUSEUM, 364th Rd., I-80 Exit 373, Goehner, NE 68364. Mailing Address: P.O. Box 188, Goehner, NE 68364-0188. Tel.: 402-643-4935 & 523-4055 (museum).
E-mail: info@sewardcountymuseum.org
Web Site: sewardcountymuseum.org
Founded: 1978.
Congressional District: 1
Key Personnel: Pres., Lynn Reinmiller; Vice Pres., Joan Kennel; Treas., Miniature Train Owner & Operator, Jim Culver; Sec., Jean Kolterman; Membership & Exhibits, Della Miers.
Personnel Profile: Full-Time Volunteers 10; Part-Time Volunteers 10.
Governing Authority: society; nonprofit organization. Tax-exempt: 501(c)(3).
Institution Type/Description: Historical Society Museum.
Collections: toys; 1910-1930 kitchen & dining room; 1900 bedroom; food preservation; furnishings; miniature farm; dolls; 2 dining rooms; butchering equipment; sewing room; locally-made single-engine plane; furnished one-room schoolhouse; 1908 Seward County two story furnished farm home; farm machinery; Civil & World Wars memorabilia; doctor's office; dentist's office; country store; bootery; beauty shop; furnished log cabin; miniature railroad system; steam engine.
Facilities: approx. 100-vol. library of books on Seward County, available for use by public on premises; 14,400 sq. ft. exhibit space; youth room. Books for sale.
Activities: train ride on miniature train 2nd & 4th Sun. Museum Sponsors: Pot Luck Dinner for Friends & Family in June; Annual Family Sunday in July; Antique Car Show in October.
Publications: Seward County Historical Society Newsletter.
Hours & Admission Prices: May-Oct. Thurs. 10-4, Sun. 1:30-5. No charge; donations accepted. &
Attendance: 1,200 (estimated)
Membership: Annual $10; Sustaining $25; Life $150.

Gordon

SCAMAHORN CHURCH MUSEUM, 200 Block of W. 5th St. in Wayland Park, Gordon, NE 69360. Mailing Address: 516 N. Elm St., Gordon, NE 69360. Tel.: 308-282-2915.
Key Personnel: Pres., Harlen Wheeler
Institution Type/Description: History Museum.
Collections: local history; quilts; obituaries; period medical equipment; office furnishings; furniture; Native American beadwork; photographs; period clothing; military artifacts from WWI & WWII; dollhouse.
Hours & Admission Prices: Memorial Day to mid-Sept. Mon.-Fri. 1-5. No charge.

SHERIDAN GALLERY, 117 N. Main St., Gordon, NE 69343. Mailing Address: P.O. Box 237, Gordon, NE 69343. Tel.: 308-282-9972.
Institution Type/Description: Art Gallery.
Collections: works by local artists including Jean Ann Curry-Hess; photographs; paintings.
Activities: workshops; classes.
Hours & Admission Prices: Call for hours.

THE TRI-STATE OLD TIME COWBOYS MEMORIAL MUSEUM, Gordon, NE 69343. Mailing Address: P.O. Box 202, Gordon, NE 69343-0202. Tel.: 308-282-0749.
Institution Type/Description: History Museum.
Collections: early chuck wagon; old saddles; chaps; spurs; tools & gear used from early ranches of the late 1880s to modern-day; artifacts & relics pertaining to ranching & cowboys.
Hours & Admission Prices: June-Sept. 15 daily 1-5; other times by appointment. No charge; donations accepted. &

Gothenburg

GOTHENBURG HISTORICAL MUSEUM, 1420 Ave. F., Gothenburg, NE 69138. Mailing Address: 1001 Lake Ave., Gothenburg, NE 69138-1947. Tel.: 308-537-4293.
E-mail: gothenburghistory@qwestoffice.com
Web Site: www.gottenburghistory.org
Founded: 1980.
Congressional District: 3
Key Personnel: Pres. (V), Dick Larson.
Personnel Profile: Part-Time Volunteers 55.
Governing Authority: Tax-exempt.
Institution Type/Description: History Museum.
Collections: local history & culture; personal artifacts; period furnishings; photographs; textiles; costumes.
Hours & Admission Prices: Call for hours. No charge; donations accepted.
Attendance: 2,282 (accurate)
Membership: Base Member $15; Family $25; Supporting $50; Contributor $100; Benefactor $500; Lifetime $1,000.

PONY EXPRESS STATION MUSEUM, 1500 Lake Ave., Gothenburg, NE 69138. Mailing Address: P.O. Box 263, Gothenburg, NE 69138. Tel.: 308-537-3505.
E-mail: chamber@gothenburgdelivers.com
Web Site: www.gothenburgdelivers.com
Founded: 1954.
Congressional District: 3
Key Personnel: Dir., Anne Anderson; Museum Shop Mgr., Joyce Kolbo.
Personnel Profile: Full-Time Paid 1; Part-Time Paid 11.
Governing Authority: Parent Institution: Gothenburg Chamber of Commerce.
Institution Type/Description: Historic Building: housed in the Sam Macchette station used by the Pony Express between 1860-1861.
Collections: local history & culture; Pony Express memorabilia; early pioneer artifacts; photographs.
Facilities: Museum-related items for sale.
Hours & Admission Prices: April & Oct. 9-3; May & Sept. daily 9-6; June-Aug. daily 8-8. No charge; donations accepted.

SOD HOUSE MUSEUM, 300 S. Lake Ave., Gothenburg, NE 69138. Tel.: 308-537-2076.
E-mail: sodlady2001@hotmail.com
Founded: 1988.
Congressional District: 39
Key Personnel: Dir., Merle Block; Museum Mgr. & Museum Shop Mgr., Linda Block.
Personnel Profile: Full-Time Volunteers 2; Part-Time Paid 2.
Governing Authority: private.
Institution Type/Description: Historic Site & Historic House: c.1800 Farmstead & Sod House.
Collections: homesteader memorabilia; photographs of sod house era; barbed wire buffalo sculpture (life size) containing 4-1/2 miles of wire; Indian & Pony barbed wire sculpture; Pony Express history; Indian artifacts.
Facilities: barn housing museum; wooden windmill. Gift items for sale.
Activities: guided tours.
Publications: annual brochure, Pony Express Times.
Hours & Admission Prices: May & Sept. daily 9-6; June-Aug. daily 8-8. Family $5, adult $1. &
Attendance: 20,000 (estimated)

Grand Island

❋ STUHR MUSEUM OF THE PRAIRIE PIONEER, (M), 3133 W. Hwy. 34, Grand Island, NE 68801-7485. Tel.: 308-385-5316. Fax: 308-385-5028. Facebook: Stuhr Museum.
E-mail: info@stuhrmuseum.org
Web Site: www.stuhrmuseum.org
Founded: 1961.
Congressional District: 3
Key Personnel: Exec. Dir., Joe Black; Chm. (V), Densel Rasmussen; Dir. Education, Tiffany Hartford; Dir. Mktg., Mike Bockoven; Finance & Human Resources Mgr., Steve Stump; Visitor Svcs. Coord. & Museum Shop Mgr., Courtney Meyer; Historian & Research Center Mgr., Jessica Waite; Historical Interpretation Mgr., Kay Cynova.
Personnel Profile: Full-Time Paid 21; Full-Time Volunteers 10; Part-Time Paid 90; Part-Time Volunteers 1,898; Interns 3.
Governing Authority: county. Hall County Museum Board. Tax-exempt: 501(c)(3).
Institution Type/Description: History Museum: Stuhr Building designed by Edward Durell Stone.

Collections: 60 restored & furnished houses & businesses; farm; clothing; household wares; pioneer crafts; 20 period railroad cars & engines; farm machinery; period autos & farm machinery; Indian & Old West artifacts; restored 19th-century Railroad Town; Gus Fonner Memorial Rotunda. Historic Buildings: 1860s Road Ranch; 1890s rural settlement & farm; Pawnee Indian Earth Lodge.

Research Fields: mid 19th to early 20th-century prairie town building; homes; business; lifestyle; material culture; agriculture; Plains Indians; steam railroading.

Facilities: 200-acres; research center; 100-seat auditorium; activity center; arboretum. Museum-related items for sale.

Activities: structured educational programs; seasonal interpretive programs; period trades on site; interpretive history exhibits; special events; guided tours; lectures; films; concerts; workshops for children & adults; volunteer program; permanent, temporary & traveling exhibitions; hands-on activities; classes.

Publications: quarterly newsletter; monthly e-newsletter.

Hours & Admission Prices: Jan.-March Tues.-Sat. 9-5, Sun. 12-5; April-Dec. Mon.-Sat. 9-5, Sun. 12-5. May-Sept. adults $8, senior citizens $7, youth 7-12 $6; discounts to AAM members; members no charge. Oct.-April: adults $6, senior citizens $5, youth 7-12 $4; discounts to AAM members; members no charge. Closed New Year's Day; Thanksgiving; Christmas. &

Attendance: 71,000 (estimated)

Membership: Individual $35; Companion $40; Family & Grandparents $45.

Grant

PERKINS COUNTY HISTORICAL SOCIETY, Central Ave. & 6th, Grant, NE 69140. Mailing Address: P.O. Box 562, Grant, NE 69140-0562. Tel.: 308-352-4977 & 4698. Fax: 308-352-4977.

Founded: 1964.

Congressional District: 44

Key Personnel: Pres. (V), Brenda Styskal.

Personnel Profile: Part-Time Volunteers 45.

Governing Authority: nonprofit organization. Tax-exempt.

Institution Type/Description: General Museum: housed in 1905 two-story frame home with wrap around porch.

Collections: archives; agriculture; costumes. Historic Buildings: 1900 Brandon Rural School; 1975 Metal Building housing farm equipment.

Facilities: 300-vol. library of school books; religious books; fiction; history available on premises.

Activities: permanent exhibitions; area high school annuals.

Publications: book, Plainscape, A Portrait of Perkins County; 125th Celebration of Perkin's County; Golden Jubilee.

Hours & Admission Prices: Sat. 1-4 & special events. No charge; donations accepted.

Attendance: 550 (estimated)

Greeley

GREELEY COUNTY HISTORICAL SOCIETY COURT-HOUSE MUSEUM, Courthouse Square, Greeley, NE 68842. Mailing Address: P.O. Box 6, Greeley, NE 68842-0288. Tel.: 308-428-3115.

Institution Type/Description: Historical Society Museum.

Collections: local history & culture; photographs; personal artifacts; military artifacts; jewelry; Native American artifacts; clothing; early books; Veterans' Wall & Memorial including over 1,000 veterans listed.

Hours & Admission Prices: Mon.-Fri. 9-4.

Greenwood

GREENWOOD HISTORICAL SOCIETY DEPOT MUSEUM, 440 Broad St., Greenwood, NE 68366. Mailing Address: P.O. Box 1, Greenwood, NE 68366-0001. Tel.: 402-430-0238. Facebook: Greenwood Depot Museum / Greenwood Historical Society.

E-mail: jarid_massa@yahoo.com

Founded: 1976.

Key Personnel: Dir. & Museum Shop Mgr., Jarid Massa; Pres. & Chm. (V), Anna Jamrog.

Personnel Profile: Part-Time Volunteers 7.

Governing Authority: Tax-exempt.

Institution Type/Description: Railroad & History Museum.

Collections: depot & railroad history; photographs.

Hours & Admission Prices: April-Oct. Mon.-Wed. 10-5; other times by appointment. No charge; donations accepted. &

Attendance: 200 (estimated)

WORLD WAR II LIBRARY AND MUSEUM, 13102 N. 238th St., Greenwood, NE 68366. Tel.: 402-944-4238.

Institution Type/Description: Military History Museum.

Collections: World War II history & artifacts; uniforms; books; photographs; documentary tapes; WWII Jeeps; military toys.

Facilities: library.

Hours & Admission Prices: Daily 9-5.

Gretna

AK-SAR-BEN AQUARIUM OUTDOOR EDUCATION CENTER, 21502 W. Hwy. 31, Gretna, NE 68028-7264. Tel.: 402-332-3901. Fax: 402-332-5853.

Web Site: outdoornebraska.ne.gov/fishing/programs/aquticed/aquarium.asp

Founded: 1979.

Congressional District: 2

Key Personnel: Dir., Tony Korth.

Personnel Profile: Full-Time Paid 4; Part-Time Paid 3.

Governing Authority: state; Parent Institution: Nebraska Game & Parks Commission, 2200 N. 33rd, Lincoln, NE 68504. Tel.: 402-471-0641. Tax-exempt.

Institution Type/Description: Aquarium: located on site of 1882, Hatch House, first fish hatchery & public picnic area in Nebraska.

Collections: fish, turtles & frogs native to Nebraska.

Facilities: aquarium; 103-seat auditorium; classrooms.

Activities: guided tours; lectures; films, hobby workshops.

Publications: informational pamphlets on facility.

Hours & Admission Prices: Spring: April-Fri. before Memorial Day Wed.-Mon. 10-4:30; Summer: Memorial Day weekend-Labor Day Mon.-Fri. 10-4:30, Sat.-Sun. 10-5; Fall: Wed. after Labor Day-Nov. 30 Wed.-Mon. 10-4:30; Winter: Dec. 1-March 31 Wed.-Sun. 10-4:30. No charge. Closed New Year's Day; Thanksgiving; Christmas. &

Attendance: 65,000 (estimated)

Harrisburg

BANNER COUNTY HISTORICAL SOCIETY, 200 N. Pennsylvania Ave., Harrisburg, NE 69345. Mailing Address: P.O. Box 74, Harrisburg, NE 69345-0074. Tel.: 308-436-7228.

E-mail: bannercountyhistoricalsociety@yahoo.com

Web Site: bannercountyhistoricalsociety.com

Founded: 1969.

Congressional District: 3

Key Personnel: C.E.O. & Cur., Judy Leafdale; Pres., Vicki Stone; Vice Pres., Steve Brown; Treas., Reta Pahl; Sec., Sherry Gifford.

Personnel Profile: Part-Time Volunteers 15.

Governing Authority: society. Tax-exempt: 501(c)(3).

Institution Type/Description: Historical Society Museum.

Collections: 1800-1940 farm & household artifacts of western Nebraska; transportation; arrowhead collection; steam engine; period vehicles. Historic Buildings: sod house; log house; frame church; one-room log schoolhouse; 1900 drug store; c.1920 horse barn; c.1930 filling station; c.1920 bank.

Facilities: library of area history books, available for use by public; 10,800 sq. ft. exhibit space.

Activities: guided tours; lectures; program on late-1800s pioneer lifestyle for fourth grade students. Annual Event: Open House in June.

Publications: Banner County Historical Newsletter, 3 times per year.

Hours & Admission Prices: June to mid-Sept. Sun. 1-5; other times by appointment. No charge; donations accepted. &

Attendance: 3,000 (estimated)

Membership: Junior $1; Adult $5.

Harrison

AGATE FOSSIL BEDS NATIONAL MONUMENT, 301 River Rd., 22 miles south of Harrison & 3 miles east of Hwy. 29, Harrison, NE 69346-2734. Tel.: 308-668-2211. Fax: 308-668-2318.

Web Site: www.nps.gov/agfo/

Founded: 1965.

Congressional District: 3

Key Personnel: Supt., James Hill; Cur., Mark Hertig; Museum Shop Mgr., Jolene S. Kaufmann.

Personnel Profile: Full-Time Paid 6; Part-Time Paid 6; Part-Time Volunteers 1; Interns 2.

Governing Authority: federal. Parent Institution: National Park Service, Washington, DC. Tax-exempt.

Institution Type/Description: Park Museum/Visitor Center: located near the site of 19-22 million year old mammal fossil remains & historic Agate Springs Ranch on the Niobrara River.

Collections: Miocene fossils; Plains Indian artifacts, 1880-1930; James H. Cook & Harold J. Cook historical papers collection; articles of 19th-century western America; interactive exhibits.
Research Fields: paleontology; prairie & fire; river water quality; Cook Indian collection; early ranching
Facilities: visitor center; movie theater. Books & postcards for sale.
Activities: self-guided tour to fossil beds; self-guided tour to Daemonelix area; guided tours for schools; slide & video programs; permanent & temporary exhibitions.
Publications: book, Agate Fossil Beds Handbook 107.
Hours & Admission Prices: Memorial Day-Labor Day daily 8-6; Sept.-May daily 8-4. No charge. Closed New Year's Day; Thanksgiving; Christmas. &
Attendance: 12,000 (estimated)

SIOUX COUNTY HISTORICAL MUSEUM, 130 Main St., Harrison, NE 69346. Tel.: 308-668-2110.
Institution Type/Description: History Museum.
Collections: local history & culture; period furnishings; personal artifacts; photographs.
Hours & Admission Prices: Memorial Day to Labor Day Mon.-Sat. 10-4, Sun. 1-4. No charge.

Hartington

CEDAR COUNTY HISTORICAL MUSEUM, 304 W. Franklin, Hartington, NE 68739. Mailing Address: P.O. Box 181, Hartington, NE 68739. Tel.: 402-254-6597.
E-mail: cchistory@hartel.net
Founded: 1964.
Institution Type/Description: History Museum.
Collections: early machinery; horse drawn buggies; Hartington's first fire engine; log cabin; carriage house; restored Model T; personal artifacts; photographs; period furnishings.
Hours & Admission Prices: May-Aug. Sun. 2-4; other times by appointment. No charge; donations accepted.

Hastings

CHILDREN'S MUSEUM OF CENTRAL NEBRASKA, Imperial Mall, 12th & Marian Rd., Hastings, NE 68901. Mailing Address: P.O. Box 1502, Hastings, NE 68902-1502. Tel.: 402-463-3300.
E-mail: director@cmocn.org
Web Site: www.cmocn.org
Key Personnel: Pres. Bd., David Bosle; Vice Pres., Patrick Cecil; Sec. & Exec. Dir., Deb Bosle; Treas., DeWayne Boesen.
Governing Authority: nonprofit. Tax-exempt: 501(c)(3).
Institution Type/Description: Children's Museum.
Collections: hands-on exhibits.
Facilities: 10,000 sq. ft. exhibit space. Museum-related items for sale.
Activities: summer science & nature programs; birthday parties; facility rental; field trips.
Publications: newsletter, Hands on Learning.
Hours & Admission Prices: Tues.-Sat. 10-5:30, Sun. 1-5. Admission $5; members & children under 1 no charge. &
Membership: Family & Grandparent $60; Membership Plus & Care Giver $75; Reciprocal Traveler $125 & up.

CLYDE SACHTLEBEN OBSERVATORY, S. Wabash Ave., Hastings, NE 68901. Mailing Address: 710 N. Turner, Hastings, NE 68901. Tel.: 402-462-7378.
Institution Type/Description: Observatory.
Collections: 14", 10" & 2 8" reflecting telescopes; space science.
Activities: astronomy presentations.
Hours & Admission Prices: Call for hours. No charge.

* **HASTINGS MUSEUM, (M),** 1330 N. Burlington, Hastings, NE 68901-3099. Mailing Address: P.O. Box 1286, Hastings, NE 68902-1286. Tel.: 402-461-2399. Fax: 402-461-2379. Facebook: Hastings Museum.
E-mail: museum@hastingsmuseum.org
Web Site: www.hastingsmuseum.org
Founded: 1926.
Congressional District: 3
Key Personnel: Dir., Rebecca Matticks; Pres., Mike Howie; Dir. Mktg. & Devel., Becky Tideman; Cur. Collections, Teresa Kreutzer-Hodson; Cur. Education, Russanne Erickson; Museum Shop Mgr., Jenny Korte.
Personnel Profile: Full-Time Paid 12; Part-Time Paid 10; Part-Time Volunteers 163.

Volunteer Hours: 2,390
Operating Expenses: 1,331,881
Operating Income: 1,355,462
Governing Authority: municipal. Parent Institution: City of Hastings. Tax-exempt: 170(b)(1)(A) & 501(c)(3).
Institution Type/Description: Natural Science & History Museum.
Collections: North American mammals & birds; fossils; rocks; minerals; firearms; clothing; Native American; regional history items; automobiles; Kool Aid.
Major Exhibits: Artic Migrations (T), 1/14-3/21/14; Hastings Public School Art Show, 3/26/14-4/23/14; Crusin the Fossil Freeway (T), 5/5/14-8/10/14; Saving Endangered Species (T), 8/29/14-11/16/14; Snapshots of D Day (T), 10/24/14-11/30/14; Infamy, 12/5/14-1/15.
Facilities: 1,000-vol. library of natural science & historical books and pamphlets available on premises; planetarium; large-format digital 3D theater. Educational items, gifts & books for sale.
Activities: guided tours, lectures & films for schools; study clubs; films; planetarium programs; permanent, temporary & traveling exhibitions; summer education program for children; holiday activities for families; cell phone tours.
Publications: newsletter, UpClose; periodical.
Hours & Admission Prices: Adult $7; discounts to military; members no charge. &
Attendance: 55,000 (accurate)
Membership: VIP Add-on Program $20; Individual $40; Companion $45; Family & Grandparent $50; Silver Premiere $150; Gold Premiere $325; Platinum Premiere $600.

Hay Springs

HERITAGE CENTER MUSEUM, 230 N. Baker St., Hay Springs, NE 69347. Mailing Address: P.O. Box 291, Hay Springs, NE 69347-0291. Tel.: 308-638-7643.
Institution Type/Description: History Museum.
Collections: local history & culture; period artifacts; photographs. Historic Buildings: 1887 Methodist Church; 1910 school house; log cabin.
Hours & Admission Prices: Memorial Day to Labor Day Mon.-Fri. 1-4; other times by appointment. No charge.

HERITAGE CENTER MUSEUM II, 133 N. Main St., Hay Springs, NE 69347. Mailing Address: P.O. Box 291, Hay Springs, NE 69347-0291. Tel.: 308-638-7643.
E-mail: dsperkins@gpcom.net
Institution Type/Description: Historic Building: housed in the former hardware store.
Collections: local history & culture; period furnishings; personal artifacts; photographs.
Hours & Admission Prices: Memorial Day to Labor Day Mon.-Fri. 1-4; other times by appointment. No charge.

Hershey

STONES AND BONES GALLERY AND EMPORIUM, 105 E. 2nd St., Hershey, NE 69143. Mailing Address: P.O. Box 85, Hershey, NE 69143. Tel.: 308-368-7400.
Founded: 2008.
Personnel Profile: Part-Time Volunteers 2.
Institution Type/Description: Art Gallery.
Collections: western & wildlife art; period artifacts; photographs
Activities: demonstrations; flint knapping.
Hours & Admission Prices: Call for hours. No charge.
Attendance: 100

Holdrege

NEBRASKA PRAIRIE MUSEUM OF THE PHELPS COUNTY HISTORICAL SOCIETY, (M), 2701 Burlington St., Holdrege, NE 68949-1347. Mailing Address: P.O. Box 164, Holdrege, NE 68949-0164. Tel.: 308-995-5015. Fax: 308-995-2241.
E-mail: prairie995@gmail.com
Web Site: www.nebraskaprairie.org
Formerly: Phelps County Historical Museum
Founded: 1966.
Congressional District: 3
Key Personnel: Dir., Dan Christensen; Vice Pres., Nancy Morse; Chm. (V), Dr. Bob Butz; Sec., Warner Carlson; Museum Shop Mgr., Keith Weaver.
Personnel Profile: Full-Time Paid 1; Part-Time Paid 2; Part-Time Volunteers 50.
Governing Authority: nonprofit organization. Parent Institution: Phelps County Historical Society. Tax-exempt: 170(b)(1)(A).

Institution Type/Description: General Museum.
Collections: agriculture; antiques; religious; art; anthropology; ethnology; Indians; World War II German prisoner of war exhibit.
Research Fields: county & state history.
Facilities: 5,000-vol. library.
Activities: guided tours; lectures; permanent & temporary exhibitions; genealogy research.
Publications: book, History of Phelps County; quarterly newsletter, Stereoscope; Prisoners on the Plains.
Hours & Admission Prices: Summer: Mon.-Fri. 9-5, Sat.-Sun. 1-5; Winter: Mon.-Fri. 9-4, Sat.-Sun. 1-4. No charge; donations requested. Closed holidays. &
Attendance: 15,000 (estimated)
Membership: Annual $35; Life $1000.

Hyannis

GRANT COUNTY MUSEUM & HISTORIC SOCIETY, Grant County Courthouse, 105 E. Harrison, Hyannis, NE 69350-9706. Mailing Address: P.O. Box 82, Hyannis, NE 69350-0082. Tel.: 308-458-2371. Fax: 308-458-2485.
Founded: 1963.
Congressional District: 3
Key Personnel: C.E.O., Harry Merrihew.
Personnel Profile: Part-Time Paid 1; Part-Time Volunteers 2.
Governing Authority: nonprofit organization. Parent Institution: Grant County Historic Society. Tax-exempt.
Institution Type/Description: General Museum: housed in the Grant County Court House.
Collections: Grant County history; cowboy artifacts & pictures; Western culture; Indian artifacts; barbed wire; John Wayne; saddles; stuntman, Chuck Hayward; genealogy.
Research Fields: Genealogy; history; local.
Activities: guided tours; permanent exhibitions; special school tours for town & country schools. Special Events: Windmill Days; Quilt Show; Arrowhead and Artifact Show; Alumni Grant Co. Fair, County Government Day.
Publications: book, History of Grant County; Nebraska, Our Towns; Grant County - It's Friends & Neighbors.
Hours & Admission Prices: Tues.-Wed. 1-4; other times by appointment. No charge; donations accepted. &
Attendance: 238 (accurate)
Membership: Annual $10; Life $50.

Kearney

FORT KEARNEY MUSEUM, 131 S. Central Ave., Kearney, NE 68847-7908. Tel.: 308-234-5200.
Founded: 1950.
Key Personnel: Dir., Marlo L. Johnson.
Personnel Profile: Full-Time Paid 2; Part-Time Paid 3.
Governing Authority: individual operation.
Institution Type/Description: General Museum.
Collections: Indian and pioneer relics, tools & arms; European and Oriental material; Egyptian & African objects; history; natural history; glass; anthropology; archaeology; mineralogy; music; geology; costumes; numismatics; paintings; sculpture; graphics; circus.
Research Fields: natural history; glass; anthropology; archaeology; mineralogy; music; geology; costumes; numismatic; history.
Facilities: Gift items for sale.
Activities: permanent exhibitions; glass bottom boat rides.
Publications: brochure, Fort Kearney Museum.
Hours & Admission Prices: Memorial Day-Labor Day Thurs.-Sat. 10:30-5, Sun. 1-5. Museum: Adults $2.50; discount to groups; children under 12 no charge. Boat Rides: Adults $3, children under 12 $2.50. &

FORT KEARNY STATE HISTORICAL PARK, 1020 V Rd., Kearney, NE 68847-8043. Tel.: 308-865-5305. Fax: 308-865-5306.
E-mail: ngpc.fort.kearny@nebraska.gov
Web Site: www.outdoornebraska.org
Founded: 1929.
Congressional District: 3
Key Personnel: Supt., Eugene A. Hunt; Asst. Supt., Joe Blazek.
Personnel Profile: Full-Time Paid 2; Part-Time Paid 8; Interns 2.
Governing Authority: state. Affiliated with Nebraska Game and Parks Commission, Statehouse, Box 30370, 2200 N. 33rd St., Lincoln, NE. 68503. 402-471-0641. Tax-exempt.
Institution Type/Description: Historic Building & Site.
Collections: 1848-1871 military records; letters; dispatches; military orders; Oregon Trail history.

Facilities: 66-seat auditorium; microfilm. Books pertaining to Western history & other museum-related items for sale.
Activities: self-guided tours; lectures; films; broadcast programs; permanent & temporary exhibitions.
Publications: pamphlet.
Hours & Admission Prices: Memorial Day-Labor Day daily 9-5. Grounds: sunrise-sunset; other times by appointment; game & parks sticker required. &
Attendance: 61,000 (estimated)

THE FRANK HOUSE AT THE UNIVERSITY OF NEBRASKA-KEARNEY, 2010 University Dr., Kearney, NE 68849. Tel.: 308-865-8284.
E-mail: frankhouse@unk.edu
Web Site: www.frankhouse.org
Founded: 1972.
Congressional District: 3
Key Personnel: Dir., KrisAnn Sulivan.
Personnel Profile: Full-Time Paid 1; Part-Time Paid 3; Part-Time Volunteers 25; Interns 1.
Governing Authority: Parent Institution: University of Nebraska. Tax-exempt.
Institution Type/Description: Historic House Museum: housed in the former home of G.W. Frank & his family; built in 1889. Listed on the National Register of Historic Places.
Collections: household artifacts; furnishings; Tiffany windows; 19th century architecture.
Research Fields: early Kearney; Frank family; hospital for tuberculosis.
Activities: various programs open to the public. Museum Sponsors: Holiday Open House; Edible Flower Tea; Titanic Diner; Saturdays at the Frank House.
Hours & Admission Prices: Mon.-Fri. 2-5; Sat. 12-5. No charge; donations accepted. Closed university holidays & breaks.
Attendance: 4,160 (accurate)

KEARNEY AREA CHILDREN'S MUSEUM, 5827 4th Ave., Kearney, NE 68845-2879. Tel.: 308-698-2228. Fax: 308-698-2229.
E-mail: kidzone@kearneykidzone.com
Web Site: www.kearneykidzone.com
Founded: 1989.
Congressional District: 3
Key Personnel: Exec. Dir., Tara Knuth; Chm., Erin Becker; Vice Chm., Ryan Kohtz; Treas., Margene Dahlstedt; Sec., Loralea Frank.
Personnel Profile: Full-Time Paid 2; Part-Time Paid 10; Part-Time Volunteers 40.
Governing Authority: nonprofit organization. Tax-exempt: 501(c)(3).
Institution Type/Description: Children's Museum.
Collections: science; hands-on exhibits; music; art; sports; trains; water.
Activities: educational programs for children & families; crafts; workshops; music; shadow room.
Publications: monthly newsletter, KACM.
Hours & Admission Prices: Wed. & Fri.-Sat. 10-5, Thurs. 10-8, Sun. 1-5. Children 1-12 $7, adults 13 & over $5; discounts to Association of Children's Museums reciprocal members; KACM members & children under one no charge. Closed all major holidays. &
Attendance: 60,000 (estimated)
Membership: Sharing & Caring $10; Family & Grandparents $72; Family Plus 2 & Grandparents Plus 2 $84; Daycare $90; Daycare Plus 2 $102; Patron $125; Patron Plus 2 $137; Daycare Center Plus $180.

MUSEUM OF NEBRASKA ART, 2401 Central Ave., Kearney, NE 68847-4501. Tel.: 308-865-8559. Fax: 308-865-8104. Facebook: Museum of Nebraska Art.
E-mail: mona@unk.edu
Web Site: mona.unk.edu
Founded: 1976.
Congressional District: 3
Key Personnel: Bd. Pres., Jeanne Salerno; Dir., Audrey S. Kauders; Cur., Teliza Rodriguez; Collections Supvr., Jean Jacobson; Dir. Education, Kelley Dachtler; Dir. Devel., Kara Foradori; Coord. ARTreach, Russ Erpelding; Coord. Mktg., Gina Garden; Membership Asst., Carol Hardesty; Museum Shop Mgr., Merilyn Anderson; Museum Shop Mgr., Janet Fox; Office Supvr., Karen Humphrey.
Personnel Profile: Full-Time Paid 4; Part-Time Paid 7; Part-Time Volunteers 120.
Governing Authority: Parent Institution: University of Nebraska Kearney. Tax-exempt: 501(c)(3).
Institution Type/Description: Art Museum: housed in neoclassical revival building. Listed on the National Register of Historic Places.

Collections: Nebraska art & artists from early 19th century to present; sculpture garden.

Major Exhibits: Cranes Taking Flight, 2/18/14-5/11/14; Spirit A Celebration of Art in the Heart, 3/15/14-4/6/14; Nebraska Now: Richard Chung, Ceramic Sculpture, 4/12/14-7/6/14; Nebraska Arts Council: Visual Artist Fellowship 2014, 4/29/14-7/27/14; Junior Curator Show, 5/9/14-7/6/14; Spotlight On: Keith Martin, 5/20/14-8/10/14; Quilts Past and Present, 5/23/14-9/14/14; Regionalist Works of Grant Reynard, 6/10/14-8/10/14; George Catlins North American Indian Portfolio, 8/19/14-12/7/14; Cut, Formed, Folded, Pressed: Paper, 9/26/14-1/25/15.

Research Fields: historic & contemporary Nebraska artists.

Facilities: archival library containing records & documents related to Nebraska art & artists for public use; sculpture garden; educational facility.

Activities: guided tours; lectures; organized education programs for children & adults; docent program; organized education programs for undergraduate or graduate college students affiliated with University of Nebraska at Kearney; traveling, temporary & loan exhibitions.

Publications: exhibition catalogues; collection catalogues.

Hours & Admission Prices: Tues.-Sat. 11-5, Sun. 1-5. No charge; donations accepted. NARM reciprocal. Closed major holidays. &

Attendance: 20,000 (accurate)

Membership: Student $20; Senior $25; Individual $30; Dual Senior $45; Household $50; Patron $125; Benefactor $250; Bronze $500; Silver $1,000; Gold $2,500; Platinum $5,000.

NEBRASKA FIREFIGHTERS MUSEUM & EDUCATION CENTER, 2434 E. 1st St., Kearney, NE 68847. Tel.: 308-338-3473.

E-mail: mail@neffm.org

Web Site: www.neffm.org

Founded: 2009.

Key Personnel: Dir., Lindsay Schluntz; Chm. (V), Norman Hoeft.

Personnel Profile: Full-Time Paid 1.

Governing Authority: Parent Institution: Nebraska Firefighters Foundation.

Institution Type/Description: Fire Fighting History Museum.

Collections: firefighting history & equipment; photographs; Firefighter & EMS Memorial; fire trucks.

Facilities: rental facilities.

Activities: educational programs; hands-on exhibitions; scout programs; group tours; temporary & permanent exhibitions.

Hours & Admission Prices: Call for hours. Adults $6, seniors $4, youth $3, children 5 & under and members no charge. &

Attendance: 10,000 (estimated)

Membership: Individual Firefighters $20; Individual $30; Firefighter Family $65; Non-Firefighter Family $75.

TRAILS & RAILS MUSEUM, (M), 710 W. 11th St., Kearney, NE 68845-7340. Mailing Address: BCHS, P.O. Box 523, Kearney, NE 68848. Tel.: 308-234-3041.

E-mail: bchs.us@hotmail.com

Web Site: www.bchs.us

Congressional District: 3

Key Personnel: Dir., Jennifer Murrish; Pres. (V), Dan Speirs; Education Coord., Lyn Hoffman.

Personnel Profile: Full-Time Paid 1; Part-Time Paid 8; Part-Time Volunteers 20; Interns 1.

Governing Authority: Parent Institution: Buffalo County Historical Society. Tax-exempt.

Institution Type/Description: History Museum.

Collections: area history; locomotive; photographs; personal artifacts; livery barn. Historic Buildings: Union Pacific Depot; 1898 church; school house; log cabin; Freighters Hotel; Boyd Ranch house.

Facilities: archives. Museum-related items for sale.

Activities: groups tours; special events; Wild Science Thursdays; Ghost Hunting with Midwest Paranormal Investigators. Annual Events: Wagons West Day in June; 1/2 Marathon; Trivia Contest; Fiddle Contest; Christmas Tree Walk in December.

Publications: Buffalo Tales.

Hours & Admission Prices: Memorial Day to Labor Day Mon.-Sat. 10-6, Sun. 1-5; Sept.-May Mon.-Fri. 1-5; groups by appointment. Adults $5, children $2; discounts to AAM members; members no charge.

Attendance: 7,589 (accurate)

Membership: Individual $35; Family $40; Institutional $50; Supporting $75.

Lewellen

ASH HOLLOW STATE HISTORICAL PARK, 4265 Hwy. 26, Lewellen, NE 69147. Mailing Address: P.O. Box 70, Lewellen, NE 69147-0070. Tel.: 308-778-5651.

E-mail: jeff.uhrich@nebraska.gov

Web Site: outdoornebraska.ne.gov

Founded: 1967.

Congressional District: 3

Key Personnel: Supt., Jeffery Uhrich.

Personnel Profile: Full-Time Paid 1; Part-Time Paid 8.

Governing Authority: state. Parent Institution: State of Nebraska. Subsidiary Institution: Nebraska Game and Parks Commission, P.O. Box 30370, Lincoln, NE 68503. Tax-exempt.

Institution Type/Description: Historic Site: camp site along the Oregon Trail.

Collections: archaeology; Indian artifacts; pioneer trail history.

Facilities: visitors center.

Hours & Admission Prices: Memorial Day to Labor Day Thurs.-Sun. 9-4. Visitor Center: adults $2, children under 13 $1; children under 3 no charge. &

Attendance: 20,562 (accurate)

Lexington

DAWSON COUNTY HISTORICAL SOCIETY MUSEUM, (M), 805 N. Taft St., Lexington, NE 68850-2029. Mailing Address: P.O. Box 369, Lexington, NE 68850-0369. Tel.: 308-324-5340.

E-mail: info@dchsmuseum.com

Web Site: www.dchsmuseum.com

Founded: 1958.

Congressional District: 3

Key Personnel: Pres., Jerry Lashley; Dir., John Woodward.

Personnel Profile: Full-Time Paid 1; Part-Time Paid 2; Part-Time Volunteers 30.

Governing Authority: nonprofit organization. Parent Institution: Dawson County Historical Society. Tax-exempt: 501(c)(3).

Institution Type/Description: Local History Museum.

Collections: local history items; pioneer household artifacts; photographs & archives; quilts; clothing; glassware; period rooms & offices; 15,000 year-old remains of a Columbian mammoth discovered in Dawson County in 1993; early history of Central Nebraska including Wheels of Progress: The Automobile in Dawson County; 1917 McCabe Baby Biplane; 1903 Baldwin steam locomotive. Historic Buildings: 1888 rural schoolhouse; 1885 Union Pacific Depot; 1865 loghouse; farm history building; 1890 Fairhaven Church.

Research Fields: local history; genealogy; historical preservation; agricultural history.

Facilities: 40-seat meeting room. Museum-related items for sale.

Activities: guided tours; inter-museum loan; special speakers on local, regional & state history.

Publications: brochures; quarterly newsletter; books.

Hours & Admission Prices: Tues.-Fri. 9-5, Sat. 10-4. No charge; donations accepted. Closed New Year's Eve & Day; Easter; Memorial Day; Thanksgiving; Christmas. &

Attendance: 2,500 (accurate)

Membership: Individual $20; Family $30; Sustaining $50; Business $100; Life $1,000; Joint Life (husband & wife) $1,750.

HEARTLAND MUSEUM OF MILITARY VEHICLES, 606 Heartland Rd., Lexington, NE 68850-5666. Tel.: 308-324-6329. Fax: 308-324-6329.

E-mail: heartlandmuseum@cozadtel.net

Web Site: www.heartlandmuseum.com

Founded: 1991.

Key Personnel: Dir., Gary Gifford

Institution Type/Description: Military Museum.

Collections: period military vehicles; equipment; uniforms; USS Lexington, CV2 artifacts; MASH.

Facilities: library.

Hours & Admission Prices: Mon.-Sat. 10-5, Sun. 1-5. No charge; donations accepted. &

Attendance: 9,000 (accurate)

Lincoln

AMERICAN HISTORICAL SOCIETY OF GERMANS FROM RUSSIA, 631 D. St., Lincoln, NE 68502-1149. Tel.: 402-474-3363. Fax: 402-474-7229.

E-mail: ahsgr@ahsgr.org

Web Site: www.ahsgr.org

Founded: 1968.

Congressional District: 1

Personnel Profile: Full-Time Paid 1; Part-Time Paid 5.

Governing Authority: society members. Tax-exempt.

Institution Type/Description: Historical Society Museum.

Collections: memorabilia brought from Russia by the Germans; summer kitchen, chapel, country store; blacksmith shop; agricultural artifacts.

Research Fields: Germans from Russia genealogy.
Facilities: research library.
Activities: annual convention.
Publications: four journals; four newsletters.
Hours & Admission Prices: Mon.-Fri. 9-4, Sat. by appointment. Outside Campus Tours: April-Oct. Mon.-Fri. 2pm; other times by appointment. No charge; donations accepted. Closed major holidays & annual convention week. &
Attendance: 3,500 (accurate)
Membership: Youth $8; Student $15; Individual & Family $35, $50, $100; Life $750, $900, $1,050.

EISENTRAGER-HOWARD GALLERY - UNIVERSITY OF NEBRASKA-LINCOLN, Dept. of Art & Art History, Richards Hall 120, Stadium Dr. & T St., Lincoln, NE 68588-0114. Tel.: 402-472-5522.
Institution Type/Description: Art Gallery.
Collections: paintings; sculpture.
Activities: special events.
Hours & Admission Prices: Call for hours. No charge. &

ELDER ART GALLERY, NEBRASKA WESLEYAN UNIVERSITY, Rogers Center for Fine Arts, 50th St. & Huntington Ave., Lincoln, NE 68504-2230. Mailing Address: Art Depart., Nebraska Wesleyan University, 5000 St. Paul Ave., Lincoln, NE 68504-2760. Tel.: 402-466-2371 & 465-2230. Fax: 402-465-2179.
E-mail: llockman@nebrwesleyan.edu
Founded: 1965.
Congressional District: 2
Key Personnel: Pres., Fred Ohles; Dir., Donald Paoletta; Gallery Asst., Regina O'Rear.
Governing Authority: university. Parent Institution: Nebraska Wesleyan University. Tax-exempt: 501(c)(3).
Institution Type/Description: Art Museum.
Collections: graphics; paintings; sculpture; ethnic & contemporary crafts.
Activities: guided tours; lectures; gallery talks; concerts; permanent, temporary & traveling exhibitions.
Hours & Admission Prices: Tues.-Fri. 10-4, Sat.-Sun. 1-4. No charge; donations accepted. Closed school holidays & between shows. &
Attendance: 1,500 (estimated)

FAIRVIEW, THE BRYAN MUSEUM, 49th St., (and Sumner St.), Lincoln, NE 68506-1299. Tel.: 402-481-8303.
E-mail: ellen.beans@bryanhealth.org
Web Site: www.bryanhealth.org
Formerly: The Bryan Museum
Key Personnel: Dir. Volunteer Svcs., Ellen Beans
Institution Type/Description: Historic House: former home of William Jennings Bryan.
Collections: displays & recordings; photographs
Hours & Admission Prices: Mon.-Fri. 10-4. Tours: by appointment. No charge. Closed New Year's Day; Thanksgiving; Christmas.

FRANK H. WOODS TELEPHONE MUSEUM, 2047 M St., Lincoln, NE 68510-1029. Mailing Address: P.O. Box 81309, Lincoln, NE 68501-1309. Tel.: 402-436-4640. Fax: 402-436-4914.
Founded: 1996.
Key Personnel: Dir., Wally Tubbs
Institution Type/Description: History Museum: named for the founder of the Lincoln Telephone Company, 1903.
Collections: history of the telephone industry; independent telephony; The Lincoln Telephone and Telegraph Company and it's founder, Frank H. Woods.
Activities: tours.
Hours & Admission Prices: Sun. 1-4; other times by appointment. No charge; donations accepted. Closed major holidays.

GREAT PLAINS ART MUSEUM, University of Nebraska-Lincoln, 1155 Q St., Hewit Place, Lincoln, NE 68588-0250. Tel.: 402-472-6220. Fax: 402-472-0463.
E-mail: amohr2@unl.edu
Web Site: www.unl.edu/plains
Formerly: Great Plains Art Collection
Founded: 1980.
Congressional District: 1
Key Personnel: Dir., Richard Edwards; Pres. (V), Thomas G. Franti; Cur., Amber Mohr.

Personnel Profile: Full-Time Paid 2; Part-Time Paid 8; Part-Time Volunteers 12; Interns 2.
Governing Authority: nonprofit organization. Parent Institution: University of Nebraska-Lincoln. Subsidiary Institution: Center for Great Plains Studies, University of Nebraska. Tax-exempt: 170(b)(1)(a).
Institution Type/Description: Art Museum.
Collections: Western art with emphasis on Great Plains; sculpture; paintings; graphics; photographs.
Research Fields: history, literature & art of the American West.
Facilities: 4,000-vol. library relating to the history of the American West available for research on premises only.
Activities: guided tours; gallery talks; changing exhibitions; artist-in-residence program.
Publications: exhibition catalogs; brochures.
Hours & Admission Prices: Tues.-Sat. 10-5, Sun. 1:30-5. No charge; donations accepted. Closed holidays; between exhibitions. &
Attendance: 17,150 (accurate)
Membership: Family $35; Donor $50; Sustaining $100; Sponsor $250.

∗ **INTERNATIONAL QUILT STUDY CENTER & MUSEUM - UNIVERSITY OF NEBRASKA-LINCOLN, (M),** 1523 N. 33rd St., Lincoln, NE 68583-0838. Tel.: 402-472-6549. Fax: 402-472-2008.
Web Site: www.quiltstudy.org
Founded: 1997.
Congressional District: 1
Key Personnel: Dir., Patricia Cox Crews.
Personnel Profile: Full-Time Paid 7; Part-Time Paid 10; Part-Time Volunteers 63.
Governing Authority: Parent Institution: University of Nebraska. Tax-exempt.
Institution Type/Description: History Museum.
Collections: quilts & related textiles.
Research Fields: quiltmaking traditions worldwide; textiles & material culture.
Facilities: classroom; reception hall.
Activities: educational programs; classes; special events.
Publications: quarterly e-newsletter for members; annual report; exhibition catalogs; quilt-of-the month email notification.
Hours & Admission Prices: Tues.-Sat. 10-4, Sun. 1-4. Adults $6; UNL faculty, staff & students and NARM & museum members no charge. Closed major holidays; university winter break. &
Attendance: 12,667 (accurate)
Membership: Student $15; Individual $45; Family $70; Contributing $100-$249; Sustaining $250 & up.

LARSEN TRACTOR TEST & POWER MUSEUM, University of Nebraska, 35th & Fair Sts., Lincoln, NE 68583. Mailing Address: P.O. Box 830833, Lincoln, NE 68583-0833. Tel.: 402-472-8389. Fax: 402-472-8367. Facebook: Larsen Tractor Museum.
E-mail: ltodd6@unl.edu
Web Site: tractormuseum.unl.edu
Founded: 1998.
Key Personnel: Dir., Dr. Mack Riley; Mgr., Lance Todd; Cur., Lou Leviticus.
Personnel Profile: Full-Time Paid 1; Full-Time Volunteers 1; Part-Time Volunteers 6.
Governing Authority: Parent Institution: University of Nebraska. Subsidiary Institution: Department of Biological Systems Engineering. Tax-exempt.
Institution Type/Description: Agriculture Museum: housed in the original Nebraska Tractor Test facility built in 1919. A Historic Landmark.
Collections: 30 historic tractors including a 1915 Ford B Tractor, Waterloo Boy, Heider, 1918 Moline Universal D, Allis Chalmers WC & Fordson; period hand tools; planters; cultivators; horse-drawn plows; sod cutter; haying tools; household implements.
Publications: newsletter, Nuts & Bolts.
Hours & Admission Prices: Tues.-Fri. 9-4, Sat. 10-2; other times by appointment. No charge, donations accepted. &
Attendance: 3,000 (estimated)

LINCOLN CHILDREN'S MUSEUM, 1420 P St., Lincoln, NE 68508-1635. Tel.: 402-477-4000. Fax: 402-477-2004.
E-mail: info@lincolnchildrensmuseum.org
Web Site: www.lincolnchildrensmuseum.org
Founded: 1989.
Key Personnel: Exec. Dir., Paul Durban; Dir. Operations, Ariel Hadwiger; Dir. Devel., Angela Smith; Dir. Mktg., Michaella Kumke.
Personnel Profile: Full-Time Paid 10; Part-Time Paid 25; Part-Time Volunteers 900; Interns 4.
Governing Authority: private; nonprofit organization. Tax-exempt.
Institution Type/Description: Children's Museum.
Collections: interactive hands-on exhibits & activities.

Activities: participatory & traveling exhibits. Annual Events: Wonderful Wednesdays; Creation Station; New Year's Eve Family Celebration; Stuff a Scarecrow workshop; Candyhouse Funshop; Adult Night; Music & Mozzarella.
Publications: quarterly newsletter, Fingerprints.
Hours & Admission Prices: Mon.-Wed. & Fri.-Sat. 9:30-5, Thurs. 9:0-7:30, Sun. 1-5. Admission $8, seniors 62 & over $7.50, children under 2 $5; children under one no charge. Closed major holidays. &
Attendance: 170,000 (accurate)
Membership: Individual $75; Caregiver $85.

LINCOLN CHILDREN'S ZOO, 1222 S. 27th St., Lincoln, NE 68502-1832. Tel.: 402-475-6741. Fax: 402-475-6742.
E-mail: jchapo@lincolnzoo.org
Web Site: www.lincolnzoo.org
Formerly: Folsom Children's Zoo and Botanical Garden
Founded: 1959.
Key Personnel: C.E.O. & Pres., John P. Chapo; Gen. Mgr., Evan Killeen; Gen. Cur., Randy Scheer; Dir. Education, Aimee Johns.
Personnel Profile: Full-Time Paid 17; Part-Time Paid 20; Part-Time Volunteers 400.
Governing Authority: nonprofit organization. Affiliated with A.R. Folsom Zoological Society. Tax-exempt: 501(c)(3).
Institution Type/Description: Children's Zoo.
Collections: zoology; horticulture; aviary; herpetology; animal sculpture; environmental education.
Facilities: theater; classrooms; concessions. Museum-related items for sale.
Activities: self-guided tour maps; education programs for children; volunteer opportunities; amphitheater presentations; critter encounter area.
Publications: four publications yearly, Zoo Tracks.
Hours & Admission Prices: April 15-May & Sept.-Oct. 15 daily 10-5; June-Aug. Wed. 10-8, Thurs.-Tues. 10-5. Adults $6.50, seniors 60 & over and children 2-11 $5.50; members & children under 2 no charge. &
Attendance: 155,000 (accurate)
Membership: Membership $75.

LINCOLN FIRE & RESCUE DEPARTMENT MUSEUM, 1801 "Q" St., Lincoln, NE 68508-1774. Tel.: 402-441-8360.
Institution Type/Description: Fire Fighting Museum.
Collections: uniforms & helmets; fire fighting equipment; fire trumpet; 1911 motorized fire engine; photographs.
Activities: group tours.
Hours & Admission Prices: Daily 9-8.

LUX CENTER FOR THE ARTS, 2601 N. 48th St., Lincoln, NE 68504-3632. Tel.: 402-466-8692. Fax: 402-466-3786.
E-mail: info@luxcenter.org
Web Site: www.luxcenter.org
Formerly: University Place Art Center
Founded: 1978.
Congressional District: 2
Key Personnel: Exec. Dir., JoAnn Emerson; Chm. (V), Carl Eskridge; Dir. Education & Gallery, Stephanie Leach; Business Mgr., Lettie Vanhemert; Cur., Susan Soriente.
Personnel Profile: Full-Time Paid 3; Part-Time Paid 2; Part-Time Volunteers 100; Interns 4.
Governing Authority: private; nonprofit organization. Tax-exempt: 501(c)(3).
Institution Type/Description: General Museum.
Collections: over 1,700 historical & collectible dolls; paperweights; 450 19th & 20th-century master prints from the United States, Europe & Asia; contemporary art.
Facilities: 225-vol. library of art & collectible books; 4,800 sq. ft. exhibit space; educational facilities.
Activities: studio art classes; lectures; tours; temporary exhibitions. Museum Sponsors: Art of Fine Craft Conference.
Publications: newsletter; class brochure; monthly, exhibition announcements.
Hours & Admission Prices: Tues.-Fri. 11-5, Sat. 10-5. No charge; donations accepted. Closed New Year's Day; Memorial Day; Independence Day; Labor Day; Thanksgiving; Christmas. &
Attendance: 8,500 (estimated)
Membership: Individual $40; Household $70; Charter $100; Benefactor $250; Philanthropist $500; Founder $1,000.

NATIONAL MUSEUM OF ROLLER SKATING, 4730 South St., Lincoln, NE 68506-1256. Tel.: 402-483-7551, ext. 16. Fax: 402-483-1465.
E-mail: directorcurator@rollerskatingmuseum.com
Web Site: www.rollerskatingmuseum.com

Founded: 1980.
Congressional District: 1
Key Personnel: C.E.O. & Pres. (V), Annelle Anderson; Dir., James Vannurden.
Personnel Profile: Full-Time Paid 1.
Governing Authority: nonprofit organization. Tax-exempt.
Institution Type/Description: Sports & Technology Museum.
Collections: roller skates, 1819-present; patents; trophies & medals; costumes; posters; photos, films & videotape; books; periodicals manuscripts.
Research Fields: all aspects of roller skating history; sports; technology; industries; roller rinks; personalities involved in the sport.
Facilities: library of books, periodicals & manuscripts. Museum-related items for sale.
Activities: films; temporary, permanent & traveling exhibitions.
Publications: quarterly, newsletter; brochures; booklets; catalogues; books, The History of Roller Skating; The Evolution of the Roller Skate; The Allure of the Rink.
Hours & Admission Prices: Mon.-Fri. 9-5. No charge; donations accepted. Closed holidays. &
Attendance: 2,200 (estimated)
Membership: Individual & Family $35; Donor $50; Associate Patron $100; Patron $250; Benefactor $500.

NEBRASKA CONFERENCE UNITED METHODIST HISTORICAL CENTER, Nebraska Wesleyan Univ., Cochrane-Woods Library, Lower Level, 5000 St. Paul Ave., Lincoln, NE 68504. Mailing Address: 3333 Landmark Circle, Lincoln, NE 68504-4760. Tel.: 402-465-2175. Fax: 402-464-6203.
E-mail: kdvorak@umcneb.org
Web Site: www.umcneb.org
Founded: 1968.
Congressional District: 1
Key Personnel: Dir., Karrie Dvorak.
Personnel Profile: Part-Time Paid 1; Part-Time Volunteers 3.
Governing Authority: church. Parent Institution: United Methodist Conference Offices. Subsidiary Institution: Commission on Archives & History, Nebraska Conference, United Methodist Church. Tax-exempt.
Institution Type/Description: Religious Archives.
Collections: artifacts related to the United Methodist Church & its predecessor denominations in Nebraska, documents & legal papers; bibles; hymnals; memorabilia from United Methodist Churches & institutions of Nebraska.
Research Fields: United Methodist persons, local churches, Conference agencies & institutions.
Facilities: 11,000-vol. library of historical material of the United Methodist Church & its predecessor denominations available on premises; archives; reading room.
Activities: permanent & temporary exhibitions; research services.
Publications: Methodist History.
Hours & Admission Prices: Tues. & Thurs. 9:30-4, Wed. 10:30-4. Historic Center: no charge. Research Fee: $20 an hour. Closed holidays. &
Attendance: 200 (estimated)

✱　NEBRASKA HISTORY MUSEUM, (M), 131 Centennial Mall N., Lincoln, NE 68508-3805. Mailing Address: P.O Box 82554, Lincoln, NE 68501-2554. Tel.: 402-471-4754. Fax: 402-471-3314.
E-mail: ann.billesbach@nebraska.gov
Web Site: www.nebraskahistory.org
Formerly: Nebraska State Historical Society's Museum of Nebraska History
Founded: 1878.
Congressional District: 1
Key Personnel: C.E.O. & Exec. Dir., Michael J. Smith; Museum Dir. & Assoc. Dir. Interpretation & Education, Ann Billesbach; Deputy Dir., Lynne Ireland; Assoc. Dir. Collection, Deborah Arenz; Coord. Exhibition Svcs., Tina Koeppe; Museum Shop Mgr., Deb McWilliams; Assoc. Dir. Publications, David Bristow; Museum Educator, Judy Keetle; Sr. Museum Cur., Laura Mooney; Senior Research Folklorist, John E. Carter.
Personnel Profile: Full-Time Paid 69; Part-Time Paid 11; Part-Time Volunteers 250.
Governing Authority: state. Parent Institution: Nebraska State Historical Society. Historic Sites: Fort Robinson Museum, Crawford; George W. Norris State Historic Site; McCook; Neligh Mill State Historic Site; Neligh; John G. Neihardt State Historic Site; Bancroft; Kennard House, Lincoln; Willa Cather State Historic Site, Red Cloud; Gerald R. Ford Conservation Center, Omaha; Chimney Rock National Historic Site; Bayard. Tax-exempt: 501(c)(3).
Institution Type/Description: History Museum.
Collections: items relating to Nebraska & the Central Plains; art; manuscript collections; anthropology; archaeology; archives; costumes; ethnology; military items; photographs.

Major Exhibits: Nebraska Cowboys: Lives, Legends & Legacies, 9/23/13-12/14.
Research Fields: anthropology; archaeology; ethnology; western history.
Facilities: auditorium. Museum-related items for sale.
Activities: guided tours; gallery talks; inter-museum loan, permanent & temporary exhibitions; films; monthly lecture series; Investigation Station, an experiential learning environment for all ages.
Publications: educational leaflets; exhibit catalogs & checklists; anthropology publications; quarterly, Nebraska History; Nebraska History newspaper for kids, Nebraska Trailblazer.
Hours & Admission Prices: Mon.-Fri. 9-4:30, Sat.-Sun. 1-4:30. Suggested Donation $2. Closed federal holidays. &
Membership: Individual $40; Household $55; Contributing $150; Supporting $250; Sustaining $500; Founder $1,000.

NEBRASKA STATE CAPITOL, 1445 K St., Lincoln, NE 68509-4696. Mailing Address: Office of the Capitol Commission, P.O. Box 94696, Lincoln, NE 68509-4696. Tel.: 402-471-6691. Fax: 402-471-6952.
Web Site: www.capitol.org
Founded: 1932.
Congressional District: 1
Key Personnel: Capitol Admin., Robert C. Ripley.
Personnel Profile: Full-Time Paid 27; Part-Time Volunteers 3; Interns 1.
Governing Authority: state. Parent Institution: State of Nebraska. Subsidiary Institution: Nebraska Capitol Commission. Tax-exempt.
Institution Type/Description: Historic Building & Site: 1922-1932, state capitol.
Collections: 5,000 drawings & blueprints, 30 ft. of specifications & correspondence; decorative arts; sculpture; murals.
Facilities: 180,000-vol. law library & public library, available for inter-library loan. Brochures, booklets & other museum-related items for sale.
Activities: guided tours; inter-museum loan & permanent exhibitions.
Publications: books, Architectural Wonder of The World; Building A Landmark; A Harmony of The Arts-The Nebraska State Capitol.
Hours & Admission Prices: Mon.-Fri. 8-5, Sat. & holidays 10-5, Sun. 1-5. No charge; donations accepted. Closed New Year's Day; Thanksgiving & day after; Christmas. &
Attendance: 100,000 (estimated)

NEBRASKA STATE HISTORICAL SOCIETY'S THOMAS P. KENNARD HOUSE, (M), 1627 H St., Lincoln, NE 68508. Mailing Address: Box 82554, Lincoln, NE 68501-2554. Tel.: 402-471-4764. Fax: 402-471-3314.
E-mail: tom.buecker@nebraska.gov
Web Site: www.nebraskahistory.org
Formerly: Thomas P. Kennard House Nebraska Statehood Memorial
Founded: 1968.
Congressional District: 1
Key Personnel: C.E.O. & Dir., Michael J. Smith; Assoc. Dir., Ann Billesbach; Site Supvr., Thomas R. Buecker; Historic Sites Mgr., Deb McWilliams.
Personnel Profile: Part-Time Paid 1; Part-Time Volunteers 2.
Governing Authority: state. Parent Institution: Nebraska State Historical Society. Tax-exempt: 501(c)(3).
Institution Type/Description: Historic House: 1869 The Kennard House.
Collections: period furnishings; 1870s period furniture.
Research Fields: 1870s Nebraska life.
Activities: guided tours; special exhibits. Annual Event: Victorian Holidays Past in December.
Publications: Nebraska History magazine; educational leaflets.
Hours & Admission Prices: Mon.-Fri. by appointment. Adults $3; discounts to groups of 20 or more; children under 18 accompanied by adults, members no charge. Closed state holidays.
Attendance: 869 (accurate)
Membership: Individual $40; Household $55; Contributing $100; Supporting $250; Sustaining $500; Founder $1,000.

PIONEERS PARK NATURE CENTER, 3201 S. Coddington, Lincoln, NE 68522-9212. Mailing Address: 2740 A St., Lincoln, NE 68502-3113. Tel.: 402-441-7895. Fax: 402-441-6468.
E-mail: nfurman@lincoln.ne.gov
Web Site: parks.lincoln.ne.gov
Founded: 1963.
Congressional District: 1
Key Personnel: Natural Resources Mgr., Terry Genrich; Nature Center Coord., Nancy Furman; Naturalist, Andrea Faas.
Personnel Profile: Full-Time Paid 5; Part-Time Paid 45; Part-Time Volunteers 200; Interns 1.

Governing Authority: municipal government. Parent Institution: City of Lincoln. Subsidiary Institution: Parks & Recreation. Tax-exempt.
Institution Type/Description: Nature Center.
Collections: indoor & outdoor exhibits emphasizing Nebraska native fauna & flora; prairie grass; school furnishings. Historic Building: one room schoolhouse.
Research Fields: prairie restoration, interpretative techniques, ecological studies.
Facilities: 668 acres of prairie, woodlands, marsh, stream & pond habitats; over 8.5 miles of trails; 3,200 sq. ft. of indoor exhibits; herb garden; interpretive facilities. Museum-related items for sale.
Activities: interpretive school programs; hikes; special interest classes; special events; teacher & parent workshops; wilderness nature camps; outreach programs & field trips; overnights; hayrack rides; pre-school program; senior programming; puppet shows.
Publications: quarterly newsletter, From The Trails.
Hours & Admission Prices: Mon.-Sat. 8:30-5, Sun. 12-5. No charge, donations accepted. Closed New Year's Day; Thanksgiving; Christmas. &
Attendance: 76,830 (accurate)

ROBERT HILLESTAD TEXTILE GALLERY - UNIVERSITY OF NEBRASKA-LINCOLN, 234 Home Economics Bldg., Lincoln, NE 68588-0802. Tel.: 402-472-2911. Fax: 402-472-0640.
E-mail: wweiss@unl.edu
Web Site: textilegallery.unl.edu
Founded: 1994.
Personnel Profile: Full-Time Paid 1.
Governing Authority: nonprofit organization. Parent Institution: University of Nebraska. Tax-exempt: 501(c)(3).
Institution Type/Description: Textile Museum.
Collections: works of art from regional, national & international artists.
Activities: educational programs.
Publications: brochures; catalogs
Hours & Admission Prices: Mon.-Fri. 8:30-4; other times by appointment. No charge. Closed university holidays; between shows. &
Attendance: 3,000 (estimated)

*** SHELDON MUSEUM OF ART AND SCULPTURE GARDEN/UNIVERSITY OF NEBRASKA-LINCOLN, (M),** 12th and R Sts., Lincoln, NE 68588-0300. Mailing Address: P.O. Box 880300, Lincoln, NE 68588-0300. Tel.: 402-472-2461. Fax: 402-472-4258. Facebook: Sheldon Museum.
E-mail: sheldon@unl.edu
Web Site: www.sheldonartmuseum.org
Founded: 1963.
Congressional District: 1
Key Personnel: Pres. (V), Lisa Smith; Dir., Jorge Daniel Veneciano; Dir. Education & Publications, Gregory Nosan; Collections Mgr., Stacey Walsh; Mktg. Mgr., Ann Gradwohl; Office Mgr., Monica Babcock; Museum Shop Mgr., Vonni Sparks.
Personnel Profile: Full-Time Paid 19; Part-Time Paid 6; Part-Time Volunteers 100; Interns 10.
Governing Authority: university. Parent Institution: University of Nebraska-Lincoln. Tax-exempt: 170(b)(1)(A).
Institution Type/Description: American Art Museum.
Collections: 19th to 21st-century American art with emphasis on 20th century; American modernism & sculpture from Rodin to Roxy Paine.
Research Fields: 19th- & 20th-century American art.
Facilities: 25,000 vol. research library; campus-wide sculpture garden; 300-seat auditorium. Art & gift items for sale.
Activities: guided tours; lectures; films; gallery talks; concerts; annual state-wide touring exhibition; inter-museum loans; formally organized education programs for children; docent program; permanent, temporary & traveling exhibitions.
Publications: exhibition catalogues, Sculpture Collection; The American Painting Collection of the Sheldon Memorial Art Gallery; quarterly newsletter.
Hours & Admission Prices: Tues. 10-8, Wed.-Sat. 10-5, Sun. 12-5. No charge, donations accepted. Closed major holidays. &
Attendance: 50,000 (accurate)
Membership: Student $15; Basic $50; Supporting $80; Contributor $150; Patron $250; Benefactor $500; Curator's Circle $1,000; Director's Circle $2,500; Trustee Circle $5,000; Sheldon Circle $10,000.

SMITH COLLECTION MUSEUM OF AMERICAN SPEED, Speedway Motors Inc. Campus, 599 Oak Creek Dr., Lincoln, NE 68528. Mailing Address: P.O. Box 81906, Lincoln, NE 68501-1906. Tel.: 402-323-3166. Fax: 402-323-3151.
E-mail: museumofamericanspeed@gmail.com

Web Site: www.museumofamericanspeed.com
Founded: 1992.
Congressional District: 1
Key Personnel: Dir., John MacKichan; C.E.O., Clay Smith.
Personnel Profile: Full-Time Paid 4; Part-Time Paid 5; Part-Time Volunteers 29.
Governing Authority: Parent Institution: Speedway Motors. Tax-exempt.
Institution Type/Description: History Museum.
Collections: cars & car engines; pedal cars; gas & oil art; license plates; air pumps; pennants; radiator caps; monkey wrenches; toys; lunch boxes.
Activities: guided tours; rental facilities.
Hours & Admission Prices: Guided Tours: May-Sept. Mon.-Fri. 2pm; Oct.-April Fri. 2pm. Admission. $10. &
Attendance: 5,364 (accurate)

UNIVERSITY OF NEBRASKA-LINCOLN BOTANICAL GARDEN & ARBORETUM, 1309 N. 17th St, Lincoln, NE 68588-0663. Tel.: 402-472-2679. Fax: 402-472-9615.
E-mail: sbudler1@unl.edu
Web Site: www.unl.edu/bga/
Institution Type/Description: Botanical Garden.
Collections: native plants & trees; perennials.
Activities: educational programs; special events; tours.
Hours & Admission Prices: Call for hours. No charge.

* **UNIVERSITY OF NEBRASKA STATE MUSEUM, (M),** 307 Morrill Hall, South of 14th and Vine Sts., Lincoln, NE 68588-0338. Tel.: 402-472-3779 & 2642 (Research Collections). Fax: 402-472-8899.
E-mail: pgrewl@unl.edu
Web Site: www.museum.unl.edu
Founded: 1871.
Congressional District: 1
Key Personnel: Chancellor, Harvey Perlman; Dir., Dr. Priscilla C. Grew; Pres. Friends of the Museum, Mark Brohman; Assoc. Dir., Mark W. Harris; Cur., Professor & Informal Science Education, Dr. Judy Diamond; Cur. Zoology, Dr. Patricia Freeman; Cur. Anthropology, Dr. Alan Osborn; Cur. Entomology, Dr. Brett C. Ratcliffe; Cur. Geology & Mineralogy, Dr. Samuel B. Treves; Cur. Invertebrate Paleontology, Robert Diffendal; Cur. Invertebrate Paleontology, Dr. David K. Watkins; Cur. Vertebrate Paleontology, Dr. Robert M. Hunt; Cur. Parasitology, Scott L. Gardner; Collection Mgr. Entomology, Matt Paulsen; Collection Mgr. Vertebrate Paleontology, R. George Corner; Graphics Specialist Exhibits, Joel Nielsen; Collection Mgr. Zoology & Botany, Thomas E. Labedz; Collection Mgr. Parasitology, Gabor Racz; Highway Salvage Paleontologist, Shane Tucker; Coord. Planetarium Programs, Jack A. Dunn; Education Coord., Kathleen A. French; Scientific Illustrator, Angie Fox; Preparator Vertebrate Paleontology, Gregory W. Brown; Preparator Vertebrate Paleontology, Robert I. Skolnick; Preparator Vertebrate Paleontology, Ellen Stepleton; Supt. Ashfall Fossil Beds State Historical Park, Rick Otto; Head Security & Public Service Assoc., Linda Beran; Museum Educator, Ann E. Cusick; Coord. Public Rels., Dana Ludvik; Sec. Research & Publications, Gail A. Littrell; Ashfall Fossil Beds State Historical Park Asst., Sandy S. Mosel; Trailside Staff Asst., Pattie Norman; Accounting Clerk, Judy Ray; Museum Educator, Cindy Loope; Museum Educator, Reservations, Ina van der Veen; Museum Educator, Annie Mumgaard; Collections Asst. Anthropology & NAGPRA Asst., Susan Curtis; Cur. Vertebrate Paleontology, Ross Secord; Collections Asst. Botany, Linda Rader; Museum Shop Mgr., Marisa Kardell.
Personnel Profile: Full-Time Paid 23; Part-Time Paid 14; Part-Time Volunteers 5; Interns 3.
Governing Authority: state; university. Parent Institution: University of Nebraska-Lincoln. Branch Museums: Trailside Natural History Museum, Ft. Robinson, Crawford, NE; Ashfall Fossil Beds State Historical Park, Royal, NE. Tax-exempt.
Institution Type/Description: Natural History Museum.
Collections: 15,000,000 specimens, primarily from the Central Plains; anthropology; archives; archaeology; botany; entomology; geology; mineralogy; invertebrate & vertebrate paleontology; parasitology; paleobotany; numismatic, philatelic, zoology.
Research Fields: anthropology; archaeology; botany; entomology; geology; mineralogy; invertebrate & vertebrate paleontology; mammalogy; parasitology; zoology.
Facilities: 60,000 sq. ft. natural science exhibit area; planetarium; 150-seat auditorium; science teacher resource center.
Activities: hands-on natural science discovery room; inquiry-based gallery programs; science outreach kits; planetarium; professional development for teachers.
Publications: scientific bulletin, Bulletin of the University of Nebraska State Museum; Mammoth; scientific reports & guidebooks.
Hours & Admission Prices: Museum: Mon.-Wed. & Fri.-Sat. 9:30-4:30, Thurs.

9:30-8, Sun. 1:30-4:30. Planetarium: Tues.-Sat. 9:30-4:30, Sun. 1:30-4:30. Show times vary. Museum (Lincoln, NE): family $10, adults $5, children 5-18 $3. Planetarium: adults $8, children $5.50. Trailside Museum of Natural History (Crawford, NE): adults $3, children 5-18 $1. Ashfall Fossil Beds State Historical Park (Royal, NE): adults $5, children 6-18 $3; discounts to AAM members; ASTC & friends members no charge. Closed New Year's Day; Easter; Independence Day; Thanksgiving; Christmas Eve & Day. &
Attendance: 100,000 (estimated)
Membership: Individual $30; Family $45; Tusker Club $60-$99; Fossil Funder $100-$249; Nautilus Club $250-$499; Scarab Society $500-$999; Mammoth Circle $1,000-$2,499; Morrill Hall Star $2,500 & up; Ashfall additional $10.

Lodgepole

LODGEPOLE DEPOT MUSEUM, 722 McCall St., Lodgepole, NE 69149. Tel.: 308-483-5620.
E-mail: museum@lodgepole.us
Founded: 1976.
Congressional District: 3
Key Personnel: Pres. (V), Laurie Abrams.
Personnel Profile: Part-Time Volunteers 7.
Volunteer Hours: 500
Governing Authority: Tax-exempt.
Institution Type/Description: History Museum: housed in the former Union Pacific Railroad Depot.
Collections: horse buggies; antique furniture; historical clothing; antique clothes irons & bells; antique machinery; pony express; arrow heads; military; Pokmy Express collectables; posters; statues.
Hours & Admission Prices: By appointment. No charge, but donations accepted. &
Attendance: 500 (estimated)

Long Pine

LONG PINE HERITAGE SOCIETY AND HERITAGE HOUSE MUSEUM, 199 W. 3rd St., Long Pine, NE 69217-0337. Mailing Address: P.O. Box 333, Long Pine, NE 69217. Tel.: 402-273-4141.
E-mail: wardene@nntc.net
Founded: 1985.
Key Personnel: Pres. (V), Wardene Roark
Governing Authority: Tax-exempt.
Institution Type/Description: History Museum.
Collections: local history & genealogy.
Publications: annual newsletter.
Hours & Admission Prices: Memorial Day to Labor Day Sat. 1-4; other times by appointment. No charge; donations accepted.
Attendance: 250 (estimated)
Membership: Individual $5; Contributing $25; Sustaining $50; Lifetime $100.

McCook

MUSEUM OF THE HIGH PLAINS, 421 Norris Ave., McCook, NE 69001-2003. Tel.: 308-345-3661.
Founded: 1969.
Congressional District: 3
Key Personnel: Chm. Bd. & Pres. (V), Russell Dowling; Dir., Karen Anderson; Vice Pres., Del Harsh; Treas., Korey Burkert.
Personnel Profile: Part-Time Paid 3; Part-Time Volunteers 40.
Governing Authority: society. Parent Institution: High Plains Historical Society.
Institution Type/Description: History Museum.
Collections: German bibles; diaries; manuscripts; fashion from 1779-present; 1800 kitchen; 1800 pharmacy; farm tools; musical instruments; piano organ; sheet music; records; doll collection; railroad memorabilia; medical equip.; pioneer memorabilia; military uniforms dating from Spanish-American War-Vietnam War; paintings done by World War II German prisoners-of-war; memorabilia from World War II's McCook Air Base; flour mill; oil industry exhibit; B.N. caboose, semaphore, way car; Hendley shop where Kool-Aid was developed; general store.
Research Fields: history of the area.
Facilities: 500-vol. library available for use on premises; reading room. Books for sale.
Activities: guided tours; lectures; permanent & temporary exhibitions.
Publications: book, McCook's First 100 Years.
Hours & Admission Prices: June-Aug. Mon. & Wed.-Sat. 9-5, Sun. 1-5; Sept.-May Mon. & Wed.-Sat. 9-5. No charge; donations accepted. Closed major holidays. &
Attendance: 10,000 (estimated)

Membership: Voting $5; Life $200.

NEBRASKA STATE HISTORICAL SOCIETY'S GEORGE NORRIS STATE HISTORIC SITE, (M), 706 Norris Ave., McCook, NE 69001-3142. Tel.: 308-345-8484. Fax: 308-345-8484.
E-mail: dawna.bates@nebraska.gov
Web Site: www.nebraskahistory.org
Formerly: Senator George Norris State Historic Site
Founded: 1969.
Congressional District: 3
Key Personnel: C.E.O. & Exec. Dir., Michael J. Smith; Assoc. Dir., Ann Billesbach; Site Supvr., Jessica Wall; Historic Sites Mgr., Deb McWilliams.
Personnel Profile: Part-Time Paid 2; Part-Time Volunteers 3.
Governing Authority: state. Parent Institution: Nebraska State Historical Society. Tax-exempt: 501(c)(3).
Institution Type/Description: Historic House Museum: 1886 home of Senator George W. Norris (1899-1944).
Collections: period furnishings & artifacts of George Norris.
Research Fields: George W. Norris; Nebraska history.
Activities: guided tours; fifteen minute video on George Norris.
Publications: pamphlets.
Hours & Admission Prices: Tues.-Sat. 1-5. Adults $3; members & children no charge. Closed state holidays.
Attendance: 474 (accurate)
Membership: Individual $40; Household $55; Contributing $100; Supporting $250; Sustaining $500; Founder $1,000.

Milford

DINOSAUR MUSEUM AND HEARTLAND HAUNT, 923 238th Rd., Milford, NE 68405. Tel.: 402-761-2129.
Institution Type/Description: Dinosaur Museum.
Collections: over 65 life-sized dinosaurs.
Facilities: theater. Museum-related items for sale.
Activities: fossil dig; mirror maze.
Hours & Admission Prices: Mon.-Sat. 10-5, Sun. 12-5. Museum or Heartland Haunt: adults $9.95, teen 13-17 $8.95, children 4-12 $7.95; children 3 & under no charge. Museum & Haunt: adults $16.95, teen 13-17 $14.95, children 4-12 $12.95; children 3 & under no charge. Closed Independence Day; Thanksgiving; Christmas.

Minden

HAROLD WARP PIONEER VILLAGE FOUNDATION, 138 E. Hwy. 6, Minden, NE 68959-2500. Tel.: 308-832-1181; 800-445-4447. Fax: 308-832-1181.
E-mail: manager@pioneervillage.com
Web Site: www.pioneervillage.org
Founded: 1953.
Congressional District: 37
Key Personnel: Pres., Harold G. Warp; Gen. Mgr., Marshall S. Nelson.
Personnel Profile: Full-Time Paid 10; Part-Time Paid 30; Part-Time Volunteers 5.
Governing Authority: board of directors. Tax-exempt.
Institution Type/Description: General Museum.
Collections: transportation; 350 cars; period rooms from 1830-present; farm machinery; musical instruments; household appliances; china; cut glass; hobbies; toys; art; general store; 50,000 historical items displayed in 26 buildings; John Rogers statuary; Jackson paintings; statuettes; 20 aircraft; 100 tractors. Historic Buildings: 1860 Pony Express station; 1869 Elm Creek Stockade.
Research Fields: 1830 to present day on all items of development.
Facilities: campground.
Activities: crafts demonstrations.
Publications: History of Man's Progress; Over the Hill & Past Our Place; 500 Fascinating Facts; Sister Clara's Letters.
Hours & Admission Prices: Memorial Day to Labor Day daily 8-6; Sept.-May daily 9-4:30. Adults $14, children $7; discount to groups, seniors, military and AAM & ICOM members; children under 6 no charge. Two day pass available. &
Attendance: 40,000 (accurate)

KEARNEY COUNTY HISTORICAL MUSEUM, 1938 E Rd., Minden, NE 68959. Tel.: 308-832-1765.
Founded: 1925.
Congressional District: 3
Key Personnel: Dir. (V), Mary Bergsten; Pres. (V), Jane Kuehn.
Personnel Profile: Part-Time Volunteers 30.
Governing Authority: society. Tax-exempt: 501(c)(3).

Institution Type/Description: Historical Society Museum: housed in 1881 first schoolhouse in Minden.
Collections: items used by Kearney County residents dating back to early 1880; family genealogy; early business, school & church records; scrapbooks. Historic Buildings: depot; grocery store & post office; rural schoolhouse.
Facilities: 25-vol. library of historical books available on premises.
Activities: guided tours; cub scouts; cemetery tours; 4th grade school tours.
Hours & Admission Prices: June-Aug. daily 1-4. No charge; donations accepted. &
Attendance: 358 (accurate)
Membership: Individual & One Year $5; Five Year $25; Life $100.

Murdock

MURDOCK HISTORICAL SOCIETY AND MUSEUM, 9014 310th St., Murdock, NE 68407. Tel.: 402-867-3331.
Institution Type/Description: Historical Society Museum.
Collections: local history & culture; period furnishings; photographs; personal artifacts; trophies; uniforms; religious & military artifacts.
Hours & Admission Prices: Sun. 1:30-5; other times by appointment.

Nebraska City

ARBOR LODGE STATE HISTORICAL PARK, 2600 Arbor Ave., Nebraska City, NE 68410-1072. Mailing Address: P.O. Box 15, Nebraska City, NE 68410-0015. Tel.: 402-873-7222. Fax: 402-874-9885. Facebook: Arbor Lodge State Park.
E-mail: ngpc.arbor.lodge@nebraska.gov
Founded: 1923.
Congressional District: 1
Key Personnel: Dir., Jim Douglas; Supt., Randall J. Fox; Asst. Supt., Mark Kemper; Asst. Dir., Jim Swenson.
Personnel Profile: Full-Time Paid 2; Part-Time Paid 17.
Governing Authority: state. Affiliated with the Nebraska Game & Parks Commission, Box 30370, 2200 N. 33rd St., Lincoln, NE 68503. Tax-exempt: 501(c)(3).
Institution Type/Description: Historic House Museum: 1855 home of J. Sterling Morton with addition made by eldest son Joy Morton in 1903-1905.
Collections: personal property & correspondence of J. Sterling Morton including carriages; furniture; Tiffany glassware; silverware; paintings & other art items. Historic Houses: 1890 log cabin; 1901 greenhouse; 1903 carriage house.
Facilities: 72-acre arboretum; formal gardens; monument square; nature trails. Museum-related items for sale.
Activities: tours; permanent & traveling exhibitions. Museum Sponsors: Living History Demonstrations in October.
Publications: brochures; pamphlets.
Hours & Admission Prices: mid-April to mid-Oct. daily 11-5. Adults $5, children 3-12 $2; children under 3 no charge. Closed Thanksgiving; Christmas.
Attendance: 75,000 (estimated)

CIVIL WAR VETERANS MUSEUM, 910 First Corso, Nebraska City, NE 68410. Tel.: 402-873-4018.
Institution Type/Description: Military History Museum: housed in the Grand Army of the Republic Hall; built in 1894.
Collections: Civil War history; photographs; military artifacts; personal artifacts.
Hours & Admission Prices: April-Oct. Fri.-Sun. 12-4.

KIMMEL-HARDING-NELSON CENTER FOR THE ARTS, 801 Third Corso, Nebraska City, NE 68410-2819. Tel.: 402-874-9600. Facebook: KHN Center.
E-mail: info@khncenterforthearts.org
Web Site: www.khncenterforthearts.org
Founded: 2001.
Key Personnel: Exec. Dir., Jenni Brant.
Personnel Profile: Full-Time Paid 1; Part-Time Paid 1.
Governing Authority: Parent Institution: Kimmel Foundation.
Institution Type/Description: Art Gallery.
Collections: works by local & regional artists; paintings; sculpture; drawings.
Major Exhibits: Heidi Bartlett & Nora Rolf, 1/6/14-2/20/14; Laurine Kimmel High School Invitational, 3/3/14-4/25/14.
Facilities: gallery.
Activities: special events; lectures; classes in art, writing & music; artist residency program.

Hours & Admission Prices: Mon.-Fri. 10-5; other times by appointment. No charge. &

Attendance: 800 (estimated)

MAYHEW CABIN & HISTORIC VILLAGE, 2012 4th Corso, Nebraska City, NE 68410. Tel.: 402-873-3115.

Key Personnel: Dir., Bill Hayes

Institution Type/Description: History Museum.

Collections: United States history; slavery; period furnishings; personal artifacts; underground cave. Historic Building: Mayhew Cabin

Activities: educational programs. Annual Event: Juneteenth Celebration in June.

Hours & Admission Prices: May-Oct. Mon.-Sat. 11-5, Sun. 12-5; other times by appointment. Adults $3, children $1.

MISSOURI RIVER BASIN LEWIS & CLARK INTERPRETIVE TRAIL & VISITOR CENTER, 100 Valmont Dr., Nebraska City, NE 68410. Mailing Address: P.O. Box 785, Nebraska City, NE 68410. Tel.: 402-874-9900. Fax: 402-874-9909.

E-mail: discover@mrb-lewisandclarkcenter.org

Web Site: www.mrb-lewisandclarkcenter.org

Founded: 2004.

Congressional District: 1

Key Personnel: Dir., Erv Friesen.

Personnel Profile: Full-Time Paid 3; Part-Time Paid 4.

Governing Authority: Tax-exempt.

Institution Type/Description: History Museum.

Collections: Lewis & Clark history; 55 ft. Keelboat replica; Native American artifacts; photographs; maps.

Facilities: theater; classrooms; trails.

Activities: educational programs.

Hours & Admission Prices: May-Sept. Mon.-Sat. 9-6, Sun. 9-5; Oct.-April Mon.-Sat. 10-4, Sun. 12-4. Adults $5.50; members no charge. Closed New Year's Day; Thanksgiving; Christmas. &

Attendance: 14,000 (accurate)

Membership: Students $15; Individual $35; Household $45; Patron $200-$499; Benefactor $500-$999; Charter (Lifetime) $1,000 & up.

NEBRASKA CITY MUSEUM OF FIREFIGHTING, 1320 Central Ave., Nebraska City, NE 68410. Mailing Address: P.O. Box 376, Nebraska City, NE 68410-0376. Tel.: 402-873-4403. Fax: 402-873-5191.

E-mail: admin@ncfire.net

Web Site: www.ncmuseumoffirefighting.net

Founded: 2008.

Congressional District: 1

Key Personnel: Pres. (V), Steven Recker.

Personnel Profile: Part-Time Paid 3; Part-Time Volunteers 40.

Governing Authority: Parent Institution: Historical Preservation Society of the NCVFD, Inc. Tax-exempt.

Institution Type/Description: Firefighting History Museum.

Collections: local volunteer fire department history; firefighting; fire safety and prevention; 1884 Button Steam Engine; 1926 & 1938 Seagrave Pumpers.

Facilities: Museum-related items for sale.

Activities: children's interactive fire engine.

Hours & Admission Prices: April-Oct. Wed.-Sat. 11-5, Sun. 12-4. Adults $3, children 4-12 $1; members no charge. &

Attendance: 3,000 (estimated)

OLD FREIGHTERS MUSEUM, 407 N. 14th St., Nebraska City, NE 68410-1947. Mailing Address: P.O. Box 175, Nebraska City, NE 68410-0175. Tel.: 402-873-9360.

Institution Type/Description: History Museum: former home of the Russell-Majors-Waddell Freighting Company in 1858.

Collections: transportation artifacts.

Hours & Admission Prices: By appointment. Adults $3, children $1.

RIVER COUNTRY NATURE CENTER, 114 S. 6th St., Nebraska City, NE 68410. Mailing Address: P.O. Box 435, Nebraska City, NE 68410. Tel.: 402-873-3411.

E-mail: rcnaturecenter@windstream.net

Web Site: www.rivercountrynaturecenter.org

Institution Type/Description: Nature Center.

Collections: mounted wildlife; Native American artifacts; photographs.

Facilities: library; classroom. Museum-related items for sale.

Activities: educational programs.

Hours & Admission Prices: By appointment.

TAYLOR-WESSEL-BICKEL (NELSON) HOUSE, 711 3rd Corso, Nebraska City, NE 68410-2817. Mailing Address: Nelson House, 806 1st Ave., Nebraska City, NE 68410-0075. Tel.: 402-873-9360.

Governing Authority: Parent Institution: Nebraska City Historical Society.

Institution Type/Description: Historic House. Built in 1857.

Collections: period furnishings; artifacts; interpretive materials; art; photographs.

Hours & Admission Prices: By appointment. Adults $2, children $1.

WILDWOOD HISTORIC CENTER, 420 S. Steinhart Park Rd., Nebraska City, NE 68410-3300. Tel.: 402-873-6340. Facebook: Wildwood House.

E-mail: wildwoodbarn@windstream.net

Web Site: wildwoodhistoriccenter.org

Founded: 1967.

Congressional District: 2

Key Personnel: Acting Chm., Pat Friedle; Museum Shop Mgr., Gail Wurtele.

Personnel Profile: Full-Time Paid 1; Part-Time Paid 8.

Governing Authority: municipal. Tax-exempt: 501(c)(3).

Institution Type/Description: Historic House Museum: c.1869 restored two-story Victorian brick home.

Collections: Victorian furnishings.

Major Exhibits: Transferware China, 9/20/14-9/21/14.

Facilities: Original art & craft items for sale.

Activities: guided tours. Center Sponsors: Arts & Craft Show; Potters in the Garden in April; Chainsaw Carving in May; Fey Family & Friends Whittling in June; Wildwood Star Party in September.

Hours & Admission Prices: mid-April to Oct. Mon.-Sat. 10-5, Sun. 1-5; other times by appointment. Adults $3, children 12 & under $1. Barn no charge. &

Attendance: 3,000 (estimated)

Neligh

JAIL MUSEUM, 509 L St., Neligh, NE 68756-1419. Tel.: 402-887-5046.

E-mail: jailmuseum@jailmuseum.net

Web Site: www.jailmuseum.net

Formerly: Antelope County Historical Museum

Founded: 1965.

Congressional District: 3

Key Personnel: Pres. & Dir., Dr. Geo. Strassler; Vice Pres., Ray Ahrens; Treas., Harlen Frasier.

Personnel Profile: Part-Time Paid 2; Part-Time Volunteers 6.

Governing Authority: society. Administered by the Antelope County Historical Society, Neligh, NE. Parent Institution: Antelope County Historical Society. Subsidiary Institution: Antelope County Museum. Tax-exempt.

Institution Type/Description: Local History Museum: housed in 1892, Gates College gymnasium; later used as county jail.

Collections: Indian artifacts of the area; pioneer-present day memorabilia; 1887 Episcopal Church; log cabin.

Research Fields: county cemetery records; genealogy research by mail.

Facilities: library of books from Gates College which was closed in 1899; newspapers of the area; books on Nebraska & Antelope County; Indian Room; furnished log cabin.

Activities: guided tours; reading room aiding genealogical research.

Publications: History of Antelope County 1868-1883; Neligh Centennial 1873-1973; Early Day Stories; Antelope County - The War Years (World War II); articles of special events; maps of county; Antelope County Pioneers; Early Pioneer Stories.

Hours & Admission Prices: Summer: Tues.-Sat. 1-5; Labor Day to Memorial Day Thurs.-Fri. 1-5. Adults $3; children 12 & under and members no charge.

Attendance: 600 (estimated)

Membership: Single $10; Family $15; Sustaining $100.

NEBRASKA STATE HISTORICAL SOCIETY'S NELIGH MILL STATE HISTORIC SITE, (M), N St & Wylie Dr., Neligh, NE 68756. Mailing Address: P.O. Box 271, Neligh, NE 68756-0271. Tel.: 402-887-4303. Fax: 402-887-4303.

E-mail: mill@gpcom.net

Web Site: www.nebraskahistory.org

Formerly: Neligh Mills

Founded: 1971.

Congressional District: 3

Key Personnel: C.E.O. & Exec. Dir., Michael J. Smith; Assoc. Dir. Education & Interpretation, Ann Billesbach; Site Supvr., Don Ofe; Historic Sites Mgr., Deb McWilliams.

Personnel Profile: Full-Time Paid 1; Part-Time Paid 1; Part-Time Volunteers 4.

Governing Authority: state. Parent Institution: Nebraska State Historical Society. Tax-exempt: 501(c)(3).
Institution Type/Description: Historic Building Museum: 1873 Neligh Mills.
Collections: 500-barrel a day flour mill with seven stands of rollers, two extra mills, thirty six elevators, two plan sifters, four purifiers, three reel sifters, four sackers, water powered turbine. Historic Building: 1883 Mill Office.
Research Fields: flour milling; water power; wheat agriculture.
Facilities: Society publications & postcards for sale.
Activities: 15 min. video on the history of the Neligh Mill & flour milling; guided tours; permanent exhibits.
Publications: Neligh Mills Cookbook; Nebraska History Magazine; Self-Guided Tour of Neligh Mill; Flour Milling in Nebraska; Water Powered Flour Mills in Nebraska.
Hours & Admission Prices: Memorial Day-Labor Day Tues.-Sat. 10-5, Sun. 1-5; Sept.-May Mon.-Fri. 10-5. Adults $3, groups of 20 $2, children $1; discount to Time Travelers; children accompanied by an adult, youth groups, Nebraska State Historical Society members & families no charge. Closed most federal holidays.
Attendance: 2,135 (accurate)
Membership: Individual $40; Household $55; Contributing $100; Supporting $250; Sustaining $500; Founder $1,000.

PIERSON WILDLIFE MUSEUM LEARNING CENTER, 205 E. 5th St., Neligh, NE 68756-1301. Tel.: 402-887-4212 & 4447.
Web Site: www.neligh.org/visit/museums
Founded: 2002.
Key Personnel: Chm. (V), Gloria Christiansen; Pres. (V), Lyle Juracek; Museum Shop Mgr., Torgus Thompson.
Personnel Profile: Part-Time Volunteers 1.
Governing Authority: Tax-exempt.
Institution Type/Description: Wildlife Museum.
Collections: African wild game; animals from around the world.
Hours & Admission Prices: Summer: Tues-Sun. 1-4; Fall: Sat.-Sun. 1-4 & by appointment; Winter: by appointment. Adults $5, seniors $4, students K-12 $3; discounts to groups of 25 or more; children under school age no charge. &
Attendance: 800 (accurate)
Membership: Single $15; Family $30.

Niobrara

NIOBRARA HISTORICAL SOCIETY MUSEUM, 89054 519 Ave., Niobrara, NE 68760-6013. Tel.: 402-857-3794.
Founded: 1977.
Key Personnel: Pres. (V), Barb Farrar.
Personnel Profile: Part-Time Volunteers 8.
Volunteer Hours: 200
Governing Authority: Tax-exempt.
Institution Type/Description: History Museum.
Collections: local history & culture; photographs; personal artifacts; hood ornaments; period wedding dresses.
Hours & Admission Prices: Memorial Day to Labor Day Fri.-Sat. 12-3, Sun. 1-3; other times by appointment. No charge; donations accepted. &
Attendance: 1,300 (estimated)
Membership: Individual $5.

PONCA TRIBAL MUSEUM, 2548 Park Ave., Niobrara, NE 68760. Tel.: 402-857-3519.
Key Personnel: Dir. Cultural Affairs., Gloria Hamilton
Institution Type/Description: Native American History Museum.
Collections: Native American history & culture; paintings; personal artifacts; photographs.
Facilities: library; archives.
Hours & Admission Prices: Mon.-Sat. 8-4. No charge. &

Norfolk

ELKHORN VALLEY MUSEUM & RESEARCH CENTER, 515 Queen City Blvd., Norfolk, NE 68701-4060. Tel.: 402-371-3886. Fax: 402-371-3886.
E-mail: evmdirector@cableone.net
Web Site: www.elkhornvalleymuseum.org
Founded: 1958.
Personnel Profile: Full-Time Paid 1; Part-Time Paid 4; Part-Time Volunteers 15.
Governing Authority: Board of Directors.
Institution Type/Description: History Museum.
Collections: history of Norfolk & Elkhorn Valley; genealogy; documents; Johnny Carson Gallery; historical exhibits; Square Turn Tractor.

Facilities: bird library. Museum-related items for sale.
Activities: special events. Museum Sponsors: Pioneer Day in the Park in August.
Publications: bimonthly newsletter.
Hours & Admission Prices: Tues.-Sat. 10-5. Adults $6, seniors over 60 $5, students $4, children $3; children under 5 no charge. Closed New Year's Day; Easter; Thanksgiving; Christmas. &
Attendance: 3,500 (accurate)
Membership: Individual $30; Family $60; Friend $125; Patron $250; Benefactor $500; Distinguished Benefactor $1,000; Pacesetter $2,500; Visioneer $5,000.

NORFOLK ARTS CENTER, 305 N. 5th St., Norfolk, NE 68701. Tel.: 402-371-7199.
E-mail: info@norfolkartscenter.org
Web Site: www.norfolkartscenter.org
Founded: 1978.
Congressional District: 1
Key Personnel: Exec. Dir., Kara Weander-Gaster; Pres. (V), Tammy Day.
Personnel Profile: Full-Time Paid 3; Part-Time Paid 2; Part-Time Volunteers 1.
Institution Type/Description: Art Gallery.
Collections: paintings; photographs; sculpture.
Facilities: classrooms; rental facility; auditorium.
Activities: classes; workshops; performances.
Publications: seasonal catalogs with classes & upcoming events; monthly newsletters.
Hours & Admission Prices: Tues.-Fri. 10-8, Sat. 10-4. No charge; donations accepted. &

North Bend

HAROLD AND LEONA'S TOY MUSEUM, 631 Chestnut St., North Bend, NE 68649. Mailing Address: 211 W. 14th St., North Bend, NE 68649. Tel.: 402-652-3481.
Founded: 1996.
Key Personnel: Dir., C.E.O. & Museum Shop Mgr., Leona Soukup.
Operating Expenses: 200,000
Operating Income: 300,000
Governing Authority: Tax-exempt.
Institution Type/Description: Toy Museum.
Collections: early toys & dolls; furniture; machinery; glassware; advertising pieces.
Activities: Tours.
Hours & Admission Prices: Adult $2. Sat. & Sun. by appointment only. &
Attendance: 812 (estimated)

North Platte

BUFFALO BILL RANCH STATE HISTORICAL PARK, 2921 Scouts Rest Ranch Rd., North Platte, NE 69101-8444. Tel.: 308-535-8035. Fax: 308-535-8070.
E-mail: ngpc.buffalo.bill@nebraska.gov
Web Site: outdoornebraska.org
Founded: 1964.
Congressional District: 3
Key Personnel: Supt., Samantha Hopkinson.
Personnel Profile: Full-Time Paid 5; Part-Time Paid 3; Part-Time Volunteers 1.
Governing Authority: state. Affiliated with the Nebraska Game and Parks Commission, Box 30370, 2200 N. 33rd St., Lincoln, NE 68503. Tax-exempt: 501(c)(3).
Institution Type/Description: History Museum: housed in 1886 home of William (Buffalo Bill) Cody.
Collections: personal property and correspondence of Buffalo Bill; showbills; photographs; film clips. Historic Structure: 1886 home of William Cody; 1887 barn.
Research Fields: Wild West Show; life & activities of Cody.
Facilities: theater. Books relating to the history of the West & Buffalo Bill & other museum-related items for sale.
Activities: guided tours; films; permanent & temporary exhibitions.
Hours & Admission Prices: May & Sept.-Oct. Mon.-Fri. 10-4; Memorial Day-Labor Day daily 9-5. Adults 13 & up $2, children 3-12 $1; children 2 & under no charge. State park permit required: daily $5 per vehicle. Times & dates of operation subject to change.
Attendance: 20,030 (accurate)

CODY PARK RAILROAD MUSEUM, 1400 N. Jeffers, North Platte, NE 69101. Tel.: 308-535-6700.
Web Site: www.ci.north-platte.ne/us/publicservices
Institution Type/Description: Railroad Museum.

Collections: Union Pacific memorabilia; railroad history; photographs; telephone & telegraph equipment; Union Pacific's steam locomotive, Challenger 3977; Union Pacific DD40AX; baggage car; mail car; caboose; depot.
Hours & Admission Prices: May-Sept. daily 10-6. No charge.

LINCOLN COUNTY HISTORICAL MUSEUM, (M), 2403 N. Buffalo Bill Ave., North Platte, NE 69101-9702. Tel.: 308-534-5640.
E-mail: lincomuseum@hamilton.net
Web Site: www.lincolncountymuseum.org
Founded: 1976.
Congressional District: 3
Key Personnel: Chm., Phyllis Slavlik; Pres. (V), Lloyd Speicher; Dir., Cur. & Museum Shop Mgr., James Griffin.
Personnel Profile: Full-Time Paid 1; Part-Time Paid 1; Part-Time Volunteers 140.
Governing Authority: private nonprofit. Tax-exempt.
Institution Type/Description: General Museum.
Collections: Lincoln County Nebraska history; WWII Canteen exhibit; village of 15 outbuildings: two-story log house, a log pony express station building, log military fort building, log cabin, church, one-room schoolhouse, large barn, railroad depot, general store, two-story frame house.
Facilities: Gift-related items for sale.
Activities: guided tours. Annual Event: Heritage Festival in June; Spirit Hunting in a Ghost Town Haunted Village in October.
Publications: quarterly newsletter.
Hours & Admission Prices: May-Sept. Mon.-Sat. 9-5, Sun. 1-5. Families $10, adults 13 & over $5, seniors 55 & over and military $4; discount to groups of 10 or more; children 12 & under no charge. &
Attendance: 12,500 (accurate)
Membership: Individual $20; Couple & Family $40; Lifetime $200.

NORTH PLATTE CHILDREN'S MUSEUM, 314 N. Jeffers St., North Platte, NE 69101. Mailing Address: P.O. Box 2088, North Platte, NE 69101. Tel.: 308-532-3512.
Institution Type/Description: Children's Museum.
Collections: hands-on exhibits.
Activities: rental facilities.
Hours & Admission Prices: Wed. & Fri. 9-3, Thurs. 9-7, Sat. 9-5, Sun. 12:30-5. Admission $4; discounts to seniors over 60; children 2 & under no charge.

O'Neill

HOLT COUNTY HISTORICAL SOCIETY MUSEUM, 401 E. Douglas, O'Neill, NE 68763. Tel.: 402-336-2344.
Founded: 1984.
Governing Authority: Tax-exempt.
Institution Type/Description: Historical Society Museum.
Collections: local history & culture; genealogy; photographs; personal artifacts.
Research Fields: family histories.
Activities: book signings; lectures; monthly meetings Feb.-Nov. Annual Events: St. Patrick's Day Celebration; Summerfest Celebration.
Publications: biannual newsletter.
Hours & Admission Prices: By appointment. No charge; donations accepted. Closed major holidays.
Attendance: 150 (accurate)
Membership: Annual & Contributing $10; Supporting, Business, Professional, & Institutional $25; Life $100; Joint Life $150.

Oakland

SWEDISH HERITAGE CENTER, 301 N. Charde Ave., Oakland, NE 68045. Tel.: 402-685-5489.
Institution Type/Description: Cultural Center.
Collections: Swedish culture & heritage; photographs; personal artifacts; period furnishings.
Activities: monthly cultural programs.
Hours & Admission Prices: May-Sept. Tues.-Sun.; Oct.-April Sat.-Sun. No charge.

Ogallala

FRONT STREET, 519 E. First, Ogallala, NE 69153-2620. Tel.: 308-284-6000. Fax: 308-284-0865.
E-mail: frontstreet@frntst.com
Web Site: www.ogallalafrontstreet.com
Founded: 1964.

Congressional District: 3
Key Personnel: Dir. & C.E.O., Darlan Rezac
Governing Authority: nonprofit organization. Parent Institution: Front Street Inc.
Institution Type/Description: History Museum.
Collections: cowboy items; saddles; farm tools; 1890s clothing; schoolhouse artifacts; Sam Bass history; medicine items; barber shop; funeral parlor; jail.
Facilities: library of books. Gift items for sale.
Publications: newspapers.
Hours & Admission Prices: Mon.-Sat. 9-9. No charge; donations accepted. Closed New Year's Day; Thanksgiving; Christmas. &
Attendance: 59,451 (accurate)

LAKE MCCONAUGHY VISITOR/WATER INTERPRETIVE CENTER, 1475 Hwy. 61 N., Ogallala, NE 69153. Tel.: 308-284-8800.
E-mail: ngpc.lake.mcconaughy@nebraska.gov
Web Site: www.outdoornebraska.org
Congressional District: 3
Institution Type/Description: History Museum.
Collections: water significance to Nebraska's past, present & future.
Facilities: aquarium; 50-seat theater; interpretive center.
Hours & Admission Prices: Summer: daily 8-5; Winter Mon.-Fri. 8-5.

MANSION ON THE HILL, 1004 N. Spruce St., Ogallala, NE 69153. Mailing Address: Keith County Historical Society, P.O. Box 5, Ogallala, NE 69153. Tel.: 308-284-4066 & 0821; 800-658-4390.
Web Site: ogallalamansiononhill.com
Governing Authority: nonprofit organization. Parent Institution: Keith County Historical Society.
Institution Type/Description: Historic House: built in 1887. Listed on the National Register of Historic Places.
Collections: local history; period furnishings; personal artifacts; photographs.
Hours & Admission Prices: Memorial Day to mid-Sept. Tues.-Sat. 9-12 & 1-4, Sun. 1-4. Adults $2, children 5-12 $1; children under 5 no charge.
Attendance: 1,500 (estimated)

PETRIFIED WOOD & ART GALLERY, 418 E. 1st St., Ogallala, NE 69153-2620. Tel.: 308-284-9996. Facebook: Petrified Wood & Art Gallery.
E-mail: hhkenfi@allophone.com
Web Site: www.petrifiedwoodgallery.com
Founded: 2000.
Congressional District: 3
Key Personnel: Dir., Kathy Zeller; Pres. (V), Doug Teaford.
Personnel Profile: Part-Time Paid 1; Part-Time Volunteers 24.
Governing Authority: Tax-exempt.
Institution Type/Description: Natural History Museum.
Collections: woods & fossils from around the world; Native American artifacts; arrowheads.
Facilities: Museum-related items for sale.
Hours & Admission Prices: Call for hours. No charge; donations accepted. &
Attendance: 14,000 (estimated)
Membership: $25; $50; $100; $250; $500; $1,000; $2,500; $5,000.

Omaha

BATCHELDER FAMILY SCOUT MUSEUM, Durham Scout Center, 12401 W. Maple Rd., Omaha, NE 68164. Tel.: 402-431-9272.
Web Site: www.mac-bsa.org
Founded: 2007.
Congressional District: 2
Key Personnel: Public Rels. Dir., Katie Godbout
Institution Type/Description: Scout Museum.
Collections: scouting history, uniforms, badges, & handbooks; photographs; paintings; personal artifacts.
Activities: tours; educational programs & activities.
Hours & Admission Prices: Call for hours. No charge.

BEMIS CENTER FOR CONTEMPORARY ARTS, (M), 724 S. 12th St., 12th and Leavenworth, Omaha, NE 68102. Tel.: 402-341-7130. Fax: 402-341-9791.
E-mail: info@bemiscenter.org
Web Site: www.bemiscenter.org
Founded: 1981.
Key Personnel: Cur., Hesse McGraw

Institution Type/Description: Contemporary Art Gallery.
Collections: works by contemporary artists; videos.
Activities: educational programs; artist-in-residence program.
Hours & Admission Prices: Tues.-Sat. 11-5; tours by appointment. No charge; donations accepted.

CATHEDRAL CULTURAL CENTER, 3900 Webster St., Omaha, NE 68131-1810. Tel.: 402-551-4888.
Web Site: www.cathedralartsproject.org
Founded: 2003.
Key Personnel: C.E.O., Bro. William Woeger, F.S.C.; Assoc. Dir., Dorothy Begley; Chm., Joni Fogarty; Treas., Charles Schultz; Public Rels., John Wees; Museum Shop Mgr., Kathy White.
Personnel Profile: Part-Time Paid 3; Part-Time Volunteers 20.
Governing Authority: church group. Parent Institution: Cathedral Arts Project. Tax-exempt.
Institution Type/Description: Religious Museum.
Collections: heritage of Saint Cecilia Cathedral & it's architect, Thomas Rogers Kimball; videos & CDs.
Major Exhibits: South Seas Images, 1/26/14-1/27/14; Landscapes, 2/16/14-3/28/14; Winnebago Images (T), 4/13/14-6/9/14; A Life of Art: Brother Mel Meyer, 6/15/14-8/14.
Research Fields: cultural history of Saint Cecilia Cathedral & interfaith heritage.
Facilities: 75-seat lecture room; 1,000 sq. ft. exhibit space. Museum-related items for sale.
Activities: concerts; docent program; films; formal education programs for children; guided tours; lectures; loan exhibitions; rental gallery. Annual Events: art shows.
Publications: book, Beauty of Thy House.
Hours & Admission Prices: Tues.-Fri. 11-4, third Sun. of each month 10:30-4:30. Suggested Donation: $2 per person. Closed religious holidays. ♿
Attendance: 5,000 (estimated)
Membership: Cornerstone $35; Pillar $50; Keystone $100; Pier $250; Rose Window $500; Arch $1,000; Portal $2,500; Capstone $5,000.

DOUGLAS COUNTY HISTORICAL SOCIETY, 5730 N. 30 St., #11B, Omaha, NE 68111-1657. Tel.: 402-451-1013 & 455-9990. Fax: 402-453-9448.
E-mail: director@omahahistory.org
Web Site: www.omahahistory.org/museum.htm
Key Personnel: Exec. Dir., Kathryn Aultz
Institution Type/Description: Historical Society Museum: housed in the General Crook House; built in 1879. Listed on the National Register of Historic Places.
Collections: local history & culture; period furnishings; personal artifacts; photographs.
Facilities: library; archives.
Activities: guided tours; rental facilities; classes; lectures; special events.
Hours & Admission Prices: Museum: Tues.-Fri. 10-4, Sat.-Sun. 1-4; groups by appointment. Jan. to mid-Nov. adults $5, students $4, children 6-12 $3; members no charge. Mid-Nov. to Dec. adults $6, students $4, children 6-12 $3; members no charge.

THE DURHAM MUSEUM, (M), 801 S. 10th St., Omaha, NE 68108-3205. Tel.: 402-444-5071. Fax: 402-444-5397.
E-mail: info@durhammuseum.org
Web Site: www.durhammuseum.org
Formerly: Durham Western Heritage Museum
Founded: 1975.
Congressional District: 2
Key Personnel: Chm., Richard Bell; Exec. Dir., Christi Janssen; Deputy Dir. & Dir. Finance, Amy Carolus; Dir. Community Rels., Shawna Forsberg; Exhibition Design Svcs., Tim Hantula; Tour Coord. & Educational Programs, Mick Hale; Cur., Carrie Wieners; Administrative Asst., Katie Moulton; Museum Shop Mgr., Diane Hileman.
Personnel Profile: Full-Time Paid 28; Part-Time Paid 10; Part-Time Volunteers 200.
Governing Authority: nonprofit organization. Tax-exempt.
Institution Type/Description: History Museum: housed in c.1930, former Union Pacific Railroad Station.
Collections: railroad memorabilia; furniture; 1930s soda fountain; clothing; decorative arts; pioneering tools; archival photography collection of 500,000 negatives & prints; Byron Reed coins, paper money & manuscripts; six restored train cars; model train layout measuring 85 feet in length with 40 train cars; Union Pacific Railroad artifacts & memorabilia.
Research Fields: Omaha, Nebraska History; Great Plains Regional History; railroad history; photographs; numismatics.

Facilities: conference rooms; catered dinners with seating for up to 1,000; 4 historic classrooms. Museum-related items for sale.
Activities: guided tours; special events; education programs for children & adults; docent programs; permanent, temporary & traveling exhibitions; offers exhibit design services to other institutions.
Publications: quarterly newsletter, Timelines.
Hours & Admission Prices: Tues. 10-8, Wed.-Sat. 10-5, Sun. 1-5. Adults $8, senior citizens $6, children 3-12 $5; children under 3 & members no charge. Closed New Year's Day, Memorial Day, Independence Day, Labor Day, Thanksgiving, Christmas. ♿
Attendance: 161,000 (estimated)
Membership: Student, Educator & Senior $25; Military Family & Individual $30; Educator Family $45; Family $55; Community Group $80; Bronze $100; Silver $250; Gold $500; Durham Society $1,000.

EL MUSEO LATINO, 4701 S. 25th St., Omaha, NE 68107-2728. Tel.: 402-731-1137. Fax: 402-733-7012.
E-mail: mgarcia@elmuseolatino.org
Web Site: elmuseolatino.org
Founded: 1993.
Congressional District: 2
Key Personnel: Exec. Dir., Magdalena A. Garcia; Pres. (V), Jim Mammel; Vice Pres., Jeffrey Keating; Treas., Maria Arbelaez; Sec., Rita Melgares.
Personnel Profile: Full-Time Paid 2; Part-Time Paid 4; Part-Time Volunteers 25; Interns 3.
Governing Authority: nonprofit organization. Tax-exempt: 501(c)(3).
Institution Type/Description: Latino Art & History Museum.
Collections: concentration of art, history, artifacts & archival material of the Latino people of the Americas.
Research Fields: art & history of Latin America; Latino & Hispanic presence in Nebraska; Latina women.
Facilities: 500-vol. bilingual reference & resource library; reading room; video & slide library.
Activities: permanent & traveling exhibits; lectures; guided tours; gallery talks; workshops; art festivals; docent program; teen, youth & bilingual docent programs; bilingual programming including films, videos, slide presentations, art & art history classes. Museum Sponsors Hispanic & Latino celebrations: Cinco de Mayo, May; Hispanic Heritage Month, mid-Sept. to mid-Oct.
Publications: exhibition catalogue, Augustin Victor Casasola: The Mexican Revolution.
Hours & Admission Prices: Mon., Wed. & Fri. 10-5, Tues. & Thurs. 1-5, Sat. 10-2. Adults $5, college students $4, senior citizens & students K-12 $3.50; discounts to AAM & ICOM members; children under 5 & members no charge. Closed New Year's Eve & Day; Independence Day; Thanksgiving; Christmas. ♿
Attendance: 50,000 (accurate)
Membership: Senior & Student $25; Individual $30; Family $50; Individual Patron & Business Associate $100; Corporate Sponsor $250.

FLORENCE MILL, 9102 N. 30th St., Omaha, NE 68112. Mailing Address: 5215 Jackson, Omaha, NE 68106-1331. Tel.: 402-551-1233. Fax: 402-561-0024.
Institution Type/Description: Historic Mill: built in 1846.
Collections: local history; photographs; newspaper clippings; early pioneer artifacts.
Hours & Admission Prices: Memorial Day to Labor Day Tues.-Sun. 1-5; other times by appointment.

FREEDOM PARK NAVY MUSEUM, 2497 Freedom Park Rd., Omaha, NE 68110-2745. Mailing Address: 1523 S. 24th St., Omaha, NE 68108. Tel.: 402-444-5955. Fax: 402-444-6838. Facebook: Freedom Park.
Web Site: www.cityofomaha.org
Founded: 1972.
Key Personnel: Dir. Parks & Recreation, Brook Bench.
Personnel Profile: Part-Time Paid 1.
Governing Authority: Parent Institution: City of Omaha Parks, Recreation & Public Property Dept. Tax-exempt.
Institution Type/Description: Military Museum.
Collections: military displays including the USS Hazard, World War II minesweeper, the USS Marlin, a training submarine; WWII anti-aircraft guns; MK5 & MK8 nuclear bomb casings; A-4C jet; A-7 jet; Coast Guard helicopter; propellers; anchors; minesweeping equipment; jeeps; crane truck; military vehicles; ambulance; radar units.
Research Fields: WWII; Korea; Vietnam; present conflicts.
Hours & Admission Prices: Temporarily closed.
Attendance: 3,300 (accurate)

GENERAL CROOK HOUSE MUSEUM AND LIBRARY/ARCHIVES CENTER, 5730 N. 30th St., 11B, Omaha, NE 68111-1658. Tel.: 402-455-9990. Fax: 402-453-9448.
E-mail: house@omahahistory.org
Web Site: www.omahahistory.org
Founded: 1956.
Congressional District: 1
Key Personnel: Exec. Dir. Historical Society, Betty J. Davis; Pres. Bd. Dirs., Mary Maxwell; Cur. General Crook House, Patricia Pixley; Chm. (V) & Museum Shop Mgr., Virgie Ward; Dir. Education & Public Programs, Liz Rea; Dir. Library Archives Center, Travis Sing; Archivist & Librarian, Don Snoddy; Research Specialist, Gary Rosenberg; Registrar, Elizabeth Krecek.
Personnel Profile: Full-Time Paid 4; Part-Time Paid 7; Part-Time Volunteers 50; Interns 1.
Governing Authority: nonprofit organization. Parent Institution: Historical Society of Douglas County. Subsidiary Institution: Crook House Guild. Tax-exempt: 501(c)(3).
Institution Type/Description: Library & Historical House Museum: 1878 General Crook House & archives center located at Historic Fort Omaha now the campus of Metropolitan Community College.
Collections: artifacts interpreting life style of a commanding officer on a major frontier fort; fort life; pieces interpreting the history of Douglas County; authentic Victorian heirloom garden; 30,000 photographs & documents dating back to the mid 1800s.
Research Fields: military history surrounding Fort Omaha & the westward movement; history & development of Omaha & Douglas County, NE; wallpaper in the Midwest; genealogy.
Facilities: 2,000-vol. library pertaining to county history available for research on premise; 6,000,000-piece archive; reading room; 30-seat auditorium; classrooms.
Activities: guided tours; lectures; study clubs; formally organized education programs for children & adults; docent program; traveling, permanent & temporary exhibitions; step-on guides for motorcoach requests.
Publications: newsletter, Banner; books, Omaha & Douglas County: A Panoramic View; Sautter House Five: Wallpapers of a German-American Homestead; A.V. Sorensen & the New Omaha; reprint of History of Omaha; oral history interviews.
Hours & Admission Prices: Museum: Mon.-Fri. 10-4, Sat.-Sun. 1-4; other times by appointment. Library: Tues.-Fri. 10-4. Adults $5, students $4, children 6-12 $3; members & library no charge. Closed holidays.
Attendance: 18,000 (estimated)
Membership: Student $20; Senior $25; Individual $30; Family $40; Historian $60; Century $100; Patron $250; Lieutenant's Council $500; General's Council & Corporate $1,000.

GERALD R. FORD CONSERVATION CENTER, 1326 S. 32nd St., Omaha, NE 68105-2044. Tel.: 402-595-1180. Fax: 402-595-1178.
E-mail: nshs.grfcc@nebraska.gov
Founded: 1995.
Governing Authority: Parent Institution: Nebraska State Historical Society. Tax-exempt.
Institution Type/Description: Conservation Center.
Collections: life & career of President Gerald R. Ford; photographs; personal artifacts.
Hours & Admission Prices: By appointment only.

GREAT PLAINS BLACK HISTORY MUSEUM, 2213 Lake St., Omaha, NE 68110. Mailing Address: 5623 Willit, Omaha, NE 68152. Tel.: 402-572-9292. Fax: 402-571-4050. Facebook: Great Plains Black History Museum.
E-mail: gpblackmuseum@aol.com
Web Site: www.gpblackmuseum.org
Key Personnel: Chm. (V) & Pres. (V), James R. Beatty; Museum Shop Mgr., Terri Sanders.
Personnel Profile: Part-Time Volunteers 20.
Governing Authority: Tax-exempt: 501(c)(3).
Institution Type/Description: History Museum: housed in the former Webster Telephone Exchange Building. Listed on the National Register of Historic Places.
Collections: Black history & culture; photographs; periodicals; manuscripts; paintings; films.
Hours & Admission Prices: Call for hours. No charge; donations accepted.
Membership: Individual $20; Family $35; Corporate $100; Lifetime $500.

HOT SHOPS ART CENTER, 1301 Nichols St., Omaha, NE 68102-4212. Tel.: 402-342-6452.
E-mail: manager@hotshopsartcenter.com

Web Site: www.hotshopsartcenter.com
Institution Type/Description: Art Gallery.
Collections: works by regional artists.
Activities: special events.
Hours & Admission Prices: Daily 8-5; other times by appointment.

✴ **JOSLYN ART MUSEUM, (M),** 2200 Dodge St., Omaha, NE 68102-1292. Tel.: 402-342-3300. Fax: 402-342-2376.
E-mail: info@joslyn.org
Web Site: www.joslyn.org
Founded: 1931.
Congressional District: 2
Key Personnel: Exec. Dir. & C.E.O., Jack F. Becker, Ph.D.; Chm. Bd., John P. Nelson; Deputy Dir. Finance & Operations, Miranda Templeman; Dir. Education, Nancy Round; Chief Cur. Richard & Mary Holland Cur. American Western Art, Toby Jurovics; Dir. Devel. & External Affairs, Sabrina Weiss; Dir. Mktg. & Public Rels., Amy Rummel; Dir. Adult Programs, Susie Severson; Asst. to Exec. Dir. & C.E.O., Brooke Masek; Campaign & Planned Giving Mgr., Sharon Latham; Corporate & Foundation Giving Mgr., Alyssa Kohler; Youth & Family Programs Mgr., Maranda Allbritten; Interpretive Media Mgr., Laura Huntimer; Studio Programs Mgr., Andrew Smith; Donor Rels. Mgr., Kenley Sturdivant-Wilson; Visitor Svcs. & Protection Mgr., Ronnie Rivers; Events Mgr., Candace Tielebein; Retail & Guest Svcs. Mgr., Jane Precella; Facility Mgr., Steven Tlsty; Asst. Cur. Exhibitions, Ruby C. Hagerbaumer; Registrar, Kay Johnson.
Personnel Profile: Full-Time Paid 60; Part-Time Paid 12; Part-Time Volunteers 153; Interns 2.
Governing Authority: nonprofit organization. Tax-exempt: 501(c)(3).
Institution Type/Description: Art Museum.
Collections: ancient through contemporary art, including European & American paintings, sculpture, graphics; art of the Western frontier; Native American art.
Research Fields: interdisciplinary study of the American West; 19th & 20th-century European & American Art.
Facilities: 25,600-vol. non-circulating research library; 1,000-seat concert hall, 140-seat lecture hall; classrooms; restaurant. Museum-related items for sale.
Activities: guided tours; lectures; films; gallery talks; concerts; formally organized education programs for children & adults; special training programs for educators; cooperative curriculum oriented school programs; inter-museum loan, permanent, temporary & traveling exhibitions.
Publications: members magazine; catalogs; brochures.
Hours & Admission Prices: Tues.-Wed. & Fri.-Sat. 10-4, Thurs. 10-8, Sun. 12-4. Adults $8, senior 62 & over and college students w/ID $6, children 5-17 $5; discounts to AAA, AAM & ICOM members; members, children 4 & under and Sat. 10am to noon no charge. Closed major holidays.
Attendance: 200,000 (estimated)
Membership: Student, Senior, Educator & Military Family $40; Individual $45; Educator & Senior Family $55; Family & Dual $60; Young Arts Patron $100; Contributor $100 & up; Patron $750 & up.

JOSLYN CASTLE TRUST, 3902 Davenport St., Omaha, NE 68131. Tel.: 402-595-2199. Fax: 402-595-1007. Facebook: Joslyn Castle.
E-mail: info@joslyncastle.com
Web Site: www.joslyncastle.com
Founded: 2010.
Congressional District: 2
Key Personnel: Exec. Dir., Julie Reilly; Events & Mktg., Emily Lasky; Education & Tours, Judy Alderman; Facilities Mgr., Peter Marion; Facilities Asst., Jake Arneson; Membership, Amy Trenolone.
Personnel Profile: Full-Time Paid 4; Part-Time Paid 2; Part-Time Volunteers 18.
Governing Authority: Tax-exempt.
Institution Type/Description: Historic Building: housed in the former home of George & Sarah Joslyn; built in 1903.
Collections: Joslyn family history; period furnishings; photographs; personal artifacts.
Activities: tours.
Publications: publication, Castle Watch.
Hours & Admission Prices: Tours: 1st & 3rd Sun. each month 1, 2, & 3; groups by appointment. Adults $6, seniors 60 & over and students $5; children under 5 no charge.
Attendance: 18,000 (estimated)

LAURITZEN GARDENS, 100 Bancroft St., Omaha, NE 68108. Tel.: 402-346-4002. Fax: 402-346-8948. Facebook: Lauritzen Gardens.
Web Site: www.lauritzengardens.org
Founded: 1983.
Key Personnel: Exec. Dir., Spencer Crews.

Governing Authority: Tax-exempt.
Institution Type/Description: Living Plant Museum.
Collections: four-season plants.
Research Fields: plant conservation.
Facilities: visitor & education center; cafe; horticulture library; rental facilities; classrooms. Gift items for sale.
Activities: educational programs; special events; seasonal floral displays; tram tours May-Oct.
Hours & Admission Prices: mid-May to mid-Sept. Mon.-Tues. 9-8, Wed.-Sun. 9-5; mid-Sept. to mid-May daily 9-5. April-Oct.: adults $7, children 6-12 $3; Nov.-March: adults $6, children 6-12 $3; members & children under 6 no charge. Tram Tours: $3 per person. Closed New Year's Day; Thanksgiving; Christmas. &
Attendance: 175,549 (accurate)
Membership: Individual $30; Dual $50; Family $60; Grandparent $70. (Plus One $20).

LOVE'S JAZZ AND ARTS CENTER, 2510 N. 24th St., Omaha, NE 68111. Tel.: 402-502-5291.
Web Site: lovesjazzartcenter.org
Institution Type/Description: Art Museum.
Collections: North Omaha Jazz history; photographs; paintings; hands-on exhibitions.
Activities: temporary exhibits; special events; educational programs.
Hours & Admission Prices: Call for hours.

MORMON TRAIL CENTER AT HISTORIC WINTER QUARTERS, 3215 State St., Omaha, NE 68112. Tel.: 402-453-9372. Fax: 402-453-1538. Facebook: Mormon Trail Center at Historic Winter Quarters.
E-mail: hswinter@ldschurch.org
Personnel Profile: Full-Time Volunteers 38; Part-Time Volunteers 2.
Governing Authority: Parent Institution: The Church of Jesus Christ of Latter-Day Saints. Tax-exempt.
Institution Type/Description: Visitors Center.
Collections: local history & culture; Mormon migration from Illinois to Utah; early pioneer life; period log cabin; ox drawn covered wagon; hand-carts; personal artifacts.
Major Exhibits: Quilt Show, 9/14; Gingerbread Festival, 11/14-12/14.
Hours & Admission Prices: Daily 9-9. No charge. Closed Thanksgiving; Christmas. &
Attendance: 41,000 (accurate)

NEBRASKA JEWISH HISTORICAL SOCIETY AND RIEKES MUSEUM, 333 S. 132nd St., Omaha, NE 68154-2106. Tel.: 402-334-6442. Fax: 402-334-6507.
E-mail: njhs@jewishomaha.org
Web Site: www.nebraskajhs.com
Founded: 1982.
Congressional District: 2
Institution Type/Description: History Museum.
Collections: artifacts, memorabilia, images & objects pertaining to Jewish life and culture since the 1880s; photographs.
Hours & Admission Prices: Mon.-Thurs. 10-4. Tours by appointment. No charge; donations accepted. &
Attendance: 450 (estimated)
Membership: Basic $25; Sponsor $50; Patron $100; Gold $250; Platinum $500; Benefactor $1,000.

OMAHA CHILDREN'S MUSEUM, 500 S. 20th St., Omaha, NE 68102-2505. Tel.: 402-342-6164. Fax: 402-342-6165.
E-mail: info@ocm.org
Web Site: www.ocm.org
Founded: 1977.
Congressional District: 2
Key Personnel: Exec. Dir., Lindy Hoyer; Pres. Bd. Dirs., Chuck Campbell; Chief Museum Officer, Jeff Barnhart; Dir. Finance, Shiree King; Dir. Guest Experience, Sara Sherman; Devel. & Donor Rels. Mgr., Joe Toppi; Mktg. & Public Rels. Coord., Cally Larsen.
Governing Authority: nonprofit organization. bd. of dirs. Tax-exempt: 501(c)(3).
Institution Type/Description: Children's Museum.
Collections: participatory exhibits in the arts, sciences & humanities.
Research Fields: science & technology; visual & performing arts; cultural & ethnic heritage; summer camps; outreach programs.
Facilities: 60,000 sq. ft. building.
Activities: art, science & humanities; special interest exhibits; theme-related workshops.

Publications: bimonthly newsletter, Fun Times; quarterly donor, Building Blocks; annual, Field Notes Program Guide.
Hours & Admission Prices: Summer: Memorial Day-Labor Day Tues., Wed. & Fri. 10-5, Thurs. 10-8, Sat. 9-5, Sun. 1-5; Winter: Tues.-Fri. 10-4, Sat. 9-5, Sun. 1-5. Children & adults 2-59 $9, seniors $8; discounts to reciprocal program, ASTC & ACM members; children under 2 & members no charge. Closed major holidays. &
Attendance: 206,164 (accurate)
Membership: Family & Grandparent $65; Family Plus $80; Patron $125.

OMAHA'S HENRY DOORLY ZOO AND AQUARIUM, 3701 S. 10th St., Omaha, NE 68107-2200. Tel.: 402-733-8401. Fax: 402-733-7868.
Web Site: www.omahazoo.com
Founded: 1965.
Congressional District: 2
Key Personnel: Dir., Dennis E. Pate; Pres. (V), John Boyer.
Personnel Profile: Full-Time Paid 253; Part-Time Paid 476; Part-Time Volunteers 408; Interns 144.
Volunteer Hours: 89,430
Operating Expenses: 27,571,688
Operating Income: 27,398,366
Governing Authority: society; nonprofit organization.
Institution Type/Description: Zoo.
Collections: approx. 5,769 live zoological specimens.
Research Fields: preventive medicine; nutrition; reproductive physiology; primate behavior.
Facilities: 200-vol. library for use on the premises; zoological park.
Activities: guided tours; lectures; films; formally organized education programs for children; docent program; tours for blind, disabled & underprivileged.
Publications: Zoo Newsletter.
Hours & Admission Prices: Daily 9:30-5. Adults 12 & over $15, senior citizens $14, children 3-11 $10; discounts to AAZPA members; members & children under 3 no charge. Closed Christmas. &
Attendance: 1,719,925 (accurate)
Membership: Individual & Military $60; Dual $88; Household & Grandparent $104.

SOKOL SOUTH OMAHA CZECHOSLOVAK MUSEUM, 2021 U St., Omaha, NE 68107. Tel.: 402-291-2893.
Institution Type/Description: Czech & Slovak Cultural History.
Collections: Czech & Slovak life, culture & history; hand-cut lead crystal; costumes; photographs.
Facilities: library. Museum-related items for sale.
Hours & Admission Prices: By appointment. &

WINTER QUARTERS MILL MUSEUM, 9102 N. 30th St., Omaha, NE 68112-1816. Mailing Address: 5215 Jackson, Omaha, NE 68106. Tel.: 402-551-1233.
Founded: 1999.
Key Personnel: Dir., Pres. (V) & Museum Shop Mgr., Linda Meigs.
Personnel Profile: Part-Time Volunteers 1.
Governing Authority: Tax-exempt.
Institution Type/Description: History Museum: founded in 1846 as the Mormon Winter Quarters gristmill; rebuilt by a Gold-Rusher; it contains some of the original hand-hewn beams & wooden pegs cut for the gristmill. Listed on the National Register of Historic Places.
Collections: historic photographs; newspaper clippings; objects from the pioneer era; agricultural artifacts; 1846 timbers; photographs; pioneer & settlement era tools & artifacts; agricultural artifacts.
Activities: tours & flour sack lunches for groups by appointment. Museum Sponsors: Sunday Summer Farmers Market.
Publications: annual, How's The Mill?.
Hours & Admission Prices: May-Oct. Wed.-Sun. 1-5; other times by appointment. No charge; donations accepted.
Attendance: 9,000 (estimated)
Membership: Friends of the Mill $20; Business Friend of the Mill $100.

Ord

VALLEY COUNTY MUSEUM, 117 S. 16th St., Ord, NE 68862. Mailing Address: P.O. Box 101, Ord, NE 68862. Tel.: 308-728-3044.
Web Site: www.ordnebraska.com/play/history
Institution Type/Description: History Museum.
Collections: local history & culture; period furnishings; personal artifacts; photographs.
Activities: special events.

Hours & Admission Prices: May-Sept. Mon.-Fri. 1-4; other times by appointment. No charge.

Oshkosh

HISTORICAL SOCIETY OF GARDEN COUNTY, West 1st & Avenue E, Oshkosh, NE 69154. Mailing Address: P.O. Box 193, Oshkosh, NE 69154-0193. Tel.: 308-772-3848.
Founded: 1969.
Congressional District: 3
Key Personnel: Chm. (V), Betty Kechley; Pres. (V), Verna Bairn.
Personnel Profile: Full-Time Volunteers 9; Part-Time Volunteers 29.
Governing Authority: nonprofit organization. Branch Museum: Rock School, W. 2nd & Avenue G, Oshkosh, NE.
Institution Type/Description: General Museum: housed in 1906-1907 Silver Hill Museum.
Collections: mounted birds; Indian artifacts; pioneer household furnishings, machinery & apparel, fossils; local history of schools; old photographs; barbed wire; other memorabilia.
Research Fields: archaeological specimens.
Facilities: 100-vol. library of textbooks, fiction & history.
Activities: guided tours; period equipment demonstrations. Annual Event: Garden County Fair.
Publications: brochures; annual newsletter.
Hours & Admission Prices: Memorial Day-Labor Day Mon.-Sat. 9-4, Sun. 2-6; call for additional hours. No charge; donations accepted. &
Attendance: 345 (accurate)
Membership: Annual $10.

Pawnee City

PAWNEE CITY HISTORICAL SOCIETY & MUSEUM, Hwy. 50/8 East, Pawnee City, NE 68420. Mailing Address: P.O. Box 33, Pawnee City, NE 68420-0033. Tel.: 402-852-3131. Facebook: Pawnee City Historical Society.
Web Site: www.pawneecountyhistory.com/pchsm/
Founded: 1968.
Congressional District: 1
Key Personnel: Chm., Roy Mullin; Vice Pres., Rita Shaw; Treas., Yvonne Dalluge.
Personnel Profile: Part-Time Volunteers 10.
Governing Authority: society; bd. of dirs. Tax-exempt.
Institution Type/Description: Regional History Museum: 22 historic buildings on grounds.
Collections: period dishes; photographs; papers; furniture; farm machinery; wire; 1,600 pairs of salt & pepper shakers; tailwind airplane; barbed wire. Historic Buildings: 1868 first schoolhouse in Pawnee City; 1862 home of first governor of Nebraska; 1857 Smithery log cabin; country store; MS Senator Kenneth Wherry Library; church; dentist office.
Activities: guided tours; arts & crafts display; motorcoach tours; living village. Annual Events: children's workshops in June; Civil War workshop in September; Civil War Reenactments in September; Country School Reunion.
Publications: biannual newsletter.
Hours & Admission Prices: Summer: Tues.-Fri. 10-3, Sat. 9-12 & 1-4, Sun. 1-4; Winter: Thurs. 9-4; other times by appointment. Adults $5; members no charge. &
Attendance: 2,500 (estimated)
Membership: K-12 Student $5; Couple $15; Family $25; Business $30; Supporting $50; Patron $100; Lifetime $200.

Pierce

PIERCE HISTORICAL SOCIETY, Gilman Park, Pierce, NE 68767. Mailing Address: Gilman Park, P.O. Box 122, Pierce, NE 68767. Tel.: 402-329-4265.
E-mail: museum@ptcnet.net
Web Site: www.ptcnet.net/museum
Congressional District: 3
Personnel Profile: Part-Time Volunteers 50.
Governing Authority: Tax-exempt.
Institution Type/Description: Historical Society Museum.
Collections: local history & culture; period furnishings; personal artifacts; movie projection equipment; printing equipment; military uniforms; farm equipment. Historic Buildings 1880 depot; school house; blacksmith shop.
Activities: special events.
Hours & Admission Prices: Memorial Day to Labor Day Sun. 1:30-4:30; other times by appointment. No charge; donations accepted. &
Attendance: 521 (accurate)
Membership: Individual & Family $10; Business $25.

Pilger

HISTORICAL SOCIETY OF STANTON COUNTY MUSEUMS AT STANTON AND PILGER, INC., 345 N. Main St., Pilger, NE 68768. Mailing Address: P.O. Box 234, Stanton, NE 68779. Tel.: 402-396-3422.
E-mail: rjensen98@cableone.net
Web Site: www.stantoncountyhistoricalsociety.org
Founded: 1965.
Congressional District: 3
Key Personnel: Pres. (V), James Duncan; Vice Pres., Rebecca Frerichs; Treas., Virgine Jensen; Sec., Frances Beck
Governing Authority: society. Parent Institution: Historical Society of Stanton County. Branch Museum: Stanton Heritage Museum, 1106 Ivy St., P.O. Box 234, Stanton, NE 68779. Tel. 402-439-2208; EUB Historic Church, 706 Kingwood, Stanton, NE 68779; Restored Rural School, 1108 Ivy St., Stanton, NE 68779. Tax-exempt.
Institution Type/Description: Historical Society Museum.
Collections: Indian artifacts; antiques. Historic Buildings: rural schoolhouse & c.1880 restored church in Stanton & Pilger.
Research Fields: Pioneer records; biographies of country Pioneers; old cemeteries.
Activities: permanent & temporary exhibitions; cataloging of county cemeteries; taped interviews with county residents.
Publications: biannual newsletter, Historical Times.
Hours & Admission Prices: Pilger Museum: Memorial Day, Pilger Days & Labor Day; other times by appointment. Stanton Heritage Museum: May-Oct. Sat. 2-4; other times by appointment. No charge; donations accepted.
Attendance: 450 (estimated)
Membership: Student $.50; Active $10; Institutional $15; Lifetime $100.

Plainview

PLAINVIEW HISTORICAL SOCIETY, 304 S. Main St., Plainview, NE 68769. Mailing Address: P.O. Box 495, Plainview, NE 68769. Tel.: 402-582-4730.
Founded: 1978.
Congressional District: 18
Governing Authority: Tax-exempt.
Institution Type/Description: Historic Building: housed in a former railroad depot; built in 1880.
Collections: local history & culture; railroad history & memorabilia; photographs; period furnishings; clothing; cooking utensils & appliances; personal artifacts.
Activities: Museum Sponsors: Antique Farm Show.
Hours & Admission Prices: May to Labor Day Tues.-Sun. 1-5; other times by appointment. No charge; donations accepted.
Attendance: 300 (estimated)
Membership: Individual $20.

PLAINVIEW KLOWN DOLL MUSEUM, Hwy. 20, Plainview, NE 68769. Mailing Address: P.O. Box 813, Plainview, NE 68769-0813. Tel.: 402-582-4433.
E-mail: clowndollmuseum@yahoo.com
Web Site: www.klowndollmuseum.com
Founded: 1980.
Congressional District: 3
Key Personnel: Chm. (V), Lee Warneke.
Personnel Profile: Part-Time Volunteers 30.
Volunteer Hours: 613
Operating Expenses: 750
Operating Income: 750
Governing Authority: Parent Institution: Plainview Chamber of Commerce. Tax-exempt.
Institution Type/Description: Clown Doll Museum.
Collections: over 7,000 clown dolls, all donated.
Hours & Admission Prices: Memorial Day-Labor Day Mon.-Sat. 1-5; Sept.-May Mon.-Fri. 1-4.30; other times by appointment. No charge; donations accepted. &
Attendance: 783 (accurate)

Plattsmouth

CASS COUNTY HISTORICAL SOCIETY MUSEUM, 646 Main St., Plattsmouth, NE 68048-1852. Tel.: 402-296-4770.
E-mail: ccohsm@windstream.net
Web Site: www.nebraskamuseums.org/casscountymuseum.htm
Founded: 1936.

Congressional District: 2
Key Personnel: Pres., Douglas V. Duey; Vice Pres., Susan Rice; Treas., Diane Berlett; Sec., Pat Meisinger; Cur., Margo Prentiss.
Personnel Profile: Full-Time Paid 1; Part-Time Paid 1; Part-Time Volunteers 67.
Volunteer Hours: 1,500
Operating Expenses: 66,000
Operating Income: 64,000
Governing Authority: society; nonprofit organization. Parent Institution: Cass County Historical Society. Branch Museums: Rock Bluffs One Room School House; Joseph Cook Log House; Burlington Northern Caboose. Tax-exempt: 501(c)(3).
Institution Type/Description: Historical Society Museum.
Collections: minerals & fossils; tools & equipment; household belongings; clothing & textiles; decorative arts; works of art; manuscripts & archives; photographs.
Major Exhibits: Don't Forget your Hat, 10/29/13-1/25/14; Saturday Evening Post Covers by John Falter, 10/29/13-1/25/14; Valentines and Teapots, 1/28-3/22/14; From Sea to Shining Sea Prints by John Falter, 3/25-6/1/14; Remembering Beautiful Merritt Beach, 6/3-8/24/14; Harvest Memories: Farm Life & Fall Festivals, 8/26-10/26/14.
Research Fields: County History; Genealogical.
Facilities: 600-vol. library of history material; Searl S. & Leila C. Davis Conference Room. Gift-related items for sale.
Activities: guided tours; lectures; workshops; brown bag programs; permanent & temporary exhibitions; open houses.
Publications: quarterly newsletter.
Hours & Admission Prices: April-Oct. Tues.-Sun. 12-4; Nov.-March Tues.-Sat. 12-4. Adults $2.50; AAM, ICOM, & museum members no charge. Closed holidays.
Attendance: 3,000 (estimated)
Membership: Friend $15; Booster $30; Promoter $50; Century $100; Investor $250; Benefactor $500 & up; Patron $1,000 & up.

Red Cloud

WEBSTER COUNTY HISTORICAL MUSEUM, 721 W. 4th Ave., Red Cloud, NE 68970-2221. Mailing Address: P.O. Box 464, Red Cloud, NE 68970-0464. Tel.: 402-746-2444.
E-mail: wchmdirector@gpcom.net
Founded: 1965.
Congressional District: 35
Key Personnel: Pres. (V), Ron Gestring; Dir., Teresa Young.
Personnel Profile: Part-Time Paid 3; Part-Time Volunteers 3.
Governing Authority: county. Tax-exempt.
Institution Type/Description: General Museum.
Collections: agriculture; period artifacts; natural history; history; medical; military; mill levee.
Facilities: library of local history available for use on the premises; nature center; reading room.
Activities: guided tours; permanent exhibitions.
Hours & Admission Prices: April-Oct. Tues.-Sun. 1-5. Adults $4, high school students $1.50, elementary school students $1; members no charge. Closed Easter; Mother's Day; Father's Day; Independence Day.
Attendance: 1,700 (estimated)
Membership: Individual $15; Family $25; Sustaining $50.

WILLA CATHER FOUNDATION, (M), 413 N. Webster St., Red Cloud, NE 68970-2466. Tel.: 402-746-2653. Fax: 402-746-2652.
E-mail: info@willacather.org
Web Site: www.willacather.org
Formerly: Willa Cather Pioneer Memorial and Educational Foundation
Founded: 1955.
Congressional District: 3
Key Personnel: Exec. Dir. & Museum Shop Mgr., Leslie C. Levy; Chm. (V), Thomas Gallagher.
Personnel Profile: Full-Time Paid 7; Part-Time Paid 6; Part-Time Volunteers 10.
Governing Authority: nonprofit organization. Subsidiary Institution: Nebraska State Historical Society. Tax-exempt.
Institution Type/Description: Art Gallery & Historic Museum.
Collections: various forms of art; letters; photographs.
Research Fields: archives.
Facilities: Literature, slides, books, stationery & prints of some original oils for sale.
Activities: guided tours; lectures.
Publications: Willa Cather Newsletter & Review.
Hours & Admission Prices: By appointment. Town & Country Tours: adults $15, children 5-12 $6; children 4 & under no charge. Closed New Year's Day; Easter; Independence Day; Thanksgiving; Christmas. &

Attendance: 10,000 (estimated)
Membership: Student $20; General $50; Sustaining $125; Friend $250; Patron $500; Benefactor $1,000; Cather Circle $2,500.

Republican City

LIGHTHOUSE MUSEUM, 103 Berrigan Rd., Republican City, NE 68971. Mailing Address: 202 Center Ave., Republican City, NE 68971-7110. Tel.: 308-799-2033.
Institution Type/Description: History Museum.
Collections: over 5,000 depression glass & glassware; McCoy pottery; beer signs; early toys; period clothing & furniture.
Hours & Admission Prices: Memorial Day to Labor Day Sat.-Sun. 10-5; other times by appointment.

Rushville

SHERIDAN COUNTY HISTORICAL SOCIETY - RUSHVILLE MUSEUM/ARMSTRONG HOUSE, 408 E. 2nd, Rushville, NE 69360-0274. Mailing Address: P.O. Box 274, Rushville, NE 69360-0274. Tel.: 308-327-2985 & 360-0299 (cell).
E-mail: nitz@gpcom.net
Formerly: Sheridan County Historical Society - Armstrong House Museum
Founded: 1962.
Congressional District: 3
Volunteer Hours: 640
Governing Authority: Tax-exempt.
Institution Type/Description: Historical Society Museum: house built in 1890.
Collections: local history & culture; period furnishings; personal artifacts; photographs.
Hours & Admission Prices: Memorial Day to Labor Day Mon.-Fri. 1-4; other times by appointment. No charge; donations accepted.
Attendance: 721 (accurate)

Schuyler

SCHUYLER/COLFAX COUNTY MUSEUM, 309 E. 11th St., Schuyler, NE 68661. Tel.: 402-352-2301 & 352-3195.
Founded: 1976.
Key Personnel: C.E.O., Nadine Beran
Institution Type/Description: History Museum.
Collections: county history & culture; photographs; personal artifacts.
Publications: annual newsletter.
Hours & Admission Prices: Jan.-March Tues. forenoons; April-Dec. Sun. 2-4, Tues. forenoons; tours by appointment. No charge; donations accepted. &
Attendance: 649 (accurate)
Membership: Individual $15; Sustaining $25; Lifetime $150.

Scotia

HAPPY JACK PEAK AND CHALK MINE, 80131 Hwy. 11, Scotia, NE 68875. Tel.: 308-245-3276.
E-mail: info@happyjackchalkmine.org
Web Site: www.happyjackchalkmine.org
Institution Type/Description: Mine Museum.
Collections: chalk room; pillar mine; local history & mining.
Facilities: over 6,000 ft. of honeycombed caverns.
Hours & Admission Prices: Memorial Day to Labor Day daily 10-5; Sept. Fri.-Sun. 10-5. Adults $6.50, seniors 60 & over and children 6-12 $5.25; children 5 & under no charge.

Scottsbluff

RIVERSIDE DISCOVERY CENTER, 1600 S. Beltline Hwy. West, Scottsbluff, NE 69361-1331. Tel.: 308-630-6236. Fax: 308-632-2953.
E-mail: anne.james@riversidediscoverycenter.org
Web Site: riversidediscoverycenter.org
Founded: 1950.
Congressional District: 3
Key Personnel: Exec. Dir., Anne James.
Personnel Profile: Full-Time Paid 9; Part-Time Paid 5; Part-Time Volunteers 50.
Operating Expenses: 674,000
Operating Income: 674,000
Governing Authority: nonprofit corporation. Parent Institution: Riverside Zoological Foundation.
Institution Type/Description: Zoo.
Collections: domestic & wild animals; Chimpanzee Conservation Center; tiger facility; lions; addax; bobcats; rainforest discovery center.

Facilities: chimpanzee conservation center; discovery center; vending area. Zoo-related gift items for sale.
Activities: guided tours; feed waterfowl on the zoo's lake; special events; playground; splash pad; petting zoo; educational programs.
Publications: newsletter, Zoo Tales.
Hours & Admission Prices: April-Oct. daily 9:30-4:30; Nov.-March daily 10:30-3:30. Adults $6, senior citizens $5, children 5-12 $4; discounts to AZA members & groups; children under 4 & members no charge. &
Attendance: 34,000 (accurate)
Membership: Family $75; Supporting $150; Contributor $250; Benefactor $500; Gift of Discovery $1,000.

WEST NEBRASKA ARTS CENTER GALLERY, 106 E. 18th St., Scottsbluff, NE 69361-2423. Tel.: 308-632-2226.
Institution Type/Description: Art Gallery.
Collections: paintings; photography; printmaking; sculpture.
Activities: special events.
Hours & Admission Prices: Tues.-Fri. 9-5, Sat.-Sun. 1-5. No charge.

Scribner

MUSBACH MUSEUM, 429 Main St., Scribner, NE 68057. Mailing Address: P.O. Box 136, Scribner, NE 68057. Tel.: 402-664-2788.
Institution Type/Description: History Museum.
Collections: German history, heritage & culture; early farm & medical equipment; kitchen china; glassware; photographs.
Hours & Admission Prices: By appointment.

Seward

BARTELS MUSEUM - CONCORDIA UNIVERSITY, Link Library, 800 N. Columbia Ave., Seward, NE 68434. Tel.: 800-535-5494; 402-643-3651. Fax: 402-643-4073.
E-mail: marvin.plamann@cune.edu
Founded: 1983.
Key Personnel: Dir., Marvin Plamann
Governing Authority: Parent Institution: Concordia University, Seward, NE. Tax-exempt.
Institution Type/Description: Geology Museum.
Collections: minerals; agate; fossils; rocks.
Hours & Admission Prices: Sept.-April Mon.-Thurs. 8am to midnight, Fri. 8-5, Sat. 1-5, Sun. 1pm to midnight; Summer: call for hours. No charge. &
Attendance: 250

MARXHAUSEN ART GALLERY, Concordia University, 800 N. Columbia Ave., Seward, NE 68434-1500. Tel.: 402-643-3651 & 7490. Fax: 402-643-4073.
E-mail: jbockelman@cune.edu
Web Site: www.cune.edu/arts/
Founded: 1951.
Congressional District: 1
Key Personnel: Cur., James E. Bockelman.
Personnel Profile: Part-Time Paid 1.
Governing Authority: denominational group. Parent Institution: Concordia College. Tax-exempt: 501(c)(3).
Institution Type/Description: Art Gallery.
Collections: over 300 works of art by national & international artists; screenprints; etchings; lithographs; prints.
Activities: gallery talks; rental gallery; temporary & traveling exhibitions.
Hours & Admission Prices: Sept.-May Mon.-Fri. 11-4, Sat.-Sun. 1-4. No charge; donations accepted. Closed Easter weekend; week of Thanksgiving; weeks before & after Christmas; school recesses including summer break.

Sidney

FORT SIDNEY MUSEUM & POST COMMANDER'S HOME, 6th Ave. & Jackson St., Sidney, NE 69162. Mailing Address: P.O. Box 596, Sidney, NE 69162-0596. Tel.: 308-254-2150 & 4030.
Founded: 1954.
Congressional District: 3
Key Personnel: Pres., Ed Bivens; Vice Pres., Brenda Blanke; Treas., Audrey Tremain; Bd. Member, Charlotte Steffens; Bd. Member, Roger Jorgensen; Bd. Member, Pat Albers; Bd. Member, Julie Gehrig; Bd. Member, Marua Elwanger; Sec., LaDonna Jung; Museum Shop Mgr., Duane Nightengale.
Personnel Profile: Full-Time Paid 1; Part-Time Paid 1; Part-Time Volunteers 10.
Volunteer Hours: 2,500
Operating Expenses: 30,000
Operating Income: 30,000

Governing Authority: nonprofit organization. Parent Institution: Cheyenne Co. Historical Assoc. Tax-exempt.
Institution Type/Description: General Museum on a Historic Building & site: 1871 Commanding Officer's Quarters; 1872 Powder House or Magazine; 1884 Officers' Quarters; Carriage House.
Collections: 1894 furnishings; artifacts; 1900 carriage; military collection; railroad room; Ladies Dressmaking and Millinery Shop; Sioux Army Depot Room; Pioneer Room; library of old & valuable books, maps & stereographs.
Research Fields: history of Fort Sidney; Cheyenne County history; past & present citizens of Cheyenne County; Sioux army depot history.
Facilities: library.
Activities: guided tours; organized education programs for children & adults; 4th of July celebration, continuous entertainment 1-4; Old Fashion Christmas, decorations & entertainment in December.
Publications: pamphlet, History of Ft. Sidney; Walking Tour of Sidney Historic Downtown District; newsletter, Cheyenne County Historical Association.
Hours & Admission Prices: Museum: June-Aug. daily 9-5; Dec. Sat.-Sun. 1-4; other times by appointment. Commander's Home: June-Aug. & Old Fashioned Christmas. No charge; donations accepted. &
Attendance: 2,808 (accurate)
Membership: Annual $5.

St. Paul

MUSEUM OF NEBRASKA MAJOR LEAGUE BASEBALL, 619 Howard Ave., St. Paul, NE 68873-2022. Tel.: 308-754-5558.
E-mail: stpaulcham@qwestoffice.net
Web Site: www.nebraskabaseballmuseum.com
Founded: 1991.
Congressional District: 3
Governing Authority: Parent Institution: St. Paul Chamber of Commerce. Tax-exempt.
Institution Type/Description: Baseball Museum.
Collections: trading cards; photographs; artifacts.
Hours & Admission Prices: Memorial Day-Labor Day Mon.-Fri. 10-4, Sat. 10-2; other times by appointment. No charge; donations accepted.
Attendance: 800 (accurate)

Stuart

WHITE HORSE MUSEUM AND HERITAGE VILLAGE, Hwy. 20, Stuart, NE 68780-5847. Mailing Address: P.O. Box 118, Stuart, NE 68780. Tel.: 402-924-3861.
E-mail: visitorinfo@stuartwhitehorsemuseum.com
Web Site: www.stuartwhitehorsemuseum.com
Key Personnel: Dir., Donald I. Schmaderer.
Personnel Profile: Part-Time Volunteers 5.
Volunteer Hours: 500
Operating Expenses: 18,000
Operating Income: 18,500
Governing Authority: township. Tax-exempt.
Institution Type/Description: History Museum.
Collections: White Horse Ranch memorabilia; period artifacts; furnishings; photographs.
Hours & Admission Prices: Memorial Day to Labor Day call for hours. Adults $3.
Attendance: 100 (estimated)

Sutton

SUTTON HISTORICAL SOCIETY, 604 S. Way Ave., Sutton, NE 68979-2142. Tel.: 402-773-0222.
Institution Type/Description: Historical Society Museum: housed in the former home of pioneers, John and Emma Gray; built in 1908.
Collections: local history & culture; period furnishings; photographs.
Hours & Admission Prices: Call for hours.

Syracuse

OTOE COUNTY MUSEUM OF MEMORIES, 1621 Thorne St., Syracuse, NE 68446-9730. Tel.: 402-269-2355.
E-mail: wittelow@windstream.net
Founded: 1972.
Congressional District: 1
Key Personnel: Pres., Phyllis Witte; Cur. & Treas., Rose Garey.
Personnel Profile: Full-Time Volunteers 6; Part-Time Volunteers 15.
Governing Authority: society. Parent Institution: Syracuse Foundation.
Institution Type/Description: History Museum.

Collections: parlor, kitchen & bedroom from the turn-of-the-century; local Indian artifacts; loom & sewing room; Thomas A. Edison exhibit; U.S. flag display; tent show exhibit; Kramer collection of big game trophies, small animals & birds; butcher shop; 1890 general store; doctor's office; carriage house & cutter; original church altar & furnishings; old fashioned herb garden; county room exhibit; tools & small farm equipment.
Research Fields: Chick Boyes tent show; financial records; play scripts; costumes & props; genealogy.
Facilities: library of history & personal memories of residents of Otoe County; meeting rooms.
Activities: tours; school tours; art exhibits; permanent exhibits. Museum Sponsors: local artists show; quilt show; crotchet show; apron show.
Hours & Admission Prices: May-Sept. Sun. 1:30-4. No charge; donations accepted. &
Attendance: 1,500 (estimated)
Membership: Individual $10; Family $15.

Table Rock

TABLE ROCK HISTORICAL SOCIETY AND MUSEUMS, 414-16 Houston St., Table Rock, NE 68447. Mailing Address: P.O. Box 194, 62676 717th Rd., Table Rock, NE 68447. Tel.: 402-839-4135 & 2985. Fax: 402-839-4135.
Founded: 1965.
Congressional District: 1
Key Personnel: Pres. (V), Gordon Clement.
Personnel Profile: Part-Time Paid 1; Part-Time Volunteers 10.
Governing Authority: society.
Institution Type/Description: General Museum.
Collections: furnishings; antiques; paintings; tools and equipment; Indian artifacts; school items; automobiles; religious exhibits; military exhibits, WWI & WWII; Table Rock Argus Print Shop. Historic Houses: 1854 Turner Log Cabin; 1877 St. John's Catholic Church; 1893 Old Opera House.
Activities: guided tours; Living History Day.
Publications: brochures.
Hours & Admission Prices: May-Sept. Sun. 1-4; other times by appointment. No charge; donations accepted. &
Attendance: 800 (accurate)
Membership: Individual $2; Sustaining $5; Life $100.

Tecumseh

JOHNSON COUNTY HISTORICAL SOCIETY & MUSEUM, 3rd & Lincoln Streets, Tecumseh, NE 68450-2116. Mailing Address: 231 Lincoln St., Tecumseh, NE 68450. Tel.: 402-335-5900.
E-mail: jc10928@windstream.net
Founded: 1962.
Congressional District: 1
Key Personnel: Pres. & Cur., Boyd C. Maddox.
Governing Authority: county. Tax-exempt: 170(b)(1)(a).
Institution Type/Description: History Museum: housed in 1889 Christian Church Building.
Collections: Indian artifacts; tools; 3,000 wooden household articles; clothing; linens; silver; china; glass; toys; World War I objects; memorabilia; pictures; scrapbooks; newspapers; books. Historic House: 1870, one-room schoolhouse.
Research Fields: genealogy.
Facilities: 75-vol. library of old atlases; records; bibles; newspapers; songs; books written by Johnson County residents; available for use on premises. Museum-related items for sale.
Activities: arts festivals; guided tours; study clubs; formally organized education programs for children & adults; docent program or council; temporary exhibitions.
Hours & Admission Prices: May-Oct. Tues.-Fri. 1-4; other times by appointment. No charge; donations accepted.
Attendance: 100 (estimated)
Membership: Individual $10; Family $20.

Tekamah

BURT COUNTY MUSEUM, 319 N. 13th St., Tekamah, NE 68061-1503. Mailing Address: P.O. Box 125, Tekamah, NE 68061-0125. Tel.: 402-374-1505.
E-mail: burtcomuseum@abbnebraska.com
Web Site: sites.google.com/site/officialburtcountymuseum
Founded: 1967.
Congressional District: 16
Key Personnel: Dir., Bonnie Newell; Pres. (V), Marlene Kaeding; Treas., Dee Bottger.

Personnel Profile: Part-Time Paid 5; Part-Time Volunteers 12.
Governing Authority: city. Subsidiary Institution: Burt County Museum, Inc. Tax-exempt.
Institution Type/Description: General Museum: housed in c.1904 home of E.C. Houston. Listed on National Register of Historic Places.
Collections: agriculture; clothing; furniture; photos; manuscript collection.
Major Exhibits: Historic Quilts, 5/14-9/14.
Research Fields: genealogy; community history events.
Facilities: meeting rooms. Museum-related items for sale.
Activities: guided tours; permanent & temporary exhibitions; tour of homes. Museum Sponsors: Christmas Candle Light Tours; Coffee on the Porch for Alumni in May; Independence Day Celebration.
Publications: Tekamah Cemetery Book; quarterly newsletters; brochures; flyers; book, Decatur & Tekamah.
Hours & Admission Prices: Tues., Thurs. & Sat. 1-5. No charge; donations accepted. Closed major holidays.
Attendance: 2,560 (estimated)
Membership: Annual $5; Five Years $10; Life $20.

Thedford

THEDFORD ART GALLERY, 509 Court St., Thedford, NE 69166. Tel.: 308-645-2586.
Institution Type/Description: Art Gallery.
Collections: paintings; photographs; pottery; sculpture.
Hours & Admission Prices: Jan.-April Tues. 1-5; May-Dec. Tues.-Sat. 10-5; other times by appointment. &

THOMAS COUNTY HISTORICAL SOCIETY MUSEUM, 609 Court St., Thedford, NE 69166. Tel.: 308-645-2477.
Institution Type/Description: Historical Society Museum.
Collections: local history & culture; period furnishings; photographs; personal artifacts.
Hours & Admission Prices: Memorial Day to Labor Day Mon., Wed. & Fri. 10-12 & 2-4; other times by appointment. No charge.

Tobias

TOBIAS COMMUNITY HISTORICAL SOCIETY, Main St., Tobias, NE 68453-2073. Mailing Address: 561 County Rd. 5, Tobias, NE 68453-2073. Tel.: 402-243-2356.
E-mail: rjrada@diodecom.net
Founded: 1968.
Congressional District: 1
Key Personnel: Pres. (V), Judith K. Rada; Sec. & Cataloging, Mary Kronhofman.
Personnel Profile: Part-Time Volunteers 1.
Governing Authority: municipal. Tax-exempt.
Institution Type/Description: General Museum: housed in original bank & Tobias Print Shop, saved in the 1891 fire which burned most of the business district.
Collections: period bottles; kitchen wares; household appliances; photographs; tools; philatelic; medical; military; Tobias High School trophies & pictures; newspapers.
Research Fields: local history.
Activities: permanent & temporary exhibitions.
Hours & Admission Prices: By appointment only. No charge; donations accepted. &
Attendance: 100 (estimated)
Membership: Individual $1; Individual Life $15; Couple Life $25.

Valentine

CHERRY COUNTY HISTORICAL SOCIETY, Main St. & Hwy. 20, Valentine, NE 69201. Mailing Address: P.O. Box 284, Valentine, NE 69201-0284. Tel.: 402-376-2015.
Founded: 1928.
Congressional District: 3
Key Personnel: C.E.O. & Pres. (V), Joyce Muirhead; Cur. & Museum Shop Mgr., Jan Howell.
Personnel Profile: Full-Time Volunteers 4; Part-Time Paid 2; Part-Time Volunteers 3.
Governing Authority: county. Tax-exempt.
Institution Type/Description: History Museum.
Collections: newspapers from 1883; county records from 1884; pictures; buttons; clothing from 1885-present; home furnishings; musical instruments; Indian artifacts.
Research Fields: history; family tree data.
Facilities: 250-vol. genealogy library of history, western stories, old school books, Nebraska & Indian history.

Activities: guided tours; microfilms; films; study clubs.
Publications: A Sandhills Century; Murder on the Plains.
Hours & Admission Prices: Memorial Day to Labor Day Thurs.-Sat. 1-4:30; other times by appointment. Adults $1, children under 12 $.50; members no charge. &

Attendance: 450 (estimated)
Membership: Individual $5; Ten Year $25; Lifetime $100.

FORT NIOBRARA NATIONAL WILDLIFE REFUGE, 39983 Refuge Rd., HC 14, Valentine, NE 69201. Tel.: 402-376-3789. Fax: 402-376-3217.
E-mail: fortniobrara@fws.gov
Web Site: fortniobrara.fws.gov
Founded: 1912.
Congressional District: 3
Key Personnel: Refuge Mgr., Alan Whited.
Personnel Profile: Full-Time Paid 10; Part-Time Paid 1.
Governing Authority: federal. Branch of Dept. of the Interior, Fish & Wildlife Service, Washington, D.C. Tax-exempt.
Institution Type/Description: Wildlife Refuge.
Collections: birds; small mammals; local historical photographs; photographs of military reservation 1879-1906.
Facilities: visitor center.
Activities: permanent exhibitions.
Publications: Fort Niobrara 1880-1906.
Hours & Admission Prices: Daily sunrise-sunset. No charge. Visitor Center: Memorial Day-Labor Day daily 8-4:30; Sept.-May Mon.-Fri. 8-4:30. No charge. &
Attendance: 50,000 (estimated)

SAWYER'S SANDHILLS MUSEUM, 440 Valentine St., Valentine, NE 69201-1942. Mailing Address: 1100 E. 10th St., Apt. 47, Valentine, NE 69201-1691. Tel.: 402-376-3293.
Founded: 1958.
Key Personnel: Owner, Dorothy Sawyer.
Governing Authority: private; nonprofit organization.
Institution Type/Description: General Museum.
Collections: period autos; guns; Indian artifacts; musical instruments; lamps; dishes; a moonshine still.
Hours & Admission Prices: Memorial Day-Labor Day by appointment only. Adults $5, children 5-13 $1.

Valley

VALLEY COMMUNITY HISTORICAL SOCIETY, INC., 218 W. Alexander St., Valley, NE 68064. Mailing Address: P.O. Box 685, Valley, NE 68064-0685. Tel.: 402-359-5544, 5323 & 2339.
Web Site: www.valleyne.org
Founded: 1966.
Congressional District: 2
Key Personnel: Pres., Max Johnson; Vice Pres., Tom Batten; Treas. & Sec., Julia A. Allen
Governing Authority: society; nonprofit. Tax-exempt: 501(c)(3).
Institution Type/Description: Local History Museum: housed in 1872 schoolhouse, then served as a Baptist Church & later a Catholic Church.
Collections: personal artifacts; manuscripts; Indian artifacts; fossils.
Activities: lectures; films; permanent exhibitions.
Publications: annual pamphlet, Yearly Report.
Hours & Admission Prices: May-Aug. Sun. 1-3; other times by appointment. No charge; donations accepted.
Attendance: 425 (accurate)
Membership: Individual $5; Life $100.

Wahoo

SAUNDERS COUNTY HISTORICAL COMPLEX, 240 N. Walnut, Wahoo, NE 68066-1858. Tel.: 402-443-3090. Fax: 402-443-3090 (call first). Facebook: Saunders County Historical Complex.
E-mail: saunderscomuseum@hotmail.com
Web Site: www.saunderscountymuseum.org
Founded: 1963.
Congressional District: 23
Key Personnel: Pres., Marlenme McDonald; Dir. & Cur., Erin Hauser; Museum Shop Mgr., Lila Zech.
Personnel Profile: Full-Time Paid 1; Part-Time Paid 1; Part-Time Volunteers 75.
Volunteer Hours: 8,000
Operating Expenses: 70,000
Operating Income: 66,000

Governing Authority: nonprofit organization. Saunders County Historical Society. Tax-exempt: 501(c)(3).
Institution Type/Description: Historical Society Museum Complex.
Collections: agricultural equipment; business machines; furniture; household equipment; costumes; decorative arts; photographs; memorabilia relating to Saunders County; materials pertaining to Czech, German & Swedish ethnic groups. Historic Houses: c.1890 Rural School; 1886 Burlington-Northern Railroad Depot; 1889 Weston Presbyterian Church; 1893 Hanson Historical House; 1873 Log House; 1950 Post Office.
Research Fields: county history; genealogy.
Facilities: 400-vol. library pertaining to state & county histories available for research on premises only.
Activities: guided tours; permanent, temporary & traveling exhibitions; educational programs. Annual Event: Christmas program.
Publications: brochure; newsletter.
Hours & Admission Prices: April-Sept. Tues.-Sat. 10-4; Oct.-March Tues.-Fri. 10-4. No charge; donations accepted. Closed major holidays. &
Attendance: 7,800 (estimated)
Membership: Individual $15; Family $20; Business $30; Lifetime $150.

Walthill

SUSAN LA FLESCHE PICOTTE CENTER, 505 Matthewson St., Walthill, NE 68067. Mailing Address: 2517 H Ave., Walthill, NE 68067-5015. Tel.: 402-406-5966.
E-mail: pmchar@hotmail.com
Founded: 1986.
Congressional District: 3
Key Personnel: Chm. (V), Keith Mahaney.
Personnel Profile: Part-Time Volunteers 1.
Institution Type/Description: Historic Building: housed in the former hospital built in 1910 by Dr. Susan La Flesche Picotte, the first Native American woman physician in the U.S.
Collections: Dr. Susan La Flesche Picotte's life; photographs; personal artifacts.
Hours & Admission Prices: By appointment. No charge. &

Weeping Water

WEEPING WATER VALLEY HISTORICAL SOCIETY, 215 W. Eldora Ave., Weeping Water, NE 68463. Mailing Address: P.O. Box 43, Weeping Water, NE 68463-0043. Tel.: 402-267-4925.
E-mail: wd85407@navix.net
Founded: 1969.
Congressional District: 2
Key Personnel: Chm. & Pres. (V), Doris Duff; Treas., Dale Nielsen.
Personnel Profile: Part-Time Volunteers 30.
Governing Authority: society; nonprofit. Parent Institution: Weeping Water Valley Historical Society. Tax-exempt: 501(c)(3).
Institution Type/Description: History & Medical Museum: located on the site of the oldest First Congregational Parsonage in state.
Collections: Dr. Lloyd N. Kunkels prehistoric Indian artifacts from 20,000 years ago to Plains Indians; fossils, period furniture, pioneer artifacts, guns & tools. Historic Buildings: 1870 chapel, 1880 original office of Dr. Jesse Fate; 1867 residence, originally the Congregational parsonage, detailing businesses housed in Weeping Water from the 1860's to the present, original soda fountain, general store, barber shop; Kevin Brack's celebrity collection.
Research Fields: archeology; excavation of prehistoric Indian houses in vicinity.
Facilities: 50-vol. library of old medical books; Nebraska history available for inter-library loan & on premises. Fossils for sale.
Activities: guided tours; lectures; permanent & temporary exhibitions.
Publications: brochure.
Hours & Admission Prices: May-Oct. Sun. 1-4; other times by appointment, call 402-267-4925. Memory Lane Mon.-Fri. 10-2. No charge; donations accepted. &
Attendance: 2,200 (estimated)
Membership: Annual $1; Life $25.

Wellfleet

DANCING LEAF CULTURAL LEARNING CENTER, 6100 E. Opal Springs Rd., Wellfleet, NE 69170. Tel.: 308-963-4233.
Web Site: dancingleaf.com
Founded: 1989.
Congressional District: 3
Key Personnel: Co-Dir., Les Hosick; Co-Dir., Jan Hosick.
Personnel Profile: Full-Time Paid 2.
Institution Type/Description: Native American History Center.

Collections: Native American history & culture; period furnishings; personal artifacts.
Facilities: Gift items for sale.
Activities: guided tours.
Hours & Admission Prices: By reservation. Adults $7, children 5-17 $5.
Attendance: 5,000 (estimated)

West Point

CUMING COUNTY HISTORICAL MUSEUM COMPLEX, 130 N. River, West Point, NE 68778. Mailing Address: 227 N. Main St., West Point, NE 68788. Tel.: 402-372-2936.
Institution Type/Description: Historical Society Museum.
Collections: local history & culture; period furnishings; photographs; personal artifacts. Historic Buildings: 1904 Chicago Northwestern Depot; Union Pacific caboose; St. Matthew's Church; 1912 District #34 one room school.
Hours & Admission Prices: Aug. during the Cuming County Fair.
Membership: Annual $5.

Wilber

WILBER CZECH MUSEUM, 224 N. High St., Wilber, NE 68465. Tel.: 402-821-2574.
E-mail: dorisourecky@windstream.net
Founded: 1965.
Congressional District: 32
Key Personnel: Gift Shop Mgr., LaVern Aksamit.
Personnel Profile: Part-Time Volunteers 12; Interns 1.
Governing Authority: nonprofit organization. Parent Institution: Nebraska Czechs of Wilber. Tax-exempt: 501(c)(3).
Institution Type/Description: General Museum.
Collections: textiles; glass; Czech dolls; Czech dishes; quilts; replicas of early immigrant homes & early Wilber businesses; agriculture; paintings; specialized laces & Czech costumes; decorative arts; medical; guns; period artifacts.
Research Fields: local history; history of Czechoslovakia.
Facilities: Hand-loomed rugs & other museum-related items for sale.
Activities: guided tours; permanent & temporary exhibitions; rug making demonstrations; Czech Festival reflecting origins of the past.
Publications: brochure; book, Poetic History of Wilber; Tales of the Czechs.
Hours & Admission Prices: March-Dec. daily 1-4; other times by appointment. No charge; donations accepted. Closed national holidays. &
Attendance: 3,900 (estimated)

Winnebago

ANGEL DECORA MUSEUM AND RESEARCH CENTER, 601 E. College Dr., Winnebago, NE 68071. Mailing Address: P.O. Box 687, Winnebago, NE 68071. Tel.: 402-878-2380, ext. 113.
E-mail: smith_deleon77@yahoo.com
Key Personnel: Dir., Emily Smith-deLeon; Chm. (V), David Smith
Institution Type/Description: History Museum.
Collections: Winnebago history & artifacts; Native American dolls, paintings, clothing, & baskets; photographs.
Hours & Admission Prices: Mon.-Fri. 9-12 & 1-4. No charge; donations accepted. Closed holidays. &
Attendance: 400 (estimated)

Wisner

WISNER HERITAGE MUSEUM, 920 Ave. "E", Wisner, NE 68791. Mailing Address: P.O. Box 842, Wisner, NE 68791-0842. Tel.: 402-529-3226.
Founded: 1995.
Key Personnel: Pres. (V), Gregg Moeller
Institution Type/Description: History Museum.
Collections: local history & culture relating to the city of Wisner.
Hours & Admission Prices: Sat.-Sun. 1-4; other times by appointment. No charge; donations accepted. &
Attendance: 200 (estimated)
Membership: Individual $10; Family $20; Lifetime $150.

York

ANNA BEMIS PALMER MUSEUM, (M), 211 E. 7th St., York, NE 68467-3022. Tel.: 402-363-2630. Fax: 402-362-0347.
E-mail: jeannec@cityofyork.net
Founded: 1967.
Congressional District: 3

Key Personnel: Dir., Kent Bedient.
Personnel Profile: Part-Time Paid 1; Part-Time Volunteers 12.
Governing Authority: municipal. Tax-exempt: 170(b)(1)(A).
Institution Type/Description: Local History Museum.
Collections: York County History 1870-1960; agricultural implements; covered wagon; 1880s living room & bedroom; Civil War memorabilia; military artifacts; household items; sod house replica
Facilities: histories of York County; atlases; newspaper artifacts; old bibles; school books.
Activities: self-guided tours; permanent & temporary exhibitions; special programs; annual open house.
Hours & Admission Prices: Daily 8-5, select evenings & weekends. No charge; donations accepted. Closed national holidays. &
Attendance: 1,500 (estimated)
Membership: Palmer Pals $10 & up.

WESSELS LIVING HISTORY FARM, 5520 S. Lincoln Ave., York, NE 68467. Tel.: 402-710-0682. Facebook: Wessels Living History Farm.
E-mail: wesselsfarm@gmail.com
Web Site: www.livinghistoryfarm.org
Founded: 2005.
Congressional District: 3
Key Personnel: Dir., Dale Clark; Museum Shop Mgr., Caroline Gaudreault.
Personnel Profile: Full-Time Paid 1; Part-Time Paid 3; Part-Time Volunteers 45.
Volunteer Hours: 2,500
Operating Expenses: 233,000
Operating Income: 210,000
Governing Authority: Parent Institution: York Community Foundation. Tax-exempt.
Institution Type/Description: Living History Farm.
Collections: farming & agricultural history; tractors; farming equipment; historic buildings.
Facilities: 145-acre farm.
Activities: special events; docent-led tours; facility rental; educational classes.
Publications: Membership newsletter.
Hours & Admission Prices: May-Oct. Mon.-Sat. 9:30-4:30, Sun. 1-4:30; Dec. daily 1-4:30; other times by appointment. Adults $5, senior citizens $4, children $2. &
Attendance: 8,000 (estimated)
Membership: Individual $15; Family & Grandparents $25; Gold $50.

NEVADA

(106 listings)

Amargosa Valley

ASH MEADOWS NATIONAL WILDLIFE REFUGE, HCR 70, Amargosa Valley, NV 89020. Tel.: 775-372-5435. Fax: 775-372-5436.
E-mail: sharon_mckelvey@fws.gov
Web Site: www.fws.gov/desertcomplex/ashmeadows
Institution Type/Description: Wildlife Refuge.
Collections: fish; plants; native wildlife.
Hours & Admission Prices: Sunrise to sunset. No charge.

Austin

AUSTIN HISTORICAL SOCIETY, 180 Main St., Austin, NV 89310. Mailing Address: P.O. Box 25, Austin, NV 89310-0025. Tel.: 775-964-1202. Fax: 775-964-2306.
Key Personnel: Pres., Nancy Gordon
Institution Type/Description: Historical Society Museum.
Collections: local history & culture; photographs; personal artifacts.
Hours & Admission Prices: Call for hours.

BERLIN-ICHTHYOSAUR STATE PARK, HC 61, Austin, NV 89310. Mailing Address: HC 61, Box 61200, Austin, NV 89310. Tel.: 775-964-2440. Fax: 775-964-2012.
Institution Type/Description: Park Museum.
Collections: plants; native wildlife; fossils; geological; 20th century mining camp.
Hours & Admission Prices: Daylight hours. Park: $3 per vehicle. Fossil House Tour: $2 per person.

Baker

GREAT BASIN NATIONAL PARK, 100 Great Basin Natl. Park, Baker, NV 89311-9701. Tel.: 775-234-7331. Fax: 775-234-7269.
Web Site: www.nps.gov/grba
Founded: 1986.
Congressional District: 2
Key Personnel: Supt., Andy Ferguson; Chief Resources, Tod Williams; Cultural Resource Mgr., Eva A. Jensen.
Governing Authority: federal. Parent Institution: National Park Service. Tax-exempt.
Institution Type/Description: Park Museum.
Collections: speleothems; historical objects found in cave; botanical specimens; photographs; archaeology.
Facilities: 750-vol. library of natural history & geology books available for use on premises; 30-seat restaurant; nature center; 35-seat theater. Books & museum-related items for sale.
Activities: guided tours; self-guiding nature trail; evening programs; films; exhibit room; formal interpretation programs for adults & children.
Hours & Admission Prices: Winter: daily 8:30-4:30; Summer: daily 7:30-6. Visitors Center: Cave Tours: 90-minute tours: Adults 12 & over $8, youth 5-11 $4; children 4 & under not allowed. 60-minute Tours: Adults $6, youth $3. 30-minute Tours: Adults $2; children 11 & under no charge. Closed New Year's Day; Thanksgiving; Christmas. &
Attendance: 86,457 (estimated)

Beatty

BEATTY MUSEUM AND HISTORICAL SOCIETY, 417 Main St., Beatty, NV 89003. Mailing Address: P.O. Box 244, Beatty, NV 89003-0244. Tel.: 775-553-2303.
E-mail: beattymuseum1@sbcglobal.net
Web Site: beattymuseum.org
Founded: 1995.
Key Personnel: Mgr., Amina Anderson; Pres., Vonnie Gray; Vice Pres., Corey Kirk.
Personnel Profile: Part-Time Paid 2; Part-Time Volunteers 4.
Governing Authority: private; nonprofit organization. Tax-exempt: 501(c)(3).
Institution Type/Description: Historical Society Museum.
Collections: Bullfrog Mining District & Southern Nye County history; photographs; books; personal artifacts.
Activities: meetings.
Publications: quarterly newsletter.
Hours & Admission Prices: Daily 10-3. No charge; donations accepted. Closed holidays. &
Attendance: 5,000 (estimated)
Membership: Senior $10; Individual $20; Family $30.

Blue Diamond

SPRING MOUNTAIN RANCH STATE PARK, State Rte. 159, Blue Diamond, NV 89004. Mailing Address: P.O. Box 124, Blue Diamond, NV 89004-0124. Tel.: 702-875-4141.
E-mail: smrrangers@parks.nv.gov
Web Site: www.parks.nv.gov/smr.htm
Institution Type/Description: Park Museum.
Collections: local history; period furnishings; photographs; historic ranch.
Activities: living history programs; guided tours; rental facilities.
Hours & Admission Prices: Daily 10-4.

Boulder City

BOULDER CITY/HOOVER DAM MUSEUM, (M), 1305 Arizona St., Boulder City, NV 89005-2613. Tel.: 702-294-1988.
E-mail: info@bcmha.org
Web Site: www.bcmha.org
Formerly: Boulder City Museum/Hoover Dam Museum
Founded: 1981.
Congressional District: 1
Key Personnel: Pres., Bret Runion; Museum Mgr., Laura Hutton; Vice Pres., Jim Breneda; Treas., Jennifer Scheldahl; Mgr. Operations, Roger Shoaff.
Personnel Profile: Full-Time Paid 1; Part-Time Paid 3; Part-Time Volunteers 7.
Governing Authority: nonprofit organization. Parent Institution: Boulder City Museum & Historical Association. Tax-exempt: 501(c)(3).
Institution Type/Description: History Museum.
Collections: historical & industrial items related to the building of Hoover (Boulder) Dam & Boulder City and lower Colorado River development.
Research Fields: oral research; audio & video memories of period when Hoover (Boulder) Dam was being built; photographs; manuscripts; personal papers; three-dimensional artifacts; serials; monographs.

Facilities: archives; 4,719 sq. ft. exhibit space. Books, videos & souvenirs regarding the construction of Hoover (Boulder) Dam & Boulder City for sale.
Activities: films; guided tours; lectures; temporary & loan exhibitions; research; 31ers educational outreach; rental facilities. Annual Events: Luncheon for Thirty-Oners; Fundraiser.
Publications: quarterly newsletter, BCMHA Newsletter.
Hours & Admission Prices: Mon.-Sat. 10-5. Adults $2, children & students $1; members no charge. Closed New Year's Day; Thanksgiving; Christmas. &
Attendance: 16,000 (accurate)
Membership: Student $15; Individual $20; Family $40; Organization $60; Business $80; Lifetime $400.

LAKE MEAD NATIONAL RECREATION AREA, 601 Nevada Hwy., Boulder City, NV 89005-2426. Tel.: 702-293-8859. Fax: 702-293-8936.
Web Site: www.nps.gov/lame/index.htm
Founded: 1968.
Congressional District: 3
Key Personnel: Acting Cultural Resource Mgr. & Park Archeologist, Steve Daron; Book Shop Mgr., Gabriel Zurn.
Governing Authority: federal. Parent Institution: National Park Service, Dept. of the Interior. Subsidiary Institution: Western National Parks Association. Tax-exempt.
Institution Type/Description: National Park & Visitor Center.
Collections: mammals; birds; archaeology; mining; geology; herbarium; entomology.
Research Fields: history; archaeology; bighorn sheep; fish; endangered plants, mammals, birds, amphibians & reptiles.
Facilities: botanic garden; 60-seat auditorium; visitor center with exhibits; theater. Books & maps for sale.
Activities: guided walks; lectures; films; environmental education programs; permanent & temporary exhibits.
Hours & Admission Prices: Visitor's Center: Wed.-Sun. 9-4:30. No charge. Closed New Year's Day; Thanksgiving; Christmas. &
Attendance: 200,000 (estimated)

Caliente

CALIENTE RAILROAD DEPOT & BOXCAR MUSEUM, 100 Depot, Caliente, NV 89008. Mailing Address: P.O. Box 1006, Caliente, NV 89008-1006. Tel.: 775-726-3129.
Institution Type/Description: Historic Building: built in 1923.
Collections: railroad history; railway station; photographs; period artifacts.
Hours & Admission Prices: Mon.-Fri. 10-2. No charge; donations accepted.

Carson City

BREWERY ARTS CENTER, 449 W. King St., Carson City, NV 89703-4205. Tel.: 775-883-1976. Fax: 775-883-1922.
Web Site: www.breweryarts.org
Founded: 1975.
Key Personnel: Dir., John Procaccini
Institution Type/Description: Arts Center.
Collections: fine arts; paintings.
Facilities: theater; classrooms. Museum-related items for sale.
Activities: performances; special events.
Hours & Admission Prices: Call for hours.

CHILDREN'S MUSEUM OF NORTHERN NEVADA, 813 N. Carson St., Carson City, NV 89701-4009. Tel.: 775-884-2226. Fax: 775-884-2179.
E-mail: info@cmnn.org
Web Site: www.cmnn.org/
Key Personnel: Exec. Dir., Luana Olsen.
Personnel Profile: Full-Time Paid 1; Part-Time Paid 6; Part-Time Volunteers 20.
Governing Authority: private; nonprofit organization. Tax-exempt.
Institution Type/Description: Children's Museum.
Collections: hands-on, virtual reality & interactive exhibits.
Facilities: Museum-related items for sale.
Activities: birthday parties; special events.
Hours & Admission Prices: Daily 10-4:30. Adults $6, seniors 55 & over $5, children 2-13 $4; discount to groups, AAM & ICOM members; children one & under no charge. ACM reciprocal memberships. Closed New Year's Day; Easter; Independence Day; Thanksgiving; Christmas. &
Attendance: 40,000 (accurate)
Membership: Children's Museum $70; Association of Children's Museum $130.

COMSTOCK GOLD MILL, F St., Carson City, NV 89440. Mailing Address: P.O. Box 1298, Carson City, NV 89702-1298. Tel.: 775-742-9694.
Institution Type/Description: Mining History Museum.
Collections: local mining history; stamp mills; mining equipment.
Activities: working mine tour; stagecoach rides.
Hours & Admission Prices: April-Oct. Wed.-Mon. 10-5. Adults $8, children 5-12 $4; children 4 & under no charge. Stagecoach Rides: Wed.-Mon. 11-5. Adults $10; children 4 & under no charge.

HISTORIC BOWERS MANSION, 4005 U.S. Hwy. 396 N., Carson City, NV 89704. Tel.: 702-849-0201.
Institution Type/Description: Historic House Museum.
Collections: local history & culture; Bowers family; photographs; period furnishings.
Hours & Admission Prices: mid-May to late Sept. daily 10-3. Closed Thanksgiving; Christmas.

NEVADA STATE LIBRARY AND ARCHIVES, 100 N. Stewart St., Carson City, NV 89701-4285. Tel.: 775-684-3360. Fax: 775-684-3330.
Institution Type/Description: Library and Archives.
Collections: books; pamphlets; manuscripts.
Facilities: library.
Hours & Admission Prices: Mon.-Fri. 8-5. No charge.

✳ **NEVADA STATE MUSEUM, (M),** 600 N. Carson Street, Carson City, NV 89701-4004. Tel.: 775-687-4810. Fax: 775-687-4168.
Web Site: www.museums.nevadaculture.org
Founded: 1939.
Congressional District: 2
Key Personnel: Chm., Robert Stoldal; Dir., Jim Barmore; Exhibits Mgr., Ray Geiser; Cur. Natural History, George Baumgardner; Cur. Anthropology, Gene Hattori; Collections Mgr. Anthropology, Rachel Malloy; Collections Mgr. Anthropology, Maggie Brown; Cur. History, Bob Nylen; Cur. Clothing & Textiles, Jan Loverin; Registrar, Sue Ann Monteleone; Cur. Education, Deborah Stevenson; Business Mgr., Holly Payson; Exhibit Preparator, Jeanette McGregor; Exhibit Preparator, David Shipman; Museum Shop Mgr., Leslie Phillips.
Personnel Profile: Full-Time Paid 17; Part-Time Paid 2; Part-Time Volunteers 90; Interns 3.
Volunteer Hours: 9,000
Operating Expenses: 1,564,052
Operating Income: 1,647,497
Governing Authority: state. Parent Institution: Nevada Department of Tourism and Cultural Affairs, Division of Museums & History. Tax-exempt.
Institution Type/Description: General Museum: located at the former U.S. Mint in Carson City; 1869.
Collections: history; natural history; anthropology.
Major Exhibits: Finding Fremont: Pathfinder of the West (T), 2/14-10/14.
Research Fields: Nevada natural and cultural history.
Facilities: 25,000-vol. library of historical, biological, geological & anthropological books & periodicals available on premises; classrooms; research center. Museum-related items for sale.
Activities: long term, temporary, & traveling exhibits; lectures; guided tours; activities for children; coin press demonstrations; special events; outreach educational trunks; research service; internships; Friends organization; virtual tours. Annual Event: Carson City Mint Coin Show.
Publications: e-newsletter, Mint Edition; Anthropological Papers.
Hours & Admission Prices: Tues.-Sun. 8:30-4:30. Adults $8; discounts to AAM & ICOM members; children under 18 & members no charge. Closed New Year's Day; Thanksgiving; Christmas. ♿
Attendance: 28,017 (accurate)
Membership: Senior Citizens $20; Individual $35; Family $60; Sustaining $100; Contributing $250; Patron $500; Benefactor $1,000; Corporate $2,500; Corporate Silver $5,000; Corporate Gold $10,000.

NEVADA STATE RAILROAD MUSEUM, 2180 S. Carson St., Carson City, NV 89701-5552. Tel.: 775-687-6953. Fax: 775-687-8294.
E-mail: gfackerman@state.nv.us
Web Site: www.nevadaculture.org
Founded: 1980.
Congressional District: 2
Key Personnel: Museum Dir., Frank Ackerman; Chm. (V), Robert Stoldal; Cur.

Education, Lara Mather; Cur. History, Wendell Huffman; Restoration Supvr., Chris DeWitt; Museum Store Mgr., John Walker; Facility Supvr., Brian Sheldon.
Personnel Profile: Part-Time Paid 11; Part-Time Volunteers 79.
Governing Authority: state. Parent Institution: State of Nevada, Dept. of Tourism and Cultural Affairs, Div. of Museum and History. Tax-exempt.
Institution Type/Description: Railroad Museum.
Collections: history; Virginia & Truckee Railway Collection of cars & locomotives; 70 pieces of railroad equipment.
Research Fields: Nevada history; history of Virginia & Truckee Railway and other Nevada railroads; restoration program.
Facilities: library of historical books available on premises; conference room; 13-acre site. Books, postcards, notepapers, arts & crafts for sale.
Activities: temporary exhibitions; annual history symposium; operation of railroad equipment; school outreach program; live steamups on holiday weekends; motor car rides weekends May to September.
Publications: triannual newsletters, The Sagebrush Headlight.
Hours & Admission Prices: Fri.-Mon. 9-5. Adults $6; members & children under 18 no charge. Closed New Year's Day; Thanksgiving; Christmas. ♿
Attendance: 17,529 (accurate)
Membership: Students & Senior Citizens $20; Regular $35; Family $50; Sustaining $100; Contributing $250; Corporate Engineer Roster $500; Corporate Conductor's Roster $1,000.

ROBERT'S HOUSE MUSEUM, 1207 N. Carson St., Carson City, NV 89701-1203. Mailing Address: P.O. Box 1864, Carson City, NV 89702. Tel.: 775-887-2174.
Governing Authority: Parent Institution: Carson City Historical Society.
Institution Type/Description: Historic House Museum.
Collections: local history & culture; photographs; period furnishings.
Activities: special events.
Hours & Admission Prices: Fri.-Sun. 1-3.

STEWART INDIAN SCHOOL MUSEUM, 5366 Snyder Ave., Carson City, NV 89701-6743. Tel.: 775-882-6929 & 687-8333. Fax: 775-882-1061.
Institution Type/Description: History Museum.
Collections: American Indian history; arrowheads; baskets; grinding rocks; Great Basin artifacts; pottery; photographs.
Facilities: Museum-related items for sale.
Hours & Admission Prices: Daily 9-5.

WARREN ENGINE COMPANY #1 MUSEUM, 777 S. Stewart St., Carson City, NV 89701-5218. Tel.: 775-887-2210. Fax: 775-887-2209.
Institution Type/Description: Fire-Fighting Museum.
Collections: Carson Fire Department history; fire equipment; lithographs; photographs; period artifacts.
Hours & Admission Prices: Mon.-Fri. 8-5. No charge; donations accepted.

Dayton

DAYTON HISTORICAL SOCIETY MUSEUM, 135 Shady Lane, at Logan Alley, Dayton, NV 89403. Mailing Address: P.O. Box 485, Dayton, NV 89403-0485. Tel.: 775-246-6316.
E-mail: virginiacitypat@sbcglobal.net
Web Site: daytonnvhistory.org
Founded: 1991.
Key Personnel: Museum Shop Mgr., Donna McElroy.
Personnel Profile: Part-Time Volunteers 10.
Governing Authority: Parent Institution: Historical Society of Dayton Valley. Tax-exempt.
Institution Type/Description: Historical Society Museum.
Collections: Dayton history; photographs; personal artifacts; mining & milling artifacts; railroad artifacts; sutro tunnel; Native American; gold discovery.
Activities: local school group tours. Museum Sponsors: Railroad Jamboree in April; Historic Preservation Month in May; Dayton Valley Days in September; Ghostly Pioneers of Old Town Dayton Return in October.
Hours & Admission Prices: March-April & June-Nov. Sat. 10-4, Sun. 1-4; May Mon.-Fri. 11-3, Sat. 10-4, Sun. 1-4. No charge; donations accepted. ♿
Attendance: 1,500 (estimated)
Membership: Single $12; Family $25; Lifetime $250.

Elko

NORTHEASTERN NEVADA MUSEUM, 1515 Idaho St., Elko, NV 89801-4021. Tel.: 775-738-3418. Fax: 775-778-9318.
E-mail: info@museumelko.org

Web Site: www.museumelko.org
Founded: 1968.
Congressional District: 2
Key Personnel: C.E.O., Kim Steninger; Dir., Claudia Wines; Registrar, Stephanie Youngquist; Archivist, Toni Mendive; Exhibit Coord., Catherine Wines; Maintenance Supvr., Ty Carrillo; Museum Shop Mgr. & Bookkeeper, Tracy Beatty.
Personnel Profile: Full-Time Paid 1; Part-Time Paid 12; Part-Time Volunteers 8; Interns 1.
Governing Authority: nonprofit organization. Parent Institution: Northeastern Nevada Historical Society. Tax-exempt: 501(c)(3).
Institution Type/Description: General Museum.
Collections: historical; Indian artifacts; natural history; prehistory; art; manuscript collections; photographs; newspapers. Historic Building: 1860 Ruby Valley Pony Express Station.
Research Fields: Northeastern Nevada history.
Facilities: 3,750-vol. library of Northeastern Nevada newspapers & census records on microfilm, & bound books publications, cataloged & cross-referenced photograph and negative file, books & manuscripts available for research on premises; reading room; 30,000 photos with negatives cataloged. Area subject publications, art, Indian-made items, stationery & postcards for sale.
Activities: lectures; films; arts festivals; formally organized education programs for children and adults; permanent & temporary exhibitions; audio & visual history productions.
Publications: Northeastern Nevada Historical Society Quarterly.
Hours & Admission Prices: Tues.-Sat. 9-5, Sun. 1-5. Adults $5, senior & students $3, children 3-12 $1; discount to groups, AAM & AASLH members; members, children under 2 & the last Sunday of every month no charge. Closed New Year's Day; Thanksgiving; Christmas. &
Attendance: 30,000 (estimated)
Membership: Individual $25; Family $35; Business & Sustaining $100; Pioneer Life $750; Life $1,000.

WESTERN FOLKLIFE CENTER, 501 Railroad St., Ste. 306, Elko, NV 89801-3785. Tel.: 775-738-7508. Fax: 775-738-2900.
E-mail: wfc@westernfolklife.org
Web Site: www.westernfolklife.org
Key Personnel: Exec. Dir., Charlie Seemann; Artistic Dir., Meg Glasser.
Governing Authority: nonprofit organization.
Institution Type/Description: Cultural Center.
Collections: American West culture.
Activities: special events; rental facility.
Hours & Admission Prices: Mon.-Sat. 10-5. Adults $5, students & seniors 60 & over $3, children 6-12 $1; 1st Sat. of month, members and children 5 & under no charge.

Ely

EAST ELY RAILROAD DEPOT MUSEUM, 1100 Avenue A, Ely, NV 89301-2486. Tel.: 775-289-1663. Fax: 775-289-1664.
E-mail: esm@mwpower.net
Founded: 1992.
Key Personnel: Cur., Sean Pitts; Program Asst., Keith Stone.
Personnel Profile: Full-Time Paid 2; Part-Time Volunteers 1.
Governing Authority: state.
Institution Type/Description: Railroad Depot Museum.
Collections: mining & transportation heritage; Nevada's industrial development; payroll ledgers; right of way maps.
Facilities: library.
Hours & Admission Prices: Daily 8-4:30. Adults $2; children under 18 no charge.
Attendance: 15,800 (accurate)

NEVADA NORTHERN RAILWAY MUSEUM, 1100 Avenue A, Ely, NV 89301 2486. Mailing Address: P.O. Box 150040, Ely, NV 89315-0040. Tel.: 775-289-2085. Fax: 775-289-6284.
E-mail: info@nnry.com
Web Site: www.nnry.com
Founded: 1985.
Key Personnel: Exec. Dir., Mark S. Bassett; Museum Shop Mgr., Maggie O'Brien.
Personnel Profile: Full-Time Paid 11; Part-Time Paid 2; Part-Time Volunteers 100.
Governing Authority: private; nonprofit organization. Tax-exempt: 501(c)(3).
Institution Type/Description: History & Transportation Museum.
Collections: railroad equipment; 3 steam locomotives; 8 diesel locomotives; Nevada Northern Railroad artifacts; White Pine County mining rail equipment.

Facilities: Museum-related items for sale.
Activities: guided tours; docent program; formal education programs; temporary exhibitions; steam & diesel train rides; Be The Engineer; caboose sleepover special events. Annual Events: Railroad Reality Week; Winter Photo Shoots; Haunted Ghost Train; Polar Express.
Publications: quarterly newsletter, Ghost Tracks.
Hours & Admission Prices: June-Aug. Wed.-Mon. 8-7; Sept.-May Wed.-Mon. 8-5. Adults $5; discounts to National Trust for Historic Preservation members; members no charge. Closed New Year's Eve & Day; Thanksgiving; Christmas Eve & Day. &
Attendance: 32,000 (estimated)
Membership: Active $30; Contributing $50; Centennial $100; Sustaining $250; Patron $500; Friend $1,000; Supporter $2,500; Benefactor $5,000; Leader $10,000.

WHITE PINE PUBLIC MUSEUM, 2000 Aultman St., Ely, NV 89301-1824. Tel.: 775-289-4710. Fax: 775-289-4710.
E-mail: wpmuseum@sbcglobal.net
Web Site: wpmuseum.org
Founded: 1959.
Congressional District: 2
Key Personnel: Chm. & C.E.O., Daniel L. Braddock; Dir., Nancy Davis.
Governing Authority: nonprofit. Branch Museum: The McGill Historical Drug Company, #11 Fourth St., Hwy. 93, McGill, NV. Tax-exempt.
Institution Type/Description: General Museum.
Collections: Indian artifacts; geology; mining items; dolls; antiques; historic costumes; natural history items; trains; railroad depot; 50,000 year old short-faced prehistoric cave bear; Shellengerger Cabin; Cherry Creek Depot. McGill Historical Drug Company: 1930 Drug Store with working soda fountains; 1950s-1970s store artifacts; business records, invoices & prescriptions dating back to 1915.
Research Fields: eastern Nevada history.
Facilities: 2,500 sq. ft. exhibit space. Books and other museum related items for sale.
Activities: school exhibitions.
Publications: book, Saving Our Heritage (Ethnic Heritage of Ely, Nevada).
Hours & Admission Prices: Daily 10-4. Donations Requested: adults $4, seniors $3, children 4-12 $2. Closed New Year's Day; Memorial Day; Thanksgiving; Christmas. &
Attendance: 11,000 (estimated)
Membership: Senior Citizen & Student $5; Regular $10; Business $25; Contributing $50; Sustaining $100; Life $500.

Eureka

EUREKA COUNTY COURTHOUSE, 10 S. Main St., Eureka, NV 89316. Mailing Address: 20 S. Main St., Eureka, NV 89316. Tel.: 775-237-5270.
Institution Type/Description: Historic Building: built in 1879.
Collections: local history; period furnishings; paintings; courtroom; walk-in vaults.
Hours & Admission Prices: Call for hours.

EUREKA OPERA HOUSE, 31 S. Main St., Eureka, NV 89316. Mailing Address: Convention & Cultural Arts Center, P.O. Box 284, Eureka, NV 89316-0289. Tel.: 775-237-6006. Fax: 775-237-6040.
Key Personnel: Dir., Andrea Rossman
Institution Type/Description: Historic Building: built in 1880.
Collections: theater history; period furnishings; fine arts.
Facilities: 300-seat auditorium; rental facilities.
Activities: rental facilities; permanent & traveling exhibits.
Hours & Admission Prices: Call for hours.

EUREKA SENTINEL MUSEUM, (M), 10 N. Monroe St., Eureka, NV 89316. Mailing Address: P.O. Box 82, Eureka, NV 89316-0082. Tel.: 775-237-5010. Fax: 775-237-6040.
E-mail: esm@eurekanv.org
Web Site: www.co.eureka.nv.us
Founded: 1982.
Key Personnel: Dir., Ree Taylor.
Personnel Profile: Full-Time Paid 1; Part-Time Paid 1; Part-Time Volunteers 1.
Institution Type/Description: Historic Building: housed in the 1879 Eureka Sentinel Newspaper Building.
Collections: 1860's press equipment; mining; ranching; kitchen; parlor; Fraternal organizations; barbershop.
Facilities: research library.
Hours & Admission Prices: May-Oct. daily 10-6; Nov.-April Tues.-Sat. 10-6. No charge; donations accepted.

Attendance: 5,000 (estimated)

Fallon

CHURCHILL COUNTY MUSEUM AND ARCHIVES, 1050 S. Maine St., Fallon, NV 89406-8815. Tel.: 775-423-3677. Fax: 775-423-3662.
E-mail: ccmuseum@phonewave.net
Web Site: ccmuseum.org
Founded: 1967.
Congressional District: 2
Key Personnel: Pres. (V), Kelly Helton; Dir. & Cur., Jane Pieplow.
Personnel Profile: Full-Time Paid 1; Part-Time Paid 9; Part-Time Volunteers 10.
Governing Authority: county. Parent Institution: Churchill County Museum Association. Tax-exempt: 501(c)(3).
Institution Type/Description: Local History Museum.
Collections: anthropology; archaeology; Churchill County & City of Fallon archives; museum archives; costumes; geology; glass; Indian artifacts; quilts; mineralogy; natural history; 1904 restored store; buggies & wagons; 1909 steam road roller; steam traction engine; horse-drawn transportation; fire fighting; telephone system.
Research Fields: Nevada and Churchill County history & prehistory.
Facilities: Books & museum-related items for sale.
Activities: guided tours; lectures; permanent & temporary exhibitions.
Publications: quarterly newsletter, Musenews; annual journal, Churchill County Museum Association, Churchill County In Focus.
Hours & Admission Prices: March-Nov. Mon.-Sat. 10-5; Dec.-Feb. Mon.-Sat. 10-4. Hidden Cave Tours: 2nd & 4th Sat. each month 9:30 a.m. weather permitting. No charge; donations accepted. Closed county holidays; Easter; Thanksgiving; Christmas. &
Attendance: 12,000 (estimated)
Membership: Junior $15; Senior 60 & over $20; Individual $25; Family $30; Wagonmaster $50; Pioneer & Business $75 & up; Homesteader $100 & up.

Fernley

WIGWAM NATIVE AMERICAN MUSEUM, 225 W. Main St., Fernley, NV 89408. Mailing Address: P.O. Box 97, Fernley, NV 89408. Tel.: 775-575-2573. Fax: 775-575-5280.
Institution Type/Description: History Museum.
Collections: Native American artifacts; arrowheads; stone tools; woven baskets; art; Kachina dolls.
Hours & Admission Prices: Daily.

Gardnerville

DOUGLAS COUNTY HISTORICAL SOCIETY - CARSON VALLEY MUSEUM & CULTURAL CENTER, 1477 Hwy. 395 N., Ste. B, Gardnerville, NV 89410-5214. Tel.: 775-782-2555. Fax: 775-783-8802. Facebook: Douglas Co. Historical Society.
E-mail: dchs@historicnevada.net
Web Site: historicnevada.net
Formerly: Carson Valley Historical Society
Founded: 1995.
Congressional District: 2
Key Personnel: Pres. (V), Grace Bower; Museum Shop Mgr., Kim Copel; Office Mgr., Cindy Rogers.
Personnel Profile: Part-Time Paid 2; Part-Time Volunteers 100.
Volunteer Hours: 5,600
Operating Expenses: 94,400
Operating Income: 138,000
Governing Authority: private; nonprofit organization. Parent Institution: DC Historical Society. Branch Museum: Courthouse Museum Genoa, 2304 Main St., Genoa, NV 89411. Tax-exempt: 501(c)(3).
Institution Type/Description: History Museum: housed in restored 1915 high school.
Collections: Washoe Indian history; Carson Valley history 1850 to present; early settlements; ranching; farming; Main Street businesses; women in history; Basque; wild horses.
Research Fields: agriculture; commerce; local towns; newspapers; Washoe culture; Basque culture; wild mustangs; local history.
Facilities: library of local history books; educational facilities; 12,000 sq. ft. exhibit space; rental room; classrooms; research library. Museum-related items for sale.
Activities: cultural activities; guided tours; lectures; film & video series; rental room; temporary exhibitions; formal education programs; woman's history program; cemetery tour. Annual Events: Young Chautauqua; Holiday Open House.
Publications: quarterly newsletter; booklets, Keepsake series; Town of Gardnerville: Then & Now.

Hours & Admission Prices: Mon.-Sat. 10-4. Adults $3, children over 6 $2; discounts to tour groups & AASLH members; Douglas County schools & members no charge. Closed New Year's Day; Carson Valley Day; Independence Day; Nevada Day; Thanksgiving; Christmas Eve & Day. &
Attendance: 6,000 (estimated)
Membership: Senior $20; Individual & Senior Couple $30; Family $40.

Genoa

DOUGLAS COUNTY HISTORICAL SOCIETY - COURTHOUSE MUSEUM GENOA, 2304 Main St., Genoa, NV 89411. Mailing Address: 1477 Hwy. 395 N, Suite B, Gardnerville, NV 89410-5570. Tel.: 775-782-4325. Fax: 775-783-8802. Facebook: Douglas County Historical Society Courthouse Museum Genoa.
E-mail: dchs@historicnevada.net
Web Site: historicnevada.net
Founded: 1961.
Congressional District: 2
Key Personnel: Chm. (V) & Museum Shop Mgr., Kim Copel; Pres. (V) & Dir., Grace Bower.
Personnel Profile: Part-Time Paid 2; Part-Time Volunteers 100.
Governing Authority: private; nonprofit organization. Parent Institution: Douglas County Historical Society, 1477 Hwy. 395 N, Gardnerville, NV 89410-5214. Tax-exempt: 501(c)(3).
Institution Type/Description: History Museum: housed in 19th-century courthouse.
Collections: original courthouse & jail; registry of early residents in Carson Valley; artifacts 1850-1900.
Research Fields: local history; Snowshoe Thompson; Pony Express; Indian culture.
Facilities: 6,000 sq. ft. exhibit space. Museum-related items for sale.
Activities: guided tours; temporary exhibitions.
Publications: quarterly newsletter.
Hours & Admission Prices: May -Oct. daily 10-4. Adults $3, children over 6 $2; discounts to groups; Douglas County schools & members no charge. &
Attendance: 4,200 (estimated)
Membership: Senior 65 & over $20; Individual & Senior Couple 65 & over $30; Family $40; Individual Lifetime $500.

MORMON STATION STATE HISTORIC PARK, 2295 Main St., Genoa, NV 89411. Mailing Address: P.O. Box 302, Genoa, NV 89411-0302. Tel.: 775-782-2590.
Web Site: www.parks.nv.gov/ms.htm
Founded: 1947.
Congressional District: 1
Key Personnel: Rgnl. Mgr., Robert Mergell; Park Supvr., Suzanne Sturtevant.
Personnel Profile: Full-Time Paid 1; Part-Time Paid 2; Part-Time Volunteers 4.
Governing Authority: state. Nevada Div. of State Parks, 901 S. Stewart St., Ste. 5005, Carson City, NV 89701-5248. Tax-exempt: 501(c)(3).
Institution Type/Description: Park, Museum & Visitor Center.
Collections: artifacts. Historic Building: replica of 1851 Mormon Station.
Facilities: picnic area; rental facilities.
Activities: permanent exhibits; rental facilities; interpretive programs.
Publications: park brochure; state parks brochure.
Hours & Admission Prices: Museum: May to mid-Oct. Wed.-Sun. 10-4. Admission $1; children 12 & under no charge. &
Attendance: 70,000 (estimated)

Hawthorne

MINERAL COUNTY MUSEUM, 400 Tenth St., Hawthorne, NV 89415. Mailing Address: P.O. Box 1584, Hawthorne, NV 89415-1584. Tel.: 775-945-5142.
E-mail: gm@mcmuseum.hawthorne.nv.us
Web Site: www.web0.greatbasin.net/~mcmuseum/
Governing Authority: nonprofit.
Institution Type/Description: History Museum.
Collections: local history; early 1800s Mission Bells; mining, fire & railroad equipment; horse drawn vehicles; turn of the century pharmacy; wildlife & fossils; rocks & minerals; Victorian furniture; cameras; vintage clothing, shoes & hats.
Facilities: Museum-related items for sale.
Hours & Admission Prices: April-Oct. Tues.-Sat. 11-5; Nov.-March Tues.-Sat. 12-4; other times by appointment. No charge.

Henderson

CLARK COUNTY MUSEUM, 1830 S. Boulder Hwy., Henderson, NV 89002-8502. Tel.: 702-455-7955. Fax: 702-455-7948. Facebook: Clark County Museum.
E-mail: mhp@clarkcountynv.gov
Web Site: www.clarkcountynv.gov/depts/parks/pages/clark-county-museum.aspx
Formerly: Clark County Heritage Museum
Founded: 1968.
Congressional District: 1
Key Personnel: Museum Admin., Mark Hall-Patton; Cur. Exhibits., Dawna Jolliff; Registrar, Cynthia Sanford; Museum Guild Pres. (V), Jeanne Brady; Museum Program Specialist, Malcom Vuksich.
Personnel Profile: Full-Time Paid 4; Part-Time Paid 21; Part-Time Volunteers 115.
Governing Authority: county government. Parent Institution: Clark County Parks & Recreation Dept. Subsidiary Institutions: Searchlight Historical Museum, Searchlight, NV; Howard W. Cannon Aviation Museum at McCarran International Airport. Tax-exempt: 170(B)A.1.
Institution Type/Description: History Museum.
Collections: prehistoric; southwest Indians; miners; 1880 ghost town; ranchers; gambling exhibits; cultural history; aviation history. Historic Buildings: 1912 Beckley House; 1930s P.S. Goumond House; 1931 Historic Boulder City Depot; 1890s Donald W. Reynolds printshop replica; 1940s Henderson Townsite house; 1900s Barcus-Giles house; 1935 Boulder City/Dam house; 1966 candlelight wedding chapel; 1911 railroad cottage; 1935 Grand Canyon Airlines ticket office; 1950 Esslinger barn.
Research Fields: Southwestern history; southern Nevada history & aviation.
Facilities: cactus garden; nature trail; Mojave outdoor classroom with 2 acres of botanical specimens. Museum-related items for sale.
Activities: permanent, temporary & traveling exhibits; special programs; historic preservation projects; video.
Publications: Desert Airways; A Short History of Clark County Aviation: 1920-1948; Asphalt Memories, Origins of Some of the Street Names in Clark County.
Hours & Admission Prices: Daily 9-4:30; bus tours by appointment; call for holiday tour information. Adults $2, children under 16 & senior citizens $1; members no charge. Closed New Year's Day; Thanksgiving; Christmas. &
Attendance: 43,807 (accurate)
Membership: Senior & Student $15; Individual $25; Family $35; Contributor $50; Sponsor $100-$149; Business Sponsor $150-$249; Donor $250-$249; Benefactor & Business Patron $500-$999; Silver Patron $1,000 & up.

HENDERSON BIRD VIEWING PRESERVE, 350 E. Galleria Dr., Henderson, NV 89011. Tel.: 702-267-4180.
Institution Type/Description: Bird Preserve.
Collections: 200 species of birds.
Facilities: 147 acre preserve; nature trails. Museum-related items for sale.
Activities: educational programs.
Hours & Admission Prices: Daily 6am-3pm; groups by appointment. &

Incline Village

THUNDERBIRD LODGE PRESERVATION SOCIETY, 5000 Hwy. 28, Incline Village, NV 89451. Mailing Address: P.O. Box 6812, Incline Village, NV 89450-6812. Tel.: 775-832-8750. Fax: 775-832-8798.
E-mail: askus@thunderbirdlodge.org
Web Site: www.thunderbirdlodge.org
Institution Type/Description: Preservation Society: listed on the National Register of Historic Places.
Collections: local history; period furnishings; personal artifacts. Historic Buildings: main house; card house; caretaker's cottage; cook/butler's house; elephant garage; admiral's house; boathouse; gatehouse.
Activities: guided tours.
Hours & Admission Prices: By appointment.

Las Vegas

THE AUTO COLLECTIONS AT THE IMPERIAL PALACE, 3535 Las Vegas Blvd., S., Las Vegas, NV 89109-8921. Tel.: 702-794-3174. Fax: 702-369-7430.
Institution Type/Description: Car Museum.
Collections: over 250 automobiles including period cars, classic cars, muscle cars, custom hot rods, movie cars, celebrity cars, restored cars, race cars, Indy 500 race & pace cars, & concept cars.
Facilities: Cars for sale.
Activities: buy, sell or trade vehicles.

Hours & Admission Prices: Daily 10-6.

BELLAGIO CONSERVATORY & BOTANICAL GARDENS, 3600 S. Las Vegas Blvd., Las Vegas, NV 89109-4303. Tel.: 702-693-7111.
Institution Type/Description: Botanical Gardens.
Collections: over 7,500 plants including orchids, ferns, lilies, & tropical flowers.
Facilities: 14,000 sq. ft. conservatory.
Activities: Seasonal Themes: Chinese New Year; Spring Celebration; Summer Garden Party; Harvest Show; Holiday Show.
Hours & Admission Prices: Daily 24 hours. No charge.

BELLAGIO GALLERY OF FINE ART, (M), 3600 Las Vegas Blvd. S., Las Vegas, NV 89109-4339. Tel.: 702-693-7871. Fax: 702-693-7872.
E-mail: fineartgallery@bellagio.com
Web Site: www.bellagio.com/bgfa
Founded: 1998.
Institution Type/Description: Art Museum.
Collections: works by national & international artists.
Facilities: 2,576 sq. ft. exhibit space. Museum-related items for sale.
Activities: intern program; school group tours; school & library outreach.
Hours & Admission Prices: Daily 10-8. Adults $15, NV residents & seniors 65 and over $12, students, teachers, & military $10; children under 12 no charge. &
Membership: Student, Teacher, & Senior $35; Individual $50; Dual $75; Family $100.

CENTAUR ART GALLERIES, Fashion Show Mall, 3200 S. Las Vegas Blvd., Ste. 1040, Las Vegas, NV 89109-0728. Tel.: 702-737-1234.
Institution Type/Description: Art Gallery.
Collections: works of art from the 16th century to modern times; paintings; sculpture.
Activities: permanent & temporary exhibits.
Hours & Admission Prices: Mon.-Sat. 10-9, Sun. 11-7. No charge.

CITY OF LAS VEGAS CHARLESTON HEIGHTS ART CENTER GALLERY, 800 S. Brush St., Las Vegas, NV 89107. Tel.: 702-229-1012. Fax: 702-383-1129.
E-mail: jvoltura@lasvegasnevada.gov
Web Site: www.artslasvegas.org
Institution Type/Description: Art Gallery.
Collections: works by contemporary artists.
Activities: temporary exhibitions; educational programs.
Hours & Admission Prices: Tues.-Fri. 11-9, Sat. 10-6.

DISCOVERY CHILDREN'S MUSEUM, (M), 360 Promenade Pl., Las Vegas, NV 89106-1470. Tel.: 702-382-3445. Fax: 702-382-0592. Facebook: Discovery Children's Museum.
E-mail: dtuller@discoverykidslv.org
Web Site: www.discoverykidslv.org
Formerly: Lied Discovery Children's Museum
Founded: 1984.
Congressional District: 1
Key Personnel: C.E.O., Linda Quinn; Pres., Mike Mathis; Chm., Troy Moser; Vice Chm., Annemarie Jones.
Personnel Profile: Full-Time Paid 44; Part-Time Paid 40; Part-Time Volunteers 60.
Governing Authority: nonprofit organization. Tax-exempt: 501(c)(3).
Institution Type/Description: Children's Museum.
Collections: hands-on exhibits in science, arts, culture & early childhood learning.
Major Exhibits: Mindbender Mansion (T), 1/30/14-4/27/14; Magic School Bus (T), 5/24/14-9/1/14; Storyland (T), 9/20/14-1/4/15.
Facilities: library; 26,000 sq. ft. exhibit space; classrooms.
Activities: guided tours; organized education programs focused on science, art & culture and early childhood development; volunteer program; intern program; multi-tiered climbing & exploring area. Museum Sponsors: Artist Residencies, Cultural Celebrations; Science Demonstrations; Early Childhood Pavilion & Programs; Youth Explainer Program.
Publications: quarterly newsletter; annual report; teacher materials including Educators Guide; family guides.
Hours & Admission Prices: June to Labor Day Mon.-Sat. 10-5, Sun. 12-5; Sept.-May Tues.-Fri. 9-4, Sat. 10-5, Sun. 12-5; call for school holiday hours. Admission $12; children under one no charge. &
Attendance: 350,000 (estimated)

Membership: Family (6 people) $125; Discovery Club (8 people) $175; Science Tower Club $275; Director's Club $500; President's Club $1,000; Museum Associates $2,500; Building Block Council $5,000; Kid's Corner $10,000; Stars for Children $25,000.

DONNA BEAM FINE ART GALLERY, UNLV, 4505 Maryland Pkwy., Las Vegas, NV 89154-5002. Mailing Address: 4505 Maryland Pkwy., Box 5002, Las Vegas, NV 89154-5002. Tel.: 702-895-3893. Fax: 702-895-3751.
E-mail: jerry.schefcik@unlv.edu
Web Site: donnabeamgallery.unlv.edu/
Founded: 1960.
Congressional District: 1
Key Personnel: Dir., Jerry A. Schefcik.
Personnel Profile: Full-Time Paid 1; Interns 3.
Governing Authority: university. Parent Institution: University of Nevada Las Vegas. Tax-exempt: 170(b)(1)(A).
Institution Type/Description: Art Museum & Center.
Collections: contemporary American art; Dorothy and Herbert Vogel collection: Fifty Works for Fifty States.
Research Fields: contemporary art.
Activities: gallery talks; lectures; formally organized education programs for undergraduate college students; temporary & traveling exhibitions.
Publications: exhibition catalogues; brochures.
Hours & Admission Prices: Mon.-Fri. 9-5, Sat. 10-2. No charge. Closed major holidays. &

Attendance: 8,000 (estimated)

HOWARD W. CANNON AVIATION MUSEUM, McCarran International Airport, 5757 Wayne Newton Blvd., Las Vegas, NV 89119. Mailing Address: 1830 S. Boulder Hwy., Henderson, NV 89002-8502. Tel.: 702-455-7968. Fax: 702-455-7948.
E-mail: mhp@clarkcountynv.gov
Web Site: www.clarkcountynv.gov/depts/parks/pages/cannon-aviation-museum.aspx
Formerly: McCarran Aviation Heritage Museum
Founded: 1993.
Key Personnel: Admin., Mark P. Hall-Patton; Registrar, Cynthia Sanford.
Personnel Profile: Full-Time Paid 2; Part-Time Paid 3; Part-Time Volunteers 3.
Volunteer Hours: 139
Operating Expenses: 118,241
Governing Authority: county/regional government. Parent Institution: Clark County Museum System. Tax-exempt: 170(b)(1)(A).
Institution Type/Description: Aviation Museum.
Collections: history of aviation of southern Nevada; commercial, general & military; uniforms; insignia; desk models; instruments; souvenirs; documents; aircraft, a Cessna 172 which holds the world endurance aloft record; automobile, 1956 Ford Thunderbird outfitted as a crashwagon used at the airport in the late 1950s; two aircraft engines an R-4360 & R-670; 38 five-minute history videos.
Research Fields: commercial & general aviation in Southern Nevada; world endurance flight 1958-1959.
Facilities: 3,500-vol. library of commercial & military aviation history and aviation history magazines available to the public; 3,000 sq. ft. exhibit space.
Publications: monographs, Desert Airways; Barnstorming the Desert, The Life of Randall Henderson; videos, History of Aviation in Southern Nevada; They Flew On and On: The 1958-59 World Endurance Flight & A View from the Top: A History of McCarran International Airport.
Hours & Admission Prices: Airport Exhibits: 24 hours daily. Office: Mon.-Fri. 8-5. No charge. &

Attendance: 440,000 (estimated)

LAS VEGAS ART MUSEUM, 3065 S. Johnes Blvd., Ste. 100, Las Vegas, NV 89146-6780. Tel.: 702-360-8000. Fax: 702-360-8080.
E-mail: info@lasvegasartmuseum.org
Web Site: www.lasvegasartmuseum.org
Founded: 1950.
Congressional District: 1
Key Personnel: Bd. Pres., Gerald Facciani; Exec. Dir., Libby Lumpkin; Cur., Dr. James Mann; Museum Shop Mgr., Don Jones.
Personnel Profile: Full-Time Paid 6; Full-Time Volunteers 1; Part-Time Paid 3; Part-Time Volunteers 220; Interns 1.
Governing Authority: nonprofit organization. Tax-exempt: 501(c)(3).
Institution Type/Description: Art Museum.
Collections: contemporary international art.
Activities: classes; lectures; art in schools.
Publications: quarterly newsletter; exhibition catalogues.

Hours & Admission Prices: Tues.-Sat. 10-5, Sun. 1-5. Adults $6, senior citizens $5, students $3; children 12 & under and members no charge. &
Attendance: 50,000 (estimated)
Membership: Student $25; Senior $35; Family $50; Supporting $100; Sustaining $250; Sponsor $500; Benefactor $1,000; Patron $2,500; Pacesetter $5,000; Leadership Contributor $10,000.

LAS VEGAS CLUB SPORTS HALL OF FAME, 18 E. Fremont St., Las Vegas, NV 89101-5678. Tel.: 702-385-1664.
Institution Type/Description: Sports Museum.
Collections: sports memorabilia; photographs; World Series bat; autographed baseballs; gloves; jerseys; Hall of Fame inductees.
Hours & Admission Prices: Daily. No charge.

LAS VEGAS INTERNATIONAL SCOUTING MUSEUM, 3025 W. Sahara Ave., Las Vegas, NV 89102. Tel.: 702-878-7268. Fax: 702-822-2020.
E-mail: lvismpresident@aol.com
Web Site: www.worldscoutingmuseum.org
Founded: 1996.
Key Personnel: Exec. Dir., Robert Lynn Horne, M.D.; Cur., James Arriola.
Personnel Profile: Part-Time Volunteers 5.
Governing Authority: private; nonprofit organization.
Institution Type/Description: Scouting Museum.
Collections: memorabilia.
Hours & Admission Prices: By appointment only. No charge; donations accepted.
Membership: Friend $25; Associate $50; Contributing $100; Sustaining $250; Patron $500; Benefactor $1,000; Life $2,500.

* **LAS VEGAS NATURAL HISTORY MUSEUM, (M),** 900 Las Vegas Blvd., N., Las Vegas, NV 89101-1112. Tel.: 702-384-3466. Fax: 702-384-5343.
E-mail: dino@lvnhm.org
Web Site: www.lvnhm.org
Founded: 1989.
Congressional District: 1
Key Personnel: Dir., Marilyn Gillespie; Chm. (V), John Good; Museum Shop Mgr., Laurie Thomas.
Personnel Profile: Full-Time Paid 6; Part-Time Paid 12; Part-Time Volunteers 300.
Volunteer Hours: 6,130
Operating Expenses: 706,000
Operating Income: 625,000
Governing Authority: nonprofit organization. Tax-exempt: 501(c)(3).
Institution Type/Description: Natural History Museum.
Collections: marine life including live sharks; ancient Egypt including King Tut's tomb; prehistoric dinosaurs & animals; world wildlife; Nevada wildlife habitat.
Major Exhibits: Hatching the Past, 5/14-9/14.
Facilities: 35,000 sq. ft. exhibit space; educational & group meeting space; children's activity area. Prints, books, toys & other gift items for sale.
Activities: guided tours; lectures; organized education programs for children; participatory, loan, traveling & temporary exhibitions; museum extension academy.
Publications: quarterly newsletter.
Hours & Admission Prices: Daily 9-4. Adults $10, senior citizens, students & military $8, children 3-11 $5; discounts to groups, ASTC, AAA & AAM members; children 2 & under and members no charge. Closed Thanksgiving; Christmas. &
Attendance: 91,774 (accurate)
Membership: Individual $35; Grandparent $45; Family $65; Honor Roll $100; Friends $250.

LAS VEGAS SPRINGS PRESERVE, 333 S. Valley View Blvd., Las Vegas, NV 89107-4372. Mailing Address: P.O. Box 98947, Las Vegas, NV 89193-8947. Tel.: 702-822-7700. Fax: 702-822-8700.
E-mail: anne.silva@springspreserve.org
Web Site: www.springspreserve.org
Founded: 2007.
Congressional District: 1
Key Personnel: Dir., Julie Wilcox; Mgr., Amy Febbo; Cur. Exhibits, Aaron Micallef.
Personnel Profile: Full-Time Paid 48; Part-Time Paid 16; Part-Time Volunteers 250.
Governing Authority: Parent Institution: Las Vegas Valley Water District. Tax-exempt.

Institution Type/Description: Historical & Cultural Complex: Origen Museum & Desert Living Center.
Collections: local history; historic buildings; botanical; archaeological.
Research Fields: restoration ecology; wildlife biology; Las Vegas history; archaeology.
Facilities: 10,000-vol. library; 180 acres; gardens; nature trails; 2,000-seat amphitheater; movie theater; cafe; meeting area. Gift items for sale.
Activities: school outreach programs; special events; permanent & temporary exhibits; intern program; research; tours; lectures.
Publications: monthly program guide; member newsletter.
Hours & Admission Prices: March 11-Nov. 3 daily 10-6; Nov. 4-March 10 daily 10-4. Nevada Residents: adults $9.95, children 5-17 $4.95. Non-Residents: adults $18.95, children 5-17 $10.95. American Horticulture Society reciprocal admission program. Closed Thanksgiving; Christmas. &
Attendance: 249,000 (accurate)
Membership: Individual $25; Family $60; Donor Bronze $100; Donor Silver $250; Donor Gold $500; Donor Platinum $1,000.

MADAME TUSSAUD'S WAX MUSEUM, LAS VEGAS, 3377 Las Vegas Blvd. S., Ste. 2001, Las Vegas, NV 89109-8910. Tel.: 702-862-7800. Fax: 702-862-7851.
E-mail: info@madametussaudslv.com
Web Site: www.mtvegas.com
Key Personnel: Head Mktg., Michele Popejoy
Institution Type/Description: Wax Museum.
Collections: over 100 wax figures of celebrities, politicians, athletes & legends.
Activities: parties; special events.
Hours & Admission Prices: Call for hours & admission prices.

THE MOB MUSEUM, (M), 300 Stewart Ave., Las Vegas, NV 89101. Tel.: 702-229-2734. Facebook: The Mob Museum.
E-mail: info@themobmuseum.org
Web Site: themobmuseum.org
Key Personnel: Exec. Dir. & C.E.O., Jonathan Ullman.
Governing Authority: nonprofit organization. Tax-exempt: 501(c)(3).
Institution Type/Description: Mob & Organized Crime History Museum: housed in a former federal courthouse & United States Post Office. Listed on the National Register of Historic Places.
Collections: organized crime & law enforcement history; photographs; hands-on exhibitions; films; historical artifacts.
Facilities: 16,800 sq. ft. exhibition space; classrooms. Museum-related items for sale.
Activities: special events; rental facilities; lectures; educational programs.
Hours & Admission Prices: Sun.-Thurs. 10-7, Fri.-Sat. 10-8. &
Membership: Individual $25; Dual $50; Family $85; Donor $150; Partner $500; Major Supporter $1,000.

NATIONAL ATOMIC TESTING MUSEUM, (M), 755 E. Flamingo Rd., Las Vegas, NV 89119-7363. Tel.: 702-794-5151. Fax: 702-794-5155. Facebook: National Atomic Testing Museum.
Web Site: atomictestingmuseum.org
Founded: 1998.
Key Personnel: Exec. Dir., Allan Palmer; Museum Store Mgr., Kathy Powell
Governing Authority: nonprofit organization. Parent Institution: Nevada Test Site Historical Foundation. Tax-exempt: 501(c)(3).
Institution Type/Description: Science & Technology Museum.
Collections: Nevada Test Site history; narratives; environmental re-creations; theatrical devices; iconic artifacts; national nuclear weapons testing program history.
Research Fields: nuclear weapons development; geopolitical conflict during the Cold War.
Facilities: 8,000 sq. ft. exhibit space. Museum-related items for sale.
Activities: multimedia presentations; environmental re-creations; narratives.
Hours & Admission Prices: Mon.-Sat. 10-5, Sun. 12-5. Adults $14, seniors, military & students $11; members and children 6 & under no charge. Closed New Year's Day; Thanksgiving; Christmas.
Attendance: 82,000 (accurate)

THE NEON MUSEUM, (M), 770 Las Vegas Blvd., N., Las Vegas, NV 89101-2030. Mailing Address: P.O. Box 147, Las Vegas, NV 89125-0147. Tel.: 702-387-6366. Fax: 702-477-7751.
E-mail: info@neonmuseum.org
Web Site: www.neonmuseum.org
Founded: 1996.
Key Personnel: Dir., Danielle Kelly; Museum Shop Mgr., Bill Lee
Institution Type/Description: General Museum.
Collections: neon signs from 1930 to present.

Hours & Admission Prices: Daily see website for hours. Day Tours: $12-$18. Night Tours: $22-$25; children under 6 no charge. &
Membership: La Concha $35; Sahara $75; Aladdin $250; Silver Slipper $500; Stardust $1,000; Corporate $2,500.

* **NEVADA STATE MUSEUM, LAS VEGAS, (M),** 309 S. Valley View/Springs Preserve, Las Vegas, NV 89107-2104. Tel.: 702-486-5205. Fax: 702-486-5172.
E-mail: dmillman@nevadaculture.org
Web Site: www.nevadaculture.org
Formerly: Nevada State Museum & Historical Society
Founded: 1982.
Congressional District: 1
Key Personnel: Exec. Dir., David Millman; Chm. (V), Robert Stohldal; Cur. History, Dennis McBride; Cur. Education, Stacy Irvin; Cur. Natural History, Barbara Adams; Dir. Exhibits, Thomas Dyer; Museum Shop Mgr., Harvey Foutz.
Personnel Profile: Full-Time Paid 17; Part-Time Paid 1; Part-Time Volunteers 35; Interns 3.
Governing Authority: state. Parent Institution: Nevada Department of Tourism. Tax-exempt: 501(c)(3).
Institution Type/Description: General Museum.
Collections: southern Nevada history, natural history & anthropology; paleontology; photographs; manuscripts; newspapers; Las Vegas history.
Research Fields: Nevada history; Las Vegas newspapers; Las Vegas neon signs; butterflies; historic preservation.
Facilities: 3,000-vol. library; 100-seat auditorium; 13,000 sq. ft. exhibit space; 4,000 sq. ft. special events room; education lab; meeting room.
Activities: guided tours; lectures; films; organized education programs for adults & children; docent program; participatory, loan, temporary & traveling exhibitions.
Publications: quarterly scholarly publication; exhibit catalogs.
Hours & Admission Prices: Fri.-Mon. 10-6. Adults $9.95; youth 17 & under and members no charge. Closed New Year's Day; Thanksgiving; Christmas. &
Attendance: 50,000 (accurate)
Membership: Student & Senior Citizen $15; Regular $25; Family $35; Sustaining $50; Contributing $100; Departmental Fellow $250; Benefactor $500; Patron $1,000. Business Support Program for Corporate & Commercial members.

OLD LAS VEGAS MORMON FORT STATE HISTORIC PARK, 500 E. Washington Ave., Las Vegas, NV 89101-1000. Tel.: 702-486-3511. Fax: 702-486-3734.
E-mail: oldfort@parks.nv.gov
Web Site: www.parks.nv.gov/olvmf.htm
Key Personnel: Park Supvr., Scott Egy
Institution Type/Description: History Museum.
Collections: pioneer settlers; personal artifacts; period furnishings; historic building.
Activities: educational programs; guided tours.
Hours & Admission Prices: Tues.-Sat. 8-4:30. Adults 12 & over $1.

RED ROCK CANYON VISITOR CENTER AND NATIONAL CONSERVATION AREA, 1000 Scenic Loop Dr., HCR 33, Las Vegas, NV 89124. Mailing Address: Box 5500, Las Vegas, NV 89124. Tel.: 702-515-5350.
Web Site: www.redrockcanyonlv.org
Institution Type/Description: National Park.
Collections: geology; botanical; zoological.
Facilities: hiking trails; picnic areas. Books for sale.
Activities: hiking.
Hours & Admission Prices: Daily 8-4:30. Adults $5.

SHARK REEF AQUARIUM AT MANDALAY BAY, 3950 Las Vegas Blvd. S., Las Vegas, NV 89119-1006. Tel.: 702-632-4555. Fax: 702-632-4553.
E-mail: sharkreef@mandalaybay.com
Web Site: www.sharkreef.com
Institution Type/Description: Aquarium.
Collections: sharks; sea turtles; crocodile; reptiles; cichlids; snakes; octopus; stingrays; fish.
Activities: touch pool; special events.
Hours & Admission Prices: Sun.-Thurs. 10-8, Fri.-Sat. 10-10. Adults $18, children 5-12 $12; discounts to Nevada residents; children 4 & under no charge. &

SHELBY AMERICAN, INC., 6755 Speedway Blvd., Unit 101, Las Vegas, NV 89115-1762. Tel.: 702-942-7325. Fax: 702-932-6272.
E-mail: terriy@shelbyamerican.com
Formerly: Shelby Automobiles
Institution Type/Description: Automobile Museum.
Collections: sports cars.
Facilities: cafe.
Hours & Admission Prices: Call for hours.

SIEGFRIED & ROY'S SECRET GARDEN & DOLPHIN HABITAT, Mirage Department of Animal Care, 3400 Las Vegas Blvd. S., Las Vegas, NV 89109-8923. Tel.: 702-791-7188. Fax: 702-792-7684.
E-mail: secretgarden@mirage.com
Web Site: www.miragehabitat.com
Institution Type/Description: Animal Garden.
Collections: tigers; lions; panthers; leopards; dolphins.
Activities: educational programs.
Hours & Admission Prices: Memorial Day to Labor day daily 10-7; Sept.-May Mon.-Fri. 11-5:30, Sat.-Sun. 10-5:30. Adults $15, children 4-12 $10; children 3 & under no charge.

SOUTHERN NEVADA MUSEUM OF FINE ART, 270450 Fremont St., Ste., Las Vegas, NV 89101. Tel.: 702-382-2926.
E-mail: snmoffa@gmail.com
Web Site: www.snmfa.com
Founded: 2003.
Key Personnel: Dir., Joseph Palermo; Museum Shop Mgr., Lynn Jones.
Governing Authority: nonprofit organization. Tax-exempt.
Institution Type/Description: Art Gallery.
Collections: works by local, regional & international artists.
Facilities: 17,000 sq. ft. exhibit space.
Activities: temporary & permanent exhibitions.
Hours & Admission Prices: Spring & Summer: Wed.-Sat. 12-5; Fall & Winter: call for hours. Admission $3; discounts to AAM members & groups of 5 or more; members no charge. &
Membership: Family $20.

SOUTHERN NEVADA ZOOLOGICAL-BOTANICAL PARK, 1775 N. Rancho Dr., Las Vegas, NV 89106-1020. Tel.: 702-647-4685. Fax: 702-648-5955.
Institution Type/Description: Zoo.
Collections: over 150 species of plants & animals; endangered cats; chimpanzee; eagles; ostriches; emus; parrots; wallabies; flamingos; reptiles; cycads; bamboos.
Facilities: 3 acre park.
Hours & Admission Prices: Daily 9-5. Adults $8, children 2-12 and seniors 62 & over $6; children under 2 no charge.

UNLV MARJORIE BARRICK MUSEUM, 4505 Maryland Pkwy., Las Vegas, NV 89154-9900. Mailing Address: P.O. Box 454012, Las Vegas, NV 89154-4012. Tel.: 702-895-3381. Fax: 702-895-5737.
E-mail: barrick.museum@unlv.edu
Web Site: barrickmuseum.unlv.edu
Founded: 1967.
Congressional District: 1
Key Personnel: Dir., Jerry Schefcik; Program Dir., Aurore Giguet; Collections Mgr., Alisha Kerlin.
Personnel Profile: Full-Time Paid 2; Part-Time Paid 4; Part-Time Volunteers 8.
Governing Authority: university. Parent Institution: University of Nevada System, Las Vegas & Reno, NE. Tax-exempt: 501(c)(3).
Institution Type/Description: Art Museum.
Collections: Pre-Hispanic, modern & contemporary art.
Research Fields: Pre-Hispanic, American & contemporary art.
Facilities: library; banquet facilities.
Activities: temporary & traveling exhibitions; lectures; gallery talks; film series; tours. Annual Events: Braunstein Symposium of Pre-Columbian Art.
Publications: newsletter.
Hours & Admission Prices: Mon.-Wed. & Fri. 9-5, Thurs. 9-8, Sat. 12-5. Suggested Donations: adults $5, seniors & children $2. Closed national & state holidays. &
Attendance: 80,000 (estimated)

THE WALKER AFRICAN AMERICAN MUSEUM & RESEARCH CENTER, 705 W. Van Buren Ave., Las Vegas, NV 89106-3042. Mailing Address: 501 Lass Circle, North Las Vegas, NV 89030. Tel.: 702-649-2238.
E-mail: walkeraamuseum1@yahoo.com
Web Site: www.churchesinlasvegas.com/walkermuseum
Founded: 1993.
Key Personnel: Pres. (V), Dir. & Cur., Gwendolyn Walker; Membership, Belinda Strong; Education, Margaret Crawford; Public Rels., Lillian McMorris; Treas., Juanita Walker; Security, Melvin Thompson; Museum Shop Mgr., Nika Sewell.
Personnel Profile: Full-Time Volunteers 8; Part-Time Volunteers 4.
Governing Authority: nonprofit. Tax-exempt: 501(c)(3).
Institution Type/Description: History Museum.
Collections: books; publications; documents; artifacts; articles on African Americans in Nevada from 1800s to present; dolls; audio; oral histories; African artifacts; local & national Black Americana; obituaries; church histories; Moulin Rouge.
Research Fields: early business owned by African Americans in Nevada.
Facilities: library; 800 sq. ft. exhibit space. Museum-related items for sale.
Activities: guided tours; films.
Publications: annual booklet, Black Pioneers of Nevada; book, From the Kitchen to the Boarder, Nevada's Black Women; Courage, Strength & Faith, Nevada's Black Men; God Is Alive and Well in Sin City, The History of Black Churches in Nevada; To Be Young, Gifted & Black In Nevada.
Hours & Admission Prices: By appointment. Adults $2. Closed Christmas.
Attendance: 750 (estimated)
Membership: Individual $25; Supporting $100.

Laughlin

DON LAUGHLIN'S CLASSIC CAR COLLECTION, Riverside Resort Hotel & Casino, 1650 S. Casino Dr., Laughlin, NV 89029-1512. Tel.: 800-227-3849; 702-298-2535. Fax: 702-298-2605.
E-mail: mosborn@riversideresort.com
Key Personnel: Cur., Mark Osborn
Institution Type/Description: Classic Car Collection.
Collections: over 80 automobiles, trucks & motorcycles.
Hours & Admission Prices: First Floor Show Room: daily 10-10. Third Floor Show Room: Sun.-Thurs. 9am-10pm, Fri.-Sat. 9am-11pm. No charge. &
Attendance: 300,000 (estimated)

Logandale

OLD LOGANDALE SCHOOL HISTORICAL AND CULTURAL SOCIETY, 3011 N. Moapa Valley Blvd., Logandale, NV 89021-0065. Mailing Address: P.O. Box 65, Logandale, NV 89021. Tel.: 702-398-7272 & 7273.
Institution Type/Description: Historic Building: housed in the former Logandale School; built in 1899.
Collections: local history & culture; period furnishings; personal artifacts; photographs.
Hours & Admission Prices: Mon.-Fri. 10-4.

Lovelock

PERSHING COUNTY COURTHOUSE, 400 Main St., Lovelock, NV 89419. Mailing Address: Pershing County Commissioners, Drawer E, Lovelock, NV 89419. Tel.: 702-273-7144. Fax: 702-273-7647.
Institution Type/Description: Historic Building: built in 1919. Listed on the National Register of Historic Places.
Collections: local history & culture; photographs; period furnishings; personal artifacts.
Hours & Admission Prices: Mon.-Fri. 10-4. No charge.

PERSHING COUNTY MARZEN HOUSE MUSEUM, 25 Marzen Lane, Lovelock, NV 89419. Mailing Address: P.O. Box 212, Lovelock, NV 89419. Tel.: 775-273-2115.
Institution Type/Description: Historic House Museum: built in 1874.
Collections: mining equipment; furnishings; Native American artifacts.
Hours & Admission Prices: May-Oct. daily 9-4; Nov.-April Mon.-Fri. 9-1:30. No charge; donations accepted.

McGill

HISTORIC MCGILL DRUG COMPANY, U.S. 93 E., McGill, NV 89318. Mailing Address: P.O. Box 757, McGill, NV 89318-0757. Tel.: 775-235-7082. Fax: 775-235-7802.
E-mail: bhaven1@sbcglobal.net
Web Site: www.mcgilldrugstoremuseum.org
Founded: 2000.
Congressional District: 35
Key Personnel: Dir., Daniel Braddock.
Personnel Profile: Part-Time Volunteers 1.
Governing Authority: Parent Institution: White Pine Public Museum, Inc. Tax-exempt.
Institution Type/Description: Historic Building: built in 1907.
Collections: drugstore history & memorabilia; period furnishings.
Research Fields: period business records; prescription records from 1915.
Hours & Admission Prices: By appointment. No charge; donations accepted. &
Attendance: 2,997 (accurate)

Mesquite

MESQUITE FINE ARTS CENTER & GALLERY, 15 W. Mesquite Blvd., Mesquite, NV 89027-4754. Tel.: 702-346-1338. Fax: 702-346-1339.
E-mail: vvarts@gmail.com
Web Site: mesquitefineartscenter.com
Founded: 1996.
Congressional District: 110
Key Personnel: Pres. (V), Linda Faas
Governing Authority: Tax-exempt.
Institution Type/Description: Art Gallery.
Collections: paintings; sculpture; drawings.
Facilities: Museum-related items for sale.
Activities: educational programs; special events. Annual Event: Student Art Exhibition & Show in May.
Hours & Admission Prices: Mon.-Sat. 10-4. No charge; donations accepted. Closed holidays. &
Attendance: 15,000 (estimated)

VIRGIN VALLEY HERITAGE MUSEUM, 35 W. Mesquite Blvd., Mesquite, NV 89027-4707. Tel.: 702-346-5705. Facebook: Mesquite NV Museum.
E-mail: emarler@mesquitenv.gov
Web Site: www.mesquitenv.gov/parkfacility/virginvalleymuseum
Founded: 1985.
Key Personnel: Museum Coord., Erika Kuta Marler
Institution Type/Description: Heritage Museum.
Collections: local history; pioneer artifacts from late 1800s to early 1900s; Skookum dolls; sewing machine; photographs; movie projector; furnishings; slot machine; quilts; wedding dresses.
Hours & Admission Prices: Tues.-Sat. 10-4. No charge; donations accepted. Closed most holidays. &

Minden

DANGBERG HOME RANCH, 1450 Hwy. 88, Minden, NV 89423. Mailing Address: 9 Charleston Court, Carson City, NV 89701. Tel.: 775-783-9417.
E-mail: dangbergcurator@xmission.com
Founded: 1857.
Congressional District: 1
Key Personnel: Cur., Mark Jensen.
Personnel Profile: Full-Time Paid 2; Part-Time Volunteers 25.
Governing Authority: Operated by the Friends of the Dangberg Home Ranch.
Institution Type/Description: Historic House Museum: built in 1857.
Collections: local history; period furnishings; personal artifacts; photographs; sculpture. Historic Structures: outbuildings.
Facilities: picnic area.
Activities: permanent & temporary exhibitions; guided tours by appointment. Museum Sponsors: outdoor speakers May to October.
Hours & Admission Prices: Tours: April to mid-Dec. Wed.-Sun. 10 & 2 by appointment. Adults $3; children 12 & under no charge. Special Exhibits: call for hours. Adults $3; children 12 & under no charge. Speaker Series: no charge.
Attendance: 3,500 (accurate)

Nixon

PYRAMID LAKE PAIUTE CULTURAL CENTER, 709 State St., Nixon, NV 89424. Mailing Address: P.O. Box 256, Nixon, NV 89424-0256. Tel.: 775-574-1088. Fax: 775-574-1090.
Key Personnel: Dir., Shannon Mandell
Institution Type/Description: Cultural Center.
Collections: tribal history & culture; photographs; personal artifacts.
Hours & Admission Prices: June-Sept. Tues.-Sat. 10-4:30; Oct.-May Mon.-Fri. 10-4:30.

North Las Vegas

LEFT OF CENTER ART GALLERY & STUDIO, 2207 W. Gowan, North Las Vegas, NV 89032-7961. Tel.: 702-647-7378. Fax: 702-647-7340.
Key Personnel: Dir., Vicki Richardson.
Governing Authority: nonprofit organization. Tax-exempt: 501(c)(3).
Institution Type/Description: Art Gallery.
Collections: works by national & local artists.
Activities: workshops; lectures; classes.
Hours & Admission Prices: Tues.-Fri. 12-5, Sat. 10-3.

THE PLANETARIUM - COLLEGE OF SOUTHERN NEVADA, 3200 E. Cheyenne Ave., North Las Vegas, NV 89030-4296. Mailing Address: Sort Code S1A, 3200 E. Cheyenne Ave., North Las Vegas, NV 89030-4296. Tel.: 702-651-4759 & 4138.
E-mail: planetarium@csn.edu
Web Site: www.csn.edu/planetarium
Founded: 1977.
Congressional District: 1
Key Personnel: Dir., Dr. Dale Etheridge; Museum Shop Mgr., Pamela Maher.
Personnel Profile: Full-Time Paid 3; Part-Time Paid 1.
Governing Authority: Tax-exempt.
Institution Type/Description: Planetarium.
Collections: Evans & Sutherland Digistar 5 projects images of the hemisphere onto the theater dome.
Facilities: theater. Museum-related items for sale.
Activities: educational programs.
Publications: monthly magazine, On Orbit.
Hours & Admission Prices: Call for hours. Adults $6; discounts to members. &
Attendance: 17,159 (accurate)
Membership: Basic $25.

Overton

LOST CITY MUSEUM, (M), 721 S. Moapa Blvd., Overton, NV 89040. Mailing Address: P.O. Box 807, Overton, NV 89040-0807. Tel.: 702-397-2193. Fax: 702-397-8987.
E-mail: lostcity@nevadaculture.org
Web Site: www.nevadaculture.org
Founded: 1935.
Congressional District: 4
Key Personnel: Dir., Jerrie Clarke; Admin. Asst., Janie Shakespear.
Personnel Profile: Full-Time Paid 6; Part-Time Paid 2; Part-Time Volunteers 15.
Governing Authority: state. Tax-exempt.
Institution Type/Description: Archaeology Museum.
Collections: Ancestral Puebloan artifacts excavated from Pueblo Grande de Nevada, Lost City; Paiute Indian artifacts; historic baskets; Southwestern Indian crafts; minerals; fossils; historic Mormon & mining artifacts.
Major Exhibits: Nevada 150, 10/25/13-11/14.
Research Fields: Ancestral Puebloan archaeology; historic archaeology; historic architecture; Paiute ethnology; Mormon history.
Facilities: 400-vol. library pertaining to archaeology, history, museum techniques & natural history available to the public. Books, Native American crafts, rocks, minerals & other museum-related items for sale.
Activities: guided tours; lectures; docent program; loan, temporary & traveling exhibitions; school loan service.
Publications: brochure, Nevada's Lost City Museum; booklet, Nevada's Lost City: A Treasure Trove of Mystery; book Nevada's Lost City.
Hours & Admission Prices: Daily 8:30-4:30. Adults over 18 $5; members & children under 18 no charge. Closed New Year's Day; Thanksgiving; Christmas. &
Attendance: 20,933 (accurate)
Membership: Senior Citizen $20; Individual $35; Family $50; Sustaining $100; Contributing $250; Patron $500; Benefactor $1,000.

Pahrump

PAHRUMP VALLEY MUSEUM, 401 E. Basin Ave., Pahrump, NV 89048. Mailing Address: P.O. Box 1510, Pahrump, NV 89041-1510. Tel.: 775-751-1970. Fax: 775-751-1970.
E-mail: pahrumpmuseum@att.net
Web Site: Facebook: Pahrump Valley Museum
Founded: 1992.
Key Personnel: Pres. (V), Philip Raneri; Treas. (V), Philip Huff; Clerk, Marilyn Davis.
Personnel Profile: Full-Time Paid 1; Part-Time Volunteers 6.
Governing Authority: Parent Institution: Nye County, NV. Tax-exempt.
Institution Type/Description: History Museum.
Collections: local history & culture; President Lincoln memorabilia; photographs; personal artifacts; Yucca Mountain nuclear exhibit; Walt Repository exhibit.
Hours & Admission Prices: Tues.-Sun. 10-4. No charge; donations accepted. &
Attendance: 3,963 (accurate)
Membership: Individual $10; Family $15; Service Organization $50; Business $100; Individual Lifetime $150.

Pioche

LINCOLN COUNTY HISTORICAL MUSEUM, 716 Main St., Pioche, NV 89043. Mailing Address: P.O. Box 515, Pioche, NV 89043-0515. Tel.: 775-962-5207.
Institution Type/Description: History Museum.
Collections: local history & culture; Chinese & Native American artifacts; mining tools; furniture; musical instruments; minerals.
Hours & Admission Prices: Daily 10-3. No charge; donations accepted. Closed New Year's Day; Thanksgiving; Christmas.

Reno

ANIMAL ARK, 1265 Deerlodge Rd., Reno, NV 89508. Tel.: 775-970-3111.
E-mail: info@vistagrille.com
Web Site: www.vistagrille.com
Institution Type/Description: Wildlife Sanctuary & Nature Center.
Collections: wildlife & their habitats.
Facilities: 38 acre site.
Activities: educational programs.
Hours & Admission Prices: March 30-Nov. 3 Tues.-Sun. 10-4:30; adults $9.50, seniors (62+) $8, children 3-12 $6,50, children 2 & under no charge.

FLEISCHMANN PLANETARIUM, 1664 N. Virginia St., Reno, NV 89503-0703. Mailing Address: Univ. of Nevada, Reno/272, Reno, NV 89557. Tel.: 775-784-4812. Fax: 775-784-4822.
Web Site: planetarium.unr.edu
Founded: 1963.
Congressional District: 1
Key Personnel: Management Asst., JoAnne Robb-Black; Assoc. Dir., Dan Ruby.
Personnel Profile: Full-Time Paid 3; Part-Time Paid 5; Part-Time Volunteers 3.
Governing Authority: state. Affiliated with University of Nevada. Reno, NV. Tax-exempt.
Institution Type/Description: Planetarium.
Collections: meteorites.
Research Fields: audio-visual techniques in planetarium.
Facilities: planetarium; theater; classroom.
Activities: lectures; films; hobby workshops; permanent, temporary & traveling exhibitions; telescope viewing; planetarium programs for school & public; rental facilities; children's birthday parties; toddler morning activities; rocket building classes.
Hours & Admission Prices: Daily call for hours. Exhibits no charge. Planetarium Show: adults 13-59 $7, children under 13 & seniors over 59 $5; discount to groups, AAM, ASTC, IPS & planetarium members. Closed New Year's Day; Thanksgiving; Christmas. &
Attendance: 20,000 (accurate)
Membership: Student & Senior $30; Individuals $40; Family $100; one-time Patron of the Planetarium (life) $2,500.

NATIONAL AUTOMOBILE MUSEUM (THE HARRAH COLLECTION), (M), 10 South Lake St., Reno, NV 89501-1558. Tel.: 775-333-9300. Fax: 775-333-9309.
E-mail: info@automuseum.org
Web Site: www.automuseum.org
Founded: 1989.
Congressional District: 2
Key Personnel: Pres. & Exec. Dir., Jackie L. Frady; Chm. (V), Ranson Webster; Mgr. Business & Membership, Lisa Panko; Sales & Mktg. Mgr., Becky Contos; Mgr. Sr. Support Svcs., Barbara Clark; Automotive Collections Mgr., Jay Hubbard; Retail Mgr., Barbara Bolenbaker; Events & Operations Supvr., Nic Olson.
Personnel Profile: Full-Time Paid 8; Part-Time Paid 4; Part-Time Volunteers 92; Interns 2.
Governing Authority: private; nonprofit organization. Tax-exempt: 501(c)(3).
Institution Type/Description: Automobile Museum.
Collections: horseless carriage; vintage, classic, special interest, & period vehicles.
Research Fields: automotive.
Facilities: library with 322,000 items pertaining to automotive history; audiovisual theater. Automotive-related gifts for sale.
Activities: rotating & permanent exhibitions.
Publications: magazine, Precious Metal.
Hours & Admission Prices: Mon.-Sat. 9:30-5:30, Sun. 10-4. Adults $10, senior citizens $8, students 6-18 $4; discounts to AAM & AMG members; staff, museum members & children under 5 no charge. Closed Thanksgiving; Christmas. &
Attendance: 60,000 (accurate)
Membership: Individual $45; Companion $60; Family $70; Contributing $150; Patron $500; Driving Force $1,000; Corporate $2,750-$10,000.

* **NEVADA HISTORICAL SOCIETY, (M),** 1650 N. Virginia St., Reno, NV 89503-1799. Tel.: 775-688-1190. Fax: 775-688-2917.
Web Site: www.nevadaculture.org
Founded: 1904.
Congressional District: 2
Key Personnel: Pres (V) NHS Docent Council, Susan Oddo; Acting Dir. & Manuscript Cur., Sheryln Hayes-Zorn; Registrar & Cur. Artifacts & Education, Christine K. Johnson; Cur. Photography, Lee Brumbaugh; Research Librarian, Michael Maher; Museum Shop Mgr., Judy Dandini.
Personnel Profile: Part-Time Paid 6; Part-Time Volunteers 60; Interns 4.
Governing Authority: state. A Division of the Department of Tourism and Cultural Affairs. Subsidiary Institution: Division of Museums & History. Tax-exempt.
Institution Type/Description: History Museum.
Collections: Nevada history from earliest cultures to 21st century; manuscripts; photography; maps; bound state newspapers; print ephemera.
Research Fields: history of Nevada, the Great Basin & the west.
Facilities: 20,000-vol. library of books, 3,000 manuscripts, 3,500 MSS collections, 450,000 photographs & periodicals, 55,000 maps, bound newspapers and government publications on Nevada & Western America available on premises.
Activities: guided tours; lectures; docent program or council; inter-museum loan exhibitions; historic outreach programs; field trips.
Publications: magazine, Nevada Historical Society Quarterly; occasional series & monographs.
Hours & Admission Prices: Museum: Wed.-Sat. 10-5. Library: Wed.-Sat. 12-4. Adults $4, children 17 & under and members no charge. Closed New Year's Day; Martin Luther King Jr. Day; Memorial Day; Independence Day; Labor Day; Veterans Day; Admissions Day; Thanksgiving; Christmas; state holidays. &
Attendance: 40,000 (estimated)
Membership: Senior & Student $20; Individual $35; Family $60; Sustaining $100; Contributing $250; Patron $500; Benefactor $1,000.

* **NEVADA MUSEUM OF ART, (M),** 160 W. Liberty, Reno, NV 89501-1916. Tel.: 775-329-3333. Fax: 775-329-1541. Facebook: Nevada Museum of Art.
E-mail: art@nevadaart.org
Web Site: www.nevadaart.org
Founded: 1931.
Congressional District: 2
Key Personnel: Pres., Nancy Fennell; Exec. Dir. & CEO, David B. Walker; Deputy Dir. & COO, Amy Oppio; Cur. & Deputy Dir., Ann Wolfe; Dir. Operations, Jes Stewart; Registrar, Brian Eyler; Volunteer Mgr., Margaret Walsh; Cur. Education, Colin Robertson; Dir. Communications, Rachel Milon; Museum Shop Mgr., Jackie Clay; Dir. Contemporary Art Initiatives, JoAnne Northrup.
Personnel Profile: Full-Time Paid 20; Part-Time Paid 20; Part-Time Volunteers 125.
Governing Authority: nonprofit organization. Tax-exempt.
Institution Type/Description: Art Museum.
Collections: over 1900 works of art & 20,000+ archival materials representing the land and environment; sculpture; paintings; photographs.
Major Exhibits: Maurice Sendak, 3/14-4/14; Doris Duke's Shangrila (T),

5/14-8/14; Alfredo Ramos Martinez, 5/14-8/14; Mayalin: What is Missing, 7/14-12/14; Stellar Axis! Lita Albuquerque, 8/14-1/15; Late Harvest, 9/14-1/15.
Research Fields: permanent collection; natural built & virtual environments.
Facilities: library; classrooms; cafe; theater. Museum-related items for sale.
Activities: docent programs; guided tours; lectures; education programs for children & adults; loan, temporary & traveling exhibitions; museum school art classes.
Publications: exhibition catalogues; quarterly newsletter; exhibition notes.
Hours & Admission Prices: Wed. & Fri.-Sun. 11-6, Thurs. 11-8. Adults $10, senior citizens & students $8, children $1; discount to AAM & NARM members; museum members no charge. Closed national holidays. &
Attendance: 90,000 (accurate)
Membership: Student $30; Senior Individual $35; Single $45; Family $65; Scholar $200; Collector $500; Sierra Circle $1,000.

SHEPPARD CONTEMPORARY AND UNIVERSITY GAL-LERIES, CFA 162, University of Nevada, Reno, NV 89557. Mailing Address: Department of Art, University of Nevada/0224, Reno, NV 89557-0224. Tel.: 775-784-6658. Fax: 775-784-6655.
E-mail: bakerprindle@unr.edu
Web Site: www.unr.edu/art/site/galleriesevents/sheppard_gallery.html
Formerly: Sheppard Fine Arts Gallery
Founded: 1960.
Key Personnel: Dir. University Galleries, Paul Baker Prindle; Preparator, Richard Jackson.
Personnel Profile: Full-Time Paid 1; Part-Time Paid 7; Part-Time Volunteers 3; Interns 16.
Governing Authority: university. Parent Institution: University of Nevada, Reno. Tax-exempt.
Institution Type/Description: Fine Arts Gallery.
Collections: print collection; drawings; sculpture; photographs; paintings.
Major Exhibits: A(MUSE), 11/26/13-1/26/14; Nancy Hom: Mandala, 2/1/14-3/28/14; Manipulated View, 2/26/14-4/13/14; Erika Harsh, 4/3/14-5/30/14; Possession, 4/23/14-6/27/14.
Research Fields: modern Nevada art; contemporary art.
Activities: guided tours; lectures; films; gallery talks; arts festivals; TV & radio programs; loan, temporary & traveling exhibitions.
Publications: exhibitions catalogs.
Hours & Admission Prices: Tues.-Wed. 11-5; Thurs. 11-8, Fri.-Sat. 10-8, Sun. 12-3. No charge; donations accepted. Closed state holidays. &
Attendance: 30,000 (estimated)

SIERRA ARTS GALLERY, 17 S. Virginia St., Ste. 120, Reno, NV 89501-2905. Tel.: 775-329-2787. Fax: 775-329-1328.
Web Site: www.sierra-arts.org
Founded: 1964.
Key Personnel: Exec. Dir., Jill Berryman
Institution Type/Description: Art Gallery.
Collections: works by local artists.
Activities: special events.
Hours & Admission Prices: Call for hours.

SIERRA SAFARI ZOO, 10200 N. Virginia St., Reno, NV 89506-9203. Tel.: 775-677-1101. Fax: 775-677-7874.
Web Site: www.sierrasafarizoo.org
Key Personnel: Gen. Mgr., Lori Acordagoitia
Institution Type/Description: Zoo.
Collections: over 150 animals of 40 species.
Facilities: Museum-related items for sale.
Hours & Admission Prices: April-Oct. daily 10-5. Adults $7, children 3-12 and seniors 55 & over $6; discounts to Public Broadcasting Members; children 2 & under no charge.

STREMMEL GALLERY, 1400 S. Virginia St., Reno, NV 89502-2806. Tel.: 775-786-0558. Fax: 775-786-0311.
E-mail: info@stremmelgallery.com
Key Personnel: Mgr., Sara Gray
Institution Type/Description: Art Gallery.
Collections: works by contemporary artists.
Hours & Admission Prices: Mon.-Fri. 9-5:30, Sat. 10-3. Closed holidays.

TERRY LEE WELLS NEVADA DISCOVERY MUSEUM, (M), 490 s. Center St., Reno, NV 89501. Tel.: 775-786-1000. Fax: 775-786-1114.
E-mail: info@nvdm.org
Web Site: www.nvdm.org
Founded: 2004.
Key Personnel: Exec. Dir., Klaus Grimm; Dir. Devel., Gretchen Kelley Bietz; Dir. Finance, Melinda Lyons; Dir. Educ., Sarah Gobbs-Hill; Dir. Exhibits, Will Durham; Mgr. Mktg., Patrick Turner; Mgr. Devel., Megan Merenda; Museum Educator, Meghan Schiedel; Volunteer Coord., Stephanie D'Arcy; Museum Shop Mgr., Emily Reid; Office Mgr., Katie Green.
Personnel Profile: Full-Time Paid 18; Full-Time Volunteers 1; Part-Time Paid 12; Part-Time Volunteers 50.
Institution Type/Description: Children's Museum.
Collections: hands-on exhibitions.
Hours & Admission Prices: Call for hours. Adults $8; members no charge. &
Membership: Family $85; Grandparent $100; Traveling $125; Giving $175; Premier $500.

W.M. KECK EARTH SCIENCE AND MINERAL ENGINEER-ING MUSEUM, Mail Stop 168, Mackay School of Mines, University of Nevada, Reno, NV 89557. Tel.: 775-784-4528. Fax: 775-784-1766. Facebook: W.M. Keck Earth Science and Mineral Engineering Museum.
E-mail: gbarmore@unr.edu
Web Site: mines.unr.edu/museum/
Formerly: MacKay Mines Museum
Founded: 1908.
Congressional District: 2
Key Personnel: Admin., Garrett J. Barmore.
Personnel Profile: Part-Time Paid 1; Interns 1.
Governing Authority: university. Affiliated with Mackay School of Mines, University of Nevada. Tax-exempt: 501(c)(3).
Institution Type/Description: Geology, Paleontology & Mineralogy Museum.
Collections: metallurgical; paleontological; geological; fossils; mineralogical; history of mining.
Major Exhibits: Extraterrestrial Rocks (T), 12/6/13-3/6/14; 100 Years of Impacting Nevadans, 2/14-2/15.
Activities: guided tours; permanent & temporary exhibitions.
Hours & Admission Prices: Mon.-Fri. 9-4. No charge; donations accepted. Closed university holidays. &

WILBUR D. MAY MUSEUM, 1595 N. Sierra St., Reno, NV 89503-2862.89503-1716. Tel.: 775-785-5961. Fax: 775-325-6891.
Web Site: www.washoecounty.us/parks/museum/museum.html
Founded: 1985.
Congressional District: 2
Key Personnel: Cur., Kristy Lide; Asst. Cur. & Public Rels., Samantha Szesciorka.
Governing Authority: county; nonprofit organization. Parent Institution: Washoe County Parks Dept. Subsidiary Institution: Rancho San Rafael Park. Tax-exempt.
Institution Type/Description: General Museum.
Collections: western art; big game trophies; silver & pewter; guns & swords; photographs; art & music material; ceramics & glass; ethnology & archaeology items.
Facilities: garden; 2 auditoriums; full kitchen; banquet facilities. Museum-related items for sale.
Activities: hands-on traveling exhibits; guided tours; lectures; gift fair; docent program; rental facilities. Special Event: Balloon Art Show.
Publications: brochures.
Hours & Admission Prices: Wed.-Sat. 10-4, Sun. 12-4. Adults $5, Seniors 62 & up and children 3-17 $3.50; children 2 & under no charge. &
Attendance: 55,000 (estimated)

WILDFLOWER VILLAGE, 4395 W. Fourth St., Reno, NV 89523-8830. Tel.: 775-827-5250 & 787-3769.
Institution Type/Description: Art Gallery.
Collections: art glass; photography; painting; ceramics.
Facilities: cafe.
Activities: art events; classes; retreats; rental facilities.
Hours & Admission Prices: By appointment.

Searchlight

SEARCHLIGHT HISTORIC MUSEUM & MINING PARK, 200 Michael Wendell Way, Searchlight, NV 89046. Mailing Address: P.O. Box 36, Searchlight, NV 89046-0036. Tel.: 702-297-1642.
Web Site: searchlighthistoricmuseum.org
Founded: 1989.
Congressional District: 3
Key Personnel: Founder, Jane Bunker Overy.
Personnel Profile: Part-Time Volunteers 1.
Governing Authority: county government; nonprofit. Parent Institution: Clark County Museum, Henderson, NV. Tax-exempt: 501(c)(3).

Institution Type/Description: History Museum.
Collections: photographs; early mining days of Searchlight; personal artifacts; furnishings; geological.
Research Fields: early Searchlight through photographs.
Facilities: 500 sq. ft. exhibit space.
Activities: rental facility for private group events. Biennial Event: Searchlight Town Founding Birthday Celebration in October.
Hours & Admission Prices: Mon.-Fri. 9-5, Sat. 9-1. No charge; donations accepted. Closed holidays. &
Attendance: 5,000 (estimated)
Membership: Founder $5; Promoter $15; Prospector $25; Surveyor $50; Miner $75; Grubstaker $100; Recorder $1,000 & up; Assessor $5,000 & up.

Silver Springs

FORT CHURCHILL STATE HISTORIC PARK, 1000 US Hwy. 95A N., Silver Springs, NV 89429. Tel.: 775-577-2345.
Institution Type/Description: Historic Site: former U.S. Army fort; built in 1861.
Collections: fort's history; building ruins.
Facilities: nature trails; visitor center.
Activities: camping; picnicking; hiking; educational programs; canoeing.
Hours & Admission Prices: Call for hours.

Sparks

SPARKS HERITAGE FOUNDATION & MUSEUM, 820 Victorian Ave., Sparks, NV 89431-5077. Tel.: 775-355-1144. Fax: 775-355-6788.
E-mail: info@sparksmuseum.org
Web Site: www.sparksmuseum.org
Founded: 1985.
Congressional District: 2
Key Personnel: Dir., Anthea Humphreys; Pres. (V), Larma Volk.
Personnel Profile: Full-Time Paid 1; Part-Time Volunteers 50.
Institution Type/Description: History Museum & Cultural Center.
Collections: local history; Native American artifacts; mining; prospecting; farming; ranching; railroad history; photographs; aerospace.
Facilities: cultural center; event & meeting room; 4,000 sq. ft. exhibition space. Gift items for sale.
Activities: monthly lecture series; cultural heritage performances; Young Chautauqua Program; train tours.
Publications: quarterly newsletter.
Hours & Admission Prices: Museum: Tues.-Fri. 11-4, Sat.-Sun. 1-4. Train: Sat.-Sun. 1-4. Adults $5; children under 12, school groups & members no charge. Closed holidays. &
Attendance: 4,800 (accurate)
Membership: Individual $25; Couple $30; Family $50; Premium $100; Lifetime $1,000; Couples Lifetime $1500.

Tonopah

CENTRAL NEVADA MUSEUM, 1900 Logan Field Rd., Tonopah, NV 89049. Mailing Address: P.O. Box 326, Tonopah, NV 89049-0326. Tel.: 775-482-9676. Fax: 775-482-5423. Facebook: CN Museum.
E-mail: cnmuseum@citlink.net
Web Site: www.tonopahnevada.com/CentralNevadaMuseum.html
Founded: 1981.
Congressional District: 4
Key Personnel: Cur., Eva La Rue.
Personnel Profile: Full-Time Paid 1; Full-Time Volunteers 1; Part-Time Paid 1.
Governing Authority: county. Parent Institution: Central Nevada Historical Society. Subsidiary Institution: Nye County, NV. Tax-exempt.
Institution Type/Description: History Museum.
Collections: central Nevada history from prehistoric times to present; including Nye & Esmeralda Counties.
Research Fields: Tonopah Army Air Field (WWII B24-P39 training base); Central Nevada history; self-guided tours.
Facilities: 4,000-vol. research library; 6,000 sq. ft. exhibit space. Museum-related items for sale.
Activities: films; school loan service; temporary exhibitions; special events.
Publications: annual, Central Nevada's Glorious Past.
Hours & Admission Prices: Tues.-Sat. 9-5. No charge; donations accepted. Closed state & federal holidays. &
Attendance: 3,816 (accurate)
Membership: Individual $20; Family $25; Business $30; Individual Life $250; Family Life $350; Business Life $400.

TONOPAH HISTORIC MINING PARK, 110 N. Burro Ave., Tonopah, NV 89049. Mailing Address: P.O. Box 965, Tonopah, NV 89049-0965. Tel.: 775-482-9274. Fax: 775-482-9327.
E-mail: tonopahminingpark@gmail.com
Web Site: www.tonopahhistoricminingpark.com
Founded: 1992.
Congressional District: 2
Governing Authority: Parent Institution: Town of Tonopah. Tax-exempt.
Institution Type/Description: Mining History Museum.
Collections: mining history & artifacts; mining equipment; mineral specimens; historic buildings.
Research Fields: mining; geology; early 1900 history in relation to Tonopah mining.
Facilities: visitor center. Museum-related items for sale.
Activities: underground tours; video; walking tour; walking bridge; Polaris tour.
Publications: member newsletter, Tailings.
Hours & Admission Prices: Daily 9-5. Walking tour: adults $5, children $4, seniors $3; members, veterans, active military & children 6 & under no charge.
Attendance: 5,000 (estimated)
Membership: Individual $20; Family $25; Business $50; Individual Life $250; Business/Family Life $350; Individual/Family Life Benefactor $1,000; Business Life Benefactor $1,500.

Virginia City

COMSTOCK HISTORY CENTER, 20 N. E St., Virginia City, NV 89440. Mailing Address: P.O. Box 128, Virginia City, NV 89440. Tel.: 775-847-0419. Fax: 775-847-0653.
E-mail: jeff.wood@shpo.nv.gov
Founded: 2005.
Congressional District: 1
Key Personnel: Mgr., Jeff Wood.
Personnel Profile: Part-Time Paid 1.
Governing Authority: Parent Institution: State of Nevada, Dept. of Conservation and Natural Resources.
Institution Type/Description: History Museum.
Collections: local history & culture; Virginia & Truckee Railroad steam engine, No. 18; photographs; period furnishings; rotating temporary exhibits featuring Comstock area mining, railroad & architectural subjects.
Facilities: library; conference room; restrooms.
Activities: traveling exhibitions.
Hours & Admission Prices: Thurs.-Sun. 11-4. No charge; donations accepted. &
Attendance: 4,700 (accurate)

FOURTH WARD SCHOOL MUSEUM, 537 S. C St., Virginia City, NV 89440. Mailing Address: P.O. Box 4, Virginia City, NV 89440-0004. Tel.: 775-847-0875. Fax: 775-847-1011.
E-mail: director@fourthwardschool.org
Web Site: www.fourthwardschool.org
Founded: 1986.
Key Personnel: Dir., Barbara Mackey; Pres. (V), Ron Gallagher.
Personnel Profile: Full-Time Paid 1; Part-Time Paid 6.
Governing Authority: Parent Institution: Historic Fourth Ward School Museum Foundation. Tax-exempt.
Institution Type/Description: History Museum.
Collections: Comstock history; mining; Mark Twain; personal artifacts; Sutro tunnel; Virginia City & Gold Hill Water Co.
Research Fields: Comstock history & genealogy.
Facilities: archives; research center.
Activities: workshops; rental facilities.
Hours & Admission Prices: May-Oct. daily 10-5. Adults $5, children 6-16 $3; discounts to Smithsonian Day, Blue Star Museum, AAM & National Trust members; members & children under 6 no charge. &
Attendance: 21,362 (accurate)
Membership: $30; $50; $75; $100; $250; $500; $1,000.

MACKAY MANSION, 129 S. D St., Virginia City, NV 89440. Mailing Address: P.O. Box 971, Virginia City, NV 89440-0971. Tel.: 775-847-0173.
Institution Type/Description: Historic House: built in 1859. Listed on the National Register of Historic Places.
Collections: family history; period furnishings; photographs.
Activities: rental facilities.
Hours & Admission Prices: Daily 10-6.

MARSHALL MINT MUSEUM, 96 N. C St., Virginia City, NV 89440. Mailing Address: P.O. Box 447, Virginia City, NV 89440-0447. Tel.: 775-847-0777; 800-321-6374. Fax: 775-847-9543.
Institution Type/Description: Mint Museum: housed in the Assay Office building; built in 1861.
Collections: minerals & gems; jewelry; coins.
Facilities: Museum-related items for sale.
Activities: view coins being minted and jewelry produced in the press & assembly rooms.
Hours & Admission Prices: Daily 10-5. Closed Thanksgiving; Christmas.

NEVADA STATE FIRE MUSEUM & COMSTOCK FIRE-MEN'S MUSEUM, (M), 125 S. C St., Virginia City, NV 89440. Mailing Address: P.O. Box 466, Virginia City, NV 89440-0466. Tel.: 775-847-0717. Fax: 775-847-9010.
Web Site: comstockfiremuseum.com
Founded: 1979.
Key Personnel: Chm. (V), Michael E. Nevin; Sec. & Treas., Joseph L. Curtis; Museum Shop Mgr., Eleanor Curtis.
Personnel Profile: Full-Time Volunteers 2; Part-Time Paid 3; Part-Time Volunteers 6.
Governing Authority: Parent Institution: Storey County Volunteer Fire Department. Subsidiary Institution: Virginia City District. Tax-exempt: 501(c)(3).
Institution Type/Description: Fire-Fighting Museum: housed in c.1870 structure.
Collections: uniforms; leather helmets; belts; trumpets; photographs; manuscripts; tools; 1860 & 1856 hand-pumped fire engines; Virginia City & Comstock Lode area history; magazines; brochures; 1839, 1860 & 1879 hand-drawn hose carriages; 1877 horse-drawn hose carriage; motorized apparatus; 1874 hand-drawn ladder truck; operational 1879 fire steamer-horse drawn.
Research Fields: history of firefighting in Nevada; background and genealogy of members of the Virginia City and Gold Hill fire departments.
Facilities: Firefighting-related items for sale.
Activities: guided tours; lectures; study clubs; media programs; participatory, loan & temporary exhibitions.
Publications: brochures; annual newsletter.
Hours & Admission Prices: late April & Nov.-Dec. Sat.-Sun. 10-5 weather permitting; May-Oct. daily 10-5. No charge; donations accepted. &
Attendance: 43,917 (accurate)
Membership: Student & Senior Citizen $5; General $10-$50; Corporate $200; Life $500-$1,000.

THE WAY IT WAS MUSEUM, 113 N. C St., Virginia City, NV 89440. Mailing Address: P.O. Box 158, Virginia City, NV 89440-0158. Tel.: 775-847-0766. Fax: 775-847-9613.
Institution Type/Description: History Museum.
Collections: local history & culture; Comstock mining artifacts; photographs; lithographs; maps.
Hours & Admission Prices: Daily 10-6. Adults $2.50; children 11 & under no charge. Closed Christmas.

Washoe Valley

BOWERS MANSION, Franktown Rd., Washoe Valley, NV 89704-9518. Mailing Address: Washoe County Parks, 2601 Plumas St., Reno, NV 89509. Tel.: 775-849-0201; 775-849-0684.
Web Site: www.washoecounty.us/parks/parkdetails~pkid=1
Founded: 1946.
Key Personnel: Cur., Mrs. Betty Hood; Park Ranger, Jerry Buzzard.
Personnel Profile: Full-Time Paid 1; Part-Time Paid 2.
Governing Authority: county; nonprofit. Affiliated with Washoe County Dept. of Parks & Recreation & Bowers Mansion Restoration Committee, 2601 Plumas St., Reno, NV 89510.
Institution Type/Description: Historic House: c.1864 restored & refurbished Mansion built by L.S. Bowers.
Collections: furnishings.
Activities: guided tours; formally organized education programs for children; permanent exhibitions.
Hours & Admission Prices: Tours: May 18 to Sept. 29 Sat.-Sun. 10-3 on the hour. Adults 18-61 $8, seniors 62 & up and children 6-17 $5; children 5 & under no charge.
Attendance: 7,000 (accurate)

Winnemucca

BUCKAROO HALL OF FAME, 30 W. Winnemucca Blvd., Winnemucca, NV 89445-3129. Mailing Address: 30774 Culp Lane, Burns, OR 97720. Tel.: 775-623-2225. Fax: 800-962-2638.
Key Personnel: Cur., Carl Hammond
Institution Type/Description: Art Museum.
Collections: paintings; sculpture; drawings.
Hours & Admission Prices: Daily 8-12 & 1-5. No charge; donations accepted.

HUMBOLDT MUSEUM, (M), 175 Museum Ln., Winnemucca, NV 89446. Mailing Address: P.O. Box 819, Winnemucca, NV 89446-0819. Tel.: 775-623-2912. Fax: 775-623-5640.
E-mail: museum@winnemucca.net
Founded: 1974.
Congressional District: 2
Key Personnel: Dir., Dana Toth; Pres. (V), Judy Adams.
Governing Authority: Parent Institution: North Central NV Historical Society. Tax-exempt.
Institution Type/Description: History Museum.
Collections: Native American artifacts; pioneer household items; tools; utensils; local history; period automobiles; historic buildings.
Research Fields: local history.
Facilities: displays; rental facility.
Activities: art sales; speakers; book signings; fundraising events.
Hours & Admission Prices: Mon.-Fri. 9-4, Sat. 1-4. No charge; donations accepted. &
Membership: Senior $15; Individual $20; Family $25; Institutional $35; Contributing $55.

Yerington

LYON COUNTY MUSEUM, 215 S. Main St., Yerington, NV 89447-2536. Tel.: 775-463-6576.
E-mail: info@lyoncomus.com
Founded: 1972.
Key Personnel: Chm. (V), Arnold S. Page; Museum Shop Mgr., Mary Page.
Personnel Profile: Part-Time Volunteers 15.
Governing Authority: Tax-exempt.
Institution Type/Description: History Museum.
Collections: local history; natural history; blacksmith shop; general store; schoolhouse; books.
Hours & Admission Prices: Thurs.-Sun. 1-4; other times by appointment. No charge; donations accepted. &
Attendance: 1,500 (estimated)
Membership: Individual $10; Family $25; Fellow $50; Patron $100; Life $500.

NEW HAMPSHIRE

(143 listings)

Allenstown

MUSEUM OF FAMILY CAMPING, Bear Brook State Park, 157 Deerfield Rd., Allenstown, NH 03275-2503. Mailing Address: 100 Athol Rd., Richmond, NH 03470-4200. Tel.: 603-239-4768.
Web Site: www.ucampnh.com/museum/welcome.html
Governing Authority: Tax-exempt.
Institution Type/Description: Camping History Museum.
Collections: family camping history; camp stove; 1935 campsite; 1895 sleeping bag; Native American artifacts; cooking equipment; camping gear; period campers; Hall of Fame inductees.
Hours & Admission Prices: Memorial Day to Sept. daily 10-4. No charge; donations accepted.

THE NEW HAMPSHIRE SNOWMOBILE MUSEUM, Bear Brook State Park, Rte. 28, Allenstown, NH 03275. Mailing Address: P.O. Box 10112, Concord, NH 03301-0112. Tel.: 603-648-2304.
E-mail: info@nhsnowmobilemuseum.com
Web Site: www.nhsnowmobilemuseum.com
Founded: 1985.
Key Personnel: Dir., Greg Lewis; Pres., Dan Lewis; Vice Pres., George Burdick.
Governing Authority: nonprofit organization.
Institution Type/Description: Snowmobile Museum.
Collections: history of snowmobiling; snowmobiles

Activities: special events. Annual Events: NHSMA 24th Annual Winter Event in February; 5th Annual Vintage Round-Up in March.
Hours & Admission Prices: Jan.-March Sat. 1-3; Memorial Day to Columbus Day by appointment. No charge; donations accepted.

Alton

THE HAROLD S. GILMAN MUSEUM, Main St. & Rte. 140, Alton, NH 03809. Mailing Address: 1 Monument Square, P.O. Box 637, Alton, NH 03809-0637. Tel.: 603-875-2161. Fax: 603-875-0207.
Web Site: www.alton.nh.gov/museum.asp
Institution Type/Description: History Museum.
Collections: guns; furniture; dolls; buttons; toys; family papers & photographs.
Hours & Admission Prices: Call for hours.

Amherst

THE CHAPEL MUSEUM, Corner of Middle and Church Sts., Amherst, NH 03031. Mailing Address: Historical Society of Amherst, P.O. Box 717, Amherst, NH 03031-0717. Tel.: 603-673-8029; 9831.
Web Site: www.hsanh.org
Key Personnel: Cur., Susan Fischer; Dir. Museums, Bonnie Struss.
Governing Authority: Parent Institution: Historical Society of Amherst.
Institution Type/Description: History Museum: built in 1858.
Collections: memorabilia relating to Amherst; maps; photographs; scrapbooks; clothing; furniture.
Hours & Admission Prices: May-Oct. 2nd Sat. each month 1-4. No charge; donations accepted.

THE WIGWAM MUSEUM, Corner of Middle and Cross Sts., Amherst, NH 03031. Mailing Address: Historical Society of Amherst, P.O. Box 717, Amherst, NH 03031-0717. Tel.: 603-672-9831; 8029.
Web Site: www.hsanh.org
Key Personnel: Genealogy Chair, Jackie Marshall; Dir. Museums, Bonnie Struss; Cur., Chris Marshall.
Governing Authority: Parent Institution: Historical Society of Amherst.
Institution Type/Description: History Museum: former Methodist chapel built in 1839.
Collections: artifacts relating to Amherst; Concord Coach; old jailhouse door.
Hours & Admission Prices: May-Oct. 2nd Sat. each month 1-4. No charge; donations accepted.

Ashland

ASHLAND RAILROAD STATION MUSEUM, 69 Depot St., Ashland, NH 03217. Mailing Address: P.O. Box 175, Ashland, NH 03217-0175.
Founded: 1999.
Congressional District: 2
Key Personnel: Pres. Historical Society, David Ruell; Vice Pres., Jane Sawyer; Treas., Philip Preston.
Personnel Profile: Part-Time Volunteers 7.
Governing Authority: private; nonprofit organization. Parent Institution: Ashland Historical Society, P.O. Box 175, Ashland, NH 03217. Tax-exempt: 501(c)(3).
Institution Type/Description: Transportation Museum: housed in c.1869 Victorian railroad station.
Collections: railroad related items & documents.
Activities: lectures.
Hours & Admission Prices: July-Aug. Sat. 1-4. Suggested Donation: Adults $2; members, school groups & children no charge. &
Attendance: 750 (estimated)
Membership: Individual $10; Family $15; Supporter $25; Patron $50; Benefactor $100.

PAULINE E. GLIDDEN TOY MUSEUM, 49 Main St., Ashland, NH 03217. Mailing Address: P.O. Box 14, Ashland, NH 03217-0014. Tel.: 603-968-7289.
Web Site: www.oldashlandnh.com
Founded: 1991.
Congressional District: 2
Key Personnel: Pres. Historical Society, David Ruell; Dir. Toy Museum, Shirley Splaine.
Personnel Profile: Part-Time Volunteers 29.
Governing Authority: nonprofit. Parent Institution: Ashland Historical Society. Tax-exempt: 501(c)(3).

Institution Type/Description: Early Toy Museum.
Collections: period toys; children's books; games; dolls; schoolroom; doll houses; penny toys.
Facilities: Museum-related items for sale.
Activities: Annual Events: Young Ladies Tea; Appraisal Day.
Publications: newsletter.
Hours & Admission Prices: July-Aug. Wed.-Fri. 1-4. Adults $2; children 12 & under no charge.
Attendance: 275 (accurate)
Membership: Individual $10; Family $20; Sponsor $50; Patron $100.

WHIPPLE HOUSE MUSEUM, ASHLAND HISTORICAL SOCIETY, 14 Pleasant St., Ashland, NH 03217. Mailing Address: P.O. Box 175, Ashland, NH 03217-0175. Tel.: 603-968-7716. Fax: 603-968-7716.
Web Site: www.oldashlandnh.org
Founded: 1970.
Congressional District: 2
Key Personnel: Pres., David Ruell; Vice Pres., Jane Sawyer; Treas., Philip Preston.
Personnel Profile: Part-Time Volunteers 20.
Governing Authority: nonprofit organization. Parent Institution: Ashland Historical Society. Subsidiary Institution: Pauline E. Glidden Toy Museum, Pleasant St., Ashland, NH 03217; Ashland Railroad Station Museum. Tax-exempt: 501(c)(3).
Institution Type/Description: Local History Museum: 1837 brick house, birthplace & childhood home of George Hoyt Whipple, winner of the Nobel Prize for medicine.
Collections: local historical items; furniture; household items; tools; agricultural items.
Facilities: library of material available to the public for use on premises only.
Activities: guided tours; lectures; temporary exhibitions.
Hours & Admission Prices: July-Aug. Wed. & Fri. 1-4. Suggested Donation: adults $2; children, school groups & members no charge.
Attendance: 100 (estimated)
Membership: Annual $10; Family $15; Supporter $25; Patron $50; Benefactor $100.

Auburn

MASSABESIC AUDUBON CENTER, 26 Audubon Way, Auburn, NH 03032-3109. Tel.: 603-668-2045.
Web Site: www.nhaudubon.org
Institution Type/Description: Nature Center.
Collections: wildlife & their habitats.
Facilities: nature trails.
Activities: educational programs; workshops.
Hours & Admission Prices: Call for hours.

Barrington

LITTLETON GRIST MILL, 79 Meadowbrook Dr., Barrington, NH 03825-7105. Tel.: 603-259-3205.
E-mail: shop@littletongristmill.com
Web Site: www.littletongristmill.com
Institution Type/Description: Historic Building: built in 1797.
Collections: mill history; period furnishings, tools & equipment.
Hours & Admission Prices: Wed.-Sat. 10:30-3:30. No charge.

Bedford

BEDFORD HISTORICAL SOCIETY, 24 N. Amherst Rd., Bedford, NH 03110-5404. Tel.: 603-471-6336.
Institution Type/Description: History Museum.
Collections: local history & culture; photographs; period furnishings.
Hours & Admission Prices: By appointment.

Berlin

THE MOFFETT HOUSE MUSEUM & GENEALOGY CENTER, 119 High St., Berlin, NH 03570-2062. Mailing Address: P.O. Box 52, Berlin, NH 03570-0052. Tel.: 603-752-7928. www.berlin-nhhistoricalsociety.org.
E-mail: bcchs@hotmail.com
Web Site: www.aannh.org/heritage/coos/moffett.php
Founded: 1997.
Congressional District: 2
Key Personnel: Pres. (V), Renney Morneau; Recording Sec., Jacklyn Nadeau.
Personnel Profile: Part-Time Paid 1; Part-Time Volunteers 7.

Governing Authority: Parent Institution: Berlin & Coos County Historical Society. Tax-exempt.
Institution Type/Description: Historic House: former home and office of Dr. and Mrs. Irving Moffett.
Collections: military memorabilia; photographs; Brown Company bulletins; souvenir china; historic barns; genealogy; logging tools.
Research Fields: industrial blueprints; genealogy; area history.
Publications: semi-annual newsletter, News From The House.
Hours & Admission Prices: Tues.-Sat. 12-4; other times by appointment. No charge; donations accepted. Closed major holidays.
Attendance: 800 (estimated)
Membership: Individual $10; Family $20.

NORTHERN FOREST HERITAGE PARK & BROWN HOUSE MUSEUM, 961 Main St., Berlin, NH 03570-3031. Tel.: 603-752-7202. Fax: 603-752-7222.
E-mail: heritage@ncia.net
Web Site: www.northernforestheritage.org
Key Personnel: Dir., Dick Huot; C.E.O., Joe Costello.
Personnel Profile: Full-Time Paid 1; Part-Time Volunteers 2.
Governing Authority: Parent Institution: Tri County Community Action Program, Inc. Tax-exempt.
Institution Type/Description: Park & Historic House.
Collections: logging history; river artifacts; works of art; period furnishings.
Facilities: outdoor amphitheater; 3 acre site. Museum-related items for sale.
Activities: lectures; seminars; concerts; cultural festivals; lumberjack competitions; blacksmith demonstrations; special events.
Hours & Admission Prices: Park: late May to early Oct. Tues.-Sat. 10-4. House: daily 9-4. Boat Tours: June-Oct. Tues.-Sat. 6pm. Tours: adult $15, children 5-11 $8; children under 5 no charge. &

Campton

CAMPTON HISTORICAL SOCIETY, Town House, Rte. 175, Campton, NH 03223. Mailing Address: P.O. Box 160, Campton, NH 03223-0160. Tel.: 603-726-3813.
E-mail: camptonhistorical@gmail.com
Institution Type/Description: Historical Society Museum: housed in the former Town Hall, library, and municipal court; built in 1855.
Collections: local history, culture & heritage; period furnishings; photographs, documents. Historic Building: 1903 carriage house.
Activities: special events; lectures; presentations.
Hours & Admission Prices: Thurs. 9-4. No charge; donations accepted.

Canaan

CANAAN HISTORICAL SOCIETY AND MUSEUM, Canaan St., Canaan, NH 03741. Tel.: 603-523-7364.
E-mail: fleethamdaniel@netzero.com
Key Personnel: Pres. (V), Daniel W. Fleetham, Sr.
Institution Type/Description: History Museum.
Collections: local history; photographs; period furniture; school desks; science equipment.
Hours & Admission Prices: June-Oct. Sat. 1-4. No charge.

Candia

FITTS MUSEUM, 185 High St., Candia, NH 03034. Mailing Address: c/o Town Office, 74 High St., Candia, NH 03034-2751. Tel.: 603-483-8881.
E-mail: fittsmuseum@comcast.net
Web Site: www.fittsmuseum.org
Personnel Profile: Part-Time Volunteers 5.
Governing Authority: Parent Institution: Town of Candia. Tax-exempt.
Institution Type/Description: Historic House Museum.
Collections: local history & culture; photographs; period furnishings; personal artifacts.
Activities: research.
Hours & Admission Prices: May-Oct. third Sat. of the month 1-4; other times by appointment. No charge; donations accepted.
Attendance: 150 (estimated)

Canterbury

CANTERBURY SHAKER VILLAGE, INC., 288 Shaker Rd., Canterbury, NH 03224-2728. Tel.: 603-783-9511, ext. 200. Fax: 603-783-9362.
Web Site: www.shakers.org
Founded: 1969.

Congressional District: 2
Key Personnel: Exec. Dir., Funi Burdick; Chm. (V), Deane Morrison; Archivist, Renee Fox; Museum Store Mgr., Dawn Demers.
Personnel Profile: Full-Time Paid 10; Part-Time Paid 40; Part-Time Volunteers 159; Interns 1.
Governing Authority: nonprofit organization. Tax-exempt: 501(c)(3).
Institution Type/Description: Historic Site & Village: 1792 preserved Shaker community consisting of 25 buildings on 694 acres.
Collections: artifacts, manuscripts & photographs related to Shaker history.
Facilities: visitor center; 140-seat restaurant; gardens; education center. Museum-related items for sale.
Activities: guided tours; lectures; craft demonstrations & workshops; organized education programs for children; special events; self-guided nature trails.
Publications: members newsletter; A Shaker Sister's Drawings; Heaven on Earth: The Art and Architecture of the Canterbury Shakers; Annual Workshops and Events brochure; cookbook; Historic Structure Reports; collection catalogs; annual report.
Hours & Admission Prices: Call for hours & admissions prices. &
Attendance: 30,000 (estimated)
Membership: Dual $55; Grandparent $65; Family $75; Steward $100; Shaker $250.

Center Sandwich

PATRICIA LADD CAREGA GALLERY, 69 Maple St., Center Sandwich, NH 03227. Mailing Address: P.O. Box 417, Center Sandwich, NH 03227. Tel.: 603-284-7728 & 6692.
E-mail: plcarega@yahoo.com
Web Site: www.patricialaddcarega.com
Institution Type/Description: Art Gallery.
Collections: works by contemporary artists; sculpture; paintings; photographs.
Hours & Admission Prices: Memorial Day to mid-Oct. Mon.-Sat. 10-5, Sun. 12-5.

SANDWICH HISTORICAL SOCIETY, 4 Maple St., Center Sandwich, NH 03227. Mailing Address: P.O. Box 244, Center Sandwich, NH 03227-0244. Tel.: 603-284-6269.
E-mail: sandwichhistory@gmail.com
Web Site: www.sandwichhistorical.org
Founded: 1917.
Congressional District: 1
Key Personnel: Pres. (V), Tom Shevenell.
Personnel Profile: Part-Time Paid 4; Part-Time Volunteers 50.
Governing Authority: society; bd. of directors. Tax-exempt.
Institution Type/Description: Local History Museum: housed in c.1850 Elisha Marston House.
Collections: vehicles; housewares; furniture; art; crafts; farm implements & tools; trade & textile tools; archives; spinning & weaving tools.
Research Fields: town, church, family & commercial records.
Facilities: local history library; genealogical research.
Activities: permanent & temporary exhibitions; lectures; tours; member receptions. Society Sponsors: annual picnic.
Publications: bulletin, Annual Excursion of the Sandwich Historical Society; Sandwich New Hampshire 1763-1900; Seven Wonders of Sandwich; Schoolhouses of Sandwich; newsletters.
Hours & Admission Prices: June-Oct. Tues.-Sat. 10-4; other times by appointment. No charge; donations accepted.
Attendance: 2,000 (estimated)
Membership: Single $20; Family & Institutional $35; Sustaining $75; Patron $125; Contributing $250; Life $500.

Charlestown

CHARLESTOWN HISTORICAL SOCIETY, Town Hall, Archives Rm., 19 Summer St., Charlestown, NH 03603. Mailing Address: P.O. Box 159, Charlestown, NH 03603. Tel.: 603-826-9943.
Key Personnel: Pres., Judi Baraly; Vice Pres., Marge Reed; Sec. & Treas., Judy Petre; Membership, Marie Hartmann; Vice Pres., Cur. & Archivist, Marge Reed.
Governing Authority: private; nonprofit organization. Tax-exempt.
Institution Type/Description: Historical Society Museum.
Collections: local history; photographs; newspapers; books; genealogy. Historic Building: 1774 one-room school house.
Major Exhibits: Our Archives: Town Treasures, 4/20/14.
Facilities: 50-vol. library.
Activities: school group tours.
Publications: The Judge's Daughter.
Hours & Admission Prices: Memorial Day weekend & Columbus Day weekend; other times by appointment. Archives: Tues. 9am to noon. No charge; donations accepted. &

Attendance: 450 (estimated)
Membership: Individual $5; Family $12; Contributing $25; Sustaining $100; Lifetime $500.

THE FORT AT NO. 4 LIVING HISTORY MUSEUM, 267

Springfield Rd., Rte. 11, Charlestown, NH 03603. Mailing Address: P.O. Box 1336, Charlestown, NH 03603-1336. Tel.: 603-826-5700.
E-mail: info@fortat4.com
Web Site: www.fortat4.org
Formerly: Old Fort Number 4 Associates
Founded: 1947.
Congressional District: 2
Key Personnel: C.E.O., Wendalyn Baker; Chm. (V), Paul Truax; Pres. (V), Wells Chandler; Vice Chm., Nicholas Westbrook.
Personnel Profile: Part-Time Volunteers 65.
Governing Authority: nonprofit organization. Tax-exempt: 501(c)(3).
Institution Type/Description: Historic Site Museum: granted in 1735 by Crown Province of Massachusetts, completed as a fortified village in 1744.
Collections: Native American artifacts; colonial furnishings & tools; weapons; canoes; 14 reconstructed buildings; two original barns; French-Indian War.
Research Fields: New England life & history, 1743-1760; King George's War; frontier fortification; families of Fort at No. 4 during 1740s-1750s.
Facilities: Eighteenth century reproductions of period pieces, books & handcrafts for sale.
Activities: lectures; films; demonstrations; recreation of daily life at the fortified village of No. 4; formally organized educational programs; online educational resource.
Publications: newsletter, The Advocate.
Hours & Admission Prices: Tours: April by appointment. Museum: May-Oct. daily 10-4:30. Adults $10, senior citizens 55 & over and children 6-12 $7, children 6-12 $5; discounts to AAM & AAA members; members & children under 6 no charge. &
Attendance: 14,000 (estimated)

Claremont

THE CLAREMONT, NEW HAMPSHIRE HISTORICAL SOCI-

ETY, INC., 26 Mulberry St., Claremont, NH 03743-2538. Tel.: 603-543-1400.
E-mail: claremont_historical@yahoo.com
Web Site: www.claremonthistoricalsociety.org
Founded: 1966.
Key Personnel: Pres. & Chm. Bd. Trustees, Colin J. Sanborn.
Governing Authority: society. Tax-exempt.
Institution Type/Description: Local History Museum.
Collections: paintings, furnishings, tools, books & manuscripts pertaining to the early history of Claremont.
Research Fields: genealogy.
Facilities: 950-vol. historical library available for use on premises by appointment.
Activities: guided tours; lectures; formally organized education programs for children; permanent & temporary exhibitions.
Publications: The Historical Papers of George Baxter Upham.
Hours & Admission Prices: mid-June to mid-Sept. Sat. 1-4; call for additional hours. Admission $1.
Membership: Student $5; Individual $10; Family $15; Business & Professional $50; Corporate $200; Life $300.

Colebrook

COLEBROOK AREA HISTORICAL MUSEUM, 17 Bridge St.,

2nd Fl., Colebrook, NH 03576. Mailing Address: P.O. Box 32, Colebrook, NH 03576-0032. Tel.: 603-237-4470.
E-mail: agoodrum@myfairpoint.net
Web Site: colebrookareahistoricalsociety.weebly.com
Key Personnel: Pres., Arnold Goodrum.
Personnel Profile: Part-Time Volunteers 5.
Institution Type/Description: History Museum.
Collections: 15,000 artifacts including fine arts; folk arts; costumes & textiles; household items; maps; photographs; manuscripts; court records; local history items.
Hours & Admission Prices: Town Hall Museum: July-Aug. Sat. 10-1, other times by appointment; Tillotson Center Annex: open during center events, other times by appointment. &
Membership: Individual $5; Family $10.

Concord

ART CENTER IN HARGATE, ST. PAUL'S SCHOOL, 325

Pleasant St., Concord, NH 03301-2552. Tel.: 603-229-4643. Fax: 603-229-5696.
E-mail: ccallahan@sps.edu
Web Site: www.sps.edu
Founded: 1967.
Congressional District: 2
Key Personnel: Dir., Colin J. Callahan.
Personnel Profile: Full-Time Paid 1; Part-Time Paid 1.
Governing Authority: nonprofit. Tax-exempt: 501(c)(3).
Institution Type/Description: Art Gallery.
Collections: works by well-known & up-and-coming artists; art donated by the school's alumni including original pieces by Alexander Calder, Henry Moore, Daniel Chester French, Rembrandt Peale, George Inness, Hans Hofmann, George Braque, Frederic Remington,& Augustus St. Gaudens.
Facilities: 2,500-vol. library; 97-seat auditorium; classrooms, labs & studios; 1,600 sq. ft. exhibit space.
Activities: guided tours; lectures; loan, temporary & traveling exhibitions.
Hours & Admission Prices: Sept.-May Tues.-Sat. 9:30-4. No charge. Closed spring break; Thanksgiving break; Christmas vacation. &
Attendance: 3,700 (estimated)

AUDUBON SOCIETY OF NEW HAMPSHIRE, 84 Silk Farm Rd.,

Concord, NH 03301-8311. Tel.: 603-224-9909. Fax: 603-226-0902.
E-mail: nha@nhaudubon.org
Web Site: www.nhaudubon.org
Founded: 1914.
Congressional District: 2
Key Personnel: C.E.O. & Pres., Michael Bartlett; Chm. (V), Paul Nickerson; Dir. Membership & Devel., Eric Berger; Museum Shop Mgr., Nancy Boisvert.
Personnel Profile: Full-Time Paid 44; Part-Time Paid 26; Part-Time Volunteers 1,500.
Governing Authority: nonprofit organization. Subsidiary Institutions: Loon Preservation committee; Amoskeag Fishways Learning Center; Massabesic Audubon Center; McLane Center; Newfound Audubon Center. Tax-exempt: 501(c)(3).
Institution Type/Description: Nature Center & Conservation Area.
Collections: stuffed birds; live wild animals.
Research Fields: endangered species & wildlife habitat.
Facilities: 2,500-vol. library on natural history available to the public; meeting room auditorium; educational facilities; field research station; nature conservation center. Museum-related items for sale.
Activities: guided tours; lectures; films; concerts; temporary exhibitions; hobby workshops; rental gallery; organized education programs; teacher workshops. Annual Events: holiday fair; library book sale; Loon Festival; backyard winter bird survey; Birdathon.
Publications: books; booklets; leaflets; quarterly newsletter, Afield; quarterly newsletter, New Hampshire Bird Records.
Hours & Admission Prices: Mon.-Fri. 9-5. No charge; donations accepted. Closed legal holidays. &
Attendance: 100,000 (estimated)
Membership: Senior & Introductory $24; Individual $39; Family $55; Contributor $100; Supporter $250; Environmentalist $500; Conservation Partner $1,000.

KIMBALL JENKINS ESTATE, 266 N. Main St., Concord, NH

03301-5053. Tel.: 603-225-3932.
Web Site: www.kimballjenkins.com
Institution Type/Description: Art Gallery.
Collections: paintings; sculpture; photographs.
Hours & Admission Prices: Mon.-Thurs. 10-4; other times by appointment.

MARY BAKER EDDY HISTORIC HOME, 62 N. State St.,

Concord, NH 03301-4330. Tel.: 603-225-3444.
Institution Type/Description: Historic House: built c.1850.
Collections: local history & culture; personal artifacts; period furnishings; photographs.
Hours & Admission Prices: May-Oct. Thurs. & Sat. 11-2; other times by appointment.

MCAULIFFE-SHEPARD DISCOVERY CENTER, 2 Institute Dr.,

Concord, NH 03301-7422. Tel.: 603-271-7827. Fax: 603-271-7832.
Web Site: www.starhop.com
Formerly: Christa McAuliffe Planetarium

Founded: 1990.
Congressional District: 2
Key Personnel: Exec. Dir., Jeanne T. Gerulskis; Chm. (V), Paul A. Burkett; Producer, Sandt Michener; Dir. Special Events & Visitor Svcs., Gina Bowler; Visitor Svcs. Asst., Cheryl Stinson; Dir. Education, David McDonald, M.Ed.; Astronomer, Kathryn Michener; Astronomy Educator, Tiffany Picard; Coord. Mktg., Jennifer Jones; Devel. Asst., Michelle Carignan; Maintenance Engineer, Mike LaRochelle; Science Store Mgr., Susan Jacques; Financial Mgr., Martha Laurie.
Personnel Profile: Full-Time Paid 10; Part-Time Paid 20; Part-Time Volunteers 43.
Governing Authority: state; Discovery Center Commission. Parent Institution: State of New Hampshire. Subsidiary Institution: New Hampshire NASA Educator Resource Center. Tax-exempt: 170(d)(1)(A).
Institution Type/Description: Science Center.
Collections: U.S. Space Program memorabilia; Christa McAuliffe memorabilia; interactive exhibits; indoor & outdoor astronomy & space science; Alan Shepard memorabilia.
Research Fields: astronomy; space science; aviation; earth science.
Facilities: planetarium; NASA Educator Resource Center; 45,000 sq. ft. exhibit space; garden. Museum-related items for sale.
Activities: docent program; formal education programs; guided tours; lectures; participatory & temporary exhibitions; indoor & outdoor exhibits on astronomy & space science; planetarium shows; monthly skywatch; lecture series; monthly Teen Night; teacher workshops & training; birthday parties; rental facilities. Annual Events: Aerospace Festival in spring; fall gala.
Publications: e-newsletter.
Hours & Admission Prices: Thurs. & Sat.-Sun. 10-5, Fri. 6:30 pm-9 pm. Adults $9, seniors & students $8, children 3-12 $6; ASTC members and children 2 & under no charge. Planetarium: additional fee charged. Closed New Year's Day; Easter; Memorial Day; Independence Day; Labor Day; Thanksgiving; Christmas. &
Attendance: 69,636 (accurate)
Membership: Student $25; Individual $40; 2 Person $60; Family $80; Family & Friends $150; Gold $250. Plus level memberships available (include unlimited admission to planetarium).

MILL BROOK GALLERY AND SCULPTURE GARDEN, 236 Hopkinton Rd., Concord, NH 03301-7911. Tel.: 603-226-2046. Fax: 603-225-9630.
E-mail: artsculpt@mindspring.com
Web Site: www.themillbrookgallery.com
Founded: 1996.
Congressional District: 1
Key Personnel: Dir., Pamela R. Tarbell.
Personnel Profile: Full-Time Volunteers 1.
Institution Type/Description: Art Gallery.
Collections: works by regional & national artists; paintings; sculpture; pottery.
Major Exhibits: 17th Annual Outdoor Sculpture Exhibit, 6/25-10/30/14.
Activities: temporary exhibitions. Museum Sponsors: Juried Outdoor Sculpture Shows.
Hours & Admission Prices: April-Dec. Tues.-Sun. 11-5. No charge; donations accepted.
Attendance: 2,000 (estimated)

* **NEW HAMPSHIRE HISTORICAL SOCIETY,** 30 Park St., Concord, NH 03301-4956. Mailing Address: 30 Park St., Concord, NH 03301-6384. Tel.: 603-228-6688. Fax: 603-228-6308.
E-mail: jdesmarais@nhhistory.org
Web Site: www.nhhistory.org
Founded: 1823.
Congressional District: 2
Key Personnel: C.E.O. & Exec. Dir., William H. Dunlap; Pres. (V), Glenn K. Currie; Asst. Exec. Dir., Joan Desmarais; Dir. Collections & Exhibitions, Wesley Balla; Dir. Publications, Donna-Belle Garvin; Dir. Devel., Anne M. Hamilton; Registrar, Douglas Copeley; Dir. Library, Sarah Hays.
Personnel Profile: Full-Time Paid 13; Part-Time Paid 28; Part-Time Volunteers 80.
Governing Authority: society. Parent Institution: New Hampshire Historical Society. Tax-exempt: 501(c)(3).
Institution Type/Description: History Museum.
Collections: decorative & fine arts; New Hampshire artifacts; furniture; silver; paintings; manuscripts; ceramics; costumes & textiles; photos; books; manuscripts; ephemera; newspapers; maps.
Research Fields: history; fine & decorative arts of New Hampshire.
Facilities: 1911 library; 125-seat auditorium. Gift items for sale.
Activities: lectures; workshops; school programs; concerts; permanent, temporary & traveling exhibitions.
Publications: semi-annual, Historical New Hampshire; quarterly newsletter;

New Hampshire Architecture, An Illustrated Guide; Plain and Elegant, Rich & Common: Documented N. H. Furniture; New Hampshire Scenery: A Dictionary of 19th-Century Artists of New Hampshire Mountain Landscapes; Instruments of Change: New Hampshire Hand Tools & Their Makers; On the Road North of Boston: New Hampshire Taverns and Turnpikes, 1700-1900; Abbot-Downing & the Concord Coach; At What Cost? Shaping the Land We Call New Hampshire; Capital Views, A Photographic History of Concord, N.H. 1850-1930; Beauty Caught and Kept: Benjamin Champney in the White Mountains; Soldiers, Sailors. Slaves and Ships: The Civil War Photographs of Henry P. Moore; The Years of the Life of Samuel Lane: A New Hampshire Man and His World; Consuming Views: Art and Tourism in The White Mountains, 1850-1900.
Hours & Admission Prices: Call for hours & admission prices. &
Attendance: 23,189 (accurate)
Membership: Individual, Library & Non-profit Organization $40; Couple $50; Family & Library with museum pass $60; Business $250-$5,000; Life $800; Couple Life $1,200.

THE PIERCE MANSE, 14 Horseshoe Pond Lane, Concord, NH 03301-5028. Mailing Address: P.O. Box 425, Concord, NH 03302-0425. Tel.: 603-225-4555. Fax: 603-225-0540.
E-mail: piercebrigade@gmail.com
Founded: 1966.
Congressional District: 1
Key Personnel: Pres. (V), Pierce Brigade.
Personnel Profile: Part-Time Volunteers 35.
Governing Authority: nonprofit organization. Operated by The Pierce Brigade Inc., Concord, NH 03301. Tax-exempt: 501(c)(3).
Institution Type/Description: Historic House Museum: 1842-1848 home of President Franklin Pierce.
Collections: furniture & memorabilia of President Pierce & his family.
Research Fields: Franklin Pierce.
Facilities: 50-seat auditorium. Souvenir items for sale.
Activities: guided tours; public programs March to Nov.
Publications: monthly members' newsletter.
Hours & Admission Prices: Pierce Manse: June 17-Sept. 1 Tues.-Sat. 11-3; other times by appointment; groups of 10 or more by appointment. Family $15, adults $7, senior citizens $6, children & students $3; members no charge. Closed holidays.
Attendance: 1,100 (accurate)
Membership: Student $10; Individual $20; Family $35.

Contoocook

THE LITTLE NATURE MUSEUM, 656 Gould Hill Rd., Contoocook, NH 03229-2809. Mailing Address: 216 Tucker Dr., Hopkinton, NH 03229-2426. Tel.: 603-746-6121.
E-mail: info@littlenaturemuseum.org
Web Site: www.littlenaturemuseum.org
Founded: 1955.
Congressional District: 2
Key Personnel: Dir., C.E.O. & Pres. (V), Sandra W. Martin; Chm. (V), Paul Basham.
Personnel Profile: Part-Time Volunteers 23.
Governing Authority: Tax-exempt.
Institution Type/Description: Natural History Museum.
Collections: fossils; rocks; minerals; mounted birds & mammals; seashells & corals; insects; fungi & galls; lichens; cones; nests.
Activities: Museum Sponsors: Nature Fest in September.
Publications: quarterly newsletter; annual report.
Hours & Admission Prices: June 23-July 22 Sat.-Sun. 1-4; July 28-Sept. 9 Sat.-Sun. 11-4; Sept. 15-Nov. 4 Sat.-Sun. 10:30-4:30; groups by appointment. Tours: $2; discounts to groups. &
Attendance: 2,666 (estimated)
Membership: Individual $10; Family $25; Friend $30; Supporter $50; Sponsor $100; Benefactor $250; Naturalist $500; Patron $1,000.

Conway

EASTMAN LORD HOUSE, 100 Main St., Conway, NH 03818. Mailing Address: Conway Historical Society, P.O. Box 1949, Conway, NH 03818-1949. Tel.: 603-447-5551. Fax: 603-447-1991.
E-mail: conwayhistory@myfairpoint.net
Web Site: www.conwayhistoricalsociety.org
Key Personnel: Pres. (V), Ken Rancourt; Acting Cur., Jim Arnold.
Personnel Profile: Part-Time Volunteers 12.
Institution Type/Description: Historic House Museum: housed in the home of Conway mill owner, William Kimball Eastman, c.1818. Listed on the National Register of Historic Places
Collections: personal artifacts; period furnishings.

Facilities: Museum-related items for sale.
Activities: monthly meetings.
Publications: quarterly newsletter.
Hours & Admission Prices: Memorial Day to Labor Day Wed. 2-4, Sat. 1-4; groups & other times by appointment. No charge; donations accepted.
Attendance: 68 (accurate)
Membership: Individual $20.

Cornish

*** SAINT-GAUDENS NATIONAL HISTORIC SITE, (M),** 139 Saint Gaudens Rd., Cornish, NH 03745-4232. Tel.: 603-675-2175, ext. 100. Fax: 603-675-2701. Facebook: Saint Gaudens National Historic Site.
E-mail: rick_kendall@nps.gov
Web Site: www.nps.gov/saga
Founded: 1926.
Congressional District: 2
Key Personnel: Supt., Rick Kendall; Administrative Officer, April May Preston; Supvr. Interpretation, Chief Ranger, Volunteer Coord. & Museum Shop Mgr., Gregory C. Schwarz; Cur., Henry J. Duffy, Ph.D.; Facility Mgr., Steven Walasewicz.
Personnel Profile: Full-Time Paid 12; Part-Time Paid 5; Part-Time Volunteers 14; Interns 2.
Governing Authority: federal. Parent Institution: U.S. Dept. of the Interior, National Park Service. Affiliated Institution: Saint-Gaudens Memorial, Inc. Affiliated Association: Trustees of the Saint-Gaudens Memorial, 34 S. Highland Ave., Ossining, NY 10562. Tax-exempt.
Institution Type/Description: Art Museum: home & studio of Augustus Saint-Gaudens, American sculptor (1848-1907).
Collections: works of American sculptor Augustus Saint-Gaudens including bronze, sculpture, marble, plaster, molds & historic furnishings; works of Cornish Colony artists; bronze sculpture.
Research Fields: sculpture, life & works of Augustus Saint-Gaudens, 1848-1907; American art & sculpture; Cornish art colony (1885-1930).
Facilities: 2,000-vol. library on American art & sculpture; historic gardens; 194 acres.
Activities: guided tours; gallery talks; education programs; temporary exhibitions; concerts.
Publications: catalogues, Catalog of Saint-Gaudens' works; 1907 U.S. Gold Coinage; The Shaw Memorial: A Celebration of an American Masterpiece; Augustus Saint-Gaudens 1848-1907: A Master of American Sculpture. Augustus Saint Gaudens American Sculptor of the Gildedage; Paul St. Gaudens Ceramic Artist.
Hours & Admission Prices: late May to Oct. daily 9-4:30. Adults $5; discounts to US fee area, America the Beautiful, Senior Passes, National Park Pass, Golden Age, Golden Eagle & annual pass; children 15 & under and school groups no charge.
Attendance: 35,000 (estimated)

Derry

ROBERT FROST FARM, 122 Rockingham Rd., Rte. 28, Derry, NH 03038. Mailing Address: P.O. Box 1075, Derry, NH 03038-1075. Tel.: 603-432-3091.
E-mail: info@robertfrostfarm.org
Web Site: www.robertfrostfarm.org
Founded: 1968.
Congressional District: 1
Key Personnel: Chm., Clair Ternam; Sec., Laura Burnham; Historic Site Specialist, Ben Wilson; Museum Shop Mgr., William Gleed.
Personnel Profile: Part-Time Paid 2; Part-Time Volunteers 1; Interns 1.
Governing Authority: state. Parent Institution: State of New Hampshire. Tax-exempt.
Institution Type/Description: Historic Site: 1900-1909 home of Robert Frost & setting for 43 of his poems.
Collections: early 1900s furniture; some family belongings.
Facilities: 30-vol. library about Robert Frost; nature-poetry walk. Robert Frost books of poetry for sale.
Activities: guided tours; summer lecture series; films; youth poet program for NH 4th graders; poetry readings.
Publications: newsletter, Friends of the Robert Frost Farm.
Hours & Admission Prices: May-June 22 & Sept. 3-Columbus Day Wed.-Sun. 10-5; June 23-Sept. 1 daily 10-5; groups by appointment. Adults $7, children 6-17 $3; NH residents and children 5 & under no charge.
Attendance: 6,500 (estimated)
Membership: Robert Frost Farm Trustees: Student $10; Individual $20; Family $35; Sponsor $50; Supporting $250; Corporate $500.

Dover

THE CHILDREN'S MUSEUM OF NEW HAMPSHIRE, (M), 6 Washington St., Dover, NH 03820-3814. Tel.: 603-742-2002. Fax: 603-742-2044. Facebook: The Children's Museum of New Hampshire.
E-mail: questions@childrens-museum.org
Web Site: www.childrens-museum.org
Formerly: The Children's Museum of Portsmouth
Founded: 1983.
Key Personnel: Exec. Dir., Justine Roberts; Chm. Bd., Eric Gregg; Exhibit Mgr., Mark Cuddy; Dir. Education, Jane Bard; Museum Shop Mgr., Carol Chambers; Dir. Visitor Svcs., Doug Tilton; Dir. Finance, Sarah Strangas; Dir. Mktg., Heidi Duncanson; Dir. Community Engagement, Paula Rais; Membership, Katie West; Dir. Devel., Stephanie Ancona; Early Childhood Coord., Xanthi Gray.
Personnel Profile: Full-Time Paid 9; Part-Time Paid 20; Part-Time Volunteers 314; Interns 3.
Governing Authority: nonprofit organization. Tax-exempt: 501(c)(3).
Institution Type/Description: Children's Museum.
Collections: custom-created hands-on exhibits emphasizing world cultures; natural science; engineering; marine research; music; art; brainwaves & paleontology.
Facilities: Books, educational & museum-related items for sale.
Activities: performances; classes; education outreach programs; curriculum themed school visits; special events; daily art projects; after school & summer programs; community outreach.
Publications: quarterly newsletter; annual report; brochure.
Hours & Admission Prices: Summer Mon.-Sat. 10-5, Sun. 12-5; Spring, Fall & Winter: Tues.-Sat. 10-5, Sun. 12-5. Adults & children over one $9, seniors 65 & over $8; discounts to groups of 10 or more; ACM & ASTC members, museum members & children under one no charge. Closed New Year's Day; Easter; Labor Day; Thanksgiving; Christmas Eve & Day. &
Attendance: 120,000 (estimated)
Membership: Grandparent $44; individual $55, $5 each additional family member; Family of 2 $70, each additional family member $10; ACM & ASTC $120; contributing membership $250.

WOODMAN INSTITUTE MUSEUM, 182 Central Ave., Dover, NH 03821-1916. Mailing Address: P.O. Box 1916, Dover, NH 03821-1916. Tel.: 603-742-1038.
Web Site: www.woodmaninstitutemuseum.org
Founded: 1916.
Congressional District: 1
Governing Authority: nonprofit organization. Tax-exempt.
Institution Type/Description: History Museum.
Collections: Indian artifacts; mineralogy; archaeology; insects; geology; herpetology; period furnishings; fire-fighting equipment; guns; toys; dolls; whaling; clothing; farm tools; local history; furniture; war memorial floor; paintings; Lincoln room. Historic Houses: c.1813 U.S. Senator John P. Hale Home; c.1818 Woodman House; c.1827 Keefe House; c.1675 Colonial Garrison House.
Research Fields: local area history.
Activities: gallery talks; permanent & temporary exhibitions.
Publications: newsletter.
Hours & Admission Prices: April-Nov. Wed.-Sun. 12-4:30. Adults $8, seniors 65 & over and students $6, children 6-15 $3; children 5 & under and members no charge. Closed holidays.
Membership: Individual $25; Family $35; Business $125.

Durham

DURHAM HISTORIC ASSOCIATION MUSEUM, Rte. 108 & Main St., Durham, NH 03824-2815. Mailing Address: 15 Newmarket Rd., Durham, NH 03824-2815. Tel.: 603-868-5436.
E-mail: durhamhistoricassn@comcast.net
Web Site: www.ci.durham.nh.us/community/historic/dha.html
Founded: 1851.
Congressional District: 1
Key Personnel: Pres., Richard Lord.
Personnel Profile: Part-Time Paid 1; Part-Time Volunteers 5.
Governing Authority: society. Tax-exempt: 501(c)(3).
Institution Type/Description: Local History Museum: housed in c.1825 Town Hall.
Collections: history pertaining to Durham; railroad benches; costumes; doctors' equipment; American Indian artifacts; lighting fixtures; handiwork of town's women; school equipment; maps; manuscripts; account books; church pewter; pew; cupola; 1835 lap organ; 1855 melodeon; 1757 Thomas Wille clock.

Research Fields: genealogy; church history; construction & history of Gundalow in the Piscataqua Basin; Durham Town History.
Facilities: approx. 300-vol. library of books on New Hampshire & local history available by special arrangement.
Activities: lectures; formally organized education programs for children; permanent & temporary exhibitions; walking tour of Durham's historic district.
Publications: quarterly newsletter, Walking Tour - Historic District; books, Landmarks in Ancient Dover, 1892; If Only Uncle Ben, 1971; History of Durham, New Hampshire in an Oystershell: 1600-1976; Durham, New Hampshire: A History 1900-1985; History of Durham, New Hampshire, 1913; Stackpole, reprinted 2 vols. in 1, 1994.
Hours & Admission Prices: By appointment. No charge; donations accepted.
Attendance: 100 (estimated)
Membership: Regular & Contributing $10; Life $50.

MUSEUM OF ART, UNH, (M), Paul Creative Arts Center, 30 Academic Way, Durham, NH 03824-2617. Tel.: 603-862-3712. Fax: 603-862-2191.
E-mail: museum.of.art@unh.edu
Web Site: www.unh.edu/moa
Founded: 1960.
Congressional District: 1
Key Personnel: Dir., Kristina Durocher; Pres. (V), Anne de Cossy; Mgr. Exhibitions & Collections, Jacqueline Finnegan; Education & Publicity Coord., Catherine A. Mazur; Administrative Asst., Cynthia Farrell.
Personnel Profile: Full-Time Paid 4; Part-Time Paid 8; Part-Time Volunteers 2; Interns 3.
Governing Authority: university. Affiliated with Univ. of New Hampshire. Tax-exempt.
Institution Type/Description: Art Museum.
Collections: 20th-century works on paper; 19th-century American landscape paintings; 19th-century Japanese woodblock prints.
Facilities: 5,350 sq. ft. exhibition space.
Activities: inter-museum loan; temporary exhibitions & permanent collection; lectures; educational programs.
Publications: booklets; exhibition catalogues, A Stern & Lovely Scene: A Visual History of the Isles of Shoals, Circle of Friends: Art Colonies of Cornish & Dublin, The White Mountains: Place & Perceptions, By Good Hands: New Hampshire Folk Art, Hyman Bloom: Paintings & Drawings; Realism & Invention in the Prints of Albrecht Durer; Deeply Rooted: New Hampshire Traditions in Wood; On Great Bay: Paintings by Christopher Cook and Arthur Di Mambro; The Simple Art: Printed Images in an Age of Magnificence; Gabriel Laderman: Unconventional Realist; Acts & Memory: Paintings by Langdon Quin; The Artists Revealed: 2010 Studio Faculty Exhibition; Legacy: Works by Distinguished Faculty; Felice Beato: Photographer in Nineteenth-Century Japan.
Hours & Admission Prices: Sept.-May Mon.-Wed. 10-4, Thurs. 10-8, Sat.-Sun. 1-5. No charge; donations accepted. Closed university holidays. &
Attendance: 8,125 (accurate)
Membership: Contributor $25; Partner $50; Donor $100; Sponsor $250; Patron $500; Benefactor $1,000.

Enfield

ENFIELD SHAKER MUSEUM, (M), 447 NH Rte. 4A, Enfield, NH 03748-3503. Tel.: 603-632-4346.
E-mail: info@shakermuseum.org
Web Site: www.shakermuseum.org
Formerly: Museum at Lower Shaker Village
Founded: 1986.
Congressional District: 2
Key Personnel: Exec. Dir., Dolores C. Struckhoff; Pres. (V), June K. Hemberger; Cur., Michael O'Connor; Coord. Events, Vreni Gust.
Personnel Profile: Full-Time Paid 2; Part-Time Paid 6; Part-Time Volunteers 95; Interns 2.
Governing Authority: private; nonprofit organization. Tax-exempt: 501(c)(3).
Institution Type/Description: Historic Village: 1793 Shaker community.
Collections: Enfield Shaker Village history from 1790s to present; agricultural equipment; furniture; photographs; clothing; display herb, vegetable, & flower gardens.
Research Fields: Shaker herb gardening; Enfield Shaker history.
Facilities: limited library of Shaker history & garden resource material available to the public by appointment; botanical garden; classrooms; 1,500 sq. ft. exhibit space; 28 acres of land. Museum-related items for sale.
Activities: guided tours; lectures; films; concerts; workshops; formal education programs for adults; docent program; loan exhibitions. Overnight programs: Shaker Forum. Museum Sponsors: Taste of the Upper Valley; Dragonfly Ball; Harvest Festival; Holiday Event.
Publications: brochure; newsletter, The Friends' Quarterly.

Hours & Admission Prices: Jan. to Memorial Day Mon.-Sat. 10-4, Sun. 12-4; late-May to Dec. Mon.-Sat. 10-5, Sun. 12-5. Adults $12, youth 12-17 $8; children under 12 no charge. Closed New Year's Day; Presidents' Day; Memorial Day; Independence Day; Labor Day; Thanksgiving; Christmas Eve & Day. &
Attendance: 5,000 (estimated)
Membership: Individual $40; Dual $50; Family $60.

LOCKEHAVEN SCHOOLHOUSE MUSEUM, Corner of Lockehaven Rd. & Ibey Rd., Enfield, NH 03748. Mailing Address: P.O. Box 612, Enfield, NH 03748-0612. Tel.: 603-632-7740.
Founded: 1947.
Congressional District: 2
Key Personnel: C.E.O., Paul Waehler; Vice Pres., Linda Jones; Treas., John P. Carr; Historian, Marjorie A. Carr
Governing Authority: Affiliated with Enfield Historical Society. Tax-exempt.
Institution Type/Description: Historic Building: 1864 Lockehaven Schoolhouse.
Collections: school textbooks & registers; town reports; scrapbooks; seats & desks; photos of former teachers back to 1854; early photos of Lockehaven & schools of surrounding districts.
Hours & Admission Prices: mid-June to Sept. Sun. 2-4. No charge.
Attendance: 150 (estimated)
Membership: Individual $10; Family $15; Contributing $25; Sustaining $50; Donor $100.

Enfield Center

ENFIELD HISTORICAL SOCIETY MUSEUM, 1047 NH Rt. 4A, Enfield Center, NH 03749. Mailing Address: P.O. Box 612, Enfield, NH 03748-0612. Tel.: 603-632-7740.
Founded: 1991.
Congressional District: 2
Key Personnel: Pres., Paul Waehler; Vice Pres., Linda Jones; Treas., John P. Carr; Historian, Marjorie A. Carr
Governing Authority: Affiliated with Enfield Historical Society. Tax-exempt.
Institution Type/Description: Historical Society Museum: housed in an 1851 structure that was used as a schoolhouse for 95 years.
Collections: local history & culture.
Facilities: 500-vol. library of newspaper clippings, private writings & photographs; 30-seat auditorium; 650 sq. ft. exhibit space.
Activities: permanent exhibitions.
Publications: newsletter.
Hours & Admission Prices: June-Sept. by appointment. No charge; donations accepted.
Attendance: 150 (estimated)
Membership: Individual $10; Family $15; Contributing $25; Sustaining $50; Patron $100.

Epping

EPPING HISTORICAL SOCIETY, 11 Water St., Epping, NH 03802. Mailing Address: P.O. Box 348, Epping, NH 03042-0348. Tel.: 603-679-2944.
E-mail: joysgarden@hotmail.com
Founded: 1971.
Key Personnel: Cur., Joy True.
Volunteer Hours: 1,000
Institution Type/Description: Historical Society Museum.
Collections: local history & culture; photographs; genealogy; period furnishings; personal artifacts.
Publications: quarterly newsletter.
Hours & Admission Prices: Mon. 8-12; other times by appointment. No charge.
Attendance: 200 (estimated)
Membership: Junior $5; Individual $10; Lifetime $50.

Exeter

AMERICAN INDEPENDENCE MUSEUM, One Governors Lane, Exeter, NH 03833-2420. Tel.: 603-772-2622. Fax: 603-772-0861.
E-mail: info@independencemuseum.org
Web Site: www.independencemuseum.org
Founded: 1991.
Congressional District: 1
Key Personnel: Dir., Gail Nessell Colglazier; Pres., Randall A. Hammond.
Personnel Profile: Full-Time Paid 1; Part-Time Paid 3; Part-Time Volunteers 40; Interns 4.
Governing Authority: nonprofit organization. Parent Institution: American Independence Center. Tax-exempt: 501(c)(3).

Institution Type/Description: Historic House Museum: c.1721 Ladd-Gilman House & c.1775 Folsom Tavern.
Collections: historic documents; Revolutionary War era artifacts; Society of the Cincinnati in NH archives; 18th-19th century decorative arts.
Research Fields: New Hampshire & national history pertaining to the Revolutionary War & the founding of the government; Ladd, Folsom & Gilman families; Society of the Cincinnati in NH.
Facilities: museum store; research library.
Activities: guided & self-guided tours; school & adult group programs; special programs for families, children & adults; rental facilities; research. Annual Event: American Independence Festival in July.
Publications: e-newsletter, The Broadside.
Hours & Admission Prices: May-Oct. Wed.-Sat. 10-4. Adults $6, seniors 65 & over $5, students 6-18 $3; discounts to AAA & AAM members; children under 6 & members no charge.
Attendance: 6,845 (accurate)
Membership: Individual $35; Family $50; Partner $125; John Adams $250; Benjamin Franklin $500; Thomas Jefferson $1,000.

EXETER HISTORICAL SOCIETY, 47 Front St., Exeter, NH 03833-2707. Mailing Address: P.O. Box 924, Exeter, NH 03833. Tel.: 603-778-2335.
E-mail: info@exeterhistory.org
Web Site: www.exeterhistory.org
Founded: 1928.
Congressional District: 1
Key Personnel: Chm. (V), Lionel Ingram; Pres. (V), Edward Rowan; Cur., Barbara Rimkunas; Program Mgr., Laura Gowing.
Personnel Profile: Part-Time Paid 2.
Governing Authority: Tax-exempt.
Institution Type/Description: Historical Society Museum: housed in the former public library; c. 1894.
Collections: local history & culture; photographs; documents; maps; period artifacts; china; dolls.
Research Fields: Exeter history; genealogy.
Activities: educational programs.
Hours & Admission Prices: Tues. & Thurs. 2-4:30, Sat. 9:30 to noon. No charge; donations accepted.
Attendance: 1,200 (estimated)
Membership: Educator & Student $10; Individual $25; Family $50; Sponsor $100; Corporate $250; Corporate Pioneer Donor $500.

*** GILMAN GARRISON HOUSE, (M),** 12 Water St., Exeter, NH 03833-2431. Mailing Address: Historic New England, 141 Cambridge St., Boston, MA 02114. Tel.: 603-436-3205. Fax: 617-227-9204.
Web Site: www.historicnewengland.org
Founded: 1965.
Congressional District: 1
Key Personnel: Pres., Carl Nold.
Governing Authority: society; nonprofit organization. Parent Institution: Historic New England, 141 Cambridge St. Boston, MA 02114. Tax-exempt: 501 (c)(3).
Institution Type/Description: Historic House: 1709 Gilman Garrison House.
Collections: furnishings; period artifacts.
Hours & Admission Prices: Call for hours. Adults $5, seniors $4, students $2.50; members no charge.
Attendance: 433 (accurate)
Membership: National $35; Individual $45; Household $55; Garden & Landscape $75; Contributing and Library & School $100; Historic New England Affiliate $100-$350; Young Friends of Historic New England $100-$1,500; Friends of the Library and Archives $125; Historic Homeowner $200; Business & Ogden Codman Design Group $250; Appleton Circle $1,750-$3,500.

THE LAMONT GALLERY, (M), Phillips Exeter Academy, 11 Tan Lane, Exeter, NH 03833. Mailing Address: 20 Main St., Exeter, NH 03383. Tel.: 603-777-3461. Fax: 603-777-4371.
E-mail: gallery@exeter.edu
Web Site: www.exeter.edu/arts/8160.aspx
Founded: 1953.
Congressional District: 1
Key Personnel: Dir., Karen Burgess Smith; Gallery Mgr., Sara Zela.
Personnel Profile: Full-Time Paid 2; Part-Time Paid 3.
Governing Authority: preparatory school. Affiliated with The Phillips Exeter Academy, Exeter, NH. Tax-exempt.
Institution Type/Description: Art Gallery.
Collections: paintings; photographs.

Facilities: 85-seat auditorium.
Activities: lectures; films; gallery talks; formally organized education program for students affiliated with The Phillips Exeter Academy; temporary & traveling exhibitions.
Hours & Admission Prices: Academic Year: Mon. 1-5, Tues.-Sat. 9-5; July Tues.-Fri. 9-4. No charge. Closed school holidays.
Attendance: 5,000 (estimated)

Fitzwilliam

FITZWILLIAM HISTORICAL SOCIETY & AMOS J. BLAKE HOUSE MUSEUM, 66 Rte. 119 - Village Green, Fitzwilliam, NH 03447. Mailing Address: P.O. Box 87, Fitzwilliam, NH 03447-0087. Tel.: 603-585-7742.
E-mail: fitzhs@peoplepc.com
Web Site: www.fitzhistoricalsociety.org/
Founded: 1966.
Key Personnel: Pres., Theresa Sillapaa.
Personnel Profile: Part-Time Volunteers 2.
Governing Authority: Parent Institution: Fitzwilliam Historical Society. Tax-exempt.
Institution Type/Description: History Museum.
Collections: local history; furnishings; period artifacts; household items.
Facilities: library.
Activities: special events.
Publications: biannual newsletter.
Hours & Admission Prices: Museum: Memorial Day to Labor Day Sat. 1-4; other times by appointment. Research: Thurs. 9-11. No charge; donations accepted.
Attendance: 35 (estimated)

Franconia

FRANCONIA HERITAGE MUSEUM AND IRON FURNACE INTERPRETIVE CENTER, 553 Main St., Franconia, NH 03580. Mailing Address: P.O. Box 169, Franconia, NH 03580-0169. Tel.: 603-823-5000.
E-mail: heritagemuseum@myfairpoint.net
Web Site: www.franconianh.org
Founded: 1998.
Congressional District: 2
Key Personnel: Pres., Nancy Heinemann; Vice Pres., Dot Wiggins.
Personnel Profile: Part-Time Volunteers 10.
Governing Authority: Parent Institution: Franconia Area Heritage Council. Tax-exempt: 501(c)(3).
Institution Type/Description: Historic House Museum.
Collections: personal artifacts; furnishings; Brooks; Parker; Dow Academy; Franconia College; Old Man of the Mountain.
Publications: semiannual newsletter.
Hours & Admission Prices: Memorial Day to Oct. Fri.-Sat. 1-4; other times by appointment. No charge; donations accepted.
Attendance: 640 (accurate)
Membership: Individual $15; Family $25; Contributing $50; Supporting $75; Sustaining $150; Plus Major Donor $250; Lifetime $1,000.

THE FROST PLACE, 158 Ridge Rd., Franconia, NH 03580. Mailing Address: P.O. Box 74, Franconia, NH 03580-0074. Tel.: 603-823-5510. Facebook: The Frost Place.
E-mail: frost@frostplace.org
Web Site: www.frostplace.org
Founded: 1976.
Key Personnel: Exec. Dir., Maudelle Driskell; Chm. Bd. (V), Deming P. Holleran; Admin., Therese Reger
Institution Type/Description: History Museum.
Collections: first editions of Robert Frost's works; memorabilia; poems; personal artifacts.
Activities: nature trail; summer conferences; school programs. Annual Events: Frost Day Celebration in July; Festival Conference in July & August.
Hours & Admission Prices: Memorial Day to July 1 Thurs.-Sun. 1-5; July-Sept. Wed.-Mon. 1-5; Sept. 15 to Columbus Day Wed.-Mon. 10-5. Adults $5, seniors $4, children $3.

NEW ENGLAND SKI MUSEUM, (M), Franconia Notch State Park, I-93 Exit 34B, Franconia, NH 03580. Mailing Address: P.O. Box 267, Franconia, NH 03580-0267. Tel.: 603-823-7177. Fax: 603-823-9505.
E-mail: staff@skimuseum.org
Web Site: www.skimuseum.org
Founded: 1977.

Congressional District: 2
Key Personnel: Exec. Dir., Jeffrey R. Leich; Pres. (V), Bo Adams; Museum Shop Mgr. & Front Desk Staffer, Linda Bradshaw; Front Desk Staffer, Kay Kerr.
Personnel Profile: Full-Time Paid 2; Part-Time Paid 3; Part-Time Volunteers 12.
Volunteer Hours: 557
Operating Expenses: 356,897
Operating Income: 390,305
Governing Authority: nonprofit. Tax-exempt: 501(c)(3).
Institution Type/Description: Ski Museum.
Collections: ski artifacts; clothing; posters; photographs; films dating to the early 1930s; trophies & awards; oral history tapes; artwork.
Major Exhibits: Skiing in Vermont, 5/31/14-3/31/15.
Research Fields: origins & development of skiing; ski technique; ski equipment.
Facilities: 1,500-vol. library pertaining to skiing, including scrapbooks, photo albums, organizational records & competition records; 35-seat auditorium; videos. Ski-related items for sale.
Activities: guided tours; lectures; loan & temporary exhibitions; film screenings on video. Annual Events: Hannes Schneider Meister Cup Race in March; Bretton Woods Nordic Marathon in March; Annual Meeting & Spirit of Skiing Award Presentation in Fall.
Publications: quarterly journal; books, Tales of the 10th, Over the Headwall; DVDs, Thrill and Spills, Ski Sentinels; reproduction vintage ski posters & photographs; postcards.
Hours & Admission Prices: Memorial Day-March daily 10-5. No charge; donations accepted. Closed Christmas. &
Attendance: 29,247 (accurate)
Membership: Individual $35; Family $50; Ski Clubs & Supporting $75; Sustaining $125; Life $1,000; Corporate $100, $250, $500, $1,000.

Franklin

DANIEL WEBSTER BIRTHPLACE, North Rd., off Rte. 127, Franklin, NH 03235. Mailing Address: P.O. Box 1856, Concord, NH 03302-1856. Tel.: 603-271-3556.
Web Site: www.nhstateparks.com/danielwebster.html
Key Personnel: Dir., Ben Wilson
Institution Type/Description: Historic Site.
Collections: local history & culture relating to Daniel Webster
Hours & Admission Prices: June 21-Sept. 1. Sat.-Sun. 9-5. Adults $7, children 6-11 $3; New Hampshire residents & children 5 and under no charge.

FRANKLIN FIREFIGHTERS MUSEUM, 59 W. Bow St., Franklin, NH 03235. Tel.: 603-934-2205.
Founded: 1987.
Institution Type/Description: Firefighters Museum.
Collections: firefighting history & equipment; early fire trucks & apparatus; photographs.
Activities: demonstrations.
Hours & Admission Prices: By appointment.

TARBIN GARDENS, 321 Salisbury Rd., Franklin, NH 03235-2503. Tel.: 603-934-3518.
E-mail: info@tarbingardens.com
Web Site: www.tarbingardens.com
Institution Type/Description: Gardens.
Collections: native & exotic plants, trees & shrubs.
Activities: special events.
Hours & Admission Prices: May to Columbus Day Tues.-Sun. 10-6; other times by appointment. Guided Tours: daily 11am. Adults $8.50, children 4-17 & seniors 62 & over $7; children 3 & under no charge.

Gilford

THOMPSON-AMES HISTORICAL SOCIETY, Gilford Village, 8 Belknap Mountain Rd., Gilford, NH 03249-6807. Tel.: 603-293-2877.
E-mail: thomames@metrocast.net
Web Site: www.gilfordhistoricalsociety.org
Key Personnel: Pres. (V), Karin Landry; Vice Pres., Donna Shinlever; Sec., Mary Jane Kwist; Treas., Rick Moses
Institution Type/Description: Historical Society Museum.
Collections: Historic Houses: The Mt. Belknap Grange/John J. Morrill Store c.1857; The Union Meetinghouse c.1834; The Benjamin Rowe House c.1838.
Hours & Admission Prices: By appointment.

Gorham

GORHAM HISTORICAL SOCIETY AND RAIL MUSEUM, 25 Railroad St., Gorham, NH 03581-1638. Mailing Address: P.O. Box 351, Gorham, NH 03581-0351. Tel.: 603-466-5338. Facebook: Gorham Historical Society and Rail Museum.
E-mail: gorhamhistoricalsociety@gmail.com
Web Site: www.gorhamnewhampshire.com/railroadmuseum.html
Founded: 1973.
Key Personnel: Pres., Reuben Rajala
Institution Type/Description: History Museum.
Collections: Gorham history; steam engine; box cars; period artifacts; clothing; photographs.
Hours & Admission Prices: Memorial Day to Columbus Day daily 10-2; other times by appointment. No charge.

Hampton

TUCK MUSEUM, 40 Park Ave., Hampton, NH 03843-1601. Mailing Address: P.O. Box 1601, Hampton, NH 03843-1601. Tel.: 603-926-2543.
E-mail: info@hamptonhistoricalsociety.org
Web Site: www.hamptonhistoricalsociety.org
Founded: 1925.
Congressional District: 1
Key Personnel: Pres. (V), Candice Stellmach; Dir., Betty Moore.
Personnel Profile: Full-Time Volunteers 1; Part-Time Volunteers 25.
Governing Authority: nonprofit organization. Parent Institution: Hampton Historical Society. Tax-exempt: 501(c)(3).
Institution Type/Description: Local History Museum: Tuck Memorial Museum, located on the site of the original settlement of Hampton in 1638.
Collections: period artifacts & documents pertaining to colonial Hampton; trolley exhibit of E. H. & A. St. Ry. Company; early fire equipment; farming & farm equipment; early industry. Historic Buildings: one-room schoolhouse; c.1796 barn; beach cottage.
Research Fields: historical research of colonial Hampton.
Facilities: 500-vol. library of old books available on premises. Historical booklets, note paper, postcards & historical maps for sale.
Activities: guided tours; permanent exhibitions.
Publications: brochure; newsletter, Gatherings
Hours & Admission Prices: Wed., Fri. & Sun. 1-4; other times by appointment. No charge; donations accepted. &
Attendance: 1,200 (estimated)
Membership: Student & Senior Citizen $10; Individual $15; Family $25; Business $100.

Hancock

HANCOCK HISTORICAL SOCIETY & MUSEUM, 7 Main St., Hancock, NH 03449-6008. Mailing Address: P.O. Box 138, Hancock, NH 03449-0138. Tel.: 603-525-9379.
E-mail: history@hancockhistoricalsociety.org
Web Site: www.hancockhistoricalsociety.org
Founded: 1903.
Congressional District: 2
Key Personnel: C.E.O. & Pres. (V), Timothy J. Lord; Vice Pres., Maribeth Sullivan; Museum Committee Chair, Roberta D. Nylander.
Personnel Profile: Part-Time Volunteers 25.
Governing Authority: private; nonprofit organization. Tax-exempt: 501(c)(3).
Institution Type/Description: General Museum: housed in c.1809 Federal-style brick building which was possibly a tavern when built.
Collections: 18th to 20th-century archives & artifacts related to Hancock.
Research Fields: collection-related research.
Facilities: archive library of photographs & unpublished papers from Hancock natives available for public viewing. Museum-related items for sale.
Activities: docent program; guided tours; lectures; loan & temporary exhibitions.
Publications: quarterly newsletter, Hancock Historical Society News.
Hours & Admission Prices: Call for hours. No charge; donations accepted.
Attendance: 150 (estimated)
Membership: Individual $15; Life $250.

Hanover

* **HOOD MUSEUM OF ART, (M),** 6 E. Wheelock St., Dartmouth College, Hanover, NH 03755. Tel.: 603-646-2808. Fax: 603-646-1400. Facebook: Hood Museum of Art.
E-mail: hood.museum@dartmouth.edu
Web Site: www.hoodmuseum.dartmouth.edu

Founded: 1772.

Congressional District: 2

Key Personnel: Dir., Michael Taylor; Assoc. Dir. & Barbara C. and Harvey P. Hood 1918 Cur. Academic Programming, Katherine Hart; Asst. Dir., Juliette Bianco; Asst. Dir. & Cur. Education, Lesley Wellman; Coord. Academic Programming, Amelia Kahl; Coord. Devel. & Membership, Julie Ann Otis; Cur. American Art, Barbara J. MacAdam; Cur. African Art, Ugochukwu-Smooth Nzewi; Curatorial Asst., Karysa Norris; Assoc. Registrar, Kathleen O'Malley; Asst. Registrar, Cynthia Gilliland; Adjunct Cur. Costumes, Margaret E. Spicer; Asst. Cur. Special Projects, Sarah Powers; Business Mgr., Nancy A. McLain; Gift Shop Mgr., Mary Ellen Rigby; Public Rels. Asst., Alison Palizzolo; Mgr. Communications & Publications, Nils Nadeau; Exhibitions Designer, Patrick Dunfey; Images & Art Start Instructor, Neely McNulty; Lead Preparator, John Reynolds; Preparator, Matthew Zayatz; Tour Coord., Adrienne Kermond; Coord. Programs & Events, Sharon Reed; Exhibitions Coord., Nicole Gilbert; Art Handler, Susan Achenbach; Museum Educator, Vivian Ladd; Asst. Cur. Education, Rebecca Karp; TMS Project/Digital Asset Mgr., Jonathan Benoit; Data Mgr., Deborah Haynes; Security & Bldg. Mgr., Gary Alafat.

Personnel Profile: Full-Time Paid 27; Part-Time Paid 9; Part-Time Volunteers 30; Interns 8.

Operating Expenses: 3,200,000

Operating Income: 3,200,000

Governing Authority: college. Parent Institution: Dartmouth College. Tax-exempt: 501(c)(3).

Institution Type/Description: Art Museum.

Collections: The Hood preserves approximately 65,000 works of art representing a broad range of cultural areas and historical periods. Selections that are always on view encompass ancient, Asian, Native American, Oceanic and African collections, European old master prints and nineteenth-century paintings, American colonial silver, portraiture, drawings, watercolors and paintings, as well as major works of modern and contemporary art.

Major Exhibits: In Residence: Contemporary Artists at Dartmouth, 1/18/14-7/6/14; Allan Houser: A Centennial Exhibition, 5/14-5/15; Witness: Art and Civil Rights in the Sixties, 8/30/14-12/21/14.

Research Fields: art; ethnography; history.

Facilities: auditorium; seminar room; collections study. Museum-related items for sale.

Activities: permanent & traveling exhibitions; inter-museum loans; gallery talks; lectures; symposia; family and school programs; tours; receptions.

Publications: exhibition catalogues; special publications; handbook, Treasures of the Hood Museum of Art; Hood Quarterly; gallery brochures; general brochures; European Art at Dartmouth: Highlights from the Hood Museum of Art; American Art at Dartmouth: Highlights from the Hood Museum of Art; Modern and Contemporary Art at Dartmouth: Highlights from the Hood Museum of Art; Native American Art at Dartmouth: Highlights from the Hood Museum of Art; Crossing Cultures: The Owen and Wagner Collection of Contemporary Aboriginal Australian Art at the Hood Museum of Art.

Hours & Admission Prices: Tues. & Thurs.-Sat. 10-5, Wed. 10-9, Sun. 12-5. No charge; donations accepted. Closed Independence Day; Thanksgiving. &

Attendance: 45,000 (accurate)

Membership: Friend $100-$499; Patron $500-$1,499; Corporate $1,000; Lathrop Fellow $1,500-$4,999; Director's Circle $5,000-$9,999; Chairman's Circle $10,000 & up.

HOPKINS CENTER FOR THE ARTS, 2 E. Wheelock St., Dartmouth College, Hanover, NH 03755. Mailing Address: 6041 Wilson Hall Lower Level, Dartmouth College, Hanover, NH 03755. Tel.: 603-646-2422.

Institution Type/Description: Art Gallery.

Collections: works by guest artists, faculty, students & alumni; paintings; sculpture.

Hours & Admission Prices: Tues.-Sat. 12:30-10, Sun. 12:30-5:30.

Hebron

PARADISE POINT NATURE CENTER, North Shore Rd., Hebron, NH 03241. Mailing Address: c/o NH Audubon, 84 Silk Farm Rd., Concord, NH 03301. Tel.: 603-744-3516 (July-Sept); 224-9909 (Sept.-July). Fax: 603-744-3516.

E-mail: nha@nhaudubon.org

Web Site: www.nhaudubon.org

Founded: 1969.

Governing Authority: nonprofit organization. Parent Institution: Audubon Society of New Hampshire. Branch Facility: Hebron Marsh Wildlife Sanctuary. Tax-exempt: 501(c)(3).

Institution Type/Description: Nature Center & Wildlife Sanctuary.

Collections: live wildlife exhibits; taxidermal bird & mammal species; wildlife artifacts; natural history materials & specimens. Historic House: late 1800's, Ash Cottage.

Facilities: resource library; educational facilities; 400 sq. ft. exhibit space; nature trails; 43 acres of grounds. Natural history related books & gifts for sale.

Activities: lectures; organized educational programs; participatory exhibits; summer festival; summer day camp.

Hours & Admission Prices: July-Sept. Mon.-Sat. 10-4; other times by appointment. No charge; donations accepted.

Membership: Individual $25; Family $35.

Henniker

NEW ENGLAND COLLEGE GALLERY, 7 Main St., Henniker, NH 03242-6225. Tel.: 603-428-2329. Fax: 603-428-2266.

E-mail: dfurtkamp@nec.edu

Web Site: www.nec.edu

Founded: 1988.

Congressional District: 2

Key Personnel: Dir., Darryl Furtkamp.

Personnel Profile: Full-Time Paid 1; Part-Time Paid 1; Interns 2.

Governing Authority: private college; nonprofit. Parent Institution: New England College. Tax-exempt.

Institution Type/Description: College Art Gallery.

Collections: concentration on contemporary art; paintings; photographs; works on paper.

Research Fields: computer imaging; children's book illustration; politically sensitive art; contemporary and post-modern art; photography.

Facilities: 1,300 sq. ft. exhibit space.

Activities: lectures; loan, participatory, temporary & traveling exhibits; training programs for professional museum workers.

Hours & Admission Prices: Academic Year: Tues.-Thurs. 11-6, Fri. 11-4. No charge. Closed Thanksgiving week; Christmas break. &

Attendance: 1,500 (estimated)

Hillsborough

THE FRANKLIN PIERCE HOMESTEAD, Rte. 31 & Rte. 9, Hillsborough, NH 03244. Mailing Address: P.O. Box 896, Hillsborough, NH 03244-0896. Tel.: 603-478-3165. Fax: 603-478-5500.

E-mail: c_chadwick@conknet.com

Web Site: www.franklinpierce.ws/homestead/homestead.html

Founded: 1804.

Congressional District: 2

Key Personnel: Chm. (V), Christina Chadwick; Pres. (V), Jane H. Waters; Architectural Historian, James Garvin, III; Museum Shop Mgr., Pat Bradley.

Personnel Profile: Part-Time Paid 2; Part-Time Volunteers 42.

Governing Authority: municipal; state; society. Parent Institution: Hillsborough Historical Society. Tax-exempt: 501(c)(3).

Institution Type/Description: Historic House Museum: 1804 Mansion, built by Benjamin Pierce; childhood home of Franklin Pierce, 14th U.S. President.

Collections: period furniture; farm & household equipment; Pierce furniture, clothing, pictures & campaign posters.

Research Fields: economic & political lives of Benjamin & Franklin Pierce; early 19th century rural life.

Facilities: 70-seat auditorium; 5,000 sq. ft. exhibit space.

Activities: guided & walking tours; lectures; organized education programs for children; loan exhibitions.

Hours & Admission Prices: June & Sept. Sat. 10-4, Sun. 1-4; July-Aug. Mon.-Sat. 10-4, Sun. 1-4; other times by appointment. Adults $5, children 6-17 $1; NH residents over 65, Hillsboro residents, members & children under 6 no charge. &

Attendance: 4,750 (estimated)

Membership: Individual $5; Family $10; Corporate $25; Patron $100.

Holderness

SQUAM LAKES NATURAL SCIENCE CENTER, 23 Science Center Rd., Holderness, NH 03245. Mailing Address: P.O. Box 173, Holderness, NH 03245-0173. Tel.: 603-968-7194. Fax: 603-968-2229. Facebook: NH Wildlife.

E-mail: info@nhnature.org

Web Site: www.nhnature.org

Formerly: Science Center of New Hampshire

Founded: 1966.

Congressional District: 2

Key Personnel: Chm. Bd. (V), David F. Martin; Exec. Dir., Iain MacLeod; Dir. Operations, Elizabeth Rowe; Mgr. Mktg. & Visitor Svcs., Amanda Gillen.

Personnel Profile: Full-Time Paid 18; Part-Time Paid 30; Part-Time Volunteers 250; Interns 4.

Volunteer Hours: 8,930
Operating Expenses: 1,766,265
Operating Income: 1,471,650
Governing Authority: nonprofit organization. Tax-exempt: 501(c)(3).
Institution Type/Description: Natural Science Museum.
Collections: 700 birds & mammals; Charles Goodhue collection.
Research Fields: environmental education.
Facilities: nature trails; classrooms.
Activities: outdoor discovery hikes; lecture-demonstrations for school groups
    during school year; natural history cruises; special member programs;
    Montessori nature preschool.
Publications: quarterly newsletter; annual report.
Hours & Admission Prices: May-Nov. daily 9:30-4:30. Adults $15; discounts
    to ANCA & AZA members; members no charge. &
Attendance: 75,000 (estimated)
Membership: Individual $40; Two Person $50; Four Person $60; Six Person
    $80; Eight Person $100; Wetlands $125; Field $250; Forest $500; Mt. Fayal
    $1,000.

Hollis

WHEELER HOUSE & ALWAYS READY ENGINE HOUSE, 20
    Main St., Hollis, NH 03079. Tel.: 603-465-3935.
Web Site: www.hollis-history.org
Founded: 1958.
Congressional District: 2
Key Personnel: Pres. (V), Dick Lates; Cur., Fredricka Olson
Institution Type/Description: Historical Society Museum.
Collections: local history & culture; period furnishings; personal artifacts;
    photographs.
Activities: genealogy; research; speakers; permanent & temporary exhibitions.
Publications: quarterly newsletter.
Hours & Admission Prices: June-Oct. Mon., Wed. and 1st & 3rd Sun. 1-4;
    Nov.-May Mon. & Wed. 1-4. No charge; donations accepted. Closed major
    holidays.
Membership: Single $15; Family $25; Supporting $50; Corporate $100.

Hopkinton

HOPKINTON HISTORICAL SOCIETY, 300 Main St., Hopkin-
    ton, NH 03229-2627. Tel.: 603-746-3825.
E-mail: nhas@tds.net
Web Site: www.hopkintonhistory.org
Formerly: New Hampshire Antiquarian Society
Founded: 1859.
Congressional District: 2
Key Personnel: Dir., Heather Mitchell; Pres. (V), Don Lane.
Personnel Profile: Part-Time Paid 2; Part-Time Volunteers 7; Interns 1.
Governing Authority: society. Tax-exempt: 501(c)(3) & 170(b).
Institution Type/Description: Local History Museum.
Collections: Hopkinton genealogies from 1737 to present; town histories; early
    music books; NH imprints; costumes; local Indian artifacts; local household
    & agricultural equipment; early textbooks; almanacs; family Bibles; por-
    traits.
Research Fields: local history; genealogy.
Facilities: 2,000-vol. library of local history & genealogy books available for
    use by appointment on premises.
Activities: programs for the public.
Publications: newsletter; books; cemetery walk DVD.
Hours & Admission Prices: Thurs.-Fri. 9-4, Sat. 9-1. Admission charge for
    special exhibits & events only. Closed holidays. &
Attendance: 1,500 (estimated)
Membership: Individual $25; Family & Couple $40; Patron $100; Benefactor
    $250; Life $1,000.

Intervale

HARTMANN MODEL R.R. & TOY MUSEUM, 15 Town Hall
    Rd., Intervale, NH 03845. Mailing Address: P.O. Box 165, Inter-
    vale, NH 03845-0165. Tel.: 603-356-9922. Fax: 603-356-9958.
E-mail: info@hartmannrr.com
Web Site: www.hartmannrr.com
Founded: 1991.
Key Personnel: Pres., Nelly Hartmann
Institution Type/Description: Model Railroad & Toy Museum.
Collections: operating model train layouts from G to Z scales; over 15,000
    items including railroad, cars, trucks, boats, planes, & transport vehicles.
Facilities: cafe. Museum-related items for sale.
Hours & Admission Prices: July-Aug. daily 10-5; Sept.-June Fri.-Mon. 10-5.
    Adults $6, seniors 60 & over $5, children $4; discounts to AAM & ICOM
    members. Closed Easter; Mother's Day; Thanksgiving; Christmas. &

Attendance: 10,000 (accurate)
Membership: Family $50.

Jaffrey

JAFFREY CIVIC CENTER, 40 Main St., Jaffrey, NH 03452-6144.
    Tel.: 603-532-6527. Fax: 603-532-6527.
E-mail: jaffreycntr@aol.com
Web Site: www.jaffreyciviccneter.com
Founded: 1966.
Key Personnel: Exec. Dir., Dion Owens; Pres. (V), Lee S. Sawyer.
Personnel Profile: Full-Time Paid 1; Part-Time Paid 1; Part-Time Volunteers
    33.
Governing Authority: nonprofit organization. Affiliated with Jaffrey Gilmore
    Foundation. Tax-exempt.
Institution Type/Description: Art Foundation & Historical Society.
Collections: paintings; photographs; sculptures; drawings.
Facilities: research library specializing in nature, art, & Americana; meeting
    room, classroom; auditorium.
Activities: meeting space for cultural & civic organizations; art exhibitions,
    classes & workshops, outreach art & history program with the schools.
Hours & Admission Prices: Tues. 10-6, Wed.-Fri. 1-5, Sat. 10-2. No charge;
    donations accepted. Closed national holidays.
Attendance: 12,000 (estimated)

MELVILLE ACADEMY MUSEUM, Thorndike Pond Rd., Jaffrey,
    NH 03452. Mailing Address: P.O. Box 722, Jaffrey, NH 03452-
    0722. Tel.: 603-532-4992.
Founded: 1920.
Key Personnel: Cur., Sarah H Larsen.
Governing Authority: Parent Institution: Village Improvement Society. Tax-
    exempt.
Institution Type/Description: History Museum.
Collections: area history; period furnishings; Hannah Davis bandboxes;
    scrapbooks; 19th century kitchen; agricultural tools.
Hours & Admission Prices: July-Aug. Sat.-Sun. 2-4; other times by appoint-
    ment. No charge; donations accepted.
Attendance: 240 (estimated)

Jefferson

JEFFERSON HISTORICAL MUSEUM, Rte. 2 900 Presidential
    Hwy., Jefferson, NH 03583. Mailing Address: P.O. Box 143,
    Jefferson, NH 03583.
Founded: 1977.
Congressional District: 2
Key Personnel: Pres., Winifred S. Ward; Sec., Marjorie Doan; Treas., Adele
    Woods
Governing Authority: Tax-exempt.
Institution Type/Description: Historic Building: housed in an 1868 church.
Collections: Jefferson history; period artifacts.
Hours & Admission Prices: June to Columbus Day Thurs. & Sun. 1-4. No
    charge; donations accepted.
Attendance: 170 (estimated)
Membership: Single $10; Family $15.

Keene

HISTORICAL SOCIETY OF CHESHIRE COUNTY, 246 Main
    St., Keene, NH 03431-4143. Mailing Address: P.O. Box 803,
    Keene, NH 03431-0803. Tel.: 603-352-1895. Fax: 603-352-9226.
    Facebook: Historical Society of Cheshire County.
E-mail: hscc@hsccnh.org
Web Site: www.hsccnh.org
Founded: 1927.
Congressional District: 2
Key Personnel: Dir. & C.E.O., Alan F. Rumrill; Pres. (V), Sydney Croteau
    Frechette; Museum Shop Mgr., Gail Currier.
Personnel Profile: Full-Time Paid 3; Part-Time Paid 2; Part-Time Volunteers
    100; Interns 1.
Governing Authority: society. Branch Museums: Wyman Tavern Museum.
    Tax-exempt: 501(c)(3).
Institution Type/Description: Historical Society: housed in c.1870 Italianate
    Mansion.
Collections: period furnishings; glass; pottery; silver; toys; photographs;
    archives.
Major Exhibits: J.A. Wright Silver Polish, 11/13-5/14.
Research Fields: local history; genealogy; local history.
Facilities: 3,000-vol. library pertaining to local history & genealogy available
    to the public; 3,500 sq. ft. exhibit space.

Activities: lectures; films; radio programs; organized education programs for children & undergraduate & graduate college students affiliated with Keene State College; temporary exhibitions.
Publications: newsletter 5 times a year.
Hours & Admission Prices: Tues. & Thurs.-Fri. 9-4, Wed. 9-9, 1st & 3rd Sat. each month 9-12. No charge; donations accepted. &
Attendance: 24,000 (estimated)
Membership: Student $10; Senior $25; Individual $35; Family $45; Sustaining $200; Contributing $100.

HORATIO COLONY HOUSE MUSEUM & NATURE PRE-SERVE, 199 Main St., Keene, NH 03431-3780. Tel.: 603-352-0460. Facebook: Horatio Colony Museum & Nature Preserve.
E-mail: colonymuseum@webryders.com
Web Site: horatiocolonymuseum.org
Founded: 1977.
Key Personnel: C.E.O., Anita Carroll-Weldon; Pres. (V), Frank Coolidge.
Personnel Profile: Full-Time Paid 1; Part-Time Paid 1; Part-Time Volunteers 8; Interns 1.
Governing Authority: private; nonprofit organization. Tax-exempt.
Institution Type/Description: Historic Houses & Historic Buildings: c.1806 family home of antique collector & author Horatio Colony, grandson of Keene's first mayor & owner of the city's woolen mill.
Collections: 19th - early 20th century furniture; furnishings; decorative arts; portraits; photos; diaries; personal artifacts; cribbage boards; ink wells; door stops; porcelain figures; pitchers; engravings; napkin rings; glass paper weights; bottles & decanters; transferware platters; Middle Eastern brass artifacts; oriental rugs; period books; Buddhist & Hindu images.
Facilities: 400 sq. ft. exhibit space; garden; 615 acre nature preserve at a noncontiguous site.
Activities: docent program; guided tours; lectures; rental gallery; temporary exhibitions; workshops; educational programs.
Hours & Admission Prices: May to mid-Oct. Tues.-Sat. 11-4; other times by appointment. No charge.
Attendance: 2,500 (accurate)

THORNE-SAGENDORPH ART GALLERY, (M), Wyman Way, Keene, NH 03435-3501. Mailing Address: 229 Main St., Keene State College, Keene, NH 03435-3501. Tel.: 603-358-2720. Fax: 603-358-2238.
E-mail: thorne@keene.edu
Web Site: www.keene.edu/tsag
Founded: 1965.
Congressional District: 2
Key Personnel: Dir., Maureen Ahern.
Personnel Profile: Part-Time Paid 4; Part-Time Volunteers 40; Interns 1.
Governing Authority: Parent Institution: Keene State College. Tax-exempt.
Institution Type/Description: Art Gallery.
Collections: paintings; prints; drawings; sculpture; crafts; regional & national artists; slide collections of all exhibits.
Major Exhibits: KSC Art Faculty Exhibit, 1/24/14-3/9/14; Emerging Art, KSC Art Studio Exhibit, 4/19/14-5/10/14.
Activities: guided tours; lectures; films; gallery talks; formally organized education programs for children; loan, temporary & traveling exhibitions.
Publications: catalogs & brochures of exhibitions.
Hours & Admission Prices: June-July Wed.-Thurs. 12-5, Fri. 3-8, Sat.-Sun. 12-5; Sept.-May Sun.-Wed. 12-5, Thurs.-Fri. 12-7, Sat. 12-8; other times by appointment. No charge; donations accepted. Closed academic holidays; semester breaks. &
Attendance: 6,000 (estimated)
Membership: Senior Citizens & Students $15; Individual $20; Family $30; Donor $50-$99; Sponsor $100-$249; Associate $250-$499; Benefactor $500 & up.

THE WYMAN TAVERN, 339 Main St., Keene, NH 03431. Mailing Address: P.O. Box 803, Keene, NH 03431-0803. Tel.: 603-352-1895. Fax: 603-352-9226. Facebook: Wyman Tavern.
E-mail: hscc@hsccnh.org
Web Site: www.hsccnh.org
Founded: 1968.
Congressional District: 2
Key Personnel: C.E.O., Alan F. Rumrill; Pres., Sydney Croteau-Frechette.
Personnel Profile: Full-Time Paid 3; Part-Time Paid 2; Part-Time Volunteers 20; Interns 1.
Governing Authority: society. Parent Institution: Historical Society of Cheshire County, Inc. Tax-exempt: 501(c)(3).
Institution Type/Description: Historic House: 1762 Wyman Tavern.
Collections: early 19th-century furnishings; portraits; objects made or owned locally.

Research Fields: local history.
Facilities: Postcards & note paper for sale.
Activities: guided tours; permanent & temporary exhibitions; craft demonstrations; school tours; special group tours; lecture series; music programs; living history.
Publications: newsletter 5 times a year.
Hours & Admission Prices: June to Labor Day Thurs.-Sat. 11-4. Adults $3; children under 12 & members no charge.
Attendance: 1,500 (estimated)
Membership: Student $10; Senior $25; Individual $35; Family $45; Sustaining $200; Contributing $100.

Laconia

THE BELKNAP MILL SOCIETY, 25 Beacon St. East, The Mill Plaza, Laconia, NH 03246-3445. Tel.: 603-524-8813. Fax: 603-528-1228. Facebook: Historic Belknap Mill.
E-mail: information@belknapmill.org
Web Site: www.belknapmill.org
Founded: 1970.
Congressional District: 1
Governing Authority: nonprofit organization. Tax-exempt: 501(c)(3).
Institution Type/Description: Art & History Museum: housed in 1823 brick textile building.
Collections: mill architecture; textile machinery and knitted products; industrial knitting; photographs; documents.
Research Fields: industrial knitting.
Facilities: library; 16,000 sq. ft. exhibit space; 260-seat theatre. Art, crafts & mill-related items for sale.
Activities: guided tours; lectures; concerts; arts festivals; theatre; participatory, loan, temporary & traveling exhibits; outdoor concerts & walking tours by the river. Society Sponsors: Book Sale; Mill Anniversary.
Publications: program booklet; quarterly newsletter; teacher packet.
Hours & Admission Prices: Mon.-Fri. 9-5, Sat. 9-3, call to confirm & for special events. Fees for special events; discounts to National Trust members. Closed New Year's Day; Memorial Day; Independence Day; Labor Day; Christmas. &
Attendance: 50,000 (accurate)
Membership: Individual $35; Family $60; Patron $100; Benefactor $200; Friend $500.

LAKE WINNIPESAUKEE MUSEUM & HISTORICAL SOCI-ETY, 503 Endicott St. N., Rte. 3, Laconia, NH 03246-1725. Mailing Address: P.O. Box 5386, Weirs, NH 03247-5386. Tel.: 603-366-5950; 603-366-7301.
E-mail: info@lwhs.us
Web Site: www.lwhs.us
Founded: 1985.
Key Personnel: Dir., Melanie Benton; Chm. (V), Robert Lawton; Pres. (V), Brian Vincent; Cur., Lynda Laflamme; Museum Shop Mgr., Vynnie Hale.
Governing Authority: Tax-exempt.
Institution Type/Description: Historic House Museum.
Collections: local history; boats; beaches; photographs; models.
Hours & Admission Prices: May-Oct. Mon.-Sat. 10-4. No charge; donations accepted. &
Attendance: 1,000 (accurate)
Membership: Single $25; Family $40; Corporate $100.

PRESCOTT FARM ENVIRONMENTAL EDUCATION CEN-TER, 928 White Oaks Rd., Laconia, NH 03246. Tel.: 603-366-5695. Fax: 603-366-5720.
E-mail: info@prescottconservancy.org
Web Site: prescottconservancy.org
Institution Type/Description: Education Center.
Collections: farm history; photographs; personal artifacts; period furnishings; farm equipment & tools; live animals; historic buildings.
Facilities: nature trails; rental facilities.
Activities: hiking trails; school groups; birthday parties; rental facilities; summer camp; educational programs; special events; workshops.
Publications: newsletter.
Hours & Admission Prices: Call for hours.

Lancaster

THE LANCASTER HISTORICAL SOCIETY, 226 Main St., Lancaster, NH 03584-3038. Mailing Address: P.O. Box 473, Lancaster, NH 03584-0473. Tel.: 603-788-3004.
Founded: 1964.
Congressional District: 1

Key Personnel: Treas., Anne Morgan.
Governing Authority: nonprofit organization. Tax-exempt: 501(c)(3).
Institution Type/Description: Local History Museum: housed in 1780 Wilder-Holton House, first two-story house built in the county and Paul F. Smith Memorial Barn.
Collections: local history; archives pertaining to people of Lancaster; manuscripts; early local farm; logging; photographs; household artifacts.
Research Fields: local manufacturing; architecture; local artists; local history.
Facilities: 50-vol. library of local history; books by local authors available for use on premises. Postcards & bicentennial items for sale.
Activities: guided tours; films; permanent & temporary exhibitions.
Hours & Admission Prices: Tours: summer by appointment. No charge; donations accepted. &
Membership: Single $10.

Lebanon

AVA GALLERY AND ART CENTER, 11 Bank St., Lebanon, NH 03766-1749. Tel.: 603-448-3117. Fax: 603-448-4827.
E-mail: info@avagallery.org
Web Site: www.avagallery.org
Key Personnel: Exec. Dir., Bente Torjusen; Dir. Education, Adam Blue; Office Mgr., Constance Creed; Publications & Web, Carrie Fradkin; Coord. Exhibition, Margaret Jacobs; Studio Mgr., Murray Ngoima; Bookkeeper, Abigail Murphy.
Governing Authority: Tax-exempt: 501(c)(3)
Institution Type/Description: Art Gallery.
Collections: paintings; photographs; sculpture.
Publications: seasonal class brochures.
Hours & Admission Prices: Tues.-Sat. 11-5; other times by appointment. No charge. &
Membership: Individual $45; Family $65; Contributing $100-$249; Patron $250-$499; Sponsor $500-$999; Benefactor $1,000 & up.

Lisbon

LISBON AREA HISTORICAL SOCIETY, 6 S. Main St., Lisbon, NH 03585. Mailing Address: P.O. Box 6, Lisbon, NH 03585. Tel.: 603-838 6146.
E-mail: info@.lisbonareahistory.org
Web Site: www.lisbonareahistory.org
Founded: 1964.
Congressional District: 1
Governing Authority: Tax-exempt.
Institution Type/Description: Historical Society Museum.
Collections: local history; period artifacts & correspondence; photographs; genealogy; documents.
Activities: research; educational programs.
Publications: books: I & II Personal Traces, An Historical Collection of Personal Accounts from Lisbon, Lyman & Landaff, NH; 1883 map; Celebrating 250 Years, A Pictorial History of Lisbon, Lyman & Landaff, NH by Andrea M. Fitzgerald.
Hours & Admission Prices: May-Oct. Fri.1-3, other times By appointment. No charge; donations accepted.

Londonderry

AVIATION MUSEUM OF NEW HAMPSHIRE, 27 Navigator Rd., Londonderry, NH 03053-4403. Tel.: 603-669-4877. Facebook: Aviation Museum of New Hampshire.
E-mail: wberthelsen@nhahs.org
Web Site: www.nhahs.org
Founded: 1995.
Key Personnel: Dir., Wendell Berthelsen.
Governing Authority: Parent Institution: NHAHS.
Institution Type/Description: Aviation History Museum.
Collections: aviation history & artifacts; photographs; personal artifacts.
Facilities: Museum-related items for sale.
Activities: special events; classes; educational programs; aviation education programs for high school juniors & seniors.
Publications: Aeronaut.
Hours & Admission Prices: Fri.-Sat. 10-4, Sun. 1-4. Adults $5; members no charge. &
Membership: $25-$500.

Manchester

ALVA DE MARS MEGAN CHAPEL ART CENTER, (M), 100 Saint Anselm Dr., #1718, Saint Anselm College, Manchester, NH 03102-1308. Tel.: 603-641-7470. Fax: 603-641-7116.
E-mail: chapelartcenter@anselm.edu
Web Site: www.anselm.edu/chapelart
Founded: 1967.
Congressional District: 1
Key Personnel: Dir., Iain MacLellan, O.S.B.; Asst. Cur., Julia Welch; Administrative Asst., Pamela Condon.
Personnel Profile: Full-Time Paid 2; Part-Time Paid 1; Interns 1.
Governing Authority: college. Parent Institution: Saint Anselm College. Tax-exempt: 501(c)(3).
Institution Type/Description: College Art Museum & Center: housed in the former college chapel; built in 1923.
Collections: works by area craftsmen; sculpture; paintings; wall & ceiling frescoes; stained glass.
Research Fields: 19th- & 20th-century art; old masters.
Activities: lectures; gallery talks; loan, temporary & traveling exhibitions; concerts.
Hours & Admission Prices: Tues.-Wed. & Fri.-Sat. 10-4, Thurs. 10-7. No charge. Closed holidays; college breaks; between exhibitions. &
Attendance: 5,000 (accurate)

AMERICA'S CREDIT UNION MUSEUM, 418-420 Notre Dame Ave., Manchester, NH 03102. Mailing Address: P.O. Box 603, Manchester, NH 03105-0603. Tel.: 603-629-1553. Fax: 603-629-1595.
Key Personnel: Exec. Dir., Peggy Powell
Institution Type/Description: History Museum.
Collections: credit union history; photographs; personal artifacts.
Hours & Admission Prices: Mon., Wed. & Fri. 10-12 & 1-4; other times by appointment.

* **CURRIER MUSEUM OF ART, (M),** 150 Ash St., Manchester, NH 03104-4347. Tel.: 603-669-6144. Fax: 603-669-7194. Facebook: Currier Museum.
E-mail: visitor@currier.org
Web Site: www.currier.org
Founded: 1929.
Congressional District: 1
Key Personnel: C.E.O. & Dir., Susan E. Strickler; Pres. (V), David Jensen; Dir. Finance, Sherry Collins; Chief Cur., P. Andrew Spahr; Assoc. Cur., Kurt Sundstrom; Head Public Programs, Leah Fox; Art Center Dir., Bruce McColl; Museum Shop Mgr., Heidi Norton.
Personnel Profile: Full-Time Paid 30; Part-Time Paid 76; Part-Time Volunteers 150.
Governing Authority: nonprofit organization. Tax-exempt: 501(c)(3).
Institution Type/Description: Art Museum.
Collections: European and American paintings & sculpture; American 17th, 18th & early 19th-century furniture, silver, glass, pewter and textiles; graphics; 15th-century Tournai tapestry. Historic House: 1950 Zimmerman House, a Usonian home designed by Frank Lloyd Wright.
Major Exhibits: American Posters 1890-1905, 5/25/13-9/2/14; Africa Interweave: Textile Diasporas, 9/13-1/14; Romare Bearden: A Black Odyssey (T), 3/29/13-6/22/14.
Research Fields: pertaining to collections.
Facilities: 7,000-vol. library of art books; periodicals; 180-seat auditorium; classrooms. Paperback books, postcards, reproductions & gallery publications for sale.
Activities: guided group tours; lectures; gallery talks; concerts; formally organized educational programs; children's studio art classes; docent program or council; inter-museum loan, permanent & temporary exhibitions.
Publications: annual report; exhibition catalogs; calendar; membership & exhibition brochures.
Hours & Admission Prices: Sun.-Mon. & Wed.-Fri. 11-5, Sat. 10-5, 1st Thurs. of month 11-8. Adults $10, seniors $9, students $8; members & children under 18 no charge. &
Attendance: 66,339 (accurate)
Membership: Student $30; Individual $45 & up.

LAWRENCE L. LEE SCOUTING MUSEUM, 40 Blondin Rd., Manchester, NH 03109-5907. Mailing Address: 571 Holt Ave., Manchester, NH 03109-5213. Tel.: 603-669-8919; 603-625-6431. Fax: 603-625-2467.
E-mail: administrator@scoutingmuseum.org

Web Site: www.scoutingmuseum.org
Founded: 1969.
Congressional District: 1
Key Personnel: Scout Exec., Donald D. Shepard, Jr.; Chm. (V), Richard Zeloski.
Personnel Profile: Part-Time Volunteers 12.
Governing Authority: nonprofit. Daniel Webster Council, Boy Scouts of America, Manchester. Tax-exempt: 501(c)(3).
Institution Type/Description: Scouting Museum.
Collections: Boy Scouts of America memorabilia; uniforms; equipment; badges & patches; books; manuals; manuscripts; periodicals; items from world Scouting; Scouts on stamps.
Research Fields: pertaining to Scouting.
Facilities: 2,800 plus-vol. library of official handbooks; training manuals; yearbooks; diaries; pamphlets; fiction; Boys' Life & Scouting magazines; manuscripts & photographs, available for use on-site by the public; 3,000 sq. ft. exhibit space. Scouting items for sale.
Activities: guided tours; scavenger hunt; Jamboree-on-the-Air.
Publications: quarterly Scouting historical journal, Scout Memorabilia.
Hours & Admission Prices: July-Aug. Sat. 10-4, call for additional hours; Sept.-June Sat. 10-4. No charge; donations accepted. Closed Independence Day; Christmas. &
Attendance: 3,037 (accurate)
Membership: Friend $25; Associate $50; Advocate $100; Patron $250; Benefactor $500; Golden Eagle $1,000.

MCININCH ART GALLERY - SOUTHERN NEW HAMP-SHIRE UNIVERSITY, Robert Frost Hall, 2500 N. River Rd., Manchester, NH 03106. Tel.: 603-629-4622. Fax: 603-668-2211, ext. 2228.
E-mail: d.disston@snhu.edu
Web Site: www.snhu.edu/art
Founded: 2001.
Key Personnel: Dir., Debbie Disston.
Personnel Profile: Full-Time Paid 1; Part-Time Paid 14.
Governing Authority: Parent Institution: Southern New Hampshire University. Tax-exempt.
Institution Type/Description: Art Gallery.
Collections: paintings; sculpture.
Activities: demonstrations; lectures; concerts; special events; educational programs.
Hours & Admission Prices: Academic Year: Mon.-Wed. & Fri.-Sat. 10-3, Thurs. 10-3 & 5-8. No charge; donations accepted. Closed university holidays & breaks. &
Attendance: 5,987 (accurate)

MILLYARD MUSEUM, (M), 200 Bedford St., Manchester, NH 03101. Mailing Address: 200 Bedford St., 1st Fl., Manchester, NH 03101-1132. Tel.: 603-622-7531. Fax: 603-641-8191.
E-mail: history@manchesterhistoric.org
Web Site: www.manchesterhistoric.org
Founded: 1896.
Congressional District: 2
Key Personnel: Exec. Dir., Aurore Eaton; Pres., Elizabeth L. La Rocca; Cur. Museum Collection, Marylou Ashooh Lazos; Cur. Library Collection, Eileen O'Brien.
Personnel Profile: Full-Time Paid 4; Part-Time Paid 4; Part-Time Volunteers 10; Interns 4.
Governing Authority: society. Parent Institution: Manchester Historic Association. Tax-exempt: 501(c)(3).
Institution Type/Description: Local History Museum.
Collections: history of greater Manchester area from prehistory to present; decorative arts; architectural illustrations & fragments; fire-fighting equipment; tools; costumes; paintings; prints; guns; toys; photographs; business records & maps; manuscripts; textile manufacturing records.
Research Fields: history of the Manchester area; early textile industry; labor history; Native Americans; immigration.
Facilities: research center.
Activities: guided tours; lectures; inter-museum loan; permanent & temporary exhibitions; special events; historic preservation workshops; concerts; walking tours; programs for children & families.
Publications: MHA News; calendar of events; Picturing Manchester; Photos from the MHA Collection; reprint of 1948 Manchester on the Merrimack; Manchester Streetcars; Valley Cemetery Walking Tour; Video: Preserving Manchester; Lake Massabesic; Images Then & Now; Remembering the Valley Cemetery; Manchester Stories.
Hours & Admission Prices: Museum: Wed.-Sat. 10-4. Research Center: Wed. & Sat. 10-4. Adults $6, seniors & college students $5, children 6-18 $2; discounts to AASLH & NEMA members; children under 6, members &

library no charge. Closed New Year's Day; Independence Day; Veterans Day; Thanksgiving; Christmas Eve & Day. &
Attendance: 6,000 (estimated)
Membership: Student & Senior $25; Individual $30; Senior Family $50; Family & Dual $60; Patron $125; Benefactor $250; Sponsor $500; F.P. Carpenter Circle $1,000; Corporate $250-$10,000.

NEW HAMPSHIRE INSTITUTE OF ART, 148 Concord St., Manchester, NH 03104-4858. Tel.: 603-623-0313. Fax: 603-641-1832.
Founded: 1898.
Congressional District: 1
Key Personnel: Pres., Roger Williams; Academic Dean, Patrick McCay; Dir. Devel., Jessica Kinsey; Dir. Admissions, Liam Sullivan; Vice Pres. Finance & Administration, Erik Gross; Shop Mgr., Joe Vivilecchia; Dir. Gallery, Alison Williams.
Personnel Profile: Full-Time Paid 48; Part-Time Paid 17.
Governing Authority: school; nonprofit educational. Tax-exempt: 501(c)(3).
Institution Type/Description: Art Institute.
Collections: contemporary regional fine arts & traditional crafts.
Research Fields: fine arts; institute; contemporary arts.
Facilities: rental facilities; 450-seat auditorium.
Activities: academic & public programs; educational services for schools & teachers; visual arts exhibitions; concerts; programs for young people.
Publications: admissions viewbook; quarterly newsletter; CE course catalogs; annual report; exhibition postings.
Hours & Admission Prices: Mon.-Fri. 9-5, Sat. 9-12. No charge. Closed legal holidays. &
Attendance: 27,500 (estimated)

SEE SCIENCE CENTER, 200 Bedford St., Manchester, NH 03101-1153. Tel.: 603-669-0400. Fax: 603-669-0400.
E-mail: info@see-sciencecenter.org
Web Site: www.see-sciencecenter.org
Founded: 1986.
Congressional District: 1
Key Personnel: Dir., Douglas Heuser; Education & Membership, Rebecca Mayhew; Operations & Design, Adele Maurier; Devel. Coord., Peter Gustafson.
Personnel Profile: Full-Time Paid 6; Part-Time Paid 15; Part-Time Volunteers 20.
Governing Authority: private; nonprofit organization. Tax-exempt: 501(c)(3).
Institution Type/Description: Science Museum.
Collections: hands-on exhibitions; Lego reproduction of Millyard.
Facilities: educational facilities. Museum-related items for sale.
Activities: formal education programs; moon walk; chem lab; momentum chair; guided tours; traveling exhibitions.
Publications: quarterly newsletter, SEE News.
Hours & Admission Prices: Mon.-Fri. 10-4, Sat.-Sun. 10-5. Admission $8; children 2 & under and members no charge. Closed New Year's Day; Easter; Memorial Day; Independence Day; Labor Day; Thanksgiving; Christmas Eve & Day. &
Attendance: 75,000 (estimated)
Membership: Family $80.

Melvin Village

TUFTONBORO HISTORICAL SOCIETY & MUSEUM, Rte. 109, Melvin Village, NH 03850. Mailing Address: P.O. Box 372, Melvin Village, NH 03850-0372. Tel.: 603-544-7225.
Key Personnel: Dir., Margaret Bashe
Institution Type/Description: History Museum.
Collections: local Tuftonboro history.
Hours & Admission Prices: July-Aug. Mon.-Fri. 2-4. No charge.

Meredith

MEREDITH CHILDREN'S MUSEUM, 28 Lang St., Meredith, NH 03253-5824. Tel.: 603-279-6307.
Institution Type/Description: Children's Museum.
Collections: hands-on exhibits.
Activities: hands-on exhibits.
Hours & Admission Prices: Mon.-Fri. 10-4, Sat.-Sun. 12-4. Admission $6. Closed Easter; Thanksgiving; Christmas.
Membership: Family $125.

Meriden

AIDRON DUCKWORTH ART MUSEUM, (M), 21 Bean Rd., Meriden, NH 03770. Mailing Address: P.O. Box 61, Meriden, NH 03770-0061. Tel.: 603-469-3444.
E-mail: info@aidronduckworthmuseum.org
Web Site: www.aidronduckworthmuseum.org
Founded: 2002.
Key Personnel: Dir. & Trustee, Grace W. Harde.
Personnel Profile: Full-Time Volunteers 1.
Volunteer Hours: 1,650
Operating Expenses: 44,710
Operating Income: 45,606
Governing Authority: Parent Institution: Aidron Duckworth Art Preservation Trust. Tax-exempt.
Institution Type/Description: Art Museum.
Collections: artwork by Aidron Duckworth & local artists.
Major Exhibits: Lorna Ritz Drawings/Paintings, 4/14-6/14; Exhibition XXIII - Duckworth, 4/14-7/14; Bhaktiziek - Weavings, 6/14-7/14; Mark Shapiro - Ceramics, 6/14-7/14; Aya Itagaki, 8/14-9/14; Exhibition XXIV - Duckworth, 8/14-10/14; Craig Stockwell, 9/14-10/14.
Research Fields: Aidron Duckworth's art & writings on teaching art.
Facilities: archives; artists studio space.
Activities: temporary exhibitions.
Publications: booklet, The Paintings and Drawings by Aidron Duckworth; "The Artists Eye" Essays by Aidron Duckworth 1981-82.
Hours & Admission Prices: Fri.-Sun. 10-5; other times by appointment. No charge; donations accepted. &
Attendance: 750 (estimated)

Milton

NEW HAMPSHIRE FARM MUSEUM, INC., 1305 White Mountain Hwy., Milton, NH 03851. Mailing Address: P.O. Box 644, Rte. 125, Milton, NH 03851-0644. Tel.: 603-652-7840. Fax: 603-652-7840 (call first).
E-mail: info@farmmuseum.org
Web Site: www.farmmuseum.org
Founded: 1970.
Congressional District: 1
Key Personnel: Exec. Dir., Kathleen Shea; Pres. (V), Otis Perry.
Personnel Profile: Full-Time Paid 2; Part-Time Paid 3; Part-Time Volunteers 6; Interns 2.
Governing Authority: bd. of trustees; nonprofit organization. Tax-exempt: 501(c)(3).
Institution Type/Description: Agriculture & Rural Life Museum: former Jones Farm with buildings dating from 1780s-1900s; the Plummer Homestead.
Collections: farm tools; machinery; implements; oral histories; photographs; books; artifacts from late 1800s to early 1900s; farm buildings; shoe shop; blacksmith shop; family farm life of the 1890's; tractors; carriages.
Research Fields: agriculture; New England farming.
Facilities: library.
Activities: house tours; country store; special events; workshops; demonstrations; permanent exhibits; farm animal tours. Annual Events: Farm Day; Children's Day.
Publications: quarterly newsletter; special events brochure & calendar.
Hours & Admission Prices: Memorial Day to mid-Oct. Wed.-Sun. 10-4. Adults $7, children 4-17 $4; discount to NEMA members & groups of 10 or more; members no charge. &
Attendance: 4,000 (estimated)
Membership: Individual $20; Couple $30; Family $45; Contributing $50; Supporting $100; Sustaining $300; Patron $500.

Moultonborough

CASTLE IN THE CLOUDS, 455 Old Mountain Rd., Moultonborough, NH 03254. Mailing Address: P.O. Box 687, Moultonborough, NH 03254-0687. Tel.: 603-476-5900. Fax: 603-476-2512. Facebook: Castle in the Clouds.
E-mail: info@castleintheclouds.org
Web Site: castleintheclouds.org
Key Personnel: Dir., Michael P. Desplaines; Pres. (V), Ann Hackl.
Personnel Profile: Full-Time Paid 4; Full-Time Volunteers 1; Part-Time Paid 50; Part-Time Volunteers 25.
Governing Authority: Parent Institution: Castle Preservation Society. Tax-exempt.
Institution Type/Description: Historic House: housed in Lucknow, the former estate of Tom & Olive Plant; built in 1914.
Collections: family history; personal artifacts; period furnishings; photographs.

Facilities: cafe; picnic area. Gift items for sale.
Activities: hiking; picnic area; horseback riding.
Publications: Castle Chronicle.
Hours & Admission Prices: May 11-June 7 Sat.-Sun. 10-4; June 8-Oct. 19 daily 10-4. Adults $16, seniors 65 & over and students 16-20 $12, children 5-15 $8; children 4 & under no charge. &
Attendance: 52,000 (accurate)
Membership: Student $30; Individual $45; Dual & Family $85; Griffin $150; The Pebble $200; Castle $500; Lucknow $1,000.

THE LOON CENTER AND MARKUS WILDLIFE SANCTUARY, 183 Lee's Mill Rd., Moultonborough, NH 03254. Mailing Address: P.O. Box 604, Moultonborough, NH 03254-0604. Tel.: 603-476-5666. Fax: 603-476-5497.
E-mail: info@loon.org
Web Site: www.loon.org
Founded: 1975.
Institution Type/Description: Wildlife Sanctuary.
Collections: New Hampshire loons & their habitats; photographs.
Facilities: nature trails. Museum-related items for sale.
Activities: presentations; educational programs; videos.
Publications: newsletter.
Hours & Admission Prices: Loon Center: Mon.-Sat. 9-5; July-Columbus Day daily 9-5; Dec.-May 15 Thurs.-Sat. 9-5. Markus Sanctuary Walking Trails: daily dawn to dusk. No charge; donations accepted. Closed winter holidays; day after Thanksgiving; afternoons of Christmas Eve & New Year's Eve. &
Attendance: 10,000 (estimated)
Membership: Senior & Student $30; Adult $40.

Nashua

JOHN M. HUNT MEMORIAL BUILDING, 6 Main St., Nashua, NH 03064-2712. Tel.: 603-594-3661.
E-mail: mortellaroc@nashuanh.gov
Web Site: www.huntbuilding.org
Founded: 1903.
Key Personnel: Chm. (V), Joy Barrett.
Personnel Profile: Part-Time Paid 1.
Governing Authority: city.
Institution Type/Description: Historic Building: housed in the former Nashua Public Library. Listed on the National Register of Historic Buildings.
Collections: local history; photographs; period furnishings.
Activities: rental facilities.
Hours & Admission Prices: Tours: by appointment. No charge; donations accepted. &
Attendance: 1,000 (estimated)

THE NASHUA HISTORICAL SOCIETY - ABBOT-SPALDING HOUSE MUSEUM AND FLORENCE SPEARE MEMORIAL MUSEUM, 5 Abbott St., Nashua, NH 03064-2119. Tel.: 603-883-0015. Fax: 603-889-8515.
E-mail: nashuahistorical@comcast.net
Web Site: www.nashuahistoricalsociety.org
Founded: 1870.
Personnel Profile: Full-Time Paid 1; Full-Time Volunteers 2; Part-Time Paid 3; Part-Time Volunteers 62.
Volunteer Hours: 1,200
Operating Expenses: 215,258
Operating Income: 191,497
Governing Authority: Parent Institution: Nashua Historical Society. Tax-exempt.
Institution Type/Description: House Museum.
Collections: Nashua history & genealogy; period furnishings; fine glass & china; paintings.
Major Exhibits: Adopt An Artifact, 12/13-12/14.
Research Fields: Nashua history & genealogy.
Facilities: library.
Activities: lecture series; school tours; scouting badge & patch support.
Publications: newsletter; book, Nashua People and Places (2011).
Hours & Admission Prices: Abbot-Spalding Museum: by appointment. Florence H. Speare Memorial Museum: March-Thanksgiving weekend Tues.-Thurs. 10-4; other times by appointment. Abbot-Spalding House Museum: $2 per person.
Attendance: 550 (estimated)
Membership: Student $5; Adult $25; Family $40; Lifetime $250.

New Ipswich

✻ BARRETT HOUSE, FOREST HALL, (M), 79 Main St., New Ipswich, NH 03071-3716. Mailing Address: Historic New England, 143 Pleasant St., Portsmouth, NH 03801. Tel.: 617-994-6675. Fax: 860-963-2208.
Web Site: www.historicnewengland.org
Founded: 1948.
Congressional District: 2
Key Personnel: Pres., Carl Nold; Regl. Site Mgr., Craig Tuminaro.
Governing Authority: society; nonprofit organization. Parent Institution: Historic New England, 141 Cambridge St., Boston, MA 02114. Tax-exempt: 501 (c)(3).
Institution Type/Description: Historic House: c.1800 Barrett House, Forest Hall, Federal residence.
Collections: 18th- & 19th-century furniture; period musical instruments; Gothic Revival summer house.
Activities: guided tours; lectures; outdoor events.
Publications: visitors guide.
Hours & Admission Prices: June-Oct. 15 2nd & 4th Sat. of month 11-5; last tour at 4. Grounds: dawn to dusk. Adults $5, seniors $4, students $2.50; discounts to seniors, AAM, AAA, ICOM, WGBH members; New Ipswich residents & Historic New England members no charge.
Attendance: 840 (accurate)
Membership: National $35; Individual $45; Household $55; Garden & Landscape $75; Contributing and Library & School $100; Historic New England Affiliate $100-$350; Young Friends of Historic New England $100-$1,500; Friends of the Library and Archives $125; Historic Homeowner $200; Business & Ogden Codman Design Group $250; Appleton Circle $1,750-$3,500.

New London

NEW LONDON HISTORICAL SOCIETY, 179 Little Sunapee Rd., New London, NH 03257. Mailing Address: P.O. Box 965, New London, NH 03257-0965. Tel.: 603-526-6564.
E-mail: info@newlondonhistoricalsociety.org
Web Site: www.nlhs.net
Founded: 1954.
Congressional District: 2
Key Personnel: Pres. (V), M.B. Ford.
Personnel Profile: Part-Time Volunteers 100.
Governing Authority: society. Tax-exempt.
Institution Type/Description: General Museum.
Collections: folklore; agriculture; archives; costumes; manuscripts; military; horse drawn vehicles; firefighting equipment. Historic Buildings: 1800 Griffin barn; 1820 Scytheville house; 1820 blacksmith shop; 1830 country store; 1830 Pleasant St. schoolhouse; 1830 carriage house; 1830 corn crib; violin & carriage painting shop; meeting house; acquisitions building; Hearse house; Concord coach; transportation building.
Facilities: 40-vol. collection of handwritten town records, school records, store ledgers, manuscripts available for inter-library use.
Activities: guided tours; lectures; formally organized education programs for children; permanent & temporary exhibitions; special programs.
Publications: triannual newsletter; irregular local history articles.
Hours & Admission Prices: Memorial Day to June & Sept.-Columbus Day Sun. & Tues. 12:30-3:30; July-Aug. Tues. 12:30-3:30; other times by appointment. Adults $6, members $4.
Attendance: 1,500 (estimated)
Membership: Regular $50; Contributing $100; Supporting $150; Sustaining $200; Scytheville Society $500; Phillips Society $1,000.

Newbury

THE FELLS, 456 Rte. 103A, Newbury, NH 03255. Mailing Address: P.O. Box 276, Newbury, NH 03255-0276. Tel.: 603-763-4789. Fax: 603-763-2452.
E-mail: info@thefells.org
Web Site: www.thefells.org
Founded: 1996.
Congressional District: 2
Key Personnel: Exec. Dir., Susan warren; Chm. (V), John Ferries; Museum Shop Mgr., Mary Lou McCrave.
Personnel Profile: Full-Time Paid 2; Part-Time Paid 12; Part-Time Volunteers 300; Interns 3.
Governing Authority: Tax-exempt.
Institution Type/Description: Historic House & Gardens: former estate of American writer and diplomat John M. Hay. Listed on the National Register of Historic Places.
Collections: local history & culture.

Research Fields: New Hampshire history; John Milton Hay; local ecology.
Facilities: formal gardens; nature trails; meeting facilities.
Activities: guided tours; hiking; classes; programs; temporary & permanent exhibits; children's camps, programs, & nature explorer backpacks.
Publications: newsletter.
Hours & Admission Prices: Gardens & Trails: dawn to dusk. Main House: Memorial Day weekend-Columbus Day Sat.-Sun. & Mon. holidays 10-4; June 18 to Labor Day Wed.-Sun. 10-4. Adults $10, seniors 65 & over and students with ID $8, children 6-17 $4; discounts to AAM, AAA & Garden Conservancy members, families & during the off season; members and children 5 & under no charge.
Attendance: 10,000 (accurate)
Membership: Individual $40; Household $60; Sponsor $125; Advocate $250; Patron $500; Benefactor $1,000; Sustainer $2,500.

North Conway

CONWAY SCENIC RAILROAD, INC., 38 Norcross Circle, North Conway, NH 03860. Mailing Address: P.O. Box 1947, North Conway, NH 03860-1947. Tel.: 603-356-5251; 800-232-5251. Fax: 603-356-7606. Facebook: Conway Scenic Railroad.
E-mail: info@conwayscenic.com
Web Site: www.conwayscenic.com
Founded: 1974.
Congressional District: 1
Key Personnel: Pres., Russell G. Seybold; Operations Mgr., Paul Hallett.
Personnel Profile: Full-Time Paid 8; Part-Time Paid 36.
Governing Authority: for profit corporation.
Institution Type/Description: Railroad Museum: housed in 1874 Victorian, wood-framed railroad station. A National Historic Landmark.
Collections: railroad artifacts & memorabilia; northern New England railroad items; period locomotives & railroad cars; 1898 Pullman sleeper-parlor-observation car; roundhouse; operating turntable; restored, operating dining car, old photographs.
Research Fields: New England railroad history.
Facilities: 250-vol. library of railroad materials; 47-seat railroad dining car. Museum-related items for sale.
Activities: organized guided tours for groups by reservation; self-guided walking tour; participatory exhibits. Museum Sponsors: Railfan's Day in September.
Hours & Admission Prices: Train Rides: Valley Train: mid-April to mid-May & Nov. to mid-Dec. Sat.-Sun.; mid-May to Oct. & Christmas vacation daily. Notch Train: late-June to early-Sept. Tues.-Sat.; early-Sept. to mid-Oct. daily. Adults $15 & up; discounts to groups & AAA members; museum & grounds no charge.
Attendance: 100,000 (accurate)

MOUNT WASHINGTON MUSEUM AND THE WEATHER DISCOVERY CENTER, 2779 White Mountain Hwy., North Conway, NH 03860-5194. Mailing Address: P.O. Box 2310, North Conway, NH 03860-2310. Tel.: 603-356-2137, ext. 203. Fax: 603-356-0307.
E-mail: email@mountwashing.org
Web Site: www.mountwashington.org
Founded: 1932.
Congressional District: 2
Key Personnel: Pres., Jack Middelton; Exec. Dir., Scot Henley; Dir. Museum Operations & Museum Shop Mgr., Bill Grenfell; Dir. Education, Michelle Cruz.
Personnel Profile: Full-Time Paid 15; Part-Time Paid 5; Part-Time Volunteers 15; Interns 4.
Governing Authority: nonprofit corporation. Parent Institution: Mount Washington Observatory. Tax-exempt: 501(c)(3).
Institution Type/Description: Park & Weather Museum: located on summit of Mount Washington.
Collections: over 100 specimens of local flora; White Mountains literature; artifacts; historical photos.
Research Fields: cloud physics; icing; cosmic rays.
Facilities: 2,600-vol. library of books on the White Mountains available for research on site; field research station. Educational material, books, pamphlets, weather instruments, photographs, slides & posters for sale.
Activities: occasional topical workshops; annual symposia.
Publications: quarterly, News Bulletin, Windswept.
Hours & Admission Prices: Mount Washington Museum: late May to mid-Oct. daily 9-6. Weather Discovery Center: daily 10-5. No charge; donations accepted.
Attendance: 40,000 (estimated)
Membership: Student $30; Senior $35; Individual $45; Family $70.

North Hampton

FULLER GARDENS, 10 Willow Ave., North Hampton, NH 03862-2228. Tel.: 603-964-5414.
E-mail: fullergrdn@aol.com
Web Site: www.fullergardens.org
Founded: 1927.
Key Personnel: C.E.O. & Dir., Jamie Colen; Museum Shop Mgr., Victoria Kaiser.
Personnel Profile: Full-Time Paid 3; Part-Time Paid 6; Part-Time Volunteers 20.
Governing Authority: nonprofit organization. Parent Institution: Fuller Foundations. Tax-exempt: 501(c)(3).
Institution Type/Description: Botanical Garden.
Collections: horticultural display gardens; 1,500 rose bushes; English perennial gardens, Japanese garden; Hosta garden; tropical & desert conservatory.
Facilities: gardens.
Activities: self-guided tour of estate & formal gardens; summer calendar of social events & seminars. Gift items for sale.
Hours & Admission Prices: mid-May to Oct. daily 10-5:30. Adults $9, senior citizens $8, children under 10 $4.50; discounts for groups of 10 or more; members & infants no charge. &

Attendance: 11,000 (estimated)
Membership: Annual $50; Perennial $65; Rose $75.

North Salem

AMERICA'S STONEHENGE, 105 Haverhill Rd., North Salem, NH 03073. Mailing Address: P.O. Box 84, North Salem, NH 03073-0084. Tel.: 603-893-8300.
E-mail: info@stonehengeusa.com
Web Site: www.stonehengeusa.com
Founded: 1958.
Key Personnel: C.E.O. & Pres., Robert E. Stone; Gen. Mgr. & Public Rels., Dennis W. Stone; Operations Mgr., Patricia Stone.
Personnel Profile: Full-Time Paid 5; Part-Time Paid 1.
Governing Authority: for-profit organization; private operation.
Institution Type/Description: Archaeological Site: located on Mystery Hill, site of 4,000 year old archaeo-astronomical site.
Collections: late 18th- & early 19th century rural household items; late Archaic, early & late Woodland tools & pottery; stone tools; carved petroglyphs; archives.
Research Fields: Mystery Hill site origins & attribution of controversial elements; archival research; preservation.
Facilities: library of research files & photographs; site related books & clippings; food service; picnic area; educational facilities; 300 sq. ft. exhibit space; 50-seat theatre; snack bar. Books & museum-related items for sale.
Activities: self-guided tours with guide maps; guided tours for groups of 20 or more.
Publications: annual bibliography of material; newsletter, Horizon; booklets, Quarry Techniques; Myth & Mythology in the Land of Academe.
Hours & Admission Prices: Daily 9-5. Adults $10, seniors $7, children 13-18 $6, children 6-12 $5. Closed Thanksgiving; Christmas. &
Attendance: 35,000 (estimated)
Membership: Single $35; Family $75.

North Sutton

MUSTER FIELD FARM MUSEUM & THE MATTHEW HARVEY HOMESTEAD, Harvey Rd., North Sutton, NH 03260. Mailing Address: P.O. Box 118, North Sutton, NH 03260-0118. Tel.: 603-927-4276.
E-mail: musterfield@tds.net
Web Site: www.musterfieldfarm.com
Founded: 1990.
Congressional District: 2
Institution Type/Description: Farm Museum: housed in an 18th century farmhouse. Listed on the National Register of Historic Places.
Collections: local agricultural history; early architecture; personal artifacts; period furnishings; photographs. Historic Buildings: barns; corn cribs; schoolhouse; blacksmith shop; houses.
Activities: special events. Museum Sponsors: Ice Day in January; June Jam; Farm Days in August; Harvest Day in October.
Hours & Admission Prices: Farm Buildings: daily. Homestead: July-Oct. 3 Sun. 1-4. June Jam: $10 per person. Farm Days & Harvest Day: $5 per person; members & children under 6 no charge. Ice Day: no charge.
Membership: See website.

Peterborough

AQUARIUS NO. 1 FIRE MUSEUM, Summer St., Peterborough, NH 03458. Mailing Address: Peterborough Fire & Rescue Assoc., Inc., P.O. Box 244, Peterborough, NH 03458. Tel.: 603-924-8090.
Institution Type/Description: Firefighting History Museum.
Collections: firefighting history & equipment; photographs; Thayer 1803 hand tub; late 1800s hose reel; 1914 American LaFrance fire truck; Remembrance Walkway.
Hours & Admission Prices: By appointment.

MARIPOSA MUSEUM & WORLD CULTURE CENTER, 26 Main St., Peterborough, NH 03458-2420. Tel.: 603-924-4555. Fax: 603-924-7893.
E-mail: info@mariposamuseum.org
Web Site: www.mariposamuseum.org
Founded: 2002.
Key Personnel: Dir., David Blair; Pres. (V), Francoise Bourdon; Admin., Mose Olenik; Museum Mgr., Nadiya Weidman.
Personnel Profile: Full-Time Paid 1; Part-Time Paid 5; Part-Time Volunteers 50.
Governing Authority: bd. of directors. Parent Institution: Journeys in Education, Inc. Tax-exempt.
Institution Type/Description: Art Museum.
Collections: cultural arts from around the world; costumes; toys; musical instruments; puppets; paintings.
Facilities: library. Museum-related items for sale.
Activities: workshops; hands-on exhibits; special events & programs; temporary exhibitions.
Hours & Admission Prices: Summer: daily 11-5; Winter: Wed.-Sun. 11-5; groups by appointment. Adults $5, children $3; first Fri. of month 5-9 & members no charge. Closed Federal holidays. &
Attendance: 5,000 (estimated)
Membership: Teacher $25; Adult & one Child $40; Family $60; Crescent $100; Admiral $150; Viceroy $250; Queen $500; Monarch $1,000.

THE MONADNOCK CENTER FOR HISTORY AND CULTURE, 19 Grove St., Peterborough, NH 03458-1422. Mailing Address: P.O. Box 58, Peterborough, NH 03458-0058. Tel.: 603-924-3235. Fax: 603-924-3200.
E-mail: info@monadnockcenter.org
Web Site: www.monadnockcenter.org
Formerly: Peterborough Historical Society
Founded: 1902.
Congressional District: 2
Key Personnel: Exec. Dir., Michelle M. Stahl.
Personnel Profile: Full-Time Paid 1; Part-Time Paid 1; Part-Time Volunteers 60; Interns 4.
Governing Authority: society. Tax-exempt: 501(c)(3).
Institution Type/Description: Historical Society Museum.
Collections: town & state history artifacts; 19th-century manuscripts pertaining to textile mills & village life.
Research Fields: small town life; 19th-century New England; 19th-century mills.
Facilities: library; archives.
Activities: guided tours; permanent & temporary exhibitions; research by appointment.
Hours & Admission Prices: Wed.-Sat. 10-4. Adults $3; members no charge. &
Attendance: 7,500 (estimated)
Membership: Educator $20; Individual $25; Family $40; Sustaining $50; Benefactor $100; Patron $500.

Plymouth

KARL DRERUP ART GALLERY, Plymouth State College, 17 High St., MSC 21B, Plymouth, NH 03264-1595. Tel.: 603-535-2614. Fax: 603-535-2938. TDD: 603-535-2679.
E-mail: camidon@plymouth.edu
Web Site: www.plymouth.edu/psc/gallery/
Founded: 1969.
Congressional District: 2
Key Personnel: Gallery Dir., Catherine Amidon.
Personnel Profile: Full-Time Paid 1; Part-Time Paid 9; Interns 1.
Governing Authority: public university; nonprofit. Tax-exempt: 501(c)(3).
Institution Type/Description: University Art Gallery.
Collections: paintings; drawings.
Activities: lectures; formal programs for undergraduate & graduate students in conjunction with certain exhibitions; traveling exhibits; poetry readings; films; discussions.

Hours & Admission Prices: Sept.-May Mon.-Tues. & Thurs.-Sat. 10-4, Wed. 10-8. No charge. Closed spring break; college holidays; Thanksgiving; Christmas. &

Attendance: 6,000 (accurate)

Membership: Associate $10; Member $25; Supporter $50; Art Appreciator $100; Art Friend $500; Art Connoisseur $1,000 & up.

Portsmouth

ALBACORE PARK SUBMARINE MUSEUM, 600 Market St., Portsmouth, NH 03801-7313. Tel.: 603-436-3680. Facebook: USS Albacore Submarine and Museum.

E-mail: albacorepark@myfairpoint.net

Web Site: www.ussalbacore.org

Founded: 1986.

Key Personnel: Dir., John Maier; Pres. (V), Phillip Munck.

Personnel Profile: Full-Time Paid 1; Part-Time Paid 2; Part-Time Volunteers 20.

Governing Authority: Parent Institution: Port of Portsmouth Maritime Museum Assoc. Tax-exempt.

Institution Type/Description: Submarine Museum.

Collections: Albacore history; photographs; submarine artifacts; crew member history.

Facilities: visitor center; memorial garden; US Navy submarine. Museum-related items for sale.

Publications: U.S.S. Albacore History.

Hours & Admission Prices: Memorial Day to Columbus Day daily 9:30-5; Oct.-May Thurs.-Mon. 9:30-4; groups by appointment. Family $12, adults $6, retired military (20 years) $4, children 7-17 $3; discounts to Historic Naval Ships Assoc. members & Tin Can Sailors; children under 7 & active duty no charge.

Attendance: 32,000 (accurate)

Membership: Annual $50.

* **GOVERNOR JOHN LANGDON HOUSE, (M),** 143 Pleasant St., Portsmouth, NH 03801-4506. Mailing Address: 141 Cambridge St., Boston, MA 02114-2702. Tel.: 603-436-3205; 617-227-3956. Fax: 603-436-4651.

Web Site: www.historicnewengland.org

Founded: 1947.

Key Personnel: Pres. & C.E.O., Carl Nold; Site Mgr., Craig Tuminaro.

Governing Authority: society; nonprofit organization. Parent Institution: Historic New England, 141 Cambridge St., Boston, MA 02114. Tel.: 617-227-3956. Tax-exempt: 501(c)(3).

Institution Type/Description: Historic House: 1784 home of Governor John Langdon.

Collections: carvings; furniture & other artifacts from the 18th & 19th century.

Facilities: rental facilities.

Activities: guided tours; lectures; special events.

Publications: magazine, Historic New England.

Hours & Admission Prices: June-Oct. 15 Fri.-Sun. 11-5. Adults $6; discounts to seniors, AAM, ICOM, AAA, WGBH members; Historic New England members no charge.

Attendance: 6,233 (accurate)

Membership: National $35; Individual $45; Household $55; Garden & Landscape $75; Contributing and Library & School $100; Young Friends of Historic New England $100-$1,500; Friends of the Library and Archives $125; Historic Homeowner $200; Business & Ogden Codman Design Group $250; Appleton Circle $2,500 & up.

HARBOR ARTS MUSEUM, 93 High St., Portsmouth, NH 03803. Tel.: 603-436-8596.

Key Personnel: Dir., Richard Smith; C.E.O., John Hansauldt.

Personnel Profile: Full-Time Volunteers 20; Interns 2.

Institution Type/Description: History Museum.

Collections: period instruments & musical memorabilia; playbills; posters; photographs.

Hours & Admission Prices: Call for hours.

Attendance: 1,000 (estimated)

* **JACKSON HOUSE, (M),** 76 Northwest St., Portsmouth, NH 03801-3556. Mailing Address: 141 Cambridge St., Boston, MA 02114-2702. Tel.: 603-436-3205.

Web Site: www.historicnewengland.org

Founded: 1924.

Key Personnel: Pres. & C.E.O., Carl Nold.

Governing Authority: private; nonprofit organization. Parent Institution: Historic New England, 141 Cambridge St., Boston, MA. Tax-exempt: 501(c)(3).

Institution Type/Description: Historic House: built c.1664 by Richard Jackson.

Collections: local history.

Publications: magazine, Historic New England.

Hours & Admission Prices: June-Oct. 15 1st & 3rd Sat. of month 11-5, last tour at 4pm. Adults $6; discounts to AAM & ICOM members; members no charge.

Attendance: 736 (accurate)

Membership: National $35; Individual $45; Household $55; Garden & Landscape $75; Contributing and Library & School $100; Young Friends of Historic New England $100-$1,500; Friends of the Library and Archives $125; Historic Homeowner $200; Business & Ogden Codman Design Group $250; Appleton Circle $2,500 & up.

JOHN PAUL JONES HOUSE MUSEUM, 43 Middle St., Portsmouth, NH 03801-4302. Mailing Address: P.O. Box 728, Portsmouth, NH 03802-0728. Tel.: 603-436-8420.

E-mail: info@portsmouthhistory.org

Web Site: www.portsmouthhistory.org

Founded: 1920.

Congressional District: 1

Key Personnel: Pres., Joshua Scott; Exec. Dir., Maryellen Burke; Sec., M. Marguerite Mathews; Cur., Sandra Rux.

Personnel Profile: Full-Time Paid 1; Part-Time Paid 10; Part-Time Volunteers 20; Interns 1.

Governing Authority: society. Parent Institution: Portsmouth Historical Society. Tax-exempt: 501(c)(3).

Institution Type/Description: History Museum: housed in 1758 John Paul Jones House.

Collections: costumes; glass; canes; documents; paintings; furniture; oriental rugs; ship models including Ranger; John Paul Jones artifacts; colonial through Victorian artifacts; 18th & 19th-century ceramics.

Facilities: welcome center; gardens.

Activities: guided & self-guided tours.

Publications: booklet, Touring John Paul Jones House Museum; newsletter.

Hours & Admission Prices: Memorial Day to Oct. daily 11-5; tours by appointment. Adults $6, Portsmouth residents & AAA members $5; discounts to groups; children 12 and under & members no charge.

Attendance: 4,000 (estimated)

Membership: Friend $25; Dual $50; Contributing $100; Patron $150; Benefactor $250; Exemplar $1,000.

MOFFATT-LADD HOUSE AND GARDEN, 154 Market St., Portsmouth, NH 03801-3730. Tel.: 603-436-8221. Fax: 603-431-9063. Facebook: Moffatt-Ladd House & Garden.

E-mail: moffattladd@gmail.com

Web Site: moffattladd.org

Founded: 1911.

Congressional District: 1

Key Personnel: Dir. & Cur., Dr. Barbara McLean Ward; Pres., Patricia S. Meyers; Museum Shop Mgr., Ms. Virginia Guy.

Personnel Profile: Full-Time Paid 1; Part-Time Paid 14; Part-Time Volunteers 50.

Governing Authority: organization. Parent Institution: The National Society of The Colonial Dames of America in the State of New Hampshire. Tax-exempt: 501(c)(3).

Institution Type/Description: Historic House Museum: c.1763 house built by Captain John Moffatt.

Collections: ancestral portraits; Portsmouth furniture & other period furnishings, early wallpaper, textiles & clothing; Georgian architecture; colonial revival garden. Historic Buildings: warehouse; counting house.

Research Fields: Portsmouth history; American decorative arts, furniture, paintings & textiles; women's history; garden history.

Facilities: 100-vol. library of genealogical material available to members only; historical colonial revival garden. Museum-related items for sale.

Activities: guided tours of house & garden; permanent exhibitions; rental facility; garden lectures; youth camps & programs. Annual Events: Festival; 100th Anniversary Events.

Publications: The Moffatt-Ladd House Cookbook II; Alexander H. Ladd's Garden Book, 1888-1895: A 19th century view of Portsmouth; The Moffatt-Ladd House and its Garden; The Moffatt-Ladd House; From Mansion to Museum (2007).

Hours & Admission Prices: mid-June to mid-Oct. Mon.-Sat. 11-5, Sun. 1-5. Adults $6, children 7-12 $2.50. Garden: $2; discounts for groups with appointment and NEMA & AAA members; children under 7, members & NSCDA no charge.

Attendance: 6,000 (accurate)

Membership: Friends of the Moffatt Ladd House: Friendly Individual $30; Friendly Family $50; Friendly Partner & Small Business $100; Good Friend

& Corporate $250; Very Good Friend $500; Best Friend & Benefactor $1,000; Whipple Society $2,500.

NEW HAMPSHIRE ART ASSOCIATION, 136 State St., Portsmouth, NH 03801-3826. Tel.: 603-431-4230. Fax: 603-431-4230.
E-mail: nhartassociation@comcast.net
Web Site: nhartassociation.org
Formerly: Robert Lincoln Levy Gallery
Founded: 1940.
Key Personnel: Dir., Billie Tooley; Pres. (V) Bd. Dir., David Hampson.
Personnel Profile: Full-Time Paid 1; Part-Time Paid 1; Part-Time Volunteers 100; Interns 1.
Institution Type/Description: Art Gallery.
Collections: paintings; photographs; sculpture; drawings; graphic arts.
Hours & Admission Prices: Wed.-Sat. 10-5, Sun. 12-4.
Attendance: 10,000 (estimated)
Membership: Senior $40; Individual $80.

PORTSMOUTH ATHENAEUM, 9 Market Square, Portsmouth, NH 03801. Mailing Address: Box 848, Portsmouth, NH 03802-0848. Tel.: 603-431-2538. Fax: 603-431-7180.
E-mail: info@portsmouthathenaeum.org
Web Site: portsmouthathenaeum.org
Founded: 1817.
Congressional District: 1
Key Personnel: Keeper, Tom Hardiman; Pres., John Shaw; Librarian, Robin Silva; Archivist, Courtney Maclachlan.
Governing Authority: corporation. Subsidiary Institution: Athenaeum Condominium Assoc. Tax-exempt.
Institution Type/Description: Library Museum: housed in 1805 Portsmouth Athenaeum building.
Collections: research library: books on local history, literature, architecture & genealogy; early New Hampshire newspapers; historic photographs & maps; library: books on art, history, theology, science, geography, and voyages & travels; over 70 large manuscripts; ephemera; postcards; Isles of Shoals material; records of many local churches & institutions; family & business papers; portraits; ship models & objects related to regional history and maritime traditions.
Research Fields: Portsmouth & Piscataqua - region history overall emphasis on 19th century; maritime & military history; architectural history.
Facilities: 40,000-vol. library; reference & research library available for use on the premises under supervision.
Activities: permanent exhibitions; temporary exhibits on historic regional culture; lectures; concerts.
Publications: newsletter.
Hours & Admission Prices: Reference Library: Tues & Thurs. 1-4, Sat. 10-4; other times by appointment. Exhibition Gallery: Tues., Thurs. & Sat. 1-4. No charge; donations accepted. Call for holiday closings. &
Attendance: 2,500 (estimated)
Membership: Annual Assessment $225; Proprietors $500 (one-time charge).

PORTSMOUTH MUSEUM OF ART, 909 Islington St., Portsmouth, NH 03801. Mailing Address: One Harbour Pl., Portsmouth, NH 03801. Tel.: 603-436-0332.
E-mail: info@portsmouthmfa.org
Web Site: www.portsmouthmfa.org
Founded: 2009.
Congressional District: 1
Key Personnel: Dir., Chm. (V) & Museum Shop Mgr., Cathy Sununu.
Personnel Profile: Full-Time Volunteers 2; Part-Time Volunteers 28.
Governing Authority: Parent Institution: New Museum of Portsmouth, Inc. Tax-exempt.
Institution Type/Description: Art Museum.
Collections: works by emerging & established artists.
Activities: studio workshops; lectures; literary readings; fine arts programming; classes; films.
Hours & Admission Prices: Temporarily closed.
Attendance: 5,500 (accurate)
Membership: Basic $25; Advisory $150.

* **RUNDLET-MAY HOUSE, (M),** 364 Middle St., Portsmouth, NH 03801-5016. Mailing Address: 141 Cambridge St., Boston, MA 02114-2702. Tel.: 603-436-3205; 617-227-3956.
Web Site: www.historicnewengland.org
Founded: 1973.
Key Personnel: Pres. & C.E.O., Carl Nold.

Governing Authority: society; nonprofit organization. Parent Institution: Historic New England, 141 Cambridge St., Boston, MA 02114. Tax-exempt: 501(c)(3).
Institution Type/Description: Historic House: 1807 three-story Federal mansion with original outbuildings, courtyard & formal gardens.
Collections: Portsmouth furnishings; Rumford roaster & range; kettle set.
Activities: guided tours.
Publications: magazine, Historic New England.
Hours & Admission Prices: June-Oct. 15 1st & 3rd Sat. of month 11-5, (last tour at 4). Adults $6; discounts to seniors, AAM, ICOM & AAA members; Historic New England members no charge.
Attendance: 718 (accurate)
Membership: National $35; Individual $45; Household $55; Garden & Landscape $75; Contributing $100, Library & School $100; Young Friends of Historic New England $100-$1,500; Friends of the Library & Archives $125; Historic Homeowner $200; Business, Ogden Codman Design Group, & Supporting $250; Appleton Circle $2,500 & up.

STRAWBERY BANKE MUSEUM, (M), 17 Hancock St., Portsmouth, NH 03801. Mailing Address: P.O. Box 300, Portsmouth, NH 03802-0300. Tel.: 603-433-1100. Fax: 603-433-1129.
E-mail: info@strawberybanke.org
Web Site: www.strawberybanke.org
Founded: 1958.
Congressional District: 1
Key Personnel: Pres., Lawrence Yerdon; Chm. Overseers Committee (V), William Wiseman; Chm. Bd. Trustees (V), Martha Fullerclark; Vice Pres. Institutional Advancement, Joanna Brode; Cur., Elizabeth Farish; Collections Mgr., Rodney Rowland; Mgr. Education, Bekki Coppola; Devel. Officer, Joanna Brode; Dir. Mktg. & Communication, Stephanie Seacord; Events & Facilities Rental Mgr., Greg Brackett; Archaeology, Alexandra Martin; Finance Mgr., Michael Walzak; Visitor Svcs. Coord. & Volunteer Coord., Jonathan Brown; Restoration, John Schnitzler; Cur. Historic Landscapes, John Forti; Human Resources Mgr., Diane Cooley; Dir. Corporate Rels., Beth Wheland; Properties Coord., Michelle Gore; Museum Shop Mgr., Gregg Keene.
Personnel Profile: Full-Time Paid 19; Part-Time Paid 5; Part-Time Volunteers 400; Interns 8.
Governing Authority: nonprofit organization. Tax-exempt: 501(c)(3).
Institution Type/Description: Historic Site: 17th-20th century Portsmouth & New England Seacoast.
Collections: decorative arts; tools; architectural specimens; archaeological; maritime; period gardens. Historic Buildings & Gardens: 42 historic buildings dating from 1695-1955.
Research Fields: material culture including decorative arts; architectural restoration; small boatbuilding; crafts; archaeological studies; historic landscaping; social and cultural history of the Piscataqua Region; ceramics; Portsmouth furniture; ceramics; Portsmouth furniture.
Facilities: period gardens. Museum-related items for sale.
Activities: guided & self-guided tours; lectures; concerts; organized education programs; permanent & temporary exhibitions; craft & teacher workshops; education programs in local & maritime history; undergraduate & graduate internships; special events; field schools in archaeology; historic landscapes preservation & restoration. Museum Sponsors: Candlelight Stroll; Brewers Festival; Bluesfest.
Publications: quarterly newsletter; booklets; Official Guide Book; Portsmouth and the Piscataqua; Social History & Material Culture; Portsmouth Historic & Picturesque; reprint, The Story of A Bad Bay.
Hours & Admission Prices: May-Oct. daily 10-5; Nov.-April Thurs.-Sat. 10-2, Sun. 12-2. Family $40, adults $15, youth 5-17 $10; discounts to groups of 10 or more; AAM, AAA, members, employees of other museums, and children 4 & under no charge. &
Attendance: 66,056 (estimated)
Membership: Student $25; Individual $35; Double $55; Family $65; Sponsor $125; Donor $250; Patron $500; Benefactor $1,000. Business Memberships: Associate $125; Sponsor $250; Donor $500; Patron $1,000; Benefactor $1,500; Leader $2,500; Joshua Drisco Club $5,000.

THREE GRACES GALLERY, 105 Market St., Portsmouth, NH 03801-3703. Tel.: 603-436-1988.
Institution Type/Description: Art Gallery.
Collections: paintings; sculpture; pottery; mixed-media.
Hours & Admission Prices: Sun. 12-5, Mon.-Tues. & Thurs.-Sat. 10-6, Wed. by appointment.

WARNER HOUSE, 150 Daniel St., Portsmouth, NH 03801-3831. Mailing Address: Box 895, Portsmouth, NH 03802-0895. Tel.: 603-436-5909.
E-mail: housemanager@warnerhouse.org

Web Site: www.warnerhouse.org
Formerly: MacPheadris/Warner House
Founded: 1931.
Congressional District: 1
Key Personnel: Co Chm., Deborah Richards; Co Chm., Ronan Donohoe; Treas., Lorn Buxton; Sec., Elizabeth Farish; Cur., Carolyn Roy; Cur., Louise Richardson.
Personnel Profile: Part-Time Paid 5; Part-Time Volunteers 10.
Governing Authority: nonprofit organization. Tax-exempt: 509(A).
Institution Type/Description: Historic House Museum: c.1716 Macpheadris-Warner House.
Collections: period furniture & decorated accessories; five Blackburn portraits; restored c.1720 murals; early 1700s books; original & other Portsmouth furnishings illustrating six generations from 1716-1930 who lived in the Macpheadris-Warner House; 18th century copper plate bed hangings.
Facilities: Plates, needlecrafts, pewterware, books & other museum-related items for sale.
Activities: guided tours; formally organized educational programs.
Publications: newsletter.
Hours & Admission Prices: mid-June to mid-Oct. Wed.-Mon. 11-4; other times by appointment. Adults $5, senior citizens $4, children under 7-12 $2.50; discounts to groups, AAA members & Time Travelers; children under 7 & members no charge.
Attendance: 2,000 (estimated)
Membership: Macpheadris $25; Penhallow $50; Wentworth $100; Wendell $250; Sherburne $500; Warner $1,000.

WENTWORTH-COOLIDGE MANSION, 375 Little Harbor Rd., Portsmouth, NH 03801. Mailing Address: 799 South St., Portsmouth, NH 03801. Tel.: 603-436-6607. Fax: 603-430-0025.
E-mail: info@wentworthcoolidge.org
Web Site: www.wentworthcoolidge.org
Founded: 1954.
Congressional District: 1
Key Personnel: Chm. (V), Noele Clews.
Personnel Profile: Part-Time Paid 1; Part-Time Volunteers 12.
Governing Authority: commission; state. Wentworth-Coolidge Commission & New Hampshire Div. of Parks & Recreation. Tax-exempt.
Institution Type/Description: Historic House: c.1750 home of first Royal Governor of New Hampshire.
Collections: 18th-20th century furniture; Wentworth & Coolidge family belongings; late 19th-century photographs; 18th-century wallpaper; lilac plantings of 18th-century origin.
Research Fields: mid-18th century wallpaper; lilac history, cultivation & preservation.
Facilities: gardens.
Activities: guided tours; lectures; art classes; picnicking. Annual Event: Lilac Festival in May.
Publications: leaflet.
Hours & Admission Prices: Tours: late June to early Sept. Wed.-Sun. 10-3; early Sept. to early Oct. Sat.-Sun. 10-3; other times by appointment.
Attendance: 7,763 (accurate)
Membership: Student $15; Individual $25; Family & Household $40; Caretaker $100; Steward $250; Overseer $500; Benefactor $1,000; Founder $2,500; Councillor $5,000.

WENTWORTH GARDNER & TOBIAS LEAR HOUSES ASSOCIATION, 50 Mechanic St., Portsmouth, NH 03801. Mailing Address: P.O. Box 563, Portsmouth, NH 03802-0563. Tel.: 603-436-4406.
E-mail: info@wentworthgardnerandlear.org
Web Site: www.wentworthgardnerandlear.org
Founded: 1940.
Key Personnel: House Mgr., Stacey Fraser-deHaan; Pres., Allan Breed; Treas., Margot Doering; Sec., Helen Rollins.
Personnel Profile: Full-Time Volunteers 1; Part-Time Paid 1; Part-Time Volunteers 10; Interns 1.
Governing Authority: nonprofit organization. Tax-exempt: 501(c)(3).
Institution Type/Description: Historic House: 1760 Wentworth Gardner House, Georgian mansion.
Collections: colonial furniture & furnishings. Historic House: 1740 Tobias Lear House.
Activities: guided tours; lectures; temporary exhibitions. Annual Event: Plant Sale.
Publications: biannual, WGTL Newsletter.
Hours & Admission Prices: mid-June to mid-Oct. Wed.-Sun. 12-4. Adults $7, children $3; discount to groups & National Trust Affiliate; members no charge.
Attendance: 450 (accurate)

Membership: Individual $25; Family $40; Contributing $100; Sustaining $250; Friends of Wallace Nutting $500; Tea With George $1,000.

Raymond

RAYMOND HISTORICAL SOCIETY, 1 Depot Rd., Raymond, NH 03077. Mailing Address: P.O. Box 94, Raymond, NH 03077-0094. Tel.: 603-895-2866.
Web Site: raymondhistoricalsociety.home.comcast.net/~raymondhistoricalsociety
Institution Type/Description: Historical Society Museum: housed in the Old Raymond Train Depot.
Collections: local history & culture; photographs; archives.
Hours & Admission Prices: Call for hours.

Rye

SEACOAST SCIENCE CENTER, 570 Ocean Blvd., Rye, NH 03870-2131. Tel.: 603-436-8043. Fax: 603-433-2235.
E-mail: w.lull@seacentr.org
Web Site: seacoastsciencecenter.org
Founded: 1992.
Congressional District: 1
Key Personnel: Pres., Wendy W. Lull; Chm. (V), John Appleton; Vice Pres., James E. Chase; Dir. Education, Perrin Chick; Dir. Devel., Nichole Rutherford; Museum Shop Mgr., Donna Johnson.
Personnel Profile: Full-Time Paid 12; Part-Time Paid 42; Part-Time Volunteers 130; Interns 4.
Governing Authority: nonprofit organization. Tax-exempt.
Institution Type/Description: Aquarium.
Collections: sea tanks & habitat exhibits; New Hampshire cultural & natural environmental history; humpback whale skeleton.
Facilities: 10,000 sq. ft. exhibit areas; distance learning studio. Museum-related items for sale.
Activities: school & teacher programs; day camps; youth & adult group programs; nature walks; lecture series; distance learning programs.
Publications: newsletter; Teacher's Guide to Rocky Shore; Teachers Guide to Salt Marsh; Footprints in Time, A Walk Where New Hampshire Began.
Hours & Admission Prices: April-Oct. daily 10-5; Nov.-March Sat.-Mon. 10-5. Park: adults $5, children 3-12 $2. Center: $1 per person; discounts to New England Museum Assoc. members; members no charge.
Attendance: 60,000 (accurate)
Membership: Individual $40; Family $65; Family Plus $100. Annual Giving Societies: Merrimack $300-$749; Piscataqua $750-$1,999; Gulf of Maine $2,000-$4,999; World Oceans $5,000 & up.

Salisbury

SALISBURY HISTORICAL SOCIETY, Salisbury Heights, Rte. 4, Salisbury, NH 03268. Mailing Address: 67 Warner Rd., Salisbury, NH 03268-5200. Tel.: 603-648-2774.
Founded: 1966.
Congressional District: 1
Key Personnel: Cur., Wendy Barrett; Pres., Patrick Walsh.
Governing Authority: Salisbury Historical Society. Tax-exempt: 501(c)(3).
Institution Type/Description: History Museum: housed in 1791 Old Baptist Church.
Collections: clothing; tools; silver; books; manuscripts; furniture.
Research Fields: genealogy; town history.
Facilities: 225-vol. library of local & state history available on premises by appointment; reading room; 200-seat auditorium. Local maps, historical books & museum-related items for sale.
Activities: guided tours; lectures; permanent & temporary exhibitions.
Publications: Salisbury Historical Newsletter.
Hours & Admission Prices: April-Oct. Sat. 1-4. No charge; donations accepted.
Attendance: 250 (estimated)
Membership: Active $5; Family $8; Life $100.

Sandown

SANDOWN HISTORICAL SOCIETY AND MUSEUM, 6 Depot Rd., Sandown, NH 03873. Mailing Address: P.O. Box 300, Sandown, NH 03873-0373.
E-mail: sandownhistoricalsociety@gmail.com
Web Site: www.sandownnh.com/history
Founded: 1977.
Congressional District: 1
Key Personnel: C.E.O. & Pres. Historical Society, Bruce Robinson; Cur. & Museum Shop Mgr., Bertha Deveau.

Personnel Profile: Part-Time Volunteers 6.
Governing Authority: society; nonprofit.
Institution Type/Description: Railroad & Local History Museum: housed in 1873 railroad station.
Collections: World War I uniform; World War II items; ration books; military buttons; handmade shoes; wagon jack; sleds; tools; 1860-1900 period adult & children clothes; telegraph key; hand-cranked telephone; lanterns; mailbag; railroad artifacts & equipment; two Flanger cars; velocipede; books; pictures; magazines; posters; victrola; graphophone; wooden tubs; foot pedal jigsaw; foot pedal sewing machine; bicycle lamp; hand carved wooden wash board; hoops for skirts; put-put gas-powered 4-wheel rail car used to bring materials to repair track; 1910 double-runner sled; 1895 Torrington Conn. Sweeper with vacuum; 1744 powder horn Jonathan Lund; Scrimshaw including ships, mermaid, animals, maze, wine glass, birds, woman; two Maine Central Flanger cars; 2-man hand car; signal light; velocipede.
Research Fields: Indian research regarding King Philip of Rhode Island; genealogy; American Indian graves; arrowheads; scraper; trail map; local history books.
Facilities: library of railroad books available to the public; picnic area. Railroad items & other museum-related items for sale.
Activities: guided tours; lectures; temporary exhibitions; hands-on exhibits. Museum Sponsors: Trip By Train; Old Home Day; Hobo Day in October.
Publications: brochure; The View from Meeting House Hill; 1921 Cookbook, Sandown, NH.
Hours & Admission Prices: May-Oct. Sat.-Sun. & holidays 1-5. No charge; donations accepted. &
Attendance: 230 (estimated)
Membership: Student $10; Individual $15; Couple $20; Family $25.

Sharon

THE SHARON ARTS CENTER, INC., 457 NH Rte. 123, Sharon, NH 03458-7116. Tel.: 603-924-7256. Fax: 603-924-6074.
E-mail: info@sharonarts.org
Web Site: www.sharonarts.org
Founded: 1947.
Congressional District: 2
Key Personnel: Exec. Dir., Keri Wiederspahn; Gallery & Store Dir., Camellia Sousa; Craft Gallery Coord., Gillie Dierauf.
Governing Authority: nonprofit corporation. Tax-exempt: 501(c)(3).
Institution Type/Description: Art Gallery.
Collections: Nora S. Unwin wood engravings; Warfield collection.
Research Fields: wood engraving.
Facilities: library of art magazines, craft magazines, art history & craft books; classrooms; handcraft shop; school of arts & crafts. Handcrafted items by New England craftsmen for sale.
Activities: guided tours; gallery talks; monthly exhibits; formally organized education programs.
Publications: 25 Wood Engravings.
Hours & Admission Prices: Call for hours. &
Attendance: 36,000 (estimated)
Membership: Individual $40; Family $60.

South Sutton

SOUTH SUTTON OLD STORE MUSEUM, 12 Meeting House Hill Rd., South Sutton, NH 03273. Mailing Address: P.O. Box 555, South Sutton, NH 03273-0555. Tel.: 603-927-4416.
Founded: 1954.
Congressional District: 2
Governing Authority: municipal. Parent Institution: Town of Sutton.
Institution Type/Description: Historic Building: c.1850 general store.
Collections: tools; dolls; bottles; patent medicines; books; linens; toys; wallpaper; photographs; costumes; store artifacts.
Activities: guided tours.
Hours & Admission Prices: Call for hours. No charge; donations accepted.
Attendance: 100 (estimated)

Stewartstown

POORE FARM MUSEUM, 629 Hollow Rd., Stewartstown, NII 03576. Mailing Address: 438 Fish Pond Rd., Colebrook, NH 03576. Tel.: 603-237-5500 & 5313.
E-mail: johnsen@moose.ncia.net
Key Personnel: Exec. Dir., Richard Johnsen.
Governing Authority: nonprofit organization. Tax-exempt: 501(c)(3).
Institution Type/Description: Historic Building: housed in the Poore family homestead; c.1830.
Collections: Poore family history; period furnishings; personal artifacts. Historic Buildings: house; barn; outbuildings.

Hours & Admission Prices: June-Sept. Mon.-Fri. 11-1, Sat.-Sun. 11-3. Suggested Donation: adults $5; children under 12 no charge.

Sugar Hill

SUGAR HILL HISTORICAL MUSEUM, Main St. (Rte. 117), Sugar Hill, NH 03586. Mailing Address: P.O. Box 591, Sugar Hill, NH 03586-0591. Tel.: 603-823-5336.
E-mail: kittyh41@gmail.com
Web Site: www.sugarhillnh.org
Founded: 1976.
Key Personnel: Chm. (V), John E. Bigelow; Dir., Katharine N. Bigelow
Governing Authority: bd. of trustees. Tax-exempt.
Institution Type/Description: History Museum.
Collections: Sugar Hill history from 1780 to present; carriage barn; blacksmith shop; sleigh shed.
Major Exhibits: Once Upon a Time..., 5/14-10/14.
Facilities: photographic library.
Publications: biannual newsletter, Sugar Hill Treasures; Sugar Hill Street; Annual guide for exhibits; Peckett's-on-Sugar Hill cookbook.
Hours & Admission Prices: June 5-Oct.17 Fri.-Sat. 11-3. No charge; donations accepted. &
Attendance: 1,789 (accurate)
Membership: $35-$500.

Sunapee

SUNAPEE HISTORICAL SOCIETY MUSEUM, 74 Main St., Sunapee Harbor, Sunapee, NH 03782. Mailing Address: P.O. Box 501, Sunapee, NH 03782-0501. Tel.: 603-763-9872.
E-mail: sunapeehistory@gmail.com
Web Site: sunapeehistoricalsociety.org
Governing Authority: Parent Institution: Sunapee NH Historical Society, Inc. Tax-exempt.
Institution Type/Description: History Museum.
Collections: local history; steamboats; boating.
Activities: programs on local history topics.
Publications: quarterly newsletter.
Hours & Admission Prices: Summer: Tues. & Thurs.-Sun 1-4, Wed. 7pm-9pm; Sept.-Oct. Sat. Sun. 1-4. No charge; donations accepted.
Membership: Individual $15; Family $25; Business $50; Life $200; Life Family $350.

Suncook

4-H NATURE CENTER, Bear Brook State Park & Campground, off Rte. 28, Allenstown, Suncook, NH 03275. Mailing Address: 157 Deerfield Rd., Allenstown, NH 03275-2503. Tel.: 603-485-9874. Fax: 603-485-4358.
Web Site: www.nhstateparks.com/bearbrook.html
Formerly: Bear Brook Nature Center
Founded: 1961.
Congressional District: 2
Key Personnel: Park Mgr., David Evans.
Governing Authority: state. Parent Institution: New Hampshire 4-H Clubs.
Institution Type/Description: Natural History Museum.
Collections: live & mounted specimens of New Hampshire plants & animals; New Hampshire rocks & minerals.
Facilities: 200-vol. library of natural history books; three self-guiding nature trails; special trail for the blind.
Activities: guided field trips; demonstrations; slide shows; school program.
Hours & Admission Prices: Memorial Day-Labor Day Fri.-Tues. 10-5. Museum: call for hours. Park Entrance Fee: adults $4, children 6-11 $2; children under 5 and NH residents 65 & over no charge. &
Attendance: 1,500

Tamworth

REMICK COUNTRY DOCTOR MUSEUM AND FARM, 58 Cleveland Hill Rd., Tamworth, NH 03886. Tel.: 603-323-7591; 800-686-6117. Fax: 603-323-8382.
E-mail: info@remickmuseum.org
Web Site: www.remickmuseum.org
Founded: 1993.
Congressional District: 1
Key Personnel: Exec. Dir., Karen Sulewski; Chm. (V), Jane Fryeburg; Museum Shop Mgr., Sharon Trott.
Personnel Profile: Full-Time Paid 10; Part-Time Paid 7; Part-Time Volunteers 12; Interns 1.

Governing Authority: private; nonprofit. Parent Institution: Edwin C. Remick Foundation. Tax-exempt: 501(c)(3).
Institution Type/Description: Medical Museum.
Collections: agricultural tools and equipment; 99 years of medical history, 1894-1993; livestock including dairy cattle, horse, pigs, sheep & chickens; historic farm house; Capt. Enoch Remich's home; barns; outbuildings.
Major Exhibits: Tamworth Mills, 1/14-12/14; Working Woods, 1/14-12/14; Tamworth History, 1/14-12/14; Medical Office, 1/14-12/14.
Research Fields: medical history, 1894-1993; agricultural history, 1750-1993; architecture, 1750-1993; harness racing, 1900-1993.
Facilities: 7,650 sq. ft. exhibit space; visitor's center; hiking trails.
Activities: guided tours; K-12 historic workshops; Foodways workshops; summer tours. Special Events: Winter Carnival & Ice Harvesting; Maple Sugaring; Fishing Derby; Capt. Enoch Remick House - Open House; Historic Thanksgiving; Victorian Christmas; Traditional Tea; Harvest Festival; Hearthside Dinners.
Publications: newsletter, The Remick Farm Journal.
Hours & Admission Prices: Memorial Day to Labor Day Mon.-Sat. 10-4; Sept.-May Mon.-Fri. 10-4. Daily Summer Activities: 12:30-1pm. Enoch Remick Tour: 1pm. Adults $5; discounts to AAM & ICOM members; children 4 & under and members no charge. Closed major holidays.
Attendance: 11,000 (estimated)
Membership: Student & Senior $15; Individual $25; Grandparent Family $45; Family $50; Patron $100; Supporting $250; Steward $500; Corporate & Benefactor $1,000.

Warner

MT. KEARSARGE INDIAN MUSEUM, (M), 18 Highlawn Rd., Warner, NH 03278. Mailing Address: P.O. Box 142, Warner, NH 03278-0142. Tel.: 603-456-2600 & 3244. Fax: 603-456-3092. Facebook: Mt. Kearsarge Indian Museum.
E-mail: info@indianmuseum.org
Web Site: www.indianmuseum.org
Founded: 1990.
Key Personnel: Dir., Lynn Clark; Chm. (V), Grace Fraser; Deputy Dir., Emmons Cobb; Education Dir., Liz Charlebois; Cur., Nancy Jo Chabot; Museum Shop Mgr., Joan Weinstein.
Personnel Profile: Full-Time Paid 3; Part-Time Paid 11; Part-Time Volunteers 70.
Governing Authority: nonprofit organization. Tax-exempt.
Institution Type/Description: Native American History Museum.
Collections: Native American heritage, dwellings, & art; garden; The Medicine Woods; archaeology.
Research Fields: archaeology; New Hampshire Native American history and material culture.
Facilities: nature trail. Museum-related items for sale.
Activities: hands-on activities; teacher workshops; school programs; lecture series; art program. Annual Event: festivals.
Hours & Admission Prices: May-Oct. Mon.-Sat. 10-5, Sun. 12-5, Nov.-Dec. Sat. 10-5, Sun. 12-5; other times by appointment. Adults $8.50, senior citizens & students $7.50, children 6-12 $6.50; discounts to AAM, AAA & AARP members; members & children under 6 no charge.
Attendance: 12,000 (estimated)
Membership: Senior & Student $25; Individual $33; Couple $44; Family $55; Library/Nonprofit $75.

NEW HAMPSHIRE TELEPHONE MUSEUM, 22 E. Main St., Warner, NH 03278-4421. Mailing Address: P.O. Box 444, Warner, NH 03278-0444. Tel.: 603-456-2234.
E-mail: info@nhtelephonemuseum.com
Web Site: www.nhtelephonemuseum.com
Key Personnel: Chm. (V), Alderico "Dick" Violette; Pres. (V), Paul E. Violette; Museum Shop Mgr., Laura French
Institution Type/Description: History Museum.
Collections: history relating to the telephone.
Hours & Admission Prices: By appointment. No charge; donations accepted.
Membership: Student & Senior $20; Individual $25; Family $40; Benefactor $100; Corporate $250.

Waterville Valley

THE MARGARET AND H.A. REY CENTER AND CURIOUS GEORGE COTTAGE, 35 Village Rd., Bldg. C, Waterville Valley, NH 03215. Mailing Address: P.O. Box 286, Waterville Valley, NH 03215-0286. Tel.: 603-236-3308.
E-mail: info@thereycenter.org
Web Site: thereycenter.org/Welcome.html

Institution Type/Description: Art Museum: housed in the former summer home of Margaret and H.A. Rey, authors of the Curious George children's book series.
Collections: works by local artists; paintings; photographs; sculpture; ceramics.
Activities: educational programs; special events.
Hours & Admission Prices: July-Sept. 5 Wed.-Sat. 10-5; Sept. 6-June Sat. 10-5. No charge; donations accepted.

Wilton

FRYE'S MEASURE MILL, 12 Frye Mill Rd., Wilton, NH 03086. Tel.: 603-654-6581.
Web Site: www.fryesmeasuremill.com
Founded: 1858.
Congressional District: 2
Key Personnel: C.E.O., Archivist & Public Rels., Harland Savage, Jr.; Pres. (V) & Gift Shop Mgr., Pamela Savage.
Governing Authority: company organized for profit.
Institution Type/Description: Industrial Museum: housed in 1858 measure mill. Listed on the National Register of Historic Places.
Collections: water-powered measure mill with associated machinery; 1850-1860, hand card machines; 19th & 20th-century colonial containers: measures, piggins, boxes.
Research Fields: colonial wooden ware; textile: hand card facing machines.
Facilities: 250-vol. library of industrial machinery & town history material. Museum-related items for sale.
Activities: guided tours; temporary exhibitions.
Publications: brochure.
Hours & Admission Prices: Jan.-March Wed.-Sat. 10-4; April to mid-Dec. Tues.-Sat. 10-5, Sun. 12-5. Tours: Jan.-March call for hours; June-Oct. Sat. at 2. Adults $5.75; discount to groups; children under 12 no charge. Closed national holidays.

Wolfeboro

THE LIBBY MUSEUM, 755 N. Main St., Wolfeboro, NH 03894. Mailing Address: P.O. Box 629, Wolfeboro, NH 03894-0629. Tel.: 603-569-1035. Fax: 603-569-2246. Facebook: The Libby Museum.
E-mail: libbymuseum@metrocast.net
Web Site: www.wolfeboronh.us
Founded: 1912.
Congressional District: 1
Key Personnel: Dir., Lauren Hammond.
Personnel Profile: Part-Time Paid 7; Part-Time Volunteers 2.
Governing Authority: municipal; nonprofit organization. Parent Institution: Town of Wolfeboro. Tax-exempt.
Institution Type/Description: History & Natural History Museum.
Collections: mounted specimens of local birds, fish, animals; skeletal exhibits; Indian relics; local history; maps.
Research Fields: history of the area.
Facilities: picnic area.
Activities: special events; children's wildlife classes; art classes; evening lectures; family adventure programs. Museum Sponsors: Celebrating 100 Years, 1912-2012.
Publications: maps, Indian Trails of the Lakes Region of New Hampshire.
Hours & Admission Prices: Memorial Day to Labor Day Tues.-Sat. 10-4. Labor Day to Columbus Day weekend hours only. Adults $2; members no charge.
Attendance: 3,000 (accurate)
Membership: Friends of the Libby Museum: Student & Senior Citizen $10; Individual $15; Family $25; Associate $100; Sponsor $250; Patron $500.

WOLFEBORO HISTORICAL SOCIETY, S. Main St., Wolfeboro, NH 03894. Mailing Address: P.O. Box 1066, Wolfeboro, NH 03894-1066. Tel.: 603-569-1683 (Sept.-Nov.); 912-961-6562 (Nov.-June).
E-mail: protopipnit@comcast.net
Web Site: www.wolfeborohistoricalsociety.org
Founded: 1925.
Key Personnel: Pres., James Rogers.
Personnel Profile: Part-Time Paid 1; Part-Time Volunteers 25.
Governing Authority: nonprofit organization; board of directors. Tax-exempt.
Institution Type/Description: Historical Society Museum.
Collections: history; pewter; china; furniture; five 1800 fire fighting equipment. Historic Houses: 1778 Clark house; c.1820 Pleasant Valley School; c.1862 Monitor Engine Co. Firehouse.
Research Fields: local genealogy.
Facilities: 1,200-vol. library of local history books.

Activities: guided tours; monthly speakers. Museum Sponsors: Pot Luck Dinner; Flea Market; 3rd Graders Day; 1820 Barn Restoration.
Publications: pamphlets; 3 volumes, History of Wolfeboro 1770-1994; annual, Clark House Crier; History of Wolfeboro Historical Society; Parker's History of Wolfeboro.
Hours & Admission Prices: July-Labor Day Wed.-Fri. 10-4, Sat. 10-2; other times by appointment. Adults $4, students $2; children under 12 & members no charge. &
Attendance: 900 (accurate)
Membership: Individual $10; Donor $25; Sponsor $50; Patron $100; Benefactor $100 & up.

WRIGHT MUSEUM OF WWII HISTORY, (M), 77 Center St., Rte. 28, Wolfeboro, NH 03894-4368. Mailing Address: P.O. Box 1212, Wolfeboro, NH 03894-1212. Tel.: 603-569-1212. Fax: 603-569-6326.
E-mail: info@wrightmuseum.org
Web Site: wrightmuseum.org
Founded: 1982.
Congressional District: 1
Key Personnel: Chm. (V), Roy Ballentine; Volunteer Coord., Donna Hamill.
Personnel Profile: Full-Time Paid 3; Part-Time Paid 4; Part-Time Volunteers 45.
Governing Authority: nonprofit organization. Tax-exempt: 501(c)(3).
Institution Type/Description: History Museum.
Collections: World War II American military & home front including artifacts, memorabilia, film, written materials, costumes & vehicles.
Activities: educational & family events; school outreach programs; speaker's bureau.
Publications: quarterly newsletter; annual report.
Hours & Admission Prices: May-Oct. Mon.-Sat. 10-4, Sun. 12-4; other times by appointment. Adults $10, senior citizens & veterans $8, students $6; children 4 & under and members no charge. &
Attendance: 12,000 (estimated)
Membership: Individual $50; Family & Household $75; Contributing $100; Sponsor $500; Patron $1,000.

Wolfeboro Falls

NEW HAMPSHIRE BOAT MUSEUM, 399 Center St., Wolfeboro Falls, NH 03896. Mailing Address: P.O. Box 1195, Wolfeboro Falls, NH 03896-1195. Tel.: 603-569-4554. Fax: 603-569-5931.
E-mail: museum@nhbm.org
Web Site: www.nhbm.org
Founded: 1992.
Key Personnel: Dir., Lisa Simpson Lutts; Chm. (V), Joe DeChiaro; Museum Shop Mgr., Allison Gamble.
Personnel Profile: Full-Time Paid 1; Part-Time Paid 16; Part-Time Volunteers 200; Interns 1.
Governing Authority: bd. of trustees. Tax-exempt.
Institution Type/Description: Boat Museum.
Collections: period boats; canoes; guide boats; sail boats; photographs; trophies.
Facilities: 5,000 sq. ft. exhibition space.
Activities: seasonal boat building school; youth & adult sailing; lecture series. Annual Events: Boat Auction; Boathouse Tour; Biannual Vintage Raceboat Regatta; Boat Show.
Publications: quarterly journal, Boathouse News.
Hours & Admission Prices: Memorial Day to Columbus Day Mon.-Sat. 10-4, Sun. 12-4. Adults $7, seniors 65 & over $4, students $3; discounts to AARP & AAA members; AAM & museum members, children 12 & under and museum staff no charge. &
Attendance: 8,476 (accurate)
Membership: Single $30; Couple $50; Family 75; Donor $100; Patron $500; Friend $1,000.

NEW JERSEY

(400 listings)

Absecon

HOWLETT HALL MUSEUM, 100 New Jersey Ave., Absecon, NJ 08201. Mailing Address: P.O. Box 1422, Absecon, NJ 08201-1422. Tel.: 609-659-9000.
Institution Type/Description: History Museum.
Collections: Absecon history & culture; period furnishings; personal artifacts; photographs.
Hours & Admission Prices: Call for hours.

Allentown

HISTORIC WALNFORD, 78 Walnford Rd., Allentown, NJ 08501. Mailing Address: Monmouth County Park System, Newmann Springs Rd., Lincroft, NJ 07738. Tel.: 609-259-6275; 732-842-4000. Fax: 609-259-0384. TDD: 732-219-9484.
E-mail: info@monmouthcountyparks.com
Web Site: www.monmouthcountyparks.com
Founded: 1985.
Congressional District: 3
Key Personnel: Park Mgr., William O'Shaughnessy; Historic Site Supvr., Sarah Bent; Cur., Cheryl Stoeber-Goff.
Personnel Profile: Full-Time Paid 3; Part-Time Paid 2; Part-Time Volunteers 5.
Governing Authority: county. Parent Institution: Monmouth County Park System, Newman Springs Rd., Lincroft, NJ 07738. Tax-exempt.
Institution Type/Description: Historic Site: 20th-century colonial revival interpretation of 19th-century milling technology; 18th-century industrial village.
Collections: 1873 grist mill tools & equipment. Historical Buildings: 1773 Georgian Plantation Home; 1873 Grist Mill; 1872 Carriage House; barns; restored sheds & outbuildings.
Research Fields: grist & sawmills; colonial revival landscapes & architecture; 18th-19th century Quakers in west NJ; 18th-19th century merchant mills & shipping.
Activities: interpretive site tours; environmental tours, changing landscapes, wetland, flora & fauna; demonstration of operating grist mill; special seasonal events; educational programs.
Publications: interpretive flyers; brochures; activity directory.
Hours & Admission Prices: Gristmill: April 15-Nov. 15 daily 8-4; other times by appointment. Park: daily 8-4. No charge; donations accepted. &
Attendance: 20,000 (accurate)

Alloway

ALLOWAY HISTORY MUSEUM, Alloway Municipal Bldg., 49 S. Greenwich St., Alloway, NJ 08001. Tel.: 856-935-4080.
Institution Type/Description: History Museum.
Collections: local history & culture; period furnishings; photographs.
Hours & Admission Prices: 1st Sat. each month 9-12; other times by appointment.

RANCH HOPE CARRIAGE MUSEUM, 45 Sawmill Rd., Alloway, NJ 08001. Mailing Address: P.O. Box 571, Alloway, NJ 08001. Tel.: 856-935-1555.
Governing Authority: Parent Institution: Ranch Hope Inc. Tax-exempt.
Institution Type/Description: Carriage Museum.
Collections: carriages from 1700s to early 1900s; sleighs.
Activities: group tours.
Hours & Admission Prices: Mon.-Fri. 8:30-4:30 by appointment. No charge; donations accepted. &
Attendance: 100 (estimated)

Asbury Park

CRANE HOUSE, 508 4th Ave., Asbury Park, NJ 07712-6010. Tel.: 732-775-5682.
Institution Type/Description: Historic House Museum: housed in the former home of author, Stephen Crane; c.1877.
Collections: Stephen Crane's personal artifacts; photographs; period furnishings.
Activities: special events.
Hours & Admission Prices: Call for hours.

Atlantic City

ABSECON LIGHTHOUSE AND KEEPER'S HOUSE MUSEUM, (M), 31 S. Rhode Island Ave., Atlantic City, NJ 08401-7760. Tel.: 609-449-1360. Fax: 609-449-1919.
Web Site: www.abseconlighthouse.org
Congressional District: 2
Governing Authority: Tax-exempt.
Institution Type/Description: History Museum & Historic Lighthouse: built in 1857.
Collections: local history; photographs; maritime artifacts; lighthouse memorabilia.
Facilities: Museum-related items for sale.
Activities: lighthouse tours; wedding venue.
Hours & Admission Prices: July-Aug. daily 10-5; Sept.-June Thurs.-Mon. 11-4. Adults $7, seniors 65 & over $5, children 4-12 $4; children 3 & under and active military no charge. Closed major holidays. &

Attendance: 28,000 (accurate)

ATLANTIC CITY AQUARIUM, 800 N. New Hampshire Ave., Atlantic City, NJ 08401-2900. Tel.: 609-348-2880. Fax: 609-345-4238.
Web Site: www.acaquarium.com
Formerly: Historic Gardner's Basin
Founded: 1976.
Congressional District: 2
Key Personnel: Chm. Bd., James L. Cooper; Vice Chm., Murray Raphel; Exec. Dir., Jack Keith.
Governing Authority: nonprofit organization. Tax-exempt: 501(c)(3).
Institution Type/Description: Maritime Village & Ocean Life Center: housed in pre-1900 waterfront homes, located on site where Atlantic City was founded.
Collections: water craft; maritime artifacts; seafaring memorabilia; working & living exhibits on lobstermen & clammers; paintings; sculpture.
Research Fields: maritime history, particularly of South Jersey; oceanographic interests; marine mammal ecological interests; documentations of the family histories of early occupants of the houses in the Basin area.
Facilities: aquarium; field research station; laboratory; 400-seat amphitheater; classrooms; 100-seat restaurant. Nautical items, shells, carvings, models & other museum-related items for sale.
Activities: guided tours; lectures; films; gallery talks; concerts; arts festivals; formally organized education programs; permanent, temporary & loan exhibitions; 3-hour sails on the square rigger Young America.
Hours & Admission Prices: Daily 10-5. Adults $8, seniors $6, children 4-12 $5; children 3 & under no charge. Closed New Year's Day; Thanksgiving; Christmas. &
Attendance: 90,000 (estimated)
Membership: Individual $35; Family $70; Ensign $150; Captain $500; Admiral $1,000.

ATLANTIC CITY ART CENTER, New Jersey Ave. & Boardwalk, Atlantic City, NJ 08401. Mailing Address: P.O. Box 1061, Atlantic City, NJ 08404-1061. Tel.: 609-347-5837 & 5838. Fax: 609-347-5844.
Institution Type/Description: Art Gallery.
Collections: works by national & regional artists; paintings; sculpture; photographs.
Activities: educational programs; special events.
Hours & Admission Prices: Summer: daily 10-4; Fall: Tues.-Sun. 10-4. No charge. Closed national holidays.

ATLANTIC CITY HISTORICAL MUSEUM, Garden Pier, New Jersey Ave. & Boardwalk, Atlantic City, NJ 08401. Mailing Address: Atlantic City Free Public Library, 1 N. Tennessee Ave., Atlantic City, NJ 08401. Tel.: 609-347-5839 & 345-2269. Fax: 609-345-5570.
E-mail: reflib@acfpl.org
Web Site: www.atlanticcityexperience.org
Founded: 1982.
Congressional District: 2
Key Personnel: Coord., Beth Ryan; Archivist, Heather Perez.
Governing Authority: government; nonprofit. Parent Institution: Atlantic City Free Public Library. Tax-exempt.
Institution Type/Description: History Museum: situated on the Atlantic City Boardwalk on Garden Pier which was built in 1912.
Collections: furnishings; personal artifacts; prints, drawings; graphic arts; costumes; textiles; photographs.
Major Exhibits: Hometown Teams (Smithsonian Moms & New Jersey Council for the Humanities (T), 3/22/14-5/4/14.
Activities: guided tours; participatory exhibits; lectures; programs; presentations.
Hours & Admission Prices: Daily 10-5. No charge; donations accepted. See website for holiday closings. &
Attendance: 50,000 (estimated)

RIPLEY'S BELIEVE IT OR NOT! MUSEUM, 1441 Boardwalk, Atlantic City, NJ 08401-7144. Tel.: 609-347-2001.
E-mail: info@ripleysatlanticcity.com
Web Site: www.ripleys.com
Key Personnel: Gen. Mgr., Chris Connelly
Institution Type/Description: General Museum.
Collections: personal artifacts & oddities from 198 countries; lock of George Washington's hair; 27 room miniature wood carved castle; videos; photographs.
Facilities: Museum-related items for sale.

Activities: tours; private parties; groups; birthdays.
Hours & Admission Prices: May-Aug. daily 10-10; Sept.-April Sun.-Fri. 10-6, Sat. 10-8. Adults 13 & over $14.99, students $12.99, senior citizen $11.99, children 5-12 $9.99; discounts to AAA members. &

Atlantic Highlands

ATLANTIC HIGHLANDS HISTORICAL SOCIETY - STRAUSS MANSION AND MUSEUM, 27 Prospect Circle, Atlantic Highlands, NJ 07716-1310. Mailing Address: P.O. Box 108, Atlantic Highlands, NJ 07716. Tel.: 732-291-1861.
E-mail: ahhs123@comcast.net
Web Site: atlantichighlandshistory.com
Founded: 1974.
Congressional District: 4
Institution Type/Description: Historical Society Museum.
Collections: local history; period furnishings; photographs; early tools; Native American artifacts.
Facilities: library. Museum-related items for sale.
Hours & Admission Prices: Sun. 1-4. No charge; donations accepted.

Avalon

AVALON HISTORICAL SOCIETY, 215 39th St., Avalon, NJ 08202-1648. Tel.: 609-967-0090.
Web Site: www.avalonmuseum.org
Key Personnel: Pres. (V), Ms. Barbara Juzaitis.
Governing Authority: nonprofit.
Institution Type/Description: History Museum.
Collections: Avalon history & culture; photographs.
Hours & Admission Prices: Mon.-Sat. 11-3. No charge; donations accepted. &

Barnegat Light

BARNEGAT LIGHT HISTORICAL SOCIETY AND MUSEUM, 501 Central Ave., Barnegat Light, NJ 08006. Mailing Address: P.O. Box 386, Barnegat Light, NJ 08006. Tel.: 609-494-8578.
Web Site: www.bl-hs.org/
Founded: 1954.
Key Personnel: Pres. (V), Karen Larson.
Personnel Profile: Part-Time Volunteers 40.
Governing Authority: nonprofit organization. Tax-exempt.
Institution Type/Description: Historical Society Museum.
Collections: Barnegat Light & Long Beach Island history; period artifacts; photographs; first order fresnel lens.
Hours & Admission Prices: June & Sept.-Oct. Sat.-Sun. 10-4; July-Aug. daily 10-4; tours by appointment. No charge; donations accepted. &
Attendance: 4,000 (estimated)
Membership: Individual $10; Lifetime $100.

Basking Ridge

ENVIRONMENTAL EDUCATION CENTER, SOMERSET COUNTY PARK COMMISSION, 190 Lord Stirling Rd., Basking Ridge, NJ 07920-1329. Mailing Address: P.O. Box 5327, North Branch, NJ 08807. Tel.: 908-766-2489. Fax: 908-766-2687.
E-mail: cschrein@scparks.org
Web Site: www.somersetcountyparks.org
Founded: 1970.
Congressional District: 5
Key Personnel: Sec. Dir., Raymond Brown; Mgr., Shawn McCrohan; Supvr., Kurt Bender; Foreman, James Bodnar; Event Specialist, Jane Parks; Admin. Asst., Jane Bodnar.
Governing Authority: county. Affiliated with Somerset County Park Commission, P.O. Box 5327, North Branch, NJ 08876. Tel. 908-722-1200. Tax-exempt.
Institution Type/Description: Environmental Education Center: located on site of Lord Stirling's Estate.
Collections: herbarium collection of vascular plants of Great Swamp; fungi of northern New Jersey; natural history photographic slides; bird nests, vertebrate & aquatic invertebrate of Great Swamp Basin.
Research Fields: field succession; management techniques relevant to county park lands; archaeological field project at c.1763 Lord Stirling Manor Site and at paleo-Indian site.
Facilities: 6,000-vol. library of natural history, environmental education, New Jersey & local history available for research on premises; trails; observation blinds; field research station; classrooms. Museum, natural history science theme items for sale.
Activities: guided tours by appointment; lectures; films; concerts; rental

gallery; formally organized education programs; docent program or council; training programs; loan & permanent exhibitions.

Publications: newsletter, Parks, Programs & People; environmental information sheets.

Hours & Admission Prices: Daily 9-5. Trails: daily sunrise to sunset. No charge; donations accepted. Closed major holidays. &

Attendance: 10,000 (estimated)

THE HISTORICAL SOCIETY OF THE SOMERSET HILLS, 15 W. Oak St., Basking Ridge, NJ 07920. Mailing Address: P.O. Box 136, Basking Ridge, NJ 07920. Tel.: 908-221-1770.

Institution Type/Description: Historical Society Museum: housed in the Brick Academy. Listed on the National Register of Historic Places.

Collections: local history & culture; period furnishings; photographs; paintings; early education.

Facilities: library.

Activities: educational programs.

Hours & Admission Prices: Museum: Sept.-June Sun. 2-4. Library: 1st & 3rd Wed. of the month 9:30 to noon; other times by appointment. Closed holidays.

Bayonne

CHIEF JOHN T. BRENNAN FIRE MUSEUM, 10 W. 47th St., Bayonne, NJ 07002-4005. Tel.: 201-858-6000.

Institution Type/Description: Firefighting History Museum: housed in the former fire department station; built in 1870. Listed on the National Register of Historic Places.

Collections: firefighting history, artifacts & equipment; hand-drawn hose carriage; photographs; personal artifacts.

Hours & Admission Prices: Call for hours.

Bayville

BERKELEY TOWNSHIP HISTORICAL MUSEUM, 630 Rte. 9, Bayville, NJ 08721. Mailing Address: P.O. Box 303, Bayville, NJ 08721-0303. Tel.: 732-269-0643.

Institution Type/Description: History Museum.

Collections: local history & culture; personal artifacts; photographs.

Hours & Admission Prices: mid-June to mid-Sept. Wed. 6pm-8pm, Sun. 2-4; other times by appointment.

Beach Haven

LONG BEACH ISLAND HISTORICAL ASSOCIATION, 129 Engleside Ave., Beach Haven, NJ 08008-1762. Mailing Address: P.O. Box 1222, Beach Haven, NJ 08008. Tel.: 609-492-0700. Fax: 609-492-3885.

E-mail: info@lbimuseum.org

Web Site: www.lbimuseum.org

Formerly: Long Beach Island Historical Society

Founded: 1976.

Key Personnel: Pres. (V), Jamie Cianoezei; Museum Shop Mgr., Kay Donelly.

Governing Authority: Tax-exempt.

Institution Type/Description: History Museum: housed in the former Holy Innocents' Episcopal Church built in 1882.

Collections: local history & culture; photographs; personal artifacts; period furnishings.

Activities: walking tours; lectures; craft shows; flea markets. Museum Sponsors: Annual Porch Party; Island Singers Concerts.

Publications: newsletters.

Hours & Admission Prices: Spring & Fall Sat.-Sun. 2-4; Summer: Tues. & Fri. 10-4, Wed.-Thurs. & Sat.-Mon. 2-4 & 7-9. Suggested Donation: $3. &

Attendance: 2,000 (estimated)

Membership: Individual $35.

NEW JERSEY MARITIME MUSEUM, 528 Dock Rd., Beach Haven, NJ 08008-1833. Tel.: 609-492-0202. Fax: 609-492-7575.

E-mail: curator@njmaritimemuseum

Web Site: www.njmaritimemuseum.org

Formerly: Museum of New Jersey Maritime History

Founded: 2007.

Congressional District: 9

Key Personnel: Chm. (V), Pres. (V) & Cur., Deb Whitcraft; Cur. & Museum Shop Mgr., Jim Vogel.

Personnel Profile: Full-Time Volunteers 2; Part-Time Volunteers 6.

Governing Authority: Tax-exempt.

Institution Type/Description: Maritime History Museum.

Collections: maritime history; photographs; historic documents; newspapers; artifacts recovered from New Jersey wreck sites.

Research Fields: New Jersey coastal history & shipwrecks.

Facilities: library. Museum-related items for sale.

Activities: research.

Hours & Admission Prices: June-Aug. daily 10-5; Sept.-May Fri.-Sun. 10-4; other times by appointment. No charge; donations accepted. &

Attendance: 7,500 (estimated)

Membership: Student & Senior $25; Surfman $50; Surfman Family $100; Keeper $250; Superintendent $500; Lifetime Individual $1,000; Lifetime Family $2,500; Lifetime Corporate $5,000 & up; Lifetime Benefactor $10,000 & up.

Beachwood

JAKES BRANCH COUNTY PARK & NATURE CENTER, 1100 Double Trouble Rd., Beachwood, NJ 08722. Tel.: 732-281-2750.

Key Personnel: Dir., Michael Mangum

Institution Type/Description: Park & Nature Center.

Collections: local history; wildlife & their habitats; hands-on exhibitions; photographs.

Facilities: nature center; nature trails; picnic area.

Activities: avian observation area; 40 ft. tall outdoor observation deck; hiking; educational programs.

Hours & Admission Prices: Call for hours.

Bedminster

JACOBUS VANDERVEER HOUSE, Rte. 206 S., Bedminster, NJ 07921. Mailing Address: P.O. Box 723, Bedminster, NJ 07921-0723. Tel.: 908-212-7000, ext. 611.

E-mail: info@jvanderveerhouse.com

Web Site: www.jvanderveerhouse.com

Key Personnel: Pres., Leslie Mole.

Governing Authority: Parent Institution: The Friends of the Jacobus Vanderveer House. Tax-exempt: 501(c)(3).

Institution Type/Description: Historic House Museum.

Collections: period artifacts & furnishings.

Hours & Admission Prices: Call for hours.

Belmar

OCEANSIDE GALLERY, 1010 Main St., Belmar, NJ 07719-2726. Tel.: 732-280-2167. Fax: 732-280-2167.

E-mail: gallery@oceansidegallery.com

Web Site: oceansidegallery.com

Institution Type/Description: Art Gallery.

Collections: paintings; sculpture.

Hours & Admission Prices: Tues. & Sat. 10-5, Wed. 10-8:30, Thurs. 10-6, Fri. 10-8; other times by appointment.

Belvidere

WARREN COUNTY HISTORICAL SOCIETY AND MUSEUM, 313 Mansfield St., Belvidere, NJ 07823. Mailing Address: P.O. Box 313, Belvidere, NJ 07823. Tel.: 908-475-4246.

Founded: 1933.

Governing Authority: nonprofit organization.

Institution Type/Description: Historical Society Museum: house built c.1848.

Collections: local history, heritage & culture; period furnishings; personal artifacts; photographs; books; maps; documents; letters.

Facilities: library.

Activities: Warren County history & genealogical research.

Publications: newsletter, Oak Leaves.

Hours & Admission Prices: Sun. 2-4; groups by appointment. No charge; donations accepted.

Membership: Student $5, Individual $15, Family $20, Life $200, Life Couple $300.

Bernardsville

SCHERMAN HOFFMAN WILDLIFE SANCTUARY, 11 Hardscrabble Rd., Bernardsville, NJ 07924. Tel.: 908-766-5787. Fax: 908-766-7775.

E-mail: scherman-hoffman@njaudubon.org

Key Personnel: Dir., Mike Anderson

Institution Type/Description: Wildlife Sanctuary.

Collections: wildlife & their habitats; ecosystem; environment; nature; hands-on exhibitions.

Facilities: 276 acres.

Activities: workshops; classes; summer camps; educational programs.

Hours & Admission Prices: Tues.-Sat. 9-5, Sun. 12-5. Closed holidays.

Bloomfield

HISTORICAL SOCIETY OF BLOOMFIELD NEW JERSEY, 90 Broad St., Bloomfield, NJ 07003-2585. Tel.: 973-566-6220 & 743-8844.
E-mail: info@hsob.org
Founded: 1966.
Congressional District: 8
Key Personnel: C.E.O. & Pres. (V), Ina Campbell.
Personnel Profile: Part-Time Volunteers 6.
Governing Authority: society. Tax-exempt: 501(c)(3).
Institution Type/Description: Local History Museum: housed in Bloomfield Public Library.
Collections: furniture; clothing & accessories; tools; household articles; paintings; toys; posters; memorabilia; dioramas; early maps & newspapers; postcards; letters; documents; books.
Research Fields: local history; genealogy.
Activities: guided tours; lectures; permanent exhibitions.
Publications: quarterly newsletter.
Hours & Admission Prices: late June to early Sept. Wed. 2-4:30; Sept.-June Wed. 2-4:30, Sat. 10-12:30; other times by appointment. No charge; donations accepted. ♿
Attendance: 380 (estimated)
Membership: Student $5; Individual & Nonprofit Organization $10; Couple $15; Commercial Organization $25.

Boonton

BOONTON HISTORICAL SOCIETY, John Taylor Bldg., 210 Main St., Boonton, NJ 07005. Tel.: 973-402-8840.
E-mail: boontonhistory@boonton.org
Web Site: www.boonton.org
Institution Type/Description: Historical Society Museum.
Collections: local history & culture; period artifacts; photographs.
Major Exhibits: Theaters of Boonton, 1/14-9/14.
Activities: educational programs.
Hours & Admission Prices: Summer: Sat. 1-4, Sun. 12-3; Winter: Sat. 1-4; other times by appointment.

NEW JERSEY FIREMAN'S HOME MUSEUM, 565 Lathrop Ave., Boonton, NJ 07005. Tel.: 201-334-0024.
Web Site: www.njfh.org
Founded: 1985.
Key Personnel: Dir., Chm. (V), L.V. Denny.
Volunteer Hours: 3,000
Governing Authority: Parent Institution: NJ Fireman's Home. Tax-exempt.
Institution Type/Description: Firefighting History Museum.
Collections: firefighting history & equipment; steamers; hose carts; fire trucks; helmets & uniforms; photographs; 911 artifacts.
Research Fields: Firematic.
Facilities: 8,000 sq. ft. exhibit space.
Hours & Admission Prices: Daily 8-4. No charge; donations accepted. ♿
Attendance: 3,500 (estimated)

Bordentown

ARTFUL DEPOSIT GALLERY, 201 Farnsworth Ave., Bordentown, NJ 08505. Tel.: 609-298-6970.
Web Site: www.theartfuldeposit.com
Institution Type/Description: Art Gallery.
Collections: works by regional, national & international artists.
Activities: permanent & temporary exhibits.
Hours & Admission Prices: mid-July to mid-Aug. Wed.-Thurs. & Sat. 1-6, Fri. 4-9, Sun. 1-5; mid-Aug to mid-July Wed.-Thurs. & Sat. 1-6, Sun. 1-5.

BORDENTOWN HISTORICAL SOCIETY, 302 Farnsworth Ave., Bordentown, NJ 08505. Mailing Address: P.O. Box 182, Bordentown, NJ 08505. Tel.: 609-298-1740 & 9181.
E-mail: bordentownhistory@hotmail.com
Institution Type/Description: Historical Society Museum.
Collections: local history & culture; period furnishings; personal artifacts; photographs.
Hours & Admission Prices: Call for hours.

Bridgeton

BRIDGETON HALL OF FAME ALL SPORTS MUSEUM, Burt Avenue Recreation Center, Bridgeton, NJ 08302. Tel.: 856-451-7300.
Institution Type/Description: Sports Museum.
Collections: local renown sports athletes; photographs; sports equipment; scrapbooks; memorabilia; trophies; medals.
Hours & Admission Prices: Call for hours. No charge.

COHANZICK ZOO, 181 E. Commerce St., Bridgeton, NJ 08302. Tel.: 856-455-3230, ext. 242.
Institution Type/Description: Zoo.
Collections: over 100 animals; monkeys; bears; big cats; crocodiles; snakes; birds.
Activities: educational programs.
Hours & Admission Prices: Winter: daily 9-4; Summer: daily 9-5; groups by appointment. No charge. Closed Christmas. ♿

NAIL MILL MUSEUM, 1 Mayor Aitken Dr., Bridgeton, NJ 08302-1347. Tel.: 856-455-4100.
Institution Type/Description: History Museum: housed in the former Cumberland Nail and Iron Company.
Collections: local history; period furnishings; photographs; personal artifacts.
Activities: special events.
Hours & Admission Prices: Temporarily closed.

WOODRUFF MUSEUM OF INDIAN ARTIFACTS, 150 E. Commerce St., Bridgeton, NJ 08302-2613. Tel.: 856-451-2620. Fax: 856-455-1049.
E-mail: bpl@bridgetonlibrary.org
Web Site: www.bridgetonlibrary.org
Founded: 1976.
Congressional District: 2
Key Personnel: Library Dir., Courtenay Reece.
Personnel Profile: Part-Time Volunteers 3.
Governing Authority: municipal. Parent Institution: Bridgeton Public Library. Tax-exempt.
Institution Type/Description: Native American History Museum.
Collections: Native American artifacts.
Facilities: 300-vol. library.
Activities: guided tours.
Hours & Admission Prices: Mon.-Fri. 1-4, Sat. 11-2; other hours by appointment. No charge; donations accepted. Closed New Year's Day; Martin Luther King Jr. Day; Presidents' Day; Good Friday; Memorial Day; Independence Day; Labor Day; Veterans Day; Columbus Day; Election Day; Thanksgiving & day after; Christmas.
Attendance: 400 (estimated)

Bridgewater

SOMERSET COUNTY HISTORICAL SOCIETY - VAN VEGHTEN HOUSE, 9 Van Veghten Dr., Bridgewater, NJ 08807. Tel.: 908-218-1281.
E-mail: schs@schsnj.com
Web Site: www.schistoryweekend.com
Institution Type/Description: Historical Society Museum: house built in 1720. Listed on the National Register of Historic Places.
Collections: local history & culture; period furnishings; personal artifacts; photographs.
Facilities: library.
Activities: research; special events.
Hours & Admission Prices: Call for hours. No charge.

Brigantine

BRIGANTINE BEACH HISTORICAL MUSEUM & SOCIETY, 3607 Brigantine Blvd., Brigantine, NJ 08203-1001. Mailing Address: P.O. Box 833, Brigantine, NJ 08203. Tel.: 609-266-9339.
Key Personnel: Dir., Andrew Solari; Chm. (V), Linda Sayers; Museum Shop Mgr., Sue Van Nest.
Personnel Profile: Part-Time Volunteers 70.
Governing Authority: Tax-exempt.
Institution Type/Description: Historical Society Museum.
Collections: local history & culture; photographs; personal artifacts.
Hours & Admission Prices: Spring-Fall Sat. 11-2, Sun. 1-4; Summer Mon.-Sat. 11-2, Sun. 1-4. No charge; donations accepted.
Attendance: 2,300 (accurate)

Membership: Annual $15; Family Lifetime $100.

MARINE MAMMAL STRANDING CENTER - SEA LIFE MU-SEUM, 3625 Brigantine Blvd., Brigantine, NJ 08203. Mailing Address: P.O. Box 773, Brigantine, NJ 08203. Tel.: 609-266-0538.
E-mail: mmsc@verizon.net
Web Site: www.marinemammalstrandingcenter.org
Key Personnel: Dir., Bob Schoelkopf.
Governing Authority: private; nonprofit organization. Tax-exempt: 501(c)(3).
Institution Type/Description: Marine Museum.
Collections: life-sized replicas of game fish, sea turtles, & marine mammals.
Facilities: Museum-related items for sale.
Hours & Admission Prices: Sept. to mid-Dec. Sat. 10-3, Sun. 11-2; Summer: call for hours.

Burlington

BURLINGTON COUNTY HISTORICAL SOCIETY, 457 High St., Burlington, NJ 08016-4514. Tel.: 609-386-4773. Fax: 609-386-4828.
E-mail: burlcohistsoc@verizon.net
Web Site: www.burlingtoncountyhistoricalsociety.org
Founded: 1915.
Congressional District: 4
Key Personnel: Exec. Dir., Lisa Fox-Pfeiffer; Pres. (V), Bernadette Boyle; Dir. Education, Jeffrey J. Macechak; Librarian, Annie Brogan.
Personnel Profile: Full-Time Paid 4; Part-Time Paid 1; Part-Time Volunteers 15.
Governing Authority: society. Branch Museums: c.1740 Capt. James Lawrence Birthplace; c.1743 Bard-How House; c.1782 James Fenimore Cooper Birthplace; Delia Biddle Pugh Library; Museum Galleries. Tax-exempt.
Institution Type/Description: Historical Society: historic house museums & local history museum.
Collections: lighting equipment; clocks; quilts; samplers; tools & equipment; decoys; costumes; period furnishings; paintings.
Research Fields: genealogical & local history research.
Facilities: local history books deeds, manuscripts & genealogical material available for research on premises.
Activities: tours; trips; educational programs.
Publications: reprint of Scott's 1876 Atlas; 1883 History of Burlington Co.; Freedom Papers: 1776-1781; After Freedom; Burlington County Remembered: Landmarks; Delaware River Decoys of Burlington County; Ingenuity & Craftsmanship: The Culture of Production in Burlington County.
Hours & Admission Prices: Tours: Tues.-Sat. 1-5, groups by appointment. Library: Tues.-Sat. 1-5. Adults $5, children under 12 $2.50; discounts to AAM members; members no charge. &
Attendance: 7,000 (accurate)
Membership: Individual $20; Household $45; Patron $60; Patron Household $75; Benefactor $120; Benefactor Household $150.

CITY OF BURLINGTON HISTORICAL SOCIETY, 202 High St., Burlington, NJ 08016. Mailing Address: c/o Philip's Furniture & Antiques, 307 High St., Burlington, NJ 08016. Tel.: 609-387-0586; 386-7125. Facebook: City of Burlington Historical Society.
Founded: 1975.
Congressional District: 4
Key Personnel: Cur., Maryanne Augustyn.
Governing Authority: municipal. Subsidiary Institution: The Carriage House.
Institution Type/Description: Historical Society Museum: housed in c.1797 Hoskins House.
Collections: period artifacts; artifacts from the Burlington area. Historic Houses: c.1797 Hoskins House; c.1876 Carriage House; c.1792 Friend School House.
Activities: guided tours; concerts. Society Sponsors: Art Show in September; Christmas Parade & Tree Lighting in December; Christmas House Tour.
Publications: newsletter.
Hours & Admission Prices: By appointment only.
Attendance: 3,000
Membership: Annual $5.

COLONIAL BURLINGTON FOUNDATION, INC., 213 Wood St., Burlington, NJ 08016. Mailing Address: P.O. Box 1552, Burlington, NJ 08016-7152. Tel.: 609-864-8152.
E-mail: applications@woodstreetfair.com
Web Site: www.woodstreetfair.com
Founded: 1970.
Congressional District: 4
Key Personnel: Pres., Matt Penisi.
Governing Authority: nonprofit organization. Tax-exempt.

Institution Type/Description: Historic House Museum: 1685 Thomas Revell House.
Collections: furnishings; herb garden; boxwood garden.
Activities: open house tours; private tours. Museum Sponsors: Annual Craft Fair; Wood Street Fair in September.
Hours & Admission Prices: first Sat. after Labor Day; other times by appointment. No charge; donations accepted.
Attendance: 12,000 (estimated)

Butler

BUTLER MUSEUM, Main St., Butler, NJ 07405. Mailing Address: One Ace Rd., Butler, NJ 07405-1348. Tel.: 973-838-7222. Fax: 973-283-9895.
E-mail: butlermuseumj@optonline.net
Web Site: www.butlermuseumnj.org
Key Personnel: Cur., Alan Bird
Institution Type/Description: Military History Museum.
Collections: military history from Civil War to Desert Storm; period artifacts; tools.
Hours & Admission Prices: Sat. 10-2. No charge. Closed holidays.

Caldwell

GROVER CLEVELAND BIRTHPLACE, 207 Bloomfield Ave., Caldwell, NJ 07006-5115. Tel.: 973-226-0001. Fax: 973-226-1810.
E-mail: gcmuseum@gmail.com
Web Site: clevelandbirthplace.org
Founded: 1913.
Key Personnel: Dir. & Cur., Sharon Farrell.
Personnel Profile: Full-Time Paid 1; Part-Time Paid 2; Part-Time Volunteers 12; Interns 3.
Governing Authority: state. Parent Institution: New Jersey Dept. of Environmental Protection, Div. of Parks & Forestry, Trenton, NJ 08625. Tax-exempt.
Institution Type/Description: Historic House Museum: 1832 Old Manse of the Caldwell First Presbyterian Church; birthplace of President Grover Cleveland, 22nd & 24th President of the United States.
Collections: furnishings & mementos of Grover Cleveland & of the period.
Research Fields: life of Grover Cleveland; local history.
Activities: guided tours.
Publications: brochure.
Hours & Admission Prices: Wed.-Sat. 10-12 & 1-4, Sun. 1-4. No charge. Closed state & federal holidays.
Attendance: 6,000 (estimated)
Membership: See Grover Cleveland Birthplace Memorial Association AAM listing.

Camden

ADVENTURE AQUARIUM, 1 Riverside Dr., Camden, NJ 08103-1060. Tel.: 856-365-3300. Fax: 856-365-3311. TDD: 856-541-8863.
E-mail: info@njaquarium.org
Web Site: www.adventureaquarium.com
Formerly: New Jersey State Aquarium
Founded: 1992.
Congressional District: 5
Key Personnel: Exec. Dir., Kevin Keppel; Husbandry Dir., Marc Kind.
Personnel Profile: Full-Time Paid 110; Part-Time Paid 53; Part-Time Volunteers 263.
Governing Authority: Tax-exempt: 501(c)(3).
Institution Type/Description: Aquarium.
Collections: over 2 million gallons of water including 8,500 aquatic animals; 36 sharks including 2 great hammerheads; outdoor seal pool.
Research Fields: biology, zoology.
Facilities: 150-seat indoor & 300-seat outdoor restaurant; 120,000 sq. ft. exhibit space. Museum-related items for sale.
Activities: seal shows; docent program; films; lectures; mobile vans; participatory exhibits; volunteer programs in many areas including SCUBA diving demonstrations in Open Ocean Tank; Drama Gills performances; 4D theater; touch tank; shark tunnel; educational programs for groups; daily shows.
Hours & Admission Prices: Daily 10-5. Adults $23.95, children 2-12 $17.95; children under 2 no charge. &
Attendance: 510,000 (accurate)
Membership: Explorer $60; Explorer Plus $90; Family Explorer $175; Family Explorer Plus $215.

BATTLESHIP NEW JERSEY MUSEUM & MEMORIAL, 62 Battleship Place, Camden, NJ 08103-3302. Tel.: 856-966-1652, ext. 126. Fax: 856-966-8228.
E-mail: j.schuck@battleshipnewjersey.org
Web Site: www.battleshipnewjersey.org
Founded: 2001.
Congressional District: 1
Key Personnel: Chm., Patricia Egan Jones; Immediate Past Chm., John Mattheusen; Museum Shop Mgr., Edwin Bishop.
Personnel Profile: Full-Time Paid 11; Part-Time Paid 8; Part-Time Volunteers 1,200; Interns 6.
Governing Authority: Parent Institution: Home Port Alliance for the USS New Jersey, Inc. Tax-exempt.
Institution Type/Description: Naval Museum: one of the largest battleships ever built, the Iowa-class ship is our Nation's most decorated.
Collections: 16-inch gun turrets; the Combat Engagement Center; Crew's quarters; BB 62's weapon's systems.
Facilities: Museum-related items for sale.
Activities: self-guided, guided & multimedia tours; special events; educational programs for students; overnight encampment; flight simulator; adult & children's audio tour.
Publications: member & donor newsletter, Scuttlebutt.
Hours & Admission Prices: Feb.-March & Nov.-Dec. Sat.-Sun. 9:30-3; April & Sept. 4-Nov. 3 daily 9:30-3; May-Sept. 3 daily 9:30-5. Fire Power Tour: adults $21.95, seniors, veterans, & children 5-11 $17; children under 5 & military no charge. Closed New Year's Day; Thanksgiving; Christmas. &
Attendance: 175,000 (estimated)
Membership: Individual $40; Family $80; Contributing $125; Sustaining $300; Benefactor $500-$999.

THE CAMDEN CHILDREN'S GARDEN, 3 Riverside Dr., Camden, NJ 08103. Tel.: 856-365-8733.
Governing Authority: nonprofit organization.
Institution Type/Description: Horticultural Garden.
Collections: hands-on exhibitions; environmental & horticultural experiences; butterfly house; tropical plants.
Activities: gardening programs; educational classes; special events; rental facilities; rides.
Hours & Admission Prices: Thurs.-Sun. 10-4. Admission $6 per person; children 2 & under no charge.

CAMDEN COUNTY HISTORICAL SOCIETY, 1900 Park Blvd., Camden, NJ 08103-3697. Mailing Address: P.O. Box 378, Collingswood, NJ 08108-0378. Tel.: 856-964-3333. Fax: 856-964-0378.
E-mail: admin@cchsnj.com
Web Site: www.cchsnj.com
Founded: 1899.
Congressional District: 1
Key Personnel: Bd. Pres., Chris Perks; Exec. Dir., Jason Allen; Interpretive Dir., William Roulett; Administrative Asst., Jacinda Williams.
Governing Authority: bd. of trustees; society. Tax-exempt: 501(c)(3).
Institution Type/Description: History Museum.
Collections: Pomona Hall: 18th & 19th-century furnishings displayed in the 18th-century restored Georgian style residence. Museum: exhibits focus on history of Camden County & Southern New Jersey, 1600 to present. Exhibits include tools of early handcrafts displayed in craft shops (blacksmith, cooper, candle-maker, spinning & weaving, carpenter, cobbler); early firefighting equip.; early American glass; Victor Talking Machine Co.; lighting devices; military, American Revolution-Civil War; toys; one room school; ethnic exhibits; regional industry exhibits, including Victor Talking Machine Co.; textiles; costumes.
Research Fields: social, cultural & economic history of Southern New Jersey; genealogy; historic preservation.
Facilities: 21,000-vol. library of the history of Camden County & Southern NJ; maps; 1800-1959 manuscripts, newspapers of area; photographs; slides. Postcards, pamphlets, books & museum-related items for sale.
Activities: guided tours; colonial open-hearth cooking; lectures; formally organized education programs for children; permanent & temporary exhibitions; spinning demonstrations; seminars. Museum Sponsors: programs pertaining to history of Southern New Jersey and region; occasional symposia; bus trips.
Publications: bulletins; newsletters; books; pamphlets; monographs.
Hours & Admission Prices: Wed.-Fri. 10:30-4:30, Sun. 12-5. Pomona Hall Tours: Thurs. & Sun. 12:30 & 2; other times by appointment. Library or Museum: adults $5, seniors & students, $4; members no charge. Pomona Hall: adults $5, seniors & students, $4; members no charge. Pomona Hall & Museum: adults $8, students $6, seniors $4; members no charge. Closed New Year's Eve & Day; Easter; Christmas; national holidays.
Attendance: 3,800 (accurate)

Membership: Seniors & Students $15; Individual $25; Family $35; Contributor $50; Benefactor & Organization $100; Sponsor $500; Partner $1,000; Patron $2,500.

STEDMAN GALLERY, (M), Rutgers-Camden Center for the Arts, 314 Linden St., Camden, NJ 08102-1403. Tel.: 856-225-6245 & 6350. Fax: 856-225-6597.
E-mail: arts@camden.rutgers.edu
Web Site: www.rutgerscamdenarts.org
Founded: 1975.
Congressional District: 1
Key Personnel: Dir., Cyril Reade; Deputy Dir. & Gallery Cur., Nancy Maguire; Cur. Arts Education & Outreach, Noreen Scott Garrity; Mgr. Community & Artist Programs, Carmen Pendleton.
Personnel Profile: Full-Time Paid 9; Part-Time Paid 2; Part-Time Volunteers 4; Interns 4.
Governing Authority: university; nonprofit. Parent Institution: Rutgers, The State University of New Jersey. Subsidiary Dept.: Rutgers-Camden Center for the Arts. Tax-exempt: 170 (b)(1)(A).
Institution Type/Description: Art Museum.
Collections: contemporary American works on paper; paintings; sculpture; 20th-century European works on paper.
Facilities: theater.
Activities: temporary exhibitions; children's education program; lectures; films; concerts; museum studies program.
Publications: exhibition catalogs.
Hours & Admission Prices: Jan. 3-Dec. 23 Mon.-Sat. 10-4; call for additional hours. No charge; donations accepted. Closed New Year's Eve & Day; Memorial Day; Independence Day; Labor Day; Thanksgiving; Christmas Eve & Day; between exhibitions. &
Attendance: 18,300 (accurate)

WALT WHITMAN HOUSE MUSEUM AND LIBRARY, 328 Mickle Blvd., Camden, NJ 08103. Tel.: 856-964-5383. Fax: 856-964-1088.
Founded: 1946.
Congressional District: 1
Key Personnel: Cur., Richard Ryan.
Personnel Profile: Full-Time Paid 1; Part-Time Paid 5; Interns 2.
Governing Authority: nonprofit organization. Parent Institution: New Jersey State Park Service. Tax-exempt: 501(c)(3).
Institution Type/Description: Historic Home Museum: 1884-1892 Walt Whitman home.
Collections: furniture; photographs; memorabilia; rare books; manuscripts; archival material.
Research Fields: literary work of Walt Whitman.
Facilities: 1,500-vol. library containing rare books, manuscripts & works of Whitman; educational facilities.
Activities: symposia, seminars & other public lectures & dialogues; educational programs & teacher workshops; poetry readings, contests, exhibitions; guided tours of Whitman's home.
Publications: biannual brochure, The Walt Whitman Assoc. Newsletter.
Hours & Admission Prices: June 15 to Labor Day Mon.-Fri. 11-4, Sat.-Sun. 12-5; Winter: Wed.-Fri. 1-4, Sat.-Sun. 11-4. Adults $4, seniors $3; members and children 18 & under no charge. Closed New Year's Day; Thanksgiving; Christmas.
Attendance: 3,000 (accurate)
Membership: Student $10; Senior $15; Individual $20; International (outside U.S) $25; Sponsor $50; Patron $100; Corporate & Benefactor $250.

Canton

LOWER ALLOWAYS CREEK HISTORIC LOG CABIN MUSEUM, 736 Smick Rd., Canton, NJ 08001. Mailing Address: 501 Locust Island Rd., P.O. Box 157, Hancocks Bridge, NJ 08038. Tel.: 856-935-1549.
Institution Type/Description: History Museum.
Collections: local history & culture; period furnishings; personal artifacts; early farm equipment; local industry; maritime history; fishing & trapping memorabilia. Historic Buildings: log cabin; outhouse; carriage shed; canning house.
Hours & Admission Prices: March-May & Sept.-Nov. 3rd Sun. each month 1-4.

Cape May

HISTORIC COLD SPRING VILLAGE, (M), 720 Rte. 9 S., Cape May, NJ 08204-4636. Tel.: 609-898-2300. Fax: 609-884-5926.
E-mail: 4info@hcsv.org

Web Site: www.hcsv.org
Founded: 1981.
Key Personnel: Dir., Anne Salvatore; Dir. Mktg., Sydney Perkins; Museum Shop Mgr., Clare Juechter
Institution Type/Description: History Museum.
Collections: early American history; 26 restored period buildings.
Activities: demonstrations; petting zoo; interactive 19th century crafts; special events.
Hours & Admission Prices: June to Labor Day call for hours. Adults $10, seniors & children 3-12 $8; discounts to AAM members; children under 3 no charge. &
Attendance: 30,000

MID-ATLANTIC CENTER FOR THE ARTS & HUMANITIES/EMLEN PHYSICK ESTATE/CAPE MAY LIGHTHOUSE/FIRE CONTROL TOWER NO. 23, (M), 1048 Washington St., Cape May, NJ 08204-1737. Mailing Address: P.O. Box 340, Cape May, NJ 08204-0340. Tel.: 609-884-5404; 800-275-4278. Fax: 609-884-2006.
E-mail: mac4arts@capemaymac.org
Web Site: www.capemaymac.org
Founded: 1970.
Congressional District: 2
Key Personnel: Dir. & C.E.O., B. Michael Zuckerman; Pres. (V), Mary McKenney; C.O.O., Melissa Zeides; Cur., Gail Capehart; Chief Outreach Officer, Mary Stewart; Mktg. & Communications Dir., Jean Barraclough; Dir. Visitor Svcs. & Special Events, Janice Coyle; C.F.O., Charles Kealy; Museum Education Coord., Robert Heinly; Cafe & Tearoom Mgr., David Corkery; Dir. Tour Operations, Nanci Coughlin; Mgr. Groups Sales, Sue Gibson; Museum Shop Mgr., Emily McLaughlin.
Personnel Profile: Full-Time Paid 25; Part-Time Paid 125; Part-Time Volunteers 300.
Governing Authority: nonprofit organization. Tax-exempt.
Institution Type/Description: Historic Building.
Collections: Victoriana, including costumes; books; ephemera; tools; furniture. Historic Buildings: 1879 Physick house, attributed to Frank Furness; Carriage House Gallery at the Physick Estate; 1859 Cape May lighthouse; 1942 Fire Control Tower No. 23.
Major Exhibits: The Way We Were...Cape May County's Once Thriving Black Business Communities, 1/18/14-4/13/14; Cape May Stage's 25th Anniversary Retrospective, 4/25/14-5/11/14; Cape May Ablaze, 5/23/14-10/14/14.
Research Fields: 1878-1916 Victorian & Edwardian periods; maritime/lighthouse history; World War II history.
Facilities: reference library of Victoriana. Physick Estate: tearoom & cafe. Museum-related items for sale.
Activities: tours. Museum Sponsors: Murder at the Physick Estate February to December; Cape May's Spring Festival April-May; Cape May Music Festival May-June; Victorian Week in October; Halloween Happenings in October; Christmas in Cape May November-December.
Publications: quarterly newsletter; This Week in Cape May, 13 issues per year; Emlen Physick Estate (illustrated booklet); Sentinel of the Jersey Cape.
Hours & Admission Prices: Physick Estate Guided Tours: daily. Lighthouse Self-guided Tours: Feb-Dec. Architectural Trolley Tours: daily. Guided Walking Tours: daily. Self-guided Historic House Tours: April-May & Oct.-Dec. audio walking tour of Cape May's historic district. Around Cape Island Boat Tours: April-Oct.; World War II Trolley Tour: April-Oct.; Delaware Bay Lighthouse Adventures (boat cruises) May-Sept. Tours: Jan.-March Sat.-Sun.; April-Dec. daily. House Museum Tours: adults $10, children $5. Lighthouse: adults $7, children $3. Trolley Tours: adults $10, children $5; discounts to AAM; museum members, writers, journalists & tour guides from other historic sites with ID no charge. &
Attendance: 260,000 (accurate)
Membership: Student $10; Individual $30; Joint $40; Grandparent & Family $50; Sponsor $100; Business $150; Benefactor $500.

NATURE CENTER OF CAPE MAY, 1600 Delaware Ave., Cape May, NJ 08204. Tel.: 609-898-8848. Fax: 609-898-8512.
Web Site: www.njaudubon.org
Key Personnel: Dir., Gretchen Whitman.
Personnel Profile: Full-Time Paid 1; Part-Time Paid 3.
Governing Authority: Parent Institution: NJ Audubon Society. Tax-exempt.
Institution Type/Description: Nature Center.
Collections: hands-on exhibitions; wildlife & their habitats.
Facilities: classrooms. Center-related items for sale.
Activities: summer programs; special events; educational programs.
Publications: quarterly, NJ Audubon.
Hours & Admission Prices: April-Oct. Mon.-Sat. 9-5; Sun. 11-4. No charge; donations accepted. &

NAVAL AIR STATION WILDWOOD AVIATION MUSEUM, 500 Forrestal Rd., Cape May Airport, Cape May, NJ 08242-2203. Tel.: 609-886-8787. Fax: 609-886-1942. Facebook: Naval Air Station Wildwood Aviation Museum.
E-mail: aviationmuseum@comcast.net
Institution Type/Description: Military Aviation Museum.
Collections: Navy airmen commemoration; 26 aircraft; military artifacts; photographs.
Facilities: library.
Activities: aviation festivals; concerts; veterans' ceremonies; lectures; school field trips; senior tours.
Hours & Admission Prices: April-Oct. 15 daily 9-5; Oct. 16-Nov. daily 9-4; Dec.-March Mon.-Fri. 9-4. Adults $10, children 3-12 $8; children under 3 no charge.
Membership: Individual $30; Family $50; Friend $75; Patron $100; Benefactor $500; Visionary $1,000.

Cape May Court House

CAPE MAY BIRD OBSERVATORY - THE CENTER FOR RESEARCH AND EDUCATION, 600 Rte. 47 N., Cape May Court House, NJ 08210. Tel.: 609-861-0700.
E-mail: cmbo2@njaudubon.org
Institution Type/Description: Bird Sanctuary.
Collections: birds & their habitats; nature; environment.
Activities: bird observation.
Hours & Admission Prices: Call for hours.

CAPE MAY COUNTY HISTORICAL MUSEUM AND GENEA-LOGICAL SOCIETY, (M), 504 Rte. 9 N., Cape May Court House, NJ 08210-1953. Tel.: 609-465-3535. Fax: 609-465-4274. Facebook: Cape May County Museum.
E-mail: museum@co.cape-may.nj.us
Web Site: www.cmcmuseum.org
Founded: 1927.
Congressional District: 1
Key Personnel: Dir. & Cur., Pary Woehlcke; Administrative Asst., Judi Davis; Genealogist, Lois Broomell; Library Coord., Sonia L. Forry.
Governing Authority: bd. of trustees. Parent Institution: Cape May County & the Cape May County Historical & Genealogical Society. Tax-exempt.
Institution Type/Description: History Museum: housed in the Cresse-Holmes House; built in 1704.
Collections: Native American artifacts; 18th-century kitchen & bedroom; Victorian artifacts; China, glass, & furniture; whaling; decoys; military & maritime; Cape May Lighthouse lens; toys; medical instruments & tools; carriages. Historic Buildings: 11 room historic house; 5 room historic barn.
Research Fields: historical; genealogical.
Facilities: library of Cape May County genealogy & local history. Gift items for sale.
Activities: guided tours; programs; fund raising activities; civic presentations; slide/tape programs; in-school presentations; lectures; museum kids club; assistance to teachers.
Publications: annual magazine, The Cape May County Magazine of History and Genealogy; quarterly newsletter, Reflections.
Hours & Admission Prices: Library: Thurs. & Fri. 10-3. Museum: Memorial Day to Labor Day Tues.-Sat. 10-3; Labor Day to Memorial Day Wed.-Sat. 10-3. No charge. &
Attendance: 3,000 (estimated)
Membership: Individual $25; Joint $35; Family $40; Business $150.

CAPE MAY COUNTY PARK AND ZOO, 707 Rte. 9 N., Cape May Court House, NJ 08210. Mailing Address: 4 Moore Rd., DN 801, Cape May Court House, NJ 08210-1645. Tel.: 609-465-5271. Fax: 609-465-5421.
Web Site: www.capemaycountygov.net
Founded: 1963.
Congressional District: 2
Key Personnel: Parks Dir., Edward Runyon; Zoo Dir., Dr. Hubert Paluch.
Personnel Profile: Full-Time Paid 41; Part-Time Paid 6; Part-Time Volunteers 10; Interns 3.
Governing Authority: county; nonprofit. Parent Institution: Cape May County Park Dept. Subsidiary Institution: Cape May County Zoological Society. Tax-exempt.
Institution Type/Description: Zoo.
Collections: over 180 animal species & endangered species.
Facilities: botanical garden; zoological park. Museum-related items for sale.
Hours & Admission Prices: Zoo: Fall & Winter: daily 10-3:45; Spring & Summer: daily 10-4:45. Park: 9am to dusk. No charge; donations accepted. Closed Christmas.

Attendance: 700,000 (estimated)

CAPE MAY NATIONAL WILDLIFE REFUGE, 24 Kimbles Beach Rd., Cape May Court House, NJ 08210. Tel.: 609-463-0994.
Institution Type/Description: Wildlife Refuge.
Collections: wildlife & their habitats.
Facilities: nature trails.
Activities: walking trails.
Hours & Admission Prices: Headquarters: Mon.-Fri. 8-4:30. Refuge: dawn to dusk.

LEAMING'S RUN GARDEN & COLONIAL FARM, 1845 Rt. 9 N., Cape May Court House, NJ 08210-1436. Tel.: 609-465-5871. Facebook: Leaming's Run Garden.
E-mail: info@leamingsrungardens.com
Web Site: www.leamingsrungardens.com
Founded: 1978.
Congressional District: 3
Key Personnel: C.E.O., Jack Aprill; Pres. (V), Emily Aprill; Dir., Gregg Aprill.
Governing Authority: individual operation; organized for profit.
Institution Type/Description: Agricultural Museum: housed in last remaining whaler's house in New Jersey.
Collections: vegetables; farm animals; historic chicken breeds.
Facilities: botanical garden; nature & conservation center. Dried flowers & gift items for sale.
Activities: guided tours.
Hours & Admission Prices: mid-May to Sept. daily 9:30-5. Adults $8, children under 18 $4; discounts to groups; children 6 & under no charge. &
Attendance: 40,000
Membership: Annual $25.

Cape May Point

CAPE MAY BIRD OBSERVATORY - THE NORTHWOOD CENTER, 701 E. Lake Dr., Cape May Point, NJ 08204. Mailing Address: P.O. Box 3, Cape May Point, NJ 08212. Tel.: 609-884-2736.
E-mail: cmbo1@njaudubon.org
Institution Type/Description: Bird Sanctuary.
Collections: birds & their habitats; environment; nature; bird conservation.
Activities: bird watching; birding activities.
Hours & Admission Prices: Call for hours.

Cedar Grove

CEDAR GROVE HISTORICAL SOCIETY, 903 Pompton Ave., Cedar Grove, NJ 07009-1225. Mailing Address: P.O. Box 461, Cedar Grove, NJ 07009-0461. Tel.: 973-239-5414.
Web Site: www.cedargrovehistoricalsociety.org
Founded: 1968.
Congressional District: 8
Personnel Profile: Part-Time Volunteers 25.
Governing Authority: private; nonprofit organization. Tax-exempt: 501(c)(3).
Institution Type/Description: Historical Society Museum: housed in mid 19th-century early Victorian vernacular Morgan family farmhouse.
Collections: letters, receipts & plans related to the farming operations on the site from 1908-1985; agricultural magazines; books; tools & equipment; sports equipment; furnishings; textiles; late 18th- to early 20th-century cemetery; barn; civic park.
Research Fields: history of agriculture in Essex County.
Facilities: 700-vol. library of books & magazines, as well as letters & receipts; 1,200 sq. ft. exhibit space; nature trails.
Activities: guided tours; lectures. Museum Sponsors: pumpkin & apple sale in Fall; Holiday Open House in December.
Publications: monthly, Cedar Grove Historical Society Newsletter.
Hours & Admission Prices: Wed. morning; other times by appointment. Volunteer work sessions: Wed. 10-1. No charge; donations accepted. &
Attendance: 500 (estimated)
Membership: Single $15; Family $20.

Chatham

CHATHAM TOWNSHIP MUSEUM - LITTLE RED SCHOOL-HOUSE, 24 Southern Blvd., Chatham, NJ 07928. Mailing Address: P.O. Box 262, Chatham, NJ 07928-0262. Tel.: 973-635-0603.
Governing Authority: Parent Institution: Chatham Historical Society.
Institution Type/Description: History Museum.

Collections: local history & culture; period furnishings; photographs.
Hours & Admission Prices: 1st Sun. of the month 2-4.

Cherry Hill

BARCLAY FARMSTEAD, 209 Barclay Ln., Cherry Hill, NJ 08034. Tel.: 856-795-6225.
Web Site: www.barclayfarmstead.org
Institution Type/Description: Historic House: built in 1816.
Collections: local history; period furnishings; personal artifacts; tool shed; corn crib; springhouse; barn; gardens.
Activities: special events; group tours; picnic area.
Hours & Admission Prices: March-Nov. Wed. 12-4, 1st Sun. each month 1-5; Dec.-Feb. Wed. 12-4; call to confirm. No charge; donations accepted.

GARDEN STATE DISCOVERY MUSEUM, (M), 2040 Springdale Rd., Ste. 100, Cherry Hill, NJ 08003-2082. Tel.: 856-424-1233. Fax: 856-424-6516.
E-mail: roree@discoverymuseum.com
Web Site: www.discoverymuseum.com
Founded: 1994.
Key Personnel: Dir., Kelly Lyons; Dir. Mktg. & Sales, Beverly Pak.
Personnel Profile: Full-Time Paid 9; Full-Time Volunteers 1; Part-Time Paid 40; Part-Time Volunteers 6.
Institution Type/Description: Children's Museum.
Collections: hands-on activities.
Facilities: Museum-related items for sale.
Activities: educational workshops & performances; rental facilities; activity rooms; birthday parties; classes.
Publications: e-newsletter.
Hours & Admission Prices: May-Sept. daily 9:30-5:30; Oct.-April Sun.-Fri. 9:30-5:30, Sat. 9:30-8:30. Adults $10.95, senior citizens $8.95; discounts to military; children under one no charge. Closed Thanksgiving; Christmas. &
Attendance: 250,000 (estimated)
Membership: Family & Grandparent $121.98; Family Gold $144.45; Family & Friends $175.48.

GOODWIN HOLOCAUST MUSEUM AND EDUCATION CEN-TER, 1301 Springdale Rd., Cherry Hill, NJ 08003. Tel.: 856-751-9500, ext. 1203.
Institution Type/Description: History Museum.
Collections: Holocaust history; photographs; personal artifacts.
Activities: workshops; educational programs; special events.
Hours & Admission Prices: Sun.-Fri.

Chester

COOPER GRISTMILL, 66 Washington Tpke., Chester, NJ 07930. Mailing Address: 66 Rte. 24, Chester, NJ 07930. Tel.: 908-879-5463.
Institution Type/Description: Historic Building: built by Nathan Cooper in 1826.
Collections: local history & heritage; mill history.
Facilities: visitor center; nature trails.
Activities: special events; demonstrations; school programs; summer camps.
Hours & Admission Prices: May-June & Sept.-Oct. Sat.-Sun. 10-5; July-Aug. Wed.-Sun. 10-5.

Clark

DR. WILLIAM ROBINSON PLANTATION & MUSEUM, 593 Madison Hill Rd., Clark, NJ 07066-3103. Mailing Address: Clark Historical Society, Municipal Bldg., 430 Westfield Ave., Clark, NJ 07066. Tel.: 732-340-1571.
E-mail: info@drrobinsonmuseum.org
Web Site: www.clarkhistoricalsociety.org
Founded: 1974.
Congressional District: 7
Key Personnel: Dir., Scott McCabe; Asst., Trish Plummer; Asst., Lisa Jo Jennings.
Personnel Profile: Part-Time Volunteers 10.
Governing Authority: nonprofit organization. Parent Institution: Clark Historical Society. Tax-exempt: 501(c)(3).
Institution Type/Description: Historic Society Museum; Historic House: c.1690, farmhouse with features of the Tudor period.
Collections: medical items; medicinal herb garden; farm equipment: milk wagon, saws, harvesting equipment, butter churns, spinning wheels, open hearth fireplace with oven.

Facilities: herb garden. Gift items for sale.
Activities: guided tours; organized education programs for children. Museum Sponsors: Open House in November; Holiday Celebration in December.
Publications: brochure; Historical Society booklet; directory of members & friends.
Hours & Admission Prices: 3rd Sun. of month 1-4; other times by appointment. No charge; donations accepted.
Attendance: 1,000 (estimated)
Membership: Annual $25.

Clayton

SCOTLAND RUN PARK NATURE CENTER, 980 Academy St., Clayton, NJ 08322. Tel.: 856-881-0845.
Institution Type/Description: Nature Center.
Collections: wildlife & their habitats; nature; environment.
Facilities: 940-acres; nature trails.
Activities: educational programs; workshops; children's activities; walking trails.
Hours & Admission Prices: Park: dawn to dusk.

Clifton

CLIFTON ARTS CENTER & SCULPTURE PARK, 900 Clifton Ave., Clifton, NJ 07013. Tel.: 973-472-5499. Fax: 973-470-8337. Facebook: Clifton Arts Center & Sculp.
E-mail: rcammilleri@cliftonnj.org
Web Site: www.cliftonnj.org
Founded: 2000.
Congressional District: 8
Key Personnel: Dir., Roxanne Cammilleri; Chm. (V), Jeff Labriola.
Personnel Profile: Full-Time Paid 1; Part-Time Volunteers 5.
Volunteer Hours: 20
Operating Expenses: 9,356
Operating Income: 9,390
Governing Authority: city. Parent Institution: Clifton Arts Center, Inc. Tax-exempt.
Institution Type/Description: Arts Center; housed in two early 20th century barns. Listed on the National Register of Historic Places.
Collections: paintings; photographs; over 30 contemporary sculptures; cultural artifacts.
Major Exhibits: Trilogy - United Nations Staff Recreation Council - Stefano Losi, Chambliss Giobbi & Marlene Luce Temblay, 2/5/14-3/8/14; In The Style of...Clifton Association of Artists, 3/19/14-4/12/14; Heirloom - Clifton High School, 5/7/14-5/24/14; An American Master Artist's Retrospect: Louis Bouche', 6/25/14-7/26/14; Louis Bouche': Part II and the Modern Artists Era, 9/17/14-10/25/14; Five Ways of Seeing - Janice Mauro, Allan Drossman, Judy Schaefer, David Soo, Elaine Lorenz, 11/5/14-12/13/14.
Activities: cultural events.
Hours & Admission Prices: Wed.-Sat. 1-4. Adults $3. &
Attendance: 30,000
Membership: Individual $15; Family $25.

HAMILTON VAN WAGONER MUSEUM, 971 Valley Rd., Clifton, NJ 07013-4028. Tel.: 973-744-5707.
E-mail: normaleeclf@aol.com
Formerly: Hamilton House Museum-van Wagoner Museum
Founded: 1974.
Congressional District: 9
Key Personnel: Pres. Bd. Trustees, Marlene Walcott; Cur. & Museum Shop Mgr., Norma Lee Smith.
Personnel Profile: Part-Time Paid 1; Part-Time Volunteers 20.
Governing Authority: municipal council. Tax-exempt: 170(b)(1)(A).
Institution Type/Description: Historic House Museum: 1815 Hamilton Van Wagoner House.
Collections: period furnishings of the 19th century.
Research Fields: local history.
Activities: guided tours; early craft workshops; formally organized education programs for children; cooking 1st Sun.
Publications: quarterly, Newsletter.
Hours & Admission Prices: Sun. 2-4; call for private tours. Adults $3. Closed major holidays.
Attendance: 1,200 (estimated)
Membership: Student $2; Individual $5; Family $10; Sustaining $25.

Clinton

HUNTERDON ART MUSEUM, 7 Lower Center St., Clinton, NJ 08809-1384. Tel.: 908-735-8415. Fax: 908-735-8416.
E-mail: info@hunterdonartmuseum.org
Web Site: www.hunterdonartmuseum.org
Founded: 1952.
Congressional District: 13
Key Personnel: Exec. Dir., Marjorie Frankel Nathanson; Pres. Bd. Trustees (V), Jim McDevitt; Dir. Exhibitions, Jonathan Greene; Dir. Education, Jennifer Brazel; Dir. Devel., Caryn Tomljanovich.
Personnel Profile: Full-Time Paid 7; Part-Time Paid 20; Part-Time Volunteers 80; Interns 1.
Governing Authority: nonprofit organization. Tax-exempt: 501(c)(3).
Institution Type/Description: Contemporary Art Museum: housed in 1836 restored stone grist mill.
Collections: contemporary works on paper.
Activities: tours; lectures; special events; gallery talks; formally organized education programs; loan, temporary, & traveling exhibitions; juried & non-juried exhibitions; art classes; docent tours; gallery talks; special events.
Publications: newsletter, Hunterdon Museum of Art; exhibition catalogues; general brochure.
Hours & Admission Prices: Tues.-Sun. 11-5. Suggested Donation: adults $5; discounts to NARM members; members no charge. &
Attendance: 33,000 (estimated)
Membership: Senior $35; College Student $35; Individual over 18 $45; Family $65; Contributor $100; Sponsor $250; Patron $500; Benefactor $1,000.

RED MILL MUSEUM VILLAGE, 56 Main St., Clinton, NJ 08809-1328. Tel.: 908-735-4101. Fax: 908-735-0914.
E-mail: director@theredmill.org
Web Site: theredmill.org
Formerly: Hunterdon Historical Museum
Founded: 1960.
Congressional District: 13
Key Personnel: Pres., Paul Muir; Exec. Dir., Eileen Morales; Collections Mgr., Elizabeth Cole.
Personnel Profile: Full-Time Paid 2; Part-Time Paid 8; Part-Time Volunteers 70; Interns 1.
Governing Authority: private; nonprofit organization. Tax-exempt: 501(c)(3).
Institution Type/Description: Historic Site & History Museum.
Collections: 18th-, 19th- & early 20th-century rural American life; industrial; agricultural; domestic. Historic Buildings: 1810 Hunt's Mill; 1860 Bunker Hill schoolhouse; log cabin; blacksmith; general store; lime kilns; quarry & stone crusher.
Research Fields: 18th-, 19th- & early 20th-century social, industrial & agricultural history.
Facilities: herb garden; outdoor concert stage. Handcrafted & other gift items for sale.
Activities: group tours; concert series; children's workshops; lecture series. Museum Sponsors: car shows.
Publications: newsletter; The Old Mill Wheel.
Hours & Admission Prices: Jan.-March Sat.-Sun. 12-4; April-Sept. & Nov.-Dec. Tues.-Sat. 10-4, Sun. 12-5; admission gate closes at 4. Adults $8, senior citizens $7, children 6-16 $6; discounts to AAM members; members & children under 6 no charge. Closed Memorial Day; Independence Day; Labor Day; Haunted Village fundraiser in October.
Attendance: 18,000 (accurate)
Membership: Individual $30; Family $45; Patron $500; Sustaining $1,000.

Closter

BELSKIE MUSEUM OF ART & SCIENCE, 280 High St., Closter, NJ 07624-1812. Tel.: 201-768-0286. Fax: 201-768-4220.
E-mail: belskiemuseum@hotmail.com
Web Site: www.belskiemuseum.com
Founded: 1994.
Personnel Profile: Part-Time Volunteers 25.
Governing Authority: Tax-exempt.
Institution Type/Description: Art Museum.
Collections: works of Abram Belskie, 1907-1988 including drawings, sculpture, medical models, metallic molds; local & international artists.
Activities: student & instructor exhibits.
Hours & Admission Prices: Sat.-Sun. 1-5; other times by appointment. No charge; donations accepted. &
Attendance: 3,000 (estimated)
Membership: $25-$100.

CLOSTER NATURE CENTER, 230 Ruckman Rd., Closter, NJ 07624. Mailing Address: P.O. Box 80, Closter, NJ 07624. Tel.: 201-750-2778.
Web Site: www.closternaturecenter.org
Institution Type/Description: Nature Center.
Collections: wildlife & their habitat.
Facilities: nature trails.
Activities: hiking; lectures; summer youth programs; school & scout programs.
Hours & Admission Prices: Call for hours.

Columbus

MANSFIELD TOWNSHIP HISTORICAL SOCIETY (1849 GEORGETOWN SCHOOL/MUSEUM), 4 Fitzgerald Ln., Columbus, NJ 08022-2383. Tel.: 609-298-4174.
Founded: 1973.
Congressional District: 8
Key Personnel: Pres. (V), Pearl J. Tusim.
Personnel Profile: Part-Time Volunteers 4.
Governing Authority: society; nonprofit. Tax-exempt: 501(c)(3).
Institution Type/Description: Historical Society Museum: housed in a rebuilt one-room school.
Collections: township history.
Facilities: 50-vol. library of 1800's school books available to the public; educational facilities.
Activities: guided tours; lectures. Annual Event: Field Day in June.
Hours & Admission Prices: By appointment. No charge; donations accepted.
Attendance: 100 (estimated)
Membership: Mansfield Twp. Historical Soc. $5 per year.

Cranbury

CRANBURY HISTORICAL & PRESERVATION SOCIETY, 4 Park Place E., Cranbury, NJ 08512-3208. Mailing Address: 6 S. Main St., Cranbury, NJ 08512-3112. Tel.: 609-860-1889.
E-mail: historycenter@comcast.net
Web Site: cranburyhistory.org
Founded: 1967.
Congressional District: 14
Key Personnel: Pres. (V) & Museum Shop Mgr., Audrey Smith; Vice Pres., Gale Scott; Dir. History Center, Virginia Swanagan; Cur., Lisa Beach; Treas., Robert Dreyling.
Personnel Profile: Part-Time Volunteers 53.
Governing Authority: society; nonprofit organization. Subsidiary Institution: Cranbury History Center, 6 S. Main St., Cranbury, NJ; Parsonage Barn, 1741 Cranbury Neck Rd., Cranbury, NJ. Tax-exempt.
Institution Type/Description: Historical Society Museum: housed in restored 1834 Dr. Garrett Voorhees house.
Collections: oral history tapes; Cranbury local history; photographs, slides & maps; genealogical data; church records; vital statistics; Revolutionary & Civil War service records; historic house, census & cemetery records. Historic Buildings: gristmiller's house, c.1860; parsonage barn, c.1741; Dr. Garrett Voorhees house, c.1834.
Research Fields: Cranbury history; data on families, houses.
Facilities: library of books, maps, photographs, newspapers, oral history tapes, genealogical records of Cranbury, New Jersey, and Revolutionary & Civil War records for use on premises by appointment only.
Activities: guided tours; lectures & cultural programming; loan & temporary exhibitions; walking tours; living histories.
Publications: newsletter; pamphlets, Walking Tour Historic Cranbury; Sara's Garden; Historic Cranbury; books, Cranbury Past & Present; Cook's Tour of Cranbury; booklet, George Washington in Cranbury.
Hours & Admission Prices: Cranbury Museum: Sun. 1-4. Cranbury History Center: Tues. 10:30-1:30; other times by appointment. No charge; donations suggested.
Attendance: 1,000 (estimated)
Membership: Individual $15; Family $25; Sponsor $35; Patron $50; Corporate $150 & up.

NEW JERSEY AUDUBON SOCIETY'S PLAINSBORO PRESERVE & NATURE CENTER, 80 Scotts Corner Rd., Cranbury, NJ 08512. Tel.: 609-897-9400.
E-mail: plainsboro@njaudubon.org
Institution Type/Description: Nature Center.
Collections: wildlife & their habitats; hands-on exhibitions.
Facilities: nature trails. Gift items for sale.
Activities: nature trails; birthday parties; school & scout programs; vacation camps.
Hours & Admission Prices: Nature Center: Tues.-Sat. 9-5, Sun. 12-5. Trails: daily sunrise to sunset. No charge. Closed holidays. &

Cranford

CRANFORD HISTORICAL SOCIETY - THE HANSON HOUSE, 38 Springfield Ave., Cranford, NJ 07016-2144. Tel.: 908-276-0082.
E-mail: cranfordhistoricalsociety@verizon.net
Web Site: cranfordhistoricalsociety.org/
Key Personnel: Pres., Patrick Paulak; Trustee, Robert Fridlington.
Governing Authority: Branch Museum: Crane Phillips Living Museum, North Union Ave., Cranford, NJ.
Institution Type/Description: Historical Society Museum.
Collections: photographs; scrapbooks; glass negatives; furniture; tools; kitchen & farm implements; Indian artifacts; costumes & textiles; books; letters; personal artifacts.
Research Fields: genealogy.
Activities: tours.
Hours & Admission Prices: Mon.-Thurs. 9-12; other times by appointment.

TOMASULO ART GALLERY - UNION COUNTY COLLEGE, Kenneth Campbell MacKay Library, 1033 Springfield Ave., Cranford, NJ 07016-1599. Tel.: 908-709-7155.
E-mail: tomasulogallery@ucc.edu
Founded: 1973.
Key Personnel: Dir., Lisa Williamson.
Personnel Profile: Part-Time Paid 3.
Governing Authority: Parent Institution: Union County College.
Institution Type/Description: Art Gallery.
Collections: contemporary paintings & sculptures.
Hours & Admission Prices: Mon. 1-4, Tues.-Thurs. 1-4 & 6-9, Sat. 10-1. No charge.
Attendance: 1,000 (estimated)

Daretown

OLD PITTSGROVE PRESBYTERIAN CHURCH AND PITTSGROVE LOG COLLEGE, 312 Daretown Rd., Daretown, NJ 08318. Tel.: 856-358-1104.
Institution Type/Description: History Museum.
Collections: Historic Buildings: 1767 church; 18th century log college.
Activities: special events.
Hours & Admission Prices: By appointment.

Demarest

THE ART SCHOOL AT OLD CHURCH, 561 Piermont Rd., Demarest, NJ 07627-1615. Tel.: 201-767-7160. Fax: 201-767-0497.
E-mail: gallery@tasoc.org
Web Site: tasoc.org
Formerly: Old Church Cultural Center
Founded: 1974.
Congressional District: 39
Key Personnel: Exec. Dir., Maria Danziger; Pres. (V), Mikhail Zakin; Gallery Dir., John J. McGurk.
Personnel Profile: Full-Time Paid 4; Part-Time Paid 45; Part-Time Volunteers 10.
Governing Authority: private; nonprofit organization. Tax-exempt: 501(c)(3).
Institution Type/Description: Art & Cultural Center.
Collections: paintings; photographs; sculpture.
Facilities: 950 sq. ft. exhibit space; 4 classrooms.
Activities: education programs. Annual Events: Small Works Show in March; Student Exhibition in April; Faculty Show in September; Pottery Show & Sale in December.
Publications: quarterly course catalog, Semester; biannual newsletter, Centerline; annual catalog, Pottery Show.
Hours & Admission Prices: Gallery & Office: Mon.-Fri. 9:30-5, Sat. 9:30-3. Please call for extended hours. No charge for exhibitions. Closed federal holidays; most Jewish holidays. &
Attendance: 2,200 (estimated)
Membership: Youth $20; Adult $40; Family $50.

Dennisville

DENNIS TOWNSHIP MUSEUM & HISTORY CENTER, 681 Petersburg Rd., Dennisville, NJ 08214. Tel.: 609-861-1899.
Institution Type/Description: History Museum: housed in a former school house; built in 1874.
Collections: local history & culture; photographs; historic documents; period furnishings.

Hours & Admission Prices: 1st & 3rd Sat. 9-1. Closed national holidays.

Dover

DOVER HISTORY MUSEUM, 55 W. Blackwell St., Dover, NJ 07801-3808. Tel.: 973-361-3525.
Institution Type/Description: Historical Society Museum.
Collections: local history & culture; period furnishings; personal artifacts; photographs.
Hours & Admission Prices: Sun. 1-4.

East Brunswick

EAST BRUNSWICK MUSEUM CORPORATION, 16 Maple St., East Brunswick, NJ 08816-4450. Mailing Address: P.O. Box 875, East Brunswick, NJ 08816-0875. Tel.: 732-257-1508. Fax: 732-257-1508.
Founded: 1978.
Congressional District: 12
Key Personnel: Pres., Karen Scott; Vice Pres., Martha Hess
Governing Authority: nonprofit organization. Subsidiary Institution: Alice Appleby Devoe Library/Harold Hoffman Memorial Archive. Tax-exempt: 501(c)(3).
Institution Type/Description: Local & Regional History Museum: housed in 1860 Simpson Methodist Church & 1850 Appleby/Devoe House.
Collections: local & regional history objects.
Facilities: 300-vol. library of reference materials; archives.
Activities: guided tours; historic cemetery tours; lectures; docent program; participatory & loan exhibitions; workshops; book sale. Museum Sponsors: Village Street Fair; haunted house tour; holiday exhibit.
Publications: periodic newsletter; exhibition catalogues.
Hours & Admission Prices: Sat.-Sun. 1:30-4; groups at other times by appointment. No charge; donations accepted. Closed New Year's Day; Easter; Mother's Day; Memorial Day; Father's Day; Independence Day; Labor Day; Thanksgiving; Christmas. &
Attendance: 8,000
Membership: Student & Senior Citizen $10; Individual $15; Corporate $100; Life $300; Corporate Life $500.

Eastampton

HISTORIC SMITHVILLE MANSION, 803 Smithville Rd., Eastampton, NJ 08060. Mailing Address: c/o Friends of the Mansion at Smithville, P.O. Box 6000, Eastampton, NJ 08060. Tel.: 609-265-5858.
E-mail: info@smithvillemansion.org
Web Site: www.smithvillemansion.org
Institution Type/Description: Historic House Museum: built in 1840.
Collections: local history & culture; period furnishings; personal artifacts; photographs.
Activities: special events.
Hours & Admission Prices: Tours: early May to late Oct. Wed. & Sun. 1, 2, & 3. Adults $5, students $3; discounts to groups of 10 or more.

Edison

THOMAS ALVA EDISON MEMORIAL TOWER AND MENLO PARK MUSEUM, 37 Christie St., Edison, NJ 08820-3860. Mailing Address: Edison Memorial Tower Corporation, P.O. Box 656, Edison, NJ 08818. Tel.: 732-494-4194. Fax: 732-494-4190.
E-mail: info@menloparkmuseum.org
Web Site: www.menloparkmuseum.com
Institution Type/Description: History Museum.
Collections: Edison's life & history; innovations including wireless transmission, the carbon button transmitter, & the Edison Effect (the foundation for the field of electronics); photographs; personal artifacts; company artifacts.
Hours & Admission Prices: Thurs.-Sat. 10-4; call to confirm. Suggested Donation: $5.
Membership: Active $25; Supporting & Family $50; Patron $100; Sponsor $250; Benefactor $500; Lifetime $1,000.

Egg Harbor City

EGG HARBOR CITY HISTORICAL SOCIETY - ROUNDHOUSE MUSEUM, 533 London Ave., Egg Harbor City, NJ 08215. Tel.: 609-965-9073.
Key Personnel: Dir., Mark Maxwell
Institution Type/Description: Historical Society Museum.

Collections: local history & culture; period artifacts; clothing; local industry; photographs; census records; city tax records; genealogy.
Hours & Admission Prices: Wed. & Sat. 1-4. No charge.

Elizabeth

BELCHER OGDEN MANSION, 1046 E. Jersey St., Elizabeth, NJ 07201-2504. Mailing Address: 1045 E. Jersey St., Ste. 101, Elizabeth, NJ 07201-2503. Tel.: 908-581-7555.
Institution Type/Description: Historic House Museum.
Collections: local history & culture; period furnishings; photographs.
Hours & Admission Prices: By appointment.

BOXWOOD HALL, 1073 E. Jersey St., Elizabeth, NJ 07201. Mailing Address: c/o NJ Dept. of Environmental Protection State Park Svc., P.O. Box 402, Mail Code 501-04, Trenton, NJ 08625-0402. Tel.: 908-282-7617.
Web Site: ucnj.org/boxwood-hall
Institution Type/Description: Historic House Museum: built in 1750s by Elizabethtown's mayor, Samuel Woodruff later home to Elias Boudinot, a United States Congressman and then President of Congress.
Collections: local history & culture; period furnishings; personal artifacts; photographs.
Hours & Admission Prices: Mon.-Fri. 9-12 & 1-5.

Englishtown

BATTLEGROUND HISTORICAL SOCIETY - THE VILLAGE INN, 2 Water St., Englishtown, NJ 07726. Mailing Address: P.O. Box 61, Tennent, NJ 07763-0061. Tel.: 732-462-4947.
E-mail: thevillageinnenglishtown@verizon.net
Web Site: www.thevillageinn.org
Founded: 1969.
Congressional District: 12
Key Personnel: Pres., Hans Kernast; Treas., Kathy Doherty.
Personnel Profile: Part-Time Volunteers 15.
Governing Authority: private; nonprofit organization. Tax-exempt: 501(c)(3).
Institution Type/Description: Historic Building.
Collections: Englishtown Village Inn is a restored 18th-century tavern which was used by American forces at the time of the Battle of Monmouth on June 28, 1778.
Facilities: Museum-related items for sale.
Activities: guided tours. Annual Events: monthly meetings of Battleground Historical Society January-May & September-October at Inn; spring luncheon.
Publications: newsletter, Matchaponix Journal.
Hours & Admission Prices: 3rd Sun. each month 1-3. No charge; donations accepted.
Attendance: 200 (estimated)
Membership: Individual $12; Family $20; Individual Life $100.

Estell Manor

ESTELL MANOR HISTORICAL SOCIETY, 134 Cape May Ave., Estell Manor, NJ 08319. Mailing Address: P.O. Box 72, Estell Manor, NJ 08319-0072. Tel.: 609-476-2884.
E-mail: estellmanorhistoricalsociety@aol.com
Web Site: estellmanorhistoricalsociety.org
Founded: 1983.
Key Personnel: Chm. (V), Robert Grant; Pres. (V), Tom Pogue; Museum Mgr., Diane Bassetti.
Governing Authority: Parent Institution: EMHS Bd. of Trustees. Tax-exempt.
Institution Type/Description: Historical Society Museum: housed in a one room schoolhouse, Risley School. Listed on the National Register of Historic Places.
Collections: local history & culture; photographs; personal artifacts; memorabilia; family history.
Research Fields: local family genealogy; local history; schoolhouse history.
Facilities: library; meeting room.
Activities: monthly historical society meetings; annual local history family program for K-8 children; fundraisers; holiday dinner.
Publications: quarterly newsletter.
Hours & Admission Prices: Call for hours. No charge; donations accepted. &
Attendance: 350 (estimated)
Membership: Student $2; Individual $10.

Ewing

BENJAMIN TEMPLE HOUSE, Drake Farm Park, 27 Federal City Rd., Ewing, NJ 08638. Tel.: 609-883-2455. Fax: 609-883-2455.
Governing Authority: Parent Institution: Township of Ewing.
Institution Type/Description: Historic House Museum: housed in the former home of Benjamin Temple, a farmer & early area settler; built c.1750.
Collections: Temple family history; period furnishings; personal artifacts; photographs.
Facilities: library.
Activities: lectures; special events; educational programs.
Hours & Admission Prices: Wed. 10-2; other times by appointment.

THE COLLEGE ART GALLERY, The College of New Jersey, Holman Hall, 2000 Pennington Rd., Ewing, NJ 08628. Mailing Address: P.O. Box 7718, Ewing, NJ 08628-0718. Tel.: 609-771-2198. Fax: 609-637-5193.
E-mail: tcag@tcnj.edu
Web Site: www.tcnj.edu/~tcag
Founded: 1855.
Governing Authority: public college. Parent Institution: The College of New Jersey. Tax-exempt.
Institution Type/Description: Art Gallery.
Collections: photography; printmaking; paintings; drawings; sculpture.
Hours & Admission Prices: Tues.-Thurs. 12-7, Sun. 1-3. No charge. &
Attendance: 5,000 (estimated)

Fair Lawn

GARRETSON FARM, 4-02 River Rd., Fair Lawn, NJ 07410-1436. Tel.: 201-797-1775.
Web Site: www.garretsonfarm.org
Institution Type/Description: Historic Homestead: built in 1719. Listed on the National Register of Historic Places.
Collections: local history & culture; period furnishings; heirloom garden.
Activities: special events; educational programs.
Hours & Admission Prices: Call for hours.

Far Hills

UNITED STATES GOLF ASSOCIATION MUSEUM, (M), 77 Liberty Corner Rd., Far Hills, NJ 07931-2570. Mailing Address: P.O. Box 708, Far Hills, NJ 07931-0708. Tel.: 908-234-2300. Fax: 908-470-5013.
E-mail: museum@usga.org
Web Site: www.usgamuseum.com
Formerly: Golf House, Museum & Library
Founded: 1935.
Congressional District: 7
Key Personnel: Dir., Robert Williams; Cur. & Historian, Mike Trostel; Asst. Dir. Museum Operations, Susan Wasser; Film & Video Archivist, Shannon Doody; Librarian, Nancy Stulack; Security, James Rau; Mgr., Mktg. & Outreach, Kim Gianetti.
Personnel Profile: Full-Time Paid 10; Part-Time Paid 3.
Governing Authority: nonprofit organization. Parent Institution: United States Golf Association. Tax-exempt: 501(c)(3).
Institution Type/Description: Golf Museum & Library: housed in 1919 Georgian style home, designed by John Russell Pope.
Collections: clubs; balls; tees; costumes; ceramics; glass; silver; gold; stamps; bags; tins; archives; photographs; books; paintings.
Research Fields: golf.
Facilities: 30,000-vol. library of golf-related material, including books, magazines & pamphlets available to the public; 16,000 sq. ft. exhibit space; research & test center. Gift items for sale.
Activities: guided tours by appointment; traveling & special exhibitions; 16,000 sq. ft. putting green.
Hours & Admission Prices: Tues.-Sun. 10-5. Adults $7, children 13-17 $3.50; discounts to veterans and AAA & museum members; children under 12 no charge. Closed major holidays. &
Attendance: 20,000 (estimated)

Farmingdale

ALLAIRE VILLAGE INC., 4263 Atlantic Ave., Farmingdale, NJ 07727. Tel.: 732-919-3500 & 938-2253. Fax: 732-938-3302.
E-mail: allairevillage@bytheshore.com
Web Site: www.allairevillage.org
Formerly: Historic Allaire Village Inc.
Founded: 1957.

Congressional District: 11
Key Personnel: Village Admin., Sharon Avale; Program Dir., Shannon Gance; Chm. (V), William Gerhanser, Ph.D.; Treas., Hance Sitkus, CPA; Group Tours, Angela Larcara.
Personnel Profile: Full-Time Paid 4; Part-Time Paid 18; Part-Time Volunteers 160; Interns 3.
Governing Authority: nonprofit organization. Subsidiary Institution: Allaire Village Auxiliary. Tax-exempt: 501(c)(3).
Institution Type/Description: Preservation Project & Museum Complex: 1830s Howell iron works.
Collections: natural science; farm tools; textiles; decorative arts; furniture; trade artifacts; letters to & from James P. & Hal Allaire; blacksmithing tax records. 12 Historic Structures: carriage house; enameling furnace; blast furnace; general store; bakery; blacksmith shop; managers house; Allaire residence; church; workers' row houses; foreman's cottage; carpenter's shop.
Research Fields: 1821-1858 cast iron manufactory business records; business & personal correspondence of James P. Allaire; genealogical information on inhabitants of the village during the 1830s decade.
Facilities: 300-vol. library of books on Allaire; snack bar. Museum-related items for sale.
Activities: educational & interpretive living history events; school & self-guided tours; workshops; lectures; antique & art shows; demonstrations; saw mill & cast iron manufactory available on premises by appointment.
Publications: book, Historic Allaire Village souvenir booklet; periodical, Calendar of Events; map, Self-Guided Tour; bimonthly volunteer newsletter, The Village Star.
Hours & Admission Prices: Village: May & Sept.-Nov. Sat.-Sun. 12-4; Memorial Day-Labor Day daily 12-4; group tours by appointment. State Park: Spring & Fall daily 8-6; Memorial Day-Labor Day daily 8-8; Winter daily 8-4:30. Adults $3, children $2; members no charge. Parking $5 per car. &
Attendance: 185,000 (estimated)
Membership: Senior Citizen, Student & Volunteer $20; Individual $25; Family $40; Sponsor $100; Contributor $150; Distinguished Donor $300; Chairman's Circle $800 & up; Chairman Circle $1,000.

Flemington

DORIC HOUSE, 114 Main St., Flemington, NJ 08822-1415. Tel.: 908-782-1091.
E-mail: hunterdonhistory@embarqmail.com
Web Site: www.hunterdonhistory.org
Founded: 1885.
Congressional District: 5
Key Personnel: Pres. (V), Richard H. Stothoff; Dir., Terry A. McNealy.
Personnel Profile: Part-Time Paid 2; Part-Time Volunteers 40.
Governing Authority: society. Affiliated with Hunterdon County Historical Society. Tax-exempt: 501(c)(3).
Institution Type/Description: General Museum: housed in the home of architect & builder, Mahlon Fisher; built in 1846.
Collections: period domestic & agricultural furnishings; Indian artifacts; manuscripts; paintings.
Research Fields: Hunterdon County local history & genealogy.
Facilities: 4,500-vol. library of state & local history; genealogy available for use on premises during regular hours or by appointment; reading room. Publications, reprints & current books of interest for sale.
Activities: guided tours; lectures.
Publications: triannual, Hunterdon Historical Newsletter.
Hours & Admission Prices: House: by appointment. Library: Thurs. 12-4, 2nd & 4th Sat. 10-4; other times by appointment. No charge; donations accepted. Closed national holidays.
Attendance: 900 (estimated)
Membership: Student 18 & under $5; Annual $20; Family $25; Contributing $50; Sustaining $100; Institutional $100 & up; Life $400; Patron $1,000 & up.

FLEMING CASTLE MUSEUM, 5 Bonnell St., Flemington, NJ 08822-1311. Mailing Address: 38 Park Ave., Flemington, NJ 08822-1321. Tel.: 908-782-4607.
E-mail: flemingcastle@yahoo.com
Web Site: www.flemingcastle.com
Key Personnel: Pres. Bd. Trustees (V), Carmen Grimes; Vice Pres., William Wachter
Institution Type/Description: Historic House Museum.
Collections: Flemington history; period furnishings.
Activities: group tours.
Hours & Admission Prices: 2nd Sun. of month 1-4. No charge; donations accepted.

Florham Park

IMAGINE THAT, A NEW JERSEY CHILDREN'S MUSEUM, 4 Vreeland Rd., Florham Park, NJ 07932-1555. Tel.: 973-966-8000. Fax: 973-966-8990.
E-mail: itmuseum@aol.com
Web Site: www.imaginethatmuseum.com
Formerly: Imagine That Children's Museum
Founded: 1992.
Key Personnel: Dir., Deborah Bodner.
Personnel Profile: Full-Time Paid 2; Part-Time Paid 6.
Institution Type/Description: Interactive Children's Museum.
Collections: over 50 hands-on exhibits including dance, science, art, & computers.
Facilities: cafe.
Activities: birthday parties; drop-off service.
Hours & Admission Prices: Daily 10-5:30; groups by appointment. Children $10.95, adults $9.95; discounts to groups & AAA members; children under one no charge. Closed Thanksgiving; Christmas. &

Forked River

LACEY HISTORICAL SOCIETY, 126 S. Main St., Rte. 9, Forked River, NJ 08731. Mailing Address: Box 412, Forked River, NJ 08731-0412. Tel.: 609-971-0467.
E-mail: laceyhistsoc@verizon.net
Formerly: Old Schoolhouse Museum
Founded: 1962.
Congressional District: 3
Key Personnel: C.E.O. & Pres. (V), Elizabeth McGrath; Museum Shop Mgr., Mary Jensen.
Personnel Profile: Part-Time Volunteers 40.
Governing Authority: society. Tax-exempt: 501(c)(3).
Institution Type/Description: General Museum: housed in 1860 old schoolhouse.
Collections: local artifacts.
Activities: permanent & temporary exhibitions.
Publications: West Jersey Under Four Flags; Lacey Township: People and Progress.
Hours & Admission Prices: June 15-Aug. Mon., Wed. & Fri. 1-3, Sat. 10-12; other times by appointment. No charge; donations accepted. &
Attendance: 1,000 (estimated)
Membership: Annual $6.

POPCORN PARK ZOO, Humane Way at Lacey Rd., Forked River, NJ 08731. Mailing Address: P.O. Box 43, Forked River, NJ 08731-0043. Tel.: 609-693-1900. Fax: 609-693-8404.
Institution Type/Description: Zoo.
Collections: refuge for wildlife, exotic & farm animals including tigers, horses, monkeys, reptiles, pigs, goats, black bear, & sheep.
Hours & Admission Prices: Daily 11-5; groups by appointment. Adults $5, children under 12 & seniors $4; discounts to groups; children under 3 no charge. &

Fort Dix

ARMY RESERVE MOBILIZATION MUSEUM, 6501 Pennsylvania Ave., Fort Dix, NJ 08640-5300. Tel.: 609-562-6983. Fax: 609-562-2164.
E-mail: noelle.a.altamirano2.civ@mail.mil
Web Site: www.dix.army.mil
Formerly: Fort Dix Museum
Founded: 1984.
Congressional District: 13
Key Personnel: Pres., Doug Hasemann; Vice Pres., John Warrick; Dir. & Cur., Noelle Altamirano.
Personnel Profile: Full-Time Paid 2.
Governing Authority: federal. Tax-exempt: 501(c)(3).
Institution Type/Description: Military Museum.
Collections: firearms; uniforms; military equipment; personal equipment; decorative arts; lithographs; archives; photographs.
Research Fields: military history.
Facilities: reference library pertaining to military history; archives.
Activities: guided tours; lectures.
Hours & Admission Prices: Mon.-Fri. 8-4. No charge. &
Attendance: 7,650 (accurate)

Fort Hancock

GATEWAY, SANDY HOOK UNIT, NRA, 58 Magruder Rd., Fort Hancock, NJ 07732. Tel.: 732-872-5970.
Web Site: www.nps.gov/gate
Formerly: Sandy Hook National Seashore
Founded: 1974.
Congressional District: 6
Key Personnel: Historian, Thomas Hoffman; Museum Cur., Mary Rasa.
Personnel Profile: Full-Time Paid 7; Part-Time Volunteers 70; Interns 1.
Governing Authority: federal. U.S. Dept. of the Interior, National Park Service, Gateway National Recreation Area.
Institution Type/Description: National Park Museum: housed in the 1899 Fort Hancock Guard House or Post Stockade. Museum is part of Sandy Hook Historic Landmark.
Collections: 8,000 historical photographs & papers; 4,000 blueprints & maps; books & Army manuals; military uniforms, weapons, accoutrements & insignia; models & dioramas; artifacts relating to U.S. Lifesaving & Lighthouse Services; personal & societal items; herbarium, insect & vertebrate collections.
Research Fields: natural, military, maritime & social history.
Facilities: reference library; photographs, maps & other archival materials available for research by appointment.
Activities: guided tours; permanent & temporary exhibitions.
Publications: pamphlets on Ft. Hancock, Sandy Hook Lighthouse; volunteer newsletter, Sandpiper.
Hours & Admission Prices: Visitor's Center: daily 10-5. Museum: Winter: Sat.-Sun. 1-5; Summer: daily 1-5. Historic House: Sat.-Sun. 1-5. Sandy Hook Lighthouse & Light Keepers Quarters: April-Oct. Mon.-Fri. 1-5, Sat.-Sun. 12-4:30; Nov. Sat.-Sun. 12-5. Parking: buses $25, cars $10; after Labor Day no charge. &
Attendance: 50,000

Fort Lee

FORT LEE HISTORIC PARK & MUSEUM, Hudson Terrace, Fort Lee, NJ 07024. Tel.: 201-461-1776. Fax: 201-461-7275.
E-mail: flhp@njpalisades.org
Web Site: www.njpalisades.org
Founded: 1976.
Key Personnel: Dir., John Muller; Pres., Jim Hall; Museum Shop Mgr., J. Muller.
Personnel Profile: Full-Time Paid 4.
Governing Authority: municipal; nonprofit. Parent Institution: Palisades Interstate Park Commission. Tax-exempt.
Institution Type/Description: History Museum: located in Historic Park.
Collections: reproductions.
Research Fields: Revolutionary War military history; life & times of Revolutionary War soldier.
Facilities: library; 204-seat theater; 1,200 sq. ft. exhibit space.
Activities: guided tours; lectures; concerts; organized education programs; training programs for professional museum workers; loan exhibitions. Museum Sponsors: encampments; musters.
Hours & Admission Prices: March-Dec. Wed.-Sun. 10-5. No charge. Parking: $5. &
Attendance: 25,998 (accurate)

Franklin

FRANKLIN MINERAL MUSEUM, 32 Evans St., Franklin, NJ 07416-1419. Mailing Address: P.O. Box 54, Franklin, NJ 07416-0054. Tel.: 973-827-3481. Fax: 973-827-0149.
E-mail: fmm1954@earthlink.net
Web Site: franklinmineralmuseum.com
Founded: 1965.
Congressional District: 24
Key Personnel: Pres. (V), Steven Phillips; Treas., A. Lee Lowell; Mgr., Doreen Longo.
Personnel Profile: Full-Time Paid 4; Full-Time Volunteers 2; Part-Time Paid 4; Part-Time Volunteers 10.
Governing Authority: private; nonprofit organization. Tax-exempt: Form 990.
Institution Type/Description: Geology & Mining Museum: adjacent to zinc mines, located in old mine engine house, built in late 19th century.
Collections: rocks & minerals from local zinc mines and from around the world; local mining artifacts; fossils; Native American relics; fluorescent minerals.
Research Fields: museum research & education fund dedicated to the science & history of the Franklin & Ogdensburg, N.J. zinc mining district.
Facilities: 2,000-vol. library relating to geology, mineralogy, mining history & crystallography; educational facilities. Museum-related items for sale.

Activities: rock collecting on mine dump; guided tours; study clubs; temporary exhibitions. Annual Events: Miner's Day; mineral shows.
Publications: newsletter, Franklin Mineral Museum.
Hours & Admission Prices: April-Nov. Mon.-Fri. 10-4, Sat.-Sun. 10-5. Adults $7, seniors $6, children $5; members no charge. Closed Easter; Thanksgiving. &
Attendance: 20,000 (accurate)
Membership: Individual $15; Family $25; Patron $50; Life $500; Benefactor $1,000; Sustaining $5,000.

Franklin Lakes

THE GALLERY AT THE PRESBYTERIAN CHURCH AT FRANKLIN LAKES, 730 Franklin Lake Rd., Franklin Lakes, NJ 07417. Tel.: 201-891-0511. Fax: 201-891-0517.
E-mail: pcflmgr@yahoo.com
Web Site: pcfl.org
Founded: 1961.
Key Personnel: Dir., Mary Guideth McColl.
Governing Authority: Parent Institution: The Presbyterian Church at Franklin Lakes. Tax-exempt.
Institution Type/Description: Art Gallery.
Collections: works by local artists; paintings; photographs; sculpture.
Activities: temporary exhibitions.
Hours & Admission Prices: Call for hours. No charge.

LORRIMER SANCTUARY, 790 Ewing Ave., Franklin Lakes, NJ 07417. Mailing Address: P.O. Box 125, Franklin Lakes, NJ 07417. Tel.: 201-891-2185.
E-mail: lorrimer@njaudubon.org
Web Site: www.njaudubon.org
Institution Type/Description: Nature Center: housed in a late 1700s house.
Collections: hands-on exhibitions; wildlife & their habitats.
Facilities: nature trails. Gift items for sale.
Activities: winter bird feeding station; educational programs; lectures; hiking trails.
Hours & Admission Prices: Wed.-Fri. 9-5, Sat. 10-5, Sun. 1-5.

Freehold

COVENHOVEN HOUSE, 150 W. Main St., Freehold, NJ 07728. Mailing Address: Monmouth County Historical Assoc., 70 Court St., Freehold, NJ 07728-1710. Tel.: 732-462-1466.
Governing Authority: Parent Institution: Monmouth County Historical Association.
Institution Type/Description: Historic Building Museum: housed in the home of William & Elizabeth Covenhoven, built in 1753; later served as headquarters for British General Henry Clinton prior to the Battle of Monmouth in 1778.
Collections: local history & culture; period furnishings; photographs; personal artifacts.
Hours & Admission Prices: May-Sept. Thurs.-Sat. 1-4. No charge; donations accepted.

METZ BICYCLE MUSEUM AND TREASURES OF YEARS GONE BY, 54 W. Main St., Freehold, NJ 07728-2136. Tel.: 732-462-7363.
E-mail: dmetz@metzbicyclemuseum.com
Web Site: www.metzbicyclemuseum.com
Institution Type/Description: History Museum.
Collections: early bicycles & children's riding toys; mouse traps; bottle openers; kitchen gadgets; farm tools.
Hours & Admission Prices: By appointment.

MONMOUTH COUNTY HISTORICAL ASSOCIATION, (M), 70 Court St., Freehold, NJ 07728-1710. Tel.: 732-462-1466. Fax: 732-462-8346. Facebook: Monmouth County Historical Association.
E-mail: emurphy@monmouthhistory.org
Web Site: www.monmouthhistory.org
Founded: 1898.
Congressional District: 12
Key Personnel: Dir., Evelyn C. Murphy, Ph.D.; Pres. (V), Claire Knopf; Devel. & Communications Dir., Laurie Bratone; Devel. & Communications Asst., Angie Sinicki; Cur., Bernadette Sigler Rogoff; Librarian/Archivist, Laura M. Poll.

Personnel Profile: Full-Time Paid 5; Part-Time Paid 4; Part-Time Volunteers 60.
Operating Expenses: 850,000
Operating Income: 850,000
Governing Authority: nonprofit organization. Branch Museums: Marlpit Hall, Middletown, NJ; Holmes-Hendrickson House, Holmdel, NJ; Covenhoven House, Freehold, NJ; Allen House, Shrewsbury, NJ. Tax-exempt: 501(c)(3).
Institution Type/Description: Historical Society Museums.
Collections: 30,000 artifacts, 17th-20th century relating to the history of Monmouth County, N.J.; decorative arts; folk art; paintings; silver; furniture; glass; ceramics; toys & household artifacts; historic houses.
Major Exhibits: Farm: Agriculture in Monmouth County 1600 through 2013, 11/16/13-11/14; Properly Dressed: Art and Reality in 18th and 19th Century Dress in Monmouth County, 2/14-8/3/14.
Research Fields: history & architecture of houses maintained; collections; county history; architecture; genealogy; New Jersey material culture.
Facilities: 5,000-vol. library manuscript materials, maps, ephemera, postcards, all relating to Monmouth County, N.J. & surrounding area.
Activities: guided tours; lectures; films; formally organized education programs; permanent & temporary exhibitions.
Publications: newsletter; The Diary of Sarah Tabitha Reid 1868-1873; Steamboats in Monmouth County: A Gazetteer; exhibition catalogue, Micah Williams: Portrait Artist.
Hours & Admission Prices: Library: Wed.-Sat. 10-4. Museum: Tues.-Sat. 10-4; groups by appointment. Adults $5, senior citizens & children 6-18 $2.50; discounts to NARM & AAM members; members & children under 6 no charge. Historic Houses: call for information. Closed New Year's Day; Thanksgiving; Christmas Eve & Day. &
Attendance: 13,046 (accurate)
Membership: Student & Senior Citizens $25; Individual $35; Family $50; Supporting $100; Patron $250; Benefactor $500.

Frenchtown

DECOYS & WILDLIFE GALLERY, 55 Bridge St., Frenchtown, NJ 08825-1229. Tel.: 908-996-6501; 888-996-6501 (Toll Free). Fax: 908-996-0807.
E-mail: decoys@decoyswildlife.com
Institution Type/Description: Art Gallery.
Collections: wildlife art; duck decoys; paintings; sculpture.
Activities: workshops; special events. Museum Sponsors: Open House in February; Original Miniature in the Fall.
Hours & Admission Prices: Daily 10-6; other times by appointment. Closed Christmas.

Galloway

EDWIN B. FORSYTHE NATIONAL WILDLIFE REFUGE BRIGANTINE DIVISION, 800 E. Great Creek Rd., Galloway, NJ 08205. Mailing Address: P.O. Box 72, Oceanville, NJ 08231. Tel.: 609-652-1665.
Institution Type/Description: Wildlife Refuge.
Collections: wildlife & their habitat; wetlands.
Activities: educational programs; school & scout groups; special events.
Hours & Admission Prices: Call for hours.

Glassboro

GLASSBORO HERITAGE GLASS MUSEUM, 25 E. High St., Glassboro, NJ 08028-2519. Tel.: 856-881-7468.
Founded: 1979.
Key Personnel: Pres. (V), Carol Schoepske; Chm. (V), Rick Granda; Museum Shop Mgr., Linda Rudisill.
Personnel Profile: Full-Time Volunteers 10; Part-Time Volunteers 10.
Governing Authority: Tax-exempt.
Institution Type/Description: Glass Museum.
Collections: bottles & jars from Whitney, Clevenger, Wheaton, & Stanger Glass Works.
Activities: educational & historical talks.
Publications: biannual newsletter.
Hours & Admission Prices: Wed. 12-3, Sat. 11-2, 4th Sun. 1-4. No charge; donations accepted.
Attendance: 1,000 (accurate)
Membership: Single $15; Family $25; Business $75; Life $250.

Greenwich

CUMBERLAND COUNTY HISTORICAL SOCIETY, 960 YeGreate St., Greenwich, NJ 08323. Mailing Address: P.O. Box 16, Greenwich, NJ 08323-0016. Tel.: 856-455-4055 & 8580. Fax: 856-455-8580.
E-mail: cchistsoc@verizon.net
Web Site: www.cchistsoc.org
Founded: 1905.
Congressional District: 2
Key Personnel: Pres., Jonathan E. Wood; Vice Pres., Linda Jones; Sec., Ruth Ann Fox; Treas., Judith Uber; Clerk, Linda Peck; Dir. Library, Warren Q. Adams; Museum Shop Mgr., Kenneth Miller.
Personnel Profile: Part-Time Paid 4; Part-Time Volunteers 50.
Governing Authority: nonprofit organization. Tax-exempt.
Institution Type/Description: General Museum: housed in furnished 1730 Gibbon house.
Collections: agriculture; archives; costumes; Civil War artifacts; furnishings. Historic Building: Prehistorical Museum (fossil & Indian artifacts); Barn Museum; 1650 Swedish Log Granary; maritime museum; manuscripts.
Research Fields: genealogical and historical research.
Facilities: Warren Lummis Library: 500-vol. library of N.J. history & family histories available on premises. Postcards, pamphlets & note paper for sale.
Activities: guided tours; lectures; films; concerts; hobby workshops; formally organized education programs for children; docent program; permanent exhibitions. Annual Events: craft fair; antique show; Farm Day; annual private home tour.
Publications: historic pamphlets; biannual newsletter, Cumberland Patriot; books.
Hours & Admission Prices: Gibbon House: April-Dec. Tues.-Sat. 1-4. Adults $3; society members no charge. Warren Lummis Gen. & Hist. Library: Wed. 10-4, Sat.-Sun. 1-4. John Dubois Maritime Museum, Matthew Potter's Tavern & Old Stone Church: by appointment only. Prehistorical Museum: Wed. & Sat.-Sun. 12-4. No charge; donations accepted.
Attendance: 5,000 (estimated)
Membership: Single $20; Couple $30.

Hackensack

EDWARD WILLIAMS GALLERY - FAIRLEIGH DICKINSON UNIVERSITY, 150 Kotte Pl., Hackensack, NJ 07601-6112. Tel.: 201-692-2449. Fax: 201-692-2503.
E-mail: geraghty@fdu.edu
Founded: 1975.
Key Personnel: Dir., Diana Soorikian.
Governing Authority: Parent Institution: Fairleigh Dickinson University. Tax-exempt.
Institution Type/Description: Art Gallery.
Collections: works by contemporary artists.
Major Exhibits: Teresa DeSalvio, 1/20/14-2/14; Nadene Grey, 3/30/14-4/25/14; Ben Georgia, 4/28/14-5/30/14.
Activities: concerts: chamber music art receptions.
Hours & Admission Prices: Mon.-Fri. 8:30-2:30, Sat. 9:30-2:30. No charge.
Attendance: 2,000 (estimated)

NEW JERSEY NAVAL MUSEUM, 78 River St., Hackensack, NJ 07601-7110. Tel.: 201-342-3268 & 873-6670. Fax: 201-342-3268.
E-mail: njnavalmuseum@yahoo.com
Web Site: www.njnm.com
Founded: 1974.
Key Personnel: C.E.O., Chris Buermeyer.
Personnel Profile: Part-Time Paid 1; Part-Time Volunteers 18.
Governing Authority: private; nonprofit organization. Parent Institution: Submarine Memorial Association. Tax-exempt: 501(c)(3).
Institution Type/Description: Naval Military Museum: located at World War II submarine USS Ling SS-297 in Hackensack River.
Collections: World War II submarine; pictures; ship models; battle flags; missiles; Vietnam river patrol boat; naval equipment; personal mementos; torpedoes; cannons; Japanese Kaiten suicide submarine; German Seehund 2 man submarine.
Facilities: library of original boat plans, prints, letters & scrapbook; 1,300 sq. ft. exhibit space. Museum-related items for sale.
Activities: guided tours; hobby workshops; loan & temporary exhibitions; children's birthday parties. Annual Events: Pearl Harbor Day, Memorial Day & Veterans Day Services.
Publications: newsletter published three times annually, Patrol Report.
Hours & Admission Prices: Sat.-Sun. 10-4. Museum & Grounds: no charge. Tours: adults $8, senior citizens $5, children under 12 $3. Closed New Year's Day; Easter; Thanksgiving; Christmas.
Attendance: 15,000 (accurate)

Membership: Junior Supporting $5; Supporting $20; Commanders Club $50; Captains Club $250; Admiral Club $500.

Hackettstown

HACKETTSTOWN HISTORICAL SOCIETY MUSEUM & LIBRARY, 106 Church St., Hackettstown, NJ 07840-2206. Tel.: 908-852-8797.
Web Site: www.hackettstownhistory.com
Founded: 1975.
Congressional District: 24
Key Personnel: Archivist, Ray Lemasters.
Personnel Profile: Part-Time Volunteers 20.
Governing Authority: society; nonprofit organization. Tax-exempt.
Institution Type/Description: Local History & Genealogy Museum: housed in 1915 Theodore G. Plate House.
Collections: Hackettstown history; regional history; archives; 500 photographs; Warren County cemetery records; genealogy.
Research Fields: genealogy; history of Hackettstown & surrounding area & buildings.
Facilities: library of local history & genealogy.
Activities: guided tours; temporary exhibitions; historical society meetings. Society Sponsors: Open House.
Hours & Admission Prices: Mon.-Tues. 9-2, Wed. & Fri. 9-4, Sun. 2-4; other times & groups by appointment. No charge; donations accepted. Closed major holidays.
Attendance: 250 (estimated)
Membership: Student under 21 $5; Adult $20; Family $35; Life $200.

Haddonfield

HISTORICAL SOCIETY OF HADDONFIELD, 343 Kings Hwy. East, Haddonfield, NJ 08033-1214. Tel.: 856-429-7375.
E-mail: info@haddonfieldhistory.org
Web Site: www.haddonfieldhistory.org
Founded: 1914.
Congressional District: 3
Key Personnel: Pres. (V), Mrs. Lee Albright; Vice Pres., Carol Smith; Treas., Michael McMullen; Sec., Barbara Hilgen; Dir. Collections, Dianne Snodgrass; Archivist & Librarian, Kenneth Cleary; Cur. Antique Tool Collection, Don Wallace.
Personnel Profile: Part-Time Paid 1; Part-Time Volunteers 20.
Governing Authority: private; nonprofit. Tax-exempt: 501(c)(3).
Institution Type/Description: Historical Society Museum: housed in Greenfield Hall, a Georgian style brick house, built in 1841. The Samuel Mickle House houses the Society's library. Both buildings listed on the National Register of Historic Places.
Collections: history of Haddonfield & its environs; furnishings including four tall case clocks, a table & mirror which was owned by Elizabeth Haddon Estaugh, founder of Haddonfield; library's manuscripts on local & southern New Jersey history; ledgers & minute books of 300 local organizations, businesses & individuals; 3,000 photographs; early tools; archives of Samuel Nicholson Rhoads, naturalist.
Research Fields: Haddonfield history; genealogy researching; architectural research about homes in Haddonfield.
Facilities: 5,000-vol. library relating to the history of New Jersey, Haddonfield & genealogy; Greenfield Hall available for rental. Museum-related items for sale.
Activities: guided tours; lectures; temporary exhibitions. Annual Events: Candlelight Dinner in March; Holiday Shop in December.
Publications: quarterly newsletter, The Bulletin.
Hours & Admission Prices: Museum: Sept.-July Wed.-Fri. 1-4, 1st Sun. each month 1-3. Adults $5. Library: June-July Mon.-Tues. 9:30-11:30; 1st Sun. of month 1-3; Sept.-May Tues. & Thurs. 9:30-11:30, 1st Sun. of month 1-3. No charge. Closed major holidays.
Attendance: 350 (estimated)
Membership: Senior Citizen $25; Contributing $35; Household $55; Patron $150; Patron Household $250; Life $1,000; Life Household $1,500.

INDIAN KING TAVERN, 233 Kings Hwy., Haddonfield, NJ 08033. Tel.: 856-429-6792.
Institution Type/Description: Historic Building: built in 1750.
Collections: local history & culture; period furnishings; photographs; personal artifacts.
Hours & Admission Prices: By appointment.

Haledon

AMERICAN LABOR MUSEUM, BOTTO HOUSE NATIONAL LANDMARK, (M), 83 Norwood St., Haledon, NJ 07508-1363. Tel.: 973-595-7953. Fax: 973-595-7291. Facebook: American Labor Museum / Botto House National Landmark.
E-mail: labormuseum@aol.com
Web Site: www.labormuseum.net
Founded: 1982.
Congressional District: 9
Key Personnel: Pres. (V), Michael Goodwin; Dir., Angelica M. Santomauro, Ed.D.; Project Mgr., Robert Leadlie; Financial Dir., Noel Christmas; Dir. Education & Museum Shop Mgr., Evelyn M. Hershey.
Personnel Profile: Full-Time Paid 2; Part-Time Paid 1; Part-Time Volunteers 16; Interns 3.
Operating Expenses: 155,000
Operating Income: 165,000
Governing Authority: nonprofit organization. Tax-exempt: 501(c)(3).
Institution Type/Description: Labor & Immigrant Studies Museum: housed in 1908 Botto House. A National Landmark.
Collections: photographs illustrating turn-of-the-century working class immigrants; textile artifacts dating from 1890s to present day; labor union memorabilia; restored period rooms; history & contemporary issues of immigrants.
Major Exhibits: Outsourced! by Robin Holder, 1/13-4/14; Mill Girls by Donna Berger, 5/13-8/14; Border Angels by Pam Calore, 9/13-12/14.
Research Fields: labor & immigrant studies.
Facilities: 950-vol. library of books on labor, labor history & ethnicity; educational facilities; gardens. Books & materials on union/labor & immigration/ethnicity for sale.
Activities: guided tours; lectures; films; school loan service; formally organized education programs; docent program; participatory, loan, temporary & traveling exhibitions; arts festivals; classes. Annual Event: Labor Day parade/picnic.
Publications: annual report; e-newsletter.
Hours & Admission Prices: Wed.-Sat. 1-4; other times by appointment. Suggested Donation: adults $5; discounts to AAM & ICOM members; children under 12, AAA & museum members no charge. Closed major holidays except Labor Day. ♿
Attendance: 16,578 (estimated)
Membership: Senior Citizen & Student $10; Individual $20; Supporting Couples $25; Family $30; Benefactor $50.

Hamilton

GROUNDS FOR SCULPTURE, (M), 18 Fairgrounds Rd., Hamilton, NJ 08619-3447. Tel.: 609-586-0616. Fax: 609-586-7303.
E-mail: info@groundsforsculpture.org
Web Site: www.groundsforsculpture.org
Founded: 1992.
Key Personnel: Cur., Ellen Landis; Mgr. Education & Grants, Christina Ely; Registrar, Faith McClellan; Community & Patron Rels. Mgr., Bonnie Brown; Event Planner, Rena Perrone; Facility Rental Mgr., Christopher Carrell; C.F.O., Robert Gross; Supvr. Admissions, Yvonne Exedaktilos; Museum Shop Mgr., Jenifer Micikas.
Personnel Profile: Full-Time Paid 70; Part-Time Paid 15; Part-Time Volunteers 100.
Governing Authority: public; nonprofit. Tax-exempt: 501(c)(3).
Institution Type/Description: Sculpture Park.
Collections: contemporary sculpture.
Facilities: 20-seat seasonal outdoor cafe; 40-seat indoor/50-seat outdoor cafe; restaurant; 20,000 sq. ft. exhibit space; 35 acre sculpture park.
Activities: permanent & temporary exhibitions; educational programs; special events.
Publications: exhibition catalogues; event guides; family guide.
Hours & Admission Prices: Tues.-Sun. 10-6. Adults $10, student & seniors $8, children 6-12 $6; members and children 5 & under no charge. Closed New Year's Day; Thanksgiving; Christmas. ♿
Attendance: 100,000 (estimated)
Membership: Seniors & Students $45; Individual $55; Dual & Family $90; Contributor $150.

HISTORICAL SOCIETY OF HAMILTON TOWNSHIP JOHN ABBOTT II HOUSE, 2200 Kuser Rd., Hamilton, NJ 08690. Mailing Address: P.O. Box 1776, Yardville, NJ 08620. Tel.: 609-585-1686.
Founded: 1976.
Congressional District: 4
Key Personnel: Pres. (V), James A. Federici; Vice Pres., Gordon Kontrath; Treas., Robert Boldt; Sec., Ruth Applegate.

Personnel Profile: Part-Time Volunteers 12.
Governing Authority: society; nonprofit. Parent Institution: Historical Society of Hamilton Township. Tax-exempt: 501(c)(3).
Institution Type/Description: Historical House: c.1730 farm house & 1840 addition.
Collections: Colonial & Victorian furnishings; textiles.
Facilities: 200-vol. library pertaining to local history. Museum-related items for sale.
Activities: guided tours; organized education programs for children. Special Events: Septemberfest; Christmas Wassail Party.
Publications: booklets, John Abbott II House; Old Nottingham; pamphlet, John Abbott II House; Narrative History of Hamilton Township.
Hours & Admission Prices: Sat.-Sun. 12-5 by appointment only. No charge; donations accepted. Closed Christmas.
Attendance: 4,000 (estimated)
Membership: Individual $10; Family $15.

KUSER FARM MANSION, 390 Newkirk Ave., Hamilton, NJ 08610-4845. Mailing Address: 2090 Greenwood Ave., P.O. 00150, Hamilton, NJ 08609-2312. Tel.: 609-890-3630. Fax: 609-586-0678.
Web Site: www.hamiltonnj.com
Founded: 1979.
Congressional District: 4
Key Personnel: Mayor, John F. Bencivengo.
Personnel Profile: Full-Time Paid 1; Part-Time Paid 8.
Governing Authority: municipal; nonprofit. Parent Institution: Township of Hamilton. Tax-exempt.
Institution Type/Description: Historic House: c.1896 Queen Anne style mansion & outbuildings, former summer home of Fred Kuser & his family.
Collections: early motion picture projection room; Mercer motor car memorabilia; carved fireplaces; 45-foot dining room. Historic Structures: coach house; 1907 clay tennis court & tennis house; windmill; barn; chicken building; pavilion; laundry house; small formal garden.
Research Fields: local history related to Kuser family & their business connections.
Facilities: 22-acre park; tourism center.
Activities: guided tours; concerts; movie nights; organized education programs; self-guided walking tours of the grounds; special lectures & demonstrations; weddings. Special Events: Summer Concerts in the Park; Winter Wonderland; Jersey Valley Model Railroad Club Open House; Christmas Holiday Open House.
Publications: brochures, Kuser Farm Mansion & Park; booklet, The Kuser Story, Self-Guided Walking Tour map.
Hours & Admission Prices: Feb.-Nov. Sat. & Sun. 11-3:15 (last tour 2:30). Dec. special Christmas tours. Call to confirm hours. No charge, donations accepted.
Attendance: 1,600 (estimated)

Hammonton

BATSTO VILLAGE, 31 Batsto Rd., Hammonton, NJ 08037-5502. Tel.: 609-561-0024. Fax: 609-567-8116.
E-mail: info@bastovillage.org
Web Site: www.batstovillage.org
Formerly: Historic Batsto Village
Founded: 1954.
Key Personnel: Supt., Rob Auermuller.
Personnel Profile: Full-Time Paid 15; Part-Time Paid 5; Part-Time Volunteers 12; Interns 1.
Governing Authority: state. Parent Institution: N.J. Div. of Parks & Forestry, Dept. of Environmental Protection, Trenton, NJ 08625. Tax-exempt.
Institution Type/Description: Historic Site Museum: Batsto Village, 33 historic buildings built in the 1800s.
Collections: 19th century life in South Jersey; archives & manuscripts; bog-iron & glass making industries; agricultural equipment & implements; transportation; decorative arts; archaeology; postal history; nature center; furnishings.
Research Fields: bog iron; glass making; agriculture; Pine Barrens; lumbering & forestry; postal history; company town; commerce; genealogy.
Facilities: 100-seat auditorium; visitor center.
Activities: guided & self-guided tours.
Publications: Batsto Village newsletter.
Hours & Admission Prices: Batsto Mansion: call for schedule. Visitor Center: daily 9-4. Grounds: daily dawn to dusk.; groups by appointment only. Tours: adults $3, children 6-11 $1; children under 6 no charge. Parking Fee: Sat.-Sun. & holidays Memorial Day to Labor Day Weekend $5. Closed New Year's Day; Thanksgiving; Christmas. ♿
Attendance: 100,000 (estimated)

Hancock's Bridge

HANCOCK HOUSE, 3 Front St., Hancock's Bridge, NJ 08038. Mailing Address: P.O. Box 139, Hancocks Bridge, NJ 08038-0139. Tel.: 856-935-4373. Fax: 856-935-2079.
E-mail: hancockhousenj@comcast.net
Web Site: www.state.nj.us/dep/parksandforests/historic/hancockhouse/hancockhouse.index.htm
Founded: 1932.
Congressional District: 2
Key Personnel: Supt., Vince Bonica; Resource Interpretive Specialist, Alicia Bjornson.
Personnel Profile: Full-Time Paid 1; Part-Time Volunteers 15.
Governing Authority: state. Administered by the Division of Parks & Forestry, New Jersey Dept. of Environmental Protection, Trenton, NJ 08625. Tax-exempt.
Institution Type/Description: Historic House Museum: 1734 Hancock House, built by Judge William Hancock & scene of Revolutionary War's British Massacre March 21, 1778.
Collections: 18th & 19th-century regional furnishings.
Activities: guided tours; concerts; craft shows; reenactments; public programs; school programs on-site & off-site.
Hours & Admission Prices: Wed.-Sat. 10-4, Sun. 1-4; call to confirm. No charge; donations accepted. Closed New Year's Day; Thanksgiving; Christmas.
Attendance: 4,000 (estimated)
Membership: Individual $10; Household $18; Silver $25.

Highlands

TWIN LIGHTS HISTORIC SITE, Lighthouse Rd., Highlands, NJ 07732. Tel.: 732-872-1814.
Web Site: twinlightslighthouse.com
Personnel Profile: Full-Time Paid 2.
Governing Authority: Parent Institution: NJ DEP Division of Parks & Forestry.
Institution Type/Description: Lighthouse: built in 1828. Listed on the National Register of Historic Places.
Collections: 9 ft. bivalve lens; lighthouse & maritime history.
Hours & Admission Prices: Memorial Day to Labor Day daily 10-4:30; Sept.-May Wed.-Sun. 10-4:30. No charge; donations accepted. Closed New Year's Day; Thanksgiving; Christmas; state holidays. &
Attendance: 80,000 (accurate)
Membership: Basic: Friends & Family $25.

Hillsborough

DUKE FARMS, 1112 Dukes Pkwy. W., Hillsborough, NJ 08844. Tel.: 908-722-3700.
Governing Authority: Parent Institution: Duke Farm Foundation.
Institution Type/Description: Historic Farm & Garden: housed in a former horse & dairy barn; built in 1906.
Collections: property history; wildlife & their habitat; orchids; ecology; horticulture; agriculture; hands-on exhibitions.
Facilities: 2,740 acres; orientation center; classroom; cafe; nature trails.
Activities: audio tours; educational programs; demonstrations; nature trails.
Hours & Admission Prices: Orientation Center & Trails: April to Dec. 1 Thurs.-Tues. 8:30-6. No charge.

Hillside

WOODRUFF HOUSE/EATON STORE MUSEUM/PHIL RIZZUTO SPORTS EXHIBIT, 111 Conant St., Hillside, NJ 07205-2801. Tel.: 908-353-8828.
Web Site: www.woodruffhouse.org
Founded: 1978.
Key Personnel: Dir., Chm. & Pres. (V), Alan D. Zimmerman; Devel., Ann Pettigrew; Treas., Helen Witting.
Personnel Profile: Part-Time Volunteers 10.
Governing Authority: private; nonprofit organization. Operated by the Hillside Historical Society. Tax-exempt.
Institution Type/Description: Historical Society Museum.
Collections: local history & culture; photographs; period artifacts; water pump; well; farm equipment; Phil Rizzuto sports memorabilia & Baseball Hall of Fame. Historic Buildings: 1735 farm house; 1900s general store; barn; privy.
Research Fields: township of Hillside history.
Facilities: archives.
Activities: guided tours.
Hours & Admission Prices: 3rd Sun. of month 2-4; other times by appointment. No charge; donations accepted.

Attendance: 900 (estimated)
Membership: Single $7.50; Family $15; Sustaining $25.

Ho-Ho-Kus

THE HERMITAGE, Friends of the Hermitage, Inc., 335 N. Franklin Turnpike, Ho-Ho-Kus, NJ 07423-1035. Tel.: 201-445-8311. Fax: 201-445-0437.
E-mail: info@thehermitage.org
Web Site: www.thehermitage.org
Formerly: Friends of the Hermitage, Inc.
Founded: 1972.
Congressional District: 5
Key Personnel: Visitor Svcs. Mgr., Linda Vreeland; Pres. (V), Richard C. Brahs; 1st Vice Pres., Patricia A. Ricci; 2nd Vice Pres., Virginia C. Bryan; Treas., Leo McManus; Sec., Carol W. Greene.
Personnel Profile: Part-Time Paid 5; Part-Time Volunteers 40; Interns 1.
Governing Authority: state. Division of Parks & Forestry, P.O. Box CN 404, Trenton, NJ 08625; & Friends of the Hermitage, Inc. Tax-exempt: 501(c)(3).
Institution Type/Description: Gothic Revival Historic House: designed by William Ranlett. Incorporates 18th century stone house that was Washington headquarters in July 1778. Estate of Rosencrantz family 1807-1970. A National Historic Landmark.
Collections: 19th century furnishing; decorative arts; recreational & everyday artifacts; costumes & textiles; Rosencrantz family archives (1807-1970); smokehouse; summer kitchen.
Major Exhibits: 350 Years at The Hermitage, 2/14-9/14.
Research Fields: 19th century cultural, social, and industrial history; Gothic Revival domestic architecture; 18th & 19th century; costumes and textiles.
Facilities: 4.9 acre property; archives of family, estate, and mill records and oral history tapes available for use by appointment; education & conference center with classroom, meeting room, catering kitchen. Museum-related items for sale.
Activities: guided tours; permanent & temporary exhibitions; school & scout programs; children & family programs; summer children's history camp; senior & educational outreach programs; reenactments; 19th century theater; history roundtable; lectures; docent programs. Special Events: antique and craft fairs.
Publications: newsletter.
Hours & Admission Prices: Wed.-Sun. 1-4; clubs & groups by special appointment. Adults $7, seniors $5, children 6-12 $4; discounts to AASCH & AAA members; children under 6 & members no charge. &
Attendance: 15,000 (estimated)
Membership: Student & Senior $20; Individual $35; Family & Business $75; Mary Elizabeth Circle $125; Aaron Burr Guild $250; Theodosia Society $500; Benefactor's Club $1,000.

Hoboken

BARSKY GALLERY, 48 Harrison St., Hoboken, NJ 07030. Tel.: 888-465-4949.
Web Site: www.barskygallery.com
Institution Type/Description: Art Gallery.
Collections: works by emerging & established artists from around the world.
Activities: temporary exhibitions.
Hours & Admission Prices: Thurs.-Sun. 11-6; other times by appointment.

HOBOKEN FIRE DEPARTMENT MUSEUM, 213 Bloomfield St., Hoboken, NJ 07030. Tel.: 201-420-2397.
Key Personnel: Cur., Bill Bergin; Asst. Cur., Joe Kennedy
Institution Type/Description: Fire Fighting History Museum: housed in the city's former firehouse; built in 1870. Listed on the National Register of Historic Places.
Collections: firefighting history & equipment; early firefighter hats & coats; period fire boxes; running board operator's desk; 1932 Ahrens-Fox fire truck; photographs.
Activities: school group tours.
Hours & Admission Prices: Spring to Fall Sat. 10-4, call to confirm; other times by appointment. No charge.

HOBOKEN HISTORICAL MUSEUM, 1301 Hudson St., Hoboken, NJ 07030-7427. Mailing Address: P.O. Box 3296, Hoboken, NJ 07030-1603. Tel.: 201-656-2240.
E-mail: info@hobokenmuseum.org
Web Site: www.hobokenmuseum.org
Key Personnel: Dir., Bob Foster
Institution Type/Description: History Museum.
Collections: Hoboken history; photographs.

Hours & Admission Prices: Tues.-Thurs. 2-7, Fri. 1-5, Sat.-Sun. 12-5. Adults $2; members & children no charge.

Holmdel

HOLMES-HENDRICKSON HOUSE, 62 Longstreet Rd., Holmdel, NJ 07733. Mailing Address: Monmouth County Historical Assoc., 70 Court St., Freehold, NJ 07728-1710. Tel.: 732-462-1466.
Institution Type/Description: Historic House Museum: built in 1754.
Collections: local history & culture; period furnishings; personal artifacts.
Activities: sheep shearing, spinning and weaving demonstrations.
Hours & Admission Prices: May-Sept. Thurs.-Sat. 1-4. No charge; donations accepted.

LONGSTREET FARM, Holmdel Park, 44 Longstreet Rd., Holmdel, NJ 07733. Mailing Address: Monmouth County Park System, 805 Newmann Springs Rd., Lincroft, NJ 07738-1628. Tel.: 732-946-3758 & 842-4000. Fax: 732-946-0750. TDD: 732-219-9484.
E-mail: info@monmouthcountyparks.com
Web Site: www.monmouthcountyparks.com
Founded: 1967.
Congressional District: 3
Key Personnel: Site Mgr., Sandra Byard; Supvr., Sean O'Herron.
Personnel Profile: Full-Time Paid 7; Part-Time Paid 10; Part-Time Volunteers 15; Interns 1.
Governing Authority: county. Parent Institution: Monmouth County Park System, Newman Springs Rd., Lincroft, NJ 07738. Tax-exempt.
Institution Type/Description: Historic Farm: 1890 Longstreet Farm.
Collections: living history: an operating farm, including the farm animals, equipment & tools necessary for its operation; 19th-century agricultural items; 19th-century household artifacts. Historic Buildings: 19 structures including 18th-century Dutch barn; 1880 carriage house; farmhouse; barns; sheds; outbuildings.
Research Fields: agriculture & rural life in central New Jersey.
Activities: interpretive farm walks; demonstration of farm practices, including sheep-shearing, making apple cider, leather work, horseshoeing, cornhusking & threshing of grain, depending on the season of the year.
Publications: interpretive flyers.
Hours & Admission Prices: June-Labor Day daily 9-5; Sept.-May daily 10-4. No charge; donations accepted. &
Attendance: 88,000 (accurate)

VIETNAM ERA MUSEUM & EDUCATIONAL CENTER, 1 Memorial Ln., Garden State Pkwy. Exit 116, Holmdel, NJ 07733. Mailing Address: P.O. Box 648, Holmdel, NJ 07733-0648. Tel.: 732-335-0033. Fax: 732-335-1107.
E-mail: shagarty@njvvmf.org
Web Site: www.njvvmf.org
Founded: 1998.
Congressional District: 12
Key Personnel: Exec. Dir., Bill Linderman; Coord. Education, Initiatives & Resources, Sarah Hagarty; Administrative Asst., Lynn Duane.
Personnel Profile: Full-Time Paid 3; Part-Time Paid 2; Part-Time Volunteers 80.
Governing Authority: NJ Vietnam Veterans' Memorial Foundation. Tax-exempt.
Institution Type/Description: Historical & Military Museum.
Collections: photos, letters, personal items, documents, objects related to the Vietnam era.
Research Fields: Vietnam War; Vietnam era; military service; political protest; 1960's & 1970's.
Facilities: library; resource room; classroom.
Activities: tours; classes; seminars; ceremonies.
Hours & Admission Prices: Tues.-Sat. 10-4. Adults $7, seniors & students $5; active & retired military and children under 10 no charge. &
Attendance: 20,000 (estimated)
Membership: Individual $40; Family $80; Bronze $125; Silver $250; Gold $500; Platinum $1,000.

Hope

HOPE HISTORICAL SOCIETY, 323 High St., Hope, NJ 07844. Mailing Address: P.O. Box 52, Hope, NJ 07844-0052. Tel.: 908-459-4277 & 4268.
Founded: 1950.
Congressional District: 5
Key Personnel: Pres., Nancy A. Treible; Vice Pres., Tessa McDonald.
Personnel Profile: Part-Time Volunteers 14.
Volunteer Hours: 210

Operating Expenses: 1,200
Operating Income: 2,300
Governing Authority: society. Tax-exempt: 170(b)(1)(A).
Institution Type/Description: General Museum: housed in early 1800s private home.
Collections: local area furniture; old photographs; store ledgers; maps.
Activities: permanent exhibitions.
Publications: booklet, The Moravian Contribution to the Town of Hope, N.J.; 200 Years of Hope, New Jersey 1769-1969; 1976 Bicentennial Picture Book of Hope; DVD, Historic Hope Town; walking tour brochure.
Hours & Admission Prices: Nov.-May Sun. 1-3; groups by appointment. No charge; donations accepted.
Attendance: 200 (estimated)
Membership: Individual $10; Family $15; Patron $25; Benefactor $50.

Hopewell

HOPEWELL MUSEUM, 28 E. Broad St., Hopewell, NJ 08525-1828. Tel.: 609-466-0103.
Founded: 1924.
Congressional District: 12
Key Personnel: Pres., David M. Mackey; Cur., Beverly Weidl.
Governing Authority: nonprofit. Tax-exempt: 501(c).
Institution Type/Description: General Museum.
Collections: history of area from Colonial period to 1900; costumes; furnishings; equipment; weapons; art; agricultural implements; documents; pictures; Native American artifacts; natural history; glass; manuscripts.
Facilities: 100-vol. library of local history & genealogy books.
Activities: guided tours; permanent & temporary exhibitions.
Publications: book, reprint of 1963 Pioneers of Old Hopewell; map, reprint of 1875 Map of Hopewell Township; book, Hopewell Valley Heritage.
Hours & Admission Prices: Mon., Wed. & Sat. 2-5. Research: Mon. & Wed. No charge; donations accepted. Closed national holidays.
Attendance: 960

Howell

HOWELL HISTORICAL SOCIETY & MACKENZIE MUSEUM, 427 Lakewood-Farmingdale Rd., Howell, NJ 07731-8723. Mailing Address: P.O. Box 694, Farmingdale, NJ 07727-0694. Tel.: 732-938-2212.
E-mail: howellhist@aol.com
Web Site: www.howellnj.com/historic/
Formerly: Howell Historical Society & Committee Museum
Founded: 1971.
Congressional District: 3
Key Personnel: Pres. (V), Steve Meyer; Corresponding Sec., Kay Coakley; Sec., Sandra Solly; Treas., Doris Howard; Museum Shop Mgr., Virginia Krzyzanowski.
Personnel Profile: Part-Time Volunteers 25.
Governing Authority: society. Subsidiary Institution: Old Ardena Schoolhouse, Old Tavern & Preventorium Rds., Howell, NJ. Tax-exempt: 501(c)(3).
Institution Type/Description: Historical Society Museum: house built c.1807.
Collections: primary & secondary source material; decorative arts; manuscripts; ceramics; glass; furniture; paintings; costumes; photographs. Historic Houses: c.1855 one room schoolhouse; c.1807-1855 Grist Miller's home; c.1870 Horse Shed.
Research Fields: local history.
Facilities: 200-vol. library of books, photographs, newspapers, postcards & maps available for research by special request. Museum-related items for sale.
Activities: guided tours; lectures; films; art festivals; reading room; docent program; permanent & temporary exhibitions.
Publications: pamphlet; calendar of events; book, History of Howell; monthly newsletter, Howell Heritage.
Hours & Admission Prices: Grist Miller's home: Sat. 9:30-12:30. Schoolhouse: last Sun. each month 1-4; other times by appointment. No charge; donations accepted.
Attendance: 1,000 (estimated)
Membership: Individual $10; Family $15; Sponsor $30; Patron $50.

MANASQUAN RESERVOIR ENVIRONMENTAL CENTER, 331 Georgia Tavern Rd., Howell, NJ 07731. Tel.: 732-751-9453.
Institution Type/Description: Environmental Center.
Collections: hands-on exhibitions; wetlands ecology; wildlife & their habitats; photographs.
Facilities: nature trails.
Activities: multi-media presentations; educational programs; observation areas; nature trails.
Hours & Admission Prices: Memorial Day to Labor Day daily Fri. 10-8:30, Sat.-Thurs. 10-5; Sept.-May daily 10-5.

OLD ARDENA SCHOOLHOUSE - HOWELL HISTORICAL SOCIETY, Old Tavern & Preventorium Rds., Howell, NJ 07731. Mailing Address: 427 Lakewood-Farmingdale Rd., Howell, NJ 07731-8723. Tel.: 732-938-2212.
Institution Type/Description: Historic Building: housed in a former schoolhouse; built c.1855.
Collections: local history & culture; period furnishings; personal artifacts; photographs.
Activities: guided tours; special events & activities.
Hours & Admission Prices: last Sun. each month 1-4.

Iselin

GARDEN FOR THE BLIND AND PHYSICALLY HANDICAPPED, 1081 Green St., Iselin, NJ 08830. Tel.: 732-283-1200.
Institution Type/Description: Garden.
Collections: sensory garden with descriptive identifications in Braille.
Activities: lectures; workshops.
Hours & Admission Prices: Mon.-Sat. 9-6.

Island Heights

OCEAN COUNTY ARTISTS' GUILD, 22 Chestnut Ave., Island Heights, NJ 08732. Mailing Address: P.O. Box 1156, Island Heights, NJ 08732. Tel.: 732-270-3111.
Web Site: www.ocartistguild.org
Institution Type/Description: Art Gallery.
Collections: works by New Jersey artists.
Facilities: Gift items for sale.
Activities: Annual Events: Monthly Solo Shows; Member & State Juried Shows.
Hours & Admission Prices: Tues.-Sun. 1-4.

Jackson

NEW JERSEY FOREST RESOURCE EDUCATION CENTER, 370 E. Veterans Hwy., Jackson, NJ 08527. Tel.: 732-928-2360.
Institution Type/Description: Education Center.
Collections: forest & environment preservation.
Facilities: nature trails.
Activities: educational programs; special events; hiking. Annual Event: Fall Forestry Festival.
Hours & Admission Prices: Center: Mon.-Fri. 8-4. Trails: dawn to dusk.

Jamesburg

HISTORIC BUCKELEW MANSION, 203 Buckelew Ave., Jamesburg, NJ 08831. Mailing Address: P.O. Box 183, Jamesburg, NJ 08831. Tel.: 732-521-2040.
Governing Authority: Parent Institution: Jamesburg Historical Association.
Institution Type/Description: Historic House Museum: built between 1685-1870.
Collections: local history & culture; period furnishings; early clothes; photographs; personal artifacts.
Hours & Admission Prices: Temporarily closed.

Jersey City

AFRO-AMERICAN HISTORICAL SOCIETY MUSEUM, 1841 Kennedy Blvd., Jersey City, NJ 07305-2106. Tel.: 201-547-5262. Fax: 201-547-5392.
Web Site: www.cityofjerseycity.org/docs/afroam.shtml
Founded: 1977.
Congressional District: 13
Key Personnel: Dir. & Pres. (V), Neal E. Brunson; Consultant, Theodore Brunson.
Governing Authority: nonprofit organization. Tax-exempt: 501(c)(3).
Institution Type/Description: History Museum.
Collections: 1800-present, New Jersey African American history; civil rights posters; musical instruments from Africa and African American communities; black dolls; black police & firemen; artifact; quilts; coverlets; Pullman porters.
Research Fields: Jersey City black history.
Facilities: 3,500-vol. library.
Activities: guided tours; lectures; films; organized education programs for children; temporary exhibitions.
Publications: Newsletter.
Hours & Admission Prices: mid-June to Aug. Mon.-Fri. 12-5; Sept. to mid-June Mon.-Sat. 10-5. No charge; donations accepted. Closed all legal holidays; election days.
Attendance: 7,500 (estimated)

CURIOUS MATTER, 272 5th St., Jersey City, NJ 07302-2304. Tel.: 201-659-5771. Facebook: Curious Matter.
E-mail: gallery@curiousmatter.org
Web Site: www.curiousmatter.org
Founded: 2007.
Key Personnel: Dir., Raymond E. Mingst; Dir., Arthur Bruso
Institution Type/Description: Art Gallery.
Collections: works by regional, national & international artists.
Publications: catalogues; artist monographs & broadsides.
Hours & Admission Prices: Sun. 12-3; other times by appointment. No charge; donations accepted.

JERSEY CITY MUSEUM, 350 Montgomery St., Jersey City, NJ 07302-4041. Tel.: 201-413-0303. Fax: 201-413-9922. TTY: 201-413-6339.
E-mail: info@jerseycitymuseum.org
Web Site: www.jerseycitymuseum.org
Founded: 1901.
Congressional District: 13
Key Personnel: Exec. Dir., Laurene Buckley; Chm. Bd. Trustees, Nathan J. Sambul; Dir. Devel. & Mktg., Nancy Shannon; Registrar, Motrja Fedorko; Museum Shop Mgr., Lady-Grace Cervantes.
Personnel Profile: Full-Time Paid 15; Part-Time Paid 8; Part-Time Volunteers 20; Interns 6.
Governing Authority: private; nonprofit organization. Tax-exempt: 501(c)(3).
Institution Type/Description: Art Museum.
Collections: 19th- & 20th-century paintings, sculptures, & works on paper; Jersey City & New Jersey related artifacts; documents & objects; emphasis on art of social content from the 1930s to present; contemporary art which reflects U.S. cultural diversity.
Research Fields: 20th-century American art with a focus on socially engaged & multiculturalism; Jersey City industrial history.
Facilities: 152-seat auditorium; 2 classrooms; video gallery. Museum-related items for sale.
Activities: lectures; guided tours; symposia, programs for teens; workshops for school children; intern programs; traveling exhibitions; family programs.
Publications: exhibition catalogs; posters; reproductions of art work on cards; quarterly newsletter.
Hours & Admission Prices: Sat. 12-3. Adults $5, students with ID and seniors 62 & over $3; members & children under 12 no charge. Closed legal holidays. &
Attendance: 21,000 (accurate)
Membership: Student, Senior Citizens & Artists $35; Friend $50; Friends & Family $80; Patron $125; Contemporary $250; Benefactor $500; August Will Circle $1,000.

LIBERTY SCIENCE CENTER, (M), Liberty State Park, 222 Jersey City Blvd., Jersey City, NJ 07305-4636. Tel.: 201-253-1201. Fax: 201-451-6949.
E-mail: phoffman@lsc.org
Web Site: www.lsc.org
Founded: 1980.
Congressional District: 31
Key Personnel: Pres. & C.E.O., Paul Hoffman; Vice Pres. Experience Integration, Jeff Sasson; C.F.O. & Vice Pres. Resource Administration, Connie Claman; Vice Pres. External Affairs, Christine Arnold-Schroeder.
Personnel Profile: Full-Time Paid 97; Part-Time Paid 165; Part-Time Volunteers 403.
Governing Authority: nonprofit organization. Tax-exempt: 501(c)(3).
Institution Type/Description: Science Center.
Collections: health, invention & environmental hands-on exhibits; photographs; geological; botanical; zoological; films.
Research Fields: entomology.
Facilities: educational facilities; 64,000 sq. ft. exhibit space; 400-seat IMAX Dome theater; 300-seat auditorium & 3D theater; 100-seat teleconferencing theater; cafe. Museum-related items for sale.
Activities: education programs for children & families; teacher professional development; theatre; participatory & traveling exhibitions; videoconferencing; camp-ins; facility rental.
Publications: E-newsletter; newsletter, Insiders Club.
Hours & Admission Prices: April-June daily 9-5; July-Aug. Mon.-Fri. 9-4, Sat.-Sun. 9-5; Sept.-March Tues.-Fri. 9-4, Sat.-Sun. 9-5. Adults $15.75, senior citizens & children $11.50; discounts to AAM members; members no charge. &
Attendance: 540,000 (accurate)

Membership: Senior Citizen $45; Individual $55; Duo (two adults) $100; Basic Family (2 adults & up to 4 children) $140; Large Family (up to 10 individuals) $240. Corporate: Scholar $2,500-$4,999; Inventor $5,000-$9,999; Pioneer $10,000-$24,999; Explorer $25,000-$49,999; Innovator $50,000 & up. Women's Leadership Council: Associate 30 & under $45; Individual $100. Luminary Society: Patron's Circle $1,000-$2,499; Leadership Circle $2,500-$4,999; President's Circle $5,000-$9,999; Chairman's Circle $10,000-$24,999; Founder's Circle $25,000-$50,000.

NEW JERSEY ASSOCIATION OF MUSEUMS, c/o Liberty Science Center, 222 Jersey City Blvd., Jersey City, NJ 07305-4636. Tel.: 201-253-1213.
E-mail: eromanaux@lsc.org
Web Site: www.njmuseums.org/index.cfm
Founded: 1973.
Key Personnel: Pres., Elizabeth Romanaux.
Governing Authority: nonprofit organization. Tax-exempt: 501(c)(3).
Institution Type/Description: Museum Service Organization: represents 124 museums in the state.
Activities: training programs for professional museum workers; annual meeting.
Publications: museum directory, Guide to New Jersey Museums; newsletter, NJAM Briefs.
Hours & Admission Prices: Mon.-Fri. 9-5.
Membership: Affiliate $20; Individual $30; Corporate $500; Institutional based on budget size.

Kearny

KEARNY MUSEUM, Kearny Public Library, 318 Kearny Ave., Kearny, NJ 07032-2505. Tel.: 201-998-2666.
Web Site: www.kearnylibrary.org/museum.htm
Institution Type/Description: History Museum.
Collections: photographs; articles of clothing; war memorabilia; Kearny High School yearbooks.
Hours & Admission Prices: Winter: Wed. 6:30pm-7:30pm, Sat. 10-12.

Keyport

KEYPORT FIRE MUSEUM AND EDUCATION CENTER, 86 Broad St., Keyport, NJ 07735-1244. Mailing Address: P.O. Box 839, Keyport, NJ 07735. Tel.: 732-739-5362.
E-mail: popsqd@msn.com
Web Site: www.keyportfd.org
Founded: 2001.
Congressional District: 13
Key Personnel: Chm. (V), John J. Kovacs; Museum Shop Mgr., Robert M. Poling.
Personnel Profile: Part-Time Volunteers 14.
Institution Type/Description: Fire-Fighting History Museum: built in 1900.
Collections: firefighting history, artifacts & equipment; photographs; steel from the World Trade Center; personal artifacts.
Activities: educational programs.
Hours & Admission Prices: April-Dec. Sat. 12-4, Sun. 11-3. No charge; donations accepted.
Attendance: 712 (accurate)
Membership: Seniors $3; Life $250.

Kingston

ROCKINGHAM, 84 Laurel Ave., Kingston, NJ 08528. Mailing Address: P.O. Box 496, Kingston, NJ 08528-0496. Tel.: 609-683-7132.
E-mail: rockingham1783@yahoo.com
Web Site: www.rockingham.net
Founded: 1896.
Key Personnel: Dir., Lisa A. Flick.
Personnel Profile: Full-Time Paid 1; Part-Time Paid 1; Part-Time Volunteers 40; Interns 2.
Governing Authority: state. Parent Institution: Division of Parks & Forestry, Department of Environmental Protection, Trenton, NJ 08625. Tax-exempt.
Institution Type/Description: Historic House: 1710 Berrien Mansion, Washington's final wartime headquarters while Continental Congress was in session in Princeton.
Collections: period furnishings; paintings; sewing samplers; prints; reverse prints on glass; 100 pieces of textiles; Washington's military reproductions; colonial kitchen garden; life-size Washington mannequin; children's hands-on exhibits; military equipment.
Facilities: 18th century kitchen garden.

Activities: guided tours; special event days; local travel baskets; lectures.
Publications: The Rockingham Story; Colonial Herbs at Rockingham; newsletter, The Sundial.
Hours & Admission Prices: Call for hours. No charge; donations accepted. &
Attendance: 4,000 (accurate)
Membership: $35.

Lakehurst

NAVY LAKEHURST HISTORICAL SOCIETY & HERITAGE CENTER, Joint Base Lakehurst, Lakehurst, NJ 08733. Mailing Address: P.O. Box 328, Lakehurst, NJ 08733-0328. Tel.: 732-818-7520. Fax: 732-244-8897.
E-mail: tours2@nlhs.com
Web Site: nlhs.com
Key Personnel: Dir. & Pres. (V), Carl Jablonski; Museum Shop Mgr., John Niedzwiecki
Institution Type/Description: History Museum: housed on the Naval Air Station and the site of the Hindenburg disaster.
Collections: station & airship history; Hindenburg control car, crash site & disaster history; airships based here included Shenandoah, Los Angeles, & Akron and U.S. Navy blimps; photographs; military artifacts; newspapers; awards. Historic Building: Hangar No. 1.
Activities: special events; group tours.
Publications: bimonthly, The Airship.
Hours & Admission Prices: By appointment 2 weeks in advance. No charge; donations accepted.

Lakewood

JEWISH HISTORICAL SOCIETY OF OCEAN COUNTY, 301 Madison Ave., Lakewood, NJ 08701. Tel.: 732-363-0530.
E-mail: ocjf@rcn.com
Web Site: www.jewishgen.org
Institution Type/Description: Jewish History Museum.
Collections: Jewish history, culture & heritage; period artifacts; photographs.
Hours & Admission Prices: Call for hours.

LAKEWOOD HERITAGE MUSEUM, 231 Third St., Municipal Bldg. 1st & 2nd Fls., Lakewood, NJ 08701-2882. Tel.: 732-276-7944.
E-mail: lakewoodmuseum@optonline.net
Web Site: www.twp.lakewood.nj.us/parkrec_cultur.htm
Institution Type/Description: History Museum.
Collections: Lakewood history & culture; paintings; postcards; photographs; sports trophies; military artifacts.
Facilities: Museum-related items for sale.
Hours & Admission Prices: July-Aug. Tues. & Thurs. 2-5; Sept.-June Tues. & Thurs. 2-5, Sun. 2-4.

M. CHRISTINA GEIS ART GALLERY - GEORGIAN COURT UNIVERSITY, Arts & Science Center, 900 Lakewood Ave., Lakewood, NJ 08701-2600. Tel.: 732-364-2200, ext. 348.
Institution Type/Description: Art Gallery.
Collections: works by professional & student artists.
Activities: special events.
Hours & Admission Prices: Mon.-Thurs. 9-8, Fri. 9-5. No charge.

Lambertville

HOLCOMBE-JIMISON FARMSTEAD MUSEUM, 1605 Daniel Bray Hwy. (Rte. 29), Lambertville, NJ 08530-2402. Tel.: 609-397-2752.
Founded: 1985.
Institution Type/Description: Historic Building.
Collections: agricultural history; farm tools & equipment; period artifacts & furnishings; household implements & appliances; outbuildings.
Activities: demonstrations; guided tours; educational programs.
Hours & Admission Prices: May-Oct. Sun. 1-4, Wed. 9 am-12 pm; groups by appointment. Suggested Donation: adults $5, students $3, children under 5 no charge. &
Attendance: 2,100 (estimated)
Membership: Annual $25.

HOWELL LIVING HISTORY FARM, 70 Woodens Lane, Lambertville, NJ 08530. Mailing Address: 101 Hunter Rd., Titusville, NJ 08560-1902. Tel.: 609-737-3299. Fax: 609-737-6524.
E-mail: pwatson@howellfarm.com

Web Site: www.howellfarm.org
Founded: 1974.
Congressional District: 12
Key Personnel: Admin., Pete Watson; Farm Mgr., Gary Houghton; Pres. (V) Friends of Howell Farm, Keith Watson; Education, Susan DeVore; Program Coord., Kathy Brilla; Cur., Margaret Newman.
Personnel Profile: Full-Time Paid 9; Part-Time Paid 10; Part-Time Volunteers 100; Interns 9.
Governing Authority: county government. Parent Institution: Mercer County Park Commission. Tax-exempt: 501(c)(3).
Institution Type/Description: Agricultural History Museum: living history farm where farm life & farming of 1900-1910 have been recreated.
Collections: 1900-1910 farm implements; period household items.
Research Fields: oral histories of local neighbors.
Facilities: 700-vol. library. Museum-related items for sale.
Activities: formal education programs; guided tours; hands-on farming activities. Annual Events: plowing match; fiddle contest; ice harvest; maple sugaring.
Publications: online newsletter, The Furrow; weekly press releases.
Hours & Admission Prices: Feb.-Nov. Tues.-Sat. 10-4, Sun. 12-4. No charge; donations accepted. Closed legal holidays. &
Attendance: 50,000 (estimated)
Membership: Basic $40; Supporting & Business $100; Benefactor $500; Patron $1,000.

JAMES WILSON MARSHALL HOUSE, 60 Bridge St., Lambertville, NJ 08530. Mailing Address: 60 Bridge St., P.O. Box 2, Lambertville, NJ 08530. Tel.: 609-397-0770.
Key Personnel: Chm. (V), Suzanne Gitomer
Institution Type/Description: Historic House Museum: built in 1816. Listed on the National Register of Historic Places.
Collections: local history & culture; period furnishings; personal artifacts; photographs.
Hours & Admission Prices: April-Oct. Sat.-Sun. 1-4; other times by appointment.

Landing

LAKE HOPATCONG HISTORICAL MUSEUM, Hopatcong State Park, Landing, NJ 07850. Mailing Address: P.O. Box 668, Landing, NJ 07850-0668. Tel.: 973-398-2616. Facebook: Lake Hopatcong Historical Museum.
E-mail: lhhistory@att.net
Web Site: www.lakehopatconghistory.com
Founded: 1955.
Congressional District: 11
Key Personnel: Pres., Martin Kane; C.E.O., Robert Kays; Museum Shop Mgr., Laurie Martin.
Personnel Profile: Full-Time Volunteers 3; Part-Time Volunteers 15.
Governing Authority: Tax-exempt.
Institution Type/Description: Local History Museum: housed in 19th century Morris Canal Lock Tender's house.
Collections: Native American; Morris Canal; Lake Hopatcong, NJ history; former residents including Lotta Crabtree, Hudson Maxim, & Joe Cook.
Facilities: 550-vol. library of books; pamphlets; maps pertaining to history of New Jersey, available for research by approval of the Trustees.
Activities: guided tours; lectures; permanent exhibitions.
Publications: quarterly newsletter.
Hours & Admission Prices: March-May & Oct.-Dec. Sun. 12-4; Summer: call for hours. No charge; donations accepted. &
Attendance: 2,500 (accurate)
Membership: Annual $25.

Lawnside

BLACK HOLOCAUST MUSEUM OF SLAVERY, 327 White Horse Pike, Lawnside, NJ 08045. Mailing Address: P.O. Box 26846, Philadelphia, PA 19134. Tel.: 888-886-5378.
Web Site: www.lestweforgetmuseumofslavery.com
Institution Type/Description: History Museum.
Collections: history of African American slavery; shackles; manacles; branding irons; photographs.
Hours & Admission Prices: Wed.-Sun. 10-6.

PETER MOTT HOUSE - LAWNSIDE HISTORICAL SOCIETY, 26 Kings Court, Lawnside, NJ 08045. Mailing Address: P.O. Box 608, Lawnside, NJ 08045-0608. Tel.: 856-546-8850.
E-mail: lhs@petermotthouse.org
Web Site: www.petermotthouse.org

Institution Type/Description: Historical Society Museum: housed in a former station on the Underground Railroad; built in 1845. Listed on the National Register of Historic Places.
Collections: local history & culture; period furnishings.
Activities: tours.
Hours & Admission Prices: Sat. 12-3; other times by appointment. Adults $5, students $2.
Membership: Youth $2; Regular $15; Contributing $25; Business & Institution $100; Lifetime Individual $250; Corporation $1,000.

Lawrenceville

NATIONAL GUARD MILITIA MUSEUM OF NEW JERSEY - LAWRENCEVILLE, Armory, 151 Eggert Crossing Rd., Lawrenceville, NJ 08648. Tel.: 609-530-6802. Fax: 609-530-6807.
E-mail: donald.kale@njdmava.state.nj.us
Web Site: www.state.nj.us/military/museum/index.html
Founded: 1989.
Congressional District: 12
Key Personnel: Co-Cur., Col. Donald Kale; Co-Cur., Col. Jon Gribbin; Co-Cur., Ltc. William Kale.
Personnel Profile: Part-Time Volunteers 15; Interns 1.
Volunteer Hours: 7,200
Governing Authority: Parent Institution: New Jersey Department of Military and Veterans Affairs. Tax-exempt.
Institution Type/Description: Military Heritage Museum.
Collections: military history & heritage; weapons; uniforms; military equipment & vehicles; tanks; cannon; photographs.
Activities: group tours.
Publications: newsletter.
Hours & Admission Prices: Call for hours. No charge, donations accepted. &
Attendance: 980 (estimated)
Membership: Annual $10; Lifetime $250.

Lincroft

* **THE MONMOUTH MUSEUM, (M),** Newman Springs Rd., Brookdale C.C. Campus, 765 Newman Springs Rd., Lincroft, NJ 07738. Mailing Address: P.O. Box 359, Lincroft, NJ 07738 0359. Tel.: 732-747-2266. Fax: 732-747-8592.
Web Site: www.monmouthmuseum.org
Formerly: Monmouth Museum & Cultural Center
Founded: 1963.
Congressional District: 12
Key Personnel: Exec. Dir., Avis H. Anderson; Chm. (V), Marion P. Becker; Public Rels., Julia Fiorino; Coord. Special Events, Mary Suszkowski; Cur. Education, Marian Kanaga; Coord. Membership, Maureen Starace; Museum Shop Mgr., Helen Brown.
Personnel Profile: Full-Time Paid 5; Part-Time Paid 17; Part-Time Volunteers 65; Interns 2.
Governing Authority: nonprofit. Tax-exempt: 501(c)(3).
Institution Type/Description: General Museum.
Collections: sewing birds.
Facilities: Museum-related items for sale.
Activities: guided tours; lectures; gallery talks; formally organized education programs; day trips; travel trunks.
Publications: quarterly newsletter; brochures; teacher education packets.
Hours & Admission Prices: Tues.-Sat. 10-4:30, Sun. 1-5. Admission $7; discounts to AAM & PBS members; members no charge. Closed New Year's Day; Easter; Memorial Day; Independence Day; Labor Day; Thanksgiving; Christmas. &
Attendance: 55,000 (accurate)
Membership: Student & Senior Citizen $25; Individual $35; Household $75; Contributing $100; Associate $150; Supporting $250; Patron $500; Benefactor $1,000.

Linwood

JAMES KIRK MARITIME MUSEUM, 301 Davis Ave., Linwood, NJ 08221. Tel.: 609-927-8293.
Institution Type/Description: Maritime Museum.
Collections: maritime history; tools; hands-on exhibits; ship models.
Hours & Admission Prices: Mon.-Thurs. 10-8, Fri. 10-5, Sat. 10-4.

LINWOOD HISTORICAL SOCIETY - LEEDSVILLE SCHOOLHOUSE, 16 Poplar Ave., Linwood, NJ 08221-1820. Tel.: 609-927-8293.
Institution Type/Description: Historical Society Museum: built in 1873. Listed on the National Register of Historic Places.

Collections: local history & culture; period artifacts; photographs.
Hours & Admission Prices: Call for hours.

Little Falls

YOGI BERRA MUSEUM AND LEARNING CENTER, Montclair State Univ., 8 Quarry Rd., Little Falls, NJ 07424-2161. Tel.: 973-655-2378. Fax: 973-655-6894.
Web Site: www.yogiberramuseum.org
Founded: 1998.
Key Personnel: Dir., Dave Kaplan; Pres. (V), Mark Markowitz; Vice Chair, Julie Jackson; Dir. Special Events, Joni Bronander; Business Mgr., Bettylou O'Dell.
Personnel Profile: Full-Time Paid 3; Part-Time Paid 3; Part-Time Volunteers 50; Interns 3.
Governing Authority: private; nonprofit organization. Tax-exempt: 501(c)(3).
Institution Type/Description: Sports Museum.
Collections: concentration on Yankees history with special interest on Yogi Berra's career; baseball history.
Facilities: 110-seat theater; 4,500 sq. ft. exhibit space. Museum-related items for sale.
Activities: arts festivals; films; formal educational programs for children; guided tours; lectures; rental gallery; loan, temporary & traveling exhibitions; broadcast programs. Annual Events: book festival; film series; symposia.
Publications: quarterly newsletter, The Yogi Berra Museum Newsletter.
Hours & Admission Prices: Wed.-Sun. 12-5. Adults $6, senior citizens $5, students & children $4; discounts to AAA, ICOM & AAM members; members no charge. Closed New Year's Eve & Day; Easter; Mother's Day; Memorial Day; Independence Day; Labor Day; Columbus Day; Christmas Eve & Day. &
Membership: Rookie $20; Individual $35; Family $65; Business $5,000.

Locust

HUBER WOODS ENVIRONMENTAL CENTER, 25 Brown's Dock Rd., Locust, NJ 07760. Tel.: 732-872-2670.
Web Site: www.monmouthcountyparks.com
Institution Type/Description: Environmental Center.
Collections: live animals & plants; weather; hands-on exhibitions; Native American artifacts.
Facilities: nature trails.
Activities: hands-on exhibitions; puppet theater; bird watching; hiking; educational programs.
Hours & Admission Prices: Mon.-Fri. 10-4, Sat.-Sun. 10-5.

Long Branch

LONG BRANCH HISTORICAL MUSEUM, 1260 Ocean Ave., Long Branch, NJ 07740-4550. Mailing Address: P.O. Box 2204, Elberon, NJ 07740-2204. Tel.: 732-223-0874.
Web Site: www.churchofthepresidents.org
Founded: 1953.
Congressional District: 10
Key Personnel: C.E.O. & Pres. (V), Karen Van Hise; 1st. Vice Pres., James Foley.
Governing Authority: society. Tax-exempt.
Institution Type/Description: History Museum: housed in 1879 Saint James Episcopal Chapel, Church of the Presidents. Worshippers included: Presidents James A. Garfield, Ulysses S. Grant, Rutherford B. Hayes, Chester A. Arthur, Woodrow Wilson, Benjamin Harrison, William McKinley.
Collections: post Civil War bronze presidential plaques; historic flags; furniture; pictures; fire engine. Historic Building: Garfield Hut.
Activities: guided tours; permanent exhibitions.
Hours & Admission Prices: Closed for restoration.
Membership: Individual $25; Family $50; Organization $100; Lifetime $250; Patron $1,000; Donor $2,500; Benefactor $5,000.

Longport

LONGPORT HISTORICAL SOCIETY, 2301 Atlantic Ave., Longport, NJ 08403. Tel.: 609-822-3770 & 823-1115.
Institution Type/Description: Historical Society Museum: housed in a 1939 Coast Guard Station. Listed on the National Register of Historic Places.
Collections: local history & culture; period furnishings; photographs; personal artifacts.
Activities: art show; concerts.
Hours & Admission Prices: Memorial Day to Labor Day Sat. 9-12; Winter: Tues. 11-1; other times by appointment.

Loveladies

LONG BEACH ISLAND FOUNDATION OF THE ARTS & SCIENCES, 120 Long Beach Blvd., Loveladies, NJ 08008-6131. Tel.: 609-494-1241. Fax: 609-494-0662.
Key Personnel: Exec. Dir., Kristy Redford
Institution Type/Description: Art, Craft, & Science Museum.
Collections: works by local & national artists.
Activities: special events; performances; lectures; adult classes & workshops; youth programs; films.
Hours & Admission Prices: Gallery: daily 9-4. Office: Mon.-Fri. 9-4, Sat.-Sun. 9-3.

Lumberton

AIR VICTORY MUSEUM, INC., 68 Stacy Haines Rd., Lumberton, NJ 08048-4106. Tel.: 609-267-4488. Fax: 609-702-1852.
Web Site: airvictorymuseum.com
Founded: 1989.
Congressional District: 3
Key Personnel: Chm. & Pres. (V) & Museum Shop Mgr., Fred Koch.
Personnel Profile: Full-Time Volunteers 1; Part-Time Volunteers 20.
Governing Authority: nonprofit organization. Tax-exempt: 501(c)(3).
Institution Type/Description: Aeronautics Museum.
Collections: aircraft from Korea to present conflicts; military artifacts; uniforms; Norden bomb sight; navigation instruments; engines; equipment; 9/11 WTC Memorial.
Facilities: research library.
Activities: educational tours with study guides; teacher educational programs. Annual Event: living history encampment WWII-Vietnam.
Publications: biannual newsletter.
Hours & Admission Prices: Wed.-Sat. 10-4, Sun. 11-4. Museum Aircraft: tours available. Adults $4; discounts to groups, military w/ID & AAA members; members and children 3 & under no charge. &
Attendance: 3,500 (accurate)
Membership: Student $10; Senior Citizen $20; Adult $25; Family $50; Sustaining $60; Patron $100; Corporate $250. Building Fund: call for further information.

Lyndhurst

NEW JERSEY MEADOWLANDS ENVIRONMENT CENTER, 2 DeKorte Park Plaza, Lyndhurst, NJ 07071. Tel.: 201-777-2431. Fax: 201-460-7836.
Web Site: www.meadowlands.state.nj.us
Formerly: Hackensack Meadowlands Development Commission Environment Center & Museum
Founded: 1969.
Key Personnel: Exec. Dir., Marcia A. Karrow; Environmental Education, Angela Cristini, Ph.D.; Operations Mgr. & Museum Shop Mgr., Donna Bocchino.
Personnel Profile: Full-Time Paid 10; Part-Time Paid 1.
Governing Authority: state. Parent Institution: New Jersey Meadowlands Commission. Tax-exempt.
Institution Type/Description: Nature Center.
Collections: native natural history species with emphasis on birds; aquarium for live fish; wetlands diorama; living natural areas to demonstrate salt marsh wetlands & native upland habitats; interactive exhibits on past, present and future of the Hackensack Meadowlands.
Facilities: library on environmental education, wetlands & solid waste; 280-seat auditorium; educational facilities; 8,031 sq. ft. exhibit space. Gift items for sale.
Activities: guided tours; hobby workshops; formal education programs for children; teacher training in environmental education; nature tours; boating; canoeing; birding. Annual Event: Children's Halloween Celebration.
Publications: Schedule of Events.
Hours & Admission Prices: Center: Mon.-Fri. 8-4, Sat.-Sun. 9-3. Store: daily 10-3. Trails: daily 8 to dusk. No charge; donations accepted. &
Attendance: 50,000 (accurate)
Membership: Individual $25; Family $50; Contributor $100; Patron $200; Corporate $500.

Madison

MUSEUM OF EARLY TRADES & CRAFTS, 9 Main St., Madison, NJ 07940-1819. Tel.: 973-377-2982, ext. 10. Fax: 973-377-7358.
E-mail: info@metc.org
Web Site: www.metc.org
Founded: 1969.

Congressional District: 25
Key Personnel: Chm., Allen Black; Vice Chm., Michele Faas; Dir., Vivian C.R. James; Mgr. Operations, April Lyzak.
Personnel Profile: Full-Time Paid 4; Part-Time Paid 10; Part-Time Volunteers 25; Interns 4.
Governing Authority: nonprofit organization. Tax-exempt.
Institution Type/Description: Trades & Crafts Museum: housed in 1900 Richardsonian Romanesque style building, the former town library.
Collections: tools & products of 18th to 19th-century New Jersey home, farm & shop crafts; woodworking & metalworking tools.
Research Fields: regional trades history; craftsmanship.
Facilities: classrooms. Museum-related items for sale.
Activities: guided tours; demonstrations; craft workshops; family, school, scout & group programs; special & rotating exhibits; lectures; gallery talks; summer camp for children; receptions.
Publications: quarterly newsletter.
Hours & Admission Prices: Summer: Tues.-Sat. 10-4; Winter: Tues.-Sat. 10-4, Sun. 12-5. Adults $5, seniors, students, & children $3; AAM members & museum members no charge. Closed major holidays. &

Attendance: 11,000 (estimated)
Membership: Individual $30; Family $50; Patron $100.

Mahwah

BERRIE CENTER FOR PERFORMING AND VISUAL ARTS, Ramapo College of NJ, 505 Ramapo Valley Rd., Mahwah, NJ 07430. Tel.: 201-684-7500.
Web Site: www.ramapo.edu/berriecenter
Institution Type/Description: Arts & Cultural Center.
Collections: works by students, faculty & international artists.
Facilities: theater.
Activities: lectures; performances.
Hours & Admission Prices: Tues. & Thurs.-Fri. 1-5, Wed. 1-7.

HINDU SAMAJ MANDIR - MUSEUM OF INDIA JOURNEY TO THE USA, 247 W. Ramapo Ave., Mahwah, NJ 07430. Tel.: 201-529-1277.
Web Site: www.hindusamajmandir.org
Institution Type/Description: India Heritage Museum: housed in a Hindu temple.
Collections: Indian history, heritage, culture & art; videos; personal artifacts.
Facilities: meditation & sculpture garden; community center.
Activities: youth programs; school groups.
Hours & Admission Prices: Call for hours.

MAHWAH MUSEUM SOCIETY, (M), 201 Franklin Tpke., Mahwah, NJ 07430. Tel.: 201-512-0099.
Institution Type/Description: History Museum.
Collections: local history & culture; photographs; period artifacts; early records & documents.
Hours & Admission Prices: Sept.-May Sat.-Sun. 2-4.

OLD STATION MUSEUM, 1871 Old Station Lane, Mahwah, NJ 07430. Mailing Address: Mahwah Museum Society, 201 Franklin Tpke., Mahwah, NJ 07430. Tel.: 201-512-0099.
Institution Type/Description: Historic Building: housed in the original station on the Erie Railroad in Mahwah; built in 1848.
Collections: local history & culture; railroad artifacts; photographs; period furnishings; personal artifacts; 1927 caboose.
Hours & Admission Prices: June-Oct. Sun. 2-4. Adults $5; members & children under 18 no charge.

Manalapan

MONMOUTH BATTLEFIELD STATE PARK, 16 Business Rte. 33, Manalapan, NJ 07726. Mailing Address: 347 Freehold - Englishtown Rd., Manalapan, NJ 07726. Tel.: 732-462-9616.
Institution Type/Description: Park Museum.
Collections: local history. Historic Building: Revolutionary War Farmhouse.
Facilities: native trails; visitor center.
Activities: interpretive programs; walking trails; picnic area. Annual Event: battle reenactments in June.
Hours & Admission Prices: Park: daily 8-8. Visitor Center: daily 9-4.

Manasquan

SQUAN VILLAGE HISTORICAL SOCIETY - BAILEY REED HOUSE, 105 South St., Manasquan, NJ 08736. Tel.: 732-223-6770. Fax: 732-223-6770.
Institution Type/Description: Historical Society Museum.
Collections: local history & culture; period furnishings; photographs; personal artifacts.
Hours & Admission Prices: By appointment.

Marlton

EVESHAM TOWNSHIP HISTORICAL SOCIETY, 10 Madison Court, Marlton, NJ 08053. Tel.: 856-988-6530.
Institution Type/Description: Historical Society Museum.
Collections: local history & culture; period furnishings; personal artifacts; photographs.
Activities: special events.
Hours & Admission Prices: Sept.-June 3rd Sat. each month 10-2.
Membership: Single $10; Family $15; Business $50; Lifetime $150.

Matawan

BURROWES MANSION MUSEUM, 94 Main St., Matawan, NJ 07747-2630. Mailing Address: 24 Monroe St., Matawan, NJ 07747-3218. Tel.: 732-566-3817 & 5605.
Founded: 1976.
Congressional District: 6
Key Personnel: Museum Shop Mgr. & Head Docent, Sarah Ellison; Historic Site Mgr., Howard Henderson.
Personnel Profile: Part-Time Volunteers 15.
Governing Authority: commission. Parent Institution: Historic Sites Commission. Subsidiary Institution: Matawan Historical Society. Tax-exempt: 501(c)(3).
Institution Type/Description: Historic House: 1723 Georgian half-house with gambrel roof, located on a Revolutionary War skirmish site.
Collections: 18th- & 19th-century furniture; postcards; costumes; photographs; archives; local history.
Research Fields: local history; cultural survey; furnishings.
Facilities: 175-vol. research library; 1,750 sq. ft. exhibit space.
Activities: guided tours; lectures; concerts; organized educational programs for children; Christmas Musical.
Publications: biannual newsletter, Matawan Historical Society Newsletter.
Hours & Admission Prices: March-Dec. 1st & 3rd Sun. of each month 2-4; other times by appointment. No charge; donations accepted. Closed Easter.
Attendance: 900 (accurate)
Membership: Student $1; Adult $7.50; Family $12; Patron $25; Benefactor $100.

CHEESEQUAKE STATE PARK INTERPRETIVE CENTER, 300 Gordon Rd., Matawan, NJ 07747. Tel.: 732-566-2161.
Institution Type/Description: Park & Interpretive Center.
Collections: local history; animals; plants; ecosystem; murals.
Facilities: visitor center; auditorium; nature trails.
Activities: bird watching; environmental programs; special events; scout groups; hiking.
Hours & Admission Prices: Memorial Day to Labor Day daily 8-4; Sept.-May Wed.-Sun. 8-4. &

Mauricetown

MAURICETOWN HISTORICAL SOCIETY - THE EDWARD COMPTON HOUSE, 1229 Front St., Mauricetown, NJ 08329. Mailing Address: P.O. Box 1, Mauricetown, NJ 08329. Tel.: 856-785-1137 & 1372.
Web Site: www.mauricetownhistoricalsociety.org
Founded: 1983.
Congressional District: 1
Key Personnel: Pres. (V), Carol D. Perrelli; Cur. & Museum Shop Mgr., Irene Ferguson.
Personnel Profile: Part-Time Volunteers 25.
Governing Authority: Tax-exempt.
Institution Type/Description: Historical Society Museum.
Collections: local history & culture; period furnishings; personal artifacts; photographs. Historic Buildings: 1864 Edward Compton house; 1830 Abraham & Ann Hoy house; cookhouse; print shop.
Facilities: library.
Activities: house tours.
Publications: brochure, Mauricetown Historical Society.

Hours & Admission Prices: 1st & 3rd Sun. of the month 1-4; other times by appointment. No charge; donations accepted. ♿
Attendance: 1,200 (estimated)
Membership: Individual $10; Life $50.

Mays Landing

WARREN E. FOX NATURE CENTER, 109 Boulevard Rte. 50, Mays Landing, NJ 08330-4323. Tel.: 609-645-5960.
Key Personnel: Dir., Jerry Barbiero
Institution Type/Description: Nature Center.
Collections: local history; plants; ecology; live animals.
Facilities: 50-seat auditorium.
Activities: public programs; lectures; school group tours; field studies; teacher workshops.
Hours & Admission Prices: Mon.-Fri. 8-4:30, Sat.-Sun. & holidays 8-4. No charge.

Medford

BARTON ARBORETUM & NATURE PRESERVE, One Medford Leas Way, Medford, NJ 08055. Tel.: 800-331-4302; 609-654-3000. Fax: 609-654-7894.
E-mail: communityrelations@medfordleas.net
Web Site: bartonarboretum.org
Institution Type/Description: Arboretum.
Collections: over 200 species of trees; plants; flowers.
Facilities: nature trails.
Activities: special events.
Hours & Admission Prices: Call for hours.

MEDFORD HISTORICAL SOCIETY, 275 Church Rd., Medford, NJ 08055. Tel.: 609-654-7767.
Web Site: www.medfordhistory.org
Founded: 1976.
Key Personnel: Pres. (V), William Stauts.
Governing Authority: Tax-exempt.
Institution Type/Description: Historical Society Museum: built in 1785.
Collections: local history & culture; photographs; period furnishings; personal artifacts.
Activities: special events.
Publications: newsletter; books.
Hours & Admission Prices: Call for hours. No charge; donations accepted.

WOODFORD CEDAR RUN WILDLIFE REFUGE, 4 Sawmill Rd., Medford, NJ 08055. Tel.: 856-983-3329.
Institution Type/Description: Wildlife Refuge.
Collections: wildlife & their habitats.
Facilities: 174-acres; nature trails. Gift items for sale.
Activities: educational programs; special events; hiking trails.
Hours & Admission Prices: Mon.-Sat. 10-4, Sun. 12-4.

Mendham

RALSTON HISTORICAL ASSOCIATION, JOHN RALSTON MUSEUM, 313 Mendham Rd., W., Mendham, NJ 07945-1000. Tel.: 973-543-6878. Fax: 973-543-1149.
E-mail: pfr14@aol.com
Web Site: www.ralstonmuseum.org
Founded: 1964.
Congressional District: 11
Key Personnel: Pres. (V), Margaret Hogun; Vice Pres., Tracy Kinsell.
Personnel Profile: Part-Time Paid 1; Part-Time Volunteers 15.
Governing Authority: nonprofit organization. Tax-exempt: 501(c)(3).
Institution Type/Description: Historical Society Museum: housed in the oldest building used as a post office in the United States, 1780s.
Collections: local artifacts; agricultural tools before 1814; hand-blown glass; industry; local papers & documents; herbarium; manuscripts. Historic Site: 1732 Wills Family cemetery.
Activities: guided tours; temporary & permanent exhibitions; lectures.
Publications: annual newsletter; book, The Mendhams; The Rock-A-Bye Baby Railroad.
Hours & Admission Prices: June-Oct. Sun. 2-5. No charge; donations accepted.
Attendance: 250 (estimated)
Membership: Senior Citizen $15; Individual $25; Family & Institution $40; Sustaining $100; Life $250; John Ralston Society $1,000.

Metuchen

METUCHEN-EDISON HISTORICAL SOCIETY, Metuchen Public Library, 480 Middlesex Ave., Metuchen, NJ 08840-1457. Mailing Address: P.O. Box 61, Metuchen, NJ 08840. Tel.: 732-632-8526.
Institution Type/Description: Historical Society Museum.
Collections: local history & culture; period furnishings; photographs; personal artifacts.
Hours & Admission Prices: Call for hours.

Middletown

DEEP CUT GARDENS, 352 Red Hill Rd., Middletown, NJ 07748. Tel.: 732-671-6050.
Institution Type/Description: Horticultural Park: housed in the former home of the Wihtol family.
Collections: native & cultivated plants; Bonsai & rose gardens.
Facilities: library; 54 acre park; horticultural center.
Hours & Admission Prices: Daily 7am - 8:30.

MARLPIT HALL MUSEUM, 137 Kings Hwy., Middletown, NJ 07748-2003. Mailing Address: Monmouth County Historical Assoc., 70 Court St., Freehold, NJ 07728-1710. Tel.: 732-462-1466.
Institution Type/Description: Historic House Museum: built c.1756.
Collections: local history & culture; period furnishings; personal artifacts; photographs.
Hours & Admission Prices: Call for hours.

MOSES D. HEATH FARM MUSEUM, 219 Harmony Rd., Middletown, NJ 07748. Tel.: 732-671-0566.
Institution Type/Description: Farm Museum.
Collections: farm history & equipment; Health family history; farming; garden; blacksmith shop; sugar cane mill; personal artifacts.
Activities: hands-on exhibitions.
Hours & Admission Prices: Call for hours.

PORICY PARK CONSERVANCY, 345 Oak Hill Rd., Middletown, NJ 07748. Mailing Address: Box 36, Middletown, NJ 07748-0036. Tel.: 732-842-5966. Fax: 732-842-6833.
Web Site: www.poricypark.org
Founded: 1978.
Congressional District: 3
Key Personnel: Exec. Dir., Joyce Ferejohn; Treas., Lee Beaumont.
Personnel Profile: Full-Time Paid 3; Part-Time Paid 12; Part-Time Volunteers 100.
Governing Authority: nonprofit organization. Poricy Park Citizen's Committee. Tax-exempt: 501(c)(3).
Institution Type/Description: Nature Center: located on 250-acre nature preserve.
Collections: mounted animals; fossils; 18th-century clothing, cooking utensils, furnishings. Historic Buildings: 18th-century farmhouse & barn.
Facilities: 150-vol. library pertaining to natural history, 18th-century life, Murray's; educational facilities; nature center. Museum-related items for sale.
Activities: guided tours; lectures; hobby workshops; organized education programs; docent program; participatory & temporary exhibitions.
Publications: newsletter & program schedule; program brochures for groups, scouts & special events.
Hours & Admission Prices: Park: daily dawn to 10pm. Nature Center: Mon.-Fri. 9-4. No charge; donations accepted. Fees for special programs. Closed national holidays. ♿
Attendance: 22,000 (estimated)
Membership: Senior $25; Individual $35; Family $50; Patron $100; Supporter $150; Champion $250; Guardian $350; Benefactor $500.

TAYLOR-BUTLER HOUSE, 127 Kings Hwy., Middletown, NJ 07748-2003. Tel.: 732-462-1466.
Web Site: www.monmouthhistory.org
Institution Type/Description: Historic House Museum: built in 1853.
Collections: local & family history; period furnishings; personal artifacts; photographs.
Hours & Admission Prices: May-Sept. Thurs.-Sat. 1-4.

Milltown

EUREKA FIRE MUSEUM, 39 Washington Ave., Milltown, NJ 08850-1219. Tel.: 732-828-7207 & 7400.
E-mail: webmaster@milltownfire.org
Web Site: www.milltownfire.org/museum.htm
Founded: 1981.
Key Personnel: Chm. (V), Brian E. Harto; Treas., Mark Steeber.
Personnel Profile: Part-Time Volunteers 2.
Governing Authority: private; nonprofit organization.
Institution Type/Description: Fire-Fighting Museum.
Collections: 200 years worth of fire fighting equipment; fire patches, badges & insignia from around the world; uniforms; helmets; fire extinguishers; hand-drawn fire apparatus; 1921 & 1947 American LaFrance pumpers; local history; photographs.
Facilities: 200-vol. library.
Hours & Admission Prices: By appointment only. No charge; donations accepted.
Attendance: 150 (estimated)

Millville

MILLVILLE ARMY AIR FIELD MUSEUM, 1 Leddon St., Millville Airport, Millville, NJ 08332-4822. Tel.: 856-327-2347. Fax: 856-327-5737.
E-mail: museum@p47millville.org
Web Site: www.p47millville.org
Founded: 1983.
Congressional District: 2
Key Personnel: Dir., Lisa Jester; Chm. (V), Russell Davis; Pres. (V), Chuck Wyble.
Personnel Profile: Full-Time Paid 1; Full-Time Volunteers 2; Part-Time Paid 10; Part-Time Volunteers 1.
Governing Authority: federal; nonprofit organization. Tax-exempt: 501(c)(3).
Institution Type/Description: Military Museum.
Collections: military aviation history; photographs; military artifacts; airfield's history; WWII.
Facilities: library; education center.
Activities: tours; programs. Annual Event: Air Show.
Publications: newsletter, Thunderbolt.
Hours & Admission Prices: Mon. by appointment only, Tues.-Sun. 10-4; guided tours by appointment. No charge; donations accepted. &
Attendance: 4,000 (estimated)
Membership: Individual $25; Family $40; Booster $100; Patron $250; Guardian Angel $500.

＊ **MUSEUM OF AMERICAN GLASS AT WHEATON ARTS AND CULTURAL CENTER, (M),** 1501 Glasstown Rd., Millville, NJ 08332-1566. Tel.: 856-825-6800. Fax: 856-825-2410.
E-mail: museum@wheatonarts.org
Web Site: www.wheatonarts.org
Formerly: Museum of American Glass at Wheaton Village
Founded: 1968.
Congressional District: 2
Key Personnel: Chm., Keith Ragone; Registrar, Elizabeth G. Wilk; Museum Shop Mgr., Catharine Nolan.
Personnel Profile: Full-Time Paid 2; Part-Time Paid 3; Part-Time Volunteers 33.
Governing Authority: nonprofit organization. Parent Institution: Wheaton Arts and Cultural Center. Tax-exempt: 501(c)(3).
Institution Type/Description: American Glass Museum.
Collections: United States glass; operating reproduction of 19th century glass factory; glass decorating; 16 restored & recreated buildings.
Research Fields: U.S. glass; U.S. & N.J. glasshouses; individual artists.
Facilities: 3,200-vol. library of material on glass, southern N.J. history, original documents & ephemera, available for use on premises; classroom; restaurant. Museum-related items & crafts for sale.
Activities: guided tours; lectures; workshops; loan, permanent & traveling exhibitions; seminars; glass artists fellowship; study clubs; concerts; performances.
Publications: catalogs, Out of The Mold; Clevenger Brothers: The Persistence of Tradition; Maximizing the Minimum: Small Glass Sculpture; Distinctively Durand: The Art Glass of Vineland, N.J.; The Wistars & Their Glass, 1739-1777; Colorful Cutting; Thousands of Flowers: American Millefiori Glass; book, Under Sail: The Dredgeboats of Delaware Bay; Gillinder Glass: Story of a Company; It Figures: American Sculptural Bottles; Flights of Fancy: The Quezal Art Glass & Decorating Company; Slivers of Light: Rich Cut Oil Lamps; Contemporary Flameworked Glass; Vanity Vessels: The Story of the American Perfume Bottle; For Show, Not Play: Glass

Chess Sets; 20/20 Vision; Glass Threads: Tiffany, Quezal, Imperial, Durand; The Fires Burn On: 200 Years of Glassmaking in Millville, New Jersey.
Hours & Admission Prices: Jan.-March call for hours; April-Dec. Tues.-Sun. 10-5. Adults $10, senior citizens $9, students $7; discounts to groups, AAM, ICOM & AAA members; members no charge. Closed New Year's Day; Easter; Thanksgiving; Christmas. &
Attendance: 70,000 (estimated)
Membership: Students & Seniors $25; Individual $35; Dual $45; Family $55.

RIVERFRONT RENAISSANCE CENTER FOR THE ARTS, 22 N. High St., Millville, NJ 08332-3830. Tel.: 856-327-4500. Fax: 856-327-9280.
E-mail: sean@rrcarts.com
Web Site: www.rrcarts.com
Founded: 2001.
Congressional District: 2
Key Personnel: Dir., Liz Nicklus.
Personnel Profile: Full-Time Paid 3; Part-Time Volunteers 2.
Institution Type/Description: Art Gallery.
Collections: paintings; sculpture; photographs.
Facilities: Museum-related items for sale.
Activities: classes; permanent & temporary exhibits; special events.
Hours & Admission Prices: Sun.-Thurs. 11-5, Fri. 11-8, Sat. 11-7. No charge; donations accepted. &
Attendance: 6,000 (accurate)
Membership: Student & Senior $35; Individual $50; Family $75; Patron $100; Benefactor $250.

Monmouth Beach

MONMOUTH BEACH CULTURAL CENTER, 128 Ocean Ave., Monmouth Beach, NJ 07750. Tel.: 732-229-4527.
Institution Type/Description: History Museum.
Collections: local history & culture; period furnishings; personal artifacts; photographs; Monmouth Beach Lifesaving Station.
Activities: special events.
Hours & Admission Prices: Call for hours.

Monroe Township

THE STONE MUSEUM, 608 Spotswood-Englishtown Rd., Monroe Township, NJ 08831-3222. Tel.: 732-521-2232. Fax: 732-521-3388.
E-mail: displayworld@erols.com
Web Site: www.thestonemuseum.com
Key Personnel: Dir., Pat Ciecko; Museum Shop Mgr., Barbara Thompson.
Personnel Profile: Full-Time Volunteers 2.
Governing Authority: private; nonprofit organization. Parent Institution: Greeks Playland.
Institution Type/Description: Geology Museum.
Collections: minerals & fossils from over 80 countries; fluorescent rock; seashells; dinosaurs & dinosaur egg from China; whale vertebra; hands-on exhibits; M60 tank; Cola helicopter.
Hours & Admission Prices: Sat.-Sun. 11-4. No charge. Closed New Year's Day; Memorial Day; Independence Day; Labor Day; Christmas. &
Attendance: 20,000 (estimated)

Montclair

＊ **MONTCLAIR ART MUSEUM, (M),** 3 S. Mountain Ave., Montclair, NJ 07042. Tel.: 973-746-5555. Fax: 973-746-9118 & 0920.
Web Site: www.montclairartmuseum.org
Founded: 1914.
Congressional District: 11
Key Personnel: Pres., Newton Schott; Dir., Lora Urbanelli; Chief Cur., Gail Stavitsky, Ph.D.; Dir. Finance, Michael Frasco; Dir. Education, Gary Schneider; Cur. Contemporary Art, Alexandra Schwartz; Dir. Mktg. & Communications, Michael Gillespie; Museum Shop Mgr., Ilana Stollman.
Personnel Profile: Full-Time Paid 28; Part-Time Paid 80; Part-Time Volunteers 314; Interns 39.
Governing Authority: nonprofit organization. Tax-exempt: 501(c)(3); 170(b).
Institution Type/Description: Art Museum: housed in Neo-classic brick & stone building.
Collections: American art; prints; drawings; sculpture; The Rand Collection of Native American art; Morgan Russell Archive & Collection; George Inness Collection.
Major Exhibits: Robert Smithson: New Jersey Earthworks, 2/16/14-6/15/14.
Research Fields: American paintings; Morgan Russell Archives.

Facilities: art school; auditorium. Museum-related items for sale.
Activities: guided tours; lectures; films; gallery talks; art classes for adults & children; concerts; permanent & temporary exhibitions; family programs.
Hours & Admission Prices: Wed.-Sun. 12-5. Adults $12, seniors & students with I.D. $10; discounts to AAM & ICOM members; members, children under 12 & Oct.-June first Thurs. of month 5-9 no charge. Closed major holidays. &
Attendance: 66,000 (accurate)
Membership: College Student $35; Individual $50; Dual & Family $70; Friend $165; Curator's Circle $325; Sustaining $750; Benefactor $1,500; Director's Circle $3,000; Inness $5,500.

MONTCLAIR HISTORICAL SOCIETY - CLARK HOUSE & LIBRARY, 108 Orange Rd., Montclair, NJ 07042-2133. Tel.: 973-744-1796. Fax: 973-783-9419.
E-mail: mail@montclairhistorical.org
Web Site: www.montclairhistorical.org
Founded: 1965.
Congressional District: 8
Key Personnel: Exec. Dir., Carlos Pomares; Pres. (V), John Weisel.
Personnel Profile: Full-Time Paid 2; Part-Time Paid 1; Part-Time Volunteers 100; Interns 2.
Governing Authority: society. Subsidiary Institution: Israel Crane Museum & Nathaniel Crane House, 110 Orange Rd., Montclair, NJ; The Charles S. Shultz House (Evergreens), 30 N. Mountain Ave., Montclair, NJ. Tax-exempt: 501(c)(3).
Institution Type/Description: Historic House Museum: housed in the former home of Dr. James Henry Clark & his wife Carrie Schenck; built in 1894. Listed on the National Register of Historic Places.
Collections: period rooms; 18th-century gardens; period costumes; period tools; history of Montclair, NJ from 1600s to present; schoolhouse contents; reference library.
Research Fields: family life of 19th-century to early 20th-century in Montclair, NJ; 18th-century open hearth cooking; 18th & early 19th-century arts & skills; genealogy materials of area families.
Facilities: 1,000-vol. library of New Jersey & Montclair history & genealogy; early American crafts; period tools.
Activities: education programs.
Publications: History of Montclair Township; Fanny Pierson Crane, Her Receipts, 1796; The Thirteen Colonies Cookbook; monthly newsletter.
Hours & Admission Prices: Call for hours & admission prices. &
Attendance: 4,800 (accurate)
Membership: Individual $25; Family $35; Sustaining & Business $60; Patron $100; Friend $250; Benefactor $500; Society Circle $1,000; Life $2,500.

MONTCLAIR HISTORICAL SOCIETY - CHARLES SHULTZ HOUSE, 30 N. Mountain Ave., Montclair, NJ 07042. Tel.: 973-744-1796.
Governing Authority: Subsidiary Institutions: Clark House & Library, 108 Orange Rd., Montclair, NJ; Israel Crane House & Nathaniel Crane House, 110 Orange Rd., Montclair, NJ.
Institution Type/Description: Historic House Museum: built in 1896. Listed on the National Register of Historic Places.
Collections: local history & culture; period furnishings; personal artifacts.
Hours & Admission Prices: April-Dec. 1st & 3rd Sun. each month 1-4. Adults $8, seniors & children $5; members no charge.

MONTCLAIR HISTORICAL SOCIETY - ISRAEL CRANE HOUSE & NATHANIEL CRANE HOUSE, 110 Orange Rd., Montclair, NJ 07042. Tel.: 973-744-1796.
E-mail: mail@montclairhistorical.org
Governing Authority: Subsidiary Institutions: Clark House & Library, 108 Orange Rd., Montclair, NJ; Charles Shultz House (Evergreens), 30 N. Mountain Ave., Montclair, NJ.
Institution Type/Description: Historic Houses: Israel Crane House built in 1796; Nathaniel Crane House built in 1818.
Collections: local history & culture; period furnishings; personal artifacts; general store.
Hours & Admission Prices: 1st & 3rd Sun. each month 1-4. Adults $8, seniors & children $5; members no charge.

VAN VLECK HOUSE & GARDENS, 21 Van Vleck St., Montclair, NJ 07042. Tel.: 973-744-4752. Fax: 973-746-1082.
E-mail: info@vanvleck.org
Web Site: www.vanvleck.org
Key Personnel: Exec. Dir., Charles Fischer
Institution Type/Description: Historic House Museum: late 19th century home.
Collections: period furnishings; paintings; gardens.
Activities: educational programs; special events.

Hours & Admission Prices: House: Mon.-Fri. 10-3. Gardens: daily dawn to dusk. No charge.

Montville

MONTVILLE TOWNSHIP HISTORICAL MUSEUM, 6 Taylortown Rd., Montville, NJ 07045. Mailing Address: P.O. Box 519, Montville, NJ 07045. Tel.: 973-334-3665.
Web Site: www.montvillenj.org
Founded: 1963.
Congressional District: 11
Key Personnel: Pres., Kathleen Fisher.
Personnel Profile: Full-Time Volunteers 6; Part-Time Volunteers 6; Interns 1.
Governing Authority: society; municipal. Parent Institution: Montville Historical Society. Tax-exempt: 501(c)(3).
Institution Type/Description: General Museum: housed in c. 1867 School House.
Collections: agriculture; children's museum; costumes; decorative arts; general; geology; history; Indian artifacts; manuscripts; photographs.
Research Fields: houses, genealogies, town records, church records.
Facilities: library containing books, tapes, letters, deeds, medical, church and account records available for use on premises.
Activities: lectures; formally organized education programs; permanent exhibitions.
Publications: book, reprints of historical maps of Montville 1853 & 1868; pamphlets, Tours of Montville.
Hours & Admission Prices: Sept.-June Sun. 1-4. No charge; donations accepted. Closed national holidays.
Attendance: 1,200 (accurate)
Membership: Student $5; Individual $15; Family $20; Corporate $100; Life $200.

Moorestown

PERKINS CENTER FOR THE ARTS, 395 Kings Hwy., Moorestown, NJ 08057-2725. Tel.: 856-235-6488; 800-387-5226. Fax: 856-235-6624.
E-mail: create@perkinscenter.org
Web Site: www.perkinscenter.org
Founded: 1977.
Key Personnel: Dir., Alan Willoughby.
Personnel Profile: Full-Time Paid 10; Part-Time Paid 7.
Governing Authority: nonprofit organization. Satellite: 30 Irvin Ave., Collingswood, NJ 08108. Tax-exempt: 501(c)(3).
Institution Type/Description: Art Center: housed in 1910 Tudor Revival House, designed by Herbert C. Wise, first editor of House & Garden magazine.
Collections: paintings; sculpture; photographs.
Facilities: visual arts studios; pottery studio; dance studio, music rooms.
Activities: guided tours; lectures; concerts; dance recitals; organized education programs; performing arts; pottery co-op; participatory & temporary exhibitions; artist in-school residencies; art, music, dance, & drama instruction.
Publications: newsletter/catalogue published quarterly per year.
Hours & Admission Prices: Thurs.-Fri. 10-4, Sat.-Sun. noon to 4. No charge; donations accepted. &
Attendance: 30,000 (estimated)
Membership: Student & Senior $30; Individual $35; Senior Family $50; Family $55; Family Plus $95.

SMITH-CADBURY MANSION, 12 High St., Moorestown, NJ 08057-3504. Mailing Address: P.O. Box 477, Moorestown, NJ 08057. Tel.: 856-235-0353.
E-mail: moorestownhistory@verizon.net
Founded: 1969.
Governing Authority: Parent Institution: Historical Society of Moorestown.
Institution Type/Description: Historic House Museum.
Collections: local history & culture; photographs; period furnishings.
Facilities: Museum-related items for sale.
Hours & Admission Prices: Sept.-June: House Tours & Gift Shop: Sun. 1-4. Library: Tues. & 2nd Sun. each month 1-4. No charge; donations accepted. &

Morganville

NEW JERSEY SCOUT MUSEUM, 705 Ginesi Dr., Morganville, NJ 07751-1235. Tel.: 732-862-1282.
E-mail: president@njsm.org
Web Site: www.njscoutmuseum.org
Founded: 2004.

Key Personnel: Chm. (V), Fred Pachman.
Personnel Profile: Part-Time Paid 1; Part-Time Volunteers 10.
Governing Authority: Tax-exempt.
Institution Type/Description: History Museum.
Collections: girl & boy scout history & memorabilia with concentration on New Jersey; personal artifacts; photographs.
Hours & Admission Prices: Sept.-June Wed. 6pm-8pm; other times by appointment. No charge; donations accepted. &
Membership: Youth $10; Family $25; Century $100; Patron $250; Life $1,000.

Morris Plains

THE STICKLEY MUSEUM AT CRAFTSMAN FARMS, 2352 Rt. 10 W., Manor Lane, Morris Plains, NJ 07950. Tel.: 973-540-0311 & 1165. Fax: 973-540-1167. Facebook: Craftsman Farms.
E-mail: info@stickleymuseum.org
Web Site: www.stickleymuseum.org
Formerly: Craftsman Farms Foundation
Founded: 1989.
Congressional District: 26
Key Personnel: C.E.O. & Exec. Dir., Heather E. Stivison, CFRE; Pres. (V), Barbara Weiskittel; Visitor Svcs. Coord. & Museum Shop Mgr., Betty Wyka; Dir. Education, Vonda Givens; Membership Coord., Linda Blume; Registrar, Bernadette Rubbo; Office Mgr., Susie Traverso; Visitor Svcs. Assoc., Kirsten McCauley.
Personnel Profile: Full-Time Paid 3; Part-Time Paid 5; Part-Time Volunteers 76; Interns 2.
Governing Authority: private; nonprofit organization. Tax-exempt: 501(c)(3).
Institution Type/Description: Historic House & Site: 30 acre National Historic Landmark includes the Craftsman Log House of Gustav Stickley, one of America's leading proponents of the Arts & Crafts movement in the early 1900s.
Collections: furniture & decorative arts from the American Arts & Crafts movement of 1890-1920; emphasis on Stickley's interiors.
Research Fields: life & career of Gustav Stickley & his family; the history of Craftsman Farms; the American Arts & Crafts movement.
Facilities: library of original Craftsman Magazines; 30 acre property. Museum-related contemporary items for sale.
Activities: guided tours; school tours; girl scout badge workshops; lectures; temporary exhibitions; volunteer programs; workshops; outdoor family days. Museum Sponsors: annual artist-in-residence; annual emerging scholars symposium; annual exhibition at Grove Park Inn, Asheville, NC; holiday program in December.
Publications: quarterly newsletter; exhibit catalogs; guidebook; books on the arts & crafts movement.
Hours & Admission Prices: Wed.-Thurs. 12-3, Fri.-Sun. 11-4. Adults $8, seniors $5, children $3; discounts to AAM & THIRTEEN members and North American Reciprocal members; members & children under 6 no charge. Closed major holidays. &
Attendance: 11,108 (accurate)
Membership: Individual $40; Dual & Family $65; Friend $150; Patron $250; Sponsor $500.

Morris Township

FOSTERFIELDS LIVING HISTORICAL FARM, 73 Kahdena Rd., Morris Township, NJ 07960-3524. Tel.: 973-326-7645. Fax: 973-631-5023.
Web Site: www.morrisparks.net
Founded: 1978.
Congressional District: 12
Key Personnel: C.E.O., Dave Helmer; Asst. Dir. Historic Sites, Lynn Laffey; Pres. Park Comm., John Sette; Pres. Friends of Fosterfields, Anne Bukata; Farm Supvr., Rob Kibbe.
Personnel Profile: Full-Time Paid 12; Part-Time Paid 21; Part-Time Volunteers 100.
Governing Authority: county; nonprofit. Parent Institution: Morris County Park Commission, Tel.: 973-326-7600. Tax-exempt: 501(c)(3).
Institution Type/Description: Living Historical Farm & Agricultural Museum: located on a 230-acre farm site.
Collections: turn-of-the-century farm buildings, machinery & tools; furniture & decorative arts; photographs; farm records; diary. Historic Buildings: 1854 Gothic Revival House; 1920s farmhouse.
Research Fields: turn-of-the-century agricultural practices; period clothing.
Facilities: 500-vol. library of agriculture-related books; 70-seat auditorium; classrooms.
Activities: guided tours; lectures; films; hobby workshops; audiovisual presentations; organized education programs; living history programs; docent program; training programs for professional museum workers; participatory & temporary exhibitions. Museum Sponsors: Harvest Activities; Holiday House Tour.

Publications: quarterly newsletter, The Farm & Mill Gazette.
Hours & Admission Prices: Farm: April-July 1 Tues.-Sat. 10-5; July 2-Oct. Wed.-Sat. 10-5, Sun. 12-5. Adults $6, seniors $5, children 4-16 $4; discounts to AAA members; members & children under 3 no charge. &
Attendance: 24,000 (estimated)
Membership: Individual $25; Family $40; Sustaining $100; Supporting $250.

THE GEORGE G. FRELINGHUYSEN ARBORETUM, 353 E. Hanover Ave., Morris Township, NJ 07960-3161. Mailing Address: P.O. Box 1295, Morristown, NJ 07962-1295. Tel.: 973-326-7600. Fax: 973-644-2726. TDD: 1-800-852-7899.
E-mail: info@parks.morris.nj.us
Web Site: www.arboretumfriends.org
Founded: 1970.
Congressional District: 11
Key Personnel: C.E.O., David Helmer; Pres. (V), Judith Schliecher; Treas., Glenn Roe; Dir. Park Maintenance, Ed Vath; Pres. Friends of Arboretum, Sue Acheson; Supt. Education, Lesley Parness; Mgr. Horticulture, John Morse; Dir. Visitors Svcs., Denise Lanza.
Personnel Profile: Full-Time Paid 12; Part-Time Paid 12; Part-Time Volunteers 200; Interns 1.
Governing Authority: county; nonprofit organization. Parent Institution: Morris County Park Commission. Branch Museums: Willowwood Arboretum, Bamboo Brook Outdoor Education Center, P.O. Box 1295, Morristown, NJ 07962-1295. Tax-exempt.
Institution Type/Description: Arboretum: located on 1891 Whippany Farm Residence, summer home of George Griswold Frelinghuysen.
Collections: labeled collections of trees & shrubs. Historic Building: c.1891 Colonial Revival House.
Research Fields: hardy ornamentals recommended for the home landscapes of Northern New Jersey.
Facilities: 2,500-vol. library of botanical & horticultural literature available for research on premises by appointment; botanical garden; education center: 180-seat auditorium; two classrooms.
Activities: guided tours; lectures; films; concerts; docent program; arts festivals; study clubs; hobby workshops; formally organized education programs for children, adults & undergraduate college students; docent program or council; permanent, temporary & traveling exhibitions. Annual Events: plant sale; Gingerbread Wonderland.
Publications: quarterly newsletter, Arboretum Leaves.
Hours & Admission Prices: Grounds: daily 8am-dusk. Education Center: daily 9-4:30. No charge; donations accepted. Closed New Year's Day; Thanksgiving; Christmas. &
Attendance: 400,000 (estimated)
Membership: Friends of Frelinghuysen Arboretum: Individual $25; Family $35; Associate $50; Organizational $75; Affiliate & Supporting $100; Contributing $250; Patron $500; Benefactor $1,000.

Morristown

HISTORIC SPEEDWELL, 333 Speedwell Ave., Ste. A, Morristown, NJ 07960-9384. Tel.: 973-285-6550. Fax: 973-285-6541.
E-mail: msutherland@morrisparks.net
Web Site: www.morrisparks.net
Formerly: Historic Speedwell - Birthplace of the Telegraph
Founded: 1966.
Congressional District: 5
Key Personnel: Pres. (V), Joseph Kane; Site Supt., Mark Sutherland; Administrative Asst., Abagail Eckert; Collections Specialist, Melanie Holster; Program Specialist, Maressa McFarlane; Program Specialist, Joshua Kandebo.
Personnel Profile: Full-Time Paid 5; Part-Time Paid 2; Part-Time Volunteers 65; Interns 2.
Governing Authority: county; nonprofit organization. Parent Institution: Morris County. Subsidiary Institution: Morris County Park Commission. Tax-exempt.
Institution Type/Description: Historic site: 19th century homestead estate.
Collections: wooden patterns from the Speedwell Iron Works; manuscripts; decorative objects; portraits by Samuel F. B. Morse; telegraph instruments; furniture. Historic Buildings: 1844 Vail house; homestead carriage house; 1849 carriage house; Granary Factory where telegraph was first demonstrated; L'Hommedieu house; Estey house; Ford cottage.
Research Fields: genealogy; architecture; decorative arts; industry; transportation; telegraph communication; iron industry.
Activities: guided tours; docent programs; permanent & temporary exhibitions; special events; educational programs.
Publications: books, At Speedwell In the 19th Century; The Speedwell Ironworks: A History of Workers & Work; student activity book.
Hours & Admission Prices: April-June Tues.-Sat. 10-5; July-Oct. Wed.-Sat.

10-5, Sun. 12-5. Adults $4, seniors $3, students & children 4-16 $2; discounts to groups & AAM members; children under 4 no charge. &
Attendance: 6,000 (accurate)
Membership: Student & Senior 65 & over $15; Individual $20; Family $50; Journeyman $100; Molder $250; Founder $500; Ironmaster $1,000.

MACCULLOCH HALL HISTORICAL MUSEUM & GARDENS, (M), 45 Macculloch Ave., Ste. 1, Morristown, NJ 07960-9374. Tel.: 973-538-2404. Fax: 973-538-9428.

E-mail: cfellows@macullochhall.org
Web Site: www.macullochhall.org
Founded: 1950.
Congressional District: 11
Key Personnel: Exec. Dir., Carrie Fellows; Pres. (V), JoAnn Burk; Treas., Sharon Cross; Cur. Collections F.M. Kirby, Ryan Hyman; Educator, Cynthia Winslow; Coord. Events & Programs, Karen Hollywood.
Personnel Profile: Full-Time Paid 3; Part-Time Paid 2; Part-Time Volunteers 45; Interns 1.
Governing Authority: nonprofit organization. Affiliated with W. Parsons Todd Foundation, New York, NY. Tax-exempt: 501(c)(3).
Institution Type/Description: Historic House: 1810 George Macculloch Home.
Collections: 18 & 19th century English & American decorative arts; oriental rugs & china; Macculloch family & Todd family archives; Thomas Nast collection; Morris Canal; archives & library.
Research Fields: Thomas Nast, Miller & Macculloch families.
Facilities: archives; garden.
Activities: guided tours; lectures; permanent & temporary exhibitions; concerts; children's programs.
Publications: brochure; catalog, newsletter.
Hours & Admission Prices: Wed.-Thurs. & Sun. 1-4; groups by appointment. Adults $8, senior citizens & students over 12 $6, children 6-12 $4; discounts to AAM members; children under 5 & members no charge. Closed New Year's Eve & Day; Memorial Day; Independence Day; Labor Day; Thanksgiving & Day after; Christmas Eve, Day & week.
Attendance: 4,000 (accurate)
Membership: Student $15; Individual $35; Family $55; Sponsor $75, Loyalist $100; Federalist $250; Whig $500; Neoclassicist $1,000.

MORRIS COUNTY HISTORICAL SOCIETY (ACORN HALL HOUSE MUSEUM), (M), 68 Morris Ave., Morristown, NJ 07960-4315. Tel.: 973-267-3465. Fax: 973-267-8773.

E-mail: acornhall@juno.com
Web Site: www.acornhall.org
Founded: 1945.
Congressional District: 11
Key Personnel: Dir., Amy Curry; Cur., Susan Orr; Outreach Coord., Karen Ann Kurlander.
Personnel Profile: Full-Time Paid 2; Part-Time Paid 2; Part-Time Volunteers 100; Interns 3.
Governing Authority: society. Tax-exempt: 501(c)(3).
Institution Type/Description: Historical Society Museum: housed in c.1853 Acorn Hall.
Collections: Crane-Hone collection of Victorian furniture; 1853 original furnishings.
Research Fields: 19th-century & mid-Victorian period; Morris County history.
Facilities: 2,500-vol. research library of books on Victorian era; garden. Historic books, maps, postcards & Victorian items for sale.
Activities: guided tours; vintage dancing; lecture series; oral history. Museum Sponsors: Vintage Bazaar; Old Fashioned Fun Day; Tea Parties; Armistice Ball; Victorian Christmas.
Publications: quarterly newsletter, Munsell's History of Morris County, 1739-1882; New Jersey's Revolutionary War Powder Mill; Tours in Historic Morris County; One Hundred Years, One Hundred Miles: The Morris Canal; DVD docudrama, Hard Winter.
Hours & Admission Prices: Mon. & Thurs. 11-4, Sun. 1-4; groups by appointment. Adults $6, senior citizens $5, students $3; discounts to AAM members & active military and their families; children under 12 & members no charge. Closed New Year's Eve & Day; Christmas Eve, Day & week.
Attendance: 3,500 (estimated)
Membership: Student $15; Senior Citizen $20; Individual $30; Family & Institution $50; Contributor $100; Sustaining $250; Sponsor $500; Patron $1,000; Life $2,500.

✳ THE MORRIS MUSEUM, (M), 6 Normandy Heights Rd., Morristown, NJ 07960-4627. Tel.: 973-971-3700. Fax: 973-538-0154. Facebook: Morris Museum.

E-mail: lmoore@morrismuseum.org
Web Site: www.morrismuseum.org
Founded: 1913.

Congressional District: 11
Key Personnel: Exec. Dir. & C.E.O., Linda S. Moore; Chm. (V), John Hemmendinger; C.F.O., Constance Reed; Mgr. Collections, Maria Ribando; Artistic Dir. Bickford Theatre, Eric Hafen; Cur. Guinness Collection, Michelle Marinelli; Museum Shop Mgr., Kathleen Haviland.
Personnel Profile: Full-Time Paid 19; Part-Time Paid 8; Part-Time Volunteers 450; Interns 6.
Volunteer Hours: 15,000
Operating Expenses: 2,380,726
Operating Income: 2,625,002
Governing Authority: nonprofit organization. Subsidiary Institution: The Morris Museum Foundation. Tax-exempt: 501(c)(3).
Institution Type/Description: General Museum.
Collections: anthropology; archaeology; zoology; geology; paleontology; natural history; cultural history; fine & decorative art; folk art; automata; Murtogh D. Guinness mechanical musical instruments.
Major Exhibits: Nano: The Science of the Super Small (T), 10/3/13-9/14; Bears - The Long & Short of It, 11/23/13-3/30/14; Egg-Zibit: The Science, Art & Culture of Eggs, 1/30/14-4/20/14; 125th Anniversary of the National Association of Women Artists, 3/20/14-7/1/14; Contemporary American Indian Pottery, 3/22/14-9/28/14; The Honey & The Hive, 5/3/14-8/15/14; Portraits of Special Olympic Athletes, 8/15/14-11/14; A Dog's Life (Art & Culture), 9/20/14-12/13/14; Bullets, Bows & Blades: Weapons from the Collection of the Morris Museum, 10/16/14-2/22/15.
Facilities: 312-seat theatre; classrooms; rental facilities.
Activities: lectures; films; gallery talks; concerts; dance performances; arts festivals; drama; extension talks to schools & hospitals; mineral astronomy & astronomy clubs; formally organized education programs for children, adults & undergraduate college students; docent program; permanent, traveling & temporary exhibitions; school loan service; inter-museum loan; rentals.
Publications: quarterly program brochures; exhibition catalogues; student & teacher program bulletin.
Hours & Admission Prices: Tues.-Sat. 11-5, 2nd & 3rd Thurs. 11-8, Sun. 12-5. Adults $10, senior citizens & children 3-12 $7; discounts to AAM members & employees of other museums with ID; Thurs. 5-8, children under 3 & members no charge. Closed major holidays. &
Attendance: 150,000 (accurate)
Membership: Student $30; Teachers & Seniors $45; Individual $50; Couple/Family/Grand Family $70; Patron $165; Benefactor $350; Sponsor $650; Director's Circle $1,500.

MORRISTOWN NATIONAL HISTORICAL PARK, 30 Washington Place, Morristown, NJ 07960-4299. Tel.: 973-539-2016. Fax: 973-539-8361. TDD: 973-539-5072.

E-mail: jude_pfister@nps.gov
Web Site: nps.gov/morr
Founded: 1933.
Congressional District: 12
Key Personnel: Supt., Jill A. Hawk; Chief Interpretation, Justin Monetti; Museum Specialist, Joni Rowe; Chief Cultural Resources, Dr. Jude M. Pfister.
Personnel Profile: Full-Time Paid 30; Part-Time Volunteers 10; Interns 6.
Governing Authority: federal. Parent Institution: National Park Service, U.S. Dept. of the Interior, Washington, DC 20240. Branch Units: Washington's Headquarters; Jockey Hollow; Fort Nonsense; New Jersey Brigade. Tax-exempt.
Institution Type/Description: Historic Site: 1777, 1779-1782 winter encampment sites of the Continental Army; 1779-1780 Washington's Headquarters.
Collections: preserved archaeological sites; 18th-century military weapons & equipment; decorative arts; paintings; furnishings; archaeology; rare books & manuscripts; reconstructed soldier huts; Revolutionary War. Historic Houses: 1772 Ford Mansion (Washington's Headquarters); 1750 Wick Farm; c.1770 Guerin House.
Research Fields: 18th-century military and domestic life, especially that relating to Morristown & the Revolutionary War in New Jersey; George Washington & the Continental Army; archaeology; early American decorative arts & 18th-century imprints; cultural & political aspects of western European history, 1550-1900.
Facilities: 50,000-vol. research library; 160-seat auditorium; visitor center; 24 miles of hiking trails & historic traces with interpretive waysides covering 1,700 acres of park land. Books for sale.
Activities: guided tours; lectures; organized educational programs for groups; audiovisual programs; living history demonstrations; nature & history walks; special events.
Publications: handbook, Morristown NHP; books, A Certain Splendid House; Morristown: The War Years, 1777-1780; booklets; special event bulletins; research monographs; guide to manuscript collections.

Hours & Admission Prices: Daily 9-5. Adults $4; children under 16 no charge. Annual Family Pass: $15 Closed New Year's Day; Thanksgiving; Christmas. &

Attendance: 200,000 (estimated)

SCHUYLER-HAMILTON HOUSE, 5 Olyphant Place, Morristown, NJ 07960-4231. Tel.: 973-267-4039. Fax: 973-539-7502.
E-mail: abren8527@aol.com
Web Site: www.co.morris.nj.us/mchc/directory-museums.html#schuyler
Founded: 1923.
Congressional District: 11
Key Personnel: Dir., Patricia Sanftner; Co-1st Vice Regent, Kathleen Cruger; Co-2nd Vice Regent, Mariane Browne.
Personnel Profile: Part-Time Volunteers 22.
Governing Authority: society. Parent Institution: National Society Daughters of the American Revolution. Subsidiary Institution: Morristown Chapter. Tax-exempt: 501(c)(3) & 216-018-192.
Institution Type/Description: Historic House Museum: 1760 home of Dr. Jabez Campfield, Revolutionary War army doctor. Site of Alexander Hamilton's engagement to Elizabeth Schuyler.
Collections: 1720-1850 period furnishings; engravings.
Research Fields: herb garden.
Facilities: 30-vol. library of Revolutionary period books & patriot's index available for use on premises; kitchenette; small garden. Dolls, toy cannons, postcards & notecards for sale.
Activities: guided tours; lectures; permanent exhibitions; society meetings open to the public; Girl Scout programs. Museum Sponsors: Free Tour Days.
Publications: house history & garden pamphlet; DAR Spirit Magazine (National Society DAR).
Hours & Admission Prices: Sun. 2-5; other times by appointment. Adults $5, children 12 to high school $2.50; children under 12 no charge.
Attendance: 1,000 (estimated)
Membership: Associate $18; D.A.R. Members $40.

Mount Holly

HISTORIC BURLINGTON COUNTY PRISON MUSEUM, 128 High St., Mount Holly, NJ 08060-1402. Mailing Address: Burlington County Division of Parks, P.O. Box 6000, Mount Holly, NJ 08060. Tel.: 609-265-5858 & 5476 (museum). Fax: 609-265-5797.
E-mail: parks@co.burlington.nj.us
Web Site: www.prisonmuseum.net
Founded: 1966.
Congressional District: 3
Key Personnel: Pres. (V) Historic Burlington County Prison Museum Assn., Janet L. Sozio; Supt. Burlington County Division of Parks, John H. Smith, Jr.; Site Attendant, Marisa Bozarth.
Personnel Profile: Full-Time Paid 12; Part-Time Paid 1; Part-Time Volunteers 7.
Governing Authority: county. A body politic & corporate of the State of New Jersey. Tax-exempt: 501(c)(3).
Institution Type/Description: Historic Building: 1810 Burlington County Prison. A National Historic Landmark.
Collections: items related to penology, prisons & Robert Mills, architect.
Facilities: library of penology books available for inter-library loan & on premises.
Activities: guided tours; lectures; permanent & temporary exhibitions.
Publications: pamphlets.
Hours & Admission Prices: Thurs.-Sat. 10-4, Sun. noon-4. Adults $4, seniors & students $2; children under 5 no charge. Closed legal holidays. &
Attendance: 5,000 (accurate)
Membership: Individual $10.

JOHN WOOLMAN MEMORIAL, 99 Branch St., Mount Holly, NJ 08060-1866. Tel.: 609 267-3226.
Web Site: www.woolmancentral.com
Founded: 1915.
Congressional District: 6
Key Personnel: Dir., Jack Walz; Dir., Carol Walz.
Governing Authority: denominational group. Tax-exempt: 501(c)(3).
Institution Type/Description: Historic Building & Site: 1783 John Woolman Memorial.
Collections: artifacts.
Activities: guided tours.
Publications: annual report.
Hours & Admission Prices: Call for hours. No charge, donations requested. &
Attendance: 2,500 (estimated)

MOUNT HOLLY HISTORICAL SOCIETY - SHINN-CURTIS LOG HOUSE, 23 Washington St., Mount Holly, NJ 08060. Tel.: 609-267-2773.
Institution Type/Description: Historical Society Museum: housed in the former home of Thomas Shinn; c.1712.
Collections: local history & culture; period furnishings; photographs.
Hours & Admission Prices: Call for hours.

THE OLD SCHOOLHOUSE, 35 Brainerd St., Mount Holly, NJ 08060. Mailing Address: 180 Burrs Rd., Westampton, NJ 08060. Tel.: 609-267-6996.
E-mail: colonialdamesnj@comcast.net
Web Site: www.colonialdamesnj.org
Founded: 1892.
Congressional District: 3
Personnel Profile: Part-Time Paid 2.
Governing Authority: Parent Institution: National Society Colonial Dames of America in the State of NJ. Tax-exempt.
Institution Type/Description: Historic Building: housed in a former schoolhouse; built in 1756; listed on National Register of Historic Places.
Collections: local history & culture; period furnishings; personal artifacts; photographs.
Activities: educational programs; school tours.
Hours & Admission Prices: By appointment. &

PEACHFIELD PLANTATION, 180 Burrs Rd., Mount Holly, NJ 08060. Tel.: 609-267-6996.
E-mail: colonialdamesnj@comcast.net
Web Site: www.colonialdamesnj.org
Founded: 1892.
Congressional District: 3
Personnel Profile: Part-Time Paid 2.
Governing Authority: Parent Institution: Colonial Dames of NJ. Tax-exempt.
Institution Type/Description: Historic House Museum: listed on the National Register of Historic Places.
Collections: local history & culture; period furnishings; personal artifacts; photographs.
Activities: special events.
Hours & Admission Prices: By appointment. &

Mount Laurel

PAWS FARM NATURE CENTER, 1105 Hainesport Rd., Mount Laurel, NJ 08054. Tel.: 856-778-8795.
Institution Type/Description: Nature Center.
Collections: local history & environment; wildlife & their habitats
Activities: educational classes & programs; story hour; birthday parties; special events; hands-on exhibitions; group tours.
Hours & Admission Prices: Wed.-Sun. 10-4. Adults $6, children $4; children under one no charge.

Mountainside

DEACON ANDREW HETFIELD HOUSE, Constitution Plaza, Mountainside, NJ 07092. Mailing Address: Mountainside Historical Preservation Committee, 1385 Rte. 22 E., Mountainside, NJ 07092-2605. Tel.: 908-687-3636.
E-mail: info@mountainsidehistory.org
Institution Type/Description: Historic House Museum: built in 1760. Listed on the National Register of Historic Places.
Collections: local history & culture; photographs; period artifacts.
Hours & Admission Prices: March-May & Sept.-Oct. 3rd Sun. of month 1-3.

TRAILSIDE NATURE AND SCIENCE CENTER, 452 New Providence Rd., Mountainside, NJ 07092-1409. Tel.: 908-789-3670. Fax: 908-789-3270.
E-mail: pbertsch@ucnj.org
Web Site: www.ucnj.org/trailside
Founded: 1941.
Congressional District: 7
Key Personnel: Dir., Patricia Bertsch; Asst. Dir., Karen Inzillo; Museum Shop Mgr., Lenore Mangan.
Personnel Profile: Full-Time Paid 7; Part-Time Paid 15; Part-Time Volunteers 5.
Governing Authority: county. Parent Institution: Union County Department of Parks & Community Renewal. Tax-exempt: 170(b)(1)(A).
Institution Type/Description: Natural Science Museum.
Collections: snakes; turtles; frogs; aquatic life; minerals; fluorescent minerals;

Watchung Reservation geology; taxidermy birds & mammals; Native American artifacts; hands-on exhibits.
Facilities: library; classrooms; 4,500 sq. ft. exhibit space; 13 miles of hiking trails; theater; children's discovery room.
Activities: guided tours; lectures; children, family & adults workshops; teacher training workshops; school & community programs; boy & girl scout programs; birthday parties.
Hours & Admission Prices: Visitor Center: daily 12-5. No charge; donations accepted. Closed New Year's Day; Easter; Independence Day; Thanksgiving & day after; Christmas Eve & Day. &
Attendance: 50,000 (estimated)

Mullica Hill

OLD TOWN HALL MUSEUM - THE HARRISON TOWNSHIP HISTORICAL SOCIETY, 62-64 S. Main St., Mullica Hill, NJ 08062. Mailing Address: P.O. Box 4, Mullica Hill, NJ 08062. Tel.: 856-478-4949.
Web Site: www.harrisonhistorical.com
Founded: 1971.
Congressional District: 2
Personnel Profile: Part-Time Volunteers 56.
Governing Authority: Tax-exempt.
Institution Type/Description: Historical Society Museum.
Collections: local history & culture; period furnishings; personal artifacts; photographs.
Publications: newsletter, Milestones.
Hours & Admission Prices: Call for hours. No charge; donations accepted.
Attendance: 750 (estimated)
Membership: Student & Senior 55 & over $5; Individual $10; Supporting $25; Family & Household $35.

National Park

RED BANK BATTLEFIELD PARK & JAMES AND ANN WHITALL HOUSE, 100 Hessian Ave., National Park, NJ 08063. Mailing Address: P.O. Box 337, Woodbury, NJ 08096. Tel.: 856-853-5120. Fax: 856-853-0950. Facebook: Whitall House.
E-mail: jjanofsky@co.gloucester.nj.us
Web Site: www.whitall.org
Personnel Profile: Part-Time Paid 1; Part-Time Volunteers 50.
Governing Authority: Parent Institution: Glocester County Parks & Recreation.
Institution Type/Description: History Museum: housed in the home of James & Ann Whitall which also served as a Revolutionary War patriot headquarters and a field hospital; built in 1748. Listed on the National Register of Historic Places.
Collections: local history & culture; period furnishings; personal artifacts; kitchen garden.
Facilities: picnic pavilions
Activities: educational programs.
Hours & Admission Prices: House: April-Sept. Wed.-Sun. 1-4; Oct.-March Wed.-Fri. 9-12 & 1-4. no charge; donations accepted. &
Attendance: 6,000 (accurate)

New Brunswick

AMERICAN HUNGARIAN FOUNDATION/MUSEUM OF THE AMERICAN HUNGARIAN FOUNDATION, 300 Somerset St., New Brunswick, NJ 08901-2248. Mailing Address: P.O. Box 1084, New Brunswick, NJ 08903-1084. Tel.: 732-846-5777. Fax: 732-249-7033.
E-mail: info@ahfoundation.org
Web Site: www.ahfoundation.org
Formerly: American Hungarian Foundation/Hungarian Heritage Center Museum
Founded: 1954.
Congressional District: 6
Key Personnel: Pres. & Acting Dir., Gergely Hajdu-Nemeth; Co Chm. Bd. (V), Zsolt Harsanyi; Co Chm. Bd. (V), August J. Molnar; Treas., Scott B. Lukacs; Librarian, Margaret Papai.
Governing Authority: nonprofit organization. Tax-exempt: 501(c)(3).
Institution Type/Description: Cultural Heritage Center.
Collections: art & artists born in Hungary who later worked in the United States; history of Hungarian immigration to the United States from colonial to modern times; Hungarian contributions to American life; art; coins; weapons; decorative arts; glass; photographs; archives; Hungarian folk art, furniture, textiles, & ceramics.
Research Fields: history of Hungarian immigration; history of Hungarian contributions to American life; Hungarian art.

Facilities: 60,000-vol. library of books dealing with Hungarian history, art & culture, available for inter-library loan & use by public. Museum-related items for sale.
Activities: guided tours; lectures; films; concerts; hobby workshops; organized education programs; loan, traveling & temporary exhibitions; school loan service.
Hours & Admission Prices: Tues.-Sat. 11-4, Sun. 1-4. Suggested Donation: $5. Closed New Year's Day; Easter; Memorial Day; Independence Day; Labor Day; Thanksgiving; Christmas. &
Attendance: 3,000 (estimated)
Membership: Associate $25; Sustaining $40; Participating $65; Friend $100; Donor $500; Sponsor $1,000; Patron $2,500; Benefactor $5,000.

BUCCLEUCH MANSION, Buccleuch Park, 800 George St. (private driveway), (enter park from Easton Ave.), New Brunswick, NJ 08901. Mailing Address: Jersey Blue Chapter NSDAR, P.O. Box 27, New Brunswick, NJ 08903. Tel.: 732-745-5094.
E-mail: jkgennaro@aol.com
Founded: 1915.
Congressional District: 6
Key Personnel: Cur., Judy Gennaro.
Governing Authority: municipal; society. Mansion & park owned by city of New Brunswick, NJ; Affiliated with Jersey Blue Chapter, Daughters of the American Revolution, New Brunswick NJ. Tax-exempt.
Institution Type/Description: Historic House: 1739 three-story Colonial house, located on the site of 1776 Revolutionary War fortifications.
Collections: 18th-century to late Victorian period rooms; c.1815 Du Four wallpaper; furniture of N.J. cabinetmakers; crafts items, including spinning wheels, carders, metal dress patterns; toys; costumes; textiles.
Research Fields: research related to house and its previous occupants.
Facilities: 80-acre park.
Activities: guided tours; permanent & temporary exhibitions.
Publications: brochure.
Hours & Admission Prices: June-Oct. last of month Sun. 1-4; groups by appointment year round. No charge; donations accepted. &
Attendance: 1,500 (estimated)

CENTER FOR LATINO ARTS & CULTURE, 122 College Ave., New Brunswick, NJ 08901-1165. Tel.: 732-932-1263. Fax: 732-932-1589.
Web Site: clac.rutgers.edu
Key Personnel: Dir., Carlos Fernandez, Ph.D.; Program Coord., Silismar Suriel
Institution Type/Description: Latino Arts & Culture.
Collections: Latino, Hispanic, Caribbean, & Latin American arts & culture.
Hours & Admission Prices: Mon.-Fri. 9-5; other times by appointment.

HENRY GUEST HOUSE, 58 Livingston Ave., New Brunswick, NJ 08901-2521. Mailing Address: 60 Livingston Ave., New Brunswick, NJ 08901-2520. Tel.: 732-745-5108. Fax: 732-846-0226.
E-mail: nbfpl@lmxac.org
Founded: 1760.
Congressional District: 17
Key Personnel: Dir., Robert Belvin.
Governing Authority: municipal. Parent Institution: New Brunswick Public Library, New Brunswick. Tax-exempt.
Institution Type/Description: Historic House: 1760 Henry Guest House. Listed on the National Register of Historic Places.
Collections: local history & culture; period furnishings; personal artifacts.
Activities: guided tours; permanent exhibitions.
Hours & Admission Prices: By appointment. No charge. &
Attendance: 2,000 (estimated)

JEWISH HISTORICAL SOCIETY OF CENTRAL JERSEY, 222 Livingston Ave., New Brunswick, NJ 08901. Tel.: 732-249-4894.
E-mail: info.jhscj@gmail.com
Web Site: www.jewishgen.org
Institution Type/Description: Jewish History Museum.
Collections: Jewish history, culture & heritage; documents; records; period artifacts; photographs; oral histories; Jewish newspapers.
Facilities: archives.
Activities: educational programs; research.
Hours & Admission Prices: Mon.-Fri. 9-1

RUTGERS GARDENS, 112 Ryders Ln., New Brunswick, NJ 08901-8519. Tel.: 732-932-8451. Fax: 732-932-7060.
E-mail: rugardens@aesop.rutgers.edu
Web Site: rutgersgardens.rutgers.edu

Founded: 1927.
Congressional District: 6
Key Personnel: Dir., Bruce Crawford; Volunteer Coord., Mary Ann McMillan; Supt., Matthew Jamicky.
Personnel Profile: Full-Time Paid 4; Part-Time Paid 1; Part-Time Volunteers 70; Interns 4.
Governing Authority: university. Parent Institution: Rutgers-The State University of NJ., New Brunswick, NJ 08903. Tel.: 908-932-9325. Subsidiary Institution: Cook College. Tax-exempt.
Institution Type/Description: Research Arboretum.
Collections: cultivars of Ilex opaca & Cornus Florida; shade tree & small tree collection; rhododendron & azalea garden; garden for sun & shade; Donald B. Lacey Display garden; Ella Quimby water conservation garden; evergreen & shrub garden; bamboo forest; Rain garden.
Research Fields: plant breeding; ornamental trees & shrubs.
Facilities: botanical garden; field research station.
Activities: guided tours; formally organized education programs for undergraduate college students; professional students; elementary, secondary & pre-school students; general public.
Publications: newsletter.
Hours & Admission Prices: Daily 8:30 am-dusk. No charge; donations accepted.
Attendance: 20,000 (estimated)
Membership: Student $10; Friend $40; Family $65; Educator $75; Sponsor $150; Benefactor $500; Patron $1,000.

RUTGERS UNIVERSITY GEOLOGY MUSEUM, Geology Hall, Old Queen Campus, 85 Somerset St., New Brunswick, NJ 08901. Tel.: 848-932-4243.

E-mail: museum@rci.rutgers.edu
Web Site: geologymuseum.rutgers.edu
Key Personnel: Assoc. Dir., Lauren Neitzke Adamo; Assoc. Dir., Patricia Irizarry-Barreto
Institution Type/Description: Geology Museum.
Collections: geology; natural history; anthropology; dinosaur trackway; mounted mastodon; rocks & minerals; florescent minerals.
Facilities: Gift items for sale.
Activities: special events.
Hours & Admission Prices: Tues.-Thurs. 10-6, Fri. 10-4, Sat. 10-2; other times by appointment. No charge.

RUTGERS UNIVERSITY - MASON GROSS ART GALLERIES, 33 Livingston Ave., New Brunswick, NJ 08901. Tel.: 732-932-2222, ext. 798. Fax: 732-932-2217.

Web Site: www.masongross.rutgers.edu/visarts/arts_gallery.html
Institution Type/Description: Art Gallery.
Collections: paintings; sculpture.
Hours & Admission Prices: Call for hours.

ZIMMERLI ART MUSEUM AT RUTGERS UNIVERSITY, (M), Rutgers, The State University of New Jersey, 71 Hamilton St., New Brunswick, NJ 08901-1248. Tel.: 848-932-7237. Fax: 732-932-8201.

E-mail: press@zimmerli.rutgers.edu
Web Site: www.zimmerlimuseum.rutgers.edu
Formerly: Jane Voorhees Zimmerli Art Museum
Founded: 1966.
Congressional District: 17
Key Personnel: Dir., Suzanne Delehanty; Chm. (V), Lisa Pretecrum; Mgr. Publications & Communications, Stacy Smith; Mgr. Business Affairs, Bernadette Clapsis; Visiting Specialist in Museum Education & Audience Engagement, Michael Norris; Operations, Security & Visitor Svcs. Mgr., Edward Schwab; Registrar, Leslie Kriff; Dir. Morse Research Center for Graphic Arts and Cur. Prints & Drawings, Marilyn Symmes; Andrew W. Mellon Liaison for Academic Programs & Cur., Donna Gustafson; Assoc. Cur. European Art, Christine Giviskos; Assoc. Cur. Russian & Soviet Nonconformist Art, Julia Tulovsky; Research Cur. Soviet Nonconformist Art, Jane A. Sharp; Dir. Devel., Whitney Prendergast.
Personnel Profile: Full-Time Paid 25; Part-Time Paid 58; Part-Time Volunteers 60; Interns 8.
Volunteer Hours: 3,163
Operating Expenses: 4,123,764
Operating Income: 4,123,764
Governing Authority: state. Parent Institution: Rutgers, The State University of New Jersey. Tax-exempt: 501(c)(3) & 170(b)(1)(A).
Institution Type/Description: Art Museum.
Collections: 18th to 21st century American art; European art from antiquities to early 20th century; 14th to 21st century Russian art including Soviet nonconformist art; rare books; Japonisme.

Major Exhibits: Norton and Nancy Dodge Collection of Nonconformist Art from the Soviet Union, 4/30/12-12/14; Meiji Photographs: A Historic Friendship between Japan and Rutgers, 9/3/13-7/14; Diane Burko: Glacial Perspectives, 9/4/13-7/14; A Gift in Honor of Tyler Clementi: Dale Chihuly's Rivera Blue Macchia Chartreuse Lip Wrap, 9/4/13-7/14; Staging Symbolism: Programs for the Theatre de l'Oeuvre in Paris, 9/14/13-2/2/14; Stars: Contemporary Prints by Derriere L'Etoile Studio Part Two, 1990s, 10/5/13-3/2/14; Artists' Portraits: Putting a Face to the Name, 10/19/13-4/6/14; Striking Resemblance: The Changing Art of Portraiture, 1/25/14-7/13/14; Never Such Innocence Again: Picturing the Great War in French Prints and Drawings, 2/8/14-7/14; Stars: Contemporary Prints by Derriere L'Etoile Studio Part Three, 2000-Present, 3/8/14-7/14; Odessa's Second Avant-Garde: City and Myth, 4/12/14-10/19/14; Norton and Nancy Dodge Collection of Nonconformist Art from the Soviet Union, 4/30/12-12/14; Maples in the Mist: Chinese Poems for Children Illustrated by Jean & Mou-sien Tseng, 7/10/13-6/22/14.
Research Fields: 19th-century French art including prints, rare books, & Japonisme; Russian & Soviet nonconformist art; American art notably works on paper; related to the collection & interdisciplinary themes.
Facilities: 3,000-vol. library of French fin de siecle illustrated books & periodicals; archives including 50,000 items created by Soviet nonconformist artists.
Activities: guided tours; programs for children & teens; K-12 teacher workshops; programs for Rutgers undergraduate & graduate students; international inter-museum loan & temporary exhibitions.
Publications: annual report; annual program guide; exhibition catalogs; ebooks.
Hours & Admission Prices: Sept.-July Tues.-Fri. 10-4:30, Sat.-Sun. 12-5, first Wed. each month 10-9. Adults 18 & over $6, seniors over 65 $5; discounts to AAM, ICOM & AAMD members; Rutgers staff & students, children under 18, members & 1st Sun. each month no charge. Closed major holidays; New Year's Day; Independence Day; Thanksgiving & day after; Christmas Eve & Day. &
Attendance: 40,000 (accurate)
Membership: Student $25; Individual $50; Dual $85; Family $100; Associate $250; Patron $500; Curator's Circle $1,000; Director's Gallery $2,500; Hamilton Society $5,000.

New Providence

SALTBOX MUSEUM, 1350 Springfield Ave., New Providence, NJ 07974. Mailing Address: Mason Rm., 377 Elkwood Ave., New Providence, NJ 07974-1837. Tel.: 908-665-1034. Facebook: New Providence Historical Society.

E-mail: newprovhistorical@gmail.com
Web Site: www.newprovidencehistorical.com
Founded: 1966.
Congressional District: 7
Key Personnel: Pres. (V), Linda J. Kale.
Personnel Profile: Part-Time Volunteers 20.
Governing Authority: Parent Institution: New Providence Historical Society. Tax-exempt.
Institution Type/Description: Historic House Museum.
Collections: period artifacts & furnishings.
Research Fields: real estate.
Activities: Museum Sponsors: Antique & Collectibles Appraisal in March; Craft Sale in Spring; trips pertaining to history.
Publications: semiannual: Turkey Tracks.
Hours & Admission Prices: March-Nov. 1st & 3rd Sun. 1-3. No charge; donations accepted.
Attendance: 250 (estimated)
Membership: Individual $15; Family $20; Contribution $30; Sustaining $50; Life $125.

New Vernon

TUNIS-ELLICKS HISTORIC HOUSE AND MUSEUM, 16 Village Rd., New Vernon, NJ 07976. Mailing Address: P.O. Box 1777, New Vernon, NJ 07976. Tel.: 973-292-3661.

E-mail: hardinghist@comcast.net
Founded: 1977.
Congressional District: 11
Personnel Profile: Part-Time Volunteers 15.
Governing Authority: Parent Institution: Harding Township Historical Society. Tax-exempt.
Institution Type/Description: Historic House Museum.
Collections: local history & culture; photographs; aerial maps.
Publications: biannual, The Review.
Hours & Admission Prices: By appointment. No charge.

Newark

ALJIRA, A CENTER FOR CONTEMPORARY ART, 591 Broad St., Newark, NJ 07102-4403. Tel.: 973-622-1600. Fax: 973-622-6526.
E-mail: info@aljira.org
Web Site: www.aljira.org
Key Personnel: Exec. Dir., Victor L. Davson
Institution Type/Description: Art Center.
Collections: works of emerging & under-represented artists.
Hours & Admission Prices: Wed.-Fri. 12-6, Sat. 11-4.

CITY WITHOUT WALLS (CWOW), 6 Crawford St., Newark, NJ 07102. Tel.: 973-622-1188. Fax: 973-622-2941.
Web Site: www.cwow.org
Key Personnel: Interim Exec. Dir., Patricia Huizing; Gallery & Prog. Mgr., Jacqueline Cruz.
Governing Authority: nonprofit organization.
Institution Type/Description: Art Gallery.
Collections: works by emerging contemporary artists.
Activities: outreach programs; temporary & traveling exhibitions.
Hours & Admission Prices: Wed.-Sat. 12-6; other times by appointment.

THE JEWISH MUSEUM OF NEW JERSEY, Congregation Ahavas Sholom, 145 Broadway, Newark, NJ 07104. Tel.: 973-485-2609.
E-mail: info@jewishmuseumnj.org
Web Site: jewishmuseumnj.org
Institution Type/Description: Jewish History Museum.
Collections: Jewish history, heritage & culture; photographs; sculptures; paintings; drawings.
Hours & Admission Prices: Sun. 1-5; other times by appointment.

THE NEW JERSEY HISTORICAL SOCIETY, 52 Park Place, Newark, NJ 07102-4302. Tel.: 973-596-8500. Fax: 973-596-6957.
E-mail: contactnjhs@jerseyhistory.org
Web Site: www.jerseyhistory.org
Founded: 1845.
Congressional District: 10
Key Personnel: Chm. (V), John G. Zinn; Pres. & C.E.O., Linda Epps; Collections Mgr., Timothy Decker; Dir. Finance & Admin, Steve Tettamanti; Cur. Manuscripts & Acting Dir. Library, Julia Telonidis; Cur. Education & Programs, Margaret Renn.
Personnel Profile: Full-Time Paid 21; Part-Time Paid 2; Part-Time Volunteers 40; Interns 9.
Governing Authority: society. Tax-exempt: 501(c)(3).
Institution Type/Description: History Museum & Library.
Collections: N.J. & American history; transportation; industry; costumes; paintings; prints; drawings; furniture; glass; ceramics; silver; textiles; household utensils; jewelry; coins; sculpture; Indian artifacts; photographs; N.J. maritime material; c.1790-1815.
Research Fields: history; art history; graphics; decorative & fine arts; architecture; genealogy.
Facilities: 65,000-vol. library of books; 1,600 manuscript groups; maps; rare books; pamphlets; broadsides & newspapers pertaining to New Jersey history & genealogy available for use on premises; reading room; 73-seat auditorium. Books, periodicals & museum-related items for sale.
Activities: guided tours; lectures; films; gallery talks; education programs; docent program; inter-museum loan, temporary exhibitions; seminars; community programs.
Publications: biannual journal, New Jersey History; newsletter, New Jersey Historical Society News; occasional books & catalogs.
Hours & Admission Prices: Museum: Tues.-Sat. 10-5. Library: Wed.-Thurs. & Sat. 12-5; school & group tours by appointment. Suggested Donation: $4. Closed major holidays.
Attendance: 20,000 (estimated)
Membership: Individual Educator $25; Individual $30; Library & Nonprofit $45; Household $60; Supporter $100; Friend $250; Patron $500; Benefactor $750; President's Circle $1,500.

✳   **THE NEWARK MUSEUM, (M),** 49 Washington St., Newark, NJ 07102-3176. Tel.: 973-596-6550. Fax: 973-642-0459. TDD: 973-596-6355.
Web Site: www.newarkmuseum.org
Founded: 1909.
Congressional District: 10
Key Personnel: Chm. Bd. (V), Arlene Lieberman; Pres., Andrew Richards; Pres. of Volunteers, Joyce Hamlin; Dir., Mary Sue Sweeney Price; Deputy Dir. Finance & Administration, Meme Omoqbai; Deputy Dir. Mktg., Mark Albin; Deputy Dir. Devel., Peggy Dougherty; Cur. Asia & the Pacific, Dr. Katherine Anne Paul; Cur. Decorative Arts, Ulysses G. Dietz; Cur. Africa & the Americas, Dr. Christa Clarke; Assoc. Cur. American Art, Dr. Holly Pyne Connor; Dir. Science Dept., Dr. Ismael Calderon; Cur. Natural Sciences, Dr. Sule Oygur; Astronomer, Kevin Conod; Deputy Dir. Collections, Rebecca A. Buck; Deputy Dir. Education, Ted Lind; Dir. School Educational Program, Shirley Thomas; Dir. Adult Education, Linda Nettleton; Dir. Innovative Education, Kim Robledo-Diga; Dir. Exhibits, Tim Wintemberg; Mgr. Visitor Svcs., David May; Mgr. Arts Workshop, Stephen J. McKenzie; Archivist, Jeffrey Moy; Librarian, Dr. William A. Peniston; Museum Shop Mgr., Lorelei Rowars; Public Rels., Allison McCartney; Dir. Special Events, Carol Blunda; Cur. American Art, Mary Kate O'Hare.
Personnel Profile: Full-Time Paid 104; Part-Time Paid 210; Part-Time Volunteers 190; Interns 15.
Governing Authority: nonprofit organization. Parent Institution: The Newark Museum Association. Tax-exempt: 501(c)(3).
Institution Type/Description: Art & Science Museum.
Collections: American art 18th- to 21st-century; 19th, 20th & 21st century decorative arts; Asian art, esp. Tibetan; ancient art: Egyptian, Greek, Coptic & Roman; numismatics; African art; Native American art; Pacific Islands art; natural sciences; planetarium. Historic Houses: 1784 schoolhouse, Ward Carriage House with Newark Fire Museum; 1885 Ballantine House (National Historic Landmark).
Research Fields: art; history & ethnology; natural sciences; numismatics.
Facilities: 50,000-vol. library of art & science reference available for inter-library loan & on premises; planetarium; garden; auditorium; cafe. Gift items for sale.
Activities: guided tours; lectures; films; gallery talks; concerts; dance recitals; organized education programs; docent program; adult art school; inter-museum loan; permanent & temporary exhibitions; volunteer organization.
Publications: quarterly magazine, Access: The Newark Museum; exhibitions catalogues.
Hours & Admission Prices: Wed.-Sun. 10-5. Suggested Donation: adults $7; discounts to AAM members; members no charge. Public planetarium performances: adults $4, children under 12 $2. Closed New Year's Eve & Day; Independence Day; Thanksgiving; Christmas.
Attendance: 948,663 (accurate)
Membership: Individual $50; Family $60; Sustaining $150; Benefactor $350; Sponsor $750; Director's Circle $1,500.

PAUL ROBESON GALLERIES-RUTGERS UNIVERSITY, Rutgers University - Newark, 350 Dr. Martin Luther King, Jr. Blvd., Paul Robeson Campus Center, Newark, NJ 07102. Tel.: 973-353-1610. Fax: 973-353-5912.
E-mail: galleryr@andromeda.rutgers.edu
Web Site: www.andromeda.rutgers.edu/artgallery
Founded: 1979.
Key Personnel: Dir. & Cur., Anonda Bell.
Personnel Profile: Full-Time Paid 1; Part-Time Paid 1; Part-Time Volunteers 2; Interns 2.
Governing Authority: Parent Institution: Rutgers University. Tax-exempt.
Institution Type/Description: Art Gallery.
Collections: photography; paintings & drawings.
Hours & Admission Prices: Mon.-Thurs. 10-5, 1st Sat. each month 12-5. No charge; donations accepted.
Attendance: 8,588 (accurate)

Newfield

MATCHBOX ROAD MUSEUM, 15 Pearl St., Newfield, NJ 08344-2603. Mailing Address: P.O. Box 977, Newfield, NJ 08344. Tel.: 856-697-6900.
E-mail: mbroad@aol.com
Web Site: mbroad.com
Founded: 1992.
Key Personnel: Everett Marshall, III
Institution Type/Description: Toy Museum.
Collections: 40,000 miniature Matchbox vehicles including cars, tractors, double-decker buses, milk wagons, garbage trucks, moving vans, horse carriers, taxis & oil trucks.
Facilities: Museum-related items for sale.
Hours & Admission Prices: By appointment. No charge; donations accepted.

Newton

NEWTON FIRE DEPARTMENT MUSEUM, 150 Spring St., Newton, NJ 07860-2009. Tel.: 973-383-0396.
E-mail: director@newtonfiremuseum.org

Web Site: www.newtonfiremuseum.org
Key Personnel: Dir., Daniel Finkle.
Governing Authority: nonprofit organization. Tax-exempt: 501(c)(3).
Institution Type/Description: Fire-Fighting Museum: housed in an historic firehouse built in 1891.
Collections: firefighting equipment from the late 1800s to present.
Hours & Admission Prices: May-Oct. Thurs. 5pm-8pm, Sat. 11-3; groups by appointment.

SUSSEX COUNTY HISTORICAL SOCIETY, 82 Main St., Newton, NJ 07860. Mailing Address: P.O. Box 913, Newton, NJ 07860-0913. Tel.: 973-383-6010. Fax: 973-383-6010.
E-mail: schs@tellurian.net; sussexcountyhs@gmail.com
Founded: 1904.
Congressional District: 12
Key Personnel: Pres., Dennis Pegg; Vice Pres., Peter Chletsos; Sec., Wayne McCabe; Treas., Ruth Ann Whitesell; Corresponding Sec., Linda Rienecker.
Personnel Profile: Full-Time Volunteers 4; Part-Time Volunteers 3.
Governing Authority: society. Tax-exempt.
Institution Type/Description: History Museum.
Collections: local history; American Indian artifacts; archaeology; mastodon; genealogy.
Facilities: 1,000-vol. library of genealogical & local history books available on premises.
Publications: Society Newsletter; books & pamphlets written by members; quarterly, Old Sussex Almanack.
Hours & Admission Prices: Fri. 9-1, 2nd Sat. each month 10-2. Museum: No charge; donations accepted. Library $15 daily.
Attendance: 400 (estimated)
Membership: Junior (under 18) $5; Individual $15; Husband & Wife $25; Business $100; Life $200; Contributing $500; Benefactor $1,000.

Newtonville

AFRICAN AMERICAN HERITAGE MUSEUM OF SOUTHERN NEW JERSEY, Dr. Martin Luther King Jr. Center, 661 Jackson Rd., Newtonville, NJ 08346. Mailing Address: P.O. Box 39, Newtonville, NJ 08346. Tel.: 609-704-5495. Fax: 609-704-5532.
E-mail: rhunter@aahmsnj.org
Web Site: www.aahmsnj.org
Key Personnel: Pres., Ralph Hunter.
Governing Authority: Branch: Mississippi & 2200 Fairmount Ave Atlantic City, NJ 08401.
Institution Type/Description: History Museum.
Collections: Martin Luther King, Jr. memorabilia; civil rights movement; African American advertisements & artwork; local African American history; photographs; period furnishings; personal artifacts; Atlantic City North Side.
Activities: traveling museum for schools & organizations.
Hours & Admission Prices: Tues.-Fri. 10-3; other times by appointment. No charge; donations accepted.

North Brunswick

THE NEW JERSEY MUSEUM OF AGRICULTURE, College Farm Rd. & Rte. 1, North Brunswick, NJ 08902. Mailing Address: P.O. Box 644, Farmingdale, NJ 07727-0644. Tel.: 732-249-2077. Fax: 732-247-1035.
E-mail: info@agriculturemuseum.org
Web Site: www.agriculturemuseum.org
Founded: 1984.
Congressional District: 6
Key Personnel: Mgr., Dana S. Vallely; Cur., Cooper Morris; Cur. Collections, Coles Roberts; Program Coord. & Admin., Kelleen Madden; Cur. Archives, George Coyne; Coord. Education, Usha Luna; Facility Coord., Nelson Seda.
Personnel Profile: Full-Time Paid 2; Part-Time Paid 5; Part-Time Volunteers 40; Interns 5.
Governing Authority: independent, nonprofit corporation; board of trustees. Tax-exempt.
Institution Type/Description: Agriculture, Educational & History Museum.
Collections: agricultural; machinery; equipment; tools; housewares; photographs; glass plate negatives; lantern slides; negatives & prints; archival documents.
Research Fields: agricultural history; household technology; scientific photography; environmental science; social change; economics.
Facilities: library of books, pamphlets, catalogs, & reports relating to agriculture & household application available for inter-library loan & by request; 130-seat meeting room.
Activities: guided tours; educational programs; seminars; field trips.
Publications: quarterly newsletter; brochures.
Hours & Admission Prices: Tues.-Sat. 10-3; Sun. for special events. Adults $4, senior citizens $3, children 4-12 $2; discounts to AAM & ICOM members; children under 4 & members no charge. Closed major holidays. &
Attendance: 40,000 (accurate)
Membership: Student $10; Seniors & Teachers $15; Individual $25; Family $35; Supporting $100; Sustaining $250; Benefactor $500; Patron $1,000.

North Plainfield

THE FLEETWOOD MUSEUM, 614 Greenbrook Rd., North Plainfield, NJ 07063-1621. Mailing Address: 135 Sandford Ave., North Plainfield, NJ 07060. Tel.: 908-756-7810.
E-mail: naciampa@aim.com
Web Site: www.fleetwoodmuseum.org
Founded: 1985.
Congressional District: 7
Key Personnel: Dir. & Chm. (V), Nicholas A. Ciampa.
Personnel Profile: Part-Time Paid 5; Part-Time Volunteers 3; Interns 1.
Governing Authority: borough. Tax-exempt.
Institution Type/Description: Art Museum: housed in the Vermeule Mansion. Listed on the National Register of Historic Places.
Collections: cameras; oil paintings; 900-vol. photo library; music boxes; musical instruments.
Research Fields: pioneers of photo technology.
Facilities: 800-vol. photo technical library.
Activities: lectures; music concerts.
Hours & Admission Prices: Sat. 10-4. No charge; donations accepted. Closed major holidays. &
Attendance: 1,200 (estimated)

North Wildwood

HEREFORD INLET LIGHTHOUSE, 111 N. Central Ave., North Wildwood, NJ 08260-5955. Mailing Address: P.O. Box 784, Rio Grande, NJ 08242-0784. Tel.: 609-522-4520. Fax: 609-522-8590.
Web Site: www.herefordlighthouse.org
Founded: 1982.
Congressional District: 1
Key Personnel: Mayor, Aldo Palombo; Chm. (V), Historical Grants, Paul DiFilippo; Vice Chm., Grounds & Exterior Bldg., Steve Murray; Mgr. & Educator, Betty Mugnier; Public Rels., Peter Harp.
Personnel Profile: Full-Time Paid 1; Full-Time Volunteers 7; Part-Time Paid 6; Part-Time Volunteers 25.
Governing Authority: municipal; nonprofit. Tax-exempt: 501(c)(3).
Institution Type/Description: Historic Site & Maritime Museum: listed in the National Directory of Historic Places.
Collections: local maritime artifacts; photographs of local history.
Research Fields: maritime history & local architecture; development of historical commission with emphasis on historic buildings.
Facilities: library; botanical garden; nature/conservation center. Museum-related items for sale.
Activities: arts festivals; craft shows; concerts; guided tours; lectures; rental gallery; study clubs; temporary & traveling exhibitions.
Publications: semiannual newsletter; annual brochure.
Hours & Admission Prices: mid-May to mid-Oct. daily 9-5; mid-Oct. to mid-May Wed.-Sun. 10-4. Tours: adults $4, children 11 & under $1; active Coast Guard & NJ Lighthouse Society members no charge.
Attendance: 30,000 (accurate)
Membership: Associate $15; Family $25.

Northfield

NORTHFIELD MUSEUM, CASTO HOUSE & WEBB'S MILITARY SHED, Birch Grove Park, Burton Ave., Northfield, NJ 08225. Mailing Address: 1600 Shore Rd., Northfield, NJ 08225-2251. Tel.: 609-383-1505. Fax: 609-641-5901.
E-mail: northfieldmuseum@aol.com
Web Site: www.cityofnorthfield.org/main/castohouse.asp
Formerly: Northfield Bicentennial Museum
Founded: 1976.
Congressional District: 2
Key Personnel: Chm. (V) & Pres. (V), Carol A. Patrick; Cur., Roy W. Clark.
Personnel Profile: Part-Time Volunteers 5.
Governing Authority: city. Parent Institution: Northfield Cultural Committee. Subsidiary Institution: Northfield Historical Society. Tax-exempt.
Institution Type/Description: Local History Museum.

Collections: Northfield history; period artifacts & furnishings; area history; Evelyn C. Ryon obituary collection; quilts; period clothing; Mill Road school photographs; military artifacts; 1905 & 1920-1930 Northfield census; obituary files; genealogy.
Research Fields: local history.
Activities: fundraisers; 2 flea markets in March & November.
Publications: Northfield Historical Society's triannual print & e-newsletter, Northfield; Northfield Arcadia published 2005.
Hours & Admission Prices: Sun. & Wed. 1-3; other times by appointment. No charge; donations accepted. Photo prints $5. Closed some holidays. &
Attendance: 275 (estimated)
Membership: Individual $10; Family $20; Lifetime $100; Lifetime Couple $150.

Nutley

NUTLEY HISTORICAL SOCIETY & MUSEUM, 65 Church St., Nutley, NJ 07110. Tel.: 973-667-1528.
E-mail: museuminfo@nutleyhistorical society
Web Site: nutleyhistoricalsociety.org
Founded: 1945.
Key Personnel: Dir., Barry Lenson; Pres. (V), Domenick Tibaldo; Asst. Dir. & Museum Shop Mgr., Nancy Greulich.
Personnel Profile: Part-Time Volunteers 10.
Institution Type/Description: Historical Society Museum.
Collections: local history & culture; postcards; toys; personal artifacts; period furnishings; Annie Oakley memorabilia; military artifacts.
Facilities: 200-vol. library.
Hours & Admission Prices: By appointment. No charge.
Attendance: 600 (estimated)
Membership: Individual $15.

Oak Ridge

JEFFERSON TOWNSHIP HISTORICAL SOCIETY, 315 Dover-Milton Rd., Oak Ridge, NJ 07438. Tel.: 973-697-0258.
E-mail: info@jthistoricalsociety.org
Key Personnel: Dir., Carol Keppel
Institution Type/Description: Historical Society Museum: housed in the former home of Amos Chamberlain; c.1870.
Collections: local history & culture; period furnishings; personal artifacts; photographs.
Hours & Admission Prices: Call for hours.

Oakland

OAKLAND HISTORICAL SOCIETY - VAN ALLEN HOUSE, Franklin Ave., Oakland, NJ 07436. Mailing Address: P.O. Box 296, Oakland, NJ 07436. Tel.: 201-405-7726.
E-mail: ohs.inc1966@verizon.net
Web Site: www.oaklandhistoricalsociety.org
Founded: 1966.
Institution Type/Description: Historical Society Museum: house built in 1777. Listed on the National Register of Historic Places.
Collections: local history & culture; period furnishings; photographs; personal artifacts; George Washington's papers.
Research Fields: Oakland history; Revolutionary War history.
Hours & Admission Prices: Feb.-June & Sept.-Oct. 3rd Sun. of the month 1-4; Dec. 1st Sun. 1-4. No charge; donations accepted.
Membership: Single $10; Family $15; Participant Contributor $100.

Ocean City

DISCOVERY SEASHELL MUSEUM, 2721 Asbury Ave., Ocean City, NJ 08226-2329. Mailing Address: 2724 Central Ave., Ocean City, NJ 08226-2334. Tel.: 609-398-2316.
Institution Type/Description: Seashell Museum.
Collections: over 10,000 varieties of seashells from around the world; coral; carved shells; air plants.
Facilities: Museum-related items for sale.
Activities: tours.
Hours & Admission Prices: Memorial Day to Dec. Mon.-Sat. 10-8, Sun. 12-6. No charge. &

OCEAN CITY ARTS CENTER, 1735 Simpson Ave., Ocean City, NJ 08226-3070. Tel.: 609-399-7628. Fax: 609-399-7089.
E-mail: info@oceancityartscenter.org
Web Site: www.oceancityartscenter.org
Key Personnel: Dir., Rosalyn Lifshin; Pres. (V), Dr. Jack Devine

Institution Type/Description: Art Gallery.
Collections: paintings; photographs; sculpture.
Activities: classes; permanent & temporary exhibits; special events; art camp. Annual Event: Juried Shows.
Hours & Admission Prices: Mon.-Fri. 9-9, Sat. 9-3. No charge; donations accepted. &
Membership: Student $15; Adult $20; Family $35; Lifetime $300.

OCEAN CITY HISTORICAL MUSEUM, INC., Cultural Community Center, 1735 Simpson Ave., Ocean City, NJ 08226-3070. Tel.: 609-399-1801. Fax: 609-399-0544.
E-mail: info@ocnjmuseum.org
Web Site: ocnjmuseum.org
Formerly: Friends of the Ocean City Historical Museum, Inc.
Founded: 1964.
Congressional District: 1
Key Personnel: Exec. Dir., Johnette H. Halpin; Pres. (V), Paul S. Anselm.
Personnel Profile: Part-Time Paid 3; Part-Time Volunteers 50; Interns 2.
Governing Authority: municipal. Tax-exempt.
Institution Type/Description: Local History Museum.
Collections: local history & heritage; shipwreck & maritime artifacts; costumes; paintings.
Facilities: 300-vol. library of material on New Jersey & Ocean City history; deeds; documents; maps available for use on premises. Period glassware, china & museum-related items for sale.
Activities: guided tours; lectures; summer weekly lecture series; educational programs; outreach.
Publications: History of Ocean City; Ocean City Memories; videotape, The Sindia: Her Final Resting Place; The Saga of the Sindia.
Hours & Admission Prices: Call for hours. No charge; donations accepted. Closed major holidays. &
Attendance: 4,000 (estimated)
Membership: Senior 62 & over and Student 16-18 $20; Individual $25; Family $40; Museum History Benefactor $100; Museum Circle $500; Sindia Society $1,000.

Ocean Grove

HISTORICAL SOCIETY OF OCEAN GROVE, NEW JERSEY, (M), 50 Pitman Ave., Ocean Grove, NJ 07756-1557. Mailing Address: P.O. Box 446, Ocean Grove, NJ 07756-0446. Tel.: 732-774-1869. Fax: 732-774-1684.
E-mail: info@oceangrovehistory.org
Web Site: oceangrovehistory.org
Founded: 1969.
Congressional District: 3
Key Personnel: Pres., Gail Shaffer; Museum Shop Mgr., Elizabeth Ogden.
Personnel Profile: Full-Time Paid 1; Part-Time Paid 1.
Governing Authority: society; nonprofit. Subsidiary Institution: Centennial Cottage. Tax-exempt: 501(c)(3).
Institution Type/Description: Historical Society Museum: located in Ocean Grove Historic District; emphasizing history of Ocean Grove & the history of the camp meeting.
Collections: local history; Victoriana; original Stokes (founder of Ocean Grove) painting, pastel, chair & desk; original trustees' desks from Ocean Grove Camp Meeting Association; pictures; postcards; religious items & costumes. Historic Building: 1870 Centennial Cottage & gardens.
Facilities: 250-vol. library of Ocean Grove history & Victoriana material, including 100 manuscripts available to the public; 500 sq. ft. exhibit space. Notepads, calendars, & postcards for sale.
Activities: guided tours; lectures; temporary exhibitions. Annual Event: house tours.
Publications: quarterly newsletter.
Hours & Admission Prices: mid-May to mid-June Fri.-Sat. 10-4; mid-June to Sept. Mon.-Thurs. 10-4, Fri.-Sat. 10-5; Oct.-Dec. Sat. 10-4. No charge; donations accepted.
Attendance: 6,000 (accurate)
Membership: Student and Senior 62 & over $12; Individual $25; Family $40; Sustaining $100; Patron & Corporate $200.

Oceanville

THE NOYES MUSEUM OF ART, (M), 733 Lily Lake Rd., Oceanville, NJ 08231. Tel.: 609-652-8848. Fax: 609-652-6166.
E-mail: info@noyesmuseum.org
Web Site: www.noyesmuseum.org
Founded: 1983.
Congressional District: 2
Key Personnel: Exec. Dir., Michael Cagno; Dir. Education & Community Programs, Saskia Schmidt; Mgr. Exhibitions, Dorrie Papademetriou.

Personnel Profile: Full-Time Paid 5; Part-Time Paid 8; Part-Time Volunteers 5; Interns 3.
Governing Authority: nonprofit organization. Parent Institution: Fred Winslow Noyes Foundation. Tax-exempt: 501(c)(3).
Institution Type/Description: Art Museum.
Collections: 20th-century American paintings & sculpture, folk art from the mid-Atlantic region including decoys, contemporary American arts & crafts
Activities: guided tours; gallery talks; Meet the Artist; environmental talks; Educators' Symposia; concerts; rotating & traveling exhibitions; Artist-in-Residence at local schools; hands-on art activities for children; trips to other institutions; craft shows & sale; facility rental.
Publications: New Jersey Arts Annual; crafts & fine arts catalogs; brochures; exhibition catalogues; quarterly newsletter.
Hours & Admission Prices: Tues.-Sat. 10-4:30, Sun. 12-5. Adults $4, seniors & students over 12 $3; discounts to AAM, & N.J. Association of Museums members; children under 6 & members no charge. Closed national holidays. &
Attendance: 17,000 (estimated)
Membership: Student & Teacher $30; Senior 60 & over $35; Individual $45; Household & Family $65; Supporting $125; Sustaining $250; Benefactor $500; Leadership $1,000.

Ogdensburg

STERLING HILL MINING MUSEUM, INC., 30 Plant St., Ogdensburg, NJ 07439-1126. Tel.: 973-209-7212. Fax: 973-209-8505.
E-mail: shmm@ptd.net
Web Site: www.sterlinghill.org
Founded: 1990.
Congressional District: 5
Key Personnel: C.E.O., Bill Kroth; Museum Shop Mgr., Rhea Cianfichi.
Personnel Profile: Full-Time Paid 5; Part-Time Paid 14; Part-Time Volunteers 10.
Governing Authority: private; nonprofit organization. Tax-exempt: 501(c)(3).
Institution Type/Description: Mining Museum: former New Jersey zinc company mine; national historic site.
Collections: geology & mineralogy of the Sterling Hill zinc ore deposit; mining technology; equipment; mining artifacts; rare minerals; mining equipment.
Research Fields: geology; minerology.
Publications: newsletter twice a year.
Hours & Admission Prices: March by appointment; April-Nov. daily 10-3; Dec. Sat.-Sun. weather permitting. Adults $11, senior citizens $10, children under 12 $8; discount to groups of 10 or more with reservations & AAA members. Closed New Year's Day; Easter; Thanksgiving; Christmas. &
Attendance: 45,000 (accurate)
Membership: Calcite-Individual $20; Calcite-Family $30; Willemite $50; Zincite $100; Lifetime & Club $500.

Old Bridge

THOMAS WARNE HISTORICAL MUSEUM & LIBRARY, 4216 Rte. 516, Old Bridge, NJ 08857. Mailing Address: Madison Twp. Historical Society, 4216 Rte. 516, Matawan, NJ 07747-7032. Tel.: 732-566-0348 & 2108. Fax: 732-566-6943.
E-mail: info@thomas-warne-museum.com
Web Site: www.thomas-warne-museum.com
Founded: 1964.
Congressional District: 6
Key Personnel: Pres., Alycia Rihacek; Treas. (V), Richard Kujawinski; Cur., Kate Philbrick.
Personnel Profile: Full-Time Volunteers 4; Part-Time Paid 1; Part-Time Volunteers 9.
Governing Authority: private; nonprofit organization. Parent Institution: Madison Township Historical Society, Inc., Old Bridge Township, NJ. Tax-exempt.
Institution Type/Description: History Museum: housed in an 1885 one room school house.
Collections: photographs; post cards; Edison phonograph & records sheet music; pottery; carpenter tools; clothing; Indian artifacts; textiles; local & NJ research files; old school books; research books of genealogy; concentration on local history, Middlesex & Monmouth Counties from fossils to World War II; agricultural tools.
Research Fields: local history & genealogy.
Facilities: library of genealogy & NJ history. Gift items for sale.
Activities: guided tours; lectures; participatory, loan & temporary exhibitions; study clubs; resource information. Annual Event: Apple Festival.
Publications: biannual, Timepiece; local history book, At the Headwaters of Cheesequake Creek; cookbook, From Groaning Board Cooks; Arcadia book, Images of America, Old Bridge.

Hours & Admission Prices: Fri. 12-4, Sat.-Sun. 12-6; other times by appointment. No charge; donations accepted. Closed New Year's Day; Christmas. &
Attendance: 560 (estimated)
Membership: Individual $15; Family $25.

Oradell

HIRAM BLAUVELT ART MUSEUM, 705 Kinderkamack Rd., Oradell, NJ 07649-1504. Tel.: 201-261-0012. Fax: 201-391-6418.
E-mail: hbam@verizon.net
Web Site: www.blauveltartmuseum.com
Founded: 1940.
Congressional District: 5
Key Personnel: Chm. Foundation, James L. Bellis, Sr.; Pres. Foundation, James Bellis, Jr.; Education Coord. & Registrar, Rosa Lara; Artist-in-Residence, Aaron Yount.
Personnel Profile: Part-Time Paid 3; Part-Time Volunteers 3.
Governing Authority: not-for-profit organization. Parent Institution: Blauvalt-Demarest Foundation. Subsidiary Institutions: 1678 Demarest House, River Edge, NJ; Old French Cemetery, New Milford, NJ. Tax-exempt: 501(c)(3).
Institution Type/Description: Art Museum: located in an 1893 shingle & turret-style carriage house.
Collections: period furnishings & artifacts; Audubon folio, extinct birds & an ivory collection; North American mammals, big game, birds & a dioramic recreation of the African Water Hole Group by Louis Paul Jonas; artists: Douglas Allen, Dennis Anderson, John James Audubon, Richard Bishop Braithman, Paul Branson, Charles Livingston Bull, John Etti, Arthur B. Frost, Dwayne Harty, Lynn Bogue Hunt, Robert Kuhn, George Edward Lodge, Richard Loffler, Roy W. Mason, Walter Matia, Carl Milles, Dino Paravano, Alexander Proctor, Frederic Remington, John Schoenherr, W.J. Schaldach, Sir Peter Scott, David Shepherd, Arthur Tait, Archibald Thorburn, Eustracz Ziegler, Geordie Millar, Robert Guelich, Rene Headings, Wayne Trimm, Lanford Monroe, John Pitcher, Pati Stajcar, John Banovich, Kent Ullberg, Charles Allmond, Nancy Glazier & Terry Miller; Matthew Hillier; Judy Chapman; Julie Chapman; Daniel Smith; Peter Zaluzec; Dale Weiler; Alan Hunt; Robert Bateman; James Coe; Guy Combes. Historic House: 1678 Demarest House.
Research Fields: wildlife artists.
Facilities: library.
Activities: arts festivals; guided tours; temporary exhibitions; educational programs for schools; demonstrations in wildlife/landscape painting; lectures; artist in residence program; artist roundtables.
Publications: Charles Livingston Bull; Douglas Allen; John Schoenherr catalogs; Art and the Animal - Society of Animal Artists 2003; catalogs; Art and the Animal - Society of Animal Artists 2004; Africa and Beyond, 2006.
Hours & Admission Prices: Wed.-Fri. 10-4, Sat.-Sun. 2-5. No charge. Closed holidays. &
Attendance: 2,000 (estimated)

Oxford

SHIPPEN MANOR, 8 Belvidere Ave., Oxford, NJ 07863-3014. Tel.: 908-453-4381. Fax: 908-453-4981.
E-mail: wcchc@nac.net
Web Site: www.wcchc.org
Key Personnel: Cur., Andy Drysdale
Institution Type/Description: Historic House: listed on the National Register of Historic Places.
Collections: period furnishings; personal artifacts.
Activities: classes.
Hours & Admission Prices: Guided Tours: 1st & 2nd Sun. each month 1-4. No charge; donations accepted. Closed holidays.

Palmyra

PALMYRA COVE NATURE PARK, 1335 Rte. 73, Palmyra, NJ 08065. Tel.: 856-829-1900. Fax: 856-786-6138.
Institution Type/Description: Park Museum & Discovery Center.
Collections: wildlife & their habitats; over 250 species of birds; trees; plants; flowers.
Facilities: nature trails; 250 acres.
Activities: special events; nature trails; educational programs.
Hours & Admission Prices: Center: Mon.-Fri. 9-4, Sat.-Sun. 10-4.

Paramus

BERGEN COUNTY ZOOLOGICAL PARK, 216 Forest Ave., Paramus, NJ 07652-5349. Tel.: 201-262-3771. Fax: 201-986-1788.
E-mail: mvella@co.bergen.nj.us

Web Site: www.co.bergen.nj.us/parks/zoo.htm
Founded: 1960.
Key Personnel: Dir., Marianne Vella; Cur. Education, Liz Carletta; Gen. Cur., Cindy Norton.
Personnel Profile: Full-Time Paid 9; Part-Time Paid 6; Part-Time Volunteers 50.
Governing Authority: county. Parent Institution: Bergen County Dept. of Parks, 21 Main St., Court Plaza S., Hackensack, NJ 07601. Tax-exempt: 501(c)(3).
Institution Type/Description: Zoo.
Collections: wild & domestic animals including 71 mammals, 111 birds & 44 reptiles; early farm equipment; 1860, Hackensack River Valley farm yard.
Research Fields: breeding endangered species.
Facilities: classrooms.
Activities: guided tours; lectures; hobby workshops; formally organized education programs for children, adults & undergraduate college students affiliated with Ramapo College; permanent & temporary exhibitions.
Hours & Admission Prices: July-Aug. Thurs. 10-8, Fri.-Wed. 10-4:30; Sept.-June daily 10-4:30. Non-Residents: adults 15 & over $8, children 3-14 $5, seniors $2; Residents: adults 15 & over $4, children 3-14 $2, seniors $1; military & Nov.-April no charge. &
Attendance: 250,000 (estimated)
Membership: Dual $40; Family $50; Sustaining $65; Contributing $125.

BUEHLER CHALLENGER & SCIENCE CENTER, 400 Paramus Rd., Parking Lot G, Paramus, NJ 07652-1508. Mailing Address: P.O. Box 647, Paramus, NJ 07653-0647. Tel.: 201-251-8589. Fax: 201-251-9049.
E-mail: missionservices@bcsc.org
Web Site: www.bcsc.org
Founded: 1994.
Key Personnel: Mission Coord., Peggy Silverman
Institution Type/Description: Science Museum.
Collections: space mission simulations; science exhibits; hands-on exhibits.
Activities: space mission simulations; public programs; space camps; tours; camp-in; teacher workshops. Museum Sponsors: Family Science Morning.
Hours & Admission Prices: Call for further information. &

THE NEW JERSEY CHILDREN'S MUSEUM, 599 Valley Health Plaza, Paramus, NJ 07652-3616. Tel.: 201-262-5151. Fax: 201-262-0560.
E-mail: nycminfo@aol.com
Web Site: www.njcm.com
Founded: 1992.
Key Personnel: Exec. Dir., Anne R. Sumers, MD; Chm., Elliott H. Sumers, MD
Personnel Profile: Full-Time Paid 12; Part-Time Paid 20.
Governing Authority: self-governing; profit.
Institution Type/Description: Children's Museum.
Collections: hands-on exhibits; 100 sq. ft. interactive train set.
Facilities: 14,000 sq. ft. exhibit space. Museum-related items for sale.
Activities: films; lectures; participatory exhibits; broadcast programs; puppet shows; story telling; skits; craft project.
Publications: quarterly newsletter.
Hours & Admission Prices: Tues.-Fri. 10-5, Sat.-Sun. 10-6. Children $11.99, adults $9.99; discount to groups; children under one no charge. Closed Thanksgiving; Christmas. &
Attendance: 150,000 (estimated)

Park Ridge

PASCACK HISTORICAL SOCIETY, 19 Ridge Ave., Park Ridge, NJ 07656-1138. Mailing Address: P.O. Box 85, Park Ridge, NJ 07656-0285. Tel.: 201-573-0307.
E-mail: info@pascackhistoricalsociety.org
Web Site: www.pascackhistoricalsociety.org
Founded: 1942.
Congressional District: 9
Key Personnel: Pres. (V), George Sherman; Museum Shop Mgr., Jackie Martin; Book Shop Mgr., Tom Norton.
Personnel Profile: Full-Time Volunteers 20; Part-Time Volunteers 40; Interns 5.
Governing Authority: nonprofit. Tax-exempt: 501(c)(3).
Institution Type/Description: General Museum: local history, housed in 1873 former Congregational Church.
Collections: history; archives; costumes; glass paperweights; kitchen, farm & country store displays; Baylor Massacre; wampum & wampum making tools; wampum drilling machine; quilts & coverlets; Lenape artifacts; Tice wedding dress quilt c.1781.

Research Fields: quilts; Dutch buildings; Pascack Valley History; Third Continental Dragoons; Lenape Indians; Jersey Dutch language; Wampum collection.
Facilities: research library. Publications, reproduction wampum & local history articles for sale.
Activities: films; permanent exhibits; special meeting programs; children's activities.
Publications: quarterly, Relics; book, Pascack Valley tales.
Hours & Admission Prices: Wed. 10 am to noon, Sun. 1-4; other times by appointment. No charge. &
Attendance: 4,000 (estimated)
Membership: Active $25; Dual $35; Contributing $50; Patron $100; Corporate $150; Life $400.

Paterson

PASSAIC COUNTY COMMUNITY COLLEGE, One College Blvd., Paterson, NJ 07505-1179. Tel.: 973-684-6555. Fax: 973-523-6085.
E-mail: jhaw@pcc.edu
Web Site: www.pccc.edu/art/gallery
Founded: 1968.
Congressional District: 35
Key Personnel: Exec. Dir. Cultural Affairs, Maria Mazziotti Gillan; Gallery Cur., Jane Haw; Assoc. Dir., Susan Balik.
Personnel Profile: Full-Time Paid 4; Part-Time Paid 3.
Governing Authority: county & state government; public college; nonprofit. Tax-exempt.
Institution Type/Description: College Museum.
Collections: contemporary, 19th & early 20th century paintings & sculpture collections.
Activities: drawing & art history courses. Annual Events: community celebrations of various ethnic groups including African American & Hispanic Heritage month.
Publications: monthly, Passaic County Arts News.
Hours & Admission Prices: June-Aug. Mon.-Thurs. 9-9; Sept.-May Mon.-Fri. 9-9, Sat. 9-5. No charge. &
Attendance: 54,000 (estimated)

PASSAIC COUNTY HISTORICAL SOCIETY & LAMBERT CASTLE MUSEUM, 3 Valley Rd., Paterson, NJ 07503-2932. Tel.: 973-247-0085. Fax: 973-881-9434.
E-mail: lambercastle@verizon.net
Web Site: lambertcastle.org
Founded: 1926.
Congressional District: 7
Key Personnel: Dir., Alison Krawiec Faubert; Pres., Lorraine Yurchak.
Personnel Profile: Full-Time Paid 2; Part-Time Paid 5; Part-Time Volunteers 50.
Governing Authority: society; nonprofit organization. Tax-exempt: 501(c)(3).
Institution Type/Description: Historical Society Museum: housed in 1893, Lambert Castle.
Collections: Passaic County & New Jersey; 18th to 19th-century decorative arts; furniture; paintings; sculpture; textiles; costumes; photographs; industrial history; genealogy; archives.
Research Fields: Passaic County history; silk industry; Victorian & Edwardian Era.
Facilities: library of books, archives, manuscripts, documents & maps on the history of Passaic County & northern N.J.; genealogy library.
Activities: guided tours; lectures; permanent, temporary & traveling exhibitions; educational programs for students; genealogy club.
Publications: newsletters, The Historic County, Castle Genie; catalogue, Gaetano Federici-The Artist as Historian; Silk & Sandstone: The Story of Catholina Lambert & His Castle; At The Sign Of The Brass Dog: Passaic County Folk Art.
Hours & Admission Prices: Museum: July-Aug. Wed.-Fri. 1-4, Sat.-Sun. 12-4; Sept.-June Wed.-Sun. 1-4. Adults $5, seniors $4, children 5-17 $3; discounts to AAM members; children under 5 & members no charge. &
Attendance: 27,500 (estimated)
Membership: Student $10; Senior $20; Individual & Regular $25; Family $35; Sustaining $50; Benefactor $100; Corporate $150.

THE PATERSON MUSEUM, 2 Market St., Ste. 202, Paterson, NJ 07501-1726. Tel.: 973-321-1260. Fax: 973-881-3435.
Founded: 1925.
Congressional District: 8
Key Personnel: Dir., Giacomo R. DeStefano; Pres. (V), Florence Bottler; Cur. History, Bruce Balistrieri; Exhibit Artist Designer, Joseph Ruffilo; Photo Archivist, Joseph Costa; Museum Attendant, Mohamed Khalil.
Personnel Profile: Full-Time Paid 4; Interns 3.

Governing Authority: municipal. Tax-exempt.
Institution Type/Description: History, Natural History & Science Museum: housed in 1871, Thomas Rogers Building, locomotive erecting shop.
Collections: mineralogy; New Jersey Indian artifacts; local history; entomological specimens; zoological specimens; John P. Holland's first & second successful submarine; Colt Paterson revolver collection 1837-1841; Silk City: textile machinery; model train layout; industrial history.
Research Fields: city atlas & local history library.
Facilities: library.
Activities: guided tours; lectures; films; gallery talks; formally organized education programs for children; permanent & temporary exhibitions.
Publications: booklets, Great Falls/S.U.M.; Historic District Walking Tour; The Colt Family; Lenni Lenape, New Jersey's Native Americans.
Hours & Admission Prices: Tues.-Fri. 10-4, Sat.-Sun. 12:30-4:30. Suggested Donation: adults $2; discount to AAM & ICOM members; children no charge. Closed major holidays. &

Attendance: 17,500 (estimated)

Paulsboro

TINICUM REAR RANGE LIGHTHOUSE, 70 2nd St., Paulsboro, NJ 08066. Mailing Address: P.O. Box 176, Paulsboro, NJ 08066. Tel.: 856-423-1505.
E-mail: info@tinicumrearrangelighthouse.org
Web Site: www.tinicumrearrangelighthouse.org
Institution Type/Description: Historic Lighthouse: listed on the National Register of Historic Places.
Collections: lighthouse history.
Facilities: Gift items for sale.
Activities: special events.
Hours & Admission Prices: April-Oct. 3rd Sun. each month 12-4.

Pemberton

NORTH PEMBERTON RAILROAD STATION MUSEUM AND RAIL TRAIL, 3 Fort Dix Rd., Pemberton, NJ 08068-1439. Mailing Address: 500 Pemberton-Brown Mills Rd., Pemberton, NJ 08068-1539. Tel.: 609-894-0546. Fax: 609-894-0568.
E-mail: pthtrust@yahoo.com
Governing Authority: nonprofit organization. Tax-exempt: 501(c)(3).
Institution Type/Description: Historic Building: housed in a former railroad station; built in 1892. Listed on the National Register of Historic Sites.
Collections: local & railroad history; period artifacts; cranberry & blueberry industry; photographs; personal artifacts.
Facilities: 2.5 mile rail trail. Museum-related items for sale.
Activities: research; walking trail; special events. Annual Events: MS Walk and Volunteer Picnic; Day at the Station & Lantern Show in June; Train, Toy & Collectible Show in November; Holiday Hayride in December.
Hours & Admission Prices: Temporarily closed. &

Pennington

STONY BROOK MILLSTONE WATERSHED ARBORETUM, 31 Titus Mill Rd., Pennington, NJ 08534. Tel.: 609-737-7592.
Key Personnel: Exec. Dir., Jim Waltman
Institution Type/Description: Arboretum.
Collections: conservation; water & environment; science; wildlife & their habitats; butterfly house.
Facilities: 930 acres; nature center; nature trails.
Activities: educational programs; special events; hiking trails.
Hours & Admission Prices: Dawn to dusk.

Penns Grove

HISTORICAL SOCIETY OF PENNS GROVE, CARNEY'S POINT, AND OLDMANS, 48 W. Main St., Penns Grove, NJ 08069. Tel.: 856-299-1556.
Institution Type/Description: Historical Society Museum.
Collections: local history & culture; personal artifacts; photographs.
Hours & Admission Prices: March-Dec. Sun. 1-3.

Pennsauken

PENNSAUKEN HISTORICAL SOCIETY, 9201 Burrough Dover Ln., Pennsauken, NJ 08110-1000. Mailing Address: P.O. Box 56, Pennsauken, NJ 08110. Tel.: 856-662-3002.
Institution Type/Description: Historical Society Museum: housed in the Burrough-Dover house; built c.1710. Listed on the National Register of Historic Places.

Collections: local history & culture; period furnishings.
Activities: Museum Sponsors: Open Houses.
Hours & Admission Prices: Call for hours.

Pennsville

CHURCH LANDING FARMHOUSE, 86 Church Landing Rd., Pennsville, NJ 08070-1203. Tel.: 856-678-4453.
E-mail: wmasten@pennsvillenb.com
Web Site: www.pvhistorical.njcool.net
Governing Authority: Parent Institution: Pennsville Historical Society.
Institution Type/Description: History Museum.
Collections: period furnishings; personal artifacts; flower & herb gardens.
Facilities: gardens.
Activities: guided tours; school group programs. Annual Event: Day at the Farm.
Hours & Admission Prices: Sun. 1-3; groups by appointment. Adults $2, seniors $1; members no charge.

Perth Amboy

THE KEARNY COTTAGE HISTORICAL ASSOCIATION, 63 Catalpa Ave., Perth Amboy, NJ 08861-4617. Tel.: 732-293-1090; 908-812-4549.
E-mail: pw857@aol.com
Web Site: www.kearnycottage.org
Formerly: The Kearny Cottage
Founded: 1928.
Key Personnel: C.E.O., Paul W. Wang.
Personnel Profile: Part-Time Paid 1; Part-Time Volunteers 25.
Governing Authority: municipal. Parent Institution: Kearny Cottage Historical Association, Inc. Tax-exempt: 501(c)(3).
Institution Type/Description: Local History Museum.
Collections: Roger Statuettes furniture; documents; works by local artists. Historic Houses: Kearny cottage; home of Captain (Commodore) Lawrence Kearny.
Research Fields: local history of Perth Amboy & New Jersey; local citizens.
Activities: public meetings; open house tours; demonstrations of colonial crafts.
Hours & Admission Prices: Mon., Thurs. & last Sun. each month 2-4. Admission by appointment. No charge; donations accepted.
Attendance: 260 (estimated)
Membership: Membership $15.

PROPRIETARY HOUSE, THE ROYAL GOVERNOR'S MAN-SION, 149 Kearny Ave., Perth Amboy, NJ 08861-4700. Tel.: 732-826-5527. Fax: 732-826-8889.
E-mail: info@proprietaryhouse.org
Web Site: www.proprietaryhouse.org
Founded: 1967.
Congressional District: 13
Key Personnel: C.E.O. & Pres. (V), Tom Ward.
Personnel Profile: Part-Time Paid 1; Part-Time Volunteers 30.
Governing Authority: private; nonprofit association. Tax-exempt.
Institution Type/Description: Historic Building.
Collections: 1764 Royal Governor's original mansion; paintings; architectural fragments; 18th- & 19th-century furnishings; ephemera.
Activities: general audience tours, school programs, special events, outreach programs.
Publications: newsletter, booklet on history of site.
Hours & Admission Prices: Mon. & Wed.-Fri. 10-4 by appointment. Closed New Year's Day; Easter; Mother's Day; Father's Day; Thanksgiving; Christmas Eve & Day. &
Attendance: 1,000 (estimated)
Membership: Student & Senior $15; Individual & Family $25-$49; Sustaining $50-$99; Supporting $100-$499; Benefactor $500-$999; Patron $1,000 & up.

Phillipsburg

PHILLIPSBURG HISTORICAL SOCIETY, Municipal Bldg., Lower Level, 675 Corliss Ave., Phillipsburg, NJ 08865. Tel.: 908-454-0816.
E-mail: pburghistory@yahoo.com
Web Site: www.pburglib.com
Institution Type/Description: Historical Society Museum.
Collections: local history & culture; books; genealogy; photographs; maps; newspapers.
Facilities: library.

Publications: newsletter.
Hours & Admission Prices: Call for hours.

Piscataway

CORNELIUS LOW HOUSE/MIDDLESEX COUNTY MU-SEUM, 1225 River Rd., Piscataway, NJ Mailing Address: 703 Jersey Ave., New Brunswick, NJ 08901-3651. Tel.: 732-745-4177 & 4489. Fax: 732-745-4507. TDD: 732-745-3888.
E-mail: info@cultureheritage.org
Web Site: co.middlesex.nj.us/culturalheritage
Founded: 1979.
Congressional District: 6
Key Personnel: Exec. Dir., Anna M. Aschkenes; Cur. Exhibitions, Katie Zavoski; Asst. Cur. Facilities & Educator, Kenneth M. Helsby.
Personnel Profile: Full-Time Paid 2.
Governing Authority: county; nonprofit. Parent Institution: Middlesex County Cultural & Heritage Commission, 703 Jersey Ave., New Brunswick. Tax-exempt.
Institution Type/Description: History Museum & Historic House: housed in a c.1741 Georgian style home featuring a wainscoted central hall, original floors & staircase & Delft-tiled fireplaces.
Collections: state & local history.
Major Exhibits: Got Work? New Deal/WPA in New Jersey, 1/14-6/14; New Jersey Diners, 9/14-12/15.
Research Fields: New Jersey history.
Activities: student workshops; public programs.
Hours & Admission Prices: Tues.-Fri. & Sun. 1-4; groups by appointment. No charge; donations accepted. Closed federal holidays. &

EAST JERSEY OLDE TOWNE, 1050 River Rd. & Old Hoes Ln., Piscataway, NJ 08855. Mailing Address: 703 Jersey Ave., New Brunswick, NJ 08901-3651. Tel.: 732-745-3030 & 4489. Fax: 732-463-1086. 732-745-3888 (TTY).
E-mail: culturalandheritage@co.middlesex.nj.us
Web Site: co.middlesex.nj.us/culturalheritage
Founded: 1971.
Key Personnel: Exec. Dir., Anna M. Aschkenes.
Personnel Profile: Full-Time Paid 3; Part-Time Volunteers 25.
Governing Authority: county; Parent Institution: Middlesex County Cultural & Heritage Commission. Tax-exempt.
Institution Type/Description: Historic Building District: 18th-century village.
Collections: 18th-century furnishings & farm equipment; 17 18th-century buildings.
Research Fields: Piscataway historic village genealogy.
Facilities: library; botanical garden.
Activities: guided tours; lectures; workshops; programs; storytelling; demonstrations.
Publications: monthly newsletter; cultural calendar; event notification; exhibit guides, large print & Braille (on request).
Hours & Admission Prices: Tues.-Fri. 8:30-4:15, Sun. 1-4. Tours: Tues.-Fri. 1:30, Sun. rotating schedule; groups by appointment. Call for holiday hours & seasonal schedule. No charge. &
Attendance: 6,000 (estimated)

Plainfield

THE DRAKE HOUSE MUSEUM, (M), 602 W. Front St., Plainfield, NJ 07060-1004. Tel.: 908-755-5831. Fax: 908-755-0132.
E-mail: thedrakehousemuseum@verizon.net
Web Site: drakehousemuseum.tripod.com
Founded: 1921.
Congressional District: 7
Key Personnel: Pres. (V), Eloise Bryant Tinley; 1st Vice Pres., Nancy Piwowar; 2nd Vice Pres., Molly Banta; Corresponding Sec., Sandy Gurshman; Office Mgr., Danielle Franklin.
Personnel Profile: Part-Time Paid 1.
Volunteer Hours: 125
Operating Expenses: 30,415
Operating Income: 40,000
Governing Authority: municipal. Parent Institution: Historical Society of Plainfield. Tax-exempt: 501(c)(3).
Institution Type/Description: History Museum: housed in 1746 Drake house.
Collections: collection of Americana; period furnishings.
Research Fields: Drake family genealogy; local history.
Facilities: Slides, postcards, stationery, books & museum-related items for sale.
Activities: guided tours; lectures; films; formally organized education programs; docent program; temporary & traveling exhibitions; luncheons.
Publications: newsletter, Communique.

Hours & Admission Prices: Mon.-Fri. 9-3, Sat. 11-1; other times by appointment. Suggested Donation $3; members no charge. Closed holiday weekends. &
Attendance: 1,800
Membership: Senior Citizen $15; Individual $25; Family $35; Contributing $50; Corporate $100; Sustaining $100 & up.

Plainsboro

PLAINSBORO MUSEUM - WICOFF HOUSE, 641 Plainsboro Rd., Plainsboro, NJ 08536-2094. Tel.: 609-799-9040.
Governing Authority: Parent Institution: Plainsboro Historical Society.
Institution Type/Description: Historical Society Museum.
Collections: local history & culture; period furnishings; Walker-Gordon Farms; Elsie the Cow; photographs.
Hours & Admission Prices: 1st & 3rd Sun. of the month 2-4:30.

Point Pleasant

NEW JERSEY MUSEUM OF BOATING, Johnson Bros. Boat Works, Bldg. #13, 1800 Bay Ave., Point Pleasant, NJ 08742-4584. Tel.: 732-606-7605.
E-mail: knnthmotz@aol.com
Web Site: njmb.org
Founded: 1999.
Key Personnel: Pres., Kenneth J. Motz; Chm., W. Birdsall.
Personnel Profile: Full-Time Volunteers 6; Part-Time Volunteers 9; Interns 1.
Governing Authority: Parent Institution: NJMB Inc.; nonprofit organization. Tax-exempt: 501(c)(3).
Institution Type/Description: Boating Museum.
Collections: boating history; photographs; boatbuilding woods; maritime arts; nautical library.
Major Exhibits: NJ Crabbing, 11/13-11/14; Shipwrecks of NJ Coast, 3/14-3/15.
Research Fields: ecology of New Jersey coast.
Activities: book signings; maritime art & photo shows; youth sailing classes. Annual Events: Spring Fling; Antique & Classic Boat Show in Sept.; Holiday Party; Kids n' boats education program.
Publications: quarterly newsletter Waterline.
Hours & Admission Prices: Daily 10-4. No charge; donations accepted. &
Attendance: 10,000 (estimated)
Membership: Individual $15; Family $20; Supporter $100; Lifetime 1,000.

Point Pleasant Beach

JENKINSON'S AQUARIUM, 300 Ocean Ave., Point Pleasant Beach, NJ 08742. Tel.: 732-892-0600, ext. 130. Fax: 732-899-1717.
E-mail: jenkinsonsaquarium@comcast.net
Web Site: www.jenkinsons.com/aquarium
Founded: 1991.
Key Personnel: Dir., Cindy Claus.
Personnel Profile: Full-Time Paid 18; Part-Time Paid 13.
Institution Type/Description: Aquarium.
Collections: sharks; penguins; seals; coral reefs; sea star; sting ray; touch tank.
Facilities: Gift items for sale.
Activities: educational programs; birthday parties; special events; animal feedings; touch tank; workshops; outreach program; adopt an animal programs.
Hours & Admission Prices: Summer: daily 10-10; Winter: Mon.-Fri. 9:30-5, Sat.-Sun. 10-5. Adults $10, seniors 62 & over and children 3-12 $6; children 2 & under no charge. Closed New Year's Day; Thanksgiving; Christmas.
Attendance: 220,000 (accurate)

Port Norris

BAYSHORE DISCOVERY PROJECT - DELAWARE BAY MUSEUM & SCHOONER A.J. MEERWALD, 2800 High St. (Bivalve), Port Norris, NJ 08349-3126. Tel.: 856-785-2060. Fax: 856-785-2893.
E-mail: info@bayshorediscoveryproject.org
Web Site: bayshorediscovery.org
Founded: 1988.
Congressional District: 2
Governing Authority: Schooner Location: High St., Port Norris.
Institution Type/Description: Maritime Museum.
Collections: maritime artifacts; shipbuilding; oystering; commercial fishing; photographs; 115 ft. schooner.
Facilities: Bivalve Center.

Activities: Schooner: education sail for school & youth groups; charter events; public sails; special events.
Hours & Admission Prices: Tues.-Sat. 11-4. &

Princeton

DRUMTHWACKET - GOVERNORS MANSION, 354 Stockton St., Princeton, NJ 08540. Tel.: 609-683-0057.
Web Site: www.drumthwacket.org
Institution Type/Description: Historic Building: housed in the official residence of New Jersey governors.
Collections: state history; period furnishings; photographs.
Activities: guided tours.
Hours & Admission Prices: By appointment.

HISTORICAL SOCIETY OF PRINCETON - BAINBRIDGE HOUSE, (M), 158 Nassau St., Princeton, NJ 08542-7006. Tel.: 609-921-6748. Fax: 609-921-6939.
E-mail: information@princetonhistory.org
Web Site: www.princetonhistory.org
Founded: 1938.
Congressional District: 12
Key Personnel: Exec. Dir., Erin Dougherty; Pres., John Dumont; Dir. Devel., Barbara Webb; Museum Shop Mgr., Julie Janokowicz.
Personnel Profile: Full-Time Paid 4; Part-Time Paid 4; Part-Time Volunteers 100; Interns 1.
Governing Authority: society; nonprofit organization. Tax-exempt: 501(c)(3).
Institution Type/Description: History Museum: housed in 1766 Bainbridge House.
Collections: arts; artifacts; documents; photographs; glass-plate negatives; historical archives; architectural research materials.
Research Fields: history of Princeton & New Jersey.
Facilities: library of Princeton and New Jersey books & documents available on premises.
Activities: exhibitions; tours; lectures; publications; educational programs for adults & school groups.
Publications: journal, Princeton History, Vol. 1-16; exhibit catalogues; guide to manuscript collections.
Hours & Admission Prices: Tues.-Sun. 12-4. No charge; donations accepted. Library: Tues. & Sat. 1-4 Adults $5; discounts to AAM members; library members no charge. Closed New Year's Day; Independence Day; Thanksgiving; Christmas. &
Attendance: 18,500 (accurate)
Membership: Senior $40; Individual $45; Family & Household $60; Contributor $85; Bainbridge Club Member $175; Bainbridge Club Patron $250; Bainbridge Club Benefactor $500; Bainbridge Club Sponsor $1,000.

MORVEN MUSEUM & GARDEN, (M), 55 Stockton St., Princeton, NJ 08540-6812. Tel.: 609-924-8144. Fax: 609-924-8331.
E-mail: info@morven.org
Web Site: www.morven.org
Formerly: Historic Morven
Founded: 1987.
Congressional District: 12
Key Personnel: Dir., Clare Michel Smith; Pres. (V), Georgia T. Schley; Cur. Exhibitions, Elizabeth Allan; Museum Shop Mgr., Kathy O'Hara.
Personnel Profile: Full-Time Paid 4; Part-Time Paid 9; Part-Time Volunteers 75; Interns 1.
Governing Authority: state; nonprofit organization. Tax-exempt: 501(c)(3).
Institution Type/Description: Historic House Museum: former home of five New Jersey governors and a signer of the Declaration of Independence.
Collections: history of the site and its occupants with special emphasis on New Jersey history; New Jersey decorative & fine arts; cultural heritage of New Jersey.
Research Fields: site & its inhabitants; garden history; central New Jersey history; New Jersey fine and decorative arts.
Facilities: guided tours; lectures; school programs; temporary exhibitions.
Publications: biannual newsletter, For the Eye of a Friend.
Hours & Admission Prices: Wed.-Fri. 11-3, Sat.-Sun. 12-4. Adults $6, seniors 60 & older and students $5; discounts to members. Closed New Year's Day; Independence Day; Thanksgiving; Christmas. &
Attendance: 8,000 (estimated)
Membership: Individual $40; Couple $50; Household $75; Patron $100; Sustainer $250; Sponsor $500.

∗ PRINCETON UNIVERSITY ART MUSEUM, (M), Princeton, NJ 08544-1018. Tel.: 609-258-3788 & 1860. Fax: 609-258-3610.
E-mail: artmuseum@princeton.edu
Web Site: artmuseum.princeton.edu
Formerly: The Art Museum, Princeton University
Founded: 1882.
Congressional District: 5
Key Personnel: Dir., James C. Steward; Exec. Asst. to the Dir., Office Mgr., Laura Hahn; Museum Asst., Henry Vega; Mgr. Campus Collections, Lisa Arcomano; Preparator, Todd Baldwin; Haskell Cur. Modern and Contemporary Art, Kelly Baum; Assoc. Cur. Prints and Drawings, Calvin Brown; Preparator, Keith Crowley; Preparator, John Franklin; Collection Technician, Molly Gibbons; Heather and Paul G. Haaga Jr., Class of 1970 Cur. Prints and Drawings, Laura Giles; Preparator, Mark Harris; Project Registrar, Janet Hawkins; Registrar, Alexia Hughes; Mgr. Exhibition Services, Michael Jacobs; Peter Jay Sharp, Class of 1952 Cur. and Lecturer in the Art of the Ancient Americas, Bryan Just; Assoc. Registrar, James Kopp; Cur. American Art, Karl Kusserow; Cur. Asian Art, Cary Liu; Preparator, Rory Mahorn; Chief Registrar, Maureen McCormick; Assoc. Registrar, Emily McVeigh; Conservator, Norman Muller; Cur. Ancient Art, J. Michael Padgett; Assoc. Registrar, Inventory Project Mgr., Virginia Pifko; Preparator, Matt Pruden; Research Cur. European Painting and Sculpture, Betsy J. Rosasco; Curatorial Research Assoc., Allison Unruh; Assoc. Dir. Finance and Operations, Karen Ohland; Business Mgr., Michael Brew; Casual Financial Asst., Linda Greiner; Mgr. Retail and Wholesale Operations, Christine Hacker; Financial Asst., Devon Hart; Museum Facilities Mgr., Craig Hoppock; Asst. Museum Facilities Mgr., Edward Murfit; Museum Store Asst., Rachel Schatz; Museum Store Asst., Laura Smith; Museum Store Asst., Samantha Smith; Assoc. Dir. Information and Technology, Janet Strohl-Morgan; Collections Data Asst., Daniel Brennan; Mgr. Collections Information and Access, Cathryn Goodwin; Collections Data Asst., Marin Lewis; Technical Support Specialist, Rebecca Purcell; Raiser's Edge Database Admin., Jill Oster; Assoc. Dir. Institutional Advancement, Nancy Stout; Mgr. Corporate, Foundation and Government Relations, Kelly Freidenfields; Mgr. Membership and Annual Support, Jennifer Fekete-Donners; Assoc. Dir. Education, Caroline Harris; Visitor Logistics Coord., Louise Barrett; Student Outreach Coord., Jessica Popkin; Mgr. School, Family and Community Programs, Brice Batchelor-Hall; Andrew W. Mellon Curatorial Fellow for Collections Engagement, Juliana Ochs Dweck; Andrew W. Mellon Curatorial Fellow for Academic Programs, Johanna Seasonwein; Assoc. Dir. Publishing and Communications, Curtis Scott; Mgr. Marketing and Public Relations, Erin Firestone; Graphic Designer, Lehze Flax; Asst. Editor, Anna Brouwer; Marketing and Public Relations Asst., Kristina Giasi; Assoc. Editor and Interpretive Mgr., Janet Rauscher; Security Operations Mgr., Albert Wise, Jr.; Head Art Security Supvr., Tracy Craig.
Personnel Profile: Full-Time Paid 58; Part-Time Paid 9; Part-Time Volunteers 120; Interns 13.
Governing Authority: university. Parent Institution: Princeton University. Tax-exempt: 501(c)(3).
Institution Type/Description: Art Museum.
Collections: ranging from ancient to modern art; Classical; Far Eastern Art especially Chinese; pre-Columbian, African, European & American paintings, sculpture, prints & drawings; European, American & Japanese photographs.
Major Exhibits: Disegno in Translation: Italian Drawings from the Princeton University Art Museum, 1/25/14-5/11/14.
Research Fields: related to collections.
Facilities: Art & museum-related items for sale.
Activities: permanent & temporary exhibitions; gallery and children's talks; guided tours for groups by appointment; viewing of print, drawing and photograph collections by appointment; Sat. & Sun. Highlights Tours.
Publications: quarterly newsletter, Record of the Princeton University Art Museum; catalogs of special exhibitions; note cards; postcards; posters; reproductions.
Hours & Admission Prices: Tues.-Wed. & Fri.-Sat. 10-5, Thurs. 10-10, Sun. 1-5. No charge; donations accepted. Closed national holidays. &
Attendance: 150,229 (accurate)
Membership: Student $40; Individual $75; Dual & Family $125; Contributor $250; Curator's Circle $500; Director's Circle $1,000; Partners $2,500.

PRINCETON UNIVERSITY MUSEUM OF NATURAL HISTORY, Princeton University, Guyot Hall, Princeton, NJ 08544. Mailing Address: 9 Eno Hall, Princeton University, Princeton, NJ 08544-0430. Tel.: 609-258-4102 & 3832. Fax: 609-258-1334.
E-mail: ehorn@princeton.edu
Founded: 1805.
Congressional District: 7
Key Personnel: Cur. Biological Collections, Elizabeth Horn.
Personnel Profile: Part-Time Paid 1.
Governing Authority: university. Affiliated with Princeton University. Tax-exempt.
Institution Type/Description: Natural History Museum.

Collections: invertebrate paleontology; geology & mineralogy; ornithology; ethnology; archaeology; osteology.
Research Fields: ecology & ornithology.
Activities: formally organized education programs for undergraduate college & graduate students affiliated with Princeton University; inter-museum loan, permanent & temporary exhibitions.
Publications: books & monographs published by Princeton University press.
Hours & Admission Prices: Call for hours. No charge. Closed holidays. &
Attendance: 500 (estimated)

THOMAS CLARKE HOUSE/PRINCETON BATTLEFIELD STATE PARK, 500 Mercer Rd., Princeton, NJ 08540-4810. Tel.: 609-921-0074. Fax: 609-921-0074.
E-mail: pbsp@aol.com
Founded: 1976.
Congressional District: 12
Key Personnel: Cur., John K. Mills.
Personnel Profile: Full-Time Paid 1; Part-Time Paid 1; Part-Time Volunteers 25.
Governing Authority: state. Parent Institution: N.J. Department of Environmental Protection, Division of Parks & Forestry, Trenton, NJ 08625. Tax-exempt.
Institution Type/Description: Historic House: c.1772 Quaker Farm, field hospital after Battle of Princeton.
Collections: Delaware Valley furniture; Revolutionary War displays; Quaker culture; 18th-century farming & domestic items; weapons.
Research Fields: Quaker culture; Battle of Princeton; Princeton & Delaware Valley history; 18th-century farming.
Facilities: 85-acre park.
Activities: guided tours; lectures; films; hobby workshops; organized education programs; docent program; participatory exhibits; demonstrations of historic skills; special events.
Hours & Admission Prices: Wed.-Sat. 10-12 & 1-4, Sun. 1-4. No charge; donations accepted. Closed New Year's Day; Thanksgiving; Christmas. &
Attendance: 90,000 (estimated)

Rahway

MERCHANT AND DROVERS TAVERN MUSEUM, 1632 St. Georges Ave., Rahway, NJ 07065-2006. Mailing Address: P.O. Box 1842, Rahway, NJ 07065-7842. Tel.: 732-381-0441.
E-mail: mdtavernmuseum@gmail.com
Web Site: www.merchantanddrovers.org
Founded: 1969.
Congressional District: 21
Key Personnel: Pres., Annette Satkowski; Vice Pres., Joseph Carpenter; Dir. Museum Operations, Alex Shipley; Program Mgr., Lisa Michaloski.
Personnel Profile: Part-Time Paid 1; Part-Time Volunteers 3.
Operating Expenses: 60,000
Operating Income: 66,250
Governing Authority: Tax-exempt.
Institution Type/Description: Historic Building: built in c.1795. Listed on the National Register of Historic Places.
Collections: early tavern life; local history & culture; photographs; period furnishings.
Facilities: Museum-related items for sale.
Activities: special events; educational programs.
Publications: quarterly newsletter.
Hours & Admission Prices: Thurs.-Fri., and 1st & 3rd Sat. of month 10-4, 2nd & 4th Sun. of month 1-4; other times by appointment. Adults $5, members $4. &
Attendance: 2,300 (accurate)
Membership: Junior $5; Individual $15; Nonprofit & Family $25; Professional & Business $50; Friend $100; Patron $250; Benefactor $500; Sponsor $1,000.

Randolph

HISTORICAL SOCIETY OF OLD RANDOLPH, 630 Millbrook Ave., Randolph, NJ 07869-3730. Mailing Address: P.O. Box 1776, Ironia, NJ 07845-1776. Tel.: 973-989-7095.
E-mail: hsor@juno.com
Web Site: www.randolphnj.org/get_to_know_us/historical_society
Founded: 1979.
Congressional District: 25
Key Personnel: Museum Shop Mgr., Joan Brembs.
Governing Authority: Tax-exempt.
Institution Type/Description: Historical Society Museum: housed in an 1860 farmhouse.
Collections: Randolph's history including industrial, agricultural, resort & suburbia eras; oral histories; school room.

Activities: colonial clothing dress-up for children.
Hours & Admission Prices: April-Nov. Sun. 1-4; tours & other times by appointment. Adults $2; discounts to AAM & ICOM members; members no charge.
Attendance: 1,500 (estimated)
Membership: Senior 65 & up $10; Individual $15; Family $25; Sustaining $100.

Readington

COLD BROOK SCHOOL, Potterstown Rd., Readington, NJ 08870. Mailing Address: P.O. Box 216, Stanton, NJ 08885-0216. Tel.: 908-236-2327. Fax: 908-236-2306.
E-mail: readingtonmuseums@gmail.com
Web Site: readingtontwp.org/ReadingtonMuseums/
Key Personnel: Program Dir., Margaret Smith
Institution Type/Description: Historic Building: housed in a one-room school house, built in 1828.
Collections: period furnishings; personal artifacts.
Activities: tours; demonstrations; 4th grade Partners in History Program. Museum Sponsors: Open House.
Hours & Admission Prices: Call for hours.

Ridgewood

SCHOOLHOUSE MUSEUM OF THE RIDGEWOOD HISTORICAL SOCIETY, INC., 650 E. Glen Ave., Ridgewood, NJ 07450-1905. Tel.: 201-447-3242.
E-mail: info@ridgewoodhistoricalsociety.org
Web Site: www.ridgewoodhistoricalsociety.org
Founded: 1949.
Congressional District: 5
Key Personnel: Pres., Sheila Brogan.
Personnel Profile: Part-Time Volunteers 20.
Governing Authority: nonprofit. Parent Institution: Ridgewood Historical Society. Subsidiary Institution: Schoolhouse Museum. Tax-exempt.
Institution Type/Description: General Museum.
Collections: Bergen County & New Jersey heirlooms from the 17th, 18th & 19th centuries; local Indian artifacts; blacksmith tools; farm tools before the machine age; military; quilts, coverlets. Historic House: 1875 one-room schoolhouse; depicts life in early America in Bergen County from Dutch Era to Victorian Age.
Research Fields: local, community & New Jersey history.
Facilities: reference library of local history, maps & genealogy.
Activities: guided tours; for school children, scouts & adult groups; program meetings; field trips; special exhibits.
Publications: pamphlets, Local History.
Hours & Admission Prices: Thurs. & Sat. 1-3, Sun. 2-4. Adults $5.
Attendance: 800 (estimated)
Membership: Single $25; Family $50; Patron $100.

Ringwood

NEW JERSEY BOTANICAL GARDEN AT SKYLANDS (NJBG), Morris Rd., Ringwood, NJ 07456. Mailing Address: P.O. Box 302, Ringwood, NJ 07456-0302. Tel.: 973-962-9534 & 7527. Fax: 973-962-1553.
E-mail: info@njbg.org
Web Site: www.njbg.org
Founded: 1966.
Congressional District: 5
Key Personnel: Pres., Frank Dyer; Treas., Schuyler Jenks; Landscape Designer, Rich Flynn; Park Supvr., Eric Pain; Museum Shop Mgr., Sonja Vieth.
Personnel Profile: Part-Time Paid 2; Part-Time Volunteers 205.
Volunteer Hours: 11,495
Governing Authority: private; nonprofit organization. Parent Institution: Skylands Association. Tax-exempt: 501(c)(3).
Institution Type/Description: Historic House & Site: 44-room Tudor-style manor house & 96-acre botanical garden.
Collections: 16th-century paneling; 42 stained glass medallions dating to Middle Ages.
Facilities: 600-vol. library on gardening & horticulture; 1,200 sq. ft. exhibit space apart from museum. Museum-related items for sale.
Activities: guided house & garden tours; botanical craft workshops; lectures. Annual Events: plant sale; Champagne and Candle-Light Evenings; Harvest Festival; Holiday Open House in December.
Publications: quarterly newsletter, The Skylands Journal; self-guiding brochures, Self-Guiding Tour-Skylands, A Guide to Skylands Manor; A Guide to Garden Ornamentation at Skylands; Solar System; Birding; Lilac; Wildflower.

Hours & Admission Prices: Gardens: daily 8-8. No charge. Manor House Tours: selected Sun. Adults $7, senior citizens $5, children 6-18 $3; children under 6 no charge.
Attendance: 98,000 (estimated)
Membership: Student & Senior $20; Individual $30; Family & Dual $55; Organization & Sponsor $100; Friend $250.

RINGWOOD MANOR, Ringwood State Park, 1304 Sloatsburg Rd., Ringwood, NJ 07456-1706. Tel.: 973-962-2240. Fax: 973-962-2247.
E-mail: rspris@verizon.net
Web Site: www.ringwoodmanor.com
Formerly: The Forges and Manor of Ringwood
Founded: 1936.
Congressional District: 5
Key Personnel: Supt., Eric Pain; Historic Preservationist, Sue Shutte; Museum Shop Mgr., Ralph Colfax.
Personnel Profile: Full-Time Paid 1; Part-Time Paid 4; Part-Time Volunteers 5; Interns 3.
Governing Authority: state. Parent Institution: New Jersey Dept. of Environmental Protection. Subsidiary Institution: Division of Parks & Forestry. Tax-exempt.
Institution Type/Description: Historic House Museum.
Collections: furniture; archives; paintings; graphics; decorative arts.
Research Fields: New Jersey iron industry 1740-1940.
Facilities: Books & other museum-related items for sale.
Activities: guided tours; permanent exhibitions. Museum Sponsors: Declaration of Independence read in July; Victorian Christmas in December.
Hours & Admission Prices: Wed.-Sun. 10-4. No charge; donations accepted. Closed New Year's Day; Good Friday; Thanksgiving; Christmas. &
Attendance: 15,000 (accurate)
Membership: Associate $10; Active $12; Patron $25, $50, $100 & up.

River Edge

BERGEN COUNTY HISTORICAL SOCIETY, 1201-1209 Main St., River Edge, NJ 07661-2026. Mailing Address: P.O. Box 55, River Edge, NJ 07661-9998. Tel.: 201-343-9492.
E-mail: contactbchs@bergencountyhistory.org
Web Site: www.bergencountyhistory.org
Founded: 1902.
Congressional District: 9
Key Personnel: Pres., Deborah Powell; 1st Vice Pres., Michael Trepicchio; Sec., Rosann Pelligrino.
Governing Authority: private; nonprofit organization. Tax-exempt.
Institution Type/Description: Local History Museum: housed in c.1713, 1752 Steuben House.
Collections: furniture; furnishings; china; glass; farm & household tools; paintings; sculptures; books; maps; manuscripts; photographs; toys; quilts & coverlets; folk art; Wolfkiel pottery; clothing.
Research Fields: history of Bergen County; genealogy; architecture.
Facilities: 1,000-vol. library of New Jersey history located at Felician College, Lodi, NJ, available for research by anyone over 18.
Activities: lectures; formally organized programs for adults; inter-museum loan, permanent & temporary exhibitions; outreach program to 4th & 5th grade school children; black studies committee; marker program for historic bldgs.
Publications: softcover, Bergen County history; hardcover, The United Churches of Hackensack & Schraalenburgh; newsletter, In Bergen's Attic.
Hours & Admission Prices: Call for hours. No charge; donations accepted. Closed New Year's Day; Thanksgiving; Christmas. &
Attendance: 15,000
Membership: Student $5; Individual, Historic Organizations & Libraries $15; Family $25; Contributing $30; Corporate $200; Benefactor $1,000.

RIVER EDGE CULTURAL CENTER, 201 Continental Ave., River Edge, NJ 07661. Mailing Address: P.O. Box 416, River Edge, NJ 07661. Tel.: 201-634-0158.
Web Site: www.recultural.org
Institution Type/Description: Cultural Center.
Collections: local history & culture; personal artifacts; photographs.
Activities: educational programs; special events; concerts; lectures; workshops; temporary exhibitions; classes.
Hours & Admission Prices: Call for hours.

Roebling

ROEBLING MUSEUM, (M), 100 Second Ave., Roebling, NJ 08554. Mailing Address: P.O. Box 9, Roebling, NJ 08554. Tel.: 609-499-7200. Fax: 609-499-7201.
E-mail: rmuseum@roeblingmuseum.org
Web Site: roeblingmuseum.org
Founded: 2007.
Congressional District: 3
Key Personnel: Exec. Dir., Patricia Millen; Pres. (V), Karl Darby; Museum Shop Mgr., Kathleen Lengel.
Personnel Profile: Full-Time Paid 1; Part-Time Paid 2.
Governing Authority: Tax-exempt.
Institution Type/Description: History Museum.
Collections: history of John A. Roebling's Sons Company, the Roebling family, & the city of Roebling; industrial artifacts; technological accomplishments; local social history.
Activities: educational programs; research; special events.
Publications: Roebling Wire, 3 times a year.
Hours & Admission Prices: April-Aug. Wed.-Sun. 11-4; Sept.-Dec. Thurs.-Sun. 11-4; other times by appointment. Adults $6, seniors & children 6-12 $5; members & children under 6 no charge. Closed New Year's Day; Thanksgiving; Christmas. &
Attendance: 5,000
Membership: Foreman $20; Shift Supervisor $50; Roebling Patron $150; Roebling Business Benefactor $250; Roebling Bridge Building Sponsor $500; Lifetime $1,000.

Roseland

ROSELAND HISTORICAL SOCIETY - THE HARRISON HOUSE, 126 Eagle Rock Ave., Roseland, NJ 07068. Mailing Address: P.O. Box 152, Roseland, NJ 07068. Tel.: 973-228-0742.
E-mail: info@roselandhistsocnj.org
Web Site: www.roselandhistsocnj.org
Institution Type/Description: Historical Society Museum: house built in 1824. Listed on the National and State Registers of Historic Places.
Collections: local history & culture; period furnishings & tools; photographs.
Hours & Admission Prices: Call for hours.

Roselle Park

ROSELLE PARK MUSEUM, 9 W. Grant Ave., Roselle Park, NJ 07204-1915. Tel.: 908-245-1776. Facebook: Roselle Park Museum.
E-mail: roselleparkmuseum@gmail.com
Web Site: www.rosellepark.org/history/museuminfo.html
Founded: 1996.
Key Personnel: C.E.O., Bill Frolich; Chm. (V), Karl Ardler; Pres. (V), Pat Butler; Historian, Pat Pagnetti.
Personnel Profile: Part-Time Volunteers 12.
Governing Authority: Parent Institution: Boro of Roselle Park, NJ. Tax-exempt.
Institution Type/Description: History Museum.
Collections: local history; photographs; personal artifacts.
Research Fields: genealogy; local historic events.
Activities: guest speakers program; slide presentations.
Publications: book, History of Roselle Park.
Hours & Admission Prices: Mon. 7pm-9pm, Wed. 10-2. No charge; donations accepted.
Attendance: 1,020 (estimated)
Membership: Individual $2.

Rutherford

MEADOWLANDS MUSEUM, 91 Crane Ave., Rutherford, NJ 07070-2539. Tel.: 201-935-1175. Fax: 201-935-9791.
E-mail: mcadowlandsmuseum@verizon.net
Web Site: www.meadowlandsmuseum.com
Founded: 1961.
Congressional District: 9
Key Personnel: Dir., Heather Hope Kuruvilla; Chm. (V), Rod Leith; Bookkeeper, Carol Eveleens.
Personnel Profile: Part-Time Paid 2; Part-Time Volunteers 16.
Governing Authority: nonprofit organization. Tax-exempt: 501(c)(3).
Institution Type/Description: Local History Museum: housed in Dutch Colonial farm house.
Collections: children's toys & games; local & fluorescent rocks & minerals; textiles; household & kitchenware; local history photographs; local history archives.
Major Exhibits: Historic Homes of the Meadowlands, 11/13-1/14; Super

Retrospective of Football in the Meadowlands, 1/14-2/14; The Great War & the Meadowlands, 3/14-4/14.
Research Fields: local history.
Facilities: 300-vol. library of local history & reference books related to exhibitions available for use on premises by appointment.
Activities: guided tours; lectures; gallery talks; craft workshops; circulating exhibits; educational work with schools; programs for Girl & Boy scouts.
Publications: e-newsletter.
Hours & Admission Prices: Sun. 10-4; groups by appointment. No charge; donations accepted. Closed New Year's Day; Memorial Day; Independence Day; Labor Day; Thanksgiving; Christmas.
Attendance: 2,500 (estimated)
Membership: Senior & Student $15; Individual $25; Family & Contributor $40; Donor $100; Sponsor $150; Patron $250; Benefactor $500.

Salem

PSEG ENERGY & ENVIRONMENTAL RESOURCE CENTER, 244 Chestnut St., Salem, NJ 08079. Tel.: 856-339-3372.
Founded: 2010.
Institution Type/Description: Energy Learning Center: housed in the former Nuclear Training Center.
Collections: energy & energy consumption; environmental challenges.
Facilities: classroom; lab.
Activities: educational programs; group tours.
Hours & Admission Prices: By appointment.

SALEM CITY FIRE MUSEUM, 166 E. Broadway, Salem, NJ 08079. Tel.: 856-362-1550.
Institution Type/Description: Fire Museum: housed in Union Fire Company's second firehouse; built in 1863.
Collections: local history; early firefighting equipment & pumper; fire service artifacts.
Activities: special events.
Hours & Admission Prices: By appointment.

SALEM COUNTY HISTORICAL SOCIETY, (M), 83 Market St., Salem, NJ 08079-1910. Tel.: 856-935-5004. Fax: 856-935-0728. Facebook: Salem County Historical Society.
E-mail: info@salemcountyhistoricalsociety.com
Web Site: www.salemcountyhistoricalsociety.com
Founded: 1884.
Congressional District: 2
Key Personnel: Pres., Maggie Maxwell Mood; Library Staff, Beverly Carr Bradway; Admin. & Cur., Andrew Coldren; Collections Asst., Kimberly Downs; Administrative Asst., Danielle Polonczyk.
Personnel Profile: Part-Time Paid 5; Part-Time Volunteers 50.
Governing Authority: society; bd. of trustees. Tax-exempt.
Institution Type/Description: Historic Building, Site & History Museum.
Collections: archaeology; archives; costumes; glass; furniture; art; Indian artifacts; military; textiles; china; silver; glass; early agricultural, manufacturing & fishing equipment. Historic Buildings: 1721 headquarters, Grant Building; 1735 brick law office; 1840s log cabin.
Research Fields: local genealogy research; local history and culture.
Facilities: 1,000-vol. library of historic deeds & records; education center.
Activities: guided tours; lectures; student programs; quarterly meetings. Society Sponsors: Salem County House & Garden Tour in May; John S. Rock Memorial Lecture in October.
Publications: quarterly newsletter.
Hours & Admission Prices: Tues.-Sat. 12-4. Adults $5; discounts to WHYY, AAA, AAM & ICOM members; members no charge. &
Attendance: 16,000 (estimated)
Membership: Student $10; Individual $25; Family & Household $40; Partner $100; Oak Tree Associate $100-$249; Benefactor $250; Pedersen Provider $250-$499; Life $500; Goodwin Provider $500-$999; Fenwick Benefactor $1,000 & up.

Scotch Plains

OSBORN CANNONBALL HOUSE, 1840 Front St., Scotch Plains, NJ 07076-1103. Mailing Address: P.O. Box 261, Scotch Plains, NJ 07076. Tel.: 908-322-6700, ext. 230.
Key Personnel: Cur., Ginger Bishop
Institution Type/Description: Historic House: housed in the former home of John & Abigail Osborn; built c.1760.
Collections: local history; period furnishings & farming equipment; photographs.
Hours & Admission Prices: March-Dec. 1st Sun. of month 2-4.

Sea Girt

NATIONAL GUARD MILITIA MUSEUM OF NEW JERSEY - SEA GIRT, (M), National Guard Training Center, Rte. 71 & Sea Girt Ave., Sea Girt, NJ 08750. Mailing Address: P.O. Box 277, Sea Girt, NJ 08750-0277. Tel.: 732-974-5966.
Personnel Profile: Part-Time Paid 2.
Governing Authority: state; nonprofit organization. Subsidiary Institution: NJ Dept. of Military & Veterans Affairs. Tax-exempt.
Institution Type/Description: Military Heritage Museum.
Collections: New Jersey military heritage; Army National Guard; Air National Guard; Naval Militia of New Jersey; photographs; uniforms; weapons; military vehicles; art; personal artifacts.
Facilities: archives.
Activities: tours.
Publications: newsletter.
Hours & Admission Prices: Tues. & Thurs. 10-3; other times by appointment. No charge; donations accepted.
Attendance: 3,141 (accurate)
Membership: Individual $10; Life $250.

SEA GIRT LIGHTHOUSE, Beacon & Ocean Ave., Sea Girt, NJ 08750. Mailing Address: Sea Girt Lighthouse Citizens Committee, P.O. Box 83, Sea Girt, NJ 08750-0083. Tel.: 732-974-0514.
Governing Authority: Tax-exempt: 501(c)(3).
Institution Type/Description: Historic Building: housed in a live-in lighthouse; built in 1896.
Collections: lighthouse history; keeper's office; Fresnel lens; photographs; documents; artifacts from Morro Castle, the cruise ship that burned offshore in 1934.
Facilities: Gifts for sale.
Activities: special events; guided tours; rental facilities. Annual Events: Summer Art Show; International Lighthouse Weekend in September.
Hours & Admission Prices: mid-April to Nov. 20 Sun. 2-4; other times by appointment.

Sea Isle City

SEA ISLE CITY HISTORICAL MUSEUM, 4800 Central Ave., Sea Isle City, NJ 08243. Mailing Address: P.O. Box 443, Sea Isle City, NJ 08243-0743. Tel.: 609-263-2992.
Founded: 1983.
Key Personnel: Acting Dir. & Pres. (V), Michael Stafford.
Personnel Profile: Part-Time Volunteers 30.
Governing Authority: Tax-exempt: 501(c)(3).
Institution Type/Description: Historical Museum.
Collections: local history; period furnishings; clothing; photographs.
Facilities: Museum-related items for sale.
Publications: newsletter, Excursion.
Hours & Admission Prices: Mon.-Sat. 10-3. Donations accepted. &
Attendance: 10,000 (accurate)
Membership: Annual $10.

Seabrook

SEABROOK EDUCATIONAL AND CULTURAL CENTER, Upper Deerfield Twsp. Municipal Bldg., 1325 Hwy. 77, Seabrook, NJ 08302-5976. Tel.: 856-451-8393.
E-mail: info@seabrookeducation.org
Web Site: www.seabrookeducation.org
Founded: 1991.
Governing Authority: Tax-exempt.
Institution Type/Description: History Museum.
Collections: local history, industry & culture; personal artifacts; 1950s Seabrook Village model; photographs; books; oral histories.
Hours & Admission Prices: Mon.-Thurs. 9-12. No charge.
Attendance: 400 (estimated)

Short Hills

CORA HARTSHORN ARBORETUM AND BIRD SANCTUARY, 324 Forest Dr. S., Short Hills, NJ 07078-2308. Tel.: 973-376-3587. Fax: 973-379-5059.
E-mail: info@hartshornarboretum.com
Web Site: www.hartshornarboretum.com
Founded: 1960.
Congressional District: 1
Key Personnel: Exec. Dir., Judy Trigg; Co Pres., Lisa Chenofsky Singer; Co Pres., Julia Rubinstein; Business Mgr., Anne Brandeis.

Personnel Profile: Full-Time Paid 2; Part-Time Paid 5; Interns 4.
Governing Authority: nonprofit. Tax-exempt.
Institution Type/Description: Nature Center.
Collections: fossils; birds; insects.
Facilities: nature conservation center; classrooms. Museum-related items for sale.
Activities: guided tours; lectures; hobby workshops; formally organized education programs; docent program or council.
Hours & Admission Prices: Gardens & Grounds: dawn to dusk. Stone House: Mon.-Fri. 9-5, Sat. 10-3. Closed New Year's Eve & Day; Christmas Eve, Day & week.
Membership: Individual $20; Family $30; Sponsor & Group $50; Patron $100; Benefactor $500; Life $1,000.

MILLBURN-SHORT HILLS HISTORICAL SOCIETY, One Station Plaza, Short Hills, NJ 07078. Mailing Address: P.O. Box 243, Short Hills, NJ 07078. Tel.: 973-564-9519. Fax: 973-564-9519 (call first).
E-mail: mshhs@comcast.net
Web Site: www.mshhistsoc.org
Founded: 1975.
Institution Type/Description: Historical Society Museum: housed in the Short Hills train station.
Collections: local history & culture; photographs; period artifacts; maps.
Publications: newsletter, The Thistle.
Hours & Admission Prices: Tues. 1-3, Wed. 3:30-5:30, Thurs. 5:30-7:30, 1st Sun. each month 2-4. No charge.
Membership: Individual $15; Family $25; Patron $50; Sponsor & Business $100; Benefactor $500.

Shrewsbury

SHREWSBURY HISTORICAL SOCIETY MUSEUM, EDUCATION & RESEARCH CENTER, 419 Sycamore Ave., Shrewsbury, NJ 07702-0333. Mailing Address: P.O. Box 333, Shrewsbury, NJ 07702-0333. Tel.: 732-530-7974.
Founded: 1976.
Congressional District: 12
Key Personnel: Pres., Donald W. Burden
Governing Authority: Tax-exempt.
Institution Type/Description: Historical Society Museum.
Collections: local history & culture; photographs; early tools; period furnishings.
Publications: biannual newsletter, Ellis: History of Monmouth County.
Hours & Admission Prices: By appointment. No charge; donations accepted.
♿
Attendance: 384 (accurate)
Membership: Individual $15; Family $25; Business $100; Lifetime $250.

Somers Point

ATLANTIC COUNTY HISTORICAL SOCIETY, (M), 907 Shore Rd., Somers Point, NJ 08244-2335. Mailing Address: P.O. Box 301, Somers Point, NJ 08244-0301. Tel.: 609-927-5218.
E-mail: ahcinfo@comcast.net
Web Site: www.atlanticheritagecenternj.org
Formerly: Atlantic Heritage Center
Founded: 1913.
Congressional District: 2
Key Personnel: Pres., Richard Squires; Cur., Joan Frankel; Librarian, Carol Raph.
Personnel Profile: Full-Time Volunteers 2; Part-Time Volunteers 30; Interns 3.
Governing Authority: exec. bd. Subsidiary Institution: Risley Homestead Committee. Tax-exempt.
Institution Type/Description: History Museum.
Collections: Victorian & 20th-century decorative, useful & fine arts; costumes; textiles; maritime including ship models & half-hulls; instruments; paintings; American Indian; images 1860-1960; military including weapons & uniforms; manuscripts from 1695; collectibles. Historic Houses: 1714 Somers Mansion: house is state owned but furnished (on loan) from our 18th & early 19th-century collections; c.1790 Risley Homestead: Oysterman's farmhouse, 19th & 20th-century furnishings.
Research Fields: genealogy; local & U.S. history.
Facilities: research library; lecture hall.
Activities: guided tours; lectures; permanent & temporary exhibitions.
Publications: yearbook; newsletter, Atlantic Heritage; books, Absegami Yesteryear; Railroading in Atlantic County; Early History of Atlantic County; Glory At Last; Our Stories.
Hours & Admission Prices: Museum: Wed.-Sat. 10-3:30. No charge; donations

accepted. Library: Wed.-Sat. 10-3:30. Research: $5; members no charge. Closed New Year's Day; Good Friday & day after; Independence Day; Thanksgiving; Christmas.
Attendance: 2,500
Membership: Junior $10; Summer $15; Individual $25; Family $35; Nonprofit Organization $50; Corporate Partner $250; Life $250-$500; Corporate Patron $500; Corporate Benefactor $1,000.

SOMERS MANSION, 1000 Shore Rd., Somers Point, NJ 08244. Mailing Address: P.O. Box 107, Cape May Point, NJ 08212. Tel.: 609-927-2212. Fax: 609-927-1827.
Founded: 1941.
Congressional District: 2
Key Personnel: Resource Interpretive Specialist-Historic Research, John Morsa.
Personnel Profile: Full-Time Paid 1.
Governing Authority: state. Parent Institution: NJ Division of Parks & Forestry 08625. Tax-exempt.
Institution Type/Description: Historic House: 1720 Richard Somers House.
Collections: period furnishings.
Activities: guided tours.
Hours & Admission Prices: Wed.-Sun. 10-12 & 1-4. No charge. Closed New Year's Day; Thanksgiving; Christmas.
Attendance: 3,500 (accurate)

Somerset

COLONIAL PARK ARBORETUM AND GARDENS, 150 Mettlers Rd., Somerset, NJ 08873. Tel.: 732-873-2459.
Institution Type/Description: Arboretum & Gardens.
Collections: flowering trees; evergreens; ornamental grass; shrubs; fragrance & sensory garden; rose garden; perennial gardens.
Facilities: rental facilities; nature trails.
Activities: special events; rental facilities; walking trails.
Hours & Admission Prices: Daily sunrise to sunset. No charge.

UKRAINIAN MUSEUM OF NEW JERSEY, INC., 135 Davidson Ave., Somerset, NJ 08873-1358. Tel.: 732-356-0090.
Formerly: Museum of the Ukrainian Orthodox Memorial Church
Key Personnel: Dir., Archbishop Antony
Institution Type/Description: Religious Museum.
Collections: Ukrainian history; religious artifacts; photographs; Ukrainian famine memorial.
Hours & Admission Prices: Call for hours.

Somerville

OLD DUTCH PARSONAGE, 71 Somerset St., Somerville, NJ 08876-2812. Tel.: 908-725-1015.
Founded: 1947.
Key Personnel: Cur., Jim Kurzenberger.
Governing Authority: state. Administered by the Division of Parks & Forestry, New Jersey, Dept. of Environmental Protection, Trenton, NJ 08625.
Institution Type/Description: Historic House: 1751 old Dutch parsonage, home of Jacob Hardenbergh, founder of Rutgers University and its first president.
Collections: Dutch furnishings; replica of period pieces.
Activities: guided tours; hands on workshops for adults & children; lectures; concerts.
Hours & Admission Prices: Wed.-Sat. 10-12 & 1-4, Sun. 1-4; groups by appointment. No charge. Closed federal & state holidays.
Attendance: 5,000

SOMERVILLE FIRE DEPARTMENT MUSEUM, 15 N. Doughty Ave., Somerville, NJ 08876. Mailing Address: 25 West End Ave., Somerville, NJ 08876. Tel.: 908-526-7098 & 4828.
Institution Type/Description: History Museum.
Collections: firefighting history & equipment; personal artifacts; photographs; early fire truck.
Hours & Admission Prices: Sat. 10am - 12pm; other times by appointment.

WALLACE HOUSE, 71 Somerset St., Somerville, NJ 08876-2812. Tel.: 908-725-1015.
Founded: 1947.
Key Personnel: Cur., Jim Kurzenberger.
Governing Authority: state. Administered by the Division of Parks & Forestry, New Jersey, Department of Environmental Protection, Trenton, NJ 08625.
Institution Type/Description: Historic House: 1778 Wallace House, Washington's headquarters during winter of 1778-79.

Collections: original architectural features; kitchen and slave quarters; furnishings; glass.
Research Fields: Revolutionary War; local history.
Activities: guided tours; interpretation programs.
Publications: newsletter, Washington Place Dispatch.
Hours & Admission Prices: Wed.-Sat. 10-12 & 1-4, Sun. 1-4. Groups of 10 or more by appointment only; school groups limited to 30 students at a time. Closed federal & state holidays.
Attendance: 5,000
Membership: Individual $7; Family $10

South Orange

WALSH GALLERY, SETON HALL UNIVERSITY, 400 S. Orange Ave., South Orange, NJ 07079-2697. Tel.: 973-275-2033. Fax: 973-761-9550.
E-mail: jeanne.brasile@shu.edu
Web Site: library.shu.edu/gallery
Key Personnel: Dir., Jeanne Brasile
Institution Type/Description: Art Gallery.
Collections: paintings; photographs; sculpture.
Hours & Admission Prices: Mon.-Fri. 10:30-4:30; call to confirm.

South River

SOUTH RIVER MUSEUM - OLD SCHOOL BAPTIST CHURCH, 64-66 Main St., South River, NJ 08882. Mailing Address: South River Historical & Preservation Society, Inc., P.O. Box 446, South River, NJ 08882. Tel.: 732-613-3078. Facebook: South River History.
E-mail: southriverhistory@gmail.com
Web Site: www.rootsweb.ancestry.com/~njsrhps/
Founded: 1988.
Congressional District: 18
Key Personnel: Pres. (V), Brian Armstrong
Governing Authority: Tax-exempt.
Institution Type/Description: Historical & Preservation Society Museum: housed in a former Baptist church; built in 1805. Listed on the National & NJ Register of Historic Places.
Collections: local history & culture; period furnishings; personal artifacts; photographs; cemetery.
Research Fields: local history & genealogy.
Activities: Museum Sponsors: Public Programs in February, April, June, September, & November.
Publications: 5 times a year, South River Historical & Preservation News; annual historical calendar.
Hours & Admission Prices: 1st Sun. of each month 1:30-3:30, (2nd Sun. is the first is a holiday). No charge; donations accepted. ⅙
Attendance: 300 (estimated)
Membership: Student $5; Adult $10; Family & Couple $15; Business & Corporate $25; Lifetime $250.

Southampton

JACK ALLEN MEMORIAL EARLY COUNTY LIVING MUSEUM, 224 Landing St., Southampton, NJ 08088-8823. Tel.: 609-267-8382. Fax: 609-267-8382.
E-mail: rallen231@comcast.net
Institution Type/Description: History Museum.
Collections: local history & culture; period furnishings; personal artifacts; photographs; early United Bank of Vincentown transaction counter; country general store artifacts; farm equipment & tools.
Hours & Admission Prices: May-Oct. Sun. 1-4. Adults $3.

Sparta

FRIAR MOUNTAIN MODEL RAILROAD MUSEUM, 240 Demarest Rd., Sparta, NJ 07871. Tel.: 973-579-9833.
E-mail: wendy@fmmrm.com
Web Site: www.fmmrm.com
Founded: 2007.
Institution Type/Description: Model Railroad Museum.
Collections: model railroad cars & memorabilia; Revolutionary War exhibits.
Activities: special events. Annual Events: St. Patricks Train in March; Easter Train in April; Thomas Train in June; Halloween Train in October; Christmas Train in December.
Hours & Admission Prices: Fri.-Sun. 10-5. Admission $6. ⅙

Spring Lake

SPRING LAKE HISTORICAL SOCIETY, 423 Warren Ave., Spring Lake, NJ 07762. Mailing Address: P.O. Box 703, Spring Lake, NJ 07762. Tel.: 732-449-0772.
E-mail: djlan34@aol.com
Web Site: springlake.org
Founded: 1978.
Key Personnel: Pres. (V), Liz Campanile; Docent, Thomas Rusoff.
Personnel Profile: Part-Time Volunteers 25.
Governing Authority: Tax-exempt.
Institution Type/Description: Historical Society Museum: housed in Spring Lake's Borough Hall; built in 1897.
Collections: local history & culture; period furnishings; personal artifacts; photographs.
Research Fields: Spring Lake history.
Facilities: library.
Activities: research; temporary exhibitions; quarterly programs; house tours.
Publications: quarterly newsletter.
Hours & Admission Prices: Thurs. & Sun. call for hours; other times by appointment. No charge; donations accepted. ⅙
Attendance: 400 (estimated)
Membership: Annual $20; Business $25; Life $200.

Springfield

SPRINGFIELD HISTORICAL SOCIETY, 126 Morris Ave., Springfield, NJ 07081. Mailing Address: 166 Milltown Rd., Springfield, NJ 07081-2313. Tel.: 973-376-4784 & 912-4464.
Web Site: www.springfield-nj.us/index.php?page=history
Founded: 1954.
Congressional District: 7
Key Personnel: Pres., Margaret Bandrowski
Institution Type/Description: Local History Museum.
Collections: local history items; drawings; prints; decorative arts; costumes; textiles. Historic House: c.1741 The Cannon Ball House.
Facilities: 500-vol. library of local history books available for use by appointment.
Activities: guided tours.
Publications: book, Battle of Springfield.
Hours & Admission Prices: By appointment. Donation Requested $1. Closed holidays.
Attendance: 50 (estimated)
Membership: Annual $7.50; Institutional $25; Life $100.

Stanton

BOUMAN-STICKNEY FARMSTEAD, 114 Dreahook Rd., Stanton, NJ 08885. Mailing Address: P.O. Box 216, Stanton, NJ 08885-0216. Tel.: 908-236-2327. Fax: 908-236-2306.
E-mail: readingtonmuseums@gmail.com
Web Site: www.readingtontwp.org/ReadingtonMuseums/
Key Personnel: Program Dir., Margaret Smith.
Governing Authority: Parent Institution: Readington Museums.
Institution Type/Description: Historic Buildings.
Collections: period furnishings. Historic Buildings: 1741 Dutch stone bank house; 19th century double corn crib; 1820 Dutch barn.
Activities: demonstrations; tours; educational programs. Museum Sponsors: Open House.
Hours & Admission Prices: Call for hours.

Stone Harbor

STONE HARBOR BIRD SANCTUARY, 11400 3rd Ave., Stone Harbor, NJ 08247. Tel.: 609-368-5102.
Institution Type/Description: Wildlife Sanctuary.
Collections: hundreds of species of Colonial wading birds & song birds.
Facilities: nature trails.
Activities: bird watching; walking trails.
Hours & Admission Prices: Call for hours.

THE WETLANDS INSTITUTE, 1075 Stone Harbor Blvd., Stone Harbor, NJ 08247-1424. Tel.: 609-368-1211. Fax: 609-368-3871.
E-mail: ltedesco@wetlandsinstitute.org
Web Site: www.wetlandsinstitute.org
Founded: 1969.
Congressional District: 1
Key Personnel: Exec. Dir., Lenore Tedesco; Chm., Raymond Burke; Museum Shop Mgr., Joyce Ferguson.

Personnel Profile: Full-Time Paid 9; Full-Time Volunteers 20; Part-Time Paid 7; Part-Time Volunteers 40; Interns 19.
Governing Authority: private; nonprofit. Tax-exempt.
Institution Type/Description: Nature Center.
Collections: aquarium, turtles; duck stamp; 6,000 acres of salt marsh; decorative carved decoy collection.
Research Fields: coastal environment.
Facilities: library of coastal wildlife & plants material; aquarium; educational facilities; nature & conservation center; birdwatching trails; viewing decks. Gift items for sale.
Activities: guided tours; lectures; films; study clubs; hobby workshops; organized education programs for children, adults, undergraduate & graduate students; docent program; annual events.
Publications: quarterly newsletter, Views From the Tower; annual report.
Hours & Admission Prices: mid-May to mid-Oct. Mon.-Sat. 9:30-4:30, Sun. 10-4; mid-Oct. to mid-May Fri.-Sun. 9:30-4:30. Adults $8, children under 12 $6; discounts to seniors, military & groups; members no charge. Closed national holidays & two weeks at Christmas. &
Attendance: 50,000 (estimated)
Membership: Individual $25; Family $40; Friend $75; Terrapin $150; Osprey Club $500; Skimmer $1,000; Tower $2,500.

Summit

THE CARTER HOUSE, 90 Butler Pkwy., Summit, NJ 07901-1617. Tel.: 908-277-1747.
E-mail: president@summitnjhistory.org
Web Site: www.summitnjhistory.org/carter.php
Key Personnel: Pres., Christine Siepert Smith.
Governing Authority: Parent Institution: The Summit Historical Society. Tax-exempt.
Institution Type/Description: Historic House: former home of Benjamin Carter; built in 1741.
Collections: local history; period furnishings; books; photographs.
Publications: brochure; newsletter, The Historian.
Hours & Admission Prices: Call for hours. No charge.
Membership: Individual $10; Family $15; Sustaining $25; Patron $50; Benefactor $100; Corporate $150; Life $200.

REEVES-REED ARBORETUM, (M), 165 Hobart Ave., Summit, NJ 07901-2908. Tel.: 908-273-8787. Fax: 908-273-6869.
E-mail: reevesreedarboretum@juno.com
Web Site: www.reeves-reedarboretum.org
Founded: 1974.
Congressional District: 12
Key Personnel: Exec. Dir., Gayle Petty-Johnson; Pres. Bd., Andy Gottesman.
Personnel Profile: Full-Time Paid 7; Part-Time Paid 3; Interns 1.
Governing Authority: nonprofit organization.
Institution Type/Description: Arboretum & Botanical Garden.
Collections: specimen trees; herbs; rose garden; daffodils; perennials; wildflowers; daylilies; wildlife habitat with pond and aquatic life; azaleas; rhododendrons; historical formal gardens. Historic Building: 1889 house.
Research Fields: environmental education; preservation; conservation; horticulture; gardening & garden design.
Facilities: 700-vol. library of botanical & horticultural books available for research to the public & members during office hours, members may check out books; formal, historical gardens; classrooms; education center. Museum-related items for sale.
Activities: guided tours; lectures; docent program; adult garden tours; gardening; workshops; field trips; guided trail walks for groups; special environmental education classes outdoors for Scouts or Brownies; children's summer nature camp; poetry readings & concerts; rental facilities. Arboretum Sponsors: Networks to Nature; Elephant Tree nature camp; family events; fund-raisers; collaborative education classes outdoors for inner-city children; Jazz on the Terrace.
Publications: newsletters; trail guides; brochures: adult education & membership information; herb garden brochure.
Hours & Admission Prices: Grounds: daily dawn-dusk. Office: Mon.-Fri. 9-5. No charge; donations accepted. &
Attendance: 46,003 (accurate)
Membership: Individual $65; Family & Club $100; Patron & Corporate $250; Benefactor $500; Sponsor $1,000. Tuition reductions for members.

∗　**VISUAL ARTS CENTER OF NEW JERSEY, (M),** 68 Elm St., Summit, NJ 07901-3472. Tel.: 908-273-9121. Fax: 908-273-1457.
E-mail: info@artcenternj.org
Web Site: www.artcenternj.org
Formerly: New Jersey Center for Visual Arts
Founded: 1933.
Congressional District: 7

Key Personnel: Exec. Dir., Marion Grzesiak; C.E.O., Ernie Palatucci; Chm. Bd. Trustees, Rachel Weinberger.
Personnel Profile: Full-Time Paid 14; Part-Time Paid 9; Part-Time Volunteers 10; Interns 2.
Governing Authority: nonprofit. Tax-exempt: 501(c)(3).
Institution Type/Description: Visual Arts Center.
Collections: works by contemporary artists.
Research Fields: artists with disabilities.
Facilities: educational facilities. Art work for sale.
Activities: lectures; concerts; temporary & permanent exhibitions; workshops; classes for children, teens, & adults; demonstrations; education programs; docent tours.
Publications: exhibition catalogues; quarterly class catalog.
Hours & Admission Prices: Gallery: Mon.-Wed. & Fri. 10-5, Thurs. 10-8, Sat.-Sun. 11-4; tours by appointment. Suggested Donation: adults $5, seniors & children $3; members, seniors & children under 12 no charge. Closed New Year's Eve & Day; Martin Luther King Jr. Day; Easter; Memorial Day; Independence Day; Labor Day; Columbus Day; Thanksgiving & day after; Christmas Eve & Day. &
Attendance: 60,000 (estimated)
Membership: Senior, Student 18 & over, & Art Educator $35; Individual $50; Dual & Family $75; Friend $125; Contributor $250; Supporter $500; Benefactor $1,000; Art Patron $5,000.

Sussex

DAR VAN BUNSCHOOTEN MUSEUM, 1097 Rte. 23, Sussex, NJ 07461-3732. Tel.: 973-875-4058 & 7634.
E-mail: bjsauve@ptd.net
Founded: 1971.
Key Personnel: Cur., Judy Smith; Regent, Wendy Wyman; Museum Shop Mgr., Diana Matties.
Personnel Profile: Full-Time Volunteers 2.
Governing Authority: Parent Institution: National Society Daughters of the American Revolution. Subsidiary Institution: Chinkchewunska Chapter. Tax-exempt.
Institution Type/Description: Historic House Museum: housed in the former home of Rev. Elias Van Bunschooten; built in 1787. Listed on the National Register of Historic Places.
Collections: personal artifacts; furnishings; clothing; quilts; china & cookware; Revolutionary War weapons; paintings; dolls; farm implements. Historic Buildings: barn; ice house; wagon house; milk room; privy.
Facilities: library.
Activities: docent-led tours; research. Annual Event: Christmas in July.
Publications: triannual Chapter newsletter.
Hours & Admission Prices: May 15 to Oct. 15 Thurs. & Sat. 1-4; groups by appointment. Adults $4, children $2; discounts to seniors on Thurs. & groups of 10 or more.
Attendance: 1,100 (estimated)

SPACE FARMS ZOO AND MUSEUM, 218 Rt. 519, Sussex, NJ 07461-2800. Tel.: 973-875-3223. Fax: 973-875-9397.
E-mail: info@spacefarms.com
Web Site: www.spacefarms.com
Founded: 1927.
Congressional District: 5
Key Personnel: C.E.O., Parker Space; Museum Shop Mgr., Jill Space.
Personnel Profile: Full-Time Paid 5; Part-Time Paid 10; Part-Time Volunteers 2.
Governing Authority: individual operation.
Institution Type/Description: General Museum and Zoo.
Collections: early American tools; household equipment; clocks; phonographs; dolls; guns; autos; wagons; sleighs; farm machinery; Indian artifacts; antiques; mounted wildlife specimens; musical instruments; live wild animals.
Facilities: restaurant; picnic area. Books & gifts for sale.
Activities: formally organized education programs for children.
Hours & Admission Prices: April-Oct. daily 9-5. Adults 13-64 $14, seniors 65 & over $13, children 3-12 $9.50; discounts to groups. &
Attendance: 80,000 (accurate)

Teaneck

PUFFIN CULTURAL FORUM, 20 Puffin Way, Teaneck, NJ 07666. Tel.: 201-836-3499.
Key Personnel: Dir., Andrew Lee.
Governing Authority: Parent Institution: The Puffin Foundation Ltd.
Institution Type/Description: Arts & Cultural Center.
Collections: works by local & national artists.
Facilities: theater.

Activities: multicultural concerts; workshops; guided tours; school groups; lectures; performances.
Hours & Admission Prices: Tues.-Thurs. 12-4; other times by appointment.

Tenafly

AFRICAN ART MUSEUM OF THE S.M.A. FATHERS, 23 Bliss Ave., Tenafly, NJ 07670-3001. Tel.: 201-894-8611. Fax: 201-541-1280.
E-mail: museum@smafathers.org
Web Site: www.smafathers.org
Founded: 1963.
Congressional District: 9
Key Personnel: Pres., Rev. Michael Moran, S.M.A.; Dir., Robert J. Koenig; Registrar & Collections Mgr., Peter H. Cade.
Personnel Profile: Full-Time Paid 1; Full-Time Volunteers 1; Part-Time Paid 4; Part-Time Volunteers 1.
Governing Authority: board of trustees. Parent Institution: Society of African Missions, Inc. (American Province), 23 Bliss Ave., Tenafly 07670. Tax-exempt: 170(b)(1)(A).
Institution Type/Description: African Art Museum: located at the Society of African Missions, American Provincialate.
Collections: traditional arts of sub-Saharan Africa; masks; sculpture; ritual objects; weapons; musical instruments; architectural elements; tools; household implements; furniture; textiles; costumes & jewelry from Liberia, Ivory Coast, Ghana, Mali, Sierra Leone, Benin, Nigeria, Cameroon, Congo, Gabon, Zaire, Kenya, Tanzania, Ethiopia & South Africa.
Research Fields: traditional visual arts of sub-Saharan Africa.
Facilities: 500-vol. library pertaining to African art, by appointment only; changing collection and special loan exhibitions; 1,300 sq. ft. exhibition space; adjacent cloister garden & chapel; extensive documentary catalogs, brochures & educational materials.
Activities: lectures; concerts; special children's programs. Museum Sponsors: Annual African Film Festival.
Publications: brochures & catalogs.
Hours & Admission Prices: Daily 10-5; group tours by appointment. No charge; donations accepted. Closed national & religious holidays. ♿
Attendance: 3,000 (estimated)
Membership: Student & Senior Citizen $6; Individual $10; Family $20; Patron $100-$500; Benefactor $600-$2,500.

Teterboro

AVIATION HALL OF FAME AND MUSEUM OF NEW JERSEY, 400 Fred Wehran Dr., Teterboro, NJ 07608-1114. Tel.: 201-288-6344. Fax: 201-288-5666.
E-mail: njahof@verizon.net
Web Site: www.njahof.org
Founded: 1972.
Key Personnel: Dir., Shea Oakley; Pres. (V), W. Timothy McSwain.
Governing Authority: Tax-exempt.
Institution Type/Description: Aviation Museum.
Collections: NJ aviation heritage; military vehicles, aircraft & artifacts; Bell 47C helicopter; Hall of Fame room; Curtiss Wright J-5 Whirlwind engine used by Charles Lindbergh; space travel; lighter-than-air flight; photographs; Martin 202 Airliner; AH-1 Cobra helicopter; Lockheed Bushmaster; OV-1A Mohawk; Convair 880 cockpit; Coast Guard HH-52A helicopter; Walter Airport Rescue and Firefighting truck.
Facilities: research library; theater. Museum-related items for sale.
Activities: lectures; conferences; group meetings; special events.
Publications: newsletter, Propwash.
Hours & Admission Prices: Tues.-Sun. 10-4. Adults $7, senior citizens & children 2 & over $5; discounts to groups of 10 or more, veterans & AAA members; members no charge. Closed New Year's Eve & Day; Easter; Independence Day; Thanksgiving; Christmas Eve & Day.
Attendance: 5,000 (accurate)
Membership: Junior $5; Senior $20; Individual $25; Family $35; Supporting $50; Sustaining $100; Patron $250; Corporate $500.

Titusville

JOHNSON FERRY HOUSE MUSEUM, Washington Crossing State Park, 355 Washington Crossing Penn Rd., Titusville, NJ 08560-1517. Tel.: 609-737-2515. TDD: 609-737-0623.
E-mail: jfhwashxing@fast.net
Founded: 1912.
Congressional District: 5
Key Personnel: JFH - Resource Interpretive Specialist, Nancy Carter Ceperley; VC - Resource Interpretive Specialist, Clay Craighead.
Personnel Profile: Full-Time Paid 1; Part-Time Paid 1; Part-Time Volunteers 7.

Governing Authority: state. Parent Institution: New Jersey Dept. of Environmental Protection, Trenton, NJ. Tax-exempt.
Institution Type/Description: Historic Building Museum & Visitor Center: housed in a c.1740, ferry house in which George Washington & his officers met after crossing the Delaware River & before the march to Trenton.
Collections: Ferry House Museum: houses colonial furniture, domestic & decorative items; recreation of 18th-century kitchen garden. Visitors Center: houses 900 American Revolutionary War artifacts.
Research Fields: domestic life on an 18th-century ferry plantation before & during the Revolution; early American spiritual heritage.
Facilities: picnic areas; playgrounds; foot paths.
Activities: guided tours of house & garden by request; living history activities & demonstrations; special musical & military events; historic cooking & food preparation; walking & outreach tours; Great Awakening Outreach Tours; lectures; discussions.
Publications: brochure, Johnson Ferry House.
Hours & Admission Prices: Ferry House: Wed.-Sat. 10-4, Sun. 1-4; groups by appointment only. No charge; donations accepted. Parking Fee: Memorial Day-Labor Day Sat.-Sun. $5 per car. Closed New Year's Day; Thanksgiving; Christmas. ♿
Attendance: 10,000 (accurate)
Membership: Friends of the Ferry House: Annual $5.

Toms River

CATTUS ISLAND COUNTY PARK - COOPER ENVIRONMENTAL CENTER, 1170 Cattus Island Blvd., Toms River, NJ 08753. Tel.: 732-270-6960.
Institution Type/Description: Environmental Center.
Collections: nature; environment; live animals; butterfly garden; hands-on exhibitions.
Facilities: nature trails.
Activities: special events; educational programs; hands-on exhibitions; hiking trails.
Hours & Admission Prices: Park: daily dawn to dusk. Center: daily 8-4:30. ♿

INSECTROPOLIS, 1761 Rte. 9, Toms River, NJ 08755-1296. Tel.: 732-349-7090.
E-mail: info@insectropolis.com
Web Site: www.insectropolis.com
Key Personnel: Pres., Thomas Koerner
Institution Type/Description: Bug Museum.
Collections: hands-on exhibits; insects.
Hours & Admission Prices: Tues.-Sat. 10-3. Admission $7; discounts to groups of 15 or more; children under 3 no charge. Closed holidays.

OCEAN COUNTY HISTORICAL SOCIETY, (M), 26 Hadley Ave., Toms River, NJ 08753-7540. Tel.: 732-341-1880. Fax: 732-341-4372. Facebook: Ocean County Historical Society.
E-mail: oceancounty.history@verizon.net
Web Site: www.oceancountyhistory.org
Founded: 1950.
Congressional District: 3
Key Personnel: Pres. & C.E.O., Cynthia Smith; Chm. (V), Kevin Neary; Museum Clerk, Donna Davis; Museum Shop Mgr., Susan Schmied; Clerk Typist, Kim Fleischer.
Personnel Profile: Part-Time Volunteers 45; Interns 7.
Volunteer Hours: 13,500
Operating Expenses: 96,000
Operating Income: 96,905
Governing Authority: society. Tax-exempt.
Institution Type/Description: History Museum.
Collections: Native American artifacts; one room schoolhouse interior; LTA history of the dirigible & Naval Air Station at Lakehurst; Revolutionary War salt works; early Ocean County industry; Victorian furniture & costumes; cranberry & blueberry industry; charcoal industry; chicken farming.
Major Exhibits: The Child Within Us, 11/4/13-2/28/14.
Research Fields: local genealogy.
Facilities: research library; archives.
Activities: lectures; films; guided school tours; programs.
Publications: picture albums; The New Jersey Coast & Pines, originally published in 1889; History of Monmouth & Ocean Counties New Jersey, originally published in 1890; Tides of Time, originally published in 1940; The Woolman & Rose Atlas of the Jersey Coast, originally published in 1878; Place Names of Ocean County New Jersey 1609-1849; Three Centuries on Island Beach; Chickaree In The Wall: a history of the one-room schools of Ocean County; Along the Toms River; Moored to the Mast: A history of lighter than air at Naval Air Station, Lakewood, NJ; From Manahawkin To New Gretna published 1997; F. Slade Dale: The Life of His

Choice published 1998; Living With the Pine Barrens, by Jack Cervetto; When Cranberries Were King by the Thursday Group 2008.
Hours & Admission Prices: Office & Polly Miller exhibit room: Mon.-Fri. 9:30-4, 1st Sat. of month 1-4; tours by appointment. Research Facilities: Tues.-Wed. & 1st Sat. of month. 1-4. Tours: no charge. Programs may require a fee; veterans no charge. &
Attendance: 5,000 (estimated)
Membership: Individual $30; Family $45; Patron $60; Sponsor $100; Benefactor $250.

ROBERT J. NOVINS PLANETARIUM - OCEAN COUNTY COLLEGE, College Dr., Rm. 100, Toms River, NJ 08754. Mailing Address: College Dr., P.O. Box 2001, Toms River, NJ 08754-2001. Tel.: 732-255-0342 & 0343.

E-mail: planetarium@ocean.edu
Web Site: www.ocean.edu/campus/planetarium/index.htm
Institution Type/Description: Planetarium.
Collections: fiber-optic star projector; astronomy.
Facilities: 119-seat theater; classroom. Museum-related items for sale.
Activities: educational programs; workshops; school programs; birthday parties; special events.
Hours & Admission Prices: Call for hours. &

TOMS RIVER SEAPORT SOCIETY & MARITIME MUSEUM, 78 E. Water St., Toms River, NJ 08753-7554. Mailing Address: P.O. Box 1111, Toms River, NJ 08754-1111. Tel.: 732-349-9209. Fax: 732-349-2498.

E-mail: tomsriverssmm@yahoo.com
Web Site: www.tomsriverseaport.org
Founded: 1976.
Key Personnel: Pres., Dan Crabbe; Dir., Tara Lueddeke; Museum Shop Mgr., Crickett Kersens.
Personnel Profile: Part-Time Paid 1; Part-Time Volunteers 30.
Governing Authority: nonprofit organization.
Institution Type/Description: Maritime Museum.
Collections: local maritime heritage & history; watercraft; ship models.
Activities: Annual Event: Wooden Boat Show in July.
Publications: biannual, Seafarer.
Hours & Admission Prices: Tues., Thurs. & Sat. 10-2. No charge; donations accepted.
Attendance: 2,000 (estimated)
Membership: Individual $20; Family $25; Business $35.

VIRGINIA PERLE ART GALLERY, 96 E. Water St., Toms River, NJ 08753. Tel.: 732-244-4300.

Founded: 2002.
Institution Type/Description: Art Gallery.
Collections: works by Virginia Perle and other local artists; paintings.
Activities: special events; permanent & temporary exhibitions.
Hours & Admission Prices: Wed.-Sat. 12-4; other times by appointment.

Trenton

MEREDITH HAVENS FIRE MUSEUM OF TRENTON, Trenton Fire Dept. Headquarters, 244 Perry St., 1st Fl., Trenton, NJ 08618-3926. Tel.: 609-989-4038. Fax: 609-989-4280.

E-mail: firemuseum@trentonnj.org
Web Site: trentonfiremuseum.org
Key Personnel: Chm. (V), Dennis M. Keenan.
Personnel Profile: Part-Time Volunteers 14.
Governing Authority: Tax-exempt.
Institution Type/Description: Fire Museum.
Collections: fire-fighting history & equipment; photographs; personal artifacts.
Hours & Admission Prices: Mon.-Fri. 9-5, Sat. 10-4; groups of 10 or more by appointment. &
Attendance: 700 (estimated)

NEW JERSEY OFFICE OF HISTORIC SITES, 501 E. State St., Trenton, NJ 08625-0420. Mailing Address: NJ Div of Parks & Forestry, Mail Code 501-04, P.O. Box 420, Trenton, NJ 08625-0420. Tel.: 609-777-0238. Fax: 609-984-0503.

E-mail: beverly.weaver@dep.state.nj.us
Web Site: www.njparksandforests.org/historic
Founded: 1903.
Key Personnel: Admin., Beverly A. Weaver.
Personnel Profile: Full-Time Paid 1; Part-Time Paid 1.
Governing Authority: state. Parent Institution: State of New Jersey Dept. of Environmental Protection, Div. of Parks & Forestry. Branch Museums: Allaire Village, Farmingdale; Batsto Village, Hammonton; Boxwood Hall, Elizabeth; Clarke House, Princeton; Craig House, Freehold; Delaware & Raritan Canal State Park, Somerset; Liberty State Park, CRRNJ Terminal, Jersey City; Grover Cleveland Birthplace, Caldwell; Hancock House, Hancock's Bridge; Indian King Tavern, Haddonfield; Johnson Ferry House, Titusville; Old Dutch Parsonage, Somerville; Ringwood Manor, Ringwood; Rockingham, Kingston; Somers Mansion, Somers Point; Steuben House, River Edge; Twin Lights, Highlands; Wallace House, Somerville; Walt Whitman House, Camden; Monmouth Battlefield, Freehold; Princeton Battlefield, Princeton; Fort Mott, Pennsville; Washington Crossing State Park, Titusville; Double Trouble State Park, Bayville. Tax-exempt.
Institution Type/Description: Historic Sites & Villages.
Collections: period furnishings, decorative art; tools; fire engines; manuscripts; transportation vehicles.
Research Fields: 18th- to 20th-century American, New Jersey & maritime history.
Facilities: library; meeting rooms. Museum-related items for sale.
Activities: guided & self-guided tours; permanent & changing exhibits; school programs; special events; craft demonstrations; lecture series; films; reenactments. Annual Event: State History Fair.
Publications: brochures; posters; guidebooks.
Hours & Admission Prices: Wed.-Sat. 10-12 & 1-4, Sun. 1-4. Program fees might apply at certain locations. Closed New Year's Day; Thanksgiving; Christmas; Wed. after a Mon. holiday. &
Attendance: 970,000 (estimated)

NEW JERSEY STATE HOUSE, 125 W. State St., Trenton, NJ 08608-1101. Mailing Address: State House Tour Office, State House Annex, P.O. Box 068, Trenton, NJ 08625-0068. Tel.: 609-633-2709. Fax: 609-292-1498.

E-mail: dapril@njleg.org
Web Site: www.njleg.state.nj.us
Founded: 1792.
Congressional District: 4
Key Personnel: Tour Program Coord., David April; Tour Program Educator, Sarah Schmidt.
Personnel Profile: Full-Time Paid 4; Part-Time Paid 2; Part-Time Volunteers 50.
Governing Authority: state; nonprofit. Parent Institution: State of New Jersey. Subsidiary Institution: Office of Legislative Services. Tax-exempt.
Institution Type/Description: Historic Building: The New Jersey State House constructed in 1792.
Collections: period furnishings & equipment associated with New Jersey history and the operations of state government from the early federal period to present day, with special emphasis on portraits of former Governors and legislative figures.
Facilities: multi-purpose space. Museum-related items for sale.
Activities: docent program, guided tours. Museum Sponsors: Heritage Days; Holiday Open House.
Publications: monthly newsletter, Volunteer News.
Hours & Admission Prices: Mon.-Fri. 10-3, Sat. 12-3. No charge. Closed state holidays. &
Attendance: 39,277 (accurate)

* NEW JERSEY STATE MUSEUM, (M), 205 W. State St., Trenton, NJ 08608-1001. Mailing Address: P.O. Box 530, Trenton, NJ 08625-0530. Tel.: 609-292-6300 & 6301. Fax: 609-292-7636.

E-mail: feedback@sos.state.nj.us
Web Site: www.newjerseystatemuseum.org
Founded: 1895.
Congressional District: 4
Key Personnel: Exec. Dir., Anthony Gardner; Pres. Bd. Trustees, Steve M. Richmond; Pres. Friends, Karen Ali; Asst. Cur. Archaeology & Ethnology, Greg Lattanzi; Business Mgr., Barbara Bower, Supvr. Exhibits, Elizabeth Beitel, Cur. Natural History, David Parris; Asst. Cur. Planetarium, Jay Schwartz; Cur. Fine Art, Margaret M. O'Reilly; Registrar, Archaeology & Ethnology, Jessie Cohen; Cur. Natural History, Jason Schein; Natural History Registrar, Rodrigo Pellegrini; Cur. Cultural History, Nicholas Ciotola; Museum Shop Mgr., Lorraine Starr-Curtin; Friends of the State Museum Exec. Dir., Nicole Jannotte; Registrar Fine Art & Cultural History, Jenny Martin-Wicoff.
Personnel Profile: Full-Time Paid 26; Part-Time Paid 10; Part-Time Volunteers 198; Interns 9.
Governing Authority: state. Parent Institution: New Jersey Dept. of State, P.O. Box 300, Trenton, NJ 08625. Tax-exempt.
Institution Type/Description: General Museum.
Collections: fine arts; decorative arts; cultural history; archaeology; ethnology; natural history.

Research Fields: art, paleontology, science, history, decorative arts, archaeology & ethnology of New Jersey.
Facilities: staff library; planetarium; auditorium; meeting rooms. Museum-related items for sale.
Activities: guided tours; lectures; films; gallery talks; concerts; science festivals; formally organized education programs; inter-museum loan, permanent, temporary & traveling exhibitions; specimen loans to schools.
Publications: exhibition catalogs; scholarly publications.
Hours & Admission Prices: Tues.-Sat. 9-4:45, Sun. 12-5. No charge; donations accepted. Closed state holidays. &
Attendance: 300,000 (estimated)
Membership: Student & Senior $20; Individual $40; Family $60; Contributor $100; Sponsor $250; Patron $500; Founder $1,000; Director's Associate $2,500; Chairman's Circle $5,000.

OLD BARRACKS MUSEUM, (M), 101 Barrack St., Trenton, NJ 08608-2007. Tel.: 609-396-1776. Fax: 609-777-4000.
E-mail: barracks@voicenet.com
Web Site: www.barracks.org
Founded: 1902.
Congressional District: 12
Key Personnel: Exec. Dir., Richard Patterson; Pres. (V), Michael Keeler; Office Mgr., Linda Mathies; Museum Shop Mgr., Vikki Bell; Chief Historical Interpreter, Gloria Bell; Tour Coord., Renee Kato.
Personnel Profile: Full-Time Paid 9; Part-Time Paid 4; Part-Time Volunteers 150.
Governing Authority: nonprofit organization. Tax-exempt: 501(c)(3).
Institution Type/Description: History Museum: housed in British military barracks built in 1758 during the French & Indian War.
Collections: Revolutionary War; period rooms; American furniture; ceramics; Chinese porcelain; American silver, firearms & accoutrements; portraits; books & manuscripts; reconstructed barracks interior; exhibits on the Battle of Trenton; 18th century beehive bake oven; Battle of Trenton; New Jersey's role in the French & Indian War; the women who saved the barracks.
Research Fields: 18th-century American military & social history; decorative arts; Delaware Valley culture.
Facilities: Museum-related items for sale.
Activities: tours; permanent & temporary exhibitions; educational programming; first-person historical interpretation; summer history day camp; re-enactment of the Battle of Trenton; special events
Publications: brochures & pamphlets; quarterly newsletter, The Parade; various publications on the barracks building, collections & the War for Independence in New Jersey.
Hours & Admission Prices: Mon.-Sat. 10-5. Adults $8, senior citizens & students $6; discounts to active military; children under 6 & members no charge. Call ahead for handicap assistance. Closed New Year's Day; Easter; Thanksgiving; Christmas Eve & Day. &
Attendance: 23,552 (estimated)
Membership: Cadet (Student w/ID) $30; Ensign $40; Lieutenant $100; Captain $150; Major $250; Washington's Company: $1,000. Seniors deduct 20% off membership levels.

THE 1719 WILLIAM TRENT HOUSE MUSEUM, 15 Market St., Trenton, NJ 08611-2147. Tel.: 609-989-3027 & 0087. Fax: 609-278-7890.
Web Site: www.williamtrenthouse.org
Formerly: William Trent House
Founded: 1939.
Congressional District: 4
Key Personnel: Admin., Trent House Association, Berverly Mills.
Personnel Profile: Part-Time Paid 5; Part-Time Volunteers 18; Interns 3.
Governing Authority: nonprofit organization; municipal. Parent Institution: City of Trenton. Subsidiary Institution: Trent House Association. Tax-exempt: 501(c)(3).
Institution Type/Description: Historic House: 1719 William Trent House.
Collections: early baroque furniture & decorative arts; ceramics; glass.
Research Fields: local history of the Colonial period; biographies of residents; foundation & development of Trenton.
Facilities: 100-vol. library of local history books; classroom; picnic area. Museum-related items & Colonial reproductions for sale.
Activities: guided tours; lectures; programs; events.
Publications: periodic newsletter; brochures; booklet, The 1719 William Trent House Museum.
Hours & Admission Prices: Daily 12:30-4; groups by appointment. Adults $4, senior citizens, students & children 12 and under $3; discounts to AAA & AAM members; members no charge. Closed holidays. &
Attendance: 15,000 (estimated)

Membership: Association: Individual $25; Family $45; Sustaining $100; Patron $250; Sponsor $500 & up; Benefactor $1,000 & up.

TRENTON CITY MUSEUM, 299 Parkside Ave., Trenton, NJ 08618. Mailing Address: 319 E. State St., Trenton, NJ 08608-1809. Tel.: 609-989-3632. Fax: 609-989-3624.
E-mail: brianohill@ellarslie.org
Web Site: www.ellarslie.org
Founded: 1971.
Congressional District: 12
Key Personnel: C.E.O. & Dir., Mr. Brian O. Hill; Pres. (V), Carolyn Sletson; Museum Shop Mgr., Carol Hall.
Personnel Profile: Full-Time Paid 2; Part-Time Paid 4; Part-Time Volunteers 33; Interns 2.
Governing Authority: municipal; nonprofit. Parent Institution: City of Trenton, NJ. Subsidiary Institution: The Trenton Museum Society. Tax-exempt.
Institution Type/Description: General Museum: housed in c.1850 Ellarslie 34-room Tuscan Villa, designed by John Notman.
Collections: Trentoniana; Trenton-made pottery; porcelain; paintings; furniture.
Research Fields: Trenton ceramics; Trenton history; Trenton industry.
Facilities: specialized reference library. Museum-related items for sale.
Activities: guided tours; lectures; films; gallery talks; concerts; docent program; inter-museum loan; permanent, temporary & traveling exhibitions.
Publications: annual, Museum Society Newsletter.
Hours & Admission Prices: Tues.-Sat. 11-3, Sun. 1-4. Suggested Donation: adults $5. Closed national holidays. &
Attendance: 22,000 (estimated)
Membership: Individual $45; Family $65; Patron $125; Donor $250; Sponsor $500; Benefactor & Corporate $1,000.

Tuckerton

GIFFORDTOWN SCHOOLHOUSE MUSEUM, 35 Leitz Blvd., Tuckerton, NJ 08087. Mailing Address: P.O. Box 43, Tuckerton, NJ 08087. Tel.: 609-294-1547.
E-mail: tuckertonhistoricalsociety@gmail.com
Web Site: tuckertonhistoricalsociety.org
Founded: 1972.
Congressional District: 9
Governing Authority: Parent Institution: Tuckerton Historical Society.
Institution Type/Description: Historic Building: housed in the former Giffordtown one-room schoolhouse.
Collections: local history & culture; photographs; documents; clothing; newspapers; Native American artifacts; period furnishings; genealogy.
Hours & Admission Prices: June-Sept. Wed. 10-4, Sat. 2-4; Oct.-May Wed. 10-4. No charge; donations accepted.
Membership: Annual $5.

TUCKERTON JUNCTION RAILROAD COMPANY, 213C E. Main St., Tuckerton, NJ 08087. Tel.: 609-812-0300.
Web Site: www.tuckertonseaport.org/exhibits
Institution Type/Description: Hobby Museum.
Collections: model railroad including Lionel trains.
Hours & Admission Prices: Wed.-Sun. call for hours.

TUCKERTON SEAPORT MUSEUM, 120 W. Main St., Tuckerton, NJ 08087-2237. Mailing Address: P.O. Box 52, Tuckerton, NJ 08087-0052. Tel.: 609-296-8868. Fax: 609-296-5810.
E-mail: info@tuckertonseaport.org
Web Site: www.tuckertonseaport.org
Formerly: Tuckerton Seaport A Project of Barnegat Bay Decoy & Baymen's Museum, Inc.
Founded: 1989.
Congressional District: 3
Key Personnel: Exec. Dir., Paul Hart; Pres. (V), Joseph Martin, Esq.; Dir. Jersey Shore Folklife Center, Jaclyn Stewart; Museum Shop, Charlene Ackerman; Administrative Asst., Dot Dow.
Personnel Profile: Full-Time Paid 4; Part-Time Paid 18; Part-Time Volunteers 400; Interns 2.
Governing Authority: private; nonprofit. Parent Institution: Barnegat Bay Decoy and Baymen's Museum.Tax-exempt: 501(c)(3).
Institution Type/Description: Maritime Museum.
Collections: baymen's tools: oyster tongs, clamming rakes, decoys, eel & clam pots, shrimp nets, snapper poles, fyke nets, bird calls, boat & sail-making tools; charter boats, sneakboxes, garvey, sloops; photographs; life car.
Research Fields: Folk life; folk art.
Activities: lectures; loan exhibitions; workshops; classes; children & adult programs; tours; boat rides (summer). Annual Events: ecotours.

Publications: quarterly newsletter.
Hours & Admission Prices: Daily 10-5. Adults $8, senior citizens 62 & over $6, children 6-12 $3; discounts to AAM members; members, children 5 & under no charge. Closed New Year's Eve & Day; Easter; Thanksgiving; Christmas. &
Attendance: 30,000 (accurate)
Membership: Individual $25; Family $35.

Union

CALDWELL PARSONAGE, 909 Caldwell Ave., Union, NJ 07083-6754. Tel.: 908-964-9047.
Formerly: Reverend James and Hannah Caldwell Parsonage
Founded: 1957.
Congressional District: 7
Key Personnel: Dir. & Treas., Michael Yesenko; Chm. (V), Thomas Beisler; Pres. (V), David Arminio; Vice Pres. & Museum Shop Mgr., Joseph Canerelli; Sec. & Museum Shop Mgr., Barbara La Mort.
Personnel Profile: Full-Time Volunteers 1; Part-Time Volunteers 1.
Governing Authority: Parent Institution: Union Township Historical Society. Tax-exempt.
Institution Type/Description: History & Culture Museum: housed in the former home of Presbyterian minister, Rev. James Caldwell; built in 1730, burned to the ground by the British Army in 1780; rebuilt in 1782. Listed on the National Register of Historic Places.
Collections: local history & culture; clothing, furniture, & paintings of Rev. James & Hannah Caldwell; Revolutionary War history & military artifacts including muskets, swords, cannon balls, & maps; household artifacts; photographs; period artifacts; tools.
Research Fields: local history; Union County & New Jersey State history; Battle of Connecticut Farms June 7, 1780; General George Washington's campaigns of 1779 & 1780.
Activities: Annual Event: Hannah Caldwell Day in June.
Publications: Township of Union: A Bicentennial History; Reverend James & Hannah Caldwell; Battle of Connecticut Farms; Brigadier General William Maxwell and the New Jersey Brigade During the American Revolutionary War.
Hours & Admission Prices: 3rd Sun. of month 1-5; other times by appointment. No charge; donations accepted. &
Attendance: 1,250 (accurate)
Membership: Individual $20; Family $30; Lifetime $100.

KARL & HELEN BURGER GALLERY, (M), Kean University, CAS Bldg., 1000 Morris Ave., Union, NJ 07083-7133. Tel.: 908-737-4452 & 0392. Fax: 908-737-4416.
E-mail: ntetkows@kean.edu
Web Site: www.kean.edu/~gallery/cas-gallery.html
Formerly: James Howe Gallery
Founded: 1971.
Congressional District: 12
Key Personnel: Dir. University Galleries, Neil Tetkowski.
Governing Authority: state. Parent Institution: Kean University. Tax-exempt.
Institution Type/Description: College Art Gallery.
Collections: 19th & 20th century artists: Audubon, Walter Darby Bannard, Werner Drewes, Homer, Nordfeldt, Rauschenberg, Rosenquest, Max Ernst, Tony Smith, Joseph Stella; art by alumni, faculty & undergraduates.
Research Fields: original exhibitions; historical & contemporary; visual communications; interior design & crafts.
Facilities: library of liberal arts & education, college archives, New Jersey history, rare books & rare art books, maps, prints; 1,000 seat auditorium; faculty dining room. Books and crafts for sale in book store.
Activities: gallery talks to community; programs for museum training, theory, practice, internship; gallery openings.
Publications: George Segal Portraits: Drawings & Sculpture.
Hours & Admission Prices: Sept.-May Mon.-Thurs. 10:30-4:30, Fri. 10:30-4. No charge. Closed college holidays. &

LIBERTY HALL MUSEUM AT KEAN UNIVERSITY, (M), 1003 Morris Ave., Union, NJ 07083-7120. Tel.: 908-527-0400. Fax: 908-352-8915.
E-mail: libertyhall@kean.edu
Web Site: kean.edu/libertyhall
Founded: 1968.
Congressional District: 7
Key Personnel: C.E.O., John Kean, Sr.; Pres. (V), Joel D. Siegel, Esq.; Exec. Dir., Richard J. O'Neill; Dir. Operations, William P. Schroh, Jr.; School & Family Programs, Lorraine Bartone; Collection Mgr., Rachel Goldberg; Museum Store Mgr., Isbett Checo.
Personnel Profile: Full-Time Paid 6; Part-Time Paid 15; Part-Time Volunteers 25; Interns 12.

Governing Authority: private; nonprofit organization. Tax-exempt: 501(c)(3).
Institution Type/Description: Historic Building: c.1772 home of Gov. William Livingston, the first elected governor of New Jersey.
Collections: Livingston & Kean family history from 1770s-1990s. Historic Buildings: carriage house; wagon shed; ice house; fire house.
Facilities: formal gardens. Museum-related items for sale.
Activities: docent program; formal education programs for children; guided tours; seasonal special events.
Hours & Admission Prices: Wed.-Sat. 10-4. Adults $10, seniors & students $6; discounts to AAM & ICOM members; children under 3 & members no charge. &
Attendance: 15,000 (accurate)
Membership: Senior & College Student $20; Individual $25; Grandparent & Single Parent $35; Family $45; Military $75; Colonist $150; Rebel $250; Patriot $500.

Upper Montclair

THE ESSEX COUNTY PRESBY MEMORIAL IRIS GARDENS, 474 Upper Mountain Ave., Upper Montclair, NJ 07043-1523. Tel.: 973-783-5974. Fax: 973-783-3833.
E-mail: presbyiris@comcast.net
Web Site: presbyirisgardens.org
Founded: 1927.
Key Personnel: Pres. (V), Nancy Skjei-Lawes; Dir., Linda Sercus
Governing Authority: Tax-exempt.
Institution Type/Description: Iris Display Garden.
Collections: 6 species & over 3,000 varieties of irises. Historic Building: 1865 house.
Activities: children's outreach programs; children's iris art exhibit; group tours; iris identification classes. Museum Sponsors: Memorial Weekend Lawn Party.
Hours & Admission Prices: mid-May to early June 10-8. Suggested Donation: $5. &
Attendance: 10,000 (estimated)
Membership: Annual $50.

MONTCLAIR STATE UNIVERSITY ART GALLERIES - GEORGE SEGAL GALLERY, (M), Valley Rd. & Normal Ave., Upper Montclair, NJ 07043. Tel.: 973-655-3382. Fax: 973-655-7665.
E-mail: artgalleries@mail.montclair.edu
Web Site: www.montclair.edu/arts/galleries
Founded: 1972.
Congressional District: 8
Key Personnel: Dir., Teresa Rodriguez.
Personnel Profile: Full-Time Paid 2; Part-Time Paid 3; Interns 2.
Governing Authority: public college; nonprofit. Tax-exempt.
Institution Type/Description: University Art Gallery.
Collections: sculpture; drawings; photos; fine art; paintings; posters; prints.
Facilities: 1,800 sq. ft. exhibit space.
Activities: temporary exhibitions.
Publications: catalogs; brochures.
Hours & Admission Prices: Sept.-July Tues.-Wed. & Fri.-Sat. 10-5, Thurs. 12:30-7:30. No charge; donations accepted. Closed New Year's Day; Good Friday; Easter; Labor Day; Thanksgiving; Christmas. &
Attendance: 5,000 (estimated)

Upper Saddle River

UPPER SADDLE RIVER HISTORICAL SOCIETY, 245 Lake St., Upper Saddle River, NJ 07458-1699. Tel.: 201-327-8644.
Web Site: usrhistoricalsociety.org
Institution Type/Description: Historical Society Museum: housed in the Hopper-Goetschius House.
Collections: area history & culture; photographs; personal artifacts.
Activities: special events; meetings. Museum Sponsors: Spring Concert; Harvest Fair in Fall; Old Time Holiday Open House in December.
Hours & Admission Prices: June-July Sun. 2-4; other times by appointment.

Vineland

VINELAND HISTORICAL AND ANTIQUARIAN SOCIETY, 108 S. Seventh St., Vineland, NJ 08362. Mailing Address: P.O. Box 35, Vineland, NJ 08362. Tel.: 856-691-1111.
E-mail: vinelandhistory@gmail.com
Web Site: www.vinelandhistory.org
Founded: 1864.
Institution Type/Description: History Museum.

Collections: local history & culture; military artifacts; Native American; musical instruments; paintings.
Facilities: library; archives.
Publications: annual magazine, Images of America: Vineland Historical Society.
Hours & Admission Prices: Tues.-Sat. 1-4. No charge; donations accepted. Closed holidays.

Wall

INFOAGE SCIENCE HISTORY LEARNING CENTER, 2201 Marconi Rd., Wall, NJ 07719. Tel.: 732-280-3000.
Key Personnel: Dir., Fred Carl
Institution Type/Description: Science & History Center: housed on the site of Camp Evans, the former Marconi Belmar Trans-Atlantic Wireless station. Listed on the National Register of Historic Places.
Collections: communications technology from spark-gap wireless technology to microwave, fiber optics & satellite technology; National Broadcasters Hall of Fame.
Hours & Admission Prices: Wed. & Sat.-Sun. 1-5. Suggested Donation: $5 per person.

NEW JERSEY MUSEUM OF TRANSPORTATION, Allaire State Park, 4265 Atlantic Ave., Wall, NJ 07727-3715. Mailing Address: P.O. Box 622, Farmingdale, NJ 07727-0622. Tel.: 732-938-5524.
E-mail: office@njmt.org
Web Site: www.njmt.org
Governing Authority: Tax-exempt: 501(c)(3).
Institution Type/Description: Transportation Museum.
Collections: transportation history; railroad artifacts & equipment.
Activities: special events; birthday parties; train rides.
Hours & Admission Prices: Call for hours.

OLD WALL HISTORICAL SOCIETY, 1701 New Bedford Rd., Wall, NJ 07719. Tel.: 732-974-1430.
Institution Type/Description: Historical Society Museum.
Collections: local history & culture; personal artifacts; photographs.
Hours & Admission Prices: Call for hours.

Waretown

WELLS MILLS COUNTY PARK AND NATURE CENTER, 905 Wells Mills Rd., Waretown, NJ 08758. Tel.: 609-971-3085.
Institution Type/Description: Park and Nature Center.
Collections: pine & oak forest; animals; plants; ecosystem.
Facilities: library.
Activities: observation deck.
Hours & Admission Prices: Call for hours.

Warren

WAGNER FARM ARBORETUM AND GARDENS, 197 Mountain Ave., Warren, NJ 07059. Tel.: 908-350-7383.
Governing Authority: nonprofit organization.
Institution Type/Description: Arboretum.
Collections: trees; shrubs; flowers; plants.
Activities: educational programs.
Hours & Admission Prices: Call for hours.

Washington

BLUE ARMY OF OUR LADY OF FATIMA AND THE IMMACULATE HEART OF MARY, 674 Mountain View Rd., Washington, NJ 07882. Tel.: 908-689-7200.
Institution Type/Description: Religious Museum.
Collections: replica of historic chapel of Fatima; gardens.
Facilities: 1,400 seat sanctuary; retreat center; shrine. Gift items for sale.
Hours & Admission Prices: Call for hours.

MERRILL CREEK RESERVOIR, 34 Merrill Creek Rd., Washington, NJ 07882. Tel.: 908-454-1213. Fax: 908-454-2747.
Web Site: www.merrillcreek.com
Key Personnel: On-site Coord., Jim Mershon
Institution Type/Description: Nature Preserve.
Collections: local history & culture; mounted wildlife.
Facilities: visitor center; 100-seat auditorium; classrooms.
Activities: educational programs; internships; bird feeding; wildlife viewing; butterfly/hummingbird demonstration garden.

Hours & Admission Prices: Visitors Center: Mon.-Fri. 8:30-4:30, Sat.-Sun. 10-4. Outdoor Areas: daily dawn to dusk. Closed New Year's Day; Easter; Thanksgiving & day after; Christmas Eve & Day.

Watchung

WATCHUNG ARTS CENTER, 18 Stirling Rd., Watchung, NJ 07069. Tel.: 908-753-0190.
E-mail: wacenter@optonline.net
Web Site: www.watchungarts.org
Key Personnel: Exec. Dir., Stacy Gannon.
Governing Authority: nonprofit organization.
Institution Type/Description: Arts Center: housed in a former schoolhouse; c.1888.
Collections: works by professional & emerging artists; paintings; photographs.
Facilities: rental facilities.
Activities: classes; workshops; performances; educational programs; special events; rental facilities.
Hours & Admission Prices: Wed.-Fri. 12-3; other times by appointment.

Wayne

DEY MANSION MUSEUM, 199 Totowa Rd., Wayne, NJ 07470-3108. Tel.: 973-696-1776. Fax: 973-696-1365; 973-523-8712 (parks).
E-mail: deymansion@passaiccounty.nj.org
Formerly: Dey Mansion/Washington's Headquarters
Founded: 1934.
Congressional District: 8
Key Personnel: Dir., Nick Roca; Coord., Arlene R. Potenzone.
Governing Authority: county. Maintained & operated by The Passaic County Dept. of Parks.
Institution Type/Description: Historic House Museum: headquarters of the Continental Army during July, October-November, 1780.
Collections: period furniture restored & furnished in 18th-century style; early American artifacts; gardens.
Research Fields: history.
Facilities: picnic area.
Activities: guided tours. Annual Events: Military Group Day; Craft Days.
Publications: quarterly newsletter.
Hours & Admission Prices: Wed.-Fri. 1-4, Sat.-Sun. 10-12 & 1-4; group tours by appointment, last tour begins at 3:30. Adults $1; children under 10 no charge. Closed major holidays.
Attendance: 3,500 (accurate)

UNIVERSITY GALLERIES - BEN SHAHN CENTER FOR THE VISUAL ARTS - WILLIAM PATERSON UNIVERSITY, (M), 300 Pompton Rd., Wayne, NJ 07470-2152. Tel.: 973-720-2654. Fax: 973-720-3290. Facebook: WPU Galleries.
E-mail: evangelistak@wpunj.edu
Web Site: www.wpunj.edu/coac/gallery
Founded: 1979.
Key Personnel: Gallery Dir., Kristen Evangelista; Public Rels., Mary Beth Zeman; Program Asst., Emily Johnsen.
Personnel Profile: Full-Time Paid 2; Part-Time Paid 5; Part-Time Volunteers 10; Interns 3.
Governing Authority: state government & public college. Parent Institution: The William Paterson University of New Jersey. Tax-exempt: 501(c)(3).
Institution Type/Description: College Art Gallery.
Collections: 19th-century landscape paintings; sculptures; 1950s-present; African & Oceanic; artists' book; outdoor sculpture.
Facilities: 5,000 sq. ft. exhibit space.
Activities: guided tours; artist talks; panel discussions; traveling exhibitions; juried exhibition; arts education workshops.
Publications: exhibition catalogs.
Hours & Admission Prices: Mon.-Fri. 10-5, Sun. call for hours; other times by appointment. No charge. Closed holidays.
Attendance: 10,000 (estimated)

VAN RIPER-HOPPER HOUSE MUSEUM AND VAN DUYNE HOUSE MUSEUM, 533 Berdan Ave., Wayne, NJ 07470-2026. Tel.: 973-694-7192, ext. 3375. Fax: 973-694-9100.
Web Site: www.waynetownship.com/
Founded: 1964.
Congressional District: 8
Key Personnel: Pres., Bob Monacelli.
Personnel Profile: Part-Time Paid 1.
Governing Authority: municipal. Affiliated with Township of Wayne. Tax-exempt.

Institution Type/Description: Historic House: 1786 Dutch Colonial Farmhouse restored.
Collections: furnishings from Colonial period to 1860; Indian artifacts; clothing; artifact display; A.P. Terhune, Sunnybank & Collie artifacts and books. Historic House: 18th century, Mead-Van Duyne House Museum.
Research Fields: early Township homes; brick making; local archaeology.
Facilities: 100-vol. library available by appointment for higher education research & local genealogy; archaeology lab. Postcards, pamphlets & local history books for sale.
Activities: guided tours; lectures; formally organized education programs; films; concerts; school loan service; historic skills workshops.
Publications: book, Under The Sign of the Eagle; The Van Riper Hopper House; Artist Coloring Book of Historical Wayne N.J; A Wondrously Beautiful Valley & Commemorative History of Wayne, NJ; The Singack and Mead's Basin Brickyards in Wayne Township; The Archaeology of Wayne; Max Schrabisch Rockshelter Archaeologist; pictorial history of Wayne.
Hours & Admission Prices: By appointment. Guided House Tour: adult $5, child $3; discounts to active military & 4 family members. Closed New Year's Day; Christmas.
Attendance: 1,800 (estimated)
Membership: Individual $15; Family $25; Lifetime $100; Corporate & Business $250.

West Long Branch

800 GALLERY & ROTARY ICE HOUSE GALLERY, Dept. Art & Design, Monmouth University, 400 Cedar Ave., West Long Branch, NJ 07764-1898. Tel.: 732-571-3428.
Web Site: www.monmouth.edu/academics/art/faculty/default.asp
Key Personnel: Dir. Galleries & Collections, Scott Knauer
Institution Type/Description: Art Museum.
Collections: paintings; photographs; sculpture; drawing; ceramics.
Activities: special events; receptions; temporary exhibits.
Hours & Admission Prices: Call for hours.

West Milford

LONG POND IRONWORKS MUSEUM, 1334 Greenwood Lake Tpke., West Milford, NJ 07421. Mailing Address: The Friends of Long Pond Ironworks, Inc., P.O. Box 809, Hewitt, NJ 07421. Tel.: 973-657-1688.
Institution Type/Description: History Museum.
Collections: local history; period furnishings; personal artifacts; photographs.
Hours & Admission Prices: April-Nov. Sat.-Sun. 1-4; other times by appointment.

WEST MILFORD MUSEUM, 1480 Union Valley Rd., West Milford, NJ 07480-1338. Tel.: 973-728-1823.
E-mail: museum@westmilford.org
Web Site: www.westmilfordmuseum.org
Key Personnel: Chairperson, Tonya Cubby; Vice Chairperson, Mary Kochka; Treas., Adrian Birdsall
Institution Type/Description: History Museum: built in 1860s, originally a Methodist Episcopal Church, later became West Milford's Town Hall 1912-1958.
Collections: town history; 19th century life; art; music; political memorabilia; weaponry; farming tools; toys; Native American artifacts.
Activities: lectures; demonstrations; special events.
Hours & Admission Prices: Closed for renovations until Spring 2014.

West Orange

ESSEX COUNTY'S TURTLE BACK ZOO, 560 Northfield Ave., West Orange, NJ 07052-2431. Tel.: 973-731-5800. Fax: 973-731-1059.
Web Site: www.turtlebackzoo.com
Founded: 1963.
Congressional District: 13
Key Personnel: C.E.O., Brint Spencer, D.V.M.; Zoological Society Pres. (V), John Doeffinger.
Personnel Profile: Full-Time Paid 40; Part-Time Paid 125; Part-Time Volunteers 80; Interns 5.
Governing Authority: county. Parent Institution: Essex County Park System. Tax-exempt.
Institution Type/Description: Zoo.
Collections: Wolf Woods; tortoises; farm animals; specimens of live wild animals; bobcats; bison; penguins; otters; cougars; reptiles; wallabies; bears; sea lions; snow leopard; gibbons.

Facilities: food service available. Museum-related items for sale.
Activities: lectures; formally organized & informal education programs for children; permanent & traveling exhibitions; rental facilities; catering.
Publications: zoo newsletter, Turtle Talk.
Hours & Admission Prices: Mon.-Sat. 10-4:30, Sun. 11-5:30. Adults $11; children $8; discounts to AZA members; children under 2 & members no charge. Closed New Year's Day; Thanksgiving; Christmas.
Attendance: 560,402 (accurate)
Membership: Individual $35; Individual Plus One $55; Family $110; Supporting $175; Patron $500; Benefactor $1,000; Director's Circle $5,000.

THOMAS EDISON NATIONAL HISTORIC PARK, 211 Main St., West Orange, NJ 07052-5612. Tel.: 973-736-0550. Fax: 973-243-7172. TDD: 973-243-9122.
E-mail: edis_superintendent@nps.gov
Web Site: www.nps.gov/edis
Founded: 1956.
Congressional District: 8
Key Personnel: Asst. Supt., Theresa Flynn Jung.
Personnel Profile: Full-Time Paid 30.
Governing Authority: federal. Administered by National Park Service, U.S. Dept. of Interior. Tax-exempt.
Institution Type/Description: Historic Site: 20 historic structures from 1880-1887 including Glenmont home & laboratory of Thomas A. Edison.
Collections: Edison archives; photographs; sound recordings; industrial & scientific machinery & equipment; late 19th-century decorative arts.
Research Fields: 19th to 20th-century American history; history of technology & science; Edison related technologies including electricity, motion picture & sound recording.
Facilities: archives of 4.5 million documents for use by appointment only. Books, postcards, & museum-related items for sale.
Activities: guided tours; films; permanent exhibitions; education programs; special events; audio tour.
Publications: The Papers of Thomas A. Edison.
Hours & Admission Prices: Estate: Fri.-Sun. 12-4. Grounds: 11:30-5. Laboratory: Wed.-Sun. 9-5; groups by appointment. Adults $7 (seven day pass); youth under 16 no charge. Annual Park Pass: $30.
Attendance: 60,000 (estimated)
Membership: Friends of Edison National Historic Site: Individual $15; Family $25; Patron $100.

West Trenton

NEW JERSEY STATE POLICE MUSEUM, 1040 River Rd., West Trenton, NJ 08628-0068. Mailing Address: P.O. Box 7068, West Trenton, NJ 08628-0068. Tel.: 609-882-2000, ext. 6401. Fax: 609-882-0321.
Web Site: www.njspmuseum.org
Founded: 1982.
Institution Type/Description: State Police History Museum.
Collections: N.J. State Police history; Trooper training; photographs; videos; Lindbergh kidnapping artifacts; early police facilities, transportation, & weaponry; criminal investigation equipment; surveillance equipment; forensic lab gear; fingerprint-lifting tools; interactive microscopes; seized firearms; police vehicles.
Activities: guided tours.
Hours & Admission Prices: Mon.-Fri. 10-4, group tours by appointment. No charge. Closed state holidays.

West Windsor

THE GALLERY, MERCER COUNTY COMMUNITY COLLEGE, Communications Center, 1200 Old Trenton Rd., West Windsor, NJ 08550-3407. Tel.: 609-586-4800, ext. 3589.
E-mail: gallery@mccc.edu
Web Site: www.mccc.edu/community-gallery.shtml
Institution Type/Description: Art Museum.
Collections: paintings; photographs; sculpture.
Facilities: 1,300 sq. ft. exhibit space.
Activities: presentations; special events; receptions.
Hours & Admission Prices: Tues. 9-4, Wed. 9-3 & 6-8, Thurs. 11-3. No charge.

Westfield

MILLER-CORY HOUSE MUSEUM, 614 Mountain Ave., Westfield, NJ 07090-3044. Mailing Address: P.O. Box 455, Westfield, NJ 07091-0455. Tel.: 908-232-1776. Fax: 908-232-1740.
E-mail: millercorymuseum@gmail.com
Web Site: www.millercoryhouse.org

Founded: 1972.
Key Personnel: C.E.O. & Chm. Bd. Governors, Lowell Schantz; Pres. (V), Patricia D'Angelo; Acquisitions, Eileen O'Shea; Museum Shop Mgr., Debbie Bailey.
Personnel Profile: Part-Time Paid 2; Part-Time Volunteers 101.
Governing Authority: nonprofit organization. Parent Institution: Westfield Historical Society, P.O. Box 613, Westfield, NJ 07091. Tel. 908-654-1794. Tax-exempt.
Institution Type/Description: Living Museum & Historic House: 1740 Miller-Cory House, two-story farmhouse.
Collections: period furniture; farm implements; cooking & craft utensils; 18th-century farmhouse & gardens.
Research Fields: 18th-century living, crafts, cooking, history & gardens; relationship of family to town & general area; New Jersey history.
Facilities: library; education center; working open-hearth kitchen; gardens. 18th-century craft items for sale.
Activities: guided tours; lecture & slide shows; open-hearth cooking & craft demonstrations during museum hours; showcase program of 18th-century skills at schools or on premises.
Publications: quarterly bulletin, The Bee Line; in-house bulletin, The Broadside; how-to sheets on 18th-century crafts & skills; study guide; cookbooks, The Groaning Board & Pleasures of 18th-Century Cooking in conjunction with New Jersey Historical Society; children's coloring book on 18th-Century Family Life.
Hours & Admission Prices: mid-Sept. to mid-June Sun. 2-4; Mon.-Fri. group tours by appointment. Adults $3, children $2; members no charge. Closed holiday weekends.
Attendance: 1,800 (estimated)
Membership: Volunteer Membership: Junior $5; Active Individual $15; Inactive Individual $25. Friends of Miller-Cory: Supporter $30; Benefactor $50; Patron $100; Life $1,000.

Westhampton

RANCOCAS NATURE CENTER, 794 Rancocas Rd., Westhampton, NJ 08060. Tel.: 609-261-2495.
Institution Type/Description: Nature Center.
Collections: wildlife & their habitats.
Facilities: 210-acres; nature trails.
Activities: walking trails; bird watching; educational programs.
Hours & Admission Prices: Tues.-Sat. 9-5, Sun. 12-5.

Whippany

JEWISH HISTORICAL SOCIETY OF METROWEST, 901 Rte. 10 E., Whippany, NJ 07981-1156. Tel.: 973-929-2995.
Institution Type/Description: Jewish History Museum.
Collections: Jewish history, culture & heritage; documents; records; period artifacts; photographs; oral histories; Jewish newspapers.
Facilities: archives.
Activities: educational programs; research.
Hours & Admission Prices: Mon.-Fri. 9-1.

WHIPPANY RAILWAY MUSEUM, 1 Railroad Plaza, Rte. 10 W. & Whippany Rd., Whippany, NJ 07981-1505. Mailing Address: P.O. Box 16, Whippany, NJ 07981-0016. Tel.: 973-887-8177.
E-mail: info@whippanyrailwaymuseum.net
Web Site: www.whippanyrailwaymuseum.net
Founded: 1973.
Congressional District: 11
Key Personnel: Pres. (V), Steven P. Hepler.
Governing Authority: Tax-exempt.
Institution Type/Description: Railway Museum.
Collections: New Jersey's railroad history & transportation heritage; railroad equipment & artifacts; Rahway Valley Locomotive #16; M&E Section Gang Car No. TC-1; delivery truck; PRR Cabin Car 477823; M&E Caboose #1; Railbus No. 10; handcars; Whitcomb switcher; 1907 steam locomotive No. 385; 1942 steam locomotive No. 4039.
Facilities: picnic area.
Activities: train excursions; carry-in picnics.
Hours & Admission Prices: April-Oct. Sun. 12-4. Adults $1, children under 12 $.50.
Attendance: 25,000 (estimated)

Whitehouse

EVERSOLE-HALL HOUSE, 511 Rte. 523 S., Whitehouse, NJ 08888. Mailing Address: P.O. Box 216, Stanton, NJ 08885-0216. Tel.: 908-236-2327.
Web Site: www.readingtontwp.org/museum_eversole-hall.html
Key Personnel: Museum Admin., Amy Hollander
Institution Type/Description: Historic Buildings.
Collections: period furnishings; personal artifacts. Historic Buildings: 1830s farmhouse; shoemaker's show; outbuildings.
Activities: tours; demonstrations; 5th grade Partners in History Program; summer camp. Annual Events: Open House; Fall Frolic Family Day.
Hours & Admission Prices: Call for hours.

Wildwood

DOO WOP EXPERIENCE MUSEUM, 4500 Ocean Ave., Wildwood, NJ 08260. Mailing Address: c/o Doo Wop Preservation League, P.O. Box 1703, Wildwood, NJ 08260. Tel.: 609-523-1958.
Web Site: www.doowopusa.org
Institution Type/Description: History Museum.
Collections: Wildwood in the 1950s & 1960s; pop culture; music; neon signs; early lamps, tables, chairs, clocks & furnishings.
Hours & Admission Prices: Call for hours.

WILDWOOD HISTORICAL SOCIETY, INC., 3907 Pacific Ave., Wildwood, NJ 08260-4722. Tel.: 609-523-0277.
E-mail: wildwoodhistoricalsociety@hotmail.com
Web Site: www.funchase.com/museum.htm
Formerly: George F. Boyer Museum
Founded: 1963.
Congressional District: 1
Key Personnel: Pres. (V) & Dir., Anna M. Vinci.
Personnel Profile: Part-Time Paid 2; Part-Time Volunteers 8.
Governing Authority: municipal. Tax-exempt.
Institution Type/Description: Historical Society Museum.
Collections: 1898-present local newspapers; 120 picture books.
Research Fields: local history.
Facilities: 1,200-vol. library of New Jersey history available for public use; reading room.
Activities: Annual Event: Postcard Show & Open House in April.
Publications: biannual newsletter.
Hours & Admission Prices: April-June 15 & Oct.-Dec. Thurs.-Sat. 9-2; June 15-Sept. Mon.-Sat. 9-2. No charge; donations accepted. Closed national holidays. &
Attendance: 2,500 (estimated)
Membership: Student $5; Individual $10; Family $15; Business $50; Life $200.

WILDWOOD INSECTARIUM, 5216 Boardwalk, Wildwood, NJ 08260. Tel.: 609-488-6939.
Web Site: www.wildwoodinsectarium.com
Institution Type/Description: Insect Museum.
Collections: live & mounted insects; hands-on exhibitions; photographs.
Facilities: Gift items for sale.
Activities: special events; educational programs.
Hours & Admission Prices: Daily 10am-11pm.

WILDWOODS VIETNAM VETERANS REMEMBRANCE WALL, 4500 Ocean Dr., Wildwood, NJ 08260. Mailing Address: P.O. Box 481, Wildwood, NJ 08260. Tel.: 609-729-4000.
Web Site: www.wildwoodswall.com
Institution Type/Description: Veterans Memorial.
Collections: half-size black granite replica of the Washington DC memorial; 58,913 names of fallen heroes of the Vietnam War.
Activities: special events.
Hours & Admission Prices: Daily. No charge; donations accepted.

Wildwood Crest

WILDWOOD CREST HISTORICAL SOCIETY, 116 E. Heather Rd., Wildwood Crest, NJ 08260-4253. Mailing Address: 2 Jefferson Ave., Somers Point, NJ 08244-1524. Tel.: 609-729-4515.
E-mail: kirkhastings5@aol.com
Web Site: cresthistory.org
Founded: 1976.

Key Personnel: Pres., Kirk Hastings; Sec., Theresa Williams
Institution Type/Description: Historical Society Museum.
Collections: local history & culture; photographs; period furnishings; personal artifacts.
Activities: monthly meetings October to May.
Hours & Admission Prices: Call for hours.

Williamstown

MONROE TOWNSHIP HISTORICAL SOCIETY - IRELAND HOFFER HOUSE, 313 S. Main St., Williamstown, NJ 08094. Tel.: 856-875-2943.
Institution Type/Description: Historical Society Museum.
Collections: local history & culture; period furnishings & clothing; photographs.
Hours & Admission Prices: By appointment.

Woodbine

THE SAM AZEEZ MUSEUM OF WOODBINE HERITAGE, 610 Washington Ave., Woodbine, NJ 08270. Mailing Address: P.O. Box 517, Woodbine, NJ 08270-0517. Tel.: 609-861-5355.
Web Site: www.thesam.org
Key Personnel: Exec. Dir., Jane Stark
Institution Type/Description: History Museum.
Collections: local history & culture; photographs; personal artifacts.
Hours & Admission Prices: Wed.-Fri. & Sun. 10-4. Closed New Year's Day; Christmas Day.

Woodbridge

BARRON ARTS CENTER & MUSEUM, 582 Rahway Ave., Woodbridge, NJ 07095-3419. Tel.: 732-634-0413. Fax: 732-634-8633.
E-mail: barronarts@twp.woodbridge.nj.us
Web Site: www.twp.woodbridge.nj.us/barronarts/
Key Personnel: Dir. Cultural Arts, Cynthia Knight.
Personnel Profile: Full-Time Paid 3.
Governing Authority: municipal. Tax-exempt.
Institution Type/Description: Art Museum.
Collections: Woodbridge history.
Hours & Admission Prices: Mon.-Fri. 11-4, Sat.-Sun. 2-4; call to confirm. No charge; donations accepted. Closed holidays. &
Attendance: 25,000 (estimated)

Woodbury

FRIENDSHIP FIRE COMPANY MUSEUM, 29 Delaware St., Woodbury, NJ 08096-5925. Tel.: 856-845-0066.
Institution Type/Description: Fire Fighting Museum.
Collections: photographs; badges & awards; wooden water main; leather, tin & high tech plastic helmets; alarm boxes; leather buckets; parade belts & hats from 1843 to present; alarm box list; 1911 Ahrens-Fox steamer; 1799 Philip Mason hand pumper.
Activities: tours.
Hours & Admission Prices: Mon.-Fri. 7:30-4.

GLOUCESTER COUNTY HISTORICAL SOCIETY, 58 N. Broad St., Woodbury, NJ 08096-4629. Mailing Address: 17 Hunter St., Woodbury, NJ 08096-4605. Tel.: 856-848-8531 (museum); 845-7881 (business office); 845-4771 (library). Fax: 856-845-4771.
E-mail: gchs@net-gate.com
Web Site: www.rootsweb.com/~njgchs
Founded: 1903.
Congressional District: 3
Key Personnel: Pres., Barbara L. Turner; Museum Coord., Kathleen Fleming; Library Coord., Barbara Price; Coord. Museum Collections, Patricia A. Hrynenko.
Personnel Profile: Part-Time Paid 4; Part-Time Volunteers 50.
Governing Authority: nonprofit organization. Tax-exempt: 501(c)(3).
Institution Type/Description: General Museum.
Collections: Colonial & Victorian furniture; archives; history; textile collection including quilts, samplers & wedding gowns; glass; children's museum; costumes; militaria; Native American artifacts; newspapers 1824-2008; deeds; maps; census; vital statistics; church records; cemetery inscriptions; court records; wills. Historic Houses: 1765 Hunter-Lawrence-Jessup House; 1786 Moravian Church.
Facilities: 10,000-vol. library of genealogical and historical books & manuscripts.

Activities: library genealogy programs & lectures; museum trunk shows.
Publications: books & pamphlets on the history & genealogy of South Jersey; quarterly bulletin.
Hours & Admission Prices: Museum: Mon., Wed. & Fri. 1-4, last Sun. of month 2-5; other times by appointment. Library: June-Sept. Tues. 6 pm-9:30 pm, Wed.-Fri. 10-4, last Sun. each month 2-5; Oct.-May Tues. 6 pm-9:30 pm, Wed.-Fri. 10-4, first Sat. each month 10-4, last Sun. each month 2-5. Museum: adults $5, children 6-18 $1; children under 5 & members no charge. Library: non-members $5. Closed national holidays. &
Attendance: 3,000 (estimated)
Membership: Student $10; Individual $30; Family $40; Business $100; Life $500.

Woodstown

PILESGROVE-WOODSTOWN HISTORICAL SOCIETY, 42 N. Main St., Woodstown, NJ 08098. Tel.: 856-769-4588.
Institution Type/Description: Historical Society Museum: housed in the 18th century Dickinson House.
Collections: local history & culture; period furnishings; personal artifacts; photographs; books. Historic Building: Eldridge's Hill School House.
Hours & Admission Prices: Sat. 10-1.

Wyckoff

JAMES A. MCFAUL ENVIRONMENTAL CENTER OF BERGEN COUNTY, 150 Crescent Ave., Wyckoff, NJ 07481-2751. Tel.: 201-891-5571. Fax: 201-891-5583. TDD: 201-343-7249.
Web Site: www.co.bergen.nj.us/parks
Founded: 1967.
Congressional District: 7
Key Personnel: C.E.O. & Dir., Raymond Dressler; Mgr., Peter Both.
Personnel Profile: Full-Time Paid 7; Part-Time Paid 2; Part-Time Volunteers 6.
Governing Authority: county. Parent Institution: Bergen County Dept. of Parks. Tax-exempt.
Institution Type/Description: Nature Center; Park Museum.
Collections: native wild animals, including birds of prey; plant material of horticultural interest; specialized gardens.
Facilities: 1,000-vol. library of environmental subjects, available for research on premises during regular hours; nature conservation center; 175-seat auditorium; classrooms; 81-acre park; nature trail; herb gardens; ornamental grass garden; observation deck; gazebo; informal gardens featuring rhododendrons, azaleas, large daffodil collection & numerous ground covers; exhibit hall; observatory overlooking small pond.
Activities: guided tours; lectures; films; formally organized education programs for children; permanent & temporary exhibitions; monthly nature art display; study workshops; various recreational activities; wildlife programs.
Publications: Calendar of Events; Nature Trail Guide; Herb Garden Guide.
Hours & Admission Prices: Programs: Mon.-Fri. 9:30-3:30. Park Grounds: Mon.-Fri. 8-sunset, Sat.-Sun. & holidays 8:30-sunset. No charge. &
Attendance: 50,000 (estimated)

NEW MEXICO

(183 listings)

Acoma Pueblo

ACOMA PUEBLO MUSEUM, NM 23 (12 miles SW of I-40), Acoma Pueblo, NM 87034. Mailing Address: P.O. Box 309, Acoma Pueblo, NM 87034-0309. Tel.: 505-252-1139.
Institution Type/Description: History Museum.
Collections: Indian pottery from the 15th century to the present; photographs & documents relating to the history of Acoma.
Hours & Admission Prices: Winter: 8-4:30. Summer: 8am-7pm.

Alamogordo

ALAMEDA PARK ZOO, 1321 N. White Sands Blvd., Alamogordo, NM 88310. Mailing Address: 1376 E. Ninth St., Alamogordo, NM 88310-5855. Tel.: 505-439-4290. Fax: 505-439-4103.
E-mail: sdiehl@ci.alamogordo.nm.us
Web Site: ci.alamogordo.nm.us/coa/communityservices/zoo.htm
Founded: 1898.
Congressional District: 2
Key Personnel: Dir., Steve Diehl; Support Asst. & Gift Shop Mgr., Kathy Chase.
Personnel Profile: Full-Time Paid 4; Part-Time Paid 4; Part-Time Volunteers 20.

Governing Authority: municipal. Parent Institution: City of Alamogordo. Subsidiary Institution: Alamogordo Friends of the Zoo. Tax-exempt.
Institution Type/Description: Zoo: located in 1898 park.
Collections: 250 animals, 90 indigenous & exotic species.
Research Fields: raptor rehabilitation; Mexican gray wolf captive management program; ring-tailed lemur species survival program.
Facilities: zoological park. Museum-related items for sale.
Activities: guided tours; formally organized education programs for children; temporary exhibitions.
Publications: AFOTZ Thoughts.
Hours & Admission Prices: Daily 9-5. Adults $2.50, senior citizens & children 3-11 $1.50; discounts to groups, AZA & AAZK members. Closed New Year's Day; Christmas. &
Attendance: 50,941 (accurate)
Membership: Alamogordo Friends of the Zoo: Junior & Senior Citizen $7.50; Individual $15; Family (inside Otero County) $25; Family (outside Otero County) & Patron $35; Benefactor $75.

ALAMOGORDO MUSEUM OF HISTORY, 1301 N. White Sands Blvd., Alamogordo, NM 88310-6659. Tel.: 575-434-4438.
E-mail: tbhs@zianet.com
Web Site: alamogordohistorymuseum.com
Formerly: Tularosa Basin Historical Society Museum
Founded: 1964.
Congressional District: 2
Key Personnel: Pres. (V), Elizabeth Padilla; Cur. & Museum Shop Mgr., Jean Ann Killer.
Personnel Profile: Part-Time Paid 1; Part-Time Volunteers 28.
Governing Authority: nonprofit. Parent Institution: Tularosa Historical Society. Tax-exempt.
Institution Type/Description: Historical Society Museum.
Collections: Indian artifacts; fossils; displays dating from early man to the atomic age, centering around the Tularosa Basin; early photographs of the Tularosa Basin & Sacramento mountains; oral histories.
Research Fields: local (Tularosa Basin, Sacramento Mountains, Otero County) history.
Facilities: research library, by appointment for large projects.
Activities: guided tours; lectures; temporary exhibitions.
Publications: quarterly newsletter; semiannual monograph; calendar.
Hours & Admission Prices: Mon.-Fri. 10-4, Sat. 10-3, Sun. 1-4. No charge; donations accepted. Closed federal holidays.
Attendance: 6,500 (accurate)
Membership: Student $10; Individual $20; Family $25; Subscriber $35; Contributing $100; Life $500. Corporate: Regular $50; Sustaining $500; Benefactor $1,000.

✻   **NEW MEXICO MUSEUM OF SPACE HISTORY, (M),** Top of New Mexico, Hwy. 2001, Alamogordo, NM 88310. Mailing Address: P.O. Box 5430, Alamogordo, NM 88311-5430. Tel.: 575-437-2840; 877-333-6589 (Toll Free). Fax: 575-437-7722. Facebook: NM Space Museum.
E-mail: cathy.harper@state.nm.us
Web Site: www.nmspacemuseum.org
Formerly: The Space Center
Founded: 1976.
Congressional District: 2
Key Personnel: Exec. Dir., Chris Orwoll; Chm. & Pres., Bob Wood; Cur., George M. House; Public Affairs, Cathy Harper; Museum Shop Mgr., Val Cory.
Personnel Profile: Full-Time Paid 28; Part-Time Paid 2; Part-Time Volunteers 35.
Governing Authority: state; nonprofit organization. Parent Institution: New Mexico Department of Cultural Affairs, State of New Mexico. Tax-exempt: 501(c)(3).
Institution Type/Description: Space Museum: located near White Sands, home of America's early space program.
Collections: objects documenting the histories of astronautics, astronomy, & space exploration; astronaut memorial garden; International Space Hall of Fame.
Research Fields: scholarly & collection-oriented historical research, astronomy materials for planetarium programs; oral history programs.
Facilities: five story museum; research library & archives; artifact restoration center; large screen dome theater & planetarium; air & space park; outdoor exhibits; education facilities; gift shop.
Activities: guided tours; lectures; summer camp; educational outreach; workshops. Annual Events: Independence Day fireworks display; New Mexico Space Academy summer camp; International Space Hall of Fame induction ceremony.

Publications: assorted brochures; schedule cards; museum guide book; scholarly journals.
Hours & Admission Prices: Daily 9-5. Museum: adults $6, senior citizens & military $5, children 4-12 $4; members & children under 4 no charge. Theatre: adults $6, senior citizens & military $5.50, children 4-12 $4.50. Closed New Year's Day; Christmas. &
Attendance: 85,000 (accurate)
Membership: Student $15; Individual $20; Family/Corporate $45, $100 & $250.

OLIVER LEE RANCH HOUSE MUSEUM, 409 Dog Canyon Rd., Alamogordo, NM 88310. Mailing Address: New Mexico State Parks, 1220 S. St. Francis Dr., Santa Fe, NM 87504. Tel.: 575-437-8284; 505-476-3355. Fax: 575-439-1290.
E-mail: wendy.justice@state_nm.us
Web Site: www.emnrd.state.nm.us
Institution Type/Description: History Museum.
Collections: local history & culture; period furnishings; photographs; desert flora.
Facilities: visitor center; nature trails.
Hours & Admission Prices: Tours: Sat.-Sun. 3pm by appointment.

TOY TRAIN DEPOT MUSEUM AND TRAIN RIDE, 1991 N. White Sands Blvd., Alamogordo, NM 88310-6200. Tel.: 888-207-3564. Fax: 575-437-8169.
E-mail: mkoral@bajabb.com
Web Site: www.toytraindepot.homestead.com
Founded: 1988.
Key Personnel: Pres., Hugh Malcolm; C.E.O., Merry Koval.
Personnel Profile: Full-Time Paid 1; Part-Time Paid 2; Part-Time Volunteers 6.
Governing Authority: nonprofit. Tax-exempt.
Institution Type/Description: History Museum.
Collections: railroad history of New Mexico & Alamogordo; model trains of all scales & gauges; train memorabilia.
Activities: 1/4 scale train ride.
Hours & Admission Prices: Museum: Winter: Wed.-Sun. 12-4; Summer: Wed.-Sun. 12-4. Train: Winter: Wed.-Sun. 12:30-4; Summer: Wed.-Sun. 12:30-5. Museum: adults $4. Train: adults $4. Combination: adults $6; discounts to groups with reservation. Closed Thanksgiving; Christmas.

WHITE SANDS NATIONAL MONUMENT, 19955 U.S. Hwy. 70, Alamogordo, NM 88310. Mailing Address: P.O. Box 1086, Holloman A.F.B., NM 88330-1086. Tel.: 505-679-2599. Fax: 505-479-4333.
E-mail: whsa_administration@nps.gov
Web Site: www.nps.gov/whsa
Founded: 1933.
Congressional District: 2
Governing Authority: federal, National Park Service. Parent Institution: U.S. National Park Service. Tax-exempt.
Institution Type/Description: Park Museum.
Collections: insects; herbs; birds; mammals; reptiles; amphibians; cultural artifacts; selenite crystals.
Research Fields: local natural history; bird observations; hydrology; geologyl arcgaeikigy,
Facilities: 500-vol. library of material on natural history and geology available on premises & on loan by special written permission of superintendent; native plant display area. General natural history items for sale.
Activities: guided tours; lectures; sound & light geology program; permanent exhibitions; full moon night events.
Publications: brochures; booklets.
Hours & Admission Prices: Visitor Center: Winter: daily 9-5; Summer: call for hours. Monument: Memorial Day to Labor Day daily 7am-10pm; Sept.-May daily 7am to sunset. Adults $3; children 16 & under no charge. Closed Christmas. &
Attendance: 470,921 (accurate)

Albuquerque

ALBUQUERQUE BIOLOGICAL PARK, 903 Tenth St., S.W., Albuquerque, NM 87102-4029. Tel.: 505-768-2000. Fax: 505-764-6281. TDD: 505-764-6297.
E-mail: biopark@cabq.gov
Web Site: www.cabq.gov/biopark
Founded: 1927.
Congressional District: 1
Key Personnel: Dir., Rick Janser; Dir. Dept., Ray D. Darnell.

Personnel Profile: Full-Time Paid 160; Part-Time Paid 20; Part-Time Volunteers 430.
Governing Authority: municipal. Tax-exempt.
Institution Type/Description: Zoo, Aquarium & Botanical Garden.
Collections: mammals; birds; reptiles; fish; plants
Facilities: 250-vol. library of medical & zoological reference material available for use by request.
Activities: guided tours; volunteer programs; education & outreach programs.
Publications: Zooscape.
Hours & Admission Prices: Summer: Mon.-Fri. 9-5, Sat.-Sun. 9-6; Winter: daily 9-5. Aquarium or Zoo: adults 13-64 $7, senior citizens over 64 & children 3-12 $3; discount to AZA members; members & children under 3 no charge. Closed New Year's Day; Thanksgiving; Christmas. &
Attendance: 975,000 (accurate)
Membership: Family $50.

✳ **ALBUQUERQUE MUSEUM OF ART & HISTORY, (M),** 2000 Mountain Rd., N.W., Albuquerque, NM 87104-1459. Tel.: 505-243-7255. Fax: 505-764-6546.
E-mail: clwright@cabq.gov
Web Site: www.cabq.gov/museum
Founded: 1967.
Congressional District: 1
Key Personnel: Dir., Cathy L. Wright; Asst. Dir., Cynthia Garcia; Cur. History, Deb Slaney; Cur. Art, Andrew Connors; Cur. Education, Elizabeth Becker; Museum Shop Mgr., Christina Godwin-Taylor.
Personnel Profile: Full-Time Paid 29; Part-Time Paid 1; Part-Time Volunteers 200; Interns 2.
Governing Authority: municipal. Parent Institution: City of Albuquerque, Albuquerque Museum Foundation. Tax-exempt: 501(c)(3).
Institution Type/Description: Art & History Museum; Sculpture Garden; Historic House (Casa San Ysidro)
Collections: general; decorative arts; regional fine arts & crafts; costumes; photography; objects & artifacts from the Middle Rio Grande Valley from 1500-present.
Major Exhibits: Smithsonian - Vivian Vance, 9/13-3/14.
Research Fields: Southwestern history & art.
Facilities: 85-seat auditorium; studio classroom; special events room; amphitheater. Books, contemporary fine art, jewelry & crafts for sale.
Activities: guided tours, lectures; films; gallery talks; inter-museum loan, permanent & temporary exhibitions.
Publications: foundation newsletter; exhibition catalogs; local history publications; gallery guides.
Hours & Admission Prices: Tues.-Sun. 9-5. Adults $4, in-state residents $3; discounts to AAM & ICOM members; members no charge. Closed holidays. &
Attendance: 121,379 (accurate)
Membership: Senior $40; General $60; Friend $100; Supporter $250; Benefactor $500; Patron's Circle Bronze $1,000-$2,499; Patron's Circle Silver $2,500-$4,999; Patron's Circle Gold $5,000.

AMERICAN INTERNATIONAL RATTLESNAKE MUSEUM, 202 San Felipe, N.W., Suite A, Albuquerque, NM 87104-1426. Tel.: 505-242-6569. Fax: 505-242-6569 (call first).
E-mail: snakemuseum@aol.com
Web Site: www.rattlesnakes.com
Founded: 1990.
Key Personnel: Dir., Bob Myers.
Personnel Profile: Full-Time Paid 2; Part-Time Volunteers 34.
Governing Authority: private; profit.
Institution Type/Description: Herpetology & Animal Conservation Museum.
Collections: American history; Native American culture & art; fine art; photography; medical artifacts; textiles; music; toys & games; anatomical; symbolism; folk art; advertising.
Activities: field trips; training programs & school field trips.
Hours & Admission Prices: Summer: Mon.-Sat. 10-6, Sun. 1-5; Sept.-May Mon.-Fri. 11:30-5:30, Sat. 10-6, Sun. 1-5. Adults $5, seniors, military & students $4, kids $3. Closed New Year's Day; Easter; Thanksgiving; Christmas. &
Attendance: 50,000 (accurate)

ANDERSON/ABRUZZO ALBUQUERQUE INTERNATIONAL BALLOON MUSEUM, 9201 Balloon Museum Dr., N.E., Albuquerque, NM 87113-2425. Tel.: 505-768-6020. Fax: 505-768-6021.
Web Site: www.cabq.gov/balloon; www.balloonmuseum.com
Founded: 2000.
Congressional District: 1
Key Personnel: Dir., Jeffrey P. Cooper-Smith; Foundation Pres. (V), Gerald Landgraf; Foundation Dir., Jennifer Taylor; Cur. Collections, Dr. Marilee Schmit Nason; Devel. Coord., Linda Hubley.

Personnel Profile: Full-Time Paid 7; Part-Time Paid 6; Part-Time Volunteers 45.
Governing Authority: nonprofit. Parent Institution: City of Albuquerque, Cultural Services Department. Tax-exempt.
Institution Type/Description: International Balloon Museum: named after Albuquerque balloon pilots Maxie Anderson and Ben Abruzzo, the first individuals to fly a balloon across the Atlantic.
Collections: art, culture, history, science & sport of lighter-than-air flight; balloons, airships & their artifacts; letters dated 1783 from the Montgolfier brothers, founders of ballooning; 6,000 objects from 1783 to present; personal artifacts; technological information; films; music; literature; stamps; coins.
Research Fields: art, culture, history, science, & sport of lighter-than-air flight (balloon airships) from 1783 to present.
Facilities: 25,000 sq. ft. exhibit space; classrooms; meeting rooms; rental facilities. Museum-related items for sale.
Activities: long-term, temporary & interactive exhibits; guided tours; formal education programs; community outreach; balloon demonstrations; arts and crafts; school field trips; community open houses; facility rental programs; balloon school. Annual Event: participation in the Albuquerque International Balloon Fiesta in October.
Publications: volunteer newsletter.
Hours & Admission Prices: Tues.-Sun. 9-5. Adults $4, seniors 65 & over $2, children 4-12 $1; discount to New Mexico residents with ID; 1st Fri. each month, Sun. 9-1, ICOM, AAM, NMAM & Foundation members, and children 3 & under no charge. Closed New Year's Day; Thanksgiving; Christmas. &
Attendance: 71,623 (accurate)
Membership: Student & Senior $20; Individual & Senior Couple $35; Family $60; Associate $100; Patron $250; Benefactor $500.

EXPLORA, 1701 Mountain Rd., N.W., Albuquerque, NM 87104-1396. Tel.: 505-224-8300. Fax: 505-224-8323.
E-mail: explora@explora.us
Web Site: www.explora.us
Formerly: Explora Science Center & Children's Museum of Albuquerque
Founded: 1996.
Congressional District: 1
Key Personnel: Exec. Dir., Patrick Lopez; Pres. (V), Tim Hendry; Assoc. Dir., Paul Tatter; Dir. Exhibits, Betsy Adamson; External Rels., Ellen Welker; Dir. Visitor Svcs., Armelle Casau; Dir. Educational Svcs., Kristen Leigh; Dir. Operations, Alfonso Romero.
Personnel Profile: Full-Time Paid 65; Part-Time Paid 16; Part-Time Volunteers 215; Interns 30.
Governing Authority: private; nonprofit organization. Tax-exempt: 501(c)(3).
Institution Type/Description: Science Center & Children's Museum.
Collections: transactive hands-on exhibits.
Activities: after-school programs; camp-in programs; classes & demonstrations; curriculum materials; field trips; school outreach; science kits; science camps; youth employment programs; workshops & institutes for teachers; homeschoolers programs; senior citizens programs.
Publications: quarterly newsletter; calendar; annual report.
Hours & Admission Prices: Mon.-Sat. 10-6, Sun. 12-6. Adults $7, senior citizens 65 & over $5, children 1-11 $3; ASTC reciprocal membership program; children under 1 & members no charge. Closed New Year's Day; week after Labor Day; Independence Day; Thanksgiving; Christmas. &
Attendance: 204,678 (accurate)
Membership: Student $25; Senior (Individual) $30; Individual $40; Senior (Couple) $50; Family, Helping Hand & Grandparents $75; Family Plus $100; Helping Hand $150.

EXPO NEW MEXICO, 300 San Pedro, N.E., Albuquerque, NM 87108-2812. Mailing Address: P.O. Box 8546, Albuquerque, NM 87198-8546. Tel.: 505-222-9700. Fax: 505-266-7784.
Web Site: exponm.com
Formerly: New Mexico State Fair Fine Arts Gallery
Key Personnel: Gen. Mgr., Dan Mourning; Sponsorship & Events Dir., Sabrina Garza
Institution Type/Description: Art Gallery.
Collections: Hispanic & Indian art; fine art; African American art.
Hours & Admission Prices: Mon.-Fri. 8-5.

516 ARTS, 516 Central Ave., S.W., Albuquerque, NM 87102. Tel.: 505-242-1445. Fax: 505-244-4101.
E-mail: info@516arts.org
Web Site: www.516arts.org
Founded: 2006.
Congressional District: 1
Key Personnel: Pres. & Exec. Dir., Suzanne Sbarge.

Personnel Profile: Full-Time Paid 4; Part-Time Volunteers 5; Interns 2.
Volunteer Hours: 650
Operating Expenses: 1,313,863
Operating Income: 1,389,070
Governing Authority: Tax-exempt.
Institution Type/Description: Art Gallery.
Major Exhibits: Heart of the City, 2/14-5/14; Digital Latin America, 6/14-8/14; Floyd D. Tunson: Son of Pop (T), 9/14-12/14.
Activities: workshops; educational programs; special events; lectures.
Hours & Admission Prices: Tues.-Sat. 12-5; other times by appointment. No charge. &
Attendance: 11,000 (accurate)
Membership: Student & Senior $25; Individual $50; Family $75; Friend $100; Contributor $250; Donor $500; Patron $1,000 & up.

HOLOCAUST & INTOLERANCE MUSEUM OF NEW MEXICO, 616 Central Ave., S.W., Albuquerque, NM 87102. Mailing Address: P.O. Box 1762, Albuquerque, NM 87103-1762. Tel.: 505-247-0606. Fax: 505-247-0606 (call first).
E-mail: info@nmholocaustmuseum.org
Web Site: www.nmholocaustmuseum.org
Founded: 2001.
Key Personnel: Admin., Lyn Berner; Pres. (V), Randee Kaiser.
Personnel Profile: Part-Time Paid 1.
Governing Authority: Tax-exempt.
Institution Type/Description: History Museum.
Collections: Holocaust history; photographs; personal artifacts.
Hours & Admission Prices: Tues.-Sat. 11-3:30. No charge; donations accepted. Closed major holidays. &
Attendance: 5,617 (accurate)

INDIAN PUEBLO CULTURAL CENTER, (M), 2401 12th St., N.W., Albuquerque, NM 87104-2397. Tel.: 505-843-7270. Fax: 505-842-6959.
Web Site: www.indianpueblo.org
Founded: 1976.
Congressional District: 1
Key Personnel: Museum Dir., Marth B. Becktell; Pres., Ronald J. Solimon; C.O.O., Dwayne Virgint; Collections Management Specialist, Amy Johnson; Gift Shop Mgr., Ira Wilson; Cultural Arts & Education Coord., Greg Analla; Volunteer & Membership Coord., Claire Canfield; Coord. Outreach Education, Felipe Estudillo-Colon.
Personnel Profile: Full-Time Paid 90; Part-Time Paid 70; Part-Time Volunteers 65; Interns 4.
Governing Authority: tribal government. Subsidiary Institution: IPMI - Indian Pueble Marketing, Inc. Tax-exempt: 501(c)(3).
Institution Type/Description: Pueblo Cultural Museum.
Collections: ethnographic; contemporary arts & crafts; pueblo culture, history, & architecture; archaeology.
Research Fields: Pueblo Indian culture & history.
Facilities: archival library; restaurant; conference & meeting rooms. Museum-related items for sale.
Activities: guided tours; lectures; films; arts festivals; theater; organized education programs for children & adults; docent program; special event planning; temporary & permanent exhibits; artist demonstration. Museum Sponsors: Arts & Crafts Fair; American Indian Week; Indian Children Art Contest; Indian Traditional Dances.
Publications: Pueblo Horizons.
Hours & Admission Prices: Daily 9-5. Adults $6, senior citizens $5.50, military $4.75, NM residents $4, students $3; discounts to groups of 10 or more; members and children 4 & under no charge. Closed New Year's Day; Memorial Day; Independence Day; Labor Day; Thanksgiving; Christmas. &
Attendance: 100,000 (estimated)
Membership: Student $20; Senior $25; Individual & Senior Plus One $40; Family $60; Coral $100; Turquoise $250.

INSTITUTE OF METEORITICS METEORITE MUSEUM, 221 Yale Blvd., N.E., Albuquerque, NM 87131. Mailing Address: MSCO3 2050, 1 University of New Mexico, Albuquerque, NM 87131. Tel.: 505-277-2747. Fax: 505-277-3577.
Web Site: epswww.unm.edu/iom/
Founded: 1944.
Key Personnel: Dir. & Cur., Dr. Carl Agee.
Governing Authority: university. Affiliated with the University of New Mexico. Tax-exempt.
Institution Type/Description: Meteorite Museum.
Collections: meteorites; tektite, impact glasses & other shocked rock from terrestrial impact craters.

Research Fields: mineralogy & petrology of meteorites, lunar rocks & terrestrial basalts.
Activities: self-guided tours; lectures; temporary & loan exhibitions.
Hours & Admission Prices: Mon.-Fri. Call for hours. No charge; donations accepted. Closed university holidays. &
Attendance: 2,500 (estimated)

*   **MAXWELL MUSEUM OF ANTHROPOLOGY, (M),** University of New Mexico, Albuquerque, NM 87131. Mailing Address: MSCO1 1050, 1 University of New Mexico, Albuquerque, NM 87131. Tel.: 505-277-4405. Fax: 505-277-1547.
E-mail: maxwell@unm.edu
Web Site: www.unm.edu/~maxwell
Founded: 1932.
Congressional District: 1
Key Personnel: Dir., E. James Dixon; Sr. Research Coord., Bruce B. Huckell; Dir. Education, Amy Grochowski; Cur. Osteology, Heather Edgar; Photo Archives, Catherine Baudoin; Admin., Amy Hathaway; Data Mgr., Dorothy Larson; Cur. Archaeology, David Phillips; Cur. Ethnology, Kathryn Klein; Cur. Exhibits, Devorah Romanek.
Personnel Profile: Full-Time Paid 14; Part-Time Volunteers 47.
Governing Authority: university. Parent Institution: University of New Mexico. Tax-exempt: 501(c)(3).
Institution Type/Description: Science & Anthropology Museum.
Collections: archaeology; ethnology; osteology; archives; library; specialized Southwestern collections; Navajo & other Southwestern weaving; silver; Mimbres & Pueblo pottery; American Indian basketry; musical instruments; Pakistani textiles & jewelry; worldwide ethnology; anthropology library; South American textiles; Native American easel art.
Major Exhibits: An Experiment in Viewing, 11/12-4/15.
Research Fields: archaeology; osteology; ethnology.
Facilities: 12,500-vol. general library with North American specialization available on premises; photo archive emphasizing the Southwest; laboratory of physical anthropology; archaeology laboratory; reading room; 3,866 sq. ft. exhibit space. Ethnic art, books, Navajo & Pueblo silver jewelry & textiles for sale.
Activities: guided tours; lectures; formally organized education programs for children, adults, undergraduate & graduate students; docent program; permanent & temporary exhibitions; school loan service; funding of public service programs.
Publications: exhibition catalogs; newsletters; brochures; occasional papers; Weaving of the Southwest; Anthropological Papers Series; General Publications including: The Fetish Carvers of Zuni; Across the Colorado Plateau: Anthropological Studies for the Transwestern Pipeline Expansion Project, Vol. VII-XX Vol. XII: Before the Sky Fell: The Pre-Eruptive Sinagua of the Flagstaff Area; Vol XIII: Excavation of Cohonina & Cerbat Sites in the Western Arizona Uplands; Vol XIV: Excavation & Interpretation of Aceramic & Archaic Sites; Vol. XV: Subsistence & Environment; Vol. XVI: Interpretation of Ceramic Artifacts; Vol. XVII: Architectural Studies, Lithic Analysis & Ancillary Studies; Vol XVIII: Human Remains & Burial Goods; Vol. XIX: Hot Nights, San Francisco Whiskey, Baking Powder & a View of the River: Life on the Southwestern Frontier; Vol. XX: Conclusions & Synthesis-Communities, Boundaries & Cultural variation; Zuni: A Village of Silversmiths; Cuando Hablan Los Santos: Contemporary Santero Traditions from Northern New Mexico; Before Pecos; Environmental Disaster & the Archaeology of Human Response; Anasazi Regional Organization.
Hours & Admission Prices: Tues.-Sat. 10-4. No charge; donations accepted. Closed holidays. &
Attendance: 86,000 (accurate)
Membership: Student $5 UNM/$10 non-UNM; Senior 60 & over $35; Individual/Family $40; Contributor $250; Curator's Circle $500 or 50 hours volunteer service; Sponsor $1,000; Patron $2,500 or above.

MUSEUM OF SOUTHWESTERN BIOLOGY, 302 Yale Blvd., N.E., CERIA 83, Rm. 204, Albuquerque, NM 87131. Mailing Address: 1 University of New Mexico/MSCO3 2020, Albuquerque, NM 87131. Tel.: 505-277-1360. Fax: 505-277-1351.
E-mail: cosborn@unm.edu
Web Site: msb.unm.edu
Founded: 1930.
Congressional District: 1
Key Personnel: C.E.O., Dir. & Cur. Mammals, Dr. Joseph A. Cook; Collections Mgr. Mammals, Dr. Jon Dunnum; Cur. Birds, Dr. Christopher Witt; Collection Mgr. Birds, Andrew B. Johnson; Cur. Amphibians & Reptiles, Dr. Howard L. Snell; Collection Mgr., Dr. Tom Giermakowski; Cur. Anthropods, Dr. Kelly Miller; Collection Mgr. Arthropods, Dr. Sandra Brantley; Collection Mgr. Anthropods, Dr. David Lightfoot; Collection Mgr. Fishes, Alexandra M. Snyder; Cur. Herbarium, Dr. Timothy K. Lowrey; Collection Mgr. Herbarium, Phil Tonne; USGS Cur., Dr. Michael

A. Bogan; USGS Collection Mgr., Cindy A. Ramotnik; Collection Mgr. Genomic Resources, Cheryl Parmenter; Collection Mgr. NHNM, Rayo McCullough; Dir. Natural Heritage New Mexico, Dr. Esteban Muldavin.
Personnel Profile: Full-Time Paid 17.
Governing Authority: university. Parent Institution: University of New Mexico. Tax-exempt.
Institution Type/Description: Research Museum.
Collections: plants; mammals; birds; reptiles; amphibians; fish; anthropods; manuscripts; slide library; frozen tissue collection.
Research Fields: systematics; biogeography; ecology; evolution; biodiversity; conservation; behavior; bioacoustics; museum science; bioinformatics & education to government & business leaders, natural resource managers; parasitology.
Activities: formally organized education programs for undergraduate & graduate students.
Publications: occasional papers and special publications of the Museum of Southwestern biology.
Hours & Admission Prices: Tours by appointment. No charge; donations accepted. &

Attendance: 1,500 (accurate)

NATIONAL HISPANIC CULTURAL CENTER, ART MUSEUM, 1701 4th St., S.W., Albuquerque, NM 87102-4508. Tel.: 505-246-2261. Fax: 505-724-4760.
Web Site: www.nhccnm.org
Founded: 1993.
Congressional District: 1
Key Personnel: Interim Exec. Dir., Gary Romero; Dir. Visual Arts, Tey Marianna Nunn; Dir. Education, Shelle Sanchez; Dir. Performing Arts, Reeve Love; Dir. History & Literary Arts, Carlos Vasquez; Dir. Technology, Ruby Williams; Exec. Admin., Rosemary Garcia.
Personnel Profile: Full-Time Paid 28; Part-Time Paid 5; Part-Time Volunteers 136; Interns 5.
Governing Authority: state; nonprofit. Parent Institution: Department of Cultural Affairs, Santa Fe, New Mexico. Tax-exempt: 501(c)(3).
Institution Type/Description: Art Museum.
Collections: contemporary & historic art from New Mexico, United States, western hemisphere, Spain & the Spanish diaspora; books; historic photographs; archival materials.
Research Fields: Latino & Hispanic visual, performing arts, history, literature & Spanish language learning resources.
Facilities: 11,000-vol. library; 11,000 sq. ft. exhibition space; 600-seat theater; 300-seat film house; lecture hall; classrooms; meeting rooms; 125-seat restaurant.
Activities: permanent exhibitions; arts festivals; concerts; dance recitals; docent program; films; formal education programs for children & University of New Mexico students; guided tours; lectures; theater; training programs. Annual Events: Dia del Nino (Day of the Child); Dia de Muertos (Day of the Dead).
Hours & Admission Prices: Tues.-Sun. 10-5. Adults $5, senior citizens $3; discounts to AAM & ICOM members; school groups, children under 16, members & Sun. no charge. Closed New Year's Day; Easter; Thanksgiving; Christmas. &

Attendance: 233,034 (accurate)

✱ NATIONAL MUSEUM OF NUCLEAR SCIENCE AND HISTORY, (M), 601 Eubank Blvd., S.E., Albuquerque, NM 87123. Tel.: 505-245-2137. Fax: 505-242-4537.
E-mail: info@nuclearmuseum.org
Web Site: www.nuclearmuseum.org
Formerly: National Atomic Museum
Founded: 1969.
Congressional District: 1
Key Personnel: Dir., Jim Walther; Deputy Dir. & Exhibits Coord., Greg Shuman; Pres. (V), John Stichman; Cur., David Hoover; Registrar, Sandra Fye; Dir. Mktg. & Public Rels., Jennifer Hayden; Dir. Education, Rachel Cutrafello; Museum Store Mgr., Molly Brunell; Volunteer Coord., Melissa Donahoo; Special Events & Facility Rental Coord., Beonka Sinclair; Dir. Devel., Charles Lowery; Devel. Assoc., Nadine Scala.
Personnel Profile: Full-Time Paid 16; Part-Time Paid 5; Part-Time Volunteers 129.
Governing Authority: federal. Parent Institution: Department of Energy; Sandia National Laboratories. Subsidiary Institutions: Lockheed Martin Corp. & National Atomic Museum Foundation. Tax-exempt: 501(c)(3).
Institution Type/Description: Nuclear Science & History Museum.
Collections: nuclear weapon cases & related materials; nuclear medicine; weapons; history of the atom; aircraft including B-29, B-52B, A7, B-47E; cruise missiles; weapon systems; civil defense; Manhattan Project history.
Research Fields: nuclear weapons history; nuclear medicine; special exhibits that portray the events & personalities of the atomic age.

Facilities: library of books & photographs on the history of nuclear weapons & various energy programs available on premises; exhibits relating to nuclear weapons history; current & future nuclear energy programs; theater. Museum-related items for sale.
Activities: guided tours; films; temporary & permanent exhibitions.
Publications: newsletter, Nuclear Times.
Hours & Admission Prices: Daily 9-5. Adults $8, senior citizens 60 & over and youth 6-17 $7; discounts to AAM members, groups, active military & veterans; members and children 5 & under no charge. Closed New Year's Day; Easter; Thanksgiving; Christmas. &
Attendance: 80,000 (accurate)
Membership: The Neutrons $30; Senior Citizen & Student $35; Individual $40; Family $75; Seaborg $150-$299; Fermi $300-$499; Oppenheimer $500-$999; Curie $1,000-$4,999; Einstein $5,000 & up; Corporate $10,000 & up.

✱ NEW MEXICO MUSEUM OF NATURAL HISTORY & SCIENCE, (M), 1801 Mountain Rd., N.W., Albuquerque, NM 87104-1375. Tel.: 505-841-2846. Fax: 505-841-2844. TDD: 505-841-2878.
E-mail: denise.hidalgo@state.nm.us
Web Site: www.NMnaturalhistory.org
Founded: 1986.
Congressional District: 1
Key Personnel: Pres. (V) Bd. Trustees, Gary Friedman; Pres. (V) Volunteer Association, Dwight Jennison; Exec. Dir., Charles H. Walter; Foundation Exec. Dir., Jotina Trussell; Deputy Dir., Alicia Borrego Pierce; Finance Mgr., Sherice Padilla; Collections Mgr., Justin Spielmann; Museum Shop Mgr., Beth Ricker; Exec. Asst. to Dir., Denise Hidalgo.
Personnel Profile: Full-Time Paid 54; Part-Time Paid 4; Part-Time Volunteers 397; Interns 7.
Governing Authority: state government. Department of Cultural Affairs, Santa Fe, NM. Tax-exempt. 501(c)(3).
Institution Type/Description: Natural History & Science Museum.
Collections: natural history materials representative of New Mexico & southwest; botany, geology, paleontology & zoology specimens; microcomputer memorabilia.
Research Fields: paleontology; botany; zoology; geology.
Facilities: 440-book & 5,000 journal library pertaining to natural history; 50,000 sq. ft. exhibit space; Extreme Screen DynaTheater; 280-seat Lockheed Martin DynaTheater; 20,000 sq. ft. Education & Research Complex; naturalist center; class rooms; educational facilities. Museum-related items for sale.
Activities: guided tours; lectures; organized education programs for children, adults, undergraduate & graduate students; docent program; training programs for professional museum workers; participatory, temporary & traveling exhibitions.
Publications: quarterly newsletter, Timetracks; brochures; exhibit catalog; posters; annual report; science bulletin.
Hours & Admission Prices: Daily 9-5. Museum: adults $7, seniors 60 & up $6, children 3-12 $4. Extreme Screen DynaTheater: adults $10, seniors 60 & up $8, children 3-12 $6. Planetarium: adults $7, seniors 60 & up $6, children 3-12 $4; discounts to AAM & ASTC members; members & children under 3 no charge. Combination packages available. Closed New Year's Day; Thanksgiving; Christmas. &
Attendance: 211,322 (accurate)
Membership: Student $25; Senior $30; Individual $45; Senior Couple $50; Family & Grandparent $75; Family Plus $100.

NORTH FOURTH ART CENTER & GALLERY, 4904 Fourth St., N.W., Albuquerque, NM 87107. Tel.: 505-344-4542. Fax: 505-345-2896.
E-mail: info@vsartsnm.org
Web Site: vsartsnm.org/gallery
Institution Type/Description: Art Gallery.
Collections: photographs; paintings; sculpture.
Hours & Admission Prices: Sat. 12-4; other times by appointment.

PETROGLYPH NATIONAL MONUMENT, 6001 Unser Blvd., N.W., Albuquerque, NM 87120-2069. Tel.: 505-899-0205, ext. 331. Fax: 505-899-0207.
Web Site: www.nps.gov/petr
Founded: 1990.
Congressional District: 1
Key Personnel: Dir., Joseph P. Sanchez; Museum Shop Mgr., Ed Dunn.
Personnel Profile: Full-Time Paid 23; Part-Time Paid 4; Part-Time Volunteers 10.
Governing Authority: Parent Institution: National Park Service.
Institution Type/Description: National Park & Landmark.

Collections: petroglyph etchings by local Native Americans.
Facilities: hiking trails.
Hours & Admission Prices: Daily 8-5. Sat.-Sun. $2, Mon.-Fri. $1. Closed holidays.
Attendance: 150,000 (estimated)

RIO GRANDE NATURE CENTER STATE PARK, 2901 Candelaria Rd., N.W., Albuquerque, NM 87107-2965. Tel.: 505-344-7240. Fax: 505-344-4505.
E-mail: friends@rgnc.org
Web Site: www.rgnc.org
Founded: 1982.
Congressional District: 2
Key Personnel: Pres., Jo Fairbanks; Office Mgr., Paula Dwall; Museum Shop Mgr., Marilyn Cimalore; Park Supt., Beth Dillingham.
Personnel Profile: Full-Time Paid 8; Part-Time Paid 3; Part-Time Volunteers 150; Interns 2.
Governing Authority: state government; nonprofit. Parent Institution: New Mexico State Parks Division, Santa Fe, NM 87504-1147. Tax-exempt.
Institution Type/Description: Nature Center: preservation & protection of wetland and riverine resources in the Rio Grande Valley & the surrounding Bosque (woods); to educate visitors on the importance of wetlands and river resources in the arid Southwest.
Collections: botanical; zoological.
Facilities: visitor center; gardens; trails. Museum-related items for sale.
Activities: classes; guided tours; special events.
Publications: quarterly member newsletter; park brochure; trail guides.
Hours & Admission Prices: Park: daily 8-5. Visitor Center: daily 10-5. Parking Fee: car $3, van & buses $15. Closed New Year's Day; Thanksgiving; Christmas. &
Attendance: 50,000 (accurate)
Membership: Student & Senior (62 & up) $15; Individual & Family $30; Donor $50; Sponsor $100; Sustaining $250; Benefactor $500; Life $1,000.

SAN FELIPE DE NERI CHURCH MUSEUM, San Felipe de Neri Parish, 2005 North Plaza N.W., Albuquerque, NM 87104. Mailing Address: P.O. Box 7007, Albuquerque, NM 87194-7007. Tel.: 505-243-4628. Fax: 505-224-9495.
E-mail: dgarcia@sanfelipedeneri.org
Web Site: www.sanfelipedeneri.org
Key Personnel: Pastor, Fr. Dennis Garcia
Institution Type/Description: Historic Church.
Collections: period furnishings; religious art.
Hours & Admission Prices: Mon.-Sat. 9:30-5. No charge.

TAMARIND INSTITUTE & GALLERY, University of New Mexico, 2500 Central Ave., S.E., Albuquerque, NM 87106. Mailing Address: University of New Mexico, MSC 04-2540, Albuquerque, NM 87131. Tel.: 505-277-3901. Fax: 505-277-3920.
E-mail: tamarind@unm.edu
Web Site: tamarind.unm.edu
Key Personnel: Dir., Marjorie Devon
Institution Type/Description: Art Gallery.
Collections: lithographs.
Activities: printing demonstration; workshops.
Hours & Admission Prices: 1st Fri. of the month 3:30. No charge.

TELEPHONE MUSEUM OF NEW MEXICO, 110 Fourth St., N.W., Albuquerque, NM 87102-3268. Mailing Address: P.O. Box 16174, Albuquerque, NM 87191-6174. Tel.: 505-842-2937. museumsusa.com.
E-mail: telmuseum@hotmail.com
Web Site: www.nmculture.org
Founded: 1961.
Key Personnel: C.E.O. & Chm. (V), Gigi Galassini; Dir., Sue Turner.
Personnel Profile: Full-Time Volunteers 8; Part-Time Volunteers 65.
Volunteer Hours: 3,500
Governing Authority: Tax-exempt.
Institution Type/Description: Telephone & Communication Museum.
Collections: telephones; switchboards; insulators; directories; hand tools; photographs; telephone equipment; telephone technical books; genealogy research.
Major Exhibits: Alexander G. Bell History, 3/14; AGB Other Forms of Communications, 10/14.
Research Fields: Growth & development of wire communications in Southwest.
Facilities: reading room.

Activities: permanent & temporary exhibitions.
Publications: books, 50 Little Known Facts About Alexander G. Bell; Theodore Vail.
Hours & Admission Prices: Mon., Wed. & Fri. 10-2; large tours by appointment. Adults $2, children under 12 $1. Tours on off-hours: $4 per person. Closed holidays. &
Attendance: 2,500 (estimated)

TURQUOISE MUSEUM, 2107 Central N.W., Albuquerque, NM 87104-1605. Mailing Address: P.O. Box 7598, Albuquerque, NM 87194-7598. Tel.: 505-247-8650. Fax: 505-247-8765.
E-mail: turquoisemuseum@yahoo.com
Web Site: turquoisemuseum.com
Founded: 1993.
Congressional District: 1
Key Personnel: Pres. & Cur., Joe Dan Lowry; Museum Shop Mgr., Katy Lowry.
Personnel Profile: Full-Time Paid 4; Part-Time Paid 2.
Governing Authority: private; for profit.
Institution Type/Description: General Museum.
Collections: J.C. Zachary, Jr. largest natural turquoise collection; educational information: history, mining, geology, cutting, marketing.
Facilities: 60-seat auditorium. Museum-related items for sale.
Activities: lapidary demonstrations.
Publications: Turquoise Unearthed.
Hours & Admission Prices: Mon.-Fri. 9:30-5, Sat. 9:30-4. Adults $4, senior citizens & children $3. Closed Independence Day; Thanksgiving; Christmas. &
Attendance: 9,375 (accurate)

UNIVERSITY OF NEW MEXICO ART MUSEUM, (M), UNM Center for the Arts, Cornell & Central N.E., Albuquerque, NM 87131. Mailing Address: UNM Art Museum, MSC04 2570, 1 University of New Mexico, Albuquerque, NM 87131. Tel.: 505-277-4001. Fax: 505-277-7315.
E-mail: artmuse@unm.edu
Web Site: www.unm.edu/~artmuse
Founded: 1963.
Congressional District: 1
Key Personnel: Interim Dir. & Cur. Academic Initiatives, Sara Otto-Diniz; Cur. Photographs & Prints, Michele Penhall; Cur. Raymond Jonson Gallery, Robert Ware; Mgr. Collections, Bonnie Verardo; Asst. Cur. Photographs & Prints, Sherri Sorensen; Curatorial Asst., Leilani Ringkvist; Preparator, Steven Hurley.
Personnel Profile: Full-Time Paid 8; Part-Time Paid 7; Part-Time Volunteers 10.
Governing Authority: university. Parent Institution: University of New Mexico. Tax-exempt: 501(c)(3).
Institution Type/Description: Art Museum.
Collections: photographs; Beaumont Newhall prints; Harold Edgerton collection; Tamarind Lithography archive; 19th- to 20th-century art; Old Master paintings & sculpture; Albert A. Anella collection; Spanish colonial art; Field collection; Raymond Jonson collection & archive.
Research Fields: photography; graphics, with emphasis on lithography; 20th-century American painting & sculpture; Spanish colonial art, with emphasis on paintings.
Facilities: library; print & photography study room.
Activities: guided tours by appointment; lectures; gallery talks; formally organized education programs for graduate students; training programs for professional museum workers; inter-museum loan, temporary & traveling exhibitions.
Publications: catalogues; gallery guides; 19th Century Photographs at the University of New Mexico Art Museum; Tamarind Lithography Workshop, Inc. 1960-1970 Catalogue Raisonne; The Potters of Mata Ortiz; Nineteenth Century Lithography in Europe; Highlights of the Collection.
Hours & Admission Prices: Tues.-Sat. 10-4. No charge; donations accepted. Closed holidays. &
Attendance: 44,374 (accurate)

THE UNSER RACING MUSEUM, (M), 1776 Montano N.W., Albuquerque, NM 87107-3245. Tel.: 505-341-1776.
E-mail: cathy@unserracingmuseum.com
Web Site: www.unserracingmuseum.com
Institution Type/Description: Racing Museum.
Collections: four generations of race cars; uniforms; period cars; trophies; pace cars; artwork.
Facilities: Museum-related items for sale.
Activities: rental facilities.

Hours & Admission Prices: Daily 10-4. Adults $10, seniors & military $6; children under 16 no charge.

WHEELS TRANSPORTATION MUSEUM, 1100 2nd S.W., Albuquerque, NM 87102-1535. Mailing Address: P.O. Box 95438, Albuquerque, NM 87199-5438. Tel.: 505-243-6269.
E-mail: info@wheelsmuseum.org
Web Site: www.wheelsmuseum.org
Key Personnel: Pres., Leba Freed.
Governing Authority: Tax-exempt: 501(c)(3).
Institution Type/Description: Transportation Museum.
Collections: historic industrial plant history; 18 structures; transportation.
Hours & Admission Prices: Call for hours. No charge.

Artesia

ARTESIA HISTORICAL MUSEUM & ART CENTER, (M), 505 Richardson Ave., Artesia, NM 88210-2062. Tel.: 575-748-2390. Fax: 575-748-7345 (attn: Museum).
E-mail: artesiamuseum@artesianm.gov
Web Site: www.artesianm.gov/index.php/visitors-main/museum-vis
Founded: 1970.
Congressional District: 2
Key Personnel: Mgr., Nancy Dunn; Registrar, Carissa Baize.
Personnel Profile: Full-Time Paid 2; Part-Time Paid 1.
Governing Authority: municipal; museum commission. Parent Institution: City of Artesia. Tax-exempt.
Institution Type/Description: History Museum: located in 1904-05 Moore-Ward House.
Collections: agriculture; ranching; archaeology; geology; oil fields; oral histories; newspapers; local artists; photographs; clothing & objects of early residents; Native American artifacts; printing press; windmill; phonograph records.
Research Fields: local & area history; local film history; Seven Rivers settlement; genealogy.
Facilities: library of books, newspapers, manuscripts; oral history tapes available for research on premises by arrangement.
Activities: guided tours; permanent, temporary, loan & participatory exhibits; educational outreach classroom programs available. Annual Events: Quilt Show; Art Show; Living Treasures Awards Ceremony; Dia de los Muertos honoring Artesia's Veterans.
Hours & Admission Prices: Tues.-Fri. 9-12 & 1-5, Sat. 1-5. No charge; donations accepted. Closed New Year's Day; Memorial Day; Independence Day; Labor Day; Thanksgiving & day after; Christmas Eve & Day. &
Attendance: 2,000 (accurate)

Aztec

AZTEC MUSEUM AND PIONEER VILLAGE, 125 N. Main Ave., Aztec, NM 87410-1923. Tel.: 505-334-9829. Fax: 505-334-9829.
E-mail: info@aztecmuseum.org
Web Site: www.aztecmuseum.org
Founded: 1963.
Congressional District: 2
Key Personnel: Dir., Dale W. Anderson.
Personnel Profile: Part-Time Paid 3; Part-Time Volunteers 2.
Governing Authority: private association. Parent Institution: Aztec Museum Association, Inc. Tax-exempt.
Institution Type/Description: General Museum: housed in 1940 City Hall building; Pioneer Village.
Collections: Lobato collection of rocks, minerals, fossils, artifacts; Abrams collection of prehistoric Anasazi pottery; 1900 barber shop; early fashions; furnished living room, bedroom & kitchen; antique farm & ranch equipment; pioneer hand tools & household items; historic oil & gas equipment including a 1920 spudder; frontier village; sheriff's office; jail; lawyer's office; doctor's office; 1880 log cabin; 1905 replica of bank; D&RG narrow gauge caboose replica; general store; post office; print shop; tin shop; replica of early 1900s farm house.
Research Fields: pioneer living & working c.1880s; oil & gas from 1920s to present.
Activities: self-guided tours; permanent & loan exhibitions.
Publications: walking tour; brochure.
Hours & Admission Prices: Summer: Mon.-Sat. 9-5. No charge; donations accepted. Closed New Year's Day; Good Friday; Memorial Day; Labor Day; Veterans Day; Independence Day; Thanksgiving; Christmas. &
Attendance: 4,000 (estimated)
Membership: Regular $25; Patron $250; Lifetime $500.

AZTEC RUINS NATIONAL MONUMENT, 84 County Rd. 2900, Aztec, NM 87410-9715. Tel.: 505-334-6174. Fax: 505-334-6372. TDD: 505-334-6174.
E-mail: azru_curatorial@nps.gov
Web Site: www.nps.gov/azru
Founded: 1923.
Congressional District: 3
Key Personnel: Supt., Dennis Carruth.
Governing Authority: federal. Parent Institution: Dept. of the Interior. Affiliated with National Park Service. Tax-exempt.
Institution Type/Description: Archaeology Museum.
Collections: archaeological materials of ancestral Puebloan people, late 1000s-1300 A.D.
Research Fields: Chaco archaeology.
Facilities: 280-vol. library on ethnology & archaeology of Southwest Indians & prehistoric Pueblo Indians. Publications, postcards, slides & guidebooks for sale.
Activities: lectures; self-guiding trail; 25 min. video.
Publications: orientation brochure; trail guide; leaflet.
Hours & Admission Prices: Memorial Day to Labor Day daily 8-6; Sept.-May daily 8-5. Adults $6; discounts for Golden Eagle, Golden Age & Golden Access passport holders; children under 15 no charge. Closed New Year's Day; Thanksgiving; Christmas. &
Attendance: 50,000 (estimated)

Belen

BELEN HARVEY HOUSE MUSEUM, 104 N. 1st St., Belen, NM 87002-4302. Tel.: 505-966-2614. Facebook: Belen Harvey House Museum.
E-mail: ronnie.torres@belen-nm.gov
Web Site: www.belen-nm.gov/community/harveyhousebelen
Founded: 1985.
Key Personnel: Museum Tech, Ronnie Torres; Museum Shop Mgr., Heide Green.
Volunteer Hours: 2,500
Operating Income: 25,000
Governing Authority: Parent Institution: Belen Public Library. Tax-exempt.
Institution Type/Description: History Museum: housed in the former Harvey House Restaurant; built in 1910.
Collections: local & New Mexico history; period furnishings; Santa Fe railroad artifacts; Harvey House & Harvey girl artifacts; local historical artifacts.
Major Exhibits: Matanza Exhibit at Harvey House, 1/18/14-1/25/14; Pilot Club Exhibit, 2/15/14-2/22/14; Belen History, 3/14; Belen Model Rail Club, 4/14-5/11/14; Railroad, 6/7/14-7/5/14; Familia, 8/1/14-8/17/14; Belen Model Rail Club, 10/14-11/9/14; Harvey House Holiday Exhibit, 11/22/14-1/4/15.
Research Fields: Harvey houses.
Activities: Museum Sponsors: Belen Model Railroad Club Winter Sale in January.
Hours & Admission Prices: Tues.-Sat. 12:30-3:30. No charge; donations accepted. Fee for groups of 25 or more.
Attendance: 10,000 (estimated)

Bernalillo

*** CORONADO STATE MONUMENT,** 485 Kuaua Rd., Bernalillo, NM 87004-7099. Tel.: 505-867-5351; 800-419-3738. Fax: 505-867-1733.
E-mail: coronadohistoricsite@gmail.com
Web Site: www.nmmonuments.org
Founded: 1940.
Key Personnel: Mgr., Angie Manning.
Personnel Profile: Full-Time Paid 3; Part-Time Paid 1; Part-Time Volunteers 5.
Governing Authority: state. Parent Institution: Museum of New Mexico. Tax-exempt: 170(c)(3).
Institution Type/Description: Historic Site: Site of Tiwa Pueblo named Kuaua dating from AD 1300.
Collections: local history & culture; Native American & Spanish artifacts; murals.
Activities: guided tours; lectures; Indian storytelling; guided tours. Museum Sponsors: Preservation Month in May; Fiesta of Cultures in October; Christmas at Kuaua in December.
Hours & Admission Prices: Wed.-Mon. 8:30-5. Adults $3; discount to AAM & ICOM members and senior groups of 10 or more; seniors on Wed. with N.M.I.D., children 16 & under, school groups by appointment & New Mexico residents on Sun. no charge. Coronado Historic Site & Jemez Historic Site: two day pass $5 each site. Annual Site Pass: $10. Closed New Year's Day; Easter; Thanksgiving; Christmas. &
Attendance: 18,260 (accurate)

Membership: Student & Teacher $15; Senior Individual (65 & up) $25; Individual $30; Senior Couple $35; Family/Dual $40; Benefactor Business $300 & up.

SANDOVAL COUNTY HISTORICAL SOCIETY MUSEUM, 161 Edmond Rd., Bernalillo, NM 87004. Mailing Address: P.O. Box 692, Bernalillo, NM 87004-0692. Tel.: 505-867-2755.
Web Site: www.sandovalhistory.com/
Founded: 1977.
Congressional District: 3
Personnel Profile: Part-Time Volunteers 6.
Governing Authority: private; nonprofit organization. Tax-exempt: 501(c)(3).
Institution Type/Description: Historical Society Museum: housed in the former home of oil painter, Edmond DeLavy.
Collections: local history & culture; photographs; local genealogical records; personal artifacts; paintings; local geology.
Facilities: 150-vol. library; 100-seat auditorium; 2,700 sq. ft. exhibit space.
Activities: society meetings; art shows; concerts; lectures; loan, participatory, temporary & traveling exhibitions.
Publications: quarterly newsletter, El Cronicon.
Hours & Admission Prices: Sept.-June Wed. 10-12, Sun. 2-4. Adults $5; members no charge. &
Attendance: 1,000 (accurate)
Membership: Single $15; Couple $25.

Bloomfield

SAN JUAN COUNTY ARCHAEOLOGICAL RESEARCH CENTER AND LIBRARY AT THE SALMON RUIN/SAN JUAN COUNTY MUSEUM ASSOCIATION, (M), 6131 U.S. Hwy. 64, Bloomfield, NM 87413. Mailing Address: P.O. Box 125, Bloomfield, NM 87413-0125. Tel.: 505-632-2013. Fax: 505-632-8633.
E-mail: sreducation@sisna.com
Web Site: www.salmonruins.com
Founded: 1973.
Congressional District: 2
Key Personnel: Pres., David Casey; Exec. Dir., Larry L. Baker; Dir. Education, Nancy Sweet Espinosa; Acting Librarian & Museum Shop Mgr., Diane Hayden.
Personnel Profile: Full-Time Paid 4; Full-Time Volunteers 1; Part-Time Paid 1; Part-Time Volunteers 17; Interns 1.
Governing Authority: nonprofit. Parent Institution: San Juan County Museum Association. Tax-exempt: 501(c)(3).
Institution Type/Description: Archaeological and Cultural Museum, Library and Research Center.
Collections: historical; archaeological; anthropological. Heritage Park: cultural habitations of the Four Corners region. Salmon Homestead Complex: 1898 adobe house & outbuilding;
Research Fields: four corners archaeology; ruins stabilization & historic structure rehabilitation; rock art; petroglyphs; local history, living cultures of the Four Corners area (Navajo, Ute, Jicarilla Apache, Hispanic).
Facilities: 7,000-vol. research library of books on archaeology, anthropology & Southwestern history; picnic area; amphitheatre; nature trails. Books, native arts & crafts and museum-related items for sale.
Activities: lectures; workshops; arts & crafts fair; permanent & temporary exhibitions; outdoor theater; Journey into the Past guided tours.
Publications: Survey & mitigation reports of sites; historic series; report, Thirty Five Years of Archaeological Research at Salmon Ruins, Volumes I, II, & III; quarterly newsletter, The Outlier.
Hours & Admission Prices: May-Oct. daily 8-5; Nov.-April Mon.-Sat. 8-5, Sun. 12-5. Adults 16 & over $3, senior citizens $2, children 6-15 $1; discounts to AAM members; children under 6, members & student groups no charge. Closed New Year's Day; Easter; Thanksgiving; Christmas. &
Attendance: 7,235 (accurate)
Membership: Student & Single Senior $10; Individual & Senior Couple $15; Family $25; Institution $50; Patron $100; Lifetime $500.

Capitan

SMOKEY BEAR MUSEUM, 118 W. Smokey Bear Blvd., Capitan, NM 88316. Mailing Address: P.O. Box 591, Capitan, NM 88316-0891. Tel.: 575-354-2748. Fax: 575-354-6012.
Web Site: smokeybearpark.com/
Institution Type/Description: General Museum.
Collections: over 500 Smokey artifacts & memorabilia from around the world from early 1940's to present; baby bottle that fed Smokey the Bear cub when he was first rescued from the fire in the Capitan Mountains; Golden Bear, a 1944 stuffed Teddy Bear photographed with President Eisenhower.
Hours & Admission Prices: Daily 9-5. Adults 13 & over $2, children 7-12 $1; children 6 & under no charge. Closed New Year's Day; Thanksgiving; Christmas.

Capulin

CAPULIN VOLCANO NATIONAL MONUMENT, 46 Volcano Rd., Capulin, NM 88414. Tel.: 505-278-2201. Fax: 505-278-2211.
E-mail: christopher_moos@nps.gov
Web Site: www.nps.gov/cavo/
Founded: 1916.
Congressional District: 3
Key Personnel: Dir., Christopher R. Moos; Park Ranger Interpretation, Joyce Umbach.
Governing Authority: federal. Parent Institution: U.S. Dept. of the Interior. Subsidiary Institution: Natl. Park Svc. Tax-exempt.
Institution Type/Description: Natural Science Museum.
Collections: herbarium; geological & zoological specimens; entomology collection; lepidopteran collection.
Research Fields: volcanism.
Facilities: 500-vol. library of natural history books & reference material available on premises; nature center; self-guided trails.
Activities: 10-min. movie of natural history of the area; permanent exhibits.
Publications: orientation brochures.
Hours & Admission Prices: Winter daily 8-4; Summer daily 7:30-6:30. Entrance fee: $10 per vehicle. All federal passes accepted. &
Attendance: 48,994 (accurate)

Carlsbad

CARLSBAD CAVERNS NATIONAL PARK, 3225 National Parks Hwy., Carlsbad, NM 88220-5254. Tel.: 575-234-6716. Fax: 575-885-4557. TDD: 575-885-8884.
E-mail: samuel_denman@nps.gov
Web Site: www.nps.gov
Founded: 1923.
Congressional District: 2
Key Personnel: Collections Mgr., Samuel W. Denman.
Personnel Profile: Full-Time Paid 2; Part-Time Paid 2; Part-Time Volunteers 3.
Volunteer Hours: 100
Governing Authority: federal. Parent Institution: the U.S. Dept. of Interior, National Park Service. Tax-exempt.
Institution Type/Description: Natural History & Archaeology Museum.
Collections: history of the caverns; botany; entomology; geology; paleontology; archaeology; herpetology; archives.
Major Exhibits: Cavern Art - Photos and Paintings of Karlsbad Caverns, 1/13-12/14.
Research Fields: archeology; caves; biological.
Facilities: 4,000-vol. library; restaurant. Gift items for sale.
Activities: cavern tours; permanent exhibitions; backcountry day & overnight hiking; bird watching; nature & plant trail.
Publications: Carlsbad Caverns National Park, Silent Chambers; Timeless Beauty; What About Bats; Bats of Carlsbad Caverns National Park.
Hours & Admission Prices: Visitor Center: daily 8-5. Cave: Labor Day-Memorial Day call for hours. Adults $10, senior citizens $5; discounts to National Parks pass holders, Golden Eagle Passport holders & Golden Age pass holders; children under 15 no charge. Ranger-guided Tours: $7-$20. Closed Christmas.
Attendance: 250,000 (accurate)
Membership: Carlsbad Caverns-Golden Age (62 & over) $20; Guadalupe Association: Annual $25; Golden Eagle $75.

CARLSBAD MUSEUM & ART CENTER, (M), 418 W. Fox St., Carlsbad, NM 88220-5743. Tel.: 575-887-0276. Fax: 575-887-7191. Facebook: Carlsbad Museum and Art Center.
E-mail: museumstaff@cityofcarlsbadnm.com
Web Site: www.cityofcarlsbadnm.com/museum.cfm
Founded: 1931.
Congressional District: 2
Key Personnel: Dir., Dave Morgan; Pres. Bd. Dir., Michael Cleary.
Personnel Profile: Full-Time Paid 2; Part-Time Paid 4; Part-Time Volunteers 2.
Governing Authority: municipal. Parent Institution: City of Carlsbad. Tax-exempt.
Institution Type/Description: General Museum and Art Center.
Collections: archaeology; regional history; New Mexico art; Pueblo pottery; Peruvian antiquities.
Major Exhibits: Bittersweet Harvest: The Bracero Program 1942-1964 (T), 11/13-1/14; Carlsbad Area Art Association Membership Show, 3/14; Carlsbad Municipal Schools Annual Show, 4/14; The Masters of the Night: The True Story of Bats (T), 5/14-8/14; Zia Quilting & Stitchery Guild, 9/14; Carlsbad Area Art Association Images 2014, 10/14.
Facilities: art classroom; lecture hall.
Activities: permanent & temporary exhibits.
Publications: newsletter, Amigos; exhibition brochures.

Hours & Admission Prices: Mon.-Sat. 10-5. No charge; donations accepted. Closed New Year's Day; Martin Luther King, Jr. Day; Memorial Day; Independence Day; Labor Day; Veterans Day; Thanksgiving & day after; Christmas Eve & Day. &
Attendance: 8,180 (accurate)
Membership: Single $15; Family $25; Patron $100; Corporate $250; Benefactor $500; Lifetime $1,000.

LIVING DESERT ZOO AND GARDENS STATE PARK, 1504 Miehls Dr., Carlsbad, NM 88220-3057. Mailing Address: P.O. Box 100, Carlsbad, NM 88221-0100. Tel.: 575-887-5516. Fax: 575-885-4478.
E-mail: adrian.stiteler@state.nm.us
Web Site: www.nmparks.com
Founded: 1971.
Congressional District: 2
Key Personnel: C.E.O. New Mexico State Parks, Tommy Mutz; Chm. (V), K.C. Sparks; Zoo Dir., Adrian Stiteler; Animal Cur., Holly Paine; Museum Shop Mgr., Tina Brummett.
Personnel Profile: Full-Time Paid 12; Part-Time Paid 3; Part-Time Volunteers 126; Interns 1.
Governing Authority: state. Parent Institution: New Mexico State Parks Div., P.O. Box 1147, Santa Fe, NM 87504 (Not affiliated or associated with The Living Desert in Palm Desert, CA). Tax-exempt.
Institution Type/Description: Zoological & Botanical State Park.
Collections: birds, mammals & reptiles of the Chihuahuan Desert; cacti & succulents, featuring native Southwestern plants & exotics; mineral collection; exhibits of plants & animals representing different habitats found within the northern Chihuahuan Desert; nocturnal exhibit.
Facilities: 500-vol. library of botanical & zoological material available for use in park only; botanical garden; zoological garden; nature & conservation center; 1.3-miles of trails. Natural history publications, curios, gifts, postcards, plants & cacti for sale.
Activities: tours; lectures; concerts; study clubs; radio & TV programs; education programs; docent environmental education programs; plant sale; Annual Events: Mescal Roast.
Publications: booklet, A Guide to Living Desert Zoological and Botanical State Park, 1989.
Hours & Admission Prices: Memorial Day-Labor Day daily 8-5 (last entrance 3.30), Sept.-May daily 9-5 (last entrance 3:30). Adults $5, children 7-12 $3, organized youth groups $.50 per person; discounts to groups & AZA members; children 6 & under no charge. &
Attendance: 44,633 (accurate)
Membership: Horticultural Society Individual & Friends of the Park Family Individual $20; Family $40; Supporting $75; Life $500; Corporate (variable).

Cedar Crest

MUSEUM OF ARCHAEOLOGY & MATERIAL CULTURE, 22 Calvary Rd., Cedar Crest, NM 87008-9314. Mailing Address: P.O. Box 582, Cedar Crest, NM 87008-0582. Tel.: 505-281-2005.
E-mail: info@museumarch.org
Web Site: www.turquoisetrail.org
Key Personnel: Dir., Bradley Bowman
Institution Type/Description: Archaeology Museum.
Collections: science of archaeology & history; Native American; Sandia Cave & turquoise mining.
Hours & Admission Prices: May-Nov. 1 daily 12-7. Adults $3, children 6-12 $1.50; children under 6 no charge.

Cerrillos

CERRILLOS TURQUOISE MINING MUSEUM, 17 Waldo St. W., Cerrillos, NM 87010. Mailing Address: Box 131, Cerrillos, NM 87010. Tel.: 505-438-3008.
E-mail: brownp52@yahoo.com
Web Site: casagrandetradingpost.com
Founded: 1979.
Congressional District: 3
Key Personnel: Dir., Todd Brown; Museum Shop Mgr., Patricia Brown
Institution Type/Description: Mining Museum.
Collections: mining history; minerals; photographs; period artifacts.
Activities: petting zoo.
Hours & Admission Prices: Daily 9-5. Admission: $2 per person; discounts to groups. &
Attendance: 10,000 (estimated)

Chimayo

CHIMAYO MUSEUM, Plaza del Cerro, Chimayo, NM 87522. Mailing Address: P.O. Box 727, Chimayo, NM 87522-0727. Tel.: 505-351-0945.
E-mail: chimayomuseo@cybermesa.com
Web Site: www.chimayomuseum.org
Founded: 1995.
Congressional District: 3
Key Personnel: Pres. (V), Brenda Romero.
Personnel Profile: Part-Time Paid 1.
Governing Authority: Parent Institution: Chimayo Cultural Preservation Association. Tax-exempt.
Institution Type/Description: History Museum: former home of Jose Ramon & Petra Mestas Ortega.
Collections: local history & culture relating to Chimayo.
Activities: monthly events during season: lectures, performances, special exhibits.
Publications: quarterly newsletter.
Hours & Admission Prices: May-June Wed.-Sat. 10-3. No charge; donations accepted.
Attendance: 2,000 (estimated)

Church Rock

RED ROCK PARK, 5757 Red Rock Park Dr., Church Rock, NM 87311. Mailing Address: P.O. Box 10, Church Rock, NM 87311-0010. Tel.: 505-722-3839. Fax: 505-905-1277.
E-mail: redrockpark@ci.gallup.nm.us
Web Site: www.gallupnm.gov
Founded: 1976.
Congressional District: 3
Key Personnel: City Mgr., Dan Dible; Recreational Dir. Red Rock Park, Special Events & Projects, Ben Welch; Parks Specialist, Beverly Lovett.
Personnel Profile: Full-Time Paid 7; Part-Time Paid 6.
Volunteer Hours: 300
Operating Expenses: 1,500,000
Operating Income: 1,500,000
Governing Authority: municipal. Parent Institution: City of Gallup, NM 87301. Subsidiary Institution: Red Rock Park. Tax-exempt: 501(c)(3).
Institution Type/Description: Anthropology & Natural History.
Collections: fine arts, crafts & artifacts of the prehistoric Anasazi & historic Navajo, Hopi, Zuni, Rio Grande Pueblos, Apache; geological & botanical artifacts.
Research Fields: community history; human & natural history of Gallup area.
Facilities: 500-seat auditorium; 148 RV sites; 7,000-seat rodeo arena.
Activities: Annual Events: Grand Canyon College Rodeo in April; All Indian Festival in May; Run For The Wall, Freedom Flight & Ride, and Mega Bucks Bullriding Championship in May; Wrangeler Gallup Invitational Rodeo in June; J.M. High School/Jr. High School Rodeo's in May; U.S. Team Roping Championship and Lions Club Rodeo in June; National Jr. High Finals Rodeo, Wild Thing Bullriding Championship, Cycle City Arena Cross Series, and Turquoise Team Roping Championship in July; Intertribal Indian Ceremonial and Turquoise Team Roping Championship in August; Animosity FMX, Good Sam's Samboree, and Jordan Circus in September; Red Rock Balloon Rally in December.
Publications: Gallup Visitor Guide; magazines, New Mexico; Roper; Super Looper.
Hours & Admission Prices: Mon.-Fri. 8-5. No charge; donations accepted. Closed national holidays. &
Attendance: 200,000 (estimated)

Cimarron

PHILMONT MUSEUMS, (M), Philmont Scout Ranch, 17 Deer Run Rd., Cimarron, NM 87714-9638. Tel.: 575-376-2281, ext. 1136. Fax: 575-376-2602.
E-mail: robin.taylor@scouting.org
Web Site: www.scouting.org/philmont/
Founded: 1967.
Congressional District: 3
Key Personnel: Dir., David Werhane; Librarian & Museum Shop Mgr., Robin Taylor.
Personnel Profile: Full-Time Paid 4; Part-Time Paid 14; Interns 1.
Governing Authority: nonprofit organization. Parent Institution: Boy Scouts of America. Tax-exempt: 170(b)(1)(A).
Institution Type/Description: History & Art Museums: housed in Philmont Museum - Seton Memorial Library & Kit Carson Museum, located on the Philmont Scout Ranch.

Collections: art; archaeology; archives; ethnology; history of the Southwest; 1910-present day artifacts of the Boy Scouts of America.
Research Fields: Ernest Thompson Seton; Boy Scouts of America; History & Art of the Southwest.
Facilities: 3,000-vol. library pertaining to art, history, & anthropology available for research on premise only.
Activities: guided tours; lectures; study clubs; inter-museum loan, permanent & temporary exhibitions.
Hours & Admission Prices: June-Aug. daily 8-5; Sept.-May Mon.-Sat. 8-12 & 1-5. Kit Carson Museum: Summer: daily 8-5. No charge; donations accepted. &

Attendance: 35,000 (estimated)

Clayton

HERZSTEIN MEMORIAL MUSEUM, S. Second & Walnut, Clayton, NM 88415. Mailing Address: P.O. Box 75, Clayton, NM 88415-0075. Tel.: 575-374-2977. Fax: 575-374-2977.
E-mail: uchs@plateautel.net
Congressional District: 3
Key Personnel: Dir., Victoria Baker; Pres. (V), David Kyea; Museum Shop Mgr., Michelle Fernandez.
Personnel Profile: Full-Time Paid 1; Part-Time Paid 2; Part-Time Volunteers 12.
Governing Authority: nonprofit organization. Parent Institution: Union County Historical Society. Tax-exempt.
Institution Type/Description: Regional Historical Museum.
Collections: history of Union County & Northeastern New Mexico; photographs.
Facilities: archives.
Activities: monthly programs including family history, music, & Santa Fe Trail; genealogy club; research.
Publications: quarterly newsletter, Union County Historical Society Review.
Hours & Admission Prices: Tues.-Sat. 10-4; other times by appointment. No charge; donations accepted. Closed major holidays. &

Attendance: 700 (estimated)
Membership: Individual $25; Family $40; Supporting $100; Business $250; Lifetime $750; President's Council $1,000.

Cleveland

CLEVELAND ROLLER MILL MUSEUM, Rte. 518, Cleveland, NM 87715. Mailing Address: P.O. Box 287, Cleveland, NM 87715-0287. Tel.: 575-387-2645. Facebook: Cleveland Roller Mill.
E-mail: dancas@nnmt.net
Web Site: www.clevelandrollermillmuseum.com
Founded: 1989.
Congressional District: 43
Key Personnel: Dir., Dan Cassidy
Governing Authority: Parent Institution: Historic Mora Valley Foundation. Tax-exempt.
Institution Type/Description: Historic Building.
Collections: mill history; photographs; log mill.
Activities: demonstrations. Museum Sponsors: Millfest in September.
Hours & Admission Prices: Memorial Day to Labor Day Sat.-Sun. 10-3. No charge; donations accepted. Millfest: fee charged.

Attendance: 3,000 (estimated)

Cloudcroft

SACRAMENTO MOUNTAINS HISTORICAL MUSEUM, 1000 U.S. Hwy. 82, Cloudcroft, NM 88317. Mailing Address: P.O. Box 435, Cloudcroft, NM 88317-0435. Tel.: 575-682-2932. Fax: 575-682-3638. Facebook: Sacramento Mountains Historical Society and Museum.
E-mail: smhsmuseum@yahoo.com
Web Site: cloudcroftmuseum.com
Founded: 1977.
Congressional District: 2
Key Personnel: Pres., Dr. Wesley Lane; Vice Pres., Jane Sutherland; Treas., Bonnie Ludwick; Sec., Charlene Basham; Dir., Lucille Swope; Dir., Mike Mills; Dir., Ed Woten; Dir., Glenn Gillett; Dir., Ed Armstrong; Dir., Melissa Turnbow.
Personnel Profile: Full-Time Paid 1; Part-Time Volunteers 10.
Governing Authority: nonprofit organization. Parent Institution: Sacramento Mountains Historical Society. Tax-exempt: 501(c)(3).
Institution Type/Description: History Museum: a 2 1/2 acre pioneer village.
Collections: micro-filmed newspapers; photographs; maps; books. Historic Buildings: 1887 Posey Cottage; 1899 J. Arthur Eddy Cottage; 1899 Chapel; 1943 log cabin; depot; school house.

Research Fields: oral history; local cemeteries.
Facilities: Books & museum-related items for sale.
Activities: lectures; guided tours. Museum Sponsors: Living History Events.
Publications: brochure, History of the Sacramento Mountains; Railroad to Cloudcroft; Logging Railroads of the Lincoln National Forest; Saga of the Sierra Blanca; Harrowing Adventures of an Old-Time Cowboy & Sheriff; book, In Pursuit of a Railroad.
Hours & Admission Prices: Summer: Mon.-Tues. & Fri.-Sat. 10-4, Sun. 1-4; Winter: Fri.-Sat. 10-4. Adults $5, children 6-12 $3; discounts to military; members & children under 6 no charge. Closed holidays. &

Attendance: 3,000 (estimated)
Membership: Adult $15; Couple $30; Family $40; Extended Family & Corporate $50.

Clovis

CLOVIS DEPOT MODEL TRAIN MUSEUM, 221 W. First St., Clovis, NM 88101-7409. Tel.: 575-762-0066; 888-762-0064. Fax: 575-762-0066 (call first).
E-mail: philipw@3lefties.com
Web Site: www.clovisdepot.com/
Founded: 1995.
Key Personnel: Dir, Cur, Phil Williams.
Personnel Profile: Full-Time Volunteers 2.
Institution Type/Description: Model Train Museum.
Collections: area railroad history; period documents & photographs; model trains; operating telegraph station.
Facilities: library. Museum-related items for sale.
Hours & Admission Prices: March-Aug. & Oct.-Jan. Wed.-Sun. 12-5; call for additional hours.

EULA MAE MUSEUM & ART GALLERY, Clovis Community College, 417 Schepps Blvd., Clovis, NM 88101-8345. Tel.: 505-769-4115.
Web Site: www.clovis.edu
Institution Type/Description: Art Gallery and Museum.
Collections: artwork from local artists.
Hours & Admission Prices: Mon.-Fri. 8-4:30; groups by appointment. No charge.

HILLCREST PARK ZOO, 1201 Sycamore, Clovis, NM 88101. Mailing Address: P.O. Box 760, Clovis, NM 88101. Tel.: 575-769-7873. Fax: 575-762-0756.
E-mail: zoo1@cityofclovis.org
Institution Type/Description: Zoo.
Collections: over 150 mammals including Siberian Tiger, Mexican Lobo's, Common Tamarins, & Gila Monster; 300 birds.
Facilities: 23 acres.
Activities: rental facilities.
Hours & Admission Prices: Winter: Tues.-Sun. 9-3:30; Summer: call for extended hours. Adults 12 & over $3, seniors 60 & over $2.50, children 3-11 $1.50; children 2 & under no charge.
Membership: Single $20; Household $50.

NORMAN & VI PETTY ROCK & ROLL MUSEUM, 105 E. Grand Ave., Clovis, NM 88101-7509. Tel.: 575-763-3435.
Web Site: www.pettymuseum.com
Institution Type/Description: History Museum.
Collections: artifacts & memorabilia on Norman & Vi Petty.
Hours & Admission Prices: Mon.-Fri. 8-12 & 1-5, Sat. & group tours by appointment. Adults $5.

Columbus

COLUMBUS HISTORICAL SOCIETY MUSEUM, Hwy. 9 & 11, Columbus, NM 88029. Mailing Address: P.O. Box 562, Columbus, NM 88029-0562. Tel.: 505-531-2620.
Key Personnel: Acting Cur., Betty Dean; Pres. (V), Richard Dean
Governing Authority: Tax-exempt.
Institution Type/Description: Historical Society Museum.
Collections: local history & culture; American & Indian collection; military & weapons collection; Pancho Villa's Raid; photographs.
Facilities: Museum-related items for sale.
Hours & Admission Prices: Daily 10-4. No charge; donations accepted. Closed Christmas.
Membership: $45-$100.

PANCHO VILLA STATE PARK, 400 W. Hwy. 9, Columbus, NM 88029. Mailing Address: P.O. Box 450, Columbus, NM 88029-0450. Tel.: 505-531-2711. Fax: 505-531-2115.
E-mail: victor.trujillo@state.nm.us
Key Personnel: Park Mgr., John Read
Institution Type/Description: State Park: located on the grounds of the former Camp Furlong.
Collections: replica of Curtiss JN-3 "Jenny" airplane; 1916 Dodge touring car; historic artifacts; military weapons & ribbons. Historic Sites: 1902 former U.S. Customs House; two adobe structures from the Camp Furlong-era; Camp Furlong Recreation Hall.
Hours & Admission Prices: Daily 8-5.

Deming

DEMING LUNA MIMBRES MUSEUM, 301 S. Silver St., Deming, NM 88030-3761. Tel.: 575-546-2382. Fax: 575-544-0121.
E-mail: director@deminglunamimbresmuseum.com
Web Site: www.deminglunamimbresmuseum.com
Founded: 1957.
Congressional District: 2
Key Personnel: Museum Shop Mgr., Joyce Peterson.
Personnel Profile: Full-Time Volunteers 1; Part-Time Volunteers 80.
Governing Authority: society; non-profit. Parent Institution: Luna County Historical Society. Tax-exempt: 501(c)(3).
Institution Type/Description: Historical Society Museum & Historic House: housed in 1916 Deming National Armory; it once housed the unit that marched against Pancho Villa after his raid on Columbus NM & in 1941 housed men who were in the Bataan Death march; The Custom House, belonged to Seaman Field, an early custom agent.
Collections: Louise Southerland Toys & Doll Collection; Gem & Mineral Society Collection; Bessie C. May Collection of Clothing from 1880; Southwestern New Mexico early photography; military artifacts from frontier period-present; Mimbres Indian artifacts; Diamond A. Chuckwagon; quilt room; 1863 concert grand piano; Indian Basket Collection; antique lace display; medical & dental room; art gallery; transportation annex; four-rooms depicting 1930s living; Indian Kiva; Old Deming Street scenes; blacksmith & forge exhibit; 1909 RH drive Ford Roadster; 1907 Red, 1912 Allison La France Fire Truck etc.; Ruth Anderson Bell Collection; A.J. Fabian Whiskey Bottle Collection; John & Mary Alice King Indian pottery exhibit; archives for research on SW New Mexico; Geode Kid collection of thunder eggs.
Research Fields: Mimbres Indian pottery; Luna County & Southwestern United States histories.
Facilities: research library on Mimbres Indian Art & Culture; archives southwestern corner of New Mexico. Handcrafted items, Indian crafts, sand paintings, turquoise and silver jewelry, books & notepaper for sale.
Activities: guided tours; formally organized education programs; permanent exhibitions; continuous art show of local artists.
Publications: book & supplement one, The History of Luna County.
Hours & Admission Prices: Spring & Summer Mon.-Sat. 9-4; Fall & Winter: Mon.-Sat. 9-4, Sun. 1:30-4; groups by appointment. Adults $2; members no charge. Closed New Year's Day; Easter; Thanksgiving; Christmas. &
Attendance: 21,000 (accurate)
Membership: Individual $3; Life $100; Endowment Fund $1,000 & up.

Edgewood

WILDLIFE WEST NATURE PARK, 87 N. Frontage Rd., Edgewood, NM 87015. Mailing Address: P.O. Box 1359, Edgewood, NM 87015-1359. Tel.: 505-281-7655; 877-981-9453. Fax: 505-281-7170.
E-mail: info@wildlifewest.org
Web Site: www.wildlifewest.org
Founded: 1992.
Governing Authority: Tax-exempt.
Institution Type/Description: Wildlife Nature Park.
Collections: wildlife.
Hours & Admission Prices: Summer: daily 10-6; Winter: daily 12-4. Adults $7, seniors $6, students $4; children under 5 no charge. &
Attendance: 25,000 (estimated)
Membership: Family $50; Life $1,000.

Espanola

BOND HOUSE MUSEUM, 706 Bond St., Espanola, NM 87532-2727. Mailing Address: 405 N. Paseo De Onate, Espanola, NM 87532-2619. Tel.: 505-747-8535.
E-mail: mail@plazadeespanola.com
Institution Type/Description: Historic House: former home of Frank Bond & his family. Listed on the National Register of Historic Places.
Collections: historical artifacts & photographs; personal artifacts; furnishings.
Hours & Admission Prices: Mon.-Fri. 12-4, Sat. 11-3. No charge.

Farmington

BOLACK MUSEUM OF FISH AND WILDLIFE AND BOLACK ELECTROMECHANICAL MUSEUM, B Square Ranch, 3901 Bloomfield Hwy., Farmington, NM 87401-2831. Tel.: 505-325-4275.
Web Site: www.bolackmuseums.com
Founded: 1990.
Key Personnel: Dir., Owner & Cur., Tommy Bolack.
Personnel Profile: Full-Time Paid 3; Part-Time Paid 2.
Governing Authority: foundation. Tax-exempt.
Institution Type/Description: Fish and Wildlife Museum.
Collections: over 2,500 specimens; over 100,000 artifacts including electric power, broadcasting, medical, agriculture, oil & gas; slate & marble switchboard; small power distribution objects.
Publications: Two Fists Full; San Juan Agate; Kiui Kiel Currents.
Hours & Admission Prices: Mon.-Sat. 9-3. No charge. Closed holidays.
Attendance: 22,000 (estimated)

E3 CHILDREN'S MUSEUM & SCIENCE CENTER, 302 N. Orchard Ave., Farmington, NM 87401-6227. Tel.: 505-599-1425.
Key Personnel: Dir., Bart Wilsey; Museum Coord., Cherie D. Powell; Museum Shop Mgr., Kandy LeMoine.
Personnel Profile: Full-Time Paid 1; Part-Time Paid 4.
Governing Authority: Parent Institution: City of Farmington. Subsidiary Institution: Farmington Museum. Tax-exempt.
Institution Type/Description: Children's Museum and Science Center.
Collections: science exhibits pertaining to dinosaurs, sound, light, magnetism & shadows art.
Hours & Admission Prices: Tues.-Sat. 10-5. No charge; donations accepted. &

FARMINGTON MUSEUM, (M), 3041 E. Main St., Farmington, NM 87402-7621. Tel.: 505-599-1174. Fax: 505-326-7572.
E-mail: bwilsey@fmtn.org
Web Site: www.farmingtonmuseum.org
Founded: 1980.
Congressional District: 3
Key Personnel: Dir., Bart Wilsey; Collections Mgr., Jon Watson; Education Coord., Adrienne Boggs; Children's Museum Coord., Cherie Powell; Cur., Don Snoddy; Nature Center Coord., Donna Thatcher; Cur. Exhibits, Tom Cunningham; Volunteer Coord., Kandy LeMoine; Hospitality Supvr., Amy Homer.
Personnel Profile: Full-Time Paid 7; Part-Time Paid 5; Part-Time Volunteers 20.
Governing Authority: municipal; nonprofit organization. Parent Institution: City of Farmington, Office of Cultural Affairs. Tax-exempt.
Institution Type/Description: General Museum.
Collections: artifacts from the history of the San Juan Basin & Four Corners area; archives; photographs; manuscripts; costumes; oil & gas industry; trading post; oral histories, oil & gas boom, trading post families; nature center; living history farm & orchards.
Research Fields: local & regional history.
Facilities: library of material on pioneer families of Farmington & records of several oil & gas booms available for research by appointment; reading room. Museum-related items for sale.
Activities: guided tours; lectures; permanent & traveling exhibitions; video & slide presentations.
Publications: bimonthly newsletter, The Confluence.
Hours & Admission Prices: Mon.-Sat. 8-5. No charge; donations accepted. Closed New Year's Day; Thanksgiving; Christmas. &
Attendance: 106,000 (accurate)
Membership: Senior Individual $15; Individual $20; Senior Couple $25; Family $35; School, Organization, Friend & Business $75; Patron $100 & up; Sponsor $250; Benefactor $1,000 & up.

HARVEST GROVE FARM & ORCHARDS EXHIBIT BARN, Animas Park off Browning Pkwy., Farmington, NM 87402. Mailing Address: 3041 E. Main St., Farmington, NM 87402-7621. Tel.: 505-599-1423.
Web Site: www.farmingtonmuseum.org
Institution Type/Description: Historic House.
Collections: period tractors; early agricultural equipment.
Activities: special events; temporary exhibits.
Hours & Admission Prices: By appointment.

RIVERSIDE NATURE CENTER, Animas Park off Browning Pkwy., Farmington, NM 87401. Mailing Address: 3041 E. Main St., Farmington, NM 87402-7621. Tel.: 505-599-1422. Fax: 505-599-1429.
Web Site: www.farmingtonmuseum.org
Institution Type/Description: Nature Center.
Collections: wildlife & their habitats.
Facilities: Museum-related items for sale.
Activities: educational programs; special events.
Hours & Admission Prices: Tues.-Sat. 1-5, Sun. 1-4; extended hours during summer. No charge; donations accepted. Closed New Year's Day; Easter; Thanksgiving; Christmas.

Folsom

FOLSOM MUSEUM, INC., Main Street, Folsom, NM 88419. Mailing Address: P.O. Box 454, Folsom, NM 88419-0454. Tel.: 505-278-3616 (Sept.-May) & 2122 (June-Aug. & Sat.-Sun. in May & Sept.).
E-mail: bkthompson@bacavalley.com
Web Site: www.folsommuseum.org
Founded: 1967.
Congressional District: 2
Key Personnel: Pres., Eleanor Krusi; Sec., MariJo Balmer; Treas., Kay Thompson.
Personnel Profile: Part-Time Paid 2; Part-Time Volunteers 7.
Governing Authority: nonprofit organization. Parent Institution: Folsom Museum. Tax-exempt.
Institution Type/Description: Archaeology Museum: housed in the 1896 Doherty Mercantile Store Building.
Collections: Folsom Man discovery; local artifacts; historic items used in the settling of the area.
Facilities: Books & museum-related items for sale.
Publications: books, The Folsom, New Mexico Story and a Pictorial Review; The Folsom Hotel Story; pamphlet, Folsom Man; books, Arrowheads & Stone Artifacts; From Martyrs to Murders; The Story of Folsom; Folsom 1888-1988, Then & Now.
Hours & Admission Prices: May & Oct. Sat.-Sun. 10-5; June-Sept. Mon.-Fri. 10-5; other times by appointment. Adults $1.50; members no charge. &
Attendance: 2,000 (estimated)
Membership: Individual $10; Family $20; Lifetime $200; Homesteaders $250; Storekeeper $500; Rancher $750; Banker $1,000.

Fort Stanton

FORT STANTON MUSEUM, 104 Kit Carson Rd., Fort Stanton, NM 88323. Mailing Address: P.O. Box 1, Fort Stanton, NM 88323-0001. Tel.: 575-354-0341.
Key Personnel: Pres. (V), Larry Auld; Museum Shop Mgr., Charlotte Rowe.
Personnel Profile: Part-Time Volunteers 40.
Institution Type/Description: Military History Museum.
Collections: military history & artifacts; photographs; personal artifacts.
Facilities: Museum-related items for sale.
Hours & Admission Prices: April-Nov. Mon. & Thurs.-Sat. 10-4, Sun. 12-4; Dec.-March Sat. 10-4, Sun. 12-4. No charge; donations accepted.
Attendance: 11,000 (estimated)

Fort Sumner

BILLY THE KID MUSEUM, 1435 E. Sumner Ave., Fort Sumner, NM 88119-9606. Tel.: 575-355-2380. Fax: 575-355-1380.
E-mail: btkmuseum@plateautel.net
Web Site: billythekidmuseumfortsumner.com
Founded: 1953.
Key Personnel: Owner, Donald E. Sweet; Museum Shop Mgr., Tim Sweet; Museum Shop Mgr., Lula Sweet.
Governing Authority: individual operation.
Institution Type/Description: Local History Museum.
Collections: Old West, local & Billy the Kid memorabilia.
Facilities: Museum-related items for sale.
Activities: tours.
Publications: brochures.
Hours & Admission Prices: Jan. 16-May 14 & Oct. 2-Dec. Mon.-Sat. 8:30-5; May 15-Oct. 1 daily 8:30-5. Adults $5, senior citizens 62 & over $4, children ages 7-15 $3; children under 7 no charge. Closed Easter; Thanksgiving; Christmas. &
Attendance: 20,000 (accurate)

BOSQUE REDONDO MEMORIAL FORT SUMNER STATE MONUMENT, 3647 Billy the Kid Rd., Fort Sumner, NM 88119-0356. Mailing Address: Box 356, Fort Sumner, NM 88119-0356. Tel.: 575-355-2573. Fax: 575-355-2575.
E-mail: josephine.lucero@state.nm.us
Web Site: www.nmmonuments.org
Formerly: Fort Sumner State Monument
Founded: 1968.
Key Personnel: Pres. (V) & Museum Shop Mgr., MaryAnn Cortese.
Personnel Profile: Full-Time Paid 3.
Governing Authority: state. Parent Institution: Museum of New Mexico. Subsidiary Institution: New Mexico State Monuments. Tax-exempt: 170(c)(1).
Institution Type/Description: Historic Site: 1862-1869 frontier fort; control & supply station for Bosque Redondo Indian Reservation; relocation site of the Mescalero, Apache & Navajo peoples in the early 1860s; destination of The Long Walk.
Collections: local history & culture; Native American artifacts; photographs; personal artifacts.
Facilities: 700 sq. ft. exhibit space.
Activities: guided tours; lectures; living history demos; film.
Publications: El Palacio.
Hours & Admission Prices: Daily 8:30-5. Adults $3; AAM members, school groups & children 17 & under no charge. Closed New Year's Day; Easter; Thanksgiving; Christmas. &
Attendance: 10,126 (accurate)
Membership: Student $15; Individual $25; Partner $50; Business $75; Corporate $500.

Gallup

REX MUSEUM, 301 W. Historic 66 Ave., Gallup, NM 87301. Tel.: 505-863-1363.
E-mail: rexmuseum@ci.gallup.nm.us
Founded: 1995.
Congressional District: 2
Key Personnel: Pres. Gallup Historical Society (V), Dale Underwood; Clerk, Virgil Smith.
Personnel Profile: Full-Time Paid 1; Part-Time Volunteers 3.
Governing Authority: city. Tax-exempt.
Institution Type/Description: History Museum.
Collections: local history & culture; photographs; newspapers; personal artifacts.
Publications: newsletter.
Hours & Admission Prices: Mon.-Fri. 8-5. No charge; donations accepted. Closed legal holidays. &
Attendance: 2,100 (estimated)
Membership: Gallup Historical Society Annual $10.

STORYTELLER MUSEUM, 201 E. Hwy. 66, Gallup, NM 87301-6126. Tel.: 505-863-4131.
E-mail: jeremy@cia-g.com
Web Site: gallupculturalcenter.org
Key Personnel: Dir., Jeremy Boucher.
Governing Authority: Parent Institution: Southwest Indian Foundation. Tax-exempt.
Institution Type/Description: Native American History.
Collections: pottery; basketry; rug weaving; Kachina dolls; model trains; exhibits on silversmithing & traditional sandpaintings.
Facilities: Museum-related items for sale.
Hours & Admission Prices: Memorial Day to Labor Day Mon.-Sat. 9-4; Winter: Mon.-Fri. 8-5. No charge. &

Grants

NEW MEXICO MINING MUSEUM, 100 N. Iron Ave., Grants, NM 87020-3657. Mailing Address: P.O. Box 297, Grants, NM 87020-0297. Tel.: 505-287-4802; 800-748-2142. Fax: 505-287-8224.
E-mail: discover@grants.org
Web Site: www.grants.org
Key Personnel: Exec. Dir., Star Gonzales
Institution Type/Description: Mining Museum.
Collections: mining history; Native American artifacts; minerals; mining machinery; drilling & blasting equipment.
Hours & Admission Prices: Mon.-Sat. 9-4. Adults 19-59 $3, seniors 60 & over and youth 7-18 $2; children 6 & under no charge. Closed holidays. &

Hillsboro

BLACK RANGE MUSEUM, Hwy. NM 152, Hillsboro, NM 88042. Mailing Address: P.O. Box 454, Hillsboro, NM 88042-0454. Tel.: 505-895-5685.
Founded: 1961.
Key Personnel: Dir., June Anders
Institution Type/Description: History Museum.
Collections: local history; mining artifacts.
Hours & Admission Prices: By appointment. No charge; donations accepted. Closed Easter; Thanksgiving; Christmas.
Attendance: 800 (estimated)

Hobbs

WESTERN HERITAGE MUSEUM COMPLEX & LEA COUNTY COWBOY HALL OF FAME, (M), 5317 Lovington Hwy., NMJC Campus, Hobbs, NM 88240-9121. Tel.: 575-392-6730. Fax: 575-492-2680. Facebook: WHM Complex.
E-mail: themuseum@nmjc.edu
Web Site: www.westernheritagemuseumcomplex.org
Founded: 1978.
Congressional District: 2
Key Personnel: Chm. Bd. (V), Ray Battaglini; Exec. Dir., Dr. Darrell Beauchamp; Cur., Erin Anderson; Dir. Education, Mary Lyle; Administrative Sec., Lupe Johnston.
Personnel Profile: Full-Time Paid 4; Part-Time Paid 6; Interns 3.
Governing Authority: college; nonprofit organization. Parent Institution: New Mexico Junior College. Tax-exempt.
Institution Type/Description: History & Archaeology Museum.
Collections: historical & archaeological artifacts from southeastern New Mexico; ranching; rodeo; archaeological & petroleum-related artifacts & archives related to Lea County, NM.
Research Fields: archaeology of southeastern NM.
Facilities: 5,000 sq. ft. exhibit space.
Activities: guided tours; lectures; temporary exhibitions; movies; performances; special events.
Publications: quarterly newsletter.
Hours & Admission Prices: Tues. Sat. 10 5, Sun. 1-5; tours by appointment. Adults $3, seniors & students 6-18 $2; discount to NMAM members; children 5 & under, NMJC students & members no charge. Closed major holidays; college holidays; Spring & Christmas break. &
Attendance: 8,000 (accurate)
Membership: Student $5; LCCHOF $20; Individual $25; Family $50; Associate, Civic & Nonprofit Organization $100; Provider $250; Investor $500; Endorser $1,000; Corporate Organization $1,000; Life $1,500.

Jemez Springs

*** JEMEZ STATE MONUMENT,** One mile North of Jemez Springs, State Hwy. 4, Jemez Springs, NM 87025. Mailing Address: P.O. Box 143, Jemez Springs, NM 87025-0143. Tel.: 505-829-3530; 800-495-1279. Fax: 505-829-3530.
E-mail: giusewa@valornet.com
Web Site: www.nmmonuments.org
Founded: 1935.
Congressional District: 65
Key Personnel: Monuments Dir., Jose Cisneros; Monument Mgr., Richard Reycraft; Ranger, Jennifer Miksula; Ranger, Joshua Madalena; Receptionist, Brenda Tafoya.
Personnel Profile: Full-Time Paid 4; Part-Time Volunteers 1.
Governing Authority: state. Parent Institution: Department of Cultural Affairs. Tax-exempt: 170(c)(1).
Institution Type/Description: Historic Site: ruins of Towa Pueblo, Giusewa, occupied 1280-1680; ruins of Spanish mission, San Jose de los Jemez.
Collections: local history & culture; Indian village ruins; Native American artifacts.
Activities: guided tours; lectures; films. Annual Events: Arts Festival; Indian Dances.
Hours & Admission Prices: Nov.-April 1 Wed.-Mon. 8:30-5, April 2-Oct. daily 8:30-4:30. Adults $3, senior residents Wed. $1; discount to AAM & ICOM members; MNM Foundation member, New Mexico seniors with I.D. on Sun. and children 16 & under no charge. Coronado & Jemez 2 Day Pass $5. Annual Site Pass $10. Annual Museum & Monument Pass $25. Closed New Year's Day; Easter; Thanksgiving; Christmas. &
Attendance: 18,000 (estimated)
Membership: Student $20; Senior Individual $35; Individual $40; Senior Couple $45; Family $50; Sponsor $75; higher levels of membership with related benefits are available.

Kingston

HISTORIC PERCHA BANK MUSEUM, 119B Main St., Kingston, NM 88042. Tel.: 575-895-5652.
E-mail: info@perchabankmuseum.org
Web Site: www.perchabankmuseum.org
Founded: 2002.
Key Personnel: Dir., Mark Nero.
Personnel Profile: Full-Time Volunteers 2; Part-Time Volunteers 1.
Governing Authority: Tax-exempt.
Institution Type/Description: Historic Bank Museum.
Collections: bank history; period furnishings; photographs; mining artifacts; newspapers.
Hours & Admission Prices: Summer: Fri.-Sun. 11-3; other times by appointment. No charge; donations accepted.
Attendance: 650 (estimated)

Lamy

LAMY RAILROAD & HISTORY MUSEUM, 151 Old Lamy Trail, Lamy, NM 87540. Tel.: 505-466-6154.
Founded: 2006.
Institution Type/Description: History Museum: housed in the former Pflueger general store; built in 1881. Listed on the National Register of Historic Places.
Collections: local history; railroad artifacts; photographs; model trains; 1950 Talladega Dining car. Historic Building: 1884 saloon.
Activities: rental facilities.
Hours & Admission Prices: By appointment.

Las Cruces

THE BRANIGAN CULTURAL CENTER, (M), 500 N. Main St., Las Cruces, NM 88001. Mailing Address: P.O. Box 20000, Las Cruces, NM 88004. Tel.: 575-541-2154. Fax: 575-541-2152.
E-mail: rslaughter@las-cruces.org
Web Site: www.las-cruces.org/public-services/museums
Formerly: Las Cruces Historical Museum & Cultural Center
Founded: 1981.
Congressional District: 3
Key Personnel: Dir., Rebecca Slaughter; Cur. Education, Andrew Albertson.
Personnel Profile: Full-Time Paid 2; Part-Time Paid 1; Part-Time Volunteers 35; Interns 2.
Governing Authority: municipal; nonprofit organization. Parent Institution: City of Las Cruces.
Institution Type/Description: Cultural Center and Art & History Museum: housed in the former city library built in 1934. Listed on the National Register of Historic Places.
Collections: local & regional art & history.
Research Fields: Mexican cultural exchange.
Facilities: 100-seat auditorium; educational facilities; 2,935 sq. ft. exhibit space; 100-seat theater.
Activities: concerts; dance recitals; docent program; films; formal education programs for adults, children, & college students; guided tours; hobby workshops; lectures; loan, participatory & temporary exhibitions. Annual Events: International Day of Dance in April; Day of the Dead in October & November.
Publications: quarterly newsletter, MuseNews.
Hours & Admission Prices: Tues.-Sat. 9-4:30. No charge; donations accepted. Closed New Year's Day; Martin Luther King Jr. Day; Presidents' Day; Memorial Day; Independence Day; Labor Day; Columbus Day; Veterans Day; Thanksgiving & day after; Christmas Eve & Day. &
Attendance: 50,650 (accurate)
Membership: Adult $5; Family $10; Institutional $20; Contributing $50; Supporting $100; Sustaining $200; Patron $500.

CITY OF LAS CRUCES MUSEUM SYSTEM ADMINISTRATION, (M), 151 N. Church St., Las Cruces, NM 88001. Mailing Address: P.O. Box 20000, Las Cruces, NM 88004. Tel.: 575-541-2296. Fax: 575-525-8587.
E-mail: wticknor@las-cruces.org
Web Site: www.las-cruces.org/public-services/museums
Founded: 1981.
Congressional District: 3
Key Personnel: C.E.O. & Dir., Will Ticknor; Education & Volunteers, Julia Hansen; Cur. Collections, Stephanie Long; Cur. Exhibitions, Carey Crane; Exec. Asst., Liz Montoya.
Personnel Profile: Full-Time Paid 17; Part-Time Paid 8; Part-Time Volunteers 107; Interns 4.
Governing Authority: municipal; nonprofit organization. Branch Museums:

Branigan Cultural Center, 500 N. Main St., Las Cruces, NM 88001; Las Cruces Museum of Art, 491 N. Main St., Las Cruces, NM 88001; Museum of Nature & Science, 411 N. Main St., Las Cruces, NM 88001; Las Cruces Railroad Museum, 351 N. Mesilla St., Las Cruces, NM 88001; Historic Rio Grande Theatre, 211 N. Main St., Las Cruces, NM 88001. Tax-exempt.

Institution Type/Description: Art & History Museums: Branigan Cultural Center is listed on the National & State Register of Historic Places. Railroad Museum: housed in the Santa Fe Depot; listed on the National Register of Historic Places. Rio Grande Theatre, built 1926, listed on the National Register of Historic Places.

Collections: Branigan Cultural Center: local history; paintings. Las Cruces Museum of Art (MoA): contemporary art. Museum of Nature & Science: natural history, science, & the Chihuahuan Desert. Railroad Museum: railroad history.

Research Fields: local history.

Publications: quarterly print & e-newsletter, MuseNews; annual report.

Hours & Admission Prices: See individual museum listings. &

Attendance: 292,603 (accurate)

Membership: Foundation $35.

DONA ANA COUNTY SHERIFF'S DEPARTMENT MUSEUM,
845 N. Motel Blvd., Las Cruces, NM 88007-8100. Tel.: 575-525-8847.

Institution Type/Description: History Museum.

Collections: law enforcement history, furnishings & artifacts; photographs; badges; guns; memorabilia; early law enforcement vehicles.

Hours & Admission Prices: Mon.-Fri. 8-5. Closed holidays.

HISTORIC RIO GRANDE THEATRE,
211 N. Main St., Las Cruces, NM 88001. Mailing Address: P.O. Box 1721, Las Cruces, NM 88004. Tel.: 575-523-6403. Fax: 575-523-4760.

E-mail: infodaac@daarts.org

Web Site: www.riograndtheatre.com

Key Personnel: Interim Exec. Dir. & Program Coord., Ceci Vasconcellos; Theatre Mgr., David Salcido.

Governing Authority: Parent Institution: City of Las Cruces.

Institution Type/Description: Historic Building: built in 1926.

Collections: theatre history; works by local artists.

Facilities: 422-seat theatre.

Activities: performances; special events; temporary exhibitions.

Hours & Admission Prices: Tours: call for hours. &

LAS CRUCES MUSEUM OF ART, (M),
491 N. Main St., Las Cruces, NM 88001. Mailing Address: P.O. Box 20000, Las Cruces, NM 88004. Tel.: 575-541-2137. Fax: 575-541-2371.

E-mail: museum-of-art@las-cruces.org

Web Site: las-cruces.org/museums

Founded: 1999.

Congressional District: 36

Key Personnel: Cur. Education, Andrew Albertson; Cur. Exhibits, Joy Miller.

Personnel Profile: Full-Time Paid 3; Part-Time Paid 3; Part-Time Volunteers 10; Interns 2.

Governing Authority: municipal; nonprofit organization. Parent Institution: City of Las Cruces.

Institution Type/Description: Art Museum.

Collections: works of contemporary art; paintings; prints; sculpture; ceramics; photographs.

Major Exhibits: The Works of Gustave Baumann, 2/7/14-3/29/14; Chicanitos, 5/2/14-7/19/14; Art From Science, 8/14-10/11/14; Fragile Waters, 10/25/14-1/15/15.

Facilities: educational facilities; 3,500 sq. ft. exhibit space. Museum-related items for sale.

Activities: arts festivals; concerts; docent program; formal education programs for adults & children; temporary exhibitions; studio program school; guided tours; lectures; loan, temporary & traveling exhibitions; study clubs.

Hours & Admission Prices: Tues.-Sat. 9-4:30. No charge; donations accepted. Closed city holidays. &

Attendance: 75,000 (accurate)

LAS CRUCES MUSEUM OF NATURAL HISTORY,
Mesilla Valley Mall, 700 S. Telshor, Las Cruces, NM 88011. Mailing Address: P.O. Box 20000, Las Cruces, NM 88004. Tel.: 575-522-3120. Fax: 575-532-3370.

E-mail: mwalczak@las-cruces.org

Web Site: www.las-cruces.org/public-services/museums

Founded: 1986.

Congressional District: 3

Key Personnel: Dir., Michael Walczak; Naturalist, Richard Quick; Cur. Education, Kim Hanson.

Personnel Profile: Full-Time Paid 4; Part-Time Paid 4; Part-Time Volunteers 30; Interns 1.

Governing Authority: municipal; nonprofit organization. Parent Institution: City of Las Cruces.

Institution Type/Description: Science & Nature Museum.

Collections: nature & science focused on the Chihuahuan Desert including reptiles, insects, amphibians & fish.

Facilities: 780 sq. ft. exhibit space; educational facilities; nature center.

Activities: docent program; formal education programs for adults & children; guided tours; lectures; temporary exhibitions.

Publications: quarterly newsletter, MuseNews.

Hours & Admission Prices: Mon.-Thurs. & Sat. 10-5, Fri. 10-8, Sun. 1-5. No charge; donations accepted. Closed New Year's Day; Martin Luther King Jr. Day; Presidents' Day; Memorial Day; Independence Day; Labor Day; Columbus Day; Veterans Day; Thanksgiving & day after; Christmas Eve & Day. &

Attendance: 169,126 (accurate)

Membership: Adult $5; Family $10; Institutional $20; Contributing $50; Supporting $100; Sustaining $200; Patron $500.

LAS CRUCES RAILROAD MUSEUM, (M),
351 Mesilla St., Las Cruces, NM 88001. Mailing Address: P.O. Box 20000, Las Cruces, NM 88004. Tel.: 575-647-4480. Fax: 575-647-4304.

E-mail: gcourts@las-cruces.org

Web Site: www.las-cruces.org/public-services/museums

Founded: 2000.

Congressional District: 3

Key Personnel: Dir., Garland Courts; Education, Joanne Beer.

Personnel Profile: Full-Time Paid 1; Part-Time Paid 1; Part-Time Volunteers 24.

Governing Authority: municipal; nonprofit organization. Parent Institution: City of Las Cruces.

Institution Type/Description: Railroad History Museum: housed in the Santa Fe Depot; built in 1909. Listed on the National Register of Historic Places.

Collections: local railroad history; railroad memorabilia; photographs; period artifacts; early 1900s wooden caboose.

Research Fields: local history, railroading, & community development.

Facilities: library; 785 sq. ft. exhibit space; educational facilities.

Activities: docent program; formal education programs for adults & children; guided tours; hobby workshops; lectures; participatory & temporary exhibits. Annual Events: Railroad Days; Holiday Light Festival in December.

Publications: quarterly newsletter, MuseNews.

Hours & Admission Prices: Thurs.-Sat. 9-4:30. No charge; donations accepted. Closed New Year's Day; Martin Luther King Jr. Day; Presidents' Day; Memorial Day; Independence Day; Labor Day; Columbus Day; Veterans Day; Thanksgiving & day after; Christmas Eve & Day. &

Attendance: 12,150 (accurate)

MERIKS AQUARIUM,
630 King James Ave., Las Cruces, NM 88007-5380.

Key Personnel: Pres. (V), Tillian Paul; COO, Margaret Deeds; Assoc. Dir., Antoinette Harvey; Volunteer Coord., Danielle Marcus.

Personnel Profile: Full-Time Paid 1; Full-Time Volunteers 3; Part-Time Volunteers 3.

Institution Type/Description: Aquarium.

Research Fields: aquaculture; fish diseases.

Publications: newsletter, Fish Tales.

NEW MEXICO FARM & RANCH HERITAGE MUSEUM,
4100 Dripping Springs Rd., Las Cruces, NM 88011-5067. Tel.: 575-522-4100. Fax: 575-522-3085.

E-mail: craig.massey@state.nm.us

Web Site: www.nmfarmandranchmuseum.org

Founded: 1991.

Congressional District: 2

Key Personnel: Dir., Mark Santiago; Foundation Pres. (V), J. Bloom; Collections Mgr., Holly Radke; Education, Scott Green; Public Rels., Craig Massey; Museum Shop Mgr., Maria Massey.

Personnel Profile: Full-Time Paid 28; Part-Time Paid 3; Part-Time Volunteers 100; Interns 2.

Governing Authority: state; nonprofit. Parent Institution: New Mexico Department of Cultural Affairs, 228 E. Palace Ave., Santa Fe 87503. Subsidiary Institution: New Mexico Farm & Ranch Heritage Foundation. Tax-exempt: 501(c)(3).

Institution Type/Description: Agriculture, History & Science Museum.

Collections: concentration on history of agriculture in New Mexico from prehistoric to modern times. Historic Buildings: blacksmith shop; barn.

Major Exhibits: Meet the Producer..., 10/13-2/14; Stitches in Time: Quilts from the Museum's Collections, 10/13-5/14; The Canada Alamosa Project: 4000 Years of Agricultural History, 10/13-5/14; The Color of Pie Town, 10/13-10/19/14; Community Art Featuring: Patricia Burnett, 12/13/13-4/6/14; Community Art Featuring: Peter Goodman, 4/18/14-8/3/14; Cowboys!, 5/14-6/14; Community Art Featuring: Bonnie Mandoe, 8/15/14-11/30/14; Community Art Featuring: Collette Wallace, 12/12/14-4/5/15.
Research Fields: oral history-institutional beginnings & rural life.
Facilities: 2,000-vol. library of books on agricultural history; 144-seat auditorium; botanical garden; 140-seat restaurant; 2 classrooms; 25,000 sq. ft. exhibit space; 47 acres consisting of crops.
Activities: arts festivals; concerts; dance recitals; docent program; formal education programs for adults, children & college students (internships, New Mexico State University); guided tours; hobby workshops; lectures; loan, participatory, temporary & traveling exhibitions; theater; training programs for professional museum workers; broadcast programs; desert gardening. Annual Events: Blessing of the Fields, May 15; Cowboy Days, music and poetry festival in March.
Publications: biannual, newsletter, New Mexico Farm & Ranch Heritage.
Hours & Admission Prices: Mon.-Sat. 9-5, Sun. 12-5. Adults $5, seniors $3, children 5-17 $2; members no charge. Closed Thanksgiving; Christmas. &
Attendance: 63,000 (accurate)
Membership: Individual $35.

PAUL W. KLIPSCH MUSEUM - NEW MEXICO STATE UNIVERSITY, College of Engineering, 1060 Frenger Mall, Engineering Complex 111, Las Cruces, NM 88033. Mailing Address: Engineering Complex 111, P.O. Box 30001, Las Cruces, NM 88033-8001. Tel.: 575-646-2913.
Institution Type/Description: History Museum.
Collections: audio industry history; Klipsch's life, career & inventions; personal artifacts; photographs; personal papers & publications; early audio equipment; awards.
Hours & Admission Prices: By appointment.

UNIVERSITY ART GALLERIES, NEW MEXICO STATE UNIVERSITY, 1390 E. University Ave., E. of Solano, Las Cruces, NM 88003-8001. Mailing Address: Box 30001, Dept. Box 3572, Las Cruces, NM 88003-8001. Tel.: 505-646-2545. Fax: 505-646-8036.
E-mail: artglry@nmsu.edu
Web Site: www.nmsu.edu/~artgal
Founded: 1973.
Congressional District: 2
Key Personnel: C.E.O., Preston Thayer.
Personnel Profile: Full-Time Paid 2; Part-Time Paid 4; Part-Time Volunteers 10; Interns 1.
Governing Authority: university; nonprofit. Parent Institution: New Mexico State University. Tax-exempt: 501(c)(3).
Institution Type/Description: University Art Gallery.
Collections: 19th-century Mexican retablos; contemporary prints; photographs; painting & works on paper; time-based media.
Research Fields: contemporary art; retablos.
Facilities: 4,600 sq. ft. exhibition gallery; lecture hall.
Activities: guided tours; lectures; gallery talks; workshops; symposia; temporary, loan & traveling exhibits.
Publications: exhibition catalogues.
Hours & Admission Prices: Tues. & Thurs.-Sat. 12-4, Wed. 12-4 & 6-8. No charge; donations accepted. Closed university holidays. &
Attendance: 23,000 (accurate)
Membership: Students & Senior Citizens $25; Individual & Family $50; Patron $150; Benefactor $400.

UNIVERSITY MUSEUM, NEW MEXICO STATE UNIVERSITY, (M), Univ. Ave. at Solano Dr., Kent Hall, Las Cruces, NM 88003. Mailing Address: P.O. Box 30001, MSC 3564, Las Cruces, NM 88003-8001. Tel.: 575-646-5161. Fax: 575-646-1419. TDD: 505-646-3739.
E-mail: museum@nmsu.edu
Web Site: www.nmsu.edu/~museum/
Founded: 1959.
Congressional District: 2
Key Personnel: Dir., Dr. Monte McCrossin; Cur. Collections & Exhibitions, Dr. Jennifer Robles; Public Program Coord., Katherine Brooks.
Personnel Profile: Full-Time Paid 2; Part-Time Paid 3; Interns 2.
Governing Authority: university. Parent Institution: New Mexico State University. Tax-exempt: 301(b).
Institution Type/Description: University Anthropology Museum.
Collections: Southwestern U.S. & Northwestern Mexico archaeology; ethnology; history; science.

Research Fields: Southwestern U.S. archaeology; ethnology; history.
Facilities: research library; archaeology labs.
Activities: guided tours; field trips; lectures; workshops; temporary exhibitions.
Publications: newsletter; Taylor, Mary Daniels (2004) A Place as Wild as the West Ever Was: Mesilla, New Mexico, 1848-1872; New Mexico State University Museum, Las Cruces.
Hours & Admission Prices: Tues.-Sat. 10-4. No charge; donations accepted. Closed major and university holidays. &
Attendance: 21,000 (estimated)
Membership: Student $10; Individual & Family $25; Contributing $50; Sustaining $100; Patron $250.

THE ZUHL MUSEUM: HOME OF THE ZUHL COLLECTION, (M), NMSU Alumni and Visitors Center, 775 College Dr., Las Cruces, NM 88003. Mailing Address: Dept. Geological Sciences, MSC 3AB, NMSU, P.O. Box 30001, Las Cruces, NM 88003-8001. Tel.: 575-646-3616 & 4714. Fax: 575-646-6123. Facebook: Zuhl Museum.
E-mail: zuhl@nmsu.edu
Web Site: www.nmsu.edu/zuhl
Founded: 2004.
Congressional District: 2
Key Personnel: Dir. & Cur., Tiffany Holder-Santos.
Personnel Profile: Full-Time Paid 1; Interns 1.
Governing Authority: Parent Institution: New Mexico State University.
Institution Type/Description: Art Gallery & Natural History Museum.
Collections: petrified wood; fossils; minerals.
Hours & Admission Prices: Mon.-Fri. 8-5. Closed holidays. &
Attendance: 1,000 (estimated)

Las Vegas

CITY OF LAS VEGAS MUSEUM AND ROUGH RIDER MEMORIAL COLLECTION, (M), 727 Grand Ave., Las Vegas, NM 87701. Mailing Address: 1700 N. Grand Ave., Las Vegas, NM 87701-4731. Tel.: 505-426-3205. Facebook: City of Las Vegas Museum and Rough Rider Memorial Collection.
E-mail: museum@desertgate.com
Web Site: lasvegasmuseum.org
Founded: 1960.
Congressional District: 3
Key Personnel: Admin., Kristin Hsueh.
Personnel Profile: Full-Time Paid 2; Part-Time Paid 2; Part-Time Volunteers 5.
Governing Authority: municipal. Tax-exempt: 170(b).
Institution Type/Description: History Museum: Santa Fe Trail interpretive site.
Collections: military artifacts; photographs; Rough Riders' Cuban campaign of the Spanish-American War; Rough Riders' reunion; 19th-20th century Las Vegas history; agriculture; ranching; industry; post 1880 Santa Fe Trail artifacts.
Facilities: Museum-related items for sale.
Activities: community involvement; volunteer program; cultural programs.
Publications: brochures.
Hours & Admission Prices: Tues.-Sat. 10-4. No charge; donations accepted. Closed most holidays. &
Attendance: 3,500 (accurate)
Membership: Student & Senior $10; Individual $15; Family $25; Supporting $30-$59; Sustaining $60-$99; Sponsor $100-$499; Patron $500 & up.

THE RAY DREW GALLERY-NEW MEXICO HIGHLANDS UNIVERSITY, National Ave., Las Vegas, NM 87701. Tel.: 505-454-3338 & 3332. Fax: 505-454-0026.
E-mail: gallery@nmhu.edu
Formerly: The Fine Arts Gallery-New Mexico Highlands University
Founded: 1982.
Key Personnel: C.E.O., Dr. Jim Fries; Dir., Bob Read.
Personnel Profile: Full-Time Paid 1.
Governing Authority: university; nonprofit. Tax-exempt: 501(c)(3).
Institution Type/Description: Art Museum.
Collections: fine art prints from 1500 to present.
Facilities: classrooms; 800 sq. ft. exhibit space.
Activities: fine art exhibitions.
Hours & Admission Prices: Mon.-Fri. 8-5. No charge. Closed major holidays.
Attendance: 9,000 (estimated)

Lincoln

LINCOLN HISTORIC SITE, Hwy. 380, Lincoln, NM 88338. Mailing Address: P.O. Box 36, Lincoln, NM 88338-0036. Tel.: 575-653-4372. Facebook: Lincoln Historic Site.
E-mail: gary.cozzens@state.nm.us
Web Site: www.nmhistoricsites.org
Formerly: Old Lincoln County Courthouse Museum & Lincoln State Monument
Founded: 1937.
Key Personnel: Monument Mgr., Gary Cozzens; Dir., Richard Sims.
Personnel Profile: Full-Time Paid 5; Part-Time Volunteers 10.
Governing Authority: state. Parent Institution: Department of Cultural Affairs. New Mexico State Historic Sites: 1880 Brent House; 1860 Convento; 1850 Torreon; 1874 Tunstall Store Museum; 1890 Fresquez House & Watson House; 1887 San Juan Mission; 1874 Courthouse Museum. Tax-exempt.
Institution Type/Description: History Museum.
Collections: local history & culture; Indian artifacts; papers; furniture; clothing; tools; art objects; historic buildings.
Research Fields: history of southeast New Mexico, Lincoln County.
Facilities: library of research files on history of area; reading room; restaurant.
Activities: permanent & temporary exhibitions; films; interpretive tours.
Hours & Admission Prices: Daily 8:30-4:30; groups by appointment. Adults $5; discounts to senior groups and AAM & ICOM members; children under 16 no charge. Closed New Year's Day; Easter; Thanksgiving; Christmas. &
Attendance: 40,000 (accurate)

Lordsburg

LORDSBURG-HIDALGO COUNTY MUSEUM, 710 E. 2nd St., Lordsburg, NM 88045. Mailing Address: 316 E. 10th St, Lordsburg, NM 88045. Tel.: 575-542-9086.
Web Site: www.lordsburghidalgocounty.net/museum/museumhome.htm
Institution Type/Description: History Museum.
Collections: Prisoner of War memorabilia; Johnson photographs; Dr. Baxter's personal artifacts; mining; military; period tools; arrowheads; minerals & rocks.
Hours & Admission Prices: Mon.-Fri. 1-5.

SHAKESPEARE GHOST TOWN, 2 1/2 Miles South of Main St., Lordsburg, NM 88045. Mailing Address: P.O. Box 253, Lordsburg, NM 88045-0253. Tel.: 505-542-9034.
Web Site: www.shakespeareghostown.com
Founded: 1970.
Key Personnel: Pres. & Dir., Emanuel D. Hough; Treas., Linda Erikson.
Personnel Profile: Full-Time Volunteers 4; Part-Time Volunteers 20.
Governing Authority: nonprofit organization. Tax-exempt: 501(c)(3).
Institution Type/Description: History Museum.
Collections: southwest history 1856-1935; furniture; period guns; custom made holsters, saddles & tack.
Facilities: 300-vol. library of southwest history books; 8,900 sq. ft. exhibit space. Books & handmade blacksmith items for sale.
Activities: guided tours; lectures. Annual Events: Living History Events in April, June, August & October.
Publications: quarterly newsletter, Shakespeare Quarterly.
Hours & Admission Prices: Tours: 2nd weekend of month 10-12 & 2-4. Adults $4, children 6-12 $3. Living History Events: 4th weekend in April, June, Aug. & Oct. 10-12 & 2-4. Adults $5, children 6-12 $4. &
Attendance: 3,000 (accurate)
Membership: Individual $10; Family $25; Business $50; Patron $100; Lifetime $500.

Los Alamos

THE ART CENTER AT FULLER LODGE, 2132 Central Ave. Front, Los Alamos, NM 87544-4013. Mailing Address: P.O. Box 1295, Espanola, NM 87532-1295. Tel.: 505-662-9331. Fax: 505-662-9334.
E-mail: director@artful.org
Web Site: artfulnm.org
Founded: 1977.
Congressional District: 3
Key Personnel: Chm. (V), Carole Rinald; Exec. Dir., John Werenko; Dir., Gloria Gilmore-House; Asst. Dir., Craig Carmer; Office Mgr., Betty Hettich; Museum Shop Mgr., Maria Theye.
Personnel Profile: Part-Time Paid 4; Part-Time Volunteers 40; Interns 1.
Governing Authority: nonprofit organization. Tax-exempt: 501(c)(3).
Institution Type/Description: Art Center & Gallery.
Collections: various forms of art media.

Activities: guided tours; lectures; films; gallery talks; traveling & invitational exhibitions; rental & sales gallery; juried exhibits; docent training programs. Annual Events: summer & autumn arts & crafts fairs; Affordable Art Benefit Sale in December.
Publications: quarterly bulletin, The Member Letter.
Hours & Admission Prices: Mon.-Sat. 10-4. No charge; donations accepted. Closed New Year's Day; Thanksgiving; Christmas. &
Attendance: 18,000 (accurate)
Membership: Senior $20; Individual & General $25; Family $35; Contributing Donor $100; Patron $250.

BANDELIER NATIONAL MONUMENT, 15 Entrance Rd., Los Alamos, NM 87544-9508. Tel.: 505-672-3861 & 0343. Fax: 505-672-9607.
E-mail: band-administration@nps.gov
Web Site: www.nps.gov/band
Founded: 1916.
Congressional District: 3
Key Personnel: Park Supt., Brad Traver; Chief Protection, Fred Patton; Chief Interpretation, Lynne Dominy; Chief Facility Management, Liza Ermelling; Park Archeologist, Rory Gauthier; Museum Technician, Gary Roybal.
Governing Authority: federal. Parent Institution: National Park Service, Dept. of Interior, Washington, DC. Tax-exempt.
Institution Type/Description: National Park & Archaeological Site Museum.
Collections: archaeological & ethnological items of Pueblo Indians of the Pajarito Plateau; plant & animal specimens of the Jemez Mountains & the Pajarito Plateau; research reports; environmental documents; maps; drawings; plans; photographs; entomology; historic structures.
Research Fields: history, archaeology, ethnographical & natural science of the site.
Facilities: campsites; visitor center. Interpretive materials for sale.
Activities: lectures; films; interpretive slide programs; campfire programs; summer night walks; culture/arts demonstrations; guided tours & self-guided tours & backcountry hiking.
Publications: quarterly guide book & trail booklet, The Tuff times.
Hours & Admission Prices: Winter: daily 8-4:30; Spring & Fall 9-5:30; Summer: 8-6. Admission $12 per car; call park for commercial vehicle fees. Annual Pass: seniors 62 & over $10; Bandelier pass $30; inter-agency pass $80. Closed New Year's Day; Christmas. &
Attendance: 300,000 (accurate)

BRADBURY SCIENCE MUSEUM, 15th & Central, Los Alamos, NM 87544. Mailing Address: MSC330, P.O. Box 1663, Los Alamos, NM 87545. Tel.: 505-667-4444. Fax: 505-665-6932.
E-mail: web-bsm@lanl.gov
Web Site: www.lanl.gov/museum
Founded: 1963.
Congressional District: 3
Key Personnel: Dir., Linda Deck; Community Programs Coord., Mary Ellen Ortiz; Exhibits Designer, Omar Juveland.
Personnel Profile: Full-Time Paid 6; Part-Time Paid 8.
Governing Authority: federal. Los Alamos National Security, LLC, on contract to the Dept. of Energy. Parent Institution: Los Alamos National Laboratory.
Institution Type/Description: Science Museum.
Collections: scientific exhibits; historical events of Manhattan Project; hands-on exhibits; history wall.
Research Fields: weapons; energy; biomedical; environmental; nature of matter; current laboratory research.
Facilities: theatre; auditorium; lab.
Activities: lectures; films; traveling exhibits; classroom programs; science drama; summer camp; science demonstrations; research.
Publications: brochure.
Hours & Admission Prices: Sun.-Mon. 1-5, Tues.-Sat. 10-5. No charge. Closed New Year's Day; Thanksgiving; Christmas. &
Attendance: 83,450 (accurate)

LOS ALAMOS HISTORICAL MUSEUM, 1050 Bathtub Row, Los Alamos, NM 87544. Mailing Address: P.O. Box 43, Los Alamos, NM 87544-0043. Tel.: 505-662-6272 (weekdays 10-4) & 4493.
E-mail: historicalsociety@losalamoshistory.org
Web Site: www.losalamoshistory.org
Founded: 1968.
Congressional District: 3
Key Personnel: Exec. Dir., Heather McClenahan; Pres., Ron Wilkins; Vice Pres., Michael Wheeler; Archivist, Rebecca Collinsworth; Museum Shop Mgr., Kathy Ankeny.
Personnel Profile: Full-Time Paid 1; Part-Time Paid 7; Part-Time Volunteers 35; Interns 1.

Governing Authority: county; historical society. Tax-exempt: 501(c)(3).
Institution Type/Description: Local History Museum.
Collections: archaeology; geology; paleontology; homesteading, history of 1918-1943 Los Alamos Ranch school; wartime atomic bomb memorabilia; slides; photos; tape recordings; manuscripts; photographs; prehistoric Pueblo Indian artifacts; documents. Historic Buildings: 1920 guest cottage of Los Alamos Ranch school; Homesteader's cabin; early Indian pueblo.
Research Fields: archaeology; land grants; wartime history; Los Alamos Ranch School history; Los Alamos Manhattan Project history; Cold War history.
Facilities: library; archives.
Activities: self-guided tours; lectures; audiovisual history programs; formally organized education programs for students; inter-museum loan; permanent & temporary exhibitions; school loan service; periodic arts & crafts shows; demonstrations; slide & film shows; field trips; docent training sessions.
Publications: books include Los Alamos: The First 50 Years; Inside Box 1663; Los Alamos Outdoors; newsletter, Los Alamos Historical Society Newsletter; A Hiker's Guide to Bandelier; Manhattan District History: Nonscientific Aspects of Los Alamos Project Y 1942-1946; Los Alamos: Beginning of an Era. Standing By Making Do: Women of Wartime Los Alamos; Non-Technical History of Project Y; Sentinels On Stone (area petroglyphs); Los Alamos Place Names. Quads, Shoeboxes and Sunken Living Rooms: A History of Los Alamos Housing, A Boy on the Hill; Savoring the Past-Recipes from 3 Cultures; Secrets! of a Los Alamos Kid 1946-1953; A Guide to the Nuclear Arms Control Treaties; Life Within Limits; Robert Oppenheimer 1904-1967; Plutonium Metallurgy at Los Alamos 1943-1945; Twilight Time - A Soldiers Role in the Manhattan Project at Los Alamos; The Secret Project Notebook, Gatekeeper to Los Alamos; Just Crazy to Ski: A 50-Year History of Skiing in Los Alamos; The Forest and the Fire; The Secret Project Notebook; Tales of Los Alamos; The Forgotten Physicist; Historic Roads of Los Alamos (2009); At Home On The Slopes of Mountains: The Story of Peggy Pond Church.
Hours & Admission Prices: Exhibit Hall: Summer Mon.-Fri. 9:30-4:30, Sat.-Sun. 11-4; Winter Mon.-Fri. 10-4, Sat.-Sun. 11-4. Archives: Mon.-Fri. 10-4. Tour: $1 donation. Museum shop discounts to members. Closed New Year's Day; Thanksgiving; Christmas. &
Attendance: 34,700 (accurate)
Membership: Student & Senior $35; Individual & Senior Couple $40; Family $50; Heritage Friend $100-$499; Heritage Contributor $500-$999; Heritage Supporter $1,000-$2,499; Heritage Benefactor $2,500 & up.

MESA PUBLIC LIBRARY ART GALLERY, 2400 Central Ave., Los Alamos, NM 87544-4014. Tel.: 505-662-8240. Fax: 505-662-8245.
Web Site: www.losalamosnm.us
Key Personnel: Dir., Charlie Kalogeros-Chattan; Dir., Carol Meine
Institution Type/Description: Art Gallery.
Collections: works by local, regional & national artists; paintings; prints; drawings; photographs; sculpture; architectural models; digital art; decorative arts.
Facilities: library.
Activities: traveling exhibitions.
Hours & Admission Prices: Mon.-Thurs. 10-9, Fri. 10-6, Sat. 10-5, Sun. 12-5.

Lovington

LEA COUNTY MUSEUM, 103 S. Love, Lovington, NM 88260-4218. Tel.: 505-396-4805. Fax: 505-396-4805.
E-mail: leacomuseum@leaco.net
Web Site: www.leacountymuseum.org
Key Personnel: Dir., Jim Harris.
Governing Authority: nonprofit organization. Tax-exempt: 501(c)(3).
Institution Type/Description: History Museum.
Collections: local history; Native Americans; ranchers; farmers; homesteaders; town builders; oil men; residents of Southeast New Mexico; wagons; buggies. Historic Buildings: Commercial Hotel; 1908 Love House; 1914 Baker School; 1913 Store and Post Office; 1908 Dugout; 1913 Caprock Store; 1950 Reed House.
Hours & Admission Prices: Tues.-Fri. 1-5, Sat. 9-5; other times by appointment. No charge. Closed New Year's Day; Independence Day; Thanksgiving; Christmas.

Madrid

OLD COAL MINE MUSEUM, 2846 State Hwy. 14, (on the Turquoise Trail), Madrid, NM 87010. Tel.: 505-438-3780; 505-473-0743 (schedule tour).
E-mail: ocm@themineshafttavern.com
Web Site: www.themineshafttavern.com
Founded: 1982.

Key Personnel: Dir. & C.E.O., Lori Lindsey
Governing Authority: Parent Institution: Mine Shaft Properties, LLC.
Institution Type/Description: Mine Museum.
Collections: mine history; 1900 locomotive; period autos & trucks; firefighting equipment; medical & office equipment; farm, homemaking & carpentry equipment; early silent & sound movie projectors; blacksmith shop.
Activities: rental facilities.
Hours & Admission Prices: Fri.-Mon. 10-5; other times by appointment weather permitting. Adults $5; discounts to seniors & children under 10. Closed Thanksgiving; Christmas.
Attendance: 3,575 (estimated)

Magdalena

BOX CAR MUSEUM, 108 N. Main St., Magdalena, NM 87825. Mailing Address: P.O. Box 145, Magdalena, NM 87825. Tel.: 575-854-2261. Fax: 575-854-2273. Facebook: Magdalena Public Library.
E-mail: mpl@gilanet.com
Web Site: www.magdalenapubliclibrary.org
Key Personnel: Dir., Jennifer Kent
Institution Type/Description: History Museum.
Collections: local history & culture; photographs; railroad history; local mining history.
Hours & Admission Prices: Tues.-Fri. 10-6, Sat. 10-4. No charge; donations accepted.
Attendance: 300 (estimated)

Mesilla

GADSDEN MUSEUM, 1875 Boutz Rd., Mesilla, NM 88046. Mailing Address: Box 147, Mesilla, NM 88046-0147. Tel.: 575-526-6293.
Web Site: www.nmohwy.com/g/gadsdemeu.htm
Founded: 1931.
Congressional District: 2
Key Personnel: Dir., Mary F. Bird; Dir. & Pres. (V), R. Eileen Betzen.
Personnel Profile: Full-Time Volunteers 1; Part-Time Volunteers 1; Interns 1.
Governing Authority: individual operation.
Institution Type/Description: History Museum.
Collections: Indian artifacts from the Southwest; pottery; baskets; Apache, Navajo, Jemez, arrowheads, peace medals, beaded leather goods; pictures painted on deerskin; artifacts representing Penitente lifestyle; Colonel Albert Jennings Fountain; Santo collection.
Activities: permanent exhibitions; tours for school students; formally organized education programs for children & college students.
Publications: pamphlet.
Hours & Admission Prices: Mon.-Sat. 9-11am & 1-5; groups & Sun. by appointment. Admission $5. Closed New Year's Day; Easter; Independence Day; Thanksgiving; Christmas. &
Attendance: 100 (estimated)

Moriarty

MORIARTY HISTORICAL SOCIETY & MUSEUM, 202 Broadway St., Moriarty, NM 87035. Mailing Address: P.O. Box 1366, Moriarty, NM 87035-1366. Tel.: 505-832-0839. Fax: 505-832-9286.
E-mail: momuseum@yahoo.com
Web Site: www.moriartymuseum.org
Founded: 1981.
Key Personnel: Pres., Sammie Pachta; Vice Pres., Tina Ortega; Past Pres. & Historian, Joseph H. "Choppo" McComb; Cur. & Archivist, Barbara Takiguchi.
Personnel Profile: Full-Time Paid 7; Full-Time Volunteers 5; Part-Time Paid 3; Part-Time Volunteers 9.
Governing Authority: municipal; private; nonprofit organization. Branch Museum: Moriarty Historical Society, Moriarty, NM. Tax-exempt: 501(c)(3).
Institution Type/Description: General Museum.
Collections: 19th to mid-20th century homesteading/ranching area history.
Research Fields: Route 66 from 1930-1960; Estancia Valley papers 1893-1950.
Facilities: 100-vol. library of state & local history books.
Activities: guided tours; lectures; school loan service; Moriarty History presentations.
Publications: book, My Life in New Mexico; Torrance County History (reprint).
Hours & Admission Prices: Tues.-Fri. 10-5, Sat. 10-2; other times by appointment. No charge; donations accepted. Closed New Year's Day; Memorial Day; Independence Day; Thanksgiving; Christmas. &

Attendance: 11,000 (accurate)
Membership: Single $7; Family $10.

U.S. SOUTHWEST SOARING MUSEUM, 918 E. Old Hwy. 66, Moriarty, NM 87035. Mailing Address: P.O. Box 3626, Moriarty, NM 87035-3626. Tel.: 505-832-0755.
E-mail: usssm1@yahoo.com
Web Site: www.swsoaringmuseum.org
Key Personnel: Pres., George Applebay.
Governing Authority: Tax-exempt: 501 (c)(3).
Institution Type/Description: History Museum.
Collections: western U.S. history; glider models; mural.
Hours & Admission Prices: Mon.-Fri. 9-4; other times by appointment. No charge; donations accepted.

Mountainair

SALINAS PUEBLO MISSIONS NATIONAL MONUMENT, 102 S. Ripley Ave., Mountainair, NM 87036. Mailing Address: P.O. Box 517, Mountainair, NM 87036-0517. Tel.: 505-847-2585. Fax: 505-847-2441.
E-mail: jeanette_wolfe@nps.gov
Web Site: www.nps.gov
Founded: 1909.
Congressional District: 1 & 3
Key Personnel: Supt., Glenn Fulfer; Chief Park Ranger & Interpretive Specialist, Norma Pineda.
Personnel Profile: Full-Time Paid 14; Part-Time Paid 6; Part-Time Volunteers 3.
Governing Authority: federal. Parent Institution: U.S. National Park Service, Interior Bldg., Washington, DC 20240. Tax-exempt.
Institution Type/Description: Archaeology Museum: located near the site of prehistoric pithouses c.800 A.D.; prehistoric Indian ruins c.1100-1670 A.D.; four Spanish mission ruins c.1622-1672.
Collections: archaeological artifacts from historic and prehistoric ruins; manuscript collection; photograph collection of approx. 5,000 prints, negatives and slides. Historic Buildings: 1627 San Isidro Mission; 1659 San Buenaventure Mission; 1300-1670 Pueblo de Las Humanas; 1630-1672 San Gregorio de Abo; 1100-1672 Pueblo de Abo; 1630-1672 Nuestra Senora de la Purisina Concepcion Mission; 1300-1672 Pueblo de Quarai.
Research Fields: archaeological; historical; botanical; environment.
Facilities: 1,000-vol. library of archaeology, history, natural science, ethnology books available for use on premises by approval of superintendent; picnic area. Books, slides & postcards for sale.
Activities: guided tours for organized groups; lectures; permanent & temporary exhibitions; 17 minute audiovisual program; 45 minute video; self-guiding trail.
Publications: orientation brochures; trail guides; excavation reports; historic structures report.
Hours & Admission Prices: Memorial Day to Labor Day daily 9-6; Sept.-May daily 9-5. No charge; donations accepted. &
Attendance: 52,099 (accurate)

Nageezi

CHACO CULTURE NATIONAL HISTORICAL PARK, 1808 County Rd. 7950, Nageezi, NM 87037. Mailing Address: P.O. Box 220, Nageezi, NM 87037-0220. Tel.: 505-786-7014 ext. 221. Fax: 505-786-7061.
Web Site: www.nps.gov/chcu
Founded: 1907.
Congressional District: 3
Key Personnel: Supt, Larry Turk.
Personnel Profile: Full-Time Paid 3; Part-Time Paid 4; Part-Time Volunteers 1.
Governing Authority: federal. Parent Institution: National Park Service. Subsidiary Institution: Southwest Region. Tax-exempt.
Institution Type/Description: Archaeology Museum.
Collections: archaeology; ethnology; prehistoric & historic sites.
Research Fields: archaeology of the Chaco Anasazi & the San Juan Basin.
Facilities: 1,300-vol. library of archaeology, ethnology, natural history, available for use by special permission; visitor center; campground. Publications for sale.
Activities: guided & self-guiding tours; lectures; films; permanent exhibitions.
Publications: orientation brochures; booklets; guide leaflets; technical reports.
Hours & Admission Prices: Park: daily 7-sunset. Visitor Center: 8-5. Weekly Pass: Vehicle $8, Individual $4. Closed New Year's Day; Thanksgiving; Christmas. &
Attendance: 100,000 (estimated)

Organ

THE SPACE MURALS MUSEUM, 12450 Hwy. 70 E., Organ, NM 88052. Mailing Address: P.O. Box 243, Organ, NM 88052-0243. Tel.: 505-382-0977.
E-mail: klin@zianet.com
Web Site: www.zianet.com
Institution Type/Description: Space Museum.
Collections: over 2,500 air & space photographs; model airplanes; replica of Space Station Freedom; astronaut-related artifacts; air & space artifacts; 1/8 scale replica of space shuttle Challenger memorial; Nike Hercules Missile; V-2 nose cone & tail piece.
Hours & Admission Prices: Mon.-Sat. 9-6, Sun. 10-6. No charge.

Pecos

PECOS NATIONAL HISTORICAL PARK, State Rd. 63, 2 mi. south of Pecos, Pecos, NM 87552. Mailing Address: P.O. Box 418, Pecos, NM 87552-0418. Tel.: 505-757-7200. Fax: 505-757-7207.
E-mail: peco_visitor_information@nps.gov
Web Site: www.nps.gov/peco
Founded: 1965.
Congressional District: 3
Key Personnel: Supt., Dennis Carruth; Chief Education, Visitor Svcs. & Cultural Resources, Christine Beekman; Cur., Heather Young.
Personnel Profile: Full-Time Paid 1.
Governing Authority: federal. Parent Institution: National Park Service. Tax-exempt.
Institution Type/Description: Cultural & Natural Historic Site.
Collections: Pecos area archaeology; 17,000 artifacts of A.V. Kidder collection 1915-1929; artifacts from Mission Churches; Kiva excavations. Historic Building: 17th & 18th-century Nuestra Senora de Los Angeles de Porciuncula missions & associated buildings. Spanish mission; Spanish homestead; Santa Fe trail; Civil War battlefield; Route 66, & 20th Century Ranching.
Research Fields: archaeology; anthropology; history.
Facilities: 300-vol. library of books on archaeology, anthropology, history, & Kidder archaeological volumes available for research on premises; visitor center including exhibit room. Books for sale.
Activities: guided tours by arrangements; permanent exhibitions; introductory film; 1.25 mile self-guided pueblo & church ruins hike; 2.25 mile self-guided Civil War Battle of Glorieta hike; seasonal fishing.
Publications: orientation brochures; trail guide; book, Pecos: Gateway to Pueblos and Plains, The Anthology. From Folsom to Fogelson: A Cultural Resources Inventory Survey of Pecos National Historical Park.
Hours & Admission Prices: Memorial Day-Labor Day: daily 8-6; Sept.-May daily 8-5. Adult $3; children under 16 no charge. Closed New Year's Day; Thanksgiving; Christmas. Senior Pass, Interagency Pass & Access Pass available.
Attendance: 40,000 (accurate)
Membership: Senior 62 & over $10; Annual $80.

Pinos Altos

PINOS ALTOS HISTORICAL MUSEUM, 33 Main St., Pinos Altos, NM 88053. Mailing Address: P.O. Box 505, Silver City, NM 88062. Tel.: 575-388-1882.
E-mail: info@pinosaltoscabins.com
Web Site: www.pinosaltos.org/museum/schaferlogcabin.html
Institution Type/Description: History Museum: housed in the Schafer Log Cabin c.1860.
Collections: 19th century Americana, including old maps & photographs; period furniture & artifacts; mining tools; weathered wood relics of the old west.
Facilities: Museum-related items for sale.
Hours & Admission Prices: Daily 10-5.

Portales

BLACKWATER DRAW MUSEUM, 42987 Hwy. 70, Portales, NM 88130. Mailing Address: Eastern NM Univ., Station 53, Portales, NM 88130. Tel.: 505-562-2202 & 1011. Fax: 505-562-2291.
E-mail: matthew.hillsman@enmu.edu
Web Site: www.enmu.edu
Founded: 1969.
Congressional District: 3
Key Personnel: Dir., Dr. John Montgomery; Archaeologist, George Crawford; Cur., Matthew Hillsman.
Personnel Profile: Full-Time Paid 2; Part-Time Paid 2; Interns 3.
Governing Authority: university. Parent Institution: Eastern New Mexico University & State of New Mexico. Tax-exempt.

Institution Type/Description: Archaeological Site: 1932 America's first multi-cultural, paleoindian archaeological site.
Collections: paleoindian archaeology & geology; paleontology; anthropology.
Research Fields: Paleoindian archaeology.
Facilities: Publications & t-shirts on early man for sale.
Activities: guided tours; lectures; formally organized education programs for children, adults, undergraduate & graduate college students; permanent & temporary exhibitions; films on anthropology, archaeology & ecology.
Publications: Eastern New Mexico University Contributions in Anthropology.
Hours & Admission Prices: Museum: Memorial Day to Labor Day Mon.-Sat. 10-5, Sun. 12-5; Sept.-May Tues.-Sat. 10-5, Sun. 12-5. Adults $3, seniors 60 & up $2, students $1; discounts to school groups; children under 6 no charge. &
Attendance: 7,000 (estimated)

MILES MINERAL MUSEUM, Eastern New Mexico University Campus, Roosevelt Hall, Portales, NM 88130. Mailing Address: Station 33, Eastern New Mexico University, Portales, NM 88130. Tel.: 575-562-2651 & 2174. Fax: 575-562-2192.
E-mail: jim.constantopoulos@enmu.edu
Web Site: w3a.enmu.edu/services/museums/miles-mineral
Founded: 1969.
Congressional District: 3
Key Personnel: Dir., Dr. Jim Constantopoulos.
Governing Authority: university. Affiliated with Eastern New Mexico University. Branch Museums: Blackwater Draw Museum & the Paleo-Indian Institute; Roosevelt Count Museum; Natural History Museum. Tax-exempt.
Institution Type/Description: Geology Museum.
Collections: minerals; rocks; fossils.
Research Fields: mineralogy.
Activities: guided tours; lectures; formally organized education programs for children, adults, undergraduate & graduate college students; permanent & temporary exhibitions.
Hours & Admission Prices: Mon., Wed. & Fri. 8-5. No charge; donations accepted. &
Attendance: 4,136 (accurate)

NATURAL HISTORY MUSEUM, Roosevelt Hall, S. Ave. K, Eastern New Mexico University, Portales, NM 88130. Tel.: 575-562-2706, 2753 & 2862. Fax: 575-562-2192.
E-mail: marv.lutnesky@enmu.edu
Web Site: www.enmu.edu/services/museums/natural-history/
Founded: 1968.
Congressional District: 66
Key Personnel: Dir. & Cur. Exhibits, Marvin Lutnesky; Head Cur. Collections, Darren A. Pollock.
Governing Authority: university. Parent Institution: Eastern New Mexico University. A unit of the Llano Estacado Center for Advanced Professional Studies & Research. Tax-exempt.
Institution Type/Description: Natural History Museum: housed in Roosevelt Hall, originally used as a men's dormitory.
Collections: 10,500 specimens of mammals; 5,000 specimens of reptiles & amphibians; 600 specimens of birds; 10,000 specimens of fish; invertebrates.
Research Fields: aquatic, & terrestrial ecology & conservation.
Activities: guided tours; permanent & temporary collections; research.
Publications: book, Studies in Natural Sciences.
Hours & Admission Prices: Mon.-Fri. 8-5. No charge. &

ROOSEVELT COUNTY MUSEUM, Eastern New Mexico Univ., Station 9, Portales, NM 88130. Mailing Address: 1200 W. University, Portales, NM 88130. Tel.: 575-562-2592. Fax: 575-562-2362.
E-mail: mark.romero@enmu.edu
Web Site: www.enmu.edu/services/museums/roosevelt-county/
Founded: 1940.
Congressional District: 3
Key Personnel: Cur., Mark Romero.
Governing Authority: university; nonprofit. Parent Institution: Eastern New Mexico University. Tax-exempt.
Institution Type/Description: History Museum.
Collections: technology; early settlers & Native American artifacts; ethnology; folklore; archives; costumes; numismatic; ranching artifacts; Roosevelt County historical artifacts.
Activities: lectures; seminars; art shows; outreach program; video & slide shows; permanent & temporary exhibitions. Museum Sponsors: musical permanent & temporary exhibitions.
Hours & Admission Prices: Mon.-Fri. 8-5, Sat. 10-4, Sun. 1-4. No charge; donations accepted. Closed university Christmas break; national holidays.

Attendance: 2,000 (estimated)

Pueblo of Acoma

SKY CITY CULTURAL CENTER/HAAK'U MUSEUM, (M), I-40 West, Exit 102, SPA 30 @ 32, Pueblo of Acoma, NM 87034. Mailing Address: P.O. Box 310, Pueblo of Acoma, NM 87034-0310. Tel.: 800-747-0181; 505-552-7861. Fax: 505-552-7883.
E-mail: friends@skycity.com
Web Site: www.acomaskycity.org
Formerly: Acoma Tourist & Visitation Center
Founded: 1978.
Congressional District: 3
Key Personnel: Dir., Emerson R. Vallo.
Personnel Profile: Full-Time Paid 23; Part-Time Paid 9.
Governing Authority: Parent Institution: Acoma Tribal Government. Subsidiary Institution: Acoma Business Enterprises, Inc. Haak'u Museum: Tax-exempt: 501(c)(3).
Institution Type/Description: Indian Historical & Cultural Museum.
Collections: photo archives & documents related to the history of Acoma; historic pottery. Historic Building: 16th century church, Old Pueblo.
Research Fields: Acoma Culture.
Facilities: library; archives.
Activities: guided tour of Old Acoma & Sky City.
Publications: catalog, One Thousand Years of Clay.
Hours & Admission Prices: Call for hours & admission prices. &
Attendance: 71,880 (accurate)
Membership: Student $25; Our Friend $50; Our Support $100; The Reserve $250; People's Council $500; Director's Council $1,000; Governor's Council $5,000; Corporate $10,000 & up.

Quemado

DIA CENTER FOR THE ARTS, Quemado, NM 87829. Mailing Address: P.O. Box 2993, Corrales, NM 87048-2993. Tel.: 505-898-3335. Fax: 505-898-3336.
E-mail: info@lightningfield.org
Web Site: www.diacenter.org
Key Personnel: Admin., Kathleen Shields
Institution Type/Description: Art Museum
Collections: 400 stainless steel poles; work of land art by Walter De Maria, The Lightning Field.
Hours & Admission Prices: May-June & Sept.-Oct. $150 per person; July-Aug. $250 per person. Call for reservations & information.

Radium Springs

FORT SELDEN STATE MONUMENT, 1280 Ft. Selden Rd., Radium Springs, NM 88054. Mailing Address: P.O. Box 2087, Santa Fe, NM 87504. Tel.: 575-526-8911.
E-mail: nathan.stone@state.nm.us
Web Site: www.nmmonuments.org/fort-selden
Founded: 1973.
Key Personnel: Ranger, Nathan Stone.
Governing Authority: state. Parent Institution Museum of New Mexico, Box 2087, Santa Fe, NM 87504. Tax-exempt: 170(c)(1).
Institution Type/Description: Military Museum: located on site of 1865-1891 army fort; 1884-1886 home of Douglas MacArthur.
Collections: artifacts found on the site; military artifacts, including uniforms, saddles & an 1864 Springfield rifle.
Research Fields: military history pertaining to the Indian Wars period.
Facilities: 100-vol. library of books on the Indian Wars available for research on premises; botanical garden.
Activities: guided tours; TV programs; formally organized education programs for children & adults; living history programs in summer or on special request; traveling & loan exhibitions; living history demos; annual Old Fort Days.
Publications: book, Fort Selden.
Hours & Admission Prices: Wed.-Sun. 8:30-5. Adults $3; children 16 and under no charge. Closed New Year's Day; Easter; Thanksgiving; Christmas. &

Ramah

EL MORRO NATIONAL MONUMENT, Hwy. 53-42 mi. S.W. of Grants, Ramah, NM 87321. Mailing Address: HC 61 Box 43, Ramah, NM 87321-9603. Tel.: 505-783-4226. Fax: 505-783-4689.
Web Site: www.nps.gov/elmo
Founded: 1906.

Congressional District: 3
Key Personnel: Supt., Kayci Cook Collins.
Governing Authority: federal. National Park Service.
Institution Type/Description: National Monument, Historical & Archaeological Site Museum: located at site of Inscription Rock, bearing inscriptions dating from 1605-1906; prehistoric Pueblo Ruins.
Collections: archaeological artifacts; pots; implements; handcrafts; weapons; religious items dating from 17th-19th centuries; prehistoric pottery; Spanish exploration artifacts; Southwest historic artifacts, inscriptions & petroglyphs.
Research Fields: archaeology, Spanish-Colonial & American-Exploratory history; southwest biology, botany & geology.
Facilities: 400-vol. library of archaeology, history, zoology & botany, available for research by personnel.
Activities: Interpretive talks and guided walks in the summer.
Publications: brochures; pamphlets.
Hours & Admission Prices: Winter: daily 9-5; Summer: daily 8-7; Spring & Fall: daily 9-6. Adults $3; under 16 no charge. NPS, Golden Eagle & Golden Age/Golden Access passes issued & honored. Closed New Year's Day; Christmas. &
Attendance: 80,000 (accurate)

WILD SPIRIT WOLF SANCTUARY, 378 Candy Kitchen Rd., Ramah, NM 87321. Mailing Address: HC 61, P.O. Box 28, Ramah, NM 87321-9601. Tel.: 505-775-3304. Fax: 505-775-3824.
E-mail: info@wildspiritwolfsanctuary.org
Web Site: www.wildspiritwolfsanctuary.org
Key Personnel: Exec. Dir., Leyton Cougar; Memberships, Education & Admin., Cheryl Vaughn; Newsletter, Advertising, Publications, Georgia Cougar
Institution Type/Description: Wildlife Sanctuary.
Collections: wolves.
Hours & Admission Prices: Tours: Tues.-Sun. 11, 12:30, 2 & 3:30. Adults $7, senior citizens $6, children $4.
Membership: Single $25; Supporting $50; Alpha $100; Yearly Sponsorship $125.

Raton

BOY SCOUT MUSEUM, 400 S. 1st St., Raton, NM 87740-4063. Tel.: 505-445-1413.
Web Site: www.santafetrailnm.org/site558.html
Key Personnel: Owner & Cur., Dennis Downing; Owner & Cur., Sue Downing
Institution Type/Description: History Museum.
Collections: boy scout history; personal artifacts.
Hours & Admission Prices: Daily 10-5; other times by appointment.

COLFAX COUNTY SOCIETY OF ART, HISTORY AND ARCHAEOLOGY, 108 S. 2nd St., Raton, NM 87740-3906.
Founded: 1938.
Congressional District: 3
Key Personnel: Pres., Kathy McQueary; Archivist, Roger Sanchez.
Personnel Profile: Full-Time Paid 1; Part-Time Volunteers 20.
Governing Authority: private; nonprofit organization.
Institution Type/Description: General Museum.
Collections: displays, artifacts & photographs of the railroad, 9 coal mining camps & ranching from 1880 to 1950; history of Raton & its early settlers.
Activities: docent program; guided tours; lectures; traveling exhibitions.
Hours & Admission Prices: Summer: May-Sept. Tues.-Sat. 9-5; Winter: Oct.-April Wed.-Sat. 10-4. No charge; donations accepted. &
Attendance: 6,000 (accurate)
Membership: Individual $30.

RATON MUSEUM, 108 S. 2nd St., Raton, NM 87740-3906. Tel.: 505-445-8979.
Founded: 1939.
Congressional District: 3
Key Personnel: C.E.O., Roger Sanchez; Pres. (V), Kathy McQueary.
Personnel Profile: Full-Time Paid 1; Part-Time Volunteers 20.
Governing Authority: board of directors/trustees; nonprofit organization. Parent Institution: Colfax County Society of Art History & Archaeology. Tax-exempt: 170(b)(1)(A).
Institution Type/Description: Historical Society Museum.
Collections: railroading; coal mining; ranching; the history of Raton & its early settlers. Historic Buildings: 1890 & 1906 buildings.
Research Fields: railroading; mining; ranching; the development of Raton & surrounding area.
Activities: lectures; docent program; traveling exhibitions; outreach program.
Publications: newsletter.

Hours & Admission Prices: May-Aug. Tues.-Sat. 9-5; Sept.-April Wed.-Sat. 10-4; will open for special groups. No charge; donations accepted. Closed New Year's Day; Memorial Day; Independence Day; Thanksgiving; Christmas. &
Attendance: 6,000 (accurate)
Membership: Individual $30.

Rio Rancho

J&R VINTAGE AUTO MUSEUM, 3650 NM Hwy. 528, Rio Rancho, NM 87144-7524. Tel.: 505-867-2881.
E-mail: info@jrvintageautos.com
Web Site: www.jrvintageautos.com
Institution Type/Description: Auto Museum.
Collections: 70 automobiles; die cast toys.
Facilities: Museum-related items for sale.
Hours & Admission Prices: Mon.-Sat. 10-5. Adults $6, senior citizens $5, children 6-12 $3; discounts to groups; children under 6 no charge.

Roswell

ANDERSON MUSEUM OF CONTEMPORARY ART, 409 E. College Blvd., Roswell, NM 88201-7524. Mailing Address: P.O. Box 1, Roswell, NM 88202. Tel.: 575-623-5600. Fax: 575-623-5603. Facebook: Roswell AMoCA.
E-mail: email@roswellamoca.org
Web Site: www.roswellamoca.org
Founded: 1994.
Congressional District: 2
Key Personnel: Dir., C.E.O. & Chm. (V), Donald B. Anderson; Exec. Vice Pres., Phelps Anderson; Education, Cymantha Liakos; Public Rels., Nancy Fleming; Museum Shop Mgr., Lanice White.
Personnel Profile: Full-Time Paid 1; Full-Time Volunteers 2; Part-Time Paid 5; Part-Time Volunteers 21.
Governing Authority: private; nonprofit organization. Tax-exempt: 501(c)(3).
Institution Type/Description: Contemporary Art Museum.
Collections: works by Roswell Artist-in-Residence program residents; paintings; prints; drawings; photographs; sculpture; digital media.
Facilities: research library; 22,000 sq. ft. exhibit space. Museum-related items for sale.
Activities: guided tours by request.
Publications: brochure; newsletter.
Hours & Admission Prices: Mon.-Fri. 9-4, Sat.-Sun. 1-5. No charge; donations accepted. Closed New Year's Day; Independence Day; Thanksgiving; Christmas. &
Attendance: 10,000 (estimated)
Membership: Fan $20-$99; Friend $100-$499; Fanatic $500 & up.

THE GENERAL DOUGLAS L. MCBRIDE MUSEUM, NEW MEXICO MILITARY INSTITUTE, 101 W. College Blvd., Roswell, NM 88201-5100. Tel.: 505-624-8220. Fax: 505-624-8258.
E-mail: klopfer@nmmi.edu
Web Site: www.nmmi.cc.nm.us/museum
Founded: 1983.
Congressional District: 2
Key Personnel: Dir., Col. Jerry Klopfer; Administrative Asst., Liz Bolin.
Personnel Profile: Full-Time Paid 1; Part-Time Volunteers 1.
Governing Authority: public college. New Mexico Military Institute. Tax-exempt.
Institution Type/Description: Military Museum: housed in c.1912 Luna Natatorium.
Collections: military & NMMI history; Civil War-present, small weapons; New Mexico Military Institute memorabilia; artifacts from 20th-century conflicts; Hall of Fame; Valor Rooms.
Research Fields: achievements of New Mexico Military Institute alumni; 20th-century general military history.
Facilities: 300-vol. library of books on weapons; uniforms; military history; 100-vols. New Mexico Military Institute yearbooks available for use by the public.
Activities: self-guided tours.
Hours & Admission Prices: Mon.-Fri. 8-4; other times by appointment. No charge; donations accepted. Closed national holidays. &
Attendance: 4,000
Membership: Annual $100.

HISTORICAL CENTER FOR SOUTHEAST NEW MEXICO, (M), 200 N. Lea Ave., Roswell, NM 88201-4655. Tel.: 575-622-8333. Fax: 575-623-8746. Facebook: Historical Center for Southeast New Mexico.
E-mail: history@dfn.com
Web Site: www.hssnm.net
Founded: 1976.
Congressional District: 2
Key Personnel: Pres., Judy Smith; Administrative Dir. & Museum Dir., Tina Williams; Museum Shop Mgr. & Administrative Asst., Amy Davis; Librarian & Archivist, Elvis E. Fleming.
Personnel Profile: Part-Time Paid 2; Part-Time Volunteers 45.
Governing Authority: nonprofit organization. Tax-exempt: 501(c)(3).
Institution Type/Description: Historical Museum & Archives Center: housed in 1910-12 J.P. White, Sr. House.
Collections: 1875-1936 period furniture & furnishings; 12,000 collection of Southeastern New Mexico, Pecos Valley & Chaves County photographs; books; audio tapes; manuscripts; periodicals.
Research Fields: Southeastern New Mexico; Pecos Valley; Chaves County & local history.
Facilities: library of photographs, manuscripts, papers, maps and books on Southeastern, New Mexico and local history, available for use with approval of archivist. Books on Southeastern Americana, photographs and other museum-related items for sale.
Activities: guided tours; lectures; docent program; permanent exhibitions.
Publications: quarterly newsletter, Facts and Traditions; books, Roundup on the Pecos; Treasures of History: Historic Buildings in Chaves County, 1870-1935; A Brief Historical Survey of the Middle Pecos River Basin; Treasures of History II: Chaves County Vignettes; The Bitter River; Treasures of History III: Southeastern New Mexico, People, Places and Events; Chaves County Schools 1881-1968; Captain Joseph C. Lea, From Confederate Guerrilla to New Mexico Patriarch; Treasures of History IV; Roundup on the Pecos II; Brief History of Roswell from Stone Age to All American City.
Hours & Admission Prices: Daily 1-4. No charge; donations accepted. Closed New Year's Day; Easter; Independence Day; Thanksgiving; Christmas Eve & Day.
Attendance: 3,500 (accurate)
Membership: Individual Bronze $30; Couple Silver $50; Gold $100; Platinum Business $150.

INTERNATIONAL UFO MUSEUM AND RESEARCH CENTER, 114 N. Main St., Roswell, NM 88203-4706. Tel.: 505-625-9495. Fax: 505-625-1907.
E-mail: mbriscoe@hotmail.com
Web Site: roswellufomuseum.com
Founded: 1991.
Governing Authority: nonprofit organization. Tax-exempt: 501(c)(3).
Institution Type/Description: UFO museum.
Collections: the Roswell Incident, 1947; history, science & research of UFO events worldwide.
Activities: special events. Museum Sponsors: International UFO Festival; Roswell UFO Festival.
Hours & Admission Prices: Daily 9-5. Adults $5; discounts to seniors & military; members no charge. ♿
Attendance: 152,000 (accurate)
Membership: Annual $20.

＊　**ROSWELL MUSEUM AND ART CENTER, (M),** 100 West 11th, Roswell, NM 88201-4998. Tel.: 505-624-6744. Fax: 505-624-6765.
E-mail: rmac@roswellmuseum.org
Web Site: www.roswellmuseum.org
Founded: 1937.
Congressional District: 2
Key Personnel: Asst. Dir., Stephen Jollmer; Pres. (V), Joyce Tucker; Cur., Sara Woodbury; Cur. Education, Meredith Bennett; Registrar, Kenna Arganbright; Museum Shop Mgr., Charles Bentley; Membership Coord., Tracy Hutcherson.
Personnel Profile: Full-Time Paid 14; Part-Time Paid 5; Part-Time Volunteers 145.
Governing Authority: municipal government. Parent Institution: City of Roswell. Tax-exempt: 501(c)(3).
Institution Type/Description: General & Art Museum.
Collections: 20th-century American paintings & sculpture; European & American prints; Southwestern painting; paintings & prints by Peter Hurd & Henriette Wyeth; ethnological & archaeological collection of Southwestern Native American art; Rogers Aston collection of the American West; Robert H. Goddard experimental liquid-propellant rocket collection.

Major Exhibits: Aria Finch, 8/17/12-2/16/14; RMAC: A to Z, 10/26/13-7/27/14.
Research Fields: Southwestern art; Federal Arts Project in New Mexico, early liquid-fueled rocketry; Native American contemporary art.
Facilities: 8,000-vol. library of art, archeology & crafts, available for research by museum students, area teachers, college students; planetarium; auditorium; classrooms; exhibition galleries. Books, reproductions, pottery & museum-related items for sale.
Activities: guided tours; lectures; films; gallery talks; formally organized education programs; inter-museum loan, permanent, temporary & traveling exhibitions.
Publications: quarterly, Roswell Museum Bulletin; monthly, catalogs of exhibitions.
Hours & Admission Prices: Mon.-Sat. 9-5, Sun. & holidays 1-5. No charge; donations accepted. Closed New Year's Day; Thanksgiving; Christmas. ♿
Attendance: 39,000 (accurate)
Membership: Senior Citizen $15; Educator $20; Individual $25; Family $35; Donor $100; Patron $250; Benefactor $500; Fellow $1,000; Corporate: I $250; II $500; III $1000.

SPRING RIVER PARK & ZOO, College & Atkinson Sts., Roswell, NM 88201-9506. Mailing Address: P.O. Drawer 1838, Roswell, NM 88202-1838. Tel.: 507-624-6760. Fax: 507-624-6941.
E-mail: roswellzoo@dfn.com
Web Site: www.roswellmysteries.com
Founded: 1966.
Congressional District: 5
Key Personnel: Dir. Parks & Recreation, Kim Elliott; Dir. Zoo, Elaine Mayfield; Pres. (V), Rita Kane-Doerhoefer.
Personnel Profile: Full-Time Paid 9; Part-Time Paid 3; Part-Time Volunteers 5; Interns 1.
Governing Authority: city; municipal government.
Institution Type/Description: Zoo.
Collections: aviary; zoo exhibits; birds & animals; Prairie Dog Town; early wooden horse carousel; miniature train; longhorn ranch exhibit; early carousel; miniature train.
Facilities: picnic area.
Activities: group meetings; youth & family recreation.
Hours & Admission Prices: Summer: 10-8; Winter: 10-5:30, weather permitting. No charge, donations accepted. ♿
Attendance: 90,000 (estimated)
Membership: Friends of the Roswell Zoo Support Group $15.

Ruidoso

RUIDOSO RIVER MUSEUM, 101 Mechem Dr., Ruidoso, NM 88345-6826. Tel.: 575-257-0296.
E-mail: elisegomber@mail2web.com
Web Site: www.ruidosorivermuseum.com
Institution Type/Description: History Museum.
Collections: local history & culture; period furnishings; clothing; guns; Lincoln County War artifacts; photographs.
Facilities: Museum-related items for sale.
Activities: permanent & temporary exhibitions.
Hours & Admission Prices: Sun.-Mon. & Thurs. 10-5. Adults $5, children 6-15 $2.50; children 5 & under no charge.

Ruidoso Downs

THE HUBBARD MUSEUM OF THE AMERICAN WEST, 841 Hwy. 70, W., Ruidoso Downs, NM 88346. Mailing Address: P.O. Box 40, Ruidoso Downs, NM 88346-0040. Tel.: 505-378-4142. Fax: 505-378-4166.
Web Site: www.hubbardmuseum.org
Formerly: Museum of the Horse
Founded: 1989.
Congressional District: 2
Key Personnel: Chm. (V), Mayor Tom Armstrong; Museum Dir., Jay Smith; Cur. Exhibits, David Mandel; Cur. Collections, Gwen McCausland; Cur. Education, Patsy Jackson; Devel., Jim Kofakis; Mktg. & Graphics, Janis Rowe; Gift Shop Mgr., Donna Franklin; Outreach, Adele Karolik; Administrative Sec., Anne Spotts; Facilities, Wayne Anzak; Custodial, Patricia Valdez.
Personnel Profile: Full-Time Paid 10; Part-Time Paid 4; Part-Time Volunteers 20; Interns 1.
Governing Authority: city; nonprofit organization. Tax-exempt: 501(c)(3).
Institution Type/Description: History Museum.
Collections: wagons; stagecoach; sleighs; saddles & other horse-related items; fine art; furniture; Indian artifacts; children's interactive area; equine monument; historic buildings.

Facilities: 2,500-vol. library relating to horses, the American West & the Lincoln County War; 12 acre site.

Activities: changing exhibits; special event programs; school programs; lectures; performing arts.

Publications: brochures; newsletter; exhibition catalogs.

Hours & Admission Prices: Daily 9-5. Adults $6, senior citizens over 65 & military $5, children 6-16 $2, Ruidoso Downs residents 6 & over $1; members and children 6 & under no charge. Closed Thanksgiving; Christmas. &

Attendance: 30,000 (accurate)

Membership: The Scout $50; The Explorer $100; The Wagon Master $150; The Wrangler $250; The Trail Boss $500; The Rancher $1,000; The Stradling Patron $1,500; The Hubbard Patron $3,000; The Free Spirits Patron $5,000.

San Antonio

EL CAMINO REAL HISTORIC TRAIL SITE, 300 E. County Rd., San Antonio, NM 87832. Mailing Address: P.O. Box 175, Socorro, NM 87801-0175. Tel.: 505-854-3600. Fax: 505-854-3609.

Web Site: www.elcaminoreal.org

Formerly: El Camino Real International Heritage Center

Key Personnel: Mgr., Chris Hanson

Institution Type/Description: History Museum.

Collections: oldest U.S. trail history; local cultural heritage; NM history; history of the first European & Spanish settlements.

Facilities: outdoor amphitheater; gardens; nature trails. Museum-related items for sale.

Activities: performances; special events.

Hours & Admission Prices: Wed.-Sun. 8:30-5. Adults $5; discounts to groups and AAM & ICOM members; mililtary veterans & children under 16 no charge. Closed New Year's Day; Easter; Thanksgiving; Christmas. &

Membership: Regular $30; Family $40; Senior $60; Nonprofit Educational Institution $200; Government & Corporate Entity $500; Charter $1,000.

Sandia Park

TINKERTOWN MUSEUM, 121 Sandia Crest Rd., Sandia Park, NM 87047. Mailing Address: P.O. Box 303, Sandia Park, NM 87047-0303. Tel.: 505-281-5233.

E-mail: tinker4u@tinkertown.com

Web Site: www.tinkertown.com

Founded: 1983.

Congressional District: 1

Key Personnel: Dir., Carla Ward.

Personnel Profile: Part-Time Paid 3.

Governing Authority: private.

Institution Type/Description: Folk Art Museum.

Collections: wood carvings; miniature animated western town & three ring circus; hand made dolls & toys; 140 wedding cake couples; swords; circus banners & memorabilia; side show giant shoes & pants; period mechanical coin operated arcade machines; western livery & mining memorabilia.

Facilities: Museum-related items for sale.

Activities: Annual Event: Festival of Tinkering.

Publications: The Tinkertown Story; Emily Finds a Dog - A Tinkertown Tale.

Hours & Admission Prices: April to Nov. 1 daily 9-6. Adults $3.50, senior citizens $3, children $1; discounts to groups, AAM & ICOM members.

Attendance: 23,725 (accurate)

Santa Fe

BATAAN MEMORIAL MILITARY MUSEUM & LIBRARY, 1050 Old Pecos Tr., Santa Fe, NM 87505-2688. Tel.: 505-474-1670.

Institution Type/Description: Military Museum.

Collections: military artifacts & memorabilia.

Hours & Admission Prices: Tues.-Fri. 10-4. No charge; donations accepted. Closed holidays.

EL RANCHO DE LAS GOLONDRINAS MUSEUM, 334 Los Pinos Rd., Santa Fe, NM 87507-4363. Tel.: 505-471-2261. Fax: 505-471-5623.

E-mail: mail@golondrinas.org

Web Site: www.golondrinas.org

Founded: 1972.

Congressional District: 3

Key Personnel: Chm. (V), Tom Simons; Exec Dir., John Berkenfield; Deputy Dir., Michael King; Cur. Collection, Daniel Goodman; Cur. Historical Interpretation, Joseph Maes; Dir. Operations, Sean Paloheimo; Cur. Agri-

culture, Julie Anna Lopez; Dir. Programs & Mktg., Amanda Crocker; Site Rental Mgr., Carol Hilgers; Museum Shop Mgr., Lolly Martin.

Personnel Profile: Full-Time Paid 12; Part-Time Paid 11; Part-Time Volunteers 180.

Governing Authority: Parent Institution: El Rancho de las Golondrinas, Inc. Tax-exempt.

Institution Type/Description: Living History Museum: housed on 1700-1885 Las Golondrinas Ranch.

Collections: Spanish, Mexican & Territorial life in New Mexico between 1710-1912; religious artifacts. Historic Houses: Leger Gristmill; Talpa Water Mill, Padilla Water Mill, Barela Water Mill; Apodaca Blacksmith Shop; Casias Wheelwright shop; Gallegos Winery; Raton Schoolhouse; Las Trampas country store; hacienda-type dwellings of various periods & styles representing northern New Mexico of the Spanish colonial period; Morada & Oratorio Chapels.

Research Fields: Spanish colonial lifestyle; agriculture & textiles; back-breeding sheep.

Facilities: lecture & film hall; food service pavilion & kitchen.

Activities: guided tours; lectures; films; formally organized education programs for children; craft workshops; permanent & temporary exhibitions. Museum Sponsors: demonstrations, workshops & hands on activities June to August; Spring Festival in June; Wine Festival in July; Summer Festival in August; Harvest Festival in October.

Publications: official guidebook, El Rancho De Las Golondrinas; photo book, Through The Seasons at El Rancho De Las Golondrinas; coloring book; The Exposition on the Province of New Mexico, 1812; El Rancho de las Golondrinas; The Ranch of the Swallows.

Hours & Admission Prices: Museum: June-Sept. Wed.-Sun. 10-4. Adults $6-$13, seniors & military $4; members & children under 13 no charge. Admission varies for weekends & special events. Closed Labor Day. &

Attendance: 55,000 (accurate)

Membership: Single $30; Family $50; Sustaining $75; Supporting $100; Patron $150; Benefactor $500; Life $1,000.

＊ **GEORGIA O'KEEFFE MUSEUM, (M),** 217 Johnson St., Santa Fe, NM 87501-1826. Tel.: 505-946-1000. Fax: 505-946-1091.

E-mail: info@okeeffemuseum.org

Web Site: www.okeeffemuseum.org

Founded: 1997.

Congressional District: 3

Key Personnel: Dir., Robert A. Kret; Pres., Lee Dirks; Exec. Asst., Debra Liggett; Mgr. Visitor Svcs. & Facility Rental, Christina Dallorso Kortz; Dir. Finance, Carl Brown; Cur. & Research Center Dir., Barbara Buhler Lynes; Dir. Education, Jackie M; Education Coord., Sarah Zurick; Registrar & Collections Mgr., Judith C. Smith; Librarian, Archivist & Asst. Dir. Research Ctr., Eumie Imm-Stroukoff; Grant Writer & Researcher, Kaaren Boullosa; Human Resources Mgr., Sylvie S. Ward; Chief Security, Gary Smith; Dir. Mktg. & Public Rels., Kristin Kautz; Mgr. Mktg., Shannon Hanson; Assoc. Cur., Carolyn Kastner; Archivist, Elizabeth Ehrnst; Museum Shop Mgr., Janice Wrhel; Conservation, Dale Kronkright.

Personnel Profile: Full-Time Paid 49; Part-Time Paid 27; Part-Time Volunteers 75.

Governing Authority: Tax-exempt.

Institution Type/Description: Art Museum.

Collections: over 3,000 works of art including 1,700 by Georgia O'Keeffe.

Facilities: 5,000-vol. non-circulating research library; research center; education facility; video orientation rooms. Museum-related items for sale.

Activities: permanent, traveling & temporary exhibitions; guided & audio tours; captioned orientation video; lectures; concert series; docent program; workshops; field trips; training programs for educators; school tours; onsite symposiums.

Publications: monographs; collection & exhibition catalogues; critical studies & theory; educational brochures; members newsletters; teachers resources.

Hours & Admission Prices: Thurs. 10-7, Fri.-Wed. 10-5. Adults $12, seniors & students $10, NM residents $6; discounts to AAM & ICOM members; reciprocal museum members, students under 18 & members no charge. Docent Led Tour: $14. Audio Guide $5. Closed New Year's Day; Easter; Thanksgiving; Christmas. &

Attendance: 172,877 (accurate)

Membership: Individual $45 NM Resident, $50 Non-Resident; Household $65 NM Resident, $80 Non-Resident; Supporter $150; Friend $250; Sustainer $350; Patron $500; Benefactor $1,000; National Council $5,000; O'Keeffe Circle $10,000.

GERALD PETERS GALLERY, 1011 Paseo de Peralta, Santa Fe, NM 87501-2735. Tel.: 505-954-5700. Fax: 505-954-5754.

E-mail: info@gpgallery.com

Web Site: www.gpgallery.com

Key Personnel: Dir., Elizabeth Hubbard Hook
Institution Type/Description: Art Gallery.
Collections: paintings; sculpture.
Hours & Admission Prices: Mon.-Sat. 10-5. &

GIB SINGLETON MUSEUM OF FINE ART, Plaza Mercado, 112 W. San Francisco St. #204, Santa Fe, NM 87501. Tel.: 505-995-9713.
E-mail: info@gibsingleton.com
Web Site: gibsingleton.com
Founded: 2008.
Institution Type/Description: Art Museum.
Collections: works by Gib Singleton & Earl Biss.
Activities: permanent & temporary exhibitions.
Hours & Admission Prices: Winter: Mon.-Fri. 9-5; Summer: Tues.-Sat. 10-5.

THE GOVERNOR'S GALLERY, State Capitol, Rm. 400, 491 Old Santa Fe Trail, Santa Fe, NM 87501-2753. Mailing Address: P.O. Box 2087, Santa Fe, NM 87504. Tel.: 505-476-5072. Fax: 505-476-5076.
E-mail: merry.scully@state.nm.us
Web Site: www.mfasantafe.org
Founded: 1975.
Congressional District: 1
Key Personnel: Cur., Merry Scully; Museum Shop Mgr., Anna Burgess.
Personnel Profile: Full-Time Paid 2; Part-Time Volunteers 5.
Governing Authority: state. Parent Institution: New Mexico Office of Cultural Affairs. Subsidiary Institution: Museum of Fine Arts. Tax-exempt: 170(b)(1)(A).
Institution Type/Description: Art Gallery: housed in State Capitol on Old Santa Fe Trail.
Collections: paintings; photographs; sculpture.
Activities: promotion of New Mexico artists & regional art forms.
Hours & Admission Prices: Mon.-Fri. 8-5. No charge. &
Attendance: 30,000 (accurate)

INSTITUTE OF AMERICAN INDIAN ARTS MUSEUM, 83 Avan Nu Po Rd., Santa Fe, NM 87508-1300. Tel.: 505 424 2300.
Institution Type/Description: Art Museum.
Collections: over 7,000 contemporary American Indian works of art from more than 120 Native American nations; paintings; sculpture; photographs; drawings; prints; textiles; clothing; baskets; pottery; ceramics.
Activities: temporary & traveling exhibitions.
Hours & Admission Prices: Call for hours.

MILL FINE ART, 530 Canyon Rd., Santa Fe, NM 87501. Tel.: 505-982-9212. Fax: 505-982-9215.
E-mail: millfineart@gmail.com
Web Site: www.millfineart.com
Key Personnel: Dir., Mary Mill; Dir., Verne Stanford
Institution Type/Description: Art Gallery.
Collections: works by national & international contemporary artists; paintings; drawings; sculpture.
Activities: temporary exhibitions.
Hours & Admission Prices: Daily 10-5.

MUNOZ WAXMAN GALLERY - CENTER FOR CONTEMPO-RARY ARTS, 1050 Old Pecos Trail, Santa Fe, NM 87505. Tel.: 505-982-1338.
Key Personnel: Exec. Dir., Craig Anderson
Institution Type/Description: Art Gallery.
Collections: works by contemporary artists; paintings; sculptures.
Activities: special events; workshops; lectures.
Hours & Admission Prices: Call for hours.

MUSEUM OF CONTEMPORARY NATIVE ARTS, A Center of the Institute of American Indian Arts, 108 Cathedral Place, Santa Fe, NM 87501-2027. Tel.: 505-983-8900. Fax: 505-983-1222.
E-mail: museum@iaia.edu
Web Site: www.iaia.edu/museum
Founded: 1972.
Congressional District: 1
Key Personnel: Dir., Patsy Phillips; Registrar, John Joe; Administrative & Finance Officer, Marcella Apodaca; Special Projects & Community Rels. Officer, Larry Phillips; Museum Shop Mgr., Laura Ellerby; Security Supvr.,

Thomas Atencio; Art Design Asst., Sallie Wesaw; Chief Cur., Ryan Rice; Cur. Collections, Tatiana Lomahaftewa Slock; Educational Program Asst., Hayes Locklear.
Personnel Profile: Full-Time Paid 18; Part-Time Paid 3; Interns 2.
Governing Authority: college. Parent Institution: Institute of American Indian & Alaska Native Culture and Arts Development. Tax-exempt.
Institution Type/Description: Culturally Specific.
Collections: over 7,000 items contemporary American Indian art; American Indian artists archive.
Research Fields: contemporary American Indian art.
Facilities: 400-vol. library of books related to American Indian art & general art history; classrooms; national Native American videotape archives; BIA Bicentennial Videotape Archives available by written request to museums & researchers. North American Indian arts & books for sale.
Activities: guided tours; lectures; films; gallery talks; arts festivals; workshops; undergraduate college students; training programs for professional museum workers; inter-museum loan; permanent, temporary & traveling exhibitions.
Publications: anthologies of student written works; exhibit catalogues & brochures; books: Creativity is Our Tradition; Gallery handouts; members newsletter, Manifestations: New Native Art Criticism.
Hours & Admission Prices: Mon. & Wed.-Sat. 10-5, Sun. 12-5. Adults $10, students & senior citizens with ID $5; discounts to AAM members; children under 16, members, & Native Americans no charge. Closed major holidays. &
Attendance: 39,761 (accurate)
Membership: Individual $40; Family $50; Sponsor $100; Patron $500; Benefactor $1,000.

✳ **MUSEUM OF INDIAN ARTS & CULTURE - LABORA-TORY OF ANTHROPOLOGY, (M),** 710 Camino Lejo, Santa Fe, NM 87505-7511. Mailing Address: P.O. Box 2087, Santa Fe, NM 87504-2087. Tel.: 505-476-1250 & 1247. Fax: 505-476-1330.
E-mail: info@miaclab.org
Web Site: www.indianartsandculture.org
Founded: 1929.
Congressional District: 3
Key Personnel: Deputy Dir., Elena Sweeney; Finance, Monica Vigil; Cur. Archaeological Research Collections, Julia Clifton; Cur. Ethnology, Antonio Chavarria; Asst. Cur. Archaeological Research Collections, Dody Fugate; Cur. Individually Cataloged Collections, Valerie Verzuh; Archivist, Rights & Reproductions, Diana Bird; Archaeological Collection Specialist, Rachel Johnson; Librarian, Allison Colborne; Dir. Living Traditions Center, Joyce Begay-Foss; Educator, Dawn Kaufmann; Museum Shop Mgr., John Stafford.
Personnel Profile: Full-Time Paid 24; Part-Time Paid 4; Part-Time Volunteers 90.
Governing Authority: state; nonprofit. Parent Institution: Department of Cultural Affairs. Tax-exempt: 170(c)(1).
Institution Type/Description: Native American Arts & Culture Museum: Laboratory of Anthropology is housed in 1931 Spanish-Pueblo revival architecture designed by John Gaw Meem.
Collections: American Indian Art, material culture & ethnology of the native peoples of the Southwest; pre-contact & historic jewelry, pottery, baskets, textiles, accessories; Kachinas; sculpture; paintings.
Major Exhibits: Woven Identities: Basketry Art from the Collection, 11/19/11-4/1/14.
Research Fields: archaeology of the Southwest; anthropology of native peoples of Southwest; contemporary arts & aesthetics.
Facilities: 45,000-vol. library on anthropology of the Southwest available to the public; educational classrooms; resource center; 30,000 sq. ft. exhibit space; field research section; amphitheater.
Activities: guided tours; lectures; films; concerts; training programs; organized education programs for children, adults, undergraduate & graduate students; internships. Annual Events: Native Treasures Indian Arts Festival in May; Native Treasures Collector's Sale in September.
Publications: popular articles in Museum of New Mexico magazine El Palacio; Secrets of Casas Grandes; A River Apart: The Pottery of Cochiti & Santo Domingo Pueblo; Huichol Art & Culture: Balancing the World; Painting the Native World: Life, Land and Animals; Spider Woman's Gift: Nineteenth-Century Dine Textiles at the Museum of Indian Arts & Culture.
Hours & Admission Prices: Memorial Day to Labor Day daily 10-5; Sept.-May Tues.-Sun. 10-5. Four-day pass to all five Museums $20, a single Museum $9; discounts to students & AAM members; NM residents with ID, Sun. & members no charge. Closed New Year's Day; Easter; Thanksgiving; Christmas. &
Attendance: 48,446 (accurate)
Membership: Student & Teacher $30; Senior $50; Senior Couple $60; Individual $65; Family & Dual $75-$124; Sponsor $125-$249; Patron

$250-$499; Benefactor $500-$1,499; Regent's Circle $1,500-$2,499; Governor's Circle $2,500-$4,999; National Council $5,000-$9,999; Chairman's Council $10,000 & up.

* **MUSEUM OF INTERNATIONAL FOLK ART, (M),** Museum Hill at Camino Lejo, Santa Fe, NM 87505. Mailing Address: P.O. Box 2087, Santa Fe, NM 87504-2087. Tel.: 505-476-1200 & 1204. Fax: 505-476-1300.
E-mail: info@moifa.org
Web Site: www.moifa.org
Founded: 1953.
Congressional District: 1
Key Personnel: Dir., Dr. Marsha C. Bol; Asst. Dir., Elena Sweeney; Collections Mgr., P.J. Smutko; Cur. Latin American Folk Art, Dr. Barbara Mauldin; Cur. U.S. Latino, Hispano, & Spanish Colonial Collections, Nicolasa Chavez; Cur. Textiles & Costumes, Dr. Bobbie Sumberg; Dir. Education, Aurelia Gomez; Librarian & Archivist, Caroline Dechert; Special Events Coord., Laura Lovejoy-May; Administrative Sec., Christine Vitagliano; Cur. European & American Folk Art, Laura Addison; Cur. Asian & Middle Eastern Folk Art, Felicia Katz Harris; Museum Educator, Patricia Sigala; Security, Andy Perea.
Personnel Profile: Full-Time Paid 28; Part-Time Paid 2; Part-Time Volunteers 115; Interns 1.
Governing Authority: state. Parent Institution: Cultural Affairs Dept. Subsidiary Institute: Museum of New Mexico. Tax-exempt.
Institution Type/Description: International Folk Art Museum.
Collections: international collection of folk art including textiles, costumes, ceramics, dolls & toys; Spanish colonial religious art; Southwest Hispanic folk art & furniture; jewelry; amulets; religious, ceremonial & ritual objects.
Major Exhibits: Brasil and Arte Popular, 11/17/13-8/10/14.
Research Fields: Spanish colonial arts, textiles, costumes; folk art.
Facilities: 16,000-vol. library of books & journals pertaining to folk art; 180-seat auditorium; 25,000 sq. ft. exhibit space; outdoor plaza & restaurant; labyrinth; gardens. Museum-related items for sale.
Activities: guided tours; lectures; films; concerts; arts festivals; organized education programs for children & adults; docent program.
Publications: magazine, El Palacio.
Hours & Admission Prices: Memorial Day to Labor Day daily 10-5; Sept.-May Tues.-Sun. 10-5. Four-day pass to five museum facilities in Santa Fe $18. Single Museum: out of state $9, in-state $6; discounts to AAM & ICOM members; children 16 & under, school groups, NM seniors 60 & over on Wed., museum foundation members & NM residents on Sun. no charge. Closed New Year's Day; Easter; Thanksgiving; Christmas. &
Attendance: 85,000 (accurate)
Membership: Student with ID $25; Senior Individual over 62 $40; Individual & Senior Couple over 62 $50; Family $60-$99; Sponsor $100-$249; Patron $250-$499; Benefactor $500-$999; Regent's Circle $1,000-$2,499; Governor's Circle $2,500-$4,999; National Council $5,000 & up.

* **MUSEUM OF NEW MEXICO, (M),** 725 Camino Lejo, Santa Fe, NM 87505-7516. Mailing Address: P.O. Box 2087, Santa Fe, NM 87504-2087. Tel.: 505-476-1125. Fax: 505-476-1127.
E-mail: steve.cantrell@state.nm.us
Web Site: www.museumofnewmexico.org
Founded: 1909.
Congressional District: 1,2 & 3
Key Personnel: Dir. Museum of International Folk Art, Dr. Marsha C. Bol; Dir. Museum of Indian Arts & Culture/Laboratory of Anthropology, Della Warrior; Dir. NM Museum of Arts, Mary Kershaw; Dir. NM History Museum/Palace of the Governors, Dr. Frances Levine; Public Rels. Mgr., Steve Cantrell.
Governing Authority: state. Parent Institution: Dept. of Cultural Affairs. Branch Museums: Palace of the Governors, Santa Fe; New Mexico Museum of Arts, Santa Fe; Museum of International Folk Art, Santa Fe; Museum of Indian Arts & Culture/Laboratory of Anthropology, Santa Fe; Fort Sumner State Monument, Fort Sumner; Coronado State Monument, Bernalillo; Jemez State Monument, Jemez Springs; Fort Selden State Monument, Radium Springs; Lincoln State Monument, Lincoln; El Camino Real Heritage Center, Socorro. Tax-exempt: 170(b)(1)(A).
Institution Type/Description: History, Art, Folk Art & Anthropology Museums.
Collections: anthropology, archaeology & ethnology of the Southwest; fine arts with emphasis on southwestern painters; international folk art; Spanish colonial history; history of New Mexico; photographic archives; American Indian art & cultures; six historic site state monuments at Lincoln, Fort Sumner, Jemez Springs, Fort Selden, Bernalillo & Socorro.
Research Fields: anthropology; folk art; history of New Mexico; Spanish Colonial history; art; archaeomagnetic dating; non-destructive trace elements dating.
Activities: guided tours; lectures; films; concerts; art competitions; docent

program; inter-museum loan, permanent, temporary & traveling exhibitions; school loan service; educational aid kits for schools.
Publications: magazine, El Palacio; 80 titles covering southwestern art, history, archaeology, ethnology & crafts; monthly newsletter.
Hours & Admission Prices: NM Museum of Art & NM History Museum: daily 10-5; no charge May-Oct. Fri. 5p.m.-8p.m. & Nov.-April first Fri. of month; closed Mon. Oct. to Memorial Day. Museum of International Folk Art & Museum of Indian Arts & Cultures: daily 10-5; no charge Memorial Day to Labor Day; closed Mon. Oct. to Memorial Day. Four-day pass to Santa Fe museums nonresidents $20, NM residents $18, single visit nonresidents $9, NM residents $6; discounts to students, groups of 10 or more, AAM & ICOM members; school groups, NM residents on Sun., NM seniors on Wed., children 16 & under no charge. Closed New Year's Day; Easter; Thanksgiving; Christmas. &
Attendance: 642,100 (accurate)
Membership: Student/Teacher $30; Senior Individual $50; Senior Couple $60; Individual $65; Family/Dual $75-$124; Sponsor $125-$249; Patron $250-$499; Benefactor $500-$1,499; The Circles $1,500 & up.

* **NEW MEXICO MUSEUM OF ART, (M),** 107 W. Palace Ave., Santa Fe, NM 87501-2014. Mailing Address: P.O. Box 2087, 107 W. Palace Ave., Santa Fe, NM 87504-2087. Tel.: 505-476-5061. Fax: 505-476-5076.
E-mail: rebecca.potance@state.nm.us
Web Site: nmartmuseum.org
Formerly: Museum of Fine Arts
Founded: 1917.
Congressional District: 1
Key Personnel: Dir., Mary Kershaw; Cur. 20th Century Art, Dr. Joe Traugott; Librarian, Rebecca Potance; Cur. Education, Ellen Zieselman; Mgr. Special Events, Martha Landry; Cur. Contemporary Art, Laura Addison; Chief Registrar, Michelle Roberts; Cur. Photography, Katherine Ware; Preparator, Sam Rykels.
Personnel Profile: Full-Time Paid 24; Part-Time Paid 1; Part-Time Volunteers 8.
Governing Authority: state; nonprofit. Parent Institution: Dept. of Cultural Affairs. Tax-exempt: 170(c)(1).
Institution Type/Description: Art Museum: housed in an architectural structure patterned after New Mexico mission churches.
Collections: over 23,000 fine art pieces emphasizing the Southwest; paintings; prints; drawings; photographs; sculpture; exhibitions on 20th-century American art.
Research Fields: American art, specializing in New Mexico.
Facilities: 6,000-vol. library relating to art in the Southwest available to the public; auditorium; three sculpture courtyards. Museum-related items for sale.
Activities: guided tours; lectures; films; concerts; organized education programs for children; traveling exhibitions.
Publications: quarterly journal, El Palacio; exhibition catalogues.
Hours & Admission Prices: Winter: Tues.-Sun. 10-5; Summer: daily 10-5. Four-day pass to 4 Santa Fe museums $20, single non-resident $9, single NM residents $6; discounts to AAM & ICOM members; school groups, NM residents on Sun., seniors 60 & over on Wed., Fri. 5-8 & museum members no charge. Closed New Year's Day; Easter; Thanksgiving; Christmas. &
Attendance: 125,000 (accurate)

* **PALACE OF THE GOVERNORS/NEW MEXICO HISTORY MUSEUM, (M),** On the Plaza, Santa Fe, NM 87501. Mailing Address: P.O. Box 2087, Santa Fe, NM 87504-2087. Tel.: 505-476-5100 (Palace) & 5200 (NMHM). Fax: 505-476-5104.
E-mail: carla.ortiz@state.nm.us
Web Site: www.palaceofthegovernors.org
Founded: 1909.
Congressional District: 3
Key Personnel: Dir., Frances Levine, Ph.D.; Deputy Dir., John J. McCarthy; Asst. Dir., Rene Harris; Finance Officer, Judy Morse; Cur., Josef Diaz; Cur. Historical Photography, Mary Anne Redding; Mgr. Collections, Wanda Edwards; Asst. Collections Mgr., Pennie McBride; Imaging Specialist, Nicholas Chiarella; Palace Press, Thomas Leech; Palace Press, James Bourland; Photo Archivist, Daniel Kosharek; Chief Librarian, Tomas Jaehn; Sr. Cataloger, Patricia Hewitt; Archivist, Lou Jaureguiberry; Coord. Public Programs, Carlotta Boettcher; Security Captain, Steve Baca; Education, Erica Garcia; Education, Emily Fijol; Exhibit Designer, Caroline Lajoie; Public Rels. & Mktg., Kate Nelson; Coord. Special Events, Inessa Williams; Graphic Design, Natalie Beca.
Personnel Profile: Full-Time Paid 38; Part-Time Paid 1; Part-Time Volunteers 70.
Governing Authority: state; nonprofit. New Mexico History Museum, 113 Lincoln, Santa Fe, NM 87501.

Institution Type/Description: History Museum: housed in c.1610 Spanish-style government building.
Collections: New Mexico & Southwest history from European contact, colonial period to the present; firearms; religious art; decorative arts; photos; archives; pottery; silver; furniture.
Research Fields: New Mexico, Southwestern, Latin American & borderlands history.
Facilities: 12,000-vol. library relating to the history of the Southwest; manuscripts; 380,000 photographs; 15,000 artifact collection.
Activities: guided walking tours of Santa Fe; foreign tours to Spain, Mexico's Central & South America; lectures; films; concerts. Museum Sponsors: Mountain Trade Fair; Christmas at the Palace; Las Posadas; Native American Dances; Portal Native American Venders Program.
Publications: Friends of the Palace Press; historical novels.
Hours & Admission Prices: Labor Day to Memorial Day Tues.-Sun. 10-5. Adults $9 nonresident, $6 residents; New Mexico residents on Sun., NM seniors on Wed., Fri. 5-8, children 16 and under, AAM & ICOM members no charge. Closed New Year's Day; Easter; Thanksgiving; Christmas. &
Attendance: 100,000

PUEBLO OF POJOAQUE, POEH MUSEUM, 78 Cities of Gold Rd., Santa Fe, NM 87506-0918. Tel.: 505-455-3334. Fax: 505-455-0174.
Web Site: poehmuseum.com
Founded: 1987.
Congressional District: 3
Key Personnel: Dir., Vernon G. Lujan; Asst. Dir., Francine Maestas; Exec. Dir., George Rivera; Cur., Melissa Talachy; Museum Shop Mgr., Andrea Tse-Pe'.
Personnel Profile: Full-Time Paid 5; Part-Time Paid 2; Interns 1.
Governing Authority: Parent Institution: Poeh Cultural Center and Museum. Tax-exempt.
Institution Type/Description: Tribal Museum.
Collections: archaeological to contemporary art from Pueblos of New Mexico; Tewa Pueblo communities; photographs; Native American; NAGPRA repository.
Research Fields: language & cultural preservation.
Facilities: classroom. Museum-related items for sale.
Activities: guided tours; temporary exhibitions.
Hours & Admission Prices: Mon.-Fri. 8-5, Sat. 9-4. No charge; donations accepted. Closed Thanksgiving; Feast Day; Christmas. &
Attendance: 750 (estimated)

SITE SANTA FE, 1606 Paseo de Peralta, Santa Fe, NM 87501-3724. Tel.: 505-989-1199. Fax: 505-989-1188. Facebook: SITE Santa Fe.
E-mail: info@sitesantafe.org
Web Site: www.sitesantafe.org
Founded: 1995.
Congressional District: 3
Key Personnel: Phillips Dir., Irene Hofmann; Deputy Dir., Catherine Putnam; Bd. Chair, Andrew Wallerstein; Bd. Vice Chair, Steve Berkowitz; Bd. Treas., William A. Miller; Bd. Sec., Courtney Finch Taylor; Pres. Emeritus, John L. Marion; Dir. Education, Joanne Lefrak; Dir. External Affairs, Anne Wrinkle; Dir. Devel., Valerie Ingram.
Personnel Profile: Full-Time Paid 13; Part-Time Paid 12; Part-Time Volunteers 19; Interns 4.
Governing Authority: private; nonprofit organization. Tax-exempt: 501(c)(3).
Institution Type/Description: Contemporary Art Museum.
Collections: works by contemporary artists.
Major Exhibits: FEAST (T), 2/14-5/17/14; SITElines: Unsettled Landscapes, 7/14-1/15; Unsuspected Possibilities, 6/15-10/15.
Facilities: 15,000 sq. ft. exhibit space. Books & catalogues for sale.
Activities: international biennial exhibition program; contemporary art exhibitions; formal education programs for adults & teens; guided tours; lectures & public programs; traveling exhibitions; artists' commissions; concerts; films.
Publications: exhibition catalogs: Looking for a Place; Roni Horn; Robert Therrien; Allan Graham; Jose Bedia; Sarah Charlesworth; Truce; Longing & Belonging; Thomas Ashcraft; Jim Hodges; Conceal/Reveal; Tom Sachs; Alan Rath; Postmark; Beau Monde: Toward a Redeemed Cosmopolitanism; Gary Simmons: Ghost House; Dara Friedman; Ernesto Neto; Teresita Fernandez; Mona Hatoum: Domestic Disturbance; Erika Wanenmacher: Grimoire; Monica Bonvicini; Doris Salcedo; Roxy Paine: Second Nature; Uta Barth: White Blind (Bright Red); Disparities and Deformations: Our Grotesque; Jim Campbell: Quantizing Effects; Paul Sarkisian; Still Points of the Turning World; Stephen Bush: Gelderland; Barry X Ball; Steina: 1970-2000 (2008); Lucky Number Seven; catalogue, SITE's 7th International Biennial; The Dissolve Catalogue, SITE's 8th International Biennial; Linda Mary Montano: You Too Are A Performance Artist; Mungo Thompson: Time People Money Crickets.

Hours & Admission Prices: Exhibition: July-Aug. Wed.-Thurs. & Sat. 10-5, Fri. 10-7, Sun. 12-5; Sept.-June Thurs & Sat. 10-5, Fri. 10-7, Sun. 12-5. Adults $10, seniors, students & children $5; discounts to AAM & ICOM members; members & Fri. 10-7 no charge. Closed national holidays. &
Attendance: 25,000 (accurate)
Membership: Student, Senior & Educator $35; Individual $50; Family $75; Supporter $125; Patron $250; Contemporary Circle $500; Director's Circle $1,000.

SAN ILDEFONSO PUEBLO MUSEUM, 02 Tunyo PO Rd., Santa Fe, NM 87506. Tel.: 505-455-3549. Fax: 505-455-7351.
Web Site: www.sanipueblo.org
Institution Type/Description: History Museum.
Collections: contemporary & traditional pottery; paintings; San Ildefonso artifacts; historic items.
Hours & Admission Prices: Mon.-Fri. 8-4:30. Vehicles $10 (includes museum admission).

SAN MIGUEL MISSION CHURCH, 401 Old Santa Fe Trail, Santa Fe, NM 87501-2746. Tel.: 505-983-3974.
E-mail: sanmiguelsantafe@yahoo.com
Founded: 1610.
Congressional District: 3
Key Personnel: C.E.O., Saundra Johnson Austin.
Personnel Profile: Full-Time Paid 1.
Governing Authority: church; nonprofit organization. Parent Institution: St. Michael's H.S. Tax-exempt.
Institution Type/Description: Historic Building: built in 1610.
Collections: Indian art; religious artifacts; pottery; oldest bell in U.S.; Spanish colonial art.
Facilities: archives; church. Religious articles for sale.
Activities: guided tours; lectures; masses.
Publications: book, Story of San Miguel.
Hours & Admission Prices: Summer: Mon.-Sat. 9-5, Sun. 10-4. Latin Mass: Sun. 2pm. Mass: Sun. 5pm. Tours: $1 per person; children under 6 no charge. &
Attendance: 65,000 (accurate)

SANTA FE CHILDREN'S MUSEUM, (M), 1050 Old Pecos Trail, Santa Fe, NM 87505-2688. Tel.: 505-989-8359. Fax: 505-989-7506.
E-mail: children@santafechildrensmuseum.org
Web Site: www.santafechildrensmuseum.org
Founded: 1985.
Congressional District: 3
Key Personnel: Exec. Dir., Ann Marie Tutera Manriquez; Pres. (V) & Chm. Bd. (V), Fidel Gutierrez; Bd. Pres., Sally Mittler; Deputy Dir., Jeff Dailey; Dir. Devel., Ginger Roherty; Museum Shop Mgr., Meri Frauwirth.
Personnel Profile: Full-Time Paid 5; Part-Time Paid 7; Part-Time Volunteers 100; Interns 6.
Governing Authority: not-for-profit organization. Tax-exempt: 501(c)(3).
Institution Type/Description: Children's Museum: located in Santa Fe's historic district.
Collections: interactive exhibits pertaining to the arts, humanities & science.
Research Fields: informal learning.
Facilities: 500-vol. library on education & life studies, available to the public; 5,000 sq. ft. exhibit space; 1 acre horticultural garden. Museum-related items for sale.
Activities: concerts; formal education programs for adults; environmental education; workshops; demonstrations; art & science education programs; interactive exhibits. Annual Events: auction; concert; holiday programming; ice cream Sunday.
Publications: quarterly newsletter; annual ad brochure, annual report; Voices of Violence, Visions of Peace; Families: Celebrating the Journey.
Hours & Admission Prices: June-Aug. Tues.-Sat. 10-6, Sun. 12-5; mid-Sept. to May Wed.-Sat. 10-5, Sun. 12-5. Adults $9, NM residents with ID $4, Sunday $2; school groups no charge. Closed New Year's Day; Easter; Independence Day; Thanksgiving; Christmas. &
Attendance: 68,818 (accurate)
Membership: Family $120; Grandparent $140; Family Plus $150; Family Plus Sponsorship $250; Dollar a Day Membership $365.

SANTUARIO DE NUESTRA SENORA DE GUADALUPE, 417 Agua Fria St., Santa Fe, NM 87501. Tel.: 505-983-8868 & 988-2027. Fax: 505-983-4304.
E-mail: santuario@ologsf.com
Web Site: www.ologsf.com
Founded: 1975.

Congressional District: 1
Key Personnel: Santuario Dir., Gail Delgado.
Personnel Profile: Full-Time Paid 1; Part-Time Volunteers 9.
Governing Authority: nonprofit organization. Tax-exempt: 501(c)(3).
Institution Type/Description: Religious Museum & Historic Site: housed in c.1780 church built by the Franciscans.
Collections: Spanish colonial mural; New Mexican Santero art.
Research Fields: living Indo-Hispanic artists & their works; adobe structures.
Facilities: plants of the Holy Land garden.
Activities: guided tours; lectures; gallery talks; concerts; arts festivals; drama.
Publications: quarterly newsletter, Noticias.
Hours & Admission Prices: Mon.-Sat. 9-12 & 1-4. No charge; donations accepted. &
Attendance: 21,000

SCHOOL FOR ADVANCED RESEARCH, INDIAN ARTS RESEARCH CENTER, 660 Garcia St., Santa Fe, NM 87505-2858. Mailing Address: P.O. Box 2188, Santa Fe, NM 87504-2188. Tel.: 505-954-7205 & 7200. Fax: 505-954-7207.
E-mail: iarc@sarsf.org
Web Site: www.sarweb.org
Formerly: School of American Research
Founded: 1907.
Congressional District: 1
Key Personnel: Interim Pres., David Sturt; Vice Pres. Business Administration, Sharon Tison; Co Dir. Publications, Lynn Baca; Dir. Indian Arts Research Center, Dr. Cynthia Chavez Lamar.
Personnel Profile: Full-Time Paid 28; Part-Time Paid 6; Part-Time Volunteers 23; Interns 2.
Governing Authority: nonprofit organization. Parent Institution: School for Advanced Research. Subsidiary Institution: Indian Arts Research Center. Tax-exempt: 501(c)(3).
Institution Type/Description: Indian Arts Research Center.
Collections: representative historic & contemporary Southwestern Pueblo Indian pottery; Navajo & Pueblo Indian textiles & jewelry collection; Native American easel paintings; Southwest Indian basketry; Katsinam & other ethnographic objects pertaining to historic & contemporary Southwest American Indian art.
Research Fields: anthropology; archaeology; Southwest Indian arts.
Facilities: library; 8 acre historic estate.
Activities: resident scholar program; advanced seminars; Southwest Indian arts research; archaeology; publications; public education.
Publications: books on Southwest Indian art, archaeology & anthropology.
Hours & Admission Prices: Jan. 2-Dec. 23 Mon.-Fri. 9-5 by appointment only. Public Tours Fri. 2 pm, call for reservations. Adults $15; members no charge. Closed Federal holidays. &
Attendance: 1,145 (accurate)
Membership: Arroyo Hondo $40; Pecos $70; Mesa Verde $150; Galisteo $250; Chaco $500; Bandelier $1,000.

SPANISH COLONIAL ARTS SOCIETY, 750 Camino Lejo, Santa Fe, NM 87505-7511. Mailing Address: P.O. Box 5378, Santa Fe, NM 87502-5378. Tel.: 505-982-2226. Fax: 505-982-4585.
E-mail: museum@spanishcolonial.org
Web Site: www.spanishcolonial.org
Founded: 1925.
Congressional District: 3
Key Personnel: Exec. Dir. Society, Donna Pedace; Cur., Robin Farwell Gavin; Pres., Jim Long; Museum Shop Mgr., Jann Phillips.
Personnel Profile: Full-Time Paid 10; Part-Time Paid 3; Interns 2.
Governing Authority: private; nonprofit organization. Parent Institution: Spanish Colonial Arts Society, Inc. Tax-exempt.
Institution Type/Description: Spanish Colonial Art Museum.
Collections: Spanish colonial art; New Mexico Hispanic art; furniture; bultos; retablos; straw applique; textiles; tinwork; jewelry; developments in Hispanic art forms.
Facilities: 6,300 sq. ft. exhibit space.
Activities: Annual Events: Celebracion Gala in March; Spanish Market in July; Matanza! in October; Winter Market in December.
Publications: newsletter, Que Pasa?
Hours & Admission Prices: Memorial Day to Labor Day daily 10-5; Sept.-May Tues.-Sun. 10-5. Adults $5; discounts to AAM members; N.M. residents on Sun. & members no charge. &
Attendance: 50,000 (estimated)
Membership: Individual $40; Family & Dual $65; Supporter $100; Sponsor $300; Sustainer $500; Patron $1,000 & up; Benefactor $2,500 & up; Chairman's Council $5,000 & up.

THE VISUAL ARTS GALLERY - SANTA FE COMMUNITY COLLEGE, 6401 Richards Ave., Santa Fe, NM 87508. Tel.: 505-428-1501.
Institution Type/Description: Art Gallery.
Collections: paintings; sculpture; photographs; ceramics.
Facilities: 1,600 sq. ft. exhibition space.
Hours & Admission Prices: Call for hours.

* **THE WHEELWRIGHT MUSEUM OF THE AMERICAN INDIAN, (M),** 704 Camino Lejo, Santa Fe, NM 87505-7511. Mailing Address: P.O. Box 5153, Santa Fe, NM 87502-5153. Tel.: 505-982-4636 & 1-800-607-4636. Fax: 505-989-7386.
E-mail: info@wheelwright.org
Web Site: www.wheelwright.org
Founded: 1937.
Congressional District: 3
Key Personnel: Dir. & C.E.O., Jonathan Batkin; Cur., Cheri Falkenstien-Doyle; Fiscal Officer, Julie Hoffman; Preparator, Arthur Esquibel; Mgr. Case Trading Post, Robb Lucas.
Personnel Profile: Full-Time Paid 6; Part-Time Paid 7; Part-Time Volunteers 130.
Governing Authority: nonprofit organization. Subsidiary Institution: Wheelwright Foundation: Tax-exempt: 501(c)(3).
Institution Type/Description: Anthropology & Art Museum.
Collections: traditional & contemporary Native American Art including Navajo, Pueblo & other Native Americans of New Mexico ethnography; art history.
Research Fields: Native American art & culture.
Facilities: 6,000-vol. library of books on Navajo studies & Native American Art. Museum-related items for sale.
Activities: guided tours; lectures; films; gallery talks; temporary exhibitions; fieldtrips; craft demonstrations; children's activities; book signings.
Hours & Admission Prices: Mon.-Sat. 10-5, Sun. 1-5. No charge; donations accepted. Closed New Year's Day; Thanksgiving; Christmas. &
Attendance: 33,000 (accurate)
Membership: Individual $40; Family $60; Behind the Scenes $150; Supporter & Business $250; Benefactor $500; Corporate $1,000.

Santa Rosa

ROUTE 66 AUTO MUSEUM, 2866 Will Rogers Ave., (Historic Rte. 66), Santa Rosa, NM 88435. Tel.: 505-472-1966.
E-mail: info@route66automuseum.com
Web Site: www.route66automuseum.com
Key Personnel: Owner, Bozo Cordova
Institution Type/Description: Car Museum.
Collections: over 30 vehicles on display.
Hours & Admission Prices: April-Oct. Mon.-Sat. 7:30am-6pm, Sun. 10-5; Nov.-March Mon.-Sat. 8-5, Sun. 10-5. Adults $5; discount to family & groups.

Santa Teresa

WAR EAGLES AIR MUSEUM, Dona Ana County Airport, Santa Teresa, NM 88008. Mailing Address: 8012 Airport Rd., Santa Teresa, NM 88008-9719. Tel.: 575-589-2000.
E-mail: mail@war-eagles-air-museum.com
Web Site: www.war-eagles-air-museum.com/
Founded: 1989.
Key Personnel: Dir., Bob Dockendorf; Operations Mgr., George Guerra; Gift Shop Mgr., Kathy Barnicoat.
Personnel Profile: Full-Time Paid 4; Part-Time Paid 4; Part-Time Volunteers 12.
Governing Authority: nonprofit organization. Parent Institution: WEAM Inc. Tax-exempt.
Institution Type/Description: Military Aviation History.
Collections: over 35 aircraft from World War II & the Korean Conflict era including P-51 Mustang, P-38 Lightning, P-40 Warhawk, F-4U-4 Corsair; twin-engine Invader bomber, DC-3 transport; German observation aircraft, the Fiesler-Storch; over 50 classic & historic automobiles.
Facilities: Museum-related items for sale.
Activities: rental facilities; educational programs; special events; rental facilities.
Publications: quarterly newsletter.
Hours & Admission Prices: Tues.-Sun. 10-4. Adults $5, seniors & military $4; students & children no charge. &
Membership: Individual $15; Family $25; Participating $100; Supporting $500; Benefactor $1,000; Life $5,000.

Silver City

FRANCIS MCCRAY GALLERY, Western NM Univ., 1000 College Ave., Silver City, NM 88062. Mailing Address: Western NM Univ., P.O. Box 680, Silver City, NM 88062-0680. Tel.: 575-538-6517. Fax: 575-538-6619.
Web Site: www.wnmu.edu
Founded: 1960.
Key Personnel: Professor, Michael Metcalf.
Personnel Profile: Full-Time Paid 1; Part-Time Paid 3.
Governing Authority: university. Parent Institution: Western New Mexico University. Tax-exempt.
Institution Type/Description: University Art Gallery
Collections: contemporary American art: paintings, prints, sculpture, pottery.
Facilities: classrooms; 1,600 sq. ft. exhibit space; 990-seat theatre.
Activities: lectures; films; concerts; dance recitals; arts festivals; rental gallery; education programs for college students affiliated with Western New Mexico University; loan, temporary & traveling exhibitions. Museum Sponsors: Scholarship Art Sale.
Publications: monthly, flyer.
Hours & Admission Prices: Sept.-May Mon.-Fri. 10-4:30; other times by appointment. No charge; donations accepted. Closed university holidays. &
Attendance: 2,400 (estimated)

✳ **SILVER CITY MUSEUM, (M),** 312 W. Broadway, Silver City, NM 88061-4921. Tel.: 575-538-5921. Fax: 575-388-1096.
E-mail: info@silvercitymuseum.org
Web Site: www.silvercitymuseum.org
Founded: 1967.
Key Personnel: Dir., Tracy Spikes; Pres. Silver City Museum Society, Liz Mikols; Cur. Collections, Jackie Becker; Volunteer Coord., Barbara Nance; Cur. Education, Jessica Tumposky; Museum Shop Mgr., Charmeine Wait.
Personnel Profile: Full-Time Paid 3; Part-Time Paid 4; Part-Time Volunteers 50.
Governing Authority: municipal; nonprofit. Parent Institution: Town of Silver City. Tax-exempt.
Institution Type/Description: History Museum: housed in the former home of H. B. Ailman; 1881.
Collections: late 19th-century objects & artifacts; Native American artifacts; items from early 20th-century mining town; collection of photographs & documents relating to the history of southwest New Mexico.
Research Fields: southwest New Mexico history; architecture.
Facilities: events courtyard; programs & research annex.
Activities: guided tours; lecture series; permanent & temporary exhibitions; outreach program; special events; family programs; school tour programs.
Publications: newsletter, The Mansardian.
Hours & Admission Prices: Tues.-Fri. 9-4:30, Sat.-Sun. 10-4. Suggested Donation: $3; members no charge. Closed New Year's Day; Thanksgiving; Christmas. &
Attendance: 11,000 (estimated)
Membership: Student & Volunteer $15; Friend & Senior $30; Couple & Family $50; Sponsor $100; Business $150; Patron $250; Benefactor $500 & up.

WESTERN NEW MEXICO UNIVERSITY MUSEUM, (M), Fleming Hall at 10th St., 1000 W. College St., Silver City, NM 88061. Mailing Address: P.O. Box 680, Silver City, NM 88062-0680. Tel.: 575-538-6386. Fax: 575-538-6385.
Web Site: www.wnmumuseum.org
Founded: 1974.
Congressional District: 2
Key Personnel: Dir., Dr. Cynthia Ann Bettison; Museum Shop Mgr., Phillip Cave.
Personnel Profile: Full-Time Paid 2; Part-Time Paid 7; Part-Time Volunteers 14.
Governing Authority: University. Parent Institution. Western New Mexico University. Tax-exempt.
Institution Type/Description: Anthropology Museum.
Collections: Eisele Collection of Southwest Native American artifacts including Mimbres painted pottery, Casas Grandes pottery, Anasazi pottery, chipped & groundstone; stone tools; basketry; The Hunter Collection containing mining implements; D.C. Hinman Photographs; archives & artwork of Editha Watson; Alvan N. White Memorial archives & memorabilia; Miller collection of baskets; c.1850 Scott Nichols single horse doctor buggies; Margaret Kelly collection of historic Navajo rugs & saddle blankets; NAN Ranch Collection of pottery and other artifacts; more than 30 years of scientific excavation led by Dr. Harry Shafer.
Research Fields: history & anthropology.
Facilities: historic classroom; antique room; short term exhibition room. Gift items for sale.

Activities: museum internships for high school juniors & seniors and Western New Mexico University undergraduates & graduates; oral history; permanent, temporary & traveling exhibits; school loan service; instruction in both archaeology & museology.
Publications: quarterly newsletter, WNMU Museum.
Hours & Admission Prices: Mon.-Fri. 9-4:30, Sat.-Sun. 10-4. No charge; donations accepted. Research: Mon.-Fri. 9-4. Closed university holidays. &
Attendance: 14,500 (estimated)
Membership: Volunteer $7; Individual & Senior $10; Family $15; Supporter & Business Patron $50; Sponsor $100; Patron $500; Benefactor $1,000.

Socorro

HAMMEL MUSEUM, 500 Sixth St., Socorro, NM 87801-4227. Mailing Address: P.O. Box 923, Socorro, NM 87801-0923. Tel.: 575-835-3183.
Web Site: socorrohistory.org
Key Personnel: Dir., Debra Dean
Institution Type/Description: History Museum: housed in the former Hammel Brewery & later operated as an ice house and soft drink bottling plant until the 1950s.
Collections: brewery history; period furnishings; equipment & machinery; industrial & commercial history.
Activities: tours.
Hours & Admission Prices: First Sat. of month 9-12. No charge.

NEW MEXICO BUREAU OF GEOLOGY MINERAL MUSEUM, New Mexico Tech, 801 Leroy Place, Socorro, NM 87801-4681. Tel.: 505-835-5140 & 5420. Fax: 505-835-6333.
E-mail: vwlueth@nmt.edu
Web Site: geoinfo.nmt.edu
Formerly: New Mexico Bureau of Mines Mineral Museum
Founded: 1889.
Congressional District: 3
Key Personnel: Cur., Virgil W. Lueth, Ph.D.
Personnel Profile: Full-Time Paid 1; Part-Time Paid 1; Interns 5.
Governing Authority: state. Parent Institution: New Mexico Institute of Mining & Technology. Subsidiary Institution: New Mexico Bureau of Mines & Mineral Resources, Campus Station, Socorro, NM 87801. Tax-exempt.
Institution Type/Description: Mineral Museum.
Collections: minerals & fossils.
Research Fields: mineralogy; petrology; economic geology; mining history; paleontology.
Facilities: X-ray diffraction & X-ray fluorescence equipment; electron microprobe.
Activities: guided tours; lectures; permanent, temporary & loan exhibitions.
Publications: bulletins, memoirs, circulars, maps & other documents relative to New Mexico's geology & mineral resources.
Hours & Admission Prices: Mon.-Fri. 8-5, Sat.-Sun. 10-3; other times by appointment. No charge; donations accepted. Closed state holidays. &
Attendance: 13,136 (accurate)

Springer

SANTA FE TRAIL MUSEUM, 516 Maxwell Ave., Springer, NM 87747. Tel.: 505-483-5554.
Web Site: www.santafetrail.nm.org/site58.html
Institution Type/Description: History Museum: housed in the former Colfax County Court House, built in 1881.
Collections: local history & culture; art; early pioneering life.
Hours & Admission Prices: Memorial Day-Labor Day daily 10-4. No charge; donations accepted.
Attendance: 400 (accurate)

Taos

GOVERNOR BENT MUSEUM, 117 Bent St., Taos, NM 87571. Mailing Address: P.O. Box 153, Taos, NM 87571-0153. Tel.: 505-758-2376. Fax: 505-758-2376.
E-mail: thn52@yahoo.com
Founded: 1958.
Congressional District: 3
Key Personnel: C.E.O. & Museum Shop Mgr., Tom Noeding.
Personnel Profile: Full-Time Paid 1.
Governing Authority: individual operation.
Institution Type/Description: History Museum: housed in c.1825 home belonging to Charles Bent, first American Governor of New Mexico territory, site of his death in 1847 during an uprising.
Collections: Indian artifacts; Bent family possessions; old Americana antiques;

Eskimo collection; guns; spinning wheels; paintings; photographs; anthropology; mineralogy; archaeology; graphics.
Facilities: paintings; prints; books; rocks. Indian artifacts, jewelry & pottery for sale.
Activities: gallery talks; permanent exhibitions.
Hours & Admission Prices: Daily 10-5. Adults $3, children 8-15 $1; discount to groups; children under 8 with parents no charge.
Attendance: 7,000 (estimated)

THE HARWOOD MUSEUM OF ART OF THE UNIVERSITY OF NEW MEXICO, (M), 238 Ledoux St., Taos, NM 87571-7009. Tel.: 505-758-9826. Fax: 505-758-1475.
E-mail: info@hartwoodmuseum.com
Web Site: www.harwoodmuseum.org
Formerly: The Harwood Foundation of the University of New Mexico
Founded: 1923.
Congressional District: 3
Key Personnel: Dir., Susan Longhenry; Pres., Linda Warning; Vice Pres., Gus Foster; Devel. Officer, Juniper Manley; Cur., Jina Brenneman; Cur. Education, Lucy Perera; Museum Shop Mgr., Carolyn Hinske; Finance, Sherry Carlton.
Personnel Profile: Full-Time Paid 10; Part-Time Paid 10; Part-Time Volunteers 12; Interns 1.
Governing Authority: university. Parent Institution: University of New Mexico. Tax-exempt.
Institution Type/Description: Art Museum.
Collections: Taos artists in all media from early 20th century to the present; 19th- & 20th-century Santos & Spanish colonial furniture of Hispanic culture; paintings; Agnes Martin Gallery.
Research Fields: history of Taos art colony.
Facilities: archives; auditorium.
Activities: permanent & changing exhibitions; education programs for children & adults; artist in residence & lecture programs.
Publications: biannual newsletter; museum collection handbook; selected exhibition catalogues; catalogs, Clemente in Taos; The Triumph of Beatrice Mandelman; Contemporary Art/Taos; Taos Pueblo Painters; Wayne Thiebaud: Country City; Jack Smith: Taos Portraits; Lilly Fenichel: Just You/Just Me; In Praise of What Persists, Images of Breast Cancer; Christine Taylor Patten; Stitched Marks; John Suazo 30 Year Retrospective; Sabra Moore: Out of the Woods; Dieberkorn in New Mexico.
Hours & Admission Prices: Tues.-Sat. 10-5, Sun. 12-5. Adults $8; discounts to members, groups & AAM members; children under 12 no charge. Closed major holidays. &
Attendance: 20,276 (accurate)
Membership: Student & Senior $25; Individual $35; Dual Family $55; Business $75; Contributor $125; Associate $250; Benefactor $500; Patron $1,000 & up.

KIT CARSON HOME & MUSEUM, INC., 113 Kit Carson Rd., Taos, NM 87571-5949. Tel.: 505-758-4945.
E-mail: directorkchm@yahoo.com
Founded: 1918.
Congressional District: 3
Personnel Profile: Full-Time Paid 1; Part-Time Paid 2.
Governing Authority: private; nonprofit organization. Tax-exempt: 501(c)(3).
Institution Type/Description: Historic House Museum: home of Kit & Josefa Carson, 1843-1865.
Collections: life & culture of Kit Carson & his family; structures.
Hours & Admission Prices: Nov. -Feb. daily 11-4; March-Oct. daily 11-5. Adults 20-61 $5, teens $3, military and seniors 62 & over $1; discounts to groups; children 12 & under and students with teachers no charge. Closed New Year's Day; Easter; Thanksgiving; Christmas.
Attendance: 23,500 (accurate)

∗ MILLICENT ROGERS MUSEUM, (M), 1504 Millicent Rogers Rd., Taos, NM 87571. Mailing Address: P.O. Box 1210, Taos, NM 87571-1210. Tel.: 505-758-2462. Fax: 575-758-2462.
E-mail: mrm@millicentrogers.org
Web Site: www.millicentrogers.org
Formerly: Millicent Rogers Museum of Northern New Mexico
Founded: 1956.
Congressional District: 3
Key Personnel: Exec. Dir., Peter S. Seibert; Bd. Pres., Philip Peralta-Ramos; Cur. Collections, Carmela Quinto; Museum Shop Mgr., Joy Jensen.
Personnel Profile: Full-Time Paid 7; Part-Time Paid 3; Part-Time Volunteers 26; Interns 2.
Operating Expenses: 970,000
Operating Income: 970,000
Governing Authority: private; nonprofit organization. Tax-exempt: 501(c)(3).

Institution Type/Description: Ethnology & Art Museum.
Collections: Southwestern Native American (Pueblo and Athabascan) jewelry, textiles, pottery, baskets, Kachinas, paintings, & works of art on paper; Maria Martinez family pottery collection; Southwestern Hispanic textiles, paintings, santos, tinwork, woodwork, Colcha embroidery, & domestic articles of the Colonial Period; historic & contemporary photography of the Southwest; jewelry of Millicent Rogers; contemporary jewelry design.
Major Exhibits: Fred Harvey and the American West, 8/13-1/14.
Research Fields: art; material culture & design of Northern New Mexico; southwestern jewelry.
Facilities: Museum-related items for sale.
Activities: guided tours; permanent & temporary exhibits; lectures; workshops; field trips; collections available for research or exhibit loans.
Publications: newsletter, Las Palabras; museum rack card & brochure; catalogues: Hebras De Vision/Threads of Vision; Retratos Nuevomexicanos, Collection of Hispanic New Mexican Photography; Oremos, Oremos; New Mexican Midwinter Masquerades; book, Padre Martinez: New Perspectives from Taos; catalogue, The Art of the Book in the Southwest.
Hours & Admission Prices: April-Oct. daily 10-5; Nov.-March Tues.-Sun. 10-5. Family $18, adults $10, seniors $8, students $6, New Mexican residents $5, children under 16 $2; discounts to AAM members; members no charge. Tours 10 & over $4 with 24 hr. notice. Closed New Year's Day; Easter; San Geronimo Day (Sept. 30, Taos Pueblo feast day); Thanksgiving; Christmas. &
Attendance: 22,244 (accurate)
Membership: Personal Memberships: Individual $35; Dual Family $50; Contributing $125; Sustaining $250; Directors Circle $500; Millicent Club 1,000 & up. Commercial Memberships: Regular $100; Supporting $250; Sustaining $500; Benefactor $1,000 & up.

TAOS ART MUSEUM, 227 Paseo del Pueblo Norte, Taos, NM 87571-7316. Mailing Address: P.O. Box 1848, Taos, NM 87571-1848. Tel.: 575-758-2690. Fax: 575-758-7320.
E-mail: director@taosartmuseum.org
Web Site: www.taosartmuseum.org
Formerly: Fechin Institute
Founded: 1992.
Key Personnel: Exec. Dir., Erion Simpson; Pres. (V), Jerry Geist; Museum Shop Mgr., Vickie Snyder.
Personnel Profile: Full-Time Paid 4; Part-Time Paid 3; Part-Time Volunteers 18; Interns 15.
Governing Authority: Tax-exempt.
Institution Type/Description: Art Museum: housed in the former home & studio of Nicolai I. Fechin; built 1927-1933.
Collections: early 20th century Taos art; paintings by the Taos Society of Artists, Nicolai I. Fechin & Taos Moderns.
Hours & Admission Prices: Tues.-Sun. 10-5. Adults $8; members & Taos County residents on Sun. no charge.
Attendance: 10,000 (estimated)

TAOS COUNTY HISTORICAL SOCIETY, INC., 121 N. Plaza, Taos, NM 87571-4110. Mailing Address: P.O. Box 2447, Taos, NM 87571-2447.
E-mail: tylerh@newmex.com
Web Site: www.taos-history.org
Founded: 1952.
Congressional District: 3
Key Personnel: Pres. (V), L.A. Lindquist.
Personnel Profile: Part-Time Volunteers 12.
Governing Authority: private; nonprofit organization. Tax-exempt: 501(c)(3).
Institution Type/Description: Historical Society.
Collections: audio & video tapes of lecture speakers sponsored by our society.
Research Fields: history of Taos area; identifying historic sites.
Facilities: 500-vol. library of news clippings & magazine articles; 150 sq. ft. exhibit space apart from museum.
Activities: guided tours; lectures. Annual Events: luncheon to honor historic preservation work.
Publications: semi-annual publication, Ayer y hoy en Taos; quarterly newsletter.
Hours & Admission Prices: Mon., Wed. & Fri. 1-3. No charge.
Attendance: 264 (estimated)
Membership: Individual $15; Family $20; Sustaining $30.

TAOS HISTORIC MUSEUMS, 222 Ledoux St., Taos, NM 87571-5944. Tel.: 575-758-0505. Fax: 575-758-0330.
E-mail: director@taoshistoricmuseums.org
Web Site: www.taoshistoricmuseums.org
Formerly: Kit Carson Historic Museums
Founded: 1949.

Congressional District: 3
Key Personnel: Exec. Dir., Carmen Zacarias; Collections, Anita McDaniel.
Personnel Profile: Full-Time Paid 3; Part-Time Paid 5; Part-Time Volunteers 20.
Governing Authority: nonprofit organization. Parent Institution: Kit Carson Historic Museums. Tax-exempt: 501(c)(3).
Institution Type/Description: History & Art Museum: 1797 Ernest L. Blumenschein Home; 1800-1827 Martinez Hacienda.
Collections: manuscript collections; Spanish colonial Taos; early fur trade period; archives; photographs; art by early Taos & Indian artists; antique furniture; furnishings of Ernest L. Blumenschein home. Historic Houses: 1797 Ernest L. Blumenschein home; 1804 La Hacienda de los Martinez; 1830 La Morada de Don Fernando de Taos.
Research Fields: Western history; fur trade; Spanish colonial period; Taos artists.
Facilities: 5,500-vol. library on fur trade, archaeology, military, Southwestern history available for use on premises; historical & anthropological research center. Books, slides, postcards, prints for sale.
Activities: guided tours; lectures; permanent & temporary exhibitions; museum of Spanish colonial & contemporary Hispanic folk arts; demonstrations.
Publications: booklets, Kit Carson, He Led the Way; Kit Carson, Patriot; The Real Kit Carson; book, Kit Carson's Own Story of his Life; A Brief History of Taos; The Short Truth About Kit Carson and the Indians.
Hours & Admission Prices: Blumenschein Home & Martinez Hacienda: May-Sept. daily 10-5; Winter: call for hours. Adults $8; discounts to groups & families; members no charge. Combination tickets for all sites available. Closed New Year's Day; Easter; Thanksgiving; Christmas.
Attendance: 31,957 (accurate)
Membership: Individual $35; Partners $60; Supporting $100; Sustaining $150; Benefactor $250; Sponsor $500; Patron $1,000.

Tome

TOME PARISH MUSEUM, 7 Church Loop, Tome, NM 87060-6001. Mailing Address: P.O. Box 100, Tome, NM 87060-0100. Tel.: 505-865-7497. Fax: 505-865-7622 (call first).
E-mail: mabaca@icchurchtome.org
Web Site: www.icchurchtome.org
Founded: 1966.
Congressional District: 2
Key Personnel: Dir., Fr. Carl Feil, O.S.M.; Museum Coord., Elena Calles.
Governing Authority: church group. Parent Institution: Immaculate Conception Parish. Tax-exempt.
Institution Type/Description: Religious Museum.
Collections: religion; paintings; santos statues.
Activities: permanent exhibitions.
Hours & Admission Prices: Tues.-Sat. 9-1. No charge; donations accepted. &

Truth or Consequences

GERONIMO SPRINGS MUSEUM, (M), 211 Main St., Truth or Consequences, NM 87901-2838. Tel.: 575-894-6600. Fax: 575-894-2888. Facebook: Geronimo Springs Museum.
E-mail: info@geronimospringsmuseum.com
Web Site: geronimospringsmuseum.com
Founded: 1972.
Congressional District: 2
Key Personnel: Dir. & Museum Shop Mgr., Marilyn Pope; Pres. (V), Cary "Jagger" Gustin.
Personnel Profile: Full-Time Paid 1; Part-Time Paid 2; Part-Time Volunteers 35.
Governing Authority: nonprofit organization. Parent Institution: Sierra County Historical Society. Tax-exempt: 501(c)(3).
Institution Type/Description: History Museum.
Collections: local history Indian artifacts; art; Mimbres & area pottery; rocks & minerals; Ralph Edwards Room; area fossils and mastodon skulls display; authentic log cabin moved log by log to museum; military history.
Facilities: library of magazine & newspaper items & books on local & New Mexico history. Jewelry, pottery & handcrafted items for sale.
Activities: films; art festivals; permanent collections; Armendaris Ranch tours. Museum Sponsors: Fiesta Art Exhibit; annual membership dinner; open houses for special or traveling displays or special community events.
Publications: museum guide booklet; members' quarterly newsletter.
Hours & Admission Prices: Mon.-Sat. 9-5, Sun. 12-5. Adults $6, members $5, students under 18 $3; discounts to senior citizens, AAM, ICOM, AARP, & AAA members, military, college students & organizations. Closed New Year's Day; Easter; Independence Day; Thanksgiving; Christmas. &
Attendance: 12,500 (accurate)
Membership: Annual $15; Couple $25; Business & Professions $30; Lifetime $500 & up; Lifetime Major Contributor $5,000 & up.

Tucumcari

MESALANDS COMMUNITY COLLEGE'S DINOSAUR MUSEUM AND NATIONAL SCIENCES LABORATORY, (M), 222 E. Laughlin, Tucumcari, NM 88401-2730. Mailing Address: 911 S. Tenth St., Tucumcari, NM 88401-3390. Tel.: 575-461-3466. Fax: 575-461-1901.
E-mail: gretcheng@mesalands.edu
Web Site: www.mesalands.edu
Formerly: Mesalands community College's Dinosaur Museum
Founded: 2000.
Congressional District: 3
Key Personnel: Dir., Gretchen Gurther; Pres. (V), Mike Lucero; Education & Cur., Axel Hungerbeuhler, Ph.D.; Museum Shop Mgr., Linda Morris.
Personnel Profile: Full-Time Paid 3; Part-Time Paid 3; Part-Time Volunteers 5; Interns 1.
Governing Authority: public college; nonprofit. Parent Institution: Mesaland Community College. Tax-exempt.
Institution Type/Description: Natural Science Museum.
Collections: life-sized bronze prehistoric skeletons; paintings; sculptures; geological; paleontological.
Research Fields: Mesozoic vertebrate paleontology & water research.
Facilities: 259-vol. library; laboratory; classrooms; 9,000 sq. ft. exhibit space; 360-acre dig site. Museum-related items for sale.
Activities: formal educational programs.
Publications: quarterly, Bare Bones.
Hours & Admission Prices: March to Labor Day Tues.-Sat. 10-6; Sept.-Feb. Tues.-Sat. 12-5. Adults $6.50, senior citizens $5.50, college students & educators $4.50, children $4; discounts to members, active military & groups. Closed New Year's Day; Thanksgiving; Christmas. &
Attendance: 14,106 (accurate)
Membership: Individual $7.50; Family $15; Business $30.

TUCUMCARI HISTORICAL MUSEUM, 416 S. Adams, Tucumcari, NM 88401-2718. Tel.: 505-461-4201.
E-mail: museum@cityoftucumcari.com
Web Site: www.cityoftucumcari.com
Founded: 1968.
Congressional District: 2
Key Personnel: Pres., Duane Moore; Vice Pres., Danny Wallace; Sec., Elizabeth Morris; Treas., Lois Sappington.
Personnel Profile: Part-Time Paid 2; Part-Time Volunteers 1.
Governing Authority: municipal; nonprofit. Tax-exempt: 501(c)(3).
Institution Type/Description: History & Folk Art Museum: housed in 1903 school.
Collections: agriculture; anthropology; archaeology; archives; art association with permanent Western collections; botany; costumes; folklore; geology; glass; herbarium; Indian artifacts; industrial; junior museum; mineralogy; military; music; natural history; paleontology; preservation project; transportation; restored 1926 fire truck; newspapers 1905 to present; Southwest Indian tipi; 1906 wood Dempster windmill.
Research Fields: archaeology; history; folklore; anthropology.
Facilities: 50-vol. library of books on geology, anthropology & history available on premises.
Activities: guided tours; permanent exhibitions.
Hours & Admission Prices: Tues.-Sat. 9-3. Adults $5, seniors 65 & over $4, children 6-15 $1; children 5 & under no charge. Closed holidays. &
Attendance: 3,000 (estimated)
Membership: Individual $10; Family $20.

Watrous

FORT UNION NATIONAL MONUMENT, I 25 Exit 366, 8 mi on Hwy. 161, Watrous, NM 87753. Mailing Address: P.O. Box 127, Watrous, NM 87753-0127. Tel.: 505-425-8025. Fax: 505-454-1155.
E-mail: foun_administration@nps.gov
Web Site: www.nps.gov/foun/
Founded: 1954.
Congressional District: 3
Key Personnel: Supt., Marie Sauter; Museum Shop Mgr., Jessica Gonzales.
Personnel Profile: Full-Time Paid 1.
Governing Authority: federal. Parent Institution: National Park Service. Tax-exempt: 501(c)(3).
Institution Type/Description: Historic Site & Building.
Collections: exhibits & objects pertaining to 1851-1891, the three Fort Unions & their significance in the development of the Santa Fe trail, the Civil War & the settling of the West; historic buildings; 1860s Fort Union.
Research Fields: military history.
Facilities: 1,300-vol. library on the history of the West. Publications for sale.

Activities: guided tours; self-guided trail with audio stations. Museum Sponsors: living history interpretive activities June-Sept.
Publications: brochures; handbook.
Hours & Admission Prices: Memorial Day to Labor Day daily 8-5; Sept.-May 8-4; tours by appointment. Adults 16 & over $3; senior citizens with Golden Age Passport & children under 16 no charge. Closed New Year's Day; Thanksgiving; Christmas. &
Attendance: 15,782 (accurate)

White Oaks

WHITE OAKS SCHOOLHOUSE MUSEUM, 974 White Oaks Rd., White Oaks, NM 88301-9000.
Institution Type/Description: Historic Building: built in 1895.
Collections: local history; period artifacts; photographs.
Hours & Admission Prices: Spring & Summer: Sat. 10-4, Sun. 12-4.

White Sands

WHITE SANDS MISSILE RANGE MUSEUM, 200 Headquarters Ave., White Sands, NM 88002. Tel.: 575-647-1116.
E-mail: darren.court@us.army.mi.
Web Site: www.wsmr-history.org
Institution Type/Description: History Museum.
Collections: missile range history & artifacts; missiles & rockets.
Facilities: Museum-related items for sale.
Hours & Admission Prices: Mon.-Fri. 8-4, Sat. 10-3. No charge. Closed holidays.

Zuni

A:SHIWI A:WAN MUSEUM AND HERITAGE CENTER, 02E Ojo Caliente Rd., Zuni, NM 87327. Mailing Address: P.O. Box 1009, Zuni, NM 87327-1009. Tel.: 505-782-4403. Fax: 505-782-4503.
Founded: 1992.
Congressional District: 6
Key Personnel: Chm. & Pres. (V), Norman Cooeyate; Exec. Dir., Jim Enote; Museum Technician, Curtis Quam; Financial Administrative Officer, Valerie Epaloose.
Personnel Profile: Full-Time Paid 3.
Governing Authority: nonprofit individual operation. Tax-exempt: 501(c)(3).
Institution Type/Description: Native American Museum.
Collections: Zuni photographs from 1880s to present; Zuni Tribe art & artifacts.
Research Fields: Zuni folk life.
Facilities: 625 sq. ft. exhibit space. Books & publications for sale.
Activities: arts festivals; guided tours; temporary exhibitions of local collections.
Publications: book, A Zuni Artist Looks at Frank H. Cushing: Cartoons by Phil Hughtie; Dialogues with Zuni Potters; Zuni: A Village of Silversmiths; Idonapshe; Zuni Cookbook; Journeys Home.
Hours & Admission Prices: Mon.-Fri. 9-5. No charge; donations accepted. Closed major holidays. &
Attendance: 7,000 (estimated)

NEW YORK

(827 listings)

Akron

KNIGHT-SUTTON MUSEUM - NEWSTEAD HISTORICAL SOCIETY, 123 Main St., Akron, NY 14001. Mailing Address: P.O. Box 222, Akron, NY 14001. Tel.: 716-542-7022.
Key Personnel: Dir., Marybeth Whiting
Institution Type/Description: Historical Society Museum.
Collections: local & national history; period furnishings & artifacts; photographs; personal artifacts.
Hours & Admission Prices: Call for hours.

RICH-TWINN OCTAGON HOUSE - NEWSTEAD HISTORICAL SOCIETY, 145 Main St., Akron, NY 14001. Mailing Address: P.O. Box 222, Akron, NY 14001-0222. Tel.: 716-542-7022.
Institution Type/Description: Historic House: housed in an octagon shaped house built by Charles B. Rich in the late 1840s. Listed on the National Register of Historic Places.

Collections: period furnishings; personal artifacts; paintings; photographs.
Activities: guided tours.
Hours & Admission Prices: March-Oct. 1st & 3rd Sun. of the month 1-3; Nov. 1st Sun. of the month 1-3; Dec. call for hours.

Albany

ALBANY HERITAGE AREA VISITORS CENTER, 25 Quackenbush Square, Albany, NY 12207-2311. Tel.: 518-434-0405. Fax: 518-434-0887.
E-mail: info@albany.org
Web Site: albany.org
Founded: 1986.
Congressional District: 21
Key Personnel: Dir. & Museum Shop Mgr., Kathy Quandt; C.E.O., Michele Vennard; Chm. (V), Todd Reichelt.
Personnel Profile: Full-Time Paid 2; Part-Time Paid 10; Part-Time Volunteers 5.
Governing Authority: Parent Institution: Albany County Convention and Visitors Bureau.
Institution Type/Description: History Museum.
Collections: Albany's history; photographs.
Facilities: planetarium. Gift items for sale.
Activities: lectures; demonstrations; educational programs; special events.
Hours & Admission Prices: Mon.-Fri. 9-4, Sat. 10-3, Sun. 11-3. No charge; donations accepted. &
Attendance: 20,000 (accurate)

*** ALBANY INSTITUTE OF HISTORY & ART, (M),** 125 Washington Ave., Albany, NY 12210-2296. Tel.: 518-463-4478. Fax: 518-462-1522.
E-mail: information@albanyinstitute.org
Web Site: www.albanyinstitute.org
Founded: 1791.
Congressional District: 21
Key Personnel: Dir. & C.E.O., Christine M. Miles; Bd. Chm. (V), George R. Hearst, III; Deputy Dir. Collections & Exhibitions, Tammis K. Groft; Public Rels. & Mktg. Mgr., Steve Ricci; Dir. Business & Finance, Lori Veshia; Dir. Education, Erika Sanger; Dir. Facilities, Robert Nilson; Museum Shop Mgr., Elizabeth Bechand.
Personnel Profile: Full-Time Paid 20; Part-Time Paid 5; Part-Time Volunteers 360; Interns 35.
Governing Authority: nonprofit organization. Tax-exempt: 501(c)(3).
Institution Type/Description: History & Art Museum.
Collections: fine arts, decorative arts, historical artifacts, photographs, manuscripts, documents, sketchbooks relating to the history, art & culture of Albany & the upper Hudson Valley, New York State & America from the 17th-century to present; 18th-century Hudson Valley portraits & scripture paintings; Hudson River School landscapes; regional folk art; 18th- to 20th-century portraits; paintings & documentary materials on Thomas Cole, Erastus Dow Palmer & Walter Launt Palmer; cast-iron stoves; 17th- to 19th-century Albany-made silver; New York State furniture; New York Central Railroad collections.
Research Fields: American, regional, New York art, history & culture.
Facilities: 13,000-vol. library, 85,000 photographs, maps & broadsides, all pertaining to the city of Albany & the Upper Hudson area, available for public reference; reading room; 150-seat multi-purpose auditorium. Gifts & books for sale.
Activities: temporary, thematic & traveling exhibitions; education programs for children & adults; lecture series; family programs; teachers' workshops; volunteer tour guide program; art & history education classes; public programs including lectures, films, gallery talks; special domestic tours. Museum Sponsors: Festival of Trees; annual Book & Ephemera Fair.
Publications: exhibit & collection catalogs; newsletter; calendars; books; annual report.
Hours & Admission Prices: Wed.-Sat. 10-5, Sun. 12-5. Adults $10, senior citizens & students $8, children 6-12 $6; discounts to AAM, ICOM, NARM, AAA, Empire State reciprocal & North American reciprocal members; children 5 & under and members no charge. Closed major holidays. &
Attendance: 90,000 (estimated)
Membership: Teacher & Student $25; Individual $50; Dual $70; Family $80; Supporter $125; Sponsor $250; Sustainer $500; Patron of the arts $1,000 & up. Seniors receive 10% discount off individual, dual or family rates.

ALBANY PINE BUSH DISCOVERY CENTER, 195 New Karner Rd., Albany, NY 12205-4605. Tel.: 518-456-0655.
E-mail: info@albanypinebush.org
Web Site: www.albanypinebush.org

Key Personnel: Exec. Dir., Christopher Hawver; Dir. Communications & Outreach, Wendy Craney.
Personnel Profile: Full-Time Paid 4.
Governing Authority: Parent Institution: Albany Pine Bush Commission.
Institution Type/Description: Nature Center.
Collections: hands-on exhibits; ecology; cultural history.
Facilities: Museum-related items for sale.
Activities: audio tours; educational programs; special events.
Hours & Admission Prices: Tues.-Fri. 9-4, Sat.-Sun. 10-4. No charge. Closed New Year's Day; Thanksgiving; Christmas.

DESTROYER ESCORT HISTORICAL MUSEUM, Broadway & Quay Sts., Albany, NY 12202. Mailing Address: P.O. Box 1926, Albany, NY 12201. Tel.: 518-431-1943. Fax: 518-432-1123. Facebook: USS Slater.
E-mail: info@ussslater.org
Web Site: www.ussslater.org
Founded: 2001.
Congressional District: 21
Key Personnel: Dir., Timothy C. Rizzuto; Chm. (V), Bartley Costello, III; Pres. (V), Anthony Esposito; Business Mgr., Rosehn M. Gipe.
Personnel Profile: Full-Time Paid 4; Part-Time Paid 4; Part-Time Volunteers 100.
Governing Authority: private; nonprofit organization. Tax-exempt: 501(c)(3).
Institution Type/Description: Military History Museum: housed on the destroyer escort, USS Slater.
Collections: destroyer escort history, equipment & furnishings; personal artifacts; photographs.
Facilities: 1,370-vol. library; educational facilities. Museum-related items for sale.
Activities: docent program; guided tours; school loan service.
Publications: quarterly newsletter, Trim But Deadly; monthly online newsletter, Slater Signals.
Hours & Admission Prices: April-Nov. Wed.-Sun. 10-4. Adults $8, senior citizens $7, children under 14 $6; discounts to groups; members no charge. Closed Easter; Thanksgiving.
Attendance: 15,000 (estimated)
Membership: Individual $25; Family $35; Lifetime $500.

ESTHER MASSRY GALLERY, (M), 1002 Madison Ave., Albany, NY 12208. Mailing Address: 432 Western Ave., Albany, NY 12203. Tel.: 518-485-3902 & 337-2390. Fax: 518-337-4967.
E-mail: flanagaj@strose.edu
Web Site: www.strose.edu/gallery
Formerly: The College of Saint Rose Art Gallery
Founded: 1972.
Key Personnel: Dir. & Cur., Jeanne Flanagan.
Personnel Profile: Full-Time Paid 1.
Governing Authority: Parent Institution: The College of Saint Rose. Tax-exempt.
Institution Type/Description: College Art Gallery.
Collections: works by regional, national & international artists.
Facilities: recital hall; classrooms; 400-seat theatre.
Activities: musical performances; art gallery receptions; lectures.
Hours & Admission Prices: Sept. 21-May 8 Mon.-Thurs. 10-8, Fri. 10-4:30, Sun. 12-4. No charge; donations accepted. Closed school holidays; winter & spring break. &
Attendance: 10,000 (accurate)

* **HISTORIC CHERRY HILL,** 523 1/2 S. Pearl St., Albany, NY 12202-1111. Tel.: 518-434-4791. Fax: 518-434-4806.
E-mail: info@historiccherryhill.org
Web Site: www.historiccherryhill.org
Founded: 1964.
Congressional District: 21
Key Personnel: C.E.O. & Dir., Liselle LaFrance; Pres., Michael R. Beiter; Dir. Education, Rebecca Watrous; Cur., Deborah Emmons-Andarawis; Business Mgr., Lauren Mastin; Program Asst. & Facilities Support Asst., Aine Leader-Nagy; Manuscript Specialist & Communications Coord., Mary Doehla.
Personnel Profile: Full-Time Paid 3; Part-Time Paid 3; Part-Time Volunteers 50; Interns 2.
Governing Authority: nonprofit organization. Tax-exempt.
Institution Type/Description: Historic House: 1787 Cherry Hill.
Collections: furnishings; textiles; ceramics; paintings; silver, personal papers & belongings of the Van Rensselaer-Rankin family from the 18th-20th centuries. Historic Building: 1787 Georgian Style house.

Research Fields: Van Rensselaer family history; Hudson Valley decorative arts; New York State; branded furniture; Colonial Revivalism.
Facilities: over 5,000-vol. library of books available for research by appointment on premises only. Books on Cherry Hill & other museum-related items for sale.
Activities: guided tours; volunteer program; lectures; workshops; special events; student intern program; NYS curriculum-related education programs.
Publications: books, Cherry Hill, The History & Collections of a Van Rensselaer Family; Murder at Cherry Hill; Different Voices, Different Truths: The 1827 Murder at Cherry Hill; Creating a Dignified Past: Museums and the Colonial Revival; Kittie Putman and the Cherry Hill Household: 1860-1884; bibliography, Selections from a Van Rensselaer Family Library, 1536-1799; On the Score of Hospitality: selected receipts of a Van Rensselaer Family, 1785-1835; The Confession of Jesse Strang.
Hours & Admission Prices: Tours: Wed. at 1, 2, & 3, Sat. at 2 & 3.
Attendance: 10,400 (accurate)
Membership: Basic $35; Family/Household $50; Copper $100; Brass $250; Silver $500; Gold $1,000.

IRISH AMERICAN HERITAGE MUSEUM, 370 Broadway, Albany, NY 12207. Tel.: 518-427-1916. Fax: 518-427-1915.
E-mail: irishamermuseum@cs.com
Web Site: irishamericanheritagemuseum.org
Founded: 1986.
Congressional District: 20
Key Personnel: C.E.O. & Chm. (V), Edward Collins; Financial Dir., David Stack.
Personnel Profile: Full-Time Paid 2; Full-Time Volunteers 1; Part-Time Volunteers 14.
Governing Authority: bd. of trustees; not-for-profit organization. Branch Museum: 2267 Rt. 145, East Durham, NY 12423. Tax-exempt: 501(c)(3).
Institution Type/Description: Heritage Museum.
Collections: objects representing the history of the Irish in America; archives.
Research Fields: Irish immigration experience; the history of the 69th NY Regiment; great hunger; Irish American women; Irish in religion; Irish in Labor; Ancient Order of Hibernians; corporate Irish.
Facilities: 2,000-vol. library of books on the history of Ireland & the Irish in America. Museum-related items for sale.
Activities: concerts; lectures; loan & traveling exhibitions; storytelling.
Publications: Irish American Heritage Museum Newsletter.
Hours & Admission Prices: Museum: Wed.-Fri. 11-4, Sat. 10-4, Sun. 12-4. Library: Mon.-Fri. 8:30-4; groups by appointment. No charge; donations accepted. East Durham Museum: Memorial Day to Labor Day Wed.-Sun. 12-4; groups by appointment. Suggested Donations: adults $3, seniors $2; children 14 & under no charge. Empire State Museum Reciprocal Program; North American Reciprocal Museum Program. &
Attendance: 25,000 (estimated)
Membership: Individual $25; Family $45; Sustaining $100; Patron $500; Corporate $1,000.

THE LITTLE GALLERY, SAGE COLLEGE OF ALBANY, 140 New Scotland Ave., Rathbone Hall, Albany, NY 12208-3491. Tel.: 518-292-8625.
Key Personnel: Faculty Advisor, Janice Medina
Institution Type/Description: Art Gallery.
Collections: works by faculty & student artists.
Hours & Admission Prices: Sun.-Fri. 1-5.

THE MUSEUM OF PRINTS AND PRINTMAKING, Albany, NY 12206-0578. Mailing Address: P.O. Box 6578, Albany, NY 12206-0578. Tel.: 518-449-4756. Fax: 518-506-6864.
E-mail: semowich@gmail.com
Web Site: pcaprint.org
Founded: 1990.
Congressional District: 23
Key Personnel: Pres. (V), Thomas Andress; Sec., Pam Williams; Cur., Charles Semowich.
Personnel Profile: Full-Time Volunteers 5; Part-Time Volunteers 20.
Governing Authority: nonprofit. Chartered by New York State Board of Regents, Print Club of Albany. Tax-exempt.
Institution Type/Description: Print Museum.
Collections: prints from the collections of the Print Club of Albany; archives of print makers including Dorothy Lathrop & Alice P. Schaefer.
Research Fields: prints; print making; print makers.
Facilities: 400-vol. library of art & print making books available to the public; 300 sq. ft. exhibit space. Prints & other museum-related items for sale.
Activities: guided tours; lectures; traveling, temporary & loan exhibitions; participatory exhibits in association with the Print Club of Albany.

Publications: 5 times annually, shares newsletter with the Print Club of Albany.
Hours & Admission Prices: By appointment only. No charge. &
Membership: Print Club of Albany.

NEW YORK STATE - EMPIRE STATE PLAZA ART COLLECTION, 2978 Corning Tower, 1 Empire State Plaza, Albany, NY 12242. Tel.: 518-473-7521. Fax: 518-474-0984.
E-mail: curatorial.services@ogs.state.ny.us
Institution Type/Description: Art Collection.
Collections: paintings; sculptures; tapestries.
Hours & Admission Prices: Concourse & Plaza: Mon.-Fri. 89:30-5. Corning Tower: Mon.-Fri. 7-6. By appointment. No charge.

NEW YORK STATE MUSEUM, (M), Rm. 3023, Cultural Education Center, Empire State Plaza, Albany, NY 12230. Tel.: 518-474-5877 & 5812. Fax: 518-486-3696.
E-mail: cryan@mail.nysed.gov
Web Site: www.nysm.nysed.gov
Founded: 1858.
Congressional District: 29
Key Personnel: Dir., Clifford A. Siegfried; Chief Geological Survey, Taury Smith; Head Education, Jeanine Grinage; Dir. Research & Collections, John Hart; Dir. Communications, Joanne Guilmette; Dir. Exhibits, Mark Schaming.
Personnel Profile: Full-Time Paid 152; Full-Time Volunteers 3; Part-Time Paid 29; Part-Time Volunteers 346; Interns 27.
Governing Authority: state. Parent Institution: New York State Education Dept., Albany, NY 12224. Tax-exempt: 170(b)(1)(A).
Institution Type/Description: General Museum.
Collections: entomology; geology; mineralogy; archaeology; New York Indian ethnology; zoology; New York historical artifacts including architecture; African American history; costume; decorative art; fire engines; militaria; musical instruments; Shakeriana; toys; transportation; agricultural, industrial & domestic technology; gas and oil; art works by New York artists; botany; tissue samples; World Trade Center.
Research Fields: entomology; geology; paleontology; mineralogy; zoology; ecology; botany; mycology; Indians of New York; history of New York, archaeology; gas and oil resources.
Facilities: 9,000-vol. library; field research station; 425,112 sq. ft. of exhibition space; lab. Museum-related items for sale.
Activities: lectures; films; formal & informal education programs; school loan service; permanent & temporary exhibits; musical performances.
Publications: various annual bulletins; map & chart series; handbooks; memoirs; circulars; educational leaflet series; exhibit catalogs; special publications.
Hours & Admission Prices: Daily 9:30-5. No charge; donations accepted. Closed New Year's Day; Thanksgiving; Christmas. &
Attendance: 719,205 (accurate)
Membership: Senior Citizen $30; Individual $35; Contributing $40; Family $50; Supporting $60; Smithsonian Affiliate $75; Sustaining $100; Discovery Club $150; Curator's Circle $250.

OPALKA GALLERY, SAGE COLLEGE OF ALBANY, (M), 140 New Scotland Ave., Albany, NY 12208-3491. Tel.: 518-292-7742. Fax: 518-292-1903. Facebook: Opalka Gallery.
E-mail: opalka@sage.edu
Web Site: www.sage.edu/opalka
Founded: 2002.
Key Personnel: Dir., Elizabeth Greenberg; Asst. to the Dir., Jacqueline Smith.
Personnel Profile: Full-Time Paid 2; Part-Time Paid 1; Part-Time Volunteers 10; Interns 1.
Governing Authority: Parent Institution: The Sage Colleges. Tax-exempt.
Institution Type/Description: Art Gallery.
Collections: paintings; photographs; sculpture; installations.
Major Exhibits: 2014 Faculty Show, 1/14-3/14; High School Regional, 3/14-4/14; BFA Exhibition, 5/14-7/14; Regis Brodie, 8/14-10/14.
Facilities: 75-seat lecture hall.
Activities: poetry readings; recitals; symposia.
Publications: publications for certain exhibitions.
Hours & Admission Prices: Winter: Mon.-Fri. 10-8, Sun. 12-4; Summer: Mon.-Fri. 10-4; other times by appointment. No charge. &

SCHUYLER MANSION STATE HISTORIC SITE, 32 Catherine St., Albany, NY 12202-1605. Tel.: 518-434-0834. Fax: 518-434-3821.
E-mail: heidi.hill@parks.ny.gov
Web Site: www.nysparks.com/hist

Founded: 1911.
Congressional District: 28
Key Personnel: Historic Site Mgr., Heidi Hill; Educator, Michelle Mavigliano.
Personnel Profile: Full-Time Paid 5; Part-Time Paid 4; Part-Time Volunteers 20; Interns 1.
Governing Authority: state. Parent Institution: New York State Office of Parks, Recreation & Historic Preservation, Saratoga & Capital District Region. Tax-exempt.
Institution Type/Description: Historic Building & Site: 1761 home of American Revolution Gen. Philip Schuyler and his family. Listed on the National Register of Historic Places.
Collections: 18th-century furnishings & decorative arts; Schuyler family possessions; books; glass; silver.
Research Fields: Gen. Philip Schuyler; Schuyler family; 18th-century Dutch in New York; colonial Albany; American Revolution in New York State; post Revolution.
Facilities: visitor center.
Activities: guided tours; lectures; gallery talks; school program; special programs & demonstrations; reenactments; museum theater productions. Museum Sponsors: Independence Day Celebration; Twelfth Night.
Publications: Schuyler Mansion: Historic Structure Report; Schuyler Genealogy: A Compendium of Sources Pertaining to the Schuyler Families in America prior to 1900.
Hours & Admission Prices: mid-May to Oct. Wed.-Sun. 11-5; Nov. to mid-May by appointment; groups by appointment. Adults $4, senior & students $3; children under 12 & members no charge. Closed New Year's Day; Thanksgiving; Christmas. &
Attendance: 10,000 (estimated)

SHAKER HERITAGE SOCIETY, Heritage Lane, Albany, NY 12211-1051. Mailing Address: 25 Meeting House Rd., Albany, NY 12211. Tel.: 518-456-7890. Fax: 518-452-7348.
E-mail: shakerdirector@gmail.com
Web Site: www.shakerheritage.org
Founded: 1977.
Congressional District: 21
Key Personnel: Exec. Dir., Starlyn D'Angelo; Pres., Stephen Iachetta; Coord. Education, Samantha Saladino; Museum Shop Mgr., Patricia Williams.
Personnel Profile: Full-Time Paid 3; Part-Time Paid 1; Part-Time Volunteers 50; Interns 3.
Governing Authority: nonprofit. Tax-exempt: 501(c)(3).
Institution Type/Description: Historical Society Museum: housed in c.1848 meeting house.
Collections: Shaker artifacts & history 1776-1938; cemetery. Historic Buildings: 1830 Trustees House; 1822 Brethrens Shop & Workshop; 1916 barn; 1856 Herb Storage & Milk House; 1820 Ministry; 1920 garage; 1848 meeting house.
Major Exhibits: Quiet Revolutionaries: Shakers in America, 11/14.
Research Fields: Shaker history; oral history; religious history.
Facilities: Shaker herb garden. Books, Shaker items & other related items for sale.
Activities: guided tours; K-12 education programs; outreach programs; programs for nursing home residents; heritage breed farm animal & herb garden program. Museum Sponsors: Outdoor craft fairs in July & September; Shaker Christmas shop in November & December.
Publications: quarterly newsletter, Watervliet Shaker Journal.
Hours & Admission Prices: Feb.-Oct. Tues.-Sat. 9:30-4; Nov. to mid-Dec. Mon.-Sat. 10-4. No charge; donations accepted. Closed major holidays. &
Attendance: 18,000 (estimated)
Membership: Individual $35; Household $50; Supporting $100; Sustaining $250 & up; Benefactor $500 & up.

TEN BROECK MANSION - ALBANY COUNTY HISTORICAL ASSOCIATION, 9 Ten Broeck Place, Albany, NY 12210-2524. Tel.: 518-436-9826. Fax: 518-436-1489.
E-mail: achadirector@onecommail.com
Web Site: www.tenbroeckmansion.org
Founded: 1942.
Congressional District: 23
Key Personnel: Exec. Dir., Wendy Burch; Pres., Matthew Kirk; First Vice Pres., Keith Bennett; Second Vice Pres., Lois Conklin; Treas., Thomas A. Devane, Jr.; Sec., Angela Markessinis.
Personnel Profile: Part-Time Paid 2; Part-Time Volunteers 20.
Governing Authority: Parent Institution: University of the State of New York Board of Regents. Tax-exempt: 501(c)(3).
Institution Type/Description: Historic House Museum: 1798 Arbor Hill, Federal style mansion, home of General Abraham Ten Broeck & later the Olcott family.
Collections: Federal & Empire furniture; herb garden; decorative arts; 19th-century wine cellar; Victorian gardens.

Research Fields: local & regional history; industrial; social; military.
Facilities: meeting rooms; gardens.
Activities: guided tours; lectures; permanent & temporary exhibitions; educational programs; concerts; bus trips; auction; wine tastings; rental space available. Annual Events: House & Garden Tour; Haunted Mansion; Holiday House.
Publications: quarterly newsletter, Prospect.
Hours & Admission Prices: May-Dec. Thurs.-Fri. 10-4, Sat.-Sun. 1-4. Adults $5, students & seniors $4, children $3; children under 5 & members no charge. Closed all major holidays.
Attendance: 5,925 (accurate)
Membership: Student & Senior Citizen $20; Individual $35; Family $50; Supporting $75; Patron $100; Sustaining $200; Corporate $500.

UNIVERSITY ART MUSEUM, UNIVERSITY AT ALBANY, (M), State University of New York, 1400 Washington Ave., Albany, NY 12222. Tel.: 518-442-4035. Fax: 518-442-5075.
E-mail: museum@albany.edu
Web Site: www.albany.edu/museum
Founded: 1967.
Congressional District: 23
Key Personnel: Dir., Janet Riker; Assoc. Dir., Corinna Ripps Schaming; Exhibit Designer, Zheng Hu; Preparator, Jeffrey Wright-Sedam; Exhibition & Outreach Coord., Naomi Lewis; Admin. Asst., Joanne Lue.
Personnel Profile: Full-Time Paid 3; Part-Time Paid 5.
Governing Authority: state. Parent Institution: University at Albany. Affiliated with State University of New York. Tax-exempt.
Institution Type/Description: Contemporary Art.
Collections: over 3,000 late modern & contemporary works including paintings, photographs & drawings.
Major Exhibits: Lamar Peterson, 2/4/14-4/5/14; American Dream(er): Selections from the Univ. at Albany Collection, 2/4/14-4/5/14; Mary Reid Kelley: Working Objects and Videos (T), 6/14-9/14.
Research Fields: Western & non-Western contemporary art.
Activities: temporary exhibitions, lectures, gallery talks, films, inter-museum loans.
Publications: exhibition catalogs.
Hours & Admission Prices: Summer: Tues.-Sat. 11-4; Sept.-June Tues. 10-8, Wed.-Fri. 10-5, Sat. 12-4. No charge. Closed major holidays & during installations. &

Attendance: 37,500 (estimated)

Albertson

CLARK BOTANIC GARDEN, 193 I.U. Willets Rd., Albertson, NY 11507-2298. Tel.: 516-484-8600 & 2208. Fax: 516-625-3718.
E-mail: darcyj@northhempstead.com
Web Site: www.clarkbotanic.org
Founded: 1966.
Congressional District: 6
Key Personnel: Bd. Chm. (V), Jerilyn Dreitlein; Commissioner, Gerard Olsen; Office Mgr., Carol Murphy.
Personnel Profile: Full-Time Paid 5; Part-Time Paid 3; Part-Time Volunteers 10; Interns 3.
Governing Authority: nonprofit. Parent Institution: Fanny Dwight Clark Memorial Garden Inc. Tax-exempt.
Institution Type/Description: Botanical Garden.
Collections: roses, wildflowers, ferns, herbs, annuals, perennials, rock garden, flowering shrubs & dwarf trees; ponds; honey bee hives; maze & children's garden; rhododendrons, azaleas, magnolias; spring bulbs.
Facilities: botanical garden; library; classrooms. Museum-related items for sale.
Activities: guided tours; lectures; concerts; formally organized education programs; special events.
Publications: biannual, Clark Garden Newsletter; Horticultural Bulletin.
Hours & Admission Prices: Summer: daily 10-4; Winter: Mon.-Fri. 10-4. No charge; donations accepted. &
Attendance: 20,000 (estimated)
Membership: Individual $20; Family $35; Group & Associate $50; Sponsor $100; Benefactor $250; Patron $500; Donor $1,000.

Alden

ALDEN HISTORICAL SOCIETY, INC., 13213 Broadway, Alden, NY 14004-1312. Mailing Address: 1594 Westcott Ave., Alden, NY 14004-1122. Tel.: 716-937-7606.
E-mail: aldenhist@gmail.com
Web Site: www.alden.erie.gov
Founded: 1965.
Congressional District: 38

Key Personnel: Pres., Roberta Vincent; Vice Pres., Sal Sardella; Cur., Karen Muchow; Treas., Ralph Davis, Jr.; Asst. Cur., Laura Airey; Sec., Janet Koelbl.
Governing Authority: nonprofit organization.
Institution Type/Description: History Museum: housed in 1859 Alden Historical Society House.
Collections: farm tools; kitchen utensils; town memorabilia; Early American bedroom; Victorian living room.
Research Fields: history of town of Alden & surrounding area.
Facilities: 125-vol. library of old books available on premises by appointment with curator.
Activities: guided tours; lectures; permanent & temporary exhibitions.
Publications: booklet, A History of Town of Alden: 125 Years & Growing; The Village of Alden; video, A Complete History of Alden, New York.
Hours & Admission Prices: Sat. 10-2; other times by appointment. No charge; donations accepted. &
Attendance: 500 (estimated)
Membership: Individual $5; Lifetime $50.

Alfred

THE SCHEIN-JOSEPH INTERNATIONAL MUSEUM OF CERAMIC ART, (M), Binns-Merrill Hall, Top Fl., Alfred University Campus, Alfred, NY 14802. Mailing Address: 2 Pine St., Alfred, NY 14802-1214. Tel.: 607-871-2421. Fax: 607-871-2615.
E-mail: ceramicsmuseum@alfred.edu
Web Site: ceramicsmuseum.alfred.edu
Founded: 1900.
Key Personnel: Collections Mgr., Susan Kowalczyk; Museum Asst., Greta Holzheimer.
Personnel Profile: Full-Time Paid 1; Part-Time Paid 1.
Governing Authority: university. Parent Institution: New York State College of Ceramics at Alfred University. Tax-exempt: 501(c)(3).
Institution Type/Description: Ceramic Museum.
Collections: Korean ceramics; Charles Fergus Binns ceramics; pottery of the Ancient Americas; Corsaw ceramics; Cybis collection; ceramic of Alfred-trained ceramists; early pottery; European porcelain; American studio ceramics from 1900-present; ceramics from around the world; technical ceramics; advanced ceramics.
Research Fields: ceramic history
Facilities: 1,500 sq. ft. temporary location in Binns-Merrill Hall on the Alfred University campus.
Activities: lectures; organized education programs for undergraduate or graduate college students; permanent & loan exhibitions.
Publications: biannual newsletter, Ceramophile; exhibition catalogues.
Hours & Admission Prices: Wed.-Fri. 10-4. Call for exhibition information. No charge; donations accepted. Closed school holidays. &
Attendance: 1,766 (accurate)
Membership: Student & Senior Citizen $15; Individual $25; Family $35; Sustaining $50-$99; Contributing & Business $100-$249; Patron $250-$499; Benefactor $500 & up.

Almond

ALMOND HISTORICAL SOCIETY/HAGADORN HOUSE MUSEUM, 7 N. Main St., Almond, NY 14804. Mailing Address: 1 Park St., Box 234, Almond, NY 14804-0234. Tel.: 607-276-6781.
E-mail: almondhistoricalsociety@gmail.com
Web Site: www.rootsweb.com/~nyahs/almondhs.html
Founded: 1965.
Congressional District: 34
Key Personnel: Pres., Lee A. Ryan; Vice Pres., Helen Spencer; Sec., Donna Ryan; Treas., Teresa Johnson.
Personnel Profile: Part-Time Volunteers 25.
Governing Authority: society; nonprofit organization. Tax-exempt: 501(c)(3).
Institution Type/Description: Historical Society Museum: housed in 1830 Hagadorn House.
Collections: genealogical files of 1500 local families beginning in 1794 with the Esterbrooks; 1800-1991 local artifacts; photographs; costumes; diaries; scrap books; amateur oil paintings; 1830-1940 furniture, glass, pottery & china; locally made tools; 1837 Jesse/Angel; 1800 cooking fireplace; 1837 6 room & stairway addition; 1794 Count Rumford Plan.
Research Fields: monograph & maps; forgotten cemeteries.
Facilities: 700-vol. library of books, magazines, 1840's records & ledgers available for research by appointment only on premises; archives; reading room. Postcards, note paper, monographs & area maps for sale.
Activities: guided tours; lectures; films; permanent exhibitions. Museum Sponsors: Strawberry Festivals; Christmas Open House music program; Trash and Treasures Sale.
Publications: quarterly newsletter, Almond Historical Society; books, School

Days Recollections of Horace Stillman 1854-1951; Forgotten Cemeteries of Almond 1975; 175th Anniversary Souvenir Program; The Cooking Fireplace; My Father's Old Fashioned Drugstore; When Grandfather Ran a General Store, 1937 reprint by permission of Christian Science Monitor; 1991 Monograph of Autographs 1885-1932.
Hours & Admission Prices: Fri. 2-4; other times by appointment. No charge. ౬
Attendance: 1,500 (estimated)
Membership: Single $10; Couple $15; Family $20; Business $30; Single Life $250; Couple Life $500.

Amagansett

EAST HAMPTON TOWN MARINE MUSEUM, 301 Bluff Rd., Amagansett, NY 11930. Mailing Address: 101 Main St., East Hampton, NY 11937-2714. Tel.: 631-324-6850. Fax: 631-324-9885.
E-mail: info@easthamptonhistory.org
Web Site: www.easthamptonhistory.org
Founded: 1966.
Congressional District: 2
Key Personnel: Pres. (V), Arthur Graham; Exec. Dir., Richard Barons.
Personnel Profile: Full-Time Paid 2; Part-Time Paid 3; Part-Time Volunteers 18.
Governing Authority: Parent Institution: East Hampton Historical Society, Main St., East Hampton, NY. Associated Museums: Clinton Academy; Mulford Farm; Jackson-Osborne House. Tax-exempt: 170(b)(1)(A); 501(c)(3).
Institution Type/Description: Marine Museum.
Collections: marine; history & technology of eastern Long Island, NY.; whaling & commercial fishing; small water craft; folklore; archaeology; photographs. Historic Building: World War II Navy Barracks.
Research Fields: folklife & technology of eastern Long Island fishermen.
Activities: guided tours; permanent exhibitions; classes in wooden boat building; classes in nautical subject matter & skills.
Hours & Admission Prices: Memorial Day to Columbus Day Sat.-Sun. 10-5. Adults $4, senior citizens $3, students $1; discounts to AAM & AAA members; members & preschoolers no charge.
Attendance: 8,000 (estimated)
Membership: Individual $35; Family & Couple $60; Business $100.

Amenia

WETHERSFIELD ESTATE AND GARDENS, (M), 214 Pugsley Hill Rd., Amenia, NY 12501-5032. Tel.: 845-373-8037.
Web Site: www.wethersfieldgarden.org
Key Personnel: Mgr., Kevin Malloy
Institution Type/Description: Historic House Museum.
Collections: sculptures; carriages; gardens.
Facilities: gardens.
Hours & Admission Prices: June-Sept. Wed. & Fri.-Sat. 12-5 by appointment. Call for admission prices. ౬
Attendance: 3,000

Amherst

BUFFALO NIAGARA HERITAGE VILLAGE, 3755 Tonawanda Creek Rd., Amherst, NY 14228-1599. Tel.: 716-689-1440. Fax: 716-689-1409. Facebook: Buffalo Niagara Heritage Village.
E-mail: info@bnhv.org
Web Site: www.bnhv.org
Formerly: Amherst Museum
Founded: 1970.
Congressional District: 27
Key Personnel: Exec. Dir., Joseph G. Weickart; Pres., Karen Vilonen; Sr. Cur., Jessica A. Johnson.
Personnel Profile: Full-Time Paid 9; Part-Time Volunteers 300.
Governing Authority: nonprofit; chartered by NYS Board of Regents; joint governing with Town of Amherst. Parent Institution: Town of Amherst. Tax-exempt: 501(c)(3).
Institution Type/Description: Historical Museum: includes 10 historic houses.
Collections: 19th-& 20th century material culture; technology; American art; photographs; archives; research material relating to the history of the Niagara Frontier & Town of Amherst, New York. Shaw Administration Building: local history; costumes; Erie Canal; Gallery of the Senses for visually impaired; emphasis on Amherst within the Niagara Frontier region of Western NY.
Major Exhibits: Body Beautiful, 1/14-12/14; Suiting Up: American Uniforms, 1/14-12/14.
Research Fields: 19th-20th century material & technological culture of Town of Amherst & Niagara Frontier.

Facilities: 5,500-vol. research library; 35-acre site; period herb gardens; wildflower gardens.
Activities: organized school & group tour programs; permanent & temporary exhibitions; living history reenactments; guided tours; local history lectures; summer camps. Special Events: Scottish Festival & Highland Games; Halloween Celebration; German Festival; Holiday Celebration & Candlelight Tours; annual Quilt & Lace seminars; Biennial International Quilt Show.
Publications: quarterly newsletter, Village Voice.
Hours & Admission Prices: April-Sept. Tues.-Fri. 9:30-4:30, Sat. 10:30-4:30, Sun. 12:30-4:30; Oct.-March Tues.-Fri. 9:30-4:30. Adults $8, seniors, active military & students $6, children $4; discounts to museum members; members no charge. Closed municipal holiday weekends. ౬
Attendance: 40,000 (estimated)
Membership: Individual $20; Family $35; Patron $50; Supporter $75; Contributor $100; Business I $250; Business II $500.

MUSEUM OF DISABILITY HISTORY, 3826 Main St., Amherst, NY 14226. Tel.: 716-629-3606. Fax: 716-629-3624.
Web Site: www.museumofdisability.org
Founded: 2003.
Congressional District: 26
Key Personnel: Dir., Theresa Fraser; Educator, Elizabeth Marotta; Cur., Douglas A. Platt; Research Specialist, Reid Dunlavey; Museum Shop Mgr., Lacy Abbott.
Personnel Profile: Full-Time Paid 5; Full-Time Volunteers 1; Part-Time Paid 3; Interns 1.
Governing Authority: private; nonprofit organization. Parent Institution: People, Inc., Williamsville, NY. Tax-exempt: 501(c)(3).
Institution Type/Description: Disability History Museum.
Collections: medical & social disability history, care & treatment; photographs; books; film.
Research Fields: history of developmental disabilities; western New York institutions; self-advocacy movement.
Facilities: 500-vol. reference library; 800 sq. ft. exhibit space.
Activities: guided tours; films; formal educational programs for adults & children; lectures; temporary & traveling exhibitions. Museum Sponsors: Kids on the Block (R) Disability Awareness Puppet Troup; Disability Awareness Scout Merit Badge Program. Annual Event: Disability Film Festival.
Hours & Admission Prices: Mon.-Fri. 10-4; other times by appointment. No charge; donations accepted. Closed municipal holidays. ౬
Attendance: 400

Amityville

LAUDER MUSEUM - AMITYVILLE HISTORICAL SOCIETY, 170 Broadway, Amityville, NY 11701-2704. Mailing Address: P.O. Box 764, Amityville, NY 11701-0764. Tel.: 631-598-1486. Fax: 631-598-1486.
Web Site: www.amityvillehistoricalsociety.com
Founded: 1973.
Congressional District: 2
Key Personnel: C.E.O., William T. Lauder; Pres., Patricia Cahaney; Treas., Martha Peterson; Cur., Seth Purdy; Devel., Caroline D'Antonio.
Personnel Profile: Part-Time Paid 1; Part-Time Volunteers 50.
Governing Authority: private; nonprofit organization. Tax exempt.
Institution Type/Description: Historical Museum: 1909 brick edifice formally used for a bank.
Collections: history of Amityville, NY.
Facilities: 1,000-vol. library. Museum-related items for sale.
Activities: docent program; guided tours; lectures; temporary exhibitions. Annual Event: Heritage Fair in June.
Publications: quarterly newsletter, The Dispatch.
Hours & Admission Prices: Tues. & Fri.-Sun. 2-4. No charge; donations accepted. Closed New Year's Day; Easter; Independence Day; Christmas. ౬
Attendance: 3,500 (estimated)
Membership: Senior $15; Adult $20; Family $25; Business & Patron $50; Friend $75; Fellow $100; Lifetime & Benefactor $1,000.

Amsterdam

WALTER ELWOOD MUSEUM, 100 Church St., Amsterdam, NY 12010-4243. Tel.: 518-843-5151. Fax: 518-843-6098. Facebook: Walter Elwood Museum.
E-mail: info@walterelwoodmuseum.org
Web Site: www.walterelwoodmuseum.org
Founded: 1933.
Congressional District: 28

Key Personnel: Exec. Dir., Ann Peconie; Pres., Susan Wollman, Esq.
Personnel Profile: Part Time Paid 2; Part-Time Volunteers 30; Interns 1.
Volunteer Hours: 500
Operating Income: 170,000
Governing Authority: bd of trustees. Tax-exempt: 501(c)(3).
Institution Type/Description: General Museum.
Collections: early American & Indian material; history; natural history; ethnology; costumes; prints & paintings; local & area historical displays; artifacts; antiques; Civil War artifacts; early lighting; fans; dolls from around the world; Victorian era.
Research Fields: local history.
Facilities: 1,500-vol. research library of books pertaining to 19th-century American children, school & literature available for research by arrangement with curator & for inter-library loan; reading room; galleries; classrooms; children's museum; junior museum. Children's books, minerals & other museum-related & gift items for sale.
Activities: guided tours; lectures; formally organized education programs for children; inter-museum loan, permanent, temporary & traveling exhibitions; adult oil-painting classes; films; gallery talks; school loan service.
Publications: brochures; membership newsletters.
Hours & Admission Prices: Mon.-Fri. 9-4; groups by appointment. Adults $3, seniors $2; discounts to AAM & ICOM members; members and children 12 & under no charge. Closed most major & school holidays ὕ
Attendance: 5,000 (accurate)
Membership: Senior Citizen & Student $15; Individual $20; Family $30; Professional $50.

Angola

EVANS HISTORICAL SOCIETY AND 1857 SCHOOL HOUSE MUSEUM, 8787 Erie Rd., Angola, NY 14006-9620. Tel.: 716-549-6139.
Institution Type/Description: Historical Society Museum: housed in a two-story brick school; built in 1857.
Collections: local history & culture; period artifacts; photographs; business & industry; toys; household artifacts; period furnishings; tools; farm equipment; early transportation; firefighting uniforms & equipment; military uniforms.
Hours & Admission Prices: July-Aug. Sun. 2-4; Sept.-June 1st Sun. of month 2-4.

Annandale-on-Hudson

CENTER FOR CURATORIAL STUDIES, BARD COLLEGE AND THE HESSEL MUSEUM OF ART, Bard College, 33 Garden Rd., Annandale-on-Hudson, NY 12504. Mailing Address: Bard College, P.O. Box 5000, Annandale-on-Hudson, NY 12504-5000. Tel.: 845-758-7598. Fax: 845-758-2442.
E-mail: ccs@bard.edu
Web Site: www.bard.edu/ccs
Founded: 1990.
Congressional District: 26
Key Personnel: Exec. Dir., Tom Eccles; Dir. Graduate Program, Paul O'Neill; Dir. Library & Archives, Ann Butler; Asst. Dir. Administration & Devel., Jaime Baird; Asst. Dir. Museum, Marcia Acita; Mgr. External Affairs, Ramona Rosenberg; Administrative & Devel. Mgr., Tracy Pollock; Registrar, Rachel von Wettberg; Preparator, Mark DeLura; Mgr. Security, George Acker; Graduate Program Coord., Sarah Higgins; Administrative Asst., Maxwell Paparella.
Personnel Profile: Full-Time Paid 11.
Governing Authority: private college; nonprofit. Parent Institution: Bard College. Tax-exempt: 501(c)(3).
Institution Type/Description: Art Museum.
Collections: international contemporary visual art focusing on the period from the mid-1960s to present including paintings, drawings, sculpture, photographs, video, installation art, performance art & artists' books.
Research Fields: contemporary visual arts; international exhibition practice & history; curatorial pedagogy & museum education.
Facilities: 21,000-vol. library of books & catalogs on contemporary art; 26,300 sq. ft. exhibit space.
Activities: two-year MA degree program in curatorial studies; lectures; temporary exhibitions.
Publications: exhibition catalogues.
Hours & Admission Prices: Center: call for hours. Museum: Thurs.-Sun. 11-6. Administration: Mon.-Fri. 9-5. Library: Sept.-June: Mon.-Wed. 10-7, Thurs.-Fri. 10-5, Sat. 1-5, Sun. 1-7. Summer: Mon.-Fri. 1-5. No charge. Closed New Year's Day; Independence Day; Thanksgiving; Christmas. ὕ
Attendance: 12,000 (accurate)

Arcade

ARCADE HISTORICAL SOCIETY, 331 W. Main St., Arcade, NY 14009-1110. Mailing Address: P.O. Box 236, Arcade, NY 14009-0236. Tel.: 585-492-4466. Facebook: Arcade Historical Society.
E-mail: office@arcadehistoricalsociety.org
Web Site: www.arcadehistoricalsociety.org
Founded: 1956.
Congressional District: 29
Key Personnel: Pres. (V), Lynn Lester; Operations Mgr., Susan Andrews.
Personnel Profile: Part-Time Paid 1; Part-Time Volunteers 20.
Volunteer Hours: 780
Governing Authority: society. Tax-exempt: 501(c)(3).
Institution Type/Description: Historical Society Museum: housed in 1903 Queen Anne-style house.
Collections: crafts; local history, 1865-present; photographs; 1903-1933 household items; toys; archives.
Research Fields: history & development of the Arcade area.
Facilities: 100-vol. library of local history & fiction books; magazines available to the public; 500 sq. ft. exhibit space. Museum-related items for sale.
Activities: guided tours; lectures; organized education programs for children; temporary exhibitions; annual meeting. Special Events: exhibit openings.
Publications: quarterly newsletter, The Arcade Historical Society News.
Hours & Admission Prices: Thurs. 1-8, Fri. 9-4; other times by appointment. No charge; donations accepted. Closed January; New Year's Day; Thanksgiving; Christmas.
Attendance: 1,883 (accurate)
Membership: Senior/Student $5; Individual $10; Family $15; Business $25; Patron $50; Sustaining $100; Life $500.

Arden

ORANGE COUNTY HISTORICAL SOCIETY, 21 Clove Furnace Dr., Arden, NY 10910. Mailing Address: P.O. Box 55, Arden, NY 10910-0055. Tel.: 845-351-4696.
Founded: 1971.
Governing Authority: Tax-exempt.
Institution Type/Description: Historical Society Museum.
Collections: local history & culture; photographs; historic buildings.
Publications: annual, OCHS Journal.
Hours & Admission Prices: Mon.-Fri. 8-4:30.

Ardsley

ARDSLEY HISTORICAL SOCIETY, 9 American Legion Dr., Library 2nd Fl., Ardsley, NY 10502. Tel.: 914-693-6027.
Institution Type/Description: Historical Society Museum.
Collections: local history & culture; period artifacts; photographs.
Activities: educational programs.
Hours & Admission Prices: Tues. 10 to noon; other times by appointment.

COUNTY OF WESTCHESTER, DEPARTMENT OF PARKS, RECREATION AND CONSERVATION, 450 Saw Mill River Rd., Ardsley, NY 10502. Tel.: 914-231-4500. Fax: 914-864-7053.
Web Site: www.westchestergov.com/parks
Founded: 1961.
Key Personnel: County Exec., Robert P. Astorino; Commissioner, Kathleen O'Connor; Naturalist, John Baker.
Governing Authority: county. Branch Facilities: Muscoot Park Interpretive Farm Tel. 914-864-7282; Trailside Nature Museum Tel. 914-864-7322; Marshlands Conservancy Tel. 914-835-4466; Kingsland Point Lighthouse Tel. 914-366-5109; Kingsland Point Park Tel. 914-366-5109; Washington Headquarters, Virginia Rd., North White Plains Tel. 914-864-PARK; Cranberry Lake Preserve Tel. 914-428-1005; Lenoir Preserve Tel. 914-968-5851; Silver Lake Preserve Tel. 914-231-4500; Mildred D. Lasdon Bird Sanctuary Tel. 914-231-4500; Edith G. Read Natural Park & Wildlife Sanctuary Tel. 914-967-8720; Croton Point Nature Center Tel. 914-862-5297; Merestead Tel. 914-231-4539; Westchester County Veterans Museum, 914-864-7268. Tax-exempt.
Institution Type/Description: Nature Centers, Sanctuaries & Historic Sites.
Collections: colonial artifacts; farm animals; natural science; 19th century paintings, prints & decorative arts; 20th century furnishings & textiles. Historic House: 1732 George Washington's headquarters, Virginia Rd., North White Plains; Historic Farm: Muscoot Park, turn-of-the-century 777-acre working farm with Manor House.
Activities: guided tours; lectures; permanent & temporary exhibitions; demonstration of colonial tools & machines.
Publications: quarterly calendar.
Hours & Admission Prices: Washington's Headquarters by appointment only.

Trailside Museum & Marshlands Conservancy: call for hours. No charge. Cranberry Lake Preserve: Tues.-Sun. 9-5. Lenoir Preserve: Tues.-Sun. 9-5. Silver Lake Preserve: daily 9-4. Lasdon Bird Sanctuary: daily dawn-dusk. Read Natural Park & Wildlife Sanctuary: Tues.-Sun. 9-5. Tarrytown Lighthouse: by appointment only.

Armonk

THE NORTH CASTLE HISTORICAL SOCIETY, 440 Bedford Rd., Armonk, NY 10504-2502. Tel.: 914-273-4510.
Founded: 1971.
Congressional District: 24
Key Personnel: Pres., Ree Shultz; Town Historian, Doris Finch Watson; Public Rels. & Publications Dir., Sharon Tomback; Museum Shop Mgr., Jodie Burns.
Personnel Profile: Part-Time Volunteers 50.
Governing Authority: board of trustees. Tax-exempt: 501(c)(3).
Institution Type/Description: Historical Society Museum: housed in c.1776 Smith's Tavern; Brundage Blacksmith Shop; East Middle Patent One Room Schoolhouse; c.1798 Quaker Meeting House.
Collections: various eras depicting home styles, fashions, tools, farm implements; blacksmith shop & forge; East Middle Patent one-room schoolhouse.
Research Fields: history of North Castle.
Facilities: library of local history, slides, photographs, movies, diaries, newspapers and manuscripts, available for use by special permission. Museum-related items for sale.
Activities: guided tours; dinners; antique shows; lectures; films; formally organized education programs for children; docent program; permanent & temporary exhibitions.
Publications: annual booklet, North Castle History; quarterly newsletter.
Hours & Admission Prices: April-Dec. Sun. 2-5, Wed. 2-4; other times by appointment. Suggested Donations: adults $5, member adults $3, children $1. &
Attendance: 750 (estimated)
Membership: Individual $30; Family $40; Friend $100; Benefactor $250; Life $500; Patron $1,000. Corporate, Business, Foundation, & Organizational $100 & up.

Astoria

✳ **MUSEUM OF THE MOVING IMAGE, (M),** 36-01 35 Ave. (at 37 St.), Astoria, NY 11106. Tel.: 718-777-6800. Fax: 718-784-4681.
E-mail: info@movingimage.us
Web Site: www.movingimage.us
Founded: 1977.
Congressional District: 9
Key Personnel: Exec. Dir., Carl Goodman; Chm. (V), Herbert S. Schlosser; Deputy Dir. Operations, Exhibitions & Design, Wendell Walker; Deputy Dir. Education & Visitor Experience, Christopher Wisniewski; Chief Cur., David Schwartz; Museum Shop Mgr., Darcie LaFarge.
Personnel Profile: Full-Time Paid 48; Part-Time Paid 29; Part-Time Volunteers 15; Interns 20.
Governing Authority: nonprofit organization. Tax-exempt: 501(c)(3).
Institution Type/Description: Art, History & Technology Museum of Film, Television & Digital Media.
Collections: 130,000 artifacts & materials related to the art, production, technology, craft, business & history of motion pictures, television & digital media; technical apparatus & installations, production design, licensed merchandise & other ancillary materials, photographs, promotional materials.
Research Fields: cinema; television; video studies; digital media.
Facilities: 267-seat theater; 68-seat screening rooms; 20,000 sq. ft. exhibition space; amphitheater; education center; laboratories; cafe. Museum-related items for sale.
Activities: film & video screenings; lectures; seminars; permanent & changing exhibitions; workshops; demonstrations; education programs in conjunction with New York City public schools; special events.
Publications: program notes; education guides; quarterly calendar of exhibitions, screenings & events.
Hours & Admission Prices: Tues.-Thurs. 10:30-5, Fri. 10:30-8, Sat.-Sun. 10:30-7. Adults $12, seniors & college students with ID $9, children 4-17 $6; members, children under 3, AAM & ICOM members, Fri. 4-8 no charge. Closed Memorial Day; Independence Day; Labor Day; Thanksgiving; Christmas. &
Attendance: 201,530 (accurate)
Membership: Individual $75; Dual $125; Family $150; Red Carpet Kids $250; Silver Screen $275; Curator's Circle $500; Director's Circle $1,000.

THE PEOPLE'S MUSEUM, (M), 22-27 Crescent St., Astoria, NY 11105-3105. Tel.: 718-204-7941 & 0031. Facebook: Peoples Museum.
E-mail: thepeoplesmuseum@hotmail.com
Web Site: thepeoplesmuseum.org
Founded: 2000.
Congressional District: 14
Key Personnel: Dir., Chm. & Pres., Mark Allen Sepanski; Devel., Education & Public Rels., April D. Sepanski; Treas., Howard S. Rose; Sec., Dr. Pamela Estelle Ransom; Registrar, Brandon Ballengee; Cur., Paul L. Sieswerda; Archivist & Artist, Fred Douglas Wilson; Trustee, Henry Galiano; Trustee, Jeffrey R. Myers; Accounting CPA Partner at Lutz & Carr, Martin Berkowitz.
Personnel Profile: Full-Time Volunteers 5; Part-Time Volunteers 60.
Governing Authority: private; board trustees; nonprofit organization. Charter by the Board of Regents of the Education Dept. of the State of New York. Tax-exempt: 501(c)(3), 509(a)(1) & 170(b)(1)(A).
Institution Type/Description: General Museum.
Collections: African; Amazonian; Americana; Native American; fossils; oceanic; Oriental; pre-Columbian; Egyptian; military including guns, swords, period arms & armor from Renaissance Europe, American Revolutionary War, Civil War, Egyptian, Sudan & 1879 Zulu War, Boer War, WWI & WWII; Whaling Harpoons 1850; dinosaur bones; coins; stamps; toys; contemporary & modern art; photographs; rare books; decorative art; fire & police department; musical instruments; archaeology; geology; zoology; New York artists art work; astronomy; early lobby cards & posters from the Mummy's Ghost 1944; werewolf of London, 1935; American Indian wars; animal heads & skeletons; railroad history; aviation history; cassettes; record albums; books; films pertaining to Native Americans; circus history over 40,000 museum items.
Research Fields: archaeology; anthropology; paleontology; Egyptian art; African art; Oceanic art; zoology; military; films; Hollywood early lobby card & posters; pre-Columbian art; American Indian; Zulu Wars 1838-1906; Boer War 1899-1902; Confederate & Union edged weapons; aviation history; railroad history; circus history.
Facilities: 40,000-vol. library.
Activities: arts festivals; docent program; films; lectures; school loan service; temporary & traveling exhibitions; broadcast programs.
Publications: newsletters, The People's Museum; The People's Museum on Art & Archaeology; The People's Museum on General History; The People's Museum on Military History; membership brochure; annual report.
Hours & Admission Prices: By appointment. No charge; donations accepted. &
Membership: Individual $25; Family $30; Contributing $75; Corporate $100; Supporting $250; Patron $500.

Auburn

CAYUGA MUSEUM & CASE RESEARCH LAB MUSEUM, 203 Genesee St., Auburn, NY 13021-3304. Tel.: 315-253-8051. Fax: 315-253-9829.
E-mail: cayugamuseum@roadrunner.com
Web Site: www.cayuganet.org/cayugamuseum
Founded: 1936.
Congressional District: 33
Key Personnel: Dir., Eileen McHugh; Pres. (V), Peter Maciulewicz; Cur., Lauren Chyle; Business Mgr., Lynn Palmieri.
Personnel Profile: Full-Time Paid 3; Part-Time Paid 3; Part-Time Volunteers 20.
Governing Authority: nonprofit organization. Tax-exempt: 501(c)(3).
Institution Type/Description: Local History Museum.
Collections: Bundy Thousand-Year Clock, inventor of the time clock; Case Research Lab; invention of sound-on-film; John Clark Collection: Civil War & Native American documents.
Research Fields: commercialization of sound film.
Facilities: 2,000-vol. library of books on history & archaeology. Books, prints & other museum-related items for sale.
Activities: guided tours; lectures; films; concerts; formally organized education programs; permanent, temporary & traveling exhibitions.
Publications: quarterly newsletter.
Hours & Admission Prices: Tues.-Sun. 12-5. No charge; donations accepted. Closed holidays. &
Attendance: 9,300 (accurate)
Membership: Individual $25; Family & Household $35; Sponsor $50; Patron $100; Benefactor $250.

THE HARRIET TUBMAN HOME, 180 South St., Auburn, NY 13021-5636. Tel.: 315-252-2081.
E-mail: khill@harriethouse.org

Web Site: www.harriethouse.org
Key Personnel: Exec. Dir., Karen V. Hill
Institution Type/Description: History Museum.
Collections: life & history of Harriet Tubman; period furniture; personal artifacts.
Activities: Museum Sponsors: special events in May.
Hours & Admission Prices: Tues.-Fri. 10-4, Sat. 10-3.

SCHWEINFURTH MEMORIAL ART CENTER, 205 Genesce St., Auburn, NY 13021-3304. Tel.: 315-255-1553. Fax: 315-255-0871.
E-mail: mail@schweinfurthartcenter.org
Web Site: www.myartcenter.org
Founded: 1975.
Congressional District: 33
Key Personnel: Exec. Dir., Donna Lamb; Pres. (V), Steven Moolin; Program Dir., Allison Graff.
Personnel Profile: Full-Time Paid 4; Part-Time Paid 3; Part-Time Volunteers 25.
Governing Authority: nonprofit organization. Tax-exempt: 501(c)(3).
Institution Type/Description: Cultural Art Center.
Collections: Julius A. Schweinfurth architectural drawings & renderings.
Facilities: cultural center. Museum-related items for sale.
Activities: tours; lectures; concerts; temporary exhibitions; special events.
Publications: monthly calendar; exhibition catalogs.
Hours & Admission Prices: Tues.-Sat. 10-5, Sun. 1-5. Suggested admission $6; children under 12 & members no charge. Closed holidays. &
Attendance: 20,000 (accurate)
Membership: Student & Senior $30; Individual $40; Senior Family $50; Family $60; Friend $125; Supporter $250; Benefactor $500; Patron $1,000.

SEWARD HOUSE MUSEUM, (M), 33 South St., Auburn, NY 13021-3929. Tel.: 315-252-1283. Fax: 315-253-3351.
E-mail: director@sewardhouse.org
Web Site: sewardhouse.org
Founded: 1951.
Congressional District: 24
Key Personnel: Pres., Daniel Fisher; Exec. Dir., Billye J. Chabot; Facilities Mgr., Andrew Roblee; Dir. Education, John Kingsley.
Personnel Profile: Full-Time Paid 5; Part-Time Paid 2; Part-Time Volunteers 60.
Governing Authority: nonprofit organization. Tax-exempt: 501(c)(3).
Institution Type/Description: Historic House: 1816 home of William H. Seward.
Collections: articles of Seward's career; original furnishings; Civil War material; paintings; uniforms; china; original costumes; Alaskan memorabilia & artifacts.
Research Fields: furniture; decorations; clothing 1816-1900; Civil War; politics, Alaska.
Facilities: 10,000-vol. library of books on Civil War & Abraham Lincoln available by appointment.
Activities: guided tours; permanent exhibitions; lectures; special topic tours
Publications: booklet; members quarterly newsletter, Diplomatic Pouch.
Hours & Admission Prices: Feb.-June & mid-Oct. to Dec. Tues.-Sat. 10-4; July to mid-Oct. Tues.-Sat. 10-4, Sun. 1-4. Adults $7, senior citizens $6, students $2; discounts to AAM & ICOM members; members no charge. Closed major holidays. &
Attendance: 16,000 (accurate)
Membership: Student $15; Individual $25; Family $40; Supporting $90; Sustaining $125.

WARD W. O'HARA AGRICULTURAL & COUNTRY LIVING MUSEUM & DR. JOSEPH F. NARPINSKI EDUCATIONAL CENTER, 6880 E. Lake Rd., Auburn, NY 13021. Tel.: 315-252-7644. Fax: 315-253-5199.
Web Site: www.cayuganet.org/agmuseum
Founded: 1975.
Congressional District: 33
Key Personnel: C.E.O., Chm. (V) & Pres. (V), Norman Riley; Dir. & Museum Shop Mgr., Timothy J. Quill.
Personnel Profile: Part-Time Paid 3; Part-Time Volunteers 10.
Governing Authority: county. Tax-exempt.
Institution Type/Description: Agricultural Museum.
Collections: agriculture machinery & implements; country kitchen; blacksmith shop; wheelwright shop; country store; photos of rural Cayuga County; dairy processing; sleighs; buggies; milking equipment; cooper's bench; early woodworking shop, tillage & planting equipment, tractors & power engines, veterinarians office & equipment; live bee colony; early 1900 Birdsal saw mill; Osburn equipment; over 500 historical photos.

Research Fields: Cayuga County businesses & farms relating to agricultural field.
Activities: guided tours; lectures; slide shows; hobby workshops; formally organized educational programs; loan, permanent, temporary & traveling exhibitions; school loan service; weaving.
Hours & Admission Prices: mid-May to mid-Sept. daily 11-4; other times by appointment. No charge; donations accepted. &
Attendance: 15,000 (estimated)

Baldwin

BALDWIN HISTORICAL SOCIETY AND MUSEUM, 1980 Grand Ave., Baldwin, NY 11510-2836. Mailing Address: P.O. Box 762, Baldwin, NY 11510-0586. Tel.: 516-546-0629.
Founded: 1971.
Congressional District: 5
Key Personnel: C.E.O. & Dir., Jack Bryck; Vice Pres. & Public Rels., Constance Grando.
Personnel Profile: Part-Time Volunteers 12.
Governing Authority: society; nonprofit organization.
Institution Type/Description: Historical Society Museum.
Collections: history of Baldwin; Baldwin photographic collection; Baldwin related artifacts & archives.
Research Fields: local history.
Facilities: library of books on local & Long Island history available by appointment only. Stock of 1910-30s Baldwin postcards & other museum-related items for sale.
Activities: guided tours; lectures; formally organized education programs.
Publications: newsletter.
Hours & Admission Prices: By appointment. No charge. Closed New Year's Day; Easter; Christmas. &
Attendance: 200 (estimated)
Membership: Student $1; Individual $10; Family $20; Life $200.

Ballston Spa

NATIONAL BOTTLE MUSEUM, 76 Milton Ave., Ballston Spa, NY 12020-1405. Tel.: 518-885-7589. Fax: 518-885-0317. Facebook: National Bottle Museum.
E-mail: nbottlemuseum@verizon.net
Web Site: www.nationalbottlemuseum.org
Founded: 1996.
Congressional District: 21
Key Personnel: Dir., Gary Moeller.
Personnel Profile: Full-Time Paid 1; Part-Time Volunteers 5.
Governing Authority: Tax-exempt: 501(c)(3).
Institution Type/Description: History Museum.
Collections: early bottle making methods; hand tools; 1800s glass furnace model; handmade bottles; stoneware.
Facilities: library.
Activities: research; classes; flameworking lessons.
Publications: newsletter, The Bottle Muse.
Hours & Admission Prices: June-Sept. Fri.-Tues. 10-4; Oct.-May Tues.-Sat. 10-4. No charge; donations accepted. Closed New Year's Day; Christmas.
Attendance: 3,750 (estimated)
Membership: Friends $10-$49; Supporters $50-$149; Benefactors $150 & up; Corporate Donors $500 & up.

SARATOGA COUNTY HISTORICAL SOCIETY, BROOKSIDE MUSEUM, (M), 6 Charlton St., Ballston Spa, NY 12020-1707. Tel.: 518-885-4000. Fax: 518-885-4055.
E-mail: info@brooksidemuseum.org
Web Site: www.brooksidemuseum.org
Formerly: Brookside, Saratoga County Historical Society
Founded: 1962.
Congressional District: 24
Key Personnel: Exec. Dir., Joy Houle; Pres. (V), Jeanne Obermayer; Dir. Education, Anne Clothier; Cur., Kathleen Coleman.
Personnel Profile: Full-Time Paid 1; Part-Time Paid 4; Part-Time Volunteers 60; Interns 7.
Volunteer Hours: 5,000
Operating Expenses: 165,000
Operating Income: 165,000
Governing Authority: society. Operated by Saratoga County Historical Society. Tax-exempt: 501(c)(3).
Institution Type/Description: General Museum: housed in 1792 Aldridge House & Resort Hotel.
Collections: Saratoga County history; items relating to area residents; art; costumes; toys & sports; agriculture; lumbering; resort life; manuscripts; Saratoga County surname files.

Major Exhibits: Here Comes the Bride, 2/14-6/14.
Research Fields: county history.
Facilities: 1,000-vol. library of local history materials; photographic collection; program space; 1,500 sq. ft. exhibit space. Museum-related items for sale.
Activities: group tours; traveling education programs; living history events; craft classes; hobby workshops; concerts; docent program; formal education programs; participatory & temporary exhibitions; lectures; school loan service.
Publications: yearly journal, The Grist Mill; bimonthly newsletter, Columns.
Hours & Admission Prices: Tues.-Fri. 10-4, Sat. 10-2. Family $5, adults $2, senior citizens $1.50, students & children $1; museum, AASLH & MANY members no charge. Closed New Year's Day; Easter; Thanksgiving; Christmas. ᶦ
Attendance: 11,000 (accurate)
Membership: Senior & Student $15; Individual $25; Family $35; Contributor $50; Supporter $75; Patron $100

Batavia

THE HOLLAND LAND OFFICE MUSEUM, 131 W. Main St., Batavia, NY 14020-2021. Tel.: 585-343-4727. Facebook: The Holland Land Office Museum.
E-mail: info@hollandlandoffice.com
Web Site: www.hollandlandoffice.com
Founded: 1894.
Congressional District: 26
Key Personnel: Dir., Jeffrey E. Donahue; Pres. (V), Robert Purk; Museum Shop Mgr., Corinne Iwanicki.
Personnel Profile: Full-Time Paid 1; Part-Time Paid 1; Part-Time Volunteers 15.
Governing Authority: county. Parent Institution: Holland Purchase Historical Society. Tax-exempt: 501(c)(3).
Institution Type/Description: General Museum: housed in 1815 cut-stone structure erected for use as the office of Holland Land Co. where land was offered for sale in 1800s.
Collections: medical; military; glass & china; early furniture; musical instruments; guns; toys; costumes; early hand tools; fireplace tools; cooking utensils; woodenware; Indian artifacts.
Research Fields: pertaining to museum collections; 19th-century rural & town life; western New York in early 1800s.
Activities: guided tours; permanent exhibitions; museum quilt guild; children's programs.
Publications: booklets, Music in Genesee County; The Sesquicentennial 1802-1952; History of Genesee County New York, 1890-1982.
Hours & Admission Prices: Jan.-May & Sept.-Nov. Tues.-Sat. 10-4; Memorial Day to Labor Day Mon.-Sat. 10-4; late-Nov. to Dec. Tues.-Sat. 10-4, Sun. 12:30-4:30. No charge; donations accepted. Closed legal holidays. ᶦ
Attendance: 6,500 (accurate)
Membership: Individual $20; Family $35; Patron $50; Sponsor $100; Benefactor $250; Life $500.

Bath

MAGEE HOUSE - STEUBEN COUNTY HISTORICAL SOCIETY, 1 Cohocton St., Bath, NY 14810. Mailing Address: P.O. Box 349, Bath, NY 14810. Tel.: 607-776-9930.
E-mail: steuben349@yahoo.com
Web Site: www.steubenhistoricalsociety.org
Founded: 1949.
Congressional District: 29
Key Personnel: Dir., Kirk House; Chm. (V), Helen K. Brink.
Personnel Profile: Part-Time Paid 1; Part-Time Volunteers 25.
Institution Type/Description: Historical Society Museum.
Collections: local history & culture; period furnishings; personal artifacts; photographs; documents; maps; genealogy.
Research Fields: Steuben County history.
Publications: quarterly, Steuben Echoes.
Hours & Admission Prices: Mon.-Fri. 10-3. No charge; donations accepted. Closed holidays.
Membership: Individual $15; Family $20; Business $50-$99; Underwriter $100 & up.

Bay Shore

SAGTIKOS MANOR HISTORICAL SOCIETY, Montauk Hwy. & Manor Ln., Bay Shore, NY 11706. Mailing Address: P.O. Box 5344, Bay Shore, NY 11706. Tel.: 631-854-0939. Facebook: Sagtikos Manor Historical Society.
E-mail: info@sagtikosmanor.com

Web Site: sagtikosmanor.com
Founded: 1963.
Key Personnel: Pres. (V), Christine Gottsch; Museum Shop Mgr., Maria Pecorale.
Personnel Profile: Part-Time Volunteers 200.
Volunteer Hours: 4,000
Operating Expenses: 45,000
Operating Income: 45,000
Governing Authority: Tax-exempt.
Institution Type/Description: Historical Society Museum: housed in the former home of the Thompson-Gardiner family; built in 1697. The estate served as the British Army's local headquarters during the Revolutionary War. President George Washington stayed in the manor during his tour of Long Island in 1790.
Collections: local & family history; Thompson-Gardiner family furnishings & personal artifacts.
Facilities: 42-room estate house museum; carriage house; cemetery; walled garden, buttery.
Activities: guided docent tours; special events.
Hours & Admission Prices: Memorial Day 1-3:30; June Sat.-Sun. 1-3:30; July-Aug. Fri.-Sun. 1-3:30; Sept. Sat. & Sun. 1-3:30; groups by appointment. Adults $7, seniors over 60 & students $5, children 3-12 $3.
Attendance: 1,500 (estimated)
Membership: Student $5; Senior $10; Individual $20; Family $40.

Bayport

BAYPORT AERODROME, Vitamin Dr., Bayport, NY 11705-1115. Mailing Address: P.O. Box 728, Bayport, NY 11705.
Key Personnel: Chm. (V), Kevin Kilroy; Pres. (V), John C. Hess; Museum Shop Mgr., Bob Mott.
Governing Authority: Tax-exempt.
Institution Type/Description: Aviation History Museum.
Collections: aviation history, aircraft & artifacts.
Activities: special events.
Hours & Admission Prices: June-Sept. 10-4. No charge; donations accepted.
Attendance: 1,800 (accurate)

Bayside

QCC ART GALLERY/CUNY, (M), 222-05 56th Ave., Bayside, NY 11364-1497. Tel.: 718-631-6396. Fax: 718-631-6620.
E-mail: qccartgallery@qcc.cuny.edu
Web Site: www.qccartgallery.org
Founded: 1966.
Congressional District: 8
Key Personnel: Dir., Mr. Faustino Quintanilla; Asst. Dir., Deanne DeNyse.
Governing Authority: nonprofit organization.
Institution Type/Description: College Art Gallery.
Collections: ethnic diversity of the College, community & the role art plays in the cultural history of people; primarily based on American artists after 1950 & women artists.
Research Fields: Trans-expressionism & spiritual behavior in art; African art; Pre-Columbian art.
Facilities: library at Queensborough Community College Art Gallery; small library of exhibition catalogues, Queens Artists Registry, & African art books available to students & community members.
Activities: guided tours; lectures; temporary, traveling & loan exhibitions; training programs for professional museum workers; formally organized education programs for adults, undergraduates & graduates affiliated with Queensborough Community College.
Publications: exhibit catalogues.
Hours & Admission Prices: Tues. & Fri. 10-5, Wed.-Thurs. 10-7, Sat.-Sun. 12-5. No charge; donations accepted. Closed major holidays. ᶦ
Attendance: 35,000 (estimated)
Membership: Senior Citizen & Student $25; Associate $50; Sustaining $100; Patron $250; Benefactor $500; President's Circle $1,000 & up.

Beacon

BEACON HISTORICAL SOCIETY, 477 Main St., Beacon, NY 12508-3819. Mailing Address: P.O. Box 89, Beacon, NY 12508-0089. Tel.: 845-831-0514.
Web Site: www.beaconhistoricalsociety.org
Founded: 1976.
Key Personnel: Pres. (V), Robert Murphy.
Personnel Profile: Part-Time Volunteers 5.
Governing Authority: private; nonprofit organization.
Institution Type/Description: Historical Museum.
Collections: paper archives; photographs; personal & Beacon government historical records; maps; pictures; news & personal items; memorabilia.

Facilities: 300-vol. library pertaining to Hudson River area; field research station.
Activities: slide collections; temporary exhibitions.
Publications: monthly newsletter.
Hours & Admission Prices: Thurs. 10-12, Sat. 1-3. Archival Research: 4th Tues. of each month. Open Meetings 7:30pm. No charge.
Attendance: 500 (estimated)
Membership: Individual $10.

DIA: BEACON, 3 Beekman St., Beacon, NY 12508. Mailing Address: 535 W. 22 St., New York, NY 10011-1119. Tel.: 212-989-5566 & 293-5518. Fax: 212-989-4055.
E-mail: info@diaart.org
Web Site: www.diaart.org
Formerly: Dia: Chelsea
Founded: 1974.
Congressional District: 17
Key Personnel: Exec. Dir., Philippe Vergne; Chm. Bd. (V), Nathalie de Gunzburg; Cur., Lynne Cooke; Dir. Operations, James P. Schaeufele; Dir. Digital Media, Sara Tucker; Publications, Karen Kelly; Asst. Dir., Steven Evans; Museum Shop Mgr., Jill Petrush Rogers.
Governing Authority: nonprofit organization. Subsidiary Institutions: The Earth Room by Walter De Maria, New York, NY; The Broken Kilometer by Walter De Maria, New York, NY; Dan Flavin Institute/Dia Center for the Arts, Bridgehampton, NY; The Lightning Field by Walter De Maria, Quemado, NM. Tax-exempt: 501(c)(3).
Institution Type/Description: Art Center.
Collections: contemporary American & European art.
Facilities: 2,800-vol. library of art books; catalogs; periodicals; 100-vols. videotapes & books on poetry; 45,000 sq. ft. exhibit space. Artists' books & exhibition catalogs for sale.
Activities: lectures; loan, temporary & traveling exhibitions; year long exhibitions of new works by individual artists; worldwide website commissions, artists web projects; Readings in Contemporary Poetry; education program; symposia; Robert Lehman Lectures on Contemporary Art.
Publications: semiannual calendar; exhibition catalogues & brochures; symposia publications; world wide web pages.
Hours & Admission Prices: Winter: Fri.-Mon. 11-4. Fall: Thurs.-Mon. 11-4. Adults $10, students & seniors $7; children under 12 no charge.
Attendance: 84,000 (estimated)
Membership: General $50; Family $100; Friend $500; Art Circle $5,000.

DIA:BEACON, RIGGIO GALLERIES, 3 Beekman St., Beacon, NY 12508-2521. Mailing Address: c/o Dia Art Foundation, 535 W. 22nd St., 4th Fl., New York, NY 10011. Tel.: 845-440-0100. Fax: 845-440-0092.
E-mail: info@diaart.org
Web Site: www.diaart.org
Founded: 2003.
Key Personnel: Chm. (V), Nathalie de Gunzburg; Dir., Philippe Vergne; Deputy Dir., Laura Raicovich; Financial Dir., Carolyn Kay Carson; Cur., Lynne Cooke; Museum Shop Mgr., Jill Petrush.
Governing Authority: private; nonprofit organization. Parent Institution: Dia Art Foundation, New York, NY. Tax-exempt: 501(c)(3).
Institution Type/Description: Art Museum.
Collections: art from 1960s to present.
Facilities: classroom; 50-seat cafeteria; 240,000 sq. ft. exhibit space. Museum-related items for sale.
Activities: guided tours; formal education programs for children.
Publications: semiannual, program calendar; exhibition publications.
Hours & Admission Prices: mid-April to mid-Oct. Thurs.-Mon. 11-6; Oct.-April Fri.-Mon. 11-4. Adults $10, senior citizens & students $7; discounts to AAM members; children under 12 no charge. Closed New Year's Eve & Day; Thanksgiving; Christmas Eve & Day.
Attendance: 80,000 (accurate)
Membership: Senior, Artist & Student $30; Individual $50; Family $100; Friend $500; Art Council $5,000.

THE MADAM BRETT HOMESTEAD, 50 Van Nydeck Ave., Beacon, NY 12508-3326. Tel.: 845-831-6533.
Founded: 1954.
Congressional District: 19
Key Personnel: Regent & Dir. Tours, Anne Thomas; Cur., Lorraine MacAulay; Cur. & Museum Shop Mgr., Evelyn Mark.
Personnel Profile: Part-Time Volunteers 20.
Governing Authority: society. Operated by the Melzingah Chapter of the Daughters of the American Revolution, 1776 D St. N.W., Washington, D.C., Melzingah Chapter, NSDAR, 50 Van Nydeck Ave. Beacon, NY 12508. Tax-exempt.
Institution Type/Description: Historic Site: c.1709 & c.1740 Madam Brett Homestead.
Collections: furnishings of seven generations of the Brett family; costumes; doll room; local history room.
Facilities: colonial garden. Museum-related items for sale.
Activities: guided tours.
Publications: brochure; booklet, Points of Historical Interest in Southern Dutchess.
Hours & Admission Prices: April-Dec. 2nd Sat. of month. Adults $5, students $2.
Attendance: 600 (estimated)

MOUNT GULIAN HISTORIC SITE, (M), 145 Sterling St., Beacon, NY 12508-1483. Tel.: 845-831-8172. Fax: 845-831-7376.
E-mail: info@mountgulian.org
Web Site: www.mountgulian.org
Founded: 1966.
Congressional District: 19
Key Personnel: Exec. Dir., Elaine Hayes.
Personnel Profile: Full-Time Paid 1; Part-Time Paid 1; Part-Time Volunteers 20.
Governing Authority: Parent Institution: Mount Gulian Society. Tax-exempt.
Institution Type/Description: Historic House Museum: housed in an 18th century Dutch Colonial Homestead.
Collections: local history; period furnishings.
Activities: weddings; rental facilities; special events; guided tours; school & group programs.
Hours & Admission Prices: Wed.-Fri. & Sun. 1-5; tours by appointment. Adults $8, seniors $6, youth 6-18 $4; AAM & Mount Gulian members no charge.
Attendance: 4,000 (estimated)

Bear Mountain

BEAR MOUNTAIN TRAILSIDE MUSEUMS AND ZOO, Bear Mountain State Park, Bear Mountain, NY 10911. Mailing Address: Palisades Interstate Park Commission, P.O. Box 427, Bear Mountain, NY 10911-0427. Tel.: 845-786-2701, ext. 263. Fax: 845-786-7157.
Web Site: www.palisadesparksconservancy.org/parks/5/
Founded: 1927.
Key Personnel: Exec. Dir., Ed McGowan.
Governing Authority: state. Palisades Interstate Park Commission.
Institution Type/Description: Natural History Museum.
Collections: local collections of animals, plants, minerals & rocks; Indian artifacts; history.
Research Fields: local history.
Facilities: approx. 3,000-vol. library of natural & local history available for use by appointment; botanical garden; zoological park; aquarium.
Activities: permanent exhibitions.
Hours & Admission Prices: Daily 10-4:30. Suggested Donations: $1 per person. Parking in state park $8. Closed Thanksgiving; Christmas.

Bedford

MUSEUM OF THE BEDFORD HISTORICAL SOCIETY, 612 Old Post Rd., Bedford, NY 10506. Mailing Address: P.O. Box 491, Bedford, NY 10506-0491. Tel.: 914-234-9751. Fax: 914-234-5461. Facebook: Museum of the Bedford Historical Society.
E-mail: info@bedfordhistoricalsociety.org
Web Site: www.bedfordhistoricalsociety.org
Founded: 1916.
Congressional District: 21
Key Personnel: Exec. Dir., Evelyne H. Ryan; Chm. (V), Kirtley Cameron; Pres. (V), Stacy E. Albanese.
Personnel Profile: Full-Time Paid 2; Part-Time Volunteers 30.
Governing Authority: society. Tax-exempt: 501(c)(3).
Institution Type/Description: Local History Museum.
Collections: archives; costumes; general; history; manuscript collections; Native American artifacts. Historic Buildings: 1806 Historical Hall; 1838 general store; 1787 Court House; 1829 schoolhouse; 1838 Post Office; 1807 Library; 1906 Stone Lounsbery Building; 1857 Stone Jackson House.
Research Fields: costumes; archives; genealogy.
Facilities: library of manuscripts, periodicals & old books available for use by appointment.
Activities: guided tours; permanent & temporary exhibitions; Images of America & Bedford; Walking Tour of Historic Bedford Village.

Publications: books, Pilgrim's Progress Bedford Version; Historical Tour of Bedford; The Burning of Bedford - July 1779.
Hours & Admission Prices: Museum: Thurs.-Sat. 12-3; other times by appointment. Adult $5, members no charge. &
Attendance: 200 (estimated)
Membership: Member $50; Founder $100-$249; Farmer $250-499; Patriot $500-999; Settler $1,000 & above.

Bedford Corners

WESTMORELAND SANCTUARY, INC., 260 Chestnut Ridge Rd., Bedford Corners, NY 10549-4812. Tel.: 914-666-8448. Fax: 914-242-1175.
E-mail: westsanc@optonline.net
Web Site: westmorelandsanctuary.org
Founded: 1957.
Congressional District: 21
Key Personnel: Dir. & Naturalist, Stephen Ricker.
Personnel Profile: Full-Time Paid 2; Part-Time Paid 1; Part-Time Volunteers 30.
Governing Authority: nonprofit organization. Tax-exempt.
Institution Type/Description: Nature Sanctuary.
Collections: rocks; minerals; mammal skulls; furs; turtle shells; colonial tools; mounted fish, mammals & birds; seasonal displays; live animals. Historic Building: 1783 church.
Research Fields: wildlife; botanical.
Facilities: 300-vol. library of natural science books available for research upon special request; nature center; 625-acre area; maple sugar house; auditorium; classroom; bird observation window.
Activities: guided tours; maple sugaring; outdoor lectures on wildlife management, ornithology, pond ecology & general ecology; evening lecture series; annual fall festival; nature programs for school groups, families & adults; workshops.
Publications: seasonal activity calendar; book, Ferris Family Cemetery in Westmoreland.
Hours & Admission Prices: Trails: daily dawn-dusk. Museum: Mon.-Sat. 9-5, Sun. 10:30-5. No charge; donations accepted.
Attendance: 40,000 (estimated)
Membership: Student & Senior $20; Individual $40; Family $75; Contributor $100; Benefactor $250; Sponsor $500; Ambassador $1,000 & up.

Bellport

BELLPORT-BROOKHAVEN HISTORICAL SOCIETY AND MUSEUM, 31 Bellport Lane, Bellport, NY 11713-2739. Tel.: 631-286-0888; 631-776-7640.
E-mail: president@bbhsmuseum.org
Web Site: www.bbhsmuseum.org
Founded: 1963.
Congressional District: 3
Key Personnel: Pres. (V), Jan Harting-McChesney; Museum Shop Mgr., Robert Duckworth.
Personnel Profile: Part-Time Volunteers 30.
Governing Authority: society. Tax-exempt: 501(c)(3).
Institution Type/Description: Local History Museum: housed in 1833 buildings.
Collections: wild fowl & shore bird decoys; memorabilia of Great South Bay & Fire Island Beach; dolls; toys; tools; guns; batteries; whaling artifacts; original scooter; textiles; Sperry gyroscope instruments; 19th-century household artifacts & costumes; tinware; early American decoration; genealogical information. Historic Buildings: c.1800's, Barn Museum; Brown Building; Post-Crowell House; Blacksmith Shop; Milk House.
Research Fields: local history.
Facilities: Museum-related items for sale.
Activities: guided tours; quarterly meetings. Museum Sponsors: Antiques Fair.
Publications: book, History of Bellport-Brookhaven; quarterly newsletter.
Hours & Admission Prices: Memorial Day to Labor Day Thurs.-Sat. 1-4; other times by appointment. Adults $5, seniors citizens 65 & over and children 13-17 $4; discounts to AAM & ICOM members; children 12 & under and members no charge. Gift Shop: May-Dec. Thurs.-Sat. 11-5. &
Attendance: 400 (estimated)
Membership: Student $15; Senior $20; Individual & Family $35; Sponsor $50; Patron $100; Benefactor $250; Individual Life $500; Couple Life $650.

Belmont

ALLEGANY COUNTY MUSEUM, 11 Wells St., Belmont, NY 14813-1052. Mailing Address: Court House Court St., Belmont, NY 14813. Tel.: 585-268-9293. Fax: 716-268-9446.
E-mail: historian@alleganyco.com

Founded: 1970.
Congressional District: 29
Key Personnel: Dir., Craig R. Braack.
Personnel Profile: Full-Time Paid 1; Part-Time Paid 1.
Governing Authority: county. Affiliated with the County Board of Legislators, Belmont, NY 14813. Tax-exempt.
Institution Type/Description: History Museum: housed in 1842 Greek Revival Church.
Collections: local land office records, maps & preservation records; impact statements; 19th century tools & furnishings.
Research Fields: genealogy; industry.
Facilities: 1,000-vol. library of books & government reports on genealogy & general history available for research on premises; 30-seat auditorium.
Activities: guided tours; lectures; formally organized education programs for children; permanent & traveling exhibitions.
Hours & Admission Prices: Mon.-Fri. 9-4:30; other times by appointment. No charge. Closed holidays. &

Bergen

BERGEN MUSEUM OF LOCAL HISTORY, 7547 Lake Rd., Bergen, NY 14416. Mailing Address: 10 Hunter St., Bergen, NY 14416. Tel.: 585-494-0080, 494-1121 & 704-4119 (cell). Fax: 585-494-1488.
Founded: 1964.
Congressional District: 35
Key Personnel: Mgr. & Museum Shop Mgr., Peggy Denton; Pres., Teresa Alexander; Vice Pres., Tracy Miller; Sec., Nancy Charcolla; Sec., Jean Stewart; Treas., Lisa Teremy
Governing Authority: municipal. Affiliated with Bergen Historical Society. Tax-exempt.
Institution Type/Description: Local History Museum: housed in 1843 schoolhouse.
Collections: photographs & documents of town residents; decorative arts; folklore; pottery; early implements & tools; farm, trade & home articles; livery.
Research Fields: history of town and people.
Activities: lecture tours for schools and organizations; formally organized education programs for children; permanent & temporary exhibitions; pioneer life in Bergen.
Publications: Bergen Historical Newsletter; 175th History Bergen.
Hours & Admission Prices: June-Oct. by appointment. No charge; donations accepted. Closed legal holidays. &
Attendance: 175 (estimated)
Membership: Single $10; Family $15; Sustaining $25.

HARFORD BARN MUSEUM, 15 S. Lake Ave., Bergen, NY 14416. Mailing Address: 10 Hunter St., Bergen, NY 14416. Tel.: 585-494-0080.
E-mail: historian@bergenny.org
Founded: 2003.
Key Personnel: Dir., Thomas M. Tiefel
Governing Authority: city. Tax-exempt.
Institution Type/Description: Historic Building: housed in a restored barn; built in the 1800s.
Collections: local history & culture; pioneer life from 1800-1900s; folk culture; sculpture; personal artifacts.
Activities: temporary & permanent exhibits.
Hours & Admission Prices: June-Oct. by appointment. No charge; donations accepted. &
Attendance: 325 (estimated)

Bethel

THE MUSEUM AT BETHEL WOODS, (M), 200 Hurd Rd., Bethel, NY 12720. Mailing Address: P.O. Box 222, Liberty, NY 12754-0222. Tel.: 845-583-2075. Fax: 845-583-4242. Facebook: Bethel Woods Center.
E-mail: wlawrence@bethelwoodscenter.org
Web Site: www.bethelwoodscenter.org
Founded: 2008.
Congressional District: 22
Key Personnel: Dir., Wade Lawrence; C.E.O., Darlene Fedun; Museum Shop Mgr., Carole Couture.
Personnel Profile: Full-Time Paid 4; Part-Time Paid 7; Part-Time Volunteers 97; Interns 2.
Volunteer Hours: 2,372
Governing Authority: Parent Institution: The Bethel Performing Arts Center, LLC. Tax-exempt.

Institution Type/Description: History Museum & Historic Site: at the site of the 1969 Woodstock Music and Art Fair.
Collections: graphic arts; photography; video; sound recordings; clothing; manuscripts; period artifacts; ephemera; oral histories.
Major Exhibits: The Beatles Invasion, 4/14-8/14; 45 Years of Woodstock, 4/14-12/14; Speak Truth to Power (T), 9/14-12/14.
Research Fields: 1960s music, fashion, politics & popular culture; Woodstock Music & Art Fair.
Facilities: permanent & temporary exhibit gallery; event gallery; classrooms; cafe; 15,000-seat outdoor concert pavilion; administrative offices; Woodstock historic site & monument.
Activities: student & adult education programs; concert, film & speaker series; seasonal festivals; community outreach.
Hours & Admission Prices: April-Dec. see website for hours. Adults $15; discounts to military, AAM & ICOM members. Closed Easter; Thanksgiving; Christmas Eve & Day. &
Attendance: 40,000 (estimated)
Membership: See website for membership information.

Big Flats

BIG FLATS HISTORICAL SOCIETY MUSEUM, 258 Hibbard Rd., Big Flats, NY 14814. Mailing Address: P.O. Box 232, Big Flats, NY 14814. Tel.: 607-562-7460.
Founded: 1971.
Institution Type/Description: Historical Society Museum.
Collections: local history & culture; period furnishings; personal artifacts; photographs.
Hours & Admission Prices: Tues. 9am to noon; other times by appointment. No charge; donations accepted. &
Membership: Individual $10.

Binghamton

BINGHAMTON ZOO AT ROSS PARK, 60 Morgan Rd., Binghamton, NY 13903-3667. Mailing Address: 185 Park Ave., Binghamton, NY 13903-3643. Tel.: 607-724-5461. Fax: 607-724-5454 & 5453.
E-mail: info@rosspark.com
Web Site: www.rossparkzoo.com
Formerly: Ross Park Zoo
Founded: 1875.
Key Personnel: Exec. Dir., David M. Conklin.
Personnel Profile: Full-Time Paid 17; Full-Time Volunteers 40.
Governing Authority: Tax-exempt.
Institution Type/Description: Zoo.
Collections: native & exotic animals including the Arctic fox & Coati Mundi; Amur Leopard; Red Panda; African Black Footed penguins; timber wolves; barred owl; African clawless otter; endangered species.
Facilities: picnic facilities; cafe. Gift items for sale.
Activities: carousel; educational programs.
Hours & Admission Prices: April-Nov. daily 10-5. Adults $7, seniors 55 & over and military with ID $6, children 3-11 $4.50; discount to groups; children under 2 & members no charge. &
Attendance: 60,000 (estimated)

THE BROOME COUNTY HISTORICAL SOCIETY, 185 Court St., Binghamton, NY 13901-3503. Tel.: 607-778-3572. Fax: 607-778-6429.
E-mail: localhistory@bclibrary.info
Web Site: www.bclibrary.info/history.htm
Founded: 1919.
Congressional District: 27
Key Personnel: Pres. (V), David J. Dixon.
Personnel Profile: Part-Time Paid 1; Part-Time Volunteers 15; Interns 1.
Operating Expenses: 22,219
Operating Income: 37,735
Governing Authority: Tax-exempt.
Institution Type/Description: History Museum: housed in Roberson Museum.
Collections: historic 1907 Renaissance Revival Residence; artifacts; manuscripts; decorative arts; regional history library maintained in Broome County Public Library.
Research Fields: local & New York state history; art; archaeology.
Facilities: 5,000-vol. library & photo archives on regional New York history available on premises.
Activities: guided tours; lectures; permanent, temporary & traveling exhibitions.
Publications: Broome County History Bulletin; exhibition catalogs.
Hours & Admission Prices: Mon. & Wed.-Fri. 9-5, Tues. 12-8. No charge; donations accepted. &

Attendance: 25,000 (estimated)
Membership: Individual $20; Family $30.

THE BUNDY MUSEUM OF HISTORY AND ART, (M), 129 Main St., Binghamton, NY 13905-2742. Mailing Address: Visitors Center, 127 Main St., Binghamton, NY 13905. Tel.: 607-772-9179. Facebook: Bundy Museum.
E-mail: info@bundymuseum.org
Web Site: www.bundymuseum.org
Formerly: The Bundy Arts & Victorian Museum
Founded: 2004.
Key Personnel: Dir., Janna Rudler; Pres. (V), Martine Barnaby.
Personnel Profile: Full-Time Paid 2; Part-Time Paid 2; Part-Time Volunteers 3.
Governing Authority: Tax-exempt.
Institution Type/Description: History Museum.
Collections: Bundy Time Recording Clocks; African artifacts; broadcasting pioneers & artifacts; period barbershop; Asian art; personal artifacts; works of modern art. Historic Buildings: Bundy house; carriage house.
Research Fields: local industrial history; Harlow E. Bundy family genealogy; architectural history.
Facilities: 70-seat theater.
Activities: permanent & temporary exhibitions; special events; educational programs; lectures; guided tours.
Hours & Admission Prices: Tues.-Sat. 11-5. Adults $7, students & senior citizens $5; discount to AAM members; children 10 & under no charge.
Attendance: 1,100 (estimated)
Membership: Senior Citizen & Student $25; Individual $35; Couple $60; Supporter $129; Charter & Lifetime $1,000.

CUTLER BOTANIC GARDEN, 840 Upper Front St., Binghamton, NY 13905-1542. Tel.: 607-772-8953. Fax: 607-723-5951.
E-mail: broome@cornell.edu
Web Site: www.cce.cornell.edu/broome
Founded: 1979.
Congressional District: 28
Key Personnel: C.E.O., David A. Bradstreet; Program Assoc., Brian Aukema.
Personnel Profile: Full-Time Paid 1; Part-Time Paid 2; Part-Time Volunteers 64; Interns 24.
Governing Authority: nonprofit. Parent Institution: Cornell Cooperative Extension of Broome County. Tax-exempt.
Institution Type/Description: Botanical Garden.
Collections: individual beds of annuals, perennials, herbs, vegetables; rock garden; heath & heather garden; antique rose border; Heritage fruit & vegetable garden; All-America display garden; seasonal theme gardens; demonstration composting site.
Research Fields: testing of hardiness & ornamental or economic value of plants.
Facilities: botanical garden.
Activities: guided tours; lectures; training programs for professional museum workers; master gardener program.
Publications: garden brochure.
Hours & Admission Prices: June-Oct. daily during daylight hours. No charge; donations accepted. &
Attendance: 15,000 (estimated)
Membership: Friend $10; Family $25; Benefactor $50.

THE DISCOVERY CENTER OF THE SOUTHERN TIER, 60 Morgan Rd., Binghamton, NY 13903-3667. Tel.: 607-773-8750 & 8661, ext. 200. Fax: 607-773-8019.
E-mail: pr@thediscoverycenter.org
Web Site: www.thediscoverycenter.org
Founded: 1984.
Congressional District: 28
Key Personnel: C.E.O., Margaret S. Crocker; Pres., Nicola R. Chanecka; Financial Dir., Catherine Fiacco; Asst. Dir., Donna Jones-Wright; Dir. Mktg., Martha J. Steed; Museum Shop Mgr., Jennifer Fiala.
Personnel Profile: Full-Time Paid 4; Part-Time Paid 33; Part-Time Volunteers 225; Interns 3.
Governing Authority: organization; nonprofit. Tax-exempt: 501(c)(3).
Institution Type/Description: Children's Museum: located in historic Ross Park. Discovery Center chartered in 1984.
Collections: fossils; minerals; natural science specimens, shells, African artifacts; textiles; contemporary & historical objects for standard exhibit or use in context; children's material culture objects of childhood.
Facilities: two classrooms; 4,800 sq. ft. exhibit space; 175-seat theater; cafe; zoological park; 1 1/2 acre garden; solar clubhouse. Educational toys, books & other museum-related items for sale.
Activities: concerts; docent program; hobby workshops; lectures; participatory

& temporary exhibits; theater; formal education programs for children & college undergraduate & graduate students; after school program, Discovery Kids; performing arts; environmental studies; paleontology; family workshops; pre-school for 3 & 4 year olds. Annual Events: Truck Day; Super Hero Day; Martin Luther King Day; Afro-American History Month; Business Day; Parlor City Boys Chorus. Museum Sponsors: early childhood programs weekly; Center Kids monthly.
Publications: quarterly newsletter, Discoverings.
Hours & Admission Prices: July-Aug. Mon.-Fri. 10-4, Sat. 10-5, Sun. 12-5; Sept.-June Tues.-Fri. 10-4, Sat. 10-5, Sun. 12-5. Children 2 & over $6, adults $5; discounts to groups, ACM members; children 1 & under and members no charge. &
Attendance: 53,420 (accurate)
Membership: Individual $45; Family $65; Family Plus $75; Contributing & ACM $125; Supporting $175; Patron $300.

❋ ROBERSON MUSEUM AND SCIENCE CENTER, 30 Front St., Binghamton, NY 13905-4779. Tel.: 607-772-0660. Fax: 607-771-8905. Facebook: Roberson Museum.
E-mail: blake@roberson.org
Web Site: www.roberson.org
Founded: 1954.
Congressional District: 28
Key Personnel: C.E.O., Terry McDonald; Pres. (V), Glenn Small; Exhibition Devel., Peter Klosky; Registrar, shannon Lindridge.
Personnel Profile: Full-Time Paid 8; Part-Time Paid 9.
Governing Authority: nonprofit. Parent Institution: Roberson Memorial, Inc. & Roberson Museum & Science Center. Tax-exempt: 501(c)(3).
Institution Type/Description: General Museum.
Collections: decorative arts; American art; mounted specimens of birds & mammals; local & regional history; ethnological & archaeological collections. Historic House: Roberson Mansion (c.1904); digital planetarium.
Major Exhibits: Home for the Holidays, 11/14-12/14.
Research Fields: regional; cultural; ethnic & social history.
Facilities: local history research library; 60-seat planetarium; 10-acre nature site. Museum-related items for sale; region's largest public model train display.
Activities: guided tours; lectures; concerts; family activities; classes; special events; school programs.
Publications: exhibit catalogues; local history monographs; newsletter.
Hours & Admission Prices: Wed.-Thurs. & Sat.-Sun. 12-5, Fri. 12-9. Adults $8, seniors & students $6, $20 family maximum (6 people maximum), reciprocal agreement with ASTC, Arnot Art Museum & 19 other New York State museums; discount to AAM & ICOM members; members no charge. Closed major holidays. &
Attendance: 35,000 (estimated)
Membership: Student $20; Senior $30; Individual $40; Family $65.

UNIVERSITY ART MUSEUM, (M), Binghamton University, Binghamton, NY 13902-6000. Mailing Address: P.O. Box 6000, Binghamton, NY 13902-6000. Tel.: 607-777-2634. Fax: 607-777-2613.
E-mail: hogan@binghamton.edu
Web Site: artmuseum.binghamton.edu
Founded: 1967.
Congressional District: 27
Key Personnel: Dir., Diane Butler; Dir. Asst., Jacqueline Hogan; Registrar, Silvia Ivanova.
Personnel Profile: Full-Time Paid 3; Part-Time Paid 3; Interns 2.
Governing Authority: university. Parent Institution: State University of New York at Binghamton & Foundation, Binghamton, NY. Tax-exempt.
Institution Type/Description: Art Museum.
Collections: paintings; sculpture; graphics; decorative arts; Wedgwood ceramics; Asian art; ancient pottery; African Art.
Facilities: 50,000-vol. fine arts library of catalogs; journals & books available for inter-library loan.
Activities: lectures; internships; inter-museum loan; temporary exhibits; faculty & student shows; educational class (Curatorial Practice) for graduate students; education outreach to public schools.
Publications: exhibition catalogs; newsletter, calendar of events.
Hours & Admission Prices: Summer: Tues.-Fri. 12-4; Sept.-June Tues.-Wed. & Fri.-Sat. 12-4, Thurs. 12-7; other times by appointment. No charge; donations accepted. Closed holidays. &

Bloomfield

A.W.A. ELECTRONIC-COMMUNICATION MUSEUM, 2 South Ave., Bloomfield, NY 14469. Mailing Address: P.O. Box 421, Bloomfield, NY 14469. Tel.: 585-657-6260. Fax: 585-257-5119.
E-mail: n2evg@arrl.net

Web Site: www.antiquewireless.org
Founded: 1952.
Congressional District: 29
Key Personnel: Dir., Thomas Peterson; Deputy Dir., Robert Hobday; Cur., Bruce Roloson; Treas., Stan Avery.
Personnel Profile: Part-Time Volunteers 30.
Governing Authority: Tax-exempt: 501(c)(3).
Institution Type/Description: Radio Museum: housed in c.1840 restored East Bloomfield Academy building.
Collections: radio & communications; electricity & electronics.
Research Fields: radio & communications.
Facilities: 20,000-vol. library of books, magazines, & catalogs on radios & electrical material available for research for members of association; 70,000 documents & photographic library; historical magnetic tape library; motion picture collection available for research by request & approval of Board.
Activities: guided tours; lectures; films; temporary exhibitions; film shows. Museum Sponsors: Annual Historical Radio Conference.
Publications: annual review; quarterly journal, Old Timers Bulletin.
Hours & Admission Prices: May-Sept. Sat. 2-4, Sun. 2-5; other times by appointment. No charge; donations accepted. Closed Independence Day.
Attendance: 2,000 (estimated)
Membership: Annual: U.S. $25; Canada & Overseas $30.

Blue Mountain Lake

❋ THE ADIRONDACK MUSEUM, 9097 State Rte. 30, Blue Mountain Lake, NY 12812-0099. Mailing Address: P.O. Box 99, Blue Mountain Lake, NY 12812-0099. Tel.: 518-352-7311, ext. 114. Fax: 518-352-7021.
E-mail: info@adkmuseum.org
Web Site: www.adirondackmuseum.org
Founded: 1952.
Congressional District: 24
Key Personnel: Dir., David M. Kahn; Chm., Kevin J. Arquit; Chief Cur., Laura S. Rice; Cur., Hallie Bond; Dir. Library, Jerold L. Pepper; Collections Mgr., Doreen Alessi; Mgr. Mktg., Kate Moore; Dir. Institutional Advancement, Sarah Lewin; Membership Coord., Michelle Bashaw; Museum Shop Mgr., Victoria Sandiford; Human Resources Mgr., Colleen Bush Sage.
Personnel Profile: Full-Time Paid 30; Part-Time Paid 6; Part-Time Volunteers 21; Interns 5.
Governing Authority: nonprofit organization. Parent Institution: The Adirondack Historical Association. Tax-exempt: 501(c)(3).
Institution Type/Description: Regional History & Art Museum.
Collections: regional lumbering & mining industries; transportation; boats; animal husbandry; agriculture; period furniture; fine art; period photographs; manuscripts; maps; films; oral histories; newspapers; tourism; decorative arts.
Research Fields: regional history; life & culture in Adirondacks; historical ecology; art history in the Adirondacks.
Facilities: research library consisting of books, pamphlets, maps, microfilms of newspapers, & photographs; cafe. Museum-related items for sale.
Activities: long-term & temporary exhibitions; research collections; school programs; continuing education programs; daily programs for children & families.
Publications: books; catalogues; monographs.
Hours & Admission Prices: Museum: Memorial Day to mid-Oct. daily 10-5. Adults $18, children 6-12 $6; discounts to groups of 15 or more, AAM & MANY members; members and children 5 & under no charge. Library: Mon.-Fri. 9-5 by appointment. &
Attendance: 59,380 (accurate)
Membership: Individual $40; Companion $75; Family $95; Benefactor $250; Patron $500.

Bolton Landing

BOLTON HISTORICAL MUSEUM, 4924 Main St., Bolton Landing, NY 12814. Mailing Address: P.O. Box 441, Bolton Landing, NY 12814-0441. Tel.: 518-644-9960.
Web Site: www.boltonhistorical.org
Personnel Profile: Part-Time Paid 2; Part-Time Volunteers 25.
Governing Authority: Parent Institution: The Historical Society of the Town of Bolton.
Institution Type/Description: History Museum: housed in a former church; built in 1890.
Collections: local history & culture; photographs; harvesting & farm equipment; clothing; personal artifacts.
Hours & Admission Prices: Call for hours. No charge; donations accepted. &
Attendance: 5,000 (estimated)
Membership: Adults $35.

THE SEMBRICH MARCELLA SEMBRICH OPERA MUSEUM, 4800 Lake Shore Dr. (Rte. 9N), Bolton Landing, NY 12814. Mailing Address: P.O. Box 417, Bolton Landing, NY 12814-0417. Tel.: 518-644-2431. Fax: 518-644-9831.
E-mail: office@thesembrich.org
Web Site: www.thesembrich.org
Founded: 1937.
Congressional District: 20
Key Personnel: Pres., William Hubert; Vice Pres., Lisa Hall; Sec., Rebecca Smith; Admin., Elizabeth Barton-Navitsky; Artistic Dir., Richard Wargo.
Personnel Profile: Full-Time Paid 2; Part-Time Paid 3; Part-Time Volunteers 30.
Governing Authority: nonprofit organization. Parent Institution: Marcella Sembrich Memorial Association, Inc., P.O. Box 417, Bolton Landing, NY 12814, 518-644-2431. Tax-exempt: 501(c)(3).
Institution Type/Description: Opera Museum & Historic Building: 1923 The Sembrich Studio, former teaching studio of Madame Marcella Sembrich, opera singer, teacher & director of the vocal departments at the Curtis Institute & the Juilliard School.
Collections: art objects, furniture, books, paintings, scores, costumes, photographs representative of the Golden Age of Opera, c.1877-1935; awards, trophies & plaques given to Madame Sembrich; opera history during the life & career period of Marcella Sembrich's lifetime.
Research Fields: opera; singing.
Facilities: 1,000-vol. library of operatic history, collection of material such as contracts, letters, reviews, available for study by scholars upon request. Biography, cassette, musical scores & CD recordings of Madame Sembrich, and postcards & books for sale.
Activities: guided tours; concerts in museum & the Lake George area; musical programs; permanent exhibitions; lectures.
Publications: biannual newsletter; book, A Recollection of Marcella Sembrich.
Hours & Admission Prices: June 15-Sept. 15 daily 10-12:30 & 2-5; May & Oct. by appointment. No charge; donations accepted. &
Attendance: 2,000 (accurate)
Membership: Individual $35; Sustaining $50; Patron $100; Benefactor $500.

Brewster

SOUTHEAST MUSEUM ASSOCIATION, INC., 67 Main St., Brewster, NY 10509-1416. Tel.: 845-279-7500. Fax: 845-279-1992.
E-mail: sem@bestweb.net
Web Site: www.southeastmuseum.org
Founded: 1963.
Congressional District: 19
Key Personnel: Exec. Dir., Amy Campanaro; Pres. (V), Elizabeth Ryder; Cur., Joan Crawford; Museum Shop Mgr., Eleanor Keefe.
Personnel Profile: Full-Time Paid 2; Part-Time Paid 4; Part-Time Volunteers 30.
Governing Authority: nonprofit organization. Tax-exempt: 501(c)(3).
Institution Type/Description: Historical Museum.
Collections: McLane Railroad Exhibit of the History of the Harlem Line; Trainer collection of minerals; Borden Condensed Milk Factory collection; costume & quilt collection; local historical memorabilia illustrating economic & social history from 19th-century to the present; early circus items; large 19th-century barn implement & tool collection.
Research Fields: Croton water system; Tilly Foster Mines; local circus history; Harlem Railroad; local Borden's Milk Condensery.
Facilities: archives of photographs & document slides available to researchers by appointment.
Activities: guided tours; permanent & changing exhibits; lectures; family programs & workshops; films.
Publications: newsletter, Southeast Sentinel; booklet accompanying permanent & temporary exhibits.
Hours & Admission Prices: April-Dec. Tues.-Sat. 10-4. No charge; donations accepted.
Attendance: 4,500 (estimated)
Membership: Senior Citizen $15; Individual $25; Family $35; Friend $50; Contributing $100; Sponsor $250; Supporter $500; Patron $1,000; Sustainer $3,000; Benefactor $5,000.

Bridgehampton

THE BRIDGEHAMPTON MUSEUM - BRIDGEHAMPTON HISTORICAL SOCIETY, 2368 Main St. & Corwith Ave., Bridgehampton, NY 11932-0977. Mailing Address: P.O. Box 977, Bridgehampton, NY 11932-0977. Tel.: 631-537-1088. Fax: 631-537-4225.
E-mail: bhhs@optonline.net
Web Site: bhmuseum.org
Founded: 1956.
Congressional District: 1
Key Personnel: Exec. Dir., John Eilertsen; Pres., Gerrit Vreeland.
Personnel Profile: Full-Time Paid 2; Full-Time Volunteers 1; Part-Time Paid 3; Part-Time Volunteers 40.
Volunteer Hours: 1,500
Operating Expenses: 300,000
Operating Income: 300,000
Governing Authority: private; nonprofit organization. Tax-exempt: 501(c)(3) & 170(b).
Institution Type/Description: General Museum.
Collections: period furniture; agriculture; costumes; tools; archival material; Hildreth Machine Shop with antique engines & early farm machines. Historic Houses: Corwith House c.1840; wheelwright shop c.1872; jail c.1902; c.1840 Nathaniel Rogers House.
Research Fields: local history.
Facilities: library of local history books and records available for research in reading room.
Activities: guided tours; permanent exhibitions; annual fair; special changing exhibits each summer; Road Rallye, Engine Run.
Publications: annual magazine, The Bridge; newsletter.
Hours & Admission Prices: mid-June to Nov. Mon.-Sat. 10-3; Dec. to mid-June Mon.-Fri. 10-3; other times by appointment. Adults $5; members no charge. &
Attendance: 5,000 (accurate)
Membership: Contributing $35; Sustaining $50; Patron $100-$499; Benefactor $500 & up.

CHILDREN'S MUSEUM OF THE EAST END, 376 Bridgehampton/Sag Harbor Turnpike, Bridgehampton, NY 11932. Mailing Address: P.O. Box 316, Bridgehampton, NY 11932-0316. Tel.: 631-537-8250. Fax: 631-537-2413.
E-mail: steve@cmee.org
Web Site: www.cmee.org
Founded: 1997.
Key Personnel: Dir., Steve Long.
Personnel Profile: Full-Time Paid 6; Part-Time Paid 4; Part-Time Volunteers 5.
Governing Authority: Tax-exempt: 501(c)(3).
Institution Type/Description: Children's Museum.
Collections: hands-on exhibits.
Activities: summer programs; classes; workshops; birthday parties.
Hours & Admission Prices: Wed.-Mon. 9-5; groups by appointment. Admission $10; members & children under one no charge. &
Membership: CMEE Be Happy 1 $60; CMEE Be Happy 2 $75; CMEE Be Grand $95; CMEE Be Delighted $125; CMEE Jump for Joy $250; CMEE Sponsor $750; CMEE Visionary $2,500; CMEE Corporate Circle $10,000.

Brockport

CAPEN HOSE COMPANY NO. 4 - FIRE MUSEUM, 237 S. Main St., Brockport, NY 14716. Tel.: 716-637-4713 & 2512.
Web Site: brockportfire.org/Museum/Museum.html
Founded: 1967.
Institution Type/Description: Firefighting History Museum.
Collections: firefighting history & equipment; photographs; fire pumps; leather fire bucket; uniforms; helmets; personal artifacts.
Activities: group tours.
Hours & Admission Prices: By appointment.

TOWER FINE ARTS GALLERY (SUNY BROCKPORT), Tower Fine Arts Bldg., 180 Holley St., Brockport, NY 14420-2985. Mailing Address: 350 New Campus Dr., Brockport, NY 14420-2985. Tel.: 585-395-2805. Fax: 585-395-2588.
E-mail: tmassey@brockport.edu
Web Site: www.brockport.edu/finearts/
Founded: 1964.
Congressional District: 32
Key Personnel: Chm., Phyllis Kloda; Dir., Timothy Massey.
Personnel Profile: Part-Time Paid 4.
Governing Authority: university; nonprofit organization. Parent Institution: SUNY College at Brockport. Tax-exempt.
Institution Type/Description: Art Museum.
Collections: works of E.E. Cummings.
Facilities: 1,900 sq. ft. exhibit space.
Activities: arts festivals; lectures; training programs for professional museum workers.
Publications: exhibit catalogs.
Hours & Admission Prices: Academic Year: Mon.-Fri. 10-5, Sun. 1-4. No charge. &

Attendance: 6,000 (estimated)

Bronx

BARTOW-PELL MANSION MUSEUM, CARRIAGE HOUSE & GARDENS, (M), 895 Shore Rd., Pelham Bay Park, Bronx, NY 10464-1030. Tel.: 718-885-1461. Fax: 718-885-9164.
E-mail: info@bpmm.org
Web Site: www.bartowpellmansionmuseum.org
Founded: 1914.
Congressional District: 19
Key Personnel: Pres. (V), Catherine Campbell Scinta; Exec. Dir., Ellen Bruzelius; Site Mgr., Mary Ellen Williamson; Mgr. Education, Margaret Highland; Dir. Garden, Frazier Holloway.
Personnel Profile: Full-Time Paid 2; Part-Time Paid 6; Part-Time Volunteers 100; Interns 5.
Governing Authority: Trustees of Bartow-Pell Landmark Fund. Parent Institution: Bartow-Pell Conservancy. Tax-exempt: 501(c)(3).
Institution Type/Description: Historic House Museum: 1842-1888 Bartow-Pell Mansion, Greek-Revival mansion.
Collections: c. 1842 Greek revival house; Empire period furnishings; paintings; c.1915 formal garden designed by Delano & Aldrich; carriage house & gardens.
Research Fields: Empire period furniture; accessories; preservation; gardens; Greek revival architecture; colonial history of Pelham Manor; development of Pelham Bay Park area of Bronx.
Facilities: 500-vol. library of books on architecture, gardening & herbs.
Activities: luncheon & tea tours; lectures; permanent exhibitions; concerts; demonstrations; public programs for adults & families; summer tunes for kids. Annual Events: Candlelight Tour; Harvest Festival.
Publications: quarterly newsletter.
Hours & Admission Prices: Wed. & Sat.-Sun. 12-4. Adults $5, seniors & students $3; discounts for AAM & ICOM members & groups; children under 6 & members no charge. Closed New Year's Eve & Day; Easter; Thanksgiving weekend; Christmas.
Attendance: 15,000 (estimated)
Membership: Friend $40; Associate $80; Family Friend $100; Heritage Individual $180; Heritage Family $300; Bartow Pell Society $500; Conservation Society $1,000; Leadership Circle $3,500.

THE BRONX COUNTY HISTORICAL SOCIETY, 3309 Bainbridge Ave., Bronx, NY 10467-2850. Tel.: 718-881-8900. Fax: 718-881-4827.
E-mail: angel@bronxhistoricalsociety.org
Web Site: www.bronxhistoricalsociety.org
Founded: 1897.
Congressional District: 7, 8, 16 & 17
Key Personnel: C.E.O., Dr. Gary D. Hermalyn; Pres. (V), Jacqueline Kutner; Cur. & Mgr. Edgar Allan Poe Cottage, Kathleen A. McAuley; Assoc. Librarian, Laura Tosi; Valentine-Varian House Mgr., Marcus Hickman; Dir. Bronx Latino History Project, Angel Hernandez; Education Outreach, Daniel Richards; Historian, Prof. Lloyd Ultan; Researcher, Dr. Stephen Stertz; Journal Editor, Dr. Elizabeth Beirne.
Personnel Profile: Full-Time Paid 8; Full-Time Volunteers 5; Part-Time Paid 5; Part-Time Volunteers 7; Interns 2.
Governing Authority: nonprofit organization. Branch Museum: Edgar Allan Poe Cottage; Museum of Bronx History. Tax-exempt: 501(c)(3).
Institution Type/Description: Historical Society.
Collections: items pertaining to the Bronx, New York City & Westchester from 1639-present.
Major Exhibits: History Woven Into Cloth (T), 10/23/13-4/1/14; Bronx Lation History (T), 2013-2014; Bronx African American History (T), 2013-2014.
Research Fields: social, colonial & urban history & geography.
Facilities: library pertaining to history for inter-library & public use; The Bronx County Archives. Books for sale.
Activities: guided tours; lectures; films; concerts; dance recitals; arts festivals; theater; expeditions, commemorations & conferences; study clubs; broadcast programs; organized education programs for children & adults; docent program; organized education programs for undergraduate or graduate college students; training programs for professional museum workers; participatory, loan & traveling exhibitions; temporary exhibitions of your own collections.
Publications: books, Geography: The Earth, The Poles & NYC; The Bronx Cookbook; The Hudson River; NYC at the Turn of the Century; biannual journal; books, History In Asphalt: The Origin of Bronx Street & Place Names Encyclopedia: The Study and Writing of History; The Beautiful Bronx, The Bronx in the Innocent Years; McNamara's Old Bronx; Elected Public Officials of the Bronx since 1898; Signers of the U.S. Constitution; Signers of The Declaration of Independence; Presidents of the U.S.; Chief Justices of the Supreme Court; Historic Landmarks & Districts; The First

Public High Schools of NYC; The Bronx: It Was Only Yesterday; Morris High School and the Creation of the New York City Public High School System; 1st Senate of the U.S.; 1st House of Representatives and the Bill of Rights; The Centennial of Greater New York; Poems & Tales of Edgar Allan Poe at Fordham; Time & the Calendar; Bronx Views: Postcards of the Bronx; Documents of the Bronx; The Student Writing of History; Yankee Stadium, The Bronx Then & Now; Publications & Other Media of the Bronx County Historical Society; Educational Culture in the Bronx; Ethnic Groups in the Bronx Bibliography; Erie Canal.
Hours & Admission Prices: Mon.-Fri. 9-5, Sat. 10-4, Sun. 1-5. Adults $5, senior citizens, children & students $3; discounts to AAM members; members no charge. ♿
Attendance: 85,000 (estimated)
Membership: Student & Senior Citizen $20; Individual, Nonprofit & Libraries $25; Family $35; Sustaining $100; Business & Fellow $250; Benefactor $2,500; Patron $5,000.

THE BRONX MUSEUM OF THE ARTS, 1040 Grand Concourse, Bronx, NY 10456-3901. Tel.: 718-681-6000. Fax: 718-681-6181.
E-mail: info@bronxmuseum.org
Web Site: www.bronxmuseum.org
Founded: 1971.
Congressional District: 21
Key Personnel: Chm., R. Douglas Rice; Exec. Dir., Holly Block; Acting Dir. Programs, Sergio Bessa; Dir. Finance, Alan Highet; Dir. Devel., Yvonne Garcia.
Personnel Profile: Full-Time Paid 20; Part-Time Paid 6; Part-Time Volunteers 2; Interns 2.
Governing Authority: nonprofit organization. Tax-exempt.
Institution Type/Description: Art Museum.
Collections: modern & contemporary works by artists of African, Latin American & Asian descent.
Major Exhibits: Urban Archives: David Bruscky, 9/15/13-2/25/14.
Research Fields: Bronx history; contemporary art; local artists; Latino & Latin American artists; African-American art; Asian American art.
Activities: formally organized education programs; internship program; film programs; curatorial & administrative apprenticeships; temporary & traveling exhibitions; demonstrations; gallery talks; live concert series; art festivals; special events.
Publications: catalogs of exhibitions; educational workbooks; brochures; gallery guides.
Hours & Admission Prices: Thurs. & Sat.-Sun. 11-6, Fri. 11-8. No charge. Closed New Year's Day; Thanksgiving; Christmas. ♿
Attendance: 40,000 (estimated)
Membership: Student & Senior Citizen $25; Individual $35; Family & Joint $50; Sustaining $100; Associate $250; Patron $500; Benefactor $1,000; Major Donor $2,500.

BRONX ZOO, (M), 2300 Southern Blvd., Bronx, NY 10460-1090. Tel.: 718-220-5100. Fax: 718-220-2685.
E-mail: PR@wcs.org
Web Site: www.wcs.org
Founded: 1895.
Congressional District: 10
Key Personnel: Pres. & C.E.O., Steven Sanderson; Dir. Operations, Ken Hutchinson; Vice Pres. Business Svcs., Robert Moskovitz; Sr. Vice Pres. & Gen. Dir. Living Institutions, Wildlife Conservation Society, Dr. Robert Cook; Exec. Vice Pres. & C.F.O., Patricia Calabrese.
Personnel Profile: Full-Time Paid 700; Part-Time Paid 675; Part-Time Volunteers 275; Interns 20.
Governing Authority: society. Parent Institution: The Wildlife Conservation Society. Tax exempt: 501(c)(3).
Institution Type/Description: Zoo.
Collections: birds; mammals; reptiles; amphibians; invertebrates.
Research Fields: ornithology; herpetology; mammalogy; natural history; animal behavior; reproductive biology.
Facilities: library by appointment or interlibrary loan. Museum-related items for sale.
Activities: guided tours; lectures; films; formally organized education programs; permanent & occasional temporary exhibitions.
Publications: bimonthly magazine, Wildlife Conservation; book; annual report; map.
Hours & Admission Prices: Zoo: April-Oct. Mon.-Fri. 10-5, Sat.-Sun. & holidays 10-5:30; Nov.-March 10-4:30. Call for rates; discounts to AAA members; members, children under 2, & Wed. no charge. Additional fees for some exhibits & rides. ♿
Attendance: 1,932,638 (accurate)
Membership: Wildlife Conservation Society: Senior $60; Senior Individual Premium $72; Individual $75; Individual Premium $90; Senior Family $96;

Family $120; Family Premium $150; Conservation Supporter $250; Conservation Fellow $500; Conservation Partner $750; Annual Patron $1,500.

CITY ISLAND NAUTICAL MUSEUM, City Island, 190 Fordham St., Bronx, NY 10464. Mailing Address: P.O. Box 82, Bronx, NY 10464-0082. Tel.: 718-885-0008 & 0507. Fax: 718-885-0507. Facebook: City Island Nautical Museum.
E-mail: cihs@cityislandmuseum.org
Web Site: www.cityislandmuseum.org
Founded: 1964.
Congressional District: 7
Key Personnel: Pres. & Cur., Tom Nye; Treas., Fred Ramftl; Vice Pres., Barbara Dolensek; Sec., Russell Schaller; Museum Shop Mgr., Barbara Hoffman.
Personnel Profile: Part-Time Paid 1; Part-Time Volunteers 15.
Governing Authority: nonprofit organization. Parent Institution: City Island Historical Society Inc., City Island, P.O. Box 82, The Bronx, NY 10464. Tax-exempt.
Institution Type/Description: Nautical Historical Museum: housed in 1897-98, P.S. 17 school building.
Collections: paintings, photographs, artifacts, documents and memorabilia from Indian times to the present, with special emphasis on the part played by City Island in the yachting industry and the America's Cup Races and the Hell Gate Pilots, archaeological collections.
Major Exhibits: Rosenfeld Photographs, 3/14-6/14.
Research Fields: local land and nautical history.
Facilities: 1,000-vol. library on local history available for use on premises.
Activities: guided tours; lectures; formally organized education programs; permanent & temporary exhibitions.
Publications: newsletter; informative brochure; City Island walking guide.
Hours & Admission Prices: Sat.-Sun. 1-5; other times by appointment. No charge; donations accepted. &
Attendance: 2,700 (estimated)
Membership: Individual $20; Family $25; Corporate $50.

EDGAR ALLAN POE COTTAGE, Poe Park, Grand Concourse at E. Kingsbridge Rd., Bronx, NY 10458. Mailing Address: Bronx County Historical Society, 3309 Bainbridge Ave., Bronx, NY 10467-2840. Tel.: 718-881-8900. Fax: 718-881-4827.
E-mail: kmcauley@bronxhistoricalsociety.org
Web Site: www.bronxhistoricalsociety.org
Founded: 1955.
Congressional District: 13
Key Personnel: C.E.O., Dr. Gary Hermalyn; Pres. (V), Jacqueline Kutner; Dir. Museums, Kathleen A. McAuley.
Personnel Profile: Full-Time Paid 4; Part-Time Paid 5; Part-Time Volunteers 33.
Governing Authority: nonprofit historical & preservation society. Parent Institutions: Bronx County Historical Society; The Historic House Trust of New York City, The Arsenal, Room 203, Central Park, New York, NY 10065; New York City Dept. of Parks & Recreation. Tax-exempt.
Institution Type/Description: Historic House: housed in the former residence of poet Edgar Allan Poe; built in 1812.
Collections: 19th-century furnishings; paintings & sculptured portraits of Poe; photographs; drawings.
Research Fields: Edgar Allan Poe's life, particularly his final years, 1846-1849.
Activities: film presentation; guided tours.
Publications: semiannual journal, Bronx County Historical Society; annual books & booklets.
Hours & Admission Prices: Mon.-Fri. 9-5 by appointment, Sat. 10-4, Sun. 1-5; group tours by appointment. Adults $5, seniors, students & children $3; discounts to AAM, ICOM members; members & Historic House Trust of NYC members no charge.
Attendance: 46,934 (estimated)
Membership: Student & Senior $20; Individual $25; Family $35; Sustaining $100; Fellow $250.

THE HALL OF FAME FOR GREAT AMERICANS, Bronx Community College, 2155 University Ave., Rm. 26, Bronx, NY 10453-2804. Tel.: 718-289-5910. Fax: 718-289-6496.
Web Site: www.bcc.cuny.edu/halloffame
Founded: 1900.
Congressional District: 22
Key Personnel: Dir., Wendell Joyner.
Governing Authority: Trustees of The Hall of Fame for Great Americans; affiliated with Bronx Community College of the City University of NY. Tax-exempt.

Institution Type/Description: Art & History Museum.
Collections: 98 original bronzes of notable Americans in the arts, sciences, humanities, government, business & labor, elected by a College of Electors.
Research Fields: American history; biography; sculpture; American architecture designed by Stanford White.
Facilities: outdoor columned arcade for display of sculptures.
Activities: guided tours plus video introduction.
Publications: brochure; tour guide.
Hours & Admission Prices: Daily 10-5. No charge; donations accepted.
Attendance: 25,000 (estimated)

LEHMAN COLLEGE ART GALLERY, (M), 250 Bedford Park Blvd. W., Bronx, NY 10468-1589. Tel.: 718-960-8731. Fax: 718-960-6991.
E-mail: susan@lehman.cuny.edu
Web Site: www.lehman.edu/gallery
Founded: 1984.
Congressional District: 19
Key Personnel: Dir., Susan Hoeltzel; Chm. (V), Virginia Cupiola; Devel. Assoc., Mary Ann Siano.
Personnel Profile: Full-Time Paid 3; Part-Time Paid 2; Part-Time Volunteers 8; Interns 10.
Governing Authority: nonprofit organization. Tax-exempt: 501(c)(3).
Institution Type/Description: College Art Museum.
Collections: works by contemporary artists.
Facilities: educational facilities.
Activities: guided tours; lectures; films; organized education programs for children; docent program; participatory & loan exhibitions. Museum Sponsors: Public Art in the Bronx.
Publications: exhibition notes; catalogues; teacher guides; Bronx Architecture.
Hours & Admission Prices: June-Aug. open by appointment; Sept.-June Tues.-Sat. 10-4. No charge. &
Attendance: 30,000 (estimated)
Membership: Senior & Student $20; Individual $30; Dual Family $50; Supporting $100; Sustaining $250; Sponsor $500; Benefactor $1,000.

LONGWOOD ART GALLERY AT HOSTOS CENTER FOR THE ARTS & CULTURE, 450 Grand Concourse, at 149th St., Bronx, NY 10451. Tel.: 718-518-4455.
Institution Type/Description: Art Gallery.
Collections: works by local, national & international artists.
Hours & Admission Prices: Mon.-Sat. 10-6.

MARITIME INDUSTRY MUSEUM AT FORT SCHUYLER, 6 Pennyfield Ave., Bronx, NY 10465-4127. Tel.: 718-409-7218. Fax: 718-409-6130.
E-mail: maritimeindustry@sunymaritime.edu
Web Site: sunymaritime.edu/maritime museum
Founded: 1986.
Key Personnel: Dir., Capt. Eric J. Johansson; Chm., Capt. James J. McNamara; Treas., Veronica Sullivan; Cur., William Sokol, Jr.; Administrator, Patricia Perez.
Personnel Profile: Part-Time Paid 3; Part-Time Volunteers 10.
Governing Authority: nonprofit.
Institution Type/Description: Maritime Museum.
Collections: over 250 ship models; Tufnell watercolor; Frank Cronican models; Maritime College history; Works Project Act (WPA) art; Ocean Liner; Victory & Liberty Ship; Robert G. Herbert ship models; Louis Weickum art; SS United States; Morro Castle; navigation instruments; underwater sea; Brooklyn Navy Yard diorama; evolution of seafaring.
Facilities: library; 7,000 sq. ft. exhibit space.
Activities: formal education programs for Maritime College students; guided tours; lectures; loan & participatory exhibits.
Publications: newsletter, Voyage Abstract.
Hours & Admission Prices: Mon.-Sat. 9-4. No charge; donations accepted. Closed on holidays. &
Attendance: 10,000 (estimated)
Membership: Senior $20; Regular $30; Chief Mate's Club $100; Captain's Club $200; Admiral Lite $1,000.

✳   THE NEW YORK BOTANICAL GARDEN, (M), 2900 Southern Blvd., Bronx, NY 10458. Tel.: 718-817-8700. Fax: 718-220-6504.
E-mail: pubrel@nybg.org
Web Site: www.nybg.org
Founded: 1891.
Congressional District: 7
Key Personnel: Chm. Bd. (V), Maureen K. Chilton; C.E.O. & Pres., Gregory

R. Long; C.O.O., J.V. Cossaboom; Dean & Vice Pres. Science, Dr. James Miller; Vice Pres. Laboratory Science, Dr. Dennis W. Stevenson; Vice Pres. Institute of Economic Botany, Dr. Michael J. Balick; Vice Pres. Mktg. & Creative Svcs., Terry Skoda; Vice Pres. Institutional Advancement & Special Asst. to Pres., Jennifer Rominiecki; Vice Pres. External Affairs, Carrie Laney; Assoc. Vice Pres. Garden Retail & Business Devel., Richard Pickett; Assoc. Vice Pres. Communications, Melinda Manning; Dir. Herbarium, Dr. Barbara M. Thiers; Cur., Dr. Scott A. Mori; Cur., Dr. Christine Padoch; Cur., Dr. William R. Buck; Dir. Museum Shop, Margaret Csala.
Personnel Profile: Full-Time Paid 434; Part-Time Paid 125; Part-Time Volunteers 1,235; Interns 99.
Governing Authority: not-for-profit corporation. Tax-exempt 501(c)(3).
Institution Type/Description: Botanical Garden & Nature Center.
Collections: 250 acres including comprehensive tree & shrub collections, specialized garden & conservatory collections of tropical & subtropical plants; manuscripts & photo collection; 7.2 million specimens in herbarium; 1,000,000 items in the LuEsther T. Mertz Library; stock culture collection of micro-organisms, principally basidiomycetes; native woodland; horticultural exhibits; 50 gardens; plant collections. Historical Landmarks: 250-acre site (national landmark); Enid A. Haupt Conservatory (NYC landmark); 1840 Stone Mill.
Research Fields: floras & monographs, including flowering plants, ferns, fungi, mosses, plant ecology; plant geography, plant nutrition, physiology & pathology of fungi; plant-insect relationships; plant evolution; problems concerned with diseases & pests of ornamental plants; economic botany; ecosystem studies; urban horticulture; molecular systematics studies & plant genomics.
Facilities: 375,000-vol. library plus over one million non-book items on botanical art, photos, manuscripts, archives, Lord & Burnham architectural plans, seed & nursery catalogs available for public reference; research laboratories; reading room. Books & publications relating to botany & gardening and botanical & plant-related items for sale.
Activities: guided tours; lectures; concerts; workshops; formal education programs for adults & school children; permanent & temporary exhibitions; horticultural therapy training; School of Professional Horticulture two-year New York state licensed program; post-secondary work-study program; symposia for plant professionals.
Publications: field notes; Garden News; serials, Botanical Review; Brittonia; Economic Botany; monographs, Memoirs of New York Botanical Garden; North American Flora; Advances in Economic Botany; Flora Neotropica; reprint series & books.
Hours & Admission Prices: Grounds: April-Oct. Tues.-Sun. 10-6; Nov.-March Tues.-Sun. 10-5. Grounds: adults $10, senior citizens & students $5, children 2-12 $2; discounts to American Horticulture Society, American Public Gardens Association, Museum Council of New York City, American Assoc. of Museums, Macauly Honors College at CUNY, Cultural Passport Program & AAM members; children under 2, Wed., & Sat. 10am-11am no charge. All Garden Pass: adults $25 (peak), $20 (off-peak); seniors $18, children 2-12 $8. Parking: $12 per vehicle. &
Attendance: 820,000 (accurate)
Membership: Individual $75; Dual $100; Family $120; Supporting $250; Contributing $600; Sustaining $1,000; discounts to seniors.

NEW YORK YANKEES MUSEUM, One E. 161st St., Bronx, NY 10451-2100. Tel.: 646-9778687.
Key Personnel: Cur., Brian Richards
Institution Type/Description: Baseball Museum.
Collections: baseball & franchise history; personal artifacts; photographs; baseball memorabilia; monuments.
Facilities: Museum-related items for sale.
Hours & Admission Prices: Museum: Game Day - from the time the gates open until the end of the 8th inning. No charge. Non-Game Days: museum access is part of the Yankee Stadium tour. Yankee Stadium Tours: May-Sept. call for hours. Tours: adults $20, children 14 & under and seniors 60 & over $15.

VALENTINE-VARIAN HOUSE/MUSEUM OF BRONX HISTORY, Varian House Park, 3266 Bainbridge Ave. at E. 208th St., Bronx, NY 10467. Mailing Address: Bronx County Historical Society, 3309 Bainbridge Ave., Bronx, NY 10467-2850. Tel.: 718-881-8900. Fax: 718-881-4827.
E-mail: kmcauley@bronxhistoricalsociety.org
Web Site: www.bronxhistoricalsociety.org
Founded: 1955.
Congressional District: 13
Key Personnel: C.E.O., Dr. G. Hermalyn; Pres. (V), Jacqueline Kutner; Dir. Museums, Kathleen McAuley; Educator, Angel Hernandez.
Personnel Profile: Full-Time Paid 4; Part-Time Paid 5; Part-Time Volunteers 33.

Governing Authority: nonprofit historical & preservation society. Parent Institution: Bronx County Historical Society. Tax-exempt.
Institution Type/Description: Historical House: second oldest house in the Bronx built by Isaac Valentine in 1758; sold to Isaac Varian in 1792.
Collections: 18th-19th century furnishings; domestic artifacts; herb garden; monument.
Research Fields: New York City history, colonial period through 20th-century.
Facilities: herb garden.
Activities: guided tours.
Publications: semi-annual journal, Bronx County Historical Society; annual books & booklets; maps.
Hours & Admission Prices: Sat. 10-4, Sun. 1-5, Mon.-Fri. tours by appointment. Adults $5, seniors, students & children $3; discounts to AAM & ICOM members; members & Historic House Trust of NYC no charge.
Attendance: 46,934 (estimated)
Membership: Student & Senior $20; Individual $25; Non-profile & Libraries $30; Family $35; Sustaining $100; Fellow & Corporate $250; Benefactor $2,500; Patron $5,000.

VAN CORTLANDT HOUSE MUSEUM, Van Cortlandt Park, Broadway at 246th St., Bronx, NY 10471. Mailing Address: 6393 Broadway, Bronx, NY 10471-2798. Tel.: 718-543-3344. Fax: 718-543-3315. TDD: 800-281-5722.
E-mail: info@vancortlandhouse.org
Web Site: www.vancortlandhouse.org
Founded: 1896.
Congressional District: 15
Key Personnel: Dir., Laura Carpenter; Pres., Ann Crawford; Museum Shop Mgr., Juana Vasquez.
Personnel Profile: Full-Time Paid 2; Part-Time Paid 2; Part-Time Volunteers 50.
Governing Authority: society. Parent Institution: National Society of Colonial Dames in the State of New York. Tax-exempt: 501(c)(3).
Institution Type/Description: Historic House: c.1748 house built by Frederick Van Cortlandt.
Collections: 17th- & 18th-century decorative & useful arts.
Research Fields: 18th-century social history; decorative arts; the American Colonial Revival.
Facilities: assembly room; Colonial Revival herb garden. Gift items for sale.
Activities: formal education programs for children; guided tours; lectures; living history demonstrations. Annual Event: Candlelight tours.
Publications: newsletter, To Absent Friends.
Hours & Admission Prices: Tues.-Fri. 10-3, Sat.-Sun. 11-4; groups by appointment. Adults $5, senior citizens & students $3; discounts to AAM members; children under 12, Wed. & members of Colonial Dames no charge. Closed major holidays.
Attendance: 8,600 (accurate)
Membership: Student & Senior $20; Friend $25; Family $35; Sustaining Friend $100; Sponsoring Friend $500.

WAVE HILL, W 249th St. & Independence Ave., Bronx, NY 10471-2899. Mailing Address: 675 W. 252nd St., Bronx, NY 10471-2899. Tel.: 718-549-3200. Fax: 718-884-8952. Facebook: Wave Hill.
E-mail: information@wavehill.org
Web Site: www.wavehill.org
Founded: 1965.
Congressional District: 17
Key Personnel: Exec. Dir. & Pres., Claudia Bonn; Chm. (V), Cathy Marks Weinroth; Deputy Dir., Michele Rossetti; Dir. Horticulture, Scott Canning; Dir. Mktg. & Communications, Mary Weitzman; Dir Visitor Svcs., Michael Wiertz; Dir. Devel., Barbara Giordano; Dir. Arts & Sr. Cur., Jennifer McGregor; Dir. Education, Debra Epstein; Dir. Facilities & Capital Projects, Frank Perrone; Shop Mgr., Jenah Barry.
Personnel Profile: Full-Time Paid 39; Part-Time Paid 22; Part-Time Volunteers 80; Interns 50.
Volunteer Hours: 1,700
Operating Income: 5,200,000
Governing Authority: nonprofit organization. Tax-exempt.
Institution Type/Description: Nature Center.
Collections: cactus & succulents; tropical plants; herbs; aquatic garden; flower garden; arboretum; historic house; contemporary art.
Major Exhibits: Sunroom Project Space 2013, 4/13-12/14.
Research Fields: horticulture; land management; inquiry-based education.
Facilities: gardens; greenhouse; conference room; 28-acre estate; 150-seat conference performing hall.
Activities: lectures; concerts; guided tours; school education programs in science & art; classes; art & literary workshops; visual & performing arts

programs; wellness programs; cooking workshops & demonstrations; beekeeping; birding; temporary exhibitions.
Publications: quarterly newsletter; trail & garden guides; exhibition catalogues.
Hours & Admission Prices: March 15-Oct. Tues.-Sun. 9-5:30; Nov.-March 14 Tues.-Sun. 9-4:30. Adults $8, seniors 65 & over and students $4, children 6-18 $2; discounts to AAM members; children under 6, Tues & Sat. 9-12 and members no charge. Closed New Year's Day; Thanksgiving; Christmas. &
Attendance: 131,000 (estimated)
Membership: Seniors 65 & over $35; Individual $40; Senior Couple $65; Family & Dual $75; Supporting $250; Patron $500; Wave Hill Partner $1,500.

Bronxville

EASTCHESTER HISTORICAL SOCIETY - 1835 MARBLE SCHOOLHOUSE, 390 California Rd., Bronxville, NY 10708. Mailing Address: P.O. Box 37, Eastchester, NY 10709-0037. Tel.: 914-793-1900. Facebook: Eastchester Historical Society.
E-mail: marbleschoolhouse@yahoo.com
Founded: 1959.
Congressional District: 23
Key Personnel: Pres. (V), Sheila Marcotte.
Governing Authority: nonprofit. Tax-exempt: 501(c)(3).
Institution Type/Description: Historical Society Museum: housed in c.1835 Marble School House & separate library building.
Collections: toys; 19th-century costumes; archives; photographs; furniture; manuscripts; 19th-century juvenile literature; costumes & textiles; school artifacts; 1881-1913 birth records. Historic Building: restored c.1835 one-room school.
Research Fields: local history; St. Paul's National Historic Site; Eastchester history; genealogy.
Facilities: 6,000-vol. research library of general history of New York State, New York City, Westchester County & Eastchester; Mt. Vernon; Tuckahoe; Bronxville; juvenile literature of 19th century.
Activities: guided tours; lectures; films; formally organized education programs for children; inter-museum loan, temporary & traveling exhibitions; school loan service.
Publications: books, Records of the Town of Eastchester, New York; Minutes of the Town of Eastchester; The Book of Strays & the Alteration of Roads; Overseers of the Poor; Minutes of the Trustees of Public Lands, Eastchester, New York; Town Property; Register of the Proceedings of St. Paul's Church at Eastchester; Burial Records of St. Paul's Church; military records of the town of Eastchester; Book of Births, Eastchester, New York 1881-1913.
Hours & Admission Prices: By appointment. Schoolhouse: adults $3; members no charge.
Attendance: 300 (estimated)
Membership: Individual/Family $30; Business $75.

Brooklyn

A.I.R. GALLERY, 111 Front St., #228, Brooklyn, NY 11201. Mailing Address: 55 Washington St., Brooklyn, NY 11201. Tel.: 212-255-6651. Fax: 212-255-6653. Facebook: A.I.R. Gallery.
E-mail: info@airgallery.org
Web Site: www.airgallery.org
Founded: 1972.
Key Personnel: Dir., Julie Lohnes; Gallery Asst., Jacqueline Ferrante
Institution Type/Description: Art Gallery.
Collections: works by local, national & international women artists.
Major Exhibits: CURRENTS - Curated by Mira Schor, 1/9/14-2/2/14.
Hours & Admission Prices: Wed.-Sun. 11-6. No charge; donations accepted. &

BAC GALLERY, 111 Front St., Ste. 218, Brooklyn, NY 11201. Tel.: 718-625-0080.
E-mail: gallery@brooklynartscouncil.org
Web Site: www.brooklynartscouncil.org
Key Personnel: Dir., Courtney J. Wendroff; Pres., Ella J. Weiss.
Governing Authority: Parent Institution: Brooklyn Arts Council. Tax-exempt.
Institution Type/Description: Art Gallery.
Collections: paintings; photographs; sculpture; drawings.
Activities: special events.
Hours & Admission Prices: Mon.-Fri. 10-5. No charge; donations accepted. Closed holidays.

BLACK & WHITE GALLERY, 483 Driggs Ave., Brooklyn, NY 11211. Tel.: 718-599-8775. Fax: 718-599-8798.
E-mail: info@blackandwhiteartgallery.com
Web Site: www.blackandwhiteartgallery.com
Founded: 2002.
Key Personnel: Dir., Tatyana Okshteyn
Institution Type/Description: Art Gallery.
Collections: works by contemporary artists; paintings; sculpture; drawings.
Activities: special events.
Hours & Admission Prices: Sept.-June Fri.-Sun. 12-6; other times by appointment.

BOSE PACIA, 163 Plymouth St., Brooklyn, NY 11201. Tel.: 212-989-7074. Fax: 212-989-6982.
E-mail: mail@bosepacia.com
Web Site: www.bosepacia.com
Key Personnel: Dir., Rebecca Davis
Institution Type/Description: Art Gallery.
Collections: works by contemporary South Asian artists; paintings; drawings.
Activities: special events.
Hours & Admission Prices: Tues.-Sat. 11-6.

BRIC CONTEMPORARY ART, 33 Clinton St., Brooklyn, NY 11201-2706. Tel.: 718-683-5604. Fax: 718-488-0609.
E-mail: contemporaryart@bricartsmedia.org
Web Site: www.bricartsmedia.org/contemporary-art
Formerly: The Rotunda Gallery
Founded: 1981.
Congressional District: 13
Key Personnel: Dir., Elizabeth Ferrer; Chm. (V), Lizanne Fontaine; Exec. Dir. BRIC Arts/Media/Bklyn, Leslie Schultz.
Personnel Profile: Full-Time Paid 4; Part-Time Paid 2; Interns 5.
Governing Authority: not-for-profit organization. Parent Institution: BRIC/Arts/Media/Bklyn (formerly BRIC/Brooklyn Information & Culture). Tax-exempt.
Institution Type/Description: Art Museum.
Collections: works in all media focusing on emerging and mid-career Brooklyn-affiliated artists; online registry of Brooklyn-affiliated artists.
Research Fields: contemporary Brooklyn-affiliated artists.
Facilities: educational facilities.
Activities: outreach educational school programs; Lori Ledis Emerging Curator program; training programs for Brooklyn-based visual artists; media fellowship for visual artists in conjunction with BRIC's community media program.
Publications: exhibition catalogs.
Hours & Admission Prices: Tues.-Sat. 12-6. No charge; donations accepted. &
Attendance: 12,000 (accurate)

BROOKLYN BOTANIC GARDEN, 1000 Washington Ave., Brooklyn, NY 11225-1099. Tel.: 718-623-7200. Fax: 718-857-2430.
E-mail: presidentsoffice@bbg.org
Web Site: www.bbg.org
Founded: 1910.
Congressional District: 13
Key Personnel: Chm. Bd., Frederick Bland; C.E.O. & Pres., Scott Medbury; Vice Pres. Mktg., Noreen Bradley; Dir. Emeritus, Elizabeth Scholtz; Vice Pres. Finance & C.F.O., Keith L. Stubblefield; Vice Pres. Science, Steven Clemants; Dir. Individual Giving & Corporate Sponsorship, Elizabeth Fallon Culp; Vice Pres. Devel., Leslie Findlen; Vice Pres. Education, Sharon Myrie; Dir. Library Svcs., Patricia Jonas; Dir. Children & Family Programs, Marilyn Smith; Dir. Security, Anthony Quarless; Dir. Capital Projects, Ralph Morgan; Dir. Continuing Education, Julie Warsowe; Dir. Government & Community Affairs, Bahia Ramos; Dir. Public Affairs, Leeann Lavin; Dir. Science, Gerry Moore; Dir. Program, BGCI (US), Dan Shepard; Dir. Computer Technology, Paul Turcotte; Dir. Institutional Funding, Kirsten Munro; Dir. Human Resources, Rochelle Cabiness; Dir. Horticulture, Jacqueline Fazio; Dir. Mktg., Marie Leahy; Dir. Facilities, Gerard Rudloff; Dir. Brooklyn Greenbridge, Ellen Kirby; Dir. Publishing, Janet Marinelli; Dir. Visitor Svcs. & Volunteers, Louis Cesario; Dir. Public Programs, Anita Jacobs; Dir. Major Gifts, Sarah Young.
Personnel Profile: Full-Time Paid 154; Part-Time Paid 96; Part-Time Volunteers 646; Interns 23.
Governing Authority: nonprofit organization. Tax-exempt.
Institution Type/Description: Arboretum & Botanical Garden.
Collections: 14,000 different Taxa; Japanese Hill & Pond Garden; rock garden; herb garden; fragrance garden; local flora section; rose garden; Shakespeare garden; children's garden; dwarf conifer collection; special collections of rhododendrons, lilacs, cherries, magnolias, conifers, water lilies; conservatories containing tropical plants, cacti & succulents, bromeliads, ferns, bonsai, orchids; insectivorous plants; house plants; aquatic plants; trail of evolution.
Research Fields: plant taxonomy.

Facilities: 55,000-vol. library of botany; horticulture; science available for inter-library loan & for use on premises; separate laboratory operation; reading room; classrooms; herbarium; four classrooms & three teaching greenhouses; discovery center for children; Palm House for special events; exhibition gallery; house museum; 44,000 sq. ft. conservatory, children's garden; garden-produced documentary films on horticultural subject & slide shows available for purchase or rent. Plants, books, horticultural supplies & other museum-related items for sale.

Activities: guided tours; lectures; films; formally organized education programs; docent program; permanent & temporary exhibitions; plant information service; plant & disease identification service.

Publications: quarterly handbook, 21st Century Gardening Series; quarterly newsletter, P & G News; monograph, Brooklyn Botanic Garden Contributions, educational booklets for parents & teachers; gardening booklet & video for children & parents.

Hours & Admission Prices: Garden: Tues.-Fri. 8-4:30, Sat.-Sun. 10-4:30. Adults $8, seniors 65 & over and students 12 & over $4; children under 12, members, school groups, & Tues. no charge. &

Attendance: 685,000 (estimated)

Membership: Subscriber & Senior Citizen Individual $35; Individual $40; International Subscriber $45; Senior Citizen Family Dual $70; Family Dual $75; Family Dual Plus $95; Signature $150; Sponsor $300; Patron $500; Gager Society $1,500.

---

* **BROOKLYN CHILDREN'S MUSEUM,** 145 Brooklyn Ave., Brooklyn, NY 11213-1900. Tel.: 718-735-4400. Fax: 718-604-7442. TDD: 718-735-4402.

Web Site: www.brooklynkids.org
Founded: 1899.
Congressional District: 12
Key Personnel: Pres. & C.E.O., Georgina Ngozi; Chm., William Rifkin; Dir. Collections, Beth Alberty.
Personnel Profile: Full-Time Paid 43; Part-Time Paid 49; Interns 20.
Governing Authority: nonprofit organization. The Brooklyn Children's Museum Corp., 145 Brooklyn Ave., Brooklyn 11213. Tax-exempt: 170(b)(1)(A), 501(c)(3) & 509(a)(1).
Institution Type/Description: General Children's Museum.
Collections: ethnology & natural history; folk crafts worldwide; hands-on exhibitions.
Facilities: children's library; outdoor theater; indoor theater; garden; cafe; party room; classrooms. Museum-related items for sale.
Activities: art, science & cultural interactions; inter-museum loan, permanent & temporary exhibitions; classes; public programs; school loan program of cultural history, natural history & science items.
Publications: educational materials; quarterly newsletter, annual report.
Hours & Admission Prices: Tues.-Sun. 10-5, 3rd Thurs. each month 10-7. Adults $9; discounts to AAM, ASTC & ACM members; members no charge. Closed New Year's Day; Thanksgiving; Christmas. &
Attendance: 250,000 (accurate)
Membership: Pioneer $85; Trailblazer $100; Globetrotter $135; Voyager $175; Explorer $275.

THE BROOKLYN HISTORICAL SOCIETY, 128 Pierrepont St., Brooklyn, NY 11201-2711. Tel.: 718-222-4111. Fax: 718-222-3794.

Web Site: www.brooklynhistory.org
Formerly: Long Island Historical Society
Founded: 1863.
Congressional District: 13
Key Personnel: Pres., Deborah Schwartz; Chm. (V), James Rossman; Vice Pres. Exhibits & Education, Kate Fermoile; Dir. Education, Emily Potter-Ndiaye; Dir. Finance & Operations, Jason Pietrangeli; Reference Librarian, Elizabeth Call; Museum Shop Mgr., Janice Monger.
Personnel Profile: Full-Time Paid 24; Part-Time Paid 6; Part-Time Volunteers 10; Interns 10.
Governing Authority: bd. trustees. Tax-exempt: 501(c)(3).
Institution Type/Description: History Museum: housed in building designed by architect, George B. Post.
Collections: paintings, graphics, sculpture, photographs, costumes; books; manuscripts; ephemera. Historic 1881 Building: houses museum & library.
Research Fields: Brooklyn history 16th-century to present.
Facilities: library of local history books.
Activities: lectures; walking & bus tours; family activities; genealogy workshops; school, after-school & summer programs; performances; readings; rental facilities.
Publications: Neighborhood History Guides.
Hours & Admission Prices: Wed.-Fri. & Sun. 12-5, Sat. 10-5. Adults $6, students & seniors $4; discounts to AAM members; members no charge. Closed New Year's Day; Independence Day; Thanksgiving; Christmas. &

Attendance: 15,000 (estimated)
Membership: Student, Senior Citizen & Teacher $35; Friend $50; Family & Partner $70; Advocate $125; Champion $275; Community Leader $550; President's Circle $1,500 & up.

---

* **BROOKLYN MUSEUM, (M),** 200 Eastern Pkwy., Brooklyn, NY 11238-6099. Tel.: 718-638-5000. Fax: 718-501-6300. TDD: 718-783-6501.

E-mail: information@brooklynmuseum.org
Web Site: www.brooklynmuseum.org
Formerly: Brooklyn Museum of Art
Founded: 1823.
Congressional District: 2
Key Personnel: Chm., John S. Tamagni; Dir., Arnold L. Lehman; Deputy Dir. Institutional Advancement, Cynthia Mayeda; Deputy Dir. Administration, Judith Frankfurt; Vice Chair, Barbara Knowles Debs; Cur. American Painting & Sculpture, Teresa Carbone; Deputy Dir. Devel., Paul Johnson; Vice Dir. Operations, Frantz Vincent; Vice Dir. Planning & Architecture, Ann Webster; Vice Dir. Merchandising, Sallie Stutz; Vice Dir. Finance & Administration, Anna Lee; Vice Dir. Education & Program, Radiah Harper; Chief Conservator & Vice Dir. Collections, Kenneth Moser; Government & Community Rels. Officer, Terri Jackson; Public Information Officer, Sally Williams; Publications & Editorial Svcs., James Leggio; Cur. John & Barbara Yogelstein, Eugenie Tsai; Cur. Arts of the Americas, Nancy Rosoff; Cur. Decorative Arts, Kevin Stayton; Chm. Egyptian, Classical & Ancient Middle Eastern Art, Edward Bleiberg; Lisa and Bernard Selz Cur., Joan Cummins; Asst. Cur. Africa & Pacific Islands, Kevin Dumouchelle; Chief Registrar, Liz Reynolds; Museum Shop Mgr., Tracy Boni; Community Involvement Mgr., Schawannah Wright; Technology, Chief, Shelley Bernstein; Community Committee Chair, Arlene Gilden; Head Librarian, Deirdre Lawrence.
Personnel Profile: Full-Time Paid 302; Part-Time Paid 52; Part-Time Volunteers 285; Interns 29.
Governing Authority: nonprofit corporation. Parent Institute: The Brooklyn Institute of Arts & Sciences. Subsidiary Institution: Brooklyn Museum Fund, Inc. Tax-exempt: 501(c)(3).
Institution Type/Description: Art Museum.
Collections: ancient Egyptian art, predynastic through Coptic; Greek & Roman art; Islamic & pre-Islamic Middle Eastern art; Asian art; pre-Columbian Central & South American collections; American Indian collections; art of Africa & of the Pacific; European & American prints, drawings & photographs; 27 American period rooms 1675-1930; American & European decorative arts; costumes & textiles; American paintings & sculpture, colonial to contemporary; European paintings & sculpture medieval to 20th century; contemporary paintings, sculpture & feminist art; outdoor sculpture garden.
Research Fields: collection-related research; archaeological expeditions to Egypt.
Facilities: 125,000-vol. library of art reference, featuring art history books; periodicals and the Wilbour Library of Egyptology; reading room; newly renovated Schapiro Wing provides an additional 30,000 sq. ft. of gallery space; museum cafe; 460-seat Iris & B. Gerald Cantor Auditorium. Books, international folk art & exclusive items based on designs from the museum's collection for sale.
Activities: guided tours; lectures; films; gallery talks; concerts; dance recitals; arts festivals; formally organized education programs; curatorial & technician apprenticeships; Wilbour fellowships in Egyptology; permanent & temporary exhibitions; late weekday events.
Publications: handbooks; guides to various collections; Wilbour monographs on ancient art; catalogs of major exhibitions.
Hours & Admission Prices: Wed. & Fri.-Sun. 11-6, Thurs. 11-10, first Sat. every month 11-11. Suggested Donations: adults $12, students & senior citizens $8; discounts to AAM & ICOM members; members no charge. Closed New Year's Day; Thanksgiving; Christmas. &
Attendance: 325,501 (accurate)
Membership: 1st Fans $20; Individual $55; Dual $85; Family $100; Contributor $150; Patron $350; Donor $600; Fellow $1,000; Curator's Circle $2,500; Benefactor's Circle $5,000.

CONEY ISLAND MUSEUM, 1208 Surf Ave., Brooklyn, NY 11224-2816. Tel.: 718-372-5159. Fax: 718-372-5101.

E-mail: info@coneyisland.com
Web Site: www.coneyisland.com/museum.shtml
Key Personnel: Museum Dir., Aaron Beebe; Mng. Dir., David Gratt
Institution Type/Description: History Museum.
Collections: history of Coney Island; the Steeplechase horse; boardwalk rolling chair; Funhouse distortion mirrors; period souvenirs & bumping cars.
Facilities: Museum-related items for sale.

Activities: lectures; films; walking tours; private parties; rental facilities; research; weddings.
Hours & Admission Prices: Sat.-Sun. 12-5. Admission $.99.

DUMBO ARTS CENTER, 30 Washington St., Ste. 2, Brooklyn, NY 11201-1007. Tel.: 718-694-0831. Fax: 718-694-0867.
E-mail: gallery@dumboartscenter.org
Web Site: www.dumboartscenter.org
Governing Authority: nonprofit organization. Tax-exempt: 501(c)(3).
Institution Type/Description: Art Gallery.
Collections: works by local & national artists.
Activities: educational programs; internships.
Hours & Admission Prices: Thurs.-Sun. 12-6. Suggested Donation: $2.

FRANKLIN FURNACE ARCHIVE, INC., 80 Arts-The James E. Davis Arts Building, 80 Hanson Pl., #301, Brooklyn, NY 11217-1506. Tel.: 718-398-7255. Fax: 718-398-7256.
E-mail: mail@franklinfurnace.org
Web Site: www.franklinfurnace.org
Founded: 1976.
Congressional District: 10
Key Personnel: C.E.O. & Dir., Martha Wilson; Deputy Dir., Harley Spiller; Program Coord., Eben Shapiro; Archivist, Michael Katchen.
Personnel Profile: Full-Time Paid 2; Part-Time Paid 2; Interns 50.
Governing Authority: nonprofit organization. Tax-exempt.
Institution Type/Description: Art Museum.
Collections: installation & performance art documentation.
Research Fields: artists' publishing from 1900 to present; performance art.
Activities: multimedia performance art presentations.
Publications: biannual calendar; CD-ROMs; DVDs; cyber art.
Hours & Admission Prices: Mon.-Fri. 10-6. No charge.
Membership: Common Denominators $34 & up; Multipliers $68 & up; Cardinals and Ordinals $100 & up; Calculators $250 & up; Statisticians $500 & up; Croupiers $1,000 & up; Supernumeraries $10,000 & up.

HARBOR DEFENSE MUSEUM, Harbor Defense Museum of Fort Hamilton, 230 Sheridan Loop, Brooklyn, NY 11252-9523. Tel.: 718-630-4349. Fax: 718-630-4888.
E-mail: info@harbordefensemuseum
Web Site: www.harbordefensemuseum.com
Founded: 1980.
Congressional District: 14
Key Personnel: Dir., Paul Morando.
Personnel Profile: Full-Time Paid 2; Part-Time Volunteers 23.
Governing Authority: federal. Parent Institution: Department of Defense. Subsidiary Institution: United States Army. Tax-exempt.
Institution Type/Description: Military Museum: housed in 1825-1831 Fort Hamilton.
Collections: 17th-century to present U.S. military; coast defense 1800-1950; artifacts reflecting Lt. Col. Rodman's career; artifacts, models, images & dioramas detailing New York's harbor defenses including an original Pattern 1844, 24-pounder flank howitzer & projectiles; operating mutoscope; NY National Guard uniforms; M1883 Gatling gun; French M1763 Charleville musket; World War II U.S. infantry weapons.
Research Fields: coast defense; fortification; New York military history.
Facilities: reference library & archives.
Activities: guided tours; temporary exhibits; lectures & presentations; classroom.
Hours & Admission Prices: Mon.-Fri. & first Sat. of the month 10-4. No charge. Closed Federal holidays. ♿
Attendance: 24,000 (estimated)

JEWISH CHILDREN'S MUSEUM, 792 Eastern Pkwy., Brooklyn, NY 11213-3502. Tel.: 718-467-0600. Fax: 718-467-1300. Facebook: Jewish Children's Museum.
E-mail: info@jcm.museum
Web Site: www.jcm.museum
Governing Authority: Tax-exempt.
Institution Type/Description: Children's Museum.
Collections: Jewish cultural heritage & history; hands-on exhibits; photographs.
Facilities: library; restaurant. Museum-related items for sale.
Activities: facilities rental; special events; shows.
Hours & Admission Prices: Mon.-Thurs. 10-4, Sun. 10-6; school & youth groups call 718-907-8888 for group bookings. Admission $13, seniors $10; children under 2 no charge.

LEFFERTS HISTORIC HOUSE, Flatbush Ave., Prospect Park, Brooklyn, NY 11215-3709. Mailing Address: Prospect Park Alliance, 95 Prospect Park W., Brooklyn, NY 11215-3709. Tel.: 718-789-2822. Fax: 718-789-4724.
E-mail: lefferts@prospectpark.org
Web Site: www.prospectpark.org
Formerly: Lefferts Homestead
Founded: 1918.
Congressional District: 10 & 12
Key Personnel: Prospect Park Admin., Emily Lloyd; Dir., Elyse Newman.
Personnel Profile: Full-Time Paid 3; Part-Time Paid 6; Part-Time Volunteers 3.
Governing Authority: Parent Institution: New York City Dept. of Parks & Recreation, Prospect Park Alliance, Inc. Tax-exempt.
Institution Type/Description: Historic House: built c.1783 Dutch-American architecture.
Collections: period furniture; historic documents; household items.
Research Fields: everyday lives of the people living in western Long Island in the early 19th-century, including European, African or Native American ancestry.
Facilities: demonstration garden.
Activities: school tours; research games; craft demonstrations; concerts; family programs; storytelling; crafts for kids; special events; games.
Publications: brochures; seasonal calendar; newsletter.
Hours & Admission Prices: April-June & Sept.-Nov. Thurs.-Sun. & holidays 12-5; July-Aug. Thurs.-Sun. 12-6; Dec.-March Sat.-Sun. & school holidays 12-4. No charge; donations accepted. Closed New Year's Day; Thanksgiving; Christmas Eve & Day. ♿
Attendance: 45,000 (accurate)

LESBIAN HERSTORY EDUCATIONAL FOUNDATION, INC. AKA LESBIAN HERSTORY ARCHIVES, 484 14th St., Brooklyn, NY 11215-5702. Tel.: 718-768-3953. Fax: 718-768-4663.
E-mail: lesbianherstoryarchives@gmail.com
Web Site: www.lesbianherstoryarchives.org
Founded: 1974.
Key Personnel: Treas., Deborah Edel.
Personnel Profile: Part-Time Volunteers 25; Interns 6.
Volunteer Hours: 1,000
Operating Expenses: 77,000
Operating Income: 67,500
Governing Authority: private; nonprofit organization. Tax-exempt: 501(c)(3).
Institution Type/Description: Lesbian History & Culture Museum.
Collections: photographs; books; videos; archival records; oral histories; subject files; ephemera.
Research Fields: preservation of lesbian history & culture.
Facilities: 15,000-vol. library available to the public.
Activities: guided tours; loan, traveling & temporary exhibitions; slide show. Annual Event: At Home Series.
Hours & Admission Prices: See website for hours. No charge; donations accepted.
Attendance: 3,000 (estimated)

MOCADA - THE MUSEUM OF CONTEMPORARY AFRICAN DIASPORAN ARTS, James E. Davis Art Bldg., 80 Hanson Place, Brooklyn, NY 11217-1506. Tel.: 718-230-0492. Fax: 718-230-0246.
E-mail: lc@mocada.org
Web Site: www.mocada.org
Key Personnel: Exec. Dir., Laurie Cumbo; Dir. Exhibitions, Kimberli Grant; Dir. Education, Regine Romain; Museum Shop Mgr., Paul Bispo
Institution Type/Description: Art Museum.
Collections: African American history & culture; paintings; sculpture; photographs.
Facilities: Museum-related items for sale.
Activities: educational programs; special events; internships.
Hours & Admission Prices: Wed.-Sun. 11-6. Adults $4, students $3; children 12 & under no charge.

NEW YORK AQUARIUM, (M), Surf Ave. & W. 8th St., Brooklyn, NY 11224-3495. Tel.: 718-265-3400 & 3405. Fax: 718-265-3482.
E-mail: brusso@wcs.org
Web Site: www.nyaquarium.org
Founded: 1896.
Congressional District: 13
Key Personnel: Pres. & C.E.O., Cristian Samper; Dir., Jon F. Dohlin; Dir. Operations, Dennis Ethier; Cur., Dave DeNardo; Mgr. Public Rels., Barbara Russo; Cur. Animal Hospital, Kate McClave.
Personnel Profile: Full-Time Paid 88; Full-Time Volunteers 20; Part-Time Paid 110; Part-Time Volunteers 270; Interns 5.

Governing Authority: nonprofit organization. Parent Institution: The Wildlife Conservation Society, Bronx Park, Bronx, NY 10460. Tax-exempt: 170(b)(1)(A).
Institution Type/Description: Aquarium.
Collections: living aquatic animals.
Research Fields: all aspects of aquatic animal biology; fish genetics.
Facilities: 2,000-vol. library of books & periodicals pertaining to aquatic environment available for use on premises; separate laboratory operation; 200-seat auditorium; 1,800-seat sea lion arena; restaurant. Aquatic & other museum-related items for sale.
Activities: lectures; films; gallery talks; formally organized education programs; permanent & temporary exhibitions; docent program; private events; public events.
Hours & Admission Prices: Memorial Day-Labor Day Mon.-Fri. 10-5, Sat.-Sun. & holidays 10-7; Sept.-May daily 10-4:30. Adults $14.95, senior citizens $11.95, children $10.95; discount to student groups. Fri. 3-5 admission by donation. &
Attendance: 767,026 (accurate)
Membership: Individual $104; Family $129.

PROSPECT PARK ZOO, (M), 450 Flatbush Ave., Brooklyn, NY 11225-3707. Tel.: 718-399-7339. Fax: 718-399-7337.
Web Site: www.prospectparkzoo.com
Formerly: Prospect Park Wildlife Center
Founded: 1993.
Congressional District: 11
Key Personnel: Chm. (V), Ward Wood; Pres., Dr. Steven Sanderson; Sr. Vice Pres. & Gen. Dir., Robert Cook; Sr. Vice Pres. & C.F.O., Patti Calabrese; Supervising Librarian & Archivist, Steve Johnson; Dir., Dr. Donald Moore; Asst. Dir. & Cur. Animals, Dr. Patricia Cole; Cur. Education, Karen Tingley; Public Rels., Barbara Russo.
Personnel Profile: Full-Time Paid 67; Part-Time Volunteers 48.
Governing Authority: private; nonprofit organization. Parent Institution: Wildlife Conservation Society, 185th St. & Southern Blvd., Bronx, NY 10460. Tax-exempt: 501(c)(3).
Institution Type/Description: Zoo.
Collections: wildlife from around the world.
Research Fields: Wyoming toads; intake & digestion in hyrax; animal cognition & behavior.
Facilities: 6,183-vol. library; 2 classrooms; 34,000 sq. ft. exhibit space; zoological park; discovery center.
Activities: docent program; formal education programs; participatory exhibits. Annual Events: Fleece Festival in April; Keeping up with Keepers in September; Boo at the Zoo in October; Presents to the Animals in December.
Publications: bimonthly magazine, Wildlife Conservation; biannual newsletter, Wildlife; quarterly newsletter, Membership News.
Hours & Admission Prices: Winter: daily 10-4:30; Summer: Mon.-Fri. 10-5, Sat.-Sun. 10-5:30. Adults $7, senior citizens $4, children 3-12 $3; members and children under 3 no charge. &
Attendance: 232,426 (accurate)
Membership: Senior $33; Individual $75; Family $120; Conservation Supporter $250; Conservation Partner $750; Patron $1,500.

ROBERT LEHMAN GALLERY AT URBANGLASS, 57 Rockwell Place, 3rd Fl., Brooklyn, NY 11217-1112. Mailing Address: 126 13th St., Brooklyn, NY 11215-4632. Tel.: 718-625-3685. Fax: 718-625-3889.
E-mail: info@urbanglass.org
Web Site: www.urbanglass.org
Founded: 1977.
Key Personnel: Exec. Dir., Dawn Bennett.
Governing Authority: Tax-exempt.
Institution Type/Description: Art Gallery.
Collections: works created with glass by aspiring & established artists.
Publications: GLASS: The Urbanglass Art Quarterly.
Hours & Admission Prices: Tues.-Fri. 10-6, Sat.-Sun. 10-5. No charge.

THE RUBELLE & NORMAN SCHAFLER GALLERY, Pratt Institute, 200 Willoughby Ave., Brooklyn, NY 11205-3802. Tel.: 718-636-3517. Fax: 718-399-4230.
E-mail: exhibits@pratt.edu
Web Site: www.pratt.edu/exhibitions
Founded: 1984.
Congressional District: 11
Key Personnel: Dir., Nick Battis; Asst. Dir., Olivia Good; Exhibit Designer, Katherine Davis.
Personnel Profile: Full-Time Paid 2; Part-Time Paid 8.

Governing Authority: nonprofit. Parent Institution: Pratt Institute. Subsidiary Institution: Pratt Manhattan Gallery, 144 W. 14th St., 2nd Fl., New York, NY 10011. Tel.: 212-647-7778. Tax-exempt: 501(c)(3).
Institution Type/Description: Art & University Museum.
Collections: 19th-21st century paintings, sculptures, prints, photography, & graphic arts by European & American artists.
Research Fields: contemporary fine arts, design & architecture.
Facilities: 185,000-vol. library of books; 135,000 pictures, 50,000 slides, 22,000 microforms, & 130,000 government documents pertaining to art & architecture; 2,300 sq. ft. exhibit space; classrooms; multi-media center; 300-seat restaurant.
Activities: lectures; performances; panel discussions & receptions; organized education programs for undergraduate or graduate students affiliated with Pratt Institute.
Publications: exhibition announcements; biannual exhibitions & events calendars.
Hours & Admission Prices: Schafler Gallery: Mon.-Fri. 9-5, Sat. 12-5. Pratt Manhattan Gallery: Sept.-July Tues.-Sat. 11-6. No charge. Closed major holidays.
Attendance: 10,000 (estimated)

WATERFRONT MUSEUM, 290 Conover St., Pier 44, Brooklyn, NY 11231-1020. Tel.: 718-624-4719.
E-mail: dsharps@waterfrontmuseum.org
Web Site: www.waterfrontmuseum.org
Founded: 1986.
Congressional District: 8
Key Personnel: Pres. & C.E.O., David Sharps; Chm. (V), Alison Tocci.
Personnel Profile: Full-Time Paid 1; Part-Time Paid 4; Part-Time Volunteers 35.
Governing Authority: private; nonprofit organization. Tax-exempt: 501(c)(3).
Institution Type/Description: Maritime Museum: housed aboard the 1914 Lehigh Valley Railroad Barge #79 used during the Lighterage Era 1860-1960, when goods were transferred from port docks to railhead terminals by tug & barge. Listed on the National Register of Historic Places.
Collections: maritime artifacts.
Research Fields: creating docking for historic vessels in NYC.
Facilities: library containing all issues of Transfer Magazine; floating classroom; 149-seat showboat. Museum-related items for sale.
Activities: maritime & environmental education; cultural programs; community meetings; special events; knot-tying; concerts; formal education programs; guided tours; theater. Museum Sponsors: CIRCUSundays.
Hours & Admission Prices: Thurs. 4-8, Sat. 1-5; call to confirm; groups by appointment. Suggested Donation: $5. &
Attendance: 18,000 (estimated)

THE WYCKOFF FARMHOUSE MUSEUM, 5816 Clarendon Rd., Brooklyn, NY 11203-5444. Tel.: 718-629-5400. Fax: 718-629-3125. Facebook: The Wyckoff Farmhouse Museum.
E-mail: info@wyckoffassociation.org
Web Site: wyckoffassociation.org
Formerly: The Pieter Claesen Wyckoff House Museum
Founded: 1982.
Congressional District: 12
Key Personnel: Chm. (V), E. Lisk Wyckoff, Jr.; Pres., Naj Wikoff; Exec. Dir., Joshua Van Kirk; Gardener & Caretaker, Jason Gaspar.
Personnel Profile: Full-Time Paid 1; Part-Time Paid 2; Part-Time Volunteers 10.
Governing Authority: nonprofit organization. Parent Institution: Wyckoff House & Association, Inc. Tax-exempt: 501(c)(3).
Institution Type/Description: Historic House: c.1652 Dutch Colonial farmhouse.
Collections: colonial & early American furnishings; documents dating from 1670-1866.
Research Fields: Dutch-Colonial; New York history; labor & immigration history.
Facilities: colonial kitchen garden; picnic area. Museum-related items for sale.
Activities: education programs for children; guided tours; lectures; children's colonial crafts workshops. Annual Events: St. Nicholas Day Celebration; Pinkster Celebration; African Lives: From Wyckoff to Weeksville; Apple Festival.
Publications: annual bulletin; annual, Calendar of Events; quarterly newsletter.
Hours & Admission Prices: April-Oct. Tues.-Sun. 10-4; Nov.-March Tues.-Sat. 10-4. Adults $5, senior citizens & children $3; discounts to AAM & ICOM members; Wyckoff Assoc. & members no charge. Closed New Year's Day; Thanksgiving; Christmas. &
Attendance: 6,000 (estimated)
Membership: Basic $40; Family $60; Supporter $70-$80; Benefactor $115-$175; Patron $550-$1,100.

Brooklyn Heights

NEW YORK TRANSIT MUSEUM, (M), Corner of Boerum Pl. & Schermerhorn St., Brooklyn Heights, NY 11201. Mailing Address: 130 Livingston St., 10th Fl., Brooklyn, NY 11201-5106. Tel.: 718-694-1600 & 1873. Fax: 718-694-1791.
E-mail: gabrielle.shubert@nyct.com
Web Site: www.mta.info/museum
Founded: 1976.
Congressional District: 13
Key Personnel: Dir., Gabrielle Shubert; Chm. (V), Susan Gilbert; Deputy Dir., Regina Asborno; Tour Coord., Luz Montano; Asst. Dir. & Devel. Officer, Marcia Ely; Admin. Mgr., Angela Agard Solomon; Operations Mgr., Timothy Keiley; Sr. Mgr. Exhibits, Robert Del Bagno; Mgr. Education, Laura Kujo; Cur., Carissa Amash; Assoc. Cur., Chandra Buie; Coord. Education, Virgil Talaid; Archivist, Carey Stumm; Mgr. Retail & MTA Products Devel., Gail Goldberg; Museum Educator, Lynette Morse; Asst. Mgr. Retail Operations, Dorla Arnold.
Personnel Profile: Full-Time Paid 26; Part-Time Paid 26; Part-Time Volunteers 2; Interns 6.
Governing Authority: municipal; nonprofit. Parent Institution: Metropolitan Transportation Authority. Subsidiary Institution: Friends of NY Transit Museum. Gallery Annex: Grand Central Terminal. Tax-exempt.
Institution Type/Description: Urban Transportation Museum: located in a decommissioned subway station c.1936.
Collections: N.Y. transportation artifacts; history of public transportation in New York including subways, buses, bridges, tunnels & commuter railroads; 1904-1968 subway cars & elevated cars; signs; signals; 250,000 photographs; architectural & engineering drawings; uniforms; mosaics; tools & equipment.
Research Fields: New York Metropolitan region transportation history and the effects these have on the development of the region.
Facilities: 3,500-vol. archive pertaining to NYC Transit history; 18,000 sq. ft. exhibit space; 75-seat theater. Museum-related items for sale.
Activities: guided tours; lectures; films; organized education programs; docent program; participatory & temporary exhibitions; readings.
Publications: books; exhibition catalogues; oral history; educational materials; tri-annual calendar of events; members' online newsletter.
Hours & Admission Prices. Tues.-Fri. 10-4, Sat.-Sun. 12-5. Adults $6, senior citizens & children under 17 $4; discounts to AAM, ICOM & ASTC members; members no charge. GCT Annex: Mon.-Fri. 8-8, Sat.-Sun. 10-6. No charge. Closed New Year's Eve & Day; Memorial Day; Independence Day; Labor Day Weekend; Thanksgiving; Christmas. &
Attendance: 458,656 (accurate)
Membership: Senior Citizen, Student & MTA Employee $30; Friend $40; Family $55; Contributing $80; Sustaining $150; Patron $250.

Brookville

HILLWOOD ART MUSEUM, C. W. Post Campus, Long Island University, 720 Northern Blvd., Brookville, NY 11548-1319. Tel.: 516-299-4073. Fax: 516-299-2787.
E-mail: museum@cwpost.liu.edu
Web Site: www.liu.edu/museum
Founded: 1973.
Congressional District: 4
Key Personnel: Dir., Barbara Applegate; Pres., Dr. David Steinberg.
Personnel Profile: Full-Time Paid 2; Part-Time Paid 8; Part-Time Volunteers 2; Interns 3.
Governing Authority: university. Parent Institution: C. W. Post, Long Island University. Tax-exempt.
Institution Type/Description: Art Museum & Public Art Program.
Collections: paintings; prints; photography; sculpture; ethnographic.
Research Fields: pre-Columbian; Chinese; African; photography; European & American prints; textiles.
Facilities: located in student union complex: restaurant; lecture hall; cinema; concert theater.
Activities: tours; lectures; performances; public art program; exhibitions for academic & surrounding community; 6 major shows per year.
Publications: catalogues for each exhibition & public art program; newsletter, MuseumNews; exhibition catalogues; exhibition guides; study guides.
Hours & Admission Prices: June-July Mon.-Fri. 9-4:30; Sept.-May Mon.-Wed. & Fri. 9:30-4:30, Thurs. 9:30-8, Sat. 11-3. No charge; donations accepted. &
Attendance: 15,000 (accurate)
Membership: Call for information.

Brownville

GENERAL JACOB BROWN HISTORICAL SOCIETY, 116 E. Main St., Brownville, NY 13615. Tel.: 315-782-4508. Fax: 315-786-1178.
Founded: 1978.
Congressional District: 30
Key Personnel: Pres. (V) & Corresponding Sec., Constance G. Hoard; Vice Pres., Randy L. McIntyre.
Governing Authority: municipal; nonprofit. Tax-exempt.
Institution Type/Description: Historical Society Museum: housed in 1811-1815 General Brown Mansion.
Collections: Brown furnishings; period pieces; dishes & cooking utensils; tools; weapons; paintings.
Research Fields: Brown Family memorabilia; local history of Brownville; War of 1812.
Facilities: Village library of books on local Jefferson County History; research materials.
Activities: guided tours upon request; lectures; films; concerts; permanent & temporary exhibitions.
Publications: pamphlet describing mansion & General Brown's achievements are available with no charge upon request.
Hours & Admission Prices: By appointment only. No charge; donations accepted.
Membership: Individual $2; Family $5.

Buffalo

*** ALBRIGHT-KNOX ART GALLERY, (M),** 1285 Elmwood Ave., Buffalo, NY 14222-1096. Tel.: 716-882-8700. Fax: 716-882-8773. Facebook: Albright-Knox Art Gallery.
E-mail: mmorreale@albrightknox.org
Web Site: www.albrightknox.org
Founded: 1862.
Congressional District: 38
Key Personnel: Dir., Dr. Janne Siren; Deputy Dir., Joe Lin-Hill; Pres. (V), Leslie Zemsky; Sr. Cur., Douglas Dreishpoon, Ph.D.; Assoc. Cur., Holly Hughes; Registrar, Laura Fleischmann; Dir. External Affairs, Maria Morreale; Dir. Advancement, Jennifer Bayles; Museum Shop Mgr., Tracey Levy.
Personnel Profile: Full-Time Paid 65; Part-Time Paid 20; Part-Time Volunteers 195; Interns 21.
Governing Authority: nonprofit organization. Parent Institution: Buffalo Fine Arts Academy. Tax-exempt: 501(c)(3).
Institution Type/Description: Art Museum.
Collections: 18th-century English paintings; 19th-century French & American paintings; 20th-21st century painting & sculpture; 3000 B.C.-present, sculpture; drawings; graphics; photography.
Facilities: 30,000-vol. art reference library by appointment; print & drawing study room; 345-seat auditorium; classrooms; restaurant. Art books, catalogues, reproductions & other museum-related items for sale.
Activities: guided tours; lectures; films; concerts; rental sales gallery; organized education programs; docent program or council; inter-museum loan, permanent, temporary & traveling exhibitions; speakers' bureau; educational exhibitions; handicap programs; artists' workshop; school resource service; special programs for high school students; participatory tours.
Publications: Annual report; exhibition catalogs; quarterly newsletter.
Hours & Admission Prices: Mon.-Fri. 10-5, first Fri. every month 10-10. Adults $12; discounts to AAM members; members no charge. Closed New Year's Day; Independence Day; Thanksgiving; Christmas. &
Attendance: 143,653 (accurate)
Membership: Students $25; Senior Citizens $30; Individual $50; Family $75; Reciprocal $150; Contributing $300; Sustaining $500; Clyfford Still Circle $1,000; Andy Warhol Circle $2,500; Jackson Pollock Circle $5,000; Georgia O'Keeffe Circle $10,000; Henri Matisse Circle $25,000.

THE BENJAMIN & DR. EDGAR R. COFELD JUDAIC MUSEUM OF TEMPLE BETH ZION, 805 Delaware Ave., Buffalo, NY 14209-2005. Mailing Address: 700 Sweet Home Rd., Buffalo, NY 14226-1444. Tel.: 716-836-6565. Fax: 716-831-1126.
Web Site: www.tbz.org
Founded: 1981.
Congressional District: 33
Personnel Profile: Part-Time Volunteers 12; Interns 6.
Governing Authority: Parent Institution: Temple Beth Zion. Tax-exempt.
Institution Type/Description: Judaic Museum.
Collections: Judaic artifacts from the 10th century to present including coins, medallions, books, holocaust remembrances, historical memorabilia, folk art and textiles; Jewish ceremonial artifacts.
Research Fields: Judaic artifacts.

Facilities: 12,000-vol. library pertaining to the Jewish faith available to the public; 500-seat auditorium.
Activities: guided tours; films; organized education programs; temporary & traveling exhibitions; mobile mini museum.
Publications: catalog; pamphlets; brochure, Contributions of Jews to America.
Hours & Admission Prices: Mon.-Fri. 9-4, Sat. 11-12; tours by appointment. No charge; donations accepted. &
Attendance: 16,000

BUFFALO AND ERIE COUNTY BOTANICAL GARDENS, **(M),** 2655 South Park Ave., Buffalo, NY 14218-1526. Tel.: 716-827-1584, ext. 204. Fax: 716-828-0091. Facebook: Buffalo and Erie County Botanical Gardens.
E-mail: egrajek@buffalogardens.com
Web Site: www.buffalogardens.com
Founded: 1898.
Congressional District: 30
Key Personnel: Pres. & C.E.O., David J. Swarts; Chm. (V), Miche Needham; Dir. Administration, Julie DeCarolis; Museum Shop Mgr., Denise Nichols.
Personnel Profile: Full-Time Paid 13; Part-Time Paid 5; Part-Time Volunteers 218; Interns 3.
Volunteer Hours: 8,860
Governing Authority: municipal government. A branch of the County of Erie Dept. of Parks, Recreation & Forestry. Tax-exempt.
Institution Type/Description: Botanical Garden.
Collections: tropical & exotic plants; hardy trees, shrubs, & flowers; orchids; ivy; herbs; palms; begonias; cacti; succulents; bromeliads; koi.
Research Fields: plant taxonomy; horticulture; botany.
Facilities: greenhouse.
Activities: guided tours; special events; concerts; holiday events; educational offerings; summer camp; school tours; private events; weddings. Museum Sponsors: Night Lights at the Gardens in January & February; Celebration of Coleus & Color in June & July; Orchid Show in February & October; Spring Flower Show in March & April; Bonsai Show in May; Succulent Show in September & October; Mum Show in October & November; Poinsettia Show in November & December.
Publications: newsletter, Under the Dome; volunteer newsletter; seasonal flyers.
Hours & Admission Prices: Daily 10-5. Adults $9, seniors 55 & over and students 13 & over with ID $8, children 3-12 $5; discount to American Horticulture Society's reciprocal program; members & children under 3 no charge. Closed Thanksgiving; Christmas. &
Attendance: 90,000 (accurate)
Membership: Senior & Student with ID $35; Individual Gardener $45; Family & Grandparent $65; Gardener Plus $100; Garden Club & Plant Society $130.

* **BUFFALO AND ERIE COUNTY HISTORICAL SOCIETY DBA THE BUFFALO HISTORY MUSEUM, (M),** One Museum Ct., Buffalo, NY 14216-3119. Tel.: 716-873-9644. Fax: 716-873-8754.
E-mail: ccaldwell@buffalohistory.org
Web Site: buffalohistory.org
Founded: 1862.
Congressional District: 41
Key Personnel: Exec. Dir., Melissa Brown; Pres., Joan Bukowski; Dir. Museum Collections, Walter Mayer; Dir. Library & Archives, Cynthia Van Ness; Sr. Dir. Admin. & Operations, Sarah E. Treanor.
Personnel Profile: Full-Time Paid 15; Part-Time Paid 12; Part-Time Volunteers 200; Interns 2.
Governing Authority: society. Tax-exempt: 501(c)3.
Institution Type/Description: Local/Regional History Museum: housed in the only permanent building constructed for the 1901 Pan American Exposition. National Historic Landmark.
Collections: archaeology; ethnology; costumes; crafts; industry; military; stamps; coins; transportation; agriculture; marine; household china, glass, furnishings; jewelry; toys; medical items; ephemera; manuscripts; photographs; communication artifacts; tools & equipment for communication, materials, science & technology; Pan-American artifacts.
Research Fields: history of the Niagara frontier; military; Great Lake maritime; genealogy.
Facilities: 20,000-vol. library of historical books of western New York & the Niagara frontier available for inter-library loan with restrictions; reading room; 250-seat auditorium.
Activities: guided tours by appointment; lectures; formally organized education programs for children; inter-museum loan, permanent, temporary & traveling exhibitions; loan service.
Publications: quarterly newsletter; books, Buffalo & Erie County Town Histories; Superior in Design & Execution: Pressed Glass catalogue;

Coming of Age in Buffalo; Second Looks-A Pictorial History of Buffalo & Erie County; Patterns in Time: Quilts of Western New York.
Hours & Admission Prices: Museum: Tues. & Thurs.-Sat. 10-5, Wed. 10-8, Sun. 12-5. Adults $6, senior citizens & students 13-21 $4, children 7-12 $2.50; discounts for groups, AAM & ICOM members; children under 7 & members no charge. Reference Library: Wed.-Sat. 1-5. Non-members $6. Closed New Year's Day; Thanksgiving; Christmas. &
Attendance: 65,000 (estimated)
Membership: Individuals: Basic $30; Teacher & Grandparent & Senior $35; Family & Household $45; Sustaining $60; Supporting $100; Collector $150; Pan-American $500.

BUFFALO AND ERIE COUNTY NAVAL & MILITARY PARK, **(M),** One Naval Park Cove, Buffalo, NY 14202-4114. Tel.: 716-847-1773. Fax: 716-847-6405.
E-mail: info@buffalonavalpark.org
Web Site: www.buffalonavalpark.org
Founded: 1979.
Congressional District: 33
Key Personnel: Chm. (V), Donald Alessi; Exec. Dir., Patrick J. Cunningham; Ship's Supt. & Staff Duty Officer, Richard Smith; Office Mgr., Anita Baril; Ship's Store Mgr., Dolores Kwiatkowski.
Personnel Profile: Full-Time Paid 14; Part-Time Paid 26; Part-Time Volunteers 67.
Governing Authority: nonprofit organization. Tax-exempt: 501(c)(3).
Institution Type/Description: Historical Ships & Military Museum.
Collections: military history & the five armed forces; photographs; archives; uniforms; 8 ship models; naval artifacts; U.S.S. The Sullivans-DD 537; M41 tank; U.S.S. Little Rock-CLG 4; U.S.S. Croaker - SS-246; P-39 Bell Airacobra; M-84 Armored Personnel Carrier; F-101 Voodoo Air Force Jet; FJ4B Navy Fury Jet; PTF-17.
Research Fields: exhibit & interpretational development.
Facilities: 4,500 sq. ft. exhibit space.
Activities: guided & self-guided tours; films; docent program; overnight encampment of scout groups.
Publications: quarterly newsletter, Ship 'N Shore.
Hours & Admission Prices: April-Oct. daily 10-5; Nov. Sat.-Sun. 10-4. Adults $9, senior citizens & children 6-16 $6; discounts to groups, active military personnel, AAM & ICOM members; children 5 & under and members no charge. &
Attendance: 55,000 (estimated)
Membership: Individual $15; Family $25; Life & Corporate $250.

BUFFALO FIRE HISTORICAL SOCIETY & MUSEUM, 1850 William St., Buffalo, NY 14206. Tel.: 716-892-8400.
Founded: 1980.
Institution Type/Description: Firefighting History Museum.
Collections: firefighting history & equipment; horse-drawn & motorized fire apparatus; uniforms; badges; patches; helmets; tools; fire alarm boxes; photographs; journals; interactive exhibits.
Activities: guided tours; school group tours; lectures; demonstrations; fire safety & prevention; hands-on exhibitions; videos.
Hours & Admission Prices: Sat. 10-4; groups by appointment. &

* **BUFFALO MUSEUM OF SCIENCE, (M),** 1020 Humboldt Pkwy., Buffalo, NY 14211-1293. Tel.: 716-896-5200; 866-291-6660. Fax: 716-897-6723.
E-mail: mmortenson@sciencebuff.org
Web Site: www.buffalomuseumofscience.org
Founded: 1861.
Congressional District: 37
Key Personnel: Pres. & C.E.O., Mark Mortenson; Chm. (V), Randall E. Burkard; Sr. Accountant, Lynn Metzger; Dir. Center for Science Learning, Karen Wallace; Mgr. Exhibits, David Cinquino; Mgr. Operations, Jerry Silvis; Dir. Science & Research, Dr. John Grehan; Dir. Devel. & External Rels., Michelle Rudnicki; Cur. Collections, Kathy Leacock; Museum Shop Mgr. - KMSSA, Lisa Cooper.
Personnel Profile: Full-Time Paid 27; Part-Time Paid 26; Part-Time Volunteers 100; Interns 4.
Governing Authority: Buffalo Society of Natural Sciences. Subsidiary Institution: Tifft Nature Preserve. Tax-exempt: 501(c)(3).
Institution Type/Description: Natural History Museum.
Collections: anthropology; botany; art equipment; botany; geology; palentology; zoology.
Research Fields: ecological management; evolutionary biology; lithics; paleontology; systematics.
Facilities: 42,000-vol. library available for research & inter-library loan; Kellogg Observatory; solar observatory; field research station; 260-acre nature preserve.

Activities: guided tours; lectures; field trip; study trips; formally organized education programs; inter-museum loan; permanent & temporary exhibitions; hands-on discovery room for children.
Publications: Bulletin of the Buffalo Society of Natural Sciences.
Hours & Admission Prices: Daily 10-4. Adults $9, senior citizens 62 & over $8, children 2-17, students & military with ID $7; discounts to AAM & WBFO members, Buffalo & Erie county Public Library cardholders, and NFTA pass cardholders; members & children under 2 no charge. &
Attendance: 111,225 (accurate)
Membership: Student $25; Senior $35; Senior Couple & Military $45; Grandparent $55; Family $60; Contributing $90; Sustaining $165.

BUFFALO TRANSPORTATION PIERCE-ARROW MUSEUM, 263 Michigan Ave., Buffalo, NY 14203-2900. Mailing Address: 24 Myrtle Ave., Buffalo, NY 14204-2048. Tel.: 716-853-0084.
E-mail: piercemuseum@roadrunner.com
Web Site: pierce-arrow.com
Founded: 1997.
Key Personnel: Dir., James T. Sandoro; Museum Shop Mgr., Tim Green.
Governing Authority: Tax-exempt.
Institution Type/Description: Transportation Museum.
Collections: period automobiles; transportation history; photographs.
Activities: rental facilities; special events; meetings; dinners.
Hours & Admission Prices: Call for hours. Adults $10. &
Attendance: 10,000 (estimated)

BUFFALO ZOO, 300 Parkside Ave., Buffalo, NY 14214-1999. Tel.: 716-837-3900, ext. 100. Fax: 716-833-3743.
Web Site: www.buffalozoo.org
Formerly: Buffalo Zoological Gardens
Founded: 1875.
Congressional District: 37
Key Personnel: Pres., Donna M. Fernandes, Ph.D.; Chm., Dorothy T. Ferguson; Dir. Admin. & Finance, Mrs. Denise Maloney; Gen. Cur., Gerald D. Aquilina; Service Systems Assoc., Jeff Blarr.
Personnel Profile: Full-Time Paid 72; Part-Time Paid 77; Part-Time Volunteers 578; Interns 135.
Volunteer Hours: 19,500
Operating Expenses: 7,149,941
Operating Income: 7,149,941
Governing Authority: society. Tax-exempt: 501(c)(3).
Institution Type/Description: Zoological Gardens & Children's Zoo.
Collections: aviary; mammals; birds; reptiles; amphibians; fish; insects; arachnids; botanical gardens; South American rainforest.
Research Fields: rare & endangered animal exhibition, propagation & management.
Facilities: 23.5 acre park; gardens; children's resource center; concession stand. Museum-related items for sale.
Activities: permanent exhibitions; interactive & educational activities; guided tours; lectures; animal demonstration; tropical rainforest; children's zoo; train rides; carousel; keeper talks; dinosaur dig; kangaroo hop trampoline; wind tunnel booth simulator; playground.
Publications: quarterly zoo magazine, ZOOLOG.
Hours & Admission Prices: Winter: Gates daily 10-4; Buildings 10-4.30; Grounds 10-5. Summer: Gates daily 10-5; Buildings 10-5.30; Grounds 10-6. Adults 15-62 $10.50, senior citizens 63 & over and student 22 & under $8.50, youth 2-14 $7.50; discounts to AZA members & groups; children under 2 & zoo members no charge. Closed Thanksgiving; Christmas. &
Attendance: 455,283 (accurate)
Membership: Individual $40; Individual Plus $55; Grandparents $60; Grandparent Plus $70; Family $75; Family Plus $90; Supporting $135; Sponsor $250; Patron $500; Benefactor $1,000.

* **BURCHFIELD-PENNEY ART CENTER, (M),** Buffalo State College, 1300 Elmwood Ave., Buffalo, NY 14222-1004. Tel.: 716-878-6012. Fax: 716-878-6003.
E-mail: burchfld@buffalostate.edu
Web Site: www.burchfieldpenney.org
Founded: 1966.
Congressional District: 140
Key Personnel: Dir. & Pres., Ted Pietrzak; C.O.O., Carolyn A. Morris Hunt, CFRE; Mktg. & Public Rels., Kathleen Heyworth; Head Collections, Nancy Weekly; Cur. Education, Kathy Gaye Shiroki; Head Programs, Don Metz; Facilities Mgr., William Menshon.
Personnel Profile: Full-Time Paid 16; Part-Time Paid 4; Part-Time Volunteers 65; Interns 3.
Governing Authority: nonprofit. Tax-exempt: 501(c)(3).

Institution Type/Description: Art Museum: housed on the State University College at Buffalo Campus.
Collections: paintings, drawings, sketches, wallpapers, prints, by Charles Burchfield; works by contemporary Western New York artists; works by historical Western New York artists; works by Charles E. Burchfield's contemporaries; archival materials relating to Charles Burchfield & Western New York artists; manuscripts; Penney collections of work by Charles Burchfield; Roycroft artists; craft artists.
Research Fields: American art; Western New York art; Charles Burchfield's art & life.
Facilities: 2,500-vol. library; archives; cafe. Museum-related items for sale.
Activities: guided tours; lectures; gallery talks; concerts; poetry reading; education programs for children, adults, undergraduate & graduate college students affiliated with State University College at Buffalo; loan, temporary & traveling exhibitions; school loan service; workshops; special events.
Publications: Burchfield Penney Art Center newsletter; exhibition catalogues; books on Charles Burchfield.
Hours & Admission Prices: Tues.-Wed. & Fri.-Sat. 10-5, Thurs. 10-9, Sun. 1-5. Adults $9, students 6-18 $5; discounts to AAM & NARM members; members & children under 6 no charge. Empire State Reciprocal membership. Closed legal holidays. &
Attendance: 70,000 (accurate)
Membership: Individual $45; Family & Dual $60; Friend $125; Curator's Circle $275; Director's Circle $500; Burchfield Circle $1,000.

CENTER FOR EXPLORATORY AND PERCEPTUAL ART, 617 Main St., #201, Buffalo, NY 14203-1400. Tel.: 716-856-2717. Fax: 716-270-0184.
E-mail: info@cepagallery.com
Web Site: www.cepagallery.org
Founded: 1974.
Congressional District: 37
Key Personnel: C.E.O., Exec. Dir. & Cur., Lawrence Brose; Chm. (V), Jim Rolls; Artistic Dir., Sean Donaher; Education, Lauren Tent; Administrative Asst., Lynda Kaszubski.
Personnel Profile: Full-Time Paid 3; Part-Time Paid 1; Part-Time Volunteers 12; Interns 12.
Governing Authority: nonprofit. Tax-exempt.
Institution Type/Description: Art Gallery.
Collections: photographs; audiovisual.
Facilities: library; 7,500 sq. ft. exhibit space.
Activities: films; dark room & computer workshops; public art. Annual Event: photography art auction.
Publications: CEPA Journal; CEPA newsletter.
Hours & Admission Prices: Mon.-Fri. 7am-9pm, Sat. 8:30-4:45, Sun. 10-3. No charge; donations accepted. &
Attendance: 1,000,000 (estimated)
Membership: Artist, Senior & Student with ID $25; Individual $35; Household & Institution $60; Patron $100; Collectors Program: Supporter $200; Sustainer $500.

DARWIN D. MARTIN HOUSE, (M), 125 Jewett Pkwy., Buffalo, NY 14214-2301. Tel.: 716-856-3858. Fax: 716-856-4009.
E-mail: info@darwinmartinhouse.org
Web Site: www.darwinmartinhouse.org
Founded: 1992.
Key Personnel: Exec. Dir., Mary F. Roberts; Chm. (V), Robert Kresse; Pres. (V), Jack Walsh; Museum Shop Mgr., Rebecca Lee.
Personnel Profile: Full-Time Paid 7; Part-Time Paid 17; Part-Time Volunteers 400.
Institution Type/Description: Historic House: housed in the former home of Darwin & Isabelle Martin; designed by Frank Lloyd Wright; built in 1904. Listed on the National Register of Historic Places.
Collections: family history; period furnishings; photographs.
Hours & Admission Prices: Call for hours. Adults $15; members no charge.
Attendance: 28,000 (accurate)

HALLWALLS CONTEMPORARY ARTS CENTER, 341 Delaware Ave., Buffalo, NY 14202. Tel.: 716-854-1694. Fax: 716-854-1696.
Web Site: www.hallwalls.org
Key Personnel: Exec. Dir., Edmund Cardoni
Institution Type/Description: Art Gallery.
Collections: works by contemporary artists.
Hours & Admission Prices: Tues.-Fri. 11-6, Sat. 11-2.

IRA G. ROSS AEROSPACE MUSEUM, (M), 1 Seymour H. Knox III Plaza, (at Perry St. in HSBC Arena), Buffalo, NY 14203-3007. Mailing Address: 2221 Niagara Falls Blvd., Ste. 7, Niagara Falls, NY 14304-1696. Tel.: 716-858-4340. Fax: 716-278-0257.
E-mail: info@wnyaerospace.org
Web Site: www.wnyaerospace.org
Formerly: Niagara Aerospace Museum
Key Personnel: Exec. Dir., Jack Wysocki
Institution Type/Description: Aviation History Museum.
Collections: Western New York aviation history; aerospace innovations.
Facilities: 140-seat theater.
Activities: group tours.
Hours & Admission Prices: Sat.-Sun. 11-4. Adults $6, senior citizens 62 & over and students $5, children 12 & under $2; discount to groups; members no charge. Closed New Year's Eve & Day; Thanksgiving; Christmas Eve & Day.

LOWER LAKES MARINE HISTORICAL SOCIETY, 66 Erie St., Buffalo, NY 14202. Tel.: 716-849-0914. Fax: 716-849-0914.
E-mail: museum@llmhs.org
Web Site: www.llmhs.org
Founded: 1987.
Personnel Profile: Part-Time Volunteers 6.
Governing Authority: Tax-exempt.
Institution Type/Description: Marine Historical Society Museum.
Collections: local history; period artifacts; photographs; Great Lakes history; Buffalo Harbor; Erie Canal; Lake Erie.
Research Fields: Great Lakes; geneology; vessel histories.
Activities: special events.
Publications: bimonthly newsletter; books.
Hours & Admission Prices: Thurs. & Sat. 10-3. No charge; donations accepted.
Attendance: 1,500 (estimated)
Membership: Student & Senior $20; Regular $30; Sustaining $50.

SHEA'S PERFORMING ARTS CENTER, 646 Main St., Buffalo, NY 14202. Mailing Address: P.O. Box 1130, Buffalo, NY 14205. Tel.: 716-847-1410. Fax: 716-847-1644.
Web Site: www.sheas.org
Institution Type/Description: Historic Building: built in 1926. A National Historic Site.
Collections: theater history & architecture; Neo-Spanish Baroque design; building restoration progress; Western New York Entertainment Hall of Fame; inductees' portraits.
Activities: guided tours.
Hours & Admission Prices: Tours: Tues., Thurs. & Sat. 10 & 1; during non-show days; groups by appointment. Adults $8, children, students & seniors $4; discounts to groups of 20 or more.

THEODORE ROOSEVELT INAUGURAL NATIONAL HISTORIC SITE, 641 Delaware Ave., Buffalo, NY 14202-1079. Tel.: 716-884-0095. Fax: 716-884-0330.
E-mail: thri_administration@nps.gov
Web Site: www.nps.gov/thri/; trsite.org
Founded: 1971.
Congressional District: 26
Key Personnel: Exec. Dir. & Site Supt., Molly Quackenbush; Pres. (V), Stanton H. Hudson, Jr.; Vice Pres. (V), Karen Gaughan Scott; Admin. Officer, Sally Goris; Cur., Lenora Henson; Education Dir., Mark Lozo; Administrative Asst., Wendy Phelps; Asst. Dir., Janice Kuzan; Maintenance Chief, Keith Krummel; Interpreter, Mark Comito.
Personnel Profile: Full-Time Paid 7; Part-Time Paid 9; Part-Time Volunteers 322; Interns 3.
Governing Authority: nonprofit organization. Parent Institution: National Park Service. Subsidiary Institution: Theodore Roosevelt Inaugural Site Foundation. Tax-exempt: 501(c)(3).
Institution Type/Description: Historic Building & Site: housed in c.1838 Buffalo Barracks, the site of the inauguration of Pres. Theodore Roosevelt in 1901.
Collections: artifacts relating to Theodore Roosevelt's inauguration in 1901, Theodore Roosevelt's life, Wilcox house 1838-1933 & 1937-1960, William McKinley's death, Pan-American Exposition & turn-of-the-20th century furnishings. Historic Buildings: Ansley Wilcox house, 1901; carriage house.
Research Fields: Theodore Roosevelt's Presidency; Buffalo history; Wilcox family history; restoration; Pan-American Exposition.
Facilities: carriage house visitor center.
Activities: guided tours; volunteer program; school programs; permanent & changing exhibitions; architectural walking tours; lecture series; naturalization ceremonies; Inaugural reenactment; lectures; walking tours. Museum

Sponsors: Teddy Bear Picnics; Public Speaking Contest; Victorian Christmas program in December.
Publications: books, Theodore Roosevelt, An American Hero in Caricature: essays explaining Puck magazine covers (1898-1909); newsletter, The Bullytin; Guidebook to the Theodore Roosevelt Inaugural National Historic Site.
Hours & Admission Prices: Tours: daily. Adults $10; discounts to National Park Pass holders; members no charge. Closed New Year's Eve & Day; Easter; Memorial Day; Independence Day; Labor Day; Thanksgiving; Christmas Eve & Day.
Attendance: 17,110 (accurate)
Membership: Individual $30; Family & Grandparent $45; Rough Rider $75; Governor $100; Vice President $250; President $500.

UNIVERSITY AT BUFFALO ART GALLERIES, (M), UB Art Gallery, Center for the Arts, 201 Center for the Arts, Buffalo, NY 14260-6000. Tel.: 716-645-6912. Fax: 716-645-6753. Facebook: UB Art Galleries.
E-mail: sholsen@buffalo.edu
Web Site: www.ubartgalleries.org
Founded: 1994.
Congressional District: 26
Key Personnel: Dir., UB Art Galleries, Sandra H. Olsen, Ph.D; Cur., UB Art Gallery, Sandra Firmin; Cur. Education, UB Anderson Gallery, Ginny Lohr; Asst. Dir., UB Anderson Gallery, Mary Moran; Asst. Dir. Collections & Exhibitions and Registrar, Robert Scalise; Staff Asst., Jim Snider; Head Preparator, Ken Short; Asst. Preparator, Tom Holt; Maintenance Supvr., Paul Wilcox.
Personnel Profile: Full-Time Paid 10; Full-Time Volunteers 11; Part-Time Paid 10; Part-Time Volunteers 12; Interns 12.
Governing Authority: public university; nonprofit. Parent Institution: University of Buffalo. Branch Gallery: UB Anderson Gallery, 1 Martha Jackson Place, Buffalo, NY 14214. Tel: 716-829-3754; Fax: 716-829-3757. Tax-exempt.
Institution Type/Description: University Art Gallery & Museum.
Collections: over 3,000 works of art including modern & contemporary paintings, prints, drawings, sculpture; cultural art from the Annette Cravens collection; Martha Jackson & David Anderson Gallery collections.
Major Exhibits: Painting Borges: Art Interpreting Literature, 9/21/13-2/23/14; Affinity in Print: Francesco Toledo and Jorge Luis Borges, 9/21/13-2/23/14; Selections of Prints by Antoni Tapies, 9/21/13-2/23/14; Legacies: Martha Jackson and David Anderson, 3/15/14-5/4/14; Art = Text = Art - Sally and Wynn Kramarsky Collection Spring to Fall 2014.
Research Fields: modern art; cultural material (global arts); contemporary art.
Facilities: archives; educational facilities; 17,800 sq. ft. exhibit space.
Activities: formal education programs for adults & college students; participatory & traveling exhibits; temporary exhibits followed by lectures, panel discussions & receptions for artists; organization & participation in conferences (national & international).
Publications: exhibition brochures & catalogs; Artpark: 1974-1984 (2010).
Hours & Admission Prices: UB Art Gallery: Tues.-Fri. 11-5, Sat. 1-5; UB Anderson Gallery: Wed.-Sat. 11-5, Sun. 1-5. No charge; donations accepted. Closed New Year's Day; Memorial Day; Independence Day; Labor Day; Thanksgiving Day; Christmas; during installations.
Attendance: 10,000 (estimated)

Burt

VAN HORN MANSION - NEWFANE HISTORICAL SOCIETY, Rt. 78, Burt, NY 14028. Mailing Address: P.O. Box 115, Newfane, NY 14108-0115. Tel.: 716-778-7197. Facebook: Town of Newfane Historical Society; The Van Horn Mansion.
Key Personnel: Dir., Jill Heck
Governing Authority: Tax-exempt.
Institution Type/Description: Historic House Museum: housed in the former home of James Van Horn; built in 1823. Listed on the National Registry of Historical Buildings.
Collections: local history; period furnishings; personal artifacts.
Activities: Annual Events: Apple Blossom Festival in May; Harvest Festival in October.
Publications: monthly newsletters.
Hours & Admission Prices: Call for hours. No charge; donations accepted.
Attendance: 3,000 (estimated)
Membership: $10.

Byron

BYRON HISTORICAL MUSEUM, 6407 Town Line Rd., Byron, NY 14422. Mailing Address: 6451 Mill Pond Rd., Byron, NY 14422-9758. Tel.: 585-548-9008.
E-mail: byron-historian@juno.com
Web Site: www.byronny.com/history.html
Founded: 1967.
Congressional District: 137
Key Personnel: Historian, Beth Wilson; Historian, Bob Wilson.
Governing Authority: society. Affiliated with Byron Historical Society. Tax-exempt.
Institution Type/Description: General Museum.
Collections: history; pulpit, organ & altar of former church; general; home & farm displays of the area.
Activities: permanent & temporary exhibitions.
Hours & Admission Prices: Memorial Day to Labor Day Sun. 2-4; other times by appointment. No charge.
Attendance: 235

Caledonia

BIG SPRINGS MUSEUM, 3095 Main St., Caledonia, NY 14423-1237. Mailing Address: Box 41, Caledonia, NY 14423-0041. Tel.: 585-538-9880.
E-mail: bigspringshistoricalsociety@yahoo.com
Web Site: www.bigspringsmuseum.org
Founded: 1936.
Congressional District: 26
Key Personnel: Pres. (V), Meg Donegan; Cur., Patty Garrett; Docent, Lois Waldron; Docent, Mike LaFave.
Personnel Profile: Part-Time Paid 3; Part-Time Volunteers 40.
Governing Authority: society. Affiliated with Big Springs Historical Society. Tax-exempt.
Institution Type/Description: General Museum.
Collections: early farm implements; Native American collection; early pictures; documents; letters; Civil War, World War I & II collections; household objects; toys; games; dolls; costumes; weaving equipment; school room articles; china; glass; medical.
Facilities: library of bound copies of local newspaper dating back to the 1900s; material on local & county histories, available for research on premises. Local historic plates, tiles, history booklets, notepaper for sale.
Activities: guided tours; lectures; formally organized education programs for children; permanent & temporary exhibitions.
Publications: members' newsletter; program flyers for members.
Hours & Admission Prices: Sun. 1-4, Mon. 9-12; call for additional hours. No charge; donations accepted. Closed Easter; Memorial Day; Independence Day; Christmas. &
Attendance: 1,800 (estimated)
Membership: Individual $20; Family $25; Silver Patron $50; Gold Patron $75; Platinum $100; Benefactor $500.

Calverton

GRUMMAN MEMORIAL PARK, Rte. 25 & 25A, Calverton, NY 11933. Mailing Address: P.O. Box 147, Calverton, NY 11933-0147. Tel.: 631-369-1826.
E-mail: gmpark@optonline.net
Web Site: www.grummanpark.org
Formerly: East End Aircraft L.I. Corp.
Founded: 1998.
Congressional District: 1
Key Personnel: Chm. (V), Joe Van de Wetering.
Personnel Profile: Part-Time Volunteers 7.
Governing Authority: private; nonprofit organization. Parent Institution: East End Aircraft L.I. Corp. Tax-exempt: 501(c)(3).
Institution Type/Description: Military Museum.
Collections: military aircraft; F14; A6E.
Hours & Admission Prices: Daily 9-5. No charge. &
Attendance: 15,000 (estimated)
Membership: One Year $10; Five Years $40; Ten Years $75.

Camden

QV HISTORICAL SOCIETY AT CARRIAGE HOUSE MUSEUM, 2 N. Park St., Camden, NY 13316-1306. Mailing Address: P.O. Box 38, Camden, NY 13316-0038. Tel.: 315-245-4652.
E-mail: historycamden@verizon.net
Founded: 1975.
Congressional District: 31

Key Personnel: Pres. (V), J.C. Kuttruff.
Personnel Profile: Part-Time Volunteers 9.
Governing Authority: society; nonprofit. Tax-exempt.
Institution Type/Description: Historical Society Museum.
Collections: tools; bottles; books; photographs; local history; china; town hall bell; maps; soil samples; barb wire; surveying instruments; jewelry; dolls; paintings; furniture; clothing; manuscripts; genealogy information; Camden's postcards; history of Camden's churches, schools, homes & Main Street stores; one room schoolhouse.
Research Fields: local history; family searches.
Facilities: library.
Activities: permanent exhibitions; research.
Publications: Camden postcard booklet; brochures.
Hours & Admission Prices: May-Oct. Wed. & Fri.-Sat. 1-4; call to confirm; other times by appointment. No charge; donations accepted. Closed holidays.
Attendance: 500 (estimated)
Membership: Student $2; Individual $10; Family $20.

Camillus

WILCOX OCTAGON HOUSE MUSEUM, 5420 W. Genesee St., Camillus, NY 13031. Mailing Address: 4600 W. Genesee St., Camillus, NY 13219. Tel.: 315-488-7800.
E-mail: octagonhouseofcamillus@gmx.com
Web Site: www.octagonhouseofcamillus.org
Institution Type/Description: Historic House Museum: built in 1856. Listed on the National Register of Historic Places.
Collections: local history; period furnishings; personal artifacts; photographs.
Activities: rental facilities; special events.
Hours & Admission Prices: April-Dec. Sun. 1-5; other times by appointment. No charge; donations accepted.

Canaan

CANAAN HISTORICAL SOCIETY, INC., 13 Warner Crossing Rd., Canaan, NY 12029. Mailing Address: 84 Old Hudson Turnpike, Canaan, NY 12029-2801. Tel.: 518-781-4228.
E-mail: nyvetcounsil@yahoo.com
Founded: 1962.
Congressional District: 19
Key Personnel: Pres., Suzanne Pemrick; Cur., Tammy Flaherty.
Personnel Profile: Part-Time Volunteers 8.
Volunteer Hours: 200
Operating Expenses: 1,797
Operating Income: 2,592
Governing Authority: society. Affiliated with University of the State of New York. Tax-exempt.
Institution Type/Description: Historic Building: 1829 former Meeting House, Presbyterian Society.
Collections: glassware; china; silver; artifacts from the Shakers; furniture; Civil War items; Canaan militia Civil War artifacts; quilts; military artifacts of Captain William Henry Warner; books by Susan & Anna Warner.
Facilities: genealogy library available for research on premise by appointment; meeting room. Museum-related items for sale.
Activities: guided tours; temporary exhibitions. Museum Sponsors: various special programs in July & August.
Publications: semi-annual newsletter, Canaan Eagle; book, Reflections: Canaan, NY 1776-1976.
Hours & Admission Prices: July-Aug. Sat. 1-4; other times by appointment. No charge; donations accepted. &
Attendance: 200 (estimated)
Membership: Junior $1; Regular $15; Family $20; Business $30; Benefactor $60; Memorial $100; Life $200.

Canajoharie

ARKELL MUSEUM, (M), 2 Erie Blvd., Canajoharie, NY 13317-1198. Tel.: 518-673-2314. Fax: 518-673-5243. Facebook: Arkell Museum.
E-mail: info@arkellmuseum.org
Web Site: www.arkellmuseum.org
Formerly: Canajoharie Library and Art Gallery
Founded: 1929.
Congressional District: 31
Key Personnel: Dir. & Chief Cur., Diane Forsberg; Pres., Charles Tallent.
Personnel Profile: Full-Time Paid 3; Part-Time Paid 5; Part-Time Volunteers 22.
Operating Expenses: 671,628
Operating Income: 587,636

Governing Authority: nonprofit organization. Tax-exempt: 501(c)(3)
Institution Type/Description: Library & Art Gallery.
Collections: American art including works by Winslow Homer, American Impressionists, the Ash Can Artists, George Innes; John Singer Sargert; Andrew Wyeth; The Beechnut archives.
Major Exhibits: Of Time and the Mohawk River, 2/15/14-5/14; Winslow Homer: The Nature and Rhythm of Life (T), 9/2/14-1/11/15.
Research Fields: American Art; Beech-Nut history.
Facilities: library of books of general interest & special collection of art of America available for inter-library loan on short term.
Activities: guided tours; gallery talks.
Hours & Admission Prices: Tues.-Fri. 10-5, Sat-Sun 1-5. Adults $7. Closed Thanksgiving, Christmas & New Year's Day. ⚹
Attendance: 44,500 (accurate)
Membership: Basic $35; Basic Dual $60; Basic Family $75; Associate $100; Sustaining $200; Benefactor $1,000; Arkell Circle $2,000.

Canandaigua

THE GRANGER HOMESTEAD SOCIETY, INC., 295 N. Main St., Canandaigua, NY 14424-1228. Tel.: 585-394-1472. Fax: 585-394-6958.
E-mail: info@grangerhomestead.org
Web Site: www.grangerhomestead.org
Founded: 1946.
Congressional District: 33
Key Personnel: Exec. Dir., Martha Herbik; Pres. (V), David Sauter; Business Mgr., Jo-el Hibbard; Administrative Asst., Libby Campbell; Museum Shop Mgr., Bonnie Kelly; Facilities Mgr., James Catalfamo.
Personnel Profile: Full-Time Paid 2; Part-Time Paid 4; Part-Time Volunteers 200.
Governing Authority: society; nonprofit. Branch Museum: The Granger Homestead Carriage Museum. Tax-exempt: 501(c)(3).
Institution Type/Description: Historic House Museum: 1816 Gideon Granger Mansion.
Collections: decorative arts; furniture; silver; china; glass & textiles; 1820-1930 horse-drawn vehicles; over 70 antique carriages; progress of light, oil lamps & brass fixtures.
Research Fields: architecture; 19th-century decorative arts & interiors; horse-drawn vehicles.
Facilities: research library; archives. Museum-related items for sale.
Activities: guided tours; lectures; concerts; workshops; classes; educational programs; docent program; special events; private rentals; society business. Museum Sponsors: carriage rides in summer; Festival of Trees; Christkindl Market; lawn sale; art lecture; art show; historical recitals & concerts.
Publications: quarterly newsletter, Gideon's Trumpet.
Hours & Admission Prices: June-Oct. Tues.-Wed. & Sat.-Sun. 1-5, Thurs.-Fri. 11-5. Adults $6, senior citizens $5, children $2; discounts to groups, AAA and AAM members & Sonnenberg Gardens visitors; members no charge. Additional charge for special events. Closed New Year's Eve & Day; Presidents' Day; Memorial Day; Independence Day; Labor Day; Columbus Day; Thanksgiving; Christmas Eve, Day & week. ⚹
Attendance: 16,000 (accurate)
Membership: Individual $35; Family $50; Friend $75; Benefactor $125; Gold $175; Patron $250; Sponsor $500; Gideon's Circle $1,000.

ONTARIO COUNTY HISTORICAL SOCIETY, 55 N. Main St., Canandaigua, NY 14424-1438. Tel.: 585-394-4975. Fax: 716-394-9351.
E-mail: director@ochs.org
Web Site: www.ochs.org
Founded: 1902.
Congressional District: 29
Key Personnel: Exec. Dir. & Museum Shop Mgr., Edward Varno; Pres. (V), Thomas Walter; Bookkeeper, Ernest Mor; Cur., Wilma T. Townsend; Educator, Nancy Parsons; Library Mgr., Linda Alexander; Computer Admin., Bruce Stewart; Museum Shop Mgr., Maureen O'Connell Baker.
Personnel Profile: Full-Time Paid 3; Full-Time Volunteers 1; Part-Time Paid 2; Part-Time Volunteers 62.
Governing Authority: society. Tax-exempt: 501(c)(3).
Institution Type/Description: Historical Society Museum.
Collections: 15,000 objects & 60,000 manuscripts & photographs relating to history of Ontario County; genealogy of the region.
Major Exhibits: Greed, 1/13-1/18; The Origins of Western NY, 1/13-1/18; The City of Canandaigua, 4/13-5/14.
Research Fields: history of western New York state; genealogy.
Facilities: 2,000-vol. library of books pertaining to western NY state, local history & genealogies available for research on premises; 2 galleries.
Activities: tours; lectures; school programs; permanent & temporary exhibitions.

Publications: newsletter, books & pamphlets on local history; photographic histories; quarterly newsletter, Chronicles; Roseland, Playground of The Finger Lakes, 1935-1985.
Hours & Admission Prices: Jan.-April Tues.-Fri. 10-4:30, Sat. 11-3; May-Dec. Tues. & Thurs.-Fri. 10-4:30, Wed. 10-9, Sat. 11-3. No charge; donations accepted. Research Room: by appointment. Research: $7.50. Research by Mail: $35 per hour. Closed national holidays. ⚹
Attendance: 10,000 (accurate)
Membership: Senior $30; Individual $35; Family $50; Friend $75; Centennial $100; Benefactor $150; Heritage Circle $250; Real Good Friend $1,000 & up.

SONNENBERG GARDENS & MANSION STATE HISTORIC PARK, (M), 151 Charlotte St., Canandaigua, NY 14424-1363. Tel.: 585-394-4922. Fax: 585-394-2192.
E-mail: director@sonnenberg.org
Web Site: www.sonnenberg.org
Founded: 1972.
Congressional District: 31
Key Personnel: Chm. (V), Dale Stell; Dir., David Hutchings; Pres. Friends, Roy Beecher.
Personnel Profile: Full-Time Paid 6; Part-Time Paid 2; Part-Time Volunteers 426; Interns 3.
Governing Authority: nonprofit organization. Parent Institution: NYS Board of Regents. Subsidiary Institution: NYSOPRHP. Tax-exempt: 501(c)(3).
Institution Type/Description: Historic House & Botanical Garden: 1887 Sonnenberg Mansion & Gardens.
Collections: period furniture & furnishings; Victorian estate garden.
Major Exhibits: Flowers & Fashion, 5/14-9/14; Vanishing Acts (T), 6/14-10/14.
Research Fields: history of mansion & original owners; historical landscape practices-early 20th century.
Facilities: 50 acres of Victorian gardens; Finger Lake Wine Center. Museum-related items for sale.
Activities: guided tours; lectures; concerts; workshops; formally organized education programs; docent program or council; loan, permanent & temporary exhibitions.
Publications: brochure.
Hours & Admission Prices: May & Sept.-Oct. 9:30-4:30; Memorial Day to Labor Day 9:30-5:30. Adults 18-59 $12, seniors 60 & over $10, military & students w/id $6, Granger Homestead combination pkg. $14; discounts to AAA members; children 12 & under and members no charge. Reciprocal admission with participating gardens, arboreta, & conservatories of the American Horticultural Society. See website for event pricing. ⚹
Attendance: 32,000 (accurate)
Membership: Students $25; Individual $40; Dual $60; Family $100; Patron $250; Rose $500; Orchid $1,000; Lily $2,500; Mary Clark Thompson Circle $5,000.

Canastota

INTERNATIONAL BOXING HALL OF FAME, 1 Hall of Fame Dr., Canastota, NY 13032. Mailing Address: 1 Hall of Fame Dr., 360 N. Peterboro St., Canastota, NY 13032. Tel.: 315-697-7095. Fax: 315-697-5356.
Institution Type/Description: Boxing Hall of Fame.
Collections: boxing history & heritage; gloves; hand wraps; fist casts; posters; life-size statues; personal artifacts; photographs; Hall of Fame inductees.
Facilities: Museum-related items for sale.
Activities: special events. Annual Event: Hall of Fame Induction Weekend.
Hours & Admission Prices: Mon.-Fri. 9-5, Sat.-Sun. 10-4.

Canisteo

KANESTIO HISTORICAL SOCIETY, 23 Main St., Canisteo, NY 14823. Mailing Address: P.O. Box 35, Canisteo, NY 14823-0035. Tel.: 607-698-2086.
E-mail: kanestio@yahoo.com
Founded: 1985.
Congressional District: 29
Key Personnel: Pres. (V), John S. Babbitt.
Personnel Profile: Part-Time Volunteers 20.
Governing Authority: Tax-exempt.
Institution Type/Description: Historical Society Museum.
Collections: local history; Canisteo Times from 1880-1956; photographs; family genealogies.
Activities: research.
Publications: biannual newsletter.
Hours & Admission Prices: Wed.-Fri. 1-3. Programs: 3rd Tues. each month 7pm. No charge.

Attendance: 192 (accurate)
Membership: Individual $7; Family $10; Life $100.

Canton

PIERREPONT MUSEUM, 864 State Hwy. 68, Canton, NY 13617-3468. Mailing Address: 872 State Hwy. 68, Canton, NY 13617-3468. Tel.: 315-386-8311 & 379-0804. Fax: 315-379-0415.
E-mail: barbara@northnet.org
Founded: 1977.
Congressional District: 26
Key Personnel: Historian, Barbara J. Daniels.
Personnel Profile: Part-Time Paid 1; Part-Time Volunteers 2.
Volunteer Hours: 100
Operating Income: 500
Governing Authority: society. Parent Institution: Pierrepont Historical Society. Tax-exempt.
Institution Type/Description: Antiques Museum: housed in early 1800 district school house.
Collections: rural school setting; farm tools of the pre-machine era; veterans exhibit; antique text books; kitchen artifacts; clothing dating from 1800-1875; paintings & biographies of local artists; minerals from local mines; old bibles.
Research Fields: genealogy.
Facilities: 100-vol. library of text books, scrap books, documentaries available for research on premises by appointment by phoning town historian.
Activities: tours for school children.
Publications: booklet, Pierrepont Memories-People, Places & Things.
Hours & Admission Prices: Memorial Day-Labor Day Sat. 10-1; other times by appointment. No charge; donations accepted.
Attendance: 150 (estimated)

✳ ST. LAWRENCE COUNTY HISTORICAL ASSOCIATION - SILAS WRIGHT HOUSE, (M), 3 E. Main, Canton, NY 13617-1416. Mailing Address: P.O. Box 8, Canton, NY 13617-0008. Tel.: 315-386-8133. Fax: 315-386-8134.
E-mail: info@slcha.org
Web Site: www.slcha.org
Founded: 1947.
Congressional District: 23
Key Personnel: C.E.O., Trent Trulock; Pres., Carlton Stickney.
Personnel Profile: Full-Time Paid 2; Part-Time Paid 1; Part-Time Volunteers 111; Interns 3.
Governing Authority: nonprofit organization. Parent Institution: St. Lawrence County Historical Association. Tax-exempt: 501(c)(3).
Institution Type/Description: Preservation Project: housed in 1834 home of Governor Silas Wright.
Collections: early 19th-century furniture; costumes; household utensils; local history artifacts & archives.
Research Fields: St. Lawrence County history & folklore.
Facilities: 1,500-vol. library of local history material; genealogical information; Wright family papers.
Activities: guided tours; lectures; temporary & permanent exhibits; special events; programs for school children.
Publications: magazine, The Quarterly; bimonthly newsletter; books: Old Hollywood; 1878 History of St. Lawrence County; The County Chronicler; Silas Wright, The Farmer Statesman; The Wright House Paper Model Book.
Hours & Admission Prices: Museum & Archives: Tues.-Thurs. & Sat. 12-4, Fri. 12-8. Museum: no charge. Archives: adults $5, college students $2.50; discounts to AAM & ICOM members; members & children no charge. Closed New Year's Eve & Day; Martin Luther King, Jr. Day; Presidents' Day; Memorial Day; Independence Day; Labor Day; Columbus Day; Thanksgiving & day after; Christmas Eve, Day and week. &
Attendance: 6,191 (accurate)
Membership: Senior Citizen & Student $25; Individual $30; Family $40; Associate $60; Contributor $100; Donor $250; Benefactor $500

ST. LAWRENCE UNIVERSITY - RICHARD F. BRUSH ART GALLERY AND PERMANENT COLLECTION, 23 Romoda Dr., Canton, NY 13617-1501. Tel.: 315-229-5174. Fax: 315-229-7445.
E-mail: ctedford@stlawu.edu
Web Site: www.stlawu.edu/gallery
Founded: 1967.
Congressional District: 30
Key Personnel: Dir., Catherine L. Tedford; Asst. Dir., Carole Mathey; Arts Programming Coord., Juli Pomainville.
Personnel Profile: Full-Time Paid 2; Part-Time Paid 1; Interns 25.
Governing Authority: college. Parent Institution: St. Lawrence University. Tax-exempt: 501(c)(3).

Institution Type/Description: University Art Gallery.
Collections: 20th-century European & American works of art on paper including prints, photographs, drawings, and artists' books and portfolios, with emphasis on American photography; 19th- & 20th-century American & European painting, sculpture, and ceramics.
Research Fields: pertaining to collections; Inuit Art; West African textiles.
Facilities: 75-seat auditorium; classrooms.
Activities: lectures; films; gallery talks; arts festivals; training programs; loan, permanent, temporary & traveling exhibitions.
Publications: exhibition brochures; book, Photographs at St. Lawrence University.
Hours & Admission Prices: Mon.-Thurs. 12-8, Fri.-Sat. 12-5. No charge. Closed college recesses. &
Attendance: 20,000 (estimated)

Cape Vincent

CAPE VINCENT HISTORICAL MUSEUM, James St., Cape Vincent, NY 13618. Mailing Address: P.O. Box 376, Cape Vincent, NY 13618-0376. Tel.: 315-654-4400.
Founded: 1968.
Congressional District: 26
Key Personnel: Pres. (V), Jeanne Thompson; Dir., Mary Hamilton; Dir. & Town Historian, Peter Margrey.
Governing Authority: municipal. Affiliated with the Cape Vincent Town Board. Tax-exempt.
Institution Type/Description: General Museum: housed in a stone building used as barracks in the War of 1812.
Collections: archives; history.
Research Fields: growth of Cape Vincent; family histories.
Facilities: library of scrapbooks, history books on early French settlers, family histories, historic buildings & activities available for use by appointment; reading room.
Activities: guided tours; permanent & temporary exhibitions.
Publications: pamphlet, History of Cape Vincent.
Hours & Admission Prices: July-Aug. Mon.-Sat. 10-4, Sun. 1-3. Historical research & family histories by appointment. No charge; donations accepted. &
Attendance: 3,000 (estimated)

Castile

WILLIAM PRYOR LETCHWORTH MUSEUM, One Letchworth State Park, Castile, NY 14427-9714. Tel.: 585-493-3600. Fax: 716-493-5272. TDD: 716-493-3070.
E-mail: brian.scriven@parks.ny.gov
Web Site: www.nysparks.state.ny.us
Founded: 1913.
Key Personnel: Park Mgr., Roland Beck; Museum Dir. & Historic Site Mgr., Brian Scriven.
Personnel Profile: Full-Time Paid 1; Part-Time Paid 5; Part-Time Volunteers 2.
Governing Authority: state. Parent Institution: Genesee State Park & Recreation Region Executive Dept. Tax-exempt.
Institution Type/Description: State Park Museum.
Collections: history; ethnology; Indian artifacts; Civilian Conservation Corps. History; pioneer artifacts. Historic Houses: pre-Revolution Council House; c.1800 Nancy Jemison cabin; William Pryor Letchworth papers.
Research Fields: Genesee Valley history; Western New York history.
Facilities: 500-vol. library of W.P. Letchworth's personal collection & his reference library of penal & epileptic institutions & Genesee Valley historical material available for research under supervision.
Activities: permanent exhibitions; interpretive historical video.
Publications: museum brochure; historical booklet.
Hours & Admission Prices: May-Oct. daily 10-5. No charge; donations accepted. &
Attendance: 25,000 (accurate)

Cattaraugus

CATTARAUGUS AREA HISTORICAL CENTER, 23 Main St., Cattaraugus, NY 14719-1032. Mailing Address: 26 S. Franklin St., Cattaraugus, NY 14719-1130. Tel.: 716-257-9500.
Founded: 1955.
Congressional District: 34
Key Personnel: Pres. (V), Robert B. Waite; Sec., Dawn D. Schmitt.
Governing Authority: Executive Council of Cattaraugus Area Historical Society. Subsidiary Institution: Medora Ball Historical Museum, Otto, NY 14766. Tax-exempt.
Institution Type/Description: Historical Society Museum: housed in historic building owned by Village of Cattaraugus.

Collections: agriculture; archives; costumes; general; Indian artifacts; manuscripts; industry; military; eight-foot, hand-carved Statue of Liberty; over 200 photos, yearbooks & historical documents of the schools of the town of New Albion, from 1800s & up. Historic Building: 1862 Otto Congregational Church.
Research Fields: local history of the Cattaraugus area: villages of Cattaraugus, Otto, East Otto and immediate rural area.
Activities: permanent & temporary exhibitions; bimonthly public programs.
Hours & Admission Prices: 2nd Sun. each month 1:30-3:30; other times by appointment. No charge; donations accepted.
Attendance: 175 (estimated)
Membership: Annual $5; Life $25.

Cazenovia

CAZENOVIA COLLEGE - ART GALLERY AT REISMAN HALL ART & DESIGN CENTER, Cazenovia College, 6 Sullivan St., Cazenovia, NY 13035-1085. Tel.: 315-655-7138. Fax: 315-655-2190.
E-mail: jpepper@cazenovia.edu
Web Site: www.cazenovia.edu
Formerly: Cazenovia College - Art & Design Gallery
Founded: 1978.
Key Personnel: Dir., Jen Pepper.
Personnel Profile: Full-Time Paid 1.
Governing Authority: private college; nonprofit. Tax-exempt.
Institution Type/Description: Art Gallery.
Collections: works by international, national & regional artists including paintings; sculpture; prints; photographs; video; digital
Facilities: 1,000 sq. ft. exhibit space.
Hours & Admission Prices: Spring, Fall & Winter: Mon.-Thurs. 1-4 & 7-9, Fri. 1-4, Sat.-Sun. 2-6. Summer hours vary. No charge; donations accepted. Closed during college holidays & breaks. &
Attendance: 1,000 (estimated)

LORENZO STATE HISTORIC SITE, 17 Rippleton Rd., Cazenovia, NY 13035-9601. Tel.: 315-655-3200. Fax: 315-655-4304.
E-mail: barbara.bartlett@parks.ny.gov
Web Site: www.lorenzony.org
Founded: 1968.
Congressional District: 32
Key Personnel: Dir., Barbara Bartlett; Cur. Asst., Jackie Vivirito.
Personnel Profile: Full-Time Paid 6; Part-Time Paid 6; Part-Time Volunteers 60.
Governing Authority: state. New York State Office of Parks, Recreation and Historic Preservation, Central Region.
Institution Type/Description: Historic House Museum: 1807 Federal-style mansion built for John Lincklaen, land agent for the Holland Land Co.
Collections: Lincklaen/Ledyard family furnishings; 19th-century decorative arts; fine arts; carriages; regional archive of family documents; furniture. Historic Buildings: c.1892 Carriage House & Barn; tool building; garage; playhouse; smokehouse; c.1814 schoolhouse.
Research Fields: Lincklaen/Ledyard family; local history; early land development & transportation.
Facilities: 4,000-vol. library of family manuscripts & 100,000 papers available for use by advance request; formal gardens; picnic area; visitor center.
Activities: guided tours; lectures; gallery talks; special events; annual horse & carriage pleasure driving competition; cross-country skiing.
Publications: newsletter; Lorenzo Guidebook.
Hours & Admission Prices: May-Oct. Wed.-Sun. 10-4:30; other times by appointment. Adults $5, tour groups & NY state senior citizens $4; Friends of Lorenzo & members no charge. &
Attendance: 32,000 (estimated)
Membership: Friends of Lorenzo: Regular Single $25; Regular Family $35; Benefactor $100; Patron $150; Sponsor $250 & up.

ROTHSCHILD PETERSEN PATENT MODEL MUSEUM, 4796 W. Lake Rd., Cazenovia, NY 13035-9670.
E-mail: museum@patentmodel.org
Web Site: www.patentmodel.org
Key Personnel: Owner, Alan Rothschild
Institution Type/Description: General Museum.
Collections: patent models.
Hours & Admission Prices: By appointment only.

STONE QUARRY HILL ART PARK, INC., 3883 Stone Quarry Rd., Cazenovia, NY 13035-8447. Mailing Address: P.O. Box 251, Cazenovia, NY 13035-0251. Tel.: 315-655-3196. Fax: 315-655-5742.
E-mail: office@stonequarryhillartpark.org
Web Site: www.stonequarryhillartpark.org
Founded: 1991.
Key Personnel: Mgr., Lesley C. Owens-Pelton, J.D.; Site Mgr., Dylan Otts.
Personnel Profile: Full-Time Paid 2; Part-Time Paid 1; Part-Time Volunteers 10.
Governing Authority: private; nonprofit organization. Tax-exempt: 501(c)(3)
Institution Type/Description: Sculpture Art Park.
Collections: outdoor sculpture consisting of site-specific installations and a permanent collection mounted in a natural setting consisting of woodlands, meadows & wetlands.
Facilities: 3,000-vol. library of art books available to members by appointment; 2,400 sq. ft. exhibit space; 80-seat classroom; 104 acre park; nature trails.
Activities: arts festivals; concerts; informal education programs; guided tours; lecture; workshops. Annual Events: Syracuse Ceramic Guild Sale; Society for New Music; Art in the Sky, kite fest; Talons: Birds of Prey; Finding Art in Nature Treasure Hunt.
Publications: biannual newsletter; exhibition catalogs; e-newsletters.
Hours & Admission Prices: Park: daily dawn to dusk. Gallery: Thurs.-Sun. 12-5. Suggested Donation: $5 per car; members no charge. &
Attendance: 30,000 (estimated)

Centerport

CENTERPORT FIRE MUSEUM, 9 Park Circle, Centerport, NY 11721. Tel.: 631-261-5916.
Institution Type/Description: Firefighting History Museum.
Collections: firefighting history & equipment; horse-drawn hand pumper; period ambulance; uniforms; helmets; photographs; department band instruments; scrapbooks; early fire alarm box; hose nozzles & fittings; siren box.
Hours & Admission Prices: By appointment.

✳ SUFFOLK COUNTY VANDERBILT MUSEUM AND PLANETARIUM, (M), 180 Little Neck Rd., Centerport, NY 11721-1145. Mailing Address: P.O. Box 0605, Centerport, NY 11721-0605. Tel.: 631-854-5579. Fax: 631-854-5594.
E-mail: info@vanderbiltmuseum.org
Web Site: www.vanderbiltmuseum.org
Founded: 1950.
Congressional District: 3
Key Personnel: Exec. Dir., Lance Reinheimer; Pres. Bd. Trustees (V), Ronald A. Beattie; Dir. Public Programming, Lorraine Vernola; Education, Beth Laxer-Limmer; Museum Shop & Business Office, Barbara Oster.
Personnel Profile: Full-Time Paid 10; Part-Time Paid 69; Part-Time Volunteers 54; Interns 1.
Governing Authority: county. Chartered by Education Dept. of University of the State of New York. Tax-exempt: 501(c)(3).
Institution Type/Description: Historical Site: Marine & Natural History Museum & Planetarium. 1910-36, Spanish Revival mansion located on 43-acre landscaped estate of William K. Vanderbilt, II.
Collections: Marine Biology & Wildlife; European, 19th-century & Late Gothic Fine Arts; medieval-19th-century decorative arts including Far & Near East, European & American furniture, metal work, ceramics & carpets; Oriental, Muslim & European firearms & swords; South Pacific, Asian, African & Persian anthropology and ethnology; original Vanderbilt natural history exhibits; marine biology, butterfly, bird & shell collections; ship models and chandlery; maritime books & paintings; documents of the Vanderbilt family; library containing the private collection of Mr. Vanderbilt including the nine books he wrote; antique & classic automobiles; Aeolian player organ.
Research Fields: Vanderbilt history & collections; astronomy & space science.
Facilities: science library available to institutions of higher learning through exchange; planetarium; specimen trees including 12 lindens, statuary & garden ornaments. Museum-related items & educational material for sale.
Activities: guided tours; lectures; workshops; concerts; classes; sky shows; planetarium shows; theater productions; car shows; music events.
Hours & Admission Prices: April 16-June 24 Tues. & Sat.-Sun. 12-5; June 25 to Labor Day Tues.-Sat. 11-5, Sun. 12-5; Sept. 10-Nov. 4 Tues. & Fri.-Sat. 11-5, Sun. 12-5; Nov. 5-April 15 Tues. & Sat.-Sun. 12-4. Adults $7, senior 62 & over and students $6, children under 12 $3; active military no charge. Planetarium: Day Shows: adults $12, seniors 62 & over and students $11, children under 12 $8; Night Shows: adults $9, students and seniors 62 & over $8, children 12 & under $7. Closed New Year's Day; Easter; Memorial Day; Independence Day; Labor Day; Thanksgiving; Christmas Eve & Day.

Attendance: 100,000 (estimated)
Membership: Student & Senior Citizen $45; Individual $60; Dual $70; Family $95; Associate $125; Director's Circle $500.

Chappaqua

NEW CASTLE HISTORICAL SOCIETY - HORACE GREELEY HOUSE, (M), 100 King St., Chappaqua, NY 10514-3433. Mailing Address: P.O. Box 55, Chappaqua, NY 10514-0055. Tel.: 914-238-4666. Fax: 914-238-1296.
E-mail: newcastlehs@aol.com
Web Site: www.newcastlehs.org
Founded: 1966.
Congressional District: 24
Key Personnel: Museum Shop Mgr., Lois Danneckar; Pres., Betsy Guardenter; Exec. Dir., Betsy Towl.
Personnel Profile: Full-Time Paid 1; Part-Time Paid 2; Part-Time Volunteers 100; Interns 1.
Governing Authority: nonprofit organization; society. Parent Institution: New Castle Historical Society. Tax-exempt.
Institution Type/Description: History Museum: housed in the Horace Greeley family country home, 1864-1872.
Collections: New Castle history; Quakers; Horace Greeley memorabilia; costumes; photographs; postcards.
Major Exhibits: Notable Neighbors in New Castle, 2/14-1/15; New Castle's Beginnings: Our Founding Farms at New Castle Town Hall, 2/14-1/15; Here Comes the Bride, 6/14-3/15.
Research Fields: genealogy & local history with emphasis on Horace Greeley & the Quakers.
Facilities: 100-vol. library of books on Horace Greeley, Westchester history & Quakers; period herb garden. Notepapers, maps of community, pamphlets for sale.
Activities: guided tours; lectures; permanent & temporary exhibitions; oral history workshops; school programs & workshops. Special Event: annual indoor antiques show.
Publications: books, The Early Quaker Hamlet of Old Chappaqua; The Chappaqua Life of Horace Greeley; Deeds & Misdeeds, 1657-1763; A Bicentennial History of the Town of New Castle: 1791-1991; The Battle of White Plains; The 1872 Nomination & Presidential Campaign of Horace Greeley; Rare Architectural Heritage - Old Houses of Chappaqua & Millwood; Annandale Farm; video, Our Quaker Heritage; newsletter; annual report; exhibition catalogues; postcards; note paper; maps; Images of America - Newcastle: Chappaqua and Millwood.
Hours & Admission Prices: Aug. Tues.-Thurs. 1-4; Sept.-July Tues.-Thurs. & Sat. 1-4. No charge; donations accepted. Closed New Year's Day; Christmas. &
Attendance: 3,899 (accurate)
Membership: Members $35; Contributor $50; Patron $100; Supporter $250; Benefactor $500; Pacesetter $1,000.

Chazy

THE ALICE T. MINER COLONIAL COLLECTION, (M), 9618 Main St., Chazy, NY 12921. Mailing Address: P.O. Box 628, Chazy, NY 12921-0628. Tel.: 518-846-7336.
E-mail: minermuseum@westelcom.com
Web Site: www.minermuseum.org
Founded: 1924.
Congressional District: 31
Key Personnel: C.E.O., Pres. (V) & Chm. (V), Mrs. Joan T. Burke; Dir. & Cur., Amanda A. Palmer.
Personnel Profile: Full-Time Paid 1; Part-Time Paid 2; Part-Time Volunteers 10.
Governing Authority: nonprofit organization. Tax-exempt: 501(c)(3).
Institution Type/Description: Colonial Revival Museum: housed in 1824 limestone building.
Collections: paintings; decorative arts; furnishings; glass; Indian; military; manuscripts.
Facilities: 900-vol. library of books of the 18th,19th & 20th centuries available on premises.
Activities: guided tours; permanent exhibitions; educational programs; entertainment events.
Hours & Admission Prices: Tues.-Sat. 10-4. Guided Tours: 10am, 12 & 2pm. Adults $3, seniors $2, students $1; student groups no charge.
Attendance: 2,500 (estimated)

Cherry Valley

CHERRY VALLEY MUSEUM, 49 Main St., Cherry Valley, NY 13320. Mailing Address: P.O. Box 115, Cherry Valley, NY 13320-0115. Tel.: 607-264-3303 & 3098. Fax: 607-264-9320.
E-mail: museum@celticart.com
Web Site: www.cherryvalleymuseum.org
Founded: 1958.
Congressional District: 31
Key Personnel: Dir., Chm. (V) & Museum Shop Mgr., Barbara Bell; Pres. (V), James Johnson.
Personnel Profile: Part-Time Paid 2; Part-Time Volunteers 30.
Governing Authority: society. Affiliated with the Cherry Valley Historical Association. Tax-exempt: 501(c)(3).
Institution Type/Description: History Museum: housed in 1832 home.
Collections: 15 rooms of exhibits pertaining to farm & village life in Cherry Valley since late 18th century.
Research Fields: genealogy.
Facilities: 100-vol. library of books & genealogies available for use on premises; reading room. Books on Cherry Valley history for sale.
Activities: guided tours; permanent exhibitions.
Publications: book, The History of Cherry Valley from 1740 to 1898; DVD & video tape, The Border Warfare of N.Y. During the Revolution; The Annals of Tryon County.
Hours & Admission Prices: Memorial Day to mid-Oct. daily 10-5. Adults $6, senior citizens 60 & over $5.50; discounts to groups of 10 or more & AAA members; children under 11 & members no charge. Closed Independence Day. &
Attendance: 700 (estimated)
Membership: Annual $10; Joint $15; Life $150; Joint Life $250.

Chester

CHESTER HISTORICAL SOCIETY, 47 Main St., Chester, NY 10918. Tel.: 845-469-2388. Facebook: Chester Historical Society.
E-mail: chester_historical@mac.com
Web Site: www.chesterhistoricalsociety.com
Founded: 1964.
Congressional District: 19
Key Personnel: Pres. (V), Debby Lu Vadala.
Personnel Profile: Part-Time Volunteers 10.
Governing Authority: Tax-exempt.
Institution Type/Description: Historical Society Museum.
Collections: local history & culture; photographs; personal artifacts; newspapers; oral histories; manuscripts; books; documents.
Research Fields: local history.
Facilities: archives.
Activities: Annual Event: Open House in December.
Publications: newsletter, four times per year.
Hours & Admission Prices: Tues. 10 to noon. No charge; donations accepted.
Membership: Individual $15; Family $20; Life $150.

CHESTER HISTORY MUSEUM - 1915 ERIE RAILROAD STATION, 19 Winkler Pl., Chester, NY 10918-1259. Mailing Address: 47 Main St., Chester, NY 10918. Tel.: 845-469-2591. Facebook: Chester Historical Society.
E-mail: chester_historical@mac.com
Web Site: www.chesterhistoricalsociety.com
Founded: 1999.
Congressional District: 19
Key Personnel: Pres. (V), Norma Stoddard.
Personnel Profile: Part-Time Volunteers 10.
Governing Authority: Tax-exempt.
Institution Type/Description: Historical Society Museum: station built in 1915.
Collections: local history & culture; railroad artifacts; photographs; Chester School alumni; agricultural artifacts.
Research Fields: local history.
Activities: special events. Annual Event: Open House in December.
Publications: quarterly newsletter.
Hours & Admission Prices: May-Oct. Sat. 9-1; groups by appointment. No charge; donations accepted. &
Attendance: 1,000 (estimated)
Membership: Individual $15; Family $20; Associate $35; Life $150.

Chittenango

CHITTENANGO LANDING CANAL BOAT MUSEUM, 717 Lakeport Rd., Chittenango, NY 13037-8584. Tel.: 315-687-3801. Facebook: Chittenango Landing Canal Boat Museum.
E-mail: info@clcbm.org
Web Site: clcbm.org
Founded: 1986.
Key Personnel: Exec. Dir., Christine O'Neil.
Governing Authority: private; nonprofit. Tax-exempt: 501(c)(3).
Institution Type/Description: Transportation Museum: housed in a 19th century structure.
Collections: artifacts & structural features recovered from the site & used for interpretation; artifacts relating to the canal boat repair building & social aspects of the canal; original canal boat drawings.
Research Fields: archaeological investigation of site to identify & delineate historic features; investigation of canal boat building practices of the 19th century; area residents related to those workers of the original & enlarged Erie Canal and to the dry docks at Chittenango Landing.
Facilities: 700-vol. library related to boat building & canal history; 1,625 sq. ft. exhibit space; 5-acre archaeological site; visitors center. Museum-related items for sale.
Activities: formal education programs; guided tours; lectures; participatory exhibits; 4 elderhostel programs yearly. Annual Events: July 4th Community Celebration; Winter Fun Day.
Publications: quarterly newsletter, From the Boatyard.
Hours & Admission Prices: May-June & Sept.-Oct. Sat.-Sun. 1-4; July-Aug. daily 11-3. Family $20, adults $6, senior citizens $5, children 4-12 $3. Closed New Year's Day; Easter; Thanksgiving; Christmas. &
Attendance: 12,000 (estimated)
Membership: Hoggee under 18 & Canaller 60 & up $10; Drydocker $15; Towpath Traveler $25; Bronze Anchor $35; Silver Anchor $50; Gold Anchor $100; Boat Builder $250; Steersman $500; Captain $1,000; Lockmaster $2,500.

Clarence

HISTORICAL SOCIETY OF THE TOWN OF CLARENCE, (M), 10465 Main St., Clarence, NY 14031-1617. Mailing Address: P.O. Box 86, Clarence, NY 14031-0086. Tel.: 716-759-8575.
Founded: 1954.
Congressional District: 38
Key Personnel: Pres., Thomas Steffan; Cur., Mrs. Alicia L. Braaten.
Personnel Profile: Part-Time Paid 1; Part-Time Volunteers 19.
Governing Authority: society; nonprofit organization. Tax-exempt.
Institution Type/Description: Historical Society & Genealogy Museum.
Collections: archaeology; genealogy of Clarence & Western New York families; Indian artifacts; the garage & workshop where Wilson Greatbatch invented the implantable pacemaker. Historic Structure: log cabin c.1825.
Research Fields: archaeology; Clarence history; Indian artifacts.
Facilities: collection of local & Western New York history available on premises; reading room.
Activities: guided tours; lectures; formally organized education programs for children; permanent & temporary exhibitions.
Publications: newsletter, A Ransom Note; Clarence bicentennial historical books; local history books.
Hours & Admission Prices: Call for hours. No charge; donations accepted. Closed holidays. &
Attendance: 2,300 (accurate)
Membership: Individual $15; Couple $25; Family $30; Life $150.

Clayton

THE ANTIQUE BOAT MUSEUM, (M), 750 Mary St., Clayton, NY 13624-1119. Tel.: 315-686-4104. Fax: 315-686-2775.
E-mail: info@abm.org
Web Site: www.abm.org
Formerly: Thousand Islands Shipyard Museum
Founded: 1964.
Congressional District: 30
Key Personnel: Exec. Dir., Frederick H. Hager; Dir. Public Programming, Lora Nadolski; Cur., Emmett Smith; Curatorial Asst., Claire Wakefield; Educator, Julie Broadbent; Watercraft Conservator, Mike Corrigan; Coord. Events, Margaret Hummel; Dir. Endowment, John MacLean; Dir. Leadership Gifts, Barbara Maddocks; Assoc. Dir. Endowment, Christine Brown; C.F.O., Dale Corsa; Accounting Clerk, Norma Zimmer; Museum Shop Mgr., Charlotte Brooks; Administrative Asst., Kirsti Touhey; Supvr. Facilities, Bud Gray; Maintenance Asst., Jim Mellowship; Maintenance Asst., Sam Hopkins.
Personnel Profile: Full-Time Paid 17; Part-Time Paid 25; Part-Time Volunteers 200; Interns 2.

Governing Authority: board of trustees; nonprofit organization. Tax-exempt: 501(c)(3).
Institution Type/Description: Antique Boat Museum.
Collections: St. Lawrence sailing & rowing skiffs; power skiffs; canoes; duckboats; outboards; outboard & inboard engines; launches; runabouts; utilities; ice boats; racing boats; photography collection; boat building molds & tools; nautical artifacts; St. Lawrence River culture.
Research Fields: freshwater recreational boating, St. Lawrence regional history.
Facilities: 700-vol. library of books pertaining to nautical & history of the Thousand Islands region available for research upon request; educational facilities. Museum-related items for sale.
Activities: guided tours; lectures; hobby workshops; loan, permanent & traveling exhibitions; charters & cruises; boat building & restoration classes; children's classes; boat rides every hour on the hour. Museum Sponsors: antique boat show, auction, small craft festival, regatta.
Publications: triannual newsletter, Gazette; brochures; pamphlets; exhibition catalogues.
Hours & Admission Prices: mid-May to mid-Oct. daily 9-5; other times by appointment. Adults $13, youth $6.50; discounts to seniors, AAA, & groups; militry and children under 6 & members no charge. &
Attendance: 33,295 (accurate)
Membership: Individual $50; Family $65; Supporting $125; Contributing $250; Patron $500; Friends of the Museum $1,000 & up.

THOUSAND ISLANDS ARTS CENTER HOME OF THE HANDWEAVING MUSEUM, 314 John St., Clayton, NY 13624-1017. Tel.: 315-686-4123.
E-mail: rebecca@tiartscenter.org
Web Site: www.tiartscenter.org
Formerly: American Handweaving Museum and Thousand Islands Craft School
Founded: 1966.
Congressional District: 26
Key Personnel: Exec. Dir., Rebecca Hopfinger; Chm. (V), Joy Rhinebeck; Events Coord., Kassandra Kittle; Education Coord., Marcia Rogers; Potter, Christin Bentley.
Personnel Profile: Full-Time Paid 3; Part-Time Paid 2; Part-Time Volunteers 21; Interns 1.
Governing Authority: nonprofit organization. Tax-exempt: 501(c)(3).
Institution Type/Description: Textile Museum: housed in late 19th-century two-story frame home.
Collections: over 1,400 textiles from ancient Egyptian to present including 20th century handwovens; archives including books, photographs & correspondence; 1,600 textile-oriented books; pattern books.
Research Fields: 20th-century American handweaving.
Facilities: Museum-related items for sale.
Activities: guided tours; arts festivals; hobby workshops; organized education programs; temporary exhibitions; classes in heritage arts. Museum Sponsors: fiber arts workshops; annual art & antique show & sale; annual juried crafts show & sale; Annual Weaving History Conference.
Publications: newsletter; collections catalog; class catalog.
Hours & Admission Prices: Museum: Mon.-Fri. 9-5, Sat. 10-4. Suggested Donation: $10. Pottery Studio: Tues. 6pm-8pm. Closed major holidays. &
Attendance: 12,000 (estimated)
Membership: Individual $35-$59; Family $60-$124; Sponsor $125-$249; Benefactor $250-$499; Patron $500-$999; Friend $1,000 & up.

THOUSAND ISLANDS MUSEUM, 312 James St., Clayton, NY 13624-1012. Mailing Address: P.O. Box 27, Clayton, NY 13624-0027. Tel.: 315-686-5794. Fax: 315-686-4867.
E-mail: info@timuseum.org
Web Site: www.timuseum.org
Founded: 1964.
Congressional District: 30
Key Personnel: Pres. (V), Thomas Humberstone; Museum Shop Mgr., April Ingerson.
Personnel Profile: Full-Time Paid 1; Full-Time Volunteers 2; Part-Time Volunteers 10.
Governing Authority: nonprofit corporation. Parent Institution: Thousand Islands Museum. Tax-exempt: 501(c)(3).
Institution Type/Description: Local & Regional History Museum.
Collections: decoy carving; muskie fishing; pictorial history of Clayton & 1000 Islands; home of the St. Lawrence Tartan; St. Lawrence River Heritage.
Major Exhibits: Country Kitchen, 5/14-12/14; One-Room Schoolhouse, 5/14-12/14; Vintage Ladies & Men's Hats, 5/14-12/14.
Research Fields: history of St. Lawrence River & Thousand Islands region.
Facilities: library. Sole supplier of St. Lawrence Tartan.
Activities: permanent & rotating exhibits. Museum Sponsors: fund-raiser,

Decoy & Wildlife Art Show; History Boat Tours in July & August; Quilt Show; History At Noon Program in July & August.
Publications: brochure; newsletter.
Hours & Admission Prices: May-Sept. Mon.-Fri. 10-5, Sat.-Sun. 10-4; Oct.-Dec. Tues.-Sun. 10-4. No charge; donations accepted. &
Attendance: 15,000 (accurate)
Membership: Individual $25; Family $35; Supporting $50; Patron $100; Benefactor (5 yrs) $250; Contributing $1,251-$4,999; Architect $5,000.

Clinton

CLINTON HISTORICAL SOCIETY, One Fountain St., Clinton, NY 13323. Mailing Address: P.O. Box 42, Clinton, NY 13323. Tel.: 315-859-1392.
Founded: 1962.
Governing Authority: nonprofit organization. Tax-exempt: 501(c)(3).
Institution Type/Description: Historical Society Museum: housed in the former Clinton Baptist Church.
Collections: local history & culture; period artifacts; photographs; paintings; maps; school yearbooks & records; Bristol-Myers Squibb Company history; business signs; railroad artifacts; scrapbooks.
Facilities: library.
Activities: educational programs.
Publications: newsletter.
Hours & Admission Prices: Wed. 1-3, Sat. 11-2. No charge. &
Attendance: 800 (estimated)

THE RUTH AND ELMER WELLIN MUSEUM OF ART, (M), Hamilton College, 198 College Hill Rd., Clinton, NY 13323-1218. Tel.: 315-859-4396. Fax: 315-859-4060. Facebook: Wellin Museum.
E-mail: wellin@hamilton.edu
Web Site: www.hamilton.edu/wellin
Formerly: The Emerson Gallery
Founded: 2012.
Congressional District: 32
Key Personnel: Dir., Tracy Adler; Assoc. Dir. & Cur., Susanna White; Public Programming & Outreach Coord., Megan Austin; Bldg. Mgr. & Preparator, William Bitter; Gen. Contact & Office Asst., Eleonore Moncheur.
Personnel Profile: Full Time Paid 3; Part-Time Paid 12; Part-Time Volunteers 4.
Governing Authority: Hamilton College. Tax-exempt.
Institution Type/Description: Art Museum.
Collections: works on paper 15th-20th century with emphasis on 19th-20th century British & American; 19th-& 20th century paintings; Native American objects; Greek vases; Roman glass; Asian painting, sculpture & ceramics; Beinecke collection of 16th-to 19th century prints, drawings & maps of West Indies.
Activities: 6-10 art exhibitions per year; lectures; films; tours; gallery talks; performances.
Hours & Admission Prices: Tues.-Fri. 10-5:30, Sat. 10-5, Sun. 12-5. No charge. Closed New Year's Day; Memorial Day; Independence Day; Christmas. &
Attendance: 8,000 (estimated)

Cohocton

COHOCTON HISTORICAL SOCIETY, Main St., Cohocton, NY 14826. Mailing Address: P.O. Box 177, Cohocton, NY 14826. Tel.: 585-384-5729.
Institution Type/Description: Historical Society Museum.
Collections: local history & culture; period artifacts; photographs.
Hours & Admission Prices: Summer: 1st Sat. of the month 9-1.

Cold Spring

PUTNAM COUNTY HISTORICAL SOCIETY & FOUNDRY SCHOOL MUSEUM, 63 Chestnut St., Cold Spring, NY 10516-2613. Tel.: 845-265-4010. Fax: 845-265-2884.
E-mail: office@pchs-fsm.org
Web Site: www.pchs-fsm.org
Founded: 1906.
Congressional District: 21
Key Personnel: Exec. Dir., Mindy Krazmien; Cur., Trudie Alexis Grace; Dir. Administration, Kara Shier; Coord. Outreach, Kendall Helbock; Office Mgr., Helen Brown; Curatorial Asst., Anne Saunders.
Personnel Profile: Part-Time Paid 6; Part-Time Volunteers 6.
Governing Authority: society. Parent Institution: Putnam County Historical Society. Subsidiary Institution: Foundry School Museum. Tax-exempt: 501(C)(3).

Institution Type/Description: Local History Museum: housed in 1828 Foundry Schoolhouse.
Collections: 19th-century paintings; items from the West Point Foundry including manufactured articles; photos; documents; records; letters; local genealogical records; manuscripts.
Research Fields: local history; genealogy; West Point Foundry.
Facilities: 1,500-vol. library of history books available for use by appointment on premises; reading room. Museum-related items for sale.
Activities: guided tours; lectures; gallery talks; historical research; inter-museum loan exhibitions; school loan service; school program.
Publications: exhibition catalogs, America the Beautiful: Women and the Flag; A Ramble Through the Hudson Highlands: A History in Pictures and the Writings of Donald H. MacDonald; High Fashion and Society in the Hudson Highlands, 1865-1914; George Pope Morris: Defining American Culture; This Perfect River View: The Hudson River School and Contemporaries in Private Collections in the Highlands.
Hours & Admission Prices: March-Dec. Wed.-Sun. 11-5. No charge; donations accepted. &
Attendance: 1,800 (estimated)
Membership: Student & Senior $30; Individual $50; Family $100; Highlander $250; Weir Circle $500; Storm King $1,000; General Putnam Brigade $2,500.

Cold Spring Harbor

SOCIETY FOR THE PRESERVATION OF LONG ISLAND ANTIQUITIES, 161 Main St., Cold Spring Harbor, NY 11724-0148. Mailing Address: P.O. Box 148, Cold Spring Harbor, NY 11724-0148. Tel.: 631-692-4664. Fax: 631-692-5265. Facebook: Society for the Preservation of Long Island Antiquities.
E-mail: info@splia.org
Web Site: splia.org
Founded: 1948.
Congressional District: 1
Key Personnel: Pres., Townsend U. Weekes, Jr.; Dir., Alexandra P. Wolfe; Educator, Sandy Branciforte; Property Mgr., Alan Tewksbury; Office Mgr., Betsy Terrizzi.
Personnel Profile: Full-Time Paid 6; Part-Time Paid 14; Part-Time Volunteers 25.
Governing Authority: nonprofit organization. Tax-exempt: 501(c)(3).
Institution Type/Description: Preservation Society.
Collections: decorative arts, textiles, ceramics, glass, photographs, paintings, prints, manuscripts, pertaining to house museums. Historic Houses: 1767 Joseph Lloyd manor, village of Lloyd Neck, Long Island; 1730-1780 Sherwood-Jayne house, E. Setauket, Long Island; c.1800 Custom house, Sag Harbor, Long Island; The Gallery, 161 Main St., Cold Spring Harbor, Long Island.
Major Exhibits: Collecting Long Island: 21st Century Acquisitions, 4/14-12/14.
Research Fields: Long Island history & material culture.
Activities: education programs; publications; preservation services; special projects.
Publications: periodical, Preservation Notes; books, A History of the Joseph Lloyd Manor House; Long Island Domestic Architecture of the Colonial & Federal Periods; Long Island is My Nation: The Decorative Arts & Craftsmen 1640-1830; Windmills of Long Island; Between Ocean & Empire: An Illustrated History of Long Island; Useful Art: Long Island Pottery; A Forgotten People: Discovering the Black Experience in Suffolk County; Long Island Country Houses & their Architects, 1860-1940; Saving Large Estates; Edward Lange's Long Island; Woven History: The Technology & Innovation of Long Island Coverlets, 1800-1860; AIA Guide to Nassau & Suffolk Counties, Long Island; Long Island Landscapes and the Women Who Designed Them; Harbor Hill - Portrait of a House; Barns of the North Fork; Anchor to Windward - The Paintings and Diaries of Annie Cooper Boyd; Long Island: An Illustrated History; Long Island Modernism - 1930-1980.
Hours & Admission Prices: Custom House: May-June & Sept.-Oct. Sat.-Sun. 10-5; July-Aug. daily 10-5; groups by appointment. Joseph Lloyd Manor: Memorial Day to Columbus Day Sat.-Sun. 1-5; groups by appointment. Sherwood-Jayne House: Memorial Day to Columbus Day by appointment only. Custom House: adults $5, seniors & children 7-14 $3. Joseph Lloyd Manor & Sherwood-Jayne Houses: adults $3, seniors & children 7-14 $2. SPLIA Gallery for Moving Exhibits: call for hours & admission prices. &
Attendance: 7,100 (estimated)
Membership: Student & Senior Citizen $20; General $30; Family $40; Sustaining $100; Corporation & Contributing $250; Life $1,000 Patron $2,500; Founder $5,000.

* **THE WHALING MUSEUM & EDUCATION CENTER OF COLD SPRING HARBOR, (M),** 301 Main St., Cold Spring Harbor, NY 11724. Mailing Address: 279 Main St., Cold Spring Harbor, NY 11724-0025. Tel.: 631-367-3418. Fax: 631-692-7037.
E-mail: kkelly@cshwhalingmuseum.org
Web Site: www.cshwhalingmuseum.org
Founded: 1936.
Congressional District: 5
Key Personnel: Exec. Dir., Nomi Dayan; Pres. (V), Art Brings; Business Mgr., Katie Kelly; Coord. Education, Leah Master.
Personnel Profile: Full-Time Paid 5; Part-Time Paid 6; Part-Time Volunteers 10; Interns 10.
Governing Authority: society. Tax-exempt: 501(c)(3).
Institution Type/Description: Whaling Museum.
Collections: scrimshaw collection; ship models; maritime paintings; manuscript collection; 19th-century whaleboat; whaling implements; permanent exhibit on Long Island Whaling industry; marine mammal conservation display.
Major Exhibits: Waves & Watercolors: Sea & Create Art (T), 9/13-9/14; Sea Ink: Tattoos and Nautical Culture, 9/14-12/14.
Research Fields: whales & whaling with emphasis on conservation.
Facilities: 2,000-vol. library on whales & whaling, film library and manuscripts by appointment. Nautical items for sale.
Activities: guided tours; lectures; films; gallery talks; formally organized education programs; permanent, rotating & changing exhibitions; maritime workshops for children & adults; walking tours; special weekend programs.
Publications: triannual newsletter, A Whaling Account; books; booklets; collection catalogues; In Their Hours of Ocean Leisure.
Hours & Admission Prices: Memorial Day-Labor Day daily 11-5; Sept.-May Tues.-Sun. 11-5; tours by appointment. Family $20, adults $6, senior citizens & students 5-18 $5; discounts to CAMM, ICOM & AAM members, & other museum professionals; active military, members & children under 5 no charge. Sun. special programs 11-1 & 3 pm. &
Attendance: 22,000 (accurate)
Membership: Individual $35; Family $50; Patron $100; Associate $250; Sponsor $500; Benefactor $750.

Commack

LONG ISLAND CULTURE HISTORY LAB & MUSEUM, Hoyt Farm Park, 200 New Hwy., Commack, NY 11725. Mailing Address: c/o Suffolk County Archaeological Assn., P.O. Box 1542, Stony Brook, NY 11790. Tel.: 631-929-8725. Fax: 631-929-8725.
E-mail: SCArchaeology@gmail.com
Web Site: www.scaa-ny.org
Founded: 1975.
Congressional District: 1
Key Personnel: Pres. (V), Douglas DeRenzo; Dir., Gaynell Stone, Ph.D.
Personnel Profile: Full-Time Paid 1; Part-Time Paid 20; Part-Time Volunteers 5.
Governing Authority: society. Parent Institution: Suffolk County Archaeological Assn. Tax-exempt: 501(c)(3).
Institution Type/Description: Historical Site: prehistoric park with an 18th-century house & 20th-century outbuildings. Site 1: Native American complex. Site 2: 18th- & 19th-century house & outbuildings; Dutch architecture.
Collections: prehistoric & historic artifacts; photographs; hands-on exhibitions.
Research Fields: Long Island archaeology; ethnohistory; technology; material culture; historic photographs.
Facilities: 300-vol. library; classrooms; cultural history center.
Activities: guided tours; lectures; organized education programs; workshops.
Publications: triannual newsletter; biannual reference series, Readings in Long Island Archaeology & Ethnohistory; biannual student series, Prehistoric Natives of Long Island; posters, Native Technology & Native Long Island; workbook, Native Life & Archaeology; Native Forts of the Long Island Sound Area; documentary film, The Sugar Connection: Holland, Barbados, Shelter Island (2011).
Hours & Admission Prices: Sept.-June Mon.-Fri. 10-2 by appointment. Call for admission prices. &
Attendance: 10,000 (accurate)
Membership: Student $10; Individual $20; Family $30; Institution $50; Contributing $100; Patron $200; Life $400.

Cooperstown

COOPERSTOWN HEROES OF BASEBALL WAX MUSEUM, 99 Main St., Cooperstown, NY 13326. Tel.: 607-547-1273.
E-mail: info@baseballwaxmuseum.com
Web Site: www.baseballwaxmuseum.com

Institution Type/Description: Wax Museum.
Collections: over 40 life-sized wax figures; videos.
Facilities: cafe; theatre. Museum-related items for sale.
Activities: photos with wax figures; batting cage; videos.
Hours & Admission Prices: Daily 9-9. Adults $9.95, seniors 65 & over and children 4-12 $7.95.

DOLLHOUSE HALL OF FAME, 635 County Rte. 33, Cooperstown, NY 13326. Tel.: 607-547-5222.
Institution Type/Description: Dollhouse Museum.
Collections: over 60 dollhouses.
Facilities: Museum-related items for sale.
Hours & Admission Prices: Mon.-Fri. 10-5, Sat.-Sun. 10-3.

* **THE FARMERS' MUSEUM, INC., (M),** 5775 State Hwy. 80, Cooperstown, NY 13326. Mailing Address: P.O. Box 30, Cooperstown, NY 13326-0030. Tel.: 607-547-1450. Fax: 607-547-1404.
E-mail: info@nysha.org
Web Site: www.farmersmuseum.org
Founded: 1943.
Congressional District: 24
Key Personnel: Pres. & C.E.O., Paul S. D'Ambrosio, Ph.D.; Chm. Bd., Jane Forbes Clark; Dir. Finance, Marnie Auld; Vice Pres. Operations, Joseph Siracusa; Sr. Dir. Human Resources, Barbara Fischer; Dir. Collections, Erin Richardson; Dir. Exhibitions, Michelle Murdock; Dir. Mktg., Todd Kenyon; Registrar, Christine Olsen; Museum Shop Mgr., Sue deBruijn.
Personnel Profile: Full-Time Paid 89; Part-Time Paid 58; Part-Time Volunteers 252; Interns 6.
Governing Authority: Tax-exempt.
Institution Type/Description: Village Museum: housed in early 19th-century Village including 29 buildings.
Collections: crafts; farm implements & technology; agriculture; transportation; textiles; 19th-century rural & village life; photography; Empire State carousel.
Research Fields: farm implements; New York rural life history; agriculture; textiles; village architecture; craft practices.
Facilities: 80,000-vol. library shared with the New York State Historical Association.
Activities: Museum Sponsors: conferences, seminars & graduate level courses of study with New York State Historical Association.
Publications: early 19th-century book reprints series; conference papers; magazine, Heritage.
Hours & Admission Prices: April to mid-May & Oct. Tues.-Sun. 10-4; mid-May to Columbus Day daily 10-5; Nov.-Dec. open for special programs only. Adults $12, seniors 65 & over $10.50, children $6; discounts to AAM & ICOM members; members, children under 7 & New York State Historical Association members no charge. &
Attendance: 60,000 (estimated)

HYDE HALL, INC., 267 Glimmerglass State Park Rd., Cooperstown, NY 13326. Mailing Address: P.O. Box 721, Cooperstown, NY 13326-0721. Tel.: 607-547-5098 & 6129. Fax: 607-547-8462. Facebook: Hyde Hall.
E-mail: jonathanmaney@hydehall.org
Web Site: www.hydehall.org
Formerly: Friends of Hyde Hall, Inc.
Founded: 1964.
Congressional District: 23
Key Personnel: Bd. Chm. (V), Andrew M. Blum; Tours & Collections Coord., Larry Smith; Operations Asst. & Special Events, Karen Sheckells.
Personnel Profile: Full-Time Paid 2; Full-Time Volunteers 1; Part-Time Paid 15; Part-Time Volunteers 2.
Governing Authority: Tax-exempt.
Institution Type/Description: Historic House Museum.
Collections: family artifacts dating from 1820.
Research Fields: family archives & collections.
Facilities: grounds available for rental.
Activities: special events.
Hours & Admission Prices: Tours: Mother's Day to Oct. daily 10-5, last tour at 4pm. Adults $10, seniors 65 & over and children 6-12 $8; children under 6 no charge. &
Attendance: 5,000 (estimated)

NATIONAL BASEBALL HALL OF FAME AND MUSEUM, 25 Main Street, Cooperstown, NY 13326. Tel.: 607-547-7200. Fax: 607-547-2044.
E-mail: jgates@baseballhalloffame.org
Web Site: www.baseballhalloffame.org

Founded: 1936.
Congressional District: 31
Key Personnel: Pres., Jeffrey Idelson; Sr. Vice Pres., William Haase; Vice Pres. Mktg., Sean Gahagan; Sr. Dir. Communications & Education, Brad Horn; Controller, Fran Althiser; Sr. Dir. Exhibits & Collections, Erik Strohl; Librarian, Jim Gates; Sec., William T. Burdick; Museum Shop Mgr., Diane Adams.
Personnel Profile: Full-Time Paid 73; Part-Time Paid 107; Part-Time Volunteers 2; Interns 30.
Governing Authority: nonprofit organization. Tax-exempt: 501(c)(3).
Institution Type/Description: Sports Museum.
Collections: plaques of members of Baseball Hall of Fame; autographed baseballs; bats; trophies; books; pictures; paintings; cigarette & gum cards; uniforms; cartoons; history time line; video tapes of World Series & All-Star Games; research library.
Research Fields: baseball statistics; history.
Facilities: 1,200-vol. library of scrapbooks, complete sets of Guides & record books, clippings, newspapers, magazines, photographs, microfilm, motion pictures, phonograph records, tapes, August Herrmann & A. G. Mills correspondence pertaining to baseball available by appointment on premises; reading room; 200-seat Baseball Experience theater; 60-seat bullpen-style theater in library. Books, baseballs & other museum-related items for sale.
Activities: films; loan & temporary exhibits; audio & visual displays; special educational programs for children grades 4-6, Nov.-April; distance learning capability. Annual Events: Baseball Symposium in June; special Sandlot Series in summer; Artifact Spotlights in summer; Hall of Fame Classic Game & Induction Ceremony.
Publications: annual yearbook, National Baseball Hall of Fame and Museum; brochure, descriptive folder of Museum; six times per year newsletter; membership magazine, Memories & Dreams; Donor brochure; membership brochure.
Hours & Admission Prices: Memorial Day to Labor Day daily 9-9; Sept.-May daily 9-5. Adults $19.50, senior citizens $12, children 7-12 $7; discount to AAM members & groups with advance reservations; members, military no charge. Closed New Year's Day; Thanksgiving; Christmas. &
Attendance: 301,755 (accurate)
Membership: Junior 12 & under $20; Individual $40; Family $70; Sustaining $100; Patron $175; President's Circle $500; Benefactor $1,000 & up.

* **NEW YORK STATE HISTORICAL ASSOCIATION - FENI-MORE ART MUSEUM, (M),** 5798 St. Hwy. 80, Cooperstown, NY 13326. Mailing Address: P.O. Box 800, Cooperstown, NY 13326-0800. Tel.: 607-547-1400. Fax: 607-547-1404.
E-mail: info@nysha.org
Web Site: www.fenimoreartmuseum.org
Founded: 1899.
Congressional District: 24
Key Personnel: Chm. Bd., Dr. Douglas E. Evelyn; Pres. & C.E.O., Paul S. D'Ambrosio, Ph.D.; Dir. Finance, Marnie Auld; Vice Pres. Operations, Joseph Siracusa; Sr. Dir. Human Resources, Barbara Fischer; Dir. Cooperstown Graduate Program, Gretchen Sorin; Dir. Collections, Erin Richardson; Dir. Exhibitions, Michelle Murdock; Dir. Mktg., Todd Kenyon; Registrar, Christine Olsen; Museum Shop Mgr., Sue deBruijn.
Personnel Profile: Full-Time Paid 85; Part-Time Paid 42; Part-Time Volunteers 171; Interns 9.
Governing Authority: Parent Institution: New York State Historical Association. Subsidiary Institution: Fenimore Art Museum. Tax-exempt: 501(c)(3).
Institution Type/Description: Art Museum & Historical Association.
Collections: academic & folk art; costumes; sculpture; graphics; history; decorative arts; manuscripts and American Indian Art.
Major Exhibits: The Adirondack World of Arthur Fitzwilliam Tait, 4/14-9/14; Winslow Homer: The Native & Rhythm of Life, 6/14-8/14; Dorothea Lange's America, 9/14-12/14.
Research Fields: folk art; American art; Native American art; U.S. & New York State history; museum studies; art history
Facilities: 80,000-vol. library of books.
Activities: formally organized education programs for children, adults & graduate students; graduate program in History Museum Studies in conjunction with State University College of New York at Oneonta.
Publications: quarterly journal, New York History; annual magazine, Heritage; monographs; various books.
Hours & Admission Prices: April to mid-May & Oct.-Dec. Tues.-Sun. 10-4; mid-May to Columbus Day daily 10-5. Adults 13-64 $12, seniors 65 & over $10.50; discounts to AAM members; members and children 12 & under no charge. Closed Thanksgiving; Christmas. &
Attendance: 40,000 (estimated)
Membership: Student & Educator $25; Individual $50; Far-Away Friends $50; Family & Household $80; Sustaining $115; Contributing $250; Benefactor $500; Fenimore Society $1,000.

Corning

* **THE CORNING MUSEUM OF GLASS, (M),** One Museum Way, Corning, NY 14830-2253. Tel.: 607-937-5371.
Web Site: www.cmog.org
Founded: 1951.
Congressional District: 39
Key Personnel: Chm. (V), James B. Flaws; Exec. Dir., Karol Wight; Pres. (V), E. Marie McKee; Sr. Dir. Creative Svcs. & Mktg., Robert K. Cassetti; Sr. Dir. Finance & Administration, Nancy J. Earley; Dir. Finance, David Togni; Head of Publication Dept., Richard Price; Dir. Mktg. & Community Rels., Beth Duane; Dir. Education & The Studio, Amy Schwartz; Dir. Human Resources, Ellen Corradini; Sr. Mgr. Communications, Yvette Sterbenk; Mgr. Hot Glass Programs, Steve Gibbs; Mgr. Operations, Dave Murray; Mgr. Glass Market, Victor Nemard; Mgr. Collections & Exhibitions, Warren Bunn; Cur., Jane S. Spillman; Cur., Tina Oldknow; Conservator, Stephen Koob; Photographer, Nicholas L. Williams; Librarian, Diane Dolbashian.
Personnel Profile: Full-Time Paid 122; Part-Time Paid 8; Part-Time Volunteers 300.
Governing Authority: nonprofit educational institution. Tax-exempt: 501(c)(3).
Institution Type/Description: Glass Art & History Museum.
Collections: comprehensive collection of glass vessels; objects; archaeological remains from every period & area of glass history; library collections of material related to glass.
Major Exhibits: Masters of Studio Glass: Richard Marquis, 2/16/13-2/2/14.
Research Fields: glass history; early technology; archaeology; conservation; chemical analysis & examination; isotope studies; physical property measurements.
Facilities: 50,000-vol. research library on art, history, technology & archaeology of glass; periodicals, trade catalogs, microfilm, microfiche, films, videotapes, slides for use on premises; cafe & snack bar; 6,000 sq. ft. studio, offering instruction & practical experience in the art & craft of glassmaking; glassworking studio. Museum publications for sale.
Activities: self-guided tours; lectures; films; gallery talks; formally organized education programs for children; paid guided tour service; inter-museum loan, permanent & temporary exhibitions; seminars on glass history; live glassmaking demonstrations; try-it-yourself glassmaking; traveling hot glass stage.
Publications: annuals, Journal of Glass Studies; New Glass Review; recent books: The Corning Museum of Glass, A Decade of Glass Collecting; Chemical Analysis of Early Glass, Vol. 1 & 2; Innovations in Glass; The American Cut Glass Industry: T.G. Hawkes and His Competitors; Designs in Miniature: The Story of Mosaic Glass (CD-ROM), produced by the Victoria and Albert Museum (London); Roman Glass in The Corning Museum of Glass Vol. 1, 2 & 3; Beyond Venice: Glass in Venetian Style 1500-1750; 25 Years of New Glass Review; European Glass Furnishings for Eastern Palaces; Glass of the Alchemists: Lead Crystal - Gold Ruby 1650-1750; Contemporary Glass Sculptures and Panels: Selections from the Corning Museum of Glass; Medieval Glass for Popes, Princess & Peasants; Voices of Contemporary Glass: The Heineman Collection; Mt. Washington & Pairpoint Glass, v. II; The Corning Museum of Glass: Notable Acquisitions 2010; DVD series, Glass Masters and Master Class.
Hours & Admission Prices: June-Aug. daily 9-8; Sept.-May daily 9-5. Adults $15; discounts to senior citizens, college students, military, AAA, AAM & ICOM; members & children 19 and under no charge. Closed New Year's Day; Thanksgiving; Christmas Eve & Day. &
Attendance: 400,000 (accurate)
Membership: See website for information.

HORNBY HISTORICAL SOCIETY, 185 South Place, Corning, NY 14830-2219. Tel.: 607-962-4471. Facebook: Hornby Historical Society.
E-mail: sjmoore@stny.rr.com
Founded: 1958.
Key Personnel: Pres. (V), Pamela Smith; Sec. & Treas., Susan J. Moore.
Personnel Profile: Part-Time Volunteers 10.
Governing Authority: society. Tax-exempt.
Institution Type/Description: Local History Museum: housed in a small country schoolhouse.
Collections: 19th-century household & farm items; journals; books; one-room schoolhouse artifacts.
Activities: permanent exhibitions; tours; scouts; school classes.
Publications: quarterly newsletter.
Hours & Admission Prices: July-Aug. Sun. 1-3; other days by appointment. No charge; donations accepted.
Attendance: 74 (accurate)
Membership: Individual $5; Family $8.

PATTERSON INN MUSEUM/CORNING - PAINTED POST HISTORICAL SOCIETY, 59 W. Pulteney St., Corning, NY 14830-2212. Mailing Address: 73 W. Pulteney St., Corning, NY 14830. Tel.: 607-937-5281. Fax: 607-937-5281.

E-mail: pattersoninnmuseum@stny.rr.com
Web Site: www.pattersonmuseum.org
Formerly: The Benjamin Patterson Inn Museum
Founded: 1946.
Congressional District: 23
Key Personnel: Dir., Sheri Golder; Pres. (V), Jane Spillman.
Personnel Profile: Full-Time Volunteers 1; Part-Time Paid 2; Part-Time Volunteers 50.
Operating Expenses: 74,000
Operating Income: 86,775
Governing Authority: board of trustees. Parent Institution: Corning-Painted Post Historical Society. Tax-exempt: 501(c)(3).
Institution Type/Description: Historic Building: 1796 Benjamin Patterson Inn.
Collections: furnishings & artifacts appropriate to the Inn & the years 1800-1850; local history items; extensive costume collection. Historic Buildings: c.1850 log cabin; c.1878 one-room schoolhouse; c.1860s four bay barn; c.1870 blacksmith shop.
Research Fields: local history.
Facilities: 300-vol. library of local & area history books available for research on premises only. Books, stationery, traditional toys & other gift items for sale.
Activities: guided tours; lectures; school programs.
Publications: bimonthly newsletter.
Hours & Admission Prices: Mon.-Fri. 10-4; other times by appointment. Family $14, adults $6, senior citizens 60 & over $4, students $2; discounts to groups over 10 & AAA members; preschool children & members no charge. Closed major holidays. ♿
Attendance: 5,000 (estimated)
Membership: Student $15; Senior Citizen $20; Individual $30; Family $50; Donor $75; Patron $100; Corporate $250; Benefactor $500.

✳ ROCKWELL MUSEUM OF WESTERN ART, (M), 111 Cedar St., Corning, NY 14830-2632. Tel.: 607-937-5386. Fax: 607-974-4536.

E-mail: info@rockwellmuseum.org
Web Site: www.rockwellmuseum.org
Founded: 1976.
Congressional District: 31
Key Personnel: Exec. Dir., Kristin Swain; C.E.O. & Chm. (V), Beth Dann; Dir. Education, Gigi Alvare; Dir. Public Programs, Visitor Svcs. & Museum Shop Mgr., Cindy Weakland; Dir. Mktg. & Communications, Beth Manwaring; Controller, Andrew Braman; Cur. Collections, James Peck.
Personnel Profile: Full-Time Paid 14; Part-Time Paid 1; Part-Time Volunteers 32; Interns 1.
Volunteer Hours: 1,252
Operating Income: 1,557,500
Governing Authority: nonprofit education institution. Tax-exempt: 501(c)(3).
Institution Type/Description: American Western & Native American Art Museum.
Collections: 19th- & 20th-century art of the North American West and Native American Art, including paintings, sculpture, works of art on paper & related materials; firearms.
Major Exhibits: Alexander Hogue Retrospective, 9/13-1/14; Painted Journeys on the Colorado Plateau, 1/24/14-5/4/14; Karl Bodmer Prints, 3/4/14-3/6/15; Artic Journeys/Ancient Memories: The Sculpture of Abraham Ruben, 5/16/14-9/7/14; Illustration & Imagination: W.H.D. Koerner's Western Paintings, 5/16/14-9/7/14; CM Russell: Harmless Hunter, 9/19/14-1/11/15; A Feeling of Humanity: Western Art from the Ken Ratner Collection, 9/19/14-1/11/15.
Research Fields: American Western Art; Native American Art.
Facilities: 2,500-vol. library. Southwestern crafts & jewelry; books on Western American art & other museum-related items for sale.
Activities: guided tours; formally organized education programs for children; docent training program; temporary & permanent exhibitions; lectures; symposiums.
Publications: exhibition catalogues; annual report; brochures; postcards; slides; prints; posters; catalog of western art collection.
Hours & Admission Prices: Memorial Day to Labor Day daily 9-8; Sept.-May daily 9-5. Adults $9, senior citizens $8; discounts to groups, NARM & Museums West Consortium members; members & youth under 19 no charge. Closed New Year's Day; Thanksgiving; Christmas Eve & Day. ♿
Attendance: 31,821 (accurate)
Membership: Friend $70; Contributor $100; Donor $250; Sustainer $500; Silver Dollar Society $1,000.

Cornwall

HUDSON HIGHLANDS NATURE MUSEUM, 120 Muser Dr., Cornwall, NY 12518. Mailing Address: P.O. Box 451, Cornwall, NY 12518. Tel.: 845-534-5506. Fax: 845-534-4581.

E-mail: mgoldin@hhnaturemuseum.org
Web Site: www.hhnaturemuseum.org
Formerly: Museum of the Hudson Highlands
Founded: 1962.
Congressional District: 97
Key Personnel: Exec. Dir., Jacqueline Grant; Chm. (V), David N. Redden; Vice Pres., Susan W. Christensen; Dir. Education, Judy Onnfer; Operations Mgr., Candace Rivera; Mgr. Mktg., Marian Goldin; Dir. Wildlife Education Center, Pam Golben; Dir. Preschool, Sandy Dixon.
Governing Authority: private. Subsidiary Institution: Outdoor Discovery Center,20 Kenridge Farm Dr., Cornwall, NY; Kenridge Farm, 120 Muser Dr., Cornwall. Tax-exempt: 501 (c)(3).
Institution Type/Description: Nature & Environmental Museum.
Collections: preserved fishes, reptiles, amphibians, live animals, Indian artifacts and geological specimens indigenous to the Hudson Valley; ichthyology collection from Hudson River tributaries.
Research Fields: exhibiting live animals; regional phenological data; field surveys; restoration of the Bald Eagle, Peregrine Falcon & Osprey populations along the Hudson River; restoration of Hudson River tidal freshwater marshes.
Facilities: nature trails. Gifts items for sale.
Activities: evening film/lecture series; summer camp; volunteer program for high school students, in-museum and in-school educational programs; creative art courses; community special events; creative art courses for children and adults; summer environmental workshops for elementary students. Kenridge Farms: bird watching, cross-country skiing; maple sugaring; seasonal events.
Publications: quarterly, Member's Bulletin; biannual, Teachers Bulletin.
Hours & Admission Prices: Wildlife Education Center: Fri.-Sat. 10-5, Sun. 1-5. Outdoor Discovery Center: Sat.-Sun. 10-4. Museum on the Hudson Highlands Kenridge Farm: daily dawn to dusk. Wildlife Education Center: $3; members no charge. Outdoor Discovery Center: $3; members no charge. ♿
Attendance: 40,000 (estimated)
Membership: Friend $40; Friend Plus $50; Family $70; Family Plus $94; Pathfinder $125; Explorer $250; Naturalists $500.

Corona

LOUIS ARMSTRONG HOUSE & ARCHIVES, (M), 34-56 107th St., Corona, NY 11368-1226. Tel.: 718-997-3670 & 478-8274.
Web Site: www.louisarmstronghouse.org
Key Personnel: Dir., Michael Cogswell
Institution Type/Description: Historic House Museum.
Collections: Armstrong history & cultural legacy; personal artifacts; photographs.
Hours & Admission Prices: Tues.-Fri. 10-5, Sat.-Sun. 12-5; groups by appointment. Adults $8, seniors, students & children $6; discounts to groups; members no charge.

Cortland

BOWERS SCIENCE MUSEUM, State Univ. of New York, Cortland Bowers Hall, Cortland, NY 13045. Tel.: 607-753-2900. Fax: 607-753-2927.
Founded: 1964.
Key Personnel: Dir. & Cur., Dr. Peter K. Ducey.
Personnel Profile: Interns 2.
Governing Authority: state; university. Parent Institution: State University College at Cortland & State University of New York, 8 Thurlow Terrace, Albany, 12201. Tax-exempt.
Institution Type/Description: Science Museum.
Collections: anatomy; local plants; gems & industrial collection; local birds & mammals; teaching skins; mounted fish & heads of game animals; local fossils; mollusks; human biology; ecology; ethology; geology & fungi.
Research Fields: conservation biology; plant, animal & fungal systematics; vertebrate evolution; ornithology; herpetology; many facets of ecology.
Facilities: nature/conservation center; field research station; planetarium; reading room; 300 seat auditorium; classrooms; nature trail; teaching units.
Activities: guided tours; lectures; films; formally organized education programs for children, undergraduate & graduate college students; permanent & temporary exhibitions; training classes for biology teachers graduate course in techniques & methods of collecting, culturing & preserving specimens.
Hours & Admission Prices: Summer: Mon.-Fri. 8-4; Sept.-June Mon.-Fri. 9-6;

special group tours by appointment with Biology Dept. or Science Bldg. Coord. No charge. Closed holidays. &

Attendance: 4,000

CORTLAND COUNTY HISTORICAL SOCIETY, 25 Homer Ave., Cortland, NY 13045-2056. Tel.: 607-756-6071.
E-mail: cortlandcountyhistoricalsociety@centralny.twcbc.com
Web Site: www.cortlandhistory.com
Founded: 1925.
Congressional District: 24
Key Personnel: Pres., Diane Ames; Dir., Mindy Leisenring; Mgr. Collections, Anita Wright.
Personnel Profile: Full-Time Paid 2; Full-Time Volunteers 1; Part-Time Paid 1; Part-Time Volunteers 50; Interns 1.
Governing Authority: nonprofit organization; board of directors. Tax-exempt: 501(c)(3).
Institution Type/Description: Historical Society Museum: housed in 1882 Suggett House built by James Suggett, holder of patents on the driven well.
Collections: clothing; military items from Revolutionary War to Desert Storm; furniture; textiles; paintings; photographs; tools; utensils; pottery; items relating to local industries; lighting devices; glass; silver; china; hair wreaths & jewelry; manuscripts; archival materials; collections related to the 200-year history of the county.
Research Fields: local history & genealogy.
Facilities: over 3,000-vol. library of books; letters, manuscripts, maps, tapes, documents & microfilms relating to Cortland County; 14,000 negatives of Brockway Motor Truck Co.; catalogs, broadsides & genealogical materials available for use in library wing. Publications of the Historical Society & other museum-related items for sale.
Activities: guided tours; walking tours; bus tours; lectures; craft classes & demonstrations; permanent & temporary exhibitions; school loan service; outreach services to clubs, organizations & schools.
Publications: newsletter, News Notes; pamphlet, Bulletin; books, Chronicles Vols. II, III, IV & V; Paris Lived in Homer; Residents of Cortland County 1800-1810; Growing up in Cortland; Index to Personal & Place Names, French's Gazetteer of NYS 1860; Rails Through Cortland; booklet, Historical Markers of Cortland County; annual pictorial calendar; A Regiment Remembered: The 157th NYV.
Hours & Admission Prices: Museum: Tues.-Sat. 1-5. Library: Tues.-Sat. 1-5. Gift Shop: Tues.-Sat. 9:30-5. Museum Tours: adults $3. Research Library $8 & $5; members no charge. Closed New Year's Day; Independence Day; Thanksgiving; Christmas. &
Attendance: 3,200 (accurate)
Membership: Senior Citizen $18; Individual $25; Family & Nonprofit $35; Supporting $60; Sustaining $125; Sponsor $300; Patron $600; Benefactor $1,200.

DOWD GALLERY, STATE UNIVERSITY OF NEW YORK COLLEGE AT CORTLAND, SUNY Cortland, Dowd Fine Arts Center, Rm. 162, Cortland, NY 13045. Mailing Address: SUNY Cortland, P.O. Box 2000, Cortland, NY 13045-0900. Tel.: 607-753-4216. Fax: 607-753-5934.
E-mail: dowd.gallery@cortland.edu
Web Site: www2.cortland.edu/departments/art/dowd-gallery/
Founded: 1967.
Congressional District: 25
Key Personnel: Interim Dir., Bryan Thomas; Gallery Mgr., Jaroslava Prihodova.
Personnel Profile: Part-Time Paid 2; Interns 5.
Governing Authority: nonprofit. Parent Institution: State University of New York College at Cortland. Tax-exempt.
Institution Type/Description: University Art Gallery.
Collections: 19th- to 20th-century European & American paintings, prints, drawings, sculpture.
Research Fields: contemporary & modern art.
Facilities: 2,500 sq. ft. exhibit space.
Activities: docent program; films; guided tours; lectures; loan & temporary exhibitions; gallery talks.
Publications: exhibition catalogs in conjunction with temporary exhibits.
Hours & Admission Prices: Mon.-Fri. 10-4; other times by appointment. No charge; donations accepted. Closed major & university holidays. &
Attendance: 3,500 (estimated)

THE 1890 HOUSE-MUSEUM & CENTER FOR ARTS, 37 Tompkins St., Cortland, NY 13045-2555. Tel.: 607-756-7551. Fax: 607-756-7551 (call first).
E-mail: the1890house@gmail.com
Web Site: www.1890house.org
Founded: 1976.

Congressional District: 32
Key Personnel: Dir., Katherine Niver.
Personnel Profile: Part-Time Volunteers 30.
Governing Authority: nonprofit organization. Tax-exempt: 501(c)(3).
Institution Type/Description: Historic House: 1890 Victorian chateauesque style mansion.
Collections: architectural & decorative features of the mansion, including parquet floors, hand carved oak & cherry woodwork, stained glass windows & skylights; 19th to early 20th century furniture; decorative arts; photographs; china; silver; glassware; textiles; clothing; paintings; wall stenciling; Victorian art.
Research Fields: family & site history.
Facilities: 500-vol. library of books on architecture & Victorian decorative arts; reading room. Museum-related items for sale.
Activities: guided tours; lectures; films; concerts; formally organized education programs; docent program; loan, permanent, temporary & traveling exhibitions; lectures; educational programs; school programs.
Publications: newsletter, Whispers Near the Inglenook.
Hours & Admission Prices: Thurs.-Sat. 12-4. Adults $8, students & senior citizens $5; discounts to groups; children under 10 no charge. Closed major holidays. &
Attendance: 5,000 (estimated)
Membership: Senior & Student $15; General $25; Contributing $26-$249; Sustaining $250-$999; Benefactor $1,000 & up.

Coxsackie

BRONCK MUSEUM, Rte. 9W, Coxsackie, NY 12051-3022. Mailing Address: P.O. Box 44, Coxsackie, NY 12051-0044. Tel.: 518-731-6490.
E-mail: gchsbm@mhcable.com
Web Site: www.gchistory.org/
Founded: 1929.
Congressional District: 29
Key Personnel: Site Mgr., Shelby Mattice; Pres., Robert Hallock; Chm., Joseph Warren; Museum Shop Mgr., Jennifer Barnhart.
Personnel Profile: Full-Time Paid 1; Part-Time Paid 2; Part-Time Volunteers 40.
Governing Authority: society. Parent Institution: The Greene County Historical Society. Tax-exempt: 501(c)(3).
Institution Type/Description: Historic House Museum: 1663 home of Pieter Bronck, cousin of the first settler north of the Harlem River for whom Bronx Borough of New York was named.
Collections: art; silver; furniture; painting; pottery; textiles; carriages; early tools; weaving paraphernalia; manuscripts; 1663 Dutch dwelling; Dutch domestic & agricultural architecture. Vedder Memorial Library; manuscripts.
Research Fields: local history; genealogy.
Facilities: 5,000-vol. library of books; newspapers; manuscripts; photographs of Greene County & the Catskill Region. Museum-related items for sale.
Activities: guided tours; lectures; permanent & temporary exhibitions.
Publications: quarterly journal; books, Out to Greenville & Beyond; Historical Sketches of Greene County; Letters from a Revolution; book, Under Three Flags: A History of the Coxsackie, Earlton & Climax Settlements; Historic Places in Greene County New York; Reports From The Homefront; Episodes From A Hudson River Town, New Baltimore, NY.
Hours & Admission Prices: Memorial Day to Oct. 15 Wed.-Fri. 12-4, Sat. 10-4, Sun. 1-4. Adults $5, youth 12-15 $3, children 5-11 $2; children under 5 & members no charge.
Attendance: 2,500 (estimated)
Membership: Student $15; Individual $20-$29; Library & Business Basic $25; Dual/Family $30-$59; Business Friend $50 & over; Supporter $60-$109; Business Supporter $100 & up; Patron $110-$249; Benefactor $250-$499; Silver Benefactor $500-$999; Gold Benefactor $1,000 & up.

Croghan

AMERICAN MAPLE MUSEUM, 9756 Main St., Croghan, NY 13327. Mailing Address: P.O. Box 81, Croghan, NY 13327-0081. Tel.: 315-346-1107.
E-mail: americanmaplemuseum@frontier.com
Web Site: www.americanmaplemuseum.org
Founded: 1977.
Congressional District: 30
Key Personnel: Pres., Donald Moser; Treas., Jane Yancey; Sec., Eleanor Allen; Site Admin., Gina Hoppel; Site Admin., Maureen Martin.
Personnel Profile: Part-Time Paid 2; Part-Time Volunteers 25; Interns 1.
Governing Authority: nonprofit organization. Tax-exempt.
Institution Type/Description: Agricultural Museum.
Collections: historic & modern materials from the maple products industry.

Research Fields: history of maple products & production systems.
Facilities: 100-vol. library of periodicals of the maple products industry available for research by appointment only; 150-seat auditorium. Maple syrup, maple products, hand-crafted items for sale.
Activities: guided tours; lectures; permanent exhibitions. Annual Events: Maple Festival in May; Concert & Ice Cream Social in July; Fall Foliage Run in October.
Hours & Admission Prices: June Mon. & Fri.-Sat. 11-4; July-Labor Day Mon.-Sat. 11-4; off season visits by appointment. Adults $4, children 5-12 $1; discounts to AAA & CAA members. Group Tours: appointment requested. &
Attendance: 1,710 (estimated)

Cropseyville

BRUNSWICK HISTORICAL SOCIETY - GARFIELD SCHOOL, 605 Brunswick Rd., Cropseyville, NY 12052. Mailing Address: P.O. Box 1776, Cropseyville, NY 12052. Tel.: 518-279-4024.
E-mail: president@bhs-ny.org
Web Site: www.bhs-ny.org
Institution Type/Description: Historic Building: housed in a two-room schoolhouse; built in 1881.
Collections: local history & culture; photographs; records; documents; personal artifacts.
Facilities: library.
Activities: educational programs; special events.
Publications: quarterly newsletter, Artifacts.
Hours & Admission Prices: Wed. 1-3, Sat. 10-3.

Cross River

TRAILSIDE NATURE MUSEUM, Ward Pound Ridge, Rte. 35 & Rte. 121, Cross River, NY 10518. Mailing Address: P.O. Box 236, Cross River, NY 10518-0236. Tel.: 914-864-7322.
Web Site: www.friendsoftrailside.org
Founded: 1937.
Key Personnel: Cur., Jason Klein.
Personnel Profile: Full-Time Paid 2; Part-Time Paid 1; Part-Time Volunteers 8.
Governing Authority: county. Parent Institution: County of Westchester Dept. of Parks, Recreation & Conservation, County Office Building, White Plains, 10601. Subsidiary Institution: Friends of Trailside Nature Museum. Tax-exempt.
Institution Type/Description: Natural History Museum.
Collections: native American library; mounted birds of prey; songbirds; mammals; insects; geology; local history.
Research Fields: natural & local history.
Facilities: 1,500-vol. library of reference books, Delaware Indian Resource Center collection of rare books, tapes, & photographs devoted to the study of the coastal Algonkian tribes available for research on premises.
Activities: lectures; arts festivals; organized educational programs; permanent & temporary exhibitions.
Publications: quarterly calendar of events; educational leaflets; members newsletter.
Hours & Admission Prices: Winter: Tues.-Sun. 9-4; Summer: Mon.-Fri. 9-4. No charge. Parking Fee: $4 with resident permit, $8 without permit. &
Attendance: 20,000 (estimated)
Membership: Individual $10; Husband & Wife $15; Family $25; Sustaining $50; Supporting $100; Sponsor $250; Patron $500-$1,000.

Crown Point

CROWN POINT STATE HISTORIC SITE, (M), 21 Grandview Dr., Crown Point, NY 12928-2852. Tel.: 518-597-4666. Fax: 518-597-4668.
Founded: 1910.
Congressional District: 23
Key Personnel: Historic Site Mgr., Tom Hughes; Pres. (V), John Freilich.
Personnel Profile: Full-Time Paid 4; Part-Time Paid 10.
Governing Authority: state. New York State Office of Parks, Recreation and Historic Preservation, and Saratoga Capital District Region. Tax-exempt: 501(c)(3).
Institution Type/Descripion: Historic Site: 1734 remains of French fort, 1759 British Fort Crown Point & Outwork fortifications controlling Lake Champlain.
Collections: military materials; archaeological artifacts; 18th-century domestic materials.
Research Fields: French & Indian Wars; American Revolution; regional history.
Facilities: 60-seat auditorium; interpretive center; picnic area.

Activities: guided tours; lectures; closed-captioned audiovisual program; gallery talks; permanent exhibitions; special programs & demonstrations.
Hours & Admission Prices: Museum: mid-May-mid-Oct. Thurs.-Mon. 9:30-5; other times by appointment. Adults $4, senior citizens & students $1. Grounds: Summer 7 am-8 pm.

Cuddebackville

NEVERSINK VALLEY MUSEUM OF HISTORY AND INNO-VATION, 26 Hoag Rd., Cuddebackville, NY 12729. Mailing Address: P.O. Box 263, Cuddebackville, NY 12729-0263. Tel.: 845-754-8870. Fax: 845-754-8870.
E-mail: nvam@frontiernet.net
Web Site: neversinkmuseum.org
Formerly: Neversink Valley Area Museum
Founded: 1967.
Congressional District: 19
Key Personnel: C.E.O. & Pres. (V), Stephen Skye; Exec. Dir., Seth Goldman; Asst. Exec. Dir., David H. Lawrence.
Personnel Profile: Full-Time Paid 3; Part-Time Volunteers 35.
Governing Authority: nonprofit organization. Tax-exempt: 501(c)(3).
Institution Type/Description: Preservation Project & History Museum: housed in 2 buildings including a 1790, salt box-type house, part of the Delaware & Hudson Canal Park Historic Landmark.
Collections: local history of the Delaware & Hudson Canal; 19th century lifestyles in the Neversink River Valley area; early motion picture history.
Research Fields: 19th-century history; early motion pictures.
Facilities: 400-vol. library of historical material & periodicals; herb garden. Museum-related items for sale.
Activities: canal tours; school & group tours; craft demonstrations; lecture series; special canal boat activity center for children; permanent & changing exhibits.
Publications: quarterly newsletter.
Hours & Admission Prices: Thurs.-Sun. 12-4. Adults $3, children $1.50; children under 6 & members no charge. &
Attendance: 7,000 (accurate)
Membership: Individual $15; Family $25; Donor $50; Business $100; Contributor $125; Sponsor $250; Sustainer $500; Founder $1,000.

Delhi

DELAWARE COUNTY HISTORICAL ASSOCIATION & FLETCHER DAVIDSON LIBRARY AND ARCHIVES, 46549 State Hwy. 10, Delhi, NY 13753. Tel.: 607-746-3849.
E-mail: dcha@delhi.net
Web Site: dcha-ny.org
Institution Type/Description: History Museum.
Collections: local history & culture; period furnishings; personal artifacts; diaries; books; maps; photographs; federal & state censuses from 1790-1905; paintings; Frisbee family cemetery. Historic Buildings: Amos Wood gunshop; Husted Hollow one-room schoolhouse; 19th century corn crib; Woodin blacksmith shop; Tollgate house; Frisbee barn; 1797 Gideon Frisbee house.
Hours & Admission Prices: Library & Archives: Tues.-Wed. 10-3; other times by appointment. Requested Donation: $5. Buildings & Galleries: Memorial Day to mid-Oct. Tues.-Sun. 11-4; Winter: Mon.-Fri. 10-3. Adults $4, children $1.50; members no charge.
Attendance: 4,500 (accurate)

Derby

GRAYCLIFF CONSERVANCY, (M), 6472 Old Lake Shore Rd., Derby, NY 14047-9731. Mailing Address: P.O. Box 823, Derby, NY 14047-0823. Tel.: 716-947-9217. Fax: 716-947-2086.
E-mail: Graycliff@verizon.net
Web Site: graycliffestate.org
Founded: 1997.
Congressional District: 27
Key Personnel: Exec. Dir., Reine Hauser; Pres. (V), Diane Schrenk; Group Tour Coord., Shannon Lyons; Museum Shop Mgr., Ryan Gravell.
Personnel Profile: Full-Time Paid 2; Part-Time Paid 6; Part-Time Volunteers 300.
Governing Authority: private; nonprofit organization. Tax-exempt: 501(c)(3).
Institution Type/Description: Historic House Museum: summer home of Isabelle and Darwin Martin; estate designed by Frank Lloyd Wright, 1926-1931.
Collections: furnishings, documents & personal artifacts of Frank Lloyd Wright, the Martin family & the Larkin Soap Company; historic gardens.
Research Fields: gardens & grounds of Graycliff; historic furnishings of Graycliff.

Facilities: 100-vol. library; botanical garden. Museum-related items for sale.
Activities: docent program; formal education programs for adults; guided tours; lectures; participatory exhibits; special events.
Publications: biannual newsletter, Graycliff; e-newsletter.
Hours & Admission Prices: See website for information. &
Attendance: 10,000 (accurate)
Membership: Individual \$35; Family \$60; Benefactor \$100; Gold \$250; Platinum \$500; Jewel \$1,000.

Dobbs Ferry

DOBBS FERRY HISTORICAL SOCIETY - THE MEAD HOUSE, 12 Elm St., Dobbs Ferry, NY 10522. Tel.: 914-674-1007.
E-mail: dfhistory@optimum.net
Web Site: www.dobbsferryhistory.org/Home.html
Founded: 1988.
Congressional District: 17
Key Personnel: Pres. (V), Mary S. Donovan.
Personnel Profile: Part-Time Volunteers 20.
Volunteer Hours: 600
Operating Expenses: 58,590
Operating Income: 50,013
Governing Authority: Tax-exempt.
Institution Type/Description: Historical Society Museum.
Collections: local history & culture; period artifacts; books; paintings; photographs; maps; oral histories.
Research Fields: local history.
Activities: special events.
Publications: books; pamphlets; newsletter, The Ferryman.
Hours & Admission Prices: Tues. 10 to noon; other times by appointment. No charge; donations accepted.
Attendance: 500 (estimated)
Membership: Senior \$10; Adult \$25; Family \$50; Patron \$100; Benefactor \$500; Historian \$1,000.

Douglaston

ALLEY POND ENVIRONMENTAL CENTER, INC., 228-06 Northern Blvd., Douglaston, NY 11362-1096. Tel.: 718-229-4000, ext. 0. Fax: 718-229-0376.
E-mail: info@alleypond.com
Web Site: www.alleypond.com
Founded: 1976.
Congressional District: 6
Key Personnel: Exec. Dir., Irene V. Scheid; Pres. (V), Rita Sherman.
Personnel Profile: Full-Time Paid 8; Part-Time Paid 5; Part-Time Volunteers 20; Interns 5.
Volunteer Hours: 2,651
Operating Expenses: 931,746
Operating Income: 894,204
Governing Authority: nonprofit organization. Tax-exempt: 501(c)(3).
Institution Type/Description: Nature & Environmental Center.
Collections: historical maps; indigenous animals & marine life.
Research Fields: local flora & fauna; impact statements on construction & development.
Facilities: auditorium; classrooms; interpretive trails.
Activities: guided tours; lectures; films; hobby workshops; formally organized education programs; conservation.
Publications: seasonal program guide & newsletter; APEC field guides; Our Wild Friends Coloring Book.
Hours & Admission Prices: Mon.-Sat. 9-4:30, Sun. 9-3:30. Environmental Center: no charge; donations accepted. Nominal fee for special programs. Closed New Year's Eve & Day; Martin Luther King Jr. Day; Presidents' Day; Easter; Mother's Day; Independence Day; Labor Day; Columbus Day; Thanksgiving & day after; Christmas Eve & Day. &
Attendance: 62,000 (accurate)
Membership: College Student & Senior Citizen \$30; Individual & Senior With Child \$40; Family \$60; Associate \$100; Sustainer \$200; Fellow \$500; Benefactor \$1,000.

Dresden

ROBERT GREEN INGERSOLL BIRTHPLACE MUSEUM, 61 Main St., Dresden, NY 14441. Mailing Address: Ingersoll Memorial Committee, P.O. Box 664, Amherst, NY 14226. Tel.: 315-536-1074.
Web Site: www.secularhumanism.org/ingersoll
Founded: 1993.
Key Personnel: Dir., Thomas Flynn; C.E.O., Ronald Lindsay; Chm. (V), Roger Greeley; Museum Shop Mgr., Frances Emerson.

Personnel Profile: Part-Time Paid 1; Part-Time Volunteers 5.
Governing Authority: Parent Institution: Council for Secular Humanism. Tax-exempt.
Institution Type/Description: Historic House Museum: housed in the birthplace of Robert Green Ingersoll; born Aug. 11, 1833.
Collections: Ingersoll's life & career; photographs; personal artifacts; period furnishings; manuscript of his famous speech; drawings.
Activities: video presentation; group tours.
Publications: annual newsletter, Ingersoll Report.
Hours & Admission Prices: Memorial Day to Oct. Sat.-Sun. 12-5. Suggested Donation: \$2 per person. &
Attendance: 450 (accurate)
Membership: Annual \$18.

East Aurora

AURORA HISTORICAL SOCIETY, 363 Oakwood Ave., East Aurora, NY 14052-2319. Mailing Address: P.O. Box 472, East Aurora, NY 14052-0472. Tel.: 716-652-4735. Facebook: Aurora Historical Society New York.
E-mail: ahs1951@verizon.net
Web Site: www.aurorahistoricalsociety.com
Founded: 1951.
Congressional District: 27
Key Personnel: Pres., Susan McBurney; Dir. & Town Historian, Robert Lowell Goller; Cur. Elbert Hubbard Museum, Tom Alcamo; Cur. Elbert Hubbard Museum, Don Meade; Cur. Millard Fillmore House, Kathy Frost.
Personnel Profile: Part-Time Paid 2; Part-Time Volunteers 32.
Governing Authority: society. Parent Institution: Aurora Historical Society. Branch Museums: Elbert Hubbard-Roycroft Museum; Millard Fillmore Presidential Site; Aurora History Museum & Town Archives. Tax-exempt.
Institution Type/Description: Museum Complex & Historic District: national landmark known as Roycroft Campus, originated c.1896 by Elbert Hubbard as a haven to promote all types of arts & crafts; still in operation.
Collections: Historic Houses: c.1825 The President Millard Fillmore House-Museum, a national historic landmark; period furnishings; rose garden; herb garden, barn with collections. The Aurora History Museum: located in the town municipal center; paintings; Indian artifacts; memorabilia; Kuster collection of Indian arrowheads; maps; atlases; photographs; genealogy files; books. Scheidemantel House, home of Elbert Hubbard Museum, a craftsman period home, housing artifacts and books of Elbert Hubbard and his Roycroft organization.
Facilities: library of history books and iconography; 115 years of East Aurora Advertiser newspapers on microfilm, available in the Town's Historian's office by request.
Activities: guided tours; self-guided tours of the campus.
Publications: books; pamphlets; brochures.
Hours & Admission Prices: June-Oct. Wed. & Sat.-Sun. 1-4. Fillmore & Hubbard museums \$10 each location; members no charge.
Attendance: 1,000 (estimated)
Membership: Student \$10; Individual \$20; Couple \$30; Family \$50; Lifetime \$250; Business sponsor levels available.

COPPER SHOP GALLERY, 31 S. Grove St., East Aurora, NY 14052-2325. Tel.: 716-655-0261. Fax: 716-655-8498.
E-mail: info@roycroftcampuscorp.com
Web Site: www.roycroftcampuscorporation.com
Key Personnel: Pres., Douglas Swift; Exec. Dir., Christine Peters.
Governing Authority: Parent Institution: Roycroft Campus Corp.
Institution Type/Description: History Museum.
Collections: local history; photographs; Roycroft artifacts; Dard Hunter window pane; early letter press; period furnishings.
Hours & Admission Prices: Daily 10-5.

ELBERT HUBBARD-ROYCROFT MUSEUM, 363 Oakwood Ave., East Aurora, NY 14052-2319. Mailing Address: P.O. Box 472, East Aurora, NY 14052-0472. Tel.: 716-652-4735. Facebook: Aurora Historical Society New York.
E-mail: ahs1951@verizon.net
Web Site: www.aurorahistoricalsociety.com
Founded: 1951.
Congressional District: 27
Key Personnel: Dir., Robert Lowell Goller; Co-Cur, Tom Alcamo; Co-Cur., Don Meade.
Personnel Profile: Part-Time Paid 2; Part-Time Volunteers 25.
Governing Authority: municipal. society; Parent Institution: The Aurora Historical Society, Inc. (see separate listing for branch museums). Tax-exempt: 501(c)(3).

Institution Type/Description: Library and History Museum: housed in 1910 Scheide Mantel Home.
Collections: history & culture of Roycroft community & Elbert Hubbard; books, furniture & craft items of Roycroft Shops.
Facilities: 1,100-vol. library of books, magazines & pamphlets printed at Roycroft Shops from 1895-1936 available for inter-library loan & for use on premises; reading room.
Activities: lectures; gallery talks; permanent, temporary & traveling exhibitions.
Publications: pamphlet, Elbert Hubbard & the Roycrofters; pamphlet, A Message to Garcia.
Hours & Admission Prices: June-Oct. Wed. & Sat.-Sun. 1-4. Adults $10, school children $5; Aurora Historical Society members no charge.
Attendance: 2,000 (estimated)
Membership: Student $10; Individual $20; Couple $30; Family $50; Lifetime $250.

EXPLORE & MORE...A CHILDREN'S MUSEUM, 300 Gleed Ave., East Aurora, NY 14052-2983. Tel.: 716-655-5131. Fax: 716-655-5466.
E-mail: mail@exploreandmore.org
Web Site: www.exploreandmore.org
Founded: 1994.
Key Personnel: C.E.O. & Dir., Barbara Park Leggett; Chm., Pete Anderson; Vice Pres., Deborah Andrews; Education, Claudia Newton; Treas., Peter Bergmann; Museum Shop Mgr., Barb Kolich.
Personnel Profile: Full-Time Paid 1; Part-Time Paid 13; Part-Time Volunteers 40.
Governing Authority: private; nonprofit organization. Tax-exempt.
Institution Type/Description: Children's Museum.
Collections: hands-on children's exhibits including building & architecture, food & nutrition, brain teasers, world cultures and infant & toddler play.
Activities: formal education programs for children; participatory exhibits; school loan service.
Publications: bimonthly newsletter, Explorations.
Hours & Admission Prices: Wed.-Sat. 10-5, Sun. 12-5, first Fri. of month 10-8, Mon.-Tues. large groups by appointment. Admission $5; ACM reciprocal; members no charge. Closed New Year's Day; Easter; Independence Day; Thanksgiving; Christmas. &
Attendance: 38,840 (accurate)
Membership: Creativity $45; Discovery & Discovery Grandparent $60; Imagination $100.

MILLARD FILLMORE PRESIDENTIAL SITE, 24 Shearer Ave., East Aurora, NY 14052. Mailing Address: Box 472, East Aurora, NY 14052-0472. Tel.: 716-652-8875 & 2432. Facebook: Aurora Historical Society New York.
E-mail: ahs1951@verizon.net
Web Site: www.aurorahistoricalsociety.com
Formerly: Millard Fillmore House
Founded: 1951.
Congressional District: 27
Key Personnel: Pres., Susan McBurney; Cur., Kathy Frost; Dir., Robert Lowell Goller.
Personnel Profile: Part-Time Paid 2; Part-Time Volunteers 45.
Governing Authority: society. Parent Institution: The Aurora Historical Society, Inc. Tax-exempt.
Institution Type/Description: Historical Society Museum: housed in 1826 home of the 13th U.S. President, Millard Fillmore.
Collections: period Fillmore furnishings; stenciled walls; barn; tools & farm implements; formal rose garden; herb garden.
Activities: tours.
Publications: brochures & postcards; books: local history.
Hours & Admission Prices: June-Oct. 15 Wed. & Sat.-Sun. 1-4; other times & tours by appointment. Adults $10; members & children under 12 no charge.
Attendance: 900 (estimated)
Membership: Student $10; Individual $20; Couple $30; Family $50; Lifetime $250; Business $500.

East Bloomfield

BLOOMFIELD ACADEMY MUSEUM, 8 South Ave., East Bloomfield, NY 14443. Mailing Address: P.O. Box 212, E. Bloomfield, NY 14443-0212. Tel.: 585-657-7244. Fax: 585-657-7244.
E-mail: director@ebhs1838.org
Web Site: www.ebhs1838.org
Founded: 1967.
Congressional District: 35

Key Personnel: C.E.O., William Burlingame; Cur., Stephen Beaulieu; Museum Shop Mgr., Richard Delong.
Personnel Profile: Part-Time Paid 1; Part-Time Volunteers 50.
Governing Authority: nonprofit organization. Parent Institution: Historical Society of the Town of East Bloomfield. Tax-exempt.
Institution Type/Description: Historic Agency: housed in 1838 Bloomfield Academy Building.
Collections: local artifacts; clothing; furniture; household furnishings; dolls; tools; shells; inkwells; agricultural tools; genealogical research materials; period textbooks; agricultural artifacts; local authors' publications; reference materials; ledgers; diaries.
Research Fields: genealogical & historical research materials & personnel.
Facilities: library of early books on town of East Bloomfield & Ontario County available for research on premises & by special request. Locally handmade crafts & museum publications for sale.
Activities: guided tours; lectures; films; arts festivals; drama; docent program; permanent & temporary exhibitions; school loan service; genealogical programs; children's programs in summer; summer concerts. Museum Sponsors: Antique Car Show in July.
Publications: monthly newsletter, Academy Chronicles; quarterly magazine; books, Bloomfield Bicentennial Historical Sketchbook; Bloomfield Schools Revisited; Bloomfield Village Walking Tour; The War Years in Bloomfield 1941-1945; 1874 town & village maps.
Hours & Admission Prices: April-Dec. Wed.-Fri. 9-2, Sat. 9-12. No charge; donations accepted. Closed major holidays.
Attendance: 1,000 (estimated)
Membership: Senior Citizen $10; Individual $12; Senior Family $15; Family $20; Friend $35; Sustaining $35-$99. Corporate: Patron $100-$499; Benefactor $500-$1,000.

East Durham

DURHAM CENTER MUSEUM/RESEARCH LIBRARY, State Rte. 145, East Durham, NY 12423. Mailing Address: P.O. Box 192, East Durham, NY 12423-0192. Tel.: 518-239-8461 & 4081. Fax: 518-239-4081.
E-mail: durhamcentermu@aol.com
Founded: 1960.
Congressional District: 102
Key Personnel: C.E.O. & Pres. (V), Bruce Hamm; Exec. Dir., Asst. Cur. & Museum Shop Mgr., Sancie Thomsen; Cur., Douglas Thomsen.
Personnel Profile: Part-Time Volunteers 4.
Governing Authority: nonprofit organization. Tax-exempt.
Institution Type/Description: General Museum.
Collections: folk art; local & Greene County artifacts; Civil War; genealogical.
Research Fields: Susquehanna Turnpike; Catskill Canajoharie Railroad.
Facilities: library of historical books & genealogy records available for research on premises; reading room.
Activities: permanent exhibitions; special exhibits of local ephemera & photography; arts & crafts workshops. Museum Sponsors: Musical Program.
Publications: Catskill-Canojaharie RR; Lyman Tremaine, Rebels at Rest; James Barker Patroon-Yesteryear Fireside Recollections.
Hours & Admission Prices: Museum: 3rd week of May to Columbus Day Thurs.-Sun. 1-4. Adults $2.50, children under 12 $1.50; discount to AAM & ICOM members; children under 5 accompanied by adult no charge. &
Attendance: 500 (estimated)
Membership: $10-$2,000.

East Hampton

CLINTON ACADEMY MUSEUM, 151 Main St., East Hampton, NY 11937-2716. Mailing Address: 101 Main St., East Hampton, NY 11937-2714. Tel.: 631-324-6850. Fax: 631-324-9885.
E-mail: info@easthamptonhistory.org
Web Site: www.easthamptonhistory.org
Founded: 1921.
Congressional District: 2
Key Personnel: Pres. (V), Arthur Graham; Exec. Dir., Richard Barons.
Personnel Profile: Full-Time Paid 4; Full-Time Volunteers 5; Part-Time Paid 5; Part-Time Volunteers 35; Interns 1.
Governing Authority: nonprofit. Parent Institution: The East Hampton Historical Society. Branch Museums: Osborn-Jackson House; Mulford Farm Complex; Town House; Marine Museum. Tax-exempt: 501(c)(3).
Institution Type/Description: History Museum: housed in a former school; built in 1784. Listed on the National Register of Historic Places.
Collections: decorative & fine arts; textiles; tools & equipment.
Research Fields: eastern Long Island history; decorative & fine arts.
Activities: temporary exhibitions; lectures.
Publications: brochures; exhibition catalogues.

Hours & Admission Prices: Memorial Day to June & Sept. to Columbus Day Sat. 10-5, Sun. 12-5; July-Aug. Fri.-Sat. 10-5, Sun. 12-5. No charge; donations accepted. &

Attendance: 1,800 (estimated)

Membership: Individual $35; Family & Couple $60; Friend $75; Business $100.

EAST HAMPTON HISTORICAL SOCIETY, INC., (M), 101 Main St., East Hampton, NY 11937-2714. Tel.: 631-324-6850. Fax: 631-324-9885.

E-mail: info@easthamptonhistory.org

Web Site: www.easthamptonhistory.org

Founded: 1921.

Congressional District: 2

Key Personnel: Exec. Dir., Richard Barons; Pres. (V), Arthur Graham.

Personnel Profile: Full-Time Paid 4; Part-Time Paid 5; Part-Time Volunteers 50; Interns 1.

Governing Authority: nonprofit organization. Branch Museums: Clinton Academy; Mulford Farm; Osborn-Jackson House; Town House; Marine Museum. Tax-exempt: 501(c)(3).

Institution Type/Description: Historical Society Museum.

Collections: decorative & fine arts; textiles; tools & equipment; historic structures & landscapes; maritime artifacts.

Research Fields: Eastern Long Island history.

Facilities: East Hampton Town Marine Museum; Clinton Academy; Town House; Mulford House & Barn; Osborn-Jackson House.

Activities: lectures; walking tours; children's programs; exhibitions; special events; school programs; historic preservation; summer camp; summer theater.

Publications: walking tour booklet; exhibition guides.

Hours & Admission Prices: Office: Tues.-Sat. 10-5. Adults $4; discounts to AAM members; members no charge. &

Attendance: 10,000 (estimated)

Membership: Individual $35; Family & Couple $60; Friend $75; Business $100.

* **GUILD HALL MUSEUM, (M),** 158 Main St., East Hampton, NY 11937-2795. Tel.: 631-324-0806. Fax: 631-324-2722.

E-mail: museum@guildhall.org

Web Site: www.guildhall.org

Founded: 1931.

Congressional District: 1

Key Personnel: Exec. Dir., Ruth Stevens Appelhof, Ph.D.; Chm. Bd. (V), Marty Cohen; Chm. (V), Sue Sylvor; Museum Dir. & Chief Cur., Christina M Strassfield; Museum Shop Mgr., Elaine Dangio.

Personnel Profile: Full-Time Paid 12; Part-Time Paid 16; Part-Time Volunteers 200; Interns 4.

Governing Authority: board of trustees. Tax-exempt.

Institution Type/Description: Art Museum.

Collections: 2,200 works by 19th-21st century regional artists, including Lynda Benglis, Ross Bleckner, Eric Fischl, Chuck Close, Audrey Flack, Willem de Kooning, Childe Hassam, Lee Krasner, Roy Lichtenstein, Thomas Moran, Robert Motherwell, Jackson Pollock, Larry Rivers; Richard Prince.

Research Fields: late 19th & 20th-century art; artists of the region.

Facilities: art library & archives; theater; classroom; meeting space. Museum-related items for sale.

Activities: temporary, permanent & traveling exhibitions; lectures; films; gallery talks; concerts; dance recitals; arts festivals; drama & poetry readings; arts & crafts workshops; educational programs; inter-museum loan. Museum Sponsors: Clothesline Art Sale in August.

Publications: exhibition catalogues; seasonal calendars.

Hours & Admission Prices: June-Labor Day daily 11-5; Sept.-May Wed.-Sat. 11-5, Sun. 12-5. Suggested Donation: $7 per person; discounts to AAM & ICOM members; members no charge. Closed New Year's Day; Thanksgiving; Christmas. &

Attendance: 50,000 (accurate)

Membership: Student & Senior Citizen $30; Individual $45; Supporting $70; Family $80; Donor $100; Sustaining $250; Gold Card $500; Sponsor $1,000; Patron $2,500; Benefactor $5,000; Chairman's Circle $10,000.

HOME SWEET HOME MUSEUM, 14 James Lane, East Hampton, NY 11937-2710. Mailing Address: East Hampton Village Hall, 86 Main St., East Hampton, NY 11937. Tel.: 631-324-0713 & 4150 & 267-6834. Fax: 631-324-0713 & 4189.

E-mail: hking@easthamptonvillage.org

Web Site: www.easthampton.com/homesweethome.org

Founded: 1928.

Congressional District: 1

Key Personnel: Historic Site Mgr., Hugh King.

Personnel Profile: Full-Time Paid 1; Part-Time Paid 2.

Governing Authority: nonprofit. Parent Institution: Inc. Village of East Hampton, 86 Main St., East Hampton, NY 11937. Tax-exempt.

Institution Type/Description: Historic House: mid-18th century saltbox dedicated to the memory of John Howard Payne, 19th-century actor, playwright & author of Home Sweet Home.

Collections: Lustreware, Staffordshire & English ceramics of late 18th & early 19th centuries; 17th-19th century American furniture including 1640 Mulliner chest; needlework & textiles.

Research Fields: 18th & early 19th-century English ceramics & textiles; Life & Times of John Howard Payne, 1791-1852.

Activities: guided tours; educational programs; temporary & permanent exhibitions.

Hours & Admission Prices: May-Sept. Mon.-Sat. 10-4, Sun. 2-4; Oct.-Nov. Fri.-Sun.; other times by appointment. Adults $4, children $2. Closed New Year's Day; Thanksgiving; Christmas. &

Attendance: 500 (accurate)

LONGHOUSE RESERVE, (M), 133 Hands Creek Rd., East Hampton, NY 11937-3808. Tel.: 631-329-3568. Fax: 631-329-4299.

E-mail: info@longhouse.org

Web Site: www.longhouse.org

Founded: 1991.

Congressional District: 1

Key Personnel: Exec. Dir., Matko Tomicic; Founder, Jack Lenor Larsen; Pres., Dianne Benson; Treas., Mark Levine; Cur., Wendy Van Deusen.

Personnel Profile: Full-Time Paid 6; Full-Time Volunteers 1; Part-Time Paid 2; Part-Time Volunteers 30.

Governing Authority: private; nonprofit organization. Tax-exempt: 501(c)(3).

Institution Type/Description: General Museum.

Collections: decorative arts; arboretum; sculpture garden.

Activities: arts festivals; concerts; dance recitals; guided tours; lectures; loan & temporary exhibitions.

Publications: quarterly newsletter, LongHouse.

Hours & Admission Prices: April 21-June & Sept. Mon., Wed. & Sat. 2-5; July-Aug. Wed-Sat. 2-5; Oct.-Dec. Sat. 2-5; Adults $10, senior citizens $8; discounts to AAM members; children & members no charge. &

Attendance: 10,000 (accurate)

Membership: Student $35; Friend $75; Friends Duo $150; Founder $300; Patron $500; Fellow $1,000; Benefactors $5,000; President's Circle $10,000; Corporate $2,500.

MULFORD FARM MUSEUM, 10 James Lane, East Hampton, NY 11937-2714. Mailing Address: East Hampton Historical Society, 101 Main St., East Hampton, NY 11937-2714. Tel.: 631-324-6850. Fax: 631-324-9885.

E-mail: info@easthamptonhistory.org

Web Site: www.easthamptonhistory.org

Founded: 1948.

Congressional District: 2

Key Personnel: Exec. Dir., Richard Barons; Pres. (V), Arthur Graham.

Personnel Profile: Full-Time Paid 2; Part-Time Paid 6; Part-Time Volunteers 35; Interns 1.

Governing Authority: nonprofit. The East Hampton Historical Society. Branch Museums: Osborn-Jackson House; Clinton Academy; Town House; Marine Museum. Tax-exempt: 501(c)(3).

Institution Type/Description: Historic Farm Complex: c.1680 frame salt box farm house and c.1720 barn with several outbuildings, continuously belonging to eight generations of the Mulford family 1710-1948 located on 3 acres.

Collections: farm implements; 17th-18th century decorative arts; crafts & domestic equipment; period of restoration 1770-1780.

Research Fields: social & economic history; agricultural; architecture; East Hampton town history.

Activities: guided tours; lectures; classes in local history & crafts; fairs; traditional skills demonstrations; living history demonstrations.

Publications: garden guide.

Hours & Admission Prices: Memorial Day-June & Sept.-Columbus Day Sat. 10-5, Sun. 12-5; July-Aug. Fri.-Sat. 10-5, Sun. 12-5. Adults $4, seniors $3, students $2; discounts to AAM & AAA members; members no charge.

Membership: Student $10; Individual $35; Family & Couple $60; Friend $75; Business $100; Corporate & Foundation $150.

OSBORN-JACKSON HOUSE, 101 Main St., East Hampton, NY 11937-2714. Mailing Address: East Hampton Historical Society, 101 Main St., East Hampton, NY 11937. Tel.: 631-324-6850. Fax: 631-324-9885.
E-mail: info@easthamptonhistory.org
Web Site: www.easthamptonhistory.org
Founded: 1979.
Congressional District: 2
Key Personnel: Exec. Dir., Richard Barons; Pres. (V), Arthur Graham.
Personnel Profile: Full-Time Paid 4; Part-Time Paid 5; Part-Time Volunteers 35; Interns 2.
Governing Authority: nonprofit organization. The East Hampton Historical Society. Branch Museums: Clinton Academy; Mulford Farm Complex; Town House; Marine Museum. Tax-exempt: 501(c)(3).
Institution Type/Description: History Museum: housed in a two-story home; built in 1740.
Collections: 18th-century decorative & fine arts; 3 furnished period rooms; East End history; Dominy furniture.
Research Fields: Long Island decorative arts; East Hampton town history; colonial & Revolutionary War history; domestic life.
Activities: lectures; guided tours; changing exhibitions; craft demonstrations; antique shows; walking tours; special events.
Publications: walking tour brochures; gallery guides.
Hours & Admission Prices: Tues.-Sat. 10-5. Adults $4, seniors $3, students $2; discounts to AAM members; members no charge.
Membership: Student $10; Individual $35; Family/Couple $60; Friend $75; Business $100; Corporate & Foundation $150.

East Islip

ISLIP ART MUSEUM, (M), 50 Irish Lane, East Islip, NY 11730-2003. Tel.: 631-224-5402. Fax: 631-224-5417.
E-mail: lynda@islipartscouncil.org
Web Site: islipartmuseum.org
Founded: 1973.
Congressional District: 2
Key Personnel: Exec. Dir., Lynda A. Moran; Dir. Public Rels., Stefanie Taylor; Cur., Karen Shaw; Cur. Permanent Collection, Janet Goleas; Admin. & Cur., Beth Giacummo; Carriage House Mgr., MaryLou Cohalan; Museum Shop Mgr. & Administrative Asst., Rosa Ramos; Educational Dir., Loretta Corbisiero.
Personnel Profile: Full-Time Paid 2; Part-Time Paid 10; Interns 3.
Governing Authority: Managing Agency: Islip Arts Council. Tax-exempt.
Institution Type/Description: Art Museum.
Collections: contemporary & avant-garde art.
Research Fields: contemporary art.
Facilities: educational facilities; 3,500 sq. ft. exhibit space. Museum-related items for sale.
Activities: arts festivals; concerts; formal education programs; guided tours; lectures; loan, traveling & participatory exhibitions.
Publications: bimonthly exhibition brochures.
Hours & Admission Prices: Wed.-Sat. 10-4, Sun. 12-4. No charge; donations accepted. Closed New Year's Day; Easter; Memorial Day; Independence Day; Labor Day; Thanksgiving & day after; Christmas. &
Attendance: 12,000 (accurate)
Membership: Senior 62 & over and Student $35; Individual $50; Family $65; Friend $125; Associate $250; Patron $500; Concert Sponsor $1,000; Benefactor $5,000.

East Meadow

NASSAU COUNTY, DIVISION OF MUSEUM SERVICES, DEPARTMENT OF RECREATION, PARKS & MUSEUMS, Eisenhower Park, East Meadow, NY 11554. Tel.: 516-572-0200. Fax: 516-572-0260.
Web Site: nassaucountyny.gov/parks
Founded: 1956.
Congressional District: 4
Key Personnel: Commissioner, Jose Lopez; Dir. Museum Svcs. Div., Herbert Mills; Sands Pt. Preserve Supvr., Gary Haglich; Old Bethpage Village Restoration Supvr., Jim McKenna; Historic Sites Cur., Harrison Hunt; Life Science Cur. & Tackapausha Museum Supvr., Lois Lindbergh; African/American Cur., David Byer-Tyre; Muttontown Preserve Supvr., Al Lindberg; Garvies Point Museum Supvr., Kathryne Natale.
Personnel Profile: Full-Time Paid 35; Part-Time Paid 54; Part-Time Volunteers 500.
Governing Authority: county; nonprofit organization, Affiliated with Friends for Long Island Heritage. Branch Museums: 1700s Saddle Rock Grist Mill, Saddle Rock; 1934 Christopher Morley Knothole, North Hills; Old Bethpage Village Restoration, Round Swamp Rd., Old Bethpage (includes:

c.1760 Schenck house; early 1840s Hewlett Farm; 1815 Cooper House; c.1850 Conklin House; c.1840 Luyster Store; 1860 Williams House & Carpentry Shop; late 1820s, Lawrence House; c.1850 Noon Inn; c.1840 Kirby House; c.1830 Ritch House & Hat Shop; 1829 Benjamin House; c.1850 Powell House; c.1857 Manetto Hill Church; c.1860 Prime Storage Building; c.1865 Layton Store-House; c.1870 Bach Blacksmith Shop; c.1845 District School #6; c.1835 Bedell House; Garvies Point Museum, Barry Drive, Glen Cove; Tackapausha Museum, Washington Ave., Seaford; Sands Point Preserve (includes: c.1923 Falaise; c.1912 Hempstead House; c.1902 Castle Gould); Muttontown Preserve (included Chelsea Cultural Art Center); African American Museum, Hempstead; Welwyn Preserve, Glen Cove; Roslyn Grist Mill, Roslyn; Cedarmere (William Cullen Bryant's Homestead), Roslyn Harbor; Cradle of Aviation Museum, Mitchel Field; Long Island Studies Institute in cooperation with Hofstra University, Hempstead. Tax-exempt.
Institution Type/Description: Preservation Project: Old Bethpage Village, farm community of pre-Civil War era, 45 historic structures; Jericho historic preserve, six structures from early 1800s, Sands Point Preserve, four early 1900s Gold Coast estate structures.
Collections: American history material before 1870; archaeological, geological & natural history specimens of local area; L.I. made aircraft & aviation artifacts.
Research Fields: history & natural history pertaining to Long Island region.
Activities: permanent & traveling exhibits: guided tours: films; formally organized education programs; inter-museum loan.
Publications: history booklet series.
Hours & Admission Prices: Call for information.
Attendance: 767,019 (accurate)

East Meredith

HANFORD MILLS MUSEUM, 73 County Hwy. 12, Corner of County Hwys. 10 & 12, East Meredith, NY 13757. Mailing Address: P.O. Box 99, East Meredith, NY 13757-0099. Tel.: 607-278-5744. Fax: 607-278-6299.
E-mail: info@hanfordmills.org
Web Site: www.hanfordmills.org
Founded: 1973.
Congressional District: 27
Key Personnel: Dir., Elizabeth Callahan; Chm. (V), Katie Boardman; Asst. Dir., Caroline de Marrais; Cur., Suzanne Soden; Mill Operations Foreman, Dawn Raudibaugh; Museum Shop Mgr., Louise Storey.
Personnel Profile: Full-Time Paid 5; Part-Time Paid 9; Part-Time Volunteers 50.
Governing Authority: nonprofit organization. Tax-exempt.
Institution Type/Description: Historic Site & Industrial Museum: housed in water powered mill & 16 historic structures.
Collections: historic structures from c.1840-1920; sawmilling, woodworking, gristmilling equipment; historical archives of Hanford Mill from 1860-1960; historic documents.
Research Fields: history of water milling in upper watershed of Delaware, Susquehanna & Schoharie River systems; vernacular engineering of mill dams & structures; social history of mill villages; history of site & village of East Meredith; power transmission technology; historic woodworking practices & production techniques; local social & cultural history.
Facilities: 250-vol. library of period catalogs of manufacturers of water, turbines, gas engines, saw mills, wagons, available for research by appointment on premises.
Activities: guided tours; formally organized education programs; permanent, temporary & traveling exhibitions; working machinery demonstrations; orientation film.
Publications: newsletter; collection catalogs, Made By Machine; The Hanford Photographs; East Meredith Memories; The Butter Business.
Hours & Admission Prices: mid-May to mid-Oct. Wed.-Sun. 10-5; other times by appointment. Adults $8.50, seniors 65 & over $6.50, military $4.25; discounts to AAA & AAM members; members and children 12 & under no charge. Buses welcome. &
Attendance: 8,200 (estimated)
Membership: Individual $15; Joint $20; Family $25.

Elizabethtown

ADIRONDACK HISTORY CENTER, (M), 7590 Court St., Elizabethtown, NY 12932. Mailing Address: P.O. Box 428, Elizabethtown, NY 12932-0428. Tel.: 518-873-6466.
E-mail: echs@adkhistorycenter.org
Web Site: adkhistorycenter.org
Founded: 1954.
Congressional District: 22
Key Personnel: Pres. (V), Carol Blakeslee-Collin; Dir., Margaret Gibbs.

Personnel Profile: Full-Time Paid 1; Part-Time Paid 5; Part-Time Volunteers 9.
Governing Authority: nonprofit. Parent Institution: Essex County Historical Society. Tax-exempt: 501(c)(3).
Institution Type/Description: History Museum.
Collections: horse-drawn vehicles, farm implements; tools; household items; dolls; mining equipment; surveying equipment; military artifacts; costumes; artifacts relating to camping, hiking & hunting; Adirondack mountain fire tower; sugar house.
Research Fields: History of Essex County.
Facilities: library; garden.
Activities: permanent, temporary & traveling exhibitions; lecture series; summer performance tour; ghost tour. Annual Events: Maple Sugar Festival; Fall Festival.
Publications: On The Trail of John Brown; A Basic Guide to Genealogical and Family History Resources for Essex County, New York.
Hours & Admission Prices: Memorial Day to Columbus Day daily 10-5. Adults $5, seniors $4, students $2; discounts to AAM members; members, school groups & children under 6 no charge. &
Attendance: 10,000 (estimated)
Membership: Individual $25; Family $40; Contributing $50; Sustaining $100; Sponsor $250; Patron $500; Benefactor $1,000.

Elma

ELMA TOWN MUSEUM - ELMA HISTORICAL SOCIETY INC., 3011 Bowen Rd., Elma, NY 14059. Mailing Address: P.O. Box 84, Elma, NY 14059. Tel.: 716-655-0046.
E-mail: elmahistory@aol.com
Web Site: www.elmanyhistory.com
Formerly: Elma Town Museum - Elma Historical Inc.
Founded: 1956.
Congressional District: 56
Key Personnel: Pres., Marlene Baumgartner.
Personnel Profile: Part-Time Volunteers 35.
Volunteer Hours: 743
Operating Expenses: 1,840
Operating Income: 2,863
Governing Authority: Parent Institution: N.Y. State Dept. of Parks, Education Dept.; Town of Elma. Tax-exempt.
Institution Type/Description: Historical Houses & Sites.
Collections: local history & culture; period artifacts; photographs; local Iroquois artifacts from 2000 BC. Historic House: former home of Clark W. Hurd, one of the founders of Elma & site of the 1st town board meeting on March 3, 1857; built in 1846. Barn Museum: Elma's dairy & farm equipment from 1830. Logging & Lumber Museum: restored 1846 sawmill with working water wheel.
Major Exhibits: Elma Center Post Office, 5/14-10/14; Jamison Rd Post Office, 5/14-10/14; Briggs Dairy, 5/14-10/14.
Research Fields: genealogy; site documents; early photographs; court records; deeds.
Facilities: Hurd House: library.
Activities: special events. Dedication & Grand Opening of Northrup Wheel Shelter; Apple Pressing Demonstration in Oct.; Pet Parade with Santa in Nov.; Town Tree Lighting Partnered with Kiwanis.
Publications: quarterly newsletter; History of East Elma; History of Elma for Children; It Is For Your Country.
Hours & Admission Prices: Thurs. 1-4, 1st & 3rd Sun. each month 1-4; other times by appointment and for special events. No charge; donations accepted. Closed holidays. &
Attendance: 3,000 (estimated)
Membership: Individual $10; Family $15; Lifetime $150.

Elmira

✳ **ARNOT ART MUSEUM, (M),** 235 Lake St., Elmira, NY 14901-3191. Tel.: 607-734-3697. Fax: 607-734-5687.
E-mail: rick@arnotartmuseum.org
Web Site: arnotartmuseum.org
Founded: 1913
Congressional District: 39
Key Personnel: Exec. Dir., Rick Pirozzolo; Business Mgr., Lynda Williams; Mgr. Education Programs, Meghan O'Loughlin; Mgr. Collections, Laura Wetmore; Dir. Tour Svcs., Wendy Taylor; Facilities Caretaker, Gregg Leavenworth.
Personnel Profile: Full-Time Paid 6; Part-Time Paid 2; Part-Time Volunteers 62; Interns 3.
Governing Authority: nonprofit organization. Tax-exempt: 509(a)(1).
Institution Type/Description: Art Museum: housed in 1833 Greek Revival style home of Matthias H. Arnot with 1985 Graham Gund Addition, and Carriage House Education Center.

Collections: restored 1890s picture gallery containing 17th-, 18th- & 19th-century European paintings; 19th- to 20th-century American Paintings; graphics; sculpture; decorative arts, emphasis on contemporary representational art.
Major Exhibits: Representing Representation VIII, 9/13-2/14.
Research Fields: 17th- & 19th-century European Salon paintings; 19th- & 20th-century American art; contemporary representation.
Activities: guided tours; lectures; films; gallery talks; concerts; formally organized education programs for children; docent program; inter-museum loan, permanent & temporary exhibitions.
Publications: catalogue, A Collector's Vision: The Bequest of Matthias H. Arnot; bimonthly members, The Column; special exhibition catalogues.
Hours & Admission Prices: Tues.-Fri. 10-5, Sat. 12-5. Adults $7; discounts to AAM & ICOM members; members no charge; reciprocal membership with 15 New York State museums. Closed national holidays. &
Attendance: 17,000 (accurate)
Membership: Single $50 & up; Family $75 & up; Friend $100 & up; Curator's Circle $250 & up; Director's Circle $500 & up; Arnot Society $1,000 & up.

✳ **CHEMUNG COUNTY HISTORICAL SOCIETY, (M),** 415 E. Water St., Elmira, NY 14901-3410. Tel.: 607-734-4167 & 4168. Fax: 607-734-1565.
E-mail: cchs@chemungvalleymuseum.org
Web Site: www.chemungvalleymuseum.org
Founded: 1923.
Congressional District: 29
Key Personnel: Pres., Russell Smith; Dir., Bruce Whitmarsh; Cur., Erin Doane; Coord. Education, Kerry Lippincott; Archivist, Rachel Dworkin; Administrative Asst., Christine Gunderson; Journal Editor, Joe Lemak.
Personnel Profile: Full-Time Paid 4; Part-Time Paid 1; Part-Time Volunteers 12; Interns 7.
Governing Authority: society. Parent Institution: Chemung County Historical Society. Tax-exempt: 501(c)(3).
Institution Type/Description: Historical Society Museum.
Collections: Civil War & Indian artifacts; Mark Twain collection; local history; archives; costumes; military; industry; agriculture; archaeology; textiles; transportation; 30,000 manuscript items; 11,500 maps & architectural records.
Research Fields: local history; archaeology. Historic Building: 1833 bank.
Facilities: 4,500-vol. library of local history books. Publications, reproductions, Mark Twain postcards & other museum-related items for sale.
Activities: guided tours; lectures; TV programs; formally organized education programs for students; permanent exhibitions; school loan service; historical excursions; veterans history project partner.
Publications: quarterly magazine, The Chemung Historical Journal; books, Chemung County, Its History; A Link in the Great Chain, A History of the Chemung Canal; booklet, Mark Twain in Elmira; newsletter, Banknotes.
Hours & Admission Prices: Museum: Mon.-Sat. 10-5. Office: Mon.-Fri. 9-5. Library & Research: Mon.-Fri. 1-5. Tours: adults $3; discounts to AAM & museum members. Personal Tours: adults $4. Closed holidays. &
Attendance: 10,000 (estimated)
Membership: Senior Citizen $25; Senior Household & Individual $35; Household $50; Research Sponsor $100; Society Patron $250; History Patron $500.

MARK TWAIN ARCHIVES - GANNETT-TRIPP LIBRARY - ELMIRA COLLEGE, One Park Place, Elmira, NY 14901. Tel.: 607-735-1869. Fax: 607-735-1712.
E-mail: mwoodhouse@elmira.edu
Institution Type/Description: Library & Archives.
Collections: Mark Twain's work including photographs & books; personal artifacts; books & articles written about him; microfilms of letters & manuscripts.
Activities: research; guided tours.
Hours & Admission Prices: By appointment.

MARK TWAIN STUDY & EXHIBITS - ELMIRA COLLEGE, One Park Place, Hamilton Hall, Elmira, NY 14901. Tel.: 607-735-1941. Fax: 607-735-1756.
E-mail: twaincenter@elmira.edu
Institution Type/Description: History Museum.
Collections: Mark Twain's life, career & writings; personal artifacts; photographs; period furnishings & clothing. Historic Structure: c.1874 study.
Activities: guided tours.
Hours & Admission Prices: Study: May to Labor Day Mon.-Sat. 9-5, Sun. 12-5; Sept. to mid-Oct. Sat. 9-5, Sun. 12-5.

NATIONAL SOARING MUSEUM, (M), Harris Hill, 51 Soaring Hill Dr., Elmira, NY 14903-9204. Tel.: 607-734-3128. Fax: 607-732-6745.
E-mail: nsm@soaringmuseum.org
Web Site: www.soaringmuseum.org
Founded: 1969.
Congressional District: 34
Key Personnel: Dir., Peter W. Smith; Pres. (V), Walter Cannon; Dir. Museum Svcs., Mary D. Flasphaler; Museum Shop Mgr., Lisa C. Bartlett; Dir. Devel. & Mktg., Ronald Ogden; Coord. Education, Caitlin Stevens.
Personnel Profile: Full-Time Paid 5; Part-Time Paid 3; Part-Time Volunteers 23; Interns 2.
Governing Authority: nonprofit organization. Tax-exempt. 501(c)(3).
Institution Type/Description: Aeronautics Museum: located on the site of the earliest recognized center for soaring flight in the United States.
Collections: motorless aircraft (gliders & sailplanes); archives including Soaring Society of America; film library; manuscripts; displays relating to motorless flight; U.S. Soaring Hall of Fame.
Research Fields: aerodynamics of low speed flight; motorless flight.
Facilities: library; archives; theater; community & special events room; classroom; 25,000 sq. ft. exhibit space. Museum-related items for sale.
Activities: guided group tours; lectures; films; special events; classes; permanent & temporary exhibitions; aviation summer camp; science & aviation programs; encampments.
Publications: NSM Journal; NSM News.
Hours & Admission Prices: Daily 10-5. Adults $7.50, senior citizens $6, youth 7-17 $4.50; discounts to groups & AAA members; members & children under 6 no charge. &
Attendance: 22,000 (accurate)
Membership: Individual $35; Family $50

TANGLEWOOD NATURE CENTER AND MUSEUM, 443 Coleman Ave., Elmira, NY 14903-9311. Tel.: 607-732-6060. Fax: 607-732-6210.
E-mail: elainems@stny.rr.com
Web Site: www.tanglewoodnaturecenter.com
Founded: 1973.
Congressional District: 31
Key Personnel: Office Mgr. & Museum Shop Mgr., Deanna Soper; Pres. Bd., Linda Hillman; Exec. Dir., Elaine Farwell; Volunteer Coord., Ian Martin; Cur., Valerie Heywood; Grounds & Bldg., Rich Gridley.
Personnel Profile: Full-Time Paid 2; Full-Time Volunteers 2; Part-Time Paid 4; Interns 2.
Governing Authority: private; nonprofit organization. Tax-exempt.
Institution Type/Description: Nature Center.
Collections: live animals; walk through immersion habitat.
Facilities: 300-vol. nature library; walk though museum; 200-seat auditorium; 400 acres of nature trails. Museum-related items for sale.
Activities: formal education programs for adults, children & interns; lectures; theatre; special events; guided hikes; astronomy club; snowshoe & kite rentals.
Publications: quarterly newsletter, Tanglewood Talk.
Hours & Admission Prices: Museum: May-Oct. Tues.-Sat. 8:30-4:30; Nov.-April Tues.-Sat. 9-4. Trails: daily dawn to dusk. No charge; donations accepted. Closed holidays. &
Attendance: 30,000 (estimated)
Membership: Individual $35; Senior Family $40; Family $45; Supporting $50-100.

Elmsford

GREATER HUDSON HERITAGE NETWORK, 2199 Saw Mill River Rd., Elmsford, NY 10523-3812. Tel.: 914-592-6726. Fax: 914-592-6946.
E-mail: info@greaterhudson.org
Web Site: www.greaterhudson.org
Formerly: Lower Hudson Conference of Historical Agencies and Museums
Founded: 1979.
Congressional District: 22
Key Personnel: C.E.O. & Exec. Dir., Priscilla Brendler; Pres. (V), Jacquetta Haley; Project Mgr., Dianne Macpherson.
Personnel Profile: Full-Time Paid 1; Part-Time Paid 2; Part-Time Volunteers 3; Interns 1.
Governing Authority: nonprofit organization. Tax-exempt: 501(c)(3).
Institution Type/Description: Museum Service Organization
Collections: museum, history and archival resource library for technical assistance and collections care.
Research Fields: management of museums & historical societies; collections care & management; programming; interpretation.
Facilities: library of museum-related technical publications.
Activities: workshops; lectures; training programs for professional & volunteer museum workers; technical assistance; conferences; museum tours; meetings. Annual Event: awards program in October.
Publications: biannual newsletter; on-line newsletter; History Keepers' Companion: Guide; Emergency Preparedness & Recovery Handbook.
Hours & Admission Prices: Mon.-Fri. 9-4. No charge. Fees for training programs. Workshop & seminar fee discount to GHHN member organizations & individuals. Closed Westchester County & New York state holidays. &
Attendance: 500 (estimated)
Membership: Student with valid ID $15; Individuals $40; Professional & Consultant $75; Sustaining $100. Organizations: Budget under $25,000 $50; Budget $25,000-$99,000 $75; Budget $100,000-$249,000 $100; Budget $250,000-$499,000 $125; Budget $500,000-$999,000 $175; Budget $1,000,000 & up $200; Business Member $500.

WESTCHESTER COUNTY HISTORICAL SOCIETY, 2199 Saw Mill River Rd., Elmsford, NY 10523-3812. Tel.: 914-592-4323. Fax: 914-592-4338.
E-mail: info@westchesterhistory.com
Web Site: www.westchesterhistory.com
Founded: 1874.
Congressional District: 23
Key Personnel: Exec. Dir., Katherine Hite; Chm. (V), Susan Morison; Librarian, Patrick Raftery.
Personnel Profile: Full-Time Paid 2; Part-Time Paid 1; Part-Time Volunteers 10.
Governing Authority: Tax-exempt.
Institution Type/Description: Historical Society Library & Research Center.
Collections: manuscripts; photographs; maps; books; pamphlets; diaries; periodicals; newspapers on Westchester County.
Research Fields: genealogy & local history.
Facilities: genealogy library; county history library.
Activities: programs featuring local history; education program; trips to historic areas in & out of the county.
Publications: quarterly, The Westchester Historian; books.
Hours & Admission Prices: Research Library: Tues.-Wed. 9-4. No charge. &
Attendance: 2,500 (estimated)
Membership: Member $35; Contributor $60; Friend $100; Sponsor $150; Benefactor $250; Patron $500; Sustainer $1,000.

Esperance

GEORGE LANDIS ARBORETUM, 174 Lape Rd., Esperance, NY 12066. Mailing Address: P.O. Box 186, Esperance, NY 12066-0186. Tel.: 518-875-6935. Fax: 518-875-6394.
E-mail: info@landisarboretum.org
Web Site: www.landisarboretum.org
Founded: 1951.
Congressional District: 21
Key Personnel: Exec. Dir., Fred Breglia; Pres. (V), James Paley; Science Educator, George Steele.
Personnel Profile: Full-Time Paid 1; Part-Time Paid 3; Part-Time Volunteers 178; Interns 1.
Governing Authority: nonprofit educational organization, chartered by Regents of New York State. Tax-exempt: 501(c)(3).
Institution Type/Description: Arboretum & Botanical Garden.
Collections: native & exotic trees & shrubs; perennial & spring bulb gardens; herbarium.
Research Fields: hardiness & performance testing of species of European & Asian origin.
Facilities: 548 acres including 40 acres planted to formal collections and over 20 acre native woodland with nature trail; visitor center; classroom; greenhouse.
Activities: guided tours; lectures; education programs for adults; horticulture & natural history programs for youth & adults; self-guided tours & trails; guided tours for groups.
Publications: quarterly newsletter; annual calendar of events.
Hours & Admission Prices: Daily dawn to dusk. No charge; donations accepted.
Attendance: 8,000 (estimated)
Membership: Student $25; Individual $35; Basic Family $50; Garden Club $100; Enhanced $125; Business $200; Patron $250; Benefactor $500; Founders Circle $1,000; Lifetime $2,500.

Fairport

FAIRPORT HISTORICAL MUSEUM, Perinton Historical Society, 18 Perrin St., Fairport, NY 14450-2122. Tel.: 585-223-3989.
E-mail: info@perintonhistoricalsociety.org
Web Site: www.angelfire.com/ny5/fairporthistmuseum
Founded: 1935.
Congressional District: 28
Key Personnel: Dir., William Keeler.
Personnel Profile: Full-Time Volunteers 40; Part-Time Paid 1; Part-Time Volunteers 30.
Governing Authority: society. Parent Institution: Perinton Historical Society. Tax-exempt.
Institution Type/Description: Local History Museum.
Collections: local history & culture; Native American artifacts; 19th-century industry; farm, home & village items.
Research Fields: local history.
Facilities: 385-vol. library of local history, historical fictions of local setting, state institution reports, genealogical reports, & books on local families available for use on premises; herb garden. Museum-related items for sale.
Activities: tours; support of school programs. Museum Sponsors: monthly meetings with speakers Sept.-April.
Publications: monthly newsletter, The Historigram.
Hours & Admission Prices: Sept.-May Sun. & Tues. 2-4, Thurs. 7-9; June-Aug. Sun. & Tues. 2-4, Thurs. 2-4 & 7-9. No charge; donations accepted. Closed major holidays. &
Attendance: 1,223 (estimated)
Membership: Seniors $5; Individual $10; Family $15.

Farmingdale

AMERICAN AIRPOWER MUSEUM, (M), 1230 New Hwy., Farmingdale, NY 11735. Mailing Address: c/o Cockpit USA, 15 W. 39th St., 12th Fl., New York, NY 10018. Tel.: 631-293-6398.
E-mail: info@americanairpowermuseum.com
Web Site: www.americanairpowermuseum.com
Founded: 1993.
Key Personnel: Pres., Jeff Clyman; Museum Shop Mgr., Jacky Clyman.
Personnel Profile: Full-Time Paid 2; Part-Time Paid 1; Part-Time Volunteers 150.
Governing Authority: Tax-exempt.
Institution Type/Description: History Museum.
Collections: military aircraft & vehicles; personal artifacts; photographs.
Activities: rental facilities; special events.
Publications: Flightlines, biennial
Hours & Admission Prices: Thurs.-Sun. 10:30-4. Adults $10, seniors & veterans $8, children 4-12 $5, children 4 & under no charge. &
Attendance: 15,000 (estimated)
Membership: Cadet $40; Wingman $75; Flight Lead $150; Mission Commander $500.

Fayetteville

THE STICKLEY MUSEUM, 300 Orchard St., Fayetteville, NY 13066-2120. Mailing Address: P.O. Box 480, Manlius, NY 13104-0480. Tel.: 315-682-5500. Fax: 315-682-6306.
Web Site: www.stickleymuseum.com
Founded: 2007.
Key Personnel: Dir., Sarah Lanigan.
Governing Authority: Parent Institution: L. & J.G. Stickley, Inc.
Institution Type/Description: Furniture Museum.
Collections: family furniture-making history; Stickley furniture & accessories.
Hours & Admission Prices: Tues. 11:30-5, Sat. 10-5; other times by appointment. No charge. &

Fineview

THE MINNA ANTHONY COMMON NATURE CENTER AT WELLESLEY STATE PARK, 44927 Cross Island Rd., Fineview, NY 13640-3105. Tel.: 315-482-2479. Fax: 315-482-2785.
Web Site: nysparks.com
Founded: 1969.
Key Personnel: Educator, Kimbrie Cullen; Pres. (V), Kerry Roberge.
Personnel Profile: Full-Time Paid 2; Part-Time Paid 7; Part-Time Volunteers 100; Interns 5.
Governing Authority: Parent Institution: New York State Office of Parks, Recreation and Historic Preservation. Subsidiary Institution: Wellesley Island State Park.
Institution Type/Description: Nature Center.
Collections: 3-dimensional displays; taxidermy; river art work; environmental exhibits.

Facilities: 8-mile hiking trail; 8-mile cross country ski trail; Butterfly House.
Activities: nature hikes; environmental education programs. Annual Events: Autumn Festival; Earth Day Activities.
Publications: newsletter, Friends of Nature Center; Geology of Wellesley Island Guide; Northfield Loop Self Guiding Booklet; bird checklist.
Hours & Admission Prices: Museum: daily call for hours. Trails: sunrise to sunset. Summer $7; Winter no charge. Closed Thanksgiving; Christmas. &
Attendance: 50,000 (estimated)
Membership: Individual $10; Family $15; Sponsor $25; Patron $50; Bluebird $100; Benefactor $250; Eagle $500; Life $1,000.

Fishkill

VAN WYCK HOMESTEAD MUSEUM - FISHKILL HISTORICAL SOCIETY, 504 Rte. 9, Fishkill, NY 12524. Mailing Address: P.O. Box 133, Fishkill, NY 12524-0133. Tel.: 845-896-9560.
E-mail: vanwyckhomestead@aol.com
Web Site: www.fishkillhistoricalsociety.org
Founded: 1962.
Congressional District: 25
Key Personnel: Pres. (V), Steve Lynch; Museum Shop Mgr., Helga Mackenzie.
Personnel Profile: Part-Time Volunteers 50.
Governing Authority: society; nonprofit organization. Parent Institution: Fishkill Historical Society. Tax-exempt: 501(c)(3).
Institution Type/Description: Historic House Museum: housed in a former Dutch Colonial Homestead built in 1732; officers' headquarters during the Revolutionary War for northern supply depot 1776-83.
Collections: early Dutch settler artifacts; Revolutionary War items; archaeology; archives; agriculture; technology; manuscripts; local genealogy; Ammi Phillips' art.
Research Fields: local history; archaeology.
Facilities: approx. 800-vol. research library of history. Gift items for sale.
Activities: guided tours; lectures; formally organized educational programs; permanent exhibitions; extensive craft program; summer archeological research project; special events. Annual Events: Craft Boutique in November; Dutch Legacy Weekend; Revolutionary War Weekend; Historic Bike Tour of Fishkill; Hudson River Valley Ramble; Christmas Open House; Candlelight Holiday Party in December.
Publications: monthly e-newsletter, Van Wyck Dispatch; Around Fishkill, Picture History of Fishkill
Hours & Admission Prices: Memorial Day to Oct. Sat.-Sun. 1-4. No charge; donations accepted.
Attendance: 2,000 (estimated)
Membership: Individual $15; Family $20; Contributing $25; Sponsoring $50; Sustaining $100; Life $500.

Flanders

THE BIG DUCK, 1012 Flanders Rd., Flanders, NY 11901. Mailing Address: P.O. Box 144, West Sayville, NY 11796. Tel.: 631-852-3377.
Governing Authority: Parent Institution: Suffolk County Parks Dept.
Institution Type/Description: Historic Building: listed on the National Register of Historic Places.
Collections: duck souvenirs; Long Island specialties; tourism information.
Facilities: tourism center. Museum-related items for sale.
Activities: special events. Annual Event: Holiday Lighting of the Big Duck.
Hours & Admission Prices: Tues.-Sat. 10-5, Sun. 2-5, call to confirm. No charge.

Floral Park

QUEENS COUNTY FARM MUSEUM, 73-50 Little Neck Pkwy., Floral Park, NY 11004-1129. Tel.: 718-347-3276, ext. 303. Fax: 718-347-3243. TDD: 800-281-5722.
E-mail: amy@queensfarm.org
Web Site: www.queensfarm.org
Founded: 1975.
Congressional District: 6
Key Personnel: Exec. Dir. & Events Coord., Amy Fischetti Boncardo; Pres. (V), James A. Trent; Chm. (V), Samuel Shapiro; Cur., Renee Tone; Dir. Education, Interpreter & Museum Shop Mgr., Diane Miller; Dir. Agriculture, Gary Mitchell; Dir. Horticultural, Annemarie Gero; Administrative Asst., Fran Erato; Interpreter, Mary Mifsud; Museum Shop Mgr., Sarah Meyer.
Personnel Profile: Full-Time Paid 7; Part-Time Paid 198; Part-Time Volunteers 108; Interns 3.
Governing Authority: nonprofit organization. Parent Institution: Colonial Farmhouse Restoration Society of Bellerose, Inc. Tax-exempt: 501(c)(3).
Institution Type/Description: Historic House: c.1772 Adriance Farmhouse.

Collections: farm animals; historic orchard; 19th-20th century farm tools & household artifacts.
Research Fields: restored farmhouse & grounds of a 20th-century urban truck farm; farming history of Queens 1772-1927; social history of farm families of Queens; truck farming in the 20th-century.
Facilities: Books, colonial reproductions & other museum-related items for sale.
Activities: guided, self-guided & school tours; concerts; fairs; festivals; docent program; quilting classes; workshops.
Publications: bimonthly newsletter, Broadside.
Hours & Admission Prices: Outdoors: daily 10-5. House Tours: Sat.-Sun. 10-5. General Admission: no charge. School tours $4-$8 per person. School Workshops $5-$8 per person. Special Events $4-$9 per person. &
Attendance: 500,000 (accurate)
Membership: Senior 65 & over or Student $20; Individual $25; Student Plus or Senior 65 & over Plus $35; Family or Individual Plus $40; Family Plus $80.

Flushing

THE BOWNE HOUSE HISTORICAL SOCIETY, 37-01 Bowne St., Flushing, NY 11354-5628. Tel.: 718-359-0528. Fax: 718-359-0873.
E-mail: office@bownehouse.org
Web Site: www.bownehouse.org
Founded: 1945.
Congressional District: 8
Key Personnel: Pres., Rosemary Vietor.
Personnel Profile: Full-Time Paid 1; Part-Time Paid 2; Part-Time Volunteers 1; Interns 2.
Governing Authority: Tax-exempt: 501(c)(3).
Institution Type/Description: Historic House Museum: 1661 home of John Bowne, religious freedom advocate.
Collections: 17th-19th century furnishings; artifacts; paintings & documents; period herb garden.
Research Fields: Bowne, Parsons & related families including the signers of the Flushing Remonstrance; early American history including religions, social, architecture & decorative arts.
Activities: school & off-site programs.
Publications: Bowne Family of Flushing Long Island.
Hours & Admission Prices: Museum is under restoration. Please call for appointment. &
Membership: Individual $25; Family $50; Sustaining $100; Corporate $500; Life $1,000.

THE GODWIN-TERNBACH MUSEUM, (M), Queens College, 405 Klapper Hall, 65-30 Kissena Blvd., Flushing, NY 11367-1575. Tel.: 718-997-4747. Fax: 718-997-4734.
E-mail: gtmuseum@qc.cuny.edu
Web Site: www.qc.cuny.edu/godwin_ternbach
Formerly: Frances Godwin & Joseph Ternbach Museum
Founded: 1957.
Congressional District: 7
Key Personnel: Dir. & Cur., Dr. Amy Winter; Chm. (V), Margaret Zeuschner.
Personnel Profile: Full-Time Paid 2; Part-Time Paid 3.
Governing Authority: Board of Trustees. Parent Institution: Queens College, CUNY. Tax-exempt.
Institution Type/Description: Art Museum.
Collections: ancient Near-Eastern, Asian, Western & contemporary art; graphics; paintings; glass; sculpture; prints.
Activities: permanent & temporary exhibitions; formally organized education programs for undergraduate & graduate students; lecture series; special events.
Publications: exhibition catalogs; Queens College Art Collection; brochure.
Hours & Admission Prices: Mon.-Thurs. 11-7, Sat. 11-5. No charge; donations accepted. &
Attendance: 9,500 (estimated)
Membership: Students & Seniors $10; Individual $25; Donor $100-$499; Sponsor $500-$999; Patron $1,000 & up.

METS HALL OF FAME & MUSEUM, Citi Field, Roosevelt Ave., Flushing, NY 11368-1699. Tel.: 718-507-8499.
Web Site: newyork.mets.mlb.com
Founded: 2010.
Institution Type/Description: Sports Museum & Hall of Fame.
Collections: Mets' history & memorabilia; signed artifacts; photographs; personal artifacts; baseballs; uniforms; bats; mitts; trophies; plaques; Hall of Fame inductees.
Facilities: 3,700 sq. ft. exhibition space. Museum-related items for sale.
Activities: interactive exhibitions; videos.

Hours & Admission Prices: During Home Games: when gates open to the end of the game. No charge. Non-game Days: access with Citi Field tour. Tickets: adults $10, children 12 & under and seniors 60 & over $7; discounts to groups; military no charge.

QUEENS BOTANICAL GARDEN, (M), 43-50 Main St., Flushing, NY 11355-4758. Tel.: 718-886-3800. Fax: 718-463-0263.
E-mail: info@queensbotanical.org
Web Site: www.queensbotanical.org
Founded: 1946.
Congressional District: 8
Key Personnel: Chm., Frank Mirovsky; Exec. Dir., Susan Lacerte; Dir. Finance & Admin., Wai Li; Maintenance Supvr., Peter Sansone; Dir. Mktg. & Devel., Darcy Hector; Sr. Attendant Guard, James Adams.
Personnel Profile: Full-Time Paid 23; Part-Time Paid 20; Part-Time Volunteers 755; Interns 3.
Governing Authority: society. Tax-exempt: 501(c)(3).
Institution Type/Description: Botanical Garden and Arboretum.
Collections: 39 acres including an herb garden; perennial garden tulip & annual displays; bee garden; rose garden; wedding garden; demonstration backyard gardens; 21-acre arboretum; compost home demonstration site. Fragrance walk; wetlands exhibit.
Research Fields: ethnobotanical.
Facilities: auditorium; classrooms; rental facilities.
Activities: guided tours; lectures; concerts; workshops; formally organized education programs; senior & children's gardening programs.
Publications: biannual newsletter, Lecture & Workshop Schedules; map & brochure printed in English, Spanish, Korean, & Chinese.
Hours & Admission Prices: April-Oct. Tues.-Sun. & Mon. holidays 8-6; Nov.-March Tues.-Sun. & Mon. holidays 8-4:30. Adults $4, seniors $3, students & children $2; APGA, American Horticultural Society & garden members no charge. &
Attendance: 205,000 (accurate)
Membership: Individual Senior $45; Individual $50; Family $85; Supporting $150; Friend $350; Patron $750; Director's Circle $1,000.

QUEENS HISTORICAL SOCIETY, (M), 143-35 37th Ave., Flushing, NY 11354-5729. Tel.: 718-939-0647, ext. 17. Fax: 718-539-9885. Facebook: Queens Historical Society.
E-mail: info@queenshistoricalsociety.org
Web Site: www.queenshistoricalsociety.org
Founded: 1968.
Congressional District: 5
Key Personnel: Dir., Ellissa Fazio; Mgr. Collections, Richard Hourahan; Coord. Education & Outreach, Danielle Hilkin; Pres., Patricia B. Sherwood; Vice Pres. History, James Driscoll; Treas., Linda Mandell; Recording Sec., Peter Byrne; Membership Sec., Catherine Williams.
Personnel Profile: Full-Time Volunteers 1; Part-Time Paid 3; Part-Time Volunteers 50; Interns 5.
Governing Authority: nonprofit organization. Subsidiary Institution: Kingsland Homestead; Moore-Jackson Cemetery. Tax-exempt: 501(c)(3).
Institution Type/Description: Historic Site: c.1785 Kingsland Homestead, located in Weeping Beech Park.
Collections: borough of Queens & Long Island history, 1600s-present; maps; memorabilia; textiles; photos; furniture; ephemera; documents; papers; decorative arts.
Major Exhibits: Practicing Equality: Quakes in Queens, 10/13-5/14; World's Fair Exhibit, 6/14-5/15.
Research Fields: Queens; New York City, Long Island.
Facilities: library of Queens related material; archives; multimedia & lecture room. Weeping Beech Park. Publications for sale.
Activities: docent program; guided tours; lectures; temporary & traveling exhibitions; workshops; outreach programs; preservation advocacy; panel discussions.
Publications: quarterly newsletter, Queens Historical Society; Angels of Deliverance: The Struggle Against Slavery in Queens and Long Island; The Road to Freedom: The Underground Railroad, New York and Beyond; 300 Years of Long Island City, 1630-1930; History of Flushing, NY; Teaching With Documents, Slavery in New York; Everything You Ever Wanted to Know About Queens - A Book of Trivia; Friends of Freedom: The Underground Railroad in Queens and on Long Island.
Hours & Admission Prices: Office: Mon.-Fri. 9:30-5. Museum: Tues. & Sat.-Sun. 2:30-4:30. Adults $5, senior citizens & students $3; members no charge. Closed major holidays. &
Attendance: 6,000 (estimated)
Membership: Senior & Student $15; Individual $20; Family $45; Business Patron $100; Corporate $250; Life Sponsor $500; Corporate Benefactor $750; Corporate Leader $1,000.

WILDLIFE CONSERVATION SOCIETY, QUEENS ZOO, (M), 53-51 111th St., Flushing, NY 11368-3301. Tel.: 718-271-1500, ext 126. Fax: 718-271-4441.
E-mail: ssilver@wcs.org
Web Site: www.queenszoo.org
Key Personnel: Dir. & Cur. Animals, Dr. Scott Silver; Asst. Cur. Animals, Craig Gibbs; Mgr. Operations, Jeffrey Blatz; Mgr. Security & Admissions, Vincent Capobianco; Cur. Education, Tom Hurtubise.
Personnel Profile: Full-Time Paid 60; Part-Time Paid 10; Part-Time Volunteers 30.
Governing Authority: Parent Institution: Wildlife Conservation Society.
Institution Type/Description: Zoo.
Collections: animals from the Americas.
Hours & Admission Prices: April-Oct. Mon.-Fri. 10-5, Sat.-Sun. & holidays 10-5:30; Nov.-March daily 10-4:30. Adults $6, senior citizens $2.25, children 3-12 $2; children under 3 no charge. &
Attendance: 208,389 (accurate)

Fonda

NATIVE AMERICAN EXHIBIT, NATIONAL KATERI SHRINE, 3636 State Hwy. 5, Fonda, NY 12068. Mailing Address: Box 627, Fonda, NY 12068-0627. Tel.: 518-853-3646.
Web Site: www.katerishrine.com
Founded: 1949.
Congressional District: 103
Key Personnel: Dir., Rev. Mark Steed, O.F.M.Conv.; Assoc. Dir. & Museum Shop Mgr., Bro. James Amrhein, O.F.M.Conv.
Personnel Profile: Full-Time Volunteers 2; Part-Time Paid 1; Part-Time Volunteers 3.
Governing Authority: church. Parent Institution: Order Minor Conventuals. Subsidiary Institution: National Kateri Shrine. Tax-exempt.
Institution Type/Description: Religious Shrine & Historic Archaeological Site: 1666-1693 staked out Mohawk Indian castle & 1666-1676 residence of Kateri Tekakwitha.
Collections: Native Americans & the shrine; Dutch Farmhouse & barn c.1782.
Facilities: chapel; 3 mile nature trail; service pavilion & grounds; Way of the Cross in wooded area. Religious & museum-related items for sale.
Activities: school tours; bus tours.
Publications: brochures, quarterly newsletter.
Hours & Admission Prices: May-Oct. daily 9-6; other times by appointment. No charge; donations accepted. Mass schedule: Sat. 4:30, Sun. 10:30. &
Attendance: 5,000 (estimated)
Membership: $7.

Fort Edward

OLD FORT HOUSE MUSEUM, 29 Lower Broadway, Fort Edward, NY 12828. Mailing Address: P.O. Box 106, Fort Edward, NY 12828-0106. Tel.: 518-747-9600. Fax: 518-747-7790.
E-mail: oldfort@localnet.com
Web Site: www.oldforthousemuseum.com
Founded: 1925.
Congressional District: 24
Key Personnel: C.E.O. & Historian, R. Paul McCarty; Pres. (V), Mary R. Smith; Dir. Education & Sec., Elizabeth O'Leary.
Personnel Profile: Part-Time Paid 3; Part-Time Volunteers 52.
Volunteer Hours: 600
Operating Expenses: 87,775
Operating Income: 87,775
Governing Authority: society. Parent Institution: Fort Edward Historical Association. Tax-exempt: 501(c)(3).
Institution Type/Description: Local History Museum: housed in 1772-73 Old Fort House.
Collections: archaeology; glass; history; colonial Victorian period furniture; Indian & colonial war artifacts; local pottery; 19th century dolls & toys; clothing; photographs; medicinal herb garden. Historic Site: Wait law office & toll house; one room schoolhouse; Cronkhite Pavilion.
Major Exhibits: 60th Anniversary Exhibit, 6/14-10/14.
Research Fields: local history; genealogy.
Facilities: research center.
Activities: guided walking tours; lectures; films; permanent & temporary exhibitions.
Hours & Admission Prices: June-Aug. daily 1-5; Sept. to Columbus Day Tues.-Sun. 1-5. Adults $5, students 13 & over $2; members no charge. Closed national holidays. &
Attendance: 9,300 (accurate)
Membership: Student & Senior Citizen $20; Individual $25; Family $35; Patron $50; Contributing $100; Supporting $250; Sustaining $500; Life $1,000. Corporation & Organization: Corporate Patron $100; Contributing $200; Supporting $300; Sustaining $500; Benefactor $750; Gold Pacesetter $1,000; Platinum Sponsor $5,000.

Fort Hunter

SCHOHARIE CROSSING STATE HISTORIC SITE, 129 Schoharie St., Fort Hunter, NY 12069-0140. Mailing Address: P.O. Box 140, Fort Hunter, NY 12069-0140. Tel.: 518-829-7516. Fax: 518-829-7491.
E-mail: janice.fontanella@parks.ny.gov
Web Site: nysparks.com
Founded: 1966.
Congressional District: 28
Key Personnel: Historic Site Mgr., Janice M. Fontanella; Supvr. Maintenance, Don Drew; Coord. Education, Tricia Shaw.
Personnel Profile: Full-Time Paid 3; Part-Time Paid 7; Part-Time Volunteers 3.
Governing Authority: state. Parent Institution: New York State Office of Parks, Recreation & Historic Preservation, & Saratoga/Capital District State Park, Recreation & Historic Preservation Commission. Tax-exempt: 501(c)(3).
Institution Type/Description: Historic Site.
Collections: Erie Canal history; archaeological remains from 18th century Fort Hunter & 18th century Native American village; parts of original locks of the Erie Canal; Schoharie Aqueduct, 1841.
Research Fields: Erie Canal; history of transportation in the Mohawk Valley; 18th century Native American, European, French & British colonial wars.
Facilities: boat landings & launching; hiking trails; bike paths; cross-country ski trails; picnic area; visitor center with exhibits on Erie Canal; restored 1850s canal store building.
Activities: guided walking tours; self guided tours; lectures; storytelling. Museum Sponsors: Canal Days Festival.
Publications: book, Erie Canal coloring book.
Hours & Admission Prices: Daily during daylight hours, weather permitting. Visitor Center: May-Oct. Wed.-Sat. 10-5, Sun. 1-5. No charge; donations accepted. &
Attendance: 80,000 (estimated)

Fort Johnson

OLD FORT JOHNSON, Rte. 5, Fort Johnson, NY 12070. Mailing Address: Fort Johnson, P.O. Box 196, Fort Johnson, NY 12070-0196. Tel.: 518-843-0300.
E-mail: museum@oldfortjohnson.org
Web Site: www.oldfortjohnson.org
Founded: 1904.
Congressional District: 30
Key Personnel: Pres., Philip V. Cortese.
Personnel Profile: Part-Time Paid 2; Part-Time Volunteers 15.
Governing Authority: Montgomery County Historical Society. Tax-exempt: 501(c)(3).
Institution Type/Description: General Museum: housed in 1749 Fort Johnson, home of Sir William Johnson, Supt. of Indian affairs & general of the Royal Militia & scene of Indian councils, military assemblies & Indian administration for British colonies.
Collections: 18th-19th century furnishings of the Johnson family & the Mohawk Valley; Mohawk Valley Indian artifacts; 19th-century textiles & costumes; Civil War & Montgomery County artifacts; manuscripts. Historic Structure: 1749 Fort Johnson.
Research Fields: Fort Johnson; William Johnson; c.1740-1850 decorative arts of Montgomery County.
Facilities: 500-vol. library on Mohawk Valley & New York Colonial, Revolutionary War & later history available by appointment for use at the museum.
Activities: guided tours; lectures; inter-museum loan, permanent, temporary & traveling exhibitions.
Publications: Quilts From Montgomery County, NY; Rufus A. Grider: Artist & Historian.
Hours & Admission Prices: May 15-Oct. 15 Wed.-Sat. 10-4, Sun. 1-5. Adults $4; members & children under 12 no charge.
Attendance: 3,500 (estimated)
Membership: Regular $15; Family $20; Sustaining $25; Corporate $100.

Fort Montgomery

FORT MONTGOMERY STATE HISTORIC SITE, 690 Rte. 9 W., Fort Montgomery, NY 10922. Mailing Address: P.O. Box 213, Fort Montgomery, NY 10922. Tel.: 845-446-2134. Fax: 845-446-2403.
Institution Type/Description: Historic Site: scene of the Revolutionary War battle for control of the Hudson River.
Collections: local history; 14-acre fortification; military artifacts & weapons;

large scale models of the fort and the attack; stone foundation of barracks; the gunpowder magazine; eroded redout walls; replica cannon.
Activities: film; living history demonstrations; special events.
Hours & Admission Prices: Call for hours.

Fort Plain

FORT PLAIN MUSEUM, 389 Canal St., Fort Plain, NY 13339-1160. Mailing Address: P.O. Box 324, Fort Plain, NY 13339-0324. Tel.: 518-993-2527.
E-mail: fortplainmuseum@yahoo.com
Web Site: www.fortplainmuseum.com
Founded: 1963.
Congressional District: 31
Key Personnel: Chm., Norm Bollen; Vice Chm., Robert Perry; Museum Shop Mgr., Patricia Perry.
Personnel Profile: Part-Time Volunteers 28.
Governing Authority: nonprofit corporation. State Education Dept. Tax-exempt: 501(c)(3).
Institution Type/Description: History Museum: housed in restored 19th-century farmhouse.
Collections: Native Americans; Revolutionary War; Erie Canal; history; anthropology; archaeology.
Research Fields: American Revolutionary period; Mohawk Valley development.
Facilities: Museum-related items for sale.
Activities: temporary exhibitions; films; junior museum; tours; seminars.
Publications: Revolutionary War Fort Plain: A Closer Look.
Hours & Admission Prices: June-Aug. daily 1-5; Sept.-May by appointment. Donations accepted. &
Attendance: 2,500 (estimated)
Membership: Sponsor $50; Patron $100; Sustainer $500.

Franklinville

ISCHUA VALLEY HISTORICAL SOCIETY, INC., 9 Pine St., Franklinville, NY 14737-1111. Mailing Address: P.O. Box 153, Franklinville, NY 14737. Tel.: 716-676-2590.
E-mail: maidlynn@aol.com
Web Site: ischuavalleyhistoricalsociety.org
Founded: 1966.
Congressional District: 34
Key Personnel: Pres., Bruce D. Fredrickson; Treas., Duane Walker; Sec., Ida Gardner; Cur., Maggie Fredrickson.
Personnel Profile: Part-Time Volunteers 50.
Governing Authority: nonprofit organization. Tax-exempt: 501(c)(3).
Institution Type/Description: Historical Society Museum: listed on State & National Historic Registers.
Collections: local artifacts & documents; photographs; family & business files. Historic Houses: 1814 Howe-Prescott - Salt Box; 1895 Miner's Cabin - Victorian.
Major Exhibits: Victorian Clothing, 6/14-8/14; Glass Photo Plates, 9/14.
Research Fields: local history; genealogy.
Facilities: library of newspapers, written local histories, scrapbooks & assorted documents available for research by appointment on site; reading room.
Activities: tours. Museum Sponsors: Harvest Dinner; Dinner Meeting in April; monthly programs March to December.
Publications: quarterly newsletter.
Hours & Admission Prices: Miner's Cabin: Memorial Day-Labor Day Sun. 2-5. Howe-Prescott House: by appointment only. No charge; donations accepted.
Attendance: 700 (estimated)
Membership: Contributing $10; Business $25; Life $100.

Fredonia

CATHY & JESSE MARION ART GALLERY, State University College, Fredonia, NY 14063. Tel.: 716-673-4897. Fax: 716-673-4990. Facebook: Cathy & Jesse Marion Art Gallery.
E-mail: tina.hastings@fredonia.edu
Web Site: www.fredonia.edu/rac
Formerly: Rockefeller Arts Center Art Gallery
Founded: 1826.
Congressional District: 34
Key Personnel: Gallery Dir., Tina Hastings.
Personnel Profile: Part-Time Paid 1; Interns 3.
Governing Authority: university. Affiliated with the State University of New York, Fredonia. Tax-exempt: 170(b)(1)(A).
Institution Type/Description: Art Gallery.
Collections: contemporary sculpture, prints & drawings.

Activities: inter-museum loan; permanent, temporary & traveling exhibitions.
Hours & Admission Prices: Academic School Year: Tues.-Thurs. & Sun. 12-4, Fri.-Sat. 12-6. No charge. Closed holidays; college breaks. &
Attendance: 3,109 (accurate)
Membership: Annual $25.

D.R. BARKER HISTORICAL MUSEUM, 20 E. Main St., Fredonia, NY 14063. Mailing Address: 7 Day St., Fredonia, NY 14063-1813. Tel.: 716-672-2114.
E-mail: barkermu@netsync.net
Web Site: www.barkermuseum.net
Formerly: Historical Museum of the D.R. Barker Library
Founded: 1884.
Congressional District: 39
Key Personnel: Pres. (V), Peter Clark; Vice Pres., Keith Sullivan; Cur., Nancy Brown.
Personnel Profile: Full-Time Paid 1; Part-Time Paid 3; Part-Time Volunteers 10; Interns 3.
Governing Authority: nonprofit organization. Parent Institution: Darwin R. Barker Library Association. Tax-exempt: 501(c)(3).
Institution Type/Description: Genealogy Library & History Museum: housed in 1821 Leverett Barker Home.
Collections: portraits; photographs; period costumes and accessories; military uniforms and records; documents and letters, including the records of the Fredonia Academy from 1821-1867; newspapers; genealogical reference works; period furniture and furnishings, tools and equipment.
Research Fields: local history; genealogy.
Facilities: 4,572-vol. genealogy & history library with material relating to the history of Western New York, the Village of Fredonia, the Town of Pomfret and Chautauqua County, and other genealogical materials available for use on premises; 1,968 sq. ft. exhibit space; meeting room.
Activities: concerts; films; hands-on activities; lectures related to exhibitions; guided tours of current exhibits on request; variety of school programs; children's programs; public programming relating to collections; permanent & temporary exhibitions.
Publications: quarterly newsletter, Barker Historical Newsletter; pamphlets, Fredonia by Gaslight, A Diary Year: 1866, The Barker Library & Museum: A History; The Dunkirk & Fredonia Telephone Company 1898-1999-A History; Photographers of Fredonia.
Hours & Admission Prices: Tues. & Thurs. 1-5 & 7-9, Wed. & Fri.-Sat. 1-5. Children's Museum: Sat. 1-5. No charge; donations accepted. Closed national holidays. &
Attendance: 3,958 (accurate)
Membership: Senior Citizens & Students $7; Individual $10; Family $15; Contributing $25; Donor $50; Sponsor $100; Patron $100 & up.

Fultonville

SHRINE OF OUR LADY OF MARTYRS AKA THE NATIONAL SHRINE OF THE NORTH AMERICAN MARTYRS, 136 Shrine Rd., Fultonville, NY 12072. Tel.: 518-853-3033. Fax: 518-853-3051.
E-mail: office@martyrshrine.org
Web Site: www.martyrshrine.org
Founded: 1885.
Congressional District: 31
Key Personnel: Dir., Rev. George H. Belgarde, S.J.; Museum Mgr. & Media Rep., Elizabeth Lynch; Supvr. Operations, Larry Steiger; Gift Shop Mgr., Joanne Wiesner.
Personnel Profile: Full-Time Paid 5; Part-Time Paid 4; Part-Time Volunteers 12.
Governing Authority: nonprofit organization; church. Parent Institution: New York Province, Society of Jesus. Subsidiary Institution: Saints of Auriesville Museum. Tax-exempt: 501(c)(3).
Institution Type/Description: Museum Complex & Historic Site: housed in c.1900 frame structure, located on the site of the 1646 martyrdom of Father Isaac Jogues, French Jesuit priest & 1656 birthplace of Kateri Tekakwitha.
Collections: anthropology; aboriginal/Catholic relations; pictorial storyboards & rare books; Catholic Sacramentals & memorabilia; Native American artifacts & handicrafts; paintings; maps.
Facilities: 300-vol. library pertaining to early history of New York & 1897, set of Jesuit Relations available for research by appointment on premises; reading room; 6,500-seat Coliseum Church; 50-seat media room. Museum-related items for sale.
Activities: guided tours; lectures; films; drama; permanent exhibitions.
Publications: weekly newsletter, Pilgrim; brochures.
Hours & Admission Prices: Museum: Mon.-Fri. 12-4, Sat.-Sun. 10-4. Visitor Center: May-Oct. Sun.-Fri. 10-5, Sat. 10-5:30. Mass Schedule: Mon.-Sat. 11:30 & 4, Sun. 9, 11 & 4. No charge; donations accepted. &
Attendance: 75,000 (estimated)

Garden City

CRADLE OF AVIATION MUSEUM, Charles Lindbergh Blvd., Garden City, NY 11530. Mailing Address: One Davis Ave., Garden City, NY 11530-6743. Tel.: 516-572-4111. Fax: 516-572-4065.
Web Site: www.cradleofaviation.org
Founded: 1979.
Congressional District: 4
Key Personnel: Exec. Dir., Andrew Parton; Chm. Bd. (V), Linda Armyn; Dir. Education, Jennifer Baxmeyer; Financial Dir., Dan Boehm; Cur., Joshua Stoff; Visitor Svcs., Gary Monti.
Personnel Profile: Full-Time Paid 20; Part-Time Paid 23; Part-Time Volunteers 250.
Governing Authority: private; nonprofit organization. Parent Institution: Museums at Mitchell. Tax-exempt: 501(c)(3).
Institution Type/Description: Air and Space Museum & Educational Center.
Collections: 73 aircraft & spacecraft significant to the aerospace heritage of Long Island.
Research Fields: history of Long Island airports.
Facilities: 6,300-vol. library; 250-seat cafeteria; classroom; 44,000 sq. ft. exhibit space; 300-seat IMAX theater. Museum-related items for sale.
Activities: concerts; docent program; formal education programs for teachers; educational programs for children; school tour programs; STEM magnet academy for high school students; club meetings; lectures.
Hours & Admission Prices: Daily 9:30-5. Museum: adult $14, seniors 62 & over and children 4-12 $12; additional fees for other attractions. Closed Thanksgiving; Christmas. &
Attendance: 225,000 (accurate)
Membership: Seniors & Veterans $40; Individual $50; Family $85; Supporter $250; Sponsor $500; IMAX Seat Sponsor $1000.

FIREHOUSE ART GALLERY, NASSAU COMMUNITY COLLEGE, One Education Dr., Garden City, NY 11530-6793. Tel.: 516-572-0619. Fax: 516-572-9673.
E-mail: gallery@ncc.edu
Web Site: firehouse.ncc.edu
Founded: 1965.
Key Personnel: Dir., Lynn Rozzi; Cur., Meg Oliveri.
Personnel Profile: Full-Time Paid 1; Part-Time Paid 4; Part-Time Volunteers 2; Interns 2.
Governing Authority: public college; nonprofit. Tax-exempt: 501(c)(3).
Institution Type/Description: College Art Gallery.
Collections: over 450 works of fine art including paintings, prints, & sculpture, 16th-century to present.
Activities: guided tours; lectures; temporary exhibitions. Annual Events: Highlights of Contemporary American Art; student & faculty shows; regional competition.
Publications: monthly mailer; events calendar.
Hours & Admission Prices: Mon., Wed.-Thurs. & Sat. 11-4, Tues. 11-7. No charge. Closed college holidays. &
Attendance: 12,000 (estimated)

LONG ISLAND CHILDREN'S MUSEUM, Charles Lindbergh Blvd., Garden City, NY 11530. Mailing Address: 11 Davis Ave., Garden City, NY 11530-6745. Tel.: 516-224-5800. Fax: 516-302-8188.
E-mail: development@licm.org
Web Site: www.licm.org
Founded: 1993.
Congressional District: 4
Key Personnel: Exec. Dir., Suzanne LeBlanc; Co Chm. (V), Robert S. Lemle; Co Chm. (V), Scott Rechler; Asst. Museum Shop Mgr., Tanya Skachinsky.
Personnel Profile: Full-Time Paid 32; Part-Time Paid 80; Part-Time Volunteers 85; Interns 5.
Volunteer Hours: 12,536
Operating Expenses: 4,726,027
Operating Income: 4,750,972
Governing Authority: nonprofit. Tax-exempt.
Institution Type/Description: Children's Museum.
Collections: Interactive; interdisciplinary.
Major Exhibits: Broken? Fix It! (T), 11/13-1/14; Secrets of Circles (T), 1/14-5/14; Alice's Wonderland (T), 5/14-8/14; Native Voices, 9/14-1/15.
Facilities: 40,000 sq. ft. exhibit space.
Activities: hands-on fun experiences; weekend multicultural workshops & performances.
Publications: newsletter; monthly calendars.
Hours & Admission Prices: July-Aug. daily 10-5; Sept.-June Tues.-Sun. 10-5.

Admission $12, seniors 65 & over $11; discounts to educators, military, AAM, AAA, ACM & ASTC members; members & children under one no charge. &
Attendance: 275,375 (accurate)
Membership: Inventor $70; Discoverer $130; Grandparent $160; Adventurer $185; Explorer $350.

NASSAU COUNTY FIREFIGHTERS MUSEUM, One Davis Ave., Garden City, NY 11530. Tel.: 516-572-4177 & 4066.
Web Site: www.ncfiremuseum.org
Founded: 2006.
Institution Type/Description: Firefighting History Museum.
Collections: firefighting history & equipment; photographs; fire trucks.
Facilities: 5,000 sq. ft. exhibition space.
Activities: fire safety & prevention programs.
Hours & Admission Prices: July-Aug. daily 10-5; Sept.-June Tues.-Sun. 10-5. Adults $5, senior citizens 62 & over, volunteer firemen and children $4.

Gardiner

LOCUST LAWN AND TERWILLIGER HOUSE, 400 Rte. 32 S., Gardiner, NY 12525. Mailing Address: Huguenot Historical Society, 88 Huguenot St., New Paltz, NY 12561-1415. Tel.: 845-255-1660. Fax: 845-255-0376.
E-mail: info@huguenotstreet.org
Web Site: www.locustlawn.org
Founded: 1894.
Congressional District: 27
Key Personnel: Exec. Dir., Eric J. Roth; Pres. (V), Eileen Crispell Ford; Coord. Patron Svcs., Laura Lucas; Cur. Education, Victoria Hughes; Cur., Leslie LeFevre Stratton; Cur. Historic Properties, Linda Pate; Museum Shop Mgr., Julianna de Grandis.
Personnel Profile: Full-Time Paid 5; Part-Time Paid 15; Part-Time Volunteers 30; Interns 2.
Governing Authority: Society. Parent Institution: The Huguenot Historical Society, 18 Brodhead Ave., New Paltz. Tax-exempt: 501(c)(3).
Institution Type/Description: Historic House & Estate: 1814 Locust Lawn house; 1738 Terwilliger house.
Collections: period furnishings from Queen Anne-Early Victorian; china; fabrics; laces; textiles; toys; paintings; coaches; tools & farm implements; paintings by Ammi Phillips. Historic Buildings: 1738 Terwilliger House; 1814 Locust Lawn house; slaughter-house; smoke house; carriage house.
Research Fields: furniture; paintings; decorative arts.
Activities: tours; children's programs; family events.
Publications: brochures.
Hours & Admission Prices: June-Oct. Sat.-Sun. 11-4; tours by appointment. Adults $7, seniors & AAA members $6, students $3; discounts to groups; members and children 5 & under no charge.
Membership: Individual $35; Household $50; Business & Contributor $100; Donor $250; Patron $500; Benefactor $1,000.

MOHONK PRESERVE, INC., (M), 3197 Rte. 44/55, Gardiner, NY 12525. Mailing Address: P.O. Box 715, New Paltz, NY 12561-0715. Tel.: 845-255-0919. Fax: 845-255-5646.
E-mail: info@mohonkpreserve.org
Web Site: mohonkpreserve.org
Founded: 1963.
Congressional District: 26
Key Personnel: Pres. Bd. Dir. (V), Ronald G. Knapp; Exec. Dir., Glenn D. Hoagland; Dep. Exec. Dir., David Toman; Dir. Communications, Gretchen Reed; Dir. Cons. Science, John Thompson; Dir. Stewardship, Hank Alicandri; Deputy Exec. Dir., Joe Alfano; Dir. Education, Kathy Ambrosini; Dir. Land Protection, Jennifer Garofalini; Dir. Devel., Sarah Kermensky.
Personnel Profile: Full-Time Paid 30; Part-Time Paid 28; Part-Time Volunteers 380; Interns 4.
Governing Authority: nonprofit organization. Tax-exempt: 501(c)(3).
Institution Type/Description: Historic Site & Nature Center: located in the Shawangunk Mountains.
Collections: natural history records; Northern Shawangunk Ridge history.
Research Fields: all aspects of ecosystem research; weather recording.
Facilities: 8,000 acres of land, visitor center; research center; teaching pavilion; 65 miles of carriage roads & trails; 4 trailheads.
Activities: public program series; K-12 education program.
Publications: newsletter; book, Time & the Mountain: A Guide to the Geology of the Northern Shawangunks; articles; monographs.
Hours & Admission Prices: Preserve: daily sunrise-sunset. Visitor Center: Mon.-Fri. 9-5. Adults: climbers & bikers $17, hikers $12; discount to ANCA members & groups; members & children under 13 no charge. Closed New Year's Day; Thanksgiving; Christmas Eve & Day. &

Attendance: 150,000 (estimated)
Membership: Seniors & Full-Time Students $45; Individual $55; Supporting $300; Sustaining $500; Sentinel $1,000; Life Endowment $3,000. Business memberships available.

Garnerville

GAGA ARTS CENTER, Garnerville Arts & Industrial Center, 55 W. Railroad Ave., Garnerville, NY 10923. Tel.: 845-947-7108.
E-mail: gaga@garnervillearts.com
Web Site: gagaartscenter.org
Key Personnel: Dir., James Tyler
Institution Type/Description: Art Gallery: housed in a 19th century textile mill.
Collections: works by established & emerging artists; paintings; sculpture; photographs; ceramics.
Facilities: 15,000 sq. ft. exhibition space.
Activities: educational programs; special events; workshops.
Hours & Admission Prices: Call for hours.

Garrison

BOSCOBEL HOUSE & GARDENS, (M), 1601 Rte. 9D, Garrison, NY 10524-4406. Tel.: 845-265-3638. Fax: 845-265-4405.
E-mail: info@boscobel.org
Web Site: www.boscobel.org
Founded: 1955.
Congressional District: 25
Key Personnel: Exec. Dir., Steven Miller; Pres. (V), Barnabas McHenry; Mktg. & Events Mgr., Donna Blaney; Museum Educator, Lisa DiMorzo; Museum Shop Mgr., Renate Smoller; Curator & Collections Mgr., Judith Pavelock; Maintenance Supvr., Richard Soedler.
Personnel Profile: Full-Time Paid 10; Part-Time Paid 28; Part-Time Volunteers 15.
Governing Authority: nonprofit organization. Tax-exempt: 501(c)(3).
Institution Type/Description: Historic House Museum: c.1808 New York Federal style home built by States Morris Dyckman.
Collections: 1800-1820 neo-classical furniture by cabinetmakers Duncan Phyfe & Michael Allison; federal period decorative arts; china, glass & silver purchased by Dyckman in London; Federal period paintings & decorative arts.
Research Fields: Dyckman Family & property history; Decorative Arts & Architecture; everyday life in America during the Federal Period.
Facilities: visitor center; gardens; one mile woodland trail. Books and museum-related items for sale.
Activities: guided tours; school tours; workshops; lectures; study kits on life in Federal Period & the Decorative Arts available to schools; changing & permanent exhibitions. Annual Events: Hudson Valley Shakespeare Festival June-August; Big Band Concert in September.
Publications: brochure; booklet, History of Boscobel; catalogue, Federal Furniture and Decorative Arts at Boscobel; summer art exhibition catalogue.
Hours & Admission Prices: April-Oct. Wed.-Mon. 9:30-5; Nov.-Dec. Wed.-Mon. 9:30-4. House & Grounds: Family of 4 $40; adults $16, senior citizens 62 & over $12; children 6-14 $7. Grounds: Family of 4 $25, adults $8, children 6-14 $5; discounts to groups of 12 or more & AAM members; museum employees of local museums in Hudson Valley with proper ID & Friends of Boscobel no charge. Closed Thanksgiving; Christmas. &
Attendance: 55,000 (accurate)
Membership: Senior Citizens 62 & over $30; Individual $40; Family $60; Loyalists $150; Federalists $300; Patriots $500; Neoclassicists $1,000.

GARRISON ART CENTER, 23 Garrison's Landing, Garrison, NY 10524-3648. Mailing Address: P.O. Box 4, Garrison, NY 10524-0004. Tel.: 845-424-3960. Fax: 8454244711.
E-mail: director@garrisonartcenter.org
Web Site: www.garrisonartcenter.org
Founded: 1964.
Congressional District: 41
Key Personnel: Exec. Dir., Carinda Swann.
Personnel Profile: Full-Time Paid 2; Part-Time Paid 3; Part-Time Volunteers 60.
Governing Authority: Tax-exempt.
Institution Type/Description: Art Center.
Collections: works by local artists.
Publications: exhibition catalogs.
Hours & Admission Prices: Tues.-Sun. 12-5. No charge; donations accepted. &
Attendance: 20,000 (estimated)

MANITOGA/THE RUSSEL WRIGHT DESIGN CENTER, (M), 584 Rte. 9D, Garrison, NY 10524. Mailing Address: P.O. Box 249, Garrison, NY 10524-0249. Tel.: 845-424-3812. Fax: 845-424-4043.
E-mail: info@russelwrightcenter.org
Founded: 1984.
Congressional District: 9
Key Personnel: Pres. (V), David McAlpin; Asst. Dir. & Museum Shop Mgr., Lori Moss.
Personnel Profile: Full-Time Paid 3; Part-Time Paid 1; Part-Time Volunteers 12; Interns 2.
Governing Authority: Tax-exempt.
Institution Type/Description: Historic House Museum: housed in the former home & studio of designer Russel Wright. A National Historic Landmark.
Collections: Wright's life & family history; personal artifacts; period furnishings; photographs.
Facilities: nature trails.
Activities: guided tours; special educational programs & events.
Publications: summer nature & design camp brochure.
Hours & Admission Prices: Tours: May-Oct. Mon.-Fri. call for hours, Sat.-Sun. 11 & 1:30; groups by appointment. Adults $15; discounts to American Horticultural Society, Long House Preserve members; members no charge.
Attendance: 3,500 (estimated)
Membership: Student $25; Individual $40; Family & Dual $70; Moss Circle $150-$300; Fern Circle $250-$500; Laurel Circle $1,000; Manitoga Circle $2,500; Dragon Rock Circle $5,000 & up.

Geneseo

BERTHA V.B. LEDERER GALLERY, SUNY Geneseo, Brodie Hall, 1 College Circle, Geneseo, NY 14454-1401. Tel.: 585-245-5814.
Founded: 1966.
Congressional District: 26
Key Personnel: Dir., Cynthia Hawkins.
Personnel Profile: Full-Time Paid 1; Part-Time Paid 12; Interns 2.
Governing Authority: Parent Institution: State University of New York, Geneseo. Tax-exempt.
Institution Type/Description: Art Gallery.
Collections: works on paper; paintings; sculpture; photographs; prints; ceramics.
Facilities: 2,000 sq. ft. exhibit space.
Activities: temporary exhibitions.
Publications: catalogs.
Hours & Admission Prices: Mon.-Thurs. 12-4, Fri.-Sat. 12-6. No charge; donations accepted. Closed Thanksgiving; Christmas, fall & spring breaks. &
Attendance: 1,000
Membership: Student $10; Basic $40; Patron $100; Benefactor $250; Bertha Lederer Society $500; President's Club $1,000.

LIVINGSTON COUNTY HISTORICAL SOCIETY MUSEUM, 30 Center St., Geneseo, NY 14454-1204. Tel.: 585-243-9147.
E-mail: lchistory@frontier.com
Web Site: www.livingstoncountyhistoricalsociety.com
Formerly: Cobblestone Museum
Founded: 1876.
Congressional District: 30
Key Personnel: Pres. (V), Liz Porter; Dir., Anna Kowalchuk.
Personnel Profile: Part-Time Paid 1; Part-Time Volunteers 20; Interns 2.
Governing Authority: society. Tax-exempt: 501(c)(3).
Institution Type/Description: Historical Society Museum: housed in 1838 cobblestone building.
Collections: agriculture; transportation; military; archaeology; archives; clothing; household items; toys; Indian artifacts; china; education; silver; period fire apparatus; 1,100 books; documents; Concord Coach; Shaker prayer stone replica. Historic Buildings: cobblestone schoolhouse; 1890 Hose Cart Firehouse.
Facilities: 1,100-vol. library.
Activities: guided tours; formally organized education programs for children; monthly programs for adults.
Publications: historic newsletter, History of Cobblestone Museum.
Hours & Admission Prices: May-June & Sept.-Oct. Thurs. & Sun. 2-5; July-Aug. Thurs. & Sun. 2-5. Call for other information or appointments. No charge; donations accepted.
Attendance: 1,500 (accurate)
Membership: Children $5; Single $10; Corporate $50; Life $100.

1941 HISTORICAL AIRCRAFT GROUP MUSEUM, 3489 Big Tree Ln., Geneseo, NY 14454. Mailing Address: P.O. Box 185, Geneseo, NY 14454. Tel.: 585-243-2100. Fax: 585-245-9802.
Web Site: www.1941hag.org/index.html
Institution Type/Description: Military Aviation History Museum.
Collections: military aviation history & aircraft; photographs; personal artifacts.
Activities: special events.
Hours & Admission Prices: April-Sept. daily 10-4; Oct.-March Mon., Wed. & Fri.-Sat. 10-4. Adults $4, children 5-12 $1; children under 5 no charge. Closed New Year's Day; Thanksgiving; Christmas.
Membership: Junior $15; Senior Citizen $30; Adult $35; Family $75; C-47 Club $250; Top Gun Club & Senior Life $500; Life $600.

Geneva

GENEVA HISTORICAL SOCIETY, 543 S. Main St., Geneva, NY 14456-3106. Tel.: 315-789-5151. Fax: 315-789-0314.
E-mail: info@genevahistoricalsociety.com
Web Site: www.genevahistoricalsociety.com
Formerly: Geneva Historical Society & Rose Hill Mansion
Founded: 1883.
Congressional District: 24
Key Personnel: Dir. Education, Anne Dealy; Museum Shop Mgr., M.J. Benda.
Personnel Profile: Full-Time Paid 5; Part-Time Paid 11; Part-Time Volunteers 40.
Governing Authority: nonprofit organization. Tax-exempt: 501(c)(3).
Institution Type/Description: Historical Society Museum & Historic Houses: Geneva History Museum, Johnson House & Rose Hill Mansion.
Collections: Empire-style period furniture: eight bedrooms; kitchen; front & rear parlors; banquet hall; library; music room; 3D objects & archival materials that chronicle the history of Geneva.
Activities: guided tours.
Publications: brochure; Rose Hill Book.
Hours & Admission Prices: Geneva History Museum: Mon.-Fri. 9:30-4:30, Sat. 1:30-4:30. No charge. Rose Hill Mansion: May-Oct. Tues.-Sat. 10-4, Sun. 1-5. Adults $7, seniors $6, children 10-18 $4. Johnston House: May-Oct. Sat. 10-4, Sun. 1-5. &
Attendance: 9,000 (estimated)
Membership: Single $25; Family $35; Friend $50; Patron $100; Sustaining $150; Sponsor $250; Benefactor $500; Life $1,000.

GENEVA HISTORY MUSEUM, 543 S. Main St., Geneva, NY 14456-3194. Tel.: 315-789-5151. Fax: 315-789-0314. Facebook: Geneva Historical Society.
E-mail: info@genevahistoricalsociety.com
Web Site: genevahistoricalsociety.com
Founded: 1883.
Congressional District: 24
Key Personnel: Dir., Kerry Lippincott; Pres. (V), Cyril Smith; Rose Hill Site Admin., Alice Askins; Cur. Collections, John C. Marks; Archivist, Karen D. Osburn; Cur. Education & Public Info, Anne F. Dealy.
Personnel Profile: Full-Time Paid 5; Part-Time Paid 10; Part-Time Volunteers 60.
Operating Expenses: 200,115
Operating Income: 132,381
Governing Authority: nonprofit educational organization. Parent Institution: Geneva Historical Society. Branch Museums: 1839 Rose Hill Mansion; Johnston House. Tax-exempt: 501(c)(3).
Institution Type/Description: Local History Museum & Historic Houses.
Collections: Federal, Empire & Victorian furnishings & accessories; material of local historical interest; archival material; agricultural; military; industrial; costumes; toys; manuscripts; paintings & graphics by local artists; extensive photograph collection.
Research Fields: local history, historic architecture; decorative arts; local artists
Facilities: 1,600-vol. library & research archive of local history & genealogy available on premises; reading room; 75-seat auditorium.
Activities: guided tours; lectures; school programs; permanent & temporary exhibits.
Publications: newsletter, published 6 times per year; annual report; local history publications: A Walking Tour-South Main Street, Geneva, NY; A Driving Tour-Architectural Landmarks; 19th-century Architecture in Geneva; Gentle Enthusiasts in Art: Geneva's Landscape Painting Families; Treasures of American Architecture; Make a Way Somehow: African-Americans in Geneva, NY 1790-1965; To Dress & Keep the Earth: The Nurseries & Nursery Men of Geneva, NY; Close to the Heart of the War: Geneva & World War II; Images of America: Geneva, 1790-1945; Images of America: Geneva, 1945-1980.
Hours & Admission Prices: Museum: July-Aug. Mon.-Fri. 9:30-4:30, Sat.-Sun. 1:30-4:30; Sept.-June Tues.-Fri. 9:30-4:30, Sat. 1:30-4:30; Rose Hill Mansion & Johnston House: May-Oct.; closed federal holidays. Museum: no charge; donations accepted. Rose Hill: Tues.-Sun. Admission charged. Johnston House: Sat. 10-4, Sun. 1-5.
Attendance: 19,172 (accurate)
Membership: Single $35; Family $50; Friend $75; Patron $100; Sustaining $150; Sponsor $250; Benefactor $500; Investor $1,000.

Germantown

CLERMONT STATE HISTORIC SITE, One Clermont Ave., Germantown, NY 12526-5632. Tel.: 518-537-4240 & 8687. Fax: 518-537-6240.
E-mail: fofc@valstar.net
Web Site: www.friendsofclermont.org
Founded: 1962.
Congressional District: 29
Key Personnel: Dir., Susan Boudreau; Pres. (V), Carl Brandt; Dir. Education, Kjirsten Gustavson; Administrative Asst., Roberta Nolan; Bd. Coord., Audrey Reifler; Historic Horticulturist, Jane Lehmuller.
Personnel Profile: Full-Time Paid 8; Part-Time Paid 16; Part-Time Volunteers 70; Interns 1.
Governing Authority: state. Parent Institution: NYS Office of Parks, Recreation & Historic Preservation. Tax-exempt: 501(c)(3).
Institution Type/Description: Historic House Museum & Estate Grounds: located on 500-acre Hudson Valley estate, with c.1730 home belonging to seven generations of the Livingston family.
Collections: 16th- through 20th-century library volumes; 18th- through 20th-century furnishings & decorative arts; Livingston family possessions; photographic & manuscripts.
Research Fields: Livingston family history; Livingston-Fulton manuscripts; American Revolution & early Federal period; landowning in the Hudson Valley.
Facilities: 3,500-vol. family library available for use by advance request; 1,100-vol. reference library; visitor center; four formal gardens; picnic area; former carriage drives for hiking; horseback riding; cross-country skiing.
Activities: guided tours; lectures; gallery talks; concerts; school programs; workshops; temporary exhibits; bird walks; Sunday garden strolls; special events & craft demonstrations; children's summer camp.
Publications: newsletters three times per year, Views from Clermont; Friends of Clermont, Inc.
Hours & Admission Prices: House: April-Oct. Tues.-Fri. 11-5 (last tour 4:30); Nov.-March Sat.-Sun. 11-4. Adults $5, seniors over 62 $4; children under 12 no charge. Visitor Center: Jan. 2-April 1 Sat.-Sun. Grounds: daily 8:30-sunset. Visitor Center & Grounds no charge. Closed holidays. &
Attendance: 100,000 (estimated)
Membership: Individual $30; Dual & Family $50; Contributor $100; Benefactor $200; Patron $500; Chancellor's Court $1,000.

Ghent

THE FIELDS SCULPTURE PARK AT OMI INTERNATIONAL ARTS CENTER, 1405 Cty. Rte. 22, Ghent, NY 12075-3809. Tel.: 518-392-4747. Fax: 518-392-4740.
E-mail: bmaynes@artomi.org
Web Site: www.artomi.org
Founded: 1991.
Congressional District: 20
Key Personnel: Dir., Bill Maynes
Institution Type/Description: Sculpture Park.
Collections: sculpture.
Hours & Admission Prices: Daily dawn to dusk. &
Attendance: 13,000 (estimated)

Glen Cove

GARVIES POINT MUSEUM & PRESERVE, 50 Barry Dr., Glen Cove, NY 11542-1765. Tel.: 516-571-8010.
E-mail: vnatale@nassaucountyny.gov
Web Site: www.garviespointmuseum.com
Founded: 1967.
Governing Authority: Parent Institution: Nassau County Department of Parks, Recreation and Museums. Tax-exempt.
Institution Type/Description: Natural History Museum.
Collections: county history & heritage; Native American artifacts; archaeology; geology; photographs.
Facilities: Museum-related items for sale.
Activities: research; educational programs; annual Thanksgiving Native American Feast (weekend before Thanksgiving); international coastal cleanup.

Hours & Admission Prices: Tues.-Sat. 10-4; closed holidays. Adults $3, children 5-12 $2 (with parent or guardian); Friends of Garevies Point Museum & Preserve & members no charge. &

Membership: Senior/Student $20; Individual $25; Family $35; Patron $100-$499; Benefactor $500-$999; Corporate $1,000 & up.

Glens Falls

THE CHAPMAN HISTORICAL MUSEUM, 348 Glen St., Glens Falls, NY 12801-3520. Tel.: 518-793-2826. Fax: 518-793-2831.

E-mail: director@chapmanmuseum.org
Web Site: chapmanmuseum.org
Founded: 1967.
Congressional District: 22
Key Personnel: Exec. Dir., Timothy Weidner; Cur., Jilian Mulder; Educator, Andrea Kinderman.
Personnel Profile: Full-Time Paid 3; Part-Time Paid 3; Part-Time Volunteers 50.
Governing Authority: nonprofit organization. Affiliated with the Glen Falls-Queensbury Historical Association, Inc. Tax-exempt.
Institution Type/Description: Regional History Museum: housed in c.1868 Zopher Isaac DeLong House & attached Carriage House Gallery.
Collections: social & industrial history of the Upper Hudson-Southeastern Adirondack area including the City of Glen Falls & the Town of Queensbury; Stoddard collection of photographs, paintings, published works & memorabilia; The Glens Falls YMCA & Glen Falls Insurance Company collections; Glen Falls City School System records; diaries & archival materials; costumes; textiles; household articles; decorative arts.
Research Fields: local social & industrial history; photography of the Greater Glens Falls-Southeastern Adirondack Area.
Facilities: 600-vol. library of primary & secondary materials on local, regional & state history available for use on premises or loan by special permission; 20,000 piece photographic library for on-site use by arrangement.
Activities: guided tours; lectures; formally organized educational programs; docent program; permanent, temporary & traveling exhibitions; school loan service; oral history projects; architectural surveys; school kits.
Publications: book, Bridging the Years; The Adirondacks Illustrated; newsletter, The Echo; 19th-century map reprints.
Hours & Admission Prices: Tues.-Sat. 10-4, Sun. 12-4. Adults $5, seniors 65 & over and students $4; children under 12 & members no charge. Closed major holidays. &
Attendance: 15,000 (accurate)
Membership: Senior $22; Individual $40; Family $50; Contributing $75; Supporting $100; Sponsor $250; Patron $500; Benefactor $1,000.

✻   **THE HYDE COLLECTION, (M), (I),** 161 Warren St., Glens Falls, NY 12801-4562. Tel.: 518-792-1761. Fax: 518-792-9197.

Web Site: www.hydecollection.org
Founded: 1952.
Congressional District: 29
Key Personnel: Dir., Charles A. Guerin; Chm. (V), Candace Wait; Financial Officer, Lynne Mason; Registrar & Collections Mgr., Barbara Bertucio; Cur., Erin M. Coe; Membership Mgr., Dede Potter; Asst. to Dir., Kathy Reed; Curatorial Asst., Susan Bishop; Cur. Education, June Leary.
Personnel Profile: Full-Time Paid 13; Part-Time Paid 15; Part-Time Volunteers 200; Interns 3.
Volunteer Hours: 2,418
Operating Expenses: 1,832,000
Operating Income: 1,866,500
Governing Authority: bd. of trustees. Tax-exempt.
Institution Type/Description: Art Museum.
Collections: European & American paintings, drawings and sculpture 14th-20th centuries including works by Botticelli, Chase, Courbet, Davies, Degas, Eakins, El Greco, Hassam, Homer, Matisse, Picasso, Raphael, Rembrandt, Renoir, Rubens, Ryder, Seurat and Whistler; Italian Renaissance, Baroque & 18th-century French furnishings; incunabula & rare first editions; Bigelow & Wadsworth designed 1912 founders' residence.
Major Exhibits: Ansel Adams: Early Works (T), 1/25/14-4/19/14; Photo-Secession: Painterly Masterworks of Turn-of-the-Century Photography (T), 1/25/14-4/19/14; Winter Light: Selections from the Thomas Clark Collection, 1/25/14-5/10/14; Larry Kagan: Lying Shadows, 6/14/14-9/14/14; Anne Diggory: Hybrid Visions, 9/27/14-1/4/15; Picturing America: Signature Works from the Westmoreland Museum of American Art (T), 9/27/14-1/4/15.
Research Fields: European & American paintings, drawings & sculpture.
Facilities: 144-seat auditorium; art studio. Museum-related items for sale.
Activities: guided tours; lectures; concerts; school & community outreach programs; special & traveling exhibitions; family activities; gallery talks; facility rentals; adult & children's art studios.
Publications: members newsletter, exhibition catalogues & brochures includ-

ing: Arthur B. Davies: Dweller on the Threshold; Family Matters: American Impressionism & Realism; FORM(ATION): Modern & Contemporary Works from the Feibes and Schmitt Collection; Painting Lake George, 1774-1900, Adolph Gottlieb: 1956; Degas & Music; George McNeil: The Late Paintings 1980-1995; An Enduring Legacy: American Impressionist Landscape Paintings from the Thomas Clark Collection; Modern Nature: Georgia O'Keeffe and Lake George.
Hours & Admission Prices: Tues.-Sat. 10-5, Sun. 12-5. Adults $8-$12, seniors 60 & up & students $6-$8; children under 12, military, members & second Sunday of the month no charge. Closed most national holidays. &
Attendance: 22,655 (accurate)
Membership: Individual $60; Dual $75; Family $85; Participating $125; Donor $250; Sustaining $500; Director's Circle $1,000-$15,000.

WORLD AWARENESS CHILDREN'S MUSEUM, (M), 89 Warren St., Glens Falls, NY 12801-4509. Tel.: 518-793-2773.

E-mail: admin@worldchildrensmuseum.org
Web Site: www.worldchildrensmuseum.org
Founded: 1995.
Key Personnel: Exec. Dir., Jacquiline Touba, Ph.D.; Chm., Paul Pontiff, Esq.; Chm. (V), Anthony Taverni.
Personnel Profile: Full-Time Paid 2; Part-Time Paid 5; Part-Time Volunteers 5.
Governing Authority: Tax-exempt.
Institution Type/Description: Children's Museum.
Collections: hands-on exhibits; international children's art; clothing & educational artifacts from around the world; international artifacts.
Activities: traveling exhibitions.
Hours & Admission Prices: Wed.-Sat. 10-4, Sun. 12-4. Admission $5; children under 3 no charge.
Membership: Senior & Student $10; Individual $25; Family $35; Business $50; Traveler $100; Explorer $250.

Glenville

EMPIRE STATE AEROSCIENCES MUSEUM, 250 Rudy Chase Dr., Glenville, NY 12302-7104. Tel.: 518-377-2191, ext. 10. Fax: 518-377-1959.

E-mail: esam@esam.org
Web Site: www.esam.org
Founded: 1984.
Congressional District: 21
Key Personnel: Pres. (V), Ralph Rosenthal; Museum Shop Mgr., Joyce Newkirk.
Personnel Profile: Full-Time Volunteers 3; Part-Time Paid 3; Part-Time Volunteers 110.
Governing Authority: nonprofit organization. Tax-exempt 501(c)(3).
Institution Type/Description: Aerosciences & Airplane Museum.
Collections: concentration on New York State aviation items with emphasis on hands-on items for children. Collection of aircraft including F-14A Tomcat; A6E Intruder; A-4 Skyhawk; F-4D Phantom II; Mig-15; Skynight & Scimitar.
Research Fields: aviation.
Facilities: research library; outdoor air park.
Activities: school & outreach programs; school tours; science & technology program for University of Albany; videos of WWII veterans' experiences; Civil Air Patrol encampment; cub scout programs; science fairs at local schools. Museum Sponsors: National Aviation Day; Dedication of F-10 Aircraft; S.T.E.M. Exposium.
Publications: monthly newsletter: Aeronotes.
Hours & Admission Prices: Sept.-June 19 Fri.-Sat. 10-4, Sun. 12-4; other times by appointment. Adults $8, seniors & military $6, youth 6-16 $5; children under 6 no charge. Closed New Year's Day; Thanksgiving; Christmas. &
Attendance: 15,000 (estimated)
Membership: Senior Citizen & Student $30; Individual $35; Family $50; School & Not-for-Profit Organization $50; Small Business $100; Business (over 10 employees) $250.

Gloversville

FULTON COUNTY MUSEUM, 237 N. Kingsboro Ave., Gloversville, NY 12078-1428. Mailing Address: P.O. Box 711, Gloversville, NY 12078-0711. Tel.: 518-725-2203.

E-mail: fultoncohist@frontier.com
Web Site: www.fultoncountymuseum.com
Founded: 1891.
Congressional District: 26
Key Personnel: Pres. (V), Mark Pollack; Suprv., Donna Terranova.
Personnel Profile: Part-Time Paid 1; Part-Time Volunteers 10; Interns 4.
Governing Authority: society. Parent Institution: Fulton County Historical Society, 237 N. Kingsboro Ave., Gloversville, NY 12078. Tax-exempt.

Institution Type/Description: Historical Society & Industrial Craft Museum: housed in Old Kingsborough School, located on the site of the original Kingsborough Academy built in 1831.
Collections: local history, with emphasis on the glove and leather industry; 18th & 19th-century weavings; 1800-present day glove machinery; leather tanning room; 18th & 19th-century muskets, pistols, swords & daggers; early school room; blacksmith shop; Indian longhouse; railroad artifacts; dioramas, photographs & documents of the Sacandaga valley from mid-1800 before the flood; Indian artifacts; model Fonda, Johnstown railroad.
Research Fields: local city & county history; glove & leather industry, early man in Fulton county, railroads.
Facilities: 500-vol. library of material on local, regional & state history, natural history, available for use on premises; reading room; classroom; 40-seat meeting room.
Activities: guided tours; lectures; permanent & temporary exhibitions; weaving demonstrations.
Publications: brochures; newsletters; program flyers.
Hours & Admission Prices: May 20 to Columbus Day Thurs.-Sun. 12-4. No charge; donations accepted. Closed Independence Day. &
Attendance: 1,470 (accurate)
Membership: Annual Single $8; Family $15; Life $125.

Goshen

THE HARNESS RACING MUSEUM AND HALL OF FAME, (M), 240 Main St., Goshen, NY 10924-2157. Tel.: 845-294-6330. Fax: 845-294-3463.
E-mail: hrm@frontiernet.net
Web Site: www.harnessmuseum.com
Formerly: Trotting Horse Museum/Hall of Fame of the Trotter
Founded: 1951.
Congressional District: 22
Key Personnel: Dir., Janet Terhune; Pres., Elbridge T. Gerry, Jr.
Personnel Profile: Full-Time Paid 8; Part-Time Paid 10; Part-Time Volunteers 4.
Governing Authority: nonprofit organization. Tax-exempt: 501(c)(3).
Institution Type/Description: Sports Museum: housed in 1913 Tudor-styled stable located next to Historic Track.
Collections: Currier & Ives lithographs; photographs; original wood carvings; dioramas; books; bronzes; statuettes; oil paintings & wash drawings of the American trotting horse; equipment used in trotting; American Standard-bred Horse history.
Research Fields: history of trotting races; breeding of the American Standard-bred; artists of the sport; genealogy of people in the sport.
Facilities: 1,000-vol. library of bound magazines, manuscripts, videos & books on the history of harness racing available on the premises; 200-seat auditorium. Books & other museum-related items for sale.
Activities: guided tours; lectures; films; gallery talks; formally organized education programs for children; permanent, temporary & traveling exhibitions; concerts; art festivals.
Publications: quarterly newsletter; gift shop catalog; souvenir journal.
Hours & Admission Prices: General Admission: no charge. Docent Guided Group Tours: $4. &
Attendance: 22,252 (accurate)
Membership: Active $35; Sustaining $50; Contributing $100; Fellow $150; Patron $500-$999; Benefactor $1,000 & up.

Gouverneur

GOUVERNEUR HISTORICAL ASSOCIATION, 30 Church St., Gouverneur, NY 13642-1416. Tel.: 315-287-0570.
E-mail: gouverneurmuseum@centralny.twcbc.com
Web Site: gouverneurmuseum.org
Founded: 1974.
Key Personnel: Pres. (V) & Cur. History & Acquisitions, Joseph Laurenza; Vice Pres., Jon Jackson; Treas., R. Joseph Weekes.
Personnel Profile: Part-Time Volunteers 40.
Volunteer Hours: 2,668
Operating Expenses: 30,980
Operating Income: 30,980
Governing Authority: nonprofit organization. Tax-exempt: 501(c)(3).
Institution Type/Description: Historical Society Museum: housed in 1890 Presbyterian Manse.
Collections: furnishings of a Victorian house; military; rocks & minerals; farm tools; fire fighting wagon; women's Victorian clothing; children's clothing & toys; Western Indians collection: clothes, pottery, paintings, dolls, sculpture.
Research Fields: local history; genealogy.
Facilities: 100-vol. library of local history; classrooms. Museum-related items for sale.

Activities: guided tours; lectures; concerts; hobby workshops; formally organized education programs for adults; school loan service; temporary & permanent exhibitions.
Hours & Admission Prices: Wed. & Sat. 1-3; other times by appointment. No charge; donations accepted. Closed New Year's Day; Christmas.
Attendance: 1,099 (accurate)
Membership: Annual $10.

Grand Island

GRAND ISLAND HISTORICAL SOCIETY - RIVER LEA, Beaver Island State Park, Grand Island, NY 14072. Mailing Address: P.O. Box 135, Grand Island, NY 14072. Tel.: 716-773-3271.
Web Site: www.isledegrande.com
Founded: 1962.
Key Personnel: Pres. (V), C. Nestark
Governing Authority: bd. trustees. Chartered by NYS Dept. of Education. Tax-exempt.
Institution Type/Description: Historical Society Museum.
Collections: Grand Island, NY & River Lea history; turn-of-the-century gowns; Grover Cleveland, Lewis F. Allen, and Edward J. Hussey; early farm tools & equipment; early kitchen utensils; early Native American artifacts; items related to Grand Island's hotels, saloons & clubs; agricultural tools. Historic Building: 1873 villa.
Research Fields: Edward J. Hussey; Lewis F. Allen; James Tillin; former owners of River Lea.
Activities: special events. Museum Sponsors: Victorian Tea in April; Paddles-Up in July.
Publications: biannual newsletter; books.
Hours & Admission Prices: By appointment. Park: entrance fee. River Lea: no charge.
Attendance: 500 (accurate)
Membership: Student $10; Single $15; Life $500.

Granville

PEMBER MUSEUM OF NATURAL HISTORY, 33 W. Main St., Granville, NY 12832-1320. Tel.: 518-642-1515.
E-mail: info@pembermuseum.com
Web Site: www.pembermuseum.com
Founded: 1909.
Congressional District: 24
Key Personnel: Dir., Patricia Bailey; Pres. (V), Dan Wilson.
Personnel Profile: Full-Time Paid 1; Part-Time Paid 3; Part-Time Volunteers 30; Interns 1.
Governing Authority: Parent Institution: Pember Library & Museum. Subsidiary Institution: Pember Nature Preserve. Affiliated with the Hebron Pember Nature Preserve. Tax-exempt: chartered under state education law.
Institution Type/Description: Natural History Museum, Nature Center & Conservation Area.
Collections: 1862-1924 private collection of Franklin Pember; 2,000 birds; 500 mammals; 860 egg sets & nests; 2,000 insects; 1,500 herbarium mounts; rocks; minerals; fossils; mollusk shells.
Facilities: 200-vol. library available on premises only; nature center; 125-acre nature preserve with trails & schoolhouse. Museum-related items for sale.
Activities: guided tours; lectures; organized educational programs; interpretive museum activities for families, adults, children, school groups & other scheduled groups; outreach & loan kits; field trips & charters; school loan service. Museum Sponsors: field trips.
Publications: book, The Pember Museum of Natural History; quarterly newsletter.
Hours & Admission Prices: Tues.-Fri. 1-5, Sat. 10-3. No charge. Closed holidays.
Attendance: 17,518 (accurate)
Membership: Individual $15; Family $25; Small Business $30; Friend $50; Supporter & Corporate $100; Sustainer $250; Patron $500; Lifetime $1,000.

SLATE VALLEY MUSEUM, (M), 17 Water St., Granville, NY 12832-1316. Tel.: 518-642-1417. Fax: 518-642-1417.
E-mail: mail@slatevalleymuseum.org
Web Site: www.slatevalleymuseum.org
Founded: 1995.
Congressional District: 22
Key Personnel: Exec. Dir., Mary Lou Willits; Bd. Pres. (V), David P. Bridges; Treas., Gladys Frustaci; Asst. Dir. & Educator, Sarah Benway.
Personnel Profile: Full-Time Paid 2; Part-Time Paid 1; Part-Time Volunteers 20.
Governing Authority: private; nonprofit organization. Tax-exempt: 501(c)(3).

Institution Type/Description: History Museum.
Collections: the history of slate quarrying & the slate industry; geological specimens; photographs; prints; paintings; artifacts; ceramics; furniture; costumes; maps; paper materials & books.
Research Fields: history of individual slate businesses in the slate valley of NY & VT 1839-present; WPA Federal Art Project mural: Men Working in Slate Quarry, Martha Levy 1939; immigration of Welsh, Irish, Eastern Europeans & Italians to the Slate regions of the U.S.
Facilities: 100-vol. library; classroom; 4,000 sq. ft. exhibit space. Museum-related items for sale.
Activities: concerts; docent program; films; formal educational programs; guided tours; lectures; loan & temporary exhibitions; school loan service. Annual Event: Holiday Open House.
Publications: newsletter 3 times a year, Slate Valley Museum Newsletter; ethnic cookbook, Cooking in the Slate Valley.
Hours & Admission Prices: Tues.-Fri. 1-5, Sat. 10-4. Adults $5; discounts to ICOM & AAM members; members, children 12 & under and Slate company employees no charge. Closed New Year's Eve & Day; Independence Day; Thanksgiving; Christmas Eve & Day. &
Attendance: 7,000 (accurate)
Membership: Individual $30; Family $55; Business Patron $70; Corporate Partner $105; Benefactor $130; Guardian $500.

Great River

BAYARD CUTTING ARBORETUM, 440 Montauk Hwy., Great River, NY 11739. Mailing Address: P.O. Box 907, Great River, NY 11739-0907. Tel.: 631-581-1002. Fax: 631-581-1031.
E-mail: nelson.sterner@parks.ny.gov
Web Site: www.bayardcuttingarboretum.com
Founded: 1952.
Congressional District: 2
Key Personnel: Dir., Nelson Sterner; Dir. Grounds, Kevin Boone; Bd. Chm., Prof. Barbara Schaedler.
Personnel Profile: Full-Time Paid 8; Part-Time Paid 12; Part-Time Volunteers 20; Interns 1.
Governing Authority: state; board of trustees. Parent Institution: New York State Parks. Subsidiary Institution: Office of Parks, Recreation & Historic Preservation. Tax-exempt.
Institution Type/Description: Arboretum: located on 1886 National Register Historic Site.
Collections: conifers; rhododendrons; oaks; dwarf evergreens; hollies; mounted birds; local Indian artifacts; Tiffany glass; antique woodwork.
Facilities: botanical garden; nature conservation center; 130-seat auditorium; classrooms; cafeteria.
Activities: guided tours; lectures; concerts; arts festivals; formally organized education programs; guided historic house tours; holiday decorations in December. Museum Sponsors: flower show.
Publications: book, Bayard Cutting Arboretum A History.
Hours & Admission Prices: April-Oct. daily 10-5; Nov.-March daily 10-4. Buses: $35-$75, Car: $8; discounts to senior citizens & handicapped persons with New York State park passes. &
Attendance: 150,000 (estimated)

Greece

GREECE HISTORICAL SOCIETY & MUSEUM, 595 Long Pond Rd., Greece, NY 14612. Tel.: 585-225-7221.
E-mail: greecehistoricalsociety@yahoo.com
Web Site: www.greecehistoricalsociety.net
Key Personnel: Pres. (V), Bill Sauers; Museum Shop Mgr., Wendy Peeck
Governing Authority: Tax-exempt.
Institution Type/Description: Local Town Historical Society Museum: house built in 1870s.
Collections: local history & culture; period furnishings; personal artifacts; photographs; Native American artifacts.
Publications: bimonthly newsletter, The Corinthian.
Hours & Admission Prices: Museum: Sun. 1:30-4; tours by appointment. Office: Mon. 9:30-12:30. No charge. &
Attendance: 1,000 (estimated)
Membership: Seniors $10; Individual $12; Family $20; Business $50; Patron $75; Life $500.

Greenport

EAST END SEAPORT MUSEUM AND MARINE FOUNDA-TION, 3rd St. at Ferry Dock, Greenport, NY 11944. Mailing Address: P.O. Box 624, Greenport, NY 11944-0624. Tel.: 631-477-2100. Fax: 631-477-0004.
E-mail: eseaport@verizon.net

Web Site: www.eastendseaport.org
Founded: 1990.
Congressional District: 1
Key Personnel: Chm. & Pres., George Peter; Museum Shop Mgr., Kelly Logsdon.
Personnel Profile: Part-Time Paid 3.
Governing Authority: private; nonprofit organization. Tax-exempt: 501(c)(3).
Institution Type/Description: Maritime Museum: housed in c.1894 historic Long Island Railroad Station depicting maritime heritage & seaport history of Eastern Long Island.
Collections: text; shipbuilding in Greenport during World War II photographs; tools; models; lighthouse artifacts; saltwater aquarium; yacht racing & maritime women highlights; Picket Patrol, story of sailboat warriors in WWII; J-Boats racing yachts of the 1930s; 1800s to 1900 maritime events of Eastern Long Island.
Facilities: aquarium. Museum-related items for sale.
Activities: lectures; guided tours; oral history of East End Maritime history. Museum Sponsors: Maritime Festival; Opening Reception.
Hours & Admission Prices: May 17-June Sat.-Sun. 11-5; July-Sept. Wed.-Mon. 11-5. Gate $2. &
Attendance: 9,500 (accurate)
Membership: Regular $25; Family $40; First Mate $100; Captain $250; Commodore $500; Admiral (life) $1,000.

RAILROAD MUSEUM OF LONG ISLAND, 440-4th St., Greenport, NY 11944-0726. Mailing Address: P.O. Box 726, Greenport, NY 11944-0726. Tel.: 631-727-7920 (Riverhead) & 477-0439 (Greenport). Fax: 631-765-2757. Facebook: Railroad Museum of Long Island.
E-mail: info@rmli.org
Web Site: www.rmli.org
Founded: 1990.
Congressional District: 1
Key Personnel: Pres., Don Fisher; Vice Pres., Dennis DeAngelis; Trustee, Dennis Harrington; Sec., James Werner; Treas., Al Schick; Museum Shop Mgr., Richard Horn.
Personnel Profile: Full-Time Volunteers 258; Part-Time Volunteers 115.
Governing Authority: bd. of trustees. Corporate office in Riverhead. Tax-exempt: 501(c)(3).
Institution Type/Description: Railroad Museum.
Collections: railroading history of Long Island; photographs; artifacts; 3 steam engines; 2 diesel engines; 8 historic railroad cars; 2 railroad speeders; 2 railroad cars in Greenport, LI. Historic Station: 1890s restored freight depot. Riverhead: period Lionel layout; rail cars & locomotives.
Facilities: Greenport: Museum-related items for sale. Riverhead: Visitor's Center. Museum-related items for sale.
Activities: guided tours; NY World's Fair Park train ride in Riverhead. Museum Sponsors: Railfest in August (Riverhead); Wooden Toy Train Play Days in October (Riverhead); Santa Claus weekend in December (Greenport & Riverhead).
Publications: quarterly newsletter, The Postboy.
Hours & Admission Prices: Riverhead: Memorial Day to Columbus Day Sat.-Sun. 10-4; Oct.-May Sat. 10-4. Greenport: Memorial Day to Oct. Sat.-Sun. 11-4. Adults $7, children 5-12 $4; children under 5 & members no charge. &
Attendance: 8,000 (estimated)
Membership: Student $25; Senior Citizen $30; Regular $50; Corporate $150.

Greenwich

WASHINGTON COUNTY FAIR FARM MUSEUM, 392 Old Schuylerville Rd., Greenwich, NY 12834-4615. Tel.: 518-692-2464. Fax: 518-692-1021.
E-mail: markwashfair@aol.com
Web Site: washingtoncountyfair.com
Founded: 1971.
Key Personnel: Fair Mgr., Mark St. Jacques, C.F.E.; Education, Joan Prouty.
Personnel Profile: Full-Time Volunteers 20; Part-Time Paid 2; Part-Time Volunteers 30.
Governing Authority: private; nonprofit organization. Parent Institution: Washington County Fair, Inc., Greenwich, NY. Tax-exempt: 501(c)(3).
Institution Type/Description: History Museum.
Collections: agricultural equipment; farm-related artifacts. Historic Buildings: 1850s schoolhouse, corn crib, milkhouse, summer kitchen.
Facilities: library. Museum-related items for sale.
Activities: guided tours. Annual Events: Planting Activities; Youth Art Show & Competition; Harvesting Activities.
Hours & Admission Prices: May-Oct. call for schedule; other times by appointment. No charge; donations accepted. &
Attendance: 25,000 (estimated)

Hamburg

HAMBURG HISTORICAL SOCIETY - THE DUNN HOUSE, 5902 Gowanda State Rd., Hamburg, NY 14075. Mailing Address: P.O. Box 400, Hamburg, NY 14075-0400. Tel.: 716-646-6460.
Institution Type/Description: Historical Society Museum.
Collections: local history; period artifacts; photographs.
Hours & Admission Prices: Wed.-Thurs. 9:30-1:30; other times by appointment.

Hamilton

THE PICKER ART GALLERY, (M), Colgate University, 13 Oak Dr., Hamilton, NY 13346-1398. Tel.: 315-228-7634. Fax: 315-228-7932.
E-mail: pickerart@colgate.edu
Web Site: www.pickerartgallery.org
Founded: 1966.
Congressional District: 22
Key Personnel: Dir., Anja Chavez; Educator, Melissa Davies; Registrar, Sarisha Guarneiri; Administrative Asst., Tammy Larson.
Personnel Profile: Full-Time Paid 3; Part-Time Paid 1.
Governing Authority: Parent Institution: Colgate University. Tax-exempt: 101(6).
Institution Type/Description: Art Gallery.
Collections: Luther W. Brady DFA, 88 20th century works on paper; Herman Collection of modern Chinese woodcuts; Gary M. Hoffer '74 Memorial photographs; Soviet photographer Evgeny Khaldei collection; Herbert Mayer '29 paintings & sculpture.
Research Fields: various forms of art media.
Facilities: graphic arts studyroom; classrooms.
Activities: lectures; films; gallery talks; inter-museum loan, temporary & traveling exhibitions.
Publications: newsletter; exhibition catalogues.
Hours & Admission Prices: See website for hours & admission prices. &
Attendance: 1,373 (accurate)
Membership: Student $10; Contributing $25; Sponsoring $50; Supporting $100; Sustaining $500; Patron $1,000; Life $5,000; Benefactors $200,000 in cash or works of art.

Hammondsport

GLENN H. CURTISS MUSEUM, 8419 State Rte. 54, Hammondsport, NY 14840-9795. Tel.: 607-569-2160. Fax: 607-569-2040.
E-mail: info@glennhcurtissmuseum.org
Web Site: www.glennhcurtissmuseum.org
Founded: 1961.
Congressional District: 39
Key Personnel: C.E.O. & Dir., Trafford L. Doherty; Pres., Richard Honeyman; Cur., Rick Leisenring; Museum Shop Mgr., Lynne Mason.
Personnel Profile: Full-Time Paid 3; Part-Time Paid 7; Part-Time Volunteers 160.
Governing Authority: local. Tax-exempt.
Institution Type/Description: Aviation & Local History Museum.
Collections: history of Glenn Curtiss aircraft & engines; local history; aviation; Curtiss Aeroplane & Motor Co.; Curtiss motorcycles; Curtiss-Wright Corp. Wine-making.
Research Fields: Glenn H. Curtiss, father of American Naval Aviation, founder of the American aircraft industry, early aviation.
Facilities: 2,600-vol. library of books & 750 periodicals devoted to Curtiss Aviation & the history of aviation.
Activities: guided tours; school tours.
Publications: quarterly, The Aerogram.
Hours & Admission Prices: June-Oct. Mon.-Sat. 9-5, Sun. 10-5; Nov.-May daily 10-4. Adults $8.50, senior citizens $7, students $5.50; discounts to groups of 10 or more & AAA members; children 6 & under and members no charge. Closed New Year's Day; Easter; Thanksgiving; Christmas. &
Attendance: 25,000 (estimated)
Membership: Basic $35; Contributing $50; Supporting $125; Donor $250; Patron $500; Benefactor $1000.

THE WINE MUSEUM OF GREYTON H. TAYLOR, 8843 Greyton H. Taylor Memorial Dr., Hammondsport, NY 14840-9635. Mailing Address: P.O. Box 458, Hammondsport, NY 14840-0458. Tel.: 607-868-4814 & 3610. Fax: 607-868-3205.
E-mail: info@bullyhill.com
Web Site: www.bullyhill.com
Founded: 1967.
Key Personnel: Pres., Lillian Taylor; Dir. & Museum Shop Mgr., Paul N. Sprague.

Personnel Profile: Full-Time Paid 1.
Governing Authority: nonprofit. Tax-exempt: 501(c)(3).
Institution Type/Description: Technology Museum: housed in 1880 Greyton H. Taylor Wine building.
Collections: old winemaking equipment; cooperage; wine bottles & labels; presidential glass.
Research Fields: viticulture; archives; paintings.
Facilities: 1,500-vol. library of books on winemaking, viticulture & wine cookery available for use on premises; reading room.
Activities: guided & self guided tours; temporary exhibitions.
Hours & Admission Prices: Daily 11-5. No charge; donations requested. &
Attendance: 6,500 (estimated)

Harpursville

COLESVILLE & WINDSOR MUSEUM AT ST. LUKE'S CHURCH, Maple St. & Monroe St., Harpursville, NY 13787. Mailing Address: P.O. Box 318, Harpursville, NY 13787-0318. Tel.: 607-693-1222.
Founded: 1971.
Congressional District: 27
Key Personnel: Pres., Eileen Rugieri; Historian-Colesville, Val La Clair.
Personnel Profile: Part-Time Volunteers 10.
Volunteer Hours: 60
Operating Income: 6,000
Governing Authority: nonprofit organization. Parent Institution: Old Onaquaga Historical Society. Tax-exempt.
Institution Type/Description: History Museum: housed in 1823 Episcopal Church.
Collections: American Indian artifacts; period furnishings; early agricultural tools & industries; period American utensils; architecture; craft tools; maps; life in the towns of Windsor & Colesville in Broome County, New York.
Research Fields: American Indian history; genealogy; public records of town & county.
Facilities: library of public records & history books available on premises by request; 100-seat sanctuary.
Activities: guided tours; lectures; formally organized education programs; permanent & temporary exhibitions.
Publications: monthly newsletter.
Hours & Admission Prices: May-Oct. 2nd Sun. 2-5; other times by appointment. No charge; donations accepted.
Attendance: 450 (estimated)
Membership: Family $15.

Hastings-on-Hudson

HASTINGS HISTORICAL SOCIETY, 407 Broadway, Hastings-on-Hudson, NY 10706. Tel.: 914-478-2249.
E-mail: hhscottage@hastingshistorical.org
Founded: 1971.
Key Personnel: Archivist, Muriel Olsson
Institution Type/Description: Historical Society Museum: housed in the former Henry Draper observatory.
Collections: local history; photographs; oral histories; videos; maps; Hastings High School yearbooks from 1918-2003; postcards; paintings; books.
Hours & Admission Prices: Mon. & Thurs. 10-2; other times by appointment. No charge; donations accepted.
Membership: General $25; Supporting $50; Business, Professional & Corporate $75.

Hempstead

*** HOFSTRA UNIVERSITY MUSEUM, (M),** 112 Hofstra University, Hempstead, NY 11549-1120. Tel.: 516-463-5672. Fax: 516-463-4743.
E-mail: beth.e.levinthal@hofstra.edu
Web Site: www.hofstra.edu/museum
Founded: 1963.
Congressional District: 4
Key Personnel: Exec. Dir., Beth E. Levinthal; Dir. Museum Education, Nancy Richner; Assoc. Dir. Exhibitions & Collections, Karen Albert; Public Rels., Ginny Greenberg; Coord. Devel. & Membership, Tiffany Jordan.
Personnel Profile: Full-Time Paid 5; Part-Time Paid 11; Part-Time Volunteers 26; Interns 3.
Governing Authority: university. Parent Institution: Hofstra University. Tax-exempt: 170(b)(1)(A).
Institution Type/Description: Art Museum.
Collections: 18th- to 21st-century paintings (European & American), sculpture, prints, drawings & photography; Asian, African, Oceanic & Pre-Columbian collections dating from 5,000 BCE; outdoor sculpture collection; blue stone 11-circuit labyrinth.

Major Exhibits: Land of the Rising Sun: Art of Japan, 9/30/13-2/2/14; David Jacobs: Sight and Sound, 2/4/14-4/27/14; Spirit and Identity: Melanesian Works from the Hum Collections, 2/18/14-8/29/14; Donald Resnick: Essence of Place, 5/13/14-8/15/14; Past Traditions: New Voices in Asian Art, 9/2/14-12/10/14; Exploring the Centuries: 7th-19th Century Asian Art, 9/16/14-2/8/15.

Research Fields: 20th- to 21st-century American painting, photography & sculpture; modern European art; African, Asian, Oceanic & Pre-Columbian art.

Facilities: Emily Lowe Gallery; David Filderman Gallery; outdoor sculpture collection.

Activities: guided tours; school programs K-12; lectures; symposia; inter-museum & temporary exhibitions; sculpture garden tours; gallery talks. Annual Events: Museum Family Festivals; Outdoor Sculpture Exploration Backpacks; Great Art Capers.

Publications: catalogs, The America of Currier & Ives; British Watercolors; Cobra; Paris: Nature in the City; Androgyny in Art; Harlem Renaissance: The Blossoming of New Promises; 1935: Jung & the Abstract Expressionists; Mother & Child: The Art of Henry Moore; People at Work: Seventeenth Century Dutch Art; The Transparent Thread: Asian Philosophy in American Art; Appeasing the Spirits: Sui & Early Tang Tomb Sculpture from the Schloss Collection; The Butcher, the Baker, the Candlestick Maker: Jan Luyken's Mirrors of 17th-Century Dutch Daily Life; Endless is the Way Leading Home: The Art of Stephen Csoka; 4 + 1 + 5 Artists in Korea: Contemporary Painting and Sculpture from Seoul, Korea; Beyond the Frame: Animation Art and Paintings; Paul Jenkins 1954-1960 The Early Years in New York and Paris; British Prints from the Hofstra Museum Collection; Fine Art and Paper Money in Jacksonian America 1820-1860; Abstract Expressionism: Then and Now; Artists After Moby Dick; Swiss Poster: The Art of Ten Masters; The Hofstra Museum at 40; Tabletop; Breaking the Walls of Bias: Survivors' Art; Places Made Sacred; Where the Island Begins; New, New Media: Recent Trends in Web Design; Bailey Musica: Preserving Hispanic Culture on Long Island; Robert Rauschenberg: Artist/Citizen; Views of Old New York City: D.T. Valentine's Manuals 1841-1870; The Photography of Ruth Orkin; Sally Gall: Selected Landscapes; Euclid to ebooks: Ideal Images, Moving Ideas; Twardowicz-Dodson: Artists in Parallel; Voiceless In The Presence of Realities: 9/11/01 Remembrances from the Long Island Studies Institute; On Location: Women Photographers from the Hofstra University Museum Collection; Stan Brodsky, The Figure: 1951-2006; Ancient Echoes in Contemporary Printmaking; Tranquil Power: The Art of Perle Fine; Andy Warhol: Recent Gifts from the Photographic Legacy Program; Children's Pleasures: American Celebrations of Childhood; Acquired Riches: Highlights of the Hofstra University Museum Collection; 75 Stories for 75 Years.

Hours & Admission Prices: Tues.-Fri. 11-4, Sat.-Sun. 1-4, Thurs. late nights 7-9. No charge; donations accepted. Closed Independence Day; Thanksgiving Recess; Christmas Recess; all university recesses. &

Attendance: 22,973 (accurate)

Membership: Student $10; Senior Citizen $25; Hofstra Alumni Individual $25; Individual $35; Hofstra Alumni Family & Senior Citizen Family $40; Family & Dual $60; Supporter $150; Supporter Alumni Hofstra $125; Sustainer $300; Hofstra Alumni Sustainer $250; Friend $500; Hofstra Alumni Friend $450; Patron $1,000.

Herkimer

HERKIMER COUNTY HISTORICAL SOCIETY, 400 N. Main St., Herkimer, NY 13350-1955. Tel.: 315-866-6413. Facebook: Herkimer County Historical Society.

E-mail: herkimerhistory@yahoo.com

Web Site: www.rootsweb.ancestry.com/~nyhchs/

Founded: 1896.

Congressional District: 31

Key Personnel: Pres. (V), Jeffrey Steele; Exec. Dir., Susan R. Perkins; Administrative Asst., Caryl Hopson.

Personnel Profile: Full-Time Paid 2; Part-Time Volunteers 10.

Volunteer Hours: 826

Operating Expenses: 105,062

Operating Income: 93,934

Governing Authority: society. Tax-exempt: 637.

Institution Type/Description: Local History Museum.

Collections: items of Herkimer County; German Palatine Settlement artifacts; memorabilia of major wars; tools & items related to industries; Herkimer County typewriter; doll house miniature collection; exhibition on Criminal Justice.

Research Fields: history of Herkimer County; genealogy.

Facilities: 2,000-vol. library of books on genealogy and history available for use on premises.

Activities: summer & fall monthly programs.

Publications: books, Genealogy of Mohawk Valley Bellingers and Allied Families; Genealogy of Mohawk Valley Caslers and Allied Families; The Petries of the Mohawk Valley; Mohawk Valley Rasbachs & Allied Families; Mohawk Valley Herkimers & Allied Families; Cookbook, Victuals & Vignettes; Prayer & Praise: Churches in County; Mohawk Valley Herkimers & Allied Families; Mohawk Valley Harters and Allied Families; A New History of Herkimer County; Herkimer County at 200; Mohawk Valley Starings and Allied Families.

Hours & Admission Prices: Mon.-Fri. 10-4. No charge; donations accepted. Closed national holidays. &

Attendance: 4,000 (estimated)

Membership: Student $15; Individual $40; Family $75; Sustaining $100; Friend $300.

Hicksville

THE HICKSVILLE GREGORY MUSEUM, 1 Heitz Place, Hicksville, NY 11801-3101. Tel.: 516-822-7505. Fax: 516-822-3227.

E-mail: mail@gregorymuseum.org

Web Site: gregorymuseum.org

Founded: 1963.

Congressional District: 4

Key Personnel: Pres., Richard Althaus; Cur. & Museum Shop Mgr., Donald Curran.

Personnel Profile: Full-Time Paid 1; Part-Time Paid 4; Part-Time Volunteers 12; Interns 4.

Governing Authority: nonprofit organization. Tax-exempt: 501(c)(3).

Institution Type/Description: Science & Technology Museum: housed in 1895 Heitz Place Court House-Jail.

Collections: minerals; fossils; Indian artifacts; sea shells; butterflies; moths; natural jewelry; nature photography; fluorescent mineral pictures; Long Island local historical artifacts.

Research Fields: local history; rocks; minerals; wild flowers; entomology; birds.

Facilities: 400-vol. library of science & nature books available on premises; field research station; lecture hall. Museum-related items for sale.

Activities: guided tours; lectures; films; slide programs; arts festivals; temporary & traveling exhibitions; formally organized educational programs; classes.

Publications: booklet, Faceting; Making A Cabochon; Maps on Mineral Locations; Hicksville-Yesterday & Today 1648-1981; annual bulletins.

Hours & Admission Prices: Tues.-Fri. 9:30-4:30, Sat.-Sun. 1-5. Adults $5, students & senior citizens $3; members no charge. Closed major holidays. &

Attendance: 10,000 (estimated)

Membership: Family $20; Sponsor $50; Patron $100; Benefactor $250.

High Falls

DELAWARE AND HUDSON CANAL HISTORICAL SOCIETY AND MUSEUM, 23 Mohonk Rd., High Falls, NY 12440. Mailing Address: P.O. Box 23, High Falls, NY 12440-0023. Tel.: 845-687-9311. Fax: 845-687-9311.

E-mail: info@canalmuseum.org

Web Site: www.canalmuseum.org

Founded: 1966.

Congressional District: 26

Key Personnel: Pres., Bill Merchant.

Personnel Profile: Part-Time Paid 2; Part-Time Volunteers 18.

Governing Authority: society. Tax-exempt.

Institution Type/Description: Transportation Museum: housed in 1885 Episcopal Church.

Collections: canal-related documents; maps; photographs; dioramas; scale models; artifacts on canal; canal operation & related industries.

Research Fields: historical development of 19th-century canals; transportation; related industries & their impact on the communities.

Facilities: 20-vol. library of books of D&H canal-related material available for research on premises only; 50-seat meeting hall. Canal & museum-related items for sale.

Activities: guided tours; lectures; temporary exhibitions; Exploring the D&H Canal & Environs. Museum Sponsors: Five Locks Walk.

Publications: newsletter.

Hours & Admission Prices: May-Oct. Sat.-Sun. 11-5. Adults $5, children under 12 $2; members no charge. &

Attendance: 1,000 (estimated)

Membership: Canawler $25; Barge Crew $40; Lock Tender $60; Tavern Keeper $100; Boat Captain $200; Paymaster $500; Financier $1,000; Tycoon $1,500.

Hogansburg

AKWESASNE MUSEUM, 321 State Rte. 37, Hogansburg, NY 13655-3114. Tel.: 518-358-2240 & 2461. Fax: 518-358-2649.
E-mail: akwmuse@northnet.org
Web Site: www.akwesasneculturalcenter.org
Founded: 1972.
Congressional District: 30
Key Personnel: C.E.O., Glory Cole; Pres. (V), Irving Papineau; Museum Shop Mgr., Sue Ellen Herne.
Personnel Profile: Full-Time Paid 1; Part-Time Paid 2; Interns 1.
Governing Authority: nonprofit organization. Parent Institution: Akwesasne Cultural Center, Inc. Subsidiary Institution: Museum & Library. Tax-exempt: 170(b)(1)(A).
Institution Type/Description: Cultural Center.
Collections: Mohawk traditional artifacts & basketry; Akwesasne Mohawk culture; contemporary Iroquoian artifacts & ethnological exhibitions.
Research Fields: local history & folklife; Mohawk Indian, Iroquois history; Kanien'kehaka (Mohawk); Haudenosaunee (Iroquois).
Facilities: library of books relating to Native Americans; reading room; classrooms. Gift items for sale.
Activities: guided tours by appointment; lectures; educational programs; training programs; loan, permanent, temporary & traveling exhibitions.
Publications: leaflets pertaining to exhibitions; films Onenhakenhra; White Seed, Music & Dance of the Mohawk; Teionkwahontasen: Basketmakers of Akwesasne; gallery guides on history, culture & traditional arts; DVD, Akwesasne Mohawk Basketry Traditions.
Hours & Admission Prices: Mon.-Fri. 9-4, Sat. by appointment; guided tours by appointment only. Adults $2, children $1; Native Americans & children 5 & under no charge. Closed all major holidays. &
Attendance: 2,034 (accurate)

Homer

HOMEVILLE ANTIQUE FIRE DEPARTMENT MUSEUM, 32 Center St., Homer, NY 13077. Tel.: 607-749-4466.
Institution Type/Description: Firefighting History Museum.
Collections: firefighting history & equipment; American LaFrance & Ward LaFrance vehicles; pumpers; alarm systems; turnout gear.
Hours & Admission Prices: Call for hours.

Hoosick Falls

BENNINGTON BATTLEFIELD STATE HISTORIC SITE AND BARNETT HOMESTEAD VISITORS' CENTER, 30 Caretaker's Rd., Hoosick Falls, NY 12090-4801. Mailing Address: PO Box 163, Grafton, NY 12082. Tel.: 518-686-8266 & 279-1155. Fax: 518-279-1902.
E-mail: melissaann.miller@oparks.ny.gov
Web Site: www.nysparks.state.ny.us
Congressional District: 29
Key Personnel: Park Mgr., Tom Conklin; Site Coord., Phyllis Chapman.
Personnel Profile: Part-Time Paid 4; Part-Time Volunteers 4; Interns 1.
Governing Authority: state. Parent Institution: New York State Parks & Recreation.
Institution Type/Description: Historic Site: 1777 Revolutionary War battle site includes monuments to participating states, map of actual battle & panorama display of the battle site.
Collections: prints; sculpture; historic reproductions; 19th century artifacts.
Research Fields: New York State, 1777 campaign; early 19th century agricultural life.
Facilities: picnic area.
Activities: interpretive signs; picnicking; hiking; game fields; self guided tours & display; costumed tour guides; demonstrations; hands-on programs.
Publications: brochure.
Hours & Admission Prices: May to Labor Day daily 10-7; Sept.-Nov. 1 Sat.-Sun. 10-7. Visitors' Center: Thurs.-Sat. 1-4. Adults $5, children & seniors $3.
Attendance: 19,066 (accurate)
Membership: Individual $10.

LOUIS MILLER MUSEUM - HOOSICK TOWNSHIP HISTORICAL SOCIETY, 166 Main St., Hoosick Falls, NY 12090. Tel.: 518-686-4682.
E-mail: staff@hoosickhistory.com
Web Site: www.hoosickhistory.com
Key Personnel: Dir., Charles Filkins
Institution Type/Description: History Museum.
Collections: local history; maps; photographs; paintings; period furnishings & clothing; portraits; wedding gowns; toys; Bob & Ray Eberle memorabilia; military uniforms.

Hours & Admission Prices: Mon.-Tues. 1-4; other times by appointment.

Hornell

HORNELL ERIE DEPOT RAILROAD MUSEUM, 82 Main St., Hornell, NY 14843. Tel.: 607-324-7421. Fax: 607-324-3150.
E-mail: historian@cityofhornell.com
Web Site: www.hornellny.us/museum.htm
Key Personnel: Coord. Collections, Collette Cornish
Institution Type/Description: History Museum.
Collections: local history; period artifacts; photographs; Erie Railroad memorabilia.
Hours & Admission Prices: Wed.-Fri. 6pm-8pm, Sat. 12:30-3.

Horseheads

HORSEHEADS HISTORICAL SOCIETY MUSEUM AT THE DEPOT, 312 W. Broad St., Horseheads, NY 14845. Tel.: 607-739-3938.
E-mail: info@horseheadshistorical.org
Web Site: www.horseheadshistorical.com
Founded: 1972.
Key Personnel: Pres. (V), Richard Margeson.
Governing Authority: Tax-exempt.
Institution Type/Description: Historical Society Museum: housed in the former railroad depot; built in 1866.
Collections: local history & culture; period furnishings; personal artifacts; photographs; railroad memorabilia.
Facilities: Museum-related items for sale.
Hours & Admission Prices: April-Dec. Tues., Thurs. & Sat. 12-3; other times by appointment. No charge. &
Attendance: 1,000 (estimated)
Membership: Senior $8; Single $10; Family $15; Patron $20; Life $125.

WINGS OF EAGLES DISCOVERY CENTER, 339 Daniel Zenker Dr., Horseheads, NY 14845-1102. Tel.: 607-358-4247. Fax: 607-358-4248. Facebook: Wings of Eagles Discovery Center.
E-mail: bbenza@wingsofeagles.com
Web Site: www.wingsofeagles.com
Formerly: National Warplane Museum
Founded: 1983.
Congressional District: 31
Key Personnel: C.E.O. & Pres., Michael S. Hall; Chm. (V) & Museum Shop Mgr., Brenda Benza; Pres. (V), Donald Keddell; Dir. Restoration, Ed Knitter; Cur., Edward Flesch.
Personnel Profile: Full-Time Paid 3; Full-Time Volunteers 1; Part-Time Paid 3; Part-Time Volunteers 200; Interns 1.
Governing Authority: nonprofit. Chartered by The NY State Board of Regents. Tax-exempt: 501(c)(3).
Institution Type/Description: Aviation & Aerospace Museum.
Collections: WWII-era flying aircraft: WWII aircraft & artifacts; training & cargo; foreign & domestic; non-flying jet & propeller driven aircraft; engines; instruments; uniforms; insignia; flight gear & clothing from WWII, Korea & Vietnam.
Research Fields: military aviation history & air power.
Facilities: 7,000-vol. library containing books, manuals, periodicals, reference files & a/v materials. Museum-related items for sale.
Activities: guided tours; organized education programs for high school groups, college students & adults; volunteer & docent programs; temporary & permanent exhibits; community events; speaker programs. Museum Sponsors: Airfest.
Publications: newsletter.
Hours & Admission Prices: Tues.-Sat. 10-4. Adults $7, senior citizens $5.50, children 6-17 $4; discounts to military, Women in Aviation, families, AAM, ICOM, ASTC & AAA members; children under 6 & members no charge. Closed New Year's Day; Easter; Thanksgiving; Christmas. &
Attendance: 80,000 (estimated)
Membership: Students $32; Active Military & Seniors $34; Individual $40; Student Family $48; Family $60; Senior Family & Military Family $51.

ZIM'S HOUSE, HORSEHEADS HISTORICAL SOCIETY, 601 Pine St., Horseheads, NY 14845. Mailing Address: Horseheads Historical Society, 312 W. Broad St., Horseheads, NY 14845. Tel.: 607-739-3938.
Web Site: www.horseheadshistorical.com
Founded: 1982.
Key Personnel: Pres. (V), Richard Margeson.
Governing Authority: corp. Tax-exempt.
Institution Type/Description: Historic House Museum: housed in the former

home of political caricaturist of the late 1800s & early 1900s, Eugene Zimmerman. Listed on the National Register of Historic Places.
Collections: Zim's life, art & career; Zim's personal artifacts & furnishings; photographs.
Hours & Admission Prices: By appointment. No charge.
Attendance: 200 (estimated)

Howes Cave

IROQUOIS INDIAN MUSEUM, (M), 324 Caverns Rd., Howes Cave, NY 12092. Mailing Address: P.O. Box 7, Howes Cave, NY 12092-0007. Tel.: 518-296-8949. Fax: 518-296-8955.
E-mail: info@iroquoismuseum.org
Web Site: www.iroquoismuseum.org
Founded: 1980.
Congressional District: 21
Key Personnel: Acting Dir., Stephanie Shultes; Chair & Pres., Christina Hanks; Educator, Mike Tarbell.
Personnel Profile: Part-Time Paid 8; Part-Time Volunteers 40.
Governing Authority: nonprofit. Tax-exempt.
Institution Type/Description: Anthropology Museum & Site: focus on Iroquois art and culture.
Collections: contemporary art & craft work of the Iroquois Indians; prehistoric materials of the Iroquois & their immediate antecedents relating to Schoharie County; color slides; black & white prints; photographic collection of Iroquois arts; living plants & their native use.
Research Fields: contemporary Iroquois arts; archaeology of Schoharie County; relationship between early settlers & the Iroquois in the county; ethnobotany of the Iroquois.
Facilities: 1,100-vol. library of books on the Iroquois & other Native American groups available for research on request for use on premises; nature trail. Art work & craftwork made by Iroquois & books for sale.
Activities: guided tours; docents; loan, permanent & temporary exhibitions; Archaeology Dept. & school programs; lectures; arts & craft demonstrations. Museum Sponsors: Iroquois Indian Festival.
Publications: quarterly newsletter, Museum Notes; exhibition catalogs.
Hours & Admission Prices: May-Nov. Tues.-Sat. 10-5, Sun. 12-5. Adults $8, seniors & youths 13-17 $6.50, children 5-12 $5; discounts to groups, AAM & ICOM members; children under 5 & members no charge. Closed Easter; Thanksgiving; Christmas Eve & Day. &
Attendance: 11,000 (accurate)
Membership: Students (full-time) & Single Seniors $20; Individuals & Senior Couples $25; Family $35; Friend $50; Donor $100; Sponsor $250; Patron $500; Benefactor $1,000.

Hudson

FASNY MUSEUM OF FIRE FIGHTING, (M), 117 Harry Howard Ave., Hudson, NY 12534-1601. Tel.: 518-822-1875, ext. 11. Fax: 518-822-8520.
E-mail: jamie@fasnyfiremuseum.com
Web Site: www.fasnyfiremuseum.org
Formerly: American Museum of Fire Fighting
Founded: 1925.
Congressional District: 24
Key Personnel: Pres. Bd. Dirs., James A. Burns; Vice Pres., Neal Van Deusen; Dir., Jamie Smith Quinn; Museum Shop Mgr., Lori Decker.
Personnel Profile: Full-Time Paid 5; Part-Time Paid 8; Part-Time Volunteers 25.
Governing Authority: nonprofit organization. Tax-exempt.
Institution Type/Description: Fire-Fighting Antiques Museum.
Collections: period fire-fighting apparatus; art gallery; tools & equipment of the trade; photographs; prints.
Research Fields: history of the Volunteer Fire Service.
Facilities: 1,700-vol. library of books, magazines & minutes available by appointment with curator. Museum-related items for sale.
Activities: guided tours; interactive exhibitions.
Hours & Admission Prices: Daily 10-5. Family $20, adults $7, children 3 & over $5; museum, FASNY, NARM & ESRP members and children 2 & under no charge. Closed major holidays. &
Attendance: 25,000 (accurate)
Membership: Individual $30; Family $50; Lieutenant $100; Captain $250; Deputy Chief $500; Chief $1,000.

OLANA STATE HISTORIC SITE, 5720 Rte. 9-G, Hudson, NY 12534. Tel.: 518-828-0135. Fax: 518-828-6742.
E-mail: linda.mclean@oprhp.state.ny.us
Web Site: www.olana.org; www.nysparks.com
Founded: 1966.
Congressional District: 29
Key Personnel: Dir. Olana State Historic Site, Linda E. McLean; Dir. Education, Carri L. Manchester; Chief Cur., Evelyn Trebilcock; Assoc. Cur., Valerie Balint; Archivist & Librarian, Ida Brier; Historic Site Asst., Roberta Bennett; Maintenance Chief, Timothy Dodge; Pres. Olana Partnership, Sara Griffen; Chm. (V) The Olana Partnership, Richard Sharp; Chm. (V), Susan Eastman; Vice Pres. Devel. of The Olana Partnership, Bob Burns; Public Rels. & Admin. Olana Partnership, Nelson Sterner; Museum Shop Mgr., Rachel Patton.
Personnel Profile: Full-Time Paid 6; Part-Time Paid 15; Part-Time Volunteers 63; Interns 2.
Governing Authority: state. Parent Institution: New York State Office of Parks, Recreation & Historic Preservation. Tax-exempt: 501(c)(3).
Institution Type/Description: Historic House: 1870 Persian style villa residence of 19th-century landscape artist, Frederic E. Church.
Collections: 19th-century decorative arts, furnishings; 19th-century American art featuring Frederic E. Church; Old Masters; 19th-century photographs; church family library; archives; sketches, papers & correspondence; 19th-century planned landscape.
Research Fields: art of the Hudson River School; Frederic E. Church; Fritz Melbye; Camille Pissarro; art, decoration & furnishings; Romantic style landscape; 19th-century art culture & American art.
Facilities: research library including family & photo archives available for research with advance written notice; visitor center; 49-seat theater; garden; 5 mile walking trail; picnic facilities. Museum-related items for sale.
Activities: guided & self-guided tours on the landscape; guided tour of furnished 19th-century home & studio; video history; concerts; children's summer arts camp; docent program; formal education programs. Annual Events: May Symposium; August Arts Weekend; holiday reception.
Publications: biannual newsletter, The Crayon.
Hours & Admission Prices: Grounds: daily 8 to sunset. House Tours: April to late Nov. Tues.-Sun. & Mon. holidays 10-5; late Nov. to March Fri.-Sun. 11-4; groups by appointment. Standard House Tour: adults $9, seniors & students $8; children under 12 no charge. Second Floor & Gallery: adults $9, seniors & students $8; children under 12 no charge. Combined House & Second Floor: adults $12, seniors & students $10; members of the Olana Partnership no charge. Vehicle Fee: May-Oct. Sat.-Sun. $5. Closed New Year's Day; Easter; Thanksgiving; Christmas. &
Attendance: 33,495 (accurate)
Membership: Individual $30; Family $60; Business $50-$150; Contributing $75; Artist's Circle $150; Sponsor $500; Patron $1,000; Corporate Patron $1,500; Benefactor $5,000.

PARKER-O'MALLEY AIR MUSEUM, 435 Old Rte. 20, Hudson, NY 12075. Mailing Address: P.O. Box 216, Ghent, NY 12075-0216. Tel.: 518-392-7200. Fax: 518-392-2227.
Founded: 1991.
Key Personnel: C.E.O., James E. McMahon.
Personnel Profile: Part-Time Volunteers 2.
Governing Authority: private. Tax-exempt: 501(c)(3).
Institution Type/Description: Aeronautics Museum.
Collections: aircraft from the 1920s-1940s; paintings of historical aviation achievements.
Facilities: 13,000 sq. ft. hangar.
Hours & Admission Prices: By appointment only. Adults $5, children $3. &

Huntington

DAVID CONKLIN FARMHOUSE, 2 High St., Huntington, NY 11743-3416. Mailing Address: 209 Main St., Huntington, NY 11743-6907. Tel.: 631-427-7045. Fax: 631-427-7056.
E-mail: rkissam@huntingtonhistoricalsociety.org
Web Site: www.huntingtonhistoricalsociety.org
Founded: 1903.
Congressional District: 3
Key Personnel: Exec. Coord., Robert "Toby" Kissam; Pres. (V), Carl Lawrence.
Personnel Profile: Full-Time Paid 1; Part-Time Paid 7; Part-Time Volunteers 110.
Governing Authority: society; nonprofit. Parent Institution: Huntington Historical Society, 209 Main St., Huntington, NY 11743. Branch Museums: Huntington Trade School; Dr. Daniel W. Kissam House; Soldiers & Sailors Memorial Building. Tax-exempt: 501(c)(3).
Institution Type/Description: Historic Site.
Collections: period rooms furnished in Colonial, Empire, & Victorian styles; coverlets; quilts; textile preparation equipment; dolls; toys; kitchen items; pottery. Historic Buildings: c.1750 shingled house with added c.1820 2 1/2 story three-quarter house & c.1900 Victorian bay window; restored 18th-century barn.
Research Fields: decorative arts; genealogy; history of Huntington & central Long Island.

Facilities: 50-capacity auditorium; classroom.
Activities: guided tours; lectures; study clubs; hobby workshops; organized education programs; docent program; loan & permanent exhibitions; school loan service.
Publications: newsletters, The New Portico; Family History Newsletter; books on regional history.
Hours & Admission Prices: Fri. & Sun. 1-4; other times by appointment. Adults $5, students & seniors $3; discounts to AAM members; children under 5 & museum members no charge. Closed holidays. ⅍
Attendance: 10,500 (estimated)
Membership: Student & Senior Citizens $25; Individual $35; Family $55; Friend $100; Supporting $200; Heritage Friend $300.

DR. DANIEL W. KISSAM HOUSE, 434 Park Ave., Huntington, NY 11743. Mailing Address: 209 Main St., Huntington, NY 11743-6907. Tel.: 631-427-7045. Fax: 631-427-7056.
E-mail: rkissam@huntingtonhistoricalsociety.org
Web Site: www.huntingtonhistoricalsociety.org
Formerly: Kissam House
Founded: 1903.
Congressional District: 3
Key Personnel: Exec. Coord., Robert "Toby" Kissam; Pres. (V), Carl Lawrence.
Personnel Profile: Full-Time Paid 1; Part-Time Paid 7; Part-Time Volunteers 110.
Governing Authority: society; nonprofit. Parent Institution: Huntington Historical Society, 209 Main St., Huntington, NY 11743. Branch Museums: David Conklin Farmhouse; Huntington Trade School; Soldiers & Sailors Memorial Building. Tax-exempt: 501(c)(3).
Institution Type/Description: Historic House: c.1795 Federal shingled house with added c.1830 kitchen wing.
Collections: home of early Huntington physicians. Historic Structures: c.1787 three-bay English style barn; sheep shed; 2 smokehouses.
Research Fields: decorative arts; genealogy; local & central Long Island history.
Facilities: Museum-related items for sale.
Activities: guided tours; organized education programs; docent program; loan & permanent exhibitions; school loan service.
Publications: monthly newsletter, The New Portico; books on regional history.
Hours & Admission Prices: Sun. 1-4. Adults $5, children under 12 $3; discounts to AAM members; members no charge. Closed New Year's Day; Easter; Thanksgiving; Christmas. ⅍
Attendance: 10,500 (estimated)
Membership: Student & Senior Citizens $25; Individual $35; Family $55; Friend $100; Supporting $200; Heritage Friend $300.

＊　**THE HECKSCHER MUSEUM OF ART, (M),** 2 Prime Ave., Huntington, NY 11743-7702. Tel.: 631-351-3250. Fax: 631-423-2145.
E-mail: info@heckscher.org
Web Site: www.heckscher.org
Founded: 1920.
Congressional District: 2
Key Personnel: Chm., Beverly J. Bell; Exec. Dir. & C.E.O., Michael W. Schantz, Ph.D.; Deputy Exec. Dir., H.E. "Skip" Show, Jr.; Cur., Lisa Chalif; Registrar, William Titus; Dir. Education, Joy Weiner; Office Mgr., Nancy O'Brien; Dir. External Affairs, Nina M. Muller.
Personnel Profile: Full-Time Paid 12; Part-Time Paid 10; Part-Time Volunteers 72; Interns 13.
Operating Expenses: 1,281,773
Operating Income: 1,543,175
Governing Authority: nonprofit educational corporation. Tax-exempt: 501(c)(3).
Institution Type/Description: Art Museum: building built in 1920.
Collections: 16th-century to present European & American paintings, sculpture & works on paper, with emphasis on American artists of the 19th-20th centuries; works by Lucas Cranach, Moran family, Arthur Dove, Helen Torr; paintings & works on paper by George Grosz including Eclipse of the Sun; works by artists associated with Long Island, including Fairfield Porter, Elaine Dekooning, Ray Johnson, & the Tile Club; photography including works by Berenice Abbott, Larry Fink, Leon Levinstein, Man Ray, Eadweard Muybridge, Garry Winogrand.
Major Exhibits: Picture Perfect: Selections from the Permanent Collection, 8/17/13-4/27/14; Off the Wall: Sculpture from the Permanent Collection (T), 12/7/13-3/16/14; Rabble-rousers: Art, Dissent, and Social Commentary (T), 12/7/13-3/16/14; Rabble-rousers: Art, Dissent, and Social Commentary (T), 3/29/14-4/20/14; Long Island's Best, 3/29/14-4/27/14; Exposed: Eadweard Muybridge and the Study of Motion, 5/3/14-8/3/14; Rhythm and Repetition in 20th Century Art, 5/3/14-8/10/14; Richard Gachot's America,

8/16/14-11/23/14; Long Island Biennial 2014, 8/16/14-11/14; From Sea to Shining Sea, 12/6/14-3/8/15; Experiments in Photography, 12/6/14-3/15/15.
Research Fields: 19th & 20th century American & European artists; Long Island artists.
Activities: guided tours; lectures; gallery talks; formally organized education programs for children; inter-museum loan exhibitions; permanent & traveling exhibitions.
Publications: quarterly programming guides; catalogue of the collection; exhibition catalogues; educational brochures.
Hours & Admission Prices: Wed.-Fri. 10-4, 1st Fri. each month 4-8:30, Sat.-Sun. 11-5. Adults $8, seniors 62 & over $6, students 10 & over $5; discount to AAM members & Huntington residents; active military, veterans & their family, children under 10 & members no charge. Closed Thanksgiving; Christmas. ⅍
Attendance: 23,369 (accurate)
Membership: Student & Senior Citizen $35; Individual (Supporter) $40; Dual $65; Family $75; Fellow $175; Patron $500; Benefactor $1,000-$2,499; Director's Circle $2,500-$4,999; Chairman's Club $5,000 & up.

HUNTINGTON SEWING & TRADE SCHOOL, 209 Main St., Huntington, NY 11743-6907. Tel.: 631-427-7045. Fax: 516-427-7056.
E-mail: cmaguire@huntingtonhistoricalsociety.org
Web Site: www.huntingtonhistoricalsociety.org
Founded: 1903.
Congressional District: 3
Key Personnel: Exec. Dir., Carol A. Maguire; Pres. (V), Kevin Arloff.
Personnel Profile: Full-Time Paid 1; Part-Time Paid 7; Part-Time Volunteers 110.
Governing Authority: society; nonprofit. Parent Institution: Huntington Historical Society. Branch Museums: David Conklin Farmhouse; Dr. Daniel W. Kissam House; Soldiers & Sailors Memorial Building. Tax-exempt: 501(c)(3).
Institution Type/Description: Historical Building: c.1905, 2 1/2 story Jacobean Revival Trade School designed by Cady, Berg & See.
Collections: maps; diaries; family & business records & papers; genealogies; local census & newspaper microfilm manuscripts; regional photographs.
Research Fields: genealogy; Huntington history & central Long Island; photography.
Facilities: 3,500-vol. library pertaining to genealogy, local & Long Island history; reading room; classrooms. Museum-related items for sale.
Activities: guided tours; lectures; concerts; study clubs; organized education programs; docent program; loan & permanent exhibitions; school loan service.
Publications: The New Portico; Family History Newsletter.
Hours & Admission Prices: Office: Mon.-Fri. 9:30-5. Archives & Library: Wed.-Thurs. 1-4. Adult $4; discounts to AAM members; members no charge. Call for special exhibit fees. Closed New Year's Day; Easter; Thanksgiving; Christmas. ⅍
Attendance: 10,500 (estimated)
Membership: Student & Senior Citizens $25; Individual $35; Family $55; Friend $100; Supporting $200; Heritage Friend $300.

SOLDIERS & SAILORS MEMORIAL BUILDING, 228 Main St., Huntington, NY 11743-6915. Tel.: 631-427-7045. Fax: 631-427-7056.
E-mail: rkissam@huntingtonhistoricalsociety.org
Web Site: www.huntingtonhistoricalsociety.org
Formerly: Huntington Historical Society
Founded: 1903.
Congressional District: 3
Key Personnel: Exec. Coord., Robert "Toby" Kissam; Pres. (V), Carl Lawrence.
Personnel Profile: Full-Time Paid 1; Part-Time Paid 7; Part-Time Volunteers 40.
Governing Authority: nonprofit organization. Parent Institution: Huntington Historical Society. Branch Museums: Huntington Trade School; David Conklin Farmhouse; Dr. Daniel Kissam House. Tax-exempt: 501(c)(3).
Institution Type/Description: History Museum: housed in 1892 Civil War Memorial building & 1st community library.
Collections: costumes; textiles; decorative arts; paintings; prints.
Research Fields: local history; genealogy & decorative arts.
Facilities: 3,500-vol. library & 25,000 photographs available in the Huntington Trade School.
Activities: guided tours; lectures; formally organized education programs; docent program; intermuseum, permanent & temporary exhibitions.
Publications: The New Portico Newsletter, Huntington/Babylon Town History, Index to the Long Islander & others.
Hours & Admission Prices: Call for hours.

Attendance: 12,000 (estimated)
Membership: Student $15; Senior Citizens $25; Individual $30; Family $50; Small Business $60; Sustaining $90; Contributing $120; Corporate $150; Sponsoring $250.

Huntington Station

WALT WHITMAN BIRTHPLACE STATE HISTORIC SITE AND INTERPRETIVE CENTER, 246 Old Walt Whitman Rd., Huntington Station, NY 11746-4148. Tel.: 631-427-5240. Fax: 631-427-5247. Facebook: Walt Whitman Birthplace Association.
E-mail: director@waltwhitman.org
Web Site: www.waltwhitman.org
Founded: 1949.
Congressional District: 5
Key Personnel: Exec. Dir. & Public Rels., Cynthia Shor; Pres. (V), William Walter; Museum Shop Mgr., Sue Anne Dennehy; Archivist, Richard A. Ryan.
Personnel Profile: Full-Time Paid 1; Part-Time Paid 12; Part-Time Volunteers 50; Interns 2.
Governing Authority: state. New York State Office of Parks, Recreation & Historic Preservation, & Long Island State Park, Recreation & Historic Preservation Commission. Subsidiary Institution: Walt Whitman Birthplace Assoc. Tax-exempt: 501(c)(3).
Institution Type/Description: Historic House: c.1819 birthplace & early home of Walt Whitman.
Collections: period furnishings; literary materials; Whitman memorabilia; 130 portraits of Walt Whitman; original letters, manuscripts & artifacts; writing desk used by Whitman; first editions of the books Leaves of Grass & Specimen Days; Whitman's voice on tape.
Research Fields: Walt Whitman; his life & works.
Facilities: 500-vol. library; visitor center; 1,500 sq. ft. exhibit space; classroom; performance space; bookstore. Museum-related items for sale.
Activities: guided tours; permanent & changing exhibits; poetry readings; poet-in-residence; education programs; lectures; audiovisual presentation. Annual Events: Marathon Reading of Leaves of Grass; Annual Walt Whitman Birthday Celebration; Walking With Whitman - Poetry in Performance.
Publications: newsletter, Starting from Paumanok; Tour to Whitmanland.
Hours & Admission Prices: June 15 to Labor Day Mon.-Fri. 11-4, Sat.-Sun. 11-5; Winter: Wed.-Fri. 1-4, Sat.-Sun. 11-4. Adults $6, senior citizens & students $5; discounts to AAM & ICOM members; children under 18, members and children 5 & under no charge. Closed New Year's Day; Easter; Thanksgiving; Christmas. &
Attendance: 16,500 (accurate)
Membership: Student & Senior over 65 $20; Individual $30; Family $40; Friend $50; Patron $100; Sponsor $250; Associate $500; Benefactor $1,000; Poets Circle $2,500; Board Circle $5,000.

Hurleyville

SULLIVAN COUNTY HISTORICAL SOCIETY, INC., 265 Main St., Hurleyville, NY 12747-0247. Mailing Address: P.O. Box 247, Hurleyville, NY 12747-0247. Tel.: 845-434-8044. Fax: 845-434-8044.
E-mail: schs@sullivancountyhistory.org
Web Site: www.sullivancountyhistory.org
Founded: 1929.
Congressional District: 27
Key Personnel: Pres. (V), William F. Burns; Finance Officer, Jacqueline Decker; Genealogy, Barbara Viele; Cur. & Archivist, Arthur Hessinger; Museum Shop Mgr., W.F. Burns.
Personnel Profile: Part-Time Volunteers 22.
Governing Authority: society; nonprofit organization. Tax-exempt: 501(c)(3).
Institution Type/Description: Historical Society Museum.
Collections: O & W Railroad; ledgers; D&H Canal; textiles & clothing; Sullivan County maps; manuscripts; Stephen Crane collection; historical Sullivan County postcards; Judge Lawrence Cooke Room; Dr. Frederick A. Cook Room-polar explorer; tanneries; personal diaries, ephemera, papers & documents; all federal & state census records & abstracts; microfilm records of all county newspapers from 1838-present; genealogy & local history.
Research Fields: genealogy.
Facilities: 300-vol. library of material pertaining to the development of Sullivan County, available for research by written request; 200-seat auditorium.
Activities: guided tours; docent program; permanent & temporary exhibitions; genealogy & local history research. Museum Sponsors: Minisink Battlefield Commemoration in July.
Publications: quarterly newsletter, Observer; History of Sullivan County.
Hours & Admission Prices: Museum: Tues.-Sat. 10-4:30, Sun. 1-4:30. No

charge; donations accepted; Archives: Wed. 10-4:30. Modest fee for extended genealogical research; veterans & spouses no charge. Closed New Year's Day; Thanksgiving; Christmas. &
Attendance: 8,500 (accurate)
Membership: Single $20; Couple $30; Corporate $100.

Hyde Park

ELEANOR ROOSEVELT NATIONAL HISTORIC SITE, 54 Val Kill Park Rd., Hyde Park, NY 12538-1917. Tel.: 877-444-6777. Fax: 845-229-0739.
Web Site: www.nps.gov/elro/
Founded: 1977.
Congressional District: 24
Key Personnel: Museum Shop Mgr., Anne Meisner.
Personnel Profile: Full-Time Paid 18; Part-Time Paid 2; Interns 1.
Governing Authority: federal. National Park Service, Dept. of the Interior, Washington, DC 20240. Tax-exempt.
Institution Type/Description: Historic House: Val-Kill, home of Eleanor Roosevelt from 1945-1962.
Collections: Eleanor Roosevelt's original & replacement furnishings.
Activities: guided tours; introductory movie. Museum Sponsors: Special Christmas exhibit in December.
Publications: site brochure.
Hours & Admission Prices: May-Oct. Thurs.-Mon. 9-5; Nov.-April Thurs.-Mon. 1 & 3. Adults $10; children under 15 no charge. Reservation for tours call 800-967-2283. Closed New Year's Day; Thanksgiving; Christmas. &
Attendance: 58,000 (accurate)

FRANKLIN D. ROOSEVELT PRESIDENTIAL LIBRARY-MUSEUM, (M), 4079 Albany Post Rd., Hyde Park, NY 12538-1999. Tel.: 800-337-8474; 845-486-7770. Fax: 845-486-1147. Facebook: FR Library.
E-mail: roosevelt.library@nara.gov
Web Site: www.fdrlibrary.marist.edu
Founded: 1940.
Congressional District: 24
Key Personnel: Dir., Lynn Bassanese; Supervisory Cur., Herman Eberhardt; Supervisory Archivist, Robert Clark; Education Specialist, Jeff Urbin; Public Affairs Specialist, Cliff Laube; Museum Store Mgr., Amy Northup.
Personnel Profile: Full-Time Paid 21; Part-Time Paid 7; Part-Time Volunteers 19; Interns 8.
Governing Authority: federal. Parent Institution: National Archives & Records Administration. Tax-exempt: 170(b)(1)(A).
Institution Type/Description: Presidential Library.
Collections: life and times of Franklin & Eleanor Roosevelt, specifically FDR's first fifty years, the Presidential Years, WWII, & life and accomplishments of Eleanor Roosevelt; President Roosevelt's White House Oval Office desk; FDR's Library Study; FDR's 1936 Ford Phaeton; personal papers; government records; photographs; motion picture films; audio & video tapes; sound recordings; head of state gifts; gifts from private citizens; political campaign items; personal & family memorabilia; President Roosevelt's collection on U.S. naval history.
Research Fields: life, career & presidential administration of President Franklin D. Roosevelt; life & career of Mrs. Roosevelt; Great Depression and New Deal; World War II; Historic Hudson Valley, New York.
Facilities: research library; visitors center; education & conference areas; theater; cafe. Museum-related items for sale.
Activities: organized educational programs for elementary, secondary school & undergraduate students; permanent, temporary & interactive exhibitions; lectures; film series; special events; elder hostels.
Publications: general information brochures; books, Franklin D. Roosevelt & Conservation; Franklin D. Roosevelt & Foreign Affairs; bibliography, The Era of Franklin D. Roosevelt, 1945-1971; bibliography, Historical Materials in the Franklin D. Roosevelt Library, 1997; microfilm, Press Conferences of Franklin D. Roosevelt; microfilm, Roosevelt-Churchill Messages, microfilm Diary of Adolf A. Berle; microfilm, Henry A. Wallace Papers, 1941-1945.
Hours & Admission Prices: April-Oct. daily 9-6; Nov.-March daily 9-5. Combination Museum & Roosevelt Home: adults $14; discounts to Golden Age & Golden Eagle passports and AAM & ICOM members; AASLH members with advance reservation, children 16 & under and school groups no charge. Closed New Year's Day; Thanksgiving; Christmas. &
Attendance: 110,000 (accurate)

HOME OF FRANKLIN D. ROOSEVELT NATIONAL HISTORIC SITE, 4097 Albany Post Rd., Hyde Park, NY 12538. Tel.: 845-229-9115 & 9116. Fax: 845-229-0739.
Web Site: www.nps.gov/hofr

Founded: 1946.
Congressional District: 24
Key Personnel: Supt., Sarah Olson; Museum Shop Mgr., Anne Meisner.
Personnel Profile: Full-Time Paid 18; Part-Time Paid 3; Part-Time Volunteers 4.
Governing Authority: federal. Unit of the National Park Service, Dept. of the Interior, Washington, DC 20240. Branch Museum: Vanderbilt Mansion. Tax-exempt.
Institution Type/Description: Historic House: Hudson River home of F.D.R.
Collections: original furnishings including ancestral portraits; naval prints; memorabilia.
Facilities: library; Rose Garden and Gravesite; The Henry A. Wallace Visitor & Education Center; cafe; auditorium. Museum-related items for sale.
Publications: pamphlets, F.D. Roosevelt and Hyde Park; Art in the Home of Franklin D. Roosevelt; Springwood: The Grounds of the Home of Franklin Delano Roosevelt National Historic Site.
Hours & Admission Prices: Daily 9-5. Adults $14; discounts to Golden Age Passport & National Park Pass holders; children under 15 & school groups no charge. Reservation required for motorcoach tours & school groups, call 800-967-2283. Closed New Year's Day; Thanksgiving; Christmas. &
Attendance: 105,000 (accurate)

VANDERBILT MANSION NATIONAL HISTORIC SITE, 119 Vanderbilt Park Rd., Hyde Park, NY 12538. Mailing Address: 4097 Albany Post Rd., Hyde Park, NY 12538-1917. Tel.: 800-337-8474. Fax: 845-229-0739.
Web Site: www.nps.gov/vama
Founded: 1940.
Congressional District: 24
Key Personnel: Supt., Sarah Olson; Museum Shop Mgr., Anne Meisner.
Personnel Profile: Full-Time Paid 18; Part-Time Paid 3; Part-Time Volunteers 18.
Governing Authority: federal. National Parks Service, Dept. of the Interior, Washington, DC 20240. Branch Museum: Home of Franklin D. Roosevelt. Tax-exempt.
Institution Type/Description: Historic House Museum: 1896-1899 Beaux-Arts style Mansion, designed by McKim, Mead, & White.
Collections: Beaux-Arts interiors by turn-of-the-century decorators include furniture; tapestries; rugs; porcelains.
Facilities: coachhouse; formal garden restoration; view of Hudson River. Slides, books & postcards for sale.
Activities: guided tours; slide video.
Publications: Vanderbilt Mansion handbook; Vanderbilt Photo Essay Book; The Grounds at the Vanderbilt Mansion; brochure, Vanderbilt Mansion NHS Site.
Hours & Admission Prices: April-Oct. call for hours; Nov.-March Fri.-Tues. 10, 12, 2, 4. Adults $10; discounts to National Park Golden Age & Access card holders; children 15 & under and school groups no charge. Closed New Year's Day; Thanksgiving; Christmas. &
Attendance: 359,000 (accurate)

Ilion

REMINGTON FIREARMS MUSEUM AND COUNTRY STORE, 14 Hoefler Ave., Ilion, NY 13357-1888. Tel.: 315-895-3200 & 3301. Fax: 315-895-3543.
E-mail: john.balio@remington.com
Web Site: www.remington.com
Founded: 1959.
Congressional District: 23
Key Personnel: Cur. & Archives Coord., Fred Supry; Museum Shop Mgr., John Balio.
Personnel Profile: Full-Time Paid 1; Part-Time Paid 2.
Governing Authority: profit-making organization. Parent Institution: Remington Arms Co. Inc.
Institution Type/Description: Firearms Museum.
Collections: Remington firearms & artifacts from 1820s to the present; wildlife & related art.
Research Fields: early & current firearms manufactured by the Remington Arms Company, Inc.; factory tour of custom gun shop.
Facilities: Custom gun shop.
Activities: custom gun shop tour.
Publications: brochure.
Hours & Admission Prices: May-Oct. Mon.-Fri. 8-5; Nov.-April Mon.-Fri. 8-4. Tours: May-Oct. Mon.-Fri. 10 & 1. No charge. Closed major holidays. &
Attendance: 50,000 (estimated)

Interlaken

INTERLAKEN HISTORICAL SOCIETY, Main St., Interlaken, NY 14847. Mailing Address: P.O. Box 270, Interlaken, NY 14847-0270. Tel.: 607-532-4213.
E-mail: orchardland@zoom-dsl.com
Web Site: www.interlakenhistory.org
Congressional District: 27
Key Personnel: Pres. (V), Diane Bassette Nelson; Vice Pres., Bill Schaffner; Sec., Ann Buddle; Treas., Karen King.
Personnel Profile: Full-Time Volunteers 9; Part-Time Volunteers 36.
Governing Authority: society. Tax-exempt: 170(b)(1)(A).
Institution Type/Description: Historical Society & Farmers Museum: housed in 1826 Lockwood Hinman House and relocated grain cradle factory.
Collections: tools; photographs; manuscripts; genealogy.
Research Fields: local history, genealogy.
Facilities: 200-vol. library of local history books available for research on premises.
Activities: guided tours; permanent & temporary exhibitions.
Publications: quarterly newsletter.
Hours & Admission Prices: July-Aug. Sun. 1-3, Sat. 10-1; other times by appointment. No charge; donations accepted. &
Attendance: 400 (estimated)
Membership: Adult & Family $10; Life $100.

Irvington

IRVINGTON HISTORICAL SOCIETY & HISTORY CENTER - THE MCVICKAR HOUSE, 131 Main St., Irvington, NY 10533. Mailing Address: P.O. Box 23, Irvington, NY 10533-0023. Tel.: 914-591-1020.
Web Site: www.irvingtonhistoricalsociety.org
Institution Type/Description: Historical Society Museum: housed in the former home of Rev. William McVickar, first rector of the Church of St. Barnabas; built in 1853. Listed on the National Register of Historic Places.
Collections: local history; period furnishings & artifacts; photographs.
Publications: newsletter, The Roost.
Hours & Admission Prices: Thurs. & Sat. 1-4. No charge. &

Ithaca

THE CORNELL PLANTATIONS, One Plantations Rd., Ithaca, NY 14850-2799. Tel.: 607-255-2400. Fax: 607-255-2404.
E-mail: plantations@cornell.edu
Web Site: www.cornellplantations.org
Founded: 1935.
Congressional District: 27
Key Personnel: Dir., Donald A. Rakow; Dir. Education, Sonja Skelly; Dir. Devel., Beth Anderson.
Personnel Profile: Full-Time Paid 30; Part-Time Paid 21; Part-Time Volunteers 85; Interns 7.
Governing Authority: university. Cornell University, Roberts Hall. Tax-exempt: 501(c)(3).
Institution Type/Description: Arboretum.
Collections: arboretum; botanical; horticultural; outdoor laboratory; natural area; interpretive trail, Plantations Path.
Research Fields: horticulture; natural sciences; education; conservation.
Facilities: Books & museum-related items for sale.
Activities: guided tours; lectures; seminars; publications; short courses; education festival; concerts.
Publications: magazine, The Cornell Plantations; informative & interpretive materials; newsletter; Path Handbook; Path video.
Hours & Admission Prices: Daily sunrise-sunset. Fee for guided tours; by appointment.
Attendance: 200,000 (estimated)
Membership: Basic Membership $50.

HANDWERKER GALLERY, 1170 Gannett Center, Ithaca College, Ithaca, NY 14850-7276. Tel.: 607-274-3018. Fax: 607-274-1774. Facebook: Handwerker Gallery.
E-mail: handwerker@ithaca.edu
Web Site: www.ithaca.edu/handwerker
Founded: 1977.
Key Personnel: Dir., Mara Baldwin.
Governing Authority: Parent Institution: Ithaca College. Tax-exempt.
Institution Type/Description: Art Gallery & College Museum.
Collections: contemporary work of faculty, students & local artists; historical shows.
Research Fields: new museology.

Activities: critical forum; video screenings; movies.
Publications: catalogues & newsletter.
Hours & Admission Prices: Mon., Wed. & Fri. 10-6, Thurs. 10-9, Sat.-Sun. 12-5. No charge. Closed vacation breaks. &
Attendance: 8,900 (estimated)

✳ HERBERT F. JOHNSON MUSEUM OF ART, (M), Cornell University, Ithaca, NY 14853. Tel.: 607-255-6464. Fax: 607-255-9940.
E-mail: museum@cornell.edu
Web Site: www.museum.cornell.edu
Founded: 1973.
Congressional District: 27
Key Personnel: Dir., Stephanie L. Wiles; Chm. (V), Ira Drukier; Assoc. Dir. Finance Admin., Peter Gould; Chief Cur. & Cur. Asian Art, Ellen Avril; Senior Cur., Nancy Green; Cur. Modern & Contemporary Art, Andrea Inselmann; Ames Assoc. Dir. & Cur. Education, Cathy Klimaszewski; Coord. School & Children's Programs, Carol Hockett; Coord. Public Programs, Sara Ferguson; Registrar, Matthew Conway; Chief of Security, Alvin Miller; Chief Preparator, Wil Millard; Dir. Devel., Matt Braun; Publications Coord., Andrea Potochniak; Cur. European Art Before 1800, Andrew Weislogel; Mellon Asst. Coord. of Univ. Programs, Kari O'Mara.
Personnel Profile: Full-Time Paid 19; Part-Time Paid 15; Part-Time Volunteers 35; Interns 19.
Governing Authority: university. Affiliated with Cornell University. Tax-exempt: 701(b)(1)(A).
Institution Type/Description: Art Museum.
Collections: Asian art; European and American painting; sculpture; prints; drawings; photographs; art from Africa, Oceania & the Americas.
Facilities: 4,000-vol. library of art books and exhibitions catalogs; 80-seat auditorium; reading room.
Activities: permanent, temporary & traveling exhibitions, educational programs; lectures; guided tours; film series.
Publications: newsletters; annual report; exhibition brochures & catalogs; educational publications; collection research.
Hours & Admission Prices: Tues.-Sun. 10-5. No charge; donations accepted. Closed Memorial Day; Independence Day; Labor Day; Thanksgiving; Christmas to New Year's Day. &
Membership: Individual $20-$44; Family $45-$99; Supporting $100-$249; Sustaining $250-$499; Charter $500-$999; Quadrangle $1,000-$4,999; Tower $5,000 & up.

THE HISTORY CENTER IN TOMPKINS COUNTY, (M), 401 E. State St., Ste. 100, Ithaca, NY 14850-4400. Tel.: 607-273-8284. Fax: 607-273-6107.
E-mail: welcome@thehistorycenter.net
Web Site: www.thehistorycenter.net
Formerly: DeWitt Historical Society of Tompkins County & Tompkins County Museum
Founded: 1935.
Congressional District: 26
Key Personnel: Exec. Dir., Scott Callan; Pres. Bd. (V), Elizabeth Bixler; Photographer, Carl Koski; Visitor Svcs. Mgr., Catherine Duffy; Archivist, Donna Eschenbrenner; Elementary Education, Carole West; Adult & Secondary Education, Paul Miller; Research Librarian, Mary Williams.
Personnel Profile: Full-Time Paid 1; Part-Time Paid 7; Part-Time Volunteers 30; Interns 5.
Governing Authority: society. Subsidiary Institution: Tompkins County Museum. Tax-exempt: 501(c)(3).
Institution Type/Description: Historical Society Museum.
Collections: 19th-20th century Tompkins County history; photographs including Verne Morton collection of 11,000 rural life images, 1897-1945; costumes; quilts; coverlets; books, documents; maps; tools; toys; sports memorabilia; art; Native American artifacts. Historic Building: 1827 octagonal school building.
Research Fields: local history; genealogy.
Facilities: 4,000-vol. research library; archives. Museum-related items for sale.
Activities: guided tours; lectures; films; formally organized education programs; inter-museum loan; permanent & temporary exhibitions; school loan service; living history program.
Publications: quarterly newsletter; books & monographs relating to area.
Hours & Admission Prices: Tues., Thurs. & Sat. 11-5. Museum: no charge; donations accepted. Research Library: call for fees. Closed New Year's Day; Thanksgiving; Christmas. &
Attendance: 6,000 (estimated)
Membership: Donors up to $99; Friends $100-$249; Researchers $250-$499; Historians $500-$999; Leadership Circle $1,000-$4,999; Director's Circle $5,000-$9,999; Chairman's Circle $10,000 & up.

L.H. BAILEY HORTORIUM, 412 Mann Library, Cornell University, Ithaca, NY 14853-4301. Tel.: 607-255-2131. Fax: 607-255-5407.
Web Site: www.plantbio.cornell.edu
Founded: 1935.
Congressional District: 27
Key Personnel: Chm. & Professor, William Crepet; Assoc. Professor, Melissa Luckow; Professor, David M. Bates; Professor, Eloy Rodriguez; Adjunct Professor, James L. Reveal; Assoc. Professor & Cur., Kevin C. Nixon; Assoc. Professor, Jerrold I. Davis; Professor, Jeffrey J. Doyle; Assoc. Cur., Anna M. Stalter; Asst. Cur. & Librarian, Peter Fraissinet.
Personnel Profile: Full-Time Paid 20; Part-Time Volunteers 2.
Governing Authority: university. Affiliated with Cornell University. Tax-exempt.
Institution Type/Description: Horticulture & Botanical Museum.
Collections: vascular plants; 860,000 sheet herbarium; 132,000 nursery & seed catalogue collection; natural history; botany; 30,000-vol. library; paleobotanical artifacts.
Research Fields: taxonomy of cultivated & economic plants; systematic botany; paleobotany.
Facilities: 30,000-vol. library of botanical books available for inter-library loan & for research.
Activities: guided tours; lectures; formally organized education programs for undergraduate & graduate college students; plant identification & information; plant locating service.
Hours & Admission Prices: Mon.-Fri. 9-5. No charge. Closed national holidays. &
Attendance: 1,000 (estimated)

McGRAW HALL MUSEUM AND ANTHROPOLOGY COLLECTIONS, (M), 150 McGraw Hall, Cornell University, Ithaca, NY 14853-4601. Tel.: 607-254-8688.
Web Site: anthropology.cornell.edu/about/collection/index.cfm
Founded: 1872.
Key Personnel: Cur., Frederic W. Gleach.
Governing Authority: Parent Institution: Cornell University.
Institution Type/Description: Anthropology Museum.
Collections: 20,000 ethnographic & archaeological objects including pre-Columbian textiles & pottery from Peru; Native American baskets; prehistoric Mississippian pottery; Ndembu ritual masks & costumes from Africa; Yir Yoront tools & weapons from Australia; Egyptian mummy cases.
Activities: teaching support.
Hours & Admission Prices: Call for hours.
Attendance: 500 (estimated)

MUSEUM OF THE EARTH, 1259 Trumansburg Rd., Ithaca, NY 14850-1398. Tel.: 607-273-6623, ext. 26. Fax: 607-273-6620.
E-mail: marketing@museumoftheearth.org
Web Site: www.museumoftheearth.org
Formerly: Museum of the Earth and Cayuga Nature Center
Founded: 1932.
Congressional District: 26
Key Personnel: Dir., Dr. Warren D. Allmon; Dir. Exhibits, Beth Stricker; Assoc. Dir. Institutional Advancement, Elizabeth Brando; Dir. Collections, Greg Dietl; Assoc. Dir. Science, Paula Mikkelsen; Assoc. Dir. Outreach, Robert Ross; Chm. (V), Linda Irany; Museum Shop Mgr., Alicia Michael.
Personnel Profile: Full-Time Paid 30; Part-Time Volunteers 175.
Volunteer Hours: 11,525
Governing Authority: nonprofit organization. Parent Institution: Paleontological Research Institution. Tax-exempt: 501(c)(3).
Institution Type/Description: Natural History Museum.
Collections: paleontology.
Research Fields: paleontology; geology; STEM education; citizen science; earth science education.
Facilities: 50,000-vol. library of science material available for use on premises by special permission; nature center; 120 acre hiking trails.
Activities: lectures; educational programs; permanent & temporary exhibitions; workshops; classes; special events; hiking trails.
Publications: Bulletins of American Paleontology; The Science Beneath the Surface: A Very Short Guide to the Marcellus Shale.
Hours & Admission Prices: Memorial Day to Labor Day Mon.-Sat. 10-5, Sun. 11-5; Sept.-May Mon. & Thurs.-Sat. 10-5, Sun. 11-5. Adults $8, seniors & college students $5, children 4-17 $3; discounts to ASTC members; members and children 3 & under no charge. Closed New Year's Day; Thanksgiving; Christmas. &
Attendance: 30,000 (estimated)
Membership: Student & Senior $30; Individual $40; Family & Household $65.

✳ **SCIENCENTER, (M),** 601 First St., Ithaca, NY 14850-3507. Tel.: 607-272-0600. Fax: 607-277-7469.
E-mail: info@sciencenter.org
Web Site: www.sciencenter.org
Founded: 1983.
Congressional District: 26
Key Personnel: Exec. Dir., Charles H. Trautmann; Chm. Bd., G. Walton Cottrell; Assoc. Dir., Lara Kimber; Dir. Visitor Svcs. & Operations, Kerry Flannery; Dir. Education, Rae Ostman; Exhibits Mgr., Robin Burlingham; Business Mgr., Brian Gold; Gift Shop Mgr. & Front Desk Coord., Susan Trask.
Personnel Profile: Full-Time Paid 17; Full-Time Volunteers 3; Part-Time Paid 4; Part-Time Volunteers 350; Interns 3.
Governing Authority: private; not-for-profit organization. Tax-exempt: 501(c)(3).
Institution Type/Description: Science & Technology Museum.
Collections: over 250 hands-on science & technology exhibits; outdoor science park.
Facilities: 9,000 sq. ft. outdoor science playground.
Activities: school classes; demonstrations; birthday parties; facility rentals; seasonal mini science golf course; 1,200-meter outdoor walking model of Solar System; school outreach; amphitheater presentation.
Publications: Passport to the Solar System $2.
Hours & Admission Prices: July-Aug. Mon.-Sat. 10-5, Sun. 12-5; Sept.-June Tues.-Sat. 10-5, Sun. 12-5. Adults $7, senior citizens $6, children 3-17 $5; discount to AAM members; members, reciprocal ASTC & ACM members, children under 3 & members no charge. Closed New Year's Day; Thanksgiving; Christmas. ♿
Attendance: 94,000 (accurate)
Membership: Senior $25; Individual $45; Family $60; Family Plus & Grandparents $75; Voyager Society $125.

Jamaica

JAMAICA CENTER FOR ARTS & LEARNING (JCAL), 161-04 Jamaica Ave., Jamaica, NY 11432-6112. Tel.: 718-658-7400, ext. 123. Fax: 718-658-7922.
E-mail: info@jcal.org
Web Site: www.jcal.org
Founded: 1972.
Congressional District: 6
Key Personnel: Acting Exec. Dir., Anita Romero-Segarra; Bd. Chm., Leilani M. Brown; Dir. Finance & Admin., Jennifer Chiang; Cur. Visual Arts, Heng-Gil Han.
Personnel Profile: Full-Time Paid 13; Part-Time Paid 4; Part-Time Volunteers 5; Interns 4.
Governing Authority: nonprofit organization. Tax-exempt: 501(c)(3).
Institution Type/Description: Multidisciplinary Art Center.
Collections: works by international artists.
Facilities: 99-seat theatre; performance studios.
Activities: arts workshops; after school performing & visual arts program; concerts; films; commissioned art work for public spaces; guided tours; temporary exhibitions.
Publications: spring & fall newsletter; exhibit announcements; catalogs; posters; brochures; rack cards.
Hours & Admission Prices: Galleries: Mon.-Sat. 10-6. Fees vary. Closed major national holidays. ♿
Attendance: 25,000 (estimated)

KING MANOR MUSEUM, (M), 150-03 Jamaica Ave., King Park, Jamaica, NY 11432. Mailing Address: 9004 161st St., Ste. 704, Jamaica, NY 11432-6101. Tel.: 718-206-0545. Fax: 718-206-0541.
E-mail: director@kingmanor.org
Web Site: www.kingmanor.org
Founded: 1900.
Congressional District: 6
Key Personnel: C.E.O., Mary Anne Mrozinski; Pres. (V), Gerald J. Caliendo.
Personnel Profile: Full-Time Paid 2; Part-Time Paid 1; Part-Time Volunteers 25; Interns 3.
Governing Authority: board of directors, King Manor Association. Building owned by city of NY; administered by department of parks. Tax-exempt: 501(c)(3).
Institution Type/Description: Historic Building/Site; History Museum: Now the centerpiece of an 11-acre historic park, King Manor was the home of Rufus King from 1805-1827. King was a signer of the U.S. Constitution, senator from NY State and ambassador to Great Britain under four presidents, and outspoken opponent of slavery. 1819 & 1820, King led the Senate debate against admission of Missouri as a slave state.
Collections: 18th & 19th-century furniture, decorative arts, memorabilia including some related to the King family, local community of Jamaica.
Research Fields: local & national roles of Rufus King & his son John Alsop King in the early anti-slavery movement; life & work at King Manor; archaeology of slavery & freedom; Jamaica in early 19th century.
Facilities: education center.
Activities: guided tours; formally organized education programs; permanent exhibits; special events; public programs; concerts; lecture series.
Hours & Admission Prices: Thurs.-Fri. 12-2, Sat.-Sun. 1-5; groups of 10 or more by appointment. Suggested Admission: adults $5, seniors & students $3; discounts to groups, AAM, ICOM and NYC Historic House Trust members; children 16 & under and members no charge.
Attendance: 6,155 (accurate)
Membership: Friend $25; Family $45; Delegate $100; Senator $250; Ambassador $500.

Jamestown

FENTON HISTORY CENTER-MUSEUM & RESEARCH CENTER, (I), 67 Washington St., Jamestown, NY 14701-6697. Tel.: 716-664-6256.
E-mail: information@fentonhistorycenter.org
Web Site: www.fentonhistorycenter.org
Founded: 1964.
Congressional District: 27
Key Personnel: Pres. (V), Michael T. Rohlin; Dir., Joni Blackman; Dir. Education, Frances Fair; Mgr. Collections, Norman P. Carlson; Archivist, Karen E. Livsey; Membership, Office & Gift Shop Mgr., Paula Bechmann.
Personnel Profile: Full-Time Paid 2; Part-Time Paid 4; Part-Time Volunteers 1.
Governing Authority: nonprofit organization. Tax-exempt: 501(c)(3).
Institution Type/Description: History Museum: housed in 1863 Italian villa style mansion of Reuben E. Fenton, U.S. Congressman, Senator, NY. State Governor & founder of NYS Republican Party.
Collections: local history.
Research Fields: genealogy; local history.
Facilities: 7,000-vol. library of local history & genealogy available on premises.
Activities: guided tours; educational programs; genealogy support group; Jr. historians (grades 1-3); history club (grades 4-6); adult history club (grades 7 & up); long-term & temporary exhibits; self-guided tours of museum & community walking tours.
Publications: book, Chautauqua Lake Steamboats; book, Trolleys & Railroads of Chautauqua County; quarterly newsletter, book, Saga from the Hills: A History of the Swedes in Jamestown; book, Electricity & Politics; book, Chautauqua Lake Hotels; book, The Civil War letters of Wm. Depledge; book, Camp James M. Brown: Jamestown's Civil War Rendezvous; A History of Chautauqua County, NY.
Hours & Admission Prices: Mon.-Sat. 10-4. Adults $8, children 5-12 $4; discounts to AAA, AAM & ICOM members; children under 5 & members no charge. Closed New Year's Day; Memorial Day; Independence Day; Labor Day; Thanksgiving; Christmas.
Attendance: 17,500 (accurate)
Membership: Student $15; Senior $30; Individual $40; Family $60; Supporting $100; Friend $200; Sustaining $500; Benefactor $1,000.

JAMES PRENDERGAST LIBRARY ASSOCIATION, ART GALLERY, 509 Cherry St., Jamestown, NY 14701-5098. Tel.: 716-484-7135. Fax: 716-487-1148.
E-mail: prendergastlibrary@yahoo.com
Web Site: www.prendergastlibrary.org
Founded: 1880.
Congressional District: 39
Key Personnel: Acting Dir., Tina A. Scott; Gallery Coord., Anne Plyler.
Governing Authority: municipal. Tax-exempt: 501(c)(3).
Institution Type/Description: Art Museum & Gallery.
Collections: art books; art reproductions; 19th-century American & European paintings.
Facilities: circulating collection of art books.
Activities: films; permanent, temporary & traveling exhibitions; recitals.
Publications: catalogue, The Mirror Up to Nature.
Hours & Admission Prices: Mon.-Fri. 9-8:30, Sat. 9-5, Sun. 1-5. No charge; donations accepted. ♿
Attendance: 5,599 (accurate)

LUCILE M. WRIGHT AIR MUSEUM, 300 N. Main St., Jamestown, NY 14701-5109. Tel.: 716-664-9500.
E-mail: lmwairmuseum@windstream.net
Web Site: www.lucilemwrightairmuseum.org
Founded: 1986.
Personnel Profile: Part-Time Paid 2.
Governing Authority: nonprofit organization. Parent Institution: NYS Education Dept. Tax-exempt: 501(c)(3).

Institution Type/Description: Aviation History Museum.
Collections: aviation history & aircraft; 1951 Piper Pacer; mini-500 experimental helicopter; 1947 North American Aviation Navion Model A; Great Lakes biplane trainer.
Activities: special events.
Hours & Admission Prices: Tues.-Sat. call for hours. No charge; donations accepted. Closed holidays. &
Attendance: 500 (accurate)

LUCILLE BALL DESI ARNAZ CENTER FOR COMEDY, (M), 2 W. Third St., Jamestown, NY 14701. Tel.: 716-484-0800. Fax: 716-484-9373.
E-mail: info@lucy-desi.com
Web Site: www.lucy-desi.com; www.lucycomdyfest.com
Formerly: Lucille Ball-Desi Arnaz Center
Founded: 1996.
Key Personnel: Exec. Dir., Journey Gunderson; Pres. (V), Tom Benson.
Personnel Profile: Full-Time Paid 3; Part-Time Paid 7.
Governing Authority: Tax-exempt.
Institution Type/Description: General Museum.
Collections: photographs; wardrobe; personal artifacts; family portraits; Wildcat memorabilia, the 1961 musical that was Lucille Ball's only Broadway Show.
Facilities: Museum-related items for sale.
Activities: special events. Museum Sponsors: Annual Festivals in May & August; Legacy of Laughter Seminars.
Publications: quarterly members newsletter, Lucy-Desi News.
Hours & Admission Prices: Mon.-Sat. 10-5, Sun. 11-4. Adults $15, seniors $14; members no charge. &
Attendance: 20,000 (estimated)
Membership: Cast $50; Director $125; Producer $250; Executive Producer $500.

ROGER TORY PETERSON INSTITUTE OF NATURAL HISTORY, (M), 311 Curtis St., Jamestown, NY 14701-9620. Tel.: 716-665-2473; 800-758-6841. Fax: 716-665-3794.
E-mail: information@rtpi.org
Web Site: www.rtpi.org
Founded: 1984.
Congressional District: 34
Key Personnel: Pres., James M. Berry; Chm., John Rappole; Treas., Bruce Rowan; Museum Shop Mgr., Linda Pierce.
Personnel Profile: Full-Time Paid 6; Part-Time Paid 6; Part-Time Volunteers 142; Interns 2.
Governing Authority: private; nonprofit organization. Tax-exempt.
Institution Type/Description: Natural History Museum & Art Gallery: building was designed by Robert A.M. Stern.
Collections: concentrate on environmental education & environmental issues; wildlife art & photography; artifacts of naturalists; the life's work of Roger Tory Peterson.
Research Fields: environmental education; nature art.
Facilities: 4,000-vol. library of natural history & environmental education; 27 acre site; nature trails. Museum-related items for sale.
Activities: lectures; artists receptions; formal educational programs; loan exhibitions. Annual Events: Birding Festival; Banff Mountain Film Festival World Tour event; forum topics related to environmental education.
Publications: bi-monthly newsletter, The Peterson Journal; The Peterson Journal for Teaching Nature Education.
Hours & Admission Prices: Tues.-Sat. 10-4, Sun. 1-5. Family $12, adults $5, children $3; members no charge. Closed national holidays. &
Attendance: 12,747 (accurate)
Membership: Friend $35 & up.

THE WEEKS GALLERY AT JAMESTOWN COMMUNITY COLLEGE, 525 Falconer St., Jamestown, NY 14701-1920. Mailing Address: P.O. Box 20, Jamestown, NY 14702-0020. Tel.: 716-338-1300. Fax: 716-338-1451.
E-mail: weeksgallery@mail.sunyjcc.edu
Web Site: weeksgallery.sunyjcc.edu
Formerly: The CCC Weeks Gallery
Founded: 1969.
Congressional District: 34
Key Personnel: Dir. & Cur., Patricia Briggs; Arts Admin., Collin Shaffer.
Personnel Profile: Full-Time Paid 1; Interns 1.
Governing Authority: nonprofit. Parent Institution: Jamestown Community College, Jamestown. Tax-exempt: 501(c)(3).

Institution Type/Description: Art Gallery.
Collections: paintings; sculpture; photographs.
Research Fields: contemporary art.
Facilities: auditorium; classroom; 1,000 sq. ft. exhibit space.
Activities: formal education programs for adults, undergraduate & graduate college students; lectures.
Publications: catalogs for selected exhibitions.
Hours & Admission Prices: Mon.-Thurs. 11-5, Fri. 11-3; Summer: Tues.-Fri. 11-4. No charge. Closed college holidays. &
Attendance: 8,500 (estimated)

Johnstown

JOHNSON HALL STATE HISTORIC SITE, 139 Hall Ave., Johnstown, NY 12095-1615. Tel.: 518-762-8712. Fax: 518-762-2330.
E-mail: wade.wells@parks.ny.gov
Web Site: www.nysparks.com
Founded: 1763.
Congressional District: 31
Key Personnel: Historic Site Mgr., Wade Wells; Pres. (V), Heidi Meka; Museum Shop Mgr., Wanda Burch.
Personnel Profile: Full-Time Paid 2; Part-Time Paid 3; Part-Time Volunteers 4.
Governing Authority: New York State Office of Parks, Recreation & Historic Preservation, Saratoga Capital District. Tax-exempt: 501(c)(3).
Institution Type/Description: Historic site; Historic House Museum: 1763 home of Sir William Johnson, Superintendent of Indian Affairs of the Six Nations Confederacy.
Collections: period furnishings; Johnson family possessions. Historic Structure: stone outbuilding.
Research Fields: Sir William Johnson; Mohawk Valley history & settlement; Loyalists in the American Revolution; 18th century Native American culture & diplomacy.
Facilities: orientation center; herb garden; picnic area.
Activities: guided tours; lectures; gallery talks; school programs; special programs & demonstrations. Annual Event: Market Fair in July.
Publications: brochures.
Hours & Admission Prices: Call for hours. Adults $4, seniors & groups $3, children 12 & under $1; members no charge. Closed most holidays. &
Attendance: 13,000
Membership: Friends of Johnson Hall: email for information at friendsofjohnsonhall@gmail.com.

JOHNSTOWN HISTORICAL SOCIETY, 17 N. William St., Johnstown, NY 12095-2115. Tel.: 518-762-7076.
Founded: 1892.
Congressional District: 31
Key Personnel: Pres. (V), Catherine Levee; Cur., James F. Morrison.
Personnel Profile: Full-Time Volunteers 2; Interns 5.
Governing Authority: society; nonprofit organization. Tax-exempt: 501(c)(3); 101-6.
Institution Type/Description: Historical Society Museum.
Collections: local history & historical notables associated with Johnstown: Sir William Johnson; Molly Brant; Fife-Major Nicholas Stoner; Capt. Silas Talbot; Lafayette; Washington Irving; Governor Enos Throop; E.L. Henry; Aaron Burr; Brigadier-Gen. Edgar S. Dudley; Grace Livingston Hill; Rose M. Knox; Elizabeth Cady Stanton; Keck Zouaves; Knox gelatin; glove industry equipment; Cooper's barrel making equipment. Historic House: 1763 restored schoolmaster's house.
Research Fields: local history; historic notables associated with Johnstown; genealogy.
Facilities: 1,600-vol. library of old books, newspaper clippings, genealogical material; 2,300 photographs, 43 maps, 105 scrapbooks & 1,350 documents for researching early information about Johnstown & vicinity, available for research on premises or can be loaned under special circumstances; reading room.
Activities: guided tours; lectures; permanent & temporary exhibitions.
Publications: quarterly, newsletter; The Johnstown Historical Society; book, Now I am Ninety.
Hours & Admission Prices: Memorial Day to Labor Day Sat.-sun. 1-4; other times by appointment. No charge; donations accepted. Closed New Year's Day; Memorial Day; Independence Day; Labor Day; Thanksgiving; Christmas.
Attendance: 794 (accurate)
Membership: Students & Senior Citizens $5; Individual $7.50; Family $15; Patron $25; Life $100; Benefactor $1,000.

Katonah

CARAMOOR CENTER FOR MUSIC & THE ARTS, INC., (M),
149 Girdle Ridge Rd., Katonah, NY 10536-3815. Mailing Address:
Box 816, Katonah, NY 10536-0816. Tel.: 914-232-5035, ext. 221.
Fax: 914-232-5521.
E-mail: museum@caramoor.org
Web Site: www.caramoor.org
Founded: 1946.
Congressional District: 21
Key Personnel: Chm. Bd. (V), Jim Attwood; C.E.O., Jeffrey P. Haydon;
Managing Dir., Paul Rosenblum; Vice Pres. & Chief Devel. Officer, Nina
Curley; Museum Mgr. & Dir. Programs, Merceds Santos-Miller; Archivist,
Hilton Bailey; Vice Pres. & C.F.O, Tammy Belanger; Dir. Education, Scott
Ellison; Exec. Asst. & Bd. Liaison, Afton Battle; Dir. Annual Giving,
Alithia Dutschke; Dir. Individual Gifts, Talia Bennick; Dir. Special Events
& Rentals, Christine Bosco; Dir. Mktg., Sal Vaccard.
Personnel Profile: Full-Time Paid 22; Part-Time Paid 14; Part-Time Volunteers
220; Interns 4.
Governing Authority: nonprofit. Subsidiary Institution: Caramoor Rosen
House. Tax-exempt: 501(c)(3).
Institution Type/Description: House Museum: housed in 54 room Mediterra-
nean style villa.
Collections: Renaissance & 17th-19th century furniture; sculpture; painting;
decorative arts; Chinese jade & porcelains; cloisonne works; Chinese hand
painted wallpapers; Oriental carpets; period textiles; chinoiserie; 16th
century Maiolica; Caldwell collection; Theremin instruments & music
scores.
Research Fields: decorative arts; 20th-century American social history, painting & sculpture; Renaissance art & furnishings; Theremin.
Facilities: theater; gardens; landscaped grounds.
Activities: guided tours; lectures; concerts; arts festivals; docent program;
temporary & permanent exhibitions; education programs for school children; afternoon teas.
Publications: biannual newsletter; brochures; pamphlets; books, A Guide to the
Collections of Caramoor; Caramoor.
Hours & Admission Prices: May-Oct. Wed.-Sun. 1-4, last tour at 3 except
during festival on Sat. 1-5, last tour at 4; Nov.-April Tues.-Fri. by
appointment only. Adults $10; discounts to AAM members; members,
children 16 & under no charge. &
Attendance: 60,000 (estimated)
Membership: Member $100; Participant $300; Donor $600; Patron $1,500;
Conductor $3,000; Composer $6,000; Director $12,000; Chairman's Circle
$25,000.

JOHN JAY HOMESTEAD STATE HISTORIC SITE, 400 Rte.
22, Katonah, NY 10536. Mailing Address: P.O. Box 832, Katonah,
NY 10536-0832. Tel.: 914-232-5651. Fax: 914-232-8085. TDD:
845-889-4100.
E-mail: heather.iannucci@oprhp.state.ny.us
Web Site: www.nysparks.com
Founded: 1958.
Congressional District: 25
Key Personnel: C.E.O. & Historic Site Mgr., Heather Iannucci; Interpretive
Programs Asst., Allan Weinreb.
Personnel Profile: Full-Time Paid 4; Part-Time Paid 2; Part-Time Volunteers
34.
Governing Authority: state. Parent Institution: New York State Office of Parks,
Recreation and Historic Preservation. Tax-exempt: 501(c)(3).
Institution Type/Description: Historic House Museum: federal style country
home reflects the residence of U.S. Chief Justice & New York state
Governor John Jay & family living here in the first third of the 19th century.
Collections: portraits; Jay family personal artifacts including c.1800-1833
American Art, furniture & decorative arts; archives & library; 12 structures
& archaeological sites; grounds & outbuildings are interpreted through the
1930's; home, parts of original farmstead, restored gardens & 62 acres of
the original 750 accessible to visitors.
Research Fields: American history; art history; Jay family; Revolutionary &
Civil War history; slavery & abolition; period farms; 19th-century social
history; New York State history; coaching.
Facilities: 4,000-vol. historic family library, personal papers & manuscripts
available for use by advance request; formal gardens & grounds with trails;
picnic area; lecture facility.
Activities: self-guided walking tour of the site & gardens; guided tours;
lectures; concerts; permanent exhibitions; lecture series; educational pro-
grams; group tours; public events & programs. Annual Event: Farm Market
June to October.
Publications: John Jay; The Jays of Bedford; The Jay Genealogy; newsletter,
Friends of John Jay Newsletter; The Herb Garden, A Walking Tour; The
Farm of Mr. Jay, A Walking Tour.

Hours & Admission Prices: April-Nov. Tues.-Sat. 10-4, Sun. 11-4. Grounds:
daily 8-dusk. Adults $7, senior citizens $5; Friends and children 12 & under
no charge. &
Attendance: 60,721 (accurate)
Membership: Friends of John Jay Homestead: Student $25; Individual $35;
Family $70; Contributing $100; Sponsor $250; Patron $500; Benefactor
$1,000; Statesman $2,500; Founder $5,000.

✱ **KATONAH MUSEUM OF ART, (M),** 134 Jay St., Katonah,
NY 10536-3737. Tel.: 914-232-9555, ext. 0. Fax: 914-232-3128.
E-mail: info@katonahmuseum.org
Web Site: www.katonahmuseum.org
Founded: 1953.
Congressional District: 25
Key Personnel: Pres. (V) Bd. Trustees, Rochelle C. Rosenberg; Exec. Dir., Neil
Watson; Dir. Education, Karen Stein; Registrar, Nancy Hitchcock; Dir.
Special Events, Allison Chernow; Coord. Learning Center, Naomi Leiser-
off.
Personnel Profile: Full-Time Paid 11; Part-Time Paid 19; Part-Time Volunteers
187; Interns 12.
Governing Authority: nonprofit organization. Tax-exempt.
Institution Type/Description: Art Museum.
Collections: paintings; photographs; sculptures.
Research Fields: pertaining to varied exhibition schedule.
Facilities: sculpture garden; children's learning center.
Activities: guided tours; lectures; films; formally organized education pro-
grams for schools; school loan service; docent training; teacher workshops;
art trips for members; children's workshops; summer jazz concerts.
Publications: catalogues, The American Eagle: Spirit & Symbol, 1782-1882;
American Painting, 1900-1976; Utopia: The Art of William Morris & His
Circle; Dreams in Motion: The Art of Windsor McCay; Le Fantastique
Reel: Graphic Works by Odilon Redon; Forever Wild: The Adirondack
Experience; George Rickey: The Art & Movement; The Intimate Eye of
Edouard Vuillard; Sticks & Stones: Ten Artists Work with Nature; The
Technological Muse; Tradewinds: The Lure of the China Trade; Watercol-
ors from the Abstract Expressionism Era; Asobi: Play in the Arts of Japan;
Dorothy Dehner: Sixty Years of Art; Friends & Family: Portraiture in the
World of Florine Stettheimer; Block/Plate/Stone: What a Print Is; Shelter &
Dreams: Playhouses by Architects & Artists; Against the Stream: Milton
Avery, Adolph Gottlieb & Mark Rothko in the 1930's; Vladimir Tytla:
Master Animator; Medieval Monsters: Dragons & Fantastic Creatures; The
Reconstructed Figure: The Human Image in Contemporary Art; At Home
with Art: Paintings in American Interiors, 1780-1920; Object as Insight:
Japanese Buddhist Art & Ritual; Toying with Architecture; Revisiting
American Art: Works from the Collections of Historically Black Colleges
and Universities; Polish Posters: Combat on Paper, 1960-1990; Pavel
Tchelitchew: The Landscape of the Body; Latin America Still Life in the
Twentieth Century; Horse Tales: American Images and Icons 1800-2000;
The Innocent Eye: American Folk Art from Colonial Williamsburg; The
Human Comedy; Edward Giobbi: Paintings; Food Matters: Explorations in
Contemporary Art; The Birth of the Banjo; Sol LeWitt: Recent Work;
catalogues, I Love The Burbs; Eternal Presence: Handprints and Footprints
in Buddhist Art; Richard Diebenkorn Prints 1948-1993.
Hours & Admission Prices: Tues.-Sat. 10-5, Sun. 12-5. Adults $5, students &
seniors $3; discounts to AAM & WNET Channel 13 members; members,
children under 12 & 10am to noon no charge. &
Attendance: 40,000 (estimated)
Membership: Individual $50; Dual & Family $85; Museum Supporter $150;
Curator's Circle $375; Friend's Circle $500; Director's Circle $1,000;
President's Circle $2,500; Exhibition Patron $5,000.

STEPPING STONES, (M), 62 Oak Rd., Katonah, NY 10536-1810.
Tel.: 914-232-4822. Fax: 914-232-2580.
E-mail: info@steppingstones.org
Web Site: www.steppingstones.org
Founded: 1979.
Congressional District: 89
Key Personnel: Dir., Annah Perch; Pres. (V), Michael Kelly.
Personnel Profile: Full-Time Paid 1; Part-Time Paid 1; Part-Time Volunteers 8.
Governing Authority: private; nonprofit organization. Tax-exempt.
Institution Type/Description: Historic House Museum: housed in the former
home of Bill & Lois Wilson, co-founders of Alcoholics Anonymous and
Al-Anon.
Collections: Bill & Lois Wilson's personal artifacts; AA & Al-Anon archives.
Facilities: 200-vol. library. Museum-related items for sale.
Activities: guided tours. Annual Event: picnic.
Publications: annual newsletter, Stepping Stones News.
Hours & Admission Prices: Daily by appointment. Suggested Donations: $10
per person. Closed Easter; Thanksgiving; Christmas.

Attendance: 1,800 (accurate)

Kinderhook

COLUMBIA COUNTY HISTORICAL SOCIETY, INC., 5 Albany Ave., Kinderhook, NY 12106-0311. Mailing Address: P.O. Box 311, Kinderhook, NY 12106-0311. Tel.: 518-758-9265. Fax: 518-758-2499.
E-mail: cchs@cchsny.org
Web Site: www.cchsny.org
Founded: 1916.
Congressional District: 16
Key Personnel: Exec. Dir., Diane Shewchuk; Pres., David Smith; Cur., Diane Shewchuk.
Personnel Profile: Full-Time Paid 2; Part-Time Paid 6; Part-Time Volunteers 50.
Governing Authority: society. Tax-exempt: 501(c)(3).
Institution Type/Description: Historic House Museums: c.1820 Vanderpoel House of History; 1737 Luykas Van Alen House; c.1850 Ichabod Crane School House (interpretation date: c.1925). History Museum: Columbia County Museum.
Collections: New York Federal furnishings, decorative & fine arts; New York Dutch furnishings, decorative & fine arts; Columbia County documents & historical artifacts; genealogical; costumes & textiles.
Research Fields: county & regional art; decorative arts; folk art; architecture; local & regional history; county architecture; genealogy.
Facilities: 3,000-vol. library of county & state historical reference & collections, manuscripts, documents & photographs.
Activities: guided tours; lectures; inter-museum loan, permanent & temporary exhibits; concerts; educational program; docent program; workshops; school programs.
Publications: magazine, Columbia County History & Heritage; books; exhibit catalogs.
Hours & Admission Prices: Museum: Thurs.-Fri. 10-4, Sat.-Sun. 12-4. Adults $5, senior citizens $2. Historic Properties: Memorial Day to Columbus Day Sat.-Sun. 12-4. Adults $5, students & senior citizens 55 & over $3; discounts to AAM & ICOM members; children under 12 & members no charge. Group tours May-Oct. by appointment.
Attendance: 24,500 (accurate)
Membership: Senior Citizen $30; Senior (dual) $40; Individual $50; Family $60; Patron $100; Sponsor $250; Benefactor $500; Second Century Circle $1,000.

MARTIN VAN BUREN NATIONAL HISTORIC SITE, Rt. 9H, Kinderhook, NY 12106. Mailing Address: 1013 Old Post Rd., Kinderhook, NY 12106-3605. Tel.: 518-758-9689. Fax: 518-758-6986.
Web Site: www.nps.gov/mava
Founded: 1974.
Congressional District: 20
Key Personnel: C.E.O., Dan Dattilio; Cur., Dr. Patricia West; Chief Interpretation, James A. McKay.
Personnel Profile: Full-Time Paid 11; Part-Time Paid 5; Part-Time Volunteers 1; Interns 1.
Governing Authority: federal. Parent Institution: National Park Service. Tax-exempt.
Institution Type/Description: Historic House: home of President Martin Van Buren.
Collections: home furnishings; personal effects of President Martin Van Buren.
Research Fields: political history; Pres. Van Buren; Van Buren Home.
Facilities: 2,000-vol. research library; microfilm material.
Activities: mansion tours & special events during summer months.
Publications: brochure.
Hours & Admission Prices: mid-May to Oct. daily 9-4. Family $12, adults $5; children under 16 & members no charge. &
Attendance: 21,000 (accurate)

Kings Point

AMERICAN MERCHANT MARINE MUSEUM, United States Merchant Marine Academy, Kings Point, NY 11024-1699. Mailing Address: US Merchant Marine Academy, 300 Steamboat Road, Kings Point, NY 11024-1699. Tel.: 516-773-5515. Fax: 516-482-5340.
E-mail: smithj@usmma.edu
Web Site: www.usmma.edu/about/museum/default.htm
Founded: 1978.
Key Personnel: Chm. (V), Capt. Warren Leback; Interim Coord., Joshua M. Smith.

Personnel Profile: Full-Time Paid 1; Part-Time Paid 1; Part-Time Volunteers 8; Interns 1.
Governing Authority: federal.
Institution Type/Description: Maritime Museum: American Merchant Marine.
Collections: ship models, paintings & artifacts relating to American Merchant Marine; National Maritime Hall of Fame; period nautical instrument collection; Sperry navigation wing; restored emery rice steam engine; re-creation of the 1945 victory ship Radio Room; Japanese sword from WWII; ship passenger lists; immigration records.
Research Fields: American Merchant Marine WWII to present; history of the American shipping companies.
Facilities: library.
Activities: guided tours; organized education programs for children; temporary exhibitions; lectures. Museum Sponsors: Annual Bowditch Award; induction ceremonies for National Maritime Hall of Fame; annual reunion of American merchant marine veterans; reunions for former employees of U.S. shipping companies.
Hours & Admission Prices: Aug. to mid-June Tues.-Fri. 10-3, Sat.-Sun. 1-4:30; groups by appointment. No charge; donations accepted. Closed federal holidays. &
Attendance: 4,000 (estimated)
Membership: Basic $30; $40; $50; $100; $500; Patron $1,000; Corporate $1,000 & up.

Kingston

FRIENDS OF HISTORIC KINGSTON, FRED J. JOHNSTON HOUSE MUSEUM, 63 Main St., Kingston, NY 12401-3801. Mailing Address: P.O. Box 3763, Kingston, NY 12402-3763. Tel.: 845-339-0720.
E-mail: fohk@verizon.net
Web Site: www.fohk.org
Founded: 1965.
Congressional District: 26
Key Personnel: Dir., Jane Kellar; Pres. (V), Patricia Murphy.
Personnel Profile: Full-Time Paid 1; Part-Time Paid 1; Part-Time Volunteers 40.
Governing Authority: private; nonprofit organization. Tax-exempt: 501(c)(3).
Institution Type/Description: History Museum.
Collections: local history; period rooms.
Research Fields: architectural history of local buildings.
Facilities: 200-vol. library; 500 sq. ft. exhibit space. Museum-related items for sale.
Activities: docent program; formal education programs for children; gallery talks; guided tours; lectures; temporary exhibitions.
Publications: newsletter, Friends of Historic Kingston; walking tour brochures.
Hours & Admission Prices: May-Oct. Sat.-Sun. 1-4; other times by appointment. Adults $5, children $2; members no charge. &
Attendance: 4,100 (estimated)
Membership: Single $30; Family $55; Patron $500; Benefactor $1,000-$5,000.

HUDSON RIVER MARITIME MUSEUM, (M), 50 Rondout Landing, Kingston, NY 12401-6092. Tel.: 845-338-0071, ext. 12. Fax: 845-338-0583.
E-mail: pmcdonough@hrmm.org
Web Site: www.hrmm.org
Founded: 1980.
Congressional District: 19
Key Personnel: Pres., Allan Bowdery; Exec. Dir., Patrick McDonough; Cur., Allynne Lange.
Personnel Profile: Full-Time Paid 1; Part-Time Paid 5; Part-Time Volunteers 30; Interns 2.
Governing Authority: nonprofit organization. Tax-exempt: 501(c)(3).
Institution Type/Description: Maritime Museum.
Collections: Hudson River steamboats & towing industry history; Hudson River lighthouses; Ruge ice boats; photographs; brickmaking & ice-cutting artifacts & documents; ship & boatbuilding industries; Anton Otto Fischer maritime paintings; boat models; recreation of Henry Hudson's aft cabin in the Half Moon.
Research Fields: Hudson River maritime history; commerce; industries; commercial & recreational vessels; Hudson River lighthouses.
Facilities: 450-vol. library on Hudson River steam & sail vessels; archives of regional river industries. Books, prints & other related items for sale.
Activities: permanent & temporary exhibits; group tours; research; educational programs for children; college intern program; waterfront events; ecology workshops; annual lecture series.
Publications: newsletter, Focs'le News; calendar of events; brochure; annual, Pilot Club Log; book, Focs'le Days; Thomas Cornell & The Cornell Steamboat Company.
Hours & Admission Prices: April-Nov. daily 11-5; other times by appointment.

Museum: adults $7, seniors & children $5; discounts to active military & CAMM members; members no charge. Blue Star Museum. &

Attendance: 20,000 (estimated)

Membership: Individual $25; Household $40; Friend $50; Sustaining $100-$249; Patron $250-$499; Fellow $500-$999; Pilot $1,000.

SENATE HOUSE STATE HISTORIC SITE, 296 Fair St., Kingston, NY 12401-3836. Tel.: 845-338-2786. Fax: 845-334-8173.
E-mail: thomas.kernan@parks.ny.gov
Web Site: www.nysparks.com
Founded: 1887.
Congressional District: 25
Key Personnel: Mgr. Historic Site, Thomas Kernan.
Personnel Profile: Full-Time Paid 4; Part-Time Paid 5; Part-Time Volunteers 25; Interns 1.
Governing Authority: state. Parent Institution: New York State Office of Parks, Recreation and Historic Preservation, & Palisades Interstate Park Commission. Subsidiary Institution: Friends of Senate House, Inc. Tax-exempt: 501(c)(3).
Institution Type/Description: Historic Site.
Collections: period furnishings; paintings by John Vanderlyn, Ammi Phillips, James Bard & other regional artists; archives; genealogy. Historic Houses: c.1676 Senate (Van Gaasbeek) House; 1873 Louhgran House.
Research Fields: state government; John Vanderlyn; Dutch culture; regional history.
Facilities: 1,000-vol. library of historical and genealogical material 40,000 item manuscript collection available for research on premises.
Activities: guided tours; lectures; films; gallery talks; concerts; special programs; permanent, traveling & temporary exhibitions.
Publications: brochures.
Hours & Admission Prices: April 15-Oct. Wed.-Sat. 10-5, Sun. 1-5; other times by appointment. Adults $4, seniors & groups $3; children under 12 no charge. Closed most holidays. &
Attendance: 25,000 (estimated)
Membership: Individual $20; Joint $30; Family $40; Supporting $50-$99; Contributor $100-$499; Patron $500-$999; Benefactor $1,000 & up.

TROLLEY MUSEUM OF NEW YORK, (M), 89 E. Strand, Kingston, NY 12401-6001. Mailing Address: P.O. Box 2291, Kingston, NY 12402-2291. Tel.: 845-331-3399.
E-mail: admin@tmny.org
Web Site: tmny.org
Founded: 1955.
Congressional District: 26
Key Personnel: Pres. (V), Jon McGrew; Archivist, Evan Jennings; Treas., William Brandt; Admin., Steve Ladin; Museum Shop Mgr., Glendon Moffett.
Personnel Profile: Part-Time Paid 1; Part-Time Volunteers 20.
Governing Authority: private; nonprofit organization. Tax-exempt: 501(c)(3).
Institution Type/Description: Transportation Museum.
Collections: trolley; subway cars; artifacts; photographs.
Facilities: 10,000 sq. ft. exhibit space; 30-seat theater. Museum-related items for sale.
Activities: Annual Events: Trolley 5K Run; National Trails Day Clean-up; Halloween Fright Train; Trolley rides with Santa in December.
Publications: bimonthly email newsletter; volunteer bulletin.
Hours & Admission Prices: early May to mid-Oct. Sat.-Sun. & holidays 12-5. Also open for special events. Adults $6, senior citizens 62 & over and children 6-12 $4; discounts to AAM members; children 6 & under no charge.
Attendance: 5,000 (estimated)
Membership: Individual $20; Family $30; Friend $60; Patron $120; Corporate $150; Fellow $250; Motorman's Club $500; President's Club $1,000.

ULSTER COUNTY HISTORICAL SOCIETY, 2682 State Rte. 209, Kingston, NY 12401. Mailing Address: P.O. Box 279, Stone Ridge, NY 12484-0279. Tel.: 845-338-5614.
E-mail: director@bevierhousemuseum.org
Web Site: www.ulstercountyhistoricalsociety.org
Founded: 1939.
Congressional District: 96
Key Personnel: Dir., Jessica Phinney; Pres., Suzanne Hauspurg.
Personnel Profile: Part-Time Paid 2; Part-Time Volunteers 5.
Governing Authority: nonprofit organization. Tax-exempt.
Institution Type/Description: Bevier House: stone farm house owned by the Bevier family from 1715-1938; registered county historic landmark & National Historic Register.
Collections: furniture; paintings; ceramics; decorative arts; Civil War items; tools.

Facilities: 250-vol. library of monographs, periodicals (on local history) & Ulster County manuscripts from 1670 to c.1880 available by appointment.
Activities: guided tours; lectures; occasional loan exhibitions.
Publications: newsletter published three times annually, Ulster County Gazette.
Hours & Admission Prices: May-Oct. Thurs.-Sun. 12-5. Adults $5, seniors $4, children $3; AAM & ICOM members no charge.
Attendance: 500 (estimated)
Membership: Student & Senior $15; Individual $25; Household $40; Patron $50.

VOLUNTEER FIREMEN'S MALL AND MUSEUM OF KINGSTON, 265 Fair St., Kingston, NY 12401-3807. Mailing Address: P.O. Box 1501, Kingston, NY 12402-1501. Tel.: 845-331-0866. Facebook: Volunteer Firemen's Mall and Museum of Kingston.
E-mail: vfmuseumofkingston@gmail.com
Founded: 1980.
Congressional District: 19
Key Personnel: Pres. (V), Billy J. Knowles.
Personnel Profile: Part-Time Volunteers 23.
Governing Authority: private; nonprofit organization. Tax-exempt: 501(c)(3).
Institution Type/Description: Fire-Fighting Museum.
Collections: personal artifacts; furnishings; firemanic artifacts; fire engines; hose carriages.
Hours & Admission Prices: April-May & Sept.-Oct. Fri. 11-3, Sat. 10-4; June-Aug. Wed.-Fri. 11-3, Sat. 10-4. No charge; donations accepted.
Attendance: 2,500 (estimated)

LaFargeville

NORTHERN NEW YORK AGRICULTURAL HISTORICAL SOCIETY, AGRICULTURAL MUSEUM AT STONE MILLS, 30950 NYS Rte. 180, LaFargeville, NY 13656. Mailing Address: P.O. Box 108, LaFargeville, NY 13656-0108. Tel.: 315-658-2353. Facebook: Stone Mills Agricultural Museum.
E-mail: agstonemills@yahoo.com
Web Site: www.stonemillsmuseum.org
Founded: 1968.
Congressional District: 30
Key Personnel: Dir., Gail Marsh; Dir., Tom Gardner; Pres. (V), Michael LaDue
Governing Authority: society. Tax-exempt: 501(c)(3).
Institution Type/Description: Agricultural History Museum Complex.
Collections: early farm & home equipment; old maps, books, records & slates; early milk handling equipment; Home Arts Exhibit Hall; Carriage House for large agriculture equipment; authentic Ice House. Historic Buildings: 1836 Stone Meeting House; Cheese Factory; 1838 one-room schoolhouse; 1800s active blacksmith shop and wood shop; 1865 Saw Mill; restored granary working wind mill; 1900s sugar shack.
Research Fields: restoration of property; history of local agriculture.
Facilities: food service available on weekends.
Activities: permanent & temporary exhibitions. Museum Sponsors: annual three-day Craft Show with over 200 craftsmen demonstrating skills; Draft Horse Show & barbecues; Old Time Gas Engines; Tractor pulls; Sheep & Wool Festival; Quilt Show; Old Time Music; Blue Grass Festival; Horse Shows; Harvest Fest in September; Family Farmer Boy Day.
Publications: brochures.
Hours & Admission Prices: May-Sept. by appointment except events. Adult $5; discounts to AAM members; children 16 & under and life members no charge. &
Attendance: 10,000 (estimated)
Membership: Individual $15; Family $25; Business $100; Life $150.

Lake George

FORT WILLIAM HENRY MUSEUM, (M), 48 Canada St., Lake George, NY 12845-1600. Tel.: 518-668-5471 & 964-6647. Fax: 518-964-6659.
E-mail: kathryn@fortwilliamhenry.com
Web Site: www.fwhmuseum.com
Founded: 1952.
Key Personnel: Pres., Robert Flacke; C.F.O., Kathy Muncil.
Personnel Profile: Full-Time Paid 3; Part-Time Paid 1.
Governing Authority: profit-making organization. Affiliated with The Fort William Henry Corp.
Institution Type/Description: Historic Site Museum: restored c.1750 Fort William Henry.
Collections: artifacts from site, c.1750-1760; Colonial documents, 1750-1790; weapons; manuscripts.
Research Fields: colonial wars of Lake George region.

Facilities: 1,500-vol. library of maps; documents; books; newspapers pertaining to Colonial period available for research by approval of director; reading room. Museum-related items for sale.

Activities: guided tours; lectures; films; gallery talks; formally organized education programs; permanent, temporary & traveling exhibitions; school loan service.

Hours & Admission Prices: May-Oct. daily 9-5. Adults $16.95, senior citizens $13.95, children 3-11 $7.95; discount to groups; children under 5 & military with ID no charge.

Attendance: 43,150 (estimated)

HOUSE OF FRANKENSTEIN WAX MUSEUM, 213 Canada St., Lake George, NY 12845-1401. Tel.: 518-668-3377.

Web Site: frankensteinwaxmuseum.com

Institution Type/Description: Wax Museum.

Collections: wax figures.

Hours & Admission Prices: Call for hours. Adults $9.30, students 13-17 $8.36, children 6-12 $4.65.

LAKE GEORGE HISTORICAL ASSOCIATION MUSEUM, Old Warren County Courthouse, 290 Canada St., Lake George, NY 12845. Mailing Address: P.O. Box 472, Lake George, NY 12845-0472. Tel.: 518-668-5044.

E-mail: lgha@verizon.net

Web Site: lakegeorgehistorical.org

Founded: 1964.

Key Personnel: Pres. Bd. Trustees (V), Alex Parrott; Dir., Maggie McClure.

Personnel Profile: Part-Time Paid 2.

Governing Authority: Lake George Historical Association. Tax-exempt: 501(c)(3).

Institution Type/Description: Historical Society Museum: housed in 1845 Warren County Courthouse, located in the region of famous battles of the French & Indian War and American Revolution.

Collections: paintings; history; natural history; photographs; Indian artifacts; tramp art; boats; local history.

Research Fields: local history.

Facilities: 150-vol. library of Adirondack history; reading room; 1845 courthouse & courtroom for presentations; bookstore.

Activities: guided tours; lectures; films; permanent & temporary exhibitions; family activities; mid-1800s jail cells.

Publications: 1889 Seneca Ray Stoddard Map of Lake George.

Hours & Admission Prices: late May to mid-June & Oct. 1 to Columbus Day Sat.-Sun. 11-4; July-Aug. Tues. & Fri.-Sat. 11-4, Wed.-Thurs. 3-8; Sept. Fri.-Sun. 11-4. No charge.

Attendance: 3,500

Membership: Senior Citizen $5; Individual $15; Family $25; Business & Contributing $50; Patron $1,000.

Lake Placid

THE HISTORY MUSEUM, LAKE PLACID-NORTH ELBA HISTORICAL SOCIETY, 242 Station St., Lake Placid, NY 12946-1949. Mailing Address: P.O. Box 189, 242 Station St., Lake Placid, NY 12946-0189. Tel.: 518-523-1608. Facebook: Lake Placid History Museum.

E-mail: thehistorymuseum@verizon.net

Web Site: lakeplacidhistory.com

Founded: 1967.

Key Personnel: Administrative Dir., Jennifer Tufano; Pres., Peter Roland, Jr.; Treas., John B. Huttlinger, Jr.

Personnel Profile: Part-Time Paid 1; Part-Time Volunteers 5.

Governing Authority: society. Tax-exempt: 501(c)(3).

Institution Type/Description: Historic House: 1904 D&H train station.

Collections: general store replica; Adirondack Mountains & Lake Placid memorabilia; Victor Herbert & Kate Smith collections; railroad memorabilia; Dewey decimal system collection; rustic furniture; Adirondack Guideboat.

Major Exhibits: Developing Our History, Finding Our Family: The Stedman and Moses Collection of Historic Photographs, 5/13-10/14.

Facilities: library of brochures, newspapers, maps, pamphlets & photographs available for research on premises only.

Activities: guided tours; permanent & temporary exhibitions.

Publications: semiannual newsletter.

Hours & Admission Prices: Memorial Day to mid-June & Labor Day to mid-Sept. Sat.-Sun. 10-4; mid-June to Sept. & mid-Sept. to Columbus Day Wed.-Sun. 10-4. No charge; donations accepted.

Attendance: 1,685 (accurate)

Membership: Friend $25; Family $50; Sponsor $100; Patron $250; Silver

Benefactor $500; Gold Benefactor $1,000. Business: Sustaining $100; Sponsor $250; Patron $500.

JOHN BROWN FARM STATE HISTORIC SITE, 115 John Brown Rd., Lake Placid, NY 12946-3248. Tel.: 518-523-3900. Fax: 518-523-3951.

E-mail: brendan.mills@oprhp.state.ny.us

Web Site: nysparks.state.ny.us/sites/info.asp?siteid=14

Founded: 1895.

Congressional District: 30

Key Personnel: Historic Site Asst., Brendan Mills.

Governing Authority: state. New York State Office of Parks, Recreation & Historic Preservation, Thousand Islands State Park, Recreation & Historic Preservation Commission & New York State Dept. of Environmental Conservation. Tax-exempt: 501(c)(3).

Institution Type/Description: Historic House Museum: 1855 frame house occupied by John Brown & his family while he carried on his anti-slavery campaigns; the burial place of John Brown.

Collections: period furnishings; personal possessions of John Brown & his family.

Research Fields: John Brown & his family; regional history.

Facilities: farm trail.

Activities: special programs.

Hours & Admission Prices: May-Oct. Wed.-Mon. 10-5. Adults $2, seniors & children $1; children under 12 no charge. &

Attendance: 37,000 (accurate)

LAKE PLACID OLYMPIC MUSEUM, (M), 2634 Main St., Lake Placid, NY 12946. Mailing Address: P.O. Box 2002, Lake Placid, NY 12946. Tel.: 518-523-1655, ext. 226. Fax: 518-523-9275.

Formerly: 1932 & 1980 Lake Placid Winter Olympics Museum

Key Personnel: Dir., Alison Haas; Chm. (V), MaryLou Brown.

Personnel Profile: Full-Time Paid 3.

Governing Authority: Parent Institution: Olympic Regional Development Authority,

Institution Type/Description: Sports Museum.

Collections: Olympic history; athletes; uniforms; equipment; videos.

Major Exhibits: Paralympic Photography, 2/1-4/14.

Hours & Admission Prices: Daily 10-5. Adults $7, seniors & juniors $5; children 6 & under no charge. Closed Thanksgiving; Christmas. &

Attendance: 28,000 (estimated)

LeRoy

LEROY HOUSE & JELL-O GALLERY, 23 E. Main St., LeRoy, NY 14482-1209. Tel.: 585-768-7433. Fax: 585-768-7579.

E-mail: info@jellogallery.org

Web Site: www.jellomuseum.com

Founded: 1940.

Congressional District: 137

Key Personnel: Dir. & Cur., Lynne J. Belluscio; Pres. (V), Jim Newkirk; Museum Shop Mgr., Carolyn Bolin.

Personnel Profile: Full-Time Paid 1; Part-Time Paid 10; Interns 1.

Governing Authority: society. Parent Institution: LeRoy Historical Society. Tax-exempt: 501(c)(3).

Institution Type/Description: Historical Society Museum.

Collections: archives; paintings; decorative arts; textiles; agriculture; graphics; children's museum; costumes; glass; history; genealogy of local families; Jell-O memorabilia.

Research Fields: archives; history.

Facilities: 1,000-vol. library of books & genealogy manuscripts available by appointment.

Activities: guided tours; lectures; study clubs; formally organized education programs.

Publications: booklets; monthly newsletters; annual report; review of pertinent historical events & year's activities; genealogical booklets; quarterly newsletter.

Hours & Admission Prices: Leroy House: call for hours. Jell-O Gallery: Jan.-March Mon.-Fri. 10-4; April-Dec. Mon.-Sat. 10-4, Sun. 1-4. Adults $4, children 6-11 $1.50; discounts to groups, tours, AAM & AAA members; children 5 & under and members no charge. Closed New Year's Day; Thanksgiving; Christmas.

Attendance: 11,000 (estimated)

Membership: Individual $20; Family $30; Sustaining $50; Supporting $100; Life $350.

Lewiston

CASTELLANI ART MUSEUM OF NIAGARA UNIVERSITY, (M), 5795 Lewiston Rd., Lewiston, NY 14109. Mailing Address: P.O. Box 1938, Niagara University, NY 14109-1938. Tel.: 716-286-8200. Fax: 716-286-8289.
E-mail: kjk@niagara.edu
Web Site: www.niagara.edu/cam/
Founded: 1978.
Congressional District: 28
Key Personnel: Dir., Kate Koperski; Gallery & Installation Mgr., Kurt Von Voetsch; Cur. Collections & Exhibitions, Michael Beam; Cur. Folk Arts, Carrie Hertz; Registrar, Kathleen Fraas; Educ. Coord., Marian Granfield; Coord. Events, Public Rels. & Membership, Susan Clements; Museum Shop Mgr., Carla Castellani; Office Coord., Daphne Wyse; Weekend & Special Events Mgr., Celia Rodino.
Personnel Profile: Full-Time Paid 7; Part-Time Paid 3; Part-Time Volunteers 68; Interns 3.
Governing Authority: Parent Institution: Niagara University. Tax-exempt.
Institution Type/Description: University & Art Museum.
Collections: 5,900 works including paintings, sculpture, prints & drawings by artists active in the 19th & 20th centuries; Underground Railroad.
Research Fields: 19th & 20th centuries.
Facilities: 4,200 sq. ft. exhibit space; classroom. Museum-related items for sale.
Activities: permanent & temporary exhibitions; art workshops for families; educational programs; docent guided tours.
Publications: exhibition catalogs; Arcadia Revisited; newsletter; In Company: Robert Creeley's Collaborations.
Hours & Admission Prices: Tues.-Sat. 11-5, Sun. 1-5. No charge; donations accepted. Closed Good Friday; Easter; Thanksgiving; Christmas; university holidays. &
Attendance: 20,000 (accurate)
Membership: Teacher $14; Senior Citizen, Student & Artist $15; Senior Citizen Couple and NU Faculty & Staff $25; Individual $30; Family $40; Contributor $100; Benefactor $500; Life $1,000; Life Fellow $5,000.

NEW YORK POWER AUTHORITY-NIAGARA POWER PROJECT VISITORS' CENTER, 5777 Lewiston Rd., Lewiston, NY 14092-2152. Tel.: 716-286-6661. Fax: 716-286-6654.
E-mail: npvista@nypa.gov
Web Site: www.nypa.gov/vc/niagara.htm
Founded: 1963.
Key Personnel: Dir. Community Affairs, Lou Paonessa; Sr. Community Rels. Rep., Teresa Martinez.
Personnel Profile: Full-Time Paid 4; Part-Time Paid 6.
Governing Authority: private; nonprofit organization. Parent Institution: New York Power Authority. Tax-exempt.
Institution Type/Description: Electricity & Technology Museum.
Collections: over 50 hands-on exhibits; hydroelectricity & local history.
Facilities: classroom; 40-seat theater; community room.
Activities: age specific educational programs by reservation; special events.
Hours & Admission Prices: Daily 9-5. No charge. Closed New Year's Eve & Day; Thanksgiving; Christmas Eve & Day. &
Attendance: 80,000 (accurate)

Lily Dale

LILY DALE MUSEUM, 16-18 Library St., Lily Dale, NY 14752. Tel.: 716-595-8721. Fax: 716-595-2442.
Institution Type/Description: History Museum: housed in a former one-room school; built in 1890.
Collections: history of spiritualism; spirit paintings & drawings; early newspapers; photographs; memorabilia; seance trumpets; spirit slates.
Hours & Admission Prices: late June to early Sept. daily 11-4.

Lindenhurst

OLD VILLAGE HALL MUSEUM, 215 S. Wellwood Ave., Lindenhurst, NY 11757-4904. Mailing Address: P.O. Box 296, Lindenhurst, NY 11757-0296. Tel.: 631-957-4385.
Founded: 1958.
Congressional District: 13
Key Personnel: Chm. (V) & Dir., Johanna Sandy.
Personnel Profile: Part-Time Paid 3.
Governing Authority: municipal. Parent Institution: Lindenhurst Historical Society and maintained by the Village of Lindenhurst, Inc. Tax-exempt.

Institution Type/Description: Local History Museum.
Collections: period artifacts; photographs; archives; industrial; costumes; architectural plans & blueprints of local buildings; music. Historic Buildings: 1901 restored railroad depot & freight house.
Research Fields: local history, architecture & music; salt hay.
Facilities: 30-vol. library of microfilms of old local newspapers available for use at Lindenhurst Memorial Library.
Activities: guided tours; lectures; concerts; formally organized education programs for children; permanent, temporary & traveling exhibitions.
Hours & Admission Prices: Wed. & Fri.-Sat. 2-4; museum relocating, call to confirm hours. No charge; donations accepted. Closed holidays.
Attendance: 1,129 (accurate)
Membership: Individual $5; Life $30.

Little Falls

HERKIMER HOME STATE HISTORIC SITE, 200 State Rte. 169, Little Falls, NY 13365-5818. Tel.: 315-823-0398. Fax: 315-823-0587.
E-mail: karen.sheckells@parks.ny.gov
Web Site: www.herkimerhomeacademy.org
Founded: 1913.
Congressional District: 31
Personnel Profile: Part-Time Paid 8; Part-Time Volunteers 60.
Governing Authority: state. Parent Institution: New York State Office of Parks, Recreation & Historic Preservation. Subsidiary Institution: The Friends of Herkimer Home, Inc. Tax-exempt: 501(c)(3).
Institution Type/Description: Historic House: housed in the former home & gravesite of Gen. Nicholas Herkimer; built c.1760.
Collections: period furnishings; items belonging to the Herkimer family, representative of early German culture in the Mohawk Valley.
Research Fields: Mohawk Valley history; Palatine German settlement in the Mohawk Valley; American Revolution.
Facilities: visitor center; picnic area; bike trail.
Activities: guided tours; lectures; gallery talks; permanent exhibitions; school programs; seasonal farm life & craft demonstrations.
Hours & Admission Prices: mid-May to mid-Oct. Wed.-Sat. 10-5, Sun. & Mon. holidays 1-5. Adults $4, seniors & students $3; children 12 & under no charge. &
Attendance: 29,316 (accurate)
Membership: Student & Senior $5; Adult $10.

LITTLE FALLS HISTORICAL SOCIETY MUSEUM, 319 S. Ann St., Little Falls, NY 13365-1362. Tel.: 315-823-0643.
E-mail: info@lfhistoricalsociety.com
Web Site: www.lfhistoricalsociety.com
Founded: 1962.
Congressional District: 30
Key Personnel: Pres. (V), Louis W. Baum, Jr.; Museum Shop Mgr., Eileen Zak.
Personnel Profile: Part-Time Volunteers 18; Interns 3.
Governing Authority: society; nonprofit organization. Tax-exempt: 170(b)(1)(A).
Institution Type/Description: Local History Museum: housed in 1833 Greek Revival bank building.
Collections: books; scrapbooks; genealogical records; pictures; artifacts; newspaper clippings.
Research Fields: local history & genealogy.
Facilities: museum housed in c.1833 bank building.
Activities: guided tours; lectures; films; permanent & temporary exhibitions.
Publications: Historic Preservation in Little Falls.
Hours & Admission Prices: June-Sept. Tues.-Fri. 1-4, Sat. 12-4. No charge, donations accepted. Closed holidays. &
Attendance: 1,200 (accurate)
Membership: Student $5; Regular $10; Family $15; Sustaining $25; Patron $50.

MOHAWK VALLEY CENTER FOR THE ARTS, INC., 401 Canal Pl., Little Falls, NY 13365. Tel.: 315-823-0808. Fax: 315-823-0805.
E-mail: director@mohawkvalleyarts.org
Web Site: ww.mohawkvalleyarts.org
Key Personnel: Dir., Barbara Boucher
Institution Type/Description: Art Gallery.
Collections: works by local artists; paintings.
Activities: workshops; classes.
Hours & Admission Prices: Tues.-Sat. 11-4

Liverpool

SAINTE MARIE AMONG THE IROQUOIS, 6680 Onondaga Lake Pkwy., Liverpool, NY 13088-5061. Mailing Address: 106 Lake Dr., Liverpool, NY 13088-5118. Tel.: 315-453-6768. Fax: 315-453-6772.
E-mail: stemarie1657@yahoo.com
Web Site: onondagacountyparks.com
Key Personnel: Park Supt., Dale Grinolds
Institution Type/Description: History Museum: housed in re-created 1657 French Mission which stood on the shores of Onondaga Lake.
Collections: 17th-century Haudenosaunee (Iroquois) culture; daily life in the 1650s.
Hours & Admission Prices: mid-May to mid-Oct. Mon.-Fri. 9-3, Sat.-Sun. 12-5; mid-Oct. to mid-May Mon.-Fri. 9-3. Mid-May to mid-Oct. adults $3, senior citizens 62 & over $2.50, children 6-17 $2; children 5 & under no charge. Mid-Oct. to mid-May no charge; donations accepted.

SALT MUSEUM, 106 Lake Dr., Onondaga Lake Park, Liverpool, NY 13088. Mailing Address: 106 Lake Dr., Liverpool, NY 13088-5118. Tel.: 315-453-6715; 6712. Fax: 315-453-6764.
E-mail: olp@ongov.net
Web Site: www.onondagacountyparks.com/parks/olp/salt-museum.php
Founded: 1934.
Congressional District: 27
Key Personnel: Park Supt., Dale Grinolds; Museum Shop Mgr., Rhoda Sikes.
Personnel Profile: Part-Time Paid 2; Part-Time Volunteers 1.
Governing Authority: county. Onondaga Co. Dept. of Parks & Recreation. Tax-exempt.
Institution Type/Description: Historic Site: salt block; reconstruction & exhibit gallery.
Collections: paintings; photographs; manuscripts; documents; tools; vehicles; technical & social history of Onondaga; central New York's 19th & 20th-century salt industry.
Research Fields: salt manufacturing industry of Central New York in the 19th & early 20th centuries.
Facilities: library. Museum-related items for sale.
Activities: guided tours; permanent exhibits; craft demonstrations; lectures.
Publications: book, Salt: A History of Salt Manufacturing in Onondaga County; teacher resource packets.
Hours & Admission Prices: mid-May to mid-Oct. weekends 1-6. No charge; donations accepted. &
Attendance: 20,000 (accurate)
Membership: Individual $15; Family $25; Patron $50; Corporate $100.

Livingston Manor

CATSKILL FLY FISHING CENTER & MUSEUM, 1031 Old Rte. 17, Livingston Manor, NY 12758. Mailing Address: P.O. Box 1295, Livingston Manor, NY 12758-1295. Tel.: 845-439-4810. Fax: 845-439-3387.
E-mail: flyfish@catskill.net
Web Site: www.cffcm.net
Founded: 1981.
Congressional District: 28
Key Personnel: Exec. Dir., Jim Krul; Pres. (V), Miriam Stone; Vice Pres., Andrew Boyar; Vice Pres., Jerry Girard; Treas., Bob Colson; Office Mgr. & Museum Shop Mgr., Erin Phelan.
Personnel Profile: Full-Time Paid 2; Full-Time Volunteers 2; Part-Time Paid 3; Part-Time Volunteers 5.
Governing Authority: nonprofit organization. Tax-exempt: 501(c)(3).
Institution Type/Description: Fly Fishing Museum.
Collections: memorabilia & history of fly fishing; Women in fly fishing; International fly fishing.
Research Fields: preserving the history of fly fishing.
Facilities: visitors center; education building; outdoor pavilion; nature trail; picnic tables; trout stream.
Activities: traveling exhibitions; on-site exhibitions; formal education programs for children, adults & handicapped persons; fly casting & rod-building demonstrations. Museum Sponsors: Fisherman's Flea Market; Casting Tournament; Annual Fund Raising Auction.
Publications: monthly newsletter, Castabout.
Hours & Admission Prices: April-Oct. daily 10-4; Nov.-March Tues.-Fri. 10-1, Sat. 10-4. Adults $3; members no charge. Closed New Year's Day; Thanksgiving; Christmas. &
Attendance: 9,000 (accurate)
Membership: Student $15; Individual $35; Family $60; Sustaining $100; "365" $365; Benefactor $1,000; Mentor $5,000; Legacy $25,000.

Lockport

NIAGARA COUNTY HISTORICAL SOCIETY DBA THE HISTORY CENTER OF NIAGARA COUNTY, (M), 215 Niagara St., Lockport, NY 14094-2605. Tel.: 716-434-7433. Fax: 716-434-3309. Niagara History Center.
Web Site: niagarahistory.org
Founded: 1921.
Congressional District: 38
Key Personnel: C.E.O. & Dir., Melissa L. Dunlap; Pres. (V), James Mucha; Dir. Erie Canal Discovery Center, Dir. Devel. Niagara County Historical Society & Museum Shop Mgr., Ray Wigle; Education Coord. & Asst. Dir., Ann Marie Linnabery; Cur., Amy Johnson; Mktg. & Public Rels., Patricia Kibler-Fries.
Personnel Profile: Full-Time Paid 3; Part-Time Paid 4; Part-Time Volunteers 42; Interns 3.
Operating Expenses: 342,107
Operating Income: 433,715
Governing Authority: nonprofit organization. Parent Institution: Niagara County Historical Society, Inc. Branch Museums: Erie Canal Discovery Center, 24 Church St., Lockport, NY; Col. William Bond/Jesse Hawley House, 143 Ontario St., Lockport, NY. Tel.: 716-439-0431. Tax-exempt: 501(c)(3).
Institution Type/Description: History Museum.
Collections: Niagara area artifacts; Civil War artifacts; Iroquois Indian culture; medical room; toys; dolls; military; pioneer kitchen; paintings; law office; fire fighting equipment; barn; furniture; glassware; Erie Canal artifacts. Historic House: 1824 Col. Bond/Jesse Hawley house furnished with Empire period furnishings.
Major Exhibits: Charles Rand Penney Collection, 2013.
Research Fields: local history of Niagara County; genealogy; business history.
Facilities: 600-vol. library of historical & genealogical material, vertical files, directories, history publications available for use on premises; reading room; auditorium; discovery center.
Activities: guided tours; monthly program meetings; permanent & temporary exhibitions; educational programs in conjunction with area schools; summer youth programs; Erie Canal programming; Erie Canal movie; local history DVDs.
Publications: bimonthly newsletter; DVDs; books.
Hours & Admission Prices: Niagara County Historical Society: Mon.-Sat. 9-5. Closed holidays. Erie Canal Discovery Center: May-Oct. daily 9-5, Nov.-April Thurs.-Sat. 10-3. Col. Bond/Jessee Hawley House: call for hours. The Penney Gallery: May-Oct. daily 9-5, Nov.-April Thurs.-Sat. 10-3. Fee varies per site.
Attendance: 33,750 (accurate)
Membership: Individual $20; Family $30; Business $50; Life $500.

Long Eddy

BASKET HISTORICAL SOCIETY MUSEUM HALL, Rte. 97, Long Eddy, NY 12760. Mailing Address: P.O. Box 199, Long Eddy, NY 12760. Tel.: 845-887-5417.
Institution Type/Description: Historical Society Museum.
Collections: local history & culture; photographs; period artifacts.
Hours & Admission Prices: Call for hours.

Long Island City

DORSKY GALLERY CURATORIAL PROGRAMS, (M), 11-03 45th Ave., at corner of 11th St., Long Island City, NY 11101-5109. Tel.: 718-937-6317. Fax: 718-937-7469.
E-mail: info@dorsky.org
Web Site: www.dorsky.org
Key Personnel: Dir., David A. Dorsky.
Governing Authority: Tax-exempt.
Institution Type/Description: Alternative Exhibition Space.
Collections: works by contemporary artists.
Facilities: 1,200 sq. ft. exhibit space.
Hours & Admission Prices: Thurs.-Mon. 11-6; other times by appointment. No charge; donations accepted. &

FISHER LANDAU CENTER FOR ART, 38-27 30th St., Long Island City, NY 11101-2716. Tel.: 718-937-0727.
E-mail: info@flcart.org
Web Site: www.flcart.org
Institution Type/Description: Art Center.
Collections: works of contemporary art.
Hours & Admission Prices: Thurs.-Mon. 12-5. No charge. &

ISAMU NOGUCHI GARDEN MUSEUM, (M), 9-01 33rd Rd., (at Vernon Blvd.), Long Island City, NY 11106. Mailing Address: 32-37 Vernon Blvd., Long Island City, NY 11106-4926. Tel.: 718-204-7088. Fax: 718-278-2348.
E-mail: museum@noguchi.org
Web Site: www.noguchi.org
Founded: 1985.
Congressional District: 9
Key Personnel: Chm., Samuel Sachs, II; Dir., Jenny Dixon; Dir. Devel., Jennifer Burlenski; Head Education, Heather Brady; Finance Mgr., Zehava Fishman; Admin. Dir., Amy Hau; Registrar, Larry Giacoletti; Dir. Collections Cur., Bonnie Rychlak; Museum Shop Mgr., Peter Scibetta.
Personnel Profile: Full-Time Paid 21; Part-Time Paid 21; Interns 2.
Governing Authority: nonprofit organization. Parent Institution: Isamu Noguchi Foundation and Garden Museum. Tax-exempt: 501(c)(3).
Institution Type/Description: Art Museum: housed in former studio of Isamu Noguchi.
Collections: sculpture; models; writings; drawings; photographic video & audio documentation of works by Noguchi.
Research Fields: life & works of Isamu Noguchi.
Facilities: outdoor sculpture garden.
Activities: education programs; public programs; tours in Japanese.
Publications: catalogue of collection, exhibition brochures & catalogues; seasonal calendar.
Hours & Admission Prices: Wed.-Fri. 10-5, Sat.-Sun. 11-6. Adults $10, students & seniors $5; discounts to AAM & ICOM members and employees of NYC Museum Council Institutions. Pay what you wish 1st Fri. of month. &
Attendance: 26,000 (estimated)
Membership: Single $75; Dual $150.

LAGUARDIA AND WAGNER ARCHIVES, LaGuardia Community College, 31-10 Thomson Ave., Rm. E-238, Long Island City, NY 11101-3007. Tel.: 718-482-5065. Fax: 718-482-5069.
E-mail: richardli@lagcc.cuny.edu
Web Site: www.laguardiawagnerarchive.lagcc.cuny.edu
Founded: 1981.
Congressional District: 9
Key Personnel: Dir., Richard K. Lieberman; Archivist, Douglas DiCarlo.
Personnel Profile: Full-Time Paid 8; Part-Time Paid 5.
Governing Authority: Parent Institution: City University of New York. Tax-exempt.
Institution Type/Description: History Museum.
Collections: 20th-century New York City; papers of Mayors Fiorello H. LaGuardia, Robert F. Wagner, Abraham D. Beame & Edward I. Koch, records of New York City Housing Authority & the piano maker Steinway & Sons; Queens local history; NY City Council.
Research Fields: 20th-century New York City political & social history.
Facilities: library relating to New York City history & social history available to the public; 18 exhibit museum on the history of New York City; archives.
Publications: brochure; annual calendar; fourth grade curriculum; Finding Aids for the Fiorello H. LaGuardia, Robert F. Wagner & New York City Housing Authority collections, Abraham D. Beame & Steinway & Sons; museum guide.
Hours & Admission Prices: Exhibits: Mon.-Fri. 7am-10pm, Sat.-Sun. 7-5. Research & Archives: Summer: Mon.-Thurs. 9:30-4:30; Fall, Winter & Spring: Mon.-Fri. 9:30-4:30. No charge. Closed major holidays. &
Attendance: 2,500 (estimated)

MOMA PS1, 22-25 Jackson Ave., Long Island City, NY 11101-4309. Tel.: 718-784-2084. Fax: 718-482-9454.
E-mail: mail_ps1@moma.org
Web Site: www.momaps1.org
Formerly. P.S. 1 Contemporary Art Center
Founded: 1971.
Congressional District: 9
Key Personnel: Dir., Klaus Biesenbach; Dir. Devel., Todd Bishop.
Personnel Profile: Full-Time Paid 20; Part-Time Paid 40; Interns 11.
Governing Authority: nonprofit. Parent Institution: The Museum of Modern Art, NY. Tax-exempt: 501(c)(3).
Institution Type/Description: Contemporary Art Museum: housed in 19th-century Romanesque Revival schoolhouse.
Collections: site-specific installation pieces by Lawrence Weiner, Abigail Lazkoz, Saul Melman, James Turrell, Alan Saret, William Kentridge, Cecily Brown, Richard Artschwager, Ernesto Calvano, Sol LeWitt, Matt Mullican, Eric Orr, Pipilotti Rist, & Alexis Rockman.
Major Exhibits: George Kuchar: Pagan Rhapsodies, 11/20/11-1/15.
Facilities: 125,000 sq. ft. exhibit space.

Activities: guided tours; lectures; films; concerts; dance; organized educational programs; traveling exhibitions; National & International Studio Programs.
Publications: exhibition catalogues; artist publications.
Hours & Admission Prices: Thurs.-Mon. 12-6. Suggested Donation: adults $10, students & senior citizens $5; discounts to AAM & ICOM members; MoMA members no charge. Closed New Year's Day; Memorial Day; Independence Day; Thanksgiving; Christmas. &
Attendance: 133,000
Membership: Individual $85; Dual $140; Family $175; Fellow $360; Supporting $600; Sustaining $1,200; Patron $1,750; Benefactor $3,000; Sustaining Benefactor $6,000; Major Benefactor $12,000.

SCULPTURE CENTER, (M), 44-19 Purves St., Long Island City, NY 11101-2907. Tel.: 718-361-1750. Fax: 718-228-6235.
E-mail: info@sculpture-center.org
Web Site: www.sculpture-center.org
Founded: 1928.
Key Personnel: Exec. Dir., Mary Ceruti; Mgr. Operations & Web Producer, Katie Bode; Visitor Svcs. & Mgr. Membership, John Emison; Cur., Fionn Meade; Devel. Assoc. & Communications Mgr., Erin Pierson.
Personnel Profile: Full-Time Paid 4; Part-Time Paid 1; Part-Time Volunteers 50; Interns 4.
Institution Type/Description: Art Gallery.
Collections: contemporary sculpture.
Activities: educational programs.
Publications: exhibit catalog, Knight's Move (2010).
Hours & Admission Prices: Thurs.-Mon. 11-6. Suggested Donation: $5.
Attendance: 8,246 (accurate)
Membership: First Dimension (Student) $35; First Dimension $$50; New Art Network $75; Second Dimension $200; Third Dimension $500; Fourth Dimension $1,000; Fifth Dimension $2,500.

SOCRATES SCULPTURE PARK, 32-01 Vernon Blvd. (at Broadway), Long Island City, NY 11106. Mailing Address: P.O. Box 6259, Long Island City, NY 11106-0259. Tel.: 718-956-1819. Fax: 718-626-1533.
E-mail: info@socratessculpturepark.org
Web Site: socratessculpturepark.org
Founded: 1986.
Congressional District: 14
Key Personnel: Dir., John Hatfield; Pres. (V), Stuart Match Sura; Dir. Public Programs & Community Rels., Shaun Leonardo; Facilities & Studio Mgr., Lars Fisk; Dir. Devel. & Communications, Katie Denay; Office Mgr. & Administrative Asst., Elissa Goldstone.
Personnel Profile: Full-Time Paid 5; Part-Time Paid 4; Part-Time Volunteers 25; Interns 2.
Governing Authority: Tax-exempt.
Institution Type/Description: Sculpture Park.
Collections: sculpture garden; contemporary art.
Facilities: working artist studio.
Activities: classes, film screenings & special events.
Publications: exhibition catalogues.
Hours & Admission Prices: Daily 8:30 to sunset. No charge. &
Attendance: 85,000 (estimated)

Lowville

LEWIS COUNTY HISTORICAL SOCIETY MUSEUM, 7552 S. State St., Lowville, NY 13367-1529. Mailing Address: 7552 S. State St., P.O. Box 446, Lowville, NY 13367-0446. Tel.: 315-376-8957.
E-mail: lewiscountyhistoricalsociety@gmail.com
Web Site: lewiscountyhistory.org
Founded: 1926.
Congressional District: 26
Key Personnel: Pres., Marian Opela; Sec., Lida Perfetto; Vice Pres., Chuck Bunke, Treas., Sharon Sears; Museum Shop Mgr., Lida Perfetto.
Personnel Profile: Part-Time Paid 2; Part-Time Volunteers 20; Interns 1.
Governing Authority: nonprofit organization. Tax-exempt: 501(c)(3).
Institution Type/Description: History Museum.
Collections: archaeology; costumes; documents; family heirlooms; geology; history; military; photographs; technology; local history; manuscripts.
Research Fields: local history.
Facilities: 500-vol. library of diaries, journals, local and general history books, historical periodicals, museum maintenance publications, antique books available for research on premises; photocopies of original documents; Lewis County Historian's Room is housed at this facility: open for general research.

Activities: guided tours; lectures; films; inter-museum loan; permanent & temporary exhibitions.
Publications: Journal of the Lewis County Historical Society; newsletter; annual calendar; books.
Hours & Admission Prices: Historical Society: June to mid-Oct. Tues.-Fri. 10-4, Sat. 10-12; mid-Oct. to June Tues.-Wed. 10-4. Lewis County Research Center: June to mid-Oct. Tues.-Thurs. 10-4; mid-Oct. to June Tues.-Wed. 10-4. No charge; donations accepted. Closed federal holidays. &
Attendance: 1,900 (estimated)
Membership: Individual $25; Family $35; Sustaining $70; Life $250.

Lyons

WAYNE COUNTY HISTORICAL SOCIETY'S MUSEUM OF WAYNE COUNTY HISTORY, (M), 21 Butternut St., Lyons, NY 14489-1124. Tel.: 315-946-4943. Fax: 315-946-0069.
E-mail: info@waynehistory.org
Web Site: www.waynehistory.org
Formerly: Wayne County Historical Society
Founded: 1946.
Congressional District: 29
Key Personnel: Exec. Dir., Larry Ann Evans; Museum Mgr., Mary O'Toole.
Personnel Profile: Full-Time Paid 2; Part-Time Paid 1; Part-Time Volunteers 10.
Governing Authority: Wayne County Historical Society. Tax-exempt: 501(c)(3).
Institution Type/Description: Historical Society Museum.
Collections: glass & ceramics; local history artifacts; early criminology; agriculture. Historic Houses: 1854 sheriff's residence; 1854 county jail; 1913 sheriff's barn.
Research Fields: local history.
Facilities: 1,500-vol. general library available on premises; reading room. Museum-related items for sale.
Activities: permanent, temporary & traveling exhibitions; school loan service; lecture programs. Museum Sponsors: holiday boutique; annual dinner; dramatic reading.
Publications: quarterly, Wayne County Historical Society Newsletter; Index to History of Pioneer Settlement of Phelps & Gorham's Purchase; Wayne County Game-The Battle of Sodus Point - War of 1812; Seth Cole Exhibit Catalog. Wayne County: The Aesthetic Heritage of a Rural Area; The Fruit Industry in Wayne County, New York 1823-1984; Wayne County: Looking Back, 1988; Annals of Arcadia; The Sodus Shaker Community; Wayne County Turning the Century; exhibit catalog; Pre-History of the Savannah, New York Area; The History of the Wayne Co. Jail, 1856-1960; Pen Pals from the Past subscription letters from 1800s on Erie Canal, Letters from the Wayne County Jail, Underground Railroad & Early Wayne County; Remembering Wayne: A Pictorial View of the People, Places & Pastimes of Wayne County, NY; ABC Book of Wayne County; Wayne County Activity Book.
Hours & Admission Prices: April-Oct. Tues.-Sat. 10-4; Nov.-March Tues.-Fri. 10-4. Adults $4, children $2; members no charge. Closed legal holidays.
Attendance: 8,000 (accurate)
Membership: Drumlin $25; Cobblestone $30; Old Jail $35; Clyde Glass & Lyons Pottery $50 & up; Peppermint $75; Chimney Bluffs $100; Erie Canal $500 & up. Business: Bronze $100; Silver $250; Gold $500; Platinum $1,000.

Macedon

MACEDON HISTORICAL SOCIETY, INC., 1185 Macedon Center Rd., Macedon, NY 14502. Mailing Address: P.O. Box 303, Macedon, NY 14502-0303. Tel.: 385-388-0629.
Web Site: www.macedonhistoricalsociety.org
Founded: 1962.
Congressional District: 29
Key Personnel: Pres. (V), David Taber; Vice Pres., Charles Packard; Cur., Sally Millick.
Governing Authority: society. Tax-exempt.
Institution Type/Description: Historical Society Museum.
Collections: Quaker schoolhouse artifacts; personal artifacts; scrapbooks. Historic Buildings: 1853 Macedon Academy; 1868 Orthodox Quaker Church, meeting house.
Activities: temporary exhibitions.
Publications: Pioneers of Macedon, Bicentennial Edition 1975-1976; Pioneers of Macedon, 1912.
Hours & Admission Prices: Summer: call for hours; other times by appointment. No charge; donations accepted. &
Membership: Individual $2; Couple $3.

Mahopac

THE PUTNAM CHILDREN'S DISCOVERY CENTER, 854 Rte. 6, Mahopac, NY 10541-1721. Mailing Address: P.O. Box 222, Carmel, NY 10512-0222. Tel.: 845-621-1260. Fax: 845-276-2078.
E-mail: info@discoveryctr.org
Web Site: www.discoveryctr.org
Key Personnel: Exec. Dir., Janice Newman
Institution Type/Description: Children's Museum.
Collections: hands-on exhibits.
Activities: outreach programs; scout month; educational programs; birthday parties; workshops.
Hours & Admission Prices: Call for hours.

Malone

FRANKLIN COUNTY HISTORICAL & MUSEUM SOCIETY, (M), 51 Milwaukee St., Malone, NY 12953-1916. Mailing Address: P.O. Box 388, Malone, NY 12953-0388. Tel.: 518-483-2750.
E-mail: fcohms@northnet.org
Web Site: www.franklinhistory.org
Founded: 1903.
Congressional District: 23
Key Personnel: Pres. (V), Cheryl Learned.
Personnel Profile: Part-Time Paid 1; Part-Time Volunteers 34.
Governing Authority: society. Parent Institution: Franklin County Historical & Museum Society. Subsidiary Institutions: Schryer Center for Historical & Genealogical Research; House of History. Tax-exempt: 501(c)(3).
Institution Type/Description: County History Museum.
Collections: headquarter papers of the 16th Civil War Regiment of New York State Volunteers; history; agriculture; period rooms: memorial room, 1877-1881 V.P. William Almon Wheeler Room; 1,000 items of clothing from 1830-present; copy of first County newspaper; reproduction of old documents; early pictures of the area; local history books.
Research Fields: Franklin County.
Facilities: 500-vol. library of general & local history; maps available for research on premises. Farmer Boy (Almanzo Wilder) books & booklets & other museum-related items for sale.
Activities: guided tours; lectures; temporary exhibitions; genealogical services; formally organized education programs for children; summer concert series. Museum Sponsors: Museum Day craft demonstrations for 4th grade students of Franklin County; craft demonstrations for groups by appointment & at Museum Open Houses.
Publications: annual periodical, Franklin Historical Review; Architecture from the Adirondack Foothills; Franklin County Family Album; 4 newsletters per year.
Hours & Admission Prices: Research: June-Aug. Tues.-Fri. 1-4; Winter: by appointment. Museum Tours: by appointment. Tours: adults $5; discounts to AASLH members; members no charge. Research: adults $10; members no charge. Closed holidays. &
Attendance: 1,200 (estimated)
Membership: Individual $20; Family $25; Patron $50; Business $100; Benefactor $500.

Mamaroneck

KOSLOWE JUDAICA GALLERY, 175 Rockland Ave., Mamaroneck, NY 10543. Tel.: 914-698-2960. Fax: 914-698-3610.
E-mail: executive@wjcenter.org
Web Site: www.wjcenter.org/community
Key Personnel: Exec. Dir., Susan Lurie
Institution Type/Description: Jewish Art & History Museum.
Collections: photographs; paintings; sculpture; folk art.
Activities: lectures; temporary exhibitions.
Hours & Admission Prices: Call for hours. No charge; donations accepted. &

LARCHMONT HISTORICAL SOCIETY, 740 W. Boston Post Rd., Ste. 301, Mamaroneck, NY 10543-3345. Mailing Address: P.O. Box 742, Larchmont, NY 10538-0742. Tel.: 914-381-2239 & 834-5136.
E-mail: archives@larchmonthistory.org
Web Site: www.larchmonthistory.org
Founded: 1980.
Congressional District: 18
Key Personnel: Pres. (V), Colette Rodbell; Treas., Jim Sweeney; Archivist, Lynne Crowley; Membership Coord., Lauren Gottfried.
Personnel Profile: Part-Time Paid 1; Part-Time Volunteers 10; Interns 2.
Governing Authority: private; nonprofit organization. Tax-exempt: 501(c)(3).
Institution Type/Description: Historical Society Museum.

Collections: history of village & postal district of Larchmont, NY.
Research Fields: architectural sites survey; June Freeman Allen costume collection (1865-1930); Flint family papers.
Facilities: 400-vol. library; 28 sq. ft. exhibit space; archives.
Activities: guided tours; lectures; temporary exhibitions; special events. Annual Event: Spring House Tour.
Publications: monthly newsletter, Gazebo Gazette; books & booklets on Larchmont subjects.
Hours & Admission Prices: Tues. & Thurs. 9-2; other times by appointment. No charge. &
Attendance: 300 (estimated)
Membership: Senior & Student $5; Individual $15; Family $25; Institutional $25; Sustaining $50; Life $150.

Manhasset

HISTORICAL SOCIETY OF THE TOWN OF NORTH HEMP-STEAD, 200 Plandome Rd., Manhasset, NY 11030-2326. Mailing Address: P.O. Box 3000, Manhasset, NY 11030-3000. Tel.: 516-869-7646.
Web Site: www.northhempstead.com
Founded: 1963.
Congressional District: 6
Key Personnel: Pres., Hon. Dolores Sedacca.
Governing Authority: society.
Institution Type/Description: Historical Society Museum.
Collections: agriculture; history; arboretum; aviary; botany; herbarium; Indian artifacts; transportation; manuscripts.
Facilities: library of books, maps, photographs & clippings available for use by appointment.
Activities: guided tours; lectures; formally organized education programs for children; inter-museum loan; permanent, temporary & traveling exhibitions.
Hours & Admission Prices: by appointment. No charge.
Membership: Student $3; Individual $5; Contributing $10; Sustaining $25; Corporate $50; Life $100; Life Corporate $1,000.

SCIENCE MUSEUM OF LONG ISLAND, Leeds Pond Preserve, 1526 N. Plandome Rd., Manhasset, NY 11030. Mailing Address: P.O. Box 908, Plandome, NY 11030-0908. Tel.: 516-627-9400, ext. 11. Fax: 516-365-8927. Facebook: Science Museum of Long Island.
E-mail: info@smli.org
Web Site: smli.org
Founded: 1963.
Congressional District: 3
Key Personnel: Dir., John T. Tanacredi, Ph.D.; Pres. (V), Carlo Manganillo.
Personnel Profile: Full-Time Paid 10; Part-Time Paid 10; Part-Time Volunteers 18.
Governing Authority: nonprofit organization. Tax-exempt: 501(c)(3).
Institution Type/Description: Science Center.
Collections: natural history artifacts; live animal exhibits; physics & technology exhibits.
Research Fields: technology; alternative energy; aquaculture, hydroponics.
Facilities: laboratories; classrooms; wildlife preserve; pond. Museum-related items for sale.
Activities: school science programs; adult education & lecture series; lecture demonstrations; workshops for pre-kindergarten-junior high school students; advanced study & research programs for high school students; barrier-free nature trails; experimental garden with greenhouse; films; summer science studies program; after-school workshops; weekend family workshops.
Publications: quarterly newsletter, workshop brochures.
Hours & Admission Prices: Office: Mon.-Fri. 9-3:30. Call for information on charges & activities; discount to museum members. &
Attendance: 15,000 (estimated)

Manlius

MANLIUS HISTORICAL SOCIETY AND MUSEUM, 109 Pleasant St., Manlius, NY 13104. Mailing Address: P.O. Box 28, Manlius, NY 13104-0028. Tel.: 315-682-6660. Fax: 315-682-6660.
E-mail: manliushistory@gmail.com
Web Site: www.manliushistory.org
Founded: 1976.
Congressional District: 33
Key Personnel: Dir., Donna L. Nortman.
Personnel Profile: Part-Time Paid 1; Part-Time Volunteers 20.
Governing Authority: society; nonprofit organization. Tax-exempt.
Institution Type/Description: Historical Society Museum.

Collections: farm & carpentry tools; wheelwright implements & tools; operating blacksmith shop; colonial ironware; herb garden; late 19th-century medical tools & doctor's pharmacy; early 20th-century toys; locally-made products; 3 locally made antique violins made between 1870 & 1900; documents related to school, church & personal histories.
Research Fields: town history.
Facilities: 600-vol. library of books & documents on local history available for research by appointment on premises; reading room.
Activities: guided tours; lectures; hobby workshops; docent program or council; loan, permanent & temporary exhibitions.
Publications: quarterly newsletter, The Seraph; People & Places: Fayetteville, Manlius, Minoa & Neighbors, Vol. I, II & III.
Hours & Admission Prices: Museum: May-Oct. Sat. 11-3; groups & other times by appointment. Research Center: Mon. & Wed.-Thurs. 10-4; other times by appointment. No charge; donations accepted. Closed Easter; Thanksgiving; Christmas.
Attendance: 2,200 (estimated)
Membership: Senior Citizen $20; Individual $30; Family $45; Sustaining $75; Partner $125; Benefactor $250; Patron $500.

Marcellus

MARCELLUS HISTORICAL SOCIETY, 18 North St., Marcellus, NY 13108. Mailing Address: P.O. Box 165, Marcellus, NY 13108-0165.
Founded: 1960.
Congressional District: 33
Key Personnel: Pres., Peg Nolan; Vice Pres., Douglas Nightingale; Treas., Hollis Abbott; Sec., Carrie Beth Pottinger.
Personnel Profile: Part-Time Volunteers 6.
Volunteer Hours: 1,684
Operating Expenses: 18,160
Operating Income: 13,320
Governing Authority: society. Tax-exempt.
Institution Type/Description: History Museum: housed in early 1830s house.
Collections: household & farm utensils; treadmill; guns; china; agricultural implements; furniture; microfilm of Marcellus Observer Newspaper 1879-present day; historic documents, photographs, business items, genealogies.
Research Fields: local history.
Facilities: meeting room.
Activities: guided tours; permanent exhibitions; monthly meetings with guest speakers.
Publications: advertising flyer with brief history; quarterly newsletter.
Hours & Admission Prices: Sun.-Mon. 1-3, Tues. 7-9, Thurs. 1-4; other times by appointment. No charge; donations accepted. Closed holidays. &
Attendance: 610 (estimated)
Membership: Senior & Student $4; Individual $5; Family $12; Senior & Student Sustaining $20; Individual Sustaining $25; Family Sustaining $60.

Marilla

MARILLA HISTORICAL SOCIETY MUSEUM, 1810 Two Rod Rd., Marilla, NY 14102. Mailing Address: P.O. Box 36, Marilla, NY 14102-0036. Tel.: 716-652-1827 & 7608.
Founded: 1960.
Congressional District: 38
Key Personnel: Pres., Mary Beth Serafin; Vice Pres., Cindy Petrinec; Treas., John Foss; Sec., Judy Mees.
Governing Authority: society. Tax-exempt.
Institution Type/Description: Local History Museum.
Collections: local history; period artifacts.
Facilities: library of books on local history available for use on premises.
Activities: informal programs for children; permanent exhibitions.
Hours & Admission Prices: Sept.-June Tues. 7pm-9pm, 3rd Sun. of month 2-4. No charge; donations accepted. &
Attendance: 200 (estimated)
Membership: Children & Senior Citizens $5; Regular $7.

Marlboro

GOMEZ MILL HOUSE MUSEUM AND HISTORIC SITE, (M), 11 Mill House Rd., Marlboro, NY 12542-6514. Mailing Address: 15 W. 16th St., New York City, NY 10011-6301. Tel.: 845-236-3126.
E-mail: gomezmillhouse@juno.com
Web Site: www.gomez.org
Founded: 1979.
Congressional District: 28
Key Personnel: Exec. Dir., Ruth K. Abrahams; Pres. (V), Robert Jacobs, Jr.; Museum Site Asst., Jill Williams; Museum Site Asst., Richard Rosencrans.

Personnel Profile: Full-Time Paid 3; Part-Time Volunteers 2.
Volunteer Hours: 453
Operating Expenses: 243,902
Operating Income: 146,981
Governing Authority: private; nonprofit organization. Parent Institution: Gomez Foundation for Mill House, 15 W. 16th St., 6th Fl., New York City, NY 10011. Tel.: 212-294-8329. Tax-exempt: ST-119-1.
Institution Type/Description: Historic House & Living History Museum.
Collections: decorative arts; furniture; artifacts; paintings; documents. Historic Buildings: Gomez Mill House; Dard Hunter Paper Mill; stone storage structure; Stone Smoke House; root cellar.
Research Fields: American history; paper history; Jewish American history; social activism.
Facilities: event space.
Activities: public programs; tours; papermaking demonstrations. Museum Sponsors: Gallery at Gomez Mill House.
Hours & Admission Prices: mid-April to Nov. Wed.-Sun. 10-4. Adults $10, seniors 55 & over $7, children & students 6-18 $4; discount to groups of 10 or more, WNET, AAM, ICOM & Channel 13 members; members & children under 6 no charge. Closed Easter; Thanksgiving; Jewish holidays; Christmas; national holidays.
Attendance: 3,000 (estimated)
Membership: Student & Senior $35; General $50; Gift Memberships: Associate $250; Sponsor $500; Patron $1,000; Benefactor $2,500; Guardian $5,000; Champion $10,000 & up.

Mastic Beach

FIRE ISLAND NATIONAL SEASHORE, William Floyd Estate, 245 Park Dr., Mastic Beach, NY 11951. Mailing Address: 120 Laurel St., Patchogue, NY 11772-3596. Tel.: 631-399-2030. Fax: 631-399-0017.
Web Site: www.nps.gov/fiis
Founded: 1964.
Congressional District: 1 & 2
Key Personnel: Cur., Steven Czarniecki.
Governing Authority: federal. Administered by the National Park Service United States Dept. of the Interior.
Institution Type/Description: National Park.
Collections: fauna & flora. Historic House: located at Mastic Beach on mainland, furnished c.1700s estate of William Floyd.
Research Fields: beach environment.
Facilities: Visitor Center: 1,000-vol. library of natural history available for research by appointment; nature trails; picnic area; campsites; wildlife refuge & bird sanctuary; boardwalk trail for handicapped. Natural history-related publications for sale.
Activities: guided tours; lectures; films; gallery talks; demonstrations; wilderness hiking; fishing.
Publications: pamphlets; brochures.
Hours & Admission Prices: Grounds: Memorial Day to Oct. Fri.-Sun. 9-6. William Floyd Estate: Memorial Day to Veterans Day Fri.-Sun. 10-4. No charge. Closed holidays.
Attendance: 8,000

Mattituck

MATTITUCK LAUREL HISTORICAL SOCIETY AND MUSEUMS, Main Rd., Mattituck, NY 11952. Mailing Address: P.O. Box 766, Mattituck, NY 11952. Tel.: 631-298-5248.
Institution Type/Description: History Museum.
Collections: local history; period furnishings & clothing; toys. Historic Buildings: 1760 Schoolhouse; Jesse Tuthill House, 1799; Ira Tuthill House, 1841; Outbuilding, c.1900; Milk House, 1880s; New Egypt School House, 1846.
Activities: special events.
Hours & Admission Prices: May-Sept. Sat.-Sun. 1-4.

Mayville

DART AIRPORT AVIATION MUSEUM, 6167 Plank Rd., Mayville, NY 14757. Mailing Address: P.O. Box 211, Mayville, NY 14757. Tel.: 716-753-2160.
Institution Type/Description: Aviation History Museum.
Collections: aviation history & aircraft; engines; models; memorabilia.
Activities: special events.
Hours & Admission Prices: May-Nov. 1 Tues.-Sun. 10-5.

Medina

MEDINA RAILROAD MUSEUM, 530 A West Ave., Medina, NY 14103-1554. Tel.: 585-798-6106. Fax: 585-798-1086.
E-mail: office@railroadmuseum.net
Web Site: www.railroadmuseum.net
Founded: 1997.
Congressional District: 27
Key Personnel: Dir., Martin C. Phelps; Pres. (V), James L. Dickinson; Treas., Hugh F. James.
Personnel Profile: Full-Time Paid 3; Full-Time Volunteers 2; Part-Time Volunteers 35.
Governing Authority: private; nonprofit organization. Tax-exempt: 501(c)(3).
Institution Type/Description: Railroad Museum.
Collections: railroad artifacts, memorabilia, photographs, models & toys; firefighting artifacts, memorabilia & toys.
Facilities: 9,300 sq. ft. exhibit space. Museum-related items for sale.
Activities: guided tours; loan exhibitions; broadcast programs; Polar Express train rides. Annual Events: Excursion Train Rides; Day Out With Thomas.
Hours & Admission Prices: Tues.-Sun. 11-5. Adults $7, senior citizens $6, students 2-17 $4, children 2-12 $3; discounts to AAM members; members no charge. Closed New Year's Day; Easter; Christmas. &
Attendance: 33,000 (estimated)
Membership: Individual $20; Family $35; Business $50; Corporate $100.

Middlesex

MIDDLESEX HERITAGE GROUP & HISTORICAL SOCIETY MUSEUM, Town Hall, Main St., Middlesex, NY 14507. Mailing Address: P.O. Box 147, Middlesex, NY 14507.
Institution Type/Description: Historical Society Museum.
Collections: local history & culture; personal artifacts; photographs.
Publications: newsletter, The Middlesex Heritage.
Hours & Admission Prices: Wed. & Sat. 9 to noon. No charge; donations accepted.
Membership: Individual $7.

Middletown

HISTORICAL SOCIETY OF MIDDLETOWN AND THE WALLKILL PRECINCT, INC., 25 East Ave., Middletown, NY 10940-5818. Mailing Address: P.O. Box 34, Middletown, NY 10940-0034. Tel.: 845-342-0941.
E-mail: enjine@aol.com
Founded: 1923.
Congressional District: 26
Key Personnel: Pres. (V), C.E.O. & Cur., Marvin H. Cohen; Sec., Dorothy Hunt-Ingrassia; Treas., Joanne Norbury.
Personnel Profile: Part-Time Volunteers 6.
Governing Authority: society. Tax-exempt: 501(c)(3).
Institution Type/Description: Local History Museum: housed in 1886 building.
Collections: photographs; Indian artifacts; clothing; china; mixed exhibits of historical objects; local genealogy.
Research Fields: city of Middletown, New York & town of Wallkill.
Facilities: 500-vol. library of genealogy & history books available for use on premises by appointment; reading room.
Activities: guided tours; lectures; permanent exhibitions; slides; pictures.
Publications: quarterly newsletter; booklets, The History of Our Society, Music in Middletown 100 Years Ago, The Clemson Story, Sally Sunflower & the Bloomer, History of Orange County Telephone Co., Clemson Park, Memoirs of Dr. Moses Ashby Stivers, Wallkill Academy, Middletown Hotels & Inn Keepers, Churches & Synagogues of Middletown and Nearby Communities, Middletown Theatres, Early Man in Orange County, New York.
Hours & Admission Prices: Wed. 1-5; other times call 845-343-4219 for appointment. No charge.
Attendance: 300 (estimated)
Membership: Regular $15; Couples $20; Institutional $25; Life $100.

THE INTERACTIVE MUSEUM, 23 Center St., Middletown, NY 10940. Tel.: 845-344-3131.
Institution Type/Description: Children's Museum.
Collections: hands-on exhibits.
Activities: permanent & temporary exhibits; educational programs; classes; performances; summer camp; special events.
Hours & Admission Prices: Thurs.-Sat. 1-4, Sun. 2-4.

Millbrook

MILLBROOK SCHOOL, TREVOR ZOO, 131 Millbrook School Rd., Millbrook, NY 12545-4932. Tel.: 845-677-3704. Fax: 845-677-3774.
E-mail: trevorzoo@millbrook.org
Web Site: www.trevorzoo.org
Founded: 1936.
Congressional District: 22
Key Personnel: Dir., Jonathan Meigs; Assoc. Dir., Alan Tousignant; Education, Jane H. Meigs; Animal Care Coord., Jessica Bennett.
Personnel Profile: Full-Time Paid 5; Part-Time Paid 1; Part-Time Volunteers 60; Interns 2.
Governing Authority: secondary school; nonprofit. Parent Institution: Millbrook School. Tax-exempt.
Institution Type/Description: Zoo.
Collections: 180 animals from around the world.
Facilities: over 6 acres.
Activities: Millbrook School students (grades 9-12) are involved with the care of the animals.
Hours & Admission Prices: Daily 8:30-5. Adults $5, children $3; discounts to groups. Summer Family Pass $40. &

Attendance: 22,000 (estimated)

Monroe

MUSEUM VILLAGE, 1010 Rte. 17 M, Monroe, NY 10950-1625. Tel.: 845-782-8248. Fax: 845-782-6432.
E-mail: info@museumvillage.org
Web Site: www.museumvillage.org
Founded: 1950.
Congressional District: 96
Key Personnel: Exec. Dir., Kate Mitchell; Chm. (V), Paul Campanella; Collections Mgr., Chris Cantrell; Education Coord., Lori Siccardi; Museum Shop Mgr., Virginia Mina.
Personnel Profile: Full-Time Paid 4; Part-Time Paid 30; Part-Time Volunteers 25.
Governing Authority: nonprofit organization.
Institution Type/Description: Living History Museum.
Collections: life & work of 19th-century Hudson Valley; recreation of 19th century village with over 25 buildings; drug store, log cabin, broom shop, natural history museum, firehouse, printshop, farm tools, schoolhouse, candleshop, livery stable, museum, wagon maker, blacksmith, dress exhibit, weave shop, pottery, store exhibit, barber shop, shoemaker, cooperage, energy exhibit.
Research Fields: mid-19th century material culture.
Facilities: 500-vol. library of research books available for use on premises; snack bar; picnic area. Books, crafts & other museum-related items for sale.
Activities: formally organized education program for children; permanent & temporary exhibitions; craft demonstrations; living history re-enactments; special events throughout season.
Publications: newsletter; calendar of special events; exhibit guide; education brochure.
Hours & Admission Prices: Call for hours. Adults $10, senior citizens & children 4-12 $8; discounts to AASLH partner institutions; members & children under 4 no charge.
Attendance: 30,757 (accurate)
Membership: Individual $40; Dual $55; Family $75; Donor $125; Sponsor $250; Patron $500; Corporate $1,000.

Montauk

MONTAUK POINT LIGHTHOUSE MUSEUM, 2000 Montauk Hwy., Montauk, NY 11954-5600. Mailing Address: P.O. Box 943, Montauk, NY 11954. Tel.: 631-668-2544. Fax: 631-668-2546.
E-mail: kccpcr@montauklighthouse.com
Web Site: www.montauklighthouse.com
Founded: 1987.
Congressional District: 2
Key Personnel: Dir. Site Management, Tricia Wood; Chm., Treas. & Museum Shop Mgr., Elizabeth L. White.
Personnel Profile: Full-Time Paid 3; Part-Time Paid 27.
Governing Authority: private; nonprofit organization. Parent Institution: Montauk Historical Society, P.O. Box 943, Montauk, NY 11954. Tax-exempt: 501(c)(3).
Institution Type/Description: Lighthouse Museum.
Collections: Fresnel lens; paintings; photographs; maritime artifacts; documents; 19th-20th century lanterns; original document signed by Thomas Jefferson commissioning the Montauk Point Lighthouse under the 2nd Congress; documents by founder, Ezra L'Honmidieu.

Research Fields: erosion control; building maintenance & preservation.
Facilities: library; 1,350 sq. ft. exhibit space; 30-seat theater.
Activities: guided tours; permanent exhibitions; films; formal education programs for adults & children; lectures. Annual Events: Montauk Point Lighthouse Sprint Triathlon & Relay; Lighthouse weekend.
Publications: newspaper, The Beacon; newsletter, Daymark.
Hours & Admission Prices: Call for hours. Adults $8.50, senior citizens $7, children $4; discounts to groups. Children must meet a minimum height requirement of 41 inches. &
Attendance: 80,000 (estimated)

Montgomery

BRICK HOUSE, 850 Rte. 17K, Montgomery, NY 12549. Mailing Address: P.O. Box 462, Montgomery, NY 12549. Tel.: 845-457-4921 & 4905. Fax: 845-457-4906.
E-mail: stucker@co.orange.ny.us
Founded: 1979.
Key Personnel: Park Commissioner, Richard Rose; Dir., Susan Tucker.
Personnel Profile: Full-Time Paid 1; Part-Time Paid 1; Part-Time Volunteers 30.
Governing Authority: county. Orange County Dept. of Parks, Recreation & Conservation, 550 Rte. 416, Montgomery, NY. Tel. 914-294-5151, ext. 1870. Branch Museum: Hill-Hold, Campbell Hall, NY. Tax-exempt.
Institution Type/Description: Historic House: 1768 Georgian style house, built by Nathaniel Hill.
Collections: family homestead furnishings, late 17th-century to present; family papers.
Activities: guided tours.
Publications: brochures; newsletter.
Hours & Admission Prices: May-Oct. Sat.-Sun. 10-4:30. Family $7, adults $3, children $2; discounts to groups. Closed Memorial Day; Independence Day; Labor Day. &
Attendance: 4,000 (accurate)
Membership: Single $15; Family $25; Patron $50.

HILL-HOLD MUSEUM, 128 Rte. 416, Montgomery, NY 12549. Mailing Address: P.O. Box 462, Montgomery, NY 12549 Tel.: 845-291-2404.
E-mail: stucker@co.orange.ny.us
Founded: 1976.
Key Personnel: Parks Commissioner, Richard Rose; Dir., Susan Tucker.
Governing Authority: county. Affiliated with Orange County Dept. of Parks, Recreation and Conservation, RD 3, Montgomery, NY. Branch: Brick House, Montgomery, NY. Tax-exempt.
Institution Type/Description: Historic House: housed in a Georgian style stone house; built in 1769.
Collections: Hudson Valley furnishings from 17th-century to 1830s.
Research Fields: 19th-century farm life & customs.
Facilities: Museum-related items for sale.
Activities: guided tours. Museum Sponsors: Candlelight Tour in December.
Publications: An Educational Guide to Hill-Hold.
Hours & Admission Prices: mid-May to mid-Oct. Wed.-Sun. 10-4:30. Family $7, adults $3, children $2; groups by appointment. Closed Memorial Day; Independence Day; Labor Day. &
Attendance: 8,000
Membership: Individual $15; Family $25; Patron $50 & up.

ORANGE COUNTY FIREFIGHTERS MUSEUM, 141 Clinton St., Montgomery, NY 12549. Mailing Address: P.O. Box 688, Montgomery, NY 12549. Tel.: 845-457-9654.
E-mail: enjine@aol.com
Web Site: www.ocfm.us
Founded: 1991.
Congressional District: 22
Key Personnel: Chm. & Pres., F. Edward Devitt; Treas., Walter Karsten; Education, John Conner; Cur., James Bair; Public Rels., Marvin H. Cohen.
Personnel Profile: Part-Time Paid 1; Part-Time Volunteers 40.
Governing Authority: private; nonprofit organization. Tax-exempt: 501(c)(3).
Institution Type/Description: Firefighting History Museum: housed in an early firehouse.
Collections: firefighting history, equipment & artifacts; fire trucks; hose carts; badges; firefighter memorial.
Research Fields: oral history of retired firefighters.
Facilities: education center.
Activities: guided tours; formal education programs for children.
Hours & Admission Prices: Sat. 1-4; other times by appointment. No charge; donations accepted.
Attendance: 1,200 (estimated)

Montour Falls

SCHUYLER COUNTY HISTORICAL SOCIETY, INC., 108 N. Catharine, Montour Falls, NY 14865. Mailing Address: P.O. Box 651, Montour Falls, NY 14865-0651. Tel.: 607-535-9741.
E-mail: info@schuylerhistory.org
Web Site: www.schuylerhistory.org
Founded: 1960.
Congressional District: 51
Key Personnel: Dir., Andrew E. Tompkins; Pres. & Journal Editor, Richard Owlett; Vice Pres., Min Clemens.
Personnel Profile: Full-Time Paid 1; Part-Time Paid 1; Part-Time Volunteers 12.
Governing Authority: society; nonprofit organization. Tax-exempt.
Institution Type/Description: Local History Museum.
Collections: local history; agriculture; artifacts; costumes; art; archives; toys; Indians; local sanitarium; quilts; music; period clothing; paintings; artifacts from the 1900s: pioneer kitchen; old tools; one-room school; research aides in library.
Research Fields: local history.
Facilities: 1,000-vol. library of local history reference books, pamphlets, scrapbooks & files; available on premises; reading room.
Activities: lectures; formally organized educational programs; temporary exhibits; 7 general meetings with historical programs; permanent exhibits.
Publications: quarterly, Schuyler County Historical Society Journal; newsletter, The Vista.
Hours & Admission Prices: April to mid-Dec. Tues.-Fri. 10-4, Sat. call for hours; other times by appointment. No charge; donations accepted. Research room: $5 per hour. Closed holidays. &

Attendance: 2,000 (accurate)
Membership: Individual $25; Couple $30; Patron $50; Benefactor $100.

Moravia

CAYUGA-OWASCO LAKES HISTORICAL SOCIETY, 14 W. Cayuga, Moravia, NY 13118. Mailing Address: P.O. Box 247, Moravia, NY 13118-0247. Tel.: 315-497-3906.
E-mail: colhs@localnet.com
Web Site: www.colhs.org
Founded: 1966.
Congressional District: 33
Key Personnel: Pres., Roger Phillips; Sec., Sandy Morehouse.
Governing Authority: N.Y. State Charter; nonprofit. Tax-exempt: 501(c)(3).
Institution Type/Description: Local History & Genealogy Museum: housed in pre-1850 History House.
Collections: photos; clothing, household & farm artifacts; family files; manuscripts; cemetery & census microfilm & reader; vital records; newspaper abstracts; Millard Fillmore articles; scrapbooks; maps; genealogies; post cards; early era fossils.
Research Fields: local historic sites; genealogy & local history.
Facilities: library of historical, church & vital records; 800 family data files; microfilm of Moravia newspaper, cemetery & census records, available for research on premises.
Activities: guided tours; lectures; films; genealogical workshops; field trips; permanent & temporary exhibitions.
Publications: newsletter.
Hours & Admission Prices: House Tours: Sat. 10-2. Research: Mon. 9-12; other times by appointment. No charge; donations accepted. Closed holidays. &
Attendance: 400 (estimated)
Membership: Individual $10; Family $15.

Mountainville

STORM KING ART CENTER, (M), Old Pleasant Hill Rd., Mountainville, NY 10953-0280. Mailing Address: P.O. Box 280, Mountainville, NY 10953-0280. Tel.: 845-534-3115. Fax: 845-534-4457.
E-mail: info@stormkingartcenter.org
Web Site: www.stormkingartcenter.org
Founded: 1960.
Congressional District: 19
Key Personnel: Pres., John P. Stern; Chm., James H. Ottaway, Jr.; Dir. & Cur., David R. Collens; Dir. Finance, Dwayne Jarvis; Dir. Devel., Rachel Coker; Dir. Education & Public Programs, Victoria Lichtendorf.
Personnel Profile: Full-Time Paid 10; Full-Time Volunteers 2; Part-Time Paid 17; Part-Time Volunteers 40; Interns 1.
Governing Authority: nonprofit organization. Tax-exempt: 501(c)(3).
Institution Type/Description: Art Museum & Sculpture Park.
Collections: post WWII modern & contemporary sculptures including: Nogu-

chi, Serra, di Suvero, Nevelson, David Smith, Armajani, Calder, Goldsworthy; Abakanowicz; von Rydingsvard; Zhang Huan.
Major Exhibits: Virginia Overton, 5/14-11/14; Zhang Huan, 5/14-11/14.
Research Fields: 20th-century American & International sculpture.
Facilities: 2,000-vol. art library; 500-acre sculpture park. Museum-related items for sale.
Activities: permanent & temporary exhibitions; site-specific sculpture; docent guided tours; self-guided handicap accessible tram tours; Acoustiguide rental; meet the artist events; lectures; music in the park; school programs; family events; hikes; docent education; outreach to community; free admission days.
Publications: exhibition catalogues; newsletter; self-guides to outdoor sculpture collection & tram tours.
Hours & Admission Prices: April 2-Nov. 30 Wed.-Sun. 10-5:30; Memorial Day to Labor Day Sat. 10-8. Adults $12, senior citizens $10, college students with valid ID & students K-12 $8; discounts to AAM & ICOM members and bus groups of 15 or more with 2-week advance reservations; children 5 & under and members no charge. Closed Thanksgiving Day & day after. &

Attendance: 76,000 (estimated)
Membership: Student & Senior $40; Individual $50; Household $75; Contributor $100; Donor $250; Sponsor $500; Patron $1,000.

Mumford

GENESEE COUNTRY VILLAGE & MUSEUM, 1410 Flint Hill Rd., Mumford, NY 14511-0310. Mailing Address: P.O. Box 310, Mumford, NY 14511-0310. Tel.: 585-538-6822. Fax: 585-538-2887 & 6927.
E-mail: info@gcv.org
Web Site: www.gcv.org
Founded: 1966.
Congressional District: 35
Key Personnel: Pres. & C.E.O., Peter S. Arnold; Chm., Philip K. Wehrheim; C.F.O., Samantha H. Nickerson; Dir. Retail & Visitor Svcs., Robin Lott; Sr. Dir. Programs & Collections, Charles A. LeCourt; Dir. Education Svcs., Jennifer Haines; Dir. Interpretation, Brian Nagel; Sr. Dir. Guest Rels. & Administration, Christine M. Rovet.
Personnel Profile: Full-Time Paid 40; Part-Time Paid 180; Part-Time Volunteers 590; Interns 2.
Volunteer Hours: 22,250
Operating Expenses: 3,166,619
Operating Income: 3,073,209
Governing Authority: nonprofit organization. Tax-exempt.
Institution Type/Description: Art Gallery, Nature Center, Recreated Village & Living History Museum.
Collections: 68 19th-century buildings including residences, shops, offices, churches, taverns, schools, stores, brewery, barn & outbuildings; manuscripts; documents; photographs; maps; furnishings & period artifacts; John L. Wehle Gallery containing 18th & 19th century paintings, sculptures & graphics; 19th century textile & clothing collection.
Research Fields: the settlement & 19th-century development of Genesee Country including its traditions, customs, occupations & professions; histories of the component buildings of Genesee Country Museum's Country Village; 19-century hunting & sporting art; 19th-century clothing.
Facilities: 6,800-vol. library of reference books & periodicals; historic village; art gallery; nature center; gardens; cafeteria & restaurant. Handcrafted objects, related reproductions & other museum-related items for sale.
Activities: 19th-century crafts; organized programs for elementary & secondary school students; internship programs with area colleges & universities; intermuseum loan & permanent exhibitions. Museum Events: Vintage Baseball; Independence Day Celebration; Civil War Battle Reenactment; Old Time Fiddler's Fair; Agriculture Society Fair, Halloween Celebration; Annual Christmas Sale & Yuletide in the Country.
Hours & Admission Prices: May 14-Oct. 16 Tues.-Fri. 10-4, Sat.-Sun. 10-5. Adults $15.50, senior citizens over 62 & students $12.50, youth 4-16 $9.50; members and children 3 & under no charge. &
Attendance: 76,418 (estimated)
Membership: Friend $60; Friend & one $80; Family $99.

Munnsville

FRYER MEMORIAL MUSEUM, William St., Munnsville, NY 13409.
E-mail: davidsadler_13043@yahoo.com
Founded: 1976.
Congressional District: 32
Key Personnel: Town Historian, David Sadler.
Personnel Profile: Full-Time Volunteers 1; Part-Time Volunteers 1.
Governing Authority: municipal; nonprofit. Parent Institution: Town of Stockbridge. Tax-exempt: 170(b)(1)(A).

Institution Type/Description: Local History Museum: housed in 1886 Munnsville Post Office. Specialize in genealogical & local history research.

Collections: genealogical materials; artifacts; manuscripts; maps; photographs; published local history books; Munnsville Plow Co. tools & manufactured items.

Research Fields: genealogy; local Indians; cemetery records; local industry.

Activities: lectures; permanent & temporary exhibitions.

Hours & Admission Prices: By appointment. No charge; donations accepted.

Attendance: 200 (estimated)

Narrowsburg

FORT DELAWARE MUSEUM OF COLONIAL HISTORY, 6615 State Rt. 97, Narrowsburg, NY 12764. Mailing Address: Sullivan County DPW, P.O. Box 5012, Monticello, NY 12701-5192. Tel.: 845-252-6660. Fax: 845-807-0335.

Web Site: co.sullivan.ny.us

Founded: 1957.

Congressional District: 27

Personnel Profile: Part-Time Paid 9; Part-Time Volunteers 12.

Governing Authority: county. Operated by the Sullivan County legislature. Tax-exempt.

Institution Type/Description: History Museum: 1755-1785 a reconstruction of Cushetunk, the first stockaded settlement in the upper Delaware Valley.

Collections: stout stockade; cabins; blockhouses; gun platform; store house; blacksmith shed; candle shed; armory; animal yard; flintlock muskets; rifles; cannon; fowling pieces; 18th-century household furnishings; farm tools & implements; early gravestones; herb & vegetable garden; Loom Shed; log cabins; Lenape Indian dwelling.

Research Fields: genealogy.

Facilities: picnic ground; vending machines. Local handcrafts & museum-related items for sale.

Activities: guided tours; lectures; daily demonstrations including candlemaking, spinning & carding wool, weaving, musket & cannon firing; slide & taped programs; 18th-century military tactics, woodworking techniques, & basketmaking; summer military encampment; workshops in spinning & quilting.

Publications: book, Cushetunk, 1754-1784 a brief history of the early settlers who called themselves, The Delaware Company; Letters to Sarah: A Year in the Life of A Settlers Family, 1769-1770.

Hours & Admission Prices: Memorial Day to June Sat. 10-5, Sun. 12-5; late June to Labor Day Mon. & Fri.-Sat. 10-5, Sun. 12-5. Adults $7, seniors $5, children 5-14 $4. &

Attendance: 5,000 (accurate)

New City

HISTORICAL SOCIETY OF ROCKLAND COUNTY, (M), 20 Zukor Rd., New City, NY 10956-4388. Tel.: 845-634-9629. Fax: 845-634-8690.

E-mail: info@rocklandhistory.org

Web Site: www.rocklandhistory.org

Founded: 1965.

Congressional District: 26

Key Personnel: Exec. Dir., Erin L. Martin; Pres. (V), Dr. Thomas F.X. Casey.

Personnel Profile: Full-Time Paid 4; Part-Time Paid 4; Part-Time Volunteers 175; Interns 2.

Governing Authority: nonprofit organization. Tax-exempt.

Institution Type/Description: History Museum.

Collections: furniture; decorative arts; paintings; prints; photographs; documents; genealogical papers; historical materials. Historic House & Structures: restored 1832 Dutch farmhouse & outbuildings belonging to Jacob Blauvelt.

Research Fields: local & regional history.

Facilities: 500-vol. library of local history books, photographs & archives for research on premises; genealogical records, available for use by appointment. Publications & reproductions for sale.

Activities: guided tours; permanent & temporary exhibitions; lecture series; educational outreach programs; folklife programs.

Publications: quarterly pamphlet, South of the Mountains; books, Gleanings from Rockland History; Now & Then & Long Ago in Rockland County; The Way it Was in North Rockland; Wine & Bitters; The Tonnetti Years At Snedens Landing; How Things Began in Rockland County; Ladies Lib; How Rockland Women Got the Vote; Bicentennial Gleanings; Politics in the Gilded Age: A Biography of Clarence Lexow; Cole's History of Rockland County; Camp Shanks & Shanks Village; Haiti on the Hudson; Green's History of Rockland County; Growing Up in New City/Rockland in the 1790s; Adventures from the Past; Blossom; Rockland County 1900-2000, Century of History: True Stories From Mine Hole; A Catch of Grand Mothers.

Hours & Admission Prices: Daily 1-5. Adults $7, children $3; members no charge. Closed major holidays. &

Attendance: 12,000 (estimated)

Membership: New Members $15; Student & Senior Citizen $25; Individual $35; Family $40; Foreign $50; Sustaining $75; Libraries $100; Centurion $100-$249; Blauvelt Fellow $250-$499; President's Circle $500-$999; Patron $1,000-$2,500.

New Hartford

NEW HARTFORD HISTORICAL SOCIETY, 2 Paris Rd., New Hartford, NY 13413. Mailing Address: P.O. Box 238, New Hartford, NY 13413-0238. Tel.: 315-724-7258.

E-mail: nhnyhistorical@yahoo.com

Founded: 1941.

Key Personnel: Pres. (V) & Museum Shop Mgr., Barbara Couture; Archivist, Elizabeth Pettingale.

Personnel Profile: Part-Time Volunteers 1.

Operating Income: 5,000

Governing Authority: Tax-exempt.

Institution Type/Description: Historical Museum.

Collections: area history; Elliott Hughes Colonial collection.

Publications: quarterly newsletter, Tally-Ho.

Hours & Admission Prices: Mon. 1-3, Sat. 10-1. No charge; donations accepted. &

Attendance: 45 (estimated)

Membership: Individual $12; Family $17; Contributing $27; Corporate $50.

New Lebanon

*** THE SHAKER MUSEUM,** 202 Shaker Rd., New Lebanon, NY 12125. Tel.: 518-794-9100, ext. 218. Fax: 518-794-8621.

E-mail: contact@shakermuseumandlibrary.org

Web Site: www.shakermuseumandlibrary.org

Founded: 1950.

Congressional District: 24

Key Personnel: Pres., David Stocks; Chm. (V), Jeff Daly; Dir. Research, Jerry Grant.

Personnel Profile: Full-Time Paid 4; Part-Time Paid 11; Part-Time Volunteers 200.

Governing Authority: nonprofit corp. Subsidiary Institution: Shaker Library, 88 Shaker Museum Rd., Old Chatham, NY 12136. Tax-exempt: 501(c)(3).

Institution Type/Description: Shaker History & Culture Museum.

Collections: furniture; baskets; textiles; woodworking tools & machinery; metal working machinery; household equipment; personal items; books; manuscripts; photographs; craft industries; clothing; tools; North Family Site at the Mount Lebanon Shaker Village.

Research Fields: Shaker history; philosophy & religion; decorative arts; folklore; industry; textiles; utopian & communal studies; material culture.

Activities: permanent exhibits; public programs for adults, children & family groups; curriculum-based school program; guided tours by appointment; docent program; membership program.

Publications: newsletter, Broadside; Shaker Series of Monographs; catalogue, Making His Mark: The Work of Shaker Craftsman Orren Haskins.

Hours & Admission Prices: June-Sept. call for hours. No charge.

Attendance: 18,000 (estimated)

Membership: Individual $40; Family & Dual $60; Sustaining $100; Sponsor $250; Patron $500; Shaker Circle $1,000 & up.

New Paltz

HISTORIC HUGUENOT STREET, (M), 88 Huguenot St., New Paltz, NY 12561-1415. Tel.: 845-255-1660. Fax: 845-255-0376.

E-mail: info@huguenotstreet.org

Web Site: www.huguenotstreet.org

Formerly: The Huguenot Historical Society

Founded: 1894.

Congressional District: 27

Key Personnel: Exec. Dir., Tracy McNally; Pres. (V), Mary Etta Schneider; Cur., Leslie LeFevre-Stratton.

Personnel Profile: Full-Time Paid 7; Part-Time Paid 3; Part-Time Volunteers 30; Interns 3.

Governing Authority: nonprofit organization. Tax-exempt: 501(c)(3).

Institution Type/Description: Historic Site & Historic House Museum: c.1680-1890, stone houses.

Collections: 17th to 18th-century French and Dutch documents; manuscripts; 17th to 18th-century furnishings. Historic Houses: 1678; 1721 Jean (Jacob) Hasbrouck House; 1721 Abraham (Daniel) Hasbrouck house; 1698-1735 Bevier-Elting; 1692-1894 Deyo house; French Church; 1694-1735 Hugo Freer house; 1799 LeFevre house; 1705 DuBois Fort.

Research Fields: 17th-century documents; items of French Huguenot & Dutch families; genealogy & history of New York state.
Facilities: 6,000-vol. library of Huguenot history, genealogy, & local history; literature available by appointment for use on the premises; reception facilities; reading room. Books & other museum-related items for sale.
Activities: guided tours; lectures; formally organized educational programs; training programs for professional museum workers; special events.
Publications: newsletter, On Huguenot Street.
Hours & Admission Prices: Grounds: year round. Visitor's Center: May-Oct. Thurs.-Mon. Tours: 11am, 1pm, 3pm. Adults $15; discounts to AAM members; children under 6 & members no charge. &
Attendance: 14,702 (estimated)
Membership: Individual $40; Household $75; Business & Contributor $100; Donor $250; Patron $500; Benefactor $1,000.

SAMUEL DORSKY MUSEUM OF ART, STATE UNIVERSITY OF NEW YORK AT NEW PALTZ, (M), 1 Hawk Dr., New Paltz, NY 12561-2447. Tel.: 845-257-3844. Fax: 845-257-3854.
E-mail: sdma@newpaltz.edu
Web Site: www.newpaltz.edu/museum
Founded: 1963.
Congressional District: 28
Key Personnel: Chm. (V), David A. Dorsky; Mgr. Art Collections, Wayne Lempka; Dir. Neil C. Trager, Sara J. Pasti; Museum Educator, Judi Esmond; Cur., Daniel Belasco; Assoc. Cur. Collections, Dr. Jaimee Uhlenbrock; Preparator, Robert Wagner; Coord. Visitor Svcs., Amy Pickering.
Personnel Profile: Full-Time Paid 3; Part-Time Paid 3; Part-Time Volunteers 2; Interns 2.
Governing Authority: university. Parent Institution: State University of New York at New Paltz. Tax-exempt.
Institution Type/Description: Art Gallery.
Collections: 19th & 20th-century American prints & paintings; regional art; Asian prints; pre-Columbian artifacts; contemporary & historical photography; metals; posters.
Activities: lectures; concerts; video screenings; temporary & traveling exhibitions; gallery talks; docent tours.
Publications: two annual catalogues.
Hours & Admission Prices: Aug. Sat.-Sun. 11-5; Sept.-July Wed.-Sun. 11-5. Suggested Donations $5. Closed legal & school holidays. &
Attendance: 16,000 (estimated)
Membership: Student $15; Seniors $25; Friend $50; Donor $100; Sponsor $250; Patron $500; Benefactor $1,000; Director's Circle $2,500.

New Rochelle

COLLEGE OF NEW ROCHELLE CASTLE GALLERY, 29 Castle Place, New Rochelle, NY 10805. Tel.: 914-654-5423. Fax: 914-654-5014.
Web Site: www.cnr.edu/cg.htm
Founded: 1980.
Key Personnel: Dir., Katrina Rhein; Gallery Mgr., Michelle Jammes
Institution Type/Description: Art Gallery: housed in Leland Castle; built in 1855. Listed on the National Register of Historic Places.
Collections: photographs; sculpture; paintings; ceramics; drawings.
Activities: lectures; temporary exhibitions; educational programs; school & group tours; outreach programs.
Hours & Admission Prices: Sept.-June Tues. & Thurs.-Fri. 10-5, Wed. 10-8, Sat.-Sun. 12-4. No charge. Closed holidays.
Membership: Individual $35; Friend $50; Patron $100; Sponsor $500; Benefactor $1,000.

New Windsor

NATIONAL PURPLE HEART HALL OF HONOR, 374 Temple Hill Rd., New Windsor, NY 12553. Mailing Address: P.O. Box 207, Vails Gate, NY 12584-0207. Tel.: 845-561-1765; 877-284-6667. Fax: 845-569-0382.
Web Site: www.thepurpleheart.com
Founded: 2006.
Key Personnel: Dir., Anita Pidala; Program Dir., Peter Bedrossian.
Personnel Profile: Full-Time Paid 3; Part-Time Paid 8; Part-Time Volunteers 2.
Governing Authority: Parent Institution: New York State, Office of Parks, Recreation & Historic Preservation. Subsidiary Institution: Palisades Interstate Park Commission.
Institution Type/Description: Military History Museum: commemorates America's military personnel that were wounded or killed by enemy action.
Collections: Purple Heart recipients; military artifacts; photographs; films; hands-on exhibits.
Research Fields: Purple Heart recipients & their stories
Facilities: theater; ceremonial grounds.

Activities: interactive exhibits; interactive computer terminals; educational programs. Annual Event: Purple Heart Award Ceremony.
Hours & Admission Prices: Mon.-Sat. 10-5, Sun. 1-5. No charge; donations accepted. Closed most holidays. &
Attendance: 23,031 (accurate)

NATIONAL TEMPLE HILL ASSOCIATION, INC., Edmonston House, Headquarters, 1042 Rte. 94, New Windsor, NY 12553. Mailing Address: Edmonston House, Headquarters, P.O. Box 315, Vails Gate, NY 12584-0315. Tel.: 845-561-5073. Fax: 845-561-5073.
Founded: 1933.
Congressional District: 21
Key Personnel: Pres. (V), Daniel S. Lucia.
Personnel Profile: Part-Time Paid 5; Part-Time Volunteers 15.
Governing Authority: nonprofit organization. Subsidiary Institution: Last Encampment of the Continental Army. Tax-exempt: 501(c)(3).
Institution Type/Description: Historic House: c.1755 Edmonston House served as headquarters during Revolutionary War for General Horatio Gates & Maj. Gen. Arthur St. Clair.
Collections: period furnishings; 18th-century architectural artifact with significant connection to the American Revolutionary events.
Research Fields: Colonial & Revolutionary Period of local area.
Facilities: meeting & research rooms.
Activities: open house; crafts workshops; programs & interpretation at the Last Encampment of the Continental Army on Rte. 300.
Publications: quarterly bulletin.
Hours & Admission Prices: Edmonston House: July-Sept. Sun. 2-5. Last Encampment of the Continental Army: late April to Oct. daylight hours; Guides Thurs.-Sun. 12-4:30. No charge; donations accepted.
Attendance: 1,200 (estimated)
Membership: Student $5; Individual $15; Couple $25; Business $50; Corporate $100; Life $1,000.

NEW WINDSOR CANTONMENT STATE HISTORIC SITE, 374 Temple Hill Rd., Rte. 300, New Windsor, NY 12553. Mailing Address: P.O. Box 207, Vails Gate, NY 12584-0207. Tel.: 845-561-1765. Fax: 845-561-6577.
E-mail: michael.mcgurty@parks.ny.gov
Web Site: www.nysparks.com
Founded: 1918.
Congressional District: 26
Key Personnel: Interpretive Programs Asst., Michael S. McGurty.
Personnel Profile: Full-Time Paid 2; Part-Time Paid 1; Part-Time Volunteers 200.
Governing Authority: state government. Parent Institution: New York State Office of Parks, Recreation & Historic Preservation. Palisades Interstate Park Commission. Tax-exempt: 501(c)(3).
Institution Type/Description: Historic Site: living history museum located on site of last encampment of Washington's northern Continental Army, 1782-1783.
Collections: Revolutionary War; original & reproduction military equipment & artifacts; local history documents from late 18th century to early 1800s; reconstruction of the Temple Building; Temple Hill Monument; Continental Army artillery; Continental Artillery Park of 1780-81; National Purple Heart Hall of Honor.
Research Fields: American Revolution encampments.
Facilities: 750-vol. library of history available by advance request; 150-seat program area; picnic area.
Activities: living history; blacksmithing & military demonstrations daily; special events; school groups.
Hours & Admission Prices: Mon.-Sat. 10-5, Sun. 1-5. Admission for group tours & educational programs. Closed New Year's Day; Columbus Day; Thanksgiving; Christmas. &
Attendance: 23,000 (accurate)
Membership: Friends of the State Historic Sites of the Hudson Highlands: Individual $15; Household $25; Contributing $50; Business Corporate $100; Sustaining $100-$499; Patron $500-$999; Benefactor $1,000 & up.

New York

ACA GALLERIES - AMERICAN CONTEMPORARY ARTISTS, 529 W. 20th St., 5th Fl., New York, NY 10011. Tel.: 212-206-8080. Fax: 212-206-8498.
E-mail: info@acagalleries.com
Web Site: www.acagalleries.com
Founded: 1932.
Personnel Profile: Full-Time Paid 6; Full-Time Volunteers 1; Part-Time Volunteers 2.

Institution Type/Description: Art Gallery.
Collections: 19th & 20th century European, American & contemporary art; paintings; sculpture; drawings.
Hours & Admission Prices: Jan. 4-Aug. 13 & Sept. 8-Dec. 23 Tues.-Sat. 10:30-6; other times by appointment. No charge. &
Attendance: 7,500 (estimated)

ABRONS ARTS CENTER/HENRY STREET SETTLEMENT, 466 Grand St., New York, NY 10002-4804. Tel.: 212-598-0400. Fax: 212-505-8329.
E-mail: jdurham@henrystreet.org
Web Site: www.abronsartscenter.org
Formerly: The Main Gallery of Henry Street Settlement/Abrons Art Center
Founded: 1893.
Key Personnel: Dir. Abrons Arts Center, Jay Wegman; C.E.O. & Dir., Verona Middleton Jeter; Chm. (V), Robert Harrison; Pres. (V), Dale Burch; House & Night Mgr., Carl Johnson; Deputy Dir. & Visual Arts Dir., Susan Flemmger; Dir. Visual Arts, Jonathan Durham; Administrative Asst., Wanda Egipciaco; Registrar, Sonia Diaz.
Personnel Profile: Full-Time Paid 3; Part-Time Volunteers 2; Interns 3.
Governing Authority: nonprofit. Parent Institution: Henry Street Settlement; Subsidiary Institution: Abrons Arts Center. Tax-exempt.
Institution Type/Description: Arts Center.
Collections: changing thematic group exhibits of contemporary art, in all media; changing solo photography exhibits.
Activities: gallery tours; workshops; family programs; instruction in all arts disciplines (visual arts, dance, theater & music); artist-in-residence workspace program.
Hours & Admission Prices: Tues.-Sat. 12-6. No charge; donations accepted. Closed national holidays. &
Attendance: 40,000 (accurate)
Membership: Balcony $55-$120; Mezzanine $200; Orchestra $500; Stage $1,000.

AMERICAN ACADEMY OF ARTS AND LETTERS, (M), 633 W. 155th St., New York, NY 10032-7501. Tel.: 212-368-5900. Fax: 212-491-4615.
E-mail: academy@artsandletters.org
Web Site: www.artsandletters.org
Founded: 1898.
Congressional District: 16
Key Personnel: Exec. Dir., Virginia Dajani; Pres., Henry N. Cobb; Music Programs, Ardith Holmgrain; Exec. Asst., Kristen Stevens; Archivist, Nancy Malloy; Library & Publications, Kathleen Dinah Trocino; Art Awards, Exhibitions & Collections, Souhad Rafey; Literature & Richard Rodgers Awards, Jane Bolster; Architecture Awards, Cody Upton; Asst. to Dir., David Clarke-Hazlett.
Personnel Profile: Full-Time Paid 7; Part-Time Paid 6.
Governing Authority: nonprofit organization. Tax-exempt: 501(c)(3).
Institution Type/Description: Art Museum & Library.
Collections: works by & about members of the American Academy of Arts & Letters; works by Childe Hassam & Eugene Speicher; art; archives; painting; sculpture; graphics; music; decorative arts.
Research Fields: art; literature; music; architecture.
Facilities: 22,000-vol. library including musical scores & books of the works of or about the artists, architects, writers & composers who have been members of the American Academy of Arts & Letters. Archives containing correspondence & manuscripts relating to members. Open to scholars by appointment.
Activities: rotating exhibitions of art and manuscripts. Organization Sponsors: yearly cash awards for painting, sculpture, graphics, architecture, music & literature; all awards by nomination of the membership with no applications accepted; annual purchase of contemporary paintings, drawings and graphics for donation to museums.
Publications: catalogs of exhibitions, Proceedings of the American Academy of Arts & Letters.
Hours & Admission Prices: Galleries: Thurs.-Sun. 1-4, when exhibitions are held. No charge. Office: Mon.-Fri. 9:30-5. Closed holidays. &
Attendance: 2,500

AMERICAN FOLK ART MUSEUM, (M), 2 Lincoln Sq., Columbus Ave. @ 66th St., New York, NY 10023. Mailing Address: 1865 Broadway, 11th Fl., New York, NY 10023-7505. Tel.: 212-595-9533.
E-mail: info@folkartmuseum.org
Web Site: www.folkartmuseum.org
Founded: 1961.
Congressional District: 17
Key Personnel: Dir., Hon. Anne-Imelda Radice; Pres., Edward V. Blanchard,

Jr.; Dir. Devel., Elizabeth Kingman; Dir. Exhibitions & Sr. Cur., Stacy Hollander; Dir. Education & Mgr. Public Rels., Rachel Rosen; Dir. Publications, Tanya Heinrich; Registrar, Ann-Marie Reilly; Cur. Art of the Self-Taught & Art Brut, Dr. Valerie Rousseau; Public Rels. Mgr., Barbara Livenstein; Dir. Institutional Giving, Karley Klopfenstein; Museum Shop Mgr., Marie DiManno.
Personnel Profile: Full-Time Paid 16; Part-Time Paid 5; Part-Time Volunteers 60; Interns 2.
Operating Expenses: 4,160,550
Operating Income: 4,122,199
Governing Authority: nonprofit organization. Tax-exempt: 501(c)(3).
Institution Type/Description: Folk Art Museum.
Collections: American 18th- to 20th-century folk sculpture & painting; art made by the self-taught; textile arts; decorative arts; environmental folk art; traditional folk art; history; furnishings; personal artifacts; recreational artifacts.
Major Exhibits: Folk Couture: Fashion and Folk Art, 1/14-4/14; Self-Taught Genius: Treasures from the American Folk Art Museum (T), 5/14-8/14.
Research Fields: folk & decorative arts; work of 20th-century self-taught artists; history.
Facilities: library; classroom. Books, gifts & reproductions for sale.
Activities: gallery talks; inter-museum loan, temporary & traveling exhibitions; classes; outreach programs to schools & interested groups, including lectures, workshops & folk art slide presentations; docent program including classes in folk art & decorative arts; internship program; tours; antique show previews.
Publications: members' journal, exhibition catalogues; permanent collection catalogues; folk art books.
Hours & Admission Prices: Tues.-Sat. 12-7:30, Sun. 12-6; guided tours by appointment. No charge. &
Attendance: 88,800 (accurate)
Membership: Student & Senior Citizen $60; Individual $70; Dual & Family $90; Contributing $150; Young Patron $150; Sustaining $250; Benefactor $500; Director's Circle $1,000; Collector's Circle $2,500.

AMERICAN IRISH HISTORICAL SOCIETY, 991 5th Ave., New York, NY 10028-0101. Tel.: 212-288-2263. Fax: 212-628-7927. Facebook: American Irish Historical Society.
E-mail: aihs@aihs.org
Web Site: www.aihs.org
Founded: 1897.
Congressional District: 14
Key Personnel: Pres. Gen., Kevin M. Cahill, M.D.; Exec. Dir., Christopher P. Cahill.
Personnel Profile: Full-Time Paid 3; Part-Time Paid 2; Part-Time Volunteers 4; Interns 4.
Governing Authority: historical society; nonprofit organization. Subsidiary Institution: AIHS, California. Tax-exempt: 501(c)(3).
Institution Type/Description: Research Library: housed in c.1900 townhouse.
Collections: Irish history & culture; history of Irish in America.
Facilities: 10,000-vol. library of Irish & American Irish references, available for use by public by appointment; reading rooms.
Activities: concerts; lectures; readings; recitals; play readings; receptions; presentations; permanent & temporary exhibits; concerts.
Publications: annual literary magazine, The Recorder; quarterly events newsletter.
Hours & Admission Prices: Mon.-Fri. 10-5.
Attendance: 3,000 (estimated)
Membership: Individual $100; Sustaining $250; Contributing $500; Sponsor $1,000; Benefactor $2,000; Patron $2,500.

AMERICAN JEWISH HISTORICAL SOCIETY, 15 W. 16th St., New York, NY 10011-6301. Tel.: 212-294-6160. Fax: 212-294-6161.
E-mail: info@ajhs.org
Web Site: www.ajhs.org
Founded: 1892.
Congressional District: 4
Key Personnel: Pres. (V), Paul Warhit; Exec. Dir., Jonathan Karp; Dir. Library & Archives, Susan Malbin, Ph.D.; Membership Sec., Rachel Lobovsky.
Personnel Profile: Full-Time Paid 12; Part-Time Paid 2; Part-Time Volunteers 7; Interns 3.
Governing Authority: nonprofit. Branch Museum: 160 Herrick Rd., Newton, MA 02459. Tel.: 617-559-8880. Tax-exempt: 501(c)(3).
Institution Type/Description: Ethnic History Museum.
Collections: archival & library holdings relating to the American Jewish experience from the 16th century to present; colonial & early Federal portraits; family materials; Yiddish theatre collections of posters, music &

actors; objects of daily life; sports memorabilia; works on paper; photographs; silhouettes; textiles; 9,000 manuscript items; 10,000 museum items.

Research Fields: American Jewish history.

Facilities: 40,000-vol. library of books, pamphlets, bulletins, periodicals & newspapers of the history of American Jews; reading room; exhibition galleries.

Activities: guided tours; lectures; films; inter-museum loan, permanent & temporary exhibits.

Publications: quarterly journal, American Jewish History; newsletter; books; tours in Europe & America on sites of Jewish interest; catalogs.

Hours & Admission Prices: Office: Mon.-Thurs. 9-5, Fri. 9-2, Sun. 11-5. Reading Room: Mon. 9:30-7:30, Tues.-Thurs. 9:30-4:30, Fri. 9:30-1:30, Sun. 11-4. No charge. Closed national holidays & major Jewish holidays.

Attendance: 2,000 (estimated)

Membership: Student $18; Library & Regular $50; Centennial $100; Preservation Fellow $360; Patron $540; American Jewish Heritage Fellow $1,000.

*** AMERICAN MUSEUM OF NATURAL HISTORY, (M),** Central Park West at 79th St., New York, NY 10024-5193. Tel.: 212-769-5100. Fax: 212-769-5018.

E-mail: communications@amnh.org

Web Site: www.amnh.org

Founded: 1869.

Congressional District: 17

Key Personnel: Pres., Ellen V. Futter; Chm. Bd. Trustees, Lewis W. Bernard; Sr. Vice Pres. & Provost, Dr. Michael J. Novacek; Sr. Vice Pres. & C.F.O., Ellen Gallagher; Sr. Vice Pres. & Gen. Counsel, Gerald R. Singer; Vice Pres. Operations & Capital Programs, Ann Siegel; Sr. Vice Pres. Exhibition, David Harvey; Sr. Vice Pres. Institutional Advancement, Education & Strategic Planning, Lisa J. Gugenheim; Sr. Vice Pres. Communications & Mktg., Anne Canty; Assoc. Dean Science Education and Exhibition, & Assoc. Cur. Div. Vertebrate Zoology, Dr. Christopher Raxworthy; Assoc. Dean Science Collections, Scott Schaefer; Chm. & Cur. Division Anthropology, Dr. Charles S. Spencer; Frederick P. Rose Dir., Hayden Planetarium, Dr. Neil deGrasse Tyson; Chm. & Cur. Division Invertebrate Zoology, Dr. Ward Wheeler; Chm. & Cur. Division Physical Sciences, Dr. Mordecai-Mark MacLow; Chm. & Cur. Division Paleontology, Dr. Mark Norell; Chm. & Cur. Division Vertebrate Zoology, Dr. Nancy B. Simmons; Dir. Center for Biodiversity & Conservation, Dr. Eleanor J. Sterling.

Personnel Profile: Full-Time Paid 996; Part-Time Paid 336; Part-Time Volunteers 1,174.

Governing Authority: board of trustees. Tax-exempt: 501(c)(3).

Institution Type/Description: Natural History Museum.

Collections: amphibians; anthropology; arthropods; birds; dinosaurs; films; fishes; frozen DNA tissue; gems; invertebrates; mammals; manuscripts; meteorites; minerals; photographs; prints; books; reptiles; rocks; shells.

Research Fields: anthropology; biodiversity conservation; comparative biology; genomics; invertebrate zoology; paleontology; physical sciences; vertebrate zoology.

Facilities: 487,000-vol. research library; The Rose Center for Earth & Space including the Hayden Planetarium; The Richard Gilder Graduate School; The Sackler Institute for Comparative Genomics; Parallel Computing Cluster; The Center for Biodiversity and Conservation; molecular laboratories; National Center for Science, Literacy, Education and Technology; Discovery room; Natural Science Center; classrooms; conservation laboratories; remote field research station; 1,000-seat LeFrak IMAX theater; 150-seat & 300-seat auditoriums; food court; cafe. Museum-related items for sale.

Activities: field expeditions; Ph.D. graduate program, undergraduate, doctoral & post-doctoral science training programs; temporary & traveling exhibitions; teacher training & professional development programs; web-based curricula & educational resources; formally organized education programs for children, families & adults; scientific conferences & symposia; inter-museum loan; AMNH Expeditions travel program; film & video festival; Moveable Museum program.

Publications: Novitates; The Bulletin of the American Museum of Natural History; Anthropological Papers of the American Museum of Natural History; quarterly calendar; scholarly books; textbooks; hall guides; teacher training material; brochures.

Hours & Admission Prices: Museum: daily 10-5:45. Rose Center: daily 10-5:45. Library: Tues.-Thurs. 2-5:30. Special Collections: by appointment. Suggested Donations: adults $19, students & seniors $14.50, children 2-12 $10.50; discounts to NYC Cultural Institutions Group members; members no charge. Additional fees for special exhibitions. Closed Thanksgiving; Christmas. &

Attendance: 5,000,000 (estimated)

Membership: Digital $60 & up; Individual $95 & up; Family $125 & up; Junior Council $375 & up; Patron Circle $1,750 & up.

AMERICAN NUMISMATIC SOCIETY, (M), 75 Varick St., Ste. 1101, New York, NY 10013-1917. Tel.: 212-571-4470. Fax: 212-571-4479.

Web Site: www.numismatics.org

Founded: 1858.

Congressional District: 8

Key Personnel: Exec. Dir., Dr. Ute Wartenberg; Pres. (V), Roger Siboni; American Cur., Robert Hoge; Greek Cur., Dr. Peter van Alfen; Collections Mgr., Dr. Elena Stolyarik; Research Scientist, Dr. Sebastian Heath; Dir. Finance & Operations, Anna Chang; Librarian, Elizabeth Hahn; Membership Assoc., Megan Fenselau; Museum Admin., Joanne Isaac; Deputy Dir., Andrew Meadows.

Personnel Profile: Full-Time Paid 15; Part-Time Paid 1; Part-Time Volunteers 6; Interns 4.

Governing Authority: society. Tax-exempt.

Institution Type/Description: Numismatics History Museum.

Collections: numismatics; history; archaeology; manuscripts; books.

Research Fields: history; art history; economics related to numismatic evidence.

Facilities: 100,000-vol. library of books on numismatics available for use on premises; reading room.

Activities: guided tours; lectures; formally organized education programs for graduate students; training programs for professional museum workers; inter-museum loan, permanent, temporary & traveling exhibitions; numismatic conversation; webcasts. Annual Events: Coinage of the Americas Conference; Krause-Mishler Forum; Archer M. Huntington Award; The Groves Forum; The Sylvia Mani Hunter Memorial Lecture; The Mark M. Salton Memorial Lecture; The Harry W. Fowler Memorial Lecture.

Publications: magazine, American Numismatic Society; American Journal of Numismatics; Numismatic Literature; Colonial Newsletter; Series, Ancient Coins in North American Collections; Numismatic Notes and Monographs; Coinage of the Americas Conference Proceedings.

Hours & Admission Prices: Library: Tues.-Fri. 9:30-12 & 1-4:30 by appointment. Coin Research: Tues.-Fri. 9:30-4:30 by appointment. No charge. Federal Reserve Bank exhibit, 33 Liberty St.: Mon.-Fri. 10-4. Closed New Year's Day; Martin Luther King Jr. Day; Presidents' Day; Memorial Day; Independence Day; Labor Day; Thanksgiving & day after; Christmas. &

Attendance: 40,000 (estimated)

Membership: Student $35; Basic Associate $55; Full Associate $80; Foreign Associate $100; Fellow $110; Library Associate $150; Corporate Associate $500; Life Fellow & Life Associate $7,500.

AMERICAS SOCIETY, (M), 680 Park Ave., (at 68th St.), New York, NY 10065-5072. Tel.: 212-249-8950. Fax: 212-249-5868.

E-mail: artgallery@as-coa.org

Web Site: www.as-coa.org/visualarts

Formerly: Center for Inter-American Relations

Founded: 1965.

Congressional District: 15

Key Personnel: Honorary Chm., David Rockefeller; Chm. (V), William Rhodes; Pres., Susan Segal; Dir. Visual Arts, Gabriela Rangel; Asst. Cur. & Exhibition Coord., Christina DeLeon.

Personnel Profile: Full-Time Paid 39; Part-Time Paid 12; Part-Time Volunteers 1.

Governing Authority: nonprofit organization. Tax-exempt: 501(c)(3).

Institution Type/Description: Art Gallery.

Collections: Latin American paintings & drawings; Latin America, Canada & the Caribbean cultures.

Research Fields: Latin American, Caribbean and Canadian art & culture.

Facilities: 3,000-vol. library of catalogues & books on Latin American art; 4,000 artists registry files with biographical information and/or slides. Exhibitions catalogs for sale.

Activities: temporary exhibits; inter-museum loans; lectures; educational programming; concerts; films; videos; panel discussions; symposia.

Publications: biannual, Review: Latin American Literature and Arts; exhibition catalogues.

Hours & Admission Prices: Wed.-Sat. 12-6. Gallery & Public Programs: no charge; donations accepted. Concerts: adults $15, students & senior citizens $10; discounts to members. Closed major holidays. &

Attendance: 12,000 (estimated)

Membership: Student & Senior Citizen $35; Individual $85; Sustaining $500; Donor $1,000; Patron $5,000.

ANTHOLOGY FILM ARCHIVES, 32 Second Ave., New York, NY 10003-8631. Tel.: 212-505-5181. Fax: 212-477-2714.

E-mail: robert@anthologyfilmarchives.org

Web Site: www.anthologyfilmarchives.org

Founded: 1970.

Congressional District: 15

Key Personnel: Pres. (V) & Dir., Jonas Mekas; Bd. Chm., Barney Oldfield; Financial Dir., John Mhiripiri; Membership, Wendy Dorsett; Public Rels., Stephanie Grey; Archivist, Andrew Lampert.
Personnel Profile: Full-Time Paid 8; Part-Time Paid 3; Part-Time Volunteers 6; Interns 5.
Governing Authority: private; nonprofit organization. Tax-exempt: 501(c)(3).
Institution Type/Description: Film Museum.
Collections: over 12,000 avant-garde, classic films & documentaries; 6,000 photographs.
Research Fields: Jim Davis; Storm DeHirsch; Marie Menken.
Facilities: 8,000-vol. library; 190-seat large screen theater. Museum-related items for sale.
Activities: films. Annual Event: Film Preservation Honors Dinner.
Publications: quarterly film exhibition schedule; Annual Film Preservation Honors Dinner Journal; catalogs; books.
Hours & Admission Prices: Jan. to 3rd week in Aug. & Sept. to 3rd week in Dec. Film Screenings: Mon.-Fri. 6-11, Sat.-Sun. 3-11. Adults $9, senior citizens & students $5; members no charge.
Attendance: 50,000 (estimated)
Membership: Student & Senior $30; Adult $50; Dual Adult $75; Donor $250; Preservation Donor $1,500.

APERTURE FOUNDATION, 547 W. 27th St., 4th Fl., New York, NY 10001-5511. Tel.: 212-505-5555. Fax: 212-979-7759.
E-mail: info@aperture.org
Web Site: www.aperture.org
Founded: 1952.
Key Personnel: Dir., Juan Garcia de Oreyza; Book Publisher, Lesley A. Martin; Chm. (V), Celso Gonzalez-Falla; Dir. Limited Edition Prints, Kelly Mclaughlin.
Personnel Profile: Full-Time Paid 35; Part-Time Paid 2; Interns 10.
Governing Authority: private; nonprofit. Tax-exempt: 501(c)(3).
Institution Type/Description: Photography Foundation.
Collections: Paul Strand archive.
Research Fields: photography; fine arts.
Facilities: Limited edition prints & books for sale.
Activities: educational programs; work scholars; portfolio reviews; traveling exhibitions; special events.
Publications: magazine, Aperture; photography books.
Hours & Admission Prices: Mon.-Sat. 10-6; call for extended hours. No charge. Closed New Year's Eve & Day; Martin Luther King Jr. Day; Presidents' Day; Memorial Day; Independence Day; Labor Day; Thanksgiving & day after; Christmas Eve, Day & week.
Attendance: 16,887 (accurate)
Membership: Art Circle $1,000; Art Net $2,500; Art Council $5,000; Art Alliance $10,000.

ARSENAL GALLERY, Arsenal Bldg., Central Park, Fifth Ave. & 64th St., New York, NY 10065. Mailing Address: New York City Dept. of Parks & Recreation, The Arsenal, Central Park, Rm. 20, New York, NY 10065. Tel.: 212-360-8163.
E-mail: artandantiquities@parks.nyc.gov
Web Site: www.nyc.gov/parks/art
Governing Authority: Parent Institution: NYC Parks & Recreation. Tax-exempt.
Institution Type/Description: Art Gallery.
Collections: paintings; drawings; sculpture.
Activities: temporary exhibitions.
Hours & Admission Prices: Mon.-Fri. 9-5. No charge.

ARTHUR A. HOUGHTON JR. GALLERY & THE GREAT HALL GALLERY, 7 E. 7th St., Foundation Bldg., New York, NY 10003-8128. Mailing Address: 30 Cooper Square, New York, NY 10003-7120. Tel.: 212-353-4200.
Founded: 1859.
Key Personnel: Dean of Art School, Saskia Bos
Institution Type/Description: Art Gallery.
Collections: graphic design; fine arts; painting; sculpture; engineering.
Hours & Admission Prices: Mon.-Fri. 12-7, Sat. 12-5. No charge.

ARTISTS SPACE, 38 Greene St., 3rd Fl., New York, NY 10013-2505. Tel.: 212-226-3970. Fax: 212-966-1434. Facebook: Artists Space.
E-mail: info@artistsspace.org
Web Site: www.artistsspace.org
Founded: 1972.
Congressional District: 8
Key Personnel: Exec. Dir., Stefan Kalmar.

Governing Authority: Subsidiary Institution: 55 Walker St., New York, NY 10013.
Institution Type/Description: Art Gallery.
Collections: works by contemporary artists.
Facilities: 7,000 sq. ft. exhibition space.
Activities: temporary exhibitions; screenings; talks; book launches; educational programs.
Publications: exhibition booklets.
Hours & Admission Prices: Wed.-Sun. 12-6. Suggested Donation $5; members no charge. Closed New Year's Eve & Day; Independence Day; Thanksgiving; Christmas Eve & Day.
Attendance: 10,000 (estimated)
Membership: Artist $40; Individual $60; Friends of Artists Space $2,500 & up.

ASIA SOCIETY MUSEUM, 725 Park Ave., New York, NY 10021-5088. Tel.: 212-288-6400. Fax: 212-517-7246 & 8315.
Web Site: www.asiasociety.org
Formerly: Asia Society Galleries
Founded: 1956.
Congressional District: 14
Key Personnel: Pres., Vishakha N. Desai; Exec. Vice Pres., Jamie Metzl; Interim Chm. (V), Charles R. Kaye; Museum Dir., Melissa Chiu; Assoc. Dir., Marion Kocot; Cur. Traditional Art, Adriana Proser; Assoc. Cur. Contemporary Art, Miwako Tezuka; Museum Shop Mgr., Anne Godshall.
Personnel Profile: Full-Time Paid 100; Part-Time Paid 5; Part-Time Volunteers 50; Interns 7.
Governing Authority: nonprofit organization. Dept. of The Asia Society, Inc. Tax-exempt: 501(c)(3).
Institution Type/Description: Asian Art Museum.
Collections: Mr. & Mrs. John D. Rockefeller 3rd collection of Asian art.
Research Fields: Asian art.
Facilities: conference & lecture rooms; 258-seat auditorium. Publications, books, catalogs & gift items for sale.
Activities: guided tours of current exhibitions; conferences; symposia; performances; lectures & films; inter-museum loan & traveling exhibitions.
Publications: Archives of Asian Art; exhibition catalogs; collection catalogue; brochures.
Hours & Admission Prices: July to Labor Day Tues.-Thurs. & Sat.-Sun. 11-6; Sept.-June Tues.-Thurs. & Sat.-Sun. 11-6, Fri. 11-9. Adults $10, seniors $7, students $5; discounts to AAM & ICOM members; members, children under 16 accompanied by a parent & Fri. 6-9 no charge. Closed New Year's Day; Independence Day; Thanksgiving; Christmas.
Attendance: 84,580 (estimated)
Membership: Student, Senior & Associate $40; Individual $65; Dual & Family $120; Asia Circle $150; Contributing $250; Sustaining $500; President's Circle $1,000; Friends of Asian Arts $1,500.

ASIAN AMERICAN ARTS CENTRE, 111 Norfolk St., Ofc 1, New York, NY 10002-3394. Tel.: 212-233-2154. Fax: 360-283-2154.
E-mail: aaacinfo@artspiral.org
Web Site: www.artspiral.org
Founded: 1974.
Congressional District: 15
Key Personnel: C.E.O. & Cur., Robert Lee; Chm. (V), Eleanor Yong.
Personnel Profile: Full-Time Paid 1; Full-Time Volunteers 1; Part-Time Paid 1; Part-Time Volunteers 5; Interns 5.
Governing Authority: nonprofit. Tax-exempt: 501(c)(3).
Institution Type/Description: Contemporary American Culture Art Museum.
Collections: June 4th Movement; artist archive; folk arts; contemporary visual art works; folk art.
Research Fields: Milieu (the Asian American artists from 1945-1965); Asian folk art.
Facilities: 500-vol. Asian American artist library; archive of Asian American art.
Activities: Museum Sponsors: folk & contemporary art exhibition; panel discussions; slide shows.
Publications: biannual newsletter, Art-Spiral.
Hours & Admission Prices: Tues.-Fri. 12:30-6:30, Sat. 1:30-4:30. No charge; donations accepted.
Attendance: 5,000 (estimated)
Membership: Individual $35.

BABCOCK GALLERIES, 525 W. 25th St., 11th Fl., New York, NY 10001-5501. Tel.: 212-767-1852. Fax: 212-767-1857.
E-mail: info@babcockgalleries.com
Web Site: www.babcockgalleries.com
Founded: 1852.
Key Personnel: Owner, John Driscoll, Ph.D.; Asst. Dir., Tess Sol Schwab;

Gallery Assoc., Alexandra V. Tagami; Cur. John Frederick Kensett Catalogue Raisonne, Huntley Plattt; Gallery Asst., Sara Gilbert Lisa Koonce
Institution Type/Description: Art Gallery.
Collections: American art specializing in 19th- & 20th-century works; works by contemporary artists.
Publications: artist catalogues & books.
Hours & Admission Prices: Mon.-Fri. 10-5, Sat. by appointment. No charge.

BARD GRADUATE CENTER: DECORATIVE ARTS, DESIGN HISTORY, MATERIAL CULTURE, (M), 18 W. 86th St., New York, NY 10024-3602. Tel.: 212-501-3000. Fax: 212-501-3079. TDD: 212-501-3012 (for public programs only).
E-mail: Mulligan@BGC.bard.edu
Web Site: www.bgc.bard.edu
Founded: 1993.
Congressional District: 6
Key Personnel: Dir., Dr. Susan Weber; Dean, Dr. Peter N. Miller; Dean Academic Admin. & Student Affairs, Elena Pinto Simon; Dir. Finance & Admin., Lorraine Bacalles; Chief Cur., Nina Stritzler-Levine; Editor West 86th: A Journal of Decorative Arts, Design History and Material Culture, Paul Stirton; Dir. Devel., Susan Wall; Dir. External Affairs, Tim Mulligan.
Personnel Profile: Full-Time Paid 73; Part-Time Paid 2.
Governing Authority: nonprofit. Parent Institution: Bard College, Annandale-on-Hudson, NY. Tax-exempt: 501(c)(3).
Institution Type/Description: Decorative Art Academic Institution & Gallery.
Collections: decorative arts, design, & material culture.
Research Fields:
Facilities: library.
Activities: educational programs; guided gallery tours; symposia; seminars; lectures; walking tours; trips; family programs; senior programs.
Publications: West 86tj Street, semiannual journal; Monograph series on the Cultural History of the Material World; exhibition catalogues.
Hours & Admission Prices: Tues.-Wed. & Fri.-Sun. 11-5, Thurs. 11-8. Adults $7, senior citizens & students $5; discounts to AAM & ICOM members and museum staff; Thurs. after 5pm & children under 12 no charge. Closed New Year's Day; Martin Luther King Jr.'s Day; Easter; Memorial Day; Independence Day; Labor Day; Thanksgiving & day after; Christmas. &
Attendance: 3,000 (estimated)

BELVEDERE CASTLE, mid-Central Park at 79th St., New York, NY 10021. Mailing Address: Central Park Conservancy, 14 E. 60th St., New York, NY 10022. Tel.: 212-772-0210. Fax: 212-772-0214.
Web Site: www.centralparknyc.org
Key Personnel: Dir. Public Programs, Terry Carta.
Governing Authority: nonprofit. New York Dept. of Parks & Recreation & Central Park Conservancy. Tax-exempt: 501(c)(3).
Institution Type/Description: Historic Site & Preservation Project: 1872 Gothic tower within 843 acre landmark park designed by Frederick Law Olmsted and Calvert Vaux.
Collections: hands-on exhibitions; science, natural history & ecology; National Weather Service station.
Activities: school group & weekend family workshops; tours.
Publications: quarterly calendar of events; map; guide.
Hours & Admission Prices: Sat.-Sun. 10-5. No charge; donations accepted. Closed New Year's Day; Thanksgiving; Christmas.
Attendance: 100,000 (accurate)
Membership: Gardener $50; Arborist $100; Protector $250; Contributor $500; Belvedere Knight $1,000; Bethesda Angel $2,500; Patron's Circle $5,000; President's Circle $10,000; Chairman's Circle $25,000.

BERNARD JUDAICA MUSEUM, CONGREGATION EMANU-EL OF THE CITY OF NEW YORK, (M), 1 E. 65th St., New York, NY 10065-6501. Tel.: 212-744-1400. Fax: 212-570-0826.
E-mail: info@emanuelnyc.org
Web Site: www.emanuelnyc.org
Formerly: The Herbert & Eileen Bernard Museum, Congregation Emanu-El of the City of New York
Founded: 1948.
Key Personnel: Pres., Marcia Waxman; Dir. Devel., Robyn W. Cimbol; Sr. Cur., Elka Deitsch.
Governing Authority: synagogue. Parent Institution: Congregation Emanu-El of the City of New York. Tax-exempt: 501(c)(3).
Institution Type/Description: Judaica Museum: housed in 1929 synagogue.
Collections: liturgical items for synagogue & home; art; commemorative items commissioned by & for the synagogue; historical documents, photographs & graphics pertaining to the congregation; collections from the Jewish communities throughout the United States, Europe, North Africa & the Near East.

Research Fields: history of the congregation & its members; Judaica.
Activities: docent guided tours by appointment; gallery lecture series.
Publications: collection catalog, A Temple Treasury: The Judaica Collection of Congregation Emanu-El of the City of New York; exhibition pamphlets; museum checklist; exhibition catalogs; Maps of the Holy Land; To Have & To Hold: Decorated Jewish Marriage Contracts.
Hours & Admission Prices: Sun.-Thurs. 10-4:30, Fri. 10-4, Sat. 1-4:30. Docent tours are available by appointment. No charge. &

BOWERY GALLERY, 530 W. 25th St., 4th Fl., New York, NY 10001-5545. Tel.: 646-230-6655.
Institution Type/Description: Art Gallery.
Collections: paintings; sculpture; drawings.
Activities: special events; temporary exhibitions.
Hours & Admission Prices: Tues.-Sat. 11-6.

BOXING HALL OF CHAMPIONS, 9 E. 40th St., Fl. 17, New York, NY 10016-0402. Mailing Address: 8022 S. Rainbow Blvd., Las Vegas, NV 89139-6477. Tel.: 212-532-1717.
E-mail: steve@bhoc.com
Web Site: www.bhoc.com
Institution Type/Description: Boxing History Museum.
Collections: boxing champions; photographs.
Hours & Admission Prices: Call for hours.

CASTLE CLINTON NATIONAL MONUMENT, Battery Park, New York, NY 10004. Mailing Address: 26 Wall St., New York, NY 10005-1996. Tel.: 212-825-6992. Fax: 212-285-6874.
E-mail: shirley_mckinney@nps.gov
Web Site: www.nps.gov/cacl
Founded: 1950.
Congressional District: 8
Key Personnel: Supt., Shirley McKinney.
Personnel Profile: Full-Time Paid 8; Part-Time Paid 4.
Governing Authority: federal. Administered by National Park Service, U.S. Dept. of the Interior. Tax-exempt.
Institution Type/Description: Historic Site.
Collections: collections & archives relating to Castle Clinton Fort; Castle Garden Entertainment Center; immigration depot; miniature dioramas depicting 19th century Manhattan.
Research Fields: 1855-1890 immigration; 1800-1825 Harbor defense.
Facilities: visitor information center; Statue of Liberty & Ellis Island ticketing.
Activities: conducted tours; special events.
Publications: site brochure.
Hours & Admission Prices: Daily 8:30-5. No charge. Closed Christmas. &
Attendance: 2,433,250 (accurate)

THE CATHEDRAL OF ST. JOHN THE DIVINE, 1047 Amsterdam Ave. & 112th St., New York, NY 10025-1798. Tel.: 212-932-7347. Fax: 212-316-7424.
E-mail: info@stjohndivine.org
Web Site: www.stjohndivine.org
Founded: 1892.
Congressional District: 19
Key Personnel: C.E.O., Very Rev. Dr. James A. Kowalski.
Personnel Profile: Full-Time Paid 60; Part-Time Paid 70; Part-Time Volunteers 50; Interns 10.
Governing Authority: church. Parent Institution: Diocese of the State of New York. Tax-exempt: 501(c)(3).
Institution Type/Description: Historic Building & Religious Museum: housed in a Romanesque & neo-Gothic Cathedral.
Collections: art; religious artifacts; tapestries; stone & masonry; stained glass; plants & herbs mentioned in the Bible.
Major Exhibits: Dog Bless You: The Photography of Mary Bloom, 9/13-11/14; Phoenix Rising: The AA of Xu Bing, 1/14-12/14; The Value of Food, 2/15.
Research Fields: Biblical plants; history; architecture; Middle Ages.
Facilities: cathedral & its 11.6 acres.
Activities: guided tours; educational programs; family programs; school programs; medieval arts workshop; temporary exhibitions; concerts; events; services.
Publications: newsletter; booklets.
Hours & Admission Prices: Daily 7-6. Visitor Center: daily 9-5. Suggested Donation: $10 per person; discounts to AAM members. &
Attendance: 1,000,000 (estimated)
Membership: Society of Regents: Young Regents 21-40 $250; Portal Circle $2,500-$4,999; Arch Circle $5,000-$9,999; Rose Window Circle $10,000 & up.

CENTER FOR BOOK ARTS, 28 W. 27th St., 3rd Fl., New York, NY 10001-6906. Tel.: 212-481-0295. Fax: 866-708-8994.
E-mail: info@centerforbookarts.org
Web Site: centerforbookarts.org
Founded: 1974.
Key Personnel: Chm., Ellen Zeifer; Exec. Dir., Alexander Campos; Mgr. Programs, Sarah Nicholls; Admin., Myongyee Jin.
Personnel Profile: Full-Time Paid 3; Part-Time Paid 40; Part-Time Volunteers 35; Interns 10.
Governing Authority: private; nonprofit organization. Tax-exempt: 501(c)(3).
Institution Type/Description: Contemporary Book Arts Museum.
Collections: over 2,000 books & prints by regional, national, and international artists & publishers.
Facilities: 500-vol. library on bookmaking & related arts; 900 sq. ft. exhibit space.
Activities: guided tours; lectures; formal education programs; participatory exhibits; workspace rentals program for book artists; book artists slide registry & book publication programs; apprentice & internship programs; outreach program for school groups. Annual Event: Open House & Sale in December.
Publications: exhibition catalogues; member newsletters; course catalogues.
Hours & Admission Prices: Mon.-Fri. 10-6, Sat. 10-4. No charge; donations accepted. Closed New Year's Day; Passover-Easter weekend; Memorial Day; Independence Day; Labor Day; Yom Kippur; Christmas. &
Attendance: 20,000 (estimated)
Membership: Associate $50; Friend $100; Patron $250: Supporter $500; Sustainer $1,000; Benefactor $2,500; Master Benefactor $5,000.

CENTRAL PARK ZOO, (M), 830 5th Ave., New York, NY 10065-7001. Tel.: 212-439-6500. Fax: 212-988-0286.
Web Site: www.centralparkzoo.com
Institution Type/Description: Zoo.
Collections: over 150 species including penguins, polar bear; snow leopards, reptiles, amphibians, mammals.
Facilities: cafe. Zoo-related items for sale.
Activities: educational programs; guided tours.
Hours & Admission Prices: April to Nov. 6 daily 10-5:30; Nov. 7-March daily 10-4:30.

CHANCELLOR ROBERT R. LIVINGSTON MASONIC LIBRARY & MUSEUM, (M), 71 W. 23rd St., New York, NY 10010-4102. Tel.: 212-337-6620. Fax: 212-633-2639.
E-mail: info@nymasoniclibrary.org
Web Site: www.nymasoniclibrary.org
Founded: 1856.
Key Personnel: Dir., Thomas M. Savini; Pres. (V), Barry Mallah; Treas., Aldo Ghirarduzzi.
Personnel Profile: Full-Time Paid 3; Part-Time Paid 1; Part-Time Volunteers 12.
Governing Authority: private; nonprofit organization. Parent Institution: Grand Lodge of New York. Subsidiary Institution: Livingston Library, Utica Branch, Utica, NY. Tax-exempt: 501(c)(3).
Institution Type/Description: History Museum: located in Masonic Hall.
Collections: concentration on Masonic history in New York state; strong holdings in Hermetic & esoteric thought; focus on New York state with items from around the world, 16th-century to present; American fraternalism history.
Facilities: 45,000-vol. library of bound books; 15,000-vol. library of Masonic periodicals.
Activities: films; lectures; study clubs; temporary exhibitions.
Publications: annual report.
Hours & Admission Prices: Mon. & Wed.-Fri. 8:30-4:30, Tues. 12-8. No charge; donations accepted. Closed legal holidays. &
Attendance: 3,200 (accurate)

CHILDREN'S MUSEUM OF MANHATTAN, The Tisch Building, 212 West 83rd St., New York, NY 10024-4901. Tel.: 212-721-1223. Fax: 212-721-1127.
Web Site: www.cmom.org
Founded: 1973.
Congressional District: 20
Key Personnel: C.E.O. & Exec. Dir., Andrew S. Ackerman; Honorary Chm., Laurie Tisch Sussman; Chm., halley k. harrisburg; Deputy Dir. Exhibits, Karen Snider; Deputy Dir. Education, Leslie Bushara; Comptroller, Candice Carnage; School & Outreach Programs Mgr., Jennifer Kozel; Museum Shop Mgr., Devon Jairam.
Personnel Profile: Full-Time Paid 40; Part-Time Paid 40; Part-Time Volunteers 75; Interns 20.
Governing Authority: nonprofit organization. Tax-exempt: 501(c)(3).

Institution Type/Description: Children's Museum: housed in former elementary school building.
Collections: hands-on & participatory exhibits: literacy; science & the environment; arts, media & communications; early childhood education.
Research Fields: education & curriculum through the arts.
Facilities: environmental center; performing arts center; early childhood center; adventure center; children's birthday party program & facilities available by reservation. Museum-related items for sale.
Activities: guided tours; lectures; films; gallery talks; concerts; dance recitals; art festivals; workshops; formally organized education programs for children, adults, undergraduate & graduate students; docent program; training programs for museum workers; loan, permanent, temporary & traveling exhibitions; summer program; research.
Publications: calendars; brochures; posters; event schedules; publicity materials.
Hours & Admission Prices: Tues.-Sun., Mon. holidays & school vacations 10-5; school groups by appointment. Admission $11, senior citizens $7; children under one & members no charge. Closed New Year's Day; Thanksgiving; Christmas. &
Attendance: 340,000 (estimated)
Membership: Family $210; Friend $295; Supporter $395; Corporate $5,000 & up.

CHILDREN'S MUSEUM OF THE ARTS, 103 Charlton St., New York, NY 10014-3645. Tel.: 212-274-0986. Fax: 212-274-1776. Facebook: Children's Museum of the Arts.
E-mail: info@cmany.org
Web Site: www.cmany.org
Founded: 1988.
Key Personnel: Exec. Dir., Barbara Hunt McLanahan; Deputy Dir., Lucy Ofiesh; Dir. Education & Curatorial Programming, Jil Weinstock; Dir. Community Programs, Rachel Rapoport; Pres. Board of Trustees, William Floyd.
Personnel Profile: Full-Time Paid 17; Part-Time Paid 50; Interns 5.
Operating Expenses: 3,000,000
Operating Income: 3,000,000
Governing Authority: private; nonprofit organization. Tax-exempt: 501(c)(3).
Institution Type/Description: Children's Museum.
Collections: 2,000 pieces of children's art from around the world; interactive exhibits.
Facilities: classrooms; art studios; media lab.
Activities: formal education programs for children; temporary exhibitions; training programs for professional museum workers & educators. Annual Event: family fundraiser in spring; adult fundraiser in winter.
Publications: quarterly newsletter.
Hours & Admission Prices: Mon. & Wed. 12-5, Thurs. & Fri. 12-6, Sat. & Sun. 10-5. Children & Adults $11; discounts to groups, corporate sponsors, cool culture members; Thurs. 4-6 pay as you wish; members, children under one & seniors over 65 no charge. Closed Memorial Day; Independence Day; Labor Day; Thanksgiving; Christmas Day. &
Attendance: 110,000 (accurate)
Membership: CMA: Family $250; Friend $375; Community Builder $750; Teen / Young Artist Kollective for Grades 6-9 free.

CHILDREN'S MUSEUM OF THE NATIVE AMERICAN, Church of the Intercession, 2nd Fl., 550 W. 155th St., New York, NY 10032-7801. Tel.: 646-330-2125. Fax: 646-707-0414.
E-mail: cccona.center@gmail.com
Founded: 1976.
Key Personnel: Dir., Norman Ernsting
Institution Type/Description: Children's Museum.
Collections: hands-on exhibits; Native American artifacts; tipi; canoe; Native American life & culture; photographs.
Activities: demonstrations; educational programs; Indian games; puppet show; workshops.
Hours & Admission Prices: Sept.-June Tues.-Fri. 8-2 by appointment. Adults $8.

CHINA INSTITUTE GALLERY, CHINA INSTITUTE IN AMERICA, 125 E. 65th St., New York, NY 10065-7088. Tel.: 212-744-8181. Fax: 212-628-4159.
E-mail: gallery@chinainstitute.org
Web Site: www.chinainstitute.org
Founded: 1926.
Key Personnel: Dir., Willow Hai Chang; Chm., Virginia A. Kamsky; Pres., Sara Judge McCalpin; Registrar & Mgr., Jennifer Lima; Gallery Coord., Eva Wen; Mgr. Art Education Program, Yue Ma.
Personnel Profile: Full-Time Paid 34; Part-Time Paid 45; Part-Time Volunteers 10; Interns 10.

Governing Authority: nonprofit organization. Parent Institution: China Institute in America. Tax-exempt: 501(c)(3).
Institution Type/Description: Art Gallery.
Collections: Chinese paintings, calligraphy & artifacts.
Facilities: 5,000-vol. library pertaining to Chinese; classrooms; 1,000 sq. ft. exhibit space. Museum-related items for sale.
Activities: concerts; films; guided tours; lectures; loan & traveling exhibitions; formal education programs.
Publications: exhibition catalogs.
Hours & Admission Prices: Tues. & Thurs. 10-8, Wed. & Fri.-Mon. 10-5. Adults $7, students & seniors $4; discount to AAM & ICOM members; Tues. & Thurs. 6pm-8pm, China Institute members, Asia Society members, THIRTEEN members & children under 12 no charge. Closed major holidays; between exhibitions.
Attendance: 12,000 (accurate)
Membership: Individual $65; Dual & Family $115; Friends $1,500.

THE CLOISTERS, 99 Margaret Corbin Dr., New York, NY 10040-1198. Tel.: 212-923-3700. Fax: 212-795-3640.
E-mail: cloisters@metmuseum.org
Web Site: www.metmuseum.org
Founded: 1938.
Congressional District: 20
Key Personnel: Cur. in Charge & Dept. Medieval Art & The Cloisters, Peter Barnet; Mgr. Administration, Christina Alphonso; Cur., Timothy Husband; Cur., Barbara Boehm; Conservator, Lucretia Kargere; Librarian, Michael Carter; Assoc. Mgr. Visitor Svcs., Keith Glutting; Museum Educator, Nancy Wu; Assoc. Security Mgr., Theodosios Kypriotis; Horticulturist, Deirdre Larkin; Museum Shop Mgr., Sheryl Ali.
Personnel Profile: Full-Time Paid 75; Part-Time Paid 20; Part-Time Volunteers 20.
Governing Authority: nonprofit organization. Parent Institution: The Metropolitan Museum of Art, 1000 5th Ave., New York, NY 10028. Tax-exempt: 501(c)(3).
Institution Type/Description: Art Museum.
Collections: medieval art; cloisters & other European architectural elements; stained-glass windows; tapestries; sculpture; decorative arts; paintings; medieval-style gardens.
Research Fields: medieval art & architecture.
Facilities: 13,500-vol. library of books, 20,000 slides & 30,000 photographs on medieval art & architecture available to scholars & accredited graduate students by appointment; papers of Summer McKnight Crosby & George Grey Barnard; archives. Books, art reproductions & other museum-related items for sale.
Activities: guided tours; lectures; gallery talks; concerts; drama; formally organized education programs for children, adults & graduate students; permanent exhibitions; garden tours.
Publications: The Unicorn Tapestries; The Cloisters Cross; Medieval Tapestries; The Cloisters: Medieval Art and Architecture.
Hours & Admission Prices: March-Oct. Tues.-Sun. 9:30-5:15; Nov.-Feb. Tues.-Sun. 9:30-4:45. Recommended Admission: adults $20, seniors $15, students $10; discounts to AAM, ICOM & NY Museum council members; members & children under 12 accompanied by an adult no charge. Touch tours & sign language interpreter available on request. &
Attendance: 240,000 (estimated)
Membership: Associate $50; Met Net $60; Individual $95; Family/Dual $190; Friend $275; Sustaining $500; Apollo Circle $1,000; Contributing $1,200; Donor $1,800; Sponsor & Met Family Circle $4,000; Patron $8,000; Patron Circle $12,000; President's Circle $20,000.

THE COLLECTORS CLUB, INC., 22 E. 35th St., New York, NY 10016-3806. Tel.: 212-683-0559. Fax: 212-481-1269.
E-mail: collectorsclub@verizon.net
Web Site: www.collectorsclub.org
Founded: 1896.
Key Personnel: Pres., Ed Grabowski; Treas., Mark Banchik; Club. Sec., Dr. David Steidley.
Governing Authority: society. Tax-exempt: 501(c)(3).
Institution Type/Description: Historic House and Library: housed in 1902, five-story brownstone rowhouse, former residence of Thomas B. Clark; neo-Georgian architecture designed by Stanford White.
Collections: medals & awards; 19th-century engraving & printing implements; early postal equipment; U.S. Classic Society's trophies; numerous philatelic collections including Harry M. Deggett's International reply coupons; Charles L. Pack's Brazil & Canada; photographic archives of Edward Knapp postal history covers; 130,000-vol. library of hard books, journals & periodicals from 18th-century to present day; maps; manuscripts; Philately: A Catalogue of The Collectors Club Library, New York City: Author Catalogue, Subject Catalogue, Title Catalogue, and Periodical Catalogue.

Research Fields: postal & philatelic history.
Facilities: library available for research & inter library loan; reading room; 65-capacity auditorium.
Activities: lectures; study clubs; hobby workshops; formally organized education programs for adults; loan & permanent exhibitions; guest speakers twice monthly. Museum Sponsors: International Philatelic Awards.
Publications: Books for philatelists, United States: The Ten Cents Stamps of 1855-59; A Census of United States Classic Plate Blocks 1851-1882. Afghanistan: Its Twentieth Century Postal Issues. Central America: Its Pre-Stamp History. Ecuador: Its Postmarks & Postal History. French Morocco: The 1943-44 Tour Hassan Issues. Honduras: The Black Air Mail. The Stamps of Jammu & Kashmir. Philatelic Handbook for Korea 1884-1905. Newfoundland Postal History. The New Hebrides: Postal Stamps & their History. New Zealand: 1898-99, Great Barrier Island Pigeon Post Stamps. Postal Stamps of Lithuania. Postal History & Postage Stamps of Serbia, 1841-1921. Catalogues, 71st Anniversary Philatelic Exhibition Commemorating the 75th Year of the Collectors Club (1971); Cumulative Index to Collectors Club Philatelist, 1922-71.
Hours & Admission Prices: Mon.-Fri. 10-5. No charge. Closed major holidays.
Membership: Non-Resident $75; Overseas $95; Resident $190.

THE CONSERVATORY GARDEN, Fifth Ave. at 105th St., New York, NY 10029. Mailing Address: Central Park Conservancy, 830 5th Ave., New York, NY 10065-7001. Tel.: 212-860-1382. Fax: 212-360-1388.
Web Site: www.centralparknyc.org
Key Personnel: Garden Dir., Lynden Miller; Garden Cur., Diane Schaub; Central Park Admin., Douglas Blonsky.
Personnel Profile: Full-Time Paid 4; Part-Time Volunteers 25.
Governing Authority: nonprofit. New York City Dept. of Parks & Recreation and Central Park Conservancy. Tax-exempt: 501(c)(3).
Institution Type/Description: Botanical & Aquatic Gardens: six acre formal garden.
Collections: 20,000 tulips; 5,000 chrysanthemums; 175 different species of English-style perennials; annuals; bulbs; Woodland garden; Italian & French style gardens with seasonal tulip & Korean mum displays; 3 fountains.
Activities: garden tours; concerts. Annual Event: art exhibit.
Publications: Conservatory Garden Highlights, seasonal guides to plantings.
Hours & Admission Prices: Daily 8-dusk. No charge. &
Attendance: 10,000 (estimated)

COOPER-HEWITT, NATIONAL DESIGN MUSEUM, SMITH-SONIAN INSTITUTION, 2 E. 91st St., New York, NY 10128-0669. Tel.: 212-849-8400. Fax: 212-849-8401.
E-mail: cooperhewitt@si.edu
Web Site: www.cooperhewitt.org
Founded: 1897.
Congressional District: 15
Key Personnel: Dir., Caroline Baumann; Chm. (V), Barbara Mandel; Pres. (V), Beth Comstock; Cur. Dir., Cara McCarty; Cur. Drawings & Prints, Gail Davidson; Cur. Product Design & Decorative Arts, Sarah Coffin; Cur. Textiles, Matilda McQuaid; Cur. Contemporary Design, Ellen Lupton; Cur. Wallcoverings, Gregory Herringshaw; Cur. Socially Responsible Design, Cynthia Smith; Dir. Communications & Mktg., Jennifer Northrop; Librarian, Stephen H. Van Dyk; Dir. Education, Caroline Payson; Museum Shop Dir., Chad Phillips.
Personnel Profile: Full-Time Paid 72; Part-Time Paid 13; Part-Time Volunteers 50; Interns 40.
Operating Expenses: 16,000,000
Governing Authority: nonprofit organization. Parent Institution: Smithsonian Institution, Washington, D.C. Tax-exempt: 501(c)(3) & 170(b)(1)(A).
Institution Type/Description: Design Museum: housed in 1901 Andrew Carnegie Mansion.
Collections: design & decorative arts of all periods & countries; architecture & design drawings; textiles; wallpaper; prints; graphics; woodwork; metalwork; ceramics; glass; furniture; graphic design; industrial design; technology; sustainable & humanitarian design; product design; landscape architecture; interior design; interactive designs.
Research Fields: permanent collection; architecture; urban & graphic design; product design & decorative arts; textiles; wall covering; prints.
Facilities: 70,000-vol. library pertaining to design, decorative arts design and architecture available for inter-library loan & research on the premises.
Activities: guided tours; lectures; gallery talks; formally organized education programs for children, adults, undergraduate & graduate college students; inter-museum loan, permanent, temporary & traveling exhibitions; home study programs; workshops; design information service. Museum Sponsors: Masters Degree program with Parson School of Design/New School in NYC & Washington, D.C.

Publications: exhibition catalogs; book, The Smithsonian Illustrated Library of Antiques; magazine; The National Design Journal.
Hours & Admission Prices: Closed for renovations until fall 2014. See website for off-site programming. &

Attendance: 200,000 (accurate)
Membership: Student & Senior Citizen $50; National & International $55; Individual $75; Dual & Family $120; Contributing $200; Design Watch $500; Patron $1,000; National Design Council $2,500.

CORTLANDT ALLEY, 2-8 Cortlandt Alley, Between Franklin St. & White St., New York, NY 10013. Mailing Address: 248 W. 35th St., 10th Fl., New York, NY 10001. Tel.: 888-763-8839.
E-mail: info@mmuseumm.com
Governing Authority: nonprofit organization. Tax-exempt: 501(c)(3).
Institution Type/Description: History Museum.
Collections: over 250 artifacts from around the world.
Facilities: cafe. Museum-related items for sale.
Activities: special events.
Hours & Admission Prices: Sat.-Sun. 12-6.

CZECH CENTER NEW YORK, 321 E. 73rd St., New York, NY 10021. Tel.: 646-422-3399. Fax: 646-422-3383.
E-mail: info@czechcenter.com
Web Site: www.czechcenter.com
Founded: 1995.
Key Personnel: Acting Dir., Mrs. Paula Niklova.
Personnel Profile: Full-Time Paid 4; Part-Time Paid 1; Part-Time Volunteers 1.
Governing Authority: foreign; nonprofit organization.
Institution Type/Description: Civic Art & Culture Center Museum.
Collections: works of Czech artists.
Facilities: 1,250-vol. of books & magazines; 500 sq. ft. exhibit space.
Activities: concerts; dance recitals; films; traveling exhibitions; theater. Annual Event: Street Fair.
Publications: bimonthly newsletter.
Hours & Admission Prices: Mon. & Wed.-Fri. 10-6, Tues. 10-7. No charge; donations accepted. Closed New Year's Day; Christmas Eve, Day & day after. &
Attendance: 10,000 (estimated)

THE DAN FLAVIN ART INSTITUTE, 535 W. 22nd St., New York, NY 10011. Tel.: 212-989-5566.
E-mail: info@diaart.org
Web Site: www.diaart.org
Institution Type/Description: Art Museum.
Collections: works by Dan Flavin & national artists.
Activities: permanent & temporary exhibitions.
Hours & Admission Prices: Call for hours.

DIEU DONNE, 315 W. 36th St., New York, NY 10018-6681. Tel.: 212-226-0573. Fax: 212-226-6088.
E-mail: kflynn@dieudonne.org
Web Site: www.dieudonne.org
Key Personnel: Exec. Dir., Kathleen Flynn.
Governing Authority: nonprofit organization.
Institution Type/Description: Art Gallery.
Collections: works made with and on handmade paper.
Hours & Admission Prices: Tues.-Fri. 10-6, Sat. 12-6; other times by appointment.

DILLON GALLERY, 555 W. 25th St., New York, NY 10001. Tel.: 212-727-8585. Fax: 212-727-8705.
E-mail: mail@dillongallery.com
Web Site: www.dillongallery.com
Institution Type/Description: Art Gallery.
Collections: works by international contemporary artists; paintings; sculpture; photographs.
Hours & Admission Prices: Call for hours.

DISCOVERY TIMES SQUARE, 226 W. 44th St., New York, NY 10036. Tel.: 866-987-9692.
E-mail: info@tsxnyc.com
Web Site: www.discoverytsx.com
Institution Type/Description: General Museum.
Collections: world cultures; art; history; hands-on exhibitions; photographs.
Facilities: cafe. Gift items for sale.
Activities: audio guides; school groups; special events.

Hours & Admission Prices: Sun.-Thurs. 10-8, Fri.-Sat. 10-9. Adults $19-$27, Seniors 65 & over $16.50-$23.50, children 4-12 $14.50-$19.50. &

THE DRAWING CENTER, 35 Wooster St., New York, NY 10013-5300. Tel.: 212-219-2166. Fax: 888-380-3362.
E-mail: info@drawingcenter.org
Web Site: www.drawingcenter.org
Founded: 1977.
Congressional District: 8
Key Personnel: Exec. Dir., Brett Littman; Co Chm., Frances Beatty Adler; Co Chm., Eric Rudin; Exec. Editor of the Drawing Center's Publications, Jonathan T.D. Neil; Dir. Education, Aimee Good; Mng. Editor, Joanna Berman Ahlberg; Cur., Claire Gilman; Cur. Open Sessions, Lisa Sigal; Communications Dir., Molly Gross; Coord. Operations, Dan Gillespie; Dir. Finance & Admin., Champ Knecht; Dir. Devel., Nicole Goldberg; Registrar, Anna Martin; Curatorial Asst., Nora Benway; Asst. Cur., Joanna Kleinberg; Special Events Assoc., Kaegan Sparks.
Personnel Profile: Full-Time Paid 10; Part-Time Paid 5; Part-Time Volunteers 10; Interns 22.
Governing Authority: nonprofit organization. Tax-exempt.
Institution Type/Description: Art Museum: housed in c.1866 historic cast iron building.
Collections: drawings.
Major Exhibits: Deborah Grant: Christ You Know It Ain't Easy!!, 1/25/14-2/28/14; Ferran Adria: Notes On Creativity, 1/25/14-2/28/14; Rashaad Newsome: Five (The Drawing Center), 3/7/14-3/12/14; Andrea Bowers and Suzanne Lacy, 3/15/14-3/26/14.
Research Fields: contemporary & historical drawing.
Facilities: 4,733 sq. ft. exhibit space. Exhibition catalogues & books for sale.
Activities: lectures; traveling & loan exhibitions; formally organized education programs for children, undergraduates & graduates; internship programs. Museum Sponsors: Artist Symposia; literary series featuring new literature & poetry; Open Sessions, a new platform for artists.
Publications: exhibition catalogues; The Drawing Papers.
Hours & Admission Prices: Wed. & Fri.-Sun. 12-6, Thurs. 12-8. Adults $5, students & seniors $3; members no charge. Closed New Year's Day; Thanksgiving; Christmas. &
Attendance: 55,000 (accurate)
Membership: Artist $35; Individual $50; Dual $85; Contributor $150; On The Mark $200; Supporter $300; Associate $500; Donor $1,000; Friend $2,500; Corporate $5,000.

DYCKMAN FARMHOUSE MUSEUM AND PARK, 4881 Broadway, (at 204th St.), New York, NY 10034-3101. Tel.: 212-304-9422. Fax: 212-304-0635.
E-mail: info@dyckmanfarmhouse.org
Web Site: www.dyckmanfarmhouse.org
Founded: 1916.
Congressional District: 16
Key Personnel: Dir., Susan DeVries.
Personnel Profile: Full-Time Paid 3; Part-Time Volunteers 10; Interns 2.
Governing Authority: municipal. Parent Institution: Historic House Trust of New York City Parks & Recreation. Tax-exempt: 170(b) & 150(c)(3).
Institution Type/Description: Historic House: 1784 Dyckman House, old Dutch-American farmhouse.
Collections: 18th-19th century furniture & furnishings; decorative art; Revolutionary War uniforms; Hessian hut; smokehouse.
Activities: guided tours; craft workshops; wine tastings.
Publications: brochures; newsletters; guide to historic sites of New York City.
Hours & Admission Prices: Fri.-Sun. 11-5. Admission 10 & over $1; children under 10 no charge.
Attendance: 12,924 (accurate)
Membership: Annual $25-$1,000.

EL MUSEO DEL BARRIO, (M), 1230 Fifth Ave., New York, NY 10029-9962. Tel.: 212-831-7272. Fax: 212-831-7927. Facebook: El Museo.
E-mail: info@elmuseo.org
Web Site: www.elmuseo.org
Founded: 1969.
Congressional District: 15
Key Personnel: Chm. (V), Tony Bechara; Pres. Bd. Trustees, Yaz Hernandez; Exec. Dir., Margarita Aguilar; Deputy Exec. Dir., Carlos Galvez; Public Rels. Officer, Von Diaz; Senior Mktg. & Communications Mgr., Rose Mary Cortes; Dir. Education & Public Programs, Gonzalo Casals; Senior Mgr. Educ. & Public Rels., Stephanie Spahr LaForcia; Registrar, Melisa Lujan; Retail & Visitor Svcs. Mgr., Monica A. Garcia; Museum Shop Mgr., Glendalys Sosa.

Personnel Profile: Full-Time Paid 35; Full-Time Volunteers 2; Part-Time Paid 1; Part-Time Volunteers 50; Interns 8.

Governing Authority: Operated by Amigos del Museo del Barrio, Inc. Tax-exempt.

Institution Type/Description: Caribbean, Latino and Latin American Art Museum.

Collections: works on paper; sculptures; prints; paintings; films; photography; Santos & pre-Colombian collection with focus on Taino culture.

Research Fields: Puerto Rican & Latin American heritage, culture, history, art, music, literature & statistics.

Facilities: cafe; theater.

Activities: guided tours; lectures; films; performances; literary readings; symposia; educational programs; inter-museum loan; permanent, temporary & traveling exhibitions. Museum Sponsors: workshop in folk art, graphic arts, painting, drawing, sculpture, installation, art history lectures.

Publications: monthly e-newsletter; annual gala journal.

Hours & Admission Prices: Tues.-Sat. 11-6, Sun. 1-5. Suggested Donation: adults $9, senior citizens & students $5; discount to AAM & ICOM members; 3rd Sat. each month, seniors on Wed., members & children under 12 accompanied by parents no charge. &

Attendance: 300,000 (estimated)

Membership: Neighbor $25; Amigo $50; Supporter $100; Young International Circle $250; Couple $350; Aficionado & Family Circle $500; Madrina & Padrino $1,000; Premium Family Circle & Patron's Circle $2,500.

FEDERAL HALL NATIONAL MEMORIAL, 26 Wall St., New York, NY 10005-1996. Tel.: 212-825-6888. Fax: 212-825-6874.

E-mail: shirley_mckinney@nps.gov

Web Site: www.nps.gov/feha/

Founded: 1939.

Congressional District: 8

Key Personnel: Supt., Shirley McKinney.

Personnel Profile: Full-Time Paid 25.

Governing Authority: federal. Administered by National Park Service, Dept. of the Interior. Tax-exempt.

Institution Type/Description: History Museum.

Collections: artifacts relating to George Washington; evolution of the Bill of Rights; historical development of New York City; evolution of Federal Hall; constitutional government; Custom House, sub-treasury.

Research Fields: 18th-19th century American history.

Facilities: Museum-related items for sale.

Activities: tours for groups; education programs; special events.

Publications: site brochure; booklet, John Peter Zenger.

Hours & Admission Prices: Mon.-Fri. 9-5. No charge. Closed New Year's Day; Martin Luther King Jr. Day; Memorial Day; Independence Day; Labor Day; Columbus Day; Veterans Day; Thanksgiving; Christmas. &

Attendance: 148,601 (accurate)

FRAUNCES TAVERN MUSEUM, (M), 54 Pearl St., New York, NY 10004-4300. Tel.: 212-425-1778, ext. 30. Fax: 212-509-3467.

E-mail: curator@frauncestavernmuseum.org

Web Site: www.frauncestavernmuseum.org

Founded: 1907.

Congressional District: 17

Key Personnel: Chm. (V), Donald Westervelt; Cur., Suzanne Prabucki; Admin., Margaret O'Shaughnessy; Accounting, Cecelia Mahnken; Educator, Jennifer Patton; Security, Lindsford Bennett.

Personnel Profile: Full-Time Paid 5; Part-Time Paid 1; Part-Time Volunteers 15; Interns 1.

Governing Authority: society. Parent Institution: Sons of the Revolution in the State of New York. Tax-exempt: 501(c)(3).

Institution Type/Description: History Museum: housed in 1907 renovated & restored 18th-century tavern & four adjacent 19th-century buildings.

Collections: 18th & 19th-century decorative arts; textiles; paintings; prints; manuscripts; Revolutionary War memorabilia; period rooms; five historic buildings.

Research Fields: early American history & culture; New York City history; historic preservation; American art, architecture & decorative arts.

Facilities: auditorium. Museum-related items for sale.

Activities: lectures; concerts; films; workshops; organized tours & programs; American history audio-visual presentations; permanent & temporary exhibitions.

Publications: monthly e-newsletter, The Patriot.

Hours & Admission Prices: Mon.-Sat. 12-5. Adults $10, seniors, students with ID & children 6-18 $5; discounts to U.S. military, NYPD, & FDNY members; children under 6 & members no charge. Closed New Year's Day; Martin Luther King Jr. Day; Good Friday; Labor Day; Christmas. &

Attendance: 7,842 (accurate)

Membership: Senior Citizen & Student $40; Individual $50; Family $70; Sustaining $150; Supporting $275; George Washington Fellow $500.

✱ **THE FRICK COLLECTION, (M), (I),** 1 East 70th St., New York, NY 10021-4981. Tel.: 212-288-0700. Fax: 212-628-4417.

E-mail: info@frick.org

Web Site: www.frick.org

Founded: 1920.

Congressional District: 14

Key Personnel: Dir., Ian Wardropper; Pres. (V), Margot Bogert; Cur., Susan Grace Galassi; Deputy Dir. & C.O.O., Robert B. Goldsmith; Registrar, Diane Farynyk; Museum Shop Mgr., Kate Gerlough; Mgr. Devel., Tia Chapman; Mgr. Operations, Dennis F. Sweeney; Librarian, Stephen J. Bury.

Personnel Profile: Full-Time Paid 137; Part-Time Paid 53; Part-Time Volunteers 74; Interns 15.

Operating Expenses: 23,235,862

Operating Income: 23,628,508

Governing Authority: nonprofit organization. Subsidiary Institution: Frick Art Reference Library. Tax-exempt: 501(c)(3).

Institution Type/Description: Art Museum: formerly private home, c.1913-1914 building designed by Carrere & Hastings.

Collections: paintings; sculpture; furniture; decorative arts; prints & drawings.

Major Exhibits: Masterpieces of Dutch Painting from the Mauritshuis, 10/13-1/14; Precision & Splendor: Clocks & Watches of the Frick Collection, 10/13-2/14; Renaissance Baroque Bronzes from the Hill Collection, 1/14-6/14.

Research Fields: areas related to works of art in the Frick Collection.

Facilities: approx. 3,000-vol. in house library of basic publications relating to works of art in The Frick Collection; also Frick Art Reference library: Founded 1920, housed in 1934 building designed by John Russell Pope, became division of the collection in 1984. 250,000 books; 78,000 sales catalogues; 3,650 serial titles; 1.2 million photographs; microforms collection. Museum-related items for sale.

Activities: lectures; concerts; symposium for graduate students affiliated with the art history departments; temporary exhibitions; seminars & gallery talks.

Publications: The Frick Collection: An Illustrated Catalogue, Vols. I-IX; Handbook of Paintings; The Frick Collection: An Introduction (video); Ingres & the Comtesse d'Haussonville; Art in the Frick Collection. A Guide to Works of Art on Exhibition; Paintings from the Frick Collection; Building The Frick Collection.

Hours & Admission Prices: Collection: Tues.-Sat. 10-6, Sun. & minor holidays 11-5. Reference Library: Sept.-May. Mon.-Fri. 10-5, Sat. 9:30-1; June-July Mon.-Fri. 10-5. Adults $20, seniors 65 & over $15, students $10; Sun. 11-1 pay what you wish; discounts to AAM & ICOM members and museum staff. Children under 10 not admitted. Closed New Year's Day; Independence Day; Thanksgiving; Christmas. &

Attendance: 325,308 (accurate)

Membership: Student $25; Non-Resident $60; Individual $75; Dual $120; Contributing $250; Supporting $600; Young Fellow $500; Sustaining $600; Non-Resident Fellow $800; Fellow $1,200; Contributing Fellow $2,500; Supporting Fellow $5,000; Sustaining Fellow $10,000; Director's Circle $35,000.

GALLERY 456, Chinese American Arts Council, 456 Broadway, 3rd Fl., New York, NY 10013. Tel.: 212-431-9740. Fax: 212-431-9789.

E-mail: info@caacarts.org

Web Site: www.caacarts.org

Founded: 1975.

Key Personnel: Dir. & C.E.O., Alan Chow; Pres. (V), Stieven S.P. Chen; Museum Shop Mgr., Ellen Chang.

Personnel Profile: Full-Time Paid 2.

Governing Authority: Parent Institution: Chinese American Planning Council. Tax-exempt.

Institution Type/Description: Art Gallery.

Collections: Chinese American art & cultural heritage; paintings; sculpture.

Hours & Admission Prices: Mon.-Fri. 12-5. No charge; donations accepted.

Attendance: 2,000 (estimated)

GENERAL GRANT NATIONAL MEMORIAL, W. 122nd St. & Riverside Dr., New York, NY 10027-2522. Mailing Address: 26 Wall St., New York, NY 10005. Tel.: 212-666-1640 & 1668. Fax: 212-932-9631.

Web Site: www.nps.gov/gegr/

Founded: 1897.

Congressional District: 17

Key Personnel: Supt., Shirley McKinney; Cur., Steve Laise.

Personnel Profile: Full-Time Paid 3.

Governing Authority: federal. Parent Institution: Manhattan Sites Unit, National Park Service, Dept. of the Interior. 212-825-6990. Tax-exempt.
Institution Type/Description: Historical Museum: housed in 1897 building.
Collections: history; sarcophagi of President Grant and his wife; memorabilia of Ulysses S. Grant; military.
Research Fields: Pres. Ulysses S. Grant; Civil War.
Facilities: 150-vol. library of biographies of Gen. Grant & Civil War history; visitor center. Civil War-related items for sale.
Activities: guided & self-guided tours; programs honoring Grant's birthday; rotating special exhibits; Civil War programs.
Publications: NPS site brochure.
Hours & Admission Prices: Thurs.-Mon. 9-5. No charge. Closed Thanksgiving & Christmas.
Attendance: 118,000 (accurate)

THE GRACIE MANSION CONSERVANCY, 88th St. & East End Ave., New York, NY 10028-8024. Tel.: 212-570-4751. Fax: 212-570-4493.
E-mail: dtoole@cityhall.nyc.gov
Web Site: www.nyc.gov/gracie
Founded: 1981.
Key Personnel: Dir., Susan Danilow; Cur., Diana Carroll Toole.
Governing Authority: municipal. Tax-exempt: 501(c)(3).
Institution Type/Description: Historic House: 1799 Archibald Gracie country house. Designated the mayoral residence of New York City in 1942.
Collections: decorative arts of New York; paintings & prints by New York artists.
Research Fields: original residents of the mansion; New York City in the Federal period.
Facilities: 200-vol. library of decorative arts, American crafts & historic architecture books; garden; conservation center. Gift items for sale.
Activities: guided tours; docent program.
Hours & Admission Prices: Tours by reservation only. School Tours: Tues. & Thurs. mornings. No charge. Special Tours: Tues. & Thurs. afternoon. Public Tours: Wed. 10-2 hourly. Adults $7, senior citizens $4. ♿
Attendance: 40,000 (estimated)

GREY ART GALLERY, NEW YORK UNIVERSITY ART COLLECTION, 100 Washington Square E., New York, NY 10003-6688. Tel.: 212-998-6780. Fax: 212-995-4024.
E-mail: greyartgallery@nyu.edu
Web Site: www.nyu.edu/greyart
Founded: 1975.
Congressional District: 17
Key Personnel: Deputy Dir., Frank Poueymirou; Head Exhibitions & Collections and Registrar, Michele Wong; Asst. to Dir. & Press Officer, Laurie Duke; Gallery Dir., Lynn Gumpert; Preparator, Reuben Lorch-Miller; Asst. Preparator, Noah Landfield; Asst. to Deputy Dir., Amber Lynn; Head Education & Programs, Lucy Oakley.
Personnel Profile: Full-Time Paid 7; Part-Time Paid 2; Part-Time Volunteers 3; Interns 1.
Governing Authority: university. Parent Institution: New York University. Tax-exempt: 501(c)(3).
Institution Type/Description: Fine Arts Museum: housed in Main Building of New York University on Washington Square, historic center of early New York City.
Collections: 19th & 20th-century paintings, sculpture & graphics; Abby Weed Grey Collection of Asian & Middle Eastern Art.
Research Fields: New acquisitions; paintings; sculpture; graphics; exhibition catalogue research.
Activities: lectures; films; gallery talks; education programs for children & adults, undergraduate & graduate college students affiliated with N.Y.U. & other Colleges in vicinity; inter-museum loan; permanent, temporary & traveling exhibitions.
Publications: exhibition catalogues.
Hours & Admission Prices: Tues. & Thurs.-Fri. 11-6, Wed. 11-8, Sat. 11-5. Suggested Donation: $3 per person. Closed holidays. ♿
Attendance: 25,000 (estimated)

GROLIER CLUB LIBRARY, 47 E. 60th St., New York, NY 10022. Tel.: 212-838-6690. Fax: 212-838-2445. Facebook: Grolier Club Library.
E-mail: jsheehan@grolierclub.org
Web Site: www.grolierclub.org
Founded: 1884.
Institution Type/Description: Library.
Collections: books; prints; drawings.
Activities: temporary exhibitions; special events; rental facilities.

Hours & Admission Prices: Sept.-July Mon.-Sat. 10-5. No charge. Closed holidays.

GROUND ZERO MUSEUM WORKSHOP, 420 W. 14th St., Fl. 2, New York, NY 10014-1064. Tel.: 212-209-3370.
E-mail: groundzeromuseum@aol.com
Founded: 2005.
Key Personnel: Dir., Marlon Suson.
Personnel Profile: Part-Time Paid 4.
Governing Authority: Tax-exempt.
Institution Type/Description: History Museum.
Collections: over 100 photographs & artifacts representing the recovery efforts at Ground Zero.
Hours & Admission Prices: Tours: Sun.-Mon. 12-2, Tues. & Thurs.-Fri. 11-1, Sat. 11, 1, & 3. Adults $25, seniors & children $19; family members of 9/11 victims & active FDNY, PAPD & NYDP no charge.
Attendance: 15,000 (estimated)

HAMILTON GRANGE NATIONAL MEMORIAL, 414 W. 141st St., New York, NY 10031. Mailing Address: c/o Federal Hall NM, 26 Wall St., New York, NY 10005. Tel.: 646-548-2310. Fax: 646-548-9366.
E-mail: shirley_mckinney@nps.gov
Web Site: www.nps.gov/hagr
Founded: 1924.
Congressional District: 15
Key Personnel: Supt., Shirley McKinney; District Ranger, Liam Strain.
Governing Authority: federal. Administered by National Park Service, U.S. Dept. of Interior. Tax-exempt.
Institution Type/Description: History Museum & Historic House: home of Alexander Hamilton.
Collections: early 19th-century furnishings; building's history; Hamilton memorabilia.
Research Fields: American history, Federal & early 19th-century life & politics.
Facilities: Books & gift items for sale.
Activities: guided tours; videos; audio tour; interpretive exhibit.
Hours & Admission Prices: Wed.-Sun. 9-5. No charge. ♿
Attendance: 11,478 (accurate)

HAMPDEN-BOOTH THEATRE LIBRARY AT THE PLAYERS, 16 Gramercy Park, New York, NY 10003-1705. Tel.: 212-228-1861. Fax: 212-253-6473.
E-mail: hampdenboo@aol.com
Web Site: hampden-booth.org
Founded: 1888.
Congressional District: 18
Key Personnel: Pres., Robert Winter-Berger; Cur. & Librarian, Raymond Wemmlinger.
Governing Authority: society. Affiliated with the Players. Tax-exempt: 501(c)(3).
Institution Type/Description: Art & Theater Museum: housed in remodeled Gothic Revival townhouse.
Collections: American & English stage; manuscripts.
Research Fields: theater.
Facilities: 8,000-vol. library of published works on the performing arts available by appointment only; reading room.
Activities: guided tours; permanent & temporary exhibitions.
Hours & Admission Prices: Mon.-Fri. 9-5. No charge; donations accepted.

THE HISPANIC SOCIETY OF AMERICA, (M), 155th St. & Broadway, New York, NY 10032. Mailing Address: 613 W. 155th St., New York, NY 10032-7597. Tel.: 212-926-2234. Fax: 212-690-0743.
E-mail: info@hispanicsociety.org
Web Site: www.hispanicsociety.org
Founded: 1904.
Congressional District: 17
Key Personnel: Exec. Dir. & Pres., Mitchell A. Codding; Asst. Dir. & Cur. Decorative Arts, Margaret E. Connors McQuade; Cur. Modern Books, Edwin Rolon; Cur. Prints & Photographs, Patrick Lenaghan; Cur. Paintings, Marcus B. Burke; Cur. Manuscripts & Rare Books, John O'Neill; Cur. Sculpture & Textiles, Constancio del Alamo; Education, Andrea Michelle Ortuno.
Personnel Profile: Full-Time Paid 28; Part-Time Paid 13; Part-Time Volunteers 2; Interns 3.
Governing Authority: nonprofit organization. Tax-exempt: 501(c)(3).
Institution Type/Description: Art Museum & Research Library.

Collections: culture of Iberian Peninsula from prehistoric times to present; paintings; sculpture; ceramics; textiles; archaeology; decorative arts; costumes; glass; art & culture from Spain, Portugal, Latin America, Philippines.

Research Fields: culture & history of Hispanic world: Spain, Portugal, Latin America & the Philippines.

Facilities: 350,000-vol. library of books, manuscripts, maps, photographs, archives, graphics & music related to the literature, language, history & arts of the Hispanic world: Spain, Portugal, Latin America and the Philippines available for reference by scholars; reading room. Books & other museum-related items for sale.

Activities: educational programs for children, adults and undergraduate & graduate students; permanent & temporary exhibitions.

Publications: books & articles on bibliography; literature & art of the Hispanic world: Spain, Portugal, Latin America & the Philippines; catalogs of library collections & exhibitions.

Hours & Admission Prices: Museum: Tues.-Sat. 10-4:30, Sun. 1-4. No charge. Donations accepted. Reading Room: Tues.-Sat. 10-4:30. Closed holidays; Thanksgiving weekend; Christmas to New Year's Day.

Attendance: 25,000 (estimated)

Membership: Friends of the Hispanic Society $50.00 and up.

HISTORIC HOUSE TRUST OF NEW YORK CITY, (M), The Arsenal, 830 Fifth Ave., Rm. 203, New York, NY 10065-7001. Tel.: 212-360-8282. Fax: 212-360-8201.

E-mail: hhtinfo@parks.nyc.gov

Web Site: www.historichousetrust.org

Founded: 1989.

Congressional District: 14

Key Personnel: Exec. Dir., Franklin D. Vagnone; Chair, Frances Eberhart; Sec., Richard Southwick; Treas., Lisa Ackerman.

Personnel Profile: Full-Time Paid 6; Part-Time Paid 2.

Governing Authority: municipal government; nonprofit. Subsidiary Institutions: Bartow-Pell Mansion Museum, 895 Shore Rd., Pelham Bay Park, Bronx, NY 10464; Edgar Allan Poe Cottage, Poe Park, 2460 Grand Concourse at Kingsbridge, Rd., Bronx, NY 10467; Valentine-Varian House, Varian Park, 3266 Bainbridge Ave. at E. 208th St., Bronx, NY 10467; Van Cortlandt House Museum, Van Cortlandt Park, Broadway at 246th St., Bronx, NY 10471; Lefferts Historic House, Prospect Park, Flatbush Ave. at Empire Blvd., Brooklyn, NY 11215; Wyckoff Farmhouse Museum, Wyckoff Park, Clarendon Rd. at Ralph Ave., Brooklyn, NY 11210; Dyckman Farmhouse Museum, Dyckman House Park, 4881 Broadway at 204th St., New York 10034; Gracie Mansion, Carl Schurz Park, 88th St. & East End Ave., New York 10128; Morris-Jumel Mansion Museum, Roger Morris Park, 17-65 Jumel Terrace at 160th St., New York 10032; King Manor Museum, King Park, 150th St. & Jamaica Ave., Jamaica, NY 11432; Kingsland Homestead, Weeping Beech Park, 143-35 37th Ave., Flushing, NY 11354; Queens County Farm Museum, 73-50 Little Neck Parkway, Floral Park, NY 11004; Alice Austen House Museum, Alice Austen Park, 2 Hylan Blvd., Staten Island, NY 10305; Conference House, Conference House Park, 7455 Hylan Blvd., Staten Island, NY 10307; Historic Richmond Town, LaTourette Park, 441 Clarke Ave., Staten Island, NY 10306; Seguine Mansion, 441 Seguine Ave., Staten Island, NY 10307; The Old Stone House, Historic Interpretive Center, J.J. Byrne Park, 3rd St. btw. 4th & 5th Ave., Brooklyn, NY 11215; Little Red Lighthouse, Fort Washington Park, New York, NY; Merchant's House Museum, 29 E. 4th St. New York, NY 10003. Swedish Cottage Marionette Theatre, 79th St. & West Dr., Central Park, New York, NY 10024; Hendrick I. Lott House, 1940 E. 36th St., Brooklyn, NY 11234; Lewis H. Latimer House Museum, 34-41 137th St., Flushing, NY 11354; Bowne House, Weeping Beech Park, 37-01 Bowne St., Flushing, NY 11354. Tax-exempt.

Institution Type/Description: Historic Agency: created in 1989 to preserve and promote 23 Historic House museums located on Park land in the five boroughs.

Collections: 23 historic houses throughout New York City; historical furnishings; personal artifacts.

Research Fields: early history of New York City; historical homes.

Activities: restoration & maintenance of 23 houses; curatorial restoration of house interiors to original styles; development of education programs & exhibitions; recreation of period gardens & landscapes.

Publications: newsletter & calendar of events, Historic House News; brochure, Historic Houses in New York City Parks.

Hours & Admission Prices: Visit website for information.

Attendance: 800,000 (accurate)

Membership: Friend of the Trust: Friend $50; Associate $75; Family Friend $125; Patron $250; Fellow $500; Guardian $1,000; Cornerstone $5,000.

HUNTER COLLEGE ART GALLERIES, 695 Park Ave., New York, NY 10065-5085. Tel.: 212-772-4991. Fax: 212-772-4554.

E-mail: tadler@hunter.cuny.edu

Web Site: www.hunter.cuny.edu/art/galleries

Founded: 1984.

Congressional District: 14

Key Personnel: Dir., Tom Weaver; Dir., Joachim Pissarro; Cur., Tracy L. Adler; Preparator, Phi Nguyen; Asst. Cur., Marc Hoberman; Studio Supvr. MFA Bldg., Tim Laun; Curatorial Asst., Jessica Gumora.

Personnel Profile: Full-Time Paid 3; Part-Time Paid 12; Interns 10.

Governing Authority: public college; nonprofit. Parent Institution: City University of New York. Subsidiary Institutions: The Bertha and Karl Leubsdorf Art Gallery at Hunter College, 68th St. & Lexington Ave. S.W. Corner, New York, NY 10021; Hunter College/Times Square Gallery, 450 W. 41st St., New York, NY 10036. Tax-exempt: 501(c)(3).

Institution Type/Description: Art Galleries.

Collections: American art since 1945.

Activities: loan exhibitions; temporary exhibitions.

Publications: exhibition catalogues; 8-10 annual catalogue publications; Exotic Representation; Doug Ohlson: Twenty Years of Painting, 1982-2002; Mark Feldstein: Recent Work, 1999-2001; Seeing Red; Vincent Longo: Reflections on Abstraction; Strange Worlds; Moved; Re-Orientations: Islamic Art and the West in the Eighteenth and Nineteenth Centuries, 2008; to-Night: Contemporary Representations of the Night, 2008; Cutters, 2009; MAs Curate MFAs, 2009; Mixing it Up, 2009; Beyond Participation, 2009-2010; Smoke & Mirrors/Shadows & Fog, 2010; Americanana, 2010; Robert Swain, 2010.

Hours & Admission Prices: Sept.-June: Bertha and Karl Leubsdorf Art Gallery: Tues.-Sat. 1-6. Times Square Gallery: Tues.-Sat. 1-6. No charge. Closed Christmas; New Year's Day. &

Attendance: 10,000 (estimated)

Membership: Associate Member $50; Friend $250; Sponsor $500; Founder $1,000; Life $10,000.

ILDIKO BUTLER GALLERY, Lincoln Center Campus, 113 W. 60th St., New York, NY 10023-7484. Tel.: 212-636-6303.

Web Site: ildikobutlergallery.com

Formerly: Center Gallery, Fordham University

Institution Type/Description: Art Gallery.

Collections: photographs; paintings; sculpture; drawings.

Hours & Admission Prices: Call for hours. No charge. &

THE INTERCHURCH CENTER, 475 Riverside Dr., Rm. 240, New York, NY 10115-0003. Tel.: 212-870-2200. Fax: 212-870-2440.

Web Site: www.interchurch-center.org

Founded: 1959.

Congressional District: 15

Key Personnel: Pres. & Exec. Dir., Paula M. Maya; Chm. (V), Louis Barbarin; Cur., Frank DeGregorie; Librarian, Tracey Del Duca.

Personnel Profile: Full-Time Paid 1; Part-Time Paid 2.

Governing Authority: nonprofit. Tax-exempt: 501(c)(3).

Institution Type/Description: Library with Exhibits.

Collections: Revised Standard Version of the Bible exhibit; temporary exhibits of contemporary art.

Facilities: 11,000-vol. library of ecumenical & denominational religious books & periodicals, general reference material; slide registry of artists who have exhibited at TIC; 350-seat auditorium; cafeteria.

Activities: internet access; guided tours upon advance request; loan & permanent exhibitions.

Hours & Admission Prices: Mon.-Fri. 10-5. No charge. Closed legal holidays; Good Friday; day after Thanksgiving; Christmas Eve. &

Attendance: 5,000

* **INTERNATIONAL CENTER OF PHOTOGRAPHY, (M),** 1133 Avenue of the Americas at 43rd St., New York, NY 10036-7703. Mailing Address: 1114 Avenue of the Americas at 43rd St., New York, NY 10036-7703. Tel.: 212-857-0000. Fax: 212-857-0090. Facebook: International Center of Photography.

E-mail: info@icp.org

Web Site: www.icp.org

Founded: 1974.

Congressional District: 14

Key Personnel: Dir., Mark Robbins; Chm. Bd. (V), Caryl S. Englander; Pres. (V), Jeffrey A. Rosen; Dir. Operations, Charles Barrett; Controller, Victor Quinones; Dir. Administration, Linda Freitag; Dir. Mktg. Communications, Kelly Heisler; Dir. Individual & Annual Giving, Becky Lipkind; Dir. Devel. Svcs., Paul Gish; Dir. Community Engagement, Gigi Loizzo; Dir. Publications, Philomena Mariani; Deputy Dir. Programs & Education, Phillip Block; Deputy Dir. Administration, Steve Rooney; Deputy Dir. Exhibitions

& Collections and Chief Cur., Brian Wallis; Cur., Christopher Phillips; Cur., Carol Squiers; Cur. Collections, Edward Earle; Cur., Cynthia Young; Cur., Kristen Lubben; Registrar, Barbara Woytowicz; Librarian, Deirdre Donohue.
Personnel Profile: Full-Time Paid 80; Part-Time Paid 50; Part-Time Volunteers 30; Interns 30.
Governing Authority: private; nonprofit organization. Tax-exempt.
Institution Type/Description: Photography Museum.
Collections: over 120,000 photographic prints, primarily 20th-century; 200,000 negatives; over 4,000 hours of original audio recordings; collection of video tapes & films related to the Center's activities, photographers & photography.
Research Fields: photography as art, communication, with concentration in documentary & photojournalism.
Facilities: 22,000-vol. library of photographic books & related materials; screening room; classrooms; archive study room available to researchers, students & museum professionals by appointment; black & white & color photo labs & digital media labs. Photographic books, portfolios, audio visuals & other museum-related items for sale.
Activities: changing exhibitions; guided tours; lectures; films; gallery talks; formally organized education programs; MFA program in conjunction with Bard College; docent program; loan & traveling exhibitions.
Publications: books; exhibition catalogs; program guides; annual report; posters; portfolios; brochures.
Hours & Admission Prices: Tues.-Wed. & Sat.-Sun. 10-6, Thurs.-Fri. 10-8. Adults $12, students & senior citizens $8; discounts to AAM & ICOM members; Fri. 5-8 admission by voluntary contribution; staff from NYC museums, children under 12, members & school group tours no charge. Closed New Year's Day; Independence Day; Thanksgiving; Christmas. &
Attendance: 125,959 (accurate)
Membership: Senior $55; Senior Citizen Double $70; Individual $75; Double $100; Supporting $200; Focus $300; Photography Circle $350; Silver Patron $650; Gold Patron $1,350; Benefactor Patron $3,500.

INTERNATIONAL PRINT CENTER NEW YORK (IPCNY), (M), 508 W. 26th St., 5th Fl., New York, NY 10001-5538. Tel.: 212-989-5090. Fax: 212-989-6069.
E-mail: contact@ipcny.org
Web Site: www.ipcny.org
Founded: 1995.
Congressional District: 8
Key Personnel: C.E.O., Pres. & Dir., Anne Coffin; Chm. (V), Joseph Goddu.
Personnel Profile: Full-Time Paid 3; Part-Time Paid 1; Part-Time Volunteers 20; Interns 9.
Governing Authority: state; nonprofit organization. Tax-exempt: 501(c)(3).
Institution Type/Description: Print Exhibition Space.
Collections: fine art prints.
Major Exhibits: New Prints 2014 - Winter, 1/16/14-3/6/14; Contemporary Brazilian Printmaking, 3/20/14-5/23/14; New Prints 2014 - Summer, 6/14-8/14.
Research Fields: national directory of print workshops; participating artists' biographies; print galleries in New York City; libraries for print research in New York area; directory of printmaking programs at US based colleges & universities.
Facilities: information desk.
Activities: internship program; lectures; curator & artist talks; loan, traveling & participatory exhibits; member events; workshops. Annual Events: Exhibition Openings; Spring Benefit Dinner.
Publications: exhibition catalogues; informational hand-outs; traveling exhibitions brochure.
Hours & Admission Prices: July Mon.-Fri. 11-6; Sept.-June Tues.-Sat. 11-6. No charge. Closed national holidays. &
Attendance: 24,000 (estimated)
Membership: Artist & Student $50; Basic $100; Contributing $250; Benefactor $500; Patron $1,000.

INTREPID SEA, AIR & SPACE MUSEUM, (M), Pier 86, W. 46th St. & 12th Ave., New York, NY 10036-4103. Tel.: 212-245-0072. Fax: 212-245-1547.
E-mail: dscialpi@intrepidmuseum.org
Web Site: www.intrepidmuseum.org
Founded: 1982.
Congressional District: 17
Key Personnel: Pres., Bill White; Exec. Dir., Susan Marenoff; Dir. Devel. & External Affairs, Lisa Yaconiello; Asst. Vice Pres. Mktg., Mike Onysko.
Personnel Profile: Full-Time Paid 126; Part-Time Paid 146; Part-Time Volunteers 50; Interns 16.
Governing Authority: nonprofit organization. Parent Institution: Intrepid Museum Foundation. Tax-exempt: 501(c)(3).

Institution Type/Description: Armed Forces Museum: 900 ft. long aircraft carrier.
Collections: carrier history; Vietnam era destroyer Edson; destroyer escort Slater; guided missile submarine Growler; Polish MIG-21; British Scimitar; French Etendard; F-14 Super Tomcast; Coast Guard cutter Tamaroa; light ship Nantucket; research ship Elizabeth M. Fisher; Soviet missile corvette Hiddensee; A-12 Blackbird; over 40 aircraft, helicopters, missiles, rockets & space vehicles; photographs.
Research Fields: 20th-century history & technology; aviation history; New York Marine history; naval history.
Facilities: 5,000-vol. library pertaining to sea, air & space history & technology; theater; educational facilities; cafe. Gift items for sale.
Activities: guided tours of smaller ships; organized education programs for children; docent program; elementary; secondary schools in science & social science; participatory, loan & temporary exhibitions; in-house volunteer programs; flight simulators; group tours; birthday parties for children; rental facilities; Growler submarine & Concorde tours.
Publications: membership newsletter, Intrepid Times.
Hours & Admission Prices: Mon.-Fri. 10-5, Sat.-Sun. 10-6. Closed Thanksgiving; Christmas. &
Attendance: 700,000 (estimated)
Membership: Intrepid Net $25; Student $30; Individual $90; Dual $120; Family $140; Patron $300; Friend $600.

ITALIAN AMERICAN MUSEUM, (M), 155 Mulberry St., New York, NY 10013-4721. Tel.: 212-965-9000. Fax: 347-810-1028.
E-mail: info@italianamericanmuseum.org
Web Site: www.italianamericanmuseum.org
Founded: 2001.
Congressional District: 14
Key Personnel: C.E.O., Chm. (V), Pres. (V) & Cur., Dr. Joseph V. Scelsa; Devel., Maria Fosco; Treas., Robert Ciofalo; Museum Shop Mgr., Dolores Jacome.
Personnel Profile: Full-Time Paid 3; Part-Time Paid 3; Part-Time Volunteers 7; Interns 2.
Governing Authority: The University of the State of New York; N.Y.S. Board of Regents. Tax-exempt: 501(c)(3).
Institution Type/Description: Cultural Heritage Museum.
Collections: Italian artifacts; cultural heritage of Italy; contributions of Italians & Italian Americans to American culture.
Facilities: 6,000-vol. library; 2,000 sq. ft. exhibit space.
Activities: lectures; temporary & traveling exhibitions.
Publications: quarterly newspaper, America Italia Review.
Hours & Admission Prices: Fri. 11-8, Sat.-Sun. 11-6; other times by appointment. No charge; donations accepted. Closed legal holidays. &
Attendance: 2,000 (estimated)
Membership: $100.

JAPAN SOCIETY GALLERY, (M), 333 E. 47th St., New York, NY 10017-2399. Tel.: 212-832-1155. Fax: 212-715-1262.
E-mail: gallery@japansociety.org
Web Site: www.japansociety.org
Founded: 1907.
Congressional District: 15
Key Personnel: Pres., Motoatsu Sakurai; Vice Pres. & Gallery Dir., Joe Earle; Chm. (V), Wilbur Ross.
Personnel Profile: Full-Time Paid 2; Part-Time Volunteers 1; Interns 1.
Governing Authority: society. A part of the Japan Society, Inc. Tax-exempt: 501(c)(3).
Institution Type/Description: Art Museum.
Collections: Japanese art.
Research Fields: Japanese art here & abroad.
Facilities: 4,500-vol. library; auditorium; garden; classrooms. Museum-related items for sale.
Activities: guided tours; lectures; films; gallery talks; concerts; dance recitals; arts festivals; drama; traveling exhibitions; language classes.
Publications: exhibition catalogs; 2 scholarly catalogues per year.
Hours & Admission Prices: Spring & Fall: Tues.-Thurs. 11-6, Fri. 11-9, Sat.-Sun. 11-5; Summer: Tues.-Fri. 11-6, Sat.-Sun. 11-5. Spring & Fall: adults $15, seniors & students $10; discounts to AAM, ICOM, NYC Council of Museums & museum members; Closed major holidays. &
Attendance: 25,000 (accurate)
Membership: Student, Senior & Assoc. (those living beyond 100 miles from NYC) $40; Individual $60; Dual $95; Contributing $150; Sustaining $250; Sponsor $500; Patron Circle $1,000 & up.

THE JAZZ MUSEUM IN HARLEM, 104 E. 126th St., Ste. 4D, New York, NY 10035. Tel.: 212-348-8300, ext. 104.
E-mail: tsmith@jmih.org

Key Personnel: Artistic Dir., Loren Schoenberg; Mng. Dir., Jasna Radonjic; Dir. Devel., Janice Shapiro
Institution Type/Description: Jazz Museum.
Collections: jazz history & artists; photographs.
Activities: group tours; educational programs by appointment; special events.
Hours & Admission Prices: Mon.-Fri. 10-4; groups of 10 or more by appointment. Closed federal holidays. &

*　　**THE JEWISH MUSEUM, (M),** 1109 Fifth Ave., New York, NY 10128-0118. Tel.: 212-423-3200. Fax: 212-423-3232. Facebook: The Jewish Museum.
E-mail: info@thejm.org
Web Site: www.thejewishmuseum.org
Founded: 1904.
Congressional District: 14
Key Personnel: Dir., Claudia Gould; Chm. (V), Robert Pruzan; Deputy Dir. Finance & Administration, Joseph Rorech; Deputy Dir. Program Adminis-tration, Ruth Beesch; Dir. Education, Nelly Silagy Benedek; Dir. Devel., Lisa Metcalf; Dir. Special Events, Linda Padawer; Chief Cur., Norman Kleeblatt; Cur. Archaeology & Judaica and Chm. Curatorial Affairs, Susan L. Braunstein; Deputy Dir. External Affairs, Ellen Salpeter; Dir. Commu-nications, Anne Scher; Deputy Dir. Exhibition & Public Programs, Jens Hoffmann; Dir. Merchandising, Stacey Zaleski; Dir. Collections & Exhibi-tions, Jane Rubin; Dir. Mktg., Colin Weil; Museum Shop Mgr., Pamela Elias.
Personnel Profile: Full-Time Paid 118; Part-Time Paid 16; Part-Time Volun-teers 101; Interns 20.
Volunteer Hours: 17,293
Operating Expenses: 18,672,328
Operating Income: 18,672,328
Governing Authority: university. Parent Institution: The Jewish Theological Seminary of America, 3080 Broadway, New York, NY 10027. Tax-exempt: 501(c)(3).
Institution Type/Description: Art Museum: housed in 1908 Felix Warburg Mansion, a seven-story French Gothic structure.
Collections: Judaica collection spanning forty centuries; ceremonial objects; paintings; sculpture; prints; drawings; textiles; antiquities; photographs; decorative arts; coins; medals; historic manuscripts; artifacts; broadcast material; multimedia.
Major Exhibits: Marc Chagall: War and Exile, 9/13-3/14; Art Spiegelman: Co-Mix, 11/13-3/14; Other Primary Structures, 4/14-8/20/14; Mel Bochner - Strong Language (T), 5/14-9/14; Lee Krasner, Norma Lewis (T), 9/14-1/15; Helena Rubinstein & Beauty's Power (T), 10/20/14-2/15.
Research Fields: archaeology; Jewish history; Jewish art.
Facilities: library of Judaica references available by special permission to research scholars; 232-seat auditorium; renovated Warburg Mansion & new annex in June 1993. Books, Israeli and other contemporary crafts, ceremo-nial objects & other museum-related items for sale.
Activities: gallery tours; interpretive guide program for special & permanent exhibitions; participatory workshops for children; lecture series; film programs; concerts; inter-museum loans, permanent & temporary exhibi-tions; community outreach program; interpreters for the hearing-impaired (24 hrs. notice).
Publications: calendars of exhibitions & events; exhibition catalogs; posters; brochures; newsletters; New Year's graphics; biennial report.
Hours & Admission Prices: Thurs. 11-8, Sat.-Tues. 11-5:45, Fri. 11-4. Adults $15 senior citizens $12, students $7.50; discounts to members & AAM members; members & children 18 & under no charge. Closed New Year's Day; Martin Luther King Jr. Day; Thanksgiving; Jewish holidays. &
Attendance: 150,000 (accurate)
Membership: Senior Citizen $65; Individual, Dual Senior Citizen & Out of Town $100; Dual $120; Family $135; Friend $150; Contributing $250; Supporting $500; Sustaining $750.

KEHILA KEDOSHA JANINA SYNAGOGUE & MUSEUM, 280 Broome St., New York, NY 10002-3702. Mailing Address: P.O. Box 72, Cooper Station, New York, NY 10276-0072. Tel.: 212-431-1619. Fax: 631-367-3905.
E-mail: kehila_kedosha_janina@netzero.net
Web Site: www.kkjsm.org
Founded: 1997.
Key Personnel: Dir., Marcia Haddad Ikonomopoulos; Chm. (V), Sol Kofinas; 2nd Vice Pres., Rose Eskononts.
Governing Authority: Tax-exempt.
Institution Type/Description: Jewish History Museum.
Collections: Jewish history & culture.
Hours & Admission Prices: Museum: Sun. 11-4; other times by appointment. Lunch Tour: Tues., Thurs. & Sun. Lunch Tour: $18 per person. Shabbat Services: Sat. 9am.

Attendance: 5,000 (accurate)

KENKELEBA GALLERY, 214 E. 2nd St., New York, NY 10009-8031. Tel.: 212-674-3939. Fax: 212-505-5080.
Key Personnel: Dir., Joe Overstreet; Dir., Corrine Jennings
Institution Type/Description: Art Gallery.
Collections: paintings; sculpture.
Hours & Admission Prices: Wed.-Sat. 11-6. No charge.

LACRASIA'S GLOVE MUSEUM, 306 W. 38th St., Rm. 8A, New York, NY 10018. Tel.: 212-686-5428.
Web Site: www.lacrasiagloves.com
Founded: 1986.
Congressional District: 8
Key Personnel: Dir., Jay G. Ruckel; Museum Shop Mgr., Lacrasia Duchein
Institution Type/Description: General Museum.
Collections: the collection of a 20 year veteran glove maker; antique gloves; vintage clothes; tools.
Research Fields: glove history; glovemaking technology.
Facilities: 2,000 sq. ft. exhibition area.
Activities: New York City museum educators' roundtable; student tour groups.
Publications: the glove letter.
Hours & Admission Prices: By appointment only. No charge; donations accepted.
Attendance: 1,000 (estimated)

LIPANI GALLERY, FORDHAM UNIVERSITY, (M), Lincoln Center Campus, 113 W. 60th St., Sub Level, Visual Arts Complex, New York, NY 10023-6594. Mailing Address: 113 W. 60th St., Rm. A23, New York, NY 10023-6594. Tel.: 212-636-6303.
Web Site: lipanigallery.com
Formerly: Push Pin Gallery Fordham University
Institution Type/Description: Art Gallery.
Collections: works by student artists including architecture, film, graphic design, painting & photography.
Hours & Admission Prices: Call for hours. No charge. &

*　　**LOWER EAST SIDE TENEMENT MUSEUM, (M),** 91 Or-chard St., New York, NY 10002-3132. Tel.: 212-431-0233, ext. 0. Fax: 212-431-0402. Facebook: Lower East Side Tenement Mu-seum; TTY: 212-431-0714.
E-mail: lestm@tenement.org
Web Site: www.tenement.org
Founded: 1988.
Congressional District: 12
Key Personnel: Pres., Morris Vogel; Chm., Paul Massey; Vice Pres. & C.O.O., Barry Roseman; Vice Pres. Education & Programming, Annie Polland; Vice Pres. Mktg. & Communications, David Eng; Dir. Museum Shop, Mary-Kate Cowell; Vice Pres. Devel., Stephanie Wilchfort; Dir. Institutional Giving, Dana Friedman; Mgr. Evening Events, Alana Rosen.
Personnel Profile: Full-Time Paid 54; Part-Time Paid 64; Part-Time Volunteers 15; Interns 5.
Operating Expenses: 7,300,000
Operating Income: 7,350,000
Governing Authority: nonprofit organization. Tax-exempt: 501(c)(3).
Institution Type/Description: Immigrant History Museum: housed in c.1863 pre Old-Law tenement building.
Collections: historical immigrant documents; artifacts.
Research Fields: immigrant & urban history.
Facilities: 500-vol. library of material on history; 60-seat lecture hall. Books on historical subjects & other museum-related items for sale.
Activities: guided tours; lectures; loan exhibitions; ESOL workshops; training programs for professional museum workers; docent program; art exhibits; digital media projects; film screenings; school group tours; teacher work-shops; evening talks; culinary programs.
Publications: weekly e-newsletter, News from the Tenement Museum; books, A Tenement Story; What Might Have Been: The Story of the Moores; 97 Orchard Street.
Hours & Admission Prices: Daily 10-6:30, Thurs. 10-8:30. Adults $22, senior citizens & students $17; discounts to Museum Council of NY, NTHP, NPS & AAM members; children under 5 (Meet Victoria Tour only) & members no charge. Closed New Year's Day; Thanksgiving; Christmas.
Attendance: 200,000 (accurate)
Membership: Friend $25; Educator $65; Individual $75; Dual $115; Family & Friends $175; Contributor $250; Patron $500.

MERCHANT'S HOUSE MUSEUM, 29 E. Fourth St., New York, NY 10003-7003. Tel.: 212-777-1089. Fax: 212-777-1104.
E-mail: pi@merchantshouse.com
Web Site: www.merchantshouse.com
Founded: 1936.
Congressional District: 12
Key Personnel: Chm., Nicholas B.A. Nicholson; Exec. Dir., Margaret Halsey Gardiner; Sec., Anne Fairfax.
Personnel Profile: Full-Time Paid 2; Part-Time Paid 1; Part-Time Volunteers 54.
Governing Authority: nonprofit organization. Parent Institution: Historic House Trust - NYC Parks & Recreation. Tax-exempt: 501(c)(3).
Institution Type/Description: Historic House: 1832 example of late Federal & Greek Revival architecture.
Collections: Seabury Tredwell family collections; 19th-century furniture, decorative arts & textiles.
Research Fields: lighting; clothing; china & porcelain; family books; garden plot.
Facilities: 350-vol. library; garden. Postcards & other museum-related items for sale.
Activities: docent program; formal education programs for adults, children, undergraduate & graduate students affiliated with NYU, FIT, Bardy, Parson's & New York School of Design; guided tours; lectures; study clubs; temporary exhibitions; training programs for professional museum workers; dramatic period readings.
Publications: quarterly newsletter, The Merchant's House Museum.
Hours & Admission Prices: Self-Guided Tours: Thurs.-Mon. 12-5. Guided Tours: Thurs.-Mon. 2pm. Adults $10, senior citizens & students $5; children under 12 & members no charge. Closed New Year's Eve & Day; Easter; Independence Day; Thanksgiving; Christmas Eve & Day.
Attendance: 6,000 (estimated)
Membership: Student & Senior $25; Good Neighbor $50; Family & Household $75; Protector $125; Cultural Hero $250; Princely Supporter $500; Leading Light $1,000; Paragon of Virtue $2,500.

✱ **THE METROPOLITAN MUSEUM OF ART, (M),** 1000 Fifth Ave., (at 82nd St.), New York, NY 10028-0113. Tel.: 212-535-7710. Fax: 212-570-3879. TTY: 212-650-2921.
E-mail: metmuseum.org/about-the-museum/contact
Web Site: www.metmuseum.org
Founded: 1870.
Congressional District: 10
Key Personnel: Dir. & C.E.O., Thomas P. Campbell; Chm. Bd., Daniel Brodsky; Pres., Emily K. Rafferty; Sr. Vice Pres., C.F.O. & Treas., Olena Paslawsky; Chief Audience Devel. Officer, Donna Williams; Sr. Vice Pres. External Affairs, Harold Holzer; Sr. Vice Pres. Sec. & Gen. Counsel, Sharon H. Cott; Vice Pres. & Gen. Mgr. Merchandise & Retail, Brad Kauffman; Vice Pres. Construction, Tom A. Javits; Assoc. Dir. Collections & Administration, Carrie Rebora Barratt; Frederick P & Sandra P. Rose Chm. of Education, Peggy Fogelman; Sr. Assoc. Counsel, Cristina Del Valle; Andrall E. Pearson Cur. in Charge, Arts of Africa, Oceania & the Americas, Julie Jones; Lawrence A. Fleischman Chm. Am. Wing, Morrison H. Heckscher; Ruth Bigelow Writson Cur., American Decorative Arts & Admin. of the American Wing, Peter M. Kenny; Anthony W. & Lulu C. Wang Cur. American Decorative Arts, Alice Cooney Frelinghuysen; Alice Pratt Brown Cur. American Paintings & Sculpture, H. Barbara Weinberg; Cur. in Charge, Ancient Near Eastern Art, Joan Aruz; Arthur Ochs Sulzberger Cur. in Charge, Arms & Armor, Stuart Pyhrr; Douglas Dillon Cur. in Charge, Asian Art, Maxwell K. Hearn; Cur. in Charge Costume Institute, Harold Koda; Drue Heinz Chm. Drawings & Prints, George R. Goldner; Lila Acheson Wallace Cur. in Charge Egyptian Art, Dorothea Arnold; John Pope-Hennessy Chm. European Paintings, Keith Christiansen; Iris & B. Gerald Cantor Cur. in Charge, European Sculpture & Decorative Arts, Luke Syson; Henry R. Kravis Cur. European Sculpture & Decorative Arts, James David Draper; Cur. in Charge Greek & Roman Art, Carlos Picon; Patty Cadby Birch Cur. in Charge Islamic Art, Sheila R. Canby; Michel David-Weill Cur. in Charge Medieval Art & The Cloisters, Peter Barnet; Frederick P. Rose Cur. in Charge, Musical Instruments, J. Kenneth Moore; Cur. in Charge, Jeff Rosenheim; Chm. of the Dept. Modern & Contemporary Art, Sheena Wagstaff; Sherman Fairchild Conservator in Charge, Paintings Conservation, Michael Gallagher; Sherman Fairchild Conservator in Charge, Objects Conservation, Lawrence Becker
Sherman Fairchild Conservator in Charge, Sherman Fairchild Center for Works on Paper and Photographs Conservation Majorie Shelley; Conservator in Charge Textile Conservation, Florica Zaharia; Arthur K. Watson Chief Librarian, Kenneth Soehner; Gen. Mgr. Retail Store Operations, Will Sullivan.
Personnel Profile: Full-Time Paid 1,760; Part-Time Paid 353; Part-Time Volunteers 1,250; Interns 200.

Governing Authority: nonprofit organization. Branch Museum: The Cloisters Museum and Gardens. Tax-exempt: 501(c)(3).
Institution Type/Description: Art Museum.
Collections: ancient through modern art from Egypt, Greece, Rome, the Near & Far East, Europe, Africa, Oceania, pre-Columbian cultures & the U.S.; painting; sculpture; architecture; prints; photographs; drawings; glass; ceramics; metalwork; manuscripts; furniture; period rooms; textiles; costumes; arms & armor; musical instruments; decorative arts.
Research Fields: American Decorative Arts; Ancient Near Eastern Art; Arms & Armor; Drawings; Egyptian Art; European Paintings & Sculpture; European Decorative Arts; Asian Art; Greek & Roman Art; Robert Lehman Collection; Medieval Art; Musical Instruments; Prints & Photographs; Arts of African, Oceania and the Americas; American Paintings & Sculpture; Islamic Art; Modern & Contemporary Art.
Facilities: 890,000-vol. library of art & related reference materials available for use by scholars & graduate students; 708-seat auditorium; classrooms & meeting rooms; restaurant & cafeteria. Books, reproductions & other museum-related items for sale.
Activities: guided tours; lectures; films; gallery talks; concerts; formally organized education programs for children & adults; docent program or council; training programs for professional museum workers; inter-museum loan, permanent, temporary & traveling exhibitions; guided tours in Spanish, French, Italian, Japanese & other languages.
Publications: quarterly bulletin; quarterly calendar; annual journal; annual report; exhibition catalogues; collection catalogues; scholarly books; popular books & calendars; brochures; educational publications.
Hours & Admission Prices: Tues.-Thurs. & Sun. 9:30-5:30, Fri.-Sat. 9:30-9. Recommended Admission: adults $25, senior citizens 65 & over $17, students $12; discount to AAM & ICOM members; children 12 & under when accompanied by an adult & members no charge. Closed New Year's Day; Thanksgiving; Christmas. &
Attendance: 6,280,000 (accurate)
Membership: Associate $60; Met Net $70; Individual $100; Family/Dual $200; Friend $275; Sustaining $550; The Apollo Circle $1,000; Contributing $1,200; Donor $2,000; The Apollo Circle Patrons $3,500; Met Family Circle $4,000; Sponsor $4,500; Patron $9,000; Patron Circle $13,000; President's Circle $20,000.

MIRIAM & IRA D. WALLACH ART GALLERY, (M), 116th St. & Broadway, Schermerhorn Hall, 8th Fl., New York, NY 10027. Mailing Address: Columbia University, 826 Schermerhorn Hall, MC 5517, 1190 Amsterdam Ave., New York, NY 10027-7054. Tel.: 212-854-7288. Fax: 212-854-7800.
E-mail: wallach@columbia.edu
Web Site: www.columbia.edu/cu/wallach
Founded: 1986.
Congressional District: 16
Key Personnel: Dir. & Cur., Deborah Cullen-Morales.
Personnel Profile: Full-Time Paid 4; Part-Time Paid 2; Interns 2.
Governing Authority: private university. Parent Institution: Columbia University. Tax-exempt: 501(c)(3).
Institution Type/Description: Art Gallery: housed in Schermerhorn Hall.
Collections: (collection is the property of Columbia University).
Major Exhibits: vector Composition No. 1, 2013 by Vargas-Suarez Universal: A Sight Specific Installation, 10/10/13-6/14; Goddess, Heroine, Beast: Sculpture by Anna Hyatt Huntington, 1/22/14-3/15/14; 2014 First Year MFA Exhibition - Columbia University School of the Arts, Visual Arts, 3/14-4/14; Multiple Occupancy: Eleanor Antin's "Selves", 3/19/14-7/6/14; MODA Curates, 4/14-6/14; The Garden Necropolis: Art and Architecture in the Landscape of New York's Woodlawn Cemetery, 9/3/14-11/1/14; Romare Bearden: A Black Odyssey, 11/15/14-12/13/14; Romare Bearden: A Black Odyssey, 1/21/15-3/28/15.
Research Fields: pertaining to student & faculty research.
Facilities: 2,300 sq. ft. exhibit space.
Activities: lectures; films; symposia; tours.
Publications: exhibition catalogs.
Hours & Admission Prices: During the academic year Wed.-Sat. 1-5. No charge. &
Attendance: 5,000 (estimated)

✱ **THE MORGAN LIBRARY & MUSEUM, (M),** 225 Madison Ave., New York, NY 10016-3405. Tel.: 212-685-0008. Fax: 212-481-3484.
E-mail: media@themorgan.org
Web Site: www.themorgan.org
Founded: 1924.
Congressional District: 15
Key Personnel: Dir., William M. Griswold; Pres., Larry Ricciardi; Deputy Dir., Brian Regan; Dir. Finance & Administration, Kristina W. Stillman; Dir.

Library & Museum Svcs., Robert E. Parks; Cur. Robert H. Taylor & Dept. Head, Declan Kiely; Cur. & Dept. Medieval & Renaissance Manuscripts, William M. Voelkle; Cur. Charles Engelhard & Head Dept. Drawings & Prints, Linda Wolk-Simon; Astor Cur. & Dept. Head, Printed Books & Bindings, John Bidwell; Dir. Member Svcs., Nadine Slowik; Assoc. Cur. Seals & Tablets, Sidney H. Babcock, IV; Controller, Loretta Greaney; Dir., Thaw Conservation Center, Margaret Holben Ellis; Mellon Conservator, Patricia Reyes; Dir. Education, Linden Chubin; Dir. Communications & Mktg., Patrick Milliman; Dir. Merchandising Svcs., Sean Hayes; Publications Mgr., Karen Banks.
Personnel Profile: Full-Time Paid 124; Part-Time Paid 41; Part-Time Volunteers 23; Interns 9.
Governing Authority: nonprofit organization. Tax-exempt: 170(b)(1)(A).
Institution Type/Description: Library & Art Museum: housed in the 1906 library built by McKim, Mead, & White for Pierpont Morgan.
Collections: paintings; art objects; ancient Near Eastern seals & tablets; medieval & Renaissance manuscripts; printed & children's books; bindings; literary, historical & music manuscripts; Gilbert & Sullivan collection; old master to 20th-century drawings & prints.
Research Fields: all fields of collections.
Facilities: library collections available for scholarly use upon written application; reading room. Museum-related items for sale.
Activities: lectures; concerts; photographic services.
Publications: books; pamphlets; exhibition catalogs; facsimiles; guide to collections.
Hours & Admission Prices: Tues.-Thurs. 10:30-5, Fri. 10:30-9, Sat. 10-6, Sun. 11-6. Adults $15; discounts to AAM & ICOM members; members no charge. Closed New Year's Day; Thanksgiving; Christmas. &
Attendance: 160,000
Membership: Intro Individual $75; Individual $100; Dual & Family $150; Contributor $250; Sustainer & Young Associate $500; Conservator $1,000; Fellow $2,000; Patron Fellow $6,000; Pierpont Fellow $12,500.

∗ **MORRIS-JUMEL MANSION, (M),** 65 Jumel Ter., New York, NY 10032-5360. Tel.: 212-923-8008. Fax: 212-923-8947.
Web Site: www.morrisjumel.org
Founded: 1904.
Congressional District: 15
Key Personnel: Pres. (V), James Kerr, Esq.; Vice Pres., Pamela Palanque North; Dir., Kenneth Moss; Asst. Dir., Sheena Brown; Dir. Education, Sarah Mellace.
Personnel Profile: Full-Time Paid 3; Full-Time Volunteers 1; Part-Time Paid 4; Part-Time Volunteers 15; Interns 2.
Governing Authority: private; nonprofit. Tax-exempt: 501(c)(3).
Institution Type/Description: Historic House: 1765 Morris-Jumel Mansion, oldest residence in Manhattan; used as Gen. Washington's headquarters during the Revolution; purchased by the Jumel family in 1810; Eliza Jumel married Aaron Burr in the front parlor.
Collections: archives; Chippendale, Federal & Empire furniture; silver, china, crystal, prints & paintings from the Colonial, Federal & Empire periods.
Research Fields: history; preservation; decorative arts; architecture.
Facilities: archives for use by appointment.
Activities: guided tours; lectures; gallery talks; concerts; educational programs; docent program; group tours by appointment only.
Publications: semi-annual newsletter; quarterly program calendar.
Hours & Admission Prices: Wed.-Sun. 10-4; other times by appointment. Adults $5, students & senior citizens $4, school groups $1.50 per child; Historic House Trust of New York City, Museum Council of New York City, museum, children under 12, AAM & ICOM members no charge. Group tours available. Closed New Year's Day; Memorial Day; Independence Day; Labor Day; Thanksgiving; Christmas.
Attendance: 35,000 (estimated)
Membership: Friend $35; Family $55; Madame Jumel Circle $100; Roger Morris Circle $250; Aaron Burr Circle $500; George Washington Circle $1,000; Octagon Society & Cornerstone Corporate Circle $2,500.

∗ **MOUNT VERNON HOTEL MUSEUM & GARDEN, (M),** 421 E. 61st St., New York, NY 10065-8736. Tel.: 212-838-6878. Fax: 212-838-7390. Facebook: Mount Vernon Hotel Museum.
E-mail: info@mvhm.org
Web Site: www.mvhm.org
Formerly: Abigail Adams Smith Museum
Founded: 1939.
Congressional District: 14
Key Personnel: Dir., Mary Anne Caton; Cur. Education, Natalia Sokolova; Cur. Collections, Ruth Osborne; Dir. Devel., Emma Wilcox.
Personnel Profile: Full-Time Paid 2; Part-Time Paid 11; Part-Time Volunteers 45; Interns 4.
Governing Authority: nonprofit organization. Parent Institution: Colonial Dames of America. Tax-exempt.

Institution Type/Description: Historic House Museum: Built in 1799 interiors from 1826.
Collections: American decorative arts; 18th- & 19th-century documents & letters; period rooms represent the c.1830 Mount Vernon Hotel.
Major Exhibits: Instructive, Moral and Entertaining: Toys & Childhood in Nineteenth Century America, 11/27/13-4/4/14; 75 Years of Women in Preservation: The Colonial Dames of America & the Mount Vernon Hotel Museum, 4/21/14-7/28/14; /The Greek Revival in America: Classical Taste & European Sources, 8/4/14-12/15/14.
Research Fields: American Decorative Arts; Jacksonian New York; history of leisure & work in NYC & America in 1820s & 30s.
Facilities: 200-seat auditorium; historic landscape. Museum-related items for sale.
Activities: school programs for K-12 students; seminars for adults; concerts; lectures & workshops for adults & families; adult group tours; monthly children's story time; monthly lunchtime lectures.
Hours & Admission Prices: Tues.-Sun. 11-4. Adults $8, students & senior citizens $7; discounts to AAM & AASLH members & employees of local museums; children under 12 & members no charge. Closed New Year's Day; Independence Day; Thanksgiving; Christmas; first weekend in May.
Attendance: 30,000 (estimated)
Membership: Senior & Student $30; Friend $40; Family $80; Supporter $120; Patron $250; Benefactor $500; Director's Circle $1,000; Leadership Circle $2,500.

THE MUNICIPAL ART SOCIETY OF NEW YORK, 111 W. 57th St., 16th Fl., New York, NY 10019-2211. Tel.: 212-935-3960, ext. 1285. Fax: 212-753-1816.
E-mail: info@mas.org
Web Site: www.mas.org
Founded: 1893.
Congressional District: 15
Key Personnel: Pres., Vin Cipolla.
Personnel Profile: Full-Time Paid 21; Part-Time Paid 8; Part-Time Volunteers 5; Interns 10.
Governing Authority: nonprofit organization. Tax-exempt: 501(c)(3).
Institution Type/Description: Advocacy, planning & preservation organization dedicated to improving livability in New York.
Collections: research materials.
Research Fields: historic preservation, planning, open space & livability issues focused on New York City.
Facilities: research library.
Activities: guided tours; films; organized education programs.
Hours & Admission Prices: By appointment. No charge; donations accepted. &
Attendance: 65,000 (estimated)
Membership: Senior $40; Individual $50; Family $75; Sustaining $125; Contributing $250; Sponsoring $500; Patron $1,000.

MUSEUM AT ELDRIDGE STREET, 12 Eldridge St., New York, NY 10002-6204. Tel.: 212-219-0888. Fax: 212-966-4782.
E-mail: contact@eldridgestreet.org
Web Site: www.eldridgestreet.org
Founded: 1986.
Congressional District: 12
Key Personnel: Exec. Dir., Bonnie Dimun; Chm. (V), Michael Weinstein; Pres. (V), Lorinda Ash Ezersky; Dir. Public Programs, Hanna Griff; Dir. Education, Judy Greenspan; Dir. Devel., Eva Brune; Deputy Dir., Amy Milford.
Personnel Profile: Full-Time Paid 7; Part-Time Paid 2; Part-Time Volunteers 40; Interns 3.
Governing Authority: private; nonprofit organization. Tax-exempt: 501(c)(3).
Institution Type/Description: Historic Site, Cultural Center & Museum: a national historic landmark.
Collections: concentration on New York Jewish history with special emphasis on the Lower East Side.
Research Fields: history of the Eldridge Street Synagogue & its congregation in the context of the development of American Judaism.
Facilities: 150-vol. library of materials relating to New York history, immigration to the United States and Judaica & Jewish history; educational facilities; 5,000 sq. ft. exhibit space. Museum-related items for sale.
Activities: concerts; docent program; formal education programs; guided tours; lectures; participatory & temporary exhibitions.
Publications: biannually, News from Eldridge Street.
Hours & Admission Prices: Synagogue Tours: Sun.-Thurs. 10-5, Fri. 10-3. Adults $10, senior citizens & students $8, children 5-18 $6; discounts to AAM members; members & children under 5 no charge. Closed Jewish & national holidays. &
Attendance: 35,000 (accurate)

✳ **THE MUSEUM AT FIT, (M),** Seventh Ave., (at 27th St.), New York, NY 10001-5992. Tel.: 212-217-4533. Fax: 212-217-4531.
E-mail: museuminfo@fitnyc.edu
Web Site: www.fitnyc.edu/museum
Founded: 1967.
Key Personnel: Dir. & Chief Cur., Dr. Valerie Steele; Head Conservator, Ann Coppinger; Deputy Dir., Patricia Mears; Registrar, Sonia Dingilian; Mgr. Exhibits, Fred Dennis.
Personnel Profile: Full-Time Paid 28; Part-Time Paid 2.
Governing Authority: university. Parent Institution: Fashion Institute of Technology (SUNY). Tax-exempt: 170(b)(1)(A).
Institution Type/Description: Fashion Museum.
Collections: over 50,000 garments & accessories dating from the mid-18th century to the present, with a focus on 20th-century fashion, including: couture & ready-to-wear women's clothing; menswear ranging from formal to activewear; swimwear; lingerie; knitwear; children's clothing; Halston designs, patterns & related records documenting designer's life work; 15,000 accessories including shoes, hats & bags; over 30,000 textiles, dating from the 6th century to the present illustrating techniques & traditions from around the world.
Activities: guided tours; organized education programs for undergraduate or graduate college students; internships; programs relating to exhibitions; loan, temporary & traveling exhibitions; fashion culture programming, tours & lectures.
Publications: Madame Gres; Gothic: Dark Glamour; Ralph Rucci: The Art of Weightlessness; Isabel Toledo: Fashion From the Inside Out; Japan Fashion Now.
Hours & Admission Prices: Tues.-Fri. 12-8, Sat. 10-5. No charge. Closed legal holidays. ♿
Attendance: 100,000 (accurate)
Membership: Couture Council $1,000.

THE MUSEUM FOR AFRICAN ART, Administrative Offices, 1280 5th Ave., Ste. 20A, New York, NY 10029-7815. Tel.: 212-444-9795. Fax: 212-444-0219.
E-mail: admin@africanart.org
Web Site: www.africanart.org
Founded: 1984.
Congressional District: 13
Key Personnel: Mgr, Publications, Carol Braide; Exec. Vice Pres. & C.E.O., Philip Conte; Community Outreach Liaison, Lawrence Ekechi; Dir. Administration, Bridget Foley; Dir. Curatorial Affairs, Donna Ghelerter; Dir. Education & Public Programs, Dan'etta Jimenez; Curatorial Fellow, Evelyn Owen; Asst. Registrar, Eve Perry; Chief Registrar & Dir. Exhibitions, Amanda Thompson; Controller, Velky Valentin; Institutional Advisor, Jerry Vogel.
Personnel Profile: Full-Time Paid 8; Full-Time Volunteers 25; Part-Time Paid 3; Part-Time Volunteers 5; Interns 4.
Governing Authority: nonprofit organization. Tax-exempt: 501(c)(3).
Institution Type/Description: Museum for African Art.
Collections: paintings; photographs; sculpture.
Research Fields: traditional & contemporary African art.
Facilities: Publications for sale.
Activities: guided tours; lectures; organized educational programs; docent program; loan & traveling exhibitions.
Publications: brochures; exhibition catalogues: African Masterpieces from the Musee de l'Homme; Yoruba: Nine Centuries of African Art & Thought; Sets, Series & Ensembles in African Art; Likeness & Beyond: Portraits from Africa & the World; Closeup: Lessons in the Art of Seeing African Sculpture; Aesthetics of African Art: Perspectives, Angles of African Art; African Masterpieces from Munich; Art/Artifact: Art of Collecting African Art; Africa & the Renaissance; Wild Spirits Strong Medicine; Africa Explores: 20th-Century African Art; Faces of the Gods: Art & Altars of Africa and the African Americans; Exhibitionism: Museums and African Art; Animals in African Art: From the Familiar to the Marvelous; Secrecy: African Art that Conceals and Reveals; Western Artists/African Art; Fusion: West African Artists at the Venice Biennale; Home & the World: Architectural Sculpture by Two Contemporary African Artists; Exhibition-ism: Museums & African Art; Animals in African Art: From the Familiar to the Marvelous; Memory: Luba Art & the Making of History; Art of the Baga: A Drama of Cultural Reinvention; Art that Heals: The Image as Medicine in Ethiopia; African Faces, African Figures: The Arman Collection; Hair in African Art and Culture; Liberated Voices: Contemporary from South Africa; A Congo Chronicle: Patrice Lumumba in Urban Art; In the Presence of Spirits: Selections from the National Museum of Ethnology, Lisbon; African Forms: Addendum; Bamana: The Art of Existence in Mali; Facing the Mask; Material Differences: Art and Identity in Africa; Looking Both Ways: Art of the Contemporary African Diaspora; Where Gods and Mortals Meet: Continuity and Renewal in Urhobo Art; Personal Affects: Power and

Poetics in Contemporary South African Art; Resonance from the Past: African Sculpture from the New Orleans Museum of Art; Grass Roots: African Origins of an American Art; Desert Jewels: North African Jewelry and Photography from the Xavier Guerrand-Hermes Collection.
Hours & Admission Prices: Office: Mon.-Fri. 10-6. See Web site for exhibition hours & locations.
Attendance: 40,000 (estimated)

MUSEUM OF AMERICAN FINANCE, (M), 48 Wall St., Lobby 2, New York, NY 10005-2910. Tel.: 212-908-4110. Fax: 212-908-4601.
E-mail: kaguilera@moaf.org
Web Site: www.moaf.org
Formerly: Museum of American Financial History
Founded: 1988.
Congressional District: 27
Key Personnel: Chm. (V), Dr. Richard Sylla; Pres. & C.E.O., David J. Cowen; Dir. Devel., Jeanne Driscoll; Deputy Dir., Kristin Aguilera; Dir. Exhibits & Archives, Becky Laughner; Dir. Visitor Svcs., Linda Rapacki; Dir. Exhibits & Educational Programs, Maura Ferguson; Business Mgr., Arturo Gomez.
Personnel Profile: Full-Time Paid 14; Part-Time Paid 5; Part-Time Volunteers 80; Interns 15.
Governing Authority: nonprofit organization. Tax-exempt: 501(c)(3).
Institution Type/Description: Financial Museum and de facto Visitor Center for the New York Stock Exchange.
Collections: American financial history from the mid-18th century to present day; stock & bond certificates; books; periodicals; associated items.
Major Exhibits: The Fed at 100, 9/13-10/14.
Facilities: library; education center; 250-seat auditorium; exhibition space; archives; rental gallery.
Activities: permanent exhibitions on the capital markets; banking & money; entrepreneurship.
Publications: quarterly magazine, Financial History; books, Scripophily; Financing the American Revolution; exhibit catalogs.
Hours & Admission Prices: Exhibits: Tues.-Sat. 10-4. Adults $8, students & seniors $5; discounts to NYC museum employees, AAM & ICOM members; children 6 & under and museum members no charge. Closed major holidays. ♿
Attendance: 50,000 (estimated)
Membership: Student & Senior $45; Individual $55; International $65; Institutional $85; Family $85; Smithsonian Affiliate $150; Hamilton Society $500; Donor $1,000.

THE MUSEUM OF AMERICAN ILLUSTRATION AT THE SOCIETY OF ILLUSTRATORS, (M), 128 E. 63rd St., New York, NY 10065-7303. Tel.: 212-838-2560. Fax: 212-838-2561.
E-mail: info@societyillustrators.org
Web Site: www.societyillustrators.org
Formerly: Society of Illustrators Museum of American Illustration
Founded: 1901.
Congressional District: 18
Key Personnel: Pres., Dennis Dittrich; Dir., Anelle Miller.
Personnel Profile: Full-Time Paid 9; Full-Time Volunteers 1; Part-Time Paid 2; Interns 5.
Governing Authority: board of directors. Tax-exempt: 501(c)(3).
Institution Type/Description: Art Museum & Gallery.
Collections: various forms of art media & books.
Research Fields: American Illustration.
Facilities: library & archives available by appointment only.
Activities: lectures; permanent & temporary exhibitions.
Publications: Annual of American Illustration.
Hours & Admission Prices: Gallery: Sept.-July Tues. 10-8, Wed.-Fri. 10-5, Sat. 12-4. No charge; donations accepted. Closed legal holidays.
Attendance: 30,000 (estimated)
Membership: Student $35; Friend & Illustrator Senior $250; Associate & Illustrator $500; Corporate $1,500.

✳ **MUSEUM OF ARTS AND DESIGN, (M),** 2 Columbus Circle, New York, NY 10019-1800. Tel.: 212-299-7777. Fax: 212-299-7701.
E-mail: marisa.bartolucci@madmuseum.org
Web Site: www.madmuseum.org
Formerly: American Craft Museum
Founded: 1956.
Congressional District: 15
Key Personnel: Dir. & Pres., Nanette L. Laitman; Chm., Lewis Kruger; Sec., Linda E. Johnson; Chief Cur., William & Mildred Lasdon, David McFadden; Cur., Ursula Ilse-Neuman; Charles Bronfman International Cur., Lowery Stokes Sims; Marcia Docter Cur., Ron T. Labaco; Registrar, Ellen

Holdorf; Gen. Mgr. Operations, Alex Berisha; Assoc. Vice Pres. Pub. Affairs, Marisa Bartolucci.
Personnel Profile: Full-Time Paid 47; Part-Time Paid 15; Interns 12.
Governing Authority: nonprofit organization. Tax-exempt: 501(c)(3).
Institution Type/Description: Art Museum.
Collections: American 20th-century crafts by artists working in ceramic, glass, fiber, metal, wood, mixed media, paper & plastic; traveling exhibition program.
Research Fields: 20th-century international craft, design & jewelry.
Facilities: theater; event space; lecture & symposia facilities. Museum-related items for sale.
Activities: tours; lectures; symposia; workshops; audiovisual programs; Meet-the-Artist program; travel opportunities for support group members; temporary exhibits.
Publications: exhibition catalogs; newsletter.
Hours & Admission Prices: Tues.-Wed. & Fri.-Sun. 11-6, Thurs. 11-9. Adults $15, students & seniors $12; Thurs.-Fri. 6pm-8:30pm pay what you wish; discounts to AAM & ICOM members; members, teenagers with ID & children under 12 no charge. &
Attendance: 500,000 (estimated)
Membership: Students $50; Individual $75; Dual $100; Family $125; 360 Young Collectors $200; Contributing $250; Supporting $500.

MUSEUM OF BIBLICAL ART, 1865 Broadway, New York, NY 10023-7505. Tel.: 212-408-1500 & 1536. Fax: 212-408-1292.
E-mail: info@mobia.org
Web Site: www.mobia.org
Founded: 2005.
Congressional District: 14
Key Personnel: Dir., Richard Townsend; Chm. (V), Roberta Ahmanson; Assoc. Cur., Adrianne Rubin; Dir. Education, Jennifer Kalter; Devel. Assoc., Bruno Nouril; Special Events & Mktg. Coord., Brittany Daulton; Visitor Svcs. Assoc., Isabella Lores-Chavez.
Personnel Profile: Full-Time Paid 10; Part-Time Paid 4; Part-Time Volunteers 1; Interns 4.
Governing Authority: board of trustees. Tax-exempt: 501(c)(3).
Institution Type/Description: Art Museum.
Collections: art in the Jewish & Christian traditions.
Major Exhibits: Sacred Visions: 19th Century Biblical Art from the Dahesh Museum Collection, 10/18/13-2/16/14; Object of Devotion: Medieval English Alabaster Sculpture from the Victoria & Albert Museum, 3/7/14-6/8/14.
Facilities: education center. Books for sale.
Activities: films; adult workshops; family workshops; walking tours; audio guided tours; family days.
Publications: scholarly books on art & religion; exhibition catalogs; educational brochures & booklets; e-newsletter.
Hours & Admission Prices: Tues.-Sun. 10-6. No charge, donations accepted. Closed New Year's Day; Independence Day; Thanksgiving; Christmas. &
Attendance: 21,932 (accurate)
Membership: Individual $125; Supporter $500; Benefactor $1,000; Patron $5,000; Director's Circle $10,000.

MUSEUM OF BUSINESS, COMMERCE & WEALTH, 177 W. 26 St., Loft 200, New York, NY 10001-6811. Tel.: 917-330-0132; 212-613-3242.
E-mail: staff@museumofbusiness-commerce-wealth.org
Web Site: www.museumofbusiness-commerce-wealth.org
Governing Authority: Tax-exempt: 501(c)(3).
Institution Type/Description: History Museum.
Collections: business & corporate history.
Hours & Admission Prices: Call for hours.

MUSEUM OF CHINESE IN THE AMERICAS, (M), 211-215 Centre St., New York, NY 10013-3601. Mailing Address: 215 Centre St., New York, NY 10013. Tel.: 212-619-4785. Fax: 212-619-4720.
E-mail: info@mocanyc.org
Web Site: www.mocanyc.org
Founded: 1980.
Congressional District: 15
Key Personnel: Exec. Dir., Helen Koh; Dir. Public Programs, Nancy A. Bulalacao; Dir. Operations, Bonnie Chin Washburn; Dep. Dir. External Affairs, Carolyn Cervantes Antonio; Assoc. Dir. External Affairs, Emily Chovanec Schappler.
Personnel Profile: Full-Time Paid 15; Part-Time Paid 6; Part-Time Volunteers 40; Interns 10.

Governing Authority: nonprofit organization. Archives, 70 Mulberry St., 2nd Fl., New York, NY 10013. Tax-exempt: 501(c)(3).
Institution Type/Description: History Museum: housed in a former public school; built in 1893.
Collections: related to Chinese-American history & culture: archives; photographs; oral histories; Cantonese opera costumes; scripts; musical instruments; Chinatown store signs; local business-related artifacts; Chinese laundry collection; World War II Chinese-American veterans collection.
Major Exhibits: Hidden Chinatown: Photographs by Annie Ling, 10/13-1/14; Tomie Arai: Portraits of New York Chinatown, 10/13-1/14.
Research Fields: history and culture of the Chinese diaspora in North and South America.
Facilities: 1,800-vol. library on Asian-American history & culture, available for use by public; 13,000 sq. ft. exhibit space. Asian American-related books for sale.
Activities: guided gallery tours; Chinatown walking tours by appointment; workshops; lectures; literary, performing & visual arts presentations; traveling & participatory exhibits; media productions for rental; multidisciplinary, multimedia public programs; family programs.
Hours & Admission Prices: Tues.-Wed., Fri.-Sun. 11-6, Thurs. 11-9. Adults $10, seniors 65 & over and students $5; discount to AAM members; children under 12 in groups less than 8 & members no charge. Closed Mon.; New Year's Day; Thanksgiving; Christmas Day. &
Attendance: 30,000 (accurate)
Membership: Student & Senior Citizens $25; Senior Dual $35; Individual $60; Dual $80; Family $125; Contributor $250; Patron $500; Sustainer $888.

MUSEUM OF COMIC AND CARTOON ART, 128 E. 63rd St, New York, NY 10065. Tel.: 212-838-2560. Fax: 212-838-2561.
E-mail: info@societyillustrators.org
Web Site: www.societyillustrators.org
Founded: 2001.
Congressional District: 8
Key Personnel: Dir., Karl Erickson; Chm. (V), Ellen S. Abramowitz.
Personnel Profile: Full-Time Paid 1; Part-Time Volunteers 10.
Governing Authority: Parent Institution: Society of Illustrators, New York, NY.
Institution Type/Description: Art Museum.
Collections: comics; cartoons; animation.
Activities: special events.
Hours & Admission Prices: Tues. 10-8, Wed.-Fri. 10-5, Sat. 12-4. Adults $5; discounts to groups; children 12 & under no charge.
Membership: Student & Senior $25; Individual $35; Individual Plus $50; Family $75; Patron $100.

MUSEUM OF JEWISH HERITAGE-A LIVING MEMORIAL TO THE HOLOCAUST, 36 Battery Place, New York, NY 10280-1502. Tel.: 646-437-4200. Fax: 646-437-4311. Facebook: Museum of Jewish Heritage.
E-mail: aspilka@mjhnyc.org
Web Site: www.mjhnyc.org
Founded: 1984.
Congressional District: 8
Key Personnel: Chm., Bruce Ratner; Vice Chm., Stephen E. Kaufman; Vice Chm., Ann Oster; Vice Chm., Ingeborg Rennert; Dir. & C.E.O., David G. Marwell, Ph.D.; Deputy Dir., Anita Kassof; Assoc. Dir., Abby R. Spilka; Dir. Operations, Michael Minerva; Dir. Education, Elizabeth Edelstein; Dir. Collections & Exhibitions, Melissa Martens Yaverbaum; Chief Devel. Officer, Clara Nyman; C.F.O., Mohad Athar; Dir. Human Resources, Tammy Chiu; Dir. Public Programs, Gabriel Sanders; Sr. Registrar & Mgr. Traveling Exhibitions, Erica Blumenfeld; Museum Shop Mgr., Warren Shalewitz.
Personnel Profile: Full-Time Paid 60; Part-Time Paid 12; Part-Time Volunteers 250; Interns 90.
Operating Expenses: 13,411,556
Operating Income: 10,856,196
Governing Authority: nonprofit organization. Parent Institution: New York Holocaust Memorial Commission. Tax-exempt: 501(c)(3).
Institution Type/Description: Jewish History Museum: 20th-century memorial to the Holocaust.
Collections: early 20th-century Jewish life; war against the Jews; Jewish renewal since 1945; multi-media installation; core exhibit; Shoah testimonies.
Major Exhibits: Against the Odds: American Jews and the Rescue of Europe's Refugees, 1933-1941, 5/21/13-4/15; Hava Nagila: A Song for the People (T), 9/13/13-4/22/14; Discovery and Recovery: Preserving Iraqi Jewish Heritage, 2/3/14-5/18/14; The Town Known as Auschwitz: Stories from a Jewish Past (T), 5/13/14-6/15.
Research Fields: the Holocaust; late 19th- to early 20th-century Jewish history; aspects of Jewish immigration to the U.S.

Facilities: 6,300-vol. library of books on Jewish history, Holocaust, reference material, memoirs; classrooms.
Activities: tours for school groups; curriculum material; family programs; outreach; speakers bureau; programs on Wed., Sun. & evenings.
Publications: quarterly, Museum Newsletter; visitor's brochure; calendar of events; annual report; program brochures; educational workbooks; exhibition brochures.
Hours & Admission Prices: Daylight Savings Time: Sun.-Tues. & Thurs. 10-5:45, Wed. 10-8, Fri. 10-5; Eastern Standard Time & Eve of Jewish Holidays: Sun.-Tues. & Thurs. 10-5:45, Wed. 10-8, Fri. 10-3. Adults $12, seniors $10, students $7; discounts to AAM & ICOM members; members, children under 12 & Wed. 4-8 no charge. Closed Jewish holidays; Thanksgiving. &
Attendance: 153,993 (accurate)
Membership: Senior, National & International $36; Young Friends $50; Individual $54; Senior Dual $70; Young Friends Dual $90; Dual $100; Contributing $150; Family $180; Circle of Hope $500; Curator's Circle $1,000; Director's Circle $2,500; Chairman's Circle $5,000.

*** THE MUSEUM OF MODERN ART,** 11 W. 53rd St., New York, NY 10019-5401. Tel.: 212-708-9400. Fax: 212-708-9889. TDD: 212-247-1230.
E-mail: comments@moma.org
Web Site: www.moma.org
Founded: 1929.
Congressional District: 14
Key Personnel: Chm., Jerry Speyer; Pres., Marie-Josee Kravis; Dir., Glenn Lowry; C.O.O., James Gara; Sr. Deputy Dir. External Affairs, Todd Bishop; Dir. Human Resources, Trish Jeffers; Sr. Deputy Dir. Exhibitions, Collections & Programs, Ramona Bannayan; Deputy Dir. Education, Wendy Woon; Chief Cur. Prints & Illustrated Books, Christophe Cherix; Gen. Counsel, Patty Lipshutz; Chief Cur. at Large, Klaus Biesenbach; Chief Cur. Painting & Sculpture, Ann Temkin; Chief Cur. Architecture & Design, Barry Bergdoll; Chief Cur. Dept. Drawings, Cornelia Butler; Chief Cur. Film, Rajendra Roy; Assoc. Dir., Kathy Halbriech.
Personnel Profile: Full-Time Paid 750; Part-Time Paid 35; Part-Time Volunteers 314; Interns 58.
Governing Authority: nonprofit corporation. Tax-exempt: 501(c)(3).
Institution Type/Description: Art Museum.
Collections: modern & contemporary art from 1880-present, painting; sculpture; drawings; prints; architecture & design; photography; film; video; posters; illustrated books; manuscripts.
Research Fields: pertaining to collections.
Facilities: 300,000-vol. library; archives; study centers; reading room; Edward John Noble Education Center; Abby Aldrich Rockefeller Sculpture Garden; 460-seat theater; 225-seat theater; restaurants; 2 cafes. Museum-related items for sale.
Activities: theaters; symposia; lecture series; films; gallery talks; conversations with contemporary artists; internship programs; inter-museum loan, permanent, temporary & traveling exhibitions; brown bag lunch lectures; courses at MoMA; weekend family programs; virtual visits & audio programs; high school programs; weekend teachers workshops; access programs; community programs.
Publications: books; exhibition catalogs; posters; cards; monthly members calendars & e-newsletter.
Hours & Admission Prices: Wed.-Thurs. & Sat.-Mon. 10:30-5:30, Fri. 10:30-8. Museum: adults $25, seniors 65 & over $18, full-time students with ID $14; discounts to AAM & ICOM members; children 16 & under, MoMA members & Fri. 4-8 no charge. Additional charge for film & media programs. Closed Thanksgiving; Christmas. &
Attendance: 2,219,554 (estimated)
Membership: Student $52; National & International $70; Individual $85; Dual $140; Family $175; Fellow $360; Supporting $600; Sustaining $1,200; Patron $1,750; Benefactor $3,000.

MUSEUM OF MOTHERHOOD - M.O.M., (M), 401 E. 84th St., New York, NY 10028. Mailing Address: P.O. Box 210, Hastings-On-Hudson, NY 10706. Tel.: 212-452-9816; 877-711-6667.
E-mail: mommuseum@gmail.com
Web Site: www.mommuseum.org
Founded: 2003.
Key Personnel: Exec. Dir., Joy Rose; Chair, Gillian Crane; Devel., Cara-Leigh Battaglia; Education, Lynn Kuechle.
Personnel Profile: Full-Time Volunteers 1; Part-Time Paid 1; Part-Time Volunteers 3.
Governing Authority: private; nonprofit organization. Tax-exempt: 501(c)(3).
Institution Type/Description: History Museum.
Collections: history of mothers; mother-art; culture of family; mothering perspectives; women in science, politics, medicine, & society.

Research Fields: mother studies.
Facilities: library; 700 sq. ft. exhibition space; educational facilities. Museum-related items for sale.
Activities: films; formal education programs; traveling exhibitions; broadcast programs; special events; rental facilities; performances. Museum Sponsors: Annual M.O.M. Conference; Mamapalooza Festival.
Publications: biannual magazine, Mamazina.
Hours & Admission Prices: Tues.-Sun. 10:45-6:30.
Attendance: 750

MUSEUM OF SEX, 233 5th Ave. @ 27th St., New York, NY 10016. Tel.: 212-689-6337.
Web Site: museumofsex.com
Institution Type/Description: History Museum.
Collections: history, evolution & cultural significance of human sexuality.
Facilities: Museum-related items for sale.
Activities: educational programs; lectures; special events.
Hours & Admission Prices: Sun.-Thurs. 10-8, Fri.-Sat. 10-9; groups by appointment. Adults $17.50, seniors and students 18 & over $15.25; discounts to groups of 10 or more; children under 18 not admitted. Closed Thanksgiving; Christmas. &
Attendance: 200,000 (accurate)

MUSEUM OF THE CITY OF NEW YORK, (M), 1220 Fifth Ave. at 103rd St., New York, NY 10029. Tel.: 212-534-1672. Fax: 212-534-0687.
E-mail: info@mcny.org
Web Site: www.mcny.org
Founded: 1923.
Congressional District: 14
Key Personnel: Dir., Susan Henshaw Jones; Chm. (V), James G. Dinan; Deputy Dir. Programs, Dr. Sarah Henry; Head of Education, Franny Kent; Vice Pres. Devel., Susan Madden; Vice Pres. Finance & Admin. & C.F.O., Carl Dreyer; Chief Registrar, Giacommo Mirabella; Dir. Collections, Lacy Schutz; Cur. Costumes, Phyllis Magidson; Mgr. Human Resources, Nancy Mercado.
Personnel Profile: Full-Time Paid 70; Part-Time Paid 10; Part-Time Volunteers 50; Interns 10.
Governing Authority: nonprofit organization. Tax-exempt: 501(c)(3).
Institution Type/Description: General Museum.
Collections: presentation relating to present & past social, economic, intellectual & political history of New York; costumes; furniture; silver; paintings; prints & photographs; decorative arts; manuscripts; archives; sculpture; graphics; textiles; toys; theatrical.
Major Exhibits: Janet Ruttenberg Paintings, 9/12/13-1/14.
Research Fields: archives; painting; sculpture; graphics; decorative arts; costumes; general history; textiles; theatre.
Facilities: 200-seat auditorium.
Activities: guided tours; lectures; gallery talks; formally organized education programs; family workshops; performances; temporary exhibitions.
Publications: biannual magazine, The Courant; exhibition catalogues.
Hours & Admission Prices: Daily 10-6. Suggested Donations: families $20, adults $10, senior citizens, students & children $6; discounts to AAM & ICOM members, Columbia University staff & students, corporate member employees & employees of NYC Museum Council Institutions; children under 12 & members no charge. Bank of American Museums on Us Program. Closed New Year's Day; Thanksgiving; Christmas. &
Attendance: 227,716 (accurate)
Membership: New Yorker at Heart $35; Senior Citizen & Student $45; Individual $60; Family $85; Young Member $100; Associate $115; Sustainers $250; Dew Witt Clinton Fellow, Women's Committee & Fellow $500; AHC $1,000; AHC Associate $2,500; AHC Fellow $5,000; AHC Patron $10,000; AHC Partner $25,000; AHC Benefactor $50,000.

MUSEUM OF TOLERANCE NEW YORK, 226 E. 42nd St., New York, NY 10017-5806. Tel.: 212-697-1180. Fax: 212-697-1314.
E-mail: arudich@wiesenthal.com
Web Site: www.mueumoftolerancenewyork.com
Founded: 2004.
Key Personnel: Dir. Operations & Community Affairs, Adam Rudich; Dir. Eastern Region, Rabbi Steven Burg.
Personnel Profile: Full-Time Paid 10; Part-Time Paid 3; Part-Time Volunteers 20; Interns 2.
Governing Authority: Parent Institution: SWC. Tax-exempt.
Institution Type/Description: Holocaust & Human Rights Museum.
Collections: Nazi Holocaust; World War II; 20th-century genocide; racism; concentration camps; Holocaust artifacts; human rights; tolerance; Civil Rights; antisemitism; art; photographs; social justice; prejudice; discrimination.

Facilities: theater. Museum-related items for sale.
Activities: workshops; videos; educational programs; group tours; videos; lectures.
Hours & Admission Prices: Mon.-Fri. 10-5, Sun. by appointment; Nov.-March Fri. 10-3:30. Adults $15, seniors $12, students $10, members no charge. &
Attendance: 15,000 (estimated)
Membership: Senior & Student $40; Individual $55; Dual $70; Family $95.

NATIONAL ACADEMY MUSEUM & SCHOOL, 1083 Fifth Ave., (& 89th St.), New York, NY 10128-0114. Tel.: 212-369-4880. Fax: 212-426-1711.
Web Site: www.nationalacademy.org
Founded: 1825.
Congressional District: 15
Key Personnel: Exec. Dir., Carmine Branagan; Cur., Bruce Weber; Asst. Cur., Marshall Price; Conservator, Lucie Kinsolving; Registrar, Athena Latocha; Artist Membership, Sei Young Kim; Controller, Michael McKay; Dir. Mktg. & Communications, Heidi Riegler; Dir. Operations & Security, Charles Biada; Dir. Devel., Ross Randall; Dir. School, Maurizio Pellegrin.
Personnel Profile: Full-Time Paid 25; Part-Time Paid 64; Part-Time Volunteers 10; Interns 6.
Governing Authority: nonprofit organization. Branch Museums: Gallery; School of Fine Arts. Tax-exempt: 501(c)(3).
Institution Type/Description: American Art Museum, School of Fine Arts, Academy of Artists & Architects.
Collections: American painting, sculpture, drawings; architectural renderings; archives pertaining to academy proceedings & members of academy.
Research Fields: 19th-21st centuries American art.
Facilities: over 3,000-vol. library.
Activities: exhibitions from permanent collection; loan exhibitions; lectures; classes & workshops in the fine arts; annual juried exhibition; public programs.
Publications: exhibition; school catalogs.
Hours & Admission Prices: Wed.-Sun. 11-6; see website for additional hours. Adults $15, senior citizens 65 & over and students with ID $10; members, children under 12 & National Academy School students no charge. Closed New Year's Day; Independence Day; Thanksgiving; Christmas. &
Attendance: 20,000 (estimated)
Membership: Students, Educator & Seniors $40; Individual $60; Dual $80; Contributor $125; Sponsor $250; Patron $500; Director's Circle $1,000.

NATIONAL ARTS CLUB, 15 Gramercy Park S., New York, NY 10003-1796. Tel.: 212-475-3424. Fax: 212-475-3692.
Web Site: www.nationalartsclub.org
Founded: 1898.
Congressional District: 18
Key Personnel: Pres. (V), O. Aldon James, Jr.; Treas., Jason de Montmorency; Cur. National Arts Club Permanent Collection, Dr. Carol Lowrey; Chm. Exhibitions Committee, Diane Bernnhard.
Governing Authority: nonprofit organization. Tax-exempt: 501(c)(3).
Institution Type/Description: Art Museum: housed in 1840s, former mansion of Governor Samuel Tilden.
Collections: late 19th & 20th-century American painting; sculpture; works on paper; decorative arts.
Research Fields: late 19th & early 20th-century American art.
Facilities: 150-vol. personal library of Robert Henri available for research upon request; public exhibitions.
Publications: brochure; Carol Lowrey, A Noble Tradition; exhibit catalogue, American Painting from the National Arts Club Permanent Collection.
Hours & Admission Prices: Call for confirmation of hours of exhibition galleries. No charge.

NATIONAL AUDUBON SOCIETY, 225 Varick St., Fl. 7, New York, NY 10014-4396. Tel.: 212-979-3000. Fax: 212-979-3188.
Web Site: www.audubon.org
Founded: 1905.
Congressional District: 8
Key Personnel: Chm., B. Holt Thrasher; Pres. & C.E.O., John Flicker; Dir. Mktg. & Communications, Nancy Severance.
Personnel Profile: Full-Time Paid 284; Part-Time Paid 44.
Governing Authority: not-for-profit organization. Tax-exempt: 501(c)(3).
Institution Type/Description: Wildlife & Wildlife Habitat Protection Society.
Collections: wildlife & their habitats.
Research Fields: energy alternatives; ancient forests of the Northwest; wildlife trade; migratory bird program with Latin America; developing Beringia Wilderness Park; Everglades & wetlands preservation; local work.
Facilities: educational facilities; field research stations; nature centers. Environmentally associated items for sale.
Activities: formal education programs for children, adults, & undergraduate or graduate college students; guided tours; lectures; slide shows; TV & cable programs; Audubon Adventures student & teacher programs; 514 national chapters.
Publications: monthly news journal, Activist; periodic wildlife identification & information guides; bimonthly magazine, Audubon; grassroots action toolkits; Audubon Adventures curriculum.
Hours & Admission Prices: Contact individual sanctuaries for hours. &
Attendance: 125,000
Membership: Introductory $20; Regular $35.

NATIONAL MUSEUM OF MATHEMATICS, (M), 11 E. 26th St., New York, NY 10010. Mailing Address: 134 W. 26th St., Ste. 4S, New York, NY 10001. Tel.: 212-542-0566.
E-mail: info@momath.org
Web Site: momath.org
Key Personnel: Exec. Dir., Glen Whitney
Institution Type/Description: Children's Museum.
Collections: hands-on exhibitions.
Facilities: Museum-related items for sale.
Activities: educational programs; special events.
Hours & Admission Prices: Daily 10-5. Adults $15, children 12 & over, students and seniors 65 & over $9; children under 2 no charge.

* **NATIONAL MUSEUM OF THE AMERICAN INDIAN - NEW YORK, GEORGE GUSTAV HEYE CENTER, (M),** Alexander Hamilton U.S. Custom House, One Bowling Green, New York, NY 10004-1415. Tel.: 212-514-3700. Fax: 212-514-3800.
E-mail: NMAI-NY@si.edu
Web Site: www.nmai.si.edu
Founded: 1916.
Congressional District: 16
Key Personnel: Dir. National Museum of the American Indian (NMAI), Kevin Gover; Assoc. Dir. NMAI Museum Programs, Tim Johnson; Dir. George Gustav Heye Center (GGHC), John Haworth; Dep. Asst. Dir. Exhibits, Programs & Public Spaces, ,GGHC, Peter Brill; Dep. Asst. Dir. Operations & Program Support, GGHC, Scott Merritt; Dir. Devel., GGHC, Lucia DeRespinis; Education Dept. Mgr., Johanna Gorelick; Resource Center Mgr., Gaetana de Gennaro; Special Events & Corp. Membership Mgr., Trey Moynihan; Visitor Svcs. Mgr., Margaret Sagan; Cultural Arts Mgr., Shawn Termin.
Governing Authority: nonprofit, federally-chartered corporation. Parent Institution: Bureau of the Smithsonian Institution, 1000 Jefferson Dr., S.W., Washington, DC 20560. Tax-exempt: 501(c)(3).
Institution Type/Description: Anthropology & Indian Museum.
Collections: American Indian archaeology & ethnology from North, Central, South America & Caribbean; artifacts; textiles; agriculture; anthropology; paintings; sculpture; decorative arts; costumes; numismatics; music; medical; literature; history; arts; language; Eskimo culture; photograph archives; manuscript collections.
Research Fields: prehistoric, historic & contemporary native people of the Western Hemisphere.
Facilities: 20,000 sq. ft. exhibition galleries; Film & Video Center; resource center; contemporary Indian arts, crafts & books sale.
Activities: guided tours; interpretive programs; Expressive Culture series; inter-museum loans; permanent, temporary & traveling exhibitions; film & video programs; off-site lectures; information services; internship program; services to Native American groups.
Publications: books; recordings; exhibition catalogues; brochures; quarterly magazine, American Indian.
Hours & Admission Prices: Mon.-Wed. & Fri.-Sun. 10-5, Thurs. 10-8. No charge; discounts & complimentary magazine subscriptions to charter members. Closed Christmas. &
Attendance: 334,506 (accurate)
Membership: Charter Circles: Golden Prairie $25; Riverbed $35; Everglades $50; Sky Meadows $100; Boundary Waters $250; Desert Sands $500.

NEUE GALERIE NEW YORK, 1048 Fifth Ave., (at 86th St.), New York, NY 10028-0111. Tel.: 212-628-6200.
Founded: 2001.
Key Personnel: Dir., Renee Price; Book Store Dir., Brunno Keusch; Design Shop Dir., Paul Landy.
Governing Authority: Tax-exempt.
Institution Type/Description: Art Museum.
Collections: paintings; sculpture; works on paper; decorative arts; photographs.
Hours & Admission Prices: Thurs.-Mon. 11-6. Adults $20, students & seniors $10; discounts to AAM & ICOM members; 1st Fri. each month 6pm-8pm no charge. Children under 12 not admitted.

Membership: Student $75; National/International $100; Individual $275; Contributing $500; Sustaining $1,000; Patron $2,500; Leadership $5,000; Benefactor $10,000.

NEW MUSEUM, 235 Bowery, New York, NY 10002-1218. Tel.: 212-219-1222, ext. 200. Fax: 212-431-5328.
E-mail: info@newmuseum.org
Web Site: www.newmuseum.org
Founded: 1977.
Congressional District: 18
Key Personnel: Dir., Lisa Phillips; Pres. Bd. of Trustees (V), Saul Dennison; Deputy Dir., Karen Wong; Chief Cur., Massimiliano Gioni; Operations Mgr., Steve Harris; Bookstore Mgr., Daniel Thiem.
Personnel Profile: Full-Time Paid 64; Part-Time Paid 40; Part-Time Volunteers 5.
Governing Authority: nonprofit organization. Tax-exempt: 501(c)(3).
Institution Type/Description: Art Museum.
Collections: contemporary art.
Research Fields: contemporary art.
Facilities: Books for sale.
Activities: guided tours; lectures; performances; temporary & traveling exhibitions; educational programs; video; family programs.
Publications: catalogs, Barry LeVa; John Baldessari: Work 1966-80; Art & Ideology; Earl Staley 1973-1983; New Work: Golub; Hans Haacke; Fake, 1987; Pat Steir; Choices: Making an Art of Everyday Life; Ana Mendieta; Markus Raetz: In the Realm of the Possible; Impressario: Malcolm McLaren & the British New Wave; Christian Boltanski; brochures; Alfred Jensen: Paintings from the Years 1957-1977; New Work/New York, 1978; Barry LeVa, The Invented Landscape; Sustained Visions, 1979; Dimensions Variable; New Work/New York, 1979; Outside New York: Ohio, 1980; Deconstruction/Reconstruction: The Transformation of Photographic Information Into Metaphor, 1980: Events: Fashion Moda, Taller Boricua, Artists Invite Artists, 1980; Al Souza; Mary Stoppert; The Reverend Howard Finster; Candace Hill-Montgomery; Joseph Hilton; Kenneth Schorr; Brad Melamed; Anne Turyn; Gary Falk; books, Art after Modernism: Rethinking Representation; Bruce Nauman, Marcus Raetz, The Decade Show: Frameworks of Identity in the 1980's; Blasted Allegories: An Anthology of Writings by Contemporary Artists; Out There: Marginalization & Contemporary Cultures; The Interrupted Life, 1991; Rhetorical Image, 1990-91; Cadences: Icon & Abstraction in Contemporary Art, 1991; homo video, Where We Are Now; Discourses: Conversations in Post-Modern Art & Culture; Bad Girls, 1994; Contemporary Art and Multicultural Education, 1996; A Labor of Love, 1996; Carolee Schneemann: Up To And Including Her Limits, 1997; Remota: Airmail Paintings by Eugenio Dittborn, 1997; Mona Hatoum, 1997; Doris Salcedo, 1998; Martin Wong, 1998; Faith Ringgold, 1998; Temporarily Possessed, 1995; Marcel Odenbach, 1998; David Wojnarowicz, 1998; Time of Our Lives, 1999; Picturing the Modern Amazon, 2000; Cildo Meireles, 1999; Pierre et Gilles, 2001; William Kentridge, 2001; Black President, The Art and Legacy of Fela Kuti, 2003; East Village, 2004; John Waters Change of Life, 2004; Younger Than Jesus catalog, 2009; Younger Than Jesus: Artist Directory.
Hours & Admission Prices: Wed. & Fri.-Sun. 11-6, Thurs.-11-9. Adult $14, seniors $12, students $10; members and children 18 & under no charge. Thurs. 7-9 no charge. Closed all major holidays. &
Attendance: 300,000
Membership: Student & Senior $35; Advocate $60; Dual $100; Deluxe $400; Premium $1,000.

NEW YORK CITY FIRE MUSEUM, (M), 278 Spring St., New York, NY 10013-1405. Tel.: 212-691-1303. Fax: 212-352-3117. Facebook: New York City Fire Museum.
E-mail: director@nycfiremuseum.org
Web Site: nycfiremuseum.org
Founded: 1987.
Congressional District: 17
Key Personnel: Pres. (V), Judith Jamison; Vice Pres., Dorothy Marks; Dir., Damon Campagna; Treas., Paul Magda; Asst. Dir., Noemi Bourdier; Museum Shop Mgr., Ashely Whelan.
Personnel Profile: Full Time Paid 6; Part-Time Paid 5; Part-Time Volunteers 9; Interns 2.
Governing Authority: public; nonprofit organization. Parent Institution: The Friends of the NYCFD Collection, Inc. Tax-exempt.
Institution Type/Description: Fire-Fighting Museum: housed in 1904 firehouse.
Collections: fire-related art & artifacts dating from 18th-century to the present including horse, hand-drawn & motorized pieces of apparatus; fire buckets; trumpets; toy & working models; helmets; parade hats; presentation silver; portraits; photographs; Currier & Ives & other prints; an important collection of fire insurance marks.
Research Fields: history of fire fighting in New York City, U.S. & world.

Facilities: audiovisual room; fire safety education Hazard House.
Activities: guided & educational tours; group tours by appointment.
Publications: member newsletter, The Housewatch.
Hours & Admission Prices: Daily 10-5. adults $8, students with ID, seniors & children 3-12 $5. AAM members with card no charge. Closed holidays. &
Attendance: 40,000 (estimated)
Membership: Individual $25; Family $40; Supporting $65; Patron $100; Contributor $250; Donor $500; Director's Circle $1,000; President's Club $2,500; Museum Sponsor $10,000.

THE NEW YORK CITY POLICE MUSEUM, 100 Old Slip, New York, NY 10005-3539. Tel.: 212-480-3100, ext. 105. Fax: 212-480-9757.
E-mail: jbose@nycpm.org
Web Site: www.nycpm.org
Founded: 1998.
Key Personnel: C.E.O., Julie Bose; Pres., Rick Friedberg; Museum Shop Mgr., Iris Stephen.
Personnel Profile: Full-Time Paid 6; Part-Time Paid 5; Part-Time Volunteers 25; Interns 4.
Governing Authority: Tax-exempt: 501(c)(3).
Institution Type/Description: Police History Museum.
Collections: police artifacts; police lanterns; carved wooden night sticks; ivory night sticks; guns; knives; other unusual weapons; handcuffs; photographs; uniform articles dating from 1870 to present day; badges & shields from 1845-present day; antiques; police vehicles.
Activities: permanent & loan exhibitions; family programs; adult education programs.
Publications: quarterly newsletter.
Hours & Admission Prices: Mon.-Sat. 10-5, Sun. 12-5; school groups by appointment. Adults $8, seniors $5; discounts to AAM members; NYPD personnel & members no charge. Closed New Year's Day; Thanksgiving; Christmas. &
Attendance: 81,500 (accurate)
Membership: Out of Town $25; Silver Shield $30; Police Protectors $30; Family $80; Gold Shield $100; Chief $300; Commissioner's Circle $1,000.

NEW-YORK HISTORICAL SOCIETY, 170 Central Park West, New York, NY 10024-5194. Tel.: 212-873-3400, ext. 273. Fax: 212-595-5253.
E-mail: lmirrer@nyhistory.org
Web Site: www.nyhistory.org
Founded: 1804.
Congressional District: 19
Key Personnel: Pres. & C.E.O., Louise Mirrer; Chm. (V), Pam B. Schafler; Dir. Library Operations, Nina Nazionale; Vice Pres. & Sr. Art Historian, Brian Allen; Chief Cur. Museum Div., Stephen Edidin; N-YHS Historian & Vice Pres. Scholarly Programs, Valerie Paley; Vice Pres. Operations, Andrew Buonpastore; Vice Pres. Communications, Laura Washington; Vice Pres. History Exhibitions, Marci Reaven; Vice Pres. Public Programs, Dale Gregory; Vice Pres. Education, Sharon Dunn; Chief Advancement Officer, Sean Lally; C.F.O., Richard Shein; Gen. Counsel & Chief Administrative Officer, Jennifer Schantz; Vice Pres. & Library Dir., Michael Ryan; Dir. Exhibitions, Gerhard Schlanzky; Dir. Merchandise Operations, Ione Saroyan; Dir. Security, Joe McGrath; Dir. Operations & Maintenance, Tony Christoforou; Mgr. Visitor Svcs., Nick Mancini; Dir. IT, Dave Lopez; Dir. Human Resources, Valerie Crane; Dir. Education, Mia Nagawiecki; Dir. Engineering Svcs., Ron Gilchrist; Dir. DiMenna Children's History Museum, Alice Stevenson; Dir. Museum Administration, Chris Catanese; Dir. Special Events, Alex Maresca.
Personnel Profile: Full-Time Paid 139; Part-Time Paid 102; Part-Time Volunteers 175; Interns 35.
Governing Authority: nonprofit organization. Subsidiary Institution: The DiMenna Children's History Museum. Tax-exempt: 501(c)(3).
Institution Type/Description: Historical Society.
Collections: Audubon watercolors; paintings; portraits; Hudson River School landscapes; genre paintings; silver; furniture; Tiffany lamps & glass; ceramics & glass; sculpture; toys; folk art; carriages; military & naval history collections; prints; photographs; architectural drawings; American imprints; broadsides; newspapers; sheet music; rare books & documents; maps; manuscripts; Henry Luce III Center for the Study of American Culture includes 46,000 artifacts spanning the 17th through the 21st centuries.
Major Exhibits: Beauty's Legacy: Gilded Age Portraits in America, 9/27/13-3/9/14; Armory Show at 100: Modern Art & Revolution, 10/11/13-2/14/14; Clarice Smith: Recollections of a Life in Art, 11/8/13-2/9/14; Bill Cunningham: The Facades Project, 3/7/14-6/15/14; The Black Fives, 3/14/14-7/20/14; Audubon's Aviary: Part II of the Complete Flock, 3/21/14-5/26/14;

Homefront & Battlefield: Quilts & Context in the American Civil War, 4/4/14-8/31/14; Chinese American: Exclusion/Inclusion, 10/14-5/15.
Research Fields: American & New York history; American fine arts & decorative arts.
Facilities: 700,000-vol. library of books pertaining to 17th- to 19th-century American history, with emphasis on New York City & State; reading room; 420-seat auditorium offering daily screenings of New York Story, orientation film; classrooms; museum store; full service restaurant Caffe Storico; conservation labs; rental space. Building is wheelchair accessible & Society exhibitions/programs are accessible to visitors with visual & hearing impairments.
Activities: guided tours; lectures; inter-museum loan, permanent & temporary exhibitions; concerts; plays; living history days; reenactments; family days; films.
Publications: books; guides to collections; catalogs of exhibitions; New-York Journal of American History.
Hours & Admission Prices: Museum: Tues.-Thurs. & Sat. 10-6, Fri. 10-8, Sun. 11-5. Library: Tues.-Fri. 9-3, Sat. 10-1. Adults $18, seniors, educators & active military $14, students $12, children 5-13 $6; members, children 4 & under & AAM members no charge. Fri. 6pm-8pm pay what you wish & no charge to use the Library. Closed Labor Day; Thanksgiving; Christmas. ♿
Attendance: 250,000 (estimated)
Membership: Individual $75; Dual $110; Family $150; Young Friend $175; Friend $250; Patron Family $500; Benefactor $1,000; Gotham Fellow $2,500.

THE NEW YORK PUBLIC LIBRARY, ASTOR, LENOX AND TILDEN FOUNDATIONS, 476 Fifth Ave., Rm. 210, New York, NY 10018-2788. Tel.: 212-930-0800. Fax: 212-930-9299.
Web Site: www.nypl.org
Founded: 1895.
Key Personnel: Chm. Bd. Trustees, Neil L. Rudenstine; Pres. & C.E.O., Anthony W. Marx; C.O.O., C.F.O. & Treas., David Offensend; Dir. & Chief Exec. The Andrew W. Mellon, The Research Libraries, Ann D. Thornton; Mgr. Special Events, Vanessa Novak; Mgr. Public & Education Programs, Paul Holdengraber; Mgr. Exhibitions, Susan Rabbiner; Mgr. Public Rels., Anne Conty; Vice Pres. External Affairs, George Mihaltses; Vice Pres. Public Svc., Anne L. Coriston; Vice Pres. Strategic Planning, Jeffrey Roth; Vice Pres. Human Resources, Louise Shea; Dir. Science, Industry, Business Library, Kristin McDonough; Museum Shop Mgr., Sara Abraham; Dir. Performing Arts Library, Jacqueline Davis; Dir. Schomburg Ctr. Research Black Culture, Khalil Gibran Muhammad.
Personnel Profile: Full-Time Paid 2,272; Part-Time Paid 267.
Governing Authority: nonprofit organization. Branch Landmarks: 1905 Mott Haven Branch; 1911 The New York Public Library's Central Building; 1832 Jefferson Market Branch; 1884 Ottendorfer Branch; 1908 115th Street Branch; 1907 Hamilton Grange Branch; 1902 Yorkville Branch. Tax-exempt: 501(c)(3).
Institution Type/Description: Public Library: housed in 1911 Central Building on the site of the Croton Reservoir.
Collections: 85 Branch Libraries & 4 Research Libraries housing over 54 million items. 1,000 public access computer terminals providing free internet access, free access to a wide range of electronic data bases, and access to the RL catalog (CATNYP) and the BL catalog (LEO). Features new Science, Industry & Business Library (SIBL), Library for the Performing Arts (LPA), Schomburg Center: The Study of Black Culture, and Center for the Humanities and Social Sciences (HSSL) with world renowned special collections as well as wide-ranging basic research & reference materials.
Research Fields: all areas except medicine, law, theology.
Facilities: library; reading room.
Activities: lectures; films; concerts; drama; formally organized educational programs; inter-museum loan, permanent & temporary exhibitions.
Publications: Biblion, Bulletin of the New York Public Library; New Technical Books; Children's Books; newsletter, Library Lines; Directory of Community Services. For a complete list of publications, contact The New York Public Library.
Hours & Admission Prices: Daily. Call for branch hours. No charge; donations accepted. Closed holidays. ♿
Attendance: 13,000,000 (accurate)
Membership: Friend $40; Participating Friend $65; Supporting Friend $100; Patron $250; Sustainer $500; Conservator $1,250.

THE NEW YORK PUBLIC LIBRARY FOR THE PERFORM-ING ARTS, 40 Lincoln Center Plaza, New York, NY 10023-7486. Tel.: 212-870-1830. Fax: 212-870-1860.
E-mail: barbaracohenstratyner@nypl.org
Web Site: www.nypl.org
Founded: 1965.
Congressional District: 17

Key Personnel: Pres., Anthony Marx; Chm. (V), Katie Marron; Exec. Dir. Performing Arts, Jacqueline Z. Davis; Sr. Designer, Caitlyn Mack; Head Exhibitions, Barbara Cohen-Stratyner; Theatre Cur., Karen Nickeson; Music Cur., George Boziwick; Cur. Recorded Sound, Jonathan Hiam; Cur. Dance, Jan Schmidt; Museum Shop Mgr., Sara Abraham.
Personnel Profile: Full-Time Paid 76; Part-Time Paid 12; Part-Time Volunteers 27; Interns 12.
Governing Authority: nonprofit organization. Parent Institution: New York Public Library, 5th Ave. at 42nd St., New York, 10016. Tel.: 212-930-0800. Tax-exempt: 170(b)(1)(A).
Institution Type/Description: History Museum of Performing Arts.
Collections: prints; letters; manuscripts; documents; photographs; posters, films; video tapes; memorabilia; dance; recordings.
Research Fields: performing arts.
Facilities: library of circulating & reference books available for inter-library loan during building hours; reading rooms; 202-seat auditorium in addition to 8 wheelchair stations.
Activities: free lectures and classes on research workshops for teachers & general public; films; concerts; dance recitals; drama; formally organized education programs for teachers; loan, temporary & traveling exhibitions.
Publications: season brochures; calendars; free exhibition brochures.
Hours & Admission Prices: Mon. & Thurs. 12-8, Tues.-Wed. & Fri. 12-6, Sat. 2-6. No charge. Closed national holidays. ♿
Attendance: 350,000 (accurate)

THE NEW YORK STUDIO SCHOOL OF DRAWING, PAINT-ING & SCULPTURE, 8 W. 8th St., New York, NY 10011-9084. Tel.: 212-673-6466. Fax: 212-777-0996.
E-mail: info@nyss.org
Web Site: www.nyss.org
Founded: 1964.
Congressional District: 17
Key Personnel: Dean, Graham Nickson; Coord. Gallery, Pamela Salisbury.
Personnel Profile: Full-Time Paid 6; Part-Time Paid 13.
Governing Authority: private; nonprofit organization. Tax-exempt: 501(c)(3).
Institution Type/Description: Art Institute: located in a federally landmarked building, site of studios of sculptors Daniel Chester French & Gertrude Vanderbilt Whitney; original site of the Whitney Museum of American Art.
Collections: paintings; sculpture; photographs; sculpture; drawings.
Facilities: 3,000-vol. library, 2,500 slide archive; classrooms; studios; 1,000 sq. ft. exhibit space.
Activities: weekly evening lecture series Oct.-May; loan exhibits; formal education programs for adults; non-affiliated certificate programs for undergraduate & graduate students.
Publications: annual newsletter, New York Studio School; illustrated brochures on selected exhibitions.
Hours & Admission Prices: Daily 10-10. Weekly Evening Lecture Series: Tues.-Wed. 6:30 pm-8 pm. No charge; donations accepted. Closed national holidays & school vacations, depending on exhibition schedule.
Attendance: 10,000 (estimated)

NICHOLAS ROERICH MUSEUM, 319 W. 107th St., New York, NY 10025-2799. Tel.: 212-864-7752. Fax: 212-864-7704.
E-mail: director@roerich.org
Web Site: www.roerich.org
Founded: 1958.
Congressional District: 15
Key Personnel: Exec. Dir., Daniel Entin; Pres. (V), Edgar Lansbury; Director's Asst., Aida Tulskaya; Dir. Cultural Programs, Jean Fletcher.
Personnel Profile: Full-Time Paid 4; Part-Time Volunteers 4; Interns 1.
Governing Authority: nonprofit organization. Tax-exempt: 501(c)(3).
Institution Type/Description: Art Museum.
Collections: paintings of Tibet, India, Himalayas by Nicholas Roerich.
Research Fields: Eastern art.
Activities: gallery talks; concerts; permanent exhibition; poetry readings.
Publications: books, The Invincible; Shambala, The Roerich Pact and Banner of Peace; Altai-Himalaya.
Hours & Admission Prices: Tues.-Sun. 2-5. No charge; donations accepted. Closed select holidays. ♿
Attendance: 10,000 (estimated)
Membership: Associate $25; Contributing $50; Sustaining $100.

9/11 TRIBUTE CENTER, (M), 120 Liberty St., New York, NY 10006-1008. Mailing Address: 22 Cortlandt St., 801, New York, NY 10007. Tel.: 866-737-1184; 212-393-9160, ext. 141.
E-mail: visitorservices@tributewtc.org
Web Site: www.tributewtc.org
Formerly: Tribute WTC Visitor Center
Congressional District: 84

Key Personnel: C.E.O., Jennifer Adams
Governing Authority: Parent Institution: September 11th Families Association. Tax-exempt.
Institution Type/Description: History Museum.
Collections: September 11th history; photographs.
Activities: special programs.
Hours & Admission Prices: Mon.-Sat. 10-6, Sun. 10-5. Adults $15; members no charge. Closed New Year's Day; Easter; Thanksgiving; Christmas Eve, Day & week.
Attendance: 500,000 (estimated)

THE PALEY CENTER FOR MEDIA, 25 W. 52nd St., New York, NY 10019-6129. Tel.: 212-621-6600. Fax: 212-621-6700.
E-mail: mreidy@paleycenter.org
Web Site: www.paleycenter.org
Formerly: The Museum of Television & Radio
Founded: 1975.
Congressional District: 18
Key Personnel: Chm. Bd., Frank A. Bennack, Jr.; Pres. & C.E.O., Pat Mitchell; Chief Mktg. Officer, Maureen Reidy; Exec. Vice Pres. Devel., Maxim Thorne; Vice Pres. Programs, Diane Lewis; Cur., Ronald Simon; Cur., David Bushman; Dir. Devel., Jennifer Juzaitis; Dir. Technical Operation & Engineering, Doug Warner; Dir. Library & Information Svcs., Douglas F. Gibbons; Mgr. Public Rels., Carrie Oman; Guest Svcs. Mgr., Bob Eng.
Personnel Profile: Full-Time Paid 75; Part-Time Paid 55; Part-Time Volunteers 6; Interns 10.
Governing Authority: nonprofit organization. Branch Location: 465 N. Beverly Dr., Beverly Hills, CA 90210. Tel.: 310-786-1000. Tax-exempt: 170(b)(1)(A).
Institution Type/Description: Digital Media Museum.
Collections: over 150,000 radio & television programs and advertisements from the 1920s to the present; Warner Bros.
Activities: NY: television & radio consoles; three screening theaters; satellite up & down links. LA: television & radio consoles; one screening room; one screening theater; satellite up & down links; programs; formally organized education programs for students; seminars on various aspects of media; programs for industry professionals.
Publications: exhibition catalogs; seminar transcripts; catalogue of events.
Hours & Admission Prices: NY: Wed. & Fri.-Sun. 12-6, Thurs. 12-8. LA: Wed. & Fri.-Sun 12-5, Thurs. 12-8. Suggested Donation: adults $10, senior citizens & students $8, children under 14 $5; discounts to AAM & ICOM members; members & California museum no charge. Closed New Year's Day; Independence Day; Thanksgiving; Christmas. &
Attendance: 120,000 (accurate)
Membership: Senior Citizen & Student $50; Individual $70; Dual & Family $100; Supporting $250. Patrons Circle: Silver $1,000; Gold $2,500; Platinum $6,000; President's Circle $15,000; Chairman's Circle $25,000.

PRATT MANHATTAN GALLERY, 144 W. 14th St., 2nd Fl., New York, NY 10011-7301. Tel.: 212-647-7778. Fax: 212-367-2484.
E-mail: exhibits@pratt.edu
Web Site: www.pratt.edu/exhibitions
Founded: 2002.
Key Personnel: Dir., Nick Battis; Asst. Dir., Olivia Good; Exhibit Designer, Katherine Davis.
Personnel Profile: Full-Time Paid 2; Part-Time Paid 8; Interns 2.
Governing Authority: nonprofit. Parent Institution: Pratt Institute. Tax-exempt: 501(c)(3).
Institution Type/Description: Art Museum & Center.
Collections: 19th- & 20th-century paintings, sculpture, prints, decorative arts & graphic arts by European & American artists.
Research Fields: contemporary fine arts, design & architecture.
Facilities: 2,200 sq. ft. exhibit space; classrooms; lecture room.
Activities: lectures; continuing education & degree programs.
Publications: exhibition announcements & catalogs; posters; newsletter; biannual exhibitions & events calendar.
Hours & Admission Prices: Tues.-Sat. 11-6. No charge. Closed major holidays. &
Attendance: 30,000

QUEEN SOFIA SPANISH INSTITUTE, 684 Park Ave., New York, NY 10065-5043. Tel.: 212-628-0420. Fax: 212-734-4177.
E-mail: info@qssi.org
Web Site: www.queensofiaspanishinstitute.org
Formerly: The Spanish Institute
Founded: 1954.
Key Personnel: Chm., Oscar de la Renta; Pres. & C.E.O., Inmaculada de Habsburgo; Dir., Pilar Vico.

Personnel Profile: Full-Time Paid 12; Part-Time Paid 30; Part-Time Volunteers 2.
Governing Authority: nonprofit organization. Tax-exempt: 170(b)(1)(A).
Institution Type/Description: Spanish Cultural Institute: housed in c.1920 building designed by McKim, Mead & White.
Collections: temporary exhibitions of art.
Research Fields: Spanish art in all media; collecting & patronage of Spanish art.
Facilities: 500-vol. library on Spanish art available to the public; 150-seat auditorium; classrooms; 2,200 sq. ft. exhibit space. Exhibition catalogs for sale.
Activities: recitals; films; videos; formal educational programs; guided tours; lectures; temporary, loan & traveling exhibitions.
Publications: exhibition catalogues, Balenciaga; Joaquin Sorolla and the Glory of Spanish Dress; Don Quixote; Gerardo Rueda: Spanish Modernist; A Life of Picasso: The Triumphant Years; From Goya to Sorolla: Queen Sofia Spanish Institute Salutes the Hispanic Society of America on Its 100th Anniversary; On Site: New Architecture in Spain; Barcelona and Modernity: Picasso, Gaudi, Miro, Dali.
Hours & Admission Prices: Mon.-Thurs. 10-6, Fri. 10-8, Sat. 11-5. Adults $15; discounts to members. Closed major holidays.
Attendance: 30,000 (estimated)
Membership: Individual $60; Friend $500; Patron $1,000; Sponsor & Corporate $5,000; Benefactor $15,000; Golden Benefactor $25,000.

THE RENEE AND CHAIM GROSS FOUNDATION, 526 LaGuardia Place, New York, NY 10012-1401. Tel.: 212-529-4906. Fax: 212-529-1966.
E-mail: info@rcgrossfoundation.org
Web Site: www.rcgrossfoundation.org
Founded: 1988.
Congressional District: 8
Key Personnel: Pres., Miriam Gross; Exec. Dir. & Cur., Susan Greenberg Fisher, Ph.D.; Archivist, Zak Vreeland.
Personnel Profile: Full-Time Paid 3; Part-Time Paid 2.
Governing Authority: private; nonprofit. Parent Institution: Renee and Chaim Gross Foundation. Tax-exempt: 501(c)(3).
Institution Type/Description: Museum: housed in a late 19th century structure with cast iron facade that served as Chaim Gross' home and studio for more than 30 years.
Collections: sculpture in wood, stone & bronze as well as drawings, watercolors and prints, 1921-1991, demonstrating the continuity of Chaim Gross' personal vision; wire armatures; clay & plaster maquettes; tools; photographs documenting the life of Chaim Gross.
Research Fields: the life & work of Chaim Gross and the historical context surrounding it.
Facilities: 2,500-vol. library of art & art history books, relating to the life of Gross and his contemporaries; photographic reference & representation of every sculpture as is known in library binders; Chaim Gross Archives for scholarly research on premises; biographies of artists; reference works, periodicals & exhibition catalogues; 2,500 sq. ft. exhibit space; skylight roofed working studio.
Activities: book & poetry readings; films; videos; guided tours; lectures; loan, temporary & traveling exhibitions.
Hours & Admission Prices: Tues.-Fri. 10-5 by appointment. No charge; donation requested. Closed national holidays. &

THE ROSE MUSEUM AT CARNEGIE HALL, 154 W. 57th St., 2nd Fl., New York, NY 10019-3321. Mailing Address: 881 7th Ave., New York, NY 10019-3210. Tel.: 212-903-9629.
E-mail: archives@carnegiehall.org
Web Site: www.carnegiehall.org
Founded: 1991.
Key Personnel: C.E.O. Carnegie Hall, Clive Gillinson; Dir. & Archivist, Gino Francesconi; Chm. (V), Sanford I. Weill; Museum Shop Mgr., Sean Morrow; Assoc. Archivist, Robert Hudson.
Governing Authority: nonprofit organization.
Institution Type/Description: Theatre Museum: housed in c.1891 Carnegie Hall building.
Collections: artifacts & photographs depicting history & development of the building; history of the studios; chronology of events from the main hall.
Facilities: 2,800-seat theatre. Souvenirs & other museum-related items for sale.
Activities: guided tours. Museum Sponsors: temporary exhibits reflecting an anniversary or festival on stage.
Hours & Admission Prices: Sept. 15-June daily 11-4:30, open to concert-goers before concerts & during intermission. No charge. &
Attendance: 25,000 (estimated)

RUBIN MUSEUM OF ART, (M), 150 W. 17th St., New York, NY 10011-5402. Tel.: 212-620-5000. Fax: 212-620-0628.
E-mail: info@rmanyc.org
Web Site: www.rmanyc.org
Founded: 1999.
Congressional District: 8
Key Personnel: Co Chm., Donald Rubin; Co Chm., Shelley Rubin; Exec. Dir., Patrick Sears; Dir. Finance & Administration, Marilena Christodoulou; Museum Shop Mgr., Prisanee Suwanwatana; Dir. Exhibitions, Jan Van Alphen; Dir. Public Programs & Performance, Tim McHenry; Dir. Education & Visitor Experience, Marcos Stafne; Dir. Museum Services & Operations, Linda Dunne; Dir. External Affairs, Cynthia Gwyer.
Personnel Profile: Full-Time Paid 62; Part-Time Paid 10; Part-Time Volunteers 210; Interns 54.
Governing Authority: Tax-exempt.
Institution Type/Description: Art Museum.
Collections: paintings, sculptures & textiles relating to the Himalayas.
Major Exhibits: From India East: Sculpture of Devotion, 5/31/13-7/7/14; Gateway to Himalayan Art, 7/3/13-6/2/14; Count Your Blessings, 8/2/13-3/24/14; The All-Knowing Buddha: A Secret Guide, 9/13/13-2/11/14.
Activities: educational programs; public programs; Brainwave; family programs; access programs; mindful connections for people with dementia & their caregivers.
Publications: exhibition catalogs.
Hours & Admission Prices: Mon. & Thurs. 11-5, Wed. 11-7, Fri. 11-10, Sat.-Sun. 11-6. Adults $10, seniors $7, students $5; discounts to AAM & ICOM members; children under 12, members & Fri. 6pm-10pm no charge. &
Attendance: 170,000 (accurate)
Membership: Artist, Educator, Neighbor, Student and Senior 60 & over $45; Individual $65; Dual & Family $85; Friend $150; Visionary Circle 21-40 $200; Visionary Circle Couple $300; Sustaining $300; Benefactor $500; Chairman's Circle $1,000; Sponsor $2,500; Collector's Circle $5,000; Donors Circle $10,000.

SALMAGUNDI MUSEUM OF AMERICAN ART, 47 5th Ave., New York, NY 10003-4396. Tel.: 212-255-7740. Fax: 212-229-0172.
Web Site: www.salmagundi.org
Founded: 1871.
Key Personnel: Chm. (V), John T. Newton; Pres. (V), Claudia Seymour; Dir., Kathleen Arffmann; Cur., Robert Mueller.
Governing Authority: nonprofit. Tax-exempt: 501(c)(3).
Institution Type/Description: Art Gallery.
Collections: American art; paintings; photographs; artist's palettes.
Research Fields: art.
Facilities: library.
Activities: temporary exhibitions.
Publications: exhibition folders.
Hours & Admission Prices: Mon.-Fri. 1-6, Sat.-Sun. 1-5. No charge; donations accepted.
Attendance: 20,000 (estimated)

THE SCHOMBURG CENTER FOR RESEARCH IN BLACK CULTURE, THE NEW YORK PUBLIC LIBRARY, The New York Public Library, 515 Malcolm X. Blvd., New York, NY 10037-1801. Tel.: 212-491-2200. Fax: 212-491-6760.
Web Site: www.schomburgcenter.org
Founded: 1925.
Congressional District: 16
Key Personnel: Dir., Khalil G. Muhammad, Ph.D.; Deputy Dir., Kara Olidge; Cur. Photographs & Prints, Mary Yearwood; Art & Artifacts, Tammi Lawson.
Governing Authority: nonprofit. Parent Institution: The New York Public Library. Research Center of The New York Public Library/Astor, Lenox & Tilden Foundations, 42nd & 5th Ave., New York, NY. 10018. Tel.: 212-790-6254. Tax-exempt: 170(b)(1)(A).
Institution Type/Description: History Museum & Reference Library.
Collections: books by & about Afro-American, African Diasporan & African life & history; periodicals; pamphlets; manuscripts; personal papers; photographs; prints & drawings; paintings; sculpture; historical artifacts; clippings; playbills; programs; broadsides sheet music; audio-video recordings; films; phonorecords.
Research Fields: black history & culture throughout the world.
Facilities: 220,000-vol. library of print and non-print materials & art objects by or about people of African descent, available for use on-site; microfilm reading area; archives area; oral & video facilities.
Activities: exhibitions mounted on premises; book parties; symposia; scholarly research; lectures & receptions for authors; concerts; film; theater.

Publications: journal, The Schomburg Center; exhibition catalog.
Hours & Admission Prices: Mon.-Wed. 12-8, Thurs.-Fri. 11-6, Sat. 10-5. Exhibits: Tues.-Sat. 10-6, Sun. 1-5. No charge; donations accepted. &
Attendance: 134,186 (accurate)
Membership: Associate $35; Friend $50; Supporter $100; Patron $250; Sustainer $500; Conservator $1,000; Heritage Circle $2,500; Chief's Circle $5,000.

SEAPORT MUSEUM NEW YORK, 12 Fulton St., New York, NY 10038-2109. Tel.: 212-748-8568. Fax: 212-748-8610.
E-mail: reservations@southstseaport.org
Web Site: www.seany.org
Formerly: South Street Seaport Museum
Founded: 1967.
Congressional District: 8
Key Personnel: Chm. Bd. (V), Frank J. Sciame, Jr.; Pres. & C.E.O., Mary Ellen Pelzer; C.F.O., Terry Polcaro; Vice Pres. Institutional Advancement, Christian Kersten.
Personnel Profile: Full-Time Paid 26; Part-Time Paid 49; Part-Time Volunteers 202; Interns 10.
Governing Authority: nonprofit organization. Satellite Facility: New York Unearthed, 17 State St., New York, NY. Tel.: 212-748-8753. Tax-exempt: 501(c)(3).
Institution Type/Description: Maritime History Museum.
Collections: documents; photographs; paintings; ship models; ship artifacts; navigational instruments; tools; printing presses; fishmarket artifacts; scrimshaw. Historic Ships: fourmasted bark, Peking; square rigged ship, Wavertree; wood fishing schooner, Lettie G. Howard; Ambrose Light Ship; schooner, Pioneer; tugboat W.O. Decker; harbor lighter, Marion M; tugboat Helen McAllister.
Research Fields: naval history; New York maritime commercial history; transportation; printing history; architectural history; 19th-century trade, economy & transportation; maritime history & technology.
Facilities: Books, prints & other museum-related items for sale.
Activities: demonstrations on ships & piers in summer; craft demonstrations; guided tours; lectures; films; classes; educational programs; children's workshops; school tours; public sailing aboard 1885 schooner Pioneer; teacher workshops.
Publications: magazine, Seaport includes Broadside Calendar of Events.
Hours & Admission Prices: Galleries & Ships: Jan.-March Thurs.-Sun. 10-5; April-Dec. Tues.-Sun. 10-6. Adults $15, seniors, students & children 2-17 $12; discounts to Council of American Maritime Museums, International Congress of Maritime Museums, groups, AAM, ICOM, AAA & CAMM members; children under 2 & members no charge. Schooner Pioneer: May-Oct. three times daily, call for reservations & fee. &
Attendance: 507,913 (accurate)
Membership: Student & Senior Citizen $45; Individual $60; Dual $75; Family $125; Boatswain $300; Captain $500; Cape Horn Society $1,000; Navigator Society $2,500.

SHEILA C. JOHNSON DESIGN CENTER, Parsons The New School for Design, 66 Fifth Ave. at 13th St., New York, NY 10011. Mailing Address: 2 W. 13th St., Rm. Z101, New York, NY 10011. Tel.: 212-229-8919.
Institution Type/Description: Art Gallery.
Collections: works by faculty, students, & outside artists and designers.
Facilities: 4,000 sq. ft. exhibition space.
Activities: special events.
Hours & Admission Prices: Thurs. 12-8, Fri.-Wed. 12-6. No charge. Closed major holidays.

SIDNEY MISHKIN GALLERY OF BARUCH COLLEGE, 135 E. 22nd St., New York, NY 10010-5505. Mailing Address: Box D-0100, New York, NY 10010-5505. Tel.: 646-660-6652. Fax: 212-802-2693.
Web Site: www.baruch.cuny.edu/mishkin/gallery.html
Founded: 1981.
Congressional District: 15
Key Personnel: Dir., Sandra Kraskin, Ph.D.
Personnel Profile: Full-Time Paid 1; Part-Time Paid 2.
Governing Authority: Parent Institution: Baruch College, The City University of New York. Tax-exempt: 501(c)(3).
Institution Type/Description: University Gallery: housed in 1939 Family Court Building erected by WPA.
Collections: 20th & 21st century paintings, sculpture, prints, & photographs.
Research Fields: modernism 1930-1960; self-taught artists; interdisciplinary; multicultural.
Activities: organized education programs for the public, undergraduate &

graduate college students affiliated with Baruch College, City University of New York; temporary exhibitions of our own collections; traveling & loan exhibitions.

Publications: exhibition catalogues; brochures.

Hours & Admission Prices: Feb.-June & Sept.-Dec. Mon.-Wed. & Fri. 12-5, Thurs. 12-7. No charge. Closed university holidays. &

THE SKYSCRAPER MUSEUM, (M), 39 Battery Place, New York, NY 10280-1501. Tel.: 212-968-1961 & 945-6324. Fax: 212-732-3039.

E-mail: info@skyscraper.org
Web Site: www.skyscraper.org
Founded: 1997.
Congressional District: 8
Key Personnel: C.O.O., William Havemeyer; C.E.O., Dir. & Pres. (V), Carol Willis; Chm. (V), Jed Marcus; Treas., Owen Gutfreund; Museum Shop Mgr., Ellie Rubin.
Personnel Profile: Full-Time Paid 2; Part-Time Paid 7; Part-Time Volunteers 1; Interns 1.
Governing Authority: private; nonprofit organization. Tax-exempt: 501(c)(3).
Institution Type/Description: Architecture Museum.
Collections: construction photographs & film; architectural & engineering drawings; contracts; builders' records; financial reports; advertising materials; periodicals; models; building tools & artifacts.
Research Fields: future exhibitions, including core exhibition, SKYSCRAPER/CITY; building document collection for virtual archive; international skyscrapers; urban density.
Facilities: 5,000 sq. ft. exhibit space; audiovisual room. Museum-related items for sale.
Activities: films; guided tours; lectures; rental gallery; walking tours outside of museum. Museum Sponsors: Making New York History award.
Publications: book, Building the Empire State; Form Follows Finance: Skyscrapers and Skylines in NY & Chicago; Lower Manhattan Plan.
Hours & Admission Prices: Wed.-Sun. 12-6. Adults $5, students & seniors $2.50; discounts to AAM & ICOM members; members no charge. &
Attendance: 25,000 (estimated)
Membership: Basic $35; Friend $50; Contributor $100; Supporter $500; Donor $1,000; Corporate $1,000 and up.

* **SOLOMON R. GUGGENHEIM MUSEUM, (M),** 1071 Fifth Ave. at 89th St., New York, NY 10128-0112. Tel.: 212-423-3500. Fax: 212-423-3787.

E-mail: info@guggenheim.org
Web Site: www.guggenheim.org
Founded: 1937.
Congressional District: 20
Key Personnel: Dir. Solomon R. Guggenheim Foundation, Richard Armstrong; C.O.O. Solomon R. Guggenheim Museum, Marc Steglitz; Chief Counsel, Sarah Austrian; Deputy Dir, External Affairs, Eleanor Goldhar; Chief Cur., Nancy Spector; Dir. Curatorial Affairs, Joan Young; Sr. Cur. Asian Art, Alexandra Monroe; Sr. Cur. Photography, Jennifer Blessing; Sr. Cur. Collections & Exhibitions, Tracy Bashkoff; Dir. Education, Kim Kanatani; Exec. Dir., Major Gifts, Adrienne Hines; Individual Giving, Major Gifts, Mary Anne Talotta; Exec. Dir. Corporate & Institutional Devel., John Wielk; Dir. Finance, Amy West; Chief Information Officer, Alexander Pasik; Dir. Media & Public Rels., Betsy Ennis; Dir. Mktg., Laura Miller; Dir. Registration, Mary Louise Napier; Project Mgr., Jessica Ludwig; Dir. Photographic Svcs. & Chief Photographer, David M. Heald; Dir. Publications, Beth Levy; Chief Graphic Designer, Marcia Fardella; Dir. Visitor Svcs., Maria Celi; Dir. & Counsel Administration, Brendan Connell; Dir. Security, Steve Ursell; Chm. Bd., William Mack; Pres. Bd., Jennifer Blei Stockman.
Personnel Profile: Full-Time Paid 289; Part-Time Paid 36.
Governing Authority: nonprofit organization. Operated by the Solomon R. Guggenheim Foundation, 527 Madison Ave., New York, NY 10022; Guggenheim Museum Bilbao, Bilbao, Spain. Sister Institutions: Peggy Guggenheim Collection in Venice, Italy; Guggenheim Museum Bilbao, Bilbao, Spain; Deutsche Guggenheim, Berlin Germany. Tax-exempt.
Institution Type/Description: Art Museum: housed in building designed by Frank Lloyd Wright.
Collections: paintings; sculpture; video; works on paper of last 100 years including concentrations of works by Picasso, Chagall, Brancusi, Delaunay, Dubuffet, Pollock, Modigliani, Max Ernst, Mondrian, Miro, Rothko, Giacometti, Leger, Marc & an extensive representation of paintings by Kandinsky & Klee; broad range of post World War II painting & sculpture from the U.S. & Europe.
Major Exhibits: Carrie Mae Weems: Three Decades of Photography and Video, 1/14-5/14; Italian Futurism 1909-1944: Reconstructing the Universe, 2/14-

9/14; Countdown to Tomorrow: The International Zero Network 1950s-60s, 10/14-1/15; V.S. Gaitonde: Painting as Process, Painting as Life, 10/14-12/15.

Research Fields: late 19th- & 20th-century art.
Facilities: 20,000-vol. library of modern art; documentation of the collection of Solomon R. Guggenheim Museum available for research by appointment only; auditorium. Catalogs, slides, posters, note cards, exhibition-related monographs for sale.
Activities: permanent, temporary & traveling exhibitions; inter-museum loan exhibitions; guided tours; lectures; films; music, dance & theater performances; symposia; panel discussions; gallery talks; internship programs for graduate & college students.
Publications: catalogues of exhibitions; handbook, The Guggenheim Museum Collection A to Z.
Hours & Admission Prices: Sun.-Wed. 10-5:45, Fri.-Sat. 10-7:45. Adults $22, seniors & students $18; discounts to AAM & ICOM members; children under 12 & members no charge. Closed Christmas Day. &
Attendance: 1,000,000 (estimated)
Membership: Individual $75; Dual $125; Family $135; Fellow Associate $250; Supporting Associate $500.

SONY WONDER TECHNOLOGY LAB, (M), 56th St. & Madison Ave., New York, NY 10022. Mailing Address: 550 Madison Ave., Annex, New York, NY 10022-3211. Tel.: 212-833-8100. Fax: 212-833-4445.

E-mail: timothy_foster@sonyusa.com
Web Site: www.sonywondertechlab.com
Founded: 1994.
Key Personnel: Senior Dir., Karen Kelso.
Personnel Profile: Full-Time Paid 22; Part-Time Paid 8; Interns 1.
Governing Authority: corporation. Parent Institution: Sony Corporation of America.
Institution Type/Description: Technology, Media & Science Museum.
Collections: technology & entertainment; animation artifacts.
Facilities: 14,000 sq. ft. interactive, hands-on science & technology center; high definition theater.
Activities: special events; screenings; tours & teacher development; workshops; early learner programs.
Hours & Admission Prices: Tues.-Sat. 9:30-5:30. No charge. Closed major holidays. &
Attendance: 220,000 (accurate)

* **STATUE OF LIBERTY NATIONAL MONUMENT & ELLIS ISLAND IMMIGRATION MUSEUM, (M),** Liberty Island, New York, NY 10004-1418. Tel.: 212-363-3200. Fax: 212-363-6304. TDD: 212-363-3301.

E-mail: stli_info@nps.gov
Web Site: www.nps.gov/stli
Founded: 1924.
Congressional District: 17
Key Personnel: Supt., Dave Luchsinger; Deputy Supt., John Hnedack; Chief Museum Svcs. Div., Diana Pardue; Cur. Exhibits & Media, Judith Giuriceo; Cur. Collections, Geraldine Santoro; Chief Maintenance, Peter O'Dougherty; Acting Administrative Officer & Contractor, Linda Deveau; Concessions Specialist, Ben Hanslin; Contract Specialist, Yeny Reyes; Chief Interpretation Div., Daniel Brown; Supervisory Archivist, George Tselos; Library, Jeff Dosik; Library, Barry Moreno; Concessioner, Brad Hill.
Personnel Profile: Full-Time Paid 177; Full-Time Volunteers 5; Part-Time Volunteers 128; Interns 10.
Governing Authority: federal. Parent Institution: U.S. Dept. of the Interior, National Park Service. Tax-exempt.
Institution Type/Description: Statue of Liberty: housed in 1886 151-ft. copper statue bearing torch of freedom was gift of French people to commemorate alliance of U.S. & France; monument includes two exhibits on the Statue of Liberty: World Heritage Site. Ellis Island: over 12 million immigrants were processed here between 1892-1954; exhibits on Ellis Island, American immigration, film, learning center, library & oral history program; access to both islands by ferry.
Collections: history; folk art; archival & other materials pertaining to the Statue of Liberty & Ellis Island; Augustus F. Sherman photographs; immigrant oral histories with transcripts; contemporary art; photographs; film & video.
Research Fields: Statue of Liberty; Ellis Island; U.S. immigration.
Facilities: research library & oral history collection; two movie theatres; oral history recording studio; learning center for groups; restaurant. Gift items for sale.
Activities: temporary exhibitions; films; ranger tours; living history presentations.

Publications: brochures; NPS research reports.
Hours & Admission Prices: Liberty Island: advanced reservations suggested. Call: 866-782-8834. Ellis Island: advanced reservations suggested. No charge; donations accepted. Circle Line Ferry Boats: call (212) 269-5755. Closed Christmas.
Attendance: 3,408,560 (accurate)

STOREFRONT FOR ART AND ARCHITECTURE, 97 Kenmare St., New York, NY 10012. Tel.: 212-431-5795.
E-mail: info@storefrontnews.org
Web Site: www.storefrontnews.org
Key Personnel: Dir., Eva Franch
Institution Type/Description: Art Gallery.
Collections: works by contemporary artists.
Hours & Admission Prices: Tues.-Sat. 11-6.

* **THE STUDIO MUSEUM IN HARLEM,** 144 W. 125th St., New York, NY 10027-4423. Tel.: 212-864-4500. Fax: 212-864-4800.
E-mail: pr@studiomuseum.org
Web Site: www.studiomuseum.org
Founded: 1967.
Congressional District: 19
Key Personnel: Dir. & Chief Cur., Thelma Golden; Asst. Cur., Lauren Haynes; Asst. Dir. Education & Public Programs, Shanta Scott.
Personnel Profile: Full-Time Paid 42; Part-Time Paid 7; Interns 6.
Governing Authority: private; nonprofit organization. Tax-exempt: 501(c)(3).
Institution Type/Description: Contemporary Art Museum: sculpture garden; collection, documentation, preservation & interpretation of the art & artifacts of Black American & the African Diaspora.
Collections: 19th-21st century African American art; traditional & contemporary African art; Caribbean art; archives.
Research Fields: 19th & 20th-century African American art; African art; art of the African diaspora.
Facilities: classrooms. Books, African textiles, exhibition catalogues, African jewelry, posters & other museum-related items for sale.
Activities: temporary & traveling exhibitions; artist-in-residence program; public programs; guided tours; workshops. Museum Sponsors: Target Free Sundays.
Publications: exhibition catalogs; studio magazine.
Hours & Admission Prices: Thurs. & Fri. 12-9, Sat. 10-6, Sun. 12-6. Suggested Donation: adults $7, students & senior citizens $3; discounts to AAM members; members & children under 12 no charge. Closed major holidays.
Attendance: 90,831 (estimated)
Membership: Student & Senior $25; Individual $50; Family & Partner $75; Supporter $125.

T.F. CHEN CULTURAL CENTER, 250 Lafayette St., Fl. 5, New York, NY 10012-4040. Tel.: 212-966-4363. Fax: 212-966-5285.
E-mail: chen@tfchen.org
Web Site: www.tfchen.org
Founded: 1996.
Key Personnel: C.E.O. & Pres. (V), Lucia Chen; Chm. (V), Dr. T.F. Chen; Financial Dir., Ted Chen; Public Rels. & Cur., Julie Chen.
Personnel Profile: Full-Time Paid 2; Full-Time Volunteers 4; Part-Time Paid 2; Interns 2.
Governing Authority: private; nonprofit organization. Tax-exempt: 501(c)(3).
Institution Type/Description: Art Museum.
Collections: art collection dating from 1951; multicultural; universal humanism; east & west.
Research Fields: artists in the Neo-Iconography style group shows.
Facilities: 500-vol. library on art history; 3,000 sq. ft. exhibit space. Museum-related items for sale.
Activities: traveling shows; art festivals; lectures; loan, participatory & temporary exhibitions.
Publications: annual newsletter.
Hours & Admission Prices: By appointment only. No charge; donations accepted. Closed New Year's Day; Christmas.
Attendance: 200 (estimated)

THEODORE ROOSEVELT BIRTHPLACE NATIONAL HISTORIC SITE, 28 E. 20th St., New York, NY 10003-1311. Mailing Address: 26 Wall St., New York, NY 10005. Tel.: 212-260-1616. Fax: 212-677-3587.
Web Site: www.nps.gov/thrb/
Founded: 1923.
Congressional District: 15

Key Personnel: Site Mgr., William Strain; Park Ranger, Michael Amato; Park Ranger, Daniel Prebutt.
Personnel Profile: Full-Time Paid 4; Part-Time Paid 1; Part-Time Volunteers 5; Interns 1.
Governing Authority: federal. Administered by National Park Service, Dept. of Interior. Tax-exempt.
Institution Type/Description: Historic Site Museum: housed in 1919-1923 reconstructed Theodore Roosevelt birthplace.
Collections: mid-19th century period rooms; Theodore Roosevelt memorabilia & historical items.
Research Fields: Theodore Roosevelt; American history.
Facilities: auditorium. Books for sale.
Activities: guided tours; films; permanent exhibitions. Chamber music concerts Sept.-May.
Publications: site brochures.
Hours & Admission Prices: Tues.-Sat. 9-5. Tours: 10-11 & 1-4 on the hour. No charge. Closed federal holidays.
Attendance: 40,000 (accurate)
Membership: National Park Service: Golden Age 62 & over $10.

TRINITY MUSEUM OF THE PARISH OF TRINITY CHURCH, Broadway & Wall St., New York, NY 10006. Mailing Address: 74 Trinity Place, Rm. 1610, New York, NY 10006-2114. Tel.: 212-602-0800 & 0872. Fax: 212-602-9648.
E-mail: archives@trinitywallstreet.org
Web Site: www.trinitywallstreet.org/history/museum
Founded: 1966.
Congressional District: 17
Key Personnel: C.E.O., Rev. Dr. James H. Cooper; Cur., Gwynedd Cannan; Museum Shop Mgr., David Jette.
Personnel Profile: Full-Time Paid 3; Part-Time Paid 2.
Governing Authority: church; nonprofit. Parent Institution: Trinity Church. Tax-exempt: 501(c)(3).
Institution Type/Description: Religious & History Museum: housed in 1846 Trinity Church, on site first used in 1697.
Collections: 1644-present, church archives including documents, books; prints; photographs; paintings; artifacts & religious objects. Historic Building: 1766 St. Paul's Chapel.
Research Fields: religious history; New York City history; English history.
Facilities: 1,000-vol. library of religious, New York City & state history & English history available for research by appointment by qualified scholars and students; parish archives. Books, religious objects & museum-related items for sale.
Activities: self-guided tours; temporary & permanent exhibits; guided tours by appointment; educational video.
Publications: brochures; videos; A Guide to Trinity Church; Trinity, A Church, A Parish, A People.
Hours & Admission Prices: Mon.-Fri. 9-11:45 & 1-5:30, Sat.-Sun. 9-3:45; St. Paul's: Mon.-Sat. 10-6, Sun. 7-6. No charge; donations accepted. Closed holidays.
Attendance: 33,000 (estimated)

THE UKRAINIAN MUSEUM, (M), 222 E. 6th St., New York, NY 10003-8201. Tel.: 212-228-0110. Fax: 212-228-1947.
E-mail: info@ukrainianmuseum.org
Web Site: www.ukrainianmuseum.org
Founded: 1976.
Congressional District: 14
Key Personnel: Dir., Maria Shust; Pres. (V), Jaroslav Leshko; Admin. Dir., Daria Bajko; Museum Shop Mgr., Chrystyna Pevny.
Personnel Profile: Full-Time Paid 4; Part-Time Paid 9; Part-Time Volunteers 50.
Governing Authority: nonprofit. Tax-exempt: 501(c)(3).
Institution Type/Description: Culturally Specific Museum.
Collections: fine art; folk art; archives; photographs; documents; Ukrainian culture, history & immigration to the U.S.; numismatics & philatelic.
Research Fields: Ukrainian ethnology; ethnography; fine arts; history.
Facilities: 3,000-vol. library; auditorium; 80-seat lecture & conference room. Museum-related items for sale.
Activities: guided tours; lectures; films; formally organized educational programs; permanent & temporary exhibitions.
Publications: exhibitions catalogues; brochures, books, annual reports.
Hours & Admission Prices: Wed.-Sun. 11:30-5. Adults $8, students & senior citizens $6; discounts to AAM & ICOM members; members & children under 12 no charge. Closed New Year's Day; Ukrainian Christmas & Easter; Easter; Independence Day; Labor Day; Thanksgiving; Christmas.
Attendance: 25,000 (estimated)
Membership: Students $10; Senior Citizens $15; Regular $40; Family $75; Contributing $100; Life $7,500 & up.

UNION FOR REFORM JUDAISM, 633 Third Ave., New York, NY 10017-6706. Tel.: 212-650-4040. Fax: 212-650-4239.
E-mail: urj@urj.org
Web Site: www.urj.org
Formerly: Union of American Hebrew Congregation
Founded: 1957.
Key Personnel: Pres., Rabbi Eric H. Yoffie; Dir. Mktg. & Comm., Emily Grotta.
Governing Authority: denominational group. Operated by the Joint Commission on Synagogue Administration Union of American Hebrew Congregations and Central Conference of American Rabbis.
Institution Type/Description: Religious Library & Museum.
Collections: books on Synagogue architecture; books on Synagogue art, ceremonial objects & works of Jewish artists; 2,500 slides on the collection.
Facilities: 200-vol. library on Synagogue architecture, art & ceremonial objects available on premises only.
Activities: reading room; slide rental service.
Publications: books, An American Synagogue for Today & Tomorrow; Contemporary Synagogue Art.
Hours & Admission Prices: Mon.-Thurs. 9-4, Fri. 9-5. Closed Jewish holidays & festivals.

VAN DOREN WAXTER GALLERY, 23 E. 73rd St., New York, NY 10021. Tel.: 212-445-0444. Fax: 212-445-0442.
E-mail: info@gvdgallery.com
Web Site: www.gvdgallery.com/gallery
Formerly: Greenberg Van Doren Gallery
Key Personnel: Exec. Dir., Dorsey Waxter
Institution Type/Description: Art Gallery.
Collections: works by regional, national & international artists.
Hours & Admission Prices: Tues.-Sat. 10-6.

THE VILCEK FOUNDATION, 167 E. 73rd St., New York, NY 10021. Tel.: 212-472-2500. Fax: 212-472-4720. Facebook: Vilcek Foundation.
E-mail: info@vilcek.org
Web Site: www.vilcek.org
Founded: 2000.
Congressional District: 14
Key Personnel: Dir., Rick Kinsel; Pres., Jan Vilcek.
Personnel Profile: Full-Time Paid 8; Part-Time Paid 1.
Governing Authority: private; nonprofit organization. Tax-exempt: 501(c)(3).
Institution Type/Description: History Museum.
Collections: works of immigrant artists, designers, & filmmakers.
Facilities: 837 sq. ft. exhibit space.
Activities: concerts; dance recitals; films; lectures. Annual Event: off-site dinner reception to celebrate the Vilcek prizes in biomedical research and the arts and humanities.
Publications: triannual e-newsletter, The Vilcek Foundation; exhibition brochures.
Hours & Admission Prices: During Exhibitions: Wed.-Sat. 12-6; other times by appointment. No charge.
Attendance: 796 (accurate)

VISUAL ARTS GALLERY, 601 W. 26th St., 15th Fl., New York, NY 10001-1138. Mailing Address: 209 E. 23rd St., New York, NY 10010-3994. Tel.: 212-592-2145. Fax: 646-638-2110.
E-mail: gallery@sva.edu
Web Site: www.schoolofvisualarts.edu
Founded: 1960.
Congressional District: 14
Key Personnel: Dir., Francis Di Tommaso; Asst. Dir., Richard Brooks.
Personnel Profile: Full-Time Paid 6; Part-Time Paid 9.
Governing Authority: private college. Parent Institution: School of Visual Arts, New York.
Institution Type/Description: Art Museum.
Collections: digital & new media; photography; fine arts; illustration & cartooning; graphic design.
Facilities: 479-seat auditorium; 266-seat auditorium; educational facilities; 4,000 sq. ft. exhibit space.
Activities: lectures; traveling exhibitions. Annual Event: The Masters Series.
Publications: exhibition brochures & catalogs.
Hours & Admission Prices: Mon.-Sat. 10-6. No charge.
Attendance: 8,158 (accurate)

✳ **WHITNEY MUSEUM OF AMERICAN ART, (M),** 945 Madison Ave., New York, NY 10021-2790. Tel.: 212-570-3600. Fax: 212-606-0207.
Web Site: www.whitney.org
Founded: 1930.
Congressional District: 18
Key Personnel: Alice Pratt Brown Dir., Adam D. Weinberg; Assoc. Dir. Human Resources, Hillary Blass; Chief Cur. & Assoc. Dir. Programs, Donna DeSalvo; Deputy Dir., John Stanley; Assoc. Dir., Kathryn Potts; Chm. Education, Helena Rubinstein; Assoc. Dir. Conservation & Research, Carol Mancusi-Ungaro; Assoc. Dir. Collections & Exhibitions Management, Christy Putnam; Chief Mktg. & Communications Officer, Jeffrey Levine; Adjunct Cur., Andy Warhol Film Project, Callie Angell; Cur. & Cur. Drawings, Carter Foster; Cur., Barbara Haskell; Anne and Joel Ehrenkranz Cur., Chrissie Iles; Cur. & Cur. Prints, David Kiehl; Cur. Permanent Collection, Dana Miller; Adjunct Cur. New Media Arts, Christiane Paul; Cur. & Sandra Gilman Cur. Photography, Elisabeth Sussman; Adjunct Cur. Performance, Limor Tomer; Dir. Devel., Alexandra Wheeler; Dir. Security, John Baliestri; Dir. Independent Study Program, Ron Clark; Dir. Foundation & Government Rels., Hillary Strong; Dir. Special Events, Gina Rogak; Dir. Corporate Partnerships, Amy Roth; Head Registrar (Collection), Barbi Spieler; Controller, John Collins; Irma & Benjamin Weiss Librarian, Carol Rusk; Head Publications, Rachel Wixom; Dir. Retail Operations, Jennifer Heslin; Mgr. Visitor Svcs., Wendy Borbee-Louvell.
Personnel Profile: Full-Time Paid 185; Part-Time Paid 27; Part-Time Volunteers 100; Interns 70.
Governing Authority: nonprofit organization. Branch Museums: Whitney Museum of American Art at Altria, 120 Park Ave., New York NY 10017. Tax-exempt.
Institution Type/Description: Art Museum.
Collections: paintings; sculpture; drawings, prints & photography; film & video.
Research Fields: 20th-21st century American art.
Facilities: 37,000-vol. library; art available for study by appointment only; 120-seat film & video gallery; restaurant.
Activities: symposia; lectures; panel discussions; film & video programs; gallery talks; performances; formally organized education programs; training programs; temporary & traveling exhibitions; public school outreach programs & teachers workshops; public programs.
Publications: exhibition catalogs, brochures & books, posters, reproductions, cards, quarterly calendar; bulletin; gallery guides; brochures; education & membership materials; museum-related websites.
Hours & Admission Prices: Wed.-Sun. 11-6. Adults $18, senior citizens 62 & over with valid ID and youth 19-25 $12; discounts to groups, ICOM & AAM members; Fri. 6-9 pay what you wish; members, NYC public high school students & youth under 19 no charge. Closed New Year's Day; Thanksgiving; Christmas.
Attendance: 650,000 (estimated)
Membership: Artist, Student & Educator $40; National & International $50; Individual $75; Dual $100; Family Contributor $150; Friend $250; Patron $500; Whitney Circle $1,000; Whitney Fellow $2,500.

YESHIVA UNIVERSITY MUSEUM AT THE CENTER FOR JEWISH HISTORY, (M), 15 W. 16th St., New York, NY 10011-6301. Tel.: 212-294-8330. Fax: 212-294-8335.
E-mail: info@yum.cjh.org
Web Site: www.yumuseum.org
Founded: 1973.
Congressional District: 8
Key Personnel: Vice Chair, Michael Jesselson; Vice Chair, Ted Mirvis; Dir., Dr. Jacob Wisse; Dir. Institutional Advancement, Rachel Lazin; Assoc. Dir. Administration, Jody Heher; Cur. Collection, Bonni-Dara Michaels; Educator, Ilana Benson; Asst. Cur., Zachary Paul Levine.
Personnel Profile: Full-Time Paid 6; Part-Time Volunteers 35; Interns 2.
Governing Authority: Parent Institution: Yeshiva University. 500 W. 185 St., New York. Tel. 212-960-5400. Tax-exempt: 501(c)(3).
Institution Type/Description: Religious & Cultural Museum: with the purpose of preserving, enriching & interpreting Jewish life as it is reflected in the arts, history & sciences.
Collections: Jewish ceremonial objects of silver & other metals; textile collection of ceremonial costumes & clothes; rare scrolls & books; manuscripts; archival materials; slides; photographs; 10 scale models of historic synagogues; photographs; fine art; sculpture & ethnographic material.
Research Fields: Jewish history, ethnography, art & culture.
Facilities: library of books on art, Jewish history & crafts available to scholars or researchers on premises; oral history archive. Ceremonial objects, books & other museum-related items for sale.
Activities: guided tours; lectures; concerts; arts festivals; craft workshops;

formally organized education programs for children, adults & undergraduate college students; docent program; loan, permanent, temporary & traveling exhibitions.

Publications: books, The Jewish Wedding; Purim, The Face & The Mask; See and Sanctify: Exploring Jewish Symbols; Daily Life in Ancient Israel; Terezin 1942-1945: Through the Eyes of Norbert Troller; Tradition and Fantasy In Jewish Needlework; Raban Remembered: Jerusalem's Forgotten Master; Ashkenaz: The German Jewish Heritage; Lights/Orot; Medieval Justice: The Trial of the Jews of Trent; Mordecai Manuel Noah: The First American Jew; The Sephardic Journey: 1492-1992; Aishet Hayil: A Woman of Valor; Sacred Realm: The Emergence of the Synagogue in the Ancient World; Theodor Herzl: If You Will It, It Is Not A Dream; Siegmund Forst: A Lifetime in Arts and Letters; Ina Golub: The Work of the Weaver In Colors; Treasures of Dubrovnik; Major Intersections; Schwebel, David's Journey; Moritz Daniel Oppenheim: Jewish Identity in 19th Century Art; Komar & Melamid: Symbols of the Big Bang; Tobi Kahn: Microcosmos; Stories Untold: Jewish Pioneer Women, 1850-1910; Art Against Forgetting: Paintings by Leonard Meiselman; Journey to No End of the World: Judaica from the Gross Family Collection, Tel Aviv; Portion of the People: 300 Years of Southern Jewish Life; Stage & Page: Jewish Theater and Book Designs of Emanuele Luzzati; Fruits of a Lifetime: The Kathryn Yochelson Collection; A Perfect Fit: The Garment Industry and American Jewelry 1860-1960; Printing the Talmud: From Bomberg to Schottenstein; Ebrei Piemontesi: The Jews of Piedmont.

Hours & Admission Prices: Mon. 5-8, Tues., Thurs. & Sun. 11-5, Wed. 11-8, Fri. 11-2:30. Tours available by appointment. Adults $8, senior citizens & students $6; discounts to AAM members; students, staff & alumni of Yeshiva University, members, Mon. & Wed. 5-8, Fri. 11-2:30 no charge. Closed Jewish holidays. &

Attendance: 40,000 (estimated)

Membership: Individual $50; Dual $72; Family $100; Supporting $150; Sustaining $250; Sponsor $500; Patron $1,000.

ZABRISKIE GALLERY, 400 E. 57th St., 19B, New York, NY 10022. Tel.: 212-752-1223. Fax: 212-752-1224.

E-mail: info@zabriskiegallery.com

Web Site: www.zabriskiegallery.com

Founded: 1954.

Key Personnel: Dir., Virginia M. Zabriskie

Institution Type/Description: Art Gallery.

Collections: Dada; Surrealism; American Modernism; photographs; contemporary art.

Hours & Admission Prices: Memorial Day-Labor Day Mon.-Fri. 10-5:30; Sept. to mid-May Tues.-Sat. 10-5:30.

Newark Valley

BEMENT-BILLINGS FARMSTEAD, 9142 Rte. 38, Newark Valley, NY 13811. Mailing Address: P.O. Box 222, Newark Valley, NY 13811-0222. Tel.: 607-642-9516. Fax: 607-642-9516.

E-mail: nvhistory@stny.rr.com

Web Site: nvhistory.org

Founded: 1977.

Congressional District: 28

Key Personnel: Bd. Trustee, Ethel Curkendall; Bd. Trustee, Marcia Kiechler; Bd. Trustee, Doug Gorsline; Pres. (V), Ross McGraw.

Personnel Profile: Full-Time Paid 1; Part-Time Paid 1; Part-Time Volunteers 220; Interns 2.

Governing Authority: nonprofit organization. Parent Institution: Newark Valley Historical Society Inc. Subsidiary Institution: Binghamton University. Tax-exempt.

Institution Type/Description: Historic Site: c.1840 Bement-Billings Farmstead.

Collections: 1840s furniture; cookware; utensils; furnishings; tools; 19th-century ledgers & daybooks of N. Tioga County.

Research Fields: early 19th-century farmers & tradesmen.

Facilities: 50-vol. library pertaining to local history, genealogy, folklife, antiques & preservation.

Activities: educational programming; interpreter training; youth program; summer workshops. Museum Sponsors: Spring Festival in June: plowing, wool & herb market, live music, craft vendors & demonstrations; Apple Fest in October; Civil War reenactment & encampment.

Publications: bimonthly newsletters; annual reports.

Hours & Admission Prices: July to early Oct. Sat.-Sun. 12-4; other times by appointment. Adults $2, students $1; members no charge.

Attendance: 10,000 (estimated)

Membership: Senior Citizen $10; Individual $15; Family $25; Contributing $50; Professional Corporate & Sponsor $100; Sustaining $200; Patron $300.

NEWARK VALLEY DEPOT MUSEUM, Depot St., Newark Valley, NY 13811. Mailing Address: P.O. Box 222, Newark Valley, NY 13811-0222. Tel.: 607-642-9516. Fax: 607-642-9516.

E-mail: info@nvhistory.org

Web Site: www.nvhistory.org

Founded: 1977.

Congressional District: 28

Key Personnel: Pres., Ross McGraw; Treas., Doug Gorline.

Personnel Profile: Full-Time Paid 1; Part-Time Paid 1; Part-Time Volunteers 100.

Governing Authority: nonprofit organization. Parent Institution: Newark Valley Historical Society Inc. Tax-exempt.

Institution Type/Description: Railroad Museum; 1869 Depot restored in 1910.

Collections: model railroad; railroad artifacts & memorabilia.

Facilities: depot for riding Tioga Scenic Railroad; snack bar. Museum-related items for sale.

Activities: Museum Sponsors: Depot Days in June to September.

Publications: bimonthly newsletter; annual report.

Hours & Admission Prices: July to 1st week of Oct. Sat.-Sun. 1:30-3. No charge; donations accepted. &

Attendance: 5,000 (estimated)

Membership: Senior Citizen $10; Individual $15; Family $25; Contributing $50; Professional $100; Corporate $100 & up.

Newburgh

HISTORICAL SOCIETY OF NEWBURGH BAY & THE HIGHLANDS - DAVID CRAWFORD HOUSE, 189 Montgomery St., Newburgh, NY 12550-3636. Tel.: 845-561-2585. Facebook: Newburgh Historical Society.

E-mail: historicalsocietynb@yahoo.com

Web Site: www.newburghhistoricalsociety.com

Founded: 1884.

Congressional District: 21

Key Personnel: Bd. Pres. (V), Warren Cahill; Vice Pres., Allynne Lange; 2nd Vice Pres., Carla Decker; Dir., Johanna Porr; Museum Shop Mgr., Anne Coon.

Personnel Profile: Full-Time Paid 1; Part-Time Paid 1; Part-Time Volunteers 25.

Governing Authority: nonprofit organization. Parent Institution: Historical Society of Newburgh Bay and the Highlands. Tax-exempt: 501(c)(3).

Institution Type/Description: Historic House: 1830 home of David Crawford.

Collections: 19th-century period furniture, decorative arts, toys & dolls; Hudson River School paintings; ship models of Hudson River crafts; local historical archives, photographs & artifacts.

Research Fields: middle 19th-century architecture, landscape design & commercial life.

Facilities: 3,000-vol. library pertaining to local & state history. Prints, crafts, reproductions & gift items for sale.

Activities: guided tours; lectures; concerts; organized education programs; loan & temporary exhibitions. Museum Sponsors: Community House tours in December; Biennial Benefit Art Auction.

Publications: newsletter; booklet, Andrew Jackson Downing. Prints, publications, book on local history & architecture & gift items for sale.

Hours & Admission Prices: April-Oct. Sun. 1-4; other times by appointment. Suggested Donation: $5. Closed major holidays. &

Attendance: 3,000 (estimated)

Membership: Senior $35; Individual $50; Family $75; Century Club $100; Crawford Club $250; 1830 Society $500; Captain's Circle $1,000.

MOTORCYCLEPEDIA MUSEUM, (M), 250 Lake St., Newburgh, NY 12550-5262. Tel.: 845-569-9065. Fax: 845-569-9063. Facebook: Motorcyclepedia Museum.

E-mail: info@motorcyclepediamuseums.com

Web Site: www.motorcyclepediamuseum.org

Founded: 2011.

Institution Type/Description: Motorcycle Museum.

Collections: 450 motorcycles; photographs; posters; memorabilia; machinery.

Activities: guided tours.

Hours & Admission Prices: Fri.-Sat. 10-5, Sun. 11-5. Adults $11, children 3-12 $5; children under 3 no charge. &

Membership: Annual $40.

WASHINGTON'S HEADQUARTERS STATE HISTORIC SITE, Corner of Liberty & Washington Sts., Newburgh, NY 12551-1476. Mailing Address: P.O. Box 1783, Newburgh, NY 12551-1783. Tel.: 845-562-1195. Fax: 845-561-1789.

Web Site: nysparks.state.ny.us/historic-sites/17/details.aspx

Founded: 1850.

Congressional District: 26
Key Personnel: Historic Site Mgr., Elyse B. Goldberg.
Personnel Profile: Full-Time Paid 5; Part-Time Paid 4.
Governing Authority: state. Parent Institution: New York State Office of Parks, Recreation & Historic Preservation; Palisades Interstate Park Commission. Tax-exempt: 501(c)(3).
Institution Type/Description: Historic House Museum: 1750-1770 Jonathan Hasbrouck House, used as Gen. George Washington's headquarters, 1782-1783.
Collections: period furnishings; firearms; documents & military artifacts of the American Revolution; local history; Martha Washington's pocket watch; portraits of the Washingtons by Asher B. Durand; statuary; audiovisuals; Hasbrouck House preservation history. Historic Monument: 1887 Centennial Monument, Tower of Victory; early museum building: 1910.
Research Fields: American Revolution in the Hudson River Valley; George & Martha Washington and their staff.
Facilities: archives of 1,500 documents including deeds from 1648-1900 of land transactions, regional history, military service & the American Revolution in the Hudson Valley available on premises by advance request; 6 1/2 acre park.
Activities: guided tours; lectures; videos; concerts; permanent & temporary exhibitions; school programs; special programs & demonstrations; three-day Washington's Birthday Celebration.
Publications: site brochures, Inside Washington's Headquarters; historical research series, no charge; newsletter to members, Friends.
Hours & Admission Prices: mid-April to Oct. Wed.-Sat. 10-5, Sun. 1-5; Nov.-March by appointment. Adults $4, seniors 62 & over $3; discount to groups; children 12 & under and FSHSHH members no charge. Braille tours available.
Attendance: 21,203 (accurate)
Membership: Friends of the State Historic Sites of the Hudson Highlands, Inc.: Individual $15; Household $25; Contributing $25-$99; Business & Corporate $100; Sustaining $100-$499; Patron $500-$999; Benefactor $1,000 & up.

Niagara Falls

AQUARIUM OF NIAGARA, 701 Whirlpool St., Niagara Falls, NY 14301-1094. Tel.: 716-285-3575, ext. 204. Fax: 716-285-8513.
E-mail: aquariumnf@aol.com
Web Site: www.aquariumofniagara.org
Founded: 1965.
Congressional District: 32
Key Personnel: Exec. Dir., Gary B. Molnar; Pres., Richard Torcasio; Cur. Education, Jeanette Brunner; Museum Shop Mgr., Simone Russell.
Personnel Profile: Full-Time Paid 20; Part-Time Paid 14; Part-Time Volunteers 10; Interns 7.
Governing Authority: nonprofit. Niagara Aquarium Foundation. Tax-exempt.
Institution Type/Description: Aquarium.
Collections: marine mammals; marine & freshwater fishes; penguins; invertebrates; outdoor seal & sea lion pool.
Research Fields: water quality; dietary supplements; marine mammal husbandry & skin properties.
Facilities: 150-vol. library of books on zoology; oceanography; natural history with emphasis on aquatic animals. Marine-related articles for sale.
Activities: marine mammal demonstrations; classes on aquatic biology for school groups, members; lectures; hands-on interactive displays for children; penguin & shark feeding demonstrations; animal interaction programs.
Publications: membership newsletter, Sea Star; teacher's manual; factsheets.
Hours & Admission Prices: Year-round daily 9-5. Adults $10, senior citizens $9, children 4-12 $6.50; discounts to AAA members & special groups; children under 4 & members no charge. State Park Pass Program. Closed Thanksgiving; Christmas. &
Attendance: 270,000 (accurate)
Membership: Senior Citizen $25; Individual & Senior Couple $35; Grandparent $45; Family $50.

NIAGARA GORGE DISCOVERY CENTER, New York State Parks, Niagara Region, Robert Moses State Pkwy. near Main St., Niagara Falls, NY 14303-0132. Mailing Address: P.O. Box 1132, Niagara Falls, NY 14303. Tel.: 716-278-1796 & 1770. Fax: 716-278-0838.
Web Site: www.niagarafallsstatepark.com
Formerly: Schoellkopf Geological Museum

Founded: 1971.
Key Personnel: Park Mgr., Tom Watt, .; Mktg. & Public Affairs, Angela P. Berti.
Governing Authority: state. Parent Institution: New York State Parks. Administered by Niagara Frontier State Park, Recreation and Historic Preservation Commission, Niagara Falls, NY. Tax-exempt.
Institution Type/Description: Natural & Local History Museum: located on the brink of the Niagara Gorge.
Collections: marine invertebrates from middle Silurian & Devonian periods; minerals of local varieties; historical artifacts from the Niagara Gorge.
Facilities: theater; live gorge & falls camera; Time Portal & Gorge Elevator Experience Interactives; trailhead building; gorge hiking trails.
Activities: permanent exhibitions; Q & A sheet.
Hours & Admission Prices: Call for hours. Adults $3, children 6-12 $2; children 5 & under no charge. Closed New Year's Day; Thanksgiving; Christmas. &
Attendance: 68,551 (accurate)

North Blenheim

LANSING MANOR HOUSE MUSEUM, 1378 State Rte. 30, North Blenheim, NY 12131. Mailing Address: P.O. Box 898, 1378 State Rte. 30, N. Blenheim, NY 12131-0898. Tel.: 800-724-0309. Fax: 518-287-6381. Facebook: Lansing Manor House Museum.
E-mail: samantha.clark@nypa.gov
Web Site: www.nypa.gov/html/vcblenhe.html
Founded: 1977.
Congressional District: 20
Key Personnel: Supvr., Samantha Clark.
Personnel Profile: Part-Time Paid 5.
Governing Authority: society; nonprofit. Parent Institution: New York Power Authority. Subsidiary Institution: Schoharie County Historical Society, P.O. Box 69, Old Stone Fort, Schoharie, NY 12157. Tax-exempt: 501(c)(3).
Institution Type/Description: Historic House: 1819 Lansing Manor House, Federal Manor house built by Chancellor John Lansing Jr., occupied by his son-in-law Jacob Sutherland as manager of the Blenheim Patent 1783-1853.
Collections: Federal & Empire furniture, textiles, ceramics, pictures & manuscripts; hydro-electric power. Historic Buildings: 1819 carriage barn; 1804-1819 outbuildings; c.1790 tenant house.
Facilities: visitors center, educational center.
Activities: guided tours.
Hours & Admission Prices: May-Oct. Wed.-Mon. 10-5 by appointment. No charge. &
Attendance: 12,400 (accurate)

North Salem

HAMMOND MUSEUM AND JAPANESE STROLL GARDEN, 28 Deveau Rd., North Salem, NY 10560-2115. Mailing Address: P.O. Box 326, 28 Deveau Rd., North Salem, NY 10560-2115. Tel.: 914-669-5033. Fax: 914-669-8221.
E-mail: gardenprogram@yahoo.com
Web Site: www.hammondmuseum.org.
Founded: 1957.
Congressional District: 2
Key Personnel: Dir., Lorraine Laken; Chm. Bd. (V), Evelyn Tapani-Rosenthal; Business Mgr., Judy Schurmacher.
Personnel Profile: Full-Time Paid 1; Part-Time Paid 2; Part-Time Volunteers 20; Interns 1.
Governing Authority: nonprofit. Tax-exempt: 170(B)(1)(A)(vi).
Institution Type/Description: Cross-Cultural Center.
Collections: Asian art; decorative arts; Japanese stroll garden.
Major Exhibits: Beush With Nature, 4/14-6/14; Photography, Tom Okada, 4/14-6/14; Installation-Inhabited-Wennie Huang, 6/14-9/14.
Facilities: botanical garden; Japanese tea room; restaurant. Art, crafts of different countries & other museum-related items for sale.
Activities: films; concerts; drama; lectures; workshops; group & school tours; formally organized education programs.
Hours & Admission Prices: Wed.-Sat. 12-4. Adults $5, seniors & students $4; discounts to AAM & AAA members & seniors; members & children under 12 no charge. &
Attendance: 5,800 (estimated)
Membership: Individual $35; Senior Family $40; Family $50; Contributing $100; Friends of Garden $125; Sustaining $250; Patron $500; Fellow $1,000; Corporate $2,000; Sponsor $2,500.

North Tonawanda

HERSCHELL CARROUSEL FACTORY MUSEUM, (M), 180 Thompson St., North Tonawanda, NY 14120-5420. Mailing Address: P.O. Box 672, North Tonawanda, NY 14120-0672. Tel.: 716-693-1885. Fax: 716-743-9018. Facebook; Herschell Carrousel Factory Museum.
E-mail: hcfm@carrouselmuseum.org
Web Site: www.carrouselmuseum.org
Founded: 1983.
Congressional District: 26
Key Personnel: Dir. (V), Raphaelle A. Proefrock; Pres. (V), Charles W. Proefrock; Museum Shop Mgr., Maureen Schumacher.
Personnel Profile: Full-Time Volunteers 1; Part-Time Paid 3; Part-Time Volunteers 43; Interns 1.
Governing Authority: nonprofit organization. Parent Institution: Carousel Society of the Niagara Frontier. Tax-exempt: 501(c)(3).
Institution Type/Description: Company Museum: housed in 1916 Allan Herschell Company factory building. Listed on the National Register of Historic Places.
Collections: history of the Herschell Company with emphasis on different types of amusement rides produced, marketing & design of rides, wood carvings, carrousel animals, historic rides, factory workers' lives, impact of factory on local economy.
Research Fields: Herschell companies history of amusement ride production.
Facilities: Museum-related items for sale.
Activities: woodcarving classes; carrousel chats; special events. Museum Sponsors: Victorian Tea in April; Renaissance Festival in June; Halloween Spooktacular in October; Lunch With Santa in November & December.
Publications: quarterly newsletter, Carrousel News.
Hours & Admission Prices: April-June 11 & Sept.-Dec. Wed.-Sun. 12-4; June 14 to Labor Day Mon.-Sat. 10-4, Sun. 12-4. Adults $5, senior citizens $4, children 2-12 $2.50; discounts to active military, AAM, AARP & AAA members; members no charge. Closed Easter; Thanksgiving; Christmas. &
Attendance: 11,000 (estimated)
Membership: Senior Citizen $15; Senior Couple $20; Individual $25; Family $45.

NORTH TONAWANDA HISTORY MUSEUM, 54-60 Webster St., North Tonawanda, NY 14120-5814. Tel.: 716-213-0554.
E-mail: nthistorymuseum@aol.com
Web Site: www.nthistorymuseum.org
Founded: 2004.
Congressional District: 26
Key Personnel: Exec. Dir., Donna Zellner Neal; Pres. (V), Robert Clark; Asst. Exec. Dir., John Zellner Neal.
Personnel Profile: Full-Time Paid 1; Full-Time Volunteers 1; Part-Time Volunteers 3.
Governing Authority: private; nonprofit organization. Tax-exempt.
Institution Type/Description: History Museum.
Collections: area ethnic heritage; Erie Canal & Niagara River influence and it's role as a lumber & industrial center during the 19th-20th centuries; photographs.
Research Fields: German, Polish, Italian, Irish, Hungarian, Lebanese, Syrian, & Slovak heritage; cemetery records; industrial & lumber heritage; Erie Canal heritage.
Facilities: library.
Activities: formal educational programs; group tours; guided tours; lectures; oral history program ghost walks; publications. Museum Sponsors: Historic Treasures Tour; Historic Homes Tour & Garden Walks; Ethnic Heritage Festival; Erie Canal music & vaudeville.
Publications: newsletter; annual report; book, North Tonawanda Ethnic Heritage Cookbook; Historic Treasures Guide 2005; North Tonawanda: The First 100 Years; North Tonawanda: The Lumber City; Historic Gardens Tour; The Rand Family Left A Lasting Imprint on North Tonawanda And The World; North Tonawanda: A Celebration of Our Diversity; North Tonawanda Families & Their Favorite Recipes; North Tonawanda: The Lumber City Tour Guide; Historic Treasures Guide 2007; Niagara Historic Trail Guide Book; Historic Treasures Tour Guide 2011; North Tonawanda: Historic Treasures.
Hours & Admission Prices: Mon.-Sat. 10-4. Adults $5, seniors & veterans $3; members no charge. Closed New Year's Eve & Day; Memorial Day; Independence Day; Labor Day; Thanksgiving; Christmas Eve & Day. &
Attendance: 20,000 (accurate)
Membership: Senior $10; Individual $15; Family $25; Business & Civic $50; Contributing $100; Life $250. (Special Veteran Rates).

Northport

NORTHPORT HISTORICAL SOCIETY, (M), 215 Main St., Northport, NY 11768. Mailing Address: P.O. Box 545, Northport, NY 11768-0545. Tel.: 631-757-9859. Fax: 631-757-9398.
E-mail: info@northporthistorical.org
Web Site: www.northporthistorical.org
Founded: 1962.
Congressional District: 3
Key Personnel: Dir., Heather Johnson; Pres. (V), Steven King; Museum Educator, Kari-Ann Carr; Museum Shop Mgr., Lois Howe.
Personnel Profile: Part-Time Paid 3; Part-Time Volunteers 85.
Governing Authority: society. Tax-exempt.
Institution Type/Description: History Museum.
Collections: ship building tools; domestic utensils; clothing; ephemera & documents; photographs; maps.
Research Fields: history of Northport; genealogy; architecture & industry.
Facilities: research library; meeting room. Museum-related items for sale.
Activities: education programs; lectures; self-guided recorded walking tour; special events; guided walking tours. Museum Sponsors: Parading Down Main Street (monthly walking tours).
Publications: books, A Light House of Stone; Faded Laurel's History of Eaton's Neck and Asharoken; Snapshots in Time: Vols I & II; seasonal member newsletter; quarterly newsletter.
Hours & Admission Prices: Tues.-Sun. 1-4:30. Suggested Donation: $5; discounts to Smithsonian Museum Day members. Closed New Year's Day; Easter; Independence Day; Thanksgiving; Christmas.
Attendance: 4,342 (accurate)
Membership: Senior Citizen & Student $20; Individual $25; Senior Family $30; Family $40; Business $60; Supporting Friend $100; Patron $250; Sustainer $500; Conservator $1,000.

Norwich

CHENANGO COUNTY HISTORICAL SOCIETY MUSEUM, 45 Rexford St., Norwich, NY 13815-1121. Tel.: 607-334-9227. Fax: 607-334-7809.
E-mail: estuscchs@roadrunner.com
Web Site: www.chenangohistorical.org
Founded: 1939.
Congressional District: 24
Key Personnel: Dir., Alan V. Estus; Pres., Linda M. Green; Sec., Mary Pat Laufair; Vice Pres., Walter O. Rogers; Treas., Howard Fogel; Cur., Meghan Molloy; Museum Shop Mgr., Liz Welch.
Personnel Profile: Full-Time Paid 2; Full-Time Volunteers 5; Part-Time Volunteers 60.
Governing Authority: private society. Tax-exempt: 501(c)(3).
Institution Type/Description: Historical Society Museum.
Collections: costumes; folklore; folkart; Native American artifacts; local early artifacts; documents; tools; Chenango County cultural history; Norwich Pharmacal Company.
Research Fields: folklore; archaeology; cemeteries of Chenango County; genealogy; manuscripts; local history publications.
Facilities: library of local history information available for use by permission of society.
Activities: guided tours; lectures; films; formally organized education programs for children; permanent & traveling exhibitions.
Publications: various booklets & books, list available upon request.
Hours & Admission Prices: Jan.-March Mon.-Fri. 1-4; April-Dec. Sun.-Fri. 1-4; other times by appointment. No charge; donations accepted. Closed legal holidays.
Attendance: 1,500 (estimated)
Membership: Senior Citizen 60 & over $15; Individual $20; Family $30; Supporting $100; Patron $200; Angel $300; Maydole $500; Chenango Circle $1,000.

NORTHEAST CLASSIC CAR MUSEUM, 24 Rexford St., Norwich, NY 13815-1172. Tel.: 607-334-2886. Fax: 607-336-6745.
E-mail: info@classiccarmuseum.org
Web Site: www.classiccarmuseum.org
Founded: 1997.
Key Personnel: Exec. Dir., Marc Michaels; Pres., Phil Giltner; Museum Shop Mgr., Brenda Rose.
Personnel Profile: Full-Time Paid 3; Part-Time Paid 2; Part-Time Volunteers 75.
Governing Authority: state; nonprofit. Tax-exempt: 501(c)(3).
Institution Type/Description: Classic Car Museum.
Collections: vintage cars; Franklin luxury cars.
Publications: biannual, Hood Release.

Hours & Admission Prices: Daily 9-5. Adults $9, children 6-18 $4; members & children under 6 no charge. Closed New Year's Day; Thanksgiving; Christmas. &

Attendance: 10,000 (estimated)

Membership: Individual $35; Family $50; Friend of Museum $150; Corporate $300; Auburn $500; Marmon $1,000; Lifetime $25,000.

Norwood

NORWOOD HISTORICAL ASSOCIATION AND MUSEUM, 39 N. Main St., Norwood, NY 13668-1123. Mailing Address: P.O. Box 163, Norwood, NY 13668-0163. Tel.: 315-353-2751.

E-mail: glacomb@twcny.rr.com

Founded: 1968.

Key Personnel: Dir., Richard Boprey; Chm. (V), Dick Boyle; Sec., Rose Valyo.

Personnel Profile: Part-Time Paid 1; Part-Time Volunteers 7.

Governing Authority: nonprofit organization. Tax-exempt: 501(c)(3).

Institution Type/Description: General Museum.

Collections: railroad articles; folklore; history; industrial; Norwood Brass Fireman; official 1984 Olympic Band; manuscript collections.

Research Fields: local history.

Facilities: 75-vol. library of assorted legal documents; pamphlets; original programs of local events; newspapers; bound ledgers; text books; atlas & maps available by special arrangement for use on the premises.

Activities: guided tours; lectures; arts festivals; formally organized education programs; inter-museum loan, permanent & temporary exhibitions. Museum Sponsors: museum interest group for adults.

Publications: handbooks, Rails into Racquetteville; The Story of Norwood, NY.

Hours & Admission Prices: May-Oct. Tues. & Thurs. 2-4, other times by appointment. No charge; donations accepted.

Attendance: 300

Membership: Junior $.50; General $2.

Nyack

EDWARD HOPPER HOUSE ART CENTER, 82 N. Broadway, Nyack, NY 10960-2628. Tel.: 845-358-0774. Fax: 845-358-0774.

E-mail: info@hopperhouse.org

Web Site: www.hopperhouse.com

Founded: 1971.

Key Personnel: Exec. Dir., Carole Perry; Pres., Victoria Hertz.

Personnel Profile: Part-Time Paid 3; Part-Time Volunteers 40; Interns 1.

Governing Authority: private; nonprofit organization. Tax-exempt: FEX-160198.

Institution Type/Description: Art Center: boyhood home of renowned American Realist painter, Edward Hopper.

Collections: photographs; Hopper memorabilia.

Research Fields: Edward Hopper.

Facilities: 950 sq. ft. exhibit space. Books, prints, postcards, notecards, exhibited art & museum-related items for sale.

Activities: summer jazz concerts; lectures; tours; educational videos on Hopper's life & work; life drawing classes. Annual Events: Book Fairs in December; National Juried Small Works Show; Figure Show; Biennial Photography Show.

Publications: pamphlet.

Hours & Admission Prices: Thurs.-Sun. 1-5. Adults $5, seniors $3; children, students, Whitney Museum & center members no charge. &

Attendance: 4,000 (estimated)

Membership: Student & Senior $25; Individual $35; Artist $50; Family $60; Sustaining $250; Patron $500; Benefactor $1,000.

Ogdensburg

* **FREDERIC REMINGTON ART MUSEUM, (M),** 303 Washington St., Ogdensburg, NY 13669-1517. Tel.: 315-393-2425. Fax: 315-393-4464.

E-mail: info@fredericremington.org

Web Site: www.fredericremington.org

Founded: 1923.

Congressional District: 26

Key Personnel: Exec. Dir., Laura A. Foster; Administrative Aide, Shannon Ghize; Pres., William Cavanaugh; Education Specialist, Lauren Gilmour; Account Clerk, Debbie Ormasen.

Personnel Profile: Full-Time Paid 4; Part-Time Paid 8; Part-Time Volunteers 16.

Governing Authority: municipal. Tax-exempt.

Institution Type/Description: Art Museum: housed in 1809-10 Mansion with modern gallery addition.

Collections: Frederic Remington's paintings, sculpture, drawings, manuscripts; archives; decorative arts.

Research Fields: Frederic Remington.

Facilities: 4,500-vol. of Remington's personal library & family books; manuscript collections; Albert and Addie Priest Newell galleries. Museum-related items for sale.

Activities: guided tours; lectures; gallery talks; inter-museum loan, permanent & temporary exhibitions.

Publications: quarterly newsletter, Remington Related; catalog, Frederic Remington Memorial Collection; special exhibition catalogs.

Hours & Admission Prices: May 15-Oct. 15 Mon.-Sat. 10-5, Sun. 1-5; Oct. 16-May 14 Wed.-Sat. 11-5, Sun. 1-5. Adults $9, seniors and students 16 & over $8; discounts to AAM members; children 15 & under and members no charge. Closed New Year's Day; Easter; Thanksgiving; Christmas. &

Attendance: 10,000 (accurate)

Membership: Sentinel $35; Pioneer $50; Dragoon $100; Sergeant $150; Trooper $250; Cheyenne $500; Ingleneuk Club $1,000.

Old Bethpage

OLD BETHPAGE VILLAGE RESTORATION, 1303 Round Swamp Rd., Old Bethpage, NY 11804-1199. Tel.: 516-572-8401. Fax: 516-572-8439.

Web Site: www.nassaucountyny.gov/parks

Founded: 1970.

Congressional District: 2

Key Personnel: Site Dir., James McKenna; Asst. Site Dir., Geraldine Jordan; Dir. Historic Interpretation, Henry Clark; Supvr. Volunteers, Judy Pockriss.

Personnel Profile: Full-Time Paid 9; Part-Time Paid 16; Part-Time Volunteers 40; Interns 6.

Governing Authority: county. Nassau County Dept. of Parks, Recreation & Museums, Museum Svcs., East Meadow, NY 11554. Tax-exempt.

Institution Type/Description: Living History Village Museum: 15 historic site units which include houses, shops, barns, outbuildings, tavern, church & schoolhouse.

Collections: Americana including tools, utensils, furnishings & decorative arts.

Research Fields: local history.

Facilities: library of books related to historic interpretations of Long Island; 200-seat auditorium; cafeteria. Museum-related items for sale.

Activities: tours; formally organized education programs; lectures; concerts; seasonal special events & activities.

Publications: interpretive leaflet, Chronicle; seasonal leaflet, Enquirer; folder, Old Bethpage Village Restoration Schedule; Old Bethpage Guidebook.

Hours & Admission Prices: April-May & Nov.-Dec. Wed.-Sun. 10-4; June-Oct. Wed.-Sun. 10-5. Adults $10, seniors & children $7; discounts to groups; children under 5 no charge. Closed New Year's Eve; Veterans Day; Thanksgiving & day after; Christmas Eve & Day.

Attendance: 89,505 (accurate)

Old Westbury

OLD WESTBURY GARDENS, 71 Old Westbury Rd., Old Westbury, NY 11568-1603. Mailing Address: P.O. Box 430, Old Westbury, NY 11568-0430. Tel.: 516-333-0048. Fax: 516-333-6807.

E-mail: emccauley@oldwestburygardens.org

Web Site: oldwestburygardens.org

Founded: 1958.

Congressional District: 4

Key Personnel: C.E.O. & Pres., John Norbeck; Chm. (V), Mary S. Phipps; Dir. Devel., Doreen Banks; Dir. Visitor Svcs., Collections Mgmt. & Museum Shop Mgr., Paul Hunchak; Dir. Public Rels., Vincent Kish; Dir. Horticulture, Maura M. Brush; Walled Garden Supvr., Kimberly Johnson; Greenhouse Supvr., Scott Lucas.

Personnel Profile: Full-Time Paid 27; Full-Time Volunteers 1; Part-Time Paid 40; Part-Time Volunteers 250; Interns 5.

Governing Authority: nonprofit organization. Parent Institution: Old Westbury Gardens, Inc. Tax-exempt: 501(c)(3).

Institution Type/Description: Historic House & Public Horticultural Display Garden: 1906 Charles II style mansion furnished with 18th-century decorative & fine arts.

Collections: English 18th-century furniture & decorations & paintings; Chinese porcelains; formal gardens; family papers & documents; photographs.

Research Fields: horticulture; decorative arts; social history; landscape design.

Facilities: library of books on architecture, horticulture, natural sciences & art; botanical garden. Books & other museum-related items for sale.

Activities: guided tours; lectures; films; concerts; hobby workshops; temporary exhibitions.

Publications: newsletter, Old Westbury Gardens; Old Westbury Gardens Guidebook; Old Westbury Gardens Picture Book; Halcyon Days; Book of Days.

Hours & Admission Prices: Grounds: late April to Oct. Wed.-Mon. 10-5.

House: late April to Oct. Wed.-Mon. 11-5; Nov.-Dec. call for hours. House & Gardens: adults $10, senior citizens over 62 $8, children 7-17 $5; members and children 6 & under no charge, except during special events. &

Attendance: 60,000 (accurate)
Membership: Individual $50; Family & Dual $75; Sustainer $100; Friend $300; Patron $500; Contributor $1,000; Benefactor $2,500.

Onchiota

SIX NATIONS INDIAN MUSEUM, 1462 County Rte. 60, Onchiota, NY 12989-2102. Tel.: 518-891-2299.
E-mail: info@sixnationsindianmuseum.com
Web Site: www.sixnationsindianmuseum.com
Founded: 1954.
Key Personnel: Dir., John Fadden; Museum Shop Mgr. & Asst., Elizabeth E. Fadden.
Governing Authority: individual operation.
Institution Type/Description: Indian Museum, with emphasis on Iroquois.
Collections: Six Nations Indians artifacts; ancient & modern Iroquois utensils, clothing, beaded record belts, wampum belts replicas; art work & dwellings; types of fires & exhibits; Iroquois culture & history; paintings reflecting Iroquois culture.
Facilities: Indian-related items, charts & pamphlets on the Iroquois for sale.
Activities: guided tours; lectures; formally organized education programs for children, adults & undergraduate and graduate college students.
Hours & Admission Prices: July-Aug. Tues.-Sun. 10-5; June & Sept. by appointment. Adults $4, children $2; Indians & special non-Indian friends no charge.
Attendance: 2,000 (estimated)

Oneida

MADISON COUNTY HISTORICAL SOCIETY-COTTAGE LAWN, 435 Main St., Oneida, NY 13421-2440. Tel.: 315-363-4136 & 361-9735. Fax: 315-361-9735. Facebook: Madison County Historical Society.
E-mail: history@mchs1900.org
Web Site: www.mchs1900.org
Founded: 1900.
Congressional District: 25
Key Personnel: C.E.O. & Exec. Dir., Sydney L. Loftus; Pres. (V), Mishell Magnusson.
Personnel Profile: Full-Time Paid 1; Full-Time Volunteers 1; Part-Time Paid 1; Part-Time Volunteers 15.
Governing Authority: nonprofit organization. Tax-exempt: 501(c)(3).
Institution Type/Description: Historical Society Museum: housed in Gothic Revival 1849 Cottage Lawn designed by Alexander Jackson Davis.
Collections: 19th-century furnishings including paintings; furniture; glassware; ceramics; textiles; 19th-century tools including agriculture; woodworking; blacksmithing; tinsmithing; 1862 stage coach, carriages, carts & wagons; archival collections; traditional crafts.
Research Fields: Madison County & central New York history; genealogy; restoration resources & folklife; 18th- & 19th-century costuming & textiles; 19th-century conservations; folk & traditional crafts.
Facilities: 2,500-vol. library of historical books available for research; traditional crafts archives containing 20,000 slides, 180 tapes & 42 8 mm movies documenting traditional crafts & folklife. Museum-related items for sale.
Activities: guided tours; lectures; formally organized education programs for children & adults; children's daycamp; membership programs. Annual Events: Craft Days; Victorian Christmas Celebration; Hop Fest; historic walking tours; historic house tours in fall.
Publications: annual journal, Madison County Heritage; County Roads Revisited; The Cultural Imprint of Madison County; annual, Studies in Traditional American Crafts; newsletters; annual report; pamphlets; videos.
Hours & Admission Prices: Mon.-Fri. 10-4. Tours: adults $5, seniors $2; discount to AAM members; children 12 & under and members no charge. Research: adults $10; members no charge.
Attendance: 12,000 (estimated)
Membership: Junior $5; Senior $15; Adult $20; Family $30; Supporting $75; Patron $125; Sustaining & Corporate $300.

ONEIDA COMMUNITY MANSION HOUSE, (M), 170 Kenwood Ave., Oneida, NY 13421-2820. Tel.: 315-363-0745. Fax: 315-361-4580.
E-mail: ocmh@oneidacommunity.org
Web Site: www.oneidacommunity.org
Founded: 1987.
Congressional District: 23
Key Personnel: Exec. Dir., Patricia A. Hoffman; Cur. Interpretation & Collections, Anthony Wonderley.
Personnel Profile: Full-Time Paid 7; Part-Time Paid 6; Part-Time Volunteers 63; Interns 1.
Governing Authority: private; nonprofit organization. Tax-exempt: 501(c)(3).
Institution Type/Description: Historic House Museum: housed in the 93,000 sq. ft. brick Mansion House, constructed between 1861 & 1914.
Collections: furniture; clothing; decorative arts; paintings; works of art on paper; approx. 2,500 utilitarian objects made in or brought to the Oneida Community; books, pamphlets, sheet music & ephemera that illustrate the intellectual, cultural & business life of the community, approx. 10,500 items; a photographic archive documenting the life of the Mansion House during the days since the Oneida Community.
Research Fields: Oneida Community; communal societies.
Facilities: 200-seat auditorium. Museum-related items for sale.
Activities: tours; educational programs; musical performances; special events.
Publications: biannual, Oneida Community Journal; The New Circular (biannual).
Hours & Admission Prices: Self-Guided Tours: Mon.-Sat. 10-4, Sun. 1-4. Guided Tours: Wed.-Sat. 10 & 2, Sun. 2; groups by appointment. Adults $5, students $3; discounts to AAM, AASLH, VHA & NARM members & groups; children & members no charge. Closed major holidays. &
Attendance: 9,843 (accurate)
Membership: Student $25; Individual $40; Family $50; Associate $100; Contributor $250; Donor $500; Benefactor $1,000.

SHAKO:WI CULTURAL CENTER, Oneida Indian Nation, 5 Territory Rd., Oneida, NY 13421-9304. Tel.: 315-829-8801. Fax: 315-829-8805.
Web Site: www.oneida-nation.net
Founded: 1993.
Congressional District: 23
Key Personnel: Dir., Kandice Watson.
Personnel Profile: Full-Time Paid 2; Part-Time Paid 2.
Governing Authority: Oneida Indian Nation. Tax-exempt.
Institution Type/Description: Tribal Museum.
Collections: Native Americans; photo archives of Oneida residents on disc; ethnographic; beadwork.
Activities: beading group; dance workshops; language instruction; painting classes.
Hours & Admission Prices: Mon.-Sat. 9-5. No charge. Closed holidays; American Indian Day.
Attendance: 3,600 (estimated)

Oneonta

SCIENCE DISCOVERY CENTER OF ONEONTA, State University College, Oneonta, NY 13820-4015. Tel.: 607-436-2011. Fax: 607-436-2654.
E-mail: scdisc@oneonta.edu
Web Site: www.oneonta.edu/academics/sdc
Founded: 1987.
Congressional District: 25
Key Personnel: Dir., Hugh Gallagher, Jr.
Personnel Profile: Full-Time Volunteers 1; Part-Time Paid 1; Part-Time Volunteers 15.
Governing Authority: nonprofit. State University of New York, College of Oneonta. Tax-exempt: 501(c)(3).
Institution Type/Description: Science Museum.
Collections: concentration on physics & physical science; interactive science exhibits.
Facilities: 200-vol. library on science experiments & activities available to the public; 3,000 sq. ft. exhibit space.
Activities: informal science activities for elementary, secondary & college classes; some formal instruction for science education majors on use of center's facilities & occasional formal science workshops for elementary teachers.
Publications: quarterly newsletter, Science Discovery Center of Oneonta Newsletter.
Hours & Admission Prices: July-Aug. Mon.-Sat. 12-4; Sept.-June Thurs.-Sat. 12-4. Groups by appointment only. No charge; donations accepted. Closed Thanksgiving; Christmas. &
Attendance: 5,720 (accurate)
Membership: Individual $25; Family $40; Benefactor $100; Life $400.

THE YAGER MUSEUM OF ART & CULTURE, (M), One Hartwick Dr., West St., Oneonta, NY 13820-4000. Mailing Address: P.O. Box 4020, Oneonta, NY 13820-4000. Tel.: 607-431-4480. Fax: 607-431-4468.
E-mail: museum@hartwick.edu
Web Site: www.hartwick.edu/museum.xml
Founded: 1929.
Congressional District: 32
Key Personnel: College Pres., Dr. Margaret L. Drugovich; Dir., Donna Anderson; Cur. Anthropological Collection, Dr. David Anthony; Collections Mgr., Gary Norman; Museum Shop Mgr., Nancy Martin-Mathewson.
Personnel Profile: Full-Time Paid 1; Part-Time Paid 25.
Governing Authority: college board of trustees. Parent Institution: Hartwick College. Branch Museums: American Indian Collection & Fine Arts Collection. Tax-exempt: 501(c)(3).
Institution Type/Description: General Museum.
Collections: Yager: Upper Susquehanna Indian artifacts, Southwest basketry & pottery; Furman collection of Mexican, Central American & South American artifacts; Sandell collection of Ecuadorian & Peruvian artifacts; Friess collection of Mexican masks; Marks collection of Pre-Columbian art. Van Ess Fine Arts collection of renaissance, baroque & American 19th century paintings, prints & sculpture; American Indians. Russian icons.
Research Fields: archaeological excavations in the Upper Susquehanna region; 19th-century fine arts; 19th & 20th century history.
Facilities: 800-vol. library of American Indian history books available on premises.
Activities: changing exhibitions; programs; seminars.
Publications: post cards; catalogues.
Hours & Admission Prices: Tues.-Sat. 12-4:30; collection research by appointment only. No charge, donations accepted. Closed school holidays. &
Attendance: 6,050 (accurate)
Membership: Student $10; Senior $25; Individual $30; Senior Couple $40; Family & Couple $50; Van Ess $500; Hassam $1,000; Yager $2,500 & up.

Ontario

HERITAGE SQUARE MUSEUM, 7147 Ontario Center Rd., Ontario, NY 14519. Mailing Address: P.O. Box 462, Ontario, NY 14519-0462. Tel.: 315-524-5356.
E-mail: jeswitz@rochester.rr.com
Web Site: heritagesquaremuseum.org
Formerly: Town of Ontario Historical & Landmark Preservation Society, Inc. Heritage Square
Founded: 1969.
Congressional District: 24
Key Personnel: Pres. (V), Jim Switzer; Vice Pres., Ann Welker; Sec., Alayna Di Santo; Treas., Jean Tsepas; Museum Shop Mgr., Vera Graves.
Personnel Profile: Part-Time Volunteers 125.
Volunteer Hours: 2,500
Operating Expenses: 25,000
Operating Income: 25,000
Governing Authority: society. Parent Institution: Town of Ontario Historical & Landmark Preservation Society. Tax-exempt. 501(c)(3).
Institution Type/Description: Historical Society Museum Complex.
Collections: iron ore mining; artifacts; furniture; agriculture; farm machinery & tools; education. Historic 1800's Buildings: c.1860 schoolhouse; c.1834 farmhouse; c.1840 meeting house; church; c.1860 log cabin; c.1890 miner's home; Ore Miner's house; barn; dry house (apples); lockup (jail); 1874 train station.
Research Fields: local Ontario history from 1807-1993.
Facilities: 100-seat auditorium.
Activities: guided tours; formally organized educational programs; video; slide shows.
Publications: newsletter, Town of Ontario Historical & Landmark Society; books, History of the Town of Ontario 1807-1993; handbook, Heritage Corners.
Hours & Admission Prices: June 6-Oct. 4 Sat.-Sun. 1:30-4; other times by appointment. School & group tours by bus are welcome. Family $10, adults $4, seniors & children $3; members no charge. Call for admission prices; members no charge. Closed Memorial Day weekend; Independence Day weekend; Labor Day weekend. &
Attendance: 1,550 (estimated)
Membership: Senior Citizens 62 & over $10; Individual $15; Senior Couple $18; Family $25; Patron $50; Business $100.

Orangeburg

ORANGETOWN HISTORICAL MUSEUM AND ARCHIVES - SALYER HOUSE AND DEPEW HOUSE, 196 Chief Bill Harris Way, Orangeburg, NY 10962-2011. Mailing Address: 26 Orangeburg Rd., Orangeburg, NY 10962-1706. Tel.: 845-398-1302. Fax: 845-398-8919.
E-mail: otownmuseum@optonline
Web Site: www.orangetownmuseum.com
Founded: 1992.
Congressional District: 20
Key Personnel: Dir., Mary R. Cardenas; Chm. (V), Catherine M. Dodge; Supvr., Thom Kleiner; Cur., Elizabeth Skrabonja; Public Rels., Laura Davie; Registrar, Joseph Barbieri; Museum Shop Mgr., Sandi Miller.
Personnel Profile: Part-Time Paid 3; Part-Time Volunteers 36; Interns 2.
Governing Authority: nonprofit. Parent Institution: Town of Orangetown. Branch Museum: DePew House, 196 Blaisdell Rd., Orangeburg, NY 10962; The Salyer House, 213 Blue Hill Rd., Pearl River, NY 10965. Tax-exempt.
Institution Type/Description: History Museum.
Collections: photographs; period artifacts; archives; books, newspaper articles & magazines containing articles pertaining to Orangetown history from the Colonial era to present. Historic Houses: Salyer House c.1779; De Pew House c.1777.
Facilities: 750-vol. library available for use on site by appointment; archives; 1,280 sq. ft. exhibit space. Museum-related items for sale.
Activities: school programs. Museum Sponsors: antique & collectible sale in May; annual dinner in June; holiday open house in December.
Publications: quarterly newsletter, Orangetown Crier.
Hours & Admission Prices: Sun. 1-4, Tues. 10-2; other times by appointment. No charge; donations accepted. Closed holidays. &
Attendance: 864 (accurate)
Membership: Student & Senior Citizen $10; Single $15; Family $20; Life $150; Corporate $500.

Orchard Park

ORCHARD PARK HISTORICAL SOCIETY, 4287 S. Buffalo St., Orchard Park, NY 14127. Mailing Address: 4100 N. Freeman Rd., Orchard Park, NY 14127-2525. Tel.: 716-662-2185.
Founded: 1951.
Congressional District: 38
Key Personnel: C.E.O. & Pres., Dennis J. Mill; Cur., Yasabel N. Gibson.
Personnel Profile: Part-Time Volunteers 10; Interns 2.
Governing Authority: society. Tax-exempt: 501(c)(3).
Institution Type/Description: Historical Society Museum: housed in the Jolls house; built in 1870.
Collections: artifacts; agricultural items; dolls; dishes; maps; deeds; clothing; utensils; toys; loom; early household items; furniture; Indian artifacts; Quaker artifacts.
Facilities: 150-vol. library of historic books available for research on premises.
Activities: lectures; permanent & temporary exhibitions; school loan service.
Hours & Admission Prices: Spring-Oct. 1st & 3rd Sat. 2-4; Nov. 28-Dec. 20 Sat.-Sun. 2-4; groups by appointment. Admission: $2. &
Attendance: 425 (estimated)
Membership: Single $10; Family $15. Life membership: Single $250; Family $350.

Orient

OYSTERPONDS HISTORICAL SOCIETY: MUSEUM & ARCHIVE OF ORIENT & EAST MARION HISTORY, (M), 1555 Village Lane, Orient, NY 11957. Mailing Address: P.O. Box 70, Orient, NY 11957-0070. Tel.: 631-323-2480. Fax: 631-323-3719.
E-mail: ohsorient@optonline.net
Web Site: www.oysterpondshistoricalsociety.org
Founded: 1944.
Congressional District: 1
Personnel Profile: Full-Time Paid 1; Part-Time Paid 2; Part-Time Volunteers 50; Interns 1.
Governing Authority: board of trustees; society. Parent Institution: Oysterponds Historical Society, Inc. Tax-exempt: 501(c)(3).
Institution Type/Description: History Village Museum Complex.
Collections: Indian artifacts; marine paintings; photographs; manuscripts & early documents; whaling & fishing artifacts; ship models & navigating instruments; spinning & weaving tools; clothing; furniture; toys; dolls; period tools; quilts; agricultural tools. Historical Houses: 1860s Village House; 1888 Old Point School House; 1860 Amanda Brown House; 1870 Seine House: sleighs & carriages; c.1780 Webb House; 1870 Vail House.

Research Fields: local history; genealogy; school books; costumes; local architecture.

Facilities: 1,000 vol. library containing 6,000 photos, 2,000 glass plate negatives; 10,000 documents available on premises for research. Museum-related items for sale.

Activities: guided tours; lectures; permanent & temporary exhibits; grade school program.

Publications: newsletter; books, Historic Orient Village; Griffin's Journal (reprint); She Went A-Whaling; Captain's Daughter; A Sense of Place.

Hours & Admission Prices: July-Oct. Thurs. & Sat.-Sun. 2-5; other times by appointment. Adults $5, children under 16 $.50; discounts to AASLH members; members no charge. &

Attendance: 5,000 (estimated)

Membership: Individual $30; Family $50; Associate $100; Sponsor $250; Patron $500; Benefactor $1,000.

Oriskany

ORISKANY BATTLEFIELD STATE HISTORIC SITE, 7801 State Rt. 69, Oriskany, NY 13424-4115. Tel.: 315-768-7224. Fax: 315-337-3081.

E-mail: nancy.demyttenaere@oprhp.state.ny.us

Web Site: www.nysparks.com

Founded: 1927.

Congressional District: 31

Key Personnel: Regl. Historic Preservation Supvr., Nancy Demyttenaere; Second in Command, Bill Acomb.

Personnel Profile: Full-Time Paid 3; Part-Time Paid 3; Part-Time Volunteers 1.

Governing Authority: state. Parent Institution: New York State Office of Parks, Recreation & Historic Preservation. Tax-exempt: 501(c)(3).

Institution Type/Description: Historic Site: Revolutionary War Battlefield & Memorial Park, site of Aug. 6, 1777 ambush of Colonial Militia & Oneida allies by Loyalist forces.

Collections: Large memorial obelisk & other smaller monuments from 1884-1929; historic landscape & archives; commemorative souvenirs.

Facilities: battlefield site; visitor center; picnic area.

Activities: guided tours; special events.

Hours & Admission Prices: April-Oct. daily 10-4. No charge; donations accepted.

Attendance: 23,000 (accurate)

Ossining

OSSINING HISTORICAL SOCIETY MUSEUM, 196 Croton Ave., Ossining, NY 10562-4504. Tel.: 914-941-0001. Fax: 914-941-0001.

E-mail: ohsm@bestweb.net

Web Site: www.ossininghistorical.org

Founded: 1931.

Congressional District: 24

Key Personnel: Pres., John C. Wunderlich; Vice Pres., Deborah Van Steen; Treas., Greg Fratianni.

Personnel Profile: Full-Time Volunteers 5; Part-Time Volunteers 5; Interns 2.

Governing Authority: society; nonprofit organization. Tax-exempt: 501(c)(3).

Institution Type/Description: Local History Museum & Fine Arts Collection.

Collections: costumes; paintings; Indian artifacts; war relics; doll collection; fine arts; books; bottles; china; photographs; portraits & paintings depicting culture of the 17th & 18th century residents; oral history collection; manuscript collections; VHS; newspaper collection, 1799 to present; photographs; microfilm; genealogy; maps; documents; toys; glass plates & negatives.

Research Fields: Sing Sing Prison; Hudson River; local & county history; genealogy; school records; Croton Dam; Croton aqueduct; architecture; local government records.

Facilities: library; local newspapers 1799-present, magazines & journals; clipping file on local history; reference material; maps; reading room; video viewing center; oral history center; research room; archives.

Activities: guided tours; lectures; gallery talks; permanent & temporary exhibitions; genealogy information.

Publications: Sing Sing Prison Electrocutions; Civil War Directory - Local; War Memorial - All Wars; Images of America: Ossining Remembered; The 1950s in Ossining, NY; The 1940s in Ossining, NY; William Dolphins Civil War Diary; book, Sparta Cemetery; The Sparta Letters, The Record of Ossining's Earliest Residents; A Primer of Ossining History; Chronicle Of A Westchester Farm.

Hours & Admission Prices: By appointment. No charge; donations accepted. &

Attendance: 1,500 (estimated)

Membership: Student & Senior Citizen $10; Individual $20; Family $30; Sustaining & Civic $50; Patron & Commercial $100.

Oswego

FORT ONTARIO STATE HISTORIC SITE, One E. 4th St., Oswego, NY 13126-1233. Mailing Address: P.O. Box 5379, Oswego, NY 13126. Tel.: 315-343-4711.

Web Site: www.fortontario.com

Founded: 1949.

Congressional District: 30

Key Personnel: Pres. (V), Charles Harrington; Site Mgr., Paul Lear; Asst., Richard LaCrosse; Cur., Jennifer Emmons; Sec., Roberta Elmer; Maintenance Supvr., Robert Clarke.

Personnel Profile: Full-Time Paid 8; Part-Time Paid 12; Part-Time Volunteers 135.

Governing Authority: state. New York State Office of Parks, Recreation & Historic Preservation, & Central New York State Parks, Recreation & Historic Preservation Commission. Tax-exempt: 501(c)(3).

Institution Type/Description: Military Museum: housed in 1839-1844 fortifications situated at the outlet of Oswego River into Lake Ontario.

Collections: military furnishings, firearms, equipment, uniforms, accoutrements, prints, paintings. Historic Buildings: 1842-1844 enlisted men's barracks, a magazine, two officers' quarters with outbuildings, post headquarters; 1863-1872 two guardhouses & five casemates; 1821 post hospital; 18th-century artifacts; photographs.

Research Fields: military & naval activities on the Great Lakes frontier from the French & Indian Wars through World War II; fortification design & construction; related political, economic & diplomatic events & personalities.

Facilities: 500-vol. collection of books & periodicals dealing with regional, military & social history available for use on premises by advance appointment; maps & plans; classroom; picnic area.

Activities: self-guided & guided tours; lectures; concerts; encampments; Civil War Artillery, Infantry and Engineer schools; kite festival & more.

Publications: folders, Fort Ontario; Fort Ontario: A Self-Guided Tour; "Welcome to Fort Ontario".

Hours & Admission Prices: May-Oct. 15 Tues.-Sun. & Mon. holidays 10-4:30; groups by appointment. Adults $4, seniors & students with I.D. $3; children 12 & under & Friends of Fort Ontario no charge.

Attendance: 15,000 (accurate)

Membership: Senior $10; Individual $15; Family $25; Small Business $50; Benefactor $100; Patron $250; Sponsor $500; Corporate $1,000.

H. LEE WHITE MARINE MUSEUM, (M), (I), W. 1st St. Pier, Oswego, NY 13126. Mailing Address: P.O. Box 101, Oswego, NY 13126-0101. Tel.: 315-342-0480. Fax: 315-343-5778. Facebook: HLWMM.

E-mail: info@hleewhitemarinemuseum.com

Web Site: hleewhitemarinemuseum.com

Founded: 1982.

Congressional District: 23

Key Personnel: Exec. Dir., Mercedes Niess; Retail & Visitor Experience Mgr., Susan Wild; Public Rels. Asst., Shelia Weldin.

Personnel Profile: Full-Time Paid 1; Part-Time Paid 4; Part-Time Volunteers 65; Interns 2.

Operating Expenses: 130,000

Operating Income: 130,000

Governing Authority: Tax-exempt.

Institution Type/Description: Maritime Museum.

Collections: maritime artifacts; photographs; boats.

Publications: newsletter.

Hours & Admission Prices: July-Aug. daily 10-5; Sept.-June daily 1-5. Adults $7, teenagers $3; discounts to SUNY students, HNSA, AAA, AAM & ICOM members, active duty military; museum members & children under 13 no charge. Closed New Year's Eve & Day; Easter; Thanksgiving; Christmas Eve, Day & week.

Attendance: 28,700 (accurate)

Membership: First Mate $25-$49; Ensign $50-$99; Commander $100-$249; Captain $250-$499; Rear Admiral $500-$999; Admiral $1,000 & up.

JOHN D. MURRAY FIREFIGHTER'S MUSEUM, 35 E. Cayuga St., Oswego, NY 13126. Mailing Address: P.O. Box 211, Oswego, NY 13126. Tel.: 315-343-2161.

Institution Type/Description: Firefighting History Museum.

Collections: firefighting history & equipment; 1925 American LaFrance pumper; 1939 American LaFrance pumper; journals; helmets; fire buckets; lanterns; fire alarm boxes; firehouse gong; photographs.

Hours & Admission Prices: Call for hours.

RICHARDSON-BATES HOUSE MUSEUM, 135 E. 3rd St., Oswego, NY 13126-2655. Tel.: 315-343-1342.
E-mail: ochs@rbhousemuseum.org
Web Site: www.rbhousemuseum.org
Founded: 1896.
Congressional District: 24
Key Personnel: Pres. (V), Justin White.
Personnel Profile: Part-Time Paid 1; Part-Time Volunteers 12.
Governing Authority: nonprofit organization; Oswego County Historical Society. Tax-exempt: 509(a)(1); 501(c)(3).
Institution Type/Description: Historic House Museum: Richardson-Bates house, built 1867-90.
Collections: Oswego County artifacts, photographs & documents; 19th-century furnishings, paintings & decorative arts; Native American artifacts.
Research Fields: local, regional & state history; Victorian decorative arts.
Facilities: 500-vol. library of material on local, regional & state history available for use upon application.
Activities: guided museum tours; permanent & changing exhibitions; outreach in-school lectures/demonstrations; lectures; special events.
Publications: Historical Journal, 1939-1977; monthly newsletter.
Hours & Admission Prices: April-Dec. Thurs.-Sat. 1-5; research library by appointment. Family $12; adults $5, students & seniors $3, children $2; discounts to American Assoc. for State & Local History, NY State Historical Assoc. & AAM members; members no charge. Closed holidays.
Attendance: 2,500 (estimated)
Membership: Senior Citizen & Student $10; Individual $20; Family $25; Order of the Sphinx $50-$99; Friends of Harriet $100-$248; Norman's Fellows $250-$499; Naomi's Circle $500-$999; Giving to the Max $1,000 & Up.

SAFE HAVEN MUSEUM AND EDUCATION CENTER, (M), 2 E. 7th St., Oswego, NY 13126-1197. Mailing Address: P.O. Box 846, Oswego, NY 13126-0846. Tel.: 315-342-3003. Fax: 315-342-1411.
E-mail: safehaven@cnymail.com
Web Site: www.oswegohaven.org
Founded: 1990.
Congressional District: 24
Key Personnel: C.E.O. & Pres., Judy Coe Rapaport.
Personnel Profile: Part-Time Paid 1.
Governing Authority: private; nonprofit organization. Tax-exempt.
Institution Type/Description: History Museum.
Collections: photographs, memorabilia, primary source documents & oral history videotapes associated with the Fort Ontario World War II Refugee Center, only facility in the U.S. to shelter European World War II Holocaust refugees.
Publications: quarterly newsletter, New Ontario Chronicle.
Hours & Admission Prices: Memorial Day to Labor Day Tues.-Sun. 11:30-4; Winter: Thurs.-Sun. 12-4. Adults $5, children $3, seniors $2, students $1; members no charge. &
Attendance: 2,500 (estimated)
Membership: Individual $20; Family $35.

TYLER ART GALLERY, State University of New York College of Arts & Science, 7060 State Rte. 104, 123 Tyler Hall, Oswego, NY 13126-3599. Tel.: 315-312-2113. Fax: 315-312-5642. Facebook: Tyler Gallery.
E-mail: michael.flanagan@oswego.edu
Web Site: www.oswego.edu/other_campus/tylerart/index.html
Congressional District: 24
Key Personnel: Dir., Michael Flanagan; Administrative Aide, Traci Terpening.
Personnel Profile: Full-Time Paid 1; Part-Time Paid 1; Interns 11.
Governing Authority: college. State University of New York, College of Arts & Science at Oswego. Tax-exempt.
Institution Type/Description: Art Gallery.
Collections: teaching collection of 19th & 20th-century European & American drawings, prints, paintings & sculpture.
Major Exhibits: Recollection: A Memory Loss Awareness Project, 1/31/14-3/2/14.
Activities: inter-museum loan exhibitions; temporary & traveling exhibitions; lectures; gallery talks.
Publications: brochures; checklists; posters.
Hours & Admission Prices: Sept.-May Tues.-Sat. 11:30-3. No charge; donations accepted. Closed during college vacation periods. &
Attendance: 18,650 (accurate)

Ovid

OVID HISTORICAL SOCIETY, 7203 Main St., Ovid, NY 14521. Mailing Address: P.O. Box 374, Ovid, NY 14521. Tel.: 607 869 5222.
Institution Type/Description: Historical Society Museum.
Collections: local history & culture; period furnishings; personal artifacts; photographs.
Hours & Admission Prices: Thurs. 10 to noon, Sat. 10-1.
Membership: Individual $5, Family $6.

Owego

TIOGA COUNTY HISTORICAL SOCIETY, (M), 110 Front St., Owego, NY 13827-1519. Tel.: 607-687-2460. Fax: 607-687-2460. Facebook: Tioga County Historical Society.
E-mail: museum@tiogahistory.org
Web Site: www.tiogahistory.org
Founded: 1914.
Congressional District: 22 & 24
Key Personnel: Exec. Dir., Kevin Lentz.
Personnel Profile: Full-Time Paid 2; Part-Time Paid 1; Part-Time Volunteers 21.
Governing Authority: nonprofit. Tax-exempt: 501(c)(3).
Institution Type/Description: History Museum.
Collections: Indian artifacts; firearms; pioneer crafts; early commerce, industry, transportation & agriculture exhibits; primitive paintings; textiles; photography.
Major Exhibits: 100 Anniversary Exhibits, 3/14-9/14.
Research Fields: genealogy; local history.
Facilities: research center; reading room.
Activities: guided tours; lectures; education programs; permanent & temporary exhibitions; summer walking tours.
Publications: monthly newsletter; books on local history; postcards.
Hours & Admission Prices: Wed.-Sat. 10-4. No charge; donations accepted. Closed New Year's Day; Independence Day; Thanksgiving; Christmas. &
Attendance: 7,168 (accurate)
Membership: Student $10; Individual $25; Family $50; Donor / Small Business $100; Corporate $200; Supporter $250; Patron $500; Gold $1,000 & up.

Oyster Bay

COE HALL, 1395 Planting Fields Rd., Oyster Bay, NY 11771-1302. Mailing Address: P.O. Box 660, Oyster Bay, NY 11771-0660. Tel.: 516-922-9210. Fax: 516-922-9226.
Web Site: plantingfields.org
Founded: 1979.
Congressional District: 3
Key Personnel: Exec. Dir., Henry Joyce; Chm. (V), Michael D. Coe; Pres. (V), G. Morgan Brown; Business Mgr., Sherley Cherenfant; Cur., Marianne Della Croce; Volunteer Coord. & Museum Shop Mgr., Katherine Sterner; Weekend Coord., Elsa Eisenberg; Coord. Membership & Dir. Devel., Patrice Panza; Field Trip Coord., Melissa Valencia; Garden Librarian, Rose Marie Papayanopulos; Asst. Cur., Kristy Caratzola; Dir. Special Events, Jennifer Lavella; Education Asst., Tracy Potavin.
Personnel Profile: Full-Time Paid 10; Part-Time Paid 2; Part-Time Volunteers 300; Interns 7.
Governing Authority: nonprofit organization. Parent Institution: Planting Fields Foundation, P.O. Box 660, Oyster Bay 11771. Tax-exempt: 501(c)(3).
Institution Type/Description: Historic House: 1918-1921 Tudor Revival Mansion.
Collections: original & period furnishings; Baroque & Renaissance paintings; archival collection of Walker & Gillette drawings; Olmsted Brothers plans & drawings; medieval stained glass; family papers & manuscripts.
Research Fields: Tudor revival architecture in England & United States with emphasis on Long Island; development of American studies programs throughout universities in the United States.
Facilities: library; 200 seat auditorium; visitor center; cafe.
Activities: guided tours; lectures; concerts; arts festivals; archives; docent program or council. Special Events: Arbor Day Festival; Dahlia Show; Winter Festival; Rhododendron Days; Fall Foliage Walks.
Publications: newsletter, Evergreen.
Hours & Admission Prices: April-Sept. daily 12-3:30; groups by appointment. Guided Tours: adults $3.50; discounts to AAM & AAA members; children 6-12 & members no charge. Arboretum: $6 per car. Closed legal holidays. &
Attendance: 200,000 (accurate)

Membership: Individual $50; Family $85; Contributor $150; Sponsor $300; Sustainer $500.

OYSTER BAY HISTORICAL SOCIETY, 20 Summit St., Oyster Bay, NY 11771-2317. Mailing Address: P.O. Box 297, Oyster Bay, NY 11771-0297. Tel.: 516-922-5032.
E-mail: obhsdirector@optonline.net
Web Site: www.oysterbayhistorical.org
Founded: 1960.
Congressional District: 3
Key Personnel: Exec. Dir., Philip Blocklyn; Pres. (V), Frank Leone; Treas., Mark De Rugeriis; Librarian & Archivist, Nicole Menchise; Museum Shop Mgr., Elizabeth Roosevelt.
Personnel Profile: Full-Time Paid 2; Part-Time Paid 3; Part-Time Volunteers 4; Interns 1.
Governing Authority: society; nonprofit. Tax-exempt: 501(c)(3).
Institution Type/Description: Historical Society Museum: housed in c.1720 Earle-Wightman House.
Collections: 18th-19th century household furnishings & decorative arts; books; manuscripts; maps; photographs; Reichman collection of early American tools & trades.
Major Exhibits: Show Days in Oyster Bar, 2/14-3/14; It's Time for Tea, 4/14-5/14.
Research Fields: Long Island history from colonial period to the 20th century with emphasis on Oyster Bay.
Facilities: 1,150-vol. library containing 2,000 manuscripts, 2,500 photographs, 75 atlases & maps pertaining to local area history available for public use; botanical garden.
Activities: guided tours; lectures; films; concerts; organized education programs; participatory & temporary exhibitions.
Publications: books, The Diary of Mary Cooper: Life on a Long Island Farm 1768-1773; The Walls Have Tongues: Oyster Bay Houses and their Stories; Walking Tour of Old Oyster Bay; quarterly magazine, Out of the Box; What Kind of Noise Annoys An Oyster: An Oyster Songster.
Hours & Admission Prices: Tues.-Fri. 10-2, Sat. 11-3, Sun. 1-4. No charge; donations requested. Closed major holidays. ⅗
Attendance: 4,000 (estimated)
Membership: Individual $35; Family $45; Contributing $75; Sponsor $100; Sustaining $250; Patron $500; Benefactor $1,000 & up. Business $75; Business Sponsor $100; Business Friend $300; Business Patron $500 & up.

OYSTER BAY RAILROAD MUSEUM, 102 Audrey Ave., Oyster Bay, NY 11771. Mailing Address: P.O. Box 335, Oyster Bay, NY 11771. Tel.: 516-558-7036.
Web Site: www.obrm.org
Governing Authority: Tax-exempt.
Institution Type/Description: Railroad Museum.
Collections: railroad history & artifacts; photographs.
Facilities: Museum-related items for sale.
Hours & Admission Prices: Call for hours.
Attendance: 2,000 (estimated)

PLANTING FIELDS ARBORETUM STATE HISTORIC PARK, 1395 Planting Fields Rd., Oyster Bay, NY 11771-1302. Mailing Address: P.O. Box 58, Oyster Bay, NY 11771-0058. Tel.: 516-922-8600. Fax: 516-922-8610.
Web Site: www.plantingfields.org
Founded: 1955.
Congressional District: 3
Key Personnel: Exec. Dir., Henry B. Joyce; Chm. Planting Fields Foundation, G. Morgan Browne; Dir. Mktg. & Special Events, Jennifer L. Lavella; Cur., Gwendolyn L. Smith.
Governing Authority: Parent Institution: New York State Office of Parks, Recreation & Historic Preservation. Subsidiary Institution: Planting Fields Foundation. Tax-exempt: 509(a)(3).
Institution Type/Description: Arboretum.
Collections: rhododendron species & cultivars including deciduous and evergreen azaleas; herbarium; specimen trees; deciduous magnolias; synoptic shrub garden; greenhouse collections of camellias, orchids, cacti & economic plants. Historic Building: Coe Hall.
Research Fields: arboretum.
Facilities: 5,000-vol. library of horticultural & botanical books; nature trails; formal gardens; tea house.
Activities: guided tours; lectures; films; formally organized educational programs; temporary exhibitions.
Publications: quarterly newsletter.
Hours & Admission Prices: Daily 9-5. $8 per car, April 1-Labor Day; no charge during winter. Coe Hall: see separate listing. Closed Christmas Day. ⅗

Attendance: 262,360 (estimated)
Membership: Individual $50; Dual/Family $85; Contributor $150; Sponsor $300; Sustainer $500; Conservator $1,000; President's Circle $5,000 & up.

✳ **RAYNHAM HALL MUSEUM, (M),** 20 W. Main St., Oyster Bay, NY 11771-2216. Tel.: 516-922-6808. Fax: 516-922-7640.
E-mail: info@raynhamhallmuseum.org
Web Site: www.raynhamhallmuseum.org
Founded: 1953.
Congressional District: 3
Key Personnel: Exec. Dir., Harriet Gerard Clark; Asst. Dir., Theresa Skvarla; Pres. (V), Kay Hutchins Sato; Coord. Education, Jessica M. Semins; Collections Mgr., Jennifer Ladd.
Personnel Profile: Full-Time Paid 2; Part-Time Paid 8; Part-Time Volunteers 30; Interns 1.
Governing Authority: nonprofit organization. Friends of Raynham Hall, Inc. Tax-exempt: 501(c)(3).
Institution Type/Description: c.1738 Historic House Museum: 1851 Gothic Revival addition.
Collections: 18th & 19th-century household furnishings & decorative arts; textiles; archives; Townsend family furnishings; United States military; costumes; toys.
Research Fields: Townsend family; life in Oyster Bay, 1738-1880.
Facilities: Museum-related items for sale.
Activities: guided tours; audio tour of house & grounds; educational programs; lecture series; temporary exhibitions; local historic preservation; cooking & craft programs for boy and girl scouts.
Publications: brochures; annual report; quarterly newsletter; historic comic book-based.
Hours & Admission Prices: Memorial Day to Labor Day Sat.-Sun. 10-5; Sept.-May Tues.-Sun. 1-5; group tours by appointment. Adults $5, students with ID & senior citizens $3; discounts to AAM members; children under 6 & members no charge. Closed New Year's Day; Thanksgiving; Christmas; most national holidays.
Attendance: 10,032 (accurate)
Membership: Junior Culper, Jr. $5; Student $10; Individual $35; Family $45; Business Friend $75; 1740 Society & Business Sponsor $100; Culper Spy Ring & Business Benefactor $250; Victorian Society & Business Patron $500; Raynham Society $1,000.

SAGAMORE HILL NATIONAL HISTORIC SITE, 20 Sagamore Hill Rd., Oyster Bay, NY 11771-1899. Tel.: 516-922-4788. Fax: 516-922-4792.
E-mail: sahi_information@nps.gov
Web Site: www.nps.gov/sahi
Founded: 1963.
Congressional District: 3
Key Personnel: Supt., Thomas Ross; Cur., Amy Verone; Museum Shop Mgr., Debbie Bulck.
Personnel Profile: Full-Time Paid 16; Part-Time Paid 10; Part-Time Volunteers 35; Interns 2.
Governing Authority: federal. Administered by the National Park Service, U.S. Dept. of Interior, Washington, DC 20240. Tax-exempt.
Institution Type/Description: Historic House Museum: 1885 Sagamore Hill, home of Theodore Roosevelt.
Collections: personal & household effects of Theodore Roosevelt & Roosevelt family.
Research Fields: life & presidency of Theodore Roosevelt.
Facilities: 350-vol. library of books written by or about Theodore Roosevelt available for study by prior arrangement with supt.
Activities: ranger-led tours of Theodore Roosevelt's home; self-guided tour of Old Orchard Museum; audiovisual programs.
Publications: guide book, Sagamore Hill.
Hours & Admission Prices: Memorial Day to Labor Day daily 9:30-5; Winter: Wed.-Sun. 9:30-5; groups by appointment. Adults $5; children under 16 & scheduled educational groups no charge. Closed New Year's Day; Thanksgiving; Christmas. ⅗
Attendance: 65,000 (accurate)
Membership: Associate $35; Family $65; Sustaining $100; Patron $250; Sponsor $500; Benefactor $1,000; Corporate $5,000.

Painted Post

PAINTED POST - ERWIN MUSEUM AT THE DEPOT, 277 Steuben St., Painted Post, NY 14870. Mailing Address: 73 W. Pulteney St., Corning, NY 14830-2212. Tel.: 607-654-7981; 937-5281 (for appts.). Fax: 607-937-5281.
E-mail: cpphs@stny.rr.com
Web Site: www.pattersonmuseum.org

Founded: 1945.
Congressional District: 23
Key Personnel: Pres., Jane Spillman; Vice Pres., Virginia Wright; Dir., Sheri Golder.
Personnel Profile: Part-Time Paid 1; Part-Time Volunteers 5.
Governing Authority: municipal. Parent Institution: Corning-Painted Post Historic Society. Subsidiary Institution: Conning Painted Post Historical Society. Tax-exempt.
Institution Type/Description: General Museum.
Collections: Indian lore; historic artifacts; newspapers; books; 1930's local history, composed of artifacts related to regional land formation, occupation, settlement, commerce, industry, transportation, recreation, society, tradition.
Research Fields: Local & Indian history; Stenben County census records 1825-1925, painted Post, Erwin & regional history.
Facilities: 400-vol. library of local, state & national history available for research on premises; reading room.
Activities: self-guided tours.
Publications: Erwin Town Cemeteries Records 1793-1979.
Hours & Admission Prices: June-Aug. Mon.-Fri. 10-4, Sat. 10-2; Sept.-May call for appointment; No charge, donations accepted. &
Attendance: 500 (estimated)
Membership: Student $15; Senior $20; Individual $30; Family $50; Donor $75; Patron $100; Corporate $250; Benefactor $500.

Palmyra

ALLING COVERLET MUSEUM, 122 William St., Palmyra, NY 14522-1030. Mailing Address: 132 Market St., Palmyra, NY 14522-1136. Tel.: 315-597-6981. Fax: 315-597-6981.
E-mail: bjfhpinc@rochester.rr.com
Web Site: www.historicpalmyrany.com
Founded: 1976.
Congressional District: 29
Key Personnel: Chm., Becke Tomkiewicz; Exec. Dir., Bonnie J. Hays; Museum Shop Mgr., Steve Hays.
Personnel Profile: Full-Time Paid 1; Part-Time Paid 2; Part-Time Volunteers 12.
Governing Authority: Parent Institution: Historic Palmyra, Inc. Tax-exempt.
Institution Type/Description: American Coverlet Museum.
Collections: coverlets; quilts; looms; weaving materials.
Major Exhibits: Gettysburg After 150 Years, 4/14-10/14.
Facilities: Textile-related books & items for sale.
Activities: tours; hands-on activities
Publications: pamphlets; books.
Hours & Admission Prices: May-Oct. Mon. 1-4, Tues.-Sat. 10:30-4:30; Nov.-April Mon.-Sat. 1-4. No charge; donations accepted. &
Attendance: 13,000 (accurate)
Membership: Students $5; Senior Citizens $10; Individual $15; Family $25; Organization $35; Supporting & Sustaining $50; Patron $100; Corporate $150; Benefactor $1,000 & over.

ERIE CANAL DEPOT MUSEUM, 140 1/2 Market St., Palmyra, NY 14522. Mailing Address: 132 Market St., Palmyra, NY 14522-1136. Tel.: 315-597-6981. Fax: 315-597-6981.
Web Site: www.historicpalmyrany.com
Formerly: Historic Palmyra's Print Shop
Founded: 1967.
Congressional District: 29
Key Personnel: C.E.O. & Chm. (V), Becke Tomkiewicz; Exec. Dir., Bonnie J. Hays; Museum Shop Mgr., Steve Hays.
Personnel Profile: Full-Time Paid 1; Part-Time Paid 2; Part-Time Volunteers 12.
Governing Authority: nonprofit. Parent Institution: Historic Palmyra. Subsidiary Institutions: Alling Coverlet Museum, Phelps Store Museum, Palmyra Historical Museum. Tax-exempt: 501(c)(3).
Institution Type/Description: History Museum: housed in an Erie Canal store & home.
Collections: textiles; decorative arts; weaving & spinning equipment; country store merchandise; home furnishings; dolls & toys; tools; quilts; 1800s printing press & cutters; printing blocks; type; advertisements; books; typewriters; comptometers; signs; lead type equipment.
Research Fields: textiles; history of Palmyra.
Facilities: 1,500-vol. library, available for use on premises. Books & items on weaving & textiles for sale.
Activities: guided tours; lectures; concerts; permanent & temporary exhibitions; school tours; cemetery tours; monthly programs.
Publications: newsletter; brochure; local information sheet.
Hours & Admission Prices: May-Oct. Tues.-Sat. 10:30-4:30; Nov.-April Tues.-Thurs. 11-4. Individual Museum: adults $3; members no charge. Trail

Pass: family $14, adults $7, students & senior citizens $5; children 10 & under and members no charge. &
Attendance: 13,000 (estimated)
Membership: Students $5; Senior Citizens $10; Individual $15; Family $25; Organization $35; Supporting & Sustaining $50; Patron $100; Benefactor $1,000 & over.

HILL CUMORAH VISITORS CENTER & HISTORIC SITES, 603 State Rte. 21, Palmyra, NY 14522-9301. Tel.: 315-597-5851.
E-mail: rsearle@ldschurch.org
Web Site: www.hillcumorah.org
Founded: 1830.
Congressional District: 33
Key Personnel: Dir., Richard Searle.
Personnel Profile: Full-Time Volunteers 44; Part-Time Volunteers 20.
Governing Authority: church; The Church of Jesus Christ of Latter-day Saints (Mormon), 50 E. North Temple, Salt Lake City, UT 84150. Tax-exempt.
Institution Type/Description: Religious Museums & Historic Sites: 5 historical sites & places of historic significance in connection with the founding of the Church of Jesus Christ of Latter-day Saints.
Collections: household furnishings typical of 1820s western New York. Historic Structures and Sites: c.1825, home of Joseph Smith, founder of the LDS church; The Sacred Grove; c.1825, The Grandin Building, 217 E. Main St.; c.1850, Martin Harris House, Maple Ave.; c.1805, Peter Whitmer log farm house, Aunkst Rd., Waterloo.
Research Fields: LDS church history.
Facilities: visitors' center.
Activities: guided tours of visitor center. Annual Event: Hill Cumorah Pageant, America's Witness For Christ in July.
Publications: books, Book of Mormon; Doctrine and Covenants; pamphlets.
Hours & Admission Prices: Mon.-Sat. 9-9, Sun. 12:30-9. No charge. Closed Christmas. &
Attendance: 150,000 (estimated)

HISTORIC PALMYRA'S - WM. PHELPS GENERAL STORE, 140 Market St., Palmyra, NY 14522-1136. Mailing Address: 132 Market St., Palymra, NY 14522-1136. Tel.: 315-597-6981. Fax: 315-597-6981.
E-mail: bjfhpinc@rochester.rr.com
Web Site: www.historicpalmyrany.com
Formerly: Historic Palmyra Inc.
Founded: 1964.
Congressional District: 29
Key Personnel: Chm. & Pres. (V), Becke Tomkiewicz; Exec. Dir., Bonnie J. Hays; Museum Shop Mgr., Steve Hays.
Personnel Profile: Full-Time Paid 1; Part-Time Paid 2; Part-Time Volunteers 12.
Governing Authority: Parent Institution: Historic Palmyra, Inc. Tax-exempt.
Institution Type/Description: Erie Canal Store & Home: housed in c.1826 building.
Collections: period fixtures & furniture; personal artifacts; archaeological. Historic Building: general store.
Research Fields: local archives; industry; religion; local residents; business.
Activities: school tours; cemetery tours; monthly programs.
Publications: newsletter; brochure; local information sheet.
Hours & Admission Prices: May-Oct. Tues.-Sat. 10:30-4:30; Nov.-April Tues.-Thurs. 11-4. Individual Pass: adults $3; members no charge. Trail Pass: family $14, adults $7, students & senior citizens $5; children 10 & under and members no charge. Closed Independence Day; Labor Day. &
Attendance: 13,000 (accurate)
Membership: Student $5; Senior Citizen $10; Individual $15; Family $25; Organization $35; Supporting & Sustaining $50; Patron $100; Corporate $150; Benefactor $100 & up.

PALMYRA HISTORICAL MUSEUM, 132 Market St., Palmyra, NY 14522-1136. Tel.: 315-597-6981. Fax: 315-597-6981.
Web Site: www.historicpalmyrany.com
Key Personnel: C.E.O. & Chm. (V), Becke Tomkiewicz; Exec. Dir., Bonnie J. Hays; Museum Shop Mgr., Steve Hays.
Personnel Profile: Full-Time Paid 1; Part-Time Paid 2; Part-Time Volunteers 12.
Governing Authority: Parent Institution: Historic Palmyra. Tax-exempt: 501(c)(3).
Institution Type/Description: Historic Building: housed in a former 26 room hotel, c.1826.
Collections: period furnishings; personal artifacts; photographs.
Activities: ghost hunts; cemetery walks; history programs; UGRR programs.
Publications: newsletters; walking brochure.
Hours & Admission Prices: May-Oct. Tues.-Sat. 10:30-4:30; Nov.-April Tues.-Thurs. 11-4. Individual Museum: adults $3; members no charge. Trail

Pass: family $14, adults $7, student & senior citizen $5; children 10 & under and members no charge.
Attendance: 13,000
Membership: Students $5; Senior Citizens $10; Individual $15; Family $25; Organization $35; Supporting & Sustaining $50; Patron $100; Corporate $150; Benefactor $1,000 & over.

Parishville

PARISHVILLE MUSEUM, 1785 Main St., Parishville, NY 13672. Mailing Address: Box 534, Parishville, NY 13672-0534. Tel.: 315-265-7619.
Founded: 1964.
Key Personnel: Pres., Joseph McGill; Vice Pres., Sherry Remington.
Personnel Profile: Part-Time Paid 1.
Governing Authority: society. Tax-exempt: 501(c)(3).
Institution Type/Description: History Museum: located in 1800 home.
Collections: items pertaining to early days of Parishville; 1850 & 1865 census; cemetery listings for each cemetery; obituary scrap books.
Research Fields: genealogy.
Facilities: library of history books available for research on premises.
Activities: guided tours; permanent exhibitions.
Publications: sketches of Parishville.
Hours & Admission Prices: July-Aug. Tues. & Thurs. 1-3; other times by appointment. No charge; donations accepted.
Attendance: 250 (estimated)
Membership: Annual $2; Life $15.

Pawling

GUNNISON MUSEUM OF NATURAL HISTORY, 378 Old Quaker Hill Rd., Pawling, NY 12564-3449. Mailing Address: P.O. Box 345, Pawling, NY 12564-0345. Tel.: 845-855-5099.
Founded: 1960.
Key Personnel: Pres., Elizabeth P. Allen; Cur., Mrs. James Mandracchia.
Personnel Profile: Part-Time Paid 1.
Governing Authority: Akin Hall Association. Tax-exempt: 501(c)(3).
Institution Type/Description: Natural History Museum.
Collections: Gunnison collection of rocks, minerals, native birds & flora; agriculture; anatomy; entomology; geology; Indian artifacts; African & Asian musical instruments; mineralogy; natural history.
Facilities: 100-vol. library of natural history reference books available for use on premises.
Hours & Admission Prices: May-Oct. Fri.-Sun. 1-4. No charge; donations accepted.
Attendance: 1,000 (accurate)

HISTORICAL SOCIETY OF QUAKER HILL & PAWLING, 126 E. Main St., Pawling, NY 12564-1428. Mailing Address: P.O. Box 99, Pawling, NY 12564-0099. Tel.: 845-855-9316.
Web Site: www.pawling-history.org
Formerly: Historical Society of Quaker Hill & Pawling/John Kane House
Founded: 1910.
Congressional District: 19
Key Personnel: Pres. (V), John Brockway; Membership, Charlotte Whaley; Museum Shop Mgr., Mrs. Jeanne Kelly.
Personnel Profile: Part-Time Volunteers 84.
Governing Authority: private; nonprofit organization. Tax-exempt: 501(c)(3).
Institution Type/Description: Historical Society.
Collections: John Kane House: George Washington's headquarters in Autumn 1778, memorabilia from Lowell Thomas, radio broadcaster, lecturer & world traveler; the 1764 Oblong Meeting House: Gen. Washington used as a hospital for his troops in 1778; Akin Free Library: containing a Quaker Museum & Natural History Museum.
Facilities: library; botanical garden. Gift items for sale.
Activities: arts festivals; docent program; lectures; temporary exhibitions; Heirloom Day, bring your antiques to be appraised; Quilt Show; Art Exhibition, local artists. Annual Events: Tag Sale; Holiday Open House.
Publications: three times a year newsletter, Historical Society Newsletter.
Hours & Admission Prices: May 15-Oct. 15 Sat.-Sun. 2-4. Special opening 2nd weekend in Dec. No charge; donations accepted.
Attendance: 418 (accurate)

Peekskill

THE PEEKSKILL MUSEUM, 124 Union Ave., Peekskill, NY 10566-3429. Mailing Address: P.O. Box 84, Peekskill, NY 10566-0084. Tel.: 914-736-0473.
Web Site: www.peekskillmuseum.com
Founded: 1946.
Congressional District: 21
Key Personnel: Pres. (V), John Curran; Vice Pres., William Stillman; Treas., Paula Connolly; Corresponding Sec., Dolores Ubben.
Personnel Profile: Part-Time Volunteers 10.
Governing Authority: nonprofit organization. Tax-exempt.
Institution Type/Description: Historic Building & History Museum: housed in 1876-77 Dwight Herrick House.
Collections: Peekskill memorabilia from Colonial period to present day; paintings; furniture; delftware; tureens; prints; period dresses; spinning wheels & looms; local newspapers; local stoves & plows; Revolutionary War cannon. Historic House: 1870s William Rutherford Mead house.
Research Fields: coal stove manufacturers; local & county history 1830-present.
Facilities: archives; research room.
Activities: lectures; formally organized education programs for children; permanent & temporary exhibitions.
Publications: quarterly newsletter, The Peekskill Museum.
Hours & Admission Prices: April-Oct. Sat. 1-4; Nov.-March Sat. 1-3. Adults $2, children $1; members no charge. &
Attendance: 500
Membership: Senior Citizen & Student $15; Family $25; Supporting $50; Benefactor $100; Corporate, Business & Contributor $150; Donor $250; Patron $500 and up.

Pelham

PELHAM ART CENTER, 155 Fifth Ave., Pelham, NY 10803-1503. Tel.: 914-738-2525. Fax: 914-738-2686.
E-mail: info@pelhamartcenter.org
Web Site: www.pelhamartcenter.org
Founded: 1970.
Congressional District: 20
Key Personnel: Dir., Lynn Honeysett; Chm., Barbara Carden; Pres., Anna Riehl.
Personnel Profile: Full-Time Paid 2; Part-Time Paid 3; Part-Time Volunteers 3; Interns 5.
Governing Authority: nonprofit. Tax-exempt: 501(c)(3).
Institution Type/Description: Art Center.
Collections: annual visual art & craft exhibitions; all forms of art media represented.
Facilities: 3 educational studios.
Activities: guided tours; lectures; concerts; art classes; outreach programs; performances.
Publications: class catalogs; summer camp catalog; calendar of events; Neighbor to Neighbor Discount Program; Birthday Flyer; Invitation to Exhibits.
Hours & Admission Prices: Tues.-Fri. 12-5, Sat. 10-4. No charge; donations accepted. Closed New Year's Day; Memorial Day; Independence Day; Columbus Day; Thanksgiving; Christmas. &
Attendance: 13,000 (estimated)
Membership: Senior & Educator $30; Individual $45; Family $70; Patron $125; Sponsor $250; Benefactor $500; Exhibition Angel $1,000.

Penn Yan

THE AGRICULTURAL MEMORIES MUSEUM, 1110 Townline Rd., Penn Yan, NY 14527-9002. Tel.: 315-536-1206.
E-mail: jrjensen@copper.net
Web Site: www.agriculturalmemoriesmuseum.com
Founded: 1997.
Key Personnel: Owner, Jennifer R. Jensen; Asst., Hilbert J. Jensen.
Governing Authority: private.
Institution Type/Description: Agricultural Museum.
Collections: horse-drawn carriages & sleighs; period tractors; gasoline engines; toys; signs; pedal tractors & cars.
Research Fields: collection.
Facilities: library; 6,500 sq. ft. exhibit space.
Activities: films; guided tours; lectures.
Hours & Admission Prices: June-Oct. by appointment. Adults $4, students & children 2-12 $1; discounts to groups; children under 2 no charge.
Attendance: 250 (estimated)

YATES COUNTY GENEALOGICAL AND HISTORICAL SOCIETY AND OLIVER HOUSE MUSEUM AND L. CAROLINE UNDERWOOD MUSEUM, 107 Chapel St., Penn Yan, NY 14527-1128. Tel.: 315-536-7318. Fax: 315-536-0976.
E-mail: ycghs@yatespast.org
Web Site: www.yatespast.org
Founded: 1860.

Congressional District: 31
Key Personnel: Exec. Dir, John Potter; Association Pres., Doug Erwin; Cur., Charles Mitchell; Administrative Asst., Lisa Harper.
Personnel Profile: Part-Time Paid 4; Part-Time Volunteers 50.
Governing Authority: society; nonprofit. Yates County Genealogical & Historical Society. Tax-exempt: 501(c)(3).
Institution Type/Description: History Museum.
Collections: 19th-century home furnishings; period rooms; costumes; portraits; photographs; manuscripts; maps & documents; tools & agricultural implements; material relating to Jemima Wilkinson; Indian artifacts; 20th century local history; genealogy. Historic Houses: 1852 Oliver House; c.1920s home.
Research Fields: historic preservation; genealogical; architecture & archaeology; cultural impact of Indians & early white settlers on area; county history.
Facilities: research center library, archival material & collections in storage available for research. Publications & gift items for sale.
Activities: guided tours; permanent & temporary exhibitions; slide shows; school outreach service; historic craft classes; genealogical classes; lectures; historical programs.
Publications: bimonthly newsletter, Yates Past.
Hours & Admission Prices: Tues.-Fri. 9-4. Museum: no charge; donations accepted. Research Center: $5 per hr.; discounts to YCGHS members. Closed major holidays. &
Attendance: 5,000 (estimated)
Membership: Individual $20; Family $45; Friend $100; Business Partner $125; Benefactor $200; Historian $500.

Piffard

TOWN OF YORK HISTORICAL SOCIETY - WARREN HOMESTEAD, 2431 Dow Rd., Piffard, NY 14533. Mailing Address: P.O. Box 464, York, NY 14592-0464. Tel.: 585-243-2027.
E-mail: gcox8@rochester.rr.com
Web Site: www.yorkwines.org
Founded: 1989.
Key Personnel: Pres. (V), Gary A. Cox.
Personnel Profile: Part-Time Volunteers 9.
Governing Authority: Tax-exempt.
Institution Type/Description: Historical Society Museum: housed in the former home of Samuel Warren, a successful commercial winemaker of the Finger Lakes & New York state; built c.1830.
Collections: local history & culture; period furnishings; personal artifacts; photographs.
Publications: newsletter.
Hours & Admission Prices: Spring to Fall 1st & 3rd Sun. each month. No charge; donations accepted.
Attendance: 200 (estimated)
Membership: Youth 13-18 $5; Individual $10; Family $20; Business $50.

Pittsford

HISTORIC PITTSFORD, 18 Monroe Ave., Pittsford, NY 14534-1928. Tel.: 585-381-2941.
E-mail: historicpittsford@gmail.com
Web Site: www.historicpittsford.com
Founded: 1966.
Congressional District: 30
Key Personnel: Dir., Peggy Brizee.
Personnel Profile: Part-Time Paid 1; Part-Time Volunteers 18.
Governing Authority: nonprofit corp. Tax-exempt: 501(c)(3).
Institution Type/Description: Historic House: 1820 lawyer's office.
Collections: local history & culture; period furnishings.
Research Fields: local architecture.
Facilities: 200-vol. library of old law books available for research by appointment. Books & museum-related items for sale.
Activities: lectures; formally organized education programs for adults; architectural consultant program; permanent exhibitions.
Publications: biannual, Historic Pittsford Newsletter; folder, A Walking Tour Map of Historic Pittsford; book, Architecture Worth Saving in Pittsford, Elegant Village; Pittsford Scrapbook (stories of early Pittsford); Echoes, Pittsford's History; Images of America Pittsford.
Hours & Admission Prices: Wed. & Sat. 9-12; other times by appointment. No charge.
Attendance: 350 (estimated)
Membership: Family $25; Business, Professional & Patron $100; Life $300.

Plattsburgh

CLINTON COUNTY HISTORICAL ASSOCIATION & MUSEUM, 98 Ohio Ave., Plattsburgh, NY 12903-4401. Tel.: 518-561-0340. Fax: 518-561-0340.
E-mail: director@clintoncountyhistorical.org
Web Site: www.clintoncountyhistorical.org
Founded: 1945.
Congressional District: 24
Key Personnel: Pres., Helen Nerska; Dir., Melissa A. Peck; Vice Pres., William Laundry; Sec., Jan Couture; Treas., Maurica Gilbert; Museum Shop Mgr., James Bailey.
Personnel Profile: Full-Time Paid 1; Part-Time Volunteers 7.
Governing Authority: nonprofit organization. Parent Institution: Clinton County Historical Association. Tax-exempt: 501(c)(3).
Institution Type/Description: Historical Society Museum.
Collections: local history; paintings; maps; firearms; glass; ceramics; textiles; photographs; furniture; costumes; period clothing; negatives of late 19th & early 20th-century citizens & scenery.
Major Exhibits: Catholic Summer School of America, 4/14-12/14.
Research Fields: Redford glass; iron mining in Clinton County; 18th & 19th-century military & naval warfare in the Champlain Valley; maritime culture of Lake Champlain.
Facilities: Local history publications for sale.
Activities: guided tours; lectures; docent program; organized education programs for adults, children & undergraduate or graduate college students affiliated with State University College at Plattsburgh; loan & temporary exhibitions.
Publications: semiannual newsletter, North Country Notes.
Hours & Admission Prices: Wed.-Sat. 10-3. Adults $5, seniors $3, students $2; discounts to AAA, AAM & ICOM members; school groups, Historical Association & members no charge. Closed legal holidays. &
Attendance: 1,000 (estimated)
Membership: Student $35; Family $55; Friend $100; Patron $165; Sponsor $350; Sustaining $650; Life $1,500.

KENT-DELORD HOUSE MUSEUM, (M), 17 Cumberland Ave., Plattsburgh, NY 12901-1849. Tel.: 518-561-1035. Facebook: Kent - Delord House Museum.
E-mail: kentdelord@primelink1.net
Web Site: www.kentdelordhouse.org
Founded: 1924.
Congressional District: 23
Key Personnel: Pres., Trevor Laughlin; Dir., Donald Wickman.
Personnel Profile: Part-Time Paid 1; Part-Time Volunteers 25.
Governing Authority: Kent-Delord Corporation. Tax-exempt.
Institution Type/Description: Historic House Museum: c.1797, Kent-Delord home.
Collections: period furnishings; paintings; 18th & 19th century decorative arts; manuscripts; photographs; textiles.
Research Fields: War of 1812; Civil War; Woman's Christian Temperance Union; 19th & early 20th century social work; domestic life in the 19th-century.
Facilities: off site access to manuscript collection.
Activities: guided tours; special exhibits; cultural events; educational programs; garden club.
Publications: book, Henry Delord & His Family; newsletter/journal, The Kent-Delord Quarterly; Love & Duty: Letters & Diaries of the Delord-Webb Women.
Hours & Admission Prices: Tues.-Fri. 12-2, Sat. call for hours; guided tours & group tours by appointment. Adults $5, students $3, children under 12 $2; discounts for groups; members no charge. &
Attendance: 7,500 (estimated)
Membership: Quartermaster Club $35; Bellevue Avenue Club $60; Carriage Club $100; Third Century Society $300.

PLATTSBURGH STATE ART MUSEUM S.U.N.Y., (M), State University of New York, 101 Broad St., Plattsburgh, NY 12901-2637. Mailing Address: State University of New York, Myers Room 235, 101 Broad St., Plattsburgh, NY 12901-2637. Tel.: 518-564-2474 & 2178. Fax: 518-564-2473.
E-mail: ceil.esposito@plattsburgh.edu
Web Site: clubs.plattsburgh.edu/museum
Founded: 1952.
Congressional District: 26
Key Personnel: C.E.O. & College Pres., John Ettling; Dir., Cecilia M. Esposito; Mgr. Collections, David Driver.
Personnel Profile: Full-Time Paid 4; Part-Time Paid 1; Part-Time Volunteers 57; Interns 4.

Governing Authority: university. Parent Institution: State University of New York. Tax-exempt: 170(b)(1)(A).

Institution Type/Description: Art Museum & Galleries.

Collections: paintings, drawings, prints & sculptures; Rockwell Kent collection including 36 paintings, 100 prints, 1,500 drawings, sketches, proofs & designs; books owned by Kent; books about the artist; a set of first editions written or illustrated by Kent; bookplates, stationery, trademarks; Christmas cards; commercial illustration proofs; exhibition catalogs; Nina Winkel sculptures; Millet Asian collection; Ackerman modern art; Regina Slatkin art; 19th-century drawing, prints & sculptures.

Research Fields: Rockwell Kent; Nina Winkel; American Art 20th & 21st centuries.

Facilities: changing exhibition gallery; permanent collection gallery; work room, curatorial room; special collections reference room where Kent material is available for research upon request; Winkel sculpture court; Regina Slatkin Art Collection study room.

Activities: guided tours; visiting artist lectures; gallery talks; formally organized education programs for undergraduate students at the university, also available to elementary, secondary schools & community groups in the geographic area; 24 exhibitions each year; antique & contemporary, all media.

Publications: exhibition catalogs; quarterly journal, Kent Collector; monthly exhibition announcements; semiannual calendar of events.

Hours & Admission Prices: Jan. 2-Dec. 23 daily 12-4. No charge; donations accepted. Closed legal holidays. &

Attendance: 20,000 (estimated)

Pocantico Hills

HISTORIC HUDSON VALLEY, 639 Bedford Rd., Pocantico Hills, NY 10591. Tel.: 914-631-8200. Fax: 914-631-0089. Historic Hudson Valley.

E-mail: info@hudsonvalley.org

Web Site: www.hudsonvalley.org

Founded: 1951.

Congressional District: 22

Key Personnel: Chm. (V), Michael Hegarty; Pres., Waddell W. Stillman; Dir. Finance & Administration, David M. Parsons; Dir. Public Rels., Rob Schweitzer; Dir. Bldgs., Grounds & Security, Brian Blaney; Dir. Retail Sales, Henri Corbacho.

Personnel Profile: Full-Time Paid 56; Part-Time Paid 144; Part-Time Volunteers 430; Interns 10.

Governing Authority: private operating foundation. Tax-exempt: 501(c)(3).

Institution Type/Description: Preservation Project.

Collections: maps; prints; photographs; architectural plans; slides & films; fine & decorative arts; manuscripts & memorabilia. Historic Houses: mid-18th century Philipsburg Manor, Upper Mills, Sleepy Hollow, NY; 1790-1812 Van Cortlandt Manor, Croton-on-Hudson; 1835-1859 Sunnyside, Tarrytown home of author Washington Irving; 1805-1985 Montgomery Place, Annandale on Hudson; Union Church of Pocantico Hills, housing stained glass works of art by Matisse & Chagall; Kykuit, the Rockefeller Estate (1900-1913).

Research Fields: life in the lower Hudson Valley.

Facilities: 30,000-vol. library of books; 100-seat auditorium. Books & museum-related items for sale.

Activities: guided tours; films; formally organized education programs; demonstrations; special events; permanent & temporary exhibitions; inter-museum loans; interpretive programming; visitation program for Kykuit, the Rockefeller house and gardens, a property of the National Trust for Historic Preservation.

Publications: books; booklets; exhibition catalogs.

Hours & Admission Prices: See website for up-to-date tour times & admission prices. &

Attendance: 243,000 (accurate)

Membership: Senior 62 & over $50; Individual $60; Dual $100; Family $160; Family Plus $220; Premier $400; Premier Plus $750; Henry Hudson Society $1,000; Pocantico Society $2,500-$5,000.

Port Jefferson

HISTORICAL SOCIETY OF GREATER PORT JEFFERSON, 115 Prospect St., Port Jefferson, NY 11777-1812. Mailing Address: P.O. Box 586, Port Jefferson, NY 11777-0586. Tel.: 631-473-2665.

E-mail: info@portjeffhistorical.org

Web Site: www.portjeffhistorical.org

Founded: 1967.

Congressional District: 1

Key Personnel: Pres. (V), Nick Acampora; Museum Shop Co-Mgr., Eileen Coen.

Personnel Profile: Part-Time Paid 3; Part-Time Volunteers 100.

Governing Authority: nonprofit organization. Tax-exempt.

Institution Type/Description: Maritime Museum: housed in c.1840 John R. Mather homestead & out buildings.

Collections: maps; books; half hulls; paintings; sailmaker's tools & loft; shipbuilding artifacts & tools; tin items; looms; c.1900 diorama of Port Jefferson Harbor; maritime art, photographs, artifacts & memorabilia depicting the shipping industry of Port Jefferson; country store, butcher shop & barber shop replicas.

Research Fields: marine history; local artists and history.

Facilities: 175-vol. library of records, deeds, logs, ships records, maps, craft house, perennial garden, books, histories & newspapers available for inter-library loan & by appointment; gardens.

Activities: permanent & temporary exhibitions; educational tours for children; regular meetings.

Publications: catalogue; booklet, Port Jefferson-Story of a Village.

Hours & Admission Prices: Memorial Day to June Sat.-Sun. 1-4; July-Aug. Tues.-Wed. & Sat.-Sun. 1-4. Family $5, adults $3; children & members no charge. &

Attendance: 1,600 (estimated)

Membership: Single $25; Family, Business, & Professional $35; Contributing $50.

Port Jervis

MINISINK VALLEY HISTORICAL SOCIETY, 125-133 W. Main St., Port Jervis, NY 12771. Mailing Address: P.O. Box 659, Port Jervis, NY 12771-0659. Tel.: 845-856-2375. Fax: 845-856-1049.

E-mail: history@minisink.org

Web Site: www.minisink.org

Founded: 1889.

Congressional District: 20

Key Personnel: Exec. Dir., Peter Osborne, III; Pres., Robert Shultz; Treas., Nancy Conod.

Personnel Profile: Full-Time Paid 1; Part-Time Volunteers 15.

Governing Authority: nonprofit organization. Tax-exempt: 501(c)(3).

Institution Type/Description: Historical Society Museum: housed in 1793 stone home.

Collections: genealogical data; family files; photographs; local history.

Research Fields: Delaware & Hudson Canal; railroads; Civil War; Indians.

Facilities: 50,000-vol. library for public use. Books and other related items for sale.

Activities: guided tours; lectures; broadcast programs; loan, temporary & traveling exhibitions. Society Sponsors: Christmas program for local groups.

Publications: triannual newsletter.

Hours & Admission Prices: Library: Thurs. 1-4; other times by appointment. Suggested Donation $4; discounts to members. Museum: July-Oct. Sat. 10-4. No charge; donations accepted. &

Attendance: 15,000

Membership: Individual $15; Family $20; Business $50.

Port Washington

COW NECK PENINSULA HISTORICAL SOCIETY SANDS-WILLETS HOUSE, 336 Port Washington Blvd., Port Washington, NY 11050-4530. Tel.: 516-365-9074.

E-mail: info@cowneck.org

Web Site: www.cowneck.org

Founded: 1962.

Congressional District: 6

Key Personnel: Pres. (V), Fred Blumlein; Treas., Richard Coyle; Cur., Harrison Hunt; Dir. Education, Mary Alice Puglise; Museum Shop Mgr., Evelyn Fitzsimmons.

Personnel Profile: Part-Time Paid 3; Part-Time Volunteers 25.

Governing Authority: nonprofit organization. Branch Museum: Thomas Dodge House, 58 Harbor Rd., Port Washington, NY 11050. Tax-exempt: 501(c)(3).

Institution Type/Description: Historical Society Museum.

Collections: Port Washington history; 18th-century costumes; 19th-century quilts; 18th to early 20th-century decorative arts; toys; dolls; manuscript archives; photographs. Historic Buildings: c.1735 Sands-Willets House; c.1721 Thomas Dodge House; c.1690 Dutch Barn.

Research Fields: Port Washington history; Long Island history; Dodge and Sands family history.

Facilities: 100-vol. library of books on Long Island available for research on premises only. Museum-related items for sale.

Activities: guided tours; lectures; hobby workshops; formally organized education programs; temporary exhibitions. Annual Events: Dodge House Day in summer; Fall Fair in September; Holiday Fair in November; Christmas Party in December.

Publications: annual journal, Cow Neck Peninsula Historical Society; publications, The Mill Pond; Lower Main Street; Sketchbook of Cow Neck Houses; How to Research Your House; Historic Mitchell Farms.
Hours & Admission Prices: Sands-Willets House: Sun. 2-4, Wed. 10-2; other times by appointment. Dodge House: April-Nov. 2nd Sat. of month 1-5; other times by appointment. Adults $3, children $1; discount to groups & AAM members; members no charge. Closed holidays.
Attendance: 3,000 (estimated)
Membership: Student $10; Individual $25; Family $35; Sustaining $50; Sponsor $100; Life $500.

POLISH AMERICAN MUSEUM, (M), 16 Belleview Ave., Port Washington, NY 11050-3607. Tel.: 516-883-6542.
Founded: 1977.
Congressional District: 3
Key Personnel: Pres., Barbara Szydlowski; Exec. Vice Pres., Julian S. Jurus; 1st Vice Pres., Steve Szachacz; 2nd Vice Pres., Irene Wierbicki; Treas., Michael Levchuck; Recording Sec., Wilma Wierbicki; Cur., Gerald Kochan; Historian, Al Novak.
Personnel Profile: Part-Time Volunteers 25.
Governing Authority: nonprofit organization. Polish American Museum Admin., 16 Belleview Ave, Port Washington, NY 11050. Tax-exempt: 501(c)(3).
Institution Type/Description: Folk Art Museum & Library.
Collections: woodcarvings; glass; art; china; tapestry all from Poland; Polish language library; Archacki archives; history display.
Major Exhibits: Jozef Pilsudski, 6/13-6/14; Folding Bicycle, 9/13-1/14; Czeslaw Wojciechowski, 10/13-10/14; Jan Frankowski, A.G. Pilot, 12/2/13-6/14.
Research Fields: Polish & Polish American history.
Facilities: 15,000-vol. library pertaining to Polish history & culture available for research on premise; photocopy available; reading room; classrooms. Items imported from Poland for sale.
Activities: lectures; films; hobby workshops; permanent exhibitions; historical & educational videotapes.
Publications: quarterly newsletter; annual journal.
Hours & Admission Prices: Wed.-Fri. 10-2, Sat.-Sun. by appointment. No charge; donations accepted. Closed legal holidays.
Attendance: 1,200 (estimated)
Membership: Active $25; Supporting $50; Life $500; Founder $1,000.

THE SALGO TRUST FOR EDUCATION, 95 Middle Neck Rd., Port Washington, NY 11050-1218. Tel.: 516-767-3654. Fax: 516-767-7881.
Founded: 1994.
Key Personnel: Trustee, Miklos Salgo; Trustee, Christina Salgo; Collection Mgr., Eileen Baral.
Governing Authority: Tax-exempt.
Institution Type/Description: Art Museum.
Collections: Hungarian paintings, sculpture & silver 15th-18th century, saddles, saddle rugs & horse covers; French & European furniture & decorations; Chinese art; game boards & draughtsmen.
Research Fields: Hungarian painting & sculpture.
Facilities: library.
Hours & Admission Prices: By appointment only. No charge.

Potsdam

POTSDAM PUBLIC MUSEUM, 2 Park St. at Civic Center, Potsdam, NY 13676. Mailing Address: P.O. Box 5168, Potsdam, NY 13676-5168. Tel.: 315-265-6910. Fax: 315-265-3149.
E-mail: museum@vi.potsdam.ny.us
Web Site: www.potsdampublicmuseum.org
Founded: 1940.
Congressional District: 30
Key Personnel: Dir. & Cur., Mimi Van Deusen; Pres., Emily Brouwer; Vice Pres., Joan Edzwald.
Personnel Profile: Full-Time Paid 1; Part-Time Paid 3; Part-Time Volunteers 2; Interns 1.
Governing Authority: municipal. Parent Institution: Village of Potsdam. Tax-exempt: 170(b)(A)(IV).
Institution Type/Description: Local History Museum: housed in 1876 sandstone church.
Collections: local historic artifacts; photographs; portraits; paintings; musical instruments; textiles; quilts; clothing; military uniforms; ceramics; tools; furniture; glass.
Major Exhibits: Toyland - 100 Yrs of Fun & Games, 12/13-9/14.
Research Fields: English pottery; glass; 19th & 20th-century costumes; American decorative arts; history of state, county, town & village; textiles.

Facilities: 2,000-vol. library including antique price guides.
Activities: guided tours; lectures; gallery talks; formally organized education programs for children, adults & undergraduate college students affiliated with State University College at Potsdam & Clarkson University; inter-museum loan, permanent & temporary exhibitions; self-guided architectural walking tour of Potsdam; school loan service.
Publications: booklets, Early History of Potsdam; Gallantry in the Field, Potsdam & the Civil War; Burnap Collection of Potsdam Museum; Clarkson Family of Potsdam; book, Images of America: Potsdam.
Hours & Admission Prices: Tues.-Sat. 10-4. Suggested Donation: $2. Closed state & national holidays. &
Attendance: 2,500 (estimated)
Membership: Basic $20; Silver $50; Gold $100.

ROLAND GIBSON GALLERY, (M), 44 Pierrepont Ave., State Univ. of New York at Potsdam, Potsdam, NY 13676-2200. Tel.: 315-267-3290. Fax: 315-267-4884.
E-mail: vasherak@potsdam.edu
Web Site: www.potsdam.edu/gibson/gibson.html
Founded: 1968.
Congressional District: 30
Key Personnel: Dir., April Vasher-Dean; Cur. Collections, Margaret Price; Registrar, Romi Sebald-Chudzinski; Dept. Sec., Claudette Fefee.
Personnel Profile: Full-Time Paid 4; Part-Time Paid 6; Interns 1.
Governing Authority: state. Affiliated with the State University of New York, University Plaza, Albany, NY 12246. Parent Institution: SUNY College at Potsdam. Tax-exempt.
Institution Type/Description: University Art Museum.
Collections: contemporary sculpture, prints, ceramics, drawings & paintings; the Roland Gibson collection of 20th-century contemporary art.
Research Fields: pertaining to exhibitions & collections.
Facilities: 300-vol. library of catalogs from previous shows & catalogs from other galleries and museums, available on request; changing exhibition galleries; work & study room.
Activities: lectures; films; gallery talks; workshops; concerts; arts festivals; formally organized education programs; museum studies courses & internships; permanent, temporary, traveling & loan exhibitions.
Publications: catalogs on individual exhibitions; newsletter; brochures; posters; monthly bulletin.
Hours & Admission Prices: Summer: Wed.-Sat. 12-4; Academic Year: Mon. & Fri. 12-5, Tues. & Thurs. 12-7, Sat. 12-4. No charge; donations accepted. Closed college recesses; public holidays. &
Attendance: 14,000 (estimated)
Membership: Senior Citizen $10; Individual $15; Family $25; Contributing $100; Supporting $250; Patron $500; Benefactor $1,000.

Poughkeepsie

ART GALLERY MARIST COLLEGE, 3399 North Rd., Poughkeepsie, NY 12601-1387. Tel.: 845-575-3000, ext. 2308. Fax: 845-471-6213.
E-mail: edward.smith@marist.edu
Web Site: www.marist.edu/commarts/art/gallery.html
Founded: 1995.
Key Personnel: Dir., Edward Smith; Chm. Art Dept., Richard Lewis.
Personnel Profile: Interns 5.
Governing Authority: Parent Institution: Marist College. Tax-exempt.
Institution Type/Description: Art Museum.
Collections: paintings; sculptures; photographs; contemporary work of artists from the Hudson Valley region & New York City.
Facilities: 3,200 sq. ft. exhibit space.
Activities: lectures; artist talks.
Publications: gallery announcements; occasional exhibition catalogues.
Hours & Admission Prices: Sept.-May Mon.-Sat. 12-5. No charge. &
Attendance: 2,000 (estimated)

DUTCHESS COUNTY HISTORICAL SOCIETY, Clinton House, 549 Main St., Poughkeepsie, NY 12601. Mailing Address: P.O. Box 88, Poughkeepsie, NY 12602-0088. Tel.: 845-471-1630. Fax: 845-471-1634.
E-mail: dchistorical@verizon.net
Web Site: www.dutchesscountyhistoricalsociety.org
Founded: 1914.
Congressional District: 25
Key Personnel: Exec. Dir., Betsy Kopstein Stuts; Pres., Dr. Candace Lewis.
Personnel Profile: Part-Time Paid 1; Part-Time Volunteers 8; Interns 4.
Volunteer Hours: 4,200
Operating Expenses: 68,000
Operating Income: 68,000

Governing Authority: society; nonprofit. Building owned by the N.Y. State Office of Parks & Recreation & Historic Preservation. Branch Museum: 1767 Glebe House Museum, Poughkeepsie, NY 12601. Tax-exempt: 501(c)(3).

Institution Type/Description: Historical Society Museum: housed in c.1765 Clinton House, used as NY state capitol during the Revolutionary War.

Collections: manuscripts & artifacts relating to the history of Dutchess County; photographs.

Research Fields: local history; oral history; genealogy; material culture.

Facilities: library of 500-volumes of printed local history, 25,000 manuscripts & 50,000 images, available for use under supervised conditions.

Activities: tours; temporary & traveling exhibitions; meetings; seminars; symposia.

Publications: Yearbook; 18th Century Documents of the Nine Partners Patent; County at Large; Family Vista; Portraits of Dutchess County; Glebe House; 150th Anniversary of the Ratification of the United States Constitution; 250th Anniversary of Poughkeepsie; Dutchess County 1778 - Year of Trial, Year of Transition; FDR at Home.

Hours & Admission Prices: Tues. & Thurs. 12-5, Wed. & Fri. 10-3 by appointment only. Library research: $20; members no charge. Closed New Year's Eve & Day; Memorial Day; Independence Day; Labor Day; Thanksgiving & day after; Christmas Eve & Day. &

Attendance: 4,236 (estimated)

Membership: Individual $50; Family & Contributor $75; Sustaining $100; Patron $250; Sponsor $500; Millennial Circle $1,000.

٭ THE FRANCES LEHMAN LOEB ART CENTER, (M), 124 Raymond Ave., Vassar College, Poughkeepsie, NY 12604. Mailing Address: Box 703, Vassar College, Poughkeepsie, NY 12604. Tel.: 845-437-5237 & LOEB (5632). Fax: 845-437-5955.

E-mail: jamundy@vassar.edu

Web Site: fllac.vassar.edu

Founded: 1864.

Congressional District: 25

Key Personnel: Ann Hendricks Bass Dir., James Mundy; The Philip & Lynn Strauss Cur. Prints & Drawings, Patricia Phagan; Registrar & Collections Mgr., Joann Potter; Asst. Registrar, Karen Hines; Preparator, Bruce Bundock; Office Specialist, Francine Brown; The Emily Hargroves Fisher '57 & Richard B. Fisher Cur., Mary-Kay Lombino; Coord. Public Education & Programs, Margaret Vetare.

Personnel Profile: Full-Time Paid 12; Part-Time Paid 2; Part-Time Volunteers 20; Interns 1.

Governing Authority: college. Parent Institution: Vassar College. Tax-exempt: 501(c)(3).

Institution Type/Description: Art Museum.

Collections: American & European painting & sculpture, medieval to modern; Magoon collection of Hudson River School & other 19th-century American paintings; Greek & Roman antiquities; Asian art; drawings & watercolors; photographs; Old Master & modern prints.

Research Fields: all fields pertaining to the collection.

Facilities: 25,000-vol. library of books pertaining to art available for interlibrary loan & for use on the premises. Catalogues, posters & other art-related items for sale.

Activities: gallery talks; permanent & temporary exhibitions.

Publications: occasional catalogues; members' quarterly newsletter.

Hours & Admission Prices: Tues.-Wed. & Fri.-Sat. 10-5, Thurs. 10-9, Sun. 1-5. No charge. Closed New Year's Eve & Day; Thanksgiving Day; Christmas Eve, Day & week. &

Attendance: 36,000 (estimated)

Membership: Friends of the Frances Lehman Loeb Art Center: Student $10; Participating $35; Contributing $100; Sustaining $250; Donor $500; Patron $1,000.

LOCUST GROVE, THE SAMUEL MORSE HISTORIC SITE, 2683 South Rd., Poughkeepsie, NY 12601-5275. Tel.: 845-454-4500, ext. 10. Fax: 845-485-7122.

E-mail: info@lgny.org

Web Site: www.lgny.org

Founded: 1979.

Congressional District: 25

Key Personnel: Exec. Dir., Kenneth Snodgrass.

Personnel Profile: Full-Time Paid 7; Part-Time Paid 20; Part-Time Volunteers 50.

Governing Authority: nonprofit organization. Private trust. Tax-exempt: 501(c)(3).

Institution Type/Description: Historic Site: 1847 residence of Samuel F.B. Morse.

Collections: furniture; china; costumes; dolls; documents; manuscripts; telegraph equipment.

Research Fields: life of Samuel F.B. Morse.

Facilities: 3,000-vol. library of history & literature books; wildlife sanctuary; hiking trails; visitor center.

Activities: guided tours; lectures; formally organized education programs; training programs for professional museum workers; temporary exhibitions.

Publications: newsletters; brochures.

Hours & Admission Prices: Visitor Center: Jan.-April 6 Mon.-Fri. 10-5; April 9-Dec. daily 10-5. Tours: April & Dec. Sat.-Sun. call for hours; May-Nov. daily 10-5; groups by appointment only. Adults $10, youth 6-18 $6; discounts to AAM members; Friends members no charge.

Attendance: 20,000 (accurate)

MID-HUDSON CHILDREN'S MUSEUM, (M), 75 N. Water St., Poughkeepsie, NY 12601-1720. Tel.: 845-471-0589. Fax: 845-471-0415.

E-mail: info@mhcm.org

Web Site: www.mhcm.org

Founded: 1989.

Congressional District: 19

Key Personnel: Dir., Lara Litchfield-Kimber; Chm. (V), Steve Loehr.

Personnel Profile: Full-Time Paid 5; Part-Time Paid 5; Part-Time Volunteers 7.

Governing Authority: private; nonprofit organization. Tax-exempt: 501(c)(3).

Institution Type/Description: Children's Museum.

Collections: hands-on exhibits.

Facilities: 12,000 sq. ft. exhibit space; planetarium.

Activities: education programs for young children; guided tours; workshops; lectures; traveling & participatory exhibitions; school loan service. Annual Event: Halloween event, New Years Eve event, summer celebration.

Publications: quarterly newsletter; calendar of events; annual report.

Hours & Admission Prices: Tues.-Sat. 9:30-5, Sun. 11-5. Admission $7.50; members & children under one no charge. Discounts to ADM & ASTC members, active military & grandparent days. &

Attendance: 30,000 (estimated)

Membership: Basic $85; Specialist $85; Grandparent $100; Explorer $100; Voyager $150; Library $250.

Pound Ridge

POUND RIDGE HISTORICAL SOCIETY - THE POUND RIDGE MUSEUM, 255 Westchester Ave., Pound Ridge, NY 10576. Mailing Address: Pound Ridge Historical Society, P.O. Box 51, Pound Ridge, NY 10576-0051. Tel.: 914-764-4333 (museum).

E-mail: prhsmuseum@earthlink.net

Founded: 1970.

Congressional District: 20

Key Personnel: Chm. (V) & Pres. (V), Joyce Butterfield; Treas., Mike Santulli.

Personnel Profile: Part-Time Volunteers 35.

Governing Authority: private; nonprofit organization. Parent Institution: Pound Ridge Historical Society. Tax-exempt: 501(c)(3).

Institution Type/Description: Historical Society & Local History Museum: housed in an 1853 wooden frame building.

Collections: Pound Ridge history & artifacts; prehistoric remains; Native American articles; early tools; archives; letters; diaries; costumes & textiles; maps; documents; furniture.

Research Fields: genealogy & history of Pound Ridge.

Facilities: 50-vol. library on local history; 600 sq. ft. exhibit space. Publications, posters, books, historical maps, & postcards for sale.

Activities: docent program; educational programs, lectures, walking tours; oral history; exhibitions & publications; bus tours. Annual Events: annual April meeting & program.

Publications: newsletter; booklets concerning Pound Ridge history; Images of America; History of the Betsey Hunt House.

Hours & Admission Prices: April-Dec. Sat.-Sun. 2-4; groups by appointment. No charge; donations accepted. Closed holidays.

Attendance: 425 (estimated)

Membership: Individual $25; Family & Business $40; Patron $50; Sponsor $100; Sustaining $250; Lifetime $500;

Prattsburgh

NARCISSA PRENTISS HOUSE, 7225 County Rte. 75, Prattsburgh, NY 14873. Mailing Address: P.O. Box 307, Prattsburgh, NY 14873-0384. Tel.: 607-522-4537.

Founded: 1940.

Congressional District: 39

Key Personnel: Dir. & Pres. (V), Charlene Wilson; Chm. (V), Joan Georgia.

Personnel Profile: Part-Time Paid 1; Part-Time Volunteers 10.

Governing Authority: nonprofit organization. Parent Institution: Committee to Preserve the Narcissa Prentiss House. Tax-exempt.

Institution Type/Description: Historic House: c.1805 birthplace of Narcissa Prentiss Whitman, missionary.

Collections: period furnishings; books on the Whitman-Spalding expedition & related historical-religious matters.
Activities: guided tours.
Publications: pamphlet; books.
Hours & Admission Prices: mid-June to Sept. Sat.-Sun. 1-4; other times by appointment. No charge; donations accepted. Closed Independence Day; Labor Day.
Attendance: 250 (estimated)
Membership: Annual $10.

Prattsville

ZADOCK PRATT MUSEUM, Main St., Rte. 23, Prattsville, NY 12468-0333. Mailing Address: P.O. Box 333, Prattsville, NY 12468-0333. Tel.: 518-299-3395.
Institution Type/Description: Historic House: housed in the former home of Zadock Pratt, Prattsville town founder; built in c.1828. Listed on the National Register of Historic Places.
Collections: Pratt family life & history; local history & culture; period furnishings; photographs; tannery artifacts; Civil War.
Facilities: Museum-related items for sale.
Hours & Admission Prices: Memorial Day to Columbus Day Thurs.-Sun. 1-5. Suggested Donation: $3.

Pultneyville

WILLIAMSON-PULTNEYVILLE HISTORICAL SOCIETY, 4130 Mill St., Pultneyville, NY 14538. Mailing Address: P.O. Box 92, Pultneyville, NY 14538. Tel.: 315-589-9892.
Web Site: www.w-phs.org
Founded: 1964.
Key Personnel: Pres. (V), Joan Carey; Museum Shop Mgr., Suzi Goodrich.
Personnel Profile: Part-Time Volunteers 30.
Volunteer Hours: 5,000
Operating Expenses: 30,000
Operating Income: 30,000
Institution Type/Description: Historical Society Museum.
Collections: local history, culture & heritage; period furnishings; personal artifacts; photographs; Captain Throop inventions & models; St. Peter shipwreck in Lake Ontario.
Major Exhibits: Pultneyville & The War of 1812, 1/12-1/14; Bicentennial Commemoration of the War of 1812: Battle of Pultneyville, May 15, 1814, 5/14-10/14.
Publications: newsletter.
Hours & Admission Prices: By appointment. No charge; donations accepted. &
Attendance: 600 (estimated)
Membership: Individual & Family $25; Business $75; Sponsor $ 250.

Purchase

ARTHUR M. BERGER ART GALLERY - MANHATTAN-VILLE COLLEGE, 2900 Purchase St., Purchase, NY 10577. Tel.: 914-694-2200.
Key Personnel: Dir., Charles McGill
Institution Type/Description: Art Gallery.
Collections: paintings; drawings; sculpture.
Hours & Admission Prices: Tues. & Thurs.-Fri. 11-6, Wed. 11-7, Sat. 12-4; other times by appointment.

BROWNSON GALLERY - MANHATTANVILLE COLLEGE, Brownson Hall, 2nd Fl., Purchase, NY 10577. Tel.: 914-694-2200.
Web Site: www.mville.edu
Institution Type/Description: Art Gallery.
Collections: works by students & alumni.
Activities: artist lectures.
Hours & Admission Prices: Call for hours. No charge. &

✳ NEUBERGER MUSEUM OF ART, PURCHASE COLLEGE, STATE UNIVERSITY OF NEW YORK, (M), 735 Anderson Hill Rd., Purchase, NY 10577-1402. Tel.: 914-251-6100. Fax: 914-251-6101.
E-mail: nma@purchase.edu
Web Site: www.neuberger.org
Founded: 1974.
Congressional District: 24
Key Personnel: Acting Dir. & Dir. Devel., Lea Emery; Chm., Helen Stambler Neuberger; Vice Chair, Wendy Gold; Vice Chair, Rachel Stern; Chief Cur.,

Helaine Posner; Cur. African Art, Marie Therese Brincard; Cur. Education, School & Family Programs, Eleanor P. Brackbill; Registrar, Patricia Magnani; Mktg., Kristi McKee; Public Rels., Carolyn Mandelker; Museum Shop Mgr., Jane Barry.
Personnel Profile: Full-Time Paid 16; Part-Time Paid 5; Part-Time Volunteers 120; Interns 6.
Governing Authority: state. Parent Institution: Purchase College, State University of New York. Tax-exempt.
Institution Type/Description: University Art Museum.
Collections: 20th & 21st century art in all media; African art.
Research Fields: 20th-century art; contemporary & African arts.
Facilities: 25,000 sq. ft. exhibit space; cafe. Museum-related items for sale.
Activities: guided tours; biennial series of Yaseen lectures; concerts; films; seminars; educational programs for elementary & secondary school students; docent council; temporary exhibitions; interdepartmental projects with other divisions of the State University of New York at Purchase; internships; lectures; dance recitals; family programs; art workshops.
Publications: exhibition catalogues; biannual calendar of events.
Hours & Admission Prices: Tues.-Sun. 12-5. Adults $5, seniors $3; discounts to AAM, ICOM, AAMD & Channel 13; members no charge. Closed major holidays. &
Attendance: 56,500 (accurate)
Membership: Senior 62 & over $40; Basic $55; Supporter $100; Patron's Circle $750; Director's Circle $2,500; Neuberger Circle $5,000.

Queens

NEW YORK HALL OF SCIENCE, 47-01 111th St., Queens, NY 11368-2999. Tel.: 718-699-0005. Fax: 718-699-1341.
E-mail: mrecord@nysci.org
Web Site: www.nysci.org
Founded: 1964.
Congressional District: 7
Key Personnel: Pres. & C.E.O., Margaret Honey; Dir. & Chief Content Officer, Eric Siegel; Exec. Vice Pres. & C.O.O., Robert Logan; Dir. Communications, Mary Record; Vice Pres. Education Svcs., Sylvia Perez.
Personnel Profile: Full-Time Paid 80; Part-Time Paid 180; Part-Time Volunteers 40; Interns 5.
Governing Authority: nonprofit organization. Tax-exempt: 501(c)(3).
Institution Type/Description: Science Museum.
Collections: over 450 hands-on science exhibits pertaining to biology, chemistry, sound, light & physics.
Research Fields: informal science education.
Facilities: library containing a multimedia interdisciplinary collection of science books; 60,000 sq. ft. outdoor science playground; space-themed miniature golf course; educational facilities. Science-related items for sale.
Activities: lectures; films; workshops; organized education programs for children & adults; teacher training; science career access program; participatory exhibits; equipment for rent. Annual Event: World Maker Faire in September.
Publications: monthly e-newsletter; e-postcards.
Hours & Admission Prices: April-June Mon.-Thurs. 9:30-2, Fri. 9:30-5, Sat.-Sun. 10-6; July-Aug. Mon.-Fri. 9:30-5, Sat.-Sun. 10-6; Sept.-March Tues.-Thurs. 9:30-2, Fri. 9:30-5, Sat.-Sun. 10-6; groups by appointment. Adults $11, children, senior citizens & college students $8; discounts to AAM & ASTC members; members no charge. Closed Labor Day; Thanksgiving; Christmas. &
Attendance: 500,000 (estimated)
Membership: Individual $40; Individual Premium $60; Senior $70; Family $85; Senior Premium $100; Family Premium $125; Corporate Partner Premium $100; Extended Family $200.

QUEENS MUSEUM OF ART, New York City Bldg., Flushing Meadows Corona Park, Queens, NY 11368-3398. Tel.: 718-592-9700. Fax: 718-592-5778.
Web Site: www.queensmuse.org
Founded: 1972.
Congressional District: 8
Key Personnel: Exec. Dir., Tom Finkelpearl; Pres., Gretchen Werwaiss; Dir. Finance, Julie Lou; Dir. Education, Lauren Schloss; Dir. Exhibitions, Hitomi Iwasaki; Museum Shop Mgr., Betty Abramowitz.
Personnel Profile: Full-Time Paid 40; Part-Time Paid 13; Part-Time Volunteers 120; Interns 4.
Governing Authority: nonprofit organization. Satellite Gallery. Queens Museum Art Bulova Corporate Center. Tax-exempt: 501(c)(3).
Institution Type/Description: Art & Cultural Center.
Collections: paintings; sculpture; prints & photographs; The Panorama of the City of New York architectural scale model; Tiffany lamps from Egon & Hildegard Neustadt Museum collection.
Research Fields: art history; American cultural history.

Facilities: theatre; workshops. Museum-related items for sale.
Activities: special exhibits, 25-30 per year; tours; lectures; films; education & community service programs for children & adults; programs for the handicapped; intern & docent volunteer programs.
Publications: catalogs & brochures of exhibitions; quarterly QMA mail.
Hours & Admission Prices: Wed.-Fri. 10-5, Sat.-Sun. 12-5. Suggested Donations: adults $5, senior citizens & students $2.50; discounts to AAM & ICOM members; members & children under 5 no charge. Closed New Year's Day; Thanksgiving; Christmas. &
Attendance: 85,216 (accurate)
Membership: Student & Senior Citizen $25; Individual $35; Family $55; Supporting $100; Friend $200; Sponsor $500; Collector's Circle $750.

Queensbury

WARREN COUNTY HISTORICAL SOCIETY, 195 Sunnyside Rd., Queensbury, NY 12804-7762. Tel.: 518-743-0734. Fax: 518-824-5861. Facebook: Warren County Historical Society.
E-mail: mail@warrencountyhistoricalsociety.org
Web Site: warrencountyhistoricalsociety.org
Founded: 1997.
Congressional District: 21
Governing Authority: Tax-exempt: 501(c)(3).
Institution Type/Description: Historical Society Museum.
Collections: local history & culture; period furnishings; personal artifacts; photographs; manuscripts; books; archaeology.
Research Fields: author Edward Eggleston; colonial military.
Facilities: archives.
Activities: educational programs; presentations; research.
Publications: quarterly, Past Times.
Hours & Admission Prices: Call for hours. No charge; donations accepted. &
Attendance: 500 (estimated)
Membership: Individual $20; Family $40; Contributing $50; Sustaining $125; Patron $300; Life $500. Corporate: Contributor $100; Donor $200; Benefactor $500.

Remsen

STEUBEN MEMORIAL STATE HISTORIC SITE, Starr Hill Rd., Remsen, NY 13438. Mailing Address: c/o Oriskany Battlefield SHS, 7801 State Rte. 69, Oriskany, NY 13424-4115. Tel.: 315-768-7224. Fax: 315-337-3081.
E-mail: nancy.demyttenaere@oprhp.state.ny.us
Web Site: www.nysparks.com
Founded: 1930.
Congressional District: 31
Key Personnel: Regl. Historic Preservation Supv., Nancy Demyttenaere; Second in Command, Bill Acomb.
Personnel Profile: Full-Time Paid 3; Part-Time Paid 2; Part-Time Volunteers 6.
Governing Authority: state. Parent Institution: New York State Office of Parks, Recreation and Historic Preservation. Tax-exempt: 501(c)(3).
Institution Type/Description: Historic Site: burial site of Baron Friederich Wilhelm von Steuben, Drillmaster of American Army, located on land granted by New York State for services in the American Revolution.
Collections: reproduction log cabin; furniture; tomb; sacred grave; historic archives.
Facilities: picnic area.
Activities: guided tours; special programs.
Hours & Admission Prices: Memorial Day to Sept. 1 daily dawn to dusk. No charge; donations accepted. &
Attendance: 6,785 (accurate)

Rensselaer

CRAILO STATE HISTORIC SITE, 9 1/2 Riverside Ave., Rensselaer, NY 12144-2927. Tel.: 518-463-8738. Fax: 518-433-1860.
E-mail: maryellen.grimaldi@oprhp.state.ny.us
Founded: 1924.
Congressional District: 29
Key Personnel: Historic Site Mgr., Heidi Hill.
Personnel Profile: Full-Time Paid 2; Part-Time Paid 2; Part-Time Volunteers 10; Interns 1.
Governing Authority: state. Parent Institution: New York State Office of Parks, Recreation & Historic Preservation; Saratoga/Capital District State Park, Recreation & Historic Preservation Commission. Tax-exempt: 501(c)(3).
Institution Type/Description: Historic Building: c.1704, brick dwelling belonging to the Van Rensselaer family; used as museum of Dutch culture in the upper Hudson Valley.
Collections: history; Dutch colonial period artifacts; 17th-century archaeological materials; models; reproductions.

Research Fields: 17th & 18th century Dutch culture in the Hudson River Valley.
Facilities: picnic area.
Activities: permanent exhibitions; school programs; summer program for children; guided tours; special programs & demonstrations. Museum Sponsors: summer concerts; preservation day open house.
Publications: Clothing the Colonists.
Hours & Admission Prices: Nov.-March Mon.-Fri. 11-4 by appointment. Adults $5, seniors & students $4; discounts to groups; children 12 & under no charge. &
Attendance: 10,000 (accurate)
Membership: Friends of Crailo: Student & Senior $5; Individual $10; Family $15; Sustaining $25; Corporate Member $100; Patron $200.

Rhinebeck

RHINEBECK AERODROME MUSEUM, 9 Norton Rd., Rhinebeck, NY 12572. Mailing Address: P.O. Box 229, Rhinebeck, NY 12572-0229. Tel.: 845-752-3200. Fax: 845-758-6481.
E-mail: info@oldrhinebeck.org
Web Site: www.oldrhinebeck.org
Founded: 1977.
Congressional District: 19
Key Personnel: Public Rels., Don Fleming; Pres. (V), Neill Harmon; Museum Shop Mgr., JoAnn DiGiocomio.
Personnel Profile: Full-Time Paid 3; Full-Time Volunteers 5; Part-Time Paid 20; Part-Time Volunteers 30.
Governing Authority: private; nonprofit organization. Subsidiary Institution: Old Rhinebeck Aerodrome Airshows, Stone Church Rd. & Norton Rd., Rhinebeck, NY 12572. Tax-exempt: 501(c)(3)
Institution Type/Description: Aeronautics & Space Museum.
Collections: period airplanes covering the pioneer, World War I & Lindbergh eras; vintage clothing; period cars & motorcycles; Spirit of St. Louis memorabilia.
Research Fields: early aircraft designs & aviation pioneers, from 1900 through 1940.
Facilities: 300-vol. set of aircraft technical manuals and early aviation books; restaurant; 28,000 sq. ft. exhibit space. Gift items for sale.
Activities: loan, temporary & traveling exhibitions; barnstorming rides. Museum Sponsors: Opening Day in May; First Air Show in June; Father's Day Special Weekend in June; Open Cockpit Weekend in June & August; Red Hook Bicentennial Celebration Air Show in June; 100 Years Ago Weekend in July; Artists-Authors-Photographers Weekend in August; Women Fly Weekend in September; Radio Controlled Model Weekend in September; World War I Weekend in September; Antique Biplane Fly-In in September; Pumpkin Festival in October; Final Air Show of Season in October; vintage car shows & antique machinery exhibits.
Publications: quarterly newsletter, Rotary Ramblings
Hours & Admission Prices: Museum: daily 10-5. Air Show: Sat.-Sun. 2pm. Mon.-Fri.: adults 18-64 $10, children 13-17 and seniors 65 & over $8, children 6-12 $3; children 5 & under no charge. Sat.-Sun.: adults $20, children 13-17 and seniors 65 & over $15, children 6-12 $5; Aviator members & above and children 5 & under no charge.
Attendance: 40,000 (estimated)
Membership: Fledgling $25; Aviator $50; Wingman $100; Flight Leader $1,000; Squadron Leader $5,000; Wing Leader $10,000 & up.

Richfield Springs

PETRIFIED CREATURES MUSEUM OF NATURAL HISTORY, U.S. Rte. 20, Richfield Springs, NY 13439. Mailing Address: P.O. Box 751, Richfield Springs, NY 13439-0751. Tel.: 315-858-2868. Fax: 315-858-2868.
E-mail: petrifiedcreaturesmuseum@yahoo.com
Web Site: www.petrifiedcreatures.com
Founded: 1934.
Congressional District: 25
Key Personnel: C.E.O., Dir. & Museum Shop Mgr., Stella C. Mlecz; Education & Cur., Richard S. Mlecz; Public Rels. & Treas., Sally E. Kennedy; Archivist, Frank Maiocco; Security, Michael Vesely.
Personnel Profile: Full-Time Paid 1; Full-Time Volunteers 3; Part-Time Paid 1; Part-Time Volunteers 15.
Volunteer Hours: 400
Operating Expenses: 35,000
Operating Income: 30,000
Governing Authority: private; nonprofit organization. Tax-exempt: 501(c)(3).
Institution Type/Description: Nature & Science Museum.
Collections: life-size restorations of dinosaurs; prehistoric animal life; petrified fossils.
Research Fields: paleontology.

Facilities: Fossils, minerals, stoneware & butterflies for sale.
Activities: digging for fossils; formally organized educational programs; narrated exhibits.
Hours & Admission Prices: May 15-June & Sept. 1-Sept. 15 Thurs.-Mon. 10-5; July-Aug. daily 10-5. Adults $9, senior citizens $7, children 5-11 $5; discount to AAM & ICOM members; children under 5 no charge. &
Attendance: 7,500 (estimated)

Ridge

BROOKHAVEN VOLUNTEER FIREFIGHTERS MUSEUM, Rte. 25 (Middle Country Rd.), Ridge, NY 11961. Mailing Address: P.O. Box 367, Ridge, NY 11961. Tel.: 631-924-8114.
E-mail: Firehistory@aol.com
Web Site: www.brookhavenfiremuseum.org
Founded: 1991.
Congressional District: 1
Key Personnel: Dir., Herb Petersen; Chm. (V), Ed Corrigan.
Governing Authority: Tax-exempt.
Institution Type/Description: Firefighting History Museum: housed in the former Center Moriches Firehouse; built in 1889.
Collections: firefighting history & equipment; early fire trucks; uniforms; hose cart; helmets.
Hours & Admission Prices: May 2nd-Oct. Tues. & Thurs. 10-3, Sat. 10-4, Sun. 12-4; other times by appointment. No charge; donations accepted. Closed holidays.
Attendance: 2,000 (estimated)
Membership: Individual $10.

Riverdale

DERFNER JUDAICA MUSEUM + THE ART COLLECTION AT THE HEBREW HOME AT RIVERDALE, (M), 5901 Palisade Ave., Riverdale, NY 10471-1253. Tel.: 718-581-1596. Fax: 718-581-1980.
E-mail: schevlowe@hebrewhome.org
Web Site: hebrewhome.org/art.asp
Founded: 1982.
Congressional District: 17
Key Personnel: Dir. & Chief Cur., Susan Chevlowe; Asst. Cur., Emily O'Leary; Educator, Elana Kaplan.
Personnel Profile: Full-Time Paid 1; Part-Time Paid 2; Part-Time Volunteers 1.
Governing Authority: nonprofit organization. Parent Institution: The Hebrew Home at Riverdale. Tax-exempt: 501(c)(3).
Institution Type/Description: Judaica and Modern/Contemporary Art Museum.
Collections: Ralph & Leuba Baum collection of ceremonial objects in silver, gold, pewter; textiles; amulet collection; modern art collection including 20th century prints, paintings, photographs, works on paper.
Facilities: Hebraica & manuscripts; educational facilities; lecture room/spaces; sculpture garden.
Activities: guided tours; lectures; concerts; organized educational programs; docent program; loan exhibitions.
Publications: exhibition brochures.
Hours & Admission Prices: Sun.-Thurs. 10:30-4:30. No charge; donations accepted. Closed federal & Jewish holidays. &
Attendance: 7,749 (accurate)
Membership: Friend $36; Associate $180; Supporting $360; Sustaining $1,000.

Riverhead

HALLOCKVILLE MUSEUM FARM, 6038 Sound Ave., Riverhead, NY 11901-5609. Tel.: 631-298-5292. Fax: 631-298-0144.
E-mail: hallockv@optonline.net
Web Site: hallockville.com
Founded: 1975.
Congressional District: 1
Key Personnel: Exec. Dir., Herbert Strobel.
Personnel Profile: Full-Time Paid 1; Part-Time Paid 3; Part Time Volunteers 60.
Governing Authority: not-for-profit organization. Hallockville, Inc., Riverhead, NY 11901. Tax-exempt: 501(c)(3).
Institution Type/Description: Historic Site: located on c.1765 Hallock Homestead.
Collections: rural & agricultural lifestyle of the Hallock family & the north fork of Long Island, NY, from 1880-1910.
Research Fields: North Fork vernacular architecture; Hallock family genealogy; North Fork agricultural history; North Fork & Northville social & community history.

Facilities: 200-vol. library of material culture reference books; Long Island local history, available to the public; classrooms. Books & museum-related items for sale.
Activities: arts festivals; formal education programs for children; guided tours; hobby workshops; lectures. Annual Events: Fall Festival & Craft Fair; Demonstration series.
Publications: quarterly newsletter, Hallockville Happenings; occasional booklets.
Hours & Admission Prices: Guided Tours: May-Dec. weekends 11-4. Adults $7, seniors & children $5; members no charge. Closed New Year's Day; Easter; Thanksgiving; Christmas Eve & Day. &
Attendance: 10,000 (estimated)
Membership: Friends $20; Supporters $50; Contributors $125; Partners $250; Stewards $500; Benefactors $1,000.

THE LONG ISLAND SCIENCE CENTER, 11 W. Main St., Ste. 101, Riverhead, NY 11901-2818. Tel.: 631-208-8000. Fax: 631-208-8304.
E-mail: programs@lisciencecenter.org
Web Site: www.lisciencecenter.org
Founded: 1992.
Key Personnel: Exec. Dir., Michelle Pelletier.
Governing Authority: Tax exempt.
Institution Type/Description: Science Center.
Collections: hands-on exhibits.
Activities: educational programs; special events.
Hours & Admission Prices: July-Aug. Tues.-Sat. 11-4; Sept.-June Sat. 11-4. Admission: $5; discounts to ASTC members; members no charge. &
Membership: Family $50.

SUFFOLK COUNTY HISTORICAL SOCIETY, (M), 300 W. Main St., Riverhead, NY 11901-2894. Tel.: 631-727-2881. Fax: 631-727-3467.
E-mail: schsociety@optonline.net
Web Site: www.suffolkcountyhistoricalsociety.org
Founded: 1886.
Congressional District: 1
Key Personnel: Pres., Bob Baraukas; Dir., Kathryn M. Curran; Museum Shop Mgr., Lee Thumser.
Personnel Profile: Full-Time Paid 1; Part-Time Paid 7; Part-Time Volunteers 6.
Governing Authority: society Tax-exempt: 501(c)(3).
Institution Type/Description: Historical Museum.
Collections: primarily Suffolk County; Long Island Indian artifacts; Revolutionary & Civil War firearms; vehicles; boat models; china & glass; whaling artifacts; textiles; costumes; ceramics & glassware; decorative arts.
Research Fields: Suffolk County history & genealogy.
Facilities: research library; archives.
Activities: guided group tours; school education program; monthly programs & seasonal workshops for children; genealogy section.
Publications: newsletter; quarterly booklet, The Register.
Hours & Admission Prices: Museum: Tues.-Sat. 12:30-4:30. Adults $5; members no charge. Library: Wed.-Sat. 12:30-4:30. Library: $2. Closed legal holidays.
Attendance: 12,000 (estimated)
Membership: Individual $35; Family $40; Corporate $100; Sustaining $500; Life $1,000.

Rochester

ARTISANWORKS, 565 Blossom Rd., Ste. L, Rochester, NY 14610. Tel.: 585-288-7170. Fax: 585-288-7186.
E-mail: victoria@artisanworks.com
Web Site: www.artisanworks.net
Key Personnel: Dir. Events, Victoria Benz-Gehrke.
Governing Authority: nonprofit organization. Tax-exempt: 501(c)(3).
Institution Type/Description: Art Space.
Collections: works by American artists; paintings; photographs; 3-dimensional & kinetic art; regional history; sculptures.
Facilities: rental facilities. Museum related items for sale.
Activities: special events; rental facilities; tours; educational programs; hands-on workshops; performances; school group tours; lectures.
Hours & Admission Prices: Fri.-Sat. 11-6, Sun. 12-5. Adults $12, students and seniors 60 & over $8; members no charge.
Membership: Student $30; Individual $45; Household $60; Friend $100; Sustaining $250; Patron $500; Corporate: $500-$10,000.

CHARLOTTE-GENESEE LIGHTHOUSE HISTORICAL SO-CIETY, 70 Lighthouse St., Rochester, NY 14612. Tel.: 585-621-6179.
E-mail: cglighthouse14612@gmail.com
Web Site: www.geneseelighthouse.org
Founded: 1984.
Key Personnel: Pres. (V), Robert L. Owens, Jr.; Chm. (V), Virginia Kobylarz; Museum Shop Mgr., Bill Briggs
Governing Authority: nonprofit organization. Tax-exempt.
Institution Type/Description: Historical Society Museum: lighthouse built in 1822.
Collections: local history & culture; period furnishings; personal artifacts; photographs.
Publications: quarterly, The Beacon.
Hours & Admission Prices: May-Nov. Fri.-Mon. 1-5. Adults $3; members no charge.
Attendance: 5,200 (estimated)
Membership: Senior 62 & over and College Student $15; Individual & Senior Couple $20; Household $30; Supporting $50; Patron $100.

DAR-HERVEY ELY HOUSE, 11 Livingston Park, Rochester, NY 14608-2047. Mailing Address: 138 Troup St., Rochester, NY 14608-2032. Tel.: 585-232-4509.
Founded: 1894.
Congressional District: 34
Key Personnel: Regent, Roberta Dreeson; Chm. House & Grounds (V), Beverly D. Henning.
Personnel Profile: Part-Time Volunteers 30.
Governing Authority: private. Owned by Irondequoit Chapter of the D.A.R. Tax-exempt.
Institution Type/Description: Historic House Museum: 1837 Hervey Ely House, example of Greek revival architecture.
Collections: furniture of early 1800s; china; glass; silver; Washington sideboard; Lafayette mirror; antiques; George Washington's drummer boy's drum; manuscripts.
Research Fields: conservation.
Facilities: 2,200-vol. library of genealogical books & periodicals available for research on premises by appointment; genealogical records; lineage research records; records of Revolutionary soldiers in area; reading room.
Activities: guided tours; lectures; films; formally organized education programs for children; permanent & temporary exhibitions.
Publications: year books, Genealogical Records; Lineage Research Records; Record Listing of All Revolutionary Soldiers; Irondequoit Chapter newsletter, Chapter Chatter.
Hours & Admission Prices: 2nd Wed. each month 10-12, 3rd Wed. each month 10:30-3:30; other times by appointment. Genealogical Library: Fri. 11-3. Adults $1.50, children $.75. &
Membership: DAR Individual $55.

GATES HISTORICAL SOCIETY - HINCHEY HOMESTEAD, 634 Hinchey Rd., Rochester, NY 14624. Tel.: 585-464-9740.
E-mail: info@gateshistory.org
Web Site: gateshistory.org
Institution Type/Description: Historical Society Museum.
Collections: local history & culture; period furnishings; personal artifacts; photographs.
Hours & Admission Prices: Call for hours.

GENESEE WARBIRDS AVIATION HISTORY MUSEUM, 16 W. Main St., Ste. 310, Rochester, NY 14614. Tel.: 585-234-5387.
E-mail: admin@geneseewarbirds.org
Web Site: www.geneseewarbirds.org
Institution Type/Description: Military Aviation History Museum.
Collections: military aviation history & aircraft.
Activities: special events.
Hours & Admission Prices: By appointment. Adults $4, senior citizens 60 & over $3, children under 12 $2; members no charge.

* **GEORGE EASTMAN HOUSE, (M),** 900 East Ave., Rochester, NY 14607-2298. Tel.: 585-271-3361. Fax: 585-271-3970.
E-mail: info@geh.org
Web Site: www.eastmanhouse.org
Founded: 1947.
Congressional District: 29
Key Personnel: Chm. (V), Thomas Jackson; C.E.O. & Mgr. Operations, Thomas Combs; Dir., Bruce Barnes, Ph.D.; Sr. Cur. Motion Picture Collection, Paolo Cherchi Usai; Cur. George Eastman Collection, Kathy Connor; Cur. Tech., Todd Gustavson; Dir. Devel., Pamela Reed Sanchez;

Publications Mgr., Amy Schelemanow; Librarian, Rachel Stuhlman; Dir. Communications & Visitor Svcs., Eliza Benington Kozlowski; Controller, Paul Piazza; Museum Shop Mgr. & Travel Exhibitions Coord., Peter Briggs; Dir. Conservation, Taina Meller; Registrar, Wataru Okada.
Personnel Profile: Full-Time Paid 64; Part-Time Paid 38; Part-Time Volunteers 200.
Governing Authority: nonprofit organization. Tax-exempt: 501(c)(3).
Institution Type/Description: Photography and Cinematography Museum: housed in 1905 George Eastman home; historic house museum & gardens.
Collections: 400,000 photographs & negatives; 56,000 books & periodicals on photography & film; world's largest collection of cinematographic equipment; 28,000 motion pictures from 1895 to present with emphasis on silent era films from all countries; over 3 million motion picture stills of both silent & sound periods; restored 1905 colonial revival mansion & gardens; 20,000 cameras & other technology.
Research Fields: photographic & cinematographic history.
Facilities: 80-seat auditorium; 535-seat theater; photographic archives by appointment. Publications for sale.
Activities: formally organized education programs; inter-museum loan; permanent & traveling exhibitions house & garden tours; symposia; senior citizen programming; photographic & film exhibitions.
Publications: newsletter; catalogues; photographic history books & portfolios, Image.
Hours & Admission Prices: Tues.-Wed. & Fri.-Sat. 10-5, Thurs. 10-8, Sun. 1-5. Adults $10, senior citizens $8, students $6, children 5-12 $4; discounts to AAM & ICOM members; children under 5 & members no charge. Closed Mondays, Christmas Day, New Year's Day & Thanksgiving. &
Attendance: 155,000 (accurate)
Membership: Student $35; Senior $45; Individual & National $50; International $60; Dual & Household $70; Contributor $100; Sustainer $150; Patron $250; Benefactor $500.

HIGHLAND BOTANICAL PARK, 171 Reservoir Ave., Rochester, NY 14620-2728. Tel.: 585-753-7275. Fax: 585-753-7284.
Web Site: www2.monroecounty.gov/parks-highland.php
Founded: 1888.
Key Personnel: Supt., Mark Quinn.
Governing Authority: Parent Institution: Monroe County Dept. of Parks, 171 Reservoir Ave., Rochester, NY 14620. Tax-exempt.
Institution Type/Description: Arboretum.
Collections: 500 varieties of flowering shrubs; trees; flowers; rock garden.
Facilities: ice-skating rink; warming shelter; softball field; hiking paths.
Activities: tours. Annual Event: Lilac Festival in May.
Hours & Admission Prices: Highland Park: April-Oct. daily 7a.m.-11p.m., Nov.-March Mon.-Thurs. 7-4. Lamberton Conservatory: daily 10-4. Adults $3, seniors 62 & over and youth 6-18 $2; children 5 & under no charge. Closed Christmas Day. &
Attendance: 50,000

THE LANDMARK SOCIETY OF WESTERN NEW YORK, 133 S. Fitzhugh St., Rochester, NY 14608-2204. Tel.: 585-546-7029, ext. 10. Fax: 585-546-4788.
E-mail: info@landmarksociety.org
Web Site: www.landmarksociety.org
Founded: 1937.
Congressional District: 37
Key Personnel: Interim Exec. Dir., David Whitaker; Dir. Museum & Education, Cindy Boyer; Pres. (V), Jerry Ludwig.
Personnel Profile: Full-Time Paid 10; Part-Time Paid 4; Part-Time Volunteers 400.
Governing Authority: nonprofit corporation. Branch Museums: Stone-Tolan House, 2370 East Ave.; 1867, Ellwanger Garden, 625 Mt. Hope Ave. Tax-exempt: 501(c)(3).
Institution Type/Description: Preservation Project: housed in 1840 Hoyt-Potter House. Historic House Museums: Stone-Tolan Houses.
Collections: art, furnishings & decorative arts of the early 19th-century; country-style architecture. Historic Houses: c.1792 Stone-Tolan House; 1867 Ellwanger Garden.
Research Fields: architecture; history; archaeology; preservation; historic landscaping & horticulture.
Facilities: 1,000-vol. library of reference books on architecture; decorative arts. Books, crafts & other museum-related items for sale.
Activities: guided tours; lectures; workshops; formally organized educational programs; docent program; inter-museum loan, permanent & temporary exhibitions; guided bus tours of historic districts.
Publications: Landmarks of Rochester and Monroe County; quarterly newsletter; booklets; brochures; guides; books, 200 Years of Rochester Architecture & Gardens, The City of Frederick Douglass: Rochester's African-American People & Places, Walking Tours of Downtown Rochester:

Images of History; Rehab Rochester: A Sensible Guide for Old-House Maintenance, Repair & Rehabilitation; Historic New York: Architectural Journeys in the Empire State; Ghost Walk: Chilling Tales from Rochester's Past.

Hours & Admission Prices: Stone-Tolan House: March-Dec. Fri.-Sat. 12-4; groups by appointment. Adults $3, children 8-18 $1; discounts to AAM members; members no charge. Ellwanger Garden: 2nd week in May (lilac festival) by appointment. Wenrich Library: by appointment; no charge. Closed national holidays. &

Attendance: 21,420 (estimated)

Membership: Individual $35; Family $45; Patron $75; Pillar $125; Cornerstone $250; Keystone $500; Corinthian $1,000.

*   **MEMORIAL ART GALLERY OF THE UNIVERSITY OF ROCHESTER, (M),** 500 University Ave., Rochester, NY 14607-1484. Tel.: 585-276-8900. Fax: 585-473-6266. TDD: 585-473-6152; Facebook MAG Rochester.

E-mail: maginfo@mag.rochester.edu

Web Site: mag.rochester.edu

Founded: 1913.

Congressional District: 34

Key Personnel: Mary W. & Donald R. Clark Dir., Grant Holcomb; Dir. Exhibitions, Marie Via; Cur. European Art, Nancy Norwood; Public Rels., Social Media & Webmaster, Meg Colombo; Permanent Collection Registrar, Monica Simpson; Exhibitions Registrar, Daniel Knerr; Dir. Gallery Advancement, Joseph Carney; McPherson Dir. Education, Marlene Hamann-Whitmore; Asst. Dir. for Admin., Kim Hallatt; Asst. Dir., Public Rels., Mktg. & Revenue, Patti Giordano; Museum Shop Mgr., Colleen Griffin-Underhill.

Governing Authority: university. Parent Institution: University of Rochester. Tax-exempt: 501(c)(3).

Institution Type/Description: Art Museum: housed in 1913 Italian Renaissance style building with additions in 1926, 1968 & 1987, located on the site of the original campus of University of Rochester.

Collections: ancient, classical, medieval, Renaissance, baroque, 18th to 20th-century American, 19th & 20th-century French, American folk, Ancient American, African, Oriental art including paintings, sculpture, prints, drawings and decorative arts.

Research Fields: pertaining to collections.

Facilities: 38,000-vol. library of art and art history books, reference works, periodicals, exhibition catalogs & museum bulletins available for inter-library loan and for use on premises; restaurant; reading room; 300-seat auditorium; classrooms. Art books, reproductions, original paintings, sculpture & other museum-related items for sale.

Activities: guided tours; lectures; films; gallery talks; concerts; rental gallery; formally organized educational programs; docent program; inter-museum loan, permanent, temporary & traveling exhibitions; school loan service.

Publications: biennial financial report; bimonthly newsletter, MAGazine; calendar; exhibition catalogs; biennial scholarly bulletin; bimonthly newsletter, ARTiculate.

Hours & Admission Prices: Wed. & Fri.-Sun. 11-5, Thurs. 11-9. Adults $12, senior citizens 62 & up & active-duty military $8, college students with ID & children 6-18 $5; discounts to AAM & New York Consortium members and Thurs. 5-9pm; members & children 5 & under no charge. Closed New Year's Day; Independence Day; Thanksgiving; Christmas. &

Attendance: 245,512 (accurate)

Membership: Basic: Individual $65, Family $80, Reciprocal $150; Contributing: Patron $300, Benefactor $600; Leadership: Director's Circle-Winslow Homer $1,500; Director's Circle-Georgia O'Keefe $3,000; Director's Circle-Claude Monet $6,000.

MONROE COMMUNITY COLLEGE, MERCER GALLERY, (M), 1000 E. Henrietta Rd., Brighton Campus, Rochester, NY 14623-5701. Tel.: 585-292-2021. Fax: 585-292-3120.

E-mail: kfarrell@monroecc.edu

Web Site: www.monroecc.edu

Congressional District: 52

Key Personnel: Dir., Ms. Kathleen M. Farrell.

Personnel Profile: Full-Time Paid 1; Part-Time Volunteers 4; Interns 3.

Governing Authority: public college; nonprofit. Tax-exempt: 501(c)(3).

Institution Type/Description: Art Gallery.

Collections: student & faculty art.

Hours & Admission Prices: Mon., Wed., Fri.10-5, Thurs. 10-5; other times by appointment. No charge.

NATIONAL SUSAN B. ANTHONY MUSEUM & HOUSE(R), (M), 17 Madison St., Rochester, NY 14608-1928. Tel.: 585-235-6124. Fax: 585-235-6212.

E-mail: czarcone@susanbanthonyhouse.org

Web Site: www.susanbanthonyhouse.org

Founded: 1946.

Congressional District: 32

Key Personnel: Chm. Bd. Directors, Jennifer Martlew; Pres. & C.E.O., Deborah L. Hughes; Dir. Public Rels., Ellen K. Wheeler; Dir. Program & Visitor Svcs., Annie Callanan; Dir. Membership, Lesia Telega; Volunteer Coord., Deborah Coffey; Sunday Coord., Breann Bresovsky; Saturday Coord., Lenny Polizzi; Museum Shop Operations, Joanne French; Administrative Asst., Claire Hawley Zarcone; Custodian, Doug Thompson.

Personnel Profile: Full-Time Volunteers 1; Part-Time Paid 4; Part-Time Volunteers 120.

Governing Authority: society. Tax-exempt: 501(c)(3).

Institution Type/Description: Historic House Museum: home of Susan B. Anthony. A National Historic Landmark, designated in 1966.

Collections: papers, writings, & personal belongings of Susan B. Anthony; pictures & writings of famous women associated with Susan B. Anthony; Woman Suffrage Campaign artifacts.

Research Fields: Women's Suffrage Movement.

Facilities: carriage house (education lecture hall).

Activities: tours, teas, and lunches; special exhibits; educational programs for students, Girl Scouts, and seniors; slide showing the 2nd & 3rd floor available for viewing. Museum Sponsors: NYS Day of Recognition & Susan B. Anthony Birthday Luncheon in February; Susan B. Anthony Amendment Festival in August; Monday Lecture Series September to June; Election Day & Arrest Day Commemorations in November.

Publications: newsletter.

Hours & Admission Prices: Tues.-Sun. 11-5. Adults $10, seniors $8, students $5; discounts to AAM & AAA members; members no charge. Closed major holidays. &

Attendance: 9,000 (accurate)

Membership: Individual $45; Family & Dual $60; 1872 League $75-$124; Red Shawl Club $125-$249; Suffragist Society $250-$499; Leadership Council $500-$999; Susan B. Anthony Circle $1,000 & up.

ROCHESTER CONTEMPORARY ART CENTER, 137 East Ave., Rochester, NY 14604. Tel.: 585-461-2222. Fax: 585-4461-2223.

E-mail: info@rochestercontemporary.org

Web Site: www.rochestercontemporary.org

Formerly: Pyramid Arts Center

Founded: 1977.

Personnel Profile: Full-Time Paid 1; Part-Time Paid 1; Part-Time Volunteers 10; Interns 4.

Institution Type/Description: Art Gallery.

Collections: works by contemporary artists.

Hours & Admission Prices: Wed.-Thurs. & Sat.-Sun. 1-5, Fri. 1-10. Adults $1; members no charge. &

Attendance: 18,000 (accurate)

Membership: $35-$500.

ROCHESTER HISTORICAL SOCIETY, 115 South Ave., Rundel Memorial Building, Rochester, NY 14604-1817. Tel.: 585-428-8470. Fax: 585-428-8478.

E-mail: mkeller@rochesterhistory.org

Web Site: www.rochesterhistory.org

Founded: 1861.

Congressional District: 34

Key Personnel: Exec. Dir., Meredith Keller; Pres. (V), Patrick Malgieri.

Personnel Profile: Full-Time Paid 1; Part-Time Paid 2; Part-Time Volunteers 62; Interns 1.

Governing Authority: society. Tax-exempt: 501(c)(3).

Institution Type/Description: Historical Society Museum.

Collections: books; manuscripts; paintings; costume collection; silver; 15,000 photographs; architectural drawings; firearms, children's toys; furnishings; ceramics; perennial garden.

Research Fields: local history.

Facilities: 10,000-vol. library of books, 20,000 photographs, over 500 portraits of Rochesterians, genealogical & local history research files.

Activities: lecture series; permanent exhibitions; publications; research.

Publications: Rochester history.

Hours & Admission Prices: Adults $3, members no charge. &

Membership: Individual $50; Family $85; Sustaining $100; Patron $250; Benefactor & Corporate $500 & up.

ROCHESTER MEDICAL MUSEUM & ARCHIVES, 333 Humboldt St., Rochester, NY 14610-1044. Tel.: 585-922-1847. Fax: 585-922-0018.

E-mail: kathleen.britton@rochestergeneral.org

Web Site: rochestergeneral.org/archives

Formerly: ViaHealth Archives Consortium

Founded: 1947.
Congressional District: 29
Key Personnel: Chm. (V), Betsy Morse; Dir., Kathleen F. Britton.
Personnel Profile: Full-Time Paid 1; Part-Time Paid 2; Part-Time Volunteers 7.
Governing Authority: nonprofit. Parent Institution: Rochester General Hospital. Tax-exempt.
Institution Type/Description: Medical Museum.
Collections: Rochester regional healthcare history including Rochester General Hospital & its affiliated agencies; nursing; medical; military medicine; costumes; archives; photographs.
Research Fields: Rochester General Hospital's contributions to the field of medicine & its development.
Facilities: 300-vol. library; 285 ft. of archival material pertaining to the hospital, available for use by public; 400 sq. ft. exhibit space.
Activities: guided tours; loan, temporary & traveling exhibitions; organized education programs for undergraduate or graduate college students affiliated with LPN Nursing School, primary & secondary school. Museum Sponsors: Century of Service Awards.
Publications: semiannual newsletter, Baker-Cederberg Notebook; occasional monographs.
Hours & Admission Prices: Research: Mon.-Fri. 9-4 ; other times by appointment. No charge; donations accepted. &

Attendance; 1,000 (estimated)
Membership: Associate $10; Contributor $25; Supporter $50; Patron $100; Benefactor $250; Legacy $500.

✳   **ROCHESTER MUSEUM & SCIENCE CENTER, (M),** 657 East Ave., Rochester, NY 14607-2177. Tel.: 585-271-4320. Fax: 585-271-5935. Facebook: Rochester Museum & Science Center.
E-mail: kate_bennett@rmsc.org
Web Site: www.rmsc.org
Founded: 1912.
Congressional District: 25
Key Personnel: C.E.O. & Pres., Kate Bennett; Chm. Bd. (V), Andrew Meloni; Dir. Mktg. & Community Affairs, Debra Jacobson; Sr. Dir. Devel., Pamela L. Jackson; Vice Pres. Finance & Administration, Candice Sheffer; Vice Pres. Operations, Joseph R. Graves; Dir. Planetarium, Steven Fentress; Dir. Nature Center, David Gotham; Dir. Collections, George McIntosh; Dir. Visitor Satisfaction, Heidi Luizzi; Dir. Education, Calvin Uzelmeier; Mgr. Welcome Center & Museum Shop, Colleen McBride.
Personnel Profile: Full-Time Paid 57; Part-Time Paid 152; Part-Time Volunteers 748; Interns 20.
Governing Authority: nonprofit educational corporation. Divisions: Rochester Museum, 657 East. Ave., Rochester, NY; Rochester Museum-Strasenburgh Planetarium-see separate listing; Rochester Museum-Cumming Nature Center, Gulick Rd., Naples, NY. Tax-exempt: 501(c)(3).
Institution Type/Description: Science & Technology, Natural Sciences & Cultural Heritage Museum.
Collections: over 1.2 million objects of regional science & technology, nature environments, & cultural heritage with emphasis on archaeological & contemporary Upstate New York Native cultures.
Research Fields: anthropology; regional culture & history; natural sciences; technology; environment.
Facilities: 30,000-vol. library of books on science, history, technology, anthropology, astronomy available for use on premises; 225-seat planetarium with large-format film; reading room; 400- & 100-seat auditoriums; classrooms. 900-acre Cumming Nature Center containing year-round trails, reconstructed pioneer homestead, log sugarhouse & outdoor exhibits; 100-seat theatre; wildlife viewing area; restaurant. Educational items pertaining to program fields for sale.
Activities: guided tours; lecture series; films; education programs; undergraduate & graduate internships; professional development for teachers; inter-museum loan; long-term & temporary exhibitions; classes; camps; scouts. Cumming Nature Center: cross country skiing & snowshoeing; maple sugaring; nature walks.
Publications: quarterly news & programs; research records; catalogs; annual report; educator guide.
Hours & Admission Prices: Rochester Museum: Mon.-Sat. 9-5, Sun. & holidays 11-5. Adults $13, senior citizens & college students $12, children 3-18 $11; discounts to AAM & ASTC members; RMSC & museum members and children under 3 no charge. Additional fee charged for some special exhibits. Strasenburgh Planetarium: call for hours & admission prices. Cumming Nature Center: Jan.-Oct. Sat.-Sun. 9-5. Suggested Donation: $3; members no charge. Museum & Planetarium: closed Thanksgiving; Christmas. &

Attendance: 388,931 (accurate)
Membership: Individual $58; Family $83; Family Plus $108; Patron $150 & up.

SENECA PARK ZOO, 2222 St. Paul St., Rochester, NY 14621-1097. Tel.: 585-336-7200. Fax: 585-342-1477.
E-mail: pcowan@senecazoo.org
Web Site: www.senecaparkzoo.org
Founded: 1894.
Congressional District: 29
Key Personnel: Dir. County Zoo, Lawrence Sorel; Exec. Dir. Seneca Park Zoo Society, Rachel Baker August; Veterinarian, Jeffrey Wyatt, D.V.M.; Museum Shop Mgr., Sue DeCaro.
Personnel Profile: Full-Time Paid 49; Part-Time Paid 32; Part-Time Volunteers 175; Interns 2.
Governing Authority: county. Parent Institution: County of Monroe, Dept. of Parks, 171 Reservoir Ave., Rochester 14620. Tax-exempt.
Institution Type/Description: Zoo.
Collections: zoology; aviary; herpetology.
Research Fields: public education on live animal behavior; pathology; herpetology.
Facilities: rental facilities; cafe; restaurant. Gift items for sale.
Activities: guided tours; lectures; films; study clubs; docent program or council.
Publications: quarterly newsletter.
Hours & Admission Prices: April-Oct. daily 10-5; Nov.-March daily 10-4. April-Oct. adults $10, seniors 63 & over $9, youth 3-11 $7; members and children 2 & under no charge. Nov.-March: adults $8, seniors 63 & over $7, youth 3-11 $5; members and children 2 & under no charge. Closed New Year's Day; Thanksgiving; Christmas. &
Attendance: 308,977 (accurate)
Membership: Individual $43; Senior Couple $54; Grandparent $70; Family $75; Grandparent Plus $89; Family Plus $94; Penguin Circle $135.

✳   **THE STRONG, (M),** One Manhattan Square, Rochester, NY 14607-3941. Tel.: 585-263-2700. Fax: 585-263-2493. TDD: 585-423-0746.
E-mail: info@thestrong.org
Web Site: www.thestrong.org
Formerly: Strong National Museum of Play
Founded: 1968.
Congressional District: 34
Key Personnel: Pres. & C.E.O., G. Rollie Adams, Ph.D.; Chm., Steve Dubnik; Sr. Vice Pres. Guest & Institutional Svcs. and Interim Vice Pres. Education, Kathleen Dengler; Vice Pres. Institutional Advancement, Lisa M. Feinstein; Vice Pres. Collections, Christopher Bensch; Vice Pres. Exhibit Devel. & Research, Jon-Paul Dyson, Ph.D.; Vice Pres. Finance & Facilities, Jeffry Flynn; Vice Pres. Play Studies, Scott G. Eberle, Ph.D.; Vice Pres. Mktg. & Communications, Suzanne Y. Seldes.
Personnel Profile: Full-Time Paid 90; Part-Time Paid 167; Part-Time Volunteers 18; Interns 11.
Governing Authority: private foundation. Tax-exempt.
Institution Type/Description: History Museum.
Collections: toys; dolls; games; electronic games; books, photographs, documents, and other materials related to play; 1918 carousel; National Toy Hall of Fame; International Center for the History of Electronic Games; Brian Sutton-Smith Library & Archives of Play.
Major Exhibits: Little Builders (T), 9/13-1/14; Animation (T), 2/14-4/27/14; Boardwalkl Arcade II, Summer 2014; Lego Castle (T), 9/27/14-2/1/15.
Facilities: 130,000-vol. library; 85,000 sq. ft. exhibit space; 160,000 sq. ft. butterfly garden; outdoor garden; 1,700 gallon coral reef aquarium; 300-seat theater; education rooms; food court. Museum-related items for sale.
Activities: interactive learning environments: family & children's programs; concerts; dramatic presentations; lectures; workshops; community forums; school lessons; curriculum materials; inner-city outreach programs; research.
Publications: three times a year; American Journal of Play.
Hours & Admission Prices: Mon.-Thurs. 10-5, Fri.-Sat. 10-8, Sun. 12-5; groups of 20 or more by appointment. Museum: $13.50; children under 2 & members no charge. Museum & Butterfly Garden: adults $17.50, senior citizens $16.50, children 2-5 $15.50; discounts to AAM members & school groups; members & children under 2 no charge. Closed Thanksgiving; Christmas. &
Attendance: 565,280 (accurate)
Membership: Family & Grandparent $119; Patron $160; Benefactor $250-$499; Sustaining $500-$999; Leading $1,000 & up.

Rome

ERIE CANAL VILLAGE, 5789 Rome-New London Rd., Rte. 46 & 49, Rome, NY 13440-8338. Tel.: 315-337-3999. Fax: 315-337-3999.
E-mail: mandm2000@twcny.rr.com
Web Site: www.eriecanalvillage.net

Founded: 1973.
Congressional District: 31
Key Personnel: Owner, Ronald Trottier; Mgr., Melody Milewski.
Personnel Profile: Part-Time Paid 5; Part-Time Volunteers 20.
Governing Authority: private.
Institution Type/Description: Outdoor Living History Museum.
Collections: agricultural equipment; domestic objects; textiles; costumes; archival material; tools; decorative art objects related to canal history & village life in 19th-century upstate NY; cheese manufacturing.
Research Fields: Erie Canal; cheesemaking; transportation; agriculture; village life.
Facilities: 550-vol. library of agriculture, canal, cheesemaking & 19th-century culture available to the public; 100-seat auditorium. Books, publications, pottery, food stuffs, village craft products, gift items & toys for sale.
Activities: guided tours; participatory exhibits.
Hours & Admission Prices: Memorial Day to Labor Day Wed.-Sat. 10-5. Adults $6.50, senior citizens $5, children $4. &
Attendance: 20,000 (estimated)

FORT STANWIX NATIONAL MONUMENT, 112 E. Park St., Rome, NY 13440-5816. Tel.: 315-338-7730. Fax: 315-334-5051.
E-mail: fost_superintendent@nps.gov
Web Site: www.nps.gov/fost
Founded: 1935.
Congressional District: 24
Key Personnel: Dir. & Supt., Debbie Conway; Chief Interpretation, Kimberly Szewczyk; Cur., Keith Routley.
Personnel Profile: Full-Time Paid 11; Part-Time Paid 10; Part-Time Volunteers 35.
Governing Authority: federal. Parent Institution: National Park Service, Dept. of the Interior. Tax-exempt.
Institution Type/Description: National Monument.
Collections: 18th, 19th and 20th-century archaeological collection including arms & accoutrements, clothing, hardware, utensils, glassware and pottery; manuscript collection.
Facilities: 400-vol. library; theater. Museum-related items for sale.
Activities: living history; guided tours; historical education; permanent exhibits.
Publications: informational brochure, historical studies.
Hours & Admission Prices: Fort: April-Nov. daily 10-4. No charge; donations accepted. Closed Thanksgiving. Visitor Center: daily 9-5. No charge. Closed New Year's Day; Thanksgiving; Christmas. &
Attendance: 71,263 (accurate)

ROME ART AND COMMUNITY CENTER, 308 W. Bloomfield St., Rome, NY 13440-4197. Tel.: 315-336-1040. Fax: 315-336-1090.
E-mail: executivedirector@romeart.org
Web Site: www.romeart.org
Founded: 1967.
Congressional District: 31
Key Personnel: Exec. Dir., Lauren Marie Getek; Chm. (V), Ann Lynch.
Personnel Profile: Full-Time Paid 2; Part-Time Paid 2; Part-Time Volunteers 15; Interns 2.
Governing Authority: nonprofit organization. Tax-exempt: 501(c)(3).
Institution Type/Description: Art Center.
Collections: all types of art media exhibits changing every month; Carpenter Home historical exhibit; historic audio tours.
Facilities: classrooms; community meeting rooms. Art-related items for sale.
Activities: guided tours; lectures; films; gallery talks; concerts; hobby workshops; formally organized educational programs.
Publications: quarterly newsletter; class schedules; quarterly, Community Cultural Calendar.
Hours & Admission Prices: Jan.-April 2 Mon.-Thurs. 10-5, Fri. 10-2; May-Dec. Mon.-Thurs. 10-5, Fri.-Sat. 10-2. No charge; donations accepted. Closed National holidays. &
Attendance: 25,000 (estimated)
Membership: Green $5; Children & Student $20; Senior $25; Individual $35; Household $60; Patron $125 & up.

ROME HISTORICAL SOCIETY MUSEUM, 200 Church St., Rome, NY 13440-5872. Tel.: 315-336-5870. Fax: 315-336-5912.
E-mail: info@romehistorical.org
Web Site: www.romehistorical.org
Founded: 1936.
Congressional District: 31
Key Personnel: Dir., Robert Avery; Pres. (V), Virginia Batchhelder; Cur., Ann Swanson; Museum Shop Mgr. & Admin., Mary Centro.
Personnel Profile: Full-Time Paid 2; Part-Time Paid 4; Part-Time Volunteers 12; Interns 2.
Governing Authority: Society. Parent Institution: Rome Historical Society. Tax-exempt.
Institution Type/Description: Historical Society Museum.
Collections: domestic objects; furniture; tools; textiles; photographs; maps; primary source documents dating from 18th century to present; Rome Turney Radiator Co. records; archives of local history and genealogical documents.
Research Fields: local history; genealogy.
Facilities: 2,000-vol. library of Rome area & Central NY materials & documents available for research on premises with authorization; reading room; auditorium. Museum-related items for sale.
Activities: guided tours; lectures; films; permanent & temporary exhibitions; outreach programs.
Publications: quarterly Annals & Recollections; quarterly newsletter, RHS News.
Hours & Admission Prices: Museum: Tues.-Thurs. 10-5, Fri. 10-3; other times by appointment. No charge; donations accepted. Library by appointment. Research Fee: $15; students no charge. Closed major holidays. &
Attendance: 22,376 (accurate)
Membership: Senior $15; Individual $25; Family $35; Supporting $50; Sustaining $100; Sponsor $250; Patron $500; Benefactor $1,000.

Rosendale

CENTURY HOUSE HISTORICAL SOCIETY - A.J. SNYDER ESTATE, 668 Rte. 213, Rosendale, NY 12472-0150. Mailing Address: P.O. Box 150, Rosendale, NY 12472-0150. Tel.: 845-658-9900.
E-mail: info@centuryhouse.org
Web Site: www.centuryhouse.org
Founded: 1988.
Congressional District: 26
Key Personnel: Pres. (V), Anne Gorrick; Museum Shop Mgr., Althea Doris Werner.
Personnel Profile: Part-Time Volunteers 24.
Governing Authority: state; New York State Education Dept. Tax-exempt.
Institution Type/Description: Historic House: 1809 old stone Century House, built by Christopher Snyder for his son, Jacob L. Snyder, pioneer in cement manufacturing. Historic Site: Snyder Estate Natural Cement Historic District, on state & national registers of historic places.
Collections: period furnishings; antiques; barn; carriage house; 20 carriages; 1820s-1940s sleighs.
Activities: history tours; special cultural events; music, art & performance events.
Publications: brochures; quarterly newsletter, Natural News.
Hours & Admission Prices: Memorial Day to Labor Day Sun. 1-4; other times by appointment. Suggested Donation $3, children $1; members no charge. &
Attendance: 4,325 (accurate)
Membership: Individual $25; Family $50; Business $250; Lifetime $500.

Roslyn

ROSLYN LANDMARK SOCIETY, 221 Main St., Roslyn, NY 11576-2168. Mailing Address: Roslyn Landmark Society, Box 234, 36 Main St., Roslyn, NY 11576. Tel.: 516-625-4363. Fax: 516-625-4363.
E-mail: info@roslynlandmarks.org
Web Site: www.roslynlandmarks.org/index.html
Formerly: Van Nostrand Starkins House
Founded: 1960.
Congressional District: 15
Key Personnel: Pres., Robert Sargent; Dir., Franklin H. Perrell.
Personnel Profile: Part-Time Paid 1; Part-Time Volunteers 2.
Governing Authority: nonprofit organization; Roslyn Landmark Society. Tax-exempt.
Institution Type/Description: Historical Building & Site: c.1680 earliest surviving building in Roslyn; site contains a 17th-century well & has had 4 archaeological investigations.
Collections: architectural exhibit showing framing & construction of a 17th-century house; American decorative arts; period rooms; artifacts.
Activities: guided tours; lectures; participatory exhibits; children's programs.
Publications: annual, House Tour Guide.
Hours & Admission Prices: June-Oct. Sat.-Sun. 1-4. Adults $4, children $2.
Attendance: 600 (estimated)
Membership: Outside Long Island $15; Contributing $25; Sustaining $50; Supporting $100; Fellowship $200; Patron $1,000; Benefactor $2,000; Life $5,000.

Roslyn Harbor

NASSAU COUNTY MUSEUM OF ART, (M), One Museum Dr., Roslyn Harbor, NY 11576-1138. Tel.: 516-484-9337. Fax: 516-484-0710.
E-mail: kwillers@nassaumuseum.org
Web Site: www.nassaumuseum.org
Founded: 1989.
Congressional District: 6
Key Personnel: Dir., Karl Emil Willers, Ph.D.; Asst. Dir. & Registrar, Fernanda Bennett; Office Mgr., Rita Mack; Pres. (V), Clarence F. Michalis; Dir. Devel., Monica Reischmann; Comptroller, Diane Roedel; Dir. Education, Laura Lynch; Asst. Cur., Rhianna Ellis; Weekend Coord., Julius Harris; Maintenance Engineer, Reynaldo Castillo; Volunteer Coord., Nancy Barone; Docent Coord., Riva Ettus; Docent Coord., Nancy Traeger; Museum Shop Mgr., Meryl Gordon.
Personnel Profile: Full-Time Paid 14; Part-Time Paid 10; Part-Time Volunteers 275; Interns 10.
Governing Authority: bd. trustees; nonprofit. Tax-exempt.
Institution Type/Description: Art Museum: housed in c.1900 three story neo-Georgian brick mansion, former estate of Childs Frick.
Collections: 19th & 20th-century American prints; drawings; 19th & 20th-century paintings; outdoor sculpture garden; formal gardens designed by Marion Cruger Coffin; architectural blueprints & drawings relating to the museum building & property.
Research Fields: all fields of art; 20th-century American sculpture.
Facilities: 145-acres of lawns, ponds & wooded areas; hiking trails; pinetum; formal gardens; studio art classrooms; bookshop. Gift items for sale.
Activities: guided tours; lectures; gallery talks; docent programs; formally organized education programs for adults, children & undergraduate college students; loan, temporary & traveling exhibitions.
Publications: quarterly, exhibition catalogs; newsletter.
Hours & Admission Prices: Tues.-Sun. 11-5. Adults $10, seniors $8, children & students $4; discounts to AAM members & corporate sponsors, Newsday, Channel 13 & radio station WLUS members; members no charge. Closed county holidays. &
Attendance: 200,000 (accurate)
Membership: Senior Citizen & Student $45; Individual $60; Family & Dual $85; Supporting $150; Contemporary Collector's Circle $200; Sustaining $275; Friend $500; Council Member $1,250; Corporate $1,000 & up; Council Circle $2,500; Director's Circle $5,000.

Rotterdam Junction

MABEE FARM HISTORIC SITE, 1100 Main St., Rotterdam Junction, NY 12150. Tel.: 518-887-5073. Fax: 518-214-0029. Facebook: Mabee Farm Historic Site.
E-mail: mabeefarm@gmail.com
Web Site: www.mabeefarm.org
Formerly: Historic Mabee Farm Site
Congressional District: 21
Key Personnel: Pres., Merritt Glennon; Chm., Marianne Blanchard.
Personnel Profile: Full-Time Paid 2; Part-Time Paid 1; Part-Time Volunteers 40; Interns 2.
Governing Authority: Tax-exempt. Parent Institution: Schenectady County Historical Society. Tax-exempt.
Institution Type/Description: History Museum.
Collections: 18th century life; furnishings; textiles; farming implements; cemetery; gardens. Historic Buildings: pre-revolutionary Dutch barn; 18th century house, inn, slave quarters, carriage shed, black smith, wood shop, & corn crib.
Research Fields: Mabee family; farming history; colonial life; enslaved peoples; documents; wars.
Facilities: education center; gardens; lecture halls; conference room; picnic area.
Activities: classes; music festival; workshops: blacksmithing; timber framing; spinning; knitting; Dutch oven cooking; soap making; make a gourd birdhouse; school programs; basket making; cornhusk doll making; tinsmithing. Museum Sponsors: Revolutionary War encampment; Community Day; Craft Fair; Farm & Foliage Day; Holiday Fair.
Publications: society newsletter.
Hours & Admission Prices: Tues.-Sat. 10-4. House Tours or Exhibits: adults $5, $8 for both tours; members no charge. Closed New Year's Eve & Day; Martin Luther King Jr. Day; Labor Day; Memorial Day; Independence Day; Labor Day; Columbus Day; Election Day; Thanksgiving & day after; Christmas Eve & Day. &
Attendance: 17,000 (estimated)
Membership: Individual $25; Family $50; Donor $75; Sponsor $100; Benefactor $250; Patron $500; Life $1,000.

Roxbury

JOHN BURROUGHS MEMORIAL STATE HISTORIC SITE, Burroughs Memorial Rd., Roxbury, NY 12474. Mailing Address: Mine Kill State Park, Rte. 30, P.O. Box 923, North Blenheim, NY 12131-0923. Tel.: 518-827-6111, ext. 91. Fax: 518-827-6782.
E-mail: john.lowe@parks.ny.gov
Web Site: www.nysparks.com
Founded: 1964.
Congressional District: 27
Key Personnel: Park Mgr., John Lowe.
Governing Authority: state. Parent Institution: New York State Office of Parks, Recreation & Historic Preservation. Tax-exempt: 501(c)(3).
Institution Type/Description: Historic Site: burial site of naturalist John Burroughs.
Collections: life & work of John Burroughs; photographs.
Hours & Admission Prices: Daily dawn to dusk. No charge.
Attendance: 2,500 (estimated)

Rye

JAY HERITAGE CENTER, 20 Boston Post Rd., Rye, NY 10580. Tel.: 914-698-9275.
Institution Type/Description: Historic House Museum: housed in the boyhood home of New York State's native founding father, John Jay; built in 1838. A National Historic Landmark.
Collections: local history & culture; period artifacts & furnishings; photographs.
Activities: educational programs; special events.
Hours & Admission Prices: Call for hours.

***   THE RYE HISTORICAL SOCIETY AND SQUARE HOUSE MUSEUM, (M),** One Purchase St., Rye, NY 10580-3002. Tel.: 914-967-7588. Fax: 914-967-6253.
Web Site: ryehistoricalsociety.org
Founded: 1964.
Congressional District: 18
Key Personnel: Exec. Dir., Dr. Ruth Smalt; Bd. Pres., Laura Brett.
Personnel Profile: Full-Time Paid 4; Part-Time Paid 1; Part-Time Volunteers 70; Interns 3.
Governing Authority: society; nonprofit organization. Parent Institution: The Rye Historical Society. Subsidiary Institution: Knapp House Library & Archives, 265 Rye Beach Ave., Rye, NY 10580. Tel.: 914-967-8657. Tax-exempt: 501(c)(3) & 170(b)(1)(A).
Institution Type/Description: History Museum: housed in c.1730 Square House.
Collections: decorative arts; furniture; costumes & textiles; 17th-20th century manuscripts; local history; photographs; archaeology. Historic Buildings: Square House c.1730, Knapp House c.1670-1750.
Research Fields: 17th-20th century manuscripts; 17th-18th century wills, inventories & deeds; regional history; photographs.
Facilities: 1,000-vol. library of local history available for research on premises; 80-seat meeting room.
Activities: guided tours; adult & family programs; school programs; Junior Volunteer summer camp for 4th-7th graders.
Publications: quarterly calendar of events; cookbook, My Grandmother Had a Woodburning Stove; Colonial Cooking at the Square House; Silent Companions: Dummy Board Figures of the 17th through 19th Centuries; The Art of Lauren Ford; Read about Rye; Estates of Grace: The Architectural Heritage of Religious Structures in Rye; 100 Years of Health Care; Father Burke's Dream to Rescue Children of the Inner City: St. Benedict's Home, Rye, New York, 1841-1941; Views of Rye: 1907-1997.
Hours & Admission Prices: Tues.-Fri. 9-4, Sat. 10-3. No charge; donations accepted.
Membership: Student $15; Individual $45; Family $75; Sustainer $125; Sponsor $250; Benefactor $500; Patron $1,000; Angel $5,000.

Sackets Harbor

SACKETS HARBOR BATTLEFIELD STATE HISTORIC SITE, 504 W. Main St., Sackets Harbor, NY 13685. Mailing Address: P.O. Box 27, Sackets Harbor, NY 13685-0027. Tel.: 315-646-3634. Fax: 315-646-1203.
E-mail: constance.barone@oprhp.state.ny.us
Web Site: www.nysparks.com
Founded: 1933.
Congressional District: 30
Key Personnel: Site Mgr., Constance B. Barone; Interpretive Programs Asst., Stephen Wallace.

Personnel Profile: Full-Time Paid 4; Part-Time Paid 16; Part-Time Volunteers 30.
Governing Authority: state. Parent Institution: New York State. Subsidiary Institution: New York State Office of Parks, Recreation & Historic Preservation and Thousand Islands State Park, Recreation & Historic Preservation Commission; Sackets Harbor Battlefield Alliance Support Group. Tax-exempt: 501(c)(3).
Institution Type/Description: Historic U.S. Navy Yard & Battlefield complex: housed in six buildings, 1818 Union Hotel; restored 1849 Commandant's & Master's houses; 1848 stable; 1850 ice house; 1832 farmhouse; located on site of 19th-century U.S. naval base, which played an important part in the War of 1812; Maritime Museum; restored 1850-60 navy yard building complex; War of 1812 battlefield in upstate N.Y.
Collections: War of 1812 weapons, armament accessories military clothing accessories; photographs; 1811-1860 microfilm on U.S. Navy; 19th-century household furniture & accessories; 1814 artifacts from Brig Jefferson, 1812-1815 Fort Tompkins. Historic Building: 1850 Commandant's house.
Research Fields: War of 1812 on Lake Ontario & in northern New York; life on a mid-19th century naval station; army life 1812-1816; ships built at Sackets Harbor; 1810-1820 military uniforms; field fortifications; 1750-1850 heirloom vegetable garden plants.
Facilities: 850-vol. library on local & military history; visitor center; reading room; picnic area.
Activities: guided tours; self-guided tours; temporary & permanent exhibits; educational programs; school lecture service; living history program; summer programs.
Publications: Archaeological Walking Tour of Sackets Harbor Battlefield; museum brochure; occasional pamphlets; Guide to 1813 Battlefield/Walking Tour.
Hours & Admission Prices: May-Sept. Thurs.-Mon.; other times for research only. Adults $3, seniors $2; Navy Yard Restoration: children under 12 no charge. &
Attendance: 112,280 (accurate)
Membership: Individual $25; Dual & Family $40; Business $50 & up; Corporate $100 & up.

Sag Harbor

SAG HARBOR WHALING & HISTORICAL MUSEUM, 200 Main St., Sag Harbor, NY 11963-3009. Mailing Address: P.O. Box 1327, Sag Harbor, NY 11963-0050. Tel.: 631-725-0770. Fax: 631-725-5638.
E-mail: info@sagharborwhalingmuseum.org
Web Site: www.sagharborwhalingmuseum.org
Founded: 1936.
Congressional District: 1
Key Personnel: Pres. (V), Barbara Lobasco; Treas., Zachary N. Studenroth; Business Mgr., Vanessa Petruccelli.
Personnel Profile: Part-Time Paid 2; Part-Time Volunteers 4.
Governing Authority: public corporation. Tax-exempt.
Institution Type/Description: Whaling Museum: housed in 1845 Greek Revival mansion, Benjamin Huntting House.
Collections: whaling tools; scrimshaw; period fishing rods, reels & lures; books; models; clocks; dinnerware; antiques; guns; toys; oil paintings; ship models; pianos & fine furniture.
Research Fields: Sag Harbor whale ships & whaling.
Facilities: children's museum; whaleboat. Maritime & museum-related items for sale.
Activities: special Sunday events; educational talks; demonstrations.
Publications: Whales & Whaling; Tales of Sag Harbor; Sag Harbor History; A Walking Tour of Sag Harbor.
Hours & Admission Prices: mid-May to mid-Oct. Mon.-Sat. 10-5, Sun. 1-5. Adults $6, senior citizens & students $5, children 3-11 $2; discount to groups; tour guides, members & bus drivers no charge.
Attendance: 8,500 (estimated)
Membership: Student & Senior $15; Individual $25; Family $30; Life $1,000.

Saint Bonaventure

THE REGINA A. QUICK CENTER FOR THE ARTS, (M), St. Bonaventure Univ., Rte. 417, Cornelius Welch Dr., Saint Bonaventure, NY 14778. Mailing Address: P.O. Drawer B.H., Saint Bonaventure, NY 14778. Tel.: 716-375-2494. Fax: 716-375-2690.
E-mail: quick@sbu.edu
Web Site: www.sbu.edu/quickcenter.aspx?id=2012
Formerly: St. Bonaventure Art Collection
Founded: 1856.
Congressional District: 39
Key Personnel: Exec. Dir., Joseph A. LoSchiavo.

Personnel Profile: Full-Time Paid 11; Part-Time Paid 5; Part-Time Volunteers 40; Interns 3.
Governing Authority: St. Bonaventure University. Tax-exempt: 501(c)(3).
Institution Type/Description: Art Museum.
Collections: paintings; sculpture; prints & drawings; Native American & Pre-Columbian pottery; Asian porcelain; Asian art; photography; creches.
Facilities: print & drawing study, & reading room (open by appointment).
Activities: guided tours; permanent & temporary exhibitions.
Publications: newsletter, QuickAccess.
Hours & Admission Prices: Summer: Tues.-Sat. 12-5. Winter: Mon.-Fri. 10-5, Sat.-Sun. 12-4. No charge; donations accepted. Closed New Year's Day; Easter; Thanksgiving; Christmas. &
Attendance: 20,000 (accurate)
Membership: Individual $45; Family/Dual $75; Contributing $100-$499; Sustaining $500-$999; Sponsor $1,000-$4,999; Patron $5,000-$9,999.

Saint Johnsville

FORT KLOCK HISTORIC RESTORATION, 7214 State Hwy. 5, Saint Johnsville, NY 13452-4502. Mailing Address: P.O. Box 42, Saint Johnsville, NY 13452-0042. Tel.: 518-568-7779.
E-mail: fortklock@gmail.com
Web Site: fortklockrestoration.org
Founded: 1954.
Congressional District: 105
Key Personnel: Chm. (V), John Case; Pres. (V), Eugene Wagner.
Personnel Profile: Part-Time Paid 1; Part-Time Volunteers 1.
Governing Authority: nonprofit organization. Tax-exempt: 501(c)(3).
Institution Type/Description: Historic House Museum: 1750 Klock Homestead.
Collections: early Palatine German architecture; farmhouse furnishings. Historic Buildings: restored Little Red Schoolhouse; Carriage House; Blacksmith Shop; Dutch Barn; 1750 Farmhouse; cheese house.
Research Fields: Dutch barns.
Facilities: picnic area. Museum-related items for sale.
Activities: guided tours; opening day colonial craft demonstrations. Museum Sponsors: Strawberry Festival in July; Stone Soup Musical Concert in July; Klock Family Reunion in August; Young Pioneer Program in August; Craft Fair in September; Interrupted Harvest in September; Fort Klock Haunted House in October; Open House for St. Nicholas Day in December.
Hours & Admission Prices: Memorial Day to mid-Oct. Tues.-Sun. 9-5; tours & special demonstration by appointment. Adults $3. &
Attendance: 2,000 (estimated)
Membership: Junior 14-18 $1; Active $10; Supporting $15; Family Supporting $20.

Salamanca

SALAMANCA RAIL MUSEUM, 170 Main St., Salamanca, NY 14779-1574. Tel.: 716-945-3133. Fax: 716-945-3133.
E-mail: salarail@verizon.net
Web Site: salamancarailmuseumassociation
Founded: 1980.
Congressional District: 31
Key Personnel: C.E.O., Chm. (V) & Cur., Gerald J. Fordham; Treas., Robert W. Irwin; Public Rels., Kevin Burleson; Museum Shop Mgr., Barbara A. Fordham.
Personnel Profile: Full-Time Volunteers 2; Part-Time Paid 2; Part-Time Volunteers 6.
Operating Expenses: 8,293
Operating Income: 9,317
Governing Authority: private; nonprofit organization. Tax-exempt: 501(c)(3).
Institution Type/Description: Transportation Museum: housed in a 1912 restored passenger depot constructed by the Buffalo, Rochester and Pittsburgh Railway.
Collections: artifacts; photographs; videos; early 20th-century office furniture; telegraph keys; railroad memorabilia; train cars; N scale model train; nostalgic railroad pieces with emphasis on the three railroads that served the region: The Erie, the Baltimore and Ohio (BR&P), and the Pennsylvania Railroads.
Research Fields: local railroad history.
Facilities: 1,200-vol. library; 30-seat theater; 2,700 sq. ft. exhibit space. Museum-related items for sale.
Activities: docent program; films; formal education programs for children; guided tours; lectures; loan exhibitions; broadcast programs.
Publications: calendar, Salamanca Rail Museum; quarterly newsletter, The Junction Express; weekly newspaper column, Tracks From The Past.
Hours & Admission Prices: April-Dec. Mon.-Sat. 10-5, Sun. 12-5. Tours: $1 donation per person. Closed Thanksgiving; Christmas. &
Attendance: 7,705 (accurate)

Membership: Senior Citizens $5; Regular $8.

SENECA-IROQUOIS NATIONAL MUSEUM, 814 Broad St., Salamanca, NY 14779-1378. Mailing Address: 252 Rochester St., Salamanca, NY 14779-1509. Tel.: 716-945-1760. Fax: 716-945-1624.
E-mail: sue.grey@sni.org
Web Site: www.senecamuseum.org
Founded: 1977.
Congressional District: 49
Key Personnel: Dir., Jare Cardinal; Public Rels. & Mktg. Mgr., Sue Grey; Museum Shop Mgr., Eva Aidman.
Personnel Profile: Full-Time Paid 7.
Governing Authority: nonprofit organization. Tax-exempt: 501(c)(3).
Institution Type/Description: Anthropology & Ethnology Museum.
Collections: period cultural ancestral artifacts; Seneca & other Iroquois Nations of the Northeast; archaeology; history; modern art.
Research Fields: Seneca-Iroquois studies.
Facilities: Native made items and artworks for sale.
Activities: guided tours; docent program; inter-museum, permanent & temporary exhibitions.
Publications: catalogue of collections; data sheets; bibliographies.
Hours & Admission Prices: May-Oct. daily 9-5; Nov.-Dec. & Feb.-April Mon.-Fri. 9-5. Adults $5, senior citizens & college students $3.50, children 7-16 $3; discounts to AAM & AAA members; children under 7 no charge. Closed SNI observed holidays. &

Attendance: 15,000 (estimated)

Sanborn

SANBORN AREA HISTORICAL SOCIETY, (M), 2822 Niagara St., Sanborn, NY 14132-9282. Mailing Address: P.O. Box 172, Sanborn, NY 14132-0172. Tel.: 716-731-9510.
E-mail: sanborngerry@frontiernet.net
Web Site: sanbornhistory.org
Founded: 1996.
Congressional District: 28
Key Personnel: Dir. & Pres. (V), Gary Townsend; Chm. (V), Hilda Snyder; Sec., Gerald E. Treichler; Public Rels. & Treas., Glenn Wienke; Archivist, Jane Schultz; Cur., Linda Jackson.
Personnel Profile: Part-Time Volunteers 110.
Governing Authority: private; nonprofit organization. Subsidiary Institution: Sanborn Area Farm Museum, 2660 Saunders Settlement Rd., Sanborn, NY. Tax-exempt: 501(c)(3).
Institution Type/Description: History Museum.
Collections: early farm equipment; kitchen; dolls; glassware; tools; toys; household artifacts; business advertising.
Facilities: Schoolhouse Museum: library; 600 sq. ft. exhibit space. Farm Museum: 6,000 sq. ft. exhibit space.
Activities: programs; school & organization groups. Annual Events: Farm Festival; Ice Cream Social; Antique Show & Sale; Kids Christmas 2 Day Event.
Publications: quarterly newsletter, Salubris.
Hours & Admission Prices: School House Museum: April-Nov. Sun. 2-4; Dec.-March 1st Sun. 2-4; other times by appointment. Farm Museum: April-Oct. Wed. 1-4, Sun. 2-4; other times by appointment. No charge; donations accepted. &

Attendance: 3,000 (estimated)
Membership: Individual $10; Family $15; Patron, Business & Professional $100.

Saranac Lake

ROBERT LOUIS STEVENSON MEMORIAL COTTAGE, 44 Stevenson Lane, Saranac Lake, NY 12983-1975. Mailing Address: P.O. Box 607, Saranac Lake, NY 12983-0607. Tel.: 518-891-1462.
E-mail: pennypiper@verizon.net
Web Site: www.robertlouisstevensonmemorialcottage.org
Founded: 1916.
Congressional District: 30
Key Personnel: Pres., William Delahant; Vice Pres., Thomas Delahant; Cur., Mike Delahant; Sec., Melinda Hadley; Treas., Les Hershhorn.
Personnel Profile: Full-Time Volunteers 1; Part-Time Volunteers 3.
Governing Authority: society. Operated by The Delahant Family. Tax-exempt.
Institution Type/Description: Literary Museum: housed in 1887 home of Robert Louis Stevenson.
Collections: personal mementos; childhood photographs; original furniture; original letters & articles of Stevenson lore.
Facilities: 125-vol. library of books on the life of Robert Louis Stevenson available for research by special request.

Activities: guided tours.
Publications: book, The Penny Piper of Saranac - An Episode in the Life of Robert Louis Stevenson.
Hours & Admission Prices: July to Columbus Day Tues.-Sun. 9:30-12 & 1-4:30; other times by appointment. Adults $5; children under 12 & members no charge.
Attendance: 500 (estimated)
Membership: Individual/Family $25; Business/Organization $50; Supporter $100 & up; Patron $500 & up; Benefactor $1,000 & up.

SARANAC LABORATORY MUSEUM, 89 Church St., Ste. 2, Saranac Lake, NY 12983. Tel.: 518-891-4606.
E-mail: mail@historicsaranaclake.org
Web Site: historicsaranaclake.org
Key Personnel: Exec. Dir., Amy Catania
Institution Type/Description: History Museum.
Collections: local history; TB research & patient care; photographs; period artifacts.
Activities: special events; group tours.
Hours & Admission Prices: Mon.-Fri. 9-3 by appointment.

Saratoga Springs

THE CHILDREN'S MUSEUM AT SARATOGA, 69 Caroline St., Saratoga Springs, NY 12866-3202. Tel.: 518-584-5540. Fax: 518-584-6059.
E-mail: info@cmssny.org
Web Site: www.cmssny.org
Founded: 1989.
Congressional District: 21 & 22
Key Personnel: Chm. (V), Michael Mihaly; Dir., Michelle Smith.
Personnel Profile: Full-Time Paid 2; Part-Time Paid 15; Part-Time Volunteers 10; Interns 4.
Governing Authority: private; nonprofit. Tax-exempt: 501(c)(3).
Institution Type/Description: Children's Museum.
Collections: interactive exhibits for children ages 2-7.
Facilities: educational facilities; 8,400 sq. ft. exhibit space. Shirts, caps & museum-related items for sale.
Activities: participatory exhibits; school outreach programs. Annual Events: Dinner Gala in March; Big Truck Day Event in August.
Publications: weekly e-newsletter.
Hours & Admission Prices: July-Labor Day Mon.-Sat. 9:30-4:30; Labor Day-June Tues.-Sat. 9:30-4:30, Sun. 12-4:30. Admission $6; children under one no charge. Closed New Year's Day; Easter; Memorial Day; Independence Day; Thanksgiving; Christmas Eve & Day. &
Attendance: 35,339 (accurate)
Membership: You & Me $70; Top Trio $85; Family Fun $100; Provider $110; Traveler Pass $125; Explorers Plus $175.

FRANCES YOUNG TANG TEACHING MUSEUM AND ART GALLERY, Skidmore College, 815 N. Broadway, Saratoga Springs, NY 12866-1632. Tel.: 518-580-8080. Fax: 518-580-5069.
E-mail: tang@skidmore.edu
Web Site: www.skidmore.edu/tang
Founded: 2000.
Congressional District: 22
Key Personnel: Education, Susi Kerr; Registrar, Elizabeth Karp; Cur., Ian Berry; Museum Shop Mgr., Barbara Schrade.
Personnel Profile: Full-Time Paid 14; Part-Time Paid 19; Part-Time Volunteers 4; Interns 4.
Governing Authority: private college; nonprofit. Parent Institution: Skidmore College. Tax-exempt: 501(c)(3).
Institution Type/Description: Art Museum.
Collections: drawings; paintings; prints; sculptures; video, audio & installation art.
Research Fields: interdisciplinary teaching with artwork and objects of material culture.
Facilities: 200-seat auditorium; educational facilities. Museum-related items for sale.
Activities: concerts; docent program; films; formal education programs for children; guided tours; lectures; participatory, traveling & temporary exhibitions.
Publications: newsletter, Tang Talks; exhibit catalogues.
Hours & Admission Prices: Tues.-Sun. 12-5. Suggested Donations: adults $5; discounts to AAM & ICOM members; members no charge. Closed New Year's Day; Thanksgiving; Christmas. &
Attendance: 50,000 (accurate)
Membership: See website for information.

HISTORICAL SOCIETY OF SARATOGA SPRINGS, The Casino, Congress Park, Saratoga Springs, NY 12866. Mailing Address: P.O. Box 216, Saratoga Springs, NY 12866-0216. Tel.: 518-584-6920. Fax: 518-581-1477.
E-mail: info@saratogahistory.org
Web Site: www.saratogahistory.org
Founded: 1883.
Congressional District: 21
Key Personnel: Dir., James D. Parillo; Pres., Lisa Millis; Cur., Becky Codner; Research Asst., John Conors; Archivist, Doris Lamont; Museum Shop Mgr., Ted Waite.
Personnel Profile: Full-Time Paid 1; Part-Time Paid 3; Part-Time Volunteers 40; Interns 1.
Governing Authority: society. Parent Institution: George S. Bolster Collection. Branch Museum: Walworth Memorial Museum. Tax-exempt: 101(6) & 501(c)(3).
Institution Type/Description: Local History Museum: housed in 1871 gambling casino, a designated National Landmark.
Collections: relating to the springs, hotels, gambling & other facets of Saratoga Springs history; Walworth Museum: furnishings of Reuben Hyde Walworth, last Chancellor of New York; papers of Frank Sullivan.
Research Fields: 19th-century resort life; spas & mineral springs; early history of the D.A.R.; Frank Sullivan; history of 19th- & 20th-century Saratoga Springs; hydrotherapy; New York State chancellors & Court of Chancery; 19th-century law & legal codes; Roman Catholic Paulist Order.
Facilities: 2,000-vol. library of books on the history of Saratoga Springs, newspapers, guidebooks, clippings, manuscripts, photographs; manuscripts, diaries, journals & correspondence of the Walworth family (1820-1950) & writer Frank Sullivan available by appointment; Ann Grey Gallery for changing exhibitions.
Activities: formally organized educational programs; monthly meetings & programs.
Publications: monthly newsletter, Chips; George S. Bolster (monograph), The Casino; exhibit catalogues.
Hours & Admission Prices: Wed.-Sun. 10-4. Adults $5, senior citizens $4; discounts to AAM members; children under 12, members, Ann Grey Gallery, Bolster Collection Archives no charge.
Attendance: 12,000 (estimated)
Membership: Senior Citizen & Student $15; Individual $25; Family $35; Sustaining $50; Patron $100.

NATIONAL MUSEUM OF DANCE & HALL OF FAME, 99 S. Broadway, Saratoga Springs, NY 12866-4557. Tel.: 518-584-2225, ext. 3001. Fax: 518-584-4515. Facebook: Dance Museum.
E-mail: info@dancemuseum.org
Web Site: www.dancemuseum.org
Founded: 1986.
Congressional District: 22
Key Personnel: Dir., Donna Skiff; Chm. Bd. Dir. (V), Michele Riggi; Design & Devel., Laura Dirado; Coord. Grants & Exhibits, Matt MacVittie; Rental Coord., Jo Ambrosio; Administrative Coord., Jessica Munson.
Personnel Profile: Full-Time Paid 4; Part-Time Paid 3; Part-Time Volunteers 30; Interns 5.
Governing Authority: private; nonprofit organization. Parent Institution: Saratoga Performing Arts Center, Saratoga, NY. Tax-exempt: 501(c)(3).
Institution Type/Description: Dance Museum.
Collections: photographs; personal items of dancers; dance & dance history; costumes; videos; audio; clippings; ephemera; 19th-20th century American professional dance artifacts; dance archives.
Research Fields: relationships between dance and socio-economical, political & cultural aspects of society at large.
Facilities: 3 dance studios; 35,000 sq. ft. exhibit space; resource room. Gift items for sale.
Activities: dance school; dance workshops; lectures; showings; interactive children's corner; live performances; arts festivals; dance concerts; guided tours; workshops; rental facilities; special events. Annual Events: Hall of Fame Induction; exhibit openings.
Publications: biannual newsletter, Foot Notes.
Hours & Admission Prices: March-Nov. Tues.-Sun. 10-4:30. Adults $6.50, senior citizens & students $5, children under 12 $3; discounts to AAM members; members no charge. Office closed New Year's Eve, day & week; Christmas Eve, Day & week.
Attendance: 9,000 (estimated)
Membership: Student & Senior $25; Enthusiast $40; Family Dance Troupe $60; Corps Dancer $100; Soloist $250; Principal $500; Choreographer $1,000; Patron $2,000; Benefactor $5,000.

NATIONAL MUSEUM OF RACING AND HALL OF FAME, (M), 191 Union Ave., Saratoga Springs, NY 12866-3556. Tel.: 518-584-0400; 800-562-5394. Fax: 518-584-4574.
E-mail: info@racingmuseum.org/nmrmedia@racingmuseum.net
Web Site: racingmuseum.org
Founded: 1950.
Congressional District: 29
Key Personnel: Dir., Christopher Dragone; Pres., Stella F. Thayer; Asst. Dir. & Membership, Cathy Maguire.
Personnel Profile: Full-Time Paid 10; Part-Time Paid 5; Part-Time Volunteers 15.
Governing Authority: nonprofit organization. Tax-exempt: 501(c)(3).
Institution Type/Description: National Thoroughbred Racing Museum.
Collections: equine paintings; sporting art; thoroughbred racing trophies; racing colors; sculpture; racing memorabilia; hall of fame.
Research Fields: thoroughbred horses; thoroughbred racing.
Facilities: library available for use by appointment; 250-seat auditorium; theater; films; video equipment. Items pertaining to thoroughbred horses & racing, books, prints & other museum-related items for sale.
Activities: guided tours for groups by advance appointment; audiovisuals; films; inter-museum & permanent exhibitions. Museum Sponsors: gala; Hall of Fame induction ceremonies.
Publications: Quarterly Member Newsletter.
Hours & Admission Prices: Racing Season daily 9-5; Off Season: Mon.-Sat. 10-4, Sun. 12-4. Adults $7, students & senior citizens with ID $5; discounts to groups, AAM & ICOM members; members & children under 5 no charge. Closed New Year's Day; Easter; Thanksgiving; Christmas.
Attendance: 60,000 (accurate)
Membership: Individual $50; Family $75; Contributor $100; Donor $250; Associate $500; Patron $1,000; Benefactor $1,500; Gold Cup $2,500.

NEW YORK STATE MILITARY MUSEUM AND VETERANS RESEARCH CENTER, (M), 61 Lake Ave., Saratoga Springs, NY 12866-2315. Tel.: 518-581-5100. Fax: 518-581-5111.
E-mail: historians@ny.ngb.army.mil
Web Site: www.nysmm.org
Founded: 1863.
Congressional District: 20
Key Personnel: Dir., Michael Aikey; Registrar, Christopher Morton; Chief Cur., Courtney Burns; Archivist, Jim Gandy.
Personnel Profile: Full-Time Paid 7; Part-Time Paid 1; Part-Time Volunteers 50; Interns 3.
Governing Authority: state.
Institution Type/Description: Military Museum.
Collections: New York State's military history from colonial times to the present; New York State's battle flag; State's veterans oral history; military equipment.
Facilities: 10,000-vol. library of military books; 70-seat auditorium; 9,000 sq. ft. exhibit space. Museum-related items for sale.
Activities: docent program; formal education programs; guided tours; lectures; temporary exhibitions.
Hours & Admission Prices: Tues.-Sat. 10-4. No charge. Closed New York state holidays.
Attendance: 12,000 (estimated)

SARATOGA AUTOMOBILE MUSEUM, (M), 110 Avenue of the Pines, Saratoga Springs, NY 12866-6220. Tel.: 518-587-1935. Fax: 518-587-4149.
Web Site: www.saratogaautomuseum.org
Founded: 2002.
Congressional District: 21
Key Personnel: Dir., Jean Hoffman; Dir. Devel., Richard Selikoff; Education & Public Rels., Alan Edstrom.
Personnel Profile: Full-Time Paid 4; Full-Time Volunteers 15; Part-Time Paid 2; Part-Time Volunteers 140; Interns 4.
Governing Authority: private; nonprofit organization. Tax-exempt: 501(c)(3).
Institution Type/Description: Automobile Museum.
Collections: automobile history; automobiles; racing cars.
Activities: formal education programs; car restoration classes for VOTEC students; guided tours; lectures; participatory exhibits. Annual Events: Auto Show in spring; Lawn Shows in summer.
Publications: quarterly newsmagazine, Horsepower; gallery guides; show program.
Hours & Admission Prices: June-Sept. daily 10-5; Columbus Day to May Tues.-Sun. 10-5. Adults $8, senior citizens, active military & students 17 & over $5, children 6-16 $3.50; discounts to groups; children under 6 no charge. Closed New Year's Eve & Day; Thanksgiving; Christmas.
Attendance: 30,000 (accurate)

Membership: Student & Senior $25; Single $35; Family $50; Patron $250; Silver Arrow $1,000.

THE SCHICK ART GALLERY, SKIDMORE COLLEGE, 815 N. Broadway, Saisselin Art Bldg., Fl. 2, Saratoga Springs, NY 12866-1698. Tel.: 518-580-5049 & 5000. Fax: 516-580-5029.
E-mail: mjablons@skidmore.edu
Web Site: www.skidmore.edu/schick
Founded: 1926.
Key Personnel: Asst. to Dir., Mary Jablonski.
Personnel Profile: Full-Time Paid 1; Part-Time Paid 4.
Governing Authority: college. Parent Institution: Skidmore College. Tax-exempt.
Institution Type/Description: Art Gallery.
Collections: paintings; graphics; ceramics; sculpture.
Facilities: gallery for changing exhibits.
Activities: lectures; films; gallery talks; temporary exhibitions.
Publications: exhibition brochures; catalogues.
Hours & Admission Prices: Sept.-May Mon.-Fri. 9-5, Sat.-Sun. 1-4:30; Summer: hours variable according to summer class schedules. No charge. &
Attendance: 25,000 (estimated)

Saugerties

OPUS 40 SCULPTURE PARK AND MUSEUM, 50 Fite Rd., Saugerties, NY 12477-3260. Tel.: 845-246-3400 & 9922. Fax: 845-246-1997.
E-mail: patopus40@hotmail.com
Web Site: www.opus40.org
Formerly: Opus 40 and the Quarryman's Museum
Founded: 1978.
Congressional District: 26
Key Personnel: Pres., Pat Richards; Sec., Tad Richards.
Personnel Profile: Part-Time Volunteers 30.
Governing Authority: nonprofit organization. Parent Institution: Opus 40, Inc. Tax-exempt: 501(c)(3).
Institution Type/Description: Art Museum & Sculpture Park.
Collections: 6.5-acre environmental bluestone sculpture; sculpture park; 19th-century quarrymen's tools & artifacts; paintings; photographs; drawings.
Facilities: 70-acres of grounds; performing arts center. Gift items for sale.
Activities: lectures; concerts; performing, visual & literary arts events; art exhibitions.
Publications: book, Opus 40: The First 20 Years; monograph, Harvey Fite's Opus 40; brochure.
Hours & Admission Prices: May-Oct. Thurs.-Sun. 11-5:30. Adults $10, senior citizens & students $7, school age children $3; discounts to AAM & ICOM members; members & children under 5 no charge.
Attendance: 15,000
Membership: Student, Senior & Veteran $25; Individual $40; Dual & Household $75; Family $100; Premiere $250; Benefactor $500; Patron $1,000.

Sayville

SAYVILLE HISTORICAL SOCIETY, Edwards St. & Collins Ave., Sayville, NY 11782. Mailing Address: P.O. Box 41, Sayville, NY 11782-0041. Tel.: 631-563-0186 & 567-1289.
Founded: 1944.
Congressional District: 2
Key Personnel: Admin., Linda Conron; Pres. (V), Constance Currie; Treas., Cathy Foudy.
Personnel Profile: Part-Time Paid 2; Part-Time Volunteers 8.
Governing Authority: board of directors; not-for-profit organization. Tax-exempt.
Institution Type/Description: Historical Society Museum.
Collections: local history from colonial times to present.
Facilities: library available to public only upon request; 2,900 sq. ft. exhibit space; herb garden.
Activities: guided tours; lectures; loan & temporary exhibits. Annual Events: Christmas Open House; Holiday House Tour; An Afternoon on the Edwards Farm.
Publications: quarterly newsletter, Homestead Happenings.
Hours & Admission Prices: Oct.-June 1st & 3rd Sun. 2-4. No charge; donations accepted. Closed when holiday falls on Open House Sunday. &
Attendance: 830 (accurate)
Membership: Junior $5; Student $10; Individual $20; Family $30; Patron $100; Sponsor $500; Benefactor $1,000.

Scarsdale

THE GREENBURGH NATURE CENTER, 99 Dromore Rd., Scarsdale, NY 10583-1705. Tel.: 914-723-3470. Fax: 914-725-6599.
E-mail: mtjimosgoldberg@greenburghnaturecenter.org
Web Site: www.greenburghnaturecenter.org
Founded: 1975.
Congressional District: 23
Key Personnel: Exec. Dir., Margaret Tjimos Goldberg; Dir. Operations & Visitor Svcs., Penny Berman; Dir. Finance & Devel., Sara Cashen; Dir. Education & Living Collections, Travis Brady; Dir. Mktg., IT & Communications, Vicki Seiden Sherman; Data Mgr., Jocelyn Lim.
Personnel Profile: Full-Time Paid 12; Part-Time Paid 2; Part-Time Volunteers 50; Interns 4.
Governing Authority: bd. of directors; nonprofit organization. Tax-exempt: 501(c)(3).
Institution Type/Description: Nature Center: located on a 33-acre greenspace, former Nunataks Estate.
Collections: wildlife; working & observational honeybees hives; pond life aquaria & brook; 25 acres of woodlands & wetland; 2-acre lawn with specimen shrubs & trees; rock outcrops of Fordham Gneiss; live animals including over 140 exotic & local fauna; glacial boulders; nature discovery room with hands-on exhibits; rock garden; apple orchard; herb garden; natural history artifacts; Native American & colonial farm artifacts.
Research Fields: bird & reptile behavior; natural history; insects; arachnids; forest ecology.
Facilities: 700-vol. nature library pertaining to zoology, botany, natural history & ecology; oasis for spring & fall migrating song birds; aquarium; 60-seat auditorium; botanical garden; nature center; zoological park; meeting rooms; nature trails. Nature-related items for sale.
Activities: guided tours; natural trails; lectures; outdoor concerts; hobby workshops; organized educational programs; outreach program; participatory exhibits; mobile van; temporary & changing exhibitions; playground. Museum Sponsors: Spring Fair; Halloween Walks; Fall Festival.
Publications: quarterly newsletter; calendar; various guides to gardens, grounds & trails.
Hours & Admission Prices: Grounds: daily dawn-dusk. Manor House: Mon.-Thurs. 9:30-4:30, Sat.-Sun. 10-4:30. Adults $7, children 2-12 $5; discounts to NYSAM members; Greenburgh residents, Westchester County Parks Pass; members no charge except for special events & programs. &
Attendance: 64,000 (estimated)
Membership: Senior Citizen $40; Individual $50; Grandparents Special $60; Family $75; Grand Family $90; Family Fun Package $150; Sustainer $250; Benefactor $500; Patron $1,000.

THE SCARSDALE HISTORICAL SOCIETY, 937 Post Rd., Scarsdale, NY 10583-5656. Mailing Address: P.O. Box 431, Scarsdale, NY 10583-0431. Tel.: 914-723-1744. Fax: 914-723-2185.
E-mail: history@cloud9.net
Web Site: scarsdalehistory.org
Founded: 1973.
Congressional District: 20
Key Personnel: Pres., Bill Doescher; Exec. Dir., Cindy Krossman; Treas., Gloria Forte; Museum Shop Mgr., Greta Fisher; Museum Shop Mgr., Etta Parker.
Personnel Profile: Full-Time Paid 1; Part-Time Paid 3; Part-Time Volunteers 20.
Governing Authority: nonprofit organization. Tax-exempt: 501(c)(3).
Institution Type/Description: Historical Society: housed in 1828 Quaker Meeting House, a 19th-century farm house of modest means.
Collections: 19th-century textiles, costumes, furniture, rugs, farm equipment, documents, paintings, kitchen utensils, maps, photographs. Historic House: 18th-century Cudner-Hyatt Farm House.
Research Fields: architecture; furnishings; gardens; cooking; costumes; decorative arts; local history; oral history; archaeology.
Facilities: 2,500-vol. library pertaining to history, architecture, costumes, furniture, cooking & gardening available on premises only; slides; tapes; maps; photographs; paintings; documents. Books & other publications for sale.
Activities: guided tours; audiovisual slide presentation; lectures; organized educational programs; docent program; temporary, loan, traveling & participating exhibitions.
Publications: brochures; booklets, Scarsdale Heritage Homes; A Celebration of Westchester Arts & Decorations of Three Hundred Years; Bronx River Retrospective; Did you Know your Roof had a Name?; The Cudner-Hyatt House-The Story of its Two Families; A Century of Ceramics in the Hudson Valley; Summer Pleasures: Suburban Leisure in the 19th Century.

Hours & Admission Prices: Mon.-Fri. 9-4; other times by appointment. Museum: adults $3, senior citizens & students $2; discounts to AAM members; members no charge. Cudner Hyatt House: adults $5, seniors & students $3; discounts to AAM members; members no charge. Closed national holidays.
Attendance: 10,000 (accurate)
Membership: Student $10; Senior Citizen $25; Business $25 and up; Senior Couple $35; Family $50; Donor $75; Sponsor $100; Associate $125; Sustainer $150; Patron $300; Benefactor $600.

WEINBERG NATURE CENTER, 455 Mamaroneck Rd., Scarsdale, NY 10583-7727. Tel.: 914-722-1289. Fax: 914-723-4784 (call first). Facebook: Weinberg Nature Center.
E-mail: info@weinbergnaturecenter.org
Web Site: www.weinbergnaturecenter.org
Founded: 1958.
Key Personnel: Pres., Dr. Melissa Grigione; Program Dir., Cindy Polera.
Personnel Profile: Part-Time Paid 2; Part-Time Volunteers 15; Interns 3.
Governing Authority: municipal. Parent Institution: The Scarsdale Parks & Recreation Dept. Tax-exempt.
Institution Type/Description: Nature Center.
Collections: rocks; shells; leaves; photographs of birds & wild flowers; live exhibits; wildlife exhibits; taxidermy specimens of native wildlife; Native American artifacts.
Facilities: 500-vol. library of books & research notes available for use by special permission; nature trails.
Activities: guided tours; lectures; formally organized educational programs; temporary exhibitions; workshops; scout programs.
Publications: newsletter format, calendar of events.
Hours & Admission Prices: Summer: Mon.-Fri. 9-5; Fall, Winter & Spring: Mon., Wed. & Fri. 10-5, Tues. & Thurs. 10-2, see website to confirm. Charge for some weekend programs. Closed village holidays. &
Attendance: 20,000 (estimated)
Membership: Friends of the Weinberg Nature Center Inc.: Individual $35; Family $50; Institutional $100; Sustaining $250; Life $1,000.

Schenectady

MANDEVILLE GALLERY, UNION COLLEGE, (M), Nott Memorial, Schenectady, NY 12308. Mailing Address: 807 Union St., Union College, Schaffer Library, Schenectady, NY 12308-3103. Tel.: 518-388-6729. Fax: 518-388-8340.
E-mail: costellm@union.edu
Web Site: www.union.edu/gallery
Founded: 1995.
Congressional District: 21
Key Personnel: Dir., Rachel Seligman.
Personnel Profile: Full-Time Paid 2.
Governing Authority: private; nonprofit organization. Parent Institution: Union College, Schenectady, NY 12308-3155. Tax-exempt: 170(b)(1)(A).
Institution Type/Description: College Museum.
Collections: 19th-21st century European & American works on paper; portraits; Asian art; early scientific & mathematical apparatus; period artifacts.
Activities: lectures; loan, participatory, traveling & temporary exhibitions.
Hours & Admission Prices: Daily 10-6. No charge; donations accepted. Closed New Year's Eve & Day; Independence Day; Thanksgiving & day after; Christmas week. &
Attendance: 5,000 (estimated)

MISCI, (M), 15 Nott Terrace Heights, Schenectady, NY 12308-3198. Tel.: 518-382-7890. Fax: 518-382-7893.
E-mail: communications@misci.org
Web Site: www.misci.org
Formerly: Schenectady Museum and Suits-Bueche Planetarium
Founded: 1934.
Congressional District: 21
Key Personnel: Exec. Dir., Dr. William "Mac" Sudduth; Pres. (V), Earl Redding; Museum Shop Coord., Lindsay Sheehan.
Personnel Profile: Full-Time Paid 9; Part-Time Paid 30; Part-Time Volunteers 153; Interns 5.
Volunteer Hours: 1,140
Operating Expenses: 785,468
Operating Income: 779,442
Governing Authority: nonprofit organization. Tax-exempt: 501(c)(3).
Institution Type/Description: History, Science & Technology Museum and Planetarium.
Collections: area science & technology including General Electric; Charles Steinmetz & American Locomotive Company photographs; WRGB television station artifacts; technology artifacts including early home appliances,

radios, televisions, experimental electric equipment, medical imaging equipment, & 1978 prototype electric car; technology-related art.
Major Exhibits: Notion of Motion: A Moving Experience from San Francisco's Exploratorium (T), 6/15/13-6/1/14.
Research Fields: history of regional science & technology.
Facilities: 30,000 sq. ft. exhibit space; 70-seat auditorium; 61-seat planetarium; classrooms; archives; radio stations WZIR & WBZCRZ; SMARA Amateur Radio Club. Museum related items for sale.
Activities: lectures; films; gallery talks; permanent & temporary exhibits in science & regional history; adult classes; workshops; education programs for children & adults; planetarium programs; docent program; outreach programs; camp-ins.
Publications: quarterly newsletters; monthly member updates; annual report.
Hours & Admission Prices: Mon.-Sat. 9-5, Sun. 12-5. Adults $9.50, seniors $8, children 3-12 $6.50; discounts to AAM, ICOM & AAA members; ASTC, Empire State Reciprocal Program members, museum members & children under 4 no charge. Planetarium Programs: Tues.-Fri. 2 pm, Sat. 1, 2 & 3. Additional $5 per person for programs. Closed New Year's Day; Independence Day; Thanksgiving; Christmas. &
Attendance: 80,128 (accurate)
Membership: Senior Couple & Individual $45; Grandparents $50; Couple $60; Family $65.

SCHENECTADY COUNTY HISTORICAL SOCIETY, 32 Washington Ave., Schenectady, NY 12305-1600. Tel.: 518-374-0263. Fax: 518-688-2825.
E-mail: curator@schist.org
Web Site: www.schist.org
Founded: 1905.
Congressional District: 21
Key Personnel: Vice Pres., Merritt Glennon; Cur., Ryan Mahoney; Librarian, Melissa Tacke; Site Mgr. Mabee Farm, Pat Barrot; Office Mgr., Jennifer Hanson.
Personnel Profile: Full-Time Paid 5; Part-Time Paid 1; Part-Time Volunteers 50; Interns 4.
Governing Authority: society. Subsidiary Institution: Mabee Farm Historic Site, 1080 Main St. (Rt. 5S), Rotterdam Junction, NY 12150. Tax-exempt: 501(c)(3).
Institution Type/Description: General Museum & Historic House Site: located within area of original Schenectady stockade built by the Dutch in 1661.
Collections: paintings; guns; toys and dolls; furniture; household goods; Indian artifacts; decorative arts; various collections pertaining to Schenectady County; books; manuscripts; genealogical data; maps; photographs; recordings.
Research Fields: genealogy; Schenectady County history; early colonial (Dutch, English & Palatine) settlement in New York, Erie Canal; American Locomotive Co.; General Electric.
Facilities: 2,500-vol. library of books available on premises; reading room; 50-seat auditorium. Books for sale.
Activities: guided tours; lectures; courses; inter-museum loan, permanent & temporary exhibitions; research guidance & training; Walkabout & Waterfront Fair.
Publications: bimonthly, Schenectady County Historical Society Newsletter. Arcadia publications: Rotterdam; Glenville; Niskayuna.
Hours & Admission Prices: Museum: Mon.-Fri. 1-5, Sat. 10-2. Library: Mon.-Fri. 9-5, Sat. 10-2. Mabee Farm: Tues.-Sat. 10-4. Museum: adults $5, children $2. Library: adults $5; members & students no charge. Closed national holidays. &
Attendance: 17,000 (accurate)
Membership: Individual $25; Family $50; Donor $75; Sponsor $100; Benefactor $250; Patron $500; Life $1,000.

Schnectady

ALCO HERITAGE MUSEUM, 1910 Maxon Rd. Ext., Schnectady, NY 12308. Tel.: 518-557-2673.
E-mail: james.cesare@ahts.org
Web Site: www.alcoheritagemuseum.org
Founded: 2012.
Key Personnel: Dir., James Cesare
Institution Type/Description: History Museum.
Collections: ALCO company history; steam & diesel locomotives; armored vehicles; oral histories; replica drafting room; photographs.
Activities: diesel locomotive simulator; special events; educational programs.
Hours & Admission Prices: Sat.-Sun. 10-5.

Schoharie

OLD STONE FORT MUSEUM COMPLEX, (M), 145 Fort Rd., Schoharie, NY 12157-4705. Tel.: 518-295-7192. Fax: 518-295-7187.
E-mail: office@schohariehistory.net
Web Site: www.theoldstonefort.org
Founded: 1889.
Congressional District: 22, 23
Key Personnel: C.E.O. & Dir., Carle J. Kopecky; Pres., Jeremy Rosenthal; Treas., Anne Hendrix; Cur. & Education, Daniel Beams; Museum Shop Mgr. & Public Rels., Laura Spickerman.
Personnel Profile: Full-Time Paid 4; Part-Time Paid 12; Part-Time Volunteers 80.
Governing Authority: society. Parent Institution: Schoharie County Historical Society. Subsidiary Institution: Lansing Manor Museum. Tax-exempt: 501(c)(3).
Institution Type/Description: General Museum: housed in 1772 church, later used as a fort.
Collections: Indian artifacts; deeds; manuscripts; documents; local genealogies; 18th-century land grants; maps; church records; cemetery records; local newspapers; tools; vehicles; furniture; household items; firearms; arts & crafts; clothing; quilts; agriculture; technology; military history; transportation; archaeology; natural history; textiles; folk art; decorative arts.
Research Fields: local history; genealogy; folklore.
Facilities: 2,500-vol. library of NY State & Schoharie County history available for use on the premises May-Oct.; reading room. Books & other museum-related items for sale.
Activities: guided tours; lectures; permanent & temporary exhibitions; living history demonstrations; concerts; formal education programs; school loan service. Museum Sponsors: Revolutionary War & Civil War reenactments; Strawberry Festival.
Publications: semiannual pamphlet, Schoharie County Historical Review.
Hours & Admission Prices: May-June & Sept.-Oct. Tues.-Sat. 10-5, Sun. 12-5; July.-Aug. Mon.-Sat. 10-5, Sun. 12-5. Adults $7, senior citizens $6, children $2; discounts to AAM & ICOM members; Schoharie County Schools & members no charge. ♿
Attendance: 4,900 (accurate)
Membership: Senior Citizens $25; Regular $30; Family $50; Sustaining $100; Benefactor $250; Life $1,000.

SCHOHARIE COLONIAL HERITAGE ASSOCIATION, Palatine House, Spring St., Schoharie, NY 12157. Mailing Address: P.O. Box 554, Schoharie, NY 12157-0554. Tel.: 518-295-7505 & 7585. Fax: 518-295-6001.
E-mail: scha@midtel.net
Web Site: www.midtel.net/~scha
Founded: 1963.
Congressional District: 31
Key Personnel: C.E.O., Sarah Sherman; Pres., Jean Harra; Vice Pres., Ruth Anne Keese; Treas., Donna McCabe.
Personnel Profile: Part-Time Paid 3; Part-Time Volunteers 5.
Governing Authority: society; nonprofit organization. Branch Museum: Depot Lane Center, Depot Lane, Schoharie, NY 12157. Tax-exempt: 501(c)(3).
Institution Type/Description: Historic Building: housed in 1743 Palatine House, old Lutheran parsonage.
Collections: 18th-century articles used by ministers at home & in church; pictures; documents; articles pertaining to 19th century Middleburgh & Schoharie Railroad. Historic Structure: 1891 railroad depot.
Research Fields: German Palatine life.
Facilities: architectural design; digs; gardens; cemetery.
Activities: guided tours; lecture series; weaving demonstrations. Association Sponsors: Wool Day, spinning & weaving lessons and workshops, dyeing & herb workshops.
Publications: pamphlet, Visit The 1743 Palatine House & 1891 Train Car Museum.
Hours & Admission Prices: May group tours only; June-Oct. Thurs.-Sun. 1-4. Donations: adults $2.50, students $1; discounts to groups; members first visit no charge. RR Museum: Memorial Day to Columbus Day Sat.-Sun. 12-4.
Attendance: 750 (estimated)
Membership: Individual $10; Family $15; Donor $25; Sustaining $50; Patron $100; Life Benefactor $1,000.

Scotia

FLINT HOUSE, 421 Reynolds St., Scotia, NY 12302-1601. Mailing Address: 13 Larkin St., Scotia, NY 12302. Tel.: 518-374-2371.
Founded: 1997.
Personnel Profile: Part-Time Volunteers 4.

Governing Authority: municipal.
Institution Type/Description: Historic House & Museum: housed in a 1735 salt box house.
Collections: village history; broom corn industry from 1860-2007.
Facilities: library.
Activities: guided tours; lectures.
Hours & Admission Prices: Call for a guided tour, 518-374-2871. No charge; donations accepted.
Attendance: 875 (accurate)

Sea Cliff

SEA CLIFF VILLAGE MUSEUM, 95 Tenth Ave., Sea Cliff, NY 11579-1127. Tel.: 516-671-0090. Fax: 516-671-2530.
E-mail: seacliffmuseum@aol.com
Founded: 1979.
Key Personnel: Dir. & Cur., Sara Reres; Museum Technician, James Reres.
Personnel Profile: Part-Time Paid 2; Part-Time Volunteers 50; Interns 2.
Volunteer Hours: 10,000
Operating Expenses: 3,000
Operating Income: 4,300
Governing Authority: Parent Institution: Village of Sea Cliff. Tax-exempt.
Institution Type/Description: Village Museum: housed in the former Sea Cliff Methodist Church, built in 1913.
Collections: Sea Cliff history, 1870 to mid-twentieth century including documents, photographs, artifacts, & costumes.
Major Exhibits: Secrets of the Sea Cliff Museum: Celebrating our 35th Year, 9/13-6/14.
Research Fields: Carpenter family; Henry Otto Korten glass negatives.
Facilities: library; 500 sq. ft. exhibit space; garden. Museum-related items for sale.
Activities: docent program; lectures. Annual Event: Friends Exhibit Opening Reception.
Publications: annual newsletter.
Hours & Admission Prices: Oct.-June Sun. 2-5. Adults $1.
Attendance: 1,000 (accurate)
Membership: Seniors $10; Individual $15; Family $25.

Seaford

SEAFORD HISTORICAL MUSEUM, 3890 Waverly Ave., Seaford, NY 11783-2614. Mailing Address: Seaford Historical Society, P.O. Box 1254, Seaford, NY 11783. Tel.: 516-781-5184.
E-mail: seafordhistoric@optonline.net
Web Site: www.seafordhistoricalsociety.org
Founded: 1968.
Congressional District: 4
Key Personnel: Pres., Charles Wroblewski.
Governing Authority: society; chartered by NY board of regents; nonprofit. Tax-exempt: 501(c)(3).
Institution Type/Description: History Museum: housed in 1893 two-room schoolhouse.
Collections: farm tools; local memorabilia; maritime artifacts associated with life of area Baymen & farmers; old photographs.
Research Fields: local history.
Activities: lectures; hobby workshops; permanent & temporary exhibitions.
Publications: semiannual newsletter, Seaford Historical Society.
Hours & Admission Prices: By appointment. No charge; donations accepted. ♿
Attendance: 2,000
Membership: Individual $15; Family $25; Individual Life & Business $50; Family Life $125.

Selden

SUFFOLK CENTER ON THE HOLOCAUST, DIVERSITY & HUMAN UNDERSTANDING, (M), Suffolk County Community College, Huntington Library, 2nd Fl., 533 College Rd., Selden, NY 11784-2851. Tel.: 631-451-4700. Fax: 631-451-4697.
E-mail: chdhu@sunysuffolk.edu
Web Site: www.chdhu.org
Institution Type/Description: History Museum.
Collections: Holocaust history; slavery; photographs; documents; period artifacts.
Activities: special events; educational programs.
Hours & Admission Prices: Mon.-Thurs. 10-2; other times by appointment.

Selkirk

BETHLEHEM HISTORICAL ASSOCIATION, 1003 River Rd., Selkirk, NY 12158-4033. Mailing Address: P.O. Box 263, Selkirk, NY 12158-0263. Tel.: 518-767-9432.
E-mail: correspondingsecretary@bethlehemhistorical.org
Web Site: bethlehemhistorical.org
Founded: 1965.
Congressional District: 28
Key Personnel: Pres., George Lenhardt; Registrar, Valerie Thompson.
Governing Authority: nonprofit organization. Tax-exempt: 501(c)(3).
Institution Type/Description: General Museum: housed in 1859 Cedar Hill school.
Collections: tools for farming; ice-harvesting; railroading; fruit-growing; dolls; toys; early school material; early items from the home; clothing from the home; clothing from Town of Bethlehem; 1851 Albany Plank Road Toll Gate.
Research Fields: local industries of past; early arts & crafts; genealogy.
Facilities: library on Bethlehem history & resource material used in setting up temporary exhibits available by request; genealogy library; herb garden.
Activities: guided tours; lectures; films; formally organized education programs for children; historic sites survey including taped interviews; permanent & temporary exhibitions.
Publications: annual membership booklet; newsletter; book, Records of People of the Town of Bethlehem, Albany County, NY 1698-1880.
Hours & Admission Prices: June-Aug. Sun. 2-5. No charge; donations accepted. ♿
Attendance: 375 (accurate)
Membership: Student $5; Single $20; Family $30; Business $50.

Seneca Falls

NATIONAL WOMEN'S HALL OF FAME, 76 Fall St., Seneca Falls, NY 13148-1451. Mailing Address: P.O. Box 335, Seneca Falls, NY 13148-0335. Tel.: 315-568-8060. Fax: 315-568-2976.
E-mail: greatwomen@greatwomen.org
Web Site: www.greatwomen.org
Founded: 1969.
Congressional District: 29
Key Personnel: Exec. Dir., Christine Moulton; Bd. Pres., Beth Quillen Thomas.
Personnel Profile: Full-Time Paid 4; Part-Time Paid 4; Part-Time Volunteers 40; Interns 2.
Governing Authority: nonprofit. Tax-exempt: 501(c)(3).
Institution Type/Description: Historic Building & Site: 1920s Victorian style bank building.
Collections: photographs; artifacts; audio loops; history related letters; personal memorabilia of honorees; suffrage items; biographies & correspondence of Inductees.
Research Fields: women nominated for induction into the hall; files on other women of accomplishment.
Facilities: 3,000-vol. library available for use pertaining to women's history. Museum-related items for sale.
Activities: guided tours; lectures; film; organized education programs for undergraduate or graduate college students; internship program; essay & new media contest; participatory, loan, temporary & traveling exhibitions; school loan service; honors ceremonies; special monthly programs.
Publications: quarterly newsletter; education kit.
Hours & Admission Prices: Feb.-April & Oct.-Dec. Wed.-Sat. 11-5; May-Sept. Mon.-Sat. 10-5, Sun. 12-5. Family $7, adults $3, students & seniors $1.50; members & children under 5 no charge. Closed New Year's Day; Easter; Memorial Day; Independence Day; Labor Day; Thanksgiving; Christmas. ♿
Attendance: 15,000 (accurate)
Membership: Students & Senior Citizens $15; Individual $25; Family $50; Sponsor $100; Friend $250; Patron $500; President's Circle $2,500.

SENECA FALLS HISTORICAL SOCIETY, 55 Cayuga St., Seneca Falls, NY 13148-1222. Tel.: 315-568-8412. Fax: 315-568-8426.
E-mail: sfhs@rochester.rr.com
Web Site: www.sfhistoricalsociety.org
Founded: 1896.
Congressional District: 24
Key Personnel: Exec. Dir., Philomena M. Cammuso; Dir. Education, Frances T. Barbieri; Pres. Bd. Trustees, Lorrilyn Bove; Collections Mgr., Kathleen Jans-Duffy.
Personnel Profile: Full-Time Paid 3; Part-Time Volunteers 35; Interns 2.
Governing Authority: society. Tax-exempt: 501(c)(3).

Institution Type/Description: Local History Museum: 23 room Victorian Mansion.
Collections: 19th-century American decorative & fine art; local history artifacts; costumes; toys; pre- & post-Industrial tools; photographs; Silsby steam fire engine; manuscripts; memorabilia of the first Women's Rights Convention held in 1848; apparel from early 1800s to the turn of the century; industry; archives; Indian artifacts; paintings of Carlos Bellows.
Research Fields: local history; fire apparatus; pump manufacturing; Woman's Rights Movement; genealogy; county & canal history.
Facilities: 2,000-vol. research library & archives for use on premises; 23 room Victorian house with period rooms. Museum-related items for sale.
Activities: guided tours; lectures; docent programs; permanent & temporary exhibitions; craft classes; school loan service.
Publications: bimonthly newsletters; maps; booklets: Cowing Story (fire-engines); Silsby Mfg. Story; Debut of Women's Rights; Sad Irons, The Seneca Falls of David Lum 1806-1875; As We Were, Vol. I & II, 19th & 20th Century Photographs of Seneca Falls; The Flats: Including the Canal and Early Industries of Seneca Falls; Finding Aids: Women's Rights; Gould's Pumps, Inc.; Stories of Seneca Falls, Vol. I.
Hours & Admission Prices: Business: Mon.-Fri. 9-4. Tours: June-Sept. Mon.-Fri. 10-3, Sat.-Sun. 10-2. Family $15, adults $7; discounts to AAA & AARP members; members no charge. Closed New Year's Day; Martin Luther King Jr. Day; Presidents' Day; Memorial Day; Independence Day; Veterans Day; Columbus Day; Thanksgiving & day after; Christmas week. ♿
Attendance: 18,000 (accurate)
Membership: Student (under 18) $10; Associate $30; Contributor (Individual & Family) $50; Supporter $75; Sustainer $125; Subscriber $250; Sponsor $500; Patron $1,000; Corporate $2,000; Benefactor $5,000.

SENECA MUSEUM OF WATERWAYS AND INDUSTRY, 89 Fall St., Seneca Falls, NY 13148. Mailing Address: P.O. Box 388, Seneca Falls, NY 13148. Tel.: 315-568-1510. Fax: 315-568-1504.
Web Site: www.senecamuseum.com
Founded: 1998.
Congressional District: 54
Key Personnel: Dir., Linda Solan; Chm. (V), Don Gentilcore; Museum Shop Mgr., Barb Dorvee.
Personnel Profile: Full-Time Paid 2; Part-Time Paid 2.
Governing Authority: Tax-exempt.
Institution Type/Description: History Museum.
Collections: industrial & canal history; water-powered grist mill model; pumps; foot-powered tools; canal boat cabin; period artifacts; dioramas.
Facilities: Museum related items for sale.
Activities: educational programs; classes; summer camps.
Hours & Admission Prices: Jan.-Feb. Tues.-Sat. 10-4; March-Dec. Mon.-Sat. 10-4, Sun. 12-1; groups by appointment. Guided Tours: adults $5; children under 14 no charge. ♿
Attendance: 17,000 (accurate)

WOMEN'S RIGHTS NATIONAL HISTORICAL PARK, 136 Fall St., Seneca Falls, NY 13148. Tel.: 315-568-0024.
Institution Type/Description: History Museum: housed in the former Village Hall; built in 1915.
Collections: women's rights history; photographs; personal artifacts; period furnishings; life-sized bronze sculptures; Declaration of Sentiments & its signers, 100 ft. long bluestone wall; quilts; statues.
Facilities: theater. Gift items for sale.
Activities: film; orientation area; special events.
Hours & Admission Prices: Daily 9-5. Closed New Year's Day; Thanksgiving; Christmas.

Setauket

GALLERY NORTH, 90 N. Country Rd., Setauket, NY 11733-1352. Tel.: 631-751-2676. Fax: 631-751-0180.
E-mail: info@gallerynorth.org
Web Site: www.gallerynorth.org
Founded: 1965.
Congressional District: 1
Key Personnel: Pres. & Chm., Paul Lamb; Dir. & Cur., Judith Levy; Business Mgr., Martha Stansbury.
Personnel Profile: Full-Time Paid 1; Part-Time Paid 4; Part-Time Volunteers 20.
Governing Authority: nonprofit organization. Tax-exempt.
Institution Type/Description: Art Gallery.
Collections: contemporary works by Long Island artists & crafts people.
Facilities: 1,200 sq. ft. exhibit space.
Activities: annual outdoor art show; lectures; trips; studio tours; adult painting

workshops; children's art workshops. Museum Sponsors: Plein Art/Wet Paint Festival in May; 48th Annual Outdoor Art Show in September; Holiday Art Show in November & December.
Publications: quarterly newsletter for members only, Gallery North News.
Hours & Admission Prices: mid-Jan. to Dec. Tues.-Sat. 10-5, Sun. 12-5. No charge. Closed Easter; Thanksgiving; Christmas. &
Attendance: 17,500 (estimated)
Membership: Friends $50-$500 & up.

THREE VILLAGE HISTORICAL SOCIETY, (M), 93 N. Country Rd., Setauket, NY 11733-1347. Tel.: 631-751-3730. Fax: 631-751-3936.
E-mail: info@tvhs.org
Web Site: threevillagehistoricalsociety.org
Founded: 1964.
Congressional District: 1
Key Personnel: Exec. Dir., Judith Estes; Pres., Patricia Yantz; Archivist, Karen Martin; Office Asst., Suzie Roberts; Office Asst., Maryanne Vigneaux.
Personnel Profile: Part-Time Paid 3; Part-Time Volunteers 350.
Governing Authority: private; nonprofit organization. Tax-exempt.
Institution Type/Description: Historical Society.
Collections: related to the history, from earliest settlement to present, of the Three Village area of Long Island's North Shore.
Research Fields: 19th-century seafaring women; 19th-century culture, folklore; 20th-century development issues; 17th- to 20th-century deed search & property development.
Facilities: Books for sale.
Activities: docent program; formal educational programs; guided tours; lectures; loan exhibitions; bicycle through history. Annual Events: Apple Festival; Spirits of Three Villages Cemetery Tour; Candlelight House Tour; Walk Through History.
Publications: annual research volume; quarterly newsletter; local history books.
Hours & Admission Prices: Mon.-Fri. 10-3, Sat.-Sun. by appointment. Adults $5, children & members $3. &
Attendance: 3,000 (estimated)
Membership: Individual $30; Family $50; Patron $100; Benefactor $250; Major Contributor $500; Founder $1,000.

Shelter Island

SHELTER ISLAND HISTORICAL SOCIETY, (M), 16 S. Ferry Rd., Shelter Island, NY 11964. Mailing Address: P.O. Box 847, Shelter Island, NY 11964. Tel.: 631-749-0025. Fax: 631-749-1825. Facebook: Shelter Island Historical Society.
E-mail: info@shelterislandhistorical.org
Web Site: shelterislandhistorical.org
Founded: 1965.
Congressional District: 1
Key Personnel: Exec. Dir., Nanette Lawrenson; Pres. (V), Elizabeth Pedersen; Sec., Belle Lareau; Museum Shop Mgr., Suzanne Louer.
Personnel Profile: Part-Time Paid 2; Part-Time Volunteers 100; Interns 2.
Governing Authority: society. Tax-exempt.
Institution Type/Description: Historic House Museum: 1743 James Havens house.
Collections: period documents, photographs, memorabilia & furniture pertaining to the town of Shelter Island. Historic Building: Havens Barn.
Research Fields: historical data concerning the town of Shelter Island.
Activities: monthly meetings; changing exhibits; lectures; educational programs; films; fairs.
Publications: biannual newsletter, Heartsease; Delftware - Gill Patterson Collection; book, History of Shelter Island: The Smallest Village; The Story of Shelter Island in the Revolution; A Chronicle of Shelter Island Churches; A Woman Named Matilda & Other True Accounts of Old Shelter Island; God's Summer Cottage; An Island Sheltered; Images of America: Shelter Island; Diary of a Little Girl in Old New York; Lords of the Soil.
Hours & Admission Prices: By appointment. No charge; donations accepted.
Attendance: 5,500 (estimated)

Sidney

SIDNEY HISTORICAL ASSOCIATION, 21 Liberty St., Rm. 218, Sidney, NY 13838-1246. Mailing Address: 21 Liberty St., Box 8, Sidney, NY 13838-1266. Tel.: 607-563-2542.
E-mail: sidneyhistorical@stny.rr.com
Web Site: www.sidneyonline.com/sha.htm
Founded: 1945.
Congressional District: 25
Key Personnel: Pres., Floyd Howard; Vice Pres., Joelene Cole; Cur., Graydon Ballard; Treas., Bonnie Curtis; Sec., Evy Avery.

Personnel Profile: Part-Time Volunteers 15.
Governing Authority: association; nonprofit organization. Tax-exempt.
Institution Type/Description: Regional History Museum.
Collections: history; archaeology; archives; photographs; Indian artifacts; Troop C, Capt. Fox picture & trophy collection 1920-1992; genealogies; manuscripts; wooden horse sleigh made in Sidney by Cortland, Cart & Carriage Co. 1890s.
Research Fields: local history; Indian artifacts & sites; genealogy.
Facilities: library of historical books, records, newspapers, diaries & microfilm; reports available for research upon request with member of association in attendance.
Activities: guided tours; lectures; films; formally organized educational programs; temporary exhibitions; antique appraisal clinics.
Publications: Organizational Histories; Sidney - Then & Now - 1772-1972; Lest We Forget; newspaper articles; community calendar; microfilming; quarterly newsletter; reproductions of historic maps.
Hours & Admission Prices: Sept.-July Wed. 4pm-6pm, Thurs. 9am-11:30am; other times by appointment. No charge. Closed holidays. &
Attendance: 630 (accurate)
Membership: Individual $10; Life $50.

Skaneateles

THE JOHN D. BARROW ART GALLERY, (M), Skaneateles Library, 49 E. Genesee St., Skaneateles, NY 13152-1314. Tel.: 315-685-5135.
Web Site: www.barrowgallery.org
Founded: 1900.
Congressional District: 25
Key Personnel: Dir., Margaret Whitehouse.
Personnel Profile: Part-Time Paid 1; Part-Time Volunteers 40.
Governing Authority: Parent Institution: Skaneateles Library Association. Tax-exempt.
Institution Type/Description: Art Gallery.
Collections: 435 paintings highlighting the historical & cultural aspects of the local community in the mid to late 1800s.
Facilities: Museum-related items for sale.
Hours & Admission Prices: Memorial Day to Labor Day Mon.-Fri. 1-4, Sat. 11-4; Sept.-May Thurs.-Fri. 11-4. Tours by appointment. No charge; donations accepted. Closed holidays.
Attendance: 5,000 (estimated)

Smithtown

SMITHTOWN HISTORICAL SOCIETY, (M), 239 Middle Country Rd., Smithtown, NY 11787-2807. Tel.: 631-265-6768. Fax: 631-979-4694.
E-mail: info@smithtownhistorical.org
Web Site: www.smithtownhistorical.org
Founded: 1955.
Congressional District: 1
Key Personnel: Dir., Kiernan Lannon; Pres. (V), Brad Harris; Dir. Education, Elizabeth Jenks.
Personnel Profile: Full-Time Paid 4; Part-Time Paid 7; Part-Time Volunteers 30; Interns 2.
Governing Authority: society. Tax-exempt.
Institution Type/Description: Historical Society Museums.
Collections: documents, books, decorative arts, costumes relating to the history of Smithtown. Historic Houses: 1700 Obadiah Smith house; 1750 Epenetus Smith Tavern; 1750 Franklin O. Arthur Farm; 1819 Caleb Smith house; 1790 Judge J. Lawrence Smith Homestead; 1870 Reading Room; 1900 Brush barn; 1918 Roseneath Cottage; 1860 Rockwell Barn & Carriage House.
Research Fields: preservation; genealogy; local history.
Facilities: library of history & genealogy books available for use on premises. Museum-related items for sale.
Activities: films; exhibit & demonstration programs; school educational programs; changing exhibits; special events.
Publications: books, Genealogy of Smith Family; Smithtown 1660-1929: Looking Back Through the Lens; Old Houses in Smithtown-Rockwell's Scrapbook; map, Early Residents of Smithtown 1660-1885.
Hours & Admission Prices: Caleb Smith House: Mon.-Fri. 9-5, Sat. by appointment. Historic Houses: by appointment. No charge; donations accepted. &
Attendance: 25,000 (estimated)
Membership: Individual $25; Family $35; Sustaining $75; Patron $100; Life $1,000.

Sodus Point

SODUS BAY HISTORICAL SOCIETY, 7606 N. Ontario St., Sodus Point, NY 14555-9536. Mailing Address: P.O. Box 94, Sodus Point, NY 14555-0094. Tel.: 315-483-4936. Fax: 315-483-1398.
E-mail: sbhslighthouse@gmail.com
Web Site: www.sodusbaylighthouse.org
Founded: 1972.
Congressional District: 27
Key Personnel: Museum Dir., Joseph O'Toole; Pres. (V), Deborah Lattime; Vice Pres., Joan Eckberg; Museum Shop Mgr., Ali Duncan.
Personnel Profile: Full-Time Paid 1; Part-Time Paid 2; Part-Time Volunteers 125.
Governing Authority: nonprofit organization. Tax-exempt: 501(c)(3).
Institution Type/Description: Lighthouse Museum: 1870 Sodus Bay Lighthouse & two outbuildings.
Collections: concentration on Great Lakes maritime history with major emphasis on Lake Ontario & Sodus Bay.
Research Fields: Great Lakes Lighthouse Service; US Lighthouse Service; US Lighthouse Board; US Coast Guard.
Facilities: 1870 Lighthouse.
Activities: Independence Day gala & 10 free concerts; guided tours; lectures.
Publications: monthly newsletter (nine issues).
Hours & Admission Prices: May-Oct. Tues.-Sun. & Mon. holidays 10-5. Adults $4, children $2. &
Attendance: 21,000 (estimated)
Membership: Friend $30; Contributing $60; Patron $120; Sustaining $300; Life $1,000.

Somers

SOMERS HISTORICAL SOCIETY, Elephant Hotel, Rte. 100 & 202, Somers, NY 10589. Mailing Address: P.O. Box 336, Somers, NY 10589-0336. Tel.: 914-277-4977.
E-mail: somershistoricalsoc@yahoo.com
Web Site: www.somershistoricalsoc.org
Formerly: Somers Historical Society Museum and Museum of the Early American Circus
Founded: 1956.
Congressional District: 25
Key Personnel: Pres. (V), Emil Antonaccio.
Personnel Profile: Part-Time Volunteers 15.
Governing Authority: society; board of trustees. Tax-exempt: 990-A.
Institution Type/Description: Circus & Local History Museum: housed in 1825 Elephant Hotel built by Hachaliah Bailey.
Collections: early American circus history; furniture, artifacts & manuscripts relating to local history; couriers, posters, lithographs, manuscripts, route books & other memorabilia relating to pioneer circus; Rowell Miniature Circus.
Research Fields: pioneer circus; genealogy; local history.
Facilities: 1,000-vol. library of local history & circus available for research by appointment. Museum-related items for sale.
Activities: guided tours; permanent & temporary exhibitions; special events.
Publications: books, The Elephant Hotel; America; reprint, History of Somers (1886); reprint, Journal of Dr. Elias Cornelius, Rev. War; Somers Remembered; Somers Confederates: James Wright & his Nephews.
Hours & Admission Prices: Thurs. 2-4; other times by appointment. No charge.
Attendance: 1,000 (estimated)
Membership: Regular: Student $5; Single $20; Family $25. Sustaining: Single $35; Family $50; Patron $100; Life $250 & $350.

Southampton

SHINNECOCK NATION CULTURAL CENTER AND MUSEUM, 100 Montauk Hwy. & W. Gate Rd., Southampton, NY 11969. Mailing Address: P.O. Box 5059, Southampton, NY 11969-5059. Tel.: 631-287-4923. Fax: 631-287-7153.
E-mail: office@shinnecockmuseum.org
Web Site: www.shinnecockmuseum.org
Institution Type/Description: Native American Museum.
Collections: Shinnecock heritage, culture & history; paintings; murals; sculptures.
Hours & Admission Prices: Thurs.-Sun. 11-5; tours by appointment. Adults $8, seniors 55 & over and students $5.50, children 5-12 $4.75; members & children under 5 no charge.

SOUTHAMPTON HISTORICAL MUSEUMS AND RESEARCH CENTER - ROGERS MANSION, 17 Meeting House Lane, Southampton, NY 11968-4911. Mailing Address: P.O. Box 303, Southampton, NY 11969-0303. Tel.: 631-283-2494. Fax: 631-283-4540.
E-mail: info@southamptonhistoricalmuseum.org
Web Site: southamptonhistoricalmuseum.org
Formerly: Southampton Colonial Society
Founded: 1898.
Congressional District: 1
Key Personnel: Exec. Dir., Tom Edmonds; Pres., Gerri MacWinnie; Asst. Dir., Diane Becker; Mgr. Research Center, Mary Cummings; Education & Events Mgr., Laurie Collins; Cur. & Registrar, Emma Ballou; Museum Asst., Diane Pinkston.
Personnel Profile: Full-Time Paid 2; Part-Time Paid 4; Part-Time Volunteers 2; Interns 2.
Volunteer Hours: 7,967
Operating Expenses: 350,000
Operating Income: 350,000
Governing Authority: Southampton Colonial Society dba Southampton Historical Museum. Subsidiary Institutions: Pelletreau Silver Shop, 80 Main St., Southampton, NY; The Thomas Halsey Homestead, 249 S. Main St., Southampton, NY; Conscience Point Historic Site and Nature Walk, N. Sea Rd., Southampton, NY; The Captain George White House, 159 Main St., Southampton, NY. Tax-exempt.
Institution Type/Description: History Museum.
Collections: local history with changing exhibits focusing on South Fork culture including folk art, Native American, whaling, rural tools, area ethnic history. Historic Sites: 1843 Roger Mansion; 1666 Thomas Halsey Homestead; 1686 Pelletreau Silver Shop; Conscience Point Historic Site & Nature Walk.
Research Fields: Southampton history.
Facilities: 4 separate properties with 14 historic buildings listed on the Natl. Register of historic places.
Activities: education programs; craft workshops; guided tours; temporary exhibitions.
Publications: newsletter; brochures.
Hours & Admission Prices: Rogers Mansion & Pelletreau Shop: Tues.-Sat. 11-4. Thomas Halsey House: July to mid-Oct. Sat.-Sun. 11-4. Adults $4; discounts to AAM, ICOM & AAA members; children 17 & under and members no charge. &
Attendance: 31,365 (accurate)
Membership: Individual $25; Family & Couple $45; Sustaining $150; Contributing $250, Sponsor $500; Patron $1,000; President's Circle $5,000.

Southold

SOUTHOLD HISTORICAL SOCIETY AND MUSEUM, (M), 54325 Main Rd., Southold, NY 11971-4646. Mailing Address: P.O. Box 1, Southold, NY 11971. Tel.: 631-765-5500 & 5551. Fax: 631-765-8510.
E-mail: sohissoc@optonline.net
Web Site: southoldhistoricalsociety.org
Founded: 1960.
Congressional District: 1
Key Personnel: Dir., Geoffrey Fleming; Pres. (V), John Barnes; Museum Shop Mgr., Deanna Walker.
Personnel Profile: Full-Time Paid 1; Part-Time Paid 4; Part-Time Volunteers 155.
Governing Authority: society. Parent Institution: Southold Historical Society. Subsidiary Institution: Horton Point Nautical Museum. Tax-exempt: 900.
Institution Type/Description: General Museum.
Collections: farm implements; china; silver; quilts; textiles; costumes; hats; furniture; maritime artifacts; archives; paintings; decorative arts; agriculture; paintings; costumes; 19th century Overton corncrib; glass; manuscripts. Historic Buildings: 1899 Hallock Currie-Bell House; middle 18th-century Pine Neck barn; c.1790 Thomas Moore House; c.1845 Irving Downs Carriage House; c.1845 Henry Cleveland blacksmith shop; 1857 Horton's Point Lighthouse & Nautical Museum; 1875 Bayview Icehouse.
Research Fields: genealogy; local history
Facilities: 1,000-vol. library of historical books available for research on the premises.
Activities: guided tours; formally organized education programs for children; permanent & temporary exhibitions.
Publications: annual report; newsletter.
Hours & Admission Prices: July-Aug. Wed. & Sat.-Sun. 1-4. Adults $2; members & children no charge. &
Attendance: 9,000 (estimated)
Membership: Annual $20; Contributing $50; Sustaining $100; Business $100-$250; Patron $750; Benefactor $1,000.

SOUTHOLD INDIAN MUSEUM, 1080 Bayview Rd., Southold, NY 11971. Mailing Address: P.O. Box 268, Southold, NY 11971-0268. Tel.: 631-765-5577. Fax: 631-765-5577.
E-mail: indianmuseum@aol.com
Web Site: southoldindianmuseum.org
Founded: 1962.
Congressional District: 1
Key Personnel: Pres. (V), Ellen Barcel; Membership, Education & Museum Shop Mgr., Martha Waide.
Personnel Profile: Part-Time Paid 1.
Governing Authority: Incorporated Long Island Chapter of the New York State Archaeological Association; nonprofit organization. Tax-exempt: 501(c)(3).
Institution Type/Description: Archaeology & Indian Museum.
Collections: spear heads; murals; Algonquin ceramic pottery; soapstone pots & bowls; arrow heads; knife blades; hoe blades; hammers; gouges; drills; aborigine vegetables, foods & herbs for medical purposes; toys; fishing tackle with net & sinkers; jewelry; ornaments; smoking pipes; clothing; religious relics; handiworks of Eskimos.
Research Fields: Long Island archaeology.
Facilities: 1,500-vol. archaeological library; 100-seat auditorium; 1,600 sq. ft. exhibit space. Publications & other museum-related items for sale.
Activities: guided tours; lectures; organized education programs for children; monthly meetings; trips; traveling exhibitions.
Publications: quarterly newsletter, Southold Indian Museum News; Archaeology leaflets; museum guide.
Hours & Admission Prices: Sun. 1:30-4:30. Adults $4; members no charge. Closed New Year's Day; Easter; Thanksgiving; Christmas.
Attendance: 3,000 (estimated)
Membership: Active $38; Active Husband-Wife $50.

Spencerport

OGDEN HISTORICAL SOCIETY, 568 Colby St., Spencerport, NY 14559. Mailing Address: P.O. Box 777, Adams Basin, NY 14410-0777. Tel.: 585-352-3672. Facebook: Ogden Historical Society.
E-mail: historian@ogdenny.com
Founded: 1958.
Congressional District: 34
Key Personnel: Town Historian, Carol Coburn.
Governing Authority: board of trustees. Tax-exempt.
Institution Type/Description: Historic House: 1811 Eastman-Colby House.
Collections: furnishings; clothing; implements of early 19th-century farm house.
Facilities: library of local history & general information on Erie Canal.
Activities: lectures; films; inter-museum loan; craft demonstrations.
Hours & Admission Prices: Sun. 2-4, group tours by appointment. No charge; donations accepted. &
Attendance: 500 (estimated)

Springville

CONCORD HISTORICAL SOCIETY, WARNER MUSEUM, 98 Main St., Springville, NY 14141-1335. Mailing Address: P.O. Box 425, Springville, NY 14141-0425. Tel.: 716-592-0094.
E-mail: lucybensley@townofconcordnyhistoricalsociety.org
Web Site: www.townofconcordnyhistoricalsociety.org
Founded: 1953.
Key Personnel: Pres. & Historian, David Batterson; Vice Pres., Joel Maul; Treas., Jeanne Fornes; Sec., Derek Otto
Governing Authority: municipal; society. Subsidiary Institution: Concord Historical Society Genealogical Library, 23 N. Buffalo St., Springville, NY 14141. Tax-exempt: 501(c)(3).
Institution Type/Description: Local History Museum.
Collections: carpenter tools; pictures; Pop Warner's Indian artifact collection; pump organ; china; furniture; local historical items.
Facilities: library of history & school books, maps & photographs available for research by appointment on premise; reading room.
Activities: temporary exhibitions; tours; genealogy research.
Publications: newsletter.
Hours & Admission Prices: 2nd & 4th Sun. each month and Mon.-Thurs. 10-2; other times by appointment. No charge; donations accepted. &
Attendance: 650 (estimated)
Membership: Senior Citizen $5; Individual $10; Life $50.

Staatsburg

STAATSBURGH STATE HISTORIC SITE, Old Post Road, Staatsburg, NY 12580-5911. Mailing Address: P.O. Box 308, Staatsburg, NY 12580-0308. Tel.: 845-889-8851. Fax: 845-889-8843. Facebook: Staatsburgh SHS.
E-mail: pam.malcolm@parks.ny.gov
Web Site: nysparks.com
Formerly: Mills Mansion State Historic Site
Founded: 1938.
Congressional District: 25
Key Personnel: Historic Site Mgr., Pamela Malcolm; Pres. (V), Caroline Carey.
Personnel Profile: Full-Time Paid 5; Part-Time Paid 8; Part-Time Volunteers 40; Interns 4.
Governing Authority: state. Parent Institution: New York State Office of Parks, Recreation & Historic Preservation Taconic Region. Tax-exempt: 501(c)(3).
Institution Type/Description: Historic House Museum: 1895 65-room Neo-classical Revival mansion designed by Stanford White.
Collections: 19th & 20th-century furnishings belonging to the Mills family; fine & decorative arts.
Research Fields: Livingston & Mills families; fine & decorative arts.
Facilities: golf course; picnic area; cross-country skiing trails; camp ground; state park.
Activities: guided tours; concerts; lectures; special events; school programs.
Publications: map of grounds; Great Estates of the Hudson Valley.
Hours & Admission Prices: Call for hours. Adults $8; seniors $6; students under 12 and members no charge. &
Attendance: 24,206 (accurate)
Membership: Student & Senior Citizen $15; Individual $20; Family $35; Contributor $50; Sustainer $100; Sponsor $250; Patron $500; Benefactor $1,000.

Staten Island

ALICE AUSTEN HOUSE MUSEUM, 2 Hylan Blvd., Staten Island, NY 10305-2002. Tel.: 718-816-4506. Fax: 718-815-3959.
E-mail: eaausten@aol.com
Web Site: www.aliceausten.org
Founded: 1979.
Congressional District: 14
Key Personnel: Exec. Dir., Carl Rutberg; Pres., David Goldfarb; Dir. Education, Annmarie McDonnell; Dir. Museum Svcs., Sara Signorelli.
Personnel Profile: Full-Time Paid 3; Full-Time Volunteers 2; Part-Time Paid 2; Part-Time Volunteers 50.
Governing Authority: nonprofit historical & preservation organization. Parent Institution: Friends of Alice Austen House, Inc. Tax-exempt.
Institution Type/Description: Historic House: c.1690 one of the oldest in New York, home of photographer Alice Austen.
Collections: Victorian furniture & furnishings; photographs by Alice Austen; archival holdings; architectural elements; gardens.
Research Fields: photography & life of Alice Austen; New York City history; aspects of Victoriana & early 20th-century social history; contemporary photos.
Facilities: garden; lawns; meeting room; video viewing room. Gift items for sale.
Activities: lectures; concerts; walking tours; gallery tours.
Publications: quarterly newsletter; museum guidebook; exhibition catalogs & checklists; walking tour booklet; Street Types of New York photograph portfolio reprint.
Hours & Admission Prices: March-Dec. Thurs.-Sun. 11-5. Adults $3; discounts to AAM members; members no charge. Closed holidays. &
Attendance: 19,000 (estimated)
Membership: Senior & Student Friend $15; Friend $25; Family Friend $40; Supporting Friend $100; Business Friend $101 & up; Sustaining Friend $250; Guardian of Clear Comfort $500.

THE CONFERENCE HOUSE, 298 Satterlee St., Staten Island, NY 10307. Tel.: 718-984-6046. Fax: 718-984-7760.
E-mail: info@conferencehouse.org
Web Site: www.conferencehouse.org
Founded: 1927.
Key Personnel: Pres., Linda Jensen; Office Operations Mgr., Colleen Siuzdak; Caretaker, Deborah Woodbridge.
Personnel Profile: Full-Time Paid 2; Part-Time Volunteers 12; Interns 1.
Governing Authority: municipal. Parent Institution: The Conference House Association. Tax-exempt.
Institution Type/Description: Historic House Museum: housed in 1675 Conference House or Billopp House, built by Capt. Christopher Billopp,

English Navy. Site of peace conference during Revolutionary War between Ben Franklin, John Adams & Edward Rutledge with Lord Admiral Richard Howe, Sept. 11, 1776.
Collections: 18th century furnishings & accessories.
Facilities: 50-seat 17th century kitchen used as classroom. Museum-related items for sale.
Activities: guided tours; permanent exhibitions; formally organized education programs for children; slide presentation; open-hearth cooking demonstrations; spinning & weaving demonstrations. Museum Sponsors: Art Show; Peace Conference Celebration; Craft Sale & Open House; Harvest Festival in October; Yuletide Celebration.
Publications: quarterly newsletter; brochures.
Hours & Admission Prices: April to mid-Dec. Fri.-Sun. 1-4. Adults $4, children & seniors $3; members no charge. Closed New Year's Day; Independence Day; Thanksgiving; Christmas.
Attendance: 3,906 (accurate)

GARIBALDI-MEUCCI MUSEUM, 420 Tompkins Ave., Staten Island, NY 10305-1704. Tel.: 718-442-1608. Fax: 718-442-8635.
E-mail: info@garibaldimeuccimuseum.org
Web Site: garibaldimeuccimuseum.org
Formerly: Garibaldi and Meucci Museum of the Order Sons of Italy In America
Founded: 1956.
Congressional District: 13
Key Personnel: Dir. & Admin., Michela Traetto; Pres., Joseph Rondinelli; Co-Chm., Joseph Sciame; Co-Chm., Luigi Squillante; Coord. Publicity, Bonnie McCourt.
Personnel Profile: Part-Time Paid 9; Part-Time Volunteers 10.
Governing Authority: nonprofit organization. Tax-exempt: 501(c)(3).
Institution Type/Description: Historic House: 1845 country home of inventor Antonio Meucci. Italian Patriot, General Giuseppe Garibaldi also occupied house 1850-53.
Collections: life & work of Meucci & Garibaldi, the Risorgimento, Italian history & culture; prints; paintings; medals; coins; military uniforms; artifacts; personal artifacts; period restored Garibaldi bedroom; contemporary Italian & Italian American culture & heritage.
Research Fields: history of house; Italian American history & culture; the Italian Risorgimento (unification of Italy); history of the telephone.
Facilities: 1,200-vol. research library; 35-seat auditorium.
Activities: guided tours; culture & history lectures; films; concerts; Italian language classes; daily school program K-12.
Publications: book, Antonio Meucci; Italian language text & work books; children's books; Anita Garibaldi Biography; Heritage Publications; Italian culture & heritage publications.
Hours & Admission Prices: Wed.-Sat. 1-5. School Programs: daily 9:30-11:30. Suggested Donation: $5; discounts to AAM & ICOM members; children under 10 & members no charge. Closed bank holidays.
Attendance: 10,000 (estimated)
Membership: Student & Senior $25; Individual $40; Household $50-$99; Sponsor $100-$499; Benefactor $500-$999; Patron $1,000 & up.

HISTORIC RICHMOND TOWN - STATEN ISLAND HISTORICAL SOCIETY, 441 Clarke Ave., Staten Island, NY 10306-1196. Tel.: 718-351-1611. Fax: 718-351-6057.
E-mail: squadrino@historicrichmondtown.org
Web Site: www.historicrichmondtown.org
Founded: 1856.
Congressional District: 14
Key Personnel: Exec. Dir., Ed Wiseman; Pres. (V), Robert Coghlan; Chief Cur., Maxine Friedman; Dir. Education, Felicity Beil; Museum Shop Mgr., Adria Scaduto.
Personnel Profile: Full-Time Paid 25; Part-Time Paid 15; Part-Time Volunteers 250; Interns 2.
Governing Authority: nonprofit. Parent Institution: Staten Island Historical Society. Tax-exempt.
Institution Type/Description: Historical Society Museum & Historic Village: society housed in housed in 1848-1917 County Clerk's & Surrogate's office, part of historic 100-acre village including 28 historic buildings.
Collections: tools; costumes; furnishings; toys; folk art; textiles; rare Stansbury press; china & glass; military; archives; manuscripts. 28 Historic Structures including oldest surviving school in U.S., Voorlezer's House c.1695; Decker Farm.
Research Fields: Staten Island history; genealogy; trades; occupations; crafts; architectural history; local history.
Facilities: 15,000-vol. library; archives; 75-seat auditorium; snack bar; cafe; rental facilities; 100 acre site. Museum-related items for sale.
Activities: guided tours; loan & permanent exhibitions; formally organized educational programs; craft demonstrations; concerts; lectures; films; summer day camp; curriculum based educational tours. Annual Events: Civil

War Encampment; Autumn Celebration; Big Apple Wine Fest in May; Richmond County Fair; October pumpkin-picking program at Decker Farm (shuttle bus available); Christmas festivities.
Publications: biannual, Staten Island Historian; calendar.
Hours & Admission Prices: July-Aug. Wed.-Sun. 10-5; Sept.-June Wed.-Sun. 1-5, call to confirm. Adults $8, senior citizens & students 12-17 $6, children 4-11 $5; discounts to AAM members; children under 3 & members no charge. Closed New Year's Day; Easter; Thanksgiving; Christmas.
Attendance: 85,000 (accurate)
Membership: Senior $30; Individual $35; Family $50; Sponsor $75; Contributing $100; Patron $500; Benefactor $1,000; Corporate $2,500.

JACQUES MARCHAIS MUSEUM OF TIBETAN ART, 338 Lighthouse Ave., Staten Island, NY 10306-1217. Tel.: 718-987-3500. Fax: 718-351-0402.
E-mail: mventrudo@tibetanmuseum.org
Web Site: www.tibetanmuseum.org
Founded: 1945.
Congressional District: 11
Key Personnel: C.E.O., Meg Ventrudo; Chm. (V), Beverly Garcia-Anderson; Bookkeeper, Jayne Catalfo; Museum Asst., Alison Baldassano.
Personnel Profile: Full-Time Paid 1; Part-Time Paid 4; Part-Time Volunteers 6; Interns 2.
Volunteer Hours: 400
Operating Expenses: 149,096
Operating Income: 155,186
Governing Authority: nonprofit. Tax-exempt.
Institution Type/Description: Tibetan Art Museum: Buddhist art, primarily Tibetan, as well as other Asian objects exhibited in a traditional Tibetan style building.
Collections: Tibetan & Buddhist art.
Major Exhibits: Freed From the Vault, 9/1/13-6/30/14.
Research Fields: oriental art & religion.
Facilities: 1,100-vol. library; gardens. Gift-related items for sale.
Activities: public lectures & performances; permanent exhibition; children's art workshops, guided tours; participatory workshops for senior citizens & students.
Publications: booklet, The Dalai Lama at the Jacques Marchais Tibetan Museum; calendar of events; brochure; posters; catalogue of collections, Treasures of Tibetan Art.
Hours & Admission Prices: Wed.-Sun. 1-5; other times by appointment. Adults $6, seniors & students $4; discounts to Museum Council of NY & Cool Culture members; members no charge. Closed major holidays.
Attendance: 5,200 (accurate)
Membership: Student & Senior $25; Individual $30; Family $45; Sponsor $75.

THE NOBLE MARITIME COLLECTION, 1000 Richmond Terr., Staten Island, NY 10301-1181. Tel.: 718-447-6490. Fax: 718-447-6056.
E-mail: erinurban@noblemaritime.org
Web Site: www.noblemaritime.org
Formerly: The John A. Noble Collection
Founded: 1986.
Congressional District: 14
Key Personnel: Exec. Dir., Erin Urban; Asst. Dir., Ciro Galeno; Dir. Programs, D.B. Lampman.
Personnel Profile: Full-Time Paid 3; Part-Time Volunteers 150.
Governing Authority: nonprofit organization. Tax-exempt: 501(c)(3).
Institution Type/Description: Art & Maritime History Museum.
Collections: 220-piece collection; lithographs; paintings; formal & plein air drawings; photographs; writings; marine artifacts; furnishings; memorabilia; John A Noble's houseboat studio; Sailors' Snug Harbor period rooms; navigation instruments; art, furnishings and documents from the Sailors' Snug Harbor collection.
Major Exhibits: Tides of 100 Years: A Noble Centennial Exhibition, 3/13-3/1/14.
Research Fields: art history; maritime history; study of the care of frail adults.
Facilities: 4,900 vol. library of material reflecting maritime concerns & history available to the public on premises. John A. Noble (1913-1983) lithographs, reproductions, books & catalogues for sale.
Activities: guided tours; lectures; films; organized education programs for children and adults; loan, temporary & traveling exhibitions.
Publications: quarterly newsletter, Hold Fast!; books, John A. Noble: The Rowboat Drawings; Hulls & Hulks in the Tide of Time: The Life & Work of John A. Noble; The Fight for Sailor's Snug Harbor; The Terrible Captain Jack Visits the Museum; Caddell Dry Dock: 100 Years Harborside; Bon A Tirer: The Prints of Herman Zaage.
Hours & Admission Prices: Thurs.-Sun. 1-5. Adults $5, seniors, students &

educators $3; discounts to AAM members; children under 10 & members no charge. Closed New Year's Day; Thanksgiving & Day after; Christmas.

Attendance: 30,000 (estimated)

Membership: Senior Citizen & Educator $25; Individual $40; Family $75; Sail $250; Mast $500; Rowboat Society $1,000; Quarterdeck $2,500; Crow's Nest $5,000.

SEQUINE MANSION, 440 Seguine Ave., Staten Island, NY 10309-3936. Mailing Address: 830 Fifth Ave., The Arsenal, Rm. 203, New York, NY 10065-7001. Tel.: 718-967-3542.

Web Site: www.historichousetrust.org
Formerly: Seguine-Burke Plantation
Founded: 1989.
Congressional District: 13
Key Personnel: Exec. Dir. Historic House Trust, Franklin D. Vagnone.
Governing Authority: nonprofit historical agency. Parent Institutions: Historic House Trust of New York City, The Arsenal, Room 203, Central Park, New York; New York City Dept. of Parks & Recreation.
Institution Type/Description: Historic House: 1837 two-story Greek Revival mansion built by businessman Joseph A. Seguine.
Collections: historic structure; stable; barn; carriage house; antique furnishings.
Research Fields: early Staten Island history.
Activities: guided tours.
Hours & Admission Prices: By appointment only. No charge.
Membership: Friend $50; Associate $75; Roof Raiser $100; Family Friend $125; Patron $250; Fellow $500; Guardian $1,000; Cornerstone $5,000.

SNUG HARBOR CULTURAL CENTER & BOTANICAL GARDEN, 1000 Richmond Ter., Staten Island, NY 10301-1114. Tel.: 718-425-3504. Fax: 718-815-0198. Facebook: Snug Harbor Cultural Center and Botanical Garden.

E-mail: info@snug-harbor.org
Web Site: snug-harbor.org
Founded: 1976.
Congressional District: 13
Key Personnel: Chm. Bd. Directors, Mark Lauria; Pres. & C.E.O., Lynn B. Kelly; C.F.O., Jefrey Manzer; C.O.O., JoAnn Mardikos; Chief Devel. Office, Cynthia Taylor.
Personnel Profile: Full-Time Paid 25; Part-Time Paid 30; Interns 1.
Governing Authority: nonprofit. Subsidiary Institution: Smithsonian affiliate. Tax-exempt: 501(c)(3).
Institution Type/Description: Cultural Center & Botanical Garden: housed in a former retirement home for sailors.
Collections: local history; contemporary art; gardens; 28 historic buildings.
Research Fields: visual & performing arts.
Facilities: botanical garden; 686-seat music hall; 150-seat auditorium; banquet hall; dance studios; classrooms; outdoor stage.
Activities: guided tours; educational programs in visual arts, horticulture, & agriculture; concerts; plays; theater performances; festivals; public programs; temporary art exhibitions.
Publications: book, Sailors Snug Harbor; art catalogues; calendar of events.
Hours & Admission Prices: Grounds: daily dawn to dusk. Visitor's Center & Newhouse Center: Wed.-Sun. 12-5. Garden: Tues.-Sun. 10-5. Adults $5, seniors & students $4; children under 12 no charge. Gardens & Galleries: adults $8, seniors & students $7; children under 12 no charge. Special Events & Performances prices may vary. Closed New Year's Day; Thanksgiving; Christmas.
Attendance: 300,000 (estimated)
Membership: Senior 65 & over and Student $35; Family $60; Individual $65; Patron $100; Snug Smithsonian $150.

STATEN ISLAND CHILDREN'S MUSEUM, (M), 1000 Richmond Terr. at Snug Harbor, Staten Island, NY 10301. Tel.: 718-273-2060. Fax: 718-273-2836.

E-mail: drosenthal@sichildrensmuseum.org
Web Site: www.statenislandkids.org
Founded: 1974.
Congressional District: 24
Key Personnel: Exec. Dir., Dina R. Rosenthal; Chm. (V), Mark Dray; Business Mgr., Barbara Brandon; Dir. Education, Addy Manipella; Volunteer & Intern Coord., Carl Jackman; Exhibits Mgr., Renee Wasser.
Personnel Profile: Full-Time Paid 13; Part-Time Paid 30; Part-Time Volunteers 400; Interns 20.
Governing Authority: nonprofit organization. Tax-exempt: 501(c)(3).
Institution Type/Description: Children's Museum.
Collections: arts & sciences; interactive exhibitions reflecting themes in arts, humanities & sciences; pre-school exhibit; insects.

Research Fields: visitor evaluation; art education; use of arts in teaching science.
Facilities: art workshop; performance space. Museum-related items for sale.
Activities: school & community youth group tours; public programs; workshop services; guest artist demonstrations & workshops; cooperative programming with the school district, community groups & other cultural organizations; outreach programs; science theater productions for museum & school audiences; teacher enhancement programs.
Publications: quarterly, activity calendars; events newsletter; brochures; exhibit catalogues.
Hours & Admission Prices: School Year: Tues.-Sun. 12-5; Summer: Tues.-Sun. 10-5. Admission: over 1 $6; discounts to AAM, AYM & ASTC members; members no charge. Closed New Year's Day; Easter; Memorial Day; Independence Day; Labor Day; Thanksgiving; Christmas.
Attendance: 230,000 (accurate)
Membership: Family $75; Companionship $100; Friendship $150 & up.

STATEN ISLAND MUSEUM, (M), 75 Stuyvesant Place, Staten Island, NY 10301-1998. Tel.: 718-727-1135, ext. 114. Fax: 718-273-5683. Facebook: Staten Island Museum.

E-mail: hbehnke@statenislandmuseum.org
Web Site: www.statenislandmuseum.org
Formerly: Staten Island Institute of Arts & Sciences (SIIAS)
Founded: 1881.
Congressional District: 17
Key Personnel: C.E.O., Elizabeth Egbert; Chm., Ralph Branca; Cur. Art., Robert Bunkin; Archivist, Cara Dellatte; Vice Pres. External Affairs & Advancement, Henryk Behnke; Vice Pres. Exhibitions & Programs, Diane Matyas; Dir. Science, Edward Johnson; Education Mgr., Claire Aniela Arthurs; School Programs Asst., Loretta Lonecke; Communications Coord., Rachel Somma; C.O.O., Cheryl Adolph; Registrar/Collections Mgr., Audrey Malachowsky; Visitor Svcs. Mgr., Renee Bushelle; Community Campaign Coord., Amanda Straniere.
Personnel Profile: Full-Time Paid 16; Part-Time Paid 20; Part-Time Volunteers 25; Interns 5.
Governing Authority: private; nonprofit organization. Tax-exempt: 501(c)(3).
Institution Type/Description: General Museum.
Collections: artworks; natural science specimens; local history archives; paintings; decorative arts; sculpture; archaeology; manuscripts; photographs; maps; prints; costumes & textiles; graphics; natural history of Staten Island & region; minerals; botany; ornithology; herpetology; entomology; Staten Island Ferry collection.
Research Fields: area art, natural science & history; decorative arts; environmental concerns; preservation of natural resources; photography; 19th-century American painting; American portraits.
Facilities: 16,000-vol. library; classroom; 80-seat auditorium. Maritime history publications, nautical gifts & ferry-related items for sale.
Activities: workshops; demonstrations; academic internships; permanent & temporary exhibitions; study clubs; formally organized educational programs for children & adults; guided Staten Island Ferry tours; nature & neighborhood walks.
Publications: quarterly newsletter; special exhibition catalogues; publications on natural science & local history of Staten Island.
Hours & Admission Prices: Mon.-Fri. 11-5, Sat. 10-5, Sun. 12-5. Adults $5, students & seniors $3; discounts to AAM members; children under 12 & members no charge. Closed New Year's Day; Memorial Day; Independence Day; Thanksgiving; Christmas.
Attendance: 75,000 (estimated)
Membership: Student, School Teacher & Senior (65 & up) $25; Individual $35; Family/Dual $50; Collector's Club $100; Connoisseur's Club $250; Curator's Circle $500; President's Circle $1,000; Trustees Circle $2,500.

STATEN ISLAND ZOO, 614 Broadway, Staten Island, NY 10310-2896. Tel.: 718-442-3100 & 3101. Fax: 718-981-8711.

E-mail: kmitchell@statenislandzoo.org
Web Site: www.statenislandzoo.org
Founded: 1936.
Congressional District: 17
Key Personnel: Exec. Dir., Kenneth C. Mitchell; Pres., William J. Frew, Jr.; Gen. Cur., Dr. Marc Valitutto, VMD
Personnel Profile: Full-Time Paid 41; Part-Time Paid 9; Part-Time Volunteers 44; Interns 2.
Governing Authority: society. Staten Island Zoological Society. Tax-exempt: 501(c)(3).
Institution Type/Description: Zoo.
Collections: herpetology; mammalogy; avian; aquarium.
Facilities: 1,000-vol. library of books on zoology and biology; snack bar. Books & other museum-related items for sale.

Activities: lectures; films; zoomobile outreach program; formally organized education programs for children.
Publications: e-newsletter.
Hours & Admission Prices: Daily 10-4:45. Adults $8, seniors 60 & over $6, children 3-14 $5, disabled $3; members, children under 3, members of reciprocal zoos & every Wed. 2-4:45 no charge. Closed New Year's Day; Thanksgiving; Christmas. &
Attendance: 157,178 (accurate)
Membership: Individual $45; Family $85; Gold Card $175; Corporate Donor $250; Life $500; Corporate Sponsor $1,000; Corporate Benefactor $2,500. Joint membership with SI Children's Museum $140.

Sterling Center

STERLING HISTORICAL SOCIETY AND LITTLE RED SCHOOLHOUSE MUSEUM, 1294 State Rte. 104A, Sterling Center, NY 13156. Mailing Address: P.O. Box 114, Sterling, NY 13156. Tel.: 315-947-5072 & 564-6721.
E-mail: sterlinghistory@lakeontario.net
Web Site: www.lakeontario.net/sterlinghistory
Founded: 1976.
Congressional District: 33
Key Personnel: Pres., Patricia Shortslef; Sec. & Trustee, Susan Parsons; Treasurer & Trustee, James Chaffee; Trustee, Mr. Marion Teachout; Trustee, Leigh Shortslef; Town Clerk, Lisa Cooper; Museum Shop Mgr., Susan Allen.
Personnel Profile: Full-Time Volunteers 1; Part-Time Volunteers 14.
Governing Authority: society. Parent Institution: Town of Sterling. Subsidiary Institution: Cayuga Museum. Branch Museum: C.C.C. (Civilian Conservation Corps.) mini museum & monuments in Fair Haven Beach State Park. Tax-exempt.
Institution Type/Description: Historical Society Museum: housed in The Little Red School House, an original wood frame, two-story structure.
Collections: personal artifacts; cameras & developing equipment; glass negatives; typewriters; Doctor's bag including medical items; Civil War; Indian artifacts; period store; barber shop; period kitchen; old blacksmith shop; 1910 NY Central/Hojack Signal tower with 85 feet of track and operating Sheffield handcar; railroad artifacts
Research Fields: genealogy.
Facilities: classroom; Raymond A. Waldren Exhibit Building.
Activities: temporary exhibitions; special crafts display; special exhibits; movie presentation; special speakers.
Publications: brochure; newsletters; books; annual community calendars.
Hours & Admission Prices: weekend after July 4th to weekend before Labor Day Sat.-Sun. 1-4. No charge; donations accepted.
Attendance: 180 (estimated)
Membership: Single $5; Family $10; Life $75.

Stillwater

SARATOGA NATIONAL HISTORICAL PARK, 648 Rte. 32, Stillwater, NY 12170-1604. Tel.: 518-664-9821, ext. 224. Fax: 518-664-3349.
E-mail: sara_info@nps.gov
Web Site: www.nps.gov/sara
Founded: 1938.
Congressional District: 24
Key Personnel: Supt., Joe Finan; Program Dir., Gina Johnson.
Personnel Profile: Full-Time Paid 16; Part-Time Paid 6; Part-Time Volunteers 20; Interns 1.
Governing Authority: federal. Administered by the Dept. of Interior, National Park Service, Interior Bldg., Washington, DC 20240.
Institution Type/Description: Military Museum: located on the site of the Battles of Saratoga, Sept. 19 and Oct. 7, 1777.
Collections: military history and the Burgoyne Campaign.
Research Fields: Burgoyne Campaign; the Battles of Saratoga.
Facilities: 600-vol. library of books, microfilm and research reports related to the American Revolution, Burgoyne Campaign and the Battles of Saratoga available for use by researchers on premises; 80-seat auditorium; theater; battlefield unit; visitor center; tour road; picnic areas; Saratoga Monument Unit; Philip Schuyler House. History-related educational items for sale.
Activities: orientation films; tours; talks & walks; period demonstrations.
Publications: brochure, Saratoga National Historical Park.
Hours & Admission Prices: Visitor Center: daily 9-5. Tour Road: April to early Dec. daily 9am to sunset. Park: no charge. Tour Road: $5. Closed New Year's Day; Thanksgiving; Christmas. &
Attendance: 150,000 (accurate)

Stony Brook

* **LONG ISLAND MUSEUM OF AMERICAN ART, HISTORY & CARRIAGES, (M),** 1200 Rte. 25A, Stony Brook, NY 11790-1992. Tel.: 631-751-0066, ext. 0. Fax: 631-751-0353.
E-mail: mail@longislandmuseum.org
Web Site: www.longislandmuseum.org
Formerly: The Museums at Stony Brook
Founded: 1935.
Congressional District: 1
Key Personnel: Pres. & C.E.O., Jacqueline Day; Chm. (V), James M. Wicks, Esq.; Chief Cur., William S. Ayres; Dir. Devel., Deirdre Doherty; Bldgs. & Grounds, Joe Esser; Bldgs. & Grounds, Matt Schery; Dir. Education, Betsy Radecki; Dir. Communications, Julie Diamond; Cur. History, Joshua Ruff; Museum Shop Mgr., Dori Portes.
Personnel Profile: Full-Time Paid 19; Part-Time Paid 4; Part-Time Volunteers 153; Interns 1.
Governing Authority: nonprofit organization. Tax-exempt: 501(c)(3).
Institution Type/Description: General Museum.
Collections: 200 carriages & horse-drawn vehicles, carriage accoutrements; 19th- & 20th-century American art including the William Sidney Mount collection of paintings, drawings, memorabilia; costumes & accessories; dolls; toys; wildfowling decoys; decorative arts; miniature rooms; local archives; early photographs; William Sidney Mount archives. Historic Buildings: 18th-century Hawkins-Mount house; 18th-century barn; 19th-century schoolhouse; 19th-century blacksmith shop; carriage shed.
Research Fields: horse-drawn transportation; 19th- & 20th-century American art; history; costumes & textiles; wildfowling decoys.
Facilities: 2,000-vol. library of American art & Long Island history; classroom; meeting room
Activities: formally organized education programs for children, adults & families; permanent & temporary exhibitions; art engagement for people with memory loss; training program for museum guides; internship programs; special events; concerts; seminars; workshops.
Publications: exhibition catalogs; brochures; books related to collections.
Hours & Admission Prices: Call for hours. Adults $9, senior citizens $7, children $4; discounts to AAM members; members no charge. Closed New Year's Day; Easter; Thanksgiving; Christmas Eve & Day. &
Attendance: 21,201 (accurate)
Membership: Students & Senior Citizens $25; Individual $40; Dual & Family $60; Contributor $100; Sponsor $250; Patron $500; President's Council $1,000.

MUSEUM OF LONG ISLAND NATURAL SCIENCES, Earth & Space Sciences Bldg.-State University of New York at Stony Brook, Stony Brook, NY 11794. Tel.: 631-632-8230. Fax: 631-632-8240. www.museumoflongislandnaturalsciences.org.
E-mail: Pamela.Stewart@sunysb.edu
Web Site: www.geosciences.stonybrook.edu/museum
Founded: 1973.
Congressional District: 1
Key Personnel: Dir., Pamela Stewart; Cur. Geology, Steven E. Englebright.
Personnel Profile: Full-Time Paid 1; Part-Time Paid 1; Part-Time Volunteers 18.
Governing Authority: university; nonprofit organization. Parent Institution: State University of New York at Stony Brook. Tax-exempt.
Institution Type/Description: Natural Science Museum.
Collections: worldwide modern & fossil marine invertebrates especially Mollusca; natural history collections of Long Island, especially insects & their host plants; general herbarium; minerals & fossils; modern marine invertebrates.
Research Fields: geology of Long Island; ecology; marine science; science education; botany.
Facilities: classrooms; meeting rooms; nature trails.
Activities: guided tours; lectures; films; gallery talks; formally organized education programs for children, adults, college students affiliated with State University of New York; loan, permanent & temporary exhibitions; field trips.
Publications: calendar of events; A Beachcombers Guide to Long Island Shores; The Pine Barrens; Our Fragile Wilderness; Tall Grass by the Sea: A Guide to Long Islands Salt Marshes; film, Long Island Wilderness...The Pine Barrens; books, The Mashomack Preserve Study, 3 vols; Long Island Water Resources Curriculum Guide, 2 vols; A Field Guide to Long Island Woodlands; A Field Guide to Long Island's Freshwater Wetlands; A Field Guide to Long Island's Seashores.
Hours & Admission Prices: Mon.-Fri. 9-5. No charge; donations accepted. Closed national holidays. &
Attendance: 20,000 (estimated)
Membership: Student $10; Individual $15; Dual & Family $25; Supporting

$30; Associate $50; Contributor $60; Sustainer $100; Affiliate $250; Patron $500; Supporter $1,000; Benefactor $2,500.

UNIVERSITY ART GALLERY, STATE UNIVERSITY OF NEW YORK AT STONY BROOK, Staller Center for the Arts, Stony Brook, NY 11794. Tel.: 516-632-7240. Fax: 516-632-1976.

E-mail: rhonda.cooper@stonybrook.edu
Founded: 1975.
Congressional District: 1
Key Personnel: Dir., Rhonda Cooper.
Personnel Profile: Full-Time Paid 2; Interns 5.
Governing Authority: university; nonprofit. Parent Institution: SUNY at Stony Brook. Tax-exempt.
Institution Type/Description: University Art Gallery.
Collections: works by students, faculty, national & international artists.
Research Fields: modern & contemporary art.
Facilities: 4,700 sq. ft. exhibit space; educational facilities.
Activities: formal education programs; traveling & temporary exhibitions.
Publications: exhibition catalogues 3 times annually.
Hours & Admission Prices: early Sept. to May Tues.-Fri. 12-4, Sat. 7-9. No charge. Closed holidays. &
Attendance: 12,000 (estimated)

Stony Point

STONY POINT BATTLEFIELD STATE HISTORIC SITE, 44 Battlefield Rd., Stony Point, NY 10980. Mailing Address: Box 182, Stony Point, NY 10980-0182. Tel.: 845-786-2521. Fax: 845-786-0463.

Web Site: www.palisadesparksconservancy.org
Founded: 1897.
Congressional District: 26
Key Personnel: Historic Site Mgr., Julia M. Warger.
Personnel Profile: Part-Time Paid 9; Part-Time Volunteers 3.
Governing Authority: state. Parent Institution: New York State Office of Parks, Recreation & Historic Preservation & Palisades Interstate Park Commission. Tax-exempt: 501(c)(3).
Institution Type/Description: Historic Site & Military Museum: located on the site of raid on British stronghold by Brigadier Gen. Anthony Wayne on July 15, 1779.
Collections: Revolutionary War equipment & artifacts. Historic Structure: c.1826 lighthouse.
Research Fields: American Revolution in Hudson Valley; lighthouses of the Hudson River Valley.
Facilities: self-guided signed trails; outdoor walking tour; picnic areas.
Activities: permanent exhibit; special programs & living history demonstrations.
Hours & Admission Prices: mid-April to Oct. Wed.-Sun. 10-5, Sun. 12-5. Museum & Lighthouse: Sat.-Sun. & Mon. holidays $5 per vehicle. Additional Fees Charged: evening tours, lantern walks, special events, & programs.
Attendance: 27,500 (estimated)

Syracuse

THE COMMUNITY FOLK ART CENTER, (M), 805 E. Genesee St., Syracuse, NY 13210-1507. Tel.: 315-442-2230. Fax: 315-442-2972.

E-mail: cfac@syr.edu
Web Site: communityfolkartcenter.org
Founded: 1972.
Congressional District: 25
Key Personnel: Exec. Dir., Kheli Willetts
Institution Type/Description: Art Center.
Collections: African American culture & art.
Activities: gallery talks; workshops; lectures; films.
Hours & Admission Prices: Tues.-Fri. 10-5, Sat. 11-5.

✱ ERIE CANAL MUSEUM, (M), 318 Erie Blvd., E., Syracuse, NY 13202-1106. Tel.: 315-471-0593. Fax: 315-471-7220.

E-mail: contactus@eriecanalmuseum.org
Web Site: www.eriecanalmuseum.org
Founded: 1962.
Congressional District: 25
Key Personnel: Exec. Dir., Diana Goodsight; Pres. (V), Richard Cordaro; Cur., Daniel Franklin Ward; Devel. & Public Rels. Dir., Natalie Stetson; Mgr. Operations, Steve Caraccuo.
Personnel Profile: Full-Time Paid 4; Part-Time Paid 4; Part-Time Volunteers 40.

Governing Authority: nonprofit organization. Tax-exempt: 501(c)(3).
Institution Type/Description: History Museum: housed in 1850 Greek Revival style Weighlock Building, a monument to Erie Canal architecture; Syracuse Heritage Area Visitors Center.
Collections: patent models; commemorative china; textiles; 19th-century household items & tools; paintings; photographs; manuscripts; maps; final account journals; engineering records; prints of the canal era. Historic Building: Weighlock Building, used for collection of tolls & administration of a division of the 19th & 20th-century Erie Canal network; the impact of Erie Canal on United States history in the 19th century; Syracuse history; 65 ft. long canal boat replica.
Research Fields: history, use & operation of the canal; economics of construction & impact on surrounding territories; life of canal era; construction of boats; development of Erie Canal Park.
Facilities: 6,000-vol. library of state legislative records, 900,000 manuscripts, 30,000 photos, negatives & slides, secondary historical works & primary sources available by appointment; towpath trail; orientation center.
Activities: guided museum tours; lectures; graduate student internships; inter-museum exhibitions; school loan service; walking tours; motor coach tours; education programs; festival for children.
Publications: quarterly newsletter, A Canal Boat Primer; books, A Novel Look at the Erie Canal; Photos From the Collection; Syracuse; An Urban Landscape; Always Know Your Pal; Children of the Erie Canal; Syracuse Architecture: A City Rises from the Banks of the Canal; Those Among Us: Uncovering the Story of Who Built the Erie Canal.
Hours & Admission Prices: Mon.-Sat. 10-5, Sun. 10-3. No charge; donations accepted. Library by appointment only. Closed New Year's Day; Independence Day; Thanksgiving; Christmas. &
Attendance: 25,000 (estimated)
Membership: General $35; Family $50; Canaller $100; Weighmaster $500 & up.

✱ EVERSON MUSEUM OF ART, (M), 401 Harrison St., Syracuse, NY 13202-3091. Tel.: 315-474-6064. Fax: 315-474-6943. Facebook: Everson Museum of Art.

E-mail: everson@everson.org
Web Site: www.everson.org
Formerly: Everson Museum of Art of Syracuse & Onondaga County
Founded: 1896.
Key Personnel: Pres. (V), Patrick Pedro; Dir., Steven Kern; Pres. (V), Cathy Winger; Cur. Education, Pam McLaughlin; Archivist & Librarian, Mary Iversen; Asst. Dir., Sarah Massett; Museum Shop Mgr., Sheila Goldie; Museum Shop Mgr., Karen Williams.
Personnel Profile: Full-Time Paid 17; Part-Time Paid 2; Part-Time Volunteers 324; Interns 7.
Governing Authority: nonprofit organization. Tax-exempt: 501(c)(3).
Institution Type/Description: Art Museum.
Collections: 18th-20th century American paintings, sculpture, drawings & graphics; American ceramics A.D. 1100-21st century; ceramics of the world, 3000 B.C.-21st century; photography; study collections of Asian, African and Oceanic art & artifacts.
Major Exhibits: African American Art: Harlem Renaissance, Civil Rights Era and Beyond (T), 10/18/13-1/4/2015; Of Heaven and Earth: 500 Years of Italian Painting from Glasgow Museums, 4/19/14-7/13/14.
Research Fields: American painting & sculpture; 19th, 20th & 21st century American ceramics.
Facilities: 10,000-vol. library of art history, art criticism, biography & exhibition catalogs, available for research on premise & for inter-library loan; 299-seat auditorium; classrooms; cafe. Museum-related items for sale.
Activities: guided tours; docent program; lectures; films; gallery talks; concerts; dance recitals; arts festivals; education programs for school children, students, families, and adults; outreach programs; training programs for professional museum workers; inter-museum loan exhibitions; permanent, traveling & temporary exhibitions.
Publications: handbooks; guides to various collections; catalogues of major exhibitions; quarterly newsletter.
Hours & Admission Prices: Tues.-Fri. & Sun. 12-5, Thurs. 12-8, Sat. 10-5. Suggested Donation: $5 per person. Closed New Year's Day; Independence Day; Thanksgiving; Christmas. &
Attendance: 116,500 (estimated)
Membership: Individual $50; Household $75; Fellow $125; Friend $250; Patron $500; Everson Circle $1,000; Everson Circle Gold $2,500; Director's Circle $5,000; Founder's Circle $10,000; 20% off any level up to Everson Circle for Seniors (65+), Educators, Students & Military.

MILTON J. RUBENSTEIN MUSEUM OF SCIENCE & TECHNOLOGY, 500 S. Franklin St., Syracuse, NY 13202-1245. Tel.: 315-425-9068, ext. 2116. Fax: 315-425-9072.

Web Site: www.most.org
Founded: 1978.

Congressional District: 25
Key Personnel: Pres., Larry R. Leatherman; Chm. (V), Richard Sykes; Vice
   Pres. (V), Edgar Galson; Treas., Patrick Dooher; Museum Shop Mgr.,
   Nancy Allison.
Personnel Profile: Full-Time Paid 23; Part-Time Paid 25; Part-Time Volunteers
   140; Interns 2.
Governing Authority: nonprofit organization. Parent Institution: Discover
   Center of Science & Technology. Tax-exempt: 501(c)(3).
Institution Type/Description: Science & Technology Museum.
Collections: concept oriented exhibits on a wide range of scientific &
   technological topics aimed at visitor participation involving hands-on
   techniques.
Research Fields: history of science.
Facilities: meeting rooms; teachers resource center; Silverman Planetarium;
   classrooms; Bristol Omnitheater (IMAX Dome). Science-related kits &
   specimens for sale.
Activities: demonstrations; lectures; films; hobby workshops; planetarium
   presentations; in-school science programs; hands-on exhibits; formally
   organized education programs; docent program or council; loan, permanent,
   temporary & traveling exhibitions; field trips.
Publications: quarterly newsletter.
Hours & Admission Prices: Wed.-Sun. 10-5. Adults $8, seniors 65 & over and
   children under 11 $7; discounts to ASTC members; members no charge. &
Attendance: 165,000 (estimated)
Membership: Senior $40; Individual $55; Grandparent $69; Family $79;
   Enhanced Family $89.

**ONONDAGA HISTORICAL ASSOCIATION MUSEUM & RE-
   SEARCH CENTER,** 321 Montgomery St., Syracuse, NY 13202-
   2098. Tel.: 315-428-1864. Fax: 315-471-2133.
E-mail: karen.cooney@cnyhistory.org
Web Site: www.cnyhistory.org
Founded: 1862.
Congressional District: 27
Key Personnel: Exec. Dir., Gregg Tripoli; Asst. Dir. & Cur. Collections,
   Thomas Hunter; Pres. (V), Lee DeAmicis; Cur. History, Dennis Connors;
   Dir. Devel., Lynne Pascale; Education Assoc., Scott Peal; Archivist, Pamela
   Priest; Administrative Asst., Karen Cooney; Museum Shop Mgr., Andrew
   Jordan.
Personnel Profile: Full-Time Paid 9; Part-Time Paid 2; Part-Time Volunteers
   40; Interns 2.
Governing Authority: society. Tax-exempt: 501(c)(3).
Institution Type/Description: Local Historical Museum.
Collections: Syracuse, Onondaga County & area history; Underground Rail-
   road artifacts; New York State canal history; landscapes & portraits by local
   artists; archaeology; archives; graphics; ethnology; transportation; stained
   glass; pottery; manuscripts; architectural drawings, newspapers, imprints,
   decorative arts, photographs; typewriters; Franklin automobiles & other
   locally manufactured products; sports history; manuscripts; maps; Syracuse
   china factory collection; clothing & textiles; Gambrinus statue; Native
   American artifacts.
Major Exhibits: Snowy Splendor, 10/9/13-3/30/14; Culture of the Cocktail
   Hour, 11/15/13-6/15/14; Fashion After Five, 11/15/13-6/15/14; Italian
   Heritage, 4/17/14-11/1/14; Crime & Punishment, 7/16/14-1/25/15.
Research Fields: Onondaga County & area history.
Facilities: library of photographs, maps, documents & other reference mate-
   rials available on premises; classroom; auditorium. Books & museum-
   related items for sale.
Activities: guided tours; lectures; workshops; craft demonstrations; temporary
   & traveling exhibitions; bus tours; Ghost Walk; National History Day
   Competition.
Publications: semi-annual news magazine, History Highlights; Syracuse
   Landmarks; Images of America: Syracuse, local photographs from 1854;
   Images of America: Greater Syracuse, local photographs from 1900;
   Crossroads In Time: An Illustrated History of Syracuse; Historic Photos of
   Syracuse; Windows Into The Past, Vol. 1.
Hours & Admission Prices: Museum: Wed.-Fri. 10-4, Sat.-Sun. 11-4. No
   charge; donations accepted. Research Center: Wed.-Fri. 10-2 Sat. 11-3:30.
   Adults $7, students $4. Closed national holidays. &
Attendance: 30,000 (estimated)
Membership: Students $25; Seniors $25; Senior Couple $30; Individual $30;
   Family $40; Researcher $50; Archivist $100; Historian $500; Scholar
   $1,000. Corporate: $250; $500; $1,000; Lifetime $2,500.

ROSAMOND GIFFORD ZOO AT BURNET PARK, 1 Conserva-
   tion Place, Syracuse, NY 13204-2590. Tel.: 315-435-8511. Fax:
   315-435-8517.
E-mail: rgzoo@ongov.net
Web Site: www.rosamondgiffordzoo.org
Formerly: Burnet Park Zoo

Founded: 1914.
Congressional District: 23
Key Personnel: Dir., Ted Fox; Cur., Tom LaBarge; Society Pres., Janet
   Agostini; Dir. Devel. & Membership, Jill Allen; Dir. Education, Nathan
   Keefe; Dir. Food & Beverage, Jim Mahler; Dir. Mktg., Maria Simmons; Dir.
   Public Rels., Jaime Alverez; Volunteer Svcs. Mgr., Ellen Vaughn; Gift Shop
   Mgr., Renee Ross.
Personnel Profile: Full-Time Paid 36; Part-Time Paid 7; Part-Time Volunteers
   376; Interns 9.
Governing Authority: county. Parent Institution: Onondaga County Parks.
   Tax-exempt: 170(b)(1)(A).
Institution Type/Description: Zoo.
Collections: invertebrates; fish; amphibians; reptiles, birds & mammals.
Research Fields: mammalian, avian, reptilian & fish behavior.
Facilities: 45 acres of grounds.
Activities: tours; seminars; film series; summer camp; internships; educational
   programs through extensive volunteer program.
Publications: teacher's guide to zoo; map; training materials; animal briefs;
   multimedia presentations; quarterly membership magazine, My Zoo; bi-
   monthly docent newsletter, Inside My Zoo.
Hours & Admission Prices: Daily 10-4:30. Jan.-Feb. adults $4, seniors 62 &
   over $2.50, youth 3-18 $2; members and children 2 & under no charge;
   March-Dec. adults $8, senior citizens 62 & over $5, youth 3-18 $4;
   members and children 2 & under no charge. Closed New Year's Day;
   Thanksgiving; Christmas. &
Attendance: 372,132 (accurate)
Membership: Individual $38; Dual Senior $42; Dual $52; Grandparents $58;
   Family $68; Family Plus One $82.

SUART GALLERIES - SYRACUSE UNIVERSITY, (M), Shaffer
   Art Bldg., Syracuse University, Syracuse, NY 13244. Tel.: 315-
   443-4097. Fax: 315-443-9225.
E-mail: suart@syr.edu
Web Site: suart.syr.edu
Formerly: Syracuse University Art Collection
Founded: 1871.
Congressional District: 27
Key Personnel: Dir., Domenic J. Iacono; Assoc. Dir., David Prince; Registrar,
   Laura Wellner; Asst. Dir. Museum Operations, Andrew J. Saluti; Coord.
   Collection & Exhibition, Emily Dittman; Administrative Asst., Joan Re-
   cuparo; Sec. & Office Coord., Alex Hahn.
Personnel Profile: Full-Time Paid 7; Part-Time Paid 5; Part-Time Volunteers 2.
Governing Authority: private university; nonprofit. Parent Institution: Syracuse
   University. Subsidiary Institution: Louise and Bernard Palitz Gallery;
   Syracuse University Lubin House, New York, NY. Tax-exempt.
Institution Type/Description: Art Gallery.
Collections: 20th-century American graphics, painting, sculpture & decorative
   arts; prehistoric to modern Western art; textiles; history of illustration.
Research Fields: art between the World Wars.
Facilities: 3,200-vol. library of catalog raisonne & general art history,
   American art monographs, indexes & listings of artists; educational
   facilities.
Activities: formal education programs for undergraduate & graduate college
   students; loan, temporary & traveling exhibitions.
Publications: highlight special collection; semi-annual collection exhibitions
   catalog; semi-annual newsletter.
Hours & Admission Prices: Tues.-Sun. 11-4:30; other times by appointment.
   No charge. Closed New Year's Eve & Day; Thanksgiving & weekend after;
   Christmas Eve, Day & week. &
Attendance: 12,000 (estimated)

Tappan

TAPPANTOWN HISTORICAL SOCIETY, Tappan, NY 10983.
   Mailing Address: Box 71, Tappan, NY 10983-0071. Tel.: 845-359-
   1149 & 2730.
Web Site: tappantown.org
Founded: 1965.
Congressional District: 47
Key Personnel: Pres., Carol LaValle; Vice Pres., John Morton; Treas., Gerri
   McCauley.
Governing Authority: nonprofit organization.
Institution Type/Description: Historical Society & Historic District: approxi-
   mately 20 18th- & 19th-century homes.
Collections: American architecture; preservation project; architectural heri-
   tage.
Research Fields: local history.
Activities: educational programs; 1-mile walking tours; guided tours; dating of
   houses.
Publications: annual booklet, Drummer Boy; quarterly newsletter.

Hours & Admission Prices: No charge.
Membership: Individual $10; Family $15.

Tarrytown

THE HISTORICAL SOCIETY, SERVING SLEEPY HOLLOW AND TARRYTOWN, One Grove St., Tarrytown, NY 10591-4122. Tel.: 914-631-8374.
E-mail: historyatgrove@aol.com
Web Site: www.thehistoricalsociety.net
Formerly: The Historical Society of Tarrytowns, Inc.
Founded: 1889.
Congressional District: 23
Key Personnel: Pres., Richard Rose; Admin. & Cur., Sara Mascia.
Personnel Profile: Full-Time Paid 1; Part-Time Volunteers 12.
Governing Authority: society. Bd. of Trustees. Tax-exempt: 170(b)(1)(A).
Institution Type/Description: General History Museum.
Collections: Native American artifacts; early Dutch history of the region; photographs; artifacts from local archaeological dig; local memorabilia; military papers; land records; wills; American Revolution: 1780 capture of Major John Andre at Tarrytown; Washington Irving & other men and women of letters; paintings; costumes; manuscripts; early military artifacts; art collections of Evart Duychinck III, Ezra Ames, John Mare, Emily N. Hatch, DeWitt C. Hay, & Edgar Mayhew Bacon; jewelry; clothing; textiles; dolls; toys; household implements.
Research Fields: genealogy; local history-social, economic, political, geographical, & architectural.
Facilities: 3,500-vol. library of material on local & New York State history & genealogy, 715 maps, records of Ward B. Burnett Post, G.A.R.; Civil War papers of Cap. Charles H. Rockwell; microfilm of local newspapers; reading room. Museum-related items for sale.
Activities: guided tours; lectures; permanent & temporary exhibitions. Society Sponsors: Annual Strawberry Festival; historic house tours.
Publications: Images of America Tarrytown and Sleepy Hollow in the 20th Century; society newsletter, The Chronicle.
Hours & Admission Prices: Wed.-Thurs. & Sat. 2-4. No charge; donations accepted. Archives: by appointment. Closed holidays.
Attendance: 2,500 (estimated)
Membership: Student $10; Individual $30; Family $40; Sustaining $50; Contributing $100; Business & Organization $100; Life $500.

LYNDHURST, 635 S. Broadway, Tarrytown, NY 10591-6499. Tel.: 914-631-4481, ext. 0. Fax: 914-631-5634.
E-mail: lyndhurst@nthp.org
Web Site: www.lyndhurst.org
Founded: 1964.
Congressional District: 22
Key Personnel: Events & Rentals, Christine Plazas; Preservation Mgr., Krystyn Hastings-Silver; Buildings & Grounds Mgr., David Ware; Visitor Svcs., Lisa Buckley.
Personnel Profile: Full-Time Paid 7; Part-Time Paid 15.
Governing Authority: nonprofit organization. Parent Institution: National Trust for Historic Preservation, 1785 Massachusetts Ave. N.W., Washington, DC 20036. Tax-exempt: 501(c)(3).
Institution Type/Description: Historic House Museum: 19th-century Gothic Revival residence of William Paulding, George Merritt & Jay Gould family, designed by Alexander Jackson Davis.
Collections: 19th- & early 20th-century decorative arts; fine arts; textiles; A.J. Davis-designed Gothic furniture; rare books; specimen trees. Historic Building: carriage house.
Research Fields: architecture; decorative arts; landscape; local history; archives consisting of family & estate records; photographs.
Facilities: meeting space. Museum-related items for sale.
Activities: guided tours; lectures; audio tour; concerts; special events; educational programs.
Hours & Admission Prices: May-Dec. see website for hours & admission prices.
Attendance: 67,510 (estimated)
Membership: Individual $50; Senior Couple $65; Family $75; Contributing $150; Supporting $250; Benefactor $500; Stewardship $1,000.

Ticonderoga

✻ **FORT TICONDEROGA, (M),** 100 Fort Ti Rd., Rte 74 E., Ticonderoga, NY 12883. Mailing Address: P.O. Box 390, Ticonderoga, NY 12883-0390. Tel.: 518-585-2821. Fax: 518-585-2210.
E-mail: fort@fort-ticonderoga.org
Web Site: www.fort-ticonderoga.org
Founded: 1909.
Congressional District: 20

Key Personnel: Exec. Dir., Beth L. Hill; Pres. (V), Peter S. Paine, Jr.; Cur. Collections, Christopher D. Fox; Cur. Landscape, Heidi teRiele Karkoski; Dir. Education, Richard Strum; Dir. Interpretation, Stuart Lilie.
Personnel Profile: Full-Time Paid 10; Part-Time Paid 53; Part-Time Volunteers 1,054; Interns 1.
Governing Authority: nonprofit organization. Parent Institution: Fort Ticonderoga Assoc.; Mt. Independence, Orwell, VT; Mt. Hope Battery, Ticonderoga; Mount Defiance, NY. Tax-exempt: 170(b)(1)(A) & 501(c)(3).
Institution Type/Description: Military History Museum: housed in 1755 barracks of reconstructed Colonial & Revolutionary fortress; restored formal garden c.1920.
Collections: 300 manuscripts; books; maps; prints; engravings; guns; cannon; swords; uniforms; engraved power horns; 18th-century military accoutrements; military manuals & biographies; paintings of 18th-century military scenes & 19th-century landscapes depicting the Fort; 1756-1920 King's Garden.
Research Fields: Colonial & Revolutionary American history; 19th-century tourism.
Facilities: 13,000-vol. library of Colonial & Revolutionary military history available for research by appointment in the research center; restaurant. Museum-related items for sale.
Activities: guided tours; lectures; formally organized educational programs; temporary exhibitions; artillery and musket demonstrations; fife & drum corps performances; garden tours; hands-on family activities; re-enactment events; historic trades workshops; corn maze.
Publications: newsletter, The Haversack; annual brochure; Robert Rogers of the Rangers; Fort Ticonderoga: Key to a Continent; Madison & Jefferson's Journey to the Northern Lakes; archives on CD-ROM, The King's Garden at Fort Ticonderoga; other books on regional military history; pictorial guide to Fort Ticonderoga.
Hours & Admission Prices: early May to late Oct. daily 9:30-5. Adults $17.50, seniors 62 & over $14, children 5-12 $8; discounts to AAM members; children 4 & under and members no charge. ♿
Attendance: 74,081 (estimated)
Membership: Sentinel $20; Individual $50; Dual $60; Family $75; Lieutenant $125; Captain $250; Colonel $1,000.

Tonawanda

HISTORICAL SOCIETY OF THE TONAWANDAS, INC., 113 Main St., Tonawanda, NY 14150-2129. Tel.: 716-694-7406.
E-mail: tonahist@gmail.com
Web Site: www.tonawandashistory.org
Founded: 1961.
Congressional District: 36
Key Personnel: Pres. (V), Patrick Barnard; Museum Shop Mgr., Robert Schweitzer.
Personnel Profile: Part-Time Volunteers 50.
Governing Authority: society. Affiliated with Board of Regents, State University of New York. Branch Museum: Long Homestead, 24 E. Niagara St. Tax-exempt: 501(c)(3).
Institution Type/Description: Historical Society Museums: housed in 1870 brick New York Central Railroad Station & 1829 Long Homestead.
Collections: museum industry; transportation; military; history; Indian artifacts; lumber; medical; costumes; archives; Erie Canal; barber shop; firemen; manuscripts; furnishings. Historic Building: Long Homestead.
Research Fields: local history.
Facilities: 15,000-vol. library of books, bound newspapers, maps, atlas, county histories, scrapbooks, diaries, business & church records, genealogies, obituary files, microfilm of early United States census & Civil War period newspapers available on premises by appointment; reading room. Booklets & other museum-related items for sale.
Activities: guided tours; permanent & temporary exhibitions.
Publications: monthly newsletter, The Lumber Shover; booklets, Civil War Veterans of the Tonawandas; Recollections of the Erie Canal; Tolerable Tales of the Tonawandas; Trolley Days in the Tonawandas; Tonawanda and North Tonawanda (Images of America).
Hours & Admission Prices: Museum: Wed. 12-4:30, Thurs. 10-6, Sat. 10-2. No charge. Long Homestead: Memorial Day to Labor Day Sun.1-4; other times by appointment. Admission $4; children under 12 & members no charge. Closed holidays.
Attendance: 4,000 (estimated)
Membership: Single $15; Family $25.

Troy

THE ARTS CENTER OF THE CAPITAL REGION, 265 River St., Troy, NY 12180. Tel.: 518-273-0552. Fax: 518-273-4591.
E-mail: info@artscenteronline.org
Web Site: www.artscenteronline.org
Institution Type/Description: Art Gallery.

Collections: works by contemporary artists; paintings; sculpture; photographs.
Activities: special events; classes; summer camps.
Hours & Admission Prices: Mon.-Thurs. 11-7, Fri.-Sat. 9-5, Sun. 12-4. No charge.

THE CHILDREN'S MUSEUM OF SCIENCE AND TECHNOL-OGY, 250 Jordan Rd., Troy, NY 12180-8394. Tel.: 518-235-2120. Fax: 518-235-6836.
Web Site: www.cmost.org
Formerly: The Junior Museum
Founded: 1954.
Congressional District: 23
Key Personnel: Interim Dir., Nancy Schultz; Chm., Paul Fahey; Museum Shop Mgr., Laurie Miedema.
Personnel Profile: Full-Time Paid 6; Part-Time Paid 6; Part-Time Volunteers 10; Interns 6.
Governing Authority: nonprofit organization. Tax-exempt: 501(c)(3).
Institution Type/Description: Children's Museum & Science Center.
Collections: live animals; science & natural history; environmental; reusable fuels; weather & meteorology; nano-science; space & astronomy.
Facilities: Digistar II planetarium. Museum-related items for sale.
Activities: interpretations of exhibits presented to school groups, other groups & individual visitors by museum teachers; weekend workshops; college workstudy program; school break workshops; summer workshops; special school programs & classes on environmental education; festivals; family gallery of multiple participatory science exhibits; summer camp; Toddler Explore More Time.
Publications: annual teacher brochure.
Hours & Admission Prices: July-Aug. Mon.-Sat. 10-5; Sept.-June Thurs.-Sun. 10-5; other times by appointment. Toddler Explore & More Time: Sept.-June Wed. 1-5. Admission 2 & over $5; discounts to ASTC members; members & children under 2 no charge. Closed New Year's Day; Easter; Memorial Day; Independence Day; Labor Day; Thanksgiving; Christmas. &
Attendance: 65,000 (accurate)
Membership: Grandparents $75-$105; Family $80-$110; Family Explorer $120; Sponsor $250; Shooting Star $500; Hendrick Hudson $1,000.

LANSINGBURGH HISTORICAL SOCIETY - THE MELVILLE HOUSE AND MUSEUM, 2 114th St., Lansingburgh Station, Troy, NY 12182-2712. Mailing Address: P.O. Box 219, Lansingburgh Station, Troy, NY 12182-0219. Tel.: 518-235-3501. Facebook: Melville House and Museum.
E-mail: lhssecretary@gmail.com
Web Site: www.lansingburghhistoricalsociety.org
Founded: 1965.
Congressional District: 20
Key Personnel: Pres. (V), David Marsh; Pres. (V), Elizabeth Van Buren.
Personnel Profile: Part-Time Volunteers 20.
Governing Authority: Tax-exempt.
Institution Type/Description: Historical Society Museum: housed in the former home of Stephan Gorham, the first postmaster of Rensselaer County; built in 1786.
Collections: local history; period furnishings; maps; photographs; diaries; town & business records; tools; local industry.
Research Fields: local history; genealogy.
Activities: meetings; guided tours; open houses; special events.
Publications: newsletter.
Hours & Admission Prices: By appointment. No charge; donations accepted. &
Attendance: 500 (estimated)
Membership: Basic $5; Sustaining $25; Patron/Life $100.

RENSSELAER COUNTY HISTORICAL SOCIETY, (M), 57 Second St., Troy, NY 12180-3928. Tel.: 518-272-7232. Fax: 518-273-1264.
E-mail: info@rchsonline.org
Web Site: www.rchsonline.org
Founded: 1927.
Congressional District: 23
Key Personnel: Pres., Joyce Chupka; Cur., Stacy Pomeroy Draper; Registrar, Kathryn Sheehan.
Personnel Profile: Part-Time Paid 3; Part-Time Volunteers 50; Interns 2.
Governing Authority: society. Subsidiary Institution: 1827 Hart-Cluett House. Tax-exempt: 501(c)(3).
Institution Type/Description: History & Art Museum: 1927 Hart-Cluett House;
Collections: early to mid-19th century furnishings; portraits; manuscripts; maps; photographs; American decorative arts; objects & archives related to the area's history; genealogy. Historic Building: 1827 Hart-Cluett House.

Research Fields: local history & American decorative arts.
Facilities: local history research library; meeting room. Museum-related items for sale.
Activities: school education program; guided tours; lectures; gallery talks; educational & family programming; inter-museum loan, permanent, temporary & traveling exhibitions.
Publications: newsletter.
Hours & Admission Prices: Feb.-Dec. 23 Tues.-Sat. 12-5. Adults $5; members no charge. &
Attendance: 8,000 (estimated)
Membership: Student $20; Individual $45; Family & Sustainer $65; Sponsor $125; Patron $250; Fellow $500; Benefactor $1,000.

Tupper Lake

NATURAL HISTORY MUSEUM OF THE ADIRONDACKS/THE WILD CENTER, (M), 45 Museum Dr., Tupper Lake, NY 12986-9712. Tel.: 518-359-7800. Fax: 518-359-3253.
Web Site: www.wildcenter.org
Founded: 1998.
Congressional District: 23
Key Personnel: Exec. Dir., Stephanie Ratcliffe; Pres. Bd. Trustees (V), Lynn S. Birdsong; Dir. Devel., Hillarie Logan-Dechene; Dir. Programs, Jennifer Kretser; Public Rels., Tracey Legat-Jolly; Cur., Dave Gross; Dir. Facilities, David St. Onge; Archivist, Betty Woods; Museum Shop, Josh Pratt.
Personnel Profile: Full-Time Paid 28; Part-Time Paid 25; Part-Time Volunteers 120; Interns 9.
Volunteer Hours: 8,500
Operating Expenses: 3,294,000
Operating Income: 3,634,000
Governing Authority: private; nonprofit organization. Tax-exempt: 501(c)(3).
Institution Type/Description: Natural History Museum.
Collections: zoological; paleontological; botanical; geological.
Facilities: 600-vol. library; aquarium; cafeteria; 13,000 sq. ft. exhibit space; 165-seat theater. Museum-related items for sale.
Activities: docent program; films; guided tours; lectures; participatory & temporary exhibits; theater; informal learning programs. Annual Event: Wildfest.
Publications: biannual newsletter, The Otter.
Hours & Admission Prices: Memorial Day-Labor Day daily 10-6; Labor Day-Columbus Day daily 10-5; Columbus Day-Memorial Day Fri.-Sun. 10-5. Adults $17, Senior Citizens $15, Children $10; discounts to groups; members & children under 3 no charge. Closed New Year's Day; Thanksgiving; Christmas. &
Attendance: 60,000 (accurate)
Membership: Individual $45; Family $85; Extended Family $125; Woodland Guide $250; River Guide $500; Mountain Guide $1,000.

Ulster Park

KLYNE ESOPUS MUSEUM, 764 Rte. 9W, Ulster Park, NY 12487. Mailing Address: P.O. Box 751, Port Ewen, NY 12466-0751. Tel.: 845-226-8221.
E-mail: klyneesopusmuseum@gmail.com
Web Site: klyneesopusmuseum.us
Founded: 1969.
Congressional District: 26
Key Personnel: Pres. (V), Alexander F. Contini; Museum Shop Mgr., Byllye Montalto.
Personnel Profile: Part-Time Paid 1; Part-Time Volunteers 30.
Governing Authority: nonprofit; state chartered educational facility. Subsidiary Institution: Perrine's Covered Bridge. Tax-exempt: 501(c)(3).
Institution Type/Description: Local History Museum: housed in c.1827 Old Dutch Church.
Collections: 10,000 BC to 20th century historical artifacts; local & regional history; genealogical & historical records; special emphasis on town of Esopus, Hudson River, Sojourner Truth, Alton B. Parker, John Burroughs & Attilio J. Contini.
Research Fields: local property history; Black & American Indian genealogy from 18th to 19th centuries.
Facilities: 161-vol. library of genealogical & historical materials available to the public; 50-seat lecture hall; 1,200 sq. ft. exhibit space. Museum-related items for sale.
Activities: arts festival; guided tours; lectures; loan & temporary exhibitions.
Publications: quarterly newsletter.
Hours & Admission Prices: last week of May to 1st week of Dec. Fri.-Tues. 1-4; groups by appointment only. No charge; donations accepted. &
Attendance: 650 (estimated)
Membership: Individual $15; Family $20; Individual Life $200; Family Life $400.

Union Springs

FRONTENAC HISTORICAL SOCIETY AND MUSEUM, 178 Cayuga St., Union Springs, NY 13160. Mailing Address: P.O. Box 338, Union Springs, NY 13160. Tel.: 315-889-7273.
Institution Type/Description: History Museum.
Collections: local history & culture; period furnishings & clothing; photographs; personal artifacts.
Hours & Admission Prices: Call for hours.

Upton

BROOKHAVEN NATIONAL LABORATORY-SCIENCE LEARNING CENTER, Brookhaven National Laboratory, Upton, NY 11973-5000. Mailing Address: P.O. Box 5000, Bldg. 400, Upton, NY 11973-5000. Tel.: 631-344-2838 & 4495. Fax: 631-344-5832.
E-mail: oep@bnl.gov
Web Site: www.bnl.gov/slc
Formerly: BNL Science Museum
Founded: 1977.
Congressional District: 1
Key Personnel: Supvr., Gail Donoghue; Coord., Bernadette Uzzi.
Personnel Profile: Full-Time Paid 3; Part-Time Paid 3.
Governing Authority: federal; nonprofit. Tax-exempt.
Institution Type/Description: Science Museum: dedicated to using inquiry methods to teach science to students in grades K-12.
Collections: objects relating to scientific research.
Research Fields: high energy physics; chemistry; medicine; biology; chemistry, environmental & applied technologies.
Facilities: educational facilities; 5,000 sq. ft. exhibit space.
Activities: onsite & outreach science programs for students; participatory exhibits; workshops. Center Sponsors: Elementary School Science Fair & Magnetic Levitation competition.
Hours & Admission Prices: Open to school groups, scouts & teachers by appointment only. No charge. &
Attendance: 37,000 (accurate)

Utica

CHILDREN'S MUSEUM OF HISTORY, SCIENCE & TECH-NOLOGY, 311 Main St., Utica, NY 13501-1282. Tel.: 315-724-6129.
E-mail: marlenebrown@roadrunner.com
Web Site: www.museum4kids.net
Formerly: Children's Museum of History, Natural History and Science at Utica, New York
Founded: 1963.
Congressional District: 31
Key Personnel: Exec. Dir., Marlene Brown.
Governing Authority: nonprofit organization. Parent Institution: New York State Board of Regents. Tax-exempt: 501(c)(3).
Institution Type/Description: Children's Museum.
Collections: Indian artifacts; historic dioramas; rocks; minerals; shells; bird & animal mounts; outdoor railroad display; wax dinosaur models & paintings; period toys; NASA exhibit.
Facilities: classrooms. Museum-related items for sale.
Activities: formally organized education programs; school loan service; permanent & temporary exhibits; music & storytelling programs for children; culturally diverse programs.
Publications: monthly newsletter; brochure.
Hours & Admission Prices: Mon., Thurs. & Sat. 9:15-12:15. Adults $8, seniors & veterans $7,children $6; discounts to groups; children under 2 & members no charge. Closed New Year's Eve & Day; Easter; Memorial Day; Independence Day; Thanksgiving Day; Christmas Eve & Day. &
Attendance: 18,000 (accurate)
Membership: Individual & Student $30; Family $45; Family Plus $60; Corporate $250 & up.

✱ **MUNSON-WILLIAMS-PROCTOR ARTS INSTITUTE MU-SEUM OF ART, (M),** 310 Genesee St., Utica, NY 13502-4799. Tel.: 315-797-0000, ext. 2168. Fax: 315-797-5608.
E-mail: rschneid@mwpai.org
Web Site: www.mwpai.org
Founded: 1919.
Congressional District: 25
Key Personnel: Dir. & Chief Cur., Anna T. D'Ambrosio; Pres., Anthony J. Spiridigloizzi; Dir. Communications, John Bach; Art Shop Mgr., Bona Starring; Cur. Modern & Contemporary Art, Mary Murray; Registrar & Exhibition Mgr., Michael D. Somple; Head Librarian, Kathryn Corcoran; Museum Educator, April Oswald.
Personnel Profile: Full-Time Paid 10; Part-Time Paid 3; Part-Time Volunteers 42; Interns 2.
Governing Authority: private; nonprofit organization. Tax-exempt.
Institution Type/Description: Art Museum.
Collections: 18th- & 19th-century American paintings; 20th-21st century European & American paintings & sculpture; European, Japanese, American prints; 19th-century American decorative arts; Proctor watch collection; manuscript collections; Historic House: 1850 Fountain Elms, restored mid-Victorian, Italian villa style house designed by William L. Woolett, Jr., four rooms in 1850 decor.
Research Fields: 18th- to 21st-century American & European Art; 19th century decorative arts.
Facilities: 25,000 books & 25,000 color slides, 500 videos plus 800 compact discs; lending library of originals; exhibition space; school of art; meetinghouse; 271-seat auditorium. Original paintings, sculpture, pottery, prints & jewelry by area artists, jewelry reproductions, art books & other museum-related items for sale.
Activities: guided tours; lectures; films; gallery talks; interactive gallery; concerts; dance recitals; arts festival; drama; rental gallery; formally organized education programs for children, adults & undergraduate students; docent program inter-museum loan, permanent, temporary & traveling exhibitions.
Publications: Order & Enigma: American Art Between the Wars; The Voyage of Life by Thomas Cole: Paintings, Drawings & Prints; 200 Years of American Art (w/AFA); Watches: The Proctor Collection; The Blue & Gray: Oneida County Stoneware; From Drawing to Dwelling: the Planning & Construction of Fountain Elms; The Art of Trenton Falls; Alex Katz: A Drawing Retrospective; John Monti: The Sculpture Court Project; The Distinction of Being Different: Joseph P. McHugh and the American Arts & Crafts Movement; Life Lines: American Master Drawings; Sculpture Space: Celebrating 20 Years; Artistry in Rosewood: Furniture by Elijah Galusha; Masterpieces of American Furniture from the Munson-Williams Proctor Institute; American Twentieth-Century Watercolors at the Munson-Williams-Proctor Arts Institute; Jewels of Time: Watches from the Munson-Williams-Proctor Collection; Collecting Modernism: European Masterworks from the MWPAI; Ferdinand Richardt: Drawings of America, 1855-1859 (2007); A Brass Menagerie: Metalwork of the Aesthetic Movement (2006); Auspicious Vision: Edward Wales Root and American Modernism (2007); James E. Freeman, 1808-1884: An American Painter in Italy; Look for Beauty: Philip Johnson and Art Museum Design.
Hours & Admission Prices: Tues.-Sat. 10-5, Sun. 1-5. No charge. Closed New Year's Day; Martin Luther King Jr. Day; Independence Day; Thanksgiving; Christmas. &
Attendance: 90,292 (accurate)
Membership: Student $25; Individual Senior 65 & over $45; Individual $50; Senior Household 65 & over $65; Family/Household $75; Contributor $150; Patron $250; Affiliate $250; President's Circle $500; Heritage Group $1,000. Business: Colleague $100; Sustainer $500; Leader $1,000.

ONEIDA COUNTY HISTORICAL SOCIETY, (M), 1608 Genesee St., Utica, NY 13502-5425. Tel.: 315-735-3642. Fax: 315-732-0806. Facebook: Oneida County Historical Society.
E-mail: ochs@oneidacountyhistory.org
Web Site: www.oneidacountyhistory.org
Founded: 1876.
Congressional District: 25
Key Personnel: Exec. Dir., Brian J. Howard; Chm. Bd. (V), Lou Parrotta; Museum Shop Mgr., George Abel.
Personnel Profile: Full-Time Paid 1; Full-Time Volunteers 1; Part-Time Volunteers 30; Interns 2.
Governing Authority: nonprofit organization. Tax-exempt: 501(c)(3).
Institution Type/Description: History Museum: housed in former First Church of Christ, Scientist.
Collections: local history books, pamphlets, photographs, maps, manuscripts & artifacts pertaining to the history of Utica, Oneida County & the Mohawk Valley.
Research Fields: local history; Mohawk Valley.
Facilities: library of books, pamphlets, documents, manuscripts, maps, photographs available for research on premises; copies or photostats may be ordered; reading room.
Activities: lectures; permanent & temporary exhibitions.
Publications: quarterly newsletter, Oniota.
Hours & Admission Prices: Tues.-Fri. 10-4, Sat. 11-3. Museum: adults $2; members no charge. Research Library: $5 per day. Closed national holidays. &
Attendance: 4,000 (accurate)
Membership: Student $15; Individual $40; Club & Organization $50; Family $70; Corporate $150.

UTICA ZOO, 99 Steele Hill Rd., Utica, NY 13501-5090. Tel.: 315-738-0472. Fax: 315-738-0475.
E-mail: info@uticazoo.org
Web Site: www.uticazoo.org
Founded: 1914.
Congressional District: 24
Key Personnel: Exec. Dir., Michael D. Wodrich; Mgr. Education, Mary Hall; Dir. Devel. & Communications, Elizabeth Irons; Cur. Animals, Michael Bates; Supt. Bldgs. & Grounds, Gary Mundschenk; Veterinarian, Ellen Hilton, D.V.M; Gift Shop Mgr., Patricia Jones.
Personnel Profile: Full-Time Paid 17; Part-Time Volunteers 50; Interns 12.
Governing Authority: nonprofit organization. Parent Institution: Utica Zoological Society. Tax-exempt: 501(c)(3).
Institution Type/Description: Zoo.
Collections: exotic & domestic animals, birds, reptiles & vertebrates.
Research Fields: veterinary medicine; animal behavior.
Facilities: library of natural history, zoology, zoo management reference material available for research on premises only; 300-seat amphitheater; 50-seat auditorium/classroom; picnic grove. Books & other items with animal themes for sale.
Activities: lectures; films; study clubs; formally organized educational programs; docent program; permanent exhibitions; travel program; zoomobile presentations; Zoo Parents program.
Publications: quarterly newsletter, Zoo News; annual report.
Hours & Admission Prices: April-Oct. daily 10-5; Nov.-March daily 10-4. April-Oct. adults $6.75, seniors 60 & up $5.75, children 4-12 $4.25; discounts to groups of 10 or more; children under 4 & members no charge. Nov.-March: $2 donation per person. ⅃
Attendance: 73,000 (accurate)
Membership: Associate $20; Individual $45; Senior Family $50; Family $55; Extended Family $65; Honor Roll $75; Silver $125; Gold $275; Platinum $500; Diamond $1,000.

Vails Gate

KNOX HEADQUARTERS STATE HISTORIC SITE, 289 Forge Hill Rd., Vails Gate, NY 12584. Mailing Address: P.O. Box 207, Vails Gate, NY 12584-0207. Tel.: 845-561-1765, ext. 22. Fax: 845-561-6577.
E-mail: michael.mcgurty@parks.ny.gov
Web Site: www.nysparks.com
Founded: 1922.
Congressional District: 26
Key Personnel: Interpretive Programs Asst., Michael S. McGurty.
Governing Authority: state government. Parent Institution: New York State Office of Parks, Recreation & Historic Preservation. Palisades Interstate Park Commission. Tax-exempt: 501(c)(3).
Institution Type/Description: Historic House Museum: 1754 John Ellison home, used as Continental Army officers' headquarters.
Collections: period furnishings; orientation exhibit; research collection. Historic Structure: c.1741 mill ruins.
Research Fields: American Revolution in Hudson Valley; Gen. Henry Knox & Horatio Gates; the Ellison family; General Nathanael Greene, Slavery in the Mid-Hudson Valley.
Facilities: library of research files available for use by appointment; nature trail.
Activities: guided tours; lectures; school programs; special programs & demonstrations; hiking trails.
Hours & Admission Prices: Tours: Memorial Day-Labor Day Wed.-Fri. 11 & 3, Sat. 10-5, Sun. 1-5; other times by appointment. No charge; donations accepted.
Attendance: 8,500 (estimated)
Membership: Friends of the State Historic Sites of the Hudson Highlands: Individual $15; Household $25; Contributing $50; Business, Corporate & Sustaining $100; Patron $500; Benefactor $1,000.

Vestal

KOPERNIK OBSERVATORY & SCIENCE CENTER, 698 Underwood Rd., Vestal, NY 13850. Tel.: 607-748-3685. Fax: 607-748-3222.
E-mail: info@kopernik.org
Web Site: www.kopernik.org
Founded: 1974.
Key Personnel: Dir., Drew Deskur, BSEE, M.B.A.
Institution Type/Description: Observatory & Science Center.
Collections: space science; astronomy; hands-on exhibitions; telescopes; amateur radio station; weather station.
Facilities: laboratories.

Activities: educational programs; special events; summer science; school programs; school & scout programs; birthday parties.
Hours & Admission Prices: March-Nov. Fri. 7:30pm; Dec.-Feb. call for hours; other times by appointment.

THE VESTAL MUSEUM, 328 Vestal Pkwy. E., Vestal, NY 13850. Mailing Address: 605 Vestal Pkwy. W., Vestal, NY 13850-1437. Tel.: 607-748-1432.
E-mail: vmuseum@vestalny.com
Founded: 1976.
Congressional District: 27
Key Personnel: Dir., Jan Roosa; Cur., Virginia Wood; Museum Shop Mgr., Cathy Roosa.
Personnel Profile: Part-Time Paid 2; Part-Time Volunteers 8; Interns 1.
Governing Authority: municipal. Parent Institution: Town of Vestal. Tax-exempt: 170(b)(1)(A).
Institution Type/Description: Historic Building: 1881 DL&W Vestal Railroad Depot.
Collections: tools of past eras; railroad memorabilia; graphic illustrations; period clothing; artifacts of past lifestyles of area.
Facilities: Gift items for sale.
Activities: guided tours; lectures; films; gallery talks; arts festivals; loan, permanent & temporary exhibitions.
Publications: monthly, Vestal Historical Society Newsletter.
Hours & Admission Prices: Feb.-March & Nov.-Dec. Tues.-Sat. 11-3; April & Oct. Tues.-Sat. 10-3; group tours by appointment. No charge; donations accepted. ⅃
Attendance: 4,000 (accurate)
Membership: Students $1; Individual $8; Family $10; Patron $25.

Victor

GANONDAGAN STATE HISTORIC SITE, 1488 Victor Bloomfield Rd. at State Rte. 444, Victor, NY 14564. Mailing Address: P.O. Box 239, Victor, NY 14564-0239. Tel.: 585-924-5848 & 5414. Fax: 585-742-1732.
Web Site: www.ganondagan.org
Founded: 1972.
Congressional District: 33
Key Personnel: Historic Site Mgr., G. Peter Jemison; Interpretive Programs, Michael Galban; Pres. Friends Ganondagan (V), Perry Groung.
Personnel Profile: Full-Time Paid 4; Part-Time Paid 1.
Governing Authority: state. New York State Office of Parks, Recreation & Historic Preservation, and Finger Lakes State Parks, Recreation & Historic Preservation Commission. Tax-exempt.
Institution Type/Description: Historic Site: late 17th-century Seneca Indian town settlement.
Collections: Seneca artifacts; trade items of European manufacture.
Research Fields: Seneca Indians.
Facilities: visitors center; 3 walking trails.
Activities: Handenosaunee Cultural events in summer & fall.
Publications: Art from Ganondagan: War Against the Seneca.
Hours & Admission Prices: Trails: 8-sunset, weather permitting. Guided Walks: Sat.-Sun. 12 & 2. Visitors Center: May-Oct. Tues.-Sun. 9-5. Adults $3, children $2; discounts to AAM members.
Attendance: 37,000 (accurate)
Membership: Friends of Ganondagan: Students & Senior Citizens $10; Single $15; Family $30; Patron $75; Longhouse Friends $100 & up; Earthkeeper $250 & up.

VALENTOWN MUSEUM, 7370 Valentown Sq., Victor, NY 14564. Mailing Address: Victor Historical Society, P.O. Box 472, Victor, NY 14564-0472. Tel.: 585-924-4170. Fax: 585-924-0523.
E-mail: info@valentown.org
Web Site: valentown.org
Founded: 1940.
Congressional District: 33
Key Personnel: Pres. (V), Carol Finch.
Personnel Profile: Full-Time Volunteers 2; Part-Time Volunteers 20.
Governing Authority: society. Parent Institution: Victor Historical Society. Tax-exempt.
Institution Type/Description: History Museum: housed in 1879 Valentown Hall, 19th century community center, located on site of camp area for War of 1812 soldiers & on Seneca Indian Trail of the 1600s used by fur traders, explorers & missionaries.
Collections: stores furnished with original material; local Civil War material; archaeological Indian artifacts; early Mormon artifacts.
Research Fields: local history; folk lore.
Activities: guided tours; lectures.

Publications: newsletter.
Hours & Admission Prices: May-Oct. Tues., Thurs. & Sat.-Sun. 12-4. Tours: by appointment. Adults $5, seniors & students $3; discount to members; children under 5 no charge.
Attendance: 4,000 (estimated)
Membership: Senior Citizens 60 & over $18; Individual $20; Family $30; Sponsor $100; Benefactor $250; Life $500.

Walden

HISTORICAL SOCIETY OF WALDEN & WALLKILL VALLEY, 34 N. Montgomery St., Walden, NY 12586-1117. Mailing Address: P.O. Box 48, Walden, NY 12586-0048. Tel.: 845-778-1173.
E-mail: hswwv@adprose.org
Web Site: www.thewaldenhouse.org
Founded: 1958.
Congressional District: 97
Key Personnel: Pres. (V), Barbara Imbasciani.
Personnel Profile: Part-Time Volunteers 15.
Governing Authority: Parent Institution: Historical Society of Walden & Wallkill Valley, Inc. Tax-exempt: 509(a)(2).
Institution Type/Description: Historic House Museum: late 18th-century Jacob Walden House.
Collections: furnishings; local history.
Research Fields: local history.
Facilities: 500-vol. library of local history available by appointment.
Activities: guided tours; lectures; slides; permanent & temporary exhibitions; monthly lectures on local history April-June & Sept.-Nov.
Publications: member newsletter.
Hours & Admission Prices: Open by appointment only. No charge; donations accepted.
Attendance: 150 (estimated)
Membership: Student K-12 $7.50; Single $15; Family $20; Business $50.

Wantagh

WANTAGH PRESERVATION SOCIETY, 1700 Wantaugh Ave., Wantagh, NY 11793. Mailing Address: P.O. Box 132, Wantagh, NY 11793-0132. Tel.: 516-826-8767. Facebook: Wantagh Preservation Society.
E-mail: wantaghmuseum@gmail.com
Web Site: wantagh.li/museum/index.htm
Founded: 1965.
Key Personnel: Pres., Robert meagher; 1st Vice Pres., Karen Chowski; 2nd Vice Pres., Bob Cook.
Personnel Profile: Part-Time Paid 15.
Governing Authority: Nassau County Department of Parks & Recreation. Tax-exempt.
Institution Type/Description: History Museum.
Collections: former railroad station built in 1885; restored turn of the century ticket agent's booth; Wantagh's first post office c.1908; Long Island railroad parlor car c.1912; telegraph apparatus; photographs; railroad station.
Activities: craft fairs; period auto displays. Museum Sponsors: Spring & Fall Craft Fairs.
Publications: The Information Window.
Hours & Admission Prices: Sun. 2-4. No charge; donations accepted.
Attendance: 3,000 (estimated)
Membership: Individual $10; Family $20; Friend $25; Patron $50; Fellow $100.

Wappinger Falls

SPORTS MUSEUM OF DUTCHESS COUNTY, Wheeler Hill Rd., Wappinger Falls, NY 12590. Mailing Address: P.O. Box 7, Poughkeepsie, NY 12602-0007. Tel.: 845-297-9308.
Web Site: www.sportsmuseumdcny.org
Founded: 1975.
Institution Type/Description: Sports Museum.
Collections: sports equipment & memorabilia; photographs; hall of fame.
Activities: special events; induction ceremony.
Hours & Admission Prices: Summer: Sat. 11-4, Sun. 1-4; see website for additional hours. No charge; donations accepted. &
Attendance: 607 (accurate)
Membership: Individual $10; Corporation $25; Life $50.

Warsaw

WARSAW HISTORICAL MUSEUM, 15 Perry Ave., Warsaw, NY 14569-1205. Tel.: 585-786-5240. Fax: 585-786-5240.
E-mail: gateshouse@basicisp.com
Web Site: warsawhistory.org
Founded: 1938.
Congressional District: 31
Personnel Profile: Part-Time Volunteers 12.
Governing Authority: society. Parent Institution: Warsaw Historical Society. Tax-exempt: 501(c)(3).
Institution Type/Description: Historic Building Museum: c.1824 building.
Collections: period furniture; local history; military items; clothing, farm & carpentry tools; dishes; manuscripts.
Research Fields: genealogy & local history.
Facilities: 500-vol. library of various old school historical books, pamphlets, old newspapers available for research upon application.
Activities: lectures & discussions; guided tours; walking tours of historic district; permanent & temporary exhibitions.
Publications: periodic newsletter.
Hours & Admission Prices: Mon.-Fri. 10-2; other times by appointment. No charge; donations accepted.
Attendance: 945 (accurate)
Membership: Individual $2; Life $25.

Warwick

THE HISTORICAL SOCIETY OF THE TOWN OF WARWICK, A.W. Buckbee Center, 2 Colonial Ave., Warwick, NY 10990. Mailing Address: P.O. Box 353, Warwick, NY 10990-0353. Tel.: 845-986-3236.
E-mail: info@warwickhistoricalsociety.org & whs@warwick.net
Web Site: www.warwickhistoricalsociety.org
Founded: 1906.
Congressional District: 19
Key Personnel: Pres., Michael Bertolini.
Personnel Profile: Part-Time Paid 2; Part-Time Volunteers 60.
Governing Authority: society. Tax-exempt: 501(c)(3).
Institution Type/Description: General Museum.
Collections: furniture from Queen Anne through Duncan Phyfe periods; examples of work done by local cabinet makers of 1810-1830; hunting & trapping equipment used by sportsman-author, William Henry Herbert (pen name Frank Forester); old carriages; sleighs; ploughs; ice-cutting equipment; old farm tools; 1890 Lehigh & Hudson River Railways four-wheeled caboose; herb garden; archaeology; costumes; history. Historic Buildings: 1810 house; 1764 old shingle house; 1810 old school Baptist meeting house; 1810 Ketchum house; 1825 barn; Baird tavern 1766, featuring ballroom.
Facilities: over 200-vol. library of history books & biographies.
Activities: guided tours; permanent exhibitions; restoration committee involved in restoring old houses for interested citizens; fundraising.
Publications: books, People of the Valleys, A History of Warwick 1700-1976; Days Gone By-A History in Pictures, Town of Warwick, NY 1827-1945; monthly newsletter.
Hours & Admission Prices: July-Oct. Thurs.-Sun. 12-4. Admission $5, family $10.
Attendance: 3,000 (estimated)
Membership: Student $10; Senior $45; Individual & Senior Dual $60; Household $120; Business $150; Donor $250; Patron $500; Benefactor $1,000; Lifetime $5,000.

PACEM IN TERRIS, 96 Covered Bridge Rd., Warwick, NY 10990-2854. Tel.: 845-986-4329. Fax: 845-986-4329.
E-mail: paceminterris@frontiernet.net
Web Site: frederickfranck.org
Founded: 1972.
Congressional District: 19
Key Personnel: Pres. & Dir., Lukas Franck; Mentor-in-Chief, Claske Berndes; Asst., Frances Jennick.
Personnel Profile: Part-Time Paid 2; Part-Time Volunteers 4.
Governing Authority: nonprofit corporation under the Education Laws of the State of New York. Tax-exempt.
Institution Type/Description: Historic House: c.1780 water mill ruin & c.1840 country inn, McCanns Hotel & Saloon.
Collections: drawings; paintings; sculptures.
Facilities: sculpture garden.
Activities: classical chamber music concerts; spiritual drama & poetry readings; temporary exhibitions.
Publications: visitors guide; newsletter, The Shoestring.

Hours & Admission Prices: May-Oct. Mon.-Fri. 11-4, Sat.-Sun. & holidays 11-6. No charge; donations accepted.
Attendance: 3,000 (estimated)

Water Mill

* **PARRISH ART MUSEUM, (M),** 279 Montauk Hwy., Water Mill, NY 11976. Tel.: 631-283-2118. Fax: 631-283-7006.
E-mail: info@parrishart.org
Web Site: www.parrishart.org
Founded: 1898.
Congressional District: 1
Key Personnel: Dir., Terrie Sultan; Public Rels. & Mktg. Dir., Mark Segal; Deputy Dir., Scott Howe; Special Events & Membership Dir., Nina Madison; Dir. Education, Cara Conklin-Wingfield; Asst. Finance, Susan Swiatocha; Chief Cur. Art & Education, Alicia Longwell; Bldg. Mgr., Walter Gallagher; Museum Shop Mgr., Damian Wolfe.
Personnel Profile: Full-Time Paid 26; Part-Time Paid 4; Part-Time Volunteers 156; Interns 2.
Governing Authority: nonprofit organization. Tax-exempt: 501(c)(3).
Institution Type/Description: Art Museum: built in 1897, under the direction of Grosvenor Atterbury, in the style of the Latin Cross.
Collections: American art from the 19th century to present; paintings & archives of William Merritt Chase; Fairfield Porter paintings.
Research Fields: William Merritt Chase; Fairfield Porter; 19th century American etchings.
Facilities: 5,300-vol. library of art & rare books, including 3,800-volume personal library of Aline B. Saarinen; 200-vols. of catalogs & periodicals; 250-seat auditorium; arboretum. Art books, publications, original prints & other museum-related items for sale.
Activities: guided tours; films; concerts; lectures; performing arts events; workshops for children; formally organized education programs for children; inter-museum loan, permanent & temporary exhibitions; school loan service.
Publications: exhibition catalogs; quarterly newsletter; interpretive brochures.
Hours & Admission Prices: May 15-Sept. 15 Fri. 11-9, Sat.-Thurs. 11-6; Sept. 16-May 14 Wed.-Thurs. & Sat.-Mon. 11-6, Fri. 11-8. Adults $10, senior citizens $8; discounts to guests of members; students, members, & children under 18 no charge. Closed New Year's Day; Easter; Independence Day; Thanksgiving; Christmas. &
Attendance: 60,000 (estimated)
Membership: Individual $55; Family & Dual $95; Associate $125; Sponsor $250; Contributor $500.

WATER MILL MUSEUM, 41 Old Mill Rd., Water Mill, NY 11976. Mailing Address: P.O. Box 63, Water Mill, NY 11976-0063. Tel.: 631-726-4625. Facebook: Water Mill Museum.
E-mail: info@watermillmuseum.org
Web Site: www.watermillmuseum.org
Founded: 1942.
Congressional District: 1
Key Personnel: Pres. (V), Ann Lombardo; Museum Shop Mgr., Joan Wilson.
Personnel Profile: Part-Time Paid 2; Part-Time Volunteers 20.
Governing Authority: nonprofit organization. Parent Institution: NYS Board of Regents. Tax-exempt.
Institution Type/Description: Historic Building Museum: oldest commercial structure on the east end of Long Island. A water-powered working grist mill.
Collections: exhibits & artifacts relating to milling and 19th-century community life; tools of the ice harvester, farmer, cooper, whaling, baymen & carpenter; extensive photographic record of early 20th-century Water Mill.
Research Fields: local history; milling; tools.
Facilities: herb garden; wetlands garden. Museum-related items for sale.
Activities: guided tours; lectures; arts festivals; children's activities. Museum Sponsors: art exhibition; school tours; quilt shows.
Publications: guide brochure; Oral History Of Water Mill; book, Celebrating Community, History of a Long Island Hamlet, 1644-1994.
Hours & Admission Prices: May 20-Oct. 10 Mon. & Thurs.-Sat. 11-5, Sun. 1-5. No charge; donations accepted.
Attendance: 1,000 (estimated)
Membership: Individual $25; Family $30.

Waterford

NEW YORK STATE OPRHP, BUREAU OF HISTORIC SITES, Peebles Island, Waterford, NY 12188. Mailing Address: P.O. Box 219, Waterford, NY 12188-0219. Tel.: 518-237-8643, ext. 3202. Fax: 518-235-4248.
Web Site: www.nysparks.state.ny.us
Founded: 1972.

Key Personnel: Dir., John Lovell; Scientist Archaeology, Michael Roets; Interpretive Programs Coord., Audrey Nieson; Restoration Coord., Chris Flagg; Cur., Susan Walker; Mgr. Collections, Ronna Dixson; Painting Conservator, Mary Bettejeski; Decorative Arts Conservator, Heidi Miksch; Textiles Conservator, Deborah Trupin; Furniture Conservator, David L. Bayne; Gilded Objects Conservator, Eric Price; Flag Conservator, Sarah Stevens; Paper Conservator, Michele Phillips; Chief Protective Svcs., Alton Malcolm.
Personnel Profile: Full-Time Paid 33; Part-Time Paid 5; Part-Time Volunteers 10; Interns 1.
Governing Authority: state. New York State Office of Parks, Recreation & Historic Preservation. Tax-exempt.
Institution Type/Description: State Agency.
Collections: The Bureau of Historic Sites, together with OPRHP park regions, administers the following state historic sites: Bennington Battlefield, Hoosick Falls; Clermont, Germantown; Crailo, Rensselaer; Crown Point, Crown Point; Fort Ontario, Oswego; Ganondagan, Victor; Herkimer Home, Little Falls; Hyde Hall, East Springfield (affiliate); John Brown Farm, Lake Placid; John Jay Homestead, Katonah; John Burroughs Memorial, Roxbury; Johnson Hall, Johnstown; Knox's Headquarters, Vails Gate; Lorenzo, Cazenovia; Darwin Martin House (affiliate); Staatsburgh, Staatsburg; New Windsor Cantonment, Vails Gate; Olana, Hudson; Oriskany Battlefield, Oriskany; Philipse Manor Hall, Yonkers; Rexford Aqueduct, Rexford; Sackets Harbor Battlefield, Sackets Harbor; Schoharie Crossing, Fort Hunter; Schuyler Mansion, Albany; Senate House, Kingston; Stony Point Battlefield, Stony Point; Steuben Memorial, Remsen; Walt Whitman Birthplace, Huntington Station (affiliate); Washington's Headquarters, Newburgh; Planting Fields Arboretum, Oyster Bay. Clinton House, Poughkeepsie (affiliate); Old Fort Niagara, Youngstown (affiliate); Grant Cottage, Wilton (affiliate); Old Erie Canal, Kirkville; Old Croton Aqueduct, Dobbs Ferry; Caumsett, Huntington; Sonnenberg Gardens & Mansion, Canandaigua (affiliate).
Major Exhibits: Frederic Church in Maine (T), 1/12-1/14.
Research Fields: state history; archaeology; historic structures; collections management; military history & technology; decorative arts; conservation of collections.
Activities: Bureau of Historic Sites provides technical program services to state historic sites & state parks; research; interpretation; restoration of buildings; archaeology; exhibit design & fabrication; collections management. Bureau Sponsors: conservation lab tours.
Publications: technical reports; papers related to activities; site-related promotional & interpretive literature.
Hours & Admission Prices: Office: Mon.-Fri. 8-5. Refer to individual site listings for hours & admission fees. &
Attendance: 500,000 (estimated)

WATERFORD HISTORICAL MUSEUM AND CULTURAL CENTER, 2 Museum Lane, Waterford, NY 12188-2639. Tel.: 518-238-0809. Fax: 518-238-0809.
E-mail: info@waterfordmuseum.com
Web Site: www.waterfordmuseum.com
Founded: 1964.
Key Personnel: Dir., Brad L. Utter; Pres. (V), Nancy Spretty.
Personnel Profile: Full-Time Paid 1; Part-Time Paid 1; Part-Time Volunteers 60; Interns 2.
Governing Authority: museum. Parent Institution: NY State Regents, Bd. of Directors. Tax-exempt: 501(c)(3).
Institution Type/Description: Local History Museum & Cultural Center: Housed in 1830 Hugh White homestead.
Collections: dresses; kitchenware; kitchen tools & implements; quilts; china; tinware; Champlain & Erie Canal photos & artifacts; medical collections; tools; objects related to history of Waterford & surrounding area.
Facilities: library of books, microfilm & newspapers pertaining to local history & barge canal; meeting room; kitchen facilities available.
Activities: guided tours; lectures; educational programs for children; permanent & temporary exhibitions; day trips; outreach programs. Museum Sponsors: Community Heritage Day; Old Fashioned Firemen's Master.
Publications: bimonthly newsletter, The Homestead; White Homestead on Wheels; Waterford to Whitehall; The Waterford Flight, Standard of the Age: A Brief History of the Button Fire Engine Works Waterford, NY.
Hours & Admission Prices: See website for current hours. No charge; donations accepted. Closed major holidays. &
Attendance: 1,000 (estimated)
Membership: Student $10; Individual $15; Family $25; Contributor $50; Sustainer $75; Patron $100.

Waterloo

M'CLILNTOCK HOUSE, 14 E. Williams St., Waterloo, NY 13165-1411. Mailing Address: 136 Fall St., Seneca Falls, NY 13148. Tel.: 315-568-0024.
Institution Type/Description: Historic House: housed in the former home of Thomas and Mary M'Clintock; the site of the drafting of the Declaration of Sentiments for the Women's Rights Convention; built in 1836.
Collections: local history; period furnishings; photographs.
Activities: guided tours.
Hours & Admission Prices: Memorial Day to Labor Day Fri.-Mon.

NATIONAL MEMORIAL DAY MUSEUM, (M), 35 E. Main St., Waterloo, NY 13165-1430. Mailing Address: 31 E. Williams, Waterloo, NY 13165-1410. Tel.: 315-539-0533 & 9611. Fax: 315-539-7798. Facebook: National Memorial Day Museum.
E-mail: terwilliger10@fltg.net
Web Site: www.wlhs-ny.org
Founded: 1966.
Congressional District: 29
Key Personnel: Pres. (V), Coreen Lowry.
Personnel Profile: Full-Time Paid 2; Part-Time Paid 1; Part-Time Volunteers 20.
Governing Authority: society. Parent Institution: Waterloo Library & Historical Society. Tax-exempt.
Institution Type/Description: Memorial Day Museum: housed in c.1836 home.
Collections: mementos of first Memorial Day & its founders; Civil War artifacts; U.S. war archives.
Activities: guided tours; historical programs; school programs; archival research.
Publications: The History and Origin of Memorial Day in Waterloo, New York.
Hours & Admission Prices: April 15-May 22 & Sept. 8-Dec. 15 Tues.-Sat. 10-5, May 23-Sept. 4 Tues.-Sun. 10-5. Tours by request. Requested Donations: family $5, adults $3, senior citizens $2, students $1; members & children under 12 no charge. Closed Labor Day.
Attendance: 1,300 (accurate)
Membership: Senior 60 & over $10; Individual $20; Family $35; Friend $100; Contributor $200; Sponsor $300; Patron $500; Benefactor $1,000.

PETER WHITMER FARM & FAYETTE CHAPEL, 1451 Aunkst Rd., Waterloo, NY 13165-9736. Mailing Address: Hill Cumorah, 603 State Rte. 21, Palmyra, NY 14522-9301. Tel.: 315-539-2552.
E-mail: info@hillcumorah.org
Web Site: www.hillcumorah.org
Founded: 1980.
Congressional District: 29
Key Personnel: Dir., Richard Searle.
Personnel Profile: Full-Time Volunteers 28; Part-Time Volunteers 5.
Governing Authority: church. Parent Institution: Church of Jesus Christ of Latter-Day Saints, 50 E. North Temple, Salt Lake City, UT 84150. Tax-exempt.
Institution Type/Description: Historic House: located on site of organization of The Church of Jesus Christ of Latter-Day Saints (Mormon), 1830.
Collections: household furnishings typical of Western New York, 1820-1830; artwork depicting events leading to church organization.
Research Fields: Latter-Day Saints Church history.
Facilities: visitor center; displays.
Activities: guided tours.
Publications: books, Book of Mormon; Doctrine & Covenants; pamphlets.
Hours & Admission Prices: Summer: Mon.-Sat. 9-7, Sun. 12:30-7; Winter: daily 9-5. No charge. Closed Christmas. &
Attendance: 70,000 (estimated)

Watertown

JEFFERSON COUNTY HISTORICAL SOCIETY, (M), 228 Washington St., Watertown, NY 13601-3379. Tel.: 315-782-3491. Fax: 315-782-2913.
E-mail: director@jeffersoncountyhistory.org
Web Site: jeffersoncountyhistory.org
Founded: 1886.
Congressional District: 30
Key Personnel: Exec. Dir., Jessica M. Phinney; Cur. Education, Melissa Widrick; Admin. Asst. & Bookkeeper, Anna Black; Office Mgr., Donna Koniz.
Personnel Profile: Full-Time Paid 3; Part-Time Paid 2; Part-Time Volunteers 60.
Governing Authority: nonprofit organization. Tax-exempt: 501(c)(3).
Institution Type/Description: History Museum: housed in 1876 Paddock Mansion.
Collections: local history & culture; period furnishings; 19th-century Americana; handmade agricultural & woodworking tools; early machinery; clothing; portraits; archives & photographs; northern New York Indian artifacts; Tyler coverlets 1834-1858; water turbines; restored 1910 Babcock automobile; Victorian garden; log cabin; one-room schoolhouse; house.
Research Fields: local history and prehistory.
Facilities: 2,200-vol. library of local & northern New York history available for research on premises; classroom; gallery; Victorian garden. Period furnishings, regional literature & museum-related items for sale.
Activities: guided tours; lectures; formally organized education programs; docent program; permanent, temporary & traveling exhibits.
Publications: annual, Jefferson County Historical Society Bulletin; newsletter, Museum Musings.
Hours & Admission Prices: Tues.-Fri. 10-5, Sat. 10-4. Adults $6; members no charge. Closed national holidays. &
Attendance: 11,000 (estimated)
Membership: Senior Citizen $25; Individual $30; Family $45; Sustaining $60; Institution $75; Patron $100; Life $500. Business: Member $100; Sustaining $250; Patron $500; Sponsor $1,000; Partner $5,000.

SCI-TECH CENTER OF NORTHERN NEW YORK, 154 Stone St., Watertown, NY 13601-3250. Tel.: 315-788-1340. Fax: 315-788-2738 (call first).
E-mail: scitech@scitechcenter.org
Web Site: www.scitechcenter.org
Founded: 1982.
Congressional District: 26
Key Personnel: C.E.O., Stephen Karon; Pres. (V), Paul Barben.
Personnel Profile: Full-Time Volunteers 1; Part-Time Volunteers 25; Interns 1.
Volunteer Hours: 1,150
Operating Expenses: 16,726
Operating Income: 25,443
Governing Authority: nonprofit organization. Tax-exempt: 501(c)(3).
Institution Type/Description: Science & Technology Museum.
Collections: interactive science & technology exhibits from the fields of physics, biology & computers, including the shadow room, giant kaleidoscope & garden of smells.
Major Exhibits: Butterflies Alive, 6/14-9/14.
Facilities: 3,500 sq. ft. exhibit space; classrooms. Gift items for sale.
Activities: guided tours; hands-on workshops for children & adults; participatory exhibits; birthday parties; chess club; lego club; regional science fair; astronomy observing sessions.
Publications: quarterly newsletter, Sci-Tech Scope; monthly update of programming.
Hours & Admission Prices: Mon. holidays & Tues.-Sat. 10-4. Adults $4, children $3, seniors $2; discounts to AAM members & groups with reservations; children under 3 & members no charge. Closed Thanksgiving; Christmas. &
Attendance: 8,306 (accurate)
Membership: Senior $15; Individual $30; Military Family $40; Family $45.

Watervliet

WATERVLIET ARSENAL MUSEUM, 1 Buffington St., Watervliet, NY 12189-4000. Tel.: 518-266-5805. Fax: 518-266-5859.
E-mail: usarmy.watervliet.tacom.list.wvacurator@mail.mil
Web Site: www.wva.army.mil/museum.php
Formerly: Watervliet Arsenal's Museum of the Big Guns
Founded: 1975.
Congressional District: 23
Key Personnel: Dir. & Cur., Robert Pfeil.
Personnel Profile: Full-Time Paid 2; Part-Time Volunteers 5.
Governing Authority: federal government. Parent Institution: Watervliet Arsenal. Subsidiary Institution: Chief of Military History. Tax-exempt.
Institution Type/Description: Military Museum: housed in 1859 cast iron warehouse for artillery factory.
Collections: military materials & equipment; 1600s-present cannon & related ordinance; early 1900 machine shop.
Research Fields: cannon development.
Facilities: 1,000-vol. library of military books & technical manuals; 5,000 photographs of the arsenal & cannon; 10,000 sq. ft. exhibit space.
Activities: videos; slide shows; permanent & temporary exhibitions; guided tours.
Publications: Watervliet Arsenal History.
Hours & Admission Prices: Sun.-Thurs. 10-3; group tours by appointment. No charge; donations accepted. Closed federal holidays. &
Attendance: 3,000 (accurate)

Watkins Glen

INTERNATIONAL MOTOR RACING RESEARCH CENTER AT WATKINS GLEN, 610 S. Decatur St., Watkins Glen, NY 14891-1613. Tel.: 607-535-9044. Fax: 607-535-9039.
E-mail: research@racingarchives.org
Web Site: www.racingarchives.org
Formerly: Watkins Glen Motor Racing Research Library
Founded: 1998.
Key Personnel: Pres., J.C. Argetsinger; Chm. (V), Syd Silverman; Dir. Admin., Max Neal.
Personnel Profile: Full-Time Paid 1; Part-Time Paid 4; Part-Time Volunteers 5.
Governing Authority: private; nonprofit organization. Tax-exempt.
Institution Type/Description: Sports Museum & Research Library.
Collections: motorsport archives; paintings; photographs; films; fine art; posters.
Research Fields: women in motorsport; West Coast racing; general motor racing history.
Facilities: 3,000-vol. library; 25-seat auditorium; 5,000 sq. ft. exhibit space.
Activities: films; lectures; broadcast programs. Annual Events: receptions; art shows.
Publications: quarterly newsletter, From the Racing Archives; promotional brochure.
Hours & Admission Prices: Mon.-Fri. 9-5, Sat. 10-2; other times by appointment. No charge; donations accepted. Closed New Year's; Independence Day; Memorial Day; Thanksgiving; Christmas. ৬
Attendance: 7,500 (estimated)
Membership: Starting Grid $25; Turn One $50; Turn Two $100; Grand Stands $250; Back Straight $500; Final Turn $1,000; Finish Line $2,500; Victory Circle $5,000.

Weedsport

HALL OF FAME & CLASSIC CAR MUSEUM, 1 Speedway Dr., Weedsport, NY 13166-9544. Mailing Address: P.O. Box 240, Weedsport, NY 13166-0240. Tel.: 315-374-3661. Fax: 315-834-9734.
E-mail: jspeno2@gmail.com
Web Site: www.dirthalloffame-classiccarmuseum.com
Founded: 1992.
Congressional District: 24
Key Personnel: Treas., Gary Spaid; Cur., Jack Speno; Museum Shop Mgr., Harry Elkema.
Personnel Profile: Part-Time Volunteers 7.
Governing Authority: municipal.
Institution Type/Description: Motorsports Museum.
Collections: race cars from 40s to present; classic cars from 30s-70's.
Facilities: 10-vol. library; 10-seat theater. Museum-related items for sale.
Activities: guided tours; broadcast programs; weekly races. Museum Sponsors: Hall of Fame Induction Ceremonies in May.
Hours & Admission Prices: May to Columbus Day Tues.-Sun. 12-5; other times by appointment. Adults $7, senior citizens & students $5; discounts to groups of 10 or more, AAM & ICOM members. Closed Easter; Mother's Day. ৬
Attendance: 5,000 (estimated)

OLD BRUTUS HISTORICAL SOCIETY, INC., 8943 N. Seneca St., Weedsport, NY 13166. Mailing Address: P.O. Box 516, Weedsport, NY 13166-0516. Tel.: 315-834-6285.
Founded: 1967.
Congressional District: 33
Key Personnel: Pres. (V), Elvin Dolph; Dir. (V), Dennis Randall; Historian, Jeanne Baker; Treas., Jean Saroodis; Sec., Barbara Ward.
Personnel Profile: Part-Time Volunteers 50.
Volunteer Hours: 1,000
Operating Expenses: 8,975
Operating Income: 7,930
Governing Authority: society; nonprofit organization. Parent Institution: Old Brutus Historical Society. Tax-exempt: 501(c)(3).
Institution Type/Description: Local Historical Society & Museum.
Collections: local history; agricultural items & implements; medical items; furnishings & memorabilia; local manufactured artifacts; maps; photographs; household items & artifacts; manuscripts; newspaper clips; 1850 1990 clothing; school room, early kitchen & canal boat cabin mock up; genealogy file on local families.
Major Exhibits: Our Hometown Around the Turn of the Century, 3/14-11/14.

Research Fields: genealogy; local history; Erie Canal.
Facilities: 2,000-vol. library of books & atlases pertaining to history & genealogy available for research on premises by appointment; 50-seat auditorium. Museum-related items for sale.
Activities: guided tours; lectures; films; formally organized education programs for children; permanent & temporary exhibitions; public programs.
Hours & Admission Prices: Mon.-Tues. 9-12; other times by appointment. No charge; donations accepted. Closed Easter; Memorial Day; Independence Day; Labor Day; Thanksgiving; Christmas. ৬
Attendance: 1,500 (estimated)
Membership: Individual $5; Life $50.

Wellsville

THE MATHER HOMESTEAD MUSEUM, LIBRARY AND MEMORIAL PARK, 343 N. Main St., Wellsville, NY 14895-1016. Mailing Address: P.O. Box 531, Wellsville, NY 14895-0531. Tel.: 585-593-1636.
Founded: 1981.
Congressional District: 23
Key Personnel: Dir., Mrs. Glenn Williams.
Personnel Profile: Part-Time Volunteers 28; Interns 1.
Governing Authority: private; nonprofit.
Institution Type/Description: Historic House Museum.
Collections: 1930s artifacts; music publications & instruments; books; games; catalogues; toys; murals by Tony Sarg.
Research Fields: 1930s artifacts.
Facilities: library; archives.
Activities: special events; concerts; films; land preserving information for local & global environmental issues. Museum Sponsors: Easy Egg Hunt; Reading Declaration of Independence in July; Halloween Paint-Out in October.
Publications: newsletter, The Homestead Hoot; music, original scores; articles; Sound Adventures.
Hours & Admission Prices: Temporarily closed. ৬
Attendance: 200 (estimated)

West Henrietta

NEW YORK MUSEUM OF TRANSPORTATION, 6393 E. River Rd., West Henrietta, NY 14586-9575. Mailing Address: P.O. Box 136, West Henrietta, NY 14586-0136. Tel.: 585-533-1113.
E-mail: info@nymtmuseum.org
Web Site: www.nymtmuseum.org
Founded: 1975.
Congressional District: 35
Key Personnel: Pres. (V), Theodore H. Strang, Jr.; Sec., James E. Dierks; Treas., Robert Nesbitt; Museum Shop Mgr., Douglas Anderson.
Personnel Profile: Part-Time Volunteers 35.
Volunteer Hours: 7,000
Operating Expenses: 31,140
Operating Income: 54,572
Governing Authority: nonprofit organization. Tax-exempt: 501(c)(3).
Institution Type/Description: Transportation Museum.
Collections: 1867-1955, elements & evidence from the transportation facilities in Western New York; transportation history; street cars; interurbans; railroad equipment; trucks; maps; printed matter; gas, steam & diesel electric locomotives; operational railroad line; horse drawn & highway vehicles; period steam engine from former Genesee River swing bridge. Historic Building: 1904 Rochester & Eastern Rapid Railway Way Station.
Research Fields: sociological impacts of changes in the transportation patterns in Western New York from 1880-1955; area trolley line history.
Facilities: 6,000-vol. library available for research by appointment; one & three-fourths mile demonstrational railway; 15,000 sq. ft. exhibit space; 25-seat theater. Rochester & Genesee Valley Railroad Museum Country Depot. Museum-related items for sale.
Activities: guided tours; lectures; permanent, temporary & traveling exhibitions; special events; off-site slide talks; interurban trolley rides on museum railroad.
Publications: quarterly newsletter, Head End.
Hours & Admission Prices: Sun. 11-5; groups and other times by appointment. Mid-May to Oct. adults $10, seniors 65 & over $8, students 3-17 $8; members no charge; Nov. to mid-May call for admission fees. Closed New Year's Day, Easter, Christmas. ৬
Attendance: 7,033 (accurate)
Membership: Student $10; Individual $25; Family $35; Sustaining $50; Sponsor $100; Benefactor $250; Patron $500.

West Park

JOHN BURROUGHS ASSOCIATION, INC., Off John Burroughs Dr., West Park, NY 12493. Mailing Address: John Burroughs Assoc., Inc., 15 W. 77th St., New York, NY 10024-5153. Tel.: 212-769-5169. Fax: 212-313-7182.
E-mail: breslof@amnh.org
Web Site: research.amnh.org/burroughs/
Formerly: Slabsides
Founded: 1921.
Key Personnel: Pres., David Liddell; Sec., Lisa Breslof.
Personnel Profile: Part-Time Volunteers 25.
Governing Authority: nonprofit organization. Parent Institution: John Burroughs Association, c/o American Museum of Natural History, Central Park West, New York, NY. 10024. Tax-exempt: 501(c)(3).
Institution Type/Description: Historic House Museum: 1895-1896 house built by naturalist John Burroughs in Nature Sanctuary.
Collections: china & housekeeping equipment used by John Burroughs.
Facilities: nature center. Museum-related items for sale.
Activities: guided walks on sanctuary trails; lectures. Museum Sponsors: Open House in May & October.
Publications: brochure, Wakerobin Newsletter published three times annually; indexes to the collected works of John Burroughs.
Hours & Admission Prices: By appointment. Trails: dawn to dusk daily. No charge; donations accepted.
Attendance: 800 (estimated)
Membership: Senior & Student $15; Annual $25; Family $35; Patron $50; Life $500.

West Point

CONSTITUTION ISLAND ASSOCIATION, (M), South Dock, West Point, NY 10996-0041. Mailing Address: P.O. Box 41, West Point, NY 10996-0041. Tel.: 845-446-8676. Fax: 845-622-6022.
E-mail: info@constitutionisland.org
Web Site: www.constitutionisland.org
Founded: 1916.
Congressional District: 21
Key Personnel: Exec. Dir., Richard de Koster; Chm. (V), Elizabeth Pugh.
Personnel Profile: Part-Time Paid 3; Part-Time Volunteers 70.
Governing Authority: nonprofit corporation. Chartered by University of the State of New York. Tax-exempt.
Institution Type/Description: Historic House Museum: c.1800 Victorian home of writers Susan & Anna Warner on Constitution Island, site of several Revolutionary War fortification ruins; traditional 19th century gardens.
Collections: Warner House collection including 15 rooms of Warner family furnishings; furniture; art; china; glass; kitchen & gardening utensils; memorabilia.
Research Fields: Warner family and books; Revolutionary War period & relationship of Constitution Island to the war.
Facilities: 2,000-vol. library of books, documents & correspondence of the Warner family; Anna B. Warner Memorial Garden. Booklets on local history for sale.
Activities: guided tours; lectures; slide program; special education program for children; permanent exhibitions; Little American Program.
Publications: annual report; local history booklet; books, Susan Warner; Gardening by Myself; Light in the Morning; Defense of the Hudson Highlands.
Hours & Admission Prices: Tours: late April to June 15 schools only, by reservation; June 21 to early Oct. Wed.-Thurs. 1 & 2. Adults $10, senior citizens & children under 16 $9; discounts to AAM & ICOM members; children 6 & under no charge.
Attendance: 6,200 (accurate)
Membership: Individual $25; Family $40; Patron $500; Benefactor $1,000.

✳ **WEST POINT MUSEUM, (M),** United States Military Academy, Bldg. 2110, West Point, NY 10996. Tel.: 845-938-2203 & 3590. Fax: 845-938-7478. Facebook: West Point Museum.
E-mail: museum@usma.edu
Web Site: www.usma.edu/museum
Founded: 1854.
Congressional District: 19
Key Personnel: Dir., David M. Reel; Cur. Arms, Leslie D. Jensen; Cur. Uniforms & Military History, Michael J. McAfee; Exhibit Specialist, Jose Cartagena; Museum Specialist Conservator, Paul R. Ackermann; Registrar, Marlana Cook; Museum Shop Mgr., Shannon Ickes; Museum Technician, Brian Rayca; Security Chief, Gloria Johnson.
Personnel Profile: Full-Time Paid 15; Full-Time Volunteers 2; Part-Time Paid 2; Part-Time Volunteers 4; Interns 4.
Governing Authority: federal. Parent Institution: Dept. of Defense, Dept. of Army & U.S. Military Academy. Tax-exempt: 401(c)(3).
Institution Type/Description: Military History Museum.
Collections: history of military events & personalities; collection of weapons, military art; artifacts, paintings, sculpture, prints, photographs, American & European uniforms, documents; Fort Putman, historic restoration of Revolutionary War fort overlooking U.S. Military Academy; arms & armor.
Research Fields: history of American & European profession of arms & uniforms; European & American paintings; West Point historical artifacts, portraits & archival material.
Facilities: 1,600-vol. library of arms, armor & military reference books available for use on premises by appointment. West Point and Army related gift items for sale.
Activities: formally organized education programs for undergraduate & graduate college students; inter-museum loan, permanent & temporary exhibitions; lectures.
Publications: The West Point Museum: A Guide to the Museum; West Point Museum: Treasure Hunt.
Hours & Admission Prices: Daily 10:30-4:15. No charge; donations accepted. Closed New Year's Day; Thanksgiving; Christmas. No charge; donations accepted. &
Attendance: 215,000 (accurate)

West Sayville

LONG ISLAND MARITIME MUSEUM, (M), 88 West Ave., West Sayville, NY 11796-1908. Tel.: 631-854-4974. Fax: 631-854-4979.
E-mail: limm@limaritime.org
Web Site: www.limaritime.org
Founded: 1966.
Congressional District: 2
Key Personnel: Dir., Stephen M. Jones; Registrar, Arlene Balcewicz; Museum Shop Mgr., Terry Blitman; Librarian, Barbara Forde; Chm. (V), Michael Eagan; Coord. Programming, Brianne Musselwhite.
Personnel Profile: Full-Time Paid 3; Part-Time Paid 8; Part-Time Volunteers 100; Interns 1.
Volunteer Hours: 10,000
Operating Expenses: 435,000
Operating Income: 470,000
Governing Authority: New York State Board of Regents; chartered by University of the State of New York Education Dept. Tax-exempt: 501(c)(3).
Institution Type/Description: Maritime History Museum.
Collections: maritime history; small craft collection; United States Life Saving service equipment; shipwreck artifacts. Historic Buildings: Penney Boatshop; Rudolph Oyster House; 1890 Bayman's Cottage. Historic Ships: Priscilla, 1888 oyster vessel; Modesty, 1923 oyster sloop; Charlotte c.1880 sandbagger converted to a tugboat 1915.
Research Fields: pertaining to collections; Long Island South Shore Boat Builders; oyster & clamming; yachting; waterfowl hunting; duck decoys.
Facilities: library of books, documents, photographs, press clippings & periodicals relating to maritime history available for research by appointment.
Activities: lectures; oral history sessions. Museum Sponsors: maritime, folkcraft demonstrations. Museum Sponsors: Nautical Festival; Seafood Festival; Art Show in June.
Publications: monthly e-newsletter, Wrack Lines.
Hours & Admission Prices: Mon.-Sat. 10-4, Sun. 12-4. Adults $6, senior citizens & children under 12 $4; discounts for AAM & ICOM members; CAMM members no charge. &
Attendance: 85,000 (estimated)
Membership: Over 65 $35; Individual $50; Family $100; Privileged $200; Organizational $200; Small Business $250; Corporate $500.

West Seneca

FIREMEN'S MEMORIAL EXHIBIT CENTER OF WESTERN NEW YORK, 4141 Seneca St., West Seneca, NY 14224-3040. Tel.: 716-712-0413.
Governing Authority: Tax-exempt: 501(c)(3).
Institution Type/Description: Firefighting History Museum.
Collections: firefighting history & equipment; photographs; buckets; helmets; badges; uniforms; hose carts; chemical carts; fire trucks.
Activities: fire prevention programs; group tours; special events.
Hours & Admission Prices: Call for hours.

WEST SENECA HISTORICAL SOCIETY, 919 Mill Rd., West Seneca, NY 14224-3038. Tel.: 716-674-5600. Fax: 716-677-4330.
E-mail: hsociety@twsny
Web Site: westsenecahistory.com

Founded: 1950.
Key Personnel: Pres. (V), Frances Deppel.
Personnel Profile: Part-Time Paid 1; Part-Time Volunteers 12.
Governing Authority: Tax-exempt.
Institution Type/Description: Historical Society Museum: housed an original Lower Ebenezer Society house; built in 1850. Listed on the Federal Registry of Historical Places.
Collections: local history; Native American heritage & culture; period furnishings; personal artifacts; tools; clothing; musical instruments; arrowheads; fossils.
Major Exhibits: Buffalo Bible Institute, 1/3/14-3/3/14.
Activities: special events; speakers; trips. Museum Sponsors: Night Out; Civil War Encampment; Christmas Dinner; Christmas Open House.
Publications: Frank Lankes history books; cookbook, Arcadia Book on West Seneca.
Hours & Admission Prices: Tues. 10-4, 1st Sun. of month 2-4. No charge; donations accepted.
Attendance: 809 (estimated)
Membership: Single $7; Family $12.

Westfield

CHAUTAUQUA COUNTY HISTORICAL SOCIETY, MC-CLURG MUSEUM, Rts. 20 & 394 (Main & Portage Sts.), Westfield, NY 14787. Mailing Address: P.O. Box 7, Westfield, NY 14787-0007. Tel.: 716-326-2977.
E-mail: mcclurg@fairpoint.net
Web Site: www.mcclurgmuseum.org
Founded: 1883.
Congressional District: 150
Key Personnel: Pres., James O'Brien; Vice Pres., Dr. David Brown; Office Mgr., Shari Golnitz.
Personnel Profile: Part-Time Paid 1; Part-Time Volunteers 18.
Governing Authority: state regents; society board of trustees. Tax-exempt.
Institution Type/Description: County History Museum: housed in 1820 Mansion, built by James McClurg, an early settler.
Collections: agriculture; history; military; Native American artifacts; archives; costumes; furniture; genealogy of Chautauqua County.
Research Fields: genealogy; Tourgee papers; Foote papers; Cushing papers; Bowman photographs.
Facilities: 2,000-vol. library of history books.
Activities: guided tours; permanent exhibits; quarterly meetings.
Publications: quarterly newsletter; books, Updated County History 1938-1978; Patriot Soldiers of 1775-1783, Update; James McClurg, Pioneer; Chautauqua County Regiments and Soldiers in the Civil War 1861-1865.
Hours & Admission Prices: Tues.-Sat. 10-4. Adults $5; discounts to groups; students & members no charge. Closed holidays.
Attendance: 2,500 (estimated)
Membership: Individual $20; Family $30; Benefactor $100; Business $250.

White Plains

WHITE PLAINS MUSEUM GALLERY, 100 Martine Ave., 2nd Fl., White Plains, NY 10601. Mailing Address: 5 Sylvan Place, New Rochelle, NY 10801. Tel.: 914-632-8226.
E-mail: david@davidobey.com
Web Site: www.davidtobey.com
Institution Type/Description: Art Gallery.
Collections: works by local & regional artists.
Hours & Admission Prices: Call for hours.

Williamsville

WILLIAMSVILLE MEETING HOUSE & MUSEUM, 5658 Main St., Williamsville, NY 14221. Mailing Address: 5565 Main St., Williamsville, NY 14221. Tel.: 716-632-4120. Fax: 716-632-6009.
Institution Type/Description: Historic Building: built in 1871.
Collections: local history & culture; period furnishings; personal artifacts; photographs.
Hours & Admission Prices: Sept.-June 2nd Sun. of the month 1-4.

Wilson

WILSON HISTORICAL MUSEUM, 645 Lake St., Wilson, NY 14172. Mailing Address: P.O. Box 830, Wilson, NY 14172-0830. Tel.: 716-751-9886. Fax: 716-751-6141.
E-mail: agaffiliat@aol.com
Web Site: www.wilsonnewyork.com/hist_society.html
Founded: 1972.

Congressional District: 36
Key Personnel: Pres. & C.E.O. (V), Wallace Goodman; Cur., Dorothy Maxfield.
Personnel Profile: Part-Time Volunteers 132.
Governing Authority: nonprofit society. Affiliated with the Wilson Historical Society. Tax-exempt: 501(c)(3).
Institution Type/Description: General & Historical Society Museum: housed in 1912 Railroad Depot.
Collections: woodworking tools; dresses; quilts; railroad artifacts; books; wood carvings; farm tools; farm implements; genealogical records; period cars; 1903 caboose.
Research Fields: local history; genealogy.
Facilities: 100-vol. library of books available for research on premises only. Ceramics, booklets & postcards for sale.
Activities: guided tours; loan, permanent & temporary exhibitions. Annual Event: Memorial Day Fair.
Publications: monthly newsletter, W.H.S. Newsletter; booklets, Story of Billy Sherman; Tall Tales and Legends; Churches of Wilson; Land of Cobblestones; The Valiant Men of Battery M; Postal Service in the Town of Wilson; Wilson Historical Society Cookbook; Wilson's Vanishing Heritage; The Albright Opera House; Story of Sunset Island; The Wilson Free Library; Wilson Sketchbook.
Hours & Admission Prices: May-Nov. Sun. 2-4. No charge; donations accepted.
Attendance: 1,200 (estimated)
Membership: Juniors under 18 $1; Single $5; Couple $10; Life $150; Honorary (85 yrs. & over) no charge.

Wilton

ULYSSES S. GRANT COTTAGE STATE HISTORIC SITE, Mount McGregor, Wilton, NY 12831. Mailing Address: P.O. Box 2294, Wilton, NY 12831-5294. Tel.: 518-584-4353.
E-mail: info@grantcottage.org
Web Site: www.grantcottage.org
Founded: 1990.
Congressional District: 22
Key Personnel: Admin. & Pres. (V), Tim Welch; Dir., Jonathan Duda; Chm. (V), Bob Conner; Museum Shop Mgr., Dave Hubbard.
Personnel Profile: Part-Time Paid 3; Part-Time Volunteers 25.
Governing Authority: Tax-exempt.
Institution Type/Description: Historic House: 1878 cottage where Gen. Ulysses S. Grant spent the last six weeks of his life in June/July 1885; completed his personal memoirs.
Collections: period furnishings; Grant memorabilia; funeral floral pieces.
Facilities: visitor center. Museum-related items for sale.
Activities: reenactments; tours; student programs; music; view of the Hudson Valley; lectures; demonstrations.
Publications: The Grant Cottage Chronicles
Hours & Admission Prices: Memorial Day-Labor Day Wed.-Sun. 10-4; Sept. to Columbus Day Sat.-Sun. 10-4. Adults $5, senior citizens & students $4; children 5 & under and members no charge. &
Attendance: 3,300 (estimated)
Membership: $15; $25; $40; $50; $100; $250 & up.

Windsor

OLD STONE HOUSE MUSEUM, 22 Chestnut St., Windsor, NY 13865-4105. Tel.: 607-655-1491.
Founded: 1970.
Congressional District: 124
Key Personnel: Dir., Luella F. English.
Personnel Profile: Part-Time Volunteers 3.
Governing Authority: individual operation.
Institution Type/Description: History Museum: housed in Federal period house of Major Jed Hotchkiss, C.S.A.
Collections: Civil War artifacts, weapons, accoutrements pertaining to men of area; historical objects of local interest from local homes, businesses & industries; local Indian relics; old pictures; books & maps; tools; implements; lamps; furniture; bottles; whips of local manufactures.
Research Fields: local military history; local history.
Facilities: 300-vol. library of books, primarily Civil War & local history, available for inter-library loan by arrangement with owner on premises only.
Activities: guided tours; permanent exhibitions.
Hours & Admission Prices: Sat.-Sun. 10-5; other times by appointment. No charge; donations accepted.
Attendance: 300 (estimated)

Woodstock

CENTER FOR PHOTOGRAPHY AT WOODSTOCK, 59 Tinker St., Woodstock, NY 12498-1236. Tel.: 845-679-7747 & 9957. Fax: 845-679-6337.
E-mail: info@cpw.org
Web Site: www.cpw.org
Founded: 1977.
Key Personnel: Exec. Dir., Ariel Shanberg; Program Assoc., Rose Wind Jerome; Operations Mgr., Larry Lewis; Coord. Education, Lindsay Stern.
Personnel Profile: Full-Time Paid 4; Part-Time Paid 1; Part-Time Volunteers 10; Interns 10.
Governing Authority: Tax-exempt: 501(c)(3).
Institution Type/Description: Photography Museum.
Collections: fine art photography.
Facilities: library.
Activities: workshops; lectures; internships available; 1 or 2-day workshops available; residencies for artists.
Publications: Photography Quarterly.
Hours & Admission Prices: Wed.-Sun. 12-5. Lecture series: adults $7; discount to members, seniors & students. Gallery: no charge; donations accepted. &
Attendance: 50,000 (estimated)
Membership: Subscribing $25; Student $30; Individual $45; Family $50; Supporting $60; Friend $100; Patron $250-$1,250.

WOODSTOCK ARTISTS ASSOCIATION & MUSEUM, (M), 28 Tinker St., Woodstock, NY 12498-1233. Tel.: 845-679-2940. Fax: 845-679-2198.
E-mail: info@woodstockart.org
Web Site: www.woodstockart.org
Founded: 1919.
Congressional District: 42
Key Personnel: Exec. Dir., Josephine Bloodgood; Dir., Carl Van Brunt; Chm., Pat Horner; Coord. Gallery & Museum Shop, Patricia Seminara.
Personnel Profile: Full-Time Paid 4; Part-Time Paid 3; Part-Time Volunteers 40; Interns 3.
Governing Authority: Tax-exempt.
Institution Type/Description: Art Association & Museum.
Collections: work of American artists associated with the Woodstock Art Colony; archives of original documents & photographs.
Major Exhibits: Mary Frank, 2/14-5/14; Wendell Jones, 6/14-9/14; Georges Malkline: Perfect Surrealist, 10/14-12/14.
Research Fields: American 20th century art.
Facilities: archives.
Activities: permanent exhibitions; adult and school programs.
Publications: exhibition catalogs, Woodstock's Art Heritage: The Permanent Collection of the WAA; The Maverich Art Colony; At Woodstock, Kuniyoshi; Peggy Bacon: Cats and Caricatures; Embracing the New: Modernism's Impact on Woodstock Artists.
Hours & Admission Prices: Thurs.-Mon. 12-5; other times by appointment. Suggested Donation: adults $5; discounts to AAM & ICOM members; members no charge. Closed New Year's Day; Labor Day; Thanksgiving; Christmas. &
Attendance: 25,000 (accurate)
Membership: Youth $40; Individual & Supporting Friend $100; Artist $135; Family Friends$150.

Wyoming

MIDDLEBURY HISTORICAL SOCIETY, 22 S. Academy St., Wyoming, NY 14591-9801. Mailing Address: P.O. Box 198, Wyoming, NY 14591-0198. Tel.: 585-495-6420. Facebook: Middlebury Historical Society.
Founded: 1941.
Congressional District: 35
Key Personnel: Pres., Mr. Douglas Norton; Historian, Doris Bannister.
Governing Authority: historical society; nonprofit. Parent Institution: Middlebury Historical Society. Tax-exempt.
Institution Type/Description: Historical Society Museum: housed in 1817 Middlebury Academy.
Collections: military equipment; musical instruments; farm tools; photographs; local history records; women's costumes & accessories; textiles.
Research Fields: local town & village history.
Facilities: 450-vol. library of material on the 18th- & 19th-centuries available for use upon application.
Activities: guided tours; permanent & temporary exhibitions; demonstrations; meetings.

Publications: A History of Middlebury; From Middlebury to Middlebury; Village of Wyoming Historic Tours-1985; Town of Middlebury's; Wyoming - The Town Where I Grew Up.
Hours & Admission Prices: Memorial Day-last weekend in Sept. Sun. 2-5. No charge; donations accepted. &
Attendance: 691 (accurate)

Yonkers

***   HUDSON RIVER MUSEUM, (M),** 511 Warburton Ave., Yonkers, NY 10701-1899. Tel.: 914-963-4550. Fax: 914-963-8558.
E-mail: info@hrm.org
Web Site: www.hrm.org
Founded: 1919.
Congressional District: 17
Key Personnel: Dir., Michael Botwinick; Trustee Chm., Jan Adelson; Asst. Dir. Finance & Administration, Jared Hammond; Cur. Collections, Laura Vookles; Cur. Exhibitions, Bartholomew Bland; Asst. Dir. Exhibitions & Programs, Jean Paul Maitinsky; Asst. Dir. Devel., Kimberly Woodward; Dir. Public Rels., Linda Locke.
Personnel Profile: Full-Time Paid 28; Part-Time Paid 41; Part-Time Volunteers 79.
Governing Authority: nonprofit organization. Tax-exempt: 501(c)(3).
Institution Type/Description: General Museum.
Collections: 19th-, 20th- & 21st-century American painting, sculpture, photography, furniture, decorative arts, historical documents, costumes, local & regional memorabilia; planetarium. Historic House: 1877 Victorian Glenview.
Research Fields: 19th, 20th & 21st-century American fine & decorative arts; architecture; astronomy; technology; Hudson River Valley art, history & natural science.
Facilities: 127-seat planetarium; education center; meeting room.
Activities: tours; planetarium shows; cultural programs for seniors; family weekend craft programs; docent programs; research by appointment; permanent, temporary & traveling exhibitions; public programs.
Publications: monthly calendar of events; newsletter; exhibition catalogs; annual camp and teacher guides.
Hours & Admission Prices: Museum: Wed.-Sun. 12-5. Planetarium Fri. 7, Sat.-Sun. 12:30, 1:30, 2:30, 3:30; school groups & tours by appointment. Adults $5, senior citizens & children 5-16 $3; discounts for Metro-North Railroad commuters, AAM, ICOM, FWMA & Channel 13 members; members no charge. Closed New Year's Day; Thanksgiving; Christmas. &
Attendance: 54,623 (estimated)
Membership: Senior Citizen $25; Dual Senior $35; Individual $40; Dual $50; Family $60; Supporter $100; Director's Circle $250.

PHILIPSE MANOR HALL STATE HISTORIC SITE, 29 Warburton Ave., (& Dock St.), Yonkers, NY 10701-2721. Tel.: 914-965-4027. Fax: 914-965-6485.
Web Site: www.nysparks.state.ny.us/historic-sites/37/details.aspx
Founded: 1911.
Congressional District: 23
Key Personnel: Dir., Kimberly Flook; Commissioner, Carol Ash; Pres. (V), Joan Jennings.
Personnel Profile: Full-Time Paid 5; Part-Time Paid 3; Part-Time Volunteers 50.
Governing Authority: state. Parent Institution: New York State Office of Parks, Recreation & Historic Preservation. Tax-exempt.
Institution Type/Description: Art & History Museum: housed in early 18th-century Georgian style manor house.
Collections: paintings; photographs; structures; architecture.
Research Fields: Philipse family; landholding in colonial New York; Loyalists in the American Revolution; American art.
Facilities: changing exhibits; 80-seat auditorium.
Activities: guided tours; concerts; lectures; gallery talks; demonstrations; craft workshops; children's theatre; permanent & temporary exhibitions; special programs; school programs; educational program.
Publications: booklet: An American Loyalist; The Ordeal of Frederick Philipse III; semiannual newsletter.
Hours & Admission Prices: April-Oct. Tues.-Sat. 10-5; Nov.-March Tues.-Sat. 10-4; other times by appointment. Adults $5, seniors & students $3; members & children under 12 no charge. Closed holidays. &
Attendance: 25,000 (estimated)
Membership: Associate $15; Individual $25; Family $50; Contributing $100; Sponsor $250; Patron $500; Benefactor $1,000.

Yorktown Heights

TOWN OF YORKTOWN MUSEUM, YCCC Building - Top Fl., 1974 Commerce St., Yorktown Heights, NY 10598-4433. Tel.: 914-962-2970. Fax: 914-962-4379.
E-mail: museum@yorktownny.org
Web Site: www.yorktownmuseum.org
Founded: 1966.
Congressional District: 19
Key Personnel: C.E.O., Alice Roker; Tour Coord., Nancy Augustowski; Asst. Cur. & Museum Shop Mgr., Adele Hobby.
Personnel Profile: Part-Time Paid 2; Part-Time Volunteers 5; Interns 1.
Volunteer Hours: 300
Governing Authority: municipal. Subsidiary Institution: Yorktown Museum Research Center, 1974 Commerce St., Yorktown, NY 10598. Tax-exempt.
Institution Type/Description: General Museum.
Collections: agricultural tools; household tools & equipment; costumes; manuscripts; textiles; dollhouses; toys & dolls; ephemera; photographs; Indian artifacts; spinning & weaving equipment; newspapers; Sylvia Newton Thorne marionettes, Old Put Line of New York Central R.R; Woodlands Room featuring items of the Mohican Indian; holiday decorations; period school books & cookbooks; period fashion magazines & ladies magazines; family genealogical files & photos; letters; documents; Bible records.
Research Fields: history of township.
Facilities: library of books of local history, antiques, architecture, collectibles & genealogy; research room. Gift & museum related items for sale.
Activities: guided tours; lectures; inter-museum loan, permanent & temporary exhibitions; school loan service; slide lectures & mini-exhibits to local school & civic organizations; genealogical & historical research; typescripts available for use to the public.
Publications: museum newsletter; art exhibit catalog; books, Yorktown History.
Hours & Admission Prices: Tues. & Thurs. 11-4, Sat. 1-4. Tours: adults $5, children $2; discounts to members. Closed Easter; Thanksgiving; Christmas. &
Attendance: 5,000 (accurate)

Youngstown

OLD FORT NIAGARA, Fort Niagara State Park, Youngstown, NY 14174. Mailing Address: P.O. Box 169, Youngstown, NY 14174-0169. Tel.: 716-745-7611. Fax: 716-745-9141.
E-mail: jbrubaker@oldfortniagara.org
Web Site: www.oldfortniagara.org
Founded: 1927.
Congressional District: 28
Key Personnel: Exec. Dir., Robert L. Emerson; Pres., David Caldwell; Asst. Dir. & Cur., Jerome Brubaker; Museum Shop Mgr., Patricia Fitzpatrick.
Personnel Profile: Full-Time Paid 9; Full-Time Volunteers 1; Part-Time Paid 45; Part-Time Volunteers 1,750.
Governing Authority: nonprofit corporation. Tax-exempt: 501(c)(3).
Institution Type/Description: Military Historic Site.
Collections: military artifacts; local history; archaeology; site-related.
Research Fields: archaeology; military; history.
Facilities: 3,000-vol. library of history books & manuscripts available for specific research. Museum-related items for sale.
Activities: guided tours; living history demonstrations; permanent exhibitions.
Publications: brochures, Old Fort Niagara; Visitor's Guide; History & Guide to Old Fort Niagara; books: Seige-1759: The Campaign Against Niagara, The Gold-Laced Coat; Green Coats & Glory; Old Fort Niagara, An Illustrated History; A New System of Domestic Cookery; The Battle of Fort George; Navy Island: Historic Treasure of the Niagara; Molly Brant: A Legacy of Her Own; 1812 Sailor's Diaries; Memoirs of Pierre Pouchot.
Hours & Admission Prices: July-Aug. daily 9-7; Sept.-June daily 9-5. Adults $12, senior citizens $11, children 6-12 $8; discounts to groups; members no charge. Closed New Year's Day; Thanksgiving; Christmas. &
Attendance: 85,000 (accurate)
Membership: Individual $30; Family $50; Life & Sustaining $1,000; Endowing $2,000 & up.

NORTH CAROLINA

(545 listings)

Aberdeen

MALCOLM BLUE FARM MUSEUM, 1177 Bethesda Rd., Aberdeen, NC 28315. Mailing Address: P.O. Box 603, Aberdeen, NC 28315-0603. Tel.: 910-944-7558. Fax: 910-944-7558.
E-mail: malcolmblue@windstream.net
Web Site: malcolmbluefarm.com
Founded: 1972.
Key Personnel: Chm. (V), Robert Martin.
Personnel Profile: Part-Time Volunteers 16.
Governing Authority: Parent Institution: Malcolm Blue Historical Society.
Institution Type/Description: Farm Museum.
Collections: local history & culture; period furnishings; personal artifacts; farm equipment & agriculture; lumbering; Aberdeen and Rockfish Railroad; pottery; Native American artifacts. Historic Buildings: 1825 farmhouse; windmill; gristmill; water well; barns.
Facilities: 7.5 acre farm.
Activities: educational programs; demonstrations. Annual Events: Bluegrass Festival in June; Malcolm Blue Historical Crafts and Farmskills Festival in September; Early American Christmas in December.
Hours & Admission Prices: Fri.-Sat. 1-4; other times by appointment. No charge; donations accepted.
Attendance: 3,000 (estimated)
Membership: Individual $25; Family $40.

Albemarle

MORROW MOUNTAIN STATE PARK, 49104 Morrow Mountain Rd., Albemarle, NC 28001-7886. Tel.: 704-982-4402. Fax: 704-982-5323.
E-mail: morrow.mountain@ncmail.net
Web Site: www.ncparks.gov
Founded: 1962.
Congressional District: 5
Key Personnel: Park Supt., Greg Schneider.
Governing Authority: state. A branch of North Carolina Dept. of Environment and Natural Resources, North Carolina Division of Parks & Recreation, P.O. Box 27687, Raleigh, NC 27611. Tax-exempt.
Institution Type/Description: Park Museum.
Collections: geological history; Indian civilization; artifacts of Morrow Mountain area; reversion to a climax forest area; mounted animals of the area; local plant species.
Facilities: 16-vol. library of nature books available for use by special request only on premises; nature center.
Activities: lectures; guided nature hikes; power point programs; environmental & educational programs.
Hours & Admission Prices: Daily 10-5. No charge. Closed Christmas. &
Attendance: 201,970 (estimated)

STANLY COUNTY HISTORIC PRESERVATION COMMISSION AND MUSEUM, 245 E. Main St., Albemarle, NC 28001-4919. Tel.: 704-986-3777. Fax: 704-986-3778.
E-mail: junderwood@co.stanly.nc.us
Web Site: www.stanlycountymuseum.com
Founded: 1973.
Congressional District: 8
Key Personnel: Dir., Jonathan A. Underwood; Chm., Christy Stoner; Cur., Lessie Huneycutt.
Personnel Profile: Full-Time Paid 2; Part-Time Paid 1; Part-Time Volunteers 45; Interns 4.
Governing Authority: county. Parent Institution: County of Stanly. Tax-exempt.
Institution Type/Description: History Museum & Visitor Center.
Collections: local history; period clothing; period furniture; textiles; photographs; documents. Historic Houses: c.1847 Freeman-Marks House; c.1852 I.W. Snuggs House.
Research Fields: local history.
Facilities: 100-vol. library of local history; visitors center. Museum-related items for sale.
Activities: guided tours; lectures; films; workshops; organized educational programs; docent program; participatory, loan & temporary exhibitions.

Publications: brochure, Badin: A Town at the Narrows; books: The Stanly County Folklife Tour - 1991, Stanly County, The Architectural Legacy of a Rural North Carolina County (1992); Images of America: Stanly County (2000).
Hours & Admission Prices: Wed.-Fri. 10-5, Sat. 10-4. No charge; donations accepted. Closed major holidays. &
Attendance: 5,881 (accurate)
Membership: Senior Citizen $10; Individual $20; Household $35; Bronze Circle $50-$99; Silver Circle $100-$199; Gold Circle $200-$499; Platinum Circle $500 & up.

Asheboro

AMERICAN CLASSIC MOTORCYCLE MUSEUM, 1170 US Hwy. 64 W., Asheboro, NC 27205. Tel.: 336-629-9564.
Institution Type/Description: Motorcycle Museum.
Collections: motorcycles; motorcycle history, clothing & helmets; posters; photographs.
Facilities: Museum-related items for sale.
Hours & Admission Prices: Mon. 6am-2pm, Tues.-Fri. 6am-5:30pm, Sat. 6am-4pm. No charge.

NORTH CAROLINA AVIATION MUSEUM, 2222-G Pilots View Rd., Asheboro, NC 27204. Tel.: 336-625-0170. Fax: 336-625-2984.
E-mail: ncam@triad.twcbc.com
Web Site: www.ncairmuseum.org
Institution Type/Description: Aviation Museum.
Collections: aviation history; aircraft; personal artifacts; photographs.
Hours & Admission Prices: April-Oct. Mon.-Sat. 10-5, Sun. 1-5; Nov. Tues.-Sat. 10-5, Sun. 1-5; Dec.-March Wed.-Sat. 10-5, Sun. 1-5. Adults $8, students under 18 $5; children 5 & under no charge.

NORTH CAROLINA ZOOLOGICAL PARK, 4401 Zoo Pkwy., Asheboro, NC 27205-1425. Tel.: 800-488-0444; 336-879-7000. Fax: 336-879-2891.
E-mail: rod.hackney@nczoo.org
Web Site: www.nczoo.org
Founded: 1972.
Congressional District: 4
Key Personnel: Chm. Zoological Park Council, Scott Reed; Dir., Dr. David M. Jones; Business Officer, Mary Joan Pugh; Human Resources Officer, Cami Bunting; Gen. Cur., Ken Reininger; Cur. Horticulture, Virginia Wall; Design Cur., Ellen Greer; Assoc. Cur. Mammals, Guy Lichty; Cur. Herpetology, John Groves; Public Rels. Mgr., Rod Hackney; Veterinarian, Michael R. Loomis; Museum Shop Mgr., David Whitaker.
Personnel Profile: Full-Time Paid 260; Full-Time Volunteers 70; Part-Time Paid 200; Part-Time Volunteers 70.
Volunteer Hours: 41,000
Operating Expenses: 17,200,000
Operating Income: 17,600,000
Governing Authority: state. North Carolina Dept. of Environment & Natural Resources, 512 Salisbury St., Archdale Bldg., Raleigh, NC 27611; Tel. 919-733-4984. Tax-exempt: 170(b)(1)(A).
Institution Type/Description: Zoo.
Collections: 800 African animals; 300 North American animals; 30,000 introduced plants.
Major Exhibits: Big Bugs, 4/1-10/31/14.
Research Fields: chimpanzee behavior; primate husbandry; rhinoceros behavior; primate reproduction; elephant migration patterns in Cameroon, Africa; Cross River gorilla behavior in Cameroon, Africa.
Facilities: 1,800 vol. library of zoo management, wildlife ecology, zoology, botany; outdoor theater; fast food facility; picnic area. Zoo-related items for sale.
Activities: formally organized education programs for children & college students affiliated with UNC-Greensboro & N.C. State Univ. School of Veterinary Medicine; docent program; special events.
Publications: brochure; society newsletter; educators tour booklet; visitor guide; self-guided aviary tour; school packets for grades K-12.
Hours & Admission Prices: Daily 9-5; groups by appointment. Adults $12, senior citizens & college students $10, children 2-12 $8; discounts for groups, AZA & selected zoological institutions; North Carolina school groups in grades K-12, members & children under 2 no charge. &
Attendance: 761,964 (accurate)
Membership: Individual $29; Individual-plus $39; Family $59; Family-plus $69; Zookeeper $100; Curator $250; Director's Guild $1,500.

Asheville

* **ASHEVILLE ART MUSEUM, (M),** 2 S. Pack Square, Asheville, NC 28801-3521. Mailing Address: P.O. Box 1717, Asheville, NC 28802-1717. Tel.: 828-253-3227. Fax: 828-257-4503.
E-mail: mailbox@ashevilleart.org
Web Site: www.ashevilleart.org
Founded: 1948.
Congressional District: 11
Key Personnel: C.E.O., Pamela L. Myers; Chm., Nancy Ferguson; Immediate Past Chm., Rob Pulleyn; Cur., Frank E. Thomson; Asst. Cur., Cole Hendrix; Mgr. Education Programs, Sharon McRorie; Membership Coord. & Financial Officer, Lindsay G. Rosson; Registrar & Preparator, Jake Ehrlund; Mgr. Communications, Jennifer Swanson; Campaign Mgr., Rebecca Lynch-Maass; Mgr. School & Family Programs, Erin Shope; Mgr. Adult Programs, Candace Reilly; Museum Shop & Visitor Svcs. Mgr., Cornelia Katchen; Museum Shop & Events Coord., Laila Boggs; Coord. Events, Amanda Bryan; Mgr. Grants, Melisa Holman.
Personnel Profile: Full-Time Paid 15; Part-Time Paid 10; Part-Time Volunteers 126; Interns 16.
Volunteer Hours: 7,549
Operating Expenses: 1,085,000
Operating Income: 1,177,000
Governing Authority: nonprofit organization. Parent Institution: Asheville Art Museum Association, Inc. Tax-exempt.
Institution Type/Description: Art Museum.
Collections: American art in all media from 1900 to present: painting, prints, drawings, sculpture photography; contemporary art & studio craft, southeastern regional works & works related to Black Mountain College.
Major Exhibits: An Electronic Medium, 1/7/14-4/27/14; Social Geographics, 1/21/14-5/18/14; Susan Weil, 2/8/14-5/25/15; Experimental Gallery, 3/14; Collectors Circle 10 Years, 3/16-14-6/8/14; Pierra Daura, 3/16/14-6/22/14; Ralph Burns, 3/29/14-7/20/14; Farm to Table, 5/14-9/28/14; Minna Citron, 6/7/14-9/22/14; Dox Trash, 6/28/14.
Research Fields: Black Mountain College; contemporary crafts; regional artists; architecture; Cherokee ceramics.
Facilities: art library; classroom & studio; art resource center; 525-seat theater; multipurpose room; children's art space. Museum-related items for sale.
Activities: guided tours; museum and travel group tours & programs; school group tours; lectures; films; gallery demonstrations; formally organized educational programs; permanent & temporary exhibitions; slide programs; rural outreach program; adult & family programs; artist programs; summer camps & workshops; children's interactive art space.
Publications: newsletters; exhibition catalogues.
Hours & Admission Prices: Tues.-Sat. 10-5, Sun. 1-5; special evening hours available. Adults $8, senior citizens & students $7; children under 5 & members no charge. Southeastern & North American reciprocal members. Closed New Year's Day; Independence Day; Labor Day; Thanksgiving; Christmas. &
Attendance: 73,364 (accurate)
Membership: Student $25; Artists, Educator & Senior $40; Individual $50; Family & Dual $70; Patron $150; Sustaining $300; Benefactor $600; Director's Forum $1,000.

BILTMORE ESTATE, One Approach Rd., Asheville, NC 28803-8900. Tel.: 828-255-1333; 800-411-3812. Fax: 828-225-6383.
E-mail: rking@biltmore.com
Web Site: www.biltmore.com
Founded: 1930.
Congressional District: 11
Key Personnel: C.E.O., William A.V. Cecil, Jr.; Exec. Vice Pres., Richard Presby; Exec. Vice Pres., Steven Miller; Vice Pres., Richard King; Sr. Vice Pres. Attraction, Tom Ruff; Exec. Vice Pres., George W. Pickering, III; Vice Pres. Finance & Controller, Stephen Watson; Vice Pres. Agricultural Svcs., Ted Katsigianis; Sr. Vice Pres. Mktg., Jerry Douglass; Vice Pres. Winemaker, Bernard Delille; Vice Pres. Sales, Jim Owens; Public Rels. Mgr., Elizabeth Sims; Group Sales Mgr., Paula Wilbur.
Personnel Profile: Full-Time Paid 584; Part-Time Paid 500.
Governing Authority: company.
Institution Type/Description: Historic House, Conservatory & Gardens: 1895 Biltmore House.
Collections: paintings; tapestries; prints; sculpture; furniture; rugs & related decorative art. Conservatory & Gardens: azaleas, roses, orchids, tropical collections.
Research Fields: 19th-century arts & furniture; tapestries; conservation; preservation; architecture; landscape design.
Facilities: gardens; three restaurants; winery; conservatory. Museum-related items for sale.

Activities: self-guided tours. Museum Sponsors: Festival of Flowers; Christmas at Biltmore; candlelight evenings (by reservation); chamber music; summer evening concerts (by reservation); fall fair.
Publications: guidebook.
Hours & Admission Prices: Welcome Center: Mon. Thurs. 8:30-3, Fri.-Sat. 8:30-4. House: Mon.-Thurs. 9-3:30, Fri.-Sat. 9-4:30. See website for admission fees. &
Attendance: 902,000 (accurate)

BILTMORE HOMESPUN SHOPS - GROVEWOOD GALLERY, 111 Grovewood Rd., Asheville, NC 28804-2858. Tel.: 828-253-7651. Fax: 828-254-2489.
E-mail: homespun@grovewood.com
Web Site: grovewood.com
Founded: 1901.
Congressional District: 11
Key Personnel: Mgr., Allison Mills; Pres., Barbara Blomberg; Pres., Marilyn Patten.
Personnel Profile: Full-Time Paid 11; Part-Time Paid 8.
Governing Authority: company, Biltmore Industries. Branch Museum: Estes-Winn Antique Automobile Museum. Tax-exempt.
Institution Type/Description: Textile Museum: complex of six buildings c.1917 operating as Biltmore handwoven homespun industries.
Collections: 1920 tools, looms, machinery & other items related to homespun.
Facilities: Local handcrafted items, woodcarvings & other related items for sale.
Activities: film demonstrating homespun transformed from fleece to finished cloth; films; permanent exhibitions.
Publications: brochure.
Hours & Admission Prices: April-Dec. Mon.-Sat. 10-5, Sun. 11-5. No charge; donations accepted. Closed Thanksgiving; Christmas. &
Attendance: 22,635 (accurate)

BLACK MOUNTAIN COLLEGE MUSEUM + ARTS CENTER, 56 Broadway, Asheville, NC 28801-2916. Mailing Address: P.O. Box 18912, Asheville, NC 28814-0912. Tel.: 828-350-8484. Fax: 828-350-8484.
E-mail: bmcmac@bellsouth.net
Web Site: www.blackmountaincollege.org
Founded: 1993.
Congressional District: 11
Key Personnel: Program Dir., Alice Sebrell.
Personnel Profile: Part-Time Paid 3; Part-Time Volunteers 50; Interns 1.
Governing Authority: private; nonprofit organization. Tax-exempt: 501(c)(3).
Institution Type/Description: Art Museum.
Collections: documents; artwork; photographs; books related to Black Mountain College & those who taught or attended it.
Major Exhibits: Cynthia Homire: Vision Quest, 1/14-5/14.
Research Fields: educational programs for public primary & elementary schools; oral histories of Black Mountain College alumni.
Activities: formal education programs for children; guided tours; lectures; participatory, temporary & traveling exhibitions.
Publications: Black Mountain College Dossier; newsletter; exhibit catalogues.
Hours & Admission Prices: Tues.-Wed. 12-4, Thurs.-Sat. 11-5. Adults $3; discounts to AAM & ICOM members; members no charge. &
Membership: Student $15; Individual $35; Family $60; Affiliate $125; Donor $500; Patron $1,000.

BLUE RIDGE PARKWAY VISITOR CENTER, 195 Hemphill Knob Rd., Asheville, NC 28803-8686. Tel.: 828-298-5330.
Web Site: www.blueridgeheritage.com
Formerly: Blue Ridge Parkway Destination Center
Founded: 2003.
Key Personnel: Exec. Dir., Angie Chandler.
Personnel Profile: Full-Time Paid 5; Part-Time Paid 5.
Governing Authority: Parent Institution: National Park Service. Tax-exempt.
Institution Type/Description: History Museum.
Collections: local history, heritage & culture; economic traditions; photographs; personal artifacts; hands-on exhibits.
Facilities: 70-seat theater; bookstore.
Activities: interactive exhibits; listening stations; family events.
Publications: e-newsletter.
Hours & Admission Prices: Daily 9-5. No charge. Closed New Year's Day; Thanksgiving, Christmas. &
Attendance: 93,434 (estimated)

BOTANICAL GARDENS AT ASHEVILLE, 151 W. T. Weaver Blvd., Asheville, NC 28804-3414. Tel.: 828-252-5190.
E-mail: bgardens@bellsouth.net

Web Site: ashevillebotanicalgardens.org
Founded: 1960.
Congressional District: 10
Key Personnel: Pres., Gwen Wisler; Vice Pres. & Museum Shop Mgr., Suzanne Wodek; Garden Mgr. & Horticulture Chm., Jay Kranyik; Administrative Asst., Heather Rayburn.
Personnel Profile: Full-Time Paid 1; Part-Time Paid 3; Part-Time Volunteers 75.
Governing Authority: nonprofit corp. Tax-exempt: 501(c)(3).
Institution Type/Description: Botanical Gardens: includes site of earthen battlements of the Civil War, Battle of Asheville.
Collections: native flora of Southern Appalachians. Historical House: The Hayes Cabin, Smoky Mountain log cabin furnished with primitive furniture of the early settlers.
Research Fields: preservation of endangered species of plants.
Facilities: botanical library & garden; visitor center; Cole Botany Library; Butler Lecture Hall.
Activities: guided tours; lectures; children's programs. Botanical Gardens Sponsors: special events for the public in the area of conservation, botany & nature study; biannual Native Plant Sales.
Publications: quarterly newsletter, The New Leaf; brochures.
Hours & Admission Prices: Visitor Center: mid-March-Dec. daily 10-4. Grounds: daily. No charge; donations accepted. Closed Thanksgiving; Christmas. &
Attendance: 35,000 (estimated)
Membership: Student $15; Individual $25; Family & Club $35; Contributor $50; Sustaining $100; Benefactor $500.

COLBURN EARTH SCIENCE MUSEUM, (M), Pack Place Education, Arts & Science Center, 2 S. Pack Square, Asheville, NC 28801. Mailing Address: P.O. Box 1617, Asheville, NC 28802-1617. Tel.: 828-254-7162. Fax: 828-257-4505.
E-mail: info@colburnmuseum.org
Web Site: www.packplace.org
Formerly: Colburn Gem & Mineral Museum, Inc.
Founded: 1960.
Congressional District: 11
Key Personnel: Dir. & Museum Shop Mgr., Vicky Ballard; Chm. (V), Sara Peacocky.
Personnel Profile: Full-Time Paid 4; Part-Time Paid 3; Part-Time Volunteers 4.
Governing Authority: nonprofit; bd. directors. Tax-exempt.
Institution Type/Description: Geology, Mineralogy & Paleontology Museum & Earth Science.
Collections: North Carolina & worldwide minerals; gems; North Carolina rocks; mining photos; mining artifacts; history of mining; landforms & Blue Ridge geology; meteorology; astronomy; oceanography; interactive computer technology.
Research Fields: mineralogy.
Facilities: library; permanent & temporary exhibit galleries; lecture hall; research lab; star lab planetarium. Gift items for sale.
Activities: self-directed & guided tours; lectures; field trips; organized educational programs; permanent & temporary exhibition.
Hours & Admission Prices: Tues.-Sat. 10-5, Sun. 1-5. Adults $6, senior citizens, students & children $5; discount to groups, AAM, ASTC & SEMC members; children under 4 & members no charge. Closed major holidays. &
Attendance: 27,232 (accurate)
Membership: Individual $35; Family $50; Patron $100; Sponsor $250; Contributor $500; Benefactor $1,000.

ESTES-WINN ANTIQUE AUTOMOBILE MUSEUM, 111 Grovewood Rd., Asheville, NC 28804-2858. Tel.: 828-253-7651. Fax: 828-254-2489.
E-mail: automuseum@grovewood.com
Web Site: www.grovewood.com
Founded: 1970.
Congressional District: 11
Key Personnel: Mgr., Sherry Masters; Pres., Marilyn Patton; Pres., Barbara Blomberg; Business Officer, Shirley Dobbs.
Personnel Profile: Part-Time Paid 2.
Governing Authority: nonprofit organization. Affiliated with Biltmore Indus. Tax-exempt: 501(c)(3).
Institution Type/Description: Transportation Museum: located on the grounds of c.1917 Biltmore Homespun Shops.
Collections: 19 rare & period cars.
Publications: brochure.
Hours & Admission Prices: April-Dec. Mon.-Sat. 10-5, Sun. 11-5. No charge; donations accepted. &
Attendance: 22,635 (accurate)

THE HEALTH ADVENTURE, 800 Brevard Rd., Ste. 620, Asheville, NC 28806. Tel.: 828-665-2217. Fax: 828-665-9436.
E-mail: info@thehealthadventure.org
Web Site: www.thehealthadventure.org
Founded: 1968.
Congressional District: 11
Key Personnel: Chm. Bd. (V), Mark Knollman, P.A.; Exec. Dir., Paige Wheeler; Dir. Operations, Mitzi Morris; Mgr. Exhibits, Kevin King.
Personnel Profile: Full-Time Paid 13; Part-Time Paid 2; Part-Time Volunteers 239.
Governing Authority: nonprofit organization. Parent Institution: Park Ridge Health. Tax-exempt.
Institution Type/Description: Science Museum.
Collections: bones, nutrition, digestion, muscles, circulation, dental, respiration; life patterns; substance abuse; brains; senses; creativity; sports health; general anatomy; transparent anatomical mannequin; astronomy; animal care.
Research Fields: hands-on science & health education.
Facilities: 20,000 sq. ft. exhibit space; theatre; classrooms.
Activities: self-directed & guided tours; lectures; films; formally organized educational programs; docent program; permanent & traveling exhibitions.
Publications: brochures, membership; programs; rack cards; e-newsletter.
Hours & Admission Prices: Mon.-Sat. 10-8, Sun. 12:30-6. Adults $9.50, seniors $7.50, children 2-11 $6.50; discount to WNC Nature Center members; ASTC & museum members & children under 2 no charge. Closed holidays. &
Attendance: 72,321 (accurate)
Membership: Individual $50; Family/Grandparent $75; Family Plus $125; Supporting $240; Sustaining $480; Adventure Society $1,000. (Caregiver & Additional Child $15)

THE NORTH CAROLINA ARBORETUM, 100 Frederick Law Olmsted Way, Asheville, NC 28806-9315. Tel.: 828-665-2492. Fax: 828-665-2371.
Institution Type/Description: Arboretum.
Collections: plants; trees; flowers.
Facilities: nature trails. Museum-related items for sale.
Activities: research; demonstrations; educational programs; workshops; tours; classes; nature walks.
Hours & Admission Prices: Mon.-Sat. 9-5, Sun. 12-5. Closed New Year's Day; Thanksgiving; Christmas.

SMITH-MCDOWELL HOUSE MUSEUM, 283 Victoria Rd., Asheville, NC 28801-4817. Tel.: 828-253-9231. Fax: 828-253-5518.
E-mail: smh@wnchistory.org
Web Site: www.wnchistory.org
Founded: 1981.
Congressional District: 11
Key Personnel: Dir., Sharon Gruber; Pres., Gwin Jones; Coord. Education, Lisa Whitfield.
Personnel Profile: Full-Time Paid 1; Part-Time Paid 2; Part-Time Volunteers 30; Interns 1.
Governing Authority: nonprofit organization. Parent Institution: Western North Carolina Historical Assn. Tax-exempt: 501(c)(3).
Institution Type/Description: Historic Building & Local History Museum: housed in 1840 Smith-McDowell House.
Collections: 19th-century decorative arts & artifacts relating to Western North Carolina.
Major Exhibits: Hillbilly Land: Myth & Reality of Appalachian Culture, 4/1/14-9/30/14.
Research Fields: Western North Carolina history.
Facilities: 500-vol. library; classroom.
Activities: guided tours; formally organized educational programs; docent program. Museum Sponsors:Heritage Festival.
Publications: quarterly newsletter.
Hours & Admission Prices: Wed.-Sat. 10-4, Sun. 12-4. Adults $8, children 8-18 $4; discounts to groups, AAA & AASLH members; members no charge. Additional fee for Christmas exhibit. &
Attendance: 2,500 (estimated)
Membership: Individual $35; Family $50; Supporting $75; Patron $100; Sponsor $250; Benefactor $500.

THE SOUTHERN APPALACHIAN RADIO MUSEUM, A-B Technical Community College, 340 Victoria Rd. Elm Bldg., Rm. 315, Asheville, NC 28801. Tel.: 828-299-1276.
Founded: 1999.

Key Personnel: Pres., John Travis, Vice Pres., Norman Harrill; Sec. & Treas., Clint Gorman.
Personnel Profile: Part-Time Volunteers 7.
Governing Authority. nonprofit. Tax exempt: 501(c)(3).
Institution Type/Description: History Museum.
Collections: radio history; test instruments; Atwater Kent; Philco; Silvertone; Crosley; Hammarlund; Harvey Wells; spark gap transmitters; keys; period QSL cards; amateur radio station, W4AFM.
Hours & Admission Prices: Feb.-Nov. Fri. 1-3; other times by appointment. No charge; donations accepted. Closed school holidays.

SOUTHERN HIGHLAND CRAFT GUILD, Milepost 382, Blue Ridge Pkwy., Asheville, NC 28805. Mailing Address: P.O. Box 9545, Asheville, NC 28815-0545. Tel.: 828-298-7928. Fax: 828-298-7962. Facebook: Folk Art Center.
E-mail: info@craftguild.org
Web Site: www.southernhighlandguild.org
Formerly: Folk Art Center
Founded: 1930.
Congressional District: 11
Key Personnel: Exec. Dir., Tom Bailey.
Personnel Profile: Full-Time Paid 20; Part-Time Paid 10; Part-Time Volunteers 60.
Governing Authority: jointly: private; nonprofit. Parent Institution: Southern Highland Handicraft Guild, in association with National Park Svcs. Tax-exempt.
Institution Type/Description: Southern Appalachian Craft Museum.
Collections: traditional & contemporary crafts & trades.
Research Fields: traditional crafts; American crafts; folk art; Southern Appalachian culture.
Facilities: library; information center. Publications & crafts for sale.
Activities: demonstration; craft related.
Publications: quarterly newsletter, Highland Highlights.
Hours & Admission Prices: Jan.-March daily 9-5; April-Dec. daily 9-6. No charge; donations accepted. Closed New Year's Day; Thanksgiving; Christmas. &
Attendance: 281,000 (accurate)
Membership: Donor Recognition Categories $50 & $150.

THOMAS WOLFE MEMORIAL, 52 N. Market St., Asheville, NC 28801-8105. Tel.: 828-253-8304. Fax: 828-252-8171.
E-mail: contactus@wolfememorial.com
Web Site: www.wolfememorial.com
Founded: 1949.
Congressional District: 11
Key Personnel: Dir., Christian Edwards; Museum Shop Mgr., Jesse Cox.
Personnel Profile: Full-Time Paid 4; Part-Time Paid 5; Part-Time Volunteers 5.
Governing Authority: state. North Carolina Dept. of Cultural Resources, 109 E. Jones St., Raleigh, NC. 27611. Parent Institution: North Carolina Historic Sites.
Institution Type/Description: Historic House: built in 1883 Thomas Wolfe boarding house, The Old Kentucky Home.
Collections: early 20th century boarding house furnishings; Thomas Wolfe memorabilia.
Research Fields: 20th-century American Literature; early 20th century Asheville, NC.
Facilities: visitors center. Museum-related items for sale.
Activities: guided tours of house; AV(isual) program in Visitors Center.
Publications: brochure; A Literary Journey; The Lost World of Thomas Wolfe; article, Historic Homes; newsletter.
Hours & Admission Prices: Tues.-Sat. 9-5, Sun. 1-5. Adults $5, students $2. &
Attendance: 18,000 (accurate)
Membership: Friend $25-$49; Boarder $50-$99; Scholar $100-$149; Angel $150 & up.

WESTERN NORTH CAROLINA NATURE CENTER, 75 Gashes Creek Rd., Asheville, NC 28805-2529. Tel.: 828-298-5600, ext. 303. Fax: 828-298-2644.
E-mail: staff@wildwnc.org
Web Site: www.wncnaturecenter.org
Founded: 1977.
Congressional District: 11
Key Personnel: Dir., Chris Gentile; Dir. Education, Keith Mastin; Cur. Animals, Allison Ballentine; Museum Shop Mgr., Mischa Trinks.
Personnel Profile: Full-Time Paid 16; Part-Time Paid 3; Part-Time Volunteers 80; Interns 6.
Governing Authority: nonprofit. Parent Institution: City of Asheville. Tax-exempt.

Institution Type/Description: Zoological Park.
Collections: living animals & plants.
Research Fields: red wolf captive breeding program.
Facilities: gardens; nature trail; zoological park; aquarium; classrooms. Gift items for sale.
Activities: guided tours; lectures; films; gallery talks; study clubs; hobby workshops; TV programs; formally organized education programs for children; docent program; permanent exhibitions.
Publications: newsletter.
Hours & Admission Prices: Daily 10-5. Adults $8, senior citizens $7, children 3-14 $4; discounts to ASTC & AZA members; members and children 2 & under no charge. Closed major winter holidays. &
Attendance: 92,408 (accurate)
Membership: Individual $30; Couple $45; Family $60; Wildlife Guardian $100; Nature Benefactor $250.

YMI CULTURAL CENTER, 39 S. Market St., Asheville, NC 28801-3726. Mailing Address: P.O. Box 7301, Asheville, NC 28802-7301. Tel.: 828-257-4540. Fax: 828-257-4539. Facebook: YMI Cultural Center.
E-mail: ymicc@att.net
Web Site: www.ymicc.org
Founded: 1893.
Congressional District: 4
Volunteer Hours: 1,000
Governing Authority: Tax-exempt.
Institution Type/Description: Art & History Gallery.
Collections: African American culture & artifacts; paintings; sculpture.
Activities: permanent & traveling exhibition; educational programs.
Hours & Admission Prices: By appointment.
Attendance: 1,000 (estimated)

Atlantic Beach

FORT MACON STATE PARK, 2303 E. Fort Macon Rd., Atlantic Beach, NC 28512-5638. Tel.: 252-726-3775. Fax: 252-726-2497.
E-mail: fort.macon@ncparks.gov
Web Site: friendsoffortmacon.org
Founded: 1924.
Congressional District: 1
Key Personnel: Park Supt., Randall Newman; Park Ranger, Kevin Bleck; Park Ranger, Paul R. Branch; Chief Maintenance, John Schell; Maintenance Mechanic, Larry Stover; Maintenance Mechanic, Robert Taber; Ranger, Benjamin Fleming; Ranger, Paul Terry, Office Asst. III, Cleta Buck.
Personnel Profile: Full-Time Paid 10; Part-Time Paid 13; Part-Time Volunteers 25.
Governing Authority: state. NC Div. of Parks & Recreation, Box 27687, Archdale Bldg., Raleigh, NC 27611. Tel. 919-733-4181.
Institution Type/Description: Museum & Historic Building: 1834 brick casemated, irregular pentagon shape, outer & inner walls with moat, Fort Macon.
Collections: 1834-1944 military artifacts; Civil War, Spanish American War & World War II artifacts & equipment; artillery projectiles; soldier & garrison life relics; two replica barbette cannons; two original mortars & a field gun.
Facilities: library pertaining to American military history, Civil War & coastal ecology available for research on premises only.
Activities: guided tours; lectures; permanent exhibitions; living history programs.
Hours & Admission Prices: Daily 9-5:30. No charge; donations accepted. Closed Christmas. &
Attendance: 1,250,000 (estimated)

Aurora

AURORA FOSSIL MUSEUM, (M), 400 Main St., Aurora, NC 27806-0352. Mailing Address: P.O. Box 352, Aurora, NC 27806-0352. Tel.: 252-322-4238. Fax: 252-322-2220.
E-mail: aurfosmus@yahoo.com
Web Site: aurorafossilmuseum.com
Founded: 1976.
Congressional District: 1
Key Personnel: Dir., Andrea W. Stilley.
Personnel Profile: Full-Time Paid 1; Part-Time Paid 4; Part-Time Volunteers 1.
Governing Authority: nonprofit. Tax-exempt.
Institution Type/Description: Geology & Paleontology Museum.
Collections: geology & paleontology from coastal plains of North Carolina; pleistocene, pliocene, & miocene marine fossils.
Facilities: picnics. Museum-related items for sale.
Activities: outreach programs; field studies; fossil pile; school groups.
Publications: AFM teacher's packet.

Hours & Admission Prices: March to Labor Day Mon.-Sat. 9-4:30, Sun. 12:30-4:30; Sept.-Feb. Mon.-Sat. 9-4:30; groups of 10 or more by appointment. No charge; donations accepted. &
Attendance: 25,000 (accurate)
Membership: Individual $50; Family $75.

Bailey

THE COUNTRY DOCTOR MUSEUM, 6642 Peele Rd., Bailey, NC 27807. Mailing Address: P.O. Box 34, Bailey, NC 27807-0034. Tel.: 252-235-4165. Fax: 252-235-2372.
Web Site: www.countrydoctormuseum.org
Founded: 1967.
Congressional District: 2
Personnel Profile: Full-Time Paid 1; Part-Time Paid 1.
Governing Authority: nonprofit organization. Parent Institution: Medical Foundation of East Carolina University. Managed by Laupus Library of ECU. Tax-exempt: 501(c)(3).
Institution Type/Description: Rural Medical & Pharmacology Museum with emphasis on Eastern North Carolina from 1850-1960.
Collections: 19th- & early 20th-century medical and pharmacy instruments, furnishings & supplies of country doctors.
Research Fields: 19th- & early 20th-century medicine and pharmacy.
Facilities: 1,000-vol. library of medical books available for inter-library loan through Laupus Library, East Carolina University Medical School and to doctors, nurses, pharmacists, nursing & research students; medicinal herb garden.
Activities: guided tours; lectures.
Publications: book, The Country Doctor Museum; Medicinal Herb Garden of the C.D. Museum; Tarheel Doctors and Patients; Patent Medicines in North Carolina; Blackberries to Fishing Worms; A Nurse's Education.
Hours & Admission Prices: Tues.-Sat. 10-4. Adults $5, senior citizens & AAA members $4, students $3. Closed holidays. &
Attendance: 2,000 (estimated)

Bakersville

PENLAND GALLERY, 2687 Conley Ridge Rd., Bakersville, NC 28705. Mailing Address: P.O. Box 37, Penland, NC 28765-0037. Tel.: 828-765-6211.
E-mail: gallery@penland.org
Web Site: www.penland.org
Institution Type/Description: Art Gallery.
Collections: works by Penland School of Crafts students; paintings; sculpture; photographs; drawings.
Activities: classes; educational programs; special events.
Hours & Admission Prices: early March to mid-Dec. Tues.-Sat. 10-5, Sun. 12-5.

Bald Head Island

OLD BALDY LIGHTHOUSE & SMITH ISLAND MUSEUM, 101 Lighthouse Wynd, Bald Head Island, NC 28461. Tel.: 910-457-7481.
Web Site: oldbaldy.org
Institution Type/Description: Historic Lighthouse: c.1850.
Collections: local history & culture; period furnishings; photographs; replica 1850s lighthouse keeper's cottage.
Hours & Admission Prices: Bald Head Island is accessible by passenger ferry from Deep Point Marina, 1301 Ferry Rd., Southport, NC. See website for ferry schedule & admission prices. Lighthouse & Museum: Mon.-Sat. 9-5, Sun. 11-5. Adults 13 & over $5, youth 3-12 $3; children 2 & under no charge.

Bath

HISTORIC BATH STATE HISTORIC SITE, 207 Carteret St., Bath, NC 27808. Mailing Address: P.O. Box 148, Bath, NC 27808-0148. Tel.: 252-923-3971. Fax: 252 923 3971.
E-mail: bath@ncdcr.gov
Web Site: www.bath.nchistoricsites.org
Founded: 1963.
Congressional District: 3
Key Personnel: Site Mgr., Leigh Swane; Gift Shop Mgr., Robyn Jackson.
Personnel Profile: Full-Time Paid 5; Part-Time Paid 4; Part-Time Volunteers 6; Interns 1.
Governing Authority: state. Parent Institution: North Carolina Dept. of Cultural Resources, Historic Sites Section 4620 Mail Service Center Raleigh, NC 27699-4620. Tax-exempt: 170(b)(1)(A).

Institution Type/Description: Visitor Center.
Collections: history. Historic Houses: 1751 Palmer-Marsh House; 1830 Bonner House; 1790 Van Der Veer House.
Research Fields: Bath & colonial North Carolina.
Facilities: Colonial publications & museum-related items for sale.
Activities: guided tours; lectures; films; formally organized education programs for children; permanent exhibitions.
Publications: guide book; brochure.
Hours & Admission Prices: Tues.-Sat. 9-5. Two house tour: adults $2, students $1; discount to groups. Closed winter holidays. &
Attendance: 30,600 (estimated)

Beaufort

BEAUFORT HISTORIC SITE, 138 Turner St., Beaufort, NC 28516-2139. Mailing Address: 150 Turner St., Beaufort, NC 28516-2139. Tel.: 252-728-5225. Fax: 252-728-4966.
E-mail: beauforthistoricsite@earthlink.net
Web Site: www.beauforthistoricsite.org
Founded: 1960.
Congressional District: 2
Key Personnel: Exec. Dir., Patricia Suggs; Chm. (V), Polly Hagle; Pres. (V), Bill Kaeser; Vice Pres., Larry Jones; Vice Pres., Dick Bierly; Treas., Lucia Stanley; Museum Shop Mgr., Diane Donovan.
Personnel Profile: Full-Time Paid 4; Part-Time Paid 11; Part-Time Volunteers 300.
Governing Authority: private volunteer board; nonprofit organization. Parent Institution: Beaufort Historical Association. Tax-exempt.
Institution Type/Description: Historical & Preservation Society.
Collections: period furnishings; historic documents; c.1709 Old Burying Ground. Historic Houses: c.1825 John C. Manson House; c.1825 Josiah Bell House; c.1796 Carteret County Courthouse; c.1829 Carteret County Jail; c.1859 Apothecary Shop; c.1778 Samuel Leffer Cottage; c.1732 Rustell House.
Research Fields: architectural history; collections research; county & North Carolina coastal history; family history; old Burying Ground of 1731 history.
Facilities: 2 acre site; herb garden; Welcome Center. Gift items for sale.
Activities: guided tours; lectures; docent program; exhibits; traditional crafts & living history; English bus tours; weaving demonstrations. Site Sponsors: Public Day in April; Old Homes Tour & Antiques show Fundraiser in June; harvest time exhibition in October; Jumble Sale & Community Thanksgiving Feast in November; Christmas Walk in December.
Publications: newsletter, Historic Times.
Hours & Admission Prices: Mon.-Sat. 9:30-5; group tours by appointment. Adults $8, children under 12 $4. English Bus Tours: Mon. & Wed.-Sat. Adults $8. Old Burying Ground Tour: Tues.-Thurs. 2:30 Adults $8; discounts to AAA members & groups of 30 or more. Closed Easter; Thanksgiving; Christmas Eve & Day. &
Attendance: 65,000 (estimated)
Membership: Senior & Student $25; Contributor $35; Donor $60; Sponsor $100; Patron $250; Benefactor $500; Heritage Club $1,000; Beaufort Circle $2,500; Somerset Circle $5,000.

*** NORTH CAROLINA MARITIME MUSEUM, (M),** 315 Front St., Beaufort, NC 28516-2124. Tel.: 252-728-7317. Fax: 252-728-2108.
E-mail: maritime@ncdcr.gov
Web Site: www.ncmaritimemuseum.org
Founded: 1975.
Congressional District: 1
Key Personnel: Dir., Joe Schwarzer; Pres. Friends of the Museum, Clarie Burdett; Cur. Education, John Hairr; Assoc. Cur., Benn Wunderly; Coord. Group Programs, Christine Brin; Registrar Collections, Frances Hayden; Collections Mgr., Lynn Anderson; Watercraft & Maritime Research Cur., Paul Fontenoy; Exhibit Designer, Mike Carraway; Exhibit Technician, Larry Copeland; Exhibit Technician, Terry Greene; Boatshop Mgr., William Prentice; Boatbuilder, Craig Wright; Coord. Cape Lookout Studies, Keith Rittmaster; Business Mgr., Randy Mana; Nautical Archaeologist, David Moore; Artist & Illustrator, Stephanie Davis; Building & Grounds Supt., Denny Hailey; Museum Shop Mgr., Sharon Resor; Southport Branch Mgr., Mary Strickland; Program Coord., Lori Sanderlin.
Personnel Profile: Full-Time Paid 21; Part-Time Paid 6; Part-Time Volunteers 100.
Governing Authority: state. Parent Institution: North Carolina Dept. of Cultural Resources, Office of Archives & History. Subsidiary Institution: Division of State History Museums. Tax-exempt.
Institution Type/Description: Natural & Maritime History Museum.
Collections: marine specimens; seashells; ship models; marine artifacts; traditional small watercraft collection; personal artifacts;

Research Fields: development & history of small craft of North Carolina.
Facilities: ship's library of plans, charts & volumes on traditional boat building; boatshop. Books, prints, pamphlets for sale.
Activities: lectures; year-round field trips; boatbuilding skills program; junior sailing program; adult learn to sail program. Museum Sponsors: the summer school of science for children; Traditional Wooden Boat Show.
Publications: newsletter, The Waterline; NC Maritime Museum newsletter; calendar; Construction Plans for Traditional North Carolina Watercraft; N.C. Traditional Work Boats; magazine, The Mari Times.
Hours & Admission Prices: Mon.-Fri. 9-5, Sat. 10-5, Sun. 1-5. No charge; donations accepted. Closed New Year's Day; Thanksgiving; Christmas Eve & Day. &
Attendance: 205,285 (accurate)
Membership: Individual $25; Family $50; Supporter $100; Patron $250; Sustaining $500; Benefactor $1,000; Lifetime $2,500.

Belhaven

BELHAVEN MEMORIAL MUSEUM, INC., 211 E. Main St., Belhaven Town Hall, 2nd Fl., Belhaven, NC 27810-1413. Mailing Address: P.O. Box 220, Belhaven, NC 27810-0220. Tel.: 252-943-6817. Fax: 252-943-2357.
Web Site: www.beaufort-county.com/Belhaven/museum/Belhaven.htm
Founded: 1965.
Congressional District: 3
Key Personnel: Pres., Ed Harris.
Personnel Profile: Part-Time Paid 2.
Governing Authority: nonprofit corporation. Tax-exempt.
Institution Type/Description: Local History Museum.
Collections: 19th- to 20th-century historical items of coastal Carolina; early phonographs; Indian artifacts; marine.
Publications: quarterly newsletter.
Hours & Admission Prices: Thurs.-Tues. 1-5. No charge; donations accepted.
Attendance: 5,000 (estimated)
Membership: Supporter $10; Explorer $25; Archivist $50; Special Contributor $100; Corporate Sponsor $200.

Belmont

DANIEL STOWE BOTANICAL GARDEN, 6500 S. New Hope Rd., Belmont, NC 28012-8788. Tel.: 704-825-4490. Fax: 704-829-1240.
Web Site: www.dsbg.org
Key Personnel: Exec. Dir., Kara Newport; Dir. Mktg. & Guest Svcs., Jim Hoffman, APR
Institution Type/Description: Botanical Garden.
Collections: themed gardens; fountains.
Facilities: gardens; nature trail. Museum-related items for sale.
Activities: special events; educational community classes.
Publications: quarterly newsletter, The Garden Path; e-news, Garden Buzz.
Hours & Admission Prices: Daily 9-5. Adults $12, seniors 60 & over $11, children 4-12 $6; discounts to AAA members; members & children under 4 no charge. Closed New Year's Day; Thanksgiving; Christmas. &
Membership: Individual $5; Household $75; Premier Household $125.

Benson

BENSON MUSEUM OF LOCAL HISTORY, 102 W. Main St., Benson, NC 27504-1504. Tel.: 919-894-3825. Facebook: Benson Museum of Local History.
E-mail: thobgood@townofbenson.com
Key Personnel: Dir., Terry Hobgood; Chm. (V), Hampton Whittington.
Personnel Profile: Full-Time Paid 1; Part-Time Paid 1; Part-Time Volunteers 3.
Governing Authority: Parent Institution: Town of Benson.
Institution Type/Description: History Museum.
Collections: local history & culture; railroad memorabilia; period furnishings; quilts; farming & hardware artifacts; newspapers; photographs.
Hours & Admission Prices: Wed.-Sat. tours vary. No charge; donations accepted. &
Attendance: 400 (estimated)

Bethania

HISTORIC BETHANIA VISITOR CENTER, 5480 Bethania Rd., Bethania, NC 27010. Mailing Address: P.O. Box 259, Bethania, NC 27010. Tel.: 336-922-0434.
E-mail: visitorcenter@townofbethania.org
Web Site: www.townofbethania.org
Institution Type/Description: Visitor Center.

Collections: local history & culture; period furnishings; personal artifacts. Historic Buildings: Moravian farmstead; c.1799 Wolff-Moser house; 1894 Alpha Chapel.
Facilities: Gift items for sale.
Hours & Admission Prices: Call for hours.

Black Mountain

SWANNANOA VALLEY MUSEUM, 223 W. State St., Black Mountain, NC 28711-3408. Mailing Address: P.O. Box 306, Black Mountain, NC 28711-0306. Tel.: 828-669-9566.
E-mail: info@swannanoavalleymuseum.org
Web Site: www.swannanoavalleymuseum.org
Founded: 1989.
Congressional District: 11
Key Personnel: Dir., Anne Chesky-Smith; Chm. (V), Wendell Begley.
Personnel Profile: Full-Time Paid 1; Part-Time Paid 1; Part-Time Volunteers 80.
Governing Authority: Parent Institution: Swannanoa Valley Historical & Preservation Assoc., Inc. Tax-exempt.
Institution Type/Description: History Museum.
Collections: area history & culture.
Activities: school, community & senior outreach; special events; guided tours. Museum Sponsors: Heritage Roundtables.
Hours & Admission Prices: April-Oct. Tues.-Sat. 10-5. Adults $2; members no charge. &
Attendance: 5,000 (estimated)
Membership: Individual $30; Family $50; Booster $100; Sponsor $250; Patron $500; Benefactor $1,000 & up.

Blowing Rock

APPALACHIAN HERITAGE MUSEUM, 175 Mystery Hill Lane, Blowing Rock, NC 28605. Tel.: 828-264-2792. Fax: 828-262-3292.
E-mail: mysteryhillnc@gmail.com
Web Site: www.mysteryhill-nc.com
Founded: 1989.
Congressional District: 5
Institution Type/Description: History Museum.
Collections: local history & culture; period artifacts; photographs.
Hours & Admission Prices: June-Aug. daily 9-8; Sept.-May daily 9-5. Adults 13-59 $9, seniors 60 & over $8, children 5-12 $7; discounts to groups; children 4 & under no charge.

BLOWING ROCK ART AND HISTORY MUSEUM (BRAHM), (M), 159 Chestnut St., Blowing Rock, NC 28605. Mailing Address: P.O. Box 828, Blowing Rock, NC 28605. Tel.: 828-295-9099. Fax: 828-295-9029.
E-mail: director@blowingrockmuseum.org
Web Site: www.blowingrockmuseum.org
Founded: 2001.
Congressional District: 5
Key Personnel: Exec. Dir., Joann Mitchell; Pres. (V), Teresa Caine.
Personnel Profile: Full-Time Paid 5; Part-Time Paid 2; Interns 2.
Governing Authority: Parent Institution: Blowing Rock Art and History Museum, Inc. Tax-exempt.
Institution Type/Description: Art & History Museum.
Collections: local visual arts, history & heritage; paintings; period artifacts.
Facilities: Museum-related items for sale.
Activities: educational programs; art & cultural events; special events; workshops.
Hours & Admission Prices: Tues.-Wed. & Fri.-Sat. 10-5, Thurs. 10-7. Adults $8, children, students & active military $5; children 4 & under no charge.
Attendance: 6,000 (accurate)

MYSTERY HILL - HERITAGE & NATIVE ARTIFACT MUSEUM, 129 Mystery Hill Lane, Blowing Rock, NC 28605-9549. Tel.: 828-264-2792.
E-mail: mysteryhillnc@gmail.com
Web Site: www.mysteryhill-nc.com
Institution Type/Description: Historic House Museum: built in 1903.
Collections: local history & culture; arrowheads; Native American artifacts; period furnishings; personal artifacts.
Activities: hands-on exhibits, experiments & illusions.
Hours & Admission Prices: June-Aug. daily 9-8; Sept.-May 9-5. Adults 13-59 $9, seniors 60 & over $8, children 5-12 $7; discounts to groups; children 4 & under no charge.

Boone

THE CHILDREN'S PLAYHOUSE, 400 Tracy Circle, Boone, NC 28607-3846. Tel.: 828-263-0011.
Web Site: www.goplayhouse.org/
Key Personnel: Exec. Dir., Kathy Parham
Institution Type/Description: Children's Museum.
Collections: hands-on exhibits.
Activities: workshops; educational programs.
Hours & Admission Prices: June-July Tues.-Sat. 10-5; Aug.-May Tues.-Fri. 10-5, Sat. 10-3. Admission $5; children one & under and members no charge.

DANIEL BOONE NATIVE GARDENS, 651 Horn in the West Dr., Boone, NC 28607. Mailing Address: P.O. Box 1705, Boone, NC 28607. Tel.: 828-264-6390.
Institution Type/Description: Native Gardens.
Collections: native Appalachian trees, shrubs, & wildflowers; plants.
Activities: rental facilities; special events.
Hours & Admission Prices: May-Oct. daily 10-6. Adults 16 & over $2.

HICKORY RIDGE LIVING HISTORY MUSEUM, 591 Horn in the West Dr., Boone, NC 28607-4283. Mailing Address: P.O. Box 295, Boone, NC 28607-0295. Tel.: 828-264-2120. Fax: 828-264-9089. Facebook: Hickory Ridge Living History Museum.
Web Site: www.hickoryridgemuseum.com
Formerly: Hickory Ridge Homestead
Founded: 1980.
Key Personnel: Cur., Dave Davis
Institution Type/Description: History Museum.
Collections: local history & culture; hands-on exhibits; period furnishings & artifacts; historic buildings.
Activities: educational programs; special events; hands-on exhibits; workshops; live-in programs; summer day camps. Museum Sponsors: Boone Heritage Festival in October.
Hours & Admission Prices: May-Oct. Sat. 9-1; late June to mid-Aug. Tues.-Sun. 5:30-8. Admission by donation; charge for private group tours.
Membership: Settler $50; Patriot $100; Hornet $250; Kings Mountain Victory $500; Boone's Trailblazers $1,000; Community Patron Producer $2,000.

TURCHIN CENTER FOR THE VISUAL ARTS, 423 W. King St., Boone, NC 28607-3523. Mailing Address: ASU Box 32139, Boone, NC 28608. Tel.: 828-262-3017. Fax: 828-262-7546. Facebook: Turchin Center.
E-mail: turchincenter@appstate.edu
Founded: 2003.
Key Personnel: Dir., Hank T. Foreman.
Personnel Profile: Full-Time Paid 5; Part-Time Paid 22; Part-Time Volunteers 5.
Institution Type/Description: Art Gallery.
Collections: works by regional, national & international artists.
Major Exhibits: Hugh Morton, 8/16/13-1/25/14.
Hours & Admission Prices: Tues.-Thurs. & Sat. 10-6, Fri. 12-8. No charge; donations accepted.
Attendance: 81,205 (accurate)

Brevard

SILVERMONT MANSION 2ND FLOOR HOUSE MUSEUM, 364 E. Main St., Brevard, NC 28712. Tel.: 828-884-3156.
Founded: 2010.
Key Personnel: Co-Chm. (V), Jan Osborne; Co-Chm. (V), Lee Stewart.
Volunteer Hours: 175
Governing Authority: Parent Institution: Transylvania County, Brevard, NC. Subsidiary Institution: Friends of Silvermont.
Institution Type/Description: Historic House Museum: built in 1917. Listed on the National Register of Historic Places.
Collections: family history; personal artifacts; period furnishings; photographs.
Activities: rental facilities.
Hours & Admission Prices: Call for hours. No charge; donations accepted.
Attendance: 500 (estimated)

SPIERS GALLERY, Sims Art Center, Brevard College, 1 Brevard College Dr., Brevard, NC 28712-4283. Tel.: 828-883-8292, ext. 8188. Fax: 828-884-3790.
E-mail: bbyers@brevard.edu
Web Site: www.Brevard.edu

Key Personnel: Dir., Bill Byers.
Personnel Profile: Full-Time Paid 4; Part-Time Paid 2; Part-Time Volunteers 1.
Governing Authority: university; nonprofit. Parent Institution: Brevard College.
Institution Type/Description: Art Gallery.
Collections: American art.
Activities: temporary exhibitions; arts festivals; lectures; poetry readings; musical performances.
Publications: exhibition notices.
Hours & Admission Prices: Sept.-May Mon.-Fri. 8-3. No charge. &

Attendance: 1,000 (estimated)

TRANSYLVANIA HERITAGE MUSEUM, 40 W. Jordan St., Brevard, NC 28712-3641. Mailing Address: P.O. Box 2347, Brevard, NC 28712-2347. Tel.: 828-884-2347.
Key Personnel: Exec. Dir., Rebecca Suddeth
Institution Type/Description: History Museum.
Collections: local history & culture; personal artifacts; photographs; period furnishings.
Facilities: Museum-related items for sale.
Activities: special events.
Hours & Admission Prices: Wed.-Sat. 10-5. No charge.

Bryson City

SMOKY MOUNTAIN TRAINS, 100 Greenlee St., Bryson City, NC 28713. Mailing Address: P.O. Box 1490, Bryson City, NC 28713-1490. Tel.: 828-488-5200; 866-914-5200. Fax: 828-488-3162.
E-mail: info@smokymountaintrains.com
Web Site: www.smokymountaintrains.com
Founded: 2002.
Key Personnel: C.E.O. & Cur., Timothy O. Cooper.
Personnel Profile: Full-Time Paid 1; Part-Time Paid 4.
Institution Type/Description: Train Museum.
Collections: 7,000 Lionel (TM) engines, cars & accessories; 1934 Blue Comet Passenger set; Joshua Lionel Cowen Challenger steam locomotives; 3,000 trains dating from 1918-2003.
Facilities: 7,000 sq. ft. exhibit space; children's activity center. Museum-related items for sale.
Activities: children's activities.
Publications: semiannual newsletter.
Hours & Admission Prices: Mon.-Sat. 9:30-5; call to confirm. Adults $9, children 3-11 $5; children under 3 no charge. Closed New Year's Day; Thanksgiving; Christmas. &

Attendance: 30,000 (accurate)

Burgaw

BURGAW TRAIN DEPOT, 115 S. Dickerson St., Burgaw, NC 28425. Mailing Address: P.O. Box 1096, Burgaw, NC 28425. Tel.: 910-259-9817. Fax: 910-300-6116.
Institution Type/Description: Historic Building: housed in the former train depot; c.1850. Listed on the National Register of Historic Places.
Collections: depot & railroad history; period furnishings; photographs.
Activities: tours.
Hours & Admission Prices: Daily 9-5.

PENDER COUNTY MUSEUM, 200 W. Bridgers St., Burgaw, NC 28425. Mailing Address: P.O. Box 1380, Burgaw, NC 28425-1380. Tel.: 910-259-8543.
E-mail: penderhist@hotmail.com
Web Site: pendercountymuseum.webs.com
Key Personnel: Chm. (V), Jeanette Jones; Pres. (V), Shelby Battle.
Personnel Profile: Part-Time Volunteers 40.
Governing Authority: private; nonprofit organization. Parent Institution: Pender County Historical Society. Tax-exempt.
Institution Type/Description: Local History Museum.
Collections: WWII veterans' oral history transcripts; farm implements; tools; furniture; genealogy; photographs.
Publications: quarterly newsletter.
Hours & Admission Prices: Thurs.-Fri. 1-4, Sat. 10-2. No charge; donations accepted.
Attendance: 500 (estimated)
Membership: Individual $20; Family $30; Sustaining $50; Benefactor $100; Corporate & Business $250; Life $300.

Burlington

ALAMANCE BATTLEGROUND STATE HISTORIC SITE, 5803 South N.C. 62, Burlington, NC 27215. Tel.: 336-227-4785. Fax: 336-227-4787. Facebook: Alamance Battleground State Historic Site.
E-mail: alamance@ncdcr.gov
Web Site: www.alamancebattleground.nchistoricsites.org
Founded: 1955.
Congressional District: 6
Key Personnel: Site Mgr., Bryan Dalton; Historic Interpreter, Bill Thompson; Museum Shop Mgr., Lisa Cox.
Personnel Profile: Full-Time Paid 3; Part-Time Paid 1; Part-Time Volunteers 6.
Governing Authority: state. Parent Institution: North Carolina Dept. of Cultural Resources, Office of Archives & History, Div. of State Historic Sites. Tax-exempt: 170(b)(1)(A).
Institution Type/Description: Historic Site.
Collections: military; area history. Historic House: c.1780 Allen House.
Research Fields: Colonial era.
Facilities: Publications relating to Colonial period & museum-related items for sale.
Activities: guided tours; lectures; films; formally organized education programs; permanent exhibitions.
Publications: brochure & newsletter, The Regulator.
Hours & Admission Prices: Mon.-Sat. 9-5. No charge; donations accepted. Closed major holidays. &

Attendance: 11,588 (accurate)
Membership: Farmer $10; Riflemen $30; Captain $50; Colonel $65; Militia $400; Governor $750; Regulator $1,771.

ALAMANCE COUNTY HISTORICAL MUSEUM, 4777 S. Hwy. 62, Burlington, NC 27215-9295. Tel.: 336-226-8254.
E-mail: achm@triad.twcbc.com
Web Site: www.alamancemuseum.org
Founded: 1976.
Key Personnel: C.E.O. & Dir., William Vincent, Ph.D.; Pres. (V), Frances Powell Barnes.
Personnel Profile: Full-Time Paid 2; Part-Time Paid 1; Part-Time Volunteers 20; Interns 1.
Governing Authority: Tax-exempt.
Institution Type/Description: Historic House Museum: housed in the former home of E.M. Holt.
Collections: period artifacts & furnishings; local history; quilts; Native American artifacts; pottery; clothing; toys; 19th century military artifacts.
Activities: temporary & permanent exhibits.
Hours & Admission Prices: Tues.-Fri. 9-4, Sat. 10:30-4, Sun. 1-4. No charge; donations accepted.
Attendance: 15,000 (accurate)
Membership: $25; $35; $50; $100; $250; $500; $1,000; $5,000.

CEDAROCK HISTORICAL FARM, 4242 Cedarock Historical Park Rd., Burlington, NC 27215. Tel.: 336-570-6759.
Institution Type/Description: Historical Farm: housed on the site of John & Polly Garrett 1830 farm.
Collections: early farming life; farm animals; early farm equipment; farmhouse.
Activities: guided & self-guided tours; special events.
Hours & Admission Prices: Call for hours.

CONSERVATORS' CENTER, 676 E. Hughes Mill Rd., Burlington, NC 27217. Mailing Address: P.O. Box 882, Mebane, NC 27302. Tel.: 336-421-0065; 800-979-3370.
Key Personnel: Co Founder, Doug Evans; Co Founder, Mindy Stinner
Institution Type/Description: Wildlife Refuge & Conservation Center.
Collections: tigers; lions; wolves; leopards; lemurs; red foxes; lynx; bobcats.
Facilities: wildlife rescue.
Activities: guided tours; Twilight Tours; educational & conservation programs.
Publications: newsletter.
Hours & Admission Prices: By appointment.

WHISTLESTOP EXHIBIT AT COMPANY SHOPS STATION, 101 N. Main St., Burlington, NC 27217. Tel.: 336-570-1444.
E-mail: raillines@ncrr.com
Web Site: www.whistlestopncrr.com
Institution Type/Description: Railroad History Museum.
Collections: railroad history & artifacts.
Hours & Admission Prices: Daily 7:30-7; other times by appointment. No charge. &

Burnsville

MOUNT MITCHELL STATE PARK & MUSEUM, 2388 Hwy. 128, Burnsville, NC 28714. Tel.: 828-675-4611.
Institution Type/Description: Park Museum.
Collections: nature, culture & history of the mountain; weather station.
Facilities: 1,385 sq. ft. exhibition space; nature trails; restaurant; observation deck; picnic area. Museum-related items for sale.
Activities: interactive weather station; hiking; nature trails; bird watching.
Hours & Admission Prices: Museum: May-Oct. daily 0-6. Park: March & Oct. daily 8-7; April & Sept. daily 8-8; May-Aug. daily 8-9; Nov.-Feb. daily 8-6. Closed Christmas.

YANCEY HISTORY ASSOCIATION - RUSH WRAY MUSEUM, McElroy House, 11 Academy St., Burnsville, NC 28714. Mailing Address: 3 Academy St., Burnsville, NC 28714-2944. Tel.: 828-678-9587.
E-mail: yhmuseum@frontier.com
Founded: 1988.
Congressional District: 11
Key Personnel: Pres. (V), Elaine Boone; Museum Shop Mgr., C. Carter.
Personnel Profile: Part-Time Paid 1; Part-Time Volunteers 14; Interns 1.
Governing Authority: Tax-exempt.
Institution Type/Description: History Museum: housed in the McElroy House; c.1840.
Collections: Yancey County history & heritage.
Research Fields: archaeology; genealogy.
Activities: educational programs; traveling trunks.
Publications: quarterly newsletter.
Hours & Admission Prices: Wed.-Sat. 10-4. Adults $3; members no charge. &
Attendance: 537 (accurate)
Membership: Individual $15; Family $30; Sponsor, Donor & Business $50.

Buxton

HATTERAS ISLAND VISITOR CENTER, Cape Hatteras Light Station, Buxton, NC 27920. Mailing Address: 1401 National Park Dr., Manteo, NC 27954-9451. Tel.: 252-473-2111 & 995-4474. Fax: 252-995-4633.
E-mail: caha_information@nps.gov
Web Site: www.nps.gov/caha
Founded: 1953.
Congressional District: 1
Key Personnel: Supt., Barclay Trimble; Archives Technician, Jami P. Lanier.
Personnel Profile: Full-Time Paid 2; Part-Time Paid 10; Part-Time Volunteers 37; Interns 2.
Governing Authority: federal. Parent Institution: National Park Service. Affiliated with Cape Hatteras National Seashore. Tax-exempt.
Institution Type/Description: Park Museum.
Collections: history; natural history. Historic Houses: 1854 Double Keeper's Quarters; 1870 Cape Hatteras Lighthouse; 1870 Cape Hatteras Lighthouse Keeper's dwellings.
Facilities: Maps, charts & books for sale.
Activities: guided tours; lectures; permanent exhibitions.
Publications: park brochure; park newspaper; site bulletins.
Hours & Admission Prices: Center: daily 9-5. No charge. Lighthouse: Easter weekend to Columbus Day. Adults $7, seniors 62 & over and children 12 & under $3.50; members no charge. Closed Christmas. &
Attendance: 4,500,000 (estimated)

Camden

CAMDEN COUNTY HERITAGE MUSEUM, 117 N. NC Hwy 343, Camden, NC 27976. Tel.: 252-771-8333; 877-771-8333.
Institution Type/Description: Historic Building: housed in the former Camden County jailhouse; c.1910.
Collections: local history & culture; period furnishings; personal artifacts; photographs.
Hours & Admission Prices: Mon.-Fri. 8-12 & 1-5; groups by appointment.

Cameron

ALOHA SAFARI ZOO, 159 Mini Lane, Cameron, NC 28348. Tel.: 919-770-7109.
E-mail: alohasafarizoonc@gmail.com
Web Site: alohasafarizoo.org
Founded: 2010.
Institution Type/Description: Zoo.
Collections: reptiles; monkeys; birds; rainforest; Watusi cattle; camels; zebras.
Activities: petting zoo; special events.
Publications: newsletter.
Hours & Admission Prices: Sat.-Sun. 10-5. Admission $5.

Canton

CANTON AREA HISTORICAL MUSEUM, 36 Park St., Canton, NC 28716-4324. Tel.: 828-646-3412.
E-mail: cantonvisitorcenter@charterinternet.com
Key Personnel: Dir., Wayne Carson
Institution Type/Description: History Museum.
Collections: area history & culture; photographs; inventor, Fillmore Christopher; author, Fred Chappell.
Hours & Admission Prices: Mon.-Fri. 8-5; other times by appointment. No charge; donations accepted.

Carrboro

THE ARTSCENTER, 300-G E. Main St., Carrboro, NC 27510-2359. Tel.: 919-929-2787, ext. 201. Fax: 919-969-8574. Facebook: ArtsCenter Live.
E-mail: info@artscenterlive.org
Web Site: www.artscenterlive.org
Formerly: The Art School
Founded: 1974.
Congressional District: 4
Personnel Profile: Full-Time Paid 10; Part-Time Paid 8.
Operating Expenses: 1,500,000
Operating Income: 1,500,000
Governing Authority: Tax-exempt.
Institution Type/Description: Art Gallery.
Collections: paintings; photographs; sculpture.
Hours & Admission Prices: Call for hours.
Attendance: 89,000 (accurate)

Carthage

BRYANT HOUSE, 3361 Mt. Carmel Rd., Carthage, NC 28327-8337. Mailing Address: P.O. Box 324, Southern Pines, NC 28388-0324. Tel.: 910-692-2051. Fax: 910-692-2051.
E-mail: moorehistory@connectnc.net
Web Site: moorehistory.com
Founded: 1946.
Governing Authority: Parent Institution: Moore County Historical Association. Tax-exempt.
Institution Type/Description: Historic House: built c.1820.
Collections: local history; period furnishings. Historic Building: McLendon Cabin, c.1760.
Hours & Admission Prices: May-Oct. 2nd & 4th Sun. of month 2-4; other times by appointment.
Attendance: 500 (estimated)

CARTHAGE MUSEUM, 202 Rockingham St., Carthage, NC 28327. Mailing Address: Town of Carthage Government, 4396 Hwy. 15501, Carthage, NC 28327. Tel.: 910-947-2331.
Institution Type/Description: History Museum.
Collections: local history, culture, & industry; military artifacts; photographs; personal artifacts.
Hours & Admission Prices: By appointment.

Cary

PAGE-WALKER ARTS & HISTORY CENTER, 119 Ambassador Loop, Cary, NC 27512. Mailing Address: P.O. Box 8005, Cary, NC 27512-8005. Tel.: 919-460-4963. Fax: 919-388-1141.
E-mail: kris.carmichael@townofcary.org
Web Site: www.townofcary.org
Founded: 1990.
Key Personnel: Dir., Kristina Carmichael; Pres., Brent Miller.
Personnel Profile: Full-Time Paid 3; Part-Time Paid 6; Part-Time Volunteers 55; Interns 1.
Governing Authority: town.
Institution Type/Description: History Museum: housed in a former hotel; built in 1868. Listed on the National Register of Historic Places.
Collections: local history & culture; photographs; period furnishings; personal artifacts; works by regional artists.
Research Fields: local history.
Facilities: classrooms; rental facilities; gardens; outdoor performance space.
Activities: classes; special events; performances; temporary & permanent exhibitions.

Hours & Admission Prices: Mon.-Thurs. 10-9:30, Fri. 10-5, Sat. 10-1; other times by appointment. No charge. &
Attendance: 25,000 (estimated)

STEVENS NATURE CENTER, 2616 Kildaire Farm Rd., Cary, NC 27518-9612. Mailing Address: P.O. Box 8005, Cary, NC 27512-8005. Tel.: 919-387-5980.
Institution Type/Description: Nature Center.
Collections: natural history; wildlife & their habitats; plants.
Facilities: classroom. Museum-related items for sale.
Activities: educational programs; special events.
Hours & Admission Prices: May-Sept. Mon.-Sat. 10-7, Sun. 1-7; Oct.-April Mon.-Sat. 10-5, Sun. 1-5. Closed holidays.

Cashiers

ZACHARY-TOLBERT HOUSE, 1940 Hwy. 107 S., Cashiers, NC 28717. Mailing Address: P.O. Box 104, Cashiers, NC 28717-0104. Tel.: 828-743-7710. Fax: 828-743-7169. Facebook: Zachary-Tolbert House.
E-mail: info@cashiershistoricalsociety.org
Web Site: www.cashiershistoricalsociety.org/
Founded: 1996.
Key Personnel: Exec. Dir., David Joy; Volunteer Chm., Gayle Eloy.
Personnel Profile: Full-Time Paid 2; Full-Time Volunteers 2; Part-Time Paid 2; Part-Time Volunteers 20; Interns 2.
Governing Authority: Tax-exempt.
Institution Type/Description: Historic House Museum.
Collections: period furnishings; personal artifacts.
Publications: newsletter, Past-Present-Future; brochure series, Valley Insights.
Hours & Admission Prices: May-Oct. Fri.-Sat. 11-3. Tours: no charge; donations accepted. &
Attendance: 1,750 (estimated)
Membership: Single $35; Dual/Family $50; Zachary Circle: Supporter $100; Friend $500; Pinnacle Circle: Patron $1,000; Benefactor $3,000; Sponsor $5,000.

Caswell Beach

OAK ISLAND LIGHTHOUSE, 300-A Caswell Beach Rd., Caswell Beach, NC 28465. Mailing Address: Friends of Oak Island Lighthouse, 1100 Caswell Beach Rd., Caswell Beach, NC 28465. Tel.: 910-278-1328.
E-mail: oakislandlighthouse@gmail.com
Web Site: oakislandlighthouse.org
Institution Type/Description: Historic Lighthouse.
Collections: 153 ft. tall lighthouse; local history.
Activities: guided tours.
Hours & Admission Prices: Tours to 2nd level of lighthouse: Memorial Day to Labor Day Wed. & Sat. 10-2. Tours to top of lighthouse: by appointment.

Cedar Mountain

GLASS FEATHER STUDIO & GARDENS, 200 Glass Feather Dr., Cedar Mountain, NC 28718-0241. Mailing Address: P.O. Box 241, Cedar Mountain, NC 28718-0241. Tel.: 828-885-8457.
E-mail: info@glassfeather.com
Web Site: glassfeather.com
Key Personnel: Owner, Patricia Travis, M.A.
Institution Type/Description: Art Gallery.
Collections: Travis collection of kiln-fired glass including plates, platters, bowls, masks, tiles, & wall hangings.
Facilities: Gift items for sale.
Activities: demonstrations; classes.
Hours & Admission Prices: April-Dec. 21 Wed.-Sat. 10-5.

Chadbourn

1910 A.C.L. DEPOT, 1st Ave. & Colony St., Chadbourn, NC 28431. Mailing Address: P.O. Box 100, Chadbourn, NC 28431-0100.
Key Personnel: Dir., Edna T. Yates; Treas., Hilda Bullard.
Personnel Profile: Part-Time Volunteers 2.
Governing Authority: private; nonprofit organization. Tax-exempt: 501(c)(3).
Institution Type/Description: History Museum: housed in a former train depot built in 1910. Listed on the Register of Historic Places.
Collections: depot & train history; paintings; period artifacts.
Activities: guided tours; temporary exhibitions; train rides; tours; school group tours. Annual Event: Strawberry Festival.

Hours & Admission Prices: Tues. 1-5, Sun. 2-5; other times by appointment. No charge; donations accepted.
Attendance: 1,000 (estimated)

Chapel Hill

* **ACKLAND ART MUSEUM, (M),** The University of North Carolina at Chapel Hill, 101 South Columbia St. at Franklin St., Chapel Hill, NC 27514. Mailing Address: Campus Box 3400, The University of North Carolina at Chapel Hill, Chapel Hill, NC 27599-3400. Tel.: 919-966-5736. Fax: 919-966-1400. TDD: 919-962-0837.
E-mail: ackland@email.unc.edu
Web Site: www.ackland.org
Founded: 1958.
Congressional District: 4
Key Personnel: Dir., Emily Kass; Chief Cur., Peter Nisbet; Cur. Collection, Timothy Riggs; Dir. Academic Programs, Carolyn Allmendinger; Dir. External Affairs, Amanda Hughes; Mgr. School & Community Programs, Jenny Marvel; Dir. Communications, Emily Bowles; Registrar, Anita Heggli-Swenson; Museum Shop Mgr., Alice Southwick.
Personnel Profile: Full-Time Paid 21; Part-Time Paid 4; Part-Time Volunteers 200; Interns 3.
Governing Authority: Parent Institution: The University of North Carolina at Chapel Hill. Tax-exempt: 501(c)(3).
Institution Type/Description: Art Museum.
Collections: European & American painting, sculpture & drawings from ancient times to 21st century; 15th- to 21st century drawings & prints; photographs; North Carolina pottery; Asian art; Indian miniatures & sculpture; contemporary art.
Research Fields: ancient, medieval, Renaissance through modern art.
Activities: guided tours; lectures; gallery talks; live music; film series; docent program; family days; children's classes & programs; drawing workshops; yoga in the galleries; curator's clinic; volunteer student guide program.
Publications: exhibition catalogues; weekly e-news.
Hours & Admission Prices: Wed. & Fri.-Sat. 10-5, Thurs. 10-8, Sun. 1-5. No charge; donations accepted. &
Attendance: 60,000 (estimated)
Membership: Student $15; Artist & Educator $30; Individual $50; Household $75; Contributor $125; Patron $250; Benefactor $500; Sponsor $1,000; Leadership $2,500; Vanguard $5,000.

CAROLINA BASKETBALL MUSEUM, 450 Skipper Bowles Dr., Chapel Hill, NC 27599. Tel.: 919-962-6000. Fax: 919-962-6002. Facebook: Carolina Basketball Museum.
E-mail: candrews@uncaa.unc.edu
Web Site: goheels.com
Founded: 2008.
Key Personnel: Dir., Clara A. Perry.
Personnel Profile: Part-Time Paid 6.
Institution Type/Description: Sports Museum.
Collections: Carolina basketball history; photographs; hands-on exhibits; uniforms; personal artifacts.
Facilities: theater.
Hours & Admission Prices: Tues.-Fri. 10-4, Sat. 9-1; groups by appointment. No charge; donations accepted.
Attendance: 60,000 (accurate)

CHARLES KURALT LEARNING CENTER & MUSEUM, Univ. of North Carolina Chapel Hill, School of Journalism, Carroll Hall, 2nd Fl., Chapel Hill, NC 27599-3365. Tel.: 919-962-1204.
Institution Type/Description: History Museum.
Collections: Kuralt's CBS News office; program tapes; awards; personal artifacts; photographs.
Hours & Admission Prices: Tues. & Thurs. 2-4; other times by appointment.

COKER ARBORETUM OF THE NORTH CAROLINA BOTANICAL GARDEN, 100 Old Mason Frm Rd., Chapel Hill, NC 27514. Mailing Address: The University of North Carolina at Chapel Hill, CB #3375 Totten Center, Chapel Hill, NC 27599-3375. Tel.: 919-962-0522. Fax: 919-962-3531.
E-mail: ncbg@unc.edu
Web Site: www.ncbg.unc.edu
Founded: 1903.
Congressional District: 4
Key Personnel: Dir., Dr. Peter White; Cur., Dan Stern.
Personnel Profile: Full-Time Paid 2; Part-Time Paid 1; Part-Time Volunteers 4.
Governing Authority: state. Parent Institution: North Carolina Botanical Garden. Tax-exempt.

Institution Type/Description: Arboretum.
Collections: dwarf conifers; southeastern U.S. native species; Asian species; daffodils; daylilies; grasses.
Activities: seasonal tours
Publications: brochure; newsletter, North Carolina Botanical Garden Newsletter.
Hours & Admission Prices: June-Aug. Mon.-Fri. 8-6, Sat. 9-6, Sun. 1-6; Sept.-May Mon.-Fri. 8-5, Sat. 9-5, Sun. 1-5. No charge; donations accepted. &
Attendance: 10,000 (estimated)
Membership: Volunteer, Student & Senior $10; Individual $35; Family $45; Organization $50.

HORACE WILLIAMS HOUSE, 610 E. Rosemary St., Chapel Hill, NC 27514-3720. Tel.: 919-942-7818. Facebook: Preservation Chapel Hill.
E-mail: info@preservationchapelhill.com
Web Site: www.preservationchapelhill.com
Founded: 1973.
Key Personnel: Dir., Cheri Szcodronski; Pres. (V), Matt Pohlman.
Personnel Profile: Full-Time Paid 1; Part-Time Paid 2; Interns 10.
Volunteer Hours: 4,000
Operating Expenses: 130,000
Operating Income: 130,000
Governing Authority: Parent Institution: Preservation Chapel Hill. Tax-exempt.
Institution Type/Description: Historic House Museum: built c.1854.
Collections: local history; period furnishings; personal artifacts; photographs.
Research Fields: local history.
Facilities: rental facilities.
Activities: educational programs; special events; monthly exhibitions.
Publications: Preservation Notes.
Hours & Admission Prices: Tues.-Fri. 10-4; other times by appointment. No charge; donations accepted.
Attendance: 6,000 (estimated)
Membership: Student no charge; Support $35; Contributor $60; Donor $150; Patron $250; Benefactor $500; Sustainer $1,000.

JORDAN LAKE EDUCATIONAL STATE FOREST, 2832 Big Woods Rd., Chapel Hill, NC 27514. Tel.: 919-542-1154.
E-mail: jordanlakeesf.ncfs@ncagr.gov
Web Site: www.ncesf.org
Institution Type/Description: State Forest.
Collections: forest ecosystems; birds of prey; deer; songbirds; flying squirrels; beavers.
Facilities: nature trails.
Activities: educational workshops; picnic facilities; ranger-conducted classes; nature trails; demonstrations.
Hours & Admission Prices: July-Aug. Tues.-Fri. 9-5, Sat.-Sun. 11-8; Sept.-June Tues.-Fri. 9-5, Sat.-Sun. 11-5.

KIDZU CHILDREN'S MUSEUM, 123 W. Franklin St., Chapel Hill, NC 27516-2524. Tel.: 919-933-1455.
E-mail: info@kidzuchildrensmuseum.org
Web Site: www.kidzuchildrensmuseum.org
Founded: 2006.
Congressional District: 4
Key Personnel: Dir., Pam Wall; Chm. (V), Dennis Schaecher.
Personnel Profile: Full-Time Paid 4; Part-Time Paid 5; Part-Time Volunteers 12; Interns 5.
Governing Authority: Tax-exempt.
Institution Type/Description: Children's Museum.
Collections: hands-on exhibits.
Activities: educational programs; birthday parties;
Hours & Admission Prices: Tues.-Sat. 10-5, Sun. 1-5. Admission 2 & over $5; children 12-23 months $2; members, children 11 months & under and Sun. 1-5 no charge.
Attendance: 32,000 (accurate)
Membership: Family $75; Family Plus $125; Supporter $300.

MOREHEAD PLANETARIUM AND SCIENCE CENTER, 250 E. Franklin St., Chapel Hill, NC 27514. Mailing Address: CB #3480 UNC, Chapel Hill, NC 27599. Tel.: 919-962-1236. Fax: 919-962-1238.
E-mail: mhplanet@unc.edu
Web Site: www.moreheadplanetarium.org
Founded: 1949.
Congressional District: 4
Key Personnel: Dir., Todd Boyette; Dir. External Rels., Jeff Hill; Dir. Education Programs, Denise Young; Reservations, Adam Phelps; Retail, Ron Risch; Mktg. Mgr., Karen Kornegay; Dir. Finance, Susan Durham; Dir. Star Theater, Richard McColman; Camp Programs, Lindley Barrow; Dir. Outreach Programs, Crystal Harden.
Operating Expenses: 4,246,515
Operating Income: 4,331,760
Governing Authority: university. Parent Institution: University of North Carolina, Chapel Hill. Tax-exempt.
Institution Type/Description: Planetarium and Science Center.
Collections: hands-on exhibits; sundial.
Facilities: 222-seat planetarium with fulldome digital projection. Science-related items for sale.
Activities: lectures; classes for children & adults; permanent & temporary exhibitions; planetarium shows; skywatching sessions; summer camps; current science programs.
Publications: e-newsletter.
Hours & Admission Prices: See website for hours. Adults $7.25, students, children & senior citizens $6; members no charge. &
Attendance: 146,000 (estimated)
Membership: See website.

NORTH CAROLINA BOTANICAL GARDEN, 100 Old Mason Farm Rd., Chapel Hill, NC 27517. Mailing Address: The University of North Carolina at Chapel Hill, CB 3375, Chapel Hill, NC 27599-3375. Tel.: 919-962-0522. Fax: 919-962-3531.
E-mail: ncbg@unc.edu
Web Site: www.ncbg.unc.edu
Founded: 1952.
Congressional District: 17
Key Personnel: C.E.O. & Dir., Dr. Peter S. White; Pres. (V), Thomas W. Earnhardt, Ph.D.; Assoc. Dir. Devel., Charlotte Jones-Roe; Asst. Dir. Conservation, John L. Randall.
Personnel Profile: Full-Time Paid 25; Part-Time Paid 6; Part-Time Volunteers 225; Interns 10.
Governing Authority: state; university. Parent Institution: The University of North Carolina at Chapel Hill. Subsidiary Institution: Botanical Garden Foundation, Inc. Tax-exempt: 501(c)(3).
Institution Type/Description: Botanical Garden.
Collections: living native plants of the Southeastern United States; nature center; woody plants; herb garden; garden of flowering plant families; native perennial borders; accessible gardening demonstration raised beds; Coker Arboretum on UNC campus; carnivorous plants of the southeastern U.S.; herbarium with over 750,000 specimens.
Major Exhibits: Following in the Bartram's Footsteps: Contemporary Botanical Artists Explore the Bartran's Legacy (T), 8/30/14-11/2/14; Sculpture in the Garden, 9/14-12/14.
Research Fields: plant taxonomy, propagation & evolution; pollination biology; ecology.
Facilities: herbarium & botanical reference library; 1,000 acres; greenhouse; classrooms; auditorium; seminar room; nature trails. Museum-related items for sale.
Activities: guided tours; horticultural therapy services; lectures; formally organized education programs; plant rescues; educational trips within US and abroad; summer nature program for children; bimonthly botanical art exhibits; certificate programs.
Publications: quarterly newsletter.
Hours & Admission Prices: June-Aug. Mon.-Fri. 8-5, Sat. 9-6, Sun. 1-6; Sept.-May Mon.-Fri. 8-5, Sat. 9-5, Sun. 1-5. No charge; donations accepted. Closed New Year's Day; Martin Luther King Jr. Day; Thanksgiving; Christmas. &
Attendance: 100,000 (estimated)
Membership: Volunteer, Teacher & Senior $30; Individual $50; Family $75; Organization $100.

NORTH CAROLINA COLLECTION GALLERY, Wilson Library, 200 South Rd., Chapel Hill, NC 27514. Mailing Address: The University of NC at Chapel Hill, North Carolina Collection, Campus Box 3930, Chapel Hill, NC 27514-8890. Tel.: 919-962-0104.
E-mail: ljacobso@email.unc.edu
Web Site: www.lib.unc.edu/ncc/gallery.html
Founded: 1989.
Congressional District: 4
Key Personnel: Dir., Linda Jacobson.
Personnel Profile: Full-Time Paid 2; Interns 1.
Governing Authority: Parent Institution: UNC - Chapel Hill. Tax-Exempt.
Institution Type/Description: History Museum.

Collections: North Carolina & University history; Hayes Plantation furnishings; late 1500s & 1600s furnishings; personal artifacts; photographs; historic rooms.
Research Fields: numismatics.
Activities: temporary exhibits; programs; tours; K-12 activities.
Hours & Admission Prices: Mon.-Fri. 9-5, Sat. 9-1, Sun. 1-5. No charge. Closed major holidays.
Attendance: 13,000 (estimated)

PATTERSON'S MILL COUNTRY STORE, 5109 Farrington Rd., Chapel Hill, NC 27517. Tel.: 919-493-8149.
Institution Type/Description: History Museum.
Collections: 1870s country store history; early doctor's office furnishings; period pharmaceutical & commercial tobacco memorabilia.
Hours & Admission Prices: Call for hours.

Charlotte

BECHTLER MUSEUM OF MODERN ART, (M), 420 S. Tryon St., Charlotte, NC 28202. Tel.: 704-353-9200. Fax: 704-353-9299.
E-mail: info@bechtler.org
Web Site: bechtler.org
Founded: 2010.
Congressional District: 12
Key Personnel: C.E.O. & Pres., John Boyer; Chm. (V), Cyndee Paterson; Museum Shop Mgr., Patty Stevens.
Personnel Profile: Full-Time Paid 12; Part-Time Paid 10; Part-Time Volunteers 70; Interns 12.
Governing Authority: Tax-exempt.
Institution Type/Description: Art Museum.
Collections: 20th century works by American & European artists.
Facilities: classrooms; video gallery; cafe. Museum-related items for sale.
Activities: permanent & temporary exhibitions; guided tours; lectures; concerts; family days; teacher training; architecture film series; school & community outreach.
Hours & Admission Prices: Mon. & Wed.-Sat. 10-5, Sun. 12-5. Adults $8, seniors 65 & over, college students, & educators $6, youth 11-18 $4; children 10 & under and members no charge. Closed major holidays. &
Attendance: 48,269 (accurate)
Membership: Senior 65 & over $45; Individual $50; Family $90; Dual Seniors 65 & over $80; Supporting $250; Sustainer $500; Patron $1,000; Modernist $2,500; Modernist Benefactor $5,000; Modernist Leader $10,000.

BILLY GRAHAM LIBRARY, 4330 Westmont Dr., Charlotte, NC 28217-1001. Tel.: 704-401-3200.
Web Site: www.billygraham.org/library
Founded: 2007.
Institution Type/Description: Library & Historic Home: childhood home of Billy Graham has been restored & moved to this location.
Collections: Graham's family history; personal artifacts; photographs; films.
Facilities: library; cafe. Museum-related items for sale.
Hours & Admission Prices: Mon.-Sat. 9:30-5; call to confirm; groups of 15 or more by appointment. No charge. Closed New Year's Day; Thanksgiving; Christmas Eve & Day. &

CAROLINAS AVIATION MUSEUM, (M), 4672 First Flight Dr., Charlotte, NC 28208-5770. Tel.: 704-997-3770. Fax: 704-469-3193.
E-mail: info@carolinasaviation.org
Web Site: www.carolinasaviation.org
Formerly: Carolinas Historic Aviation Compassion
Founded: 1991.
Congressional District: 8
Key Personnel: Exec. Dir., Wally Coppinger; Pres. (V), Shawn Dorsch; Museum Shop Mgr. & Office Mgr., Donna Auer.
Personnel Profile: Full-Time Paid 5; Full-Time Volunteers 1; Part-Time Paid 2; Part-Time Volunteers 150.
Governing Authority: private; nonprofit organization. Tax-exempt: 501(c)(3).
Institution Type/Description: Aviation Museum.
Collections: focus on aviation history & artifacts pertaining to North Carolina and South Carolina from Wright Brothers first flight at Kitty Hawk, North Carolina, to present; World War II military aircraft; Piedmont Airlines DC-3 & two OV-1D Mohawk turbo-prop aircraft, all flyable & taken to airshow & aviation events; restored KC-97 cockpit; restored F-4 cockpit & fuselage; Airbus A320 Flight 1549 - Miracle on the Hudson; World War II aircraft service manuals & aircraft around the world; 10,000 aviation magazines; 1,000 pilot biographies.
Research Fields: restoration of aircrafts using research of military records that

indicate whether originally was Navy, Marine or Air Force, original paint scheme, nose art & tail numbers
Facilities: aviation library; 39,900 sq. ft. exhibit space. Aviation-related items for sale.
Activities: guided tours; temporary exhibitions; birthday packages; scout programs; view aircraft restoration; events in hangar.
Publications: monthly newsletter, Contact; hotel rack cards.
Hours & Admission Prices: Mon.-Fri. 10-4, Sat. 10-5, Sun. 1-5. Adults $11, senior citizens $9, students $7; discounts to groups & military; children 5 & over no charge. Closed Easter; Thanksgiving; Christmas. &
Attendance: 56,000 (accurate)
Membership: Individual $45; Family $70; Silver Wings $200; Gold Wings $500; Eagle Wings $1,000.

CHARLOTTE MUSEUM OF HISTORY, 200 E. 7th St., Charlotte, NC 28202-2508. Tel.: 704-568-1774. Fax: 704-566-1817.
E-mail: info@charlottemuseum.org
Web Site: www.charlottemuseum.org
Founded: 1976.
Congressional District: 9
Key Personnel: C.E.O. & Pres., Mary Davis Smart; Chm. (V), Mark Henriques; Exec. Dir., Kay Peninger; Museum Shop Mgr., Meredith Olan.
Personnel Profile: Full-Time Paid 16; Part-Time Paid 6; Part-Time Volunteers 50; Interns 7.
Governing Authority: Parent Institution: Charlotte Museum of History, Inc. Tax-exempt.
Institution Type/Description: History Museum.
Collections: furnishings from 1774-1820; regional artifacts & colonial to modern objects; archives; American Freedom Bell; hands-on exhibits Historic Building: 1774 rock house.
Research Fields: early history of the North Carolina Piedmont; genealogy of Alexander family; ancillary genealogy of Charlotte-Mecklenberg area.
Facilities: library. Gift items for sale.
Activities: guided tours; lectures; films; docent programs, loan, traveling & changing exhibitions. Annual Events: Revolutionary Charlotte Haunted Homesite; Twelfth Night; Colonial Fair; Independence Day Celebration.
Publications: newsletter.
Hours & Admission Prices: Museum: Tues.-Sat. 10-5, Sun. 1-5. Homesite Tours: Tues.-Sun. 1:15 & 3:15. Adult $6; discounts to senior citizens, AAA, AAM, ICOM, Time Travelers, Smithsonian Affiliates & groups with advance reservations; members no charge. &
Attendance: 54,928 (accurate)
Membership: Individual $35; Dual Senior $55; Household $65; Keystone $125; Patriot $250; Patron $500.

CHARLOTTE NATURE MUSEUM, (M), 1658 Sterling Rd., Charlotte, NC 28209-1599. Mailing Address: 301 N. Tryon St., Charlotte, NC 28202-2138. Tel.: 704-372-6261, ext. 300; 800-935-0553. Fax: 704-333-8948.
Web Site: www.discoveryplace.org
Founded: 1947.
Congressional District: 9
Key Personnel: C.E.O. & Pres., John L. Mackay; Chm. (V), Chris Perri; Exec. Vice Pres & C.O.O., Dean Briere; Vice Pres. Education, Robert Corbin.
Personnel Profile: Full-Time Paid 4; Part-Time Paid 6; Part-Time Volunteers 15; Interns 1.
Governing Authority: nonprofit organization; board of trustees. Parent Institution: Discovery Place, Inc., 301 N. Tryon St., Charlotte, NC 28202. Tax-exempt: 501(c)(3) & 170(b)(1)(A).
Institution Type/Description: Nature Museum.
Collections: teaching collection of biological material used with school classes; live collection of indigenous animals; live butterfly exhibit.
Facilities: early childhood teaching facility; nature center; puppet theater; classrooms, including outside teaching deck; 5 acres of nature trails; planetarium. Museum-related items for sale.
Activities: lectures; hobby workshops; formally organized educational programs; docent program or council; permanent & temporary exhibitions.
Publications: quarterly workshop bulletin.
Hours & Admission Prices: Tues.-Fri. 9-5, Sat. 10-5, Sun. 12-5. Admission $6; discounts to groups of 15 or more; children 2 & under and members no charge. Closed New Year's Day; Easter; Independence Day; Thanksgiving; Christmas Eve & Day. &
Attendance: 63,200 (accurate)
Membership: Individual $50; Family $100; Adventurer $250; Voyager $500; Explorer $1,000; Pioneer $2,500; Discoverer $5,000; Innovator $10,000.

CLAYWORKS GALLERY, 4506 Monroe Rd., Charlotte, NC 28205-7716. Tel.: 704-344-0795.
E-mail: adellinger@clayworksinc.org

Web Site: clayworksinc.org
Institution Type/Description: Art Gallery.
Collections: sculpture; pottery.
Activities: special events; classes; outreach programs; workshops. Annual
   Event: Open House in December.
Hours & Admission Prices: By appointment.

✻   **DISCOVERY PLACE, INC., (M),** 301 N. Tryon St., Charlotte,
   NC 28202-2138. Tel.: 704-372-6261. Fax: 704-337-2670.
E-mail: johnm@discoveryplace.org
Web Site: www.discoveryplace.org
Founded: 1981.
Congressional District: 12
Key Personnel: Pres. & C.E.O., John L. Mackay; Chm. (V), Chris Cecil; Vice
   Pres. Education & Programs, Robert Corbin; Vice Pres. Mktg. & Commu-
   nications, Debra Small; Dir. Devel., Julie Allen; Exec. Vice Pres. & C.O.O.,
   Dean Briere; Dir. Omnimax Theatre, Brian Hester; Dir. Human Resources
   & Volunteer Svcs., Ervin Gourdine; Dir. Sales, Christie Hussey; Collections
   Mgr., Dawn Cobb.
Personnel Profile: Full-Time Paid 81; Part-Time Paid 63; Part-Time Volunteers
   324.
Governing Authority: board of trustees; nonprofit organization. Tax-exempt:
   501(c)(3) & 170(b)(1)(A).
Institution Type/Description: Science & Technology Center, Omnimax Theater
   & Planetarium.
Collections: ethnological; geological; technological; archaeology; entomol-
   ogy; ethnology; herpetology; mineralogy; paleontology; malacology; ornithology; rainforest with tropical bird species.
Research Fields: entomology; ichthyology; archaeology; paleontology; history
   of technology.
Facilities: restaurant; aquarium; Omnimax; planetarium. Museum-related
   items for sale.
Activities: lectures; films; science festivals; formally organized education
   programs for children & adults; docent program; internship program; school
   classes offered for kindergarten through college; permanent & temporary
   exhibitions; school loan service; outreach program; touch pool.
Publications: quarterly workshop bulletin; quarterly newsletter; annual teach-
   er's guide; annual report.
Hours & Admission Prices: Mon.-Fri. 9-4, Sat. 10-6, Sun. 12-5. Adults $10,
   senior citizens 60 & over and youth 2-13 $8; children under 2 no charge.
   Closed Easter; Thanksgiving; Christmas Eve & Day. ♿
Attendance: 6,882,656 (accurate)
Membership: Individual $35; Family & Grandparent $75; Sponsor $125;
   Adventurer $250-$500; Voyager $501-$1,000; Explorer $1,001-$2,500;
   Pioneer $2,501-$5,000.

ELDER GALLERY, 1520 S. Tryon St., Charlotte, NC 28203. Tel.:
   704-370-6337.
E-mail: lelder@mindspring.com
Web Site: www.elderart.com
Key Personnel: Founder, Larry Elder
Institution Type/Description: Art Gallery.
Collections: works by national & international artists; paintings; sculptures.
Facilities: 6,500 sq. ft. exhibition space.
Activities: permanent & temporary artists.
Hours & Admission Prices: Wed.-Fri. 10-5:30, Sat. 10-2; other times by
   appointment.

**HARVEY B. GANTT CENTER FOR AFRICAN AMERICAN
   ARTS & CULTURE, (M),** 551 S. Tryon St., Charlotte, NC 28202.
   Tel.: 704-547-3700. Fax: 704-547-3770.
Web Site: www.ganttcenter.org
Formerly: Afro-American Cultural Center
Founded: 1974.
Congressional District: 12
Key Personnel: Pres. & C.E.O., David R. Taylor; Museum Shop Mgr., Priscilla
   Bivens.
Personnel Profile: Full-Time Paid 12; Part-Time Paid 1; Part-Time Volunteers
   200; Interns 2.
Governing Authority: private; nonprofit organization. Tax-exempt: 501(c)(3).
Institution Type/Description: Cultural Center: housed in an AIA award winning
   facility.
Collections: African & African American art.
Facilities: classroom; performance area; rental facilities.
Activities: visual arts; performance arts; arts education.
Publications: exhibition catalogs.
Hours & Admission Prices: Tues.-Sat. 10-5, Sun. 1-5. Adults $8, students,
   teachers, seniors & military $6; discounts to groups of 10 or more; children

5 & under and members no charge. Closed New Year's Day; Easter;
   Independence Day; Thanksgiving; Christmas. ♿
Attendance: 50,000 (estimated)
Membership: Teacher, Student & Senior Citizen $35; Individual $50; Profes-
   sional Investor $55; Family $100; Community Investor $150; Ebony
   Society Bronze $250; Ebony Society Silver $500; Organizational Investor
   $600; Corporate & Ebony Society Gold $1,000.

HISTORIC DOWD HOUSE, 2216 Monument St., Charlotte, NC
   28208. Tel.: 704-506-8036.
Institution Type/Description: Historic House: housed in the former Headquarters for Camp Greene Army Training Base during WWI; built in 1879.
Collections: period furnishings; WWI artifacts & equipment; photographs;
   personal artifacts.
Hours & Admission Prices: 3rd day each month by appointment.

HISTORIC ROSEDALE PLANTATION, 3427 N. Tryon St.,
   Charlotte, NC 28206-2052. Tel.: 704-335-0325. Fax: 704-335-
   0384.
E-mail: dhunter@historicrosedale.org
Web Site: historicrosedale.org
Founded: 1989.
Congressional District: 12
Key Personnel: Exec. Dir., Deborah A. Hunter; Pres., Pete Heynen; Vice Pres.,
   Woods Potts; Education Cur., Beth Harris.
Governing Authority: private; nonprofit organization. Parent Institution: His-
   toric Rosedale Foundation. Tax-exempt.
Institution Type/Description: General Museum.
Collections: period furniture; family items.
Research Fields: decorative arts of the Catawba River Valley; family history;
   African-American slaves on the plantation; mapping of area; use of herbs in
   medicinal practices; regional history.
Facilities: 150-vol. library; 400 sq. ft. exhibit space.
Activities: docent program; formal education programs for children; guided
   tours; lectures. Museum Sponsors: Spring Frolic With Dancing in the
   Gardens; Christmas at Rosedale; major guest speaker fundraiser.
Publications: quarterly newsletter, Historic Rosedale.
Hours & Admission Prices: Tours: Thurs.-Sun. 1:30 & 3; groups of 15 or more
   by appointment. Adults $10, senior citizens 65 & up and children 4-18 $8;
   members and children 3 & under with guardian no charge. Closed major
   holidays.
Attendance: 7,500 (accurate)
Membership: Colonial $35-$74; Federal $75 $149; Antebellum $150-$249;
   Reconstruction $250-$499; Millennium $500-$999; 21st Century $1,000 &
   up; 1815 Society (annually for 3 Years): Alice & Mary Louise Circle
   $1,000-$2,499; Caldwell Circle $2,500-$4,999; Davidson Circle $5,000 &
   up.

IMAGINON: THE JOE & JOAN MARTIN CENTER, 300 E.
   Seventh St., Charlotte, NC 28202. Tel.: 704-416-4600 & 973-2780.
Web Site: www.imaginon.org
Institution Type/Description: Children's Museum.
Collections: hands-on exhibitions.
Facilities: library; theater.
Activities: special events; educational programs; children's theater classes.
Hours & Admission Prices: Library: Tues.-Thurs. 10-7, Fri.-Sat. 10-5. Chil-
   dren's Theater: Mon.-Fri. 10-5.

JERALD MELBERG GALLERY, 625 S. Sharon Amity Rd.,
   Charlotte, NC 28211-2811. Tel.: 704-365-3000. Fax: 704-365-
   3016.
E-mail: gallery@jeraldmelberg.com
Web Site: www.jeraldmelberg.com
Founded: 1983.
Key Personnel: Dir., Jerald Melberg
Institution Type/Description: Art Gallery.
Collections: works by artists from the US, Argentina & Spain.
Major Exhibits: Thirty Years at Jerald Melberg Gallery, 1/14-2/14; Wolf Kahn,
   4/14-5/14; Brian Rutenberg, 9/14-10/14.
Hours & Admission Prices: Mon.-Sat. 10-6; other times by appointment. ♿

JOIE LASSITER GALLERY, 1440 S. Tryon St., Ste. 104, Char-
   lotte, NC 28203. Tel.: 704-373-1464.
E-mail: info@lassitergallery.com
Web Site: www.lassitergallery.com
Institution Type/Description: Art Gallery.
Collections: works by regional, national & international artists.

Hours & Admission Prices: Tues.-Fri. 10-5:30, Sat. 11-4.

LEVINE MUSEUM OF THE NEW SOUTH, (M), 200 E. Seventh St., Charlotte, NC 28202-2508. Tel.: 704-333-1887. Fax: 704-333-1896.
E-mail: ezimmern@museumofthenewsouth.org
Web Site: www.museumofthenewsouth.org
Founded: 1991.
Congressional District: 12
Key Personnel: Pres. & C.E.O., Emily F. Zimmern; Chm., Jamie McLawhorn; Treas., Carl Mallio.
Personnel Profile: Full-Time Paid 15; Part-Time Paid 1; Part-Time Volunteers 20; Interns 4.
Governing Authority: nonprofit organization. Tax-exempt: 501(c)(3).
Institution Type/Description: History Museum.
Collections: history of American South from the Civil War to present.
Research Fields: local history.
Facilities: education & social program areas.
Activities: formal educational programs for adults, children & students; guided tours; lectures; loan & participatory exhibitions; training programs for professional museum workers; oral & multi-media history presentation.
Publications: newsletter; book.
Hours & Admission Prices: Mon.-Sat. 10-5, Sun. 12-5. Adults $6; discounts to AAM & ICOM members; members no charge. Closed New Year's Day; Memorial Day; Independence Day; Labor Day; Thanksgiving; Christmas.
Attendance: 57,000 (estimated)
Membership: Individual $35; Household $60; Donor $150; Sustainer $250; Benefactor $500; Visionary $1,000; Connector $2,500; Innovator $5,000; Catalyst $10,000.

THE LIGHT FACTORY - CONTEMPORARY MUSEUM OF PHOTOGRAPHY & FILM, 345 N. College St., Charlotte, NC 28202. Tel.: 704-333-9755. Fax: 704-333-5910.
E-mail: info@lightfactory.org
Web Site: www.lightfactory.org
Founded: 1972.
Congressional District: 12
Key Personnel: Exec. Dir., Marcie Kelso; Chm., Liam Stokes; Treas., Charles Moore; Chief Cur., Dennis Kiel.
Personnel Profile: Full-Time Paid 8; Part-Time Paid 1; Part-Time Volunteers 150; Interns 5.
Governing Authority: private; nonprofit organization. Tax-exempt: 501(c)(3).
Institution Type/Description: Art Museum.
Collections: light-generated media including photography, video & film.
Research Fields: contemporary photography; interactive new media.
Facilities: 500-vol. library of art & photography books available to the public; 3,500 sq. ft. exhibit space; classroom; darkroom.
Activities: educational programs; films; guided tours; hobby workshops; lectures; traveling exhibitions; community outreach; special events.
Publications: quarterly newsletter; quarterly calendar; catalogues.
Hours & Admission Prices: Mon.-Sat. 9-6, Sun. 1-6. No charge; donations accepted. Closed New Year's Day; Easter; Independence Day; Thanksgiving; Christmas; major holidays.
Attendance: 70,000 (estimated)
Membership: Basic: Students $25; Individual $50; Paparazzi $70; Household $75. Supporting: Snapshot $100-$249; Aerial $250-$499; Panoramic $500-$999. Lumiere Patron: $1,000-$2,499; Silver Lumiere $2,500-$4,999; Gold Lumiere $5,000-$9,999; Grand Lumiere $10,000-$24,999.

MCCOLL FINE ART GALLERY, 126 Cottage Pl., Charlotte, NC 28207-2210. Tel.: 704-333-5983. Fax: 704-333-5816.
E-mail: info@mccollfineart.com
Web Site: www.mccollfineart.com
Institution Type/Description: Art Gallery.
Collections: paintings.
Activities: special events.
Hours & Admission Prices: Tues.-Fri. 10-6, Sat. 10-3; other times by appointment. Call to confirm hours.

MCDOWELL NATURE CENTER AND PRESERVE, 15222 York Rd., Charlotte, NC 28278. Tel.: 704-588-5224. Fax: 704-588-5226.
Institution Type/Description: Nature Center.
Collections: over 119 species of birds; 21 species of mammals; 21 species of reptiles; 14 species of amphibians; plants; flowers; trees.
Facilities: 1,115 acres; nature trails.
Activities: educational programs; hiking trails.
Hours & Admission Prices: Nature Center: Mon.-Sat. 9-5, Sun. 1-5. Nature Preserve: daily 7am to sunset. No charge.

MINT MUSEUM OF RANDOLPH, 2730 Randolph Rd., Charlotte, NC 28207-2031. Tel.: 704-337-2000. Fax: 704-337-2101.
E-mail: info@mintmuseum.org
Web Site: www.mintmuseum.org
Formerly: Mint Museum of Art
Founded: 1936.
Congressional District: 9
Key Personnel: Exec. Dir., Kathleen V. Jameson; Chm., Richard T. Williams; C.F.O., Mike Smith; Chief Cur. Fine Art, Charles Mo; Dir. Education, Cheryl Palmer; Chief Cur. Craft & Design, Annie Carlano; Cur. American Art, Jonathan Stuhlman; Cur. Contemporary Art, Carla Hanzal; Cur. Decorative Arts, Brian D. Gallagher; Head Design & Installation, Kurt Warnke; Registrar, Martha T. Mayberry; Librarian, Joyce Weaver; Museum Shop Mgr., Sandy Fisher; Facility Mgr., Hank McKiernan.
Personnel Profile: Full-Time Paid 44; Part-Time Paid 35; Part-Time Volunteers 2,000; Interns 20.
Governing Authority: nonprofit organization. Parent Institution: The Mint Museums. Tax-exempt.
Institution Type/Description: Art Museum: housed in 1835 first branch of the U.S. Mint, expanded in 1967 & 1985.
Collections: decorative arts; art of the ancient Americas; historic costumes & fashionable dress.
Research Fields: European, American & pre-Columbian art; Delhom Gallery devoted to research & study in ceramics.
Facilities: 15,000-vol. library of art books & periodicals available for use on premises; 180-seat auditorium. Museum-related items for sale.
Activities: guided tours; lectures; gallery talks; concerts; formally organized educational programs; docent program; inter-museum loan, permanent, temporary & traveling exhibitions.
Publications: catalogs of major exhibitions; books, North Carolina Pottery; Experience Art at the Mint Museums: A Look at the Collections; Experiencing Art of The Mint Museums: A Look at the Collections; North Carolina Pottery: The Collection of The Mint Museum; Passionate Journey: The Grice Collection of Native American Art; Lois Mailou Jones: A Life in Vibrant Color.
Hours & Admission Prices: Tues. 10-9, Wed.-Sat. 10-6, Sun. 1-5. Adults $10, seniors & college students $8, children 5-17 $5; members and children 4 & under no charge. Closed major holidays.
Attendance: 118,000 (estimated)
Membership: Individual $60; Dual $80; Family $100; Sustainer $250; Benefactor $500; Silver Circle $1,200.

✴ MINT MUSEUM UPTOWN, (M), 500 S. Tryon St., Charlotte, NC 28202. Tel.: 704-337-2000. Fax: 704-337-2101.
E-mail: info@mintmuseum.org
Web Site: www.mintmuseum.org
Formerly: Mint Museum of Craft + Design
Founded: 2010.
Congressional District: 9
Key Personnel: Exec. Dir., Kathleen V. Jameson; Chm. (V), Richard T. Williams; Chief Cur. Craft + Design, Annie Carlano; Chief Cur. Fine Art, Charles Mo; Cur. Contemporary Art, Carla Hanzal; Cur. American Art, Jonathan Stuhlman; Cur. Decorative Arts, Brian D. Gallagher; Dir. Education, Cheryl Palmer; Facility Mgr., Hank McKiernan; Head Design & Installation, Kurt Warnke; Registrar, Martha T. Mayberry; Librarian, Joyce Weaver; Museum Shop Mgr., Sandy Fisher.
Personnel Profile: Full-Time Paid 44; Part-Time Paid 35; Part-Time Volunteers 2,000; Interns 20.
Governing Authority: nonprofit organization. Parent Institution: The Mint Museum. Tax-exempt.
Institution Type/Description: Arts & Crafts Museum.
Collections: American art; contemporary art; craft & design, European art.
Facilities: 240-seat auditorium; cafe. Museum-related items for sale.
Activities: guided tours; gallery talks; lectures; artist demonstrations; temporary & traveling exhibitions.
Publications: Allan Chasanoff Ceramic Collection; Constructing Elozua: A Retrospective in 2004; The Nature of Craft and the Penland Experience; Spectrum: The Sculpture of John Kuhn; Contemporary British Studio Ceramics: The Grainer Collections; American Quilt Classics, 1800-1980: The Bresler Collection.
Hours & Admission Prices: Tues. 10-9, Wed.-Sat. 10-6, Sun. 1-5. Adults $10, seniors & college students $8, children 5-17 $5; children 4 & under and members no charge. Closed major holidays.
Membership: Individual $60; Dual $80; Family $100; Sustainer $250; Benefactor $500; Silver Circle $1,200.

NASCAR HALL OF FAME, 400 E. Martin Luther King, Jr. Blvd., Charlotte, NC 28202. Tel.: 704-654-4400. Facebook: NASCAR Hall of Fame.
Web Site: www.nascarhall.com
Founded: 2010.
Key Personnel: Dir., Winston Kelley; C.E.O., Tom Murray.
Personnel Profile: Full-Time Paid 26; Part-Time Paid 80; Interns 5.
Governing Authority: Parent Institution: Charlotte Regional Visitors Authority.
Institution Type/Description: Sports Museum & Hall of Fame.
Collections: racing history & heritage; race cars & memorabilia; photographs; personal artifacts; Hall of Fame inductees.
Facilities: 278- seat theater; restaurant. Museum-related items for sale.
Activities: racing simulator; special events.
Hours & Admission Prices: Daily 10-6. Adult $19.95, military & seniors $17.95, children $12.95. &
Attendance: 274,000 (accurate)
Membership: $25-$1,000.

PROVIDENCE GALLERY, 601-A Providence Rd., Charlotte, NC 28207. Tel.: 704-333-4535.
E-mail: providenceframes@bellsouth.net
Web Site: www.providencegallery.net
Institution Type/Description: Art Gallery.
Collections: works by local, regional, & national artists; paintings.
Activities: special events.
Hours & Admission Prices: Call for hours.

PUBLIC LIBRARY OF CHARLOTTE AND MECKLENBURG COUNTY, 310 N. Tryon St., Charlotte, NC 28202-2139. Tel.: 704-416-0100.
Web Site: www.plcmc.org
Founded: 1903.
Congressional District: 9
Key Personnel: Mgr., Patrice Ebert.
Governing Authority: nonprofit. Tax-exempt: 170(b)(1)(A).
Institution Type/Description: Public Library.
Collections: books; films.
Research Fields: local history; genealogy.
Facilities: library; reading room; 200-seat auditorium.
Activities: guided tours; films; local artist gallery.
Publications: Friends of Library newsletter; Business Information newsletter; bibliographies; PLCMC News; Annual Report.
Hours & Admission Prices: Tues.-Thurs. 10-7, Fri.-Sat. 10-5, Sun. 1-5. No charge. Closed national holidays. &

RENEE GEORGE GALLERY, 2839 Selwyn Ave., Ste. Z, Charlotte, NC 28209. Mailing Address: 401 Eastover Rd., Charlotte, NC 28207-2351. Tel.: 704-332-3278. Fax: 704-332-3277.
E-mail: info@reneegeorgegallery.com
Web Site: www.reneegeorgegallery.com
Institution Type/Description: Art Gallery.
Collections: works by national & international artists; paintings; sculpture.
Hours & Admission Prices: Tues.-Fri. 10-3, Sat. 11-4.

SHAIN GALLERY, 2823 Selwyn Ave., Charlotte, NC 28209. Tel.: 704-334-7744.
E-mail: shainart@earthlink.net
Web Site: www.shaingallery.com
Institution Type/Description: Art Gallery.
Collections: works by regional & national artists; paintings; sculpture.
Hours & Admission Prices: Tues.-Sat. 10-5.

SPIRIT SQUARE CENTER FOR ARTS AND EDUCATION, 345 N. College St., Charlotte, NC 28202-2113. Tel.: 704-372-1000.
Web Site: www.blumenthalcenter.org
Key Personnel: Pres., Tom Gabbard; Public Rels. & Media, Kathy Scott Rummage.
Personnel Profile: Full-Time Paid 84; Part-Time Paid 17; Part-Time Volunteers 447; Interns 9.
Institution Type/Description: Art Gallery.
Collections: paintings.
Facilities: 720-seat theater.
Activities: classes.
Hours & Admission Prices: Call for hours.
Attendance: 298,729 (estimated)

UNC CHARLOTTE BOTANICAL GARDENS, 9201 University City Blvd., Charlotte, NC 28223. Tel.: 704-687-0720 & 0721.
Founded: 1966.
Personnel Profile: Full-Time Paid 4; Part-Time Paid 3; Part-Time Volunteers 4.
Governing Authority: Parent Institution: UNC Charlotte. Tax-exempt.
Institution Type/Description: Botanical Gardens.
Collections: native wildflowers; trees; rhododendrons; Winter garden; Asian garden; Water garden; orchids; desert plants.
Activities: workshops; guided tours; educational programs.
Hours & Admission Prices: Gardens: daily dawn to dusk. Greenhouse: Mon.-Sat. 10-3, Sun. 1-4. No charge; donations accepted.
Attendance: 20,000 (estimated)
Membership: Individual $25; Family $40.

WING HAVEN, 248 Ridgewood Ave., Charlotte, NC 28209-1632. Tel.: 704-331-0664. Fax: 704-331-9368.
E-mail: winghavengardens@carolina.rr.com
Web Site: www.winghavengardens.com
Founded: 1927.
Congressional District: 9
Key Personnel: Interim Dir., Sherry Hall; Cur. Garden, Jeffrey Drum.
Personnel Profile: Full-Time Paid 5; Part-Time Paid 8; Part-Time Volunteers 400.
Governing Authority: Parent Institution: Wing Haven Foundation, Inc. Tax-exempt.
Institution Type/Description: Arboretum.
Collections: living plants; garden statuary.
Facilities: nature conservation center; gardens. Plants & garden-related items for sale.
Activities: guided tours; educational programs; musical performances.
Publications: newsletter: Wing Haven; Birds of Charlotte and Mecklenburg County; Verses from the Garden; Guide to Wing Haven; A Bird in the House.
Hours & Admission Prices: Tues. 3-5, Wed. 10-12, Sat. 10-5; groups by appointment. Adults $6; members no charge. Closed major holidays. &
Attendance: 11,500 (accurate)
Membership: Individual $45; Family $75; Friend $150; Patron $250; Sponsor $500; Benefactor $1,000; Founder $2,500; Clarkson Society $5,000.

Cherokee

CHEROKEE HISTORICAL ASSOCIATION, 564 Tsali Blvd., Cherokee, NC 28719. Mailing Address: P.O. Box 398, Cherokee, NC 28719. Tel.: 828-497-2111; 866-554-4557. Fax: 828-497-6987.
E-mail: cherokeehistorical.info@gmail.com
Web Site: www.cherokeehistorical.org
Key Personnel: Exec. Dir., John Tissue; Village Dir., Laura Blythe; Opers. Mgr., Christopher McCoy; Program Coord., Philenia Walkingstick
Institution Type/Description: History Museum: re-created 18th century Oconaluftee Indian Village.
Collections: Cherokee history, heritage & culture; early tools and artifacts; replica Cherokee homes, council house, & square ground.
Facilities: Museum-related items for sale.
Activities: live reenactments; interactive demonstrations; children's pottery classes.
Hours & Admission Prices: Tours: May to mid-Oct. Mon.-Sat. 9-4. Adults $18, children $14; children 5 & under no charge.

MOUNTAIN FARM MUSEUM-GREAT SMOKY MOUNTAINS NATIONAL PARK, 1194 Newfound Gap Rd., Cherokee, NC 28719-8249. Tel.: 828-497-1900 & 1904. Fax: 828-497-1910.
E-mail: grsm_smokies_information@nps.gov
Web Site: nps.gov/grsm
Founded: 1953.
Congressional District: 11
Key Personnel: Site Supvr., Lynda Doucette.
Personnel Profile: Full-Time Paid 1; Part-Time Volunteers 6.
Governing Authority: federal. Parent Institution: Great Smoky Mountains National Park Service, Gatlinburg, TN 37738; Tel. 865-436-1256; 436-1200 (automated information). Tax-exempt.
Institution Type/Description: Park Visitor Center.
Collections: 10 historic farm buildings, including dwellings moved to current location from throughout Great Smoky Mountains National Park.
Facilities: Postcards, slides & books for sale.
Activities: permanent exhibitions.
Publications: self guiding booklet, Mountain Farm Museum.
Hours & Admission Prices: Call for hours. No charge; donations accepted. Closed Christmas. &
Attendance: 100,000 (estimated)

MUSEUM OF THE CHEROKEE INDIAN, 589 Tsali Blvd., Cherokee, NC 28719. Mailing Address: P.O. Box 1599, Cherokee, NC 28719-1599. Tel.: 828-497-3481. Fax: 828-497-4985.
E-mail: littlejohn@cherokeemuseum.org
Web Site: www.cherokeemuseum.org
Founded: 1948.
Congressional District: 11
Key Personnel: Dir., Ken Blankenship; Administrative Mgr. & Museum Shop Mgr., Sharon Littlejohn; Facilities Mgr., Driver Pheasant, Jr.; Financial, Wendy McRay.
Personnel Profile: Full-Time Paid 12.
Governing Authority: nonprofit organization. Owned by the Eastern Band of Cherokee Indians. Tax-exempt: 501(c)(3).
Institution Type/Description: Cherokee Indian Museum.
Collections: Indian artifacts; relics; archives; 16th century Cherokee homestead exhibit.
Research Fields: Cherokee Culture; history; language.
Facilities: 3,000-vol. rare document reference library.
Activities: self-guided tours; interactive exhibit.
Publications: annual, Journal of Cherokee Studies.
Hours & Admission Prices: Memorial Day to Labor Day Mon.-Sat. 9-7, Sun. 9-5; Sept.-May daily 9-5. Adults $9, children 6-14 $6; complimentary subscription to Journal for members; discounts for AAA & AARP members Jan.-May, groups & AAM members; members & children 5 & under no charge. Closed New Year's Day; Thanksgiving; Christmas. &
Attendance: 150,000
Membership: Individual & Institution $30; Family $50; Friend $100; Clan $250; Tsali $500; Chief $1,000.

Cherryville

C. GRIER BEAM TRUCK MUSEUM, 111 N. Mountain St., Cherryville, NC 28021-2940. Mailing Address: P.O. Box 238, Cherryville, NC 28021-0238. Tel.: 704-435-3072.
E-mail: info@beamtruckmuseum.com
Web Site: www.beamtruckmuseum.com
Founded: 1982.
Congressional District: 10
Key Personnel: C.E.O. & Chm. (V), Michael N. Beam; Pres., Sandra B. Dismukes; Museum Shop Mgr., Joseph C. Dismukes.
Personnel Profile: Full-Time Paid 2; Part-Time Paid 4; Part-Time Volunteers 6.
Governing Authority: board of directors. Tax-exempt.
Institution Type/Description: Transportation Museum.
Collections: 14 trucks; history of Carolina trucking company.
Facilities: Museum-related items for sale.
Activities: Museum Sponsors: Annual Spring Antique Car Show; Antique Farm Tractor Show in August; Open House during Christmas.
Hours & Admission Prices: Thurs. & Sat. 10-3, Fri. 10-5. No charge; donations accepted. Closed New Year's Day; Independence Day; Christmas Eve & Day. &
Attendance: 3,000 (estimated)
Membership: Annual $15.

CHERRYVILLE HISTORICAL MUSEUM, 109 E. Main St., Cherryville, NC 28021-3406. Mailing Address: P.O. Box 307, Cherryville, NC 28021-0307. Tel.: 704-435-8011.
Founded: 1992.
Key Personnel: Dir., Pat Sherrill; Pres. (V), Darrell Carpenter; Museum Shop Mgr., Katherine Goins.
Personnel Profile: Part-Time Volunteers 10.
Governing Authority: Parent Institution: Cherryville Historical Assoc., Inc. Tax-exempt.
Institution Type/Description: Historic Building: built in 1911.
Collections: local history & culture; fire truck; photographs; art; medical instruments; farm equipment; recreational artifacts; hobby items; toys; games; period artifacts.
Facilities: library. Museum-related items for sale.
Hours & Admission Prices: Sat. 10-2. No charge; donations accepted.
Attendance: 1,200 (estimated)

Clayton

CLEMMONS EDUCATIONAL STATE FOREST, 2411 Old U.S. 70 W., Clayton, NC 27520. Tel.: 919-553-5651.
E-mail: clemmonsesf.ncfs@ncagr.gov
Web Site: www.ncesf.org
Institution Type/Description: State Forest.
Collections: forest ecology.
Facilities: nature trails; camp sites.

Activities: nature trails; video; educational programs; classes; Talking Trees; picnic area; camping; workshops.
Hours & Admission Prices: mid-March to June & Sept. to mid-Nov. Tues.-Fri. 9-5, Sat.-Sun. 11-5; July-Aug. Tues.-Fri. 9-5, Sat.-Sun. 11-8.

Clemmons

ARBORETUM AT TANGLEWOOD, 4201 Manor House Cir., Clemmons, NC 27012. Mailing Address: 4061 Clemmons Rd., Clemmons, NC 27012. Tel.: 336-703-6400.
Institution Type/Description: Arboretum.
Collections: native flora & herbs; rose garden; AIDS Memorial Garden.
Facilities: rental facilities; nature trails.
Activities: rental facility; special events; nature education programs; nature trails.
Hours & Admission Prices: 7am to sunset. Entrance Fee: $2 per person. Closed Christmas.

Clinton

SAMPSON COUNTY HISTORY MUSEUM, 313 Lisbon St., Clinton, NC 28328. Mailing Address: P.O. Box 786, Clinton, NC 28329-0786. Tel.: 910-590-0007. Fax: 910-590-0007.
E-mail: schm@intrstar.net
Web Site: www.sampsonhmc.com
Founded: 1997.
Congressional District: 2
Key Personnel: Dir., Jeannie King; Pres. (V), David King.
Personnel Profile: Full-Time Paid 1; Full-Time Volunteers 1; Part-Time Volunteers 10.
Governing Authority: Tax-exempt.
Institution Type/Description: History Museum.
Collections: local history, heritage & culture; period furnishings; personal artifacts; photographs; Confederate artifacts. Historic Buildings: Holmes House; Bunting Log Cabin; 1920s Wooten Country Store; blacksmith shop; Law Enforcement Museum; Agriculture Grange Museum; Military Museum; Doctors Museum.
Activities: Museum Sponsors: Craft Demonstration Day in October.
Hours & Admission Prices: Tues.-Sat. 10-4. No charge; donations accepted. &
Attendance: 5,000 (accurate)

Columbia

COLUMBIA THEATER CULTURAL RESOURCES CENTER, 304 Main St., Columbia, NC 27925. Mailing Address: P.O. Box 55, Columbia, NC 27925. Tel.: 252-766-0200; 888-737-0437. Fax: 252-766-0202.
E-mail: ctcrcenter@embarqmail.com
Founded: 1998.
Institution Type/Description: Historic Building; housed in the former Columbia Theater built in 1938 by German immigrant Fred Schlez.
Collections: theater history & memorabilia; environmental & cultural history; fishing; farming; forestry.
Facilities: rental facilities.
Hours & Admission Prices: Wed.-Fri. 10-4.

POCOSIN LAKES NATIONAL WILDLIFE REFUGE - WALTER B. JONES, SR. CENTER FOR THE SOUNDS, 205 S. Ludington Dr., Columbia, NC 27925. Mailing Address: P.O. Box 55, Columbia, NC 27925. Tel.: 252-796-3004 & 3008. Fax: 252-796-3196.
E-mail: pocosinlakes@fws.gov
Web Site: www.fws.gov/pocosinlakes
Institution Type/Description: Wildlife Refuge & Visitor Center.
Collections: wildlife & their habitats; photographs; local history.
Facilities: nature trails; visitor center. Gift items for sale.
Activities: special events; film; nature trails. Annual Event: Christmas at The Bear's Den.
Hours & Admission Prices: Refuge: daily daylight hours. Center: May-Sept. Tues.-Sat. 10-4; Oct.-April Wed.-Sat. 10-4.
Attendance: 63,000

Columbus

HOUSE OF FLAGS MUSEUM, 363 Green Creek Dr., Columbus, NC 28722. Mailing Address: P.O. Box 1090, Columbus, NC 28722. Tel.: 828-894-5640.
E-mail: flagmuseum@gmail.com

Institution Type/Description: Flag History Museum.
Collections: over 300 flags including all 27 official US flags; state & territory flags; US military flags, Veteran's flags; religious flags; pre-Colonial & Colonial flags.
Activities: educational video programs.
Hours & Admission Prices: March-Nov. Sat. 2-4; other times by appointment. No charge; donations accepted.

POLK COUNTY HISTORICAL ASSOCIATION MUSEUM, 60 Walker St., Columbus, NC 28722. Mailing Address: P.O. Box 503, Columbus, NC 28722-0503. Tel.: 828-894-3351.
Web Site: www.polkcounty.org
Founded: 1977.
Key Personnel: Pres. (V), Anne Conner.
Governing Authority: Tax-exempt.
Institution Type/Description: History Museum.
Collections: Cherokee Indian artifacts; early settlers; Revolutionary War; Civil War; WWI & WWII; settlers' tools & clothing; railroad memorabilia; personal artifacts; photographs; paintings.
Activities: local & regional history programs on 1st Tues. of the month.
Publications: biannual newsletter.
Hours & Admission Prices: Tues. & Thurs. 10-1, Sat. 10-4; other times by appointment. No charge; donations accepted. &
Attendance: 1,500 (estimated)
Membership: Individual $15-$25; Corporate $50; Lifetime $150.

Concord

BACKING UP CLASSICS AUTO MUSEUM, 4545 Concord Pkwy. S., Concord, NC 28027-4618. Tel.: 704-788-9500.
E-mail: info@morrisonmotorco.com
Web Site: www.morrisonmotorco.com
Key Personnel: Dir., Lindsay Morrison Hartman
Institution Type/Description: Classic Car Museum.
Collections: over 50 vehicles including classics, 50s, 60s, & period cars; photographs.
Facilities: 18,000 sq. ft. exhibit space. Museum-related items for sale.
Activities: rental facilities.
Hours & Admission Prices: Mon.-Tues. & Thurs.-Sat. 10-5.

BOST GRIST MILL, 4701 Hwy. 200, Concord, NC 28025-8170. Tel.: 704 782 1600.
E-mail: bostgristmill@netzero.com
Web Site: www.bostgristmill.com
Institution Type/Description: Historic Grist Mill.
Collections: mill history; working grist mill.
Facilities: Mill-related items for sale.
Activities: guided tours; special events.
Hours & Admission Prices: By appointment. &

CHARLOTTE MOTOR SPEEDWAY, 5555 Concord Pkwy., S., Concord, NC 28027. Mailing Address: P.O. Box 600, Concord, NC 28026. Tel.: 704-455-3200; 800-455-3267.
E-mail: tickets@charlottemotorspeedway.com
Web Site: www.charlottemotorspeedway.com
Institution Type/Description: Sports Museum.
Collections: Charlotte Motor Speedway history; NASCAR Sprint Cup Series.
Facilities: restaurant. Gift items for sale.
Activities: guided tours; navigate through two infield race tracks & Pit Road; take a photograph in Victory Circle; visit zMAX Dragway & The Dirt Track.
Hours & Admission Prices: Feel the Thrill Speedway Tour: Mon.-Sat. 9:30, 10:30, 11:30, 1:30, 2:30 & 3:30, Sun. 1:30, 2:30 & 3:30. Tours not available on race days. Adults $12, seniors 55 & over, military, EMS/Fire/Police personnel and children 9 & under $10. Over the Wall Tour: Mon.-Sat. 10:30 & 1:30, Sun. groups only. Adults $20, seniors 55 & over, military, EMS/Fire/Police personnel and children 9 & under $18.

HENDRICK MOTORSPORTS MUSEUM, 4400 Papa Joe Hendrick Blvd., Concord, NC 28262. Tel.: 704 455 3400; 877 467 4890 (Toll Free).
Institution Type/Description: Sports Museum.
Collections: motorsports history; Mark Martin, Jeff Gordon, Jimmie Johnson, & Dale Earnhardt, Jr. memorabilia; photographs; Chevrolets; racing competition memorabilia; personal artifacts.
Hours & Admission Prices: Mon.-Fri. 9-5, Sat. 10-4. No charge.

ROUSH FENWAY RACING MUSEUM, 4600 Roush Pl., N.W., Concord, NC 28027.
E-mail: roushracing@roushracing.com
Institution Type/Description: Auto Racing Museum.
Collections: Jack Roush's life & career; photographs; trophies; race vehicles; personal artifacts.
Facilities: Museum-related items for sale.
Hours & Admission Prices: Mon.-Fri. 8-5.

SAM BASS GALLERY, 6104 Performance Dr., Concord, NC 28027-3435. Tel.: 800-556-5464. Fax: 704-455-6916.
E-mail: info@sambass.com
Web Site: www.sambass.com
Institution Type/Description: Art Gallery.
Collections: over 300 works of art; NASCAR art; paintings; posters; prints.
Hours & Admission Prices: Tues.-Fri. 10-12:30 & 1:30-5; groups by appointment.

Conover

CATAWBA COUNTY FIREFIGHTER'S MUSEUM, 3957 Herman Sipe Rd., Conover, NC 28613. Mailing Address: 100 A S.W. Blvd., Newton, NC 28658. Tel.: 828-466-0911 & 465-8238.
E-mail: mpettit@catawbacountync.gov
Institution Type/Description: Fire-Fighting Museum.
Collections: fire-fighting history & equipment; photographs; plaques; six fire trucks; communication equipment; 1906 alarm system; lights & sirens; uniforms; helmets; fire extinguishers; personal artifacts; Firefighter Memorial.
Facilities: classroom.
Activities: lectures; videos.
Hours & Admission Prices: Sat. 11-4, Sun. 1-4; other times by appointment.

Cooleemee

MILL HOUSE MUSEUM, 163 Cross St., Cooleemee, NC 27014. Mailing Address: P.O. Box 667, Cooleemee, NC 27014-0667. Tel.: 336-284-6040.
E-mail: blinky1@yadtel.net
Founded: 2006.
Congressional District: 5
Key Personnel: Dir., Lynn Rumley; Pres. (V), Tony Steele.
Personnel Profile: Full-Time Paid 1; Part-Time Volunteers 16.
Governing Authority: Parent Institution: Cooleemee Historical Assoc. Tax-exempt.
Institution Type/Description: History Museum.
Collections: local history & culture; period furnishings; photographs; personal artifacts.
Research Fields: local history; southern cultures.
Activities: guided tours; training; children's programs; festival.
Publications: Cooleemee History Loom.
Hours & Admission Prices: Guided Tours by appointment. Adults $4, seniors $3. &
Attendance: 3,000 (estimated)
Membership: Individual $15.

TEXTILE HERITAGE MUSEUM, 131 Church St., Cooleemee, NC 27014. Mailing Address: P.O. Box 667, Cooleemee, NC 27014-0667. Tel.: 336-284-6040.
E-mail: blinky1@yadtel.net
Web Site: www.textileheritage.org
Founded: 1995.
Congressional District: 5
Key Personnel: Dir., Lynn Rumley; Pres. (V), Tony Steele.
Personnel Profile: Full-Time Paid 1; Part-Time Volunteers 16.
Governing Authority: Parent Institution: Cooleemee Historical Assoc. Tax-exempt.
Institution Type/Description: History Museum.
Collections: local history & culture; period furnishings; photographs; archives.
Research Fields: local history; textiles; southern cultures.
Facilities: archives.
Activities: guided tours; training; children's programs; festival.
Publications: Cooleemee History Loom.
Hours & Admission Prices: Wed.-Sat. 10-4. No charge; donations accepted. &
Attendance: 5,000 (estimated)
Membership: Individual $15.

Cornelius

THE COMMUNITY ARTS PROJECT GALLERY, 20700 N. Main St., Unit 112, Cornelius, NC 28031. Tel.: 704-896-8980.
Institution Type/Description: Art Gallery.
Collections: ceramics; sculptures; paintings.
Activities: special events; birthday parties; classes; outreach programs.
Hours & Admission Prices: Mon.-Thurs. 9-5, Fri. 9-12.

MICHAEL WALTRIP RACING HEADQUARTERS, 20310 Chartwell Center Dr., Cornelius, NC 28031. Tel.: 704-897-5555 & 655-9550.
Web Site: michaelwaltripracing.com
Institution Type/Description: Sports Museum.
Collections: 3 NASCAR Nextel Cup teams; one Nascar Busch Series team.
Facilities: Gift items for sale.
Activities: tours; view car assembly & fabrication areas and pit crew practices.
Hours & Admission Prices: Mon.-Fri. 9-5.

Corolla

CURRITUCK BEACH LIGHTHOUSE, 1101 Corolla Village Rd., Corolla, NC 27927. Mailing Address: P.O. Box 58, Corolla, NC 27927. Tel.: 252-453-4939.
E-mail: info@currituckbeachlight.com
Web Site: www.currituckbeachlight.com
Institution Type/Description: Historic Lighthouse: built in 1875.
Collections: lighthouse history & artifacts; Fresnel lens; shipwrecks; lighthouse keepers.
Facilities: Museum-related items for sale.
Activities: tours.
Hours & Admission Prices: late March to May & Sept. to late Nov. daily 9-5; Memorial Day to Labor Day Wed.-Thurs. 9-8, Fri.-Tues. 9-5. Admission $7; children 7 & under no charge. Closed Thanksgiving.

THE WHALEHEAD CLUB, (M), 1100 Club Rd., Corolla, NC 27927. Mailing Address: P.O. Box 307, Corolla, NC 27927-0307. Tel.: 252-453-9040. Fax: 252-457-0129.
E-mail: director@whaleheadclub.com
Web Site: whaleheadclub.org
Founded: 1992.
Congressional District: 3
Key Personnel: Pres. (V), Jeanne Melggs; Public Rels., Shannon O'Sullivan; Treas., Kimberley Hoey; Cur., Jill Landen; Museum Shop Mgr., Donna Strartak.
Personnel Profile: Full-Time Paid 6; Part-Time Paid 16; Part-Time Volunteers 9; Interns 2.
Governing Authority: private; nonprofit organization. Tax-exempt: 501(c)(3).
Institution Type/Description: Historic House Museum: listed on the National Register of Historic Sites.
Collections: Art Nouveau; 1920s-1930s decorative arts; Hunt Club history. Historic Structures: house; boathouse; bridge.
Research Fields: history of the Knight family & their servants; Art Nouveau; Tiffany; Hunt Club's on the Northern Outer Banks of North Carolina.
Facilities: 21,000 sq. ft. exhibit space. Museum-related items for sale.
Activities: arts festivals; concerts; docent program; guided tours. Annual Events: Independence Day; Haunted Village; Wine Festival.
Publications: seasonal newsletter, The Whalehead Window.
Hours & Admission Prices: Mon.-Sat. 9-5. Adults $9; discounts to AAM members, groups of 15 or more and children 8 & under no charge. Closed New Year's Day; Martin Luther King Jr. Day; Easter; Thanksgiving; Christmas Eve, Day & day after.
Attendance: 20,000 (estimated)
Membership: Senior $20; Individual $50; Family $100; Contributor $250; Donor $500; Sustainer $1,000; Knight's Circle $5,000.

WILD HORSE MUSEUM, , 1129 Corolla Village Rd., Corolla, NC 27927. Tel.: 252-453-8002. Fax: 252-453-8073.
E-mail: info@corollawildhorses.com
Web Site: www.corollawildhorses.com
Institution Type/Description: Horse Museum.
Collections: wild Spanish Mustangs; hands-on exhibitions; aerial map of horse's range.
Facilities: Gift items for sale.
Activities: live mustang viewings; horse painting; mustang rides.
Hours & Admission Prices: Call for hours.

Creswell

PETTIGREW STATE PARK, 2252 Lake Shore Rd., Creswell, NC 27928. Tel.: 252-797-4475.
Institution Type/Description: Park & American Indian History Museum.
Collections: American Indian culture; local history; photographs; personal artifacts.
Facilities: 16,600 acre natural lake; camp sites; nature trails; picnic area.
Activities: hiking; camping.
Hours & Admission Prices: Call for hours.

SOMERSET PLACE STATE HISTORIC SITE, 2572 Lake Shore Rd., Creswell, NC 27928-9174. Tel.: 252-797-4560. Fax: 252-797-4171.
E-mail: somerset@ncdcr.gov
Web Site: www.nchistoricsites.org/somerset/somerset.htm
Founded: 1969.
Congressional District: 1
Key Personnel: Site Mgr., Karen Hayes.
Governing Authority: state. Parent Institution: North Carolina Department of Cultural Resources, 109 E. Jones St., Raleigh, NC. 27601-2807. Tax-exempt: 501(c)(3).
Institution Type/Description: Plantation.
Collections: agriculture; dairy; ice house; smoke house; kitchen storehouse; kitchen-laundry; 1790-1910 plantation slave records & federal population census schedules for Chowan, Tyrrell & Washington counties. Historic Houses include c.1820 Colony House, 1830 Collins Mansion House, excavated dwellings in slave community; two-story four room slave quarter; one room slave quarter; slave hospital.
Research Fields: African-American history; agriculture; archaeology; 19th-century decorative arts.
Facilities: visitor center.
Activities: guided tours; lectures; permanent exhibitions.
Publications: brochure; newsletter.
Hours & Admission Prices: Tues.-Sat. 9-5. No charge, donations accepted. Closed holidays. ♿
Attendance: 23,407 (accurate)
Membership: Individual $20; Family $30; Donor $50; Patron $100; Friend $500; Benefactor $1,000.

Crossnore

CROSSNORE FINE ARTS GALLERY, Crossnore School, 205 Johnson Lane, Crossnore, NC 28616. Mailing Address: P.O. Box 249, Crossnore, NC 28616. Tel.: 828-733-3144 & 387-1695.
E-mail: artgallery@crossnoreschool.org
Web Site: www.crossnoregallery.org
Key Personnel: Mgr., Heidi Fisher
Institution Type/Description: Art Gallery.
Collections: paintings; sculpture.
Activities: temporary exhibitions; special events.
Hours & Admission Prices: Mon.-Sat. 9-5.

CROSSNORE WEAVERS AND GALLERY: A WORKING MUSEUM, 205 Johnson Ln., Crossnore, NC 28616. Mailing Address: P.O. Box 249, Crossnore, NC 28616-0249. Tel.: 828-733-4660. Fax: 828-733-3250.
E-mail: weavingroom@crossnoreschool.org
Web Site: www.crossnoreschool.org
Formerly: The Weaving Room
Founded: 1920.
Congressional District: 11
Key Personnel: Dir., Martha Hill; Chm. (V), Freda Nichols.
Personnel Profile: Full-Time Paid 1; Part-Time Paid 2; Part-Time Volunteers 4.
Governing Authority: bd. of trustees. Parent Institution: Crossnore School. Subsidiary Institution: Crossnore Weavers. Tax-exempt.
Institution Type/Description: History Museum: listed on the National Register of Historical Places.
Collections: early American life; period furnishings & looms.
Activities: handweaving demonstrations.
Publications: newsletter, The Crossnore School; catalogue, Crossnore Weavers: A Working Museum.
Hours & Admission Prices: Mon.-Sat. 9-5. No charge; donations accepted. Closed Easter; Thanksgiving; Christmas. ♿
Attendance: 5,500 (estimated)

Cullowhee

FINE ARTS MUSEUM - JOHN W. BARDO FINE AND PER-FORMING ARTS CENTER, 199 Centennial Dr., Cullowhee, NC 28723. Tel.: 828-227-3591.
E-mail: ddrury@wcu.edu
Web Site: www.fineartmuseum.wcu.edu
Formerly: Fine Arts Museum - Western Carolina University Fine & Performing Arts Center
Founded: 2005.
Congressional District: 11
Key Personnel: Interim Dir., Denis Drury; Museum Technician, Kevin Kirkpatrick; Museum Attendant, Dawn Behling
Institution Type/Description: Fine Arts Museum.
Collections: works by contemporary artists.
Facilities: 10,000 sq. ft. exhibit space.
Activities: third Thurs. receptions; visiting artist lectures; FAM film series. Museum Sponsors: Family Days.
Hours & Admission Prices: Mon.-Wed. & Fri. 10-4, Thurs. 10-7. No charge; donations accepted. Closed university holidays. &
Attendance: 7,627 (accurate)
Membership: Recent Alumni $25; WCU Faculty & Staff $35; Supporter $50-$99; Contributor $100-$249; Partner $250-$499; Sponsor $500-$999; Patron $1,000-$2,499; Connoisseur $2,500-$4,999; Benefactor $5,000-$9,999; Visionary $10,000 & up.

MOUNTAIN HERITAGE CENTER, (M), Western Carolina Univ., 150 H. F. Robinson Bldg., Cullowhee, NC 28723. Tel.: 828-227-7129.
Web Site: www.wcu.edu/mhc
Founded: 1975.
Congressional District: 11
Key Personnel: Dir., L. Scott Philyaw; Cur., Pam Meister; Education Specialist, Peter Koch; Coord. Events, Trina Royar; Office Mgr., Anne Lane.
Personnel Profile: Full-Time Paid 2; Part-Time Paid 3.
Governing Authority: university. Parent Institution: Western Carolina University, Cullowhee. Tel.: 828-227-7211. Tax-exempt: 170(b)(1)(A).
Institution Type/Description: History Museum.
Collections: Center: Cherokees & the European pioneer artifacts; tools; photographs; oral history; folklore recordings. Library: manuscripts & records.
Research Fields: Scotch-Irish migration; Appalachian culture; folklore; regional history; Cherokee Indian history.
Facilities: research room; 91-seat auditorium.
Activities: guided tours; lectures; films; permanent exhibits; formally organized educational programs; permanent, temporary & traveling exhibitions.
Publications: book, Our Western North Carolina Mountain Heritage; Leave-taking: The Scotch Irish Come to Western North Carolina.
Hours & Admission Prices: Mon.-Wed. & Fri. 8-5, Thurs. 8-7. No charge. Closed Christmas to New Year's Day. &
Attendance: 30,000 (estimated)
Membership: Annual $25.

Currie

MOORES CREEK NATIONAL BATTLEFIELD, 40 Patriots Hall Dr., Currie, NC 28435. Tel.: 910-283-5591. Fax: 910-283-5351.
E-mail: jonathan_grubbs@nps.gov
Web Site: www.nps.gov/mocr
Founded: 1926.
Congressional District: 1 & 3
Key Personnel: Park Ranger, Jonathan Grubbs.
Personnel Profile: Full-Time Paid 4; Part-Time Paid 1; Part-Time Volunteers 8; Interns 4.
Governing Authority: federal. Parent Institution: U.S. Dept. of the Interior, National Park Service. Tax-exempt.
Institution Type/Description: Military Museum: located on the site of the Feb. 27, 1776 Battle of Moores Creek Bridge.
Collections: weapons; diorama.
Research Fields: history of the park, including period before park establishment in 1926.
Facilities: library of North Carolina, Revolutionary War history, parks, & environment books available for use on premises. Historical booklets, books, cards & museum-related items for sale.
Activities: ranger-guided trail tours.
Publications: brochure, Moores Creek National Battlefield; Moores Creek Bridge Campaign; Roster of the Loyalists at the Battle of Moores Creek; Roster of the Patriots at the Battle of Moores Creek; poster, Colonial Militia; video, Moores Creek Bridge.

Hours & Admission Prices: Daily 9-5. No charge. Closed New Year's Day; Thanksgiving; Christmas. &
Attendance: 48,988 (accurate)
Membership: Student $5; Individual $10; Family & Group $25; Corporate $100; Patron $125; Life Individual $150; Life Family $200; Life Benefactor $500.

Dallas

* **GASTON COUNTY MUSEUM OF ART AND HISTORY, (M),** 131 W. Main St., Dallas, NC 28034-2021. Mailing Address: P.O. Box 429, Dallas, NC 28034-0429. Tel.: 704-922-7681. Fax: 704-922-7683.
E-mail: elaine.jackson@co.gaston.nc.us
Web Site: www.gastoncountymuseum.org
Founded: 1975.
Congressional District: 10
Key Personnel: Dir., Jeff Pruett; Trustee Chm., Vann Noblett; Administrative Asst., Elaine Jackson; Cur., Stephanie Elliott; Registrar, Regan Brooks.
Personnel Profile: Full-Time Paid 3; Part-Time Paid 7; Part-Time Volunteers 25; Interns 2.
Governing Authority: county; nonprofit. Tax-exempt: 501(c)(3).
Institution Type/Description: Art and History Museum: housed in 1852 Hoffman Hotel.
Collections: art; regional history; carriages & sleighs; textiles; 19th-century parlors, including hands-on parlor. Historic Buildings: 1901 Dallas Depot; 1852 Hoffman Hotel; 1848 Gaston County jail.
Research Fields: regional history; textile history.
Facilities: classrooms. Museum-related items for sale.
Activities: guided tours; lectures; workshops; educational programs; loan, permanent, temporary & traveling exhibitions.
Publications: quarterly newsletter; walking tour guide for historic district, Historic Dallas: A Stroll Through 19th Century America.
Hours & Admission Prices: Tues.-Fri. 10-5, Sat. 10-3; other times by appointment. No charge. Closed holidays. &
Attendance: 27,000 (estimated)
Membership: Student & Senior $25; Gaston County Individual $30; Non-resident Individual $40; Gaston County Family $50; Non-resident Family $60; Benefactor $125; Sustainer $250; Patron $500; Gaston Gold Society $1,000.

Danbury

J.E. PRIDDY'S GENERAL STORE, 2121 Sheppard Mill Rd., Danbury, NC 27016. Tel.: 336-593-8786.
Web Site: priddysgeneralstore.com
Institution Type/Description: Historic Building Museum: built in 1888.
Collections: early furnishings & merchandise; local history.
Activities: Annual Events: live Bluegrass in February, May & October; Christmas Celebration.
Hours & Admission Prices: Mon.-Sat. 8-6.

Davidson

TOM CLARK MUSEUM, 131 N. Main St., 2nd Fl., Davidson, NC 28036. Tel.: 704-894-6246.
Institution Type/Description: Art Museum.
Collections: sculptures.
Hours & Admission Prices: Mon.-Sat. 10-6. No charge.

VAN EVERY/SMITH GALLERIES, 315 N. Main St., Davidson College Visual Arts Center, Davidson, NC 28036-9404. Mailing Address: P.O. Box 7117, Davidson, NC 28035-7117. Tel.: 704-894-2520 & 2519. Fax: 704-894-2691.
E-mail: brthomas@davidson.edu
Web Site: www.davidson.edu
Founded: 1962.
Congressional District: 9
Key Personnel: Dir., Brad Thomas
Personnel Profile: Full-Time Paid 1; Part-Time Paid 11.
Governing Authority: college. Parent Institution: Davidson College. Subsidiary Institution: William H. Van Every & Edward M. Smith Galleries. Tax-exempt.
Institution Type/Description: College Art Gallery.
Collections: 15th- to 20th-century graphics; painting; sculpture; photography.
Activities: lectures; gallery talks; temporary exhibitions.
Publications: exhibition brochures with essay & posters.
Hours & Admission Prices: Sept.-May Mon.-Fri. 10-5, Sat.-Sun. 12-4. No charge. Closed college & national holidays. &

Attendance: 4,000 (accurate)

Dunn

AVERASBORO BATTLEFIELD AND MUSEUM, 3300 Hwy. 82, Dunn, NC 28334-6571. Mailing Address: P.O. Box 1811, Dunn, NC 28335. Tel.: 910-891-5019.
E-mail: bpearce7@nc.rr.com
Governing Authority: Parent Institution: Averasboro Battlefield Commission. Tax-exempt.
Institution Type/Description: History Museum.
Collections: local history; military artifacts, weapons, & equipment; personal artifacts; period furnishings; photographs.
Facilities: Museum-related items for sale.
Activities: special events. Annual Events: World War I Encampment in September; Battle of Averasboro Reenactment in October.
Publications: Averasboro Advocate.
Hours & Admission Prices: Call for hours. No charge; donations accepted. &
Attendance: 12,000 (estimated)
Membership: Individual $35.

GENERAL WILLIAM C. LEE AIRBORNE MUSEUM, 209 W. Divine St., Dunn, NC 28334. Mailing Address: P.O. Box 1111, Dunn, NC 28334-1111. Tel.: 910-892-1947. Facebook: General William C. Lee Airborne Museum.
E-mail: info@generalleeairbornemuseum.org
Web Site: generalleeairbornemuseum.org
Founded: 1986.
Key Personnel: Pres., Oscar Harris.
Personnel Profile: Part-Time Paid 1; Part-Time Volunteers 3.
Governing Authority: Parent Institution: General William C. Lee Memorial Commission. Tax-exempt.
Institution Type/Description: Military History Museum.
Collections: General Lee's life & career; personal artifacts; period furnishings; military artifacts; photographs.
Hours & Admission Prices: Mon.-Fri. 10-4, Sat. 11-4; groups by appointment. No charge; donations accepted. Closed holidays.
Attendance: 1,500 (estimated)

Durham

BENNETT PLACE STATE HISTORIC SITE, 4409 Bennett Memorial Road, Durham, NC 27705-2307. Tel.: 919-383-4345. Fax: 919-383-4349.
E-mail: bennett@ncdcr.gov
Web Site: www.bennettplacehistoricsite.com
Formerly: Bennett Farm
Founded: 1961.
Congressional District: 2
Key Personnel: Site Mgr., John Guss; Pres. Support Group, Karen Edwards.
Personnel Profile: Full-Time Paid 3; Part-Time Paid 1; Part-Time Volunteers 12; Interns 1.
Volunteer Hours: 5,554
Governing Authority: state. Parent Institution: North Carolina Dept. of Cultural Resources, 109 E. Jones St., Raleigh, NC 27611. Tax-exempt: 170(b).
Institution Type/Description: Military Museum & Historic Farm: reconstructed c.1850 Bennett House.
Collections: log kitchen; history; Civil War; military uniforms & civilian clothing; historic structures; weaponry; farm implements; paintings.
Research Fields: Civil War & Agriculture in North Carolina.
Facilities: library; visitor center; theater; nature trail; picnic area. Gift-related items for sale.
Activities: guided tours; costumed interpretation; special events; theater presentation; school programs.
Publications: brochure; newsletter; Bennett Place Courier.
Hours & Admission Prices: Tues.-Sat. 9-5. No charge; donations accepted. Closed major holidays. &
Attendance: 22,000 (estimated)

DUKE HOMESTEAD STATE HISTORIC SITE, 2828 Duke Homestead Rd., Durham, NC 27705-2726. Tel.: 919-477-5498. Fax: 919-479-7092.
E-mail: duke@ncdcr.gov
Web Site: www.nchistoricsites.org/duke/
Founded: 1974.
Congressional District: 2
Key Personnel: Site Mgr., Jennifer Farley; Pres. (V), Mary Bell; Cur., Martha Jackson.

Personnel Profile: Full-Time Paid 4; Part-Time Paid 4; Part-Time Volunteers 50.
Governing Authority: state. North Carolina Dept. of Cultural Resources, 109 E. Jones St., Raleigh 27611. Tax-exempt: 501(c)(3) & 170(b)(1)(A).
Institution Type/Description: Historic House: 1852 homestead of Washington Duke, founder of the American Tobacco Co.
Collections: furnished house of early 1870s; tobacco barn; well house; packhouse; 1870 third factory; reconstructed first factory; tobacco history.
Research Fields: tobacco history.
Facilities: visitor center.
Activities: guided tours; lectures; audiovisual program; craft workshops; participatory demonstrations; special events; formally organized education program for children.
Publications: guide book; biannual publication.
Hours & Admission Prices: Tues.-Sat. 9-5. No charge; donations accepted. Closed Thanksgiving; Christmas Eve & Day. &
Attendance: 18,000 (accurate)
Membership: Friend $20-$49; Hander $50-$99; Stringer $100-$249; Primer $250-$499; Grower $500-$999; Corporate & Charter $1,000 & up.

HISTORIC STAGVILLE, 5828 Old Oxford Hwy., Durham, NC 27712-9758. Tel.: 919-620-0120.
E-mail: info@stagville.org
Web Site: www.stagville.org
Founded: 1977.
Congressional District: 4
Key Personnel: Site Mgr., Frachele Scott.
Personnel Profile: Full-Time Paid 3; Part-Time Paid 2; Part-Time Volunteers 3; Interns 5.
Governing Authority: state; nonprofit corporation. Parent Institution: North Carolina, Div. of Archives & History, Dept. of Cultural Resources, 4610 Mail Service Center, Raleigh, NC 27699-4610. Tax-exempt.
Institution Type/Description: Historic House.
Collections: furniture & furnishings of the period; farm implements; tools. Historic Buildings: 1787 Plantation House; 1860 Slave Cabins; 1860 Barn; late 18th-century Cottage.
Research Fields: African-American history; oral history; preservation education; southern & plantation history.
Facilities: classrooms.
Activities: guided tours; classes in historic architecture, preservation, African-American history; historic landscapes; restoration.
Publications: newsletter, The Key.
Hours & Admission Prices: Tours: Tues.-Sat. 10-3. No charge; donations accepted. Closed New Year's Day; Martin Luther King Jr. Day; Good Friday; Memorial Day; Independence Day; Labor Day; Veterans Day; Thanksgiving; Christmas. &
Attendance: 14,000 (estimated)
Membership: Individual $25; Joint $30; Contributing $50; Sustaining $100; Patron $200; Benefactor $500.

HISTORY OF MEDICINE COLLECTIONS, Duke University Medical Center Library, Durham, NC 27710. Mailing Address: DUMC 3702, Durham, NC 27710. Tel.: 919-660-1143 & 1144. Fax: 919-681-7599.
E-mail: mclhistory@mc.duke.edu
Web Site: www.mclibrary.duke.edu/hmc
Founded: 1930.
Key Personnel: Cur., Suzanne Porter.
Personnel Profile: Full-Time Paid 1; Part-Time Paid 1.
Governing Authority: university. Parent Institution: Duke University. Tax-exempt: 501(c)(3).
Institution Type/Description: University Library.
Collections: books; journals; manuscripts; medical instruments & artifacts; prints; photographs.
Research Fields: history of medicine.
Facilities: 33,000-vol. library of medical & scientific books, journals, manuscripts, reference material on the history of medicine & biology available for research on premises; reading room; medicinal herb garden.
Activities: lectures; formally organized education programs for graduate students; temporary exhibitions.
Publications: newsletter, Trent Associates Report.
Hours & Admission Prices: Tues.-Fri. 10-4; appointments encouraged. No charge. &
Attendance: 2,000 (estimated)
Membership: Trent Associates: Member $25; Contributing $50; Sustaining $100.

LYDA MOORE MERRICK GALLERY - THE HAYTI HERITAGE CENTER, 804 Old Fayetteville St., Durham, NC 27707. Tel.: 919-683-1709. Fax: 919-682-5869.
E-mail: info@hayti.org
Web Site: www.hayti.org
Institution Type/Description: Art Gallery.
Collections: works by local, regional & national artists.
Hours & Admission Prices: By appointment.

MUSEUM OF LIFE AND SCIENCE, 433 Murray Ave., Durham, NC 27704-3101. Tel.: 919-220-5429. Fax: 919-220-5575.
Web Site: lifeandscience.org
Formerly: North Carolina Museum of Life and Science
Founded: 1946.
Congressional District: 2
Key Personnel: Pres. & C.E.O., Barry Van Deman; Bd. Chm., Tracey Martin.
Governing Authority: nonprofit organization. Tax-exempt: 501(c)(3).
Institution Type/Description: Science & Technology Center.
Collections: hands-on science exhibits; wildlife & flora indigenous to North Carolina; small railway; weather; communications; physics; geology; paleontology; aerospace; animal habitats; biology; children's exhibits; physical & natural science discovery rooms; classrooms; farmyard & animals; moonrook; meteorite; large animals including bears & red wolves; butterfly house includes 1,000 exotic butterflies from Africa, Asia, Central & South America; insectarium includes live insects & their predators, specimens, interactive exhibits, & murals; insect environments.
Research Fields: science & technology education.
Facilities: butterfly house & insectarium; discovery rooms; lab & classrooms; meeting room with A/V booth; catering kitchen; 70-acre outdoor campus features outdoor exhibit areas, farmyard, small railway, nature park with animals & outdoor maze; cafe. Museum-related items for sale.
Activities: classes; lectures; science-in-suitcase outreach program; after-school programs; teacher training; special events; demonstrations. Annual Event: science camp in summer.
Publications: bimonthly newsletter; teacher's guide; school guide; visitor's guide; corporate newsletter.
Hours & Admission Prices: Tues.-Sat. 10-5, Sun. 12-5. Adults $14, senior citizens 65 & over and military with ID $11, children 3-12 $10; discounts to AAA members; members and children 2 & under no charge. Train Rides $3. Closed New Year's Day; Thanksgiving; Christmas. &
Attendance: 300,000 (accurate)
Membership: Explorer: Couple $95, Four People $135, Six People $155, Eight People $175; Supporting: Sustainer $300, Patron $600, Chrysalis Society $1,000 & up.

NCCU ART MUSEUM, North Carolina Central University, Lawson St. (Btw. Fine Arts Bldg. & Music Bldg.), Durham, NC 27707. Mailing Address: P.O. Box 19555, Durham, NC 27703. Tel.: 919-530-6211. Fax: 919-560-5649.
Web Site: web.nccu.edu/artmuseum
Founded: 1971.
Congressional District: 4
Key Personnel: Dir., Kenneth G. Rodgers; Registrar, Pat Jones.
Governing Authority: state; nonprofit. Parent Institution: University of North Carolina. Subsidiary Institution: North Carolina Central University. Tax-exempt: 501(c)(3); 170(b)(1)(A); 509(A)(1).
Institution Type/Description: University Art Museum.
Collections: contemporary painting, sculpture, & original prints with a focus on the works by African American artists; traditional African sculpture & artifacts.
Research Fields: minority artists; exhibition subjects.
Activities: lectures; gallery talks; formally organized education programs for undergraduate college students; loan, permanent, temporary & traveling exhibitions.
Publications: catalogue; American Landscape East & West: 1820-1920; Duncanson: A British-American Connection; Geoffrey Holder, Painter; Gullah Life Reflections: A Traveling Exhibition of the Paintings of Jonathan Green; Joy of Living: Romare Bearden's Late Work.
Hours & Admission Prices: Tues.-Fri. 9-4:30, Sun. 2-4. No charge. &

＊　**NASHER MUSEUM OF ART AT DUKE UNIVERSITY, (M),** 2001 Campus Dr., Durham, NC 27705-1003. Mailing Address: P.O. Box 90732, Durham, NC 27708-0732. Tel.: 919-684-5135. Fax: 919-681-8624.
E-mail: nasherinfo@duke.edu
Web Site: www.nasher.duke.edu
Formerly: Duke University Museum of Art
Founded: 1969.
Congressional District: 2

Key Personnel: C.E.O., Prcs. & Pres. of Duke University, Richard H. Brodhead; Dir., Dr. Kimerly Rorschach; Pres., Angela O. Terry; Coord. Membership, Amy Weaver; Cur., Dr. Sarah W. Schroth; Dir. Devel., Kristen Greenaway; Mgr. Mktg. & Communications, Wendy Hower Livingston; Registrar, Charles Carroll; Coord. Special Events, Kathleen Wright; Cur., Trevor Schoonmaker; Cur., Anne Schroder; Deputy Dir. Operations, Dorothy Clark; Museum Shop Mgr., Arienne Cheek.
Personnel Profile: Full-Time Paid 29; Part-Time Paid 20; Part-Time Volunteers 100; Interns 12.
Governing Authority: university. Parent Institution: Duke University. Tax-exempt: 501(c)(3).
Institution Type/Description: Art Museum.
Collections: modern & contemporary art; African art; Greek & Roman period artifacts; early American art; American & European paintings, sculpture & works on paper; medieval sculpture.
Research Fields: pertaining to collections.
Activities: guided tours; gallery talks; poetry readings; concerts; plays; film series; docent program; inter-museum loan, permanent & temporary exhibitions; lectures; symposia; student art volunteer program; available for rental to nonprofit organizations, member corporations & Duke Univ. departments.
Publications: exhibition catalogs; newsletter; annual report.
Hours & Admission Prices: Tues.-Wed. & Fri.-Sat. 10-5, Thurs. 10-9, Sun. 12-5. Suggested Donation: adults $5, seniors & Duke Alumni $4, non-Duke students $3; children under 16, members, Duke students, faculty & staff no charge. &
Attendance: 115,000 (accurate)
Membership: Friends of the Art Museum: Student $20; Individual $50; Duke Faculty & Staff $40; Family $60; Sponsor $100; Patron $50; Director's Circle $500; Brummer Society Bronze $1,000; Brummer Society Silver $2,500; Brummer Society Gold $5,000.

SARAH P. DUKE GARDENS, Duke University West Campus, 420 Anderson St., Durham, NC 27708. Mailing Address: Duke University, P.O. Box 90341, Durham, NC 27708-0341. Tel.: 919-684-3698.
Key Personnel: Exec. Dir., William M. LeFevre
Institution Type/Description: Gardens.
Collections: over 2,000 plant species.
Facilities: 55 acres; visitor center; cafe; walking trails; rental facilities. Gift items for sale.
Activities: special events; rental facilities; classes; tours; summer camps; educational programs. Annual Event: Duke Performances' Music in the Gardens summer outdoor concert series.
Hours & Admission Prices: Daily. No charge; donations accepted.
Attendance: 300,000

Edenton

CHOWAN ARTS COUNCIL GALLERY - THE GALLERY, 504 S. Broad St., Edenton, NC 27932. Tel.: 252-482-8005.
Web Site: www.chowanarts.com
Key Personnel: Dir., Murielle Harmon; Gallery Liaison, Mary Altman.
Governing Authority: nonprofit organization.
Institution Type/Description: Art Gallery.
Collections: works by local, state & regional artists; paintings; sculpture.
Activities: outreach programs; temporary exhibitions. Annual Event: 5 County Student Art Show in February; 2 Man Show in March; Member's Exhibition in October; Confection Perfection in December.
Hours & Admission Prices: Mon.-Fri. 11-4, Sat.-Sun. 10-2. No charge; donations accepted. &
Membership: Single $35; Double $60; Sponsor $100; Sustainer $250; Benefactor $500.

EDENTON NATIONAL FISH HATCHERY, 1102 W. Queen St., Edenton, NC 27932. Tel.: 252-482-4118. Fax: 252-482-2106.
Key Personnel: Hatchery Mgr., Stephen C. Jackson
Institution Type/Description: Aquarium.
Collections: hatchery history; fish & fishery management.
Facilities: trails; aquarium; 63 acres.
Activities: educational programs; walking trails; guided tours.
Hours & Admission Prices: By appointment.

HISTORIC EDENTON STATE HISTORIC SITE, 108 N. Broad St., Edenton, NC 27932. Tel.: 252-482-2637. Fax: 252-482-3499. www.harrietjacobs.org.
E-mail: edenton@ncdcr.gov
Web Site: www.edenton.nchistoricsites.org
Formerly: James Iredell House State Historic Site

Founded: 1951.
Congressional District: 3
Key Personnel: Site Mgr., Linda Jordan Eure; Operations Mgr., Judith W. Chilcoat; Historic Interpreter, Keith Furlough; Historic Interpreter, Sharon K. Keeter; Historic Interpreter, Charles Boyette; Historic Interpreter, Carolyn A. Owens; Bldg. & Environmental Svcs., George S. Lassiter; Maintenance Mechanic, Blake S. Harmon.
Personnel Profile: Full-Time Paid 8; Part-Time Paid 4; Part-Time Volunteers 35.
Governing Authority: state. Parent Institution: North Carolina Office of Archives & History, 4610 Mail Service Center, Raleigh, NC 27699-4610. Subsidiary Institution: James Iredell Historical Association, Inc. Tax-exempt.
Institution Type/Description: Historic House Museum.
Collections: period furnishings; kitchen utensils; formal gardens; 1756 kitchen; c.1827 dairy. Historic Buildings: c.1827 Bandon Schoolhouse; c.1827 Bandon Smokehouse; carriage house; 1800/1827 James Iredell House; 18th century necessary house; 1886 Roanoke River Lighthouse; 1767 Chowan County Courthouse, a National Historic Landmark.
Research Fields: James Iredell; period furnishings; architecture; 18th century history of U.S. Supreme Court; 18th century legal & social history; maritime history.
Facilities: visitor center; information services. Local history books, local handcrafted items & guidebooks for sale.
Activities: guided tours; formally organized education programs; audiovisual program; permanent exhibitions; special events & programs.
Publications: brochure; town map; Harriet Jacobs Self-Guided Brochure; Edenton Architecture Self-Guided Brochure.
Hours & Admission Prices: Daily 9-5. Guided walking and trolley tours: $10, $6, or $4 per person; discounts to seniors. Visitor Center: no charge. Closed New Year's Day; Martin Luther King Jr. Day; Veterans Day; Thanksgiving; Christmas Eve & Day. &
Attendance: 23,954 (accurate)

Elizabeth City

THE JAQUELIN JENKINS GALLERY, 516 E. Main St., Elizabeth City, NC 27907. Tel.: 252-338-6455. Fax: 252-338-3156.
E-mail: info@artsaoa.com
Web Site: www.artsaoa.com
Institution Type/Description: Art Gallery.
Collections: works of over 250 artists, craftsmen & photographers.
Activities: workshops; temporary exhibitions; opening receptions.
Hours & Admission Prices: Call for hours.

MUSEUM OF THE ALBEMARLE, (M), 501 S. Water St., Elizabeth City, NC 27909-4863. Tel.: 252-335-1453. Fax: 252-335-0637.
E-mail: moa@ncdcr.gov
Web Site: www.museumofthealbemarle.com
Founded: 1967.
Congressional District: 1
Key Personnel: Admin., William J. McCrea; Administrative Officer, Mary Cherry Tirak; Cur., Wanda Stiles; Coord. Education, Charlotte Patterson; Facilities Mgr., Wayne Mathews; Lighting, Electronics & Interactive Technician, Lynette Sawyer; Educator, Lori Meads; Utility Worker, William Seymore; Museum Shop Mgr., Mary Temple; Office Asst., Gina Cappellano; Public Information Asst., Lisa Doepker; Exhibit Designer, Jamie McCargo; Carpenter, Matthew Ferrell; Utility Worker, Ben Shipley; Collections Specialist, Clay Swindell.
Personnel Profile: Full-Time Paid 13; Part-Time Paid 6; Part-Time Volunteers 40.
Governing Authority: Parent Institution: North Carolina Div. of State History Museum Dept. of Cultural Resources, 5 E. Edenton St., Raleigh, NC 27601-1011. Tax-exempt.
Institution Type/Description: History Museum.
Collections: regional history; Indian artifacts; agricultural exhibits; lumbering items; decoys; farming; fishing; military featuring U.S. Coast Guard.
Major Exhibits: Out of the Blue: Coast Guard Aviation, 11/12-12/14; Al Norte Al Norte, 10/13-6/14; Reminiscing the 1980s (T), 11/13-2/14; Across Three Centuries: Art from the Edwin T. & Diana D. Hardison Collection, 10/13-5/14; Under Both Flags: Civil War in the Albemarle, 10/13-12/15; Out of the Blue: Coast Guard Aviation, 1/14-7/15; Al Norte al Norte: Latino Life in North Carolina (T), 2/14-6/14; Tea Time!, 7/14-7/15.
Research Fields: regional history; area industries & lifestyle.
Facilities: 1,000-vol. library of historical information available for use on premises by appointment. Museum-related items for sale.
Activities: guided tours; changing exhibitions; formally organized educational programs; living history; hands-on history programs.

Publications: brochure; quarterly newsletter; annual school programs brochure.
Hours & Admission Prices: Tues.-Sat. 10-4. No charge; donations accepted. Closed state holidays. &
Attendance: 42,627 (accurate)
Membership: Student $15; Individual $30; Grandparents & Family $50; Patron $100-$499; Lifetime $500 & up.

PORT DISCOVER: NORTHEASTERN NORTH CAROLINA'S CENTER FOR HANDS-ON SCIENCE, 611 E. Main St., Elizabeth City, NC 27909. Tel.: 252-338-6117.
Web Site: portdiscover.wildapricot.org
Key Personnel: Dir., Robin Kelly-Goss; Mgr. Mktg. & Membership, Chrissy Benton; Dir. Education & Exhibits, Michelle Donahue; Visitor Svcs. Mgr., Judi Stuart; Science Educator, Loren Cartwright; Science Educator, Adrienne Cole
Institution Type/Description: Science Center.
Collections: hands-on exhibitions.
Activities: educational programs; special events.
Hours & Admission Prices: Tues.-Fri. 1-5, Sat. 10-4; groups by appointment.

Ellerbe

RANKIN MUSEUM OF AMERICAN HERITAGE, 131 W. Church St., Ellerbe, NC 28338. Mailing Address: P.O. Box 499, Ellerbe, NC 28338-0499. Tel.: 910-652-6378. Fax: 910-652-6130.
E-mail: rankinmuseum@etinternet.net
Web Site: www.rankinmuseum.org
Founded: 1986.
Congressional District: 19
Key Personnel: Art Cur., Supvr., Museum Shop Mgr., Gail Benson; Pres. (V), Julian Carter; Vice Pres., Judy Richardson; Financial Dir., Jim Chavis.
Personnel Profile: Full-Time Paid 1; Part-Time Paid 3; Part-Time Volunteers 2.
Governing Authority: private; nonprofit organization. Parent Institution: Rankin Museum of American Heritage, Box 499, Ellerbe, NC 28338. Tax-exempt: 501(c)(3).
Institution Type/Description: General Museum.
Collections: Native Americans of North, Central & South America; natural history including animal studies, geology & paleontology; heritage & history of North Carolina, South Carolina, & Richmond County; Civil War artifacts; paintings; medical artifacts from 1800 to early 1900s.
Facilities: research library; classroom; 6,500 sq. ft. exhibit space. Museum-related items for sale.
Activities: adult & student workshops; education programs with schools; annual fundraisers. Museum Sponsors: Geology Dig and Fossil Fair for students; Christmas Open House.
Publications: annual newsletter.
Hours & Admission Prices: Mon.-Tues. & Thurs.-Fri. 10-4, Sat. 1-5, Sun. 2-5. Adults $4, students $1; discounts to AAA members, groups of 15 or more & school groups; members and children 4 & under no charge. Closed New Year's Day; Thanksgiving; Christmas. &
Attendance: 4,000 (estimated)
Membership: Student $5; Adult $15; Family $25; Donor $50; Sustainer $100; Benefactor $250; Sponsor & Memorial $500; Patron $1,000.

Farmville

MAY MUSEUM & PARK, 3802 S. Main, Farmville, NC 27828-8548. Mailing Address: P.O. Box 623, Farmville, NC 27828-0086. Tel.: 252-753-6725. Cell: 252-327-8859.
E-mail: maymuseum@farmville-nc.com
Founded: 1991.
Congressional District: 1
Key Personnel: Dir., Deb Higgins.
Personnel Profile: Part-Time Paid 1; Part-Time Volunteers 8.
Governing Authority: municipal; nonprofit. Parent Institution: Town of Farmville. Tax-exempt.
Institution Type/Description: History Museum.
Collections: quilts; furniture; photographs; books; documents; sketches & artwork.
Major Exhibits: The Personal Side of Walter B. Jones, Sr., 12/8/13-4/11/14; Clothing Thru the Ages, 4/20/14-12/1/14.
Research Fields: May family genealogy.
Facilities: botanical garden.
Activities: Annual Events: Garden Party in April; Friends of the May Museum Membership Party in October; Christmas Tea in December.
Hours & Admission Prices: Mon.-Fri. 9-5. No charge; donations accepted. &
Attendance: 2,000 (estimated)
Membership: Student $10; Individual $25; Family $40.

Fayetteville

THE AIRBORNE & SPECIAL OPERATIONS MUSEUM, 100 Bragg Blvd., Fayetteville, NC 28301-4806. Tel.: 910-643-2766. Fax: 910-643-2792.
E-mail: jamesh.huggins@us.army.mil
Web Site: www.asomf.org
Founded: 2000.
Congressional District: 8
Key Personnel: C.E.O. & Chm., Paul Galloway; Dir., James H. Huggins; Chm. (V), Gen. James J. Lindsay, (U.S.A.) (R); Pres. (V), Henry Holt; Cur. Collections, Mary Dennings; Collections, George Stefanski; Museum Shop Mgr., Amanda Swan.
Personnel Profile: Full-Time Paid 17; Part-Time Paid 4; Part-Time Volunteers 97.
Governing Authority: federal; nonprofit organization. Parent Institution: U.S. Army. Tax-exempt: 501(c)(3).
Institution Type/Description: Military Museum.
Collections: 1940-present military artifacts; uniforms; equipment; weapons; soldering equipment; C-47 WWII airplane; CG4A WACO glider; Huey helicopter; AH-6 helicopter; HUMMV.
Facilities: 22,000 sq. ft. exhibit space; 200-seat theater. Museum-related items for sale.
Activities: docent program; films; rental gallery; temporary exhibitions; theater; 24-seat motion simulator; paver stones available to memorialized veterans. Annual Event: National Airborne Day in August.
Publications: newsletter; annual report; gift shop catalog.
Hours & Admission Prices: Mon. Federal holidays & Tues.-Sat. 10-5, Sun. 12-5. No charge; donations accepted. Closed New Year's Day; Easter; Thanksgiving; Christmas. &
Attendance: 164,639 (accurate)
Membership: Friend: Student under 18 & military; Individual $35; Student & Military Family $40; Family $50. Contributor $100-$499; Donor $500-$599; Patron $1,000-$2,499; Sponsor $2,500-$4,999.

ARTS COUNCIL OF FAYETTEVILLE/CUMBERLAND COUNTY, 301 Hay St., Fayetteville, NC 28301-5535. Mailing Address: P.O. Box 318, Fayetteville, NC 28302-0318. Tel.: 910-323-1776. Fax: 910-323-1727.
E-mail: admin@theartscouncil.com
Web Site: www.theartscouncil.com
Founded: 1974.
Congressional District: 7
Key Personnel: Exec. Dir., Deborah Martin Mintz; Gen. Mgr., Nancy Silver; Dir. Opers., Robert Pinson; Dir. Mktg., Mary Kinney; Exec. Asst., Jennifer Gilbertson.
Governing Authority: nonprofit organization. Tax-exempt: 501(c)(3).
Institution Type/Description: Arts Center: housed in c.1910 former Post Office & Library.
Collections: paintings.
Activities: self-guided tours; concerts; dance recitals; arts festivals; theatrical productions; temporary exhibitions. Gallery Sponsors: exhibits by & competitions for local & trade area artists.
Publications: monthly newsletter.
Hours & Admission Prices: Mon.-Thurs. 8:30-5, Fri. 8:30-Noon; other times by appointment. No charge. Closed New Year's Day; Good Friday; Memorial Day; Independence Day; Labor Day; Thanksgiving; Christmas. &
Attendance: 85,000 (estimated)
Membership: Apprentice $35-$49; Coontributor $50-$99; Patron $100-$249; Supporter $250-$499; Associate Director $500-$999; Director $1,000-$2,499; Producer $2,500-$4,999; Executive Producer $5,000 & up.

CAPE FEAR BOTANICAL GARDEN, 536 N. Eastern Blvd., Fayetteville, NC 28301. Mailing Address: P.O. Box 53485, Fayetteville, NC 28305. Tel.: 910-486-0221. Fax: 910-486-4209.
E-mail: info@capefearbg.org
Web Site: www.capefear.org
Institution Type/Description: Botanical Garden.
Collections: over 2,000 varieties of plants; gardens
Facilities: 79 acres.
Activities: special events; facility rental; educational programs; docent-led tours.
Hours & Admission Prices: March to mid-Dec. Mon.-Sat. 10-5, Sun. 12-5; mid-Dec. to Feb. Mon.-Sat. 10-5. Adults $6, children 6-12 $1; children 5 & under and members no charge.

DAVID MCCUNE INTERNATIONAL ART GALLERY - METHODIST UNIVERSITY, William F. Bethune Center for Visual Arts, 5400 Ramsey St., Fayetteville, NC 28311. Tel.: 910-425-5379.
E-mail: sfoti@davidmccunegallery.org
Web Site: www.davidmccunegallery.org
Key Personnel: Exec. Dir., Silvana M. Foti
Institution Type/Description: Art Gallery.
Collections: works by students, regional, national & international artists.
Hours & Admission Prices: Mon.-Fri. 10-5. No charge. Closed holidays.

FASCINATE-U CHILDREN'S MUSEUM, 116 Green St., Fayetteville, NC 28301-5024. Mailing Address: P.O. Box 2671, Fayetteville, NC 28302-2671. Tel.: 910-829-9171. Fax: 910-433-1639.
E-mail: webmail@fascinate-u.com
Web Site: www.fascinate-u.com
Institution Type/Description: Children's Museum.
Collections: hands-on exhibits.
Hours & Admission Prices: Tues. & Thurs.-Fri. 9-5, Wed. 9-7, Sat. 10-5, Sun. 12-5. Children $4, adults $3. Closed New Year's Day; Easter; Thanksgiving; Christmas Eve & Day.

FAYETTEVILLE AREA TRANSPORTATION AND LOCAL HISTORY MUSEUM, 325 Franklin St., Fayetteville, NC 28301. Mailing Address: 121 Lamon St., Fayetteville, NC 28301-4953. Tel.: 910-433-1457.
Web Site: www.fayettevillenc.net/sites/st_trainstation2.htm
Institution Type/Description: Local History Museum.
Collections: local history & culture; transportation history; photographs; period artifacts.
Hours & Admission Prices: Call for hours.

FAYETTEVILLE INDEPENDENT LIGHT INFANTRY ARMORY & MUSEUM, 210 Burgess St., Fayetteville, NC 28301. Mailing Address: 227 Hillside Ave., Fayetteville, NC 28301. Tel.: 910-433-1457; 800-255-8217.
Institution Type/Description: Military History Museum.
Collections: local history; weapons; uniforms; military & personal artifacts; period furnishings; photographs.
Hours & Admission Prices: By appointment. No charge.

MUSEUM OF THE CAPE FEAR HISTORICAL COMPLEX, (M), 801 Arsenal Ave., Fayetteville, NC 28305. Mailing Address: P.O. Box 53693, Fayetteville, NC 28305-3693. Tel.: 910-486-1330. Fax: 910-486-1585.
E-mail: david.reid@ncdcr.gov
Web Site: museumofthecapefear.ncdcr.gov
Founded: 1985.
Congressional District: 7
Key Personnel: Admin., David E. Reid; Assoc. Cur. Education, Leisa Greathouse; Exhibit Designer, Margaret Shearin; Carpenter & Exhibit Builder, Jim Frederickson; Assoc. Cur. Research, Kathryn A. Beach; 1897 Poe House Educator, Megan Maxwell; Arsenal Park Educator, Chris Woodson; Historic Interpreter, Jim Brisson; Collections Asst., Bill Surface.
Personnel Profile: Full-Time Paid 13; Part-Time Paid 4; Part-Time Volunteers 80.
Governing Authority: Division of State History Museums, N.C. Dept. of Cultural Resources, 5 E. Edenton St., Raleigh, NC 27601. Tax-exempt.
Institution Type/Description: History Museum: octagonal structure & building foundations from the federal arsenal in Arsenal Park.
Collections: artifacts pertaining to history & culture of southern North Carolina. Historic Building: 1897 historic house.
Research Fields: regional history.
Facilities: reference & slide library; 6,900 sq. ft. exhibit galleries; classroom. Museum-related items for sale.
Activities: changing exhibitions; educational programs; guided tours; lectures & film series; workshops; special events.
Publications: brochure; quarterly newsletter, the Longleaf; quarterly calendar of events.
Hours & Admission Prices: Tues.-Sat. 10-5, Sun. 1-5. No charge; donations accepted. Closed New Year's Day; Easter; Independence Day; Thanksgiving; Christmas Eve & Day. &
Attendance: 25,000 (accurate)
Membership: Individual $25; Family & Patron $100; Benefactor $500; Sustainer $1,000. Corporate: Bronze $250; Silver $500; Gold $1,000.

**NORTH CAROLINA VETERANS PARK & VISITORS CEN-
TER,** 300 Bragg Blvd., Fayetteville, NC 28301. Tel.: 910-433-
1547.
Web Site: www.ncveteranspark.org
Institution Type/Description: Military Veterans Memorial.
Collections: military veterans memorial for all branches of the armed services;
  Service Ribbon Wall; chandelier made from 33,500 dog tags; personal
  stories.
Facilities: 3,500 sq. ft. visitors center.
Hours & Admission Prices: Tues.-Sat. 10-5, Sun. 12-5.

Ferguson

WHIPPOORWILL ACADEMY AND VILLAGE, 11928 NC Hwy.
  268 W., Ferguson, NC 28624. Mailing Address: P.O. Box 458,
  Ferguson, NC 28624. Tel.: 336-973-3237.
E-mail: whippoorwillacademy@hotmail.com
Founded: 1985.
Congressional District: 5
Key Personnel: Dir. & Museum Shop Mgr., Edith F. Carter; Chm. (V),
  Margaret Martine.
Personnel Profile: Part-Time Volunteers 1.
Governing Authority: Tax-exempt.
Institution Type/Description: History Museum.
Collections: local history & culture; period furnishings; crafts; one-room
  schoolhouse; Daniel Boone's replica cabin; Chapel of Peace; blacksmith
  shop; country store; jail; tavern.
Major Exhibits: Lenoir Students Art Exhibit, 5/14-6/14; Ferguson Family Art
  Exhibit, 6/14-8/14; Eddie Tallant Art Exhibit, 8/14-10/14; Brush: Palette Art
  Club Exhibit, 10/15/14-11/15/14.
Facilities: Museum-related items for sale.
Activities: Museum Sponsors: Gospel Singing in September; Daniel Boone
  Day; Christmas Open House.
Hours & Admission Prices: April-Dec. Sat.-Sun. 3-5; other times by appoint-
  ment. Adults $3, children $1. &
Attendance: 6,000 (estimated)

Flat Rock

CARL SANDBURG HOME NATIONAL HISTORIC SITE, 81
  Carl Sandburg Lane, Flat Rock, NC 28731-8635. Tel.: 828-693-
  4178. Fax: 828-693-4179.
E-mail: carl_administration@nps.gov
Web Site: www.nps.gov/carl
Founded: 1968.
Congressional District: 11
Key Personnel: Park Supt., Mr. Tyrone Brandyburg.
Personnel Profile: Full-Time Paid 1; Part-Time Volunteers 3.
Governing Authority: federal. Parent Institution: National Park Service, U.S.
  Dept. of Interior. Tax-exempt.
Institution Type/Description: Historic House: 1838 home of Confederate Sec.
  of Treasury C. G. Memminger & later acquired by Carl Sandburg in 1945.
Collections: Carl Sandburg's working library; books; letters; papers; photo-
  graphs; furnishings of the Sandburg family; Lilian Sandburg's goat files &
  farming equipment.
Research Fields: Carl Sandburg; American literature; goat breeding; 1950s
  material culture.
Facilities: 10,000-vol. library of American literature, history, biography;
  summer outdoor auditorium. Carl Sandburg works for sale.
Activities: guided tours; films; permanent exhibitions; hiking trails.
Hours & Admission Prices: Daily 9-5. Adults $5; children under 17 no charge.
  &
Attendance: 100,000 (estimated)

Fontana Dam

FONTANA DAM & VISITORS CENTER, Hwy. 28 S., Fontana
  Dam, NC 28733-9700. Mailing Address: Tennessee Valley Author-
  ity, 400 W. Summit Hill Dr., Knoxville, TN 37902. Tel.: 800-467-
  1388; 828-498-2234.
Web Site: www.tva.com/sites/fontana.htm
Key Personnel: Site Mgr., Laura Smith
Institution Type/Description: History Museum.
Collections: history of the dam; hydroelectric power; photographs.
Facilities: visitor center.
Activities: guided tours.
Hours & Admission Prices: Visitor Center: May-Nov. daily 9-7.

Forest City

RUTHERFORD COUNTY FARM MUSEUM, 240 Depot St.,
  Forest City, NC 28043-3654. Tel.: 828-248-1248.
Founded: 1994.
Congressional District: 11
Key Personnel: Dir., Wilbur Burgin.
Personnel Profile: Full-Time Volunteers 1.
Governing Authority: Tax-exempt.
Institution Type/Description: Farm Museum.
Collections: farm equipment; period artifacts; murals.
Hours & Admission Prices: Wed.-Sat. 10-4. Adults $2; children no charge.
Attendance: 500 (estimated)

Fort Bragg

82ND AIRBORNE DIVISION MUSEUM, Bldg. C-6841, 5108
  Ardennes St., Fort Bragg, NC 28310. Mailing Address: P.O. Box
  70119, Fort Bragg, NC 28307-0119. Tel.: 910-432-3443 & 5307.
  Fax: 910-432-1642.
Web Site: www.bragg.army.mil/18abn/museums.htm
Founded: 1946.
Congressional District: 7
Key Personnel: Dir. & Cur., John W. Aarsen; Chm. (V), Richard O'Hore;
  Registrar, Betty J. Rucker; Archivist, Jimmie Hallis; Museum Shop Mgr.,
  Ami Cooper.
Personnel Profile: Full-Time Paid 4; Part-Time Paid 1; Part-Time Volunteers
  20.
Governing Authority: federal government. Parent Institution: Museum Branch,
  Training Division. Affiliated with the 82nd Airborne Historical Society.
  Tax-exempt: 501(c)(3).
Institution Type/Description: Military History Museum.
Collections: history of World War I & II, history of 82nd Division from World
  War I to present, Desert Storm & Desert Shield, Panama, Dominican
  Republic, Grenada, Vietnam; U.S. & foreign materials; weapons; uniforms;
  equipment; flags; military vehicles & aircraft; art; photographs; personal
  artifacts.
Research Fields: history of airborne units & warfare.
Facilities: library; 5,000 sq. ft. exhibit space; educational facilities; 65-seat
  theater. Museum-related gifts for sale.
Activities: lectures; films; permanent & temporary exhibitions. Annual Event:
  All American Week in May.
Publications: 82nd Airborne Division History: 1917 to Present.
Hours & Admission Prices: Tues.-Sat. 10-4:30 (valid photo ID required). No
  charge; donations accepted. Closed New Year's Day; Thanksgiving; Christ-
  mas. &
Attendance: 65,000 (accurate)
Membership: Lifetime $5.

JFK SPECIAL WARFARE MUSEUM, (M), Ardennes & Marion
  Sts., Bldg. D-2502, Fort Bragg, NC 28307. Mailing Address: JFK
  SW/SF Branch Museum, P.O. Box 70060, Fort Bragg, NC 28307-
  5000. Tel.: 910-432-1533 & 4272. Fax: 910-432-4062.
E-mail: merrittr@soc.mil
Web Site: www.soc.mil/swcs/museum/html
Founded: 1963.
Key Personnel: Cur., Roxanne M. Merritt; Museum Assn. Pres., Col. William
  Palmer, (USA Ret.); Gift Shop Mgr., Betty Amaker.
Governing Authority: federal. Parent Institution: U.S. Army Special Warfare
  Center & School. Subsidiary Institution: JFK Special Warfare, SF Branch
  Historical & Memorial Association. Tax-exempt: 501(c)(3).
Institution Type/Description: U.S. Army Military Museum.
Collections: unconventional warfare, military & ethnographic collection; U.S.
  Army JFK Special Warfare Center; U.S. Army Special Operations Com-
  mand; U.S. Army Special Forces; U.S. Army Psychological Operations
  Groups; U.S Army Ranger & Civil Affairs units; Hall of Heroes.
Research Fields: unconventional warfare; unit histories of special groups
  including Darby's Rangers, Merrill's Marauders; 1st Special Service Force,
  OSS & Alamo Scouts; Special Forces involvement in Vietnam conflict;
  Airborne Rangers; Special Operations in Latin America & Desert Storm.
Facilities: Items related to Special Forces, Rangers, 4th Psyops, Civil Affairs,
  Airborne & Ft. Bragg for sale.
Activities: permanent & traveling exhibitions.
Publications: quarterly newsletter, Museum Musings; gift shop catalog;
  brochure.
Hours & Admission Prices: Tues.-Sun. 11-4. No charge; donations accepted.
  Closed most federal holidays. &
Attendance: 56,750 (accurate)
Membership: US Army JFK Special Warfare Museum Association: Annual

$10; In Memory/Honor of Donation $50; Life or Century $100; Life & Century $200; Benefactor $1,000; Patron $2,000; Leadership $5,000; Honor $10,000; Distinguished Donor $20,000; Pacesetter $50,000; Special Donor $100,000.

Four Oaks

BENTONVILLE BATTLEFIELD STATE HISTORIC SITE, 5466 Harper House Rd., Four Oaks, NC 27524-9125. Tel.: 910-594-0789. Fax: 910-594-0027.
E-mail: bentonville@ncdcr.gov
Web Site: www.bentonvillebattlefield.nchistoricsites.org
Founded: 1961.
Congressional District: 3
Key Personnel: Site Mgr., Donald B. Taylor; Site Asst. Mgr., Derrick Brown; Site Interpreter, Jeff Fritzinger.
Personnel Profile: Full-Time Paid 4; Part-Time Paid 6.
Governing Authority: state. North Carolina Dept. of Cultural Resources, 109 E. Jones St., Raleigh 27611. Parent Institution: North Carolina Division of Archives & History. Subsidiary Institution: North Carolina Historic Sites Section. Tax-exempt: 170(b).
Institution Type/Description: Historic Site.
Collections: military artifacts; history items. Historic House: c.1855 Harper House, used by Union army as field hospital during Battle of Bentonville.
Facilities: visitor center; audiovisual room; picnic areas.
Activities: guided tours; lectures; walking trail to union trenches; artillery demonstrations; audiovisual programs; seasonal living history; self guided driving tour; fiber optic map of the first day's fighting (6 minute presentation). Annual Events: living history events.
Publications: brochure.
Hours & Admission Prices: Tues.-Sat. 9-5. No charge; donations accepted. Closed most major holidays.
Attendance: 28,000 (accurate)
Membership: Bentonville Battleground Historical Association: Annual $20.

HOWELL WOODS ENVIRONMENTAL LEARNING CENTER, 6601 Devil's Racetrack Rd., Four Oaks, NC 27524. Tel.: 919-938-0115.
Web Site: www.johnstoncc.edu/howellwoods
Key Personnel: Dir., Jordan Astoske
Institution Type/Description: Wildlife & Nature Center.
Collections: native birds, butterflies, reptiles, amphibians & their habitats; freshwater fishes; plants.
Facilities: library; 2,800 acre center; nature trails; camp sites.
Activities: special events; educational programs; hiking trails; live animal exhibits.
Hours & Admission Prices: Call for hours.

Franklin

FRANKLIN GEM & MINERAL MUSEUM, 25 Phillips St., Franklin, NC 28734-3029. Tel.: 828-369-7831 & 342-6360.
E-mail: franklingemsociety@fastmail.fm
Founded: 1974.
Congressional District: 11
Key Personnel: Pres. (V), Lon Peden; Museum Shop Mgr., Wally Smith.
Personnel Profile: Part-Time Volunteers 110.
Institution Type/Description: Gem & Mineral Museum: housed in the former Macon County Public Jail; built in 1850.
Collections: gems; minerals; mining; fossils; Native American artifacts; sea shells.
Activities: special events. Annual Events: Gemboree in July & October.
Publications: The Mountain Gem.
Hours & Admission Prices: May-Oct. Mon.-Sat. 12-4; Nov.-April Sat. 12-4; groups by appointment. No charge. Closed Independence Day.
Attendance: 9,000 (estimated)
Membership: Adult $15, $10 each additional adult, $3 additional per child under 18.

MACON COUNTY HISTORICAL SOCIETY & MUSEUM, 36 W. Main, Franklin, NC 28734. Tel.: 828-524-9758.
E-mail: fund43@aol.com
Founded: 1947.
Congressional District: 11
Key Personnel: Dir., R. Steven Rice; Chm. (V), Robert Poindexter.
Personnel Profile: Full-Time Paid 1; Part-Time Paid 4; Part-Time Volunteers 18.
Governing Authority: Tax-exempt.
Institution Type/Description: History Museum: housed in the Pendergrass Building, c.1904.

Collections: period artifacts; photographs; county history.
Research Fields: genealogical & historical; precontact America.
Facilities: Civil War Letters DVD, histories of Mason County, local history, historical fiction & other museum-related items for sale.
Activities: public programs; school local history & Civil War programs.
Publications: quarterly newsletter; books; Echoes.
Hours & Admission Prices: May-Oct. Tues.-Fri. 10-5, Sat. 1-5; Nov.-April Tues.-Fri. 10-4, Sat. 1-4; other times by appointment. No charge; donations accepted.
Attendance: 5,000 (estimated)
Membership: Student $10; Single $25; Family $35; Sponsor $100; Patron $200; Sutton Society $1,000; Founder Summit $5,000.

THE SCOTTISH TARTANS MUSEUM, 86 E. Main St., Franklin, NC 28734-3026. Tel.: 828-524-7472.
E-mail: tartans@scottishtartans.org
Web Site: www.scottishtartans.org/
Institution Type/Description: General Museum.
Collections: origins, history & development of tartans; Scottish culture, history, dress, migration & military; over 500 tartans.
Facilities: Museum-related items for sale.
Activities: special events; tours.
Hours & Admission Prices: Mon.-Sat. 10-5. Adults $2, children $1.

UPTOWN GALLERY, 30 E. Main St., Franklin, NC 28734. Tel.: 828-349-4607.
E-mail: info@uptowngalleryoffranklin.com
Web Site: www.uptowngalleryoffranklin.com
Institution Type/Description: Art Gallery.
Collections: works by local artists.
Activities: classes; workshops; educational programs; special events; temporary exhibitions.
Hours & Admission Prices: Jan.-April Mon.-Sat. 11-3; May-Dec. Mon.-Sat. 10-5; other times by appointment.

WILDERNESS TAXIDERMY & OUTFITTERS MUSEUM, 5040 Highlands Rd., Franklin, NC 28734-4009. Tel.: 828-524-3677. Fax: 828-349-4200.
Institution Type/Description: Taxidermy Museum.
Collections: animals from around the world; wildlife art.
Hours & Admission Prices: Mon.-Tues. & Thurs.-Fri. 8-5, Sat. 8 am-12 pm.

Fremont

CHARLES B. AYCOCK BIRTHPLACE STATE HISTORIC SITE, 264 Governor Aycock Rd., Fremont, NC 27830-7906. Tel.: 919-242-5581. Fax: 919-242-6668.
E-mail: aycock@ncdcr.gov
Web Site: www.ah.dcr.state.nc.us/hs/Aycock/Aycock.htm
Founded: 1959.
Congressional District: 3
Key Personnel: Site Mgr., Leigh V. Strickland; Museum Shop Mgr., Sarah Pittman.
Personnel Profile: Full-Time Paid 3; Part-Time Paid 1; Part-Time Volunteers 20.
Governing Authority: state. Parent Institution: North Carolina Dept. of Cultural Resources, 4601 Mail Service Center, Raleigh, NC 27699-4601. Tax-exempt: 170(b).
Institution Type/Description: History Museum.
Collections: history. Historic Houses: c.1846 Charles B. Aycock Birthplace; 1893 schoolhouse.
Research Fields: C.B. Aycock; 19th-century farm life; 19th-century education in North Carolina.
Facilities: visitor center.
Activities: guided tours; lectures; films; formally organized education programs for children; permanent exhibitions; special living history program.
Publications: brochure; semi-annual newsletter, The Orator.
Hours & Admission Prices: Mon.-Sat. 9-5. No charge; donations accepted. Closed Martin Luther King Jr. Day; Memorial Day; Independence Day; Labor Day; Veterans Day; Thanksgiving; Christmas Eve & Day. &
Attendance: 19,198 (accurate)

Frisco

FRISCO NATIVE AMERICAN MUSEUM AND NATURAL HISTORY CENTER, 53536 Hwy. 12, Frisco, NC 27936. Mailing Address: P.O. Box 399, Frisco, NC 27936-0399. Tel.: 252-995-4440. Fax: 252-995-4030. Facebook: Frisco Native American Museum.
E-mail: admin@nativeamericanmuseum.org
Web Site: www.nativeamericanmuseum.org
Founded: 1986.
Congressional District: 3
Key Personnel: Exec. Dir., Carl Bornfriend; Chm. & Pres., James Goes; Chm. (V), Elvin Hooper; Education & Public Rels., Joyce Bornfriend; Museum Shop Mgr., Amber Roth; Maintenance, Charles Donald Carmen, III
Personnel Profile: Full-Time Paid 2; Full-Time Volunteers 2; Part-Time Volunteers 45.
Governing Authority: private; nonprofit organization. Tax-exempt: 501(c)(3).
Institution Type/Description: History Museum: located on Hatteras Island, the central building is more than 100 years old with a history of use as a village post office & general store.
Collections: concentration on Native American artifacts from North America including the Algonquian Tribe of Hatteras Island.
Research Fields: Chiricahua, Apache Photo Collection; women's roles in Native American culture; representative samples of Native American dress.
Facilities: 4,000-vol. library; educational facilities; 2,445 sq. ft. exhibit space; nature center. Museum-related items for sale.
Activities: docent program; formal educational programs; guided tours; lectures; participatory exhibits; study clubs. Museum Sponsors: seminars & workshops on Native American culture; Annual Inter-Tribal PowWow, Journey Home; summer programs including beginning archaeology & beginning birding classes every Friday June-August.
Publications: annual brochure; semiannual newsletter, Museum Update.
Hours & Admission Prices: Summer: Tues.-Sun. 10:30-5, Mon. by appointment; Winter: call for hours. Family $15, adults $5, senior citizens $3; discount to AAM members. Closed Thanksgiving; Christmas. &
Attendance: 35,000 (estimated)
Membership: Student $5; Individual $10; Family $25; Charter $100; Sponsoring $200; Life $1,000.

Gastonia

AMERICAN MILITARY MUSEUM, 109 W. Second Ave., Gastonia, NC 28052. Tel.: 704-866-6068.
E-mail: jimboi812@carolina.rr.com
Institution Type/Description: Military Museum.
Collections: military history & memorabilia; personal artifacts; photographs; uniforms; weapons; military ships, planes, tanks & vehicle models; medals; Drum and Bugle Corps memorabilia.
Hours & Admission Prices: Sun. 1-5; other times by appointment.

✻ **SCHIELE MUSEUM OF NATURAL HISTORY AND LYNN PLANETARIUM, (M),** 1500 E. Garrison Blvd., Gastonia, NC 28054-5133. Tel.: 704-866-6908. Fax: 704-866-6041.
E-mail: carried@cityofgastonia.com
Web Site: www.schielemuseum.org
Founded: 1960.
Congressional District: 9
Key Personnel: Dir., Dr. V. Ann Tippitt; Chm. (V), Margaret Mackie; Asst. Dir., Karl McKinnon; Dir. Advancement, Debbie Windley; Dir. Mktg., Amy Ballard; Collections Mgr., Carrie V. Duran; Cur. Life Sciences, Dawn Flynn; Research Coord., Dr. J. Alan May; Head Education, Tony Pasour; Dir. Planetarium, Jim Craig; Security, Mark Rudisill.
Personnel Profile: Full-Time Paid 24; Part-Time Paid 16; Part-Time Volunteers 86; Interns 5.
Volunteer Hours: 5,300
Operating Expenses: 2,690,000
Operating Income: 2,690,000
Governing Authority: municipal. Parent Institution: City of Gastonia. Subsidiary Institution: Board of Trustees of Schiele Museum, Inc. Tax-exempt: 501(c)(3).
Institution Type/Description: Natural History Museum.
Collections: North American mammals; birds; reptiles & amphibians; fishes; invertebrate; fossils; rocks, minerals & gems; anthropology; insects; bird eggs; natural history art; Native American objects; late 18th-century farm; Catawba Indian village.
Major Exhibits: Mammal Safari, 1/14-4/14; Farmers, Warriors, Builders: Secret Life of Ants (T), 1/14-8/14; 50 Greatest Photographs (T), 6/14-10/14.
Research Fields: anthropology; archaeology; mycology; Colonial & Native American lifestyles; entomology; regional ecosystem change; malacology.
Facilities: auditorium; environmental classroom; planetarium; nature trail.

Activities: lectures; films; formally organized educational programs; environment program; docent programs; traveling exhibits; travel programs; education expeditions; museum internship; summer workshops in astronomy, geology, ecology & conservation; Native American studies; Aboriginal studies program; outdoor environmental studies; living history programs in Native American studies & Colonial lifestyles. Museum Sponsors: Colonial Christmas & Thanksgiving Outdoor learning site program.
Publications: annual report; quarterly newsletter; exhibition guides; catalogues; brochures.
Hours & Admission Prices: Mon.-Sat. 9-5, Sun. 1-5. Museum: City Residents: adults $5, students & seniors $4. Non-Residents: adults $7, students & seniors $6. Planetarium: $3. Members & ASTC members no charge. Closed Easter; Thanksgiving; Christmas Eve & Day. &
Attendance: 89,356 (accurate)
Membership: Individual $50; Family $75; Business Partner $150; Corporate Partner & Benefactor $250; Executive Partner & Schiele Society $500; President's Partner $1,000; Chairman's Partner $2,500.

Glencoe

TEXTILE HERITAGE MUSEUM, 2406 Glencoe St., Glencoe, NC 27217. Mailing Address: 1403 E. Webb Ave., Burlington, NC 27217. Tel.: 336-260-0038.
E-mail: textileheritage@triad.rr.com
Web Site: www.textileheritagemuseum.org
Founded: 2001.
Key Personnel: Dir., Jerrie Nall
Governing Authority: Tax-exempt.
Institution Type/Description: Historic Building Museum: housed in the former Company Store and Office Building in the Glencoe Mill Village; built in 1880.
Collections: local textile history; machinery & equipment; mill village life; family labor system; period furnishings; photographs.
Research Fields: textile manufacturers; textiles.
Publications: triannual newsletters; booklets, Spools of Gold; Machines of Industrial Revolution; calendars; annual report.
Hours & Admission Prices: Sat.-Sun. 1-4; groups by appointment. No charge; donations accepted. Closed New Year's Eve & Day; Easter; Thanksgiving; Christmas. &
Attendance: 9,000 (estimated)
Membership: Student $5; Individual $15; Family $25; Friend $50; Century Donor $100; Textile Sponsor $250; Heritage Leader $500; Benefactor $1,000; Museum Patron $5,000; Glencoe Citizen $10,000.

Gold Hill

HISTORIC GOLD HILL, 735 St. Stephens Church Rd., Gold Hill, NC 28071. Mailing Address: P.O. Box 206, Gold Hill, NC 28071. Tel.: 704-960-6457.
Web Site: historicgoldhill.com
Institution Type/Description: Historic Buildings.
Collections: local history & culture; photographs. Historic Buildings: Rock Jail; Mauney's 1840 Store; E.H. Montgomery Store.
Hours & Admission Prices: Call for hours.

Goldsboro

ARTS COUNCIL OF WAYNE COUNTY, 102 N. John St., Goldsboro, NC 27530. Tel.: 919-736-3300.
E-mail: sarah@artsinwayne.org
Web Site: artsinwayne.org
Formerly: Community Arts Council, Inc.
Founded: 1963.
Congressional District: 3
Key Personnel: Exec. Dir., Sarah Merritt; Gallery & Education Dir., Becca Scott Reynolds.
Personnel Profile: Full-Time Paid 2; Part-Time Paid 1.
Governing Authority: nonprofit. Tax-exempt.
Institution Type/Description: Art Museum.
Collections: permanent collection works by contemporary artists.
Facilities: studios; classrooms. Art & fine crafts for sale.
Activities: adult & children's art classes; art demonstrations; workshops; film showings; lectures; tours; monthly exhibits; county-wide outreach programs.
Publications: quarterly newsletter.
Hours & Admission Prices: Mon.-Wed. 9-5, Thurs.-Fri. 9-7, Sat. 4-7. No charge; donations accepted. Closed New Years Day & day after; Good Friday; Easter; Memorial Day; Independence Day; Labor Day; Thanksgiving & day after; Christmas Eve, Day & week. &

Membership: Friends of the Arts: Individual $10; Family $15; Contributing $25; Supporting $50; Patron $100; Sustaining Patron $250; Benefactor $500; Sustaining Benefactor $1,000.

CHERRY HOSPITAL MUSEUM, 201 Stevens Mill Rd., Goldsboro, NC 27530. Tel.: 919-580-2936. Fax: 919-731-3418.
E-mail: tanya.rollins@dhhs.nc.gov
Founded: 1981.
Institution Type/Description: History Museum: housed in a hospital building that was opened by the state in 1880 for African Americans with mental illness.
Collections: African American history; medical & farming equipment; photographs; log books.
Hours & Admission Prices: Mon.-Fri. 8-12 & 1-5. No charge; donations accepted. &
Attendance: 500 (estimated)

GOLDSBOROUGH BRIDGE BATTLEFIELD, Old Mt. Olive Hwy., Goldsboro, NC 27530. Mailing Address: 103 S. George St., Goldsboro, NC 27530. Tel.: 919-736-4423.
E-mail: info@goldsborough.bridge.com
Web Site: www.goldsboroughbridge.com
Institution Type/Description: Historic Site: battlefield where 15,000 soldiers fought in Dec. 1862.
Collections: battlefield history.
Activities: self-guided & guided tours. Annual Event: Battle of Goldsborough Bridge Reenactment in December.
Hours & Admission Prices: Daily sunrise to Sunset. No charge; donations accepted.
Membership: Individual $10; Family $25; Business $50.

SEYMOUR JOHNSON AIR FORCE BASE, 1510 Wright Brothers Ave., Ste. 200, Goldsboro, NC 27534. Tel.: 919-734-2241 & 722-0027.
Web Site: seymourjohnson.af.mil
Institution Type/Description: Military Museum.
Collections: military aircraft; F-15E Strike Eagle; KC-135R Stratotanker.
Activities: tours.
Hours & Admission Prices: last Thurs. each month by appointment.

WAYNE COUNTY MUSEUM, 116 N. William St., Goldsboro, NC 27530. Mailing Address: P.O. Box 665, Goldsboro, NC 27533-0665. Tel.: 919-734-5023.
E-mail: waynecountymuseu@bellsouth.net
Web Site: waynecountyhistoricalnc.org
Founded: 1988.
Congressional District: 1
Key Personnel: Dir., Teresa Williams; Pres. (V), Jane Rustin.
Personnel Profile: Part-Time Paid 2; Part-Time Volunteers 2.
Governing Authority: Parent Institution: Wayne County Historical Association.
Institution Type/Description: History Museum.
Collections: local history, cultural heritage, & science; Civil War artifacts; personal artifacts; period furnishings; photographs.
Facilities: rental facilities.
Activities: talks. Museum Sponsors: Scavenger Hunt.
Publications: quarterly newsletter, Reflections.
Hours & Admission Prices: Tues. 11-8, Wed.-Fri. 11-4, Sat. 12-4; other times by appointment. No charge; donations accepted. &
Attendance: 2,500 (accurate)

WAYNESBOROUGH HISTORICAL VILLAGE & VISITOR CENTER, 801 US Hwy. 117 Bypass S., Goldsboro, NC 27530. Tel.: 919-731-1653.
Web Site: www.waynesboroughhistoricalvillage.com
Institution Type/Description: Historical Village.
Collections: local history & culture; period furnishings; personal artifacts; Native American.
Hours & Admission Prices: Summer: Mon.-Sat. 11-5, Sun. 1-5; Winter: Mon.-Sat. 10-4, Sun. 1-4. No charge.

Graham

CHILDREN'S MUSEUM OF ALAMANCE COUNTY, 217 S. Main St., Graham, NC 07253. Tel.: 336-228-7997.
E-mail: info@childrensmuseumofalamance.org
Web Site: www.childrensmuseumofalamance.org
Founded: 2012.
Congressional District: 6 & 13

Personnel Profile: Full-Time Paid 4; Part-Time Paid 16.
Governing Authority: Tax-exempt.
Institution Type/Description: Children's Museum.
Collections: hands-on exhibitions.
Activities: educational programs.
Hours & Admission Prices: Sat. 10-5, Sun. 1-5. Adults $5; discounts to ACM reciprocal members; members no charge. &
Membership: Single Parent $75; Grandparent $135; Family $150.

GRAHAM HISTORICAL MUSEUM, 135 W. Elm St., Graham, NC 27253. Mailing Address: 637 Johnson Ave., Graham, NC 27253. Tel.: 336-513-4773 & 226-4794.
Institution Type/Description: History Museum.
Collections: Jeanne Robertson, former Miss North Carolina, pageant memorabilia; local history; photographs; personal artifacts; 1930 Seagraves fire truck & fire fighting equipment; Graham High School memorabilia & Sports Hall of Fame; period military artifacts & weapons; Elon College printing press; pottery.
Hours & Admission Prices: Sun. 2-5; other times by appointment. No charge; donations accepted.

Grantsboro

PAMLICO COUNTY MUSEUM & HERITAGE CENTER, 10642 Hwy. 55 E., Grantsboro, NC 28529. Mailing Address: P.O. Box 33, Grantsboro, NC 28529. Tel.: 252-745-2239.
E-mail: pcha@pamlico.net
Web Site: pamlicohistory.com
Institution Type/Description: History Museum.
Collections: county history & heritage; farming; fishing; forestry; photographs; genealogy; newspaper articles; periodicals; blacksmith shop; one-room school house; barbershop; historical crop field.
Activities: 6,000 sq. ft. exhibit space; educational programs; demonstrations. Annual Event: Heritage Day in October.
Hours & Admission Prices: Tues.-Thurs. 10-4. No charge. &

Greensboro

ACC HALL OF CHAMPIONS, 1921 W. Lee St., Greensboro, NC 27403. Tel.: 336-315-8411.
Web Site: acchallofchampions.net
Institution Type/Description: Sports Museum.
Collections: past & present Atlantic Coast Conference coaches & athletes; hands-on exhibitions; game & player memorabilia; personal artifacts; photographs; trophies.
Activities: tours.
Hours & Admission Prices: Thurs.-Sat. 10-4. Adults $5, children 2-12 $3; discounts to groups of 4 or more.

AFRICAN AMERICAN ATELIER, INC., Greensboro Cultural Center, 200 N. Davie St., Ste. 14, Greensboro, NC 27401-2865. Tel.: 336-333-6885. Fax: 336-373-4826.
E-mail: info@africanamericanatelier.org
Web Site: www.africanamericanatelier.org
Founded: 1991.
Key Personnel: Bd. Pres., Eresterine Guidry; Asst. Gallery Mgr., Lou Mecia Koonce; Asst. Gallery Mgr., Angela Fitzgerald.
Governing Authority: Tax-exempt.
Institution Type/Description: Art Gallery.
Collections: African American history & culture; paintings; photographs.
Major Exhibits: 23rd Founding Members Invitational, 1/14-2/14; Gordon James, 3/14-4/14; William "Bill" Gamble, 4/14-5/14; Visual & Performing Arts Program featuring Guilford County Students, 5/14-6/14; Youth Expremiers Exhibit, 6/14-7/14.
Facilities: 2,000 sq. ft. exhibit space.
Activities: youth programs; educational programs; Atelier Around the World Youth Program (AAW); Alberta W. Cuthbertson Visual & Performing Arts Program.
Hours & Admission Prices: Aug.-June Tues.-Sat. 10-5, Sun. 2-5. No charge; donations accepted. &
Attendance: 20,000 (estimated)
Membership: Full-Time Student $10; Senior Citizen $20; Individual $35; Supporter $50-$249; Patron $250-$499; Benefactor $500-$1,000; Sponsor $1,000 & up.

ANNE RUDD GALYON & IRENE CULLIS GALLERIES, GREENSBORO COLLEGE, College Pl., Cowan Humanities Bldg., Greensboro, NC 27401. Mailing Address: 815 W. Market St., Greensboro, NC 27401. Tel.: 336-272-7102. Fax: 336-217-7245.
E-mail: langerj@greensboro.edu
Web Site: www.greensboro.edu/about/galleries
Founded: 1838.
Congressional District: 6
Key Personnel: Gallery Dir., James Langer.
Governing Authority: college. Greensboro College.
Institution Type/Description: College Art Gallery.
Collections: paintings; photographs; sculpture.
Activities: monthly exhibitions: professional, faculty & student shows.
Hours & Admission Prices: Mon.-Fri. 9-5, Sat. 9 to noon. No charge. Closed college holidays. &
Attendance: 3,000 (estimated)

ARTQUEST AT GREENHILL, 200 N. Davie St., Greensboro, NC 27401. Tel.: 336-333-7460. Fax: 336-333-2612.
Web Site: greenhillnc.org
Key Personnel: Exec. Dir. & C.E.O., Laura Way
Institution Type/Description: Art Studio.
Collections: hands-on art studio.
Facilities: restaurant; outdoor amphitheatre; sculpture garden.
Activities: hands-on activities including painting, mold clay, weave, & collage.
Hours & Admission Prices: Tues., Thurs. & Sat. 12:30-5, Wed. 12:30-7, Fri. 10-5. Admission $5; children 1 & under no charge.

BLANDWOOD MANSION, (M), 447 W. Washington St., Greensboro, NC 27401-2348. Mailing Address: P.O. Box 13136, Greensboro, NC 27415-3136. Tel.: 336-272-5003. Fax: 336-271-8049.
E-mail: bbriggs@blandwood.org
Web Site: www.blandwood.org/
Key Personnel: Exec. Dir., Benjamin Briggs
Institution Type/Description: Historic House: former home of North Carolina Governor John Motley Moorehead, c.1790. A National Historic Landmark.
Collections: period furnishings; personal artifacts; photographs.
Hours & Admission Prices: Tues.-Sat. 11-4, Sun. 2-5; groups of 20 or more by appointment. Adults $8, seniors $7, children under 12 $5; discounts to AAA members & groups; school groups no charge. Closed holidays.

BROCK HISTORICAL MUSEUM, Greensboro College, Main Bldg., 815 W. Market St., Greensboro, NC 27401-1875. Tel.: 336-272-7102, ext. 283. Fax: 336-271-6634.
Institution Type/Description: History Museum.
Collections: college history; life at the college; college events; sculpture; photographs.
Hours & Admission Prices: Mon.-Fri. 9-4 by appointment.

CENTER FOR VISUAL ARTISTS - GREENSBORO, 200 N. Davie St., Box 13, Greensboro, NC 27401-2819. Tel.: 336-333-7485. Fax: 336-333-7477.
E-mail: info@greensboroart.org
Web Site: www.greensboroart.org
Formerly: Greensboro Artists' League
Founded: 1956.
Congressional District: 6
Key Personnel: Pres., Martha Mason; Dir. Education, Katie Lank; Education Asst., Christie Gulley; Cur., Kristy Thomas; Programming Asst., Melanie Greene.
Personnel Profile: Full-Time Paid 1; Part-Time Paid 2; Part-Time Volunteers 1; Interns 1.
Governing Authority: nonprofit organization. Parent Institution: United Arts Council of Greensboro. Tax-exempt: 501(c)(3).
Institution Type/Description: Art Gallery: located in the Greensboro Cultural Center.
Collections: works by emerging local artists.
Activities: lectures; gallery talks; juried competition; workshops; special projects.
Publications: monthly member newsletter.
Hours & Admission Prices: Tues. & Thurs.-Sat. 10-5, Wed. 10-7, Sun. 2-5. No charge; donations accepted. Closed legal holidays. &
Attendance: 90,000 (estimated)
Membership: Seniors 55 & over and Students 18 & over $30; Individual 18 & over $40; Family & Supportive $75; Patron $100; Benefactor $250; Sustaining Benefactor $500; Business Sponsorships $1,000.

COLONIAL HERITAGE CENTER AT GUILFORD COURT-HOUSE NATIONAL MILITARY PARK, 2200 New Garden Rd., Greensboro, NC 27410-2354. Tel.: 336-545-5315. Fax: 336-545-5314.
E-mail: guco_administration@nps.gov
Web Site: www.nps.gov/guco
Formerly: Tannenbaum Historic Park
Founded: 1988.
Congressional District: 6
Key Personnel: Dir., Charles Cranfield.
Personnel Profile: Full-Time Paid 3; Part-Time Paid 6; Part-Time Volunteers 20; Interns 2.
Governing Authority: municipal. Parent Institution: City of Greensboro, NC. Tax-exempt.
Institution Type/Description: Historic Site.
Collections: graphic arts; furnishings; historic structures; 18th century maps.
Research Fields: colonial backcountry.
Facilities: picnic facilities.
Activities: living history programs.
Hours & Admission Prices: Fri.-Sat. 8:30-5. No charge.
Attendance: 20,000 (estimated)
Membership: Guilford Battleground Company: Student $10; Individual $20; Family $35; Contributor $50; Donor $100; Benefactor $1,000.

GREEN HILL CENTER FOR NORTH CAROLINA ART, 200 N. Davie St., Greensboro, NC 27401. Mailing Address: 200 N. Davie St., Box 4, Greensboro, NC 27401. Tel.: 336-333-7460. Fax: 336-333-2612. Facebook: Greenhill Art.
E-mail: laura.way@greenhillnc.org
Web Site: www.greenhillnc.org
Formerly: Green Hill Art Gallery
Founded: 1974.
Key Personnel: Exec. Dir. & C.E.O., Laura Way; Dir., Cutorial & Artistic Programs, Edie Carpenter; Dir., Youth & Outreach Programs, Jaymie Meyer; Dir., Education & Community Programs, Mary Young; Devel. & Membership Assoc., Courtney Whittington; ArtQuest Program Coord. & Educator, Verna Fricke; Curatorial Asst., Michelle Janine Lanteri; Receptionist, Delois Bynum.
Governing Authority: nonprofit organization. Tax-exempt: 501(c)(3).
Institution Type/Description: Art Gallery.
Collections: visual arts of North Carolina; North Carolina art including solo & group shows; interactive children's gallery focusing on process of making art; hands-on exhibits.
Research Fields: contemporary North Carolina Art.
Facilities: Sales gallery with North Carolina art & fine crafts for sale.
Activities: guided tours; lectures; gallery talks; concerts; formally organized educational programs.
Publications: newsletter; calendar; catalogs of exhibitions.
Hours & Admission Prices: Gallery & Shop: Tues. & Thurs.-Sat. 10-5, Wed. 10-7, Sun. 2-5. Suggested Donation: $5. ArtQuest: Tues., Thurs., & Sat. 12:30-5, Wed. 12:30-7, Fri. 10-5. Admission $5; children 1 & under, Wed. 5 p.m. to 7 p.m., Household members & up no charge. Closed legal holidays. &
Attendance: 68,652 (accurate)
Membership: Friend $35; Household $65; Contributor $125; Supporter $250; Donor $500; Patron $1,000; Benefactor $2,500; Leadership $5,000; Visionary $10,000; Presidential $20,000.

GREENSBORO CHILDREN'S MUSEUM, 220 N. Church St., Greensboro, NC 27401-2918. Tel.: 336-574-2898. Fax: 336-574-3810.
E-mail: info@gcmuseum.com
Web Site: www.gcmuseum.com
Founded: 1999.
Key Personnel: C.E.O., Betsy Grant; Chm. (V), John Cross; Mktg., Steffany Reeve.
Personnel Profile: Full-Time Paid 9; Part-Time Paid 5; Part-Time Volunteers 20; Interns 8.
Governing Authority: Tax-exempt: 501(c)(3).
Institution Type/Description: Children's Museum.
Collections: hands-on exhibits.
Facilities: Museum-related items for sale.
Activities: rental facilities; educational programs; special events; birthday parties; teacher workshops.
Hours & Admission Prices: Mon. 9-12 (members only), Tues.-Thurs. & Sat. 9-5, Fri. 9-8, Sun. 1-5. Adults $6, seniors $5; discounts to groups of 10 or more, Fri. 5-8 & Sun. 1-5; members & children under one no charge. Closed

New Year's Day; Easter; Memorial Day; Independence Day; Labor Day; Thanksgiving; Christmas Eve & Day. &
Attendance: 140,000 (accurate)
Membership: Family $95; Grandparent $80; ACM $150.

* **GREENSBORO HISTORICAL MUSEUM, INC., (M),** 130 Summit Ave., Greensboro, NC 27401-3016. Tel.: 336-373-2043 & 2982 (before 10am & Mon.). Fax: 336-373-2204.
Web Site: www.greensborohistory.org
Founded: 1924.
Congressional District: 6
Key Personnel: Pres., Margaret Benjamin; Cur. & Registrar, Susan Webster; Cur. Collections, Jon Zachman; Community Historian, Linda Evans; Archivist, Elise Allison; Museum Shop Mgr., Cynthia Kennard.
Personnel Profile: Full-Time Paid 6; Part-Time Paid 9; Part-Time Volunteers 150.
Governing Authority: municipal & historical society. Parent Institution: City of Greensboro. Tax-exempt: 501(c)(3).
Institution Type/Description: Local History Museum.
Collections: local history; transportation; military; Confederate longarms; Dolley Madison; O. Henry; pharmacy & medical artifacts; decorative arts including large collection of American historical glass; local ethnic & religious group exhibits including the Greensboro lunch counter civil rights sit-ins; room interiors; early 1900's Greensboro. Historic Buildings: Christian Isley House; Francis McNairy House; Hockett blacksmith; woodworking shops.
Research Fields: county history.
Facilities: 150-seat auditorium; archives. Museum-related items for sale.
Activities: guided tours; lectures; films; gallery talks; education programs for children; docent program; permanent & temporary exhibitions. Museum Sponsors: intern programs with local universities; school outreach.
Publications: quarterly newsletter.
Hours & Admission Prices: Tues.-Sat. 10-5, Sun. 2-5. No charge; donations accepted. Closed city of Greensboro holidays. &
Membership: Senior Citizen $25; Individual $30; Family $50; Sponsor $100; Sustainer $250; Patron $500; Benefactor $1,000 & up.

* **GREENSBORO SCIENCE CENTER, (M),** 4301 Lawndale Dr., Greensboro, NC 27455-1899. Tel.: 336-288-3769. Fax: 336-288-2531.
E-mail: marketing@greensboroscience.org
Web Site: www.greensboroscience.org
Formerly: The Natural Science Center of Greensboro
Founded: 1957.
Congressional District: 6
Key Personnel: Exec. Dir., Glenn Dobrogosz; Chm. (V), Gary Brown; Dir. Finance & Human Resources, Beth Hamphill; Cur. Herps & Invertebrates, Rick Bolling; Cur. Mammals & Birds, Jessica Hoffman; Cur. Collections & Volunteers, Kelli Landing; Cur. Fisheries, Dennis Law; Cur. Omnisphere, Roger D. Joyner; Dir. Exhibits & Programs, Rick Betton; Retail & Admissions Mgr., Mitch Inman; Cur. Programs, Ron Settle; Cur. Programs, Martha Regester; Mktg. & Membership Mgr., Erica Brown; Dir. Operations, Lindsey Zarecky.
Personnel Profile: Full-Time Paid 31; Part-Time Paid 77; Part-Time Volunteers 620.
Governing Authority: bd. of trustees; nonprofit organization. Tax-exempt: 501(c)(3).
Institution Type/Description: Natural History, Science Museum, Aquarium, Planetarium & Zoo.
Collections: geology; ornithology; entomology; mammals; reptiles; fish; paleontology; conchology; period scientific documents & instruments; living collections.
Major Exhibits: A T-Rex Named Sue (T), 1/18/14-5/4/14.
Facilities: science museum; 3D, full dome & laser dome theater; live theater; zoological park; aquarium; cafe; retail shop.
Activities: theater shows; interactive science exhibits; aquarium dive shows; animal encounters; live shows; zoo.
Publications: brochures; newsletters; educators guides.
Hours & Admission Prices: Museum & Aquarium: daily 9-5. Zoo: daily 10-4. Adults $12.50, seniors 65 & over and children 3-13 $11.50; discounts to Greensboro residents, military, AAM & AZA members; members and children 2 & under no charge. OmniSphere Theater: $3-$5. Closed Thanksgiving; Christmas. &
Attendance: 319,076 (accurate)
Membership: Individual $60; Grandparent $70; Family $80; Extended Family $100; Keeper $150; Curator $250; Director $500.

GUILFORD COLLEGE ART GALLERY, (M), 5800 W. Friendly Ave., Greensboro, NC 27410-4108. Tel.: 336-316-2438. Fax: 336-316-2950.
E-mail: thammond@guilford.edu
Web Site: www.library.guilford.edu/art-gallery
Founded: 1990.
Congressional District: 6
Key Personnel: Dir. & Cur., Theresa N. Hammond.
Personnel Profile: Full-Time Paid 1; Part-Time Paid 1; Interns 1.
Governing Authority: private college. Parent Institution: Guilford College. Tax-exempt: 501(c)(3).
Institution Type/Description: College Art Gallery: located in Hege Library.
Collections: fine art & crafts representing a variety of periods, styles & cultures.
Major Exhibits: Adele Wayman Retrospective, 2/14-5/14.
Facilities: 5,000 sq. ft. exhibit space; permanent collections.
Activities: guided tours; temporary exhibitions; student internships.
Hours & Admission Prices: Main Gallery: Mon.-Fri. 9-5, Sun. 2-5. Atrium Galleries: Mon.-Thurs. 8:30-2am, Fri. 8:30-6, Sat. 10-9, Sun. 12-2am. No charge. Closed college holidays. &
Attendance: 7,500 (estimated)

GUILFORD COURTHOUSE NATIONAL MILITARY PARK, 2332 New Garden Rd., Greensboro, NC 27410-2355. Tel.: 336-288-1776. Fax: 336-282-2296.
E-mail: guco_administration@nps.gov
Web Site: www.nps.gov/guco
Founded: 1917.
Congressional District: 13
Key Personnel: Supt., Charles Cranfield; Museum Shop Mgr., Nancy Stewart.
Personnel Profile: Full-Time Paid 10; Part-Time Paid 1; Part-Time Volunteers 20; Interns 1.
Governing Authority: federal. Parent Institution: Dept. of Interior. Subsidiary Institution: National Park Service. Tax-exempt.
Institution Type/Description: Military Museum.
Collections: Revolutionary War weapons & equipment; soldier mannequins; military artifacts; introductory movie; battle map.
Research Fields: battle & Southern campaign of Revolutionary War.
Facilities: 2,000-vol. library of microfilm, military reports, books pertaining to the American Revolution, books & pamphlets concerning the state of North Carolina available for research on premises by appointment. Booklets pertaining to the Revolutionary War, reproductions of the Declaration of Independence, postcards & reproductions of 1781 Tarleton Map of the Battle of Guilford Courthouse for sale.
Activities: special programs & lectures by appointment; audiovisual program of the Battle of Guilford Courthouse.
Publications: books, Another Such Victory, The Monuments at Guilford Courthouse National Military Park; Battlemap.
Hours & Admission Prices: Visitor Center open Tues.-Sat. 8:30-5. Tour road closed to vehicles Sun. No charge, donations accepted. Closed New Year's Day; Thanksgiving & Christmas. &
Attendance: 70,000 (accurate)

GUILFORD NATIVE AMERICAN ART GALLERY, 200 N. Davie St., Greensboro, NC 27402. Tel.: 336-273-6605.
Web Site: greensboro.nc.gov
Institution Type/Description: Art Gallery.
Collections: contemporary Native American art & crafts.
Facilities: Gift items for sale.
Activities: temporary exhibitions.
Hours & Admission Prices: Tues.-Sat. 10-5, Sun. 2-5.

INTERNATIONAL CIVIL RIGHTS CENTER & MUSEUM, 134 S. Elm St., Greensboro, NC 27401. Tel.: 336-274-9199; 800-748-7116. Fax: 336-274-6244.
E-mail: info@sitinmovement.org
Web Site: www.sitinmovement.org
Key Personnel: Exec. Dir., Bamidele Demerson; CFO, John L. Swaine; Coord. Exhibition Tours, Anita Johnson; Museum Shop Assoc. & Docent, Jean Dulin; Curatorial Program Assoc., Lolita Watkins; Special Events & Membership Coord., Pamela Glass; Museum Shop Assoc., Shirley Tate; Facilities Coord., Tyrone Cook.
Personnel Profile: Full-Time Paid 5; Part-Time Paid 17; Interns 2.
Volunteer Hours: 3,840
Governing Authority: Parent Institution: Sit-in Movement, Inc. Tax-exempt.
Institution Type/Description: History Museum.
Collections: civil & human rights history; photographs; period artifacts; F.W. Woolworth lunch counter; Greensboro Rail Depot reproduction; ballot boxes; voter registration forms & questionnaires.

Facilities: 30,000 sq. ft. exhibit space; 172-seat auditorium; archival center. Museum-related items for sale.
Activities: guided tours; special events; children's story hour; film; permanent & traveling exhibition.
Hours & Admission Prices: Guided Tours: April-Sept. Mon.-Thurs. 9-6, Fri.-Sat. 9-7; Oct.-March Mon.-Sat. 10-6. Adults $10, seniors 65 & over and students $8, children 6-12 $6; children under 6 & AAM members no charge. &
Attendance: 70,000 (accurate)
Membership: Senior Individual $30; Individual $40; Senior Family $60; Family $100.

NORTH CAROLINA A&T STATE UNIVERSITY GALLERIES, Corner of Bluford and Dudley Sts., Greensboro, NC 27411. Mailing Address: Dudley Bldg., 1601 E. Market St., Greensboro, NC 27411-0002. Tel.: 336-334-3209. Fax: 336-334-4378.
E-mail: sharris@ncat.edu
Web Site: www.ncat.edu/~museum/
Formerly: Mattye Reed African Heritage Center
Founded: 1968.
Congressional District: 6
Key Personnel: Cur., Christi Pemberton; Administrative Asst., Lisa Phillips.
Governing Authority: university. Tax-exempt.
Institution Type/Description: University Museum.
Collections: arts & history from over 31 African nations, Mattye Reed African collection; H.C. Taylor contemporary art.
Research Fields: African art history, culture & heritage.
Facilities: education room; artist studio; resource room.
Activities: guided tours; lectures; education programs for students; permanent & traveling exhibitions; community & university departmental partnership programs; artist in residency program; Friends of the Galleries Committee.
Publications: brochures; exhibition catalogues; museum & cultural sites directories.
Hours & Admission Prices: Mon.-Fri. 10-5; other times by appointment. No charge; donations accepted. Closed university holidays.
Attendance: 10,000 (estimated)

* **WEATHERSPOON ART MUSEUM, (M),** Spring Garden & Tate St., Univ. of NC at Greensboro, Greensboro, NC 27402-6170. Mailing Address: P.O. Box 26170, Greensboro, NC 27402-6170. Tel.: 336-334-5770. Fax: 336-334-5907.
E-mail: weatherspoon@uncg.edu
Web Site: weatherspoon.uncg.edu
Formerly: Weatherspoon Art Gallery
Founded: 1941.
Congressional District: 6
Key Personnel: Dir., Nancy Doll; C.E.O., Dr. David Perrin; Pres., Cheryl Stewart; Public & Community Rels. Officer, Loring Martensen; Cur. Collections, Elaine D. Gustafson; Cur. Education, Ann Grimaldi; Cur. Exhibitions, Xandra Eden; Assoc. Cur. Education, Terri Dowell-Dennis; Museum Shop Mgr., Kate Hill.
Personnel Profile: Full-Time Paid 14; Part-Time Paid 2; Part-Time Volunteers 110; Interns 6.
Governing Authority: state; university. Parent Institution: The University of North Carolina at Greensboro. Tax-exempt: 501(c)(3).
Institution Type/Description: University Art Museum.
Collections: American modern & contemporary paintings; sculpture; works on paper; Dillard collection of art on paper; Claribel & Etta Cone collection including Matisse lithographs & bronzes; Lenoir C. Wright Japanese prints; video & digital.
Research Fields: American modern & contemporary art.
Facilities: sculpture garden; auditorium; meeting rooms. Gift items for sale.
Activities: guided tours; lectures; gallery talks; docent program; temporary, traveling & permanent exhibitions; youth programs; membership activities including Contemporary Collectors trips to art venues; performances; film screenings.
Publications: exhibition catalogs; biennial bulletin; newsletters; gallery guides.
Hours & Admission Prices: Tues.-Wed. & Fri. 10-5, Thurs. 10-9, Sat.-Sun. 1-5. No charge; donations accepted. Closed university holidays. &
Attendance: 31,737 (accurate)
Membership: Student $15; Artist $20; Senior Citizen $25; Individual $35; Family $50; Supporter $100; Contributor $250; Friend; $500; Benefactor $1,000; Patron $2,500.

Greenville

* **GREENVILLE MUSEUM OF ART, INC., (M),** 802 S. Evans St., Greenville, NC 27834-3268. Tel.: 252-758-1946.
E-mail: info@gmoa.org
Web Site: www.gmoa.org
Founded: 1956.
Congressional District: 1
Key Personnel: C.E.O., Dir. & Dir. Education, Charlotte Fitz; Bd. Pres., Heather Stepp; Exhibit Designer, Christopher Daniels.
Personnel Profile: Full-Time Paid 2; Part-Time Paid 3; Part-Time Volunteers 10; Interns 1.
Governing Authority: nonprofit organization. Tax-exempt: 501(c)(3).
Institution Type/Description: 20th-century American Art & Visual Arts Museum.
Collections: oils; water colors; sculpture; graphic arts; ceramics; education & children's collections; 20th-century American art.
Research Fields: North Carolina visual art; 1900-1940, American drawings & paintings.
Facilities: classrooms; studios.
Activities: guest artists' series; gallery talks; seminars; docent program; museum tour trips; Artist Association programs; art classes & workshops for adults & children; multicultural programs for children 6 & over.
Publications: quarterly newsletter; monthly exhibition announcements.
Hours & Admission Prices: Tues.-Fri. 10-4:30, Sat. 1-4. Tours: Wed.-Fri. between 10 & 12. No charge; donations accepted. Closed major holidays. &
Attendance: 12,000 (accurate)
Membership: Individual $45; Family $60; Sponsor $100; Donor $150; Patron $250; Benefactor $600; Honorary Trustee $1,200.

WALTER L. STASAVICH SCIENCE AND NATURE CENTER, River Park N., 1000 Mumford Rd., Greenville, NC 27858. Tel.: 252-329-4560. Fax: 252-329-4547.
Institution Type/Description: Science & Nature Center.
Collections: wildlife & their habitats; hands-on exhibits; snakes; fish.
Facilities: 70-seat theater; 10,000 gallon freshwater aquarium.
Activities: turtle touch tank.
Hours & Admission Prices: Tues.-Sat. 9:30-5, Sun. 1-5.

WELLINGTON B. GRAY GALLERY, East Carolina Univ., Jenkins Fine Arts Cntr., Greenville, NC 27858-4353. Tel.: 252-328-6336. Fax: 252-328-6441.
E-mail: braswellg@ecu.edu
Web Site: www.ecu.edu/art/home/html
Founded: 1978.
Congressional District: 1
Key Personnel: Dir., Tom Braswell.
Personnel Profile: Full-Time Paid 3; Part-Time Paid 14; Part-Time Volunteers 10; Interns 1.
Governing Authority: university. Parent Institution: East Carolina University. Subsidiary Institution: East Carolina University School of Art. Tax-exempt.
Institution Type/Description: Art Gallery.
Collections: contemporary & African art.
Research Fields: contemporary art.
Facilities: 6,000 sq. ft. exhibit space.
Activities: lectures; films; gallery talks; temporary & traveling exhibitions; international, national & regional exhibitions of contemporary works, installations & site-specific work.
Publications: catalogs for selected exhibitions; Fiber: Fabrication/Revelation; Jacob Lawrence: An American Master; Minnie Evans: Artist; Anders Knutsson: A Retrospective; Baltic Ceramist: 1996; Robert Lee Humber: A Collector Creates; International Photography & Digital Image Exhibition.
Hours & Admission Prices: Mon.-Fri. 10-4, Sat. 10-2. No charge; donations accepted. Closed for university holidays. &
Attendance: 20,102 (accurate)

Grifton

GRIFTON MUSEUM AND CATECHNA INDIAN VILLAGE, 437 A Creekshore Dr., Grifton, NC 28530. Mailing Address: P.O. Box 85, Grifton, NC 28530. Tel.: 252-524-0190.
Institution Type/Description: American Indian Village & Museum.
Collections: Native American artifacts; early century doctor's office; Civil War artifacts; photographs; personal artifacts; furnishings.
Activities: demonstrations.
Hours & Admission Prices: 1st & 3rd Sun. 1-5; other times by appointment.

Grover

PRESIDENTIAL CULINARY MUSEUM, 301 Cleveland Ave.,
Grover, NC 28073. Tel.: 704-937-2940. Fax: 888-217-3470.
E-mail: curator@presidentialculinarymuseum.org
Web Site: www.presidentialculinarymuseum.org
Founded: 2007.
Congressional District: 110
Key Personnel: Dir. & Pres., Martin C.J. Mongiello; C.E.O. & Chm., Michael
　C. Mongiello, Jr.; Chm. (V), Allan B. Miller; Pres. (V), Rick E. Scott;
　Museum Shop Mgr., Stormy L. Mongiello.
Personnel Profile: Full-Time Paid 4; Part-Time Paid 8; Part-Time Volunteers 5;
　Interns 1.
Volunteer Hours: 411
Operating Income: 191,000
Governing Authority: Parent Institution: Mongiello Holding, LLC. Subsidiary
　Institution: American Revolutionary War Living History Center & Presi-
　dential Service Center. Supported by The National Archives and several
　presidential foundations.
Institution Type/Description: History Museum.
Collections: private & public collections of Chef Martin CJ Mongiello, a
　former Executive Chef to the President of the United States; photographs;
　personal artifacts; art & menus from around the world; letters from
　Presidents; Presidential china; Oil paintings recently commissioned.
Research Fields: Colonial War of 1776-1783; presidential memorabilia;
　presidents & first families eating habits, foods & recipes.
Facilities: 2 gift shops.
Activities: bus tours; special tours; live stage shows; cooking classes.
Publications: monthly e-newsletter; books on private estate management;
　cookbooks & hotel legal guide
Hours & Admission Prices: Guided Tours: daily 9:30am; bus tours & specialty
　groups by appointment. Adults $7, senior citizens & military $5; discounts
　to AAM & ICOM members; children under 12 no charge.
Attendance: 3,220 (accurate)

Halifax

HISTORIC HALIFAX STATE HISTORIC SITE, 25 St. David
　St., Halifax, NC 27839. Mailing Address: P.O. Box 406, Halifax,
　NC 27839-0406. Tel.: 252-583-7191. Fax: 252-583-9421.
E-mail: halifax@ncdcr.gov
Web Site: www.halifax.nchistoricsites.org
Founded: 1955.
Congressional District: 2
Key Personnel: Sites Mgr., Monica Moody; Chm., Wrenn Phillips; Museum
　Shop Mgr., Sarah Hill.
Personnel Profile: Full-Time Paid 5; Part-Time Volunteers 6.
Governing Authority: state. Parent Institution: North Carolina Historic Sites,
　109 E. Jones St., Raleigh, NC 27611. Tax-exempt: 170(b)(1)(A).
Institution Type/Description: Preservation Project & Visitor Center.
Collections: archaeology; history. Historic Houses: 1760 Owens House; 1808
　Burgess House; 1833 Clerk's office; 1838 Jail; 1790 Tap Room; 1790 Eagle
　Tavern; 1808 Sally-Billy House; 1783 William R. Davie House.
Research Fields: archaeology; history; preservation project.
Facilities: visitor center; picnic area.
Activities: guided tours; lectures; permanent exhibitions.
Publications: brochure.
Hours & Admission Prices: Tues.-Sat. 9-5. No charge; donations accepted.
　Closed holidays. &
Attendance: 30,000 (estimated)
Membership: Individual $15; Contributors $30; Donor $75; Patron $125.

Hamilton

**FORT BRANCH CONFEDERATE EARTHEN FORT CIVIL
　WAR SITE,** 2883 Fort Branch Rd., Hamilton, NC 27840. Mailing
　Address: P.O. Box 355, Hamilton, NC 27840. Tel.: 800-776-8566.
Web Site: www.fortbranchcivilwarsite.com
Institution Type/Description: Civil War Site: housed on the site of the
　Confederate Army in 1861.
Collections: Civil War history; cannons; Native American pottery; military
　artifacts.
Activities: Annual Events: Battle Re-enactments in November; Fort Branch
　Christmas in December.
Hours & Admission Prices: April to early Nov. Sat.-Sun. 1:30-5:30; other
　times by appointment.

Hamlet

**NATIONAL RAILROAD MUSEUM AND HALL OF FAME,
　INC.,** 120 E. Spring St., Hamlet, NC 28345. Mailing Address: P.O.
　Box 1583, Hamlet, NC 28345-1583. Tel.: 910-582-2383.
E-mail: nrrmhof@yahoo.com
Founded: 1976.
Congressional District: 8
Key Personnel: Pres., Tim Nevinger; Sec. & Treas., Kay Cavendish; Public
　Rels., Bobbie Williams.
Personnel Profile: Part-Time Volunteers 9.
Governing Authority: private; nonprofit organization. Tax-exempt: 501(c)(3).
Institution Type/Description: Railroad Museum.
Collections: photographs; maps; model railroad layout; four pieces of rolling
　stock; recreated telegraph office; SAL locomotive 1114 SDP 35 & caboose
　SAL 5241; 1892 Tornado replica.
Facilities: library. Museum-related items for sale.
Activities: formal education programs for children; guided tours; lectures;
　facility rental for special groups. Annual Event: Seaboard Festival Day in
　October.
Publications: brochures.
Hours & Admission Prices: Sat.-Sun. 1-4; other times by appointment. No
　charge; donations accepted. Closed Christmas. &
Attendance: 6,500 (estimated)
Membership: Student $5; Adult $10.

Harkers Island

**CAPE LOOKOUT LIGHTHOUSE & LIGHT STATION VISI-
　TOR CENTER AND KEEPERS QUARTERS MUSEUM,** 131
　Charles St., Harkers Island, NC 28516. Tel.: 252-728-2250. Face-
　book: Cape Lookout.
Institution Type/Description: Historic Lighthouse: built in 1859.
Collections: lighthouse history; photographs; nautical artifacts; keepers quar-
　ters.
Facilities: visitors center.
Activities: group tours.
Hours & Admission Prices: Lighthouse: Wed.-Sat. 10-3:45; groups by appoint-
　ment; call for holiday hours. Adults $8, children 12 & over and seniors 62
　& over $4. Ferry transportation not included in fees.

**CORE SOUND WATERFOWL MUSEUM & HERITAGE CEN-
　TER,** 1785 Island Rd., Harkers Island, NC 28531-9670. Mailing
　Address: P.O. Box 556, Harkers Island, NC 28531-0556. Tel.:
　919-728-1500. Fax: 919-728-1742.
E-mail: museum@coresound.com
Web Site: www.coresound.com
Founded: 1992.
Congressional District: 3
Key Personnel: Dir., Karen Willis Amspacher; Chm. (V), Charles S. Jones;
　Museum Shop Mgr., Jennifer Taylor.
Personnel Profile: Full-Time Paid 2; Part-Time Paid 8; Part-Time Volunteers
　425.
Governing Authority: private; nonprofit organization. Tax-exempt.
Institution Type/Description: Heritage Center.
Collections: local history & culture; waterfowling traditions.
Research Fields: local history; coastal culture.
Facilities: library; archives; community center. Museum-related items for sale.
Activities: carving demonstrations; boatbuilding & restoration; oral histories;
　community documentary projects; foodways.
Hours & Admission Prices: Mon.-Sat. 10-5, Sun. 2-5. No charge; donations
　accepted. Closed New Year's Day; Easter; Thanksgiving; Christmas. &
Attendance: 25,000 (estimated)
Membership: Individual $30; Family $50; Businesses $75.

Harrisburg

U.S. LEGEND CARS INTERNATIONAL, 5245 Hwy. 49 S.,
　Harrisburg, NC 28075. Tel.: 704-455-3896. Fax: 704-455-3820.
Web Site: 600racing.com
Institution Type/Description: Race Car Manufacturer.
Collections: manufacturer of Legends Car, Bandolero, Thunder Roadsters, &
　Legends Dirt Modified cars.
Activities: facility tours.
Hours & Admission Prices: Call for hours.

WOOD BROTHERS RACING, 7201 Caldwell Rd., Harrisburg, NC
　28075. Tel.: 704-456-1221.
E-mail: 21team@woodbrothersracing.com

Web Site: woodbrothersracing.com
Institution Type/Description: Racing Company.
Collections: photographs; race cars.
Facilities: Gift items for sale.
Activities: viewing windows.
Hours & Admission Prices: Call for hours.

Hatteras

GRAVEYARD OF THE ATLANTIC MUSEUM, (M), 59200
Museum Dr., Hatteras, NC 27943. Mailing Address: P.O. Box 191,
Hatteras, NC 27943-0191. Tel.: 252-986-2995. Fax: 252-986-1212.
E-mail: museum@graveyardoftheatlantic.com
Web Site: www.graveyardoftheatlantic.com
Governing Authority: Parent Institution: State of North Carolina Museum of
History. Tax-exempt.
Institution Type/Description: Maritime History Museum.
Collections: maritime history; North Carolina Outer Banks shipwrecks;
seafaring history.
Hours & Admission Prices: Mon.-Fri. 10-4. No charge; donations accepted. &
Attendance: 50,000 (accurate)

Havelock

**EASTERN CAROLINA AVIATION HERITAGE FOUNDA-
TION,** 201 Tourist Center Dr., Havelock, NC 28532. Mailing
Address: P.O. Box 368, Havelock, NC 28532. Tel.: 252-444-7260.
E-mail: dmiller@havelocknc.us
Web Site: www.ecaviationheritage.com/Index.html
Institution Type/Description: Military Aviation History Museum.
Collections: Marine Corps aviation history, aircraft, equipment & artifacts;
photographs; uniforms.
Activities: educational programs; special events.
Hours & Admission Prices: Mon.-Sat. 9-5.

Haw River

**CHILDREY HOUSE, WORLD WAR II, HOME FRONT MU-
SEUM,** 309 E. Main St., Haw River, NC 27258. Mailing Address:
c/o Haw River Historical Association, P.O. Box 936, Haw River,
NC 27258. Tel.: 336-578-0784 & 684-1002.
Founded: 2009.
Congressional District: 6 & 13
Key Personnel: Dir., Gail Knauft; Dir., Buddy Boggs.
Personnel Profile: Part-Time Volunteers 50.
Governing Authority: Parent Institution: Haw River Historical Association.
Tax-exempt.
Institution Type/Description: History Museum.
Collections: period furnishings; personal artifacts; photographs; interpretive
notes; 1940's posters & magazines.
Publications: quarterly newsletter, The Shuttle.
Hours & Admission Prices: Veterans Day weekend & Memorial Day weekend;
other times by appointment. No charge; donations accepted.
Attendance: 150 (estimated)

HAW RIVER HISTORICAL MUSEUM, 201 E. Main St., Haw
River, NC 27258. Mailing Address: P.O. Box 936, Haw River, NC
27258-0936. Tel.: 336-578-0784 & 684-1002. Facebook: Haw
River Museum.
Founded: 1997.
Key Personnel: Dir., Gail Knauff; Dir., Cathy Wilson; Treas., Buddy Boggs.
Personnel Profile: Part-Time Volunteers 100.
Governing Authority: Parent Institution: Haw River Historical Association.
Tax-exempt.
Institution Type/Description: History Museum.
Collections: local history & culture; period furnishings; personal artifacts;
photographs.
Activities: educational programs; lectures; special events.
Publications: quarterly newsletter, The Shuttle.
Hours & Admission Prices: Sat.-Sun. 1-4; other times by appointment. No
charge; donations accepted.
Attendance: 600 (estimated)
Membership: Individual $10; Family $15.

Hawk

DELLINGER GRIST MILL ON CANE CREEK, 4020 Cane
Creek Rd., Hawk, NC 27949. Mailing Address: 52 Maple St.,
Bakersville, NC 28705. Tel.: 828-688-1009.
E-mail: jackdellinger@bellsouth.net
Web Site: www.dellingermill.com
Institution Type/Description: Grist Mill: built in 1867. Listed on The National
Register of Historic Places.
Collections: mill history; period equipment.
Facilities: Mill-related items for sale.
Activities: demonstrations; special events.
Hours & Admission Prices: Call for hours; groups by appointment.

Hayesville

**CLAY COUNTY HISTORICAL AND ARTS COUNCIL MU-
SEUM,** 21 Davis Loop, Hayesville, NC 28904. Mailing Address:
P.O. Box 5, Hayesville, NC 28904-0005. Tel.: 828-389-6814.
Institution Type/Description: History Museum: housed in a former county jail;
built in 1912.
Collections: Native American artifacts; photographs; early farm kitchen; 1800s
loom & clothing; 1916 switchboard; 1838 quilt.
Hours & Admission Prices: Memorial Day to Labor Day Tues.-Sat. 10-4;
Sept.-Oct. Fri.-Sat. 10-4. No charge; donations accepted.
Attendance: 800

Henderson

HENDERSON INSTITUTE HISTORICAL MUSEUM, 629 W.
Rockspring St., Henderson, NC 27536. Mailing Address: P.O. Box
2081, Henderson, NC 27536. Tel.: 252-430-0616.
Institution Type/Description: Historic Building: housed in the former second-
ary school for African Americans in Vance County; built in 1887.
Collections: school history & records; photographs; period furnishings.
Activities: permanent exhibitions.
Hours & Admission Prices: Wed.-Sat. 1-4; other times by appointment.

Hendersonville

HENDERSON COUNTY HERITAGE MUSEUM, 1 Historic
Courthouse Sq., Main St., Hendersonville, NC 28792. Tel.: 828-
694-1619.
Web Site: www.hendersoncountymuseum.com
Founded: 2008.
Key Personnel: Exec. Dir., Bette Carter; Supvr., Gale Henry
Institution Type/Description: Heritage Museum.
Collections: county history, heritage, & culture; personal artifacts; photo-
graphs; period furnishings; military artifacts.
Activities: special events; demonstrations.
Hours & Admission Prices: Wed.-Sat. 10-5, Sun. 1-5.

HISTORIC JOHNSON FARM, 3346 Haywood Rd., Henderson-
ville, NC 28791-9721. Tel.: 828-891-6585. Fax: 828-890-7001.
Institution Type/Description: Historic Farm: housed on a late 19th-century
tobacco farm. Listed on the National Register of Historic Places.
Collections: family & farm history; personal artifacts; period furnishings.
Historic Buildings: 10 farm buildings.
Facilities: nature trails.
Activities: guided tours.
Hours & Admission Prices: Tours: June-Aug. Mon.-Thurs. 10:30am; Sept.-
May Tues.-Fri. 10:30am. Adults $5, students K-12 $3; preschool & under
no charge.

HOLMES EDUCATIONAL STATE FOREST, 1299 Crab Creek
Rd., Hendersonville, NC 28739-8440. Tel.: 828-692-0100.
Institution Type/Description: State Forest.
Collections: plants; trees; flowers; ecology.
Facilities: nature trails.
Activities: audio stations; classes; picnic area; hiking trails; educational
programs.
Hours & Admission Prices: mid-March to late Nov. Tues.-Sun.

**MINERAL AND LAPIDARY MUSEUM OF HENDERSON
COUNTY INC.,** 400 N. Main St., Hendersonville, NC 28792-
4901. Tel.: 828-698-1977. Fax: 828-698-1977.
E-mail: info@mineralmuseum.org
Web Site: www.mineralmuseum.org

Founded: 1996.
Congressional District: 11
Key Personnel: C.E.O. & Pres., Helen Hauser; Treas., Zeb Palmer; Museum Shop Mgr., Diane Lapp.
Personnel Profile: Part-Time Volunteers 65.
Governing Authority: private; nonprofit. Tax-exempt.
Institution Type/Description: Mineral & Lapidary Museum.
Collections: minerals from USA & world, mostly from North Carolina.
Facilities: 1,100 sq. ft. exhibit space. Items related to minerals, gems, fossils, fluorescents, jewelry for sale.
Activities: formal education programs for children; hobby workshops.
Hours & Admission Prices: Mon.-Fri. 1-5, Sat. 10-5. No charge; donations accepted. Closed New Year's Day; Thanksgiving; Christmas. &
Attendance: 31,865 (accurate)

MOUNTAIN FARM AND HOME MUSEUM, 101 Brookside Camp Rd., Hendersonville, NC 28792-1101. Tel.: 828-697-8846.
Founded: 1998.
Congressional District: 11
Key Personnel: Pres. (V), A.B. Wexler
Governing Authority: Tax-exempt.
Institution Type/Description: History Museum.
Collections: period farm machinery & tools.
Major Exhibits: N. C. Mt. State Fair (T).
Research Fields: N
Activities: educational programs. Annual Event: North Carolina Mountain State Fair.
Hours & Admission Prices: Mon.-Fri. 9-3. No charge; donations accepted. &
Attendance: 500 (estimated)

WESTERN NORTH CAROLINA AIR MUSEUM, 1340 E. Gilbert St., Hendersonville, NC 28792. Mailing Address: P.O. Box 2343, Hendersonville, NC 28793-2343. Tel.: 828-698-2482. Facebook: Western North Carolina Air Museum.
Founded: 1990.
Key Personnel: Dir. Docents, Robert Field; Pres. (V) & Museum Shop Mgr., Donald Buck.
Personnel Profile: Part-Time Volunteers 230.
Governing Authority: private. Tax-exempt: 501(c)(3).
Institution Type/Description: Aviation Museum.
Collections: aviation heritage of North Carolina; period aircraft; flight manuals; engines; photographs; models; books.
Facilities: Museum related items for sale.
Activities: special events. Annual Event: Air Fair first weekend in June; Apple Festival in September.
Publications: monthly newsletter.
Hours & Admission Prices: April-Oct. Sun. & Wed. 12-5, Sat. 10-5; Nov.-March Wed. & Sat.-Sun. 12-5. No charge; donations accepted. &
Membership: Individual $30; Family $40; Life $350; Family Life $450.

Hertford

THE JIM "CATFISH" HUNTER BASEBALL MUSEUM, 118 W. Market St., Hertford, NC 27944-1151. Tel.: 252-426-5657.
Key Personnel: Exec. Dir. Chamber of Commerce, J. Sid Eley
Institution Type/Description: Sports Museum.
Collections: baseball history & memorabilia; personal artifacts; photographs; books; documents; bronze plaques; Hall of Fame Inductees.
Hours & Admission Prices: Mon.-Fri. 9:30-4:30, Sat. 10-12; other times by appointment. No charge; donations accepted.

NEWBOLD WHITE HOUSE HISTORIC SITE, 151 Newbold-White Rd., Hertford, NC 27944-8240. Mailing Address: P.O. Box 103, Hertford, NC 27944-0103. Tel.: 252-426-7567.
E-mail: nbwh1730@embarqmail.com
Web Site: www.perquimansrestoration.org
Founded: 1971.
Congressional District: 3
Key Personnel: Dir. & Pres. (V), Phil McMullen; Museum Shop Mgr., Sylvia Wyatt.
Personnel Profile: Part-Time Volunteers 130.
Volunteer Hours: 2,000
Operating Expenses: 34,825
Operating Income: 61,381
Governing Authority: Parent Institution: The Perguimans County Restoration Association, Inc. Tax-exempt.
Institution Type/Description: Historic Site.
Collections: period furnishings; 18th-19th century artifacts; reproduction

Periauger work boat; 18th century smokehouse; kitchen garden; 17th century Quaker cemetery. Historic House: c.1730 Sanders' home.
Facilities: nature trail; picnic area. Museum-related items for sale.
Hours & Admission Prices: April-Oct. Thurs.-Sat. 10-4; other times by appointment. Adults $5, child $3; discounts to groups; members & children under 6 no charge.
Attendance: 1,100 (accurate)
Membership: Individual $30; Family $40; Sponsor $100; Donor $250; Friend $500; Patron $1,000; Abraham Sanders $5,000.

Hickory

CATAWBA SCIENCE CENTER, 243 3rd Ave., N.E., Hickory, NC 28601-5168. Mailing Address: P.O. Box 2431, Hickory, NC 28603-2431. Tel.: 828-322-8169, ext. 300. Fax: 828-322-1585.
E-mail: info@catawbascience.org
Web Site: www.catawbascience.org
Founded: 1975.
Congressional District: 10
Key Personnel: Dir., Alan Barnhardt; Asst. Dir., Tricia Little; Pres. (V), Roger Young; Dir. Exhibits, Phil Hawn; Dir. Programs, Erin Graves; Museum Shop Mgr., Nadia Scopes.
Personnel Profile: Full-Time Paid 13; Part-Time Paid 30; Part-Time Volunteers 50.
Governing Authority: nonprofit organization. Tax-exempt: 501(c)(3).
Institution Type/Description: Science & Technology Center.
Collections: saltwater & freshwater marine habitats; astronomy.
Facilities: digital planetarium.
Activities: informal education programs for children; permanent & temporary exhibitions; pre-school outreach program.
Publications: newsletter; brochure.
Hours & Admission Prices: Tues.-Fri. 10-5, Sat. 10-4, Sun. 1-4. Adults $7, senior citizens 62 and over, military with ID & youth 3-18 $5; children under 3, members & ASTC Passport Program Participants no charge. Closed major holidays. &
Attendance: 101,000 (accurate)
Membership: Senior Citizen $40; Individual $55; Family/Grandparents $70; Patron $100; Benefactor $250; Stellar Society $500; Angel $1,000; President's Circle $2,500; Philanthropist $5,000.

HARPER HOUSE/HICKORY HISTORY CENTER, (M), 310 N. Center St., Hickory, NC 28601-5031. Tel.: 828-324-7294. Fax: 828-465-9813.
E mail: palmoreccha@gmail.com
Web Site: catawbahistory.org
Founded: 2006.
Key Personnel: Dir., Michelle Palmore; C.E.O., Melinda L. Herzog.
Personnel Profile: Full-Time Paid 1; Part-Time Paid 1.
Governing Authority: Parent Institution: Catawba County Historical Association. Tax-exempt.
Institution Type/Description: History Museum.
Collections: local history & culture; period furnishings; personal artifacts; photographs. Historic Houses: 1887 Harper House; 1912 Bonniwell-Lyerly House.
Research Fields: Hickory, NC history; textile history; furniture industry.
Hours & Admission Prices: Thurs.-Sat. 9-4, Sun. 1:30-4:30; other times by appointment. Tours: Harper House: $5; discounts to AAM members; members no charge. Bonniwell-Lyerly House: no charge.
Attendance: 12,250 (accurate)

HICKORY AVIATION MUSEUM, Hickory Regional Airport, 3101 9th Ave. Dr., N.W., Hickory, NC 28601-8646. Tel.: 828-323-7408.
E-mail: lindajhill@bellsouth.net
Web Site: hickoryaviationmuseum.org
Institution Type/Description: Aviation History Museum.
Collections: aviation history, equipment & artifacts; aircraft; photographs; personal artifacts; flight suits; paintings.
Hours & Admission Prices: Sat. 10-5, Sun. 1-5; other times by appointment.

HICKORY FURNITURE MART MUSEUM, 2220 Hwy. 70 S.E., Level 1, West Entrance, Hickory, NC 28602. Tel.: 800-462-6278.
Web Site: hickoryfurniture.com
Institution Type/Description: Furniture Museum.
Collections: early furnishings & workshop including furniture-making tools; area furniture manufacturers' history.
Facilities: cafe; visitors center.
Activities: group tours; videos.
Hours & Admission Prices: Mon.-Sat. 9-6.

HICKORY LANDMARKS SOCIETY-PROPST HOUSE AND MAPLE GROVE MUSEUMS, 542 2nd Ave., Hickory, NC 28601. Mailing Address: P.O. Box 2341, Hickory, NC 28603-2341. Tel.: 828-322-4731. Fax: 828-327-9096.
E-mail: info@hickorylandmarks.org
Web Site: www.hickorylandmarks.org/
Founded: 1968.
Congressional District: 10
Key Personnel: Exec. Dir. & C.E.O., Patrick T. Daily; Cur. Collections, Leslie Keller.
Personnel Profile: Full-Time Paid 1; Part-Time Paid 2; Part-Time Volunteers 40.
Governing Authority: nonprofit organization. Affiliated of the Catawba County Council for the Arts. Tax-exempt: 501(c)(3).
Institution Type/Description: Three Victorian House museums (1882-1895).
Collections: Victorian period furniture.
Research Fields: 19th century local history.
Activities: house tours, special events.
Publications: newsletter, Landmarkings.
Hours & Admission Prices: Propst St. House Museum: March 15-Dec. 15 Thurs. & Sun. 1:30-4:30. Maple Grove Museum: Mon.-Fri. 9-5. No charge; donations accepted.
Attendance: 6,300 (estimated)
Membership: Individual $25; Family $35.

✱ **THE HICKORY MUSEUM OF ART, (M),** 243 Third Ave., N.E., Hickory, NC 28601-5168. Mailing Address: P.O. Box 2572, Hickory, NC 28603-2572. Tel.: 828-327-8576. Fax: 828-327-7281.
E-mail: info@hickorymuseumofart.org
Web Site: www.hickorymuseumofart.org
Founded: 1944.
Congressional District: 10
Key Personnel: Exec. Dir., Lise C. Swensson; Pres. (V), Kit Cannon; Dir. Education, Virginia Zellmer; Dir. Communications & Mktg., Kristina Anthony; Dir. Membership, Chrissy Schramm; Museum Shop Mgr., Ronni Smith.
Personnel Profile: Full-Time Paid 5; Part-Time Paid 13; Part-Time Volunteers 125; Interns 2.
Volunteer Hours: 5,740
Operating Expenses: 564,184
Operating Income: 536,903
Governing Authority: board of trustees. Tax-exempt: 501(c)(3).
Institution Type/Description: Art Museum.
Collections: 1850 to present American art; American art pottery; NC glass & pottery; Southern contemporary folk art.
Major Exhibits: Juie Ratley III, 10/13-1/14; Elizabeth Catlett: Artist, Activist, 1/14-3/14; Discover Folk Art, 1/14-12/14; Gordon Wetmore & Friends, 1/14-3/14; 7 Painters Launch 70 Years, 3/14-8/14; Qualities of Light: Works from the Permanent Collection, 4/14-9/14; Wolf Kahn, 6/14-10-14; Artistic Influences on Paul Whitener, 9/14-2/15; In Time We Shall Know Ourselves: Photographs by Raymond Smith, 10/14-1/15.
Research Fields: American art.
Facilities: 1200-vol. library of art; research books; classrooms.
Activities: guided tours; lectures; films; gallery talks; permanent & traveling exhibitions; art classes.
Publications: quarterly newsletter; monthly announcements; catalogues on Museum collections; exhibition catalogues.
Hours & Admission Prices: Tues.-Sat. 10-4, Sun. 1-4. No charge; donations accepted. Closed New Year's Day; Easter; Independence Day; Thanksgiving; Christmas. ♿
Attendance: 36,124 (accurate)
Membership: Individual $40; Family $50; Contributor $100; Patron $150; Sustainer $250; Benefactor $500; Sponsor $1,000; President's Circle $2,500.

Hiddenite

HIDDENITE CENTER, INC., (M), 316 Church St., Hiddenite, NC 28636. Mailing Address: P.O. Box 311, Hiddenite, NC 28636-0311. Tel.: 828-632-6966. Fax: 828-632-5756.
E-mail: info@hiddenitecenter.com
Web Site: www.hiddenitecenter.com
Founded: 1981.
Congressional District: 5
Key Personnel: Asst. Dir., Karen B. Walker; Dir. Education, Allison S. Houchins.
Personnel Profile: Part-Time Paid 8; Part-Time Volunteers 60.
Governing Authority: state. Tax-exempt: 501(c)(3).

Institution Type/Description: General Museum & Art Center: housed in 1900 James Paul Lucas Mansion.
Collections: dolls; gem & mineral display; pottery; basketry; quilts; visual arts; paintings; woven works; multicultural exhibits; house furnishings.
Research Fields: quilt documentation; historic architecture; folklorist research of local tradition.
Facilities: 250-seat auditorium; educational facilities; theatre. Handcrafted items, books & pens for sale.
Activities: lectures; guided tours; concerts; dance recitals; arts festivals; theatre; organized educational programs; docent program; participatory, loan & temporary exhibitions; hobby workshops. Center Sponsors: Celebration of Arts; Local Art Competition; Heritage Fair; Christmas Show; Quilt Symposium.
Publications: bi-monthly newsletter, Friendsletter.
Hours & Admission Prices: Tues.-Fri. 10-4:30, Sat. 10-3. Museum: adults $2.50, seniors & students $1.50; discounts to AAM members & NC Museums Council; children under 6, members, Art Center, gallery & doll collection no charge. Closed Easter & day after; Independence Day; Labor Day; Thanksgiving; Christmas. ♿
Attendance: 25,000 (estimated)
Membership: Individual, Senior $25; Family $40; Patron $50; Grand Patron $100; Sponsor $500; Endowment Donor $1,000; Angel $1,500.

High Point

ALL-A-FLUTTER BUTTERFLY FARM, 7850-B Clinard Farms Rd., High Point, NC 27265. Tel.: 336-454-5651. Facebook: All A Flutter and All-A-Flutter Butterfly Farm.
E-mail: buttersare@aol.com
Web Site: www.all-a-flutter.com
Founded: 2001.
Congressional District: 6
Institution Type/Description: Butterfly Farm
Collections: butterflies & their lifecycle.
Facilities: picnic area.
Activities: feed the butterflies; educational programs; birthday parties; family shows. Annual Event: Release Day at the Farm in October.
Hours & Admission Prices: By appointment. Admission $6; children under 2 no charge. ♿

BERNICE BIENENSTOCK FURNITURE LIBRARY, 1009 N. Main St., High Point, NC 27262. Tel.: 336-883-4011.
Institution Type/Description: Historic Building: housed in the Grayson House; built in 1923. Listed on the National Register of Historic Places.
Collections: over 4,000 furniture & design-related books; drawings.
Facilities: library.
Activities: classes; lectures; art events; research.
Publications: newsletter.
Hours & Admission Prices: Daily 9-5.

HIGH POINT MUSEUM, (M), 1859 E. Lexington Ave., High Point, NC 27262-3499. Tel.: 336-885-1859. Fax: 336-883-3284. Facebook: High Point Museum.
E-mail: hpmuseum@highpointnc.gov
Web Site: www.highpointmuseum.org
Founded: 1966.
Congressional District: 6
Key Personnel: Dir., Edith Brady; Chm. (V), Donna Kaiser; Registrar, Corinne Midgett; Community Rel., Teresa Loflin; Cur. Collections, Marian Inabinett; Museum Shop Mgr., Mary Barnett.
Personnel Profile: Full-Time Paid 5; Part-Time Paid 5; Part-Time Volunteers 40.
Governing Authority: society; nonprofit organization. Parent Institution: High Point Historical Society, Inc. Tax-exempt: 501(c)(3).
Institution Type/Description: History Museum.
Collections: 18th-19th century tools, textiles, ceramics, furniture; industrial, social, civic, military; local area history; photographs; history of High Point, including Native Americans, Quakers, development of furniture & textile industries from settlement to present. Historic Buildings: 1786 John and Phebe Haley House; 1824 Richard Mendenhall Store; 1819 Jamestown Friends Meeting House; blacksmith shop c.1841; 1801 Hoggatt house; High Point, home of John Coltrane & Fantasia Barrino: personal artifacts including his piano; The Little Red Schoolhouse; 1929.
Research Fields: history of Piedmont; Quaker settlement; High Point local history; economic & technological history of local industries.
Facilities: reference library; 150-seat lecture hall; 50-seat educational classroom; 40-seat meeting room. Museum-related items for sale.
Activities: guided tours; lectures; films; temporary exhibitions; school &

community educational outreach programs; 18th-century skills & crafts hands-on exhibits; museum classes.

Publications: quarterly newsletter; When Racing was Racing; exhibit related pamphlets; African-American Gallery Guide.

Hours & Admission Prices: Museum: Wed.-Sat. 10-4:30. Historical Park: Sat. 10-4. No charge; donations accepted. Closed New Year's Day; Martin Luther King Jr. Day; Easter; Memorial Day; Independence Day; Labor Day; Thanksgiving & day after; Christmas Eve & Day. &

Attendance: 16,000 (estimated)

Membership: Individual $30; Household $60; 1859 Club Member $150; Friend of History $300; Founder $500; Collector $1,000.

MILLIS REGIONAL HEALTH EDUCATION CENTER, 600 N. Elm St., High Point, NC 27261. Mailing Address: P.O. Box HP-5, High Point, NC 27261-1899. Tel.: 336-878-6713.

Institution Type/Description: Health Education Center.

Collections: health & the human body; teeth & dental care; senses; hands-on exhibits.

Activities: educational programs.

Hours & Admission Prices: Mon.-Fri. 8:30-5 by appointment.

PIEDMONT ENVIRONMENTAL CENTER, 1220 Penny Rd., High Point, NC 27265-9182. Tel.: 336-883-8531. Fax: 336-883-8537.

E-mail: dick.thomas@highpointnc.gov

Web Site: www.piedmontenvironmental.com

Founded: 1973.

Congressional District: 6

Key Personnel: Exec. Dir., Richard Thomas.

Personnel Profile: Full-Time Paid 4; Part-Time Volunteers 10; Interns 1.

Governing Authority: city; nonprofit organization. Tax-exempt.

Institution Type/Description: Environmental & Nature Center.

Collections: rocks & minerals of Eastern U.S.; insects; native animals; area reptiles; astronomical; biological; gardens.

Facilities: educational facilities.

Activities: public programming; local & extended field trips; school programming; international eco tours.

Publications: bimonthly newsletter.

Hours & Admission Prices: Mon.-Fri. 9-5. No charge; donations accepted. Closed New Year's Day; Easter; Independence Day; Thanksgiving; Christmas. &

Attendance: 80,000 (estimated)

Membership: Students $15; Seniors $20; Individual $25; Family $35; Club $50; Leadership $200.

ROSETTA C. BALDWIN MUSEUM, 1408 R.C. Baldwin Ave., High Point, NC 27260. Tel.: 336-289-1942.

E-mail: hpafmuseum@yahoo.com

Web Site: rosettacbaldwinfoundation.com

Key Personnel: Dir., C.E.O. & Chm. (V), Julius Clark.

Governing Authority: Tax-exempt: 501(c)(3).

Institution Type/Description: Historic House: housed in the former home of Rosetta Baldwin who turned her home into a school for African American children in 1942.

Collections: African American history; personal artifacts; photographs; period furnishings.

Activities: guided tours.

Publications: African American Heritage Guide.

Hours & Admission Prices: Tues. & Thurs. 10-4; other times by appointment. No charge; donations accepted.

Attendance: 150 (estimated)

SPRINGFIELD MUSEUM OF OLD DOMESTIC LIFE, 555 E. Springfield Rd., High Point, NC 27263-1843. Mailing Address: 803 Kingston Dr., High Point, NC 27262-7047. Tel.: 910-889-4911; 336-882-3054.

E-mail: bgh@northstate.net

Founded: 1935.

Key Personnel: Dir. & Chief Cur., Brenda Haworth; Asst. Cur., Dan Warren.

Governing Authority: church. Parent Institution: Springfield Memorial Assoc. Tax-exempt.

Institution Type/Description: General Museum: housed in 1858 3rd meeting house located on the site of one of the first Normal schools in North Carolina.

Collections: items used by early settlers: cooking utensils, woodworking tools, shoe making items, old plank road material; loom; Indian artifacts.

Activities: guided tours.

Publications: booklet; book of photographs; cemetery & genealogical listing book.

Hours & Admission Prices: By appointment. No charge; donations accepted.

Attendance: 100 (estimated)

THEATRE ART GALLERIES - TAG, 220 E. Commerce Ave., High Point, NC 27260. Tel.: 336-887-2137. Fax: 336-887-3415.

Web Site: www.tagart.org

Key Personnel: Exec. Dir., Jeff Horney

Institution Type/Description: Art Gallery.

Collections: works by regional artists.

Activities: summer art camps; adults workshops; special events; classes; temporary exhibitions; educational & outreach programs.

Hours & Admission Prices: Call for hours.

Highlands

THE BASCOM, 323 Franklin Rd., Highlands, NC 28741. Mailing Address: P.O. Box 766, Highlands, NC 28741-0766. Tel.: 828-526-4949. Fax: 828-526-0277.

Key Personnel: Exec. Dir., Linda Steigleder

Institution Type/Description: Art Gallery.

Collections: works by student artists.

Activities: workshops; special events; educational programs.

Hours & Admission Prices: Call for hours. No charge.

MUSEUM OF AMERICAN CUT AND ENGRAVED GLASS, 472 Chestnut St., Highlands, NC 28741. Mailing Address: 218 Whiteside Mountain Rd., Highlands, NC 28741-7357. Tel.: 828-526-3415 & 3427.

E-mail: geobon@hcgexpress.net

Founded: 1996.

Congressional District: 11

Key Personnel: Dir., David Cashion; Dir., C.E.O. & Museum Shop (V), George E. Siek, Sr.

Personnel Profile: Full-Time Volunteers 1; Part-Time Volunteers 5.

Governing Authority: Tax-exempt.

Institution Type/Description: Glass Museum.

Collections: American Brilliant Period cut & engraved glass from 1876-1916.

Hours & Admission Prices: May-Oct. Tues., Thurs. & Sat. 1-4; Dec.-April by appointment. No charge; donations accepted. Closed Independence Day unless it falls on Tues., Thurs., or Sat. No charge; donations accepted.

Attendance: 1,000 (estimated)

Hillsborough

AYR MOUNT AND POET'S WALK, 376 St. Mary's Rd., Hillsborough, NC 27278-2523. Tel.: 919-732-6886.

Governing Authority: Parent Institution: Classical American Homes Preservation Trust.

Institution Type/Description: Historic House Museum: housed in the former home of William Kirkland of Ayr, Scotland; c.1815.

Collections: family history; period furnishings; portraits; personal artifacts; photographs; decorative arts.

Facilities: one-mile walking trail.

Activities: walking trail; guided tours.

Hours & Admission Prices: House Guided Tours: March 21-Dec. 20 Wed. 11am, Thurs.-Sat. 11am & 2 pm, Sun. 2 pm; other times by appointment. $10 per person. Poet's Walk: March-April & Sept.-Oct. daily 9-6; May & Aug. daily 9-7; June-July daily 9-8; Nov.-Feb. daily 9-5. No charge.

BURWELL SCHOOL HISTORIC SITE, 319 North Churton St., Hillsborough, NC 27278. Mailing Address: P.O. Box 922, Hillsborough, NC 27278-0922. Tel.: 919-732-7451. Fax: 919-644-7577.

E-mail: info@burwellschool.org

Web Site: burwellschool.org

Governing Authority: Parent Institution: Historic Hillsborough Commission. Tax-exempt: 501(c)(3).

Institution Type/Description: Historic Building: housed in the former home of Robert & Anna Burwell, c.1821.

Collections: local history & culture; period furnishings; personal artifacts. Historic Building: 1837 Burwell School.

Activities: educational programs; lectures; special events.

Hours & Admission Prices: Wed.-Sat. 11-4, Sun. 1-4. No charge.

Attendance: 7,200 (estimated)

THE GARDENS OF MONTROSE, St. Marys Rd., Hillsborough, NC 27278. Mailing Address: P.O. Box 957, Hillsborough, NC 27278-0957. Tel.: 919-732-7787.
Institution Type/Description: Garden Conservancy.
Collections: plants; trees; flowers; historic buildings.
Activities: tours; seminars. Annual Event: Garden Open Day.
Hours & Admission Prices: Guided Tours: Tues. & Thurs. 10am, Sat. 10am & 2pm, by appointment. Tours: adults 12 & over $10; children 6-11$5; children under 6 no charge.

HILLSBOROUGH VISITOR'S CENTER - ALEXANDER DICKSON HOUSE, 150 E. King St., Hillsborough, NC 27278-2685. Tel.: 919-732-7741.
E-mail: office@historichillsborough.org
Web Site: www.visithillsboroughnc.com
Formerly: Orange County Visitor's Center - Alexander Dickson House
Founded: 1993.
Key Personnel: Dir., Sarah DeGennaro.
Governing Authority: Tax-exempt.
Institution Type/Description: Visitor's Center & Historic House: site includes an office used by Confederate Gen. Joseph E. Johnston before he surrendered his troops to Union Gen. William T. Sherman in April 1865.
Collections: local history; period furnishings; photographs; 18th-19th century garden.
Hours & Admission Prices: Center: Mon.-Sat. 10-4, Sun. 12-4. Tours: 2nd Sat. of month 10am & 2pm; other times by request. Adults $5. Closed major holidays.
Attendance: 6,000 (estimated)

MOOREFIELDS, 2201 Moorefields Rd., Hillsborough, NC 27278. Tel.: 919-732-4941.
E-mail: moorefields@earthlink.net
Web Site: www.moorefields.org
Governing Authority: nonprofit organization. Administered by the Moorefields Foundation. Tax-exempt.
Institution Type/Description: Historic House Museum: housed in the former summer home of U.S. Supreme Court Justice Alfred Moore; built in 1785. Listed on the National Register of Historic Places.
Collections: period furnishings & 19th century portraits belonging to Draper-Savage.
Facilities: rental facilities.
Hours & Admission Prices: By appointment.

ORANGE COUNTY HISTORICAL MUSEUM, 201 N. Churton St., Hillsborough, NC 27278-2535. Tel.: 919-732-2201. Facebook: Orange County Historical Museum.
E-mail: info@orangenchistory.org
Web Site: www.orangecountymuseum.org
Founded: 1957.
Congressional District: 4
Key Personnel: Exec. Dir., Brandie Fields; Historic Interpreter, Carol Yavelak.
Personnel Profile: Part-Time Paid 3; Part-Time Volunteers 15; Interns 1.
Governing Authority: board of directors. Tax-exempt.
Institution Type/Description: Local History Museum.
Collections: crafts of early settlers; Indian artifacts; 20-piece set of King's Standard weights & measures; china; silver; loom, spinning wheels & home spun bed spreads & costumes; guns; pump organ; medical instruments.
Major Exhibits: From Horses to Horsepower - History of the Occaneechee Speedway, 5/14-9/14.
Research Fields: Orange County history.
Facilities: Gift items for sale.
Activities: guided tours; permanent & temporary exhibitions; heritage education program.
Hours & Admission Prices: Jan. Tues.-Sun. 1-4; Feb.-Dec. Tues.-Sat. 11-4, Sun. 1-4; group tours by appointment. No charge; donations accepted. Blue Star Museum.
Attendance: 10,766 (accurate)
Membership: Contact for further information.

RUFFIN-ROUHLAC HOUSE, 101 E. Orange St., Hillsborough, NC 27278. Tel.: 919-732-1270. Fax: 919-644-2390.
Web Site: www.historichillsborough.org
Institution Type/Description: Historic Building: housed in the former home of Chief Justice Thomas Ruffin; presently houses the Town Hall.
Collections: local & house history; photographs; period furnishings.
Activities: tours.
Hours & Admission Prices: Mon.-Fri. 9-5. No charge.

Huntersville

CAROLINA RAPTOR CENTER, 6000 Sample Rd., Huntersville, NC 28078. Mailing Address: P.O. Box 16443, Charlotte, NC 28297-6443. Tel.: 704-875-6521. Fax: 704-875-8814. Facebook: Carolina Raptor Center.
E-mail: raptors@carolinaraptorcenter.org
Web Site: www.carolinaraptorcenter.org
Founded: 1981.
Congressional District: 9
Key Personnel: Exec. Dir., Tim Warren; Chm. Bd. (V), Doug Bowman; Assoc. Exec. Dir., Michele Miller Houck; Museum Shop Mgr., Kelsey Hoke.
Personnel Profile: Full-Time Paid 12; Full-Time Volunteers 7; Part-Time Volunteers 300; Interns 5.
Governing Authority: Tax-exempt.
Institution Type/Description: Living Museum.
Collections: over 25 species of native & exotic raptors.
Research Fields: veterinary medicine conservation-environmental science.
Facilities: nature trail; amphitheatre; picnic area; event space. Museum-related items for sale.
Activities: self guided tours; nature trail; flight shows; live presentations.
Hours & Admission Prices: Mon.-Sat. 10-5, Sun. 12-5. Adults $10, seniors, military & teachers $8, students 5 & over $6; discounts to groups of 15 or more; members and children 4 & under no charge. Closed New Year's Day; Thanksgiving; Christmas.
Attendance: 40,000 (estimated)
Membership: Individual $35; Dual $45; Family $50; Extended Family $100; Family Plus $250; Small Business $500; Corporate $1,000.

DISCOVERY PLACE KIDS - HUNTERSVILLE, 105 Gilead Rd., Huntersville, NC 28078. Tel.: 704-372-6261.
Governing Authority: private; nonprofit organization. Tax-exempt: 501(c)(3).
Institution Type/Description: Children's Museum.
Collections: hands-on exhibitions; science; technology; engineering; mathematics.
Activities: educational programs; special events; classes; birthday parties; rental facilities.
Publications: newsletter.
Hours & Admission Prices: Tues.-Sat. 9-5, Sun. 12-5. Admission one & over $8; children under one & members no charge. Closed Easter; Labor Day; Thanksgiving; Christmas Eve & Day.

ENERGY EXPLORIUM, 13339 Hagers Ferry Rd., Huntersville, NC 28078. Tel.: 980-875-5600; 800-777-0003.
Web Site: www.duke-energy.com
Institution Type/Description: Science Museum.
Collections: hands-on exhibitions; electricity & Lake Norman; nuclear energy; weather; environment; butterfly garden.
Facilities: nature trail; picnic area.
Activities: educational programs; hands-on exhibitions; special events; videos.
Hours & Admission Prices: Mon.-Fri. 9-5, Sat. 12-5; groups of 10 or more by appointment. Closed New Year's Day; Easter; Thanksgiving; Christmas Eve & Day. ♿

HISTORIC LATTA PLANTATION, 5225 Sample Rd., Huntersville, NC 28078-9107. Tel.: 704-875-2312. Fax: 704-875-1724.
Web Site: www.lattaplantation.org
Founded: 1972.
Key Personnel: Exec. Dir., Kristin Toler; Bd. Pres., Lawrence Kimbrough; Dir. Education, Blair Elder; Visitor Svcs. Mgr., Nicole Glinski.
Personnel Profile: Full-Time Paid 7; Part-Time Paid 2; Part-Time Volunteers 35.
Governing Authority: nonprofit educational institution. Tax-exempt.
Institution Type/Description: Historic House Museum: 1800 house.
Collections: federal period furniture; decorative arts; agricultural artifacts; back country items; plantation outbuildings; working kitchen; farm animals; period cooking utensils. Historic Buildings: 1800 house; barns; log house.
Research Fields: slavery in Mecklenburg County; 19th century clothing; Latta family; 19th century agriculture; Yeoman farmers.
Facilities: 200-vol. library available to volunteers, Latta descendents & history scholars; nature center; nature trails. Museum-related items for sale.
Activities: self guided tours; educational programs for children; historic craft workshops; living history demonstrations; docent program; membership program; Civil War & Revolutionary War encampments; Back of the Big House program; special events; living history summer camp for children. Museum Sponsors: Folk Life Festival; Backcountry Christmas.

Publications: quarterly newsletter; The Latta Journal; monthly volunteer newsletter, Around the Plantation.
Hours & Admission Prices: Tues.-Sat. 10-5, Sun. 1-5. Adults $6, seniors 62 & over and students $5; discount to groups of 15 or more & AAM members; members & children under 5 no charge. Closed major holidays.
Attendance: 30,000 (accurate)
Membership: Individual $35; Family $60; Patron $100; Sponsor $250; Sustainer $500; Benefactor $1,000.

LATTA PLANTATION NATURE PRESERVE, 6211 Sample Rd., Huntersville, NC 28078. Tel.: 704-875-1391. Fax: 704-875-1394.
Institution Type/Description: Nature Preserve.
Collections: wildlife and their habitats.
Facilities: 1,351 acre preserve; nature trails.
Activities: educational programs; children's camps; hiking; picnic area.
Hours & Admission Prices: Mon.-Sat. 9-5, Sun. 1-5. No charge.

METROLINA GREENHOUSES, 16400 Huntersville Concord Rd., Huntersville, NC 28078. Tel.: 704-875-1371; 800-543-3915. Fax: 704-875-6741.
Web Site: www.metrolinagreenhouses.com
Institution Type/Description: Agricultural Museum.
Collections: greenhouse production, packaging & distribution; spring bedding plants; summer annuals; fall mums & pansies; winter poinsettias.
Facilities: 192 acres.
Activities: guided tours.
Hours & Admission Prices: By appointment.

RURAL HILL, 4431 Neck Rd., Huntersville, NC 28078-8342. Mailing Address: P.O. Box 1009, Huntersville, NC 28070-1009. Tel.: 704-875-3113. Fax: 704-875-3193.
E-mail: office@ruralhill.net
Web Site: www.ruralhill.net
Key Personnel: Exec. Dir., Jeff Fissel; Dir. Education, Zac Vinson
Institution Type/Description: Historic Site: housed on the former farm of Major John & Violet Davidson, one of the original signers of the Mecklenburg Declaration of Independence.
Collections: local history & culture; period furnishing; photographs. Historic Buildings: smoke house; ash house; well house; barn; chicken shed & granary; one-room schoolhouse.
Activities: educational programs; nature trails.
Hours & Admission Prices: Mon.-Fri. 9-4, Sat. call for hours. Adults 13 & over $6, children 5-12 $4; children 4 & under no charge.

Jacksonville

LYNNWOOD PARK ZOO, 1071 Wells Rd., Jacksonville, NC 28540. Tel.: 910-938-5848.
Web Site: www.lynnwoodparkzoo.com
Institution Type/Description: Zoo.
Collections: over 80 animals including reptiles, birds, mammals.
Facilities: 10 acres.
Activities: guided tours.
Hours & Admission Prices: Fri.-Mon. 10-5; groups by appointment. Adults $9, children 2-12 $7.

MONTFORD POINT MARINES MUSEUM, Bldg. M101 East Wing, Marine Corps Base, Camp Gilbert H. Johnson, Jacksonville, NC 28540. Tel.: 910-450-1340.
Web Site: montfordpointmarines.com
Institution Type/Description: Military History Museum: housed on the former training facility for black Marines from 1842-1949.
Collections: Camp Johnson history; Black American military heroes; photographs; personal artifacts.
Hours & Admission Prices: Tues. & Thurs. 11-2 & 4-7, Sat. 11-4; groups by appointment.

30 ACRES AND A MULE FARM, 125 McGowan Rd., Jacksonville, NC 28540. Tel.: 910-324-4499.
Institution Type/Description: Agricultural Museum.
Collections: local history; farm animals.
Facilities: Gift items for sale.
Activities: rental facilities; birthday parties; gem mining; children's activities; tours.
Hours & Admission Prices: Thurs.-Sat. 10-6.

Jamestown

MENDENHALL PLANTATION, (M), 603 W. Main St., Jamestown, NC 27282. Mailing Address: P.O. Box 512, Jamestown, NC 27282-0512. Tel.: 336-454-3819. Facebook: Mendenhall Plantation.
E-mail: director@mendenhallplantation.org
Web Site: www.mendenhallplantation.org
Key Personnel: Dir., Shawn M. Rogers; Pres. (V), Shirley Haworth.
Governing Authority: Parent Institution: Historic Jamestown Society Inc.
Institution Type/Description: Historic House Museum: home built in 1811. Listed on the National Register of Historic Places.
Collections: local history & culture; period furnishings; early tools & equipment; personal artifacts; photographs; historic buildings.
Activities: special events.
Hours & Admission Prices: early Jan.-Feb. Fri.-Sat. 1-4; March to late Dec. Tues.-Fri. 11-3, Sat. 1-4, Sun. 2-4. Adults $2, seniors, students & children $1.

Kenansville

COWAN MUSEUM, 411 S. Main St., Kenansville, NC 28349. Mailing Address: P.O. Box 950, Kenansville, NC 28349-0950. Tel.: 910-296-2149.
Web Site: www.cowanmuseum.com
Key Personnel: Cur., Donna Cowan
Institution Type/Description: History Museum.
Collections: local history & culture; photographs; period artifacts.
Hours & Admission Prices: Tues.-Sat. 10-4. No charge; donations accepted. Closed holidays; holiday weekend.

LIBERTY HALL PLANTATION, 409 S. Main St., Kenansville, NC 28349. Mailing Address: P.O. Box 634, Kenansville, NC 28349. Tel.: 910-296-2175.
E-mail: libertyhalltours@embarqmail.com
Web Site: www.libertyhallnc.org
Institution Type/Description: Historic House Museum: housed in an 11 room Southern plantation; built in the early 1800s.
Collections: Kenan family history; early furnishings; photographs; personal artifacts; wine cellar; carriage house; servants quarters; chicken house; smokehouse; woodshed; tool house; wash shed; overseer's cottage.
Facilities: Gift items for sale.
Activities: guided tours.
Hours & Admission Prices: Tues.-Sat. 10-4. Admission $5; groups by appointment. Closed major holidays.

Kenly

TOBACCO FARM LIFE MUSEUM, INC., Hwy. 301 N., 709 Church St., Kenly, NC 27542. Mailing Address: P.O. Box 88, Kenly, NC 27542-0088. Tel.: 919-284-3431; 800-965-1437. Fax: 919-284-9788.
E-mail: gayle@tobaccofarmlifemuseum.org
Web Site: www.tobaccofarmlifemuseum.org.
Founded: 1983.
Congressional District: 2
Key Personnel: Exec. Dir., Gayle Kildoyle; Cur., Melody Worthington.
Personnel Profile: Full-Time Paid 2; Part-Time Paid 1; Part-Time Volunteers 20; Interns 1.
Volunteer Hours: 1,039
Operating Expenses: 162,702
Operating Income: 160,641
Governing Authority: Tax-exempt: 501(c)(3).
Institution Type/Description: Agriculture Museum.
Collections: farm tools, implements & equipment used to produce, harvest & market flue-cured tobacco and other native NC products; agricultural tools & items; rural household furnishings; textiles; personal papers & documents. Historic Buildings: c.1900 farmhouse; detached kitchen; smoke house; log tobacco curing barn; blacksmith shop; one room schoolhouse.
Research Fields: eastern North Carolina farming families.
Facilities: 6,000 sq. ft. exhibit space; 50-seat theatre. Gifts, books & museum related items for sale.
Activities: tours; films; organized educational programs; hands-on children's exhibit; intern program; farmers market; special events; demonstrations; weekly Saturday interactive demonstrations.
Publications: annual report; newsletter.
Hours & Admission Prices: mid-Jan. to Dec. Tues.-Sat. 9:30-5. Adults $8, senior citizens $7, students $6; discounts to AAM, AAA & NCMC

members; members & children 2 & under no charge. Closed Easter; Thanksgiving; Christmas & day after. &

Attendance: 6,000 (accurate)

Membership: Individuals $25; Family $50. Business & Corporates available.

Kernersville

KORNER'S FOLLY, 413 S. Main St., Kernersville, NC 27284-2737. Mailing Address: P.O. Box 2091, Kernersville, NC 27285. Tel.: 336-996-7922. Fax: 336-996-1199. Facebook: Korner's Folly.

E-mail: director@kornersfolly.org

Web Site: www.kornersfolly.org

Founded: 1996.

Congressional District: 5

Key Personnel: Exec. Dir., Dale Pennington; Pres., Duane Long; Treas., Laurie McDaniel.

Personnel Profile: Full-Time Paid 1; Part-Time Paid 3; Part-Time Volunteers 25; Interns 1.

Governing Authority: private; nonprofit organization. Parent Institution: Korner's Folly Foundation. Tax-exempt: 501(c)(3).

Institution Type/Description: Historic House: c.1880 Victorian home.

Collections: furnishings; smoke house; Aunt Dealy's house.

Research Fields: construction & restoration 1880-1900.

Activities: formal education programs & tours for school children; guided tours; theater; monthly children's puppet show. Annual Events: Christmas at Korner's Folly; Kenersville Little Theater's Annual Fall Production; Annual Kernersville Oktoberfest.

Publications: quarterly newsletter, The Kornerstone; monthly newsletter, The Reubin Rink Observer.

Hours & Admission Prices: Thurs.-Sat. 10-4, Sun. 1-4. Adults $10, children $6; discounts to groups of 20 or more & Big Brothers/Big Sisters; children under 6 & members no charge. Closed New Year's Day; Easter; Thanksgiving; Christmas.

Attendance: 8,500 (accurate)

Membership: Individual $35; Couple $50; Family $100; J.K. Society $300.

PAUL J. CIENER BOTANICAL GARDEN, 215 S. Main St., Kernersville, NC 27284. Tel.: 336-996-7888.

Web Site: pjcbg.org

Key Personnel: Exec. Dir., Kitty Lyon; Prog. Coord., Toni Hays; Museum Shop Mgr., Kim Babyak; Cur., Adrienne Roethling

Institution Type/Description: Botanical Garden.

Collections: 15 gardens including 1,300 different plants.

Facilities: visitor center. Museum-related items for sale.

Activities: educational programs; lectures; guided tours.

Publications: newsletter.

Hours & Admission Prices: Gardens: daily dawn to dusk. Welcome Center: Mon.-Fri. 9-5. No charge.

Membership: Individual $35; Family $60.

Kill Devil Hills

WRIGHT BROTHERS NATIONAL MEMORIAL, 1000 Croatan Hwy., Kill Devil Hills, NC 27948. Mailing Address: 1401 National Park Dr., Manteo, NC 27954. Tel.: 252-441-7430, ext. 0. Fax: 252-441-7730.

Web Site: www.nps.gov/wrbr

Founded: 1927.

Congressional District: 3

Key Personnel: Supt., Barclay Trimble; Archives Technician, Jami P. Lanier.

Personnel Profile: Full-Time Paid 6; Part-Time Volunteers 12.

Governing Authority: federal. Parent Institution: Dept. of the Interior. Subsidiary Institution: U.S. National Park Service, Interior Building, Washington, DC 20240. Tax-exempt.

Institution Type/Description: Located on the site of the Wright Brothers first powered flight in 1903. Centennial Pavilion: reconstructed Wright camp buildings & 60 ft. memorial shaft atop Big Kill Devil Hill.

Collections: replicas of 1903 flyer; 1902 glider & wind tunnel moon cloth that Neil Armstrong took to the moon in 1969; original parts of the Wright Flyer, tools used by the Wright brothers; Wright Brothers sculpture.

Research Fields: aviation library; photographs.

Facilities: Books & museum-related items for sale.

Activities: guided tours; lectures. Annual Events: celebration of first successful powered flight by Orville & Wilbur Wright; Wilbur Wright's Birthday in April; Annual Wright Kite Festival in July; National Aviation Day & Orville Wright's Birthday in August; Anniversary of the First Flight in December.

Hours & Admission Prices: mid-June to Labor Day daily 9-6; Sept. to mid-June daily 9-5. Adults $4; seniors with Golden Age Card Federal Pass & children under 16 no charge. Closed Christmas. &

Attendance: 496,500 (accurate)

Kings Mountain

CROWDERS MOUNTAIN STATE PARK, 522 Park Office Lane, Kings Mountain, NC 28086-7902. Tel.: 704-853-5375. Fax: 704-853-5391.

E-mail: crowders.mountain@ncdenr.gov

Founded: 1973.

Congressional District: 10

Key Personnel: Park Supt., Larry Hyde.

Personnel Profile: Full-Time Paid 11; Part-Time Paid 5; Part-Time Volunteers 10.

Governing Authority: Parent Institution: NC State Parks. Tax-exempt.

Institution Type/Description: Park & Visitors Center.

Collections: wildlife & their habitats; mounted wildlife; plants; environment.

Facilities: classroom; auditorium; nature trails.

Activities: hiking; canoeing; backpack camping; nature observation & study; rock climbing.

Hours & Admission Prices: Park: March-April & Sept.-Oct. daily 8-8; May-Aug. daily 8-9; Nov.-Feb. daily 8-6. No charge. Closed Christmas. &

Attendance: 349,000 (accurate)

KINGS MOUNTAIN HISTORICAL FIRE MUSEUM, 211 Cleveland Ave., Kings Mountain, NC 28086. Mailing Address: c/o Kings Mountain Fire Dept., P.O. Box 429, Kings Mountain, NC 28086-0429. Tel.: 704-734-0555.

Web Site: kmfire.com

Founded: 1973.

Key Personnel: Chm. (V), Frank Burns

Governing Authority: city.

Institution Type/Description: Fire Museum.

Collections: firefighting equipment & uniforms; 1938 Ford fire truck; hose cart; fire extinguishers; photographs; personal artifacts; fire patches from around the world.

Hours & Admission Prices: Call for hours. No charge.

Attendance: 350 (estimated)

KINGS MOUNTAIN HISTORICAL MUSEUM, (M), 100 E. Mountain St., Kings Mountain, NC 28086. Mailing Address: P.O. Box 552, Kings Mountain, NC 28086. Tel.: 704-739-1019. Fax: 704-734-4537.

E-mail: director@kingsmountainmuseum.org

Web Site: www.kingsmountainmuseum.org

Founded: 1986.

Key Personnel: Dir. & Cur., Adria L. Focht.

Governing Authority: Tax-exempt.

Institution Type/Description: History Museum.

Collections: local history; military artifacts; early clothing; WWI & II memorabilia; textiles; archives; photographs.

Major Exhibits: Common Threads: Kings Mountain's Textile Heritage from Prehistory to Today, 2/14-5/24/14; Say Ahh! The Incredible Medical History of Kings Mountain, 6/14-10/14.

Facilities: Museum-related items for sale.

Activities: special events. Museum Sponsors: Annual Reverse Raffle & Auction in September.

Hours & Admission Prices: Tues.-Sat. 10-4; groups by appointment. No charge; donations accepted. Closed New Year's Day; Independence Day; Thanksgiving; Christmas Eve & Day.

Kinston

CSS NEUSE STATE HISTORIC SITE AND GOV. RICHARD CASWELL MEMORIAL, U.S. Hwy. 70 Business, 2612 W. Vernon Ave., Kinston, NC 28504. Mailing Address: P.O. Box 3043, Kinston, NC 28502. Tel.: 252-522-2091. Fax: 252-527-7036.

E-mail: cssneuse@ncdcr.gov

Web Site: www.cssneuse.nchistoricsites.org

Founded: 1965.

Congressional District: 4

Key Personnel: Historic Site Mgr., Guy Smith; Historic Interpreter III, Morris Bass; Historic Interpreter II & Museum Shop Mgr., Holly Weaver; Historic Site Asst., Thomas R. Dawson; Office Asst. III, Sharon Clements; Maintenance Mechanic II, Gaston Davis.

Personnel Profile: Full-Time Paid 6; Part-Time Paid 3.

Governing Authority: state. North Carolina Dept. of Cultural Resources, 4601 Mail Service Center, Raleigh, NC 27699-4601. Tax-exempt: 170(b).

Institution Type/Description: Historic Site.

Collections: artifacts from the ram Neuse; sunken Confederate iron clad gunboat c.1862-65; hull of ship; items depicting life of Richard Caswell, first elected governor of N.C., 1776-1780 & 1784-1787.

Facilities: picnic facility.
Activities: guided tours; lectures; audiovisual program; formally organized education programs; living history demonstrations; hands-on ropemaking & demonstrations.
Publications: brochure; The CSS NEUSE: A Question of Iron & Time.
Hours & Admission Prices: Mon.-Sat. call for hours. No charge; donations accepted. Closed major state holidays. &

Attendance: 13,000 (estimated)

C.S.S. NEUSE II, 118 W. Heritage St., Kinston, NC 28502. Mailing Address: C.S.S. Neuse Foundation, Inc., P.O. Box 251, Kinston, NC 28502-0251. Tel.: 252-560-2150.
E-mail: info@cssneuseii.org
Web Site: www.cssneuseii.org
Institution Type/Description: Military History Museum.
Collections: full-scale Civil War replica; Civil War sailor history.
Activities: tours.
Hours & Admission Prices: Sat. 10-4; other times by appointment.

CASWELL CENTER MUSEUM AND VISITOR CENTER, 2415 W. Vernon Ave., Kinston, NC 28504. Tel.: 252-208-3780. Fax: 252-208-3771.
E-mail: caswell.center@ncmail.net
Institution Type/Description: History Museum.
Collections: institutional life from the early 1900s to present; photographs; period furnishings.
Hours & Admission Prices: No charge; donations accepted.

CASWELL NO. 1 FIRE STATION MUSEUM, 118 S. Queen St., Kinston, NC 28501. Mailing Address: 1005 Oriental Ave., Kinston, NC 28504. Tel.: 252-522-4676 & 527-1566. Facebook: Caswell No. 1 Fire Station Museum.
Institution Type/Description: Historic Building: housed in the city's original fire station; built in 1895. Listed on the National Register of Historic Places.
Collections: 19th century fire equipment & truck; photographs; station & town history.
Hours & Admission Prices: Tues., Thurs. & Sat. 10-4. No charge.

COMMUNITY COUNCIL FOR THE ARTS, 400 N. Queen St., Kinston, NC 28501-4328. Tel.: 252-527-2517. Fax: 252-527-8280.
E-mail: slandis@kinstoncca.com
Web Site: www.kinstoncca.com
Formerly: Kinston Arts Council, Inc.
Founded: 1965.
Congressional District: 1
Key Personnel: Exec. Dir., Sandy Landis; Pres., Vickie Robinson; Dir. Gallery, Niki Litts; Financial Svcs. Dir., Elaine Carmon.
Personnel Profile: Full-Time Paid 2; Part-Time Paid 3; Part-Time Volunteers 15.
Governing Authority: nonprofit. Affiliated with North Carolina Arts Council, Raleigh, NC 27611. Tel.: 919-733-7897. Tax-exempt.
Institution Type/Description: Arts Council: housed in historic building.
Collections: works by North Carolina & international artists; model trains.
Facilities: meeting room; children's gallery; classrooms. Museum-related items for sale.
Activities: guided tours; arts festivals; competitive art exhibition; permanent & temporary exhibitions; classes; workshops; special projects & events; performing arts programs.
Publications: monthly newsletter, Kaleidoscope.
Hours & Admission Prices: Tues.-Fri. 10-6, Sat. 10-2. No charge, donations accepted. Closed state holidays. &

Attendance: 10,000
Membership: Individual $50; Family $100; Donor $150; Sponsor $250; Patron $500; Sustainer $1,000; Renaissance $5,000.

EXCHANGE NATURE CENTER, 401 W. Caswell St., Kinston, NC 28501. Tel.: 252-939-3367.
Institution Type/Description: Nature Center.
Collections: live & mounted wildlife and their habitats; salt water touch tank; minerals.
Facilities: nature trails.
Activities: miniature train ride; touch tank; picnic area.
Hours & Admission Prices: Tues.-Sat. 9:30-5, Sun. 1-5.

LENOIR MEMORIAL HOSPITAL HEALTH & SCIENCE MUSEUM - NEUSEWAY PLANETARIUM, 401 W. Caswell St., Kinston, NC 28501. Tel.: 252-939-3302.
Institution Type/Description: Science Museum & Planetarium.
Collections: hands-on health & science exhibitions; NASA videos.
Facilities: 52-seat planetarium.
Activities: videos; school groups; day camps; birthday parties.
Hours & Admission Prices: Museum: Tues.-Sat. 9:30-5, Sun. 1-5. Planetarium: Tues.-Fri. 1 & 4, Sat. 11am, 1 & 4, Sun. 2 & 4.

Kitty Hawk

OUTER BANKS CHILDREN AT PLAY, 3810 N. Croatan Hwy., Kitty Hawk, NC 27949. Tel.: 252-261-0290.
E-mail: info@childrenatplayobx.com
Web Site: www.childrenatplayobx.com
Founded: 2008.
Key Personnel: Dir., Alyssa Hannon; Pres. (V), Diane Wehner.
Personnel Profile: Full-Time Paid 1; Part-Time Paid 3; Interns 1.
Volunteer Hours: 200
Operating Expenses: 62,000
Governing Authority: Tax exempt.
Institution Type/Description: Children's Museum.
Collections: hands-on exhibitions.
Activities: educational programs; special events; art Wednesdays, science Thursdays, music Fridays.
Hours & Admission Prices: Tues.-Sat. 10-5. Children $7, adults $5, 12 months & under no charge. Closed Thanksgiving; Christmas Eve, Day & day after. &

Attendance: 8,000 (accurate)
Membership: two people $65; three people $75; four people $85; five people $95; reciprocal network $130; supporter $300.

Kure Beach

FORT FISHER STATE HISTORIC SITE, U.S. Hwy. 421-1610 Fort Fisher Blvd., S. of Kure Beach, Kure Beach, NC 28449. Tel.: 910-458-5538. Fax: 910-458-0477.
E-mail: fisher@ncdcr.gov
Web Site: www.nchistoricsites.org/fisher
Founded: 1961.
Congressional District: 7
Key Personnel: Dir., Jim Steele; Museum Shop Mgr., Becky Sawyer.
Personnel Profile: Full-Time Paid 8; Part-Time Paid 5; Part-Time Volunteers 3; Interns 1.
Governing Authority: state. North Carolina Dept. of Cultural Resources, 109 E. Jones St., Raleigh, 27611. Tax-exempt: 170(b).
Institution Type/Description: Historic Site & Visitor Center.
Collections: underwater archaeology; military accoutrements; memorabilia; Civil War fort ruins & remains; Civil War artifacts; findings from sunken blockade runners.
Research Fields: Civil War Coastal Defenses.
Facilities: visitor center; nature trails.
Activities: guided tours; lectures; formally organized education programs; living history events.
Publications: brochure; newsletter.
Hours & Admission Prices: Tues-Sat 9-5. No charge, donations accepted. Closed most state holidays. &

Attendance: 600,000 (accurate)
Membership: Military & Student $20; Individual $25; Family $40; Business $100 & up; Sustaining, Armstrong Society $500 & up.

NORTH CAROLINA AQUARIUM AT FORT FISHER, 900 Loggerhead Rd., Kure Beach, NC 28449-3786. Tel.: 910-458-8257. Fax: 910-458-6812.
E-mail: casey.davis@ncaquariums.com
Web Site: www.ncaquariums.com
Founded: 1976.
Congressional District: 7
Key Personnel: Dir., Peggy Sloan; Chm. Aquarium Society, Neal Conoley; Dir. Operations & Husbandry, Paul Barrington; Div. Dir. Aquariums, David Griffin; Cur. Exhibits, David Barney; Cur. Education, Jennifer Metzler-Fiorino; Aquariology Cur., Hap Fatzinger; Dive Coord., Brian Germick; Aquarist, Monica Dudley; Aquarist, Michael Suchy; Aquarist, Rich Bamberger; Aquarist, Keith Farmer; Aquarist, Julie Johnson; Aquarist, Marc Neill; Bldg. Supt., Tom Coit; Business Mgr., Angie Leary; Visitor Svcs. Coord., Joanna Zazzali; Media Technician, Bob Griffin; Events Coord., Emily Clark; Publicity & Mktg., Robin Nalepa; Museum Shop Mgr., Kathy Roper.

Personnel Profile: Full-Time Paid 40; Part-Time Paid 33; Part-Time Volunteers 230; Interns 2.

Governing Authority: state. Parent Institution: North Carolina Dept. of Environment & Natural Resources, Aquariums Division, 3125 Poplarwood Ct., Ste. 160, Raleigh, NC 27604. Tel. 919-877-5500. Subsidiary Institution: NC Aquarium Society. Other Aquariums: North Carolina Aquarium-Roanoke Island, P.O. Box 967, Manteo, NC 27954. Tel. 252-473-3493; North Carolina Aquarium-Pine Knoll Shores, 1 Roosevelt Blvd., Atlantic Beach, NC 28512. Tel. 252-247-4003. Tax-exempt.

Institution Type/Description: Aquarium: aquatic life & habitats of North Carolina.

Collections: live saltwater & freshwater fishes; sea turtles; sharks; stingrays; alligators; lion fish.

Research Fields: marine sciences, with emphasis on marine biology.

Facilities: freshwater conservatory; gardens; aquarium; auditorium; classrooms. Educational items & aquarium-related items for sale.

Activities: lectures; films; formally organized education programs by appointment; docent programs or council; field trips; summer camps; outreach programs.

Publications: monthly online Calendar of Events; teachers guide, The Aquarium News; various information folders.

Hours & Admission Prices: Daily 9-5. Adults 13-61 $8, seniors 62 & over $7, children 3-12 $6; discounts to AAM members; children 2 & under, registered school groups & NCAS members no charge. Closed New Year's Day; Thanksgiving; Christmas. ♿

Attendance: 420,176 (accurate)

Membership: Individual $40; Family $60; Donor $100; Patron $300; Benefactor $1,000.

Lake Junaluska

WORLD METHODIST MUSEUM, 575 N. Lakeshore Dr., Lake Junaluska, NC 28745-9742. Mailing Address: P.O. Box 518, Lake Junaluska, NC 28745-0518. Tel.: 828-456-9432, ext. 4. Fax: 828-456-9433.

Web Site: www.worldmethodistcouncil.org

Founded: 1954.

Congressional District: 11

Key Personnel: Gen. Sec. World Methodist Council, Bishop Juan Abrahams; Asst. to Gen. Sec., Roma Wyatt; Museum Dir., Amanda Riera-Gomez.

Personnel Profile: Full-Time Paid 1; Interns 1.

Governing Authority: church. Parent Institution: World Methodist Council. Tax-exempt.

Institution Type/Description: Religious History Museum.

Collections: items of early Methodism from 18th century England including original letters by John Wesley, Francis Asbury & Thomas Coke; traveling pulpit used by Wesley; pottery busts of John Wesley & other Staffordshire artisans; John Hurst Watercolors of Wesley's England; Frank O. Salisbury portraits.

Research Fields: Methodist Church history & the 74 member churches of Council.

Facilities: library of rare books.

Activities: guided & self-guided tours; DVD introduction.

Publications: newsletter of World Methodist Council, World Parish; newsletter, WMC First Friday Newsletter.

Hours & Admission Prices: Summer: Mon.-Fri. 9-5, Sat. 10-3; Fall, Winter, & Spring Mon.-Fri. 9-5. No charge; donations accepted. Closed holidays. ♿

Attendance: 6,000 (accurate)

Membership: Friends of the Museum: Contributor $25; Sustainer $50; Sponsor $100; Patron $500; Benefactor $1,000; Grand Benefactor $3,000.

Lake Waccamaw

LAKE WACCAMAW DEPOT MUSEUM, 201 Flemington Ave., Lake Waccamaw, NC 28450. Mailing Address: P.O. Box 386, Lake Waccamaw, NC 28450-0386. Tel.: 910-646-1992.

E-mail: info@lakewaccamawdepotmuseum.com

Web Site: lakewaccamawdepotmuseum.com

Founded: 1977.

Congressional District: 7

Key Personnel: Chm. (V), Nancy Sigmon; Treas., Martha Lowe; Cur., Ginger Littrell; Museum Shop Mgr., Lynn Cain.

Personnel Profile: Part-Time Paid 1.

Governing Authority: nonprofit organization. Tax-exempt: 501(c)(3).

Institution Type/Description: Marine Science & History Museum: housed in c.1904 railroad station.

Collections: photographs; deeds; artifacts; fossils; tools; mounted birds; railroad & logging industry artifacts; Waccamaw-Siouan Indians from Archaic Period (8,000-500 B.C.).

Activities: guided tours; lectures; films; formally organized educational programs; permanent exhibitions; slide presentation; canoeing; camping; hiking; fossil hunts; field trips. Museum Sponsors: The Way It Was; chartered bus tours; Indian dance teams.

Publications: brochures; quarterly program, Calendar of Events.

Hours & Admission Prices: Wed.-Fri. 10-3, Sun. 3-5. No charge; donations accepted. ♿

Attendance: 3,600 (estimated)

Membership: Individual $15; Family $25; Associate $35; Patron $50; Sustaining $100; Benefactor $500 & up.

Laurinburg

JOHN BLUE HOUSE AND MUSEUM, 13040 X-Way Rd., Laurinburg, NC 28352. Mailing Address: P.O. Box 152, Laurinburg, NC 28353. Tel.: 910-280-0435.

E-mail: walkeramj@bellsouth.net

Web Site: johnbluecottonfestival.com

Key Personnel: Museum Shop Mgr., Jim Walker.

Governing Authority: Parent Institution: Scotland County Historic Properties Commission. Tax-exempt.

Institution Type/Description: Historic House: built in the early 1890s by businessman, farmer & inventor, John Blue Sr.

Collections: local & family history; personal artifacts; period furnishings; mule powered cotton gin.

Activities: Annual Event: John Blue Cotton Festival; Highland Games; Storytelling Festival.

Hours & Admission Prices: Call for hours. No charge; donations accepted. ♿

Attendance: 5,000 (estimated)

LUMBEE INDIAN MUSEUM OF THE CAROLINAS, 607 Turnpike Rd., Laurinburg, NC 28352. Tel.: 910-276-2496.

Web Site: lumbee-indianmuseum.com

Institution Type/Description: Native American History Museum.

Collections: Native American artifacts; pottery; jewelry; clothing; weapons; tools.

Hours & Admission Prices: Call for hours.

MUSEUM OF SCOTLAND COUNTY, 13043 X Way Rd., Laurinburg, NC 28353. Tel.: 910-276-2496.

Web Site: www.scotlandcounty-museum.com

Institution Type/Description: History Museum.

Collections: local history, farming & industry; agricultural & textile machinery; early farm implements; period cars & fire engine; local art; model trains; Civil War artifacts; personal artifacts; books.

Hours & Admission Prices: Call for hours.

ST. ANDREWS PRESBYTERIAN COLLEGE ART GALLERY, 1700 Dogwood Mile, Laurinburg, NC 28352-5521. Tel.: 910-277-5555. Fax: 910-277-5020.

Institution Type/Description: Art Gallery.

Collections: paintings; photographs; sculpture.

Hours & Admission Prices: Mon.-Fri. 9-4:30. No charge.

SCOTTISH HERITAGE CENTER, St. Andrews Presbyterian College, DeTamble Library, 1700 Dogwood Mile, Laurinburg, NC 28352. Tel.: 910-277-5236. Fax: 910-277-5020.

E-mail: bagpipe@sapc.edu

Web Site: www.sapc.edu

Key Personnel: Dir., Bill Caudill

Institution Type/Description: Heritage Center.

Collections: Scottish history, culture, music & traditions; early books; genealogy.

Activities: Annual Event: Scottish Heritage Weekend in March.

Hours & Admission Prices: Call for hours.

Lenoir

CALDWELL ARTS COUNCIL, 601 College Ave., Lenoir, NC 28645-5406. Mailing Address: P.O. Box 1613, Lenoir, NC 28645-1613. Tel.: 828-754-2486. Fax: 828-754-2440.

E-mail: info@caldwellarts.com

Web Site: www.caldwellarts.com

Founded: 1976.

Key Personnel: Dir., Lee Carol Giduz

Institution Type/Description: Cultural Arts Museum.

Collections: paintings; sculpture.

Activities: art classes; artist competitions; grants to community; temporary exhibitions.

Hours & Admission Prices: Tues.-Fri. 9-5, Sat. by appointment. No charge.

CALDWELL HERITAGE MUSEUM, INC., 112 Vaiden St., S.W., Lenoir, NC 28645-5670. Tel.: 828-758-4004. Fax: 828 758 4242.
E-mail: caldheritmus@aol.com
Web Site: www.caldwellheritagemuseum.org
Founded: 1991.
Key Personnel: Dir., John O. Hawkins; Chm. (V), Mike Gibbons.
Personnel Profile: Full-Time Volunteers 1; Part-Time Paid 1; Part-Time Volunteers 40.
Governing Authority: Tax-exempt.
Institution Type/Description: History Museum.
Collections: Caldwell County history; photographs.
Publications: quarterly newsletter.
Hours & Admission Prices: Tues.-Fri. 10-4:30, Sat. 10-3. No charge; donations accepted. &
Attendance: 2,500 (estimated)

FORT DEFIANCE, 1792 Fort Defiance Dr., Lenoir, NC 28645-6606. Tel.: 828-758-1671.
Institution Type/Description: Historic Building: former home of Revolutionary War hero, Gen. William Lenoir; built in 1792.
Collections: military history; personal artifacts; period furnishings; military uniforms; clothing.
Hours & Admission Prices: April-Oct. Thurs.-Sat. 10-5, Sun. 1-5; Nov.-March Sat.-Sun. by appointment.

TUTTLE EDUCATIONAL STATE FOREST, 3420 Playmore Beach Rd., Lenoir, NC 28655. Tel.: 828-757-5608.
Governing Authority: nonprofit organization.
Institution Type/Description: Park Museum.
Collections: local history; ecosystems; plants; trees; flowers.
Facilities: nature trail.
Activities: classes; workshops; picnic area; hiking.
Hours & Admission Prices: mid-March to mid-Nov. Tues.-Sun. call for hours.

Lexington

DAVIDSON COUNTY HISTORICAL MUSEUM, (M), 2 S. Main St., Old Courthouse on the Square, Lexington, NC 27292-3320. Tel.: 336-242-2035. Fax: 336-242-2871.
E-mail: choffmann@co.davidson.nc.us
Web Site: www.co.davidson.nc.us/HistoricalMuseum
Founded: 1976.
Congressional District: 6
Key Personnel: Library Dir., Ruth Ann Copley; Cur., Catherine Hoffmann.
Governing Authority: county. Parent Institution: Davidson County. Subsidiary Institution: Davidson County Historical Association. Tax-exempt: 501(c)(3).
Institution Type/Description: Local History Museum: housed in the county's oldest existing courthouse; built in 1858. Listed on the National Register of Historic Places.
Collections: local history; courtroom; jury room; holding room.
Research Fields: local history.
Activities: guided tours; lectures; formally organized education programs; temporary & traveling exhibitions.
Publications: quarterly newsletter.
Hours & Admission Prices: Tues.-Fri. 10-4, first Sun. of month 2-4. No charge; donations accepted. Closed holidays. &
Attendance: 18,000 (accurate)

Lincolnton

LINCOLN COUNTY MUSEUM OF HISTORY, 403 E. Main St., Lincolnton, NC 28092-3305. Tel.: 704-748-9090. Fax: 704-732-9057.
E-mail: lcmh@bellsouth.net
Web Site: www.lincolncountyhistory.com
Founded: 1955.
Congressional District: 10
Key Personnel: Exec. Dir., Jason L. Harpe; Cur. Archaeology & Collections, January W. Porter; Administrative Asst., Tina Guffey.
Personnel Profile: Full-Time Paid 1; Part-Time Paid 2; Part-Time Volunteers 20.
Governing Authority: Tax-exempt.
Institution Type/Description: History Museum.
Collections: Lincoln County heritage & history; Native American artifacts; early immigrants; the Battle of Ramsour's Mill; Civil War; archaeological artifacts; ceramics; archives.

Activities: seminars & programs; cemetery preservation & restoration; Tarheel Junior Historian Camp; digital photography courses; archaeology camp; historic preservation.
Publications: The Lincoln Sentinel.
Hours & Admission Prices: Tues. & Thurs. 1-5, Sun. 2-5. No charge; donations accepted. &
Attendance: 4,000 (estimated)

Linville

GRANDFATHER MOUNTAIN, 2050 Blowing Rock Hwy., Linville, NC 28646. Mailing Address: P.O. Box 129, Linville, NC 28646. Tel.: 800-468-7325. Fax: 828-733-2608. Facebook: Grandfather Mountain.
E-mail: nature@grandfather.com
Web Site: www.grandfather.com
Founded: 1952.
Institution Type/Description: Nature Center.
Collections: black bears; river otters; cougars; golden eagles; white-tailed deer.
Facilities: nature trails. Museum-related items for sale.
Activities: hiking.
Hours & Admission Prices: Spring & Fall daily 8-6, Summer: daily 8-7; Winter: daily 9-5. Adults 13-59 $18, seniors 60 & over $15, children 4-12 $8; discounts to AAA members; children under 4 no charge.
Attendance: 250,000 (estimated)

Louisburg

LOUISBURG COLLEGE ART GALLERY, 501 N. Main St., Louisburg, NC 27549-2399. Tel.: 919-497-3238. Fax: 919-496-1788.
Web Site: www.louisburg.edu/news/art.html
Founded: 1957.
Key Personnel: Dir. & Cur., William Hinton; Business Officer, Belinda Faulkner; Public Rels. & Publications Dir., Amy McManus.
Governing Authority: college; nonprofit organization. Affiliated with Louisburg College. Tax-exempt.
Institution Type/Description: Art Gallery: housed in 1787 Louisburg College auditorium theatre complex.
Collections: primitive, American impressionist & contemporary art.
Facilities: classrooms.
Activities: guided tours; lectures; gallery talks; arts festivals; loan, permanent, & temporary exhibitions.
Publications: Alumni Review
Hours & Admission Prices: Mon.-Fri. 9-5. No charge. Closed holidays. &
Attendance: 10,000

Lucama

WHIRLIGIGS, Wiggins Mill Rd., Lucama, NC 27542. Tel.: 800-497-7398.
Institution Type/Description: Sculpture Garden.
Collections: Vollis Simpson's wind-powered sculptures.
Hours & Admission Prices: Call for hours.

Lumberton

EXPLORATION STATION, 104 N. Chestnut St., Lumberton, NC 28358. Tel.: 910-738-1114.
E-mail: fun@explorationkids.com
Web Site: www.explorationkids.com
Institution Type/Description: Children's Museum.
Collections: hands-on exhibitions.
Activities: educational programs; special events. Museum Sponsors: Monthly Kids Kamps; Weekly Storytimes; St. Patrick's Day Party; Annual Prissy Polka Dilly Tea Party.
Hours & Admission Prices: Tues.-Wed. & Fri. 10-5, Thurs. 10-8, Sat. 10-4. Children $3, adults $1.

ROBESON COUNTY HISTORY MUSEUM, 101 S. Elm St., Lumberton, NC 28358. Tel.: 910-738-7979.
Web Site: robesoncountyhistory.org
Founded: 1987.
Institution Type/Description: History Museum: housed in the Southern Express Building.
Collections: local history & culture; period furnishings; personal artifacts; photographs; transportation; agriculture; military artifacts; uniforms; weapons.
Hours & Admission Prices: Tues. & Thurs. 10 to noon, Sun. 2-4. No charge; donations accepted. Closed holidays. &

Membership: Individual $25; Family $50; Sponsor $100; Patron $200; Benefactor $1,000.

ROBESON PLANETARIUM AND SCIENCE CENTER, 410 Caton Rd., Lumberton, NC 28358. Tel.: 910-735-2147.
E-mail: brandt@uncp.edu
Web Site: www.robesonsky.com
Key Personnel: Dir., Ken Brandt
Institution Type/Description: Planetarium & Science Center.
Collections: geology; space science.
Activities: educational programs & shows.
Hours & Admission Prices: Call for hours & admission.

Maggie Valley

WHEELS THROUGH TIME MUSEUM, INC., 62 Vintage Ln., Maggie Valley, NC 28751. Mailing Address: P.O. Box 790, Maggie Valley, NC 28751-0790. Tel.: 828-926-6266. Fax: 828-926-9158.
E-mail: info@wheelsthroughtime.com
Web Site: www.wheelsthroughtime.com
Founded: 2002.
Congressional District: 11
Governing Authority: Tax-exempt.
Institution Type/Description: Motorcycle & Automobile Museum.
Collections: over 320 period motorcycles & automobiles from 1903 to present; factory lithographs; photographs; posters; clothing; trophies; autographed photos.
Research Fields: Blue Ridge transportation history.
Facilities: Museum-related items for sale.
Hours & Admission Prices: April-Nov. Thurs.-Mon. 9-5. Adults $12, seniors 65 & over $10, children 6-12 $6. &
Attendance: 50,000 (estimated)
Membership: Annual $100; Lifetime $300.

Manteo

ALLIGATOR RIVER NATIONAL WILDLIFE REFUGE & VISITOR CENTER, 100 Conservation Way, Manteo, NC 27954. Mailing Address: P.O. Box 1969, Manteo, NC 27954. Tel.: 252-473-1131. Fax: 252-473-1668.
E-mail: alligatorriver@fws.gov
Key Personnel: Refuge Mgr., Mike Bryant; Deputy Refuge Mgr., Scott Lanier; Visitor Svcs. Mgr., Bonnie Strawser; Visitor Ctr. Mgr., Susie Kowlok
Institution Type/Description: Wildlife Refuge.
Collections: wildlife & their habitats including alligators, black bears, & red wolves; plants.
Facilities: 150,000 acres of wetland habitats; nature trails; theater. Gift items for sale.
Activities: hiking; observation platforms.
Hours & Admission Prices: Refuge: daily daylight hours. Visitor Center: Mon.-Sat. 9-4, Sun. 12-4.

DARE COUNTY REGIONAL AIRPORT MUSEUM, 410 Airport Rd., Manteo, NC 27954. Mailing Address: P.O. Box 429, Manteo, NC 27954. Tel.: 252-475-5570. Fax: 252-473-1196.
E-mail: museum@darenc.com
Web Site: www.darenc.com/airport/museum.htm
Key Personnel: Dir. Airport, Robert A. Benson
Institution Type/Description: Aviation History Museum.
Collections: aviation history; model airplanes; photographs; Civil Air Patrol monument.
Hours & Admission Prices: Daily 8-7.

ELIZABETHAN GARDENS, 1411 National Park Dr., Manteo, NC 27954. Tel.: 252-473-3234. Fax: 252-473-3244.
E-mail: tours@elizabethangardens.org
Web Site: www.elizabethangardens.org
Key Personnel: Exec. Dir., Carl Curnutte, III
Personnel Profile: Full-Time Paid 10; Part-Time Paid 8; Part-Time Volunteers 54; Interns 3.
Institution Type/Description: Gardens.
Collections: flowering plants, shrubs and trees including azaleas, dogwoods, rhododendrons, herbs, roses, magnolias, crape myrtle, hydrangeas & bedding plants; statues.
Facilities: Garden-related items for sale.
Activities: special events; educational programs; workshops.
Hours & Admission Prices: March & Oct.-Nov. daily 9-5; April-May & Sept. daily 9-6; June-Aug. daily 9-7; Dec.-Feb. daily 10-4. Adults $8, youth 6-17 $5; children 5 & under no charge. Closed New Year's Day; Thanksgiving; Christmas Eve & Day. &

FORT RALEIGH NATIONAL HISTORIC SITE, 1401 National Park Dr., Manteo, NC 27954. Tel.: 252-473-5772. Fax: 252-473-1049.
E-mail: caha_information@nps.gov
Web Site: www.nps.gov/fora
Founded: 1941.
Congressional District: 3
Key Personnel: Supt., Barclay Trimble; Archives Technician, Jami P. Lanier; Volunteer Coord., Mary Doll; Museum Shop Mgr., Rulaine Kegerris.
Personnel Profile: Full-Time Paid 3; Part-Time Paid 2; Part-Time Volunteers 3.
Governing Authority: federal. Parent Institution: National Park Service. Tax-exempt.
Institution Type/Description: Historic Park Museum: located on the site of the 1585-1587 Roanoke Island Colony attempts.
Collections: relics from period of first colony; history; archaeology; Indian artifacts.
Facilities: 70-seat auditorium. Museum-related items for sale.
Activities: guided tours; lectures; films; permanent exhibitions. Park Sponsors: The Lost Colony in summer.
Publications: handbook, Fort Raleigh.
Hours & Admission Prices: Park: daily sunrise to sunset. Visitor Center: daily 9-5. Closed Christmas. &
Attendance: 170,780 (estimated)

NORTH CAROLINA AQUARIUM ON ROANOKE ISLAND, 374 Airport Rd., Manteo, NC 27954-9485. Mailing Address: P.O. Box 967, Manteo, NC 27954. Tel.: 252-473-3494. Fax: 252-473-1980.
E-mail: rimail@ncaquariums.com
Web Site: www.ncaquariums.com
Founded: 1976.
Congressional District: 3
Key Personnel: Dir., Maylon White; C.E.O. & Business Mgr., Deborah Edelman; Dir. Operations & Husbandry, Brian Dorn; Cur. Education, Pat Raves.
Personnel Profile: Full-Time Paid 38; Part-Time Paid 32; Part-Time Volunteers 65; Interns 7.
Governing Authority: state government. Parent Institution: NCDENR. Tax-exempt.
Institution Type/Description: Aquarium.
Collections: over 300 species of native/regional fish, reptiles & amphibians and invertebrates; over 2000 specimens; concentration in coastal northeastern North Carolina habitats & aquatic life.
Research Fields: aquatic animal (sea turtle & marine mammal) rehabilitation.
Facilities: 800-vol. library of books & reports of coastal nature; extensive vertical files of coastal North Carolina topics covering over 20 years. Educational items of marine & aquatic nature for sale.
Activities: docent program; films; formal education programs for adults, children & college students; lectures; participatory & temporary exhibitions; training programs; rental gallery. Annual Events: Wild Foods Weekend; Shark Discovery Day; Earth Day Festival.
Publications: Events Calendar & Newsletter.
Hours & Admission Prices: Daily 9-5. Adults $8, senior citizens $7, children 3-12 $6; children 2 & under, members, NC Aquarium society, NC school groups, Martin Luther King Day, & Veterans Day no charge. Closed New Year's Day; Thanksgiving; Christmas. &
Attendance: 288,000 (accurate)
Membership: Individual $40; Family $60; Curator $100; Patron $300; Director $500; Benefactor $1,000.

ROANOKE ISLAND FESTIVAL PARK, One Festival Park, Manteo, NC 27954-9396. Tel.: 252-475-1500. Fax: 252-475-1507.
E-mail: festivalparkinformation@ncdcr.gov
Web Site: www.roanokeisland.com
Founded: 1983.
Congressional District: 2
Key Personnel: Exec. Dir., Kim Sawyer; Chm. (V), Ellen Newbold; Pres. (V), Friends of Elizabeth II, Tod Clissold; Operations Mgr., Amy Hinnant; Facilities Mgr., Carroll Williams; Communications Mgr., Tanya Young.
Personnel Profile: Full-Time Paid 26; Part-Time Paid 10; Part-Time Volunteers 46; Interns 1.
Governing Authority: state. Parent Institution: North Carolina Dept. of Cultural Resources, Raleigh, NC 27611. Subsidiary Institution: Roanoke Island Commission. Tax-exempt: 170(b)(1)(A).
Institution Type/Description: History Museum: living history 1585 settlement site & replica 16th century ship.

Collections: reproductions of 16th-century artifacts; tools; equipment; personal artifacts & furnishings; Outer Banks history; working English sailing vessel representative of 16th-century.
Research Fields: Elizabethan England & exploration; 16th-century life in coastal North America; Roanoke Island Civil War; Roanoke Island Freedman's Colony; Outer Banks; maritime history; national cultures of Roanoke Island.
Facilities: 242-seat theater; 3,500-seat outdoor pavilion; visitor center; nature trails. Museum-related items for sale.
Activities: tours; audiovisual program; living history presentation; volunteer sailing program.
Publications: brochures.
Hours & Admission Prices: March 2-Dec. daily 9-5. Adults $10, students $7; discounts to groups; members & children under 5 no charge. &
Attendance: 135,000 (accurate)
Membership: Individual $30; Dual $50; Family $75; Participating $125; Corporate $250; Patron $500; Benefactor $1,000.

ROANOKE ISLAND MARITIME MUSEUM, 104 Fernando St., On the Manteo Waterfront, Manteo, NC 27954. Mailing Address: c/o Town of Manteo, P.O. Box 246, Manteo, NC 27954. Tel.: 252-475-1750.
Institution Type/Description: Maritime Museum.
Collections: local & regional maritime history & heritage; watercraft; photographs; period artifacts; 1883 Shad boat, Ella View; boat building; commercial fishing.
Activities: educational programs; workshops.
Hours & Admission Prices: Temporarily closed.

ROANOKE MARSHES LIGHTHOUSE, Queen Elizabeth Ave., Manteo, NC 27954. Mailing Address: P.O. Box 246, Manteo, NC 27954. Tel.: 252-475-1750.
Institution Type/Description: History Museum.
Collections: lighthouse & maritime history; photographs; period artifacts; replica 1877 screwpile lighthouse.
Hours & Admission Prices: Temporarily closed.

Marion

HISTORIC CARSON HOUSE, 1805 US Hwy. 70 W., Marion, NC 28752. Tel.: 828-724-4948.
E-mail: info@historiccarsonhouse.com
Web Site: www.historiccarsonhouse.com
Founded: 1964.
Congressional District: 11
Key Personnel: Chm., Dr. James Haney; Exec. Dir., Sara Bryant; Treas., Leslie Morgan; Sec., Ann McNutt.
Personnel Profile: Part-Time Paid 1; Part-Time Volunteers 30.
Governing Authority: nonprofit organization. Parent Institution: Carson House Restoration Inc. Tax-exempt: 501(c)(3); 170(b)(1)(A).
Institution Type/Description: History Museum: housed in 1793 Col. John Carson House.
Collections: 350 artifacts dating back to pioneer times; manuscripts; hand woven coverlets; 4 quilts made & quilted by slaves; quilt woven by a Princess slave; hand woven towels; clocks; period furniture; grand piano; organ; musical instruments.
Research Fields: family histories; Civil War; Revolutionary War.
Facilities: library of genealogy, family histories, national, state & local books available for use on premises; reading room.
Activities: guided tours; study clubs; formally organized educational programs; permanent & temporary exhibitions.
Publications: book, Stories Not Told in History Books; Where Early Settlers Made Their Homes; Life of Samuel Price Carson; Research Study of Carson House; McDowell County Pictorial History; book, Everything That's All; Shining Dust.
Hours & Admission Prices: April-Nov. Wed.-Sat. 10-4, Sun. 2-5; groups by appointment. Adults $5; members & children under 12 no charge. &
Attendance: 3,000 (estimated)
Membership: Individual $25; Family $50; Sponsor $100; Patron $500; Benefactor $1,000.

LINVILLE CAVERNS, 19929 US 221 N., Marion, NC 28752. Tel.: 800-419-0540. Fax: 828-756-4171.
E-mail: info@linvillecaverns.com
Institution Type/Description: Geology Museum.
Collections: geology; local history; science.
Facilities: visitors center.
Activities: guided tours.
Hours & Admission Prices: March & Nov. daily 9-4:30; April-May &

Sept.-Oct. daily 9-5; June to Labor Day daily 9-6; Dec.-Feb. Sat.-Sun. 9-4:30. Adults $7.50, seniors 62 & over $6.50, children 5-12 $5.50; children under 5 no charge.

Mars Hill

RURAL LIFE MUSEUM, (M), Mars Hill College, Montague Bldg., Cascade & College Sts., Mars Hill, NC 28754. Mailing Address: Mars Hill College, P.O. Box 6706, Mars Hill, NC 28754-5000. Tel.: 828-689-1262.
Web Site: www.mhc.edu/ramsey-center/rural-life-museum
Institution Type/Description: History Museum.
Collections: Southern Appalachian culture & history.
Hours & Admission Prices: By appointment. No charge.

Mayodan

MAYO RIVER STATE PARK, 500 Old Mayo Park Rd., Mayodan, NC 27027. Tel.: 336-427-2530.
Web Site: ncparks.gov/visit/parks/mari/main.php
Institution Type/Description: Park Museum.
Collections: environment; ecology; natural resources.
Facilities: nature trails.
Activities: hiking trails; educational programs; picnic area; catch-and-release fishing.
Hours & Admission Prices: Call for hours.

McLeansville

REPLACEMENTS, LTD. MUSEUM, 1089 Knox Rd., McLeansville, NC 27301. Tel.: 800-737-5223.
Institution Type/Description: China Museum.
Collections: over 2,000 tableware pieces.
Activities: guided tours of showrooms, museum, warehouse & restoration facility.
Hours & Admission Prices: Tours: daily 9:30-6.

Midland

REED GOLD MINE STATE HISTORIC SITE, 9621 Reed Mine Rd., Midland, NC 28107-9673. Tel.: 704-721-4653 Fax: 704-721-4657.
E-mail: larry.neal@ncdcr.gov
Web Site: www.nchistoricsites.org/reed
Founded: 1971.
Congressional District: 8
Key Personnel: Site Mgr., Larry K. Neal, Jr.; Museum Shop Mgr., Susan Smith.
Personnel Profile: Full-Time Paid 6; Part-Time Paid 8; Part-Time Volunteers 6.
Governing Authority: state. Parent Institution: North Carolina Department of Cultural Resources, 109 E. Jones St., Raleigh, NC 27611. Tax-exempt: 501(c)(3); 170(b)(1)(A).
Institution Type/Description: Historic Site: 1799 site of the first documented discovery of gold in the United States.
Collections: 19th- to 20th-century mining machinery; coins; steam engines; period artifacts; underground tunnels. Historic Building: c.1895 operating stamp mill.
Research Fields: local history; gold mining.
Facilities: library; visitor center; nature trails; picnic area.
Activities: guided tours; lectures; formally organized educational programs for children; instruction in gold panning; film; underground workings.
Publications: guidebook; brochure; The Story of John Reed.
Hours & Admission Prices: Tues.-Sat. 9-5. No charge; donations accepted. Gold panning: $3 per person. Closed state holidays. &
Attendance: 48,000 (accurate)

Montreat

PRESBYTERIAN HERITAGE CENTER, (M), 318 Georgia Ter., Montreat, NC 28757. Mailing Address: P.O. Box 207, Montreat, NC 28757-0207. Tel.: 828-669-6556.
Key Personnel: Pres., Frank L. Arnold
Institution Type/Description: Heritage Center.
Collections: Presbyterian history; photographs; period furnishings; religious artifacts.
Hours & Admission Prices: Tues.-Fri. 10-4, Sat. 1-4, Sun. 1:30-4.

Mooresville

DALE EARNHARDT INC. MUSEUM, 1675 Dale Earnhardt Hwy. 3, Mooresville, NC 28115. Tel.: 704-662-8000. Fax: 704-663-7945.
E-mail: mlucas@dei-zone.com
Institution Type/Description: Sports Museum.
Collections: Dale Earnhardt's life & career; photographs; personal artifacts; trophies; awards; race cars.
Facilities: Museum-related items for sale.
Hours & Admission Prices: Mon.-Fri. 9-5. No charge.

JR MOTORSPORTS, 349 Cayuga Dr., Mooresville, NC 28117. Tel.: 866-576-8883.
E-mail: shoptours@jrmracing.com
Web Site: jrmotorsport.com
Institution Type/Description: Dale Earnhardt Jr's Race Shop.
Collections: cars that compete in NASCAR Nationwide Series & local weekly late model division; life-size wax figure of Dale Earnhardt Jr.
Facilities: 66,000 sq. ft. race shop.
Activities: viewing window.
Hours & Admission Prices: Tours: Thurs.-Fri. 9-4:30 by appointment. (email request for tour).

KYLE BUSCH MOTORSPORTS, 351 Mazeppa Rd., Mooresville, NC 28115.
Web Site: www.kylebuschmotorsports.com
Institution Type/Description: Motorsports Museum.
Collections: fleet of Toyota Camry's & Tundras.
Facilities: 77,000 sq. ft. shop. Gift items for sale.
Activities: view race preparations.
Hours & Admission Prices: Mon.-Fri. 9-4.

LAZY 5 RANCH, 15100 Mooresville Rd., Mooresville, NC 28115-7245. Tel.: 704-663-5100. Fax: 704-664-1549.
E-mail: lazy5ranch@aol.com
Web Site: www.lazy5ranch.com
Institution Type/Description: Zoo.
Collections: over 750 animals from around the world.
Activities: drive thru park.
Hours & Admission Prices: Mon.-Sat. 9 to dusk. Adults $8.50, children 2-11 and seniors 60 & over $5.50; discounts to groups. Wagon Rides: adults $5, children 2-11 and seniors 60 & over $3; discounts to groups.

MEMORY LANE MUSEUM, 769 River Hwy., Mooresville, NC 28117. Tel.: 704-662-3673; 877-716-6723. Facebook: Memory Lane Museum.
E-mail: memorylanemuseum@gmail.com
Web Site: www.memorylaneautomuseum.com
Founded: 2001.
Key Personnel: Dir., Sam Alex Beam; Chm. (V), Sam A. Beam, III; Museum Shop Mgr., Michele Tilton
Institution Type/Description: Automobile & Racing History Museum.
Collections: racing history; automobiles; race cars; motorcycles; period artifacts; toys.
Facilities: banquet facilities. Museum-related items for sale.
Activities: group tours.
Hours & Admission Prices: Summer: Mon.-Sat. 10-5; Winter: Thurs.-Sat. 10-5. Race Week: daily 9-5. Adults $10, children 6-12 $6; children under 6 no charge. &

NORTH CAROLINA AUTO RACING HALL OF FAME, 119 Knob Hill Rd., Lakeside Park, Mooresville, NC 28117-6847. Tel.: 704-663-5331.
E-mail: donna@ncarhof.com
Web Site: www.ncarhof.com
Key Personnel: Mgr., Donna DeNardo
Institution Type/Description: Auto Racing Museum.
Collections: heritage of motorsports; over 35 race cars; Hall of Fame Inductees.
Facilities: theater. Museum-related items for sale.
Activities: Museum Sponsors: Induction Ceremony.
Hours & Admission Prices: Mon.-Fri. 10-5, Sat. 10-3. Adults $6, seniors 55 & over and children 6-12 $4.

PENSKE RACING, 200 Penske Way, Mooresville, NC 28115. Tel.: 704-664-2300.
Web Site: www.raceshops.com

Institution Type/Description: Racing Facility.
Collections: three NASCAR teams; Indy Car and ALMS teams.
Facilities: Gift items for sale.
Activities: watch the crews prepare for upcoming NASCAR races from fan walkway.
Hours & Admission Prices: Mon.-Fri. 9-5; call to confirm. No charge.

Morehead City

THE HISTORY PLACE, (M), The History Place, 1008 Arendell St., Morehead City, NC 28557-4143. Tel.: 252-247-7533. Fax: 252-247-2756.
E-mail: historyplace@thehistoryplace.org
Web Site: www.thehistoryplace.org
Formerly: Carteret County Museum of History & Art
Founded: 1971.
Key Personnel: Pres. (V), Janet Eshleman; Librarian, Pat Edwards; Exec. Dir., Cindi Hamilton; Museum Shop Mgr., Stacey Veros.
Personnel Profile: Full-Time Paid 1; Part-Time Volunteers 95.
Governing Authority: Tax-exempt.
Institution Type/Description: History & Art Museum.
Collections: local genealogy; memorabilia of Carteret County; pictures & artifacts.
Facilities: research library; tea room; auditorium; banquet facility; conference room. Gift items for sale.
Activities: walking tours; programs; bus tours; genealogy seminars; Sunday supplement programs; concerts; children's programs; antique shows; concerts; lunch with Rodney - history lectures.
Publications: history journal, The Researcher; local books; bimonthly newsletter.
Hours & Admission Prices: Summer: Tues.-Sat. 10-4; Winter: Tues.-Fri. & 1st Sat. each month 10-4. No charge; donations accepted. Closed holidays. &
Attendance: 38,000 (accurate)
Membership: $10; $25; $30; $35; $250; Business $100.

Morganton

HERITAGE MUSEUM, 102 E. Union St., Morganton, NC 28655-3448. Mailing Address: P.O. Box 915, Morganton, NC 28680. Tel.: 828-437-4104. Facebook: Historic Burke Foundation.
E-mail: historicburke@compascable.net
Web Site: www.historicburke.org
Institution Type/Description: History Museum: housed in the Old Burke County Courthouse; built in 1837.
Collections: local & courthouse history; period furnishings; photographs; North Carolina Supreme Justices from 1847-1862; video.
Activities: video presentation.
Hours & Admission Prices: Mon.-Fri. 9-4. No charge; donations accepted. &
Attendance: 2,000 (estimated)

THE HISTORY MUSEUM OF BURKE COUNTY, 201 W. Meeting St., Morganton, NC 28655. Tel.: 828-437-1777.
E-mail: thehistorymuseum@directus.net
Web Site: www.thehistorymuseumofburke.org
Institution Type/Description: History Museum.
Collections: Burke County history, heritage & culture; photographs; personal artifacts; period furnishing; early beauty shop, doctor & dentist's office; military artifacts; housewares.
Activities: docent tours; special events; permanent & temporary exhibitions.
Publications: quarterly newsletter.
Hours & Admission Prices: Call for hours.

MCDOWELL HOUSE AT QUAKER MEADOWS, 119 St. Mary's Church Rd., Morganton, NC 28680. Mailing Address: P.O. Box 915, Morganton, NC 28680-0915. Tel.: 828-437-4104. Facebook: Historic Burke Foundation.
E-mail: historicburke@compascable.net
Web Site: www.historicburke.org
Institution Type/Description: Historic House Museum: built in 1812 by Captain Charles McDowell, Jr.
Collections: area history; 19th century life.
Hours & Admission Prices: April-Oct. Sun. 2-4; other times by appointment. No charge; donations accepted. &

SENATOR SAM J. ERVIN, JR. LIBRARY AND MUSEUM, Western Piedmont Community College, 1001 Burkemont Ave., Morganton, NC 28655. Tel.: 828-438-6152.
E-mail: library@wpcc.edu

Web Site: www.samervinlibrary.org
Founded: 1990.
Congressional District: 10
Key Personnel: Cur., Daniel R. Smith; Asst. Cur., Nancy Daniel.
Personnel Profile: Part-Time Paid 4.
Governing Authority: Parent Institution: Western Piedmont Foundation. Subsidiary Institution: Ervin Library Fund. Tax-exempt.
Institution Type/Description: History Museum.
Collections: replica of Senator Ervin's home library; books; correspondence; photographs; furniture; memorabilia.
Hours & Admission Prices: Mon.-Fri. 8-5. No charge.

Mount Airy

THE ANDY GRIFFITH MUSEUM, 218 Rockford St., Mount Airy, NC 27030. Tel.: 336-786-1604.
E-mail: arts@surryarts.org
Web Site: www.andygriffithmuseum.com
Institution Type/Description: History Museum.
Collections: Andy Griffith's life & career; personal artifacts; photographs; memorabilia; TV show props.
Hours & Admission Prices: Daily call for hours; groups by appointment. Closed Thanksgiving.

GERTRUDE SMITH HOUSE, 708 N. Main St., Mount Airy, NC 27030. Mailing Address: 615 N. Main St., Mount Airy, NC 27030-3723. Tel.: 336-786-6856; 336-755-3283.
E-mail: gilmersmith708@gmail.com
Founded: 1984.
Congressional District: 5
Key Personnel: Chm., David Beal; Supvr. Buildings & Grounds, Cindy Puckett; Sec. Foundation, Tom Fawcett.
Personnel Profile: Full-Time Paid 1; Part-Time Paid 2; Part-Time Volunteers 1.
Governing Authority: nonprofit organization. Parent Institution: Gilmer-Smith Foundation. Tax-exempt: 501(c)(3).
Institution Type/Description: Cultural & Enrichment Center: housed in the former Jefferson Davis Smith family home; built in 1903. Listed on the National Register of Historic Places.
Collections: family history & personal artifacts; period furnishings; paintings.
Facilities: 20-seat lecture room; park. Garden for the Senses, (for the visually impaired) with Braille labels.
Activities: guided tours; lectures; concerts; outreach programs. Open House: Thanksgiving to Christmas Eve.
Hours & Admission Prices: April-Dec. Mon., Wed. & Fri.-Sat. 11-4; other times by appointment, call 336-786-6856. No charge; Closed holidays. ♿
Attendance: 8,947 (accurate)

MOORE HOUSE, 202 Moore Ave., Mount Airy, NC 27030. Mailing Address: c/o Mount Airy Restoration Foundation, PO Box 447, Mount Airy, NC 27030. Tel.: 336-786-6116; 800-948-0949.
Institution Type/Description: Historic House: built in 1862.
Collections: local history; period furnishings. Historic Structure: c.1865 gazebo.
Hours & Admission Prices: Call for hours.

MOUNT AIRY MUSEUM OF REGIONAL HISTORY, (M), 301 N. Main St., Mount Airy, NC 27030-3811. Tel.: 336-786-4478; 336-786-1666.
E-mail: mamrh@northcarolinamuseum.org
Web Site: www.northcarolinamuseum.org
Key Personnel: Dir., Matthew J. Edwards; Cur. Collections, Amy Snyder; Museum Shop Mgr., Nancy Davis.
Personnel Profile: Full-Time Paid 1; Part-Time Paid 5; Part-Time Volunteers 100.
Institution Type/Description: History Museum.
Collections: regional history; personal artifacts.
Hours & Admission Prices: April-Oct. Tues.-Sat. 10-4; Nov.-March Tues.-Sat. 10-2. Adults $5, senior citizens & students $3; discounts to groups. ♿
Attendance: 15,000 (estimated)
Membership: Student & Senior $25; Individual $35; Family $50; Friends Circle $100; Historians Circle $250; Patriots Circle $500; Benefactors Circle $1,000; Sterling Benefactor $2,500; Granite Circle $5,000.

Mount Gilead

TOWN CREEK INDIAN MOUND STATE HISTORIC SITE, 509 Town Creek Mound Rd., Mount Gilead, NC 27306-8506. Tel.: 910-439-6802. Fax: 910-439-6441.
E-mail: towncreek@ncdcr.gov

Web Site: www.towncreekindianmound.com
Founded: 1937.
Congressional District: 8
Key Personnel: Site Mgr., Rich Thompson.
Personnel Profile: Full-Time Paid 4; Part-Time Paid 4; Part-Time Volunteers 6; Interns 1.
Governing Authority: state. North Carolina Dept. of Cultural Resources, 4620 Mail Service Center, Raleigh, NC 27699-4620. Tax-exempt: 170(b).
Institution Type/Description: Historic Site & Visitor Center: Mississippian period Ceremonial Center restored; temple on top of earth mound, priest's dwelling, burial house & mud-plastered palisade surrounding temple.
Collections: anthropology; archaeology; Indian artifacts; Indian mound; reconstructed Indian structures.
Research Fields: Woodland Indians of North Carolina; archaeology.
Facilities: visitor center.
Activities: guided tours; lectures; formally organized education programs.
Publications: brochure; guidebook.
Hours & Admission Prices: Tues.-Sat. 9-5, Sun. 1-5. No charge; donations accepted. Closed holidays. ♿
Attendance: 22,000 (accurate)

Mount Ulla

KERR MILL AT SLOAN PARK, 550 Sloan Rd., Mount Ulla, NC 28125. Tel.: 704-637-7776.
Web Site: co.rowan.nc.us
Institution Type/Description: Historic Building: housed in a former grist mill; built in 1823.
Collections: mill history & equipment.
Hours & Admission Prices: Sat.-Sun. 1-7; other times by appointment.

Murfreesboro

BRADY C. JEFCOAT MUSEUM OF AMERICANA, 201 W. High St., Murfreesboro, NC 27855-1819. Mailing Address: P.O. Box 3, Murfreesboro, NC 27855. Tel.: 252-398-5922; 910-358-1202. Fax: 252-398-5871.
E-mail: mha@murfreesboronc.org
Web Site: murfreesboronc.org
Founded: 1997.
Congressional District: 1
Key Personnel: Dir., Colon Ballance.
Personnel Profile: Part-Time Paid 1; Part-Time Volunteers 14; Interns 1.
Volunteer Hours: 1,500
Operating Expenses: 35,000
Operating Income: 25,000
Governing Authority: Parent Institution: Murfreesboro Historical Association. Tax-exempt.
Institution Type/Description: History Museum.
Collections: early furnishings including washing machines, irons, churns, music boxes, radios, & phonographs; glassware; toasters; farm tolls & equipment; photographs; personal artifacts.
Hours & Admission Prices: Sat. 11-4, Sun. 2-5; groups by appointment. Adults $8, seniors 60 & over $7, students $6. ♿
Attendance: 2,000 (estimated)

MURFREESBORO HISTORICAL ASSOCIATION, INC., 116 E. Main St., Murfreesboro, NC 27855-1407. Mailing Address: P.O. Box 3, Murfreesboro, NC 27855-0003. Tel.: 252-398-5922. Fax: 252-398-5871.
E-mail: mha@murfreesboronc.org
Web Site: www.murfreesboronc.org
Founded: 1967.
Congressional District: 1
Key Personnel: Pres. (V), Carol B. Lassiter.
Personnel Profile: Full-Time Paid 2; Part-Time Volunteers 150.
Governing Authority: society. Parent Institution: Murfreesboro Historical Assoc. Tax-exempt: 501(c)(3).
Institution Type/Description: Historical Society: housed in the Roberts-Vaughan House; built in 1790.
Collections: agriculture; history; American Indian artifacts; early educational materials; manuscripts & photographs; Richard J. Gatling Room; Gatling artifacts; Gatling family. Historic Building: 1790 William Rea Store; 1814 Wheeler House, 1870 Winborne County Store & Law Office.
Research Fields: religion; agriculture; education; American Indian artifacts.
Activities: guided tours; lectures; permanent & temporary exhibitions.
Publications: brochures; The Tuscaroras Vol. I & II; books, Legends & Myths of North Carolina's Roanoke-Chowan Area; The Peanut Story; The Fabled Doctor Jim Jordan; The Tuscaroras Mythology, Medicine, Culture, Vol. 1; The Tuscaroras History, Tradition, Cuture, Vol. 1; Murfreesboro and the

Founding of the American Republic; Murfreesboro, NC and the Great Intercoastal Waterway; Murfreesboro, NC Cradle of Titans; Murfreesboro, NC and the Roots of Nat Turner's Revolt; The Gatling Aeroplane of 1873, America's First Airplane; Trial Separation: Murfreesboro, NC and the Civil War; A History of the Riddick PLantation Hertford County, North Carolina; Captain Jack: The Life of Dragoon Captain John Henry King Burgwin on the Great Plains.
Hours & Admission Prices: Mon.-Fri. 9-4. Roberts-Vaughan House no charge. Tours: Sat. 11-4. William Rea Store, Wheeler House & Winborne Country Store & Law Office: adults $8, students $5; children under 6 no charge. Jefcoat Museum $8.
Attendance: 1,500 (estimated)
Membership: Individual $25; Family $35; Sponsor $100; Patron $300; Life $1,000; Benefactor $1,500.

Murphy

CHEROKEE COUNTY HISTORICAL MUSEUM, INC., 87 Peachtree St., Murphy, NC 28906-2940. Tel.: 828-837-6792. Fax: 828-837-6792.
E-mail: cchistoricalmuseum@gmail.com
Founded: 1977.
Congressional District: 11
Key Personnel: Dir. & Museum Shop Mgr., Wanda Stalcup; Chm. (V), Glenda Fisher; Vice Chm., Mary Ann Thompson.
Personnel Profile: Full-Time Paid 1; Part-Time Volunteers 7.
Governing Authority: nonprofit organization; museum council. Tax-exempt: 501(c)(3) & 170(b)(1)(A).
Institution Type/Description: History Museum.
Collections: Indian artifacts; early American housewares; primitives; minerals; firearms; musical instruments; local history items; manuscripts; over 500 dolls; Cherokee culture. Historic Building: c.1800 Cherokee house.
Research Fields: local history.
Facilities: library of books about the Cherokee Indian Nation & local history; reading room. Postcards & notepaper for sale.
Activities: guided tours; lectures; permanent & temporary exhibitions.
Publications: brochures, The History & Architecture of Cherokee County, NC.; books, Cherokee County Heritage, Vol. I & II, Pictorial History of Cherokee County; Heritage & Cherokee County, Vol. III; Cherokee County History Nuggets; The Civil War in Cherokee County; Clay and Graham Counties; Cherokee County Oral Histories.
Hours & Admission Prices: Mon.-Fri. 9-5; group tours by appointment. Adults $3, children $1; students no charge. Closed New Year's Day; Good Friday through Easter Monday; Memorial Day; Independence Day; Labor Day; Thanksgiving; Christmas. &
Attendance: 7,000 (estimated)

FIELDS OF THE WOOD, 10000 Hwy. 294, Murphy, NC 28906. Mailing Address: P.O. Box 2910, Cleveland, TN 37320. Tel.: 828-494-7855. Facebook: COGOP Heritage.
E-mail: fieldsofthewood@cogop.org
Web Site: fieldsofthewood.net
Founded: 1939.
Key Personnel: Dir., Paul Holt.
Governing Authority: Parent Institution: Church of God of Prophecy International Offices.
Institution Type/Description: History Museum.
Collections: religious artifacts; the Ten Commandments written in stone; replica of Jesus' tomb.
Facilities: 200-acre Bible park; nature trails; cafe. Gift items for sale.
Hours & Admission Prices: Sunrise to sunset. No charge; donations accepted.
Membership: Heritage Ministries Association $20.

New Bern

ATTMORE-OLIVER HOUSE MUSEUM, 510 Pollock St., New Bern, NC 28560. Mailing Address: 511 Broad St., New Bern, NC 28560. Tel.: 252-638-8558. Fax: 252-638-5773.
E-mail: lynne@newbernhistorical.org
Web Site: www.newbernhistorical.org
Key Personnel: Exec. Dir., Lynne Harakal.
Governing Authority: Parent Institution: New Bern Historical Society.
Institution Type/Description: Historic House Museum: c.1790.
Collections: local history; 18th-19th century furnishings; period artifacts; photographs; Civil War artifacts; dolls.
Activities: Museum Sponsors: Haunted Evening; Spirited Through Time Dramas.
Hours & Admission Prices: By appointment. Adults $4; students & children no charge.

THE BIRTHPLACE OF PEPSI-COLA STORE, 256 Middle St., New Bern, NC 28560. Tel.: 252-636-5898.
Institution Type/Description: History Museum: housed in the birthplace of Pepsi-Cola; invented by Caleb Bradham in his pharmacy in 1898.
Collections: Caleb Bradham's life & career; Pepsi history; photographs; period furnishings; personal artifacts.
Facilities: Museum-related items for sale.
Hours & Admission Prices: Jan.-Feb. Mon.-Sat. 10-6; March-Dec. Mon.-Sat. 10-6, Sun. 12-4.

FRIENDS OF FIREMEN'S MUSEUM, 408 Hancock St., New Bern, NC 28560-4923. Tel.: 252-636-4087 & 4020. Fax: 252-636-4087.
E-mail: firechief-nb@admin.ci.new-bern.nc.us
Formerly: New Bern Firemen's Museum
Founded: 1955.
Key Personnel: C.E.O., Charles Williams; Sec. & Treas., Richard M. Register; Museum Shop Mgr., Ben Gaskill.
Personnel Profile: Full-Time Paid 1; Part-Time Paid 1; Part-Time Volunteers 1.
Governing Authority: municipal. Tax-exempt.
Institution Type/Description: Fire-Fighting Museum.
Collections: history; firematic display; Civil War display; preservation project; world record 1884 horse-drawn steamer; horse-drawn hose wagons & hand reel; steam engines; 1914 fire engine; 1800s leather fire helmets; preserved 1900s fire horse; period fire fighting equipment; photographs.
Research Fields: permanent exhibits.
Facilities: Firematic-related materials for sale.
Activities: guided tours; permanent exhibitions.
Publications: newsletter, Museum Sparks.
Hours & Admission Prices: Mon.-Sat. 10-4. Adults $5, children $2.50; children under 6 no charge. Closed New Year's Day; Thanksgiving; Christmas.
Attendance: 14,000 (estimated)

✱ TRYON PALACE, (M), 529 S. Front St., New Bern, NC 28562. Mailing Address: P.O. Box 1007, New Bern, NC 28563-1007. Tel.: 252-639-3500; 800-767-1560. Fax: 252-514-4876. Facebook: Tryon Palace.
E-mail: info@tryonpalace.org
Web Site: www.tryonpalace.org
Formerly: Tryon Palace Historic Sites & Gardens
Founded: 1945.
Congressional District: 1
Key Personnel: Chm., William C. Cannon, Jr.; Deputy Dir., Philippe Lafargue; Coord. Human Resources & Volunteer, Laurie Bowles; Dir. Collections, Nancy Packer; Controller, Susan L. Flowers; Conservator, Richard Baker; Foundation Coord., Gwen Parish; African American Outreach Coord., Sharon C. Bryant; Security Coord., Sean Creamer; Security Coord., Orlando Venters; Museum Store Mgr., Susan Briley.
Personnel Profile: Full-Time Paid 55; Part-Time Paid 35; Part-Time Volunteers 351.
Governing Authority: state; nonprofit organization. Affiliated with The Tryon Palace Commission, Office of Archives & History, North Carolina Dept. of Cultural Resources. Tax-exempt.
Institution Type/Description: History Museum.
Collections: 18th, 19th, & 20th century English & American objects, including furniture, paintings, books, porcelains, textiles, sculptures & metalwork; heritage plants; architectural fragments; prints; currency; maps. Historic Buildings: c.1830 George W. Dixon House; c.1779 John Wright Stanly House; c.1810 Daves House; c.1809 Jones House; c.1890 Commission House; c.1880 Disosway House; c.1809 New Bern Academy; c.1785 Gaston House; c.1804 Robert Hay House; c.1842 William Hollister House.
Research Fields: North Carolina history; 18th- &19th-century material culture; African-American history of the central North Carolina coast; African American artisans of 18th & 19th centuries, New Bern.
Facilities: 4,000-vol. library of reference books & books on history & decorative arts, available for research by scholars by advance appointment; 200-seat auditorium; 14 period gardens; classrooms; 2 theatres; cafe. Museum-related items for sale.
Activities: guided tours; lectures; conservation, preservation & educational programs; domestic skills & craft demonstrations; garden workshops & events; daily living history programs; character interpretation program; African American history program; interactive children's exhibits. Museum Sponsors: New Bern's 300th Anniversary Celebration; Independence Day Celebration; Tryon Palace Decorative Arts Symposium in March; Garden events; Fife & Drum Corps events; recreating African American Celebration of Christmas in eastern North Carolina; Christmas Candlelight Tours.
Publications: guidebook; periodic pamphlets & leaflets; books, History of the New Bern Academy; A Candlelight Christmas; Tryon Treasury; A New

Bern Album; Profile of a Patriot; Historic Architecture of New Bern & Craven County; Singleton Slave Narrative; magazine, Palace.
Hours & Admission Prices: Jan.-June Mon.-Sat. 9-5, Sun. 12-5, last tour at 4. Gardens: June-Aug. Mon.-Sat. 9-7, Sun. 12-5, last tour at 4. One Day Pass: adults $20, students grades 1-12 $10; Two Day Pass: adults $26, students $13; discounts to seniors 62 & over, military, AAM & AAA members, college students and groups. Galleries: adults $12, students $4. Gardens: adults $6, students $3. Closed New Year's Day; Thanksgiving; Christmas Eve, Day & day after. &
Attendance: 150,000 (accurate)

New Hill

HARRIS ENERGY AND ENVIRONMENTAL CENTER, 3932 New Hill, New Hill, NC 27562. Tel.: 919-362-3261. Fax: 919-362-3446.
Web Site: progress-energy.com/harris
Institution Type/Description: Environmental Center.
Collections: electricity generation & transmission; alternative energy; energy efficiency; nuclear power; hands-on exhibits.
Activities: educational programs.
Hours & Admission Prices: By appointment. Closed holidays.

NORTH CAROLINA RAILWAY MUSEUM, 5121 Daisy St., New Hill, NC 27562. Mailing Address: P.O. Box 40, New Hill, NC 27562-0040. Tel.: 919-362-5416.
Institution Type/Description: Railway Museum.
Collections: railroad history & equipment; photographs; tools; signs.
Facilities: Museum-related items for sale.
Hours & Admission Prices: Sat.-Sun. 9-5. No charge.

Newland

AVERY COUNTY MUSEUM, 1829 Schultz Circle, Newland, NC 28657. Mailing Address: P.O. Box 266, Newland, NC 28657-0266. Tel.: 828-733-7111.
E-mail: averymuseum@interlink-cafe.com
Web Site: averymuseum.com
Founded: 1977.
Congressional District: 10
Key Personnel: Exec. Dir., Cindy Peters.
Personnel Profile: Part-Time Volunteers 10.
Governing Authority: nonprofit organization. Subsidiary Institution: Avery County Historical Society. Tax-exempt: 501(c)(3).
Institution Type/Description: County Museum.
Collections: local genealogies; artifacts of prominent citizens of the past; local minerals.
Activities: guided tours; lectures; formally organized education programs for children; permanent & temporary exhibitions; group tours.
Publications: Avery County Heritage: Biographies, Genealogies, Church Histories, Vols. I, II & III; Avery County Heritage: Historic Sites, Vol. IV.
Hours & Admission Prices: May-Oct. Fri. 10-4, Sat. 11-3; other times by appointment. No charge; donations accepted.
Attendance: 1,000 (estimated)
Membership: Student $2; Adults $15; Family $20; Business $50.

Newton

CATAWBA COUNTY MUSEUM OF HISTORY, (M), 30 N. College Ave., Newton, NC 28658. Mailing Address: P.O. Box 73, Newton, NC 28658-0073. Tel.: 828-465-0383. Fax: 828-465-9813.
E-mail: mherzognc@gmail.com
Web Site: www.catawbahistory.org
Founded: 1949.
Congressional District: 10
Key Personnel: C.E.O. & Dir., Melinda Herzog; Pres. (V) & Chm. (V), Shuford Abernethy, III; Cur., Kendall Reese; Mktg. Dir., Joshua Cummings; Registrar & Site Mgr., Jennifer Marquart-Leach; Business Officer, Doris Teague; Registrar, John Powers; Librarian, Marian Stearns.
Personnel Profile: Full-Time Paid 3; Part-Time Paid 6; Part-Time Volunteers 123.
Volunteer Hours: 13,882
Operating Expenses: 449,814
Operating Income: 450,830
Governing Authority: nonprofit. Parent Institution: Catawba County Historical Association. Subsidiary Institution: Harper House/Hickory History Center; Bunker Hill Covered Bridge, Claremont, NC; Murray's Mill Historic Site, Catawba, NC. Tax-exempt: 501(c)(3).
Institution Type/Description: History Museum: housed in former 1924 Catawba Courthouse.

Collections: Upper Piedmont history including household utensils, tools, weapons, textiles, folk art; agricultural implements; furniture; objects relating to the development of furniture industry; manuscripts; manufacturing. Historic Buildings: 1900 Grist Mill Complex; 1894 Bunker Hill Covered Bridge; 1887 Historic Harper House; 1790 Huffman Plantation House; Bonniwell Lyerly House 1912; 1913 overshot waterwheel-powered grist mill; 1890s country store; 1880 wheat house; 1913 miller's residence.
Major Exhibits: Elegantly Crafted, 11/13-4/14; Women of the River, 3/14; Qualla Arts, 3/14-5/14.
Research Fields: local, family & architectural history.
Facilities: 7,800-vol. library of rare books & books on local history available for use on premises only; reading room.
Activities: guided tours; lectures; films; gallery talks; formally organized educational programs; docent program or council; permanent & temporary exhibitions; school loan service.
Publications: books, A History of Catawba County; Tombstone Inscriptions of Old St. Paul's Church; Grandmother's House; A History of the Yount Family; Blackburn Family History; Catawba County: An Architectural History; The Catawbans, Vol. I & II.
Hours & Admission Prices: Museum of History: Tues.-Sat. 9-4. Guided Tour $3.50. Murray's Mill: Thurs.-Sat. 9-5, Sun. 1-5. Admission $5, guided tour $6. Bunker Hill Covered Bridge: daily sunrise to sunset. No charge. Harper House: Thurs.-Sat. 10-5, Sun. 1-5. Admission $5. Closed major holidays. &
Attendance: 43,480 (estimated)
Membership: Senior $25; Individual $30; Senior Family $40; Family $50; Friend $100; Sponsor $150; Donor $250; Patron $500; Corporate $2,500; Institutional $5,000.

North Wilkesboro

WILKES ART GALLERY, 913 C St., North Wilkesboro, NC 28659-4119. Tel.: 336-667-2841. Fax: 336-667-9264.
E-mail: info@wilkesartgallery.org
Web Site: wilkesartgallery.org
Founded: 1962.
Congressional District: 5
Key Personnel: C.E.O., Kara Minton-Elmore; Pres., Madeline Johnson; Vice Pres., John Harwell; Office Mgr., Eric Blahnik.
Personnel Profile: Full-Time Paid 1; Full-Time Volunteers 2; Part-Time Paid 3; Part-Time Volunteers 2; Interns 1.
Governing Authority: nonprofit organization. Tax-exempt: 501(c)(3).
Institution Type/Description: Art Gallery.
Collections: contemporary paintings, graphics, sculpture, primarily of North Carolina artists.
Facilities: classrooms; 3,500 sq. ft. exhibition space. Original crafts & artwork for sale.
Activities: guided tours; lectures; films; gallery talks; arts festivals; formally organized education programs; docent program or council; permanent, temporary & traveling exhibitions; school & community loan service.
Publications: monthly newsletter, Wilkes Art Gallery Newsletter; monthly brochures & catalogues, Title of Exhibition.
Hours & Admission Prices: Tues.-Fri. 10-5, Sat. 10-2. No charge; donations accepted. Closed New Year's Day; Easter & day after; Memorial Day; Independence Day; Labor Day; Thanksgiving; Christmas Eve & Day. &
Attendance: 12,000 (estimated)
Membership: Student & Senior Citizen $25; Individual $35; Family $50; Friend $125-$299; Advocate $300-$499; Benefactor $500 & up. Call for information about higher levels of membership.

Oak Island

OAK ISLAND NATURE CENTER, 52nd St., N.E. & Yacht Dr., Oak Island, NC 28465. Mailing Address: 4601 E. Oak Island Dr., Oak Island, NC 28465. Tel.: 910-278-5011.
Institution Type/Description: Nature Center.
Collections: live animals; botanical.
Facilities: nature trails.
Activities: educational programs; nature trails.
Hours & Admission Prices: Memorial Day to Labor Day Wed.-Sun. 12-3; Sept.-May Fri.-Sun. 12-3.

Oak Ridge

OLD MILL OF GUILFORD, 1340 NC Hwy. 68 N., Oak Ridge, NC 27310. Tel.: 336-643-4783.
E-mail: olmill@triad.rr.com
Web Site: www.oldmillofguilford.com
Institution Type/Description: Historic Building: housed in a working gristmill; built in 1767. Listed on the National Register of Historic Places.
Collections: gristmill history.

Facilities: Gift items for sale.
Hours & Admission Prices: Daily 9-5.

Oakboro

OAKBORO REGIONAL MUSEUM OF HISTORY, 231 N. Main St., Oakboro, NC 28129. Mailing Address: P.O. Box 565, Oakboro, NC 28129. Tel.: 704-485-4222.
E-mail: oakboromuseum@aol.com
Founded: 1999.
Key Personnel: Chm. (V), Bob Barbee.
Volunteer Hours: 974
Governing Authority: Tax-exempt.
Institution Type/Description: History Museum.
Collections: local history & culture; period furnishings; personal artifacts; early 1900 Stanly County map; photographs.
Activities: guided tours.
Hours & Admission Prices: Sun.-Mon. 2-4, Thurs. 10am-12pm; groups of 8 or more by appointment; call for additional hours. &
Attendance: 825 (estimated)
Membership: Individual $20; Family $30; Patron $60; Donor $100; Benefactor $150.

Ocean Isle Beach

MUSEUM OF COASTAL CAROLINA, 21 E. Second St., Ocean Isle Beach, NC 28469. Tel.: 910-579-1016. Fax: 910-575-4770.
E-mail: minfo@museumplanetarium.org
Web Site: www.museumplanetarium.org
Founded: 1991.
Key Personnel: Exec. Dir., Terry Bryant; Dir. Planetarium, Mark Jankowski; Mkgt., Susan Silk; Membership Svcs., Deb Boyce; Educator, Ed Orsenik; Museum Shop Mgr., Lynn Wiedman; Asst. Mgr. Museum Shop, Rita Ashley.
Governing Authority: nonprofit organization. Parent Institution: Ocean Isle Museum Foundation. Subsidiary Institution: Ingram Planetarium, 7625, High Market St., Sunset Beach, NC 28468.
Institution Type/Description: Natural History Museum.
Collections: wildlife & marine dioramas; shells; shark jaws; touch tank & aquarium with live animals; hands-on exhibits; early maritime artifacts; Civil War memorabilia.
Facilities: Museum-related items for sale.
Activities: educational programs; special events; lectures; bird walks; fossil pit; hands-on exhibitions; touch tank. Annual Event: Home Tour in December.
Hours & Admission Prices: Memorial Day to Labor Day Mon.-Fri. 10-5, Sat. 10-2; Sept.-May Fri.-Sat. 10-4. Adults $8, seniors 61 & over $7, children 3-12 $6; children 2 & under no charge.

Ocracoke

DAVID WILLIAMS HOUSE MUSEUM, 49 Water Plant Rd., Ocracoke, NC 27960. Mailing Address: P.O. Box 1240, Ocracoke, NC 27960-1240. Tel.: 252-928-7375. Facebook: Ocracoke Preservation Society.
E-mail: info@ocracokepreservation.org
Web Site: www.ocracokepreservation.org
Founded: 1983.
Key Personnel: Dir., Amy Howard; Museum Shop Mgr., Ann Borland.
Personnel Profile: Full-Time Paid 1; Part-Time Paid 2; Part-Time Volunteers 30.
Governing Authority: nonprofit organization. Tax-exempt.
Institution Type/Description: Historic House Museum: c.1900. Listed on the National Register of Historic Places.
Collections: period furnishings; photographs.
Major Exhibits: Island Faces, 3/14-12/14; Hurricanes, 3/14-12/14.
Facilities: library. Museum-related items for sale.
Activities: porch talks. Museum Sponsors: Membership Meeting in May & November; Educational Programs June, July & August; Wassail Party in December.
Publications: biannual membership newsletter, The Mullet Wrapper.
Hours & Admission Prices: Call for hours. No charge; donations accepted.
Attendance: 15,000 (estimated)
Membership: Student $20; Individual $25; Family $45; Business $100; Sustaining $300-$1,000.

OCRACOKE ISLAND VISITOR CENTER, Hwy. 12, Ocracoke, NC 27960. Mailing Address: 1401 National Park Dr., Manteo, NC 27954-9451. Tel.: 252-473-2111 & 928-4531. Fax: 252-473-2595.
E-mail: caha_information@nps.gov

Web Site: www.nps.gov/caha
Founded: 1956.
Congressional District: 1
Key Personnel: Supt., Barclay Trimble; Archives Technician, Jami P. Lanier.
Personnel Profile: Full-Time Paid 1; Part-Time Paid 3.
Governing Authority: federal. Affiliated with Cape Hatteras National Seashore. Parent Institution: Dept. of the Interior. Subsidiary Institution: U.S. National Park Service. Tax-exempt.
Institution Type/Description: Park Museum & Visitor Center.
Collections: wall panel exhibits; history; natural history.
Facilities: Maps & books for sale.
Activities: programs by interpretive park rangers June-Aug.
Publications: park brochure; park newspaper; site bulletins.
Hours & Admission Prices: mid-June to Labor Day daily 9-6; Sept.-June daily 9-5. No charge. &
Attendance: 54,600 (accurate)

Old Fort

MOUNTAIN GATEWAY MUSEUM, 102 Water St., Old Fort, NC 28762. Mailing Address: P.O. Box 1286, Old Fort, NC 28762-1286. Tel.: 828-668-9259. Fax: 828-668-0041.
E-mail: mgm@ncmail.net
Founded: 1971.
Congressional District: 11
Key Personnel: C.E.O., Dir. & Cur., Sam Gray.
Personnel Profile: Full-Time Paid 3; Part-Time Paid 5.
Governing Authority: Parent Institution: N.C. Dept. of Cultural Resources. Branch: North Carolina Museum of History, North Carolina Dept. of Cultural Resources, Div. of Archives & History, 109 E. Jones St., Raleigh, NC 27611. Tax-exempt.
Institution Type/Description: History Museum.
Collections: mountain life artifacts; pottery; farm implements; history & culture of Western North Carolina; log cabins.
Research Fields: local history.
Facilities: Museum-related items for sale.
Activities: guided tours; education programs.
Hours & Admission Prices: Mon. 12-5, Tues.-Sat. 9-5, Sun. 2-5. No charge. Closed state holidays.
Attendance: 20,000 (accurate)

OLD FORT RAILROAD MUSEUM, 25 Hwy. 70 W., Old Fort, NC 28762. Mailing Address: 38 Catawba Ave., Old Fort, NC 28762. Tel.: 828-668-4282.
Institution Type/Description: Historic Building: housed in a former deport; built in 1890.
Collections: local history; railroad memorabilia & artifacts; hand tools; signal lights; period furnishings; signs; caboose.
Hours & Admission Prices: Call for hours.

Oriental

ORIENTAL'S HISTORY MUSEUM, 802 Broad St., Oriental, NC 28571. Mailing Address: P.O. Box 103, Oriental, NC 28571. Tel.: 252-670-9318.
Key Personnel: Cur., Marsha Shirk
Collections: steamship Orental's porthole; dugout canoe; local history & artifacts; photographs; drawings; nautical artifacts.
Hours & Admission Prices: Fri. 11-3, Sat. 12-4, Sun. 1-4. No charge; donations accepted.

Oxford

GRANVILLE COUNTY HISTORICAL SOCIETY MUSEUM, 1 Museum Ln., Oxford, NC 27565. Mailing Address: P.O. Box 1433, Oxford, NC 27565-1433. Tel.: 919-693-9706. Fax: 919-693-9706.
E-mail: webmail@granvillemuseumnc.org
Web Site: www.granvillemuseumnc.org
Founded: 1996.
Congressional District: 78
Key Personnel: Dir. & Museum Shop Mgr., Pam Thornton; Asst. Dir., Valerie Heinssen; Pres. (V), Dr. Richard Taylor.
Personnel Profile: Full-Time Paid 1; Part-Time Paid 5; Part-Time Volunteers 70.
Governing Authority: private; nonprofit. Parent Institution: Granville County Historical Society. Branch Museum: Harris Exhibit Hall, 110 Count St., Oxford, NC. Tax-exempt: 501(c)(3).
Institution Type/Description: History & Science Museum.
Collections: Granville County history, art & science. Harris Exhibit Hall: science, arts, cultural topics & history exhibits.

Facilities: 75-vol. library on local history & historic preservation; 3,000 sq. ft. exhibit spaces. Harris Hall: museum-related items for sale.
Activities: docent program; films; formal education program for children; guided tours; lectures; loan, temporary & traveling exhibitions.
Publications: quarterly, Granville County Historical Society Newsletter; monthly, Docent Diary.
Hours & Admission Prices: Wed.-Fri. 10-4, Sat. 11-3. No charge; donations accepted. Closed Easter; Thanksgiving; Christmas. &
Attendance: 6,932 (accurate)
Membership: Individual $25; Double $40; Business $75; Lifetime $300.

Pembroke

NATIVE AMERICAN RESOURCE CENTER, University of North Carolina at Pembroke, Pembroke, NC 28372. Mailing Address: P.O. Box 1510, Pembroke, NC 28372-1510. Tel.: 910-521-6282.
E-mail: nativemuseum@uncp.edu
Web Site: www.uncp.edu/nativemuseum/
Founded: 1979.
Congressional District: 7
Key Personnel: Dir., Dr. Stanley Knick; Business Officer, Neil R. Hawk; Publications & Public Rels. Dir., Amber Rach.
Governing Authority: college. University of North Carolina at Pembroke. Tax-exempt: 170 (b)(1)(A).
Institution Type/Description: Native American Museum: housed in 1923 Old Main Building.
Collections: Indian artifacts; handicrafts; art, books, cassettes, record albums & films pertaining to Native Americans & Lumbee Indians.
Research Fields: Lumbee history & culture; prehistory of southeastern North Carolina; Indian health & educational issues.
Facilities: 90-seat auditorium.
Activities: guided tours; lectures; films; traveling exhibitions.
Publications: book, Robeson Trails Archaeological Survey: Reconnaissance in Robeson County; reader, Along the Trail: A Reader About Native Americans; brochure; The Lumbee in Context; River Spirits: A Collection of Lumbee Writings; Fine in the World: Lumbee Language in Time and Place; DVDs: Indian By Birth; Lumbee By Grace; In The Heart of Tradition; Our People: The Sappony; Dancing In The Gardens of the Lord; Listen to the Drum; DVDs: Our People: The Lumbee; Our People: The Coharie; Our People: The Occaneechi Band of Saponi Nation; VHS: A Healing Faith.
Hours & Admission Prices: Mon.-Sat. 8-5; groups by appointment. No charge. Closed state holidays. &
Attendance: 10,000 (accurate)
Membership: Student $5; Subscriber $15; Teacher $20, Sustainer $30; Sponsor $60; Benefactor $100; Patron $250; Angel $500.

Penland

PENLAND GALLERY & VISITORS CENTER, 3135 Conley Rdige Rd., Penland, NC 28765. Mailing Address: P.O. Box 37, Penland, NC 28765. Tel.: 828-765-6211.
E-mail: gallery@penland.org
Web Site: penland.org/gallery
Key Personnel: Dir., Kathryn Gremley; Gallery Asst., Sarah McClary.
Personnel Profile: Full-Time Paid 2; Part-Time Paid 3.
Governing Authority: Parent Institution: Penland School of Crafts.
Institution Type/Description: Art Museums.
Collections: works by current and former Penland instructors, resident artists, & students from around the country.
Facilities: visitor center.
Activities: temporary exhibitions.
Hours & Admission Prices: March to early Dec. Tues.-Sat. 10-5, Sun. 12-5. No charge. &
Attendance: 10,750 (estimated)

Pine Knoll Shores

NORTH CAROLINA AQUARIUM AT PINE KNOLL SHORES, One Roosevelt Dr., Pine Knoll Shores, NC 28512. Mailing Address: One Roosevelt Blvd., Pine Knoll Shores, NC 28512. Tel.: 252-247-4003. Fax: 252-247-0663.
E-mail: pksmail@ncaquariums.com
Web Site: www.ncaquariums.com
Founded: 1976.
Key Personnel: Dir., Allen Monroe; Dir. Husbandry & Operations, Stuart May; Cur. Education, Windy Arey Kent; Coord. Public Rels., Julie Powers.
Personnel Profile: Full-Time Paid 45; Part-Time Paid 45; Part-Time Volunteers 230; Interns 6.
Governing Authority: state; nonprofit. Parent Institution: North Carolina

Aquariums, 3125 Poplarwood Ct., Ste. 160, Raleigh, NC 27604. Tax-exempt: 501(c)(3).
Institution Type/Description: Aquarium.
Collections: live displays & supportive artifacts that focus on the natural history of North Carolina's aquatic life.
Facilities: 300-vol. library of books related to coastal & aquatic natural history; 150-seat auditorium; educational facilities; nature trails. Books, educational games, shirts & museum-related items for sale.
Activities: films; formal educational programs; lectures; day camps; family programs; field trips; participatory exhibits; conservation special events. Museum Sponsors: Surf Fishing School.
Publications: magazine, Aquarium News; seasonal calendar of activities & events.
Hours & Admission Prices: Adults $8, seniors 62 & over $7, children 3-12 $6; children under 3 & members no charge. Closed New Year's Day; Thanksgiving; Christmas. &
Attendance: 403,769 (accurate)
Membership: Individual $40; Family $60; Curator $100; Patron $300; Benefactor $1,000.

Pinehurst

GIVEN MEMORIAL LIBRARY AND TUFTS ARCHIVES, 150 Cherokee Rd., Pinehurst, NC 28374. Mailing Address: P.O. Box 159, Pinehurst, NC 28370. Tel.: 910-295-6022 & 3642.
Governing Authority: nonprofit organization. Tax-exempt: 501(c)(3).
Institution Type/Description: Library & Archives.
Collections: Pinehurst history; books; photographs.
Activities: research.
Hours & Admission Prices: Mon.-Fri. 9:30-5, Sat. 9:30-12:30. No charge; donations accepted.

SANDHILLS HORTICULTURAL GARDENS, 3395 Airport Rd., Pinehurst, NC 28374-8778. Tel.: 910-695-3882. Fax: 910-695-3894.
E-mail: johnsond@sandhills.edu
Web Site: www.sandhillshorticulturalgardens.com
Founded: 1980.
Key Personnel: Dir., Dee Johnson; Chm. (V), Andy Auman.
Personnel Profile: Full-Time Paid 3; Part-Time Paid 2; Part-Time Volunteers 18.
Governing Authority: Parent Institution: Sandhills Community College. Tax-exempt.
Institution Type/Description: Botanical Garden.
Collections: gardens including plants, trees, & flowers; design composition; nature.
Facilities: visitor center; 32 acre garden.
Activities: educational programs; lectures; special events; rental facilities; seminars.
Publications: quarterly newsletter, Bloomin' News.
Hours & Admission Prices: Visitor Center: daily 8-5. Gardens: dawn to sunset. No charge; donations accepted. &
Attendance: 7,500 (estimated)
Membership: Individual $25; Family $45; Lifetime $350.

Pineville

PRESIDENT JAMES K. POLK STATE HISTORIC SITE, (M), 12031 Lancaster Hwy., Pineville, NC 28134. Mailing Address: P.O. Box 475, Pineville, NC 28134-0475. Tel.: 704-889-7145. Fax: 704-889-3057.
E-mail: polk@ncdcr.gov
Web Site: www.polk.nchistoricsites.org
Founded: 1968.
Congressional District: 9
Key Personnel: Site Mgr., Scott Warren; Pres. (V), Sharon Van Kuren.
Personnel Profile: Full-Time Paid 3; Part-Time Paid 2; Part-Time Volunteers 30; Interns 1.
Governing Authority: state. Parent Institution: North Carolina Dept. of Cultural Resources, 109 E. Jones St., Raleigh, NC 27611. Subsidiary Institution: J.K. Polk Memorial Support Fund, Inc. Tax-exempt: 170(b)(1)(A).
Institution Type/Description: Historic Buildings & Visitor Center.
Collections: life & times of James K. Polk; late 18th- & early 19th-century furnishings. Historic Buildings: main house, barn, separate kitchen.
Research Fields: life & times of James K. Polk 1795-1849.
Facilities: visitor center.
Activities: guided tours; films; formally organized education programs for students; permanent exhibitions; audiovisual programs; lectures & demonstrations for organized school groups & special events.

Publications: brochure (available in Braille); exhibit text in Braille; The Young Hickory News.
Hours & Admission Prices: Tues.-Sat. 9-5. No charge; donations accepted. Closed New Year's Day; Memorial Day; Independence Day; Labor Day; Thanksgiving; Christmas Eve & Day. &
Attendance: 16,048 (accurate)
Membership: Congressman $20; Governor $25; Speaker of the House $40; President $100.

Pinnacle

HORNE CREEK LIVING HISTORICAL FARM, 308 Horne Creek Farm Rd., Pinnacle, NC 27043. Tel.: 336-325-2298. Fax: 336-325-3150.
E-mail: hornecreek@ncmail.net
Institution Type/Description: History Museum.
Collections: North Carolina's rural heritage & history; photographs; historic buildings.
Facilities: nature trails. Museum-related items for sale.
Activities: special events; hands-on working farm.
Hours & Admission Prices: Tues.-Sat. 9-5. No charge; donations accepted. Closed most major holidays.

Pisgah Forest

ALLISON-DEAVER HOUSE MUSEUM, 200 Hwy. 280, Pisgah Forest, NC 28768. Mailing Address: Transylvania County Historical Society, P.O. Box 2061, Brevard, NC 28712-2061. Tel.: 828-884-5137 & 8570.
Web Site: www.preservingourpast.org
Institution Type/Description: Historic House Museum: listed on the National Register of Historic Places.
Collections: period furnishings; personal artifacts.
Facilities: Museum-related items for sale.
Hours & Admission Prices: May-Oct. by appointment.

FOREST DISCOVERY CENTER AND CRADLE OF FORESTRY HISTORIC SITE, 11250 Pisgah Hwy., Pisgah Forest, NC 28768. Mailing Address: 1600 Pisgah Hwy., Pisgah Forest, NC 28768. Tel.: 828-877-3130.
Institution Type/Description: History Museum.
Collections: forest conservation history; school of forestry; 1915 logging locomotive; helicopter; hands-on exhibitions.
Facilities: Gift items for sale.
Activities: video; hands-on exhibitions; fire fighting helicopter simulator; nature-based scavenger hunt; special events.
Hours & Admission Prices: mid-April to early Nov. daily 9-5. Adults 16 & over $5; children 15 & under and Tues. no charge.

PISGAH CENTER FOR WILDLIFE EDUCATION, 1401 Fish Hatchery Rd., Pisgah Forest, NC 28768. Mailing Address: P.O. Box 1600, Pisgah Forest, NC 28768. Tel.: 828-877-4423. Fax: 828-877-4792.
E-mail: melinda.patterson@ncwildlife.org
Founded: 1998.
Key Personnel: Dir., Melinda Patterson; Program Coord., Lee Sherrill.
Personnel Profile: Full-Time Paid 7.
Operating Expenses: 300,000
Governing Authority: Parent Institution: North Carolina Wildlife Resources Commission.
Institution Type/Description: Wildlife Center.
Collections: North Carolina's wildlife & their habitats; trout raceways; aquariums including turtles, salamanders, & fish.
Facilities: indoor exhibit; outdoor exhibit; hatchery raceways; restrooms; video viewing area.
Activities: educational programs; workshops; video; feeding of the brook, brown & rainbow trout.
Publications: Wildlife in North Carolina.
Hours & Admission Prices: April-Nov. Mon.-Sat. 8-4:45; Dec.-March Mon.-Fri. 8-4:45. No charge; donations accepted. &
Attendance: 125,000 (accurate)

Pittsboro

CAROLINA TIGER RESCUE, 1940 Hanks Chapel Rd., Pittsboro, NC 27312-9794. Tel.: 919-542-4684. Fax: 919-542-4454.
E-mail: info@cptigers.org
Web Site: www.carolinatigerrescue.org

Institution Type/Description: Wildlife Refuge.
Collections: tigers; caracals; servals; ocelots; binturongs; kinkajous.
Activities: guide tours.
Hours & Admission Prices: By appointment. Adults $12, youth 4-12 $7; children 3 & under no charge.

Plymouth

GOD'S WILDLIFE CREATIONS, 111 W. Water St., Plymouth, NC 27962. Mailing Address: P.O. Box 706, Plymouth, NC 27962. Tel.: 252-793-6600.
E-mail: tourism@washingtoncountygov.com
Web Site: www.visitwashingtoncountync.com
Formerly: Wildlife Museum & Gallery
Founded: 2007.
Institution Type/Description: General Museum.
Collections: mounted wildlife; wildlife habitat; photographs.
Activities: touch tables.
Hours & Admission Prices: Mon.-Fri. 9-4. Adults $3. &
Attendance: 200 (estimated)

PORT-O-PLYMOUTH MUSEUM, 302 E. Water St., Plymouth, NC 27962. Mailing Address: P.O. Box 296, Plymouth, NC 27962-0296. Tel.: 252-793-1377. Fax: 252-741-9501. Facebook: Washington County NC Historical Society / Port O' Plymouth Museum.
E-mail: headquarters@livinghistoryweekend.com
Web Site: www.livinghistoryweekend.com
Founded: 1988.
Key Personnel: Cur., Kimberly McCray.
Governing Authority: private; nonprofit organization. Parent Institution: Washington County Historical Society. Tax exempt: 501(c)(3).
Institution Type/Description: History Museum.
Collections: local history & culture; photographs; Civil War memorabilia & artifacts; paintings.
Major Exhibits: Local Interpretations of the Civil War, 1/14-3/14.
Facilities: Museum-related items for sale.
Activities: Museum Sponsors: Living History Weekend; Odyssey of an Ironclad.
Hours & Admission Prices: Tues.-Sat. 9-4. Adults $3. &
Membership: Individual $15; Family $25; Business $100.

ROANOKE RIVER LIGHTHOUSE & MARITIME MUSEUM, 206 W. Water St., Plymouth, NC 27962. Tel.: 252-217-2204.
E-mail: info@roanoakeriverlighthouse.org
Web Site: www.roanokeriverlighthouse.org
Institution Type/Description: History Museum.
Collections: maritime history; photographs; personal artifacts; replica lighthouse.
Hours & Admission Prices: Tues.-Sat. 11-3; other times by appointment.

Pollocksville

FOSCUE PLANTATION HOUSE, 7509 N. U.S. 17 Hwy., Pollocksville, NC 28573. Mailing Address: 909 Rockford Rd., Highpoint, NC 27262. Tel.: 252-224-1803.
Web Site: www.foscueplantation.com/
Institution Type/Description: Historic House Museum: housed in the former home of Simon Foscue, Jr.; built in 1824.
Collections: local history & culture; period furnishings; personal artifacts; photographs.
Hours & Admission Prices: Thurs. 10-4; other times by appointment.

Princeton

POWELL'S GARDEN, 9468 U.S. Hwy. 70, Princeton, NC 27569. Tel.: 919-936-4421.
Institution Type/Description: Garden.
Collections: over 600 varieties of bearded irises & daylilies.
Hours & Admission Prices: Mon.-Sat. 10-6.

Purlear

RENDEZVOUS MOUNTAIN EDUCATIONAL STATE FOREST, 1956 Rendezvous Mountain Rd., Purlear, NC 28665. Tel.: 336-667-5072.
E-mail: rendezvousmountainesf.ncfs@ncagr.gov
Web Site: www.ncesf.org/rmesf/home.htm
Institution Type/Description: Park Museum.

Collections: forest ecology & history; logging.
Facilities: nature trails; covered amphitheater; picnic facilities.
Activities: talking tree trail; demonstrations; outdoor workshops; classes; nature trails; special events; educational programs.
Hours & Admission Prices: March to late Nov. call for hours.

Raleigh

AFRICAN AMERICAN CULTURAL COMPLEX, 119 Sunnybrook Rd., Raleigh, NC 27610-1827. Tel.: 919-250-9336. Fax: 919-212-3598.
Web Site: www.aaccmuseum.com
Governing Authority: nonprofit organization. Tax-exempt.
Institution Type/Description: History Museum.
Collections: African American history, culture, & inventions; documents; personal artifacts; period furnishings; Hall of Fame inductees.
Facilities: amphitheater; nature trails.
Hours & Admission Prices: By appointment.

ARTSPACE, 201 E. Davie St., Ste. 101, Raleigh, NC 27601-1830. Tel.: 919-821-2787. Fax: 919-821-0383. Facebook: Artspace.
E-mail: info@artspacenc.org
Web Site: www.artspacenc.org
Founded: 1986.
Key Personnel: Exec. Dir., Mary Poole; Dir. Devel., Rae Marie Czuhai; Dir. Programs & Exhibitions, Shana Dumont-Garr.
Personnel Profile: Full-Time Paid 7; Part-Time Paid 5; Part-Time Volunteers 350; Interns 1.
Governing Authority: private; nonprofit organization. Tax-exempt: 501(c)(3).
Institution Type/Description: Art Museum & Gallery.
Collections: works by regional, national & international artists.
Facilities: educational facilities. Museum-related items for sale.
Activities: formal education programs; docent program; guided tours; lectures; rental gallery; children's summer camp program.
Publications: biannual newsletter; monthly e-newsletter; course catalogs; exhibition brochures.
Hours & Admission Prices: Tues.-Sat. 10-6. No charge; donations accepted. Closed New Year's Day; Christmas. &
Attendance: 150,000 (estimated)
Membership: Students & Seniors $30; Individual $45; Family $75; Art Lovers $100-$249; Art Advocates $250-$499; Art Collectors $500-$999; Sponsors $1,000 & up.

CAM RALEIGH - CONTEMPORARY ART MUSEUM RALEIGH, 409 W. Martin St., Raleigh, NC 27603. Tel.: 919-513-0946. Fax: 919-515-7330. Facebook: CAM Raleigh.
Web Site: www.camraleigh.org
Founded: 1983.
Congressional District: 4
Key Personnel: Exec. Dir., Elysia Borowy-Reeder; Cur. Education, Nicole Welch; Mgr. Gallery & Exhibitions, Kate Shafer.
Personnel Profile: Full-Time Paid 7; Part-Time Paid 6; Part-Time Volunteers 4; Interns 3.
Governing Authority: nonprofit organization. Parent Institution: Contemporary Art Foundation.
Institution Type/Description: Contemporary Art.
Collections: national, regional & international contemporary art & design; performance & public art projects.
Research Fields:
Activities: guided tours; lectures; films; concerts; performances; broadcast programs; organized education programs; docent program; traveling exhibitions.
Publications: exhibition catalogues.
Hours & Admission Prices: Call for hours. Adults $5; children under 10, NARM, MODCO & museum members no charge. Closed New Year's Day; Christmas. &
Attendance: 30,000 (estimated)
Membership: Student $25; Out of Town & Educator $30; Senior $40; Individual $50; Dual & Family $75; Friend $150; Art Advocate $500; Cam/Silver & Business Benefactor $1,000; Cam/Gold & Business Supporter $2,500; Director's Vision Circle & Corporate Circle/Silver $5,000; Chair's Circle & Corporate Circle/Gold $10,000; Corporate Circle/Platinum $25,000.

DR. MARTIN LUTHER KING JR. MEMORIAL GARDENS, 900 Rock Quarry Rd., Raleigh, NC 27610. Mailing Address: c/o Raleigh Parks, Recreation and Cultural Resources Dept., P.O. Box 590, Raleigh, NC 27601. Tel.: 919-996-3000.
Founded: 1990.
Institution Type/Description: Memorial Garden.
Collections: life-sized bronze statue of Dr. King; memorial wall; 12-ton granite water monument honoring the civil rights movement.
Hours & Admission Prices: Dawn to dusk.

GREGG MUSEUM OF ART & DESIGN AT NORTH CAROLINA STATE UNIVERSITY, (M), 516 Brickhaven Dr., Ste. 200, Raleigh, NC 27606. Mailing Address: Box 7306, Raleigh, NC 27695. Tel.: 919-515-3503. Fax: 919-515-6163.
E-mail: gregg@ncsu.edu
Web Site: www.ncsu.edu/gregg
Formerly: North Carolina State University Gallery of Art & Design
Founded: 1979.
Congressional District: 4
Key Personnel: Dir., Roger Manley; Pres. (V), Bernard J. Hyman; Cur. Education & Resources, Zoe Starling; Art Preparator, Matt Gay; Registrar, Mary Hauser; Museum Ops. Mgr., Hilary Kinlaw.
Personnel Profile: Full-Time Paid 7; Full-Time Volunteers 2; Part-Time Paid 2; Part-Time Volunteers 4; Interns 15.
Governing Authority: university. Parent Institution: North Carolina State University. Tax-exempt: 170(b)(1)(A).
Institution Type/Description: Art Museum.
Collections: contemporary & historic textiles, ceramics, ethnographic art, photography, glass, furniture, metals, outsider art as well as work by faculty of the NCSU College of Design.
Major Exhibits: Theater of Belief: Afro-Atlantic Costuming and Masking in Large-Format Color Photographs (T), 1/23/14-5/11/14; Remnants of the Floating World: Japanese Art from the Permanent Collection, 2/6/14-5/23/14.
Research Fields: textiles; ceramics pertaining to North Carolina & the South; product design; furniture; material culture; photography.
Facilities: auditorium; theater; classrooms; cafeteria.
Activities: formally organized education programs for undergraduate & graduate college students; loan, permanent, temporary & traveling exhibitions.
Publications: newsletter, Friends of the Gregg; exhibition catalogues.
Hours & Admission Prices: Hours vary. No charge; donations accepted. Closed university holidays. &
Attendance: 47,323 (estimated)
Membership: Friends of the Gregg: NC State Faculty & Staff $35; Supporter $50-$99; Contributor $100-$249; Collector's Circle $250-$499; Sponsor $500-$999; Patron $1,000-$2,499; Connoisseur $2,500-$4,999; Benefactor $5,000-$9,999; Visionary $10,000 & up.

HAYWOOD HALL HOUSE AND GARDEN, 211 New Bern Place, Raleigh, NC 27601. Tel.: 919-832-8357 & 4158.
Governing Authority: Parent Institution: The National Society of the Colonial Dames of America.
Institution Type/Description: Historic House: housed in the former home of State Treasurer John Haywood; built in 1799.
Collections: Haywood family history; personal artifacts; period furnishings; photographs; paintings.
Facilities: gardens.
Activities: rental facilities.
Hours & Admission Prices: By appointment.

HISTORIC OAK VIEW COUNTY PARK, 4028 Carya Dr., Raleigh, NC 27610-2913. Tel.: 919-250-1013. Fax: 919-250-1119.
E-mail: oakview@wakegov.com
Web Site: www.wakegov/parks/oakview
Personnel Profile: Full-Time Paid 2; Part-Time Paid 5.
Governing Authority: Parent Institution: Wake County Parks, Recreation and Open Space.
Institution Type/Description: Historic Farmstead.
Collections: agricultural heritage; local history. Historic Structures: 1855 farmhouse; plank kitchen; c.1900 cotton gin house; livestock barn; carriage house.
Facilities: visitor's center.
Activities: educational programs; special events; permanent exhibits.
Hours & Admission Prices: Park: Mon.-Sat. 8:30-5, Sun. 1-5. No charge. Closed New Year's Day; Thanksgiving, Christmas Eve & Day. &
Attendance: 100,000 (estimated)

HISTORIC YATES MILL COUNTY PARK, 4620 Lake Wheeler Rd., Raleigh, NC 27603. Tel.: 919-856-6675.
Institution Type/Description: History Museum: housed in a 1756 gristmill.
Collections: mill history; period furnishings & equipment; personal artifacts; photographs.
Facilities: nature trails; visitor center.
Activities: hiking trails; educational programs; guided tours.

Hours & Admission Prices: Mill Tours: March-Nov. call for hours. Visitor Center: daily 8:30-5. Tours: adults $5, seniors 60 & over $4, children 7-16 $3; children 6 & under no charge.

JC RAULSTON ARBORETUM AT NC STATE UNIVERSITY,
4415 Beryl Rd., Raleigh, NC 27695. Mailing Address: Dept. Horticultural Science, Box 7522, Raleigh, NC 27695. Tel.: 919-515-3132. Fax: 919-515-5361.
E-mail: ted.bilderback@ncsu.edu
Web Site: www.ncsu.edu/jcraulstonarboretum
Founded: 1976.
Key Personnel: Dir., Ted Bilderback; Chm. (V), Dr. Bobby Ward.
Personnel Profile: Full-Time Paid 7; Part-Time Paid 8; Part-Time Volunteers 210; Interns 4.
Governing Authority: Parent Institution: N.C. State University. Tax-exempt.
Institution Type/Description: Arboretum.
Collections: over 6,000 plants.
Activities: teaching & research programs.
Hours & Admission Prices: April-Oct. daily 8-8; Nov.-March daily 8-5. No charge; donations accepted. &
Attendance: 35,000 (estimated)
Membership: Individual $50; Family & Dual $75; Sponsor $300; Patron $600; Founder $1,250; Benefactor $2,500; Philanthropist $5,000.

JOEL LANE HOUSE MUSEUM, (M), 160 S. Saint Mary's St.,
Raleigh, NC 27603-1618. Mailing Address: P.O. Box 10884, Raleigh, NC 27605-0884. Tel.: 919-833-3431. Fax: 919-833-9431.
E-mail: joellane@bellsouth.net
Web Site: www.joellane.org
Founded: 1972.
Key Personnel: Pres. (V), Fair Pickel; Cur., Isabella Long.
Personnel Profile: Full-Time Paid 1; Part-Time Paid 2; Part-Time Volunteers 50.
Governing Authority: Parent Institution: NSCDA in North Carolina. Tax-exempt.
Institution Type/Description: Raleigh Historic Site: housed in the former home of Colonel Joel Lane. Listed on the National Register of Historic Places.
Collections: Lane family history; personal artifacts; period furnishings; hands-on exhibits; local history.
Research Fields: African Americans 18th century in Wake County, NC.
Facilities: Museum-related items for sale.
Activities: educational programs; hands-on activities.
Publications: Chameleon on the Crabtree: The Story of Joel Lane; newsletter, The Parshment Press.
Hours & Admission Prices: Tours: Sat. 1, 2 & 3; other times by appointment. Adults $5, seniors 65 & over $4, students $3; discounts to Time Travelers & AAA members; members & children under 6 no charge.
Attendance: 6,000 (accurate)
Membership: Individual $40; Family $50; Colonial Patron $100.

MARBLES KIDS MUSEUM, 201 E. Hargett St., Raleigh, NC
27601-1437. Tel.: 919-834-4040 & 857-1085. Fax: 919-834-3516.
E-mail: info@marbleskidsmuseum.org
Web Site: www.marbleskidsmuseum.org
Formerly: Exploris
Founded: 2007.
Key Personnel: Chm. Bd. (V), David Fountain; Pres., Sally Edwards; Vice Pres. Finance & Human Resources, Ben Blankenship; Dir. Mktg., Katy Burgwyn; Vice Pres. Exhibits, Pam Hartley; Vice Pres. Operations, Tim Hazlehurst; Vice Pres. External Affairs, Britt Thomas; Museum Shop Mgr., Toni Strickland.
Personnel Profile: Full-Time Paid 35; Part-Time Paid 50; Part-Time Volunteers 65; Interns 2.
Governing Authority: private; nonprofit organization. Tax-exempt: 501(c)(3).
Institution Type/Description: Children's Museum.
Collections: interactive exhibits.
Facilities: 272-seat IMAX theatre; classrooms; rental spaces.
Activities: summer camps; school's out camps; daily & weekly programming; special events; birthday parties; IMAX films; traveling exhibits.
Publications: monthly newsletter; annual report; educators guide; brochures; e-blasts; calendars.
Hours & Admission Prices: Tues.-Sat. 9-5, Sun. 12-5. Adults $5; discounts to ACM members; members no charge. Closed Easter; Thanksgiving; Christmas. &
Attendance: 546,000 (estimated)
Membership: Marbles Grand $90; Family $100; Family Plus $150.

MORDECAI HISTORIC PARK, 1 Mimosa St., Raleigh, NC
27604-1203. Mailing Address: P.O. Box 28072, Raleigh, NC 27611-8072. Tel.: 919-857-4364. Fax: 919-834-7314.
E-mail: info@cappresinc.org
Web Site: www.raleighnc.gov/mordecai
Founded: 1972.
Congressional District: 2
Key Personnel: Pres. & C.E.O., Gary G. Roth.
Personnel Profile: Full-Time Paid 3; Part-Time Paid 7; Part-Time Volunteers 225.
Governing Authority: Parent Institution: Capital Area Preservation, Inc. Tax-exempt.
Institution Type/Description: Historic Houses: housed in the birthplace of Andrew Johnson.
Collections: period furnishings; period linens. Historic Houses: early Raleigh office; Badger/Iredell law office; c.1785 Mordecai house; 2 original plantation dependencies; 1830 Ellen Mordecai garden; c.1842 Allen kitchen.
Facilities: Museum-related items for sale.
Activities: guided tours; special events.
Publications: interpretive brochure, Gleanings From Long Ago.
Hours & Admission Prices: Tues.-Sat. 10-3, Sun. 1-3, last tour 3. Adults $5, youth 7-17 & seniors $3; children 6 & under no charge. &
Attendance: 11,000 (accurate)
Membership: Annual Membership: Individual $30; Family $50. President's Circle Membership: Sustainer $100; Contributor $250; Patron $500; Benefactor $1,000.

NORTH CAROLINA EXECUTIVE MANSION, 200 N. Blount
St., Raleigh, NC 27601. Tel.: 866-724-8687.
Web Site: www.nchistoricsites.org/capitol/exec/exectour.htm
Institution Type/Description: Historic Mansion: housed in the home of North Carolina's governors since 1891.
Collections: local history; period furnishings; personal artifacts; photographs.
Facilities: gardens.
Activities: guided tours; special events. Annual Event: Holiday Open House.
Hours & Admission Prices: Tours: by appointment.

* NORTH CAROLINA MUSEUM OF ART, (M), 2110 Blue
Ridge Rd., Raleigh, NC 27607-6494. Mailing Address: 4630 Mail Service Center, Raleigh, NC 27699-4600. Tel.: 919-839-6262. Fax: 919-733-8034.
E-mail: cwoodrum@ncmamail.dcr.state.nc.us
Web Site: www.ncartmuseum.org
Founded: 1956.
Congressional District: 4
Key Personnel: Dir., Lawrence J. Wheeler; Deputy Dir. Art & Cur. Modern Art, John Coffey; Cur. Ancient Art, David H. Steel; Chief Deputy Dir. & C.F.O., Caterri Woodrum; Cur. Northern European, Dennis P. Weller; Cur. Ancient Art, Mary Ellen Soles; Dir. Mktg., Melanie Davis-Jones; Dir. Education, Susan Glasser; Dir. Devel. & Membership, Ellen Stone; Chief Registrar, Maggie Gregory; Librarian, Natalia Lonchyna; Mgr. Communications, Jennifer Bahus; Chief Conservator, Bill Brown; Chief Cur. & Cur. Contemporary Art, Linda Dougherty; Assoc. Cur. Contemporary Art, Kinsey Katchka; Mgr. Exhibitions, Tiara L. Paris; Dir. Planning & Design, Dan Gottlieb.
Personnel Profile: Full-Time Paid 140; Full-Time Volunteers 500; Part-Time Paid 8; Part-Time Volunteers 320; Interns 4.
Governing Authority: state. Parent Institution N.C. Dept. of Cultural Resources. Subsidiary Institution: N.C. Museum of Art Foundation, Inc. Tax-exempt.
Institution Type/Description: Art Museum.
Collections: European & American art; ancient art; African, Oceanic & pre-Columbian art; Kress collection; Mary Duke Biddle Education Gallery; North Carolina art; Judaic art.
Facilities: 34,000-vol. art reference library; restaurant. Museum-related items for sale.
Activities: guided tours; lectures; films; concerts; workshops; teacher seminars; art consultation; art reference library; state extension program; permanent & temporary exhibitions.
Publications: bulletin; magazine; exhibition & permanent collection catalogues & brochures; calendar of events.
Hours & Admission Prices: Tues.-Thurs. & Sat.-Sun. 10-5, Fri. 10-9. &
Attendance: 400,000 (estimated)
Membership: Student $24; Senior Citizen $30; Individual $40; Senior Couple $50; Dual $65; Family $75; Patron $125; Sustainer $250; Benefactor $500; Fellow $1,000; Director's Circle $2,500; Humber Society $5,000; Collectors Cabinet/Gold $10,000; Collectors Cabinet & Platinum $25,000.

*** NORTH CAROLINA MUSEUM OF HISTORY, (M),** 5 E. Edenton St., Raleigh, NC 27601-1011. Tel.: 919-807-7900. Fax: 919-733-8655. Facebook: North Carolina Museum of History.
E-mail: ken.howard@ncdcr.gov
Web Site: ncmuseumofhistory.org
Founded: 1902.
Congressional District: 4
Key Personnel: Dir. N.C. Museum of History & Div. of State History Museums, Kenneth B. Howard; Assoc. Dir., Jackson Marshall; Exec. Dir. No. Carolina Museum of History Assoc., William Wilson; C.O.O., Heyward McKinney; Admin. Mountain Gateway Museum & Heritage Center, Terrell Finley; Admin. Museum of the Cape Fear Historical Complex, David Reid; Dir. Museum of the Albemarle, Bill McCrea; Chief Cur., Dr. Joseph Porter; Chief Collections Management, John Campbell; Chief Administrative Svcs., Thom Swindell; Museum Store Mgr., Lynn Brower; Public Information Officer, Susan Friday Lamb; Dir. NC Maritime Museums & The Graveyard of the Atlantic Museum, Joseph K. Schwarzer.
Personnel Profile: Full-Time Paid 144; Part-Time Paid 5; Part-Time Volunteers 134; Interns 10.
Volunteer Hours: 9,774
Governing Authority: state. Parent Institution: North Carolina Dept. Cultural Resources, Office of Archives and History. Subsidiary Institutions: Mountain Gateway Museum and Heritage Center; Museum of the Albemarle; Museum of the Cape Fear Historical Complex; North Carolina Maritime Museum in Beaufort; N.C. Maritime Museum in Southport; Graveyard of the Atlantic Museum. Affiliate of the Smithsonian Institution. Tax-exempt.
Institution Type/Description: North Carolina History Museum.
Collections: decorative arts; furnishings; costumes; uniforms; tools & equipment; industry; folklife; numismatics; currency; weapons; textiles; anthropology; paintings; graphics; military; medicine; transportation; toys; tobacco; North Carolina history & material culture.
Major Exhibits: Watergate: Political Scandal and the Presidency, 5/13-8/14; Formed, Fired & Finished: Art Pottery from James Farmer, 11/13-8/14; Civil War 1864, 1/14-12/14; Cedars in the Pines - Lebanese Life in N.C., 2/21/14-8/14; Stagville Plantation Photographs, 3/14-1/15; N.C. Decorative Arts, 4/14-8/14; Spanish Exploration, 9/14-2/15; Starring N.C. - Filmmaking in N.C., 11/14-1/16.
Research Fields: culture, social, military, economic, political history & folk life of North Carolina.
Facilities: 8,000-vol. library; 315-seat auditorium; classrooms; conference rooms. Museum-related items for sale.
Activities: guided tours; hands-on museum classroom program; lecture series; workshops for children, adults & teachers; films; traditional music presentations; festivals; symposia; permanent & temporary exhibitions; media center; hands-on classroom kits; online workshops; distance learning classes; interactive website; virtual field trips; video presentations & podcasts on web site. Museum Sponsors: Tar Heel Junior Historian Association; consultant services for in-state history museums.
Publications: docent newsletter; program-related educational leaflets & pamphlets; magazine & newsletter, Tar Heel Junior Historian Association; exhibition-related publications; e-newsletter; biennial magazine, Circa; bimonthly program calendar.
Hours & Admission Prices: Mon.-Sat. 9-5, Sun. 12-5. No charge; donations accepted. Closed New Year's Day; Good Friday; Easter; Memorial Day; Independence Day; Labor Day; Veterans Day; Thanksgiving; Christmas Eve & Day. &
Attendance: 342,000 (accurate)
Membership: Scribe $50; Archivist $75; Patriot $100; Enthusiast $250; Curator's Assembly $500; Historian $750; Benefactor $1,000; Sterling Benefactor $1,500; Fred Olds Society $2,500; Executive Gallery $5,000.

*** NORTH CAROLINA MUSEUM OF NATURAL SCIENCES, (M),** 11 W. Jones St., Raleigh, NC 27601-1029. Mailing Address: Mail Service Center 1626, Raleigh, NC 27699-1600. Tel.: 919-733-7450. Fax: 919-733-1573.
Web Site: www.naturalsciences.org
Founded: 1879.
Congressional District: 4
Key Personnel: Dir., Dr. Betsy Bennett; Deputy Dir. Museum Operations, Alvin L. Braswell; Dir. Devel., Bonnie L. Smith; Dir. Exhibits, Roy G. Campbell; Deputy Dir. Research & Strategic Planning and Interim Dir. Nature Museum Research Center, Dr. Karen Giroux, Dir. School Programs, Liz Baird; Dir. Prairie Ridge, Mary Ann Brittain; Dir. Public Programs, Jesse P. Perry, III; Mktg. Dir., Jeff Williford; Dir. Communications, Jonathan Pishney; Dir., Friends of the Museum, Angela Baker-James; Cur. Mammals, Lisa J. Gatens; Cur. Birds, John Gerwin; Cur. Terrestrial Invertebrates, Dr. Rowland Shelley; Cur. Paleontology, Vince Schneider; Cur. Herpetology, Dr. Bryan Stuart; Cur. Crustaceans, Dr. John Cooper; Cur. Fishes, Dr. Wayne Starnes; Cur. Geology, Dr. Christopher Tacker; Cur. Aquatic Invertebrates, Dr. Arthur Bogan; Museum Shop Mgr., Heather Heath.
Personnel Profile: Full-Time Paid 111; Part-Time Paid 55; Part-Time Volunteers 297; Interns 28.
Governing Authority: state. Division of North Carolina Department of Environment Natural Resources, 412 Salisbury St., Raleigh. Subsidiary Institution: N.C. Museum of Forestry, Whiteville, NC. Tax-exempt: 170(c)(1).
Institution Type/Description: Natural Science Museum.
Collections: zoology, geology, paleontology, with emphasis on North Carolina & the southeastern United States.
Research Fields: mammalogy; ornithology; herpetology; ichthyology; invertebrate zoology; ecology; endangered species; systematics.
Facilities: 7,000-vol. library of natural history books available for use on premises; 262-seat auditorium; naturalist center; cafe; laboratory; rental facilities; 38 acre outdoor education facility. Nature study & natural history books for sale.
Activities: permanent, temporary & traveling exhibits; statewide teacher workshops & outreach programs; classes; trips to local & international natural areas; adult, children & family events; high definition films; resource box loans; self-guided tours; lectures; school & group programs; presentations; internships; meeting place for natural history groups; faunal surveys; advisory services for government agencies. Annual Events: Astronomy Day; A Natural History Halloween; Bugfest; Groundhog Day; Reptile & Amphibian Day; Planet Earth Celebration.
Publications: quarterly popular natural history magazine & program calendar; information leaflets; special publications; educator guide.
Hours & Admission Prices: Mon.-Sat. 9-5, Sun. 12-5, 1st Fri. of month 9-9. No charge; donations accepted. &
Attendance: 724,000 (accurate)
Membership: Individual $45; Family $55; Discovery Club $100; Naturalist Society $250; Explorers Society $500; Adventure Society $1,000.

NORTH CAROLINA OFFICE OF ARCHIVES AND HISTORY, 109 E. Jones St., Raleigh, NC 27601-1023. Mailing Address: 4610 Mail Service Center, Raleigh, NC 27699-4610. Tel.: 919-807-7280. Fax: 919-733-8807.
E-mail: webmaster@ncmail.net
Web Site: www.ah.dcr.state.nc.us/
Founded: 1903.
Congressional District: 4
Key Personnel: Deputy Sec., Dr. Jeffrey J. Crow; Chief Division of Historical Resources, Dr. David Brook; Chief Historic Sites Section, Keith Hardison; Chief State Capitol Visitor Svcs. Section, Deanna Kerrigan; Chief Historical Publications Section, Donna Kelly; Chief Tryon Palace Section, Kay P. Williams; Chief Archives & Records, Jesse Lankford; Chief North Carolina Maritime Museum, Dr. Joseph Schwarzer; Chief Archaeology, Steve Claggett.
Personnel Profile: Full-Time Paid 407.
Governing Authority: state. Parent Institution: North Carolina Dept. of Cultural Resources. Subsidiary Institutions: NC Transportation Museum; Tobacco Museum at Duke Homestead; North Carolina Maritime Museum. Tax-exempt: 170(b)(1)(A).
Institution Type/Description: Historic House & Site.
Collections: decorative arts & furnishings; costumes; uniforms; manuscripts; photography; industry; archaeology; folklore; numismatics; weapons; textiles; anthropology; paintings; graphics; military; medicine; transportation. Sites & Historic Houses: 1744 Palmer-Marsh House, Bath; 1759 James Iredell House, Edenton; 1760 Owens House, Halifax; 1770 Constitution House, Halifax; 1770 Tryon Palace, New Bern; 1770 House in the Horseshoe, Carthage; 1782 Allen House, Burlington; 1799 Stagville Plantation, Durham; 1825 Bonner House, Bath; 1830 Collins House, Creswell; 1833 Clerk of Court's office, Halifax; 1838 colonial jail, Halifax; 1840 State Capitol, Raleigh; 1840 Aycock birthplace, Fremont; 1850 Harper House, Newton Grove; 1852 Duke Homestead, Durham; 1883 Thomas Wolfe House, Asheville; reconstructed 1960 Vance birthplace, Weaverville; reconstructed 1962 Bennett House, Durham; reconstructed 1968 Polk Homestead Pineville; 1880 Hauser Farmhouse Pinnacle; 1924 Roundhouse, Spencer; 1926 Canary cottage (Palmer Memorial Institute), Sedalia; 1726-1830 Brunswick town remains Winnabow; 1864 confederate seaboat wooden hull, Kinston; site of 1756 log Fort Dobbs, Statesville; 1862-1865 Ft. Fisher earthworks Kure Beach; 1850-1912 restored underground workings of Reed Gold Mine, Stanfield; 1300-1500 Town Creek Indian mound with reconstructed temples, Mt. Gilead.
Research Fields: museology; decorative arts; architecture; archaeology; historic preservation; genealogy; North Carolina history; maritime history.
Facilities: archives.
Activities: guided tours; lectures; films; special events: such as encampments, living history & reenactments; docent programs; inter-museum loan, permanent, temporary & traveling exhibitions.

Publications: quarterly scholarly journal, NC. Historical Review; bimonthly newsletter, Carolina Comments; Federation Bulletin.
Hours & Admission Prices: Archives Research Room: Tues.-Fri. 8-5, Sat. 9-4. State Historic Sites: call for hours. State Capitol: Mon.-Fri. 8-5, Sat. 9-5, Sun. 1-5. Admission at most sites no charge. Closed national holidays. &

NORTH CAROLINA SPORTS HALL OF FAME, 5 E. Edenton St., Raleigh, NC 27622. Mailing Address: P.O. Box 31524, Raleigh, NC 27622. Tel.: 919-845-3455.
E-mail: dfish@ncsportshalloffame.org
Web Site: www.ncshof.org
Founded: 1962.
Key Personnel: Exec. Dir., Don Fish; Assoc. Dir., Kevin Brafford
Institution Type/Description: Sports Museum.
Collections: sports history & athletes; Hall of Fame Inductees. personal artifacts; photographs; fitness; nutrition; sports medicine.
Facilities: 3,000 sq. ft. exhibition hall.
Activities: traveling exhibitions. Annual Event: Induction Ceremony.
Hours & Admission Prices: Mon.-Sat. 9-5, Sun. 12-5. No charge.

NORTH CAROLINA STATE CAPITOL, 1 E. Edenton St., Raleigh, NC 27601. Mailing Address: 4624 Mail Service Center, Raleigh, NC 27699-4624. Tel.: 919-733-4994 & 807-7950. Fax: 919-715-4030.
E-mail: state.capitol@ncdcr.gov
Web Site: www.nchistoricsites.org/capitol
Institution Type/Description: Historic Building: built in 1840. A National Historic Landmark.
Collections: state government & history; photographs.
Facilities: library.
Activities: research; guided tours.
Hours & Admission Prices: Self-Guided Tours: Mon.-Fri. Guided Tours: Sat. 11-2; groups by appointment. No charge; donations accepted.

NORTH CAROLINA STATE LEGISLATIVE BUILDING, 16 W. Jones St., Raleigh, NC 27601. Tel.: 919-733-7929.
Institution Type/Description: State Government Building.
Collections: North Carolina General Assembly; North Carolina history.
Facilities: legislative library; cafeteria.
Activities: guided tours; observe legislative procedures.
Hours & Admission Prices: Mon.-Fri. 8-5 Sat. 9-5, Sun. 1-5. No charge.

RALEIGH CITY MUSEUM, (M), Briggs Bldg., Ste. 100, 220 Fayetteville St., Raleigh, NC 27601-1358. Tel.: 919-832-3775. Fax: 919-832-3085.
E-mail: rcm@raleighcitymuseum.org
Web Site: www.raleighcitymuseum.org
Founded: 1993.
Key Personnel: Mgr. Operations, Dianne Davidson; Collections & Membership, Elizabeth Weichel; Exhibits, Jennifer Carpenter; Education & Outreach, Stormi Souter; Visitor Svcs., Grayson Walton.
Personnel Profile: Part-Time Paid 5; Part-Time Volunteers 5; Interns 4.
Governing Authority: private; nonprofit organization. Tax-exempt.
Institution Type/Description: History Museum.
Collections: artifacts & archives related to the city of Raleigh.
Facilities: 5,000 sq. ft. exhibit space; classroom. Museum-related items for sale.
Activities: docent program; guided tours; lectures; rental gallery; school loan service; temporary exhibitions; broadcast programs.
Publications: exhibit catalogs, Let Us March On: Raleigh's Journey Toward Civil Rights; Businesses That Built Raleigh; Historic Raleigh; semi-annual newsletter, Baliwick.
Hours & Admission Prices: Tues.-Fri. 10-4, 1st Fri. each month 10-4 & 6-9, Sat. 1-4. No charge; donations accepted. Closed New Year's Day; Thanksgiving; Christmas Eve & Day. &
Attendance: 23,918 (accurate)
Membership: Student & Senior $25; Individual $50; Family $100; Benefactor $500; Heritage $1,000; Founder $2,500; Golden Oak $5,000; Capital Society $10,000.

RAY PRICE LEGENDS OF HARLEY DRAG RACING MUSEUM, 1126 S. Saunders St., Raleigh, NC 27603-2204. Tel.: 919-832-2261.
Institution Type/Description: Drag Racing Museum.
Collections: motorcycles; racing memorabilia; photographs; personal artifacts.
Hours & Admission Prices: Mon.-Tues. & Thurs.-Fri. 8-6, Sat. 8-5, Sun. 12-5.

WILLIAM B. UMSTEAD STATE PARK, 8801 Glenwood Ave., Raleigh, NC 27617. Tel.: 919-571-4170.
E-mail: william.umstead@ncparks.gov
Web Site: www.ncparks.gov
Institution Type/Description: State Park & Visitor Center.
Collections: natural history & heritage.
Facilities: 22 miles of nature trails; visitor center.
Activities: hiking trails; camping; picnic area; nature study; educational programs; hands-on exhibits; workshops.
Hours & Admission Prices: Call for hours.

Ramseur

MILLSTONE CREEK ORCHARDS, 506 Parks Crossroad Church Rd., Ramseur, NC 27316. Tel.: 336-824-5263.
E-mail: info@millstonecreekorchards.com
Web Site: millstonecreekorchards.com
Institution Type/Description: Working Orchard.
Collections: fruit-growing industry; orchard history.
Facilities: picnic area. Orchard-related items for sale.
Activities: special events; tours; hands-on exhibitions; pick your own fruit; hayrides. Annual Events: Fruity Pickin Storytime; Family Fun Days.
Hours & Admission Prices: mid-May to Nov. Mon.-Sat. 9-6, Sun. 1-5:30; groups by appointment.

Randleman

RICHARD PETTY MUSEUM, 142 W. Academy St., Randleman, NC 27317-1502. Mailing Address: 311 Branson Mill Rd., Randleman, NC 27317. Tel.: 336-495-1143. Fax: 336-495-1543.
E-mail: jwilson@pettys-garage.com
Web Site: www.rpmuseum.com
Founded: 1988.
Key Personnel: Exec. Dir., Jean Wilson; C.E.O., Richard Petty; Museum Shop Mgr., Bonnie Davis.
Personnel Profile: Full-Time Paid 2; Part-Time Paid 3.
Governing Authority: Tax-exempt.
Institution Type/Description: Petty Family & History Museum.
Collections: race cars; awards; photographs; fan memorabilia; videos; personal artifacts; family history.
Facilities: Museum-related items for sale.
Activities: Annual Events: Capital Campaign; Annual Fund; Petty Fest.
Hours & Admission Prices: Wed.-Sat. 9-5. Adults $10, seniors $8, students $5; discounts to groups & military families; children 6 & under and military no charge. Closed Thanksgiving; Christmas. &
Attendance: 3,500 (estimated)
Membership: Individual $25; Family $60.

Richlands

ONSLOW COUNTY MUSEUM, 301 S. Wilmington St., Richlands, NC 28574-8326. Tel.: 910-324-5008. Fax: 910-324-2897.
E-mail: museum@onslowcountync.gov
Web Site: www.onslowcountync.gov/museum
Founded: 1976.
Congressional District: 3
Key Personnel: Division Head, Lisa Whitman-Grice; Pres., Arthine Thomas; Chm., John Chandler; Collections Mgr., Patricia Hughey; Education Coord., Chancellor Mellman; Exhibits Facilitator, Kenneth Barbee; Office Asst., Cacelia Davis.
Personnel Profile: Full-Time Paid 3; Part-Time Paid 2; Part-Time Volunteers 63.
Governing Authority: nonprofit organization. Parent Institution: Onslow County Government & Museum Foundation Board, Inc.
Institution Type/Description: General Museum.
Collections: 19th- & early 20th-century tools & farm implements, woodworking tools, 19th-century costumes; Native American artifacts & quilts.
Research Fields: Onslow County history, architectural history in area, surrounding counties; genealogical research.
Facilities: multi-purpose classroom & meeting room; research room.
Activities: guided tours; lectures; musical performances; changing exhibits. Annual Events: Living History Weekend in April; Summer Art Program for Youth June & July; Annual Art Craft Festival in November.
Publications: brochures: A Brief History of Onslow; Historic Sites in Onslow; Richlands Historic District: Walking & Driving Tour; books, The Architectural History of Onslow County, North Carolina; The Water and the Wood, the history of Onslow County.
Hours & Admission Prices: Tues.-Fri. 10-4:30, Sat. 10-4; school groups by appointment. Adults $2, students & youth 3-18 $1; children 3 & under no charge. Closed New Year's Day; Martin Luther King Jr. Day; Good Friday;

Easter; Independence Day; Labor Day; Veterans Day; Thanksgiving & day after; Christmas. &

Attendance: 10,000 (estimated)

Membership: Individual $10; Family $25; Contributor $50; Sponsor $100; Patron $500 & up.

Roanoke Rapids

ROANOKE CANAL MUSEUM AND TRAIL, 15 Jackson St. Ext., Roanoke Rapids, NC 27870-1901. Tel.: 252-537-2769.

E-mail: canalmuseum@roanokerapidsnc.com

Web Site: www.roanokecanal.com/museum

Key Personnel: Mgr., Harold Jacobson; Cultural Resources Leader, Rodney D. Pierce

Institution Type/Description: History Museum: listed on the National Register of Historic Places.

Collections: Roanoke River Valley history; engineering of the canal; hydroelectric power; photographs.

Facilities: nature trails. Museum-related items for sale.

Activities: educational programs; guided tours; hiking.

Hours & Admission Prices: Museum: Tues.-Sat. 10-4. Trail: daily dawn to dusk. Admission: $4; children 8 & under no charge.

ROANOKE VALLEY VETERANS MUSEUM, 1620 E. 10th St., Roanoke Rapids, NC 27870. Tel.: 252-537-2514; 800-522-4282.

Web Site: roanokevalleyveteransmuseum.com

Governing Authority: Tax-exempt: 501(c)(3).

Institution Type/Description: Military History Museum.

Collections: military veterans history, memorabilia & medals; photographs; personal artifacts.

Hours & Admission Prices: Tues.-Sat. 10-4. No charge; donations accepted.

Robbinsville

JUNALUSKA MEMORIAL AND MUSEUM, 1 Junaluska Dr., Hwy. 143, Robbinsville, NC 28771. Mailing Address: P.O. Box 1209, Robbinsville, NC 28771. Tel.: 828-479-4727.

E-mail: friendsofjuno@dnet.net

Web Site: www.junaluska.com

Key Personnel: Mgr., Thomas Holland

Institution Type/Description: History Museum.

Collections: Cheoah Valley artifacts, art & crafts; arrowheads & spearheads; gravesite of Cherokee Warrior Junaluska.

Facilities: Museum-related items for sale.

Hours & Admission Prices: Mon.-Fri. 8-4. No charge.

Rockingham

DISCOVERY PLACE KIDS - ROCKINGHAM, 233 E. Washington St., Rockingham, NC 28379. Tel.: 910-997-5266.

Governing Authority: private; nonprofit organization. Tax-exempt: 501(c)(3).

Institution Type/Description: Children's Museum.

Collections: hands-on exhibition; science; technology; engineering; mathematics.

Activities: educational events; school groups; birthday parties.

Publications: newsletter.

Hours & Admission Prices: Tues.-Sat. 9-5, Sun. 12-5. Admission one & over $8; children under one & members no charge. Closed Easter; Labor Day; Thanksgiving; Christmas Eve & Day.

Rocky Mount

MARIA V. HOWARD ARTS CENTER AT THE IMPERIAL CENTRE, (M), 270 Gay St., Rocky Mount, NC 27804-5442. Mailing Address: P.O. Box 1180, Rocky Mount, NC 27802-1180. Tel.: 252-972-1163. Fax: 252-972-1563. Facebook: MVH Arts Center.

E-mail: kelvin.yarrell@rockymountnc.gov

Web Site: www.imperialcentre.org

Formerly: Rocky Mount Arts Center

Founded: 1956.

Congressional District: 4

Key Personnel: Cultural Arts Admin., David Griffin; Arts Education Specialist, Jennifer Rankin; Administrative Sec., Felicia Murphy; Box Office Mgr., Adrienne Lynch; Dir. Theatre, David Nields; Preparator, Brad Grooms; Program Asst., Andre Jenkins; Theatre Costumer, Pat Allen; Volunteer Coord., Nancy Stetz; Registrar, Marion Clark Weathers; Museum Shop Mgr., Emma Lynn Wheeler.

Personnel Profile: Full-Time Paid 7; Part-Time Paid 12; Part-Time Volunteers 24; Interns 1.

Governing Authority: municipal; nonprofit organization. Parent Institution: City of Rocky Mount, NC. Tax-exempt: 501(c)(3).

Institution Type/Description: Civic Art & Cultural Center.

Collections: 2 & 3 dimensional works predominantly by North Carolina artists 1950 present.

Major Exhibits: Sculpture Salmagundi XVII, 11/13-8/15/14; Nancy Cook: Seed Play, 2/14-5/11/14; 2014 Handcrafted Juried Exhibition, 2/14-5/18/14; Pocosin: Cabin Fever Relieved, 2/14-5/18/14; 57th National Juried Art Show, 6/14-9/22/14.

Activities: guided tours; lectures; films; gallery talks; arts festivals; community theatre & other theatrical productions; concerts; formally organized educational programs; loan, permanent & traveling exhibitions.

Publications: newsletter, Arty Facts; brochure, The Arts Center Season; exhibition catalogs; ceramic programs.

Hours & Admission Prices: Tues.-Sat. 10-5, Sun. 1-5. Gallery: no charge; donations accepted. Productions: call for prices. Closed Easter; Thanksgiving; Christmas. &

Attendance: 45,000 (accurate)

Membership: Friends of the Arts Center, Inc.: Jefferson $25; Shakespeare $50; Davinci $100; Einstein $500.

ROCKY MOUNT CHILDREN'S MUSEUM AND SCIENCE CENTER, INC., 270 Gay St., Rocky Mount, NC 27804-5442. Tel.: 252-972-1167. Fax: 252-972-1535.

E-mail: museum@imperialcentre.org

Web Site: museum.imperialcentre.org

Founded: 1952.

Congressional District: 2

Key Personnel: Dir., Candy L. Madrid; Space Science Educator, Steve Schmidt; Cur. Education, Leigh White; Cur. Exhibits, Frank Armstrong; Sec., Tabitha Richardson.

Personnel Profile: Full-Time Paid 7; Part-Time Paid 14; Part-Time Volunteers 50.

Governing Authority: municipal. Parent Institution: City of Rocky Mount. Tax-exempt: 501(c)(3).

Institution Type/Description: Science Museum.

Collections: hands-on science & technology exhibits; health exhibits; natural history; early childhood; digital planetarium; live animal habitats.

Activities: hands-on classes; on-floor science demonstrations; puppet shows for school & day care groups; permanent & temporary exhibitions; science camps; field trips; docent training & outreach programs; digital planetarium; laser light shows.

Publications: newsletter, Star Stuff.

Hours & Admission Prices: Tues.-Sat. 10-5, Sun. 1-5. Adults $5, children 2-15 $4, senior citizens 60 & over $3; museum & ASTC members, children one & under and residents on Sun. 1-5 no charge. Planetarium: $5. Combination pricing available. Closed Thanksgiving; Christmas. &

Attendance: 100,000 (accurate)

Membership: Individual $25; Family & Grandparent $50; Out-of-State $100; Patron $500; Guardian Angel $1,000.

ROCKY MOUNT FIRE MUSEUM, 404 S. Church St., Rocky Mount, NC 27804-5809. Tel.: 252-972-1376.

Institution Type/Description: Fire Museum: housed in the original Fire Station No. 2; built in 1924.

Collections: local history & culture; fire fighting equipment; photographs; period artifacts.

Hours & Admission Prices: Mon.-Fri. 8:30-5; other times by appointment.

STONEWALL MANOR, 1331 Stonewall Lane, Rocky Mount, NC 27804. Mailing Address: P.O. Box 9028, Rocky Mount, NC 27804. Tel.: 252-442-0063. Fax: 252-443-0137.

E-mail: stonewalllf@embarqmail.com

Web Site: www.stonewallmanor.org

Founded: 1970.

Congressional District: 2

Key Personnel: Pres. (V), Morris Wilder; Nash County Historical Coord., Lauren Fillicttaz; Tour Guide, Mary Dyer; Tour Guide, Barbara Hardisen Privette; Treas., Stewart Gibson.

Personnel Profile: Part-Time Volunteers 8.

Governing Authority: nonprofit organization. Parent Institution: Nash County Historical Association, Inc. Tax-exempt: 501(c)(3).

Institution Type/Description: Historic House & Site.

Collections: local historical memorabilia. Historic House: c.1830 Stonewall, late Federal Manor House.

Facilities: Museum-related items for sale.

Activities: guided tours; lectures; films; permanent exhibitions.

Hours & Admission Prices: Call for hours. Adults $5, seniors $3, children under 12 $2; discounts to groups of 6 or more; children 4 & under no charge. &

Attendance: 1,000 (estimated)

Membership: Individual $10; Family $15; Associate $25; Corporate $35; Sustaining $50; Life $100.

Rodanthe

CHICAMACOMICO LIFE-SAVING STATION HISTORIC SITE AND MUSEUM, 23645 N.C. Hwy. 12, Rodanthe, NC 27968. Mailing Address: P.O. Box 5, Rodanthe, NC 27968. Tel.: 252-987-1552. Fax: 252-987-1559.

Governing Authority: nonprofit organization.

Institution Type/Description: History Museum.

Collections: local history; life-saving service history; period artifacts; photographs.

Facilities: Museum-related items for sale.

Activities: special events; reenactments; educational programs; guided tours.

Hours & Admission Prices: mid-April to Nov. Mon.-Fri. 10-5. Adults $6, seniors 65 & over and students $4; discounts to groups of 20 or more.

PEA ISLAND NATIONAL WILDLIFE REFUGE, NC 12 S., Rodanthe, NC 27968. Mailing Address: P.O. Box 1969, Manteo, NC 27954. Tel.: 252-473-1131.

E-mail: peaisland@fws.gov

Web Site: www.fws.gov/peaisland

Key Personnel: Refuge Mgr., Mike Bryant

Institution Type/Description: Wildlife Refuge.

Collections: refuge & breeding ground for migratory birds, mammals, reptiles & amphibians.

Facilities: Gift items for sale.

Activities: wildlife observation; environmental education; guided tours.

Hours & Admission Prices: Daily 9-4.

Attendance: 2,500,000

Roxboro

PERSON COUNTY MUSEUM OF HISTORY, 309 N. Main St., Roxboro, NC 27573-5326. Mailing Address: P.O. Box 1792, Roxboro, NC 27573-1792. Tel.: 336-597-2884.

E-mail: pcmuseum@roxboro.net

Web Site: www.visitroxboronc.com/heritage/museum.htm

Institution Type/Description: History Museum.

Collections: local history; dolls; china; military uniforms; Native American artifacts; period furnishings; photographs.

Hours & Admission Prices: Wed.-Fri. 10-4, Sat. 10-2; other times by appointment.

Rutherford

GREEN RIVER PLANTATION, 6333 Coxe Rd., Rutherford, NC 28139. Tel.: 828-286-1461.

Institution Type/Description: Historic House Museum: housed in a 42 room mansion; built in 1807. Listed on the National Register of Historic Places.

Collections: local history; period furnishings; photographs.

Activities: guided tours.

Hours & Admission Prices: Daily call for hours; groups by appointment. Adults 12 & over $20, seniors 65 & over $18, children under 12 $10; discounts to groups; children under 4 no charge.

Rutherfordton

BROAD RIVER GEMS AND MINING COMPANY, 218 River Landing, Rutherfordton, NC 28139-7390. Tel.: 828-286-1220.

Web Site: broadrivergems.com

Institution Type/Description: Mining Museum.

Collections: minerals; fossils.

Facilities: Gift items for sale.

Activities: gem mining; fossil dig.

Hours & Admission Prices: Daily 10-6.

KIDSENSES CHILDREN'S MUSEUM, 172 N. Main St., Rutherfordton, NC 28139-2502. Mailing Address: P.O. Box 150, Rutherfordton, NC 28139-0150. Tel.: 828-286-2120.

E-mail: info@kidsenses.com

Web Site: www.kidsenses.com

Institution Type/Description: Children's Museum.

Collections: hands-on exhibits.

Activities: workshops.

Hours & Admission Prices: Tues.-Thurs. & Sat. 9-5, Fri. 9-8. Adults $5, seniors 55 & over and Mon.-Fri. after 3 pm $3.

Salisbury

DAN NICHOLAS PARK NATURE CENTER, 6800 Bringle Ferry Rd., Salisbury, NC 28146-7144. Tel.: 704-216-7803. Fax: 704-639-0947.

E-mail: bringled@co.rowan.nc.us

Web Site: www.dannicholas.net

Founded: 1975.

Congressional District: 7

Key Personnel: Parks & Recreation Dir., Dan Bringle; County Mgr., Gary Page; Nature Center Dir., Bob Pendergrass; Asst. Naturalist, David Jones.

Personnel Profile: Full-Time Paid 3; Part-Time Paid 4.

Governing Authority: county. Parent Institution: County of Rowan. Tax-exempt.

Institution Type/Description: Zoo & Nature Center Museum.

Collections: over 50 species of native wildlife; marine room with salt water aquarium tank. Historic Building: 1858 log cabin.

Facilities: 100-seat auditorium & classroom; 500-seat amphitheater; petting barn; outdoor animal area; nature trails; 13 acre lake.

Activities: lectures; films; arts festivals; study clubs; formally organized education programs; permanent & temporary exhibitions.

Hours & Admission Prices: April-Oct. Mon.-Thurs. 8-5, Fri.-Sun. 8-8; Nov.-March daily 9-5. Petting Barn $.50. &

Attendance: 100,000

DR. JOSEPHUS HALL HOUSE, 226 S. Jackson St., Salisbury, NC 28144-4838. Mailing Address: P.O. Box 4221, Salisbury, NC 28145-4221. Tel.: 704-636-0103. Facebook: Historic Salisbury Foundation.

Web Site: www.historicsalisbury.org

Founded: 1972.

Governing Authority: Parent Institution: Historic Salisbury Foundation. Tax-exempt.

Institution Type/Description: Historic House Museum: housed in a former school; built in 1820. Listed on the National Register of Historic Places.

Collections: local history & culture; period furnishings; personal artifacts.

Activities: group & school tours.

Publications: quarterly newsletter, The Salisbury Watchman.

Hours & Admission Prices: Sat.-Sun. 1-4; other times by appointment. Adults $5, children 6-12 $1; children under 6 no charge.

HORIZONS UNLIMITED SUPPLEMENTARY EDUCATIONAL CENTER, 1636 Parkview Circle, Salisbury, NC 28144-2461. Tel.: 704-639-3004. Fax: 704-639-3015.

Web Site: www.rss.k12.nc.us/horizons/HU1/home.html

Founded: 1967.

Congressional District: 8

Key Personnel: Dir. & Space Science Specialist, Lisa Wear; Science Specialist & Dir. Planetarium, Patsy Wilson; History Specialist, Theresa Pierce.

Governing Authority: public school district. Tax-exempt.

Institution Type/Description: General Museum.

Collections: natural science; astronomy; wildlife; Apollo lunar lander (1/3 scale model); Indian artifacts; 1850 Chilean Ore Mill. Historic Building: 1842 one room log school.

Facilities: 300-vol. library of material on visual art, space science, natural science, history available on premises or to be checked out by local residents, students or teachers; nature center; planetarium; classrooms; 24-acre wetlands natural area; aquariums; touch tanks; rain forest aviary; health museum.

Activities: guided tours; lectures; films; gallery talks; formally organized education programs for children, adults & undergraduate students; temporary & traveling exhibitions; school loan service.

Publications: monthly bulletins.

Hours & Admission Prices: By appointment. &

Attendance: 21,850

ROWAN MUSEUM, INC., 202 N. Main St., Salisbury, NC 28144-4356. Tel.: 704-633-5946. Fax: 704-633-9858.

E-mail: rowanmuseum@carolina.rr.com

Web Site: rowanmuseum.org

Founded: 1953.

Congressional District: 12

Key Personnel: Exec. Dir., Kaye Brown Hirst; Pres., Paul Brown.

Personnel Profile: Full-Time Paid 2; Full-Time Volunteers 50; Part-Time Paid 12; Part-Time Volunteers 6; Interns 1.
Governing Authority: corporate trustees. Tax-exempt: 501(c)(3).
Institution Type/Description: History Museum.
Collections: artifacts of region from 1700s-1900s. Historic Houses: 1766 Old Stone House, residence of Michael Braun; 1819 Utzman-Chambers House.
Activities: guided tours. Museum Sponsors: History Summer Day Camp; Heritage Celebration at Old Stone House; Colonial Christmas; Antiques Show.
Publications: biannual newsletter.
Hours & Admission Prices: Utzman-Chambers House: April-Nov. Sat.-Sun. 1-4. Rowan Museum: Mon.-Fri. 10-4, Sat.-Sun.1-4; other times by appointment. Old Stone House: April-Dec. Thurs.-Sun. 1-4. Adults $3, students $1.50, children $1; discounts to AAM members. &
Attendance: 5,000 (estimated)
Membership: Individual $35-$59; Family $60-$100; Sustaining $100-$249; Donor $250-$499; Business $250 & up; Patron $500-$999; Maxwell Chambers Society $1,000 & up.

* **WATERWORKS VISUAL ARTS CENTER, (M),** 123 E. Liberty St., Salisbury, NC 28144-5038. Tel.: 704-636-1882. Fax: 704-636-1895.
E-mail: info@waterworks.org
Web Site: waterworks.org
Founded: 1959.
Congressional District: 12
Key Personnel: Pres. (V), Tim Proper; Exec. Dir., Anne Scott Clement.
Personnel Profile: Full-Time Paid 3; Part-Time Paid 20; Part-Time Volunteers 110; Interns 2.
Governing Authority: nonprofit organization. Waterworks Visual Arts Center, Inc. Tax-exempt: 501(c)(3).
Institution Type/Description: Art Museum.
Collections: paintings; photographs; sculpture.
Facilities: art library; conference room; sensory gardens.
Activities: guided tours; lectures; gallery talks; formally organized education programs for children, adults & special populations; docent program.
Publications: Exhibition catalogue.
Hours & Admission Prices: Mon.-Wed. & Fri. 10-5, Thurs. 10-7, Sat. 10-4. No charge; donations accepted. Closed major holidays. &
Attendance: 13,106 (accurate)
Membership: Individual & Family $50 & up; Friend $200; Affiliate $500 & up; Patron $1,000 & up; Fellow $2,500 & up; Benefactor $5,000 & up.

Saluda

PEARSON'S FALLS, 2748 Pearson Falls Rd., Saluda, NC 28773. Tel.: 828-749-3031.
E-mail: info@pearsonsfalls.org
Web Site: www.pearsonsfalls.org
Institution Type/Description: Nature & Wildlife Preserve.
Collections: natural heritage; wildlife preserve; 90 ft. falls.
Facilities: 320 acre nature & wildlife preserve; nature trails.
Activities: nature trails.
Hours & Admission Prices: Feb. & Nov.-Dec. Mon.-Sat. 10-5, Sun. 12-5; March-Oct. Mon.-Sat. 10-6, Sun. 12-6. Adults 13 & up $5, youth 6-12 $1; children under 6 no charge. Closed Thanksgiving; Christmas.

Sanford

HOUSE IN THE HORSESHOE STATE HISTORIC SITE, 288 Alston House Rd., Sanford, NC 27330-8712. Tel.: 910-947-2051. Fax: 910-947-2051.
E-mail: horseshoe@ncdcr.gov
Web Site: www.houseinthehorseshoe.nchistoricsites.org
Founded: 1955.
Congressional District: 8
Key Personnel: Site Mgr., Alex Cameron.
Personnel Profile: Full-Time Paid 3; Part-Time Paid 1; Part-Time Volunteers 12.
Governing Authority: state. North Carolina Dept. of Cultural Resources, Historic Sites Section, 532 N. Wilmington St., 4620 Mail Service Center, Raleigh, NC 27699-4620. Tax-exempt.
Institution Type/Description: Historic House: housed in the former home of North Carolina Governor Benjamin Williams; built in 1772; site of Revolutionary War militia skirmish.
Collections: colonial artifacts & furnishings.
Research Fields: North Carolina agricultural history; Revolutionary War & local history; War of 1812.
Facilities: picnic area. Gift items for sale.
Activities: guided tours; lectures; formally organized education programs;

musket demonstrations for organized groups; special events. Museum Sponsors: Battle Reenactment in August; Christmas Open House in December.
Publications: brochure; electronic newsletter.
Hours & Admission Prices: Tues.-Sat. 9-5. No charge; donations accepted. Closed state holidays. &
Attendance: 17,064 (accurate)

MUSEUM OF RAILROAD HOUSE HISTORICAL ASSOCIATION MUSEUM INC., 110 Charlotte Ave., Sanford, NC 27330-4304. Mailing Address: P.O. Box 1023, Sanford, NC 27331-1023. Tel.: 919-776-7479.
Web Site: www.railroadhouse.org
Founded: 1962.
Congressional District: 3
Key Personnel: Pres. (V), Rebekah S. Harvey.
Personnel Profile: Part-Time Volunteers 40.
Governing Authority: bd. of directors; nonprofit. Tax-exempt: 501(c)(3).
Institution Type/Description: Historic House: 1872 Railroad House, Gothic Revival, oldest house in Sanford, home of the first mayor & site of first school.
Collections: local historical artifacts & documents; railroad history; artifacts from Endor Iron Furnace; Cole pottery, collection of one of the last hand kiln potters; soil exhibit; fossils; uniforms from WW I & II.
Research Fields: restoration of the c.1859 stack of the Endor Iron Furnace, a Civil War site for smelting pig iron; archaeological studies at Endor Site.
Activities: reading room; permanent & temporary exhibitions.
Publications: newsletter; The History & Architecture of Lee Co.; Men of Endor; Sanford and Lee County, North Carolina; Images of America Series; In Celebration of the 2007 Centennial of Lee County, North Carolina.
Hours & Admission Prices: Sat.-Sun. 1-5; other times by appointment. No charge; donations accepted.
Attendance: 2,000 (estimated)
Membership: Family $25; Corporate By Donation.

Scotland Neck

SYLVAN HEIGHTS BIRD PARK, 500 Sylvan Heights Pkwy., Scotland Neck, NC 27874. Mailing Address: P.O. Box 368, Scotland Neck, NC 27874. Tel.: 252-826-3186. Fax: 252-826-3273.
E-mail: info@shwpark.com
Web Site: shwpark.com
Key Personnel: Exec. Dir., Mike Lubbock; Cur., Brad Hazelton; Opers. Mgr., Ali Lubbock; Retail Mgr., JoAnn Josey; Education Coord., Lee Peoples; Membership & Devel., Brent Lubbock.
Governing Authority: nonprofit organization. Tax-exempt: 501(c)(3).
Institution Type/Description: Bird Park.
Collections: over 2,000 ducks, geese, swans, & other exotic birds from around the world; waterfowl, wildlife & wetland conservation.
Facilities: 18 acre park; gardens; nature trails.
Activities: walking trails; educational programs.
Hours & Admission Prices: April-Oct. Tues.-Sun. 9-5; Nov.-March Tues.-Sun. 9-4. Adults $9, senior citizens 62 & over $7, children 3-12 $5; children under 3 no charge. Closed Thanksgiving; Christmas.

Seagrove

MUSEUM OF NC TRADITIONAL POTTERY, 127 E. Main St., Seagrove, NC 27341. Mailing Address: P.O. Box 500, Seagrove, NC 27341. Tel.: 336-873-7887.
E-mail: ncpottery122@embarqmail.com
Web Site: www.seagrovepotteryheritage.com
Founded: 1983.
Institution Type/Description: Pottery History Museum.
Collections: pottery history; pottery; photographs.
Activities: Annual Event: Seagrove Pottery Festival.
Hours & Admission Prices: No charge; donations accepted.

NORTH CAROLINA POTTERY CENTER, 233 East Ave., Seagrove, NC 27341-0531. Mailing Address: P.O. Box 531, Seagrove, NC 27341-0531. Tel.: 336-873-8430. Fax: 336-873-8530.
E-mail: manager@ncpotterycenter.org
Web Site: www.ncpotterycenter.org
Founded: 1998.
Congressional District: 6
Key Personnel: Pres., Linda Carnes-McNaughton; Mgr., Paulett Badgett.
Personnel Profile: Full-Time Paid 1; Part-Time Paid 4; Part-Time Volunteers 45; Interns 0.

Governing Authority: private; nonprofit organization. Tax-exempt: 501(c)(3).
Institution Type/Description: Pottery Center.
Collections: North Carolina pottery from prehistoric to contemporary; regional, national, & international clay traditions; southern pottery & related tools, equipment & documentation; photographs.
Research Fields: North Carolina & Southern pottery; pottery techniques & technology.
Facilities: 100-vol. library; classrooms; 2,300 sq. ft. exhibit space. Museum-related items for sale.
Activities: lectures; workshops; demonstrations; outreach programs. Annual Events: Catawba Valley Pottery Festival; Seagrove Pottery Festival.
Publications: newsletter, North Carolina Pottery Center Newsletter.
Hours & Admission Prices: Tues.-Sat. 10-4. Adults $2, students 9th-12th grade $1; students K-8th grade & members no charge. &
Attendance: 12,000 (accurate)
Membership: Individual $35; Family $50; Contributor $100; Donor $500; Benefactor $1,000.

SEAGROVE ORCHIDS, 3451 Brower Mill Rd., Seagrove, NC 27341. Tel.: 336-879-6677.
Institution Type/Description: Working Nursery.
Collections: over 200 species of orchids.
Activities: guided tours.
Hours & Admission Prices: Tues.-Sat. 10-5; other times by appointment.

Sedalia

CHARLOTTE HAWKINS BROWN MUSEUM, 6136 Burlington Rd., Sedalia, NC 27249. Mailing Address: P.O. Box B, Sedalia, NC 27342-0190. Tel.: 336-449-4846. Fax: 336-449-0176.
Web Site: www.nchistoricsites.org/chb/chb.htm
Personnel Profile: Full-Time Paid 5; Part-Time Paid 1; Part-Time Volunteers 5; Interns 5.
Governing Authority: Parent Institution: North Carolina Department of Cultural Resources. Subsidiary Institution: Division of State Historic Sites.
Institution Type/Description: History Museum.
Collections: African American history & culture; women's history; period furnishings; personal artifacts. Historic Buildings: house; school.
Activities: Museum Sponsors: Heritage Day in July.
Hours & Admission Prices: Mon.-Sat. 9-5. No charge; donations accepted. Closed most major state holidays. &
Attendance: 13,000 (accurate)

Seven Springs

CLIFFS OF THE NEUSE STATE PARK, 240 Park Entrance Rd., Seven Springs, NC 28578-8968. Tel.: 919-778-6234. Fax: 919-778-7447.
E-mail: cliffs.neuse@ncmail.net
Web Site: www.ncparks.gov/Visit/parks/clne/main.php
Founded: 1979.
Key Personnel: Park Supt., Lyden Sutton; Dist. Naturalist, Jeanne Peacock.
Governing Authority: state. Affiliated with the North Carolina Div. of Parks & Recreation, 1615 Mail Service Center, Raleigh, NC 27699-1615. Tax-exempt.
Institution Type/Description: Park Museum.
Collections: American Indian pottery & arrowheads; herbarium; insects; mammals.
Research Fields: geology; botany; paleontology; archaeology; zoology.
Facilities: guide books.
Activities: guided tours; lectures; films; permanent exhibitions; audiovisuals.
Publications: environmental educational learning experience: The Cliffs of Time, designed for grades 6-8.
Hours & Admission Prices: March-May & Sept.-Nov. daily 8-8; June-Aug. daily 8-9; Dec.-Feb. daily 8-6. No charge. Closed Christmas.
Attendance: 70,000 (estimated)

Shelby

CLEVELAND COUNTY ARTS CENTER, 111 S. Washington St., Shelby, NC 28150. Tel.: 704-484-2787. Fax: 704-481-1822.
E-mail: info@ccartscouncil.org
Web Site: www.ccartscouncil.org
Key Personnel: Pres., Shearra Miller; Mktg. Coord., Violet Arth.
Governing Authority: nonprofit organization.
Institution Type/Description: Arts Center: housed in the former post office building.
Collections: works by regional artists.
Facilities: Gift items for sale.

Activities: classes; special events; educational programs; temporary exhibitions; performing arts; classes; summer art camp. Annual Event: Student Art Competition.
Hours & Admission Prices: Mon.-Fri. 9-5:30.

CLEVELAND COUNTY HISTORICAL MUSEUM, Court Square, Shelby, NC 28150. Mailing Address: P.O. Box 1335, Shelby, NC 28151-1210. Tel.: 704-482-8186. Fax: 704-482-8186.
Founded: 1976.
Congressional District: 10
Key Personnel: Dir. & Cur., Lamar Wilson.
Governing Authority: nonprofit organization. Parent Institution: Cleveland County Historical Association, Inc. Tax-exempt: 501(c)(3).
Institution Type/Description: Historical Society Museum: housed in 1907 classical revival courthouse.
Collections: photographs; documents; clothing; textiles; tools; furniture; housewares; medical; communication; military; agricultural; religious; manuscripts; live living history displays at Peeler Mill cabin & Broadriver cabin.
Research Fields: local history.
Facilities: historical books available for research on premises; reading room. Historical, general & hand-crafted items for sale.
Activities: guided tours; lectures; films; gallery talks; arts festivals; study clubs; hobby workshops; docent program; formally organized education programs for undergraduate & graduate college students of Gaston College; training programs; permanent & temporary exhibitions; living history cabin.
Publications: quarterly newsletters.
Hours & Admission Prices: Tues.-Fri. 9-4. Closed all national & county holidays. &
Attendance: 25,000
Membership: Individual $15; Family $25; Patron $50; Professional $100; Corporate $200; Individual Life $500.

INTERNATIONAL LINEMANS MUSEUM, 529 Caleb Rd., Shelby, NC 28152. Mailing Address: P.O. Box 1740, Shelby, NC 28152. Tel.: 704-482-7638. Facebook: Lineman Museum.
E-mail: ilm@linemanmuseum.com
Web Site: www.linemanmuseum.com
Founded: 1997.
Key Personnel: C.E.O., Andy Price.
Governing Authority: Tax-exempt: 501(c)(3).
Institution Type/Description: History Museum.
Collections: lineman history; electrical utility industry; photographs; personal artifacts; period furnishings.
Hours & Admission Prices: Mon.-Fri. 8-5; other times by appointment. No charge.
Attendance: 200 (estimated)

Siler City

THE FARM AT CELEBRITY DAIRY, 2106 Mt. Vernon - Hickory Mountain Rd., Siler City, NC 27344. Tel.: 877-742-5176.
Web Site: www.celebritydairy.com
Institution Type/Description: Dairy Farm.
Collections: 100 Alpine & Saanen goats; farm history & equipment.
Activities: daily milking; farmstead goat cheese production; special events; school group tours.
Hours & Admission Prices: By appointment.

Smithfield

ARBORETUM AT JOHNSTON COMMUNITY COLLEGE, 1240 E. Market St., Smithfield, NC 27577. Mailing Address: P.O. Box 2350, Smithfield, NC 27577. Tel.: 919-209-2052.
Web Site: www.johnstoncc.edu/arboretum
Institution Type/Description: Arboretum.
Collections: trees; flowers; ornamental shrubs; dogwoods; azaleas; lilies; roses; irises.
Facilities: nature trails.
Activities: educational programs; walking trails; special events.
Hours & Admission Prices: Call for hours.

AVA GARDNER MUSEUM, 325 E. Market St., Smithfield, NC 27577-3919. Tel.: 919-934-5830. Fax: 919-934-6998.
E-mail: avainfo@avagardner.org
Web Site: www.avagardner.org
Founded: 1996.
Key Personnel: Dir., K. Todd Johnson; Chm. (V), Rick Lotz.
Personnel Profile: Full-Time Paid 1; Part-Time Paid 4; Part-Time Volunteers 2.

Governing Authority: Tax-exempt.
Institution Type/Description: Film Museum.
Collections: artifacts reflecting the movie career & private life of Ava Gardner including photographs, newspaper articles, personal artifacts, & portraits.
Activities: Annual Event: Ava Gardner Festival in October.
Hours & Admission Prices: Mon.-Sat. 9-5, Sun. 2-5. Adults $7, seniors 65 & over, military and children 13-16 $6, children 6-12 $5; Ava Advocates & children under 6 no charge. &
Attendance: 12,000 (accurate)
Membership: Student $20; Individual $25; Family $50; Military & Senior Citizen Family $40.

Snow Camp

SNOW CAMP HISTORICAL SOCIETY, 301 Drama Rd., Snow Camp, NC 27349. Mailing Address: P.O. Box 535, Snow Camp, NC 27349-0535. Tel.: 336-376-6948.
E-mail: snowcampot@aol.com
Web Site: www.snowcampdrama.com
Founded: 1971.
Congressional District: 26
Key Personnel: Chm. (V), James Shields.
Personnel Profile: Full-Time Volunteers 2; Part-Time Volunteers 6.
Institution Type/Description: History Museum.
Collections: local history; personal artifacts; photographs.
Hours & Admission Prices: May-Aug. Mon.-Fri. 9-5. No charge. &
Attendance: 10,000 (estimated)

South Mills

DISMAL SWAMP STATE PARK & VISITORS CENTER, 2294 US 17 N., South Mills, NC 27976. Tel.: 252-771-6593 & 6582.
E-mail: dismal.swamp@ncparks.gov
Web Site: www.ncparks.gov/visit/parks/disw/main.php
Institution Type/Description: State Park & Visitors Center.
Collections: local history; Underground Railroad history.
Facilities: nature trails.
Activities: hiking.
Hours & Admission Prices: Park: March-Oct. daily 8-6; Nov.-Feb. daily 8-5. Visitors Center: Mon.-Fri. 8-4:30, Sat.-Sun. 9:30-4:30. Closed Christmas.

South Nags Head

BODIE ISLAND LIGHT STATION, Bodie Island Lighthouse, South Nags Head, NC 27959. Mailing Address: 1401 National Park Dr., Manteo, NC 27954-9451. Tel.: 252-473-2111 & 441-5711. Fax: 252-449-0788.
E-mail: caha_information@nps.gov
Web Site: www.nps.gov/caha
Formerly: Bodie Island Visitor Center
Founded: 1956.
Congressional District: 3
Key Personnel: Supt., Barclay Trimble; Archives Technician, Jami P. Lanier.
Personnel Profile: Full-Time Paid 1; Part-Time Paid 3; Part-Time Volunteers 8.
Governing Authority: federal. Affiliated with Cape Hatteras National Seashore. Parent Institution: National Park Service. Tax-exempt.
Institution Type/Description: Park Museum & Visitor Center.
Collections: Historic Building: 1872 Bodie Island Light Station includes lighthouse & keeper dwelling.
Facilities: Books & theme-related items for sale.
Publications: park brochure; various related site bulletins.
Hours & Admission Prices: Daily 9-5. No charge; donations accepted. &

Southern Pines

ARTS COUNCIL OF MOORE COUNTY - CAMPBELL HOUSE, 482 E. Connecticut Ave., Southern Pines, NC 28387-5624. Mailing Address: P.O. Box 405, Southern Pines, NC 28388-0405. Tel.: 910-692-2787. Fax: 910-693-1217.
E-mail: acmc@mooreart.net
Web Site: www.mooreart.net
Founded: 1973.
Institution Type/Description: Art Museum.
Collections: works by local, regional & national artists.
Hours & Admission Prices: Mon.-Fri. 9-5, 3rd Sat.-Sun. of month 2-4. No charge.

SHAW HOUSE, 110 Morganton Rd., Southern Pines, NC 28387. Mailing Address: P.O. Box 324, Southern Pines, NC 28388-0324. Tel.: 910-692-2051. Fax: 910-692-2051.
E mail: moorehistory@connectnc.net
Web Site: moorehistory.com
Founded: 1946.
Governing Authority: Parent Institution: Moore County Historical Association. Tax-exempt.
Institution Type/Description: Historic House Museum: built c.1820.
Collections: local history; period furnishings. Historic Buildings: Sanders Cabin, c.1770; Garner House, c.1790; corn crib; tobacco barn.
Hours & Admission Prices: Tours: Tues.-Fri. 1-4; other times by appointment. No charge.
Attendance: 1,500 (estimated)

TAXIDERMY HALL OF FAME OF NORTH CAROLINA/ANTIQUE TOOL MUSEUM/CREATION MUSEUM, 156 N.W. Broad St., Southern Pines, NC 28387. Tel.: 910-692-3471.
Web Site: thecreationmuseum.org
Institution Type/Description: History Museum.
Collections: North Carolina wildlife; early tools & equipment; religious artifacts.
Hours & Admission Prices: Call for hours.

WEYMOUTH WOODS-SANDHILLS NATURE PRESERVE MUSEUM, 1024 Fort Bragg Rd., Southern Pines, NC 28387-7319. Tel.: 910-692-2167. Fax: 910-692-8042.
E-mail: weymouth.woods@ncparks.gov
Web Site: www.ncparks.gov/Visit/parks/wewo/main.php
Founded: 1969.
Congressional District: 8
Key Personnel: Ranger, Kim Hyre.
Governing Authority: state. Parent Institution: N.C. Dept. of Environment, Natural Resources, Division of Parks & Recreation, P.O. Box 27687, Raleigh, NC 27611. Tax-exempt.
Institution Type/Description: Natural History & Science Museum.
Collections: study specimens; Indian artifacts; turpentine industry artifacts; herbarium; wildlife refuge & bird sanctuary.
Research Fields: archaeological; ethnological; botanical; zoological; interpretive planning.
Facilities: 800-vol. library of natural history available on premises; nature center; field search station; auditorium.
Activities: guided tours; lectures; films; formally organized education programs for children; permanent & temporary exhibits.
Publications: brochure; bird & wildflower checklist.
Hours & Admission Prices: Daily 9-5. No charge; donations accepted. Closed Christmas.
Attendance: 22,295 (estimated)

Southport

DUKE ENERGY BRUNSWICK NUCLEAR PLANT VISITOR CENTER, 8470 River Rd. S.E., Southport, NC 28461. Tel.: 910-457-2418.
Institution Type/Description: Energy Museum.
Collections: over 30 energy exhibitions including electricity production, energy conservation, radiation & nuclear power.
Activities: video.
Hours & Admission Prices: Call for hours. No charge. Closed major holidays.

FORT JOHNSON - SOUTHPORT MUSEUM, 219 E. Bay St., Southport, NC 28461. Tel.: 800-388-9635; 910-457-7927.
Institution Type/Description: Military History Museum: housed on the site of Fort Johnston; built in 1748.
Collections: Fort Johnson history, artifacts & documents.
Activities: docent led or self-guided tours.
Hours & Admission Prices: Mon.-Sat. 9:30-4:30, Sun. 1-4.

NORTH CAROLINA MARITIME MUSEUM AT SOUTHPORT, 204 E. Moore St., Southport, NC 28461-3928. Tel.: 910-457-0003.
Institution Type/Description: Maritime History Museum.
Collections: maritime history; local commerce & fishing; photographs; period artifacts.
Hours & Admission Prices: Temporarily closed.

Sparta

DOUGHTON PARK - BRINEGAR CABIN AND CAUDILL FAMILY HOMESTEAD, Blue Ridge Pkwy., MP 240, Sparta, NC 28675. Tel.: 336-372-8877.
Institution Type/Description: Historic House Museum & Park.
Collections: history of isolated mountain life; period furnishings & artifacts. Historic Buildings: c.1885 Brinegar Cabin; 1895 Caudill Cabin.
Facilities: 7,000 acre park; nature trails; camping.
Activities: craft demonstrations; hiking; nature trails; camping.
Hours & Admission Prices: May-Oct. call for hours. No charge.

Spencer

NORTH CAROLINA TRANSPORTATION MUSEUM, 411 S. Salisbury Ave., Spencer, NC 28159-2238. Tel.: 704-636-2889. Fax: 704-639-1881.
E-mail: nctrans@nctrans.org
Web Site: www.nctrans.org
Founded: 1977.
Congressional District: 12
Key Personnel: Exec. Dir., Samuel Wegner; Museum Foundation Pres. (V), Roy Johnson; Facility Mgr., Brian Howell; Public Information Officer, Mark Brown; C.F.O., Marlene Minshew; Education Programming Coord., Brian Moffitt; Volunteer Coord., LeAnne Johnson; Volunteer Coord., Vickie Peacock; Exhibits Coord., Bob Hopkins; Historian, Walter Turner.
Personnel Profile: Full-Time Paid 11; Part-Time Paid 5; Part-Time Volunteers 100; Interns 1.
Governing Authority: state. Parent Institution: North Carolina Department of Cultural Resources, Division of Historic Sites. Tax-exempt.
Institution Type/Description: Historic Transportation Museum: housed in 1896, Southern Railway steam primary staging & repair facility complex containing 20 structures, 37 bay roundhouse, turntable & 90,000 feet back shop.
Collections: 8 steam locomotives; 10 diesel locomotives (6 operating); 50 assorted rolling stock; transportation artifacts & memorabilia; Conestoga wagon; 1896 mail buggy. Historic Cars: 1902 Loretto Rail Car; 1907 Ford Model R Roadster; 1917 Doris private rail car; 1919 Dodge Brothers Roadster; 1959 Edsel Corsair.
Research Fields: transportation.
Facilities: 500-vol. library pertaining to transportation; 30-seat auditorium. Books, Transportation items & other museum-related items for sale.
Activities: guided tours; lectures; films; docent program; formally organized education for adults & undergraduate students; permanent exhibitions. Museum Sponsors: car shows; special events. Annual Events: Rail Days; A Day Out with Thomas the Tank Engine; Santa Train.
Publications: brochures; history publications; semiannual magazine, Shop Talk.
Hours & Admission Prices: Jan.-March Tues.-Sat. 9-5; April-Aug. Mon.-Sat. 9-5, Sun. 1-5; Sept.-Dec. Tues.-Sat. 9-5, Sun. 1-5. Museum: adults $5, seniors & active military $4, children 3-12 $3. Museum & Train Ride: adults $10, children 3-12 $6, seniors & active military $4; discounts to AAA members & groups; children 2 & under no charge. Closed New Year's Day; Easter; Veterans Day; Thanksgiving; Christmas. ⅌
Attendance: 80,000 (estimated)
Membership: Senior & Youth 3-17 $35; Individual $45; Two People $60; Two Adults & up to 4 children $80.

Spruce Pine

EMERALD VILLAGE - NORTH CAROLINA MINING MUSEUM, 331 McKinney Mine Rd., Spruce Pine, NC 28777. Mailing Address: P.O. Box 98, Little Switzerland, NC 28749. Tel.: 828-765-6463.
Institution Type/Description: Mining Museum.
Collections: mining; mining history & equipment.
Facilities: Museum-related items for sale.
Activities: underground mine tours; pan for gold; emerald dig.
Hours & Admission Prices: April-May & Sept.-Oct. daily 9-5; Memorial Day to Labor Day daily 9-6. Mine Tour: adults $7, senior citizens 60 & over $6, students $5; discounts to groups; preschool children no charge. Gold Panning: $10 & up.

MUSEUM OF NORTH CAROLINA MINERALS, Milepost 331, Blue Ridge Pkwy. at Hwy. 226, Spruce Pine, NC 28777. Mailing Address: Blue Ridge Pkwy. Foundation, 717 S. Marshall St., Ste. 105 B, Winston-Salem, NC 27101-5865. Tel.: 828-765-2761. Fax: 828-765-0202.
Founded: 1955.

Congressional District: 11
Key Personnel: District Ranger, Tim Francis.
Personnel Profile: Full-Time Paid 2; Part-Time Paid 4; Part-Time Volunteers 5; Interns 2.
Governing Authority: federal. Parent Institution: National Park Service. Subsidiary Institution: Blue Ridge Parkway. Tax-exempt.
Institution Type/Description: Mineral & Mineral Industry Museum.
Collections: geology; study collection; industry.
Research Fields: mineralogy; mineral industries.
Activities: permanent exhibitions; lapidary demonstrations. Museum Sponsors: revolutionary military encampment in September.
Hours & Admission Prices: Daily 9-5. No charge. Closed New Year's Day; Thanksgiving; Christmas. ⅌
Attendance: 254,000 (estimated)

Stanley

BREVARD STATION MUSEUM, 112 S. Main St., Stanley, NC 28164-1750. Tel.: 704-263-9801.
E-mail: brevardstation@gmail.com
Web Site: www.brevardstation.com
Founded: 1991.
Congressional District: 9
Key Personnel: Admin., Joyce J. Handsel; Admin., Ruth Wood; Bd. Pres. (V), W. Barry Smith.
Personnel Profile: Full-Time Volunteers 2; Part-Time Volunteers 2.
Governing Authority: Tax-exempt.
Institution Type/Description: History Museum.
Collections: local history & culture; railroad; textile industry; town government; farm life; period clothing & furnishings; personal artifacts; military uniforms & equipment; sculpture; photographs; research literature; sports artifacts.
Activities: research.
Hours & Admission Prices: Tues.-Thurs. 10-4; other times by appointment. No charge; donations accepted.
Membership: Associate $20-$49; Annual $25; Friend $50-$74; Donor $75-$124; Patron $125-$499; Benefactor $500-$999; Magnolia Circle $1,000 & up.

Statesville

ALLISON WOODS OUTDOOR LEARNING CENTER, 2106 Turnersburg Hwy., Statesville, NC 28625. Mailing Address: P.O. Box 211, Statesville, NC 28687-0211. Tel.: 704-873-5976.
E-mail: selena@allisonwoodslivinghistory.org
Web Site: www.allisonwoodslivinghistory.org
Institution Type/Description: Historic Buildings & Gardens: listed on the National Register of Historic Places.
Collections: local history; natural & cultural heritage; historic structures & gardens.
Facilities: 3,000 acres of historic gardens.
Activities: demonstrations; living history events; Civil War reenactments; special events; educational programs.
Hours & Admission Prices: Call for hours.

FORT DOBBS STATE HISTORIC SITE, 438 Fort Dobbs Rd., Statesville, NC 28625-1915. Tel.: 704-873-5882. Fax: 704-873-5995.
E-mail: richard.douglas.brown@ncdcr.gov
Web Site: www.fortdobbs.org
Founded: 1969.
Congressional District: 9
Key Personnel: Historic Site Mgr., Doug Brown; Chm. (V), Ralph Bentley; Historic Interpreter, Scott Douglas.
Personnel Profile: Full-Time Paid 3; Full-Time Volunteers 20; Part-Time Paid 5; Part-Time Volunteers 100; Interns 2.
Governing Authority: state. North Carolina Dept. of Cultural Resources, 109 E. Jones St., Raleigh, NC 27611. Tax-exempt: 501(c)(3); 170(b)(1)(A).
Institution Type/Description: Historic Site.
Collections: mid-18th century frontier fort; French & Indian War.
Research Fields: French & Indian War archaeology; NC frontier; back country settlement.
Facilities: picnic shelter; visitor center; nature trail.
Activities: special events; living history programming; daily interpretive programs.
Publications: brochure; Fort Dobbs Gazette.
Hours & Admission Prices: Tues.-Sat. 9-5. No charge; donations accepted. ⅌
Attendance: 32,000 (accurate)
Membership: Cadet $10; Sentinel $20; Corporal $50; Sergeant $60; Ensign $150; Lieutenant $200; Captain $500; Major $1,000; Colonel $5,000.

IREDELL MUSEUMS, (M), 134 Court St., Statesville, NC 28677. Mailing Address: P.O. Box 223, Statesville, NC 28687-0223. Tel.: 704-873-4734. Facebook: Iredell Museums.
E-mail: info@iredellmuseums.org
Web Site: www.iredellmuseums.org/
Founded: 1956.
Congressional District: 9
Personnel Profile: Full-Time Paid 2; Full-Time Volunteers 1; Part-Time Volunteers 45.
Governing Authority: nonprofit organization. Parent Institution: Tax-exempt.
Institution Type/Description: General Museum.
Collections: local history & culture; period artifacts & furnishings. Historic Buildings: Pioneer Log Cabin & Farmstead; c.1750 smoke house; 1800 pioneer cabin; WWI, WWII, Korea, Vietnam war artifacts; 1750's-1800 furniture; textiles.
Research Fields: Statesville history; art; culture.
Facilities: nature trails.
Activities: guided tours; lectures; films; temporary exhibitions; educational programs; classes; nature trails. Annual Events: Art on the Green; Art in Bloom; Christmas at the cabins.
Publications: newsletter; pamphlets.
Hours & Admission Prices: By appointment & special events. &
Attendance: 25,000 (estimated)
Membership: Student & Senior Citizen $15; Individual $25: Family $40; Patron $100; Benefactor $500.

Sunset Beach

INGRAM PLANETARIUM, 7625 High Market St., Sunset Beach, NC 28468. Tel.: 910-575-0033. Fax: 910-575-0031.
E-mail: pinfo@museumplanetarium.org
Web Site: www.museumplanetarium.org
Key Personnel: Exec. Dir., Terry Bryant; Planetarium Dir., Mark Jankowski; Mktg., Susan Silk; Educational Svcs. Coord., Allison Smith; Membership Svcs., Deb Boyce; Gift Shop Mgr., Lynn Wiedman.
Governing Authority: nonprofit organization. Parent Institution: Ocean Isle Museum Foundation. Subsidiary Institution: Museum of Coastal Carolina, 21 E. Second St., Ocean Isle Beach, NC 28469.
Institution Type/Description: Planetarium.
Collections: astronomy; science; photographs; hands-on exhibits.
Facilities: Museum-related items for sale.
Activities: educational programs; special events; laser shows.
Hours & Admission Prices: Hours & programs are seasonal, check website. Adults $8, seniors 60 & over and students $6, children 3-4 $4, children 2 & under no charge.

Tarboro

BLOUNT-BRIDGERS HOUSE/HOBSON PITTMAN MEMO-RIAL GALLERY, 130 Bridgers St., Tarboro, NC 27886-3868. Tel.: 252-823-4159. Fax: 252-823-6190.
E-mail: edgecombearts@embarqmail.com
Web Site: edgecombearts.com
Formerly: Blount-Bridgers House/Edgecombe Country Art Museum
Founded: 1982.
Congressional District: 2
Key Personnel: Dir., Joyce Turner; Asst. Dir., Carol Banks; Chm. Bd., Barbara Campbell Davis.
Personnel Profile: Full-Time Paid 1; Part-Time Paid 1; Interns 4.
Volunteer Hours: 60
Governing Authority: nonprofit organization. Parent Institution: Edgecombe County Arts Council. Tax-exempt: 501(c)(3).
Institution Type/Description: Art Museum: housed in c.1808 Blount-Bridgers plantation house.
Collections: Hobson Pittman 20th-century art; 20th-century American art; 19th-century silver; 19th-century ceramics; 19th-century American furniture.
Research Fields: Hobson Pittman; Blount Family history; Edgecombe County history.
Facilities: 2,000-vol. library; 75-seat gallery; 3,400 sq. ft. exhibit space. Arts & crafts by local artists for sale.
Activities: guided tours; lectures; arts festivals; organized educational programs; docent program; loan, traveling & temporary exhibitions; walking tours of historic district.
Publications: newsletter; annual report; brochure; book, The Poet's Pallette, Selected Works by Hobson Pittman.
Hours & Admission Prices: Summer: Tues.-Fri. 10-4, By appointment only for weekends. Groups by appointment. Adults $5, children under 12 $2; members no charge. Closed New Year's Day; Good Friday; Easter; Memorial Day; Independence Day; Labor Day; Thanksgiving; Christmas. &

Attendance: 4,000 (estimated)
Membership: Student & Teacher $15; Individual $25; Family & Couple $45; Donor $50-$249; Sponsor $250-$499; Patron $500-$999; Benefactor $1,000 & up.

Thomasville

1870 TRAIN DEPORT, 44 W. Main St., Thomasville, NC 27360. Tel.: 800-611-9907.
Web Site: thomasvilletourism.com
Institution Type/Description: Historic Building: built in 1870. Listed on the National Register of Historic Places.
Collections: local history & culture; period furnishings; photographs.
Facilities: visitor center.
Activities: temporary exhibitions.
Hours & Admission Prices: Mon.-Fri. 9-5, Sat. 9-1. No charge. &

Thurmond

PRECIOUS ALPACA FARM, 2930 S. Center Church Rd., Thurmond, NC 28683. Tel.: 336-957-3581.
Institution Type/Description: Farm.
Collections: alpaca history from the origin to the different colors & breeds.
Facilities: Farm-related items for sale.
Activities: guided tours; video; weaving demonstrations; petting zoo.
Hours & Admission Prices: Sat. 9-5, Sun. 1-5; groups by appointment.

Tillery

TILLERY HISTORY HOUSE MUSEUM, 321 Community Center Rd., Tillery, NC 27887. Mailing Address: P.O. Box 61, Tillery, NC 27887. Tel.: 252-826-3017. Fax: 252-826-3244.
E-mail: tillery@aol.com
Web Site: www.cct78.org
Founded: 1995.
Congressional District: 1
Key Personnel: Dir. & Chm. (V), Gary R. Grant.
Personnel Profile: Part-Time Volunteers 1.
Volunteer Hours: 1,000
Governing Authority: Parent Institution: Concerned Citizens of Tillery (CCT).
Institution Type/Description: Historic House Museum: housed on former plantation land worked by generations of African-American slaves.
Collections: local history; African American slave history; photographs; personal artifacts; farm equipment & tools.
Activities: guided tours.
Hours & Admission Prices: By appointment. No charge; donations accepted. &
Attendance: 1,500 (accurate)

Topsail Island

MISSILES & MORE MUSEUM, 720 Channel Blvd., Topsail Island, NC 28460. Tel.: 800-626-2780; 910-328-2488.
Institution Type/Description: History Museum: housed in the Assembly Building; built in 1946.
Collections: local history & culture; military artifacts; photographs.
Activities: group tours.
Hours & Admission Prices: April to mid-May & Sept. to mid-Oct. Mon.-Fri. 2-5; Memorial Day to Labor Day Mon.-Sat. 2-5.

Trinity

LINBROOK HERITAGE ESTATE, 5507 Snyder Country Rd., Trinity, NC 27370. Tel.: 336-861-6959.
Web Site: www.linbrookheritageestate.com
Institution Type/Description: History Museum: estate comprises Linbrook Hall, the Historic Hoover House, and the Neal Agricultural and Industrial Museum.
Collections: Neal family's personal artifacts & furnishings; early John Deere tractors & farm equipment.
Activities: guided tours; demonstrations.
Hours & Admission Prices: Tues.-Sat. 9-5, Sun. 2-5; groups by appointment. Linbrook Hall: adults $20, seniors 65 & over and youth 12-17 $15; children under 12 no charge. Historic Hoover House, Neal Agricultural & Industrial Museum: adults $5, seniors 65 & over and youth 12-17 $3; children under 12 no charge.

NEAL JOHN DEERE TRACTOR MUSEUM, 5507 Snyder Country Rd., Trinity, NC 27370. Tel.: 336-861-6959.
E-mail: johndeer@northstate.net
Web Site: nealsjohndeeretractors.com
Institution Type/Description: Tractor Museum.
Collections: early John Deere tractors.
Hours & Admission Prices: Sat. 9-5, Sun. 2-5.

Tryon

FOOTHILLS EQUESTRIAN NATURE CENTER (FENCE), 3381 Hunting County Rd., Tryon, NC 28782. Tel.: 828-859-9021.
Web Site: www.fence.org
Governing Authority: nonprofit organization.
Institution Type/Description: Nature Center.
Collections: natural history; wildfowl; native plants & trees; preservation & conservation.
Facilities: 320-acre nature center; 5 mile nature trails; equestrian center; classrooms; meeting rooms.
Activities: educational programs; outdoor concerts; summer camp; school & scout programs; birthday parties.
Hours & Admission Prices: Call for hours.

TOY MAKERS HOUSE MUSEUM, 43 E. Howard St., Tryon, NC 28782-2400. Tel.: 828-290-6600.
Institution Type/Description: Toy History Museum.
Collections: toy history; wood working; toys; furniture; tools; personal artifacts; photographs.
Facilities: Museum-related items for sale.
Activities: art classes.
Hours & Admission Prices: Call for hours.

Valdese

WALDENSIAN HERITAGE MUSEUM, 208 Rodoret St., Valdese, NC 28690-2841. Mailing Address: P.O. Box 111, Valdese, NC 28690-0111. Tel.: 828-874-1111& 879-2531. Fax: 828-874-1111.
E-mail: museum@waldensianpresbyterian.org
Web Site: www.waldensianpresbyterian.org
Formerly: Museum of Waldensian History
Founded: 1955.
Congressional District: 10
Key Personnel: Exec. Dir. & Museum Shop Mgr., Gretchen Costner; Pres. (V), Jewell Bounous.
Personnel Profile: Full-Time Paid 1; Part-Time Volunteers 15.
Governing Authority: church. Parent Institution: Waldensian Presbyterian Church. Tax-exempt: 170(b)(1)(A).
Institution Type/Description: Religious Museum.
Collections: late 19th & early 20th-century clothing & household furnishings; farm implements & construction tools related to the Waldensians; church furnishings & religious items related to the Waldensian Presbyterian Church; photographs of church people, events & early homes; World War I & II uniforms & artifacts.
Facilities: 500-vol. library of books related to Waldensian history, including Bibles & services dating back to the 16th century, available for research on premises; reading room. Books on Waldensian history & language & gift items for sale.
Activities: guided tours; lectures; docent program or council; outdoor theatre.
Publications: books, The Waldenses of Valdese; The Provencal Speech of The Waldensian Colonists of Valdese, NC; Genealogy of the Waldensian Settlers in Valdese, NC 1893-1990; booklets, The Waldenses of Burke County; The History of the Waldenses; The History & Heritage of the Waldensian Presbyterian Church in Valdese (first 100 years).
Hours & Admission Prices: Summer during outdoor drama: Fri.-Sat. 4-6. Tours: Tues.-Fri. 11 & 2; other times by appointment. Adults $2, students $1. Closed New Year's Day; Easter; Memorial Day; Labor Day; Thanksgiving; Christmas. &
Attendance: 2,000 (estimated)

WALDENSIAN TRAIL OF FAITH, 401 Church St., N.W., Valdese, NC 28690. Mailing Address: P.O. Box 1256, Valdese, NC 28690. Tel.: 828-874-1893; 800-635-4778. Facebook: Waldensian Trail of Faith.
E-mail: trailoffaith1893@embarqmail.com
Web Site: www.waldensiantrailoffaith.org
Founded: 1993.
Key Personnel: Dir., Anthony Collins; Pres. (V), Jim Jacumin.
Personnel Profile: Part-Time Paid 2; Part-Time Volunteers 24.
Volunteer Hours: 1,600

Governing Authority: Tax-exempt.
Institution Type/Description: History Museum.
Collections: Waldensian history & religious heritage.
Activities: guided tours; special events.
Hours & Admission Prices: Mon.-Fri. 9-5, Sat.-Sun. 2-5. Guided Tours: adults $9, seniors $8, teachers & seniors $6, teachers $5, students $3. Closed New Year's Day; Thanksgiving; Christmas. &
Attendance: 5,000 (accurate)

Wadesboro

ANSON COUNTY HISTORICAL SOCIETY, INC., 206 E. Wade St., Wadesboro, NC 28170-2229. Tel.: 704-694-6694. Fax: 704-694-3763.
E-mail: ansonhistorical@windstream.net
Web Site: www.ansonhistoricalsociety.org
Founded: 1962.
Congressional District: 8
Key Personnel: Dir., John Jennings Dunlap, III
Personnel Profile: Part-Time Volunteers 10.
Governing Authority: society. Tax-exempt: 170(b)(1)(A), 501(c)(3).
Institution Type/Description: General Museum.
Collections: period furniture; agriculture; paintings; colonial garden; Indian artifacts; Ashe-Covington Medical Museum (medical artifacts). Historic Houses: 1783 Boggan-Hammond House; 1839 Alexander Little Wing.
Research Fields: paintings; colonial garden.
Activities: guided tours; lectures; arts festivals; permanent & temporary exhibitions.
Publications: quarterly newsletter; pamphlets; books, A Pictorial Tribute; History of Anson County 1750-1976; Cemeteries of Anson County, Volume I; Cemeteries of Anson County, Volume II; Eastview Cemetery, Wadesboro, NC.
Hours & Admission Prices: Mon.-Fri. 9-1; groups by appointment only. Mon.-Fri. 9-1. Adults $5; members no charge. &
Attendance: 850 (estimated)
Membership: Individual $35; Family $50; Business $100; Life $2,000.

PEE DEE NATIONAL WILDLIFE REFUGE, 5770 U.S. Hwy. 52 N., Wadesboro, NC 28170. Tel.: 704-694-4424. Fax: 704-694-6570.
Web Site: www.fws.gov/peedee
Founded: 1963.
Key Personnel: Refuge Mgr., J.D. Bricken; Asst. Refuge Mgr., Greg Walmsley
Institution Type/Description: Wildlife Refuge.
Collections: natural history; over 180 bird species; 49 amphibian & reptile species; 28 mammals; 20 fish species.
Facilities: nature trails.
Activities: special events; educational programs; hiking trails; wildlife observation. Annual Events: Friends of the Pee Dee Barbeque in March; Youth Turkey Hunt in April; Youth Fishing Day in April; International Migratory Bird Day in May; National Fishing Week in June.
Hours & Admission Prices: Call for hours.
Attendance: 35,000

WADESBORO ROTARY PLANETARIUM AND SCIENCE CENTER, 320 Camden Rd., Wadesboro, NC 28170. Tel.: 704-694-7016.
E-mail: phillips.lanette@anson.k12.nc.us
Web Site: www.ansonschools.org
Institution Type/Description: Planetarium & Science Center.
Collections: space, physical, life & earth sciences; astronomy; wildlife dioramas; NASA; photographs; hands-on exhibits.
Activities: educational programs.
Hours & Admission Prices: Mon.-Fri. 8-10:45. Adults $5, students $3.

Wake Forest

WAKE FOREST COLLEGE BIRTHPLACE SOCIETY, INC., (M), 414 N. Main St., Wake Forest, NC 27587. Mailing Address: P.O. Box 494, Wake Forest, NC 27588-0494. Tel.: 919-556-2911. Fax: 919-556-2991.
E-mail: morrisce@wfu.edu
Web Site: www.wakeforestmuseum.org
Founded: 1956.
Congressional District: 4
Key Personnel: Exec. Dir., Ed Morris; Pres. (V), Tom Parrish; Vice Pres., Durward Matheny; Asst. Dir., Jennifer Smart.
Personnel Profile: Full-Time Paid 1; Part-Time Paid 1; Part-Time Volunteers 50; Interns 1.

Governing Authority: nonprofit organization. Parent Institution: Wake Forest University. Tax-exempt: 501(c)(3).
Institution Type/Description: Historic Society Museum: housed in the first home of Wake Forest University; built in 1820.
Collections: memorabilia of Wake Forest College & town of Wake Forest; period furniture; publications; archives; sports memorabilia.
Research Fields: college & town history; history of education in NC; Baptist heritage; sports; medical school.
Facilities: 75-seat auditorium; gardens.
Activities: formally organized educational programs for children.
Publications: brochure describing house; newsletter; research journal.
Hours & Admission Prices: Tues.-Fri. 9-12 & 1:30-4:30, Sun. 2-5; tours by appointment. No charge. Closed major holidays. &

Attendance: 7,000 (estimated)
Membership: Individual $35; Family $50; Patron $250; Samuel Wait Society $500.

Warrenton

BUCK SPRING - NATHANIEL MACON HOMEPLACE, 1348 State Rd., Warrenton, NC 27589. Tel.: 252-257-3640. Fax: 252-257-5616.
Institution Type/Description: History Museum.
Collections: local history & culture; period furnishings; personal artifacts; replica of Nathaniel Macon's home (1758-1837); outbuildings; gravesite.
Facilities: nature trails.
Hours & Admission Prices: Call for hours.

Warsaw

DUPLIN COUNTY VETERANS MEMORIAL MUSEUM, 119 E. Hill St., Warsaw, NC 28398-1917. Mailing Address: P.O. Box 137, Warsaw, NC 28398. Tel.: 910-293-2190.
Web Site: duplincountyveteransmuseum.com
Founded: 1998.
Congressional District: 3
Key Personnel: Cur., Earl Rouse; Chm. (V), Jene Thompson.
Personnel Profile: Part-Time Paid 1.
Governing Authority: Parent Institution: Duplin County Veterans Foundation.
Institution Type/Description: Military Museum: housed in the L. P. Best house; built in 1894. Listed on the National Register of Historic Places.
Collections: Best family furnishings & personal artifacts; photographs; portraits; Victrola; Duplin County servicemen & women memorial; military artifacts & memorabilia; POW/MIA memorial.
Hours & Admission Prices: Thurs.-Fri. 1-4, Sat. by appointment. Adults $2; members no charge. &

Attendance: 100 (estimated)
Membership: Individual $25.

Washington

NORTH CAROLINA ESTUARIUM, 223 E. Water St., Washington, NC 27889. Tel.: 252-948-0000.
E-mail: estuarium@embarqmail.com
Institution Type/Description: Environmental Center.
Collections: hands-on exhibitions; science; period artifacts; seashells; works by regional artists; aquariums; terrariums; photographs.
Facilities: Museum-related items for sale.
Activities: audio-visual presentation; educational programs; classes; special events; rental facilities.
Hours & Admission Prices: Tues.-Sat. 10-4. Adults $4, students K-12 $2; children 4 & under no charge.

Waxhaw

MEXICO CARDENAS MUSEUM, 6403 Davis Rd., Waxhaw, NC 28173. Tel.: 704-843-6045.
Web Site: www.jaars.org
Institution Type/Description: History Museum.
Collections: history of the people of Mexico, their former president Lazaro Cardenas, & Wycliffe Bible Translators; personal artifacts; paintings; photographs; period artifacts; clothing.
Hours & Admission Prices: Mon. Sat. 9 12 & 1 4.

THE MUSEUM OF THE ALPHABET, 6409 Davis Rd., The JAARS Center, Waxhaw, NC 28173. Mailing Address: P.O. Box 248, Waxhaw, NC 28173-0248. Tel.: 704-843-6066. Fax: 704-843-6200.
E-mail: info@jaars.org

Web Site: www.jaars.org/museum/alphabet/index.htm
Founded: 1990.
Congressional District: 8
Key Personnel: Dir., LaDonna Mann; Financial Dir., Roy Self; Public Rels., Mike Osborn.
Personnel Profile: Full-Time Volunteers 1; Part-Time Volunteers 12.
Governing Authority: nonprofit organization. Parent Institution: JAARS. Tax-exempt: 501(c)(3) & 170(b)(1)(A).
Institution Type/Description: Alphabet Museum.
Collections:
Facilities: 750-vol. library of art; history; linguistics books; 5,000 sq. ft. exhibit space. Museum-related items for sale.
Activities: films; guided tours; participatory exhibits.
Publications: booklet, The Alphabet Makers, Alphabet Roots; Alphabet Account, Past Masters, Old Lamp Lighters; exhibition brochures; Faculty of Letters; Medley of Alphabets.
Hours & Admission Prices: Mon.-Sat. 9-12 & 1-4. No charge; donations accepted. Closed government holidays. &

Attendance: 5,500 (accurate)

MUSEUM OF THE WAXHAWS & ANDREW JACKSON MEMORIAL, 8215 Waxhaw Hwy. - Hwy. 75, Waxhaw, NC 28173. Mailing Address: P.O. Box 7, Waxhaw, NC 28173-1038. Tel.: 704-843-1832. Fax: 704-843-1832.
E-mail: mwaxhaw@museumofthewaxhaws.com
Web Site: www.museumofthewaxhaws.com/
Founded: 1996.
Congressional District: 8
Key Personnel: Dir., Sharon Murrer; Pres. (V), Mary Alice Wilson.
Personnel Profile: Part-Time Paid 1; Part-Time Volunteers 25.
Governing Authority: private; nonprofit organization. Parent Institution: Andrew Jackson Historical Foundation. Tax-exempt: 501(c)(3).
Institution Type/Description: Regional History Museum focus on settlement period, American Revolution & Andrew Jackson's life.
Collections: civilian & military related artifacts associated with the Waxhaw Settlement & President Andrew Jackson; time line 1650-1900 tells the story of The Waxhaws a border region along NC/SC line; Andrew Jackson, 7th US President was born in the area (1767), serves as a living memorial to President Jackson; areas of focus are the American Revolution & the Civil War.
Research Fields: area study of Scots-Irish immigration & culture; Andrew Jackson.
Facilities: 8,500 sq. ft. exhibit space; 66-seat theater. Museum-related items for sale.
Activities: guided tours; lectures; theater & outdoor drama company located with museum; craft workshops; youth reenactment days. Annual Events: two living history days; Outdoor drama in June.
Publications: quarterly newsletter, Museum Monitor.
Hours & Admission Prices: Fri.-Sat. 10-5, Sun. 2-5. Adults $5, seniors 60 & over $4, children 6-12 $2; members and children 5 & under no charge. Closed New Year's Day; Thanksgiving; Christmas. &

Attendance: 3,000 (accurate)
Membership: Patron's Society $1-$99; Director's Society $100-$249; Scotch-Irish Society $250-$499; William R. Davie Society $500-$749; Waxhaw Heritage Society $750-$999; "Old Hickory" Society $1,000-$2,499; Elizabeth Jackson Society $2,500-$4,999; President Andrew Jackson Society $5,000 & up.

Waynesville

MUSEUM OF NORTH CAROLINA HANDICRAFTS IN THE HISTORIC SHELTON HOUSE, 49 Shelton St., Waynesville, NC 28786-5795. Mailing Address: P.O. Box 145, Waynesville, NC 28786. Tel.: 828-452-1551.
E-mail: museumnc@bellsouth.net
Web Site: www.sheltonhouse.org
Founded: 1977.
Governing Authority: Tax-exempt.
Institution Type/Description: Handicraft Museum: housed in the former home of Stephen John Shelton and then his son, William Taylor Shelton, founder of the Shiprock New Mexico Navajo Indian Reservation & School; built in 1875. Listed on the National Register of Historic Places.
Collections: 18th-21st century handicrafts & furniture; Cherokee & Navajo artifacts.
Hours & Admission Prices: May-Oct. Tues.-Sat. 10-4; Winter: call for hours. Adults $5, students $3; discounts to groups of 10 or more; children under 5 no charge.
Attendance: 700 (estimated)
Membership: Bronze Patron $1-$299; Individual $25; Couple $40; Family $75; Silver Patron $300-$499; Gold Patron $500-$999; Platinum Patron $1,000 & up.

Weaverville

ZEBULON B. VANCE BIRTHPLACE STATE HISTORIC SITE, 911 Reems Creek Rd., Weaverville, NC 28787-8710. Tel.: 828-645-6706. Fax: 828-645-0936.
Web Site: www.nchistoricsites.org/vance/vance.htm
Founded: 1961.
Congressional District: 11
Key Personnel: Site Mgr., Chris Morton.
Personnel Profile: Full-Time Paid 3; Part-Time Paid 3; Part-Time Volunteers 1.
Governing Authority: state. Parent Institution: North Carolina Department of Cultural Resources, 109 E. Jones St., Raleigh, NC 27611. Tax-exempt: 170(b).
Institution Type/Description: Park Museum Visitor Center.
Collections: Historic House: 1795 reconstructed log house & out buildings.
Research Fields: Zebulon B. Vance.
Facilities: visitor center.
Activities: guided tours; lectures; formally organized education programs for children; permanent exhibitions; participatory demonstrations.
Publications: brochure.
Hours & Admission Prices: Tues.-Sat. 9-5. No charge; donations accepted. Closed major holidays. &
Attendance: 11,764 (accurate)

Welcome

RICHARD CHILDRESS RACING MUSEUM, 180 Industrial Dr., Welcome, NC 27374. Mailing Address: P.O. Box 360, Welcome, NC 27374. Tel.: 366-731-3389; 800-476-3389.
Institution Type/Description: Sports Museum.
Collections: race cars; hot rods; trophies; awards; racing memorabilia; photographs.
Facilities: Museum-related items for sale.
Hours & Admission Prices: Mon.-Fri. 9-5. Guided Tours: Mon.-Fri. 10am & 2pm.

West Jefferson

ASHE ARTS CENTER, 303 School Ave., West Jefferson, NC 28694. Tel.: 336-846-2787.
E-mail: jane@ashecountyarts.org
Web Site: www.ashecountyarts.org/galleryshop.htm
Key Personnel: Exec. Dir., Jane Lonon; Dir. Opers., Linda Dreyer; Dir. Programs, Rebecca Williams
Institution Type/Description: Art Gallery.
Collections: works by local artists; paintings; photographs; drawings.
Activities: performances; classes; educational programs; temporary exhibits; special events.
Hours & Admission Prices: Jan.-March Mon.-Fri. 9-4; April-Dec. Mon.-Fri. 9-4, Sat. 10-4.

ASHE COUNTY CHEESE PLANT & STORE, 106 E. Main St., West Jefferson, NC 28694. Mailing Address: P.O. Box 447, West Jefferson, NC 28694. Tel.: 800-445-1378; 336-246-2501.
E-mail: info@ashecountycheese.com
Web Site: www.ashecountycheese.com
Institution Type/Description: Cheese Plant.
Collections: over 20 varieties of cheese; butter; fudge; cheese curd.
Facilities: Plant-related items for sale.
Activities: cheese making viewing room.
Hours & Admission Prices: Mon.-Sat. 8:30-5. &

BLUFF MOUNTAIN NATURE PRESERVE, Edwards Rd., West Jefferson, NC 28694. Mailing Address: 334 Blackwell St., Ste. 300, Durham, NC 27701. Tel.: 336-497-1972.
E-mail: bluffmountainpreserve@gmail.com
Web Site: www.blueridgeheritage.com
Key Personnel: Guide, Kim Hadley.
Governing Authority: Parent Institution: The Nature Conservancy.
Institution Type/Description: Nature Preserve.
Collections: natural heritage; plants; flowers; trees; birds; bobcats.
Activities: guided tours.
Hours & Admission Prices: By appointment.

Whiteville

NORTH CAROLINA MUSEUM OF FORESTRY, 415 S. Madison St., Whiteville, NC 28472-4125. Tel.: 910-914-4185. Fax: 910-641-0385.
E-mail: forestry.museum@naturalsciences.org
Founded: 2000.
Congressional District: 7
Key Personnel: Pres. Bd., Harold Blanchard; Exhibit Coord., Sara Capps; Educator, Kellie Lewis; Administrative Asst., Rhonda Billeaud.
Personnel Profile: Full-Time Paid 4; Part-Time Paid 5; Part-Time Volunteers 33; Interns 1.
Governing Authority: state. Parent Institution: North Carolina Museum of Natural Sciences, Raleigh, NC. Tax-exempt: 501(c)(3).
Institution Type/Description: Natural History Museum.
Collections: North Carolina forest history & natural history.
Facilities: 164-vol. library; educational facilities; 8,000 sq. ft. exhibit space.
Activities: docent program; guided tours; lectures; loan, participatory & traveling exhibitions; rental gallery. Annual Events: Festival of Trees; Wildlife Encounters; Pecan Festival.
Hours & Admission Prices: Mon.-Fri. 9-5, Sat. 1-4, Sun. 2-5. No charge. Closed state holidays. &
Attendance: 13,421 (accurate)

Wilkesboro

WILKES HERITAGE MUSEUM, 100 E. Main St., Wilkesboro, NC 28697. Mailing Address: P.O. Box 935, Wilkesboro, NC 28697-0935. Tel.: 336-667-3171.
E-mail: info@wilkesheritagemuseum.com
Web Site: www.wilkesheritagemuseum.com
Key Personnel: Dir., Jennifer Furr
Institution Type/Description: History Museum.
Collections: early settlers; pottery; education; religion; NASCAR; moonshine; industry; military. Historic Buildings: Old Wilkes Jail; Robert Cleveland Log Home.
Hours & Admission Prices: Mon.-Fri. 10-4. Admission $6; children 5 & under no charge. Closed major holidays.

Willard

PENDERLEA HOMESTEAD MUSEUM, 284 Garden Rd., Willard, NC 28478-6780. Tel.: 910-285-3490.
E-mail: info@penderleahomesteadmuseum.org
Web Site: www.penderleahomesteadmuseum.org
Key Personnel: Pres. Bd. Dirs., Allison Rankin; Vice Pres., Michael Booth; Cur., Ann Southerland Cottle
Institution Type/Description: History Museum.
Collections: local history & culture; early settlers; farming; photographs; period artifacts.
Hours & Admission Prices: Sat. 1-4. No charge; donations accepted. &
Membership: Individual $15; Family $20; Lifetime $250.

Williamston

ASA BIGGS HOUSE - MARTIN COUNTY HISTORICAL SOCIETY, 100 E. Church St., Williamston, NC 27892. Mailing Address: P.O. Box 851, Williamston, NC 27892. Tel.: 252-792-6605. Fax: 252-792-8710.
E-mail: tourism@visitmartincounty.com
Institution Type/Description: Historical Society Museum: housed in the former home of attorney, federal judge & U.S. Senator Asa Biggs; built in 1831. Listed on the National Register of Historic Places.
Collections: Asa Biggs' life & career; personal artifacts; photographs; period furnishings.
Hours & Admission Prices: Call for hours.

Wilmington

AIRLIE GARDENS, 300 Airlie Rd., Wilmington, NC 28403-3706. Tel.: 910-798-7700.
Institution Type/Description: Gardens.
Collections: plants; trees; flowers; sculpture; over 130 bird species; works of Minnie Evans; mosaics.
Facilities: nature trails.
Activities: hiking; summer camp.
Hours & Admission Prices: Jan. 2-March 19 Mon.-Sat. 9-5; March 20-April 2 & May 18-Dec. daily 9-5; April 3-May 17 Sun.-Wed. 9-5, Thurs.-Sat. 9-7. Adults $8, county residents & military $5, children 4-12 $3.

BATTLESHIP NORTH CAROLINA, Eagles Island, #1 Battleship Rd., Wilmington, NC 28401. Mailing Address: P.O. Box 480, Wilmington, NC 28402-0480. Tel.: 910-251-5797. Fax: 910-251-5807.
E-mail: museum@battleshipnc.com
Web Site: www.battleshipnc.com
Founded: 1961.
Congressional District: 7
Key Personnel: Exec. Dir., Capt. Terry Bragg; Chm. (V), Samuel Southern; Asst. Dir. Operations, Chris Vargo; Comptroller, Elizabeth Rollinson; Dir. Promotions, Heather Loftin; Maintenance Supvr., Terry Kuhn; Dir. Museum Svcs., Kim Robinson Sincox; Cur., Mary Ames Booker; Dir. Sales, Leesa McFarlane; Dir. Programs, Danielle Wallace.
Personnel Profile: Full-Time Paid 25; Part-Time Paid 15; Part-Time Volunteers 50.
Governing Authority: state. Subsidiary Institution: Friends of the Battleship NC. Tax-exempt.
Institution Type/Description: Historic Ship Museum.
Collections: World War II & U.S. Navy paintings & photographs; Kingfisher float plane; naval artifacts of the World War II era; artifacts & archival materials from ships named North Carolina, 1818-1947; SSN777 attack submarine North Carolina (2008-); archival material from U.S.S. NORTH CAROLINA (BB-55); early 20th century U.S. Navy archival material.
Research Fields: ships named NORTH CAROLINA.
Facilities: auditorium; visitor's orientation center; picnic grounds; rental spaces. Gifts, postcards & books for sale.
Activities: self-guided tours; permanent & changing exhibitions; school group presentations; volunteer program; life long learning programs. Annual Events: Memorial Day service; Independence Day; Living History Weekends.
Publications: books, revised & expanded second edition - Battleship North Carolina; USS North Carolina Ship's Data I; newsletter.
Hours & Admission Prices: Memorial Day to Labor Day daily 8-8; Sept.-May daily 8-5. Adults $12, seniors & military $10, children 6-11 $6; discounts to groups; Friends of the Battleship & children under 5 no charge.
Attendance: 200,000 (accurate)
Membership: Chief $15; Lieutenant (jg) $35; Lieutenant $45; Lieutenant Commander $75; Commander $150; Captain $300; Commodore $500.

BELLAMY MANSION MUSEUM OF HISTORY AND DESIGN ARTS, 503 Market St., Wilmington, NC 28401-4634. Tel.: 910-251-3700, ext. 102. Fax: 910-763-8154.
E-mail: info@bellamymansion.org
Web Site: www.bellamymansion.org
Founded: 1993.
Congressional District: 1
Key Personnel: Exec. Dir., Beverly Ayscue; Chm., Sharon Stone; Dir. Mktg. & Facilities, Gene Ayscue; Dir. Public Education, Madeline Flagler.
Personnel Profile: Full-Time Paid 2; Part-Time Paid 3; Part-Time Volunteers 75.
Governing Authority: private; nonprofit organization. Parent Institution: The Historic Preservation Foundation of North Carolina. Tax-exempt: 501(c)(3).
Institution Type/Description: Historic House Museum: housed in c.1861 Bellamy Mansion, a 22-room Greek Revival and Italianate residence built by free and enslaved African Americans that includes original slave quarters.
Collections: original furnishings; textiles; family material; archeological finds; structures.
Research Fields: history and design arts.
Facilities: 5,000 sq. ft. exhibit space.
Activities: guided tours; temporary & traveling exhibitions.
Publications: quarterly newsletter, Bellamy Mansion News.
Hours & Admission Prices: Tues.-Sat. 10-5, Sun. 1-5. Adults $10, children 5-12 $4; discounts to groups, Preservation North Carolina & Friends of Bellamy Mansion Museum members; members & National Trust for Historic Preservation members no charge. Closed New Year's Day; Easter; Memorial Day; Independence Day; Thanksgiving; Christmas.
Attendance: 15,000 (accurate)
Membership: Individual $35; Contributor $50; Family $60; Friend $125; Supporter $150; Sponsor $250; Benefactor $500; Cornerstone $1,000; Heritage Leader $2,500.

THE BURGWIN-WRIGHT MUSEUM, 224 Market St., Wilmington, NC 28401-4444. Tel.: 910-762-0570. Fax: 910-762-8650.
Facebook: The Burgwin-Wright Museum.
E-mail: info@burgwinwrighthouse.com
Web Site: www.burgwinwrighthouse.com
Formerly: The Burgwin-Wright Museum and Gardens
Founded: 1770.

Congressional District: 7
Key Personnel: Pres. (V), Mary Eggleston; Dir., Christine Lamberton; Exec. Dir. & C.E.O., Joy Allen; Museum Mgr., Jackie Pastis Margoles.
Personnel Profile: Full-Time Paid 2; Part-Time Paid 6; Interns 1.
Governing Authority: society. Parent Institution: The National Society of the Colonial Dames of America in the State of North Carolina, Wilmington, NC 28401. Tax-exempt: 501(c)(3).
Institution Type/Description: Historic House Museum: housed in 1771 Burgwin-Wright House & Garden.
Collections: 17th, 18th & early 19th-century furnishings, gardens & orchard; colonial kitchen; 1700's shards from tunnel under house.
Activities: guided tours; open hearth cooking demonstrations in colonial kitchen; 12/9/12 Christmas Stroll Through the Past.
Hours & Admission Prices: Tues.-Sat. 10-4, last tour at 3. Adults $10, children 5-12 & students $5; discount for multiple house ticket; children under 5, military & members no charge. Closed national holidays; Christmas week & Jan.
Attendance: 4,500 (estimated)

CAMERON ART MUSEUM, (M), 3201 S. 17th St., Wilmington, NC 28412-6554. Tel.: 910-395-5999. Fax: 910-395-5030.
Web Site: www.cameronartmuseum.org
Formerly: St. John's Museum of Art
Founded: 1962.
Congressional District: 7
Key Personnel: Dir., Anne Brennan; Chm. (V), Frances Goodman; Cur. Public Programs, Daphne Holmes; Property Mgr., Johnnie McKoy; Cur. Education, Georgia Mastroieni; Resident Master Artist, Hiroshi Sueyoshi; Registrar, Holly Tripman; Museum Shop Mgr., Nan Pope.
Personnel Profile: Full-Time Paid 17; Part-Time Paid 14; Part-Time Volunteers 100; Interns 3.
Governing Authority: nonprofit organization. Tax-exempt: 501(c)(3).
Institution Type/Description: Art Museum.
Collections: fine art; design & crafts.
Research Fields: folk art (Minnie Evans Study Center).
Facilities: art library; cafe; auditorium; reception hall. Museum-related items for sale.
Activities: permanent & temporary exhibitions; artist & gallery talks; interdisciplinary programs; classes; workshops; museum school; tours; monthly children's programs; child docent program; artist-in-residence program (Clay Studio).
Publications: member bulletin; exhibition catalogues.
Hours & Admission Prices: Tues.-Wed. & Fri.-Sun. 10-5, Thurs. 10-9. Adults $8; discount to students, senior citizens, NARM & AAM members; members no charge. Closed holidays. &
Attendance: 40,000 (accurate)
Membership: Students & Seniors $35; Individual $50; Household $100; Friend & CAM Contemporary Household $150; CAM Contemporary $175; Associate $240; Patron $500; Donor $1,000.

* **CAPE FEAR MUSEUM OF HISTORY AND SCIENCE, (M),** 814 Market St., Wilmington, NC 28401-4752. Tel.: 910-798-4350. Fax: 910-798-4382.
E-mail: amangus@nhcgov.com
Web Site: www.CapeFearMuseum.com
Founded: 1898.
Congressional District: 7
Key Personnel: Dir., Ruth Haas; Chair, Allen Trask, III; Pres. (V) CFM Associates, Inc., Mike Ryan; Administrative Asst., Lynda Danley; Volunteer Coord. & Museum Shop Mgr., Karen Smith; Cur., Barbara L. Rowe; Mgr. Education, Amy Thornton; Public Rels. Specialist, Amy Mangus; Educator, Tom Osborne; Educator, Jameson McDermott; Educator, Pepper Hill; Mgr. Exhibits, Adrienne Garwood; Exhibits Designer, John Timmerman; Registrar, Terri Hudgins; Historian, Janet Davidson; Donor Relations Dir., Cindy Anzalotti.
Personnel Profile: Full-Time Paid 12; Part-Time Paid 5; Part-Time Volunteers 80.
Volunteer Hours: 5,915
Operating Expenses: 1,700,151
Operating Income: 1,700,151
Governing Authority: county-appointed museum advisory bd.; administrative. Parent Institution: New Hanover County. Subsidiary Institution: Cape Fear Museum Associates, Inc. Tax-exempt.
Institution Type/Description: History & Science Museum.
Collections: objects, documents & photographs representing the history, science & cultures of lower Cape Fear, including business & industry; natural history; household items; costumes & textiles; decorative arts; agriculture; military; maritime; forestry & lumbering; collection from former Blockade Runner Museum; model of 1863 Wilmington Water Front.

Major Exhibits: Nano, 12/13-3/14; Handbags, 1/14-7/14; A View from Space (OMSI) (T), 2/14-9/14; Communication, 7/14-1/15.
Research Fields: Lower Cape Fear history & natural history.
Facilities: library of books, pamphlets, research files & media on local, state, regional history, natural history & museum practice available on museum premises by appointment; 10,000 sq. ft. exhibition space.
Activities: long-term & changing exhibitions; K-12 school programs; outreach programs & classroom kits; public programs & events for children, family, adult & community groups; collection-based research by appointment.
Publications: brochures & rack cards; calendars; educational program guides; publicity materials; semi-annual magazine, Cape Fear Stories.
Hours & Admission Prices: May-Sept. Mon.-Sat. 9-5, Sun. 1-5; Sept.-May Tues.-Sat. 9-5, Sun. 1-5. Adults $7, seniors, college students & military $6, children 6-17 $4; discounts to ASTC & SEMC members; children under 6 & members no charge. Closed major holidays ⓗ
Attendance: 42,877 (accurate)
Membership: Student $25; Individual $45; Family $60; Donor $100; Patron $250; Director $500; Benefactor & Corporate Contributor $1,000; Corporate Patron $2,500; Corporate Benefactor $5,000.

CAPE FEAR SERPENTARIUM, 20 Orange St., Wilmington, NC 28401-4419. Tel.: 910-762-1669. Fax: 910-762-1669.
E-mail: reptileeducation@gmail.com
Web Site: www.capefearserpentarium.com
Founded: 2001.
Key Personnel: Dir., Dean Ripa
Institution Type/Description: Herpetology Museum.
Collections: reptiles from around the world including snakes, lizards, & crocodiles.
Research Fields: herpetology.
Activities: educational programs; birthday parties; class field trips; feeding shows.
Publications: occasional papers.
Hours & Admission Prices: Summer: Mon.-Fri. 11-5, Sat. 11-6, Sun. 11-5; Winter: call for hours. Admission $8; discounts to groups; children under 2 no charge.

THE CHILDREN'S MUSEUM OF WILMINGTON, 116 Orange St., Wilmington, NC 28401-4421. Tel.: 910-254-3534. Fax: 910-254-3565. Facebook: The Children's Museum of Wilmington.
E-mail: info@playwilmington.org
Web Site: www.playwilmington.org
Founded: 1991.
Congressional District: 7
Key Personnel: Dir., Rick Lawson; Pres. (V), Emily Reiniche.
Personnel Profile: Full-Time Paid 6; Part-Time Paid 10; Part-Time Volunteers 10; Interns 5.
Volunteer Hours: 300
Operating Expenses: 659,123
Operating Income: 549,063
Governing Authority: Tax-exempt.
Institution Type/Description: Children's Museum.
Collections: hands-on exhibits.
Activities: rental facilities; birthday parties; school field trips; rental facilities.
Hours & Admission Prices: Mon.-Sat. 9-5, Sun. 1-5. Admission $8; discounts to educators, military & seniors; ACM reciprocal program; members & children under one no charge. Closed Easter; Thanksgiving; Christmas Eve & Day. ⓗ
Attendance: 60,000 (accurate)
Membership: Weekday $90; Anytime $125; ACM Reciprocal $150; Contributing $225.

LOWER CAPE FEAR HISTORICAL SOCIETY, INC., 126 S. Third St., Wilmington, NC 28401-4556. Tel.: 910-762-0492 & 2976. Fax: 910-763-5869. Facebook: Latimer House Lower Cape Fear Historical Society.
E-mail: info@latimerhouse.org
Web Site: www.hslcf.org
Founded: 1956.
Congressional District: 7
Key Personnel: Chm. (V), Pat Hardee; Pres. (V), John Golden; Vice Pres., Clauston Jenkins; Office Mgr., Brittany Bennett; Archives Representative, James Rush Beeler; Archivist, Candace McGreevy.
Personnel Profile: Full-Time Volunteers 40; Part-Time Paid 3; Part-Time Volunteers 50.
Volunteer Hours: 1,000
Operating Expenses: 116,000
Operating Income: 116,000

Governing Authority: nonprofit corporation. Tax-exempt.
Institution Type/Description: Decorative Arts Museum: housed in 1852 Latimer House, on National Register of Historic Places.
Collections: original furnishings, artifacts & portraits; Cape Fear region lifestyle in 1850s.
Research Fields: archives of the Lower Cape Fear region.
Facilities: books & material relating to Wilmington archives available for research by application.
Activities: guided tours; walking tour of historic district. Museum Sponsors: three annual formal lectures.
Publications: newsletter; journal; bulletins; brochure; short stories on Cape Fear area.
Hours & Admission Prices: Mon.-Sat. 10-3. Adults $10, children & students $5; discounts to AAA members; members no charge.
Attendance: 3,000 (accurate)
Membership: Individual $40; Family $75.

MUSEUM OF WORLD CULTURES/UNIVERSITY OF NORTH CAROLINA AT WILMINGTON, William M. Randall Library, UNC at Wilmington, 601 S. College Rd., Wilmington, NC 28403-5649. Tel.: 910-962-3276.
E-mail: parnellg@uncw.edu
Web Site: library.uncw.edu/museum
Key Personnel: Coord. Special Collections, Jerry Parnell
Institution Type/Description: University Museum.
Collections: artifacts from around the world; clothing; textiles; jewelry; pottery; furniture; figures; drawings; photographs; prints; scrolls.
Hours & Admission Prices: Call for hours.

POPLAR GROVE HISTORIC PLANTATION, 10200 U.S. Hwy. 17 N., Wilmington, NC 28411-6854. Tel.: 910-686-4868, ext. 26. Fax: 910-686-4309.
E-mail: pgp@poplargrove.com
Web Site: www.poplargrove.com
Founded: 1980.
Congressional District: 7
Key Personnel: Pres. (V), Chris Wilcox; Dir., Nancy Simon; Museum Shop Mgr., Nancy Kroeger; Volunteer Coord., Jeanne Walker.
Personnel Profile: Full-Time Paid 9; Part-Time Paid 6; Part-Time Volunteers 100; Interns 1.
Governing Authority: nonprofit organization. Tax-exempt.
Institution Type/Description: Historic House & Site: housed in 1850 Greek Revival Plantation Manor House.
Collections: 19th-century furnishings; agricultural implements. Historic Buildings: 1850 manor house; kitchen; smoke house; 1875 tenant house.
Research Fields: agriculture; archaeology; Black history; genealogy; 19th-century trades & crafts.
Facilities: cultural arts center; picnic area; restaurant. Gift items for sale.
Activities: guided tours; organized educational programs; classes; craft workshops. Special Events: Summer Fair; Antique Fair in May; Halloween Festival; Christmas Open House.
Publications: newsletter.
Hours & Admission Prices: Feb.-Dec. Mon.-Sat. 9-5, Sun. 12-5. Adults $10, senior citizens $9, students 6-16 $5; discount to military, groups of 15 or more & AAA members; members & children under 5 no charge. Closed Easter; Thanksgiving; Christmas.
Attendance: 40,000 (estimated)
Membership: Patron $25; Dozier Society $50; Goober Society $100; Mumford Society $500; Foy Society $500 and up.

WILMINGTON RAILROAD MUSEUM, (M), 505 Nutt St., Ste. 6, Wilmington, NC 28401-3316. Tel.: 910-763-2634.
E-mail: wrrmnc@bellsouth.net
Web Site: www.wrrm.org
Founded: 1979.
Key Personnel: Exec. Dir., Mark W. Koenig; Pres. (V), William Bryden.
Personnel Profile: Full-Time Paid 1; Part-Time Paid 5; Part-Time Volunteers 40.
Governing Authority: not-for-profit organization. Parent Institution: Wilmington Railroad Museum Foundation Inc. Tax-exempt: 501(c)(3).
Institution Type/Description: Railroad Museum: housed in 1883 freight warehouse building.
Collections: late 19th-century to middle 20th-century railroad artifacts; photographs; tools; manuals; textiles; china; silver; railroad memorabilia; rolling stock. Historic Buildings: c.1882 & 1883 freight warehouses.
Facilities: 1,000-vol. library of railroad related material; 6,500 sq.ft. exhibit space; archival material from Atlantic Coast Line RR.
Activities: guided tours; programs for children & adults.
Publications: quarterly newsletter, The Dispatcher.

Hours & Admission Prices: Call or visit website for confirmation of hours. Adults $8.50, military & seniors 60 & over $7.50, children 2-12 $4.50; discounts to groups, AAM & ICOM members; members & children under 2 no charge. Closed New Year's Eve & Day; Easter; Thanksgiving; Christmas Eve & Day. &
Attendance: 21,675 (accurate)
Membership: Coach $50; Club Car $75; Pullman $125; Private Car $250; RR Baron $500.

Wilson

ARTS COUNCIL OF WILSON, 124 Nash St., S.W., Wilson, NC 27893. Tel.: 252-291-4329. Fax: 252-234-0049.
Web Site: www.wilsonarts.com
Institution Type/Description: Art Gallery.
Collections: works by local, regional & national artists.
Facilities: theater. Museum-related items for sale.
Activities: temporary exhibitions; performances; special events. Annual Event: Holiday Invitational Show & Sale.
Publications: newsletter.
Hours & Admission Prices: Call for hours.

BARTON ART GALLERIES, Whitehead & Gold St., Wilson, NC 27893. Mailing Address: Art Dept., Barton College, P.O. Box 5000, 704A College St., Wilson, NC 27893-7000. Tel.: 252-399-6477 & 6300.
E-mail: artgalleries@barton.edu
Web Site: www.barton.edu
Founded: 1965.
Congressional District: 2
Key Personnel: Dir., Susan Fecho.
Personnel Profile: Full-Time Paid 1; Part-Time Paid 3; Interns 3.
Governing Authority: college. Parent Institution: Barton College. Tax-exempt.
Institution Type/Description: Art Gallery.
Collections: various art media & other works donated to the college or purchased by the museum.
Major Exhibits: Textile International, 3/16/14-4/12/14.
Facilities: 4,000-vol. library pertaining to art history available for research & for inter-library loan.
Activities: lectures; gallery talks; arts festivals; formally organized education programs for undergraduate college students; temporary exhibitions.
Publications: gallery programs; exhibition catalogues.
Hours & Admission Prices: mid-Aug. to mid-May Mon.-Fri. 10-4. No charge. Closed New Year's Day; Martin Luther King; Good Friday; Thanksgiving; Christmas; fall & spring breaks. &
Attendance: 800 (estimated)
Membership: Student, Faculty, & Staff $10; Individual $25; Family $50; Hirsham $100; Guggenheim $250; Metropolitan $500; Louvre $1,000 & up.

IMAGINATION STATION SCIENCE MUSEUM, 224 E. Nash St., Wilson, NC 27893. Mailing Address: P.O. Box 2127, Wilson, NC 27894-2127. Tel.: 252-291-5113. Fax: 252-291-2968.
E-mail: mail@imaginescience.org
Web Site: www.imaginescience.org
Founded: 1989.
Congressional District: 1
Key Personnel: C.E.O., Nancy Van Dolsen; Pres., Richard Smith.
Personnel Profile: Full-Time Paid 3; Part-Time Paid 7; Interns 1.
Governing Authority: private; nonprofit organization. Tax-exempt: 501(c)(3)
Institution Type/Description: Science & History Museum.
Collections: local history; interactive science exhibits; reptiles; bees.
Major Exhibits: Sharkabet, 1/14-5/14.
Facilities: 14,000 sq. ft. exhibit space; 100-seat auditorium; education facilities; planetarium. Museum-related items for sale.
Activities: mobile vans; school loan service; traveling exhibition; training for teachers.
Publications: quarterly newsletter, Lab Notes.
Hours & Admission Prices: Mon.-Sat. 9-5. Adults $5, seniors & students $4; discounts to AAM & ASTC members; children under 4 & members no charge. Closed Thanksgiving; Christmas. &
Attendance: 20,000 (accurate)
Membership: Family $60.

NORTH CAROLINA BASEBALL MUSEUM, Fleming Stadium, 300 Stadium St., Wilson, NC 27893. Tel.: 252-296-3048. Facebook: North Carolina Baseball Museum.
Web Site: ncbaseballmuseum.com
Founded: 2004.

Volunteer Hours: 1,500
Governing Authority: Tax-exempt.
Institution Type/Description: Baseball Museum.
Collections: baseball history; players' memorabilia; personal artifacts; bats; balls; uniforms; gloves; photographs; early stadium locker room stalls; baseball cards; magazine & newspaper articles; Walk of Fame.
Hours & Admission Prices: During home games & by appointment. Adults $3, children 17 & under and seniors 65 & over $1; discounts to groups.
Attendance: 8,000 (estimated)

OLIVER NESTUS FREEMAN ROUND HOUSE MUSEUM, 1202 Nash St., Wilson, NC 27893. Tel.: 252-296-3056.
Founded: 2001.
Congressional District: 1
Key Personnel: Dir., Bill Myers
Institution Type/Description: History Museum: housed in the former home of Freeman; built in 1946.
Collections: African American history & culture; period furnishings & artifacts.
Major Exhibits: George Washington Carver, 10/13-1/15.
Hours & Admission Prices: Tues.-Sat. 9-4. No charge; donations accepted. &
Attendance: 3,000 (estimated)

WILSON BOTANICAL GARDENS, 1806 S.W. Goldsboro St., Wilson, NC 27893. Tel.: 252-237-0113. Fax: 252-237-0114. Facebook: Wilson Botanical Gardens.
E-mail: info@wilsonbotanicalgardens
Web Site: www.wilsonbotanicalgardens.com
Founded: 2005.
Key Personnel: Pres. (V), Dr. Roger Thurman
Institution Type/Description: Botanical Garden.
Collections: trees; turf grass; perennial beds; native plants; bird & butterfly gardens.
Facilities: picnic area.
Activities: educational programs.
Hours & Admission Prices: Daily dawn to dusk. No charge; donations accepted. &
Membership: Individual $25; Family $40; Sponsor $100-$500; Patron $500-$1,000; Benefactor $1,000-$9,999; Angel $10,000 & up.

WILSON ROSE GARDEN, 1800 Herring Ave., Wilson, NC 27893-6727. Tel.: 252-399-2261. Fax: 252-399-2196.
E-mail: hbass@wilsonnc.org
Institution Type/Description: Gardens.
Collections: over 1,200 rose plants consisting of 180 different varieties including grandifloras, old garden, English, hybrid teas, & miniature.
Hours & Admission Prices: Dawn to dusk. No charge.

Windsor

HISTORIC HOPE FOUNDATION, INC., 132 Hope House Rd., Windsor, NC 27983-7458. Tel.: 252-794-3140. Fax: 252-794-5583.
E-mail: hopeplantation@coastalnet.com
Web Site: www.hopeplantation.org
Founded: 1965.
Congressional District: 1
Key Personnel: Pres. (V), Dr. John L. Hill; Administrative Asst., Belinda Winborne; Cur., Gregory Tyler.
Personnel Profile: Full-Time Paid 1; Part-Time Paid 10; Part-Time Volunteers 50; Interns 1.
Governing Authority: nonprofit organization. Tax-exempt: 501(c)(3).
Institution Type/Description: Historic House Museum: located on the Hope Plantation.
Collections: period furniture; decorative arts; fine art. Historic Houses: 1763 King-Bazemore House; 1800 Samuel Cox House; 1803 Hope Mansion; Roanoke-Chowan Heritage Center: history; art; African American; Native American.
Research Fields: gardening; African American heritage.
Facilities: research library; nature trails.
Activities: guided tours; lectures; films; decorative arts symposium; permanent exhibitions.
Publications: semi-annual newsletter.
Hours & Admission Prices: April-Dec. 20 Mon.-Sat. 10-4, Sun. 2-5; other times by appointment. Adults $8, seniors $7, children & students $3; discount to AAA, AAM & ICOM members; members & Bertie County students no charge. Closed Thanksgiving; Christmas.
Attendance: 17,500 (estimated)
Membership: Individual $30; Family $50; Donor $75; Contributing $100; Corporate/Business $125; Patron $500; Benefactor $1,000.

ROANOKE/CASHIE RIVER CENTER, 112 W. Water St., Windsor, NC 27983. Tel.: 252-794-2001. Fax: 252-794-5202.
E-mail: roanoke_cashierc@embarqmail.com
Founded: 2000.
Institution Type/Description: History Museum.
Collections: natural & cultural history; neotropical migratory songbirds; local bottomland habitats; native wild turkeys; period artifacts; photographs.
Facilities: Museum-related items for sale.
Activities: rental facilities.
Hours & Admission Prices: Wed.-Fri. 10-4, Sat. 10-2.

Wingate

JESSE HELMS CENTER, 3910 U.S. 74 E., Wingate, NC 28174. Mailing Address: P.O. Box 247, Wingate, NC 28174-0247. Tel.: 704-233-1776. Fax: 704-233-1787.
Web Site: jessehelmscenter.org
Founded: 2001.
Key Personnel: Facility Coord., Ladonna Snodgrass
Institution Type/Description: History Museum.
Collections: hands-on exhibitions; Helms' personal correspondence; photographs; film; United Nations model; office replica of Helm's Washington, D.C. Senate office.
Activities: hands-on exhibitions; self-guided tours.
Hours & Admission Prices: Mon.-Fri. 9-5. No charge.

Winnabow

BRUNSWICK TOWN/FORT ANDERSON STATE HISTORIC SITE, 8884 St. Philips Rd., S.E., Winnabow, NC 28479-5035. Tel.: 910-371-6613. Fax: 910-383-3806.
E-mail: brunswick@ncdcr.gov
Web Site: www.nchistoricsites.org/brunswic/brunswic.htm
Founded: 1958.
Congressional District: 7
Key Personnel: Regl. Supvr. East Region, James A. Bartley.
Personnel Profile: Full-Time Paid 5; Part-Time Paid 1; Part-Time Volunteers 8.
Governing Authority: state. Parent Institution: North Carolina Division of Archives & History. Subsidiary Institution: North Carolina Dept. of Cultural Resources, 109 E. Jones St., Raleigh, NC 27611. Tax-exempt: 170(b)(1)(A).
Institution Type/Description: Historic Site: 1726-1776 excavated foundations of port town; earthen Confederate Fort Anderson.
Collections: 18th-century English & Civil War artifacts; colonial.
Research Fields: local history.
Facilities: visitor center.
Activities: guided tours; lectures; school programs.
Publications: brochure.
Hours & Admission Prices: Tues.-Sat. 9-5. No charge; donations accepted. Closed Thanksgiving; Christmas Eve & Day.
Attendance: 54,680 (accurate)

Winston-Salem

CHARLOTTE AND PHILIP HANES ART GALLERY, WAKE FOREST UNIVERSITY, Art Dept., 1834 Wake Forest Rd., Winston-Salem, NC 27106. Mailing Address: P.O. Box 7232, Winston-Salem, NC 27109. Tel.: 336-758-5795 & 5585. Fax: 336-758-6014.
E-mail: faccinto@wfu.edu
Web Site: www.wfu.edu/Academic-departments/Art/gall_index.html
Formerly: Wake Forest University Fine Arts Gallery
Founded: 1976.
Key Personnel: Dir., Victor Faccinto; Asst. Dir., Paul Bright.
Personnel Profile: Full-Time Paid 2; Part-Time Paid 6.
Governing Authority: university; nonprofit. Tax-exempt.
Institution Type/Description: Art Museum.
Collections: paintings; drawings; sculpture.
Facilities: educational facilities.
Activities: lectures; loan & traveling exhibitions; curating contemporary & historical exhibitions in various media.
Publications: catalogues for exhibitions.
Hours & Admission Prices: Sept.-May Mon.-Fri. 10-5, Sat.-Sun. 1-5. No charge. Closed university holidays.
Attendance: 6,500

CHILDREN'S MUSEUM OF WINSTON-SALEM, 390 S. Liberty St., Winston-Salem, NC 27101-5260. Tel.: 336-723-9111. Fax: 336-723-9469.
E-mail: info@childrensmuseumofws.org
Web Site: www.childrensmuseumofws.org
Founded: 2004.
Key Personnel: Exec. Dir., Elizabeth Dampier; Dir. Mktg., Brandy Hall; Dir. Programming, Christine Simonson; Dir. Guest Svcs., Lesa Pierce.
Governing Authority: nonprofit organization. Tax-exempt: 501(c)(3).
Institution Type/Description: Children's Museum.
Collections: hands-on exhibits.
Facilities: library; meeting room. Museum-related items for sale.
Activities: special events; bean stalk climber. Museum Sponsors: Annual Gala; Storybook Luncheon in May; Storybook Soirree in October.
Hours & Admission Prices: Memorial Day to Labor Day Mon.-Sat. 10-4, Sun. 1-5; Sept.-May Tues.-Sat. 10-4, Sun. 1-5. Admission one & over $7, seniors 62 & over $6; discounts to groups of 15 or more; members, educators & children under one no charge. ACM reciprocal membership. Closed New Year's Day; Easter; Thanksgiving; Christmas.
Attendance: 70,000 (estimated)
Membership: Summer $50; Grandparents $90; Family $105; Family Plus $135.

DAVIS GALLERY - SAWTOOTH SCHOOL FOR VISUAL ART, 251 N. Spruce Street, Winston-Salem, NC 27101. Tel.: 336-723-7395. Fax: 336-773-0132. Facebook: Sawtooth School for Visual Art.
E-mail: info@sawtooth.org
Web Site: www.sawtooth.org
Founded: 1945.
Congressional District: 12
Key Personnel: Exec. Dir., JoAnne Vernon
Institution Type/Description: Art Gallery.
Collections: works by students, faculty, regional & national artists.
Major Exhibits: Fiberworks, 2/14-3/14; Word as Art, 4/14-5/14; Toward Spirituality, 6/14.
Facilities: Gallery-related items for sale.
Activities: classes; special events.
Hours & Admission Prices: Mon.-Fri. 9-5. No charge.
Membership: Student and Senior 65 & over $35; Individual $50; Family $125.

DELTA ARTS CENTER, 2611 New Walkertown Rd., Winston-Salem, NC 27101-1948. Tel.: 336-722-2625.
E-mail: delta2611@bellsouth.net
Web Site: www.deltafinearts.org/
Key Personnel: Interim Exec. Dir., Daphne Holmes-Johnson
Institution Type/Description: Arts Center.
Collections: paintings & sculpture by African-American artists from North Carolina.
Hours & Admission Prices: Tues.-Fri. 10-5, Sat. 11-3; groups by appointment. No charge. Closed New Year's Eve, Day & day after; Thanksgiving & day after; Christmas Eve, Day & week.

DIGGS GALLERY AT WINSTON-SALEM STATE UNIVERSITY, 601 Martin Luther King Jr. Dr., Winston-Salem, NC 27110-0003. Tel.: 336-750-2458. Fax: 336-750-2463.
E-mail: diggsinfo@wssu.edu
Web Site: www.wssu.edu
Founded: 1990.
Congressional District: 12
Key Personnel: Dir., Cur. & Devel., Belinda Tate; Cur. Education, Dara Silver; Office Asst., Monica Scott.
Personnel Profile: Full-Time Paid 2; Part-Time Paid 1; Part-Time Volunteers 8; Interns 2.
Governing Authority: public university; nonprofit. Parent Institution: Winston-Salem State University. Tax-exempt: 501(c)(3).
Institution Type/Description: University Art Gallery.
Collections: African-American art; sculptures by Mel Edwards, Tyrone Mitchell, Beverly Buchanan & Dennis Peacock; John Biggers murals; paintings, prints & sculptures by Romare Bearden, Stephanie Pogue, Samuel Brown, Selma Burke & William Artis; emphasis on North Carolina & southeastern African-American artists.
Research Fields: African art; African-American art (historical & contemporary); art of the African diaspora; memory jugs: African-American grave markers.
Facilities: 100-vol. library of African & African-American related books.
Activities: arts festivals; concerts; films; formal education programs for adults, children & college students at Winston-Salem State University; guided

tours; lectures; loan, traveling & participatory exhibits; temporary exhibits of our own collection; training programs for professional museum workers.
Publications: exhibition catalogs.
Hours & Admission Prices: Gallery: Tues.-Sat. 11-5. Office: Mon.-Fri. 8-5. No charge. Closed New Year's Eve & Day; Martin Luther King Jr. Day; Good Friday; Memorial Day; Independence Day; Labor Day; Veterans Day; Thanksgiving; Christmas Eve, Day & week. &
Attendance: 14,700 (accurate)
Membership: Annual $25.

HISTORIC BETHABARA PARK, 2147 Bethabara Rd., Winston-Salem, NC 27106-2701. Tel.: 336-924-8191. Fax: 336-924-0535.
E-mail: ekutcher@triadbiz.rr.com
Web Site: www.bethabarapark.org
Founded: 1966.
Congressional District: 5
Key Personnel: Dir., Ellen M. Kutcher; Chm. (V), Olan Beam, Jr.; Museum Shop Mgr., Diana B. Overbey.
Personnel Profile: Full-Time Paid 4; Part-Time Paid 21; Part-Time Volunteers 300; Interns 4.
Governing Authority: bd. of trustees. Parent Institution: City of Winston-Salem Recreation & Parks Dept. Subsidiary Institution: Historic Bethabara Park, Inc. Tax-exempt: 501(c)(3).
Institution Type/Description: Historic Site & Wilderness Preserve: 1753 site of the first Moravian Settlement in North Carolina.
Collections: Moravian pottery, furniture. Historic Buildings: 1782 Potter's House; 1788 Gemeinhaus (church); 1803 Herman Buttner Distiller's House; 1816 log house; 1756-1763 Reconstructed Palisade Fort; stabilized archaeological foundations of original buildings in 1753 community; reconstructed 1759 community garden; 1761 medical garden; reconstructed 1754 colonial village.
Research Fields: Moravian history; archaeology.
Facilities: wildlife preserve; restored colonial gardens; theater; 15 km. of nature trails; picnic area; visitor center.
Activities: guided tours; lectures; special events; audiovisual presentation; volunteers wildlife preserve trails.
Publications: brochures; Historic Bethabara Park Field Guide.
Hours & Admission Prices: Visitor Center & Buildings: April to mid-Dec. Tues.-Fri. 10:30-4:30, Sat.-Sun. 1:30-4:30. Grounds: daily. Adults $4, children $1. &
Attendance: 130,829 (estimated)

HISTORIC OAK GROVE SCHOOL MUSEUM, 2637 Oak Grove Circle, Winston-Salem, NC 27106. Tel.: 336-722-5138, ext. 225.
Institution Type/Description: Historic Building Museum: housed in a former school built in the early 1900s for African American students. Listed on the National Register of Historic Places.
Collections: school history; period furnishings & fixtures; school-related artifacts.
Hours & Admission Prices: By appointment.

MUSEUM OF ANTHROPOLOGY, (M), Wake Forest University, Wingate Rd., Winston-Salem, NC 27106. Mailing Address: P.O. Box 7267, Winston-Salem, NC 27109-7267. Tel.: 336-758-5282. Fax: 336-758-5116. Facebook: WFUMOA.
E-mail: moa@wfu.edu
Web Site: moa.wfu.edu
Formerly: Museum of Man
Founded: 1963.
Congressional District: 5
Key Personnel: Dir., Dr. Stephen Whittington; Pres. (V), Evelyn Pursley; Educator, Tina Smith; Registrar & Collections Mgr., Kyle Elizabeth Bryner; Public Rels., Mktg. & Membership Coord., Sara Cromwell.
Personnel Profile: Full-Time Paid 4; Part-Time Paid 4; Part-Time Volunteers 18; Interns 4.
Volunteer Hours: 264
Operating Expenses: 308,470
Operating Income: 377,603
Governing Authority: university. Parent Institution: Wake Forest University. Tax-exempt.
Institution Type/Description: Anthropology Museum.
Collections: anthropological and archaeological collections from the Americas, Africa, Asia & the Pacific.
Research Fields: archaeology; ethnohistory; Native American art; cultural ecology.
Facilities: library.
Activities: formally organized education programs for kindergarten-12th grade; museum studies courses & internships for undergraduate students; perma-

nent & temporary exhibitions, lecture series; college classes; tours & outreach programs; summer camps; family days.
Publications: MOA News.
Hours & Admission Prices: Tues.-Sat. 10-4:30; groups by appointment only. No charge; donations accepted. &
Attendance: 6,571 (accurate)
Membership: Educator $25; Individual $30; Family $50; Supporting $75; Patron $100; Sustainer $250; Benefactor $500; Banks Founder's Circle $1,000 & up.

* **MUSEUM OF EARLY SOUTHERN DECORATIVE ARTS (MESDA),** 924 S. Main St., Winston-Salem, NC 27101-5335. Tel.: 336-721-7360. Fax: 336-721-7367.
E-mail: research@oldsalem.org
Web Site: www.mesda.org
Founded: 1965.
Congressional District: 5
Key Personnel: C.E.O. & Pres., Lee French; C.F.O., Eric Hoyle; Vice Pres. & Chief Cur., Robert Leath; Dir. Research, June Lucas; Dir. Education & Special Programs, Sally Gant; Assoc. Cur., Daniel Ackermann; Dir. Devel., Frances Beasley; Dir. Collections & Cur., Johanna M. Brown; Collections Mgr., Abigail Linville; Office Admin., Martha Ashley; Photographer, Wes Stewart; Librarian, Michele Doyle; Vice Pres. Publications, Gary Albert.
Personnel Profile: Full-Time Paid 14; Part-Time Paid 40.
Governing Authority: nonprofit organization. Parent Institution: Old Salem, Inc., 600 S. Main St., Winston-Salem, NC 27101. Tax-exempt: 501(c)(3).
Institution Type/Description: Decorative Arts Museum.
Collections: architecture; furniture; paintings; ceramics; textiles; prints; metalwares used or made in the South through 1820.
Research Fields: southern decorative arts through 1820 in Maryland, Virginia, Kentucky, Tennessee, the Carolinas, & Georgia.
Facilities: library; research center; auditorium. Museum-related items for sale.
Activities: guided tours; lectures; formally organized programs for adults & undergraduate college students; annual graduate Summer Institute in coordination with the University of Virginia; temporary exhibitions.
Publications: magazine, Old Salem Museum & Gardens.
Hours & Admission Prices: Tues.-Sat. 9:30-4:30, Sun. 1-5. Adults $21, children 6-16 $10; discounts to AAM & ICOM members. Closed Easter; Thanksgiving; Christmas Eve & Day. &
Attendance: 19,176 (accurate)
Membership: Annual $50 & up.

* **OLD SALEM MUSEUMS & GARDENS, (M),** 900 Old Salem Rd., Winston-Salem, NC 27101-5329. Mailing Address: 600 S. Main St., Winston-Salem, NC 27101-5329. Tel.: 336-721-7300; 888-653-7253. Fax: 336-721-7335.
E-mail: info@oldsalem.org
Web Site: www.oldsalem.org
Formerly: Historic Town of Salem
Founded: 1950.
Congressional District: 5
Key Personnel: Pres., Ragan Folan; Chm. (V), Judy Lambeth; C.F.O., Eric Hoyle; Vice Pres., Horton Center, Paula Locklair; Vice Pres. & Dir. Restoration, John Larson; Vice Pres. Devel., Frances Beasley; Chief Cur. & Vice Pres. Collections & Research, Robert Leath; Cur. Collections, Johanna Brown; Coord. Crafts, Nat Norwood; Museum Shop Mgr., Ann Johnson.
Personnel Profile: Full-Time Paid 100; Part-Time Paid 100; Part-Time Volunteers 40; Interns 2.
Governing Authority: nonprofit organization. Branch Museum: The Museum of Early Southern Decorative Arts. Tax-exempt: 501(c)(3).
Institution Type/Description: Historic Restoration Village: 1766 Moravian Congregation Town.
Collections: 18th, early 19th-century Moravian artifacts; authentic furnishings; tools for craft shops in Single Brothers House; history; gardens; music; decorative arts. Historic Buildings: 1769-1786 Single Brothers House; 1771 Miksch Tobacco Shop; 1784 Salem Tavern; 1794 Boys School; 1800 Winkler Bakery; 1802 Vierling House; 1803 Market Fire House; 1819 John Vogler House; 1827 Shultz Shoemaker Shop; 1833 Vogler Gunsmith Shop; 1861 St. Philips Church.
Major Exhibits: Needlework of Salem, 8/13-8/14; A Large and Handsome Assortment of Earthenware, 8/13-8/14; Art in Clay (T), 9/13-9/14.
Research Fields: architecture; crafts; community life of the early Moravians & enslaved African Americans.
Facilities: 2,000-vol. library of books, articles, & periodicals; reading room; auditoriums; restaurants. Reproductions, gifts & bakery goods for sale.
Activities: guided tours; lectures; films; concerts; formally organized education programs for children, adults & undergraduate college students; permanent exhibitions.
Publications: books, Candle Lovefeast; An Adventure in Historic Preservation; The Moravian Potters in North Carolina; The Quiet People of the Land; The

Three Forks of Muddy Creek; Johann Ludwig Eberhardt & His Salem Clocks; Moravian Decorative Arts in North Carolina; Journal of Early Southern Decorative Arts.
Hours & Admission Prices: Visitor Center: Tues.-Sat. 9-5, Sun. 12:30-5. Town: Tues.-Sat. 9:30-4:30, Sun. 1-4:30. Adults $23, children $11; discounts to AAM & ICOM members. Closed Easter; Thanksgiving; Christmas Eve & Day. ♿
Attendance: 100,000 (estimated)
Membership: Student & Senior Citizen $35; New Benefactor $40; Individual $50; Family $99; Patron $250; Sustaining $500; Marshall Society $1,000 & up.

REYNOLDA GARDENS OF WAKE FOREST UNIVERSITY,
100 Reynolda Village, Winston-Salem, NC 27106-5123. Tel.: 336-758-5593. Fax: 336-758-4132.
E-mail: gardens@wfu.edu
Web Site: www.reynoldagardens.org
Founded: 1962.
Congressional District: 5
Key Personnel: Mgr., Preston Stockton; Asst. Mgr., John Kiger; Cur. Education, Camilla Wilcox.
Personnel Profile: Full-Time Paid 6; Part-Time Volunteers 30.
Governing Authority: nonprofit. Reynolda Gardens Committee of Wake; Forest University. Tax-exempt.
Institution Type/Description: Conservatory: housed in 1912 building.
Collections: tropical plants; roses; cactus; orchids.
Research Fields: botany.
Facilities: garden; nature trails; 60-seat auditorium.
Activities: lectures; concerts; formally organized educational programs; volunteer program.
Publications: biannual gardeners journal; biannual calendar; web publication, Naturalist's Notebook for K-5 educators.
Hours & Admission Prices: Greenhouses: Jan. & July Mon.-Fri. 10-5; Feb.-June & Aug.-Dec. Mon.-Sat. 10-5. Grounds: daily sunrise-sunset. No charge. ♿
Attendance: 100,000 (estimated)
Membership: Friend $30; Donor $50; Family $100; Sponsor $250; Patron $500; Benefactor $1,000.

✳ REYNOLDA HOUSE MUSEUM OF AMERICAN ART, (M),
2250 Reynolda Rd., Winston-Salem, NC 27106-5117. Mailing Address: P.O. Box 7287, Winston-Salem, NC 27109-7287. Tel.: 336-758-5150; 888-663-1149. Fax: 336-758-5704. Facebook: RHMAA.
E-mail: reynolda@reynoldahouse.org
Web Site: reynoldahouse.org
Founded: 1967.
Congressional District: 5
Key Personnel: Dir., Allison C. Perkins; Chm. Bd., Dianne Blixt; Pres. (V), Louise Bazemore; Founding Dir., Barbara B. Millhouse; Business Mgr., Kim Hampton; Dir. Devel., Stephan Dragisic; Dir. Curatorial & Education Division, Elizabeth Chew, Ph.D.; Dir. Education, Kathleen F.G. Hutton; Dir. Collections Management, Rebecca Eddins; Dir. Mktg. & Communications, Sarah R. Smith; Dir. Public Programs, Philip Archer; Museum Shop Mgr., Cindy Bird.
Personnel Profile: Full-Time Paid 25; Part-Time Paid 32; Part-Time Volunteers 183; Interns 8.
Volunteer Hours: 6,745
Operating Expenses: 3,642,145
Operating Income: 3,642,145
Governing Authority: nonprofit charitable organization. Governing Authority: 24 member Board of Directors. Affiliated with Wake Forest University. Tax-exempt: 501(c)(3).
Institution Type/Description: Art Museum: housed in 1917 home of R.J. Reynolds, founder of R.J. Reynolds Tobacco Company.
Collections: American paintings, sculpture & prints; costumes; decorative arts; archives.
Major Exhibits: American Moderns, 1910-1960: From O'Keeffe to Rockwell, 2/7/14-5/4/14; Color and Space and Light: Watercolors from Reynolda's Collection, 4/12/14-9/13/14; The Art of Seating: Two Hundred Years of Modern Design (T), 8/23/14-12/31/14; Partisans: Social Realism in American Art, 10/5/14-3/16/14; Love and Loss, 10/11/14-3/8/15.
Research Fields: American art; American Country House; Country Place Era; American history.
Facilities: library of books & pamphlets pertaining to American art available for research; botanical gardens. Museum-related items for sale.
Activities: guided tours; lectures; films; concerts; arts festivals; drama; formally organized education programs children & adults as well as under-

graduate & graduate college students; continuing education classes; January college intern program; docent program; inter-museum loan & permanent exhibitions.
Publications: catalog, American Originals: Selections from Reynolda House, Museum of American Art; Reynolda: A History of an American Country House, 1997; The Reynolda House Aeolian Organ; The Paris Gowns in the Reynolda House Collection; A World of Her Own Making Katharine Smith Reynolds & the Landscape of Reynolda; Log of Aeroplane NR-898W; Reynolda Farm; American Wilderness: The Story of The Hudson River School of Painting; Reynolda House: An American Country Home Becomes a Home for American Art.
Hours & Admission Prices: Feb.-Dec. Tues.-Sat. 9:30-4:30, Sun. 1:30-4:30; see website for additional hours. Adults $14; children, students with current ID, Wake Forest Univ. faculty & staff & members no charge. Closed New Year's Day; Thanksgiving; Christmas Eve & Day. ♿
Attendance: 37,934 (accurate)
Membership: Educator $40; Individual $60; Dual & Family $85; Patron $150; Benefactor $250; Sustainer $500; Reynolda Society $1,000 & up.

SCIWORKS, (M),
400 Hanes-Mill Rd., Winston-Salem, NC 27105-9667. Tel.: 336-767-6730. Fax: 336-661-1777.
E-mail: bssanford@sciworks.org
Web Site: www.sciworks.org
Founded: 1964.
Congressional District: 5
Key Personnel: C.E.O., Dr. Beverly Sanford; Bd. Chm., Michael Myers; Vice Pres. Finance, Sam Hancock; Vice Pres. Programs & Education, Kelli Isenhour; Vice Pres. Exhibits, Tom Wilson; Vice Pres. Facilities, Carl Nisbet; Science Shop Mgr., Bobbie Tucker.
Personnel Profile: Full-Time Paid 13; Part-Time Paid 15; Part-Time Volunteers 88; Interns 10.
Governing Authority: nonprofit organization. Tax-exempt: 501(c)(3).
Institution Type/Description: Science & Technology Museum.
Collections: hands-on exhibits relating to the physical sciences, natural sciences, health sciences; natural history includes rocks, minerals, sea shells & mounts; planetarium; 15-acre environmental park.
Facilities: 30,000 sq. ft. exhibit space; 150-seat science theatre; 120-seat planetarium; classrooms; 32 acres of grounds; 15-acre environmental park; picnic area. Science museum-related items for sale.
Activities: education science programs for walk-in visitors & organized groups of children, students & adults; outreach programs; permanent & traveling exhibits; workshops for children & teachers; volunteer program; annual special events & festivals; laser shows.
Publications: quarterly newsletter.
Hours & Admission Prices: Mon.-Fri. 10-4, Sat. 11-5. Exhibits: adults $11, students & seniors $9, children 3-5 $6; discounts for ASTC & AAM members; children under 3 & members no charge. Closed New Year's Day; Thanksgiving; Christmas. ♿
Attendance: 79,000 (accurate)
Membership: Cardholder $50; Cardholder +1 $65; Cardholder +2 $80; Cardholder +3 $95; Cardholder +4 $110; Large Family $120; supersize any level $15.

SINGLE SISTERS HOUSE MUSEUM,
601 S. Church St., Winston-Salem, NC 27101. Tel.: 336-721-2600.
Web Site: www.salem.edu
Institution Type/Description: Historic Building: established in 1772 by early Moravian settlers, dedicated to the education of women; built in 1785.
Collections: education of girls & women since 1772; photographs.
Activities: tours.
Hours & Admission Prices: Mon.-Fri. 9-4:30, Sat. 9-1. ♿

SOUTHEASTERN CENTER FOR CONTEMPORARY ART (SECCA), (M),
750 Marguerite Dr., Winston-Salem, NC 27106-5861. Tel.: 336-725-1904. Fax: 336-722-9142.
E-mail: general@secca.org
Web Site: www.secca.org
Founded: 1956.
Congressional District: 5
Key Personnel: Dir. & Mktg., Public Rels., Mark R. Leach; Bd. Advisory, Dr. Wesley Davis; Cur. Art, Steven Matijcio.
Personnel Profile: Full-Time Paid 8; Part-Time Paid 3; Interns 1.
Governing Authority: nonprofit; bd. of directors. Parent Institution: North Carolina Dept. of Cultural Resources. Subsidiary Institution: North Carolina Museum of Art. Tax-exempt: 501(c)(3).
Institution Type/Description: Contemporary Art.
Collections: .
Research Fields: contemporary art.
Facilities: library; auditorium; 32-acres of wooded grounds & sculpture sites.

Activities: lectures; panels; films; gallery talks; education programs for children & adults in arts & humanities; music, dance, drama performances; traveling & temporary exhibitions; special events. Museum Sponsors: Artists in the Community II.
Publications: gallery notes; exhibition catalogs; e-newsletter; Twitter; Facebook.
Hours & Admission Prices: Tues.-Wed. & Fri.-Sat. 10-5, Thurs. 10-8, Sun. 1-5. No charge. Closed major holidays. &
Attendance: 30,000 (estimated)
Membership: Contributing: Individual $50; Household $75; Collaborator $150. Supporting: $250-$500. Director's Circle $5,000; Sustaining $1,000-$2,500.

WINSTON CUP MUSEUM, 1355 N. Martin Luther King Jr. Dr., Winston-Salem, NC 27101. Tel.: 336-724-4557. Fax: 336-724-4558.
E-mail: wcminfo@winstoncupmuseum.com
Web Site: www.winstoncupmuseum.com
Founded: 2005.
Key Personnel: Dir., Bill Soper; C.E.O., William Spencer; Museum Shop Mgr., Kathleen Allen
Institution Type/Description: History Museum.
Collections: NASCAR history R.J. Reynolds Tobacco Company's NASCAR sponsorship; mural; race cars; trophies; driver uniforms; helmets; winner' checks; photographs racing posters.
Activities: facilities rental.
Hours & Admission Prices: Tues.-Sat. 10-5. Adults $8, children 5-12 $4; children under 5 no charge.
Attendance: 2,000 (estimated)

Winton

C.S. BROWN CULTURAL ARTS CENTER AND MUSEUM, 511 S. Main St., Winton, NC 27986. Mailing Address: P.O. Box 435, Winton, NC 27986. Tel.: 252-506-9480.
E-mail: csbrownculturalartscenter@gmail.com
Web Site: csbrownculturalartscenter.weebly.com
Founded: 1986.
Personnel Profile: Part-Time Volunteers 20.
Volunteer Hours: 1,000
Operating Expenses: 10,000
Operating Income: 10,000
Governing Authority: Tax-exempt.
Institution Type/Description: History Museum: housed in Brown Hall; built in 1926. Listed on the National Register of Historic Places.
Collections: local history & culture; period furnishings; personal artifacts; photographs.
Facilities: 10,000sq. ft. exhibition space; auditorium.
Activities: film festival; teen forum. Annual Events: Founders Day Holiday Extravaganza; Harvest Day, Veterans Day.
Hours & Admission Prices: Call for hours. No charge; donations accepted. &
Membership: Students $15; Adults $25; Family $50; Life $200.

WINTON CENTURY POST OFFICE MUSEUM, 404 N. Main St., Winton, NC 27986. Mailing Address: c/o Historic Winston, P.O. Box 627, WInston, NC 27986. Tel.: 252-358-5788.
Institution Type/Description: Historic Building Museum: housed in the former town post office building.
Collections: local history & culture; period furnishing; photographs; postal service history.
Hours & Admission Prices: Call for hours.

Wrightsville Beach

WRIGHTSVILLE BEACH MUSEUM OF HISTORY, 303 W. Salisbury St., Wrightsville Beach, NC 28480-1819. Mailing Address: P.O. Box 584, Wrightsville Beach, NC 28480-0584. Tel.: 910-256-2569. Fax: 910-256-2569.
E-mail: info@wbmuseum.com
Web Site: www.wbmuseum.com
Founded: 1995.
Key Personnel: Dir., Madeline Flagler
Personnel Profile: Full-Time Paid 1; Full-Time Volunteers 1; Part-Time Paid 20; Part-Time Volunteers 3.
Governing Authority: private; nonprofit organization. Tax-exempt.
Institution Type/Description: History Museum: housed in Myer's Cottage.
Collections: local history & culture; personal artifacts; photographs.
Activities: book, Tide & Tim: A History of Wrightsville Beach; newsletter, Beach Breeze.

Hours & Admission Prices: Tues.-Fri. 10-4, Sat. 12-5, Sun. 1-5. No charge; donations accepted.
Attendance: 5,426 (accurate)
Membership: Sand Daisy $25; Jelly Jumper $50; Beachcomber $100; Sandpiper $250; Lifeguard $500; Lumina Club $1,000.

Yanceyville

RICHMOND-MILES HISTORY MUSEUM, 15 Main St., Yanceyville, NC 27379. Mailing Address: c/o CCHA, P.O. Box 278, Yanceyville, NC 27379. Tel.: 336-694-4965.
Governing Authority: Parent Institution: Caswell County Historical Association.
Institution Type/Description: History Museum: housed in the Graves-Florance-Gatewood House; built in 1822.
Collections: local history & culture; period furnishings; personal artifacts; photographs; Maud Gatewood art; Native American tools & arrow heads; uniforms; portraits; 1840s tobacco & slave ledger; genealogical records.
Facilities: Museum-related items for sale.
Hours & Admission Prices: Wed.-Fri. 1-4; other times by appointment.

NORTH DAKOTA

(151 listings)

Abercrombie

FORT ABERCROMBIE STATE HISTORIC SITE, 935 Broadway, Abercrombie, ND 58001. Mailing Address: P.O. Box 148, Abercrombie, ND 58001-0148. Tel.: 701-553-8513 & 328-2666. Fax: 701-328-3710. Facebopok: Fort Abercrombie State Historic Site.
E-mail: histsoc@nd.us
Web Site: www.history.nd.gov/historicsites/abercrombie
Founded: 1905.
Key Personnel: Site Supvr., Thomas Casler.
Governing Authority: society. Parent Institution: State Historical Society of North Dakota. Subsidiary Institution: Friends of Fort Abercrombie. Tax-exempt: 501(c)(3).
Institution Type/Description: Historic Site: This site preserves the military post that served from 1857 to 1877 as the gateway to the Dakota frontier. It was besieged by the Sioux during the Dakota conflict of 1862.
Collections: Red River ox cart; relics from pioneer days & early history of Fort; flagpole. Historic Buildings: 2 reconstructed blockhouses; original guardhouse (restored); ghosted building sites & palisade.
Facilities: Museum-related items for sale.
Activities: summer programs: Dutch Oven demos; quilting demos; hide tanning demos; Memorial Day & Independence Day programs; living history demos Saturday & Sunday May-September; children's story time; arts & crafts for children. Museum Sponsors: Aber Days community celebration in June.
Publications: brochure.
Hours & Admission Prices: May 16-Sept. 15 daily 8-5. Adults $5, children 6-14 $2.50, student in groups $1; discounts to groups of 20 or more. &
Attendance: 10,000 (estimated)

Adams

KNUDT SALLE LOG CABIN, Rt. 1, Adams, ND 58210. Mailing Address: 323 3rd St., Minto, ND 58261. Tel.: 701-944-2792.
Founded: 1970.
Congressional District: 11
Governing Authority: society. Affiliated with The Walsh Co. Historical Society & Adams Community Club. Tax-exempt.
Institution Type/Description: Preservation Project: housed in 1884 log cabin located in city park.
Collections: cast iron kitchen range; cast iron cookware; 1880 furnishings; working plow; trip hammer.
Research Fields: 1880s household furnishings & utensils.
Activities: permanent exhibitions.
Publications: brochure, History of Cabin.
Hours & Admission Prices: Memorial Day-Labor Day Sun. 2-5; other times by appointment. No charge; donations accepted. &
Attendance: 60 (estimated)

Alexander

LEWIS AND CLARK TRAIL MUSEUM, US Hwy. 85, Alexander, ND 58831. Mailing Address: P.O. Box 343, Alexander, ND 58831-0343. Tel.: 701-828-3157.
Institution Type/Description: History Museum: housed in 1914 school building.
Collections: scale model of Fort Mandan; historical artifacts on North Dakota's homestead days; photographs.
Hours & Admission Prices: Call for hours. No charge; donations accepted.

Almont

ALMONT HERITAGE PARK AND MUSEUM, Main St., Almont, ND 58520. Tel.: 701-843-7927.
Key Personnel: Chm. (V) & Pres. (V), Tracy Larson; Co-Chm. (V), Nancy Doll.
Governing Authority: Parent Institution: Almont Historical Society.
Institution Type/Description: History Museum.
Collections: local history; photographs; historic buildings.
Hours & Admission Prices: Memorial Day to Labor Day. No charge.

Ashley

MCINTOSH COUNTY HISTORICAL SOCIETY, 615 Center Ave. N., Ashley, ND 58413-7011. Tel.: 701-288-3374.
Founded: 1977.
Key Personnel: Pres. (V), Ronald J. Meidinger.
Personnel Profile: Part-Time Volunteers 15.
Governing Authority: county.
Institution Type/Description: Historical Museum.
Collections: Lutheran Church; sod house; outdoor baking oven; schoolhouse; domestic equipment; furniture; clothing; farm machinery; blacksmith equipment; 500-line railroad depot; 500-line railroad caboose; 500-line railroad baggage & freight wagons.
Hours & Admission Prices: June-Sept. Sun. 2-4; other times by appointment. No charge; donations accepted.
Attendance: 250 (estimated)
Membership: Single $10; Family $15; Business $35; Contributing $50; Patron $100; Life $1,000.

Beach

GOLDEN VALLEY COUNTY MUSEUM, 185 1st Ave., S.E., Beach, ND 58621. Mailing Address: P.O. Box 384, Beach, ND 58621-0384. Tel.: 701-872-3938.
Founded: 1970.
Congressional District: 39
Key Personnel: Pres. (V), Judy M. Ridenhower.
Personnel Profile: Part-Time Paid 1.
Governing Authority: county. Parent Institution: Golden Valley County Historical Society. Tax-exempt.
Institution Type/Description: History Museum.
Collections: area history; personal artifacts; period tractors & farm equipment; fossils; Indian arrowheads; wagons. Historic Building: 1909 schoolhouse.
Hours & Admission Prices: May 30 to Labor Day Tues.-Fri. 1-4; other times by appointment. No charge; donations accepted.
Attendance: 200
Membership: 3 Years $15; 5 Years $20; Life $100.

Berthhold

UPPER SOURIS NATIONAL WILDLIFE REFUGE, 17705 212th Ave., N.W., Berthhold, ND 58718-9666. Tel.: 701-468-5467. Facebook: Upper Souris National Wildlife Refuge.
E-mail: uppersouris@fws.gov
Web Site: www.fws.gov/uppersouris
Founded: 1935.
Institution Type/Description: Wildlife Refuge.
Collections: birds; bird feathers; nests; waterfowl; nesting cormorants & great blue herons; white-tailed deer; animals hides; antlers; skulls & bones.
Hours & Admission Prices: Refuge: daily 5am-10pm. Visitor Center: Mon.-Fri. 8-4:30. No charge.

Beulah

HELMUTH PFENNIG WILDLIFE MUSEUM, 6148 3rd St., N.W., Beulah, ND 58523-9488. Tel.: 701-873-4889.
Institution Type/Description: Natural History Museum.

Collections: over 175 animal specimens from around the world.
Hours & Admission Prices: Call for hours.

MERCER COUNTY MUSEUM, 108 Seventh St., N.E., Beulah, ND 58523. Mailing Address: P.O. Box 1134, Beulah, ND 58523-1134. Tel.: 701-873-5070.
Founded: 1979.
Key Personnel: Pres. (V), Ron Kessler.
Personnel Profile: Part-Time Volunteers 7.
Governing Authority: Parent Institution: State Historical Society of ND. Tax-exempt.
Institution Type/Description: General Museum.
Collections: local history & culture; frontier life; personal artifacts; period furnishings; military artifacts; cars; tractors.
Publications: quarterly newsletter, Mercer County Chronicles.
Hours & Admission Prices: Memorial Day to Labor Day Sun. 1-4; other times by appointment. No charge; donations accepted.
Attendance: 500 (estimated)
Membership: Individual $5; Business $25; Life $100; Business Life $500.

Bismarck

BISMARCK ART & GALLERIES ASSOCIATION, 422 E. Front Ave., Bismarck, ND 58504-5641. Tel.: 701-223-5986. Fax: 701-223-8960.
E-mail: baga@midconetwork.com
Web Site: www.bismarck-art.org
Key Personnel: Exec. Dir., Linda Christman; Program Dir., Sherry Niesar
Institution Type/Description: Art Gallery.
Collections: local history & culture; art exhibits.
Hours & Admission Prices: Tues.-Fri. 10-5, Sat. 1-3.
Membership: Artist & Senior $30; Individual $40; Household $50; Contributor & Business $100-$499; Donor $500-$999; Supporter $1,000-$4,499; Sustainer $2,500-$4,999; Benefactor $5,000-$9,999; Patron $10,000 & up.

BUCKSTOP JUNCTION, E. Bismarck Expwy., Bismarck, ND 58501. Mailing Address: P.O. Box 941, Bismarck, ND 58502-0941. Tel.: 701-250-8575.
E-mail: marlette@bis.midco.net
Web Site: www.BuckstopJunction.org
Founded: 1992.
Key Personnel: Pres. (V), Darrel Pittman; Museum Shop Mgr., Marlette Pittman.
Personnel Profile: Part-Time Volunteers 40.
Governing Authority: Parent Institution: Missouri Valley Historical Society. Tax-exempt.
Institution Type/Description: Historic Village.
Collections: town history; period furnishings; historic buildings.
Activities: rental facilities. Annual Events: Corn Feed & Old Settlers Day in August; Annual Bismarck Cancer Center's Applefest in September; December Hay Ride; Vintage Shoppe Christmas Open House.
Publications: quarterly newsletter, The Sentinel.
Hours & Admission Prices: Office & Shoppe: May to mid-Sept. Fri.-Sat. 12-4. Guided Tours: June-Aug. Sat. 1pm; other times by appointment. Adults $5; members no charge.
Attendance: 8,000 (estimated)
Membership: Single $25; Family $35; Donor $50; Business Sponsorship $250-$500.

CAMP HANCOCK STATE HISTORIC SITE, 101 E. Main Ave., Bismarck, ND 58301. Tel.: 701-328-9528.
E-mail: histsoc@nd.gov
Web Site: history.nd.gov/historicsites/hancock/
Founded: 1951.
Key Personnel: Site Supvr., Johnathan Campbell.
Personnel Profile: Full-Time Paid 1.
Governing Authority: state. Parent Institution: State Historical Society of North Dakota.
Institution Type/Description: Historic Site: preserves part of military installation originally established as Camp Greeley in 1872 to protect work gangs building the Northern Pacific Railroad.
Collections: local history; railroad steam engine. Historic Buildings: 1872 log house; 1881 church.
Publications: quarterly, North Dakota History and Plains Talk.
Hours & Admission Prices: May 15-Sept. 16, Fri.-Sun. 1-5; other times by appointment. No charge; donations accepted.
Attendance: 2,000 (accurate)

THE CLELL AND RUTH GANNON GALLERY AT BISMARCK STATE COLLEGE, Library Bldg., 1500 Edwards Ave., Bismarck, ND 58501-1276. Tel.: 701-391-9840.
E-mail: andrea.fagerstrom@bsc.nodak.edu
Web Site: www.bismarckstate.edu/faculty/art/gallery
Key Personnel: Dir., Andrea Fagerstrom.
Personnel Profile: Part-Time Paid 1.
Institution Type/Description: Art Museum.
Collections: works by local, regional & national artists.
Hours & Admission Prices: Mon.-Thurs. 7am-9pm, Fri. 7-4, Sun. 4-8. No charge.

DAKOTA ZOO, 602 Riverside Park Rd., Bismarck, ND 58502. Mailing Address: P.O. Box 711, Bismarck, ND 58502-0711. Tel.: 701-223-7543. Fax: 701-258-8350. Facebook: Dakota Zoo.
E-mail: director@dakotazoo.org
Web Site: www.dakotazoo.org
Founded: 1961.
Key Personnel: Dir., Terry Lincoln; Pres. (V), Kristy Entzi; Museum Shop Mgr., Jace Schacher.
Personnel Profile: Full-Time Paid 9; Part-Time Paid 16; Part-Time Volunteers 75.
Volunteer Hours: 5,600
Operating Expenses: 1,069,749
Operating Income: 1,201,863
Governing Authority: private. Parent Institution: Dakota Zoological Society. Tax-exempt.
Institution Type/Description: Zoo.
Collections: mammals; reptiles; birds; insects; amphibians, fish.
Research Fields: mountain lions; wild horses & elk.
Facilities: meeting room; discovery center.
Activities: adopted animal program; volunteer docent program; lectures; school visits; slide presentations.
Publications: Zoo guides.
Hours & Admission Prices: May-Sept. daily 10-7. Adults $7, children 12 & under $4; discounts to AZA members; members no charge. &
Attendance: 147,600 (accurate)
Membership: Individual $40; Family & Grandparent $60; Family Plus $90.

THE ELSE FORDE GALLERY AT BISMARCK STATE COLLEGE, Schafer Hall, 1500 Edwards Ave., Bismarck, ND 58501-1276. Tel.: 701-224-5601.
E-mail: barbara.jirges@bsc.nodak.edu
Web Site: www.ndga.org/galleries/bscg.html
Key Personnel: Dir., Barbara Jirges
Institution Type/Description: Art Gallery.
Collections: student artwork; works by local, regional & national artists.
Hours & Admission Prices: Mon.-Thurs. 7am-9pm, Fri. 7-4, Sun. 6pm-9pm.

FORMER GOVERNORS' MANSION STATE HISTORIC SITE, 320 E. Ave. B, Bismarck, ND 58501-3676. Mailing Address: State Historical Society of North Dakota, North Dakota Heritage Center, 612 E. Boulevard Ave., Bismarck, ND 58505. Tel.: 701-328-9528.
E-mail: histsoc@nd.gov
Web Site: histroy.nd.gov/historicsites/fgm/
Founded: 1895.
Key Personnel: Site Supvr., Johnathan Campbell, Jr.
Personnel Profile: Full-Time Paid 2; Part-Time Paid 1; Part-Time Volunteers 20.
Governing Authority: state. Parent Institution: State Historical Society. Subsidiary Institution: Society for the Preservation of the Former Governors' Mansion. Tax-exempt: 501(c)(3).
Institution Type/Description: Historic Site: housed in the restored Victorian house which served as residence for 21 governors of North Dakota from 1893 to 1960; built in 1884
Collections: original furnishings; gubernatorial memorabilia. Historic Buildings: 1884 residence & carriage house.
Activities: summer programs. Museum Sponsors: Annual Lawn Party every August.
Publications: brochures.
Hours & Admission Prices: Mid-May to mid-Sept. Mon.-Fri. 10-5, Sat.-Sun. 12-4; Oct.-May 2nd Fri. & Sat. of the month 1-5; other times by appointment. No charge; donations accepted.
Attendance: 3,500
Membership: State Historical Society of North Dakota Foundation: Individual $35; Family $45; Sustaining $100; Patron $250; Coporate $500; Founder $1,000; Director $5,000; Benefactor $10,000.

GARY'S GALLERY, 305 E. Broadway, Bismarck, ND 58501-4007. Tel.: 701-258-0060.
E-mail: info@garypmillerart.com
Web Site: www.garymillerart.com
Founded: 1974.
Institution Type/Description: Art Museum.
Collections: paintings; illustrations; prints.
Hours & Admission Prices: Call for hours.

GATEWAY TO SCIENCE, 1810 Schafer St., Ste. 1, Bismarck, ND 58501-1218. Tel.: 701-258-1975. Fax: 701-222-7515.
E-mail: gscience@gscience.org
Web Site: www.gatewaytoscience.org
Founded: 1994.
Congressional District: 1
Key Personnel: Exec. Dir., Elisabeth Demke; Pres. (V), Randy Binegar.
Personnel Profile: Part-Time Paid 9; Part-Time Volunteers 100.
Governing Authority: private; nonprofit organization. Tax-exempt: 501(c)(3).
Institution Type/Description: Science Museum.
Collections: hands-on science exhibits.
Facilities: 2,200 sq. ft. exhibit space.
Activities: formal education programs for children; workshops; guided tours; traveling & participatory exhibits; school loan service; summer science camps. Annual Events: Einstein on Wine in fall; Environmental Festival in spring; What's the Weather in June; Family Rocket Day in July; Bubble Bliss in August; Spooktacular Science in October.
Publications: newsletter, Gateway to Science News.
Hours & Admission Prices: Mon.-Thurs. 12-7, Fri.-Sat. 10-5. Adults $6, students 4-17 $3; discount to ASTC members; members, Gateway to Science members & children under 4 no charge. Closed New Year's Eve & Day; Easter; Independence Day; Thanksgiving; Christmas Eve & Day. &
Attendance: 15,900 (accurate)
Membership: Student & Senior Citizen (62 & Up) $20; Individual $35; Grandparents $60; Family $60; Family Plus $80; Sustaining $150.

NORTH DAKOTA HERITAGE CENTER, (M), 612 East Blvd. Ave., Bismarck, ND 58505-0660. Tel.: 701-328-2666. Fax: 701-328-3710.
Web Site: www.history.nd.gov
Founded: 1895.
Congressional District: 1
Key Personnel: Dir., Merlan E. Paaverud, Jr.; Pres. (V), Gereld Gerntholz; Pres. (V), Calvin Grinnell; Museum Shop Mgr., Rhonda Brown.
Personnel Profile: Full-Time Paid 62; Full-Time Volunteers 12; Part-Time Paid 2; Part-Time Volunteers 215; Interns 8.
Governing Authority: state. Tax-exempt.
Institution Type/Description: History Museum.
Collections: local history & culture; photographs; artifacts; paleontology.
Research Fields: North Dakota history.
Facilities: state archives.
Publications: North Dakota History; Plains Talk.
Hours & Admission Prices: Exhibit Galleries: Mon.-Fri. 8-5, Sat.-Sun. 10-5. No charge; donations accepted. Closed New Year's Day; Easter; Thanksgiving; Christmas. &
Attendance: 100,000
Membership: Individual $35; Family $45; Sustaining $100; Patron $250; Corporate $500; Founder $1,000; Trustee $2,002; Director $5,000; Benefactor $10,000.

NORTH DAKOTA STATE CAPITOL, 600 E. Boulevard Ave., Bismarck, ND 58505. Tel.: 701-328-2480 & 2471. Fax: 701-328-0121.
Institution Type/Description: Historic Building: built in 1933.
Collections: U.S. history & culture; photographs; paintings; period furnishings.
Hours & Admission Prices: Tours: Memorial Day to Labor Day Mon.-Fri. 8-4, Sat. 9-4, Sun. 1-4; Sept.-May Mon.-Fri. 8-4. No charge.

✴ STATE HISTORICAL SOCIETY OF NORTH DAKOTA, (M), North Dakota Heritage Center, 612 E. Blvd., Bismarck, ND 58505. Tel.: 701-328-2666. Fax: 701-328-3710. TDD: 800-366-6888.
E-mail: histsoc@state.nd.us
Web Site: www.history.nd.gov
Founded: 1895.
Key Personnel: Dir., Merlan E. Paaverud, Jr.; Asst. Dir., David C. Skalsky; Dir. Communications, Kimberly Jondahl; Dir. Museum & Education, Chris Johnson; Dir. State Archives, Ann Jenks; Dir. Historic Preservation &

Deputy State Historic Preservation Officer, Fern Swenson; Museum Store Mgr., Rhonda Brown.
Personnel Profile: Full-Time Paid 69; Part-Time Paid 40; Part-Time Volunteers 200; Interns 6.
Governing Authority: State Historical Board. Branch Museums: Fort Abercrombie, Abercrombie, ND; Chateau de Mores, Medora, ND; Camp Hancock, Bismarck, ND; Fort Buford, Buford, ND; De Mores State Historic Site, Medora, ND; Former Governors' Mansion, Bismarck, ND; Fort Totten, near Devils Lake, ND; Gingras Trading Post, Walhalla, ND; Whitestone Hill, Kulm, ND; Missouri-Yellowstone Confluence Interpretive Center, Buford, ND; Pembina State Museum, Pembina, ND; Ronald Regan Minuteman Missile Site, Cooperstown, ND. Tax-exempt.
Institution Type/Description: History Museum.
Collections: 50,000 history, natural history & ethnology artifacts; 1.5 million archeological items; 150,000 photographic images; 1,650 historical manuscripts; 2,800 archival records series; 1,400 newspaper titles; 1,200 recorded oral histories; 115,000 books & periodicals; 10,000 maps; 3.5 million ft. of film.
Research Fields: archaeology; architectural; railroad; North Dakota history.
Facilities: 100,000-vol. library with reading room, archives, manuscripts, maps & photographs available for use on premises.
Activities: permanent & temporary exhibits; gallery demonstrations; public programming for adults & children; lectures; film workshops; docent & volunteer programs; traveling exhibition program; youth, family & school programs including suitcase exhibits; annual workshops for county & local historical societies; annual history conference; teacher workshops; outdoor adventures.
Publications: quarterly journal, North Dakota History; quarterly newsletter, Plains Talk; books, Sacred Beauty: Quillwork of Plains Women (1998); A Traveler's Companion to North Dakota's State Historic Sites (2003); Lewis and Clark in North Dakota (2003); Fort Totten (2nd Ed. 2004); No Two Horns: A Gallery of Art and Exploits (2003).
Hours & Admission Prices: Heritage Center: Mon.-Fri. 8-5, Sat.-Sun. 10-5. Research Library: Mon.-Fri. 8-4:30, 2nd Sat. every month. No charge; donations accepted. Closed New Year's Day; Easter; Thanksgiving; Christmas. &
Attendance: 100,000 (estimated)
Membership: Individual $35; Family $45; Sustaining $100; Patron $250; Corporate $500; Founder $1,000; Director $5,000; Benefactor $10,000.

UTTC CULTURAL ARTS INTERPRETIVE CENTER, United Tribes Technical College, Bismarck, ND 58504. Tel.: 701-255-3285.
Institution Type/Description: Native American Museum.
Collections: Native American history & culture; personal artifacts; photographs; art.
Hours & Admission Prices: Mon.-Fri. 8-5 by appointment.

Bowdon

BOWDON CENTENNIAL MUSEUM, 232 40th Ave., N.E., Bowdon, ND 58418. Tel.: 701-962-3736.
E-mail: lindawidicker@daktel.com
Founded: 1989.
Key Personnel: Treas., Vivian Miller; Pres. (V) & Public Rels., Rod L. Widicker; Security, Laurel Jones.
Governing Authority: private; nonprofit organization. Tax-exempt.
Institution Type/Description: History Museum.
Collections: local history & culture; period furnishings; personal artifacts; photographs.
Facilities: 5,000-vol. library.
Activities: guided tours; fundraisers.
Hours & Admission Prices: late May to late Oct. Wed.-Sun. 1-5; other times by appointment. No charge; donations accepted. &
Attendance: 350 (estimated)

Bowman

PIONEER TRAILS REGIONAL MUSEUM, 12 First Ave., N.E., Bowman, ND 58623-4010. Mailing Address: P.O. Box 78, Bowman, ND 58623-0078. Tel.: 701-523-3600. Fax: 701-523-3600.
E-mail: ptrm@ptrm.org
Web Site: ptrm.org
Founded: 1992.
Key Personnel: Pres. (V), Dean Pearson; Dir., Megan Miller.
Personnel Profile: Full-Time Paid 1; Part-Time Paid 1; Interns 1.
Governing Authority: Parent Institution: Bowman County Historical and Genealogical Society. Tax-exempt.
Institution Type/Description: History Museum.

Collections: area history; ranching; Native American artifacts; military; fossils; rodeo; paleontology.
Facilities: garden. Museum-related items for sale.
Activities: Museum Sponsors: Halloween Spooktacular; Easter Eggstravaganza.
Hours & Admission Prices: May-Sept. Mon.-Sat. 9-5, Sun. 1-5; Labor Day-Memorial Day Mon.-Fri. 10-4. Adults 14 & over $3; children under 14 & members no charge. Closed New Year's Eve & Day; Easter; Memorial Day; Independence Day; Labor Day; Thanksgiving; Christmas Eve & Day.
Attendance: 2,000 (estimated)
Membership: Single $20; Family $30.

Cando

CANDO ARTS CENTER, 502 4th Ave., Cando, ND 58324-6161. Mailing Address: Cando Arts Council, P.O. Box 368, Cando, ND 58324. Tel.: 701-968-3655.
E-mail: candoarts@gmail.com
Web Site: candoarts.weebly.com
Key Personnel: Dir., Shelley Lord
Institution Type/Description: Art Center.
Collections: student artwork.
Hours & Admission Prices: Tues. 1-6:30, Wed.-Fri. 1-4 and by appointment.

CANDO PIONEER FOUNDATION, INC., 502 Main St., Cando, ND 58324-0142. Mailing Address: P.O. Box 142, Cando, ND 58324-0142. Tel.: 701-968-3943 & 3490.
E-mail: slarson@gondtc.com
Formerly: Pioneer Museum
Founded: 1976.
Volunteer Hours: 600
Operating Expenses: 2,000
Operating Income: 600
Institution Type/Description: History Museum.
Collections: local history & culture; newspapers; photographs. Historic Building: one room school.
Hours & Admission Prices: By appointment. No charge; donations accepted. &
Attendance: 750 (estimated)

Carrington

FOSTER COUNTY MUSEUM, 2nd St. S. & 16th Ave., Carrington, ND 58421. Tel.: 701-652-1313; 701-652-3363 & 674-3270 (tours).
Key Personnel: Cur., Ralph Harmon; Cur., Henry Gussiaas
Institution Type/Description: History Museum.
Collections: photographs; display cases; agricultural equipment.
Hours & Admission Prices: Sun. afternoons & by appointment. No charge; donations accepted.

Cavalier

PEMBINA COUNTY HISTORICAL SOCIETY AND MUSEUM, 13572 Hwy. 5, Cavalier, ND 58220. Tel.: 701-265-4941.
E-mail: pchsm@polarcomm.com
Web Site: ndpchs.com/museum.htm
Founded: 1968.
Key Personnel: Admin., Zelda Hartje
Institution Type/Description: Historical Society Museum.
Collections: Pembina County history; pioneer farm machinery; historic buildings.
Activities: Museum Sponsors: Annual Pioneer Machinery Show and Tractor Pull in September.
Publications: newsletter.
Hours & Admission Prices: By appointment. No charge.
Membership: Individual $10.

PIONEER HERITAGE CENTER, 13571 Hwy. 5, Cavalier, ND 58220-9545. Tel.: 701-265-4561. Fax: 701-265-4443.
E-mail: isp@nd.gov
Web Site: www.parkrec.nd.gov
Founded: 1989.
Congressional District: 1
Key Personnel: Pres. (V), Rosemarie Myrdal; Park Mgr., Justin Robinson; Park Ranger & Interpretive Coord., Dennis Clark; Museum Shop Mgr., Lorraine Schroeder.
Personnel Profile: Full-Time Paid 2; Part-Time Paid 1; Part-Time Volunteers 25.

Governing Authority: Parent Institution: Northeastern North Dakota Heritage Association. Subsidiary Institution: Icelandic State Park. Tax-exempt.
Institution Type/Description: State Park Museum.
Collections: historical, dealing with settlement period 1870-1920.
Facilities: homestead; nature preserve; nature trail; interpretive center. Pioneer Heritage Buildings: church; town hall; one room school; log cabin.
Activities: self-guided tours; guided group tours by appointment; special events.
Publications: Home Quarter Newsletter.
Hours & Admission Prices: Center: Memorial Day-Labor Day daily 9-5; Fall, Winter & Spring Mon.-Fri. 9-5, Sun. 1-5. $5 per vehicle; $25 per year; members no charge. Closed New Year's Day; Easter; Mother's Day; Thanksgiving; Christmas. &
Attendance: 4,000 (accurate)
Membership: Individual $15; Organization $20.

Cayuga

TEWAUKON NATIONAL WILDLIFE REFUGE, 9754 143 1/2 Ave., S.E., Cayuga, ND 58013-9764. Tel.: 701-724-3598. Fax: 701-724-3683.
E-mail: tewaukon@fws.gov
Web Site: www.fws.gov/tewaukon
Founded: 1945.
Institution Type/Description: Wildlife Refuge.
Collections: migratory birds & other wildlife.
Facilities: 8,363-acre refuge.
Activities: recreational activities.
Hours & Admission Prices: Refuge: daily 5am-10pm. Office & Visitor Center: Mon.-Fri. 8-4:30.

Center

FORT CLARK TRADING POST STATE HISTORIC SITE, 1074 27th Ave., S.W., Center, ND 58530-9429. Mailing Address: 612 E. Blvd. Ave., Bismarck, ND 58505. Tel.: 701-328-2666.
Web Site: www.state.nd.us/hist
Founded: 1965.
Key Personnel: Agency Dir., Merlan E. Paaverud, Jr.; Dir. Historic Preservation, Fern Swenson; Historic Sites Mgr., Diane Rogness.
Personnel Profile: Full-Time Paid 1; Part-Time Paid 1.
Governing Authority: state. Parent Institution: State Historical Society of North Dakota.
Institution Type/Description: Historic Site Museum: built in 1830-1831, the fort was burned down in 1861.
Collections: archeological Mandan Indian villages; Fort Clark fur trading post eras covering the period 1822-1862.
Publications: quarterly, North Dakota History and Plains Talk.
Hours & Admission Prices: mid-May to mid-Sept. daily 8-5. No charge; donations accepted. Society's Museum Stores: discounts to members.
Attendance: 10,000 (accurate)
Membership: For membership information, please call (701) 222-1966.

Coleharbor

AUDUBON NATIONAL WILDLIFE REFUGE, 3275 11th St., N.W., Coleharbor, ND 58531-9419. Tel.: 701-442-5474. Fax: 701-442-5546.
E-mail: audubon@fws.gov
Web Site: www.fws.gov/audubon/
Key Personnel: Dist. Mgr., Kathy Baer
Institution Type/Description: Wildlife Refuge.
Collections: birds; mammals; reptiles; amphibian; fish.
Facilities: 14,735 acres.
Hours & Admission Prices: Mon.-Fri. 8-4:30. No charge. Closed federal holidays.
Membership: Single $10; Family $15.

Cooperstown

GRIGGS COUNTY HISTORICAL MUSEUM, 203 12th St., S.E., Cooperstown, ND 58425. Mailing Address: P.O. Box 242, Cooperstown, ND 58425-0242.
Institution Type/Description: History Museum.
Collections: Griggs County history; period artifacts; newspapers; blacksmith shop; one-room school; Rhodes animated collection.
Activities: Museum Sponsors: Pioneer Days in July; Old Fashioned Christmas in December.
Hours & Admission Prices: May-Sept. Sun. 1-4:30; other times by appointment. Adults $5, students $2; children under 5 no charge.

RONALD REAGAN MINUTEMAN MISSILE STATE HISTORIC SITE, 555 113 1/2 Ave., N.W., Hwy. 45, Cooperstown, ND 58425-0006. Mailing Address: P.O. Box 6, Cooperstown, ND 58425-0006. Tel.: 701-797-3691. Fax: 701-797-3693.
E-mail: histsoc@nd.gov
Web Site: www.history.nd.gov
Founded: 2009.
Key Personnel: Mark Sundlov.
Personnel Profile: Full-Time Paid 1; Part-Time Paid 7.
Governing Authority: state. Parent Institution: State Historical Society of North Dakota, Bismarck, ND. Tax-exempt: 501(c)(3).
Institution Type/Description: Historic Sites: housed on the former Oscar-Zero Missile Alert Facility and November-33 Launch Facility which were part of the nation's minuteman missile force.
Collections: site history; weapons; furnishings; armament; paintings; photographs.
Research Fields: Minuteman missile weapon system; Cold War; economic & social impact of nuclear weapons.
Facilities: 50-vol. library; theater. Museum-related items for sale.
Activities: films; guided tours; participatory exhibits; theater.
Hours & Admission Prices: March to mid-May & mid-Sept. to Oct. Mon. & Wed.-Sat. 10-6, Sun. 1-5; mid-May to mid-Sept. daily 10-6; Nov.-Feb. by appointment. Adults $10, students $3; discounts to groups of 20 or more; children & members no charge. &
Attendance: 8,500 (estimated)

Crosby

DIVIDE COUNTY HISTORICAL SOCIETY MUSEUM, 300 Second Ave., N.E., Crosby, ND 58730. Mailing Address: P.O. Box 130, Crosby, ND 58730-0130. Tel.: 701-339-0059.
E-mail: dctthreshingbee@gmail.com
Founded: 1969.
Key Personnel: Pres. (V), John Tysse.
Governing Authority: nonprofit organization. Tax-exempt.
Institution Type/Description: Village Museum.
Collections: pioneer artifacts; farm implements: vehicles; stationary gas engines; miniature trains. Historic Buildings: bank; Lutheran church; homestead claim shack; schools; stores; blacksmith shop; saw mill; car club displays.
Activities: temporary exhibitions. Annual Event: Threshing Bee & parades in July.
Hours & Admission Prices: Tours by appointment, call for information. No charge; donations accepted. Facilities: $12 for 3 days.
Attendance: 3,600 (estimated)

Devils Lake

LAKE REGION HERITAGE CENTER, 502 4th St., Devils Lake, ND 58301-0245. Mailing Address: P.O. Box 245, Devils Lake, ND 58301. Tel.: 701-662-3701. Fax: 701-662-2810.
Web Site: lrhs.homestead.com
Key Personnel: Cur., Jim Schiele
Institution Type/Description: Heritage Center.
Collections: local history & culture; personal artifacts; photographs; paintings; automobiles.
Activities: guided tours.
Hours & Admission Prices: Summer: Mon.-Fri. 7-5, Sat.-Sun. 12-5; Sept.-June Mon.-Fri. 8:30-5.

LAKE REGION HERITAGE HOUSE MUSEUM, 416 Sixth St., Devils Lake, ND 58301-0626. Mailing Address: P.O. Box 245, Devils Lake, ND 58301-0245. Tel.: 701-662-3701 & 7080.
Institution Type/Description: Historic House Museum.
Collections: local history; period furnishings; photographs; personal artifacts.
Hours & Admission Prices: Wed.-Sun. 1-4; other times by appointment. Adults $3, seniors & students $2; children under 6 no charge.

NORTH DAKOTA MARITIME MUSEUM, Fifth St. & Fourth Ave., Devils Lake, ND 58301-0626. Mailing Address: P.O. Box 626, Devils Lake, ND 58301-0626. Tel.: 701-662-7031. Fax: 701-662-7049.
Institution Type/Description: Maritime Museum.
Collections: maritime history; photographs; uniforms; posters; books; military artifacts.
Hours & Admission Prices: Sun. 2-5 by appointment.

SULLYS HILL NATIONAL GAME PRESERVE, 221 2nd St., W., Devils Lake, ND 58301-2963. Tel.: 701-662-8612.
Web Site: sullyshill.fws.gov
Institution Type/Description: Wildlife Preserve.
Collections: migratory birds & big game including elk, deer, prairie dogs; waterfowl; foxes; raccoons; skunks; weasels; mink; beaver; rabbits, wild turkey.
Facilities: theater; classroom; visitor center; aquarium. Museum-related items for sale.
Activities: waterfowl observation; birdwatching; special events; educational programs.
Hours & Admission Prices: Auto Tour: May-Oct. daily 8 am to sunset. $2 per vehicle.

Dickinson

DSU ART GALLERY, Klinefelter Hall, Dickinson State Univ., Dickinson, ND 58601-4896. Mailing Address: Klinefelter Hall, Box 28, Dickinson State Univ., Dickinson, ND 58601-4896. Tel.: 800-279-HAWK.
Institution Type/Description: Art Gallery.
Collections: paintings; sculpture; drawings.
Hours & Admission Prices: Mon.-Fri. 8-5. No charge.

DAKOTA DINOSAUR MUSEUM, 200 E. Museum Dr., Dickinson, ND 58601-4000. Tel.: 701-225-3466. Fax: 701-227-0534.
E-mail: info@dakotadino.com
Web Site: www.dakotadino.com
Founded: 1991.
Key Personnel: C.E.O. & Museum Shop Mgr., Alice League; Pres. (V), Richard Johnson.
Personnel Profile: Full-Time Paid 2; Part-Time Volunteers 2.
Volunteer Hours: 100
Operating Expenses: 88,594
Operating Income: 82,426
Governing Authority: private; nonprofit organization. Tax-exempt: 501(c)(3).
Institution Type/Description: Paleontology, Mineral & Geology Museum.
Collections: 12,000 paleontology & mineral specimens; complete triceratops skeleton collected in the Hell Creek formation; seashells; 14 full scale dinosaurs (11 inside and 3 outside); 10,000-year old bison; rhinoceros.
Hours & Admission Prices: May to Labor Day daily 9-5. Adults $8, children 3-12 $5; discounts to groups. &
Attendance: 9,700 (accurate)
Membership: Individual $25; Family $55; Contributor $100; Sponsor $250.

DICKINSON MUSEUM CENTER, (M), 188 Museum Dr. E., Dickinson, ND 58601-4088. Tel.: 701-456-6225.
E-mail: info@dickinsonmuseumcenter.org
Web Site: www.dickinsonmuseumcenter.org
Formerly: Joachim Regional Museum
Founded: 1983.
Key Personnel: Dir., Daniel Ingram; Pres. (V), Kristin Lacher; Museum Asst., Emily Bradbury.
Personnel Profile: Full-Time Paid 2; Part-Time Paid 3; Part-Time Volunteers 30; Interns 2.
Governing Authority: Parent Institution: Southwestern North Dakota Museum Foundation, Inc. Tax-exempt: 501(c)(3).
Institution Type/Description: History Museum.
Collections: history; art archives.
Research Fields: local history.
Activities: summer kids program. Annual Event: Holiday Event in December.
Hours & Admission Prices: Memorial Day to Labor Day daily 9-5; Winter: Mon.-Fri. 9-5. No charge; donations accepted. &
Attendance: 7,000 (estimated)

UKRAINIAN CULTURAL INSTITUTE, 1221 W. Villard, Dickinson, ND 58601-4849. Mailing Address: P.O. Box 6, Dickinson, ND 58602-0006. Tel.: 701-483-1486. Fax: 701-483-4366.
Institution Type/Description: Ukrainian History Museum.
Collections: Ukrainian history & culture; hand-crafted Ukrainian eggs; folk art; religious artifacts.
Facilities: Museum-related items for sale.
Hours & Admission Prices: Mon.-Thurs. 8-4:30, Fri.-Sun. by appointment. No charge; donations accepted.

Drayton

DRAYTON UNITED METHODIST CHURCH, 203 N. Main St., Drayton, ND 58225-0327. Mailing Address: P.O. Box 327, Drayton, ND 58225-0327. Tel.: 701-454-3880.
E-mail: umc@polarcomm.com
Formerly: Methodist Episcopal Church
Founded: 1879.
Institution Type/Description: Historic Building: housed in a church built in 1905. Listed on the National Register of Historic Places.
Collections: religious artifacts & furnishings; paintings; church history.
Hours & Admission Prices: Tours: by appointment. No charge.

Dresden

CAVALIER COUNTY MUSEUM, 10123 95th St., Dresden, ND 58249. Mailing Address: 324 8th Ave., Langdon, ND 58249. Tel.: 701-283-5417.
Institution Type/Description: History Museum.
Collections: local history & culture; period furnishings; personal artifacts; dolls; household artifacts; photographs; newspapers; books; magazines; early industry & business; log cabin; period machinery; fire truck; Model T.
Hours & Admission Prices: Memorial Day to Labor Day Sun.-Mon., Wed. & Fri. 1-5. No charge; donations accepted.

Dunn Center

DUNN COUNTY HISTORICAL SOCIETY & MUSEUM, 153 Museum Trail, Dunn Center, ND 58626. Mailing Address: P.O. Box 145, Killdeer, ND 58640-0145. Tel.: 701-548-8111.
E-mail: dunncountymuseum@ndsupernet.com
Web Site: www.dunncountymuseum.org
Founded: 1977.
Governing Authority: Tax-exempt.
Institution Type/Description: Historical Society Museum
Collections: local history & culture; period kitchen; Western display; photographs. Historic Building: country school house.
Publications: Newsletter: Tales & Trails 3 issues per year.
Hours & Admission Prices: By appointment. Call for admission prices. &
Attendance: 500 (estimated)
Membership: Single $10; Couple $15; Single Lifetime $100; Couple Lifetime $150.

Dunseith

INTERNATIONAL PEACE GARDEN, 10939 Hwy. 281, Dunseith, ND 58329-9445. Mailing Address: R.R. 1 Box 116, Dunseith, ND 58329-9445. Tel.: 701-263-4390; 888-432-6733. Fax: 701-263-3169.
E-mail: kathy@peacegarden.com
Web Site: www.peacegarden.com
Founded: 1932.
Key Personnel: C.E.O. & C.O.O., Doug Hevenor; Pres. (V), Ed Anderson; Vice Pres., Tyrone Langager; Finance Mgr., Leonard Richard.
Personnel Profile: Full-Time Paid 5; Part-Time Paid 40.
Governing Authority: nonprofit. Tax-exempt.
Institution Type/Description: Arboretum.
Collections: garden history; Carillon Bell Tower; floral clock; gardens; photographs; 911 memorial.
Research Fields: botanical.
Facilities: garden; chapel; camp grounds; concessions; interpretive center; convention & wedding reception facilities. Gift-related items for sale.
Activities: tours; International Music camp; Legion Athletic camp; concerts; bird watching. Annual Event: International Country Gospel Fest.
Publications: Annual Visitors Guide.
Hours & Admission Prices: Garden: daily; Peak time for flowers: July 15-Aug. 15. Gate: late May to mid-Sept. daily 9-7. Office: Mon.-Fri. 80-4. Donations: $125 per tour bus, $25 season pass, $20 per vehicle a season, $10 per vehicle a day, $5 pedestrian. &
Attendance: 150,000
Membership: Foundation $150.

Edmore

WHEATLAND MANOR, 405 S. Grant St., Edmore, ND 58330. Mailing Address: P.O. Box 8, Edmore, ND 58330-0008. Tel.: 701-644-2291 & 2453.
Institution Type/Description: History Museum.
Collections: local history & culture; furnishings of early 1900s; photographs.

Hours & Admission Prices: Memorial Day-Labor Day Sun. 1-3; other times by appointment.

Egeland

TOWNER COUNTY HISTORICAL MUSEUM, Main St., Egeland, ND 58331. Mailing Address: 108 4th Ave., Munich, ND 58352. Tel.: 701-682-5106.
Founded: 1974.
Key Personnel: Pres. (V), Anita Barrett; Museum Shop Mgr., David Barrett.
Personnel Profile: Part-Time Paid 3; Part-Time Volunteers 10.
Volunteer Hours: 320
Institution Type/Description: Historical Society Museums.
Collections: horse-drawn hearse; period automobiles; Bavarian crystal; dolls; Soo Line Railroad depot including period furnishings.
Hours & Admission Prices: By appointment. No charge, except for special events; donations accepted.
Attendance: 60 (estimated)
Membership: Individual $5; Life $100.

Ellendale

COLEMAN MUSEUM, Southeast Corner of Main St. & Railroad Ave., Ellendale, ND 58436. Mailing Address: 8836 92nd St., S.E., Ellendale, ND 58436.
Founded: 1985.
Governing Authority: Parent Institution: Ellendale Historical Society.
Institution Type/Description: Historical Society Museum.
Collections: artifacts & memorabilia of the Ellendale area; photographs.
Hours & Admission Prices: June-Sept. Tues. & Fri. 1-5. No charge; donations accepted.

Epping

BUFFALO TRAILS MUSEUM, Main St., Epping, ND 58843. Mailing Address: P.O. Box 22, Epping, ND 58843-0022. Tel.: 701-859-4361 (June-Aug.).
E-mail: buffalotrailmuseum@yahoo.com
Web Site: epping.govoffice.com
Founded: 1966.
Key Personnel: Pres., Duane Syverson.
Personnel Profile: Part-Time Paid 2; Part-Time Volunteers 6.
Governing Authority: nonprofit organization. Parent Institution: Williams County Historical Society. Tax-exempt: 501(c)(3).
Institution Type/Description: Regional History Museum.
Collections: regional history; Indian artifacts; geology; mineralogy; archaeology; natural history. Historic Buildings: 1903 Fosse Barber shop; 1906 O. E. Ellingson general merchandise store; 1906 C. F. Carpenter harness & hardware store; 1908 homestead log cabin; 1926 one-room rural school.
Research Fields: regional history; archaeology.
Facilities: cafe; picnic area.
Activities: guided tours; permanent & temporary exhibitions; camping.
Publications: museum brochure; tour guide.
Hours & Admission Prices: May-July Tues.-Sat. 10-4, Sun. 12-5. Adults $5, students & groups over 10 $2; discounts to groups; life members no charge.
Attendance: 275 (estimated)
Membership: Life $200.

Fargo

THE CHILDREN'S MUSEUM AT YUNKER FARM, 1201 28th Ave. N., Fargo, ND 58102-1337. Tel.: 701-232-6102. Fax: 701-232-4605.
E-mail: info@childrensmuseum-yunker.org
Web Site: childrensmuseum-yunker.org
Founded: 1989.
Congressional District: 45
Key Personnel: Exec. Dir., Yvette Nasset.
Personnel Profile: Full-Time Paid 3; Full-Time Volunteers 1; Part-Time Paid 5; Part-Time Volunteers 300; Interns 4.
Governing Authority: nonprofit. Tax-exempt: 501(c)(3).
Institution Type/Description: Children's Museum
Collections: hands-on exhibits.
Facilities: nature trail; picnic shelter. Museum-related items for sale.
Activities: classes; workshops; train rides; special events. Annual Events: Party in the Pumpkin Patch, Halloween Party, Easter Celebration; Kidcology Camp; Yunkie Club; Think Thursday; Children's Musical Theater.
Publications: quarterly newsletter, Museum Mania.
Hours & Admission Prices: Summer: Mon.-Sat. 10-5, Sun. 1-5; School Year: Tues.-Sat. 10-5, Sun. 1-5. Admission $5; children under one & members no charge. &

Attendance: 46,000 (estimated)
Membership: Individual $40; Family $70; Contributor $100; Patron $250.

FARGO AIR MUSEUM, 1609 19th Ave. N., Fargo, ND 58102-1886. Tel.: 701-293-8043. Fax: 701-293-8103.
Web Site: fargoairmuseum.org
Founded: 2001.
Key Personnel: Exec. Dir., Fran Brummund; Chm., Rex Hammarback.
Personnel Profile: Full-Time Paid 1; Part-Time Paid 1; Part-Time Volunteers 29; Interns 1.
Governing Authority: private; nonprofit organization. Tax-exempt: 501(c)(3).
Institution Type/Description: Air & Space Museum.
Collections: 20 civilian & military aircraft; 4 murals depicting a century of aviation & agricultural aviation; aviation artifacts & history; air racing; Happy Hooligan.
Facilities: 2,000-vol. library; 1,800 sq. ft. exhibit space. Museum-related items for sale.
Activities: films; swing dances. Annual Events: Amelia Earhart; Santa Fly-In; Air Racing.
Publications: newsletter, Legends.
Hours & Admission Prices: Memorial Day to Labor Day Mon.-Sat. 9-5, Sun. 12-4; Sept.-May Tues.-Sat. 9-5, Sun. 12-4. Adults $6, senior citizens $5, children 5-12 $4; discounts to groups; members no charge. Closed New Year's Day; Easter; Thanksgiving; Christmas.
Attendance: 30,000 (estimated)
Membership: Senior, Student & Military $15; Individual $25; Dual $35; Family $55.

FARGO THEATRE, 314 Broadway, Fargo, ND 58102-4715. Tel.: 701-239-8385.
Web Site: fargotheatre.org
Founded: 1926.
Institution Type/Description: Historic Building: housed in a fully-restored art deco theatre built in 1926 to host Vaudeville & silent movies.
Collections: theatre history; Wurlitzer pipe organ; statue commemorating the film Fargo.
Facilities: 870-seat main screen theatre; 76-seat theatre.
Activities: tours; independent & art films; live stage events; special events & performances; concerts; guest speakers. Theatre Events: Midnight Movie Series; Classic Film Series; Silent Movie nights; Fargo Film Festival.
Hours & Admission Prices: Tours: call for hours. Films: Mon.-Fri. 5, 7 & 9pm, Sat.-Sun. 1, 3, 5, 7, & 9pm. Closed Christmas Eve. &

GALLERY 4, LTD, 114 Broadway N., Ste. G2, Fargo, ND 58102-4942. Tel.: 701-237-6867.
Web Site: www.gallery4fargo.com
Founded: 1975.
Institution Type/Description: Art Gallery.
Collections: paintings; drawings; sculpture; glass; wood.
Hours & Admission Prices: Tues.-Wed. & Fri.-Sat. 11-5, Thurs. 11-7. &

MAURY WILLS MUSEUM, 1515 15th Ave. N., Ground Fl., Fargo, ND 58102-5701. Tel.: 701-235-6161. Fax: 701-297-9247.
E-mail: redhawks@fmredhawks.com
Founded: 2000.
Institution Type/Description: Sports Museum.
Collections: Maury Wills' life & career; personal artifacts; baseball history & artifacts; photographs.
Hours & Admission Prices: Mon.-Fri. 9-5. No charge. &

MEMORIAL UNION GALLERY, (M), 258 Memorial Union, North Dakota State Univ., Fargo, ND 58105-5476. Mailing Address: North Dakota State University, Dept. 5340, P.O. Box 6050, Fargo, ND 58108-6050. Tel.: 701-231-8239. Fax: 701-231-7866. Facebook: Memorial Union Gallery.
E-mail: ndsu.mugallery@ndsu.edu
Web Site: www.mu.ndsu.edu/gallery/exhibits_and_artists
Founded: 1975.
Key Personnel: Gallery Coord., Netha Cloeter.
Personnel Profile: Full-Time Paid 1; Part-Time Paid 12.
Governing Authority: university. Parent Institution: North Dakota State University. Tax-exempt.
Institution Type/Description: Art Gallery.
Collections: contemporary works of art by regional & local American artists; works by prominent artists including Andy Warhol, Pablo Picasso, & Salvador Dali; 1977 edition of prints from Lakeside Studios.
Major Exhibits: Put A Bird on It: Ali LaRock & Paul Noot (T), 1/4/14-1/31/14;

The Art of Theater: Master-Mentor-Medium, 2/6/14-4/11/14; NDSU Visual Arts - BFA Show, 4/15/14-5/2/14; Rosenquist Visiting Artist Show, 5/4/14-6/7/14; PEARS - Printmaking Exchange, 6/8/14-7/14; Shane Blakowitsch (T), 8/4/14-8/31/14; Oil & Water: Printmaking Exchange (T), 10/15/14-11/15/14.
Activities: guided tours; lectures; gallery talks; loan, permanent, temporary & traveling exhibitions; student exhibits.
Publications: exhibit mailers & posters.
Hours & Admission Prices: mid-May to mid-Aug. Tues.-Fri. 11-4; Aug. 15 to mid-May Tues.-Wed. & Fri.-Sat. 11-5, Thurs. 11-8. No charge. Closed holidays. &
Attendance: 2,300 (accurate)

* **PLAINS ART MUSEUM, (M),** 704 First Ave. N., Fargo, ND 58102-4904. Tel.: 701-232-3821. Fax: 701-293-1082.
E-mail: museum@plainsart.org
Web Site: plainsart.org
Founded: 1973.
Key Personnel: Chairperson, Julie Burgum; Dir. & C.E.O., Colleen Sheehy; Dep. Dir. & C.F.O., Mark Henze; Devel. Dir., Laurie Wigtil; Education Dir., Kristin Bergquist; Communicataions Dir., Amy Richardson; Collections & Opers. Dir., Mark Ryan; Cur., Becky Dunham.
Governing Authority: private; nonprofit organization. Tax-exempt: 501(c)(3).
Institution Type/Description: Art Museum.
Collections: over 3,000 objects including fine art, Native American & ethnographic with an emphasis on art of the upper Midwest.
Facilities: 7,400 sq. ft. exhibit space; cafe. Museum-related items for sale.
Activities: outreach art education programs to North Dakota & Minnesota; concerts; dance recitals; docent program; guided tours; lectures; mobile gallery; participatory, loan, temporary & traveling exhibitions. Annual Events: Trash or Treasure in April; Spring Gala Fundraiser in May.
Publications: quarterly newsletter, Museum News; triannual education newsletter, Experiences for Life!
Hours & Admission Prices: Mon., Wed. & Fri. 11-5, Tues. & Thurs. 11-9, Sat. 10-5, Sun. 12-5. Adults $5, seniors & educators w/ID $4; discounts to AAM & MPMA members; members, children & students with ID no charge. Closed New Year's Day; Memorial Day; Labor Day; Thanksgiving; Christmas. &
Attendance: 52,837 (accurate)
Membership: Senior & Student $35; Individual $50; Senior Household $70; Household $100; Enthusiast $250; Advocate $500; Patron $1,000; Curator $2,500; Director $5,000.

RED RIVER ZOO, 4255 23rd Ave., S., Fargo, ND 58104-8603. Tel.: 701-277-9240. Fax: 701-277-9238.
Founded: 1996.
Key Personnel: Dir., Lisa Tate; Pres. (V), Bruce Furness; Museum Shop Mgr., Anna Miller.
Personnel Profile: Full-Time Paid 12; Part-Time Paid 15; Part-Time Volunteers 30; Interns 30.
Governing Authority: Parent Institution: Red River Zoological Society.
Institution Type/Description: Zoo.
Collections: over 300 animals representing 75 species; exotic plants.
Activities: special events; rental facilities; educational programs.
Hours & Admission Prices: May-Sept. daily 10-7; Oct.-April Sat.-Sun. 10-5. Adults $8.75, seniors over 60 $7.75, children 2-14 $5.75; members & children under 2 no charge.
Attendance: 75,000 (accurate)
Membership: Individual $45; Household $65.

ROGER MARIS MUSEUM, West Acres Shopping Center, 3902 13th Ave. S., Fargo, ND 58103-3357. Mailing Address: c/o West Acres Mall, 3902 13th Ave. S., Ste. 3717, Fargo, ND 58103. Tel.: 701-282-2222; 800-783-6450.
E-mail: westacres@westacres.com
Web Site: www.rogermarismuseum.com
Founded: 1984.
Institution Type/Description: Sports Museum.
Collections: personal artifacts; replica of Roger's Yankee Stadium monument & 1961 locker; video.
Hours & Admission Prices: Mon.-Sat. 10-9, Sun. 12-6. No Charge. Closed Easter; Thanksgiving; Christmas.

Fessenden

WELLS COUNTY HISTORICAL SOCIETY MUSEUM, 903 4th St. N.E., Fessenden, ND 58438. Tel.: 701-547-3684.
Formerly: Wells County Museum
Founded: 1972.

Congressional District: 1
Key Personnel: Pres., Carol Beck; Vice Pres., Pat Leny; Sec., Judy Martin; Treas., Betty Hirschkorn.
Personnel Profile: Part-Time Paid 1; Part-Time Volunteers 10.
Governing Authority: county; nonprofit society. Affiliated with Wells County Historical Society. Branch Museum: Hurd Round House. Tax-exempt: 501(c)(3).
Institution Type/Description: Historic Building: housed in a 1919 2-room school house.
Collections: period agriculture implements; period automobiles; horse drawn vehicles; ethnic artifacts; pioneer tools & utensils; manuscripts; clothing; photographs; early home & church furnishings; early school textbooks & furnishings; complete county newspaper archives; various documents & ledgers of early businesses in the county.
Research Fields: local history.
Activities: guided tours; lectures; films; permanent exhibitions. Annual Event: Fair Days in June. Museum Sponsors: business meetings in May & Oct.
Publications: monthly newsletter, Wells County History.
Hours & Admission Prices: By appointment. No charge.
Attendance: 350 (estimated)

Forman

SARGENT COUNTY MUSEUM, 8987 Hwy. 32, Forman, ND 58032. Mailing Address: 13443 88th St., S.E., Forman, ND 58032-9709. Tel.: 701-724-3194.
E-mail: sargentcountymuseum@yahoo.com
Web Site: sargentcountymuseum.org
Personnel Profile: Part-Time Volunteers 6.
Governing Authority: Tax-exempt.
Institution Type/Description: History Museum.
Collections: local history & culture; photographs; personal artifacts; newspapers from mid 1880s to 1970s on microfilm.
Hours & Admission Prices: Memorial Day to Labor Day Sun.-Fri. 1-4; other times by appointment. No charge; donations accepted. &
Attendance: 500 (accurate)
Membership: Single $7.50; Family $15.

Fort Ransom

RANSOM COUNTY HISTORICAL SOCIETY, 101 Mill Road, S.E., Fort Ransom, ND 58033-9740. Mailing Address: P.O. Box 5, Fort Ransom, ND 58033-0005. Tel.: 701-678-2045.
Web Site: rchsmuseum.tripod.com
Founded: 1972.
Key Personnel: Pres. (V), Richard Birklid.
Personnel Profile: Part-Time Paid 3; Part-Time Volunteers 10; Interns 1.
Governing Authority: society; nonprofit organization. Parent Institution: State Historical Society of North Dakota, Bismarck, ND 58505. Tax-exempt: 501(c)(3).
Institution Type/Description: History Museum: housed in 1867-1872 U.S. Military Fort.
Collections: pictures; books; records; clothing; tools; household items. Historic Building: 1881 T.J. Walker Flour Mill.
Activities: tours.
Publications: Starting a Cookbook.
Hours & Admission Prices: Memorial Day-Oct. daily 1-5; other times by appointment. Adults $1; members & children no charge.
Attendance: 5,000 (estimated)
Membership: Active $1; Individual $5; Family $10; Business Institutional $25; Contributing & Sustaining $50; Lifetime over 60 $50; Lifetime $100.

Fort Totten

FORT TOTTEN STATE HISTORIC SITE, 417 Cavalry Cir., Fort Totten, ND 58335. Mailing Address: P.O. Box 224, Fort Totten, ND 58335-0224. Tel.: 701-766-4441. Fax: 701-766-1382.
E-mail: ngronseth@nd.gov
Web Site: www.discovernd.com/hist
Founded: 1960.
Key Personnel: Dir. State Historical Society of North Dakota, Merl Paaverud; Div. Dir. Historic Sites, Fern Swenson; Eastern Rgnl. Site Mgr., Diane Rogness; Site Supvr. & Museum Store Mgr., Jack Mattson.
Personnel Profile: Full-Time Paid 2; Part-Time Paid 8; Part-Time Volunteers 15.
Governing Authority: state. Parent Institution: State Historical Society of ND. Subsidiary Institution: Fort Totten State Historic Site Foundation & Pioneer Daughters. Tax-exempt: 501(c)(3).
Institution Type/Description: Historic Site & Outdoor Museum: 1868-1890 Military Post; Pioneer Daughters Museum; 1891-1959 Indian School, consisting of 17 structures.

Collections: artifacts pertaining to Native American & pioneer heritage of Devils Lake & North Dakota region. Historic Structures: officers row; enlisted men's barracks; commissary; quartermaster storehouse; powder magazine; hospital; bakery; adjutant's office.

Research Fields: military history; Indian School history; pioneers of Devil's Lake region history.

Facilities: interpretive center; 300-seat auditorium. Books, postcards & other Native American & military-related items for sale.

Activities: self-guided tours; organized group guided tours; organized educational programs for children & adults; tours with catered meals on request. Annual Event: Living History Field Day in September.

Publications: booklet, Fort Totten: Military Post & Indian School, 1867-1959; video, Images of Fort Totten.

Hours & Admission Prices: Visitor Center & P.W. Museum Facilities: mid-May to mid-Sept. daily 8-5; mid-Sept. to mid-May Mon.-Fri. by appointment. Adults $5, children 6-15 $1.50; discounts to school groups, NDHF & FFTHS members; members & children under 6 no charge. Bus: $40. Season Passes: Family $20, Individual $10.

Attendance: 15,000 (estimated)

Membership: Friends of Ft. Totten Historic Site: Raccoon $15; Star $26; Turtle $201; Deer $501; Buffalo $1,001; Storyteller $5,001.

Fullerton

ROSEBUD SCHOOL MUSEUM, Main St., Fullerton, ND 58441. Mailing Address: P.O. Box 27, Fullerton, ND 58441-0027. Tel.: 701-375-7521.

Institution Type/Description: Historic Building: housed in a former one-room schoolhouse; built in 1901.

Collections: local history; period furnishings; personal artifacts; photographs; local school artifacts.

Hours & Admission Prices: May-Oct. by appointment.

Garrison

HERITAGE PARK MUSEUM, First St. & First Ave., N.W., Garrison, ND 58540. Mailing Address: Garrison Convention & Visitors' Bureau, P.O. Box 1000, Garrison, ND 58540-0850. Tel.: 701-463-2546.

Web Site: www.garrisonnd.com/?id=81

Key Personnel: Dir., Garrison Convention & Visitors' Bureau, McKaila Matteson

Institution Type/Description: History Museum.

Collections: local history & culture; railroad depot; homestead house & blacksmith shop.

Hours & Admission Prices: Memorial Day to Labor Day Mon. 11-4 or by appointment.

NORTH DAKOTA FIREFIGHTER'S MUSEUM & HALL OF FAME, 52 N. Main St., Garrison, ND 58540. Mailing Address: P.O. Box 1000, Garrison, ND 58540. Tel.: 701-463-2345; 800-799-4242.

E-mail: ndfm@restel.com

Web Site: www.ndfm.org

Founded: 2006.

Institution Type/Description: Fire Fighters Museum.

Collections: fire fighting history; firefighters records, personal artifacts & memorabilia; uniforms; photographs; fire apparatus & equipment.

Activities: educational programs; induction ceremony; Hall of Fame. Museum Sponsors: Fire Prevention Week.

Publications: ND Firefighter's Journal.

Hours & Admission Prices: Mon.-Fri. 8-5. No charge.

NORTH DAKOTA FISHING HALL OF FAME AND MUSEUM, 4034 Hwy. 37 Bypass, Garrison, ND 58540. Mailing Address: 4034 Hwy. 37 Bypass, Garrison, ND 58540. Tel.: 701-463-2600.

Personnel Profile: Part-Time Paid 1.

Institution Type/Description: Fishing Museum.

Collections: fishing memorabilia; native fishermen.

Hours & Admission Prices: Mon.-Fri. 1-6, Sat.-Sun. 11-5.

Glen Ullin

GLEN ULLIN MUSEUM, 207 S. 10th St., Glen Ullin, ND 58631. Mailing Address: 6315 46th St., Glen Ullin, ND 58631-9734. Tel.: 701-348-3295.

Founded: 1983.

Key Personnel: Pres. (V), Lance Gartner; Dir., Mike Schirado; Dir., Laura Gartner; Dir., Viola Weindhart.

Personnel Profile: Part-Time Volunteers 6.

Governing Authority: Tax-exempt.

Institution Type/Description: Historic Buildings.

Collections: local history & culture; period furnishings; personal artifacts; photographs. Historic Buildings: house, 1884; Bethany church, 1896, schoolhouse, 1920.

Hours & Admission Prices: Sept. 3rd Sat.-Sun.; other times by appointment. No charge; donations accepted.

Attendance: 750 (estimated)

Membership: Individual $2; Life $50.

Grafton

HERITAGE VILLAGE & JUGVILLE MUSEUM, 695 W. 12th St., Grafton, ND 58237-2115. Tel.: 701-360-0088.

E-mail: vernaaasand@gmail.com

Web Site: www.walshcountyhistorical.com

Key Personnel: Pres., Ken Hoffmann; Office Mgr., Verna Aasand.

Personnel Profile: Part-Time Paid 1.

Volunteer Hours: 200

Operating Expenses: 27,613

Operating Income: 39,627

Governing Authority: Parent Institution: Walsh County Heritage Village, 695 W. 12th St., Hwy. 17 W., Grafton, ND 58237. Tax-exempt.

Institution Type/Description: History Museum.

Collections: village & farm implements; monuments; gazebos.

Hours & Admission Prices: Memorial Day to Labor Day Sun. 1-4; other times by appointment. No charge; donations accepted.

Attendance: 100 (estimated)

HISTORIC ELMWOOD HOUSE, Stephen Ave. & 2nd St., Grafton, ND 58237. Mailing Address: Chamber of Commerce, 432 Hill Ave., Grafton, ND 58237-1002. Tel.: 701-352-0152.

Institution Type/Description: Historic House Museum: housed in the former home of North Dakota's second Attorney General, Cam Spencer; built in 1895. Listed on the National Register of Historic Places.

Collections: family history; personal artifacts; period furnishings; photographs.

Activities: rental facilities.

Hours & Admission Prices: By appointment. No charge; donations accepted.

Grand Forks

THE ARTSPLACE, 1110 2nd Ave. N., Grand Forks, ND 58203-3620. Tel.: 701-746-6479.

Founded: 1988.

Institution Type/Description: Art Gallery.

Collections: works by local, regional & national artists; prints; paintings; photographs.

Hours & Admission Prices: Mon.-Sat. 10-5; other times by appointment. No charge; donations accepted.

Attendance: 1,173 (accurate)

BROWNING ARTS, 23 S. Fourth St., Grand Forks, ND 58201-4733. Tel.: 701-746-5090.

Institution Type/Description: Art Gallery.

Collections: ceramics; painting; drawings; photography; computer art; sculpture; jewelry.

Facilities: Museum-related items for sale.

Hours & Admission Prices: Jan. to late Nov. Mon.-Fri. 9-5:30; Thanksgiving to Christmas Mon.-Sat. 9-5:30.

GRAND FORKS COUNTY HISTORICAL SOCIETY, 2405 Belmont Rd., Grand Forks, ND 58201-7505. Tel.: 701-775-2216.

E-mail: gfhistory@midconetwork.com

Web Site: grandforkshistory.com

Founded: 1970.

Congressional District: 17 & 18

Key Personnel: Pres., Greg Vettel; Dir., Leah Byzewski; Treas., Wallace Bloom.

Personnel Profile: Full-Time Paid 1; Part-Time Paid 3; Part-Time Volunteers 2.

Governing Authority: society; nonprofit organization. Branch Museums: Campbell House; Myra Museum; Grand Forks Post Office; one-room schoolhouse; Carriage House; Myra Centennial Pavilion. Tax-exempt: 501(c)(3).

Institution Type/Description: Historical Society Museum: located on the former farmsite belonging to Tom Campbell, the agriculturalist.

Collections: historic objects mostly from Grand Forks County area. Buildings: Campbell House; Myra Museum; Grand Forks Post Office; one-room

schoolhouse; Carriage House; Myra Centennial Pavilion; 1950 Lustron House; 1930s era gas station replica.
Research Fields: local history.
Facilities: browsing library; meeting room.
Activities: guided tours; permanent & temporary exhibitions. Museum Sponsors: ice cream social; craft sale; Antique Show & Sale.
Publications: newsletter, The Paddlewheel Press.
Hours & Admission Prices: Grounds: May 15-Sept. 15 daily 1-5. Office: Summer Mon.-Fri. 9-12; Winter Sat.-Sun. 1-5. Adults 16 & over $5, children 10-15 $3; children under 10 & members no charge. Closed Independence Day. &
Attendance: 5,000 (estimated)
Membership: Individual $25; Family $50; Corporate $100 & up.

NORTH DAKOTA MUSEUM OF ART, 261 Centennial Dr., Stop 7305, Grand Forks, ND 58202-6003. Tel.: 701-777-4195. Fax: 701-777-4425. www.facebook.com/ndmoa.
E-mail: ndmoa@ndmoa.com
Web Site: www.ndmoa.com
Founded: 1970.
Congressional District: 1
Key Personnel: Dir. & Chief Cur., Laurel J. Reuter; Pres. Bd. Trustees, Wayne Zimmerman; Exhibition Coord. & Registrar, Greg Vettel; Dir. Education, Sue Fink; Dir. Rural Art, Matthew Wallace; Asst. to Dir., Brian Lofthus; Accountant, Brad Werner; Office Mgr., Becca Grandstrand; Collection Care Specialist, Danielle Masters; Mgr. Cafe, Justin Welsh; Museum Shop Mgr., Sheila Dalgliesh.
Personnel Profile: Full-Time Paid 10; Part-Time Paid 10; Part-Time Volunteers 80.
Governing Authority: private cultural institution. Tax-exempt: 501(c)(3).
Institution Type/Description: Art Museum.
Collections: national & international contemporary art; Northern Plains art; national American Indian since 1970.
Research Fields: contemporary art; American Indian art, contemporary & historical.
Facilities: outdoor sculpture garden; video viewing room; lecture room; cafe; gift shop.
Activities: concerts; readings; performances; films, as well as special events for both children & adults.
Publications: quarterly newsletter; occasional catalogs of specific exhibitions.
Hours & Admission Prices: Mon.-Fri. 9-5, Sat.-Sun. 1-5. No charge; donations accepted. &
Attendance: 50,000 (estimated)
Membership: Student $10; Individual $35; Household $50; Sustaining $100; Supporting $250; Sponsor $500; Patron $1,000; Benefactor $5,000; Barton Benes Donor Wall $10,000.

UNIVERSITY OF NORTH DAKOTA ART COLLECTIONS GALLERY AT THE EMPIRE ARTS CENTER, 415 Demers Ave., Grand Forks, ND 58201. Mailing Address: Hughes Fine Arts Center, 3350 Campus Rd., Stop 7099, Grand Forks, ND 58202-7099. Tel.: 701-777-2257. Fax: 701-777-2903.
E-mail: art.jones@und.edu
Founded: 2012.
Key Personnel: Dir., Dr. Arthur Jones
Institution Type/Description: University Art Collection Gallery.
Collections: paintings; photographs; ceramics; sculpture; prints; Native American art; folk art.
Hours & Admission Prices: Mon.-Fri. 11-4. No charge; donations accepted. &

UNIVERSITY OF NORTH DAKOTA ZOOLOGY MUSEUM, Dept. of Biology, Stop 9019, Grand Forks, ND 58202. Tel.: 701-777-2621. Fax: 701-777-2623.
E-mail: susan.felege@email.und.edu
Web Site: www.und.edu/dept/biology/undergrad
Founded: 1883.
Congressional District: 1
Key Personnel: Cur. Vertebrates, Susan Ellis-Felege; Cur. Invertebrates, Dr. Jefferson Vaughan.
Governing Authority: university. Affiliated with University of North Dakota. Tax-exempt: 170(b)(1)(A).
Institution Type/Description: Natural History Museum.
Collections: birds; fishes; mammals; insects; reptiles & amphibians; parasite of Aquatic Invertebrates.
Research Fields: pertaining to collections.
Facilities: 3,000-vol. library of taxonomy & history books available for inter-library loan; field research station; laboratories; classrooms.
Activities: guided tours; formally organized education programs for under-

graduate & graduate students affiliated with University of North Dakota; permanent & temporary exhibits.
Hours & Admission Prices: Mon.-Fri. 8-4:30 by appointment. No charge. Closed national holidays.
Attendance: 200 (estimated)

Grand Rapids

LAMOURE COUNTY MUSEUM, Memorial Park, Grand Rapids, ND 58458. Mailing Address: P.O. Box 128, LaMoure, ND 58458-0128. Tel.: 701-883-5301.
Institution Type/Description: Historical Building.
Collections: personal artifacts; period furnishings; Senator Milton Young.
Hours & Admission Prices: Memorial Day to Labor Day Sat.-Sun. 1-8; other times by appointment.

Grassy Butte

GRASSY BUTTE HISTORIC POST OFFICE, 101 Museum Dr., Grassy Butte, ND 58634. Mailing Address: 581 Highway 85 S, Grassy Butte, ND 58634. Tel.: 701-863-6769 & 6604.
E-mail: xrranch@ndsupernet.com
Formerly: Old Sod Post Office
Founded: 1912.
Key Personnel: Pres. (V) & Museum Shop Mgr., Gail Chinn; Chm. (V), Lois Flick; Dir., Rose Eschenko.
Personnel Profile: Full-Time Volunteers 2; Part-Time Paid 2; Part-Time Volunteers 2.
Governing Authority: Parent Institution: McKenzie County Historic Society. Tax-exempt.
Institution Type/Description: History Museum: built in 1912 of logs & sod, housed Grassy Butte Post Office from 1914 until 1964. Listed on the National Register of Historic Places.
Collections: period artifacts from 1800s & 1900s.
Hours & Admission Prices: Memorial Day to June Sat.-Sun. 9-4; July to Labor Day daily 9-4. No charge; donations accepted.
Attendance: 535 (accurate)

Hanks

PIONEER TRAILS MUSEUM, 9 miles west junction Hwy. 85 & Hwy. 50, Hanks, ND 58856. Mailing Address: 2310 University Ave., Williston, ND 58801. Tel.: 701-572-4759.
Founded: 1970.
Institution Type/Description: History Museum.
Collections: local history & culture; photographs; historical area artifacts from the early settlement days.
Hours & Admission Prices: June-Sept. Sun. 1-5, Mon.-Fri. by appointment. No charge; donations accepted.
Attendance: 85 (estimated)

Hatton

HATTON-EIELSON MUSEUM & HISTORICAL ASSOCIATION, 405 Eielson St., Hatton, ND 58240. Mailing Address: P.O. Box 278, Hatton, ND 58240-0278. Tel.: 701-543-3828. Fax: 701-543-4013.
E-mail: mlc@gra.midco.net
Web Site: www.hattonmuseum.org
Founded: 1973.
Congressional District: 20
Key Personnel: Pres., Gary Lillemoen; Vice Pres., Jerry Pederson; Asst. Cur., Eileen Mork; Public Rels., Treas. & Museum Shop Mgr., Eileen Holt.
Personnel Profile: Part-Time Volunteers 20.
Volunteer Hours: 95
Operating Expenses: 25,563
Operating Income: 26,183
Governing Authority: society; nonprofit organization. Tax-exempt.
Institution Type/Description: Historic Building: c.1900 Victorian home of aviator-explorer Carl Ben Eielson.
Collections: photographs; local history items; pioneer artifacts; mail service items; aviation memorabilia pertaining to Alaska & Antarctica; newspapers; Victorian furnishings; glassware; manuscripts; 1921 Fokker airplane fuselage that Eielson flew in Alaska.
Research Fields: local history.
Facilities: 500-vol. library of historical books available for research on premises. Museum-related items for sale.
Activities: guided tours; formally organized education programs for children & adults; permanent exhibitions. Annual Events: Barbeque; Spring Meatball Dinner; Christmas Tour.

Hours & Admission Prices: June–Aug. Sun. 1–4 or by appointment. Adults $5, children $1; preschool no charge.
Attendance: 1,000 (estimated)
Membership: Individual $5; Family $10; Business $25; Booster $100; Top of the World $1,000.

Hebron

HEBRON HISTORICAL AND ART SOCIETY, 606 Lincoln Ave., Hebron, ND 58638. Tel.: 701-878-4644.
Web Site: www.hebronnd.org/live_historical.html
Founded: 1979.
Congressional District: 32
Key Personnel: Dir. & Pres. (V), Robyn Renalds; Chm., Lyle Hoerauf; Vice Pres., Claudia Meberg; Museum Shop Mgr., Henry Mische
Governing Authority: Tax-exempt.
Institution Type/Description: History Museum.
Collections: area history; over 2,000 artifacts & memorabilia; 1929 Model A Ford Snowmobile; hand drawn horse cart; period stationary engines; photographs; 900 dolls from 57 countries.
Hours & Admission Prices: By appointment. No charge; donations accepted. ♿
Attendance: 200 (estimated)
Membership: Annual $3.

Hillsboro

TRAILL COUNTY HISTORICAL SOCIETY, 306 Caledonia W. Ave., Hillsboro, ND 58045. Mailing Address: Box 173, Hillsboro, ND 58045-0173. Tel.: 701-636-5571.
Founded: 1965.
Congressional District: 1
Key Personnel: Pres. (V), Shirley Nysveen; Cur., Marilu Person.
Personnel Profile: Part-Time Volunteers 200.
Governing Authority: Traill County. Affiliated with North Dakota State Historical Society, Bismarck. Tax-exempt.
Institution Type/Description: General Museum: housed in 1897 three-story brick mansion.
Collections: household furnishings of the 1800s; Indian artifact room & basement of early agricultural implements; Norwegian immigrant items; rare Red River fur trader's cart; old farm equipment & machinery. Historic Buildings: Pioneer Agricultural building; St. Olaf's Chapel; 1870 log cabin; 1920s rural one-room school.
Research Fields: local Indian archaeology.
Hours & Admission Prices: June–Aug. Sat.-Sun. 2-5; other times by appointment. Adults $2; members & children no charge when accompanied by an adult.
Attendance: 1,000 (estimated)
Membership: Individual $5; Family $10.

Hope

STEELE COUNTY HISTORICAL SOCIETY, 301 Steele Ave., Hope, ND 58046. Mailing Address: P.O. Box 144, Hope, ND 58046-0144. Tel.: 701-945-2394. Fax: 701-945-2394.
E-mail: scmuseum@invisimax.com
Web Site: www.steelecomuseum.com
Formerly: Steele County Museum
Founded: 1966.
Congressional District: 23
Key Personnel: Pres. (V), Homer Wennerston; Dir., Sue Johnson.
Personnel Profile: Full-Time Paid 1; Part-Time Paid 2; Part-Time Volunteers 15.
Governing Authority: nonprofit organization. Parent Institution: North Dakota State University. Tax-exempt: 501(c)(3).
Institution Type/Description: Historical Society Museum.
Collections: clothing; furniture; archival materials; photographs; newspapers; farm machinery; Indian artifacts; manuscripts; quilts; looms; folk art; cook car. Historic Buildings: 1882 Baldwin Arcade; one-room schoolhouse c.1893; Enger log house c.1872 ; Mitchell house c.1882; St. Petri Church.
Research Fields: local history; agriculture; rural life.
Activities: guided tours; lectures; films; arts & holiday festivals; permanent & temporary exhibitions; permanent weaving studio; traditional arts workshops; costume rentals.
Publications: annual report.
Hours & Admission Prices: May-Sept. Tues.-Fri. 9-5, Sun. 2-5; Oct.-April Tues.-Fri. 9-5. Suggested Donation: adults $5, seniors & children $3; students & members no charge. ♿
Attendance: 1,800 (estimated)
Membership: Individual $10; Family $15 & up; Friend $50-$99; Supporter $100-$499; Sustaining $500-$999; Benefactor $1,000 & up.

Jamestown

THE ARTS CENTER, 115 2nd St., S.W., Jamestown, ND 58401-4114. Mailing Address: P.O. Box 363, Jamestown, ND 58402-0363. Tel.: 701-251-2496. Fax: 701-251-1749.
E-mail: tbarnes@jamestownarts.com
Web Site: www.jamestownarts.com
Founded: 1982.
Congressional District: 47
Key Personnel: Dir., Taylor Barnes; Mgr. Gallery, Sally Jeppson.
Personnel Profile: Full-Time Paid 2; Part-Time Paid 2.
Governing Authority: Parent Institution: Jamestown Fine Arts Association. Tax-exempt.
Institution Type/Description: Art Museum.
Collections: works by regional artists.
Facilities: theater.
Activities: classes; performances; workshops; art shows; temporary exhibitions.
Hours & Admission Prices: Mon.-Fri. 9-5, Sat. 10-2. No charge; donations accepted. ♿
Attendance: 9,000 (estimated)
Membership: Senior $20; Individual $35; Family $50.

FORT SEWARD MUSEUM, 605 10th Ave., N.W., Jamestown, ND 58401-2027. Mailing Address: 4145 91st Ave., S.E., Ypsilanti, ND 58497. Tel.: 701-251-1875.
Institution Type/Description: Military History Museum: housed on the original site of the military fort, 1872-1877.
Collections: local history; military artifacts; photographs; memorial wall; personal artifacts; 30 x 60 ft. U.S. flag.
Hours & Admission Prices: Museum: Memorial Day to Labor Day daily 10-6. Grounds: daily.

FRONTIER VILLAGE ASSOCIATION, INC., 17th St., S.E., Jamestown, ND 58401. Mailing Address: P.O. Box 324, Jamestown, ND 58402-0324. Tel.: 701-252-6307. Fax: 701-252-5455.
E-mail: mhager@unisonbank.com
Founded: 1959.
Congressional District: 48
Key Personnel: Pres. (V), Mitzi Hagar; Pres. (V), Darcy Herman; Vice Pres., Charles Tanata.
Personnel Profile: Part-Time Paid 5; Part-Time Volunteers 4.
Governing Authority: nonprofit organization. Parent Institution: City of Jamestown. Tax-exempt: 501(c)(3).
Institution Type/Description: History Museum.
Collections: local history & culture; early Pioneer artifacts.
Activities: guided tours; arts festival; theater; stagecoach rides, pony rides, stagecoach hold-up (acting).
Publications: brochure.
Hours & Admission Prices: Memorial Day-Labor Day daily 9-9. No charge; donations accepted. ♿
Attendance: 150,000 (estimated)

NATIONAL BUFFALO MUSEUM, (M), 500 17th St., S.E., Jamestown, ND 58401-6456. Tel.: 701-252-8648; 800-807-1511. Fax: 701-253-5803.
E-mail: director@buffalomuseum.com
Web Site: www.buffalomuseum.com/
Founded: 1993.
Key Personnel: Exec. Dir., Kim Penrod; Museum Sho Mgr., Jessica Manson.
Personnel Profile: Full-Time Paid 2; Part-Time Paid 6; Part-Time Volunteers 15.
Governing Authority: nonprofit organization. Parent Institution: North Dakota Buffalo Foundation. Tax-exempt (c)(3).
Institution Type/Description: Buffalo Museum.
Collections: cultural & natural history of bison and the Great Plains; Plains Indian artifacts; paintings; sculpture; Native American art; 19th century firearms; North Dakota wildlife; Lewis & Clark; live bison herd.
Facilities: Museum-related items for sale.
Activities: educational programs; video presentation; community events; school & youth programs. Annual Events: White Cloud's Birthday Celebration; Potholes and Prairies Birding Festival.
Hours & Admission Prices: May & Sept.-Oct. Mon.-Fri. 9-5, Sat. 10-5, Sun. 12-5; Memorial Day to Labor Day daily 8-8; Nov.-April Mon.-Fri. 9-5, Sat. 10-5. Adults $5, seniors $4, students 7-18 $1; discounts to groups of 15 or more, AAA, AAM & ICOM members; members, military & children 6 & under no charge. National Park members no charge, donations accepted. ♿

Attendance: 18,898 (accurate)
Membership: Individual $15; Family $25; Lifetime $350. Business: Pewter $50; Silver $100; Gold $150; Diamond $200.

NORTH DAKOTA SPORTS HALL OF FAME, 212 3rd Ave., N.E., Jamestown, ND 58401. Tel.: 701-252-4835. Fax: 701-252-8089. Facebook: North Dakota Sports Hall of Fame.
E-mail: cdiestler@daktel.com
Institution Type/Description: Sports Hall of Fame.
Collections: sports history & athletes; personal artifacts; photographs; Hall of Fame inductees.
Activities: special events.
Publications: booklet, North Dakota Sports Hall of Fame, Inc.
Hours & Admission Prices: Mon.-Fri. 8-5. No charge.

STUTSMAN COUNTY MEMORIAL MUSEUM, 321 3rd Ave., S.E., Jamestown, ND 58401-4208. Mailing Address: P.O. Box 1002, Jamestown, ND 58402-1002. Tel.: 701-252-6741 & 4809.
Web Site: jamestownnd.com
Founded: 1964.
Congressional District: 12
Key Personnel: C.E.O. & Pres. (V), Harold Sahr; Cur., Alden Kollman.
Personnel Profile: Part-Time Paid 3; Part-Time Volunteers 10.
Governing Authority: nonprofit organization. Tax-exempt: 501(c)(3).
Institution Type/Description: Historic Building: built in 1907.
Collections: local history & culture; period artifacts; agriculture; costumes; dolls; dishes; equipment; Indian artifacts; medical; military; music; railroad room; country store; chapel; homesteaders shanty.
Facilities: 100-vol. library of local and state history books & 35 scrap books available for historical research; reading room.
Activities: Museum Sponsors: annual ice cream social fun night.
Publications: biannual newsletter.
Hours & Admission Prices: Memorial Day to Sept. Mon.-Fri. 10-5, Sat.-Sun. 1-5; tours by appointment. No charge; donations accepted. &
Attendance: 1,700 (estimated)
Membership: Individual $20, $50, $100, $250, $500.

Kulm

WHITESTONE HILL BATTLEFIELD STATE HISTORIC SITE, 7310 86th St., S.E., Kulm, ND 58456-9555. Mailing Address: 612 E. Boulevard Ave., Bismarck, ND 58505-0660. Tel.: 701-328-3508.
E-mail: histsoc@nd.gov
Web Site: history.nd.gov/historicsites/whitestone/
Founded: 1904.
Key Personnel: Site Supvr., James E. Hill; Historic Sites Mgr., Diane Rogness.
Governing Authority: state. Parent Institution: State Historical Society of North Dakota. Subsidiary Institution: Friends of Whitestone Hill Battlefield State Historic Site. Tax-exempt: 501(c)(3).
Institution Type/Description: Historic Site Museum: army troops under General Alfred Sully battled Sioux warriors in 1863.
Collections: period artifacts; graves; monuments commemorating the Battle of Whitestone Hill.
Facilities: 2 monuments; park, picnic & recreational sites.
Publications: quarterly, North Dakota History and Plains Talk.
Hours & Admission Prices: May 16 to Sept. 15 Thurs.-Mon. 10-5. No charge; donations accepted.
Attendance: 2,000 (accurate)

LaMoure

TOY FARMER MUSEUM, 7496 106th Ave., SE, LaMoure, ND 58458-9404. Tel.: 701-883-5206; 800-533-8293. Fax: 701-883-5209.
E-mail: info@toyfarmer.com
Web Site: www.toyfarmer.com/museum
Key Personnel: C.E.O., Cathy Scheibe
Institution Type/Description: Toy Farmer Museum.
Collections: toy memorabilia.
Facilities: Museum-related items for sale.
Hours & Admission Prices: May-Sept. Mon.-Fri. 10-6, Sat.-Sun. 12-5; Oct-April Mon.-Fri. 10-5, Sat.-Sun. 12-5. No charge. Closed New Year's Day; Easter; Thanksgiving; Christmas.

Lakota

A.M. TOFTHAGEN LIBRARY & MUSEUM AKA LAKOTA CITY LIBRARY, 116 W. B Ave., Lakota, ND 58344. Mailing Address: P.O. Box 307, Lakota, ND 58344-0307. Tel.: 701-247-2543.
E-mail: gerry.wagness@sendit.nodak.edu
Founded: 1927.
Congressional District: 23
Institution Type/Description: Library & History Museum: listed on the National Register of Historic Places.
Collections: books; paintings; sculpture; personal artifacts; local history & culture; scrapbooks.
Facilities: library.
Activities: story hour.
Hours & Admission Prices: June-Aug. Tues. 2-5 & 7-9, Thurs. 2-5, Sat. 1-5; Sept.-May Tues. & Thurs. 2-5, Sat. 1-5. No charge; donations accepted.

Larimore

LARIMORE COMMUNITY MUSEUM, 310 Towner Ave., Larimore, ND 58251. Mailing Address: P.O. Box 524, Larimore, ND 58251-0524. Tel.: 701-397-5723.
Key Personnel: Pres., Helen Welte; Treas., B. Jean Swanson
Institution Type/Description: History Museum.
Collections: local history & culture; photographs; personal artifacts.
Activities: fundraising event.
Hours & Admission Prices: Memorial Day to Labor Day. Suggested Donations: adults $1, students $.50.
Membership: Lifetime $25.

Lidgerwood

LIDGERWOOD COMMUNITY MUSEUM, 10 Third Ave., SE, Lidgerwood, ND 58053. Mailing Address: P.O. Box 36, Lidgerwood, ND 58053. Tel.: 701-538-4466.
E-mail: lmuseum@rrt.net
Key Personnel: Dir., Annette Smykowski
Institution Type/Description: History Museum.
Collections: local history & culture; antique furnishings; works of North Dakota native sculptor & artist Ida Prokop.
Hours & Admission Prices: First & third Sun. of the month 1-5. No charge; donations accepted.

Linton

EMMONS COUNTY MUSEUM, NW First and Oak, Linton, ND 58552. Mailing Address: P.O. Box 862, Linton, ND 58552-0862. Tel.: 701-254-4399.
Key Personnel: Pres. (V), Mary Ann Gefroh.
Personnel Profile: Full-Time Volunteers 2; Part-Time Volunteers 7.
Governing Authority: county. Parent Institution: Emmons County Historical Society. Tax-exempt.
Institution Type/Description: History Museum: housed in St. James Episcopal Church.
Collections: household items; military items; period buggy, sleigh & tools; leather tack.
Hours & Admission Prices: Fri. & Sun. 2-4; other times by appointment. No charge; donations accepted.
Attendance: 225 (estimated)
Membership: Annual $5; Life $50.

Makoti

MAKOTI THRESHERS' MUSEUM, 106 7th Ave. W., Makoti, ND 58756. Mailing Address: 30000 338th St. S.W., Ryder, ND 58779-9522. Tel.: 701-726-5656.
Web Site: www.makoti.net
Founded: 1961.
Key Personnel: Pres. (V), Merle Dreher
Governing Authority: Tax-exempt.
Institution Type/Description: History Museum.
Collections: over 250 stationary engines; 150 period farm tractors; prairie community including post office, church, blacksmith shop, & school; cars; trucks.
Hours & Admission Prices: June-Sept. by appointment. No charge; donations accepted.
Attendance: 3,000 (accurate)
Membership: Lifetime $10.

Mandan

FORT ABRAHAM LINCOLN STATE PARK, 4480 Ft. Lincoln Rd., Mandan, ND 58554-7947. Tel.: 701-667-6340. Fax: 701-667-6349.
E-mail: falsp@nd.gov
Web Site: www.parkrec.nd.gov
Founded: 1936.
Key Personnel: Park Mgr, Dan Schelske.
Personnel Profile: Full-Time Paid 3; Part-Time Paid 20.
Governing Authority: state. Affiliated with North Dakota Parks & Recreation Dept., 1600 E. Century Ave. #3, Bismarck, ND 58503. Tax-exempt.
Institution Type/Description: State Park Museum.
Collections: Mandan Indian artifacts; 1870s military; 7th Cavalry & Custer. Historic Sites: Slant Indian village; cavalry post; blockhouses.
Facilities: Books & other museum-related items for sale.
Activities: slide-tape program; movies on area; outdoor interpretive programs; guided & self-guided tours; interpretive markers.
Publications: brochures.
Hours & Admission Prices: May-Oct. daily 9-5; other times by appointment. Park Vehicle Fee: $5. Custer House Tours: adults $6, students $4. &
Attendance: 150,000 (estimated)

NORTH DAKOTA STATE RAILROAD MUSEUM, 3102 37th St., N.W., Mandan, ND 58554-7001. Mailing Address: P.O. Box 1001, Mandan, ND 58554-7001. Tel.: 701-663-9322.
Founded: 1985.
Personnel Profile: Part-Time Volunteers 15.
Governing Authority: Tax-exempt.
Institution Type/Description: Railroad History Museum.
Collections: train cars; HO models; photographs; timetables.
Facilities: Museum-related items for sale.
Activities: Annual Event: Railroad Day in August.
Hours & Admission Prices: Memorial Day to Labor Day daily 1-5. No charge; donations accepted. &
Attendance: 2,550 (estimated)

Manvel

MANVEL MUSEUM, Main St., Manvel, ND 58256. Mailing Address: 3286 Hwy. 81, Ardoch, ND 58261-9510. Tel.: 701-696-2279.
Institution Type/Description: History Museum.
Collections: personal items of former state representative Dagne Olson; antique furnishings; books.
Hours & Admission Prices: By appointment.

Marmarth

UNIQUE ANTIQUE AUTO MUSEUM, 305 1st St., W., Marmarth, ND 58643. Tel.: 701-279-5904.
E-mail: vanhorn1159@n.d.supernet.com
Web Site: www.uniqueantiquemuseum.com
Founded: 2002.
Key Personnel: Owner, J.D. VanHorn; Museum Shop Mgr., Janice M Abraham.
Governing Authority: private.
Institution Type/Description: Auto Museum.
Collections: 50 period automobiles; license plates from 1916 to present; early toys & dolls; hand-made wagons, coaches, & tractors; photographs.
Hours & Admission Prices: May 7-Sept. 7 daily 9-5. Adults $7; Lifetime members no charge. &
Membership: Lifetime $100.

Mayville

GOOSE RIVER HERITAGE CENTER, Main St. & 1st Ave., S.E., Mayville, ND 58257. Mailing Address: 320 1st St NW, Mayville, ND 58257-1107. Tel.: 701-788-4115.
Key Personnel: Pres., Betty Karaim.
Personnel Profile: Part-Time Volunteers 15.
Governing Authority: Tax-exempt.
Institution Type/Description: History Museum.
Collections: local history & culture; photographs; personal artifacts.
Hours & Admission Prices: Memorial Day to Labor Day Sat.-Sun. 1-4; other times by appointment. No charge; donations accepted.

Medora

CHATEAU DE MORES STATE HISTORIC SITE, 3448 Chateau Rd., Medora, ND 58645. Mailing Address: P.O. Box 106, Medora, ND 58645-0106. Tel.: 701-623-4355. Fax: 701-623-4921.
E-mail: shschateau@nd.gov
Web Site: www.nd.gov/list/chateau; www.history.nd.gov
Founded: 1936.
Key Personnel: Dir. State Historical Society of ND, Merl Paaverud; Dir. Historic Sites, Fern Swenson; Mgr. State Sites, Diane Rogness; Museum Shop Mgr., Rhonda Brown.
Personnel Profile: Full-Time Paid 1; Part-Time Paid 24; Part-Time Volunteers 12; Interns 2.
Governing Authority: state. Parent Institution: State Historical Society of North Dakota.
Institution Type/Description: Historic Site: housed in a 26 two-story chateau built as the summer residence for the Marquis de Mores family.
Collections: personal items belonging to the Marquis de Mores, French nobleman and entrepreneur; original furnishings; riding tack, coaches & clothing.
Facilities: visitor center; interpretive center. Video & museum-related items for sale.
Activities: self-guided tours; summer programs, including De Mores Day & History Alive.
Publications: books, Aristocracy on the Western Frontier: The Legacy of the Marquis de Mores (1994); The Career of the Marquis de Mores in the Badlands of North Dakota (1994).
Hours & Admission Prices: mid-May to mid-Sept. daily 8:30-6:30. Interpretive Center: Sept. 16-May 15 Wed.-Sun. 9-5. Adults $7, children 6-15 $3; discounts to groups of 20 or more; active military & children under 6 no charge. Season Pass: $20. Closed Easter; Thanksgiving; Christmas. &
Attendance: 85,000 (accurate)
Membership: Individual $35; Family $45.

THEODORE ROOSEVELT NATIONAL PARK-VISITOR CENTER, 315 2nd Ave., Medora, ND 58645. Mailing Address: P.O. Box 7, Medora, ND 58645-0007. Tel.: 701-623-4466. Fax: 701-623-4840. Facebook: Theodore Roosevelt National Park.
E-mail: eileen_andes@nps.gov
Web Site: www.nps.gov/thro
Founded: 1959.
Congressional District: 1
Key Personnel: Park Supt., Val Naylor; Chief Interpretation, Eileen Andes.
Personnel Profile: Full-Time Paid 3; Part-Time Paid 7; Part-Time Volunteers 3; Interns 4.
Governing Authority: federal government. Parent Institution: Theodore Roosevelt National Park. Tax-exempt.
Institution Type/Description: Visitor Center & National Park.
Collections: partial collection of Theodore Roosevelt's ranching effects; herbarium; anthropology; archaeology; botany; geology; Indian artifacts; industry; natural history, as it relates to Theodore Roosevelt National Park. Historic Building: 1883 Maltese Cross cabin.
Facilities: 2,500-vol. library of history & natural history books available for use on premises; visitors center. Publications for sale.
Activities: guided tours; permanent exhibitions; evening campfire programs.
Publications: Natural history handbooks; book, Theodore Roosevelt in the Dakota Badlands.
Hours & Admission Prices: Park: daily. Visitor Centers: daily 8-4:30; call or see website to confirm hours. Park: Vehicle $10, Individual $5. Annual Pass $80. Closed New Year's Day; Thanksgiving; Christmas. &
Attendance: 630,319 (accurate)

VON HOFFMAN HOUSE, 485 Broadway, Medora, ND 58645. Mailing Address: P.O. Box 198, Medora, ND 58645-0198. Tel.: 800-633-6721.
E-mail: medora@medora.com
Web Site: www.medora.com
Formerly: Medora Doll House
Founded: 2012.
Key Personnel: Dir., Kinley R. Slauter
Institution Type/Description: Historic House Museum: built in 1884.
Collections: period furnishings.
Hours & Admission Prices: Memorial Day to Labor Day daily 10-7. No charge; donations accepted.

Minot

DAKOTA TERRITORY AIR MUSEUM, 100 34th Ave., N.E., Minot, ND 58703. Mailing Address: P.O. Box 195, Minot, ND 58702-0195. Tel.: 701-852-8500.
E-mail: airmuseum@minot.com
Web Site: dakotaterritoryairmuseum.com/
Founded: 1986.
Congressional District: 1
Key Personnel: Chm. (V), Don Larson; Cur., Glenn Blackaby.
Personnel Profile: Full-Time Paid 1; Part-Time Paid 5; Part-Time Volunteers 40.
Institution Type/Description: Aviation History Museum.
Collections: Lockheed T-33 Jet Trainer (U.S. Air Force's first jet trainer); Douglas C-47 (DC-3) World War II Transport.; Douglas C-47 Cockpit; L-T-V (Chance Vought) A-7 Corsair 2; '34 Stinson Reliant; North American T-28; '34 Fairchild 24; '32 Monocoupe 110; '29 Waco 10; '40 Waco UPF-7; 1972 Pietenpol Aircamper; '37 J-2 Cub; '79 Vari-eze; '67 Volksplane; '46 Luscombe; '89 Rotorway Exec helicopter; '29 Arrow Sport; AT-6 Texan; Twin Beech D-18; '31 Waco QCF-2; '67 Starduster; '45 Taylorcraft; '46 Ercoupe; '42 Piper J-3 Cub; '47 A-3 Callair; '47 Cessna 195; '67 Breezy; '68 BU-133 Jungmeister; '38 Monocoupe 110 special racer; 2/3rds Cub Experimental; Cherokee II Sailplane; Wolf Boredom biplane; Rally 2B Ultralight; 1917 Buick D45; Minot Fire Department trucks from 1920s & 1930s; Arrow Sport Model F; Wright Flyer (replica); 1910 Curtiss Pusher (replica); Interstate Cadet L-6; 1951 Tripacer; cessna 170; 1958 Mooney M20A; Link Trainer; 1930 Taper Wing Waco; F-15A Eagle; F-106 Delta Dart; literature & periodicals; area military aviation service flight gear & equipment; photographs; aviation's role in the region's agriculture and transportation needs; collection of flying WWII warbirds from the Houston-based Texas Flying Legends Museum on display.
Activities: special events.
Publications: members' newsletter, Dakota Territory Air Museum.
Hours & Admission Prices: mid-May to mid-Oct. Mon.-Sat. 10-5, Sun. 1-5; other times by appointment. Family $15, adults $5, children 6-17 $3; members no charge. &
Attendance: 5,079 (estimated)
Membership: Individual $35; Family $45; Individual Lifetime $200; Family Lifetime & Business Lifetime $300.

NORTH DAKOTA ART GALLERY ASSOCIATION, (M), #2 11th Ave., N.W., Minot, ND 58701-6420. Tel.: 701-858-3242. Fax: 701-858-3894.
E-mail: ndaga@ndaga.org
Web Site: www.ndaga.org
Founded: 1977.
Key Personnel: Dir., Linda A. Olson.
Personnel Profile: Part-Time Paid 2; Part-Time Volunteers 2; Interns 2.
Institution Type/Description: Art Gallery.
Collections: works by local & regional artists.
Activities: children's programs; performances; workshops; student art show; student traveling show; arts trunks; cultural encounters & games; story telling; arts trunks.
Hours & Admission Prices: Mon.-Fri. 9-5.

NORTHWEST ART CENTER, Minot State University, 11th Ave. N.W., Minot, ND 58707. Mailing Address: 500 University Ave. W., Minot, ND 58707. Tel.: 701-858-3264.
E-mail: nac@minotstateu.edu
Web Site: www.minotstateu.edu/nac
Formerly: Hartnett Hall Gallery
Founded: 1976.
Key Personnel: Dir., Avis Veikley.
Personnel Profile: Part-Time Paid 4; Interns 2.
Governing Authority: Parent Institution: Minot State University. Tax-exempt.
Institution Type/Description: Art Center.
Collections: contemporary & traditional art by local, regional & national artists.
Facilities: library.
Activities: art exhibitions; lecture series.
Publications: Americas 2000.
Hours & Admission Prices: Mon.-Fri. 8-4:30; other times by appointment. No charge; donations accepted. &
Attendance: 5,000 (estimated)
Membership: Student & Senior Citizen $10; Individual $20; Household $25; Sponsor $50-$99; Patron $100-$999; Benefactor $1,000.

OLD SOO DEPOT TRANSPORTATION MUSEUM, 15 N. Main St., Minot, ND 58703-3103. Mailing Address: P.O. Box 2148, Minot, ND 58702. Tel.: 701-852-2234.
E-mail: soodepot@srt.com
Founded: 2000.
Institution Type/Description: Historic Building: housed in the restored 1912 Soo Line Depot.
Collections: American West transportation history including railroads, automobiles, buses, & aviation.
Hours & Admission Prices: Call for hours. No charge, donations accepted.

RAILROAD MUSEUM OF MINOT, 19 First St., N.E., Minot, ND 58701-3960. Mailing Address: P.O. Box 74, Minot, ND 58702-0074. Tel.: 701-852-7091. Facebook: Railroad Museum of Minot.
E-mail: railroadmuseum@srt.com
Founded: 1986.
Key Personnel: Dir., James Huston; Museum Shop Mgr., Roger Burchill
Institution Type/Description: Railroad Museum.
Collections: over 100 years of regional rail history; photographs; memorabilia; communication systems.
Facilities: library. Gift items for sale.
Activities: train rides.
Hours & Admission Prices: Sat. 10-2; other times by appointment. No charge; donations accepted. &
Attendance: 2,000 (estimated)
Membership: Annual $25.

ROOSEVELT PARK ZOO, 1219 Burdick Expwy., E., Minot, ND 58701. Mailing Address: P.O. Box 549, Minot, ND 58702-0549. Tel.: 701-857-4166. Fax: 701-857-4169. Facebook: RP Zoo.
E-mail: info@srt.com
Web Site: www.rpzoo.com
Founded: 1921.
Key Personnel: Chm., Bob Petry; Dir., David Merritt; Pres. (V), Jenny Steckler; Museum Shop Mgr., Jennifer Fry.
Personnel Profile: Full-Time Paid 7; Full-Time Volunteers 2; Part-Time Paid 20; Part-Time Volunteers 100.
Governing Authority: municipal. Parent Institution: Minot Park District. Subsidiary Institution: Greater Minot Zoological Society. Tax-exempt: 501(c)(3).
Institution Type/Description: Zoo.
Collections: mammals; birds; reptiles.
Activities: guided tours; organized education programs for children & adults.
Publications: quarterly newsletter, Inside Tracks.
Hours & Admission Prices: May-Sept. daily 10-8. Adults $7, Adults 55 & over $6.50, children 4-12 $3.75; discounts to groups; AZA members and children 3 & under no charge. Season passes available. &
Attendance: 75,000 (accurate)
Membership: Senior Citizen 55 & over $15; Individual $35; Family Plus $80; Patron $105.

SCANDINAVIAN HERITAGE PARK, 1020 S. Broadway, Minot, ND 58701-4660. Mailing Address: P.O. Box 862, Minot, ND 58702-0862. Tel.: 701-852-9161.
E-mail: scandha@srt.com
Web Site: www.scandinavianheritage.org
Key Personnel: Pres., Gail S. Peterson
Institution Type/Description: Cultural Heritage Museum.
Collections: Scandinavian heritage & culture; early area settlers; statues; historic house.
Activities: Annual Event: Norsk Hostfest in Oct.
Hours & Admission Prices: Mid-May to first weekend in Oct. Mon.-Fri. 10-4. No charge.
Membership: Sustaining $35; Sponsor $100-$499; Benefactor $500 & up.

TAUBE MUSEUM OF ART, 2 N. Main St., Minot, ND 58703-3104. Tel.: 701-838-4445. Fax: 701-838-6471.
E-mail: taube@srt.com
Web Site: www.taubemuseum.org
Formerly: Taube Museum of Art and Minot Art Association
Founded: 1970.
Congressional District: 2
Key Personnel: Exec. Dir. & Museum Shop Mgr., Nancy F. Walter; Pres. (V), Jim Hochhalter.
Personnel Profile: Full-Time Paid 1; Part-Time Paid 2; Part-Time Volunteers 25; Interns 1.
Governing Authority: nonprofit organization. Tax-exempt: 501(c)(3).
Institution Type/Description: Art Museum: housed in renovated bank building.

Collections: paintings; sculptures; photos.
Facilities: 2,346 sq. ft. main gallery. Original artworks & museum-related items for sale.
Activities: art auction; North Dakota student art show; art festivals; art classes for all ages; participation in local festivals & fund-raisers.
Publications: quarterly newsletter.
Hours & Admission Prices: Tues.-Fri. 10:30-5:30, Sat. 11-4; other times by appointment. No charge; donations accepted. Closed holidays. &
Attendance: 77,500 (estimated)
Membership: Students & Senior Citizens $25; Adult $45; Family $60; Sustainer $100; Business $200; Business Patron $500; Benefactor $1,000; Corporate $1,500.

WARD COUNTY HISTORICAL SOCIETY, 2005 Burdick Expwy. E., Minot, ND 58702. Mailing Address: P.O. Box 994, Minot, ND 58702-0994. Tel.: 701-839-0785. Facebook: Ward County Historical Society.
E-mail: wchs@wchsnd.org
Web Site: www.wchsnd.org
Founded: 1951.
Key Personnel: Dir., Sue Bergan; Pres., Glynn Brewer.
Personnel Profile: Part-Time Paid 1; Part-Time Volunteers 8; Interns 1.
Governing Authority: county. Tax-exempt.
Institution Type/Description: Historic Building & Site.
Collections: farm implements; church; blacksmith shop; county courthouse; dental facility; rail depot & artifacts; homesteader cabin; schoolhouse; post office; bank.
Research Fields: genealogical.
Facilities: research facility.
Activities: various seasonal programs.
Publications: bimonthly newsletter, Prairie Perspectives.
Hours & Admission Prices: Memorial Day-Labor Day Tues., Thurs. & Sat. 11-5; extended hours during ND State Fair. No charge; donations accepted. &
Attendance: 6,000 (accurate)
Membership: Single $20; Family $30; Lifetime $250.

Minto

WALSH COUNTY HISTORICAL MUSEUM, 323 3rd St., Minto, ND 58261. Tel.: 701-248-3237 & 3414.
Institution Type/Description: History Museum.
Collections: local history & culture; trading post; log cabin; area wildlife; photographs.
Hours & Admission Prices: Memorial Day to Labor Day Sun. 1-4; other times by appointment. Adults 18 & over $5, students $3; preschool no charge.

Mohall

RENVILLE COUNTY HISTORICAL SOCIETY, 504 First St., N.E., Mohall, ND 58761-4200. Mailing Address: 204 Central Ave., Apt. 2D, Mohall, ND 58761-0163. Tel.: 701-240-7015. Facebook: Renville County Historical Society.
E-mail: trevor_hoyt@hotmail.com
Founded: 1978.
Congressional District: 1
Key Personnel: Pres., Trevor Hoyt; Sec., Betty Johnson; Treas., Joyce Lunde.
Personnel Profile: Part-Time Paid 1; Part-Time Volunteers 4.
Volunteer Hours: 49
Governing Authority: nonprofit organization. Tax-exempt: 170(b)(1)(A).
Institution Type/Description: General Museum: housed in a pioneer church from Norma, ND & first depot built in Mohall, ND.
Collections: local items used in the area during homestead days; historic buildings.
Research Fields: history of governmental subdivisions.
Facilities: depot; church; quonsets.
Activities: guided tours; permanent exhibitions.
Publications: facebook page.
Hours & Admission Prices: Mid-June-mid-Sept. Thurs. 2-5. No charge; donations accepted. &
Attendance: 38 (estimated)
Membership: Annual $5; Life $100.

Mooreton

BAGG BONANZA HISTORICAL FARM, 8025 169th Ave., S.E., Mooreton, ND 58061. Mailing Address: P.O. Box 702, Mooreton, ND 58061. Tel.: 701-274-8989 & 642-2411.
Web Site: www.baggfarm.com

Founded: 1986.
Key Personnel: Chm. (V), Norma Nosek; Museum Shop Mgr., Pat Ward.
Personnel Profile: Part-Time Volunteers 40.
Governing Authority: Tax-exempt.
Institution Type/Description: History Museum.
Collections: farming from the 1800s to early 1900s; restored 21-bedroom main house.
Facilities: 15-acre farm.
Hours & Admission Prices: Summer: Fri.-Sun. 12-5; other times by appointment. Adults $5, children under 12 $3.50; members & children under 6 no charge. &
Attendance: 5,000 (estimated)
Membership: Single $25; Family $45; Business $100.

Morton

NORTH DAKOTA COWBOY HALL OF FAME, 105 3rd Ave., N.W., Morton, ND 58554. Tel.: 701-623-2000. Fax: 701-623-2001. Facebook; North Dakota Cowboy Hall of Fame.
E-mail: info@northdakotacowboy.com
Web Site: www.northdakotacowboy
Founded: 1995.
Key Personnel: Exec. Dir., Raymond Morrell
Institution Type/Description: History Museum.
Collections: Native American, ranching, & rodeo history; photographs; paintings; personal artifacts; Hall of Fame inductees.
Facilities: library; archives; 45-seat theater. Museum-related items for sale.
Activities: special events.
Hours & Admission Prices: May 15-Sept. 15 daily 9-6; off-season by appointment. Discounts to military members. &
Membership: See website.

Mott

MOTT GALLERY OF HISTORY & ART, Brown Ave., Mott, ND 58646. Mailing Address: 508 Minn Ave., Mott, ND 58646. Tel.: 701-824-2552.
Founded: 2003.
Personnel Profile: Part-Time Volunteers 18.
Governing Authority: city. Tax-exempt.
Institution Type/Description: Art Gallery & History Museum: housed in the historic bank building.
Collections: works by local artists; area history & culture; personal artifacts; period furnishings; photographs.
Activities: special events.
Hours & Admission Prices: Summer: Sun. & Thurs. 1-4. No charge; donations accepted. &
Attendance: 586 (accurate)

Napoleon

LOGAN COUNTY MUSEUM, 207 Lake St. W., Napoleon, ND 58561. Mailing Address: 208 E. 5th St., Napoleon, ND 58561-7217. Tel.: 701-754-2640 & 2221.
Founded: 1984.
Congressional District: 1
Key Personnel: Dir., Charles Weigel.
Personnel Profile: Part-Time Volunteers 20.
Governing Authority: county. Parent Institution: Logan County Historical Society. Tax-exempt.
Institution Type/Description: Historical Society Museum.
Collections: local history & culture; photographs.
Publications: annual newsletter.
Hours & Admission Prices: Sun., Memorial Day, Independence Day & Labor Day 1-4; other times by appointment. No charge; donations accepted.
Attendance: 150 (estimated)
Membership: Annual $5; Life $50.

New Rockford

EDDY COUNTY MUSEUM, 1115 1st Ave N, New Rockford, ND 58356. Mailing Address: P.O. Box 135, New Rockford, ND 58356. Tel.: 701-947-2205. Facebook: Eddy County Museum.
E-mail: dillon1348@gmail.com
Founded: 1964.
Key Personnel: Pres. (V), Jessica Dillon.
Institution Type/Description: History Museum.
Collections: local history & culture; photographs; historic furnishings & exhibits.

Hours & Admission Prices: Labor Day-Memorial Day Sun. 1-4. No charge; donations accepted. &

Attendance: 400 (estimated)

New Town

THREE TRIBES MUSEUM, 302 Frontage Rd., New Town, ND 58763. Mailing Address: P.O. Box 147, New Town, ND 58763-0147. Tel.: 701-627-4477. Fax: 701-627-3805.

E-mail: tatmuseum@restel.net

Founded: 1964.

Personnel Profile: Part-Time Paid 2; Part-Time Volunteers 2.

Governing Authority: Tax-exempt.

Institution Type/Description: Native American Museum.

Collections: Mandan, Hidatsa & Arikara Indian history & culture; photographs; personal artifacts.

Facilities: Museum-related items for sale.

Activities: tours.

Hours & Admission Prices: May-Oct. Mon.-Fri. 10-4. Adults $3, seniors & children 12-18 $2; children 11 & under no charge.

Attendance: 5,000 (accurate)

Oakes

DICKEY COUNTY HISTORICAL SOCIETY, 5th & Main, Oakes, ND 58474. Mailing Address: 9225 104th St., S.E., Oakes, ND 58474-9449. Tel.: 701-785-4361. Fax: 701-783-4485.

E-mail: mkunrath@drtel.net

Founded: 1975.

Key Personnel: Chm. (V), Lane Bredeson; Pres. (V), Mary Ann Kunrath.

Personnel Profile: Part-Time Volunteers 10.

Volunteer Hours: 150

Governing Authority: Tax-exempt.

Institution Type/Description: History Museum.

Collections: local history & culture; Burlington Northern caboose & depot; photographs.

Hours & Admission Prices: By appointment. No charge; donations accepted. &

Attendance: 400 (estimated)

Membership: Annual $10.

Parshall

PAUL BROSTE ROCK MUSEUM, Main St., Parshall, ND 58770. Mailing Address: P.O. Box 184, Parshall, ND 58770-0184. Tel.: 701-862-3264.

Founded: 1964.

Institution Type/Description: Rock Museum.

Collections: rocks; crystals.

Hours & Admission Prices: May to Labor Day Wed.-Sat. 12-4. Adults $4, students $2. &

Membership: Life $100.

Pembina

PEMBINA STATE MUSEUM, 805 State Hwy. 59, Pembina, ND 58271-0456. Mailing Address: P.O. Box 456, Pembina, ND 58271-0456. Tel.: 701-825-6840. Fax: 701-825-6383.

E-mail: jblanchard@nd.gov

Web Site: history.nd.gov

Founded: 1996.

Congressional District: 1

Key Personnel: Museum Site Supvr., Jeff Blanchard.

Personnel Profile: Full-Time Paid 1; Part-Time Paid 5.

Governing Authority: state; nonprofit organization. Parent Institution: State Historical Society of North Dakota. Tax-exempt.

Institution Type/Description: History Museum.

Collections: history of the Red River Valley of the North.

Major Exhibits: How Does Your Garden Grow? Gardening in North Dakota (T), 4/6/12-2/1/14.

Research Fields: fur trade; agricultural history; settlement; transportation history; frontier military-Fort Pembina; American Indian: Metis.

Facilities: public meeting room; 110 ft. observation tower; travel information center. Museum-related items for sale.

Activities: bimonthly interpretive programming.

Publications: quarterly, North Dakota History; quarterly, Plains Talk.

Hours & Admission Prices: Summer: May 16-Sept. 15 Mon.-Sat. 9-6, Sun. 1-6; Winter: Sept. 16-May 15 Mon.-Sat. 9-5, Sun. 1-5. Museum: no charge; donations accepted. Tower: $2; SHSND Foundation no charge. Closed New Year's Day; Easter; Thanksgiving; Christmas. &

Attendance: 5,042 (accurate)

Membership: State Historical Society of ND Foundation: Individual $35; Family $45.

Plaza

PLAZA COMMUNITY MUSEUM, 502 5th Ave., Plaza, ND 58771. Mailing Address: P.O. Box 188, Plaza, ND 58771-0188. Tel.: 701-497-3724.

E-mail: snowwhite@restel.com

Institution Type/Description: History Museum: housed in a church.

Collections: historic items from the Plaza community; photographs.

Hours & Admission Prices: By appointment. No charge; donations accepted.

Ray

RAY OPERA HOUSE MUSEUM, 119 Main St., Ray, ND 58849. Mailing Address: 420 4th Ave. E., Ray, ND 58849. Tel.: 701-568-3578.

E-mail: tricodev@nccray.com

Founded: 1989.

Congressional District: 2

Key Personnel: Pres., Gordon Lokken; Dir., Jerry Engel.

Personnel Profile: Part-Time Volunteers 1.

Governing Authority: Tax-exempt.

Institution Type/Description: History Museum.

Collections: local history; period furnishings; school memorabilia.

Hours & Admission Prices: June-Oct. Sun. 2-4; other times by appointment. No charge; donations accepted. &

Attendance: 50 (estimated)

Membership: Annual $1.

Regent

HETTINGER COUNTY HISTORICAL SOCIETY, Main Street, Regent, ND 58650. Mailing Address: P.O. Box 151, Regent, ND 58650-0151. Tel.: 701-563-4643.

Founded: 1962.

Congressional District: 31

Key Personnel: Chm. (V), Jess Kouba; Pres. (V), Gary Greff; Sec. & Treas., Paula Anderson.

Personnel Profile: Part-Time Paid 1; Part-Time Volunteers 7.

Governing Authority: society. Affiliated with State Historical Society of North Dakota. Bismarck, ND 58501. Tax-exempt.

Institution Type/Description: Local History Museum.

Collections: Pioneer Street: blacksmith shop; barber shop; rural church; print shop; schoolhouse; harness shop; bar; bank; insurance co.; general merchandise store; post office; hotel; jail; meat market; pioneer machinery & farm tools; written histories; genealogical sheets of county residents; personal artifacts; collectibles; 1911-1965 Dr. S. W. Hill Drug Store including equipment & tools; Hettinger County area furnishings; two period cars; home-built airplane; American Indian artifacts; antique store & merchandise; shells; stuffed birds & animals; army uniforms, WWI, WWII & Gulf War; artifacts from wars; two-year old German-Hungarian building full of authentic Banat (Germany) artifacts; genealogies of Hettinger County residents.

Research Fields: pioneer antiques; Indian relics; genealogical study.

Facilities: approx. 300-vol. library of medical books, newspapers, old books, maps & periodicals.

Activities: guided tours; annual meetings.

Publications: brochure; postcard.

Hours & Admission Prices: Memorial Day to Labor Day daily 9-5; other times by appointment. Adults $5, children $3; discount to groups. &

Attendance: 400 (estimated)

Membership: Individual $5.

Riverdale

GARRISON DAM NATIONAL FISH HATCHERY & VISITOR CENTER, Hatchery Rd., Riverdale, ND 58565. Mailing Address: P.O. Box 530, Riverdale, ND 58565-0530. Tel.: 701-654-7451.

Key Personnel: Mgr., Rob Holm

Institution Type/Description: Fish Hatchery & Visitor Center.

Collections: northern pike; walleye; smallmouth bass; crappie; rainbow trout; lake trout; brown trout; five 400 gallon aquariums; waterfowl; birds; mammals.

Facilities: nature trails.

Activities: observation area; educational programs; tours.

Hours & Admission Prices: Visitor Center: Labor Day to Memorial Day daily 8-3:30. Hatchery: daily.

Rugby

PRAIRIE VILLAGE MUSEUM, 102 Hwy. 2, S.E., Rugby, ND 58368-2424. Mailing Address: P.O. Box 232, Rugby, ND 58368-0232. Tel.: 701-776-6414 & 7606.
E-mail: prairievillagemuseum@gmail.com
Web Site: www.prairievillagemuseum.com
Founded: 1964.
Key Personnel: Exec. Dir., Catherine Jelsing; Pres. (V), Randy Myers.
Personnel Profile: Full-Time Paid 2; Part-Time Paid 6; Part-Time Volunteers 30.
Governing Authority: Parent Institution: Geographical Center Historical Society. Tax-exempt.
Institution Type/Description: History Museum.
Collections: late 1800s to present including turn of century household artifacts; farm machinery; 23 historic buildings; automobiles; Native American artifacts.
Activities: Annual Event: Village Fair Festival in August.
Publications: quarterly newsletter, Prairie Times.
Hours & Admission Prices: May 15-Sept. 16 daily 8-6; groups by appointment. Adults $7, students and seniors 65 & over $6, children 7-17 $3; active military & reserves and their families, children 6 & under no charge. &
Attendance: 3,900 (accurate)
Membership: Single $15; Household $30.

VICTORIAN DRESS MUSEUM, 312 Second Ave., S.W., Rugby, ND 58368-1708. Tel.: 701-776-2189.
Institution Type/Description: Dress Museum: housed in the former Saint Paul's Episcopal Church. Listed on the National Register of Historic Places.
Collections: 23 Victorian-style dresses; accessories.
Activities: tours.
Hours & Admission Prices: By appointment.

Ryder

RYDER HISTORICAL SOCIETY MUSEUM, 20510 184th St., S.W., Ryder, ND 58779-9547. Tel.: 701-758-2527.
Founded: 1978.
Key Personnel: Pres. (V), David Kraft; Museum Shop Mgr., Faye Karna.
Personnel Profile: Part-Time Volunteers 2.
Governing Authority: Tax-exempt.
Institution Type/Description: Historical Society Museum.
Collections: local history & culture; photographs; dolls; salt-n-pepper shakers; pens; military uniforms; school room & clothing; old irons; license plates; cameras; dental equipment; dishes; old tools; barb wire; period glasses; bottles; railroad items; creamery items; blacksmith shop items; store artifacts; church artifacts; fireman artifacts; souvenirs; dental office supplies; barber shop supplies; early toys; dolls; soo line train artifacts.
Hours & Admission Prices: June-Sept. call for hours; other times by appointment. No charge; donations accepted. &
Attendance: 150 (estimated)
Membership: Life $35.

Saint John

ROLETTE COUNTY HISTORICAL SOCIETY MUSEUM, Main St., Saint John, ND 58369. Mailing Address: P.O. Box 377, Saint John, ND 58369-0377. Tel.: 701-477-3026 & 244-5814.
Founded: 1974.
Governing Authority: county. Parent Institution: Rolette County Historical Society.
Institution Type/Description: Historical Society Museum.
Collections: local & regional exhibits; clothing; photography; household & agricultural tools; machinery.
Hours & Admission Prices: Memorial Day to Labor Day Sun. 2-4.

Stanley

FLICKERTAIL VILLAGE AND MUSEUM, 5th St., S.E. off U.S. 2, Stanley, ND 58784. Mailing Address: 1012 E. LaSalle Dr., Bismarck, ND 58503-8895. Tel.: 701-628-3335 & 2802.
Key Personnel: Dir., Robert G. Liebl
Institution Type/Description: History Museum.
Collections: 18 buildings including depot, jail, school, church, homestead, & country store; Girl Scout artifacts; restored dolls.
Hours & Admission Prices: June-Aug. Wed. 6:30pm-8:30pm, Sun. 2-4; other times by appointment. Adults $3, children 11 & under $1.

Stanton

KNIFE RIVER INDIAN VILLAGES NATIONAL HISTORIC SITE, 564 County Rd. 37, Stanton, ND 58571-9422. Mailing Address: P.O. Box 9, Stanton, ND 58571-0009. Tel.: 701-745-3300. Fax: 701-745-3708.
E-mail: knri_information@nps.gov
Web Site: www.nps.gov/knri
Founded: 1974.
Congressional District: 1
Key Personnel: Supt., Wendy Ross.
Governing Authority: federal. Parent Institution: Dept. of the Interior. Subsidiary Institution: National Park Service, Washington, D.C. Tax-exempt: 501(c)(3).
Institution Type/Description: Historic Site.
Collections: Hidatsa & Mandan Cultural Artifacts; Native Americans of the Northern Plains.
Facilities: visitor center. Books for sale.
Activities: tours; special events; self-guided nature & cross-country ski trails. Museum Sponsors: Northern Plains Indian Culture Festival last weekend in July.
Hours & Admission Prices: Memorial Day to Labor Day daily 8-6, tours available; Sept.-May 8-4:30. No charge. Closed New Year's Day; Thanksgiving; Christmas. &
Attendance: 24,704 (accurate)

Steele

KIDDER COUNTY MUSEUM, 103 W. Broadway, Steele, ND 58482-7110. Mailing Address: P.O. Box 383, Steele, ND 58482. Tel.: 701-475-2133 & 2741.
Key Personnel: Pres. (V), Glen DeKrey; Vice Pres., Eleanore Wolbaum; Sec., Rachel DeKrey; Treas., Bev Johnson.
Governing Authority: Parent Institution: Kidder county Historical Society.
Institution Type/Description: History Museum.
Collections: local history & culture; period furnishings; personal artifacts; photographs; household artifacts; medical equipment.
Hours & Admission Prices: Mon.-Fri. 9-12; other times by appointment. No charge.

Strasburg

PIONEER HERITAGE, INC. - LAWRENCE WELK HOUSE, 845 88th St., S.E., Strasburg, ND 58573. Mailing Address: P.O. Box 52, Strasburg, ND 58573-0052. Tel.: 701-336-7777; 701-254-4439.
Formerly: Ludwig Welk Farmstead
Founded: 1988.
Congressional District: 28
Key Personnel: Pres. (V), Adam Baumstarck.
Personnel Profile: Part-Time Paid 6.
Institution Type/Description: Historic House Museum: housed in the boyhood home of Lawrence Welk.
Collections: personal artifacts; period furnishings; farm machinery.
Hours & Admission Prices: Memorial Day to Labor Day Thurs.-Sun. 10-5. Adults $5, children 6-12 $3; children under 6 no charge. &
Attendance: 898 (accurate)

Tioga

NORSEMAN MUSEUM, Corner of Second St. N. & Welo, Tioga, ND 58852. Mailing Address: P.O. Box 699, Tioga, ND 58852-0699. Tel.: 701-664-2702.
Institution Type/Description: History Museum.
Collections: local history & culture; photographs & artifacts of early settlers homes & farms.
Hours & Admission Prices: June-Aug. Sun. 2-4; other times by appointment.

Valley City

BARNES COUNTY HISTORICAL MUSEUM, 315 Central Ave. N., Valley City, ND 58072-2954. Tel.: 701-845-0966. Fax: 701-845-5223 (Attn: Wes).
E-mail: bchistoricalsociety@hotmail.com
Web Site: www.hellovalley.com
Founded: 1930.
Congressional District: 1
Key Personnel: Cur., Wes Anderson.
Personnel Profile: Full-Time Paid 1; Part-Time Volunteers 40.

Governing Authority: society. Parent Institution: Barnes County Historical Society. Tax-exempt.
Institution Type/Description: History Museum.
Collections: historical exhibits; military artifacts; costumes; tools; glasswares; dolls; Far Eastern artifacts.
Research Fields: genealogy; county history.
Facilities: newspaper research files.
Activities: historical tours; member study of county history; lectures.
Publications: book, History of Barnes County.
Hours & Admission Prices: Mon.-Sat. 10-4, Sun. by appointment. No charge; donations accepted. Closed national holidays. &
Attendance: 10,000 (accurate)
Membership: Individual $20; Family $25; Friends Club $100-$1,000.

Wahpeton

CHAHINKAPA ZOO, 1004 R.J. Hughes Dr., Wahpeton, ND 58075. Mailing Address: P.O. Box 1325, Wahpeton, ND 58074-1325. Tel.: 701-642-8709. Fax: 701-642-9285. Facebook: Chahinkapa Zoo.
E-mail: administration@chahinkapazoo.org
Web Site: www.chahinkapazoo.org
Key Personnel: Zoo Dir., Kathy Diekman; Cur., Tom Schmaltz; Lead Keeper/Trainer & Registrar, Addy Paul
Institution Type/Description: Zoo.
Collections: over 200 animals & birds representing 60 species including fossa, otters, bison, monkeys, gibbon apes, camels, snow leopards, grizzly bears, wallabies, cougars, gemsbok, llamas, elk & Bengal tigers.
Hours & Admission Prices: April-Aug. daily 10-7; Sept.-Oct. daily 10-5. Adults $8, seniors 62 & up and military with ID $7, children 4-12 $4; Prairie Rose Carousel $2; members and children 3 & under no charge.
Membership: Individual $30; Family & Grandparent $55; Donor $100; Patron $250; Benefactor $1,000.

RICHLAND COUNTY HISTORICAL MUSEUM, 2nd St. and 7th Ave. N., Wahpeton, ND 58075. Mailing Address: P.O. Box 1292, Wahpeton, ND 58074-1292. Tel.: 701-642-3075.
E-mail: richcomuseum@702.com.net
Web Site: richco.m702com.net
Founded: 1946.
Congressional District: 25
Key Personnel: Pres. (V), Lois Berndt; Museum Shop Mgr., Marjo Johnson.
Personnel Profile: Full-Time Paid 4; Part-Time Paid 2; Part-Time Volunteers 5.
Governing Authority: county. Affiliated with Richland County Historical Society. Tax-exempt.
Institution Type/Description: Historical Society Museum.
Collections: archives; Indian artifacts; Rosemeade pottery; Roger's statues; manuscripts; genealogical; land records; cemetery records; 1885 census.
Research Fields: local history.
Facilities: Museum-related items for sale.
Activities: guided tours; permanent exhibitions; historical society meets monthly.
Publications: History of Richland County; Wahpeton Daily News; Hankinson Monitor.
Hours & Admission Prices: mid-April to Oct. Tues., Thurs. & Sat.-Sun. 1-4. No charge; donations accepted. Closed Easter; Independence Day; Labor Day. &
Attendance: 1,300 (estimated)
Membership: Life $25 & up.

Walhalla

GINGRAS TRADING POST STATE HISTORIC SITE, 12882 105th St., N.E., Walhalla, ND 58282-9757. Mailing Address: C/o Pembina State Museum, 805 Hwy. 59, P.O. Box 456, Pembina, ND 58271. Tel.: 701-549-2775. Fax: 701-825-6383.
E-mail: jblanchard@nd.gov
Web Site: history.nd.gov
Founded: 1956.
Congressional District: 1
Key Personnel: Site Supvr., Jeff Blanchard.
Personnel Profile: Full-Time Paid 1; Part-Time Paid 3.
Governing Authority: state. Parent Institution: State Historical Society of North Dakota. Tax-exempt.
Institution Type/Description: Historic Site Museum.
Collections: fur trade hand-hewn oak log store & home of Antoine Gingras.
Facilities: Fur trade store.
Activities: monthly interpretive programming.
Publications: quarterly, North Dakota History and Plains Talk.
Hours & Admission Prices: May 16-Sept. 15 daily 10-5. No charge; donations accepted.

Attendance: 793 (accurate)
Membership: State Historical Society of ND Foundation: Individual $35; Family $45.

Washburn

MCLEAN COUNTY HISTORICAL SOCIETY MUSEUM, 610 Main St., Washburn, ND 58577. Mailing Address: P.O. Box 345, Washburn, ND 58577-0345. Tel.: 701-462-3744.
E-mail: vmerkel@westriv.com
Web Site: www.wrtc.com/vmerkel/McleanCountyMuseum
Founded: 1967.
Congressional District: 4
Key Personnel: Dir. & Pres. (V), Jenell Olson; Cur., Vivian Merkel.
Personnel Profile: Part-Time Paid 4; Part-Time Volunteers 6.
Governing Authority: society. Affiliated with the McLean County Historical Society. Tax-exempt.
Institution Type/Description: Historical Society Museum: housed in 1905, County Courthouse & separate building adjacent.
Collections: pioneer & Indian artifacts; office machine; threshing machine; steam engine; birds & small animals; audiovisual & film; photo equipment & prints; keelboat; medical equipment; geological; farm tools; fishing equipment.
Research Fields: newspapers; pioneer books; town centennial books; family histories.
Facilities: 300-vol. library of old books, diaries, papers, historical materials, available for research on premises. Books for sale.
Activities: guided tours; research.
Publications: reprints, Fifty Pioneer Mothers; Pioneer Days of Washburn; Centennial Book of Towns in McLean County.
Hours & Admission Prices: June-Aug. Tues.-Sat. 1-4; other times by appointment. No charge; donations accepted. &
Attendance: 800 (estimated)
Membership: Individual $5; Family $10; Life $100.

THE NORTH DAKOTA LEWIS & CLARK INTERPRETIVE CENTER, 2876 N. 8th St., S.E., Washburn, ND 58577. Mailing Address: P.O. Box 607, Washburn, ND 58577-0607. Tel.: 701-462-8535; 877-462-8535. Fax: 701-462-3316.
E-mail: info@fortmandan.org
Web Site: www.fortmandan.org
Founded: 1997.
Key Personnel: Pres., David Borlaug; Dir., Clay Jenkinson; Membership Dir. & Exec. Asst., Nancy Krebsbach; Mktg. Coord., Nicolette Borlaug; Museum Store Mgr., Sarah Trandahl
Institution Type/Description: History Museum.
Collections: Lewis & Clark expedition; Native American artifacts; wood canoe; Karl Bodmer watercolors.
Activities: facility rental.
Hours & Admission Prices: Memorial Day to Labor Day daily 9-5; Sept.-May Mon.-Sat. 9-5, Sun. 12-5; groups by appointment. Adults $7.50, students $5; members no charge. &
Membership: Corps $50-$99; Captain $100-$249; Gallatin $250-$499; Madison $500-$999; Jefferson $1,000 & up.

Watford City

PIONEER MUSEUM OF MCKENZIE COUNTY, 100 2nd Ave., S.W., Watford City, ND 58854. Mailing Address: P.O. Box 126, Watford City, ND 58854-0126. Tel.: 701-444-2990. Fax: 701-444-5804.
E-mail: museum@ruggedwest.com
Web Site: www.4eyes.net
Founded: 1968.
Congressional District: 39
Key Personnel: Pres. (V), Jennifer Sorenson; Dir., Charlotte Schilke; Sec. & Treas., Jan Dodge.
Personnel Profile: Full-Time Volunteers 1; Part-Time Paid 3; Part-Time Volunteers 3.
Governing Authority: nonprofit organization. Parent Institution: McKenzie County Historical Society, Watford City, ND 58854. Tax-exempt: 501(c)(3).
Institution Type/Description: Pioneer Museum.
Collections: pioneer furnishings & clothing; photographs.
Activities: permanent exhibitions.
Hours & Admission Prices: Mon.-Sat. 10-6. Adults $2, students $1; pre-school children no charge. &
Attendance: 1,000 (estimated)

West Fargo

CASS COUNTY HISTORICAL SOCIETY AT BONANZA-VILLE, 1351 W. Main Ave., West Fargo, ND 58078-1321. Mailing Address: P.O. Box 719, West Fargo, ND 58078-0719. Tel.: 701-282-2822. Fax: 701-282-7606.
E-mail: info@bonanzaville.com
Web Site: www.bonanzaville.com
Formerly: Red River & Northern Plains Regional Museum
Founded: 1954.
Key Personnel: Pres., Brian Stavenger; Exec. Dir., Troy White; Cur., Andrew Nielsen; Facilities Mgr., Joe Vasek.
Personnel Profile: Full-Time Paid 4; Full-Time Volunteers 2; Part-Time Paid 1; Part-Time Volunteers 100; Interns 1.
Governing Authority: society. Tax-exempt: 501(c)(3) & 170(b).
Institution Type/Description: Pioneer Village & Museum Complex: consisting of 44 buildings.
Collections: Indian artifacts; pioneer home equipment; crafts; textiles; maps & atlas; city directories; historical North Dakota history from 1869-1920s; publications; limited genealogical material; minerals; period cars; cameras; dolls; farm machinery; Rosemeade pottery. Historic Structures: 1920 log cabin; 1890 Arthur Town Hall; 1887 saw mill; 1880 grainery; 1897 Hagen House; 1881 Houston House; 1884 farm house; 1895 one-room school; 1898 St. John's Church; 1890 homestead; 1900 barber shop.
Research Fields: local & regional history.
Facilities: banquet & wedding facilities. Museum-related items for sale.
Activities: self-guided tours; special events; rotating & permanent exhibitions. Annual Events: Pioneer Days in August; Living History events in summer; Christmas on the Prairie.
Publications: quarterly newsletter; bimonthly newsletter, Bonanzaville Times.
Hours & Admission Prices: Village: call for seasonal hours. Museum: Mon.-Sat. 10-5, Sun. 12-5. Adults $10, children $5; members no charge.
Attendance: 19,414 (accurate)
Membership: Pioneer $30; Explorer $50; Pathfinder $100; Trailblazer $250; Settler $500; Discoverer $1,000; Harvester $2,500; Homesteader $5,000.

Williston

FORT BUFORD STATE HISTORIC SITE, 15349 39th Lane, N.W., Williston, ND 58801-8677. Tel.: 701-572-9034. Fax: 701-572-9033.
E-mail: shsbuford@nd.gov
Web Site: www.nd.gov/hist
Founded: 1931.
Key Personnel: Dir. State Historical Society of ND, Merl Paaverud; Div. Dir. Historic Sites, Fern Swenson; Mgr. Historic Sites, Diane Rogness; Site Supvr., Steven Reidburn; Asst. Site Supvr., Kerry Finsaas; Museum Shop Mgr., Rhonda Brown.
Personnel Profile: Full-Time Paid 2; Part-Time Paid 10; Part-Time Volunteers 3.
Governing Authority: state. nonprofit organization. Parent Institution: State Historical Society of North Dakota. Subsidiary Institution: Friends of Fort Union/Fort Buford. Tax-exempt: 170(b)(1)(A).
Institution Type/Description: Historic Site: housed in 1871 officers quarters of Fort Buford, at confluence of Yellowstone and Missouri Rivers; site of Sitting Bull's surrender July 20, 1881.
Collections: military uniforms, weapons, furnishings used at the site during this era; history of confluence region & North Dakota; reconstructed barracks; officers' quarters; magazine buildings; post cemetery sites.
Facilities: picnic area; campground; visitor center; meeting room. Museum-related items for sale.
Activities: permanent & temporary exhibits.
Publications: book, The Last Years of Sitting Bull, by Herbert T. Hoover & Robert C. Hollow (second printing, 1985); North Dakota history journal.
Hours & Admission Prices: May 16-Sept. 15 daily 8-6. Confluence Center: Sept. 16 to May 15 Wed.-Sun. 9-4. Adults $5, children 6-15 $2.50, school groups $1 each; discounts to SHSND foundation members; members and children 5 & under no charge. Tour Bus: $40 per bus. Annual Site Pass: $20. Closed New Year's Day, Easter, Thanksgiving, Christmas. &
Attendance: 14,753 (accurate)
Membership: Individual $35; Family $45.

FORT UNION TRADING POST NATIONAL HISTORIC SITE, 15550 Hwy. 1804, Williston, ND 58801-8680. Tel.: 701-572-9083. Fax: 701-572-7321.
E-mail: lisa.sander@nps.gov
Web Site: www.nps.gov/fous
Founded: 1966.
Key Personnel: Supt., Andrew Banta.
Personnel Profile: Full-Time Paid 7; Part-Time Paid 1; Part-Time Volunteers 1.
Governing Authority: federal. Affiliated with U.S. National Park Service, Dept. of the Interior. Tax-exempt.
Institution Type/Description: Historic Site: site of American Fur Company reconstructed fur trading fort at the historic confluence of the Missouri & Yellowstone Rivers.
Collections: archaeological specimens & field records; fur trade related artifacts; Ben Innis book & manuscripts; Union-Buford Council papers.
Research Fields: Northern Plains fur trade era including the Anglo European influence & the Native American influence; Lewis & Clark.
Facilities: 3,000-vol. library of books & microfilms on Northern Plains Indians, military, & fur trade of the upper Missouri Region including North Dakota, Montana & Saskatchewan, available on site. Books on the Upper Missouri River Region & its history for sale.
Activities: guided tours; lectures; summer living history programs.
Publications: newsletter, Confluence News.
Hours & Admission Prices: Winter: daily 9-5:30; Summer: daily 8-6:30. No charge; donations accepted. Closed New Year's Day; Thanksgiving; Christmas. &
Attendance: 20,000 (accurate)
Membership: Individual $10.

FRONTIER MUSEUM, 6330 2nd Ave. W., Williston, ND 58801. Mailing Address: c/o Williams County Historical Society, 519 11th Ave., W., Williston, ND 58801. Tel.: 701-577-4504.
E-mail: jimr@co.williams.nd.us
Founded: 1958.
Congressional District: 1
Key Personnel: Pres., Jim Ryen.
Governing Authority: historical society. Parent Institution: Williams County Historical Society; nonprofit organization. Tax-exempt: 501(c)(3).
Institution Type/Description: General Museum.
Collections: history; paintings; sculpture; graphics; decorative arts; archaeology; military; numismatic; philatelic; textiles; transportation. Historic Buildings: 1890 Log Cabin; 1906 old judge's home; 1906 country church; 1906 grocery store; Great Northern caboose; 1914 country school; old time doctor's & dentist's office.
Research Fields: frontier living; Indian artifacts; rock identification.
Facilities: 300-vol. library of material on history, biography, & exploration available for use on premises. Locally-made crafts & other museum-related items for sale.
Activities: guided tours; lectures; tapes; permanent exhibitions.
Hours & Admission Prices: By appointment. Adults $3, children $1.50. &
Attendance: 2,000 (estimated)
Membership: Annual $3; Patron $10; Sustaining $25; Life $100.

Wolford

DALE & MARTHA HAWK MUSEUM, 4839 78 St., Wolford, ND 58385-9402. Tel.: 701-583-2381.
E-mail: dmhawk@gondtc.com
Web Site: www.hawkmuseum.org
Founded: 1981.
Key Personnel: Dir. & Pres. (V), Lowell Johnson; Cur., Gordon Thingvold.
Governing Authority: Tax-exempt.
Institution Type/Description: History Museum.
Collections: period cars, farm equipment & threshers; personal artifacts; steam engines; period tractors; early furniture; general store; church; one-room school; black smith shop; motorcycles.
Activities: camping. Annual Events: Farm Show in June.
Hours & Admission Prices: May-Oct. daily 9-5. Adults $5, children under 12 $1.50. &
Attendance: 3,500 (estimated)
Membership: Lifetime $500.

Woodworth

MELZER MUSEUM, Main St., Woodworth, ND 58476. Tel.: 701-752-4119.
Institution Type/Description: History Museum.
Collections: local history & culture; American Indian photography; office of Dr. Melzer, an early settlement physician.
Hours & Admission Prices: By appointment.

OHIO
(472 listings)

Akron

✳ **AKRON ART MUSEUM, (M),** One South High, Akron, OH 44308-1801. Tel.: 330-376-9185. Fax: 330-376-1180.
E-mail: mail@akronartmuseum.org
Web Site: www.akronartmuseum.org
Founded: 1922.
Congressional District: 14
Key Personnel: Dir. & C.E.O., Mitchell Kahan, Ph.D.; Pres. (V), Fred Bidwell; Dir. Education, Melissa Higgins-Linder; Sr. Educator, Alison Caplan; Dir. Devel., Jon Trainor; Dir. Mktg. Communication, Elizabeth Wilson; Collections Mgr., Arnold Tunstall; Interim Chief Curator, Ellen Rudolph; C.O.O., Gail Wild; Museum Shop Mgr., Laura Firestone; Asst. to Dir., Cristina Alexander.
Personnel Profile: Full-Time Paid 25; Part-Time Paid 40; Part-Time Volunteers 150; Interns 10.
Governing Authority: nonprofit organization. Tax-exempt: 501(c)(3).
Institution Type/Description: Art Museum.
Collections: paintings & sculpture by Philip Guston, Frank Stella, Alma Thomas, Claes Oldenburg, Donald Judd, David Salle, Richard Deacon, Lari Pittman, Sol LeWitt, El Anatsui & Doris Salcedo; works by photographers Lewis Hine, Robert Frank, Harry Callahan, Lee Friedlander, Gilbert & George, Hiroshi Sugimoto, Sophie Calle & Carrie Mae Weems.
Research Fields: modern art; contemporary art & photography.
Facilities: 12,493-vol. library of technical art books & publications and 5,500 artist files available to students & teachers from area universities.
Activities: lectures; concerts; video programs; formally organized educational programs; docent program.
Publications: biennial report; quarterly magazine; gallery guides; exhibition catalogues.
Hours & Admission Prices: Administrative Office: Mon.-Fri. 9-5. Gallery: Wed. & Fri.-Sun. 11-5, Thurs. 11-9. Adults $7, students and seniors 65 & over $5; discounts to AAM, NARM, & Metro RTA members; 1st Sun., children 12 & under and members no charge. Ohio Reciprocal Program. Closed New Year's Day; Memorial Day; Independence Day; Labor Day; Thanksgiving; Christmas Eve & Day. &
Membership: Individual $50; Dual & Household $65; Contributor $100; Contributor Plus $150; Sponsor $250; Sustainer $500; Director's Circle $1,000 & up.

AKRON POLICE MUSEUM, Harold K. Stubbs Justice Center, 217 S. High St., #508A, Akron, OH 44308-1636. Tel.: 330-375-2390.
Institution Type/Description: Police History Museum.
Collections: confiscated weapons; gambling & narcotics paraphernalia; counterfeit money; police equipment, uniforms & weapons; photographs; 1965 Harley-Davidson police motorcycle.
Hours & Admission Prices: Mon.-Fri. 8-3:30; groups by appointment.

AKRON ZOOLOGICAL PARK, 504 Euclid Ave., Akron, OH 44307. Mailing Address: 500 Edgewood Ave., Akron, OH 44307. Tel.: 330-375-2550. Fax: 330-375-2575.
E-mail: info@akronzoo.org
Web Site: www.akronzoo.org
Founded: 1953.
Congressional District: 14
Key Personnel: Pres. & C.E.O., L. Patricia Simmons; Dir. Mktg. & Guest Svcs., David Barnhardt; Dir. Devel., Pamela Webb.
Governing Authority: bd. of trustees; nonprofit. Tax-exempt: 501(c)(3).
Institution Type/Description: Zoo.
Collections: lions; tigers; jaguars; snow leopards; condors; bats; penguins; Komodo dragons; flamingos; Chinese alligators; Galapagos tortoises; eagles; river otters; bears; gibbons; waterfowl; Ohio farmland; parrots.
Research Fields: animal behavior.
Facilities: food concession; picnic grounds. Zoo-related items for sale.
Activities: school programs & orientations; docent program; train rides; temporary, traveling & permanent exhibits; guided tours; formally organized education programs. Zoo Sponsors: Animal Show in summer; Halloween Trick or Treat adventure.
Publications: quarterly newsletter, Zoo Tales; brochure.
Hours & Admission Prices: May-Oct. daily 10-5. Adults $11, senior citizens 62 & up $9, children 2-14 $8. Nov.-April daily 11-4. Admission $7. Discounts to groups & AZA members; members & children under 2 no charge. Closed New Year's Day; Thanksgiving; Christmas Eve & Day. &
Attendance: 255,000 (estimated)
Membership: Individual $55; Companion $65; Family & Grandparent $75; Family Deluxe $125; Explorer $250; Adventurer $500; Discoverer $1,000.

BETHLEHEM CAVE AND NATIVITY MUSEUM, Nativity of The Lord Jesus Catholic Church, 2425 Myersville Rd., Akron, OH 44312. Tel.: 330-699-5086. Fax: 330-699-4299.
E-mail: nativity@neohio.twcbc.com
Web Site: www.nativityofthelord.org/museum.asp
Institution Type/Description: Religious Museum.
Collections: Bethlehem Cave replica; Altar of the Nativity; Altar of the Magi; Manger of the Infant Jesus; Nativity scenes from around the world; 16th century church bell; stained glass windows depicting the Christmas gospels.
Activities: pilgrimages; guided tours.
Hours & Admission Prices: Mon.-Fri. 9-4, Sat. 4-6, Sun. 8-12:30. No charge.

EMILY DAVIS GALLERY - UNIVERSITY OF AKRON, Myers School of Art, 150 E. Exchange St., Akron, OH 44325-7801. Mailing Address: 302 Buchtel Commons, Akron, OH 44325. Tel.: 330-972-5950.
Web Site: www.uakron.edu/art/galleries
Institution Type/Description: Art Gallery.
Collections: paintings; sculpture; drawings.
Activities: special events.
Hours & Admission Prices: Mon.-Tues. & Fri.-Sat. 10-5, Wed.-Thurs. 10-9.

HOWER HOUSE, University of Akron, 60 Fir Hill, Akron, OH 44325-2401. Tel.: 330-972-6909. Fax: 330-384-2635.
E-mail: howerhouse@uakron.edu
Web Site: howerhouse.org
Key Personnel: Dir., Cindy Bussey
Institution Type/Description: Historic House Museum: former home of John Henry Hower, a leading Akron industrialist; c.1871.
Collections: Howard family artifacts; period furnishings; personal artifacts.
Activities: summer programs; holiday festivities; speakers; special events.
Hours & Admission Prices: Feb.-Dec. Wed.-Sat. 12-3:30; groups by appointment. Adults $8, senior citizens 65 & over $6, students & children $2; discounts to groups; children 6 & under no charge. Closed major holidays.
Membership: Individual $25; Family $35; Contributing $50; Sustaining $100; Sponsor $250; Patron $500; Benefactor $1,000; Golden Benefactor $5,000.

✳ **STAN HYWET HALL AND GARDENS, INC., (M),** 714 N. Portage Path, Akron, OH 44303-1399. Tel.: 330-836-5533 & 315-3284. Fax: 330-836-2680.
E-mail: mheppner@stanhywet.org
Web Site: www.stanhywet.org
Founded: 1957.
Congressional District: 14
Key Personnel: Pres. & Exec. Dir., Linda Conrad; Vice Pres. Outreach & Communications, Gailmarie Fort; Vice Pres. Finance & Operations, Sean Joyce; Museum Shop Mgr., Kristi Woodill.
Personnel Profile: Full-Time Paid 50; Part-Time Paid 70; Part-Time Volunteers 1,300; Interns 7.
Governing Authority: nonprofit. Affiliated with Stan Hywet Hall Foundation, Inc. Tax-exempt: 501(c)(3).
Institution Type/Description: Historic House Museum: 1912-15 65-room Tudor Revival: manor house and country estate of Frank A. Seiberling, founder of Goodyear Tire & Rubber Co.
Collections: English & American period artifacts; 16th- to 18th-century European tapestries; 18th- to 20th-century American & British fine art; 19th- to 20th century glass, ceramics, textiles & other decorative arts.
Research Fields: American landscape architecture, 20th-century architecture & interior design; 20th-century social history; decorative arts; historic preservation.
Facilities: 260-seat auditorium; conservatory & greenhouses; 70 acres of landscaped gardens & grounds; cafe; rentals. Museum-related items for sale.
Publications: Stan Hywet Hall & Gardens Magazine; Calendar of Events; Annual Report & Honor Roll of Donors; volunteer newsletter.
Hours & Admission Prices: April-Dec. Tues.-Sun. 11-4:30. Adults $18-$20, seniors $11, youth 13-17 $6; discounts to members, groups & AAM members; children 12 & under no charge. Corbin Conservatory Exhibit: additional fee. Closed New Year's Eve & Day; Easter; Nov. 13; Thanksgiving; Christmas Eve & Day. &
Attendance: 200,000 (estimated)

THE SUMMIT COUNTY HISTORICAL SOCIETY OF AKRON, OHIO, (M), 550 Copley Rd., Akron, OH 44320-2398. Tel.: 330-535-1120. Fax: 330-535-0250.
E-mail: schs@summithistory.org
Web Site: summithistory.org
Founded: 1924.

Congressional District: 14
Key Personnel: Exec. Dir., Leianne Neff Heppner; Pres. (V), Richard Comstock; Business Mgr., Sandra Pecimon; Education Coord., Alison First; Record Mgr., Charlotte Gintert.
Personnel Profile: Full-Time Paid 4; Part-Time Paid 2; Part-Time Volunteers 60; Interns 4.
Governing Authority: society. Tax-exempt: 501(c)(3).
Institution Type/Description: History Museum.
Collections: early Americana; transportation; pottery; costumes; glass; manuscripts. Historic Houses: 1830 John Brown House; 1840 Old Stone School; 1837 Simon Perkins Mansion.
Research Fields: transportation; pottery; industrial & urban history.
Facilities: Books for sale.
Activities: guided tours; lectures; films; study clubs; permanent & temporary exhibitions; preservation efforts; outreach programs for elementary schools & older adults.
Publications: quarterly bulletin, Old Portage Trail Review; books; pamphlets.
Hours & Admission Prices: Call for hours. Adults $6, seniors and children 18 & under $4; members no charge. Closed national holidays.
Attendance: 5,000 (estimated)
Membership: Individual $40; Family $50; Summit Sponsor $100; John Brown Benefactor $250; Simon Perkins Benefactor $500.

Alliance

MABEL HARTZELL HISTORICAL HOME, 840 N. Park Ave., Alliance, OH 44601-1728. Mailing Address: P.O. Box 2044, Alliance, OH 44601-0044. Tel.: 330-823-4115. Facebook: Alliance History.
E-mail: alliancehistory@yahoo.com
Web Site: www.alliancehistory.org
Founded: 1939.
Congressional District: 16
Key Personnel: Pres., Karen Perone; Vice Pres., Joseph Zelasko; Treas., Lucy Harrison; Sec., Jennifer Crist.
Personnel Profile: Part-Time Volunteers 30.
Governing Authority: private; nonprofit. Parent Institution: Alliance Historical Society. Tax-exempt: 501(c)(3).
Institution Type/Description: Historic House: an 1867 Italianate built by Matthew Earley, a prominent businessman & politician; restored to 1880s period with original furnishings.
Collections: life in the Victorian era; Renaissance revival furniture; curved staircase; sitting room parlor; home entertainment artifacts from late 1800s to early 1900s.
Research Fields: D.W. Crist, composer & music publisher; history of Alliance.
Facilities: library of photographs related to the Alliance community & Morgan Engineering.
Activities: docent program; guided tours; lectures; local history programs.
Publications: biannual society newsletter, Then and Now; books, Alliance As I Knew It by William H. Magrath; What's Cooking? Recipes and Stories from Alliance and Beyond.
Hours & Admission Prices: By appointment only. Admission $3; children under 12 no charge.
Attendance: 1,500 (estimated)
Membership: Regular $15; Family $20; Patron $30; Mabel Hartzell Club $100; Life $300.

Amherst

AMHERST HISTORICAL SOCIETY, 113 S. Lake St., Amherst, OH 44001. Tel.: 440-988-7255. Fax: 440-988-2951.
E-mail: amhersthistory@centurytel.net
Web Site: amhersthistoricalsociety.org
Personnel Profile: Full-Time Paid 2; Part-Time Paid 1.
Institution Type/Description: Historical Society Museum.
Collections: local history & culture; period furnishings; personal artifacts; photographs.
Hours & Admission Prices: May to early Dec. Wed. 10-12, Sun. 2-4.
Attendance: 1,000 (estimated)

Archbold

SAUDER VILLAGE, 22611 State Rte. 2, Archbold, OH 43502-9452. Mailing Address: P.O. Box 235, Archbold, OH 43502-0235. Tel.: 419-446-2541; 800-590-9755. Fax: 419-445-5251. TDD: 419-445-9610.
E-mail: info@saudervillage.org
Web Site: www.saudervillage.org
Founded: 1971.
Congressional District: 5

Key Personnel: Exec. Dir., Debbie Sauder David; Dir. Historic Operations, Kris Jemmott; Cur. Education, Andrea Erbskorn; Cur. Collections, Tracie Evans; Business Officer, Corey Smeltzer; Dir. Mktg., Jeanette Smith; Museum Shop Mgr., Leslie Hartman.
Personnel Profile: Full-Time Paid 45; Part-Time Paid 350; Part-Time Volunteers 400.
Governing Authority: nonprofit organization. Tax-exempt.
Institution Type/Description: Living History Village: a 37-building pioneer village.
Collections: farm equipment; woodworking tools; household items; blacksmith; potter; glassblower; spinning & weaving; broom making; cooper; tinsmith; basket maker; grist mill; herbalist.
Major Exhibits: Blood, Sweat & Tears: How the Civil War Touched Northwest Ohio, 4/11-10/14.
Facilities: visitors center; 350-seat restaurant; cafe; 47-site campground; 98 room country inn. Gift items for sale.
Activities: tours; lectures; arts & music festivals; workshops; working mill; permanent exhibitions; pioneer craft demonstrations; educational programming. Museum Sponsors: monthly festivals April-Oct.
Publications: quarterly member's magazine
Hours & Admission Prices: May & Sept.-Oct. Tues.-Fri. 10-3:30, Sat. 10-5, Sun. 12-4; Memorial Day to Labor Day Tues.-Sat. 10-5, Sun. 12-4. Adults $16, students 6-16 $10; discount to seniors, AAA members & groups; members and children 5 & under no charge. &
Attendance: 100,000 (accurate)
Membership: Single $50; Single Plus & Couple $75; Family & Grandparent $85. Expanded level: Single $100; Couple $150; Family & Grandparent $170.

Ashland

ASHLAND COUNTY HISTORICAL SOCIETY, 420 Center St., Ashland, OH 44805-3247. Tel.: 419-289-3111. Fax: 419-207-8153.
Key Personnel: Dir., Chris Box.
Personnel Profile: Part-Time Paid 4.
Governing Authority: Tax-exempt.
Institution Type/Description: Historical Society Museum.
Collections: county history & culture; photographs; personal artifacts; quill baskets; art glass tumblers; pressed glass goblets.
Hours & Admission Prices: Tours: April-Dec. Wed. & Fri. 1-4; other times by appointment. No charge.
Attendance: 4,000 (estimated)

Ashtabula

ASHTABULA ARTS CENTER, 2928 West 13th St., Ashtabula, OH 44004-2498. Tel.: 440-964-3396. Fax: 440-964-3396.
E-mail: aac@suite224.net
Web Site: www.artscenternews.com
Founded: 1953.
Congressional District: 11
Key Personnel: Exec. Dir., Elizabeth Koski; Pres. (V), Judy Robson; Art Coord., Meeghan Humphrey; Business Mgr., Cindy Rimplela; Theater Coord., Kimberly Godfrey; Dance Dept. Coord., Shelagh Dubsky.
Personnel Profile: Full-Time Paid 8; Part-Time Paid 3; Part-Time Volunteers 125.
Governing Authority: nonprofit organization. Tax-exempt: 501(c)(3).
Institution Type/Description: Art Museum Center.
Collections: contemporary art.
Research Fields: local history.
Facilities: classrooms; performance area; theater.
Activities: lectures; gallery talks; dance recitals; arts festivals; hobby workshops; formally organized education programs for children & adults in visual arts, dance, drama, music; outreach program in schools; permanent, temporary & traveling exhibitions; performance ensembles & repertory groups.
Publications: bimonthly newsletter; annual report; general brochure.
Hours & Admission Prices: Mon.-Thurs. 9-9, Fri.-Sat. 9-5. No charge. Closed holidays. &
Attendance: 15,000 (estimated)
Membership: Individual $25; Family $50; Supporters $100; Friends $200; Patrons $500; Donors $1,000; Benefactors $2,000; Golden Investors $5,000 & up.

ASHTABULA MARITIME AND SURFACE TRANSPORTATION MUSEUM, 1071 Walnut Blvd., Ashtabula, OH 44004-3249. Mailing Address: Ashtabula Maritime Museum, P.O. Box 1546, Ashtabula, OH 44047-0036. Tel.: 440-997-5370; 964-6847.
Web Site: www.ashtabulamarinemuseum.org
Key Personnel: Exec. Dir., Robert Frisbie

Institution Type/Description: Marine Museum: housed in the former residence of the Lighthouse Keepers and the Coast Guard Chief, built in 1871/1898.
Collections: models; paintings; marine artifacts; photographs of early Ashtabula Harbor, ore boats & tugs; miniature hand-made brass tools; scale model of a Hulett Ore Unloading Machine.
Hours & Admission Prices: Memorial Day to Aug. Fri.-Sun. & 4th of July 12-5; Sept. Sat.-Sun. 12-5. Adults $5, children 6-16 $3; children under 6 no charge.
Membership: Individual $20; Family $30; Sustaining $50; Century $100; Corporate $1,000; Corporate Lifetime $5,000.

Athens

ATHENS COUNTY HISTORICAL SOCIETY AND MUSEUM, 65 N. Court St., Athens, OH 45701-2506. Tel.: 740-592-2280. Fax: 740-594-8352.
Founded: 1980.
Key Personnel: Dir., Ron Luce; Pres. (V), Tom O'Grady; Museum Shop Mgr., Laura Farrell.
Personnel Profile: Full-Time Paid 1; Part-Time Paid 3; Part-Time Volunteers 20; Interns 15.
Governing Authority: private; nonprofit organization. Tax-exempt.
Institution Type/Description: Historical Society Museum.
Collections: county history & culture; photographs; personal artifacts; hands-on exhibits.
Facilities: library; meeting & lecture hall.
Activities: school programs; hands-on exhibits; history camp; movie night; history lectures; genealogy classes.
Publications: birth, death, marriage, cemetery, & township history records; local history.
Hours & Admission Prices: Tues.-Sat. 12-4. No charge; donations accepted. &
Attendance: 3,220 (accurate)
Membership: Individual $15; Family $30; Friend $100; Business Club $175; Sponsoring $250; Life $500; Patron $1,000.

THE DAIRY BARN ARTS CENTER, 8000 Dairy Lane, Athens, OH 45701-9393. Mailing Address: P.O. Box 747, Athens, OH 45701-0747. Tel.: 740-592-4981. Fax: 740-592-5090.
E-mail: artsinfo@dairybarn.org
Web Site: www.dairybarn.org
Formerly: The Dairy Barn Southeastern Ohio's Cultural Arts Center
Founded: 1978.
Congressional District: 6
Key Personnel: Dir., Andrea Lewis; Museum Shop Mgr., Claire White.
Personnel Profile: Full-Time Paid 5; Part-Time Paid 2; Interns 1.
Governing Authority: Tax-exempt.
Institution Type/Description: Arts Center: housed in a former dairy barn; built in 1914. Listed on the National Register of Historic Places.
Collections: works by contemporary artists.
Facilities: performance area; classroom.
Activities: festivals; educational programs; special events.
Publications: exhibition catalogs, Quilt National; Bead International; Beyond Basketry; Contemporary Ceramics; Quilting Traditions - The Art of the Amish.
Hours & Admission Prices: Winter: Tues.-Wed. & Fri.-Sat. 12-5; Summer: call for extended hours. Adults $7, students & seniors $5; members & children under 12 no charge. &
Attendance: 22,000 (accurate)
Membership: Student & Senior $25; Individual $50; Family $100; Sustainer $250.

❋ **KENNEDY MUSEUM OF ART, (M),** Ohio University, Lin Hall, Athens, OH 45701-2979. Tel.: 740-593-1304. Fax: 740-593-1305.
E-mail: kennedymuseum@ohio.edu
Web Site: www.ohiou.edu/museum
Founded: 1993.
Congressional District: 15
Key Personnel: Dir., Edward Pauley; Cur. Education, Sally Delgado; Cur., Petra Kralickova; Registrar, Jeffrey Carr; Mktg., Public Rels. & Guest Svcs., Maryann Gunderson.
Personnel Profile: Full-Time Paid 5; Part-Time Paid 2; Part-Time Volunteers 150; Interns 12.
Governing Authority: university. Parent Institution: Ohio University. Tax-exempt.
Institution Type/Description: University Art Museum.
Collections: southwest Native American Collection of Edwin L. & Ruth E. Kennedy; works on paper; American painting, sculpture; photography; ceramics; non western art.

Research Fields: American art; contemporary prints; southwest Native American textiles.
Facilities: cafe.
Activities: guided tours; lectures; symposiums; gallery talks; performances; organized education programs; docent programs; school programs; teacher training; internships; permanent, temporary & traveling exhibitions.
Publications: newsletter, Kennedy Museum of Art Magazine.
Hours & Admission Prices: Gallery: Mon.-Wed. & Fri. 10-5, Thurs. 10-8, Sat.-Sun. 1-5. Cafe: Mon.-Fri. 8-3. No charge; donations accepted. &
Attendance: 10,000 (accurate)
Membership: Individual $25; Household $35; Patron $50; Benefactor $100; Sustaining $250; Partner $500; Director's Circle $1,000.

OHIO UNIVERSITY ART GALLERY, 528 Seigfred Hall, Athens, OH 45701. Tel.: 740-593-0796. Fax: 740-593-1305.
E-mail: kralicko@ohio.edu
Web Site: www.ohiou.edu/art/ougallery.html
Key Personnel: Dir. Exhibition, Petra Kralickova
Institution Type/Description: University Art Gallery.
Collections: works by national & regional artists.
Facilities: 2,500 sq. ft. exhibit space.
Activities: Annual Events: Graphic Design Bachelor of Fine Arts exhibition; juried show of undergraduate artwork.
Hours & Admission Prices: Mon.-Sat. 10-4. No charge.

Aurora

AURORA HISTORICAL SOCIETY, (M), 115 E. Pioneer Trail, Aurora Memorial Library Bldg., Aurora, OH 44202-7922. Tel.: 330-995-3336.
E-mail: aurorahist@windstream
Web Site: aurorahistorical.org
Founded: 1968.
Congressional District: 14
Key Personnel: Pres., John Kudley; Dir., Marcelle R. Wilson, Ph.D.; Museum Shop Mgr., Josephine Smalley.
Personnel Profile: Part-Time Paid 1; Part-Time Volunteers 20.
Governing Authority: society; nonprofit organization. Affiliated with Library Trust-Aurora Memorial Library. Tax-exempt: 501(c)(3).
Institution Type/Description: General Museum.
Collections: church, family & business records; cheese-making equipment; farm implements; home furnishings; photo records; genealogies; WWII; Geauga Lake Park & Sea World exhibit.
Major Exhibits: Journey Stories: The Tales of How We & Our Ancestors Came to America (T), 3/11/14-4/5/14.
Research Fields: century homes; general stores; genealogy.
Facilities: library.
Activities: guided tours; lectures; gallery talks; permanent exhibitions; local research; century homes recognition; antique show.
Publications: Aurora Story 1 and 2, The Pioneer; Images of America: Aurora; Tastes of Aurora: Favorite Recipes From The Past & From Today.
Hours & Admission Prices: Mon. & Wed. 2-4; other times by appointment. No charge. &
Attendance: 1,400 (accurate)
Membership: Individual $25; Family $35; Contributing $50; Benefactor $100; Life $1,000.

Bainbridge

DR. JOHN HARRIS DENTAL MUSEUM, 208 West Main St., Bainbridge, OH 45612. Mailing Address: Bainbridge Historical Society, P.O. Box 424, Bainbridge, OH 45612-0424. Tel.: 740-634-2228 & 2246.
Web Site: www.bainbridgedentalmuseum.com
Founded: 1939.
Key Personnel: Chm. (V), Mrs. Cherry Miller.
Personnel Profile: Part-Time Volunteers 10.
Governing Authority: Parent Institution: Bainbridge Historical Society. Tax-exempt: 501(c)(3).
Institution Type/Description: Dental History Museum: housed in 1827 former office of Dr. John Harris.
Collections: dental artifacts from 1627-present.
Facilities: library of dental literature.
Activities: guided tours.
Hours & Admission Prices: April-May & Sept.-Oct. Sat.-Sun. 12-4; June-Aug. Tues.-Sun. 12-4. Adults $2; children under 12 & members no charge. &
Attendance: 350 (estimated)
Membership: Bainbridge Historical Society: Individual $5.

Barnesville

BELMONT COUNTY VICTORIAN MANSION MUSEUM, 532
N. Chestnut St., Barnesville, OH 43713-1274. Mailing Address:
P.O. Box 434, Barnesville, OH 43713-0434. Tel.: 740-425-2343 &
2926.
Formerly: Gay 90's Mansion Museum
Founded: 1966.
Congressional District: 20
Key Personnel: C.E.O. & Pres., Sheryl McClellan; Dir. & Treas., Rebecca J.
Thomas.
Personnel Profile: Part-Time Volunteers 35.
Governing Authority: county; society. Operated by Belmont County Historical
Society. Tax-exempt: 501(c)(3).
Institution Type/Description: General Museum: housed in 1890 Richardsonian
mansion.
Collections: glass; china; utensils; furniture; clothing; manuscripts; quilts.
Facilities: 150-vol. library of old books available for use on premises; reading
room.
Activities: guided tours; formally organized education programs for children.
Annual Event: Victorian Christmas Show.
Publications: annual brochures & fact sheets.
Hours & Admission Prices: May-Sept. Wed.-Sun. & holidays 1-4; special tours
by appointment. Adults $5, children 6-18 $2; discounts to groups.
Attendance: 1,058 (accurate)
Membership: Single $15; Couple $25; Sustaining $25; Life $150; Life Couple
$200.

Batavia

TRI-STATE WARBIRD MUSEUM, 4021 Borman Ave., Batavia,
OH 45103. Tel.: 513-735-4500.
Web Site: www.tri-statewarbirdmuseum.org/museum.html
Governing Authority: nonprofit organization. Tax-exempt: 501(c)(3).
Institution Type/Description: Military Aviation Museum.
Collections: military aviation history; aircraft.
Facilities: library; classroom; 5,000 sq. ft. exhibition space.
Activities: school group tours.
Hours & Admission Prices: Wed. 4-7, Sat. 10-3. Adults $12, student & veteran
$7.

Bath

HALE FARM AND VILLAGE, 2686 Oak Hill Rd., Bath, OH
44210. Mailing Address: P.O. Box 296, Bath, OH 44210-0296.
Tel.: 330-666-3711. Fax: 330-666-9497.
E-mail: kfalcone@wrhs.org
Web Site: www.halefarm.org
Founded: 1956.
Congressional District: 14
Key Personnel: C.E.O. & Pres., Dr. Gainor B. Davis, WRHS; Bd. Chm., Don
Dailey, WRHS; Sr. Vice Pres. Interpretation & C.O.O., Kelly Falconc;
Property Mgr., Jason Klein; Education Program Mgr., Lisa Leaman.
Personnel Profile: Full-Time Paid 8; Part-Time Paid 20; Part-Time Volunteers
50; Interns 15.
Governing Authority: Parent Institution: The Western Reserve Historical
Society, 10825 East Blvd., Cleveland, OH. 44106. Tax-exempt:
170(b)(1)(A).
Institution Type/Description: Village Museum.19th century agrarian & village
communities in the Western Reserve.
Collections: agriculture; crafts. Historic Buildings: 1816 log schoolhouse;
1825 Hale House; 1825 Wade law office; 1825 Franklin glassworks; steam
sawmill; 1830 salt box house; 1832 Goldsmith House; 1832 Brown Land
Office; 1844 Greek revival house; 1845 stone Herrick house; carriage
museum; 1850 Stow House; 1852 meetinghouse; 1804 log cabin; 1850 mill.
Research Fields: Midwestern glass; mid-19th century rural costumes; agricultural history & research; architecture.
Facilities: cafe; visitor center. Gift items for sale.
Activities: guided & self-guided tours; lectures; permanent exhibitions; craft
demonstrations; educational programs for K-12 students.
Publications: The Jonathan Hale Farm.
Hours & Admission Prices: June-Oct. Wed.-Sun. 10-5; Nov.-Dec. Wed.-Fri.
9:30-2, Sat.-Sun. 10-5. Adults $10, children 3-12 $5; discounts to groups
and AAM & ICOM members; members no charge except for special events.
Closed New Year's Day; Thanksgiving; Christmas. &
Attendance: 60,000 (accurate)
Membership: Individual $50; Family & Couple $70; Sustaining $150; Fellow
$250; Special Fellow $500.

Bay Village

BAYARTS, 28795 Lake Rd., Bay Village, OH 44140-1399. Tel.:
440-871-6543 & 5678. Fax: 440-871-0452. Facebook: BAYarts.
E-mail: info@bayarts.net
Web Site: www.bayarts.net
Formerly: Baycrafters, Inc.
Founded: 1948.
Congressional District: 23
Key Personnel: Exec. Dir., Nancy Heaton; Education Dir., Erin Stack; Gallery
Dir., Eileen Stockdale.
Governing Authority: nonprofit organization. Tax-exempt.
Institution Type/Description: Art Association: housed in original Huntington
Estate. Visual arts organization with consignment shop for members in 1882
Nickel Plate Bay Village Railway Station with a Norfolk & Western
Caboose.
Collections: works by regional artists.
Facilities: 200-vol. library of arts & crafts reference & instruction books &
materials available for research with membership in BAYarts; classrooms.
Arts & crafts by regional artists & craftsmen for sale.
Activities: gallery talks; arts festivals; workshops; formally organized educa-
tion programs; consignment shop.
Publications: periodical, Brochure of Classes & Events; Artists Entry Form for
Juried Competitions.
Hours & Admission Prices: Mon.-Sat. 10-5, Sun. 12-5. No charge; donations
accepted. Closed holidays.
Attendance: 100,000
Membership: Individual $30; Family $50; Super Supporter $100.

LAKE ERIE NATURE & SCIENCE CENTER, 28728 Wolf Rd.,
Bay Village, OH 44140-1350. Tel.: 440-871-2900. Fax: 440-871-
2901.
E-mail: info@lensc.org
Web Site: www.lensc.org
Formerly: Lake Erie Junior Museum
Founded: 1945.
Congressional District: 9
Key Personnel: Exec. Dir., Catherine Timko; Pres. (V), Colleen Lowmiller;
Museum Shop Mgr., Sheryl Caine.
Personnel Profile: Full-Time Paid 11; Part-Time Paid 21; Part-Time Volunteers
307; Interns 4.
Volunteer Hours: 10,000
Operating Expenses: 1,161,222
Operating Income: 1,255,522
Governing Authority: nonprofit organization. Tax-exempt: 501(c)(3).
Institution Type/Description: Nature Center.
Collections: astronomy; botany; geology; physical science; zoology; live
wildlife; ecological & environmental.
Facilities: library; wild flower & teaching garden; classrooms; planetarium;
aquariums.
Activities: natural & physical science classes for school children; weekend
family programs; adult workshops; nature films; nature art shows; travel-
ogues. Center Sponsors: Nature classes, traveling programs for hospitalized
children, the handicapped & senior citizens.
Publications: annual report; program guide; newsletter; informational brochures.
Hours & Admission Prices: Daily 10-5; groups by appointment only. No
charge; donations accepted. Closed New Year's Day; Easter; Memorial
Day; Independence Day; Labor Day; Thanksgiving; Christmas. &
Attendance: 180,000 (estimated)
Membership: Individual $25; Family $50; Contributing $200; Supporting
$500.

ROSE HILL MUSEUM, 27715 Lake Rd., Bay Village, OH 44140.
Mailing Address: P.O. Box 40187, Bay Village, OH 44140-0187.
Tel.: 440-871-7338 & 835-2718.
E-mail: mail@bayhistorical.com
Web Site: www.bayhistorical.com
Founded: 1960.
Congressional District: 23
Key Personnel: Pres. (V), Cynthia Eakin; Vice Pres., Doug Gertz; Correspond-
ing Sec., Cindi Lindgren; Treas., Abigail Sammon; Accessions & Museum
Shop Mgr., Janet Zvara.
Personnel Profile: Part-Time Volunteers 12.
Governing Authority: society. Parent Institution: Bay Village Historical Soci-
ety. Tax-exempt: 501(c)(3).
Institution Type/Description: History Museum: housed in 1818 Western
Reserve farmhouse belonging to the Cahoon Family, first settlers in Dover
Township.

Collections: clothing; children's toys; Wischmeyer boats; primitive furniture of the earliest settlers; Empire & Victorian furniture belonging to the Cahoon family; 1818 bedroom; roped bed; cornhusk mattress; Victorian bedroom; cellar containing a furnished summer kitchen & food storerooms; Cleveland history; local genealogy; 1810 replica log cabin with furnishings.
Research Fields: history of the Dover Township-Western Reserve area.
Facilities: 200-vol. library of genealogy, Cahoon & local history books available for research on premises only; smoke house; cabin-replica of Cahoons; Reuben Osborn Learning Center. Books & other museum-related items for sale.
Activities: outreach to school & other organizations; guided tours; hobby workshops; formally organized education programs for children; temporary exhibitions; school tour service by appointment. Museum Sponsors: Victorian Tea in April, Cahoon in June crafts & antiques fair in June.
Publications: book, Bay Village, A Way Of Life; Bay Village Historical Society Cook Book; Photo History book, Images of America: Bay Village.
Hours & Admission Prices: Sun. 2-4:30, call for special tours. No charge; donations accepted. &
Attendance: 1,500
Membership: Single $15; Family $25; Patron $50.

Beachwood

MALTZ MUSEUM OF JEWISH HERITAGE, (M), 2929 Richmond Rd., Beachwood, OH 44122-3270. Tel.: 216-593-0575. Fax: 216-593-0576.
E-mail: info@mmjh.org
Web Site: maltzmuseum.org
Founded: 2005.
Congressional District: 16
Key Personnel: Chm. Bd., Milton Maltz; Pres., Tamar Maltz; Exec. Dir., Judith Feniger; Devel., Laura Whay Klein; Education, Lynda Bender; Media Rels., Adam Teresi; Registrar, Amber Anderson; Mgr. Special Exhibits, Stacy Singerman; Admin., Laurie Hughes; Archivist, Sean Martin; Museum Shop Mgr., Martha Sivertson.
Personnel Profile: Full-Time Paid 10; Part-Time Paid 5.
Governing Authority: private; nonprofit organization. Tax-exempt: 501(c)(3).
Institution Type/Description: Jewish Heritage Museum.
Collections: Jewish heritage; films; personal artifacts; oral histories; art; photographs.
Facilities: 60-seat theater; 12,000 sq. ft. exhibit space.
Hours & Admission Prices: Tues., Thurs.-Fri. & Sun. 11-5, Wed. 11-9, Sat. 12-5. Adults $7, students & seniors $5; children under 12 no charge. &

Bedford

BEDFORD HISTORICAL SOCIETY MUSEUM AND LIBRARY, (M), 30 S. Park St. (Squire Place), Bedford, OH 44146. Mailing Address: P.O. Box 46282, Bedford, OH 44146-0282. Tel.: 440-232-0796.
E-mail: museum@bedfordohiohistory.org
Web Site: www.bedfordohiohistory.org
Formerly: Bedford Museum
Founded: 1955.
Congressional District: 21
Key Personnel: C.E.O., Janet Caldwell; Pres., Robert Schroeter, Sr.; Archivist, Debra Grubb; Librarian, Paul Pojman; Mgr., Doris Shriver.
Personnel Profile: Full-Time Paid 1; Part-Time Paid 3; Part-Time Volunteers 12.
Governing Authority: society; Bedford Historical Society. Tax-exempt: 501(c)(3).
Institution Type/Description: History Museum.
Collections: Americana; archaeology; archives; period furniture; glass; local & Ohio history; Indian artifacts; Jacka 1876 Centennial collection; Siegel railway collection; numismatic; manuscripts; the Squire Lincoln collection; the Barnum Civil War library. The art of Archibald Willard & Richard Sedlon. Historic Building: 1874 Bedford Township Hall; 1832 Hezekiah Dunham House; 1882 Bedford Railroad Depot; 1893 Gothic Church.
Research Fields: local history; genealogy.
Facilities: 15,000-vol. library of local & state history books available for use on premises. Used books & other museum-related items for sale.
Activities: guided tours; permanent & temporary exhibitions; lectures; educational programs. Museum Sponsors: Annual Strawberry Festival.
Publications: quarterly bulletin; book, Bedford Vignettes; Bedford Village Views-1992; children's activity book, All About Bedford, Ohio; Images of America: Bedford & Bedford Township Ohio (2012); Bedford Cemetery: Past & Present.
Hours & Admission Prices: Mon. & Wed. 7:30pm-10pm, Thurs. 10am-4pm; Second Sun. of the month 2-5. No charge; donations accepted. &
Attendance: 10,000 (accurate)

Membership: Junior (17 & under) $5; Annual $10; Family $20; Contributing $25; Life $150; Corporate $250.

Bellaire

THE NATIONAL GLASS IMPERIAL MUSEUM, 3200 Belmont St., Bellaire, OH 43906-1521. Mailing Address: P.O. Box 534, Bellaire, OH 43906-0534. Tel.: 740-671-3971.
E-mail: info@imperialglass.org
Web Site: www.imperialglass.org/museum.htm
Founded: 2003.
Congressional District: 6
Key Personnel: Admin., Rosalie Wenckoski; Chm. (V), Paul Douglas.
Personnel Profile: Part-Time Paid 1; Part-Time Volunteers 30.
Governing Authority: Parent Institution: National Imperial Glass Collectors' Society, Inc. Tax-exempt.
Institution Type/Description: Glass Museum.
Collections: Imperial glassware; tools; documents; photos.
Research Fields: glass.
Facilities: research room.
Activities: school groups; civic groups; business groups.
Publications: The Imperial Collectors Glasszette.
Hours & Admission Prices: April-Oct. Thurs.-Sat. 11-3. Admission: $3; NIGCS members no charge. Closed Independence Day. &
Attendance: 480 (estimated)
Membership: Associate $5; Primary $20; Patron $50; President's Club $100; Sponsor's Club $250; Benefactor's Club $500.

UNOFFICIAL LEGO & TOY MUSEUM, 4597 Noble St., Bellaire, OH 43906. Tel.: 740-671-8890.
E-mail: toymuseum@hotmail.com
Web Site: danstoymuseum.blogspot.com
Institution Type/Description: Toy Museum.
Collections: LEGO structures.
Hours & Admission Prices: Wed.-Sun. 12-7. Adults $8, children $6; children 4 & under no charge.

Bellbrook

BELLBROOK HISTORICAL MUSEUM, 42 N. Main St., Bellbrook, OH 45305-2009. Mailing Address: 3640 Ferry Rd., Bellbrook, OH 45305-0285. Tel.: 937-848-4666.
Key Personnel: Chm. (V), Dwight W. Bartlett.
Personnel Profile: Part-Time Paid 1; Part-Time Volunteers 6.
Governing Authority: city. Tax-exempt.
Institution Type/Description: History Museum: housed in The Crowl Building, a former mortuary.
Collections: artifacts that portray the story of Bellbrook from 1816; 19th century funeral business.
Hours & Admission Prices: Sat. 12-5. No charge; donations accepted. Closed holidays. &
Attendance: 500 (estimated)

Bellefontaine

LOGAN COUNTY HISTORICAL SOCIETY MUSEUM, 521 E. Columbus Ave., Bellefontaine, OH 43311-2401. Tel.: 937-593-7557.
E-mail: historycenter@loganhistory.org
Web Site: www.logancountymuseum.org
Founded: 1945.
Congressional District: 7
Key Personnel: Dir. & Cur., Todd McCormick; Pres. (V), David Wagner.
Personnel Profile: Full-Time Paid 2; Part-Time Paid 2; Part-Time Volunteers 62.
Volunteer Hours: 3,000
Operating Expenses: 265,000
Operating Income: 265,000
Governing Authority: nonprofit. Affiliated with Logan County Historical Society, 521 E. Columbus Ave., Bellefontaine. Tax-exempt.
Institution Type/Description: Historical Society Museum.
Collections: household appliances; tools; clothing; cameras; communications; music; firearms; uniforms & other insignia; railroad artifacts; autos; carriages; aviation; AJ Miller Co. hearses; military & pioneer costumes; geology; Warren Cushman art; American Indian relics; manuscripts; agriculture; modern day.
Research Fields: local history; genealogy.
Facilities: 300-vol. library available for use on premises; 40-seat auditorium; reading room.

Activities: guided tours; lectures; permanent & temporary exhibitions.
Publications: brochure, Tour Guide of the County; booklets, First Concrete Street in America; leaflet, Welcome to Our Museum; Pictorial History Book; Historic Glimpses of Logan County, Ohio; Life in Rebel Prisons; quarterly newsletter.
Hours & Admission Prices: May-Oct. Wed. & Fri.-Sun. 1-4; Nov.-April Fri.-Sat. 1-4. No charge; donations accepted. &
Attendance: 5,000 (estimated)
Membership: Young Historians $2; Adult $15; Family $20; Business $30.

Bellevue

HISTORIC LYME VILLAGE ASSOCIATION, 5001 State Rte. 4, Bellevue, OH 44811. Mailing Address: P.O. Box 342, Bellevue, OH 44811-0342. Tel.: 419-483-4949.
E-mail: info@lymevillage.org
Web Site: www.lymevillage.com
Founded: 1972.
Congressional District: 4
Key Personnel: Pres., Neil Leimbach; Vice Pres., Bill Drown; Treas., Dennis Bauer; Dir. & Museum Shop Mgr., Raymond Parker.
Personnel Profile: Part-Time Volunteers 40; Interns 0.
Volunteer Hours: 1,500
Operating Expenses: 63,000
Operating Income: 65,000
Governing Authority: private; nonprofit organization. Tax-exempt: 501(c)(3).
Institution Type/Description: History Museum.
Collections: early-1800s to early-1900s houses, buildings & furnishings.
Activities: guided tours of village; Pioneer Days in September; Victorian Dinners and Christmas tours in December.
Publications: newsletter, Lyme Lines.
Hours & Admission Prices: June-Aug. Tues.-Sat. 11-4, Sun. 12-4; Sept. Sun. 12-4. Adults $9, senior citizens 60 & over $8, children 6-12 $5. &
Attendance: 7,000 (estimated)
Membership: Student & Senior Citizen 60 & over $10; Individual $20; Family $45; Corporate $200 & up; Individual Life $300.

MAD RIVER & NKP RAILROAD MUSEUM, 253 Southwest St., Bellevue, OH 44811-1377. Mailing Address: 233 York St., Bellevue, OH 44811-1377. Tel.: 419-483-2222.
E-mail: madriver@onebellevue.com
Web Site: madrivermuseum.org
Founded: 1973.
Key Personnel: Pres. (V), Christopher Beamer.
Governing Authority: Tax-exempt.
Institution Type/Description: Railroad History Museum.
Collections: Mad River & Lake Erie Railroad history; railroad artifacts; photographs; tools; china; linens; badges; uniforms; lanterns; drawings; rolling stock.
Facilities: picnic area. Museum-related items for sale.
Activities: hands-on exhibits; group tours.
Publications: quarterly newsletter, Caboose Cable.
Hours & Admission Prices: May & Sept.-Oct. Sat.-Sun. 12-4; Memorial Day to Labor Day daily 12-4. Adults $7, seniors 60 & over $6, children 3-12 $4; members & children under 3 no charge.
Attendance: 4,500 (estimated)
Membership: Senior over 60 $15; Adult under 60 $20; Family $30; Sustaining $75; Life $350; Golden Life $500.

Belpre

BELPRE HISTORICAL SOCIETY/FARMER'S CASTLE MUSEUM, 509 Ridge St., Belpre, OH 45714. Mailing Address: P.O. Box 731, Belpre, OH 45714-0731. Tel.: 740-423-7588.
E-mail: info@belprehistory.com
Web Site: belprehistory.com
Key Personnel: Pres., Nancy Sams
Institution Type/Description: History Museum.
Collections: local history & culture; period furnishings; photographs; personal artifacts.
Hours & Admission Prices: April-Oct. Wed. & Sat. 1-4. Adults $4, students $1.
Membership: Student $5; Adult $15; Family $20; Business $25; Partner $50.

Berea

FAWICK ART GALLERY, Kleist Center for Art & Drama, 95 E. Bagley Rd., Berea, OH 44017. Tel.: 440-826-2152.
Institution Type/Description: Art Gallery.
Collections: ceramics; paintings; sculpture.

Activities: temporary exhibitions.
Hours & Admission Prices: Mon.-Fri. 2-5; other times by appointment.

Berlin

AMISH & MENNONITE HERITAGE CENTER, 5798 County Rd. 77, Berlin, OH 44610. Mailing Address: P.O. Box 324, Berlin, OH 44610. Tel.: 877-858-4634.
E-mail: director@amheritagecenter.com
Web Site: www.behalt.com
Founded: 1981.
Personnel Profile: Full-Time Paid 2; Part-Time Paid 4; Part-Time Volunteers 10.
Governing Authority: Tax-exempt.
Institution Type/Description: Heritage Center.
Collections: Amish & Mennonite history, faith & culture; personal artifacts; period furnishings; photographs.
Facilities: Museum-related items for sale.
Activities: video.
Hours & Admission Prices: Mon.-Sat. 9-5. Adults $8.75-$12.50; members no charge. &
Attendance: 23,000 (estimated)

Beverly

THE OLIVER TUCKER MUSEUM, 441 5th St., Beverly, OH 45715. Mailing Address: 70 Maple Circle, Waterford, OH 45786-5321. Tel.: 740-984-2489.
Web Site: olivertuckermuseum.com
Founded: 1971.
Congressional District: 11
Key Personnel: Pres. (V), Susan Trotter; Bd. Trustee, Wayne Fansworth; Bd. Trustee & Vice Pres., Francis M. Sampson.
Personnel Profile: Part-Time Volunteers 10.
Governing Authority: society; nonprofit organization. Parent Institution: Lower Muskingum Historical Society, P.O. Box 191, Waterford, OH 45786. Tax-exempt.
Institution Type/Description: Historic Buildings.
Collections: Mary Tucker Townsend artifacts; photographs; documents; doctor's office furnishings & instruments; early furniture & clothing; period church alter c.1800; early settlers of the Lower Muskingum Valley area; riverboat items; log cabin containing period furnishings including working loom, weaving items, spinning wheels; hardware; letter distribution box from late 1700 post office Waterford, OH 45786; Civil War artifacts. Historic Buildings: 1835 John Dodge Home & 1886 Oliver Tucker Home; barn.
Research Fields: genealogy.
Facilities: Museum-related items for sale.
Activities: guided tours; permanent exhibitions.
Publications: magazine, Reflections Along The Muskingum; historical booklets on cemetery lots; Bicentennial magazines.
Hours & Admission Prices: June-Aug. Sat.-Sun. 1-4; groups & other times by appointment. No charge; donations accepted. &
Attendance: 425 (estimated)
Membership: Lower Muskingum Historical Society: Student $1; Individual $3; Family $5; Life $100.

Bexley

BEXLEY HISTORICAL SOCIETY, 2080 Clifton Ave., Bexley, OH 43209-1405. Mailing Address: P.O. Box 9285, Bexley, OH 43209-0285. Tel.: 614-559-4360. Fax: 614-235-3420.
E-mail: info@bexleyhistory.org
Web Site: www.bexleyhistory.org
Founded: 1974.
Congressional District: 12
Key Personnel: Pres., Mark Epstein; Recording Sec., Nancy Beck; Treas., Doug Wittig; Museum Shop Mgr., Edie Mae Herrel.
Personnel Profile: Part-Time Volunteers 17; Interns 1.
Governing Authority: nonprofit organization. Tax exempt.
Institution Type/Description: Local History Museum.
Collections: local memorabilia.
Research Fields: local history.
Facilities: 150-vol. library of historical books available for research on premises.
Activities: formally organized education programs for children; loan, permanent & temporary exhibitions.
Hours & Admission Prices: 2nd Mon. of month 6:30pm-8pm; other times by appointment. No charge; donations accepted. Closed holidays. &
Attendance: 500 (estimated)

Membership: Senior Citizens $15; General $25; Supporter $50; Patron $100.

Blanchester

BLANCHESTER AREA HISTORICAL SOCIETY MUSEUM,
206 W. Main St., Blanchester, OH 45107. Tel.: 937-382-1965.
E-mail: info@clintoncountyohio.com
Institution Type/Description: Historical Society Museum.
Collections: local history & culture; period furnishings; personal artifacts; photographs; religious artifacts; local industry; school memorabilia; genealogy.
Facilities: library.
Hours & Admission Prices: Wed. & Sat. 1-4. Adults $5; members and children 13 & under no charge.

Bluffton

THE LION AND LAMB PEACE ARTS CENTER, Bluffton University, Spring St., Bluffton, OH 45817-2104. Mailing Address: Bluffton University, 1 University Dr. - 50, Bluffton, OH 45817-2104. Tel.: 419-358-3207.
E-mail: lionlamb@bluffton.edu
Web Site: www.bluffton.edu/lionlamb
Founded: 1987.
Key Personnel: Dir., Louise Matthews.
Personnel Profile: Part-Time Paid 1.
Governing Authority: Parent institution: Bluffton University. Tax-exempt.
Institution Type/Description: Arts & Literature Center.
Collections: peace & peace-related items; children's literature; book illustrations; children's art; music; videos & DVDs; fine art.
Publications: yearly, 2 newsletters.
Hours & Admission Prices: Academic Year: Tues.-Thurs. 9-5, Fri. 9-12; other times by appointment. No charge. &

Attendance: 4,000 (estimated)

Bolivar

FORT LAURENS STATE MEMORIAL, (M), 11067 Fort Laurens Rd., N.W., Bolivar, OH 44612. Mailing Address: P.O. Box 508, Zoar, OH 44697-0508. Tel.: 330-874-2059. Fax: 330-874-2936.
E-mail: fortlaurens@wilkshire.net
Web Site: ohsweb.ohiohistory.org/places/ne02/index.shtml
Founded: 1972.
Congressional District: 18
Key Personnel: Dir., Victoria Branson.
Personnel Profile: Full-Time Paid 1; Part-Time Paid 2.
Governing Authority: nonprofit. Parent Institution: The Ohio Historical Society, Ohio Historical Center, 1982 Velma Ave., Columbus, OH 43211. Tax-exempt.
Institution Type/Description: Military Fort Museum: site of the only Revolutionary War fort in Ohio; the site of the Tomb of the Unknown Patriot of the Revolutionary War.
Collections: archaeological artifacts from the fort; costumed mannequins; weapons; accoutrements of the American Revolution; video of fort's history.
Facilities: theater; picnic area; 3-mile nature trail.
Activities: audiovisual program.
Hours & Admission Prices: Memorial Day-Labor Day Wed.-Sat. 9:30-5, Sun. & holidays 12-5; Sept.-Oct. Sat. 9:30-5, Sun. 12-5; groups by appointment April-Oct. Adults $3, students $2; discounts to groups, AAM & ICOM members; members & Ohio Historical Society members no charge. &
Attendance: 3,873 (accurate)
Membership: Senior Individual $55; Individual $60; Senior Family $70; Family $75.

Bowling Green

BOWLING GREEN STATE UNIVERSITY FINE ARTS CENTER GALLERIES, Fine Arts Center, Bowling Green State University, Bowling Green, OH 43403. Mailing Address: 1303 Fine Arts Center, Bowling Green State University, Bowling Green, OH 43403-0204. Tel.: 419-372-8525. Fax: 419-372-2544. Facebook: BGSU Art Galleries.
E-mail: galleries@bgsu.edu
Web Site: gallery.bgsu.edu
Founded: 1960.
Congressional District: 5
Key Personnel: Gallery Dir., Jacqueline S. Nathan.
Governing Authority: public university; nonprofit. Tax-exempt.
Institution Type/Description: Art Gallery.
Collections: prints.
Facilities: glass, sculpture, printmaking, ceramics, photography, design & computer art studios; drawing and painting, fibers & art history lecture room.
Activities: student & faculty exhibits; guided tours; lectures; school loan service; temporary exhibitions of gallery collections; loan & traveling exhibition; National & Regional Invitational Exhibitions. Annual Event: New Music & Art Festival.
Publications: catalogs once or twice a year.
Hours & Admission Prices: School Year: Tues.-Wed. & Fri.-Sat. 11-4, Thurs. 11-4 & 6pm-9pm, Sun. 1-4; Summer: call for hours. No charge; donations accepted. Closed university holidays. &
Attendance: 9,000 (estimated)
Membership: Medici Circle $25 & up.

NATIONAL CONSTRUCTION EQUIPMENT MUSEUM, 16623 Liberty Hi Rd., Bowling Green, OH 43402-9309. Tel.: 419-352-5616.
Web Site: www.hcea.net
Founded: 1986.
Congressional District: 5
Key Personnel: Chm. (V), Dare Geis; Pres. (V), Larry Kotkowski; Museum Shop Mgr., Dan Frantz.
Personnel Profile: Full-Time Paid 2; Part-Time Volunteers 12.
Governing Authority: Tax-exempt.
Institution Type/Description: History Museum.
Collections: construction equipment & history; photographs; videos; documents.
Research Fields: science & technology; genealogy; public works; community history.
Facilities: archives.
Activities: individual & group tours; rental facilities. Museum Sponsors: International Convention & Old Equipment Exposition in 2012.
Publications: quarterly magazine, Equipment Echoes.
Hours & Admission Prices: Mon.-Fri. 1-5. Adults $7; discounts to seniors & groups; members no charge.
Attendance: 200 (estimated)
Membership: Individual $32-$40; Corporate $500; Life $1,000.

WOOD COUNTY HISTORICAL CENTER AND MUSEUM, (M), 13660 County Home Rd., Bowling Green, OH 43402-9281. Tel.: 419-352-0967. Fax: 419-352-6220.
E-mail: director@woodcountyhistory.org
Web Site: www.woodcountyhistory.org
Founded: 1955.
Congressional District: 5
Key Personnel: Dir., Dana Nemeth; Coord. Education, Michael McMaster; Coord. Mktg. & Events, Kelli Kling; Cur., Holly Hartlerode; Volunteer Coord., Tim Gaddie.
Personnel Profile: Full-Time Paid 4; Part-Time Paid 1; Part-Time Volunteers 100; Interns 4.
Governing Authority: society. Parent Institutions: Wood County Commissioners. Tax-exempt: 501(c)(3).
Institution Type/Description: Regional History Museum: housed in a former county infirmary or poor farm on 50-acre site.
Collections: regional history; poor farm history; photography; oil boom history; working derrick; drilling equipment; farm & community life; period furniture; decorative arts; farm machinery; medical instruments; crime evidence; Native American artifacts; log cabin.
Research Fields: Wood County History & Infirmary.
Facilities: arboretum; herb garden; nature trails; picnic areas; meeting rooms.
Activities: tours; programs on local history; historical displays; monthly tea series. Museum Sponsors: Spring Open House in April; Power of Yesteryear Antique Tractor Show in June; Halloween Folklore & Funfest in October; Old Home Holiday Tours in December.
Publications: quarterly newsletter, The Black Swamp Chanticleer; book, The Wood County Atlases, 1875-1912, A Century or So of Wood County Weather, 1787-1991.
Hours & Admission Prices: Feb.-March Tues.-Fri. 9:30-4:30; April-Oct. Tues.-Fri. 9:30-4:30, Sat.-Sun. 1-4; Dec. Holiday Tours: Mon.-Wed. 10-5, Thurs.-Fri. 10-8, Sat.-Sun. 1-4. Suggested Donation: adults $4; discounts to AAM & AAA members; members no charge. Closed holidays. &
Attendance: 25,000 (estimated)
Membership: Seniors & Students $10; Individual $15; Grand $20; Family $25; Business $45; Sustaining $60; Patron $100; Life Individual $300; Life Couple $500.

Bradford

BRADFORD OHIO RAILROAD MUSEUM, (M), 200 N. Miami Ave., Bradford, OH 45308-1164. Mailing Address: P.O. Box 101, Bradford, OH 45308-0101. Tel.: 740-653-2220. Fax: 740-654-0505.
E-mail: mkosier@brhio.com
Web Site: bradfordrrmuseum.org
Founded: 2002.
Congressional District: 49
Key Personnel: C.E.O. & Pres. (V), Sandy Edminson; Devel., Marilyn Kosier; Chm. (V), Don Wick; Education, Jeremy Martin; Public Rels., Sue Vickroy; Treas., Jordon Ingle; Archivist, Bill Haines; Museum Shop Mgr., Gail Shafer.
Governing Authority: private; nonprofit organization. Tax-exempt: 501(c)(3).
Institution Type/Description: History & Railroad Museum.
Collections: historic railyards; railroad artifacts & tower; photographs.
Activities: walking tour. Annual Events: Salute to the Railroaders; Railroad Days; Hobo Days; Dining by Rail; Santa at the Museum.
Hours & Admission Prices: April-Dec. Sat. 10-4, Sun. 1-4. Adults $4, seniors $2; members no charge. &
Membership: Individual $20.

Brecksville

CUYAHOGA VALLEY NATIONAL PARK, 15610 Vaughn Rd., Brecksville, OH 44141-3097. Tel.: 216-524-1497. Fax: 216-524-2604.
Web Site: www.nps.gov/cuva
Formerly: Cuyahoga Valley National Recreation Area
Founded: 1974.
Congressional District: 10, 11, 13, 14 & 17
Key Personnel: Deputy Supt., Paul J. Stoehr; Chief, Interpretation & Visitor Svcs., Jennie Vasarhelyi.
Governing Authority: federal. National Park Service, Dept. of the Interior, Washington, D.C. Tax-exempt.
Institution Type/Description: Park Museum.
Collections: historic archaeology; historic structures.
Research Fields: zoology; biology; aquatic ecosystems; history; archeology; oral history.
Facilities: reference library; environmental education center; visitor centers.
Activities: interpretive programs; special cultural art events.
Publications: brochures; quarterly, calendar of events.
Hours & Admission Prices: Park: daily. Towpath Trail daily 24 hours; other trails dawn to dusk. Park Visitor Centers: call for hours. No charge; donations accepted. &
Attendance: 2,800,000 (accurate)

Brimfield

THE KELSO HOUSE MUSEUM, 4158 State Route 43, Brimfield, OH 44240. Tel.: 330-673-1058.
E-mail: curator@kelsohouse.org
Web Site: www.kelsohouse.org
Founded: 1963.
Congressional District: 11
Key Personnel: Cur., Judi Allen.
Personnel Profile: Part-Time Paid 1; Part-Time Volunteers 18; Interns 1.
Governing Authority: society; nonprofit organization. Affiliated with Brimfield Memorial House Association, Inc. Tax-exempt: 501(c)(3).
Institution Type/Description: Regional History Museum: housed in the former home of William R. Kelso originally the Union Inn & Tavern; built in 1833.
Collections: 1833 Kelso House: furniture; household items; tools; photographs. Grounds: 1845 New England style barn; corn crib; granary; 1870 Sylvester B. Jones House. The Edgar L. McCormack local history collection, located in special collections, Kent State University Library includes manuscripts of local origin; records of Brimfield churches & schools; diaries; account books.
Research Fields: local history.
Facilities: auditorium.
Activities: guided tours; lectures; permanent & temporary exhibitions.
Publications: newsletter, The Kelso Courier.
Hours & Admission Prices: Thurs. & Sat. 12-4; other times by appointment. No charge; donations accepted.
Attendance: 300 (estimated)
Membership: Student $10; Individual $15; Family $25; Sponsoring $50; Life (one name) $250.

Brooklyn

BROOKLYN HISTORICAL SOCIETY, (M), 4442 Ridge Rd., Brooklyn, OH 44144-3353. Mailing Address: P.O. Box 44422, Brooklyn, OH 44144-0422. Tel.: 216-941-0160. Facebook: Brooklyn Historical Society.
E-mail: groundhogsgarden@wowway.com
Founded: 1970.
Congressional District: 10
Key Personnel: Pres. (V), Barbara Stepic; Vice Pres., Elaine Schmidt.
Personnel Profile: Part-Time Volunteers 30; Interns 2.
Governing Authority: society; nonprofit organization. Tax-exempt.
Institution Type/Description: Local History Museum.
Collections: clothes; glass; hair receivers; tools; furniture; period artifacts & memorabilia of Victorian era; clothes; furniture & memorabilia of 1920s; perennials; herb garden including over 60 varieties; World War I; early schools in Brooklyn; Brooklyn High School yearbooks; area high school & college yearbooks; period city records; 1930s-1940s toys; medical artifacts; 1955 reprints - Elvis appeared at Brooklyn High School.
Research Fields: genealogy; early school history; city records; voting records; yearbooks.
Facilities: research library. Crafts, herbs & museum-related items for sale.
Activities: guided tours; permanent & temporary exhibitions; school tours; school 25th & 50th reunion tours; talks on history; quilting & rug loom demonstrations. Museum Sponsors: Old-Time Craft Sale in September; Power Point program early school in Brooklyn & early pictures/images of area.
Publications: brochure; tour guide; annual newsletter; docent guide; History of Schools in Brooklyn.
Hours & Admission Prices: Tues. 10-2. No charge; donations accepted. Closed holiday weekends.
Attendance: 1,200 (estimated)
Membership: Student $1.50; Single $5; Couple $7; Life $100.

Bryan

SPANGLER STORE & MUSEUM, 400 N. Portland St., Bryan, OH 43506-1200. Mailing Address: 71P.O. Box, Bryan, OH 43506-0071. Tel.: 419-633-6439; 888-636-4221.
Founded: 2006.
Institution Type/Description: Company Museum.
Collections: company & family history; photographs; machinery; equipment; tools; candy industry.
Facilities: Museum-related items for sale.
Activities: guided tours.
Hours & Admission Prices: Memorial Day to Labor Day Mon.-Fri. 10-3; Winter: Wed.-Fri. 10-3; other times by appointment. Museum: no charge. Tours: adults $5, seniors $4, children 6-18 $3; children 5 & under no charge. Closed New Year's Eve & Day; Good Friday; Memorial Day; Independence Day; Thanksgiving & day after; Christmas Eve & Day.

Bucyrus

BUCYRUS HISTORICAL SOCIETY, (M), 202 S. Walnut St., Bucyrus, OH 44820-2326. Mailing Address: P.O. Box 493, Bucyrus, OH 44820-0493. Tel.: 419-562-6386 & 9073.
Web Site: www.bucyrusonline.com/bhs
Founded: 1969.
Congressional District: 4
Key Personnel: C.E.O., Dr. John K. Kurtz; Cur., Mary Ellen Lust; Cur., Don Lust.
Governing Authority: society; nonprofit. Tax-exempt.
Institution Type/Description: Historical Society Museum: housed in 1839 Scroggs Family Home.
Collections: furniture & furnishings of the period; restored Niagara horse-drawn steam engine c.1826 & Silby hose cart c.1869.
Facilities: library of books available for research on premises.
Activities: guided tours; permanent exhibitions.
Hours & Admission Prices: Mon. 1-4; tours by appointment. No charge; donations accepted. Closed holidays.
Attendance: 600 (estimated)
Membership: Single $7; Life $100; Patron $500.

CRAWFORD AGRICULTURAL MUSEUM, 610 Whetstone St., Bucyrus, OH 44820. Mailing Address: c/o Crawford Antique Farm Machinery Assn., P.O. Box 105, Bucyrus, OH 44820. Tel.: 419-562-1123.
E-mail: rhaas5@columbus.rr.com
Key Personnel: Treas., Dorothy E. Haas

Institution Type/Description: Agricultural Museum.
Collections: local history & culture; period farm equipment & machinery; farm life.
Activities: guided tours.
Hours & Admission Prices: May-Sept. 1st & 3rd Sun. 1-4. No charge; donations accepted. &

Membership: Individual $10.

Burton

CENTURY VILLAGE MUSEUM, 14653 E. Park St., Burton, OH 44021. Mailing Address: P.O. Box 153, Burton, OH 44021-0153. Tel.: 440-834-1492. Fax: 440-834-4012.
Web Site: geaugahistorical.org
Founded: 1938.
Congressional District: 11
Key Personnel: Pres. (V), Kurt Updegraff; Site Mgr. & Museum Shop Mgr., Cheryl McNulty; Office Mgr., Terry Kwasniewski; Dir. Tours, Rosemary Kneale; Maintenance Supvr., William Troyer.
Personnel Profile: Full-Time Paid 4; Part-Time Paid 12; Part-Time Volunteers 150.
Governing Authority: private; nonprofit organization. Parent Institution: Geauga County Historical Society. Tax-exempt: 501(c)(3).
Institution Type/Description: Historic Village Museum.
Collections: history & development of Geauga County within the Western Reserve; furniture & artifacts of Geauga County; historic homes & buildings forming a recreated 19th-century village; railroad station; caboose; 19th-century farming operation; archival material of local & genealogical interest. Historic Buildings: 1817 William Law House; 1834 Boughton House; 1824 Hitchcock House; 1838 Hickox Brick; 1872 one room schoolhouse; 1850 country store; 1825 blacksmith shop; 1806 Cook House; 1856 red barn; 1850-1889 white barn; 1846 church; 1870 ladies' apparel shop.
Research Fields: genealogy & local history.
Facilities: research library. Country store, gift & craft items for sale.
Activities: guided tours; formally organized education programs for youth. Museum Sponsors: 12 special event weekends during the summer; Apple Butter Festival in October.
Publications: Geauga County Historical Society Quarterly; brochures; pamphlets.
Hours & Admission Prices: April-Nov. Public Tours: Sat.-Sun. 1 & 3. School Tours: by appointment. Adults $7, children 6-12 $4; discounts to AAA & Golden Buckeye members; children under 6 no charge.
Attendance: 30,000 (estimated)
Membership: Senior Citizens $25; Individual $35; Family $55; Silver $100; Gold $250; Platinum $500.

Cadiz

CLARK GABLE MUSEUM, 138 Charleston St., Cadiz, OH 43907. Mailing Address: P.O. Box 65, 138 Charleston St., Cadiz, OH 43907. Tel.: 740-942-4989. Fax: 740-942-4989.
Web Site: clarkgablefoundation.com
Key Personnel: Exec. Dir. & Museum Shop Mgr., Nan Mattern; Pres. (V), Dr. Gary Barber.
Personnel Profile: Part-Time Paid 1.
Governing Authority: Tax-exempt.
Institution Type/Description: Historic House Museum: housed in the birthplace of Clark Gable.
Collections: Clark Gable's personal artifacts; period furnishings; photographs; 1954 Cadillac.
Facilities: Museum-related items for sale.
Hours & Admission Prices: May & Sept. Wed.-Sat. 10-4; June to Labor Day Wed.-Sat. 10-4, Sun. 1:30-4; Oct.-Nov. Wed.-Fri. 10-4. Adults $5.50, seniors $4.75, children $3; discounts to AAM members & groups of 15 or more; members no charge. Closed major holidays. &
Attendance: 1,200 (estimated)
Membership: Annual $20, $60 & $235; Lifetime $500; Benefactor $1,000.

HARRISON COUNTY HISTORY OF COAL MUSEUM, Puskarich Public Library, 200 E. Market St., Cadiz, OH 43907-1214. Tel.: 740-942-2623.
Institution Type/Description: History Museum.
Collections: early mining tools; heavy machinery; working scale model of a dragline; historic photographs.
Hours & Admission Prices: Mon.-Sat. 9-5, Sun. by appointment. No charge.

Cambridge

GUERNSEY COUNTY MUSEUM, 218 N. 8th St., Cambridge, OH 43725-1840. Mailing Address: P.O. Box 741, Cambridge, OH 43725-0741. Tel.: 740-439-5884.
E-mail: curator@gcohmuseum.org
Web Site: www.gcohmuseum.org
Founded: 1963.
Congressional District: 18
Key Personnel: Pres., Madelyn Joseph; Vice Pres., Diane Krall; Sec., Diana Wetzel; Treas., Mary Jane Downerd; Cur., Judy Clay.
Personnel Profile: Part-Time Paid 1.
Governing Authority: society. Guernsey County Historical Society. Tax-exempt.
Institution Type/Description: General Museum: housed in 1831 McFarland Home.
Collections: glass; military; china; pottery; hand tools; furniture; kitchen ware; Victorian parlor, Hall of Fame; ladies boutique; John Glenn artifacts; one-room schoolhouse.
Research Fields: genealogy.
Activities: quilt shows; class tours; historical slide programs throughout the county; portrayals of historical men, The Legend of Fighting Bill Reed; Chaplain McFarland.
Publications: brochures.
Hours & Admission Prices: mid-March to mid-Dec. Tues., Thurs. & Sat. 12-3; tours by appointment. Adults $5, children $2; members no charge.
Attendance: 189 (accurate)
Membership: Individual $5; Family $10; Business $25; Life $100.

HOPALONG CASSIDY MUSEUM, 127 S. 10th St., Cambridge, OH 43725. Tel.: 740-432-3364.
E-mail: lbates1205@frontier.com
Founded: 1992.
Congressional District: 18
Key Personnel: Dir., Owner, Laura Bates.
Personnel Profile: Full-Time Volunteers 1; Part-Time Volunteers 3.
Institution Type/Description: History Museum: housed in the home town of William Boyd aka Hopalong Cassidy.
Collections: Cassidy's life & career; photographs; Hoppy memorabilia; cowboy artifacts.
Hours & Admission Prices: Mon.-Sat. 10-4. Donation: adults $1. &
Attendance: 1,000 (estimated)

NATIONAL MUSEUM OF CAMBRIDGE GLASS, 136 S. Ninth St., Cambridge, OH 43725-2453. Mailing Address: P.O. Box 416, Cambridge, OH 43725-0416. Tel.: 740-432-4245.
Web Site: www.cambridgeglass.org
Founded: 1973.
Key Personnel: Chm. (V), Cynthia Arent; Pres. (V), David Ray.
Governing Authority: Tax-exempt.
Institution Type/Description: Glass Museum.
Collections: glassmaking history; Cambridge glass; period furniture.
Facilities: Museum-related items for sale.
Hours & Admission Prices: April-Oct. Wed.-Sat. 9-4, Sun. 12-4. Adults $5, senior citizens & AAA members; children under 12 no charge. Closed Easter; Independence Day. &

Canal Fulton

CANAL FULTON HERITAGE SOCIETY, 116 S. Canal St., Canal Fulton, OH 44614-1317. Tel.: 1-800-854-3808; 1-800-Helena 3. Fax: 330-854-2013.
E-mail: cfhs@discovercanalfulton.com
Founded: 1968.
Congressional District: 16
Key Personnel: Pres., John Hatfield.
Personnel Profile: Part-Time Paid 1; Part-Time Volunteers 12.
Governing Authority: bd. of trustees; nonprofit organization. Tax-exempt: 501(c)(3).
Institution Type/Description: Historical & Preservation Society.
Collections: furnishings; canal artifacts; photographs; maps. Historic Houses: 1847 Oberlin House; 1870s Heritage House; 1900 Blank House.
Research Fields: local history; Ohio canals.
Facilities: Canal-related items & books and historical books for sale.
Activities: guided tours; lectures; films; docent program or council; canal boat ride on St. Helena III; permanent exhibitions.
Hours & Admission Prices: Museum: May-Oct. Sat.-Sun. 10-4, Oct.-May by appointment only. No charge; donations accepted.
Attendance: 10,000 (estimated)

Membership: Student & Senior $10; Individual & Senior Couple $15; Family $25; Business $50 & up; Sustaining $50; Contributing $100; Life $1,000.

Canal Winchester

BARBERS HALL OF FAME MUSEUM, 2 S. High St., Canal Winchester, OH 43110. Mailing Address: P.O. Box 15, Canal Winchester, OH 43110. Tel.: 614-833-1846 & 837-8400.
E-mail: maippoliti@insight.rr.com
Web Site: www.nationalbarbermuseum.org
Founded: 1968.
Congressional District: 20
Key Personnel: Dir., Mike Ippoliti
Governing Authority: Parent Institution: Canal Winchester Area Historical Society. Tax-exempt.
Institution Type/Description: History Museum.
Collections: 71 barber poles; barber chairs; re-created period barber shop; mugs; razors; medical tools.
Hours & Admission Prices: By appointment. Adults $5, seniors $4, students $3; discounts to groups and AAM & ICOM members.
Attendance: 500 (estimated)

SLATE RUN LIVING HISTORICAL FARM, METRO PARKS, 1375 State Rte. 674 N., Canal Winchester, OH 43110-9406. Tel.: 614-833-1880. Fax: 614-834-1220. TDD: 614-895-6240.
Web Site: www.metroparks.net/ParksSlateRunFarm.aspx
Founded: 1981.
Key Personnel: C.E.O., John O'Meara; Dir. Farm Program, Ann Culek.
Personnel Profile: Full-Time Paid 5; Part-Time Paid 2; Part-Time Volunteers 60; Interns 1.
Governing Authority: Franklin County Metro Parks. Tax-exempt.
Institution Type/Description: Historic Site: 1856 house; 1881 barn.
Collections: agricultural implements & machinery; household goods; buildings; livestock.
Research Fields: 1880-1900 mid-Ohio agriculture & social life; local history.
Activities: guided & self-guided tours; organized education programs; docent program.
Publications: monthly newsletter, Slate.
Hours & Admission Prices: April-May & Sept.-Oct. Tues.-Sat. 9-4, Sun. 11-4; June-Aug. Tues.-Thurs. 9-4, Fri.-Sat. 9-6, Sun. 11-6; Nov.-March Wed.-Sat. 9-4, Sun. 11-4. No charge. Closed New Year's Day; Thanksgiving; Christmas. &
Attendance: 40,000 (estimated)

Canfield

THE WAR VET MUSEUM, 23 E. Main St., Canfield, OH 44406-1360. Tel.: 330-533-6311.
E-mail: warvetmuseum@gmail.com
Web Site: www.warvetmuseum.org
Key Personnel: Owner & Operator, Lew Speece
Institution Type/Description: Military Museum.
Collections: over 36,000 artifacts from all periods of U.S. history; model trains; photographs.
Hours & Admission Prices: Call for hours. No charge; donations accepted.

Canton

CANTON CLASSIC CAR MUSEUM, 555 Market Ave. S., Canton, OH 44702-2111. Mailing Address: 612 Market Ave. S., Canton, OH 44702-2114. Tel.: 330-455-3603. Fax: 330-455-0363.
E-mail: char@cantonclassiccar.org
Web Site: www.cantonclassiccar.org
Founded: 1978.
Congressional District: 14
Key Personnel: Dir., Char Lautzenheiser; Chm. (V), Florence Belden; Pres. (V), Marshall Belden, Jr.; Financial Dir., Timothy Belden; Operations Mgr., Gary Pelger; Sr. Cur., Al Parsons; Cur., Norman Munson; Admissions, Dennis Dickey.
Personnel Profile: Full-Time Paid 2; Part-Time Paid 5; Part-Time Volunteers 6; Interns 1.
Governing Authority: private; nonprofit organization. Tax-exempt.
Institution Type/Description: Automobile Museum.
Collections: pre-World War II domestic vehicles; memorabilia; post World War II vehicles.
Research Fields: study of Lincoln Highway.
Facilities: 1,000-vol. library of automotive history; 20,000 sq. ft. exhibit space. Gift items for sale.
Activities: formal education programs; guided tours.

Publications: quarterly newsletter, CCCM News.
Hours & Admission Prices: Daily 10-5. Adults $7.50, senior citizens $6, children 6-18 $5; discounts to AAA, tour & family groups; members & children under 6 no charge. Closed New Year's Day; Easter; Thanksgiving; Christmas. &
Attendance: 10,000 (estimated)
Membership: Individual $25; w/Spouse $35; Family $50; Sustaining $250; Life $1,000; Corporate $5,000.

*   **THE CANTON MUSEUM OF ART, (M),** 1001 Market Ave. N., Canton, OH 44702-1075. Tel.: 330-453-7666. Fax: 330-453-1034.
E-mail: staff@cantonart.org
Web Site: www.cantonart.org
Founded: 1935.
Congressional District: 16
Key Personnel: Exec. Dir., M. J. Albacete; Pres. (V), Donald Jensen; Business Mgr., Sonja Herwick; Mktg. Mgr., Mary Byrne; Coord. Education, Lauren Kuntzman.
Personnel Profile: Full-Time Paid 6; Full-Time Volunteers 1; Part-Time Paid 7; Part-Time Volunteers 125; Interns 2.
Governing Authority: nonprofit organization. Tax-exempt: 501(c)(3).
Institution Type/Description: Art Center and Museum.
Collections: 19th-20th century American art; watercolors; works on paper; contemporary ceramics.
Research Fields: American watercolor; contemporary ceramics.
Facilities: 2,500-vol. library; education wing; sculpture courtyard.
Activities: lectures; films; formally organized education programs; docent program; permanent, temporary & traveling exhibits. Museum Sponsors: arts & craft festival.
Publications: class schedules; exhibition catalogs; annual report.
Hours & Admission Prices: Tues.-Thurs. 10-8, Fri. 10-5, Sat. 10-3, Sun. 1-5. Adults $6, senior citizens & college students $4; discounts to AAM members; children 12 & under and members no charge. Closed New Year's Day; Memorial Day; Independence Day; Labor Day; Thanksgiving; Christmas. &
Attendance: 35,121 (accurate)
Membership: Student $15; Individual $30; Family $55; Gallery Circle $100; Director's Circle $300; Life $2,500.

MCKINLEY PRESIDENTIAL LIBRARY & MUSEUM, 800 McKinley Monument Dr., N.W., Canton, OH 44708-4800. Tel.: 330-455-7043. Fax: 330-455-1137. Facebook: McKinley Presidential Library & Museum.
E-mail: mmuseum@neo.rr.com
Web Site: www.mckinleymuseum.org
Founded: 1946.
Congressional District: 16
Key Personnel: Dir., Joyce Yut; Pres. (V), Robert F. Belden; Vice Pres., Don Deitemyer; Cur., Kimberly Kenney; Sec., Robert Leibensperger; Treas., Mark Wright; Asst. Sec., Treas. & Museum Shop Mgr., Cindy Sober; Dir. Planetarium, David Richards; Dir. Education, Chris Kenney; Dir. Science, Lynette Reiner; Volunteer Dir., Stephanie Span; Librarian, Mark Holland; Office Mgr., Rita Zwick.
Personnel Profile: Full-Time Paid 6; Full-Time Volunteers 1; Part-Time Paid 22; Part-Time Volunteers 140.
Volunteer Hours: 8,800
Operating Expenses: 1,103,458
Operating Income: 1,127,723
Governing Authority: nonprofit organization. Parent Institution: Stark County Historical Society. Tax-exempt: 501(c)(3).
Institution Type/Description: History, Science & Comprehensive Family Museum.
Collections: archives; astronomy; glass; natural and physical science; architecture; ceramics; clocks & watches; decorative arts; industrial & natural history; transportation; Street of Shops; 96 ft. replica of HO train; Foucault Pendulum; dolls; toys; Dueber-Hampden watches; McKinley memorabilia & history; Discover World: natural history; current ecosystem; Space Ship Earth.
Major Exhibits: A Secret Gift, 11/15/13-2/2/14; Black Wings (T), 2/7/14-4/27/14; Peanuts Naturally (T), 5/23/14-8/17/14; Underwear, 9/5/14-11/14; Legacy of Ferdinand Brader, 12/4/14-3/15/15.
Research Fields: astronomy; Stark County history; Pres. William McKinley; Paleo-Indians.
Facilities: 5,000-vol. library of books on McKinley, Stark County, Ohio, Civil War, Spanish American War available on premises; planetarium; science center; 150-seat auditorium; classrooms; McKinley National Memorial & grounds. Museum-related items for sale.

Activities: guided tours; lectures; films; formally organized education programs for children; interactive science exhibits,

Publications: bimonthly newsletter; brochure, Discover World; planetarium brochure; teacher's guide; schedule of classes; books, The McKinley Monument A Tribute to a Fallen President; Canton's West Lawn Cemetery; Canton A Journey Through Time; The Wm. McKinley Presidential Library & Museum; Canton Pioneers in Flight; Canton Entertainment.

Hours & Admission Prices: Mon.-Sat. 9-4, Sun. 12-4. Adults $8, senior citizens $7, children 3-18 $6; discounts to adult groups over 20 & ASTC reciprocity members; museum members & children under 3 no charge. Closed New Year's Day; Easter; Memorial Day; Labor Day; Thanksgiving; Christmas. &

Attendance: 47,061 (accurate)

Membership: Individual $39; Family $52; Celestial $85; Silver $140; Gold $300; Diamond $575; William McKinley Circle $1,700; Individual Lifetime $5,000.

NATIONAL FIRST LADIES' LIBRARY, (M), 331 S. Market Ave., Canton, OH 44702-2107. Tel.: 330-452-0876. Fax: 330-445-2008.

E-mail: pkrider@firstladies.org

Web Site: www.firstladies.org

Founded: 1998.

Congressional District: 16

Key Personnel: Exec. Dir., Patricia Krider; Pres. (V), Mary Regula; Education, Lucinda Frailly; Museum Shop Mgr., Mary Rhodes.

Personnel Profile: Full-Time Paid 6; Part-Time Paid 5; Part-Time Volunteers 28; Interns 1.

Governing Authority: private; nonprofit organization. Tax-exempt: 501(c)(3).

Institution Type/Description: History Museum.

Collections: contributions by First Ladies to our nation; clothing; photographs; letters; audiovisual.

Facilities: 3,000-vol. library; 90-seat auditorium; educational facilities; 10,000 sq. ft. exhibit space. Museum-related items for sale.

Activities: formal education programs; guided tours; lectures; participatory & temporary exhibits; auditorium.

Publications: biannual newsletter.

Hours & Admission Prices: Guided Tours: June-Aug. Tues.-Sat. 9:30, 10:30, 12:30, 1:30 & 2:30, Sun. 12:30, 1:30 & 2:30; Sept.-May Tues.-Sat. 9:30, 10:30, 12:30, 1:30 & 2:30. Adults $7, senior citizens $6, students $5. Closed major holidays. &

Attendance: 8,500 (estimated)

Membership: Individual $50; Inaugural Circle $100; Ida Saxton McKinley Circle $250; Executive Circle $500; First Ladies' Inner Circle $1,000; Heritage Circle $5,000.

NATIONAL FOOTBALL MUSEUM, INC., (M), 2121 George Halas Dr., N.W., Canton, OH 44708-2630. Tel.: 330-456-8207. Fax: 330-588-3801.

Web Site: www.profootballhof.com

Founded: 1963.

Congressional District: 7

Key Personnel: Pres. & Exec. Dir., Stephen Perry; Chm., Thomas W. Schervish; Vice Pres. Retail Sales & Licensing, Steve Strawbridge; Vice Pres. & C.F.O., D. William Allen; Vice Pres. Communications & Exhibits, Joe Horrigan; Vice Pres. Mktg. & Operations, Dave Motts; Registrar, Christy Davis; Information Svcs. Mgr., Pete Fierle; Operations Mgr., Kevin Shiplet; Education Program Coord., Jerry Csaki; Cur. Collections, Jason Aikens; Researcher, Saleem Choudhry; Researcher, Jon Kendle; Museum Shop Mgr., Michelle Hunt; Communications Asst., Chris Schilling; Information Svcs. Specialist, Chad Reese; Event Mktg. Specialist, Gail McLaughlin.

Personnel Profile: Full-Time Paid 38; Part-Time Paid 40; Part-Time Volunteers 300; Interns 6.

Volunteer Hours: 14,877

Governing Authority: nonprofit organization. Tax-exempt: 501(c)(3).

Institution Type/Description: Professional Football Museum.

Collections: mementoes from the games & players including equipment, photos, films; videotapes; material pertinent to the development of pro football in the U.S.; printed game accounts; files on players, games & circumstances contributing to football's growth; artists' works, illustrations, photos & paintings; Hall of Fame Enshrinement; enshrinee mementoes room; historical displays.

Research Fields: professional football history.

Facilities: 3,000-vol. library of material on professional football, available for use on premises or through correspondence with personnel by mail; 350-seat theater; turnable theater featuring NFL action in Cinemascope; snack shop. Officially licensed National Football League items for sale.

Activities: films; fan activated videos; interactive exhibits.

Publications: annual magazine, Pro Football Hall of Fame Yearbook; newsletter 3 times a year, Insider.

Hours & Admission Prices: Memorial Day-Labor Day daily 9-8; Sept.-May daily 9-5. Adults $22, senior citizens $18, children 6-12 $16; discount to groups & AAM members; children 5 & under, members and Insider's Club members no charge. Closed Christmas. &

Attendance: 186,429 (accurate)

Carroll

HISTORICAL AIRCRAFT SQUADRON MUSEUM, Fairfield County Airport, 3266 Old Columbus Rd. N.W., Carroll, OH 43112-9723. Tel.: 740-653-4778. Fax: 740-653-2387.

E-mail: rdebevoise@columbus.rr.com

Web Site: www.historicalaircraftsquadron.com

Institution Type/Description: Aviation History Museum.

Collections: aviation history; military aircraft & vehicles.

Hours & Admission Prices: Wed.-Sat. 10-4, Sun. 12-5.

Carrollton

ASHTON HOUSE MUSEUM, (M), 120 3rd St., N.W., Carrollton, OH 44615. Mailing Address: c/o John H. & Evelyn L. Ashton Preservation Assoc., 60 W. Main St., Carrollton, OH 44615. Tel.: 330-627-2682.

E-mail: curator@ashtonhousemuseum.com

Web Site: ashtonhousemuseum.com

Key Personnel: Cur., Jim Painting

Institution Type/Description: History Museum.

Collections: John Ashton's life & career; personal artifacts; period furnishings; photographs.

Hours & Admission Prices: April-Dec. Wed.-Sat. 10-5, Sun. 1-5; other times by appointment. Adults $3, children 4-12 $2; children 3 & under no charge. Blue Star Museum.

BLUEBIRD FARM TOY MUSEUM, 190 Alamo Rd., Carrollton, OH 44615-9581. Tel.: 330-627-7980.

Institution Type/Description: Toy Museum.

Collections: toys from the 1700s to present; wooden, wax, china, French & German bisque, mechanical papier-mache, & cloth dolls; Teddy bears; stuffed animals; Raggedy Ann & Andy; Shirley Temple; Mickey & Minnie Mouse; toy china sets; period dollhouses.

Facilities: Museum-related items for sale.

Hours & Admission Prices: Call for hours.

MCCOOK HOUSE, CIVIL WAR MUSEUM, (M), 15 S. Lisbon St., Carrollton, OH 44615. Mailing Address: c/o Carrol County Historical Society, P.O. Box 174, Carrollton, OH 44615-0174. Tel.: 330-627-3345; 800-600-7172. Fax: 330-627-5366.

Web Site: www.ohiohistory.org/museums-and-historic-sites/museum--historic-sites-by-name/mccook-house

Founded: 1963.

Congressional District: 18

Key Personnel: Mgr., Cur. & Museum Shop Mgr., Shirley Anderson; Pres. (V), David McMahon; Vice Pres., Ann Myers; Sec., Diane George; Treas., Jennifer Cramer.

Governing Authority: state; nonprofit. Parent Institution: Ohio Historical Society. Subsidiary Institution: Carroll County Historical Society. Tax-exempt: 501(c)(3).

Institution Type/Description: History Museum.

Collections: local history; military items; period medical instruments and textiles; local pottery. Historic House: 1837 McCook house.

Research Fields: military.

Facilities: 35-vol. collection of local newspapers.

Activities: guided tours.

Publications: quarterly newsletter, Carroll County Historical Society News.

Hours & Admission Prices: Memorial Day to Labor Day & 2nd weekend in Oct. Fri.-Sat. 10-5, Sun. 1-5; Sept. to 1st weekend in Oct. Sat. 10-5, Sun. 1-5; tours by appointment. Adults $3, children $1; Carroll Co. Historical Society & Ohio Historical Society members no charge.

Attendance: 1,195 (accurate)

Membership: Carroll County, Ohio Historical Society: Student $5; Individual $12; Family $15; Business/Institutional $50; Life Individual $150; Life Couple $200; Life Business $250.

SUSIE'S MUSEUM OF CHILDHOOD, Bluebird Farm Park, 190 Alamo Rd., Carrollton, OH 44615. Tel.: 330-627-8046. Fax: 330-627-8046.
E-mail: info@ccparkdistrict.org
Web Site: www.ccparkdistrict.org
Institution Type/Description: History Museum.
Collections: toys from the 1700s to present; dolls; stuffed animals; Teddy bears; dollhouses; child-size furniture.
Hours & Admission Prices: April-Dec. Tues.-Sun. 11-4.

Celina

MERCER COUNTY HISTORICAL MUSEUM, THE RILEY HOME, 130 E. Market, Celina, OH 45822-1731. Mailing Address: P.O. Box 512, Celina, OH 45822-1731. Tel.: 419-678-2614.
E-mail: histalig@bright.net
Founded: 1959.
Congressional District: 8
Key Personnel: Pres. (V), Joyce L. Alig.
Personnel Profile: Part-Time Volunteers 125.
Governing Authority: society; nonprofit. Parent Institution: Mercer County Historical Society. Tax-exempt: 501(c)(3).
Institution Type/Description: Historical Society Museum: housed in 1896 the Riley home.
Collections: archives of Mercer County history; agriculture; carpenter's & blacksmith's tools; period furniture; school display; costume collections; glassware; guns; Indian artifacts; medical, dental & health display; art; archival collections of Mercer County history & Capt. James Riley heritage; 1898-1900 Ohioans in Alaska gold rush.
Research Fields: Mercer County history & genealogy; Captain James Riley & the Riley family; James Zura Riley, Alaska & Yukon Gold rush, 1898-1900; Coldwater Ohio New Idea Co.
Facilities: restricted library; Mercer County & Ohio history books.
Activities: guided tours; lectures; films; gallery talks; study clubs; TV programs; formally organized education programs; temporary & permanent exhibitions. Annual Events: Prehistoric Artifacts in June; Military Artifacts in October; Postcards & Post Offices in November.
Publications: quarterly, Mercer County Historical Society Newsletter; museum brochure; books: 1978 Mercer County History; History of Celina, Ohio, 1834-1984; History of St. Henry, Ohio, 1837-1987; Coldwater Sesquicentennial 1838-1988; Native Americans & Early Settlers; The Meeting of Cultures, 1780's-1980's; Those Magnificent Big Barns of Mercer County of Western Ohio; Mercer County Centennial Buildings; Ohio's Last Frontiersman, Captain James Riley; Mercer County, Ohio's Courthouses 1824-1998; Mercer County, Ohio: History of the Land Between the St. Marys & Wabash River Valleys; Old Gold Rush to Alaska Diaries, 1898-1900, A True Story; Mercer County, Ohio Fair, 1852-2002; German-American Death Card Collection, 2 Vols.; Celina, Ohio, Our Post Card Past; Rockford, Ohio, Our Post Card Past; Mercer County, Ohio, Our Post Card Past; Grand Lake St. Marys, Ohio, Our Post Card Past; Passport to Mercer County, Ohio History: Granville Township; St. Henry, Burkettsville, Cranberry Prairie, Himmelgarten Convent & St. Marys Novitiate (2007); Biographical historical resources of Mercer County Townships, Ohio History (8 volumes completed out of 14 Townships); New Idea Co. Horse-Drawn Manure Spreaders, Maria Stein & Coldwater, Ohio.
Hours & Admission Prices: by appointment only. No charge.
Attendance: 6,000 (accurate)
Membership: Membership fee voluntary.

Centerville

CENTERVILLE - WASHINGTON TOWNSHIP HISTORICAL SOCIETY & ASAHEL WRIGHT COMMUNITY CENTER, 26 N. Main St., Centerville, OH 45459-4619. Tel.: 937-291-2223. Fax: 937-432-9296.
Web Site: www.mvcc.net/centerville/histsoc
Founded: 1966.
Key Personnel: Pres., Ray Turton; Admin., Vickie Bondi; Administrative Asst., Peggy Brooker; Coord. Education, Marcia Rouse.
Personnel Profile: Part-Time Paid 4; Part-Time Volunteers 30.
Governing Authority: private; nonprofit corporation. Parent Institution: Centerville - Washington Township Historical Society, 89 W. Franklin St., Centerville, OH 45459. Tax exempt: 501(c)(3).
Institution Type/Description: Historic House Museum: housed in the former home of the Wright Brothers great uncle.
Collections: furnishings & personal artifacts depicting life in the early 1800s.
Facilities: library; classroom. Museum-related items for sale.
Activities: formal education programs; guided tours; lectures; broadcast programs. Annual Events: Victorian Tea; July 4th Americana Festival; Autumn Ball.

Publications: monthly newsletter, The Curator.
Hours & Admission Prices: Tues.-Fri. 12-4. Gift Shop: Tues.-Sat. 12-4. No charge; donations accepted. Closed New Year's; Thanksgiving; Christmas.
Attendance: 1,500 (estimated)
Membership: Individual Senior over 60 $20; Individual: $25; Family $35; Business: $45; Sustaining $50; Patron: $100; Life $500; Angel $1,000; Benefactor $5,000.

CENTERVILLE - WASHINGTON TOWNSHIP HISTORICAL SOCIETY & WALTON HOUSE MUSEUM, 89 W. Franklin St., Centerville, OH 45459-4735. Tel.: 937-433-0123. Fax: 937-424-4629.
E-mail: cwths@sbcglobal.net
Web Site: www.mvcc.net/centerville/histsoc
Founded: 1966.
Congressional District: 3
Key Personnel: Dir., Vickie Bondi; Pres., Susan Ross.
Personnel Profile: Part-Time Paid 6; Part-Time Volunteers 50.
Governing Authority: private; nonprofit organization. Branch Museums: Walton House Museum, Centerville, OH; Asahel Wright Community & Visitor Center, Centerville, OH. Tax-exempt.
Institution Type/Description: History Museum.
Collections: area artifacts, data & information; family histories; landmark files; society archives.
Facilities: library; research center.
Activities: summer youth classes; monthly speaker series; spring and fall fundraising events.
Publications: local historical books & publications; monthly newsletter.
Hours & Admission Prices: Tues.-Fri. 12-4, Sat. by appointment. No charge; donations accepted.
Attendance: 5,000 (estimated)
Membership: Individual $25; Family $35; Business $50; Friend $100; Patron $250; Supporter $500; Benefactor $1,000.

Chagrin Falls

CHAGRIN FALLS HISTORICAL SOCIETY, 21 Walnut St., Chagrin Falls, OH 44022-3125. Tel.: 440-247-4695.
E-mail: chaghist@sbcglobal.net
Web Site: www.chagrinhistorical.org
Founded: 1949.
Congressional District: 22
Key Personnel: Pres., John Bourisseau; Cur., Pat E. Zalba; Photograph Collection, Zo Sykora; Librarian, Laura Gorretta.
Personnel Profile: Part-Time Volunteers 50.
Governing Authority: society. Tax-exempt: 501(c)(3).
Institution Type/Description: Historical Society Museum.
Collections: doll collection of Marian Jencick; costumes; household articles; china; Bullard butter molds; glassware; regional art & artifacts; Ober, Williams & Chagrin Falls Manufacturing Co. sad irons & stands; paintings & information on primitive artist Henry Church, Jr. paintings & history, native of Chagrin Falls; Max Barnard paintings, birdhouses & history; photographs of early C.F. industries; village; homes; schools; transportation; people; organizations (1870s to present day); Williams cast iron banks.
Research Fields: village history; genealogy; relating to items in museum; Civil War related to village residents.
Facilities: library of c.1874-1964 newspapers, The Chagrin Falls Exponent available.
Activities: guided tours; formally organized education programs; permanent & temporary exhibitions; monthly program meeting open to public; old-time baseball games played by 1860s rules, featuring The Forest Citys versus other historical society sponsored teams throughout the region. Annual Event: Christmas Season Open House.
Publications: monthly newsletter; pamphlet, Village Victorian; books, Annie's Anecdotes; The Drawings of Max Barnard; cookbook, Taste of the Past; Chagrin Falls; Whence the Name; pamphlet, How to Care for Your Historic Home; video, Chagrin Falls: A Look at Our Past; book, Chagrin Falls: An Ohio Village History; books, The Stranahan Folding Canvas Boat Company of Chagrin Falls; The History of the Ober Manufacturing Company of Chagrin Falls, 1862-1959; History of the Chagrin Hardware Company; Cleveland & Chagrin Falls Electric Company, 1897-1925; Mill Town on the Chagrin River: The History of the Early Mills and Foundries in Chagrin Falls, Ohio.
Hours & Admission Prices: Thurs. 2-4; other times by appointment. No charge; donations accepted. Closed holidays.
Attendance: 2,500 (estimated)
Membership: Senior $20; Individual $35; Family $40; Corporate $100; Corporate Life $500.

Chillicothe

✻ ADENA MANSION AND GARDENS, (M), 847 Adena Rd., Chillicothe, OH 45601-1380. Tel.: 740-772-1500; 800-319-7248. Fax: 740-775-2746.
E-mail: info@adenamansion.com
Web Site: www.adenamansion.com
Formerly: Adena State Memorial, The Home of Thomas Worthington
Founded: 1946.
Congressional District: 6
Key Personnel: C.E.O., Ed Behanna; Dir., Kathy Styer.
Personnel Profile: Full-Time Paid 3; Part-Time Paid 7; Part-Time Volunteers 100.
Governing Authority: society; nonprofit. Parent Institution: Ohio Historical Society, Ohio Historical Center, Columbus, OH 43211-2497. Tax-exempt.
Institution Type/Description: Historic House Restoration: 1806-1807 original mansion of Ohio's sixth governor & first senator, Thomas Worthington.
Collections: American furniture & furnishings, 1760-1830; settlement of Ohio; Ohio's path to statehood; Thomas Worthington family artifacts; outbuildings.
Research Fields: history of Worthington & Swearingen families.
Facilities: education center; gardens.
Activities: guided tours; public programs. Annual Event: Heirloom Plant Sale.
Hours & Admission Prices: April-Oct. Wed.-Sat. 9-5, Sun. 12-5; Nov.-March for special events only. Adults $8, children 6-12 $4; discounts to senior citizens & AAM members; OHS members and children 5 & under no charge. Closed Easter. ♿
Attendance: 11,000 (accurate)
Membership: Family $30.

THE FRIENDS OF LUCY HAYES HERITAGE CENTER, 90 W. Sixth St., Chillicothe, OH 45601-3838. Mailing Address: P.O. Box 1790, Chillicothe, OH 45601-5790. Tel.: 740-775-5829 (center). Fax: 740-775-5829.
E-mail: lucy@lucyhayes.org
Web Site: www.lucyhayes.org
Founded: 1996.
Congressional District: 6
Key Personnel: Pres. (V), Paul Thacker; Treas., Linda Barrett; Sec., Melody Smith.
Personnel Profile: Part-Time Volunteers 15.
Governing Authority: private; nonprofit organization. Tax-exempt: 501(c)(3).
Institution Type/Description: General Museum: restored birthplace of First Lady Lucy Hayes.
Collections: pictures; pamphlets of speeches; White House invitations; civil war memorabilia; postcards; letters; Hayes family personal artifacts; Empire Style furniture; replica of the birthplace; 1876 newspapers; hooked rug, Lucy doll.
Activities: guided tours; loan exhibitions; meeting room. Annual Events: Ohio Statehood Day; open house & program; bus tours to historical sites; ice cream & cake social in August; Annual Holiday Dinner; Hayes Anniversary.
Publications: quarterly newsletter, News From The Board - Lucy Hayes Heritage Center.
Hours & Admission Prices: April-Sept. Mon. 10-2, Sat. 1-4; other times by appointment. Adults $4, children $2; members no charge. Closed Good Friday; Memorial Day weekend.
Attendance: 502 (accurate)
Membership: Individual $10; Family $20; Organization $25.

HOPEWELL CULTURE NATIONAL HISTORICAL PARK, 16062 State Rte. 104, Chillicothe, OH 45601-9701. Tel.: 740-774-1126. Fax: 740-774-1140. Facebook: Hopewell Culture National Historical Park.
E-mail: hocu_superintendent@nps.gov
Web Site: www.nps.gov/hocu
Formerly: Mound City Group National Monument
Founded: 1923.
Congressional District: 7
Governing Authority: federal. Parent Institution: U.S. Dept. of the Interior, National Park Service. Tax-exempt.
Institution Type/Description: Park Museum: 200 B.C.-500 A.D., Hopewell Indian earthwork.
Collections: archaeological artifacts from Hopewell culture who built earthen mounds & embankments between 200 B.C. and A.D. 500; items from World War I army training facility, Camp Sherman.
Research Fields: archaeological; local history.
Facilities: 1,000-vol. library on Native American culture, particularly the Hopewell culture, and National Park Service; Visitor Center; environmental study area. Printed material, postcards, booklets, replicas of artifacts for sale.
Activities: guided tours; lectures; formally organized education programs for children; permanent & temporary exhibitions.
Publications: biannual newsletter, Hopewell Happenings; newsletter, Hopewell Archeology.
Hours & Admission Prices: Visitor's Center: Memorial Day to Labor Day daily 8:30-6; Sept.-May daily 8:30-5. No charge; donations accepted. Closed New Year's Day; Thanksgiving; Christmas. ♿
Attendance: 30,000 (accurate)
Membership: Senior $10; Annual $80.

JAMES M. THOMAS TELEPHONE MUSEUM, 68 E. Main St., Chillicothe, OH 45601. Tel.: 740-772-8200.
Institution Type/Description: History Museum.
Collections: telephone industry history; early switchboard; phone directories; telephone equipment.
Hours & Admission Prices: Mon.-Fri. 8:30-4:30.

ROSS COUNTY HISTORICAL SOCIETY, INC., (M), 45 W. 5th St., Chillicothe, OH 45601-3227. Tel.: 740-772-1936.
E-mail: info@rosscountyhistorical.org
Web Site: www.rosscountyhistorical.org
Founded: 1896.
Congressional District: 6
Key Personnel: C.E.O., Dir. & Museum Shop Mgr., Thomas G. Kuhn; Pres. (V), Robert Nelson.
Personnel Profile: Full-Time Paid 2; Part-Time Paid 6; Part-Time Volunteers 25.
Governing Authority: society; nonprofit. Parent Institution: Ross County Historical Society. Tax-exempt: 501(c)(3).
Institution Type/Description: General Museum.
Collections: prehistoric artifacts; manuscripts; pioneer tools; household furnishings; 19th to 20th-century toys & costumes; naval paintings of Admiral Henry W. Walke; Ohio's Constitution Table from 1800-1816 when Chillicothe was capital of the Northwest Territory & first capital of Ohio; pioneer crafts; early state & local history; Civil War, Camp Sherman & WWI; portrait gallery. Historic Houses: 1838 William T. & Elizabeth A. McClintick Home; 1838 David McCandless McKell Library; 1907 Marianne & Charles Franklin Home devoted to the role of the 19th-century woman; 1827 Knoles Log House.
Research Fields: Indian artifacts; prehistoric & historic Native Americans; Ohio history documents; War of 1812; Civil War; Camp Sherman; WWI.
Facilities: 12,000-vol. library of 19th-century books & manuscripts with emphasis on Ohio History; McKell collection of early children's books, illuminated manuscripts, incunabula & prints available for use by request. Postcards & other museum-related items for sale.
Activities: guided tours; lectures; films; permanent & temporary exhibitions.
Publications: postcards; booklets; newsletter.
Hours & Admission Prices: Jan.-March by appointment only; April-Dec. Tues.-Sat. 1-5. Adults $4, senior citizens & students $2; discounts to AAM, ICOM & AAA members; members & Ross County School classes no charge. Closed major holidays. ♿
Attendance: 5,000 (accurate)
Membership: Youth $10; Individual $20; Family $30; Contributing $75; Business $100; Life $500; Patron $750; Benefactor $1,000.

Cincinnati

AMERICAN CLASSICAL MUSIC HALL OF FAME, 1225 Elm St., Cincinnati, OH 45202-7531. Mailing Address: P.O. Box 42295, Cincinnati, OH 45242-0295. Tel.: 800-499-3263. Fax: 513-621-1563.
E-mail: admin@americanclassicalmusic.org
Web Site: classicalwalkoffame.org
Founded: 1995.
Congressional District: 1
Key Personnel: Exec. Dir., Nina Perlove.
Personnel Profile: Full-Time Paid 2.
Governing Authority: private; nonprofit organization. Tax-exempt: 501(c)(3).
Institution Type/Description: Music History Museum.
Collections: artifacts reflecting the many facets of classical music in U.S.
Facilities: 100-seat auditorium.
Activities: interactive exhibits & displays.
Publications: annual commemorative book of current inductees.
Hours & Admission Prices: Mon.-Fri. 9-4:30. No charge. Closed New Year's Day; Independence Day; Thanksgiving; Christmas; major holidays.
Attendance: 2,500 (estimated)

Membership: Ambassador $50 & up.

AMERICAN SIGN MUSEUM, 1330 Monmouth St., Cincinnati, OH 45225. Tel.: 513-258-4020. Fax: 513-421-5144.
E-mail: info@signmuseum.org
Web Site: www.signmuseum.org
Founded: 1999.
Key Personnel: Dir. & Pres., Tod Swormstedt; Devel., John Johnson; Treas., Brian Foos.
Personnel Profile: Full-Time Paid 1; Part-Time Paid 1; Part-Time Volunteers 4.
Governing Authority: private; nonprofit organization. Tax-exempt: 501(c)(3).
Institution Type/Description: Sign Museum.
Collections: signs from all over the United States; history of technology & advertising in America; photographs.
Facilities: library; 5,500 sq. ft. exhibit space. Museum-related items for sale.
Activities: guided tours; participatory & temporary exhibitions; receptions.
Publications: quarterly newsletter, American Sign Museum.
Hours & Admission Prices: By appointment. Adults $10; discounts to groups. Closed Thanksgiving week. &
Attendance: 2,196 (accurate)
Membership: Basic $35; Family $50; Friend $100; Supporter $500; Corporate $1,000; Sustaining $2,500.

BETTS HOUSE RESEARCH CENTER, 416 Clark St., Cincinnati, OH 45203-1423. Tel.: 513-651-0734.
E-mail: info@thebettshouse.org
Web Site: thebettshouse.org
Founded: 1995.
Congressional District: 1
Key Personnel: Exec. Dir., Julie Carpenter; Pres. (V), Sally Connelly; Treas., Matt Carley; Sec., Suzanne Herdeg.
Personnel Profile: Part-Time Paid 1; Part-Time Volunteers 14.
Governing Authority: private; nonprofit. Tax-exempt: 501(c)(3).
Institution Type/Description: Historic Building: 1804 farm house with mid- & late-19th-century additions.
Collections: items relating to the history of the building & building materials.
Major Exhibits: Bricks, Barrel Vaults & Beer (T), 10/13-2/14.
Research Fields: the built environment, including architecture, historic preservation, building trades, construction technologies, and building materials.
Activities: guided tours; lectures. Museum Sponsors: Bond At The Betts House, a taste of trades mini-camp for youth.
Hours & Admission Prices: Tues.-Thurs. 11-2, 2nd & 4th Sat. 12:30-5; other times by appointment. Adults $2, discounts to AAM & ICOM members; members no charge. Closed federal holidays.
Attendance: 1,431 (accurate)
Membership: Student $10; Individual $35; Dual $50; Family $75.

CARY COTTAGE, Clovernook Center for the Blind and Visually Impaired, 7000 Hamilton Ave., Cincinnati, OH 45231-5240. Tel.: 513-522-3860. Fax: 513-728-3946. TDD: 513-522-3860.
E-mail: clovernook@clovernook.org
Web Site: www.clovernook.org
Founded: 1903.
Congressional District: 1
Key Personnel: Pres. & C.E.O., Robin Usalis; Dir., Betsy Baugh.
Personnel Profile: Full-Time Paid 105; Part-Time Paid 28.
Governing Authority: nonprofit organization. Parent Institution: The Clovernook Center for the Blind. Tax-exempt.
Institution Type/Description: Historic House: 1832 home of poets Alice & Phoebe Cary.
Collections: 1832-1850 period furnishings; kitchen herb garden; literature pertaining to Cary sisters.
Activities: guided tours by appt.; permanent exhibitions; special events.
Publications: brochures; The Perspective newsletter; Impact Report.
Hours & Admission Prices: Tours by appointment. No charge; donations accepted.
Attendance: 150 (estimated)

CINCINNATI ART GALLERIES, 225 E. Sixth St., Cincinnati, OH 45202-3209. Tel.: 513-381-2128. Fax: 513-381-7527.
E-mail: sandlers@cincyart.com
Web Site: www.cincinnatiartgalleries.com
Institution Type/Description: Art Gallery.
Collections: 19th & 20th century American and European paintings; sculpture; pottery; art glass.
Hours & Admission Prices: Winter: Mon.-Fri. 9-5, Sat. 10-4. Summer: Mon.-Fri. 9-4, Sat. 10-4.

* **CINCINNATI ART MUSEUM, (M),** 953 Eden Park Dr., Cincinnati, OH 45202-1596. Tel.: 513-639-2954. Fax: 513-721-0129.
E-mail: information@cincyart.org
Web Site: www.cincinnatiartmuseum.org
Founded: 1881.
Congressional District: 1
Key Personnel: Dir., Aaron Betsky; Deputy Dir. Institutional Advancement, Dave Linnenberg; Deputy Dir. Major Gifts & Planned Giving, Debbie Bowman; Deputy Dir. Collections, Anita Ellis; Cur. Prints, Kristin Spangenberg; Cur. Contemporary Art, Jessica Garcia Flores; Cur. American Art, Julie Aronson; Cur. Decorative Arts & Designs, Amy Dehan; Cur. Fashion Arts & Textiles, Cynthia Amneus; Cur. Asian Art, Hou-mei Sung; Museum Shop Mgr., Connie Newman.
Personnel Profile: Full-Time Paid 97; Part-Time Paid 90; Part-Time Volunteers 243; Interns 10.
Governing Authority: nonprofit organization. Parent Institution: Cincinnati Museum Assoc. Tax-exempt: 501(c)(3).
Institution Type/Description: Art Museum.
Collections: ancient art; Egyptian, Greek, Roman, Indian, Chinese, Japanese, Islamic, Nabatean, near & far Eastern and medieval art; 16th- to 20th-century European painting & sculptures; 18th- to 20th-century American paintings; 18th- to 19th-century portrait miniatures; decorative arts; costumes; textiles; musical instruments; contemporary art; African & Native American art.
Facilities: 100,000-vol. library of material pertaining to art available for inter-library loan & use on premises; auditorium; lecture hall; restaurant. Museum-related items for sale.
Activities: guided tours; lectures; gallery talks; formally organized education programs; inter-museum loan, permanent & special exhibitions; evenings for educators; speakers bureau.
Publications: exhibition catalogues; permanent collection catalogues; publications brochure.
Hours & Admission Prices: Tues.-Sun. 11-5. No charge; donations accepted. Parking: non members $4; members no charge. Closed New Year's Day; Independence Day; Thanksgiving; Christmas. &
Attendance: 272,352 (accurate)
Membership: Student $30; Individual $45; Family $75; Collector $150; Associate $300; Fellow $600; Founders Society $1,500.

CINCINNATI AVIATION HERITAGE SOCIETY & MUSEUM, 262 Wilmer Ave., Rm. 26, Cincinnati, OH 45226. Tel.: 859-442-7334.
Institution Type/Description: History Museum.
Collections: aviation history & artifacts; photographs; personal artifacts.
Hours & Admission Prices: Mon. & Fri. 10-2.

CINCINNATI CHRISTIAN UNIVERSITY, 2700 Glenway Ave., Cincinnati, OH 45204-1738. Tel.: 513-244-8100 & 8445. Fax: 513-244-8140.
E-mail: rick.bullard@ccuniversity.edu
Key Personnel: Dir., Rick Bullard.
Governing Authority: private; nonprofit organization.
Institution Type/Description: Religious Museum.
Collections: period Near Eastern, Greek & Roman artifacts; minerals; rocks.
Facilities: library; educational facilities; 100 sq. ft. exhibit space.
Activities: temporary exhibitions.
Hours & Admission Prices: Mon.-Fri. 8-4:30. No charge. Closed Memorial Day; Labor Day; Thanksgiving; Christmas & week after. &
Attendance: 300 (estimated)

CINCINNATI FIRE MUSEUM, 315 W. Court St., Cincinnati, OH 45202-1073. Tel.: 513-621-5553. Fax: 513-621-1456.
Web Site: www.cincyfiremuseum.com
Founded: 1979.
Congressional District: 2
Key Personnel: Chief Exec. Dir., Barbara M. Hammond; Pres., Tom Hardy.
Personnel Profile: Full-Time Paid 1; Part-Time Paid 5; Part-Time Volunteers 40.
Governing Authority: nonprofit organization. Tax-exempt: 501(c)(3).
Institution Type/Description: History & Fire-Fighting Museum: housed in 1906 firehouse.
Collections: fire-fighting artifacts dating from mid 19th century to present; photographs; documents; manuscripts.
Research Fields: history of fire-fighting.
Facilities: 90-seat auditorium; 18-seat theater. Museum-related items for sale.
Activities: guided tours; lectures; films; formally organized education programs for children; docent program; permanent & temporary exhibitions; Safe House fire prevention & safety education programs; hands-on exhibits.

Publications: A Guide to Cincinnati's Historic Firehouses.
Hours & Admission Prices: Tues.-Fri. 10-4, Sat.-Sun. 12-4. Adults $7, senior citizens $6, children 6-12 $5; discounts to AAM members & groups; children 5 & under and members no charge. Closed holidays. &
Attendance: 20,000 (accurate)
Membership: Individual $30; 2 Alarm $40; 3 Alarm $50; 4 Alarm $75; 5 Alarm $150. Firefighters: Individual $25; Family $35. Corporate: 2 Alarm $100; 3 Alarm $250; 4 Alarm $500; 5 Alarm $1,000.

* **CINCINNATI MUSEUM CENTER AT UNION TERMINAL, (M),** 1301 Western Ave., Cincinnati, OH 45203-1123. Tel.: 513-287-7000. Fax: 513-287-7011.
E-mail: mknight@cincymuseum.org
Web Site: www.cincymuseum.org
Founded: 1818.
Congressional District: 1
Key Personnel: C.E.O. & Pres., Douglass W. McDonald; Chm., Francie S. Hiltz; Vice Pres. Administration & C.F.O., A. Larry Sisk; Vice Pres. Featured Experiences, David Duszynski; Vice Pres. Mktg. & Public Rels., Elizabeth Wiecher Pierce; Vice Pres. Institutional Advancement, Elizabeth Lee Hoffheimer; Asst. Vice Pres. Collections & Research, Dr. Glenn Storrs; Asst. Vice Pres. Community Engagement, Terry Dickey; Asst. Vice Pres. Education, Tony Lawson; Dir. Volunteer Svcs., Angie Smorey; Dir. Facilities, Steve Terheiden; Registrar, Jane MacKnight; Dir. Exhibits, Chris Novy; Dir. Nature Preserve, Christopher Bedel; Sr. Dir. Tourism & Group Sales, Violet Rae Downey; Museum Shop Mgr., Barbara Witschger.
Personnel Profile: Full-Time Paid 181; Part-Time Paid 161; Part-Time Volunteers 764.
Volunteer Hours: 74,272
Operating Expenses: 20,974,350
Operating Income: 21,926,908
Governing Authority: private; nonprofit organization. Subsidiary Institution: Cincinnati History Museum; The Cincinnati Historical Library and Archives; Robert D. Lindner Family OMNIMAX Theater; Museum of Natural History & Science; Duke Energy Children's Museum; Geier Collections and Research Center; Richard and Lucile Durrell Eulett Center at the Edge of Appalachia Preserve. Tax-exempt: 501(c)(3).
Institution Type/Description: Natural & Regional History Museum: housed in a 1933 Art Deco building. A National Historic Landmark.
Collections: historical regional ephemera; invertebrate paleontology; conchology; entomology; ethnology; geology; mineralogy; vertebrate paleontology; prehistoric & historic archaeology; ornithology; mammology; herpetology; photographs; manuscripts; broadcast archives; regional business history archives.
Major Exhibits: Diana: A Celebration, 2/14-8/14.
Research Fields: archaeology; zoology; paleontology; Cincinnati area history.
Facilities: library; 225-seat auditorium; 13,000-acre nature preserve; research center; theater. Museum-related items for sale.
Activities: lectures; films; formally organized education programs; docent program; permanent, traveling & temporary exhibitions; field trips; outreach (school-site visits); producer of touring exhibits & Omnimax films; audiovisual aids.
Publications: Celebrating Cincinnati Museum Center at Union Terminal (2007); quarterly historical journal, Ohio Valley History; A Cincinnati Story: Emil Steinmann, E.C. Landberg and Associated Families (2008); Cincinnati Locomotive Builders: 1845-1868 (2004). Science: Vol. 1 - Natural History of the Cincinnati Region (2006); Vol. 2 - Stratigraphic Renaissance in the Cincinnati Arch-Implications for upper Ordovician Paleontology and Paleoecology (2008); Vol. 3 - 9th North American Paleontological Convention Abstracts (2009); Vol. 4 - A Naturalist Afield - Reflections on Cincinnati Nature, 1937-1984; Cincinnati Museum Center Scientific Contributions Number 5; On the Edge: A History of the Richard & Lucile Durrell Edge of Appalachia Preserve System, Adams County, Ohio.
Hours & Admission Prices: Mon.-Sat. 10-5, Sun. 11-6. Adults $8.50, senior citizens 60 & over $7.50, children 3-12 $6.50; members no charge. Closed Thanksgiving; Christmas. &
Attendance: 1,300,000 (accurate)
Membership: Duke Energy Children's Museum, Cincinnati History Museum or Museum of Natural History & Science: Individual $55; Dual $95; Family $130; Family Premium $160; Concourse Club $325; Whispering Fountains Club $575.

CINCINNATI REDS HALL OF FAME & MUSEUM, 100 Joe Nuxhall Way, Cincinnati, OH 45202. Tel.: 513-765-7000.
Key Personnel: Exec. Dir., Rick Walls
Institution Type/Description: Sports Museum.
Collections: Cincinnati Reds history & artifacts; baseball uniforms & equipment; photographs; Hall of Fame inductees.
Activities: Annual Event: Hall of Fame Induction Ceremony.

Hours & Admission Prices: Tues.-Sun. 10-5. Adults $10, seniors 60 & over and students $8, active military & veterans $6; children 4 & under no charge. Closed New Year's Day; Thanksgiving; Christmas.

CINCINNATI SKIRBALL MUSEUM HEBREW UNION COLLEGE-JEWISH INSTITUTE OF RELIGION, 3101 Clifton Ave., Cincinnati, OH 45220-2488. Tel.: 513-487-3053. Fax: 513-221-0321.
E-mail: aschwartz@huc.edu
Web Site: www.huc.edu/museums/cn
Founded: 1913.
Congressional District: 1
Key Personnel: Dir., Abby Schwartz; Dir., Rabbi Jean Eglinton.
Personnel Profile: Part-Time Paid 2; Part-Time Volunteers 25; Interns 1.
Governing Authority: college. Parent Institution: Hebrew Union College. Tax-exempt: 501(c)(3).
Institution Type/Description: Art and Jewish History Museum.
Collections: ceremonial objects relating to Jewish customs, rituals & life cycle events; biblical archaeology; textiles; decorative arts; prints; paintings; photographs.
Major Exhibits: The Boris Schatz Collection at Hebrew Union College, 12/13-2/14; Spotlight on Moses Ezekiel, 2/14-3/14.
Facilities: Archaeology Center; conference room; auditorium.
Activities: guided tours; docent program; permanent & temporary exhibitions; special exhibitions.
Hours & Admission Prices: Tues. & Thurs. 11-4, Sun. 1-5; other times by appointment. No charge; donations accepted. &
Attendance: 1,000 (estimated)
Membership: Participating $35; Supporting $100; Sustaining $250; Patron $500.

* **CINCINNATI ZOO & BOTANICAL GARDEN,** 3400 Vine St., Cincinnati, OH 45220-1333. Tel.: 513-281-4700, ext. 0 & 559-7724. Fax: 513-487-3336. TDD: 513-559-2730.
E-mail: info@cincinnatizoo.org
Web Site: www.cincinnatizoo.org
Founded: 1875.
Congressional District: 01
Key Personnel: Exec. Dir., Thane Maynard; C.O.O., Dave Jenike; Dir. Mktg. & Public Rels., Chad Yelton; Veterinarian, Mark Campbell, D.V.M.; Museum Shop Mgr., Lisa Sparks; Dir. Human Resources, Jeff Walton; Mgr. Guest Rels., T.R. Amrine; Mktg., Tiffany Barnes.
Personnel Profile: Full-Time Paid 196; Part-Time Paid 430; Part-Time Volunteers 1,100; Interns 35.
Governing Authority: society. Tax-exempt: 501(c)(3).
Institution Type/Description: Zoo.
Collections: aviary; aquarium; herpetology; arboretum; 3,000 taxa of plants; botanical garden; mammals; invertebrates; carnivore house.
Research Fields: aviary; aquarium; herpetology-amphibian research; mammals; reproductive research.
Facilities: 2,500-vol. library of books available for inter-library loan; CREW (Center for Reproduction of Endangered Wildlife) Research Dept. dedicated to preservation & propagation of endangered wildlife; children's zoo; jungle trials; Asian & African rainforest.
Activities: guided tours; lectures; films; concerts; study clubs; formally organized education programs for children & undergraduate college students; inter-museum loan, permanent & temporary exhibitions.
Publications: book, Cincinnati Zoo Official Guide Book; annual report; newsletter, Wildlife Explorer.
Hours & Admission Prices: Winter: daily 9-5; Summer: daily 9-6. General Admission: adults $15, seniors 62 & over and children 2-12 $10. Parking $8; discounts to groups of 25 or more, AAM & ICOM members; museum members & AZA members no charge. &
Attendance: 1,300,000 (estimated)
Membership: Basic: Individual $57; Single Parent Family $75; Family $89. Premium: Individual $87; Single Parent Family $105; Family $119. Gold: Individual $146; Single Parent Family $167; Family $189.

CIVIC GARDEN CENTER OF GREATER CINCINNATI, 2715 Reading Rd., Cincinnati, OH 45206-1617. Tel.: 513-221-0981. Fax: 513-221-0961.
E-mail: vciotti@civicgardencenter.org
Web Site: www.civicgardencenter.org
Founded: 1942.
Key Personnel: Exec. Dir., Vickie Ciotti; Volunteer Coord., Connie Booth; Coord. Youth Education, Corina Bullock; Horticulturist I, Cara Hague; Exec. Asst., Terry Houston; Coord. Neighborhood Gardens, Peter Huttinger; Horticulturist III, Paul Koloszar; Finance Mgr., Judy Rahm.

Personnel Profile: Full-Time Paid 7; Part-Time Paid 2; Part-Time Volunteers 550; Interns 2.
Governing Authority: private; nonprofit organization. Tax-exempt.
Institution Type/Description: Horticultural Center.
Collections: gardening; botanical; horticulture.
Facilities: library.
Activities: formal education programs; gardening classes; Community Garden training & programming for inner city youths & adults.
Publications: Gardener's Quarterly.
Hours & Admission Prices: Mon.-Fri. 9-4. No charge; donations accepted. &
Attendance: 10,000 (estimated)
Membership: Student, Teacher & Senior Citizens $25; Individual $35; Family $50.

THE CONTEMPORARY ARTS CENTER, (M), 44 E. Sixth St., Cincinnati, OH 45202-3998. Tel.: 513-345-8400. Fax: 513-721-7418.
E-mail: admin@cacmail.org
Web Site: www.contemporaryartscenter.org
Founded: 1939.
Congressional District: 1
Key Personnel: Dir., Raphaela Platow; Asst. to Dir., Bettina Bellucci; Chm., Otto M. Budig, Jr.; Pres., Jennie Rosenthal Berliant; Dir. External Affairs, Melodee DuBois; Dir. Finance, Margaux Higgins; Sr. Cur., Toby Kamps; Curatorial Asst., Clare Norwood; Curatorial Asst., Maiza Hixson; Dir. Facilities, Dave Gearding; Cur. Education, Scott Boberg; Dir. Devel., Kathryn Brass; Dir. Mktg., Peggy Shannon; Mgr. Membership, Merrilee Luke-Ebbeler; Visitor Svcs. Mgr., Melanie Derrick.
Personnel Profile: Full-Time Paid 28; Part-Time Paid 15; Part-Time Volunteers 59; Interns 10.
Governing Authority: nonprofit organization. Tax-exempt.
Institution Type/Description: Art Museum.
Collections: works by contemporary artists.
Research Fields: contemporary art.
Facilities: 1,000-vol. library. Catalogues, booklets & other museum-related items for sale.
Activities: guided tours; lectures; films; gallery talks; dance performances; docent & educational programs; traveling exhibitions; video programs; artists' residences.
Publications: newsletters; gallery guides; exhibition catalogues; brochures; audio guides.
Hours & Admission Prices: Mon. 10-9, Wed.-Fri. 10-6, Sat.-Sun. 11-6. Adults $7.50, senior citizens $6.50, students $5.50, children 3-13 $4.50; discounts to AAM members; children under 3, Mon. after 5pm & members no charge. &
Attendance: 177,000 (accurate)
Membership: Student $25; Individual $45; Individual Plus One $60; Family $65; Collector's Connection $125; Patron's Circle $250; Curator's Circle $500; Director's Circle $1,000; Trustee Circle $2,500.

DAAP GALLERIES, University of Cincinnati, 2624 Clifton Ave., Cincinnati, OH 45221. Mailing Address: Univ. of Cincinnati, P.O. Box 210016, Cincinnati, OH 45221-0016. Tel.: 513-556-2839. Fax: 513-556-3288.
E-mail: daapgalleries@us.edu
Web Site: www.daap.uc.edu/Gallery/gallery.htm
Founded: 1993.
Key Personnel: Collections Asst., Jonathan Nolting; Curatorial Asst., Sandra Geiser; Gallery Asst., Forest Harman; Mgr. Meyers Gallery, Rob Anderson; Mgr. Reed Gallery, Tijana Antonic; Mgr. University Galleries on Sycamore, Maria Seda-Reeder.
Governing Authority: public university; nonprofit. Parent Institution: University of Cincinnati, Cincinnati 45221. Tax-exempt: 170(b)(1)(A).
Institution Type/Description: Art Gallery.
Collections: art of ancient Greece, Europe, Asia & the Americas; art of the United States emphasizing late 19th- & early 20th-century paintings, sculpture & works on paper.
Research Fields: American art, especially the late 19th & early 20th centuries.
Activities: formal education programs for undergraduate & graduate students affiliated with the Univ. of Cincinnati; lectures, loan, temporary & traveling exhibitions.
Publications: exhibition announcements; exhibition catalogues.
Hours & Admission Prices: Reed Gallery & Meyers Gallery: Mon.-Fri. 10-5. University Galleries on Sycamore Tues.-Fri. 11-5, Sat. 11-4. No charge. Closed university holidays. &
Attendance: 10,000 (estimated)

THE DELHI HISTORICAL SOCIETY MUSEUM, 468 Anderson Ferry Rd., Cincinnati, OH 45238-5281. Tel.: 513-451-4313. Fax: 513-451-4300.
Web Site: delhihistoricalsociety.org
Institution Type/Description: Historical Society Museum: formerly the Joe Witterstaetter Homestead, built in the 1880s.
Collections: local history & culture; exhibits pertaining to Delhi history.
Hours & Admission Prices: Tues., Thurs. & Sun. 12-3.

GREATER CINCINNATI POLICE MUSEUM, 959 W. Eighth St., Cincinnati, OH 45203-1203. Tel.: 513-300-3664.
E-mail: director@gcphs.com
Web Site: www.gcphs.com
Founded: 2006.
Congressional District: 1
Key Personnel: Dir. & C.E.O., Lt. Stephen R. Kramer, (Ret.); Pres. (V), Lt. Alan C. March; Museum Shop Mgr., Det. Richard W. Gross.
Personnel Profile: Part-Time Volunteers 40.
Governing Authority: Parent Institution: Greater Cincinnati Police Historical Society. Tax-exempt.
Institution Type/Description: History Museum.
Collections: local police & regional law enforcement history; uniforms; equipment; badges; memorials; photographs; personal artifacts.
Research Fields: personnel research.
Facilities: library.
Publications: annual report.
Hours & Admission Prices: Tues., Thurs. & Sat. 10-4. No charge; donations accepted. &
Attendance: 1,200 (estimated)
Membership: Annual $25; Life $1,000.

HARRIET BEECHER STOWE HOUSE, (M), 2950 Gilbert Ave., Cincinnati, OH 45206. Tel.: 513-751-0651. www.stowehousecincy.org.
E-mail: stowehouse@zoomtown.com
Web Site: www.ohiohistory.org/places/stowe
Personnel Profile: Part-Time Volunteers 25.
Governing Authority: Tax-exempt.
Institution Type/Description: Historic House Museum: housed in the former of Harriet Beecher Stowe, author of Uncle Tom's Cabin.
Collections: Beecher-Stowe family history; photographs; personal artifacts; period furnishings; Underground Railroad; African American history.
Major Exhibits. Harriet Tubman Sculpture, 2013; Uncle Tom's Cabin, 2013.
Facilities: Meeting rooms.
Activities: special events; educational programs.
Hours & Admission Prices: Feb.-April & Dec. Wed. 10-12, Sat. 10-2; May to Labor Day Tues.-Thurs. & Sat. 10-2; Sept.-Nov. Tues.-Wed. & Sat. 10-2; other times by appointment. No charge; donations accepted. Closed Federal holidays.
Attendance: 2,600 (accurate)
Membership: Individual $25.

INDIAN HILL HISTORICAL SOCIETY, 8100 Given Rd., Cincinnati, OH 45243-1520. Tel.: 513-891-1873. Fax: 513-891-1873.
E-mail: ihhist@cinci.rr.com
Web Site: www.indianhill.org
Founded: 1973.
Congressional District: 26
Key Personnel: Pres. (V), Barbara Hauck; Admin., Helen Verkamp.
Personnel Profile: Part-Time Paid 3; Part-Time Volunteers 35.
Governing Authority: society; nonprofit. Tax-exempt.
Institution Type/Description: Local History Museum & Genealogies; housed in 1873 one-room school building & 1860 house.
Collections: monthly rotating collections.
Research Fields: local history.
Facilities: 500-vol. library of books available for inter-library loan.
Activities: monthly programs; lectures; concerts; arts festivals; study clubs; hobby workshops; formally organized education programs.
Publications: monthly newsletter, The Sampler.
Hours & Admission Prices: By appointment only. No charge. &
Attendance: 1,500 (estimated)
Membership: Individual $40; Family $50; Sustaining $100; Patron $175.

KROHN CONSERVATORY, 1501 Eden Park Dr., Cincinnati, OH 45202-6030. Tel.: 513-421-4086 & 5707. Fax: 513-421-6007. TDD: 513-352-3380.
E-mail: andrea.schepmann@cincinnati.oh.gov
Web Site: www.cineypark.com

Founded: 1933.
Congressional District: 2
Key Personnel: Dir. & Museum Shop Mgr., Andrea Schepmann; District Supvr., Ruth Spears; Volunteer Coord., Christyl Johnson.
Personnel Profile: Full-Time Paid 6; Part-Time Paid 8; Part-Time Volunteers 270.
Governing Authority: municipal. Affiliated with City of Cincinnati, Board of Park Commissioners. Tax-exempt.
Institution Type/Description: Botanical Garden & Conservatory.
Collections: tropical & warm climate plants; cacti; succulents; palms; ferns; aroids; orchids; crotons; carnivorous display - Bonsai.
Research Fields: horticulture.
Facilities: 50-vol. library on horticulture available for inter-library loan; 30-seat meeting room. Museum-related items for sale.
Activities: guided tours; lectures; study clubs; permanent & temporary exhibitions.
Publications: map; self-guiding walking tour; tour brochure.
Hours & Admission Prices: Daily 10-5; special evening hours during Christmas Show. Charge for special events and shows only. &
Attendance: 185,000 (accurate)
Membership: Individual $25; Family $40; Patron $100; Founder $1,000.

MOUNT AIRY ARBORETUM, 5083 Colerain Ave., Cincinnati, OH 45223-1072. Tel.: 513-541-8176. Fax: 513-541-8176.
E-mail: paula.miller@cincinnati-oh.gov
Web Site: www.cincinnati-oh.gov/parks
Founded: 1932.
Congressional District: 23
Key Personnel: Dir. Parks, William Carden; Supvr. Mt. Airy, Larry Parker.
Personnel Profile: Full-Time Paid 1; Part-Time Volunteers 2.
Governing Authority: municipal. Affiliated with Cincinnati Bd. of Park Commissioners, 950 Eden Park Dr., 45202. Tax-exempt.
Institution Type/Description: Botanical Garden: located on the site of Mount Airy Forest, the first municipal reforestation project in the U.S., started in 1911.
Collections: native hardwoods; evergreens; lilacs; azaleas; viburnums; perennial gardens; dwarf conifer garden; quince; crabapples; ground covers; clematis vine display garden.
Research Fields: plant trials.
Facilities: campsites available for Scout troops; 2 rustic lodges-75, 150 or 200 person capacity, available for meetings or receptions; reserved picnic area.
Activities: guided tours; lectures.
Publications: guide, Mount Airy Arboretum Plant Guide Map.
Hours & Admission Prices: Forest: daily 6am-10pm. Arboretum daily 7:30-dark. No charge. All facilities require reservations; call 513-352-4080.

NATIONAL UNDERGROUND RAILROAD FREEDOM CENTER, (M), 1301 Western Ave., Cincinnati, OH 45203-1138. Tel.: 513-333-7500; 877-648-4838 (toll free).
Web Site: www.freedomcenter.org
Founded: 1995.
Key Personnel: Pres., Donald Murphy; Chm. (V), John Pepper; Museum Shop Mgr., James Tecco.
Personnel Profile: Full-Time Paid 52; Part-Time Paid 29; Interns 3.
Institution Type/Description: History Museum.
Collections: slavery; resistance movements; achievement of freedom.
Facilities: cafe. Museum-related items for sale.
Activities: interactive educational programs; research.
Hours & Admission Prices: Tues.-Sun. 11-5. Adults $12, students and seniors 60 & over $10, children 6-12 $8; discounts to groups of 10 or more with reservation. Closed Thanksgiving; Christmas. &

STUDIO SAN GIUSEPPE ART GALLERY, College of Mt. St. Joseph, 5701 Delhi Rd., Cincinnati, OH 45233-1670. Tel.: 513-244-4314. Fax: 513-244-4942.
E-mail: Jerry_Bellas@mail.msj.edu
Web Site: www.msj.edu
Founded: 1962.
Congressional District: 2
Key Personnel: Dir., Gerald M. Bellas.
Personnel Profile: Part-Time Paid 1; Interns 2.
Governing Authority: private college; nonprofit. Parent Institution: College of Mount St. Joseph. Tax-exempt: 501(c)(3).
Institution Type/Description: Art Gallery: housed in the Dorothy Meyer Ziv Art Building.
Collections: art works & crafts donated to the school over the years.
Facilities: 1,500 sq. ft. exhibit space.
Activities: temporary exhibits relating to art curriculum; guided tours; lectures; participatory exhibit; films.

Hours & Admission Prices: Academic Year: Mon.-Fri. 10-5, Sat.-Sun. 1-5. No charge. Closed major holidays. &
Attendance: 5,000 (estimated)
Membership: Individual $10; Associate $25; Sponsor $50; Patron $100.

*** TAFT MUSEUM OF ART, (M),** 316 Pike St., Cincinnati, OH 45202-4293. Tel.: 513-241-0343. Fax: 513-241-7762.
E-mail: taftmuseum@taftmuseum.org
Web Site: www.taftmuseum.org
Founded: 1932.
Congressional District: 2
Key Personnel: Dir. & C.E.O., Deborah Emont Scott; Chm. (V), Gerald H. Greene; Chief Cur., Lynne D. Ambrosini, Ph.D.; Dir. Organizational Resources & Planning, Cynthia M. Kearns; Dir. Institutional Advancement, Natalie Mathis; Mgr. School & Docent Programs, Lisa Morrisette; Asst. Cur., Tammy Muente; Dir. Security, Greg Brock; Mgr. Public Programs, Mary Ladrick; Chief Preparator & Exhibition Designer, Mark Rohling; Museum Shop Mgr. & Visitor Svcs., Brooke Sherritt; Mgr. Facility Rentals, Kitty Paschall; Registrar & Collections Mgr., Joan C. Hendricks; Mktg. Mgr., Emma Caro; Exec. Asst. to Dir., Christine Miller; Dir. Finance, Patricia B. Hassel, CPA, CEMA
Personnel Profile: Full-Time Paid 19; Part-Time Paid 20; Part-Time Volunteers 165; Interns 8.
Volunteer Hours: 10,692
Operating Expenses: 3,649,000
Operating Income: 3,649,000
Governing Authority: nonprofit organization. Subsidiary Institution: Taft Publications, Inc. Tax-exempt: 170(b)(1)(A).
Institution Type/Description: Art Museum: housed in 1820 Baum-Longworth-Taft House.
Collections: Dutch, English, Spanish & French Old Master paintings; Kangxi, Yongzheng, Qianlong Chinese porcelains; Limoges enamels; Italian 16th-century Maiolica & engraved rock crystals; European & English 17th-& 18th-century watches; 19th-century European & American paintings; decorative arts; Federal furniture.
Major Exhibits: Threads of Heaven: Silken Legacy of China's Last Dynasty, 2/7/14-5/18/14; Charles-Francois Doubigny: The Birth of Impressionism (T), 2/13/14-5/17/14; Small Paintings from the Taft Collection, 2/21/14-6/15/14; America's Eden: Thomas Cole & the Voyage of Life, 6/13/14-9/14/14; Kehinde Wiley: Memling, 7/11/14-10/15/14; French Twist: Masterworks of Photography from Atget to Man Ray (T), 10/3/14-1/11/15.
Research Fields: pertaining to collections.
Facilities: 2,000-vol. library of books available for use on premises by appointment. Catalogs, booklets & other museum-related items for sale.
Activities: guided tours; lectures; gallery talks; concerts; formally organized education programs for children, students & adults; docent program or council; inter-museum loan, permanent & temporary exhibitions; Duncanson Artist-in-Residence; film series.
Publications: books, J.M.W. Turner: The Foundations of Genius; Louise Bourgeois; The Taft Museum: Its History & Collections; Night Lights; Skating in the Arts of 17th Century Holland; The American Weigh; Masterworks/Enamel/87; Paul Ashbrook; Nicholas Longworth-Art Patron of Cincinnati; China in 1700: Kangxi Porcelains at the Taft Museum; Christmas in Naples; At the Table; William Wegman: History of Travel; Tyrone Geter, Artist Face to Face: Two Centuries of Self-Portraits; Patterns in a Revolution: French Printed Textiles, 1759-1821; Oliver Newberry Chaffee (1881-1944); Fin de Siecle, Prints, Posters & Prose; An Exhibition of Tang Sancai Pottery from the collection of Alan & Simone Hartman; Tributes to the Tafts; Cavaliers & Cardinals: 19th Century French Anecdotal Painting; Master Dutch Drawings & Watercolors of the 19th Century from Haags Gemeentemuseum; Tell Me A Story; Tarleton Blackwell; Looking for Leonardo: Native & Folk Art Objects Found in America by Bates & Isabel Lowry; Dutch Drawings & Watercolors from the Kharkiv Art Museum; The Vanishing Frontier: Henry F. Varny, 1847-1916; East Meets West: Chinese Export and Design; The Great Migration: The Evolution of African American Art, 1790-1945; TAFT Museum of Art, An Illustrated Guide; Hiram Powers: Genius in Marble; Brush, Clay & Wood: The Nancy & Ed Rosenthal Collection of Chinese Art Portico; members newsletter, Portico.
Hours & Admission Prices: Wed.-Fri. 11-4, Sat.-Sun. 11-5. Adults $10, students & senior citizens $8, children 12-17 $4; discounts to AAM & ICOM members; members no charge. No charge on Sunday. Closed New Year's Day; Independence Day; Thanksgiving; Christmas. &
Attendance: 50,999 (accurate)
Membership: Individual $45; Dual $60; Family $70; Patron $135; Sustaining $275; Fellow $500; Contributor $800; Collector's Circle $1,250; Trustee's Circle $2,500; Director's Circle $5,000; Sponsor $10,000.

TRAILSIDE NATURE CENTER, Brookline Dr., Burnet Woods Park, Cincinnati, OH 45220. Mailing Address: 4 Beech Lane, Cincinnati, OH 45208-2614. Tel.: 513-321-6070 & 751-3679. Fax: 513-321-6218.
E-mail: vivian.wagner@cincinnati-oh.gov
Web Site: www.cinci-parks.org
Formerly: Trailside Nature Center & Museum
Founded: 1930.
Congressional District: 1
Key Personnel: C.E.O. & Dir. Parks, Willie F. Carden, Jr.; Dir. Outdoor Centers, Erin Morris; Coord. Education, Vivian Wagner.
Personnel Profile: Full-Time Paid 1; Part-Time Paid 2.
Governing Authority: municipal. Parent Institution: City of Cincinnati, Bd. of Park Commissioners, Interpretive Services Div. Subsidiary Institution: Wolff Plantation. Tax-exempt.
Institution Type/Description: Park Museum.
Collections: local birds, insects & other animals; tree & plant specimens; rocks & fossils; surface, underground & water habitats; exhibits displaying components of the park; sky chart exhibit.
Facilities: 50-seat auditorium; 30-seat auditorium; classrooms; planetarium; nature library.
Activities: lectures; field trips; nature clubs; summer day camp; adult education classes; handicraft classes; weather & night sky information; planetarium programs; teacher & youth-leader education.
Publications: calendar of Nature Activities.
Hours & Admission Prices: No charge; donations accepted. Closed holidays. &
Attendance: 20,000 (estimated)

WILLIAM HOWARD TAFT NATIONAL HISTORIC SITE, 2038 Auburn Ave., Cincinnati, OH 45219-3025. Tel.: 513-684-3262. Fax: 513-684-3627.
Web Site: www.nps.gov/wiho/index.html
Founded: 1969.
Congressional District: 2
Key Personnel: Supt., Reggie Tiller; Museum Shop Mgr. & Chief of Interpretation & Resources Mgmt., Ray Henderson.
Personnel Profile: Full-Time Paid 8; Part-Time Paid 4; Part-Time Volunteers 20.
Governing Authority: federal government. Parent Institution: National Park Service/Dept. of Interior. Tax-exempt.
Institution Type/Description: Historic House Museum: c.1857 birthplace & boyhood home of William Howard Taft, 27th President of the U.S. & 10th Chief Justice. Taft Education Center, 2038 Auburn Ave., Cincinnati, OH 45219-3025.
Collections: furnishings; memorabilia of President Taft & his family.
Research Fields: history of William Howard Taft & his birthplace.
Facilities: library related to William Howard Taft, his family, his birthplace.
Activities: guided tours; lectures; temporary exhibitions; educational programs for school groups.
Publications: brochure; newsletter.
Hours & Admission Prices: Daily 8-4. No charge; donations accepted. Closed New Year's Day; Thanksgiving; Christmas. &
Attendance: 17,000 (accurate)
Membership: Friends of the William Howard Taft Birthplace: Individual $25.

XAVIER UNIVERSITY ART GALLERY, 3800 Victory Pkwy., Cincinnati, OH 45207-1035. Tel.: 513-745-3811. Fax: 513-745-1098. TDD: 513-745-3811.
Web Site: xavier.edu/art/gallery.cfm
Founded: 1831.
Key Personnel: Dir., Katherine Uetz.
Governing Authority: private university; nonprofit. Parent Institution: Xavier University. Tax-exempt.
Institution Type/Description: Art Gallery.
Collections: paintings; sculpture; photographs.
Activities: student & professional art exhibitions; films.
Hours & Admission Prices: Academic year: Mon.-Fri. 10-4. No charge. Closed official & university holidays. &
Attendance: 1,800

Cleveland

THE CHILDREN'S MUSEUM OF CLEVELAND, 10730 Euclid Ave., Cleveland, OH 44106-2200. Tel.: 216-791-7114 & 791-KIDS. Fax: 216-791-8838.
E-mail: info@clevelandchildrensmuseum.org
Web Site: clevelandchildrensmuseum.org
Formerly: Rainbow Children's Museum and TRW Early Learning Center

Founded: 1981.
Congressional District: 21
Key Personnel: Exec. Dir., Maria Camparnelli; Chm. Bd., Jannifer Lesny-Fleming; Coord. Membership & Group Tours, Lisa Merk; Mgr. Guest Svcs. & Coord. Volunteers, Kelley McClelland; Facility & Exhibit Maintenance, Chris Beal; Dir. Operations, Leland Merk; Dir. Education, Dr. Sandra Redmond; Program Mgr., Alys Hileman; Museum Shop Mgr., Betty Rosskamm.
Personnel Profile: Full-Time Paid 8; Part-Time Paid 9; Part-Time Volunteers 3; Interns 10.
Operating Expenses: 1,003,792
Operating Income: 1,218,146
Governing Authority: nonprofit organization. Tax-exempt: 501(c)(3).
Institution Type/Description: Children's Museum.
Collections: hands-on participatory exhibits.
Research Fields: pertaining to exhibits.
Facilities: Museum-related items for sale.
Activities: organized education programs for children & volunteers; loan exhibitions; hands-on family learning.
Publications: monthly calendar; quarterly newsletter; brochures; rack cards; curriculum guides.
Hours & Admission Prices: Daily 10-4:45. Admission $7; members & children under 1 no charge. Closed major holidays. &
Attendance: 96,331 (accurate)
Membership: Individual Plus (1 adult/1 child) $60; Additional members $15.

CLEVELAND BOTANICAL GARDEN, 11030 East Blvd., Cleveland, OH 44106-1706. Tel.: 216-721-1600, ext. 194. Fax: 216-721-2056.
E-mail: nronayne@cbgarden.org
Web Site: www.cbgarden.org
Founded: 1930.
Congressional District: 11
Key Personnel: Dir., Natalie Ronayne; Pres., Jeffrey Biggar; Vice Pres., Matthew V. Crawford; Devel. Officer, Sara Stone; Museum Shop Mgr., Kate Fox.
Personnel Profile: Full-Time Paid 49; Part-Time Paid 29; Part-Time Volunteers 350; Interns 12.
Governing Authority: nonprofit organization. Tax-exempt: 501(c)(3).
Institution Type/Description: Botanical Garden.
Collections: Japanese, herb, rose, perennial & wild flower gardens; reading garden; Hershey children's garden. Glasshouse: Costa Rica & Madagascar ecosystems.
Facilities: 16,000-vol. library of books on gardening & horticulture, 125 videos on gardening & landscape design, nursery catalogs, available for research in library & circulation to members; reading room; 325-seat auditorium; classrooms; garden. Garden-related items for sale.
Activities: lectures; classes; workshops; formally organized education programs for adults, children & special populations. Annual Events: Orchid Mania in February; Ripe! Food & Garden Festival in September; Winter Show in December.
Publications: quarterly magazine, The Bulletin; annual report; schedule of events; visitor brochure; visitor map.
Hours & Admission Prices: Tues.-Sat. 10-5, Sun. 12-5. Adults $8.50, children $3; members no charge. Hershey Children's Garden: children 2-12 Closed New Year's Day; Thanksgiving; Christmas. &
Attendance: 140,000 (accurate)
Membership: Individual $55; Family $65; Friend $100; Partner $250; Sponsor $500; Director's Circle $1,000; Founder's Society $5,000.

CLEVELAND GRAYS ARMORY MUSEUM, (M), 1234 Bolivar Rd., Cleveland, OH 44115. Tel.: 216-621-5938. Fax: 216-621-5941. Facebook: Cleveland Grays Armory Museum.
E-mail: grays1837@yahoo.com
Web Site: www.graysarmory.com
Founded: 1837.
Congressional District: 11
Key Personnel: Pres. (V), William A. Roediger.
Governing Authority: Tax-exempt.
Institution Type/Description: Historic Building: built in 1893. Listed on the National Register of Historic Places.
Collections: military history & heritage; personal artifacts; military equipment & artifacts.
Activities: educational programs; special events; guided tours.
Publications: The Shako, online membership newsletter.
Hours & Admission Prices: By appointment. Donations accepted. &
Membership: Seniors $37.50; Annual $75.

CLEVELAND METROPARKS ZOO, 3900 Wildlife Way, Cleveland, OH 44109-3132. Tel.: 216-661-6500, ext. 0. Fax: 216-661-3312.

E-mail: zooinfo@clevelandmetroparks.com

Web Site: www.clemetzoo.com

Founded: 1882.

Congressional District: 10

Key Personnel: Facilities Mgr., Elizabeth Geith; Guest Svcs. Mgr., Edith Ricchiuto; Mktg. Mgr., Susan Allen; Cur. Education, Victoria Searles; Exec. Dir. Zoo Society, Elizabeth Fowler.

Personnel Profile: Full-Time Paid 170; Part-Time Paid 42; Part-Time Volunteers 673; Interns 6.

Governing Authority: Cleveland Metroparks Board, 4101 Fulton Pkwy., Cleveland, OH 44144. Parent Institution: Cleveland Metroparks.

Institution Type/Description: Zoo.

Collections: 3,000 animals representing over 600 species; rainforest wildlife; Africa; Australia; bears; primates.

Research Fields: animal behavior & social structures of primates.

Facilities: 7,105-vol. library of zoology books available on premises during office hours; aquarium. Museum-related items for sale.

Activities: lectures; formally organized education programs for children; permanent & traveling exhibitions; runs; walks; simulator ride.

Publications: quarterly magazine, Z to U; semiannual magazine, Z; monthly magazine, The Emerald Necklace.

Hours & Admission Prices: Memorial Day-Labor Day Mon.-Fri. 10-5, Sat.-Sun. & holidays 10-7; Sept.-May daily 10-5. Zoo & Rainforest: adults $12.25, children 2-11 $8.25; Mon. residents of Cuyahoga County & Hinckley Township, children under 2 no charge. Closed New Year's Day; Christmas. &

Attendance: 1,318,458 (accurate)

Membership: Senior Plus $40; Individual Plus $57; Family $72; Family Plus $95; Family Select $127; VIP ZooKeepers Circle $250 & up.

---

*   **THE CLEVELAND MUSEUM OF ART, (M),** 11150 East Blvd., Cleveland, OH 44106-1797. Tel.: 216-421-7340. Fax: 216-421-0411. TDD: 216-421-0018.

E-mail: info@clevelandart.org

Web Site: www.clevelandart.org

Founded: 1913.

Congressional District: 21

Key Personnel: Chm., R. Steven Krestner; Advisory Chm., Alfred M. Rankin; Pres. & C.E.O. Sarah S. & Alexander M. Cutler Dir., David Franklin; Sarah S. Cutler Vice Pres. & Sec., Ellen Stirn Mavec; Vice Pres., James A. Ratner; Deputy Dir. Admin. & Treas., Janet G. Ashe; Deputy Dir. & Chief Cur., C. Griffith Mann; Editor Publications, Barbara Bradley; Dir. Collections Mgmt., Mary Suzor; Deputy Dir. Devel. & Chief Advancement Officer, August Napoli; Dir. Library & Archives, Elizabeth Lantz; Dir. Design & Architecture, Jeffrey W. Strean; Cur. Chinese Art, Anita Chung; Cur. Greek & Roman Art, Michael J. Bennett; Cur. Prints, Jane Glaubinger; Cur. Modern European Art, William H. Robinson; Cur. Medieval Art, Stephen N. Fliegel; Cur. European Painting & Sculpture, 1500-1800, Jon Seydl; Cur. Textiles & Islamic Art, Louise W. Mackie; Cur. Decorative Art & Design, Stephen A. Harrison; Cur. Pre-Columbian & Native North American Art, Susan Bergh; Cur. African Art, Constantine Petridis; Cur. Drawings, Heather Lemonedes; Cur. Photography, Barbara L. Tannenbaum; Assoc. Cur. American Painting & Sculpture, Mark Cole; Chief Conservator, Marcia Steele; Dir. Education & Public Programs, Caroline Goeser; Dir. Exhibitions, Heidi Domine Strean; Dir. Performing Arts, Music & Film, Massoud Saidpour; Dir. Human Resources, Sharon Reaves; Dir. Auxiliary Svcs.& Museum Shop Mgr., Catherine Surratt.

Personnel Profile: Full-Time Paid 282; Part-Time Paid 95; Part-Time Volunteers 397; Interns 23.

Governing Authority: nonprofit organization. Tax-exempt: 501(c)(3).

Institution Type/Description: Art Museum.

Collections: art from all cultures & periods; paintings; sculpture; graphics; decorative arts; music; numismatic; textiles; photography; digital images; microfilm; slides.

Research Fields: paintings; sculpture; graphics; decorative arts; music; textiles; photography.

Facilities: 442,039-vol. library of art books & periodicals; 765-seat auditorium; two 160-seat halls; restaurant. Books, postcards, print reproductions, color slides, paper goods, small decorative objects & jewelry for sale.

Activities: Annual Events: Parade the Circle in June; Chalk Fest in September.

Publications: monthly, Members Magazine; annual report; Handbook of the Cleveland Museum of Art; exhibition catalogs; books relating to museum collections; guide to the galleries, collection catalogues, Masterpieces in Asian Art 19th-Century European Paintings; Knockouts: A Pocket Guide to the CMA (2002).

Hours & Admission Prices: Tues., Thurs. & Sat.-Sun. 10-5, Wed. & Fri. 10-9.

No charge; donations accepted. Closed New Year's Day; Independence Day; Thanksgiving; Christmas. &

Attendance: 308,000 (accurate)

Membership: CMA Annual: Senior Citizen $35; Senior Couple $45; Individual $50; Family $65; Classic $100; Fellow $250; Patron $300; Contributing $500; Director's Circle $1,000; President's Circle $2,500; Founder's Society $5,000; Collector's Circle $10,000; Patron Sponsor $25,000.

---

*   **THE CLEVELAND MUSEUM OF NATURAL HISTORY, (M),** 1 Wade Oval Dr., University Circle, Cleveland, OH 44106-1767. Tel.: 216-231-4600. Fax: 216-231-5919. TDD: 216-231-7777; Facebook: Cleveland Museum of Natural History.

Web Site: www.cmnh.org

Founded: 1920.

Congressional District: 21

Key Personnel: Exec. Dir. & C.E.O., Evalyn Gates, Ph.D.; Dir. Communications & Mktg., Kim Gillan-Shafron; C.F.O.O., Todd Welki; Registrar, Amber Anderson; Cur. Invertebrate Paleontology, Dr. Joseph Hannibal; Cur. Vertebrate Zoology, Dr. Timothy Matson; Cur. Paleobotany & Paleoecology, Dr. Denise Su; Cur. Vertebrate Paleontology, Dr. Michael Ryan; Cur. Mineralogy, Dr. David Saja; Cur. Ornithology, Dr. Andrew Jones; Dir. Education, Carin Miller; Dir. Conservation & Cur. Botany, James Bissell, L.H.D.; Dir. Wildlife Resources, Harvey Webster; Cur. Archaeology, Dr. Brian Redmond; Dir. GreenCityBlueLake Institute, David Beach.

Personnel Profile: Full-Time Paid 103; Part-Time Paid 87; Part-Time Volunteers 300; Interns 30.

Governing Authority: nonprofit corporation. Branch Museum: Mentor Marsh Nature Center. Tax-exempt: 501(c)(3).

Institution Type/Description: Natural History Museum.

Collections: archives; archaeology; ethnography; physical anthropology; natural history arts; entomology; herbarium; herpetology; mammalogy; ornithology; ichthyology; mineralogy; invertebrate paleontology; vertebrate paleontology; paleobotany.

Major Exhibits: Nature's Mating Game: Beyond the Birds & the Bees (T), 10/26/13-4/27/14; Be the Dinosaur (T), 2/1/14-5/4/14; Traveling the Silk Road (T), 5/31/14-10/5/14; Animal Secrets (T), 5/17/14-9/14/14; Mammoths and Mastodons: Titans of the Ice Age (T), 11/28/14-4/19/14.

Research Fields: archaeology; cultural anthropology; physical anthropology; botany; entomology; herpetology; mammalogy; invertebrate paleontology; mineralogy; astronomy; vertebrate paleontology; paleobotany; ornithology; wildlife research.

Facilities: 60,000-vol. library of natural history; 500-seat auditorium; Discovery Center; observatory; planetarium; conservation lab; casting lab; environmental garden; captive wildlife area; 27 off-site natural areas; classrooms; meeting & banquet room; cafe. Museum-related items for sale.

Activities: guided tours; lectures; films; TV & radio programs; formally organized education programs for children, adults, undergraduate & graduate college students; permanent, temporary & traveling exhibitions; astronomy programming; live animal programs; Science Resource Center; Future Scientists Program; summer internships.

Publications: members magazine, Explore; scientific papers; annual report; program guides; school programs & teacher resources catalog; natural areas calendar.

Hours & Admission Prices: Mon.-Tues. & Thurs.-Sat. 10-5, Wed. 10-10, Sun. 12-5. Adults $12, youth, students & senior citizens $10; discount to AAM members; children 2 & under & members no charge. Planetarium: $4. Closed New Year's Day; Memorial Day; Independence Day; Labor Day; Thanksgiving; Christmas. &

Attendance: 263,000 (accurate)

Membership: Senior $40; Individual $55; Couple Senior $55; Family $75; Family Plus $95; Friend $150; Associate $250; Mentor $500.

CLEVELAND POLICE HISTORICAL SOCIETY & MUSEUM, (M), 1300 Ontario St., Cleveland, OH 44113-1600. Tel.: 216-623-5055 & 5056. Fax: 216-623-5145.

E-mail: clevelandpolicemus@roadrunner.com

Web Site: www.clevelandpolicemuseum.org

Founded: 1983.

Congressional District: 11

Key Personnel: Museum Shop Mgr., Geraldine Diemert; Office Mgr., Marilyn Jech; Pres. (V), Tom Armelli.

Personnel Profile: Part-Time Paid 2; Part-Time Volunteers 4.

Governing Authority: nonprofit organization. Tax exempt: 501(c)(3).

Institution Type/Description: Police Museum: presently housed in Cleveland police headquarters.

Collections: history of the Cleveland Police Dept. from 1866-present; incomplete files of annual reports, case files, city & state codes, criminology texts, directories, manuals, police blotters & orders, scrapbooks; over 13,000 photographs including police personnel, buildings, equipment, selected

criminal cases & Cleveland events; memorabilia includes badges, patches, uniforms, equipment, firearms, call boxes, police motorcycles, a jail, fingerprint table.
Research Fields: Cleveland Ohio law enforcement history.
Activities: guided tours; traveling museums in boxes; living history outreach program; slide programs; preservation research.
Publications: newsletter, CPHS; newsletter, The Hot Sheet.
Hours & Admission Prices: Mon.-Fri. 10-4. No charge; donations accepted. Closed New Year's Day; Memorial Day; Independence Day; Labor Day; Thanksgiving; Christmas. &
Attendance: 8,878 (accurate)
Membership: CPD Retired $25; Individual $78.

CRAWFORD AUTO AVIATION COLLECTION OF WESTERN RESERVE HISTORICAL SOCIETY, 10825 East Blvd., Cleveland, OH 44106-1703. Tel.: 216-721-5722. Fax: 216-721-0891.
E-mail: alowrie@wrhs.org
Web Site: www.wrhs.org
Founded: 1963.
Congressional District: 21
Key Personnel: Pres., Gainor B. Davis, Ph.D.; Dir. Community Engagement, Becky Carlino; C.F.O., Mary Thoburn; Vice Pres. Museums, Dr. Edward Jay Pershey; Vice Pres. Institutional Advancement, Shari Kochman; Education, Janice Ziegler; Dir. Mktg. & Sales, Angie Lowrie.
Governing Authority: private; nonprofit organization. Parent Institution: Western Reserve Historical Society, Cleveland, OH. Tax-exempt: 501(c)(3).
Institution Type/Description: Automobile & Aircraft Museum.
Collections: automobiles, aircraft, motorcycles, carriages, bicycles.
Facilities: library; 75,000 sq. ft. exhibit space.
Activities: guided tours; hobby workshops; lectures; rental gallery; participatory & temporary exhibitions.
Publications: quarterly newsletter; monthly eBlasts.
Hours & Admission Prices: Tues.-Sat. 10-5, Sun. 12-5. Adults $8.50, senior citizens $7.50, students 3-17 $5; discounts to AAM members, tours & groups; members no charge. Closed New Year's Day; Independence Day; Thanksgiving; Christmas Eve & Day. &
Attendance: 205,000 (accurate)
Membership: Individual $50; Family & Couple $70; Sustaining $150; Fellow $500.

DITTRICK MUSEUM OF MEDICAL HISTORY, 11000 Euclid Ave., Cleveland, OH 44106-1714. Tel.: 216-368-3648. Fax: 216-368-0165.
E-mail: james.edmonson@case.edu
Web Site: www.case.edu/artsci/dittrick/museum
Founded: 1926.
Congressional District: 21
Key Personnel: Chief Cur., James M. Edmonson; Museum Registrar & Archivist, Jennifer K. Nieves; Website Design & Photography, Laura Travis.
Personnel Profile: Full-Time Paid 3; Part-Time Volunteers 2.
Governing Authority: Case Western Reserve University. Affiliated with College of Arts & Sciences. Tax-exempt: 509(a)(1).
Institution Type/Description: Medical History Museum.
Collections: over 75,000 historical objects related to the practice of medicine, dentistry, pharmacy & nursing; exhibits emphasize medicine in the 19th- & early 20th-centuries including recreation of doctors' offices & a 19th-century pharmacy; contraception history; manuscript collections.
Research Fields: history of the health sciences.
Facilities: 10,000-vol. library of medical history available for inter-library loan; 450-seat auditorium.
Activities: guided tours; education programs for undergraduate & graduate students; inter-museum loan, temporary exhibits.
Publications: biannual newsletter, Newsletter of the Cleveland Medical Library Association.
Hours & Admission Prices: Mon.-Fri. 10-5. No charge. Closed New Year's Day; Memorial Day weekend; Independence Day; Labor Day weekend; Thanksgiving; Christmas Eve & Day. &
Membership: Friends of the Dittrick: Associate Partner $30, Friend $55; Sustaining $100;

DUNHAM TAVERN MUSEUM, 6709 Euclid Ave., Cleveland, OH 44103-3913. Tel.: 216-431-1060.
E-mail: dunhamtavern@sbcglobal.net
Web Site: www.dunhamtavern.org
Founded: 1939.
Congressional District: 22
Key Personnel: Pres. (V), Marsha French; Treas., Garrit Wamelink.

Personnel Profile: Part-Time Paid 1; Part-Time Volunteers 40.
Governing Authority: society. Tax-exempt: 501(c)(3).
Institution Type/Description: General Museum: housed in 1824 Dunham Tavern.
Collections: folklore; glass; textiles; mocha ware; pewter; lustre ware; Ohio room; early American artifacts; herb garden; Queen Ann & Chippendale furniture; McIntire mantel; historic log cabin located on heritage trail.
Research Fields: Cleveland & Western Reserve history taverns; transportation.
Facilities: 2,000-vol. library of Ohio & Cleveland history books available for use by application to the Board of Trustees or librarian; outdoor trail of 19th-century life; picnic area; English garden.
Activities: guided tours; lectures; films; permanent & temporary exhibitions.
Publications: monthly bulletin.
Hours & Admission Prices: Wed. & Sun. 1-4. Adults $3, children under 12 $2; discounts to AAM members; members no charge; group tours call for appointment. Closed New Year's Day; Easter; Thanksgiving; Christmas. &
Attendance: 4,200 (estimated)
Membership: Single $30; Family $50; Corporate $150; Life $1,000.

GALLERIES AT CLEVELAND STATE UNIVERSITY, 1307 Euclid Ave., Cleveland, OH 44115. Mailing Address: 2121 Euclid Ave., AG 116, Cleveland, OH 44115-2226. Tel.: 216-687-2103 & 2000. Fax: 216-687-9340. TDD: 216-687-2000.
E-mail: t.knapp@csuohio.edu
Web Site: www.csuohio.edu/artgallery
Founded: 1973.
Congressional District: 10
Key Personnel: Dir. & Cur., Robert Thurmer; Chm., Jennifer Visocky-O'Grady; Chm. (V), Mark Slankard; Pres., Ronald Berkman; Asst. Dir., Tim Knapp; Devel., Craig Zullig; Public Rels., Mary Grodek.
Personnel Profile: Full-Time Paid 2; Full-Time Volunteers 1; Part-Time Paid 10; Part-Time Volunteers 18; Interns 4.
Governing Authority: state government & public university; nonprofit. Parent Institution: Cleveland State University. Subsidiary Institution: Art Gallery. Tax-exempt: 501(c)(3).
Institution Type/Description: University Art Gallery.
Collections: works by students, national & international artists.
Research Fields: critical analysis of the relationship of art & society.
Facilities: educational facilities; 4,500 sq. ft. exhibit space.
Activities: formal education programs for undergraduate & graduate college students; guided tours; lectures; workshops by artists, performance art presentations.
Publications: exhibition catalogs published 3-6 times annually.
Hours & Admission Prices: Sept.-May Mon.-Fri. 10-5, Sat. 12-4. No charge; donations accepted. Closed federal, state & university holidays. &
Attendance: 18,231 (accurate)
Membership: Students $10; $25; $40; $200; $500; $1,000.

GREAT LAKES SCIENCE CENTER & STEAMSHIP MATHER MUSEUM, (M), 601 Erieside Ave., Cleveland, OH 44114-1021. Tel.: 216-694-2000. Fax: 216-696-2140. TDD: 216-696-3823.
Web Site: www.greatscience.com
Founded: 1988.
Congressional District: 21
Key Personnel: Chm. Bd., Paul Dolan; Interim Pres. & C.F.O., Don Paterson; Dir. Exhibits, Valence Davillier; Vice Pres. Mktg. & Guest Svcs., Jamie Simoneau; Vice Pres. Devel., Laura Rayburn; Dir. Facilities & Security, Gordon Milne; Mgr. Special Events, Alyssa Hennings; Vice Pres. Education, Whitney Owens; Dir. Human Resources, Renee Jones.
Personnel Profile: Full-Time Paid 41; Part-Time Paid 57; Part-Time Volunteers 236.
Governing Authority: private; nonprofit organization. Tax-exempt: 501(c)(3).
Institution Type/Description: Technology & Science Museum.
Collections: hands-on exhibits; science; biomedical technology; 1925 Great Lakes freighter.
Facilities: 200-seat auditorium; 157,458 sq. ft. exhibit space; visitor center; cafe; 320-seat OMNIMAX theater.
Activities: hands-on exhibits; omnimax theater; education programs; traveling exhibitions; workshops; summer, spring, & winter science camps; birthday parties; space rentals; funhouse.
Hours & Admission Prices: Daily 10-5. Adults $14, senior $13, youth 2-18 $12; discounts to AAM members. Closed Thanksgiving; Christmas. &
Attendance: 382,454 (accurate)
Membership: Individual Max $90; Family $95; Family Plus $110; Family Max $160; Sustaining Family Max $250; Patron Family Max $500; President's Circle $1,000.

INTERNATIONAL WOMEN'S AIR & SPACE MUSEUM, INC., (M), 1501 N. Marginal Rd., Rm. 165, Cleveland, OH 44114-3726. Tel.: 216-623-1111. Fax: 216-623-1113.
E-mail: cluhta@iwasm.org
Web Site: www.iwasm.org
Founded: 1976.
Congressional District: 10
Key Personnel: Pres. (V), Caroline Luhta; Interim Exec. Dir., Barbara Williams; Office Mgr., Heather Alexander.
Personnel Profile: Full-Time Paid 2; Part-Time Paid 2; Part-Time Volunteers 6.
Governing Authority: nonprofit organization. Tax-exempt: 501(c)(3).
Institution Type/Description: Women's Aeronautics Museum: housed at an airport.
Collections: photos; military uniforms; personal papers; clothing & costumes; trophies & plaques; aviation art; posters; aircraft components; First Day covers; Ruth Nichols collection of papers; Smith mini-plane; history of women in aviation & space.
Research Fields: All areas of women in aviation & aerospace.
Facilities: 1,000-vol. library of aviation & women in aeronautics material available to the public; study area available for researchers. Books & other museum-related items for sale.
Activities: loaned exhibits; tours; local speakers. Annual Events: career fair; pre-air show event; wine tasting event.
Publications: quarterly newsletter, IWASM Quarterly.
Hours & Admission Prices: Mon.-Sat. 10-4; exhibits accessible daily. No charge; donations accepted. Closed holidays. &
Attendance: 10,000 (estimated)
Membership: Student $10; Senior 65 & over $25; Individual $30; International $35; Family $50; Individual Gold $100; Corporation $500.

MUSEUM OF CONTEMPORARY ART CLEVELAND, 8501 Carnegie Ave., Cleveland, OH 44106-2919. Tel.: 216-421-8671. Fax: 216-421-0737.
E-mail: info@mocacleveland.org
Web Site: www.mocacleveland.org
Formerly: Cleveland Center for Contemporary Art
Founded: 1968.
Congressional District: 21
Key Personnel: Dir., Jill Snyder; Bd. Pres., Harriet Warm; Community Rels. Mgr., Jude Goergen; Emily Hall Tremaine Curatorial Fellow, Megan Lykins; Dir. Mktg. & Communications, Kelly Bird; Dir. Finance, Grace Garver; Sr. Cur., Margo Crutchfield; Graphic Designer, Danielle Rini Uva; Visitor Svcs. Mgr. & Museum Shop Mgr., Heather Young; School Programs & Tour Coord., Dara Sepkoski; Dir. Devel., John Grayson; Registrar & Cur. Coord., Ann Albano; Mgr. Member Rels., Rob Sikora; Exhibitions Mgr., Ray Juaire; Asst. Preparator, Paul Sydorenko; Finance Asst. & IT Mgr., Terri Tokar; Asst. Cur., Ana Vejzovic; Administrative Asst., Andrea Kormos.
Personnel Profile: Full-Time Paid 13; Part-Time Paid 5; Part-Time Volunteers 10; Interns 10.
Governing Authority: nonprofit organization. Tax-exempt: 501(c)(3).
Institution Type/Description: Art Museum.
Collections: contemporary visual art.
Research Fields: contemporary visual arts.
Facilities: Artspace. Museum-related items for sale.
Activities: lectures; tours; multi-arts programs; films; loan & traveling exhibitions.
Publications: catalogues; newsletter, invitations; brochures.
Hours & Admission Prices: Tues. & Thurs.-Sun. 11-5, Wed. 11-8. Adults $4, students & seniors $3; discounts to AAM & ICOM members; children under 12, members and MOCA members no charge. Closed New Year's Day; Easter; Independence Day; Thanksgiving; Christmas. &
Attendance: 20,000
Membership: Student, Senior & Artist $25; Individual $40; Household $50; Plus $75; MOCA Society $150; MOCA Fellow $250; Patron $500-$10,000.

NASA GLENN RESEARCH & VISITOR CENTER, Lewis Field, 21000 Brookpark Rd., Cleveland, OH 44135-3191. Tel.: 216-433-9653.
E-mail: mack.g.thomas@nasa.gov
Web Site: www.nasa.gov/centers/glenn/events/index.html
Institution Type/Description: Space Science.
Collections: aeronautics, aerospace & space science and technology.
Facilities: Gift items for sale.
Activities: guided tours.
Hours & Admission Prices: April-Oct. by appointment. Government-issued photo ID required for entry.

THE NATURE CENTER AT SHAKER LAKES, 2600 S. Park Blvd., Cleveland, OH 44120-1699. Tel.: 216-321-5935, ext. 227. Fax: 216-321-1869.
E-mail: naturecenter@shakerlakes.org
Web Site: www.shakerlakes.org
Founded: 1966.
Congressional District: 22
Key Personnel: Exec. Dir., Kay Carlson; Bd. Pres., Cynthia M. Klug; Dir. Advancement, Stephen Sedam; Financial Officer, Christopher Hall; Mgr. Visitor & Administrative Svcs., Brittany Coffin.
Personnel Profile: Full-Time Paid 8; Part-Time Paid 12; Part-Time Volunteers 10.
Governing Authority: nonprofit organization. Tax-exempt: 501(c)(3).
Institution Type/Description: Nature Center.
Collections: marsh & stream habitat; plants; trees; geology.
Facilities: nature center; stewardship center; nature trails; 2,000 sq. ft. exhibit space; 90-seat auditorium; classrooms; wildflower & butterfly gardens; outdoor pavilion. Museum-related items for sale.
Activities: lectures; films; arts festivals; study clubs; hobby workshops; TV programs; formally organized earth education programs for a variety of children & adult audiences; docent program; temporary exhibitions; self-guided trails; wildlife viewing area.
Publications: bimonthly letter, The Rookery; quarterly, program schedule; annual report; School Program Topics Guide.
Hours & Admission Prices: Mon.-Sat. 10-5, Sun. 1-5. No charge; donations accepted. Closed New Year's Day; Easter; Memorial Day; Independence Day; Labor Day; Thanksgiving; Christmas. &
Attendance: 30,000 (estimated)
Membership: Student $25; Senior $40; Individual $45; Senior Couple & Supporter $60; Family $65; Sustainer $100; Sponsor $250; Teal $500; Heron $1,000; Green Heron $2,500; Blue Heron $5,000; Great Blue Heron $10,000.

REINBERGER GALLERIES AT THE CLEVELAND INSTITUTE OF ART, 11141 East Blvd., Gund Bldg., Cleveland, OH 44106-1710. Tel.: 216-421-7000.
Web Site: www.cia.edu/galleries/galleries.php
Key Personnel: Dir., Bruce Checefsky
Institution Type/Description: Art Gallery.
Collections: art exhibitions.
Hours & Admission Prices: Tues.-Sat. 10-6, Sun. 12-6.

THE ROCK AND ROLL HALL OF FAME AND MUSEUM, 751 Erieside Ave., Cleveland, OH 44114-1023. Mailing Address: 1100 Rock and Roll Blvd., Cleveland, OH 44114. Tel.: 216-781-7625. Fax: 216-781-1832.
E-mail: marketing@rockhall.org
Web Site: www.rockhall.com
Founded: 1985.
Congressional District: 10
Key Personnel: C.E.O. & Pres., Greg Harris; Pres. & C.E.O. Rock and Roll Hall of Fame Foundation, Joel Peresman; Co Chm., Jann Wenner; Co Chm., Alec Wightman; Exec. Vice Pres. & C.F.O., Brian Kenyon; Vice Pres. Education & Public Programs, Lauren Onkey; Vice Pres. Devel., Caprice Bragg; Vice Pres. Mktg. & Communications, Todd Mesek.
Personnel Profile: Full-Time Paid 100; Part-Time Paid 19; Part-Time Volunteers 60; Interns 4.
Governing Authority: private; nonprofit. Parent Institution: Rock and Roll Hall of Fame Foundation, Cleveland Rock & Roll Inc. Tax-exempt: 501(c)(3).
Institution Type/Description: Rock and Roll Music Museum.
Collections: musical instruments; stage props; costumes; vehicles; promotional artifacts; artwork; documents; ephemera; photographs; sound recordings; periodicals.
Research Fields: history of rock and roll from its earliest roots in folk, blues, country, R&B, & gospel music to present; Hall of Fame inductees; music business; rock music journalism & criticism.
Facilities: research library & archives; 164-seat 3D theater; rental facilities; 50,000 sq. ft. exhibit space; cafe. Museum-related items for sale.
Activities: lectures; concerts; films; formally organized education programs for children & adults including distance learning & teacher education; permanent, temporary & traveling exhibitions; community festivals; facility rentals; docent & volunteer program. Annual Events: Induction Ceremony; American Music Masters (TM).
Publications: biannual member newsletter; exhibit guidebook; annual report; catalogs.
Hours & Admission Prices: Memorial Day to Labor Day Wed. & Sat. 10-9, Thurs.-Tues. 10-5:30; Sept.-May Wed. 10-9, Thurs.-Tues. 10-5:30. Adults $22, senior citizens 65 & over $17, children 9-12 $13; members no charge. Closed Thanksgiving; Christmas. &

Attendance: 460,600 (accurate)
Membership: Solo Rocker $50; Duet Rocker $75; Roller $140; Inductee $250; Headliner $500; Platinum $1,000; Chairman's Club $2,500; Rock Star $5,000; Legend $10,000.

ROMANIAN ETHNIC ART MUSEUM, St. Mary's Romanian Orthodox Cathedral, 3256 Warren Rd., Cleveland, OH 44111-1144. Tel.: 216-521-8449. Fax: 216-941-3068.
E-mail: st.mary.cathedral@sbcglobal.net
Web Site: www.smroc.org/culture.php
Founded: 1960.
Congressional District: 20
Key Personnel: Pres., George Dobrea.
Personnel Profile: Part-Time Volunteers 25.
Governing Authority: church; nonprofit. Affiliated with Romanian Orthodox Episcopate, 2522 Grey Tower Rd., Jackson, MI 49201. Tax-exempt.
Institution Type/Description: Folk Art Museum.
Collections: Anisoara Stan folk art; Dr. O.K. Cosla Romanian art; Gunther books, art & costume; Romanian artist paintings.
Research Fields: Romanian folk art & culture.
Facilities: 3,000-vol. library of books on Romanian culture, available for use on premises; reading room; classrooms. Religious & Romanian articles for sale.
Activities: guided tours; lectures; films; reading room; study clubs; permanent exhibitions.
Hours & Admission Prices: Museum: by appointment. Office: Mon.-Fri. 9-5. No charge; donations accepted.
Attendance: 1,500 (estimated)

SPACES GALLERY, 2220 Superior Viaduct, Cleveland, OH 44113. Tel.: 216-621-2314.
E-mail: contact@spacesgallery.org
Web Site: spacesgallery.org
Key Personnel: Exec. Dir., Christopher Lynn
Institution Type/Description: Art Gallery.
Collections: works by contemporary artists; paintings; sculpture.
Activities: public programs.
Hours & Admission Prices: Tues.-Thurs. & Sat. 11-5:30, Fri. 11-7, Sun. 1-5.

THE TEMPLE MUSEUM OF RELIGIOUS ART, (M), University Circle at Silver Park, 1855 Ansel Rd., Cleveland, OH 44106. Mailing Address: 26000 Shaker Blvd., Beachwood, OH 44122-7199. Tel.: 216-831-3233. Fax: 216-831-4216.
Web Site: www.ttti.org
Founded: 1950.
Congressional District: 21
Key Personnel: Dir., Bob Allenick; Museum Dir., Sue Koletsky.
Personnel Profile: Full-Time Paid 1; Part-Time Paid 3; Part-Time Volunteers 10; Interns 1.
Governing Authority: denominational group. Parent Institution: The Temple - Tifereth Israel, 26000 Shaker Blvd., Beachwood, OH 44122. Branch gallery: The Maltz Museum of Jewish Heritage, 2929 Richmond Rd., Beachwood, OH 44122. Tax-exempt.
Institution Type/Description: Religious Judaica Museum.
Collections: religious objects; ritual silver; paintings; sculpture; graphics; decorative arts; archaeology.
Research Fields: Judaic ritual; decorative & folk art.
Facilities: 45,000-vol. library.
Activities: guided tours; special exhibits.
Publications: bulletin, The Temple.
Hours & Admission Prices: Museum tours available by appointment Mon.-Fri. 9-4. No charge; donations accepted. Closed Jewish & legal holidays.
Attendance: 4,000 (estimated)

UKRAINIAN MUSEUM-ARCHIVES, INC., (M), 1202 Kenilworth Ave., Cleveland, OH 44113-4417. Tel.: 216-781-4329. Fax: 216-781-5844.
E-mail: staff@umacleveland.org
Web Site: www.umacleveland.org
Founded: 1952.
Congressional District: 10
Key Personnel: Dir., Taras Szmagala, Sr.; Chm. Bd. Dirs., Daria Kowcz Jakubowycz, Sr.; Sec. Bd. Dirs., Taras Szmagala, Jr.; Treas. Bd. Dirs., Zenon Holubec; Cur., Aniza Kraus.
Personnel Profile: Full-Time Paid 1; Part-Time Paid 2; Part-Time Volunteers 35.
Governing Authority: nonprofit organization. Tax-exempt: 501(c)(3).
Institution Type/Description: Ukrainian History, Culture & Art Museum.

Collections: artifacts, documents, books, sculptures, newspapers, pictures, Easter eggs & other items pertaining to Ukrainian culture & history, particularly in the United States.
Research Fields: Ukrainian press outside the Ukraine.
Facilities: 35,000-vol. library of Ukrainian books, books on the Ukraine in other languages & periodical titles including many historical publications, available for use on premises only; reading room.
Activities: community discussion; music; lectures.
Publications: Ukrainian books; bibliographical indexes; pamphlets; chronicles; private stamps.
Hours & Admission Prices: Tues.-Sat. 10-3. No charge; donations accepted. Closed holidays.
Attendance: 2,500 (estimated)
Membership: Regular Senior & Student $10; Regular $25; Sustaining $50; Sponsor $100; Silver $250; Gold $500.

WESTERN RESERVE HISTORICAL SOCIETY, 10825 East Blvd., Cleveland, OH 44106-1788. Tel.: 216-721-5722, ext. 1407. Fax: 216-721-0891.
E-mail: apurvis@wrhs.org
Web Site: www.wrhs.org
Founded: 1867.
Congressional District: 21
Key Personnel: Chm. (V), Donald Dailey; Sr. Vice Pres. Research & Publications, John Grabowski; Sr. Vice Pres. Interpretation & C.O.O., Kelly Falcone; Sr. Vice Pres. Finance, Administration & C.F.O., Mary Thoburn; Museum Shop Mgr., Angie Lowrie.
Personnel Profile: Full-Time Paid 35; Part-Time Paid 35; Part-Time Volunteers 100; Interns 25.
Governing Authority: society. Branch Museums: History Center, 10825 E. Blvd., Cleveland, OH 44106, Tel. 216-721-5722; Hale Farm & Village, 2686 Oakhill Rd., Bath, OH 44210, Tel. 330-666-3711; Shandy Hall, 6333 S. Ridge W., Geneva, OH 44041, Tel. 440-466-3680; Loghurst, 3067 Boardman-Canfield, Canfield, OH 44406, Tel. 330-533-4330. Tax-exempt: 501(c)(3).
Institution Type/Description: General Museum.
Collections: costumes; textiles; glassware; porcelain; early aircraft; 1895-1976 automobiles; 1770-1920 furnished rooms. Historic Buildings: 1810-1938 Hale Farm & Village, Bath, Ohio, working farm & restored village buildings; 1815 Shandy Hall, pioneer home in Unionville, Ohio; 1803 Loghurst.
Major Exhibits: Tying the Knot: Cleveland Wedding Fashions, 1830-1980, 6/12-2/14; Dior & More - For the Love of Fashion, 1/14-3/14.
Research Fields: United States, Ohio & local history; automotive history; regional aviation history.
Facilities: research library covering national & regional history, genealogical sources.
Activities: guided tours; lectures; special exhibitions; musical concerts; receptions; special events; craft demonstrations.
Publications: books, Merging Traditions; Birth of Modern Cleveland; Town to Tower; Cleveland Architecture, 1876-1976; Treasures by the Bay; James E. Taylor Sketchbook; Guide to Archives; The Jonathan Hale Farm; A Tour to New Connecticut in 1811; The Narrative of Henry Leavitt Ellsworth; Golden Wheels; If Elected; Balanced in the Wind; The Frances Payne Bolton Papers; The Terminal Tower: Tower City Center-A Historical Perspective; The Frederick C. Crawford Collection: The Automobile in American Culture; Showplace of America: Cleveland's Euclid Avenue 1850-1910; Cleveland-The Making of a City; The Western Reserve: The Story of New Connecticut in Ohio; The Encyclopedia of Cleveland History 2nd edition (in cooperation with Case Western Reserve University) & The Dictionary of Cleveland Biographies (in cooperation with Case Western Reserve University); There Are No Strangers At the Feast: Catholicism & Community in Northeastern Ohio; Cleveland A-Z.
Hours & Admission Prices: Tues.-Sat. 10-5, Sun. 12-5. Adults $10, children 3-12 $5; members & children 2 no charge. &
Attendance: 70,500 (accurate)
Membership: Individual $50; Family & Couple $70; Sustaining $150; Fellow $500.

Clyde

CLYDE HISTORICAL MUSEUM, 124 W. Buckeye St., Clyde, OH 43410-1934. Mailing Address: PO Box 97, Clyde, OH 43410-0097. Tel.: 419-547-7946.
E-mail: clydeheritageleague@yahoo.com
Institution Type/Description: History Museum.
Collections: local history & culture; Native American artifacts; military; personal artifacts; photographs.
Activities: summer family programs.

Hours & Admission Prices: April-Sept. Thurs. 1-4; other times by appointment.

Columbiana

THE LOG HOUSE MUSEUM OF THE HISTORICAL SOCIETY OF COLUMBIANA-FAIRFIELD TOWNSHIP, 10 E. Park Ave., Columbiana, OH 44408-1350. Tel.: 330-482-2983.
Formerly: The Historical Society of Columbiana-Fairfield Township
Founded: 1953.
Congressional District: 19
Key Personnel: Cur. & Historian, Nora Salmen; Genealogist, Joan Beatty.
Governing Authority: society; nonprofit organization. Tax-exempt.
Institution Type/Description: Local History Museum: housed in original log house on site of 1807 post office.
Collections: Civil War & World War I & II artifacts; 8 bound volumes of letters written by General E. Holloway; bound volumes of early Columbiana newspapers; dolls; wool coverlets with dates; quilts; mastodon bones; period furniture; clothing; wooden tools; books; pottery; Harvey S. Firestone; local industry.
Major Exhibits: Indian Artifacts, 11/13-12/14; Funeral Parlor Exhibit, 1/14-12/14.
Research Fields: genealogy; local history.
Facilities: 300-vol. library of Bibles, ledgers, early local newspapers, local histories available for research on premises; reading room. Museum-related items for sale.
Activities: guided tours; permanent & temporary exhibitions.
Publications: History of Columbiana & Fairfield Township, 1805-2005; brochure; DVD on Columbiana, 1805-2005; Bob Clark's Cartoons, 2010.
Hours & Admission Prices: Memorial Day to early Sept. Sat.-Sun. 2-4. Tours by appointment. No charge; donations accepted.
Attendance: 1,000 (estimated)
Membership: Individual $10; Family $15; Contributing $20; Lifetime $250; Patron $350.

Columbus

BUNTE GALLERY, Franklin University-Alumni Hall, 301 E. Rich St., Columbus, OH 43215-4960.
Institution Type/Description: Art Gallery. Built in honor of Dr. Frederick J. Bunte, Franklin University's second president.
Collections: artwork of local & regional artists; Emerson Burkhart Collection; Elijah Pierce Piece sculpture.
Hours & Admission Prices: Mon.-Thurs. 8-5.

COSI, 333 W. Broad St., Columbus, OH 43215-2738. Tel.: 614-228-2674. Fax: 614-629-3226.
E-mail: call_center@mail.cosi.org
Web Site: www.cosi.org
Founded: 1964.
Congressional District: 15
Key Personnel: Pres. & C.E.O., David E. Chesebrough, Ed.D.; Chm. Bd., Pablo Vegas; Vice Pres. Admin. & Finance & C.F.O., Rick Dodsworth; Vice Pres. Experience Div., Andy Zakrasjek.
Personnel Profile: Full-Time Paid 138; Part-Time Paid 96; Part-Time Volunteers 2,287.
Governing Authority: board of trustees. Tax-exempt: 501(c)(3).
Institution Type/Description: Science/Technology Center.
Collections: over 300 hands-on exhibits; science; technology; industry; history.
Major Exhibits: The International Exhibition of Sherlock Holmes (T), 2/8/14-9/1/14.
Research Fields: science; technology; industry; history; early childhood development; learning in informal settings.
Facilities: 300,000 sq. ft. exhibit space; rental facilities; 7-story digital extreme screen theater; cafe. Museum-related items for sale.
Activities: hands-on exhibits; curriculum oriented programs for schools, teachers & groups; workshops; formally organized education programs; permanent, traveling & temporary exhibits; rental facilities.
Publications: quarterly stakeholder newsletter & annual report, COSI News; programming brochures; teacher newsletter, Teacher e-news; member newsletter & e-news.
Hours & Admission Prices: Please visit website for hours and admission pricing.
Attendance: 627,833 (accurate)
Membership: Please visit website for membership pricing.

CENTRAL OHIO FIRE MUSEUM, 260 N. 4th St., Columbus, OH 43215-2511. Tel.: 614-464-4099.
E-mail: cofmuseum@aol.com
Web Site: www.centralohiofiremuseum.com
Founded: 1982.
Personnel Profile: Full-Time Paid 4; Part-Time Volunteers 6.
Governing Authority: Tax-exempt.
Institution Type/Description: Fire Museum: housed in restored 1908 engine house. Listed on the National Register of Historic Places.
Collections: hand-drawn, horse-drawn & early motorized fire apparatus; hands-on exhibits.
Activities: fire safety education programs for all ages.
Hours & Admission Prices: Tues.-Sat. 10-4; groups by appointment. Adults $6; members no charge. Closed holidays.
Attendance: 10,000 (estimated)

COLUMBUS CULTURAL ARTS CENTER, 139 W. Main St., Columbus, OH 43215-5044. Tel.: 614-645-7047. Fax: 614-645-5862. TDD: 614-645-3317.
Web Site: www.culturalartscenteronline.org
Founded: 1978.
Congressional District: 15
Key Personnel: Arts Admin., Geoffrey Martin; Asst. Arts Admin., Todd Camp; Programming & Event Coord., Eric Rausch; Mktg. Mgr., Veda Gilp.
Governing Authority: municipal. Parent Institution: A Division of Columbus Recreation & Parks Dept. Tax-exempt.
Institution Type/Description: Center for Cultural & Visual Arts.
Collections: paintings.
Facilities: meeting rooms & studios for jewelry, enameling, stone carving, ceramics, sculpture, weaving, bronze casting, painting, drawing & surface design.
Activities: organized classes; temporary & touring exhibits; concerts; poetry readings; workshops; community outreach through public art.
Publications: quarterly catalogue of class & program offerings.
Hours & Admission Prices: Mon. 1-4 & 7pm-10pm, Tues.-Thurs. 9-4, & 7pm-10pm, Fri.-Sat. 9-4. No charge; donations accepted.
Attendance: 50,000

*** COLUMBUS MUSEUM OF ART, (M),** 480 E. Broad St., Columbus, OH 43215-3886. Tel.: 614-221-6801. Fax: 614-221-0226.
E-mail: info@cmaohio.org
Web Site: www.columbusmuseum.org
Founded: 1878.
Congressional District: 12
Key Personnel: Pres., John C. Vorys; 1st Vice Pres., Steve English; 2nd Vice Pres., Joy Gonsiorowski; Exec. Dir., Nannette V. Maciejunes; Dir. Curatorial Admin., Dominique Vasseur; Assoc. Cur. American Art, Melissa Wolfe; Dir. Education, Cynthia Foley; Chief Registrar & Exhibitions Mgr., Melinda Knapp; Exhibition Designer, Greg Jones; Deputy Dir. Operations, Rod Bouc; Dir. Devel., Norma Sexton; Dir. Mktg. & Communications, Melissa Ferguson; Assoc. Cur. Contemporary Art, Tyler Cann; Dir. Visitor Svcs. & Volunteers & Museum Shop Mgr., Pam Edwards.
Personnel Profile: Full-Time Paid 74; Part-Time Paid 111; Part-Time Volunteers 1,000; Interns 50.
Operating Expenses: 8,100,000
Operating Income: 8,100,000
Governing Authority: nonprofit organization. Tax-exempt: 501(c)(3).
Institution Type/Description: Art Museum.
Collections: late 19th & early 20th-century European & American paintings, sculpture and works on paper; photography; contemporary art; 20th-century folk art; 19th-century American textiles; 16th to 18th-century European paintings; Philip & Suzanne Schiller collection of American Social Commentary Art, 1930-1970.
Major Exhibits: Toulouse-Lautrec & La Vic Moderne: Paris 1800-1910 (T), 2/8/14-5/18/14; The Art of Matrimony: Splendid Marriage Contracts from the Jewish Seminary Library, 4/14-6/15/14; Imagine! Picturebook Art from the Collection of Carol & Guy Wolfenbarger, 6/20/14-9/28/14; Modern Dialect: American Paintings from the John & Susan Horseman Collection, 6/29/14-9/21/14; Paul Bourginon: A 50th Anniversary Retrospective, 9/5/14-1/18/15; In--We Trust: Art & Money, 10/3/14-4/22/15.
Research Fields: pertaining to collection.
Facilities: 284-seat auditorium; community lecture room; sculpture garden; Derby Court; The Palette Cafe. Museum-related items for sale.
Activities: guided tours; lectures; seminars; gallery talks; films; concerts; workshops; self-guided tours; permanent, temporary & traveling exhibitions; family programs; studio classes; interactive exhibition for children & families.
Publications: bimonthly members' magazine, ArtSpeaks; annual report; exhibition & permanent collection catalogs; gallery guides.
Hours & Admission Prices: Tues.-Wed. & Fri.-Sun. 10-5:30 & Thurs.10-8:30. Adults $12, senior citizens 60+ & students 18+ $8, students 6-17 $5;

discounts to AAM members; children 5 & under, members and Sun. no charge. Closed New Year's Day; Independence Day; Thanksgiving; Christmas. &

Attendance: 209,000 (accurate)

Membership: Student $25; Friend $60; Household $75; Reciprocal $120; Supporter $250; Patron $500; Benefactor $1,000; Directors Circle $2,500.

FRANKLIN PARK CONSERVATORY AND BOTANICAL GARDENS, 1777 E. Broad St., Columbus, OH 43203-2040. Tel.: 614-715-8000; 800-214-7275. Fax: 614-645-5921.

E-mail: marketing@fpconservatory.org
Web Site: www.fpconservatory.org
Founded: 1895.
Key Personnel: Dir., Bruce Harkey; Museum Shop Mgr., Kathy Steedman.
Personnel Profile: Full-Time Paid 70; Part-Time Paid 27; Part-Time Volunteers 215.
Institution Type/Description: Conservatory.
Collections: botanical garden; plants; flowers; trees; butterflies; Chihuly artwork; contemporary art.
Facilities: cafe. Gift items for sale.
Activities: butterfly emergence center; glass blowing demonstrations; performances; family activities & crafts; educational programs & classes in gardening, visual arts, cooking, & wellness; tours.
Hours & Admission Prices: Wed. 10-8, Thurs.-Tues. 10-5. Adults $11, students and seniors 60 & over $9, children 3-17 $6; children under 2 & members no charge. Closed Thanksgiving; Christmas. &
Attendance: 275,000 (estimated)
Membership: Senior Individual & Student $30; Individual $40; Add a Guest $55; Household & Grandparent $70; Supporting $95; Centennial $125; Patron $250.

GRANGE INSURANCE AUDUBON CENTER, 505 W. Whittier St., Columbus, OH 43215. Tel.: 614-545-5475.

Key Personnel: Facilities Operations Mgr., Josh Cherubini
Institution Type/Description: Audubon Center.
Collections: wildlife & birds and their habitats; hands-on exhibitions.
Facilities: library; classrooms. Center-related items for sale.
Activities: demonstrations; observation deck; temporary & permanent exhibitions; special events; educational programs; school programs.
Hours & Admission Prices: Tues.-Fri. 10-5, Sat. 9-3, Sun. 11-5.

IIISTORIC COSTUME & TEXTILES COLLECTION, (M), 1787 Neil Ave., Columbus, OH 43210-1220. Tel.: 614-292-3090. Fax: 614-688-8133.

E-mail: strege.2@osu.edu
Web Site: costume.osu.edu
Congressional District: 15
Key Personnel: Cur., Gayle Strege; Asst. Cur., Marlise Schoeny.
Personnel Profile: Full-Time Paid 1; Part-Time Paid 1; Part-Time Volunteers 2; Interns 2.
Governing Authority: Parent Institution: The Ohio State University. Tax-exempt.
Institution Type/Description: Costume & Textile Museum.
Collections: costumes & textiles.
Research Fields: costume history.
Activities: tours; lectures; research.
Publications: past & present exhibit catalogs.
Hours & Admission Prices: Wed.-Thurs. 11-6, Fri.-Sat. 12-4; other times by appointment. No charge, donations accepted. Closed university holidays. &
Attendance: 1,000 (estimated)
Membership: Senior $25; Active $35; Sustaining $50; Patron $100; Trendsetter $250; Couture $500; Corporate $1,000.

JACK NICKLAUS MUSEUM, 2355 Olentangy River Rd., Columbus, OH 43210-1032. Tel.: 614-247-5959. Fax: 614-247-5906.

E-mall: info@nicklausmuseum.org
Web Site: www.nicklausmuseum.org
Key Personnel: Cur., Steve Auch; Events Mgr., Barbara Hartley
Institution Type/Description: Sports Museum.
Collections: Jack Nicklaus' life & career including trophies, photographs, & mementos from his 20 major championships & 100 worldwide professional victories; golf history.
Facilities: 24,000 sq. ft. exhibit space; theater. Museum-related items for sale.
Activities: corporate functions; banquets; receptions; rehearsal dinners; special events.
Hours & Admission Prices: Tues.-Sat. 9-5. Adults $10, students $5.

JUBILEE MUSEUM AND CATHOLIC CULTURAL CENTER, 57 S. Grubb St., Columbus, OH 43215-2747. Tel.: 614-221-4323. Fax: 614-221-9818.

E-mail: jubileemuseum@columbus.rr.com
Web Site: www.jubileemuseum.org
Formerly: Jubilee Museum at Holy Family
Founded: 1998.
Key Personnel: Cur., Rev. Kevin F. Lutz.
Personnel Profile: Full-Time Paid 2; Part-Time Paid 2; Part-Time Volunteers 10.
Governing Authority: Parent Institution: Diocese of Columbus.
Institution Type/Description: Religious Museum: housed in the former Holy Family High School.
Collections: Holy Land & Catholic religious artifacts; paintings; statues; vestments; altars; personal artifacts.
Publications: newsletter.
Hours & Admission Prices: Guided Tours: Sat. 11am; other times by appointment. Suggested Donation: $7 per person.
Attendance: 3,500 (estimated)

KELTON HOUSE MUSEUM & GARDEN, (M), 586 E. Town St., Columbus, OH 43215-4888. Tel.: 614-464-2022.

E-mail: keltonhouse@cs.com
Web Site: www.keltonhouse.com
Founded: 1976.
Congressional District: 15
Key Personnel: Chm., Liz Zuercher; Pres., Karlye Martin; Dir., Georgeanne Reuter.
Personnel Profile: Part-Time Paid 3; Part-Time Volunteers 35.
Governing Authority: nonprofit organization. Parent Institution: Junior League of Columbus. Tax-exempt: 501(c)(3).
Institution Type/Description: Historic House Museum.
Collections: period furniture; decorative arts; china & ceramics; Victorian toys & clothing; music boxes; Civil War items; items emphasizing Columbus history.
Research Fields: local history.
Facilities: 200-vol. library; archives; rental meeting rooms & banquet facilities. Gift items for sale.
Activities: guided tours; lectures; underground railroad learning station; docent program; organized education programs; intern program for college students; audio tours. Museum Sponsors: Victorian Christmas.
Publications: biannual newsletter, Keltonian.
Hours & Admission Prices: Sun. 1-4, Mon.-Fri. 10-4 (audio tour). Adults $6, senior citizens $5, children over 6 $2; members no charge. Closed Easter; Christmas. &
Attendance: 7,081 (accurate)
Membership: Friend $35; Educator $50; Preservationist $100; Ketlon Society $250 & up.

THE OHIO CRAFT MUSEUM, 1665 W. Fifth Ave., Columbus, OH 43212-2315. Tel.: 614-486-4402. Facebook: Ohio Designer Craftsmen.

E-mail: btalbott@ohiocraft.org
Web Site: www.ohiocraft.org
Key Personnel: Exec. Dir., Sharon Kokot; Museum/Artistic Dir., Betty Talbott; Education Coord., Megan Moriarty
Institution Type/Description: Craft Museum.
Collections: exhibits on fine craft.
Hours & Admission Prices: Mon.-Fri. 10-5, Sat.-Sun. 1-4. No charge. Closed holidays.
Membership: Individual $20; Family $25.

✱ OHIO HISTORICAL SOCIETY, (M), 800 E. 17th Ave., Columbus, OH 43211-2474. Tel.: 614-297-2300; 800-686-6124. Fax: 614-297-2352. TDD: 800-750-0750; FacebooK: Ohio Historical Society.

E-mail: executivedirector@ohiohistory.org
Web Site: www.ohiohistory.org
Formerly: Ohio Historical Center
Founded: 1885.
Congressional District: 15
Key Personnel: Dir. & C.E.O., Burt Logan; Pres. Bd., Glenda S. Greenwood; Cur., Brad Lepper; Cur., Robert Glotzhober; Cur., Cliff Eckle; Dir. Museum & Library Svcs., Sharon Dean; Chief Devel. Officer, Jim Walker; Timeline Editor, David Simmons; Dir. Historic Sites & Facilities, George Kane; Dir. Collections, Angela O'Neal; Dir. Communications, Jane Mason; Dir. Education & Outreach, Jackie Barton; Mgr. Visitor Experience, Megan

Wood; C.F.O., Finance & Administration, Jeff Ward; Dir. Community & Govt. Rels., Todd Kleismit; Museum Shop Mgr., Jim Riley.

Personnel Profile: Full-Time Paid 162; Part-Time Paid 24; Part-Time Volunteers 1,485.

Governing Authority: nonprofit organization. Branch Museums: Adena, Chillicothe; Armstrong Air & Space Museum, Wapakoneta; Buckeye Furnace, Jackson County; Campus Martius Museum of the Northwest Territory, Marietta; Wahkeena Nature Preserve, Sugar Grove; Cedar Bog, Urbana; Paul L. Dunbar House, Dayton; Flint Ridge Museum, Brownsville; Fort Ancient, Lebanon; Fort Hill, Hillsboro; Fort Laurens, Bolivar; Fort Meigs, Perrysburg; Fort Recovery, Fort Recovery; U.S. Grant Birthplace, Point Pleasant; U.S. Grant Boyhood Home and Schoolhouse, Georgetown; Benjamin R. Hanby House, Westerville; Rutherford B. Hayes Presidential Center, Fremont; Indian Mill Museum, Upper Sandusky; McCook House, Carrollton; National Road-Zane Grey Museum, Norwich; Our House Tavern, Gallipolis; Johnston Farm and Indian Agency, Piqua; Museum of Ceramics, East Liverpool; Schoenbrunn Village, New Philadelphia; Serpent Mound, Peebles; Zoar Village, Zoar; Ohio Village, Columbus; Warren G. Harding Home & Memorial, Marion. Youngstown Center of Industry and Labor, Youngstown; National Afro-American Museum & Cultural Center, Wilberforce; Cooke House, Sandusky; Shaker Historical Museum, Shaker Heights; Newark Earthworks Great Circle Museum, Newark; Ohio River Museum, Marietta; Quaker Meeting House, Mount Pleasant; Rev. John Rankin House, Ripley; Harriet Beecher Stowe House, Cincinnati; Tallmadge Church, Tallmadge. Tax-exempt: 501(c)(3).

Institution Type/Description: History Museum.

Collections: prehistoric archaeology of Midwest; pre-Columbian Indian artifacts & art objects; natural history: invertebrates, insects, fish, reptiles, birds, mammals, minerals; historical objects: paintings, decorative arts, drawings, prints, craft tools & products, textiles, clothing, glass & ceramics of Ohio; industrial & military artifacts; library & manuscript collections; state archives; artifacts from the Zoar & Schoenbrunn religious groups.

Major Exhibits: The 1950s: Building the American Dream, 7/12/13-6/18.

Research Fields: archaeology; natural history & environmental studies; architectural; business; industrial; labor; military; political & social history.

Facilities: 142,000-vol. library on Ohio history, prehistory, & natural history available for inter-library loan & on premises; reading room; 280-seat auditorium. Reproductions, publications & other museum-related items for sale.

Activities: permanent & temporary exhibitions; guided tours; gallery talks; films; lectures; school loan service; TV & radio programs; volunteer program & council; formally organized education programs for undergraduate & graduate college students; state historic preservation program; training programs for professional museum workers; field services to other historical organizations; teacher professional development; school outreach programs.

Publications: quarterly magazine, Timeline; bimonthly newsletters, Echoes, The Local Historian & Ohio Preservation.

Hours & Admission Prices: Ohio Historical Center & Museum Store: Wed.-Sat. 10-5, Sun. 12-5. Archives & Library: Wed.-Sat. 10-5. Ohio Village & Branch Sites: call 800-686-6124 for hours. Adults $10, seniors $9, children 6-12 $5, students $4; discounts to Golden Buckeye, OHLA, AAA, AARP, AAM & ICOM members; OHS members & children under 6 no charge. Closed New Year's Eve & Day; Thanksgiving; Christmas Eve & Day; Monday holidays. &

Attendance: 440,000 (accurate)

Membership: Individual $35; Family $55; History Lover $85; Supporter $235.

OHIO WOMEN'S HALL OF FAME, 274 E. First Ave., Ste. 300, Columbus, OH 43201. Tel.: 614-466-3847. Fax: 614-728-6974.

E-mail: ohioana@ohioana.org

Founded: 1978.

Key Personnel: Exec. Dir., David Weaver

Institution Type/Description: History Museum.

Collections: contributions by Ohio's women; Hall of Fame inductees; photographs.

Hours & Admission Prices: Mon.-Fri. 9-4:30. No charge.

ORTON GEOLOGICAL MUSEUM, OHIO STATE UNIVERSITY, (M), 155 S. Oval Mall, Columbus, OH 43210-1308. Tel.: 614-292-6896. Fax: 614-292-1496.

E-mail: gnidovec.1@osu.edu

Web Site: ortongeologicalmuseum.osu.edu

Founded: 1892.

Congressional District: 11

Key Personnel: Dir. & Professor, Bill Ausich; Mgr. Collection & Cur., Dale Gnidovec.

Personnel Profile: Full-Time Paid 2; Part-Time Volunteers 1.

Governing Authority: university; nonprofit organization. Parent Institution: Ohio State University, School of Earth Sciences. Tax-exempt.

Institution Type/Description: College Geology Museum: housed in Orton Hall.

Collections: fossils; minerals; rocks; meteorites.

Research Fields: paleontology; mineralogy.

Facilities: Fossils, minerals & rocks for sale.

Activities: guided tours; permanent & temporary exhibitions.

Hours & Admission Prices: Mon.-Fri. 9-5. No charge; donations accepted. &

Attendance: 11,927 (accurate)

THE SCHUMACHER GALLERY, CAPITAL UNIVERSITY, 1 College and Main, Columbus, OH 43209-2394. Tel.: 614-236-6319. Fax: 614-236-6490.

Web Site: www.schumachergallery.org

Founded: 1964.

Congressional District: 12

Key Personnel: Dir., Dr. Cassandra Tellier; Asst. to Dir., David Gentilini.

Personnel Profile: Full-Time Paid 2; Part-Time Paid 12; Part-Time Volunteers 6; Interns 2.

Governing Authority: university. Parent Institution: Capital University. Tax-exempt: 170(b)(1)(A).

Institution Type/Description: University Art Museum.

Collections: Ohio artists; graphics; contemporary painting & sculpture; period works, 16th-19th century; ethnic art; Asian art; Inuit art; African & Oceanic Art.

Facilities: library.

Activities: guided tours; lectures; films; gallery talks; concerts; formally organized education programs for undergraduate & graduate college students; permanent, temporary & traveling exhibitions.

Hours & Admission Prices: Sept.-May Mon.-Sat. 1-5. No charge; donations accepted. &

Attendance: 10,000

THE SNOWDEN-GRAY HOUSE, (M), 530 E. Town St., Columbus, OH 43215-4820. Mailing Address: P.O. Box 38, Columbus, OH 43216-0038. Tel.: 614-341-2129; 866-554-1870. Fax: 614-228-6303.

E-mail: ksmith@kkg.org

Web Site: www.kappa.org/heritagemuseum

Formerly: Heritage Museum of Kappa Kappa Gamma

Founded: 1980.

Key Personnel: Archivist & Cur., Kylie Towers.

Personnel Profile: Full-Time Paid 1; Part-Time Volunteers 12; Interns 1.

Governing Authority: Parent Institution: Kappa Kappa Gamma Foundation. Tax-exempt: 501(c)(3).

Institution Type/Description: History Museum: former home of an Ohio governor, built in 1852. Listed on the National Register of Historic Places.

Collections: period furnishings; photographs.

Hours & Admission Prices: Wed.-Thurs. 1-4; other times by appointment. No charge; donations accepted.

Attendance: 1,000 (estimated)

Membership: Donor $100; Partner $250; Sponsor $500; Patron $1,000; Benefactor $5,000.

WEXNER CENTER FOR THE ARTS, The Ohio State University, 1871 N. High St., Columbus, OH 43210-1105. Tel.: 614-292-0330 & 3535. Fax: 614-292-3369.

E-mail: info@wexarts.org

Web Site: www.wexarts.org

Founded: 1989.

Congressional District: 15

Key Personnel: Dir., Sherri Geldin; Pres., James Lyski; Deputy Dir., Jack Jackson; Dir. Mktg. & Communications, Jerry Dannemiller; Dir. Exhibitions Mgmt., Jill Davis; Cur. at Large, William Horrigan; Dir. Performing Arts, Charles Helm; Dir. Education, Shelly Casto; Dir. Design, M. Christopher Jones; Cur. Film/Video Studio Prog., Jennifer Lange; Dir. Devel., Christy Rosenthal; Dir. Patron Svcs., Megan Cavanaugh; Chm. (V), Leslie H. Wexner.

Governing Authority: Ohio State University. Tax-exempt.

Institution Type/Description: Contemporary Arts Center.

Collections: painting, sculpture, photography & graphic arts of 1960-1970s; sub-collections of graphic & Asian arts; multi-disciplinary, contemporary arts center.

Facilities: film & video auditorium; 2 performing arts theaters; video & audio production & editing; print viewing room; cafe. Museum-related items for sale.

Activities: performing arts; film & video programs; lectures; gallery talks; formally organized education programs; loan, permanent, & temporary circulating exhibitions.

Publications: monthly calendar; catalogues; gallery guides.

Hours & Admission Prices: Tues.-Wed. & Sun. 11-6, Thurs.-Sat. 11-8. No charge. Closed New Year's Day; Martin Luther King Day; Memorial Day; Independence Day; Labor Day; Veterans' Day; Thanksgiving; Christmas. &
Attendance: 190,380 (accurate)
Membership: Friend $50; Household $75; Patron $125; Sponsor $250; Fellow $500.

Conneaut

CONNEAUT RAILROAD MUSEUM, Depot St., Conneaut, OH 44030. Mailing Address: P.O. Box 643, Conneaut, OH 44030-0643. Tel.: 440-599-7878.
E-mail: ronbgrumpy@suite224.net
Founded: 1962.
Congressional District: 11
Key Personnel: Natl. Dir., Ronald Brundage, Jr.; Vice Pres., Ed Trenn; Sec., Michelle Jewel; Treas., Norman Gross.
Personnel Profile: Part-Time Paid 10; Part-Time Volunteers 5.
Governing Authority: municipal. Affiliated with the National Railway Historical Society. Tax-exempt: 501(c)(3) & 170(b)(1)(A).
Institution Type/Description: Antique Railroad Museum: housed in 1900 former New York Central depot.
Collections: engines; hopper car; wood caboose; 1866 stock certificate of Red River Line, N.Y.C.; relics of Ashtabula disaster 1876; scale models of locomotives & equipment; lanterns; photos; timetables; watches; passes; steam era display.
Activities: guided tours; formally organized education programs for children; temporary & traveling exhibitions.
Publications: monthly newsletter, The Semaphore.
Hours & Admission Prices: Memorial Day-Labor Day daily 12-5. No charge; donations accepted. &
Attendance: 10,000 (estimated)
Membership: Sustaining $10; International $35.

Copley

AKRON FOSSILS & SCIENCE CENTER, 2080 S. Cleveland-Massillon Rd., Copley, OH 44321. Tel.: 330-665-3466. Fax: 330-666-9801.
E-mail: info@akronfossils.com
Web Site: www.akronfossils.com
Founded: 2005.
Governing Authority: nonprofit organization. Tax-exempt: 501(c)(3).
Institution Type/Description: Science Center.
Collections: hands-on science-related exhibitions; fossils; astronomy; earth science.
Activities: guided tours; birthday parties; camps; classes; educational programs; lectures; workshops.
Publications: quarterly newsletter.
Hours & Admission Prices: May-Sept. Tues.-Sat. 10-5; Sept.-May Fri.-Sat. 10-5. Adults $8, children 3-12 $6; children 2 & under no charge. Closed New Year's Eve & Day; Memorial Day; Independence Day; Christmas Eve & Day. &
Membership: Family $65.

Coshocton

✳ JOHNSON-HUMRICKHOUSE MUSEUM, (M), Roscoe Village, 300 N. Whitewoman St., Coshocton, OH 43812-1061. Tel.: 740-622-8710. Fax: 740-622-8710 *51.
E-mail: jhmuseum@jhmuseum.org
Web Site: www.jhmuseum.org/default.htm
Founded: 1931.
Congressional District: 18
Key Personnel: Dir., Patti Malenke.
Personnel Profile: Full-Time Paid 1; Part-Time Paid 5; Part-Time Volunteers 4.
Governing Authority: county. Administered by Board of Trustees, Coshocton County Public Library. Tax-exempt: 170(b)(1)(A).
Institution Type/Description: General Museum.
Collections: American Indian basketry, beadwork, pottery, Inuit carvings & artifacts; Ohio prehistoric points & tools; Japanese & Chinese fine & decorative arts, porcelains, lacquer ware, metal & wood sculptures, cloisonne, textiles, and Japanese prints, armor & weaponry; 19th- & 20th-century American & European textiles, period firearms, 19th-century American; cut glass, ceramics & decorative arts; tools, numismatics & items of local history; Newark Holy Stones.
Facilities: community meeting room. Museum-related items for sale.
Activities: guided tours; lectures; gallery talks; formally organized education programs for children; inter-museum loan, permanent & temporary exhibitions.

Publications: quarterly newsletter.
Hours & Admission Prices: May-Oct. daily 12-5; Nov.-April Tues.-Sun. 1-4:30. Adults $3, youth $2; members no charge. Closed New Year's Day; Easter; Thanksgiving; Christmas Eve & Day. &
Attendance: 9,831 (accurate)
Membership: Individual $30; Family $50; Patron $100; Sustaining $200; Benefactor $350; Founder $500.

POMERENE CENTER FOR THE ARTS, 317 Mulberry St., Coshocton, OH 43812-2037. Tel.: 740-622-0326.
E-mail: acornell@pomerenearts.org
Web Site: www.pomerenearts.org/index.htm
Key Personnel: Dir., Anne Cornell
Institution Type/Description: Art Museum: housed in an 1836 Greek Revival home.
Collections: works by local, regional & national artists.
Hours & Admission Prices: Tues.-Fri. 1-5; Sat.-Sun. by appointment. No Charge.

ROSCOE VILLAGE FOUNDATION, 600 N. Whitewoman St., Coshocton, OH 43812. Tel.: 740-622-7644; 800-877-1830. Fax: 740-623-6555.
E-mail: rvhistorian@roscoevillage.com
Web Site: www.roscoevillage.com
Founded: 1968.
Congressional District: 18
Key Personnel: C.E.O., James McClure; Dir. Landscaping, Connie Miller; Dir. Facilities, Chad A. Miller; Dir. Operations, Rhonda Hurt; Historian, Chris Hart.
Personnel Profile: Full-Time Paid 7; Part-Time Paid 15; Part-Time Volunteers 30; Interns 1.
Governing Authority: nonprofit organization. Parent Institution: Roscoe Village Foundation, Inc. Tax-exempt: 501(c)(3).
Institution Type/Description: Historic Site: more than 30 restored buildings; Greek Revival structures.
Collections: living history buildings with emphasis on Ohio's canal-era period; period furniture; clothing; crafts; decorative arts; photos; household utensils; canal-era artifacts; maps; deeds; tools; tin smith shop; spinning wheel; interactive display for children: water wheel & woodworking.
Research Fields: Ohio & Erie Canal history.
Facilities: 800-vol. library; 65-seat auditorium; educational facilities; meeting rooms; 3 restaurants. Museum-related items for sale.
Activities: self-guided tours; lecture series; festivals; hobby workshops; craft demonstrations; organized education programs; participatory exhibits. Village Sponsors: Apple Butter Stirrin'; Christmas Candle Lightings.
Publications: brochures; Roscoe Village News; books, The Big Ditch; Twenty-Five Miles to Nowhere; Around the Stove in Roscoe's General Store; Plan & Profile Map of Ohio & Erie Canal; map, canals of Ohio.
Hours & Admission Prices: Visitor's Center: daily 10-5. Building Tour: family $29.95, adults $9.95, student $4.95; discounts to seniors, AAA members & groups. Closed New Year's Day, Easter, Thanksgiving, Christmas. &
Attendance: 100,000 (estimated)
Membership: Individual $30; Family $50; Patron $75; Sustaining $100; Contributing $250; Fellow $500; William Roscoe Society $10,000; Corporate $500, $750, $1,000, $2,500, $4,000, $5,000.

Covington

FORT ROWDY MUSEUM, 101 Spring St., Covington, OH 45318. Tel.: 937-473-2270.
Governing Authority: Parent Institution: Covington-Newberry Historical Society.
Institution Type/Description: Historical Society Museum: built in 1850.
Collections: local history & culture; period furnishings; personal artifacts; photographs; local industry; early tools, toys, clothing, & kitchen artifacts.
Activities: special events. Museum Sponsors: Fort Rowdy Days Festival.
Hours & Admission Prices: By appointment.

Crestline

CRESTLINE SHUNK MUSEUM, 211 N. Thoman St., Crestline, OH 44827-1444. Mailing Address: P.O. Box 456, Crestline, OH 44827-0456. Tel.: 419-683-3410.
Founded: 1947.
Key Personnel: Pres. (V), Ray Holland.
Governing Authority: society; nonprofit organization. Parent Institution: Crestline Historical Society. Tax-exempt: 501(c)(3).
Institution Type/Description: Local History Museum: housed in 1860 Victorian home.

Collections: railroad artifacts; summer kitchen; Victorian parlor & bedroom; glassware; silver; china; toys; dolls; early school memorabilia; Crawford Indian room.
Activities: guided tours; lectures.
Publications: annual booklet, Crestline Historical Society.
Hours & Admission Prices: Memorial Day-Labor Day Sat.-Sun. 2-4; other times by appointment. No charge; donations accepted.
Attendance: 200 (estimated)
Membership: Individual $6; Family $10; Life $150.

LOWE-VOLK PARK NATURE CENTER, Crawford Park District, 2401 State Rte. 598, Crestline, OH 44827. Tel.: 419-683-9000. Fax: 419-683-6281. Facebook: Crawford Park District.
E-mail: bfisher@crawfordparkdistrict.org
Web Site: www.crawfordparkdistrict.org
Key Personnel: Park Dir., Bill Fisher
Institution Type/Description: Nature Center.
Collections: local natural history & culture; murals; wildlife; photographs; stained glass.
Facilities: Museum-related items for sale.
Activities: educational programs; special events.
Hours & Admission Prices: Mon.-Sat. 8-4.
Membership: Friends of the Crawford Park District: Kinglet $10; Cardinal $20; Flock $30; Bur Oak Circle $50; Bald Eagle Club $100; Tallgrass Prairie Club $250; Sycamore Club $500; Natural Resource Partner $1,000.

Dalton

DALTON COMMUNITY HISTORICAL SOCIETY, 115 E. Main St., Dalton, OH 44618. Mailing Address: P.O. Box 273, Dalton, OH 44618. Tel.: 330-828-2221.
E-mail: daltonhistoricalsociety@gmail.com
Founded: 1978.
Congressional District: 16
Institution Type/Description: Historical Society Museum: housed in the former Old Eagle Hotel and Tavern; built in 1821.
Collections: local history & culture; period furnishings; personal artifacts; photographs.
Publications: annual calendar.
Hours & Admission Prices: By appointment. No charge; donations accepted.
Membership: Student $1; Individual $15; Couple $25.

Dayton

AULLWOOD AUDUBON CENTER AND FARM, (M), 1000 Aullwood Rd., Dayton, OH 45414-1129. Tel.: 937-890-7360. Fax: 937-890-2382.
E-mail: ckrueger@audubon.org
Web Site: aullwood.center.audubon.org
Founded: 1957.
Congressional District: 3
Key Personnel: Exec. Dir., Charity Krueger; Pres. (V), Josh Lounsbury; Education Coord., Tom Hissong; Environmental Education Specialist, Mikell Kloeters; Environmental Education Specialist, Nicole Conrad; Environmental Education Specialist, Sarah Alverson; Environmental Education Specialist, Tara Pitstick; Volunteer Coord., Nina Lapitan; Environmental Education Specialist, Bev Holland; Farm Mgr., John Stedman; Maintenance Mgr., Pat Rice; Resource Technician, Ken Fasimpaur; Devel. Coord., Ardith Hamilton; Office Mgr., Barb Trick; Museum Shop Mgr., Wendy Jacoby; Outreach Environmental Education Specialist, Chris Rowlands; Farm Asst., Bill Heilman; Housekeeper, Melissa Nicely.
Personnel Profile: Full-Time Paid 13; Part-Time Paid 25; Part-Time Volunteers 1,425; Interns 5.
Volunteer Hours: 17,838
Operating Expenses: 1,273,198
Operating Income: 1,273,198
Governing Authority: nonprofit organization. Parent Institution: National Audubon Society, 225 Varick St., 7th Fl., New York, NY 10014. Subsidiary Institution: Friends of Aullwood. Tax-exempt: 501(c)(3).
Institution Type/Description: Environmental Education Facility & Working Educational Farm.
Collections: natural history; energy conservation; agricultural material; working farm includes buildings, pasture, croplands, livestock & machinery; 200 acres of natural & agricultural land, interpretative buildings & exhibits including stores, restored & created tall-grass prairie.
Major Exhibits: Aullwood's Art Quilt Exhibit - Wild Lands - Wild Life, 6/8/14-8/18/14; Aullwood's Amish Quilt Exhibit, 9/14-10/11/14; Aullwood's Avian Art Contest Exhibit, 4/12/14-6/15/14.
Research Fields: energy; agriculture; sustainable agriculture; birds; butterflies; plants; environmental education techniques.

Facilities: 2,000-vol. library of natural history, conservation & environmental books available for staff & selected public use; nature center. Museum-related items for sale.
Activities: guided tours; lectures; films; consulting services to schools; formally organized education programs; professional teachers; permanent & temporary exhibitions; special events.
Publications: Aullwood Audubon Center & Farm Newsletter.
Hours & Admission Prices: Mon.-Sat. 9-5, Sun. 1-5. Adults $5, children $3; ANCA, National Audubon Society & Friends of the Aullwood members and children under 2 no charge. Closed winter holidays. &
Attendance: 115,000 (accurate)
Membership: Student $20; Individual & Senior Couple $40; Family & Household or Grandparent & Grandchild $50-$124; Sustainer $125-$249; Sponsor $250-$499; Patron $500 & up.

✻   **BOONSHOFT MUSEUM OF DISCOVERY, (M),** 2600 DeWeese Pkwy., Dayton, OH 45414-5499. Tel.: 937-275-7431. Fax: 937-275-5811. Facebook: Boonshoft Museum of Discovery.
E-mail: info@boonshoftmuseum.org
Web Site: www.boonshoftmuseum.org
Founded: 1893.
Congressional District: 10
Key Personnel: Pres. & C.E.O., Mark J. Meister; Chm. (V), William J. Williams; C.F.O., Peter Klosterman; Vice Pres. Collections & Research, Lynn Hanson; Vice Pres. Education, Susan Pion; Dir. Astronomy, Cheri Adams; Vice Pres. External Rels., Dona Vella; Museum Shop Mgr., Angela Shaffer; Dir. Facilities Mgmt., Mike McFann; Cur. Anthropology, William Kennedy; Sun Watch Indian Village/Archaeological Park Site Mgr. & Site Anthropologist, Andrew Sawyer; Cur. Live Animals, Mark Mazzei; Site Mgr. Fort Ancient, Jack Blosser.
Personnel Profile: Full-Time Paid 55; Part-Time Paid 29; Part-Time Volunteers 400; Interns 10.
Governing Authority: nonprofit organization. Parent Institution: Dayton Society of Natural History. Subsidiary Institutions: Sunwatch Indian Village/Archaeological Park; Fort Ancient; Boonshoft Museum of Discovery, Springfield. Tax-exempt: 501(c)(3) & 170(b)(1)(A).
Institution Type/Description: Natural History Museum & Archaeological Park.
Collections: archaeology; ornithology; ethnology; entomology; geology; herpetology; invertebrates; mammals; paleontology; live animals; archives.
Major Exhibits: Amazon Voyage (T), 2/14-4/14; Glow (T), 6/14-8/14; Robots & Us (T), 10/14-1/15.
Research Fields: biology; paleontology; archaeology; astronomy.
Facilities: theater; observatory; cafe; computer center. Museum-related items for sale.
Activities: guided school tours; lectures; films; study clubs; formally organized education programs; weekend demonstrations; permanent & temporary exhibitions; daily planetary programs; weekend laser shows.
Publications: annual report; monthly calendar, Quick E Bytes; monthly newsletter, Smart-E-News; biannual newsletter.
Hours & Admission Prices: Tues.-Sat. 10-5, Sun. 12-5. Adults $6.50, children 3-16 $4.50; members, ACM, AZA & ASTC members no charge. Call for show information. Laser programs available on some weekends, call for information. Closed New Year's Eve & Day; Easter; Thanksgiving; Christmas Eve & Day. &
Attendance: 247,128 (accurate)
Membership: Military $65; Military $75; Family $85; Adventurer $100; Adventure VIP $150; Explorer $250; Discoverer $500; John B. Greene Society $1,000 & up.

COX ARBORETUM METROPARK, 6733 Springboro Pike, Dayton, OH 45449-3496. Tel.: 937-434-9005. Fax: 937-438-1221. TDD: 937-275-PARK.
E-mail: diane.hart@metroparks.org
Web Site: www.metroparks.org/parks/coxarboretum
Formerly: Cox Arboretum & Gardens MetroPark
Founded: 1963.
Congressional District: 6
Key Personnel: Acting Dir., Rosie Melia; Horticulturist, Richmond Pearson; Family & Children's Education, Katrina Arnold; Garden Store Mgr., Diane Hart.
Personnel Profile: Full-Time Paid 15; Part-Time Paid 7; Part-Time Volunteers 378; Interns 3.
Governing Authority: county. Parent Institution: Five Rivers Metro Park, 1375 E. Siebenthaler, Dayton, OH 45414. Tel: 937-275-7275. Branch Museum: Aullwood Gardens. Tax-exempt.
Institution Type/Description: Arboretum & Gardens.
Collections: rock garden; shrub garden; crabapple allee; water garden; wildflower garden; clematis arbor; conifer knoll; edible landscape garden; herb

garden; conservation corner - prairie; butterfly meadow; magnolia collection; children's boxwood maze; meditation garden; nature trails; butterfly house.

Research Fields: magnolia, lilacs.

Facilities: 750-vol. library of books on horticultural & gardening available for research by staff or public on premises; botanical garden; reading room; 50-seat auditorium; 175-seat conference center; edible landscape demonstration garden; herb garden; propagation greenhouses; rock garden. Horticultural-related items, gifts & books for sale.

Activities: guided tours; lectures; group talks; study clubs; formally organized education programs for children & adults; docent programs; permanent exhibitions.

Publications: newsletter, Arb News; program schedule; brochures & mailers.

Hours & Admission Prices: Grounds: April-Oct. daily 8am-10pm; Nov.-March daily 8am-8pm. Buildings: Mon.-Fri. 8-5, Sat.-Sun. 11-4. No charge; donations accepted. Closed New Year's Day; Christmas. &

Attendance: 360,000 (estimated)

Membership: Crabapple $40; Dogwood $75; Lilac $100; Cherry $250; Magnolia $500; Founders Society $1,000.

DAYTON ART INSTITUTE, (M), 456 Belmonte Park N., Dayton, OH 45405-4700. Tel.: 937-223-5277. Fax: 937-223-3140.

E-mail: info@daytonartinstitute.org

Web Site: www.daytonartinstitute.org

Founded: 1919.

Congressional District: 3

Key Personnel: Dir. & C.E.O., Janice Driesbach; Chm., Michael Gretizer; Dir. Planned Gifts & Membership Coord., Laura Letton; Dir. Public Rels. & Mktg., Dona Vella; Dir. Educational Resources & Svcs., Susan Anable; Chief Cur., Will South; Museum Shop Mgr., Diane Haskell.

Personnel Profile: Full-Time Paid 22; Part-Time Paid 55; Part-Time Volunteers 355; Interns 7.

Governing Authority: nonprofit organization. Tax-exempt.

Institution Type/Description: Art Museum.

Collections: European & American paintings & sculpture, classical, Asian, pre-Columbian primitive objects; prints; graphic arts; ceramics & decorative arts; contemporary art collection.

Research Fields: all fields of art history.

Facilities: 500-seat auditorium; conference & meeting space; classrooms; cafe. Books, reproductions, jewelry, sculpture & postcards for sale.

Activities: guided tours; lectures; gallery talks; concerts; formally organized education programs; docent program; participatory gallery; permanent, temporary & traveling exhibitions. Museum Sponsors: annual Oktoberfest; annual Art Ball.

Publications: quarterly members magazine; annual report; exhibition catalogs.

Hours & Admission Prices: Tues.-Wed. & Fri. 10-4, Thurs. 10-8, Sat.-Sun. 12-5. Special exhibition fee charged. Closed major holidays. &

Attendance: 147,000 (accurate)

Membership: Student $25; Senior $30; Individual $40; Senior Couple 65 & over $45; Family $60; Sponsoring $125; Sustaining $250; Cantilever Society $500; Jefferson Patterson Society $1,300 & up.

DAYTON HISTORY AT CARILLON PARK, (M), 1000 Carillon Blvd., Dayton, OH 45409-2023. Tel.: 937-293-2841, ext. 100. Fax: 937-293-5798.

E-mail: bkress@carillonpark.org

Web Site: www.carillonpark.org

Formerly: Carillon Historical Park

Founded: 1950.

Congressional District: 3

Key Personnel: Chm. (V), Rob Connelly; Pres. & C.E.O., Brady Kress; Mgr. Community Collections, Nancy Horlacher; Dir. Education & Program Svcs., Alex Heckman; Mgr. Events, Teresa Beachler; Dir. Grounds & Maintenance, Lloyd Miller; Lead Interpreter, Danny Schlegal; Lead Interpreter, Angela Neimie; Communications Mgr., Christopher Jones; Exec. Asst., Pamela Ribic; Museum Shop Mgr. & Visitors Svcs. Mgr., Linda Vanover; Dir. Business Operations, Gail Hamer; Mgr. Devel. Technology, Joy McMeekin.

Personnel Profile: Full-Time Paid 15; Part-Time Paid 12; Part-Time Volunteers 175.

Governing Authority: nonprofit organization. Parent Institution; Dayton Foundation. Tax-exempt.

Institution Type/Description: Historical Museum Complex.

Collections: Wright brothers' 1905 airplane; camera; tools; bicycles; engine; printing items; locally made antique automobiles & bicycles; train car; bridges; canal lock; 1835 locomotive; interurban car; trolley bus; caboose; Conestoga wagon; Concord coach; 1776 Newcom tavern; 1815 stone cottage; 1895 canal superintendent's office; 1896 railway station; 1896

one-room school; 1924 gas station; 1930s print shop equipment; Deeds Carillon bell tower.

Research Fields: Wright brothers; local printing; industry companies.

Facilities: 23 buildings with exhibits information. Museum-related items for sale.

Activities: interpreters in buildings; special events; concerts; one-room schoolhouse program for fourth graders; three videos; student workshops.

Publications: booklets; pamphlets.

Hours & Admission Prices: Mon.-Sat. 9:30-5, Sun. 12-5. Adults $8, senior citizens $7, students 3-16 $5; discounts to AASLH members; children under 3 & members no charge. &

Attendance: 160,000 (estimated)

Membership: Individual $30; Family $40; Supporter $50; Sponsor $100; Patron $200.

DAYTON INTERNATIONAL PEACE MUSEUM, 208 W. Monument Ave., Dayton, OH 45402-3015. Tel.: 937-227-3223. Fax: 937-224-2713.

E-mail: info@daytonpeacemuseum.org

Web Site: www.daytonpeacemuseum.org

Formerly: Dayton Peace Museum

Founded: 2004.

Congressional District: 3

Key Personnel: Chm. (V), Bill Shaw.

Personnel Profile: Part-Time Volunteers 35; Interns 3.

Governing Authority: nonprofit organization. Tax-exempt: 501(c)(3).

Institution Type/Description: Peace Museum: house listed on the National Register of Historic Places.

Collections: history of peace, nonviolence, & peaceful cultures; cultural & religious tolerance; photographs.

Major Exhibits: Gun Violence (T), 1/14-3/14; Peace Corps, 4/14-6/14.

Research Fields: Non-violence.

Activities: educational programs; special events; traveling Peace Mobile; monthly children's hour.

Publications: monthly newsletter, Peace Connection.

Hours & Admission Prices: Tues.-Sat. 10-5, Sun. 1-5. No charge; donations accepted. Closed major holidays.

Membership: Student $25; Individual $45; Family $65; Peace Promoter $100; Organization $150; Peace Advocate $500; 1000 Cranes Society $1,000; Peace Patron $5,000; Founder's Club $10,000.

DAYTON VISUAL ARTS CENTER, 118 N. Jefferson St., Dayton, OH 45402. Tel.: 937-224-3822.

Web Site: www.daytonvisualarts.org

Key Personnel: Exec. Dir., Eva Buttacavoli

Institution Type/Description: Art Gallery.

Collections: works by contemporary artists.

Activities: special events; permanent & temporary exhibitions.

Hours & Admission Prices: Tues.-Sat. 11-6, 1st Fri. each month 11-8.

PATTERSON HOMESTEAD, 1815 Brown St., Dayton, OH 45409-2414. Mailing Address: c/o Dayton Histroy, 1000 Carillon Blvd. #D, Dayton, OH 45409-2023. Tel.: 937-222-9724; 973-293-2841. Fax: 937-222-0345. FacebooK: Patterson Homestead.

E-mail: info@daytonhistroy.org

Web Site: www.daytonhistroy.org

Founded: 1953.

Congressional District: 3

Key Personnel: Dir., Denise L. Darling.

Governing Authority: municipal. Affiliated with Montgomery County Historical Society, 224 N. St. Clair St., Dayton, OH 45402. Tax-exempt.

Institution Type/Description: Historic House: Patterson Homestead, a vernacular Ohio Federal style farmhouse built between 1816-1850.

Collections: period furniture, ranging from the hand-made to machine produced products including Queen Anne, Chippendale, Federal, American Empire, Eastlake & the classical revival style of the Victorian period, several oil portraits of members of the Patterson family; manuscript; furniture textiles, & books belonging to Pattersons.

Research Fields: Kentucky & Ohio pioneers: Col. Robert Patterson, founder of Lexington, KY; Frank & John H. Patterson, founders of National Cash Register Co.

Facilities: 110-seat rental facility available for public & private events.

Activities: guided tours; lectures; formally organized education programs for children & graduate students affiliated with Wright State University; docent program or council; permanent & temporary exhibitions; rental facilities.

Publications: Montgomery County Historical Society newsletter, Ionic Columns.

Hours & Admission Prices: Thurs.-Sat. 10-4. Suggested Donation: $2 per person. Closed legal holidays. &

Attendance: 1,280 (accurate)

Membership: Seniors $15; Individual $25; Family $40; Sustaining $100; Patron $500; Benefactor $1,000 and up.

PAUL LAURENCE DUNBAR STATE MEMORIAL, (M), 219 N. Paul Laurence Dunbar St., Dayton, OH 45402-6502. Mailing Address: 1000 Carillon Blvd., Dayton, OH 45409. Tel.: 937-313-2010.

E-mail: education@daytonhistory.org
Web Site: www.ohiohistory.org/places/dunbar
Founded: 1936.
Congressional District: 3
Key Personnel: Pres. (V), Benette DeCoux; Program Mgr., Josh Cain.
Personnel Profile: Full-Time Paid 1; Part-Time Paid 12; Part-Time Volunteers 15; Interns 1.
Governing Authority: society. Parent Institution: The Ohio Historical Society, 1985 Velma Ave., Columbus, OH. 43211. Tax-exempt.
Institution Type/Description: Historic House: housed in the home of African American poet, Paul Laurence Dunbar.
Collections: Paul Laurence Dunbar's personal artifacts including clothing worn by Dunbar, family paintings, photographs, Bible, his typewriter, publications & personal library.
Research Fields: American history.
Facilities: rental facilities; auditorium.
Activities: Annual Events: Dunbar's Birthday Commemoration in June; Gravesite Tribute, A Commemoration of Dunbar's Life and Legacy.
Hours & Admission Prices: Fri.-Sat. 10-4, Sun. 12-4. Adults $6, children 3-17 $3; discounts to active military, AAA members, seniors 65 & over & groups; Dayton History & Ohio Historical Society members & National Park Service Pass holders no charge. Closed New Year's Eve & Day; Thanksgiving; Christmas Eve & Day. &
Attendance: 1,200 (accurate)
Membership: Dayton History: Individual $40; Family & Grandparent $50; Carousel $75; Supporting $100; Sponsor $250; Patron $500.

ROBERT AND ELAINE STEIN GALLERIES, 3640 Colonel Glenn Hwy., 128-CAC, Dayton, OH 45435. Tel.: 937-775-2978. Fax: 937-775-4082.

E-mail: tess.cortes@wright.edu
Web Site: www.wright.edu/artgalleries/
Formerly: Wright State University Art Galleries
Founded: 1974.
Congressional District: 7
Key Personnel: Gallery Coord., Tess Cortes.
Personnel Profile: Full-Time Paid 1; Part-Time Paid 12; Part-Time Volunteers 10; Interns 1.
Governing Authority: university. Parent Institution: Wright State University. Tax-exempt.
Institution Type/Description: University Art Gallery.
Collections: contemporary works of art; works on paper; paintings; sculpture; visual documentations.
Research Fields: all areas of contemporary art.
Facilities: preparation areas.
Activities: contemporary art exhibitions; lectures; films; gallery talks; formally organized education programs for undergraduate & graduate students affiliated with Wright State Univ.
Publications: exhibition catalogs; artists' books; print editions; CD-ROM multimedia catalogues.
Hours & Admission Prices: Tues.-Fri. 10-4, Sat.-Sun. 12-5, special hours on WSU Theatre nights. No charge. Closed holidays. &
Attendance: 12,000 (accurate)
Membership: Student $25; Alumni $50; Artist $100; Educator $150; Friend $500.

*** SUNWATCH INDIAN VILLAGE/ARCHAEOLOGICAL PARK, (M),** 2301 W. River Rd., Dayton, OH 45417. Tel.: 937-268-8199. Fax: 937-268-1760.

E-mail: sunwatch@sunwatch.org
Web Site: www.sunwatch.org
Formerly: Sunwatch Prehistoric Indian Village
Founded: 1988.
Congressional District: 3
Key Personnel: Pres. & C.E.O., Mark Meister; Bd. Chm., Bill Williams; Dir. Education, Jean Copas; Museum Shop Mgr., Janet Williams; Site Mgr., Andrew Sawyer.
Personnel Profile: Full-Time Paid 3; Part-Time Paid 1; Part-Time Volunteers 20.

Governing Authority: private; nonprofit organization. Parent Institution: Dayton Society of Natural History, Dayton. Tax-exempt: 501(c)(3).
Institution Type/Description: Historic Site: reconstructed 800 year old village built by the Fort Ancient Indians. A National Historic Landmark.
Collections: more than one million artifacts discovered from the Fort Ancient Indian culture of the AD 1200s; restored native prairie.
Research Fields: Archaeology.
Facilities: visitor center; lecture rooms.
Activities: family days; scout days; overnights; pow wow; nature flute gathering; annual lecture series. Museum Sponsors: Keeping The Tradition Pow Wow; Sun Watch Flute & Art Gathering; Summer Lore.
Publications: Adventures.
Hours & Admission Prices: Tues.-Sat. 9-5, Sun. 12-5. Adults $5, senior citizens 60 & over and students 6-16 $3; discounts to AAM members; members no charge. Closed New Year's Eve & Day; Easter; Thanksgiving; Christmas Eve & Day. &
Attendance: 20,000 (accurate)
Membership: Ambassador $50; Military $75; Family & Grandparent $85; Adventurer $100; Explorer $250; Discoverer $500; John B. Greene Society $1,000 & up.

Defiance

AU GLAIZE VILLAGE, 12296 Krouse Rd., Defiance, OH 43512. Mailing Address: P.O. Box 801, Defiance, OH 43512-0801.
Web Site: www.auglaizevillage.com
Founded: 1966.
Congressional District: 5
Key Personnel: Pres. (V), Scott Lanton; Vice Pres., Jerry DeLong; Museum Shop Mgr., Deb Kutzli.
Personnel Profile: Full-Time Volunteers 4; Part-Time Volunteers 12.
Governing Authority: county; nonprofit. Parent Institution: Defiance County Historical Society. Branch Museums: William Bensinger Military Museum; Charles Slocum Natural History Museum. Tax-exempt.
Institution Type/Description: Village Museum.
Collections: local history; farm implements & equipment; household items; clothing; military uniforms; natural history; archaeology; model trains & equipment; dentist; printing. Historic Buildings: 1875 Mark Center Post Office; 1854 Jewell R.R. Station; 1875 St. John Lutheran Church; Ayersville Telephone Office; 1850 Sherry School House; Vaughn's Lockkeepers House; Kinner cabin; Kieffer cabin; 1875 Chapel of the Crosses; Meyers Cider Mill.
Research Fields: local history.
Facilities: food service available. Museum-related items for sale.
Activities: docent program or council; permanent & temporary exhibitions; special events. Museum Sponsors: Days of Yesteryear in May; Black Swamp Show in June; Car Show in July; Johnny Appleseed Festival in October.
Publications: annual newsletter, Village Guide; newspaper listing buildings; Justice in Defiance.
Hours & Admission Prices: May-Oct. event weekends & by appointment. Adults $3, children 6-16 $1; discounts for Ohio Pass, senior citizens $2 on event days; children 6 and under & members no charge. &
Attendance: 6,000 (estimated)
Membership: Individual $5; Family $10; Supporting $25; Sustaining $50; Life $100.

Delaware

DELAWARE COUNTY HISTORICAL SOCIETY - NASH HOUSE MUSEUM, 157 E. William St., Delaware, OH 43015-2165. Tel.: 740-369-3831.
Founded: 1955.
Institution Type/Description: Historic House. Built in the 1878.
Collections: local history & culture; photographs; furnishings; period artifacts.
Publications: semiannual newsletter, Delaware Historian.
Hours & Admission Prices: mid-March to mid-Nov. Wed. & Sat.-Sun. 2-4:30, Thurs. 10-4:30; mid-Nov. to mid-March. Sun. 2-4:30. No charge; donations accepted. Closed major holidays.
Membership: Individual $10; Household $15; Patron $35; Sponsor $100; Silver Sponsor $250; Gold Sponsor $500; Platinum Sponsor $1,000.

RICHARD M. ROSS ART MUSEUM, (M), Ohio Wesleyan University, 60 S. Sandusky St., Delaware, OH 43015-2333. Tel.: 740-368-3606. Fax: 740-368-3515.
E-mail: ramuseum@owu.edu
Founded: 2002.
Key Personnel: Dir., Justin Kronewetter; Sr. Asst., Tammy Wallace; Gallery Asst., Stephen Perakis
Institution Type/Description: Art Museum.

Collections: paintings; sculpture; photographs.
Activities: permanent, temporary & traveling exhibitions; lectures; special events.
Hours & Admission Prices: Tues.-Wed. & Fri. 10-5, Thurs. 10-9, Sun. 1-5. No charge.

Delphos

DELPHOS CANAL COMMISSION MUSEUM, 241 Main St., Delphos, OH 45833-1764. Mailing Address: P.O. Box 256, Delphos, OH 45833-0256. Tel.: 419-695-7737. Facebook: Delphos Canal Commission Museum.
E-mail: info@delphoscanalcommission.com
Web Site: www.delphoscanalcommission.com
Volunteer Hours: 7,500
Governing Authority: Tax-exempt.
Institution Type/Description: History Museum.
Collections: canal & canal boat history; local history & culture; photographs; personal artifacts; period furnishings; 1902 Sears Buggy Roadster; period tools & manufacturing equipment; business; industry; schools; churches.
Hours & Admission Prices: Thurs. 9-12, Sat.-Sun. 1-3; groups by appointment. No charge.
Attendance: 2,000 (estimated)
Membership: Student & Senior $10; Individual $20; Family $30; Supporting $50; Partner $125; Patron $250. Corporate: Not for Profit $75; For Profit (under 250 employees) $250; For Profit (over 250 employees) $500.

DELPHOS POSTAL MUSEUM, 339 N. Main St., Delphos, OH 45833-0174. Mailing Address: P.O. Box 174, Delphos, OH 45833-0174. Tel.: 419-303-5482.
E-mail: mphdelphos@gmail.com
Web Site: www.postalhistorymuseum.org
Founded: 1995.
Congressional District: 4
Key Personnel: Dir., Gary S. Levitt; Pres. (V), Rev. David Howell.
Personnel Profile: Full-Time Volunteers 1; Part-Time Volunteers 35.
Volunteer Hours: 4,800
Operating Expenses: 32,000
Operating Income: 24,000
Governing Authority: Tax-exempt.
Institution Type/Description: History Museum: housed in a former horse avery; built in 1902.
Collections: postal history; period artifacts; mail processing; stamps; 1906 Harrington coach; mail sled; jeep; 1963 Westcoaster; 1957 Cushman mailster; replica railway mail service car including original equipment.
Major Exhibits: AAA Spectacular Columbus (T), 1/14; Audio Tour, 3/14.
Research Fields: postal history; philately; transportation; RPO, HPO.
Publications: quarterly newsletter; bulletins; article, Curator's Corner.
Hours & Admission Prices: Call for hours. No charge; donations accepted.
Attendance: 4,000 (estimated)
Membership: Individual $30; Family $50; Corporate $250.

Dennison

THE DENNISON RAILROAD DEPOT MUSEUM, (M), 400 Center St., Dennison, OH 44621-1402. Mailing Address: P.O. Box 11, Dennison, OH 44621-0011. Tel.: 740-922-6776; 877-278-8020 (toll free). Fax: 740-922-4929.
E-mail: director@dennisondepot.org
Web Site: www.dennisondepot.org
Founded: 1984.
Congressional District: 18
Key Personnel: Dir., Wendy R. Zucal; Bd. Pres., Melissa Stocker; Event Coord., Jacob Masters; Cur., Kim Turkovic; Membership Coord., Kim Johnson; Experience Coord., Olivia Marsh; Experience Coord., Alicia Miller; Museum Shop Mgr., Mary Galbreath; Trax Diner Inc., Sandy Armitt; Trax Diner Inc., Michael Smith.
Personnel Profile: Full-Time Paid 2; Full-Time Volunteers 25; Part-Time Paid 6; Part-Time Volunteers 150; Interns 3.
Governing Authority: municipal; private; nonprofit organization. Tax-exempt: 501(c)(3).
Institution Type/Description: National Historic Landmark: housed in 1873 Pennsylvania Railroad Depot.
Collections: artifacts; railroad; World War II Canteen Site; local history; model train display; Keystone exhibition hall; rolling stock; railroad cars.
Facilities: library; 70-seat restaurant; 2,564 sq. ft. exhibit space. Museum-related items for sale.
Activities: docent program; guided tours; lectures; loan exhibitions; broadcast programs; Polar Express train rides; children's interactive train car. Museum Sponsors: Railroad Festival; Railroad Symposium; Spring Ball; Garden Tour; Ghost Tours.
Publications: quarterly newsletter, The Timetable.
Hours & Admission Prices: Tues.-Fri. 10-5, Sat. 11-4, Sun. 11-3. Adults $8, senior citizens $6, students 7-17 $4; discounts to AAM & AAA members; children under 7 & members no charge. Closed New Year's Day; Easter; Independence Day; Thanksgiving; Christmas.
Attendance: 85,000 (estimated)
Membership: Student $5; Individual $10; Family $20; Nonprofit $30; Corporate $40; Individual Life $125; Married Life $175; Corporate Life & Memorial $200. (Supersize for additional $10).

Dover

J.E. REEVES HOME & MUSEUM, 325 E. Iron Ave., Dover, OH 44622-2105. Tel.: 330-343-7040; 800-815-2794. Fax: 330-343-6290.
E-mail: reeves@tusco.net
Web Site: www.doverhistory.org
Founded: 1958.
Congressional District: 18
Key Personnel: Chm. Bd., Zach Feller; Dir., Matt Lautzenheiser; Pres., Greg Bair; Vice Pres., Patti Feller.
Personnel Profile: Full-Time Paid 1; Part-Time Paid 10; Part-Time Volunteers 50.
Governing Authority: nonprofit. Affiliated with the Dover Historical Society. Tax-exempt.
Institution Type/Description: General Historical Society Museum: housed in 19th-century restored Victorian mansion, home of J.E. Reeves.
Collections: historical maps; agriculture; archaeology; archives; paintings; costumes; geology; glass; music players; Civil War artifacts; original draperies; chandeliers; carved & tufted period furniture; decorative arts; mineralogy; vehicles; carriages; 1922 Rauch & Lang electric car; restored 19th-century Victorian Mansion with original furnishings & Carriage House Museum holding historic artifacts, including the family sleigh and carriage, and memorabilia surrounding Dover born guerrilla Civil War leader, William Clark Quantrill.
Research Fields: birthplace of William Clark Quantrill.
Facilities: over 500-vol. library of books available by permission; reading room.
Activities: guided tours; lectures; permanent & temporary exhibitions.
Publications: quarterly newsletter; books.
Hours & Admission Prices: June-Oct. Tues.-Sun. noon-4; Nov. 11-Dec. 22 daily 1-7. Adults $8, seniors $7, children 5-18 $3; discounts to AAA members; members no charge.
Attendance: 4,000 (estimated)
Membership: Individual $12; Couple $18; Family $20; Supporting $35.

WARTHER MUSEUM INC., 331 Karl Ave., Dover, OH 44622-2767. Tel.: 330-343-7513.
E-mail: info@warthers.com
Web Site: www.warthers.com
Founded: 1936.
Key Personnel: Pres., Mark Warther; Museum Shop Mgr., Carol Moreland.
Personnel Profile: Full-Time Paid 2; Part-Time Paid 11.
Governing Authority: individual operation.
Institution Type/Description: Arts & Crafts Museum.
Collections: hand-carved works of art by Ernest Warther; history of steam engines from 250 B.C.-present day in ivory, ebony & pearl; 8 ft. ebony & ivory replica of the Lincoln funeral train; 50,000 buttons arranged in designs on walls & ceiling; telegraph station; railroad park.
Facilities: picnic facilities; Swiss gardens; 3-acre landscaped grounds.
Activities: guided tours; formally organized education programs for children.
Hours & Admission Prices: Daily 9-5. Adults $13, students $5; discounts to groups. Closed New Year's Day; Easter; Thanksgiving; Christmas.
Attendance: 100,000 (estimated)
Membership: Single $16; Couple $32.

Doylestown

CHIPPEWA-ROGUES HOLLOW HISTORICAL SOCIETY, 17500 Galehouse Rd., Doylestown, OH 44230-9773. Tel.: 330-882-3375.
E-mail: marymertic@juno.com
Web Site: www.chippewarogueshollow.org
Key Personnel: Pres., Doug Zook; Vice Pres., Mary Mertic; Treas., Earl Kerr
Institution Type/Description: Historical Society Museum.
Collections: local history & culture; period furnishings; personal artifacts; replica of mill.
Hours & Admission Prices: May to Labor Day Sun. 1-4.

Dublin

WORLD'S FIRST WENDY'S RESTAURANT, One Dave Thomas Blvd., Dublin, OH 43017-5452. Tel.: 614-764-3100.
Institution Type/Description: General Museum: housed in the first Wendy's Restaurant opened in 1969; named after his daughter Melinda (Wendy) Thomas, a nickname given to her by her siblings.
Collections: original test griddle; ad campaigns; photographs; personal artifacts; trophies; grand opening memorabilia.
Facilities: restaurant.
Hours & Admission Prices: Mon.-Fri. 10-8, Sat. 10-7, Sun. 11-6. No charge.

East Liverpool

LOU HOLTZ/UPPER OHIO VALLEY HALL OF FAME, 120 E. Fifth St., East Liverpool, OH 43920-3031. Tel.: 330-386-5443. Fax: 330-382-0244.
E-mail: director@louholtzhalloffame.com
Web Site: www.louholtzhalloffame.com
Founded: 1998.
Congressional District: 6
Key Personnel: Pres., Frank C. Dawson; Treas., Jackman Vodrey; Dir., Robin Webster.
Personnel Profile: Full-Time Paid 1; Part-Time Paid 1; Part-Time Volunteers 2.
Governing Authority: private; nonprofit organization. Tax-exempt: 501(c)(3).
Institution Type/Description: Hall of Fame.
Collections: memorabilia related to Coach Lou Holtz & other inductees; communities of the Upper Ohio Valley.
Facilities: 3,500 sq. ft. exhibit space. Books & sports-related items for sale.
Activities: films; guided tours. Museum Sponsors: Hall of Fame weekend in summer; Induction banquet; celebrity golf tournament.
Publications: quarterly newsletter.
Hours & Admission Prices: Mon.-Fri. 10-5, Sat. 10-1. No charge; donations accepted. Closed legal holidays. &
Attendance: 2,500 (estimated)
Membership: Varsity $15; MVP $50; All-Conference $100; All-State $250; All-American $500; Coaches Club $1,000; Hall of Fame $5,000; Leadership $10,000.

MUSEUM OF CERAMICS, (M), 400 E. 5th St. at Broadway, East Liverpool, OH 43920-3134. Tel.: 330-386-6001; 800-600-7180. Facebook: Museum of Ceramics.
E-mail: museumofceramics@gmail.com
Web Site: www.TheMuseumOfCeramics.org
Founded: 1980.
Congressional District: 18
Key Personnel: Dir. & Museum Shop Mgr., S.W. Vodrey; Historic Site Technician, Philip L. Rickerd.
Personnel Profile: Part-Time Paid 3; Part-Time Volunteers 6.
Governing Authority: Parent Institution: Museum of Ceramics Foundation. Tax-exempt.
Institution Type/Description: History Museum: housed in 1909 Old City Post Office.
Collections: East Liverpool area ceramics; 1930s paintings; WPA era mural; Lotus ware; art pottery; porcelain.
Research Fields: pertaining to collections.
Facilities: Museum-related items for sale.
Activities: guided tours; special events; lectures; temporary exhibitions; audiovisual shows; school outreach programs; kids' clay camps; art classes; ceramics classes.
Hours & Admission Prices: Tues.-Sat. 9:30-3:30. Adults $4, students $2; discounts to AAA members & Time Travelers; children 5 & under and members no charge. Closed federal holidays. &
Attendance: 1,900 (estimated)
Membership: Ironstone $25; Rockingham $50; Parian $100; Majolica $250; Belleek $500; Lotus Ware $5,000.

THOMPSON HOUSE MANSION, 305 Walnut St., East Liverpool, OH 43920. Mailing Address: East Liverpool Historical Society, P.O. Box 476, East Liverpool, OH 43920. Tel.: 330-386-5964. Fax: 330-386-9279.
Web Site: www.eastliverpoolhistoricalsociety.net
Founded: 1907.
Key Personnel: President, Tim Brookes; Treas., Wm. Gray
Governing Authority: Parent Institution: East Liverpool Historical Society. Tax-exempt.
Institution Type/Description: Historic House Museum: housed in the former home of C.C. Thompson, built in 1876. Listed on the National Register of Historical Places.

Collections: local history & culture; period furnishings; personal artifacts; photographs; pottery.
Activities: special events.
Hours & Admission Prices: Call for hours. No charge; donations accepted.
Attendance: 350
Membership: Individual & Family $20.

Eastlake

CROATIAN HERITAGE MUSEUM & LIBRARY, 34900 Lakeshore Blvd., Eastlake, OH 44095-3575. Tel.: 440-946-2044 (museum) & 316-7211. Fax: 216-991-2310.
E-mail: croatianmuseum@sbcglobal.net
Web Site: www.croatianmuseum.com
Founded: 1983.
Congressional District: 19
Key Personnel: Pres. (V) & Cur., Branka M. Malinar; Treas., Kathy Kuhar; Archivist & Librarian, Jerry Malinar; Devel., Judith Zivic; Public Rels., Suzanne Jerin.
Personnel Profile: Part-Time Volunteers 15.
Governing Authority: nonprofit. Parent Institution: American Croatian Lodge. Tax-exempt: 501(c)(3).
Institution Type/Description: Folk Art Museum.
Collections: Croatian artifacts; folk art & dress; textiles; folk costumes; wood carvings; sculpture; metal work; leather work; paintings.
Research Fields: Croatian textiles, customs, & architecture; folk art; wedding customs.
Facilities: 13,500 library of books & Croatian journals; restaurant; educational facilities. Museum-related items for sale.
Activities: guided tours; school loan service; traveling exhibitions. Museum Sponsors: Easter event; Christmas event.
Publications: brochures for weaving exhibit; Croatian Castles of Zagorje, Croatia; Croatian Christmas Customs; Croatian Easter Customs Croatian Wedding Customs.
Hours & Admission Prices: Fri. 12-6; other times by appointment. No charge; donations accepted. Closed New Year's Day; Christmas. &
Attendance: 9,000 (estimated)
Membership: Individual $20; Family $30; Supporting $50; Supporter Family $100; Patron $101-$499; Benefactor $500 & up.

Elmore

SCHEDEL ARBORETUM & GARDENS, 19255 W. Portage River S. Rd., Elmore, OH 43416-9743. Tel.: 419-862-3182. Facebook: Schedel Arboretum Gardens.
E-mail: info@schedel-gardens.org
Web Site: www.schedel-gardens.org
Key Personnel: Dir., Rodney Noble; Grounds Supt., Jeff Saffran; Events Coord. & Exec. Asst., Veronica Sheets; Asst. Grounds Supt., Kendra Schwartz.
Governing Authority: nonprofit organization.
Institution Type/Description: Arboretum.
Collections: 17 acre gardens include 15,000 annuals; perennials; flowering trees; shrubs. Schedel Home: period Persian rugs; carved jade; bronze; personal artifacts; paintings; photographs; furnishings.
Facilities: botanical garden. Museum-related items for sale.
Activities: arts festivals; guided tours; temporary exhibits; weddings; receptions; workshops; concerts.
Publications: quarterly newsletter, The Torii.
Hours & Admission Prices: May-Oct. Tues.-Sat. 10-4, Sun. 12-4. Adults $10, seniors 60 & up $9, children 6-12 $6; discounts to AAA members; children 5 & under and members no charge. Closed Memorial Day; Independence Day; Labor Day.
Attendance: 15,000 (accurate)
Membership: Friend's Circle: Buckeye $35, Redbud & Cedar $50, Lilac $100, Bald Cypress $150, Dawn Redwood $200, Dogwood $250, Magnolia $500; Director's Circle: Lily $1,000, Dahlia $2,500, Rose $5,000, Life $10,000.

Elyria

THE LORAIN COUNTY HISTORICAL SOCIETY - THE HICKORIES, 509 Washington Ave., Elyria, OH 44035-5128. Mailing Address: 284 Washington Ave., Elyria, OH 44035. Tel.: 440-322-3341. Fax: 440-322-2817.
E-mail: lchs@lchs.org
Web Site: lchs.org
Founded: 1889.
Congressional District: 13
Key Personnel: Exec. Dir. & C.E.O., William Bird; Pres. (V), Steve Chavez;

Office Mgr., Linda Greenaway; Collections & Research Asst., Bill Kies; Education Coord., Janet Bird; Education Coord., Jim Smith; Archivist, Eric Greenly.
Personnel Profile: Full-Time Paid 2; Part-Time Paid 4; Part-Time Volunteers 70.
Governing Authority: society. Parent Institution: Lorain County Historical Society, Inc. Subsidiary Institution: The Loran County History Center, 284 Washington Ave., Elyria, OH 44035. Tax-exempt: 501(c)(3).
Institution Type/Description: History Museum.
Collections: photographic images 1842-present; costumes & textiles 1860-1950; original artwork representing local people & scenes and by local artists; artifact collection, 1800-1950; late 19th- & early 20th-century household furnishings, woodworking & textile production tools & equipment, glassware; toys. Historic Buildings: 1894-95 The Hickories, Elyria; 1857 Starr House, Elyria.
Research Fields: genealogy; local history; family history; social history (family & home) 1890-1920.
Facilities: library & local history archives.
Activities: museum tours; permanent & temporary exhibitions; school programs on & off site; workshops & seminars for members, guests & area historical groups; special events & fundraisers.
Publications: quarterly newsletter, Hickory Leaves.
Hours & Admission Prices: Tours: Tues.-Fri. 1-4, Sat. 1-3; group & school tours by appointment. Adults $5, youth 13-18 $3, children 6-12 $2; discounts to groups; members of Northeastern Ohio Inter Museum Council, LCHS members and employees & their immediate family no charge. Closed holidays.
Attendance: 4,000 (estimated)
Membership: Student (with ID) $15; Individual $25; Family $35; Sustaining $60; Patron $125; Heritage $250; Garford Club $500.

THE LORAIN COUNTY HISTORICAL SOCIETY - THE LORAIN COUNTY HISTORY CENTER, 284 Washington Ave., Elyria, OH 44035. Tel.: 440-322-3341.
E-mail: lchs@lchs.org
Web Site: lchs.org
Governing Authority: Parent Institution: Lorain County Historical Society. Subsidiary Institution: The Hickories, 509 Washington Ave., Elyria, OH.
Institution Type/Description: History Center: housed in an Italianate-style mansion; built in 1857. Listed on the National Register of Historic Places.
Collections: local history & culture; period furnishings & artifacts;
Activities: guided tours.
Hours & Admission Prices: Tues.-Fri. 10-4, Sat. 1-3; other times by appointment. Adults $5, youth 13-18 $3, children 6-12 $2; children under 6 & members no charge. &
Membership: Student (with ID) $15; Individual $25; Family $35; Sustaining $60; Patron $125; Heritage $250; Garford Club $500.

Euclid

NATIONAL CLEVELAND-STYLE POLKA HALL OF FAME AND MUSEUM, 605 E. 222nd St., Euclid, OH 44123-2031. Tel.: 216-261-3263. Fax: 216-261-4134.
E-mail: polkashop@aol.com
Web Site: www.polkafame.com
Founded: 1987.
Key Personnel: C.O.O. & Pres., Joseph Valencic.
Personnel Profile: Part-Time Paid 4; Interns 1.
Governing Authority: Parent Institution: American Slovenia Polka Foundation. Tax-exempt.
Institution Type/Description: Music History Museum.
Collections: Polka history; Frank Yankovic's costumes & accordion; Johnny Vadnal's accordion; personal artifacts; photographs; videos.
Major Exhibits: The Women of Polka, 3/13-3/14; All Star Accordions: Celebrity Squeezeboxes and Their Stories, 10/13-9/14.
Research Fields: music history; American immigration history; popular culture.
Activities: Museum Sponsors: 11th Slovenian Sausage Festival in September; 27th Annual Awards Show Gala in November; Thanksgiving Polka Party Weekend in November.
Publications: quarterly.
Hours & Admission Prices: Tues.-Wed. & Fri. 12-5, Sat. 10-3. No charge; donations accepted.
Membership: Annual $18.

Fairport Harbor

FAIRPORT MARINE MUSEUM, (M), 129 Second St., Fairport Harbor, OH 44077-5816. Tel.: 440-354-4825.
E-mail: fhhs@ncweb.com
Web Site: www.fairportlighthouse.com
Founded: 1945.
Congressional District: 11
Key Personnel: Pres. (V), Dan Maxson; Museum Shop Mgr. (V), Mary A. Gladding.
Personnel Profile: Part-Time Volunteers 20.
Volunteer Hours: 4,000
Operating Expenses: 21,000
Operating Income: 20,000
Governing Authority: society. Affiliated with Fairport Harbor Historical Society. Tax-exempt.
Institution Type/Description: Marine Museum: housed in the 1871 Fairport Lighthouse & keeper's residence.
Collections: navigation instruments; marine charts; manuscripts; pictures & paintings of ships; lanterns; lighthouse lens; ship carpenters' tools; models & half-hulls of ships; iron ore; Native American artifacts; pilot house from Laker Frontenac.
Research Fields: local history; Great Lakes merchant vessels.
Facilities: 300-vol. library of marine books available for use by appointment.
Activities: guided tours; permanent exhibitions.
Publications: newsletter; brochure; blog, www.news-herald.com/medialab or www.fairportlighthouse.com
Hours & Admission Prices: Memorial Day to mid-Sept. Wed., Sat.-Sun. & legal holidays 1-6; group tours by appointment. Adults $4, senior citizens & children 6-12 $3; discounts to military, AAM, ICOM & Intermuseum Council members; members & children under 6 no charge. &
Attendance: 4,000 (accurate)
Membership: Regular $12; Couple $20; Benefactor $25; Life $250.

FINNISH HERITAGE MUSEUM, 301 High St., Fairport Harbor, OH 44077. Mailing Address: P.O. Box 1121, Fairport Harbor, OH 44077. Tel.: 440-352-8301. Facebook: Finnish Heritage Museum.
E-mail: info@finnishheritagemuseum.org
Web Site: finnishheritagemuseum.org
Founded: 2002.
Congressional District: 14
Key Personnel: Pres. (V), Lasse Hiltunen; Museum Shop Mgr., Sue Troutman.
Governing Authority: nonprofit organization. Tax-exempt.
Institution Type/Description: Heritage Museum.
Collections: Finnish heritage, culture, & history; photographs; personal artifacts; period furnishing.
Hours & Admission Prices: Spring, Summer & Fall Wed. & Sun. 1-4, Sat. 9-3; Winter Sat. 9-3. Adults $3, youth 13-17 $1.50; children 12 & under no charge.
Attendance: 1,500 (estimated)

Fairview Park

CLEVELAND METROPARKS OUTDOOR EDUCATION DIVISION, 4500 Valley Parkway, Fairview Park, OH 44126. Tel.: 440-331-8517. Fax: 440-331-8555.
E-mail: outdooreduc@clevelandmetroparks.com
Web Site: www.clevelandmetroparks.com
Founded: 1917.
Congressional District: 20
Key Personnel: Exec. Dir., Brian M. Zimmerman; Dir. Outdoor Education, Wendy Weirich; Div. Sec., Maryellen Dombek; Naturalist & Artist, Jennifer Brumfield; Mgr. Nature Center (Brecksville), Sharon Hosko; Naturalist-Brecksville, Jenny McClain; Naturalist-Brecksville, Kelly McGinnis; Naturalist-Brecksville, John Miller; Naturalist-Brecksville, Pam Taylor; Naturalist (Canal Way), Jill Hauger; Cultural History Interpreter, Karen Lakus; Historical Interpreter (Canal Way), Doug Kusak; Mgr. Nature Center (Garfield Park), Carl Casavecchia; Naturalist (Garfield Park), Stacey Allen; Naturalist (Garfield Park), LaDonna Sifford; Naturalist (Garfield Park), Beth Whiteley; Mgr. Outdoor Recreation (OR), Dana Smith; Outdoor Recreation Inclusion Specialist (OR), Rachel Nagle; Mgr. Look About Lodge, Barb Holtz; Naturalist (Look About Lodge), Carly Martin; Naturalist (Look About Lodge), Stefanie Verish; Outreach Mgr., Nature Tracks, Demetrius Lambert-Falconer; Education Specialist (Nature Tracks), Mario Jackson; Education Specialist (Nature Tracks), Rachel Mazzola; Mgr. Nature Center (North Chagrin), Barb Holtz; Naturalist (North Chagrin), Angelec Hillsman; Naturalist (North Chagrin), Deborah Marcinski; Naturalist (North Chagrin), Mindy Murdock; Naturalist (North Chagrin), Jeffrey Riebe; Naturalist (North Chagrin), Traci Williams; Mgr. Nature Center (Rocky River), Valerie Fetzer; Naturalist (Rocky River), Jen Bromfield; Naturalist (Rocky River), Min Sui Keung; Naturalist (Rocky River), Gretchen Motts; Naturalist (Rocky River), Joni Norris; Naturalist (Rocky River), Kathleen Schmidt; Mgr. Youth Outdoors, John Rode; Recreation Specialist (Youth Outdoors), Brian Fyfe; Recreation Specialist (Youth Outdoors), Valerie Hearst; Recreation Specialist (Youth Outdoors), Andy

Hudak; Recreation Specialist (Youth Outdoors), Faruq Abdul-Khaliq; Recreation Specialist (Youth Outdoors), Ashley Rossetti.
Personnel Profile: Full-Time Paid 32; Part-Time Paid 39; Part-Time Volunteers 918.
Governing Authority: political subdivision, State of Ohio. Cleveland Metroparks System. Nature Centers & Services: Brecksville Nature Center; Canal Way Visitor Center; North Chagrin Nature Center at North Chagrin; Rocky River Nature Center; Garfield Park Nature Center; Look-About Lodge; Nature Tracks mobile nature center; Watershed Stewardship Center. Tax-exempt.
Institution Type/Description: Nature Center & Conservation Area.
Collections: botanical; zoological.
Research Fields: white-tailed deer population dynamics; oak-hickory forest restoration; prairie restoration; marsh restoration.
Facilities: six nature & visitor centers located in regional parks; education complex at Cleveland Metroparks Zoo & Watershed Stewardship Center.
Activities: guided tours; lectures; gallery talks; study clubs; TV & radio programs; formally organized education programs; temporary exhibitions.
Publications: monthly newsletter, Emerald Necklace; trail maps.
Hours & Admission Prices: Trails: daily 6am-11pm. Nature Centers: daily 9:30-5. No charge. Nature Centers: closed New Year's Day; Easter; Thanksgiving; Christmas. &
Attendance: 473,508 (estimated)

Findlay

HANCOCK HISTORICAL MUSEUM, 422 W. Sandusky St., Findlay, OH 45840-3222. Tel.: 419-423-4433. Fax: 419-423-2154.
E-mail: reference@hancockhistoricalmuseum.org
Web Site: www.hancockhistoricalmuseum.org
Founded: 1970.
Congressional District: 4
Key Personnel: Exec. Dir., Neil Allen; Pres., Roger Criblez; Cur., Renee Smith; Archivist, Joy Bennett; Accountant, Carrie Glass.
Personnel Profile: Full-Time Paid 1; Part-Time Paid 4; Part-Time Volunteers 95; Interns 5.
Governing Authority: nonprofit organization. Branch Museums: Log Cabin; Red School House; De Wald-Funk House. Tax-exempt.
Institution Type/Description: Local History Museum.
Collections: 1875-1900, art glass; 1886-1889, pattern glass made in Findlay, Ohio; 1916 Grant car; camera & dolls; miniature train. Historic Buildings: 1880 Hull House; one room schoolhouse; log house; De Wald-Funk House; Little Red Schoolhouse.
Research Fields: Findlay Glass, Andrews Raiders (Civil War), women of Hancock County; Hancock County history; ethnic roots of Hancock County.
Activities: guided tours; permanent & temporary exhibitions; miniature train rides.
Publications: newsletter; reprints of local histories, atlases; Museum Herstory: Voices From the Past Women of Hancock County, Ohio; pictorial history of Hancock Co.; Ethnic Roots of Hancock County, Ohio; Families and Facades, Divided By A River; Porches & Parlors; Back to Sandusky Street.
Hours & Admission Prices: Wed.-Sat. 10-4, Sun. 1-4; tours available by appointment. Adults $5, seniors $3; children $2; members no charge. &
Attendance: 40,000 (accurate)
Membership: Individual $40-$74; Contributing $75-$249; Sustaining Friends $250-$499; Supporting Friends $500-$999; Heritage $1,000.

MAZZA MUSEUM, Gardner Fine Arts Pavilion, University of Findlay, Findlay, OH 45840-3653. Mailing Address: The University of Findlay, 1000 N. Main St., Findlay, OH 45840-3653. Tel.: 800-472-9502, ext. 5521. Fax: 419-434-6480.
E-mail: teeple@findlay.edu
Web Site: www.mazzamuseum.org
Founded: 1982.
Key Personnel: Dir., Benjamin E. Sapp; Museum Shop Mgr., Lee Myers.
Personnel Profile: Full-Time Paid 4.
Governing Authority: Parent Institution: The University of Findlay. Tax-exempt.
Institution Type/Description: Art Museum.
Collections: over 7,500 works of art; children's picture book art; literacy.
Activities: guided tours; special events.
Hours & Admission Prices: Wed.-Fri. 12-5, Sun. 1-4; other times by appointment. No charge; donations accepted. Closed major holidays. &
Attendance: 10,000
Membership: Individual $25.

Fort Recovery

FORT RECOVERY MUSEUM, (M), 1 Fortsite St., Fort Recovery, OH 45846. Mailing Address: P.O. Box 533, Fort Recovery, OH 45846. Tel.: 419-375-4649.
Web Site: www.fortrecoverymuseum.com
Founded: 1982.
Congressional District: 8
Key Personnel: Pres. (V), Helen LeFevre.
Personnel Profile: Part-Time Paid 2; Part-Time Volunteers 10.
Governing Authority: society. Parent Institution: Ohio Historical Society, Columbus, 43211. Tel.: 614-466-1500. Tax-exempt: 170(b)(1)(A).
Institution Type/Description: Military & Indian Museum: housed in partially reconstructed Anthony Wayne fort with two blockhouses.
Collections: relics from 1791, battle of St. Clair's defeat; relics of 1794, Wayne's victory; accoutrements; military artifacts; Indian artifacts. Historic Structures: two blockhouses; partial fort buildings; blacksmith shop.
Activities: school tours. Museum Sponsors: Indian Artifacts Show in June.
Publications: Ft. Recovery Historical Sketch; VHS documentary, St. Clair's Defeat.
Hours & Admission Prices: May & Sept. Sat.-Sun. 12-5; June-Aug. daily 12-5. Adults $3, students $1; members, OHS members, children 5 & under no charge.
Attendance: 3,346 (estimated)
Membership: Richard Butler Patron $10; Anthony Wayne Patron $25; St. Clair Patron $50; Presidential Patron $100.

Fostoria

FOSTORIA AREA HISTORICAL MUSEUM, 123 W. North St., Fostoria, OH 44830-2232. Mailing Address: P.O. Box 142, Fostoria, OH 44830-0142.
Founded: 1972.
Congressional District: 26
Key Personnel: C.E.O. & Treas., Leonard Skonecki; Pres. (V), George A. Gray.
Personnel Profile: Part-Time Volunteers 13.
Governing Authority: private; nonprofit organization. Parent Institution: Fostoria Area Historical Society. Tax-exempt.
Institution Type/Description: General Museum.
Collections: furniture; personal artifacts; fire equipment.
Facilities: 14,000 sq. ft. exhibit space. Museum-related items for sale.
Activities: films; guided tours; lectures; broadcast programs. Annual Events: 8 monthly meetings open to members & public.
Publications: monthly bulletin, Historical Activities Summary.
Hours & Admission Prices: Temporarily closed. &
Attendance: 1,000 (estimated)
Membership: Individual $15; Family $25; Corporate $30; Life $200.

GLASS HERITAGE GALLERY, 109 N. Main St., Fostoria, OH 44830-2215. Tel.: 419-435-5077.
E-mail: museum@fostoriaglass.com
Web Site: fostoriaglass.com
Founded: 1992.
Key Personnel: Pres. (V), William King.
Personnel Profile: Part-Time Volunteers 20.
Volunteer Hours: 1,320
Governing Authority: Parent Institution: Fostoria Ohio Glass Assoc. Tax-exempt.
Institution Type/Description: Glass Museum.
Collections: glassmaking history & industries; local industry glass, 1887-1920; vases; lamps; pitchers.
Research Fields: Fostoria Ohio Glass manufacturers.
Publications: quarterly newsletter, Victoria Views.
Hours & Admission Prices: March Thurs.-Sat. 10-3; April-Dec. Wed.-Sat. 10-4; other times by appointment. No charge; donations accepted. &
Attendance: 300 (estimated)
Membership: Single $20; Couple $32; Life Single $300; Life Couple $500.

Franklin

HARDING MUSEUM OF THE FRANKLIN AREA HISTORI-CAL SOCIETY, 302 Park Ave., Franklin, OH 45005-3549. Tel.: 937-746-8295.
E-mail: mnenninger@aaa-alliedgroup.com
Web Site: www.franklinmuseums.org
Founded: 1965.
Congressional District: 3
Key Personnel: Pres., Robert Bowman; Museum Mgr., Mary Nenninger.
Personnel Profile: Part-Time Paid 1; Part-Time Volunteers 15.

Operating Expenses: 14,050
Operating Income: 15,000
Governing Authority: society; nonprofit. Parent Institution: Franklin Area Historical Society. Subsidiary Institution: 1804 Log Post Office, Franklin, OH. Tax-exempt: 501(c)(3).
Institution Type/Description: Local History Museum: housed in 1901 home of Major General E.F. Harding located in Franklin's Mackinaw Historic District.
Collections: Franklin area history; Franklin in 1913 Flood; Gen. E. Forrest Harding's personal artifacts.
Major Exhibits: St. Mary Church: 100 Years, 10/13-3/14.
Research Fields: history of Franklin & Clear Creek Townships; Franklin in 1913 Flood; military history; World War II; Nuremberg Trials.
Facilities: library of local history & military books available for research on premises.
Activities: guided tours; permanent exhibitions; ghost classes; overnight tours of 1804 log post office. Annual Event: Christmas Ornament Sale.
Publications: FAHS newsletter; books, History of Franklin in the Great Miami Valley, Franklin, Images of America, Guide of Franklin in the 1913 Flood; tour pamphlets for city and Mackinaw Historic District; walking tour booklet.
Hours & Admission Prices: April-Nov. Sun. 2-5; other times by appointment. Adults $3; members no charge.
Attendance: 1,000 (estimated)
Membership: Student $5; Individual $25; Family $40; Business $50; Bronze up to $99; Silver $100-$199; Gold $200-$499; Diamond $500 & up.

Fremont

✱ RUTHERFORD B. HAYES PRESIDENTIAL CENTER, (M), Spiegel Grove, Fremont, OH 43420-2796. Tel.: 419-332-2081. Fax: 419-332-5424.
E-mail: tculbertson@rbhayes.org
Web Site: www.rbhayes.org
Founded: 1916.
Congressional District: 5
Key Personnel: Exec. Dir., Thomas J. Culbertson; Pres. (V), Stephen A. Hayes; Cur. Manuscript, Nan Card; Head Librarian, Rebecca Hill; Head Photographic Resources, Gilbert Gonzalez; Dir. Devel., Kathy Boukissen; Museum Shop Mgr., Merry May.
Personnel Profile: Full-Time Paid 16; Part-Time Paid 22; Part-Time Volunteers 145.
Governing Authority: nonprofit organization. Affiliated with The Ohio Historical Society, Columbus, OH 43211. Tax-exempt: 501(c)(3).
Institution Type/Description: U.S. Library & Museum.
Collections: Hayes family memorabilia including Mrs. Hayes' wedding gown & White House reception gowns; the President's carriage; Civil War relics; Fanny Hayes's two doll houses; Hayes family portraits; 5,000 linear ft. of manuscripts; Indian relics; weapons; White House dining room sideboard; White House china; 75,000 photographs from 1840-present; memorabilia collected by Colonel Webb C. Hayes; 240 periodicals & newsletters; 6,300 newspaper volumes; 6,100 reels of microfilm. Historic Buildings: 1859 Hayes residence; 1873 Dillon House; 1870s carriage house, service buildings; White House gates 1873; tomb of Lucy & Rutherford B. Hayes.
Research Fields: American history 1865-1914, emphasis on Gilded Age; Ohio & local history; Spanish-American War; monetary & prison reform; African American history.
Facilities: 70,000-vol. library of material on American history available for inter-library loan, including 12,000-vol. of President Hayes' personal library for use under staff supervision; reading room; auditorium; 25-acre estate; gardens; walking & jogging paths; Indian portage & Harrison military trail. Museum-related items for sale.
Activities: guided tours; lecture series; school programs; slide presentations; music programs; nature walks.
Publications: quarterly newsletter, The Statesman; monthly newsletter for members & volunteers, Billet.
Hours & Admission Prices: Museum & Residence: Tues.-Sat. 9-5, Sun. & holidays 12-5. Adults $13, children 6-12 $5; discounts to groups with appointment, senior citizens, active military & their families, AAA members, Civil War Trust & Natl. Historical Society; children under 6, AAM, AASLH, OHS & HPC members & grounds no charge. Closed New Year's Day; Easter; Thanksgiving; Christmas. Library: Tues.-Sat. 9-5. No charge. Closed holidays. ♿
Attendance: 40,137 (accurate)
Membership: Student $20; Individual $30; Individual Plus $45; Family & Grandparent $50; Patron $100; Representative $250; Cabinet $500; Advisor $1,000; Executive $1,500.

SANDUSKY COUNTY HISTORICAL SOCIETY - HOLDER-MAN HOUSE, 514 Birchard Ave., Fremont, OH 43420. Tel.: 419-332-0303.
Formerly: Sandusky County Pioneer and Historical Association
Congressional District: 88 & 26
Key Personnel: Pres. (V), Earl Wammes.
Governing Authority: Tax-exempt.
Institution Type/Description: Historical Society Museum.
Collections: local history & culture; period furnishings; personal artifacts; photographs.
Activities: special events.
Hours & Admission Prices: May-Nov. Wed. & Sun. 1-4; other times by appointment. No charge; donations accepted. Closed Thanksgiving. ♿
Membership: Individual $10; Lifetime $250.

Gahanna

THE CENTER FOR CIVIL WAR PHOTOGRAPHY, (M), Gahanna, OH 43230. Mailing Address: 947 E. Johnstown Rd. #161, Gahanna, OH 43230-1851. Tel.: 614-656-2297.
E-mail: info@civilwarphotography.org
Web Site: www.civilwarphotography.org
Founded: 1999.
Key Personnel: Pres., Bob Zeller; Vice Pres., Garry Adelman; Exec. Dir., Jennifer Kon; Dir., Justin A. Shaw; Dir. Devel., Charles Morrongiello; Dir. Imaging, John Richter; Sec. & Treas., Ronald Perisho.
Personnel Profile: Part-Time Paid 1; Part-Time Volunteers 10.
Volunteer Hours: 550
Operating Expenses: 77,623
Operating Income: 98,154
Governing Authority: private; nonprofit organization. Tax-exempt: 501(c)(3).
Institution Type/Description: Virtual Museum.
Collections: 1,000 negatives of Gettysburg Battlefield, 1930s-1950s; slide-mounted Civil War stereographs; Civil War books, artwork, video & audiotape; photographs.
Research Fields: Civil War photography.
Activities: school outreach program; lectures; temporary exhibitions; video & audiotape presentations. Annual Event: Civil War Photography Seminar in October.
Publications: semiannual newsletter, The Battlefield Photographer; 99 Historic Photographs of Culp's Hill; 99 Historic Images of Fredericksburg and Spotsylvania Civil War Sites; 99 Historic Images of Richmond Civil War Sites, 99 Historic Images of Civil War Washington, 99 Historic Images of Civil War Petersburg; 99 Historic Images of Harpers Ferry; 99 Historic Images of Civil War Charleston; 99 Historic Images of Gettysburg; 99 Historic Images of the Civil War in the West; Manassas Battlefields Then & Now; Antietam in 3-D; Gettysburg in 3-D.
Hours & Admission Prices: Online museum.
Membership: Student $25; Individual $35; Family $50; Ambrotype $100; Daguerrotype $250; Folio $500.

Galion

GALION HISTORICAL MUSEUM, 132 S. Union St., Galion, OH 44833-2524. Mailing Address: P.O. Box 125, Galion, OH 44833-0125. Tel.: 419-468-9338. Facebook: Galion Historical Society.
E-mail: galionhistory@gmail.com
Web Site: www.galionhistory.com
Founded: 1955.
Congressional District: 8
Key Personnel: Dir., Kristen Hoffert; Pres., Mary Court.
Personnel Profile: Part-Time Paid 1.
Governing Authority: nonprofit. Parent Institution: Galion Historical Society, Inc. Tax-exempt: 501(c)(3).
Institution Type/Description: History Museum.
Collections: Galion & Crawford county history from prehistoric to modern times; Indian artifacts; pioneer items; local industrial items; 1906 horse-drawn fire engine.
Facilities: Museum-related items for sale.
Activities: guided tours; temporary exhibitions of your own collections; regularly changing exhibitions.
Publications: weekly newspaper columns on local history; quarterly newsletter.
Hours & Admission Prices: June-Oct. Sun. 1-4. House Tour: $5; discount to Blue Start Museum members. ♿
Attendance: 500 (estimated)
Membership: Students $5, Individual $10, Family $15, Sustaining $25, Contributing $50, Patron & Corporate $100.

GALION HISTORICAL SOCIETY & BROWNELLA COTTAGE, 201 S. Union St., Galion, OH 44833-2524. Mailing Address: P.O. Box 125, Galion, OH 44833-0125. Tel.: 419-468-9338.
E-mail: galionhistory@gmail.com
Web Site: www.galionhistory.com
Founded: 1955.
Congressional District: 8
Key Personnel: Pres., Craig Clinger; Dir., Amber L. Wertman; Vice Pres., Dan Brown; Treas., Brenda Treisch; Sec., Thomas Palmer.
Personnel Profile: Full-Time Paid 1.
Governing Authority: society; nonprofit. Branch Museum: Galion Historical Society, Inc. Tax-exempt: 501(c)(3).
Institution Type/Description: Historic House: Home & Study of Bishop William Montgomery Brown; Galion Historical Museum.
Collections: artifacts; personal belongings; furniture; area historical items.
Facilities: study containing printed works of Bishop William Montgomery Brown.
Activities: guided tours; lectures.
Publications: quarterly newsletter, The Historian.
Hours & Admission Prices: June-Oct. Fri. & Sun. 1-4; groups year-round by appointment. Adults $5, students $3.
Attendance: 500 (accurate)
Membership: Students $5, Individual $10, Family $15, Sustaining $25, Contributing $50, Patron & Corporate $100.

Gallipolis

FRENCH ART COLONY, 530 First Ave., Gallipolis, OH 45631-1245. Mailing Address: P.O. Box 472, Gallipolis, OH 45631-0472. Tel.: 740-446-3834. Fax: 740-446-3834.
E-mail: info@frenchartcolony.org
Web Site: www.frenchartcolony.org
Founded: 1971.
Congressional District: 6
Key Personnel: Bd. Member, Jan Thaler; Bd. Member, Peggy Evans; Exec. Dir., Joseph Wright; Chm. (V), Amy Weaver.
Personnel Profile: Full-Time Paid 2; Full-Time Volunteers 4; Part-Time Paid 1; Part-Time Volunteers 80.
Governing Authority: nonprofit organization. Tax-exempt: 501(c)(3).
Institution Type/Description: Art Gallery: housed in 1855 Holzer Family Home.
Collections: private collections of fine art & antiques.
Research Fields: design arts; OAC pilot study.
Facilities: library of visual art; classrooms; gardens.
Activities: guided tours; lectures; gallery talks; concerts; classes; arts festivals; hobby workshops; educational programs; rental facilities; live theater; music performance.
Publications: bimonthly newsletter, Currents.
Hours & Admission Prices: Galleries: Tues.-Fri. 9-4, Sat. 10-3, Sun. 1-5. No charge; donations accepted. &
Attendance: 7,000 (estimated)
Membership: Student $15; Senior Citizen $27; Individual $30; Family $50; Supporter $100; Donor $500; Benefactor $1,000.

OUR HOUSE STATE MEMORIAL, (M), 432 First Ave., Gallipolis, OH 45631. Mailing Address: P.O. Box 607, Gallipolis, OH 45631-0607. Tel.: 740-446-0586.
E-mail: amasara@suddenlink.net
Web Site: ohiohistory.org/places/ourhouse
Founded: 1933.
Congressional District: 10
Key Personnel: Pres., Sara Sheets; Chm. (V), Carol Warren; Site Mgr., Becky Pasquale.
Personnel Profile: Full-Time Paid 1; Part-Time Paid 1; Part-Time Volunteers 5.
Governing Authority: society. Parent Institution: Ohio Historical Society, Columbus, OH 43211. Subsidiary Institution: Friends of Our House. Tax-exempt.
Institution Type/Description: History Museum: housed in 1819 restored Ohio River tavern.
Collections: period furnishings; Native American artifacts & arrowheads.
Research Fields: history of Gallipolis; Ohio French settlement.
Activities: guided tours; special events; Victorian teas; birthday parties; luncheons & dinners; children's day camp. Museum Sponsors: History Day; Colonial Days; Founder's Day.
Hours & Admission Prices: Memorial Day-Labor Day Wed.-Sat. 10-4, Sun. 1-4. No charge; donations accepted.
Attendance: 500 (estimated)

Gambier

KENYON COLLEGE OLIN ART GALLERY, (M), Olin Library, Gambier, OH 43022. Tel.: 740-427-5346.
E-mail: youngerd@kenyan.edu
Web Site: www2.kenyon.edu/artgallery/info/general.htm
Key Personnel: Dir., Dan Younger
Institution Type/Description: Art Gallery.
Collections: works by regional, national & international artists.
Activities: permanent & temporary exhibits; educational programs; lectures.
Publications: brochures; exhibition catalogues.
Hours & Admission Prices: Mon.-Fri. 10-8, Sat.-Sun. 10-5.

Gates Mills

GATES MILLS HISTORICAL SOCIETY, 7580 Old Mill Rd., Gates Mills, OH 44040. Mailing Address: P.O. Box 191, Gates Mills, OH 44040-0191. Tel.: 440-423-4808.
E-mail: mganselmo@roadrunner.com
Web Site: gatesmillshistoricalsociety
Founded: 1946.
Congressional District: 22
Key Personnel: Pres., Marcia Anselmo; Treas., Helen Gelbach; Sec., Sherry Levering.
Governing Authority: society. Tax-exempt: 501(c)(3).
Institution Type/Description: History Museum.
Collections: interior furnishings; kitchen utensils; Native Indian.
Research Fields: Local genealogy & Western Reserve.
Facilities: library. Decorative maps, postcards & books for sale.
Activities: permanent exhibitions.
Publications: books, 1826-1976 A Pictorial History of Gates Mills; 1920-1970, George Brown of Gates Mills; Bill Henderson's Gates Mills.
Hours & Admission Prices: By appointment. No charge; donations accepted.
Attendance: 1,500
Membership: Regular $25; Family $60; Archivist $100; Benefactor $250 & up.

Geneva

PLATT R. SPENCER MEMORIAL ARCHIVES AND SPECIAL COLLECTIONS AREA AT THE GENEVA PUBLIC LIBRARY, Geneva Public Library, 860 Sherman St., Geneva, OH 44041-9101. Tel.: 440-466-4521, ext. 107 & 109. Fax: 440-466-0162.
E-mail: legezalo@oplin.org
Web Site: www.acdl.info/archives
Founded: 1988.
Key Personnel: Dir., William Tokarczyk; Archivist, Louise Legeza.
Personnel Profile: Part-Time Paid 3; Part-Time Volunteers 6.
Governing Authority: county government. Parent Institution: Ashtabula Co. District Library.
Institution Type/Description: Archive.
Collections: Ashtabula County history & genealogy including over 3,200 family files; postcards; phonetic spelling movement books; Platt R. Spencer family memorabilia, penmanship items & business artifacts; Archie Bell manuscripts & published travel books; letters & photos of turn-of-the-century actors, opera singers & authors; Edith M. Thomas poetry; children's texts; newspapers; court house records; obituaries; The Panny Magazine (published 1830s); Godey's Lady's book & magazine (published 1860s); The Penny encyclopedia (1850s); The Times History of the War (WWI) (1917); 105 vols. The Century Illustrated Monthly Magazine.
Research Fields: local history; genealogy; local authored poetry & books; phonetic spelling movement; Platt R. Spencer & penmanship: family memorabilia.
Facilities: 2,500-vol. collection of selected U.S., foreign history & biographical/history books, bound periodicals, 1723-1774.
Activities: participatory exhibits.
Publications: brochures: Ashtabula County Genealogical Society quarterly newsletter.
Hours & Admission Prices: Call for hours. Private tours on Tues. by appointment. No charge; donations accepted. Closed New Year's Eve & Day; Easter; Memorial Day; Independence Day; Labor Day; Thanksgiving; Christmas Eve & Day. &
Attendance: 1,000 (estimated)

SHANDY HALL, 6333 S. Ridge Rd., Geneva, OH 44041-8377. Mailing Address: Western Reserve Historical Society, 10825 East Blvd., Cleveland, OH 44106. Tel.: 216-721-5722.
Web Site: www.wrhs.org
Founded: 1937.

Congressional District: 11
Key Personnel: Pres. & C.E.O. WRHS, Gainor Davis; Chm. WRHS, Donald Dailey; Sr. Vice Pres. Interpretation & C.O.O., Kelly Falcone.
Personnel Profile: Full-Time Paid 1.
Governing Authority: society. Owned & Operated by Western Reserve Historical Society, 10825 East Blvd., Cleveland 44106. Tax-exempt: 170(b)(1)(A).
Institution Type/Description: Historic House: 1815 Shandy Hall, early Western Reserve home.
Collections: original Harper family furnishings; clothes; books; toys.
Activities: guided tours.
Hours & Admission Prices: By appointment.
Membership: Individual $50; Family $70; Sustaining $150; Partner $250; Fellow $500.

Geneva-on-the-Lake

ASHTABULA COUNTY HISTORICAL SOCIETY, 5865 Lake Rd., Geneva-on-the-Lake, OH 44041-9427. Mailing Address: P.O. Box 36, Jefferson, OH 44047-0036. Tel.: 440-466-7337.
E-mail: nan@alltel.net
Web Site: www.ashtcohs.com
Founded: 1838.
Congressional District: 11
Governing Authority: society. Tax-exempt: 501(c)(3).
Institution Type/Description: House Museum & Landmark: 1811 Blakeslee Log Cabin, 1823 Jennie Munger Gregory Memorial Museum, Geneva-on-the-Lake; 1823 Joshua R. Giddings law office.
Collections: period furnishings 1850-1900; office furnishings & library; manuscripts; documents; pictures; postcards; archival papers.
Research Fields: historical.
Facilities: library of historical books & 800-1860 Congressional documents.
Activities: guided tours; lectures; permanent & temporary exhibitions; dedications of century plaques on century homes, buildings & businesses. Museum Sponsors: Log Cabin days in September.
Publications: quarterly bulletin, Ashtabula County History.
Hours & Admission Prices: May & July-Sept. Wed.-Fri. 12-4; June Wed.-Sat. 12-4. Adults $4, children $2.
Attendance: 1,000 (estimated)
Membership: Individual $15; Family $20; Business $25; Patron $50; Life $200.

Georgetown

GRANT SCHOOLHOUSE, 508 S. Water St., Georgetown, OH 45121. Mailing Address: US Grant Homestead Association, 112 N. Water St., Georgetown, OH 45121. Tel.: 937-378-3087.
E-mail: baileyho@frontier.com
Founded: 1972.
Congressional District: 2
Key Personnel: Pres. (V), Stan Purdy
Governing Authority: Parent Institution: Ohio Historical Society.
Institution Type/Description: Historic Building: housed in the one room schoolhouse that Ulysses attended from the ages of 6 to 13; built in 1829.
Collections: local history & culture; period furnishings; photographs.
Hours & Admission Prices: Memorial Day to Labor Day Wed.-Sun. 12-5; Sept.-Oct. Sat.-Sun. 12-5; groups by appointment. Adults $3, children 6-12 $1. Admission covers both Grant Schoolhouse & Grant Boyhood home.
Attendance: 500 (estimated)

U.S. GRANT BOYHOOD HOME, 219 E. Grant Ave., Georgetown, OH 45121. Mailing Address: US Grant Homestead Association, 112 N Water St., Georgetown, OH 45121. Tel.: 937-378-3087; 877-372-8177.
E-mail: baileyho@frontier.com
Web Site: usgrantboyhoodhome.org
Founded: 1972.
Congressional District: 2
Key Personnel: Dir., Chm. (V) & Pres. (V), Stan Purdy; Museum Shop Mgr., Nancy Purdy.
Personnel Profile: Part-Time Paid 4; Part-Time Volunteers 1.
Volunteer Hours: 200
Governing Authority: Parent Institution: Ohio Historical Society. Subsidiary Institution: U.S. Grant Boyhood Home Assoc. Tax-exempt.
Institution Type/Description: Historic House Museum: housed in the boyhood home of Ulysses S. Grant from 1823-1839. Listed on the National Register of Historic Places.
Collections: Grant family history; period furnishings; Grant & Georgetown memorabilia; photographs.
Hours & Admission Prices: Memorial Day to Labor Day Wed.-Sun. 12-5;

Sept. Oct. Sat.-Sun. 12-5; other times & groups by appointment. Adults $3, children 6-12 $1; members no charge. Admission covers both Grant Schoolhouse & Grant Boyhood Home. &
Attendance: 500 (estimated)
Membership: Individual $10.

Glenford

FLINT RIDGE STATE MEMORIAL MUSEUM, (M), 7091 Brownsville Rd., S.E., Glenford, OH 43739-9639. Tel.: 740-787-2476; 800-283-8707.
E-mail: meweingartner@ohiohistory.org
Web Site: ohsweb.ohiohistory.org/places/c01/index.shtml
Founded: 1933.
Congressional District: 10
Key Personnel: Mgr., M.E. Weingartner; Education Specialist, Hapi Cummons.
Personnel Profile: Full-Time Paid 4.
Governing Authority: society. Parent Institution: The Ohio Historical Society, Ohio Historical Center Site Operations, 1985 Velma Ave., Columbus, OH 43211. Tax-exempt.
Institution Type/Description: Natural History Museum: located on the site of prehistoric flint pit.
Collections: geology exhibits; flint deposits; geologic time scale; flint relics & artifacts; displays on prehistoric man.
Facilities: nature trails; facilities & trail for visually impaired & disabled; picnic area. Museum-related items for sale.
Publications: pamphlet.
Hours & Admission Prices: Memorial Day-Labor Day Sat.-Sun. 12-5. Adults $4, children 6-12 $3; discounts to school groups, seniors & AAA members; members and children 5 & under 6 no charge. &
Attendance: 7,158 (accurate)
Membership: Student $25; Senior 60 & over $35; Individual $40; Full Senior 60 & over and up to 8 guests $42; Full $47; Life $1,000.

Gnadenhutten

GNADENHUTTEN HISTORICAL PARK & MUSEUM, 352 S. Cherry St., Gnadenhutten, OH 44629. Tel.: 740-254-4143. Fax: 740-254-4992.
E-mail: gnadmuse@gnaden.com
Founded: 1963.
Congressional District: 18
Governing Authority: society. Parent Institution: Gnadenhutten Historical Society. Tax-exempt.
Institution Type/Description: Historical Park Museum.
Collections: c.1772-1875 Indian artifacts; historical manuscripts of missions & village; memorial monument; replica of log cabin church & Indian leader's cabin; mannikin, John Heckwelder.
Activities: lectures; permanent exhibitions.
Publications: Massacre At Gnadenhutten.
Hours & Admission Prices: May Tues. & Sun. 1-5, Sat. 10-5; June-Oct. Tues.-Sat. 10-5, Sun. 1-5; other times by appointment. No charge, donations accepted. &
Attendance: 10,000 (estimated)

Gomer

GOMER WELSH COMMUNITY MUSEUM, 7365 Gomer Rd., Gomer, OH 45809. Tel.: 419-999-5820 & 642-5911.
Institution Type/Description: History Museum.
Collections: Welsh family history, heritage, & culture; personal artifacts; oral histories; photographs.
Hours & Admission Prices: 2nd & 4th Sun. of month 1:30-4. No charge.

Granville

DENISON MUSEUM, (M), 240 W. Broadway, Granville, OH 43023-1120. Mailing Address: P.O. Box 810, Granville, OH 43023-0810. Tel.: 740-587-6255. Fax: 740-587-5628.
E-mail: harlachers@denison.edu
Web Site: denisonmuseum.org
Formerly: Denison University Art Gallery
Founded: 2006.
Congressional District: 17
Key Personnel: Dir., Sherry Harlacher; Cur. Collections, Anna Cannizzo; Lead Data Specialist, Sarah Baker.
Personnel Profile: Full-Time Paid 3; Part-Time Paid 1; Interns 12.
Operating Expenses: 172,757

Governing Authority: university. Parent Institution: Denison University. Tax-exempt.
Institution Type/Description: University Museum.
Collections: Asian textiles, paintings, sculpture, lacquerware & ceramics; European and American prints, paintings & drawings; Kuna Indian artifacts; Burmese art.
Major Exhibits: Personal Space, 9/13-12/14; Green Revolution, 2/14-5/14.
Research Fields: Asian, European, American & Kuna Indian art.
Facilities: museum laboratory.
Activities: traveling & permanent exhibitions; national & state arts councils & exhibiting artists; lectures; guided tours.
Publications: exhibition catalogs.
Hours & Admission Prices: Thurs. 12-7, Fri.-Wed. 12-5. No charge; donations accepted. Closed university holidays. &

Attendance: 1,495 (accurate)

GRANVILLE HISTORICAL SOCIETY MUSEUM, 115 E. Broadway, Granville, OH 43023-1303. Mailing Address: P.O. Box 129, Granville, OH 43023-0129. Tel.: 740-587-3951.
E-mail: granvillehistorical@gmail.com
Web Site: www.granvillehistory.org
Founded: 1885.
Congressional District: 12
Key Personnel: Pres. (V), Kevin Bennett; Vice Pres. (V), Cynthia Cort.
Personnel Profile: Part-Time Paid 2; Part-Time Volunteers 40.
Governing Authority: nonprofit. Parent Institution: Granville Historical Society. Tax-exempt: 501(c)(3).
Institution Type/Description: History Museum: housed in 1816 Alexandrian Bank.
Collections: village records; artifacts of local nature; family & business papers; carpenter tools prior to 1850; history; industry; agriculture; archaeology; costumes; glass; decorative arts; geology; military; textiles.
Research Fields: limited archival research.
Facilities: Publications for sale.
Activities: guided tours; permanent exhibitions.
Publications: pamphlets & books on local subjects of interest; quarterly newsletter, Historical Times; monthly electronic newsletter, Modern Times.
Hours & Admission Prices: Mid-April to Dec. Fri. 12-3, Sat. 10-4, Sun. 1-4; other times by appointment. No charge; donations accepted. &
Attendance: 1,500 (estimated)
Membership: Family $40.

ROBBINS HUNTER MUSEUM, AVERY-DOWNER HOUSE, (M), 221 E. Broadway, Granville, OH 43023-1305. Mailing Address: P.O. Box 183, Granville, OH 43023-0183. Tel.: 740-587-0430. Fax: 740-587-0430.
E-mail: annlowder@robbinshunter.org
Web Site: www.robbinshunter.org
Founded: 1981.
Congressional District: 12
Key Personnel: Pres., Don DeSapri; Dir., Ann K. Lowder.
Personnel Profile: Part-Time Paid 2; Part-Time Volunteers 50.
Governing Authority: bd. of governors. Parent Institution: Licking County Historical Society, P.O. Box 785, 6th St. Park, Newark, OH 43055. Tel.: 614-345-4898. Subsidiary Institution: Board of Governors of Robbins Hunter Museum. Tax-exempt: 501(c)(3).
Institution Type/Description: Historic House: 1842 American Greek Revival house, designed by Minard Lafever.
Collections: period furnishings; artifacts pertaining to the 19th century, life & times of Granville, Ohio 1840-1870.
Research Fields: 19th-century decorative arts & furnishings; American Greek Revival architecture & interior design.
Facilities: banquet facilities.
Activities: guided tours; loan, permanent & traveling exhibitions; rental facilities.
Publications: brochures; local history newsletter, exhibit catalogs; monthly newsletter.
Hours & Admission Prices: April 3-Dec. 20 Wed.-Sat. 1-3. No charge; donations accepted. Closed Independence Day; Thanksgiving; Christmas. &
Attendance: 12,000 (estimated)
Membership: Senior & Student $25; Individual $35; Family & Grandparent $50; Contributor $100; Supporter $200; Patron $500.

Greenville

GARST MUSEUM, (M), 205 N. Broadway, Greenville, OH 45331-2222. Tel.: 937-548-5250. Fax: 937-548-7645.
E-mail: information@garstmuseum.org
Web Site: www.garstmuseum.org

Formerly: Darke County Historical Society
Founded: 1903.
Congressional District: 8
Key Personnel: Pres. (V), John F. Marchal; Vice Pres., Richard Brown; Dir., Clay Johnson, Ph.D.; Treas., Allen Hauberg; Museum Shop Mgr., Brenda Arnett.
Personnel Profile: Full-Time Paid 2; Part-Time Paid 12; Part-Time Volunteers 60; Interns 1.
Governing Authority: society. Parent Institution: The Darke County Historical Society, Inc. Tax-exempt: 2954.
Institution Type/Description: History Museum: housed in 1852 Inn.
Collections: Annie Oakley; Native American; Treaty of Greenville; Lowell Thomas; American antiques; military history; 1890s to 1940s shops; farm equipment.
Research Fields: Darke County history; Annie Oakley & Treaty of Greenville; genealogy; Native Americans.
Facilities: library.
Activities: guided tours; films; arts festivals; permanent & temporary exhibitions; educational field trips; docent tours; speakers. Annual Events: Veterans Day Panel; Annie Oakley Days; Christmas Open House; Gathering at Garst.
Publications: books, newsletters & pamphlets relating to Annie Oakley; Anthony Wayne & Treaty of Greenville; The Autobiography of Annie Oakley; Frank Butler, The Man Behind the Woman.
Hours & Admission Prices: Feb.-Dec. Tues.-Sat. 10-4, Sun. 1-4. Adults $8, senior citizens 60 & over $7, youth 6-17 $5; discounts to AAM & AAA members; members no charge. Closed New Year's Eve & Day; Easter; Independence Day; Thanksgiving; Christmas Eve & Day. &
Attendance: 19,000 (accurate)
Membership: Individual $30; Family $50; Copper $100; Bronze $250; Silver $500; Gold $1,000; Platinum $2,000.

Groveport

MOTTS MILITARY MUSEUM, (M), 5075 S. Hamilton Rd., Groveport, OH 43125-9336. Tel.: 614-836-1500.
E-mail: info@mottsmilitarymuseum.org
Web Site: www.mottsmilitarymuseum.org
Founded: 1988.
Congressional District: 12
Key Personnel: Dir. & Archivist, Warren E. Motts; Chm. (V), Bo Hindall; Financial Dir., Ronald Albers; Devel., Gerrit Vanstraten; Public Rels. & Museum Shop Mgr., Daisy Motts; Museum Shop Mgr., Ken Macklin.
Personnel Profile: Full-Time Paid 1; Full-Time Volunteers 4; Part-Time Volunteers 60.
Governing Authority: nonprofit. Tax-exempt: 501(c)(3).
Institution Type/Description: Military Museum.
Collections: historical military items from the Civil War, WWI, WWII, Korea, Vietnam & Desert Storm.
Research Fields: Vietnam soldiers' graves to place grave stones.
Facilities: 1,000-vol. library of military history books; video histories, research classrooms. Museum-related items for sale.
Activities: guided tours; temporary, traveling & loan exhibitions. Annual Events: Community, Veteran & Educational programs; Christmas Party.
Publications: quarterly newsletter, From the Trenches.
Hours & Admission Prices: Tues.-Sat. 9-5, Sun. 1-5. Adults $5, senior citizens $4, students $3; discounts to groups; children 5 & under no charge. Closed national holidays. &
Attendance: 10,000 (estimated)
Membership: Student $15; Individual $35; Family $45; Corporate $100; Individual Lifetime $500; Family Lifetime $750; Veterans Corporate Life $1,000; Corporate Life $1,500.

Hamilton

BUTLER COUNTY HISTORICAL SOCIETY, (M), 327 N. 2nd St., Hamilton, OH 45011-1651. Tel.: 513-896-9930. Fax: 513-896-9936.
E-mail: bcomuseum@fuse.net
Web Site: www.bchistoricalsociety.com
Formerly: Butler County Museum
Founded: 1934.
Congressional District: 8
Key Personnel: Pres. (V), Greg Young; Dir., Kathy Creighton.
Personnel Profile: Full-Time Paid 1; Part-Time Paid 1; Part-Time Volunteers 15.
Governing Authority: nonprofit organization. Parent Institution: Butler County Historical Society. Tax-exempt.
Institution Type/Description: History Museum: housed in Benninghofen mansion; built 1861.

Collections: glassware; china; musical instruments; toys; dolls; Indian artifacts; original interior architectural elements; period furnishings from mid-19th century through the 1900s; local & regional history; industry; military artifacts.
Research Fields: genealogy; Southwest Ohio & Butler County history.
Facilities: 300-vol. research library pertaining to history & genealogy; auditorium.
Activities: guided tours; permanent & temporary exhibitions; local history programs.
Publications: newsletter, Benninghofen Post.
Hours & Admission Prices: Tues.-Wed. 11-4, Fri. 9-4, Sat. 9-2; groups by appointment. No charge; donations accepted. &
Attendance: 1,500 (estimated)
Membership: Senior $15; Individual $20; Household $30; Corporate Sponsorship $350.

PYRAMID HILL SCULPTURE PARK & MUSEUM, 1763 Hamilton Cleves Rd., St. Rd. 128, Hamilton, OH 45013-9601. Tel.: 513-868-8336. Fax: 513-868-3585.
E-mail: pyramid@pyramidhill.org
Web Site: www.pyramidhill.org
Founded: 1997.
Congressional District: 8
Key Personnel: Dir., H.T. Wilks.
Personnel Profile: Full-Time Paid 6; Full-Time Volunteers 1; Part-Time Paid 8; Part-Time Volunteers 15; Interns 1.
Governing Authority: Tax-exempt.
Institution Type/Description: Sculpture Park & Museum.
Collections: large monumental works of sculpture; over 90 pieces of ancient sculpture.
Facilities: amphitheater; pavilion; tea room.
Activities: concert series; children's programs.
Publications: quarterly newsletter.
Hours & Admission Prices: April-Oct. Mon.-Fri. 8-5, Sat.-Sun. 8-6; Nov.-March Mon.-Fri. 8-5, Sat.-Sun. 10-5. Adults $8, children $1.50; members no charge. &
Attendance: 112,366 (accurate)
Membership: Individual $40; Family $45; Contributor $125; Patron $250; Sponsor $500; Benefactor $1,000; Ambassador $2,500; Founder's Society $5,000.

Harrison

AMERICAN WATCHMAKERS-CLOCKMAKERS INSTITUTE, 701 Enterprise Dr., Harrison, OH 45030-2164. Tel.: 513-367-9800. Fax: 513-367-1414.
E-mail: jordan@awci.com
Web Site: www.awci.com
Founded: 1960.
Congressional District: 2
Key Personnel: Exec. Dir., James E. Lubic.
Personnel Profile: Part-Time Volunteers 2.
Governing Authority: nonprofit organization. Subsidiary Institution: Education Library Charitable Trust. Tax-exempt: 501(c)(6).
Institution Type/Description: Horological Display.
Collections: watches; clocks; tools of the trade; chronometers.
Facilities: 3,000-vol. library pertaining to horology.
Activities: guided tours by appointment only.
Publications: Horological Times.
Hours & Admission Prices: Mon.-Fri. 8-5 by appointment only. Adults $3, senior citizens $2; discounts to AAM & ICOM members; school groups, children 12 & under, AWCI members no charge. Closed national holidays. &
Attendance: 300 (estimated)
Membership: Individual $169; Industry $500.

VILLAGE HISTORICAL SOCIETY OF HARRISON, INC., Governor Othniel Looker Home, 10580 Marvin Rd., Harrison, OH 45030. Mailing Address: P.O. Box 419, Harrison, OH 45030-0419. Tel.: 513-367-9285.
E-mail: mlsmith6@cinci.rr.com
Founded: 1962.
Congressional District: 2
Key Personnel: Pres. (V), Mary Lou Smith; Chm. (V), Russell Radcliffe; Vice Pres., John Anthony; Recording Sec., John Mohr; Corresponding Sec., Ann L. Woelfel; Corresponding Sec., Nancy Gibson; Treas., Linda Losekamp.
Personnel Profile: Part-Time Volunteers 10.
Governing Authority: society. Tax-exempt: 501(c)(3).
Institution Type/Description: Historic House: 1804 Othniel Looker Home. Listed on the National Register of Historic Places.

Collections: local history & culture; period furnishings; photographs.
Facilities: Museum-related items for sale.
Activities: monthly speakers; displays; films; readings.
Hours & Admission Prices: May-Sept. third Sun. of each month 1:30-4; other times & tours by appointment. No charge; donations accepted.
Membership: Single $10; Family $15.

Heath

THE GREAT CIRCLE EARTHWORKS, 455 Hebron Rd., Heath, OH 43056. Mailing Address: 1982 Velma Ave., Columbus, OH 43211-2453. Tel.: 800-589-8224. Fax: 740-345-4403.
E-mail: communications@ohiohistory.org
Formerly: Moundbuilders State Memorial & Museum
Congressional District: 10
Personnel Profile: Full-Time Paid 4; Part-Time Volunteers 4.
Governing Authority: society. The Ohio Historical Society. Ohio Historical Center, 1982 Velma Ave., Columbus, OH 43211. Tax-exempt.
Institution Type/Description: Prehistoric Indian Art Museum & Historical Site: embankment 1,200 feet in diameter with earthen walls 8-14 feet in height enclosing 26 acres; comprise The Great Circle Earthworks, ceremonial grounds of prehistoric Hopewell Indians, 1000 B.C.-700 A.D.
Collections: art objects & other media representing achievements of the Adena & Hopewell cultures 1000 B.C.-700 A.D; prehistoric knives, pottery, pipes & jewelry.
Facilities: picnic area.
Hours & Admission Prices: Mon.-Fri. 8:30-5, Sat. 10-5, Sun. 1-5. No charge; donations accepted. Closed New Year's Eve & Day; Thanksgiving & day after; Christmas Eve & Day. &
Attendance: 7,000 (estimated)
Membership: Ohio Historical Society: Individual $35; Family $55.

Hilliard

EARLY TELEVISION FOUNDATION AND MUSEUM, 5396 Franklin St., Hilliard, OH 43026. Tel.: 614-771-0510.
E-mail: info@earlytelevision.org
Institution Type/Description: Communication Museum.
Collections: television history from 1920s to 1950s; over 150 TV sets; picture tubes; TV studio equipment; flying spot scanner TV camera.
Hours & Admission Prices: Sat. 10-6, Sun. 12-5. Suggested Donations: adults $5, children over 6 $2.

Hillsboro

HIGHLAND HOUSE MUSEUM, 151 E. Main St., Hillsboro, OH 45133-1450. Tel.: 937-393-3392.
Founded: 1965.
Governing Authority: society. Parent Institution: Highland County Historical Society. Tax-exempt.
Institution Type/Description: Historic House: 1844 Highland House.
Collections: furnished typically of the era; furniture; china; glassware; tools; documents; newspapers; Indian relics; early memorabilia from the 1874, Crusade Against Intoxicating Liquid; receipts of trade oddities.
Facilities: rental facilities.
Activities: rental facilities.
Publications: books, Blackburns: Today & Yesterday; Bearers of the Pioneer Spirit, The McAnallys; Highland County Pictorial History.
Hours & Admission Prices: Fri. 1-5, Sun. 1-4; other times by appointment. No charge; donations accepted.
Attendance: 1,500 (estimated)
Membership: Single $10; Family $25; Sustaining $25.

Holland

J.H. FENTRESS ANTIQUE POPCORN MUSEUM, 7922 Hill Ave., Holland, OH 43528. Tel.: 419-308-4812.
E-mail: info@antiquepopcornmuseum.com
Web Site: www.antiquepopcornmuseum.com
Key Personnel: Dir., Jim Fentress
Institution Type/Description: History Museum.
Collections: Holcombe & Hoke Manufacturing Company history, advertising, & artifacts; Butter-Kist popcorn & peanut machines; popcorn-related memorabilia, tins & boxes.
Hours & Admission Prices: By appointment. No charge.

Hudson

HUDSON LIBRARY AND HISTORICAL SOCIETY, 96 Library St., Hudson, OH 44236-5122. Tel.: 330-653-6658. Fax: 330-650-3373.
E-mail: archives3@hudson.lib.oh.us
Web Site: www.hudson.lib.oh.us
Founded: 1910.
Congressional District: 14
Key Personnel: Dir. & Cur., E. Leslie Polott; Pres. Bd. Trustees (V), Deborah Baker-Hall; Archivist, Gwendolyn Mayer.
Personnel Profile: Full-Time Paid 26; Part-Time Paid 20; Part-Time Volunteers 15; Interns 1.
Governing Authority: society. Tax-exempt.
Institution Type/Description: Regional History Museum.
Collections: historical objects & documents relating to the history of Hudson Township; archives; manuscripts; John Brown family papers. Historic House: 1833 Frederick Baldwin House.
Facilities: 4,000-vol. library of history & genealogy books available for use on premises only.
Activities: guided local history tours.
Publications: Books to Bytes; Yesterday In Hudson.
Hours & Admission Prices: Mon.-Thurs. 9-9, Fri.-Sat. 9-5, Sun. 12-5. No charge. Closed New Year's Eve & Day; Easter; Memorial Day; Independence Day; Labor Day; Thanksgiving; Christmas Eve & Day.

Ironton

LAWRENCE COUNTY GRAY HOUSE MUSEUM, 506 S. 6th St., Ironton, OH 45638-1825. Mailing Address: P.O. Box 73, Ironton, OH 45638-0073. Tel.: 740-532-1222.
Founded: 1988.
Congressional District: 10
Key Personnel: Pres. (V), Patricia Arrington; Treas., Herbert Brown; Museum Shop Mgr., Peggy Karshner.
Personnel Profile: Part-Time Volunteers 30.
Governing Authority: nonprofit organization. Parent Institution: Lawrence County Historical Society. Tax-exempt.
Institution Type/Description: General Museum.
Collections: charcoal iron furnaces; Hanging Rock Iron region artifacts; stone cutting tools; Indian cooking utensils; vintage clothing; period furniture; Lyons railroad photos; Lawrence Co. sports legends; Capt. Brotherton riverboats; Nannie Kelly Wright collection; Oros sheet music; John Rankin room; Col. & Mrs. George Gray furniture; vintage photographs.
Research Fields: old country cemeteries.
Activities: auctions; appraisal fairs. Annual Events: Easter Egg Hunt; Spring Tea in April; Christmas Walk in Dec.
Publications: monthly newsletters; books, 1892 in Ironton; Folk Lore & Legends; local poetry.
Hours & Admission Prices: mid-April to mid-Dec. Fri.-Sun. 1-4. No charge, donations accepted. &
Attendance: 1,723 (accurate)
Membership: Single $10; Family $15; Friend $50; Patron $100; Life $1,000; Joint Life $1,100.

Jackson

LILLIAN JONES MUSEUM, (M), 75 Broadway St., Jackson, OH 45640-1610. Tel.: 740-286-2556.
E-mail: lejmuseum@yahoo.com
Web Site: lillianjones.museum.com/home.html
Founded: 1995.
Key Personnel: Dir., Megan Malone; Cur., Rhonda Woolum.
Personnel Profile: Full-Time Paid 1; Part-Time Paid 1; Part-Time Volunteers 1.
Governing Authority: Parent Institution: City of Jackson, OH; Subsidiary Institution: Jackson Museum Board. Tax-exempt.
Institution Type/Description: History Museum.
Collections: period artifacts; furnishings; Jackson County history.
Facilities: main museum; Carriage House Genealogical Center.
Activities: special events; research.
Hours & Admission Prices: Tues.-Thurs. 10-3; other times by appointment. No charge; donations accepted. Closed major holidays. &
Attendance: 500 (accurate)
Membership: Youth $5; Individual $15; Family $25; Contributing $50; Preservationist $100; Business/Corporate $200; Life $500; Benefactor $1,000.

Kalida

PUTNAM COUNTY HISTORICAL SOCIETY MUSEUM, 201 E. Main St., Kalida, OH 45853. Mailing Address: P.O. Box 264, Kalida, OH 45853-0264. Tel.: 419-532-3008. Fax: 419-532-2944.
E-mail: pchs@bright.net
Web Site: www.bright.net/~pchs/
Founded: 1873.
Congressional District: 5
Key Personnel: Pres., Joe Balbaugh; Vice Pres., Janis Lentz; Chm. (V) & Cur., Carol Wise; Recording Sec., Ruth Oglesbee; Corresponding Sec., Lori Ann Hemenway; Treas., Shirley Berelsman; Membership, Deb Carder.
Personnel Profile: Part-Time Paid 2; Part-Time Volunteers 35.
Governing Authority: nonprofit organization. Parent Institution: Putnam County Historical Society. Tax-exempt.
Institution Type/Description: Historical Building: housed in 1901 old Methodist church.
Collections: c.1900 local artifacts; items for genealogical research.
Research Fields: genealogy; historical.
Facilities: 250-vol. library of local history & adjoining counties available for research on premises.
Activities: guided tours; speeches. Society Sponsors: Christmas Open House; work shops; fair booth; bicentennial program.
Publications: quarterly newsletter, Putnam County Heritage; books, Centennial History 1873-1973; Blizzard of 1978 in Putnam County, Ohio; Historical History 1880 & Historical Atlas of Putnam County 1895; Pioneer Reminiscences 1878 & 1887; 1896 Portrait & Biographical Record of Putnam County; One-Room Schools in Putnam County 1985; Oral History; 1834-1934 Centennial of Putnam County.
Hours & Admission Prices: Sun. 1-4, Wed. 9-12. No charge; donations accepted. &
Attendance: 2,000 (estimated)
Membership: Individual $10; Life $150.

Kent

*** KENT STATE UNIVERSITY MUSEUM,** 515 Hilltop Dr., Kent, OH 44242. Mailing Address: P.O. Box 5190, Kent, OH 44242. Tel.: 330-672-3450. Fax: 330-672-3218.
E-mail: museum@kent.edu
Web Site: www.kent.edu/museum/
Founded: 1982.
Congressional District: 17
Key Personnel: Dir., Jean Druesedow; Mgr. Collections & Museum Registrar, Joanne Fenn; Administrative Asst., Stephanie Roach.
Personnel Profile: Full-Time Paid 6; Part-Time Paid 20; Part-Time Volunteers 27.
Governing Authority: university. Parent Institution: Kent State University. Subsidiary Institution: KSU Foundation. Tax-exempt: 501(c)(3).
Institution Type/Description: Costume & Decorative Arts Museum: housed in 1927 building, first library of the University.
Collections: history of Western dress, 1750 to present; 19th-20th century American glass; Asian & African regional dress; 18th-20th century ceramics; 17th-20th century furniture; 17th-century to modern textiles; 19th-century English, French & German fashion periodicals.
Research Fields: costume history; textiles.
Facilities: 2,500-vol. costume & decorative art reference library; 120-seat auditorium; eight galleries. Museum-related items for sale.
Activities: guided tours; docent program; loan & temporary exhibitions; organized educational programs for undergraduate college students affiliated with Kent State University.
Publications: brochures; catalogues.
Hours & Admission Prices: Wed. & Fri.-Sat. 10-4:45, Thurs. 10-8:45, Sun. 12-4:45. Adults $5, senior citizen $4, children $3; discounts to AAM & ICOM members; children under 7 no charge. Annual pass $25. Closed university & national holidays. &
Attendance: 12,439 (accurate)
Membership: Annual $25; Supporter $100; Platinum Circle $1,000.

KENT STATE UNIVERSITY, SCHOOL OF ART GALLERIES, Kent, OH 44242. Mailing Address: P.O. Box 5190, Kent, OH 44242. Tel.: 330-672-7853. Fax: 330-672-4729.
E-mail: galleries@kent.edu
Web Site: galleries.kent.edu
Founded: 1950.
Congressional District: 11
Key Personnel: Dir., Christine Havice.
Personnel Profile: Full-Time Paid 1; Part-Time Paid 12; Part-Time Volunteers 1; Interns 2.
Governing Authority: state. Parent Institution: Kent State University. Branch

Galleries: Michener Gallery; Downtown Gallery; William H. Eells Art Gallery; Collection Gallery; Student Galleries. Tax-exempt: 501(c)(3).
Institution Type/Description: Art School and Gallery.
Collections: paintings; sculpture; prints; graduate theses; decorative arts; Hazel Janicki & William Schock collection; James A. Michener collection; Milton Adams collection.
Research Fields: contemporary American art.
Facilities: lectures; guided tours.
Activities: performances; lectures; special events; formally organized education programs; undergraduate & graduate students exhibitions.
Publications: booklets: Contemporary American, Canadian & European Enamelists; Leadership Artifacts of West Africa; Objects in Clay, Fiber, Glass & Metal; Indonesian Textiles; Contemporary Platinum Prints & Photographs; An Interview with Patrick Ireland; West African Textiles & Dress; A New Generation of Ohio Artists; Contemporary Woven Work: America & Abroad; European Glass; Akron & Kent Painters: 1940-1970; The Cleveland Enamelists: 1930-1955; Response to the City: Photography & Sculpture; Robert Smithson's Partially Buried Woodshed; 15th Annual Collage National Exhibit; Print in Enamel; West African Men's Wearing: Design & Technique.
Hours & Admission Prices: Tues.-Fri. 11-5. No charge; donations accepted. Closed school holidays. &
Attendance: 26,000 (accurate)
Membership: Student & Senior Citizen $10; Individual $15; Family $25; Sponsor $50; Benefactor $100; Patron $500.

Kenton

HARDIN COUNTY HISTORICAL MUSEUMS, INC., 223 N. Main St., Kenton, OH 43326-1505. Tel.: 419-673-7147.
E-mail: hardincountymuseums@windstream.net
Web Site: www.hardinmuseums.org
Founded: 1991.
Congressional District: 4
Key Personnel: Dir., Linda Iams.
Personnel Profile: Part-Time Paid 1; Part-Time Volunteers 100.
Governing Authority: nonprofit. Parent Institution: Hardin County Historical Museums, Inc. Tax-exempt.
Institution Type/Description: History Museum.
Collections: local history; Native American artifacts; military artifacts; Kenton Hardware toys; 1st Medal of Honor winner Jacob Parrott militaria; pioneer crafts & tools; D.A.R. memorabilia; Charles Shanafelt Collection of relics & natural curiosities; farming implements; one-room schoolhouse; Fred Machetanz Alaskan art & literature.
Research Fields: local history.
Facilities: archives.
Activities: guided tours; lectures; audiovisual programs; walking tours; permanent & temporary exhibitions; school visitation.
Hours & Admission Prices: Tues.-Fri. 1-4. No charge; donations accepted. Closed holidays. &
Attendance: 500 (estimated)
Membership: Individual $25; Family $35; Business $100; Life $500.

Kettering

SOCIETY FOR THE PRESERVATION OF BRITISH TRANSPORTATION IN AMERICA, INC. DBA BRITISH TRANSPORTATION MUSEUM, (M), 2304 Wrenside Lane, Kettering, OH 45440-2324. Tel.: 937-985-7204; 434-1750.
E-mail: britcarmuseum@aol.com
Web Site: www.britishcarmuseum.org
Founded: 1998.
Congressional District: 3
Key Personnel: Chm. (V), Chuck White; Pres. (V), Peter Stroble; Vice Pres. & Public Rels., Paul Rich; Treas., Archivist & Museum Shop Mgr., Richard R. Smith; Educ., Amanda Hawker.
Personnel Profile: Full-Time Volunteers 5; Part-Time Volunteers 59.
Governing Authority: private; nonprofit organization. Parent Institution: Society for the Preservation of British Transportation in America, Inc. Subsidiary Institution: British Transportation Museum. Tax-exempt: 501(c)(3).
Institution Type/Description: Transportation Museum.
Collections: British transportation history & modes; 22 cars; 9 British bicycles; British outboard motor; British vehicle models; British automotive art work; photographs; paintings; memorabilia; periodicals; repair & parts manuals; books; posters.
Research Fields: British transportation vehicles, their drivers & manufacturers.
Facilities: library of books, manuals, & periodicals; 2,000 sq. ft. exhibit space.
Activities: guided tours; temporary exhibitions. Annual Events: British Car Meet in May; British Car Day in August; British Car Day at Dayton Concours d'Elegance in September.

Publications: bimonthly newsletter, British Automobile; annual corporate report; members subscription, British Marque Car Club News.
Hours & Admission Prices: Dec.-March call for hours. No charge; donations accepted.
Attendance: 300 (estimated)
Membership: Individual $25; Family $45; Contributing $100; Sustaining $250; Bronze $500; Silver $1,000; Gold $1,500.

Kidron

KIDRON SONNENBERG HERITAGE CENTER, 13153 Emerson Rd., Kidron, OH 44636-0234. Mailing Address: P.O. Box 234, Kidron, OH 44636. Tel.: 330-857-9111.
E-mail: kidronheritagecenter@hotmail.com
Web Site: kidronhistoricalsociety.org
Founded: 1994.
Key Personnel: Dir., Prudy Steiner.
Governing Authority: Parent Institution: Kidron Community Historical Society. Tax-exempt.
Institution Type/Description: History Museum.
Collections: local history & culture; period furnishings & tools; travel documents; personal artifacts.
Facilities: Gift items for sale.
Hours & Admission Prices: April-Dec. Thurs. & Sat. 11-3. No charge; donations accepted.

Kirtland

* **THE HOLDEN ARBORETUM,** 9500 Sperry Rd., Kirtland, OH 44094-5172. Tel.: 440-946-4400. Fax: 440-602-3857.
E-mail: holden@holdenarb.org
Web Site: www.holdenarb.org
Founded: 1931.
Congressional District: 11
Key Personnel: Dir., Clem Hamilton; Chm. Bd. (V), Joe Mahovlic; Dir. Finance, Jim Ansberry; Dir. Horticulture & Conservation, Roger Gettig; Dir. Education, Paul C. Spector; Dir. Devel., Stephen Sedam; Dir. Guest Svcs., David A. Desimone; Dir. Human Resources, Nancy Spellman; Museum Shop Mgr., Kristie Hawley.
Personnel Profile: Full-Time Paid 52; Part-Time Paid 10; Part-Time Volunteers 700; Interns 7.
Governing Authority: nonprofit organization. Subsidiary Institution: David G. Leach Research Station. Tax-exempt: 501(c)(3).
Institution Type/Description: Arborctum.
Collections: native & cultivated woody plants, wildflowers, rare books & nutcrackers.
Research Fields: stress biology of urban forests.
Facilities: 9,000-vol. library on horticulture & botany available on premises by request; nature center; bird observation blind; classrooms. Educational literature & museum-related items for sale.
Activities: guided tours; lectures; formally organized education programs.
Publications: bimonthly magazine, Leaves, The Holden Arboretum Class & Events Magazine.
Hours & Admission Prices: Daily 9-5. Adults $6, senior citizens $5, children 6-12 $3; discounts to groups & AAM members; seniors on Tues., members & children under 6 no charge. Closed Thanksgiving; Christmas. &
Attendance: 87,437 (accurate)
Membership: Aspen & Buckeye $40; Maple $50.

N. K. WHITNEY STORE MUSEUM, 7800 Kirtland Chardon Rd., Kirtland, OH 44094. Tel.: 440-256-9805. Fax: 440-256-2692.
E-mail: vckirtland@ldschurch.org
Governing Authority: Parent Institution: LDS Church.
Institution Type/Description: History Museum: housed in a restored 1830s country store & post office.
Collections: period furnishings & artifacts; replicas of merchandise from more than 150 years ago.
Hours & Admission Prices: Mon.-Sat. 9am to dusk, Sun. 11:30 to dusk. No charge.
Attendance: 75,000 (accurate)

Lakewood

CLEVELAND ARTISTS FOUNDATION AT BECK CENTER FOR THE ARTS, 17801 Detroit Ave., Lakewood, OH 44107-3413. Tel.: 216-227-9507.
E-mail: laurenhansgen@clevelandartists.org
Web Site: www.clevelandartists.org
Founded: 1984.

Congressional District: 23
Key Personnel: Exec. Dir., Lauren Hansgen; Pres., James D. Gibans.
Personnel Profile: Full-Time Paid 1; Part-Time Paid 2; Part-Time Volunteers 40; Interns 2.
Governing Authority: nonprofit. Tax-exempt: 501(c)(3).
Institution Type/Description: Art Museum.
Collections: contemporary & historic art from Northeast Ohio.
Facilities: auditorium; theater; classrooms; gallery for Cleveland artists foundation.
Activities: gallery talks; workshops; lectures.
Hours & Admission Prices: Summer: Tues.-Fri. 1-5; Sept.-May Tues.-Thurs. 1-5, Fri.-Sat. 1-8. No charge; donations accepted. Closed major holidays. &
Attendance: 30,000 (estimated)
Membership: Individual $35; Family $60; Premier $100; Sustaining $250; Patron $500; Lifetime $2,500.

OLDEST STONE HOUSE MUSEUM, (M), 14710 Lake Ave., Lakewood, OH 44107-1353. Tel.: 216-221-7343. Fax: 216-221-0320.
E-mail: museum@lakewoodhistory.org
Web Site: www.lakewoodhistory.org
Founded: 1952.
Congressional District: 19
Key Personnel: Pres. (V), Kathy Haber, Jr.; Dir., Greg Palumbo.
Personnel Profile: Full-Time Paid 1; Part-Time Paid 2; Part-Time Volunteers 50.
Governing Authority: board of trustees; municipal; society; Lakewood Historical Society; nonprofit organization. Tax-exempt: 501(c)(3).
Institution Type/Description: Historic House Museum: 1838 Old Stone House.
Collections: period furnishings of the 1830s & 40s; slide collection of early photographs.
Research Fields: local landmark houses & early settlers.
Facilities: 300-vol. library of the early history of Ohio, research books relating to antiques & early textbooks available for use by members & on premises. Museum-related items for sale.
Activities: guided tours; lectures; formally organized education programs for children; permanent & temporary exhibitions; school loan service.
Publications: books, Romance in Lakewood Streets; A Child's Journal of Early Lakewood; Lakewood, The First Hundred Years; quarterly, Lakewood Historical Society Newsletter; Lakewood Lore.
Hours & Admission Prices: Feb.-Nov. Wed. 1-4, Sun. 2-5. No charge; donations accepted. Closed national holidays.
Attendance: 2,500 (accurate)
Membership: Senior Citizens $8; Family & Individual $15; Contributing $25; Sustaining $50; Patron $100; Benefactor $250.

Lancaster

DECORATIVE ARTS CENTER OF OHIO, (M), 145 E. Main St., Lancaster, OH 43130-3713. Tel.: 740-681-1423. Fax: 740-681-2713.
Web Site: www.decartsohio.org
Founded: 1997.
Congressional District: 7
Key Personnel: C.E.O., Julia C. Parke; Pres. (V), Sheila Heath; Dir. Education, Trisha Clifford-Sprouse; Dir. Operations, Andrea Brookover; Museum Shop Mgr., Barbara Jupin; Weekend Mgr., Maggie Conrad.
Personnel Profile: Full-Time Paid 3; Part-Time Paid 5; Part-Time Volunteers 50; Interns 2.
Governing Authority: private; nonprofit organization. Tax-exempt.
Institution Type/Description: Decorative Arts Museum.
Collections: 19th & 20th-centuries decorative arts.
Research Fields: decorative arts; crafts; domestic architecture.
Facilities: library; art studios & classrooms.
Activities: lectures; concerts; workshops; classes; temporary exhibitions.
Publications: semiannual member newsletter, News From the Center.
Hours & Admission Prices: Tues.-Sat. 10-4, Sun. 1-4. General Admission: no charge; donations accepted. Groups of 10 or more $3. Closed major holidays. &
Attendance: 10,463 (accurate)
Membership: Student K-12 and Senior 55 & over $25; Individual $35; Grandparent, Family & Household $55; Contributor $100; Supporter $250; Patron $500.

THE GEORGIAN MUSEUM, 105 E. Wheeling St., Lancaster, OH 43130-3706. Tel.: 740-654-9923. Fax: 740-654-9121.
E-mail: info@fairfieldheritage.org
Web Site: www.fairfieldheritage.org
Founded: 1976.

Congressional District: 10
Key Personnel: Dir., Myrna Figgins; Pres. (V), Joyce Harvey; Office Mgr., Karen S. Smith; Museum Shop Mgr., Janet McCafferty; Museum Shop Mgr., Delores Troup.
Personnel Profile: Part-Time Paid 6; Part-Time Volunteers 125.
Governing Authority: nonprofit organization. Owned & Operated by The Fairfield Heritage Association, Inc., 105 E. Wheeling St. Tax-exempt: 107(b)(1)(A).
Institution Type/Description: Historical Society Museum: housed in 1830-1832 The Georgian, Federal-style house with Regency features.
Collections: furniture; silver; costumes; tools; glass; Indian artifacts found in Fairfield county; Fairfield County historical items depicting lifestyle of affluent families from 1820-1850.
Research Fields: county survey of historic structures over 100 years old; preservation.
Facilities: reference library available for use on premises. Gift items including books pertaining to Fairfield County for sale.
Activities: lectures; films; gallery talks; tours of the museum, Square 13 historical district & early churches; reading room; hobby workshops; docent program; permanent, temporary & traveling exhibitions. Museum Sponsors: oral histories; fifth grade week; Christmas Candlelight Tour.
Publications: reprints of early histories; quarterly, Fairfield Heritage; books, Cross Roads & Fence Corners; Architecture & Arts of Fairfield County; Campfire to Courthouse; Sherman Family Chronicle; Covered Bridges of Fairfield County.
Hours & Admission Prices: Jan.-March by appointment only; April to mid-Dec. Tues.-Sun. 1-4. Adults $6, students 6-18 $1; discount to groups & seniors; children under 6 & members no charge. Closed holidays. &
Attendance: 7,500 (estimated)
Membership: Student 6-18 $10; Senior Citizen $15; Individual $20; Family $40; Sustaining $50; Fellow $75; Patron $100; Historian $150; Supporter $250; Curator $350; Partner $500; Archivist $650; Preservationist $1,000; Executive $1,500.

THE OHIO GLASS MUSEUM & GLASS BLOWING STUDIO, (M), 124 W. Main St., Lancaster, OH 43130-3756. Tel.: 740-687-0101. Fax: 740-687-0273.
E-mail: ohioglassmuseum@gmail.com
Web Site: ohioglassmuseum.org
Founded: 2002.
Congressional District: 7
Key Personnel: Dir., Douglas Ingram; Chm. (V), Margie Burket.
Personnel Profile: Full-Time Paid 1.
Governing Authority: Tax-exempt.
Institution Type/Description: Glass Museum.
Collections: glass industry & history; glass from local & national artists and companies including Heisey, Fenton, Imperial, Cambridge, Anchor-Hocking/Lancaster Glass, Lancaster Lens/Lancaster Glass, Gay-Fad, and Erickson.
Major Exhibits: Birds of a Feather...Captured in Glass, 9/14/13-3/14.
Facilities: theater.
Activities: film.
Hours & Admission Prices: March-Oct. Tues.-Sun. 1-4; Nov.-Feb. Tues.-Sat. 1-4; other times by appointment. Adults $6, seniors $5, students 6-18 $3; members no charge. &
Attendance: 6,820 (accurate)
Membership: Individual $40; Family $70; Patron $100; Sponsor $250; Benefactor $500.

SHERMAN HOUSE MUSEUM, 137 E. Main St., Lancaster, OH 43130-3713. Mailing Address: 105 E. Wheeling St., Lancaster, OH 43130-3706. Tel.: 740-687-5891. Fax: 740-654-9121.
E-mail: fairheritage@greenapple.com
Web Site: www.shermanhouse.com
Congressional District: 7
Key Personnel: Dir., Laura Bullock; Office Mgr., Karen S. Smith; Exec. Sec., Betty Ann Boone; Museum Shop Mgr., Janet McCafferty; Museum Shop Mgr., Delores Troup.
Personnel Profile: Part-Time Paid 1; Part-Time Volunteers 100.
Governing Authority: nonprofit organization. Parent Institution: The Fairfield Heritage Association, 105 E. Wheeling St. Tax-exempt: 107(b)(1)(A).
Institution Type/Description: Historical House.
Collections: furniture; silver; costumes; tools; glass; Indian artifacts found in Fairfield County; Fairfield County historical items; Civil War artifacts; Sherman family items; Civil War era art; quilts; coverlets; veteran & GAR artifacts; garden history.
Activities: lectures; films; gallery talks; tours of the museum; Square 13 historical district & early churches; oral histories; reading room; hobby workshops; docent program; permanent, temporary & traveling exhibitions.

Museum Sponsors: Civil War Roundtable; Annual Garden Party; fifth grade week; Christmas Candlelight tour.
Publications: reprints of early histories; quarterly, Fairfield Heritage; books, Cross Roads & Fence Corners; Architecture & Arts of Fairfield County; Campfire to Courthouse; Covered Bridges of Fairfield County; combination atlas map of Fairfield County, OH 1875; Fifty Six Miles into the Hills, The story of Lancaster Lateral & Hocking Canals; William T. Sherman Activity & Coloring Book.
Hours & Admission Prices: April to mid-Dec. Tues.-Sun. 1-4. Adults $6; discount to groups, seniors, students and AAA & AAM members; children under 6 & members no charge. Closed holidays. &
Attendance: 11,000 (estimated)
Membership: Senior Citizen $15; Individual $20; Family $40; Sustaining $50; Fellow $75; Patron $100; Historian $150; Supporter $250; Curator $350; Partner $500; Archivist $650; Preservationist $1,000; Executive $1,500.

Lebanon

GLENDOWER HISTORIC MANSION, 105 Cincinnati Ave., Lebanon, OH 45036-2117. Mailing Address: 105 S. Broadway, Lebanon, OH 45036-1707. Tel.: 513-932-1817. Fax: 513-932-8560.
E-mail: wchs@wchsmuseum.org
Web Site: wchsmuseum.org
Formerly: Glendower State Memorial
Founded: 1944.
Congressional District: 7
Key Personnel: Dir., Victoria V.H. Tappy.
Personnel Profile: Part-Time Paid 1; Part-Time Volunteers 12.
Governing Authority: Parent Institution: Warren County Historical Society. Tax-exempt.
Institution Type/Description: Historic Building Museum: restored Greek revival house.
Collections: 19th-century furniture & furnishings.
Activities: Museum Sponsors: Civil War Encampments; Concerts on Lawn; Christmas Programs.
Hours & Admission Prices: June to Labor Day Wed.-Sat. 12-4; call for Christmas hours; groups by appointment. Adults $5, children $3.50; WCHS members no charge.
Attendance: 1,384 (estimated)

WARREN COUNTY HISTORY CENTER, 105 S. Broadway, Lebanon, OH 45036-1707 Tel.: 513-932-1817. Fax: 513-932-8560.
E-mail: wchs@wchsmuseum.org
Web Site: wchsmuseum.org
Founded: 1940.
Congressional District: 7
Key Personnel: Exec. Dir., Victoria V.H. Tappy; Pres. (V), J. William Duning; Asst. Dir. & Historian, John J. Zimkus; Cur., Mary Klei.
Personnel Profile: Full-Time Paid 1; Full-Time Volunteers 1; Part-Time Paid 4; Part-Time Volunteers 40; Interns 2.
Governing Authority: society; nonprofit organization. Parent Institution: Warren County Historical Society. Tax-exempt: 501(c)(3).
Institution Type/Description: Local History Museum.
Collections: early artifacts from southwestern Ohio; paleontology; Shaker furniture and household articles; Native American artifacts; genealogy library.
Research Fields: genealogy; Shaker history.
Facilities: 2,000-vol. library of local history, biographies and genealogy books available for use on premises; reading room. Books, museum publication & hand-crafted items for sale.
Activities: guided tours; lectures; temporary & traveling exhibitions.
Publications: quarterly newsletter, Historicalog; local history booklets.
Hours & Admission Prices: Tues.-Sat. 9-4, Sun. 12-4. Adults $5, senior citizens 65 & over $4.50, children 5-18 $3.50; members no charge. Closed national holidays. &
Attendance: 3,057 (estimated)
Membership: Individual $35; Family $50.

Lexington

RICHLAND COUNTY MUSEUM, 51 Church St., Lexington, OH 44904-1258. Mailing Address: P.O. Box 3153, Lexington, OH 44904-0153. Tel.: 419-884-2230.
E-mail: woodsiewoman@yahoo.com
Web Site: www.richlandcountymuseum.org
Founded: 1966.
Congressional District: 4

Key Personnel: Pres., Jeffrey Mandeville; Treas., Loretta Hilliard; Sec., Shirley Addlesperger.
Personnel Profile: Part-Time Volunteers 20; Interns 1.
Governing Authority: society. Tax-exempt: 501(c)(3).
Institution Type/Description: General Museum: housed in 1800 school building.
Collections: history; agriculture; art; children's toys; Indian; medical; Jacquard coverlets; Richland County weavers.
Activities: guided tours; permanent exhibitions.
Publications: one-room schools booklet.
Hours & Admission Prices: May-Oct. Sun. 1:30-4:30; group tours by appointment. No charge; donations accepted. &
Attendance: 2,000 (estimated)
Membership: Individual $5; Family $10; Associate $25-$49; Benefactor $50-$99; Distinguished $100 & up; Corporate $1,000.

Lima

*** ALLEN COUNTY MUSEUM, (M),** 620 W. Market St., Lima, OH 45801-4665. Tel.: 419-222-9426. Fax: 419-222-0649.
Web Site: www.allencountymuseum.org
Founded: 1908.
Congressional District: 4
Key Personnel: Dir., Patricia F. Smith; Pres., William C. Timmermeister; Chm. (V), Sue Clover; Vice Pres., Richard Boehr; Cur. Manuscripts & Archives, Anna B. Selfridge; Cur. Collections, John Carnes; Cur. Education, Sarah Rish; Asst. Cur., Charles Bates; Museum Shop Mgr., Joann Park; Admin. Asst. & Bookkeeper, Donna Collins.
Personnel Profile: Full-Time Paid 4; Part-Time Paid 16; Part-Time Volunteers 60.
Governing Authority: society. Parent Institution: Allen County Historical Society. Tax-exempt: 501(c)(3).
Institution Type/Description: History Museum.
Collections: Indian relics; minerals; fossils; documents; manuscripts; photographs; drawings pertaining to steam & electric railroads; files from old Lima locomotive works; pioneer rooms; tools; furniture; 19th century fire-fighting equipment; Lincoln Park railroad exhibit; furnished miniature model of Mount Vernon. Historic Houses: 1848 log house; 1890 MacDonell house.
Research Fields: local history; railroad history.
Facilities: 5,000-vol. general library of books available for use on premises; reading room; 150-seat auditorium; 250-seat changing gallery.
Activities: guided tours; lectures; films; gallery talks; concerts; arts festivals; formally organized education programs for children; permanent & temporary exhibitions.
Publications: biannual local history periodical, The Allen County Reporter.
Hours & Admission Prices: Museum: Tues.-Sat. 1-5, Sun. 1-4. Children's Discovery Center: Tues.-Sun. 1-5. Museum: no charge; donations accepted. MacDonell House: admission 11 & over $3; members no charge. Closed holidays. &
Attendance: 64,000 (estimated)
Membership: Individual $20; Family $30; Patron $50; Centennial $100; Trustee's Circle $500; Founders $1,000.

ARTSPACE/LIMA, 65 Town Square, Lima, OH 45801-4950. Tel.: 419-222-1721.
Key Personnel: Operations Mgr., Bill Sullivan
Institution Type/Description: Art Gallery.
Collections: paintings; photographs; sculpture.
Facilities: Museum-related items for sale.
Activities: classes; permanent & temporary exhibits.
Hours & Admission Prices: Tues.-Fri. 10-5, Sat. 10-2.

LIMA FIRE FIGHTERS MEMORIAL MUSEUM, Lima Municipal Center, 50 Town Sq., Lima, OH 45801. Tel.: 419-221-5164.
Institution Type/Description: Fire Fighters Museum.
Collections: Lima's fire fighting history & heroes; 19th century horse-drawn steam pumper; photographs; equipment; uniforms; newspaper clippings.
Hours & Admission Prices: By appointment. No charge.

Lisbon

LISBON HISTORICAL SOCIETY, 117/119 E. Washington St, Lisbon, OH 44432. Mailing Address: P.O. Box 191, Lisbon, OH 44432-0191. Tel.: 330-424-1861.
E-mail: lisbonhs@epohi.com
Web Site: www.lisbonhistory.org
Founded: 1938.
Congressional District: 6

Key Personnel: Co Chm., Kenneth Everett; Pres. (V), Wendell Cole; Co Chm. & Head Cur., Gene Krotky.
Personnel Profile: Full-Time Volunteers 1; Part-Time Volunteers 8.
Governing Authority: nonprofit organization. Subsidiary Institutions: Old Stone House, 117 E. Washington St., Lisbon 44432; Erie Railroad Station Museum, 119 E. Washington St., Lisbon, OH; Bye Barn. Tax-exempt: 501(c)(3).
Institution Type/Description: General Museum.
Collections: early items of local history; Lisbon & New Lisbon history; Center Township history; Civil War artifacts; abolitionism; early records; genealogical materials.
Major Exhibits: Abolitionism/Underground Railroad Participants and Places, 4/11-12/15; Civil War - Who, What, Where in New Lisbon, 4/11-12/15; The Boys Come Home Heroes, 4/15-12/15.
Facilities: 100-vol. library of early Ohio & local history, including probate records 1803-1920, available on request.
Activities: guided tours; permanent & temporary exhibitions; research; children's programs; educational programs; living history programs.
Publications: quarterly newsletter; books, Reflections of a Village: Lisbon 1803-2003; The First Baptist Church of New Lisbon 1806-1825; A Proud Heritage; Mack's History of Columbiana County, OH 1879 (reprint).
Hours & Admission Prices: Tues. 10-3; other times by appointment. No charge; donations accepted. Closed holidays.
Attendance: 1,500 (estimated)
Membership: Individual $15; Family $25; Patron $75; Life $500.

Lithopolis

THE WAGNALLS MEMORIAL, 150 E. Columbus St., Lithopolis, OH 43136. Mailing Address: P.O. Box 217, Lithopolis, OH 43136. Tel.: 614-837-4765. Fax: 614-833-4767. FaceBook: The Wagnalls Memorial.
E-mail: emcclay@wagnalls.org
Web Site: www.wagnalls.org
Founded: 1924.
Congressional District: 10
Key Personnel: Dir., M. Ellen Gruber; Chm. (V), Jared McGill.
Personnel Profile: Part-Time Paid 2.
Governing Authority: nonprofit organization. Tax-exempt: 501(c)(3).
Institution Type/Description: Library & Community Center; Foundation.
Collections: paintings of John Ward Dunsmore; poems of Edwin Markham; books & personal items of Mabel Wagnalls Jones; letters from O'Henry
Research Fields: Ohio history.
Facilities: 35,000-vol. library; reading room; 400-seat auditorium; dining hall; recreation room.
Activities: guided tours; lectures; permanent exhibits; rotating & special exhibits; displays; musical programs; education class; children's acting classes; theatre productions.
Publications: newsletter, Wagnalls Digest.
Hours & Admission Prices: Mon.-Thurs. 10:30-7:30, Sat. 10-2. No charge; donations accepted. Closed national holidays.

Logan

PENCIL SHARPENER MUSEUM - HOCKING HILLS REGIONAL WELCOME CENTER, 13178 State Rte. 664 S., Logan, OH 43138-8560. Tel.: 740-753-4634.
Key Personnel: Owner, Charlotte A. Johnson; Dir., Karen Raymore
Institution Type/Description: General Museum.
Collections: over 3,000 pencil sharpeners.
Hours & Admission Prices: By appointment. No charge.

Lorain

BLACK RIVER HISTORICAL SOCIETY OF LORAIN, 309 W. 5th St., Lorain, OH 44052-1611. Tel.: 440-245-2563. Fax: 440-245-3591.
E-mail: brhsmoore@centurytel.net
Web Site: www.loraincityhistory.org
Founded: 1981.
Congressional District: 13
Key Personnel: Pres., Pat Morrisson; Vice Pres., Rosemary Balchak; Chm. (V), Carmalene Januzzi; Treas., Karen Shawver; Cur., Ron Sauer; Education, Rodney Beals; Museum Shop Mgr., Carolyn Sipkovsky.
Personnel Profile: Full-Time Paid 1; Part-Time Volunteers 25.
Governing Authority: municipal; nonprofit. Tax-exempt: 501(c)(3).
Institution Type/Description: General Museum.
Collections: artifacts; photographs; stories; histories of businesses & industries of Lorain.
Facilities: 100-vol. library. Museum-related items for sale.

Activities: guided tours; school service.
Publications: quarterly newsletter, Black River Historical Society Newsletter; Moore Memos.
Hours & Admission Prices: Mon.-Fri. 10-4:30, Sun. by appointment. Adults $3, students $1; discounts to NEOIMC; children under 5 & members no charge. Closed major holidays.
Attendance: 300 (estimated)
Membership: Youth $5; Historian $15; Archivist $50; Benefactor $100; Life $400; Corporate Benefactor $150; Corporate Life $3,000.

Loudonville

CLEO REDD FISHER MUSEUM, 203 E. Main St., Loudonville, OH 44842-1214. Tel.: 419-994-4050. Facebook: Cleo Redd Fisher Museum.
E-mail: mohicanhistoricalsociety@gmail.com
Founded: 1973.
Congressional District: 17
Key Personnel: Pres., Jeanne Coriffin; Cur., Kenny Libben.
Personnel Profile: Part-Time Volunteers 15.
Governing Authority: society. Mohican Historical Society. Tax-exempt.
Institution Type/Description: Local History Museum.
Collections: Flxible Bus Company; inventors Hugo Young & Charles Kettering; Indian artifacts; pioneer & Victorian artifacts; local history; Peter Reinhard animated Mechanical village; 19th century printing press.
Research Fields: Flxible Company; Charles Kettering; Mohican area history.
Facilities: 400-vol. library; 150-seat meeting room.
Activities: guided tours; lectures; loan exhibitions.
Hours & Admission Prices: By appointment; call for holiday weekend hours. No charge; donations accepted.
Attendance: 5,000 (estimated)
Membership: Mohican Historical Society: Individual $10, $5 for additional family members.

Loveland

GREATER LOVELAND HISTORICAL SOCIETY MUSEUM, 201 Riverside Dr., Loveland, OH 45140-2303. Tel.: 513-683-5692. Fax: 513-683-7409.
E-mail: glhsm@fuse.net
Web Site: www.lovelandmuseum.org
Founded: 1975.
Congressional District: 6
Key Personnel: Dir., Janet Beller; Pres., Robert Bauer; Librarian, Jo Funke; Museum Shop Co-Mgr., Nancy Garfinkel.
Personnel Profile: Part-Time Paid 1; Part-Time Volunteers 75.
Governing Authority: society; nonprofit organization. Parent Institution: Greater Loveland Historical Society. Tax-exempt: 501(c)(3).
Institution Type/Description: Local History Museum: housed in c.1861 two-story frame structure, 1797 log house & herb garden.
Collections: dolls; costumes; photographs; genealogy library; 35-volumes of Loveland newspapers; local history from Indian habitation to present time; Victorian furnishings; 1920's kitchen; textiles; Nancy Ford Cones photographs. Historic Structures: 1797 log cabin & herb garden; 1897 Bishop-Coleman Gazebo.
Research Fields: homes; genealogy; Loveland High School annuals; bound local newspapers; local history.
Facilities: library; archives, 5,400 sq. ft. exhibit space. Museum-related items for sale.
Activities: school & group guided tours; membership & community programs; rotating, loan & permanent exhibits; workshops; special events; lectures; rental facilities; Christmas programs.
Publications: quarterly newsletter, Reflections; books: Passages Through Time: A Loveland History; 25th Anniversary Cookbook, a collection of favorite recipes from members; Loveland Memorabilia.
Hours & Admission Prices: Sat.-Sun. 1-4:30; groups & other times by appointment. No charge; donations accepted. Closed Christmas.
Attendance: 1,000 (estimated)
Membership: Individual $10; Family $15; Contributing $25; Sustaining $50; Supporting $100; Life $1,000 & up; Corporate $50-$1,000 & up.

Lucas

MALABAR FARM STATE PARK, 4050 Bromfield Rd., Lucas, OH 44843-9745. Tel.: 419-892-2784. Fax: 419-892-3988.
E-mail: malabar.farm.parks@dnr.state.oh.us
Web Site: www.malabarfarm.org
Founded: 1939.
Congressional District: 4
Key Personnel: C.E.O. & Museum Shop Mgr., Jason Wesley; Chm. (V) & Museum Shop Mgr., Sybil Burskey.

Personnel Profile: Full-Time Paid 6; Full-Time Volunteers 1; Part-Time Paid 9; Part-Time Volunteers 150; Interns 1.
Governing Authority: state. Parent Institution: Malabar Farm Foundation. Subsidiary Institution: Columbus Community Foundation. Tax-exempt.
Institution Type/Description: Park Museum: housed in The Big House, former home of author Louis Bromfield.
Collections: original furnishings; 2 Grandma Moses paintings. Historic Building: 1820 Malabar Inn.
Facilities: 6,500-vol. library of ecology books; campground; restaurant; domestic animal zoo; picnic area.
Activities: guided tours; nature programs; seasonal special events.
Publications: guide, Malabar Farm Tour; maps; Ohio State Park brochures.
Hours & Admission Prices: Nov.-April Sat.11-5, Sun. 12-5; call for additional hours. Adults $4, seniors $3.60, children 6-18 $2; Farm Wagon Tours: $2 per person; discounts to Golden Buckeye card holders & Malabar Farm Foundation members; children under 6 & members no charge. Closed New Year's Day; Martin Luther King Jr. Day; Presidents' Day; Columbus Day; Thanksgiving; Christmas. &
Attendance: 352,600 (estimated)
Membership: Individual $50.

Mansfield

BIBLE WALK, 500 Tingley Ave., Mansfield, OH 44905-1234. Tel.: 800-222-0139. Fax: 419-524-2002.
E-mail: lbmjulia@richnet.net
Web Site: www.biblewalk.us
Formerly: Living Bible Museum
Key Personnel: Dir., Julia Mott-Hardin
Institution Type/Description: Religious Museum.
Collections: life-sized Bible re-creations; woodcarvings; Bibles; paintings; American votive folk art.
Activities: tours.
Hours & Admission Prices: Mon.-Sat. 9-6, Sun. 3-7. Various tours available. Call for prices. &

ELEMENT OF ART STUDIO & GALLERY, 96 N. Main St., Mansfield, OH 44902. Tel.: 419-522-2965. Facebook: Element of Art Studio & Gallery.
Web Site: www.eoastudiogallery.com
Governing Authority: nonprofit organization.
Institution Type/Description: Art Gallery.
Collections: works by artists with developmental disabilities; paintings; photographs; sculptures.
Hours & Admission Prices: Tues.-Fri. 10-5, Sat. 10-3. No charge.

KINGWOOD CENTER, 900 Park Ave., W., Mansfield, OH 44906-2999. Tel.: 419-522-0211. Fax: 419-522-0211.
E-mail: info@kingwoodcenter.org
Web Site: www.kingwoodcenter.org
Founded: 1953.
Congressional District: 4
Key Personnel: C.E.O., Charles T. Gleaves.
Personnel Profile: Full-Time Paid 14; Part-Time Paid 12; Part-Time Volunteers 250.
Governing Authority: nonprofit organization. Tax-exempt: 501(c)(3).
Institution Type/Description: Botanical Garden.
Collections: display gardens; greenhouse.
Research Fields: horticulture.
Facilities: 8,500-vol. library on botany, horticulture, gardening and nature; reading room; classrooms; meeting rooms.
Activities: guided tours; lectures; films; concerts; study clubs; hobby workshops; formally organized education programs; docent program; special events.
Publications: newsletter, Kingwood Center News.
Hours & Admission Prices: Grounds & Gardens: March & Nov. daily 8-5; April-Oct. daily 8-7. Kingwood Hall: April-Oct. Sat. 1-5; Dec. call for extended hours. Greenhouse: March-Oct. daily 8-4:20. Parking: April 14-Oct. 14 $5; members no charge. American Horticultural Society reciprocal program. Closed holidays.
Attendance: 250,000 (estimated)
Membership: Member $40; Family & Dual $50; Donor $100; Patron $500; Benefactor $1,000.

LITTLE BUCKEYE CHILDREN'S MUSEUM, 44 W. Fourth St., Mansfield, OH 44902. Tel.: 419-522-2332.
E-mail: info@littlebuckeye.org
Web Site: www.littlebuckeye.org
Institution Type/Description: Children's Museum.

Collections: hands-on exhibitions.
Activities: birthday parties; educational programs; classes; camps.
Hours & Admission Prices: Wed.-Sat. 10-6, Sun. 1-6. Admission 2 & over $7. Closed New Year's Eve & Day; Easter; Mother's Day; Father's Day; Thanksgiving; Christmas Eve & Day.

THE MANSFIELD ART CENTER, 700 Marion Ave., Mansfield, OH 44906-5006. Tel.: 419-756-1700. Fax: 419-756-0860.
E-mail: info@mansfieldartcenter.org
Web Site: www.mansfieldartcenter.org
Founded: 1946.
Congressional District: 17
Key Personnel: Exec. Dir., Tracy Graziani; Museum Shop Mgr., Roger Coulton.
Governing Authority: nonprofit. Tax-exempt: 501(c)(3).
Institution Type/Description: Art Gallery.
Collections: changing exhibitions.
Facilities: 500-vol. library of fine arts books; classrooms. Fine arts & crafts by artists for sale.
Activities: guided tours; lectures; films; gallery talks; group & theme shows; competitive exhibitions; workshops; formally organized education programs; docent program; inter-museum loan & traveling exhibitions.
Publications: monthly newsletter; quarterly class schedules; annual members report; annual report; special exhibit guides.
Hours & Admission Prices: Tues.-Sun. 11-5. No charge; donations accepted. Closed national holidays. &
Attendance: 80,000
Membership: Student $25; Associate $50; Friend $100; Fellow $175; Supporting $250; Sustaining $500; Patron $1,000 & up.

MANSFIELD FIRE MUSEUM, 1265 W. Fourth St., Mansfield, OH 44906. Tel.: 419-529-2573.
Institution Type/Description: Fire Museum.
Collections: firefighting history, tools & equipment; replica turn-of-the-century firehouse; photographs.
Hours & Admission Prices: June-Sept. 1 Sat.-Wed. 1-4; Sept.-May Sat.-Sun. 1-4.

MANSFIELD MEMORIAL MUSEUM, 34 Park Ave., W., Mansfield, OH 44902-1603. Tel.: 419-525-2491.
Key Personnel: Owner, Julia Hardin; Gen. Mgr., Scott Cater
Institution Type/Description: History Museum.
Collections: local history & culture; military history; photographs; period artifacts.
Hours & Admission Prices: Feb.-Dec. Sat. 10-4, Sun. 12-4. No charge; donations accepted.

Marblehead

JOHNSON'S ISLAND MUSEUM & INFORMATION CENTER, 414 W. Main St., Marblehead, OH 43440. Mailing Address: P.O. Box 1865, Marblehead, OH 43440-0495.
E-mail: jipres@johnsonsisland.org
Institution Type/Description: History Museum.
Collections: island's history; Civil War POW Depot; Pleasure Resort; quarry operations; photographs; personal artifacts.
Hours & Admission Prices: Memorial Day to Labor Day Sat.-Sun. & holidays 12-5; other times by appointment.

Marietta

＊ **CAMPUS MARTIUS MUSEUM, (M),** 601 2nd St., Marietta, OH 45750-2122. Tel.: 740-373-3750; 800-860-0145. Fax: 740-373-3680.
E-mail: info@campusmartiusmuseum.org
Web Site: campusmartiusmuseum.org
Founded: 1929.
Congressional District: 6
Key Personnel: Mgr., Le Ann Hendershot.
Personnel Profile: Full-Time Paid 4; Part-Time Paid 3; Part-Time Volunteers 60.
Governing Authority: private; nonprofit; educational organization. Parent Institution: Friends of the Museums, 601 Second St., Marietta, OH 45750. Tax-exempt: 501(c)(3); 509(a).
Institution Type/Description: History Museum.
Collections: items pertaining to history of early Northwest Territory & Marietta, Ohio; 19th- & early to mid-20th century farming, transportation & industry; steamboat photographs; area genealogical information; migration of farmers to cities of Ohio from 1880-1920; migration of Appalachian

people to Ohio's urban centers from 1917-1970. Historic Houses: 1788 Ohio Company Land Office; 1789 Rufus Putnam House.

Major Exhibits: Touched by Conflict: Southeastern OH & The Civil War, 7/12-7/15; Imagining Marietta Mural, 5/13-12/15.

Research Fields: history of Northwest Territory; pioneer life and settlement; the Ohio River.

Facilities: 3,000-vol. library of historical, genealogical & river material available for use on premises; auditorium; audiovisual programs; interactive video programs. Ohio-made crafts & other museum-related items for sale.

Activities: guided tours; permanent & temporary exhibits; special educational programs, including school workshops; summer day camp.

Hours & Admission Prices: Mon. & Wed.-Sat. 9:30-5, Sun. 12-5; tours by appointment. Adults $7, students $4; OHS members and children 5 & under no charge. Closed New Year's Day; Easter; Thanksgiving; Christmas. &

Attendance: 9,500 (accurate)

Membership: See separate listing for Ohio Historical Society. Friends of the Museum: Student $10; Individual $15; Family $35; Special Friend $75.

THE CASTLE, 418 Fourth St., Marietta, OH 45750-2003. Tel.: 740-373-4180. Fax: 740-373-4233.

E-mail: castle@mariettacastle.org

Web Site: www.mariettacastle.org

Founded: 1992.

Key Personnel: Dir., Scott Britton; Pres., Judy Grize; Treas., Molly Frye.

Personnel Profile: Full-Time Paid 3; Part-Time Paid 4; Part-Time Volunteers 40.

Governing Authority: private; nonprofit organization. Parent Institution: Betsey Mills Corp. Tax-exempt: 501(c)(3).

Institution Type/Description: Historic House: c.1855 Gothic Revival-style home.

Collections: Victorian furnishings.

Research Fields: history of site & families that occupied the house.

Activities: guided tours; workshops; lectures; history camp; concerts; docent program; ghost tours. Annual Events: Herb Sale; storytelling; Victorian Christmas; Victorian Funeral.

Publications: quarterly newsletter, The Calling Card.

Hours & Admission Prices: Jan.-March group tours only; April-May & Sept.-Dec. Mon. & Thurs.-Sat. 10-4, Sun. 1-4; June-Aug. Mon.-Tues. & Thurs.-Sat. 10-4, Sun. 1-4. Adults $7, senior citizens $6.50, students $4; discounts to groups, AARP, AAM & AAA members; children under 6 & members no charge. Closed New Year's Day; Easter; Thanksgiving; Christmas. &

Attendance: 6,023 (accurate)

Membership: Individual $25; Family $40; Contributing $50; Patron $100.

THE CHILDREN'S TOY & DOLL MUSEUM, 206 Gilman St., Marietta, OH 45750-2837. Mailing Address: P.O. Box 4034, Marietta, OH 45750-7034. Tel.: 740-373-8820.

Founded: 1989.

Key Personnel: Chm. (V), Phyllis Wells; Pres. (V), Donna Kern.

Personnel Profile: Part-Time Volunteers 15.

Governing Authority: Tax-exempt.

Institution Type/Description: Children's Museum.

Collections: international dolls; period toys; doll houses; miniature circus; toy transportation.

Activities: children's playroom.

Hours & Admission Prices: May-Nov. 1 Sat.-Sun. 1-4; other times by appointment. Family $10, adults $4, children $2; discount to families; members no charge.

Membership: Regular $10; Family $15; Bronze Patron $25; Silver Patron $50; Gold Patron $100; Platinum Patron $200.

MARIETTA SODA MUSEUM, 109 Maple St., Marietta, OH 45750-2777. Tel.: 740-376-2653.

Formerly: Butch's Coca-Cola Museum

Institution Type/Description: History Museum.

Collections: Coca-Cola history & artifacts from 1920s to present; Coke memorabilia; dolls; coolers; billfolds; aluminum bottle carrier; signs; tins; dinnerware; machines; clothing.

Hours & Admission Prices: Tues.-Sun. 10-5. No charge.

OHIO RIVER MUSEUM, (M), 601 Front St., Marietta, OH 45750. Mailing Address: 601 Second St., Marietta, OH 45750-2122. Tel.: 740-373-3750; 800-860-0145. Fax: 614-373-3680.

E-mail: info@campusmartiusmuseum.org

Web Site: campusmartiusmuseum.org

Founded: 1941.

Congressional District: 6

Key Personnel: Mgr., Le Ann Hendershot.

Personnel Profile: Full-Time Paid 1; Part-Time Paid 2; Part-Time Volunteers 12.

Governing Authority: private; nonprofit educational organization. Parent Institution: Friends of the Museums, 601 Second St., Marietta, OH 45750. Tax-exempt: 501(c)(3); 509(a).

Institution Type/Description: History Museum & Historic Ship: 1918 sternwheel Steamer, W.P. Snyder Jr., a National Historic Landmark.

Collections: ecological, recreational & commercial history of the Ohio River from its origin-present day; river history exhibits.

Research Fields: river history.

Facilities: library of river information material; auditorium. Museum-related items for sale.

Activities: guided tours; lectures; permanent exhibits; summer day camp; special educational programs; audiovisual program.

Hours & Admission Prices: April to Labor Day Mon. & Wed.-Sat. 9:30-5, Sun. 12-5; Sept.-Oct. Sat. 9:30-5, Sun. 12-5; tours by appointment. Adults $7, students $4; OHS members & children 5 and under no charge. &

Attendance: 4,200 (estimated)

Membership: see separate listing for Ohio Historical Society. Friends of the Museum: Student $10; Individual $15; Family $35; Special Friend $75.

PEOPLES MORTUARY MUSEUM, 408 Front St., Marietta, OH 45750. Tel.: 740-373-1111.

E-mail: brian@cawleyandpeoples.com

Institution Type/Description: History Museum.

Collections: early 1900s funeral history & memorabilia; 1938 Packard; 1940 Packard; 1948 Packard Deluxe; 1927 Henney; 1895 horse-drawn hearse.

Hours & Admission Prices: By appointment. No charge.

Marion

BUCKEYE TELEPHONE MUSEUM AKA THE CLARE E. WILLIAMS TELEPHONE MUSEUM ASSOCIATION, CWA Union Hall, 581 Bellefontaine Ave., Marion, OH 43302. Mailing Address: 6075 Marion Edison Rd., Caledonia, OH 43314-9413. Tel.: 419-947-8676; 800-371-6688; 740-389-7990.

E-mail: jekosto@frontier.com

Founded: 1985.

Key Personnel: Dir., John E. Kosto; C.E.O., Jerron Reeder.

Personnel Profile: Part-Time Volunteers 4.

Governing Authority: Tax-exempt.

Institution Type/Description: History Museum.

Collections: telephone industry history & equipment; crank phones; maps; phonebooks; testing gear.

Hours & Admission Prices: By appointment. Suggested Donation: $2. &

Attendance: 250 (estimated)

HARDING HOME AND MUSEUM, (M), 380 Mt. Vernon Ave., Marion, OH 43302-4120. Tel.: 740-387-9630; 800-600-6894. Fax: 740-387-9630 (call first).

E-mail: shall@hardinghome.org

Web Site: www.hardinghome.org

Founded: 1925.

Congressional District: 7

Key Personnel: Site Mgr., S. Hall.

Personnel Profile: Full-Time Paid 2; Part-Time Paid 2; Part-Time Volunteers 1.

Governing Authority: nonprofit organization. Parent Institution: Ohio Historical Society, Inc. I-71 & 17th Ave., Columbus, OH 43211. Tax-exempt: 501(c)(3).

Institution Type/Description: History Museum: housed in 1920 cottage, press corps center for Harding's Campaign.

Collections: mementos of Harding's life from boyhood to presidency; presidential memorabilia; original furniture & furnishings. Historic House: 1890 Harding Home.

Activities: guided tours; formally organized education programs for children; permanent exhibitions.

Publications: Ohio Historical Society, Time Line.

Hours & Admission Prices: Memorial Day weekend to Sept. Wed.-Sun. 12-5; Sept.-Oct. weekends 12-5 and by appointment on weekdays; Nov.-May by appointment only. Adults $7, seniors 60 & up $6, students 12-17 $4, children 6-11 $3; discounts AAA members; Ohio Historical Society, Marion Technical College student with ID and active military & families with ID no charge. &

Attendance: 6,000 (estimated)

HUBER MACHINERY MUSEUM, 220 E. Fairground St., Marion, OH 43302. Mailing Address: P.O. Box 6010, Marion, OH 43301-6010. Tel.: 740-389-1098.
Governing Authority: bd. of trustees. Tax-exempt.
Institution Type/Description: Machinery Museum.
Collections: local history; early steam & gasoline tractors; road-building equipment; steam shovel.
Hours & Admission Prices: March-Dec. Sat. 1-4; other times by appointment. No charge; donations accepted. &
Attendance: 2,500 (accurate)
Membership: Annual $7.50, $15, $25; Life $150.

THE MARION COUNTY HISTORICAL SOCIETY MUSEUM, 169 E. Church St., Ste. A, Marion, OH 43302-3826. Tel.: 740-387-4255. Fax: 740-387-0117.
E-mail: mchs@marionhistory.com
Web Site: www.marionhistory.com
Founded: 1969.
Congressional District: 4
Key Personnel: Dir., C.E.O. & Museum Shop Mgr., Gale E. Martin; Pres., Fred Malone; Treas., Diane Mault; Museum Shop Mgr., Meredithe Predmore.
Personnel Profile: Full-Time Paid 1; Part-Time Paid 4; Part-Time Volunteers 36.
Governing Authority: private; nonprofit organization. Tax-exempt: 501(c)(3).
Institution Type/Description: History Museum.
Collections: Marion County area history & artifacts; Warren G. Harding presidential artifacts & documents.
Research Fields: history of individual townships in Marion County, Ohio; genealogy; Warren G. Harding.
Facilities: 1,000-vol. library of history & genealogy books; 50-seat auditorium. Museum-related items for sale.
Activities: guided tours; lectures; temporary exhibitions; in-school programs. Annual Event: Marion County Fair; Night at Heritage Hall; Saturday in the Park; Blast From the Past.
Publications: quarterly newsletter, Hallmarks.
Hours & Admission Prices: March-April & Nov.-Dec. Sat.-Sun. 1-4; May-Oct. Wed.-Sun. 1-4; group tours by appointment. Adults $4, seniors $3, children $1.50; discounts to AAA & Golden Buckeye members; members no charge. &
Attendance: 8,000 (accurate)
Membership: Student $10; Regular $25; Family $35; Sustaining $60; Pride $100-$3,000.

STENGEL-TRUE MUSEUM, 504 S. State St., Marion, OH 43302-5036. Tel.: 740-387-6140.
Founded: 1973.
Congressional District: 4
Key Personnel: Chm. (V), Kevin Hall.
Governing Authority: nonprofit organization. Tax-exempt: 501(c)(3).
Institution Type/Description: Historical Society Museum: housed in c.1860 Judge Ozias Bowen home.
Collections: period artifacts; glass; Victorian bric-a-brac; table zither; John Miller family paintings; china; guns; Indian artifacts; pewter; copper; silver; timepieces; eyeglasses.
Activities: guided tours.
Hours & Admission Prices: By appointment. Requested Donation: $1. Closed holidays.
Attendance: 2,400 (estimated)

WYANDOT POPCORN MUSEUM, Heritage Hall, 169 E. Church St., Ste. B, Marion, OH 43302-3826. Tel.: 740-387-4255. Fax: 7404-387-0117.
E-mail: mchs@marionhistory.com
Web Site: www.marionhistory.com
Founded: 1981.
Congressional District: 4
Key Personnel: Dir., Gale E. Martin; Pres. (V), Brooks Brown; Museum Shop Mgr., Meredithe Predmore.
Personnel Profile: Full-Time Paid 1; Part-Time Paid 1.
Governing Authority: Tax-exempt.
Institution Type/Description: Popcorn Museum.
Collections: early popcorn poppers; peanut roasters.
Hours & Admission Prices: May-Oct. Wed.-Sun. 1-4; Nov.-April Sat.-Sun. 1-4. Adults $4, seniors $3, children 6-12 $1.50; discounts to AAA & Golden Buckeye members; members no charge. Closed New Year's Day; Easter; Memorial Day; Independence Day; Labor Day; Thanksgiving; Christmas Eve & Day. &
Attendance: 6,000 (estimated)

Marshallville

MARSHALLVILLE HISTORICAL SOCIETY, 4 E. Church St., Marshallville, OH 44645. Tel.: 330-855-4041.
Institution Type/Description: Historical Society Museum.
Collections: local history & culture; period furnishings; personal artifacts; Gerstenslager buggies.
Hours & Admission Prices: By appointment.

Marysville

THE UNION COUNTY HISTORICAL SOCIETY, 246 W. Sixth St., Marysville, OH 43040-1531. Mailing Address: P.O. Box 303, Marysville, OH 43040-0303. Tel.: 937-644-0568.
Founded: 1949.
Congressional District: 15
Key Personnel: Pres., Robert W. Parrott; Vice Pres., Stephen W. Badenhop; Treas., John Woerner.
Personnel Profile: Part-Time Paid 1; Part-Time Volunteers 12.
Governing Authority: society. Tax-exempt.
Institution Type/Description: Historical Society Museum: housed in the Henry W. Morey home; built in 1891.
Collections: farm tools; furniture; china; glass; silver; clothing; manuscripts; war relics pertaining to Union County; Indian arrowheads, chisels and hammers.
Facilities: reading room.
Activities: guided tours; gallery talks; formally organized education programs for children; permanent, temporary & traveling exhibitions.
Hours & Admission Prices: April-Nov. Wed. 1-4; other times by appointment. No charge. Adults $5; members no charge. Closed New Year's Day; Thanksgiving; Christmas.
Membership: Individual $12; Life $150.

Mason

MASON HISTORICAL SOCIETY - ALVERTA GREEN MUSEUM, 207 W. Church St., Mason, OH 45040. Tel.: 513-398-6750.
E-mail: masonhistorical@yahoo.com
Web Site: www.masonhistoricalsociety.org
Key Personnel: Cur., Letha Hendrickson
Institution Type/Description: History Museum.
Collections: local history & culture; period furnishings; personal artifacts; photographs.
Hours & Admission Prices: Thurs. Fri. 1-4. No charge; donations accepted.

Massillon

* **MASSILLON MUSEUM, (M),** 121 Lincoln Way, E., Massillon, OH 44646-6633. Tel.: 330-833-4061. Fax: 330-833-2925.
E-mail: ancoon@massillonmuseum.org
Web Site: www.massillonmuseum.org
Founded: 1933.
Congressional District: 16
Key Personnel: Interim Exec. Dir. & Cur., Alexandra Nicholis Coon; Chm. (V), Elizabeth Pruitt; Archivist, Mandy Pond; Museum Shop Mgr., Scot Phillips.
Personnel Profile: Full-Time Paid 6; Part-Time Paid 5; Part-Time Volunteers 150; Interns 4.
Volunteer Hours: 4,500
Operating Expenses: 750,000
Operating Income: 750,000
Governing Authority: private; nonprofit organization. Tax-exempt: 501(c)(3).
Institution Type/Description: History & Art Museum.
Collections: American folk art; glass; china; pottery; metal; American & European fine & decorative arts; costumes; Indian artifacts; military artifacts; paintings; tools & utensils; photographs; quilts & coverlets.
Major Exhibits: Imagining a Better World: The Artwork of Nelly Toll (T), 3/8/14-5/18/14; Fragile Waters (T), 6/7/14-9/14/14.
Research Fields: Ohio artists; local history.
Facilities: library of local history manuscripts & books available to qualified graduate students & adults for use on premises; classrooms; research room; coffee shoppe. Arts & crafts for sale.
Activities: guided tours; lectures; arts festivals; workshops; docent program; permanent, temporary & traveling exhibitions; formally organized education programs; lecture series; music series; research; vintage operational photo booth.
Publications: annual report; semi-annual newsletter; exhibit catalogs.
Hours & Admission Prices: Tues.-Sat. 9:30-5, Sun. 2-5; evenings by appointment. No charge; donations accepted. Closed legal holidays. &
Attendance: 22,000 (accurate)

Membership: Student $10; Individual $25; Family $50; Contributing $100-$249; Sustaining $250-$499; Benefactor $500-$999; Director's Circle $1,000-$1,499; James & Eliza Duncan Society $1,500 & up.

OHIO SOCIETY OF MILITARY HISTORY, 316 Lincoln Way E., Massillon, OH 44646. Tel.: 330-832-5553.
E-mail: osmh@sssnet.com
Founded: 1983.
Congressional District: 16
Volunteer Hours: 1,400
Governing Authority: Tax-exempt.
Institution Type/Description: Military History Museum.
Collections: military history; period uniforms & documents; medals; photographs; 13 Ohio Medal of Honor groupings.
Hours & Admission Prices: March 15-Dec. 15 Wed.-Fri. 10-5, Sat. 10-3. No charge; donations accepted. &
Attendance: 884 (accurate)
Membership: Annual $10.

Maumee

WOLCOTT HOUSE MUSEUM COMPLEX, 1031 River Rd., Maumee, OH 43537-3460. Tel.: 419-893-9602. Fax: 419-893-3108.
E-mail: mvhs@buckeye-access.com
Web Site: www.wolcotthouse.org
Founded: 1961.
Congressional District: 9
Key Personnel: Pres., Paul Sullivan; Exec. Dir., Jack Hiles; Museum Shop Mgr., Judy Walrod.
Personnel Profile: Part-Time Paid 6; Part-Time Volunteers 50.
Governing Authority: society. Parent Institution: Maumee Valley Historical Society. Tax-exempt.
Institution Type/Description: Historic Structures: 1836 Wolcott House; c.1850 log cabin; c.1840 saltbox style farmhouse; c.1880 train depot; c.1901 country church; c.1840 Greek Revival townhouse; c.1850 one room county schoolhouse.
Collections: mid-19th century items pertaining to Maumee Valley History; house furnishings; clothing.
Research Fields: local history; historic preservation.
Facilities: Museum-related items for sale.
Activities: guided tours; lectures; special events. Museum Sponsors: Family Fall Festival; Annual Antiques Show & Sale; Annual Lawn Sale.
Publications: Northwest Ohio History Journal.
Hours & Admission Prices: Museum Guided Tours: Thurs.-Sun. 12:30 & 2:30. Adults $5, seniors $4, students $2.50; discounts to AAA members; members no charge. Gift Shop: April-Dec. Thurs.-Sun. 12-4:30.
Attendance: 10,000 (estimated)
Membership: Senior, Student & Teacher $15; Individual $25; Family $35; Legacy Club $50; Founder's Club $100; Heritage Club $150; Curator's Club $250.

McCutchenville

MCCUTCHEN OVERLAND INN, 283 State Hwy. 53 N., McCutchenville, OH 44844. Mailing Address: P.O. Box 372, Upper Sandusky, OH 43351-0372. Tel.: 419-981-2052.
Founded: 1967.
Personnel Profile: Part-Time Paid 2.
Governing Authority: nonprofit. Tax-exempt: 501(c)(3).
Institution Type/Description: Historic House: housed in a former stage coach stop built by Col. Joseph McCutchen in 1829.
Collections: Victorian furniture; kitchen items; quilts; coverlets; tea leaf ironstone; primitive furniture; china; rope beds; toys.
Facilities: Gift items for sale.
Activities: guided tours.
Publications: bimonthly newsletter.
Hours & Admission Prices: June-Oct. Sat.-Sun. 1-4:30. Adults $2, children 13 & under $1. &
Membership: Annual $15.

Medina

AMERICA'S ICE CREAM & DAIRY MUSEUM, 1050 Lafayette Rd., Medina, OH 44256-3549. Mailing Address: 317 Forest Meadows Dr., Medina, OH 44256-1611. Tel.: 330-722-3839.
E-mail: elmfarm1934@aol.com
Web Site: www.elmfarm.com
Formerly: Elm Farm Ice Cream & Dairy Museum

Founded: 1999.
Key Personnel: C.E.O., Sherry S. Abell; Cur., Carl T. Abell.
Governing Authority: private corporation.
Institution Type/Description: Ice Cream & Dairy Museum.
Collections: ice cream memorabilia; dispensers; signs; advertising; 1905 soda fountain; photographs; prints; period toys; prototype scooper; full-size rare trucks & bottles; cream separators; butter churns; advertising; milk wagon; milk trucks.
Research Fields: historical facts on ice cream.
Facilities: reference library on ice cream history & manufacturing; ice cream parlor; outside courtyard seating. Gift items for sale.
Activities: guided tours.
Hours & Admission Prices: March-Dec. Tues.-Sun. 12-5. Adults $4.50, senior citizens $4, students $3; discounts to groups, AAM members; children under 5 no charge. &
Attendance: 5,000 (estimated)

MEDINA COUNTY HISTORICAL SOCIETY, THE JOHN SMART HOUSE, 206 N. Elmwood St., Medina, OH 44256-1829. Mailing Address: P.O. Box 306, Medina, OH 44258-0306. Tel.: 330-722-1341.
E-mail: info@medinahistorical.org
Web Site: www.medinahistorical.com
Founded: 1922.
Congressional District: 17
Key Personnel: C.E.O., Brian Feron; Pres. (V), Judy Davanzo; Cur., Thomas D. Hilberg.
Personnel Profile: Part-Time Paid 1; Part-Time Volunteers 20.
Governing Authority: independent; nonprofit. Tax-exempt.
Institution Type/Description: Historical Society Museum: located in 14-room Victorian home.
Collections: Levitt & Wolbach collections; Indian & Civil War artifacts; Victorian furnishings.
Research Fields: genealogy & history of Medina County.
Facilities: library available for use on premise only.
Activities: guided tours; seminars; field trips for members.
Publications: monthly, report of events; quarterly, newsletter.
Hours & Admission Prices: Tues. & Thurs. 9-5, 1st Sun. of the month 1-4; other times by appointment. No charge; donations accepted.
Attendance: 2,300 (estimated)
Membership: Seniors & Students $8; Individual $15; Family $20; Organization $25; Patron $35; Corporate $75.

MEDINA TOY AND TRAIN MUSEUM, 7 Public Square, Medina, OH 44256-2203. Tel.: 330-764-4455. Fax: 330-722-2205.
Founded: 2000.
Institution Type/Description: Toy & Train Museum.
Collections: railroad memorabilia; model trains; hands-on exhibits; model cars & airplanes; toys; dolls; 500 train engines.
Facilities: library.
Hours & Admission Prices: Mon.-Sat. 10-5, Sun. 12-5. Adults $2.
Attendance: 5,500 (estimated)
Membership: Individual $20.

Mentor

JAMES A. GARFIELD NATIONAL HISTORIC SITE, 8095 Mentor Ave., Mentor, OH 44060-5753. Tel.: 440-255-8722. Fax: 440-974-2045. Facebook: James A. Garfield National Historic Site.
E-mail: sherda_williams@nps.gov
Web Site: www.nps.gov/jaga
Founded: 1936.
Congressional District: 14
Key Personnel: Site Mgr., Sherda Williams.
Personnel Profile: Full-Time Paid 6; Part-Time Paid 7; Part-Time Volunteers 20.
Governing Authority: Parent Institution: National Park Service.
Institution Type/Description: Historic House: c.1880 home of former Pres. James Garfield.
Collections: Garfield memorabilia; furniture; photographs; campaign office; presidential library.
Facilities: picnic area; visitors center. Horse Barn: education facility.
Activities: guided tours; organized education programs for children; video; special events.
Publications: quarterly newsletter, The Garfield Telegraph.
Hours & Admission Prices: May-Oct. Mon.-Sat. 10-5, Sun. 12-5; Nov.-April Sat.-Sun. 12-5. Admission $5; children 15 & under and NPS Pass holders no charge. Closed New Year's Day; Thanksgiving; Christmas. &
Attendance: 40,000 (estimated)

MENTOR FIRE MUSEUM, 7262 Jackson St., Mentor, OH 44060.
Tel.: 440-299-0202.
E-mail: museum@mentorsafetyvillage.com
Web Site: www.mentorsafetyvillage.com
Key Personnel: Dir. & Cur., Don Zimmerman; Treas., Gabe Zelinka; Archivist,
Jeremy Szydlowski.
Personnel Profile: Full-Time Volunteers 10.
Governing Authority: private; nonprofit organization.
Institution Type/Description: Firefighting History Museum: housed in a former
fire station; built in 1940s.
Collections: firefighting history & equipment; personal artifacts; photographs;
EMS equipment; 1942 Ford fire truck; 1986 Pierce Arrow.
Hours & Admission Prices: By appointment. No charge; donations accepted.
Attendance: 50 (estimated)

Miamisburg

MIAMISBURG HISTORICAL SOCIETY, 4 N. Main St., Miamis-
burg, OH 45342-2313. Mailing Address: P.O. Box 774, Miamis-
burg, OH 45343-0774. Tel.: 937-859-5000.
E-mail: mhsociety@att.net
Web Site: www.miamisburg.org
Founded: 1963.
Congressional District: 3
Key Personnel: Dir., Randy Staley; Pres., Paul Schultz; Cur., Larry Suttman;
Museum Shop Mgr., Judy Frizzell.
Personnel Profile: Part-Time Volunteers 40.
Governing Authority: Branch Museum: Daniel Gebhart Tavern Museum, Lock
St. & Old Main St., Miamisburg, OH. Tax-exempt.
Institution Type/Description: Historical Society Museum.
Collections: local history & culture; photographs.
Research Fields: genealogy.
Facilities: Museum-related items for sale.
Activities: rental facilities; research.
Publications: bimonthly newsletter.
Hours & Admission Prices: Wed. & Sat. 1-4; other times by appointment. No
charge; donations accepted. &
Attendance: 800 (estimated)
Membership: Student $5; Individual $15; Family $25; Sponsor $50; Patron
$100; Lifetime $500.

WRIGHT ”B“ FLYER, INC., 10550 Springboro Pike, Miamisburg,
OH 45342. Tel.: 937-885-2327. Fax: 937-885-3310.
E-mail: wbflyer@dayton.net
Web Site: www.wright-b-flyer.org/about-us.html
Formerly: Wright ”B“ Flyer's Hangar and Museum
Institution Type/Description: Aviation History Museum.
Collections: aviation history; replica Model B airplanes.
Activities: educational programs; flight simulator.
Hours & Admission Prices: Tues., Thurs. & Sat. 9-2:30; other times by
appointment. No charge.

Middletown

CANAL MUSEUM, 1605 N. Verity Pkwy., Middletown, OH 45042.
Mailing Address: c/o Middletown Historical Society, P.O. Box 312,
Middletown, OH 45042. Tel.: 513-422-5539.
Founded: 1982.
Institution Type/Description: History Museum.
Collections: local history & industry; Miami-Erie Canal history.
Hours & Admission Prices: Sun. 2-4. No charge.

MIDDLETOWN ARTS CENTER, 130 N. Verity Pkwy., Middle-
town, OH 45042. Mailing Address: P.O. Box 441, Middletown, OH
45042. Tel.: 513-424-2417.
E-mail: pattbelisle@middletownartscenter.com
Web Site: www.middletownartscenter.com
Founded: 1957.
Key Personnel: Dir., Patt Belisle; Pres. (V), Jackie Philips; Asst. Dir., Kim
Minor; Program Coord., Leslie Pinto; Administrative Asst. & Volunteer
Coord., Cheryl Landen.
Personnel Profile: Full-Time Paid 1; Part-Time Paid 3; Part-Time Volunteers
45.
Governing Authority: Tax-exempt.
Institution Type/Description: Art Gallery.
Collections: paintings; sculpture; drawings.
Major Exhibits: Ohio Watercolor Society (T), 1/11/14-2/8/14; Tomorrow's

Artist Today, 2/14/14-3/15/14; Photography Competition Exhibit, 3/21/14-
4/19/14; Area Art Show, 9/26/14-1025/14; Annual Reunion Exhibit,
11/7/14-12/13/14.
Activities: visual arts classes; mini workshops; temporary exhibitions.
Hours & Admission Prices: Mon. 9-12pm, Tues& Thurs. 1-9, Wed. 9am-9pm,
Sat. 9am-12pm. No charge; donations accepted. Closed President's Day;
Memorial Day; 4th of July; Labor Day; Nov. 26-Nov. 30; Dec. 9-Jan. 4,
2015. &
Attendance: 4,750 (estimated)
Membership: Individual $25; Household $60; Patron $100-$249; Benefactor
$250-$499; Master's Circle $500-$999; Double Nickel Club $550;
Founder's Circle $1,000 & up.

**MIDDLETOWN HISTORICAL SOCIETY - PICKWICK
BUILDING MUSEUM,** 56 S. Main St., Middletown, OH 45042.
Mailing Address: P.O. Box 312, Middletown, OH 45042. Tel.:
513-424-5539.
E-mail: vivianmoon@juno.com
Institution Type/Description: Historical Society Museum.
Collections: local history & culture; period furnishings; photographs; personal
artifacts.
Hours & Admission Prices: Wed.-Thurs. 10-2. No charge; donations accepted.

Milan

MILAN HISTORICAL MUSEUM, INC., (M), 10 Edison Dr.,
Milan, OH 44846-9319. Mailing Address: P.O. Box 308, Milan,
OH 44846-0308. Tel.: 419-499-2968. Fax: 419-499-9004.
E-mail: museum@milanhistory.org
Web Site: www.milanhistory.org
Founded: 1955.
Congressional District: 2
Key Personnel: Pres. (V), Mr. Sparky Weilnam; C.E.O. & Dir., Ann Basilone-
Jones.
Personnel Profile: Full-Time Paid 2; Part-Time Paid 20.
Governing Authority: nonprofit. Tax-exempt: 170(c).
Institution Type/Description: General Museum: housed in 1846 home of Dr.
Lehman Galpin who assisted in the birth of Thomas A. Edison.
Collections: folklore; textiles; history; costumes; decorative arts; Indian
artifacts; marine; natural history; outdoor museum; transportation; 1,500
pieces of art glass; 400 period dolls. Historic House & Structures: 1843,
Sayles House; Newton Arts Building; general store; blacksmith shop.
Research Fields: art glass.
Facilities: Museum-related items for sale.
Activities: guided tours; permanent exhibitions.
Publications: self-guided walking tours; newsletter, New Milan Ledger.
Hours & Admission Prices: April-May & Sept.-Oct. Sat.-Sun. 1-5; June-Aug.
Tues.-Sat. 10-5, Sun. 1-5; other times by appointment. Adults $7, Senior
Citizens $6, children 6-12 $4; discount to AAM members & groups;
members no charge. Closed Labor Day weekend. &
Attendance: 5,700
Membership: Active $20; Family $40; Sponsor $50; Patron $100; Business
$200; Lifetime $1,000.

THOMAS EDISON BIRTHPLACE MUSEUM, 9 N. Edison Dr.,
Milan, OH 44846-9321. Mailing Address: P.O. Box 451, Milan,
OH 44846-0451. Tel.: 419-499-2135. Fax: 419-499-2135 (call
first).
E-mail: director@tomedison.org
Web Site: www.tomedison.org
Founded: 1947.
Congressional District: 9
Key Personnel: C.E.O. & Pres. (V), Robert K.L. Wheeler; Dir. & Museum
Shop Mgr., Lois J. Wolf; Cur., Laurence J. Russell.
Personnel Profile: Full-Time Paid 1; Part-Time Paid 11; Part-Time Volunteers
2.
Governing Authority: nonprofit organization. Parent Institution: Thomas Edi-
son Birthplace Assoc. Tax exempt.
Institution Type/Description: Historic House Museum: 1841 birthplace &
home of Thomas A. Edison.
Collections: family furnishings from 1780-1870; inventions; personal memo-
rabilia; photographs; documents.
Facilities: Books & postcards for sale.
Activities: guided tours; formally organized education programs for children.
Publications: book, Edison-Inspiration to Youth.
Hours & Admission Prices: Feb. & Dec. Sat.-Sun. 1-4 (last tour 3:30); Feb. 11
(Edison's Birthday) 1-4; March & Nov. Wed.-Sun. 1-4 (last tour 3:30);
April-May & Sept.-Oct. Tues.-Sun. 1-5 (last tour 4:30); June-Aug. Tues.-
Sat. 10-5, Sun. 1-5 (last tour 4:30). Adults $7, senior citizens $6, children

$4; discount to groups of 10 or more; military & their family no charge. Blue Star Museums Program. Closed New Year's Day; Easter; Labor Day; Thanksgiving; Christmas.
Attendance: 7,968 (accurate)
Membership: Single $25; Family $50; Special Friends $100; Lifetime $1,000; Corporate Sponsor $5,000.

Milford

PROMONT HOUSE MUSEUM, Greater Milford Area Historical Society Inc., 906 Main St., Milford, OH 45150-1767. Tel.: 513-248-0324. Fax: 513-248-2304.
E-mail: info@milfordhistory.net
Web Site: www.milfordhistory.net
Founded: 1967.
Congressional District: 6
Key Personnel: C.E.O. & Pres. (V), Dick Nordloh; Museum Shop Mgr., Tracy Lanham.
Personnel Profile: Part-Time Paid 1; Part-Time Volunteers 80.
Governing Authority: society; nonprofit organization. The Greater Milford Area Historical Society, Inc. Tax-exempt: 501(c)(3).
Institution Type/Description: Historical Society Museum: housed in 1865, Italianate 3-story brick home.
Collections: artifacts; documents; furnishings of the area; black history vintage wedding gowns 1864-1995; Indian items.
Facilities: library of papers & documents, miscellaneous subject matter pertaining to the area.
Activities: guided tours; special exhibits.
Publications: newsletter.
Hours & Admission Prices: Fri.-Sun. 1:30-4:30; special groups by appointment. Adults $5, children $1; members no charge.
Attendance: 1,800 (estimated)
Membership: Single $25; Family $35; Patron Single $50; Patron Family $100; Lifetime $1,000. Business: Small $500; Large $1,000.

Millersburg

HOLMES COUNTY HISTORICAL SOCIETY, 484 Wooster Rd., Millersburg, OH 44654. Mailing Address: P.O. Box 126, Millersburg, OH 44654-0126. Tel.: 330-674-0022; 888-201-0022. Facebook: Victorian House Museum.
E-mail: info@holmeshistory.com
Web Site: www.victorianhouse.org, www.holmeshistory.com
Founded: 1958.
Congressional District: 17
Key Personnel: Exec. Dir., Mark Boley; Pres., Camille Nowels; Vice Pres., Susan Helal; Sec., Pat Shrock; Treas., Bonnie Self; Cur., Candi Barnhart.
Personnel Profile: Full-Time Paid 1; Part-Time Volunteers 50.
Governing Authority: society; nonprofit organization. Tax-exempt.
Institution Type/Description: Historical Society Museum: housed in 1902 Victorian-style house.
Collections: period furniture, furnishings & artifacts; early medical equipment; war mementoes; early law office; pioneer tools.
Research Fields: oral history project of Holmes County.
Facilities: 1,000-vol. library of antique books.
Activities: guided tours; lectures.
Hours & Admission Prices: March Sat.-Sun. 1-4; April-Oct. Tues.-Sun. 1-4; group tours by appointment. Adults $8, seniors 65 & over $7, students $3; discounts to groups; children under 12 & members no charge.
Attendance: 5,000 (estimated)
Membership: Single $20; Supporting $20-$99; Family $30; Life $500.

Minster

LAKE LORAMIE HERITAGE MUSEUM, Lake Loramie State Park, State Rte. 362, 4401 Fort Loramie Swanders Rd., Minster, OH 45865-9306. Mailing Address: Lake Loramie State Park, 834 Edgewater Dr., Saint Marys, OH 45885-1132. Tel.: 937-295-2011.
Institution Type/Description: History Museum.
Collections: local history, heritage, & culture; photographs; harvesting ice tools; documents; wood & canvas kayak; personal artifacts; period furnishings.
Hours & Admission Prices: Summer: Sat. 10-5, Sun. 10-2.

Monroe

MONROE HISTORICAL SOCIETY, 10 E. Elm St., Monroe, OH 45050. Mailing Address: P.O. Box 82, Monroe, OH 45050. Tel.: 513-539-2270.
E-mail: info@monroeoh-historicalsociety.org
Web Site: www.monroeoh-historicalsociety.org
Formerly: Chickahominy House Museum
Founded: 1967.
Key Personnel: C.E.O. & Pres. (V), Norman Hayes; Museum Shop Mgr., James Price.
Governing Authority: Parent Institution: Monroe Ohio Historical Society. Tax-exempt.
Institution Type/Description: Historical Society Museum.
Collections: local history & culture; period furnishings; personal artifacts; photographs.
Hours & Admission Prices: By appointment. No charge; donations accepted.
Attendance: 450 (estimated)
Membership: Individual $12.

1910 GENERAL STORE MUSEUM, 2 E. Elm St., Monroe, OH 45050. Mailing Address: Monroe Historical Society, P.O. Box 82, Monroe, OH 45050. Tel.: 513-539-2270.
E-mail: info@monroeoh-historicalsociety.com
Web Site: www.monroeoh-historicalsociety.org
Founded: 1967.
Key Personnel: Dir. & Museum Shop Mgr., James Price.
Governing Authority: Parent Institution: Monroe Historical Society. Tax-exempt.
Institution Type/Description: Historical Society Museum.
Collections: local history & culture; period furnishings; personal artifacts; photographs.
Hours & Admission Prices: By appointment. No charge.
Attendance: 400 (estimated)

Montpelier

WILLIAMS COUNTY HISTORICAL MUSEUM, 611 E. Main St., Williams County Fairgrounds, Montpelier, OH 43543. Mailing Address: P.O. Box 415, Montpelier, OH 43543-0415. Tel.: 419-485-8200.
E-mail: wchs@williams-net.com
Web Site: williamscountyhistory.org
Founded: 1956.
Congressional District: 5
Key Personnel: Exec. Dir. & Cur., Kara Custar.
Personnel Profile: Full-Time Paid 1; Part-Time Volunteers 15.
Governing Authority: county; society. Tax-exempt: 501(c)(3).
Institution Type/Description: General Museum.
Collections: American prehistoric & historic Indian artifacts; agricultural items; anthropology; archaeology; archives; history; Indian artifacts; Hopewell Indian Mounds; railroad memorabilia. Historic Houses: 1850 Lett log cabin; c.1845 Kunkle Log House; 1883 Edon; 1880 Wabash Caboose.
Research Fields: local history.
Facilities: 300-vol. library of local history books available for use only on premises. Museum-related items for sale.
Activities: guided tours; lectures; formally organized education programs for adults; docent program & council; Jr. Historian Summer Program.
Publications: quarterly newsletter, Northwest Historian.
Hours & Admission Prices: May-Oct. Mon.-Thurs. 1-4; Nov.-April Mon.-Thurs. 10-4; other times & tours by appointment. Adults 2, children 6-18 $1; members no charge. Closed holidays. &
Attendance: 7,200 (accurate)
Membership: Student $7; Senior $24; Individual $29; Senior Couple $34; Family $39; Life $399; Professional $299; Life $399; Life Couple $549.

Mount Pleasant

FRIENDS (QUAKER) YEARLY MEETING HOUSE STATE MEMORIAL, (M), 298 Market St., Mount Pleasant, OH 43939. Mailing Address: P.O. Box 35, Mount Pleasant, OH 43939-0035. Tel.: 740-769-2893; 800-752-2631.
Founded: 1814.
Congressional District: 18
Key Personnel: Site Operations Mgr., Sherry Sawchuk.
Personnel Profile: Part-Time Volunteers 20.

Governing Authority: nonprofit. Parent Institution: The Ohio Historical Society, I71 & 17th Ave., Columbus, 43211. Subsidiary Institution: Mt. Pleasant Historical Society.
Institution Type/Description: Historic House Museum: built in 1814.
Collections: local history & culture; Quaker's history & artifacts; tools; period furnishings; archives; religious items.
Activities: guided tours. Annual Event: Garden Tour in August; Pilgrimage Tour in August.
Publications: biannual newsletter.
Hours & Admission Prices: May-Oct. by appointment. Meeting House: adults $6, children 6-12 $3; discounts to groups; OHS members & children under 5 no charge. Six Building Tour: adults $10, students 6-12 $5; discounts to groups; children under 5 no charge.
Attendance: 500 (estimated)

MOUNT PLEASANT HISTORICAL SOCIETY, 342 Union St., Mount Pleasant, OH 43939-9800. Mailing Address: P.O. Box 35, Mount Pleasant, OH 43939. Tel.: 740-769-2893; 800-752-2631.
Web Site: users.lst.net./gudzent/
Founded: 1948.
Congressional District: 18
Key Personnel: C.E.O. & Pres. (V), Sherry Sawchuk.
Personnel Profile: Full-Time Volunteers 15; Part-Time Volunteers 20.
Governing Authority: society. Branch Museums: Friends (Quaker) Meeting House, 298 Market St., Mount Pleasant, OH; P. L. Bone Store Museum; The Elizabeth House Mansion Museum; Historical Center: The Burriss Store Museum; The Tin Shop Museum. Tax-exempt: 501(c)(3).
Institution Type/Description: General Museum.
Collections: books; tools; documents; records; period clothing; pictures; photos; dolls; bottles & jars; historic store.
Research Fields: local history; Quaker genealogy; underground railroad.
Facilities: library of local & Quaker history books available for research on premises; reading room.
Activities: guided tours; permanent & temporary exhibitions.
Publications: newsletter, The Town Crier; educational workbook, Discover Historic Mount Pleasant, Ohio.
Hours & Admission Prices: May-Oct. by appointment. Adults $15, children 6-12 $7; discount to groups of 20 or more; children under 5 no charge.
Attendance: 400 (estimated)
Membership: Individual $15.

Mount Vernon

KNOX COUNTY AGRICULTURAL MUSEUM, Knox County Fairgrounds, State Rte. 3, Mount Vernon, OH 43050. Mailing Address: Knox County Agriculture Association, P.O. Box 171, Mount Vernon, OH 43050. Tel.: 740-397-1423.
Institution Type/Description: Agriculture Museum.
Collections: Ohio farm-life from 1800s to early 1900s; household artifacts; farming tools & equipment; one-room schoolhouse; log house.
Hours & Admission Prices: By appointment.

KNOX COUNTY HISTORICAL SOCIETY, 875 Harcourt Rd., Mount Vernon, OH 43050-4325. Mailing Address: P.O. Box 522, Mount Vernon, OH 43050-0522. Tel.: 740-393-5247.
E-mail: kchs@knoxhistory.org
Web Site: www.knoxhistory.org/about.htm
Key Personnel: Dir., James K. Gibson
Institution Type/Description: Historical Society Museum.
Collections: local history & culture; photographs.
Activities: educational programs.
Publications: newsletter.
Hours & Admission Prices: Wed. 6pm-8pm, Thurs.-Sun. 2-4.

Napoleon

DR. BLOOMFIELD HOME AND CARRIAGE HOUSE, West Clinton & Webster Sts., Napoleon, OH 43545. Mailing Address: Henry County Historical Society, P.O. Box 443, Napoleon, OH 43545. Tel.: 419-758-3262.
Institution Type/Description: Historic House Museum: c.1879.
Collections: family history; personal artifacts; period furnishings; photographs.
Activities: special events. Annual Events: Victorian Days; Christmas Open House.
Hours & Admission Prices: May-Oct. Sun. 2-4; other times by appointment.

HENRY COUNTY HISTORICAL SOCIETY, Henry County Fairgrounds, Napoleon, OH 43545. Mailing Address: P.O. Box 443, Napoleon, OH 43545. Tel.: 419-758-3262.
Institution Type/Description: Historical Society Museum.
Collections: Historic Buildings: c.1860 log cabin; 1897 Immanuel Lutheran one-room schoolhouse.
Activities: Annual Events: Henry County Fair; Napoleon Fall Festival.
Hours & Admission Prices: By appointment.

Nelsonville

STUART'S OPERA HOUSE, 52 Public Sq., Nelsonville, OH 45764-1133. Mailing Address: P.O. Box 217, Nelsonville, OH 45764-0217. Tel.: 740-753-1924. Fax: 740-753-1982.
Web Site: www.stuartsoperahouse.org
Formerly: Hocking Valley Museum of Theatrical History, Inc.
Founded: 1978.
Congressional District: 10
Key Personnel: Pres., Miki Brooks; Dir., Tim Peacock; C.O.O., Adam Fischer.
Personnel Profile: Full-Time Paid 3; Full-Time Volunteers 1; Part-Time Volunteers 20; Interns 1.
Governing Authority: nonprofit organization. Tax-exempt: 501(c)(3).
Institution Type/Description: Theater Museum: housed in 1879 Stuart's Opera House.
Collections: scenery; chairs; dressing rooms.
Activities: guided tours; lectures.
Publications: brochures.
Hours & Admission Prices: Mon.-Fri. 10-5. No charge; donations accepted. &
Attendance: 20,000 (estimated)

New Athens

FRANKLIN MUSEUM, 187 N. Main St., New Athens, OH 43981. Tel.: 740-968-1042 & 4066.
Institution Type/Description: History Museum: housed in the former Franklin College building. Listed on the National Register of Historic Places.
Collections: local history & culture; period furnishings; personal artifacts.
Facilities: Museum-related items for sale.
Hours & Admission Prices: May-Sept. Tues.-Thurs. 12-4; other times by appointment. Adults $5.

New Bremen

THE BICYCLE MUSEUM OF AMERICA, 7 W. Monroe St., New Bremen, OH 45869-1146. Tel.: 419-629-9249. Fax: 419-629-3256. Facebook: Bicycle Museum.
E-mail: jessica.howison@crown.com
Web Site: www.bicyclemuseum.com
Founded: 1997.
Key Personnel: Cur., Public Rels. & Museum Shop Mgr., Jessica Howison.
Personnel Profile: Full-Time Paid 1; Part-Time Paid 2.
Governing Authority: private; nonprofit. Tax-exempt: 501(c)(3).
Institution Type/Description: Bicycle Museum.
Collections: over 1,000 bicycles from 1816 to present.
Facilities: library; educational facilities; 13,000 sq. ft. exhibit space. Museum-related items for sale.
Activities: films; guided tours; lectures; loan & traveling exhibitions; training programs for professional museum workers; study clubs.
Hours & Admission Prices: June-Sept. Mon.-Fri. 9-7, Sat. 10-2; Oct.-May Mon.-Fri. 9-5, Sat. 10-2. Adults $3, senior citizens $2, students $1; drivers of groups & children under 6 no charge.
Attendance: 10,000 (estimated)
Membership: Single $10; Family $20.

New Carlisle

HONEY CREEK FIRE MUSEUM, 315 N. Adams St., New Carlisle, OH 45344. Tel.: 937-845-0480.
Founded: 1986.
Institution Type/Description: Firefighting History Museum.
Collections: firefighting history & equipment; hand-operated equipment; c.1875 hook & ladder; fire pump; leather fire hoses.
Activities: educational programs.
Hours & Admission Prices: By appointment.

New Philadelphia

HISTORIC SCHOENBRUNN VILLAGE, (M), 1984 W. High Ave., New Philadelphia, OH 44663. Mailing Address: P.O. Box 11, Dennison, OH 44621-0011. Tel.: 740-922-6776; 800-752-2711. Fax: 740-922-4929.
E-mail: director@dennisondepot.org
Web Site: www.ohiosfirstvillage.com
Formerly: Schoenbrunn Village State Memorial
Founded: 1928.
Congressional District: 18
Key Personnel: Dir., Wendy Zucal; Garden & Volunteer Coord., Michelle Hallman; Gift Shop Mgr., Joan Beorn; Event Coord., Jacob Masters.
Personnel Profile: Part-Time Paid 6; Part-Time Volunteers 30; Interns 5.
Governing Authority: society. Parent Institution: Ohio Historical Society Dennison Railroad Depot Museum. Subsidiary Institution: Uhrichsville Clay Museum. Tax-exempt: 170(b)(1)(A).
Institution Type/Description: Museum Village Complex.
Collections: 17 reconstructed log structures of Moravian Indian Mission, Ohio's first Christian settlement.
Research Fields: Native American history; American mission & religious history.
Facilities: picnic area. Museum-related items for sale.
Activities: Special Events: June-Oct.
Publications: calendar of special events.
Hours & Admission Prices: Memorial Day-Labor Day Tues.-Sat. 9:30-5, Sun. 12-5; Sept.-Oct. Sat. 9:30-5, Sun. 12-5. Adults $7, seniors $5, children $4; discounts to groups & AAM members; members no charge. &
Attendance: 12,500 (estimated)
Membership: Student $5; Individual $10; Family $20; Nonprofit $30; Corporate $40; Memorial $200; Benefactor $500.

Newark

THE DAWES ARBORETUM, (I), 7770 Jacksontown Rd., S.E., Newark, OH 43056-9380. Tel.: 740-323-2355 & 800-44-DAWES.
E-mail: laappleman@dawesarb.org
Web Site: www.dawesarb.org
Founded: 1929.
Congressional District: 17
Key Personnel: Co Chm., Josephine Jacobsmeyer; Co Chm., Teresa Young; Dir., Luke Messinger; Public Information Officer, Laura Appleman; Horticulturist, Michael E. Ecker; Naturalist, Lori A. Totman; Supt. Grounds, Darrell Romine; Dir. Devel., Erin Neeb; Historian, Leslie Wagner; Business Mgr., Beverly Telepchak; Museum Shop Mgr., Brandon Clayton.
Personnel Profile: Full-Time Paid 36; Part-Time Paid 12; Part-Time Volunteers 312; Interns 5.
Governing Authority: private; nonprofit foundation. Tax-exempt: 501(c)(3); 509(a).
Institution Type/Description: Arboretum & Nature Center.
Collections: over 16,000 labeled trees, shrubs & vines; gardens; natural history. Historic Building: Daweswood House.
Research Fields: botanical; zoological.
Facilities: 2,000-vol. library of books on horticulture, history & natural history available for use on premises; 175-seat auditorium; 1,790 acres; nature center; classrooms; history center. Museum-related items for sale.
Activities: guided tours; lectures; classes; films; workshops; formally organized education programs for children, adults and undergraduate college students; temporary exhibitions.
Publications: monthly newsletter; annual report; various brochures, pamphlets & information sheets; web site materials.
Hours & Admission Prices: Grounds: daily dawn-dusk. No charge; donations accepted. Daweswood House Museum: Sat.-Sun. 1:30 & 3. Museum: adults $2, children 6-12 $1. Closed New Year's Day; Thanksgiving; Christmas. &
Attendance: 275,000 (estimated)
Membership: Sugar Maple (Individual) $30; Hickory (Household) $50; Buckeye (Supporting) $100.

LICKING COUNTY ARTS, 50 S. 2nd St., Newark, OH 43055. Tel.: 740-349-8031 & 350-7490.
E-mail: lcagalleryonsecond@gmail.com
Web Site: www.lickingcountyarts.com
Founded: 1959.
Congressional District: 10
Key Personnel: Museum Shop Cur., Mary Helen Fernandez Stewart.
Personnel Profile: Part-Time Paid 1; Part-Time Volunteers 30.
Governing Authority: federal; nonprofit organization. Parent Institution: Licking County Art Assoc. Tax-exempt: 501(c)(3).
Institution Type/Description: Art Gallery; Exhibition Gallery.

Collections: 19th- & 20th-century prints & paintings; ceramics; fiber pieces; hand-blown glass; wood sculpture.
Facilities: studio; class rooms. Gallery-related items for sale.
Activities: traveling & temporary exhibitions; guided tours; art workshops; art class series; gallery talks.
Publications: monthly newsletter, Art Print; monthly docent news.
Hours & Admission Prices: Tues.-Sat. 11-4; school tours by appointment. No charge; donations accepted. Closed holidays. &
Attendance: 2,500
Membership: Individual $20; Family $30; Patron $50; Business & Sustaining $100; Life $500; Benefactor $1,000 & up.

LICKING COUNTY HISTORICAL SOCIETY, Veterans Park, N. 6th St., Newark, OH 43058. Mailing Address: P.O. Box 785, Newark, OH 43058-0785. Tel.: 740-345-4898. Fax: 740-345-4898 (call first).
Web Site: www.lchsohio.org
Founded: 1947.
Congressional District: 17
Key Personnel: Pres., Linda Leffel; Cur., Emily Larson.
Personnel Profile: Part-Time Paid 5; Part-Time Volunteers 10.
Governing Authority: nonprofit organization. Subsidiary Institutions: Sherwood Davidson House, Webb House Museum, Buckingham Meeting House, Newark, OH; Robbins-Hunter Museum, Granville, OH. Tax-exempt: 501(c)(3).
Institution Type/Description: Historic Buildings.
Collections: 18th- & 19th-century artifacts; early 20th-century items. Historic Buildings: c.1820 Sherwood-Davidson House; 1835 Buckingham Meeting House; 1842 Avery Downer House; 1907 Webb House.
Research Fields: local artists.
Facilities: 1,000-vol. library of books & newspapers, available to the public; 100-seat restaurant; 5,000 sq. ft. exhibit space. Locally made items for sale.
Activities: docent program; films; formal education programs for children; guided tours; lectures; participatory & temporary exhibitions; school loan service; trips; 4 traveling trunk museums. Annual Events: John Clem Drummer Boy Breakfast; Graveyard Walk at Cedar Hill Cemetery; Christmas Open House.
Publications: quarterly newsletter, Licking County Historical Society Quarterly.
Hours & Admission Prices: Office & Library: by appointment. Library: no charge.
Attendance: 3,200 (estimated)
Membership: Student $10; Senior Citizen $20; Individual $25; Joint Senior $30; Family $35; Contributing $60; Sustaining $125; Life $1,000.

NATIONAL HEISEY GLASS MUSEUM, 169 W. Church St., Newark, OH 43055-4945. Tel.: 740-345-2932. Fax: 740-345-9638.
E-mail: business@heiseymuseum.org
Web Site: www.heiseymuseum.org
Founded: 1974.
Congressional District: 18
Key Personnel: Dir., Larry Burge.
Personnel Profile: Part-Time Paid 3.
Governing Authority: nonprofit organization; private collectors' club. Parent Institution: Heisey Collectors of America, Inc., 169 W. Church St., Newark. Tax-exempt: 501(c)(3).
Institution Type/Description: Glass Museum: housed in 1831 King House.
Collections: Heisey glassware; glassmakers' tools; 4,000 Heisey glass moulds; 700 Heisey etching plates; 7,000 Heisey mould drawings; Heisey factory memorabilia; original Heisey factory records & catalogs.
Research Fields: history, production & preservation of Heisey Glassware; history of glassmaking.
Facilities: library of original catalogs & factory records of the A.H. Heisey & Co.; materials on other glass factories, available for private use by members of the Heisey Collectors of America; media center; conference room; reading room. Museum-related items for sale.
Activities: guided tours; lectures; films; study clubs; TV programs; docent program; permanent, temporary & loan exhibitions; film & slide shows available on rental basis.
Publications: monthly newsletter, Heisey News.
Hours & Admission Prices: Jan.-Feb. Tues.-Sat. 10-4; March-Dec. Tues.-Sat. 10-4, Sun. 1-4. Adults $4; discount to groups of 10 or more; members & children under 18 accompanied by adult no charge. Blue Star Museum. Closed holidays. &
Attendance: 3,500 (estimated)
Membership: Associate & Voting Member $30; Individual Contributing $50; Joint Contributing $60 (two people in the same household); Family Contributing $75 (parents & children under 18); Patron $125; Sponsor $250; Benefactor $500.

SHERWOOD-DAVIDSON HOUSE, Veterans Park 6th St., Newark, OH 43058-0785. Mailing Address: P.O. Box 785, Newark, OH 43058-0785. Tel.: 740-345-4898. Fax: 740-345-4898 (call first).
E-mail: sherwooddavidson@yahoo.com
Web Site: www.lchsohio.org
Founded: 1947.
Congressional District: 17
Key Personnel: Pres., Linda Leffel; Cur., Emily Larson.
Personnel Profile: Part-Time Paid 1; Part-Time Volunteers 15.
Governing Authority: society. Parent Institution: Licking County Historical Society. Tax-exempt: 501(c)(3).
Institution Type/Description: Historic House Museum: c.1820 Sherwood-Davidson house-a federal style home built by Buckingham Sherwood.
Collections: Federal & early Victorian furnishings; historical costumes; toys; antique books; glass.
Research Fields: Licking County history.
Facilities: 1835 Buckingham House, available for meetings & receptions.
Activities: guided tours; lectures; films; concerts; arts festivals; TV & radio programs; docent program; permanent, temporary & traveling exhibitions.
Publications: brochures; quarterly newsletter, The Quarterly.
Hours & Admission Prices: May to mid-Dec. Tues., Thurs. & Sat. 1-4. Adults $3; members no charge. Closed holidays.
Attendance: 900 (estimated)
Membership: Student $10; Senior Citizens 65 & over $20; Individual $25; Joint Senior Citizens $30; Family $35; Contributing $60; Sustaining $125; Life $1,000.

WEBB HOUSE MUSEUM, 303 Granville St., Newark, OH 43055-4480. Tel.: 740-345-8540.
E-mail: webbhouse@windstream.net
Web Site: lchsohio.org
Founded: 1978.
Congressional District: 17
Key Personnel: Cur., Mindy Honey Nelson.
Personnel Profile: Part-Time Paid 3; Part-Time Volunteers 2.
Governing Authority: society; nonprofit. Licking County Historical Society, P.O. 785, 6th St. Park, Newark, OH 43055. Tel. 740-345-4898. Tax-exempt: 501(c)(3).
Institution Type/Description: Historic House Museum & Gardens.
Collections: period furnishings, art, silver, china, glass, photographs by Clarence White & Ema Spencer; carriage house.
Research Fields: Arts & Crafts Movement.
Facilities: lawns & gardens with a 1900 atmosphere.
Activities: guided tours; craft classes. Museum Sponsors: Herbal Luncheon in summer; Midsummer Twilight Garden Tour in July; Holiday Open House in December.
Publications: newsletter, Licking County Historical Society Quarterly.
Hours & Admission Prices: Thurs.-Fri. 1-4; other times by appointment. No charge; donations accepted. Tour groups $2 per person.
Attendance: 200 (estimated)

THE WORKS: OHIO CENTER FOR HISTORY, ART & TECHNOLOGY, 55 S. First St., Newark, OH 43055-5429. Mailing Address: P.O. Box 721, Newark, OH 43058-0721. Tel.: 740-349-9277. Fax: 740-345-7252.
Web Site: www.attheworks.org
Formerly: Institute of Industrial Technology
Founded: 1996.
Congressional District: 18
Key Personnel: C.E.O., Marcia W. Downes; Chm., John Hinderer; Museum Shop Mgr., Tawna England.
Personnel Profile: Full-Time Paid 12; Part-Time Paid 6; Part-Time Volunteers 90; Interns 1.
Governing Authority: Parent Institution: Howard E. Lefevre Foundation. Subsidiary Institution: Ohio State University at Newark. Tax-exempt: 501(c)(3).
Institution Type/Description: Science & History Museum.
Collections: Licking County Ohio history; glass blowing studio; working letterpress; interurban train; 1881 steam engine factory; wood turning shop; car barn; wood working shop.
Facilities: Museum-related items for sale.
Activities: tours; glass blowing demonstrations; school field trips; home school programs; artist-in-residence program in glass & print.
Publications: quarterly newsletter, The Works; quarterly class schedules; annual report.
Hours & Admission Prices: Jan.-March Tues.-Sat. 10-6, Sun. 12-4; April-Dec. Tues.-Sat. 10-6. Adults $8, senior citizens $6, children $4; discounts to ASTC & Smithsonian Affiliate members; members no charge. Closed Labor Day; Memorial Day; Thanksgiving; Christmas. &

Attendance: 63,305 (accurate)
Membership: Individual $40; Family $65; Partner $100; Sustaining $200; Sponsor $500; Patron $1,000.

Newcomerstown

OLDE MAIN STREET MUSEUM & SOCIAL CENTER, 213 W. Canal St., Newcomerstown, OH 43832-1101. Mailing Address: P.O. Box 443, Newcomerstown, OH 43832-0443. Tel.: 740-498-7735.
Founded: 2007.
Congressional District: 16
Key Personnel: Pres., Mary Watts; Vice Pres., Ray McFadden; Sec., Ellen Pickrell; Treas., Don Fenton; Coord. Social Events, Marlene Ross; Historian, Elaine Mayenschien.
Governing Authority: society. Parent Institution: Newcomerstown Historical Society. Tax-exempt.
Institution Type/Description: History Museum.
Collections: local history; period furnishings; personal artifacts.
Facilities: banquet facilities.
Activities: rental facilities.
Hours & Admission Prices: Memorial Day to Oct. Tues.-Sat. 10-4, Sun. 1-4; tours by appointment. Adults $3. &
Membership: Regular $10; Business $20.

TEMPERANCE TAVERN, 221 W. Canal St., Newcomerstown, OH 43832-1101. Mailing Address: P.O. Box 443, Newcomerstown, OH 43832-0443. Tel.: 614-498-7735.
Founded: 1974.
Congressional District: 16
Key Personnel: Pres., Mary Watts; Vice Pres., Ray McFadden; Sec., Ellen Pickrell; Treas., Don Fenton; Coord. Social Events, Marlene Ross; Historian, Elaine Mayenschien.
Personnel Profile: Part-Time Volunteers 30.
Governing Authority: society. Newcomerstown Historical Society. Tax-exempt.
Institution Type/Description: Historical Society Museum: housed in c.1841 Temperance Tavern.
Collections: local history; Indian artifacts; period furniture; archaeology display; 1827-1930 costumes; Cy Young monuments; Woody Hayes.
Research Fields: local history; Indian artifacts.
Facilities: library of local history books available for inter-library loan & for research.
Activities: guided tours; films.
Hours & Admission Prices: Memorial Day to Labor Day Tues.-Sat. 10-3, Sun. 1-4; tours by appointment. Adults $3; donations accepted.
Attendance: 800 (estimated)
Membership: Regular $10; Business $20.

Niles

NATIONAL MCKINLEY BIRTHPLACE MEMORIAL & MCKINLEY BIRTHPLACE HOME, 40 N. Main St., Niles, OH 44446-5012. Tel.: 330-652-1704. Fax: 330-652-5788.
E-mail: mckinley@mcklib.org
Web Site: www.mckinley.lib.oh.us
Founded: 1911.
Congressional District: 13
Key Personnel: Pres. (V), J. Terrance Dull; Vice Pres., Carol Smaltz; Dir. Library, Patrick Finan.
Personnel Profile: Full-Time Paid 9; Part-Time Paid 7; Part-Time Volunteers 1.
Governing Authority: nonprofit organization. McKinley Birthplace Home, 40 S. Main St., Niles, OH 44446. Tel. 330-652-1704, ext. 203; Fax: 330-652-5788.
Institution Type/Description: History Museum: housed in Memorial Building located on the site formerly occupied by Little White School House attended by Pres. McKinley.
Collections: William McKinley memorabilia & artifacts.
Facilities: 500-seat auditorium. Museum-related items for sale.
Activities: guided tours; gallery talks; permanent exhibitions.
Hours & Admission Prices: Museum: Mon.-Sat. 9-5. Home: May-Oct. Sat. 9-5; other times by appointment. No charge. Closed holidays. &
Attendance: 8,000 (estimated)
Membership: Associate $10; Sustaining $25; Patron $50; Life $100.

WARD-THOMAS MUSEUM, 503 Brown St., Niles, OH 44446. Mailing Address: P.O. Box 368, Niles, OH 44446. Tel.: 330-544-2143.
E-mail: curator@nileshistoricalsociety.org

Founded: 1976.
Key Personnel: Pres. (V), Nancy Malone; Cur., Audrey John.
Governing Authority: Parent Institution: Niles Historical Society. Tax-exempt.
Institution Type/Description: Historical Society Museum: housed in the former home of James Ward and John & Margaret Thomas, two prominent Niles industrial families; built in 1862.
Collections: Ward & Thomas family history; period furnishings; personal artifacts; photographs; clothing; replica First Ladies gowns. Historic Buildings: barn; 1925 greenhouse.
Research Fields: family history.
Publications: newsletter; books.
Hours & Admission Prices: 1st Sun. each month; other times by appointment. No charge; donations accepted.
Membership: Individual $15; Family $25; Business $50-$100.

North Canton

HOOVER HISTORICAL CENTER, 1875 E. Maple St., North Canton, OH 44720-3331. Tel.: 330-499-0287. Fax: 330-494-4725.
E-mail: ahaines@walsh.edu
Web Site: www.walsh.edu/hoover-historical-center
Founded: 1978.
Congressional District: 16
Key Personnel: Operations Coord. & Museum Shop Mgr., Ann Haines; Center Rep., Patty Garber.
Personnel Profile: Full-Time Paid 1; Part-Time Paid 6; Part-Time Volunteers 75; Interns 2.
Governing Authority: company. Parent Institution: Walsh University. Tax-exempt.
Institution Type/Description: Company & History Museum: housed in 1853 Hoover family home.
Collections: tanning & leather craft tools; vacuum cleaners; Victorian furniture, decor, clothing & ABC plates; vintage clothing & accessories; photographs; Hoover Company memorabilia; vintage advertisements; Hoover family; company; advertising; World War II artifacts by Hoover Company; herb gardens; 1800-1980s ladies fashion. Historic Buildings: c.1840 Tannery; 1853 farmhouse with 1870 Victorian Italianate style addition.
Research Fields: Victorian customs 1870-1900; Hoover Company history; vacuum cleaner technology; Victorian architecture & decor; herbs; North Canton Community; Walsh University history.
Facilities: company archives; herb gardens.
Activities: docent program; videos; guided tours; Christmas caroling & horse-drawn wagon rides. Annual Events: 1860s baseball matches May-September; Summer Garden Tea & Tours; Tales in Thyme (outdoor storytelling) in August; Christmas Open House in December.
Publications: quarterly newsletter, Center News.
Hours & Admission Prices: March to mid-Dec. Wed.-Sat. 1-4; groups of 8 or more by appointment. Adults $5; discounts to veterans with military ID during veterans' week; children under 12 no charge. Closed major holidays. &

Attendance: 2,000 (estimated)

MILITARY AVIATION PRESERVATION SOCIETY AIR MU-SEUM, 2260 International Pkwy., North Canton, OH 44720. Tel.: 330-896-6332.
E-mail: kovesci.kim@mapsairmuseum.org
Web Site: mapsairmuseum.org
Founded: 1990.
Key Personnel: Dir., Kim Kovesci; Chm. (V), Robert Schwartz; Museum Shop Mgr., Robert Johnston.
Personnel Profile: Full-Time Paid 2.
Governing Authority: nonprofit organization.
Institution Type/Description: Military History Museum.
Collections: military aviation history & artifacts; aircraft.
Activities: special events; groups tours.
Hours & Admission Prices: Tues.-Sat. 9-4:30, Sun. 11:30-4. Closed New Year's Day; Easter; Memorial Day; Independence Day; Labor Day; Thanksgiving; Christmas. &
Attendance: 27,000 (accurate)
Membership: Student $25; Senior 60 & over $30; Individual and Senior Family 60 & over $40; Family $50; Life $400.

Northfield

PALMER HOUSE, HISTORICAL SOCIETY OF OLDE NORTHFIELD, 9390 Olde Eight Rd., Northfield, OH 44067. Mailing Address: P.O. Box 99, Northfield, OH 44067. Tel.: 330-468-0909. Fax: 330-468-1163.
E-mail: hson@worldnet.att.net

Web Site: www.hson.info
Founded: 1956.
Congressional District: 13
Key Personnel: Treas., Jill Potter.
Personnel Profile: Part-Time Volunteers 5.
Governing Authority: society. Parent Institution: Historical Society of Olde Northfield. Tax-exempt: 501(c)(3).
Institution Type/Description: Historical Society Museum.
Collections: agriculture; costumes; folklore; glass; Indian artifacts; kitchen ware; toys; tools; records; pictures. Historical House: 1844 Vertical Plank Western Reserve House.
Research Fields: history of Western Reserve lands.
Facilities: 100-vol. library of local & county history books available for use on premises. Surplus books & museum-related items for sale.
Activities: guided tours; lectures; films; formally organized education programs for adults; temporary exhibitions.
Publications: quarterly tertiary newsletter.
Hours & Admission Prices: 4th Sun. each month 2-4; other times by appointment. No charge.
Attendance: 20 (estimated)
Membership: Individual $7; Husband & Wife $10; Contributing $25; Life $100.

Norwalk

FIRELANDS HISTORICAL SOCIETY MUSEUM & LANING-YOUNG RESEARCH CENTER, 4 Case Ave., Norwalk, OH 44857-1404. Mailing Address: P.O. Box 572, Norwalk, OH 44857-0572. Tel.: 419-668-6038. Facebook: Firelands Historical Society & Laning-Young Research Center.
E-mail: director@firelandsmuseum.org
Web Site: www.firelandsmuseum.org
Founded: 1857.
Congressional District: 13
Key Personnel: Dir. & Cur., Kristie Bilger.
Personnel Profile: Part-Time Paid 1; Part-Time Volunteers 30; Interns 1.
Governing Authority: society; nonprofit. Parent Institution: Firelands Historical Society. Subsidiary Institution: Laning-Young Research Center, 9 Case Ave., Norwalk, OH 44857. Tax-exempt: 501(c)(3).
Institution Type/Description: History Museum: housed in 1836 Preston-Wickham House.
Collections: archaeology; archives; primitive paintings; costumes; decorative arts; folklore; geology; glass; Indian artifacts; numismatic collections; manuscripts; paleontology; transportation; weapons.
Major Exhibits: Norwalk Businesses: Past & Present, Summer 2013 to Fall 2014; American Indian: The Firelands Perspective, Fall 2013 to Fall 2016.
Facilities: 4,000-vol. library of historical and genealogical books available for research on premises; research center; reading room. Museum-related items for sale.
Activities: guided tours; lectures; study clubs; permanent & temporary exhibitions; Firelands Steward Program. Museum Sponsors: Holiday Home Tour; Garden Tour.
Publications: society newsletter, Firelands Pioneer.
Hours & Admission Prices: May & Sept.-Oct. Sat. 10-3, Sun. 12-4; June-Aug. Tues.-Sat. 10-3, Sun. 12-4; other times by appointment. Adults $5, senior $4, youth 12-18 $3; discounts to groups of 10 or more; members & children under 12 no charge.
Attendance: 550 (estimated)
Membership: Individual $20; Couple $35; Business $100. Life: Individual $500; Business $1,000.

Norwich

NATIONAL ROAD/ZANE GREY MUSEUM, (M), 8850 E. Pike, Norwich, OH 43767-9785. Mailing Address: c/o John & Annie Glenn Museum Foundation, P.O. Box 107, New Concord, OH 43762. Tel.: 740-872-3143. Fax: 740-872-3510.
E-mail: jmuseum@newconcord-oh.gov
Web Site: johnglenn.org
Founded: 1973.
Congressional District: 10
Key Personnel: Dir. Operatioins, Debbie Allender; Exec. Dir., Keith Eberly; Mgr., Kathryn Miller; Educ. Spec., JoAnna Duncan; Admin., Laura Holmes.
Personnel Profile: Full-Time Paid 1; Part-Time Paid 4; Part-Time Volunteers 20.
Governing Authority: society. Parent Institution: The Ohio Historical Society, Ohio Historical Center, 1985 Velma Ave., Columbus, OH 43211. Subsidiary Institution: John & Annie Glenn Foundation. Tax-exempt: 170(b)(1)(A).
Institution Type/Description: General Museum.

Collections: items pertaining to transportation; vehicles; Ohio art pottery; personal items of Zane Grey.
Facilities: Publications & other items relating to transportation for sale.
Activities: guided tours; permanent exhibitions.
Hours & Admission Prices: May-Sept. Wed.-Sat. 10-4, Sun. 1-4. Adult $7, seniors $6, students $3; discounts to groups and AAA, & Blue Star Museum member; children 5 & under and member no charge. Closed holidays. &
Attendance: 5,112 (accurate)
Membership: Individual $35; Family $55; History Lover $85; Supporter $235.

Norwood

DRAKE PLANETARIUM, 2020 Sherman Ave., Norwood, OH 45212-2616. Tel.: 513-396-5578. Fax: 513-396-6486.
E-mail: csteger@drakeplanetarium.org
Web Site: www.drakeplanetarium.org
Key Personnel: Dir., Pamela Bowers; Business Mgr., Carolyn Steger.
Governing Authority: nonprofit organization.
Institution Type/Description: Planetarium.
Collections: astronomy-related exhibits.
Activities: educational programs.
Hours & Admission Prices: Call for hours.

DRAKE SCIENCE CENTER, (M), 2020 Sherman Ave., Norwood, OH 45212-3100. Tel.: 513-396-5578. Fax: 513-396-6486.
E-mail: csteger@drakeplanetarium.org
Web Site: www.drakeplanetarium.org/contact.html
Key Personnel: Business Mgr., Carolyn Steger
Institution Type/Description: Science Center.
Collections: hands-on exhibits.
Activities: summer camp; educational programs.
Hours & Admission Prices: Call for hours. Discounts to AAM members.

Oak Hill

WELSH-AMERICAN HERITAGE MUSEUM, 412 E. Main St., Oak Hill, OH 45656-1229. Tel.: 740-682-7057.
Web Site: jacksonohio.org/welshmuseum.htm
Founded: 1971.
Key Personnel: Cur., Mildred Bangert.
Governing Authority: nonprofit organization.
Institution Type/Description: History Museum.
Collections: Welsh culture & traditions; records; artifacts; books; photographs.
Hours & Admission Prices: By appointment. No charge; donations accepted.

Oberlin

* **ALLEN MEMORIAL ART MUSEUM - OBERLIN COLLEGE, (M),** 87 N. Main St., Oberlin, OH 44074-1161. Tel.: 440-775-8665. Fax: 440-775-6841.
E-mail: mharding@oberlin.edu
Web Site: www.oberlin.edu/amam
Founded: 1917.
Congressional District: 13
Key Personnel: Dir. The John G.W. Cowles, Andria Derstine; Publications, Membership & Media, Megan Harding; Registrar, Lucille Stiger; Cur. Education, Jason Trimmer; Cur. Academic Programs, Liliana Milkova; Administrative Asst., Sally Moffitt.
Personnel Profile: Full-Time Paid 11; Part-Time Volunteers 15; Interns 1.
Governing Authority: college; board of trustees. Parent Institution: Oberlin College. Tax-exempt: 501(c)(3).
Institution Type/Description: Art Museum.
Collections: Egyptian, Greek & Roman sculpture & decorative arts; East Asian & South Asian sculpture, painting & prints, including the Mary A. Ainsworth collection of Japanese woodblock prints; Islamic carpets; medieval art; 14th to 20th-century European paintings, sculpture & decorative arts; 19th to 20th-century American paintings, sculpture & decorative arts; African & pre-Columbian art; contemporary painting, sculpture & video; European & American prints, drawings & photographs; textiles.
Major Exhibits: Latin American Art at the Allen, 9/14-5/15.
Research Fields: all fields related to collections; Eva Hesse Archives; Frantisek Kupka.
Facilities: 82,000-vol. Clarence Ward Art Library; Wolfgang Stechow Print Study Room. Slides, catalogues & AMAM Bulletin publications for sale.
Activities: guided tours; lectures; gallery talks; adult & children's classes; temporary exhibitions; concerts; domestic & international loans.
Publications: biannual newsletter; Allen Memorial Art Museum News; annual bulletin, Allen Memorial Art Museum Bulletin; exhibition catalogues & brochures; permanent collection catalogue on CD-Rom; cards; postcards.

Hours & Admission Prices: Tues.-Sat. 10-5, Sun. 1-5. No charge; donations accepted. Closed major holidays. &
Attendance: 22,533 (accurate)
Membership: Senior Citizen & Student $20; Individual $50; Family $75; Contributing $150; Supporting $500; Director's Circle $1,000 & up.

NEW UNION CENTER FOR THE ARTS, 39 S. Main St., Oberlin, OH 44074-1662. Tel.: 440-774-7158. Fax: 440-775-1107.
E-mail: favagallery@oberlin.net
Key Personnel: Exec. Dir., Betsy Manderen
Institution Type/Description: Art Gallery.
Collections: sculpture; paintings; photographs.
Hours & Admission Prices: Tues.-Sat. 11-5, Sun. 1-5.

* **OBERLIN HERITAGE CENTER, (M), (I),** 73-1/2 S. Professor St., Oberlin, OH 44074. Mailing Address: P.O. Box 0455, Oberlin, OH 44074-0455. Tel.: 440-774-1700. Fax: 440-774-8061. Facebook: Oberlin Heritage Center.
E-mail: history@oberlinheritage.org
Web Site: www.oberlinheritage.org
Formerly: Oberlin Heritage Center/O.H.I.O. (Oberlin Historical and Improvement Organization)
Founded: 1903.
Congressional District: 13
Key Personnel: Dir. & C.E.O., Patricia Murphy; Pres. (V), Walter Edling; Business Mgr., Bethany Hobbs.
Personnel Profile: Full-Time Paid 2; Part-Time Paid 4; Part-Time Volunteers 250; Interns 5.
Volunteer Hours: 4,328
Operating Expenses: 365,566
Operating Income: 459,637
Governing Authority: nonprofit organization. Parent Institution: The Oberlin Historical & Improvement Organization. Tax-exempt.
Institution Type/Description: Historic House: 1866 Italianate villa, originally the home of Giles Shurtleff, leader of the first African American regiment from Ohio in the Civil War & the home of James Monroe, occupied by Monroe family 1870-1930, a college professor, important abolitionist, U.S. Congressman & colleague of Frederick Douglass and President Lincoln's ambassador to Brazil; part of the National Park Service's Underground Railroad Network to Freedom program.
Collections: 19th-century furniture & furnishings; local history materials. Historic Buildings: The 1836 Little Red Schoolhouse was Oberlin's first public school, now restored as a pioneer era one-room school. The 1884 Jewett House, a brick Victorian house, was the home of Oberlin College chemistry professor Frank Fanning Jewett & his wife Frances, an author of school textbooks on public health. The Jewetts rented out second floor rooms to male Oberlin college students. It is listed on the National Register of Historic Places.
Research Fields: abolitionists; Oberlin College; underground railroad; temperance movement; women's history; Civil War; building inventories; Oberlin family histories.
Activities: public programs on aspects of local history; heritage preservations & civic affairs; children's summer camps; school field trips; speaker's bureau programs.
Publications: Newsletter, Oberlin Heritage Center Gazette; e-newsletter, E-Gazette.
Hours & Admission Prices: Guided Tours: Tues., Thurs. & Sat. 10:30 & 1:30; other times by appointment. Adults $6; discounts to AAM, ICOM, AAA, AASLH & Time Traveler's members; members & children accompanied by adult no charge. Closed Christmas-New Year's Day; all major holidays. Gift Shop Office (at Monroe House): Tues.-Sat. 10-3.
Attendance: 8,996 (accurate)
Membership: Senior & Student $15; Individual $25; Family & Dual $40; Heritage Collector $50; Heritage Rescuer $100; Heritage Leader $250; Heritage Ambassador $500; Heritage Champion $1,000. Business or Organization Membership: Active $25; Heritage Collector $50; Heritage Rescuer $100; Heritage Leader $250; Heritage Ambassador $500; Heritage Champion $1,000.

Oregon

OREGON-JERUSALEM HISTORICAL SOCIETY, 1133 Grasser St., Oregon, OH 43616-7632. Mailing Address: P.O. Box 167632, Oregon, OH 43616-7632. Tel.: 419-693-7052. Fax: 419-693-7052.
Web Site: www.ojhs.org
Founded: 1963.
Congressional District: 9
Key Personnel: Pres. (V), Betty Metz; Museum Shop Mgr., Theresa Berry.
Personnel Profile: Part-Time Volunteers 30.

Governing Authority: society. Tax-exempt.
Institution Type/Description: General Museum.
Collections: school records; maps; old farm equipment; musical instruments; office equipment; old household furniture; Indian artifacts; china; glass; costumes; country store; 1900 one-room school; barber shop; doctor's office; Civil War artifacts: oil paintings, clothes, swords, drums, medical instruments, flags, cannon balls, shells, bayonets; Spanish-American War; World War I; World War II; Korean; Vietnam military. Historic Building: Red Brick Schoolhouse.
Research Fields: historical sites for markers.
Facilities: elementary school.
Activities: guided tours; genealogy assistance available to visitors; history day camp; tour of historic homes; series of First Lady's and event teas.
Publications: bimonthly newsletter.
Hours & Admission Prices: Thurs. 10-2; other times by appointment. No charge; donations accepted. &
Attendance: 1,500 (accurate)
Membership: Individual $15; Family $30; Contributing $100 & up & Individual Life; Family Life $200.

Oregonia

* **FORT ANCIENT MUSEUM, (M),** 6123 State Rt. 350, (Exit 32 & 36 off I-71), Oregonia, OH 45054-9708. Tel.: 513-932-4421. Fax: 513-932-4843; 800-860-0141.
E-mail: jblosser@fortancient.org
Web Site: fortancient.org
Founded: 1891.
Congressional District: 6
Key Personnel: Dir., Mark Meister; Area Mgr., Jack K. Blosser.
Personnel Profile: Full-Time Paid 2; Part-Time Paid 3; Part-Time Volunteers 18.
Governing Authority: nonprofit organization. Parent Institution: Ohio Historical Society, I-71 & 17th Ave., Columbus, OH. 43211. Subsidiary Institution: Dayton Society of Natural History. Tax-exempt.
Institution Type/Description: Prehistoric Site Museum.
Collections: artifacts of prehistoric Indian life & culture.
Research Fields: ceremonial sites; habitation areas; Ohio Valley prehistory; hilltop enclosure.
Facilities: 9,000 sq. ft. exhibit space; garden; picnic area; classroom. Literature & other museum-related items for sale.
Activities: trails; school programs; outdoor nature programs. Museum Sponsors: Annual American Indian Gathering in June; Bluegrass Festival.
Hours & Admission Prices: April-Nov. Tues.-Sat. 10-5, Sun. 12-5; Dec.-March Sat. 10-5, Sun. 12-5; other times by appointment. Adults $6, seniors 60 & over and students $5; discounts to AAA, AAM & ICOM members; OHS & Dayton Society of Natural History members and children under 5 no charge. &
Attendance: 23,900 (estimated)
Membership: Ambassador $50; Military $75; Family & Grandparents $85.

Orrville

ORRVILLE HISTORICAL MUSEUM, 142 Depot St., Orrville, OH 44667. Mailing Address: P.O. Box 437, Orrville, OH 44667. Tel.: 330-930-0113.
Governing Authority: Parent Institution: Orrville Historical Society. Subsidiary Institution: Smith Orr Homestead, 365 W. Market St., Orrville, OH 44667.
Institution Type/Description: History Museum: housed in the former Manhattan Restaurant.
Collections: local history & culture; period furnishings; personal artifacts; J.M. Smucker Company memorabilia; 1876 Reed Organ.
Publications: quarterly newsletter.
Hours & Admission Prices: Feb.-Nov. Wed. 9-12, 2nd & 4th Sat. each month 1-4; other times by appointment.
Membership: Individual $15; Family $30.

Oxford

HEFNER MUSEUM OF NATURAL HISTORY, (M), 100 Upham Hall, Miami Univ., Oxford, OH 45056. Tel.: 513-529-4617. Fax: 513-529-6900.
E-mail: kaufmadg@miamioh.edu
Web Site: www.environmentaleducationohio.org
Formerly: Hefner Zoology Museum
Founded: 1951.
Key Personnel: C.E.O., Donald G. Kaufman; Education, Julie Robinson; Public Rels., Cecilia Berg.
Personnel Profile: Part-Time Paid 4.
Governing Authority: public university; nonprofit. Parent Institution: Miami University, Oxford, OH. Tax-exempt: 501(c)(3).

Institution Type/Description: Natural History Museum.
Collections: fauna of southwest Ohio.
Facilities: 350-vol. library; 1,800 sq. ft. exhibit space; educational facilities.
Activities: formal education programs; guided tours; lectures; temporary exhibitions.
Hours & Admission Prices: June-Aug. Mon.-Fri. 9-5; Sept.-May Mon.-Fri. 9-5. No charge; donations accepted. Closed university holidays. &
Attendance: 5,000 (estimated)

HIESTAND GALLERIES, 124 Art Bldg., Oxford, OH 45056. Tel.: 513-529-2900.
E-mail: art@muohio.edu
Web Site: arts.muohio.edu/art/facilities/hiestand-galleries
Key Personnel: Dir., Ann Taulbee
Institution Type/Description: Art Gallery.
Collections: paintings; sculpture; drawings.
Activities: special events.
Hours & Admission Prices: Mon.-Fri. 9-4:30; other times by appointment.

* **MIAMI UNIVERSITY ART MUSEUM, (M),** 801 S. Patterson Ave., Oxford, OH 45056-3435. Tel.: 513-529-2232. Fax: 513-529-6555. TDD: 513-529-1541.
E-mail: wicksrs@miamioh.edu
Web Site: www.miamioh.edu/art-museum
Founded: 1978.
Congressional District: 8
Key Personnel: Dir., Robert S. Wicks, Ph.D.; Cur. Exhibitions, Jason Shaiman; Cur. Education, Cynthia C. Collins; Registrar, Laura Stewart; Program Assoc., Susan V. Gambrell; Preparator & Operations Mgr., Mark De-Gennaro; Mktg. & Commun. Coord., Sherri Krazl.
Personnel Profile: Full-Time Paid 8; Part-Time Paid 2; Part-Time Volunteers 120; Interns 3.
Governing Authority: university. Parent Institution: Miami University. Subsidiary Institution: McGuffey Museum. Tax-exempt: 501(c)(3).
Institution Type/Description: University Museum.
Collections: ancient; Islamic; Native American; pre-Columbian; 19th-20th century American & European painting, sculpture & works on paper; African; European & American decorative arts; Leica cameras; international folk art & textiles; contemporary American folk art & outsider art; Chinese, oceanic & ancient art.
Research Fields: pertaining to collection.
Facilities: library of exhibition catalogues & reference books related to permanent collection available for research by appointment; 150-seat auditorium. Museum-related items for sale.
Activities: guided tours; lectures; films; gallery talks; concerts; docent program; formally organized education programs for undergraduate & graduate college students affiliated with Miami University; art education program for school children; loan, permanent, temporary & traveling exhibitions; educational outreach program; diverse volunteer & intern program; membership association activities.
Publications: brochure, Calendar of Events; occasional exhibition catalogues, Major Exhibitions; newsletter; annual report.
Hours & Admission Prices: Tues.-Fri. 10-5, Sat. 12-5; additional hours during evening programs & events. No charge. Closed national & university holidays. &
Attendance: 30,000 (estimated)
Membership: Miami Individual $35; Individual $40; Miami Family $60; Family $70; Reciprocal Individual $100; Reciprocal Family $120; Distinguished $500; Director's Circle $1,000; Exhibitor $2,500.

WILLIAM HOLMES MCGUFFEY MUSEUM, 401 E. Spring St., Oxford, OH 45056-3646. Tel.: 513-529-8380. Fax: 513-529-2637.
E-mail: mcguffeymuseum@miamioh.edu
Web Site: www.units.muohio.edu/mcguffeymuseum
Founded: 1960.
Congressional District: 8
Key Personnel: Pres. (V), Dr. Sue Jones; Dir., Robert Wicks; Cur., Stephen C. Gordon.
Personnel Profile: Part-Time Paid 1; Part-Time Volunteers 25; Interns 2.
Governing Authority: university. Parent Institution: Miami University. Tax-exempt: 501(c)(3).
Institution Type/Description: Historic House Museum: 1833 home of William Holmes McGuffey.
Collections: McGuffey Eclectic Readers; McGuffey memorabilia, furniture & other decorative arts of the 19th century.
Major Exhibits: Miami University & the Stantons, 3/14-12/14.
Facilities: 19th century restored home.
Activities: guided tours; permanent exhibitions; special volunteer group.

Publications: leaflets, William Holmes McGuffey; The McGuffey Museum of Miami University, Oxford, Ohio; newsletter, The McGuffey Report Card.
Hours & Admission Prices: Thurs.-Sat. 1-5. No charge; donations accepted. Closed university holidays. ♿
Attendance: 1,500 (accurate)

Painesville Township

LAKE COUNTY HISTORICAL SOCIETY, (M), 415 Riverside Dr., Painesville Township, OH 44077-5321. Tel.: 440--639-2945. Fax: 440-639-2947. Facebook: Lake County History Center.
Web Site: www.lakehistory.org
Founded: 1936.
Congressional District: 19
Key Personnel: Exec. Dir., Kathie Purmal; Pres., Morris W. Beverage, III; Pres. (V) & Museum Shop Mgr., Annie Hitchcock.
Personnel Profile: Full-Time Paid 3; Full-Time Volunteers 2; Part-Time Paid 2; Part-Time Volunteers 225.
Governing Authority: nonprofit organization. Tax-exempt.
Institution Type/Description: History Museum.
Collections: 19th- & 20th-century decorative arts, costumes, photographs, archival material, tools, implements & manuscripts.
Research Fields: local history; architecture; genealogy; family history.
Facilities: 2,350-vol. research library of history & genealogy books available to researchers. Museum-related items, books, quarterly journal & pamphlets for sale.
Activities: Historic Home register; plaques for structures at least 50 years old; in-school & on-site education programs.
Publications: quarterly newsletter.
Hours & Admission Prices: Museum: May-Oct. Tues.-Sat. 10-2, Sun. 1-4; Nov.-April Tues.-Sat. 10-2. Research Library: Tues.-Sat. 10-2, other times by appointment. Museum Self-Guided Tour: $3. Research Library: adults $7; discount to AAM & ICOM members; members no charge.
Attendance: 28,000 (estimated)
Membership: Senior Citizen (over 60) $25; Senior Couple (over 60) $40; Individual $35; Family $50; Gold Patron $100; Platinum $500.

Parma

STEARNS HOMESTEAD, 6975 Ridge Rd., Parma, OH 44129. Mailing Address: P.O. Box 29002, Parma, OH 44129. Tel.: 440-845-9770 & 5522.
Institution Type/Description: Historic Homestead.
Collections: local history & culture; period furnishings; farm equipment & tools; farm animals; historic houses.
Facilities: Gift items for sale.
Activities: special events; demonstrations; school & group tours.
Hours & Admission Prices: May to mid-Oct. Sat.-Sun. 12-4. No charge.

Peebles

SERPENT MOUND MUSEUM, 3850 State Rte. 73, Peebles, OH 45660-9128. Mailing Address: c/o Arc of Appalacia Preserve System, 7660 Cave Rd., Bainbridge, OH 45612. Tel.: 937-587-2796. Fax: 937-587-1116. Facebook: Arc of Appalacia Preserve System.
E-mail: serpentmound@arcofappalachia.org
Web Site: www.arcofappalachia.org
Founded: 1900.
Congressional District: 6
Key Personnel: Dir., Nancy Stranahan.
Governing Authority: Parent Institution: Ohio Historical Society, 1985 Velma Ave., Columbus, OH. 43211. Subsidiary Institution: Arc of Appalachia Preserve System (AAPS). Tax-exempt.
Institution Type/Description: Indian Museum.
Collections: Adena Indian culture.
Hours & Admission Prices: Park & Earthworks: daily dawn-dusk. Museum & Gift Shop: March & Nov.-Dec. Sat.-Sun. 10-4, April daily 10-4; May & Sept.-Oct. Thurs.-Mon. 10-4, Fri.-Sun. 9-5; June-Aug. Mon.-Thurs. 10-4, Fri.-Sun. 10-6. Parking $8; OHS members no charge.
Attendance: 40,000 (accurate)
Membership: Annual $35 & up.

Perrysburg

✷ FORT MEIGS: OHIO'S WAR OF 1812 BATTLEFIELD, (M), 29100 W. River Rd., Perrysburg, OH 43551-6019. Perrysburg, OH 43552. Tel.: 419-874-4121. Facebook: Fort Meigs: Ohio's War of 1812 Battlefield.
E-mail: rfinch@ohiohistory.org
Web Site: www.fortmeigs.org/
Formerly: Fort Meigs State Memorial
Founded: 1950.
Congressional District: 5
Key Personnel: Dir., Rick Finch; Chm. (V), George Jones, III; Museum Shop Mgr., Barb Brauer.
Personnel Profile: Full-Time Paid 4; Part-Time Paid 2; Part-Time Volunteers 35; Interns 1.
Governing Authority: The Ohio Historical Society. Ohio Historical Center Site Operations, 1985 Velma Ave., Columbus, OH. 43211. Subsidiary Institution: Fort Meigs Association, P.O. Box 3, Perrysburg, OH 43552. Tax-exempt.
Institution Type/Description: Military Fort Museum: reconstructed 1813 fort from the War of 1812; History Museum.
Collections: cannon batteries; earthen traverses; seven block houses; dioramas; artifacts & weapons of the war; history displays.
Research Fields: War of 1812; early Ohio statehood.
Facilities: 14,000 sq. ft. exhibit space; visitor center.
Activities: tours; lectures; living history demonstrations; permanent exhibitions; hands-on programming; simulated reality.
Publications: books, Men of Patriotism, Courage, and Enterprise; Fort Meigs in the War of 1812.
Hours & Admission Prices: Museum & Visitors Center: daily. Fort: April-Oct. Wed.-Sat. 9:30-5, Sun. 12-5. Adults $8, seniors 60 & over $7, students $4; members & children 5 and under no charge.
Attendance: 30,000 (accurate)

OWENS COMMUNITY COLLEGE/WALTER E. TERHUNE GALLERY, Center for Fine & Performing Arts, 30335 Oregon Rd., Perrysburg, OH 43551-4593. Mailing Address: P.O. Box 10000, Toledo, OH 43699-1947. Tel.: 567-661-2721. Fax: 567-661-7011.
Web Site: www.owens.edu/
Founded: 2003.
Personnel Profile: Part-Time Paid 1; Interns 3.
Governing Authority: public college.
Institution Type/Description: Art Museum.
Collections: works by students & faculty including paintings, photography, ceramics & sculpture.
Major Exhibits: Forming a Visual History, 2/3/14.
Facilities: 50-seat cafeteria; 1,880 sq. ft. exhibit space; 513-seat theater.
Activities: concerts; dance recitals; films; formal education programs for adults & college students; guided tours; hobby workshops; lectures; loan, participatory, traveling & temporary exhibitions.
Hours & Admission Prices: Mon.-Tues. & Fri. 10-4, Wed.-Thurs. 10-7, Sat. 10-4. No charge. Closed major holidays; school breaks. ♿
Attendance: 7,000 (accurate)

Pickerington

MOTORCYCLE HALL OF FAME MUSEUM, (M), 13515 Yarmouth Dr., Pickerington, OH 43147-8214. Tel.: 614-856-2222, ext. 1234. Fax: 614-856-2221.
E-mail: info@motorcyclemuseum.org
Web Site: www.motorcyclemuseum.org
Formerly: Motorcycle Heritage Museum.
Founded: 1990.
Congressional District: 12
Key Personnel: Operations Mgr., Katy Wood; Chm., Jeffrey V. Heininger; Sec./Treas., Rob Dingman; Dir. Operations, Jack Penton; Museum Shop Mgr., Beth Myers.
Personnel Profile: Full-Time Paid 3; Part-Time Paid 4; Part-Time Volunteers 6.
Governing Authority: society; nonprofit. Parent Institution: American Motorcyclist Association. Subsidiary Institution: American Motorcycle Heritage Foundation. Tax-exempt: 501(c)(3).
Institution Type/Description: Sports Museum & Hall of Fame
Collections: motorcycles; riding clothing & accessories; racing gear; competition memorabilia; archives; photographs; film; video; Hall of Fame; paintings; sculpture.
Major Exhibits: Motorcycle Art, 3/13-3/14.

Research Fields: motorcycle history.
Facilities: 8,000 sq. ft. exhibit space. Museum & motorcycle-related items for sale.
Activities: theme, loan & temporary exhibitions; guided tours. Museum Sponsors: Vintage Motorcycle Days (fundraiser)
Publications: annual report.
Hours & Admission Prices: Daily 9-5. Adults $10, seniors $8, students 12-17 $3; discounts to members, American Motorcyclist Association, AMCA & AHRMA members. Closed New Year's Day; Easter; Thanksgiving; Christmas. &
Attendance: 15,000 (accurate)
Membership: AMA $49.

Piqua

∗ **JOHNSTON FARM & INDIAN AGENCY, (M),** 9845 N. Hardin Rd., Piqua, OH 45356-9707. Tel.: 937-773-2522. Fax: 937-773-4311.
E-mail: ahite@ohiohistory.org
Web Site: www.johnstonfarmohio.com
Formerly: Piqua Historical Area State Memorial
Founded: 1972.
Congressional District: 8
Key Personnel: Site Mgr., Andy Hite; Groundskeeper, Rob Cline; Museum Shop Mgr., Diana Jacobs; Lead Interpreter, Marla Fair.
Personnel Profile: Full-Time Paid 2; Part-Time Paid 8; Part-Time Volunteers 60.
Governing Authority: society. Parent Institution: Ohio Historical Society, 1982 Velma Ave., Columbus, OH 43211. Tax-exempt: 170(b)(1)(A).
Institution Type/Description: History Museum.
Collections: early 18th- & 19th-century furnishings & tools; historic American Indian tools; weapons; costumes; art; canoes; trade items; graphic art; historic & theme-related material; historic woodland culture in Ohio; agriculture from Native American to present European style farms; the Eastern Woodland Indians of Ohio. Historic Buildings: Johnston home & farm; outbuildings.
Facilities: picnic area. Museum-related items for sale.
Activities: demonstrations of early 19th-century crafts; guided tours for school groups; canal boat ride; permanent exhibits.
Publications: Ohio Agricultural History 1800-1900; Native American History 1700-1900; Transportation-Canal History 1830-1900.
Hours & Admission Prices: April-May. & Sept.-Oct. Mon.-Fri. 9-2 by appointment; June-Aug. Thurs.-Fri. 10-5, Sat.-Sun. 12-5. Adults $8, senior citizens $7, children 6-12 $4; discounts to groups with advanced reservation; OHS, members no charge. &
Attendance: 20,000 (accurate)
Membership: Family (1 yr.) $25; Family (2 yrs.) $40; Senior, Student & Individual $10.

Point Pleasant

GRANT'S BIRTHPLACE STATE MEMORIAL, (M), 1551 State Rte. 232, Point Pleasant, OH 45153-9301. Mailing Address: P.O. Box 2, New Richmond, OH 45157-0002. Tel.: 513-553-4911; 800-283-8932. Fax: 614-297-2352.
Web Site: www.ohiohistory.org/places/grantbir
Congressional District: 2
Key Personnel: Museum Attendant, Greg Roberts.
Governing Authority: society. Affiliated with Ohio Historical Society, Columbus, OH. 43211. Tax-exempt: 170(b)(1)(A).
Institution Type/Description: Historic House: 1821 restored cottage, birthplace of Ulysses S. Grant.
Collections: artifacts relative to Ulysses S. Grant.
Hours & Admission Prices: April-Oct. Wed.-Sat. 9:30-12 & 1-5, Sun. 1-5; groups by appointment. Adults $2.50, seniors $2, children 6-12 $1.50; children 5 & under and OHS members no charge. &
Attendance: 6,000

Pomeroy

MEIGS COUNTY MUSEUM, 144 Butternut Ave., Pomeroy, OH 45769-1260. Mailing Address: P.O. Box 145, Pomeroy, OH 45769-0145. Tel.: 740-992-3810. Fax: 740-992-3810. Facebook: Meigs County Historical Society and Museum.
E-mail: meigscohistorical@frontier.net
Web Site: meigscohistorical.org
Founded: 1960.
Congressional District: 6
Key Personnel: Pres., Margaret Parker.

Governing Authority: society. Parent Institution: Meigs County Pioneer & Historical Society. Tax-exempt.
Institution Type/Description: Local History Museum.
Collections: history exhibits; artifacts & memorabilia; history items of Meigs County.
Research Fields: Ohio & river history; genealogy.
Facilities: 3,500-vol. library of material on local history; theater; meeting rooms.
Activities: guided tours; hobby workshops; seminars; oral history project.
Publications: books, Meigs County History 1979; reprints, Wilkesville-Salem 1874; Hardesty Meigs History 1883; Pioneer History-Larkin 1908; Poll Book Records 1800s; Pioneer Annual Meetings; Bedford Township; Meigs County History 1987; Index to Meigs County, OH death records: 1867-1908; Meigs County Birth and Death Records: 1867-1908; Pictorial History of Meigs County; Meigs County History Vol. II & Vol. III; Pictorial History of Portland, Ohio.
Hours & Admission Prices: Tues.-Fri. 10-3. Closed federal holidays. &
Attendance: 3,000
Membership: Individual $12; Family $20; Business $100; Life $200.

Port Clinton

AFRICAN SAFARI WILDLIFE PARK, 267 S Lightner Rd., Port Clinton, OH 43452. Tel.: 419-732-3606. Fax: 419-734-1919. Facebook: African Safari Wildlife Park.
E-mail: Info@africansafariwildlifepark.com
Web Site: www.africansafariwildlifepark.com
Institution Type/Description: Wildlife Park.
Collections: zebra; giraffe; camel; alpaca; gibbons; ocelot; tortoise; bison; llama.
Facilities: cafe. Gift items for sale.
Activities: pony & camel rides; pig races; animal shows; drive-thru safari; feed the animals.
Hours & Admission Prices: late Feb. to May & Sept.-Nov. daily 10-5; Memorial Day to Labor Day daily 9-7. Summer: admission 7 & up $21.95, children 4-6 $12.95; discounts to seniors 55 & over and veterans; children 3 & under no charge. Spring & Fall: admission 7 & up $15.95, children 4-6 $9.95; discounts to seniors 55 & over and veterans; children 3 & under no charge. &

OTTAWA COUNTY HISTORICAL MUSEUM, 126 W. 3rd. St., Port Clinton, OH 43452-1842. Mailing Address: P.O. Box 845, Port Clinton, OH 43452-0845. Tel.: 419-732-2273.
E-mail: ochm@cros.net
Founded: 1932.
Congressional District: 5
Key Personnel: Cur., Peggy Debien.
Personnel Profile: Part-Time Paid 1; Part-Time Volunteers 8.
Governing Authority: nonprofit. Parent Institution: Area Heritage Foundation Inc. Tax-exempt.
Institution Type/Description: History Museum.
Collections: Indian relics; early household items; hand machine equipment; clothing; guns; rocks; toys; china; cameras; historical photos; local photographs; newspaper clippings.
Research Fields: local history.
Facilities: library of genealogical & historical books.
Activities: permanent exhibitions.
Publications: book, The Heritage of Port Clinton, Ohio.
Hours & Admission Prices: Winter: Wed. 12-3; Summer: Tues.-Thurs. 12-3; other times by appointment. No charge; donations accepted. &
Attendance: 600 (estimated)
Membership: Annual $15; Family $20; Life $1,000.

Portsmouth

SOUTHERN OHIO MUSEUM, (M), 825 Gallia St., Portsmouth, OH 45662-0990. Mailing Address: 825 Gallia St., P.O. Box 990, Portsmouth, OH 45662-0990. Tel.: 740-354-5629. Fax: 740-354-4090.
E-mail: info@somacc.com
Web Site: www.somacc.com
Founded: 1979.
Congressional District: 2
Key Personnel: Dir., Pegi Wilkes; Visual Arts Dir., Darren Baker; Museum Shop, Sara Johnson.
Personnel Profile: Full-Time Paid 4; Part-Time Paid 6; Part-Time Volunteers 65.
Governing Authority: nonprofit organization. Parent Institution: Southern Ohio Museum Corp. Tax-exempt: 501(c)(3).
Institution Type/Description: Museum & Cultural Center.

Collections: American Art; folk art; decorative arts; prehistoric Native American artifacts.
Research Fields: local history; American Art; Contemporary Art; prehistoric American artifacts
Facilities: theater; reading room; classrooms. Crafts, toys & museum publications for sale.
Activities: guided tours; lectures; films; concerts; drama; formally organized education programs; docent program; temporary, traveling & loan exhibitions.
Publications: quarterly newsletter; exhibition catalogues.
Hours & Admission Prices: Tues.-Fri. 10-5, Sat. 1-5. Adults $2, students & children under 12 $1; members & Fri. no charge; gift shop discount to AAM & ICOM members; Closed national holidays. ⅊
Attendance: 18,000 (estimated)
Membership: Student & Senior $20; Individual $25; Family $35; Supporting $250; Donor $500; Patron $1,000.

Powell

COLUMBUS ZOO AND AQUARIUM, 4850 W. Powell Rd., Powell, OH 43065-0400. Mailing Address: P.O. Box 400, Powell, OH 43065-0400. Tel.: 614-645-3400. Fax: 614-645-3465.
E-mail: info@columbuszoo.org
Web Site: www.columbuszoo.org
Formerly: Columbus Zoological Park Association, Inc.
Founded: 1927.
Congressional District: 12
Key Personnel: Pres. & C.E.O., Tom Stalf; Chm., Phil Pikelny; Sr. Vice Pres. Animal Care & Conservation, Lewis Greene; Vice Pres. Mktg., Pete Fingerhut; Museum Shop Mgr., Lisa Jones.
Personnel Profile: Full-Time Paid 229; Part-Time Paid 15; Part-Time Volunteers 717.
Volunteer Hours: 39,177
Operating Expenses: 56,689,899
Operating Income: 64,597,256
Governing Authority: association. Tax-exempt.
Institution Type/Description: Zoology Museum.
Collections: aquarium; aviary; herpetology; mammals.
Research Fields: behavioral, reproductive & physiology behavior, artificial breeding; education on endangered & exotic animals.
Facilities: library of zoology books available for use by staff or volunteers.
Activities: lectures; docent tours; train rides; formally organized education programs; permanent & traveling exhibitions.
Publications: quarterly magazine, Beastly Banner; annual report.
Hours & Admission Prices: Summer daily 9-7; Winter daily 10-5. Adults $14.99, seniors 60 & over $10.99, children 2-9 $9.99; members & children under 2 no charge. Parking: $7. Closed Thanksgiving; Christmas Eve & Day. ⅊
Attendance: 2,322,934 (accurate)
Membership: Individual $49; Individual Plus One $79; Family $99; Family Plus $129; Explorer's Club $150; Keeper's Club $250; Curator's Club $500; Colo Club Bronze $1,000; Colo Club Silver $2,500; Colo Club Gold $5,000; Colo Club Platinum $10,000.

Put-in-Bay

ANTIQUE CAR MUSEUM, 979 Catawba Ave., Put-in-Bay, OH 43456-0708. Mailing Address: P.O. Box 708, Put-in-Bay, OH 43456-0708. Tel.: 419-285-2283.
Institution Type/Description: Car Museum.
Collections: period cars including Model Ts; snow mobiles; gasoline memorabilia.
Hours & Admission Prices: April & Oct. Sat.-Sun. 11-5; May-Sept. daily 10:30-6. No charge.

CHOCOLATE CAFE & MUSEUM, 820 Catawba Ave., Put-in-Bay, OH 43456. Mailing Address: Matthew James, Inc., 337 N. Miami Beach Dr., Port Clinton, OH 43452. Tel.: 419-734-7114.
Institution Type/Description: History Museum.
Collections: history of chocolate & chocolate making; period chocolate collectibles.
Hours & Admission Prices: May-Sept. daily.

LAKE ERIE ISLANDS HISTORICAL SOCIETY, 25 Town Hall Place, Put-in-Bay, OH 43456. Mailing Address: P.O. Box 25, Put-in-Bay, OH 43456-0025. Tel.: 419-285-2804.
Institution Type/Description: Historical Society Museum.
Collections: local history & culture; photographs; personal artifacts; period furnishings.

Activities: educational programs; special events.
Hours & Admission Prices: mid-May to June & Sept. daily 11-5; July-Aug. daily 10-6; Oct. Sat.-Sun. 11-5; groups by appointment. Adults $2, youth 12 & over $1; children 11 & under no charge.

PERRY'S CAVE, 979 Catawba Ave., Put-in-Bay, OH 43456. Mailing Address: P.O. Box 708, Put-in-Bay, OH 43456-0708. Tel.: 419-285-2283.
Institution Type/Description: Geology Museum: an Ohio Natural Landmark.
Collections: local history & geology; natural limestone cave; underground lake.
Facilities: Museum-related items for sale.
Activities: guided tours; private lantern tours.
Hours & Admission Prices: April & Oct. Sat.-Sun. 11-5; May-Sept. daily 10-6. Adults $7.50, children 6-12 $4.50; children under 6 no charge.

PERRY'S VICTORY & INTERNATIONAL PEACE MEMORIAL, 93 Delaware Ave., Put-in-Bay, OH 43456. Mailing Address: P.O. Box 549, Put-in-Bay, OH 43456-0549. Tel.: 419-285-2184. Fax: 419-285-2516.
E-mail: jef_helmer@nps.gov
Web Site: www.nps.gov/pevi/
Founded: 1936.
Congressional District: 9
Key Personnel: Supt., Blanca Alvarez Stransky; Chief of Interpretation, Jeff Helmer.
Governing Authority: federal. Parent Institution: National Park Service, Dept. of the Interior Washington D.C. Tax-exempt.
Institution Type/Description: Park Museum; History Museum.
Collections: weapons & equipment from the War of 1812; War of 1812 naval ordnance; paintings; engravings; lithographs; items relating to the construction & early history of Perry's Victory & International Peace Memorial.
Research Fields: War of 1812 in the Northwest; Battle of Lake Erie; Memorial Column; international peace between U.S. & Canada.
Facilities: 1,000-vol. library of books & pamphlets pertaining to the War of 1812 & the Battle of Lake Erie; papers of the Perry Centennial Commission available for research; 317 ft. observation deck; visitor contact station. Books & other museum-related items for sale.
Activities: interpretive talks; living history demonstrations; tour of Memorial Column.
Publications: Deep Water Sailors-Shallow Water Soldiers: Manning the U.S. Fleet on Lake Erie, 1913; Oliver Hazard Perry & The Battle of Lake Erie; Amongst My Best Men; African Americans & the War of 1812.
Hours & Admission Prices: late April to mid-May & late Sept. to mid-Oct. daily 10-5; mid-May to mid-June & early Sept. daily 10-6; mid-June-Aug. daily 10-7; late Oct. to mid-April by appointment. Adults $3; National Park Service Day, senior citizens over 62 w/Golden Age Passport & children under 16 no charge. ⅊
Attendance: 204,161 (accurate)

STONEHENGE ESTATE, 808 Langram Rd., Put-in-Bay, OH 43456. Tel.: 419-285-6134 & 2585.
Institution Type/Description: Historic House Museum: housed in a stone farmhouse. Listed on the National Register of Historic Places.
Collections: local history & culture; photographs; period furnishings; personal artifacts; wine cellar. Historic Building: Wine Press Cottage.
Facilities: Museum-related items for sale.
Hours & Admission Prices: Memorial Day to Labor Day daily 11-5. Adults 16 & over $7, children 6-15 $4; children 5 & under no charge.

Ravenna

PORTAGE COUNTY HISTORICAL SOCIETY, 6549 N. Chestnut St., Ravenna, OH 44266-3907. Tel.: 330-296-3523.
E-mail: pchsohio@neo.rr.com
Web Site: www.portagecountyhistoricalsociety.org
Founded: 1951.
Congressional District: 11
Key Personnel: Pres. (V), Wayne Enders; Museum Shop Mgr., Barbara Petroski.
Personnel Profile: Part-Time Volunteers 15; Interns 3.
Governing Authority: trustees & society; nonprofit organization. Tax-exempt: 501(c)(3).
Institution Type/Description: General Museum.
Collections: history artifacts; archives; archaeology finds; glassware; books of Portage history. Historic Homes: c.1820, 1832, 1869.
Research Fields: Portage county history; archaeology; genealogy; glass.
Facilities: 150-vol. library of genealogy; over 1,000 history & genealogy

books; court records available for use by appointment; reading room; rental facilities. amphitheater. Museum-related items for sale.
Activities: guided tours; monthly lectures; temporary exhibitions; scout campouts; musical programs; rental facilities.
Publications: quarterly, newsletter; Ravenna, A Bicentennial Album of 19th Century Photographs; History of Portage County Ohio Illustrated 1885; over 100 books.
Hours & Admission Prices: Thurs. & Sat. 2-6. Museum Tours: $3. Grounds: $5. Closed holidays. &
Attendance: 1,200 (estimated)
Membership: Student $5; Senior Citizen $8; Individual $15; Family $20; Sustaining & Patron $30; Contributing $100; Life $300; Portage History Gold $1,000.

Reading

READING HISTORICAL SOCIETY MUSEUM, 22 W. Benson St., Reading, OH 45215-3202. Tel.: 513-761-8535.
Founded: 1988.
Congressional District: 2
Key Personnel: Pres. (V), James J. Lichtenberg.
Personnel Profile: Part-Time Volunteers 9.
Governing Authority: nonprofit organization. Tax-exempt.
Institution Type/Description: Historical Society Museum; 1905 house built by local tinsmith.
Collections: furnishings; personal artifacts; marching band uniforms; military uniforms; replica of Doughboy statue from WWI monument; 2 period pianos; photographs.
Facilities: library; reading room.
Activities: lectures to seniors & school groups. Museum Sponsors: Memorial Day open house; Annual Settlement Day Dinner; Reading Cross Roads Celebration.
Publications: quarterly newsletter, Bridging Time; 2001 History of Reading - Sesquicentennial of Incorporation; 1994 History of Reading - Bicentennial of Founding.
Hours & Admission Prices: first Sun. of each month 1-3; other times by appointment. No charge; donations accepted.
Attendance: 100 (estimated)
Membership: Annual $5; Patron $25, $50, $100, $500

Ripley

RANKIN HOUSE STATE MEMORIAL, (M), 6152 Rankin Hill Rd., Ripley, OH 45167-1044. Mailing Address: P.O. Box 176, Ripley, OH 45167-0176. Tel.: 937-392-1627.
E-mail: ripleyohio@aol.com
Web Site: www.ripleyohio.net
Founded: 1938.
Congressional District: 2
Key Personnel: Dir., Betty Campbell.
Governing Authority: Affiliated with Ohio Historical Society, Columbus, OH. Tel.: 614-297-2610. Tax-exempt: 170(b)(1)(A).
Institution Type/Description: Historic Site: restored home of abolitionist Rev. John Rankin, 1828; underground railroad site.
Collections: period furnishings.
Hours & Admission Prices: May to mid-Dec. Tues.-Sat. 10-5, Sun. 12-5; groups by appointment. Adults $3, students 6-18 $1; members and children 5 & under no charge.
Attendance: 5,000
Membership: Annual $47.

Rittman

RITTMAN HISTORICAL SOCIETY, 393 W. Sunset Dr., Rittman, OH 44270-1054. Mailing Address: P.O. Box 583, Rittman, OH 44270-0583. Tel.: 330-715-7629. Facebook: Rittman Historical Society / Pioneer Cemetery.
Founded: 1960.
Congressional District: 16
Key Personnel: Pres., Jack Rice; Vice Pres., Mike Burg; Treas., Fred Winkler; Sec., Cindy Ferguson.
Governing Authority: society. Tax-exempt.
Institution Type/Description: Historic House Museum.
Collections: Historic Houses: 1817 Old Knupp Church; 1859 Gish house, built on old tobacco farm.
Facilities: 200-vol. library of historical books available to society members for use on premises.
Activities: geology study of pioneer families.
Hours & Admission Prices: By appointment only. No charge; donations accepted.

Membership: Student $5; Single $7; Couple $10; Business $25; Lifetime $100.

Roseville

NATIONAL CERAMIC MUSEUM AND HERITAGE CENTER, 7327 Ceramic Rd., N.E., Roseville, OH 43777-9694. Mailing Address: P.O. Box 200, Crooksville, OH 43731-0200. Tel.: 740-697-7021. Fax: 740-697-0171.
E-mail: natceramicmuseum@yahoo.com
Formerly: Ohio Ceramic Center
Founded: 1970.
Key Personnel: Dir., Kathy Campbell; Pres. (V), Kent Papageorge; Vice Pres., Joan Spring; Museum Shop Mgr., Betty Larabee.
Personnel Profile: Full-Time Paid 1; Part-Time Paid 1.
Governing Authority:
Institution Type/Description: Arts & Crafts Museum.
Collections: development of pottery from Ohio Bluebird potteries from 1750 to modern ceramic industry; Roseville, Crooksville & Zanesville, Ohio manufacturers.
Research Fields: pottery in Crooksville, Roseville & surrounding areas.
Facilities: Pottery & other museum-related items for sale.
Activities: pottery demonstrations; research; ceramic classes. Annual Events: Pottery Festival in July; Theme Is Pottery Judged Show in September.
Hours & Admission Prices: March-Dec. Wed.-Sat. 10-5, Sun. 12-5; other times by appointment. Adults $4, senior citizens $3.50, students $2; discounts to AAM & ICOM members; children under 5 & members no charge. Closed Memorial Day; Independence Day; Labor Day. &
Attendance: 10,000
Membership: Seniors & Students $10; Individual $15; Family $25; Business $50.

Saint Marys

AUGLAIZE COUNTY HISTORICAL SOCIETY, Daniel Mooney Museum, 223 S. Main St., Saint Marys, OH 45885-2208. Tel.: 419-393-8532 & 738-9328.
Web Site: www.auglaizecountyhistory.org/
Founded: 1963.
Key Personnel: Pres. (V), Karen Dietz; Vice Pres., George Neargarder; Treas., James Heinrich.
Personnel Profile: Part-Time Paid 1; Part-Time Volunteers 15.
Governing Authority: county; nonprofit organization. Subsidiary Institution: Wapakoneta Local History Museum, 206 W. Main St., Wapakoneta, OH. Tax-exempt.
Institution Type/Description: General Museum: housed in c.1876 home of Civil War Officer Major Charles Hipp.
Collections: Fort St. Marys, Fort Amanda & Fort Barbee artifacts; American Indian artifacts; Miami Erie Canal; oil wells; Gordon State Park; early industry; Civil War; local manufacturing. Historic Building: Gary Log House - Auglaize County Fair Grounds; churns; military exhibits.
Research Fields: archaeological dig at Fort St. Marys.
Activities: guided tours; traveling exhibitions by mobile home to schools. Museum Sponsors: Log House at County Fair.
Publications: newsletter, Auglaize County.
Hours & Admission Prices: Mon. & Wed.-Fri. 8-5, Tues. 8-7, Sat. 1-4. No charge; donations accepted. Closed holidays.
Attendance: 1,500 (estimated)
Membership: Basic $25.

Salem

SALEM HISTORICAL SOCIETY AND MUSEUM, 208 S. Broadway Ave., Salem, OH 44460-3004. Tel.: 330-337-8514. Facebook: The Salem Historical Society and Museum.
E-mail: thesalemhistoricalsociety@gmail.com
Web Site: www.salemhistoricalsociety.org
Founded: 1947.
Congressional District: 18
Key Personnel: Pres. (V), David J. Shivers; Museum Dir., David C. Stratton; Cur., Janice Lesher; Museum Shop Mgr., Dixie Gordon.
Personnel Profile: Part-Time Volunteers 10.
Governing Authority: nonprofit organization. Parent Institution: Salem Historical Society. Tax-exempt: 501(c)(3).
Institution Type/Description: Local History Museum.
Collections: abolitionist artifacts & exhibits; Ohio Women's Rights exhibit; Civil War collection; carpenter shop; woodworking tools & artifacts; coal mine exhibit; Fire Department exhibit; Quaker exhibit; Romanian & Saxon exhibits; Salem China Co.; 50s barbershop display; early china & glass; Mullin's Manufacturing history; photographs; valentines; postcards; toy room; kitchen utensils; loom; spinning wheels; anti-slavery history, pictures; one-room schoolhouse; local artifacts.

Research Fields: local history; early businesses.
Facilities: library of general, history & rare books available for use on premises; reading room. Local publications & other museum-related items for sale.
Activities: guided tours; formally organized education programs for children; permanent exhibitions. Museum Sponsors: Founders Day Dinner; City Celebration in the Fall; participates in community sponsored events, ie I Fest, Fall Fun Day.
Publications: quarterly newsletter, Bugle.
Hours & Admission Prices: May-Oct. Sun. 1-4; other times call for appointment. Adults $4, groups of 25 or more $3, children $2; members no charge. Closed major holidays. ♿
Attendance: 3,500 (estimated)
Membership: Single $20; Family $40; Life $1,000.

Sandusky

ELEUTHEROS COOKE HOUSE AND GARDEN, 1415 Columbus Ave., Sandusky, OH 44870. Tel.: 419-627-0640.
E-mail: tour@oldhouseguild.org
Founded: 1995.
Key Personnel: C.E.O., Richard Keller; Chm. (V), Ann Marie Flood.
Governing Authority: Parent Institution: Ohio Historical Society. Tax-exempt.
Institution Type/Description: Historic House Museum: housed in the former home of Eleutherus Cooke, Sandusky's first lawyer & politician serving in the Ohio Legislature and U.S. Congress; built c.1840.
Collections: Cooke family history; period furnishings; personal artifacts; photographs.
Publications: quarterly members' newsletter.
Hours & Admission Prices: April-Dec. Tues.-Fri. 12-3, Sat. 10-1; other times by appointment. No charge; donations accepted.
Attendance: 626 (accurate)

MARITIME MUSEUM OF SANDUSKY, 125 Meigs St., Sandusky, OH 44870-2834. Tel.: 419-624-0274.
E-mail: sanduskymaritime@bex.net
Web Site: www.sanduskymaritime.org
Founded: 1995.
Congressional District: 9
Personnel Profile: Full-Time Paid 1; Part-Time Paid 4; Part-Time Volunteers 150; Interns 1.
Governing Authority: Parent Institution: Sandusky Area Maritime Association. Tax-exempt.
Institution Type/Description: Maritime History Museum.
Collections: maritime history; ship models; tools; boat building; photographs.
Facilities: Museum-related items for sale.
Activities: educational programs; model making; outreach programs; birthday parties; classes; boat building & restoration.
Publications: newsletter, The Messenger.
Hours & Admission Prices: June-Aug. Tues.-Sat. 10-4, Sun. 12-4; Sept.-May Fri.-Sat. 10-4, Sun. 12-4. Adults $4, senior citizens & children under 12 $3. Closed major holidays. ♿
Attendance: 24,417 (accurate)
Membership: Student & Senior $20; Individual $25; Family $50; Contributor $100; Life $1,000.

THE MERRY-GO-ROUND MUSEUM, 301 Jackson St., Sandusky, OH 44870. Tel.: 419-626-6111. Fax: 419-626-1297.
E-mail: merrygoround39@peoplepc.com
Web Site: merrygoroundmuseum.org
Founded: 1989.
Key Personnel: Dir., Veronica VandenBout.
Personnel Profile: Full-Time Paid 3; Part-Time Volunteers 118.
Institution Type/Description: Carousel Museum.
Collections: carousel history, carving styles & animals; 1939 Allan Herschell carousel.
Facilities: Museum-related items for sale.
Activities: rental facilities; guided tours; special events.
Hours & Admission Prices: Feb. Sat. 11-5, Sun. 12-5; March-May & Sept.-Dec. Wed.-Sat. 11-5, Sun. 12-5; June to Labor Day Mon.-Sat. 10-5, Sun. 12-5. Adults $6, seniors 60 & over $5, children 4-14 $4; children 3 & under no charge.
Attendance: 25,000 (estimated)
Membership: Senior Citizens & Students $20; Individual $25; Family $40; Jumper $100; Prancer $250; Standers $500; Chariot $1,000; Brass Ring $5,000.

MUSEUM OF CAROUSEL ART & HISTORY, 301 Jackson St., Sandusky, OH 44870-2621. Tel.: 419-626-6111. Fax: 419-626-1297.
E-mail: merrygoround39@peoplepc.com
Web Site: www.merrygoroundmuseum.org
Formerly: Merry Go Round Museum
Founded: 1990.
Key Personnel: C.E.O., Veronica Vanden Bout; Pres. (V) & Chm. (V), Gary Mortus; Financial Dir., Bridget Castle; Museum Shop Mgr., Carol Brown.
Personnel Profile: Full-Time Paid 2; Part-Time Paid 3; Part-Time Volunteers 105; Interns 1.
Governing Authority: private; nonprofit organization. Tax-exempt: 501(c)(3).
Institution Type/Description: Carousel Art & History Museum.
Collections: American & International carousel art & history; tools & partially carved pieces from G.A. Dentzel carving shop; fully restored & operational Herschell carousel.
Research Fields: carousel art & history.
Facilities: 200-vol. library on carousel art & history; 10,000 sq. ft. exhibit space. Gift items for sale.
Activities: wood carving classes; public art program; local artist exhibits. Annual Events: New Year's Eve Gala; Follies; Carving Weekend; Toast of Ohio Wine Festival.
Publications: quarterly newsletter, Stargazer; monthly volunteer newsletter, Ponytales.
Hours & Admission Prices: Jan.-Feb. Sat. 11-5, Sun. 12-5; March-May & Sept.-Dec. Wed.-Sat. 11-5, Sun. 12-5; Memorial Day-Labor Day Mon.-Sat. 10-5, Sun. 12-5. Adults $5, senior citizens $4, children $3, discounts to groups of 10 or more; member adults no charge. Closed New Year's Eve & Day; Easter; Thanksgiving; Christmas Eve & Day. ♿
Attendance: 21,232 (estimated)
Membership: Senior & Student $20; Individual $25; Family $40; Jumpers $100; Prancers $250; Standers $500; Chariot $1,000; Brass Ring $5,000; Lead Horse $150,000.

OHIO VETERANS HOMES MUSEUM, 3416 Columbus Ave., Sandusky, OH 44870. Tel.: 419-625-2454, ext. 1447.
Key Personnel: Dir., Naomi Twine
Institution Type/Description: Military Museum.
Collections: military history & artifacts; photographs; personal artifacts.
Hours & Admission Prices: Fri.-Sat. 12-4; other times by appointment. No charge.

SANDUSKY LIBRARY FOLLETT HOUSE MUSEUM, 404 Wayne St., Sandusky, OH 44870-2751. Mailing Address: Sandusky Library, 114 W. Adams St., Sandusky, OH 44870-2751. Tel.: 419-625-3834. Fax: 419-625-4574. Facebook: Follett House Museum.
E-mail: museumservices@sandusky.lib.oh.us
Web Site: www.sandusky.lib.oh.us/
Founded: 1902.
Congressional District: 5
Key Personnel: Museum Admin., Maggie Marconi; Dep. Dir., Dennis McMullen; Archives Librarian, Ron Davidson.
Personnel Profile: Full-Time Paid 3; Part-Time Paid 3; Part-Time Volunteers 1; Interns 1.
Governing Authority: nonprofit association. Parent Institution: Sandusky Library Association, 114 W. Adams. Tax-exempt.
Institution Type/Description: Local History Museum: housed in the 1834-37 Greek Revival home of Oran Follett.
Collections: household objects; furniture; toys; artifacts of Sandusky, Erie County & Johnson's Island Civil War Confederate Officers Prison; 1834-37 Greek Revival home of Oran Follett.
Research Fields: Sandusky & Erie county history; Johnson's Island prison.
Facilities: library; archives.
Activities: guided tours by appointment; traveling exhibits to local schools; speakers available.
Publications: books, At Home in Early Sandusky; From The Widow's Walk: A View of Sandusky Vol. I & II; Images of America: Sandusky, Ohio; Erie County and The Erie Isles: A Pictorial History of The Early Years; Erie County & The Erie Isles: A Pictoral History 1940-1975; Erie County & The Erie Isles: A Pictoral History 1975-Today; Face of the Firelands: The Early Years.
Hours & Admission Prices: April-May & Oct.-Dec. Sat. 12-4; June-Aug. Wed. & Fri. 12-4, Sat. 10-1; Sept. Sat. 10-1; group tours by appointment. No charge; donations accepted. Closed Easter; Thanksgiving; Christmas.
Attendance: 3,000 (estimated)

Seville

NORTHERN OHIO RAILWAY MUSEUM, 5515 Buffham Rd., Seville, OH 44273. Mailing Address: P.O. Box 458, Chippewa Lake, OH 44215-0458. Tel.: 330-769-5501.
Institution Type/Description: Railway Museum.
Collections: railroad history, equipment & artifacts; photographs; street cars; rapid transit cars; freight cars.
Hours & Admission Prices: May 15 to Oct. Sat. 10-4. No charge; donations accepted.

Shaker Heights

THE SHAKER HISTORICAL SOCIETY, (M), 16740 S. Park Blvd., Shaker Heights, OH 44120-1641. Tel.: 216-921-1201. Fax: 216-921-2615.
E-mail: shakerhistory@shakerhistory.org
Web Site: www.shakerhistory.org
Founded: 1947.
Congressional District: 22
Key Personnel: Pres. (V), Keith Arian; Exec. Dir., Ann Cicarella.
Personnel Profile: Full-Time Paid 1; Part-Time Paid 1; Part-Time Volunteers 26; Interns 7.
Governing Authority: society; nonprofit. Tax-exempt: 501(c)(3).
Institution Type/Description: Historical Society Museum: located on land once owned by North Union Colony of Shakers, in a historic 1910 house.
Collections: furniture & artifacts of North Union & other Shaker communities; materials; books; maps; memorabilia related to Shaker Heights; an early 'garden city' community; the Shaker Rapid Transit and Warrensville Township.
Major Exhibits: A Miniature Wonderland, 1/14.
Research Fields: local history.
Facilities: 2,000-vol. library of Shakers, Northern Ohio history with emphasis on Shaker Heights & the Van Sweringen Brothers available for research by appointment and fee. Miniature furniture & other items related to Shakers for sale.
Activities: guided tours; video; lectures and programs; educational program for schools by arrangement; permanent and special exhibits; special events.
Publications: quarterly journal, The JOURNAL.
Hours & Admission Prices: Tues.-Fri. 11-5, Sun. 2-5; other times by appointment. Adults $4, children 6-18 $2; discounts to Ohio Historical Society members; members no charge. Closed holidays.
Attendance: 2,500 (accurate)
Membership: Individual $35; Family $60; Bronze $75; Silver $150; Gold $250; Platinum $500 & up.

Sharonville

HISTORIC SOUTHWEST OHIO, INC., Heritage Village, 11450 Lebanon Pike, Rte. 42, Sharonville, OH 45241. Mailing Address: P.O. Box 62475, Cincinnati, OH 45262-0475. Tel.: 513-563-9484. Fax: 513-563-0914.
E-mail: wdichtl@heritagevillagecincinnati.org
Web Site: www.heritagevillagecincinnati.org
Founded: 1964.
Congressional District: 1
Key Personnel: Pres. (V), Wayne Puritun; Dir., William J. Dichtl; Dir. Education, Steve Preston.
Personnel Profile: Full-Time Paid 3; Part-Time Volunteers 70; Interns 5.
Governing Authority: private; nonprofit corporation. Parent Institution: Historic Southwest Ohio, Inc. Subsidiary Institution: Heritage Village Museum. Tax-exempt: 501(c)(3).
Institution Type/Description: Historic Village Museum.
Collections: Historic Structures. Heritage Village: Medical Office; 1852 Hayner House; 1815 Gatch Barn; 1818 Elk Lick House & dependencies; 1872 Chester Park Railroad Station; 1825 Vorches House; 1804 Kemper Log House; 1860 Owensville Store. 19th century furniture & decorative arts; farm equipment & vehicles.
Research Fields: history; decorative arts of the 19th-century southwestern Ohio.
Facilities: resource center. Gifts & books for sale in Heritage Village.
Activities: tours; lectures; continuing education; school tours, outreach, kids camp, special events.
Publications: quarterly newsletter; survey booklets; walking tour brochures.
Hours & Admission Prices: Heritage Village: May-Sept. Tues.-Sat. 10-5, Sun. 1-5. Adults $5, children $3; members no charge. ♿
Attendance: 15,000 (accurate)
Membership: Individual $30; Family $50; Business $100; Hayner Society $1,000 & up.

Sheffield Lake

103RD OHIO VOLUNTEER INFANTRY CIVIL WAR MUSEUM, 5501 E. Lake Rd., Sheffield Lake, OH 44054-1900. Tel.: 440-949-2790.
Web Site: 103ovi.com
Formerly: 103rd Ohio Volunteer Infantry Memorial Foundation
Founded: 1972.
Key Personnel: Pres., Connie Parker; Cur., Deborah Wagner; Museum Shop Mgr., Julie Piazza.
Personnel Profile: Part-Time Volunteers 10.
Governing Authority: nonprofit organization. Tax-exempt: 501(c)(3).
Institution Type/Description: Military Museum: housed in c.1900 Elfordilno frame structure located on 4-acre site purchased in 1866 by the 103rd Ohio Volunteer Infantry.
Collections: historic Civil War relics; furniture; camp lamps; uniforms; fife & drum; flags; photographs; guns; trade items; books; monographs; biographies; manuscripts.
Research Fields: the Civil War; 103rd Ohio Volunteer Infantry.
Facilities: 300-vol. library of documented Civil War histories; 300-seat hall. Museum-related items for sale.
Activities: guided tours; lectures; temporary exhibitions.
Publications: quarterly newsletter, Assembly Call.
Hours & Admission Prices: Call for appointment. Suggested Donation: adults $2, students $1; members no charge.
Attendance: 500 (estimated)
Membership: Individual $10; Family $25.

Sidney

SHELBY COUNTY HISTORICAL SOCIETY, 201 N. Main Ave., Sidney, OH 45365. Mailing Address: P.O. Box 376, Sidney, OH 45365-0376. Tel.: 937-498-1653. Facebook: Shelby County Historical Society.
E-mail: info@shelbycountyhistory.org
Web Site: www.shelbycountyhistory.org
Founded: 1948.
Key Personnel: Dir., Tilda Phlipot
Institution Type/Description: Historical Society Museum.
Collections: local history & culture; period furnishings; personal artifacts; photographs.
Publications: monthly, Historical Highlights.
Hours & Admission Prices: Mon.-Fri. 1-5, Sat. 9-12. No charge. Closed holidays.
Attendance: 15,000
Membership: Individual $20; Family $25; Friends $150; Legacy Partner $250.

Smithville

SMITHVILLE COMMUNITY HISTORICAL SOCIETY, 381 E. Main St., Smithville, OH 44677. Mailing Address: P.O. Box 12, Smithville, OH 44677. Tel.: 330-669-9308.
Institution Type/Description: Historical Society Museum.
Collections: local history; period furnishings; quilts. Historic Buildings: 1880s Mishler Mill; 1830s Pioneer Log Cabin; Sheller House; blacksmith shop.
Activities: special events.
Hours & Admission Prices: Mill: Wed. 1:30-4. Buildings: June-Oct. 2nd & 4th Sun. 1:30-4.

Springfield

CLARK COUNTY HISTORICAL SOCIETY - HERITAGE CENTER OF CLARK COUNTY, (M), 117 S. Fountain Ave., Springfield, OH 45502-1207. Tel.: 937-324-0657. Fax: 937-324-1992.
E-mail: rsherrock@heritagecenter.us
Web Site: www.heritagecenter.us
Founded: 1897.
Congressional District: 7
Key Personnel: C.E.O., Roger Sherrock; Pres., William A. Kinnison; Cur., Kasey Eichensehr.
Personnel Profile: Full-Time Paid 7; Part-Time Paid 2; Part-Time Volunteers 75; Interns 2.
Governing Authority: county; society; nonprofit. Restoration Site: 1856, Clark County, OH. Tax-exempt.
Institution Type/Description: History Museum.
Collections: 1829-1970 newspaper files; furniture; pioneer utensils; early tools & farm equipment; documents, government records & periodicals.
Research Fields: local & area history.

Facilities: 2,500-vol. library of early newspapers & other material pertaining to 19th-century Clark County & general Americana available for use on premises; archives; reading room.
Activities: gallcry talks; permanent & temporary exhibitions; lectures.
Publications: annual monographs; quarterly newsletter.
Hours & Admission Prices: Gallery: Tues.-Sat. 9-5. Library & Archives: Wed.-Sat. 10-5. Suggested Donations: Gallery & Museum $10 per family, $5 per person. Library & Archives: $4 per day; members no charge. Closed national holidays. &
Attendance: 28,401 (accurate)
Membership: Individual $35; Family $50.

PENNSYLVANIA HOUSE MUSEUM, 1311 W. Main St., Springfield, OH 45504-2815. Tel.: 937-322-7668.
Web Site: www.pennsylvaniahousemuseum.info
Founded: 1839.
Key Personnel: Chm. (V), Martha O'Connor.
Personnel Profile: Part-Time Volunteers 50.
Volunteer Hours: 1,016
Operating Expenses: 20,000
Operating Income: 22,000
Governing Authority: Parent Institution: Lagonda Chapter of the Daughters of the American Revolution. Tax-exempt.
Institution Type/Description: Historic House Museum: housed in a Federal-style home built by David Snively in 1839; former home of Dr. Isaac K. Funk of Funk & Wagnalls.
Collections: period furnishings; personal artifacts; photographs; buttons.
Hours & Admission Prices: March-Dec. Sat.-Sun. 1-3; groups by appointment. Adults $5, students $2; discounts to AAA members. Closed holidays.
Attendance: 800 (estimated)

✳ **SPRINGFIELD MUSEUM OF ART, (M),** 107 Cliff Park Rd., Springfield, OH 45504-2501. Tel.: 937-325-4673. Fax: 937-325-4674.
E-mail: smoa@main-net.com
Web Site: www.springfieldart.museum
Founded: 1946.
Congressional District: 7
Key Personnel: Exec. Dir., Angus Randolph; Bd. Pres., Andy Inck; Cur., Charlotte Gordon; Dir. Mktg., Katherine Denney; Facilities Mgr., James Brewer; Systems Mgr., Ken Pinkham.
Personnel Profile: Full-Time Paid 3; Part-Time Paid 5; Part-Time Volunteers 95.
Governing Authority: nonprofit organization. Parent Institution: Springfield Art Association. Tax-exempt: 501(c)(3) & 170(b)(1)(A).
Institution Type/Description: Art Museum.
Collections: 19th & 20th-century American art.
Research Fields:
Facilities: 4,300-vol. library of fine arts books available for use on premises. Original works, paintings, sculpture, drawings & museum-related items for sale.
Activities: lectures; gallery talks; formally organized education programs; art workshops; studio art classes.
Publications: bimonthly newsletters; exhibition catalogues; gallery handouts; quarterly class schedules.
Hours & Admission Prices: Tues.-Sat. 9-5, Sun. 12:30-4:30. Adults $5; discounts to AAM members; members & Sun. no charge. Closed New Year's Day; Memorial Day; Independence Day; Labor Day; Christmas. &
Attendance: 39,989 (accurate)
Membership: Student $25; Individual $40; Family $60; Supporter $100; Sustainer $250; Patron $500; Gallery Circle $1,000.

THE WESTCOTT HOUSE FOUNDATION, 1340 E. High St., Springfield, OH 45505-1166. Tel.: 937-327-9291. Fax: 937-327-9074.
E-mail: info@westcotthouse.org
Web Site: www.westcotthouse.org
Founded: 2001.
Key Personnel: Chm., Mark Chepp; Dir. Devel., Jenny Montgomery; Cur., Marta Wojcik; Volunteer Coord., Erik Lindsjo; Facilities Mgr., Tom Fyffe.
Personnel Profile: Full-Time Paid 4; Full-Time Volunteers 1; Part-Time Paid 1; Part-Time Volunteers 100.
Governing Authority: private; nonprofit organization. Tax-exempt: 501(c)(3).
Institution Type/Description: Historic House Museum: housed in Frank Lloyd Wright's only Prairie Style home in Ohio.
Collections: period furnishings; personal artifacts.
Research Fields: history & architecture of the house and community.
Activities: concerts; docent program; films; school programs; guided tours; lectures; temporary exhibitions; summer arts camp. Annual Events: Annual Fundraiser; Donor Recognition Night; Volunteer Recognition Night.
Publications: quarterly, the Westcott House Foundation Newsletter.
Hours & Admission Prices: House Guided Tours: May-Oct. Wed.-Sat. 11, 12, 1, 2, 3, 4, Sun. 1, 2, 3, 4; Nov.-April Wed.-Fri. 11, 1, 3, Sat. 11, 12, 1, 2, 3, 4, Sun. 1, 2, 3, 4. Adults $15, seniors 65 & over and students $12; discounts to groups; members no charge. Frank Lloyd Wright reciprocal membership. &
Attendance: 9,003 (accurate)
Membership: Student $25; Single $35; Family $55; Family Plus $85; Apprentice $150; Journeyman $250; Craftsman $500; Masterbuilder $1,000; Designer $2,500.

Steubenville

THE JEFFERSON COUNTY HISTORICAL ASSOC., 426 Franklin Ave., Steubenville, OH 43952-1818. Mailing Address: Box 4268, Steubenville, OH 43952-8268. Tel.: 740-283-1133 & 282-9776. Fax: 740-282-9161.
E-mail: jmusm@att.net
Web Site: rootsweb.com/~ohjcha
Founded: 1973.
Congressional District: 18
Key Personnel: Pres., Judy Brancazio; 1st Vice Pres., Eleanor Naylor; 2nd Vice Pres. & Library Dir., Charles Green.
Personnel Profile: Part-Time Volunteers 18.
Governing Authority: society; nonprofit organization. Subsidiary Institution: Vivian Snyder Genealogical Library. Tax-exempt: 501(c)(3).
Institution Type/Description: Historical Society Museum: housed in 1918 mansion.
Collections: Jefferson County history; transportation gallery; riverboat room; trains; bridal room; bridal gowns, some 200 years old; president's room; feature memorabilia & photos, books, presidents born in Ohio; Civil War; WWI & WWII; Steubenville Pottery Company history; Russell Wright.
Research Fields: genealogy; history.
Facilities: 5,000-vol. library of genealogical & history books available for research on premises. Area history & museum related items for sale.
Activities: guided tours; lectures; permanent exhibitions.
Publications: quarterly newsletter.
Hours & Admission Prices: April-Nov. by appointment. Library: March-Dec. Tues.-Fri. 10-3. Office: April-Nov. Tues.-Fri. 10-3. Suggested Donation: adults $2; members no charge. Closed holidays.
Attendance: 1,000 (estimated)
Membership: Annual $15; Sponsor $50; Life $200.

Strongsville

GARDENVIEW HORTICULTURAL PARK, 16711 Pearl Rd. Rte. 42, 1 1/2 miles S. of Rte. 82, Strongsville, OH 44136-6048. Tel.: 440-238-6653.
E-mail: gardenviewhp@gmail.com
Web Site: sites.google.com/site/gvhortpk
Founded: 1949.
Key Personnel: Dir., Henry A. Ross.
Personnel Profile: Full-Time Volunteers 2.
Governing Authority: nonprofit corporation. Parent Institution: Gardenview Horticultural Park, Inc. Tax-exempt: 501(c)(3).
Institution Type/Description: Public Horticultural Park.
Collections: 6 acres of English Cottage gardens; 10-acre arboretum containing 2,500 flowering trees, 100,000 daffodils, underplanted with spring bulbs & wildflowers.
Facilities: 6,000-vol. library of reference books on gardening, birds, animals, travel; crafts available for use by members only; 10 acre arboretum; 6 acre English cottage gardens.
Activities: guided tours by appointment for groups.
Publications: annual report to members.
Hours & Admission Prices: March-Nov. 1 Sat.-Sun. 12-6; groups & members year-round by appointment. Adults $5, children $3; members no charge.
Membership: Annual $25; Sustaining $50-$99; Supporting $100-$999; Lifetime $1,000.

Sugar Grove

WAHKEENA NATURE PRESERVE, (M), 2200 Pump Station Rd., Sugar Grove, OH 43155-9665. Tel.: 740-746-8695; 800-297-1883.
E-mail: wahkeena@att.net
Web Site: ohsweb.ohiohistory.org/places/c13/
Founded: 1957.
Congressional District: 10
Key Personnel: Site Mgr., Thomas Shisler.

Personnel Profile: Full-Time Paid 1; Part-Time Paid 1; Interns 2.
Governing Authority: society. The Ohio Historical Society. Site Operations 1982 Velma Ave., Columbus, OH. 43211. Tax-exempt.
Institution Type/Description: Nature Center.
Collections: 100 species of birds; eight species native orchids; 15 species mammals; rhododendron & mountain laurel; 30 types of ferns.
Research Fields: botany; zoology; ecology.
Facilities: nature center; hiking trails.
Activities: nature guides & trails; interpretation area; seminars; walks; formally organized education program.
Publications: Timeline.
Hours & Admission Prices: April-Oct. Sat.-Sun. 8-4:30. $2 per car; AAM & ICOM members no charge.
Attendance: 3,500
Membership: Ohio Historical Society: Family $50.

Sugarcreek

ALPINE HILLS HISTORICAL MUSEUM, 106 W. Main St., Sugarcreek, OH 44681. Mailing Address: P.O. Box 293, Sugarcreek, OH 44681-0293. Tel.: 888-609-7592.
E-mail: ldyoungen@frontier.com
Web Site: www.villageofsugarcreek.com
Founded: 1977.
Key Personnel: Pres. (V), Lowell Youngen.
Personnel Profile: Part-Time Volunteers 9.
Volunteer Hours: 1,680
Governing Authority: Tax-exempt.
Institution Type/Description: History Museum.
Collections: local history & culture; Swiss & Amish heritage; Amish kitchen; cheese making; photographs; period furnishings; audio-visuals.
Activities: audio-visual presentations.
Publications: brochure.
Hours & Admission Prices: April 2 to mid-Nov. Mon.-Sat. 9:30-4:30. No charge; donations accepted.
Attendance: 8,000 (accurate)
Membership: Individual $20; Family $35; Business & Lifetime Individual $100; Lifetime Family $150; Lifetime Business & Memorial (one time fee) $200.

JOHN S. YODER HOME, 116 Andreas Dr., N.E., Sugarcreek, OH 44681. Mailing Address: P.O. Box 508, Sugarcreek, OH 44681. Tel.: 330-852-4644.
Institution Type/Description: Historic House Museum: housed in an Amish home built in 1869.
Collections: Amish & Mennonite culture in Ohio; personal artifacts; furnishings.
Hours & Admission Prices: May-Oct. Fri.-Sat. 12-5. No charge; donations accepted.

Sylvania

SYLVANIA HISTORICAL VILLAGE & HERITAGE CENTER MUSEUM, 5717 N. Main St., Sylvania, OH 43560. Tel.: 419-882-4865.
E-mail: hist.village@sev.org
Key Personnel: Cur., Joyce Armstrong
Institution Type/Description: History Museum: housed in the former home of Dr. Uriah A. Cooke.
Collections: local history & culture; period furnishings; personal artifacts; photographs; medical equipment. Historic Buildings: 1840s log home; 1844 Stone Academy; 1858 train depot.
Activities: demonstrations; educational programs.
Hours & Admission Prices: Museum: Wed. 3-7, Sat.-Sun. 1-4. Village: by appointment.

Tiffin

AMERICAN CIVIL WAR MUSEUM OF OHIO, 217 S. Washington, Tiffin, OH 44883. Tel.: 419-455-9551.
E-mail: info@acwmo.org
Web Site: www.acwmo.org
Institution Type/Description: Military History Museum.
Collections: military history & artifacts; personal artifacts; photographs.
Hours & Admission Prices: Wed.-Sat. 12-4; other times by appointment. Adults $6, seniors 55 & over $5, students $3; children under 6 & members no charge.

ENCHANTED MOMENT DOLL MUSEUM, 74 Jefferson St., Tiffin, OH 44883. Tel.: 419-443-0038.
Web Site: www.angelfire.com/oh3/dollcollectors
Institution Type/Description: Doll Museum.
Collections: over 3,000 dolls from around the world.
Hours & Admission Prices: Wed.-Fri. 10-4:45, Sat. 10-4; other times by appointment. Admission $6.

SENECA COUNTY MUSEUM, 28 Clay St., Tiffin, OH 44883-2259. Tel.: 419-447-5955. Fax: 419-443-7940.
Founded: 1942.
Congressional District: 5
Key Personnel: Dir., Rosalie Adams; Pres. (V), Barry Porter.
Personnel Profile: Full-Time Paid 1; Part-Time Paid 1; Part-Time Volunteers 25; Interns 1.
Governing Authority: museum foundation. Parent Institution: Seneca County. Subsidiary Institution: Seneca County Museum Foundation Inc. Tax-exempt.
Institution Type/Description: Historic House: 1853 Rezin W. Shawhan residence & carriage house.
Collections: costumes; glass; porcelain history; pressed glass; art glass; Tiffin glassware; weapons; folk craft pieces; historical items; primitive farm tools; period fire equipment.
Research Fields: state & local history.
Facilities: library of local history; educational center.
Activities: guided tours; lectures; films.
Hours & Admission Prices: Wed.-Thurs. 1-4; other times by appointment. Admission $1. Closed holidays.
Attendance: 7,000
Membership: Individual $10; Family $15.

TIFFIN GLASS MUSEUM, 25 S. Washington St., Tiffin, OH 44883-2347. Tel.: 419-448-0200.
E-mail: museum@tiffinglass.org
Web Site: www.tiffinglass.org
Founded: 1998.
Key Personnel: Dir. & Museum Shop Mgr., Ruth Hemminger.
Personnel Profile: Full-Time Volunteers 2; Part-Time Volunteers 10.
Institution Type/Description: Glass History Museum.
Collections: over 2,000 pieces of glass including stemware, lamps, vases, & bowls; factory history & documents.
Major Exhibits: Christmas Glass Display, 12/13-1/14.
Facilities: Books & glassware items for sale.
Activities: Annual Events: Glass Shows in June; Christmas Open House in December.
Publications: members' newsletter.
Hours & Admission Prices: Tues.-Sat. 1-5; other times by appointment. No charge; donations accepted. Closed New Year's Day; Easter; Independence Day; Thanksgiving; Christmas. &
Attendance: 2,100 (accurate)
Membership: Individual $20.

Toledo

BLAIR MUSEUM OF LITHOPHANES, 5403 Elmer Dr., Toledo, OH 43615-2803. Tel.: 419-245-1356. Fax: 419-535-5770.
E-mail: margaretcarney@sbcglobal.net
Web Site: www.lithophanemuseum.org
Founded: 1966.
Key Personnel: Dir. & Cur., Margaret Carney, Ph.D.
Governing Authority: municipal. Parent Institution: Friends of the Blair Museum, Inc. Tax-exempt: 501(c)(3).
Institution Type/Description: Decorative Arts Museum.
Collections: over 2,340 19th century porcelain lithophanes; engravings.
Research Fields: lithophanes; 19th century engravings.
Publications: biannual membership newsletter, The Blair Museum of Lithophanes Bulletin.
Hours & Admission Prices: May-Oct. Sat.-Sun. 1-4; other times by appointment. No charge; donations accepted. Special Tours: $5 per person. &
Attendance: 2,500 (accurate)

IMAGINATION STATION, One Discovery Way, Toledo, OH 43604. Tel.: 419-244-2674. Fax: 419-255-2674.
Key Personnel: Exec. Dir., Lori Hauser; Dir. Devel., Karen George; Sr. Programs Officer, Sloan Eberly; Outreach Mgr., Jamie Pafford; Chief Scientist & Dir. Exhibits, Carl Nelson; Exhibits Mgr., Jim Repolesk; Information Technology Mgr., John Hambro; Sr. Mktg. Officer, Sara Young
Institution Type/Description: Children's Museum.
Collections: hands-on exhibitions.

Activities: camp-in; workshops; scout programs; summer camps.
Hours & Admission Prices: Tues.-Sat. 10-5, Sun. 12-5. Adults 13 & over $9, seniors 65 & over $8, children 3-12 $7; discounts to military & groups of 15 or more; children 2 & under no charge. Closed New Year's Day; Easter; Thanksgiving; Christmas Eve & Day.

IMAGINATION STATION, (M), 1 Discovery Way, Toledo, OH 43604-1579. Tel.: 419-244-2674. Fax: 419-255-2674.
E-mail: pmorin@imaginationstationtoledo.org
Web Site: imaginationstationtoledo.org
Formerly: COSI Toledo
Founded: 1997.
Key Personnel: C.E.O., Lori Hauser; Opers. Dir., Amy Hering; Chief Scientist & Exhibits Dir., Carl Nelson; Corporate Devel. Officer, Karen George; Asst. Dir. STEM Education, Sloan Mann; Sr. Mktg. Officer, Sara Young; P.R. & Communications Coord., Paul Morin.
Governing Authority: private; nonprofit organization. Tax-exempt: 501(c)(3).
Institution Type/Description: Science Museum.
Collections: hands-on science exhibits.
Facilities: restaurant; labs. Museum-related items for sale.
Activities: formal education programs; participatory exhibits; broadcast programs; temporary exhibitions. Annual Events: Bash (fundraiser); various week long events.
Publications: quarterly newsletter, Discover.
Hours & Admission Prices: Tues.-Sat. 10-5, Sun. 12-5. Adults 13 & over $8.50, seniors 65 & over $7.50, children 3-12 $6.50; children 2 & under no charge. &
Attendance: 209,430 (accurate)
Membership: Individual $55; Grandparent $65; Family $70; Family & Guest $80; Family Plus $100; Copper $250; Silver $500; Gold $1,000.

NATIONAL MUSEUM OF GREAT LAKES, 1701 Front St., Toledo, OH 43605. Mailing Address: P.O. Box 8218, Toledo, OH 43605. Tel.: 440-967-3467.
Web Site: www.inlandseas.org
Institution Type/Description: History Museum.
Collections: local history; photographs; geological.
Hours & Admission Prices: Call for hours.

TOLEDO BOTANICAL GARDEN, 5403 Elmer Dr., Toledo, OH 43615-2803. Tel.: 419-536-5566. Fax: 419-536-5574.
E-mail: membership@toledogarden.org
Web Site: www.toledogarden.org
Formerly: Crosby Gardens
Founded: 1982.
Congressional District: 5
Key Personnel: Dir., Karen Ranney Wolkins; Pres. (V), Dale Theis.
Personnel Profile: Full-Time Paid 14; Part-Time Paid 12; Part-Time Volunteers 110; Interns 6.
Governing Authority: Tax-exempt.
Institution Type/Description: Botanical Garden.
Collections: Gardens: herbs; perennial; pioneer; vegetable; hosta; hemerocallis; NAPCC North American Plant Collection Consortium Accredited Hosta; rhododenron & azaleas; beech; rose; peony; iris; oak; conifer. Historic Building: Peter Navarre pioneer cabin, c.1831.
Activities: special events; weddings; rental facilities. Museum Sponsors: Crosby Festival of the Arts; Jazz in the Garden; Plant Sale; Night Fall Festival; Seed Swap; Crosby Award Luncheon; Arts in the Garden; Heralding the Holidays.
Publications: magazine, Cultivation.
Hours & Admission Prices: Daily dawn to dusk. No charge. &
Attendance: 130,000 (estimated)
Membership: Senior $30; Individual $40; Family & Grandparent $50; Friend $75; Sustaining $100; Supporting $250; Patron $500; Toledo Botanical Society $1,000.

TOLEDO FIREFIGHTERS MUSEUM, (M), 918 Sylvania Ave., Toledo, OH 43612-1343. Tel.: 419-478-3473.
E-mail: toledofiremuseum@bex.net
Founded: 1976.
Personnel Profile: Part-Time Volunteers 11; Interns 1.
Volunteer Hours: 3,300
Governing Authority: nonprofit organization. Tax-exempt: 501(c)(3).
Institution Type/Description: Firefighters History Museum: housed in Old Number 18 Fire House.
Collections: over 175 years of Toledo fire department & fire fighting history; period fire fighting equipment & uniforms; advancements in fire fighting technology; 1837 Neptune, Toledo's first fire pumper; 1927 American-LaFrance pumper; 1929 Pirsch pumper; 1936 Schacht service ladder truck; 1968 Willy's Jeep; firehouse gongs; helmets; period fire toys; photographs; 1933 Federal ladder truck; 1919 Model T Ford chemical truck; station journals; c.1880 Silsby steamer; model fire trucks.
Research Fields: local history.
Facilities: library.
Activities: children's educational activities & safety programs; guided tours; fire safety education programs. Museum Sponsors: Fire Prevention Week in October.
Publications: newsletter, Hook & Letter.
Hours & Admission Prices: Sat. 12-4; other times by appointment. No charge; donations accepted. &
Attendance: 2,500 (estimated)

* **THE TOLEDO MUSEUM OF ART, (M),** 2445 Monroe St., Toledo, OH 43620-1500. Mailing Address: P.O. Box 1013, Toledo, OH 43697-1013. Tel.: 419-255-8000. Fax: 419-255-5638. TDD: 419-255-8000; Facebook: The Toledo Museum of Art.
E-mail: info@toledomuseum.org
Web Site: www.toledomuseum.org
Founded: 1901.
Congressional District: 9
Key Personnel: Chm. Bd., David K. Welles, Jr.; Dir., Brian Kennedy, Ph.D.; C.O.O., Carol Bintz; Chief Cur., Carolyn Putney; Cur. William Hutton European Paintings & Sculpture before 1900, Lawrence W. Nichols; Cur. Glass, Jutta-Annette Page; Assoc. Dir., Amy Gilman; Registrar, Andrea Mall; Head Librarian, Alison Huftalen; Controller, Tim Szymanski; Dir. Devel., Susan Palmer; Dir. Communications, Kelly Fritz Garrow; Museum Shop Mgr., Heather Blankenship.
Personnel Profile: Full-Time Paid 93; Part-Time Paid 201; Part-Time Volunteers 426; Interns 10.
Volunteer Hours: 12,753
Governing Authority: nonprofit organization. Tax-exempt: 501(c)(3).
Institution Type/Description: Art Museum.
Collections: European & American painting, sculpture, decorative arts; ancient, European & American glass; early ancient glass; ancient & medieval art; books & manuscripts; prints; photography; jewelry; modern & contemporary art.
Major Exhibits: Varujan Boghosian, 12/13-4/14; The Art of the Louvre's Tuileries Garden (T), 2/14-5/14; In Fine Feather: Birds, Art & Science, 4/14-6/14; The Art of Video Games (T), 6/14-9/14; The Great War: Art on the Front Lines, 7/14-9/14.
Research Fields: pertaining to collections.
Facilities: over 60,000-vol. library of books, slides & periodicals on art & music available for use on premises; reading room; 165-seat lecture hall; classrooms; restaurant; 1,750-seat concert hall; sculpture garden. Books on art, postcards, reproductions of art & original works by area artists for sale.
Activities: guided tours; lectures; films; gallery talks; concerts; formally organized education programs for children, young adults, adults; docent program or council; inter-museum loan, permanent, temporary & traveling exhibitions; rental facilities.
Publications: quarterly members magazine; exhibition & collection catalogues; information guides for campus architecture.
Hours & Admission Prices: Tues.-Wed. 10-4, Thurs.-Fri. 10-9, Sat. 10-5, Sun. 12-5. No charge. Closed occasional holidays. &
Attendance: 400,000 (accurate)
Membership: Senior Citizens $35; Individual $55; Senior Couple $60; Family $75; Contributing $125; Reciprocal $250; Supporting $500; President's Council and Business Council $1,000 & up; Apollo Society $5,000.

THE TOLEDO ZOO, 2700 Broadway St., Toledo, OH 43609-3100. Mailing Address: P.O. Box 140130, Toledo, OH 43614-0130. Tel.: 419-385-5721. Fax: 419-389-8670.
Web Site: www.toledozoo.org
Founded: 1900.
Congressional District: 9
Key Personnel: Exec. Dir., Jeff Sailer; Pres. (V), Mary Ellen Pisanelli; Veterinarian, Dr. Chris Hanley; Dir. Human Resources, Nancy Foley; Conservation Biologist, Dr. Peter Tolson; Dir. Horticulture, Nancy Bucher; Cur. Fishes, Jay F. Hemdal; Cur. Interpretive Svcs., Alex DeBeukelaer; Cur. Herpetology, Andy Odum; Coord. (V), Bill Davis; Registrar, Glenous Favata; Merchandise Buyer, Deborah L. Noward; Cur. Birds, Robert Webster; Cur. Education, Mitchell Magdich; Cur. Mammals, Randi Meyerson; Construction Mgr., Rick Payeff.
Personnel Profile: Full-Time Paid 158; Part-Time Paid 353; Part-Time Volunteers 541; Interns 15.
Governing Authority: society; nonprofit organization. Parent Institution: Toledo Zoological Society. Tax-exempt: 501(c)(3).
Institution Type/Description: Zoo.

Collections: natural history; greenhouse; botanical gardens; mammals, birds; fish; reptiles; herptiles; polar bears; seals.

Research Fields: SSP programs; Aruba Island rattlesnake; Bali starling; cheetah; chimpanzee; Dumeril's ground boa; elephants; lowland gorilla; orangutan; Puerto Rican crested toad; radiated tortoise; snow leopard; Virgin Islands boa; white rhinoceros; Wyoming toad, cinereous vulture; Karner blue butterfly; Mitchell's Satyr butterfly; Purplish Copper butterfly; Swamp Metalmark butterfly.

Facilities: botanical garden; aquarium; outdoor amphitheater; classrooms; theaters; children's petting zoo; cafe. Zoo-related items for sale.

Activities: summer concerts; formally organized education programs; docent program or council; educational outreach (schools, nursing homes & libraries); distance learning; summer camp; children's zoo; teacher workshops; interpretive programming; home school; gifted & talented; rental facilities.

Publications: quarterly magazine, Safari!; souvenir guidebook.

Hours & Admission Prices: May-Sept. daily 10-6; Oct.-April daily 10-5. Adults $11, seniors over 60 & children 2-11 $8; discounts to groups & AZA members; members with card & children under 2 no charge. Parking: $6. Closed New Year's Day; Thanksgiving; Christmas. &

Attendance: 900,000 (estimated)

Membership: Single $44; Individual Plus $58; Grandparent $67; Family $73; Family Plus $101; Family Deluxe $127.

UNIVERSITY OF TOLEDO STRANAHAN ARBORETUM, 4131 Tantara Dr., Toledo, OH 43623. Tel.: 419-841-1007.

Key Personnel: Dir., Dr. Daryl Dwyer

Institution Type/Description: Arboretum.

Collections: trees; plants; flowers; ponds; wetlands.

Facilities: 47-acre site.

Hours & Admission Prices: Mon.-Fri. 8-5.

WILDWOOD MANOR HOUSE, 5100 W. Central Ave., Toledo, OH 43615-2106. Tel.: 419-461-0520; 419-277-0107.

Web Site: www.metroparkstoledo.com

Founded: 1975.

Congressional District: 9

Key Personnel: Dir. Programming, Heather Norris; Program/Manor House Staff, Susan Roberts-McGlade; Rental Facilities Coord., Grace Peoples

Institution Type/Description: Historic House: c.1938.

Collections: period furnishings; photographs.

Major Exhibits: Champion Spark Plug Co., 1/14-12/14; Frank Stranahan/Amateur & Professional Golfer, 1/14-12/14.

Activities: special events; educational programs. Annual Event: Holidays in the Manor House in December.

Hours & Admission Prices: Jan.-March Sat.-Sun. 12-5; April-June Thurs.-Sat. 12-5; July-Sept Tues.-Fri. 12-5; Oct. to mid-Nov. Thurs.-Fri. & Sun. 12-5; First full week of Dec. 10-8. No charge; donations accepted. Closed holidays. &

Attendance: 30,000 (accurate)

Trenton

CHRISHOLM HISTORIC FARMSTEAD, 2070 Woodsdale Rd., Trenton, OH 45067-9752. Mailing Address: Metroparks of Butler County, 2051 Timberman Rd., Hamilton, OH 45013. Tel.: 513-867-5835.

E-mail: friendsofchrisholm@yahoo.com

Web Site: www.chrisholmhistoricfarmstead.org

Founded: 1995.

Key Personnel: Pres. (V) Friends of Chrisholm, Bill McKnight.

Governing Authority: Parent Institution: Metro Parks of Butler County. Tax-exempt.

Institution Type/Description: Historic Site: housed in the Samuel Augspurger farmhouse, built in 1874. Listed on the National Register of Historic Places.

Collections: period furnishings; Amish-Mennonite settlers; county history.

Facilities: picnic area.

Activities: educational programs; special events; rental facilities.

Publications: Friends of Chrisholm newsletter.

Hours & Admission Prices: Park: daily 8am to dusk. Home: Summer call for hours; other times by appointment. Park Permit: Non-Residents $10 annual, $5 daily; Butler County Residents no charge. Home: no charge; donations accepted.

Attendance: 1,360 (estimated)

Membership: Friend $30; Harvester $100; Visionary $100-$499; Preservationist $500 & up.

Troy

BRUKNER NATURE CENTER, 5995 Horseshoe Bend Rd., Troy, OH 45373-9485. Tel.: 937-698-6493.

E-mail: info@bruknernaturecenter.com

Web Site: www.bruknernaturecenter.com

Governing Authority: nonprofit organization.

Institution Type/Description: Nature Center.

Collections: wetland; forest; prairie; wildlife. Historic Building: 1804 Iddings log home.

Facilities: nature trails; 165 acre nature preserve; auditorium.

Activities: educational programs.

Hours & Admission Prices: Mon.-Sat. 9-5, Sun. 12:30-5. Adults $1, children $.25.

MIAMI VALLEY VETERANS MUSEUM, 107 W. Main St., Troy, OH 45373. Mailing Address: P.O. Box 154, Troy, OH 45373. Tel.: 937-451-1455.

E-mail: director@theyshallnotbeforgotten.org

Web Site: www.theyshallnotbeforgotten.org

Institution Type/Description: Military History Museum.

Collections: military history & artifacts; photographs; personal artifacts.

Hours & Admission Prices: Tues. 1-4, Wed. & Sat. 9-1, Thurs. 1-5.

MUSEUM OF TROY HISTORY, 124 E. Water St., Troy, OH 45373. Tel.: 937-339-5155.

Web Site: museumoftroyhistory.org

Governing Authority: Tax-exempt.

Institution Type/Description: History Museum: housed in the former home of John Kitchen.

Collections: local history & culture; period furnishings; personal artifacts; sports; religion; industry; healthcare.

Activities: Museum Sponsors: June Ice Cream Social; Annual Tour of Homes in October; Christmas Open House in December.

Hours & Admission Prices: Sat.-Sun. 2-4. No charge.

OVERFIELD TAVERN MUSEUM, 201 E. Water St., Troy, OH 45373-3438. Tel.: 937-335-4019.

E-mail: info@overfieldtavernmuseum.com

Web Site: www.overfieldtavernmuseum.com/index.htm

Founded: 1966.

Congressional District: 4

Key Personnel: Dir., Robert Patton; Asst. Cur., Busser Howell; Cur., Kelly Smith.

Governing Authority: nonprofit organization. Tax-exempt: 501(c)(3).

Institution Type/Description: Historic Building Museum: housed in 1808 Overfield Tavern a 2-story, hewed-log building, which served as the first courthouse in Troy, Ohio.

Collections: late 18th & early 19th century artifacts; pewter; redware; slipware; wrought-iron accessories; brass; copper & tin primitives; fireplace equipment; tavern tables; Windsor chairs; lighting devices; old leather-bound books.

Research Fields: period pioneer artifacts; the pioneer tavern & how it was operated; genealogy.

Activities: guided tours; lectures; arts festivals; formally organized education programs for children. Museum Sponsors: arts & crafts demonstrations, such as spinning, weaving, fireplace cookery; historic displays at Troy-Hayner Cultural Center.

Hours & Admission Prices: April-Oct. Sat.-Sun. 1-5; other times by appointment. No charge; donations accepted. Closed New Year's Eve & Day; Easter; Independence Day; Thanksgiving.

Membership: Individual $5; Family $10; Life $100.

TROY-HAYNER CULTURAL CENTER, 301 W. Main St., Troy, OH 45373-3241. Tel.: 937-339-0457. Fax: 937-335-6373.

E-mail: troyhaynercenter@troyhayner.org

Web Site: www.troyhayner.org

Founded: 1976.

Congressional District: 8

Key Personnel: Pres., Carol Jackson; Dir., Linda Lee Jolly; Asst. Dir., Terri Boehringer.

Personnel Profile: Full-Time Paid 4; Part-Time Paid 8; Part-Time Volunteers 119; Interns 1.

Governing Authority: Troy-Hayner Board of Governors. Parent Institution: Troy City School District, 500 N. Market St., Troy, OH 45373. Tel.: 513-332-6700. Tax-exempt.

Institution Type/Description: Cultural Center: housed in c.1914 Mary Jane Hayner House, built in the Norman-Romanesque Revival style of architecture.
Collections: 1800's Hayner Distillery exhibit & Mary Jane Hayner Family memorabilia; Mary Coleman Allen miniatures; distillery artifacts.
Facilities: rental facilities.
Activities: guided tours; lectures; concerts; recitals; classes & workshops; temporary exhibits; formally organized education programs; community events; private parties.
Publications: newsletter; brochures; annual report.
Hours & Admission Prices: Mon. 7pm-9pm, Tues.-Thurs. 9-5 & 7-9, Fri.-Sat. 9-5, Sun. 1-5. No charge; donations accepted. Closed holidays. &
Attendance: 41,534 (estimated)
Membership: Friends of Hayner: Annual $30.

WACO HISTORICAL SOCIETY & LEARNING CENTER, 1865 S. County Rd. 25A, Troy, OH 45373. Tel.: 937-335-9226. Fax: 937-335-4357.
E-mail: admin@wacoairmuseum.org
Web Site: www.wacoairmuseum.org/index.html
Founded: 1997.
Key Personnel: Dir., Gretchen Hawk; Dir. Learning Center, Lisa Hokky; Pres. (V), Dave Bucher; Museum Shop Mgr., Patty Wagner.
Personnel Profile: Part-Time Paid 4; Part-Time Volunteers 6; Interns 2.
Governing Authority: Tax-exempt.
Institution Type/Description: Historical Society Museum.
Collections: WACO Aircraft Company history; aviation artifacts; photographs; drawings; airplanes; videos.
Facilities: rental facilities; picnic shelter.
Activities: WACO simulator; robotics teams; adult lectures; aviation camp; bi-plane rides. Annual Event: WACO Fly-In in September.
Publications: quarterly newsletter, WACO Word.
Hours & Admission Prices: Mon.-Fri. 9-12, Sat.-Sun. 12-5. Adults $6, military with ID $5, children 7-17 $3; children under 7 no charge. &
Attendance: 3,500 (estimated)
Membership: Student $15; Individual $30; Family $45; Partner $60; Supporter $100; Contributor $250; Benefactor $500; Life $1,000.

Twinsburg

TWINSBURG HISTORICAL SOCIETY, 8996 Darrow Rd., Twinsburg, OH 44087-2127. Mailing Address: P.O. Box 7, Twinsburg, OH 44087-0007. Tel.: 330-487-5565.
E-mail: akancler@windstream.net
Web Site: lwkweb.com/twinsburghistoricalsociety/
Founded: 1963.
Congressional District: 2
Key Personnel: Pres., Audrey Kancler; Vice Pres., Sue Graham; Sec., Bonnie Williams; Corresponding Sec., Lea Bissell; Treas., Dan Simecek.
Personnel Profile: Part-Time Volunteers 20.
Governing Authority: nonprofit organization. Tax-exempt: 501(c)(3).
Institution Type/Description: General Museum: housed in 1865 school.
Collections: 19th-century furniture; tools; clothes; toys; local artifacts; photographs. Historic Building: 1870 barn.
Activities: guided tours; demonstrations; permanent exhibitions; information programs; special programs. Museum Sponsors: Ice Cream Social in July; old time music program & kids day, toys without batteries in summer; Olde Thyme Fayre in September; Cemetery Walk & Halloween story telling in the barn partner with Twinsburg Public Library in October.
Publications: monthly newsletter.
Hours & Admission Prices: Feb.-Dec. last Sun. each month 2-5; other times by appointment. No charge; donations accepted. &
Attendance: 500 (estimated)
Membership: Active Single $8; Active Family & Contributing Single $15; Contributing Couple $25; Professional $35; Life $100.

Uhrichsville

THE UHRICHSVILLE MUSEUM OF CLAY INDUSTRY & FOLKART, 330 N. Main St., Uhrichsville, OH 44683. Mailing Address: P.O. Box 11, Dennison, OH 44621. Tel.: 740-922-6776 & 5455. Fax: 740-922-4929.
E-mail: claymuseum@dennisondepot.org
Web Site: claymuseum.com
Formerly: Uhrichsville Clay Museum
Founded: 2008.
Congressional District: 18
Key Personnel: Dir., Wendy Zucal
Governing Authority: Parent Institution: Dennison RR Depot Museum. Subsidiary Institution: Historic Schoenbrunn Village. Tax-exempt.

Institution Type/Description: History Museum.
Collections: clay history; clay brick; industrial history; clay sewer pipes; folk art; sewer pipes; period tools; photographs.
Facilities: facility rental.
Activities: Museum Sponsors: Clay Workers Picnic; Clay Queen's Tea.
Hours & Admission Prices: Tues.-Fri. 9-5, Sat. 11-4. Adults $3, seniors $2, students $1; discounts to groups of 20 or more. Closed major holidays.
Attendance: 2,500 (estimated)
Membership: Student $5; Individual $10; Family $20; Nonprofit $30; Corporate $40; Individual Life $125; Married Life $175; Corporate Life & Memorial $200; Benefactor $500; (additional $10 to Supersize).

Upper Sandusky

INDIAN MILL MUSEUM STATE MEMORIAL, (M), 7417 Co. Hwy. 47, Upper Sandusky, OH 43351-1430. Mailing Address: c/o Wyandot County Museum, P.O. Box 372, Upper Sandusky, OH 43351-0372. Tel.: 419-294-3857.
Web Site: www.ohiohistory.org/indianmill
Founded: 1967.
Congressional District: 4
Governing Authority: society. Affiliated with Ohio Historical Society, 1985 Velma Ave., Columbus, OH. 43211.
Institution Type/Description: Historic Site.
Collections: grist mill.
Hours & Admission Prices: Memorial Day to Labor Day Fri.-Sun. 1-6; Sept.-Oct. Sat.-Sun. 1-6; groups by appointment. Admission 13 & over $1, children 6-12 $.50; children under 6 & Ohio Historical Society members no charge. &
Attendance: 2,245 (accurate)

WYANDOT COUNTY HISTORICAL SOCIETY, 130 S. 7th St., Upper Sandusky, OH 43351-1339. Mailing Address: P.O. Box 372, Upper Sandusky, OH 43351-0372. Tel.: 419-294-3857. Facebook: Wyandot County Historical Society.
E-mail: curator@wyandothistory.org
Web Site: www.wyandothistory.org
Founded: 1929.
Key Personnel: Dir. & Cur., Ronald I. Marvin, Jr.
Governing Authority: nonprofit organization. Tax-exempt: 501(c)(3).
Institution Type/Description: General Museum.
Collections: agriculture; arboretum; archives; paintings; children's museum; manuscripts; costumes; glass; history; Indian artifacts; marine; medical; musical instruments. Historic House: 1852 McCutchen Overland Inn.
Activities: lectures; formally organized education programs for children; permanent & temporary exhibitions.
Publications: bimonthly newsletter.
Hours & Admission Prices: Museum: May-Oct. Fri.-Sun. 1-4:30. Adults $2, students & children $1. &
Membership: Annual $15.

Urbana

CEDAR BOG NATURE PRESERVE, (M), 980 Woodburn Rd., Urbana, OH 43078-9417. Mailing Address: P.O. Box 510, Urbana, OH 43078-0510. Tel.: 937-484-3744; 800-860-0147.
Web Site: www.ohiohistory.org/places/cedarbog
Founded: 1942.
Congressional District: 1
Key Personnel: Site Mgr., E. Doerzbacher.
Governing Authority: society. The Ohio Historical Society, Ohio Historical Center Museums Division, 1982 Velma Ave., Columbus, OH 43211. Tax-exempt.
Institution Type/Description: Nature Preserve: post-glacial alkaline bog in Ohio.
Collections: plants typical of a boreal fen; fish; endangered spotted turtles; endangered massasauga rattlesnakes; butterflies; birds; trees in the bog from the Ice Age.
Research Fields: geology; botany, ecology.
Facilities: nature & observation area.
Publications: books, Ohio Journal of Science; Cedar Bog Symposium.
Hours & Admission Prices: April-Oct. Wed.-Sun. 9-4:30; Nov.-March by appointment. Adults $4, children 6-12 & school groups $3; children 5 & under and members no charge. &
Attendance: 5,000 (accurate)
Membership: Cedar Bog Association: $10 & up; Ohio Historical Society: $30 & up.

CHAMPAIGN AVIATION MUSEUM, Urbana Grimes Airport I74, 1652 N. Main St., Urbana, OH 43078. Tel.: 937-652-4710.
Founded: 2008.
Congressional District: 4
Institution Type/Description: Aviation Museum.
Collections: aviation history; aircraft.
Hours & Admission Prices: Tues.-Sat. 9-4; other times by appointment. No charge; donations accepted. &
Attendance: 15,000 (estimated)

CHAMPAIGN COUNTY HISTORICAL MUSEUM, 809 E. Lawn Ave., Urbana, OH 43078. Mailing Address: P.O. Box 65, Urbana, OH 43078-0065. Tel.: 937-653-6721. Facebook: Champaign County Historical Museum.
E-mail: champhistmus@ctcn.net
Web Site: www.champaigncountyhistoricalmuseum.org
Founded: 1934.
Congressional District: 7
Key Personnel: Pres. (V), Ward Lutz; Treas., Howard Brust.
Personnel Profile: Part-Time Volunteers 7.
Governing Authority: nonprofit organization. Parent Institution: Champaign County Historical Society. Tax-exempt: 501(c)(3).
Institution Type/Description: Historical Society Museum: housed in 1912 school for Champaign County Children's Home.
Collections: Civil War; Brand Whitlock; manuscripts; period artifacts; Simon Kenton collection; local photographs; early farm implements; Native American diorama.
Facilities: 100-vol. library of historical books available upon request.
Activities: guided tours; lectures; films; arts festivals; formally organized education programs for adults; permanent & temporary exhibitions. Annual Event: Octoberfest & craft show in October
Publications: A Brief History of Simon Kenton; A Brief History of Richard Stanhope.
Hours & Admission Prices: Mon.-Tues. 10-4; other times by appointment. No charge; donations accepted. &
Attendance: 1,500 (accurate)
Membership: Family $10.

GRIMES FLYING LABORATORY MUSEUM, Grimes Field, 1636 N. Main St., Urbana, OH 43078. Tel.: 937-873-5764 & 652-4319.
Institution Type/Description: History Museum.
Collections: Warren G. Grimes' life & career in designing aircraft lighting; Flying Lab test vehicle; period & modern aircraft lighting.
Hours & Admission Prices: Sat. 9-1; other times by appointment.

JOHNNY APPLESEED SOCIETY MUSEUM, (M), 579 College Way, Urbana University, Urbana, OH 43078-2081. Tel.: 937-484-1303. Fax: 937-484-1322.
E-mail: jbesecker@urbana.edu
Web   Site: www.urbana.edu/index.php/alumni_and_friends/appleseed_society/museum/
Founded: 1998.
Key Personnel: Dir. & Museum Shop Mgr., Joe D. Besecker; Chm (V), Dr. Francis Hazard.
Personnel Profile: Full-Time Paid 1; Part-Time Volunteers 5.
Governing Authority: Parent Institution: Urbana University. Tax-exempt.
Institution Type/Description: History Museum.
Collections: Johnny (Appleseed) Chapman memorabilia & written materials.
Publications: 3 books.
Hours & Admission Prices: Tues.-Fri. 10-2, Sat. 12-4; other times by appointment. No charge; donations accepted. &
Attendance: 1,350 (estimated)
Membership: Appleseed $25; Seedling $100; Apple Nursery $250; Apple Tree $500.

Utica

VELVET ICE CREAM COMPANY MUSEUM & MILLING MUSEUM, 11324 Mt. Vernon Rd., Utica, OH 43080. Mailing Address: P.O. Box 588, Utica, OH 43080. Tel.: 800-589-500; 740-892-3921. Fax: 740-892-4339.
E-mail: info@velveticecream.com
Web Site: www.velveticecream.com
Founded: 1970.
Institution Type/Description: Ice Cream & Milling Museums.
Collections: company history; 1817 ice cream parlor; 1817 grist mill.

Facilities: visitors center; nature trails; restaurant. Museum-related items for sale.
Activities: ice cream production viewing area; petting zoo; rental facilities; nature trails.
Hours & Admission Prices: Museum & Parlor: May, June-Aug. & Sept. daily 11-7; Oct. daily 11-6. Mill & Factory Tours: Mon.-Fri. 11-3. &
Attendance: 150,000 (estimated)

Van Wert

CENTRAL INSURANCE FIRE MUSEUM, 800 S. Washington St., Van Wert, OH 45891-2381. Mailing Address: P.O. Box 351, Ven Wert, OH 45891. Tel.: 419-238-1010.
Web Site: www.central-insurance.com/docs/museum.htm
Institution Type/Description: Fire-Fighting Museum.
Collections: leather fire buckets dating back to the 1700s; over 600 antique fire toys; fire extinguishers & glass fire "grenades" from the 1850's; antique fireman helmets & uniforms; hand-drawn pumper used in 1871; Ahrens horse-drawn steam pumper; 1926 Ahrens-Fox pumper.
Hours & Admission Prices: 3rd Fri. of each month 1-3; other times by appointment.

VAN WERT COUNTY HISTORICAL SOCIETY, 602 N. Washington St., Van Wert, OH 45891-1265. Mailing Address: P.O. Box 621, Van Wert, OH 45891-0621. Tel.: 419-771-9851.
E-mail: vwc.historicalsociety@gmail.com
Web Site: www.vanwert.com/museum
Founded: 1955.
Congressional District: 5
Key Personnel: Pres. (V), Jon Amundson.
Personnel Profile: Part-Time Volunteers 40; Interns 57.
Governing Authority: Tax-exempt.
Institution Type/Description: Historic House Museum: housed in the former home of John O. & Tacey Viella Clark; built in 1895.
Collections: local history & culture; photographs; period furnishings; personal artifacts; Native American artifacts. Historic Buildings: 1906 one-room school; 1860 log house; 1875 gazebo.
Activities: special events.
Hours & Admission Prices: March-Nov. Sun. 2-4:30; groups by appointment. No charge; donations accepted. &
Attendance: 3,500 (estimated)

Vandalia

TRAPSHOOTING HALL OF FAME & MUSEUM, 601 W. National Rd., Vandalia, OH 45377-1036. Mailing Address: P.O. Box 281, Vandalia, OH 45377. Tel.: 937-660-5663. Fax: 937-660-5664.
E-mail: staff@traphof.org
Web Site: www.traphof.org
Institution Type/Description: Trapshooting Museum.
Collections: trapshooting artifacts & memorabilia; photographs.
Hours & Admission Prices: Mon.-Fri. 9-3. No charge. Closed New Year's Day; Memorial Day; Independence Day; Labor Day; Thanksgiving; Christmas.

Wapakoneta

❋ **ARMSTRONG AIR & SPACE MUSEUM, (M),** 500 Apollo Dr., Wapakoneta, OH 45895. Mailing Address: P.O. Box 1978, Wapakoneta, OH 45895-0978. Tel.: 419-738-8811.
E-mail: info@armstrongmuseum.org
Web Site: www.armstrongmuseum.org
Founded: 1972.
Congressional District: 4
Key Personnel: Dir., Chris Burton; Pres. (V), Tom Finkelmeier, Jr.
Personnel Profile: Full-Time Paid 2; Part-Time Paid 5; Part-Time Volunteers 14.
Governing Authority: Tax-exempt.
Institution Type/Description: Aeronautics & Space Museum.
Collections: symbols & relics of Ohio air & space achievements; moon rock display; space shuttle display; space flight suits; memorabilia from Neil Armstrong's boyhood interest in aviation; space shuttle landing simulator; lunar landing simulator. Historic Aircrafts: 1946 Aeronca 7AC Champion; 1966 Gemini VIII; c.1960 F5D Skylancer.
Facilities: 80-seat theater. Museum-related items for sale.
Activities: film & art exhibits; audiovisual; infinity cube; permanent & loan exhibitions. Museum Sponsors: Summer Moon Festival in July.
Hours & Admission Prices: Tues.-Sat. 9:30-5, Sun. 12-5. Adults $8, children

$4, discount to groups, Golden Buckeye & AAA members; members & children under 5 no charge. Closed New Year's Day, Thanksgiving & Christmas. &
Attendance: 50,000 (accurate)
Membership: Student $10; Individual $20; Family $35.

Warren

JOHN STARK EDWARDS HOUSE & MUSEUM, (M), 303 Monroe St., N.W., Warren, OH 44483. Mailing Address: P.O. Box 1907, Warren, OH 44482-1907. Tel.: 330-394-4653. Fax: 330-394-4653.
E-mail: museum@trumbullcountyhistory.org
Web Site: www.trumbullcountyhistory.org
Founded: 1938.
Congressional District: 13
Key Personnel: Pres. (V), Robt. Smith; Cur., Eileen Blaney; Cur., Trish Scarmuzzi.
Personnel Profile: Part-Time Volunteers 30; Interns 2.
Governing Authority: society. Trumbull County Historical Society. Tax-exempt.
Institution Type/Description: Historic House: 1807 John Stark Edwards house.
Collections: 1800 quilts; old tools; 1800s clothing; musical instruments; old books; arrowheads; postcards; furniture; kitchen items; toys.
Research Fields: Trumbull County history.
Facilities: 100-vol. library of historical data on the Connecticut Western Reserve & Trumbull County, Warren, Ohio, available for research on premises. History books & other museum-related items for sale.
Activities: guided tours.
Publications: quarterly newsletter, Trump of Fame.
Hours & Admission Prices: First Sun. each month except Jan., Feb. & March 2-5; groups by appointment. No charge; donations accepted. Closed holidays.
Attendance: 2,100 (accurate)
Membership: Students $10; Individual $25; Family $40; Corporation $100; Life $250.

NATIONAL PACKARD MUSEUM, (M), 1899 Mahoning Ave., N.W., Warren, OH 44483-2081. Tel.: 330-394-1899. Fax: 330-394-7796.
E-mail: national@packardmuseum.org
Web Site: www.packardmuseum.org
Founded: 1989.
Key Personnel: C.E.O., Cur. & Archivist, Mary Ann Porinchak; Pres. (V), Dr. Robert A. Walton, D.D.S.; Treas., Roger Phillips; Dir. Operations, Charlie Ohlin, Esq.; Event Mgr., Christine Bobco.
Personnel Profile: Full-Time Paid 1; Full-Time Volunteers 2; Part-Time Paid 3; Part-Time Volunteers 190; Interns 4.
Governing Authority: private; nonprofit organization. Tax-exempt: 501(c)(3).
Institution Type/Description: History Museum.
Collections: Packard family history; motor car memorabilia; Packard Electric, Packard Motor Car & Ohio Lamp history; Packard motor cars 1900-1958.
Research Fields: Packard Motor Car Company & their involvement in the war effort; building of the PT boat, fighter plane & war machine engines.
Facilities: archives; classroom. Museum-related items for sale.
Activities: docent program; guided tours; hobby workshops; lectures; loan, temporary & participatory exhibitions; internships with Kent State & YSU; school tours with scavenger hunt; facility rental. Annual Events: All Packard Car Show; All Makes & Models Car Show; Antique Motorcycle Exhibit.
Publications: quarterly newsletter, Time Machine.
Hours & Admission Prices: Tues.-Sat. 12-5, Sun. 1-5; varied hours for special events & prearranged tours. Adults $8, senior citizens & students $5; discounts to groups, AARP & AAA members; members & children under 7 no charge. Closed New Year's Eve & Day; Easter; Memorial Day; Independence Day; Labor Day; Thanksgiving; Christmas Eve & Day. &
Attendance: 20,000 (estimated)
Membership: Senior Citizen 65 & over $18; Individual $30; Family $60; Supporter $120. Life: Packard Live Wire $500; Packard Old Pacific $750; Packard No One $1,000; Packard Phaeton $5,000.

SUTLIFF MUSEUM, (M), 444 Mahoning Ave., N.W., Warren, OH 44483. Tel.: 330-399-8807, ext. 121.
E-mail: info@sutliffmuseum.org
Founded: 1971.
Key Personnel: Cur., Sally Thomas; Asst. Cur., Samantha Basile.
Personnel Profile: Part-Time Paid 2; Part-Time Volunteers 50.
Governing Authority: Parent Institution: Warren Library Association. Tax-exempt.

Institution Type/Description: History Museum.
Collections: local history & culture; period furnishings; personal artifacts; photographs; underground railroad documents & memorabilia.
Activities: school & group tours; one day bus tours to underground railroad sites & Victorian houses; lectures on Victorian arts & artifacts.
Hours & Admission Prices: Wed.-Sat. 2-4; other times by appointment. No charge. &
Membership: Students $5; Individual $25-$99; Family $40-$99; Bronze $100-$249; Silver $250-$499; Gold $500 & up.

Washington Court House

FAYETTE COUNTY MUSEUM, 517 Columbus Ave., Washington Court House, OH 43160-1427. Tel.: 740-335-2953.
Founded: 1948.
Key Personnel: Pres. (V), Warren Craig; Financial Dir., Craig Breedlove; Sec. Bd. Trustees, Donald J. Moore.
Personnel Profile: Part-Time Paid 1; Part-Time Volunteers 25.
Governing Authority: private; nonprofit organization. Tax-exempt: 501(c)(3).
Institution Type/Description: History Museum: housed in a 14 room Victorian mansion.
Collections: period furnishings; artifacts; memorabilia depicting history & heritage of Fayette County, Ohio.
Research Fields: history of local schools.
Facilities: 200-vol. library of County & Ohio history; 1,400 sq. ft. exhibit space.
Activities: lectures; school loan service; temporary exhibitions. Museum Sponsors: Ice Cream Social; Holiday Open House.
Publications: county-wide tour map; 2010 calendar.
Hours & Admission Prices: May-Aug. Sat.-Sun. 1-4. No charge; donations accepted.
Attendance: 1,350 (accurate)
Membership: Student $1; Adult $10; Business $50; Life $100.

Wauseon

FULTON COUNTY HISTORICAL MUSEUM, 229 Monroe St., Wauseon, OH 43567-1127. Mailing Address: P.O. Box 104, Wauseon, OH 43567-0104. Tel.: 419-337-7922. Facebook: Fulton County Historical Museum.
E-mail: museum@fultoncountyhs.org
Web Site: www.fultoncountyhs.org
Founded: 1883.
Congressional District: 5
Key Personnel: Pres. (V), Carl Buehrer; Dir., John D. Swearingen, Jr.
Personnel Profile: Part-Time Paid 1; Part-Time Volunteers 10.
Volunteer Hours: 2,000
Operating Expenses: 34,000
Operating Income: 34,000
Governing Authority: society. Fulton County Historical Society, Inc. Tax-exempt.
Institution Type/Description: History Museum.
Collections: artifacts of Fulton County from 1830-1976. Historic Buildings: 1838 Log Cabin; 1896 Lake Shore & Michigan Southern Railroad Depot; 1861 blacksmith shop; Swan Creek Township building.
Major Exhibits: Hell & Homefront: Civil War through Fulton County Eyes, 1/14-12/31/15.
Research Fields: archives, free use general public.
Facilities: museum; 4 historic structures.
Publications: quarterly newsletter, Fulton County Pioneer.
Hours & Admission Prices: April-Nov. Tues. - Fri. 10-4, Sat. 10-2; other times by appointment. Adults $4, senior $3, student $2. Discounts to members. Closed holidays. &
Attendance: 850 (estimated)
Membership: Individual $20; Senior Couple $25; Family $30.

Waynesville

CAESAR'S CREEK PIONEER VILLAGE, Caesar's Creek State Park, 3999 Pioneer Village Rd., Waynesville, OH 45068-8630. Tel.: 513-897-1120.
E-mail: ccpv@embarqmail.com
Web Site: www.caesarscreekpioneervillage.org/CCPV/Home.html
Key Personnel: Dir., Kathy Sewell.
Governing Authority: nonprofit organization.
Institution Type/Description: History Museum.
Collections: Southwestern Ohio frontier history from 1793-1812; Ohio's heritage.
Activities: school programs; rental facilities. Museum Sponsors: festivals.
Hours & Admission Prices: Call for hours.

Wellington

SPIRIT OF '76 MUSEUM, (M), 201 N. Main St., Wellington, OH 44090. Mailing Address: P.O. Box 76, Wellington, OH 44090-0076. Tel.: 440-647-4367 & 4576.
Founded: 1968.
Congressional District: 5
Key Personnel: Pres., John Perry.
Personnel Profile: Part-Time Volunteers 14.
Governing Authority: nonprofit. Parent Institution: Southern Lorain County Historical Society. Tax-exempt: 501(c)(3).
Institution Type/Description: Historical Society Museum: housed in 1872 building.
Collections: items associated with Archibald M. Willard's painting The Spirit of '76; 16 Willard oils, water colors and murals; artifacts associated with the history of Southern Lorain County.
Facilities: reading room; 100-seat auditorium; classrooms. Gift items for sale.
Activities: guided tours; lectures; films; formally organized education programs for adults.
Publications: newsletters.
Hours & Admission Prices: April-Oct. Sat.-Sun. 2:30-5 by appointment; groups of 10 or more by appointment. No charge; donations accepted. &
Attendance: 950 (accurate)
Membership: 18 & Under $1; Single $5; Couple & Patron $10; Life $50.

Wellston

BUCKEYE FURNACE STATE MEMORIAL, (M), 123 Buckeye Park Rd., T167, Wellston, OH 45692-9511. Mailing Address: Friends of Buckeye Furnace, Inc., P.O. Box 475, Jackson, OH 45640-0475. Tel.: 740-384-3537.
Web Site: ohsweb.ohiohistory.org/places/se02
Founded: 1936.
Congressional District: 18
Key Personnel: Pres. (V), J. Michael Stroth; Museum Attendant, Tammie Mash; Museum Attendant, Jan McKibben.
Personnel Profile: Part-Time Paid 2; Part-Time Volunteers 20.
Governing Authority: society. The Ohio Historical Society. Ohio Historical Center Site Operations Office, 1985 Velma Ave., Columbus, OH. 43211.
Institution Type/Description: History Museum: 1851 restored Iron Furnace Complex.
Collections: casting shed; company store; office; charging house; engine house; iron masters house.
Facilities: picnic area; nature trails.
Activities: multimedia presentation; Spring wildflower walks.
Publications: site brochures; iron technology books.
Hours & Admission Prices: Memorial Day to Oct. call for hours; other times by appointment. Adults $4, children 6-12 $3; children 5 & under and members no charge.
Attendance: 2,300 (accurate)

Wellsville

RIVER MUSEUM, 1003 Riverside Ave., Wellsville, OH 43968-1374. Mailing Address: P.O. Box 13, Wellsville, OH 43968-0013. Facebook: Wellsville Historical Society.
E-mail: wellsvillerivermuseum@hotmail.com
Founded: 1955.
Congressional District: 18
Key Personnel: Pres., Robert Lloyd; Vice Pres., Tom Davidson; Treas, Jeff Weekley; Sec., Janet Taggart.
Personnel Profile: Part-Time Volunteers 8; Interns 7.
Governing Authority: private; nonprofit organization. Tax-exempt.
Institution Type/Description: History Museum: housed in c.1870 home of Dr. John Hammond.
Collections: period furnishings; 1953 Pennsylvania Railroad cabin car; Civil War memorabilia, saber; Indian artifacts; pottery; period kitchen utensils; railroad displays; boat display room; military artifacts; French Indian WWI & II memorabilia; vintage clothing; hatpins & holders; quilts; original Pretty Boy Death Mask; firehouse artifacts & engine.
Facilities: 100-vol. library of genealogy.
Activities: guided tours.
Hours & Admission Prices: June-Sept. Sun. 1-4:30; private tours available all year. Charge for special tours only. &
Attendance: 1,200 (estimated)
Membership: Individual $10.

West Liberty

PIATT CASTLES, 10051 Township Rd. 47, West Liberty, OH 43357. Mailing Address: P.O. Box 497, West Liberty, OH 43357-0497. Tel.: 937-465-2821 & 844-3480. Fax: 937-465-7774. Facebook: Piatt Castles.
E-mail: mpcastle@aol.com
Web Site: www.piattcastles.org
Founded: 1912.
Congressional District: 7
Key Personnel: C.E.O. & Pres., Margaret Piatt; Vice Pres. & Dir. Opers., James White; Dir. Mktg. & Communications, Kate Piatt-Eckert.
Governing Authority: private.
Institution Type/Description: Historic House Museums: Mac-A-Cheek completed in 1871; Norman-French style home with several generations of Piatt Family furnishings & objects; Mac-O-Chee Castle completed in 1881 Flemish style home with American furnishings & objects.
Collections: Piatt family furnishings & objects; firearms; archival & library holdings; American furnishings & objects.
Research Fields: Piatt family; Donn Piatt; General A. Sanders Piatt.
Facilities: library; archives.
Activities: guided tours; school & adult programs; summer evening performances. Annual Event: Christmas Open House.
Publications: brochures.
Hours & Admission Prices: Spring & Fall weekends 11-4; Memorial Day-Labor Day daily 11-5. Single Castle: adults $12, seniors 65 & up $11, youth 5-15 $7; Two Castles: adults $20, seniors 65 & up $18, youth 5-15 $12; discounts to AAA members & groups.
Attendance: 1,200 (accurate)
Membership: Friend $30; Patron $50; Contributor $100; Supporter $250.

Westerville

ANTI-SALOON LEAGUE MUSEUM, Westerville Public Library, 126 S. State St., Westerville, OH 43081-2029. Tel.: 614-259-5028. Fax: 614-882-5369.
E-mail: bweinhar@westervillelibrary.org
Web Site: www.wpl.lib.oh.us/antisaloon/
Founded: 1990.
Key Personnel: Archivist, Beth Weinhardt.
Personnel Profile: Full-Time Paid 2; Part-Time Volunteers 3.
Governing Authority: nonprofit. Parent Institution: Westerville Public Library, 126 S. State St., Westerville, OH 43081.
Institution Type/Description: History Museum: housed in the building that served as League headquarters.
Collections: publications; fliers; posters; song books; microfilm; archives of community history from the first settlement to modern times.
Facilities: 225-vol. library.
Activities: formal education programs; guided tours; lectures.
Hours & Admission Prices: Mon.-Fri. 9-6. No charge. &
Attendance: 3,000 (estimated)

HANBY HOUSE, (M), 160 W. Main St., Westerville, OH 43081. Mailing Address: P.O. Box 1063, Westerville, OH 43086-7063. Tel.: 614-891-6289; 800-600-6843.
Web Site: www.hanbyhouse.org
Founded: 1937.
Congressional District: 12
Key Personnel: Chm. (V) & Museum Shop Mgr., Pam Allen; Pres. (V), Bill Merriman; Sec., Ann Pryfogle.
Personnel Profile: Part-Time Volunteers 50.
Governing Authority: state. Parent Institution: Ohio Historical Society, 800 E. 17th Ave., Columbus, OH 43211. Subsidiary Institution: Westerville Historical Society, P.O. Box 1063, Westerville, OH 43081. Tax-exempt.
Institution Type/Description: Historic House: 1853 pre-Civil War home of Bishop William Hanby, father of composer & author Benjamin R. Hanby. Listed on the Underground Railroad Network to Freedom of the National Park Service.
Collections: music; furniture; household items of the Hanby family; underground railroad site.
Research Fields: life in Civil War times; underground railroad in Ohio; genealogy of the Hanby Family.
Facilities: library of Hanby music & family history books available for use on premises.
Activities: permanent exhibitions; audiovisual; demonstrations of household crafts of 1860s; tour guides in costume; presentation of life of the Hanby family.
Publications: books, Choose You This Day, The Legacy of The Hanbys; A

Hundred Years with Dacia Custer Shoemaker; The House of Brotherhood, Story of Hanby House.
Hours & Admission Prices: May-Sept. Sat.-Sun. 1-4; other times by appointment. Adults $3, children 6-17 $1; discounts to senior citizens, AAA, AAM, Golden Buckeye & Ohio Historical Society members; Hanby Club members & children under 6 no charge. &
Attendance: 1,400 (estimated)
Membership: Annual $10.

INNISWOOD METRO GARDENS, 940 S. Hempstead Rd., Westerville, OH 43081-3612. Tel.: 614-895-6216. Fax: 614-895-6352.
Web Site: www.inniswood.org
Institution Type/Description: Botanical Garden: former estate of Grace and Mary Innis.
Collections: more than 2,000 species of plants; specialty collections of hostas, daffodils & daylilies.
Hours & Admission Prices: Daily 7-dusk. No charge.

WEITKAMP OBSERVATORY AND PLANETARIUM, Otterbein College, 155 W. Main St., Westerville, OH 43081-1430. Tel.: 614-823-1316. Fax: 614-823-1968.
E-mail: ewerwa@otterbein.edu
Web Site: www.otterbein.edu/physics/weitkamp-observatory.asp
Founded: 1955.
Congressional District: 15
Key Personnel: Dir., Uwe Trittmann.
Personnel Profile: Part-Time Volunteers 1.
Governing Authority: college. Affiliated with the Otterbein College. Tax-exempt: 501(c)(3).
Institution Type/Description: Observatory & Planetarium.
Collections: astronomy; 14 & 8 inch Celestron Schmidt-Cassegrain telescopes; Meade CCD camera & guidance system; Spitz planetarium projector.
Facilities: planetarium; observatory deck.
Activities: lectures; formally organized education programs.
Hours & Admission Prices: Call for hours.
Attendance: 200 (estimated)

Whitehouse

THE BUTTERFLY HOUSE, 11455 Obee Rd., Whitehouse, OH 43571. Tel.: 419-285-4855.
Web Site: butterfly-house.com
Key Personnel: Owner, Duke Wheeler
Institution Type/Description: Butterfly Museum.
Collections: over 500 species of butterflies from around the world; flowers; plants; trees; gardens.
Facilities: Museum-related items for sale.
Activities: educational program.
Hours & Admission Prices: May-Aug. Mon.-Sat. 10-5, Sun. 12-5; Sept. Thurs.-Sat. 10-5, Sun. 12-5; Oct. Sat.-Sun. 12-5. Adults $9, seniors 65 & up $8, children 4-11 $7; children 3 & under no charge.

THE BUTTERFLY HOUSE, 11455 Obee Rd., Whitehouse, OH 43571-9205. Tel.: 419-877-2733. Facebook; Wheeler Farms.
Institution Type/Description: Butterfly House.
Collections: over 1,000 butterflies from around the world; butterfly life cycle.
Hours & Admission Prices: May-Aug. Mon.-Sat. 10-5, Sun. 12-5; Sept. Thurs.-Sat. 10-5, Sun. 12-5; Oct. Sat. & Sun. 12-5. Adults 13-64 $8, seniors 65 & over $7, children 4-12 $6.50; children 3 & under no charge.

Wilberforce

NATIONAL AFRO-AMERICAN MUSEUM & CULTURAL CENTER, (M), 1350 Brush Row Rd., Wilberforce, OH 45384-0578. Mailing Address: P.O. Box 578, Wilberforce, OH 45384-0578. Tel.: 937-376-4944, ext. 122. Fax: 937-376-2007.
E-mail: wbillingsley@ohiohistory.org
Web Site: www.blackohio.org
Founded: 1972.
Congressional District: 7
Key Personnel: Site Mgr., William Billingsley; Chm. (V), Dr. Kenneth Goings; Cur., Dr. Floyd R. Thomas; Registrar, Wendy Felder.
Personnel Profile: Full-Time Paid 7; Part-Time Paid 1.
Governing Authority: nonprofit organization. Parent Institution: Ohio Historical Society. Tax-exempt: 501(c)(3).
Institution Type/Description: African American History Museum: located on grounds of c.1856 Wilberforce University campus.
Collections: African American life between 1945 & the passage of the Voting Rights Act of 1965; objects & images reflecting African-American families,

jobs, schools, churches, organizations, music; African-American published serials; 2,147 slides; 1,551 recordings.
Facilities: 6,481-vol. library of books on all aspects of African-American experience; 40-seat theatre. Gift items for sale.
Activities: guided tours; rental gallery; theatre; traveling exhibitions.
Hours & Admission Prices: Tues.-Sat. 9-5. Adults $4, seniors $3.60, children & students $1.50; discounts to AAM & Assoc. of African American Museum members; members no charge. Closed major holidays. &
Attendance: 2,500 (accurate)
Membership: Basic $30; Family $50; Contributing $75; Basic Plus $130; Supporting $125; Family Plus $150.

PAUL ROBESON CULTURAL AND PERFORMING ARTS CENTER, Central State University, 1400 Brush Row Rd., Wilberforce, OH 45384. Mailing Address: P.O. Box 1004, Wilberforce, OH 45384-1004. Tel.: 937-376-6403.
Web Site: www.centralstate.edu/academics/art_science/fine_performing_arts
Key Personnel: Chm., James Smith; Asst. Prof., Kenneth Pointer; Assoc. Prof., Dr. Ronald Claxton; Assoc. Prof., Dwayne Daniel
Institution Type/Description: Art Gallery.
Collections: paintings.
Facilities: 850-seat auditorium; recital hall.
Activities: convocations; lectures; recitals; workshops; temporary exhibitions.
Hours & Admission Prices: Call for hours. No charge.

Willoughby

INDIAN MUSEUM OF LAKE COUNTY, OHIO, 25 Public Sq.-Technical Center, Bldg. B, Willoughby, OH 44094. Mailing Address: P.O. Box 883, Willoughby, OH 44096-0883. Tel.: 440-951-3813.
Web Site: indianmuseumoflakecounty.org
Founded: 1980.
Congressional District: 11
Key Personnel: Dir. & Museum Shop Mgr., Ann L. Dewald; Pres. (V), Douglas R. Divish; Treas., John Brewster.
Personnel Profile: Full-Time Volunteers 1; Part-Time Paid 2; Part-Time Volunteers 21; Interns 2.
Governing Authority: nonprofit organization. Parent Institution: Lake County Chapter of Archaeological Society of Ohio. Tax-exempt: 501(c)(3).
Institution Type/Description: Native American Museum.
Collections: 10,000 B.C.-1650 A.D. pre-contact artifacts of Ohio; 1800 A.D.-present crafts & art of North American Native Americans.
Research Fields: pre-contact Native Americans of northeastern Ohio; Tarahumara & Huichol of Mexico; The Southwest: (Hopi, Navaho, Apache, Papago); Natives of Alaska; Natives of the Pacific Northwest; Plains Indians; Native Americans of the Eastern Woodlands; Navajo weaving; pictographs & petroglyphs.
Facilities: 900-vol. library pertaining to anthropology, art, history of Native Americans for public use. Books for sale.
Activities: guided tours; lectures; organized education programs; docent program.
Publications: newsletter.
Hours & Admission Prices: Mon.-Fri. 10-4, Sat.-Sun. 1-4. Adults $2, seniors $1.50; members no charge. Closed major holiday weekends; winter & spring breaks. &
Attendance: 3,000 (estimated)
Membership: Regular $20-$29; Contributing $30-$49; Donor $50-$99; Sustainer $100-$149; Patron $150-$249; Benefactor $250-$999; Honors $1,000 & up.

KIRTLAND TEMPLE VISITOR CENTER, 7809 Joseph St., Willoughby, OH 44094-9255. Tel.: 440-256-1830. Fax: 440-256-1929.
E-mail: info@kirtlandtemple.org
Web Site: www.kirtlandtemple.org
Key Personnel: C.E.O., Ronald Romig; Pres., Stephen Veazey.
Personnel Profile: Full-Time Paid 3; Full-Time Volunteers 3; Part-Time Paid 2; Part-Time Volunteers 2; Interns 6.
Governing Authority: Community of Christ. Tax-exempt: 501(c)(3).
Institution Type/Description: Historic Site. house of worship; 1833-1836.
Collections: 1830s tools & furnishings; museum display space with period publications.
Facilities: Visitor Center.
Activities: guided tours; school programs; reserved services.
Publications: Restoration Trail Forum.
Hours & Admission Prices: Feb. by appointment; March-April Mon.-Sat. 10-4, Sun. 1-5; May-Oct. Mon.-Sat. 9-5, Sun. 1-5; Nov.-Dec. Mon.-Sat. 10-4, Sun. 1-4; groups by appointment. Tours: $3. Closed New Year's Day; Easter; Thanksgiving; Christmas Eve & Day.

Attendance: 30,000 (accurate)

Wilmington

CLINTON COUNTY HISTORICAL SOCIETY, (M), 149 E. Locust St., Wilmington, OH 45177-2338. Mailing Address: P.O. Box 529, Wilmington, OH 45177-0529. Tel.: 937-382-4684. Fax: 937-382-5634.
E-mail: directorcchs@frontier.com
Web Site: clintoncountyhistory.org
Founded: 1948.
Key Personnel: Dir., Kay Fisher.
Personnel Profile: Full-Time Paid 1; Part-Time Paid 1; Part-Time Volunteers 25.
Governing Authority: private; nonprofit organization. Tax-exempt: 501(c)(3).
Institution Type/Description: Historical Society Museum: housed in an 1835 Greek Revival mansion.
Collections: Quaker heritage & settlement of the county; sculpture; paintings; textiles.
Facilities: 700-vol. library. Museum-related items for sale.
Activities: guided tours; lectures; formal education programs.
Publications: newsletter, History Center News.
Hours & Admission Prices: Museum: March-Dec. Wed.-Fri. 1-4. Library: March-Nov. Wed.-Fri. 1-4. Adults $5; members no charge. &
Attendance: 1,200 (estimated)
Membership: Single $25; Family $35; Patron & Business $100; Benefactor $1,000.

QUAKER HERITAGE CENTER OF WILMINGTON COL-LEGE, (M), College St. & Douglas St., Wilmington, OH 45177. Mailing Address: Wilmington College of Ohio, 1870 Quaker Way, Wilmington, OH 45177. Tel.: 937-382-6661, ext. 719.
E-mail: qhc@wilmington.edu
Web Site: www2.wilmington.edu
Institution Type/Description: History Museum.
Collections: local history & culture; period furnishings; photographs; personal artifacts.
Hours & Admission Prices: Mon.-Fri. 9-4; other times by appointment. No charge.

Wilmot

THE WILDERNESS CENTER INC., 9877 Alabama Ave., S.W., Wilmot, OH 44689. Mailing Address: P.O. Box 202, Wilmot, OH 44689-0202. Tel.: 330-359-5235. Fax: 330-359-7898.
E-mail: gordon@wildernesscenter.org
Web Site: www.wildernesscenter.org
Founded: 1964.
Congressional District: 16
Key Personnel: Exec. Dir., Gordon T. Maupin; Pres., C. Andrew Haag; Dir. Education, Joann L. Ballbach; Caretaker / Naturalist, Kenneth R. Schlegel, Jr.; Naturalist, Carrie Elvey; Naturalist, Lynda A. Price; Dir. Devel., Barbara Vitcosky; Dir. Land Stewardship, Gary Popotnik; Bookstore & Office Mgr., Rebecca Cyphert; Staff Accountant, Gina Mast; Mktg. & Public Rels., Vicki L. Capps; Custodian, Paul Lyon.
Personnel Profile: Full-Time Paid 11; Full-Time Volunteers 5; Part-Time Paid 6; Part-Time Volunteers 721; Interns 1.
Volunteer Hours: 17,041
Operating Expenses: 900,000
Operating Income: 900,000
Governing Authority: board of trustees; nonprofit organization. Tax-exempt: 501(c)(3).
Institution Type/Description: Nature Center & Land Trust.
Collections: land-living communities; natural-living biological communities; Eastern deciduous forest; tallgrass prairie.
Research Fields: bird banding; insects; endangered plants.
Facilities: 1,926-acre nature center & nature preserve; seven nature trails; 7.5 acre lake; planetarium; observation tower; picnic area; auditorium; meeting rooms; astronomical observatory; 1820s stone house.
Activities: guided tours; lectures; films; hobby & natural history workshops; formally organized education programs; interactive exhibits; bird viewing room. Special Interest Clubs: astronomy, birding, backpacking, nature photography, storytellers, woodcarvers, fly-fishing; geology, cavers & climbers; botanizers artistic endeavors; land conservancy work; ecopreneurial ventures; Foxfield Preserve; nature preserve cemetery; consulting forestry; ecotours; wetland mitigation.
Publications: weekly podcast, Wild Ideas...the Podcast; monthly, Members' Newsletter; annual report.
Hours & Admission Prices: Interpretive Building: Tues.-Sat. 9-5, Sun. 1-5.

Nature Center: daily dawn-dusk. No charge; donations accepted; discounts to ANCA members. &
Attendance: 80,000 (accurate)
Membership: Individual $30; Family $40; Supporting $75; Sustaining $100; Fellow $200; Life $1,250; Endowment Benefactor $3,000.

Woodsfield

PARRY PARK MUSEUMS, 217 Eastern Ave., Woodsfield, OH 43793. Mailing Address: Monroe County Historical Society, P.O. Box 538, Woodsfield, OH 43793. Tel.: 740-472-1933. Facebook: Friend Monroe County Historical Society.
E-mail: moncohissoc@att.net
Web Site: www.rootsweb.ancestry.com/~ohmchs
Founded: 1976.
Key Personnel: Dir. & Pres. (V), Joyce Wiggins; Museum Shop Mgr., Bart Wiggins.
Personnel Profile: Full-Time Volunteers 2; Part-Time Volunteers 10; Interns 1.
Volunteer Hours: 2,400
Operating Expenses: 12,000
Operating Income: 12,000
Governing Authority: Parent Institution: Monroe County Park District. Subsidiary Institution: Monroe County Historical Society. Tax-exempt.
Institution Type/Description: History Museum.
Collections: local history & culture; period furnishings; personal artifacts; photographs; archives. Historic Buildings: Hollister-Parry House, c.1800; Byers-Oak Ridge Stone school house; Yaussey-Winkler Dairy barn.
Major Exhibits: Woodsfield Bicentennial Relic Room, 7/4/14-7/5/14.
Research Fields: genealogy; local history.
Facilities: records room; senior center.
Activities: children's historical literature program in summer.
Publications: annual calendar; occasional newsletters.
Hours & Admission Prices: Museums: June-Oct. call for hours or appointment. No charge; donations accepted. Senior Center Office: Mon.-Tues. 9-4, Fri. 10-2. Member discount for publications. &
Attendance: 350 (estimated)
Membership: Single $10; Family $15; Life $100; Life Sustaining $100 annually; Patron $2,500.

Woodville

WOODVILLE HISTORICAL SOCIETY MUSEUM, 107 E. Main St., Woodville, OH 43469. Tel.: 419-849-2349.
E-mail: m.k.o@embarqmail.com
Founded: 1970.
Key Personnel: Pres. (V), Michael K. O'Connor.
Governing Authority: Parent Institution: Woodville Historical Society. Tax-exempt.
Institution Type/Description: Historical Society Museum.
Collections: local history & culture; period furnishings; personal artifacts; photographs; household tools; glassware; fossils.
Hours & Admission Prices: Call for hours. No charge; donations accepted.
Attendance: 350 (estimated)
Membership: Single $2; Couples $3.

Wooster

THE COLLEGE OF WOOSTER ART MUSEUM, Ebert Art Center, 1220 Beall Ave., Wooster, OH 44691. Tel.: 330-263-2495. Fax: 330-263-2633. Facebook: College of Wooster Art Museum.
E-mail: kzurko@wooster.edu
Web Site: wooster.edu/cwam
Founded: 1930.
Congressional District: 16
Key Personnel: Dir., Kitty McManus Zurko; Preparator, Douglas McGlumphy.
Personnel Profile: Full-Time Paid 1; Part-Time Paid 2.
Governing Authority: The College of Wooster. Tax-exempt: 101(A).
Institution Type/Description: College Art Gallery.
Collections: John Taylor Arms Collection of European & American prints; Persian decorative arts; Chinese bronzes; African art; ancient Middle Eastern, Chinese & Japanese ceramics.
Major Exhibits: Willie Cole: Complex Conversations (T), 1/14-3/2/14; Chinese & Japanese Ceramics, 10/1-11/14.
Research Fields: American, African, Asian & European graphics; decorative arts.
Facilities: 2,450 sq. ft. exhibit space.
Activities: lectures; films; gallery talks; programs for undergraduate college students; temporary exhibitions; music.
Publications: exhibition brochures & catalogues.
Hours & Admission Prices: Tues.-Fri. 10:30-4:30, Sat.-Sun. 1-5. No charge. Closed college breaks. &

Attendance: 9,500 (accurate)

WAYNE CENTER FOR THE ARTS, 237 S. Walnut St., Wooster, OH 44691-4753. Tel.: 330-264-2787. Fax: 330-264-9314.
E-mail: waynectr@wayneartscenter.org
Web Site: www.wayneartscenter.org
Founded: 1973.
Congressional District: 16
Key Personnel: Exec. Dir., Robb Hyde; Pres. (V), Matthew Long; Bookkeeper, Marge Yochheim; Education Coord., Liza Zemancik; Museum Shop Mgr., Kristin Larson.
Personnel Profile: Full-Time Paid 2; Part-Time Paid 5; Part-Time Volunteers 30; Interns 1.
Governing Authority: Tax-exempt.
Institution Type/Description: Art Center.
Collections: works by regional & national artists; pottery by international potters.
Hours & Admission Prices: Mon. 12-9, Tues.-Thurs. 9-9, Fri. 9-7, Sat. 9-2. No charge; donations accepted. &
Attendance: 9,850 (estimated)
Membership: Annual $75.

WAYNE COUNTY HISTORICAL SOCIETY, 546 E. Bowman St., Wooster, OH 44691-3110. Tel.: 330-264-8856.
E-mail: host@waynehistoricalohio.org
Web Site: www.waynehistorical.org
Founded: 1954.
Congressional District: 16
Key Personnel: Pres., David Broehl; Office Admin., Lucy Drabenstott.
Personnel Profile: Part-Time Paid 1; Part-Time Volunteers 100.
Governing Authority: society. Tax-exempt.
Institution Type/Description: General Museum: Housed in the original home of General Reasin Beall built between 1815 & 1817 on a land grant signed by Pres. James Madison.
Collections: early lustre ware; pressed glass; china; mounted birds & animals; guns & other militaria; portraits of Wayne County & Wooster citizens; Indian artifacts from Wayne County; period tools, household artifacts, furniture, dolls & toys; fire equipment; transportation; military; farm tools; taxidermy. Historic buildings: General Reasin Beall Homestead; 1873 one-room schoolhouse; Kister Building containing 1880's carpenter; blacksmith; farm; tools; general store including early 1900s furnishings; general mercantile store; ladies dress shop; Carriage House, including original Gerstenslager carriages; Relief Co. No.4 firehouse.
Research Fields: archaeological investigations of local sites.
Facilities: 1,000-vol. documents, letters & manuscripts pertaining to Ohio & Wayne County local history & pioneer families.
Activities: guided tours; formally organized education programs for children; permanent exhibitions; special quarterly exhibits.
Publications: maps, abstracts of court records; quarterly newsletter.
Hours & Admission Prices: Feb.-Oct. Wed.-Sun. 2-4:30; Nov.-Jan. by appointment. Adults $5; members & students no charge. Closed national holidays; Christmas. &
Attendance: 10,000 (estimated)
Membership: Seniors 60 & over $15; Individual $20; Family $25; Patron $50; Benefactor $100; Life - Single Payment $350. Business & Corporate Memberships: Silver $100; Gold $250; Platinum $500.

Worthington

OHIO RAILWAY MUSEUM, 990 Proprietors Rd., Worthington, OH 43085. Tel.: 614-885-7345. Facebook: Ohio Railway Museum.
E-mail: info@ohiorailwaymuseum.net
Web Site: www.ohiorailwaymuseum.org
Founded: 1945.
Congressional District: 15
Key Personnel: Pres., Chris Howell.
Governing Authority: nonprofit organization. Tax-exempt: 501(c)(3).
Institution Type/Description: Transportation Museum.
Collections: historic steam & electric railway equipment operating on 1 mile of track; artifacts; manuscripts; Ohio interurbans; Columbus Streetcar; electric streetcars & interurbans; Pullmans, 1929 The John Greenleaf Whittier; 1925 Time Square; 1928 Williamsport business car; 1920 mail car; 1924 Vulcan steam engine; 1910 Norfolk & Western #578 steam engine; 1918 passenger & baggage cars.
Facilities: library. Museum-related items for sale.
Activities: guided tours; organized education programs for children; permanent exhibitions; electric train, streetcar & interurban rides. Museum Sponsors: State Fair in August; Twilight Trolleys; Old Worthington Market Day; Ghost Train; Santa Trolley.
Publications: newsletter, Rail Fax.

Hours & Admission Prices: Sun. 12-4; groups by appointment. Adults $8, seniors 65 & over $7, children 4-12 $6; children 3 & under no charge.
Attendance: 4,000

WORTHINGTON HISTORICAL SOCIETY, 50 W. New England Ave., Worthington, OH 43085-3536. Tel.: 614-885-1247. Fax: 614-885-1040. Facebook: Worthington Historical Society.
E-mail: info@worthingtonhistory.org
Web Site: www.worthingtonhistory.org
Founded: 1955.
Congressional District: 15
Key Personnel: Dir., Kate LaLonde; Pres. Worthington Historical Society (V), Jutta C. Pegues; Cur., Sue Whitaker; Museum Shop Mgr., Karen Cantlon; Membership, Henrietta Nichols.
Personnel Profile: Part-Time Paid 3; Part-Time Volunteers 100.
Governing Authority: society; nonprofit organization. Branch Museums: Orange Johnson House, 956 High St., Worthington, OH 43085. Tel.: 614-885-1274; The Doll Museum, 50 W. New England Ave., Worthington 43085. Tax-exempt.
Institution Type/Description: Historic House Museums: 1811-1819, Orange Johnson House & Garden; The Doll Museum, 1845 Classical Revival Manse & Society Headquarters.
Collections: Worthington Indian Mound; 19th & 20th century dolls; lace; early community memorabilia; manuscripts; period clothing.
Facilities: 300-vol. library of books on local history available for inter-library loan by special arrangement. Craft items for sale.
Activities: guided tours; lectures; study clubs; formally organized education programs; docent training; permanent & special exhibitions.
Publications: monthly newsletter, The Intelligencer.
Hours & Admission Prices: Orange Johnson House: April-Dec. Sun. 2-5; other times by appointment. Adults $3, seniors & children $2; members no charge. The Doll Museum: Tues.-Fri. 1-4, Sat. 10-2. Library: 1st & 3rd Wed. 1:30-4. Gift Shop: Tues.-Fri. 1-4, Sat. 10-2. Closed holidays.
Attendance: 3,900 (estimated)
Membership: Senior Citizen $20; Individual & senior couple $25; Family $30; Contributing $50; Patron $100; Golden Patron $200. Business: Pioneer $35; Settler $50; Town Crier $100; Town Marshal $200; Village Founder $500.

Wright-Patterson Air Force Base

*** NATIONAL MUSEUM OF THE UNITED STATES AIR FORCE, (M),** 1100 Spaatz St., Wright-Patterson Air Force Base, OH 45433-7102. Tel.: 937-255-3286. Fax: 937-255-0523.
E-mail: nationalmuseum.usaf@wpafb.af.mil
Web Site: www.nationalmuseum.af.mil
Founded: 1923.
Congressional District: 3
Key Personnel: Dir., Lt. Gen. John L. "Jack" Hudson, USAF Ret.; Senior Cur., Terry Aitken; Education, Judith Wehn; Public Rels., Diana Bachert; Collection Mgr., Krista Strider; Research/Archives, Wes Henry; Operations, John Marang; Air Force Museum Foundation, Col. Larry Cooper, USAF Ret.; Museum Shop Mgr., Kim Pierre.
Personnel Profile: Full-Time Paid 96; Part-Time Volunteers 500.
Governing Authority: federal. Parent Institution: United States Air Force, The Pentagon, Washington, DC. Tax-exempt.
Institution Type/Description: Military Aviation Museum: located at historic Wright Field, site of early aviation pioneering.
Collections: 360 aerospace vehicles & missiles; 400,000 photos & 8.9 million documents, books, test reports, operational manuals, aircraft drawings & unit histories related to the history of military aviation; video & audio tapes; uniforms; personal military memorabilia; aviation guns & instruments; aircraft squadron insignia; military badges; space hardware & foods; models; WWII Control Tower.
Research Fields: aeronautical technology; military apparel & air force history.
Facilities: 1.2 million sq. ft. of exhibit space on 450 acre campus; outdoor airpark & memorial park; 500-seat auditorium; 500-seat IMAX theater; restaurant. Museum-related items for sale.
Activities: guest lectures; in house & outreach educational programs; school tours for children grades 4-12; self-guided tours; daily heritage tours; limited behind the scenes public tours for adults; special exhibit openings; concerts; flying events.
Publications: brochures; illustrated exhibit guide; quarterly journal, Friends Journal; annual report; e newsletter; calendar of events.
Hours & Admission Prices: Daily 9-5. Museum: no charge. IMAX Theatre: call for admission prices. Closed New Year's Day; Thanksgiving; Christmas. &
Attendance: 1,000,000 (estimated)
Membership: Aviator $30-$80; Wingman $50-$140; Flight Lead $100-$275; Falcon $250; Eagle $500; Ace $1,000; Lifetime $2,500.

Xenia

GREENE COUNTY HISTORICAL SOCIETY, 74 W. Church St., Xenia, OH 45385-2902. Tel.: 937-372-4606. Fax: 372-372-5660 (call first).
E-mail: GCHSXO@sbcglobal.net
Founded: 1929.
Congressional District: 7
Key Personnel: Pres., Ben Thompson; Exec. Dir., Catherine Wilson.
Personnel Profile: Part-Time Paid 2; Part-Time Volunteers 30.
Governing Authority: society. Tax-exempt: 501(c)(3).
Institution Type/Description: Historical Society Museum.
Collections: Cosley coverlets; period furniture; paintings; agricultural implements; dolls. Historic Houses: 1798 Galloway Log House; 1877 Victorian style home; Carriage House Museum: medical, farming, railroad, archaeological exhibits.
Research Fields: local historical topics.
Facilities: Books pertaining to Greene County history for sale.
Activities: guest speakers; special events. Museum Sponsors: Model Railroad workshops; Old Fashioned Days open house; annual tour of private homes.
Publications: monthly newsletter, Our Heritage; books on Greene County History.
Hours & Admission Prices: Tues.-Fri. 9-3:30; call for weekend hours. Adults $3, youth under 18 $2. Closed New Year's Day; Independence Day; Thanksgiving; Christmas Eve, Day & week. &
Attendance: 3,000 (estimated)
Membership: Student $5; Senior Citizen Single $10; Regular Single $15; Senior Citizen Double $20; Family $30; Lifetime $250.

Yellow Springs

GLEN HELEN ECOLOGY INSTITUTE TRAILSIDE MUSEUM, 405 Corry St., Yellow Springs, OH 45387-1843. Mailing Address: 1075 State Rt. 343, Yellow Springs, OH 45387. Tel.: 937-769-1902. Fax: 937-767-6655.
E-mail: rjaramillo@antioch-college.edu
Web Site: www.glenhelen.org
Founded: 1951.
Congressional District: 7
Key Personnel: Exec. Dir., Nick Boutis.
Governing Authority: college. Antioch University. Operated by Glen Helen Ecology Institute. Tax-exempt.
Institution Type/Description: Natural History Museum & Nature Center.
Collections: natural history. Historic Buildings: 1814 Grinnell Mill; covered bridge.
Facilities: 500-vol. library of natural history books available on premises; nature preserve. Publications for sale.
Activities: guided tours; lectures; formally organized education programs for children.
Publications: Guide to Historical Sites in Glen Helen; Map of Glen Helen; quarterly newsletter, In the Glen.
Hours & Admission Prices: Call for hours.
Attendance: 12,000
Membership: Contributing $25; Supporting $40; Sustaining $65; Sponsoring $125; Benefactor $500; Patron & Corporate $1,000.

Youngstown

✱ **THE ARMS FAMILY MUSEUM OF LOCAL HISTORY,** 648 Wick Ave., Youngstown, OH 44502-1289. Tel.: 330-743-2589. Fax: 330-743-7210.
E-mail: mvhs@mahoninghistory.org
Web Site: www.mahoninghistory.org
Founded: 1961.
Congressional District: 17
Key Personnel: Pres., William J. Cleary, Jr., M.D.; Dir., H. William Lawson; Mgr. Education & External Rels. and Museum Shop Mgr., Leann Rich; Registrar, T. Lea Mollman; Mgr. Collections & Curatorial Svcs., Jessica D. Trickett; Archivist, Pamela Speis.
Personnel Profile: Full-Time Paid 9; Part-Time Paid 17; Part-Time Volunteers 15; Interns 2.
Governing Authority: nonprofit organization; society. Parent Institution: Mahoning Valley Historical Society. Tax-exempt: 501(c)(3) & 509(a)(1).
Institution Type/Description: Local History Museum.
Collections: Arms family house & contents; Native American history; furnishings; costumes; archives; B.F. Wirt Eclectic collection includes books, manuscripts, art works & antiquities from a Youngstown attorney, public official & world traveler of the late 19th & early 20th centuries; business & media archives of the Mahoning Valley collection includes business records, broadcasting equipment & film & videotape library from WKBN, Youngstown's CBS network affiliate & radio & television pioneers.

Research Fields: local history.
Facilities: local history museum in preserved arts & crafts mansion; archives collections housed in carriage house on site; primary & secondary local history resources especially strong on family history, cultural history, development of built environment, local steel manufacturing material, local newspapers.
Activities: guided tours; programs; slide presentations; lectures; traveling suitcase museums for elementary schools.
Publications: pamphlet, Discover Greystone; quarterly newsletter, Historical Happenings; A Read & Color Book about the Mahoning Valley.
Hours & Admission Prices: Tues.-Sun. 1-5. Adults $4, college students & senior citizens $3, students under 18 $2; discounts for AAM, ICOM; members no charge. Closed national holidays. &
Attendance: 15,000 (accurate)
Membership: Student $10; Individual $30; Family $50; Sustaining $75; Preservation Society $100 & up.

✱ **THE BUTLER INSTITUTE OF AMERICAN ART, (M),** 524 Wick Ave., Youngstown, OH 44502-1286. Tel.: 330-743-1711. Fax: 330-743-9567.
E-mail: info@butlerart.com
Web Site: www.butlerart.com
Founded: 1919.
Congressional District: 19
Key Personnel: Dir., Louis A. Zona; Pres. (V), Nicholas J. Zennasio; Asst. Dir., M. Susan Carfano; Dir. Education, Carole O'Brien; Business Mgr., Amy Kaufman; Dir. Information, Kathy Earnhart; Museum Shop Mgr., Renee Sheakoski.
Personnel Profile: Full-Time Paid 21; Part-Time Paid 11; Part-Time Volunteers 150; Interns 5.
Governing Authority: nonprofit. Tax-exempt: 501(c)(3).
Institution Type/Description: American Art Museum.
Collections: American art from Colonial to present times; 5,000 works including major paintings by Copley, Earl, Homer, Eakins, Heade, Henri, Sheeler, Hopper, Koch, Gottlieb, Mitchell, Warhol, Nevelson, Rockwell, Rauschenberg & Leslie; Western & Marine collections; American watercolor paintings & works on paper; Americana ceramics, sculpture & contemporary photographs; holograms & other technological work.
Major Exhibits: 300 Years of American Painting from the Butler Collection, 1/14-12/14; Paul Jenkins, 3/30/14-6/15/14; Maurice Sendak (T), 5/18/14-7/6/14; Bob Guccione, 6/29/14-8/14.
Research Fields: American art.
Facilities: library of American art books available on premises; The Beecher Center Wing for Technology Art; electronic archive center. Art publications for sale.
Activities: guided tours; lectures; films; gallery talks; concerts; permanent & temporary exhibitions; children & adult art classes.
Publications: annual, National Midyear Show Catalog; catalogues of permanent collection; brochures; postcards; individual exhibition catalogues.
Hours & Admission Prices: Tues.-Sat. 11-4, Sun. 12-4. No charge; donations accepted. Closed New Year's Day; Easter; Independence Day; Thanksgiving; Christmas. &
Attendance: 140,000 (estimated)
Membership: Student $10; Individual $35; Family $45; Sustaining $60; Donor $100; Trustees Circle: Sponsor $300; Patron $500; Collector $1,000; Connoisseur $3,000 & up.

THE BUTLER INSTITUTE OF AMERICAN ART - SALEM BRANCH, 524 Wick Ave., Youngstown, OH 44502-1213. Tel.: 330-332-8213. Fax: 330-337-8286.
E-mail: info@butlerart.com
Key Personnel: Exec. Dir. & Chief Cur., Dr. Louis A. Zona
Institution Type/Description: Art Museum.
Collections: works by regional artists.
Facilities: classroom. Museum-related items for sale.
Hours & Admission Prices: Temporarily closed.

FORD NATURE EDUCATION CENTER, 840 Old Furnace Rd., Youngstown, OH 44511-1470. Mailing Address: Mill Creek MetroParks, P.O. Box 596, Canfield, OH 44406. Tel.: 330-740-7107. Fax: 330-740-7133.
E-mail: generalinfo@millcreekmetroparks.com
Web Site: www.millcreekmetroparks.com/visit/places/mill-creek-park/ford-nature-center/
Founded: 1972.
Congressional District: 17
Key Personnel: Exec. Dir., Dennis Miller; Outdoor Education Mgr., Raymond Novotny; Pres., Louis Schiavoni.

Governing Authority: Parent Institution: Mill Creek Metropolitan Park District. Tel. 330-702-3000. Tax-exempt: 501(c)(3).
Institution Type/Description: Nature Center: housed in 1912, 13-room stone mansion.
Collections: zoology, botany & geology specimens from Northeast Ohio.
Facilities: 2,000-vol. library pertaining to natural history material available to the public; classroom & meeting room; nature & conservation center; nature trail. Nature book & supplies for sale.
Activities: guided tours; lectures; films; hobby workshops; organized education programs for children and adults; docent program; participatory exhibits; organized education programs for undergraduate or graduate college students affiliated with Youngstown State University. Annual Event: Photography Competition.
Hours & Admission Prices: Tues.-Sat. 10-5, Sun. 12-5. No charge; donations accepted. Closed New Year's Day; Thanksgiving; Christmas. &
Attendance: 20,000 (accurate)

MCDONOUGH MUSEUM OF ART, (M), Youngstown State University, 525 Wick Ave., Youngstown, OH 44555. Mailing Address: One University Plaza, Youngstown, OH 44555. Tel.: 330-941-1400. Fax: 330-941-1492.
E-mail: labrothers@ysu.edu
Web Site: www.mcdonoughmuseum.ysu.edu
Founded: 1991.
Key Personnel: Dir., Leslie A. Brothers; Asst. Dir., Angela DeLucia; Exhibition Design & Production Mgr., Robyn Maas.
Governing Authority: Parent Institution: Youngstown State University. Tax-exempt.
Institution Type/Description: Contemporary Art Center.
Research Fields: contemporary art; education & community.
Facilities: 8,000 sq. ft. exhibition space.
Activities: temporary exhibitions; performances; lectures; community events.
Publications: annual poster; catalogues.
Hours & Admission Prices: Tues.-Sat. 11-4. No charge; donations accepted. Closed all major holidays. &
Attendance: 18,000 (estimated)
Membership: Student/Senior Citizen $20; Individual/Family $35; Patorn $50; Guardian $100; Partner $500-$1,000.

OH WOW! THE ROGER & GLORIA JONES CHILDREN'S CENTER FOR SCIENCE & TECHNOLOGY, 11 W. Federal St., Youngstown, OH 44503. Tel.: 330-744-5914. Fax: 330-259-0258.
E-mail: info@ohwowkids.org
Web Site: ohwowkids.org
Formerly: Children's Museum of the Valley
Founded: 2001.
Congressional District: 17
Key Personnel: Dir., Suzanne Barbati.
Personnel Profile: Full-Time Paid 4; Part-Time Paid 3; Part-Time Volunteers 20.
Governing Authority: private; nonprofit organization. Tax-exempt: 501(c)(3).
Institution Type/Description: Children's Museum.
Collections: hands-on exhibits.
Facilities: classroom; 10,000 sq. ft. exhibit space; cafeteria. Museum-related items for sale.
Activities: special events; weekly programs; participatory exhibits; storytime programs. Annual Events: Halloween Spooktacular; Spring Fling; CMV Birthday Party.
Publications: quarterly newsletter, Play Matters.
Hours & Admission Prices: Tues.-Fri. 10-4:30, Sat. 12-4:30. Adults $5, senior citizens $4; discounts to groups & ACM reciprocal members; first Wed. of month & children under 3 no charge. Closed New Year's Day; Independence Day; Christmas Eve & Day. &
Attendance: 18,000 (estimated)
Membership: Individual $35; Family $50; Grandparent $60; Sponsor $100.

YOUNGSTOWN HISTORICAL CENTER OF INDUSTRY & LABOR, (M), 151 W. Wood St., Youngstown, OH 44503-1034. Mailing Address: History Dept. - YSU, One University Plaza, Youngstown, OH 44555. Tel.: 330-941-1314; 800-262-6137. Fax: 330-941-2301.
E-mail: m.pallante@ysu.edu
Founded: 1992.
Congressional District: 17
Key Personnel: Site Mgr., Martha Pallante.
Personnel Profile: Part-Time Paid 1; Interns 2.
Governing Authority: Parent Institution: Ohio Historical Society. Subsidiary Institution: Youngstown State University. Tax-exempt.

Institution Type/Description: History Museum.
Collections: workers' tools & clothing; "last heats", the last batches of steel; photographs.
Research Fields: industry; labor; ethnicity.
Facilities: meeting rooms; rental facilities.
Hours & Admission Prices: Wed.-Fri. 10-4, Sat. 12-4. Adults $7, children 6-12 $3; discount to AAA members, seniors & active military; children 5 & under, YSU students & employees, and Ohio Historical Society members no charge. Closed New Year's Day; Independence Day; Labor Day; Thanksgiving; Christmas. &
Membership: Individual $25; Family $35; Sponsor $50.

Zanesville

DR. INCREASE MATHEWS HOUSE, 304 Woodlawn Ave., Zanesville, OH 43701-4940. Mailing Address: 115 Jefferson St., Zanesville, OH 43701-4905. Tel.: 740-454-9500.
Web Site: www.muskingumhistory.org
Founded: 1970.
Congressional District: 18
Key Personnel: Dir., Jim Geyer; Pres., Bob Jenkins.
Personnel Profile: Part-Time Paid 1; Part-Time Volunteers 20.
Governing Authority: nonprofit organization. Parent Institution: Pioneer & Historical Society of Muskingum County. Tax-exempt.
Institution Type/Description: Historical Society Museum: housed in 1805 Dr. Increase Matthews House.
Collections: material owned by early settlers & local townspeople; costumes; furniture; military; portraits; kitchen implements; art pottery; textiles; primitive paintings.
Research Fields: Muskingum County furniture, history, geography, & archeology.
Facilities: library of local history books, papers & pamphlets available for research by special request. Books & local history items for sale.
Activities: guided tours; lecture series; formally organized education programs for children; temporary exhibitions.
Publications: quarterly, Muskingum Journal; booklets, Muskingum County Courthouse 1877-1977; Dr. Increase Mathews House; Zanesville & Muskingum County Bicentennial Military History; Historic Homes of Zanesville; Muskingum Annuals, Vols. I-V.
Hours & Admission Prices: April-May & Oct. 1st & 3rd Sun. 1-5; June-Sept. Sat.-Sun. 1-4; other times by appointment. Adults $2, students $1; discounts to groups, AAM & ICOM members; members no charge.
Attendance: 534 (estimated)
Membership: Senior Citizen $10; Individual $15; Family $30; Life $300.

PUTNAM UNDERGROUND RAILROAD EDUCATION CENTER, 522 Woodlawn Ave., Zanesville, OH 43701. Tel.: 740-450-3100.
Institution Type/Description: History Museum.
Collections: local history & culture; period furnishings; personal artifacts; photographs.
Activities: educational programs; special events.
Hours & Admission Prices: Feb. Mon.-Fri. 10-2; Sun. 1-4; March-Jan. Mon.-Fri. 10-2. Adults $3, seniors $2; discounts to groups; children under 5 no charge.

THE STONE ACADEMY, 115 Jefferson St., Zanesville, OH 43701. Tel.: 740-454-9500.
Governing Authority: Parent Institution: Pioneer and Historical Society Office.
Institution Type/Description: Historic House Museum: housed in the childhood home of Elizabeth Robins, late 19th century actress, author & activist; built in 1809.
Collections: local history & culture; period documents, clothing, & furnishings; fine art; personal artifacts; photographs.
Facilities: archives.
Hours & Admission Prices: Tues.-Fri. 12-4.

ZANESVILLE MUSEUM OF ART, (M), 620 Military Rd., Zanesville, OH 43701-1533. Tel.: 740-452-0741. Fax: 740-452-0797.
E-mail: vanessa@zanesvilleart.org
Web Site: www.zanesvilleart.org
Formerly: Zanesville Art Center
Founded: 1936.
Congressional District: 10
Key Personnel: Pres. (V), Richard Duncan; Operations Technician, Fred Orr; Temporary Operations Mgr., Andrew Near; Administrative Asst., Vanessa Brosie.
Personnel Profile: Full-Time Paid 4; Part-Time Paid 1; Part-Time Volunteers 70.

Governing Authority: nonprofit. Tax-exempt: 501(c)(3).
Institution Type/Description: Art Museum.
Collections: paintings; drawings; prints; photography; sculpture; American, European, Asian, Pre-Columbian, & African art; Ohio, national & international ceramics & glass.
Major Exhibits: Michael McEwen: Ohio Landscapes, 11/13-1/14; Water, Color, Paper, 1/18/14-3/22/14; Sung Soo Kim Glass, 3/14-5/17/14; Without a Word (T), 5/3/14-7/24/14; Owens Art Pottery, 6/14-9/20/14; 70th Ohio Annual Exhibition, 10/11-1/6/15.
Research Fields: American paintings & pottery.
Facilities: 6,000-vol. library of art books; reading room; 100-seat auditorium; classrooms.
Activities: guided tours; lectures; gallery talks; workshops; education programs; permanent, temporary & traveling exhibitions.
Publications: quarterly bulletin; brochures.
Hours & Admission Prices: Wed. & Fri.-Sat. 10-5, Thurs. 10-7:30. Adults $6, seniors $4; members no charge. Reciprocity with Ohio Art Museums. Closed holidays. &
Attendance: 24,000 (estimated)
Membership: Individual $30; Household $55 & up; Sustaining $75 & up; Fellow $100 & up; Contributing $200 & up; Benefactor $500 & up; Masterpiece Society $1,000 & up.

Zoar

ZOAR COMMUNITY ASSOCIATION, (M), 198 Main St., Zoar, OH 44697. Mailing Address: P.O. Box 621, Zoar, OH 44697-0508. Tel.: 330-874-3011 & 2646; 800-262-6195. Fax: 330-874-3211.
E-mail: zoarinfo@zca.org
Web Site: www.zca.org
Formerly: Zoar State Memorial
Founded: 1930.
Congressional District: 18
Key Personnel: Dir., Jennifer Donato; Pres. (V), Jon Elsasser.
Personnel Profile: Full-Time Paid 2; Part-Time Paid 5; Part-Time Volunteers 60.
Governing Authority: nonprofit organization. Parent Institution: Ohio Historical Society, 1982 Velma Ave., Columbus, OH 43211. Tax-exempt.
Institution Type/Description: Historic Site: 12 restored buildings of 1817 German religious communal sect.
Collections: Germanic-American folk arts & crafts; tools; furniture; greenhouse; textiles.
Research Fields: communal societies; German-American studies; religious history; 19th-century agrarian community.
Facilities: meeting room; banquet rooms.
Activities: guided tours; demonstrations & interpretive staff.
Publications: book, Zoar: An Ohio Experiment in Communalism.
Hours & Admission Prices: April-May & Oct. Sat. 11-4, Sun. 12-4; June-Sept. Wed.-Sat. 11-4, Sun. 12-4. Adults $8, children 4-17 $4; Time Travelers, ZCA & OHS members no charge. &
Attendance: 23,956 (accurate)
Membership: Individual $20; Family & Corporate $30; Patron $50; Sustaining $100; Lifetime $250.

OKLAHOMA

(296 listings)

Ada

ADA ARTS AND HERITAGE CENTER, 400 S. Rennie, Ada, OK 74820. Tel.: 580-332-7302.
E-mail: adaarts_heritage@yahoo.com
Web Site: adaartsandheritagecenter.com
Key Personnel: Dir., Sarah Hilton
Institution Type/Description: Historic Building: housed in the former Ada Public Library; built in 1939. Listed on the National Register of Historic Places.
Collections: building history; early furnishings.
Activities: special events; rental facilities.
Hours & Admission Prices: Mon.-Fri. 10-2.

CHICKASAW NATIONAL HEADQUARTERS & CULTURAL CENTER, 520 E. Arlington, Ada, OK 74821. Mailing Address: P.O. Box 1548, Ada, OK 74821. Tel.: 580-436-2603.
Institution Type/Description: Native American Museum.
Collections: Chickasaw cultural heritage & history; photographs; personal artifacts; tribal artifacts.
Hours & Admission Prices: Tues.-Fri. 9-5. No charge.

Afton

NATIONAL ROD & CUSTOM CAR HALL OF FAME MUSEUM, 55251 E. Hwy. 85A, Afton, OK 74331-2774. Tel.: 918-257-4234. Fax: 918-257-4234.
E-mail: dstarbird@wavelinx.net
Web Site: darrylstarbird.com
Founded: 1995.
Key Personnel: Dir., Donna Starbird; C.E.O., Darryl Starbird.
Personnel Profile: Full-Time Volunteers 1.
Governing Authority: Tax-exempt: 501(c)(3).
Institution Type/Description: Automobile Museum.
Collections: over 50 custom built cars.
Hours & Admission Prices: March-Nov. Wed.-Mon. 10-5; other times by appointment. Adults 13 & over $8, seniors 65 & over $7, children 8-12 $5. &

Aline

SOD HOUSE MUSEUM, 4628 State Hwy. 8, Aline, OK 73716-0821. Tel.: 580-463-2441.
E-mail: sodhouse@okhistory.org
Web Site: www.okhistory.org
Founded: 1968.
Congressional District: 3
Key Personnel: Historic Properties Mgr., Renee Trindle
Governing Authority: state. Parent Institution: Oklahoma Historical Society. (See separate listing). Tax-exempt.
Institution Type/Description: Historic House: built 1894, following the opening of the Cherokee Outlet.
Collections: original homesteader sod house with furnishings; pioneer artifacts; photographs; root cellar; horse-drawn equipment; farm implements.
Facilities: picnic tables.
Activities: monthly public programs; trunk programs & scavenger hunt for students.
Hours & Admission Prices: Tues.-Sat. 9-5. Adults $4, students 6-18 $2; discounts to senior citizens & groups of 25; members & children 5 and under no charge. Closed legal holidays. &
Attendance: 6,500 (accurate)
Membership: Annual $15; Lifetime $100.

Altus

MORGAN DOLL MUSEUM, 909 E. Broadway, Altus, OK 73521. Tel.: 580-482-2387.
Founded: 2006.
Key Personnel: Dir., Mary Morgan.
Operating Expenses: 2,500
Institution Type/Description: Doll Museum.
Collections: over 5,000 period & collectible dolls including 500 Madam Alexander dolls; doll buggies; doll houses.
Activities: guided tours.
Hours & Admission Prices: Tues., Thurs. & Sat.-Sun. 1-5; other times by appointment. No charge. &
Attendance: 700 (estimated)

MUSEUM OF THE WESTERN PRAIRIE, 1100 Memorial Dr., Altus, OK 73521-2600. Tel.: 580-482-1044. Fax: 580-482-0128.
E-mail: jbuchanan@okhistory.org
Web Site: www.okhistory.org/sites/westernprairie
Founded: 1970.
Congressional District: 4
Key Personnel: Pres. (V), Dennis Vernon; Dir., Jennie Buchanan; Museum Shop Mgr., Mary Jane Winsett.
Personnel Profile: Full-Time Paid 1; Part-Time Paid 1; Part-Time Volunteers 47.
Governing Authority: state. Parent Institution: Oklahoma Historical Society. (see separate listing). Tax-exempt: 501(c)(3) & 170(b)(1)(A).
Institution Type/Description: General & Historical Society Museum.
Collections: agricultural & ranching tools; fossils; Native American artifacts; local history; archives.
Research Fields: prairie lifestyles; local history.
Facilities: library.
Activities: self-guided tours; guided tours by appointment; permanent & temporary exhibitions.
Hours & Admission Prices: Tues.-Sat. 10-5. Adults $4, seniors & military $3, children 6-18 $1; society members no charge. Closed state holidays. &
Attendance: 4,037 (accurate)
Membership: Oklahoma Historical Society: Individual $25; Family & Institutional $40; Supporting $75; Life $500; Benefactor $1,000.

Alva

CHEROKEE STRIP MUSEUM, 901 14th St., Alva, OK 73717-2500. Tel.: 580-327-2030.
Founded: 1976.
Congressional District: 6
Key Personnel: C.E.O. & Chm. (V), Brodie Bush; Dir., Beth Smith; Museum Shop Mgr., JoRetta Rhodes.
Personnel Profile: Full-Time Paid 1; Part-Time Volunteers 10; Interns 5.
Governing Authority: society; nonprofit organization. Tax-exempt.
Institution Type/Description: General Museum.
Collections: costumes; Indian sculptures with bow & headdress; post office; washing machine room; rock & natural science; farm machinery & equipment; guns; prisoner-of-war artifacts; American Indian; nursery; telephones; newspapers; flags of Oklahoma; dolls; dishes; furniture; war uniforms; clocks; spinning wheel; general store & kitchen. Historic Structures: wooden windmills; 1932 schoolhouse.
Research Fields: genealogy & history of the area; schools; churches; professions; founding of neighboring towns.
Facilities: 50-vol. library of Abraham Lincoln historical publications, historical & biographical books on Cherokee Strip, early settler poems & cattle brands available for research on premises.
Activities: guided tours; docent program or council; permanent & temporary exhibitions; special tours for children's classes, scout groups & handicapped persons. Museum Sponsors: Annual Festival of Trees in December.
Publications: brochures; books on local history.
Hours & Admission Prices: Tues.-Sun. 2-5; groups by appointment. Adults $3; members no charge. Closed Easter; Christmas. &
Attendance: 3,000 (estimated)
Membership: Single $25; Family $50; Organization $100; Lifetime $200; Business $300.

NORTHWESTERN OKLAHOMA STATE UNIVERSITY MUSEUM, Jesse Dunn Bldg., 709 Oklahoma Blvd., Alva, OK 73717-2749. Tel.: 580-327-1700, ext. 8513 & 8564. Fax: 580-327-8556.
E-mail: sdthompson@nwosu.edu
Web Site: www.nwosu.edu
Founded: 1902.
Congressional District: 6
Key Personnel: Co Dir., Dr. Steven Thompson; Co Dir., Dr. Aaron Place.
Personnel Profile: Part-Time Paid 6; Part-Time Volunteers 2.
Governing Authority: university. Parent Institution: Northwestern Oklahoma State University. Tax-exempt: 170(b)(1)(A).
Institution Type/Description: Natural History Museum.
Collections: mounted bird & mammal specimens; study skins of birds & mammals; Pleistocene fossils; Indian artifacts; natural science; paleontology; mineralogy; geology; entomology; ichthyology; vascular plants; historical & archival material about NWOSU.
Facilities: 115,000-vol. library; 220,000 microfiche units; 5,000 microfilm reels.
Activities: occasional guided tours; permanent exhibitions.
Hours & Admission Prices: Call for hours. No charge; donations accepted. Closed college holidays.
Attendance: 600 (estimated)

Ames

AMES ASTROBLEME MUSEUM, 109 E. Main St., Ames, OK 73718. Mailing Address: P.O. Box 568, Ames, OK 73718. Tel.: 580-753-4624. Fax: 580-753-4221.
Founded: 2007.
Institution Type/Description: History Museum: housed on the site of a meteor crash that occurred 450 million years ago.
Collections: town history; computer animations, graphs & charts of the asteroid's impact; photographs.
Activities: videos.
Hours & Admission Prices: Daily 24 hours.
Attendance: 400 (estimated)

HAJEK MOTORSPORTS MUSEUM, RR 1, Ames, OK 73718. Mailing Address: P.O. Box 106, Ames, OK 73718. Tel.: 580-753-4611. Fax: 580-753-4266.
Institution Type/Description: Motorsports Museum.
Collections: racing cars & memorabilia; photographs.
Activities: special events.
Hours & Admission Prices: Call for hours.

Anadarko

ANADARKO HERITAGE MUSEUM, 311 E. Main St., Anadarko, OK 73005-3023. Tel.: 405-247-3240. Fax: 405-247-3240.
E-mail: anadarkomuseum@netride.net
Formerly: Philomathic Museum
Founded: 1936.
Congressional District: 6
Key Personnel: C.E.O. & Pres. (V), Betty Bell; Co-Cur., Wilson Daingkau; Co-Cur., Cathy Crowell.
Personnel Profile: Part-Time Paid 1.
Governing Authority: society. Parent Institution: Philomathic Club. Subsidiary Institution: City of Anadarko. Tax-exempt: 501(c)(3).
Institution Type/Description: General Museum.
Collections: toys & dolls; doctor's office; country store; Indian Agency; Caddo County history; railroad memorabilia; military equipment & uniforms; American Indian dolls, paintings, clothing & artifacts; glassware & period household items; photographs including photographers Horrace Poolaw & Mrs. Annette Ross Hume.
Activities: off-site presentations for schools, clubs, & organizations by request.
Hours & Admission Prices: Mon.-Fri. 10-4, Sat. by appointment. No charge; donations accepted. Closed legal holidays. &
Attendance: 1,000 (estimated)

DELAWARE NATION MUSEUM, 31064 State Hwy. 281, Anadarko, OK 73005. Tel.: 405-247-2448, ext. 1181. Fax: 405-247-8905.
E-mail: jross@delawarenation.com
Web Site: www.delawarenation.com
Formerly: Delaware Tribal Museum
Key Personnel: Museum Shop Mgr., Gina Wooster
Institution Type/Description: Native American Museum.
Collections: Native American artifacts; beadwork; clothing.
Facilities: Museum-related items for sale.
Hours & Admission Prices: Mon.-Fri. 8-5. No charge; donations accepted. Closed holidays. &

INDIAN CITY U.S.A. CULTURAL CENTER, 2 1/2 Miles S. on Hwy. 8, Anadarko, OK 73005. Mailing Address: Kiowa Tribe, P.O. Box 369, Carnegie, OK 73015. Tel.: 580-654-2300.
Founded: 1955.
Personnel Profile: Full-Time Paid 12.
Governing Authority: Parent Institution: Kiowa Tribe of Oklahoma.
Institution Type/Description: Natural History Museum: located on 1887 site of Tonkawa Massacre.
Collections: Indian artifacts; pottery; dance costumes; cradles, bags; moccasins; leggings; rugs; Indian dolls; paintings; arrowheads; Indian dressed mannequins; early American Indian articles.
Activities: guided tours; dance recitals.
Publications: brochures.
Hours & Admission Prices: Temporarily closed. &
Attendance: 38,471 (estimated)

NATIONAL HALL OF FAME FOR FAMOUS AMERICAN INDIANS, Hwy. 62, E., Anadarko, OK 73005. Mailing Address: P.O. Box 548, Anadarko, OK 73005-0548. Tel.: 405-247-5555 & 3331. Fax: 405-247-5571.
Founded: 1952.
Congressional District: 6
Key Personnel: C.E.O. (V), Joe McBride, Jr.; Sec., Carolyn N. McBride.
Personnel Profile: Part-Time Volunteers 4.
Governing Authority: bd. of directors. Parent Institution: State Historical Society of Oklahoma. Tax-exempt: 501(c)(3).
Institution Type/Description: Native American Museum.
Collections: sculptured bronze busts of famous Native Americans in outdoor landscaped area.
Research Fields: historians.
Facilities: Visitor's Information Center.
Activities: dedication ceremonies for honorees.
Publications: self-tour brochure, The National Hall of Fame for Famous American Indians.
Hours & Admission Prices: Mon.-Sat. 9-5, Sun. 1-5. No charge; donations accepted. Closed New Year's Day; Thanksgiving; Christmas. &
Attendance: 20,000 (estimated)
Membership: Annual Individual $10; Family $25; Life $100.

PHILOMATHIC PIONEER MUSEUM, 311 E. Main St., Anadarko, OK 73005-3023. Tel.: 405-247-3240.
Institution Type/Description: History Museum.
Collections: local history; railroad memorabilia; American Indian dolls; military artifacts; replica country store & doctor's office.
Hours & Admission Prices: Tues.-Fri. 10-5, Sat.-Sun. 1-5. No charge.

SOUTHERN PLAINS INDIAN MUSEUM, 801 E. Central Blvd., Anadarko, OK 73005-4437. Mailing Address: P.O. Box 749, Hwy. 62, E., Anadarko, OK 73005-0749. Tel.: 405-247-6221. Fax: 405-247-7593.
Founded: 1947.
Congressional District: 4
Key Personnel: Cur., Bambi Allen.
Personnel Profile: Full-Time Paid 3.
Governing Authority: federal. Parent Institution: Indian Arts & Crafts Board, Rm. 4004, U.S. Department of the Interior, Washington, DC 20240. Tax-exempt.
Institution Type/Description: Indian Art Museum.
Collections: historic & contemporary arts of the southern Plains Indian peoples; paintings; sculpture.
Research Fields: historic & contemporary Native American art.
Facilities: Paintings, jewelry, suede handbags, beaded jewelry, accessories, books & audio cassettes for sale (operated by Oklahoma Indian Arts & Crafts Coop).
Activities: lectures; gallery talks; permanent & temporary exhibitions.
Publications: exhibition brochures; catalogs.
Hours & Admission Prices: Tues.-Sat. 9-5. No charge. Closed New Year's Day; Thanksgiving; Christmas. &
Attendance: 25,000 (estimated)
Membership: Student $2; Indian Artist & Craftsman $5; Individual $6; Family $10; Business $100.

Antlers

WILDLIFE HERITAGE CENTER MUSEUM, 610 S.W. D St., Antlers, OK 74523. Tel.: 580-298-9933.
E-mail: antlersdeerfestival@yahoo.com
Web Site: www.wildlifeheritagecenter.org
Founded: 2006.
Governing Authority: Parent Institution: Deer Capital Tourism Association. Tax-exempt.
Institution Type/Description: Wildlife Museum.
Collections: live & mounted wildlife and their habitats; hands-on exhibitions.
Activities: educational programs; special events; workshops.
Hours & Admission Prices: Mon.-Fri. 9-4, Sat. 10-2. No charge; donations accepted. Closed holidays. &
Attendance: 4,000 (accurate)
Membership: Deer Capital Tourism Association: Individual $25.

Apache

APACHE HISTORICAL SOCIETY MUSEUM, 101 W. Evans, Apache, OK 73006. Mailing Address: P.O. Box 101, Apache, OK 73006-0101. Tel.: 580-588-3392. Fax: 580-588-3393.
Founded: 1971.
Congressional District: 3
Key Personnel: Chm. & Pres. (V), Danny Swanda; Mgr., Mary Joyce Swanda.
Personnel Profile: Part-Time Paid 1; Part-Time Volunteers 1.
Volunteer Hours: 1,000
Operating Expenses: 2,113
Operating Income: 1,555
Governing Authority: Tax-exempt.
Institution Type/Description: Historical Society Museum.
Collections: period furnishings; records; photographs; town history; Indian art. Historic Building: 1901 State Bank.
Publications: weekly newspaper column.
Hours & Admission Prices: Mon.-Fri. 12-4. No charge; donations accepted. Closed holidays. &
Attendance: 400
Membership: Annual $10.

Ardmore

CHARLES B. GODDARD CENTER FOR VISUAL AND PERFORMING ARTS, 401 First Ave., S.W., Ardmore, OK 73401-4725. Mailing Address: P.O. Box 1624, Ardmore, OK 73402-1624. Tel.: 580-226-0909. Fax: 580-226-8891. Facebook: Charles B. Goddard The Goddard Center.
E-mail: goddardcenter@yahoo.com
Web Site: www.goddardcenter.org
Founded: 1969.
Congressional District: 3
Key Personnel: Exec. Dir., Leila Lenore; Chm. (V), Stan Kittrell; Treas. (V), Charles Williams; Cur. & Preparator, Rudy Ellis.
Personnel Profile: Full-Time Paid 7; Part-Time Paid 3; Part-Time Volunteers 50.
Governing Authority: nonprofit. Tax-exempt: 501(c)(3).
Institution Type/Description: Center for Visual & Performing Arts.
Collections: contemporary paintings; lithographs; graphics; sculptures; photographs; contemporary printmaking; Native American & western painting; western & contemporary sculpture.
Facilities: 313-seat theatre; classroom/studio; art studio; dance studio.
Activities: plays; concerts; dance performances; education programs in art & dance; art classes; temporary & permanent art exhibits; theatre.
Publications: quarterly newsletter; Annual Season Review Guide.
Hours & Admission Prices: Tues.-Fri. 9-5, Sat. 1-4. No charge; donations accepted. Closed national holidays. &
Attendance: 10,000 (estimated)
Membership: Student $25; Single $50; Couple $100; Patron $285; Personal Benefactor $550; Business Benefactor $800.

ELIZA CRUCE HALL DOLL COLLECTION, 320 E. St., N.W., Ardmore, OK 73401-4304. Tel.: 580-223-8290.
Founded: 1971.
Congressional District: 3
Key Personnel: Dir., Daniel R. Gibbs.
Governing Authority: municipal. Parent Institution: The Ardmore Public Library. Tax-exempt: 170(b)(1)(A).
Institution Type/Description: Antique Doll Museum.
Collections: 300 dolls dating back to early 1700s; three original French Court dolls depicting members of the court of Marie Antoinette; English peddlers; French fashion dolls; Lenci; Kruse, Kestner, Ravca, Klumpe, Montanari, Schoenhut & Bye-lo dolls; miniature tea services.
Facilities: 50-vol. library of books on antique dolls, toys & costuming.
Activities: guided tours; lectures.
Hours & Admission Prices: Mon.-Thurs. 10-8, Fri. 10-6, Sat.-Sun. 1-5. No charge; donations accepted. Closed legal holidays. &
Attendance: 10,000 (estimated)

GREATER SOUTHWEST HISTORICAL MUSEUM, (M), 35 Sunset Dr., Ardmore, OK 73401-2852. Tel.: 580-226-3857. Fax: 580-226-3357.
Web Site: www.gshm.org
Key Personnel: Dir., Michael W. Anderson; Museum Shop Mgr., Barbara Salamone.
Personnel Profile: Full-Time Paid 3; Part-Time Paid 2; Interns 2.
Institution Type/Description: History Museum.
Collections: South-Central Oklahoma history & culture; Native American artifacts; photographs; military artifacts.
Facilities: research library. Museum-related items for sale.
Activities: educational programs.
Publications: quarterly newsletter, Frontier Tales & Trails.
Hours & Admission Prices: Tues.-Sat. 10-5. No charge; donations accepted. &
Attendance: 10,000 (estimated)
Membership: Student $10; Senior $15; Individual $25; Couple $35; Family $50; Business & Sustaining $100; Patron $250; Benefactor $500.

TUCKER TOWER NATURE CENTER, c/o Lake Murray State Park, 13528 Scenic State Hwy. 77, Ardmore, OK 73401-7083. Tel.: 580-223-2109. Fax: 580-223-4052. Facebook: Lake Murray State Park.
Web Site: www.oklahomaparks.com
Founded: 1952.
Congressional District: 3
Key Personnel: Park Naturalist, Mark Teders.

Personnel Profile: Full-Time Paid 1; Part-Time Paid 3; Part-Time Volunteers 10; Interns 1.
Volunteer Hours: 440
Operating Expenses: 180,000
Operating Income: 9,600
Governing Authority: state. Parent Institution: Oklahoma Tourism & Recreation. Subsidiary Institution: Lake Murray State Park. Tax-exempt: 170(b)(1)(A).
Institution Type/Description: Park Museum: built 1933-35 by WPA & CCC for a Governor's Retreat.
Collections: fossils; minerals; meteorite; nature exhibits; WPA/CCC exhibits.
Research Fields: pertaining to collection.
Activities: guided walks; lectures; audiovisual programs; outdoor activities; permanent & temporary exhibitions.
Hours & Admission Prices: Feb.-May & Sept.-Nov. Wed.-Sun. 9-5; Memorial Day-Labor Day daily 9-7. Any donation for admission. &
Attendance: 35,000 (accurate)

Atoka

CONFEDERATE MEMORIAL MUSEUM, US Hwy. 69 N., Atoka, OK 74525. Mailing Address: P.O. Box 245, Atoka, OK 74525-0245. Tel.: 580-889-7192. Fax: 580-889-7192.
E-mail: atokamuseum@yahoo.com
Web Site: www.civilwaralbum.com/atoka
Founded: 1986.
Congressional District: 2
Key Personnel: Site Mgr., Gwen Walker; Museum Dir., Cindy Wallis; Pres. (V), Paula Hardman.
Personnel Profile: Full-Time Paid 1; Full-Time Volunteers 1; Part-Time Volunteers 1.
Governing Authority: Parent Institution: Atoka County Historical Society. Tax-exempt.
Institution Type/Description: Military Museum.
Collections: military artifacts; personal artifacts; uniforms; photographs; U.S. & Confederate Battle Flags.
Facilities: Museum-related items for sale.
Activities: reenactments. Annual Event: Living History Presentations.
Publications: quarterly newsletter; Civil War in the Western Choctaw Nation; Old Boggy Depot, 1838-1883; Father Murrow, The Life of J.S. Murrow; Confederate Volunteers of Atoka County; Down the Texas Road.
Hours & Admission Prices: Mon.-Fri. 9-4. No charge; donations accepted. Closed national holidays. &
Attendance: 9,000 (estimated)
Membership: Individual $10; Family $15.

Barnsdall

BIGHEART MUSEUM, 616 W. Main, Barnsdall, OK 74002. Mailing Address: P.O. Box 475, Barnsdall, OK 74002-0475. Tel.: 918-847-2397.
Key Personnel: Cur., Joe Williams; Cur., Faye Wickware
Institution Type/Description: History Museum: named after Osage Chief James Bigheart.
Collections: Native American history & culture; town history.
Hours & Admission Prices: Tues.-Fri. 12-4, Sat. 9-1. No charge.

Bartlesville

BARTLESVILLE AREA HISTORY MUSEUM, 401 S. Johnstone Ave., City Bldg., 5th Fl., Bartlesville, OK 74003-6619. Tel.: 918-338-4290. Fax: 918-338-4264.
E-mail: history@cityofbartlesville.org
Web Site: www.bartlesvillehistorycom
Founded: 1964.
Congressional District: 5
Key Personnel: Dir., Joan Singleton; Registrar, Matthew Clapper; Volunteer Coord. & Public Rels., Jo Crabtree; Collections Mgr., Debbie Neece.
Personnel Profile: Full-Time Paid 4; Part-Time Volunteers 10.
Governing Authority: municipal government. Tax-exempt
Institution Type/Description: History Museum.
Collections: early settlers & Native Indian tribes living in the western frontier from late 1800s to present with major focus on early 1900s.
Research Fields: commerce & trade; early city residents.
Facilities: 8,000 sq. ft. exhibit space.
Activities: guided tours; lectures; traveling exhibitions; school presentations; civic organization speaker.
Hours & Admission Prices: Mon.-Fri. 10-4. No charge; donations accepted. Closed major holidays. &
Attendance: 5,000 (estimated)

FRANK PHILLIPS HOME, 1107 S. Cherokee, Bartlesville, OK 74003-5027. Tel.: 918-336-2491. Fax: 918-336-3529. Facebook: Frank Phillips Home.
E-mail: jgoss@okhistory.org
Web Site: www.frankphillipshome.org
Founded: 1973.
Congressional District: 2
Key Personnel: Dir. & Cur., Jim L. Goss.
Personnel Profile: Full-Time Paid 1; Part-Time Paid 3; Part-Time Volunteers 25.
Governing Authority: state. Parent Institution: Oklahoma Historical Society. (See separate listing). Tax-exempt.
Institution Type/Description: Historic House: 1909 Frank Phillips home.
Collections: 1930s period furniture.
Research Fields: Phillips family history; history of F. Phillips, founder of Phillips Petroleum Co.; Oklahoma oil industry.
Facilities: 2,000-vol. library of books.
Activities: guided tours; arts festivals; temporary & permanent exhibitions.
Hours & Admission Prices: Tours: Wed.-Fri. 10, 11, 2, 3, & 4, Sat. 10, 11, 1, 2, 3, 4. Adults $5. Behind the Scenes Director's Tour: Wed.-Fri. 9-10:30. Adults $10. Closed major holidays. &
Attendance: 7,941 (accurate)
Membership: Friend of Frank & Jane $100; Driller $101-$250; Wildcatter $251-$500; Gusher $501-$999; Oil Baron/Baroness $1,000 & up.

PHILLIPS PETROLEUM COMPANY MUSEUM, 410 Keeler, Bartlesville, OK 74004. Tel.: 918-661-8687.
E-mail: lorrie.l.rockman@conocophillips.com
Web Site: www.phillips66museum.com
Key Personnel: Dir., Lorrie Rockman
Institution Type/Description: History and Technology Museum.
Collections: artifacts & memorabilia pertaining to the Phillips Petroleum Company.
Hours & Admission Prices: Mon.-Sat. 10-4. No charge. Closed holidays.

PRICE TOWER ARTS CENTER, (M), 510 S. Dewey, Bartlesville, OK 74003-3560. Mailing Address: P.O. Box 2464, Bartlesville, OK 74005-2464. Tel.: 918-336-4949. Fax: 918-336-7117.
E-mail: info@pricetower.org
Web Site: www.pricetower.org
Founded: 1985.
Congressional District: 5
Key Personnel: Exec. Dir., Timothy Boruff; Chm. (V), C.J. Silas; Pres. (V), Robbie A. Morris; Cur. Collections & Exhibitions, Scott Perkins; Docent Coord., Judy DuVall; Dir. Mktg. & Public Rels., Lori Esser.
Personnel Profile: Full-Time Paid 12; Part-Time Paid 2; Part-Time Volunteers 40; Interns 1.
Governing Authority: nonprofit organization. Subsidiary Institution: Inn at Price Tower, Inc. Tax-exempt: 501(c)(3).
Institution Type/Description: Art, Architecture & Design Museum: housed in Price Tower, designed by Frank Lloyd Wright.
Collections: Frank Lloyd Wright, Bruce Goff, Dennis Oppenheim, & Karim Rashid; art, architectural & design objects of the 20th & 21st centuries.
Research Fields: permanent collection; 20th-century & contemporary art, architecture and design.
Facilities: library; 7,000 sq. ft. exhibit space; restaurant; rental facilities. Museum-related items for sale.
Activities: guided tours; lectures; docent program; traveling & permanent exhibitions.
Publications: newsletter, The View.
Hours & Admission Prices: Gallery: Tues.-Sat. 10-5, Sun. 12-5. Adults $4, seniors $3; discounts to AAM, ICOM, NARM members, & Frank Lloyd Wright public sites; members and students 16 & under no charge. Tower Tours: Tues.-Sat. 11 & 2, Sun. 2; reservations recommended. Tours: adults $10, seniors $8, students $5. Closed New Year's Day; Thanksgiving; Christmas. &
Attendance: 25,000 (accurate)
Membership: Artist, Educator & Student $30; Personal $40; Family $66; Contributing $100; Supporter $250; Associate $1,000. Giving Societies: Director's Circle $2,500; Chairman's Council $5,000.

WOOLAROC MUSEUM, (M), 1925 Woolaroc Ranch Rd., Bartlesville, OK 74003-7171. Tel.: 918-336-0307, ext. 10; 888-WOOLAROC. Fax: 918-336-0084.
E-mail: lstone@woolaroc.org
Web Site: www.woolaroc.org
Founded: 1929.
Congressional District: 5

Key Personnel: Dir., Kenneth Meek; C.E.O., Bob Fraser; Cur. Art, Linda Stone; Bldg. Supt., Tim Sydebotham; Museum Shop Mgr., Beth Greene.
Personnel Profile: Full-Time Paid 5; Part-Time Paid 1; Part-Time Volunteers 70.
Governing Authority: nonprofit organization. Parent Institution: The Frank Phillips Foundation, Inc., Woolaroc Ranch, P.O. Box 1647, Bartlesville, OK 74005. Tax-exempt.
Institution Type/Description: Art & History Museum.
Collections: Indian artifacts; paintings; sculpture; ethnology; archaeology; anthropology; guns.
Facilities: nature trails; snack bar. Books, reproductions & Indian handicrafts for sale.
Activities: docent program or council; permanent exhibitions; school loan service of slides & film about Woolaroc; provides color transparencies for educational purposes.
Publications: guidebook, Woolaroc; coffee table book, Woolaroc.
Hours & Admission Prices: Memorial Day to Labor Day Tues.-Sun. 10-5; Sept.-May Wed.-Sun. 10-5. Adults $10, senior citizens 65 & up $8; discounts to AAM members, special needs groups & organized school groups; members & children under 12 no charge. Closed Thanksgiving; Christmas. &
Attendance: 100,000 (accurate)
Membership: Partner $100; Associate Sponsor $250; Sponsor $500; Sustaining $1,000; Benefactor $2,500; Patron $5,000; Grand Patron $10,000 & up.

Beaver

JONES AND PLUMMER TRAIL MUSEUM, Fairgrounds, S. Douglas St., Beaver, OK 73932. Tel.: 580-625-4439. Fax: 580-625-3265.
E-mail: jplummermuseum@ptsi.net
Institution Type/Description: History Museum.
Collections: local history & culture; period furnishings; personal artifacts; early dishes & clothing; office equipment; tools & farming equipment; wagons & buggies; piano; windmill; period newspapers & printed records; one-room schoolhouse.
Hours & Admission Prices: Tues.-Sat. 11-3, Sun. 1-4. No charge; donations accepted. &
Membership: Single $15; Family $25.

Bethany

OKLAHOMA MUSEUM OF FLYING, 7110 Millionaire Dr., Bethany, OK 73008. Tel.: 405-535-9565.
Institution Type/Description: History Museum.
Collections: aviation history; aircraft; personal artifacts; photographs; paintings; uniforms.
Activities: scout groups; educational programs.
Hours & Admission Prices: Mon.-Fri. 9-4, Sat. 10-2. No charge; donations accepted.

Billings

DR. RENFROW-MILLER MUSEUM, 201 S. Broadway St., Billings, OK 74630. Mailing Address: 2100 Ranch, Billings, OK 74630. Tel.: 580-725-3258 & 3487.
Institution Type/Description: Historic House Museum: housed in the former home and office of pioneer doctor, Thomas F. Renfrow; built in 1901. Listed on the National Register of Historic Places.
Collections: Renfrow's life, family & career; personal artifacts; period furnishings; photographs.
Activities: guided tours.
Hours & Admission Prices: By appointment. No charge; donations accepted.

HENRY AND SHIRLEY BELLMON LIBRARY & MUSEUM, Main & Broadway, Billings, OK 74630. Mailing Address: 2500 Zig Zag, Billings, OK 74630-2003. Tel.: 580-725-3487.
Institution Type/Description: Library & Museum: c.1900 building. Listed on the National Register of Historic Places.
Collections: local history; photographs; documents; campaign materials; Henry Bellmon's personal artifacts & memorabilia.
Activities: guided tours.
Hours & Admission Prices: By appointment. No charge.

Binger

CADDO HERITAGE MUSEUM, Caddo Nation Complex, Hwy. 281 & 152, Binger, OK 73009. Mailing Address: P.O. Box 478, Binger, OK 73009. Tel.: 405-658-2344.
Institution Type/Description: History Museum.
Collections: local history & culture; archives; Caddo artifacts; contemporary art.
Facilities: Museum-related items for sale.
Activities: guided tours.
Hours & Admission Prices: Mon.-Fri. 9-4. No charge. &

Bixby

BIXBY HISTORICAL SOCIETY MUSEUM, 24 E. McKennon Ave., Bixby, OK 74008-4332. Mailing Address: P.O. Box 1046, Bixby, OK 74008-1046. Tel.: 918-366-1200.
E-mail: bixby.okhs@cox.net
Web Site: www.rootsweb.ancestry.com/~okbhs/index.html
Institution Type/Description: Historical Society Museum.
Collections: artifacts & memorabilia pertaining to the Bixby area.
Hours & Admission Prices: Wed. & 1st Sat. of each month 11-2.

Blackwell

TOP OF OKLAHOMA HISTORICAL MUSEUM, 303 S. Main St., Blackwell, OK 74631-3347. Tel.: 580-363-0209.
E-mail: blackwellmuseum@gmail.com
Founded: 1972.
Congressional District: 6
Key Personnel: Dir., Joanna Campbell; Pres. (V), Ralph Gose; Sec., Fredda Ganer.
Personnel Profile: Part-Time Paid 3; Part-Time Volunteers 25.
Governing Authority: society. Tax-exempt.
Institution Type/Description: Local History Museum: housed in c.1912 Electric Park Pavilion.
Collections: personal artifacts; period pioneer equipment & house furnishings; local school history of Maroon Spirit; church; military artifacts; industrial exhibits; historical documents; period photographs; articulture; early 1900s bedroom, parlor & kitchen; audiovisual room; seminar room; military room; 1893 Cherokee outlet land run to 1920; period school artifacts; Indian artifacts; music room.
Facilities: library of land records & historical documents; club room.
Activities: lectures; films; traveling exhibitions.
Publications: quarterly newsletter.
Hours & Admission Prices: Mon.-Sat. 10-4, Sun. 1-4. No charge; donations accepted. Closed major holidays. &
Attendance: 1,150 (estimated)
Membership: Children $1; Adults $5; Lifetime $100.

Boise City

CIMARRON HERITAGE CENTER MUSEUM, 1300 N. Cimarron, Boise City, OK 73933. Mailing Address: P.O. Box 214, Boise City, OK 73933-0441. Tel.: 580-544-3479. Facebook: Cimarron Heritage Center Museum.
E-mail: museum@ptsi.net
Web Site: www.chicmuseum.com
Key Personnel: Dir., Jody Risley; Pres. (V), Beverly Baker.
Governing Authority: Tax-exempt.
Institution Type/Description: History Museum.
Collections: local history & culture; dinosaur artifacts; the Santa Fe Trail; blacksmith shop; well house; wash house; early machinery, wagons & buggies; one room schoolhouse; Dust Bowl artifacts.
Hours & Admission Prices: Mon.-Sat. 10-12 & 1-4. No charge; donations accepted. Closed major holidays.
Attendance: 5,700 (estimated)
Membership: Individual $5; Lifetime $100.

Bokchito

MUSEUM OF CREATION TRUTH, 3290 State Rd. 22, Bokchito, OK 74726. Tel.: 580-924-0803.
Founded: 2006.
Key Personnel: Dir., Billy Gordon; Dir., Delie Gordon
Institution Type/Description: Natural History Museum.
Collections: natural history; fossils; period artifacts; mounted native wildlife; early tools & weapons; 16 ft. wall chart from creation to present.

Facilities: 30-seat theater.
Activities: documentaries; guided tours.
Hours & Admission Prices: Sat. 10-4; other times by appointment. No charge.
  &
Attendance: 300 (estimated)

Boley

BOLEY HISTORICAL MUSEUM, 10 W. Grant St., Boley, OK 74829. Mailing Address: P.O. Box 158, Boley, OK 74829-0158. Tel.: 918-667-9790.
Key Personnel: Chm. (V), Henrietta Hicks.
Governing Authority: town.
Institution Type/Description: History Museum.
Collections: local history & culture.
Hours & Admission Prices: By appointment. No charge; donations accepted.
Attendance: 1,100

Bristow

BRISTOW HISTORICAL MUSEUM, 1 Railroad Place, Bristow, OK 74010-3040. Mailing Address: P.O. Box 127, Bristow, OK 74010-0127. Tel.: 918-367-5151.
Web Site: www.visitbristowok.com/museums.htm
Institution Type/Description: History Museum: housed in 1923 restored depot.
Collections: local history & culture; railroad memorabilia.
Hours & Admission Prices: Mon.-Fri. 9-4.

Broken Arrow

BROKEN ARROW HISTORICAL SOCIETY, 400 S. Main, Broken Arrow, OK 74012. Tel.: 918-258-2616.
E-mail: llewis.bamuseum@yahoo.com
Web Site: www.bahistoricalsociety.com
Founded: 1975.
Congressional District: 1
Key Personnel: Dir., Lori Lewis
Institution Type/Description: Historical Society Museum.
Collections: local history & culture; Native American artifacts; railroad depot furnishings; photographs; military artifacts. Historic Building: 1861 log cabin.
Hours & Admission Prices: Tues.-Fri. 10-4, Sat. 10-2. Adults $5; children under 18 no charge. &

Broken Bow

BEAVERS BEND WILDLIFE MUSEUM, RR 4, Broken Bow, OK 74728. Mailing Address: P.O. Box 25, Broken Bow, OK 74728. Tel.: 580-494-6193.
E-mail: nature@pine-net.com
Web Site: www.pine-net.com
Institution Type/Description: Wildlife Museum.
Collections: mounted wildlife & their habitats.
Activities: guided tours; videos; educational programs.
Hours & Admission Prices: Mon.-Sat. 9-5. Adults & children over 12 $6, senior citizens 65 & over $5, children 12 & under $3.50; discounts to groups.

GARDNER MANSION & MUSEUM, Hwy. 70 E., Broken Bow, OK 74728. Tel.: 580-584-6588. Fax: 580-584-6588.
Institution Type/Description: Historic House Museum: housed in the former home of the Choctaw Indian Chief; built in 1884.
Collections: local history & culture; Native American artifacts; fossils.
Facilities: Museum-related items for sale.
Activities: guided tours.
Hours & Admission Prices: Summer: Mon.-Sat. 8-6, Sun. 2-6; Winter: Mon.-Sat. 10-5, Sun. 2-5.

INDIAN MEMORIAL MUSEUM, 402 E. 2nd St., Broken Bow, OK 74728. Tel.: 580-584-6531.
Institution Type/Description: History Museum.
Collections: Native American pottery from 900-1500 AD; arrowheads; stone artifacts; axes; modern pottery; ceramics; rugs; period glass; quartz crystal; fossils.
Facilities: Museum-related items for sale.
Hours & Admission Prices: Mon.-Fri. 9-6.

OKLAHOMA FOREST HERITAGE CENTER, (M), Beavers Bend State Park, US-259A, Broken Bow, OK 74728. Mailing Address: P.O. Box 157, Broken Bow, OK 74728-0157. Tel.: 580-494-6497 Fax: 580-494-6689. Facebook: Forest Heritage Center.
E-mail: fhc@beaversbend.com
Web Site: www.forestry.ok.gov/fhc
Founded: 1976.
Congressional District: 2
Key Personnel: Dir., Doug Zook; Chm. (V), Rick Harder; Museum Shop Mgr., Vicki Taylor; Asst., Hanna Anderson.
Personnel Profile: Full-Time Paid 2.
Governing Authority: state.
Institution Type/Description: Forestry Museum.
Collections: historical documents; forestry tools; wood art; homestead memorabilia; historic photo murals; papermaking; 1940s lumbering; dioramas.
Major Exhibits: Woodworking Artist of the Year, 3/14-5/14; Masters at Work: Competition & Exhibit, 9/14-10/14; Folk Festival, 11/14.
Activities: formal education programs.
Publications: newsletter, Forest Heritage; book, Traveling Timber Towns; Civilian Conservation Corps of McCurtain County.
Hours & Admission Prices: Daily 8-8. No charge; donations accepted. &
Attendance: 150,000 (estimated)
Membership: Forest Friend $25; Forest Patron $100; Forest Benefactor $500.

Buffalo

BUFFALO MUSEUM, 108 S. Hoy, Buffalo, OK 73834. Mailing Address: P.O. Box 224, Buffalo, OK 73834. Tel.: 580-735-2628.
Key Personnel: Pres. (V), Sydney B. Malt.
Personnel Profile: Interns 5.
Institution Type/Description: History Museum.
Collections: local history & culture; period furnishings; personal artifacts; photographs.
Facilities: Museum-related items for sale.
Hours & Admission Prices: May-Oct. Fri.-Sat. 1-4, Sun. 2-4; other times by appointment. No charge; donations accepted. &

Cache

QUANAH PARKER STAR HOUSE/EAGLE PARK GHOST TOWN, Rte. 2 (SH-115 at US-62), Cache, OK 73527. Mailing Address: 810 N. 8th St., Cache, OK 73527-9630. Tel.: 580-429-3420.
Key Personnel: Mgr., Kathy Threadwell; Mgr., Wayne Gipson.
Governing Authority: Parent Institution: Trading Post, Inc.
Institution Type/Description: History Museum.
Collections: period furnishings. Historic Buildings: 1884 home of Comanche Chief Quanah Parker; 1869 log building; Saddle Mountain Indian Baptist Church.
Hours & Admission Prices: By appointment only.

Caddo

CADDO INDIAN TERRITORY MUSEUM AND LIBRARY, 110 Buffalo St., Caddo, OK 74729. Mailing Address: P.O. Box 274, Caddo, OK 74729. Tel.: 580-367-2787.
Key Personnel: Cur., Bonnie Chaffin
Institution Type/Description: History Museum.
Collections: local history & culture; period furnishings; photographs; personal artifacts; horse-drawn fire cart; blacksmith shop; arrowheads.
Hours & Admission Prices: Tues.-Fri. 10-3, Sat. 1-3; other times by appointment. No charge.

Carnegie

KIOWA CULTURE PRESERVATION AUTHORITY, Hwy. 9 W., Carnegie, OK 73015. Mailing Address: P.O. Box 369, Carnegie, OK 73015. Tel.: 580-654-2300, ext. 370.
Formerly: Kiowa Tribe Museum and Resource Center
Key Personnel: Chm. (V), David Sullivan; Sec. & Treas., Tommie Louise Doyebi.
Personnel Profile: Part-Time Volunteers 2.
Governing Authority: Parent Institution: Kiowa Tribe of Oklahoma. Tax-exempt.
Institution Type/Description: American Indian Museum.
Collections: Kiowa art, language & cultural heritage; paintings; native crafts; Kiowa Sundance teepee; warriors memorial.
Hours & Admission Prices: Mon.-Fri. 8-4:30. No charge; donations accepted. &

Attendance: 1,500 (accurate)
Membership: Sponsor $5,000; Associate $1,000; Friend $2,500.

Catoosa

ARKANSAS RIVER HISTORICAL SOCIETY MUSEUM, 5350
Cimarron Rd., Catoosa, OK 74015-3027. Tel.: 918-266-2291. Fax:
918-226-7678.
Institution Type/Description: Historical Society Museum.
Collections: area navigation system history; photographs; waterway memora-
bilia; Native American artifacts; Lock and Dam motorized model.
Hours & Admission Prices: Mon.-Fri. 8-4:30. No charge; donations accepted.

CATOOSA HISTORICAL SOCIETY MUSEUM, 207 N. Chero-
kee, Catoosa, OK 74015. Mailing Address: c/o Catoosa Historical
Society, P.O. Box 738, Catoosa, OK 74015. Tel.: 918-266-3296.
Fax: 918-266-7156.
E-mail: georgiamcafee@yahoo.com
Founded: 1994.
Congressional District: 1
Institution Type/Description: Historical Society Museum.
Collections: local history & culture; period furnishings; personal artifacts;
photographs.
Hours & Admission Prices: Tues. & Fri. 10-3. No charge; donations accepted.
&
Attendance: 250 (estimated)

D.W. CORRELL MUSEUM, 19934 E. Pine St., Catoosa, OK 74015.
Mailing Address: P.O. Box 190, Catoosa, OK 74015. Tel.: 918-
266-3612.
Institution Type/Description: History Museum.
Collections: local history & culture; early automobiles including 1898 Locomobile; 1902 Oldsmobile Run About; 1906 Cadillac; 1914 Dodge Touring
Car; 1914 Oldsmobile; 1917 Twin 6 Packard; 1919 Franklin; 1927 Stutz 8;
1930 Model A Ford Convertible; 1944 Ford 4-door; 1948 Dodge Coupe;
bottles; decanters; rocks; gems; minerals; sea shells; period newspapers &
toys.
Activities: group tours.
Hours & Admission Prices: Tues. & Thurs. 11-7, Wed. & Fri.-Sat. 9-5. Adults
18 & over $3, seniors 62 & over $2; military, school & nursing home groups
and youth 17 & under no charge. Closed government holidays.

Chandler

**LINCOLN COUNTY HISTORICAL SOCIETY AND MUSEUM
OF PIONEER HISTORY,** 717-719 Manvel Ave., Chandler, OK
74834-2842. Tel.: 405-258-2425. Fax: 405-258-2809.
E-mail: lincolncountyhs@sbcglobal.net
Web Site: pioneermuseumolh.org
Founded: 1954.
Congressional District: 6
Key Personnel: Dir. & Pres. (V), Diana Kinzey; Dir., Pam Anderson; Chm.
(V), Dee Douglas; Museum Shop Mgr., Carol Beckman.
Personnel Profile: Full-Time Paid 1; Part-Time Paid 1; Part-Time Volunteers 7.
Governing Authority: society. Tax-exempt: 501(c)(3).
Institution Type/Description: Local History Museum.
Collections: local artifacts, relics & history of pioneer families in 1897
building; cemetery records; county census on microfilm; land record;
county papers on microfilm; family files; old telephone books; school
yearbooks; manuscripts; memorabilia of Sheriff Bill Tilghman, the last of
the frontier Marshalls & Times man, Benny Kent, frontier movie news
photographer; hand operated printing press; military artifacts; mural depict-
ing growth in Lincoln County; doctor's office; school room.
Research Fields: genealogy.
Facilities: microfilm & microfilm reader & printer for county newspapers;
auditorium; fiche reader.
Activities: guided tours; lectures; permanent & temporary exhibitions.
Publications: quarterly newsletter; booklet, History of Lincoln County.
Hours & Admission Prices: Tues.-Sat. 10-4. No charge; donations accepted. &
Attendance: 3,954 (accurate)
Membership: Individual $15; Family $25; Patron $50; Sustaining & Corporate
$100.

Cheyenne

PIONEER MUSEUM, CITY PARK, 107 Pioneer Park Way, Chey-
enne, OK 73628. Mailing Address: P.O. Box 34, Cheyenne, OK
73628. Tel.: 580-497-3882.
E-mail: tdale55@aol.com; lmad43@hotmail.com
Web Site: www.rogermills.org
Founded: 1990.
Key Personnel: Dir., Barbara Little.
Personnel Profile: Full-Time Paid 1.
Institution Type/Description: Park & Museum Complex: consists of a 1903
One Room School, Pioneer Museum, Minnie Slief Community Museum,
Veterans Display, Santa Fe Depot, Chapel, & early 1900s Kendall House
Log Cabin.
Collections: Pioneer Museum: local history; period furnishings & clothing;
early farm equipment & machinery. School House: early furnishings;
books; period artifacts. Minnie Slief Community Museum: county history
& memorabilia; Veterans' artifacts including uniforms, photographs & war
memorabilia. Depot: local history; railroad furnishings & artifacts. Chapel:
church furnishings; stained glass windows; memorial path. Log Cabin:
early furnishings; local history.
Facilities: handicap & large vehicle parking.
Activities: Annual Event: Pioneer Day in September.
Hours & Admission Prices: Museum: March-Nov. Tues.-Sat. 10-4; I-Room
School: April, May, Sept. Oct.; call for hours. No charge; donations
accepted. &
Attendance: 2,000 (estimated)

Chickasha

GRADY COUNTY HISTORICAL SOCIETY, (M), 415 W. Chick-
asha Ave., Chickasha, OK 73018. Mailing Address: P.O. Box 495,
Chickasha, OK 73018. Tel.: 405-222-6480.
E-mail: gchistorical@att.net
Web Site: www.gradycountyhistorical.org
Founded: 1973.
Congressional District: 4
Governing Authority: Tax-exempt.
Institution Type/Description: Historical Society Museum: housed in the former
department store, The Dixie; built in 1907.
Collections: local history & culture; period furnishings; photographs; personal
artifacts; Geronimo Hotel murals by Peyraud; Mary H. Bailey Oklahoma
history books; military artifacts; medical equipment.
Facilities: Museum-related items for sale.
Activities: educational programs; groups tours.
Hours & Admission Prices: Mon.-Fri. 10-3. No charge. &
Membership: Student $10; Individual $20; Family $30; Contributing $75;
Business $150.

Claremore

J.M. DAVIS ARMS & HISTORICAL MUSEUM, 330 N. J.M.
Davis Blvd., Claremore, OK 74017-7066. Tel.: 918-341-5707. Fax:
918-341-5771.
E-mail: info@thegunmuseum.com
Web Site: www.thegunmuseum.com
Founded: 1965.
Congressional District: 2
Key Personnel: Exec. Dir., Wayne McCombs; Cur., Jason Schubert; Chm. (V),
William Higgins; Tourism Coord., Kimberly Thompson.
Personnel Profile: Full-Time Paid 5; Part-Time Paid 5.
Governing Authority: state. Parent Institution: State of Oklahoma. Tax-exempt:
170(b)(1)(A).
Institution Type/Description: Firearms & Historical Museum.
Collections: 35,000 artifacts; 13,000 firearms; western memorabilia; Native
American artifacts; statuaries; knives and swords; World War I posters;
music boxes from 1880's; steins from around the world; replica 1800's
gunsmith shop & gun store.
Research Fields: firearms; edged weapons; steins.
Facilities: 2,050-vol. library of research books available for use on premises or
removed by special permission; reception room. Indian-made items, books
& other museum-related items for sale.
Activities: guided tours; lectures; special events. Children's Museum: teaching
gun safety; classic western movies; arcade.
Publications: brochures & museum legends.
Hours & Admission Prices: Mon.-Sat. 8:30-5, Sun. 10-5. No charge; donations
requested. Closed Thanksgiving; Christmas. &

WILL ROGERS MEMORIAL MUSEUM, (M), 1720 W. Will Rogers Blvd., Claremore, OK 74017-3208. Mailing Address: P.O. Box 157, Claremore, OK 74018-0157. Tel.: 918-341-0719. Fax: 918-343-8119.
E-mail: wrinfo@willrogers.org
Web Site: www.willrogers.org
Founded: 1938.
Congressional District: 2
Key Personnel: Dir., Steven K. Gragert; Museum Shop Mgr., Julie Luna.
Personnel Profile: Full-Time Paid 7; Part-Time Paid 11; Part-Time Volunteers 88; Interns 1.
Governing Authority: state. Parent Institution: Will Rogers Memorial Commission. Subsidiary Institution: Will Rogers Birthplace. Tax-exempt: 170(b)(1)(A).
Institution Type/Description: History Museum.
Collections: statue of Will Rogers by Jo Davidson; personal items; documents; saddle collection; dioramas; archives; sculpture; history; manuscripts; paintings by Charles Banks Wilson, Torres Rojas, Count Tambourini & Wayne Cooper; Will Rogers films.
Research Fields: life & career of Will Rogers; Will Rogers' genealogy.
Facilities: 2,500-vol. library of books, films, manuscripts available for research by appointment only; 175-seat theater. Copies of Will Rogers' writing & other museum-related items for sale.
Activities: guided tours; lectures; films; permanent exhibitions; gallery talks by appointment.
Publications: booklet, Will Rogers Memorial Booklet; copies of all Will Rogers' writings; books, Will Rogers & Wiley Post in Alaska; Radio Broadcasts of Will Rogers; There's Not a Bathing Suit in Russia; Ether and Me; Never Met a Man I Didn't Like; The Life and Writings of Will Rogers; Will Rogers at the Ziegfeld Follies; the Genealogy of Will Rogers; The Papers of Will Rogers.
Hours & Admission Prices: Daily 8-5. Adults $5, seniors 62 & over and military $4; members and youth 17 & under. &
Attendance: 152,026 (accurate)
Membership: Student $20; Senior $25; Individual $30; Family & Senior Couple $45; Roper $100; Performer $250; Communicator $500; Ambassador $750; Will's Lariat of Friends $1,000.

Clinton

CHEYENNE CULTURAL CENTER, 2250 N.E. Rte. 66, Clinton, OK 73601. Tel.: 580-323-6224.
Institution Type/Description: Native American History Museum.
Collections: Cheyenne culture & history; personal artifacts; Native American artifacts; photographs; hands-on exhibits.
Activities: hands-on exhibits; educational programs.
Hours & Admission Prices: Call for hours.

MOHAWK LODGE INDIAN STORE, 22702 Rte. 66 N., Clinton, OK 73601-7526. Tel.: 580-323-2360.
Founded: 1892.
Key Personnel: Owner, Pat Henry
Institution Type/Description: American Indian Museum: trading post opened in 1892.
Collections: Plains & western tribe history; 1890s Native American artifacts; Indian art & crafts supplies.
Hours & Admission Prices: Mon.-Sat. 9-5. No charge; donations accepted. &

OKLAHOMA ROUTE 66 MUSEUM, 2229 W. Gary Blvd., Clinton, OK 73601-5305. Tel.: 580-323-7866. Fax: 580-323-2870.
E-mail: rt66mus@okhistory.org
Web Site: www.route66.org
Founded: 1967.
Congressional District: 6
Key Personnel: C.E.O., Bob Blackburn; Dir., Pat Smith; Cur., Andy Watson.
Personnel Profile: Full-Time Paid 4; Part-Time Paid 2; Part-Time Volunteers 20.
Governing Authority: Parent Institution: Oklahoma Historical Society. Tax-exempt.
Institution Type/Description: History Museum.
Collections: historical artifacts.
Research Fields: history of Rte. 66 in Oklahoma.
Facilities: library & archives; park with picnic facilities; meeting room. Museum-related items for sale.
Activities: permanent & temporary exhibitions.
Hours & Admission Prices: Mon.-Sat. 9-5, Sun. 1-5. Adults $4, senior citizens $3, students 6-18 $1. &
Attendance: 34,000 (accurate)
Membership: Friends of Oklahoma Rte. 66 Museum: Individual $15; Family

$25; Student 18 & under, Seniors 55 & over $10; Organizations & Clubs $50; Corporate $100.

Coalgate

COAL COUNTY HISTORICAL & MINING MUSEUM, 212 S. Broadway, Coalgate, OK 74538. Tel.: 580-927-2360.
Institution Type/Description: History Museum.
Collections: mining history, tools & equipment; photographs; period artifacts; early cemetery & funeral home records; tax rolls from 1910-1930.
Facilities: Museum-related items for sale.
Activities: guided tours.
Hours & Admission Prices: Tues.-Thurs. 10-3:30, Sat. 10-1. No charge. &

Colbert

COLBERT HISTORICAL MUSEUM, 100 N. Burney, Colbert, OK 74733. Mailing Address: P.O. Box 1299, Colbert, OK 74733. Tel.: 580-296-2385.
E-mail: wyotahannan@brightok.net
Founded: 1979.
Congressional District: 2
Volunteer Hours: 200
Governing Authority: Tax-exempt.
Institution Type/Description: History Museum.
Collections: local history & culture; early pioneers; photographs; documents.
Hours & Admission Prices: By appointment. No charge; donations accepted. &
Attendance: 100 (estimated)
Membership: Individual $5.

Colcord

TALBOT LIBRARY AND MUSEUM, 500 S. Colcord Ave., Colcord, OK 74338. Mailing Address: P.O. Box 349, Colcord, OK 74338. Tel.: 918-326-4532.
E-mail: talbotlibrary@earthlink.net
Web Site: talbotlibrary.org
Governing Authority: nonprofit organization. Tax-exempt: 501(c)(3).
Institution Type/Description: Library.
Collections: books; period furnishings; personal artifacts. Historic Building: Springtown School, c.1920.
Hours & Admission Prices: Wed.-Sat. 10-4.

Collinsville

COLLINSVILLE DEPOT MUSEUM, 115 S. 10th St., Collinsville, OK 74021-3124. Mailing Address: 337 N. 20th St., Collinsville, OK 74021. Tel.: 918-371-3540.
E-mail: wttkitty@sbcglobal.net
Formerly: Santa Fe Railroad Depot & Caboose
Founded: 1975.
Congressional District: 1
Key Personnel: Pres. Historical Society, William Terrill Thomas; Dir., Ronald G. Evans; Owner Collinsville Newspaper, Ted Wright; Museum Shop Mgr., Thomas Evans Wright.
Governing Authority: nonprofit organization. Collinsville Historical Society. Tax-exempt.
Institution Type/Description: Antiques Museum: housed in 90-year-old depot.
Collections: early 1800 living room & kitchen; local & railroad artifacts; World War I & II uniforms, photos; caboose; memorial brick sidewalk; time capsule in burial vault; city court docket; bank ledger; lodge photographs; steam clothing press; city attorney's organ; newspaper articles; lead & zinc smelters; player piano; farm tools; military uniforms.
Activities: guided tours; arts festivals; permanent exhibitions; scout, senior & nonprofit groups.
Hours & Admission Prices: by appointment only. No charge; donations accepted. &
Attendance: 300 (estimated)
Membership: $15 per person or family; Life $100.

NEWSPAPER MUSEUM, 1110 W. Main St., Collinsville, OK 74021-3113. Tel.: 918-371-1901.
Web Site: www.cvilleok.com/museum.html
Institution Type/Description: History Museum.
Collections: artifacts & memorabilia pertaining to The Collinsville News.
Hours & Admission Prices: By appointment. Tours: adults $1.

Cordell

WASHITA COUNTY MUSEUM, 115 E. 1st St., P.O. Box 153, Cordell, OK 73632. Tel.: 580-832-3681.
Founded: 1968.
Congressional District: 3
Key Personnel: C.E.O., Wayne Boothe; Chm. (V), Landon Jones; Pres. (V), Lavern Berry.
Governing Authority: Parent Institution: Washita County Historical Society. Tax-exempt 501 (c)(5)
Institution Type/Description: History Museum: housed in the Carnegie Library; built in 1911.
Collections: local history & culture; horse-drawn implements; 1890 Chuck wagon; period furnishings; household artifacts. Historic Buildings: 1893 house; barn.
Activities: guided tours.
Hours & Admission Prices: Fri. 2-4; other times by appointment. No charge.
Attendance: 800 (estimated)
Membership: Annual $9; Life $50.

Coweta

MISSION BELL MUSEUM, 204 S. Bristow Ave., Coweta, OK 74429-2301. Mailing Address: P.O. Box 850, Coweta, OK 74429-0850. Tel.: 918-486-2189. Fax: 918-486-2513.
Founded: 1977.
Congressional District: 2
Personnel Profile: Part-Time Volunteers 1.
Governing Authority: nonprofit organization. Tax-exempt.
Institution Type/Description: Historical Society Museum: housed in a former Presbyterian Church; built in 1907.
Collections: local history & culture; photographs; period furnishings.
Activities: guided tours.
Hours & Admission Prices: Tues. & Thurs. open four hours & by appointment only. No charge; donations accepted.
Attendance: 200 (estimated)
Membership: Historical Society $4.

Crescent

FRONTIER COUNTRY MUSEUM, 500 N. Grand St., Crescent, OK 73028. Mailing Address: P.O. Box 856, Crescent, OK 73028. Tel.: 405-969-3660. Fax: 405-969-3660.
Web Site: www.frontiercountrymuseum.org
Founded: 2000.
Congressional District: 3
Key Personnel: Pres. (V), Sandra Voskuhl; Museum Shop Mgr., Carol Oltmanns.
Governing Authority: Parent Institution: Frontier Country Historical Society. Tax-exempt.
Institution Type/Description: History Museum.
Collections: local history & culture; photographs; personal artifacts; period furnishings; military artifacts; medical equipment; barbershop; general store; one room schoolhouse; hands-on exhibits.
Facilities: Museum-related items for sale.
Activities: group tours.
Hours & Admission Prices: Tues.-Sat. 10-4, Sun. 1-4. Adults $5, seniors 55 & over $3, children 1st to 12 grade $2; members no charge. &
Membership: Junior $5; Regular $25; Family $50; 89er $100-$999; Pioneer $1,000-$4,999; Homesteader $5,000-$9,999; Oil Baron $10,000 & up.

Cushing

RICHARD O. DODRILL'S MUSEUM OF ROCKS, MINERALS & FOSSILS, 123 S. Cleveland, Cushing, OK 74023. Tel.: 918-225-0662.
Institution Type/Description: History Museum.
Collections: rocks; minerals; fossils; period artifacts; local history; Native American artifacts; photographs.
Facilities: Museum-related items for sale.
Activities: guided tours.
Hours & Admission Prices: Tues.-Sat. 10-4. No charge. &

Cyril

CYRIL MUSEUM, Main St. & Hwy. 277, Cyril, OK 73029. Mailing Address: P.O. Box 346, Cyril, OK 73029. Tel.: 580-464-2547.
Institution Type/Description: History Museum.
Collections: local history & culture; period furnishings; personal artifacts; photographs; bank & school memorabilia; military uniforms; barber shop; household items; china & crystal; city records. Historic Building: 1910 house.
Hours & Admission Prices: By appointment. No charge.

Davis

ARBUCKLE HISTORICAL SOCIETY MUSEUM, 12 Main St., Davis, OK 73030. Tel.: 580-369-2518.
Institution Type/Description: Historical Society Museum.
Collections: local history & culture; period furnishings; personal artifacts; Native American artifacts; early clothing; family histories; military artifacts; photographs.
Hours & Admission Prices: Daily 10-4. No charge.

Dewey

DEWEY HOTEL MUSEUM, 801 N. Delaware, Dewey, OK 74029-1609. Mailing Address: P.O. Box 255, Bartlesville, OK 74005-0255. Tel.: 918-534-0215.
Founded: 1899.
Congressional District: 2
Key Personnel: Pres. (V), Gary Mackey.
Personnel Profile: Full-Time Paid 1; Part-Time Paid 1; Part-Time Volunteers 12; Interns 5.
Governing Authority: nonprofit; society. Parent Institution: Washington County Historical Society, Inc.
Institution Type/Description: Historic Building: 1899 Victorian-style wood frame hotel built for Jacob H. Bartles.
Collections: furniture & furnishings from 1890-1910; photographs from 1880-1930; Delaware, Cherokee & Osage Indian artifacts; white furniture & furnishings; western articles including saddles, chaps, & spurs.
Research Fields: period clothing, furniture & furnishing.
Activities: guided tours; lectures; meeting room.
Publications: newsletter.
Hours & Admission Prices: April-Nov. daily 10-4, Sun. 1-4. Adults $1; members & children under 13 with an adult no charge. Closed holidays.
Attendance: 6,000 (accurate)
Membership: Annual $10; Life $250.

TOM MIX MUSEUM, 721 N. Delaware, Dewey, OK 74029-2307. Mailing Address: P.O. Box 190, Dewey, OK 74029-0190. Tel.: 918-534-1555.
E-mail: tommix@cableone.net
Web Site: tommixmuseum.com
Founded: 1968.
Congressional District: 2
Key Personnel: Museum Shop Mgr., Fawn Lassiter; Museum Shop Asst. Mgr., Karen Michno.
Personnel Profile: Part-Time Paid 2.
Governing Authority: state. Parent Institution: Oklahoma Historical Society (see separate listing). Tax-exempt.
Institution Type/Description: History Museum.
Collections: personal belongings of silent movie star, Tom Mix; his career & horse including clothing, saddles, & guns.
Research Fields: life of Tom Mix.
Facilities: small movie theater.
Activities: tours; permanent exhibitions; Tom Mix films; cowboy program; scavenger hunts. Museum Sponsors: Annual Festival.
Publications: quarterly newsletter.
Hours & Admission Prices: Jan.-Feb. Thurs.-Sat. 10-4:30; March-Dec. Tues.-Sat. 10-4:30. Suggested Donation: adults $3, children $.50. Closed legal holidays. &
Attendance: 5,500 (estimated)
Membership: Trail Rider (student) $15; Range Rider (individual) $20; Posse (family) $25; Deputy Marshal (silver membership) $50; Marshal (gold membership $100; Lifetime Membership $500.

Drumright

DRUMRIGHT COMMUNITY HISTORICAL MUSEUM, 301 E. Broadway, Drumright, OK 74030-3805. Tel.: 918-352-3002.
Web Site: www.drumrighthistoricalsociety.org
Founded: 1969.
Congressional District: 3
Personnel Profile: Part-Time Paid 1.
Governing Authority: Tax-exempt.
Institution Type/Description: History Museum: housed in a 1916 Santa Fe Depot. Listed on the National Register of Historic Places.

Collections: local history; oil industry; equipment, photographs; period artifacts.
Hours & Admission Prices: Wed.-Fri. 12-4, Sat. 9-5; groups by appointment. No charge; donations accepted.
Attendance: 1,000 (estimated)
Membership: Member $10; Lifetime $500; Corporate Sponsor $250.

Duncan

CHISHOLM TRAIL HERITAGE CENTER, 1000 N. Chisholm Trail Pkwy., Duncan, OK 73533-1539. Tel.: 580-252-6692 & 6563. Fax: 580-252-6567. Facebook: On the Chisholm Trail.
E-mail: info@onthechisholmtrail.com
Web Site: www.onthechisholmtrail.com
Formerly: On The Chisholm Trail Statue & Museum
Founded: 1998.
Congressional District: 4
Key Personnel: Exec. Dir., Stacy Cramer Moore; Bd. Pres., Marilyn Hugon; Museum Shop Mgr., Syvonna Davis.
Personnel Profile: Full-Time Paid 3; Part-Time Paid 14; Part-Time Volunteers 1.
Governing Authority: private; nonprofit organization. Parent Institution: On the Chisholm Trail Association. Tax-exempt: 501(c)(3).
Institution Type/Description: History Museum.
Collections: hands-on exhibits; cowboy, cavalry & Native American clothing & personal artifacts; photographs; artifacts & memorabilia of the Chisholm & other cattle trails; sculptures including a full-scale bronze cattle drive scene.
Facilities: 3,000 sq. ft. exhibit space; 45-seat theater; Statuary Park. Museum-related items for sale.
Activities: films; guided tours; lectures; loan exhibitions. Annual Event: Western Spirit Celebration in September.
Publications: newsletter.
Hours & Admission Prices: Mon.-Sat. 10-5, Sun. 1-5. Adults $6, seniors 55 & over $5, youth 5-17 $4; discount to groups. Closed Thanksgiving; Christmas.
Attendance: 20,000 (estimated)
Membership: Individual $30; Family $50; Family & Friends $100; Sustaining $250; Patron $500; Corporate $500 & up; Benefactor $1,000.

STEPHENS COUNTY HISTORICAL SOCIETY AND MUSEUM, Hwy. 81 & Beech, Fuqua Park, Duncan, OK 73533. Mailing Address: P.O. Box 1294, Duncan, OK 73534-1294. Tel.: 580-252-0717; 800-782-7167. Fax: 580-251-3195.
Founded: 1971.
Congressional District: 3
Key Personnel: C.E.O., John Jennings; Chm. & Asst. Dir. (V), Vickie Zimmerman; Dir., Pee Wee Cary; Assoc. Dir., Louise Elliott; Assoc. Dir., Marge Rigdon; Assoc. Dir., Sharleen Johns; Museum Shop Mgr., Patty Woolf.
Personnel Profile: Full-Time Paid 1; Part-Time Volunteers 7.
Governing Authority: municipal & county. Tax-exempt: 501(c)(3) & 170(b).
Institution Type/Description: History Museum.
Collections: anthropology; archaeology; paintings; sculpture; graphics; decorative art; costumes; ethnology; geology; American Indian artifacts; mineralogy; numismatics; panorama of the oil industry depicting exploration, drilling, completion & refining of oil; manuscript collection; dioramas, 3 dimensional models of research & development 1920-1980; philatelic; technical; transportation; gems & lapidary.
Research Fields: oil industry, Halliburton Services; Sun Petroleum; Mack Oil Co.
Facilities: library of material on Oklahoma history; genealogical research; Halliburton Cementers available on premises only.
Activities: guided tours; lectures; films; gallery talks; custom hand quilting by special order; arts festivals; study clubs; formally organized education programs for adults, children, undergraduate & graduate students; intermuseum, permanent & temporary exhibitions; archives of pioneer historical research.
Publications: Pictorial history of Stephens County.
Hours & Admission Prices: Tues. & Thurs.-Sat. 1-5. No charge; donations accepted. Closed New Year's Eve & Day; Easter; Thanksgiving; Christmas.
Attendance: 6,894 (accurate)
Membership: Individual $15; Family $25; Organizations $35; Reunion $100; Sustaining $250-$500; Life $500.

W.T. FOREMAN PRAIRIE HOUSE FOUNDATION, 814 W. Oak, Duncan, OK 73533. Mailing Address: P.O. Box 2094, Duncan, OK 73534. Tel.: 580-252-1780 & 251-0027. Facebook: The Prairie House.
E-mail: gloafman@cableone.net
Web Site: theprairiehouse.com
Founded: 2002.
Key Personnel: Dir., Kathy Smith; C.E.O., Gail Loafman.
Personnel Profile: Full-Time Volunteers 12; Part-Time Volunteers 20.
Governing Authority: Tax-exempt: 501(c)(3).
Institution Type/Description: Historic House Museum: built in 1918. Listed on the National Register of Historic Places.
Collections: local history; period furnishings; personal artifacts; photographs.
Activities: group tours; special events.
Hours & Admission Prices: Tues. & Thurs. 1-4; other times by appointment. No charge; donations accepted.
Membership: Friend $35-$50; Patron $51-$100; Sponsor $101-$499; Founder $500 & up.

Durant

FORT WASHITA, 3348 State Rd. 199, Durant, OK 74701-9443. Tel.: 580-924-6502.
E-mail: ftwashita@okhistory.org
Web Site: www.okhistory.org/sites/fortwashita
Founded: 1967.
Congressional District: 3
Key Personnel: Site Mgr., Jim Argo; Historic Interpreter, Ron Petty.
Governing Authority: state. Parent Institution: Oklahoma Historical Society. (See separate listing). Tax-exempt.
Institution Type/Description: Historic Site: 1842-1865 frontier military fort.
Collections: frontier maps; ruins; restoration of historical building.
Activities: military & civilian living history demonstrations.
Publications: quarterly newsletter, Chronicles of Oklahoma; monthly newsletter, Mistletoe Leaves.
Hours & Admission Prices: Tues.-Sat. 9-4:30, Sun. 1-4:30. No charge; donations accepted. Closed national holidays.
Attendance: 25,000 (accurate)
Membership: Annual $10; Lifetime $200.

THREE VALLEY MUSEUM, 401 W. Main St., Durant, OK 74701-5026. Tel.: 580-920-1907.
E-mail: tvm@netcommander.com
Founded: 1976.
Key Personnel: Pres., Greg Phillips.
Governing Authority: Parent Institution: Durant Historical Society. Tax-exempt.
Institution Type/Description: History Museum.
Collections: Choctaw, Chickasaw & Cadolo Nations; Southeastern Oklahoma specifically Bryan County history. Historic Buildings: machine shop; Sinclair Station c.1940.
Hours & Admission Prices: Tues.-Fri. 1-5, Sat. 11-3; other times by appointment. No charge; donations accepted. Closed Thanksgiving; Christmas.
Attendance: 2,000 (estimated)
Membership: Students & Seniors $10; Singles & Couples $20.

Durham

BREAK O' DAY FARM AND METCALFE MUSEUM, EW 86, Durham, OK 73642. Mailing Address: 8647 N. 1745 Rd., Durham, OK 73642. Tel.: 580-655-4467. Fax: 580-655-4654.
E-mail: metcalfe@dobsonteleco.com
Web Site: www.metcalfemuseum.org
Congressional District: 3
Key Personnel: Dir., Roger Lester; Dir., Lloydelle Lester; Pres. (V), Janna Montgomery; Museum Shop Mgr., Becky Buster
Institution Type/Description: Art Museum.
Collections: paintings; wildlife habitat.
Facilities: nature trails. Museum-related items for sale.
Hours & Admission Prices: March-Nov. Tues.-Sat. 10-5. No charge; donations accepted. Closed holidays.

Edmond

EDMOND HISTORICAL SOCIETY MUSEUM, (M), 431 S. Boulevard, Edmond, OK 73034-3873. Tel.: 405-340-0078. Fax: 405-340-2771.
E-mail: edmondhistory@edmondhistory.org
Web Site: edmondhistory.org

Founded: 1983.
Congressional District: 5
Key Personnel: Exec. Dir., Jena Mottola; Pres. (V), Amy Bailey; Treas., Steve Foskin.
Personnel Profile: Full-Time Paid 4; Full-Time Volunteers 1; Part-Time Paid 3; Part-Time Volunteers 100; Interns 1.
Operating Expenses: 244,133
Operating Income: 266,841
Governing Authority: private; nonprofit organization. Tax-exempt: 501(c)(3).
Institution Type/Description: Historical Society Museum: housed in 1936 native stone armory.
Collections: concentration on the history of Edmond & the surrounding area from prehistoric to modern times.
Research Fields: Edmond 1889ers who made the land run into Oklahoma in 1889; Route 66 in Edmond.
Facilities: research library; 7,500 sq. ft. exhibit space; children's hands-on center.
Activities: docent program; formal education program; scholarly lectures; guided tours; workshops; temporary & traveling exhibitions. Annual Events: 1889er Homestead Fair; Heritage Dinner, Preservation Awards & Roll of Honor.
Publications: triannual newsletter.
Hours & Admission Prices: Tues.-Fri. 10-5, Sat. 1-4. No charge; donations accepted. Closed major holidays. &
Attendance: 15,000 (accurate)
Membership: Student & Super Senior 75 & over $10; Senior 62-74 $20; Individual $35; Family $50; Contributing $90; Business $175.

LABORATORY OF HISTORY MUSEUM, (M), University of Central Oklahoma, Department of History & Geography, 100 N. University Dr., Edmond, OK 73034-5207. Tel.: 405-974-4669.
Founded: 1915.
Key Personnel: Dir., Heidi Vaughn.
Governing Authority: Parent Institution: University of Central Oklahoma. Tax-exempt.
Institution Type/Description: History Museum.
Collections: local history & culture.
Hours & Admission Prices: Call for hours. No charge; donations accepted.

El Reno

CANADIAN COUNTY HISTORICAL MUSEUM, 300 S. Grand, El Reno, OK 73036-3610. Tel.: 405-262-5121. Fax: 405-262-9397.
E-mail: vptrishane@aol.com
Web Site: www.elreno.org
Founded: 1969.
Congressional District: 5
Key Personnel: Pres., Vicki Proctor; Cur., Pat Reuter; Gift Shop Mgr., Marguerite Stoakes.
Personnel Profile: Full-Time Paid 1; Full-Time Volunteers 1; Part-Time Paid 2; Part-Time Volunteers 5.
Governing Authority: historical society; nonprofit organization. Tax-exempt.
Institution Type/Description: Historical Society Museum: housed in 1906 Rock Island Railway Station on 98th Meridian.
Collections: railway artifacts; Native American artifacts & history; Ft. Reno; Darlington Indian Agency; pioneer furniture & kitchen; barbed wire; tools; vehicles; fire fighting equipment; U.S. Reformatory; Red Cross Canteen; farm machinery; business equipment; clothing; rural school; mounted wildlife; Canadian County cowboys.
Research Fields: county history; rural schools; local towns; ghost towns.
Facilities: 200-vol. library of historical material for use on premises. Indian-related gift items for sale (gift shop located in El Reno Hotel).
Activities: guided tours; lectures; temporary exhibits; historical programs; trolley rides.
Publications: letters.
Hours & Admission Prices: Wed.-Sat. 9-5, Sun. 1-5. No charge; donations accepted. Closed New Year's Day; Independence Day; Thanksgiving; Christmas. &

Membership: Youth Under 12 $2; Annual $10; Life $100.

FORT RENO VISITOR CENTER, 7107 W. Cheyenne St., El Reno, OK 73036. Tel.: 405-262-3987. Fax: 405-262-0133.
E-mail: info@fortreno.org
Web Site: www.fortreno.org
Institution Type/Description: Visitor Center.
Collections: local history; photographs; personal artifacts; period furnishings.
Hours & Admission Prices: Daily 10-4.

Elk City

ANADARKO BASIN MUSEUM OF NATURAL HISTORY, 204 N. Main St., Elk City, OK 73644-4754. Tel.: 580-243-0437.
Institution Type/Description: Natural History Museum: housed in the former Casa Grande Hotel; built in the late 1920s. Listed on the National Register of Historic Places.
Collections: local history; minerals; fossils; natural history; rocks; petrified wood; Native American artifacts; photographs; paintings; beadwork; oil & gas memorabilia; oil & gas exploration and production equipment.
Hours & Admission Prices: Temporarily closed.

NATIONAL ROUTE 66 MUSEUM AND OLD TOWN COMPLEX, 2717 W. Hwy. 66, Elk City, OK 73644. Mailing Address: P.O. Box 5, Elk City, OK 73648-0005. Tel.: 405-225-6266. Fax: 580-225-3234.
E-mail: jacksom@elkcity.com
Web Site: www.elkcity.com
Founded: 1966.
Congressional District: 6
Key Personnel: Chm., L.V. Baker, Jr.; Dir., Basil Weatherly; Cur., Wanda Queenan; Museum Shop Mgr., Maxine Jackson.
Personnel Profile: Full-Time Paid 5; Part-Time Paid 3; Part-Time Volunteers 25.
Governing Authority: municipal; nonprofit organization. Tax-exempt.
Institution Type/Description: General Museum.
Collections: pioneer artifacts; costumes; glass; medical equipment; Rock Bluff School; caboose & depot; wagon yard; pioneer chapel; schoolhouse; Western rodeo; Indian artifacts; period cars; motorcycles; farming & ranching; windmills; blacksmith shop; walking plows & cultivators; hand tools; barbed wire; 1919 Rumely tractor; 1936 John Deere tractor Model D; broomcorn. Historic Buildings: 1912 museum building; grist mill; bank. Store Fronts: mercantile, furniture, grocery, jewelry, barber shop, land office, post office, doctor's office, opera house.
Facilities: 400-vol. library of school & history books; 200-seat Opera House.
Activities: guided tours; summer series. Museum Sponsors: old-fashioned Christmas.
Publications: newspaper, weekly feature in local newspaper.
Hours & Admission Prices: Memorial Day to Labor Day Mon.-Sat. 9-5, Sun. 2-5; Sept.-May Mon.-Sat. 9-5, Sun. 2-5. Adults $5, seniors & children over 6 $4; discounts to groups, AAA members, bus tours, members, senior & school groups. Closed New Year's Day; Easter; Thanksgiving; Christmas Eve & Day. &
Attendance: 35,000 (accurate)

TRANSPORTATION MUSEUM, Old Hwy. 66 N., Elk City, OK 73648. Mailing Address: P.O. Box 5, Elk City, OK 73648. Tel.: 580-225-6266.
Institution Type/Description: Transportation History Museum.
Collections: transportation history; hands-on exhibitions; cars; motorcycles; 1917 RIO fire truck.
Hours & Admission Prices: Mon.-Sat. 9-5, Sun. 2-5; other times by appointment.

Enid

CHEROKEE STRIP REGIONAL HERITAGE CENTER, (M), 507 S. 4th St., Enid, OK 73701-5835. Tel.: 580-237-1907. Fax: 580-234-1055.
E-mail: aholland@okhistory.org
Web Site: csrhc.org
Formerly: Museum of the Cherokee Strip
Founded: 1951.
Congressional District: 3
Key Personnel: Pres., Andrea Holland.
Personnel Profile: Full-Time Paid 6; Part-Time Paid 5; Part-Time Volunteers 64; Interns 1.
Governing Authority: state. Parent Institution: Oklahoma Historical Society.
Institution Type/Description: Regional History Museum.
Collections: farm implements; Native American artifacts; regional history; 1893 land run & development of Northwest Oklahoma.
Research Fields: regional history & opening of Cherokee Outlet.
Facilities: library; auditorium; theater; research center.
Activities: guided tours; permanent & traveling exhibitions; special events.
Publications: newsletter, Deeds of '93; Journal of the Cherokee Strip Pioneers.
Hours & Admission Prices: Museum: Tues.-Sat. 10-5, Sun. 1-5. Families $13, adults $5, seniors 62 & over, students $3; discounts to groups; children 5 & under, members and active duty military & veterans no charge. Closed New Year's Day; Easter; Thanksgiving; Christmas. &

Attendance: 15,000 (accurate)
Membership: International $30; Family $50.

HUMPHREY HERITAGE VILLAGE, 507 S. 4th St., Enid, OK 73701. Tel.: 580-237-1907.
Web Site: regionalheritagecenter.org
Institution Type/Description: Historic Village Museum.
Collections: local history & culture; period furnishings. Historic Buildings: Victorian home; 1893 office; schoolhouse; church.
Hours & Admission Prices: Tues.-Fri. 9-5, Sat.-Sun. 2-5; other times by appointment. No charge. Closed state holidays.

LEONA MITCHELL SOUTHERN HEIGHTS HERITAGE CENTER AND MUSEUM, 616 Leona Mitchell Blvd., Enid, OK 73701. Tel.: 580-237-6989.
E-mail: achoctaw1866@aol.com
Founded: 2001.
Key Personnel: Exec. Dir., Barbara Finley.
Governing Authority: Tax-exempt: 501(c)(3).
Institution Type/Description: Heritage Center.
Collections: local history & culture; Native American artifacts; African American artifacts; photographs; personal artifacts.
Activities: educational programs.
Hours & Admission Prices: Tues.-Fri. 10-4, Sat. 10-2; other times by appointment. Adults $8, children 5-17 $5.

LEONARDO'S CHILDREN'S MUSEUM, 200 E. Maple, Enid, OK 73701. Mailing Address: P.O. Box 348, Enid, OK 73702. Tel.: 580-233-2787. Fax: 580-237-7574.
Web Site: www.leonardos.org
Formerly: Leonardo's Discovery Warehouse
Founded: 1992.
Congressional District: 3
Key Personnel: Dir., Julie Baird; Chm (V), Jill Phillips; Museum Shop Mgr., Shelly Conrad
Governing Authority: Parent Institution: Enid Arts & Science Foundation. Tax-exempt: 501(c)(3).
Institution Type/Description: Children's Museum.
Collections: hands-on art & science exhibits; live animals.
Facilities: Museum-related items for sale.
Activities: birthday parties; facility rentals; internships; educational programs; school visits; summer camp.
Hours & Admission Prices: Mon.-Sat. 10-5, Sun. 1-5. Admission $7; discounts to groups; children under 2 no charge.
Attendance: 96,000 (accurate)

MIDGLEY MUSEUM, 1001 Sequoyah Dr., Enid, OK 73703. Tel.: 580-234-7265.
Institution Type/Description: Historic House: housed in a home built of petrified wood & rock.
Collections: rocks; crystals; minerals; big-game trophy heads; photographs.
Facilities: Museum-related items for sale.
Activities: guided tours.
Hours & Admission Prices: Wed.-Fri. 1-5, Sat. 2-5. No charge; donations accepted.

RAILROAD MUSEUM OF OKLAHOMA, 702 N. Washington, Enid, OK 73701-3138. Tel.: 580-233-3051.
Governing Authority: nonprofit organization. Tax-exempt: 501(c)(3).
Institution Type/Description: Railroad Museum.
Collections: railroading history, equipment & memorabilia; circus train; cabooses; train cars; model trains; railroad dining car china & silverware; photographs; videos.
Facilities: library. Museum-related items for sale.
Hours & Admission Prices: Tues.-Fri. 1-4, Sat. 9-1, Sun. 2-5. Suggested Donations: $3 per person. Closed holidays.

Erick

ROGER MILLER MUSEUM, Corner of Roger Miller Blvd. & Sheb Wooley Ave., Erick, OK 73645. Mailing Address: P.O. Box 464, Erick, OK 73645-0464. Tel.: 580-526-3833.
E-mail: rsnowden55@yahoo.com
Web Site: www.rogermillermuseum.com
Institution Type/Description: History Museum.
Collections: artifacts & memorabilia pertaining to singer & songwriter Roger Miller.

Hours & Admission Prices: Wed.-Sat. 10-5, Sun. 1-5; other times by appointment.

Fairview

MAJOR COUNTY HISTORICAL SOCIETY, State Hwy. 58, Fairview, OK 73737. Mailing Address: P.O. Box 555, Fairview, OK 73737. Tel.: 580-227-2265.
E-mail: office@mchsok.net
Web Site: www.mchsok.net
Institution Type/Description: Historical Society Museum.
Collections: local history & culture; early farming; farm machinery & equipment; period furnishings; personal artifacts; photographs; 2522 locomotive; caboose; pioneer life. Historic Buildings: Fairview train depot; Wohlgemuth house; two churches; sawmill; blacksmith shop; one room schoolhouse.
Activities: special events.
Hours & Admission Prices: Tues.-Fri. 1-5, Sat. 10-2. No charge; donations accepted. Closed holidays.
Membership: Single $15; couples $25.

Fort Gibson

FORT GIBSON HISTORIC SITE, 907 N. Garrison Ave., Fort Gibson, OK 74434. Mailing Address: P.O. Box 457, Fort Gibson, OK 74434-0457. Tel.: 918-478-4088. Fax: 918-478-4089.
E-mail: fortgibson@okhistory.org
Web Site: www.fortgibson.com/historical_sites.htm
Founded: 1936.
Congressional District: 2
Key Personnel: Dir., David Fowler; Museum Shop Mgr., Omar Reed.
Personnel Profile: Full-Time Paid 4; Full-Time Volunteers 1; Part-Time Paid 1; Part-Time Volunteers 42.
Governing Authority: state. Parent Institution: Oklahoma Historical Society, Historical Bldg., Oklahoma City 73105. Tax-exempt.
Institution Type/Description: Military Museum Complex.
Collections: photographs; military tools, equipment & artifacts; accoutrements; weapons; replica of 1824 log fort; historic fortifications. Historic Structures: 1845 barracks; 1871 post hospital; 1867 adjutant's office; 1867 blacksmith's shop; 1863 powder magazine; 1846 commissary.
Research Fields: Oklahoma military history; Indian removal; Mexican War; Trail of Tears; American military history.
Facilities: visitor center. Books & museum-related items for sale.
Activities: tours; education programs. Special Events: Spring Bake Day in March; Spring Encampment in April; Fourth at the Fort in July; Fall Encampment in October; Fall Bake Day in November; Candlelight Tours in December.
Publications: Mistletoe Leaves.
Hours & Admission Prices: Tues.-Sat. 10-5. Adults $3, senior citizens $2.50, students 6-18 $1; discounts to AAM members; children 5 & under & veterans no charge.
Attendance: 60,432 (accurate)
Membership: Friends of Fort Gibson $25.

Fort Sill

FORT SILL NATIONAL HISTORIC LANDMARK AND MUSEUM, 437 Quanah Rd., Fort Sill, OK 73503-5100. Tel.: 580-442-5123. Fax: 580-442-8120. Facebook: Fort Sill National Historic Landmark and Museum.
Web Site: http://sill-www.army.mil/museum
Founded: 1934.
Congressional District: 4
Key Personnel: Dir., Scott A. Neel, Ph.D.
Personnel Profile: Full-Time Paid 4; Part-Time Volunteers 20.
Governing Authority: federal government. Parent Institution: U.S. Army. Affiliated with U.S. Army Center of Military History. Tax-exempt: 501(c)(3).
Institution Type/Description: Military Museum: housed in the 1869-75 original stone buildings of Fort Sill's Indian Territory.
Collections: U.S. Cavalry, Infantry & Indian items from Western frontier period; military ordnance; uniforms; equipment; horse furnishings; vehicles; Native American artifacts; paintings; photographs & archives; African American Buffalo Soldiers in the west; 27 historic buildings of frontier army posts including guard house & cavalry barracks.
Research Fields: history of frontier army; South Plains frontier; Kiowa, Apache & Comanche Indians; frontier law enforcement.
Facilities: 10,000-vol. reference library & 60,000 photo archive on history & development of U.S. Field Artillery, Fort Sill & the South Plains frontier available for research by appointment; interpretive center.

Activities: guided tours; education programs; permanent exhibitions. Museum Sponsors: Annual Vintage Baseball Game; Candlelight Stroll; Frontier Army Days.
Hours & Admission Prices: Tues.-Sat. 8:30-5. No charge; donations accepted. Closed New Year's Eve & Day; Thanksgiving; Christmas Eve & Day. &
Attendance: 35,000 (estimated)

Fort Towson

FORT TOWSON HISTORIC SITE, HC 63, Fort Towson, OK 74735-9273. Mailing Address: P.O. Box 1580, Fort Towson, OK 74735-9273. Tel.: 580-873-2634. Fax: 580-873-9385.
E-mail: fttowson@okhistory.org
Web Site: www.okhistory.org/military/forttowson.html
Formerly: Fort Towson Military Park
Founded: 1972.
Congressional District: 3
Key Personnel: Site Mgr. & Dir., John Davis; Interim Cur., Keith Reese.
Personnel Profile: Full-Time Paid 3; Part-Time Volunteers 1.
Governing Authority: state. Parent Institution: Oklahoma Historical Society, Oklahoma City, OK (see separate listing). Tax-exempt.
Institution Type/Description: Historic Site & Ruins.
Collections: local history; personal artifacts; period furnishings; photographs.
Activities: tours & lectures by appointment.
Hours & Admission Prices: Mon.-Fri. 9-5, Sat.-Sun. 1-5. No charge; donations accepted. Closed state holidays.
Attendance: 29,454 (accurate)
Membership: Please see separate listing for Oklahoma Historical Society, Oklahoma City.

Frederick

CRAWFORD COLLECTION, 115 W. Main St., Frederick, OK 73542. Tel.: 580-335-3211.
Institution Type/Description: General Museum.
Collections: over 170 mounted animals including lions, bears, rhinoceros, python, & giraffe.
Activities: guided tours.
Hours & Admission Prices: 2nd Sat. each month 11-2. No charge. &

PIONEER HERITAGE TOWNSITE CENTER, 201 N. 9th St., Frederick, OK 73542. Tel.: 580-335-5844.
E-mail: pioneer@okhistory.org
Founded: 1977.
Institution Type/Description: History Museum: housed in a pioneer village including 8 structures.
Collections: local history, heritage & agriculture; period furnishings; personal artifacts; farm machinery & equipment; tractors; plows; grain separators; general store; railroad artifacts. Historic Buildings: 1902 Horse Creek schoolhouse; c.1900 Nill house; 1924 farm house; 1924 St. Paul Church; Frisco Depot.
Hours & Admission Prices: Mon.-Fri. 9-3, 2nd Sat. 10-1; other times by appointment. No charge.

Gage

JIM'S SCRAP METAL ART MUSEUM, Hwys. 46 & 15 S., Gage, OK 73843. Mailing Address: 205 2nd, Gage, OK 73843. Tel.: 580-923-7935.
Institution Type/Description: Art Museum.
Collections: eclectic scrap metal folk art & sculptures including giant insects, buffaloes, elephants, dinosaurs, & birds.
Hours & Admission Prices: By appointment.

Gate

GATEWAY TO THE PANHANDLE, Main St., Gate, OK 73844. Mailing Address: P.O. Box 27, Gate, OK 73844-0027. Tel.: 580-934-2004.
E-mail: emaphet@ptsi.net
Founded: 1975.
Congressional District: 6
Key Personnel: Dir., Pres. (V) & Museum Shop Mgr., L. Ernestine Maphet; Cur. & Librarian, Peggy Whiteman; Deputy Dir., Karen Bond; Deputy Dir., Valerie Spurgeon; Asst. Dir., Louise Hein.
Personnel Profile: Part-Time Paid 1; Part-Time Volunteers 9.
Governing Authority: nonprofit organization. Parent Institution: Gateway to the Panhandle Museum Association & Library. Tax-exempt: 501(c)(3).
Institution Type/Description: Historic Buildings.

Collections: farm implements & equipment; household items; dishes; Civil War artifacts; prehistoric bones; painted murals; historical Gate School & Gym; school artifacts; decorative arts. Historic Buildings: 1912 depot; 1912 school (listed on the National Register of Historic Places).
Research Fields: local history; genealogy.
Facilities: library of books available for the town & community use. Museum-related items for sale.
Activities: guided tours; permanent & temporary exhibitions.
Publications: book, Gate History & Vigilantes; Family Lost; Climb a Mountain; You Can't Buy a Home; Gate School History 1892-1992; Hangman's Tree; historical calendars.
Hours & Admission Prices: Mon.-Sat. 11-4, Sun. & holidays by appointment. No charge; donations accepted. Closed Thanksgiving; Christmas. &
Attendance: 450 (estimated)
Membership: Annual donations requested.

Geary

CANADIAN RIVERS HISTORICAL SOCIETY MUSEUM, 717 S. Broadway, Geary, OK 73040. Tel.: 405-884-2608.
Institution Type/Description: Historical Society Museum.
Collections: local history & culture; Rock Island caboose; early 1900 log jail; period furnishings; Native American artifacts; photographs; personal artifacts.
Hours & Admission Prices: By appointment. No charge.

Gene Autry

GENE AUTRY OKLAHOMA MUSEUM, 47 Prairie St., Gene Autry, OK 73436. Mailing Address: P.O. Box 44, Gene Autry, OK 73436-0044. Tel.: 580-294-3047. Fax: 580-294-3454.
E-mail: townofgeneautry@brightok.net
Web Site: geneautryokmuseum.com
Founded: 1990.
Key Personnel: Dir., Elvin R. Sweeten; Museum Shop Mgr., Flora R. Sweeten.
Personnel Profile: Part-Time Volunteers 2.
Governing Authority: Tax-exempt.
Institution Type/Description: General Museum.
Collections: Gene Autry, Roy Rogers, Rex Allen, Tex Ritter, Jimmy Wakely, Eddie Dean, & others who appeared in musical Western movies of the 1930s & 1940s.
Facilities: Museum-related items for sale.
Activities: Annual Event: Film & Music Festival.
Hours & Admission Prices: March-Nov. call for hours. No charge; donations accepted. &
Attendance: 15,000 (estimated)
Membership: $25; $50; $75; $100; $200; $500; $700.

Goodwell

NO MAN'S LAND HISTORICAL SOCIETY, 214 E. Ave., Goodwell, OK 73939. Mailing Address: P.O. Box 278, Goodwell, OK 73939-0278. Tel.: 580-349-2670. Fax: 580-349-2670.
E-mail: nmlhs@ptsi.net
Web Site: www.nmlhs.org
Founded: 1934.
Congressional District: 6
Key Personnel: Dir., Sue Weissinger; Pres. (V), Ron Kincannon.
Personnel Profile: Part-Time Paid 2.
Governing Authority: society. Parent Institution: Panhandle State University & Oklahoma Historical Society. Tax-exempt: 501(c)(3).
Institution Type/Description: Historical Society Museum.
Collections: Hal Clark & William B. Baker Indian artifacts; Duckett Alabaster carvings; anthropology; archaeology; geology; mineralogy; paleontology; archives; zoology; agriculture; costumes; art; No Man's Land history; pioneer history; Dust Bowl history; natural history; flora & fauna.
Research Fields: Indian artifacts; Western history.
Facilities: 500-vol. library of Western & local history documents available for use on premises; reading room.
Activities: guided tours; lectures.
Publications: newsletter.
Hours & Admission Prices: May-Aug. Tues.-Fri. 10-12 & 1-4, Sat. 10-4; Sept.-April Tues.-Fri. 10-12 & 1-3. No charge; donations accepted. Closed legal holidays. &
Attendance: 4,000 (estimated)
Membership: Individual $15; Family $25; Life $200.

Grove

HAR-BER VILLAGE MUSEUM, (M), 4404 W. 20th St., Grove, OK 74344-5136. Tel.: 918-786-6446 & 3488. Fax: 918-787-6213. Facebook: Har-Ber Village Museum.
E-mail: director@har-bervillage.com
Web Site: www.har-bervillage.com
Founded: 1968.
Congressional District: 2
Key Personnel: Exec. Dir., Amelia Chamberlain; Bd. Trustees, Pete Churchwell.
Personnel Profile: Full-Time Paid 8; Part-Time Paid 7; Part-Time Volunteers 3; Interns 1.
Volunteer Hours: 200
Operating Expenses: 519,196
Operating Income: 528,225
Governing Authority: individual operation. Tax-exempt.
Institution Type/Description: Historic Village Museum.
Collections: pottery; china; toys; natural history; furniture; lamps; dolls; farm machinery; stagecoaches; prairie schooner; steam engines; wagons; buggies; guns; musical instruments; clothing. Historic Buildings: approx. 102 buildings on Grand Lake O' the Cherokees, including doctor's offices, drug store & bank; Sweet Annie Herb Garden; Ecology Center; nature trails.
Facilities: visitor center; picnic pavilion; nature trails; event tent. Gift items for sale.
Activities: self-guided tours; permanent exhibitions; interactive stations; demonstrations; events; workshops.
Publications: newsletters, Har-Ber Village Happenings; brochures; annual report; visitor map guide.
Hours & Admission Prices: March 15-Oct. 14 Mon.-Sat. 9-6, Sun. 12:30-6; Oct. 15-Nov.15 Mon.-Sat. 9-5, Sun. 12:30-5. Adults $10, seniors 62 & over $7.50, children 6-14 $5; discounts to veterans, AAM members & groups of 10 or more; members and children 5 & under no charge. School Tours: $1 per student. &
Attendance: 15,000 (accurate)
Membership: Individual $25; Dual $20; Family $60; Friend $100; Supporter $150; Bronze Patron $250; Silver Patron $500; Gold Patron $1,000; Platinum Patron $2,500; Diamond Patron $5,000.

Guthrie

NATIONAL LIGHTER MUSEUM, 5715 S. Sooner Rd., Guthrie, OK 73044-6739. Tel.: 405-282-3025.
E-mail: tballard8@cox.net
Web Site: www.nationallightermuseum.com
Key Personnel: Owner & Cur., Ted C. Ballard
Institution Type/Description: History Museum.
Collections: fire-making devices.
Hours & Admission Prices: By appointment. No charge.

OKLAHOMA FRONTIER DRUG STORE MUSEUM, 214 W. Oklahoma, Guthrie, OK 73044-3132. Tel.: 405-282-1895.
E-mail: drugstoremuseum@aol.com
Web Site: www.drugmuseum.org
Founded: 1992.
Key Personnel: Dir., G. Mark Ekiss, R.PH.
Governing Authority: Parent Institution: Oklahoma Pharmacy Heritage Foundation. Tax-exempt.
Institution Type/Description: Drugstore & Pharmacy Museum.
Collections: period artifacts; 1923 soda fountain; 1890s pharmaceutical memorabilia.
Facilities: Museum-related items for sale.
Hours & Admission Prices: Tues.-Sat. 10-5. No charge; donations accepted. Closed New Year's Day; Christmas.
Attendance: 10,000 (estimated)

OKLAHOMA TERRITORIAL MUSEUM, (M), 406 E. Oklahoma Ave., Guthrie, OK 73044-3317. Tel.: 405 282 1889.
E-mail: guthriecomplex@okhistory.org
Web Site: www.oklahomaterritorialmuseum.org
Founded: 1970.
Congressional District: 6
Key Personnel: Dir., Nathan Turner; Cur. Collections, Erin Brown; Main Tech, James Ray; Clerk, Sharen Bowers.
Personnel Profile: Full-Time Paid 4; Part-Time Volunteers 20.
Governing Authority: state. Parent Institution: Oklahoma Historical Society (see separate listing). Tax-exempt.
Institution Type/Description: History Museum: housed in the Fred Pfeiffer Memorial Museum building & the Carnegie Library.

Collections: items relating to Oklahoma 1889 Land run & territorial urban period.
Research Fields: territorial Oklahoma.
Facilities: 4,000-vol. library from 1900-1910.
Activities: arts festivals; permanent & temporary exhibitions; special activities; educational programs.
Hours & Admission Prices: By appointment. Closed holidays. &
Attendance: 16,685 (accurate)
Membership: Student & Senior Citizen $10; Individual $15; Family & Institutional $25; Supporting $50; Life $300; Benefactor $500.

OWENS ARTS PLACE MUSEUM, 1202 E. Harrison Ave., Guthrie, OK 73044. Tel.: 405-260-0204.
E-mail: mail@owensmuseum.com
Web Site: www.owensmuseum.com
Founded: 2005.
Key Personnel: Dir. & C.E.O., Wallace Owens.
Personnel Profile: Part-Time Volunteers 400.
Volunteer Hours: 38
Operating Expenses: 600
Operating Income: 700
Governing Authority: Tax-exempt: 501(c)(3).
Institution Type/Description: Art Museum.
Collections: works of fine art; paintings; sculpture.
Activities: special events.
Hours & Admission Prices: Tues.-Thurs. 10-4, Sun. 1-4. No charge; donations accepted. Closed holidays. &
Attendance: 1,300 (estimated)
Membership: Individual $30.

STATE CAPITAL PUBLISHING MUSEUM, 301 W. Harrison Ave., Guthrie, OK 73044-4414. Mailing Address: 800 Nazih Zuhdi Dr., Oklahoma City, OK 73105. Tel.: 405-282-4123. Fax: 405-282-1081.
E-mail: publishingmuseum@yahoo.com
Web Site: www.guthrieok.com/MUSEUM.html
Founded: 1976.
Congressional District: 5
Key Personnel: Dir., Jeff Hirzel; C.E.O., Ed Wood; Dir. & Museum Shop Mgr., Tim Poindexter; Cur., Melissa Fusler.
Personnel Profile: Part-Time Paid 2; Part-Time Volunteers 20.
Governing Authority: state. Parent Institution: Oklahoma Historical Society. (See separate listing). Tax-exempt.
Institution Type/Description: Publishing Museum: housed in 1902 State Capital Co. Building, originally constructed by frontier editor & publisher Frank Hilton Greer.
Collections: original printing & publishing artifacts; company records; working vintage printing presses; 1889-1910 newspapers published by the State Capital Co. on microfilm; manuscript collection.
Research Fields: history of printing & newspaper publishing with emphasis on the Oklahoma territorial 1890-1907 & early statehood period.
Facilities: library of 1889-1910, microfilm newspapers.
Activities: printing; permanent & temporary exhibitions; educational programs.
Hours & Admission Prices: Temporarily closed. &
Attendance: 14,000 (estimated)
Membership: Student & Senior Citizen $10; Individual $15; Family & Institutional $25; Supporting $50; Life $300; Benefactor $500.

TERRITORIAL CAPITAL SPORTS MUSEUM, 315 W. Oklahoma Ave., Guthrie, OK 73044-3107. Tel.: 405-260-1342. Fax: 405-260-1342. Facebook: Territorial Capital Sports Museum.
E-mail: oklahomasportsmuseum@sbcglobal.net
Web Site: territorialcapitalsportsmuseum.org
Formerly: Oklahoma Sports HOF Museum-Guthrie
Founded: 1993.
Key Personnel: Dir., Richard Hendricks.
Personnel Profile: Full-Time Paid 1; Part-Time Paid 2.
Governing Authority: Parent Institution: Jim Thorpe Assoc., Inc. Tax-exempt.
Institution Type/Description: Sports Museum.
Collections: artifacts & memorabilia of professional & Olympic athletes from Oklahoma.
Activities: rental facility; interactive exhibit.
Hours & Admission Prices: Tues.-Thurs. 1-4, Fri.-Sat. 10-4, Sun. by appointment. No charge; donations accepted. &
Attendance: 1,500 (estimated)

Harrah

HARRAH HISTORY CENTER AND RAILROAD DEPOT MU-SEUM, 20881 E. Main St., Harrah, OK 73045. Mailing Address: P.O. Box 846, Harrah, OK 73045. Tel.: 405-454-6911.
E-mail: harrahhsitorycente@att.net
Web Site: www.harrahhistorycenter.com
Formerly: Harrah Historical Society
Founded: 1977.
Key Personnel: Pres. (V), Tom Barron.
Personnel Profile: Part-Time Volunteers 9.
Volunteer Hours: 3,431
Governing Authority: nonprofit. Tax-exempt 501(c)(3).
Institution Type/Description: Historical Society Museum.
Collections: local history & culture; photographs; personal artifacts; period furnishings; local industry, organizations & churches from 1896 to present; map; military; railroad artifacts; medical instruments; dolls; family research files.
Publications: quarterly newsletter.
Hours & Admission Prices: Tues.-Thurs. 10-4; other times by appointment. No charge; donations accepted. &

Attendance: 400 (estimated)
Membership: Family $25.

Healdton

HEALDTON OIL MUSEUM, 315 E. Main St., Hwy. 76, Healdton, OK 73438-1714. Mailing Address: c/o Oklahoma Historical Society, 800 Nazih Zuhdi Dr., Oklahoma City, OK 73105. Tel.: 580-229-0900.
Web Site: www.okhistory.org/sites/healdtonoil
Founded: 1973.
Congressional District: 3
Key Personnel: Mgr., Melanie Williams.
Personnel Profile: Part-Time Paid 1.
Governing Authority: state. Branch of Oklahoma Historical Society. Tax-exempt.
Institution Type/Description: Technology Museum.
Collections: oil field rigs; equipment; tools, photographs; documents related to Oklahoma oil industry.
Research Fields: oil industry.
Activities: tours; permanent exhibitions.
Hours & Admission Prices: Mon.-Fri. 9-4. No charge; donations accepted. Closed legal holidays. &
Attendance: 2,411 (estimated)
Membership: Individual $25; Family & Institutional $40; Supporting $75; Life $500; Benefactor $1,000.

Heavener

HEAVENER RUNESTONE STATE PARK, 18365 Runestone Rd., Heavener, OK 74937-7493. Mailing Address: 103 E. Ave. B, Heavener, OK 74937-2601. Tel.: 918-653-2241. Fax: 918-653-3435.
E-mail: heavener@oklahomaparks.com
Founded: 1967.
Congressional District: 3
Key Personnel: Park Mgr., Rick Sanders; Maintenance & Repair Tech., William Rowland.
Personnel Profile: Full-Time Paid 3.
Governing Authority: state. Affiliated with Oklahoma Tourism and Recreation Dept., 500 Will Rogers Building, Oklahoma City, OK 73105. Tax-exempt.
Institution Type/Description: Historic Site: Runestone inscription, Glome Valley.
Collections: local history & culture; photographs.
Facilities: visitor center; interpretive center & trail; nature trail; playground & picnic area; community building.
Activities: Annual Events: Easter Sunrise Service; Easter Egg Hunt; Trash Off Day; July 4th Musical; Car Show in August.
Publications: brochure.
Hours & Admission Prices: March-Oct. daily 8-dusk; Nov.-Feb. daily 8-dusk. No charge; donations accepted. &
Attendance: 90,000 (accurate)

PETER CONSER HOME, 47114 Conser Creek Rd., Heavener, OK 74937-9022. Mailing Address: Oklahoma Historical Center, 800 Nazih Zuhdi Dr., Oklahoma City, OK 73105. Tel.: 918-653-2493.
Founded: 1970.
Congressional District: 3

Key Personnel: C.E.O., Dr. Bob Blackburn; Site Mgr., A.G. Hembree.
Governing Authority: state. Parent Institution: Oklahoma Historical Society. Tax-exempt.
Institution Type/Description: Historic House: 1894 Peter Conser House; a captain of the Lighthorsemen of the Moshulatubbe District, a noted law enforcement group of the Choctaw nation.
Collections: furnishings & artifacts from 1894-1910.
Research Fields: Choctaw history, particularly that of the Lighthorsemen.
Facilities: library.
Activities: tours.
Publications: Mistletoe Leaves; Chronicles.
Hours & Admission Prices: Wed.-Sat. 10-5, Sun. 1-5. No charge; donations accepted. Closed national holidays. &
Attendance: 6,000 (estimated)
Membership: Student & Retired $10; Individual $15; Family & Institution $25; Supporting $50; Life $300; Benefactor $500.

Henryetta

HENRYETTA TERRITORIAL MUSEUM, (M), 410 W. Moore, Henryetta, OK 74437-5255. Mailing Address: P.O. Box 220, Henryetta, OK 74437-0220. Tel.: 918-652-7112. Fax: 918-652-7112.
Web Site: www.territorialmuseum.net
Founded: 1982.
Congressional District: 2
Key Personnel: Pres., Marsha Smith; Pres. (V), Mike Doak.
Personnel Profile: Part-Time Paid 1.
Governing Authority: Parent Institution: Henryetta Historical Society. Tax-exempt.
Institution Type/Description: History Museum.
Collections: area history; photographs.
Hours & Admission Prices: Wed.-Sat. 11-3; other times by appointment. No charge; donations accepted. Closed holidays.
Attendance: 1,500 (estimated)
Membership: Individual $20; Family $30; Business $50, $100, $250.

Hinton

HINTON HISTORICAL MUSEUM & PARKER HOUSE, 801 S. Broadway, Hinton, OK 73047. Tel.: 405-542-3181.
Institution Type/Description: History Museum.
Collections: local & state history; horse carriages; barbwire; furnishings; clothing; photographs; early phones & switchboards; radios; Native American artifacts.
Hours & Admission Prices: Mon.-Sat. 10-4.

Hobart

GENERAL TOMMY FRANKS LEADERSHIP INSTITUTE AND MUSEUM, (M), 507 S. Main, Hobart, OK 73651. Mailing Address: P.O. Box 222, Hobart, OK 73651-0222. Tel.: 580-726-5900. Fax: 580-726-5901.
E-mail: museum@tommyfranksmuseum.org
Web Site: tommyfranksmuseum.org
Founded: 2006.
Key Personnel: Dir., Warren Martin; Mgr., Scott Cumm; Office Mgr., Nikki Macy; Mgr. Finance, Michelle Gather
Governing Authority: Tax-exempt.
Institution Type/Description: History Museum.
Collections: General Tommy Franks life; military history; photographs; uniforms; personal artifacts; medals; challenge coins.
Facilities: Museum-related items for sale.
Activities: special events; educational programs. Museum Sponsors: Four Star Leadership with Gen. Tommy Frank; Celebration of Freedom; Road Show.
Hours & Admission Prices: Mon.-Sat. 10-12 & 1-5. No charge; donations accepted. Closed holidays. &
Attendance: 4,500 (estimated)

KIOWA COUNTY HISTORICAL MUSEUM, 518 S. Main, Hobart, OK 73651. Tel.: 580-726-6202.
Founded: 1993.
Institution Type/Description: History Museum: housed in the former Rock Island Depot; built in 1909. Listed on the National Register of Historic Places.
Collections: county history; period furniture; personal artifacts; memorabilia; early farm equipment; one-room schoolhouse; dental chairs; office equipment; Native America artifacts; buggies; vehicles.
Hours & Admission Prices: Call for hours.

Hominy

DRUMMOND HOME, 305 N. Price, Hominy, OK 74035-1007. Tel.: 918-885-2374. Facebook: Drummond Home of the Oklahoma Historical Society.
E-mail: bwhitcomb@okhistory.org
Founded: 1986.
Key Personnel: Site Mgr., Beverly Whitcomb.
Personnel Profile: Full-Time Paid 1.
Governing Authority: state. Parent Institution: Oklahoma Historical Society (see separate listing). Tax-exempt.
Institution Type/Description: Historic House: 1905 Drummond Home.
Collections: original furnishings; clothing; documents; photographs; decorative arts.
Major Exhibits: Antique Dolls, 2/14; Quilts, 4/14; Drummond Family Heirloom, 8/14.
Research Fields: Drummond family & their relations with the Osage; cattle industry; period furnishings.
Activities: tours; Christmas decoration. Annual Events: Ice Cream Social; Christmas Open House in December.
Publications: Mistletoe Leaves (Oklahoma Historical Society); Friend's newsletter.
Hours & Admission Prices: Wed.-Sat. 9-5, Sun. 1-5. Tours: 10-4. Adults $3, seniors $2.50, children 6-18 $1; discounts to groups; children 5 & under no charge. Closed state holidays.
Attendance: 1,800 (estimated)
Membership: Please see separate listing for Oklahoma Historical Society.

FIELD HISTORICAL PRINTING MUSEUM, 109 W. Main St., Hominy, OK 74035-1031. Tel.: 918-885-2688.
Institution Type/Description: History Museum.
Collections: two 1930 linotypes; 12 printing presses; engraver; ludlow; strip caster; cameras; brass; foundry & wood type; engravings; teletypesetter; Western Union equipment; Dow Jones machines.
Hours & Admission Prices: By appointment. No charge.

Hugo

CHOCTAW COUNTY HISTORICAL SOCIETY, 309 North B St., Hugo, OK 74743-3325. Mailing Address: P.O. Box 577, Hugo, OK 74743-0577. Tel.: 580-326-6630.
E-mail: friscodepot@live.com
Web Site: www.friscodepot.org
Founded: 1978.
Congressional District: 3
Key Personnel: Chm. (V), Noel Pence; Treas. & Museum Mgr., Norman Pence.
Personnel Profile: Part-Time Volunteers 12.
Governing Authority: society; nonprofit organization. Parent Institution: Choctaw County Historical Society. Subsidiary Institution: Frisco Depot Museum. Tax-exempt.
Institution Type/Description: Railroad Museum: housed in 1912 Frisco Railroad Depot; original Harvey House restaurant.
Collections: railroad items; local history; newspaper stories of early day events; two early fire trucks. Historic Structure: restored Harvey House restaurant, dormitory rooms & manager's apartment; working miniature trains with various engines of Frisco models; early day farm equipment display; hand made miniature circus display; large collection of early day photos of area, building & surrounding towns; restored kitchen, includes early day gas, kerosene & wood cook stoves; first long distance telephone switchboard in area; 250 year-old spinning wheel & loom; telegraph equipment; hand crank wooden washing machines; ladies turn-of-the-century apparel & accessories; country school room; doctor's office; old barber shop c.1900.
Research Fields: local history of towns in surrounding area.
Facilities: library of books on history of towns & early day events before statehood; 150-seat auditorium. Books on local history for sale.
Activities: guided tours; arts festivals. Annual Event: Homecoming days in June.
Publications: book, Hugo, 1916; book, Smoke Signals from Indian Territory.
Hours & Admission Prices: Tues.-Sat. 10-4; tours by appointment. Museum: no charge; donations accepted. &
Attendance: 2,900 (accurate)
Membership: $25.

Idabel

BARNES-STEVENSON HOUSE - MCCURTAIN COUNTY HISTORICAL SOCIETY, 302 S.E. Adams Ave., Idabel, OK 74745. Tel.: 580-212-3639.
Institution Type/Description: Historical Society Museum.
Collections: local history; period furnishings; photographs.
Hours & Admission Prices: By appointment.

MUSEUM OF THE RED RIVER, 812 E. Lincoln Rd., Idabel, OK 74745-7815. Tel.: 508-286-3616. Fax: 508-286-3616.
E-mail: motrr@hotmail.com
Web Site: www.museumoftheredriver.org
Founded: 1974.
Congressional District: 3
Key Personnel: Dir., Henry Moy; Pres. (V) Herron Foundation, Donald A. Herron; Pres. (V) Idabel Museum Society, Judy Petre; Volunteer Coord., Sallie Webb; Keeper of Collections, Daniel Vick; Head Programs, Jeanette Bohanan; Cur. Assoc., Mario Rivera; Cur. Asst., Christina Eastep; Museum Asst., John Malin; Business Mgr., Vickie Smith; Museum Shop Mgr., Sherron Mitchell.
Personnel Profile: Full-Time Paid 6; Full-Time Volunteers 2; Part-Time Paid 3; Part-Time Volunteers 10; Interns 3.
Governing Authority: municipal; owned and maintained by Herron Foundation on behalf of the city of Idabel; operated by Idabel Museum Society, Inc. Tax-exempt.
Institution Type/Description: General Museum.
Collections: prehistoric to contemporary American Indian collections, emphasis on local Indian history, Caddo, Choctaw; interpretive exhibits material culture of the Americas; Native American art & archaeology; regional natural history; regional paleontological & geological specimens.
Major Exhibits: Recent Acquisitions, 11/26/13-2/23/14; Russell Whiting & Donald Swineford, 2/25/14-4/13/14; Choco (Panama), 4/22/14-6/29/14; Southeast Traditions, 7/8/14-8/31/14; Jeralyn Lujan Lucero and Raymond Wiger, 9/9/14-11/2/14; Recent Acquisitions, 11/11/14-2/22/15.
Research Fields: Caddoan archaeology.
Facilities: library & archaeological study collections available for research on premises. Indian & folk crafts for sale.
Activities: rotating & permanent exhibits & occasional special exhibitions; gallery lectures; films; children's programs.
Publications: archeological survey reports; educational leaflets.
Hours & Admission Prices: Tues.-Sat. 10-5, Sun. 1-5. No charge; donations accepted. Closed New Year's Day; Memorial Day; Independence Day; Thanksgiving; Christmas. &
Attendance: 11,900 (accurate)

Indiahoma

WICHITA MOUNTAINS WILDLIFE REFUGE, 32 Refuge Headquarters, Indiahoma, OK 73552-2478. Tel.: 580-429-3222.
Institution Type/Description: Wildlife Refuge.
Collections: wildlife & their habitats; plants; ecology.
Facilities: nature trails.
Activities: hiking; camping; fishing.
Hours & Admission Prices: Mon.-Fri. 8-4:30.

Jay

DELAWARE COUNTY HISTORICAL SOCIETY & MARIEE WALLACE MUSEUM, 538 Krause St., Jay, OK 74346. Mailing Address: P.O. Box 855, Jay, OK 74346-0855. Tel.: 918-253-4345.
E-mail: jaymuseum@brightok.net
Founded: 1976.
Key Personnel: Dir. & Museum Shop Mgr., Jackie Coatney; Pres. (V), Becki Farley.
Personnel Profile: Full-Time Paid 1.
Governing Authority: Tax-exempt.
Institution Type/Description: Historical Society Museum.
Collections: local history & culture; toy trains; buggies; wagons; artifacts donated by families within Delaware County.
Research Fields: genealogy.
Publications: Heritage of the Hills - A Delaware County History; biannual magazine, Heritage of the Hills.
Hours & Admission Prices: Summer: Mon.-Tues. & Thurs.-Fri. 9-5, Wed. 1:30-5; Winter: Mon.-Fri. 9-2. No charge; donations accepted.
Attendance: 3,000 (estimated)
Membership: Individual $20; Family $25; Silver $100; Gold $200; Platinum $500; Double Platinum $1,000.

Jenks

OKLAHOMA AQUARIUM AND THE KARL AND BEVERLY WHITE NATIONAL FISHING TACKLE MUSEUM, 300 Aquarium Dr., Jenks, OK 74037-4148. Mailing Address: P.O. Box 910, Jenks, OK 74037-0910. Tel.: 918-296-3474. Fax: 918-296-3467.
Web Site: okaquarium.org
Key Personnel: Exec. Dir., Teri Bowers
Institution Type/Description: Aquarium.
Collections: over 200 salt & fresh water species and mammals; 20,000 pieces of fishing tackle.
Facilities: deli. Museum-related items for sale.
Activities: summer camp; educational programs; special events.
Hours & Admission Prices: Tues. 10-9, Wed.-Mon. 10-6. Adults $13.95, senior citizens & military $11.95, children 3-12 $9.95; children 2 & under no charge. Closed Christmas.

Jones

JONES HISTORICAL SOCIETY MUSEUM, 100 Main St., Jones, OK 73049. Tel.: 405-399-2228.
Institution Type/Description: Historical Society Museum.
Collections: local history & culture; photographs; personal & military artifacts.
Hours & Admission Prices: By appointment.

Kaw City

KANZA MUSEUM, 698 Grandview Dr., Kaw City, OK 74641. Mailing Address: Drawer 50, Kaw City, OK 74641. Tel.: 580-269-2552. Fax: 580-269-2301.
E-mail: cdouglas@kawnation.com
Web Site: www.kawnation.com
Founded: 1995.
Key Personnel: Dir., Crystal Douglas; C.E.O., Guy Monroe.
Governing Authority: Kaw Nation.
Institution Type/Description: Native American Museum.
Collections: Native American history, culture & heritage; personal artifacts; photographs; paleontology.
Publications: book, Kanza Reader: Learning Literacy through Historical Texts
Hours & Admission Prices: Mon.-Fri. 8-5. No charge. &

KAW CITY MUSEUM, 910 Washunga Dr., Kaw City, OK 74641. Mailing Address: P.O. Box 56, Ponca City, OK 74602. Tel.: 580-269-2366.
Institution Type/Description: History Museum: housed in a former train depot; built in 1902.
Collections: local history & culture; photographs; period furnishings; personal artifacts.
Hours & Admission Prices: May to Labor Day Mon.-Tues. & Thurs.-Fri. 1-5, Sat.-Sun. 2-5. No charge. &

Keota

OVERSTREET-KERR HISTORICAL FARM, 28878 Kerr-Overstreet Rd., Keota, OK 74941-6817. Tel.: 918-966-3396. Fax: 918-966-3396.
E-mail: okhfarm@crosstel.net
Web Site: www.kerrcenter.com
Founded: 1990.
Congressional District: 2
Key Personnel: Dir., Alan Ware; Education, Registrar & Museum Shop Mgr., Jeremy Henson; Pres. (V), Jim Horne; Public Rels., Maura McDermott; Treas., Ann Ware; Farm Devel. Mgr., Jim Combs.
Personnel Profile: Full-Time Paid 2; Part-Time Paid 1; Part-Time Volunteers 2.
Governing Authority: private; nonprofit organization. Parent Institution: Kerr Center, P.O. Box 588, Poteau, OK 74953.
Institution Type/Description: Farm Museum.
Collections: lifestyle of the Native American & Pioneer farmer from 1871 to 1952; endangered livestock breeds; 1890-1940 farm equipment; outbuildings.
Research Fields: John Deere hay making & grain production equipment from 1890-1940.
Facilities: 324 sq. ft. exhibit space; nature trail; heirloom orchard. Museum-related items for sale.
Activities: formal education programs for children; guided tours; hobby workshops; demonstrations. Annual Event: Fall Farm Fest in October.

Publications: monthly newsletter, Field Notes; promotional brochures.
Hours & Admission Prices: Mon.-Sat. by appointment. Adults $3; children under 6 & students no charge. &
Attendance: 5,000 (estimated)

Kingfisher

CHISHOLM TRAIL MUSEUM, 605 Zellers Ave., Kingfisher, OK 73750-4228. Tel.: 405-375-5176.
E-mail: ctmus@pldi.net
Web Site: www.chisholmandseay.com
Founded: 1970.
Congressional District: 6
Key Personnel: Pres., Jeremy Ingle; Dir., Adam Lynn; Museum Shop Mgr., Marvin Reames.
Personnel Profile: Full-Time Paid 2; Part-Time Paid 1.
Governing Authority: state. Parent Institution: Oklahoma Historical Society, Wiley Post Bldg., 2401 N. Laird Ave., Oklahoma City, OK 73105. Tax-exempt.
Institution Type/Description: General Museum.
Collections: agriculture; archaeology; arrowheads; barbed wire; photographs; newspapers; cowboy & old farm equipment; parts of restored frontier town; early days stores; Indian artifacts; Dalton Cabin; First Bank of Kingfisher; Gant schoolhouse (one room); Harmony Church; log cabin (two room).
Research Fields: Pioneer Kingfisher; Chisholm Trail; Jesse Chisholm; J.V. Admire.
Facilities: wildlife diorama; meeting room. Museum-related items for sale.
Activities: guided tours. Museum Sponsors: Living History in Spring; Christmas at Governor Seay Mansion & Pioneer Village in December.
Publications: brochures.
Hours & Admission Prices: Tues.-Sat. 10-5. Adults $4, children 6-18 $2; children 5 & under no charge. Closed state holidays. &
Attendance: 8,500 (estimated)
Membership: Annual $10; Life $100.

GOVERNOR SEAY MANSION, 605 Zellers Ave., Kingfisher, OK 73750-4228. Tel.: 405-375-5176. Fax: 405-375-5176.
E-mail: chisholmtrail@okhistory.org
Web Site: www.chisholmandseay.com
Founded: 1967.
Congressional District: 6
Key Personnel: C.E.O., Bob Blackburn; Dir., Adam Lynn; Pres., Jeremy Ingle; Cur., Renee Mitchell; Museum Shop Mgr., Marvin Reames.
Personnel Profile: Full-Time Paid 2; Part-Time Volunteers 30.
Governing Authority: state. Parent Institution: Oklahoma Historical Society, Wiley Post Bldg., 2401 N. Laird Ave., Oklahoma City, OK 73105-7914. Subsidiary Institution: Chisholm Trail Museum, Inc. Tax-exempt.
Institution Type/Description: Historic House: Built-in 1892 restored mansion of second territorial Gov. A.J. Seay.
Collections: period furniture.
Research Fields: Governor A.J. Seay.
Activities: guided tours.
Publications: brochure.
Hours & Admission Prices: Tues.-Sat. 10-5. Adults $4, children 6-18 $2; children under 5 no charge. Closed state holidays. &
Attendance: 7,000 (estimated)
Membership: Annual $10; Life $100.

Krebs

KREBS HERITAGE MUSEUM, 85 S. Main St., Krebs, OK 74554. Mailing Address: P.O. Box 1493, Krebs, OK 74554-1493. Tel.: 918-426-0377.
E-mail: krebsmuseum@yahoo.com
Web Site: krebsmuseum.com
Founded: 1997.
Congressional District: 2
Key Personnel: C.E.O., Dir. & Chm. (V), Steve DeFrange; Museum Shop Mgr., Jack Cordell.
Personnel Profile: Part-Time Volunteers 1.
Governing Authority: bd. dirs. Tax-exempt.
Institution Type/Description: Historical Museum.
Collections: Civil War; coal mining; general military; Krebs City artifacts; area family booths; horse-drawn vehicles; early American tools.
Hours & Admission Prices: Call for hours. No charge; donations accepted. &
Attendance: 875 (estimated)

Langston

MELVIN B. TOLSON BLACK HERITAGE CENTER, Langston University, Sanford Hall, Langston, OK 73050. Mailing Address: P.O. Box 652, Langston, OK 73050. Tel.: 405-466-3346. Fax: 405-466-2979.
Founded: 1970.
Key Personnel: Dir., Jameka B. Lewis.
Personnel Profile: Full-Time Paid 2.
Governing Authority: Parent Institution: Langston University.
Institution Type/Description: Black Heritage Center.
Collections: African American books, newspapers, videos & film; paintings; photographs; sculpture.
Hours & Admission Prices: Mon.-Fri. 8-5. No charge.

Lawton

CAMERON UNIVERSITY ART GALLERY, Louise D. McMahon Fine Arts Complex, Rm. 125, Lawton, OK 73505-6377. Tel.: 580-581-2211.
Institution Type/Description: Art Gallery.
Collections: paintings; photographs; sculpture.
Hours & Admission Prices: No charge.

COMANCHE NATIONAL MUSEUM AND CULTURAL CENTER, (M), 701 N.W. Ferris Ave., Lawton, OK 73507-5442. Tel.: 580-353-0404.
E-mail: info@comanchemuseum.com
Web Site: comanchemuseum.com
Founded: 2007.
Key Personnel: Dir., Phyllis Wahahrockah-Tasi; Collections Mgr., Nicki Hise; Program Asst., Candy Morgan; Cultural Specialist, Bambi Allen.
Governing Authority: Tax-exempt.
Institution Type/Description: Native American Museum.
Collections: local history & culture relating to the Comanche Indians.
Hours & Admission Prices: Mon.-Fri. 8-5, Sat. 10-2. No charge; donations accepted. Closed Thanksgiving; Christmas. ら
Attendance: 10,000 (estimated)

LESLIE POWELL FOUNDATION AND GALLERY, 620 S.W. D Ave., Lawton, OK 73501-4508. Tel.: 580-357-9526. Fax: 580-357-9526.
E-mail: lpartgallery@sbcglobal.net
Web Site: www.lpgallery.org
Founded: 1986.
Key Personnel: Exec. Dir. & Cur., Nancy P. Anderson.
Personnel Profile: Full-Time Paid 1; Part-Time Paid 1; Interns 1.
Governing Authority: Tax-exempt.
Institution Type/Description: Art Gallery.
Collections: works by noted artists; paintings; sculptures; 44 works of Leslie Powell Japanese wood cut collection on display.
Facilities: 80-seat lecture hall.
Activities: concerts; lectures; openings.
Publications: online newsletter; biennial show catalog, even years in November.
Hours & Admission Prices: Mon.-Sat. 12-4. No charge. ら
Attendance: 3,000 (accurate)

＊　**MUSEUM OF THE GREAT PLAINS, (M),** 601 N.W. Ferris Ave., Lawton, OK 73507-5443. Tel.: 580-581-3460. Fax: 580-581-3458.
E-mail: info@museumgreatplains.org
Web Site: www.museumgreatplains.org
Founded: 1960.
Congressional District: 4
Key Personnel: Dir. & Cur. Exhibits, John Hernandez; Chm. (V), Janice Bell; Sr. Cur., Librarian & Archivist, Deborah Baroff; Living History Interpreter, Tim Poteete; Cur. Collections & Acting Registrar, Jim Whiteley; Cur. Educator, Jana Brown; Dir. Technical Svcs., Brian Smith; Deputy Dir. & Dir. Devel., Bart McClenny; Museum Shop Mgr., Rebecca Royal; Exec. Asst., Mary Owensby; Bldg. Mgr., Larry Holland; Receptionist & Weekend Clerk, Dean Keiser; Office Asst., Lora Moffet.
Personnel Profile: Full-Time Paid 10; Part-Time Paid 3; Part-Time Volunteers 20; Interns 1.
Governing Authority: Museum of the Great Plains Authority. Tax-exempt: 501(c)(3).
Institution Type/Description: Regional History Museum.
Collections: primary documents pertaining to Great Plains region; agricultural & hardware catalogs; 1869-2000 photographs and microfilm of blacksmith & carriage periodicals and wagon catalogs; archives; photographs of Plains Indians & white settlement; exhibits and artifacts representing the material culture of man from prehistoric times to the present; outdoor exhibits of agriculture machinery; 300 ton Baldwin steam locomotive. Historic Buildings: 1902 depot & Blue Beaver one room schoolhouse; 1830s Red River Trading Post & living history program; Tingley Indian store collections from Anadarko, OK are described in Vol. 17 (1978) Great Plains Journal, published by the Institute of the Great Plains.
Research Fields: history & archeology of Great Plains region.
Facilities: 20,000-vol. library, 200,000 documents, 30,000 photographs for research; 150-seat auditorium; outdoor amphitheater; laboratories.
Activities: guided tours; art shows; special films; lectures; demonstrations.
Publications: annual, Great Plains Journal; The Writings of J. Frank Dobie: A Bibliography; Domebo: A Paleo-Indian Mammoth Kill in the Prairie Plains; Early Indian Trade Guns 1625-1775; A Historical Guide to Wagon Hardware & Blacksmith Supplies; Test Excavations in the Mangum Reservoir Area of Southwestern Oklahoma; An Archaeological Reconnaissance of Fort Sill, Oklahoma; Archaeological Investigations of the Kiowa & Comanche Indian Agency Commissaries; an Archaeological Survey in the Gypsum Breaks on the Elm Fork of Red River; an Archaeological Reconnaissance of the Salt Plains Areas of Northwestern Oklahoma; Archaeological Investigations Along the Waurika Pipeline; MGP Record Newsletter.
Hours & Admission Prices: Mon.-Sat. 10-5, Sun. 1-5. Adults $6, senior citizens $5, children 7-11 $2.50; members & children under 7 no charge. Closed New Year's Day; Thanksgiving; Christmas. ら
Attendance: 27,049 (accurate)
Membership: Institutional $25; Individual $30; Family $50; Contributing $100; Bison $200-$1,000. Oklahoma Museum Network upgrade $40.

Lindsay

MURRAY-LINDSAY MANSION & PIKES PEAK SCHOOL, Hwy. 76 S., Lindsay, OK 73052-0282. Mailing Address: P.O. Box 282, Lindsay, OK 73052-0282. Tel.: 405-756-2121.
Founded: 1971.
Congressional District: 4
Key Personnel: Pres. (V), Shawn Bridwell.
Personnel Profile: Part-Time Volunteers 30.
Governing Authority: society. Parent Institution: City of Lindsay. Subsidiary Institution: Lindsay Community Historical Society. Tax-exempt.
Institution Type/Description: Historic Buildings.
Collections: local history & culture; Chickasaw setters; period school furnishings; 187-piece teapot collection; photographs; personal artifacts. Historic Buildings: 1879 Murray-Lindsay Mansion; 1908 Pikes school.
Research Fields: period furnishings & Murray family history; one & two room schools of Garvin County.
Activities: guided tours; replicate 1908 school classes.
Publications: From Pioneers to Progress; Pikes Peak School History 1908 to Present; Military History - Lindsay History Book, 1902-2002.
Hours & Admission Prices: Wed.-Fri. & Sun. 1-4, Sat. 10-2; other times by appointment. No charge; donations accepted. Closed all holidays. ら
Attendance: 2,500 (estimated)
Membership: Annual $10; Lifetime $200.

Madill

MARSHALL COUNTY OF OKLAHOMA GENEALOGICAL & HISTORICAL SOCIETY & THE MUSEUM OF SOUTHERN OKLAHOMA, 400 W. Overton, Rm. #3, Madill, OK 73446. Tel.: 580-795-5060.
E-mail: genealogical@att.net
Web Site: www.themoso.com
Founded: 1975.
Congressional District: 2
Key Personnel: Vice Pres. & Cur., Doris Albin; Pres. (V) & Security, Royce Bartee; Treas., Devel. & Registrar, Marcia Jones.
Personnel Profile: Part-Time Volunteers 10.
Governing Authority: private; nonprofit organization. Tax-exempt: 501(c)(3).
Institution Type/Description: History Museum.
Collections: local & family history; military, school, tax & cemetery records; personal artifacts; photographs; Native American artifacts; period furnishings.
Research Fields: military history; family history; historical sites in Marshall County; Chickasaw Indian history.
Facilities: 218-vol. library.
Activities: films; guided tours; temporary exhibitions; special events in spring & fall. Annual Events: Veteran Parade; Christmas Parade.
Publications: quarterly newsletter, Marshall County Genealogical & Historical Society.

Hours & Admission Prices: Mon.-Fri. 10-4, Sat. 10-2 by appointment. No charge; donations accepted. Closed New Year's Day; Independence Day; Memorial Day; Labor Day; Piomingo Day; Thanksgiving; Christmas. &

Attendance: 750 (estimated)

Membership: Individual $12; Family $18; Lifetime $100; Sponsorship $100-$10,000.

Mangum

OLD GREER COUNTY MUSEUM & HALL OF FAME, 222 W. Jefferson St., Mangum, OK 73554-4000. Mailing Address: P.O. Box 2, Mangum, OK 73554-0002. Tel.: 580-782-2851. Facebook: Old Greer County Museum & Hall of Fame.

E-mail: museum222@att.net

Web Site: oldgreercountymuseum.com

Founded: 1972.

Congressional District: 3

Key Personnel: Pres. (V), Dick Stickle; Museum Office Mgr., Judy Forehand.

Personnel Profile: Part-Time Paid 2; Part-Time Volunteers 3.

Volunteer Hours: 300

Operating Expenses: 17,000

Operating Income: 25,000

Governing Authority: nonprofit. Tax-exempt: 501(c)(3).

Institution Type/Description: History Museum.

Collections: quilts; replica of rooms including a millinery store, barber shop, optometrist office, Masonic Hall, civic room; mementoes of schools; photographs; dolls; salt & pepper shakers; Indian artifacts; farming items; horse drawn plow; barbed wire; music; military artifacts; cowboy memorabilia; medical equipment; local art; bird collection; replica of half-dugout; Fire Station; 1928 Seagrave Fire Truck; tax records from 1896-1950.

Research Fields: local artifacts.

Facilities: 60 rooms of Old Greer County history; tax record books from 1890's;1928 restored fire truck; replica of dugout house; 50-vol. library of history books available for research on premises; conference room; reading room.

Activities: guided tours; school projects; reenactments; monument dedications.

Publications: History of Old Greer County & Its Pioneers; weekly newspaper column; quarterly newsletter.

Hours & Admission Prices: Mon.-Sat. 9-12 & 1-4. Adults $3, children 6-12 $1; Blue Star discount for active duty military; OMA & museum members no charge. Closed legal holidays. &

Attendance: 600 (estimated)

Membership: Individual $25; Sustaining $100 & up.

Marietta

LOVE COUNTY MILITARY MUSEUM, 408 1/2 W. Chickasaw St., Marietta, OK 73448. Tel.: 580-276-3192.

Institution Type/Description: Military History Museum.

Collections: military history & artifacts; over 700 photographs; memorial tribute; 1910 jailhouse.

Activities: guided tours.

Hours & Admission Prices: Thurs.-Sat. 9-2. No charge. &

LOVE COUNTY PIONEER MUSEUM, 409 W. Chickasaw, Marietta, OK 73448. Mailing Address: P.O. Box 134, Marietta, OK 73448. Tel.: 580-276-9020.

Founded: 1979.

Governing Authority: bd. of trustees.

Institution Type/Description: History Museum.

Collections: early pioneer life; local history; rocks; fossils; Native American artifacts; farm equipment & tools; guns; period furnishings.

Facilities: Museum-related items for sale.

Activities: guided tours; research.

Publications: quarterly newsletter; historical calendar.

Hours & Admission Prices: Thurs.-Sat. 10-2; groups by appointment. No charge; donations accepted. &

Membership: Individual $10; Life $150.

NORTON'S INDIAN TERRITORY MUSEUM, 115 W. Main, Marietta, OK 73448. Mailing Address: P.O. Box 23, Marietta, OK 73448. Tel.: 580-276-2568. Fax: 580-276-5249.

Institution Type/Description: History Museum.

Collections: local history & culture; over 500 early pop bottles; photographs; letterheads; national currency; medicine bottles; arrowheads.

Hours & Admission Prices: Mon.-Fri. 9-5:30, Sat. 9:30-1. No charge.

Marlow

MARLOW AREA MUSEUM, 127 W. Main St., Marlow, OK 73055. Mailing Address: 223 W. Main, Marlow, OK 73055. Tel.: 580-658-2212.

E-mail: marlowchamber@cableone.net

Web Site: marlowchamber.org

Founded: 1998.

Congressional District: 4

Institution Type/Description: History Museum.

Collections: local history; Marlow brothers & family artifacts; photographs; personal artifacts.

Facilities: Museum-related items for sale.

Hours & Admission Prices: Call for hours. No charge; donations accepted.

Maud

MAUD HISTORICAL MUSEUM, 130 E. Main, Maud, OK 74854. Mailing Address: P.O. Box A, E. Main, Maud, OK 74854. Tel.: 405-374-2880.

Institution Type/Description: History Museum: housed in the former Irby Drug Building; built in 1928.

Collections: local history & culture; period drug store furnishings; soda fountain; early school, organizations, & business artifacts; clothing; singer Wanda Jackson memorabilia; photographs; personal artifacts.

Hours & Admission Prices: June-Dec. Sat.-Sun. 2-4; other times by appointment. No charge.

McAlester

J.G. PUTERBAUGH HOUSE & GARRARD ARDENEUM, 345 E. Adams, McAlester, OK 74501-4651. Mailing Address: P.O. Box 759, McAlester, OK 74502. Tel.: 918-423-8555.

Institution Type/Description: Historic House Museum: housed in the former home of J.G. Puterbaugh, one of the founding fathers of McAlester's coal business.

Collections: furnishings; maps; photographs; period artifacts.

Hours & Admission Prices: By appointment.

J. J. MCALESTER MANSION, 14 E. Smith, McAlester, OK 74501-2648. Tel.: 918-423-8620.

Institution Type/Description: Historic House Museum: built in 1870. Listed on the National Register of Historic Places.

Collections: period furnishings; personal artifacts; photographs.

Hours & Admission Prices: Fri. by appointment. $2 per person.

MCALESTER SCOTTISH RITE MASONIC CENTER, 305 N. 2nd St., McAlester, OK 74501-4648. Mailing Address: P.O. Box 609, McAlester, OK 74502-0609. Tel.: 918-423-6360. Fax: 918-423-6362.

E-mail: djones@mcalesterscottishrite.org

Web Site: www.mcalesterscottishrite.org

Founded: 1901.

Key Personnel: Dir., Don G. Jones; Chm. (V), Bill Cox; Pres. (V), Bob Bartheld; Museum Shop Mgr., Bill Erkin.

Personnel Profile: Full-Time Paid 3; Part-Time Paid 2; Part-Time Volunteers 200.

Governing Authority: Parent Institution: A&A Scottish Rite of Freemasonry. Tax-exempt.

Institution Type/Description: History Museum.

Collections: Masonic history; photographs; costumes; 1930 custom-built Kimball organ.

Facilities: library.

Publications: biannually, McAlester Scottish Rite News.

Hours & Admission Prices: Masonic Center: Mon.-Fri. 8-12 & 1-5. Tours: Mon.-Fri. 9-12 & 1-4. No charge; donations accepted. &

Attendance: 150 (estimated)

OKLAHOMA PRISONS' HISTORICAL MUSEUM, Stonewall & West St., McAlester, OK 74501-4651. Mailing Address: P.O. Box 97, McAlester, OK 74502-0097. Tel.: 918-423-4700.

E-mail: dale.cantrell@doc.state.ok.us

Founded: 1983.

Key Personnel: Pres. (V), Dale Cantrell.

Personnel Profile: Part-Time Volunteers 1.

Governing Authority: Tax-exempt.

Institution Type/Description: Prison History Museum.

Collections: prison history; photographs; Oklahoma's electric chair.

Hours & Admission Prices: Call for hours. No charge.
Attendance: 200 (estimated)

OKLAHOMA TROLLEY MUSEUM, 21 E. Monroe, McAlester, OK 74501-4651. Mailing Address: P.O. Box 145, McAlester, OK 74501. Tel.: 918-423-2446.
Institution Type/Description: Trolley Museum.
Collections: restored trolley cars from 1907 to 1933; local history; photographs.
Hours & Admission Prices: Call for hours.

PITTSBURGH COUNTY GENEALOGICAL AND HISTORICAL SOCIETY, INC., 113 E. Carl Albert Pkwy., McAlester, OK 74501-5039. Tel.: 918-426-0388.
Web Site: www.pittsburghcogenealogical.org
Formerly: Pittsburgh County Historical Museum
Founded: 1979.
Key Personnel: Pres. (V), Christina Thurber
Institution Type/Description: History Museum.
Collections: books; genealogy; mining; Native American artifacts; local history.
Facilities: library.
Activities: monthly meetings.
Publications: triannual, Tobucksy News.
Hours & Admission Prices: Mon.-Fri. 9-3. No charge; donations accepted. Closed major holidays.
Membership: Individual $15.

TANNEHILL MUSEUM, 500 W. Stonewall, McAlester, OK 74501-2346. Mailing Address: P.O. Box 3215, McAlester, OK 74502. Tel.: 918-470-5755. Fax: 918-423-7720.
Founded: 1966.
Personnel Profile: Part-Time Volunteers 5; Interns 1.
Governing Authority: private. Tax-exempt.
Institution Type/Description: Gun Museum.
Collections: guns; weapons; local history.
Hours & Admission Prices: By appointment. No charge.
Attendance: 500 (estimated)

McLoud

MCLOUD HISTORICAL SOCIETY MUSEUM AND HERITAGE CENTER, 421 W. Broadway, McLoud, OK 74851. Mailing Address: P.O. Box 1292, McLoud, OK 74851. Tel.: 405-964-5169.
E-mail: glendakuhn@yahoo.com
Web Site: www.mcloudhistoricalsociety.org
Founded: 2004.
Congressional District: 5
Key Personnel: Dir., Glenda Kuhn; Treas., Jim Metcalf; Cur., Sylvia Metcalf.
Personnel Profile: Part-Time Volunteers 5.
Governing Authority: nonprofit. Tax-exempt: 501(c)(3).
Institution Type/Description: Historical Society Museum: housed in c.1932 Ford dealership building.
Collections: central Oklahoma historical relics & records; local history; Native American.
Major Exhibits: The History of Writing, 11/12-12/14; Physicians of the Area, 1/1/13-12/14.
Research Fields: veterans from McLoud, Dale & Newalla killed in combat.
Facilities: 10 vol. library of newspapers. Locally handcrafted items for sale.
Activities: education programs for adults & children. Museum Sponsors: Community Sunday Dinner; Arts Festival in October; Christmas Historical Home Tours.
Publications: quarterly newsletter, McLoud Historical Society Newsletter.
Hours & Admission Prices: Feb.-June & Aug.-Dec. Tues. & Fri. 10-3. No charge; donations accepted.
Attendance: 500 (estimated)
Membership: Student $5; Individual $15; Family $25; Patron $50; Sustaining $100.

Medford

GRANT COUNTY MUSEUM, Main & Cherokee Sts., Medford, OK 73759. Mailing Address: P.O. Box 241, Medford, OK 73759. Tel.: 580-395-2342. Fax: 580-395-2343.
Founded: 1965.
Congressional District: 6
Governing Authority: county. Parent Institution: Oklahoma State Historical Society. Tax-exempt.

Institution Type/Description: General Museum.
Collections: artifacts; archives; sod plow; wheel; wheat cradle; churns; pictures; old fashioned kitchen; local history & genealogy.
Research Fields: memorabilia; family histories; genealogies.
Activities: lectures; monthly programs; quilt & antique show.
Publications: books, Vol. I, Family Histories of Grant County; Early Legends of Osage Creek; Vol. 2, Family Histories, Land Marks, Schools, Churches; Historical Tales of the Cherokee Strip and the Rhubarb Farm; Glimmer on the Hill; Chisholm Trail.
Hours & Admission Prices: By appointment only. No charge; donations accepted. ර්
Membership: Active $10; Life $150.

Meeker

CARL HUBBELL MUSEUM, 510 W. Carl Hubbell Blvd., Meeker, OK 74855. Mailing Address: P.O. Box 186, Meeker, OK 74855. Tel.: 405-279-3813.
Founded: 1981.
Key Personnel: Chm. (V), Vernon Markwell
Governing Authority: Tax-exempt.
Institution Type/Description: Baseball Museum.
Collections: Carl Hubbell's life & career as a NY Giant's pitcher from 1928-1943; photographs; personal artifacts; signed baseballs; awards; news articles.
Hours & Admission Prices: Mon.-Fri. 8-5. No charge; donations accepted.
Attendance: 550 (estimated)

Miami

DOBSON MUSEUM, 110 A St. S.W., Miami, OK 74354-6806. Mailing Address: P.O. Box 242, Miami, OK 74354. Tel.: 918-542-5388.
E-mail: ochs@dobsonmuseum.com
Web Site: www.dobsonmuseum.com
Institution Type/Description: History Museum.
Collections: local history & culture; Native American artifacts; photographs; paintings; mining; woodworking tools; furniture; period newspapers.
Hours & Admission Prices: Sun., Wed. & Fri. 1-4. No charge.

Midwest City

ATKINSON HERITAGE CENTER, 1001 N. Midwest Blvd., Midwest City, OK 73110-2799. Mailing Address: c/o Rose State College Foundation, 6420 S.E. 15th St., Midwest City, OK 73110. Tel.: 405-732-5832.
E-mail: atkinson@rose.edu
Web Site: www.rose.edu/atkinson-heritage-center
Institution Type/Description: Historic House Museum: housed in the former home of Midwest city founder, W.P. "Bill" Atkinson; built in 1955.
Collections: Atkinson family history; bound volumes of the Oklahoma Journal; a pony barn.
Activities: special events; rental facilities.
Hours & Admission Prices: By appointment.

Minco

MINCO HISTORICAL SOCIETY AND MUSEUM, 304 N.W. Main St., Minco, OK 73059. Tel.: 405-352-4480.
Founded: 2007.
Governing Authority: Tax-exempt: 501(c)(3).
Institution Type/Description: Historical Society Museum.
Collections: local history & culture; period furnishings; personal artifacts; photographs.
Hours & Admission Prices: Thurs.-Fri. 10-3; other times by appointment.

Muskogee

ATALOA LODGE MUSEUM, 2299 Old Bacone Rd., Muskogee, OK 74403-1568. Tel.: 918-683-4581, ext. 283; 888-682-5514, ext. 7283. Fax: 918-687-5913.
Web Site: www.bacone.edu/ataloa
Founded: 1932.
Congressional District: 2
Key Personnel: Museum Dir., John Timothy.
Personnel Profile: Interns 4.
Governing Authority: private; college. Parent Institution: Bacone College. Tax-exempt: 501(c)(3).
Institution Type/Description: Indian Artifacts Museum.

Collections: Native American material, stone artifacts; rugs; blankets; basketry; pottery; beadwork; quillwork. Historic Building: 1932 Ataloa Lodge.
Facilities: Bacone College Research Library available for use by appointment.
Activities: guided tours; lectures; permanent & rotating exhibitions.
Publications: booklets, Baconian; Smoke Signals.
Hours & Admission Prices: Wed.-Sat. 8:30-5, Sun. 1-5. No charge; donations accepted. Closed national holidays except by appointment. &
Attendance: 6,000

THE FIVE CIVILIZED TRIBES MUSEUM, 1101 Honor Heights Dr., Muskogee, OK 74401-1321. Tel.: 918-683-1701. Fax: 918-683-3070.
E-mail: 5tribesdirector@sbcglobal.net
Web Site: www.fivetribes.org
Founded: 1966.
Congressional District: 2
Key Personnel: Dir., Mary Robinson.
Governing Authority: nonprofit organization; board of directors. Tax-exempt.
Institution Type/Description: American Indian Museum & Art Gallery: housed in 1875 Union Indian Agency Building; pertains to the Cherokee, Chickasaw, Choctaw, Creek & Seminole tribal histories & American Indian Territory History.
Collections: paintings in traditional American Indian art style & sculpture by Cherokee, Choctaw, Chickasaw, Creek & Seminole tribe artists on tribal subjects; artifacts; photographs; books; documents; manuscripts; maps.
Research Fields: history; literature; traditions; cultures; genealogy; theater; music; biography; artifacts; source correspondence; sculpture.
Facilities: library; art gallery. Items created by members of the Five Civilized Tribes for sale.
Activities: guided tours; gallery talks; formally organized education programs for children; annual art show, student art competition, craft show, competitive art show, masters art show.
Publications: newsletter; book, Poems of Alexander Lawrence Posey-Creek Indian Bard, Art Prints, Cherokee Book, Muskogee Book, Pow Wow Chow Cookbook.
Hours & Admission Prices: Mon.-Fri. 10-5, Sat. 10-2. Adults $3, seniors 62 & up $2, students $1.50. &
Attendance: 28,728 (accurate)
Membership: Senior Citizens $20; Associate $30; Sustaining $50; DA-CO-TAH $100; Sequoyah $300; Osceola $500; The Golden Eagle $1,000; The Council $2,500.

MUSKOGEE WAR MEMORIAL PARK & MILITARY MUSEUM, (M), 3500 Batfish Rd., Muskogee, OK 74401. Mailing Address: P.O. Box 253, Muskogee, OK 74402-0253. Tel.: 918-682-6294. Fax: 918-682-1642.
E-mail: ussbatfish@sbcglobal.net
Web Site: www.ussbatfish.com
Formerly: USS Batfish
Founded: 1972.
Congressional District: 2
Key Personnel: Park Mgr., Rick Dennis; Chm. Bd., James Gully; Sec., Pam Bush.
Personnel Profile: Full-Time Paid 1; Full-Time Volunteers 8; Part-Time Volunteers 20; Interns 2.
Governing Authority: municipal; nonprofit. Tax-exempt.
Institution Type/Description: Maritime Naval Museum: War Memorial & WWII Submarine, U.S.S. Batfish.
Collections: artifacts from all branches of service: Army, Navy, Marine, & Air Force; memorial to 52 subs lost in World War II; Civil War to present; USS Batfish submarine.
Facilities: meeting room. Museum-related items for sale.
Activities: self-guided submarine tours; picnic area; overnight program.
Publications: tour guide, Batfish.
Hours & Admission Prices: mid-March to mid-Oct. Wed.-Sat. 10-6, Sun. 1-6. Adult $6, senior citizens $4, children $3; discounts available for groups, military & AAA and AAM members. Closed Easter; Thanksgiving; Christmas.
Attendance: 11,000 (accurate)
Membership: Senior, Student & Veteran $20; Individual $30; Couples $35; Family $40.

THOMAS-FOREMAN HOME, 1419 W. Okmulgee, Muskogee, OK 74401-6740. Tel.: 918-686-6624. Fax: 918-682-3477. Facebook; Thomas-Foreman Historic Home.
E-mail: 3riversmuseum@sbcglobal.net
Web Site: www.thomas-foremanhistorichome.com
Founded: 1970.
Congressional District: 2

Key Personnel: Chm. (V), Sue Tolbert; Pres. (V), Judy Dotson.
Personnel Profile: Part-Time Volunteers 8.
Governing Authority: Parent Institution: Three Rivers Museum, 220 Elgin, Muskogee, OK 74401. Supporting Organization: Friends of Thomas-Foreman Home. Tax-exempt.
Institution Type/Description: Historic House: 1898 Home of Judge John R. Thomas and Grant & Carolyn Thomas Foreman.
Collections: original furnishings; private collections of Thomas & Foreman families.
Research Fields: pre-statehood Oklahoma.
Activities: guided tours. Museum Sponsors: Azalea Festival in April.
Hours & Admission Prices: Fri.-Sat. 10-5; tours by appointment. Adults $2, seniors & students $1. Closed national holidays.
Attendance: 1,200
Membership: Individual $20; Family $30; Builder $100; Friend $250; Sustainer $500.

THREE RIVERS MUSEUM, 220 Elgin, Muskogee, OK 74401-7019. Tel.: 918-686-6624. Fax: 918-682-3477. Facebook: Three Rivers Museum.
E-mail: 3riversmuseum@sbcglobal.net
Web Site: www.3riversmuseum.com
Founded: 1985.
Congressional District: 2
Key Personnel: Chm., Roger Bell; Dir., Sue Tolbert.
Personnel Profile: Full-Time Paid 1; Part-Time Paid 2; Part-Time Volunteers 20.
Governing Authority: Parent Institution: Three Rivers Museum of Muskogee, OK, Inc. Tax-exempt.
Institution Type/Description: History Museum: housed in restored 1916 Midland Valley Railroad Depot.
Collections: photographs; letters & documents; furnishings; clothing; military; railroad; area memorabilia.
Research Fields: Three Rivers region.
Facilities: conference room. Museum-related items for sale.
Activities: tours; special programs.
Publications: quarterly newsletter, 3 Rivers Historian; e-news.
Hours & Admission Prices: Wed.-Sat. 10-5. Adults $3, students $1.50; children under 6 no charge. &
Attendance: 2,000 (estimated)
Membership: Individual $25; Family $35; Builder $100; Sponsor $250; Sustainer $500.

Mustang

MUSTANG HISTORICAL MUSEUM, 470 W. State Hwy. 152, Mustang, OK 73064. Mailing Address: P.O. Box 464, Mustang, OK 73064. Tel.: 405-745-3365.
Founded: 1999.
Key Personnel: Pres. (V), Glen Muse.
Personnel Profile: Part-Time Volunteers 1.
Governing Authority: Tax-exempt.
Institution Type/Description: History Museum.
Collections: local history & culture; personal artifacts; photographs; period furnishings; agricultural farming equipment; Czech dance costumes; early hardware.
Activities: Annual Events: Antique Critique; Western Days.
Publications: History of Mustang, 1901-1914; Mustang Township Oklahoma.
Hours & Admission Prices: Sat. 10-2; other times by appointment. No charge; donations accepted.
Attendance: 90 (accurate)
Membership: Junior $5; Individual $10; Family $15; Lifetime $100.

Newkirk

NEWKIRK COMMUNITY HISTORICAL MUSEUM, 101 S. Maple, Newkirk, OK 74647-4026. Tel.: 580-362-2377. Fax: 580-362-3390.
Institution Type/Description: History Museum.
Collections: Kay County artifacts from 1893 to 1940; local history & culture; restored mail buggy; period furnishings; photographs; Native American artifacts.
Facilities: Museum-related items for sale.
Hours & Admission Prices: Sun. 2-4; other times by appointment. No charge.

Noble

TIMBERLAKE ROSE ROCK MUSEUM, 419 S. Hwy. 77, Noble, OK 73068-0663. Mailing Address: P.O. Box 663, Noble, OK 73068-0663. Tel.: 405-872-9838.
Key Personnel: Museum Shop Mgr., Nancy Stine
Institution Type/Description: Natural History Museum.
Collections: barite rose rocks; quartz crystals.
Activities: Museum Sponsors: Rose Rock Festival in May.
Hours & Admission Prices: Tues.-Fri. 10-6, Sat. 10-4. No charge.
Attendance: 4,000 (estimated)

Norman

FIREHOUSE ART CENTER, 444 S. Flood Ave., Norman, OK 73069-5513. Tel.: 405-329-4523. Fax: 405-292-9763.
E-mail: info@normanfirehouse.com
Web Site: www.normanfirehouse.com
Founded: 1971.
Key Personnel: Exec. Dir., Douglas Shaw Elder; Pres. (V), James Schwartz; Museum Shop Mgr., Sally Frech.
Personnel Profile: Full-Time Paid 3; Part-Time Paid 4.
Governing Authority: nonprofit organization. Tax-exempt.
Institution Type/Description: Art Center.
Collections: paintings; clay; sculpture; photographs.
Hours & Admission Prices: Mon.-Fri. 9:30-5:30, Sat. 10-4. No charge; donations accepted. Closed federal holidays. &
Membership: Basic $45; Family $70; Patron $120; Associate $300; Benefactor $500; Silver $1,000; Gold $1,500; Platinum $2,500; Sponsor $5,000.

*   **THE FRED JONES JR. MUSEUM OF ART, (M),** University of Oklahoma, 555 Elm Ave., Norman, OK 73019. Tel.: 405-325-3272 & 0843. Fax: 405-325-7696.
Web Site: www.ou.edu/fjjma
Founded: 1936.
Congressional District: 4
Key Personnel: Interim Dir., The Wylodean & Bill Saxon, Mark White, Ph.D.; Deputy Dir., Gail Kana Anderson; Mgr. Administration & Operations, Becky Z. Trumble; Registrar, Tracy Bidwell; Dir. Education, Susan G. Baley; Mgr. Audience Devel., Chelsea Julian; Cur. Schools & Family Programs, Karen McWilliams; Cur. Eugene B. Adkins & Chief Cur., Mark A. White; Asst. Cur. James T. Bialac Native American Art & Non-Western Art, Heather Ahtone; Head Preparator, Brad Stevens; Preparator, Ian Carrig; Museum Shop Mgr., Jin Jo Garton; Asst. Museum Shop Mgr., Kati Kain; Administrative Asst., Tanya Denton; Dir. Museum Security & Facilities & Special Events, Joyce Cummins; Asst. Security Supvr., Josh Puckett; Asst. Security Supvr., Robert Santa Cruz; Asst. Security Supvr., John Paul, iV; Asst. Registrar, Selena Capraro; Customer Rels., Lacy Jo Burgess; Customer Rels., Melissa Verville; Coord. Academic Programs, Jessica Farling.
Personnel Profile: Full-Time Paid 32; Part-Time Paid 5; Part-Time Volunteers 100; Interns 6.
Governing Authority: state; university. Parent Institution: University of Oklahoma. Tax-exempt: 501(c)(3).
Institution Type/Description: Art Museum.
Collections: French impressionism; American art; Native American art; photography; contemporary art; Asian art; European graphics from the 16th century to present.
Research Fields: French impressionism; 20th century American art; contemporary art; Native American art; art of the American West; photography.
Facilities: Museum-related items for sale.
Activities: guided tours; lectures; films; gallery talks; concerts; inter-museum loan; permanent, temporary & traveling exhibitions.
Publications: exhibition calendar; posters; catalogs; brochures; postcards.
Hours & Admission Prices: Tues.-Thurs. & Sat. 10-5, Fri. 10-9, Sun. 1-5. No charge. &
Attendance: 31,546 (accurate)
Membership: OU Faculty/Staff Individual $25; OU Faculty/Staff Family & Individual $40; Metro Arts Circle $50; Family $60; Associate $100; Supporter $250; Patron $500; Benefactor $1,000; Friend of the Arts $2,000; Director's Circle $5,000.

MOORE-LINDSAY HOUSE HISTORICAL MUSEUM, 508 N. Peters, Norman, OK 73069-7251. Tel.: 405-321-0156.
E-mail: cchs@coxinet.net
Web Site: www.normanhistorichouse.com
Formerly: The Norman Cleveland County Historical Museum
Founded: 1973.
Congressional District: 4
Key Personnel: Dir., Stephen Martin; Pres., Vernon Maddux.
Personnel Profile: Full-Time Paid 1; Part-Time Paid 2; Part-Time Volunteers 6; Interns 1.
Governing Authority: nonprofit organization. Tax-exempt: 501(c)(3).
Institution Type/Description: Historical Society Museum: housed in 1899 Queen Anne style urban house.
Collections: furniture; furnishings; decorative arts; textiles; toys; books; photographs; documents relating to Cleveland County history, particularly during the territorial period; Victorian artifacts.
Research Fields: documentation of collections.
Facilities: library of books, photos, documents focusing on Cleveland county history c.1900, available for use on premises.
Activities: guided tours; formally organized education programs for children; adult evening programs; local history slide programs; temporary exhibitions.
Publications: quarterly newsletter, The Round Tower.
Hours & Admission Prices: Wed.-Sat. 10-4; other times by appointment. No charge; donations accepted. Closed holidays.
Attendance: 2,000 (estimated)
Membership: Individual $20; Family $25; Sustaining $50; Corporate $100.

ROBERT BEBB HERBARIUM, 770 Van Vleet Oval, Rm. 206, Norman, OK 73019-6155. Mailing Address: Department of Botany & Microbiology, Univ. of Oklahoma, 206 Cross Hall, Norman, OK 73019. Tel.: 405-325-7533. Fax: 405-325-7619.
E-mail: bebbherbarium@ou.edu
Web Site: www.biosurvey.ou.edu/bebb/bebbhome.html
Founded: 1920.
Congressional District: 4
Key Personnel: Cur., Wayne J. Elisens; Collections Mgr., Amy Buthod.
Personnel Profile: Full-Time Paid 2; Part-Time Paid 1.
Governing Authority: university. Parent Institution: University of Oklahoma. Tax-exempt.
Institution Type/Description: Herbarium.
Collections: botany; herbarium with emphasis on vascular plants of the Great Plains, Southwestern & Southeastern U.S. & Oklahoma.
Research Fields: molecular systematics; pollination ecology; floristics; phylogenetics; palynology; orchid phylogeny; flora of Oklahoma.
Facilities: associated with Noble Laboratory of Electron Microscopy & Oklahoma Natural Heritage Inventory.
Activities: inter-institutional loans for study; exchange of specimens.
Hours & Admission Prices: Mon.-Fri. 9-5. No charge. Closed university holidays. &

*   **SAM NOBLE OKLAHOMA MUSEUM OF NATURAL HISTORY, (M),** University of Oklahoma, 2401 Chautauqua, Norman, OK 73072-7029. Tel.: 405-325-4712. Fax: 405-325-7699.
E-mail: snomnh@ou.edu
Web Site: www.snomnh.ou.edu
Founded: 1899.
Congressional District: 4
Key Personnel: Dir. & Cur. Mammals, Dr. Michael A. Mares; Cur. Vertebrate Paleontology, Dr. R.L. Cifelli; Cur. Herpetology, Dr. Cameron Siler; Cur. Archeology, Dr. Marc Levine; Cur. Invertebrate Paleontology, Dr. S. Westrop; Asst. Cur. Mammals, Dr. J.K. Braun; Cur. Vertebrate Paleontology, Dr. N.J. Czaplewski; Coord. Dept. Computing Systems, Paul King; Head Operations, Melanie Davidson; Registrar & Repatriation Specialist, Elsbeth Dowd; Security & Facilities Operations, David Dagg; Cur. Ichthyology, Dr. E. Marsh-Matthews; Cur. Native American Languages, Dr. M. Linn; Head Education, Dr. Jes Cole; Museum Shop Mgr., Ksenia Goode.
Personnel Profile: Full-Time Paid 93; Part-Time Paid 53; Part-Time Volunteers 304.
Governing Authority: university. Parent Institution: University of Oklahoma. Tax-exempt: 501(c)(3).
Institution Type/Description: General Museum.
Collections: mammals; birds; fish; amphibians; reptiles; insects; invertebrates; botany; anthropology; ethnology; archaeology; history; classical art; paleontology; paleobotany; minerals; textiles; Indian artifacts; Native American languages.
Research Fields: zoology; mammalogy; anthropology; ornithology; ichthyology; geology; entomology; ethnology; plants; herpetology; history; mineralogy; classical art; archaeology; paleontology; textiles; museology.
Facilities: research labs; 174-seat auditorium; discovery room; teaching lab; classroom; cafe. Museum-related items for sale.
Activities: films, & guided tours of exhibits for school & other groups; summer workshops; public lectures; field trips for members & other adults; school programs & educational loans; inter-museum loan, traveling, temporary & permanent exhibits; special events; volunteer program.
Publications: newsletter, Sam Noble Oklahoma Museum of Natural History; miscellaneous publications; Occasional Papers of the Sam Noble Oklahoma Museum of Natural History.

Hours & Admission Prices: Museum: Mon.-Sat. 10-5, Sun. 1-5. Office: Mon.-Fri. 8-5. Adults $5, seniors $4, children $3; discounts for AAM & ICOM members, military and AAA members; members & children under 5 no charge. Closed major holidays. &

Attendance: 170,431 (accurate)

Membership: Senior $20; Individual $30; Family $45; Supporter $250; Patron $500; Benefactor $1,000.

Nowata

NOWATA COUNTY HISTORICAL SOCIETY MUSEUM, 121 S. Pine St., Nowata, OK 74048-3413. Tel.: 918-273-1191.

Founded: 1969.

Congressional District: 2

Governing Authority: society; nonprofit. Parent Institution: Nowata County Historical Society Board of Trustees. Tax-exempt: 501(c)(3).

Institution Type/Description: Historical Society Museum: housed in former Nowata Clinic Hospital Building.

Collections: cowboy room; tools; art; old dentist office; dining room with dishes; school rooms; 3 bedrooms with quilts, chambers cradle & old clothing; trophies & pictures; furniture; equipment; Indian Suite; Army & Navy display; wood carvings from all over the world; salt & pepper collections; doll room; oil field equipment.

Research Fields: history of the county & surrounding area.

Facilities: library of historical, fiction & non-fiction books available for research on premises. Museum-related items for sale.

Activities: tours.

Publications: quarterly newsletter, Nowata Historical Society Newsletter.

Hours & Admission Prices: Tues.-Sat. 1-4; other times by appointment. No charge; donations accepted. &

Attendance: 2,050 (estimated)

Membership: Annual $5; Lifetime $50.

Okemah

OKFUSKEE COUNTY HISTORY CENTER, 407 W. Broadway St., Okemah, OK 74859. Mailing Address: P.O. Box 409, Okemah, OK 74859. Tel.: 918-623-2440. Fax: 918-623-2440.

E-mail: ron.gott@vosdic.com

Web Site: okemahok.com

Formerly: Okfuskee County Historical Society

Founded: 1926.

Congressional District: 4

Key Personnel: Dir. & Pres. (V), Ron Gott; Museum Shop Mgr., Wayland Bishop.

Personnel Profile: Full-Time Paid 2; Part-Time Volunteers 2.

Governing Authority: Tax-exempt.

Institution Type/Description: Historical Society Museum: housed in a Masonic temple; built in 1926.

Collections: Okfuskee County history & culture; period furnishings; early schoolroom artifacts; Woody Guthrie.

Activities: school tours. Museum Sponsors: Woody Guthrie Festival.

Hours & Admission Prices: Call for hours. No charge; donations accepted. &

Attendance: 6,000 (estimated)

Oklahoma City

ASA NATIONAL SOFTBALL HALL OF FAME AND MUSEUM COMPLEX, 2801 N.E. 50th, Oklahoma City, OK 73111-7200. Tel.: 405-424-5266. Fax: 405-424-3855. Facebook: ASA USA Softball.

E-mail: cwarren@softball.org

Web Site: www.usasoftball.com

Founded: 1957.

Congressional District: 6

Key Personnel: Exec. Dir., Craig Cress; Pres. (V), Phil Gutierrez; Hall of Fame Svcs. Mgr., Codi Warren.

Personnel Profile: Full-Time Paid 28.

Governing Authority: nonprofit organization. Parent Institution: Amateur Softball Assoc. of America. Tax-exempt.

Institution Type/Description: Sports Museum.

Collections: plaques; uniforms; softball registry; memorabilia pertinent to softball; Hall of Fame inductees & memorabilia; Olympic team displays; softball video displays; USA softball memorabilia; NCAA Women's College World Series artifacts.

Research Fields: sports; history of ASA & USA softball.

Facilities: research center.

Activities: guided tours; films; video cassettes; permanent exhibitions.

Publications: brochures; flyers; pamphlets; monthly newsletter; quarterly online newsletter; rule books; catalogues; scorebooks.

Hours & Admission Prices: May-Sept. Mon.-Fri. 8-4:30, Sat. 10-4, Sun. 1-4; Oct.-April Mon.-Fri. 8-5. No charge; donations accepted. Closed New Year's Day; Independence Day; Thanksgiving & day after; Christmas. &

Attendance: 150,000 (estimated)

Membership: Life $150.

AMERICAN BANJO MUSEUM, (M), 9 E. Sheridan Ave., Oklahoma City, OK 73104-2424. Mailing Address: 116 E. Oklahoma Ave., Guthrie, OK 73044. Tel.: 405-604-2793.

E-mail: info@banjomuseum.org

Formerly: National Four-String Banjo Hall of Fame Museum

Key Personnel: Exec. Dir., Johnny Baier

Institution Type/Description: Musical Instrument Museum.

Collections: Banjo history & music; instruments; recordings; film; video; printed music; instructional materials; photographs.

Hours & Admission Prices: Tues.-Sat. 11-6, Sun. 12-5. Family $15, adults $6, seniors $5, children 6-17 $4.

[ARTSPACE] AT UNTITLED, 1 N.E. 3rd St., Oklahoma City, OK 73104. Tel.: 405-815-9995.

E-mail: info@artspaceatuntitled.org

Web Site: www.artspaceatuntitled.org

Key Personnel: Exec. Dir., Jon Burris.

Governing Authority: nonprofit organization. Tax-exempt: 501(c)(3).

Institution Type/Description: Contemporary Art Center.

Collections: works by contemporary national & international artists.

Hours & Admission Prices: Tues.-Fri. 10-5, Sat. 10-4. No charge. &

BETTY PRICE GALLERY, Oklahoma State Capitol, 2300 N. Lincoln Blvd., Oklahoma City, OK 73105. Mailing Address: P.O. Box 52001-2001, Oklahoma City, OK 73152-2001. Tel.: 405-521-2039. Fax: 405-521-6418. Facebook: Oklahoma Arts Council.

E-mail: clint.stone@arts.ok.gov

Web Site: www.arts.ok.gov/Art_at_the_Capitol/Betty_Price_Gallery.html

Founded: 2007.

Congressional District: 5

Governing Authority: Parent Institution: State of Oklahoma.

Institution Type/Description: Art Gallery.

Collections: basketry; sculpture; metal relief; painting; printmaking; ceramics; mixed media; modern and contemporary art & sculpture.

Major Exhibits: Skip Thompson, 2/3/14-3/14; J. Don Cook, 3/24/14-5/18/14; Virginia Stroud, 3/31/14-5/25/14; Brett Debring, 5/26/14-7/20/14; Mark Zimmerman, 7/28/14-9/21/14; Virgil Lampton, 8/4/14-9/28/14; Eyakem Gulilat, 9/29/14-11/14; Dana Tiger, 10/6/14-12/7/14; Angela Riehl, 12/8/14-1/15; David Holland, 12/22/14-1/15.

Activities: special events.

Hours & Admission Prices: Mon.-Fri. 8-6, Sat.-Sun. 9-4. No charge. &

CITY ARTS CENTER, (M), 3000 General Pershing Blvd., Oklahoma City, OK 73107-6202. Tel.: 405-951-0000. Fax: 405-951-0003.

E-mail: sherry@cityartscenter.org

Web Site: www.cityartscenter.org

Founded: 1989.

Congressional District: 5

Key Personnel: Chm. (V), Christian K. Keesee; Pres. (V), Lori Tyler; Exec. Dir., Mary Ann Prior; Administrative Dir., Sherry Fair; Artistic Dir., Clint Stone.

Personnel Profile: Full-Time Paid 9; Part-Time Paid 2; Part-Time Volunteers 100.

Institution Type/Description: Arts Center.

Collections: contemporary works.

Activities: performances; classes; special events; demonstrations; meetings; social events.

Hours & Admission Prices: Mon.-Thurs. 9am-10pm, Fri.-Sat. 9-5. No charge; donations accepted. Closed major holidays. &

Attendance: 8,000 (estimated)

CLASSEN HIGH SCHOOL MUSEUM, 1901 N. Ellison, Oklahoma City, OK 73137. Mailing Address: P.O. Box 270905, Oklahoma City, OK 73137. Tel.: 405-525-3936.

E-mail: info@classen.org

Web Site: classen.org

Institution Type/Description: History Museum: built in 1920.

Collections: Classen High School history, trophies, records & photographs from 1920 to present.

Hours & Admission Prices: Temporarily closed.

45TH INFANTRY DIVISION MUSEUM, (M), 2145 N.E. 36th, Oklahoma City, OK 73111-5396. Tel.: 405-424-5313 & 5393. Fax: 405-424-3748.
E-mail: curator@45thdivisionmuseum.com
Web Site: www.45thdivisionmuseum.com
Founded: 1976.
Congressional District: 5
Key Personnel: Dir., Col. Dave Brown, (Ret.); Cur., Michael E. Gonzales; Historian, Mike Alen Beckett; Cur. & Museum Shop Mgr., Peggy Covey; Vehicle & Weapons Specialist, SSG Clark Briand, (Ret.)
Personnel Profile: Full-Time Paid 5; Part-Time Paid 2; Part-Time Volunteers 33.
Governing Authority: state. Branch of Oklahoma Military Dept., 3501 Military Circle, Oklahoma City, OK 73111. Tel.: 405-425-8000. Tax-exempt: 501(c)(3).
Institution Type/Description: Military History Museum.
Collections: military arms including the American Civil War; personal property of Adolf Hitler; original Willie & Joe cartoons created by Bill Mauldin for which he received the Pulitzer prize; uniforms; accoutrements; vehicles; photos; maps; drawings; manuscripts; technical manuals; field manuals; charts; sound recordings; film; video tape.
Research Fields: military history of Oklahoma & the Southwestern United States.
Facilities: 7,000-vol. library of general military history; vehicle identification, technical manuals available for research; auditorium; theater; classrooms. Museum-related items for sale.
Activities: lectures; films; loan, permanent & temporary exhibitions.
Publications: occasional pamphlet, Thunderbird Imprints, Historical Monographs; unit histories; authorized Bill Mauldin prints.
Hours & Admission Prices: Tues.-Fri. 9-4:15, Sat. 10-4:15, Sun. 1-4:15. No entry within 45 minutes of closing, park gates locked no later than 5. No charge; donations accepted. Closed holidays; open patriotic holidays. &
Attendance: 45,000 (estimated)

GAYLORD-PICKENS OKLAHOMA HERITAGE MUSEUM, (M), 1400 Classen Dr., Oklahoma City, OK 73106-6614. Tel.: 405-235-4458.　　Fax:　　405-235-2714. www.facebook.com/okheritage.
E-mail: oha@oklahomaheritage.com
Web Site: www.oklahomaheritage.com
Founded: 1971.
Congressional District: 5
Key Personnel: Pres., Shanon L. Rich; Chm. (V), Neyle R. Cable.
Personnel Profile: Full-Time Paid 12; Part-Time Paid 6.
Operating Expenses: 2,515,090
Operating Income: 2,651,554
Governing Authority: private. Tax-exempt. Parent institution: Oklahoma Heritage Association.
Institution Type/Description: Heritage Museum.
Collections: Oklahoma Hall of Fame portraits, busts, & archives.
Activities: special exhibits; tours of center. Museum Sponsors: statewide scholarship history contest; essay & poster contest; Oklahoma Hall of Fame in November; 5K Run.
Publications: magazine, Oklahoma; Oklahoma Trackmakers Series; Oklahoma Horizons Series.
Hours & Admission Prices: Tues.-Fri. 9-5, Sat. 10-5. Adults $7, seniors citizens 62 & up and students 6-17 $5; discounts to groups of 15 or more; members no charge. Closed New Year's Day; Thanksgiving; Christmas. &
Attendance: 5,000
Membership: Regular $35; Centennial $100; Sponsor $250; Pioneer $500; Honor $1,000; Corporate Circle $2,500; President's Circle $5,000; Chairman's Circle $10,000.

HARN HOMESTEAD AND 1889ER MUSEUM, 1721 N. Lincoln Blvd., Oklahoma City, OK 73105-4911. Tel.: 405-235-4058. Fax: 405-235-4041.
E-mail: mgregg@harnhomestead.com
Web Site: www.harnhomestead.com
Founded: 1986.
Congressional District: 6
Key Personnel: Pres., Rob Saunders; Exec. Dir., Melessa Gregg.
Personnel Profile: Full-Time Paid 3; Part-Time Volunteers 3; Interns 1.
Governing Authority: Parent Institution: William Fremont Harn Gardens, Inc. Tax-exempt: 170(b)(1)(A).
Institution Type/Description: Historic House: housed in 1904 Victorian house & barn located on the site of 10-acre tract which was part of 1889 Land Run.
Collections: pre-1907 furnishings; 6 historic buildings; exhibit barn.
Facilities: 10-acre facility; picnic grounds.

Activities: guided tours; lectures; formally organized education programs for children & adults; docent program & council; permanent & temporary exhibits.
Publications: Children Today, Oklahoma Yesterday; Teachers Guide to Territorial History Activities.
Hours & Admission Prices: Mon.-Fri. 10-4. Adults $5, seniors & military $4; members and children 3 & under no charge. Closed Federal holidays. &
Attendance: 25,000 (accurate)
Membership: Boomer $35-$49; Stake Holder $50-$99; Homesteader $100-$249; Wagon Master $250-$749; Marshal $750-$1,499; Territorial Governor $1,500 & up.

INTERNATIONAL PHOTOGRAPHY HALL OF FAME & MUSEUM, 2100 NE 52nd St., Oklahoma City, OK 73111-7107. Tel.: 405-424-4055. Fax: 405-424-4058.
E-mail: info@iphf.org
Web Site: www.iphf.org
Founded: 1963.
Key Personnel: Exec. Dir., Michael Scalf, Sr.; Pres. (V), John Nagel.
Governing Authority: Tax-exempt.
Institution Type/Description: Photography Museum.
Collections: cameras, darkroom & studio equipment from the 20th century; 20th century photographic images.
Hours & Admission Prices: Mon.-Fri. 9-5, Sat. 9-6, Sun. 11-6. Adults $9.50, seniors 65 & over and children 3-12 $8.25. Closed Thanksgiving; Christmas Eve & Day.
Membership: Student $20; Individual $28; Supporting $50; Exhibitor $150; Elector $250; Patron $500.

MELTON ART REFERENCE LIBRARY, 4300 N. Sewell, Oklahoma City, OK 73118-8010. Tel.: 405-525-3603. Fax: 405-525-0396.
E-mail: MeltonArt@aol.com
Web Site: www.marl-okc.org
Founded: 1979.
Congressional District: 6
Key Personnel: C.E.O. & Pres., Robynne Mulcahy; Dir., Suzanne Silvester.
Personnel Profile: Part-Time Paid 1; Part-Time Volunteers 1; Interns 1.
Governing Authority: nonprofit organization. Tax-exempt: 501(c)(3).
Institution Type/Description: Art Resource Library.
Collections: monographs, raisonnes & biographies of artists; original art.
Research Fields: lesser known artists & their works.
Facilities: 6,000-vol. library of raisonnes, biographies & autobiographies of artists, art reference & resource books & catalogs available for use by the public; educational facilities.
Activities: guided tours; lectures; organized education programs for children & adults; loan, temporary & traveling exhibitions; study access for art students.
Publications: resource book; Artists and Their Museums - Vol. I & II; Progress on the Land; Legacy Collection; Directory of Oklahoma Artists; History of the Art Renaissance Club; Directory of Oklahoma Artists.
Hours & Admission Prices: Mon.-Fri. 10-5. No charge; donations accepted. Closed holidays. &
Attendance: 1,500

MUSEUM OF OSTEOLOGY, (M), 10301 S. Sunnylane Rd., Oklahoma City, OK 73160. Tel.: 405-814-0006. Fax: 405-794-6985.
E-mail: info@museumofosteology.org
Web Site: www.museumofosteology.org
Founded: 2010.
Key Personnel: Founder, Jay Villemarette; Dir., Joey Williams; Mktg., Josh Villemarette; Education, Kristi Carlucci.
Personnel Profile: Full-Time Paid 4; Part-Time Paid 1.
Governing Authority: private; nonprofit organization. Tax-exempt: 501(c)(3).
Institution Type/Description: Science Museum.
Collections: over 10,000 skeletal artifacts; over 300 human & animal skeletons.
Facilities: classrooms. Museum-related items for sale.
Activities: temporary exhibitions.
Hours & Admission Prices: Mon.-Fri. 8-5, Sat. 11-5, Sun. 1-5. Adults $6; discounts to groups of 20 or more; children under 3 no charge. Closed New Year's Day; Memorial Day; Labor Day; Thanksgiving; Christmas.
Attendance: 75,000 (accurate)

MYRIAD BOTANICAL GARDENS, 301 W. Reno Ave., Oklahoma City, OK 73102. Tel.: 405-297-3995. Fax: 405-297-3620.
Web Site: www.oklahomacitybotanicalgardens.com
Founded: 1975.

Key Personnel: Exec. Dir., Maureen Heffernan; Dir. Facilities, Matthew Maly; Dir. Devel., Kelley Barnes; Dir. Myriad Gardens Foundation Community Bd., Debora Morey; Dir. Mktg. & Communications, Christine Eddington; Dir. Horticulture, Casey Sharber; Chief Financial Officer, Tammie Lowe.
Governing Authority: municipal. Parent Institution: City of Oklahoma City Parks & Recreation Dept.
Institution Type/Description: Botanical Gardens.
Collections: over 1,000 species of plants.
Facilities: library; 17 acres; botanical garden. Museum-related items for sale.
Activities: arts festival; concert; docent program; formal education programs for children; guided tours; hobby workshops; lectures; theater; temporary & traveling exhibitions; radio programs; plant sales; rentals; meetings. Annual Events: Holiday Lights; Independence Day; Mother & Father's Day.
Publications: quarterly newsletter, Crystal Connection.
Hours & Admission Prices: June to Labor Day Mon.-Sat. 9-5, Sun. 11-7; Sept.-May Mon.-Sat. 9-5, Sun. 11-5. Adults $7, seniors 62 & over and students $6, children 4-12 $4; children 3 & under no charge.
Attendance: 1,000,000
Membership: Individual $25; Family & Grandparents $35; Contributor $50; Sustainer $100.

✳ **NATIONAL COWBOY & WESTERN HERITAGE MUSEUM,** 1700 N.E. 63rd St., Oklahoma City, OK 73111-7997. Tel.: 405-478-2250. Fax: 405-478-4714.
E-mail: askarchives@nationalcowboymuseum.org
Web Site: www.nationalcowboymuseum.org
Formerly: National Cowboy Hall of Fame and Western Heritage Center
Founded: 1955.
Congressional District: 6
Key Personnel: Exec. Dir., Chuck Schroeder; Asst. Dir. & Cur. Native American Collections, Mike Leslie; C.F.O., Gary Moore; Dir. Devel., Craig Clemons; Dir. Mktg. & Editor-in-Chief, Leslie Baker; Dir. Research Center, Gerrianne Schaad; Dir. Education, Gretchen Jeane; Dir. Operations, Doug Lane; Mgr. Information Systems, Sharon Kasper; Cur. Art, Anne Morand; Cur. History, Richard Rattenbury; McCasland Chair Cowboy Culture, Don Reeves; Dir. Tourism Mktg., Aaron Martin; Dir. Publications, Judy Hilovsky; Dir. Public Rels. & Museum Events, Catherine Page; Registrar, Melissa Owens; Museum Shop Mgr., Laney Carey; Facility Sales Mgr., Charlene Ferris.
Personnel Profile: Full-Time Paid 108; Part-Time Paid 17; Part-Time Volunteers 300; Interns 3.
Governing Authority: nonprofit organization. Tax-exempt: 501(c)(3).
Institution Type/Description: Art & History Museum.
Collections: contemporary Western & Native American fine arts including works by Russell, Remington, Schreyvogel & Fechin; Arthur & Shifra Silberman Native American Fine Arts Collection; Joe Grandee Collection of the Frontier West; cowboys & ranching; rodeo history; western popular culture; frontier military history materials; Native American material cultural; Weitzenhoffer Fine American firearms; Glenn D. Shirley Western Americana Collection.
Research Fields: fine arts (Western & Native American); Native American material culture; cowboy culture & ranching; rodeo history; western popular culture; frontier military; social history; firearms; Western Americana; popular western imagery.
Facilities: 31,000-vol. library & archives at research center; restaurant; gardens; education center; theatre; banquet facilities. Museum-related items for sale.
Activities: guided tours; Prix de West seminars & demonstrations; after school program; docent program; chuckwagon gathering; art education workshops; Western Heritage Awards; rental facilities. Museum Sponsors: Tuesdays at Sundown; Cowboy Christmas Ball; Traditional Cowboy Arts Association Seminar.
Publications: quarterly magazine, Persimmon Hill; quarterly, The Ketchpen.
Hours & Admission Prices: Daily 10-5. Adults $12.50, senior citizens & students $9.75, children $5.75; discount to groups & AAM members; members no charge. Closed New Year's Day; Thanksgiving; Christmas. &
Attendance: 203,529 (accurate)
Membership: Premium Family $75; Golden Spike Society $250; Pony Express Society $500; Grand Canyon Society $1,000; Prix De West Society $3,500; American West Society $5,000; Remington & Russell Society $10,000.

NINETY-NINES MUSEUM OF WOMEN PILOTS, (M), 4300 Amelia Earhart Rd., Oklahoma City, OK 73159-1106. Tel.: 405-685-9990.
E-mail: museum@ninety-nines.org
Web Site: museumofwomenpilots.com
Institution Type/Description: History Museum.
Collections: women pilots & aviation history; personal artifacts; photographs; aviation equipment.
Activities: special events.

Hours & Admission Prices: Call for hours.

THE OKLAHOMA BLACK MUSEUM AND PERFORMING ARTS CENTER, 4701 N. Lincoln Blvd., Oklahoma City, OK 73105-3318. Tel.: 405-213-8077.
Institution Type/Description: Black History Museum.
Collections: Black history, heritage & culture; photographs; sculpture; paintings.
Activities: special events; educational programs.
Hours & Admission Prices: Call for hours.

✳ **OKLAHOMA CITY MUSEUM OF ART, (M), (I),** 415 Couch Dr., Oklahoma City, OK 73102-2214. Tel.: 405-236-3100, ext. 200. Fax: 405-236-3122.
E-mail: emwhittington@okcmoa.com
Web Site: www.okcmoa.com
Founded: 1945.
Key Personnel: C.E.O. & Pres., E. Michael Whittington; Chm., Frank W. Merrick; Exec. Asst. to Pres. & C.E.O., Michelle Lory; Exec. Office Coord., Ashley Jimenez; Dir. Finance, Rodney L. Lee; Mgr. Human Resources & Accountant, Jennifer Johnson; Dir. Devel., Sandy Cotton; Cur., Alison Amick; Cur., Jennifer Klos; Cur. Film, Brian Hearn; Project & Film Asst., Dudley Marshall; Curatorial Asst., Catherine Shotick; Education & Community Outreach Coord., Amanda Harmer; Sr. Assoc. Cur. Educator, Chandra Boyd; Asst. Cur. Education, Bryon Chambers; Education Admin. Asst., Neely Simms; Dir. Facilities Operations, Jack Madden; IT Mgr., Patrick Fisher; Sr. Devel. Officer, Whitney Moore; Mgr. Mktg. & Communications, Ralph Cornelius; Head Design & Installation, Ernesto Sanchez; Preparator, Trent Lawson; Registrar, Maury Ford; Mgr. Database, Sandy Stewart; Finance Asst., Diane Glenn; Visitor Svcs. Assoc., Marsha Jones; Chief Safety & Security, Josh Silverhorn; Museum Shop Mgr., Wendy Neer; Mgr. Museum Cafe, Ahmad Farnia; Asst. Mgr. Cafe, Lauren Cates; Sous Chef, Henry Boudrewaux.
Personnel Profile: Full-Time Paid 45; Part-Time Paid 20; Part-Time Volunteers 200; Interns 8.
Governing Authority: nonprofit organization. Tax-exempt: 501(c)(3).
Institution Type/Description: Art Museum.
Collections: 19th-21st century European & American art; contemporary art & photography; glass sculpture by Dale Chihuly.
Research Fields: 19th & 20th-century American & European art.
Facilities: library; theatre; cafe; restaurant; education center. Museum-related items for sale.
Activities: guided tours; lectures; films; education programs for children & adults; docent program.
Publications: quarterly newsletter; exhibition brochures & catalogues.
Hours & Admission Prices: Tues.-Wed. & Fri.-Sat. 10-5, Thurs. 10-9, Sun. 12-5. Adults $12, senior citizens & students $10, military $5; discounts to groups of 15 or more & AAM members; members and children 5 & under no charge. Closed New Year's Day; Easter; Independence Day; Thanksgiving; Christmas. &
Attendance: 135,000 (accurate)
Membership: Student $20; Individual $50; Dual $75; Family $85; Fellow $100; Friend $500; Sustainer $1,000.

✳ **OKLAHOMA CITY NATIONAL MEMORIAL & MUSEUM, (M),** 620 N. Harvey, Oklahoma City, OK 73102-3032. Mailing Address: P.O. Box 323, Oklahoma City, OK 73101-0323. Tel.: 405-235-3313. Fax: 405-235-3315.
E-mail: kariwatkins@oklahomacitynationalmemorial.org
Web Site: www.oklahomacitynationalmemorial.org
Founded: 1995.
Congressional District: 5
Key Personnel: Exec. Dir., Kari F. Watkins; Dir. Operations, Joanne Riley; Chm. (V), Gary Pierson; Visitor Svcs. Mgr., Joyce Andrews.
Personnel Profile: Full-Time Paid 22; Part-Time Paid 3; Part-Time Volunteers 65.
Governing Authority: private; nonprofit organization. Parent Institution: Oklahoma City National Memorial Foundation. Tax-exempt: 501(c)(3).
Institution Type/Description: Historic Foundation.
Collections: terrorism & the bombing of Oklahoma City's Alfred P. Murrah Federal Building; photographs; audio recordings; damaged furnishings & building pieces; personal artifacts; videos; memorial tribute.
Research Fields: terrorism; justice; historical; memorialization; archives; community response.
Facilities: archives; reading room; center for education & outreach.
Publications: Walking Tour of the Oklahoma National Memorial & Museum.
Hours & Admission Prices: Mon.-Sat. 9-6, Sun. 12-6. Adults $12, seniors, students & military $10; discounts to groups. Closed New Year's Day; Easter; Thanksgiving; Christmas Eve & Day. &

Attendance: 389,246 (accurate)

＊ OKLAHOMA CITY ZOO, 2000 Remington Pl., Oklahoma City, OK 73111-7199. Mailing Address: 2101 N.E. 50th St., Oklahoma City, OK 73111. Tel.: 405-424-3344. Fax: 405-425-0207.
Web Site: www.okczoo.com
Founded: 1904.
Congressional District: 97
Key Personnel: C.E.O. & Exec. Dir., Dwight Scott; Exec. Dir., Oklahoma Zoological Society, Dana McCrory; Dep. Dir., Alan Varsik; Dir. Veterinary Services, Brian Aucoin; Dir. Public Rels. & Mktg., Tara Henson; Dir. Education, Teresa Randall; Dir. Bldgs. & Grounds, Ernest Wilson; Dir. Finance, Sandra Hover; Cur., Darcy Henthorn; Cur., Pearl Pearson.
Governing Authority: municipal; nonprofit organization. Parent Institution: Oklahoma City Zoological Trust. Tax-exempt.
Institution Type/Description: Zoo.
Collections: over 2,100 animals representing 500 species; exotic horticultural collection.
Research Fields: ethology; reproduction; camouflage; husbandry.
Facilities: aquarium; aquatics center; 300-seat auditorium; classrooms; food concessions. Gift items for sale.
Activities: guided tours; lectures; concerts; formally organized education programs for children, adults, undergraduate & graduate students; docent program; zoomobiles.
Publications: weekly newsletter, Gnusweek.
Hours & Admission Prices: Daily 9-5. Adults $8, children 3-11 & senior citizens $5; discounts to active-duty & retired military and their families; members and children 2 & under no charge. Closed New Year's Day; Thanksgiving; Christmas. &
Attendance: 787,717 (accurate)
Membership: Junior $15; Wildcard $45; Family $65; Family Plus $90; ZooPride $145-$995.

OKLAHOMA FIREFIGHTERS MUSEUM, 2716 N.E. 50th St., Oklahoma City, OK 73111-7299. Tel.: 405-424-3440; 800-308-5336. Fax: 405-425-1032.
E-mail: jims@osfa.info
Web Site: osfa.info
Founded: 1970.
Congressional District: 5
Key Personnel: Dir., Jim Sanders; Mgr., Mike Billingsley.
Personnel Profile: Full-Time Paid 2; Part-Time Paid 2; Part Time Volunteers 6.
Governing Authority: nonprofit organization. Parent Institution: The Oklahoma State Firefighters Association. Tax-exempt: 501(c)(3).
Institution Type/Description: Fire-Fighting History Museum.
Collections: fire-fighting apparatus & appliances. Historic Building: 1869 first fire station in Oklahoma.
Facilities: Fire service items & gifts for sale.
Activities: audio tours; guided tours; safety DVD; lectures; films; formally organized education programs for children; permanent exhibitions; rental facilities.
Publications: Oklahoma Firefighter; Oklahoma Chiefs Association.
Hours & Admission Prices: Mon.-Sat. 9-4:30, Sun. 1-4:30. Adults $5, senior citizens $4, children 6-12 $2; discounts to AAM, ICOM & AAA members; members & children under 6 no charge. Closed major holidays. &
Attendance: 7,000 (accurate)
Membership: Retired $22; Chief $45; Firefighter $56.

OKLAHOMA HISTORICAL SOCIETY, 800 Nazih Zuhdi Dr., Oklahoma City, OK 73105-7917. Tel.: 405-521-2491. Fax: 405-521-2492.
Web Site: www.okhistory.org
Founded: 1893.
Congressional District: 5
Key Personnel: Exec. Dir., Dr. Bob Blackburn; Pres., Emmy Stidham; Deputy Dir., Dr. Tim Zwink; Dir. Research Div., Chad Williams; Oklahoma Museum of History Dir., Dan Provo; Dir. Museums & Historic Sites Div., Kathy Dickson; Dir. Publications, Elizabeth Bass; Dir. Finance, Terry Howard; Dir. Personnel, Sherri Henderson; Deputy State Historic Preservation Officer, Melvena Thurman Heisch; Museum Shop Mgr., Jera Winters.
Personnel Profile: Full-Time Paid 144; Part-Time Paid 45; Part-Time Volunteers 1,500; Interns 4.
Volunteer Hours: 9,000
Operating Expenses: 5,200,000

Operating Income: 632,000
Governing Authority: state. Branch Museums: Oklahoma Territorial Museum, Guthrie; Museum of the Western Prairie, Altus; Tom Mix Museum, Dewey; Fort Gibson; Cherokee Strip Regional Heritage Center, Enid; Cherokee Strip Museum, Perry; No Man's Land Museum, Goodwell; Oklahoma History Center, Oklahoma City; White Hair Memorial, Ralston Historic Sites & Houses: 1830 Fort Towson; 1842 Fort Washita; 1920's Jim Thorpe Home, Yale; 1898 Peter Conser House, Heavener; 1828 Sequoyah's Cabin, Sallisaw; 1894 Sod House, Aline; 1930s Frank Phillips' Mansion, Bartlesville; 1903 Overholser Mansion, Oklahoma City; 1905 Drummond Home, Hominy; Chisholm Trail & Seay Mansion, Kingfisher; Pioneer Woman Museum, Ponca City; Oklahoma Rte. 66 Museum, Clinton; Fort Supply, Fort Supply; George M. Murrell Home Site, Park Hill; Pawnee Bill Ranch Site, Pawnee; Spiro Mounds Archaeological Park, Spiro; T.B. Ferguson Home Site, Watonga. Tax-exempt.
Institution Type/Description: State Historical Society Museums.
Collections: approx. 1,000,000 objects; Native American collections covering indigenous & removal tribes, Spiro Mounds objects, archeological material from early historic sites; military collections; political campaign material; clothing; decorative arts; printing equipment; farm & ranch implements; paintings; toys; textiles; business machines. For more specific information see individual listings.
Major Exhibits: Century Chest, 1/14-12/14; Alan Houser: Born to Freedom, 4/14-12/14.
Research Fields: Oklahoma history; U.S. Indian policy; Native American tribes in Oklahoma; genealogy; historic architecture; military in Oklahoma.
Facilities: 80,000-vol. library of history & genealogy material available for research on premises; newspapers; microfilm; three million manuscripts & documents. Museum-related items for sale.
Activities: lectures; films; gallery talks; formally organized education programs for children; permanent, temporary & traveling exhibitions.
Publications: quarterly journal, The Chronicles of Oklahoma; brochures; the Oklahoma Series; exhibition catalogs; monthly newsletter, Mistletoe Leaves.
Hours & Admission Prices: Oklahoma History Center: Mon.-Sat. 10-5. Administration Offices: Mon.-Fri. 8-5. Field Facilities: Tues.-Fri. 9-5, Sat.-Sun. 2-5. Center: adults $7; discounts to AAM & OMA members; members no charge. Closed New Year's Day; Thanksgiving; Christmas. &
Attendance: 297,543 (accurate)
Membership: Individual $35; Institutional $50; Friend $100; Associate $250; Fellow $500; Director's Circle $1,000; Benefactor $5,000.

OKLAHOMA MUSEUM OF HISTORY, (M), 800 Nazih Zuhdi Dr., Oklahoma City, OK 73105. Tel.: 405-522-5248. Fax: 405-521-5402.
Web Site: www.okhistory.org
Formerly: The State Museum of History
Founded: 1913.
Congressional District: 5
Key Personnel: Exec. Dir. & C.E.O., Dr. Bob Blackburn; Bd. Pres. & Chm. (V), Emmy Scott Stidham; Museum Dir., Dan Provo; Asst. Museum Dir., Jeff Briley; Dir. Education, Jason Harris; Dir. Exhibits, David Davis; Research Dir., Bill Welge; Outreach Dir., Kathy Dickson; Museum Shop Mgr., Russ Haynes.
Personnel Profile: Full-Time Paid 29; Part-Time Paid 8; Part-Time Volunteers 99; Interns 2.
Governing Authority: board of directors. Parent Institution: Oklahoma Historical Society (see separate listing). Tax-exempt.
Institution Type/Description: State History Museum.
Collections: material pertaining to state history; decorative arts; textiles; costumes; Native American artifacts; research collections; oil & gas; fine art; popular culture.
Research Fields: Oklahoma history; American Indian; firearms; furniture; textiles; photography; genealogy.
Facilities: library; classroom; archives; research center; special event center; cafe. Gift items for sale.
Activities: special exhibits; extension & group programs; student education; school programs; tours; evening programming; traveling trunks.
Publications: newsletter, Mistletoe Leaves; Chronicles of Oklahoma.
Hours & Admission Prices: Mon.-Sat. 10-5. Families $18, adults $7, seniors $5, students $4; discount to AAM members, Time Travelers Network and Smithsonian Institute & affiliate members; children 5 & under and members no charge. Closed New Year's Day; Thanksgiving; Christmas. &
Attendance: 185,000 (accurate)
Membership: Single/Individual $35; OHS Institutional, Family & Dual $50; Friend $100; Associate $250; Fellow $500; Director's Circle $1,000; Benefactor $5,000.

OKLAHOMA MUSEUM OF TELEPHONE HISTORY, 111 Dean A McGee Ave., Rm. 178, Oklahoma City, OK 73102. Tel.: 405-236-6153.
E-mail: oktelmuseum@yahoo.com
Founded: 1997.
Congressional District: 5
Governing Authority: Parent Institution: Telephone Pioneers of America, Oklahoma Chapter. Tax-exempt.
Institution Type/Description: History Museum.
Collections: telephones from 1900 to present; switchboards; magneto wall phones; decorator phones; tools; hands-on exhibits.
Hours & Admission Prices: Mon. & Wed. 10-2. No charge. &
Attendance: 200 (accurate)

OKLAHOMA MUSEUMS ASSOCIATION, 2100 NE 52nd St., Oklahoma City, OK 73111-7107. Tel.: 405-424-7757. Fax: 405-427-5068.
E-mail: info@okmuseums.org
Web Site: www.okmuseums.org
Founded: 1972.
Congressional District: 5
Key Personnel: Exec. Dir., Brenda Granger; Pres., Gena Timberman.
Personnel Profile: Full-Time Paid 2; Part-Time Paid 1; Part-Time Volunteers 65; Interns 1.
Volunteer Hours: 2,936
Operating Expenses: 250,042
Operating Income: 250,679
Governing Authority: not-for-profit organization. Tax-exempt: 501(c)(3).
Institution Type/Description: Museum Service Organization.
Collections: museum service organization.
Facilities: library.
Activities: conferences; seminars; traveling exhibitions; consultations; workshops; online discussions.
Publications: quarterly newsletter, Musenews.
Hours & Admission Prices: Call for hours. &
Membership: Individual: Student $15; Retired Professional $35; Individual $50; Friend $100; Patron $200. Institutional (based on staff size): Elevated $100-$1,000; Pinnacle $125-$1,125. Affiliated Organization & Corporate: Level I $250; Level II $500.

OKLAHOMA RAILWAY MUSEUM, 3400 N.E. Grand Blvd., Oklahoma City, OK 73111-4417. Tel.: 405-424-8222. Fax: 405-424-0504.
Web Site: www.oklahomarailwaymuseum.org
Founded: 1999.
Congressional District: 5
Key Personnel: Pres. (V), Stan Hall; Sec., Drake Rice.
Governing Authority: Tax-exempt.
Institution Type/Description: Railway Museum.
Collections: Oklahoma railroad history.
Activities: seasonal train rides.
Hours & Admission Prices: Thurs.-Sat. 9-4. Museum: no charge; donations accepted. Train Rides: April-Sept. first & third Sat. adults 13 & over $8, children 3-15 $5; discounts to NRHS, ARM & TRAIN members; children under 3 no charge.
Attendance: 20,368 (estimated)
Membership: Student & Senior $30; Individual $36.

OKLAHOMA VISUAL ARTS COALITION, 730 W. Wilshire Blvd., Ste. 104, Oklahoma City, OK 73116-7738. Tel.: 405-232-6991. Fax: 405-316-5611.
E-mail: director@ovac-ok.org
Web Site: www.ovac-ok.org
Key Personnel: Exec. Dir., Julia Kirt; Prog. Asst., Stephanie Ruggles Winter; Publications & Mktg. Mgr., Kelsey Karper
Institution Type/Description: Art Museum.
Collections: works by Oklahoma artists.
Hours & Admission Prices: Call for hours.

OVERHOLSER MANSION, 405 N.W. 15th, Oklahoma City, OK 73103-3503. Tel.: 405-525-5325.
E-mail: overholsermansion@preservationok.org
Web Site: overholsermansion.org
Founded: 1972.
Congressional District: 5
Key Personnel: Exec. Dir., Preservation Oklahoma, David Pettyjohn; Museum Coord., Lisa Escalon.
Personnel Profile: Full-Time Paid 2.

Governing Authority: state. Parent Institution: Oklahoma Historical Society (See separate listing.) Tax-exempt.
Institution Type/Description: Historic House: 1902-04 Overholser Mansion.
Collections: original furnishings; costumes; articles; documents.
Research Fields: Urban history of Oklahoma Territory; social, cultural & economic.
Activities: guided tours; permanent exhibitions; lectures; workshops; education programs.
Hours & Admission Prices: Feb.-Dec. Tues.-Sat. 10-3. Adults $7, senior citizens $5, students & children $4; discounts to AAM & ICOM members; children under 6 no charge. Closed legal holidays.
Attendance: 1,000 (estimated)
Membership: See separate listing for Oklahoma Historical Society.

RED EARTH MUSEUM, 6 Santa Fe Plaza, Oklahoma City, OK 73102-9027. Tel.: 405-427-5228. Fax: 405-427-8079.
E-mail: info@redearth.org
Web Site: www.redearth.org
Founded: 1978.
Congressional District: 5
Key Personnel: Chm. (V), Lou C. Kerr; Pres. (V), Janet Dyke; Dir. Communications, Eric Oesch; Dir. Devel., Christy Alcox.
Personnel Profile: Full-Time Paid 2; Part-Time Paid 3; Part-Time Volunteers 3.
Institution Type/Description: Native American Museum.
Collections: American Indian cultures & history; cradleboards; contemporary & traditional art; hands-on exhibits.
Facilities: library. Museum-related items for sale.
Activities: educational programs; special events.
Hours & Admission Prices: Mon.-Fri. 10-5. No charge. &
Attendance: 50,000 (estimated)

* **SCIENCE MUSEUM OKLAHOMA, (M),** 2100 N.E. 52 St., Oklahoma City, OK 73111-7107. Tel.: 405-602-6664. Fax: 405-602-3767. Facebook: Science Museum Oklahoma.
E-mail: otto@sciencemuseumok.org
Web Site: sciencemuseumok.org
Formerly: Omniplex Science Museum
Founded: 1958.
Congressional District: 6
Key Personnel: C.E.O. & Pres., Don Otto; Chm. Bd. of Trustees, Colin Fitzsimons; Dir. Human Resources, Jennifer Friend; Dir. Devel., Linda Maisch; Dir. Oklahoma Museum Network, Sherry Marshall; Vice Pres. Operations & Finance, Kevin Wilson; Vice Pres. Programs & Interpretation, Suzette Ellison; Accounting, Kathryn Anderson; Dir. Primary Guest Interaction, Melody Muniz; Museum Shop Mgr., Kristen Hammett.
Personnel Profile: Full-Time Paid 62; Part-Time Paid 73; Part-Time Volunteers 30.
Operating Expenses: 5,742,000
Operating Income: 63
Governing Authority: nonprofit organization. Tax-exempt: 501(c)(3).
Institution Type/Description: Science & Technology Museum.
Collections: space science; aircraft; bicycles; photographs; personal artifacts; medals; apparatus; awards; WWI; WWII; Navy; Air Force; works of art.
Facilities: planetarium; classrooms; auditorium; domed IMAX; gardens; cafe. Gift items for sale.
Activities: demonstrations; merit badge classes; field trips; science; traveling labs; overnight bright nights; bright night science; live science shows; education programs for children & adults.
Publications: general museum brochures; educational brochures; email member newsletter.
Hours & Admission Prices: Mon.-Fri. 9-5, Sat. 9-6, Sun. 11-6. Adults $12.95 senior citizens & children 3-12 $10.95; discounts to ASTC & AAA members and military; members no charge. Closed Thanksgiving; Christmas Eve & Day. &
Attendance: 347,401 (accurate)
Membership: Silver 2 $75; Gold 2 $95; Silver 6 $115; Gold 6 $175.

WORLD OF WINGS PIGEON CENTER, (M), 2300 N.E. 63rd, Oklahoma City, OK 73111-8208. Tel.: 405-478-5155; 866-570-2473. Fax: 405-478-4552.
E-mail: pigeoncenter@aol.com
Web Site: www.pigeoncenter.org
Founded: 1973.
Congressional District: 5
Key Personnel: Cur., Lorrie Monteiro.
Personnel Profile: Part-Time Paid 1.
Governing Authority: private; nonprofit organization. Parent Institution: American Homing Pigeon Institute, Oklahoma City, OK. Tax-exempt: 501(c)(3).

Institution Type/Description: Pigeon Center Museum.
Collections: paintings; sketches; books; lithographs; scientific research papers & equipment; World War I & II and European use of pigeons.
Facilities: 1,000-vol. library books on pigeon flying, breeding etc.
Activities: guided tours; temporary exhibitions. Annual Event: annual races.
Hours & Admission Prices: Museum/Library: Mon.-Fri. 8-5. Administrative Offices: Mon.-Fri. 8-5. No charge; donations accepted. Closed holidays. &
Membership: Individual $20; Family $35; Sustaining $100; Benefactor $500; Patron $1,000.

THE WORLD ORGANIZATION OF CHINA PAINTERS, 2641 N.W. 10th St., Oklahoma City, OK 73107-5407. Tel.: 405-521-1234. Fax: 405-521-1265.
E-mail: wocporg@theshop.net
Web Site: www.theshop.net/wocporg
Founded: 1967.
Key Personnel: Exec. Dir., Patricia Dickerson; World Show Pres., Paige Gray; World Show Vice Pres., Mary Ann Clarin; Museum Shop, Office & Property Mgr., Mary Early; Exhibit Clerk, Michelle Richardson.
Personnel Profile: Full-Time Paid 3; Part-Time Volunteers 20.
Governing Authority: not-for-profit organization. Tax-exempt: 501(c)(3).
Institution Type/Description: Porcelain Museum.
Collections: 19th-20th century hand painted porcelain from around the world; hand-painted porcelain from 20th century artists.
Facilities: 500-vol. library; 4,000 sq. ft. exhibit space; educational facilities. Hand painted porcelain items for sale.
Activities: guided tours; formal education programs for children & adults; painting seminars & classes; temporary exhibitions; international guest artist teachers.
Publications: bimonthly porcelain art magazine, The China Painter.
Hours & Admission Prices: Mon.-Thurs. 9-5, Sat.-Sun. call for hours. No charge; donations accepted. Closed New Year's Eve & Day; Memorial Day; Independence Day; Labor Day; Thanksgiving; Christmas. &
Attendance: 2,500
Membership: Organization $27; Worldwide $34; International $39.

Okmulgee

CREEK COUNCIL HOUSE MUSEUM, 106 W. 6th St., Okmulgee, OK 74447-5014. Tel.: 918-756-2324. Fax: 918-756-3671. Facebook: Creek Council House Museum.
E-mail: creekmuseum@sbcglobal.net
Founded: 1923.
Congressional District: 2
Key Personnel: Cur., Clint Sago.
Governing Authority: nonprofit organization. Tax-exempt.
Institution Type/Description: Muscogee Creek History Museum: housed in 1878 Creek Council House.
Collections: Indian history; archives; archaeology; history of Muscogee Creek Nation.
Research Fields: Creek trade items in Indian territory; types of Creek ornamentation.
Facilities: Creek & Oklahoma history books available for research on premises. Indian arts & crafts for sale.
Activities: guided tours.
Publications: Creek/English Dictionary, 1st & 2nd Creek readers.
Hours & Admission Prices: Tues.-Fri. 10-4, Sat. 10-3. No charge; donations accepted. Closed legal holidays. &
Attendance: 7,843 (accurate)

Oologah

OOLOGAH HISTORICAL SOCIETY, 202 W. Cooweescoowee Ave., Oologah, OK 74053. Mailing Address: P.O. Box 185, Oologah, OK 74053-0185. Tel.: 918-443-2934.
E-mail: oologahhistorical@atlasok.com
Founded: 1987.
Congressional District: 2
Key Personnel: Pres., Marian Clark.
Personnel Profile: Part-Time Paid 2; Part-Time Volunteers 10.
Operating Expenses: 13,800
Operating Income: 15,000
Governing Authority: Tax-exempt.
Institution Type/Description: Historical Society Museum.
Collections: local history & culture; photographs; personal artifacts.
Hours & Admission Prices: Mon.-Fri. 11-4. No charge; donations accepted. &
Attendance: 450 (estimated)
Membership: Regular $10 & up; Contribution $25 & up; Friends $100 & up; Heritage Club $500 & up; Lifetime $1,000 & up.

Owasso

OWASSO HISTORICAL MUSEUM, 26 S. Main St., Owasso, OK 74055-3109. Mailing Address: Owasso Historical Society, P.O. Box 1481, Owasso, OK 74055. Tel.: 918-272-4966.
E-mail: mhinkle@cityofowasso.com
Web Site: cityofowasso.com/museum/index.html
Key Personnel: Dir., Marcia Boutwell
Institution Type/Description: Historical Society Museum: housed in the Komma Building, built in 1928.
Collections: historical artifacts & memorabilia pertaining to the city of Owasso.
Hours & Admission Prices: Tues.-Fri. 12-4, Sat. 10-4. No charge; donations accepted.

Park Hill

CHEROKEE NATIONAL MUSEUM, 21192 S. Keeler Dr., Park Hill, OK 74451. Mailing Address: P.O. Box 515, Tahlequah, OK 74465-0515. Tel.: 918-456-6007. Fax: 918-456-6165.
E-mail: info@cherokeeheritage.org
Web Site: www.cherokeeheritage.org
Founded: 1963.
Congressional District: 2
Key Personnel: Exec. Dir., Carey Tilley; Cur., I. Mickel Yantz; Archivist, Tom Mooney; Dir. Devel., Penny L. Moore; Museum Shop Mgr., Kathryn Roastingear.
Personnel Profile: Full-Time Paid 40; Part-Time Volunteers 10.
Governing Authority: board of directors; nonprofit organization. Parent Institution: Cherokee National Historical Society. Subsidiary Institution: Cherokee Heritage Center. Tax-exempt: 501(c)(3).
Institution Type/Description: History Museum: site of 1851 Cherokee Female Seminary, burned in 1887.
Collections: artifacts; paintings; 42-structure ancient village; 9-structure rural village; Cherokee pottery & heritage arts.
Research Fields: Cherokee history and culture; Cherokee genealogy; Ancient Village.
Facilities: 2,500-vol. library on Cherokee heritage, archives, manuscripts, photographs; microfilm; 9,000 sq. ft. exhibit space; 1,800-seat outdoor theater; arboretum. Museum-related items for sale.
Activities: traditional art classes. Museum Sponsors: Indian Territory Days; Cherokee Heritage Gospel Sing; Trail of Tears Art Show; Cherokee Genealogy Conferences; Ancient Cherokee Days; Cherokee Homecoming Art Show.
Publications: quarterly, The Columns; E-Newsletter, Cherokee Connection.
Hours & Admission Prices: Feb.-April & Sept.-Dec. Mon.-Sat. 9-5; May-Labor Day daily 9-5. Adults $8.50, seniors 55 & over and college students $7.50, youth 5-18 $5; discounts to groups of 10 or more; children 5 & under and Cherokee National Historical Society, Inc. members no charge. Closed New Year's Eve; Easter; Thanksgiving; Christmas Eve & Day. &
Attendance: 21,000 (estimated)
Membership: Learner $25; Elder $30; Adult $40; Family $75; Mentor $250; Provider Patron $1,000; John Ross Patron $2,500; Sequoyah Patron $5,000.

MURRELL HOME, 19479 E. Murrell Home Rd., Park Hill, OK 74451-2001. Tel.: 918-456-2751. Fax: 918-456-2751. Facebook: Murrell Home.
E-mail: murrellhome@okhistory.org
Web Site: www.okhistory.org/murrellhome
Founded: 1948.
Congressional District: 2
Key Personnel: Site Mgr., David Fowler; Museum Shop Mgr., Amanda Pritchett.
Personnel Profile: Full-Time Paid 3; Part-Time Paid 2.
Volunteer Hours: 1,191
Governing Authority: society. Parent Institution: Oklahoma Historical Society, 800 Nazih Zuhdi Dr., Oklahoma City, OK 73105-7917. Tax-exempt.
Institution Type/Description: Historic House: Murrell Home built c.1845.
Collections: out-buildings; Indian artifacts; archives; costumes; pre-Civil War period furnishings.
Research Fields: Cherokee-American history.
Facilities: 200-vol. library of Cherokee Indian history books; picnic area; nature trail.
Activities: guided tours; living history education program. Gift-related items for sale.
Publications: Friends of Murrell Home Newsletter, quarterly
Hours & Admission Prices: Tues.-Sat. 10-5. No charge; donations accepted. Closed state legal holidays. &
Attendance: 45,533 (accurate)
Membership: Individual $10; Family $15; Business & Organization $25.

Pauls Valley

PAULS VALLEY HISTORICAL SOCIETY - SANTA FE DE-POT MUSEUM, 204 S. Santa Fe, Pauls Valley, OK 73075. Tel.: 405-238-2244 & 2779.
Founded: 1977.
Key Personnel: Dir., Adrienne Grimmett.
Personnel Profile: Part-Time Paid 1; Part-Time Volunteers 2.
Governing Authority: Tax-exempt.
Institution Type/Description: Historic Building Museum: depot built in 1905.
Collections: local history & culture; photographs; period furnishings; personal artifacts; school memorabilia; early Frisco caboose; 1905 Santa Fe locomotive & coal tender.
Publications: quarterly members' newsletter.
Hours & Admission Prices: Tues.-Sat. 9-4:30; other times by appointment. No charge; donations accepted. Closed holidays. &
Membership: Individual $10; Lifetime $100.

TOY & ACTION FIGURE MUSEUM, 111 S. Chickasaw St., Pauls Valley, OK 73075. Mailing Address: P.O. Box 314, Pauls Valley, OK 73075. Tel.: 405-238-6300. Fax: 405-238-6301.
E-mail: director@actionfiguremuseum.com
Web Site: actionfiguremuseum.com
Founded: 2000.
Key Personnel: Dir., Ryan Peacock; Cur., Kevin Stark
Institution Type/Description: Toy Museum.
Collections: over 13,000 action figures; cartoonists' drawings.
Activities: summer reading program. Annual Events: Hot Wheels Double Dog Dare Derby in May; International Superhero Day in August.
Hours & Admission Prices: Memorial Day to Labor Day Mon.-Thurs. 10-5, Fri.-Sat. 10-7, Sun. 1-5; Sept.-May Mon.-Sat. 10-5, Sun. 1-5. Adults $6; discounts to seniors & military.
Membership: Individual $50; Family $100.

Pawhuska

OSAGE COUNTY HISTORICAL SOCIETY MUSEUM, 700 N. Lynn Ave., Pawhuska, OK 74056-3238. Tel.: 918-287-9119.
E-mail: ochs@att.net
Web Site: www.osagecohistoricalmuseum.com
Founded: 1964.
Congressional District: 2
Key Personnel: C.E.O., Dir. & Cur., Mrs. J.B. Smith; Museum Shop Mgr., Judy Taylor.
Personnel Profile: Full-Time Paid 1; Full-Time Volunteers 1; Part-Time Paid 2; Part-Time Volunteers 3.
Governing Authority: society. Tax-exempt.
Institution Type/Description: Historic Building: 1923 Santa Fe Depot Building.
Collections: pioneer; Indian, Western & military artifacts; items pertaining to oil; memorabilia of the first Boy Scout troop in America, 1909, organized by Rev. Mitchell of London, England with British charter; Old Santa Fe depot; early oil and American Indian exhibits & pictures; stainless steel Santa Fe; combination passenger, mail & freight car; Santa Fe cattle freight car; 1900 era one-room furnished schoolhouse.
Facilities: Books & other museum-related items for sale.
Activities: guided tours; permanent & inter-museum loan exhibitions.
Publications: History book, Osage County Profiles; booklet, Pioneer Days with the Osage Indians.
Hours & Admission Prices: Mon.-Fri. 9-5. No charge; donations accepted. Closed Thanksgiving; Christmas. &
Attendance: 9,000 (estimated)
Membership: Individual $5; Sustaining $12; Lifetime $100; Business $250.

OSAGE TRIBAL MUSEUM, LIBRARY, AND ARCHIVES, 819 Grandview Ave., Pawhuska, OK 74056-3203. Mailing Address: P.O. Box 779, Pawhuska, OK 74056-0779. Tel.: 918-287-5441. Fax: 918-287-5227.
E-mail: kredkorn@osagetribe.org
Web Site: www.osagetribalmuseum.com
Founded: 1938.
Key Personnel: Dir., Kathryn Red Corn.
Personnel Profile: Full-Time Paid 3; Full-Time Volunteers 2; Part-Time Volunteers 6.
Institution Type/Description: Native American Museum: listed on the National Register of Historic Places.
Collections: Osage culture & history; photographs from 1800s to present; John L. bird collection; documents; maps; paintings.
Facilities: Museum-related items for sale.
Activities: Osage heritage classes; videos; lectures; public programs.

Hours & Admission Prices: Tues.-Sat. 8:30-5. No charge; donations accepted. Closed federal holidays.
Attendance: 5,000 (estimated)

Pawnee

PAWNEE BILL RANCH AND MUSEUM, (M), 1141 Pawnee Bill Rd., Pawnee, OK 74058-3563. Mailing Address: P.O. Box 493, Pawnee, OK 74058-0493. Tel.: 918-762-2513. Fax: 918-762-2514. Facebook: Pawnee Bill Ranch and Museum.
E-mail: pawneebill@okhistory.org
Web Site: www.pawneebillranch.org
Founded: 1962.
Congressional District: 1
Key Personnel: Dir., Ron Brown; C.E.O., Dr. Bob Blackburn; Pres. (V), Larry Troxell; Chm. (V) & Museum Shop Mgr., Anna Davis; Historical Collections Specialist, Erin Brown; Maintenance Construction Technician, Kevin Webb.
Personnel Profile: Full-Time Paid 6; Part-Time Paid 5; Part-Time Volunteers 16; Interns 1.
Governing Authority: state. Parent Institution: Oklahoma Historical Society, 800 Nazih Zuhdi Dr., Oklahoma City, OK 73105-7917. Subsidiary Institution: Friends of Pawnee Bill Ranch Association. Tax-exempt.
Institution Type/Description: Historic House: 1910 restored house. Home of Wild West showman and bison ranch.
Collections: 1910-1940, original furnishings of Pawnee Bill; photos & exhibits of Wild West Show days; calliope; stagecoach; 1900 billboard advertising Pawnee Bill's Wild West Show; Indian artifacts; early day farming, ranching equipment.
Research Fields: Pawnee Bill and his Wild West.
Facilities: drive through Longhorn & Buffalo pasture; picnic shelters; small meeting room; activity & camping area for organized youth groups.
Activities: seminars; workshops; WW show reenactment; mansion tours & special tours. Museum Sponsors: weekend Wild West show in June.
Publications: newsletters; brochures; fact sheets.
Hours & Admission Prices: April-Oct. Sun.-Mon. 1-4, Tues.-Sat. 10-5 (last tour 4:15); Nov.-March Wed.-Sat. 10-5, Sun. 1-4. Adults $3, senior citizens 65 & over $2.50, students 6-18 $1.50; discounts to groups; OHS Ranch & Association members and children 5 & under no charge. Closed state holidays. &
Attendance: 45,681 (accurate)
Membership: Individual $15; Family $25; Institutional $75.

PAWNEE COUNTY HISTORICAL SOCIETY MUSEUM AND DICK TRACY HEADQUARTERS, 513 6th St., Pawnee, OK 74058. Tel.: 918-762-4681. Facebook: Pawnee County Historical Society.
E-mail: pawnee@pawneechs.org
Web Site: www.pawneechs.org
Founded: 1979.
Congressional District: 3
Governing Authority: Parent Institution: Pawnee County Historical Society. Tax-exempt.
Institution Type/Description: Historical Society Museum.
Collections: local history & culture; period furnishings; personal artifacts; photographs; Native American artifacts; artifacts of Pawnee native Chester Gould, creator of the Dick Tracy comic strip.
Activities: special events.
Publications: bimonthly members' newsletter.
Hours & Admission Prices: Wed.-Sat. 10-2; call to confirm. No charge; donations accepted.
Membership: Junior $5; Individual $15; Husband & Wife $25; Family & Business $30; Sustaining $50; Lifetime $200.

Perkins

OKLAHOMA TERRITORIAL PLAZA, 750 N. Main St., Perkins, OK 74059. Mailing Address: P.O. Box 667, Perkins, OK 74059. Tel.: 405-547-2777.
E-mail: info@okterritory.org
Web Site: www.okterritory.org
Formerly: Old Church Center and Museum
Founded: 2007.
Congressional District: 3
Key Personnel: Dir., W. David Sasser.
Personnel Profile: Full-Time Volunteers 1; Part-Time Volunteers 23.
Governing Authority: Parent Institution: Oklahoma Territorial Plaza Trust. Tax-exempt.
Institution Type/Description: Historic Buildings.

Collections: Oklahoma history; period furnishings & artifacts; Cimarron Valley history. Historic Buildings: 1891 church; 1896 one-room schoolhouse; 1901 log cabin; 1907 barn; 1916 Santa Fe railroad depot; 1960s service station; restored home of Frank "Pistol Pete" Eaton.
Hours & Admission Prices: Buildings: Sat. 1-4; other times by appointment. Park: daily. No charge; donations accepted. &
Attendance: 7,300 (estimated)

Perry

CHEROKEE STRIP MUSEUM AND ROSE HILL SCHOOL, 2617 W. Fir St., Perry, OK 73077-7903. Tel.: 580-336-2405. Fax: 580-336-2064.
E-mail: csmuseum@okhistory.org
Web Site: cherokee-strip-museum.org
Formerly: Cherokee Strip Museum
Founded: 1965.
Congressional District: 6
Key Personnel: Dir., Peggy Haxton.
Personnel Profile: Full-Time Paid 2; Part-Time Paid 4; Part-Time Volunteers 20.
Governing Authority: state. Parent Institution: Oklahoma Historical Society (see separate listing). Tax-exempt.
Institution Type/Description: History Museum.
Collections: agricultural equipment, medical & historical artifacts; Otoe-Missouri collections; costumes; glass. Historic Building: Rose Hill School 1895.
Facilities: library of books, newspapers & Noble County family history.
Activities: tours; traveling & temporary displays; permanent exhibits.
Hours & Admission Prices: mid-Jan. to Dec. Tues.-Fri. 9-5, Sat. 10-4. Adults $3, senior citizens 62 & over $2.50, children 6-17 $1; children 5 & under, Oklahoma Historical Society members & Cherokee Strip Historical Society members no charge. Closed legal holidays. &
Attendance: 8,000 (estimated)
Membership: Annual $25; Lifetime $150.

HERITAGE CENTER & DITCH WITCH MUSEUM, 6th & Cedar St., Perry, OK 73077. Mailing Address: 1959 W. Fir St., Perry, OK 73077. Tel.: 580-336-4402 & 4684.
Institution Type/Description: Industrial Museum: housed in the original Charles Machine Works, Inc. Company building, a manufacturer of Ditch Witch underground construction equipment.
Collections: company history, Ditch Witch equipment; photographs.
Activities: group tours.
Hours & Admission Prices: By appointment. No charge.

Ponca City

CANN MEMORIAL BOTANICAL GARDEN, 1500 E. Grand, Ponca City, OK 74604-5209. Mailing Address: 905 W. Hartford, Ponca City, OK 74601-1162. Tel.: 580-767-0430.
Institution Type/Description: Botanical Garden.
Collections: wisteria arbors; sundials; water garden; perennials; herbs; annuals. Historic House: c.1908 home.
Facilities: 10 acre gardens; nature trails.
Activities: rental facilities.
Hours & Admission Prices: Daily dawn to dusk. No charge; donations accepted.

CONOCO MUSEUM, 501 W. South Ave., Ponca City, OK 74601-6105. Tel.: 580-765-8687. Fax: 580-767-2147.
E-mail: oneilcm@p66.com
Web Site: www.conocomuseum.com
Founded: 2007.
Key Personnel: Dir., Carla O'Neill
Governing Authority: Parent Institution: Phillips 66. Tax-exempt.
Institution Type/Description: History Museum.
Collections: artifacts & memorabilia pertaining to Marland Oil, Continental Oil Company and Conoco.
Hours & Admission Prices: Mon.-Sat. 10-5, Sun. 1-5. No charge. Closed holidays. &
Attendance: 5,000 (accurate)

MARLAND MANSION ESTATE, 901 Monument Rd., Bldg. 2, Ponca City, OK 74604-3600. Tel.: 580-767-0420; 800-422-8340. Fax: 580-763-8054.
E-mail: marlandmansion@poncacityok.gov
Web Site: www.marlandmansion.com

Key Personnel: Exec. Dir., David Keathly
Institution Type/Description: Historic House Museum: housed in a 55-room Italian Renaissance villa, c.1925.
Collections: murals; mosaic ceiling; period furnishings; personal artifacts.
Activities: guided tours.
Hours & Admission Prices: Mon.-Sat. 10-5, Sun. 1-5. Adults $7, seniors 65 & over and students 12-17 $5, students 6-11 $4.

MARLAND'S GRAND HOME, 1000 E. Grand, Ponca City, OK 74601-5607. Tel.: 580-767-0427.
Web Site: marlandgrandhome.com
Formerly: Ponca City Cultural Center Museum
Founded: 1968.
Congressional District: 38
Key Personnel: Dir., David Keathly.
Personnel Profile: Full-Time Paid 1; Part-Time Paid 1; Part-Time Volunteers 20.
Governing Authority: municipal. Tax-exempt.
Institution Type/Description: Cultural Center & Ethnology Museum.
Collections: cultural artifacts of five neighboring Native American tribes: Ponca, Kaw, Otoe, Osage & Tonkawa; 101 ranch & Wild West Show memorabilia; D.A.R. display; restored historic furnishings & decor.
Research Fields: American Indians; cowboys; ranch life; E.W. Marland.
Publications: museum brochure.
Hours & Admission Prices: Tues.-Sat. 10-5. Adults $3, children 6-16 $1. Closed holidays.
Attendance: 8,500 (accurate)
Membership: Pioneers $10; Land Seekers $25; Claim Stakers $100; Sod Busters $250; Homesteaders $500; Builders $1,000.

PIONEER WOMAN MUSEUM & STATUE, (M), 701 Monument Rd., Ponca City, OK 74604-3910. Tel.: 580-765-6108. Fax: 580-762-2498. Facebook: Pioneer Woman Museum.
E-mail: piown@okhistory.org
Web Site: www.pioneerwomanmuseum.com
Founded: 1958.
Congressional District: 6
Key Personnel: Dir., Robbin Davis; Pres. (V), Angela Kennedy; Historical Interpreter, Keith Fagan.
Personnel Profile: Full-Time Paid 2; Part-Time Volunteers 2.
Governing Authority: state. Affiliated with the Oklahoma Historical Society, 800 Nazih Zuhdi Dr., Oklahoma City, OK 73105. Tax-exempt.
Institution Type/Description: Women's History Museum.
Collections: 17' bronze statue representing a confident woman by Bryant Baker; Oklahoma's pioneering women; material culture of Oklahoma including toys, home furnishings, farm equipment, textiles, art, dolls, photographs, & 20th century artifacts; important women of Oklahoma.
Major Exhibits: Breaking News: Oklahoma Women Journalists, 1/14-12/14; Bound to Please: A History of Corsets (T), 1/14-12/14; Bending the Rules, 1/14-12/14.
Research Fields: E. W. Marland; 101 Ranch & Miller brothers; pioneering women of Oklahoma; Cherokee strip of Oklahoma.
Facilities: 10,000 sq. ft. exhibition space; education room; 14 acres. Museum-related items for sale.
Activities: guided tours; weaving demonstrations; traveling exhibitions; educational programming.
Publications: museum exhibits & Oklahoma women booklet.
Hours & Admission Prices: Tues.-Sat. 10-5. Adults $4, senior citizens $3, students 6-18 $1; discounts to Oklahoma Museums Assoc., Oklahoma Historical Society, Mountain Plains Museums Assoc., AAM & AAA members. Closed state holidays. &
Attendance: 8,500 (accurate)

PONCA CITY ART CENTER, 819 E. Central, Ponca City, OK 74601-5506. Mailing Address: P.O. Box 1394, Ponca City, OK 74602-1394. Tel.: 580-765-9746.
E-mail: pcartcenter@sbcglobal.net
Web Site: www.poncacityartcenter.com
Founded: 1966.
Key Personnel: Dir. & Office Mgr., Jerry Cathey.
Governing Authority: Tax-exempt.
Institution Type/Description: Art Museum: housed in Soldani Mansion. Listed on the National Register of Historic Places.
Collections: paintings; regional artists.
Facilities: Museum-related items for sale.
Activities: art classes; workshops. Annual Event: Fine Arts Festival.
Hours & Admission Prices: Wed.-Sun. 1-5. No charge; donations accepted. Closed Independence Day; Thanksgiving, Christmas.

Membership: Individual & Family $20; Sponsor $30; Donor $50; Fellow of the Arts $75; Founder $150; Life $1,000; Benefactor $5,000.

STANDING BEAR MUSEUM AND EDUCATION CENTER, Standing Bear Park, 601 Standing Bear Pkwy., Ponca City, OK 74601. Mailing Address: P.O. Box 247, Ponca City, OK 74602. Tel.: 580-762-1514.
E-mail: info@standingbearpark.com
Founded: 1994.
Institution Type/Description: Native American History Museum.
Collections: Native American history, culture, heritage & artifacts; paintings; 22 ft. bronze sculpture.
Facilities: nature trail.
Hours & Admission Prices: Mon.-Fri. 9-5, Sat. 10-2. No charge. &

Poteau

LEFLORE COUNTY HISTORICAL SOCIETY, 303 Dewey Ave., Poteau, OK 74953. Mailing Address: P.O. Box 457, Poteau, OK 74953. Tel.: 918-647-9330.
E-mail: leflorecountyhistoricalsociety@windstream.net
Web Site: leflorecountyhistoricalsocietymuseum.webs.com
Institution Type/Description: Historical Society Museum: housed in the former Hotel Lowrey; built in 1922.
Collections: local history & culture; period furnishings; photographs; personal artifacts.
Hours & Admission Prices: Wed.-Thurs. 9-6; other times by appointment.

ROBERT S. KERR MUSEUM, 23009 Kerr Mansion Rd., Poteau, OK 74953-8119. Tel.: 918-647-8221. Fax: 918-647-3952.
E-mail: cburleigh@carlalbert.edu
Web Site: www.casc.cc.ok.us/kerr_center
Founded: 1968.
Congressional District: 3
Key Personnel: Dir., Cheryl Burleigh.
Personnel Profile: Full-Time Paid 1; Part-Time Paid 3.
Governing Authority: society. Parent Institution: Carl Albert State College. Tax-exempt: 501(c)(3).
Institution Type/Description: Oklahoma History Museum.
Collections: Senator Robert S. Kerr pictures & mementoes; Spiro Mounds artifacts; farm & home implements; geology of east Oklahoma; barbed wire; pioneer artifacts; Choctaw Indian artifacts; Viking runestones.
Research Fields: prehistory.
Facilities: Kerr Country Mansion; conference rooms.
Activities: lectures; art & quilt exhibits.
Publications: newsletter; booklets; Butterfield Stage; Pre-Historic People; The Choctaw Story; The Edward's Store; When Coal Was King.
Hours & Admission Prices: Mon.-Sun. upon request or by appointment. No charge; donations accepted. Closed Jan-March & major holidays. &
Attendance: 3,556 (accurate)
Membership: Annual $10; Contributing $25; Lifetime $500.

Prague

PRAGUE HISTORICAL MUSEUM, 1008 N. Jim Thorpe Blvd., Prague, OK 74864. Mailing Address: 8601 NBU, Prague, OK 74864. Tel.: 405-567-4750.
Institution Type/Description: History Museum.
Collections: local history & culture; photographs; Olympian Jim Thorpe artifacts.
Activities: Annual Event: Kolache Festival in May.
Hours & Admission Prices: Mon., Wed. & Fri. 1-4; groups by appointment. No charge; donations accepted.

Purcell

MCCLAIN COUNTY MUSEUM, 203 W. Washington, Purcell, OK 73080. Tel.: 405-527-5894. Facebook: McClain County Museum.
E-mail: mcmuseum@mail.com
Founded: 1973.
Congressional District: 4
Key Personnel: Chm. (V) & Dir., Pam Ellis-Hobbs; Pres. (V), Mark Brady.
Governing Authority: Tax-exempt.
Institution Type/Description: History Museum.
Collections: local history & culture; period furnishings; personal artifacts; photographs; genealogy; early records.
Hours & Admission Prices: Mon.-Fri. 12-4; other times by appointment. No charge; donations accepted. Closed holidays.
Attendance: 250 (estimated)

Membership: Annual $25.

Ripley

WASHINGTON IRVING TRAIL & MUSEUM, 3918 S. Mehan Rd., Ripley, OK 74062-6278. Mailing Address: P.O. Box 1852, Stillwater, OK 74076. Tel.: 405-624-9130.
E-mail: cchlouber@aol.com
Web Site: www.washingtonirvingtrailmuseum.com
Founded: 1994.
Congressional District: 3
Key Personnel: Dir., Cur. & Museum Shop Mgr., Dale Chlouber; Chm. (V) & Pres. (V), John Wilson.
Personnel Profile: Full-Time Volunteers 1; Part-Time Volunteers 12.
Volunteer Hours: 2,000
Operating Expenses: 3,000
Operating Income: 6,000
Governing Authority: Parent Institution: Payne Co. Museum Assoc. Tax-exempt.
Institution Type/Description: History Museum.
Collections: local, state & regional history; Civil War in Oklahoma; Southwest Indian artifacts.
Publications: The Otto Gray Cowboy Band.
Hours & Admission Prices: Wed.-Sat. 11-5, Sun. 1-5. No charge. &
Attendance: 7,500 (estimated)

Sallisaw

14 FLAGS MUSEUM, 400 E. Cherokee, Sallisaw, OK 74955. Mailing Address: Echo Rider, 105274 S. 4690 Rd., Sallisaw, OK 74955. Tel.: 918-775-2608. Fax: 918-775-9550.
E-mail: chamber@sallisawok.org
Founded: 1986.
Institution Type/Description: History Museum: housed in a log cabin; built in 1845.
Collections: Oklahoma's history & culture; period furnishings; general store; cattle brands; caboose.
Hours & Admission Prices: Daily 9-5. No charge.

SEQUOYAH CABIN, 470288 Hwy. 101, Sallisaw, OK 74955-9744. Tel.: 918-775-2413.
E-mail: seqcabin@okhistory.org
Formerly: Sequoyah Home Site
Founded: 1936.
Congressional District: 2
Key Personnel: C.E.O., Bob Blackburn; Cur. & Museum Shop Mgr., Jerry Dobbs.
Personnel Profile: Full-Time Paid 2; Full-Time Volunteers 1; Part-Time Volunteers 5.
Volunteer Hours: 300
Operating Expenses: 10,000
Operating Income: 10,000
Governing Authority: state. Parent Institution: Oklahoma Historical Society. Tax-exempt.
Institution Type/Description: Historic Building & Site: housed in 1829, Sequoyah Log Cabin.
Collections: artifacts; house furnishings & exhibits relating to Native Americans (primarily Cherokee) & the inventor of the Cherokee Syllabary (Sequoyah); interpretative exhibit center.
Major Exhibits: Native American Games, 12/13-2/14.
Research Fields: Cherokee history.
Facilities: visitor center; picnic area.
Activities: various cultural & educational activities.
Publications: booklets: Mistletoe Leaves; quarterly newsletter, Chronicles of Oklahoma.
Hours & Admission Prices: Tues.-Fri. 9-5, Sat.-Sun. 2-5. No charge; donations accepted. Closed holidays. &
Attendance: 20,000 (estimated)
Membership: Student & Retired over 64 $15; Annual Individual $25; Institutional & Family $40; Supporting $75; Individual Life $500; Benefactor $1,000.

Sand Springs

SAND SPRINGS CULTURAL & HISTORICAL MUSEUM, 9 E. Broadway St., Sand Springs, OK 74063. Tel.: 918-246-2509. Fax: 918-245-7101.
E-mail: museum@sandspringsok.org
Web Site: www.sandspringsmuseum.org

Founded: 1991.
Congressional District: 1
Key Personnel: C.E.O., Dir. & Museum Shop Mgr., Dr. Stacy Reaves; Chm. (V), Jerry Hanner; Chm. (V), Ed Dubie; Pres. (V), Cynthia Phillips; Public Events, Ruth Ellen Henry.
Personnel Profile: Part-Time Volunteers 14.
Governing Authority: private; nonprofit organization. Parent Institution: City of Sand Springs. Tax-exempt.
Institution Type/Description: Cultural and Historical Museum.
Collections: cultural heritage artifacts.
Facilities: Page Memorial Library.
Activities: genealogy workshops; children's art shows; art classes; tours.
Publications: quarterly, Sands Springs Reflections; annual, Sand Springs Cultural & Historical Museum History Calendar.
Hours & Admission Prices: Tues.-Fri. 1-5; other times by appointment. No charge; donations accepted. &
Attendance: 5,000 (estimated)
Membership: Senior $10; Individual $15; Family $20; Business $50; Contributing $100; Corporate $250; Sponsor $500; Patron $1,000.

Sapulpa

SAPULPA HISTORICAL MUSEUM, 100 E. Lee, Sapulpa, OK 74066-4216. Tel.: 918-224-4871. Fax: 918-224-7765. Facebook: Sapulpa Historical Society Museum.
E-mail: sapulpahistsoc@tulsacoxmail.com
Web Site: sapulpahistoricalsociety.com
Founded: 1968.
Congressional District: 1
Key Personnel: Dir., Mike Jeffries; Pres. (V), Rick Woolery; Vice Pres., Larry White; Treas., Russell Crosby; Sec., Belinda Crosby; Gift Shop Mgr. & Administrative Asst., Shanna Rutledge.
Personnel Profile: Full-Time Volunteers 1; Part-Time Paid 1; Part-Time Volunteers 40.
Governing Authority: nonprofit. Tax-exempt: 501(c)(3).
Institution Type/Description: History Museum: housed in three-story, 1910 Wills Building, which was renovated in 1982.
Collections: early 1900s kitchen equipment; parlor furniture; hand-pulled ladder wagon; 1900s-1920s clothes; militaria; Frisco railroad items; glass industry items; medical items; Native American artifacts; Native American and African-American photographs; arrowheads; 1939 Fire Engine; Catfish String Band instruments; early oil field display; early schoolroom; general store; war artifacts.
Research Fields: local history.
Activities: guided tours.
Publications: biannual members' newsletter, Our Heritage; Sapulpa History Books, Vols. I & II; book, Sapulpa OK The Greatest City in the Known World; Sapulpa OK 74066 Vols. I & II.
Hours & Admission Prices: Jan. to late Aug. & Sept. to late Dec. Tues.-Sat. 10-3; group tours by appointment. No charge; donations accepted. Closed major holidays. &
Attendance: 1,800 (accurate)
Membership: Annual $15; Family $25; Life $125.

Sayre

RS & K RAILROAD MUSEUM, 411 N. 6th St., Sayre, OK 73662-263. Tel.: 580-928-3525.
Institution Type/Description: Toy Museum.
Collections: over 250 model trains; railroad artifacts; local history.
Activities: guided tours.
Hours & Admission Prices: Daily 9-9 by appointment. &

SHORTGRASS COUNTRY MUSEUM SOCIETY, 106 E. Poplar Ave., Sayre, OK 73662-2933. Mailing Address: P.O. Box 260, Sayre, OK 73662-0260.
Founded: 1992.
Key Personnel: Dir. & Museum Shop Mgr., Bob Carey; Pres. (V), Kenny Bibb; Business Advisor & Dir., Bunny Neff.
Personnel Profile: Part-Time Paid 1; Part-Time Volunteers 7.
Governing Authority: Parent Institution: City of Sayre. Tax-exempt.
Institution Type/Description: Historic Building Museum: housed in the former Rock Island Depot; built in 1901.
Collections: western Oklahoma history; railroad history; personal artifacts.
Research Fields: local family histories.
Activities: reading program with school; outreach presentations. Museum Sponsors: Open Houses.
Publications: newsletter.
Hours & Admission Prices: Feb.-Nov. Tues.-Fri. 9-12; Dec.-Jan. call for hours; other times by appointment. No charge; donations accepted. &

Attendance: 350 (estimated)
Membership: Retired over 65 $10; Single $15; Family $20; Business $30; Donor $250; Support $500; Endowment $1,000; Patron $2,500.

Seiling

REDINGER FUNERAL MUSEUM, 105 E. Gary England Ave., Seiling, OK 73663. Mailing Address: P.O. Box 236, Seiling, OK 73663-0236. Tel.: 580-922-4226. Fax: 580-922-4228.
E-mail: rredinger@pldi.net
Web Site: www.redingerfuneralhome.com
Founded: 2000.
Institution Type/Description: History Museum.
Collections: funeral history & artifacts; horse-drawn hearse.
Hours & Admission Prices: By appointment. No charge.

Seminole

JASMINE MORAN CHILDREN'S MUSEUM, 1714 Hwy. 9 W., Seminole, OK 74868. Tel.: 405-382-0950. Fax: 405-382-3707.
Web Site: www.jasminemoran.com/frameset.html
Key Personnel: Exec. Dir., Marci Donato
Institution Type/Description: Children's Museum.
Collections: hands-on exhibits.
Activities: train ride.
Hours & Admission Prices: Tues.-Sat. 10-5, Sun. 1-5. Admission 3-60 $8, seniors over 60 $7; children under 3 no charge. Closed major holidays.

OKLAHOMA OIL MUSEUM, 1800 Hwy. 9 West, (Wrangler Blvd.), Seminole, OK 74868. Mailing Address: P.O. Box 202, Seminole, OK 74818-0202.
Formerly: Seminole Historical & Oil Museum
Founded: 1999.
Key Personnel: C.E.O. & Pres. (V), Chuck Chadick; Museum Shop Mgr., Barbara Ross.
Personnel Profile: Part-Time Paid 2.
Governing Authority: Parent Institution: Seminole Historical Society. Tax-exempt.
Institution Type/Description: Seminole County History Museum.
Collections: local history & culture; early woodworking tools; oil machinery & equipment; 1926 dental x-ray machine; photographs; personal artifacts.
Publications: members' newsletter; calendars.
Hours & Admission Prices: Mon.-Thurs. 10-4, Fri.-Sat. 10-2, Sun. 1-5. Closed holidays. &
Attendance: 2,000 (estimated)
Membership: Student $5; Senior $10; Individual $15; Family $25; Patron $500; Benefactor $750; Lifetime $1,000.

Shattuck

SHATTUCK WINDMILL MUSEUM & PARK, 1100 S. Main, Shattuck, OK 73858. Mailing Address: P.O. Box 227, Shattuck, OK 73858-0227. Tel.: 580-938-5291.
Web Site: www.shattuckwindmillmuseum.org
Founded: 1994.
Congressional District: 3
Key Personnel: Dir., Phillis Ballew; Pres. (V), Edgar Longhofer; Museum Shop Mgr., Naomi Bradley.
Personnel Profile: Part-Time Volunteers 19.
Governing Authority: Tax-exempt.
Institution Type/Description: History Museum.
Collections: 63 restored mills from 1870 to 1970; wooden wheels; 18' railroad eclipse. Historic Buildings: 1901 homestead; Halladay standard mill; half dugout soddy, 1904.
Hours & Admission Prices: Daily. No charge; donations accepted. &
Attendance: 1,500 (accurate)

Shawnee

CITIZEN POTAWATOMI NATION CULTURAL HERITAGE CENTER, 1899 S. Gordon Cooper Dr., Shawnee, OK 74801-9004. Tel.: 405-878-5830; 800-880-9880. Fax: 405-878-5840.
Web Site: www.potawatomi.org
Formerly: Citizen Potawatomi Museum
Founded: 2006.
Key Personnel: Chm., John A. "Rocky" Barrett; Mgr. Collections, Stacy S. Coon; Facilities & Operations Mgr., Cindy Stewart
Institution Type/Description: Native American Museum.
Collections: Potawatomi history & cultural heritage; art; crafts; language; textiles.

Hours & Admission Prices: Tues.-Fri. 8-5, Sat. 10-3. No charge; donations accepted. Closed holidays. &
Attendance: 21,000 (estimated)

MABEE-GERRER MUSEUM OF ART, (M), 1900 W. MacArthur, Shawnee, OK 74804-2403. Tel.: 405-878-5300. Fax: 405-878-5133.
E-mail: info@mgmoa.org
Web Site: www.mgmoa.org
Founded: 1914.
Congressional District: 5
Key Personnel: Dir., Dane Pollei; Preparator, Daniel J. Lay; Cur. Education, Donna Merkt; Cur. Collections & Museum Shop Mgr., Delaynna Trim; Dir. Devel., Tonya Ricks.
Personnel Profile: Full-Time Paid 5; Part-Time Paid 3; Part-Time Volunteers 100; Interns 4.
Volunteer Hours: 2,000
Operating Expenses: 475,000
Operating Income: 475,000
Governing Authority: nonprofit corporation. Parent Institution: St. Gregory's Abbey. Tax-exempt: 501(c)(3).
Institution Type/Description: Art Museum.
Collections: artifacts from ancient civilizations including: Egyptian, Babylonian, pre-Columbian, North, South and Central American Indians, African, South Pacific, Asian, European & American oil paintings; prints; drawings; watercolors; Persian & Chinese oriental rugs; bronze, ivory, marble and Romanesque wood sculpture.
Major Exhibits: Voices: A Sculptural Book, inspired by the Dictionary of American Regional English, 2/1/14-3/23/14; Barbara Cleary (T), 5/3/14-6/22/14; Organic: Exploring Nature Through Art, 7/12/14-8/24/14; From Tusk to Treasure: Ivory from the Milligan-Kirkpatrick Collection (T), 9/13/14-10/26/14; Regional Exhibit, 11/8/14-11/23/14.
Activities: guided tours; lectures; organized education programs for children & adults, including art classes, art camp & programs with local schools; docent program; loan exhibitions.
Publications: catalogs with exhibitions.
Hours & Admission Prices: Tues.-Sat. 10-5, Sun. 1-4. Adults $5, Seniors $4, children 6-17 $3; discount to AAM members; members & children under 5 no charge. Closed major holidays. &
Attendance; 30,000 (estimated)
Membership: Student $15; Individual $35; Family $50; Hudson River League $100; Chase Studio League $250; Papyrus Guild $500; School of Raphael $1,000.

SANTA FE DEPOT MUSEUM, 614 E. Main St., Shawnee, OK 74801. Mailing Address: P.O. Box 114, Shawnee, OK 74801. Tel.: 405-275-8412.
Institution Type/Description: History Museum.
Collections: local history & culture; period furnishings; personal artifacts; photographs; one-room schoolhouse; Native American artifacts. Historic House: 1891 Beard cabin.
Facilities: Museum-related items for sale.
Hours & Admission Prices: Tues.-Fri. 10-4, Sat.-Sun. 2-4. Suggested Donation: adults $2, children & students $1; discounts to groups. Closed holidays.

Skiatook

SKIATOOK MUSEUM, 115 S. Broadway, Skiatook, OK 74070-1540. Tel.: 918-396-7558.
Key Personnel: Pres., Donna Sue Jones; Vice Pres., John Reynolds
Institution Type/Description: History Museum: housed in a former doctor's home, built in 1912.
Collections: Cherokee & Osage artifacts including clothing & drums; local history; personal artifacts; documents; furniture; photographs; newspapers; military uniforms & medals; WWI, WWII & Korean War; period doctor's equipment.
Hours & Admission Prices: Tues.-Fri. 1-4. No charge; donations accepted. Closed legal holidays.

Spiro

SPIRO MOUNDS ARCHAEOLOGICAL CENTER, 18154 1st St., Spiro, OK 74959-4463. Tel.: 918-962-2062. Fax: 918-962-2062.
E-mail: spiro@okhistory.org
Web Site: www.okhistory.org
Founded: 1978.
Congressional District: 3
Key Personnel: Historic Property Mgr., Dennis Peterson.

Personnel Profile: Full-Time Paid 1; Part-Time Paid 1.
Volunteer Hours: 200
Operating Expenses: 20,000
Operating Income: 28,000
Governing Authority: state. Parent Institution: Oklahoma Historical Society. Tax-exempt.
Institution Type/Description: Archaeological Site.
Collections: original & reproductions of artifacts; 1250-1450, Spiro Phase period.
Research Fields: archaeology; Mississippian culture & art.
Facilities: interpretive center.
Activities: trail; group tours. Annual Events: Kite Flite Day; Winter Solstice, Summer Solstice, Vernal Equinox & Autumnal Equinox tours & programs.
Publications: brochures.
Hours & Admission Prices: Wed.-Sat. 9-5, Sun. 12-5. Adults $4, senior citizens $3, children $1; discounts to Oklahoma Historical Society members. Closed state holidays. &
Attendance: 10,000 (estimated)
Membership: Spiro Mounds Development Associates: Individual $10; Organization $25; Lifetime $500.

Stigler

HASKELL COUNTY HISTORICAL SOCIETY MUSEUM, American Legion Hut, 204 E. Main St., Stigler, OK 74462. Mailing Address: P.O. Box 481, Stigler, OK 74462-0481. Tel.: 918-967-2161.
E-mail: stiglermuseum@yahoo.com
Institution Type/Description: Historical Society Museum.
Collections: local history & culture; period furnishings; personal artifacts; photographs.
Hours & Admission Prices: Mon., Wed. & Fri. 9:30-2.

Stillwater

GALLAGHER-IBA ARENA - HERITAGE HALL SPORTS MUSEUM, Oklahoma State University, Stillwater, OK 74078. Tel.: 405-744-3864. Fax: 405-744-4535.
Founded: 2001.
Institution Type/Description: Sports Museum.
Collections: Oklahoma State University's athletic programs & athletes; personal artifacts; photographs; sports memorabilia; uniforms; trophies; Hall of Honor; OSU Olympians; national & conference championships.
Facilities: Museum-related items for sale.
Activities: guided tours.
Hours & Admission Prices: Mon.-Fri. 9-5. No charge. &

GARDINER ART GALLERY, (M), Oklahoma State University, Stillwater, OK 74078. Mailing Address: 108 Bartlett Center, Stillwater, OK 74078-4084. Tel.: 405-744-6016. Fax: 405-744-5767.
E-mail: osumuseum@okstate.edu
Web Site: art.okstate.edu
Founded: 1965.
Congressional District: 6
Key Personnel: Gallery Dir., Shiyuan "Shawn" Yuan.
Personnel Profile: Full-Time Paid 2; Part-Time Paid 1; Interns 1.
Governing Authority: university. Parent Institution: Oklahoma State University, Dept. of Art. Tax-exempt.
Institution Type/Description: Art Museum.
Collections: graphics.
Activities: rotating exhibitions.
Publications: calendar; occasional catalog.
Hours & Admission Prices: Mon.-Fri. 8-5. No charge. Closed national holidays. &
Attendance: 10,000 (estimated)

NATIONAL WRESTLING HALL OF FAME & MUSEUM, 405 W. Hall of Fame Ave., Stillwater, OK 74075-5025. Tel.: 405-377-5243. Fax: 405-377-5244.
E-mail: info@nwhof.org
Web Site: www.nwhof.org
Founded: 1976.
Congressional District: 3
Key Personnel: C.E.O., Lee Roy Smith; Chm. (V), James Keen; Office Mgr. & Museum Shop Mgr., Maghan Cawlfield.
Personnel Profile: Full-Time Paid 3; Part-Time Paid 2.
Governing Authority: nonprofit corporation. Tax-exempt: 501(c)(3).
Institution Type/Description: Sports Museum: located near campus of Oklahoma State University.

Collections: Wall of Champions display; Hall of Outstanding Americans; Hall of Distinguished Members; memorabilia including medals, trophies, scrapbooks, statues, photos & clothing pertaining to amateur wrestling.
Facilities: 10,000-vol. library available for research on premises; reading room; auditorium; theater. Wrestling-related gift items for sale.
Activities: films; lectures.
Publications: newsletter.
Hours & Admission Prices: Mon.-Fri. 9-4; other times by appointment. Adults $5, students $2 Closed holidays. &
Attendance: 10,000 (estimated)

OKLAHOMA STATE UNIVERSITY BOTANICAL GARDEN, 360 Agricultural Hall, Stillwater, OK 74078-6025. Tel.: 405-744-5414. Fax: 405-744-9709.
Web Site: hortla.okstate.edu
Founded: 1935.
Congressional District: 3
Key Personnel: Dir., Dr. Dale M. Maronek; Education, Mr. David Hillock; Cur., Dr. Mike Schnelle.
Personnel Profile: Full-Time Paid 4; Part-Time Paid 6; Part-Time Volunteers 43; Interns 1.
Governing Authority: public university.
Institution Type/Description: Botanical Garden.
Collections: theme gardens.
Research Fields: turfgrass; woody & herbaceous ornamental plants.
Facilities: botanical garden; educational facilities; field research station.
Activities: formal education programs for adults & university students. Annual Event: Garden Fest.
Hours & Admission Prices: Mon.-Fri. 8-5. No charge; donations accepted.
Attendance: 12,000 (estimated)
Membership: Student $20; Corporate Partner $125; Sustaining Partner $250; Garden Partner $500; Founding Partner $1,000.

SHEERAR MUSEUM OF STILLWATER HISTORY, (M), 702 S. Duncan St., Stillwater, OK 74074-4443. Tel.: 405-377-0359. Facebook: Sheerar Museum.
E-mail: thesheerarmuseum@sbcglobal.net
Web Site: www.sheerarmuseum.org
Formerly: The Sheerar and Cultural Heritage Center
Founded: 1974.
Congressional District: 3
Key Personnel: Dir., Ammie Bryant; Pres. (V), Gladeen Allred.
Personnel Profile: Part-Time Paid 4; Part-Time Volunteers 35; Interns 4.
Volunteer Hours: 2,175
Operating Expenses: 66,000
Operating Income: 66,000
Governing Authority: Parent Institution: Stillwater Museum Association. Tax-exempt.
Institution Type/Description: History Museum.
Collections: local history & culture; household articles; clothing; tools; 3,450 buttons from 1740-1930.
Major Exhibits: Old Central at 120 Years, 2/14-11/14; Notorious Payne County, 9/14-12/14; AQSG: Colonial Revival Quilts (T), 11/14-2/15.
Publications: quarterly member newsletter, Stillwater Muse.
Hours & Admission Prices: Tues.-Fri. 11-5, Sat.-Sun. 1-4. No charge; donations accepted. Closed major holidays. &
Attendance: 10,000 (estimated)
Membership: Supporter under 25 $25; Single & Family $25-$49; Sustainer $50-$99; Donor $100-$499; Benefactor $500-$999; Patron $1,000 & up.

STILLWATER AIRPORT MEMORIAL MUSEUM, 2020 W. Airport Rd., Stillwater, OK 74075. Tel.: 405-372-7881. Fax: 405-372-8460.
Institution Type/Description: Airport Museum.
Collections: airport history from 1918 to present; photographs; International Flying Farmers; OSU Flying Service; Stillwater Air Force, OSU Flying Aggies; Wiley Post's visit.
Hours & Admission Prices: Museum: by appointment. Lobby: daily 7am-10pm. No charge. &

Stroud

THE SAC AND FOX NATIONAL PUBLIC LIBRARY, 920883 S. Hwy. 99, Stroud, OK 74079. Mailing Address: Administration Bldg., 920883 S. Hwy. 99 Bldg. A, Stroud, OK 74079. Tel.: 918-968-3526, ext. 2020. Fax: 918-968-4837.
Founded: 1988.
Congressional District: 3

Key Personnel: Dir., Kathy Platt; Historical Researcher, Catherine Walker; Library & Archives Tech., Marlena Starr.
Personnel Profile: Full-Time Paid 2.
Institution Type/Description: Library.
Collections: books; DVDs; Native American.
Facilities: Cultural Center; two archive rooms; education & newspaper office.
Activities: Annual Events: Anniversary Open House Cookout in April; Book Sale in July.
Hours & Admission Prices: Mon.-Fri. 8-6, Sat. 9-1. No charge; donations accepted. Closed National holidays.

Sulphur

ARBUCKLE HISTORICAL SOCIETY MUSEUM, 402 W. Muskogee Ave., Sulphur, OK 73086-4614. Tel.: 580-622-5593.
Institution Type/Description: Historical Society Museum: housed in the former Sulphur City Hall; built in 1917.
Collections: local history, culture & heritage; Chickasaw Indian Nation artifacts; personal artifacts; photographs; period furnishings; courtroom; jail cell.
Hours & Admission Prices: Fri.-Sun. 12-5; groups by appointment.

CHICKASAW NATIONAL RECREATION AREA, 1008 W. Second St., Sulphur, OK 73086-4814. Tel.: 580-622-7220. Fax: 580-622-2296. TDD: 580-622-3165.
Web Site: www.nps.gov/chic
Formerly: Platt National Park
Founded: 1906.
Congressional District: 4
Key Personnel: Supt., Bruce Noble; Chief Interpreter, Ron Parker.
Personnel Profile: Full-Time Paid 50; Part-Time Paid 1.
Governing Authority: federal. Operated by National Park Service, U.S. Dept. of Interior. Tax-exempt.
Institution Type/Description: Nature Center.
Collections: snakes; frogs; turtles; insects; plants; fish; mammals; birds; herbarium.
Research Fields: natural history
Facilities: 700-vol. library of books on natural history, conservation, ecology, environmental problems, history of the area, interpretation & education available for use on premises; reading room; 140-seat auditorium. Books for sale.
Activities: nature walks; films; children's programs; environmental education; evening interpretive programs; Junior Ranger program; living history program c.1906.
Publications: orientation brochures, trail guide, visitors guide.
Hours & Admission Prices: Park: daily. Nature Center: Memorial Day-Labor Day daily 9-5:30; Sept.-May daily 9-4:30. No charge. Donations accepted. Closed New Year's Day; Thanksgiving; Christmas. &
Attendance: 95,167 (accurate)

NATIONAL MUSEUM OF HORSE SHOEING TOOLS AND HALL OF HONOR, 7781 U.S. Hwy. 177, Sulphur, OK 73086. Tel.: 580-622-4644. Fax: 580-622-4669.
E-mail: carousel@brightok.net
Web Site: www.horseshoeingmuseum.com
Key Personnel: Dir., Lee Liles
Institution Type/Description: History Museum.
Collections: horseshoeing history & heritage; blacksmith tools & equipment; shoes; photographs.
Hours & Admission Prices: By appointment. No charge.

Tecumseh

TECUMSEH HISTORICAL SOCIETY MUSEUM, 114 S. Broadway, Tecumseh, OK 74873-3206. Tel.: 405-598-8666.
Institution Type/Description: Historical Society Museum: housed in the former Dixon Millinery.
Collections: local history & culture; photographs; period ledgers; Dixon Millinery company artifacts including hats.
Hours & Admission Prices: Sat. 10-2. No charge.

Tishomingo

CHICKASAW BANK MUSEUM & JOHNSTON COUNTY MUSEUM, 413 E. Main St., Tishomingo, OK 73460. Mailing Address: P.O. Box 804, Tishomingo, OK 73460-0804. Tel.: 580-371-3141. Fax: 580-371-3141.
Founded: 1971.
Key Personnel: Pres. (V), Jackie Baker; Museum Shop Mgr., Letha Clark.

Personnel Profile: Part-Time Paid 2.
Governing Authority: Tax-exempt.
Institution Type/Description: History Museum: housed in a Chickasaw Tribe bank building; built in 1902. Listed on the National Register of Historic Places.
Collections: local history & culture; personal artifacts; period furnishings; photographs; genealogy.
Activities: research.
Hours & Admission Prices: Tues.-Fri. 9-5, Sat. 9:30-4:30; other times by appointment. No charge; donations accepted. &
Attendance: 3,024 (accurate)
Membership: Individual $15; Family $25; Sponsor $50.

CHICKASAW COUNCIL HOUSE MUSEUM, 209 N. Fisher, Tishomingo, OK 73460-1717. Mailing Address: P.O. Box 1548, Ada, OK 74821-1548. Tel.: 580-371-3351.
E-mail: museum@chickasaw.net
Web Site: www.chickasaw.net
Founded: 1970.
Congressional District: 3
Key Personnel: Mgr., Flora Fink.
Personnel Profile: Full-Time Paid 4; Interns 1.
Governing Authority: state. Parent Institution: Chickasaw Nation. Tax-exempt.
Institution Type/Description: Native American Museum.
Collections: articles relating to the Chickasaws life in Oklahoma Indian territory; first Council House built & used by Chickasaws; Mississippi migration; Chickasaw works.
Research Fields: genealogy; local history; court records; Chickasaw Dawes Rolls cemetery records.
Facilities: library of Indian Territory Court records & material pertaining to history of the Chickasaws available for use on premises.
Activities: guided tours; lectures; study clubs; permanent exhibitions.
Hours & Admission Prices: Mon.-Fri. 9-6, Sat. 10-4. No charge; donations accepted. Closed legal holidays. &

Tonkawa

A.D. BUCK HISTORY & WELCOME CENTER, 1220 E. Grand, Tonkawa, OK 74653. Mailing Address: P.O. Box 310, Tonkawa, OK 74653-0310. Tel.: 580-628-3318. Fax: 405-628-6209.
E-mail: kirby.tickel@noc.edu
Web Site: www.noc.edu
Formerly: The A.D. Buck Museum of Natural History & Science
Founded: 1913.
Congressional District: 6
Key Personnel: Vice Pres. Devel. & Community Rels., Sheri Snyder; Dir. Alumni Rels., Kirby Tickel-Hill.
Personnel Profile: Part-Time Paid 1.
Governing Authority: college. North Central Accreditations Association. Parent Institution: Northern Oklahoma College. Tax-exempt: 170(b)(1)(A).
Institution Type/Description: History Museum.
Collections: Northern Oklahoma College history.
Research Fields: local history.
Activities: guided tours; permanent & temporary exhibitions.
Hours & Admission Prices: Sept.-May Mon.-Thurs.1-4. No charge. &
Attendance: 117 (accurate)

MCCARTER MUSEUM, 220 E. Grand Ave., Tonkawa, OK 74653. Mailing Address: P.O. Box 27, Tonkawa, OK 74653. Tel.: 580-628-2895 & 2898.
E-mail: marillehelton@att.net
Institution Type/Description: History Museum.
Collections: local history & culture; Native American artifacts; Three Sands Oil fields; WWII Prisoner of War Camp; Northern Oklahoma College; photographs; period furnishings.
Hours & Admission Prices: Tues.-Sat. 1-3; other times by appointment. No charge.

TONKAWA TRIBAL MUSEUM, 10951 Allen Dr., Tonkawa, OK 74653. Mailing Address: 1 Rush Buffalo Rd., Tonkawa, OK 74653. Tel.: 580-628-2561. Fax: 580-628-2279. Facebook: Tonkawa Tribe.
E-mail: info@tonkawatribe.com
Web Site: www.tonkawatribe.com
Congressional District: 3
Key Personnel: Dir., Don L. Patterson; Museum Shop Mgr., Miranda Allen
Institution Type/Description: Native American Museum.
Collections: Tonkawa tribe history & culture; personal artifacts; photographs.
Activities: Annual Event: Pow-Wow in June
Publications: annual brochure.

Hours & Admission Prices: Mon.-Fri. 8:30-4:30. No charge. Closed all federal holidays. &

Tulsa

ALEXANDRE HOGUE GALLERY - UNIVERSITY OF TULSA, 2930 E. 5th St., Tulsa, OK 74104. Mailing Address: 500 S. Tucker Dr., Tulsa, OK 74104. Tel.: 918-631-2739. Fax: 918-631-3423.
Institution Type/Description: Art Gallery.
Collections: works by local, national & international artists.
Major Exhibits: Glenn Godsey - Digits and Doodies, 1/16/14-2/20/14; Libby Williams - MFA Thesis Exhibitions, 2/27/14-3/20/14; 46th Annual Gussman Juried Student Exhibition, 3/27/14-4/18/14; Senior Exhibition, 4/24/14-5/9/14; Ken Kewley - Concepts, Drawings & Painting, 8/28/14-9/25/14; Red Heat - Cermaics Exhibition, 10/2/14-10/30/14; Dennis Oppenheim, 11/6/14-12/5/14.
Activities: special events; lectures.
Hours & Admission Prices: Mon.-Fri. No charge.

ELSING MUSEUM, LRC 137B, 7777 S. Lewis Ave., Tulsa, OK 74171. Tel.: 918-495-6262.
E-mail: rbush@oru.edu
Web Site: elsing.oru.edu
Founded: 1975.
Key Personnel: Dir., Cur. & Museum Shop Mgr., Roger Bush; Educational Liaison, Dr. Catherine Klehm; Geologist, Dr. Steve Herr.
Personnel Profile: Part-Time Paid 6; Part-Time Volunteers 2; Interns 1.
Governing Authority: Parent Institution: Oral Roberts University. Tax-exempt.
Institution Type/Description: Geology Museum.
Collections: gems; minerals; natural art; Indian artifacts; oriental artifacts.
Research Fields: minerals; Oriental art.
Hours & Admission Prices: Wed.-Sat. 1:30-4:30; call to confirm during holiday season. No charge; donations accepted. Closed holidays. &
Attendance: 600 (estimated)

* **GILCREASE MUSEUM, (M),** 1400 N. Gilcrease Museum Rd., Tulsa, OK 74127-2100. Tel.: 918-596-2700; 888-655-2278 (Toll Free). Fax: 918-596-2770.
Web Site: gilcrease.utulsa.edu
Founded: 1949.
Congressional District: 1
Key Personnel: Exec. Dir., Duane H. King, Ph.D.; Chm. (V), Cindy Field; Exec. Asst. to the Dir., Sandra Freeman; Rights & Reproduction, Michelle Maxwell; Dir. Membership, Mary Barnes; Dir. Corporate & Business Philanthropy & Art Sales, Linda Galbraith; Asst. Registrar, Alicia Perkins; Dir. Museum Shop, Amanda Burns; Event Coord., Ken Barton; Mgr. Communications, Melani Hamilton; Coord. Volunteer Svcs., Mark Dolph; Dir. Curatorial Affairs & Public Programs and Dir. Museum Science & Management, Robert B. Pickering, Ph.D.; Head Collection Digitization, Diana Folsom.
Personnel Profile: Full-Time Paid 57; Part-Time Paid 10; Part-Time Volunteers 160; Interns 10.
Volunteer Hours: 25,760
Governing Authority: Parent Institution: City of Tulsa. Subsidiary Institution: University of Tulsa. Tax-exempt: 501(c)(3).
Institution Type/Description: American History & Art Museum.
Collections: American sculpture & painting; manuscripts; cultures of Five Civilized Tribes; American Indian artifacts from the Arctic to Mexico; the westward movement in U.S.; documents; graphics; Central & South American cultural artifacts.
Major Exhibits: Gilcrease Museum Folio's, 7/28/13-3/30/14; Folio Editions: Art in the Service of Science, 8/4/13-3/23/14; Buffalo Bill's Wild West Warriors: Photographs by Gertrude Kasebier, 11/24/13-2/2/14; Form and Line: Allan Houser's Sculpture and Drawings, 2/23/14-6/29/14; Rendezvous 2014: Ross Matteson and Greg Beecham, 4/10/14-7/13/14; Alexandre Hogue: An American Visionary Paintings and Works on Papr, 8/24/14-11/30/14; 2014 Collectors' Reserve, 10/25/14-11/9/14; From Frontier to Foundry: The Making of Small Bronze Sculpture in the Gilcrease Collection, 12/21/14-3/23/15.
Research Fields: archaeology; ethnology; art history.
Facilities: 111,000-vol. library of documents, manuscripts, maps, books & photographs on trans-Mississippi west, Five Civilized Tribes, surveys & Western movement, colonial, Indian & Hispanic history available for use by appointment; reading room; 220-seat auditorium; restaurant; 23 acres landscaped with historic theme gardens. Books, prints of paintings, Indian jewelry & other museum-related items for sale.
Activities: guided tours; lectures; films; gallery talks; rental facilities; formal education programs for children, adults & undergraduate college students; docent program; permanent & temporary exhibitions.
Publications: biannual, Gilcrease Journal; bimonthly, Gilcrease Newsletter;

books, Thomas Moran: The Field Sketches; A Guidebook to Manuscripts; First Artist of the West: George Catlin Paintings & Watercolors from the Collection of Gilcrease Museum; Catlin Catalogue; Visitor's Guide to Collections; Treasures of Gilcrease; Alfred Jacob Miller: Watercolors of the American West; George Catlin's Souvenir of the North American Indians; After Lewis and Clark: The Forces of Change 1806-1871; The Many Faces of Edward Sherriff Curtis; Charles Banks Wilson: An Oklahoma Life in Art; Thomas Gilcrease, Willard Stone, Art of the Oklahoma State Capitol, The Senate Collection; Perfectly American, The Art-Union & Its Artists; To Capture the Sun, Gold of Ancient Panama; Forging a Nation; Peace Medals, Negotiating Power in Early America.

Hours & Admission Prices: Tues.-Sun. 10-5. Adults $8, seniors 62 & over and active military $6, college students $5; discounts to groups of 10 or more; University of Tulsa students, members, children 18 & under, 1st Tues each month, Western Reciprocal, Museums West Consortium & North American Reciprocal members no charge. Closed Christmas. &

Attendance: 113,561 (accurate)

Membership: Member Levels: Student (w/ID) $25; Individual $50; Family/Dual $65; Friend $125; Supporter $250; Patron $500. Director's Society: Bierstadt $1,000; Moran $1,500; Catlin $2,000; Russell $3,000; Remington $5,000; Taos $7,500-$9,999. Business Art Alliance: Collector Level $1,000; Partner Level $2,000; Investor Level $3,000; Benefactor Level $5,000; Coouncil Level $10,000 & above. Gilcrease Coouncil $5,000 & above.

MABLE B. LITTLE HERITAGE HOUSE MUSEUM, 322 N. Greenwood Ave., Tulsa, OK 74120-1026. Tel.: 918-596-1006. Fax: 918-583-2770.

E-mail: mlmackey@sbcglobal.net

Institution Type/Description: Historic House.

Collections: period artifacts; early furnishings; personal artifacts.

Hours & Admission Prices: Mon.-Fri. 9-4:30, Sat. by appointment.

PHILBROOK DOWNTOWN, 116 E. Brady St., Tulsa, OK 74103. Tel.: 918-749-7941.

E-mail: mbrown@philbrook.org

Web Site: www.philbrook.org

Founded: 2013.

Institution Type/Description: Art Museum.

Collections: Native American art; modern & contemporary art.

Major Exhibits: In a Glorious Light, 10/13-3/16/14; Unexpected, 10/13-5/18/14; Beauty Within, 3/23/14-9/7/14; Allan Houser, 5/25/14-11/2/14.

Facilities: 30,000 sq. ft. exhibition space; study center.

Activities: permanent & temporary exhibitions.

Hours & Admission Prices: Wed.-Sat. 12-7, Sun. 12-5. Adults $9, students & senior citizens $7; youth under 17 & members no charge. Closed New Year's Day; Independence Day; Thanksgiving; Christmas.

* **THE PHILBROOK MUSEUM OF ART, INC., (M),** 2727 S. Rockford Rd., Tulsa, OK 74114-4104. Mailing Address: P.O. Box 52510, Tulsa, OK 74152-0510. Tel.: 918-749-7941. Fax: 918-743-4230. TDD: 918-749-7941 (public info. line).

E-mail: mbrown@philbrook.org

Web Site: www.philbrook.org

Founded: 1938.

Congressional District: 1

Key Personnel: Exec. Dir., Pres. & C.E.O., Randall Suffolk; Chm., Holbrook Lawson; Deputy Dir., David Singleton; Facility Mgr., Charisse Cooper; Dir. Exhibitions & Collections, Christine Knop Kallenberger; Librarian, Tom Young; Museum Shop Mgr., Susan Shrewder; Dir. Communications, Tricia Milford-Hoyt; Dir. Finance, Donna Durrin; Cur. Native American & Non-Western Art, Christina Burke; Chief Cur. & Cur. American Art, Catherine Whitney; Nancy E. Meinig Cur. Modern & Contemporary Art, Lauren Ross.

Personnel Profile: Full-Time Paid 69; Part-Time Paid 15; Part-Time Volunteers 229; Interns 2.

Volunteer Hours: 9,473

Operating Expenses: 7,727,594

Operating Income: 8,959,909

Governing Authority: state. nonprofit organization. Tax-exempt: 501(c)(3).

Institution Type/Description: Art Museum & Gardens.

Collections: Samuel H. Kress Italian Renaissance paintings & sculptures; American & European paintings & works on paper; American Indian baskets, pottery & paintings; African sculpture; Japanese screens & scrolls; Southeast Asian ceramics; modern & contemporary art & design; Eugene B. Adkins Native American & western art.

Major Exhibits: Collective Future, 11/13-1/14; Collective Future: Gifts Celebrating Philbrook's 75th Anniversary, 11/3/13-1/26/14; Georges Rouault:

Through a Glass, Darkly, 1/19/14-4/20/14; Ronen &Erwan Bouroullec: album, 3/2/14-5/11/14; Collective Past: Works on Paper, 4/27/14-7/20/14; Hard Times in Oklahoma, 1939-40: The Documentary Photography of Russell Lee, 7/27/14-10/26/14; Monet's Mornings on the Seine: Impressions of a River, 6/29/14-9/21/14; Impact. Philbrook's Indian Annual, 10/19/14-1/11/15; Howard Cook and Barbara Latham on Paper: A Creative Union, 11/2/14-1/4/15.

Research Fields: American Indian art.

Facilities: 18,900-vol. library; classrooms; 23 acres of formal & natural gardens.

Activities: guided tours; lectures; concerts; special trips; films; gallery talks; docent program; speakers bureau; inter-museum loan, permanent, temporary & traveling exhibitions; yoga; tai chi; summer camps; adult classes; children classes.

Publications: quarterly bulletins to members; catalogs of temporary exhibitions.

Hours & Admission Prices: Tues.-Wed. & Fri.-Sun. 10-5, Thurs. 10-8. Adults $9, senior citizens & higher education students $7; discounts to military, chaperones, groups of 10 or more & AAM members; children 18 & under, members, and 2nd Sat. each month no charge. &

Attendance: 128,762 (accurate)

Membership: Individual $55; Family & Dual $80; Associate $125; Supporter $250; Sponsor $500; Contributor $750; Masters Society (under 40) $1,000; Masters Society $2,000; Masters Society Patron $5,000; Masters Society Benefactor $10,000.

* **SHERWIN MILLER MUSEUM OF JEWISH ART, (M),** 2021 E. 71st St., Tulsa, OK 74136-5408. Tel.: 918-492-1818. Fax: 918-492-1888.

E-mail: info@jewishmuseum.net

Web Site: www.jewishmuseum.net

Formerly: Fenster Museum of Jewish Art

Founded: 1966.

Congressional District: 1

Key Personnel: Exec. Dir., Drew Diamond; Devel. & Mktg., Melissa Schnur; Cur., Dr. Karen York; Admin. Asst., Tracey Herst-Woods.

Personnel Profile: Full-Time Paid 3; Part-Time Paid 1; Part-Time Volunteers 25.

Governing Authority: nonprofit organization. Tax-exempt: 501(c)(3).

Institution Type/Description: Judaica Art Museum.

Collections: Jewish history, culture & art from biblical times to present; ritual objects; synagogue textiles; ethnographic artifacts; costumes; archaeological artifacts; documents; archival materials; photographs; prints; paintings; sculptures; architectural elements; Holocaust era artifacts, documents & photographs.

Research Fields: Jewish arts & crafts; regional & state Jewish history.

Facilities: 1,000-vol. library containing reference works for the study of Jewish art available to the public; educational facilities.

Activities: guided tours; lectures; organized education programs for children; docent program; participatory & loan exhibitions.

Publications: newsletter, Emuse; exhibition catalogues.

Hours & Admission Prices: Adults $6.50, senior citizens $5.50, students $3.50; discounts to law enforcement, uniform services, groups of 10 or more, military, teachers, ICOM & AAM members; Blue Star members & museum members no charge. &

Attendance: 10,000 (estimated)

Membership: Student $25; Senior $35; General $50; Family $75; Friend $125; Contributing $250; Supporting $500; Sustaining $750; Patron $1,000; Sponsor $1,500; Benefactor $2,500; Director's Circle $5,000; Life $25,000.

SOCIETY OF EXPLORATION GEOPHYSICISTS - GEOSCIENCE CENTER, 8801 S. Yale, Tulsa, OK 74137-3573. Mailing Address: P.O. Box 702740, Tulsa, OK 74170-2740. Tel.: 918-497-5555. Fax: 918-497-5557.

Institution Type/Description: Earth Science Center.

Collections: earth science; geology; rocks; minerals; fossils.

Activities: educational programs.

Hours & Admission Prices: Call for hours.

TULSA AIR AND SPACE MUSEUM & PLANETARIUM, 3624 N. 74th E. Ave., Tulsa, OK 74115-3622. Tel.: 918-834-9900. Fax: 918-834-6723.

E-mail: kjones@tulsamuseum.org

Web Site: www.tulsaairandspacemuseum.org

Formerly: Tulsa Air and Space Center

Founded: 1998.

Congressional District: 1

Key Personnel: Exec. Dir., Glenn Wright; Deputy Dir., Kim Jones; Chm. (V), Carmine Romano; Museum Shop Mgr., Pam Gunn.

Personnel Profile: Full-Time Paid 8; Part-Time Paid 4; Part-Time Volunteers 65.
Governing Authority: Tax-exempt: 501(c)(3).
Institution Type/Description: Space Museum & Planetarium.
Collections: history & future of aerospace; science & technology; hands-on exhibits; photographs; military & civil aircraft including 1931 Spartan C-2, the Rockwell Ranger 2000, the HuGo Craft, & the Grumman F-14A Tomcat; Spartan 12.
Facilities: 25,000 sq. ft. exhibit space; planetarium.
Activities: special events; educational programs; hands-on activities.
Publications: quarterly news, Flight Path.
Hours & Admission Prices: Tues.-Sat. 10-5, Sun. 1-5. Adults $12, seniors 62 & over, students and military $10, youth 4-12 $7; children 3 & under and members no charge. Closed holidays.
Attendance: 62,000 (accurate)
Membership: Pilot $65; Aviator $80; Captain $120; Colonel $200; Astronaut $1,000.

TULSA HISTORICAL SOCIETY AND MUSEUM, 2445 S. Peoria, Tulsa, OK 74114-1326. Tel.: 918-712-9484. Fax: 918-712-1939. Facebook: Tulsa History.
E-mail: ths@tulsahistory.org
Web Site: tulsahistory.org
Key Personnel: Exec. Dir., Michelle Place; Communications & Facilities Mgr., Britni Worley; Dir. Education & Exhibits, Maggie Brown; Cur. Collections & Archivist, Ian Swart; Dir. Devel., Meredith Miers.
Personnel Profile: Full-Time Paid 5; Full-Time Volunteers 1; Part-Time Volunteers 40; Interns 4.
Governing Authority: Tax-exempt.
Institution Type/Description: Historical Society Museum: housed in the Samuel Travis Mansion.
Collections: 15,000 photographs; rare books; film & video archives; maps; historical costumes; decorative arts.
Major Exhibits: Tulsa in the 1930s, 2/14-1/15.
Hours & Admission Prices: Tues.-Sat. 10-4. Adults $5, seniors $3, members, students & children no charge. Closed holidays.
Membership: Individual $35; Family $50; Supporting $100; Sustaining $500; Recognition $1,000.

TULSA ZOO & LIVING MUSEUM, 6421 E. 36th St., N., Tulsa, OK 74115-2100. Tel.: 918-669-6600 & 6202. Fax: 918-669-6260.
E-mail: info@tulsazoo.org
Web Site: tulsazoo.org
Founded: 1927.
Congressional District: 1
Key Personnel: C.E.O., Terrie Correll.
Governing Authority: municipal. Parent Institution: Tulsa Park & Recreation Board, 1712 W. Charles Page Blvd. Tax-exempt.
Institution Type/Description: Zoo & Natural History Museum.
Collections: live animals & plants; North American fossils, geological specimens; Native American, African American, tropical American, Asian cultural artifacts & displays.
Research Fields: cooperative animal behavior research; archaeological; paleontological.
Facilities: 1,730-vol. library of zoological books available upon request; restaurant. Books & educational novelties for sale.
Activities: guided tours for schools, civic groups; lectures; TV & radio programs; semi-annual training programs for docents & zooteens; summer workshops in zoology, art, photography; in-school programs for grades 4 & 6; work experience in zoo management for college credit.
Publications: newsletter, Zoo News; Volunteer Newsletter; annual reports of zoo.
Hours & Admission Prices: Daily 9-5. Adults $10, senior citizens 65 & over $8, children 3-11 $6; discounts to AZA members; children 2 & under & members no charge. Closed Christmas; third Fri. in June. &
Attendance: 630,000 (estimated)
Membership: Associate $55; Family $75; Friends & Family $150; Adventure Pass $230.

Vinita

EASTERN TRAILS MUSEUM, 215 W. Illinois, Vinita, OK 74301-3129. Tel.: 918-256-2115.
Web Site: www.vinitapl.okpls.org/museum.htm
Founded: 1964.
Key Personnel: Cur., Wanda Norton.
Personnel Profile: Full-Time Volunteers 1; Part-Time Volunteers 7.
Governing Authority: Tax-exempt.
Institution Type/Description: History Museum.

Collections: items pertaining to the history of the Cherokee, Shawnee & Delaware Indians.
Hours & Admission Prices: Mon.-Fri. 1-4, Sat. 1-3. No charge; donations accepted. &

Wagoner

CITY OF WAGONER HISTORICAL MUSEUM, 122 S. Main, Wagoner, OK 74467-5221. Mailing Address: P.O. Box 406, Wagoner, OK 74477-0406. Tel.: 918-485-9111.
Institution Type/Description: History Museum.
Collections: local history & culture; period clothing; Civil War artifacts; photographs.
Activities: guided tours.
Hours & Admission Prices: Tues.-Sat. 10-3. No charge. &

Wakita

TWISTER THE MOVIE MUSEUM, 101 W. Main St., Wakita, OK 73771. Mailing Address: P.O. Box 285, Wakita, OK 73771-0285. Tel.: 580-594-2312.
Web Site: www.twistercountry.com
Institution Type/Description: Movie Museum: located in the town where the movie Twister was filmed.
Collections: film memorabilia; props; pictures; Twister pinball machine; original Dorothy machine.
Facilities: Museum-related items for sale.
Activities: tours.
Hours & Admission Prices: Tues.-Sat. 1-5. No charge.

Walters

COTTON COUNTY MUSEUM, 116 N. Broadway, Walters, OK 73572. Mailing Address: P.O. Box 244, Walters, OK 73572. Tel.: 580-875-3335.
Governing Authority: city.
Institution Type/Description: History Museum: housed in the former Grand Theatre.
Collections: local history & culture; period furnishings; personal artifacts; photographs; military artifacts; barbed wire; tools.
Hours & Admission Prices: Mon.-Thurs. 9-12 & 1-3; other times by appointment.

Warner

WALLIS MUSEUM, 1000 College Rd., Warner, OK 74469-9700. Mailing Address: Rte. 1, Box 1000, Warner, OK 74469-9700. Tel.: 918-463-6236. Fax: 918-463-6314.
E-mail: mrigney@connorsstate.edu
Web Site: www.connorsstate.edu
Formerly: Rural Farming & Agriculture Museum
Founded: 1908.
Congressional District: 2
Key Personnel: C.E.O., Dr. Tim Faltyn.
Personnel Profile: Part-Time Volunteers 1.
Governing Authority: college. Parent Institution: Connors State College. Tax-exempt: 501(c)(3).
Institution Type/Description: University Museum.
Collections: agriculture & farming artifacts; college history.
Activities: permanent exhibitions.
Hours & Admission Prices: Closed for relocation. &

Watonga

T.B. FERGUSON HOME, 519 N. Weigel, Watonga, OK 73772. Tel.: 580-623-5069.
Founded: 1972.
Congressional District: 6
Key Personnel: Cur., Mary Deane.
Personnel Profile: Part-Time Paid 1; Part-Time Volunteers 6.
Governing Authority: state. Parent Institution: Oklahoma Historical Society, Oklahoma History Center, 2401 N. Laird Ave., Oklahoma City, OK 73105-7914. Tax-exempt.
Institution Type/Description: Historic House: c.1901 T.B. Ferguson Home, home of sixth territorial Governor.
Collections: period furnishings; 1892-1921 early-day pioneer items; personal items of T.B. & Elva Ferguson; tax rolls; maps; books. Historical Buildings: c.1893 City Jail; c.1870 Federal Remount Station.
Research Fields: Governor Ferguson.

Activities: guided tours; political re-enactment. Museum Sponsors: Cheese Festival.
Publications: brochures; weekly news column in Watonga Republican Newspaper.
Hours & Admission Prices: Wed.-Sat. 10-5, Sun. 2-5. No charge; donations accepted. Closed state holidays.
Attendance: 4,320 (estimated)
Membership: Friends $10; Individual $15; Family $25; Institutional $35; Life $300; Benefactor $500.

Waurika

ROCK ISLAND DEPOT MUSEUM, 105 S. Meridian, Waurika, OK 73573. Mailing Address: 122 S. Main St., Waurika, OK 73573-3054. Tel.: 590-228-3274. Fax: 590-228-2907.
E-mail: museum@waurika.lib.ok.us
Congressional District: 4
Key Personnel: Museum Bd. Pres., Nancy Way; Librarian & Cur., Cathy Dumas.
Governing Authority: city of Waurika.
Institution Type/Description: History Museum.
Collections: Rock Island Railroad memorabilia; authentic period caboose.
Facilities: library; meeting room.
Publications: Oklahoma flyer, Waurika; Post Offices in and near Jefferson County, Oklahoma.
Hours & Admission Prices: Mon.-Fri. 9-5.

Weatherford

HEARTLAND OF AMERICA MUSEUM, (M), 1600 S. Frontage Rd., Weatherford, OK 73096-6119. Tel.: 580-774-2212.
E-mail: heartland@oklahomaheartlandmuseum.com
Web Site: oklahomaheartlandmuseum.com
Founded: 2007.
Key Personnel: Dir., Rick Lundquist; C.E.O. & Pres. (V), Jana Lou Scott; Museum Shop Mgr., Jewrell Crall.
Personnel Profile: Part-Time Paid 3; Part-Time Volunteers 30.
Governing Authority: Parent Institution: Heartland of America Heritage Foundation. Tax-exempt.
Institution Type/Description: History Museum.
Collections: local history & culture; period furnishings; personal artifacts; photographs. Historic Buildings: one-room schoolhouse; 1930 era blacksmith shop.
Hours & Admission Prices: Tues.-Fri. 9-5, Sun. 1-5; other times by appointment. Adults 19 & over $6, students 6-18 $2; discount on Veteran's Day & to groups of 10 or more. Closed New Year's Day; Easter; Memorial Day; Independence Day; Labor Day; Thanksgiving; Christmas. &
Attendance: 1,000 (estimated)
Membership: Keeper $100.

STAFFORD AIR & SPACE MUSEUM, 3000 Logan Rd., Weatherford, OK 73096-2681. Tel.: 580-772-5871.
E-mail: director@staffordmuseum.org
Web Site: www.staffordmuseum.org
Founded: 1981.
Governing Authority: Tax-exempt: 501(c)(3).
Institution Type/Description: Air & Space Museum.
Collections: history of air & space flight; jet engines & equipment; model aircraft; General Staffords missions; rocket engines & equipment.
Hours & Admission Prices: Mon.-Sat. 9-5, Sun. 1-5. Adults 19-54 $7, seniors 55 & over $5, children & students $2. Closed New Year's Day; Easter; Memorial Day; Independence Day; Labor Day; Thanksgiving; Christmas. &
Attendance: 25,000 (estimated)
Membership: Individual $35; Family $65; Supporter $500; Benefactor $1,000; Corporate Sponsorship $3,000.

Webbers Falls

WEBBERS FALLS HISTORICAL SOCIETY AND MUSEUM, Commercial & Main, Webbers Falls, OK 74470. Mailing Address: P.O. Box 5, Webbers Falls, OK 74470. Tel.: 918-464-2728.
Institution Type/Description: Historical Society Museum: town named in honor of Chief Walter Webber who settled here in 1828.
Collections: local history & culture; Cherokee settlers; photographs; personal artifacts.
Hours & Admission Prices: Call for hours.

Weleetka

WELEETKA TOWN HALL AND JAIL, State Hwy. 75 & Seminole, Weleetka, OK 74880. Mailing Address: P.O. Box 733, Weleetka, OK 74880. Tel.: 405-786-2501 & 3251.
Governing Authority: Parent Institution: Weleetka Art & Historical Society.
Institution Type/Description: Historic Building: housed in the former City Hall building; built in 1910. Listed on the National Register of Historic Places.
Collections: local history & culture; period furnishings; personal artifacts; local newspapers; Native American artifacts.
Hours & Admission Prices: By appointment.

Wewoka

SEMINOLE NATION MUSEUM, 524 S. Wewoka Ave., Wewoka, OK 74884-3239. Mailing Address: Box 1532, Wewoka, OK 74884-1532. Tel.: 405-257-5580. Fax: 405-257-5580.
E-mail: director@theseminolenationmuseum.org
Web Site: www.theseminolenationmuseum.org
Founded: 1974.
Congressional District: 4
Key Personnel: C.E.O. & Pres. (V), William Wantland; Exec. Dir. & Chief Cur., Richard Ellwanger; Asst. Cur., Lewis Johnson; Physical Plant, Noah Hail; Registrar, Karen Smith.
Personnel Profile: Full-Time Paid 3; Part-Time Paid 1; Part-Time Volunteers 27.
Governing Authority: society. Parent Institution: The Seminole Nation Historical Society. Tax-exempt.
Institution Type/Description: History Museum.
Collections: Seminoles; freedmen Blacks; pioneer history; Native American art gallery; Oklahoma Oil Boom & railroad history.
Research Fields: Seminole Indians; Freedman; Genealogy.
Facilities: reading room. Arts & crafts of the Seminole and Creek Indians, beadwork, needlepoint, Seminole dolls & paintings for sale.
Activities: guided tours; school programs.
Publications: quarterly newsletter.
Hours & Admission Prices: Mon.-Sat. 10-5. No charge; donations accepted. Closed Federal holidays. &
Attendance: 10,000 (estimated)
Membership: Sustaining $25; Sponsor $50; Century Club $100 & up.

Wilburton

LUTIE COAL MINER'S MUSEUM, 2307 E. Main St., Wilburton, OK 74578. Tel.: 918-465-2216.
Institution Type/Description: Mining Museum: housed in the former Hailey Ola Coal Company house; built in 1901.
Collections: coal mining history; miners memorial; coal mining tools & equipment.
Hours & Admission Prices: By appointment. No charge.

ROBBERS CAVE NATURE CENTER, Hwy. 2 N., Wilburton, OK 74578. Mailing Address: P.O. Box 9, Wilburton, OK 74578-0009. Tel.: 918-465-2562; 800-654-8240. Fax: 918-465-2781. Facebook: Robbers Cave State Park.
E-mail: robberscave@travelok.com
Web Site: www.TravelOK.com/RobbersCave
Key Personnel: Park Mgr., Merle Cox; Park Naturalist, Jacque Martin
Institution Type/Description: Nature Center.
Collections: Native American history; natural history. Historic Building: 1930s natural stone bathhouse.
Facilities: nature trails. Museum-related items for sale.
Activities: haunted cave tours; hiking.
Hours & Admission Prices: Mon.-Thurs. & Sun. 8-6, Fri.-Sat. 8-8. No charge.

Wilson

WILSON HISTORICAL MUSEUM, 1270 8th St., Wilson, OK 73463. Tel.: 580-668-2505.
E-mail: whm@wilsonhistoricalmuseum.org
Web Site: www.wilsonhistoricalmuseum.org
Founded: 2001.
Governing Authority: Tax-exempt.
Institution Type/Description: History Museum: housed in the Dr. Darling/Wilson Post-Democrat building; built in 1926.
Collections: local history & culture; period furnishings; personal artifacts; photographs.
Hours & Admission Prices: Tues. & Thurs.-Sat. 10-4; other times by appointment. No charge. &

Woodward

PLAINS INDIANS & PIONEERS MUSEUM, (M), 2009 Williams Ave., Woodward, OK 73801-5717. Tel.: 580-256-6136. Fax: 580-256-2577.
E-mail: contact@pipm1.org
Web Site: www.pipm1.org
Founded: 1966.
Congressional District: 5
Key Personnel: Dir., Robert Roberson; Cur., Tammy Hawbaker; Gift Shop Mgr., Jo Simmons.
Personnel Profile: Full-Time Paid 3; Full-Time Volunteers 20; Part-Time Paid 1; Part-Time Volunteers 20.
Governing Authority: foundation; nonprofit organization. Parent Institution: Plains Indians & Pioneers Historical Foundation. Tax-exempt.
Institution Type/Description: Regional History Museum.
Collections: early day pioneer life & Indian artifacts; bank from Fargo; Fort Supply fire house & post office; Woodward's centennial; historic photographs & textiles; agricultural development in northwest Oklahoma featuring murals by Fred Olds & Jana Sol; Lee & Lienemann restored pioneer cabin; stable.
Research Fields: northwest Oklahoma local & regional history; Woodward tornado April 9, 1947.
Facilities: Books, art prints & museum-related items for sale.
Activities: permanent & changing exhibitions; lectures; tours; shows on hobbies & art; children's workshops & programming.
Publications: newsletter, Northwest Winds; books, Woodward County Pioneer Families Before 1915; Woodward County Family Histories; Below Devil's Gap; Northwest Oklahoma Territory Map; map, Woodward County Schools 1910.
Hours & Admission Prices: Tues.-Sat. 10-5. No charge; donations accepted. Closed major holidays. &
Attendance: 9,000 (accurate)
Membership: Individual $35; Family $50; Pioneer $100; Heritage $250; Paul Laune $500; Temple Houston $1,000; C.E. Williams Legacy $5,000.

Wynnewood

ESKRIDGE HOTEL MUSEUM, 114 E. Robert S. Kerr Blvd., Wynnewood, OK 73098-6621. Tel.: 405-238-4567. Fax: 405-665-4619. Facebook: The Eskridge Hotel Museum.
E-mail: theeskridgehotel@yahoo.com
Web Site: www.eskridgehotelmuseum.org
Founded: 1973.
Key Personnel: Dir., Sandy Campbell.
Governing Authority: Parent Institution: Wynnewood Historical Society. Tax-exempt.
Institution Type/Description: History Museum: built by Pinckney Reid Eskridge in 1907.
Collections: Oklahoma history; period artifacts; photographs; furnishings; Native American artifacts & arrowheads; dolls; old hats, shoes & clothing; personal toiletries; musical instruments; old doctor's room with medical items; Victrolas & old phonograph players; old barbershop chair with barber equipment; shoeshine chair; old safes; military uniforms & items; old-time permanent-wave machine.
Facilities: Museum-related items for sale.
Activities: rental facilities. Annual Theatrical Production: Tales from the Eskridge.
Hours & Admission Prices: Mon.-Thurs. 11-4; other times by appointment. Adults $4, seniors 55 & over $3, students $2; discounts to groups; children 5 & under and school groups & military veterans no charge.
Membership: Individual $25; Family $40; Business/Corporate: Gold $150, Platinum $300.

Yale

JIM THORPE HOME, 706 E. Boston Ave., Yale, OK 74085-4004. Tel.: 918-387-2815.
E-mail: frick.linda@yahoo.com
Founded: 1973.
Congressional District: 33
Key Personnel: Cur., Linda C. Frick; Cur., Virginia Stanford.
Governing Authority: state. Parent Institution: Oklahoma Historical Society (see separate listing). Subsidiary Institution: Jim Thorpe Memorial Foundation. Tax-exempt.
Institution Type/Description: History Museum.
Collections: 1920s furnishings & memorabilia; Jim Thorpe family photographs; replica of Jim Thorpe's boyhood cabin. Historic House: 1876 cabin.
Research Fields: Life of Jim Thorpe.

Hours & Admission Prices: Wed.-Sat. 10-5. No charge. Closed major holidays. &
Attendance: 1,200 (estimated)
Membership: Annual $10; Lifetime $200.

Yukon

YUKON HISTORICAL SOCIETY MUSEUM & ART CENTER, 601 Oak, Yukon, OK 73099-2538. Tel.: 405-354-5079.
Web Site: www.thestagedoorinc.org
Key Personnel: Pres., John Knuppel
Institution Type/Description: History & Art Museum: housed in a 1910 school building.
Collections: period artifacts; Czech history; flour mill history.
Facilities: theater.
Activities: theater productions; fundraiser.
Hours & Admission Prices: By appointment. Museum: no charge. Art Center: call for admission prices.

YUKON'S BEST RAILROAD MUSEUM, 410 Oak Ave., Yukon, OK 73099-2640. Tel.: 405-354-5079.
Key Personnel: Pres. & Cur., John Knuppel; Cur., Jack Austerman
Institution Type/Description: Railroad Museum.
Collections: model trains; Rock Island artifacts; railroad memorabilia.
Hours & Admission Prices: By appointment. No charge; donations accepted.

OREGON

(199 listings)

Agness

AGNESS-ILLAHE MUSEUM, INC., 34470 Agness-Illaha Rd., P.O. Box 36, Agness, OR 97406-9701. Mailing Address: 29419 Ellensburg, P.O. Box 1598, Gold Beach, OR 97444. Tel.: 541-247-2014.
E-mail: ptl@dishmail.net
Web Site: www.agnessmuseum.com
Founded: 1993.
Key Personnel: Dir., Linda Graves; C.E.O., Dennis Graves.
Personnel Profile: Full-Time Volunteers 2; Part-Time Volunteers 4.
Governing Authority: Parent Institution: Gold Beach Historical Museum.
Institution Type/Description: History Museum.
Collections: Native American artifacts; historical items.
Activities: Museum Sponsors: Big Bend historical talk/walk in September.
Hours & Admission Prices: May-Sept. daily 11-2. No charge; donations accepted. &
Attendance: 1,000 (estimated)

Albany

ALBANY REGIONAL MUSEUM, 136 Lyon St., S., Albany, OR 97321-2703. Tel.: 541-967-7122. Facebook; Albany Regional Museum.
E-mail: armuseum@peak.org
Web Site: www.armuseum.com
Founded: 1980.
Congressional District: 5
Key Personnel: Chm. (V), Kristen Shuttpelz; Exec. Dir., Judie Weissert; Collections & Exhibits Coord., Megan Lallier-Barron; Clerk, Peggy Kowal.
Personnel Profile: Part-Time Paid 3; Part-Time Volunteers 20.
Governing Authority: private; nonprofit organization.
Institution Type/Description: Historical Museum.
Collections: cultural artifacts; memorabilia; photos.
Research Fields: historical; historic homes; genealogical; local history & culture.
Facilities: 200-vol. library; 4,000 sq. ft. exhibit space. Museum-related items for sale.
Activities: guided tours; temporary exhibitions; public business meetings; monthly events; members only events; public events.
Publications: quarterly newsletter, Albany Old Times.
Hours & Admission Prices: Mon.-Fri. 12-4, Sat.10-2. No charge; donations accepted. Closed New Year's Day; Labor Day; Thanksgiving; Christmas. &
Attendance: 4,156 (accurate)
Membership: Friends $15; Business Friends $50; Patrons $115; Business Patrons $150; History Circle $250; Chautauqua Circle $1,000-$9,999 in lifetime gifts; Brenneman Society $10,000 or more in lifetime gifts.

THE MONTEITH HOUSE MUSEUM, 518 Second Ave., S.W., Albany, OR 97321-2239. Mailing Address: Monteith Historic Society, P.O. Box 965, Albany, OR 97321-0362. Tel.: 541-928-0911. Facebook: Monteith House Museum.
Web Site: www.monteithhouse.com
Key Personnel: Pres. (V), Rusty van Rossmann
Institution Type/Description: Historic House: built in 1849. Listed on the National Register of Historic Places.
Collections: period artifacts & memorabilia dedicated to the pioneer ancestors.
Hours & Admission Prices: June 15-Sept. 15 Wed.-Sat. 12-4. No charge, donations accepted.
Attendance: 2,000 (estimated)

Ashland

HANSON HOWARD GALLERY, 89 Oak St., Ashland, OR 97520-1802. Tel.: 541-488-2562.
E-mail: hhgall@mind.net
Web Site: hansonhowardgallery.com
Institution Type/Description: Art Gallery.
Collections: works by contemporary northwest artists; paintings; sculpture; ceramics; fine art prints.
Activities: temporary exhibitions.
Hours & Admission Prices: Call for hours.

SCHNEIDER MUSEUM OF ART, (M), Southern Oregon University, 1250 Siskiyou Blvd., Ashland, OR 97520-5001. Tel.: 541-552-6245. Fax: 541-552-8241.
E-mail: sma@sou.edu
Web Site: www.sou.edu/sma
Founded: 1986.
Congressional District: 2
Key Personnel: Acting Dir., Erika Leppmann.
Personnel Profile: Full-Time Paid 2; Part-Time Paid 2; Part-Time Volunteers 48; Interns 8.
Governing Authority: nonprofit. Parent Institution: Southern Oregon University. Tax-exempt.
Institution Type/Description: Art Museum.
Collections: contemporary art; Native American baskets; pre-Columbia artifacts; works on paper.
Activities: workshops; visiting artists; lectures; major fundraisers.
Publications: quarterly bulletin; catalogs.
Hours & Admission Prices: Mon.-Sat. 10-4. No charges, donations accepted. Closed state holidays; university holidays. &
Attendance: 15,000 (estimated)
Membership: Student $15; Individual $30; Family $60; Benefactor $100; Patron $250.

SCIENCEWORKS HANDS-ON MUSEUM, 1500 E. Main St., Ashland, OR 97520-1312. Tel.: 541-482-6767. Fax: 541-482-5716.
E-mail: info@scienceworksmuseum.org
Web Site: www.scienceworksmuseum.org
Founded: 2002.
Congressional District: 2
Key Personnel: Exec. Dir., Chip Lindsey; Chm. (V), Elaine Sweet; Museum Shop Mgr., Rachel Cardillo.
Personnel Profile: Full-Time Paid 6; Part-Time Paid 10; Part-Time Volunteers 120.
Governing Authority: Tax-exempt.
Institution Type/Description: Science Museum.
Collections: hands-on exhibits.
Activities: special events; school groups; summer science camps; festivals; birthday parties; summer internship.
Hours & Admission Prices: Wed.-Sat. 10-4, Sun. 12-4. Adults $7.50, children 2-12 & senior 65 & over $5; discounts to groups of 10 or more & ASTEC members; children under 2 & members no charge. Closed New Year's Day; Memorial Day; Independence Day; Labor Day; Thanksgiving; Christmas. &
Attendance: 45,000 (accurate)
Membership: Family $60.

SOUTHERN OREGON UNIVERSITY MUSEUM OF VERTEBRATE NATURAL HISTORY, 1250 Siskiyou Blvd., Ashland, OR 97520-5001. Mailing Address: Southern Oregon University, Dept. of Biology, Ashland, OR 97520. Tel.: 541-552-6749 & 6341. Fax: 541-552-6415.
E-mail: stonek@sou.edu
Founded: 1969.
Congressional District: 52
Key Personnel: Cur., Dr. Karen Stone; Sec. Biology Dept., Colleen Martin.
Governing Authority: university. Parent Institution: Southern Oregon University.
Institution Type/Description: Vertebrate Biology Museum.
Collections: vertebrate specimens; bird skins; mammal skins; reptiles; amphibians; fish.
Research Fields: vertebrate biology.
Activities: research; teaching.
Hours & Admission Prices: By appointment only. No charge; donations accepted.

Astoria

* **COLUMBIA RIVER MARITIME MUSEUM, (M),** 1792 Marine Dr., Astoria, OR 97103-3525. Tel.: 503-325-2323. Fax: 503-325-2331.
E-mail: information@crmm.org
Web Site: www.crmm.org
Founded: 1962.
Congressional District: 1
Key Personnel: Exec. Dir., Sam Johnson; Deputy Dir., David Pearson; Chm. (V), Roger Qualman; Cur., Jeff Smith; Museum Shop Mgr., Blue Anderson.
Personnel Profile: Full-Time Paid 14; Part-Time Paid 10; Part-Time Volunteers 100.
Governing Authority: nonprofit organization. Tax-exempt: 501(c)(3).
Institution Type/Description: History & Maritime Museum.
Collections: lightship (afloat); small craft; marine engines; naval weapons; nautical instruments; tools; miscellaneous marine artifacts; ship models; paintings; prints; photographs; maps; books; manuscripts; navigational aids; bridge from USS Knapp; charts commercial fishing craft and gear; Coast Guard craft and gear; cannery tools and gear.
Research Fields: Pacific Northwest maritime history.
Facilities: 8,000-vol. library of nautical books available for use on the premises & by appointment only.
Activities: permanent & temporary exhibitions; films; lectures; demonstrations of maritime skills; school programs; summer family programs; summer day camp.
Publications: quarterly newsletter, Quarterdeck.
Hours & Admission Prices: Daily 9:30-5. Adults $12, senior citizens $10, children $5; members & children under 6 no charge. Reciprocal admission with participating Council of American Maritime Museums. Closed Thanksgiving; Christmas. &
Attendance: 100,463 (accurate)
Membership: Statesman $25; Ensign $30; Crew $50; Helmsman $75; Boatswain $125; Pilot $250; Navigator $500; Captain $1,000; Commodore $2,500; Admiral $5,000.

FLAVEL HOUSE MUSEUM, (M), 441 8th St., Astoria, OR 97103-4620. Mailing Address: P.O. Box 88, Astoria, OR 97103-0088. Tel.: 503-325-2203. Fax: 503-325-7727.
E-mail: cchs@cumtux.org
Web Site: www.cumtux.org
Formerly: Captain George Flavel House Museum
Founded: 1951.
Congressional District: 1
Key Personnel: Exec. Dir., McAndrew Burns; Pres. (V), Kent Easom; Cur. & Archivist, Liisa Penner; Business Mgr., Martha L. Dahl; Dir. Mktg. & Devel., W. Sam Rascoe; Cur., Lisa Studts; Museum Maintenance, Chuck Bean.
Personnel Profile: Full-Time Paid 6; Part-Time Paid 2; Part-Time Volunteers 50.
Governing Authority: private; nonprofit organization. Parent Institution: Clatsop County Historical Society. Subsidiary Institution: Heritage Museum; Uppertown Fire Fighters Museum. Tax-exempt.
Institution Type/Description: Historic House Museum.
Collections: furniture; art; Victorian era furnishings; household objects.
Research Fields: Victorian art; local & regional history; Flavel family history.
Facilities: Museum-related items for sale in the Carriage House.
Activities: guided tours by appointment.
Publications: quarterly journal, CUMTUX; newsletter; catalogue: John H. Trullinger Paintings; Historic Flavel House.
Hours & Admission Prices: May-Sept. daily 10-5; Oct.-April daily 11-4. Family $15, adults $5, seniors citizens $4, youth 6-17 $2; discounts to AAA members; children under 6 & members no charge. Closed New Year's Day; Thanksgiving; Christmas Eve & Day.
Attendance: 23,000 (estimated)
Membership: Individual $35; Family & Foreign $50; Contributing $100-$499; Patron $500-$999; Benefactor & Corporate $1,000 & up.

THE HERITAGE MUSEUM, 1618 Exchange St., Astoria, OR 97103-3615. Mailing Address: P.O. Box 88, Astoria, OR 97103-0088. Tel.: 503-325-2203.
E-mail: cchs@cumtux.org
Web Site: www.cumtux.org
Founded: 1985.
Congressional District: 1
Key Personnel: Exec. Dir., McAndrew Burns; Bd. Pres., Paul Mitchell; Business Mgr., Martha L. Dahl; Dir. Mktg., Sam Rascoe; Archivist, Liisa Penner; Cur., Amber Glen.
Governing Authority: private; nonprofit. Parent Institution: Clatsop County Historical Society. Subsidiary Institution: Flavel House, Uppertown Fire Fighters Museum. Tax-exempt.
Institution Type/Description: Historical Society Museum.
Collections: natural history; geology; Native American artifacts; early immigrants & settlers of the region; nautical events; commerce in Clatsop County & along the Columbia River; logging; lumber; fish packing; ethnic exhibit.
Research Fields: Clatsop county history.
Facilities: research library; temporary & permanent archives; rental facility. Museum-related items for sale.
Activities: special events; lectures; guided tours.
Publications: quarterly, CUMTUX; newsletter; oral history videos, Steam Whistle Logging; Remembering Uniontown; art catalogues, Legacy of John H. Trillinger.
Hours & Admission Prices: Tues.-Fri. 10-12 & 1-3. Adults $4, senior citizens & AAA members $3, children 6-17 $2; discounts to AAA members & seniors; members no charge. Closed New Year's Eve & Day; Thanksgiving; Christmas Eve & Day. &
Attendance: 6,500 (estimated)
Membership: Student $15; Individual $35; Family/Dual $55; Friend $100; Business $200; Contributor $250; Patron $500; Corporate $1,000.

LEWIS & CLARK NATIONAL HISTORICAL PARK, 92343 Fort Clatsop Rd., Astoria, OR 97103. Tel.: 503-861-2471. Fax: 503-861-2585. TDD: 503-861-1620.
Web Site: www.nps.gov/lewi
Formerly: Fort Clatsop National Memorial
Founded: 1958.
Congressional District: 1
Key Personnel: Supt., Scott Tucker; Chief Visitor Svcs., Jill Harding; Chief Resource Mgmt., Chris Clatterbuck; Chief Facility Mgmt., Andrew Rasmussen.
Governing Authority: federal. Parent Institution: National Park Service, Washington, DC. Tax-exempt: 170(b).
Institution Type/Description: Historic Fort: replica of the 1805-1806, winter encampment of the Lewis & Clark Expedition; objects relating to the Lewis & Clark expedition.
Collections: Lewis & Clark Expedition; research library; cultural & natural objects & specimens; Fort Clatsop replica.
Research Fields: Lewis & Clark Expedition.
Facilities: 2,000-vol. reference library; visitor center; canoe landing & fresh water spring site; hiking trails. Museum-related books for sale.
Activities: living history programs in summer; visitor center programs; audiovisual program; on & off-site school programs; environmental programs.
Publications: 75 titles on Lewis & Clark, natural & cultural history; 3 titles on the Charbonneau Family; Plants of Fort Clatsop & the Clatsop Indians.
Hours & Admission Prices: Mid-June to Labor Day daily 9-6; after Labor Day to mid-June daily 9-5. 7-day entrance fee 16 & up $3; children 15 & under no charge. Closed Christmas. &
Attendance: 226,000 (accurate)

UPPERTOWN FIREFIGHTERS MUSEUM, 2986 Marine Dr., Astoria, OR 97103. Mailing Address: Clatsop County Historical Society, P.O. Box 88, Astoria, OR 97103-0088. Tel.: 503-325-2203. Fax: 503-325-7727.
E-mail: cchs@cumtux.org
Web Site: www.cumtux.org
Founded: 1990.
Congressional District: 1
Key Personnel: Exec. Dir., McAndrew Burns; Pres. (V), Kent Easom; Business Mgr., Martha L. Dahl; Dir. Mktg. & Devel., W. Sam Rascoe; Archivist, Liisa Penner; Museum Maintenance, Chuck Bean; Cur., Lisa Studts.
Personnel Profile: Full-Time Paid 6; Part-Time Paid 2; Part-Time Volunteers 50.
Governing Authority: private; nonprofit organization. Parent Institution: Clatsop County Historical Society. Subsidiary Institutions: Heritage Museum; Flavel House. Tax-exempt.

Institution Type/Description: Firefighters Museum.
Collections: fire-fighting equipment & memorabilia; photographs; fire pole; 1912 American LaFrance chemical wagon; 1921 Stutz; 1878 Hayes ladder wagon.
Facilities: meeting room.
Activities: children's activities.
Publications: Uppertown Fire Fighters' Museum.
Hours & Admission Prices: By appointment. Closed New Year's Eve & Day; Thanksgiving; Christmas Eve & Day. &
Attendance: 3,000 (estimated)
Membership: Individual $35; Family $55; Contributing $100-$499; Patron $500-$999; Benefactor $1,000 and up.

Aurora

AURORA COLONY HISTORICAL SOCIETY, (M), 15018 2nd St., N.E., Aurora, OR 97002. Mailing Address: P.O. Box 202, Aurora, OR 97002-0202. Tel.: 503-678-5754. Fax: 503-678-5756.
E-mail: info@auroracolony.org
Web Site: www.auroracolony.org
Founded: 1963.
Congressional District: 18
Key Personnel: Pres. (V), Reg Keddie.
Personnel Profile: Full-Time Paid 2; Part-Time Paid 5; Part-Time Volunteers 75.
Governing Authority: Parent Institution: Aurora Colony Historical Society. Subsidiary Institution: Old Aurora Colony Museum. Tax-exempt.
Institution Type/Description: Historical Society Museum.
Collections: local 19th century American communal society history; personal artifacts; photographs; furniture; tools; music; instruments; textiles; quilts; clothing; legal documents; maps; letters; books; ledgers; household items; historic buildings.
Activities: lectures; tours; living history demonstrations; 4th grade educational program. Annual Events: Quilt Show; Strawberry Social; Antique Spinning Wheel Showcase.
Publications: newsletter.
Hours & Admission Prices: Feb.-Dec. Tues.-Sat. 11-4, Sun. 12-4. Adults $6, seniors 60 & over $5, students $2; discounts to AAA & AAM members; children 5 & under and members no charge. Closed major holidays. &
Attendance: 4,500 (accurate)
Membership: Senior & Student $20; Individual $30; Individual Plus One $35; Family $40; Educational Supporter $50; Willie Keil Society $100; Colony Society $250; Aurora Society $500.

Baker City

ADLER HOUSE MUSEUM, 2305 Main St., Baker City, OR 97814. Mailing Address: 2480 Grove St., Baker City, OR 97814. Tel.: 541-523-9308.
E-mail: museum@bakercounty.org
Web Site: www.bakerheritagemuseum.com/adler-house.html
Founded: 1998.
Congressional District: 2
Governing Authority: Parent Institution: Baker County Museum Commission. Tax-exempt.
Institution Type/Description: Historic House Museum: housed in the former home of philanthropist, Leo Adler; built in 1889.
Collections: Adler family history; personal artifacts; period furnishings; paintings; photographs.
Activities: guided tours.
Hours & Admission Prices: Memorial Day to Labor Day Fri.-Mon. 10-2; other times by appointment. Adults $6; children 12 & under no charge.

BAKER HERITAGE MUSEUM, 2480 Grove St., Baker City, OR 97814-2719. Tel.: 541-523-9308. Fax: 541-523-9308.
E-mail: ccantrell@bakercounty.org
Web Site: www.bakerheritagemuseum.com
Formerly: Oregon Trail Regional Museum
Founded: 1982.
Congressional District: 59
Key Personnel: Dir., Chris Cantrell; Chm. (V), Steve Bogart.
Personnel Profile: Full-Time Volunteers 40; Part-Time Paid 3; Part-Time Volunteers 35.
Governing Authority: county; nonprofit organization. Subsidiary Institution: Adler House Museum, Baker City, OR. Tax-exempt: 170(b)(1)(A).
Institution Type/Description: History Museum.
Collections: artifacts of the original area settlers from 1800 to 1920s; photo archives; historical research materials, texts & papers; rocks & minerals. Historic House: Adler House, a turn-of-the-century, Victorian-Italian style home.

Facilities: 300-vol. library; 25,000 sq. ft. exhibit space. Museum-related items for sale.
Activities: formal education programs for children; lectures.
Publications: quarterly newsletter, Museum Memo.
Hours & Admission Prices: mid-March to Oct. daily 9-4; see website for additional winter hours. Family $20, adults $6, senior citizens 60 & over and youth 13-17 $5; discounts to groups; children 12 & under and members no charge. &
Attendance: 20,000 (accurate)
Membership: Individual $20; Family $30; Partner $60; Patrons $100; Contributor $250; Steward $500; Life $1,000.

NATIONAL HISTORIC OREGON TRAIL INTERPRETIVE CENTER, 22267 Oregon Hwy. 86, Baker City, OR 97814. Mailing Address: P.O. Box 987, Baker City, OR 97814-0987. Tel.: 541-523-1843. Fax: 541-523-1834.
E-mail: or_nhotic_mail@blm.gov
Web Site: www.blm.gov/or/oregontrail/
Founded: 1992.
Congressional District: 2
Key Personnel: Center Dir., Sarah LeCompte; Museum Shop Mgr., Phoebe Charbonneau; Maintenance Lead, Zack Freiwald; Park Ranger, Kelly Burns; Visitor Information Asst., Cheri Garver; Visitor Information Asst., Calvin Henshaw; Exhibit Specialist, Gypsy McFelter.
Personnel Profile: Full-Time Paid 10; Full-Time Volunteers 40; Part-Time Paid 5; Part-Time Volunteers 60; Interns 2.
Governing Authority: federal government; nonprofit organization. Parent Institution: Bureau of Land Management. Tax-exempt: 501(c)(3).
Institution Type/Description: History Museum & Interpretive Center: interpretive center along Oregon Trail; historic gold mining site.
Collections: personal, transportation, recreational & archaeological artifacts; historical photographs.
Research Fields: Oregon Trail 1840-1880.
Facilities: 750-vol. library; 150-seat auditorium; outdoor wagon encampment & lode mine site; interpretive overlook; outdoor lighted amphitheater; 4 mile hiking trail leading to Oregon Trail Ruts; picnic pavilion. Books & museum-related items for sale.
Activities: lectures; loan, participatory & temporary exhibitions; living history. Annual Events: Holiday Open House; Labor Day Wagon Encampment.
Publications: newsletter, Trail Mix; Oregon Trail Education Resource Guide.
Hours & Admission Prices: Feb. 16-March & Nov. daily 9-4; April-Oct. daily 9-6; Dec.-Feb. 15 Tues. & Thurs.-Sun. 9-4. April-Oct. adults $8, senior citizens 62 & over $4.50; discounts to groups & tours; Inter-Agency Pass, NIIOTIC Individual & Family members, schools and children under 16 no charge; Nov.-March adults $5, seniors $3.50; discounts to groups & tours; Inter-Agency Pass, NHOTIC Individual & family members, schools and children under 16 no charge. Closed New Year's Day; Thanksgiving; Christmas. &
Attendance: 69,852 (accurate)
Membership: Individual $25; Family $45; Trail Scout $85; Wagon Master $200; Business $100-$1,000.

Bandon

BANDON HISTORICAL SOCIETY MUSEUM, 270 Fillmore & Hwy. 101, Bandon, OR 97411. Mailing Address: P.O. Box 737, Bandon, OR 97411-0737. Tel.: 541-347-2164. Fax: 541-347-2164.
E-mail: bandonhistoricalmuseum@yahoo.com
Web Site: bandonhistoricalmuseum.org
Formerly: Coquille River Museum/Bandon Historical Society
Founded: 1977.
Congressional District: 24
Key Personnel: Exec. Dir. & Museum Shop Mgr., Judy Knox; Pres. (V) & Museum Shop Mgr., Kathy Dornath; Treas., John Gamble; Vice Pres. & Sec., Betty Hiley; Education, Jim Proehl.
Personnel Profile: Part-Time Paid 2; Part-Time Volunteers 40.
Governing Authority: private; nonprofit organization. Tax-exempt: 501(c)(3).
Institution Type/Description: History Museum.
Collections: area history & artifacts; local Native American artifacts; over 1,500 photographs; Bandon's two fires 1914 & 1936; 1920-1960s period clothing; early businesses; local area schools; pioneer family stories & memorabilia; natural history; local military; fishing; farming; cheese making; timber industry; maritime boat building; riverboats; sculling vessels & commerce of the area from 1800s-1960.
Research Fields: local cemeteries & obituaries; newspaper; school students; old business; Pioneer families genealogy; Bandon School class annuals; alumni.
Facilities: library & research center; classroom; 5,000 sq. ft. exhibit space. Gift-related items for sale.
Activities: docent program; educational programs; tours; lectures; training for

volunteers; book signings; special displays. Museum Sponsors: Veterans Day; Cranberry Festival activities; Christmas activity; special exhibits annually.
Publications: newsletter 3 times annually, The Bandon Light; annual historic photograph calendar; books, Woodenships & Master Craftsmen; Shipbuilding & Their Builders; Bandon Then and Now; Bandon by the Sea; Bandon Burns: Personal Stories By Bandon Fire Survivors.
Hours & Admission Prices: Feb.-May & Oct.-Dec. Mon.-Sat. 10-4; June-Sept. Mon.-Sun. 10-4. Adults $3; children under 12 & society members no charge. Closed New Year's Day; Thanksgiving; Christmas Eve & Day. &
Attendance: 3,500 (estimated)
Membership: Individual $15; Family $25; Business $35; Life $250; Benefactor $500; Patron $1,000.

COQUILLE RIVER LIGHTHOUSE/BULLARDS BEACH STATE PARK, 2 miles N. of Bandon, Hwy. 101, Bandon, OR 97411. Mailing Address: P.O. Box 569, Bandon, OR 97411-0569. Tel.: 541-347-2209. Fax: 541-347-4656.
Web Site: www.oregonstateparks.org/park_71.php
Founded: 1896.
Key Personnel: Park Mgr., Ben Fisher; Park Ranger, Eric Cook.
Personnel Profile: Full-Time Paid 2; Part-Time Volunteers 22.
Governing Authority: state. Tax-exempt.
Institution Type/Description: Historic Building Museum: housed in 1896 lighthouse.
Collections: interpretive display.
Facilities: 300 sq. ft. exhibit space. Gift items for sale.
Activities: formal education programs for children; guided tours; lectures; temporary exhibitions. Museum Sponsors: OPRD Day in June.
Publications: brochure, Coquille River Lighthouse.
Hours & Admission Prices: Coquille River Lighthouse: May-Oct. daily 11-5. No charge; donations accepted.
Attendance: 28,380 (accurate)

Beaverton

OREGON SPORTS HALL OF FAME AND MUSEUM, 4840 S.W. Western Ave. Ste. 600, Beaverton, OR 97005. Tel.: 503-227-7466. Fax: 503-235-5688.
E-mail: info@oregonsportshall.org
Web Site: www.oregonsportshall.org
Formerly: State of Oregon Sports Hall of Fame
Founded: 1978.
Key Personnel: Pres. (V), Chuck Richards; Exec. Dir., Mike Rose.
Personnel Profile: Full-Time Paid 2.
Governing Authority: nonprofit organization. Tax-exempt: 501(c)(3).
Institution Type/Description: Sports Museum.
Collections: athletic memorabilia; photos; implements; posters; written descriptions; uniforms; videos; shoes.
Research Fields: sports & athletic achievement.
Hours & Admission Prices: Temporarily closed for relocation. &
Membership: Individual $35; Family $50; Contributor $100; All League $500; All State $1,000; All American $2,500; MVP $5,000.

Bend

DES CHUTES HISTORICAL MUSEUM, 129 N.W. Idaho Ave., Bend, OR 97701-2602. Tel.: 541-389-1813. Fax: 541-317-9345.
E-mail: info@deschuteshistory.org
Web Site: www.deschuteshistory.org
Founded: 1975.
Congressional District: 5
Key Personnel: Exec. Dir., Kelly Cannon-Miller; Pres. (V), Les Joslin.
Personnel Profile: Part-Time Paid 4; Part-Time Volunteers 40.
Governing Authority: private; nonprofit organization. Parent Institution: Deschutes County Historical Society. Tax-exempt: 501(c)(3).
Institution Type/Description: History Museum: housed in 1914 Reid School.
Collections: county history; veterans & soldiers of central Oregon; photographs; Bend Bulletin archives; county history.
Facilities: 1,000-vol. library; meeting facilities. Museum-related items for sale.
Activities: arts festivals; films; formal education programs; guided tours; loan exhibitions; school loan service; study clubs.
Publications: monthly newsletter, The Homesteader; books on Central Oregon History.
Hours & Admission Prices: Tues.-Sat. 10-4:30; other times by appointment. Adults $5, children 13-17 $2; Independence Day, children 12 & under and members no charge. Closed New Year's Day; Thanksgiving; Christmas. &
Attendance: 8,000 (estimated)
Membership: Single $15; Family $25; Business $40; Donor $50; Patron & Business Plus $100; Benefactor & President's Club $500.

✳ **HIGH DESERT MUSEUM, (M),** 59800 S. Hwy. 97, Bend, OR 97702-7963. Tel.: 541-382-4754. Fax: 541-382-5256.
Web Site: www.highdesertmuseum.org
Founded: 1974.
Congressional District: 2
Key Personnel: Pres., Janeanne A. Upp; Human Resources & Volunteer Mgr., Tracy Suckow; Vice Pres. Programs, Dana Whitelaw, Ph.D.; Cur. Western History, Margaret Lee; Cur. Natural History, John Goodell; Cur. Living History, Linda Evans; Facilities Mgr., Arnie Tronson; Vice Pres. Communications & Visitor Svcs., Melissa Hochschild.
Personnel Profile: Full-Time Paid 42; Part-Time Paid 2; Part-Time Volunteers 250; Interns 7.
Governing Authority: nonprofit organization. Tax-exempt: 501(c)(3).
Institution Type/Description: Natural & Cultural History Museum.
Collections: over 120 live animals representing regional wildlife; local & natural history; Great Basin prehistory & cultural history; historic artifacts; Native American ethnographic objects, especially Columbia River plateau clothing & bags; historic, landscape & wildlife art & photographs; Plateau Native Americans; forest ecosystem; turn-of-the-century sawmill; U.S. Forest Service.
Research Fields: onsite monitoring of free-roaming wildlife.
Facilities: library; classrooms; cafe. Museum-related items for sale.
Activities: lectures; educational programs for schools & special events; adult & teen volunteer program; field excursions; permanent & temporary exhibitions; daily presentations & demonstrations.
Publications: quarterly Desert Sage Newsletter; brochures; educational materials.
Hours & Admission Prices: May-Oct. daily 9-5; Nov.-Sept. daily 10-4. May-Oct. adults 13-64 $15, seniors 65 & over $12, children 5-12 $9; discounts to groups & AAM members; members and children 4 & under no charge. Nov.-Sept. adults 13-64 $10, seniors 65 & over $9, children 5-12 $6; discounts to AAM members; children 4 & under no charge. Closed Independence Day; Thanksgiving; Christmas. ♿
Attendance: 155,000 (accurate)
Membership: Individual $60; Individual Plus $72; Household or Grandparent $90; Household Plus $144; Silver Spur $300; Golden Spur $500; Fellow/Business $1,000; Curator's Circle $2,500; Museum Associate $5,000; President's Council $10,000; Founder's Circle $25,000.

Brooks

ANTIQUE CATERPILLAR MACHINERY MUSEUM, 3995 Brooklake Rd., N.E., Brooks, OR 97303-9732. Tel.: 503-538-3935.
Web Site: antiquecaterpillarmuseum.org
Key Personnel: Pres., Don Leffler.
Governing Authority: nonprofit organization. Tax-exempt: 501(c)(3).
Institution Type/Description: Machinery Museum.
Collections: caterpillar history & equipment; period trucks, tractors & steamrollers.
Activities: demonstrations; educational programs.
Hours & Admission Prices: Call for hours.

PACIFIC NORTHWEST TRUCK MUSEUM, 3995 Brooklake Rd., N.E., Brooks, OR 97305. Mailing Address: P.O. Box 9087, Brooks, OR 97305-0087. Tel.: 503-312-0039 & 463-8701.
E-mail: office@pacificnwtruckmuseum.org
Web Site: www.pacificnwtruckmuseum.org
Founded: 1989.
Key Personnel: Pres. (V), Terry Dovre; Chm. (V), Doug Delano; Treas., Craig Vogel; Archivist, Ken Goudy, Jr.; Museum Shop Mgr., Jerry Crume.
Personnel Profile: Part-Time Volunteers 100.
Volunteer Hours: 2,700
Governing Authority: nonprofit. Tax-exempt: 501(c)(3).
Institution Type/Description: Truck Museum.
Collections: 1912-2009 Class 6 & larger trucks; truck manufacturing & transportation industries; artifacts; photographs; drawings; manuals; patents; literature; biographies; monkey wrenches.
Facilities: library; 30,500 sq. ft. exhibit space. Museum-related items for sale.
Activities: guided tours; hobby workshops; lectures; loan exhibitions. Museum Sponsors: Truck Show in August; parades; festivals; road tours.
Publications: bimonthly newsletter, Pacific Northwest Truck Museum.
Hours & Admission Prices: Museum: April-Sept. Sat.-Sun. 10-4:30; other times by appointment. Donation: adults $5; members no charge. Truck Show: admission $10; children under 12 no charge. Closed New Year's Day; Christmas. ♿
Attendance: 18,000 (estimated)
Membership: Individual $30; Household $40.

Brownsville

LINN COUNTY HISTORICAL MUSEUM AND MOYER HOUSE, 101 Park Ave., Brownsville, OR 97327. Mailing Address: Box 607, Brownsville, OR 97327-0607. Tel.: 541-466-3390 & 3070. Fax: 541-466-3390. Facebook: Linn County Historical Museum.
E-mail: lchm@centurytel.net
Web Site: linnparks.com
Founded: 1962.
Congressional District: 37
Key Personnel: C.E.O., Brian Carrol; Chm. Trust (V), Glenn Harrison; Pres. Friends (V), Myrna Baughman; Museum Coord., Ashley Toutain; Museum Shop Mgr., Tricia Thompson.
Personnel Profile: Part-Time Paid 4; Part-Time Volunteers 15.
Governing Authority: county; nonprofit. Parent Institution: Linn County Parks & Recreation. Subsidiary Institution: Linn County Museum Friends. Tax-exempt.
Institution Type/Description: Local History Museum: arranged to represent 1800s city with exhibits of businesses operating in the community showing life-style pursuits in other fields such as agriculture, mining & timber; early travel 1850-1920s. Historic House: 1881 Italianate Victorian, Moyer House.
Collections: Linn County history; industry; commerce; domestic arts; furnishings; documents; photographs; genealogical records; family books; Linn County files; reference books; pioneer artifacts; Native American artifacts.
Research Fields: genealogy; architectural; local history; early pioneer life.
Facilities: 9,290 sq. ft. exhibit space; 30-seat theater. Historical pamphlets, books & other museum-related items for sale.
Activities: guided tours; lectures; organized educational programs for children; participatory, loan & temporary exhibitions. Museum Sponsors: Living History drama at Moyer House.
Publications: biannual newsletter: guidebook: Take a Walk; semi-annual newsletter published by Friends of the Museum.
Hours & Admission Prices: Mon.-Sat. 11-4, Sun. 1-5. Guided Tours $2. Closed New Year's Day; Easter; Thanksgiving; Christmas. ♿
Attendance: 8,298 (estimated)
Membership: Pioneer Descendant $10; Individual $15; Family $25; Patron $40; Benefactor $100.

Burns

HARNEY COUNTY HISTORICAL MUSEUM, (M), 18 W. D St., Burns, OR 97720-1226. Mailing Address: P.O. Box 388, Burns, OR 97720-0388. Tel.: 541-573-5618.
E-mail: harneymuseum@centurytel.net
Web Site: hchistoricalsociety.com
Founded: 1960.
Congressional District: 2
Key Personnel: Pres., Bill Renwick; Vice Pres., Mildred Fine.
Personnel Profile: Full-Time Volunteers 24; Part-Time Volunteers 25.
Governing Authority: society. Parent Institution: Harney County Historical Society. Tax-exempt.
Institution Type/Description: General Museum.
Collections: archaeology; history; natural history; pioneer kitchen; geology; Indian artifacts; mineralogy; zoology; costumes; Indian headdress; butterfly collection; paintings; clothing; pioneer kitchen; Ilda Mae Hayes room; wagon shed room; bird showcase; Hanley room; salt & pepper shakers; guns; Dan Opie quarter horse trophy collection; early historical photographs; civilian conservation corps display; wildlife diorama; mounted displays; ancestor trade and skills.
Research Fields: agriculture; Indian artifacts; timber; genealogy.
Activities: guided tours; occasional demonstrations of pioneer crafts; permanent & temporary exhibitions.
Publications: quarterly newsletter, Harney County Historical Highlights; Harney County, Oregon and Its Rangeland, George Francis Brimlow; Harney County: An Historical Inventory, Royal Jackson & Jennifer Lee; Pieces of the Past: Harney County Historical Cemeteries - Andrews-Happy Valley-Riverside and silver Creek Area; Fort Harney Cemetery; Boat Ford, Cameron-Crane-Saddle Butte and Windy Point; Drewsey-Van-Muller; Burns Cemetery.
Hours & Admission Prices: April -Sept. Mon.-Sat. 10-4. Suggested Donation: Family of 5 $15 (additional children $.50 each), couples $8, single $5, senior citizens & youth 13-17 $3, children 6-12 $1; discounts to groups; members no charge. ♿
Attendance: 1,024 (accurate)
Membership: Individual $15; Single Lifetime $150; Couple Lifetime $250.

Canby

CANBY HISTORICAL SOCIETY, 888 N.E. 4th Ave., Canby, OR 97013-2300. Mailing Address: P.O. Box 160, Canby, OR 97013-0160. Tel.: 503-266-6712.
E-mail: depotmuseum@canby.com
Web Site: www.canbyhistoricalsociety.org
Formerly: Canby Depot Museum
Founded: 1968.
Key Personnel: Pres., Nora Clark; Vice Pres., Cheryl Rahn.
Personnel Profile: Part-Time Paid 1; Part-Time Volunteers 50.
Governing Authority: private; nonprofit organization. Parent Institution: Canby Historical Society. Tax-exempt: 501(c)(3).
Institution Type/Description: History Museum: housed in an 1891 railroad depot.
Collections: Oregon history from prehistoric to modern; history of Canby & the surrounding area; caboose; speeder car; railroading, agriculture & daily living of early Oregon pioneers.
Research Fields: genealogical.
Facilities: 100-vol. library; 2,000 sq. ft. exhibit space. Museum-related items for sale.
Activities: formal education programs for children; guided tours; temporary exhibitions. Museum Sponsors: Pioneer Breakfast in July; semiannual Flea Market.
Publications: quarterly newsletter, Canby Historical Society; annual historical calendar.
Hours & Admission Prices: March-Dec. Thurs.-Sun. 1-4. Suggested Donation: adults $2. Closed holidays; holiday weekends. &
Attendance: 1,100 (estimated)
Membership: Individual $15; Family $25; Supporter $100-$249; Contributor $250-$499; Sponsor $500-$999; Depot Patron $1,000 & up.

Cannon Beach

CANNON BEACH HISTORY CENTER AND MUSEUM, 1387 S. Spruce St., Cannon Beach, OR 97110. Mailing Address: P.O. Box 1005, Cannon Beach, OR 97110-1005. Tel.: 503-436-9301. Fax: 503-436-0490.
E-mail: info@cbhistory.org
Web Site: www.cbhistory.org
Founded: 1969.
Congressional District: 1
Key Personnel: Dir., Elaine Murdy; Pres. (V), Robert Mushen.
Personnel Profile: Part-Time Paid 2; Part-Time Volunteers 30.
Governing Authority: Tax-exempt.
Institution Type/Description: History Museum.
Collections: historical artifacts & memorabilia pertaining to Cannon Beach.
Activities: Museum Sponsors: Acoustic Folk! concert series last Saturday of month; Lecture Series 2nd Saturday of month; Annual Cottage Tour 2nd Saturday in September.
Hours & Admission Prices: Wed.-Mon. 1-5. No charge; donations accepted. &
Attendance: 3,500 (accurate)
Membership: Basic $30; Family $50; Supporter $125; Patron $250; Sponsor $500 & up.

Canyon City

GRANT COUNTY HISTORICAL MUSEUM, 101 S. Canyon City Blvd., Hwy. 395, Canyon City, OR 97820. Mailing Address: P.O. Box 464, Canyon City, OR 97820-0464. Tel.: 541-575-0362. Fax: 541-575-0515.
E-mail: museum@ortelco.net
Web Site: www.gchistoricalmuseum.com
Founded: 1953.
Congressional District: 2
Key Personnel: Dir., Jayne Primrose.
Personnel Profile: Part-Time Paid 1; Part-Time Volunteers 11.
Governing Authority: municipal. Tax-exempt.
Institution Type/Description: Local History Museum: located in Canyon City, site of active gold mining area in 1862.
Collections: pioneer artifacts & relics; native rock display; Indian artifacts; gold mining; newspapers 1884-1903. Historic Buildings: 1864 Home of Joaquin Miller; 1910 Greenhorn City Jail; 1966 Orient Buckboard automobile.
Facilities: library of books & pictures on county history available for research on premises.
Activities: permanent exhibitions.
Hours & Admission Prices: May-Sept. Mon.-Sat. 9-4:30; other times by appointment. Adults $4, senior citizens $3.50, children 7-17 $2; children 6 & under no charge.

Attendance: 1,046 (accurate)

Cascade Locks

CASCADE LOCKS HISTORICAL MUSEUM, 1 Marine Dr., Cascade Locks, OR 97014. Mailing Address: P.O. Box 307, Cascade Locks, OR 97014-0307. Tel.: 541-374-8619.
Web Site: www.portofcascadelocks.org/marinepark.htm
Institution Type/Description: History Museum.
Collections: local transportation history including the Oregon Pony, the first steam engine in the Pacific Northwest.
Hours & Admission Prices: May-Oct. daily 12-5.

Central Point

CRATER ROCK MUSEUM, (M), 2002 Scenic Ave., Central Point, OR 97502-2185. Mailing Address: P.O. Box 3999, Central Point, OR 97502-0041. Tel.: 541-664-6081. Fax: 541-664-3848.
E-mail: roxyanngems@msn.com
Institution Type/Description: Earth Sciences Museum.
Collections: rocks; fossils; petrified wood; arrowheads.
Hours & Admission Prices: Tues.-Sat. 10-4. Adults $4, seniors, students & children $2.

Chiloquin

COLLIER MEMORIAL STATE PARK & LOGGING MUSEUM, 46000 Hwy. 97 N., Chiloquin, OR 97624-9631. Tel.: 541-783-2471. Fax: 541-783-2707.
E-mail: kirk.barham@state.or.us
Web Site: collierloggingmuseum.org
Founded: 1945.
Congressional District: 4
Key Personnel: Park Mgr., Jonathan Moll.
Personnel Profile: Full-Time Paid 4; Part-Time Paid 6; Interns 1.
Governing Authority: state. Parent Institution: Oregon State Parks. Tax exempt.
Institution Type/Description: Logging, Pioneer & Lumber Museum & Village.
Collections: logging & lumbering artifacts; Dolbeer Donkey Engine; 1929 Hines sawmill; Beloit Tree Harvester; Romeo & Juliet Bridge; Sumner Sash gang saw; photographs; locomotives; track layer; trout spawning beds; giant Corliss twin steam engine; Indian artifacts; mid-1800 pioneer log houses; ox shoeing blacksmith shop; wooden wheeled wagons; caterpillars; high wheels; lumber trucks; logging trucks; McVay Log loader; snag pushers; 85 ft. steel spar poles simulating log transfer; steam logging equipment; railroad logging equipment; logging arches; 1880-1980 logging equipment & artifacts; pioneer surveying instruments.
Research Fields: millwright, machine & logging tools.
Facilities: camp & picnic grounds; Spring Creek; 650-acre Pine Forest; overnight trailer camp.
Activities: fishing; beaver aspen harvest; permanent exhibitions; hiking; canoeing; equestrian.
Publications: pamphlets; Collier State Park-Logging Museum-Walking Tour Guide.
Hours & Admission Prices: Museum: daily 8-4. Park: daily dawn-dusk. Office: Mon.-Fri. 8-4:30. No charge; donations accepted. &
Attendance: 88,164
Membership: Friends of Collier Memorial State Park $5.

Clackamas

BRIGADIER GENERAL JAMES B. THAYER OREGON MILITARY MUSEUM, Camp Withycombe, 15300 S.E. Minuteman Way, Clackamas, OR 97015. Tel.: 503-683-5359. Fax: 503-683-4913.
E-mail: tracy.m.thoennes.civ@mail.mil
Web Site: www.oregonmilitarymuseum.org
Founded: 1974.
Congressional District: 5
Key Personnel: Cur., Tracy Thoennes; Museum Technician, Kathleen Daly.
Personnel Profile: Full-Time Paid 2; Part-Time Volunteers 15; Interns 1.
Governing Authority: State of Oregon Military Department, National Guard. Tax-exempt.
Institution Type/Description: Military History Museum.
Collections: Oregon's military history from mid-19th century to present; aircraft; artwork; equipment; insignia; ordnance; tanks; uniforms; vehicles. Historic Buildings: WWII Quonset Hut; 1911 Artillery Barn.
Research Fields: Oregon National Guard in major American Wars 1898-present; Oregon casualties & necrology in WWI & WWII.

Facilities: 20,000-vol. reference library by appointment.
Activities: docent tours; educational activities; research; outreach programs for community & National Guard events. Museum Sponsors: Annual Living History Day.
Publications: Annual Report; brochure.
Hours & Admission Prices: Temporarily closed for renovation. &

Attendance: 3,500 (accurate)

Columbia City

CAPLES HOUSE MUSEUM, 1915 1st St., Columbia City, OR 97018. Mailing Address: Caretaker, Caples Museum, P.O. Box 263, Columbia City, OR 97018-0263. Tel.: 503-397-5390.
Web Site: www.rootsweb.com/~orossdar/Caples.htm
Institution Type/Description: Historic House Museum.
Collections: personal artifacts; period furnishings.
Hours & Admission Prices: March-Oct. Fri.-Sun. & holidays 1-5; other times by appointment. Adults $4, children $2; discounts to members.

Condon

GILLIAM COUNTY HISTORICAL SOCIETY & MUSEUM COMPLEX, Hwy. 19 at Burns Park, Condon, OR 97823. Mailing Address: P.O. Box 377, Condon, OR 97823-0377. Tel.: 541-384-4233.
E-mail: catlee1999@hotmail.com
Web Site: gilliamcountyhistory.org
Founded: 1985.
Congressional District: 59
Key Personnel: Pres. (V), Karen Wilde; Treas., Cheryl Baker.
Personnel Profile: Part-Time Paid 1; Part-Time Volunteers 30.
Governing Authority: private; nonprofit organization. Tax-exempt: 501(c)(3).
Institution Type/Description: History Museum.
Collections: Gilliam County history; farm machinery.
Research Fields: genealogy.
Facilities: library; 10,000 sq. ft. exhibit space.
Activities: films; guided tours; loan & temporary exhibitions. Annual Events: Independence Day parade float; County Fair booth; banquet; Appreciation Day.
Publications: biannual newsletter.
Hours & Admission Prices: May-Oct. Wed.-Sun. 1-5; call for holiday hours. No charge; donations accepted. &

Attendance: 882 (accurate)
Membership: Single $15; Couple & Family $25.

Coos Bay

COOS ART MUSEUM, (M), 235 Anderson, Coos Bay, OR 97420-1610. Tel.: 541-267-3901 & 4877. Fax: 541-267-4877 (please call 267-3901). Facebook: Coos Art Museum.
E-mail: adavenport@coosart.org
Web Site: www.coosart.org
Founded: 1966.
Congressional District: 4
Key Personnel: Dir., Steven Broocks; Pres. (V), Bill Delimont.
Personnel Profile: Full-Time Paid 1; Part-Time Paid 7; Part-Time Volunteers 83.
Governing Authority: nonprofit organization. Tax-exempt: 501(c)(3).
Institution Type/Description: Art Museum: housed in 1935 Post Office building.
Collections: contemporary American prints; historical maritime photography; Northwest art in various media.
Major Exhibits: Playing With Fire, 12/13/13-2/15/14; Biennial Student Art Exhibit, 3/7/14-4/12/14; Expressions West, 4/18/14-6/28/14; 21st Annual Maritime Art, 7/12/14-9/27/14; CAM Biennial, 10/10/14-12/6/14; Pacific Shores Photo Competition, 12/12/14-2/15.
Research Fields: 20th-century art.
Facilities: classrooms.
Activities: classes; workshops; lectures; films; tours; special art events; rental & sales program. Museum Sponsors: Fundraiser in July & September.
Publications: exhibition announcements; class schedules.
Hours & Admission Prices: Tues.-Fri. 10-4, Sat. 1-4. Adults $5, seniors & students $2; discount to veterans; members no charge. &

Attendance: 10,574 (accurate)
Membership: Student $15; Individual $45; Family $70; Business/Active Supporter $150; Gallery Club $250; Curator's Circle $500; President's Circle $1,000; Benefactor $2,500.

MARSHFIELD SUN PRINTING MUSEUM, 1049 Front St., Coos Bay, OR 97420. Mailing Address: P.O. Box 783, Coos Bay, OR 97420-0148. Tel.: 541-266-0901.
E-mail: gtinker1@frontier.com
Founded: 1975.
Key Personnel: Pres. (V), George Tinker.
Personnel Profile: Part-Time Volunteers 6.
Governing Authority: Parent Institution: Marshfield Sun Association. Tax-exempt.
Institution Type/Description: Historic Building Museum: housed in the former Marshfield Sun newspaper building.
Collections: printing & local history; printing equipment.
Hours & Admission Prices: Memorial Day to Labor Day Tues.-Sat. 10-4; other times by appointment. No charge; donations accepted. &

Attendance: 747 (accurate)
Membership: Regular $35; Donor $50; Benefactor $150; Life $200; Business $500.

Corbett

FRIENDS OF VISTA HOUSE, Crown Point State Park, 40700 E. Historic Columbia River Hwy., Corbett, OR 97019. Mailing Address: Friends of Vista House, P.O. Box 204, Corbett, OR 97019-0204. Tel.: 503-695-2230 & 2240.
E-mail: friends@vistahouse.com
Web Site: www.vistahouse.com
Founded: 1982.
Key Personnel: Exec. Dir., Marguerite Perry.
Personnel Profile: Full-Time Paid 1; Part-Time Paid 12.
Governing Authority: Parent Institution: Oregon State Parks & Recreation Dept. Tax-exempt.
Institution Type/Description: Historic Building Museum: built in 1916. Listed on the National Register of Historic Places.
Collections: Oregon pioneer history; photographs; sculpture; observatory.
Facilities: Gift items for sale.
Publications: newsletter.
Hours & Admission Prices: March-Oct. daily; Nov.-Feb. Sat.-Sun. call for hours. No charge; donations accepted. &

Attendance: 1,000,000 (estimated)
Membership: Individual $25; Family $50; Donor $100; Sponsor $250-$5,000.

Corvallis

CORVALLIS ARTS CENTER, 700 S.W. Madison Ave., Corvallis, OR 97333. Tel.: 541-754-1551.
Institution Type/Description: Art Gallery.
Collections: works by local, regional & national artists.
Facilities: Gallery-related items for sale.
Hours & Admission Prices: Tues.-Sat. 12-5.

OREGON STATE UNIVERSITY MEMORIAL UNION CONCOURSE GALLERY, Jefferson St., OSU, Corvallis, OR 97331. Mailing Address: 10 Memorial Union E., Corvallis, OR 97331-8592. Tel.: 541-737-6371. Fax: 541-737-1565.
E-mail: susan.bourque@oregonstate.edu
Web Site: www.osumu.org/about_art.htm
Founded: 1927.
Congressional District: 1
Key Personnel: Dir., Kent Sumner; Coord. Exhibits, Susan Bourque.
Personnel Profile: Full-Time Paid 1; Part-Time Paid 1; Part-Time Volunteers 2.
Governing Authority: state; university. Affiliated with the Oregon State Board of Higher Education. Tax-exempt.
Institution Type/Description: Art Institute.
Collections: paintings; sculpture; prints.
Major Exhibits: OSU Womens' Center: Call to women Artists, 2/14; Four Portland Artists: A. O'Donoghue, R. Fortney, M. Hensley & T. Murino-Brault, 3/14-4/14; OSU Art Department Printmaking Exhibit, 5/14-6/14; Handmade in Oregon XI, 11/14-12/14; I Want the Wide American Earth: An Asian Pacific American Story (T), 12/15-2/16.
Activities: guided tours; lectures; films; gallery talks; concerts; arts festivals; hobby workshops; permanent & temporary exhibitions.
Hours & Admission Prices: June-Aug. Mon.-Fri. 8-5; Sept.-May Mon.-Thurs. 7-11, Fri. 7-12, Sat. 7:30-12, Sun. 8:30-11. No charge. Closed national holidays. &

Attendance: 36,000 (estimated)

Cottage Grove

BOHEMIA GOLD MINING MUSEUM, 737 Main St., Cottage Grove, OR 97424. Mailing Address: P.O. Box 1658, Cottage Grove, OR 97424. Tel.: 541-942-5022.
Founded: 2003.
Key Personnel: Dir., Sara Smith.
Personnel Profile: Part-Time Volunteers 12.
Governing Authority: Tax-exempt.
Institution Type/Description: Mining History Museum.
Collections: gold mining history; tools; maps; photographs; murals; rocks & minerals.
Activities: group tours; panning for gold.
Publications: books.
Hours & Admission Prices: Wed.-Sat. 1-4. Suggested Donation: $2.
Attendance: 3,200 (estimated)

COTTAGE GROVE MUSEUM & ANNEX, 147 H St. & Birch Ave., Cottage Grove, OR 97424. Mailing Address: P.O. Box 142, Cottage Grove, OR 97424-0005. Tel.: 541-942-5658. Fax: 541-942-9804.
E-mail: longfellow.marie@gmail.com
Founded: 1961.
Congressional District: 4
Key Personnel: Chm. (V), Becky Venice; Dir., Marie Longfellow; Museum Shop Mgr., Joanne Skelton.
Personnel Profile: Part-Time Paid 1; Part-Time Volunteers 3.
Governing Authority: C.G. Museum Perpetuation Corp. Tax-exempt.
Institution Type/Description: General Museum: housed in former 1896 octagonal Roman Catholic Church.
Collections: pioneer articles; farm & industry artifacts; Indian artifacts; mining tools; firearms; Civil War items; Italian stain glass windows; original Oregon covered bridge prints; Titanic artifacts; quilts; military uniforms.
Facilities: 130-vol. library pertaining to the War of the Rebellion available to the public.
Hours & Admission Prices: Summer: Wed.-Sun. 1-4; Winter: Sat.-Sun. 1-4. Summer: no charge; Winter: donations requested.
Attendance: 1,500 (estimated)

OREGON AVIATION HISTORICAL SOCIETY, 2475 Jim Wright Way, Cottage Grove, OR 97424. Mailing Address: P.O. Box 553, Cottage Grove, OR 97424. Tel.: 541-767-0244. Facebook: Oregon Aviation Historical Society.
E-mail: oregonaviation.org@gmail.com
Web Site: www.oregonaviation.org
Founded: 1985.
Congressional District: 4
Key Personnel: Pres. (V), Douglas Kindred.
Personnel Profile: Full-Time Volunteers 2; Part-Time Paid 1; Part-Time Volunteers 12.
Governing Authority: Tax-exempt.
Institution Type/Description: Historical Society Museum.
Collections: aviation history, heritage & artifacts; photographs.
Research Fields: History of aviation in Oregon.
Facilities: library; archives.
Publications: newsletter 3 times per yr.
Hours & Admission Prices: May-Sept. Sat. 10-4; other times by appointment. No charge; donations accepted. &
Membership: Associate $25; Family $35; Sustaining $50; Contributing $100.

Crater Lake

CRATER LAKE NATIONAL PARK, Hwy. 62, Crater Lake, OR 97604. Mailing Address: P.O. Box 7, Crater Lake, OR 97604-0007. Tel.: 541-594-3095. Fax: 541-594-3010.
E-mail: mary_merryman@nps.gov
Founded: 1902.
Congressional District: 2
Key Personnel: Supt., Craig Ackerman; Park Cur., Mary Merryman.
Personnel Profile: Full-Time Paid 1.
Governing Authority: federal. Parent Institution: U.S. Dept. of the Interior. Subsidiary Institution: National Park Service. Tax-exempt.
Institution Type/Description: Park Museum.
Collections: natural history; period artifacts; herbarium; zoology; geology.
Research Fields: Crater Lake, geology; botany; zoology.
Facilities: 1,500-vol. library of natural science & history & 8,000 natural resource specimens available for research by appointment on premises. Books & slides for sale.
Activities: guided tours June-September; lectures.

Publications: leaflet, Nature Trail; Birds of Crater Lake; Crater Lake Trails; A Guide To Crater Lake.
Hours & Admission Prices: Visitor Center: daily 10-4:30. Park entry fee. &
Attendance: 500,000
Membership: Individual $10; Family $15; Supporting $100; Benefactor $250; Life $500; Corporate $500 & up.

Depoe Bay

OREGON COAST SPORTS MUSEUM & INTERNATIONAL SPORTS HALL OF FAME AND OLYMPIC MUSEUM, 110 N.E. Hwy. 101, Depoe Bay, OR 97341. Mailing Address: P.O. Box 166, Depoe Bay, OR 97341-0166. Tel.: 541-765-2923; 702-346-1776.
Web Site: www.olympicsource.org
Founded: 1980.
Key Personnel: Exec. Dir. & Museum Shop Mgr., Heidi E. Nash
Institution Type/Description: Sports Museum.
Collections: Olympic memorabilia 1896 to present; pins; medals; posters; stamps; uniforms; porcelain; torches; books; photographs; personal artifacts.
Research Fields: Olympic history.
Hours & Admission Prices: By appointment. Adults $8, children under 12 $5; discounts to groups, AAM & ICOM members; members no charge.

Echo

CHINESE HOUSE/O.R. & N. MUSEUM & ST. PETER'S CATHOLIC CHURCH, 230 W. Bridge & 33208 Marble, Echo, OR 97826. Mailing Address: P.O. Box 426, Echo, OR 97826-0426. Tel.: 541-376-8411. Fax: 541-376-8218.
E-mail: ecpl@centurytel.net
Web Site: www.echo-oregon.com
Founded: 1987.
Congressional District: 2
Key Personnel: C.E.O., Diane Berry.
Personnel Profile: Part-Time Paid 1; Part-Time Volunteers 7.
Governing Authority: private; nonprofit organization. Tax-exempt: 501(c)(3).
Institution Type/Description: History Museum.
Collections: artifacts; photographs; documents; tools; early Catholic relics; statuary.
Activities: guided tours; temporary exhibitions.
Publications: Echo Story: Part I Early Days to Arrival of the Railroad.
Hours & Admission Prices: Call for hours & appointment. No charge; donations accepted.
Attendance: 75 (estimated)
Membership: Individual/Library $20; Contributing $50; Patron $100.

FORT HENRIETTA PARK, 10 W. Main St., Echo, OR 97826. Mailing Address: P.O. Box 9, Echo, OR 97826-0009. Tel.: 541-571-3597.
Web Site: www.echo-oregon.com/fort.html
Founded: 1985.
Congressional District: 2
Governing Authority: municipal; nonprofit. Tax-exempt.
Institution Type/Description: History Museum & Interpretive Center.
Collections: replica of Fort Block House; interpretive panels; covered wagon museum; fire equipment c.1905-1915; hook & ladder; chemical wagon; 1st Umatilla County jail.
Activities: self-guided tours.
Publications: Echo Cultural Inventory; The Echo Story - Part I Early Days to Arrival of the Railroad.
Hours & Admission Prices: Park: Mon.-Fri. dawn-dusk. Library: Mon.-Fri. 8:30-4:30. No charge; donations accepted. &

OREGON TRAIL ARBORETUM, 1 Neely Lane, Echo, OR 97826. Mailing Address: 20 S. Bonanza, P.O. Box 9, Echo, OR 97826-0009. Tel.: 541-376-8411. Fax: 541-376-8218.
E-mail: ecpl@centurytel.net
Web Site: www.echo-oregon.com
Founded: 1993.
Congressional District: 2
Key Personnel: C.E.O. & Records, Diane Berry; Mayor, Richard Winter, Ph.D.
Personnel Profile: Part-Time Volunteers 2.
Governing Authority: municipal. Parent Institution: City of Echo. Tax-exempt.
Institution Type/Description: Arboretum.
Collections: over 130 trees & shrubs; history of trees in Echo.
Research Fields: northeastern Oregon ornamental nursery trade.
Facilities: nature trails.

Activities: self guided tours; educational programs for children. Annual Events: Tree Fair in April; Arbor Day Poster contest.
Publications: brochure, Echo Historical & Cultural Inventory.
Hours & Admission Prices: Mon.-Fri. dawn-dusk. No charge; donations accepted. &

Eugene

CONGER STREET CLOCK MUSEUM, 730 Conger St., Eugene, OR 97402. Tel.: 541-344-6359. Fax: 541-338-0869.
Institution Type/Description: Clock History Museum.
Collections: over 1,000 period clocks; cameras; miniature pedal cars.
Hours & Admission Prices: Mon.-Sat. 10-5:30. No charge.

＊ JORDAN SCHNITZER MUSEUM OF ART, (M), 1430 Johnson Lane, Eugene, OR 97403. Mailing Address: 1223 University of Oregon, Eugene, OR 97403-1223. Tel.: 541-346-3027 & 0973. Fax: 541-346-0976.
E-mail: jschnitz@uoregon.edu
Web Site: jsma.uoregon.edu
Formerly: University of Oregon Museum of Art
Founded: 1932.
Congressional District: 4
Key Personnel: Exec. Dir., Jill Hartz; Chief Cur. Collections, Asian Art & Academic Support, Anne Rose Kitagawa; Assoc. Dir. Operations & Exhibits, Kurt Neugebauer; Dir. Education, Lisa Abia-Smith; Mgr. Collections, Chris White; Chief Preparator, Joey Capadona; Mgr. Communications, Debbie Williamson-Smith; Visitor Svcs., Jamie Leaf.
Personnel Profile: Full-Time Paid 24; Part-Time Paid 5; Part-Time Volunteers 107; Interns 25.
Operating Expenses: 3,102,008
Operating Income: 3,176,656
Governing Authority: university. Parent Institution: University of Oregon. Tax-exempt: 170(b)(1)(A).
Institution Type/Description: Art Museum.
Collections: American & regional art; European, Korean, Chinese & Japanese; Russian icons; Latin America.
Major Exhibits: Art of Traditional Japanese Theater, 10/13-7/6/14; We Tell Ourselves Stories in Order to Live, 1/18/14-3/16/14; Emancipating the Past: Kara Walker's Tales of Slavery & Power (T), 1/2514-4/6/14; The Human Touch: Selections from the RBC Wealth Management Art Collection, 4/26/14-12/7/14; Ryo Toyanaga: Awakening, 9/27/14-12/28/14.
Research Fields: Asian, European, Latin American & Pacific Northwest American art.
Facilities: garden & sculpture court; cafe. Museum-related items for sale.
Activities: guided tours; gallery talks & demonstrations; student practica; inter-museum loan exhibitions; rental space available; exhibition interpreter program; studio classes; interactive exhibits; art-making studio.
Publications: collections catalog; brochure; newsletters; annual report; exhibition catalogs.
Hours & Admission Prices: Tues. & Thurs.-Sun. 11-5, Wed. 11-8. Adults $5, seniors $3; discounts to AAM & NARM members; Univ. of Oregon students, faculty & staff; high school & non-UO college students, children 13 & under and members no charge. &
Attendance: 50,233 (estimated)
Membership: Individual $45; Family $55; Member $100-$1,000. (Discounts to seniors & educators)

LANE COMMUNITY COLLEGE ART GALLERY, 4000 E. 30th Ave., Eugene, OR 97405-0640. Tel.: 541-463-5409. Fax: 541-463-4185.
Web Site: www.lanecc.edu/artgallery/
Founded: 1970.
Key Personnel: Gallery Dir., Jennifer Salzman.
Personnel Profile: Part-Time Paid 1.
Governing Authority: public college; nonprofit.
Institution Type/Description: Art Gallery.
Collections: paintings; photographs; sculpture; pottery.
Facilities: educational facilities.
Activities: lectures.
Hours & Admission Prices: Sept.-June Mon.-Thurs. 8-9, Fri. 8-4. No charge. Closed New Year's Day; Presidents' Day; Memorial Day; Labor Day; Thanksgiving; Christmas. &
Attendance: 12,500 (accurate)

LANE COUNTY HISTORICAL SOCIETY & MUSEUM, (M), 740 W. 13th Ave., Eugene, OR 97402-4010. Mailing Address: P.O. Box 5407, Eugene, OR 97405. Tel.: 541-682-4242. Fax: 541-682-7361.
E-mail: info@lanecountyhistoricalsociety.org
Web Site: www.lanecountyhistoricalsociety.org
Founded: 1951.
Congressional District: 4
Key Personnel: Dir., Robert L. Hart; Office Mgr., Molly Hoisington.
Personnel Profile: Full-Time Paid 1; Part-Time Paid 9; Part-Time Volunteers 40.
Governing Authority: nonprofit organization. Tax-exempt: 501(c)(3).
Institution Type/Description: History Museum.
Collections: 1850s-1950s textiles, tools, vehicles, household furnishings; 14,000 photographs; early settlement period; transportation artifacts; 1920s & 1930s trades; children's exhibits. Historic Building: 1853 county clerk's building.
Research Fields: Lane County history.
Facilities: 2,500-vol. library & archives on Lane County history.
Activities: formally organized education programs for children; permanent & temporary exhibitions; school trunk loan service; scheduled programs; annual picnic; long term color slide digitization project. Museum Sponsors: Annual Quilt Exhibition in April.
Publications: booklet, Oregon Trail & 19th Century Lane County; books on Lane County & Oregon history; Oregon Trail published material & diary reprints; triannual newsletter, The Lane County Historian; quarterly newsletter, The Artifact.
Hours & Admission Prices: Tues.-Sat. 10-4. Adults $3, seniors $2, youth 15-17 $.75; children under 14 & members no charge. Closed national holidays.
Attendance: 10,755 (accurate)
Membership: Senior Citizen $15; Individual $25; Supporter & Family $40; Sponsor $75-$199; Business & Patron $200-$999; Benefactor $1,000 & up.

MAUDE I. KERNS ART CENTER, 1910 E. 15th Ave., Eugene, OR 97403-2094. Tel.: 541-345-1571. Fax: 541-345-6248.
E-mail: staff@mkartcenter.org
Web Site: www.mkartcenter.org
Founded: 1962.
Key Personnel: Bd. Pres. (V), David Wade; Exec. Dir., Karen Pavelec; Assoc. Dir., Sabrina Hershey; Exhibits Coord., Michael Fisher; Publicity Coord., Marsha Shankman; Administrative Asst., Kelly McCormick.
Personnel Profile: Full-Time Paid 2; Part-Time Paid 3; Part-Time Volunteers 200; Interns 20.
Governing Authority: nonprofit organization. Tax-exempt: 501(c)(3).
Institution Type/Description: Art Museum: housed in 1895 historic landmark church building.
Collections: work of Maude Kerns.
Facilities: classroom; printmakers studio; ceramics cooperative.
Activities: arts festivals; Art & the Vineyard; formal education programs for adults & children; guided tours; lectures. Museum Sponsors: Art and the Vineyard; 24th Annual Jello Art Show in March; Independence Day Weekend in July.
Publications: quarterly newsletter; quarterly class schedules.
Hours & Admission Prices: Mon.-Fri. 10-5:30, Sat. 12-4. Suggested Donations: family $5, individual $2. &
Attendance: 60,000 (estimated)
Membership: Students & Seniors $35; Adult $45; Family $65; Business $85; Patron $125; Benefactor $250 and up.

MOUNT PISGAH ARBORETUM, 34901 Frank Parrish Rd., Eugene, OR 97405-9673. Tel.: 541-747-3817. Fax: 541-741-4904.
E-mail: office@mountpisgaharboretum.org
Web Site: www.mountpisgaharboretum.org
Founded: 1973.
Congressional District: 4
Key Personnel: Exec. Dir., Brad van Appel; Pres. Bd., Anne Forrestel; Operations Mgr., Tom LoCascio; Education Mgr., Fran Rosenthal.
Personnel Profile: Full-Time Paid 7; Full-Time Volunteers 10; Part-Time Paid 1; Part-Time Volunteers 500.
Governing Authority: nonprofit organization. Parent Institution: Friends of Mount Pisgah Arboretum. Tax-exempt: 501(c)(3).
Institution Type/Description: Arboretum-Botanical Garden.
Collections: native vegetation; 200 species of exotics.
Research Fields: natural habitat rehabilitation.
Facilities: 500-vol. library pertaining to botany & horticulture available for research on premises; nature trails & guides; 200-acre site.
Activities: guided tours; formally organized education programs for children; permanent exhibitions. Arboretum Sponsors: Wildflower Festival in Spring; Arbor Week celebrations in April; Mushroom Festival in Fall.

Publications: newsletter; pamphlets; brochure; mushroom cookbook; nature guide; newsletter, Tree Time.
Hours & Admission Prices: Daily dawn-dusk. No charge; donations accepted.
　&
Attendance: 177,000 (estimated)
Membership: Student & Senior Citizen $40; Individual & Nonprofit Organization $50; Family $60; Business $100; Life $1,000.

OREGON AIR & SPACE MUSEUM, 90377 Boeing Dr., Eugene, OR 97402-9536. Tel.: 541-461-1101. Fax: 541-461-1101.
E-mail: charalyn@charink.com
Web Site: oasm.info
Founded: 1991.
Congressional District: 4
Key Personnel: Pres. (V), Bruce Lamont; Dir. Operations & Museum Shop Mgr., Charalyn Glade.
Personnel Profile: Part-Time Volunteers 30; Interns 1.
Volunteer Hours: 19,500
Governing Authority: nonprofit organization. Tax-exempt: 501(c)(3).
Institution Type/Description: Aviation & Space Museum.
Collections: aircraft; engines; uniforms; flight suits; photographs; history of flight; Oregon aviation history; space program; display of over 1,000 scale model aircraft; military honors; experimental aircraft; engines; gliders; ultra-lights.
Facilities: two hangers; library; class seating; lecture space; small kitchen; gift shop.
Activities: guided tours; lectures; organized education programs; temporary & traveling exhibitions. Museum Sponsors: banquet; air show; USO style hangar dances.
Publications: quarterly newsletter, Wings of Oregon.
Hours & Admission Prices: April-Oct. Wed.-Sun. 12-4; Nov.-March Wed.-Sat. 12-4. Adults $7, senior citizens 62 & up $6, children 6-17 $3; children 5 & under, members, Pearson Air Museums, Seattle Museums of Flight, Lane Community College, Civil Air Patrol & LEC Aviation Academy members no charge; group rates available; rental rates negotiable. Closed some holidays. &
Attendance: 4,000 (estimated)
Membership: Individual & Senior Family $35; Family $45; Sustaining $60; Contributing $150.

SCIENCE FACTORY CHILDREN'S MUSEUM & EXPLORATION DOME, 2300 Leo Harris Pkwy., Eugene, OR 97401-8834. Mailing Address: P.O. Box 1518, Eugene, OR 97440-1518. Tel.: 541-682-7888. Fax: 541-484-9027. Facebook: Science Factory.
E-mail: info@sciencefactory.org
Web Site: www.sciencefactory.org
Formerly: Science Factory Children's Museum & Planetarium
Founded: 1961.
Congressional District: 4
Key Personnel: Exec. Dir., Carolyn Rebbert; Chm. (V), Catherine Rainwater; Dir. Operations, Kim Miller.
Personnel Profile: Full-Time Paid 6; Part-Time Paid 6; Part-Time Volunteers 100.
Governing Authority: nonprofit organization. Tax-exempt: 501(c)(3).
Institution Type/Description: Science Museum.
Collections: interactive science exhibits.
Major Exhibits: Eat Well, Play Well (T), 1/14-6/14.
Facilities: dome theater/planetarium; classrooms; computer labs. Museum-related items for sale.
Activities: interactive exhibits; planetarium shows; full dome movies; holiday science camp; no school day camp; summer science camp; special events; birthday parties; science workshops; formally organized education programs; permanent & temporary exhibitions; preschool discovery days; science clubs; speakers.
Publications: newsletter, monthly e-news.
Hours & Admission Prices: July-Aug. daily 10-4; Sept.-June Wed.-Sun. 10-4. Exhibits: adults $4, seniors $3; children 2 & under and members no charge. Exploration Dome: adults $4, seniors $3; discounts for members to Live Shows & dome movies; discounts to ASTC members; members to Seasonal Star Shows and children 2 & under no charge. Combination: adults $7, seniors $6; children 2 & under no charge. Closed New Year's Day; Easter; Independence Day; Thanksgiving; Christmas Eve & Day. &
Attendance: 35,000 (accurate)
Membership: Individual $30; Individual Plus One $40; Individual Plus Two $50; Family Two Plus Two $60; Family Two Plus Four $80; Supporting $150.

UNIVERSITY OF OREGON MUSEUM OF NATURAL AND CULTURAL HISTORY, (M), 1680 E. 15th Ave., Eugene, OR 97403-1224. Mailing Address: 1224 University of Oregon, Eugene, OR 97403-1224. Tel.: 541-346-3024. Fax: 541-346-5334. Facebook: Oregon Natural History.
E-mail: mnch@uoregon.edu
Web Site: natural-history.uoregon.edu
Founded: 1936.
Congressional District: 4
Key Personnel: Dir., Jon Erlandson; Dir. Prog., Patricia Krier; Dir. Research, Thomas Connolly; Dir. Collections, Pamela Endzweig; Asst. Dir. Visitors Svcs., Judith Pruitt; Asst. Dir. Public Programs, Ann Craig.
Personnel Profile: Full-Time Paid 22; Part-Time Paid 22; Part-Time Volunteers 52; Interns 4.
Operating Expenses: 2,314,000
Operating Income: 3,093,000
Governing Authority: state; university. Parent Institution: University of Oregon. Subsidiary Institution: Oregon State Museum of Anthropology. Tax-exempt: 170(c)1.
Institution Type/Description: Natural History Museum. (Association of Science - Associated member with ASTC Technology Centers).
Collections: Oregon natural & cultural history & archaeology; worldwide ethnology.
Major Exhibits: Cruisin' the Fossil Freeway with Artist Ray Troll and Paleontologist Kirk Johnson (T), 10/13-2/14; Site Seeing: Snapshots of Historical Archaeology in Oregon, 10/13-5/14.
Research Fields: Oregon archaeology & anthropology; paleontology; geology; paleoecology.
Facilities: 1,000-vol. library of technical reports & documentation; 3,500 sq. ft. exhibit space; research area. Museum-related items for sale.
Activities: guided tours; lectures; organized education programs; temporary exhibitions; docent program.
Publications: quarterly newsletter, Fieldnotes; scholarly special publications; annual report.
Hours & Admission Prices: Tues.-Sun. 11-5. Family $10, adults $5, seniors & youth $3; UO students, faculty & staff and MNCH & museum members no charge. ASTC reciprocal benefits. Closed New Year's Day; Independence Day; Thanksgiving; Christmas. &
Attendance: 22,000 (estimated)
Membership: Individual $40; Family $50; Supporter $100; Director's $500 & up.

Fairview

FAIRVIEW-ROCKWOOD-WILKES HISTORICAL SOCIETY, 60 Main St., Fairview, OR 97024. Mailing Address: P.O. Box 946, Fairview, OR 97024-0946. Tel.: 503-261-8078.
Web Site: www.frwhs.org
Formerly: Zimmerman House and Heslin House Museums
Founded: 1987.
Key Personnel: Pres. (V), Dodi Davies.
Personnel Profile: Part-Time Volunteers 25.
Governing Authority: Parent Institution: Fairview-Rockwood-Wilkes Historical Society. Branch Museum: Zimerman House, 17111 N.E. Sandy Blvd., Gresham, OR 97030. Tax-exempt.
Institution Type/Description: Historic Houses: housed in the Heslin house; built in 1893.
Collections: local history & culture; period furnishings; personal artifacts; photographs.
Activities: educational programs; group tours.
Publications: Reflections.
Hours & Admission Prices: Heslin House: 3rd Sat. of month 10-3. Donation: adults 12 & over $3; children no charge. &

Florence

SIUSLAW PIONEER MUSEUM, 278 Maple St., Florence, OR 97439. Mailing Address: P.O. Box 2637, Florence, OR 97439-0164. Tel.: 541-997-7884.
E-mail: museum@winfinity.com
Web Site: www.siuslawpioneermuseum.com
Founded: 1970.
Key Personnel: Chm. (v), Del Phelps; Pres. (V), Stephen Skidmore; Museum Shop Mgr., Jean Chapman.
Personnel Profile: Part-Time Paid 1; Part-Time Volunteers 40.
Governing Authority: nonprofit organization. Tax-exempt.
Institution Type/Description: History Museum.
Collections: local history & culture; photographs.
Hours & Admission Prices: Feb.-Nov. Tues.-Sun. 12-4. Adults $3; children under 16 & members no charge. Closed Easter; Thanksgiving; Christmas.

Attendance: 2,458 (accurate)

Forest Grove

PACIFIC UNIVERSITY MUSEUM, Pacific University, 2043 College Way, Forest Grove, OR 97116-1756. Tel.: 503-352-2211. Fax: 503-352-2252.
E-mail: deni@pacificu.edu
Web Site: www.pacificu.edu
Founded: 1949.
Congressional District: 1
Key Personnel: Pres., Dr. Lesley M. Hallick.
Personnel Profile: Part-Time Volunteers 12.
Governing Authority: university. Tax-exempt: 101(6).
Institution Type/Description: History Museum housed in: 1850 Old College Hall.
Collections: artifacts relating to the history of Pacific University & Tualatin Academy; Asian cultures & missionary activity.
Research Fields: biographical; institutional history of Pacific University; development of higher education in the Pacific Northwest.
Facilities: meeting & reception room; chapel.
Activities: guided tours.
Publications: Splendid Audacity: The Story of Pacific University; On Your Own Two Feet: A Self Guided Tour of Our Historic Campus.
Hours & Admission Prices: First Wed. 1-4, or by appointment. No charge; donations accepted. Closed university vacations & holidays.
Attendance: 1,500 (estimated)

VALLEY ART GALLERY, 2022 Main St., Forest Grove, OR 97116. Mailing Address: P.O. Box 333, Forest Grove, OR 97116. Tel.: 503-357-3703. Facebook: Valley Art Association.
E-mail: office@valleyart.org
Web Site: www.valleyart.org
Founded: 1966.
Key Personnel: Pres., Dana Zurcher.
Personnel Profile: Part-Time Volunteers 50.
Institution Type/Description: Art Gallery.
Collections: works by established & emerging artists; paintings; pottery; sculpture.
Facilities: classrooms.
Activities: workshops; classes; special events.
Publications: bimonthly newsletter.
Hours & Admission Prices: Mon.-Sat. 11-5:30. No charge. Closed New Year's Day; Thanksgiving; Christmas. &
Attendance: 1,000 (estimated)
Membership: Individual $15; Family $25.

Fort Rock

FORT ROCK VALLEY HISTORICAL HOMESTEAD MUSEUM, 64696 Fort Rock Rd., Fort Rock, OR 97735. Mailing Address: Fort Rock Valley Historical Society, P.O. Box 84, Fort Rock, OR 97735-0084. Tel.: 541-576-2207.
Institution Type/Description: Historic Buildings: eleven buildings in a village setting.
Collections: local history & culture; photographs; period furnishings. Historic Buildings: homestead houses; church; school buildings.
Facilities: rental facilities. Museum-related items for sale.
Hours & Admission Prices: mid-May to mid-Sept. Fri.-Sun. 10-4. Admission 12 & over $1.

Fossil

FOSSIL MUSEUM, First & Main, Fossil, OR 97830. Mailing Address: P.O. Box 465, Fossil, OR 97830-0465. Tel.: 541-763-2113. Fax: 541-763-2026. Facebook: Fossil Museum, Fossil, Oregon.
Founded: 1966.
Congressional District: 2
Key Personnel: Chm. (V), Donna D. Hopper; Pres. (V), Marilyn G. Garcia.
Personnel Profile: Part-Time Volunteers 25.
Volunteer Hours: 1,136
Operating Expenses: 3,300
Operating Income: 5,613
Governing Authority: city; nonprofit. Parent Institution: City of Fossil, OR. Tax-exempt.
Institution Type/Description: History Museum.
Collections: photographs; local newspapers; household & personal items from 1840-present.

Facilities: 2,400 sq. ft. exhibit space.
Activities: guided tours; loan exhibitions; local artists exhibitions; school loan service; community schools partner. Museum Sponsors: Quilt Show; Music in the Park.
Publications: annual newsletter; Fossil Museum; Days of Yore and Then Some More; local history book.
Hours & Admission Prices: Memorial Day to Labor Day Mon. 9-4, Wed.-Sun. 1-4. No charge; donations accepted &
Attendance: 1,431 (accurate)

Garibaldi

GARIBALDI MUSEUM, (M), 112 Garibaldi Ave., Garibaldi, OR 97118. Mailing Address: P.O. Box 5, Garibaldi, OR 97118. Tel.: 503-322-8411. Fax: 5033228411.
E-mail: info@garibaldimuseum.com
Web Site: www.garibaldimuseum.com
Founded: 1986.
Congressional District: 5
Key Personnel: Museum Shop Mgr., Anna Rzuczek.
Personnel Profile: Full-Time Paid 1; Part-Time Volunteers 12.
Governing Authority: bd. of directors. Tax-exempt.
Institution Type/Description: Maritime Museum.
Collections: local maritime heritage & history; Captain Robert Gray's life & historic vessels including models of the Columbia Rediviva & Lady Washington; reproductions of seafarers clothing & musical instruments; photographs; Native American (Tillamook Peoples) replica hunting & cooking tools.
Publications: biannual newsletter.
Hours & Admission Prices: April-Oct. Thurs.-Mon. 10-4. Adults $3, seniors & children 5-17 $1.50; discounts to groups; children under 5 no charge. &
Attendance: 4,000 (accurate)
Membership: Honored Senior $10; Deck Cadet $15; Ordinary Seaman & Mate $20; Captain & Family $25; Fleet Commander $100; Commodore $500. Corporate: Maritime Pilot $100; additional crew members $10 each.

Gold Beach

ROGUE RIVER MUSEUM, 29880 Harbor Dr., Gold Beach, OR 97444. Mailing Address: P.O. Box 1011, Gold Beach, OR 97444-1011. Tel.: 541-247-4571.
Web Site: www.roguejets.com/museum.php
Institution Type/Description: History Museum.
Collections: historical artifacts; photographs; taxidermy.
Hours & Admission Prices: Call for hours.

Government Camp

MT. HOOD CULTURAL CENTER & MUSEUM, 88900 E. Hwy. 26, Business Loop, Government Camp, OR 97028. Mailing Address: P.O. Box 55, Government Camp, OR 97028-0055. Tel.: 503-272-3301.
E-mail: info@mthoodmuseum.org
Web Site: www.mthoodmuseum.org
Founded: 1998.
Congressional District: 3
Key Personnel: Pres. (V), Bing Sheldon; Museum Shop Mgr., Cheryl Maki.
Personnel Profile: Full-Time Paid 1; Part-Time Paid 2; Part-Time Volunteers 80; Interns 2.
Governing Authority: Tax-exempt.
Institution Type/Description: History Museum.
Collections: history of winter sports; early exploration; settlement & natural history of Mt. Hood.
Facilities: community meeting room; art sales gallery. Museum-related items for sale.
Activities: arts & crafts school; historic home tours.
Publications: quarterly newsletter.
Hours & Admission Prices: Daily 9-5. No charge; donations accepted. Closed Thanksgiving; Christmas. &
Attendance: 20,000 (accurate)
Membership: Individual $20; Family $35; Patron $100; Benefactor $250; Summit Club $1,000.

Grand Ronde

CHACHALU TRIBAL MUSEUM & CULTURAL CENTER, 9615 Grand Ronde Rd., Grand Ronde, OR 97347-9712. Tel.: 503-879-2248. Fax: 503-879-2126.
E-mail: david.lewis@grandronde.org

Web Site: www.grandronde.org
Formerly: Confederated Tribes of Grand Ronde Cultural Resources Department Museum & Cultural Center
Founded: 1995,
Congressional District: 5
Key Personnel: Education & Outreach, Kathy Cole; Coord. Site Protection & Tribal Historic Preservation Officer, Eirik Thorsgard; Language & Cultural Specialist, Bobby Mercier; Interpretive Specialist, Julie Brown; Cultural Resources Specialist, Brian Krehbiel; Tribal Historian, David Lewis; Cultural Collections Specialist, Veronica Montano; Site Protection Specialist, David Harrelson; Archaeologist, Briece Edwards; Land & Culture Mgr., Jan Rubach; Sec., Flicka Lucero.
Personnel Profile: Full-Time Paid 18; Interns 1.
Governing Authority: The Confederated Tribes of the Grand Ronde Community of Oregon. Tax-exempt.
Institution Type/Description: Tribal Museum.
Collections: Indian artifacts; manuscripts; oral histories; traditional Indian crafts & art; archival materials; photographs & language materials related to Grand Ronde tribal community; digital database of source documents.
Major Exhibits: Kuri-tsaqw Tilixam: River People, 4/14-5/26/14; We Were Here First...And We Are Here to Stay: Grand Ronde Termination & Restoration (T), 4/12/14-5/27/14; Phase I Grand Ronde People, 6/14-9/14.
Research Fields: Indian culture & history, specifically of the Confederated Tribes of the Grand Ronde Community of Oregon & related tribes; U.S. Government management of American Indians; Indian Treaties; Indian removal; early history of tribe.
Facilities: 100-vol. library of culture & history of Grand Ronde Tribes & related tribes; tribal repository.
Activities: cultural protection & interpretation of tribal history & culture for tribal community; cultural education classes; tribal signs project.
Publications: Tribal newspapers, Smoke Signals; virtual gallery, Ntsayka Ikanum: Our Story.
Hours & Admission Prices: Offices: Mon.-Fri. 8-5. Closed New Year's Day; Presidents' Day; Independence Day; Labor Day; Veterans Day; Thanksgiving; Christmas; National Indian Day; Restoration Day.

Grants Pass

GRANTS PASS MUSEUM OF ART, 229 S.W. G St., Grants Pass, OR 97526-2415. Mailing Address: P.O. Box 966, Grants Pass, OR 97528-0081. Tel.: 541-479-3290. Fax: 541-479-1218.
E-mail: museum@gpmuseum.com
Web Site: www.gpmuseum.com
Founded: 1979.
Congressional District: 2
Key Personnel: Exec. Dir., Chris Pondelick; Pres. (V), Susan E. Burnes; Museum Shop Mgr., Cindy Kahoun.
Personnel Profile: Part-Time Paid 2; Part-Time Volunteers 32; Interns 2.
Governing Authority: nonprofit organization.
Institution Type/Description: Art Museum.
Collections: contemporary regional art.
Facilities: classrooms.
Activities: guided tours; lectures; gallery talks; formally organized education programs for children; permanent & traveling exhibitions; workshops.
Publications: quarterly newsletter; exhibit announcements.
Hours & Admission Prices: Summer: Tues.-Sat. 12-4; Winter: Tues.-Fri. 12-4, Sat. 10-1:30. No charge; donations accepted. Closed Easter; Memorial Day; Independence Day; Labor Day; Thanksgiving weekend. &
Attendance: 21,346 (accurate)
Membership: Student & Teacher $20; Individual $30; Family $60; Guild $100; Patron $250; Benefactor $500 & up.

SCHMIDT HOUSE MUSEUM & RESEARCH LIBRARY, 508 & 512 S.W. 5th St., Grants Pass, OR 97526-2804. Mailing Address: 512 S.W. 5th St., Grants Pass, OR 97526-2804. Tel.: 541-479-7827. Fax: 888-488-5410.
E-mail: josephine@historicalsociety.us
Web Site: www.jocohistorical.org
Founded: 1960.
Congressional District: 2
Key Personnel: C.E.O., Dir. & Museum Shop Mgr., Rose Scott; Pres. (V), Joan Momsen; Museum Shop Mgr., Martha Metcalf.
Personnel Profile: Part-Time Paid 2; Part-Time Volunteers 50.
Governing Authority: private; nonprofit organization. Parent Institution: Josephine County Historical Society, Grants Pass, OR. Tax-exempt: 501(c)(3).
Institution Type/Description: Historic House & Research Library.
Collections: 1885-1925 furniture; toys; clothes; photos; documents; books; microfilm; magazines; newspapers.
Research Fields: genealogy; local history.
Facilities: library. Museum-related items for sale.

Activities: films; guided tours; lectures; temporary exhibitions. Museum Sponsors: Pie Social; Heritage Events.
Publications: quarterly newsletter, The Oldtimer.
Hours & Admission Prices: Tues.-Fri. 10-4. Admission $5; discount to AAA members.
Attendance: 1,500 (estimated)
Membership: Individual $15; Family $25; Business $55; Sustaining $100; Newsletter $225; Lifetime $1,000.

WISEMAN & FIREHOUSE GALLERIES, ROGUE COMMUNITY COLLEGE, 3345 Redwood Hwy., Grants Pass, OR 97527-9291. Tel.: 541-956-7241. Fax: 503-471-3588.
E-mail: kbrake@roguecc.edu
Web Site: www.roguecc.edu/galleries
Founded: 1985.
Key Personnel: Dir., Karl Brake; Pres. Rogue Community College, Peter Angsadt; Gallery Coord., Heather Green.
Personnel Profile: Full-Time Paid 1; Part-Time Paid 1; Part-Time Volunteers 15.
Governing Authority: municipal; nonprofit. Parent Institution: Rogue Community College. Tax-exempt.
Institution Type/Description: Art Galleries: the FireHouse Gallery is housed in the firehouse portion of historic city hall; the Wiseman Gallery is located on the Redwood Campus & Rogue Community College campus in the Wiseman Center.
Collections: contemporary art including paintings, prints, drawings, sculpture & photographs.
Major Exhibits: Ronnie Cramer "Valentines", 1/14; Ming Zhou "Glorious Life", 2/14; Nishiki Tayui, 3/14; Scott Mayberry "Collapse", 4/14; Joana Salska 2009-2013, 5/14; Sarah Fagen "Time Together", 6/14; Joseph Lastomirsky, 7/14; Southern Oregon Art Show, 8/14; James Lilly "Currents", 9/14; Braeden Cox, 11/14.
Facilities: educational facilities; 250-seat theater.
Activities: volunteer program; formal education programs for adults & children; temporary & traveling exhibitions. Annual Events: Art Auction & Dance; Black, White and the Blues.
Publications: annual directory exhibits listing for each gallery; catalogues that accompany exhibits.
Hours & Admission Prices: Wiseman Gallery: Mon.-Thurs. 8-7, Fri. 8-3. FireHouse Gallery: Tues.-Fri. 11:30-4:30. No charge. Closed national holidays & between college quarters. Call for summer hours. &
Attendance: 19,000 (estimated)

Gresham

FAIRVIEW-ROCKWOOD-WILKES HISTORICAL SOCIETY - THE ZIMMERMAN HOUSE, 17111 N.E. Sandy Blvd., Gresham, OR 97030. Mailing Address: P.O. Box 946, Fairview, OR 97024-0946. Tel.: 503-261-8078.
Web Site: www.frwhs.org
Founded: 1987.
Key Personnel: Pres. (V), Dodi Davies.
Personnel Profile: Part-Time Volunteers 30.
Governing Authority: Parent Institution: Fairview-Rockwood-Wilkes Historical Society. Branch Museum: Heslin House Museum, 60 Main St,. Fairview, OR 97024. Tax-exempt.
Institution Type/Description: Historic House Museum: built in 1874.
Collections: local history & culture; period furnishings; personal artifacts; photographs.
Activities: educational programs; group tours.
Publications: Reflections.
Hours & Admission Prices: 3rd Sat. each month 10-3. Donation Requested: adults 12 & over $3; children no charge.
Membership: Individual $15; Family $25.

GRESHAM HISTORY MUSEUM, 410 N. Main Ave., Gresham, OR 97030-7212. Mailing Address: P.O. Box 65, Gresham, OR 97030-0011. Tel.: 503-661-0347.
Web Site: community.gorge.net/ghs/
Formerly: Gresham Pioneer Museum
Founded: 1976.
Key Personnel: Pres. (V), Dorothy Douglas; Security, Larry Kerr; Education, Utahna Kerr.
Personnel Profile: Part-Time Paid 1; Part-Time Volunteers 35.
Governing Authority: nonprofit organization. Parent Institution: Gresham Historical Society. Tax-exempt: 501(c)(3).
Institution Type/Description: Historical Society Museum: housed in 1913 Carnegie library.
Collections: photographs; historical artifacts; ephemera.

Activities: lectures; organized education programs; school loan service; loan, temporary & traveling exhibitions.
Publications: quarterly, Gresham Historical Society Newsletter.
Hours & Admission Prices: Tues. & Thurs. 10-4, Sat. 12-4. No charge; donations accepted. &

Attendance: 400 (estimated)
Membership: Student $10; Senior $15; Regular $25; Family, Business & Organization $50; Business & Organization with Business Card $100.

Haines

EASTERN OREGON MUSEUM, 610 Third St., Haines, OR 97833-6388. Mailing Address: 14514 Muddy Creek Lane, Haines, OR 97833-6388. Tel.: 541-856-3233. Facebook: Eastern Oregon Museum.
E-mail: riderm622@gmail.com
Formerly: Eastern Oregon Museum on the Old Oregon Trail
Founded: 1958.
Congressional District: 2
Key Personnel: Pres., Linda Smith; Vice Pres., Viola Perkins; Treas., Mary Rider; Sec., Teri Johnson.
Personnel Profile: Part-Time Volunteers 20.
Governing Authority: board of directors; nonprofit organization. Tax-exempt: 501(c)(3).
Institution Type/Description: General Museum: housed in a 1931 old school gym.
Collections: over 10,000 artifacts; farm implements; blacksmith shop; authentic parlor; buggies; cutters; hacks & a surrey; musical instruments; bells; dolls & toys; replica kitchen; washing & sewing machines; old newspapers; old school books; uniforms; guns; arrowheads; Oregon Cattlemen Assoc. history. Historic Building: late 1880s railroad depot.
Facilities: approx. 100-vol. library of local & Oregon history, available for use on the premises by appointment. Craft items, jewelry, handmade items for sale.
Activities: guided tours for schools, club groups & senior citizens groups.
Publications: brochures.
Hours & Admission Prices: May 15-Sept. 15 Thurs.-Sun. & holidays; other times by appointment. Family $5, adult $2. &
Attendance: 4,000 (accurate)
Membership: Individual $3; Family $5; Life $100.

Hammond

FORT STEVENS HISTORIC MILITARY SITE & MUSEUM, Fort Stevens State Park, 100 Peter Iredale Rd., Hammond, OR 97121-9712. Tel.: 503-861-1671. Facebook: Fort Stevens Historic Military Site.
E-mail: john.koch@state.or.us
Key Personnel: Pres. (V), Ron Kinsley; Museum Shop Mgr., Jim Forst.
Personnel Profile: Full-Time Paid 1; Part-Time Paid 2; Part-Time Volunteers 30.
Governing Authority: Parent Institution: Oregon Parks and Recreation Dept. Subsidiary Institution: The Friends of Old Fort Stevens.
Institution Type/Description: Historic Site: listed on the National Register of Historic Places.
Collections: artifacts & interpretive displays pertaining to the history of Fort Stevens.
Hours & Admission Prices: June-Sept. daily 10-6; Oct.-May daily 10-4. No charge; donations accepted. Closed New Year's Day; Christmas.
Attendance: 46,296 (accurate)

Harbor

CHETCO VALLEY HISTORICAL MUSEUM, 15461 Museum Rd., Harbor, OR 97415. Mailing Address: P.O. Box 2004, Harbor, OR 97415-0303. Tel.: 541-469-6651.
E-mail: julie@ross-insurance.com
Web Site: www.chetcomuseum.org
Congressional District: 4
Governing Authority: Tax-exempt.
Institution Type/Description: Historic House Museum: housed in the Blake house; built in 1857.
Collections: early pioneer life; local history & culture; Native American artifacts; photographs; period furnishings.
Hours & Admission Prices: Winter: Sat.-Sun. 12-4; Summer: call for extended hours. No charge; donations accepted.
Membership: Annual $25.

Heppner

AGRICULTURAL EQUIPMENT MUSEUM, Riverside & A Sts., Heppner, OR 97836. Mailing Address: P.O. Box 515, Heppner, OR 97836. Tel.: 541-676-5546.
Institution Type/Description: History Museum.
Collections: local history; period agricultural equipment; photographs.
Hours & Admission Prices: May-Sept. Tues.-Fri. 1-5, Sat. 11-3.

MORROW COUNTY MUSEUM, 444 N. Main St., Heppner, OR 97836. Mailing Address: P.O. Box 1153, Heppner, OR 97836. Tel.: 541-676-5524.
Institution Type/Description: History Museum.
Collections: local history; period furnishings & clothing; 1903 Heppner flood.
Activities: research.
Hours & Admission Prices: May-Sept. Tues.-Fri. 1-5, Sat. 11-3; other times by appointment. No charge; donations accepted.

Hillsboro

CLASSIC AIRCRAFT AVIATION MUSEUM, 3005 N.E. Cornell Rd., Hillsboro, OR 97124-6316. Mailing Address: P.O. Box 91430, Portland, OR 97291-0008. Tel.: 503-693-1414.
E-mail: donkel@classicaircraft.org
Web Site: www.classicaircraft.org
Key Personnel: Dir., Doug Donkel, Ph.D.
Governing Authority: nonprofit organization. Tax-exempt: 501(c)(3).
Institution Type/Description: Aviation Museum.
Collections: aviation history, technology, & engineering; aircraft.
Activities: air shows; historical reunions; aviation events; flight demonstrations.
Hours & Admission Prices: Mon.-Thurs. 9-4; other times by appointment. No charge.

RICE NORTHWEST MUSEUM OF ROCKS AND MINERALS, 26385 N.W. Groveland Dr., Hillsboro, OR 97124-9351. Tel.: 503-647-2418.
E-mail: info@ricenorthwestmuseum.org
Web Site: www.ricenorthwestmuseum.org
Founded: 1996.
Key Personnel: Exec. Dir., Melena Wallace.
Personnel Profile: Full-Time Paid 3; Part-Time Paid 9; Part-Time Volunteers 2.
Governing Authority: Tax-exempt.
Institution Type/Description: Geology Museum.
Collections: northwest minerals; crystallized minerals; meteorites; petrified wood; fossils; agates; lapidary arts; synthetics; fluorescents; gemstones.
Facilities: banquet facilities.
Activities: research; rental facilities.
Hours & Admission Prices: Wed.-Sun. 1-5; groups by appointment. Adults $7, seniors $6, students 5-17 $5; members & children under 5 no charge. Closed New Year's Day; Independence Day; Thanksgiving; Christmas. &
Attendance: 30,000 (accurate)
Membership: Individual $50; One + One $75; Family $90; Family Plus $100; Friend $150; Patron $250; Benefactor $500.

Hood River

THE HISTORY MUSEUM OF HOOD RIVER COUNTY, 300 E. Marina Dr., Hood River, OR 97031-1198. Mailing Address: P.O. Box 781, Hood River, OR 97031-0026. Tel.: 541-386-6772. Fax: 541-386-6772.
E-mail: thehistorymuseum@hrecn.net
Web Site: www.co.hood-river.or.us/museum
Formerly: Hood River County Historical Museum
Founded: 1907.
Congressional District: 2
Key Personnel: C.E.O., Dave Meriwether; Chm. (v), Carol Faull; Dir. & Museum Shop Mgr., Connie Nice.
Personnel Profile: Full-Time Paid 1; Full-Time Volunteers 15; Part-Time Volunteers 50.
Governing Authority: county; nonprofit organization. Parent Institution: Hood River County. Subsidiary Institution: Hood River County Historical Society. Tax-exempt.
Institution Type/Description: General Museum.
Collections: biographical files dating from early pioneers; photographic history of Hood River.
Research Fields: early pioneer; family; Hood River & area.
Facilities: library; picnic area. Museum-related items for sale.
Activities: Museum Sponsors: Cemetery Tales in September.

Hours & Admission Prices: Mon.-Sat. 10-5, Sun. 1-5. Admission $3; military & their families, children under 10 and members no charge. Closed major holidays. &
Attendance: 5,300 (estimated)
Membership: Student $20; Senior $25; Individual $35; Family $50.

WESTERN ANTIQUE AEROPLANE & AUTOMOBILE MU-SEUM, Ken Jernstedt Airfield 4S2, 1600 Air Museum Rd., Hood River, OR 97031-9800. Tel.: 541-308-1600. Fax: 541-308-1601.
E-mail: info@waaamuseum.org
Web Site: www.waaamuseum.org
Founded: 2006.
Key Personnel: Pres. & Founder, Terry Brandt; Dir., Judy Newman; Dir. Restorations, Thomas Murphy.
Governing Authority: Tax-exempt: 501(c)(3).
Institution Type/Description: Transportation History Museum.
Collections: transportation artifacts & memorabilia.
Hours & Admission Prices: Daily 9-5. Adults $12, senior citizens & veterans $10, students 5-18 $6; children 4 & under and active military with ID no charge. Closed New Year's Day; Thanksgiving; Christmas. &
Attendance: 25,000 (accurate)
Membership: Solo $50; Family $100.

Independence

HERITAGE MUSEUM SOCIETY, (M), 112 S. 3rd St., Independence, OR 97351. Mailing Address: P.O. Box 7, Independence, OR 97351-0007. Tel.: 503-838-4989. Fax: 503-606-3282.
E-mail: orheritage@minetfiber.com
Web Site: orheritage.org
Founded: 1976.
Congressional District: 5
Key Personnel: Dir., Peggy Smith; Society Pres., Robert Richards; Aide, Andrea Pittman.
Personnel Profile: Part-Time Paid 2; Part-Time Volunteers 12; Interns 4.
Governing Authority: city. Tax-exempt: 501(c)(3).
Institution Type/Description: Historical Society Museum & Historic Site.
Collections: local history & culture; period furnishings; documents; photographs.
Research Fields: genealogy; agriculture; pioneer history.
Activities: group tours; internships. Annual Events: Member Picnic; Volunteer Recognition; Spring Clean Up Day; Hop & Heritage Festival; Host Students with Sr. Projects; Holiday Open House.
Publications: Historic Walking Tour Maps; Early History of Independence, OR; cookbooks, On to Oregon Cavalcade; historic house notecards; diary of the 1959 Oregon Trail re-enactment, Moving West.
Hours & Admission Prices: Wed.-Sat. 1-5; groups of 5 or more by appointment. Suggested Donation: $3; members and children 12 & under no charge.
Attendance: 1,837 (accurate)
Membership: Senior $10; Individual $20; Family $35; Corporate $50; Life $500.

John Day

KAM WAH CHUNG STATE HERITAGE SITE, Ing Hay Way, John Day, OR 97845. Mailing Address: P.O. Box 115, John Day, OR 97845-0115. Tel.: 541-575-2800; 800-551-6949.
Formerly: Kam Wah Chung & Co. Museum
Key Personnel: Cur., Christina Sweet
Governing Authority: Parent Institution: Oregon Parks & Recreation Dept.
Institution Type/Description: Chinese Heritage Museum.
Collections: Oregon's Chinese heritage, history & culture.
Hours & Admission Prices: May-Oct. daily 9-5. No charge.
Attendance: 6,000 (accurate)

Joseph

WALLOWA COUNTY MUSEUM, 110 S. Main, Joseph, OR 97846. Mailing Address: P.O. Box 430, Joseph, OR 97846-0430. Tel.: 541-432-6095.
Key Personnel: Bd. Chm., Caryl Coppin; Cur., Ann Hayes
Institution Type/Description: History Museum.
Collections: county history; Native American artifacts; early settlers.
Hours & Admission Prices: Memorial Day to late Sept. daily 10-5; other times by appointment. Adults $4, seniors 65 & over $3, children 7-17 $2; children 6 & under no charge.

Junction City

JUNCTION CITY HISTORICAL SOCIETY, 655 Holly St., Junction City, OR 97448-1631. Tel.: 541-952-0900.
E-mail: cgoodin253@msn.com
Web Site: www.junctioncity.com/history
Founded: 1971.
Congressional District: 4
Key Personnel: Pres., Linda VanOrden; Vice Pres., Dale Rowe; Cur., Kitty Goodin.
Personnel Profile: Part-Time Volunteers 30.
Governing Authority: nonprofit organization. Branch Museum: 1874 vintage Mary Pitney House, 289 W. 4th St., Junction City, OR. Tax-exempt: 170(b)(1)(3).
Institution Type/Description: History Museum & House: housed in 1872 Dr. Norman Lee house, first doctor in Junction City.
Collections: furnishings belonging to Dr. Lee; medical tools; dental tools; Indian artifacts, including the war bonnet of Chief Red Cloud; clothing & household items belonging to Clarence Pitney; artifacts donated by local citizens; Junction City's first jail; railroad diorama; hands-on exhibitions.
Activities: guided tours; formal & informal education programs; oral history; walking tour. Special Event: Scandinavian Festival in August.
Publications: quarterly newsletter.
Hours & Admission Prices: Thurs. 3-5, first Sat. of month 1-4. No charge; donations accepted.
Attendance: 1,424 (accurate)
Membership: Individual $15; Family $30; Institutional $40.

Keizer

KEIZER HERITAGE MUSEUM, 980 Chemawa Rd., N.E., Keizer, OR 97307-3716. Mailing Address: P.O. Box 20845, Keizer, OR 97307-0845. Tel.: 503-393-9660. Fax: 503-393-0209.
E-mail: heritage@wvi.com
Web Site: www.keizerheritage.org
Founded: 1988.
Key Personnel: Co Dir., Evelyn Franz; Co Dir., Ray Hanson; Pres. (V), Judy Peterson.
Personnel Profile: Part-Time Volunteers 50.
Governing Authority: Parent Institution: Keizer Heritage Foundation, Inc. Tax-exempt: 501(c)(3).
Institution Type/Description: History Museum.
Collections: historical artifacts & memorabilia pertaining to the city of Keizer. Historic Building: 1916 restored craftsman-style elementary school.
Research Fields: local Keizer area history.
Facilities: rental facilities.
Activities: special events.
Publications: Looking Back; More Looking Back; Keizer Story; One Man's Journey Through Keizer's History.
Hours & Admission Prices: Tues. & Thurs. 2-4, Sat. 10-4. No charge; donations accepted. &
Attendance: 1,317 (accurate)
Membership: Senior $15; Individual $20; Family & Business $35.

Kerby

KERBYVILLE MUSEUM, 24195 Redwood Hwy., Kerby, OR 97531. Mailing Address: P.O. Box 3003, Kerby, OR 97531-3003. Tel.: 541-592-5252.
Founded: 1959.
Congressional District: 4
Key Personnel: Pres., Dennis Strayer; Vice Pres., Lloydeen K. Davis; Sec., Donna Tellyer; Treas., Chuck Rigby.
Personnel Profile: Part-Time Volunteers 9.
Governing Authority: nonprofit. Parent Institution: Kerbyville Museum & History Center. Subsidiary Institution: Kerbyville Museum board of directors. Tax-exempt: 501(c)(3).
Institution Type/Description: History Museum: part of which is housed in 1880s Old Naucke Residence located on the site of the Old Town of Kerbyville.
Collections: military items of World War I & II; Indian artifacts; pioneer home furnishings; mining, farming, logging equipment; drugs; medicines; bottles; glassware; dishes; costumes; country store; American Express Office; old Post Office facade, picture gallery; dolls; tools; musical instruments. Historic Building: 1898 Grimmett School House.
Facilities: research library.
Activities: local tours, temporary exhibitions.
Hours & Admission Prices: mid-May to mid-Sept. Thurs.-Tues. 11-3. Adults $4, seniors $3, children 6-16 $2; discount to school groups & families; children under 6 no charge. &

Attendance: 1,000 (estimated)

Kimberly

JOHN DAY FOSSIL BEDS NATIONAL MONUMENT, 32651
Hwy. 19, Kimberly, OR 97848-6228. Tel.: 541-987-2333 (admin-
istrative). Fax: 541-987-2336. TDD: 541-987-2334.
E-mail: joda_paleontology@nps.gov
Web Site: www.nps.gov/joda
Founded: 1975.
Congressional District: 2
Key Personnel: Supt., Sheley Hall; Collections Mgr., Chris Schierup; Chief
Interpretation, Jeff Axel; Cur., Joshua Samuels; Preparator, Jennifer Cavin.
Personnel Profile: Full-Time Paid 3; Part-Time Paid 2.
Governing Authority: federal. Parent Institution: National Park Service.
Branch Museum: Visitor Information Station. Tax-exempt.
Institution Type/Description: Park Museum.
Collections: fossils from Clarno, John Day, Mascall & Rattlesnake geologic
formations of Eastern Oregon; rock samples. Historic Building: ranch house
1917-18.
Research Fields: geology; paleontology; fossil preservation; paleo-climates,
paleoenvironments of related formations.
Facilities: 500-vol. library of books, numerous research papers on geology,
paleontology, modern natural history & history of Eastern Oregon &
National Parks available for research on premises; visitor center. Books on
general geology & natural history of area for sale.
Activities: guided tours; formally organized education programs for children;
permanent exhibitions.
Hours & Admission Prices: Daily 9-5. No charge; donations accepted. Closed
federal holidays. &

Attendance: 30,000 (estimated)

Klamath Falls

FAVELL MUSEUM INC., 125 W. Main St., Klamath Falls, OR
97601-4287. Tel.: 541-882-9996.
E-mail: favellmuseum@gmail.com
Web Site: www.favellmuseum.org
Formerly: Favell Museum of Western Art and Indian Artifacts
Founded: 1972.
Congressional District: 53
Key Personnel: Dir., Janann Loetscher; Cur., Patsy McMillan; Museum Shop
Mgr., Missy Hadwick.
Personnel Profile: Full-Time Paid 1; Part-Time Paid 3.
Governing Authority: nonprofit foundation. Tax-exempt.
Institution Type/Description: Indian History & Fine Arts Museum: located on
an old Indian camp.
Collections: Indian artifacts including 60,000 arrowheads, stonework, bone-
work, pottery, beadwork & quillwork; contemporary Western art; paintings;
bronzes; dioramas; woodcarvings; 125 miniature firearms; guns; rocks &
minerals including fire opal arrowhead artifacts.
Research Fields: Oregon archaeology.
Facilities: banquet & meeting room. Museum-related items for sale.
Activities: lectures; gallery talks; school programs.
Publications: The Favell Museum: A Treasury of Our Western Heritage.
Hours & Admission Prices: Tues. Sat. 11-4. Family $22, adults $8, children
6-16 $5; members no charge.
Attendance: 4,000 (estimated)
Membership: Single $25; Family $50; Patron $100; Small Business $500;
Business $1,000; Corporate $5,000.

KLAMATH COUNTY MUSEUM, (M), 1451 Main St., Klamath
Falls, OR 97601-5989. Tel.: 541-883-4208. Fax: 541-883-5710.
E-mail: tkepple@co.klamath.or.us
Web Site: www.co.klamath.or.us/museum/index.htm
Founded: 1954.
Congressional District: 53
Key Personnel: Mgr., Todd Kepple; Cur., Lynn Jeche; Museum Shop Mgr.,
Nancy Sieverts; Museum Asst., Susan Rambo.
Personnel Profile: Full-Time Paid 2; Part-Time Paid 4; Part-Time Volunteers
20.
Governing Authority: county. Parent Institution: Klamath County, Oregon.
Branch Museums: Baldwin Hotel Museum, 31 Main St.; Ft. Klamath
Museum, Ft. Klamath, OR. Tax-exempt.
Institution Type/Description: History & Natural History Museum.
Collections: anthropology; local history; natural history.
Research Fields: pertaining to local history & items in collection.
Facilities: 7,000-vol. library of history books, maps, manuscripts, microfilms
& documents available on premises. Books & museum-related items for
sale.

Activities: tours; lectures; films; workshops; inter-museum, permanent, tem-
porary & traveling exhibitions.
Publications: research papers; book, Guardhouses, Gallows, & Graves; His-
torical coloring book.
Hours & Admission Prices: Klamath County Museum: Tues.-Sat. 9-5. Adults
13 & over $5, seniors & students $4; children 12 & under no charge.
Baldwin Hotel Museum: June-Sept. Wed.-Sat. 10-4. Adults $10. Ft.
Klamath Museum: June to Labor Day Thurs.-Mon. 10-6. No charge;
donations accepted. Closed major holidays.
Attendance: 15,000 (estimated)

SENATOR GEORGE BALDWIN HOTEL, 31 Main St., Klamath
Falls, OR 97601-3174. Mailing Address: 1451 Main St., Klamath
Falls, OR 97601-5915. Tel.: 541-883-4208. Fax: 541-883-5170.
Founded: 1954.
Congressional District: 53
Key Personnel: Mgr., Todd Kepple; Cur., Lynn Jeche.
Personnel Profile: Full-Time Paid 2; Part-Time Paid 3; Part-Time Volunteers
20.
Governing Authority: county; nonprofit.
Institution Type/Description: Historic Houses & Historic Buildings Museum:
Victorian 3 story structure built in 1906.
Collections: furnishings; structures; personal artifacts.
Facilities: 250-vol. library. Museum-related items for sale.
Activities: school loan service; monthly antique appraisals. Annual Event:
Antique Toy Show.
Hours & Admission Prices: June-Sept. Tues.-Sat. 10-4. Adults $2, seniors &
students $1; children 6 and under & members no charge.
Attendance: 4,500 (estimated)

La Grande

EASTERN OREGON FIRE MUSEUM & LEARNING CENTER,
207 Depot St., La Grande, OR 97850-2618. Tel.: 541-963-8588.
Fax: 541-963-3936.
Institution Type/Description: Fire Museum: housed in the former fire station of
downtown La Grande. Listed on the National Register of Historic Places.
Collections: firefighting history; period fire trucks; firefighting equipment;
photographs.
Hours & Admission Prices: Memorial Day to Labor Day Mon.-Fri. 9-5, Sat.
9-3; Sept.-May Mon.-Fri. 9-5; other times by appointment. No charge;
donations accepted.

Lakeview

SCHMINCK MEMORIAL MUSEUM, 128 S. E St., Lakeview, OR
97630-1721. Tel.: 541-947-3134.
E-mail: dandlschminck@centurytel.net
Founded: 1938.
Congressional District: 2
Key Personnel: Dir., Monica Lawson.
Personnel Profile: Full-Time Paid 1; Part-Time Volunteers 1.
Governing Authority: society. Administered by Oregon State Society Daugh-
ters of American Revolution. Tax-exempt.
Institution Type/Description: History Museum: housed in the former home of
pioneers, Dalph & Lula Schminck; home is a Sears Craftsmen Built house
ordered from Sears Catalog 1922.
Collections: clothing; household implements; furniture; tools; quilts; Native
American artifacts; toys; saddles; ranching implements; personal artifacts
including belonging that came over with Lula Schminck's mother on the
Applegate Wagon Train in 1846; rose garden.
Facilities: 750-vol. library; rose garden.
Activities: guided tours; permanent & temporary exhibitions; tours for classes
& related children's groups.
Publications: Tri & fold brochure.
Hours & Admission Prices: March-Oct. Wed.-Sat. 11-4. Adults & teens $4,
seniors $3; children under 13 no charge. Closed holidays.
Attendance: 800 (estimated)

Lincoln City

NORTH LINCOLN COUNTY HISTORICAL MUSEUM, 4907
S.W. Hwy. 101, Lincoln City, OR 97367-1417. Tel.: 541-996-6614.
Fax: 541-996-1244.
E-mail: mlchmdirector@gmail.com
Web Site: northlincolncountyhistoricalmuseum.org
Founded: 1987.
Congressional District: 5
Key Personnel: Dir., Anne Hall; Museum Shop Mgr., Ann Murdock.

Personnel Profile: Full-Time Paid 1; Part-Time Paid 1; Part-Time Volunteers 19.
Governing Authority: Tax-exempt.
Institution Type/Description: History Museum.
Collections: county history & culture; personal artifacts; photographs; Native American artifacts; fishing; logging; tools; period artifacts.
Research Fields: North Lincoln County.
Facilities: research library; meeting room.
Activities: historical programs. Annual Event: Pioneer Picnic.
Publications: quarterly newsletter; monograph, Lincoln City and the Twenty Miracle Miles.
Hours & Admission Prices: Feb.-May 15 & Oct. 16-Dec. 15 Wed.-Sat. 12-5; May 16-Oct. 15 Wed.-Sun. 12-5. No charge. Closed holidays. &
Attendance: 5,600 (accurate)
Membership: Individual $15; Sustaining $50; Supporter & Business $100; Benefactor $100 & up.

Marylhurst

THE ART GYM, Marylhurst University, 17600 Pacific Hwy. 43, Marylhurst, OR 97036-0261. Mailing Address: P.O. Box 261, Marylhurst, OR 97036-0261. Tel.: 503-699-6243 & 636-8141. Fax: 503-636-9526.
E-mail: artgym@marylhurst.edu
Web Site: www.marylhurst.edu
Formerly: The Art Gym at Marylhurst University
Founded: 1980.
Congressional District: 1
Key Personnel: Dir. & Cur., Terri M. Hopkins.
Personnel Profile: Part-Time Paid 3; Part-Time Volunteers 15; Interns 1.
Governing Authority: private; university; nonprofit. Parent Institution: Marylhurst University. Tax-exempt.
Institution Type/Description: University Museum.
Collections: contemporary Northwest art.
Research Fields: contemporary Northwest art & artists; art in the Northwest since 1900.
Facilities: 2,500 sq. ft. exhibit space.
Activities: guided tours; lectures; loan exhibitions; formal education programs for undergraduate & graduate students affiliated with Marylhurst University.
Publications: exhibition catalogs; exhibition announcements published six times annually.
Hours & Admission Prices: Jan.-June & Sept.-Nov. Tues.-Sun. 12-4. No charge; donations accepted. Closed on holidays. &
Attendance: 5,000 (accurate)
Membership: Artists & Students $15-$34; General $35-$74; Sponsor $75 & up; Patron $150 & up; Guarantor $250 & up; Visionary $500 & up.

McMinnville

EVERGREEN AVIATION & SPACE MUSEUM, 500 N.E. Captain Michael King Smith Way, McMinnville, OR 97128-8877. Tel.: 503-434-4180 & 4185. Fax: 503-434-4188.
E-mail: publicity@sprucegoose.org
Web Site: www.evergreenmuseum.org
Formerly: Evergreen Airventure Museum
Founded: 1992.
Congressional District: 1
Key Personnel: Exec. Dir., Larry Wood; Dir. Operations, Philip Jaeger; Dir. Public Rels. & Mktg., Kasey Richter; Chm. (V), Delford M. Smith; Dir. Special Events, Melissa Grace; Dir. Finance, Olga Mery; Cur., Stewart Bailey.
Personnel Profile: Full-Time Paid 30; Part-Time Paid 4; Part-Time Volunteers 300; Interns 4.
Governing Authority: private; nonprofit, bd. of trustees. Parent Institution: Michael King Smith Education Foundation. Tax-exempt: 501(c)(3).
Institution Type/Description: Aeronautics & Space Museum.
Collections: over 200 aircraft, spacecraft & exhibits; period commercial, military & general aviation with emphasis on warbirds (fighters, bombers, trainers, recon) restored and flown regularly; historic HK 1 Hughes Flying Boat; SR71; children's exhibits.
Research Fields: restoration techniques for period furnishings; HK-1 assembly; HK-1 disassembly & move from Long Beach, CA to McMinnville, OR.
Facilities: 232-seat digital 3D theater; over 200,000 sq. ft. exhibit space; cafe. Museum-related items for sale.
Activities: films; guided tours; hobby & restoration workshops; lectures; temporary exhibitions. Annual Events: 3D digital movies; fly-in events; military & patriotic celebrations.
Publications: quarterly newsletter.
Hours & Admission Prices: Daily 9-5. Museum: adults 17-64 $20, seniors 65 & over $19, youth 5-16 $18; members no charge. Museum & 3D Theater:

adults $27, seniors 65 & over $26, youth 5-16 $25; discounts to ASTC members; members no charge. Closed holidays. &
Attendance: 250,000 (estimated)
Membership: Individual $80; Dual $100; Family $120; Patron $145; Flight Leader $250; Squadron Commander $500; Group Commander $1,000; Wing Commander $2,500.

Medford

KID TIME!, 106 N. Central Ave., Medford, OR 97501. Tel.: 541-772-9922. Fax: 541-773-5713.
Web Site: www.kid-time.org
Formerly: Kids' Imagination Discovery Space
Institution Type/Description: Children's Museum.
Collections: hands-on exhibits.
Activities: birthday parties; special events.
Hours & Admission Prices: Mon.-Sat. 10-5, Sun. 12-5. Children $6, adults $3; children under 1 & members no charge.
Membership: Family: 3 month $40; 6 month $65; 1 Year $95.

ROGUE GALLERY & ART CENTER, 40 S. Bartlett, Medford, OR 97501-7216. Tel.: 541-772-8118. Fax: 541-772-0294.
E-mail: director@roguegallery.org
Web Site: www.roguegallery.org
Founded: 1959.
Congressional District: 2
Key Personnel: Exec. Dir., Kim Hearon; Administrative Asst., Rachel Barrett; Dir. Education, Brooke Nuckles Genteko.
Personnel Profile: Full-Time Paid 1; Full-Time Volunteers 2; Part-Time Paid 2; Part-Time Volunteers 65; Interns 3.
Governing Authority: nonprofit organization. Parent Institution: Rogue Valley Art Association. Tax-exempt: 501(c)(3).
Institution Type/Description: Visual Art Center & Art Gallery.
Collections: regional prints, printmakers, Oregon sculpture; regional paintings & artworks.
Facilities: classroom; boardroom. Museum-related items for sale.
Activities: guided tours; lectures; films; gallery talks; rental gallery; formally organized education programs for children; docent program or council; temporary & traveling exhibitions.
Publications: bimonthly newsletters.
Hours & Admission Prices: Tues.-Fri. 10-5, Sat. 11-3. No charge; donations accepted. Closed national holidays. &
Attendance: 20,000 (estimated)
Membership: Youth (under 21) $25; Individual $35, Family & Dual $50; Patron $100; Sustainer $250; Benefactor $500. Business $100-$1,000.

SOUTHERN OREGON HISTORICAL SOCIETY, 106 N. Central Ave., Medford, OR 97501. Tel.: 541-773-6536. Fax: 541-776-7994.
E-mail: communicate@sohs.org
Web Site: www.sohs.org
Founded: 1946.
Congressional District: 2
Key Personnel: Exec. Dir., John Enders; Dir. Finance & Operations, Maureen Smith; Education & Programs Coord., Stephanie Butler; Devel. Coord., Richard Seidman; Cur. Collections, Suzanne M.M. Warner; Library Mgr., Carol Samuelson.
Personnel Profile: Full-Time Paid 17; Part-Time Paid 10; Part-Time Volunteers 250.
Governing Authority: society. Parent Institution: Southern Oregon Historical Society, Inc. Branch Institutions: Research Center, Medford; Jacksonville Museum of Southern Oregon History, Children's Museum, Beekman House, Beekman Bank, Catholic Rectory, U.S. Hotel, History Store, Jacksonville; Hanley Farm, Central Point. Tax-exempt: 501(c)(3).
Institution Type/Description: History Museum & Historic Sites.
Collections: historical artifacts; textiles; clothing; Peter Britt collection; 19th- & 20th-century photographic equipment; furnishings; firearms; Dorland Robinson artwork; tools & equipment from regional businesses & industries; manuscripts; maps; oral histories; photographs. Historic Buildings: 1883 Jackson County courthouse; 1910 Children's Museum; 1873 CC Beekman house; 1863 Beekman bank; 1860s Hanley Farm.
Research Fields: local & southwestern Oregon history.
Facilities: library of books on local & Southwestern Oregon history; meeting rooms. Museum-related items for sale.
Activities: tours of permanent & temporary exhibits; walking, cemetery tours; adult, family, youth education programs; educational outreach programs; living history program.
Publications: quarterly popular history magazine, Southern Oregon Heritage Today; A Century of the Photographic Arts in Southern Oregon: A Directory

of Jackson County Photographers, 1856-1956; brochures, Historic Discovery Drives; Spirit of Ashland.
Hours & Admission Prices: Call 541-773-6536 for hours & admission prices. Members no charge. Closed New Year's Day; Thanksgiving; Christmas. ♿
Attendance: 269,867 (accurate)
Membership: Individual $35; Family $50; Patron $100; Curator $200; Business $250; Director $500; Historian's Circle $1,000; Lifetime $2,500.

Milton Freewater

FRAZIER FARMSTEAD MUSEUM, 1403 Chestnut St., Milton Freewater, OR 97862. Tel.: 541-938-4636 & 3480.
Governing Authority: Parent Institution: Milton Freewater Area Historical Society.
Institution Type/Description: Historic Farmstead: housed in the former home of community founder, W.S. Frazier; built in 1892. Listed on the National Register of Historic Places.
Collections: Frazier family history; personal artifacts; period furnishings; photographs. Historic Structure: 1918 barn.
Activities: rental facilities.
Hours & Admission Prices: April-Dec. Thurs.-Sat. 11-4; other times by appointment. No charge; donations accepted.

Milwaukie

MILWAUKIE MUSEUM, 3737 S.E. Adams St., Milwaukie, OR 97222-5917. Tel.: 503-659-5780.
E-mail: tutumom@msn.com
Web Site: milwaukiemuseum.tripod.com
Founded: 1975.
Key Personnel: Pres. (V), Adele Wilder.
Personnel Profile: Full-Time Volunteers 1; Part-Time Volunteers 15.
Governing Authority: Parent Institution: Milwaukie Historical Society. Tax-exempt.
Institution Type/Description: Historic House Museum: housed in the former home of the George Wise family, built in 1865.
Collections: period furnishings; personal artifacts.
Publications: Milwaukie history.
Hours & Admission Prices: Sat. 1-5. No charge; donations accepted. Closed Easter; Christmas. ♿
Attendance: 650 (estimated)
Membership: Individual $15; Lifetime $150.

Moro

SHERMAN COUNTY HISTORICAL MUSEUM, (M), 200 Dewey St., Moro, OR 97039. Mailing Address: P.O. Box 173, Moro, OR 97039. Tel.: 541-565-3232.
E-mail: info@shermanmuseum.org
Web Site: www.shermanmuseum.org
Founded: 1983.
Personnel Profile: Full-Time Paid 1; Part-Time Volunteers 75.
Governing Authority: Parent Institution: Sherman County Historical Society. Tax-exempt.
Institution Type/Description: Historical Society Museum.
Collections: local history & culture; photographs; personal artifacts; Native American artifacts; farm machinery; household items; printing press.
Publications: biannual historical anthology, Sherman County: For the Record; biannual newsletter, The Plow.
Hours & Admission Prices: May-Oct. daily 10-5. Adults $3; discounts to groups. ♿
Attendance: 3,050 (accurate)
Membership: Household $30.

Myrtle Point

COOS COUNTY LOGGING MUSEUM, 705 Maple St., Myrtle Point, OR 97458. Mailing Address: P.O. Box 325, Myrtle Point, OR 97458-0325. Tel.: 541-572-1014.
Key Personnel: Pres. (V), Chas. King.
Governing Authority: nonprofit organization. Tax-exempt.
Institution Type/Description: Logging Museum.
Collections: period logging equipment; chain saws; spring boards; rigging equipment; axes; railroad artifacts; photographs; myrtlewood carvings.
Facilities: Museum-related items for sale.
Hours & Admission Prices: Mon.-Sat. 10-4, Sun. 1-4. No charge; donations accepted. ♿
Attendance: 4,800 (estimated)

Newberg

HOOVER-MINTHORN HOUSE MUSEUM, 115 S. River St., Newberg, OR 97132-3153. Tel.: 503-538-6629.
E-mail: hooverminthornhousemuseum@gmail.org
Web Site: www.hooverminthornhousemuseum.com
Founded: 1955.
Congressional District: 1
Key Personnel: Dir., Sarah B. Munro; Chm. (V), Elizabeth Lilley; Pres. (V), Anita Barbey.
Personnel Profile: Full-Time Volunteers 2; Part-Time Volunteers 5; Interns 3.
Governing Authority: Parent Institution: The National Society of The Colonial Dames of America in the State of Oregon. Tax-exempt.
Institution Type/Description: Historic House Museum: housed in the boyhood home of President Herbert Hoover, 1885-1889; built in 1881. Listed on the National Register of Historic Places.
Collections: local & family history; Hoover's bedroom furniture; personal artifacts; period furnishings; photographs.
Major Exhibits: All the Live Long Day: Work in the Valley, 1/14-3/14.
Activities: Herbert Hoover's 140 birthday celebration.
Hours & Admission Prices: March-Nov. Wed.-Sun. 1-4; Dec.-Feb. Sat.-Sun. 1-4. Closed holidays & Jan.
Attendance: 1,200 (estimated)

Newport

LINCOLN COUNTY HISTORICAL SOCIETY, 545 S.W. Ninth St., Newport, OR 97365-4726. Tel.: 541-265-7509. Fax: 541-265-3992. Facebook: Lincoln County Historical Society Newport Oregon.
E-mail: coasthistory@newportnet.com
Web Site: www.oregoncoasthistory.org
Formerly: Oregon Coast History Center
Founded: 1948.
Congressional District: 1
Key Personnel: Dir., Steve Wyatt; Pres. (V), Robert Olson; Museum Shop Mgr., Brenda Baker.
Personnel Profile: Full-Time Paid 2; Part-Time Paid 7; Part-Time Volunteers 12; Interns 1.
Governing Authority: society; bd. of directors. Parent Institution: Lincoln County Historical Society. Affiliated with Oregon Historical Society. Tax-exempt: 501(c)(3).
Institution Type/Description: History Museum.
Collections: Pacific Maritime & Heritage Center: maritime art & history; fishing tourism; surfing. Burrows House: county history including Siletz Reservtion artifacts. Log Cabin: archival research center.
Major Exhibits: Cars: Motoring the Coast, 1/13-12/14.
Research Fields: Oregon coast history; Lincoln County history; Siletz Reservation & Tribe; biographies & genealogies of local inhabitants & families; local industries & railroads.
Facilities: 150-vol. library of Oregon historical publications & books; reading room. Research Center: photographs, publications, manuscripts, & family history.
Activities: guided tours; gallery talks; permanent, temporary & traveling exhibitions.
Publications: books, The Land That Kept Its Promise; Pacific Spruce Corp.; Pictorial History of Otter Rock; Steam Towards The Sunset; Siletz Indian Reservation 1855-1900; School District #61; At Rest in Lincoln County; Yaquina Bay 1778-1978; Lincoln County Anthology; Pictorial Toledo, OR; reprint of 1923 Pacific Spruce Corp. & Subsidiaries; Lincoln County Kitchen Memories; Pathfinder: The First Automobile Trip from Newport to Siletz Bay; Siletz-Survival for an Artifact; When Time Seemed to Pause; Tragedy on Yaquina Bar; The Voyage of the Prairie Schooner; book, The Bay Front; DVDs, The Road from Mud to Glory; Rooms With A View; Yaquina Rails: From Tree to Sea.
Hours & Admission Prices: Pacific Maritime & Heritage Center and Burrows House: Thurs.-Sun. 11-4. Center: adults $5, children under 12 $3. Burrows House: by donation. Closed New Year's Day; Independence Day; Thanksgiving; Christmas. ♿
Attendance: 15,000 (accurate)
Membership: Pioneer $20 & up; Family $30 & up; Organization & Business $30 & up; Patron $100 & up; Corporate & Sponsor $250 & up; Pathfinders Circle $500 & up.

NEWPORT VISUAL ARTS CENTER, 777 N.W. Beach Dr., Newport, OR 97365. Tel.: 541-265-6540.
E-mail: shouck@coastarts.org
Web Site: www.coastarts.org
Institution Type/Description: Art Gallery.
Collections: paintings; sculpture.

Activities: educational programs; classes; workshops; rental facilities.
Hours & Admission Prices: Runyan Gallery: Tues.-Sun. 11-6. Upstairs Gallery: Tues.-Sat. 12-4. No charge.

OREGON COAST AQUARIUM, 2820 S.E. Ferry Slip Rd., Newport, OR 97365-5269. Tel.: 541-867-3474. Fax: 541-867-6846.
E-mail: info@aquarium.org
Web Site: www.aquarium.org
Founded: 1992.
Congressional District: 5
Key Personnel: Pres. & C.E.O., Carrie E. Lewis; Chm. Bd., Brent Denham.
Personnel Profile: Full-Time Paid 67; Full-Time Volunteers 350; Part-Time Paid 3; Interns 3.
Governing Authority: Tax-exempt.
Institution Type/Description: Aquarium.
Collections: sea otters; seals & sea lions; sea bird aviary.
Research Fields: rockfish reproduction; non-native species monitoring.
Publications: biannual, Waterlines.
Hours & Admission Prices: Memorial Day to Labor Day daily 9-6; Sept.-May daily 10-5. Adult 18-64 $18.95, seniors 65 & over $16.95, young adult 13-17 $16.95; children 3-12 $12.95; children 2 & under no charge. &
Attendance: 448,579 (accurate)
Membership: Individual $40; One Tone $65; Dual Plus One $75; Family $110; Family Plus $150; Sponsor $300; Patron $600.

YAQUINA BAY LIGHTHOUSE, 842 Government St., Newport, OR 97365. Mailing Address: Friends of Yaquina Lighthouses, P.O. Box 410, Newport, OR 97365. Tel.: 541-270-0131.
Web Site: www.yaquinalights.org
Institution Type/Description: Historic Lighthouse: built in 1871. Listed on the National Register of Historic Places.
Collections: lighthouse & area history; period furnishings.
Activities: tours; rental facilities.
Hours & Admission Prices: Summer: daily 11-5; Winter: daily 12-4. No charge; donations accepted.

YAQUINA HEAD LIGHTHOUSE AND INTERPRETIVE CENTER, 750 Lighthouse Dr. #7, Newport, OR 97365. Tel.: 541-574-3129.
Web Site: www.yaquinalights.org
Institution Type/Description: Historic Lighthouse & Museum.
Collections: lighthouse history; local & natural history; photographs; period artifacts. Historic Building: 93 ft. lighthouse built in 1872.
Activities: lighthouse tours.
Hours & Admission Prices: Center: Summer: 9-5; Winter 10-4. Lighthouse: call for hours.

North Bend

COOS HISTORICAL & MARITIME MUSEUM, 1220 Sherman, North Bend, OR 97459-3666. Tel.: 541-756-6320. Fax: 541-756-6320.
E-mail: info@cooshistory.org
Web Site: www.cooshistory.org
Formerly: Coos County Historical Society Museum
Founded: 1891.
Congressional District: 47
Key Personnel: C.E.O. & Dir., Annie Donnelly; Pres., Steve Greif; Pres. (V), Jennifer Groth.
Personnel Profile: Full-Time Paid 2; Part-Time Paid 2; Part-Time Volunteers 50.
Governing Authority: nonprofit organization. Owned & operated by Coos County Historical Society. Tax-exempt: 501(c)(3).
Institution Type/Description: General History Museum.
Collections: artifacts relating to the tidewater highways of Coos County & Southwestern Coastal Oregon; the Fahy collection of Indian artifacts; the Magee collection of international artifacts; steam donkey engine used for logging in Coos County; historic photographs & negatives
Research Fields: Coos County history.
Facilities: 1,100-vol. library of regional print and manuscript material, photographs & maps.
Activities: permanent & temporary exhibits; educational programs; book signings; research assistance; group tours.
Publications: Waterways newsletter quarterly; books, A Century of Coos and Curry; Coos Bay Region.
Hours & Admission Prices: Tues.-Sat. 10-4. Adults $4, senior citizens & students $2; members no charge. Closed Thanksgiving; Christmas. &
Attendance: 3,800 (accurate)

Membership: Individual $25; Family $35; Business & Corporate $100; Benefactor $250.

Nyssa

OREGON TRAIL AGRICULTURAL MUSEUM, 117 Good Ave., Nyssa, OR 97913-3833. Mailing Address: P.O. Box 2303, Nyssa, OR 97913-0303. Tel.: 541-372-3712.
Key Personnel: Chm. (V), Harry Flock; Chm. (V), Betty Holcomb; Dir., Nancy DeBoer
Operating Expenses: 1,067
Operating Income: 76
Governing Authority: Parent Institution: Nyssa Historical Society, Inc. Tax-exempt.
Institution Type/Description: Agriculture Museum.
Collections: local history; agricultural heritage; early farm & ranch equipment; restored sheep wagons; photographs; Oregon Trail history.
Activities: demonstrations.
Hours & Admission Prices: Summer: Sat. 10-4, Sun. 1-4. No charge; donations accepted. &
Attendance: 129 (accurate)
Membership: Individual $5; Couple $7.50; Family $10; Business & Organization $25.

Oakland

OAKLAND MUSEUM, 130 Locust St., Oakland, OR 97462. Mailing Address: P.O. Box 624, Oakland, OR 97462-0624. Tel.: 541-459-3087.
Web Site: www.historicoaklandoregon.com
Founded: 1969.
Key Personnel: Dir., Louise J. Stearns.
Personnel Profile: Part-Time Volunteers 30.
Governing Authority: Tax-exempt.
Institution Type/Description: History Museum.
Collections: historical artifacts & memorabilia pertaining to the city of Oakland & surrounding area.
Hours & Admission Prices: Daily 12:30-3:30. No charge; donations accepted.
Attendance: 2,535 (accurate)
Membership: $10 & up.

Oakridge

OAKRIDGE-WESTFIR PIONEER MUSEUM, 76433 Pine St., Oakridge, OR 97463. Mailing Address: P.O. Box 807, Oakridge, OR 97463-0807. Tel.: 541-782-2402.
Institution Type/Description: History Museum.
Collections: local history & culture; photographs; period artifacts.
Hours & Admission Prices: Call for hours.

Ontario

FOUR RIVERS CULTURAL CENTER & MUSEUM, 676 S.W. 5th Ave., Ontario, OR 97914-3436. Tel.: 541-889-8191. Fax: 541-889-7628.
E-mail: info@4rcc.com
Web Site: www.4rcc.com
Founded: 1987.
Key Personnel: Chm. (V), John Gaskill; Pres. (V), Nancy Bent; Exec. Dir., Matthew Stringer; Museum Shop Mgr., Shawn Maggar.
Personnel Profile: Full-Time Paid 1; Part-Time Paid 5; Part-Time Volunteers 60; Interns 3.
Governing Authority: private; nonprofit organization. Tax-exempt: 501(c)(3).
Institution Type/Description: History Museum.
Collections: historical artifacts; clothing; first settlers to the eastern Oregon area including Basque, Northern Paiutes, Japanese-American, Hispanic & Euro American.
Research Fields: Western Treasure Valley (east Oregon/southwest Idaho) history with special emphasis on the history of local Japanese-Americans, Basque-Americans, Hispanic Americans, Native Americans & other cultural and ethnic groups represented in the area & the relationship of these people to the land over time; agriculture.
Facilities: botanical garden; 542-seat theater, 16,000 sq. ft. exhibit space; conference center. Museum-related items for sale.
Activities: lectures; temporary exhibits; classes.
Publications: quarterly newsletter, Four Rivers Cultural Center; Shining Water News.
Hours & Admission Prices: Mon.-Sat. 10-5. Adults $4, seniors $3; discounts to groups over 20; members no charge. &

Attendance: 7,693 (accurate)
Membership: Students $10; Individual $30; Family $50; Contributor $125; Corporate $275.

Oregon City

MCLOUGHLIN HOUSE, McLoughlin Park, 713 Center St., Oregon City, OR 97045-1948. Tel.: 503-656-5146.
Founded: 1909.
Governing Authority: Parent Institution: National Park Service.
Institution Type/Description: Historic House Museum: housed in the former home of Dr. John McLoughlin, Chief Factor of the Hudson Bay Company; built in 1846. Listed on the National Register of Historic Places.
Collections: family & local history; period furnishings; photographs; personal artifacts.
Activities: special events; demonstrations.
Hours & Admission Prices: Feb.-Dec. Wed.-Sat. 10-4. No charge.

MUSEUM OF THE OREGON TERRITORY - CLACKAMAS COUNTY HISTORICAL SOCIETY, (M), 211 Tumwater Dr., Oregon City, OR 97045-2900. Mailing Address: P.O. Box 2211, Oregon City, OR 97045-0054. Tel.: 503-655-5574. Fax: 503-655-0035.
Governing Authority: Subsidiary Institution: Stevens-Crawford Historical House, 603 6th St., Oregon City, OR 97045.
Institution Type/Description: Oregon Territory Museum.
Collections: local history & culture; photographs; Lady Justice statue. Historic House: 1908 Stevens-Crawford house.
Facilities: rental facilities.
Hours & Admission Prices: Daily 11-4. Adults $7, children 5-17 $5; children under 5 no charge.

ROSE FARM MUSEUM, Holmes Ln. at Rilance St., Oregon City, OR 97045. Mailing Address: 713 Center St., Oregon City, OR 97045. Tel.: 503-656-5146.
Institution Type/Description: Historic House Museum: housed in the former home of William & Louisa Holmes; built in 1847.
Collections: local & family history; furnishings; photographs.
Hours & Admission Prices: May-Sept. Sat. 10-4, Sun. 1-5. Adults $3, seniors & youth 6-17 $2; children 6 & under no charge.

STEVENS CRAWFORD MUSEUM, 603 6th St., Oregon City, OR 97045-2232. Tel.: 503-655-2866.
Institution Type/Description: Historic House Museum: built in 1908.
Collections: family & local history; period furnishings & toys; photographs.
Hours & Admission Prices: Wed.-Fri. 12-4, Sat. 1-4. Adults $4, seniors $3, students $2; children under 5 no charge.

Parkdale

INTERNATIONAL MUSEUM OF CAROUSEL ART, 4976 Alexander Dr., Parkdale, OR 97041-7604. Mailing Address: P.O. Box 1522, Hood River, OR 97031. Tel.: 541-352-7663. Fax: 541-387-8797.
E-mail: osperron@gmail.com
Web Site: www.carouselmuseum.com
Founded: 1982.
Key Personnel: Dir., Mark Reed; Chm. (V), Duane Perron.
Governing Authority: nonprofit organization. Tax-exempt: 501(c)(3).
Institution Type/Description: General Museum.
Collections: period carousels; carousel art & history.
Activities: classes.
Hours & Admission Prices: Temporarily closed. Adults $5, students & seniors $4, children 5-10 $2; members and children 4 & under no charge. &

Pendleton

CHILDREN'S MUSEUM OF EASTERN OREGON, 400 S. Main St., Pendleton, OR 97801-2248. Tel.: 541-276-1066.
Web Site: www.cmeo.org
Founded: 1996.
Key Personnel: Exec. Dir., Kim Talarico; Pres., Sarah Haug
Institution Type/Description: Children's Museum.
Collections: hands-on exhibits.
Activities: birthday parties; special events; educational programs; classes.
Hours & Admission Prices: Tues.-Sat. 10-5; groups by appointment. Admission $3; discounts to groups of 15 or more; children under one no charge.
Membership: Family $50; Family + 2 $80.

HERITAGE STATION, UMATILLA COUNTY HISTORICAL SOCIETY MUSEUM, (M), 108 S.W. Frazer, Pendleton, OR 97801-2138. Mailing Address: P.O. Box 253, Pendleton, OR 97801-0253. Tel.: 541-276-0012. Fax: 541-276-7989. Facebook: Heritage Station.
E-mail: info@heritagestationmuseum.org
Web Site: www.heritagestationmuseum.org
Founded: 1974.
Congressional District: 2
Key Personnel: Exec. Dir., Barbara Lund-Jones; Pres., Keith May; Treas., Ryan Levy; Museum Shop Mgr., Rebecca Lynn Foster.
Personnel Profile: Full-Time Paid 2; Part-Time Volunteers 75; Interns 1.
Governing Authority: private; nonprofit. Parent Institution: Umatilla County Historical Society. Tax-exempt: 501(c)(3).
Institution Type/Description: Historical Society Museum: located in a 1909 train depot.
Collections: prehistory & history of Umatilla County & its population; refurbished 1879 one-room school; hand-hewn log cabin.
Research Fields: Umatilla County history.
Facilities: library; 4,200 sq. ft. exhibit space. Museum-related items for sale.
Activities: guided tours; lectures; temporary & traveling exhibitions. Museum Sponsors: spring & fall series of informal educational events.
Publications: triannual periodical, Pioneer Trails; biannual, The UCHS Newsletter.
Hours & Admission Prices: Call for hours. Family $10, adults $5, seniors $4, students $2; discount to AAM members & groups; children under 5 & members no charge. Closed New Year's Day; Thanksgiving; Christmas. &
Attendance: 5,500 (estimated)
Membership: Individual $40; Family $50; Business $75.

PENDLETON CENTER FOR THE ARTS, 214 N. Main St., Pendleton, OR 97801. Tel.: 541-278-9201.
E-mail: info@pendletonarts.org
Web Site: www.pendletonarts.org
Key Personnel: Exec. Dir., Roberta Lavadour
Institution Type/Description: Art Gallery: housed in the former Carnegie Library; built in 1916.
Collections: paintings; sculpture; photographs.
Activities: classes; rental facilities.
Hours & Admission Prices: Call for hours.

PENDLETON ROUND-UP AND HAPPY CANYON HALL OF FAME, 1114 S.W. Court Ave., Pendleton, OR 97801. Mailing Address: 1205 S.W. Court Ave., Pendleton, OR 97801. Tel.: 541-278-0815. Fax: 541-276-9776.
Founded: 1969.
Institution Type/Description: History Museum & Hall of Fame.
Collections: Round-Up history; photographs; personal artifacts; clothing; Hall of Fame inductees; firearms; trophies; wagons.
Hours & Admission Prices: Sat. 10-4. Adults $5, seniors $4, children 10 & under $2.

TAMASTSLIKT CULTURAL INSTITUTE, (M), 47106 Wildhorse Blvd., Pendleton, OR 97801-3379. Tel.: 541-429-7700. Fax: 541-429-7709. Facebook: Tamastslikt Cultural Institute.
E-mail: bobbie.conner@tamastslikt.org
Web Site: www.tamastslikt.org
Founded: 1998.
Congressional District: 2
Key Personnel: Dir., Roberta Conner; Education, Susan Sheoships; Mktg., Michelle Liberty; Registrar & Cur., Randall Melton; Mgr. Research & Collections, Malissa Minthorn Winks; Museum Shop Mgr., Hilda Alexander.
Personnel Profile: Full-Time Paid 20; Part-Time Paid 10.
Governing Authority: Tribal Government. Parent Institution: Confederated Tribes of the Umatilla Indian Reservation. Tax-exempt.
Institution Type/Description: History Museum.
Collections: history & culture of American Indian Tribes, the Cayuse, Umatilla & Walla Walla; their relationship with Lewis & Clark, early fur traders, missionaries, Oregon Trail immigrants & modern members of the surrounding areas; artifacts; photographs; video & interactive multimedia.
Facilities: 14,000 sq. ft. exhibit space. Museum-related items for sale.
Activities: Living Culture Village: special events in summer.
Publications: As Days Go By; Our History, Our Land, and Our People; The Cayuse, Umatilla and Walla Walla.
Hours & Admission Prices: April-Sept. daily 9-5; Oct.-March Mon.-Sat. 9-5. Adults $8; discounts to AAM & AAA members; Blue Star families no charge. Closed New Year's Day; Thanksgiving; Christmas. &

Attendance: 34,000 (estimated)
Membership: Friend $50; Advocate $75; Patron $150; Sponsor $250; Benefactor $500.

Philomath

BENTON COUNTY HISTORICAL MUSEUM, (M), 1101 Main St., Philomath, OR 97370. Mailing Address: P.O. Box 35, Philomath, OR 97370-0035. Tel.: 541-929-6230. Fax: 541-929-6261.
E-mail: info@bentoncountymuseum.org
Web Site: www.bentoncountymuseum.org
Founded: 1980.
Congressional District: 5
Key Personnel: Exec. Dir., Irene Zenev.
Personnel Profile: Full-Time Paid 3; Part-Time Paid 5; Part-Time Volunteers 60; Interns 3.
Volunteer Hours: 4,000
Operating Expenses: 395,000
Operating Income: 397,000
Governing Authority: nonprofit organization. Parent Institution: Benton County Historical Society. Tax-exempt: 501(c)(3).
Institution Type/Description: History Museum & Art Gallery: housed in c.1867 brick building, originally used as the Philomath College building.
Collections: local history & culture; photographs; archives.
Major Exhibits: Cool Tools, 11/13-11/14.
Research Fields: local history.
Facilities: 300-vol. library of local history; 100-seat auditorium, museum store; 13,500 sq. ft. collections storage facility.
Activities: lectures; organized education programs for children & adults; participatory, loan & traveling exhibitions; temporary exhibitions of your own collections; docent guild in class presentation.
Publications: quarterly newsletter, The Society Record.
Hours & Admission Prices: Philomath: Tues.-Sat. 10-4:30. No charge; donations accepted. Closed major holidays. &

Attendance: 8,000 (estimated)
Membership: Individual $25; Family $40; Patron $100; Benefactor $375; Sustainer $500; History-Maker $1,000. (Seniors 65 & over receive 10% discount on any level). Business: Basic $200; Corporate $1,000.

Port Orford

CAPE BLANCO HERITAGE SOCIETY, 92331 Coast Guard Hill Rd., Port Orford, OR 97465. Mailing Address: Cape Blanco Heritage Society, P.O. Box 1132, Port Orford, OR 97465-1132. Tel.: 541-332-0521.
E-mail: president@capeblancoheritagesociety.com
Web Site: www.capeblancoheritagesociety.com
Formerly: Point Orford Heritage Society
Founded: 2000.
Congressional District: 4
Key Personnel: Pres., Dave Holman; Museum Shop Mgr., Steve Roemen.
Personnel Profile: Part-Time Volunteers 12.
Governing Authority: Tax-exempt.
Institution Type/Description: Coast Guard History Museum.
Collections: local history; Coast Guard artifacts; photographs; model ships; newspapers; nautical artifacts.
Publications: biannual newsletter.
Hours & Admission Prices: April-Oct. Wed.-Mon. 10-3:30; other times by appointment. No charge; donations accepted.
Attendance: 3,400 (estimated)
Membership: Individual $15; Family $25; Business & Associate $50; Benefactor $100; Lifetime $1,000.

Portland

BLUE SKY, THE OREGON CENTER FOR THE PHOTOGRAPHIC ARTS, 122 N.W. 8th Ave., Portland, OR 97209. Tel.: 503-225-0210. Fax: 503-225-2990.
E-mail: bluesky@blueskygallery.org
Web Site: www.blueskygallery.org
Founded: 1975.
Congressional District: 1
Key Personnel: Exec. Dir., Todd J. Tubutis.
Personnel Profile: Full-Time Paid 3; Part-Time Volunteers 50.
Governing Authority: Tax-exempt.
Institution Type/Description: Art Gallery.
Collections: works by local, national & international photographic artists.
Facilities: resource library.
Activities: educational programs.
Publications: annual catalogue, Yearbook.

Hours & Admission Prices: Tues.-Sun. 12-5; groups of 12 or more by appointment. No charge; donations accepted. Closed New Year's Day; Independence Day; Thanksgiving; Christmas. &
Attendance: 20,000 (estimated)
Membership: Basic $40; Book Level $75; Print Level 1 $150; Print Level 2 $275; Print Level 3 $500.

THE DOUGLAS F. COOLEY MEMORIAL ART GALLERY, REED COLLEGE, (M), 3203 S.E. Woodstock Blvd., Portland, OR 97202-8138. Tel.: 503-771-1112 & 7251. Fax: 503-788-6691 (Reed College).
E-mail: snyders@reed.edu
Web Site: web.reed.edu/gallery
Founded: 1989.
Congressional District: 3
Key Personnel: Dir. & Cur., Stephanie Snyder; Registrar & Program Coord., Colleen Gotze; Coord. Education Outreach, Gregory MacNaughton.
Personnel Profile: Full-Time Paid 2; Part-Time Paid 1.
Governing Authority: college. Parent Institution: Reed College. Tax-exempt: 501(c)(3).
Institution Type/Description: Exhibition Gallery.
Collections: 20th-century American, 19th-century European & other works on paper.
Activities: lectures; gallery talks; symposia; temporary & traveling exhibitions.
Publications: exhibition catalog, The Serpentine Lattice; Ree Morton; Modern Art in America; Documenting a Myth; Robert Adams; What is a Man; Bruce Nauman: Going Solo; David Reed: Lives of Paintings; ABSTRACT; Lloyd Reynolds: A Life of Forms in Art.
Hours & Admission Prices: Sept.-June Tues.-Sun. 12-5. No charge. &
Attendance: 5,500
Membership: Cooley Gallery Art Assoc. $100; $250.

HELLENIC AMERICAN CULTURAL CENTER & MUSEUM, (M), 3131 N.E. Glisan, Portland, OR 97232-2501. Tel.: 503-858-8567.
E-mail: haccmpdx@gmail.com
Web Site: www.hellenicamericancc.org
Founded: 2006.
Key Personnel: Chm. (V), Stefanos Vertopoulos; Pres. (V), Alexandra Andronikos; Administrative Asst., Katherine Karafotias.
Personnel Profile: Part-Time Paid 1; Part-Time Volunteers 70; Interns 1.
Governing Authority: nonprofit organization. Tax-exempt: 501(c)(3).
Institution Type/Description: Cultural Center.
Collections: Hellenic (Greek) history & culture in Oregon & SW Washington; photographs; personal artifacts.
Publications: quarterly, Museletter.
Hours & Admission Prices: Call for hours. No charge; donations accepted. &
Attendance: 1,500 (estimated)

HOYT ARBORETUM, 4000 S.W. Fairview Blvd., Portland, OR 97221-2706. Tel.: 503-865-8733. Fax: 508-823-4213.
E-mail: info@hoytarboretum.org
Web Site: www.hoytarboretum.org
Founded: 1928.
Congressional District: 1
Key Personnel: Exec. Dir. Hoyt Arboretum Friends Foundation, Peggie Schwarz; Pres. Bd., David Ellis; City Nature West Natural Resources Supvr. Portland Parks & Recreation, James Allison.
Personnel Profile: Full-Time Paid 2; Part-Time Paid 1; Part-Time Volunteers 250; Interns 1.
Governing Authority: municipal. Parent Institution: City of Portland. Subsidiary Institution: Hoyt Arboretum Friends Foundation. Tax-exempt.
Institution Type/Description: Arboretum.
Collections: 1,100 species of trees & shrubs.
Research Fields: taxonomy; plant systematics; arboriculture.
Facilities: library; self-guided trails. Museum-related items for sale.
Activities: guided & self-guided tours; lectures; workshops & education sessions; international seed exchange.
Publications: trail guides; maps; educational brochures.
Hours & Admission Prices: Park: daily 6am-10pm. Visitor Center: Mon.-Fri. 9-4, Sat. 11-3. No charge; donations accepted. &
Attendance: 350,000 (estimated)
Membership: K9 $25; Maple $35; Oak $50; Fir $100; Duncan Society $1,000.

LITTMAN & WHITE GALLERIES, 1825 S.W. Broadway #250, Portland, OR 97201-3256. Mailing Address: P.O. Box 751-SD, Portland, OR 97207. Tel.: 503-725-5656. Fax: 503-725-5680.
E-mail: artcom@pdx.edu
Formerly: Portland State University Galleries

Founded: 1976.
Key Personnel: C.E.O. & Coord., Emilie Gerber; Coord., Theresa Tate.
Personnel Profile: Full-Time Paid 2.
Institution Type/Description: Art Gallery.
Collections: original prints; paintings; sculpture.
Facilities: Littman Gallery: 1,300 sq. ft. exhibit space.
Activities: lectures; gallery talks; loan exhibitions.
Hours & Admission Prices: Littman Gallery: Mon.-Fri. 12-4. No charge. White Gallery: Mon.-Fri. 8-10, Sat. 9-7. No Charge. &

MUSEUM OF CONTEMPORARY CRAFT, (M), 724 Northwest Davis St., Portland, OR 97209-3663. Tel.: 503-223-2654. Fax: 503-223-0190. Facebook: Museum of Contemporary Craft.
E-mail: info@museumofcontemporarycraft.org
Web Site: www.museumofcontemporarycraft.org
Formerly: Contemporary Crafts Museum & Gallery
Founded: 1937.
Congressional District: 5
Key Personnel: Cur., Namita Gupta Wiggers.
Personnel Profile: Full-Time Paid 5; Part-Time Paid 3; Part-Time Volunteers 65; Interns 12.
Governing Authority: private; nonprofit corporation. Tax-exempt.
Institution Type/Description: Contemporary Craft Museum.
Collections: 20th century arts & crafts; wood; clay; glass; metal; textiles.
Research Fields: craft; museology; contemporary practice.
Facilities: library; event facilities. Museum-related items for sale.
Activities: guided tours; artist & curator talks; panel discussions; films; workshops; hands-on art activities; special events.
Publications: Unpacking the Collection: Selections from the Museum of Contemporary Craft; exhibition catalogues.
Hours & Admission Prices: Tues.-Wed. & Fri.-Sat. 11-6, 1st Thurs. of month 11-8. Adults $4, members no charge; discounts to NARM. Closed New Year's Day; Independence Day; Thanksgiving; Christmas. &
Attendance: 75,000 (accurate)
Membership: Artist, Student & Senior $35; Individual $40; Household $65; Friend $125; Supporter $250; Sustainer $500; Patron $1,000; Benefactor $2,500.

THE NATIONAL HAT MUSEUM, 1928 S.E. Ladd Ave., Portland, OR 97214-4737. Tel.: 503-232-0433.
E-mail: justalyce@usa.net
Web Site: www.thehatmuseum.com
Founded: 2005.
Key Personnel: Dir., Alyce Cornyn-Selby; Museum Shop Mgr., Gracie Zusman
Institution Type/Description: Hat Museum: housed in the Ladd-Reingold House; built in 1910. Listed on the National Register of Historic Places.
Collections: over 1,300 hats; photographs; period furnishings.
Facilities: Museum-related items for sale.
Hours & Admission Prices: Reservations required. Admission $15 per person. Closed Christmas.
Attendance: 1,100 (estimated)

NEWSPACE CENTER FOR PHOTOGRAPHY, 1632 S.E. 10th Ave., Portland, OR 97214. Tel.: 503-963-1935.
E-mail: info@newspacephoto.org
Web Site: www.newspacephoto.org
Institution Type/Description: Art Gallery.
Collections: photographs.
Activities: classes; workshops; special events.
Hours & Admission Prices: Mon.-Thurs. 10am-10:30, Fri.-Sun. 10-6.

THE OLD CHURCH SOCIETY, INC., 1422 S.W. 11th Ave., Portland, OR 97201-3304. Tel.: 503-222-2031. Fax: 503-222-2981.
E-mail: staff@oldchurch.org
Web Site: www.oldchurch.org
Founded: 1968.
Key Personnel: Pres., Kelli Fields; Gen. Mgr., Amanda Stark; Restoration Architect, Bill Hawkins, III
Personnel Profile: Part-Time Paid 6; Part-Time Volunteers 3.
Governing Authority: society; nonprofit organization. Tax-exempt: 501(c)(3).
Institution Type/Description: Historical Society: housed in 1882 Calvary Presbyterian Church.
Collections: Lannie Hurst Parlor restored & furnished with Victorian furnishings; 1883 Hook & Hastings tracker action pipe organ.
Facilities: 350-seat auditorium. Note paper & posters for sale.
Activities: Gallery Sponsors: Wed. sack lunch recitals.
Publications: quarterly newsletter, The Old Church Organ.

Hours & Admission Prices: Mon.-Fri. 11-3, Sat. by appointment. No charge; donations accepted. Closed New Year's Day; Memorial Day; Independence Day; Labor Day; Thanksgiving; Christmas. &
Attendance: 50,000 (estimated)
Membership: Individual $20; Family $35; Sustaining $50; Guarantor $100 Patron $500; Life $1,000; Benefactor $5,000.

* **OREGON HISTORICAL SOCIETY, (M),** 1200 S.W. Park Ave., Portland, OR 97205-2483. Tel.: 503-222-1741. Fax: 503-221-2035.
E-mail: orhist@ohs.org
Web Site: www.ohs.org
Founded: 1898.
Congressional District: 1
Key Personnel: Pres., Dr. Jerry Hudson; Exec. Dir., Kerry Tymchuk; Dir. Devel. & Mktg., Sue Metzler; Dir. Public Svcs., Marsha Matthews; Editor Oregon Historical Quarterly, Eliza Canty-Jones; Museum Shop Mgr., Kell Smith.
Personnel Profile: Full-Time Paid 32; Part-Time Paid 6; Part-Time Volunteers 240; Interns 2.
Governing Authority: society; nonprofit organization. Parent Institution: Oregon Historical Society. Tax-exempt: 501(c)(3).
Institution Type/Description: History Museum.
Collections: anthropological; ethnographic; archaeological artifacts of Oregon Country & Pacific Rim; maritime collection; paintings; costumes; textiles; furnishings; tools; political memorabilia; advertising; manuscripts; rare books; oral histories; films; videos; photographs; maps; periodicals; microfilm & fiche; architectural drawings; government documents.
Research Fields: Oregon history; Pacific Rim exploration & settlement; native American history; genealogical; prehistory to present; Oregon business history; Northwest artists; Oregon quilts & textiles.
Facilities: research library of books, maps, pamphlets, manuscripts, sound recordings, statewide newspaper microfilm collection, two million plus photographs, diaries & journals, early films. Books & other museum-related items for sale.
Activities: lectures; guided tours; organized educational programs for children; films; reading room; permanent, temporary & traveling exhibits; Oregon Geographic Names Board; OHS Affiliate program.
Publications: Oregon Historical Quarterly.
Hours & Admission Prices: Museum: Tues.-Sat. 10-5, Sun. 12-5. Library: Tues. 1-5, Wed.-Sat. 10-5. Adults $11, students & seniors $9, youth 6-18 $5; members, Multhoman County residents and children 5 & under no charge. &
Attendance: 40,000 (estimated)
Membership: Student $25; Individual $60; Family $80; Contributor $100; Grantor $250; Steward $500.

OREGON JEWISH MUSEUM, (M), 1953 N.W. Kearney St., Portland, OR 97209-1414. Tel.: 503-226-3600. Fax: 503-226-1800.
E-mail: info@ojm.org
Web Site: www.ojm.org
Founded: 1989.
Congressional District: 1
Key Personnel: Dir., Judith Margles; Pres. (V), David Newman; Museum Shop Chair, Carol Dan.
Personnel Profile: Full-Time Paid 2; Part-Time Paid 4; Part-Time Volunteers 75; Interns 5.
Governing Authority: Tax-exempt.
Institution Type/Description: Jewish Museum.
Collections: Jewish life & culture; documents; family & business histories; photographs.
Major Exhibits: Bat Mitzvah Comes of Age, 10/9/13-1/26/14; Illuminated Letters: Threads of Connection, 2/14-4/14; Vida Seafood: A Century of Sephardic Life in Oregon, 6/14-9/14; Shirley Gittelsohn, 10/14-12/14.
Research Fields: Oregon Jewish history.
Facilities: library; archives; 50-seat theater. Gift items for sale.
Activities: movies; lectures; music events; holiday-related events; children's events; facilities rental.
Publications: e-newsletters.
Hours & Admission Prices: Tues.-Thurs. 10:30-4, Fri. 10:30-3, Sat.-Sun. 12-4. Adults $6, seniors & students $4; discounts to groups; members & children under 12 no charge. Closed holidays. &
Attendance: 5,500 (estimated)
Membership: Student $20; Educator $30; Senior Individual 62 & over $35; Individual and Senior Couple 62 & over $45; Family & Household $55; Friend $100; Sponsor $150; Donor $250; Patron $500; Angel $1,000.

OREGON MARITIME MUSEUM, (M), S.W. Pine St. & Naito Pkwy., Portland, OR 97204. Mailing Address: 115 S.W. Ash St., Ste. 400C, Portland, OR 97204-3568. Tel.: 503-224-7724.
E-mail: info@oregonmaritimemuseum.org
Founded: 1980.
Personnel Profile: Part-Time Paid 2; Part-Time Volunteers 100.
Institution Type/Description: Maritime Museum.
Collections: maritime heritage; river history; maritime records & artifacts; paintings; nautical instruments; photographs; local shipyard history.
Activities: educational programs.
Hours & Admission Prices: Wed.-Sat. 11-4, Sun. 12:30-4:30. Adults 13 & over $7, seniors 62 & over $5, students under 12 $3; members & children under 6 no charge. ‎
Attendance: 2,500 (estimated)
Membership: Seniors $20; Adults $25; Family $35; Supporting $50; Sustaining $100; Grantor $250; Steward $500; Life $1,000.

OREGON MUSEUM OF SCIENCE AND INDUSTRY, (M), 1945 S.E. Water Ave., Portland, OR 97214-3356. Tel.: 503-797-4000. Fax: 503-797-4566. Facebook: OMSI Museum.
E-mail: rhuynh@omsi.edu
Web Site: www.omsi.edu
Founded: 1944.
Congressional District: 3
Key Personnel: Pres., Nancy Stueber; Bd. Chm., Don Vollum; Sr. Vice Pres. Support & Administration, Paul Carlson; Vice Pres. Mktg., Retail & Sales, Mark Patel; Vice Pres. Programs, Janie Hurd; Vice Pres. Finance, Human Resources & Volunteer Svcs., Tim Mack; Dir. Retail Operations, Russ Repp; Museum Shop Buyer & New Product Designer, Arlana Burke.
Personnel Profile: Full-Time Paid 152; Part-Time Paid 112; Part-Time Volunteers 725; Interns 10.
Volunteer Hours: 113,963
Operating Expenses: 20,037,323
Operating Income: 19,602,080
Governing Authority: nonprofit organization. Tax-exempt: 501(c)(3).
Institution Type/Description: Science & Technology Center.
Collections: mineralogy; paleontology; zoology; natural history; archaeology; computers; electricity; health; fine art; historical objects; 219-ft. submarine, USS Blueback.
Research Fields: physical & natural sciences; education.
Facilities: field research station; 300-seat theater; 200-seat planetarium; 300-seat auditorium; classrooms; nature center; restaurant. Books, science supplies, telescopes, science kits & souvenirs for sale.
Activities: guided tours; lectures; films; study clubs; hobby workshops; organized education classes for children & adults; school loan service; outdoor education programs; temporary exhibits; submarine; residential camp programs throughout Oregon & Pacific Northwest; 15-person motion simulator. Museum Sponsors: research facilities for high school students.
Publications: quarterly, OMSI Member Magazine; annual catalogs: Fieldtrip Planner, Outreach; Camps; Classes; OMSI Member Elements.
Hours & Admission Prices: See website for hours. Adults $12, youth 3-13 and seniors 63 & over $9; discounts to AAA members; children under 3 & ASTC members no charge. Closed Thanksgiving; Christmas. Call 503-797-4661 for group information & reservations. Special pricing for premier featured exhibits. ‎
Attendance: 950,000 (estimated)
Membership: OMSI After Dark for One $50; OMSI After Dark for Two & OMSI for Two $85; Family $100; Family Plus $125; Friend $150; Patron $210.

OREGON NIKKEI ENDOWMENT/OREGON NIKKEI LEGACY CENTER, 121 N.W. 2nd Ave., Portland, OR 97209-3903. Tel.: 503-224-1458.
E-mail: info@oregonnikkei.org
Web Site: www.oregonnikkei.org
Founded: 1989.
Congressional District: 1
Institution Type/Description: Japanese Heritage Museum.
Collections: Japanese history & culture; photographs; art; personal artifacts.
Facilities: library. Museum-related items for sale.
Activities: summer internships; special events; educational programs.
Hours & Admission Prices: Tues.-Sat. 11-3, Sun. 12-3. Adults $3; members no charge. ‎
Membership: Senior & Student $20; Individual $35; Senior Plus $55; Individual Plus $60; Family $60-$99; Family Plus $80; Patron $100-$499; Benefactor $500-$999; Sustainer $1,000 & up.

OREGON ZOO, 4001 S.W. Canyon Rd., Portland, OR 97221-2799. Tel.: 503-226-1561. Fax: 503-226-6836.
Web Site: www.oregonzoo.org
Founded: 1887.
Congressional District: 1
Key Personnel: Council Pres., David Bragdon; Deputy Dir. Living Collections, Mike Keele; Finance Mgr., Craig Stroud; Gen. Cur., Chris Pfefferkorn; Asst. Zoological Cur., Michael Illig; Asst. Zoological Cur., Gilbert Gomez; Asst. Zoological Cur., Shawn St. Michael; Mng. Dir. Oregon Zoo Foundation, Kregg Hanson; Dir. Devel. Oregon Zoo Foundation, Karen Lloyd; Mktg. Mgr., Jane Hartline; Veterinarian, Mitch Finnegan, D.V.M.; Veterinarian, Lisa Harrenstien, D.V.M.; Deputy Dir. Operations, Carmen Hannold; Mgr. Operations, Ivan Ratcliff; Exhibits Mgr., Brent Shelby; Media Rels. Officer, Bill LaMarche; Construction & Maintenance Mgr., Steve Chaney; Volunteer Mgr., Jennifer Payne; Education Mgr. Programs, Charis Henrie; Conservation Mgr., Anne Warner; Guest Svcs. Mgr., Jim Gilbert; Retail Shop Mgr., Terri Pelham; Security Mgr., Dan Lorenzen.
Personnel Profile: Full-Time Paid 145; Part-Time Paid 13; Part-Time Volunteers 2,100; Interns 20.
Governing Authority: regional government. Parent Institution: Metro.
Institution Type/Description: Zoo.
Collections: mammals; birds; reptiles; amphibians; insects; fish; invertebrates.
Research Fields: studies of ecology of zoo animals; captive breeding & insemination; animal behavior.
Facilities: library; classrooms; research center.
Activities: formally organized education programs; volunteer programs; permanent & traveling exhibitions; mobile vans; field trips; classes for pre-school thru college age level; programs for disabled & senior citizens; birds of prey.
Publications: brochures; quarterly member newsletter; research publications.
Hours & Admission Prices: April 15-Sept. 15 daily 9-6; Sept. 16-April 14 daily 9-4. Adults $10.50, senior citizens 65 & over $9, youth 3-11 $7.50; 2nd Tues. each month $2; discounts to groups & AZA members; children 2 & under no charge. Closed Christmas. ‎
Attendance: 1,621,521 (accurate)
Membership: Individual $44; Zoo4Two $54; Individual Plus $59; Family $69; Zoo4Two Plus $74; Family Plus $94; Patron $125; Benefactor $250; Sponsor $500; Conservation Circle $1,000.

PITTOCK MANSION, 3229 N.W. Pittock Dr., Portland, OR 97210-5099. Tel.: 503-823-3623. Fax: 503-823-3626.
Web Site: www.pittockmansion.org
Founded: 1965.
Key Personnel: Exec. Dir., Marta Bones; Mgr. Visitor Svcs., Elizabeth Marcum; Cur. & Program Mgr., Patricia Larkin.
Personnel Profile: Full-Time Paid 6; Part-Time Paid 10; Part-Time Volunteers 122.
Governing Authority: Parent Institution: Pittock Mansion Society. Subsidiary Institution: City of Portland, Bureau of Parks, 1120 S.W. Fifth Ave., Portland, OR 97204. Tax-exempt: 501(c)(3) & 170(b)(1)(A).
Institution Type/Description: Historic House: 1909-1914 French Renaissance, Pittock Mansion.
Collections: period furnishings; fine art; decorative art.
Facilities: Museum-related items for sale.
Activities: architectural tours; artists after hours events; behind-the-scenes tours; groups tours; kids spring & summer day camps; museum store artists showcase.
Publications: society membership newsletter, Pittock Papers; visitor brochures.
Hours & Admission Prices: Feb.-June & Sept.-Jan. 1 daily 11-4; July-Aug. daily 10-5. Adults $8.50, senior citizens $7.50, children 6-18 $5.50; discounts to military & AAA members; members no charge. Closed Thanksgiving; Christmas. ‎
Attendance: 72,000 (accurate)
Membership: Individual $30; Dual $45; Family $60; Supporter $100-$249; Contributor $250-$499; Sponsor $500-$999; Patron $1,000 & up. Corporate: Bronze $250; Silver $750; Gold $1,500.

❋ **PORTLAND ART MUSEUM, (M),** 1219 S.W. Park Ave., Portland, OR 97205-2486. Tel.: 503-226-2811. Fax: 503-226-4842.
E-mail: info@pam.org
Web Site: www.portlandartmuseum.org
Founded: 1892.
Congressional District: 1
Key Personnel: Dir., Brian Ferriso; C.F.O., Gareth Nevitt; Dir. Mktg., Beth Heinrich; Dir. Devel., J.S. May; Chief Cur., Bruce Guenther; Dir. Northwest Film Center, Bill Foster; Cur. Native American Art, Anna Strankman; Cur. Prints & Drawings, Dr. Annette Dixon; Dir. Education, Tina Olsen; Dir.

Collections Mgmt., Donald Urquhart; Conservator, Elizabeth Chambers; Librarian, Debra Royer; Museum Shop Mgr., Michelle Betcone.
Personnel Profile: Full-Time Paid 127; Part-Time Paid 67; Part-Time Volunteers 865; Interns 31.
Governing Authority: nonprofit organization. Subsidiary Institution: Northwest Film Center. Tax-exempt: 501(c)(3).
Institution Type/Description: Art Museum.
Collections: American: 19th & 20th century paintings, sculpture, decorative arts; Asian: sculptures, paintings, bronzes, ceramics, decorative arts; European: paintings, sculpture, silver, decorative arts; Graphics: Vivian & Gordon Gilkey Graphics Art, Japanese prints; Photography: 19th & 20th-century, Northwest regional collection; Modern & Contemporary Art: 20th-century paintings, sculpture, decorative arts, Clement Greenberg collection; Native American: Elizabeth Cole Butler collection of Native American art, Rasmussen collection of Northwest Coast Indian & Eskimo arts, Cameroon art; pre-Columbian art; Northwest Art: paintings, sculptures, decorative arts.
Facilities: art reference library; 115,000 sq. ft. gallery space; conservation laboratory; film study center; rental sales gallery; multimedia resource center; 380-seat auditorium; classrooms; 2 ballrooms. Museum-related items for sale.
Activities: guided tours; lectures & symposia; films; gallery talks & activities; permanent, temporary & traveling exhibitions; classes for adults, youth & families.
Publications: quarterly newsletter.
Hours & Admission Prices: Mon. holidays, Tues.-Wed. & Sat. 10-5, Thurs.-Fri. 10-8, Sun. 12-5. Adults $10, seniors & students 18 & over $9; discounts for groups, AAM & Western Reciprocal members; children 17 & under and members no charge. Additional fee for special exhibitions. Closed Christmas. &
Attendance: 350,000 (estimated)
Membership: Individual $45; Family & Dual $75; Friend $100; Young & Art $150; Sponsor $250; Sustainer $500; Patron $1,000 & up.

PORTLAND CHILDREN'S MUSEUM, 4015 S.W. Canyon Rd., Portland, OR 97221-2759. Tel.: 503-223-6500. Fax: 503-223-6600. Facebook: Portland CM.
E-mail: mbridges@portlandcm.org
Web Site: www.portlandcm.org
Formerly: CM2-Children's Museum 2nd Generation
Founded: 1949.
Congressional District: 1
Key Personnel: Pres. Bd., Shawn DuBurg; Dir. Exhibits & Operations, JJ Rivera; Dir. Education & Research, Susan MacKay; Dir. Programs, Anne Kluesner; Dir. Devel., Sue Dixon; Dir. Mktg., Communications, & Sales, Melody Bridges; Dir. Finance, Brenda Smith; Museum Shop Mgr., Johnny Youngblood.
Personnel Profile: Full-Time Paid 35; Part-Time Paid 28; Part-Time Volunteers 35; Interns 6.
Governing Authority: municipal; nonprofit corporation. Parent Institution: Portland Bureau of Parks & Recreation, 1120 S.W. Fifth Ave., Portland 97204. Jointly administered by Friends of the Children's Museum. Tax-exempt: 501(c)(3) & 170(c)(1).
Institution Type/Description: Children's Museum.
Collections: natural history; paleontology; cultural history; transportation; toys; dolls; stuffed animals of North America; hands-on exhibit; children's art.
Major Exhibits: Dora and Diego: Let's Explore (T), 10/13-1/14.
Research Fields: early childhood education; playground games; multicultural artifacts of childhood.
Facilities: ceramic & arts classrooms; hands-on museum for children; presentation facilities.
Activities: drop-in art activities; presentations on ceramic & multicultural subjects; birthday parties; presentations to school groups.
Publications: weekly e-newsletter; eBlasts; annual report; tourist guides & flyers.
Hours & Admission Prices: March to Labor Day Thurs. 9-8, Fri.-Wed. 9-5; Sept.-Feb. Tues.-Wed. & Fri.-Sun. 9-5, Thurs. 9-8. Admission $9, military and seniors 55 & over $8; discount to AAM members; members & children under one no charge. &
Attendance: 320,000 (estimated)
Membership: 1 + 1 $65; Family $85; Family Plus $100; Premier $135; Premier Plus $250.

PORTLAND POLICE MUSEUM, 1111 S.W. 2nd Ave., Portland, OR 97204-3231. Tel.: 503-823-0019. Facebook: Portland Police Museum.
E-mail: info@portlandpolicemuseum.com
Web Site: www.portlandpolicemuseum.com

Founded: 1976.
Key Personnel: Dir., Jim Huff; Pres. (V), David Simpson; Museum Shop Mgr., Leslie Poole.
Personnel Profile: Part-Time Paid 2; Part-Time Volunteers 2; Interns 2.
Governing Authority: Parent Institution: Portland Police Historical Society. Tax-exempt.
Institution Type/Description: Police Museum.
Collections: police history & memorabilia; photographs; uniforms; equipment; documents; police officers memorial.
Research Fields: genealogy history.
Facilities: Museum-related items for sale.
Hours & Admission Prices: Tues.-Fri. 10-3. No charge; donations accepted. &
Attendance: 4,000 (accurate)
Membership: $35 per yr.

RONNA AND ERIC HOFFMAN GALLERY OF CONTEMPORARY ART, 0615 S.W. Palatine Hill Rd., MSC 95, Portland, OR 97219. Tel.: 503-768-7687.
E-mail: gallery@lclark.edu
Web Site: www.lclark.edu/hoffman_gallery
Founded: 1997.
Key Personnel: Dir., Linda Tesner
Institution Type/Description: Art Gallery.
Collections: works by contemporary artists; paintings; sculpture; drawings.
Hours & Admission Prices: Tues.-Sun. 11-4. No charge.

SAFETY LEARNING CENTER & FIRE MUSEUM, 900 S.E. 35th Ave., Portland, OR 97214. Tel.: 503-823-3615.
Web Site: www.jeffmorrisfoundation.org
Institution Type/Description: Firefighting History Museum.
Collections: firefighting history & equipment; photographs; hands-on exhibitions; 1859 Jeffers sidestroke handpump fire engine; 1879 Amoskeag steam pumper; 1860 hose cart.
Hours & Admission Prices: By appointment. No charge; donations accepted.

3D CENTER OF ART AND PHOTOGRAPHY, 1928 N.W. Lovejoy St., Portland, OR 97209. Mailing Address: P.O. Box 700, Gladstone, OR 97027-0700. Tel.: 503-227-6667.
Institution Type/Description: Art Gallery.
Collections: works by contemporary artists.
Hours & Admission Prices: Thurs. 11-9, Fri.-Sat. 11-5, Sun. 1-5. Adults 15 & over $5; children under 15 & Thurs. no charge. Closed New Year's Day; Easter; Thanksgiving; Christmas.

WASHINGTON COUNTY HISTORICAL SOCIETY, 17677 N.W. Springville Rd., Portland, OR 97229-1743. Tel.: 503-645-5353. Fax: 503-645-5650 (call first).
E-mail: info@washingtoncountymuseum.org
Web Site: www.washingtoncountymuseum.org
Founded: 1956.
Congressional District: 1
Key Personnel: Exec. Dir., Samuel Shogren; Pres., Gary J. Imbrie; Treas., Devon Reese; Collections & Programs Mgr., Jennifer Kozik; Museum Research Asst., Winnifred Herrschaft.
Personnel Profile: Full-Time Paid 2; Part-Time Paid 2; Part-Time Volunteers 75; Interns 1.
Governing Authority: private; nonprofit organization. Tax-exempt: 501(c)(3).
Institution Type/Description: History Museum.
Collections: artifacts; photo images; manuscripts; documents; Washington County, Oregon maps: Native American through high technology.
Research Fields: Washington County history; census; genealogy; Oregon trail; pioneer life; schools.
Facilities: 550-vol. library; 2,000 sq. ft. exhibit space. Museum-related items for sale.
Activities: Mobile Museum visits schools; guided tours; traveling exhibits; workshops; monthly lecture series; Scout Saturdays program; summer camp. Museum Sponsors: Annual Draft Horse Plowing Exhibition.
Publications: semi-annual, The Washington County Historian; pictorial history of Washington County, This Far Off Sunset Land; guide to historic sites & attractions of Washington County, Hidden Treasures.
Hours & Admission Prices: Museum: Mon.-Sat. 10-4:30. Adults $3, senior citizens & children 6-17 $2; members, faculty, staff, students of PCC & children under 6 no charge. Research Library: Fri.-Sat. 10-4:30, call first. Closed major holidays. &
Attendance: 4,839 (accurate)
Membership: Individual $20; Couples $30; Family & Grandparent $40; George Ebbert Society $50 & up; Joseph Gale Society $100 & up; Joseph Meek Society $250 & up; Tabitha Brown Society $500 & up; Director's Circle $1,000 & up.

WELLS FARGO HISTORY MUSEUM, 1300 S.W. Fifth Ave., Portland, OR 97201-5688. Mailing Address: Wells Fargo Historical Services, 420 Montgomery St., MAC-A0101-106, San Francisco, CA 94163. Tel.: 503-886-1102.
Founded: 2001.
Key Personnel: Cur., Steve Greenwood; Asst. Cur., William Taylor.
Governing Authority: profit-making organization. Affiliated with Wells Fargo Bank.
Institution Type/Description: Company History Exhibit.
Collections: Concord Stagecoach; Wells Fargo banking & express history; mining; staging.
Activities: guided group tours; audiovisual programs.
Publications: scholarly pamphlets.
Hours & Admission Prices: Mon.-Fri. 9-6. No charge. Closed bank holidays. &

WORLD FORESTRY CENTER DISCOVERY MUSEUM, (M), 4033 S.W. Canyon Rd., Portland, OR 97221-2798. Tel.: 503-228-1367. Fax: 503-228-4608.
E-mail: info@worldforestry.org
Web Site: www.worldforestry.org
Founded: 1964.
Congressional District: 1
Key Personnel: C.E.O. & Pres., Gary Hartshorn; Chm., John Warjone; Operations Dir., Mark Reed; Museum Shop Mgr., Louise George.
Personnel Profile: Full-Time Paid 32; Part-Time Paid 9; Part-Time Volunteers 1,500; Interns 5.
Governing Authority: nonprofit organization. Parent Institution: World Forestry Center. Subsidiary Institution: World Forest Institute. Tax-exempt: 501(c)(3).
Institution Type/Description: Global Forestry Museum.
Collections: forest & forestry artifacts including the Pacific NW & forests of the world; global petrified woods; books; photographs; interactive exhibits.
Research Fields: world forestry.
Facilities: 215 acres demonstration forests & nature center; visitor center; classrooms; 250-seat auditorium; 400-seat auditorium; information center.
Activities: guided tours; lectures; permanent & temporary exhibits; forestry classes for youths; ecology classes; tree planting lessons; Boy Scout campouts; interpretive hikes; international fellowship program; computer simulations; canopy lift ride; summer camp; teachers' summer camp; international educators institute; tree farm guided hikes. Annual Events: Chocolate Fest; Coffee Fest; Veterans Day Week; Smithsonian Museum Day; Girl Scout's Day; Day of the African Child.
Publications: WFI newsletter, Northwest Woodlands.
Hours & Admission Prices: Daily 10-5. Adult $8, senior citizens $7, youth 5-18 $5; discounts to groups; children 4 & under and members no charge. Closed Thanksgiving; Christmas Eve & Day. &
Attendance: 70,000 (accurate)
Membership: Individual $35; Individual Plus $40; Family $45; Family Plus $50.

YU CONTEMPORARY ART CENTER, 800 S.E. 10th Ave., Portland, OR 97214. Tel.: 503-236-7996. Fax: 503-235-3664.
E-mail: yu@yucontemporary.org
Web Site: www.yucontemporary.org
Founded: 2010.
Institution Type/Description: Art Gallery.
Collections: works by national & international contemporary artists
Facilities: library; cafe.
Activities: lectures; temporary exhibitions; educational programs.
Publications: magazine, Veneer.
Hours & Admission Prices: Call for hours.

Prairie City

DEWITT MUSEUM, 425 S. Main St., Prairie City, OR 97869. Mailing Address: P.O. Box 370, Prairie City, OR 97869. Tel.: 541-820-3330 & 3605.
E-mail: pchall@ortelco.net
Web Site: www.prairiecityoregon.com
Founded: 1974.
Key Personnel: Dir., Kathy Smith; Chm., Jay Waterhouse.
Volunteer Hours: 407
Institution Type/Description: History Museum: housed in the former Sumpter Valley Railway Station; built in 1910. Listed on the National Register of Historic Places.
Collections: depot history; railroad artifacts; station agent's living quarters;

period furnishings; personal artifacts; photographs; rocks & minerals; mining artifacts; DeWitt family history; narrow gauge railroad artifacts; early documents.
Hours & Admission Prices: mid-May to mid-Oct. Wed.-Sun. 10-5. &

Princeton

BENSON MEMORIAL MUSEUM, MALHEUR NATIONAL WILDLIFE REFUGE, 36391 Sodhouse Lane, Princeton, OR 97721-9523. Tel.: 541-493-2612. Fax: 541-493-2405.
Web Site: www.fws.gov/malheur
Founded: 1954.
Congressional District: 2
Key Personnel: Visitor Services Mgr., Carey Goss.
Governing Authority: federal. Tax-exempt.
Institution Type/Description: Wildlife Refuge & Bird Sanctuary.
Collections: representative specimens of species of birds & small mammals; zoology.
Activities: self-guided tours.
Publications: general leaflet, bird list.
Hours & Admission Prices: Daily sunrise-sunset. Visitor Center: Mon.-Fri. 8-4, Sat.-Sun. call for hours. No charge; donations accepted. &
Attendance: 60,000 (estimated)

Prineville

A.R. BOWMAN MEMORIAL MUSEUM, (M), 246 N. Main St., Prineville, OR 97754-1852. Tel.: 541-447-3715. Fax: 541-447-3715 (call first).
E-mail: bowmuse@netscape.net
Web Site: www.bowmanmuseum.org
Founded: 1971.
Key Personnel: Dir. & C.E.O., Gordon Gillespie; Pres., Jan Anderson.
Personnel Profile: Full-Time Paid 1; Full-Time Volunteers 2; Part-Time Paid 2; Part-Time Volunteers 11.
Governing Authority: county; society. Parent Institution: Crook County Historical Society. Tax-exempt.
Institution Type/Description: Local History Museum: housed in 1910 bank building.
Collections: personal artifacts of local pioneers; American Indian artifacts; local mineral society's rocks; genealogy; newspaper index; photographs.
Research Fields: history of area.
Facilities: Local historical books & other museum-related articles for sale.
Activities: local history field trips; quarterly meetings; historical programs.
Publications: quarterly newsletter.
Hours & Admission Prices: Feb.-May & Sept.-Dec. Tues.-Fri. 10-5, Sat. 11-4; June-Aug. Mon.-Fri. 10-5, Sat.-Sun. 11-4. No charge; donations accepted. Closed major holidays.
Attendance: 10,882 (estimated)
Membership: Senior 65 & over $6; Adult $10; Family $15; Business $25; Sustaining $50.

Reedsport

UMPQUA DISCOVERY CENTER, 409 Riverfront Way, Reedsport, OR 97467-1495. Tel.: 541-271-4816. Fax: 541-271-4816. Facebook: Umpqua Discovery Center.
E-mail: info@umpquadiscoverycenter.com
Web Site: www.umpquadiscoverycenter.com
Founded: 1993.
Congressional District: 4
Key Personnel: Dir. & Museum Shop Mgr., Diane Novak.
Personnel Profile: Full-Time Paid 1; Part-Time Volunteers 25.
Volunteer Hours: 3,000
Governing Authority: municipal; nonprofit organization. Tax-exempt: 501(c)(3).
Institution Type/Description: History Museum.
Collections: Oregon coast history with emphasis on the city of Reedsport and the Lower Umpqua area; dioramas & natural sounds of tidewater towns; history of towns where life, commerce & transportation depended on the ebb & flow of the tides; paintings; simulated indoor trail of the natural history of the area.
Research Fields: local Umpqua Valley area history.
Facilities: 7,000 sq. ft. interpretive center, 50-seat theater. Museum-related items for sale.
Activities: permanent & temporary exhibits; concerts; films; formal education programs; guided tours; lectures; rental gallery; simulated indoor trail. Center Sponsors: Riverfront Rhythms Concerts in Summer; Tsalila Education Days in September.
Publications: quarterly newsletter, News of Discovery.

Hours & Admission Prices: March 15-Oct. 14 Mon.-Sat. 9:30-5, Sun. 11-4; Oct. 15-March 14 Mon.-Sat. 10-4, Sun. 11-4. Adults $8, children 5-16 $4; group & school rates available; members no charge. Closed New Year's Day; Thanksgiving; Christmas. &

Attendance: 17,000 (estimated)

Membership: Individual $25; Family (one household) & Grandparent (4 grandchildren per visit) $45; Business $125, $250, $500 &, $1,000.

UMPQUA RIVER LIGHTHOUSE MUSEUM, 1020 Lighthouse Rd., Reedsport, OR 97467. Tel.: 541-271-4361. Facebook: Umpqua River Light.

E-mail: mjkoreiv@co.douglas.or.us

Founded: 1980.

Congressional District: 4

Key Personnel: Museum Mgr. & Lighthouse Keeper, MJ Koreiva.

Personnel Profile: Full-Time Paid 1; Part-Time Volunteers 42.

Volunteer Hours: 2,000

Operating Expenses: 104,000

Operating Income: 120,000

Governing Authority: Parent Institution: Douglas County, OR. Tax-exempt.

Institution Type/Description: History Museum: housed in a 4-story former US Coast Guard Station; built in 1939.

Collections: history of the 19th & early 20th century US Lighthouse Keepers Service and the US Lifesaving Service; life & work of the newly formed US Coast Guard in 1939; 1st Order Fresnel Lens. Historic Building: 1894 active lighthouse.

Hours & Admission Prices: March-April & Nov.-Dec. Fri.-Sun. 10-4; May-Oct. daily 10-4:30. Museum: no charge; donations accepted. Guided Lighthouse Tours: adults $5, seniors & students (no age limit on students) $3; free to active duty military & military families; children 3-5 no charge.

Attendance: 23,412 (accurate)

Rickreall

POLK COUNTY MUSEUM, 560 S.W. Pacific Hwy., Rickreall, OR 97371. Mailing Address: P.O. Box 67, Monmouth, OR 97361-0067. Tel.: 503-623-6251.

E-mail: pchsoregon@gmail.com

Web Site: polkcountyhistoricalsociety.com

Founded: 1991.

Governing Authority: Parent Institution: Polk County Historical Society. Tax-exempt.

Institution Type/Description: History Museum.

Collections: local history & culture; photographs; Native American artifacts; natural history; personal artifacts; period furnishings.

Research Fields: genealogy.

Facilities: library. Books for sale.

Activities: educational programs.

Publications: quarterly newsletter, Poker; bi-annual Historically Speaking; annual, historical calendar; family & county histories.

Hours & Admission Prices: Mon. & Wed.-Sat. 1-5. Adults $3, seniors $2, students $1; members & children under 5 no charge. &

Attendance: 1,965 (accurate)

Membership: Individual $10; Family $15; Business $20; Supporting $30; Life 65 & over $150; Life $250.

Rogue River

WOODVILLE MUSEUM, INC., 199 First St., Rogue River, OR 97537. Mailing Address: P.O. Box 128, Rogue River, OR 97537-0128. Tel.: 541-582-3088.

Founded: 1986.

Congressional District: 2

Key Personnel: Chm., Samuel D. Evensizer; Treas., Shirley O. Allen.

Personnel Profile: Full-Time Volunteers 12; Part-Time Volunteers 3.

Governing Authority: private; nonprofit organization.

Institution Type/Description: History Museum: housed in Hatch House, c.1909.

Collections: local history & culture; photographs; personal artifacts.

Facilities: 2,500 sq. ft. exhibit space.

Activities: concerts; temporary exhibitions. Museum Sponsors: Rooster Crow.

Hours & Admission Prices: Tues.-Sat. 12-4. No charge; donations accepted. Closed legal holidays.

Attendance: 2,000 (estimated)

Membership: Regular $10.

Roseburg

DOUGLAS COUNTY HISTORICAL SOCIETY - LANE HOUSE, 533 S.E. Douglas Ave., Roseburg, OR 97470. Mailing Address: P.O. Box 2534, Roseburg, OR 97470-0430. Tel.: 541-673-0466.

Web Site: douglascountyhistoricalsociety.org

Founded: 1953.

Key Personnel: Pres. (V), John Robertson; Chm. (V), Gwen Bates.

Personnel Profile: Full-Time Volunteers 30; Part-Time Volunteers 10.

Governing Authority: society. Parent Institution: Douglas County Historical Society. Tax-exempt: 501(c)(3).

Institution Type/Description: Historic House: Floed-Lane House.

Collections: period furnishings; history of house, General Joseph Lane & Lane family.

Facilities: 200-vol. library of old books available for use on premises.

Activities: Museum Sponsors: XMAS Open House in December.

Publications: quarterly magazine, Umpqua Trapper; book, Historic Douglas County; 620 stories of families.

Hours & Admission Prices: Sat.-Sun. 1-4; other times by appointment. No charge; donations accepted. Closed Easter; Independence Day; Labor Day; Thanksgiving; Christmas. &

Attendance: 320 (accurate)

Membership: Individual $25.

DOUGLAS COUNTY MUSEUM OF HISTORY AND NATURAL HISTORY, 123 Museum Dr., Roseburg, OR 97471-5308. Tel.: 541-957-7007. Fax: 541-957-7017.

E-mail: museum@co.douglas.or.us

Web Site: www.co.douglas.or.us/museum

Founded: 1968.

Congressional District: 4

Key Personnel: C.E.O., Gardner Chappell; Cur., Jena Mitchell; Exhibit Preparator, James Davis; Research Librarian & Museum Shop Mgr., Karen Bratton.

Personnel Profile: Full-Time Paid 4; Part-Time Volunteers 75.

Governing Authority: county. Parent Institution: Douglas County. Tax-exempt: 501(c)(3).

Institution Type/Description: General Museum.

Collections: Native Indian artifacts; manuscripts; pioneer tools; guns; utensils; railroad equipment; agricultural equipment; marine artifacts; vehicles; machinery; 20,000 historic photographs; native birds, animals & herbarium.

Research Fields: county history of Indians; settlers; agriculture; logging & milling; railroads & forest service Oregon & California grant lands; transportation; fur trade; mining; natural science; exploration.

Facilities: 2,600-vol. library of Oregon & Pacific Northwest history available for use on premises only; microfilm & oral histories; reading room.

Activities: guided tours; lectures; films; TV & radio programs; permanent, temporary & traveling exhibitions; genealogy. Museum Sponsors: slide programs presented to groups at any outside location on request.

Publications: newsletter.

Hours & Admission Prices: Jan.-Oct. Mon.-Fri. 9-5, Sat. 10-5, Sun. 12-5; Nov.-Dec. Tues.-Fri. 9-5, Sat. 10-5. Adults $4, seniors $3, children 4-17 & college students $2; children under 4 & members no charge. Closed legal holidays. &

Attendance: 12,000 (accurate)

Membership: Individual $15; Couple & Family $30; Patron $50; Contributor $100; Sponsor $250.

Saint Benedict

MOUNT ANGEL ABBEY MUSEUM, One Abbey Dr., Saint Benedict, OR 97373. Tel.: 503-845-3030. Fax: 503-845-3594.

E-mail: info@mountangelabbey.org

Web Site: www.mountangelabbey.org/monastery/museum.html

Institution Type/Description: Religious Museum.

Collections: religious history & artifacts; early liturgical vestments; American Civil War memorabilia; North American mounted mammals; natural history; paintings; mosaics; photographs; sculpture.

Hours & Admission Prices: Daily 10-11:30 & 1-5.

Saint Helens

HISTORICAL SOCIETY OF COLUMBIA COUNTY, 2194 Columbia Blvd., Saint Helens, OR 97051-1739. Tel.: 503-366-3650.

E-mail: rj@historyofcc.org

Web Site: www.historyofcc.org

Formerly: Columbia County Historical Society Museum

Founded: 1930.

Congressional District: 1

Key Personnel: Pres. (V), Cathy Taylor.
Personnel Profile: Part-Time Paid 1; Part-Time Volunteers 5; Interns 1.
Governing Authority: nonprofit historical society; affiliated with Oregon Historical Society, 1230 S.W. Park, Portland, OR 97205. Tax-exempt 501(c)(3).
Institution Type/Description: General Museum: housed in historic train depot.
Collections: period furniture; books; records: Columbia County marriages 1856-1900, alphabetical listing of tombstones in Columbia County: 1850-1984; photos; Indian artifacts; historical relics; pottery; bottles; quilts; pioneer equipment.
Research Fields: family records; early history; cemetery records.
Facilities: 22-vol. library of history & scrapbooks available for use at the museum; reading space. Postcards & booklets for sale.
Activities: self-guided tours.
Publications: Columbia County Historical Booklets; monthly newsletters; periodicals.
Hours & Admission Prices: Tues.-Sat. 10-2. No charge; donations accepted. &
Attendance: 250 (estimated)
Membership: Single $15; Couple $21; Family $27.

Saint Paul

CHAMPOEG STATE HERITAGE AREA VISITOR CENTER, 8239 Champoeg Rd., N.E., Saint Paul, OR 97137-9796. Mailing Address: 7679 Champoeg Rd. NE, St. Paul, OR 97137-9525. Tel.: 503-678-1251, ext. 221. Fax: 503-678-6142.
Web Site: www.oregonstateparks.org/park_113.php
Founded: 1901.
Key Personnel: Park Mgr., Bryan Nielsen.
Personnel Profile: Full-Time Paid 8; Part-Time Paid 10.
Governing Authority: Parent Institution: Oregon Parks & Recreation Dept. Tax-exempt.
Institution Type/Description: Park Museum Visitor Center: at location of first provisional government in the Pacific Northwest (1843).
Collections: tools; period artifacts; furniture; implements; dishes; transportation; riverboats; portraits of pioneers; American Indian artifacts; Champoeg significance in the settlement of the Oregon Territory.
Research Fields: history; archeology.
Facilities: visitor center.
Activities: summer living history demonstrations; videos; Champoeg Promise school program.
Publications: brochure, Champoeg State Park.
Hours & Admission Prices: Summer: daily 9-5; Fall & Winter: Mon.-Fri. 11-4, Sat.-Sun. 10-5. Car parking $5. No charge; donations accepted. Closed Thanksgiving; Christmas. &
Attendance: 17,250

PIONEER MOTHERS MEMORIAL CABIN, 8035 Champoeg Rd., N.E., Saint Paul, OR 97137-9709. Tel.: 503-633-2237.
Web Site: www.newellhouse.com/pioneercabin.htm#DAR
Institution Type/Description: History Museum.
Collections: period pioneer furnishings; collapsible Hudson's Bay heating stove; guns & muskets from 1777-1853; hair wreath; feather & yarn wreaths; fife play at Lincoln's funeral; china & glassware.
Activities: group tours; special events.
Hours & Admission Prices: March-Oct. Fri.-Sun. & holidays 1-5; other times by appointment. Adults $4, seniors & DAR members $3, children $2.

ROBERT NEWELL HOUSE, DAR MUSEUM, 8089 Champoeg Rd., N.E., Saint Paul, OR 97137-9709. Tel.: 503-266-3944.
E-mail: newellhousemuseum@centurytel.net
Web Site: www.newellhouse.com/index.htm
Founded: 1959.
Key Personnel: Dir., Judy Van Atta; State Cur. and DAR State Chm. Buildings & Grounds, Barbara Kieffer; Chm. Grounds, Judith Gardner.
Personnel Profile: Part-Time Paid 1; Part-Time Volunteers 3.
Governing Authority: society. Parent Institution: Oregon State Society DAR. Subsidiary Institution: Pioneer Mothers Memorial Cabin Museum. Tax-exempt.
Institution Type/Description: Historic House: 1959 restoration of 1852 Robert E. Newell House and 1931 pioneer cabin.
Collections: Indian artifacts; quilts & hand work; Oregon's governors wives' inaugural gowns; furniture & furnishings; Masonic items; guns & muskets, 1777-1853. Historic Buildings: 1850 Butteville jail; two-room schoolhouse.
Facilities: Stationery, books & periodicals for sale.
Activities: guided tours including school groups (at Newell House); hands-on demonstrations.
Publications: books, Men of Champoeg; Historic Houses in Oregon; Oregon Landmark Books; brochure on museum.

Hours & Admission Prices: March-Sept. Fri.-Sun. 1-5; private tours by appointment. Adults $4, children under 12 $2; discounts to members & seniors.
Attendance: 4,500 (estimated)

Salem

A.C. GILBERT'S DISCOVERY VILLAGE, 116 Marion St., N.E., Salem, OR 97301-3437. Tel.: 503-371-3631; 800-208-9514. Fax: 503-316-3485.
E-mail: info@acgilbert.org
Web Site: www.acgilbert.org
Formerly: Gilbert House Children's Museum
Founded: 1987.
Congressional District: 5
Key Personnel: Exec. Dir., Pamela Vorachek; Pres. (V), Shannon Martinez.
Personnel Profile: Full-Time Paid 5; Part-Time Paid 20; Part-Time Volunteers 35.
Governing Authority: nonprofit. Tax-exempt: 501(c)(3).
Institution Type/Description: Historic Houses: home of Andrew T. Gilbert, banker; home of C.S. Rockenfield, nurseryman; home of J.L. Parrish, businessman & minister.
Collections: toys & inventions of Alfred Carlton Gilbert; contemporary hands-on exhibits in arts, sciences & humanities.
Facilities: classroom; resource center; gardens; outdoor discovery center.
Activities: tours; arts festivals; hobby workshops; organized education programs for children; outreach educational programs; participatory & traveling exhibitions; organized education programs with students affiliated with Western Oregon University & Willamette University.
Publications: quarterly newsletter, Discovery.
Hours & Admission Prices: Mon.-Sat. 10-5, Sun. 12-5. Adults 3-59 $6, seniors 60 & over $4.50, children 1-2 $3; discounts to ASTC & AAA members; members no charge. Closed New Year's Day; Easter; Thanksgiving; Christmas. &
Attendance: 82,000 (accurate)
Membership: You & Me $45; You and Me $50; Grandparent $60; Family $70.

HALLIE FORD MUSEUM OF ART, (M), Willamette University, 700 State St., Salem, OR 97301. Mailing Address: Willamette University, 900 State St., Salem, OR 97301-3922. Tel.: 503-370-6855. Fax: 503-375-5458. Facebook: Hallie Ford Museum of Art.
E-mail: museum-art@willamette.edu
Web Site: www.willamette.edu/arts/hfma
Founded: 1998.
Congressional District: 5
Key Personnel: Dir. & C.E.O., John Olbrantz; Cur. Collections, Jonathan Bucci; Cur. Education, Elizabeth Garrison; Designer & Preparator, David Andersen; Membership & Public Rels. Mgr., Andrea Foust.
Personnel Profile: Full-Time Paid 8; Part-Time Paid 2; Part-Time Volunteers 12; Interns 4.
Volunteer Hours: 2,080
Operating Expenses: 530,197
Operating Income: 530,197
Governing Authority: private university; nonprofit. Parent Institution: Willamette University. Tax-exempt: 501(c)(3).
Institution Type/Description: Art Museum.
Collections: archives; American, European, Asian, Native American & regional art.
Research Fields: University's art collection with emphasis on historic & contemporary art.
Facilities: print study center; lecture hall; 9,000 sq. ft. exhibit space. Museum-related items for sale.
Activities: concerts; docent program; films; formal education programs for Willamette University students; guided tours; lectures; loan, temporary & traveling exhibitions; poetry readings; artist demonstrations. Museum Sponsors: Symposia.
Publications: biannual calendar of events, Brushstrokes; quarterly alumni magazine, Willamette Scene; temporary exhibitions catalogues.
Hours & Admission Prices: Tues.-Sat. 10-5, Sun. 1-5. Adults $6, senior citizens 55 & over $4 students 18 & over with ID $3; discounts to groups and AAM, AIA, AAA, & NARM members; members and children & students 17 & under no charge. Closed New Year's Eve & Day; Easter; Independence Day; Thanksgiving & day after; Christmas Eve, Day & week. &
Attendance: 30,000 (estimated)
Membership: Individual $25; Family & Dual $50; Sponsor $100; Donor $250; Benefactor $500; Patron $1,000.

HISTORIC DEEPWOOD ESTATE, 1116 Mission St., S.E., Salem, OR 97302-6207. Tel.: 503-363-1825. Fax: 503-363-3586.
E-mail: historicdeepwoodestate@yahoo.com
Web Site: www.historicdeepwoodestate.org
Founded: 1974.
Key Personnel: Exec. Dir., Lois M. Cole.
Personnel Profile: Full-Time Paid 1; Part-Time Paid 2; Part-Time Volunteers 60; Interns 1.
Governing Authority: private; nonprofit organization.
Institution Type/Description: Historical Museum.
Collections: personal artifacts; furnishings.
Hours & Admission Prices: May-Oct. 15 Wed.-Mon. 9 to noon; Oct. 16-April Wed.-Thurs. & Sat. 11-3. Adults $4, senior citizens & students $3, children $2; members & children under 5 no charge.
Attendance: 17,106
Membership: Student and Senior 55 & over $20; Individual $30; Family $50; Patron $50; Business & Photographer $100; Corporate $250; Sustaining $500; Benefactor $1,000.

MARION COUNTY HISTORICAL SOCIETY, 260 12th St., S.E., Salem, OR 97301-2287. Tel.: 503-364-2128. Fax: 503-391-5356.
E-mail: mchs@marionhistory.org
Web Site: www.marionhistory.org
Founded: 1950.
Congressional District: 5
Key Personnel: C.E.O., Ross Sutherland; Pres. (V), Coburn Grabenhorst; Dir., Amy Vandegrift.
Personnel Profile: Full-Time Paid 2; Part-Time Volunteers 50.
Governing Authority: private; nonprofit organization. Tax-exempt: 501(c)(3).
Institution Type/Description: General Museum.
Collections: history of Marion County from the Kalapuya Indians to present.
Facilities: 1,500-vol. library; 1,500 sq. ft. exhibit space. Museum-related items for sale.
Activities: formal education programs for adults; lectures; school loan service; traveling exhibitions.
Publications: quarterly historical journal, Historic Marion; quarterly program newsletter, Member Matters.
Hours & Admission Prices: Tues.-Sat. 12-4. Adults $4, senior citizens $3.50, children $2.50; discount to military, AAA & AAM members; members no charge.
Attendance: 2,500 (estimated)
Membership: Seniors & Student $35; Family $50; Kalapuya $100; Pudding $250; Santiam $500; Willamette $1,000; Heritage $2,500.

OREGON ELECTRIC RAILWAY HISTORICAL SOCIETY, INC., 3395 Brooklake Rd., N.E., Salem, OR 97303. Tel.: 503-393-2424.
E-mail: apmaoffice@eschelon.com
Web Site: www.oerhs.org
Founded: 1957.
Congressional District: 1
Key Personnel: Museum Dir., Greg Bonn; Pres., Charles Philpot; Chm. (V), John Nagy.
Personnel Profile: Full-Time Volunteers 3; Part-Time Volunteers 25.
Governing Authority: society. Parent Institution: Oregon Electric Railway Historical Society. Trolley Operations: Willamette Shore Trolley, 311 N. State St., Lake Oswego, OR, 97034. Tax-exempt: 501(c)(3).
Institution Type/Description: Electric Railway Museum.
Collections: light electric streetcars of 1899-1977; interurban cars; railroad lore; operating tramway with all related equipment & facilities themed to 1910; videos of historic street railway.
Research Fields: tramway technology.
Facilities: 100-vol. library of historical & technical street railway subjects available by request. Gift items for sale.
Activities: guided tours; trolley car rides.
Publications: bulletin, The Transfer.
Hours & Admission Prices: Museum: adults $6. Williamette Shore: adults $10; members no charge. &
Attendance: 6,856 (accurate)
Membership: Active $30; Family $40.

SALEM ART ASSOCIATION-BUSH HOUSE MUSEUM AND BUSH BARN ART CENTER, 600 Mission St., S.E., Salem, OR 97302-6203. Tel.: 503-581-2228. Fax: 503-371-3342.
E-mail: info@salemart.org
Web Site: www.salemart.org
Founded: 1919.
Congressional District: 5
Key Personnel: Exec. Dir., Sandra Burnett; Chm. Salem Art Assoc. Bd. of

Directors, Paula Kanarek; Dir. Bush Barn Art Center, Catherine Alexander; Dir. Community Arts Education, Kathy Dinges-Rice; Dir. Bush House Museum, Ross Sutherland; Dir. Fundraising, Debbie Leahy.
Personnel Profile: Full-Time Paid 6; Part-Time Paid 8; Part-Time Volunteers 125; Interns 6.
Governing Authority: society. Parent Institution: Salem Art Association. Tax-exempt: 501(c)(3).
Institution Type/Description: Historic House Museum: housed in 1878 Asahel Bush, II house. Historic Barn: home of the Salem Art Association.
Collections: period furnishings; fine art; decorative arts; period artifacts.
Research Fields: Oregon art, furnishings & decorative arts, 1850s-1950s; Oregon craft traditions from the early 1800s; Oregon residential architectural & household technology, 1830s-1950s.
Facilities: conservatory; art center; 90 acre park.
Activities: guided tours; NW art exhibitions & lectures; in-school art education programs. Annual Event: Salem Art Fair & Festival in July.
Publications: quarterly newsletter; annual report; exhibition catalogues.
Hours & Admission Prices: Bush House Tours: March-Dec. 23 Wed.-Sun. 1, 2, 3, & 4pm. Adults $6, seniors 62 & over $5, students 16 & over $4, youth 6-15 $3; SAA members and children 5 & under no charge. Bush Barn Art Center: Tues.-Fri. 10-5, Sat.-Sun. 12-5. &
Attendance: 85,000 (estimated)
Membership: Artist $25; Individual $30; Family $50; Patron $100; Business $150; Sponsor $500.

WILLAMETTE HERITAGE CENTER AT THE MILL, 1313 Mill St., S.E., Salem, OR 97301-6591. Tel.: 503-585-7012. Fax: 503-588-9902. Facebook: Willamette Heritage Center.
E-mail: info@willametteheritage.org
Web Site: www.willametteheritage.org
Founded: 2010.
Congressional District: 5
Key Personnel: Exec. Dir., Sam J. Wegner; Pres. (V), Ted Stang; Devel. Dir., Amy Vandergrift; Cur. & Museum Dir., Keni Sturgeon; Operations Dir., Sean O'Hara; Rental & Events Coord., Sharon Osbon; Bookkeeper, Cathie Hawkins; Museum Shop Mgr., Kylie Pine.
Personnel Profile: Full-Time Paid 8; Part-Time Paid 6; Part-Time Volunteers 200; Interns 4.
Volunteer Hours: 11,500
Operating Expenses: 970,244
Operating Income: 1,184,348
Governing Authority: nonprofit organization. Tax-exempt: 501(c)(3).
Institution Type/Description: History Museum.
Collections: artifacts, photographs, archives & books that document & preserve the history of Oregon's Mid-Willamette Valley region, specifically Marion County & the greater Salem area.
Major Exhibits: All the Live Long Day: Work in the Valley, 1/17/14-3/8/14; Chinookan People of the Willamette River, 4/11/14-5/26/14; Boys of Summer: Mid-Valley Baseball, 6/20/14-8/18/14; From the Sheep's Back to Yours: Pendleton Blankets, an Oregon Tradition, 9/26/14-12/24/14.
Research Fields: the The Methodist Mission to Oregon (1834-44); Oregon Trail & early settlement (1841-69); provincial & territorial government; early commerce & industry; religious organizations; Salem & Marion County history; Oregon textile history & industry (1856-); agricultural history in the Willamette Valley; industry significant in the Mid-Valley including fur trade, canning, forestry & milling; lifeways including foodways, clothing, religion, transportation, leisure, entertainment and cultural & ethnic groups; education in the Valley over time; commercial activities, businesses & professions.
Facilities: 1896 Thomas Kay Woolen mill with attendant machine shop, turbine & other outbuildings; 1841 Jason Lee House; 1841 Methodist Parsonage; 1847 John Boon House; 1858 Pleasant Grove Church; research library; rental facilities; picnic grounds.
Activities: guided & self-guided tours; Summer History Pub speaker series; docent program; internships; textile art classes; special exhibitions; Sheep to Shawl, Magic at the Mill holiday celebration of lights & Oregon Trail Live family festivals; heritage bus trips; Quiltopia & Handweavers Sale; Spooky Mill Tour; Heritage Awards; DIY heritage workshops;
Publications: weekly e-bulletin; quarterly newsletter; biennial journal, Willamette Valley Voices; monthly heritage-focused articles in Salem's Statesman Journal daily newspaper.
Hours & Admission Prices: Mon.-Sat. 10-5. Adults $6, seniors $5, students $4, children 6-17 $3; discounts to AAM & AAA members; members no charge. Closed New Year's Day; Martin Luther King Jr. Day; Thanksgiving; Christmas. &
Attendance: 73,540 (accurate)
Membership: Student Scholar $15; Senior Citizen 55 & up $30; Individual $40; Household $50; Millwright $100; Dye Master $250; Willamette Mission $500; Heritage Club $1,000.

Scio

SCIO HISTORICAL SOCIETY & DEPOT MUSEUM, 39004
N.E. 1st Ave., Scio, OR 97374. Mailing Address: P.O. Box 226,
Scio, OR 97374-0226. Tel.: 503-394-2199.
E-mail: maintrain1800@smt-net.com
Founded: 1984.
Key Personnel: Pres. (V), Stephanie Bates.
Governing Authority: Tax-exempt.
Institution Type/Description: History Museum.
Collections: historical items, printed material & memorabilia.
Publications: Scio Historical Society newsletter.
Hours & Admission Prices: May-Oct. Sat.-Sun. 1-4. No charge; donations
accepted.
Membership: Annual $10; Lifetime $100.

Seaside

SEASIDE AQUARIUM, 200 N. Prom, Seaside, OR 97138-5945.
Tel.: 503-738-6211.
E-mail: aquarium@seasideaquarium.com
Web Site: www.seasideaquarium.com
Institution Type/Description: Aquarium.
Collections: marine wildlife.
Hours & Admission Prices: Call for hours. Adults $7.50, senior citizens 64 &
over $6.25, children 6-13 $3.75; children 5 & under no charge.

SEASIDE MUSEUM & HISTORICAL SOCIETY, 570 Necani-
cum Dr., Seaside, OR 97138-6040. Tel.: 503-738-7065. Fax:
503-738-7065.
E-mail: seasidemuseum@hotmail.com
Web Site: www.seasidemuseum.org/
Founded: 1974.
Congressional District: 1
Key Personnel: Pres., Chris Gonzalez; Chm. (V), Roger Waller; Museum Shop
Mgr., Val Smith.
Personnel Profile: Part-Time Paid 1; Part-Time Volunteers 25.
Governing Authority: Tax-exempt.
Institution Type/Description: History Museum.
Collections: history of Seaside; period artifacts. Historic House: c.1912
Butterfield Cottage.
Facilities: research library; gardens. Museum-related items for sale.
Activities: special events.
Publications: quarterly newsletter.
Hours & Admission Prices: Mon.-Sat. 10-3; groups by appointment. Adults $3,
senior citizens $2, students $1; members and children 6 & under no charge.
Closed New Year's Day; Easter; Mother's Day; Memorial Day; Father's
Day; Labor Day; Thanksgiving; Christmas.
Attendance: 1,777 (accurate)
Membership: Students $2; Adult $10; Couple $15; Family $18.75; Business
$25-$49; Supporting $50-$99; Corporate $100-$299; Patron $300-$499;
Benefactor $500-$999; Life $1,000.

Springfield

SPRINGFIELD MUSEUM, (M), 590 Main St., Springfield, OR
97477-5469. Tel.: 541-726-3677. Fax: 541-726-3688.
E-mail: museumdirector@springfield.or.us
Web Site: www.springfieldmuseum.com
Founded: 1981.
Congressional District: 4
Key Personnel: Exec. Dir., Jim Cupples; Pres. of Board, Mike Walker; Cur. &
Registrar, Jan McKee.
Personnel Profile: Full-Time Paid 1; Part-Time Volunteers 45.
Governing Authority: municipal; nonprofit organization. Parent Institution:
City of Springfield. Tax-exempt.
Institution Type/Description: General Museum: housed in 1911 building.
Collections: costumes; photographs; documents; tools.
Research Fields: Springfield, Oregon's history.
Facilities: 2,300 sq. ft. exhibit space. Museum-related items for sale.
Activities: guided tours; organized education programs for children; temporary
exhibitions; Historic Springfield Interpretive Center.
Publications: quarterly newsletter, Museum Notes; annual exhibit schedule;
Historic Springfield Interpretive Center pamphlet.
Hours & Admission Prices: Tues.-Fri. 10-4, Sat. 12-4. Adults $2; discounts to
OMA & AAM members; members no charge. Closed legal holidays.
Attendance: 8,978 (accurate)
Membership: Individual & Senior Family $25; Family $50; Sponsor $75;
Donor $100; Benefactor $250; Patron $500; Sustaining $1,000.

Sweet Home

EAST LINN MUSEUM, 746 Long St., Sweet Home, OR 97386-
3303. Tel.: 541-367-4580.
E-mail: eastlinnmuseum@yahoo.com
Founded: 1976.
Congressional District: 4
Key Personnel: Pres. (V), Gail Gregory; Museum Shop Mgr., Glenda Hopkins.
Personnel Profile: Part-Time Volunteers 20.
Volunteer Hours: 1,800
Operating Expenses: 24,000
Operating Income: 17,000
Governing Authority: Tax-exempt.
Institution Type/Description: History Museum.
Collections: local history & culture; photographs; personal artifacts; tools &
equipment.
Publications: quarterly newsletter.
Hours & Admission Prices: Feb.- Nov. Thurs.-Sat. 11-4, Sun. 1-4. Closed
Easter, Memorial Day & July 4. No charge, donations accepted.
Attendance: 2,500 (estimated)
Membership: Family $10; Supporting $25; Contributing $50.

The Dalles

* **COLUMBIA GORGE DISCOVERY CENTER AND
WASCO COUNTY MUSEUM, (M),** 5000 Discovery Dr., The
Dalles, OR 97058-9755. Tel.: 541-296-8600. Fax: 541-298-8660.
E-mail: cpurcell@gorgediscovery.org
Web Site: www.gorgediscovery.org
Founded: 1997.
Congressional District: 2
Key Personnel: Chm. (V), William G. Dick, II; Exec. Dir., Carolyn Purcell;
Visitor Svcs. Mgr., Kristen May; Dir. Education, Steve Thompson.
Personnel Profile: Full-Time Paid 8; Part-Time Paid 10; Part-Time Volunteers
38.
Volunteer Hours: 3,311
Operating Expenses: 850,111
Operating Income: 928,615
Governing Authority: private; nonprofit. Tax-exempt: 501(c)(3).
Institution Type/Description: Interpretive Center & Museum.
Collections: geologic history of the Columbia River Gorge; wildlife & natural
history; Native American artifacts; Lewis and Clark; local industry; pioneer
artifacts; photographs; documents; manuscripts; textiles; furnishings; petro-
glyphs.
Major Exhibits: Sasquatch in Myth & Legland, 1/1/14-3/1/14; Smithsonian
Traveling Exhibitt: "Bittersweet Harvest: The Bracero Program" (T),
12/1/14-3/1/15.
Research Fields: cultural, historical & natural resources of the north & central
Oregon region and the Columbia River Gorge; settlement history; Lewis
and Clark archaeology.
Facilities: research library; cafe; classroom; picnic area; nature trails.
Museum-related items for sale.
Activities: organized educational programs; permanent, temporary & traveling
exhibits; special guest lectures; volunteer program & training; live raptor
programs.
Publications: quarterly newsletter; books, Celilo Falls: Remembering Thunder
by Wilma Roberts; A Road, a Railroad, and a Country Store: The Story of
Boyd, Oregon by Nancy Ward; annual report; Cargo Exhibit Catalog; I Am
All Alone: The Diary of Mary Evans; Win-Quatt: A Brief History of the
Dalles.
Hours & Admission Prices: Daily 9-5. Adults $9, senior citizen $6.50, children
$5; discounts to groups, active military & AAA members; members no
charge. Closed New Year's Day; Thanksgiving; Christmas.
Attendance: 23,066 (accurate)
Membership: College Student, Teacher & Senior $25; Individual $35; Family
Dual $50; Business $125-$500; Discovery $150.

THE DALLES ART ASSOCIATION, 220 E. 4th St., The Dalles,
OR 97058-2206. Tel.: 541-296-4759.
E-mail: thedallesart@earthlink.net
Web Site: thedallesartcenter.org
Founded: 1967.
Congressional District: 2
Key Personnel: Dir., Carolyn Wright.
Personnel Profile: Part-Time Paid 1; Part-Time Volunteers 50.
Governing Authority: private; nonprofit organization. Tax-exempt: 501(c)(3).
Institution Type/Description: Art Association Gallery.
Collections: paintings; photographs; sculpture.
Facilities: classroom.
Activities: student show; art auctions; monthly exhibits.

Publications: bimonthly newsletter; monthly showcard.
Hours & Admission Prices: mid-Jan. to Dec. Tues.-Sat. 11-5. No charge; donations accepted. Closed New Year's Day; Independence Day; Thanksgiving; Christmas. &
Attendance: 5,000 (estimated)
Membership: Student & Senior $25; Individual $35; Family $45; Business & Friend $75; Sponsor $100; Corporate & Sustaining $250; Benefactor $500; Associate $1,000.

FORT DALLES MUSEUM/ANDERSON HOMESTEAD, 500 W. 15th St., The Dalles, OR 97058-1527. Mailing Address: Box 806, The Dalles, OR 97058-0806. Tel.: 541-296-4547.
E-mail: fortdallesmuseum@gmail.com
Web Site: www.historicthedalles.org/fort_dalles
Founded: 1951.
Congressional District: 2
Key Personnel: Chm. (V), Sam Woolsey; Museum Shop Mgr., Hilary Hines.
Personnel Profile: Part-Time Paid 7; Part-Time Volunteers 4.
Governing Authority: municipal; county. Parent Institution: Wasco County - City of The Dalles. Tax-exempt.
Institution Type/Description: Historic Building Museum: 1856 Fort Dalles Surgeon's quarters.
Collections: history; military; vehicles.
Facilities: Gift items for sale.
Activities: permanent exhibitions.
Hours & Admission Prices: May 15-Sept. 15 daily 10-5; other times by appointment. Adults $5, seniors 55 & over $4, students 7-17 $1; discounts to groups; members no charge. Closed winter holidays.
Attendance: 2,750 (accurate)
Membership: Individual $15; Family $35; Rifle Regiment $100; Medical Officer $250; Commanding Officer $500; Inspector General $1,000 & up.

Tillamook

LATIMER QUILT AND TEXTILE CENTER, (M), 2105 Wilson River Loop Rd., Tillamook, OR 97141. Tel.: 503-842-8622.
Institution Type/Description: Textile Museum.
Collections: quilts; quilt blocks & templates; fabric samples; clothing; hand-woven coverlets; looms; spinning wheels; quilting & weaving textile tools & implements.
Facilities: library. Museum-related items for sale.
Activities: demonstrations; educational programs. Museum Sponsors: Tidal Treasures Quilt and Fiber Arts Festival in June.
Hours & Admission Prices: May-Sept. daily 10-5; Oct. & April Tues.-Sat. 10-4, Sun. 12-4; Nov.-March Tues.-Sat. 10-4. Admission $3; discounts to groups; members & children under 6 no charge. Closed New Year's Day; Independence Day; Thanksgiving; Christmas.

TILLAMOOK AIR MUSEUM, 6030 Hangar Rd., Tillamook, OR 97141-9641. Tel.: 503-842-1130. Fax: 503-842-3054.
E-mail: info@tillamookair.com
Web Site: www.tillamookair.com
Formerly: Tillamook Naval Air Station Museum
Founded: 1994.
Key Personnel: C.E.O., Mike Oliver; Cur., Christian Gurling; Museum Shop Mgr., Michelle Forster.
Personnel Profile: Full-Time Paid 6; Part-Time Paid 12.
Institution Type/Description: General Museum.
Collections: top 5 U.S. war birds; aircraft 1917 to present; use of blimps in WWII.
Facilities: 47-seat restaurant; 300,000 sq. ft. exhibit space; 50-seat theater. Museum-related items for sale.
Activities: docent program; films.
Hours & Admission Prices: Daily 9-5. Adults $9, senior citizens & military with active ID $8, youth 6-17 $5; children 5 & under no charge. Closed Thanksgiving; Christmas. &
Attendance: 80,000 (accurate)

TILLAMOOK CHEESE FACTORY & VISITOR'S CENTER, 4175 Hwy. 101 N., Tillamook, OR 97141-7770. Tel.: 503-815-1300. Fax: 503-815-1305.
Web Site: www.tillamook.com
Founded: 1909.
Institution Type/Description: Visitor Center.
Collections: artifacts & memorabilia pertaining to the history of the cheese-making process.
Hours & Admission Prices: mid-June to Labor Day daily 8-8; Sept.-June daily 8-6. No charge. Closed Thanksgiving; Christmas. &
Attendance: 950,000

TILLAMOOK COUNTY PIONEER MUSEUM, (M), 2106 Second St., Tillamook, OR 97141-2399. Tel.: 503-842-4553. Fax: 503-842-4553. Facebook: Tillamook County Pioneer Museum.
E-mail: director@tcpm.org
Web Site: www.tcpm.org
Founded: 1935.
Congressional District: 1
Key Personnel: Dir., Gary E. Albright; Chm. (V), Phyllis Wustenberg; Museum Shop Mgr., Ruby Fry-Matson.
Personnel Profile: Full-Time Paid 1; Part-Time Paid 7; Part-Time Volunteers 73; Interns 1.
Volunteer Hours: 2,183
Governing Authority: nonprofit association. Tax-exempt.
Institution Type/Description: Historical Museum: housed in 1905 Old Courthouse.
Collections: pioneer artifacts; archaeology; natural history; dishes; rocks; minerals; zoology; mounted animals, birds & fish; wildlife & natural history dioramas; logging displays; great grandmother's kitchen; Victorian parlor; blacksmith shop; 1870s photos; genealogical files.
Major Exhibits: Art of the Pioneer, 12/17/13-3/16/14; Women in the Arts Redoux, 12/18/13-2/18/14; Bayocean Park: A Grand Notion, 3/4-8/1/14; Tillamook County Arts Network (TCAN) Juried Art Show, 8/5-9/28/14; Our Collection, 10/1-11/14; Festival of Trees, 12/5-12/14/14.
Facilities: library of county & state history books available for use on premises.
Activities: permanent & temporary exhibitions; fundraisers for museum & nonprofits; speakers series; educational & environmental programs.
Hours & Admission Prices: Tues.-Sun. 10-4. Research Library: Tues.-Fri. 10-4 by appointment. Adults $4, seniors 62 & over $3, students 10-17 $1; discounts to active military members & their families; members & children under 10 no charge. Closed major holidays. &
Attendance: 11,000 (accurate)
Membership: Senior & Student $10; Individual $15; Couple $25; Family $30; Corporate $100; Sustaining $250; Patron $500; Life $1,000.

Toledo

TOLEDO HISTORY CENTER, 208 S. Main St., Toledo, OR 97391-1203. Mailing Address: P. O. Box 213, Toledo, OR 97391-0213. Tel.: 541-336-1203.
Formerly: Toledo Historical Museum at Toledo City Hall
Founded: 2005.
Key Personnel: Pres. (V), Ann Edmondson.
Personnel Profile: Part-Time Volunteers 5.
Governing Authority: Tax-exempt.
Institution Type/Description: History Museum.
Collections: historical photographic exhibits pertaining to logging, railroad & settlement history of the area.
Hours & Admission Prices: Wed.-Sun. 12-4. No charge; donations accepted. Closed holidays. &
Attendance: 1,500 (estimated)

YAQUINA PACIFIC RAILROAD HISTORICAL SOCIETY, (M), 100 N.W. A St., Toledo, OR 97391-1570. Mailing Address: P.O. Box 119, Toledo, OR 97391-0119. Tel.: 541-336-5256.
E-mail: yprhs@peak.org
Web Site: www.yaquinapacificrr.org
Founded: 1993.
Congressional District: 5
Key Personnel: Pres. (V), Bill Bain.
Personnel Profile: Part-Time Paid 1.
Governing Authority: Tax-exempt.
Institution Type/Description: Railroad History Museum.
Collections: railroad & timber history; period railroad rolling stock & equipment; 1907 wooden caboose; 1922 Baldwin steam engine. Historic Building: 1923 Railway Post Office.
Publications: bimonthly, Yaquina Shortline.
Hours & Admission Prices: Summer: Tues.-Sat. 10-4; Winter: Tues.-Sat. 10-2. No charge. &
Attendance: 1,875 (accurate)
Membership: Participating $52; Sustaining $100; Bronze Spike $250; Silver Spike $500; Golden Spike $1,000; Platinum Spike $3,000.

YAQUINA RIVER MUSEUM OF ART, (M), 151 N.E. Alder St., Toledo, OR 97391-1521. Tel.: 541-336-1907. Fax: 541-336-1907. Facebook: Yaquina River Museum of Art.
E-mail: yrmaoffice@qwestoffice.net
Web Site: www.michaelgibbons.net/museum.htm
Founded: 2002.

Key Personnel: Cur., Michael Gibbons; Chm. (V), Jill Lyon; Museum Shop Mgr., Judy Gibbons.
Personnel Profile: Part-Time Paid 1; Part-Time Volunteers 7.
Governing Authority: Tax-exempt.
Institution Type/Description: Art Museum.
Collections: artwork by artists working in the Yaquina Watershed area & Michael Gibbons, founder of the museum.
Major Exhibits: Promise, 1/14; Permanent Collection, 1/14-4/14.
Hours & Admission Prices: Wed.-Sun. 12-4. No charge; donations accepted.
Attendance: 4,000 (estimated)
Membership: Basic $30; Contributing $50; Supporting $100; Friend $500; Sponsor $1,000; Patron $5,000 & up.

Trail

TRAIL TAVERN MUSEUM, 144 Old Hwy. 62, Trail, OR 97541. Mailing Address: P.O. Box 245, Trail, OR 97541-0245. Tel.: 541-621-4462.
E-mail: trapperjackv@gmail.com
Founded: 1994.
Key Personnel: Chm. (V), Jim Colleir.
Personnel Profile: Interns 10.
Operating Expenses: 3,042
Operating Income: 2,278
Governing Authority: Parent Institution: Upper Rogue Historical Society. Tax-exempt.
Institution Type/Description: History Museum: housed in a former tavern.
Collections: historical artifacts & memorabilia pertaining to the life on the Upper Rogue River.
Activities: blacksmithing demos.
Publications: quarterly newsletter.
Hours & Admission Prices: April 15-Sept. Thurs.-Sun. 12:30-4:30, Oct. Sat.-Sun. 12:30-4:30; tours available by appointment, call 541-621-4462. No charge; donations accepted. Closed Nov. to mid-April.
Attendance: 476 (accurate)
Membership: Annual Individual $10; Life $100; Supporter $250; Benefactor $500.

Troutdale

TROUTDALE HISTORICAL SOCIETY, 726 E. Historic Columbia River Hwy., Troutdale, OR 97060-2061. Mailing Address: 104 S.E. Kibling, Troutdale, OR 97060-2012. Tel.: 503-661-2164. Fax: 503-674-2995.
E-mail: info@troutdalehistory.org
Web Site: www.troutdalehistory.org
Founded: 1968.
Key Personnel: Pres., Scott Cuningham; Vice Pres. & Museum Shop Mgr., Mona Mitchoff; Cur., Mary Bryson; Historian, Sharon Nesbit.
Personnel Profile: Full-Time Volunteers 2; Part-Time Paid 1; Part-Time Volunteers 45.
Governing Authority: private; nonprofit organization. Branch Museums: Rail Depot Museum, 473 E. Historic Columbia River Hwy., Troutdale, OR; Harlow House Museum, 726 E. Historic Columbia River Hwy., Troutdale, OR; Barn Museum, 732 E. Historic Columbia River Hwy., Troutdale, OR. Tax-exempt: 501(c)(3).
Institution Type/Description: Historical Society Museum.
Collections: photographs; furnishings; structures. Historic Buildings: Harlow House, 1900 historic home containing period furnishings; Rail Depot Museum, 1907 Depot (Union Pacific) containing local railroad related artifacts; Barn Museum.
Research Fields: Lewis & Clark.
Facilities: library. Museum-related items for sale.
Activities: monthly education programs. Annual Events: Historic Tea & Tour in May; Cemetery Tour in May; Ice Cream Social in June; Trek in September.
Publications: monthly newsletter, Bygone Times.
Hours & Admission Prices: Depot: Tues.-Fri. 10-4. Harlow House & Barn: call for hours. No charge; donations accepted.
Attendance: 2,500 (estimated)
Membership: Individual $20; Family $30. Lovers Oak Club: Individual $100; Family $150. Business Sponsor $250.

Umatilla

UMATILLA MUSEUM & HISTORICAL FOUNDATION, 911 Sixth St., Umatilla, OR 97882. Mailing Address: P.O. Box 975, Umatilla, OR 97882-0975. Tel.: 541-922-0209.
E-mail: yakka1@msn.com
Founded: 1992.
Congressional District: 2

Key Personnel: Pres. (V), Keith Harding; Dir. & Chm. (V), Marge Nelson; Museum Shop Mgr., Larry Nelson.
Personnel Profile: Part-Time Volunteers 12.
Volunteer Hours: 255
Governing Authority: Tax-exempt.
Institution Type/Description: History Museum: housed in the former city hall building.
Collections: Umatilla history & culture; period furnishings; military artifacts; photographs.
Research Fields: Native American.
Activities: special events; educational programs.
Publications: monthly newsletter; handout brochures.
Hours & Admission Prices: Call for hours. No charge; donations accepted.
Attendance: 1,000 (estimated)
Membership: Student $5; Senior 60 & over $10; Senior Couple $18; Individual $20; Couple $35; Business $50; Patron $100; Donor $250.

Union

UNION COUNTY MUSEUM, 311 S. Main St., Union, OR 97883. Mailing Address: P.O. Box 190, Union, OR 97883-0190. Tel.: 541-562-6003.
Web Site: www.ucmuseumoregon.com
Founded: 1969.
Congressional District: 2
Key Personnel: Pres. (V), Sharon Hohstadt; Museum Shop Mgr., Carol Mulvany.
Personnel Profile: Part-Time Volunteers 42.
Governing Authority: society; nonprofit. Tax-exempt: 501(c)(3).
Institution Type/Description: History Museum: housed in 1881 former First National Bank of Union building.
Collections: settlement & development of Union County, including artifacts, photographs & manuscripts dating from 1830-present.
Research Fields: development of Union County oral history & historic buildings & sites.
Facilities: 100-vol. library of local and regional history & environment, maps, photos & unpublished research, available for use upon application to the curator; reading room; 50-seat auditorium.
Activities: guided tours; lectures; reading room; educational programs for schools.
Publications: quarterly newsletter, The Scout.
Hours & Admission Prices: Mother's Day-Columbus Day Mon.-Sat. 10-4; other times by appointment. Adults $5, senior citizens $4, students $3; members no charge.
Attendance: 300 (estimated)
Membership: Annual $10; Family $20; Business $50; Life $300.

Vale

MALHEUR HISTORICAL PROJECT, STONEHOUSE MUSEUM, 255 Main St., Vale, OR 97918. Mailing Address: P.O. Box 413, Vale, OR 97918-0413. Tel.: 541-473-2070.
Founded: 1995.
Congressional District: 2
Key Personnel: Pres., Gary Fugate; Treas., Charlotte Fugate.
Personnel Profile: Part-Time Volunteers 25.
Governing Authority: private; nonprofit organization. Tax-exempt: 501(c)(3).
Institution Type/Description: Historic House: 1872 stone house built as a wayside house on the Oregon Trail.
Collections: ancient Indian artifacts & local historical artifacts, documents & photographs of Malheur County & the Oregon Trail and the Northern Piautes & Bannock Indians.
Research Fields: survey of historical sites in Malheur County.
Facilities: library; botanical garden; 800 sq. ft. exhibit space; research facility.
Activities: docent programs; guided tours; lectures; temporary exhibitions. Annual Event: Independence Day celebrations.
Publications: quarterly newsletter, Stone House.
Hours & Admission Prices: May-Oct. Tues.-Sat. 12-4. No charge; donations accepted.
Attendance: 2,000 (estimated)
Membership: Seniors & Students $10; Individuals $12; Family $25.

Vernonia

VERNONIA PIONEER MUSEUM, 511 E. Bridge St., Vernonia, OR 97064-1406. Mailing Address: P.O. Box 26, Vernonia, OR 97064-0026. Tel.: 503-429-3713; 503-705-2173. Facebook: Vernonia Pioneer Museum.
E-mail: veronicamuseum@gmail.com
Web Site: vernonia-or.gov

Formerly: Vernonia Historical Museum
Founded: 1962.
Congressional District: 1
Key Personnel: Pres. (V), Jay Anderson; Vice Pres., Ralph Keasey; Sec., Barbara Larsen.
Personnel Profile: Part-Time Volunteers 12.
Volunteer Hours: 600
Operating Expenses: 2,000
Operating Income: 2,000
Governing Authority: county. Affiliated with Oregon State Historical Society. Parent Institution: Vernonia Hands-on Art Center. Tax-exempt.
Institution Type/Description: Historical Museum: housed in the old Oregon American Mill office built in the late 1920s; listed on the National Register of Historic Places.
Collections: history of Nehalem Valley pioneers; logging; photographs; 1930s sawmill.
Publications: quarterly newsletter.
Hours & Admission Prices: June-Aug. Fri.-Sun. 1-4; Sept.-May Sat.-Sun. 1-4. No charge; donations accepted. Closed New Year's Day; Easter; Mother's Day; Independence Day; Independence Day; Christmas. &
Attendance: 1,000 (accurate)
Membership: Annual $5.

Waldport

WALDPORT HERITAGE MUSEUM, 320 N.E. Grant St., Waldport, OR 97394. Mailing Address: P.O. Box 822, Waldport, OR 97394-0822. Tel.: 541-563-7092.
E-mail: waldportmuseum@peak.org
Web Site: waldportmuseum.org
Founded: 1997.
Key Personnel: Pres., Colleen Nickerson; Pres. (V), Judy Gibbs.
Personnel Profile: Part-Time Paid 1.
Volunteer Hours: 356
Governing Authority: Managed by the Alsi Historical & Genealogical Society. Tax-exempt.
Institution Type/Description: History Museum.
Collections: historical artifacts & memorabilia; genealogy.
Major Exhibits: Pioneer of the Year, 9/14-1/15.
Hours & Admission Prices: Wed.-Fri. 12-4, Sat.-Sun. 10-4. No charge; donations accepted. &
Attendance: 849 (accurate)
Membership: Single $5; Family $10; Business $25; Lifetime $50.

Warm Springs

THE MUSEUM AT WARM SPRINGS, (M), 2189 Hwy. 26, Warm Springs, OR 97761. Mailing Address: P.O. Box 909, Warm Springs, OR 97761-0909. Tel.: 541-553-3331. Fax: 541-553-3338.
E-mail: maws@museumatwarmsprings.org
Web Site: www.museumatwarmsprings.org
Founded: 1993.
Congressional District: 2
Key Personnel: Exec. Dir., Carol Leone; Office Mgr. & Bd. Sec., Evaline Patt; Pres. (V), Olney Patt, Jr.; Vice Pres., Hon. Victor Atiyeh; Devel. Officer, Debra Stacona; Receptionist, Leminnie Smith; Cur., Natalie Kirk; Museum Shop Mgr., Sonmiet Maben.
Personnel Profile: Full-Time Paid 10; Part-Time Volunteers 87.
Governing Authority: private; nonprofit organization. Parent Institutions: The Confederated Tribes of the Warm Springs Reservation of Oregon. Tax-exempt: 501(c)(3).
Institution Type/Description: Tribal Museum.
Collections: artifacts & archival material that explain the culture & heritage of the Warm Springs, Wasco & Paiute Tribes that comprise the Confederated Tribes of the Warm Springs Reservation of Oregon.
Facilities: library; photo archives; interpretive trail. Museum-related items for sale.
Activities: educational programs; cultural workshops; storytelling for school tours; demonstrations; crafts; lectures; permanent & temporary exhibitions. Museum Sponsors: living traditions all summer; community craft fairs, Memorial & Labor Day weekends.
Publications: online newsletter for members.
Hours & Admission Prices: Summer: May-Sept. daily 9-5; Winter: Oct.-April Tues.-Sat. 9-5. Adults $7, seniors $6, children $3; discount to groups, AAA, AAM, NARMA, Western Museums Assoc., OMA & AASLH members; members no charge. Closed New Year's Day; Thanksgiving; Christmas. &
Attendance: 14,000 (estimated)
Membership: Senior Citizen $25; Individual $35; Family $45; Fellow $65; Sponsor $100; Sustainer $250; Contributor $500; Patron $1,000; Benefactor $2,500; Twanat $10,000. Corporate & Business memberships available.

Winston

WILDLIFE SAFARI, 1790 Safari Rd., Winston, OR 97496. Mailing Address: P.O. Box 1600, Winston, OR 97496-1600. Tel.: 541-679-6761. Fax: 541-679-1148.
Web Site: www.wildlifesafari.net
Founded: 1972.
Key Personnel: Dir., Dan Van Slyke; Chm. (V), Pat Merkham.
Personnel Profile: Full-Time Paid 50; Part-Time Paid 20; Part-Time Volunteers 50.
Governing Authority: nonprofit organization. Parent Institution: Safari Game Search Foundation. Tax-exempt: 501(c)(3).
Institution Type/Description: Zoo.
Collections: over 600 animals representing more than 80 species; bear; tigers; deer; camel; eagles; elephants; lion; zebra; giraffe; alligator; bob cat; cougar; monkeys; reptiles.
Facilities: cafe. Gift items for sale.
Activities: petting zoo; special events; animal feedings.
Hours & Admission Prices: Drive-thru daily 9-5. Adults $17.99, seniors 60 & over $14.99, children 4-12 $11.99.
Attendance: 150,000 (estimated)

Woodburn

WOODBURN ART CENTER - GLATT HOUSE GALLERY, 2551 N. Boones Ferry Rd., Woodburn, OR 97071-9669. Tel.: 503-982-6450.
Institution Type/Description: Art Gallery.
Collections: paintings; sculpture.
Hours & Admission Prices: Call for hours.

WOODBURN HISTORICAL MUSEUM, 270 Montgomery St., Woodburn, OR 97071-4730. Tel.: 503-982-9531.
E-mail: recreation@ci.woodbum.or.us
Web Site: www.woodburn-or.gov/communitydevelopment/history/default.aspx
Formerly: Woodburn Museum
Key Personnel: Dir., Stu Spence.
Governing Authority: Parent Institution: City of Woodburn, OR.
Institution Type/Description: History Museum.
Collections: Woodburn's history, agriculture, farming, railroad, social, educational & civic heritage.
Hours & Admission Prices: March-Dec. Fri.-Sat. 11-3; other times by appointment.
Attendance: 500 (estimated)

Yachats

LITTLE LOG CHURCH & MUSEUM, 328 W. Third St., Yachats, OR 97498. Mailing Address: P.O. Box 712, Yachats, OR 97498-0712. Tel.: 541-547-3976.
Key Personnel: Dir., Karl Christianson
Institution Type/Description: Historic Building Museum: housed in the former Little Log Church; built in the shape of a cross.
Collections: local history; books; photographs; sculptures; personal artifacts; drawings; paintings.
Activities: weddings; memorials; special events.
Hours & Admission Prices: Fri.-Wed. 12-3.

PENNSYLVANIA

(529 listings)

Abington

BRIAR BUSH NATURE CENTER, 1212 Edge Hill Rd., Abington, PA 19001-3203. Tel.: 215-887-6603. Fax: 215-887-9079.
E-mail: greta@briarbush.org
Web Site: www.briarbush.org
Founded: 1962.
Congressional District: 13
Key Personnel: Exec. Dir., Greta Brunschwyler; Chm. (V), Stan Lexow; Dir. Devel., Karen Serfass; Dir. Public Programs, Tesha Omeis; Dir., Greta Brunschwyler; Environmental Educator, Ehren Gross; Environmental Educator, Katie Fisk; Treas., Michele Kaczalek; Sr. Naturalist, Mark Fallon; Museum Shop Mgr., Patti Platt.
Personnel Profile: Full-Time Paid 8; Part-Time Paid 10; Part-Time Volunteers 150.
Governing Authority: municipal; nonprofit organization. Parent Institution: Abington Township. Subsidiary Institution: Friends of Briar Bush. Tax-exempt: 501(c)(3).

Institution Type/Description: Nature Center.
Collections: ecology of the mid-Atlantic states; live animals; active beehive; pond fed by windmill pump; wildflowers; plants of woodland; seeds; photos; bird observatory.
Facilities: 1,000-vol. library pertaining to science & nature; nature & conservation center; bird observatory; 60-seat meeting room; nature trails; nature plays cape pond. Museum-related items for sale.
Activities: guided tours; lectures; films; organized environmental education programs; docent program; participatory exhibits; school programs; after school programs; birding trips; wellness programs; senior programs; seasonal butterfly house; Nature Playscape.
Publications: newsletter, Briar Flyer; blog, The Briar Blog.
Hours & Admission Prices: Observatory & Museum: Mon.-Sat. 9-5, Sun. 1-5. Nature Trails sunrise-sunset. Adults $3, children $2; township residents, and AAM & museum members no charge. Closed New Year's Day; Independence Day; Thanksgiving; Christmas. ᶜ
Attendance: 65,000 (accurate)
Membership: Senior Citizen $25; Individual $35; Family $50.

Academia

TUSCARORA ACADEMY MUSEUM - JUNIATA COUNTY HISTORICAL SOCIETY, (M), Academia Rd. & Academy Rd., Academia, PA 17082. Mailing Address: 498 Jefferson St., Ste. B, Mifflintown, PA 17059. Tel.: 717-436-5152.
E-mail: jchs1931@juniatacountyhistoricalsociety
Web Site: www.juniatacountyhistoricalsociety.org
Founded: 1931.
Congressional District: 9
Key Personnel: Pres., Audrey R. Sizelove; Vice Pres., Nancy Crago.
Personnel Profile: Part-Time Volunteers 7.
Governing Authority: state. Operated by the Juniata County Historical Society. Tax-exempt.
Institution Type/Description: Historic Building: housed in a former boarding school; built in 1816.
Collections: boarding school dormitory rooms; WWI & II military artifacts; Indian artifacts; geology; early home furnishings; country store; post office.
Activities: guided tours; special programs. Annual Event: Civil War Artifacts & Civil War Encampment with Re-enactors in August.
Publications: quarterly newsletter, Juniata Jottings.
Hours & Admission Prices: June-Aug. Sun. 1:30-4; other times by appointment. No charge; donations accepted.
Attendance: 200 (estimated)
Membership: Juniata County Historical Society: Individual $15; Family $20; Life $100.

Alexandria

HARTSLOG HERITAGE MUSEUM, Alexandria Public Library, 2nd Fl., Main St., Alexandria, PA 16611. Mailing Address: P.O. Box 3, Alexandria, PA 16611-0003. Tel.: 814-669-4313.
Institution Type/Description: History Museum.
Collections: area history; carpentry & blacksmith tools; period furniture & clothing; paintings; tableware; photographs; business records.
Hours & Admission Prices: 1st Sun. of month 2-4; other times by appointment. No charge; donations accepted.

Allentown

✳ **ALLENTOWN ART MUSEUM, (M),** 31 N. 5th St., Allentown, PA 18101-1605. Tel.: 610-432-4333. Fax: 610-434-7409.
E-mail: info@allentownartmuseum.org
Web Site: allentownartmuseum.org
Founded: 1934.
Congressional District: 15
Key Personnel: Pres. & C.E.O., J. Brooks Joyner; Chm. Bd. Trustees, Dolores A. Laputka, Esq.; Chief Cur., Diane P. Fischer, Ph.D.; Dir. Devel. & Mktg., Elsbeth Haymon; Mgr. Government & Foundations Rels., Rhonda K. Hudak; Museum Shop Mgr., Sharon Yurkanin; Bldg. Operations Mgr., Douglas Bowerman; Mgr. Collections, Nathan Marzen.
Personnel Profile: Full-Time Paid 20; Part-Time Paid 22; Part-Time Volunteers 300.
Governing Authority: nonprofit organization. Tax-exempt.
Institution Type/Description: Art Museum.
Collections: European & American paintings & sculpture; prints; drawings; textiles; decorative arts.
Research Fields: pertinent to collection.
Facilities: 15,000-vol. library; 150-seat auditorium; cafe. Catalogs, reproductions & postcards for sale.
Activities: guided tours; lectures; films; gallery talks; concerts; permanent, temporary & traveling exhibitions; school loan service.

Publications: quarterly newsletter; exhibition catalogues.
Hours & Admission Prices: Wed.-Sat. 11-5, Sun. 12-5. Adults $12, students, seniors and children 6 & over $10; children 5 & under, members, & active military no charge. NARM reciprocal. Closed national holidays. ᶜ
Attendance: 65,000 (accurate)
Membership: Student $25; Senior $35; Individual $45; Dual Grandparent $55; Household $75; Contributor $125; Sustainer $225; Patron $550; Kress Society Associate $650; Kress $1,250.

AMERICA ON WHEELS, (M), 5 North Front St., Allentown, PA 18102-5303. Tel.: 610-432-4200. Fax: 610-432-3670.
E-mail: director@americaonwheels.org
Web Site: www.americaonwheels.org
Founded: 1994.
Congressional District: 15
Key Personnel: Exec. Dir., Linda Merkel; Pres. (V), Jack Curcio.
Personnel Profile: Full-Time Paid 3; Part-Time Paid 2; Part-Time Volunteers 60.
Volunteer Hours: 11,100
Operating Expenses: 475,000
Operating Income: 500,000
Governing Authority: private; nonprofit organization. Tax-exempt: 501(c)(3).
Institution Type/Description: Transportation Museum.
Collections: vehicles; transportation history; artifacts; cars; trucks; motorcycles & bikes; pedal cars.
Major Exhibits: Brilliant Brass Beauties, 11/13-4/14; Orphan Cars Debut, 4/15/14-9/14.
Facilities: 300-vol. library; classroom; 23,000 sq. ft. exhibit space; cafe; 40-seat theater; rental facilities; restoration learning center. Museum-related items for sale.
Activities: group tours; educational programs; workshops; corporate events & symposiums; birthday parties; anniversary & wedding receptions.
Publications: newsletter, Spoke & Word.
Hours & Admission Prices: Jan.-Feb. Wed.-Sat. 10-5, Sun. 12-5; March-Dec. Tues.-Sat. 10-5, Sun. 12-5. Adults $8, seniors 62 & over $6, children 6-16 $4; discount to AOW & AAA members. Closed New Year's Day; Independence Day; Thanksgiving; Christmas. ᶜ
Attendance: 50,000 (accurate)
Membership: Annual $35-$2,500.

DA VINCI SCIENCE CENTER, 3145 Hamilton Blvd. Bypass, Allentown, PA 18103-3686. Tel.: 484-664-1002. Fax: 484-664-1022. Facebook: Da Vinci Science Center.
E-mail: ask@davincisciencecenter.org
Web Site: www.davincisciencecenter.org
Founded: 1992.
Key Personnel: Exec. Dir. & C.E.O., Lin Erickson
Institution Type/Description: Science Center.
Collections: hands-on science exhibits; photographs; videos; Hall of Fame.
Major Exhibits: Space: A Journey to Our Future (T), 5/24/14-9/7/14.
Activities: special events; educational programs.
Hours & Admission Prices: Mon.-Sat. 10-5, Sun. 12-5. Adults $12.95, seniors 62 & over, military, and children 4-12 $9.95; children 3 & under & members no charge. Group rate ten or more $9.50. Closed New Year's Day; Easter; Thanksgiving; Christmas Day.
Attendance: 99,193 (accurate)

LEHIGH VALLEY HERITAGE MUSEUM, 432 W. Walnut St., Allentown, PA 18102-5428. Tel.: 610-435-1074, ext. 19. Fax: 610-435-9812.
Web Site: www.lchs.museum
Formerly: Lehigh County Historical Society
Founded: 1904.
Congressional District: 15
Key Personnel: Exec. Dir., Joseph Garrera; Museum Cur., Dir. Library & Archives, Jill Youngken; Reference Librarian, Carol Herrity; Dir. Education, Sarah Thayer Nelson.
Personnel Profile: Full-Time Paid 10; Full-Time Volunteers 5; Part-Time Paid 32; Part-Time Volunteers 125; Interns 2.
Governing Authority: society; nonprofit corporation. Branch Museums & Historic Sites: (1756) Troxell-Steckel House & Farm Museum, Egypt, PA; (1768) George Taylor House, Catasauqua, PA; (1770) Trout Hall, Allentown, PA; (1760) Haines Mill, Allentown, PA; (1817) Lehigh County Museum, Allentown, PA; (1892) David O. Saylor Cement Industry Museum, Coplay, PA; (1868) Lock Ridge Furnace Museum, Alburtis, PA; (1893) One-Room Schoolhouse, Claussville. Tax-exempt: 501(c)(3).
Institution Type/Description: Historical Society.
Collections: 35,000 objects of decorative arts; furniture; textiles; architectural elements; agricultural & industrial tools & equipment; fine & folk art;

domestic equipment; Lenni Lenape Indian materials; horse-drawn & motorized vehicles; 3 million documents, maps, government records, personal papers, business & commercial records; religious & social manuscripts & printed texts; genealogical materials; Pennsylvania German decorative texts & fraktur; 65,000 photographs & negatives; 10,000 books. Historic Buildings: Reninger House (c.1860); Gruber House (c.1912).

Research Fields: Lehigh Valley social, economic, cultural, architectural history; genealogy.

Facilities: 10,000-vol. research library; classrooms; 250-seat lecture hall; 13,000 sq. ft. exhibit space. Museum-related items for sale.

Activities: permanent & temporary exhibitions; inter-museum loans; guided tours; formally organized education programs for students K-12; undergraduate & graduate internships; lectures; films; workshops; bus tour travel program; special events; holiday programs; outdoor festivals & concerts.

Publications: quarterly newsletter; semiannual bulletins; biennial book, Proceedings of the Lehigh County Historical Society; occasional papers; newsletter, Town Crier; books, Allentown 1762-1987: A 225-Year History; Yuscht fer Schee: Architectural Ornament in Allentown; Hidden From History: The Latino Community of Allentown, Pennsylvania.

Hours & Admission Prices: Museum: Tues.-Sat. 10-4, Sun. 11-4. Adults $6, children $3; members no charge. Taylor House & Troxell-Steckel House & Barn: June-Oct. Sat.-Sun. 1-4. Adults $5. Trout Hall: April-Nov. Tues.-Sat. 12-3, Sun. 1-4. Adults $5; discounts to groups. Lock Ridge Furnace Museum & Haines Mill: May-Sept. Sat.-Sun. 1-4. No charge. Claussville Schoolhouse: by appointment. Adults $5. Saylor Cement Museum: daily. No charge. Research $6. Closed major holidays. &

Attendance: 36,500 (estimated)

Membership: Student $15; Individual $35; Family $50; Patron $100; Contributing $250; Sustaining $500; Benefactor $1,000.

MACK TRUCKS HISTORICAL MUSEUM, 2402 Lehigh Pkwy. S., Allentown, PA 18103. Tel.: 610-351-8999. Fax: 610-351-8756.

Key Personnel: Cur., Don Schumaker

Institution Type/Description: Company History Museum.

Collections: Mack company history; documents; trucks including a 1911 Mack Jr. & 1918 AC truck; buses; photographs.

Hours & Admission Prices: Call for hours.

MARTIN ART GALLERY, Baker Center for the Arts, Muhlenberg College, Allentown, PA 18104. Mailing Address: 2400 Chew St., Allentown, PA 18104. Tel.: 484-664-3467. Fax: 484-664-3113.

E-mail: kburke@muhlenberg.edu

Web Site: www.muhlenberg.edu/

Founded: 1976.

Personnel Profile: Full-Time Paid 1; Part-Time Paid 3; Interns 1.

Governing Authority: Parent Institution: Muhlenberg College. Tax-exempt.

Institution Type/Description: College Art Gallery.

Collections: paintings; works on paper; sculpture.

Major Exhibits: Girl Band, 1/15/14-2/22/14; Veit Stratmann: A Muhlenberg Floor, 3/12/14-4/12/14; Lehigh Art Alliance 79th Juried Exhibition, 6/8/14-8/1/14; American Abstract Artists 75th Anniversary Print Portfolio (T), 9/14; Lydia Panas Photography, 10-14-11/14.

Publications: exhibition booklets.

Hours & Admission Prices: Tues.-Sat. 12-8. No charge. Closed all major holidays; semester breaks. &

Attendance: 7,000 (estimated)

MUSEUM OF INDIAN CULTURE, (M), 2825 Fish Hatchery Rd., Allentown, PA 18103-9214. Tel.: 610-797-2121. Fax: 610-797-2801. Facebook: Museum Indian Culture.

E-mail: info@museumofindianculture.org

Web Site: www.museumofindianculture.org

Formerly: Lenni Lenape Historical Society

Founded: 1980.

Key Personnel: Exec. Dir., Pat Rivera; Cur., Archivist & Registrar, Lee Hallman.

Personnel Profile: Full-Time Volunteers 5; Part-Time Volunteers 12; Interns 2.

Governing Authority: private; nonprofit organization. Tax-exempt: 501(c)(3).

Institution Type/Description: History Museum.

Collections: concentration on Northeast Woodland Indians & Inter Tribal with some early Pennsylvania German artifacts. Historic House: 18th-century farm house.

Research Fields: Northeast Woodland Indians.

Facilities: 3,100-vol. library. Gift items for sale.

Activities: arts festivals; formal educational programs; guided tours; hobby workshops; lectures; participatory, loan & traveling exhibitions. Annual Events: Roasting Ears of Corn Festival in August; Fright Night in October; Open House in November.

Publications: quarterly newsletter, Indian Culture Quarterly.

Hours & Admission Prices: June 22-Sept. 8 Thurs.-Sun. 10-4, Sept. 8-June 22 Fri.-Sun. 10-4. Adults $5, seniors & children 12-17 $4; members no charge. Closed national holidays. &

Attendance: 14,000 (estimated)

Membership: Senior Citizen, Students & Teachers $20; Individual $30; Family $50; Sponsor $100-$499; Patron $500-$999; Benefactor $1,000 & up.

Allenwood

CLYDE PEELING'S REPTILAND, 18628 U.S. Route 15, Allenwood, PA 17810-9731. Tel.: 570-538-1869. Fax: 570-538-1714.

E-mail: info@reptiland.com

Web Site: www.reptiland.com

Founded: 1964.

Key Personnel: Dir., Clyde Peeling; Operations Mgr., Chad Peeling; Office Mgr. & Museum Shop Mgr., Chris Peeling; Museum Shop Mgr., Melody Drick; Exhibit Designer, Elliot Peeling.

Personnel Profile: Full-Time Paid 35; Part-Time Paid 10; Part-Time Volunteers 3; Interns 2.

Governing Authority: individual operation organized for profit.

Institution Type/Description: Zoo.

Collections: reptiles; amphibians; birds; Komodo dragons.

Major Exhibits: Dinosaurs Come to Life, 4/19/14-9/1/14; Butterflies, 4/19/14-10/14.

Facilities: botanical garden; 250-seat theatre; zoological park. Jewelry, books & other zoo-related items for sale.

Activities: lectures; theatre; formally organized education programs for children; traveling exhibitions; TV & radio programs.

Publications: newsletter, Reptiletter.

Hours & Admission Prices: April-May & Sept.-Oct. daily 10-6; Memorial Day-Labor Day daily 9-7; Nov.-March daily 10-5. Adults 12 & over $16, children 3-11 $14; children 2 & under and members no charge. Closed New Year's Day; Thanksgiving; Christmas. &

Attendance: 60,000 (estimated)

Membership: Individual $39; Family of Two $72 (each additional person $20).

Allison Park

DEPRECIATION LANDS MUSEUM, 4743 S. Pioneer Rd., Allison Park, PA 15101-2400. Mailing Address: P.O. Box 174, Allison Park, PA 15101-0174. Tel.: 412-486-0563.

E-mail: depreciationlands@gmail.com

Web Site: www.depreciationlandsmuseum.org

Founded: 1974.

Congressional District: 14

Personnel Profile: Part-Time Volunteers 50.

Governing Authority: nonprofit; municipal.

Institution Type/Description: Living History Museum, village & cemetery: Includes c. 1837 Covenanter Church, c. 1803 log house, one room school, blacksmith shop, carriage house & 18th c. style mercantile.

Collections: The Depreciation Lands history; log house; herb & dye gardens; cemetery; wagon (Kramer); tool barn; 1-room schoolhouse, furnishings including pot-bellied stove; old desks-different sizes for children; teacher's desk; books; slates; blacksmith shop; gunsmithing shop; forge; tools; old time sleigh; period bake oven; smoke house.

Major Exhibits: Civil War in Pennsylvania (T), 4/12-5/6/14.

Research Fields: Indian Wars of 1784-1795; causes of the financial problems of the Revolutionary War soldiers; history of the soldiers who took up tracts in these lands.

Facilities: library includes reports related to the Depreciation Lands, old school books & township histories and maps, and the Indian Wars; nature & conservation center. Museum-related items for sale.

Activities: tours; lectures; education programs; permanent & temporary exhibitions. Annual Events: Colonial Teas in February; Fall Festival in October; Halloween Tours in October.

Publications: newsletter.

Hours & Admission Prices: May-Oct. Sun. 1-4; Nov.-April by appointment. Adults $3. &

Attendance: 5,000 (estimated)

Membership: Single $15; Family $25; Group $50.

Altoona

QUAINT CORNER CHILDREN'S MUSEUM, 2000 Union Ave., Altoona, PA 16601-2059. Tel.: 814-944-6830.

Web Site: www.quaintcorner.org

Governing Authority: nonprofit organization. Tax-exempt: 501(c)(3).

Institution Type/Description: Children's Museum: housed in the former home of Daniel O'Rorke; built 1893.

Collections: hands-on exhibits.

Activities: birthday parties; classes; programs.
Hours & Admission Prices: Fri.-Sat. 10-5. Family $5.

RAILROADERS MEMORIAL MUSEUM, 1300 9th Ave., Altoona, PA 16602-2487. Tel.: 814-946-0834. Fax: 814-946-9457.
E-mail: admin@railroadcity.com
Web Site: www.railroadcity.com
Founded: 1972.
Congressional District: 9
Key Personnel: Exec. Dir., Larry Salone; Chm. (V), Dr. Andy Mulhollen; C.O.O., Cynthia Hershey.
Personnel Profile: Full-Time Paid 5; Part-Time Paid 10; Part-Time Volunteers 60.
Governing Authority: nonprofit organization. Subsidiary Institution: Horseshoe Curve National Historic Landmark. Tax-exempt: 501(c)(3).
Institution Type/Description: Railroad Museum.
Collections: railroading artifacts; paintings; photographs; documents; communications equipment; uniforms; models; tools; rolling stock; GG-1; RPO car; dining car; K-4 & 0-4-0 steam locomotives.
Research Fields: Walter L. Main circus wreck; oral history research; ethnicity.
Facilities: 45,000 sq. ft. interpretive center; 80-seat auditorium. Museum-related items for sale.
Activities: lectures; films; docent program; loan & temporary exhibitions; school loan service.
Publications: books, The Wreck of the Red Arrow; Great Circus Train Wreck of 1893; quarterly newsletter, The Standard.
Hours & Admission Prices: Museum: April 4-April 27 Fri.-Sat. 9-4, Sun. 12-4; May 2-Oct. 26 Mon.-Sat. 9-5, Sun. 12-5; Oct. 31-Dec. 21 Fri.-Sat. 9-4, Sun. 12-4. Adults 12-61 $10, seniors 62 & over $9, children 2-11 $8; children under 2 no charge. Horseshoe Curve: April 4-May 1 Mon.-Sat. 9-4, Sun. 12-6; May 2-Oct. 26 Mon.-Sat. 9-6, Sun. 12-6; Oct. 31-Nov. 23 Fri.-Sat. 9-4, Sun. 12-6. Admission 2 & over $7; children under 2 no charge. Closed Easter; Thanksgiving. &
Attendance: 113,356 (accurate)
Membership: Associate $35; Contributing $70; Trustee $150; Benefactor $500; Patron $1,000.

SOUTHERN ALLEGHENIES MUSEUM OF ART AT ALTOONA, 1210 11th Ave., Altoona, PA 16601. Tel.: 814-946-4464. Fax: 814-946-3131.
E-mail: altoona@sama-art.org
Web Site: www.sama-art.org
Founded: 1976.
Key Personnel: Dir., Gary Moyer
Institution Type/Description: Art Museum.
Collections: photographs; prints; paintings.
Major Exhibits: Art in Common, 9/13-1/1/19.
Activities: special events; permanent & temporary exhibitions; exhibition luncheons & receptions. Museum Sponsors: Summer Art Camp in July; The Art of Wine in November.
Publications: catalogues, Colleen Browning; Donald Robinson.
Hours & Admission Prices: Tues.-Fri. 10-5, Sat. 1-5. No charge; donations accepted. &
Membership: Student & Senior Citizen $25; Individual & Artist $35; Family $50-$99; Sponsor $100-$249; Sustaining $250-$499; Benefactor $500-$999; Connoisseur $1,000-$1,499; Exhibition Sponsor $1,500; Education Sponsor $2,500; Museum Associate $5,000; Director's Circle $10,000.

Ambler

THE STOOGEUM, 904 Sheble Ln., Ambler, PA 19002. Mailing Address: P.O. Box 747, Gwynedd Valley, PA 19437-0747. Tel.: 267-468-0810.
E-mail: garystooge@aol.com
Web Site: www.stoogeum.com
Founded: 2004.
Key Personnel: Cur., Gary Lassin
Institution Type/Description: Comedy Museum.
Collections: Three Stooges memorabilia from 1918 to present; personal artifacts; movie props & costumes; photographs; movie posters; toys; games; artwork; hands-on exhibits
Facilities: library; 85-seat theater.
Activities: films; lectures; special presentations. Annual Event: Three Stooges Fan Club meeting.
Publications: journal, The Three Stooges.
Hours & Admission Prices: See website for hours. Admission $10; children 12 & under no charge. &
Attendance: 2,500 (estimated)

Ambridge

OLD ECONOMY VILLAGE, (M), 270 Sixteenth St., Ambridge, PA 15003-2225. Tel.: 724-266-4500. Fax: 724-266-3010.
Web Site: www.oldeconomyvillage.org
Founded: 1919.
Congressional District: 4
Key Personnel: Site Dir., Michael Knecht; Pres. (V), Greg Gleason; Office Coord., Elaine Voss.
Personnel Profile: Full-Time Paid 11; Part-Time Paid 2; Part-Time Volunteers 150.
Governing Authority: state. Commonwealth of Pennsylvania. Parent Institution: Pennsylvania Historical & Museum Commission, Box 1026, Harrisburg, PA 17108. Tax-exempt.
Institution Type/Description: Historic Village Museum: 18 buildings of the original town of Economy (now Ambridge), PA, built between 1824 & 1831.
Collections: history; communitarian society; textiles; technology; manuscript collections; industrial; decorative arts; archives. Historic Buildings: George & Frederick Rapp Houses; 1826 grotto; 1826 Baker House (family dwelling); 1826 cabinet shop; 1827 store; 1827 granary; 1828 feast hall, mechanics' building; 1831 pavilion.
Research Fields: textiles & textile manufacturing; religion; furniture; early 19th century methods of life in communal society; 19th century business history of Western PA.
Facilities: 5,000-vol. library of the Harmony Society available by appointment on premises; classrooms; outdoor museum. Handicrafts & museum-related items for sale.
Activities: guided tours; lectures; films; concerts; formally organized educational programs; docent program; permanent & temporary exhibitions. Museum Sponsors: Young Harmonists.
Publications: newsletter, Bibliography of Harmony Society; guidebook, manuals for guides & docents.
Hours & Admission Prices: Tues.-Sat. 10-5, Sun. 12-5. Adults 12-64 $10, senior citizens 65 & over $9, children 3-11 $6; discount to AAA members; children under 2 & members no charge. Closed New Year's Day; Martin Luther King Jr. Day; Presidents Day; Columbus Day; Veterans Day; Thanksgiving; Christmas. &
Attendance: 15,999 (accurate)
Membership: Youth 12-17 $15; Student 19-22 $25; Individual $35; Family & Grandparent $40; Friends $125; Patron $250; Benefactor $500; Sustaining $1,000.

Annville

SUZANNE H. ARNOLD ART GALLERY, LEBANON VALLEY COLLEGE, (M), 101 N. College Ave., Annville, PA 17003-1404. Tel.: 717-867-6445. Fax: 717-867-6124.
E-mail: mcnulty@lvc.edu
Web Site: www.lvc.edu/gallery
Founded: 1994.
Congressional District: 17
Key Personnel: Dir., Barbara McNulty; Co Chm., Richard Charles; Registrar & Cur., Crista Detweiler.
Personnel Profile: Full-Time Paid 2; Part-Time Volunteers 6.
Governing Authority: private college. Parent Institution: Lebanon Valley College. Tax-exempt: 501(c)(3).
Institution Type/Description: Art Museum.
Collections: Pennsylvania Fraktur, Chinese, African & Inuit art; prints; paintings; sculpture; photographs.
Facilities: 200-seat auditorium; 1,000 sq. ft. exhibit space.
Activities: concerts; formal education programs for college students; guided tours; lectures; loan exhibitions. Annual Event: Artist's Demonstrations.
Publications: biannual newsletter, Friends of the Gallery; exhibition catalogs & brochures.
Hours & Admission Prices: Wed. 5-8pm, Thurs.-Fri. 1-4:30, Sat.-Sun. 11-5. No charge; donations accepted. Closed during college holidays. &
Attendance: 4,500 (estimated)
Membership: Individual $35-$59; Family $60-$99; Sustainer $100-$249; Sponsor $250-$499; Patron $500-$999; Director $1,000-$4,999; Benefactor $5,000 & up.

Apollo

W.C.T.U. BUILDING - APOLLO AREA HISTORICAL SOCIETY, 317 N. 2nd St., Apollo, PA 15613. Mailing Address: P.O. Box 434, Apollo, PA 15613. Tel.: 724-478-3037, 2899 & 1217. Facebook: Apollo Area Historical Society.
E-mail: apollo.history@yahoo.com
Founded: 1970.

Governing Authority: Tax-exempt.
Institution Type/Description: Historical Society Museum: housed in the former home of the Women's Christian Temperance Union and the home of the first public library in Armstrong County; built in 1909.
Collections: local history & culture; period furnishings; photographs. Historic Building: 1816 Drake's log cabin.
Publications: newsletter, Our Heritage; books, Windows to Apollo's Past.
Hours & Admission Prices: April-Dec. Wed. 11-2. No charge; donations accepted. &
Membership: Student $2; Adult $10; Family $25; Life $125; Business & Corporation Life $300.

Ashland

PIONEER TUNNEL COAL MINE & STEAM TRAIN, 19th & Oak Sts., Ashland, PA 17921. Tel.: 570-875-3850. Fax: 570-875-3301. Facebook: Official Pioneer Tunnel Coal Mine.
E-mail: ashpa@ptd.net
Web Site: www.pioneertunnel.com
Founded: 1962.
Key Personnel: Gen. Mgr., Keith Neidig; Business Mgr., Kathy Lattis; Pres. (V), Dennis Kane.
Personnel Profile: Full-Time Paid 2; Part-Time Paid 15.
Governing Authority: Tax-exempt.
Institution Type/Description: Mining Museum.
Collections: history of mining; 1920s steam locomotive.
Activities: coal mine tour; train ride; festivals. Annual Event: Pioneer Day in August.
Hours & Admission Prices: Mine Tours: April daily 11, 12:30 & 2; May & Sept.-Oct. Mon.-Fri. 11, 12:30 & 2, Sat.-Sun. 10-5; Memorial Day to Labor Day: daily 10-5. Coal Mine Tours: adult $10, children 2-11 $7; discounts to groups. Steam Train Ride: adult $8, children 2-11 $6; discounts to groups. Combination Ticket: adults $16.20, children 2-11 $12.60. &
Attendance: 30,102 (accurate)

Athens

TIOGA POINT MUSEUM, 724 S. Main St., Athens, PA 18810-1000. Mailing Address: P.O. Box 143, Athens, PA 18810-0143. Tel.: 570-888-7225. Facebook: Tioga Point Museum.
E-mail: tpointmuseum@stny.rr.com
Web Site: www.tiogapointmuseum.com
Founded: 1895.
Congressional District: 23
Key Personnel: Dir., Margaret Boritz; Pres., Mark Orshaw.
Personnel Profile: Full-Time Paid 1; Part-Time Paid 1; Part-Time Volunteers 5.
Governing Authority: nonprofit organization. Tax-exempt: 101.
Institution Type/Description: General Museum.
Collections: local & natural history; fine arts; early canals; railroads; Revolutionary War; Civil War; Native American archaeology & ethnography; local history archives; rare books with decorated bindings.
Research Fields: local genealogy; local history; French Azilum; Native American archaeology.
Facilities: 500-vol. library of rare books available by appointment on premises; reading room. Pamphlets for sale.
Activities: guided tours; lectures; permanent & temporary exhibitions.
Publications: newsletter.
Hours & Admission Prices: Tues. & Thurs. 12-8, Sat. 10-1; other times by appointment. No charge; donations accepted. Closed national holidays. &
Attendance: 1,423 (accurate)
Membership: Junior $5; Student & Senior Individual $15; Individual & Senior Family $25; Family $40; Research Sponsor $60; Contributing $100; Supporting $500; Sustaining $1,000.

Audubon

JOHN JAMES AUDUBON CENTER AT MILL GROVE, 1201 Pawlings Rd., Audubon, PA 19403-2242. Tel.: 610-666-5593. Fax: 484-831-5305.
E-mail: millgrove@audubon.org
Web Site: pa.audubon.org/centers_mill_grove.html
Formerly: Mill Grove, The Audubon Wildlife Sanctuary
Founded: 1951.
Congressional District: 13
Key Personnel: Dir., Jean Bochnowski; Chm. (V), Leigh Altadonna; Senior Cur., Nancy S. Powell; Facilities Coord., Susannah Conard; Administrative Coord., Linda Ridgway; Education Coord., Carrie Ashley.
Personnel Profile: Full-Time Paid 5; Part-Time Paid 1; Part-Time Volunteers 149.
Governing Authority: Partnership between Montgomery County, PA and the National Audubon Society. Tax-exempt.

Institution Type/Description: Art Museum, Wildlife Refuge & Historic Site: built in 1762; first home in America of artist/naturalist John James Audubon (1803-1806). A National Historic Landmark.
Collections: artwork published by John James Audubon; original 19th-century editions; natural history; feeding stations; nesting boxes.
Research Fields: J.J. Audubon; ornithology; natural history.
Facilities: nature trails. Museum-related items for sale.
Activities: formal educational programs for children & adults; seasonal guided nature walks; seasonal special events; permanent & temporary exhibits; self-guided tours.
Publications: newsletter, calendar of events.
Hours & Admission Prices: Tues.-Sat. 10-4, Sun. 1-4; guided tours by appointment. Adults $4, seniors 60 & over $3, children 4-17 $2; discounts to AAM & ICOM members; members no charge. Grounds: 7 am to dusk, no charge. Closed New Year's Eve & Day; Easter; Independence Day; Thanksgiving; Christmas Eve & Day; major holidays.
Attendance: 20,000 (accurate)
Membership: Senior $35; Individual $40; Senior Family $50; Family $55; Supporter $100.

Avella

MEADOWCROFT ROCKSHELTER AND HISTORIC VILLAGE, (M), 401 Meadowcroft Rd., Avella, PA 15312-2759. Tel.: 724-587-3412. Fax: 724-587-3414.
Web Site: www.heinzhistorycenter.org
Formerly: Meadowcroft Rockshelter and Museum of Rural Life
Founded: 1969.
Congressional District: 18
Key Personnel: Dir., David R. Scofield; Cur., Bonnie Reese; Dir. Education, Dr. John Boback; Visitor Svcs. Mgr., Fran Skariot.
Personnel Profile: Full-Time Paid 5; Part-Time Paid 12; Part-Time Volunteers 22; Interns 2.
Governing Authority: Parent Institution: Senator John Heinz History Center. Tax-exempt.
Institution Type/Description: History Museum: 275-acre history village with 19th century rural village recreation; 17th century eastern woodlands Indian village recreation. A National Historic Landmark-Meadowcroft Rockshelter.
Collections: local history & culture; photographs; personal artifacts.
Activities: educational programming; public programming.
Publications: quarterly, Western PA History.
Hours & Admission Prices: May & Sept.-Oct. Sat. 12-5, Sun. 1-5; Memorial Day to Labor Day Wed.-Sat. 12-5, Sun. 1-5. Adults $12, children 6-17 $6; discounts to AAM members; members & children under 6 no charge. &
Attendance: 15,608 (accurate)
Membership: Individual $57; Grandparents $80; Family $85; Contributor $125; Patron $250; Benefactor $500; President's Circle $1,000; 1879 Society $1,879.

Beaver Falls

AIR HERITAGE, INC., 35 Piper St., Beaver Falls, PA 15010-1043. Tel.: 724-843-2820. Fax: 724-847-4581.
E-mail: airheritage1@verizon.net
Web Site: www.airheritage.org
Founded: 1986.
Key Personnel: Pres. (V), Bill Schillig; Museum Shop Mgr., Virgil Wylie.
Personnel Profile: Full-Time Paid 30; Part-Time Volunteers 20.
Governing Authority: Tax-exempt.
Institution Type/Description: Aviation History Museum.
Collections: civilian, commercial & military aviation history; aviation artifacts; WWII & Vietnam era aircraft.
Research Fields: aeronautical.
Facilities: Museum-related items for sale.
Activities: air shows.
Hours & Admission Prices: Mon.-Sat. 10-5, Sun. by appointment. No charge; donations accepted. Closed major holidays. &
Attendance: 1,400 (estimated)
Membership: Individual $30.

BEAVER FALLS HISTORICAL SOCIETY AND MUSEUM, 1301 7th Ave., Beaver Falls, PA 15010-4219. Mailing Address: P.O. Box 493, Beaver Falls, PA 15010. Tel.: 724-494-2439.
Founded: 1945.
Key Personnel: Pres. (V), Kenneth Britten.
Personnel Profile: Part-Time Paid 1; Part-Time Volunteers 2.
Volunteer Hours: 1,250
Operating Expenses: 1,235
Operating Income: 1,785

Governing Authority: Tax-exempt.
Institution Type/Description: Historical Society Museum.
Collections: local history & culture; photographs; genealogy; personal artifacts.
Activities: research.
Hours & Admission Prices: 10-3 Mon.-Wed. & Fri. 10-3. No charge; donations accepted. &

Attendance: 995 (accurate)
Membership: Individual $10; Family $15.

Bedford

BEDFORD COUNTY HISTORICAL SOCIETY, 6441 Lincoln Hwy., Bedford, PA 15522. Tel.: 814-623-2011. Facebook: Bedford County Historical Society.
E-mail: bedfordhistory@embarqmail.com
Web Site: www.bedfordpahistory.com
Founded: 1937.
Key Personnel: Dir., Gillian K. Leach; Pres. (V), Glenden G. Casteel.
Personnel Profile: Full-Time Paid 1; Part-Time Paid 1; Part-Time Volunteers 30.
Governing Authority: Tax-exempt.
Institution Type/Description: Historical Society Museum.
Collections: local history & culture; photographs; personal artifacts; Civil War artifacts.
Research Fields: genealogy; Bedford County history.
Facilities: library.
Activities: research; lectures; tours. Annual Events: History Banquet and Quilt Show & Sale in October.
Publications: The Pioneer Magazine.
Hours & Admission Prices: Mon.-Fri. 9-4, 3rd Sat. each month 9-2. &
Attendance: 1,500 (accurate)
Membership: Individual $25; Family $35; Donor $50; Life $500.

FORT BEDFORD MUSEUM, 110 Fort Bedford Dr., Bedford, PA 15522. Mailing Address: 244 W. Penn St., Bedford, PA 15522. Tel.: 814-623-8891.
E-mail: info@fortbedfordmuseum.org
Web Site: www.fortbedfordmuseum.org
Founded: 1958.
Congressional District: 9
Key Personnel: Mgr. & Museum Shop Mgr., Lisa Merritt; Cur., Larry Yantz.
Personnel Profile: Full-Time Paid 1; Part-Time Paid 1; Part-Time Volunteers 1.
Governing Authority: municipal. Affiliated with the Bedford Boro Council, West Penn St. Tax-exempt.
Institution Type/Description: History Museum.
Collections: Indian artifacts; period vehicles & furnishings.
Research Fields: local area artifacts.
Facilities: Museum-related items for sale.
Activities: guided tours; temporary special exhibits.
Publications: descriptive brochures.
Hours & Admission Prices: April-Oct. Wed.-Sun. 11-5; other times by appointment. Adults $5, senior citizens 65 & over $4.50, students 6-18 $3; children under 6 no charge.
Attendance: 2,870 (accurate)

THE NATIONAL MUSEUM OF THE AMERICAN COVERLET, 322 S. Juliana St., Bedford, PA 15522-1734. Tel.: 814-623-1588. Facebook: National Museum of the American Coverlet.
E-mail: info@coverletmuseum.org
Web Site: www.coverletmuseum.org
Founded: 2006.
Key Personnel: Museum Dir., Melinda Zongor.
Governing Authority: nonprofit organization. Tax-exempt.
Institution Type/Description: Coverlet Museum.
Collections: American woven coverlet history; period American woven coverlets.
Facilities: Museum-related items for sale.
Activities: special programs; temporary exhibitions. Museum Sponsors: Fiber Retreat; Coverlet College; Fall Foliage.
Publications: membership newsletter, Yarns.
Hours & Admission Prices: Mon.-Sat. 10-5, Sun. 12-4. Adults $8, senior citizens 60 & over $6; members no charge.
Membership: Individual $40; Family $50; Sponsor $100; Corporate $250; Patron $500; Benefactor $1,000.

Bellefonte

CENTRE COUNTY LIBRARY AND HISTORICAL MUSEUM, 203 N. Allegheny St., Bellefonte, PA 16823-1601. Tel.: 814-355-1516. Fax: 814-355-2700.
E-mail: paroom@centrecountylibrary.org
Web Site: www.centrecountylibrary.org
Founded: 1939.
Congressional District: 23
Key Personnel: Mgr. Historical Collections, Alissa Zawoyski; Information Svcs. Librarian, Karen Kanipe.
Personnel Profile: Full-Time Paid 1; Part-Time Paid 1; Part-Time Volunteers 8.
Governing Authority: board of directors. Parent Institution: Centre County Library. Tax-exempt.
Institution Type/Description: Historic Library & Local History Museum.
Collections: furniture; china; county artifacts. Historic House: 1814-1816 Miles-Humes House.
Research Fields: local history; genealogy.
Facilities: 3,000-vol. library of Pennsylvania historical & genealogical books, records, letters & manuscripts available for use by public.
Activities: guided tours; temporary exhibitions.
Publications: books, Centre County Marriages, 1800-1850; Centre County Marriages, 1851-1873; Centre County Marriages 1874-1885; Deaths of Centre County 1821-1869.
Hours & Admission Prices: Mon., Wed., Fri. 10-5, Tues. & Thurs. 12-5, 3rd Sat. each month 10-2. No charge; donations accepted. Closed national holidays.
Attendance: 962 (accurate)

Bethel

GOLDEN AGE AIR MUSEUM, Grimes Airfield, 371 Airport Rd., Bethel, PA 19507. Tel.: 717-933-9566.
E-mail: info@goldenageair.org
Web Site: www.goldenageair.org
Founded: 1997.
Governing Authority: nonprofit organization. Tax-exempt: 501(c)(3).
Institution Type/Description: Aviation History.
Collections: aviation history; aircraft; automobiles.
Hours & Admission Prices: May-Oct. Sat. 10-4, Sun. 11-4. Adults $5, children 6-12 $3; discounts to groups; members no charge.
Membership: Individual $25; Family $35; Lifetime $500; Corporate $1,000.

Bethlehem

BANANA FACTORY, 25 W. Third St., Bethlehem, PA 18015-1238. Tel.: 610-332-1300.
E-mail: info@fest.org
Web Site: www.artsquest.org
Founded: 1998.
Key Personnel: Dir., Janice Lipzin.
Governing Authority: Parent Institution: ArtsQuest. Tax-exempt.
Institution Type/Description: Art Gallery.
Collections: paintings; photographs; sculpture.
Activities: guided tours; art programs; adult & children's classes.
Hours & Admission Prices: Gallery: daily 11-4. Building: Mon.-Fri. 8am-9:30pm, Sat.-Sun. 8:30-5. No charge. Closed New Year's Day; Thanksgiving; Christmas. &

BURNSIDE PLANTATION, INC., 1461 Schoenersville Rd., Bethlehem, PA 18018-1889. Mailing Address: 74 W. Broad St., Ste. 260, Bethlehem, PA 18018. Tel.: 610-882-0450. Fax: 610-882-0460.
E-mail: info@historicbethlehem.org
Web Site: www.historicbethlehem.org
Founded: 1986.
Congressional District: 15
Key Personnel: Pres., Charlene Donchez Mowers; Vice Pres. & Mng. Dir., LoriAnn Wukitsch; Finance, Tom Homanick; Chm., Joseph Kochanasz; Cur. Collections & Exhibits, Amy C. Frey.
Personnel Profile: Full-Time Paid 6; Part-Time Paid 3; Part-Time Volunteers 50.
Governing Authority: Parent Institution: Historic Bethlehem Partnership. Tax-exempt: 501(c)(3).
Institution Type/Description: Historic Site.
Collections: farming practices, domestic crafts, & decorative arts from 1748-1848. Historic Buildings: c.1748 farm house with 1818 addition; c.1825 summer kitchen; mid 19th-century bank barn with horsepower wheel, corn crib & wagon shed.

Research Fields: Moravian agricultural practices; Pennsylvania German rural furnishings.
Facilities: educational facilities; nature center.
Activities: docent program; formal education programs for children; lectures. Annual Events: Blueberry Festival in July; Harvest Festival in Fall; Colonial Winter Festival in December.
Publications: Historic Bethlehem Partnership Newsletter.
Hours & Admission Prices: July-Aug. Sat. 12-4; call for additional hours; tours by appointment. Self-guided walking tours: daily. No charge; donations accepted.
Attendance: 4,000 (estimated)
Membership: Student $20; John Adams Travel Partner $25; Individual $50; Family $75; deSchweinitz Club $125; J.S. Goundie Club $250; James Burnside Society $500; Annie Kemerer Society $1,000; 1741 Society $2,000-$5,000.

COLONIAL INDUSTRIAL QUARTER, Main St. and Church St., Bethlehem, PA 18018. Mailing Address: 74 W. Broad St., Ste. 260, Bethlehem, PA 18018. Tel.: 610-882-0450. Fax: 610-882-0460.
E-mail: info@historicbethlehem.org
Web Site: www.historicbethlehem.org
Formerly: Historic Bethlehem Inc.
Founded: 1957.
Congressional District: 15
Key Personnel: Pres., Charlene Donchez Mowers; Vice Pres. & Mng. Dir., LoriAnn Wukitsch; Finance, Thomas Homanick; Chm., Joseph Kochanasz; Cur. Collections & Exhibits, Amy C. Frey.
Personnel Profile: Full-Time Paid 6; Part-Time Paid 20; Part-Time Volunteers 35.
Governing Authority: Parent Institution: Historic Bethlehem Partnership. Subsidiary Institutions: Burnside Plantation, Bethlehem, PA; Kemerer Museum of Decorative Arts, Bethlehem, PA; Moravian Museum of Bethlehem, Bethlehem, PA; Historic Bethlehem, Inc. Tax-exempt: 501(c)(3).
Institution Type/Description: Historical Site.
Collections: historic artifacts from 1700-1885 with special emphasis on Moravian industrial history. Historic Buildings: 1761 Tannery; 1762 Waterworks; 1869 Luckenbach Mill; 1782-1831 Miller's House; reconstructed Springhouse; 1750 Smithy. Historic House: 1810 Goundie House.
Research Fields: Bethlehem history; Moravian history; 18th-century industry & trades; crafts.
Activities: Museum Sponsors: demonstrations from June to September.
Publications: Historic Bethlehem Partnership Newsletter.
Hours & Admission Prices: Colonial Industrial Quarter Smithy: Summer & Fall Fri.-Sun. 12-4. Grounds: daily. Goundie House: Tues.-Sat. 10-5, Sun. 12-5. Visitors Center: Jan.-Nov. Tues.-Sat. 10-5, Sun. 12-5; Nov. 25-Dec. 23 Sun.-Wed. 10-6, Thurs.-Sat. 10-8; Dec. 26-Dec. 31 Mon.-Fri. 10-8, Sat. 10-3. &
Attendance: 10,000 (estimated)
Membership: Student $20; John Adams Travel Partner $25; Individual $50; Family & deSchweinitz Club $125; J.S. Goundie Club $250; James Burnside Society $500; Annie Kemerer Society $1,000; 1741 Society $2,000-$5,000.

KEMERER MUSEUM OF DECORATIVE ARTS, 427 N. New St., Bethlehem, PA 18018-5802. Mailing Address: c/o Historic Bethlehem, 74 W. Broad St., Ste. 260, Bethlehem, PA 18018. Tel.: 610-882-0450. Fax: 610-882-0460.
E-mail: info@historicbethlehem.org
Web Site: www.historicbethlehem.org
Founded: 1954.
Congressional District: 15
Key Personnel: Pres., Charlene Donchez Mowers; Vice Pres. & Mng. Dir., LoriAnn Wukitsch; Finance, Thomas Homanick; Chm., Joseph Kochanasz; Cur. Collections & Exhibits, Amy C. Frey; Museum Site Coord., Sara Mercer.
Personnel Profile: Full-Time Paid 6; Part-Time Paid 3; Part-Time Volunteers 50.
Governing Authority: nonprofit organization. Parent Institution: Historic Bethlehem Partnership. Tax-exempt: 501(c)(3).
Institution Type/Description: Regional Decorative Arts Museum.
Collections: 18th- & 19th-century American decorative arts; photographs, prints & stereographic views of regional interest; regional 19th-century oil landscape paintings.
Research Fields: design influences on regional decorative arts, local craftsman & artists life & collecting activities of Annie S. Kemerer.
Facilities: 750-vol. library of history books by appointment.
Activities: guided tours; lectures; educational programs, permanent & temporary exhibitions.
Publications: newsletter, Historic Bethlehem Partnership.

Hours & Admission Prices: Fri.-Sun. 12-4, holiday Thurs.-Fri. & Sun. 12-5, Sat. 10-5. Adults $5, senior citizens $4, youth 6-12 $2.50; discount to AAA members; HBP members no charge. &
Attendance: 10,000 (estimated)
Membership: Student $20; John Adams Travel Partner $25; Individual $50; Family $75; deSchweinitz Club $125; J.S. Goundie Club $250; James Burnside Society $500; Annie Kemerer Society $1,000; 1741 Society $2,000-$5,000.

LEHIGH UNIVERSITY ART GALLERIES/MUSEUM OPERA-TIONS, (M), Zoellner Arts Center, 420 E. Packer Ave., Bethlehem, PA 18015-3010. Tel.: 610-758-3615. Fax: 610-758-4580.
E-mail: db01@lehigh.edu
Web Site: www.luag.org
Founded: 1864.
Congressional District: 15
Key Personnel: Dir. & Cur., Ricardo Viera; Coord. Collections & Exhibitions, Mark Wonsidler; Asst. Dir. & Museum Shop Mgr., Denise Stangl; Editor, Patricia Kandianis; Collections Asst., Vasti DeEsch; Asst. Collections Mgr. & Preparator, Jeffrey W. Ludwig; Preparator, Khalil Allaik; Volunteer Coord., Patricia McAndrew.
Personnel Profile: Full-Time Paid 5; Part-Time Paid 4; Part-Time Volunteers 35.
Volunteer Hours: 1,146
Governing Authority: university. Parent Institution: Lehigh University. Subsidiary Institution: Art Galleries/Museum Operation. Tax-exempt.
Institution Type/Description: University Museum.
Collections: 18th- & 19th-century American, English & French paintings; prints & photography; 20th-century American, Ashcan & contemporary paintings; contemporary Latin American photography; 20th-century contemporary prints; European, American & Oriental prints & artist's books 19th- & 20th-century & contemporary photography; Etruscan Bronzes; 19th- & 20th-century African gold weights; pre-Columbian artifacts; contemporary American folk art; public contemporary sculpture garden collection.
Research Fields: methodology; Hispanic American art; contemporary art & photography.
Facilities: library of books & catalogs, available for staff & museum studies reference only.
Activities: guided tours; lectures; gallery talks; formally organized education programs for undergraduate college students; museum studies; inter-museum loan; permanent, traveling & temporary exhibitions; interactive children's workshop.
Publications: exhibition catalogs; current exhibitions calendar; announcements; posters.
Hours & Admission Prices: Zoellner Arts Center Galleries: Wed.-Sat. 11-5, Sun. 1-5. DuBois Gallery & Maginnes Hall: Mon.-Fri. 9-10, Sat. 9-12. Siegel Gallery & Iacocca Hall: Mon.-Thurs. 9-10, Fri. 9-5. Gallery at Rauch Business Center: Mon.-Fri. 8-10, Sat. 8-5. Call for Summer hours. No charge; donations accepted. Closed national holidays; school holidays. &
Attendance: 5,402 (accurate)

MORAVIAN MUSEUM OF BETHLEHEM, INC., (M), 66 W. Church St., Bethlehem, PA 18018. Mailing Address: 74 W. Broad St., Ste. 260, Bethlehem, PA 18018. Tel.: 610-882-0450. Fax: 610-882-0460.
Web Site: www.historicbethlehem.org
Founded: 1938.
Congressional District: 15
Key Personnel: Pres., Charlene Donchez Mowers; Finance, Thomas Homanick; Chm., Anne Zug; Cur. Collections & Exhibitions, Amy C. Frey; Site Coord., Melanie Depeinski.
Personnel Profile: Full-Time Paid 6; Part-Time Paid 21; Part-Time Volunteers 70; Interns 2.
Governing Authority: private; nonprofit. Parent Institution: Historic Bethlehem Partnership. Tax-exempt: 501(c)(3).
Institution Type/Description: Historic Site Museum: housed in 1741 Gemeinhaus. A National Historic Landmark.
Collections: domestic objects; religious artifacts; tools; textiles that document the Moravian community of Bethlehem. Historic Buildings: 1741 Gemeinhaus; 1752 Apothecary; 1758 Nain House.
Research Fields: Moravian history & decorative arts.
Activities: guided & self-guided tours; Moravian Community walking tour; docent program; lectures; musical programs.
Publications: book, Gemeinhaus; Historic Bethlehem Partnership e-newsletter.
Hours & Admission Prices: Jan. to late Nov. Thurs.-Sun. 12-4; late Nov. to Dec. Thurs.-Fri. & Sun. 12-5, Sat. 10-5. Adults $7, youth 6-12 $5; children under 6 no charge. Closed New Year's Day; Thanksgiving; Christmas Eve & Day.
Attendance: 5,000 (estimated)

Membership: Student $20; Teachers $40; Individual $50; Family $75; deSch-weinitz Club $125; J.S. Goundie Club $250; James Burnside Society $500; Annie Kemerer Society $1,000; 1741 Society $2,000-$5,000.

NATIONAL MUSEUM OF INDUSTRIAL HISTORY, (M), 530 E. Third St., Bethlehem, PA 18015-1314. Tel.: 610-694-6644. Fax: 610-694-6641.
E-mail: nmih@fast.net
Web Site: www.nmih.org
Founded: 1997.
Congressional District: 15
Key Personnel: C.E.O. & Pres., Stephen G. Donches; Chm. (V), Priscilla Payne Hurd; Treas., Theodore W. Harlan.
Personnel Profile: Full-Time Paid 2; Part-Time Volunteers 7; Interns 2.
Governing Authority: private; nonprofit organization. Tax-exempt: 501(c)(3).
Institution Type/Description: History Museum: housed on the site of the former Bethlehem Steel plant.
Collections: steam engines; slate industry equipment; short-line railroad memorabilia; machine tools.
Hours & Admission Prices: Closed for construction.

SUN INN PRESERVATION ASSOCIATION INC., 556 Main St., 2nd Fl., Bethlehem, PA 18018-5861. Tel.: 610-866-1758. Fax: 610-866-3360.
E-mail: suninnbethlehem@aol.com
Web Site: www.suninnbethlehem.org/
Founded: 1971.
Congressional District: 15
Key Personnel: Pres., John Howard; Chm. (V), Jean Kessler; Vice Pres., Seth Cornish; Treas., Forrest O'Brien; Innkeeper, Bucky Szwborski.
Personnel Profile: Part-Time Paid 3; Part-Time Volunteers 20; Interns 2.
Governing Authority: private; nonprofit organization. Tax-exempt: 501(c)(3).
Institution Type/Description: Historical Museum: housed in a Germanic stone building which hosted the military leaders, statesmen and the Founding Fathers during the American Revolution.
Collections: guest registers & inventory records of The Inn; furnishings; photographs.
Activities: school tours; guided tours; lectures. Annual Events: Strawberry Festival; Chocolate Fest; Christmas Tour & Film.
Publications: brochure three times a year, Sonnenschein; newsletter, Sonnen-schein.
Hours & Admission Prices: Fri.-Sun. 12-4. No charge; donations accepted. &
Attendance: 11,500 (accurate)
Membership: Senior Citizen $25; Individual $40; Family $55; Benefactor $125; Preservationist $250; Sponsor $500; Presidential $1,000.

Biglerville

NATIONAL APPLE MUSEUM, 154 W. Hanover St., Biglerville, PA 17307. Mailing Address: P.O. Box 656, Biglerville, PA 17307-0656. Tel.: 717-677-4556.
E-mail: info@nationalapplemuseum.com
Web Site: nationalapplemuseum.com
Founded: 1990.
Congressional District: 19
Key Personnel: Pres. (V), Harold L. Griffie; Archivist, Tim Smith.
Personnel Profile: Part-Time Volunteers 30.
Governing Authority: society; nonprofit. Owned & managed by Biglerville Historical and Preservation Society. Tax-exempt: 501(c)(3).
Institution Type/Description: Local History & Agriculture Museum: housed in restored Civil War barn.
Collections: apple production, processing & utilization history; associated cultural life artifacts; home furnishings; community memorabilia; photo collections; genealogy; land records; maps; farm equipment.
Research Fields: apple & related fruit production; processing & utilization history; land ownership patterns; genealogy.
Facilities: library of research material available to the public; 150-seat auditorium; 9,500 sq. ft. exhibit space; banquet facilities. Fruit gift packs, books, jewelry & other novelty items for sale.
Activities: guided tours; lectures; films; concerts; temporary exhibitions. Museum Sponsors: Apple Blossom Festival; Apple Harvest Festival; Flea Market Day; Founders Day Festival.
Hours & Admission Prices: May-Oct. Sat. 11-4, Sun. 1-4; groups & other times by appointment. Adults $3, seniors 60 & over $2, children 6-16 $1.50; discounts to AAM & ICOM members. &
Attendance: 3,500 (estimated)
Membership: Annual $5; Associate $10; Sustaining $25; Life $100.

Birdsboro

DANIEL BOONE HOMESTEAD, (M), 400 Daniel Boone Rd., Birdsboro, PA 19508-8735. Tel.: 610-582-4900. Fax: 610-582-1744.
E-mail: info@danielboonehomestead.org
Web Site: www.danielboonehomestead.org
Founded: 1937.
Congressional District: 6
Key Personnel: Pres. (V), Brad Kissam; Dir., Amanda Machik.
Personnel Profile: Full-Time Paid 1; Part-Time Paid 3; Part-Time Volunteers 100; Interns 1.
Governing Authority: state. Administered by the Friends of the Daniel Boone Homestead.
Institution Type/Description: Open Air Museum: located on site of Daniel Boone's birth.
Collections: rural decorative arts; furniture; agricultural & blacksmithing tools. Historic Buildings: 1730-1779 Boone House; smokehouse; barn; blacksmith shop; Bertolet log house; bake house; Bertolet Sawmill.
Research Fields: Pennsylvania History.
Facilities: wildlife sanctuary; visitor center; 579-acres of grounds; nature trails. Publications relating to Daniel Boone & Pennsylvania history for sale.
Activities: tours; education programs; living history programs; lectures; environmental education; camping facilities for youth groups.
Publications: Pennsylvania Historical & Museum Commission publications; site brochures; site booklet; newsletter, Friends.
Hours & Admission Prices: Recreation: Tues.-Fri. 9-4:45, Sat. 10-4:45, Sun. 11-4:45. Visitor Center & Historic Area: Fri.-Sat. 10-4, Sun. 12-4; see website for seasonal hours. Adults $6, senior citizens $5.50, children 5-11 $4; discounts to AAM & AAA members; members and children 4 & under no charge. Closed New Year's Day; Martin Luther King Jr. Day; Presidents' Day; Easter; Columbus Day; Veterans Day; Thanksgiving; Christmas.
Attendance: 41,000 (estimated)
Membership: Junior $5; Individual $20; Family $30; Patron $75; Sustaining $150; Benefactor $500.

Blairsville

HISTORICAL SOCIETY OF BLAIRSVILLE, 116 E. Campbell St., Blairsville, PA 15717-1310. Tel.: 724-459-0580.
Institution Type/Description: Historical Society Museum: housed in the former home of Nellie Stitt; built in 1909.
Collections: local history & culture; period furnishings; personal artifacts; photographs.
Facilities: Museum-related items for sale.
Hours & Admission Prices: Tues.-Sat. 10-2; other times by appointment.

Bloomsburg

BILL'S OLD BIKE BARN, 7145 Columbia Blvd., Bloomsburg, PA 17815-8635. Tel.: 570-759-7030. Fax: 570-759-9684.
E-mail: billsbikebarn@uplink.net
Web Site: www.billsbikebarn.com
Founded: 2000.
Key Personnel: Dir., William Morris; Museum Shop Mgr., Judy E. Laubach
Institution Type/Description: History Museum.
Collections: over 200 motorcycles; European carousel horses; 1939 New York World's Fair memorabilia; military artifacts; period clothing & shoes; posters; photographs; barber shop; music shop; smoke shop; train shop; post office; toys; radio repair shop.
Facilities: 40,000 sq. ft. exhibition space; banquet facilities.
Publications: museum brochure.
Hours & Admission Prices: Thurs.-Fri. 10-6, Sat. 9:30-3, Sun. 1-5. Adults $5.

THE CHILDREN'S MUSEUM, INC., (M), 2 W. Seventh St., Bloomsburg, PA 17815-2603. Tel.: 570-389-9206.
E-mail: info@the-childrens-museum.org
Web Site: www.the-childrens-museum.org
Founded: 1985.
Personnel Profile: Full-Time Paid 2; Part-Time Paid 4; Part-Time Volunteers 100.
Governing Authority: nonprofit organization. Tax-exempt.
Institution Type/Description: Children's Museum.
Collections: hands-on exhibits.
Facilities: Museum-related items for sale.
Hours & Admission Prices: Jan.-Feb. Fri.-Sat. 10-4; March 6 to June 9 Tues.-Fri. 12-4, Sat. 10-4; June 12 to Dec. 20 Tues.-Sat. 10-4. Adults $5; discounts to AAM members; children under 2 no charge. Closed major holidays. &

Attendance: 14,000 (accurate)
Membership: Individual $35; Just the 2 of Us $60; Family & Grandparents $75; Contributing $100.

COLUMBIA COUNTY HISTORICAL AND GENEALOGICAL SOCIETY, 225 Market St., Bloomsburg, PA 17815-0360. Mailing Address: P.O. Box 360, Bloomsburg, PA 17815-0360. Tel.: 570-784-1600.

E-mail: research@colcohist-gensoc.org
Web Site: www.colcohist-gensoc.org
Founded: 1914.
Congressional District: 11
Key Personnel: Pres., John R. Thomas; Exec. Dir., Bonnie Farver; Museum Dir., Barbara Parker.
Personnel Profile: Part-Time Volunteers 20.
Governing Authority: society; board of directors. Tax-exempt: 501(c)(3).
Institution Type/Description: Historical Society Museum.
Collections: local & state history; Indian artifacts; agricultural implements; 1870-1915 household items; newspapers; genealogies; photographs; microfilm.
Research Fields: genealogy; local county history & photos.
Facilities: 2,000-vol. library; reading room.
Activities: lectures; research services; courthouse research.
Publications: quarterly newsletter series on local subjects.
Hours & Admission Prices: Tues. & Fri. 9-3, Thurs. 9-7:30, Sat. 9am-11:30am. Library Research: $2 per hour; members no charge. ♿
Attendance: 2,877 (accurate)
Membership: Individual $20; Husband & Wife $25; Family $30; Life $200.

HAAS GALLERY OF ART - BLOOMSBURG UNIVERSITY, Haas Center for the Arts & Mitrani Hall, 2nd Fl., Bloomsburg, PA 17815. Mailing Address: Dept. of Art & Art History, 400 E. Second St., Bloomsburg, PA 17815. Tel.: 570-389-4708.

E-mail: rmorgan@bloomu.edu
Key Personnel: Gallery Assoc., Rebecca Rugg
Institution Type/Description: Art Gallery.
Collections: paintings; photographs; drawings; sculpture.
Activities: temporary exhibits.
Hours & Admission Prices: Mon.-Fri. 9-4, Sat. 12-2. No charge. Closed university holidays. ♿

Blue Bell

WISSAHICKON VALLEY HISTORICAL SOCIETY, 799 Skippack Pike, Blue Bell, PA 19422. Mailing Address: P.O. Box 96, Ambler, PA 19002-0096. Tel.: 215-646-6541.

E-mail: info@wvalleyhs.org
Web Site: www.wvalleyhs.org
Founded: 1976.
Governing Authority: Tax-exempt.
Institution Type/Description: Historical Society Museum: housed in the former Whitpain Public School; built 1895. Listed on the National Register of Historic Places.
Collections: local history & culture; photographs; personal artifacts; early furnishings.
Activities: Annual Event: Fall Market Day in October.
Hours & Admission Prices: Call for hours.

Boalsburg

BOALSBURG HERITAGE MUSEUM, 304 E. Main St., Boalsburg, PA 16827. Mailing Address: P.O. Box 346, Boalsburg, PA 16827. Tel.: 814-466-3035.

Institution Type/Description: History Museum.
Collections: local history & heritage; period furnishings; personal artifacts; photographs.
Hours & Admission Prices: Tues. & Sat. 2-4; other times by appointment. No charge; donations accepted.

COLUMBUS CHAPEL, BOAL MANSION MUSEUM, 163 Boal Estate Dr., Boalsburg, PA 16827. Mailing Address: P.O. Box 116, Boalsburg, PA 16827-0116. Tel.: 814-466-6210 & 9266. Fax: 814-466-6210.

E-mail: office@boalmuseum.com
Web Site: boalmuseum.com
Founded: 1952.
Congressional District: 5
Key Personnel: C.E.O. & Pres. (V), Christopher Lee.

Personnel Profile: Full-Time Paid 1; Part-Time Paid 4; Part-Time Volunteers 15; Interns 1.
Governing Authority: society; nonprofit organization. Tax-exempt: 501(c)(3).
Institution Type/Description: Historic House Museum: 1809, Boal Mansion.
Collections: 16th-century Christopher Columbus Chapel interior brought from Spain in 1909; history; art; weapons from Colonial period through World War I; period furnishings; Columbus relics; decorative arts; farm implements; carriages.
Research Fields: Centre County 1764-present; Christopher Columbus family 1451-present; American cultural history 1764-present.
Facilities: 750-vol. library of historical books available for research by written permission of director; Columbus & related family archives (1451-1902) on micro-film. Color slides & historical pamphlets for sale.
Activities: concerts; docent program; lectures; guided tours; formal education programs for children; stations tour for school students; Action Learning Experience for grades 4-6; rental facilities. Museum Sponsors: Memorial Day Festival; Columbus Day Ball; annual benefit musicale.
Publications: semi-annual, Columbus Chapel & Boal Mansion Museum Newsletter.
Hours & Admission Prices: May-Oct. Tues.-Sun. 1:30-5; Summer: Tues.-Sat. 10-5, Sun. 12-5. Adults $10, children $6; discount to prearranged group tours over 10. ♿
Attendance: 25,000 (estimated)
Membership: Patron $150; Sponsor $300; Benefactor $600; Grand Benefactor $1,500.

PENNSYLVANIA MILITARY MUSEUM AND 28TH DIVISION SHRINE, (M), 602 Boalsburg Pike, Boalsburg, PA 16827-1251. Mailing Address: P.O. Box 160A, Boalsburg, PA 16827-0660. Tel.: 814-466-6263. Fax: 814-466-6618.

E-mail: karlsmith@pa.gov
Web Site: www.pamilmuseum.org
Founded: 1969.
Congressional District: 5
Key Personnel: Dir. & Cur., Chuck Smith; Pres., Steve Schroder; Business Mgr. & Museum Shop Mgr., Danielle Hughs; Museum Educator, Joseph Horvath.
Personnel Profile: Full-Time Paid 4; Part-Time Paid 2; Part-Time Volunteers 15; Interns 2.
Governing Authority: state. Parent Institution: Pennsylvania Historical and Museum Commission, Box 1026; William Penn Memorial Museum, Harrisburg, PA 17108. Tax-exempt: 170(c)(1).
Institution Type/Description: Military Museum.
Collections: Pennsylvania military history & artifacts from 1747 to present; personal histories of Pennsylvania veterans.
Research Fields: military equipment, clothing, arms & related items; Pennsylvania militia units since 1747; Civil War regiment histories with unit rosters.
Facilities: 400-vol. library of books & pamphlets on military history available for inter-library loan & for research on premises.
Activities: guided tours; lectures; formally organized education programs for children; permanent & temporary exhibitions.
Publications: brochures.
Hours & Admission Prices: Museum: call for hours. Shrine & Museum Grounds: Summer Tues.-Sat. 9-5, Sun. 12-5; Winter call for hours. Adults 12-64 $6, seniors & groups $5.50, youths 3-11 $4; discounts to AAM members; children under 3 & members no charge. Closed New Year's Day; Thanksgiving; Christmas. ♿
Attendance: 15,000 (estimated)
Membership: Individual $25; Family $40.

Boyertown

BOYERTOWN AREA HISTORICAL SOCIETY, 43 S. Chestnut St., Boyertown, PA 19512-1508. Tel.: 610-367-5255.

E-mail: boyertownhistory@windstream.net
Web Site: www.boyertownhistory.org
Founded: 1972.
Congressional District: 130
Key Personnel: Pres. (V), Brian Quigley; Dir. Collections, Lindsay Dierolf.
Personnel Profile: Part-Time Paid 1; Part-Time Volunteers 20.
Governing Authority: nonprofit organization. Tax-exempt: 501(c)(3).
Institution Type/Description: Historical Society Museum: housed in the former home of George Unger & later the St. Columbkill Roman Catholic Church; built in 1902.
Collections: local history & culture; period furnishings; personal artifacts; photographs.
Research Fields: genealogy; local history.
Facilities: research library; meeting room.
Activities: walking tour. Museum Sponsors: Holiday House Tour; Craft Fair.

Publications: newsletter, The Gazette.
Hours & Admission Prices: Thurs. 1-9; other times by appointment. Adults $5; discounts to students under 18; members no charge. &
Attendance: 1,000 (estimated)
Membership: Individual $25; Family $40; Life $200; Corporate $250.

BOYERTOWN MUSEUM OF HISTORIC VEHICLES, (M), 85 S. Walnut St., Boyertown, PA 19512-1462. Tel.: 610-367-2090. Fax: 610-367-9712.
E-mail: mail@boyertownmuseum.org
Web Site: boyertownmuseum.org
Founded: 1965.
Congressional District: 6
Key Personnel: Exec. Dir., David V. Beard; Chm. Exec. Committee (V), Robert H. Dare; Pres. (V), Bernard Hofmann; Operations Mgr., Loretta Wolf; Facilities Supvr., Roderick Reinert; Cur., Kendra Cook; Museum Receptionist, Darlene Brunner.
Personnel Profile: Full-Time Paid 6; Part-Time Paid 1; Part-Time Volunteers 3.
Governing Authority: nonprofit organization. Tax-exempt: 501(c)(3).
Institution Type/Description: Transportation Museum.
Collections: horse drawn, gas, steam, & electric powered autos, trucks; fire apparatus; tools used by vehicle builders; art; models; S.E. Pennsylvania memorabilia; restored 1938 roadside diner; Antique Track Club of America archives.
Research Fields: horse drawn & mechanized vehicles and their builders.
Facilities: meeting room; banquet facilities. Museum-related items for sale.
Activities: self-guided & directed group tours; education program in schools; lectures; special exhibits; production of a one hour TV program seen on cable. Museum Sponsors: annual Duryea Days, antique & classic auto show; & other events.
Publications: book, A Century of Vehicle Craftsmanship; newsletter for members.
Hours & Admission Prices: Tues.-Sun. 9:30-4. Adults $6, seniors $5, students $4; discounts to AAA & AAM members; children under 6 no charge. Closed major holidays. &
Attendance: 4,900 (accurate)
Membership: Individual $20; Family $40; Associate $50; Sponsor $100; Sustaining $250; Patron $500; Benefactor $1,000.

Bradford

ZIPPO/CASE VISITORS CENTER, 1932 Zippo Dr., Bradford, PA 16701-5414. Mailing Address: 33 Barbour St., Bradford, PA 16701-1998. Tel.: 814-368-1932 & 2711. Fax: 814-368-2874.
E-mail: lmeabon@zippo.com
Web Site: www.zippo.com
Founded: 1994.
Congressional District: 5
Key Personnel: C.E.O. & Pres. Zippo Mfg. Co., Gregory W. Booth; Dir., Patrick Grandy; Cur. & Archivist, Linda Meabon; Museum Shop Mgr., Jesse Noga.
Personnel Profile: Full-Time Paid 4; Part-Time Paid 4.
Governing Authority: private organization. Parent Institution: Zippo Manufacturing Co., 33 Barbour St., Bradford, PA 16701.
Institution Type/Description: Company Museum.
Collections: Zippo lighters and other Zippo products & company history from the founding of the company in 1932 to present day; Case knives & Case company history from 1889 to present day; interactive exhibits; re-creation of the 1947 Chrysler Saratoga Zippo Car.
Facilities: Zippo & Case Cutlery products for sale.
Activities: guided tours-motorcoach.
Publications: annual reference guide, Zippo Lighter Collectors Guide.
Hours & Admission Prices: Mon.-Sat. 9-5, Sun. 11-4. No charge. Closed New Year's Day; Easter; Thanksgiving; Christmas. &
Attendance: 50,000 (estimated)

Bridgeville

WOODVILLE PLANTATION, 1375 Washington Pike, Bridgeville, PA 15017-2821. Tel.: 412-221-0348.
Institution Type/Description: History Museum: housed in the former home of John & Presley Neville; c.1780.
Collections: local history & culture; period furnishings; photographs; personal artifacts.
Activities: special events.
Hours & Admission Prices: Self-Guided Tours: Wed.-Sat. 10-6. Guided Tours: Sun. 1-4.

Bristol

MARGARET R. GRUNDY MEMORIAL MUSEUM, (M), 610 Radcliffe St., Bristol, PA 19007. Tel.: 215-788-9432 & 7891.
Web Site: www.grundyfoundation.com
Institution Type/Description: History Museum.
Collections: Grundy family history; period furnishings; personal artifacts; photographs.
Hours & Admission Prices: Tues.-Sat. 1-4; groups by appointment. No charge. Closed major holidays.

Bryn Athyn

GLENCAIRN MUSEUM: ACADEMY OF THE NEW CHURCH, (M), 1001 Cathedral Rd., Bryn Athyn, PA 19009-0757. Mailing Address: Box 757, Bryn Athyn, PA 19009-0757. Tel.: 267-502-2600 & 2990. Fax: 267-502-2986.
E-mail: info@glencairnmuseum.org
Web Site: www.glencairnmuseum.org
Founded: 1878.
Congressional District: 8
Key Personnel: Dir., Stephen H. Morley; Cur., C. Edward Gyllenhaal; Registrar & Collection Mgr., Bret Bostock; Operations Admin., Doreen F. Carey; Coord. Education, Diane M. Fehon; Outreach & Public Rels. Coord., Joralyn Echols; Asst. Operations Admin., Edwin Steiner; Collections Asst. & Concert Coord., Peter Childs.
Personnel Profile: Full-Time Paid 8; Part-Time Paid 17; Part-Time Volunteers 20; Interns 8.
Governing Authority: Parent Institution: The Academy of the New Church. Subsidiary Institutions: Glencairn Archives. Tax-exempt: 170(b)(1)(A).
Institution Type/Description: Religious Art Museum: housed in a Romanesque style building crafted in stained glass, mosaic, & sculptured granite; c.1939.
Collections: Native American artifacts; Egyptian sculpture & artifacts; Far Eastern; furniture; Greek & Roman sculpture, jewelry & artifacts; medieval stained glass, ivories, enamels, sculpture, arms & armor; 19th- & 20th-century art; oriental rugs; coins; tapestries; Mesopotamian sculpture & artifacts; model of tabernacle of Israel.
Research Fields: medieval art; classical archeology.
Facilities: 1,000-vol. library pertaining to art & medieval architecture; education facilities.
Activities: guided tours; concerts; organized education programs for schools, undergraduate & graduate students; interpreter program; temporary exhibitions; lecture series; workshops. Museum Sponsors: quarterly Open Houses.
Publications: newsletter; collection catalogs; exhibition catalogs.
Hours & Admission Prices: Mon.-Fri. 9-5 by appointment. Guided Tours: Mon.-Fri. 2:30, Sat. 1, 1:30, 2:30 & 3; reservations suggested. Adults $8, seniors & students $6; discounts for members; children 4 & under no charge. Fees subject to change. Closed national holidays.
Attendance: 19,600 (estimated)
Membership: Single $30; Family $40; Frequent Visitor $80.

Butler

BUTLER COUNTY HISTORICAL SOCIETY - SENATOR WALTER LOWRIE HOUSE, 123 W. Diamond St., Butler, PA 16001. Mailing Address: P.O. Box 414, Butler, PA 16003-0414. Tel.: 724-283-8116. Fax: 724-283-2505.
E-mail: society@butlerhistory.com
Web Site: www.butlerhistory.com
Founded: 1924.
Key Personnel: Dir., Patricia M. Collins; Pres. (V), Brett W. Ligo.
Personnel Profile: Part-Time Paid 3; Part-Time Volunteers 12; Interns 2.
Governing Authority: Tax-exempt.
Institution Type/Description: Historic House Museum: home built in 1828.
Collections: local history; period furnishings; personal artifacts; photographs; genealogy.
Research Fields: Butler County history & genealogy.
Hours & Admission Prices: Tours: by appointment. Office: Tues.-Fri. 10-4. Adults $5, children 16 & under $2. Closed legal holidays.
Attendance: 3,000 (estimated)

MARIDON MUSEUM, (M), 322 N. McKean St., Butler, PA 16001-4913. Fax: 724-282-0567.
Web Site: www.maridon.org
Founded: 2004.
Congressional District: 3
Key Personnel: Pres. Bd., Kenneth Bronder; Mktg. & Public Rels., Larry Berg.
Personnel Profile: Part-Time Paid 3.
Governing Authority: private; nonprofit organization. Tax-exempt: 501(c)(3).

Institution Type/Description: History Museum.
Collections: Chinese & Japanese history including jade; ivory; paintings; precious stones; bronzes; 18th-19th century German Meissen porcelain figures.
Facilities: library; classrooms. Museum-related items for sale.
Activities: guided tours; educational workshops; lectures; temporary exhibits. Annual Events: Chinese New Year Celebration; Meissen Fest; Hinamatsuri Girl's Festival; 753 Shichi Go San Festival for Children.
Publications: brochures; booklet, The Meridon Museum: A New Window on the World.
Hours & Admission Prices: Tues.-Sat. 11-5, Sun. 12-5. Adults $4, senior citizens & students $3; members and children 8 & under no charge. Closed New Year's Day; Easter; Memorial Day; Independence Day; Labor Day; Thanksgiving; Christmas. &
Attendance: 2,100 (estimated)
Membership: Student & Senior $20; Individual $25; Family $35; Contributor $100; Friend $250; Donor $500; Sustaining Patron $1,000; Museum Circle $2,500 & up; Lifetime $25,000.

California

THE GALLAGHER HOUSE - CALIFORNIA AREA HISTORI- CAL SOCIETY, 429 Wood St. @ 5th St., California, PA 15419. Mailing Address: P.O. Box 624, California, PA 15419-0624. Tel.: 724-938-3250.
Key Personnel: Pres. (V), Patricia Cowen
Governing Authority: Parent Institution: California Area Historical Society. Tax-exempt.
Institution Type/Description: Historic House: built in 1903.
Collections: local history & culture; photographs; period artifacts; obituaries 1900 to present; local funeral home & cemetery records; tax records; maps.
Research Fields: genealogy; local history.
Activities: research.
Publications: newsletter, The California Crier.
Hours & Admission Prices: Tues.-Thurs. 9-4; other times by appointment. No charge; donations accepted. Closed holidays.
Membership: Single $20; Couple $25; Corporate $100; Life $300.

Camp Hill

THE FRIENDS OF PEACE CHURCH, (M), St. John's and Trindle Roads, Camp Hill, PA 17011. Mailing Address: P.O. Box 3034, Shiremanstown, PA 17011-3034. Tel.: 717-737-6492.
Web Site: www.historicpeacechurch.org
Founded: 1979.
Congressional District: 19
Key Personnel: Pres., Vernon Cleary; Vice Pres., James Bower; Treas., Earnest Kepner; Sec., Charlene Cleary; Property Placement Officer, Janice Mullen; Museum Shop Mgr., Betty O'Neill.
Personnel Profile: Part-Time Paid 3; Part-Time Volunteers 25.
Governing Authority: state. Administered by the Pennsylvania Historical and Museum Commission, P.O. Box 1026, Harrisburg, PA 17120. Tax-exempt.
Institution Type/Description: Historic Site & Building: 1799 Georgian style stone church.
Collections: pewter communion service; 1807 Conrad Doll Organ made with wooden & pewter pipes; 19th-century German Bible & hymnbook; funeral bier; whale oil lamps; hanging oil lamps; 10 plate stoves; original pews; hourglass pulpit.
Facilities: Museum-related items for sale.
Activities: organ concerts; site interpretation; craft demonstrations; concerts; lectures; weddings; Good Friday service; Christmas Carol Sing; Easter Sunrise Candlelight Service.
Publications: quarterly newsletter, The Sounding Board.
Hours & Admission Prices: June-Sept. Sun. 2-5. No charge; donations accepted. &
Attendance: 9,290 (accurate)
Membership: Senior $10; Regular $15; Patron $25.

Carbondale

CARBONDALE HISTORICAL SOCIETY AND MUSEUM, One N. Main St., 3rd Fl., Carbondale, PA 18407. Mailing Address: P.O. Box 151, Carbondale, PA 18407-0151. Tel.: 570-282-0385.
E-mail: info@carbondalehistorical.org
Founded: 1983.
Institution Type/Description: Historical Society Museum: housed in the Carbondale City Hall building. Listed on the National Register of Historic Places.
Collections: local history, heritage & culture; photographs; period furnishings; dolls; personal artifacts.

Activities: lectures; educational programs; ceremonies; permanent & temporary exhibitions.
Hours & Admission Prices: Mon.-Fri. 12-5. No charge; donations accepted. &

Carlisle

CUMBERLAND COUNTY HISTORICAL SOCIETY, THE HAMILTON LIBRARY, THE TWO MILE HOUSE AND HISTORY ON HIGH THE SHOP, (M), 21 N. Pitt St., Carlisle, PA 17013-2945. Mailing Address: P.O. Box 626, Carlisle, PA 17013-0626. Tel.: 717-249-7610. Fax: 717-258-9332.
E-mail: info@historicalsociety.com
Web Site: www.historicalsociety.com
Founded: 1874.
Congressional District: 19
Key Personnel: Exec. Dir., Linda F. Witmer; Librarian, Cara Holtry; Museum Shop Mgr., Kim Laidler.
Personnel Profile: Full-Time Paid 8; Part-Time Paid 15; Part-Time Volunteers 250; Interns 8.
Governing Authority: society; nonprofit. Subsidiary Institution: Two Mile House, 1189 Walnut Bottom Rd., Carlisle, PA 17015; History on High The Shop, 33 W. High St., Carlisle, PA 17013. Tax-exempt.
Institution Type/Description: Local History Library & Museum.
Collections: Library: papers; books; newspapers; county records; genealogy; photo archives; publications; Indian School materials. Museum: provides visitors with a glimpse of early life & crafts of county; children's touch gallery; Indian School gallery.
Research Fields: Cumberland County history; Indian School materials; genealogy.
Facilities: 7,000-vol. library; genealogical files; reading room; visitors center.
Activities: guided tours; lectures; permanent & special exhibitions; AV aids; school educational programs.
Publications: quarterly newsletter; annual journal; annual photo essay books on county topic.
Hours & Admission Prices: Mon. 4-8, Tues.-Fri. 10-4, Sat. 10-2. Library: adults $5. Museum & Exhibits no charge; donations accepted. Closed New Year's Day; Memorial Day; Independence Day; Labor Day; Martin Luther King, Jr. Day; Presidents' Day; Thanksgiving; Christmas & day after. &
Attendance: 60,000 (accurate)
Membership: Individual $40; Family $45; Supporter $50; Patron $150; Business $250; Corporate $500; James Hamilton $500; Heritage Circle $1,000.

THE TROUT GALLERY, (M), 240 W. High St., Carlisle, PA 17013. Mailing Address: Dickinson College, P.O. Box 1773, Carlisle, PA 17013-2896. Tel.: 717-245-1344. Fax: 717-254-8929.
E-mail: trout@dickinson.edu
Web Site: www.troutgallery.org
Founded: 1983.
Congressional District: 19
Key Personnel: Dir., Dr. Phillip Earenfight; Registrar & Exhibition Preparator, James Bowman; Asst. Dir. Design Svcs., Patricia Pohlman; Dir. Design Svcs., Kimberley Nichols.
Personnel Profile: Full-Time Paid 4; Part-Time Paid 10; Interns 1.
Governing Authority: college; nonprofit. Parent Institution: Dickinson College. Tax-exempt.
Institution Type/Description: Art Museum.
Collections: art history; Gerofsky African art; Potamkin collection; Cole Oriental & decorative arts; Carnegie prints; 5,000 Old Master & modern prints.
Major Exhibits: Sue Coe Exhibition, 11/13-2/8/14.
Research Fields: pertaining to collections.
Facilities: 230-capacity auditorium; educational facilities; print study.
Activities: guided tours; lectures; films; concerts; organized education programs for area school and adult groups in addition to undergraduate college students affiliated with course work towards a Bachelor's of Art in Art history & studio art; internships; loan, temporary & traveling exhibitions; symposia.
Publications: catalogues, Toshiko Takaezu Ceramics Textiles & Bronzes; Wayne Thiebaud, Paintings, Drawings & Graphics 1961-1983; Homage to Alumni & Friends, Recent Gifts to Dickinson College's Fine Arts Collection; Etruscan Pottery: The Meeting of Greece & Etruria; 11 Contemporary Latin American Artists: Works On Paper; Rauschenberg's Surface Series; The Bible & 20th-Century Artists; Joseph Priestley in America, 1794-1804; A Historical Sculpture of Africa; The Grand Tradition: Nineteenth-Century Landscapes from the Permanent Collection; Objects of Diversity: Process and Interpretation; Fields of Vision: Selected Works by Contemporary Irish Artists; African Objects of Prestige & Personal Adornment from the Permanent Collection; A Decade of Giving, The Trout Gallery 1983-1993; Unraveling the Mask: Portraits of Twentieth-Century Experience; Visions

of Home: American Impressionist Images of Suburban Leisure and Country Comfort; brochures, Calendar of Shows per semester; 20th century American women artists: Selections from the Permanent Collection at Dickinson College; Vessels: Selections from the Permanent Collection; Writing on Hands: Memory and Knowledge in Early Modern European; Selective Visions: The Art of Ralston Crawford; 19th Century Life: A Closer Look; Images of Transience: Nature and Culture in Art; Grace Hartigan: Painting Art History; Woodcuts to Wrapping Paper: Concepts of Originality in Contemporary Prints; Quincy: Selected Paintings; Within the Landscape: Essays on Nineteenth-Century American Art and Culture; Inked Impressions: Ellen Day Hale and the Painter-Etcher Movement; A Kiowa's Odyssey: A Sketchbook from Fort Marion; America en plein air: Impressions by Henry Ryan MacGinnis; New Lives for Asian Images; Handentanden: Senior Studio Art Majors Exhibition; Joyce Kozloff: Co+Ordinates; Through the Lens: Studies in Photography; LIMINAL; Elusive Imprints: Translating the Unseen in the Twentieth Century; Found and Lost and Found: Senior Studio Art Majors Exhibition.
Hours & Admission Prices: Tues.-Sat. 10-4. No charge. Closed national holidays; Thanksgiving break; Christmas break. &
Attendance: 7,634 (accurate)
Membership: Supporter: Student $15; Individual $30; Dual $45; Family $60. Sustainer $250; Partner $500; Benefactor $1,000; The John Dickinson Society $2,500.

UNION FIRE COMPANY MUSEUM, 35 W. Louther St., Carlisle, PA 17013. Tel.: 717-243-2123.
E-mail: mrugh4128@aol.com
Web Site: www.unionfireco.org/museum
Governing Authority: Parent Institution: Union Fire Company No. 1.
Institution Type/Description: Firefighting History Museum: housed in the former fire station built in 1888.
Collections: firefighting history & equipment; leather fire buckets; helmets; meeting documents; ribbons.
Activities: group tours.
Hours & Admission Prices: Daily 10-8; groups by appointment. No charge; donations accepted.
Attendance: 500 (estimated)

UNITED STATES ARMY HERITAGE AND EDUCATION CENTER - ARMY HERITAGE MUSEUM, 950 Soldiers Dr., Carlisle, PA 17013-5021. Tel.: 717-245-3972. Fax: 717-245-4370 (COM) & 242-4370 (DSN).
E-mail: usamhi@carlisle.army.mil
Web Site: www.usahec.org
Founded: 1967.
Congressional District: 17
Key Personnel: Dir., Col. Mathew Dawson; Exec. Officer, Ltc. Christopher Lelgetol; Dir. US Army Military History Institute, Mr. Thomas Hendrix; Dir. Army Heritage Museum, John Leighow; Dir. Visitor & Education Svcs., John Giblin.
Personnel Profile: Full-Time Paid 73; Part-Time Paid 1; Part-Time Volunteers 39; Interns 3.
Governing Authority: federal. Parent Institution: U.S. Army War College. Tax-exempt.
Institution Type/Description: Military Museum.
Collections: US Army history, culture & personal artifacts; books; manuscripts; diaries & letters; photographs; Hessian Guard House.
Major Exhibits: The Soldier Experience, 11/12-9/20; A Great Civil War: 1863, 5/13-2/14; Veterabs: Artist Nina Talbot, 10/13-9/14.
Research Fields: history of U.S. army; world military history.
Facilities: military research facility; army heritage trail; visitor & education center.
Activities: self-guided tours; special events.
Publications: brochures, descriptions of main collection & special programs.
Hours & Admission Prices: Hessian: March-Nov. Mon.-Sat. 9-5, Sun. 11-5; Dec.-Feb. Tues.-Sat. 9-5, Sun. 11-5. No charge. Closed some federal holidays. &
Attendance: 123,000 (accurate)
Membership: Foundation: Minuteman $25; Continental $50; Regulars $100; Rough Rider $250; Doughboy $500; Ranger $1,000; Quartermaster $2,500; GI Lifetime $5,000.

Catawissa

CATAWISSA RAILROAD CO., 119 Pine St., Catawissa, PA 17820-1239. Tel.: 570-356-2345. Fax: 570-356-7876.
E-mail: walt@caboosenut.com
Web Site: caboosenut.com
Founded: 1980.
Institution Type/Description: Railroad Museum.

Collections: 14 railroad cabooses; 1931 Davenport steam engine & tender; 400 ft. of track; railroad memorabilia.
Hours & Admission Prices: Daily 8-5. No charge.
Attendance: 2,000 (estimated)

Centre Hall

PENN'S CAVE & WILDLIFE PARK, 222 Penns Cave Rd., Centre Hall, PA 16828-8103. Tel.: 814-364-1664. Fax: 814-364-8778. Facebook: Penn's Cave 1885.
E-mail: info@pennscave.com
Web Site: www.pennscave.com
Founded: 1885.
Congressional District: 5
Institution Type/Description: Cavern & Wildlife Park.
Collections: North American wildlife including bears, wolves, elk, deer, bobcats, bison, longhorn cattle, mustangs, & cougar; local natural history; geology; biology.
Facilities: Museum-related items for sale.
Activities: motorboat cave & wildlife tours; special events; guided tours.
Hours & Admission Prices: Call for hours. Wildlife Park: adults $19.95, senior citizens 65 & over $18.95, children 2-12 $11.95; discounts to military; children under 2 no charge. Cave: adults $16.95, senior citizens 65 & over $15.95, children 2-12 $8.95; discounts to military. Cave & Wildlife Park: adults $30.95, seniors citizens 65 & over $29.95, children 2-12 $16.95; discounts to military; children under 2 no charge. &

Chadds Ford

BRANDYWINE BATTLEFIELD PARK, (M), 1491 Baltimore Pike, Chadds Ford, PA 19317-7369. Mailing Address: Brandywine Battlefield Park Assoc., P.O. Box 202, Chadds Ford, PA 19317. Tel.: 610-459-3342, ext. 3001. Fax: 610-459-9586.
E-mail: judythorpe@verizon.net
Web Site: www.brandywinebattlefield.org
Founded: 1947.
Congressional District: 5
Key Personnel: Pres. (V), Richard Bowers; Site Admin., Michael Bertheaud.
Personnel Profile: Part-Time Paid 5; Part-Time Volunteers 75; Interns 2.
Governing Authority: state. Parent Institution: Pennsylvania Historical & Museum Commission. Tax-exempt.
Institution Type/Description: Historic Site: site of the Battle of the Brandywine.
Collections: history; costumes; period furnishings; prints; Revolutionary War equipment & firearms. Historic Houses: Gen. Washington's headquarters; Lafayette's quarters.
Research Fields: Revolutionary War; Battle of Brandywine, 1777; Quakers' Lifestyles.
Facilities: picnic areas; Visitor Center with exhibits. Museum-related items for sale.
Activities: guided tours; inter-museum loan & permanent exhibitions; military reenactments; organized educational programs; special events.
Publications: booklet, Driving Tour of Brandywine Battlefield.
Hours & Admission Prices: March 12-March 31 Fri.-Sat. 9-4, Sun. 12-4; April-Nov. Wed.-Sat. 9-4, Sun. 12-4. Adults $8, senior citizens & youth 6-14 $5; children under 6 & military no charge.
Attendance: 20,000 (estimated)
Membership: Brandywine Battlefield Associates: Individual $25; Family $35; Contributor $60; Benefactor $125; Patron $250; Guardian $500.

* **BRANDYWINE RIVER MUSEUM, (M),** U.S. Rte. 1, at Hoffman's Mill Rd., Chadds Ford, PA 19317. Mailing Address: P.O. Box 141, Chadds Ford, PA 19317-0141. Tel.: 610-388-2700. Fax: 610-388-1197.
E-mail: inquiries@brandywine.org
Web Site: www.brandywinemuseum.org.
Founded: 1971.
Congressional District: 7
Key Personnel: C.E.O. & Dir., Virginia A. Logan; Chm. (V), George A. Weymouth; Pres. (V), Wendell Fenton, Esq.; Cur. Collections, Virginia H. O'Hara; Assoc. Cur., Audrey Lewis; Assoc. Cur. N.C. Wyeth Collections, Christine B. Podmaniczky; Asst. Cur., Amanda C. Burdan; Registrar, Jean A. Gilmore, Dir. Finance & Administration, Joel E. Necowitz; Dir. Public Rels., Hillary Holland; Museum Shop Mgr., Erika G. Bucino; Supvr. Education, Mary W. Cronin; Asst. Educator, Jane V. Flitner; Chief of Security, Robert Booker; Dir. Devel., Suzanne M. Regnier; Dir. Volunteers, Donna M. Gormel.
Personnel Profile: Full-Time Paid 68; Part-Time Paid 63; Part-Time Volunteers 398; Interns 3.

Governing Authority: nonprofit organization. Parent Institution: Brandywine Conservancy. Branch Museum: N.C. Wyeth House & Studio; Kuerner Farm; Andrew Wyeth Studio. Tax-exempt: 501(c)(3).
Institution Type/Description: Art Museum: housed in 1864 Hoffman's Mill located on the Brandywine River.
Collections: American art with emphasis on art of Brandywine region; still-life painting landscape illustration; artist studios; art of N.C., Andrew & Jamie Wyeth.
Research Fields: art history of the region; American illustration & still-life painting.
Facilities: 11,000-vol. library on American art available by appointment; 150-seat lecture room; 120-seat restaurant. Books, reproductions & museum-related items for sale.
Activities: permanent & temporary exhibitions; guided tours; gallery talks; formally organized education programs; docent program; concerts.
Publications: exhibition catalogs; quarterly newsletter, Catalyst; Catalogue of the Permanent Collection.
Hours & Admission Prices: Daily 9:30-4:30. Adults $12, senior citizens $8, students & children 6-12 $6; discounts to AAM, ICOM & AAMD members; children under 6 & members no charge. Closed Christmas. &
Attendance: 101,000 (accurate)
Membership: Senior & Educator $40; Individual $50; Dual $70; Family $75; Donor $125; Patron $250; Sponsor $500; Benefactor $1,000; Preservation Fellow $5,000; Trustee Circle $10,000.

CHADDS FORD HISTORICAL SOCIETY, (M), 1736 Creek Rd., Chadds Ford, PA 19317. Mailing Address: P.O. Box 27, Chadds Ford, PA 19317-0027. Tel.: 610-388-7376.
E-mail: info@chaddsfordhistory.org
Web Site: www.chaddsfordhistory.org
Founded: 1968.
Key Personnel: Exec. Dir., Ginger Tucker; Pres. (V), Dr. George Franz; Education Coord. & Collections Mgr., Lynda Gillow; Office Mgr., Aleida Diaz.
Personnel Profile: Part-Time Paid 3; Part-Time Volunteers 3; Interns 3.
Institution Type/Description: Historical Society Museum.
Collections: local history; photographs.
Activities: permanent exhibitions; young adult & adult educational programs; school programs; demonstrations; special events; interpretive videos; youth educational tours.
Publications: newsletter, Notes From the Ford.
Hours & Admission Prices: House Tours: May-Sept. Sat.-Sun. 1-5; other times by appointment. Barn Visitor's Center: Mon.-Fri. 9-2, Sat.-Sun. 1-5. Barn: no charge. Each House: $5 per person; members no charge.
Attendance: 20,000 (estimated)
Membership: Individual $25; Family $40; Contributor $75; Patron $125; Sustaining $250; Benefactor $500; Lifetime $3,000.

THE CHRISTIAN C. SANDERSON MUSEUM, 1755 Creek Rd., Chadds Ford, PA 19317. Mailing Address: P.O. Box 153, Chadds Ford, PA 19317-0153. Tel.: 610-388-6545.
E-mail: info@sandersonmuseum.org
Web Site: www.sandersonmuseum.org
Founded: 1967.
Congressional District: 5
Key Personnel: Pres. (V), Susan M. Minarchi; Cur., Charles E. Ulmann.
Personnel Profile: Part-Time Paid 1; Part-Time Volunteers 26.
Governing Authority: nonprofit organization. Tax-exempt: 501(c)(3).
Institution Type/Description: History Museum.
Collections: American Revolution to WWII; personal & family items; historical artifacts; early Wyeth family items.
Hours & Admission Prices: March-Nov. Thurs.-Sun. 12-4; groups by appointment. Adults $5; members no charge. Closed Easter; Thanksgiving; Christmas.
Attendance: 1,000 (estimated)
Membership: Individual $20; Family $30; Patron $100; Benefactor $500.

Chambersburg

CHAMBERSBURG VOLUNTEER FIREMAN'S MUSEUM, 441 Broad St., Chambersburg, PA 17201. Tel.: 717-263-6215.
Web Site: www.chambersburgfire.com
Institution Type/Description: Firefighting History Museum.
Collections: firefighting history & equipment; fire trucks; trophies; marching uniforms; musical instruments; photographs.
Hours & Admission Prices: Call for hours.

Cheltenham

CHELTENHAM CENTER FOR THE ARTS, 439 Ashbourne Rd., Cheltenham, PA 19012. Tel.: 215-379-4660. Fax: 215-663-1946.
E-mail: info@cheltenhamarts.org
Web Site: cheltenhamarts.org
Institution Type/Description: Art Gallery.
Collections: works by artists of Cheltenham & Philadelphia areas.
Activities: special events; temporary exhibitions.
Hours & Admission Prices: Call for hours.

Chester

WIDENER UNIVERSITY ART COLLECTION AND GALLERY, (M), 14th & Chestnut Sts. (University Center), Chester, PA 19013. Mailing Address: One University Pl., Chester, PA 19013-5792. Tel.: 610-499-1189 & 4000. Fax: 610-499-4425.
E-mail: rmwarda@widener.edu
Web Site: www.widener.edu/artgallery
Founded: 1970.
Congressional District: 7
Key Personnel: Collections Mgr., Rebecca M. Warda.
Governing Authority: university. Tax-exempt: 501(c)(3).
Institution Type/Description: University Art Museum.
Collections: 19th- & 20th-century American paintings & sculpture; Alfred O. Deshong Collection: 18th- & 19th-century European genre paintings; 18th- & 19th-century Oriental art objects; 19th-century American paintings.
Research Fields: American impressionism; Deshong genealogy.
Activities: lectures; temporary exhibitions.
Publications: pamphlets.
Hours & Admission Prices: Sept.-May Tues. 10-7, Wed.-Sat. 10-4:30; June-Aug. call for hours. No charge. Closed major holidays. &
Attendance: 4,000 (estimated)

Clarion

CLARION COUNTY HISTORICAL SOCIETY, SUTTON-DITZ HOUSE MUSEUM, 17 S. Fifth Ave., Clarion, PA 16214-1501. Tel.: 814-226-4450. Fax: 814-226-7106.
E-mail: clarionhistory@comcast.net
Founded: 1955.
Congressional District: 23
Key Personnel: Dir. & Cur., Mary Lea Lucas; Chm. (V), Berlie Etzel, Jr.; Museum Shop Mgr., Karen Slack.
Personnel Profile: Full-Time Volunteers 2; Part-Time Volunteers 26; Interns 1.
Governing Authority: society; nonprofit organization. Tax-exempt: 501(c)(3).
Institution Type/Description: County History Museum: located in c.1850 home, renovated in 1909.
Collections: local history; manuscripts; 1800-present historical photographs.
Research Fields: Western Pennsylvania history.
Facilities: 3,000-vol. library of books of genealogical & local historical interest available on premises.
Activities: guided tours of the Museum; lectures; & historical programs; temporary & permanent exhibits; bus tours.
Publications: Century Homes & Buildings of Clarion County, Vols. I, II & III; quarterly newsletter, Clarion County & Its Beginnings; cemetery listings & obituaries of Clarion County.
Hours & Admission Prices: Museum: Tues.-Fri. 10-4; other times by appointment. No charge; donations accepted. Closed major holidays. &
Attendance: 1,800 (estimated)
Membership: Individual $15; Family $25; Sustaining $35; Patron $50; Life $300.

UNIVERSITY GALLERIES, (M), Clarion Univ. of Pennsylvania, Carlson Library A-4, Clarion, PA 16214. Tel.: 814-393-2523. Fax: 814-393-2168.
E-mail: gallery@clarion.edu
Formerly: Sandford Gallery
Founded: 1982.
Congressional District: 23
Key Personnel: Dir., Vicky A. Clark.
Personnel Profile: Part-Time Paid 7.
Governing Authority: university. Tax-exempt: 501(c)(3).
Institution Type/Description: Art Gallery.
Collections: 19th- & 20th-century American art; Kuba textiles.
Facilities: 1,120 sq. ft. exhibit space.
Activities: lectures; organized educations programs for students affiliated with Clarion University; participatory, temporary & traveling exhibitions.
Publications: show brochures.

Hours & Admission Prices: Mon.-Thurs. 10-3. No charge. Closed university holidays.
Attendance: 1,500 (estimated)
Membership: Annual $25; University Club $1,000.

Clearfield

CLEARFIELD COUNTY HISTORICAL SOCIETY, 104 E. Pine St., Clearfield, PA 16830-2517. Tel.: 814-765-6125.
Web Site: www.clfdhistory.com
Founded: 1955.
Congressional District: 23
Key Personnel: Pres., Denny Shaffner; 1st Vice Pres., Susan Williams; Sec., Mary Kay Royer; Membership, Cathie Hughes; Treas., Brent Thomas; Museum Shop Mgr., Warren Fox.
Personnel Profile: Part-Time Volunteers 21.
Governing Authority: society. Tax-exempt.
Institution Type/Description: Local History Museum: housed in c.1880, 3-story brick building.
Collections: Indian flints; built-in mine mouth; lumbering display; PA primitives; Central PA history; Clearfield County genealogy records; carriage house; Bloody Knox log cabin.
Research Fields: family genealogy; local history.
Activities: guided tours; lectures; annual dinner.
Publications: semi-annual bulletin; pamphlets; reprints; fall & spring bulletin.
Hours & Admission Prices: May-Oct. Thurs. & Sun. 1:30-4:30; other times by appointment. No charge; donations accepted.
Attendance: 1,300 (accurate)
Membership: Student $2; Individual $10; Business, Club, Industry $35; Life $150.

Coalport

COALPORT AREA COAL MUSEUM, 961 Forest St., Coalport, PA 16627. Mailing Address: P.O. Box 248, Coalport, PA 16627-0248. Tel.: 814-672-4378 & 312-8620.
E-mail: rwsnyder@windstream.net
Web Site: www.coalportmuseum.org
Formerly: Coalport Area Museum Commission
Founded: 1991.
Congressional District: 9
Key Personnel: Pres. (V), Robert J. Counsman, Jr.; Cur., Richard W. Snyder, II
Governing Authority: Parent Institution: Coalport Area Museum Commission.
Institution Type/Description: History Museum.
Collections: coal mining industry, heritage & history; photographs; personal artifacts; tools; company documents; genealogy; local obituaries; family tree files.
Research Fields: coal mining history; local genealogy.
Facilities: library; reference room; community room.
Publications: annual newsletter.
Hours & Admission Prices: Tues.-Wed. 10-4, Thurs. 12-8, Fri. 9-12, Sun. 1-4. No charge; donations accepted.
Membership: Life $25.

Collegeville

* **PHILIP AND MURIEL BERMAN MUSEUM OF ART AT URSINUS COLLEGE, (M),** 601 E. Main St., Collegeville, PA 19426-1000. Mailing Address: P.O. Box 1000, Collegeville, PA 19426-1000. Tel.: 610-409-3500. Fax: 610-409-3664.
E-mail: lhanover@ursinus.edu
Web Site: www.ursinus.edu/berman
Founded: 1987.
Congressional District: 13
Key Personnel: Pres., Dr. John Strassburger; Chm. Art Advisory Bd., Robert L. Brant, Esq.; Dir., Lisa Tremper Hanover; Administrative Asst., Suzanne Calvin; Collections Mgr., Julie Choma; Assoc. Dir. Education, Susan Shifrin.
Personnel Profile: Full-Time Paid 4; Part-Time Volunteers 10; Interns 12.
Governing Authority: private university; nonprofit. Parent Institution: Ursinus College. Tax-exempt: 501(c)(3).
Institution Type/Description: Art Museum.
Collections: 19th- & early 20th century American landscape, portrait & genre paintings, prints & watercolors, emphasis on Pennsylvania Impressionist & regional schools; Old Master & contemporary Japanese woodcut prints, scrolls & artifacts; 18th- & 19th-century European portraits; 1950-1990 American painting & graphics; Pennsylvania German Fraktur; documents; artifacts; eastern European paintings; contemporary sculpture; private collection of British sculptor Lynn Chadwick.

Research Fields: 19th-century Pennsylvania regional painting (New Hope school); Pennsylvania German history & culture.
Facilities: 1,200-vol. library of art material available for inter-library loan & to the public; educational facilities. Museum-related items for sale.
Activities: guided tours; lectures; films; loan, temporary & traveling exhibitions; organized education programs for undergraduate & graduate students affiliated with Ursinus College. Museum Sponsors: Friends Annual Event.
Publications: quarterly calendar of events; exhibition catalogues; Friends of the Museum Newsletter.
Hours & Admission Prices: Tues.-Fri. 10-4, Sat.-Sun. 12-4:30. No charge; donations accepted. Closed college holidays. &
Attendance: 32,000 (estimated)
Membership: Patron $50; Sponsor $100; Collector Club $250; Connoisseur Club $750; American Impressionists $1,500; Chadwick Society $2,500; Old Master Association $5,000.

Columbia

COLUMBIA MUSEUM OF HISTORY - COLUMBIA HISTORIC PRESERVATION SOCIETY, 19-21 N. Second St., Columbia, PA 17512. Mailing Address: P.O. Box 578, Columbia, PA 17512.
Founded: 1984.
Key Personnel: Dir. & Pres. (V), Christopher Vera.
Personnel Profile: Part-Time Volunteers 10.
Governing Authority: Tax-exempt.
Institution Type/Description: History Museum: housed in the former First Evangelical Lutheran Church; built in 1853.
Collections: local history & culture; photographs; personal artifacts; period furnishings. Historic Buildings: 1853 church; 1890 Parsonage English Evangelical-Lutheran.
Research Fields: Columbia newspaper microfilm; Columbia directories 1887-1963; Columbia census; Crawford index card file.
Activities: genealogy research; educational programs; lectures; kids programs. Museum Sponsors: Underground Railroad in October.
Hours & Admission Prices: Sun. & Sun. 1-4; other times by appointment. No charge; donations accepted.
Attendance: 2,000 (estimated)
Membership: Student $10; Individual $15; Family $20; Lifetime $200.

FIRST NATIONAL BANK MUSEUM, 170 Locust St., Columbia, PA 17512-1109. Tel.: 717-684-8864; 717-341-7229. Fax: 717-684-8048. Facebook: First National Bank Museum.
E-mail: bankmuseum@yahoo.com
Key Personnel: Owner & Cur., Nora Motter Stark; Owner & Cur., Michael Stark
Institution Type/Description: History Museum.
Collections: bank history; period artifacts & furnishings including teller cages, the President's office, walk-in vault, & check canceller.
Hours & Admission Prices: Call for hours. Suggested donation $8-$10.

* **THE NATIONAL WATCH & CLOCK MUSEUM, (M),** 514 Poplar St., Columbia, PA 17512-2124. Tel.: 717-684-8261. Fax: 717-684-0878.
Web Site: www.museumoftime.org
Founded: 1971.
Congressional District: 16
Key Personnel: Exec. Dir., J. Steven Humphrey; Museum Dir., Noel Poirier; Cur., Carter Harris; Library Dir., Sharon Gordon; Librarian, Nancy Dyer.
Personnel Profile: Full-Time Paid 20; Part-Time Paid 13; Part-Time Volunteers 27.
Governing Authority: nonprofit organization. Parent Institution: NAWCC, Inc. Tax-exempt: 501(c)(3).
Institution Type/Description: Horological Museum.
Collections: American & foreign clocks, watches, horological tools; timekeeping from the sundial to GPS; NYU & James Arthur collection.
Research Fields: horology; watch & clock industry.
Facilities: 5,000-vol. library of reference books; 2,000-vol. lending library for members only; pamphlets, catalogues & ephemera on horology & file of 40,000 U.S. & foreign horological patents; rare book collection of horological material including some formerly in the Franklin Institute; Hamilton Watch Company records.
Activities: special exhibitions. School of Horology: clock repair, watch repair, reverse painting on glass techniques; school programs; public programs.
Publications: bimonthly Bulletin; bimonthly Mart.
Hours & Admission Prices: April-Nov. Tues.-Sat. 10-5, Sun. 12-4; Dec.-March Tues.-Sat. 10-4. Adults $7; discounts to children, senior citizens, groups, AAA & AAM members; ASTC passport & NAWCC members no charge. Closed holidays. &

Attendance: 15,000 (estimated)
Membership: Annual $70; Lifetime under 40 $3,000; Lifetime 40 & over $2,000.

WRIGHT'S FERRY MANSION, (M), 38 S. Second St., Columbia, PA 17512-1402. Mailing Address: P.O. Box 68, Columbia, PA 17512-0068. Tel.: 717-684-4325.
Founded: 1974.
Congressional District: 16
Key Personnel: Exec. Dir., Thomas Cook; Cur., Elizabeth Meg Schaefer.
Personnel Profile: Full-Time Paid 2; Part-Time Paid 2.
Governing Authority: nonprofit organization. Parent Institution: The von Hess Foundation. Tax-exempt: 501(c)(3) & 509(a).
Institution Type/Description: Historic House: 1738 Susanna Wright House.
Collections: early 18th-century Philadelphia furniture, English ceramics, glass, textiles & metals.
Research Fields: early 18th-century decorative arts, architecture & interiors; early 18th-century Philadelphia; Susanna Wright; Quaker social & cultural life.
Activities: guided tours; lectures.
Publications: brochure; two-vol. set of books, Wright's Ferry Mansion: The House and the Collection.
Hours & Admission Prices: May-Oct. Tues.-Wed. & Fri.-Sat. 10-3 (last tour begins at 3). Adults $5, children 6-18 $2.50; discount to groups. Closed holidays.
Attendance: 2,000

Concordville

PIERCE-WILLITS MAIN HOUSE, Smithbridge Rd., Concordville, PA 19331. Mailing Address: Concord Township Historical Society, P.O. Box 152, Concordville, PA 19331. Tel.: 610-459-8911.
Institution Type/Description: Historical Society Museum.
Collections: local history & culture; period furnishings; documents; personal artifacts.
Hours & Admission Prices: By appointment.

Conshohocken

MONTGOMERY COUNTY GUILD OF PROFESSIONAL ART-ISTS (MCGOPA), Inquirer Bldg., Rte. 23, Conshohocken, PA 19428. Mailing Address: P.O. Box 60736, King of Prussia, PA 19406. Tel.: 610-803-3248.
E-mail: mcgopa@comcast.net
Key Personnel: Dir., Maria Lourdes Solomon.
Governing Authority: nonprofit organization. Tax-exempt: 501(c)(3).
Institution Type/Description: Art Gallery.
Collections: paintings; sculpture; drawings.
Activities: special events.
Hours & Admission Prices: Call for hours.

Coolspring

COOLSPRING POWER MUSEUM, 179 Coolspring Rd., Coolspring, PA 15730. Mailing Address: P.O. Box 19, Coolspring, PA 15730-0019. Tel.: 814-849-6883.
E-mail: cpm@coolspringpowermuseum.org
Web Site: www.coolspringpowermuseum.org
Founded: 1985.
Congressional District: 5
Key Personnel: Dir., Vance Packard; Dir., Paul Harvey, M.D.; Dir., Chris Austin; Dir., Douglas Fye, Jr.; Dir. & Financial Dir., Jennifer Fye; Pres., Clark Colby; Dir., John Hanley; Dir., Kim Himes; Dir., Kevin Kusel; Dir., Lee Caylor; Dir., Glenn Anthony; Dir., Michael Murphy; Dir., Tommy Turner; Chm. (V) & Membership, Gail Lavender; Museum Shop Mgr., Fran Colby.
Personnel Profile: Part-Time Paid 1; Part-Time Volunteers 171.
Governing Authority: nonprofit organization. Tax-exempt: 501(c)(3).
Institution Type/Description: Historical Museum.
Collections: internal combustion engines; pumps; electrical generation equipment; oil field artifacts; oil pipeline equipment; machine tools.
Research Fields: early builders of internal combustion engines; oil line & petroleum production companies; developmental technology of the internal combustion engine.
Facilities: 1,490-vol. library of patent drawings, engineering drawings for several early engine manufactures, internal combustion texts, equipment catalogs & related periodicals, available for use by the public; 25,100 sq. ft. exhibit space; 41-acres of outside display area. Museum-related items for sale.

Activities: guided tours; lectures; films, docent program; participatory & loan exhibitions. Annual Event: Antique Power Exhibition & Show.
Publications: annual technical journal, Bores & Strokes; bimonthly newsletters, CPM News, Iron & Oil Review.
Hours & Admission Prices: April-Oct. 3rd. full weekend of month 10-5; other times by appointment. Adult $6, senior citizen, children under 12 & students $1; discounts to groups; children & members no charge. &
Attendance: 19,125 (estimated)
Membership: Individual $30; Family $40; Patron $1,000; Cornerstone Patron $5,000; Endowment Patron $10,000.

Cornwall

CORNWALL IRON FURNACE, (M), 94 Rexmont Rd., Cornwall, PA 17016. Mailing Address: P.O. Box 251, Cornwall, PA 17016-0251. Tel.: 717-272-9711. Fax: 717-272-0450.
E-mail: ssomers@pa.gov
Web Site: www.cornwallironfurnace.org
Founded: 1932.
Congressional District: 16
Key Personnel: Historic Site Admin., Stephen G. Somers.
Personnel Profile: Full-Time Paid 2; Part-Time Paid 1; Part-Time Volunteers 15.
Governing Authority: state. Parent Institution: Pennsylvania Historical & Museum Commission, 300 North St., Harrisburg, PA 17120. Subsidiary Institution: Cornwall Iron Furnace Associates, Inc. Tax-exempt.
Institution Type/Description: Industrial Museum.
Collections: charcoal process; ironmaking artifacts; geology; 19th-century horse-drawn vehicles; charcoal house; preserved 19th-century furnace complex; roasting oven; blacksmith shop; smokehouse.
Facilities: visitor center; picnic area.
Activities: guided tours; film series; lectures; craft demonstrations; changing exhibits.
Publications: books, booklets on industrial state & local history.
Hours & Admission Prices: Tues.-Sat. 9-5, Sun. 12-5. Adults $6, senior citizens 65 & over $5.50, youth 3-12 $4; discounts to AAM members; children under 6 no charge. Closed New Year's Day; Martin Luther King Jr. Day; Presidents' Day; Columbus Day; Veterans Day; Thanksgiving & day after; Christmas. &
Attendance: 9,000 (estimated)
Membership: Cornwall Iron Furnace Associates: Student $15; Senior Citizen $18; Individual $20; Family $30.

Corry

CORRY AREA HISTORICAL SOCIETY, 945 Mead Ave., Corry, PA 16407. Mailing Address: P.O. Box 107, Corry, PA 16407-0107. Tel.: 814-664-4749.
E-mail: admin@corryareahistoricalsociety.org
Web Site: www.corryareahistoricalsociety.org
Founded: 1965.
Congressional District: 24
Key Personnel: C.E.O., Ann Clark; Vice Pres., Jan Bemis; Sec., Loretta Beckerink; Treas. & Cur., Alisa Puckly; Treas., Connie Burkhart.
Personnel Profile: Part-Time Volunteers 20.
Governing Authority: Parent Institution: City of Corry. Tax-exempt.
Institution Type/Description: Local History Museum.
Collections: industry; transportation; costumes; agriculture; music.
Research Fields: Corry area.
Facilities: library of books, newspapers & deeds available on premises.
Activities: permanent exhibitions; Mead Park Days; monthly exhibit meetings.
Publications: book, Everyone A Hero.
Hours & Admission Prices: Memorial Day-Labor Day Sat.-Sun. & holidays 2-4, groups & tours at other times by appointment. No charge; donations accepted. &
Attendance: 5,500 (accurate)
Membership: Senior 62 & over $3.00; Individual $10; Life $150.

Coudersport

POTTER COUNTY HISTORICAL SOCIETY, 308 N. Main St., Coudersport, PA 16915-1626. Mailing Address: P.O. Box 605, Coudersport, PA 16915-0605. Tel.: 814-274-4410. Fax: 814-274-4411.
E-mail: pottercohist@zitomedia.net
Web Site: www.paintedhills.org
Founded: 1919.
Congressional District: 23
Key Personnel: Pres., David Castano; 1st Vice Pres., Charles Nelson; 2nd Vice Pres., Leon Reed; Sec. & Treas., James Centanni; Librarian, Kay Reed; Cur., Robert K. Currin.

Personnel Profile: Part-Time Volunteers 14.
Governing Authority: society. Tax-exempt.
Institution Type/Description: Historical Society Museum.
Collections: pioneer artifacts; prints; paintings; Bliss Indian collection; newspapers; books; materials covering local area & the industrial, political and cultural development of Potter County from pre-history to the present; photograph collection Civil War to present.
Research Fields: local history & genealogy specializing in indexed census, vital statistics, family files, CCC, Civil War, railroads & lumbering.
Facilities: 1,200-vol. library on history, genealogy and census records available for use on premises; reading room.
Activities: guided tours; research.
Publications: quarterly bulletin.
Hours & Admission Prices: Museum Tours & Research: Mon. & Fri. 10:30-4. No charge; donations accepted.
Attendance: 1,800 (estimated)
Membership: Individual $10; Family $15.

Cresson

ADMIRAL PEARY MONUMENT, 7468 Admiral Peary Hwy., Rte. 2014, Cresson, PA 16630-1717. Mailing Address: Bureau of Historic Sites & Museums-Commonwealth Keystone Building, 400 North St., Harrisburg, PA 17120-0053. Tel.: 717-705-0559.
Founded: 1945.
Congressional District: 12
Key Personnel: Property Placement Div. Chief, Robert N. Sieber.
Governing Authority: state. Administered by Pennsylvania Historical and Museum Commission, Box 1026, Harrisburg, PA. 17108. Tel.: 717-783-5406.
Institution Type/Description: State Monument: dedicated to Admiral Robert E. Peary, discoverer of the North Pole in 1909.
Collections: sculpture.
Facilities: picnic area.
Hours & Admission Prices: Daily dawn-dusk. No charge.

Dallas

PAULY FRIEDMAN ART GALLERY, Misericordia University, 301 Lake St., Dallas, PA 18612-1008. Tel.: 570-674-6250. Fax: 570-674-6416.
E-mail: dposatko@misericordia.edu
Web Site: www.misericordia.edu
Formerly: MacDonald Art Gallery
Founded: 2009.
Key Personnel: Dir., Brian J. Benedetti; Cur. & Gallery Asst., Dona Posatko.
Personnel Profile: Full-Time Paid 1; Part-Time Paid 1.
Governing Authority: Parent Institution: Misericordia University. Tax-exempt.
Institution Type/Description: Art Gallery.
Collections: paintings; photography; sculpture; decorative arts; prints; drawings.
Hours & Admission Prices: Mon.-Thurs. 10:30-8, Sat.-Sun. 1-5. No charge. &

Dayton

MARSHALL HOUSE MUSEUM, 107 N. State St., Dayton, PA 16222. Mailing Address: P.O. Box 447, Dayton, PA 16222-0447. Tel.: 814-257-8846.
Institution Type/Description: Historic House Museum: built in 1868. Listed on the National Register of Historic Places.
Collections: local history & culture; period furnishings; personal artifacts; photographs.
Activities: special events.
Hours & Admission Prices: Memorial Day to mid-Oct. Sat. 2-4; other times by appointment. &
Attendance: 200 (estimated)

Delaware Water Gap

ANTOINE DUTOT MUSEUM & GALLERY, 24 Main St., (Rte. 611), Delaware Water Gap, PA 18327. Mailing Address: P.O. Box 484, Delaware Water Gap, PA 18327-0484. Tel.: 570-476-4240.
Institution Type/Description: History Museum & Art Gallery: housed in a former school house; c.1850.
Collections: local history & culture; period furnishings; personal artifacts; paintings; photographs.
Hours & Admission Prices: Memorial Day to Columbus Day Sat.-Sun. 1-5. Suggested Donation: adults $2; children under 12 no charge.

Devault

GREAT VALLEY NATURE CENTER, 4251 State Rd., Devault, PA 19432. Mailing Address: P.O. Box 82, Devault, PA 19432. Tel.: 610-935-9777. Fax: 610-935-0130.
E-mail: gvnature@gmail.com
Web Site: www.gvnc.org/home
Founded: 1974.
Key Personnel: Exec. Dir., Tom Pascocello
Institution Type/Description: Nature Center.
Collections: environment & nature; birds of prey; Native American Lenape village replica; maple sugar house; hands-on exhibitions.
Facilities: 10 1/2 acre site.
Activities: summer camps; educational programs; special events; internships. Museum Sponsors: Maple Sugaring Festival in March; Easter Egg Hunt; Halloween Costume Party; Great Valley Nature Center Annual Dinner and Silent Auction in December.
Hours & Admission Prices: By appointment.

Devon

JENKINS ARBORETUM & GARDENS, 631 Berwyn Baptist Rd., Devon, PA 19333-1001. Tel.: 610-647-8870.
Web Site: www.jenkinsarboretum.org
Founded: 1976.
Congressional District: 5
Key Personnel: Dir., Dr. Harold E. Sweetman.
Personnel Profile: Full-Time Paid 7; Part-Time Paid 4; Part-Time Volunteers 25; Interns 2.
Governing Authority: nonprofit organization. Tax-exempt.
Institution Type/Description: Arboretum: on grounds of former home of H. Lawrence Jenkins & Elisabeth Phillippe Jenkins.
Collections: living collection of azaleas, rhododendrons & wildflowers; naturalistic landscape design emphasizing native flora.
Facilities: botanical garden.
Publications: annual newsletter.
Hours & Admission Prices: Daily 8am to sunset. No charge; donations accepted. &
Attendance: 30,000 (estimated)
Membership: Regular $45; Donor $100; Patron $250; Circle of Friends $1,000.

Dillsburg

DILL'S TAVERN, 227 N. Baltimore St., Dillsburg, PA 17019. Mailing Address: c/o Northern York County Historical Society, 35 Greenbrier Ln., Dillsburg, PA 17019. Tel.: 717-502-1440.
Web Site: www.dillstavern.org
Institution Type/Description: Historic Building: housed in a former tavern built between 1794-1819. Listed on the National Register of Historic Places.
Collections: local history; tavern life in early 19th century; period furnishings.
Activities: rental facilities; hearth cooking; educational programs. Museum Sponsors: Open Hearth Dinners.
Hours & Admission Prices: By appointment.

Doylestown

BUCKS COUNTY CIVIL WAR LIBRARY & MUSEUM, (M), 32 N. Broad St., Doylestown, PA 18901-4317. Mailing Address: 197 W. Court St., Doylestown, PA 18901-4144. Tel.: 215-348-8293 & 4870.
E-mail: bettystrecker@comcast.net
Web Site: buckscivilwar.org
Founded: 2003.
Congressional District: 8
Key Personnel: Exec. Dir., Treas. & Public Rels., Betty J. Strecker; Pres. (V), Harry Sotan.
Personnel Profile: Full-Time Paid 1; Part-Time Paid 1; Part-Time Volunteers 10; Interns 1.
Governing Authority: private; nonprofit organization. Parent Institution: Bucks County Civil War Round Table, Inc. Tax-exempt: 501(c)(3).
Institution Type/Description: History Museum.
Collections: Civil War & 104th Volunteer PA Regiment artifacts; manuscripts; books; articles.
Facilities: library. Museum-related items for sale.
Activities: guided tours; lectures; loan, participatory & temporary exhibits; study clubs. Annual Events: Community Open House in May.
Publications: quarterly newsletter, Swamp Angel II News; Civil War Veterans in Bucks County PA Cemeteries; 1862 War Diary of William Worthington.
Hours & Admission Prices: Sat. 10:30-2; other times by appointment. No charge; donations accepted.

Attendance: 820 (estimated)
Membership: Student $5; Individual $20; Family $30; Friend $50; Corporate $100.

DOYLESTOWN HISTORICAL SOCIETY, (M), 56 S. Main St., Doylestown, PA 18901-0550. Tel.: 215-345-9430. Fax: 215-345-5119.
E-mail: info@doylestownhistorical.org
Web Site: www.doylestownhistorical.org
Founded: 1995.
Congressional District: 8
Key Personnel: C.E.O., Chm. (V) & Pres. (V), Stu Abramson; Admin. & Museum Shop Mgr., Fletcher Walls.
Personnel Profile: Part-Time Paid 1; Part-Time Volunteers 37; Interns 5.
Governing Authority: Tax-exempt.
Institution Type/Description: Historical Society Museum.
Collections: local history & culture; photographs; personal artifacts.
Activities: educational programs; video histories.
Publications: quarterly newsletter, Membership News; biannual, Doylestown Correspondent.
Hours & Admission Prices: Jan.-March 15 Sat. 10-4; March 16-Dec. Sat. 10-4, Sun. 12-4; other times by appointment. No charge.
Attendance: 5,000 (estimated)
Membership: Senior $20; Senior Couple $30; Family $40; Business $100; Lifetime $250.

＊ FONTHILL MUSEUM, (M), E. Court St. & Rte. 313, Doylestown, PA 18901. Mailing Address: 84 S. Pine St., Doylestown, PA 18901-4930. Tel.: 215-348-9461. Fax: 215-348-9462.
E-mail: fhmail@fonthillmuseum.org
Web Site: www.fonthillmuseum.org
Formerly: Fonthill Museum of the Bucks County Historical Society
Founded: 1930.
Congressional District: 8
Key Personnel: C.E.O., Doug Dolan; Chm., William Maeghin; Site Admin., Edward L. Reidell.
Personnel Profile: Full-Time Paid 2; Part-Time Paid 25; Part-Time Volunteers 30; Interns 4.
Volunteer Hours: 2,615
Operating Expenses: 313,588
Operating Income: 295,665
Governing Authority: nonprofit organization. Governed by Fonthill Trust. Administered by Bucks County Historical Society. Tax-exempt: 501(c)(3).
Institution Type/Description: Historic House Museum & National Historic Landmark: Fonthill, a 44-room concrete castle-like building built 1908-1912, was the home of Henry C. Mercer (1856-1930), noted archaeologist, collector and arts & crafts movement tilemaker.
Collections: Mercer's Moravian tiles; Persian, Spanish & Delft tiles; 7,000 prints & engravings; ceramics; anthropology; photographs; archives.
Research Fields: arts & crafts movement; tile manufacturing; early concrete construction.
Facilities: 6,000-vol. library of Henry Mercer with limited availability to the public through Bucks County Historical Society Library. Publications, tile related gifts & postcards for sale.
Activities: guided tours; organized education programs for children; guide program. Museum Sponsors: special events; rental program.
Hours & Admission Prices: Mon.-Sat. 10-5, Sun. 12-5; guided tour only, reservations suggested. Adults $12. See website for current admission fees. Discounts to AAM & ICOM members. Closed New Year's Day; Thanksgiving; Christmas.
Attendance: 30,000 (estimated)
Membership: Student $35; Senior $35; Individual $40; Family & Grandparent $60; Sustaining $90; Sponsor $175; Contributor $250.

HERITAGE CONSERVANCY, 85 Old Dublin Pike, Doylestown, PA 18901-2468. Tel.: 215-345-7020. Fax: 215-345-4328. Facebook: Heritage Conservancy.
E-mail: info@heritageconservancy.org
Web Site: www.heritageconservancy.org
Founded: 1958.
Congressional District: 8
Key Personnel: C.O.O., Linda Cacossa; Chm., Marvin L. Woodall; Pres., Jeffrey Marshall.
Governing Authority: nonprofit membership organization.
Institution Type/Description: Headquarters housed in 1927, Tudor Revival Style Aldie Mansion; emphasis on land conservation and historic preservation.

Collections: artwork of local artists; decorative arts; photographs
Research Fields: Bucks County; Pennsylvania history & architecture.
Facilities: archives on Bucks County architecture & nature areas.
Activities: guided tours; lectures; films; organized education programs.
Publications: newsletter, Environs.
Hours & Admission Prices: By appointment only. Adults $4; children 12 & under no charge.
Attendance: 23,000 (estimated)
Membership: Individual $35; Family $50; Advocate $100; Benefactor $250; Conservator $500; Guardian $1,000. Business membership categories available.

＊ JAMES A. MICHENER ART MUSEUM, (M), 138 S. Pine St., Doylestown, PA 18901-4931. Tel.: 215-340-9800, ext. 113. Fax: 215-340-9807.
E-mail: jamam1@michenerartmuseum.org
Web Site: www.michenerartmuseum.org
Founded: 1987.
Congressional District: 8
Key Personnel: Dir. & C.E.O., Lisa Tremper Hanover; Chm. Bd. Trustees (V), Kevin Putman; Pres. Bd. Trustees (V), Louis Della Penna; Senior Cur., Brian Peterson; Accounting Mgr., Dorothy Landes; Exhibitions Registrar, Sean Wells; Librarian & Database Coord., Birgitta Bond; Preparator, Bryan Brems; Membership Coord. & Systems Network Admin., Joan Welcker; Dir. Mktg., Kathleen McSherry; Exec. Asst., Candace Clarke; Dir. Programs, Zoriana Siokalo; Cur. Education, Adrienne Romano; Cur. Collections, Constance Kimmerle, Ph.D.; Facility Mgr., Gilbert Winner; Dir. Operations & Visitor Svcs., Hollie Brown.
Personnel Profile: Full-Time Paid 20; Part-Time Paid 22; Part-Time Volunteers 229; Interns 2.
Institution Type/Description: Art Museum.
Collections: Impressionists artists including Edward Redfield, Daniel Garber, William Lathrop & John Folinsbee; American abstract expressionists; 19th & 20th-century American paintings.
Research Fields: 19th & 20th-century Bucks County artists.
Facilities: 48,907 sq. ft. exhibit space; educational facilities; sculpture gardens; performance venue.
Activities: guided tours; lectures; films; concerts; arts festivals; organized educational programs; docent program; permanent & temporary exhibitions.
Publications: exhibition catalogues.
Hours & Admission Prices: Tues.-Fri. 10-4:30, Sat. 10-5, Sun. 12-5. Adults $15; discounts to AAM members; members no charge.
Attendance: 150,000 (accurate)
Membership: Full Time Student & Senior (60 & over) $35; Individual $40; Grandparent, Dual & Family $60; Key & Contributor $100; Sponsor $250; Donor $500; Michener Circle $1,000.

＊ MERCER MUSEUM OF THE BUCKS COUNTY HISTORICAL SOCIETY, (M), 84 S. Pine St., Doylestown, PA 18901-4930. Tel.: 215-345-0210. Fax: 215-230-0823.
E-mail: info@mercermuseum.org
Web Site: www.mercermuseum.org
Founded: 1916.
Congressional District: 8
Key Personnel: C.E.O. & Pres., Douglas C. Dolan; Chm., William Maeghin; Exec. Vice Pres., Molly W. Lowell; Vice Pres. Collections & Interpretation, Cory Amsler.
Personnel Profile: Full-Time Paid 1; Part-Time Paid 25; Part-Time Volunteers 125; Interns 4.
Volunteer Hours: 5,300
Operating Expenses: 1,388,285
Operating Income: 1,371,572
Governing Authority: society. Parent Institution: Bucks County Historical Society. Tax-exempt: 501(c)(3).
Institution Type/Description: History, Historical Technology & Folk Art Museum: housed in 1916 concrete castle-like building. A National Historic Landmark.
Collections: 40,000 tools & products of early handcraft & pre-industrial America; artifacts of everyday life; Pennsylvania German material culture; folk art; local history items.
Major Exhibits: Step Right Up! Behind the Scenes of the Circus Big Top, 1890-1965 (T), 1/28/14-3/16/14; Playing Together: Games (T), 1/28/14-5/11/14; Rt. 66: America's Road (T), 5/24/14-8/17/14; For All the World to See: Civil Rights (T), 9/1/14-10/20/14; Under the Tree: A Century of Holiday Toys, 11/14-1/20/15.
Research Fields: pre-industrial technology; material culture; local history; genealogy.
Facilities: 20,000-vol. research library of Bucks Countiana genealogical

materials & historical technology books, maps, periodicals, photographs and pamphlets available for use during library hours.

Activities: workshops; lectures; education programs; family programs; permanent & temporary exhibitions; digital audio guide; summer camp.

Publications: newsletter, Penny Lots; various books & pamphlcts.

Hours & Admission Prices: Museum: Sun. 12-5, Mon.-Sat. 10-5. Library: call for hours. Adults $12, senior citizens $10, students $6; discounts to groups, AAA, AAM & ICOM members; children under 5 & members no charge. Closed New Year's Day; Thanksgiving; Christmas. &

Attendance: 65,000 (estimated)

Membership: Student $35; Senior $35; Individual $40; Family & Grandparent $60; Sustaining $90; Sponsor $175; Contributor $250; Distinguished Donor $500; President's Circle $1,000.

MORAVIAN POTTERY & TILE WORKS, 130 Swamp Rd., Doylestown, PA 18901-2451. Tel.: 215-348-6098. Fax: 215-345-1361.

E-mail: mptw@buckscounty.org
Web Site: www.buckscounty.org/visitors
Founded: 1969.
Congressional District: 8
Key Personnel: Dir., Charles Yeske.
Personnel Profile: Full-Time Paid 12; Part-Time Paid 4.
Governing Authority: county. Parent Institution: County of Bucks. Subsidiary Institution: Bucks County Parks Dept. Tax-exempt.
Institution Type/Description: Historic Building: a National Historic Landmark.
Collections: tile industry, production, & installation; tiles; mosaics.
Activities: factory tours.
Hours & Admission Prices: Tours: daily 10-4:45. Adults $5, senior citizens $4, children 7-17 $3; discounts to AAM members. Tile Shop: no charge. Closed major holidays. &
Attendance: 26,387 (accurate)
Membership: Individual $35.

DuBois

WINKLER GALLERY OF FINE ART, 36 N. Brady St., DuBois, PA 15801-2256. Tel.: 814-375-5834.
Institution Type/Description: Art Gallery.
Collections: paintings; photography; mosaics; blown glass; sculpture.
Hours & Admission Prices: Tues.-Thurs. 11-5, Fri.-Sat. 11-8.

Eagles Mere

EAGLES MERE MUSEUM, 288 Eagles Mere Ave., Eagles Mere, PA 17731. Mailing Address: P.O. Box 440, Eagles Mere, PA 17731. Tel.: 570-525-3155.
Institution Type/Description: History Museum.
Collections: local history & culture; period furnishings; personal artifacts; photographs.
Facilities: Museum-related items for sale.
Activities: educational programs.
Hours & Admission Prices: Summer: daily 1-5. Donations requested.

East Brady

BRADY'S BEND HISTORICAL SOCIETY, St. Stephen's Old Stone Church, Rte. 68, East Brady, PA 16029. Mailing Address: P.O. Box 451, East Brady, PA 16029. Tel.: 724-526-5693.
E-mail: dmcc1018@aol.com
Web Site: www.bradysbendhistoricalsociety.org
Institution Type/Description: Historical Society Museum.
Collections: local history & culture; period furnishings; personal artifacts; oral histories; cemetery records; photographs.
Hours & Admission Prices: By appointment.

East Stroudsburg

POCONO INDIAN MUSEUM, 5905 Milford Rd., Rte. 209, East Stroudsburg, PA 18302. Mailing Address: P.O. Box 1222, Bushkill, PA 18324. Tel.: 570-588-9338.
E-mail: dream38@ptd.net
Web Site: poconoindianmuseum.com
Founded: 1976.
Institution Type/Description: American Indian Museum.
Collections: Delaware Indian history & culture; period artifacts; weapons; tools; sculpture.
Facilities: Museum-related items for sale.
Hours & Admission Prices: Memorial Day to Labor Day 10-7; Winter: 10-5:30. Adults $5, children 6-12 $2.50; children under six no charge.

Easton

LAFAYETTE COLLEGE ART GALLERIES, (M), 317 Hamilton St., Easton, PA 18042-1768. Mailing Address: 111 Quad Dr., Easton, PA 18042-1768. Tel.: 610-330-5361. Fax: 610-330-5642.
E-mail: okayam@lafayette.edu
Web Site: galleries.lafayette.edu
Founded: 1983.
Congressional District: 136
Key Personnel: Pres. Lafayette College, Dan Weiss; Dir. Williams Center for the Arts, H. Ellis Finger; Vice Pres. & Treas., Mitchell Wein; Dir. Lafayette Art Galleries, Michiko Okaya; Skillman Library Archivist, Diane Windham Shaw; Dir. Security, Hugh Harris.
Personnel Profile: Full-Time Paid 1; Part-Time Paid 1.
Governing Authority: private college; not-for-profit. Parent Institution: Lafayette College. Tax-exempt.
Institution Type/Description: College Art Gallery.
Collections: 19th- & 20th-century American & British portrait & historical paintings; photographs; Abraham Lincoln & Marquis de Lafayette memorabilia & artwork; 20th-century contemporary prints; Kirby paintings.
Research Fields: pertaining to collections.
Facilities: 2,400 sq. ft. exhibit space.
Activities: lectures; in-house loan & temporary exhibitions.
Publications: triannual, exhibit catalogs & brochures; newsletter.
Hours & Admission Prices: Sept.-May Mon.-Tues.-Wed. & Fri. 11-5, Thurs. 11-8, Sat.-Sun. 12-5. No charge. Closed school holidays. &
Attendance: 7,000 (estimated)

*** NATIONAL CANAL MUSEUM, DELAWARE AND LEHIGH NATIONAL HERITAGE CORRIDOR, INC.,** 2750 Hugh Moore Park Rd., Easton, PA 18042-7120. Tel.: 610-923-3548. Fax: 610-250-6554.
E-mail: ncm@canals.org
Web Site: canals.org
Formerly: National Canal Museum, Hugh Moore Historical Park and Museums
Founded: 1970.
Congressional District: 15
Key Personnel: Chm., Nick Forte; C.E.O., Elissa Garofalo.
Personnel Profile: Full-Time Paid 12; Part-Time Paid 12; Part-Time Volunteers 4; Interns 1.
Volunteer Hours: 4,000
Operating Expenses: 900,000
Operating Income: 950,000
Governing Authority: municipal & nonprofit corporation. Parent Institution: Delaware and Lehigh National Heritage Corridor, Inc. Tax-exempt.
Institution Type/Description: Transportation and Industrial Museum.
Collections: canal related artifacts; manuscripts; photo archives; industrial artifacts & records. Historical House: 1923 Locktenders House Museum.
Research Fields: Towpath canals; canal-related industries; early railroads; canal folklore; oral history; recreational use of canals; anthracite coal mining; anthracite iron production; industrial history, technology, steel production & research; communication; technology.
Facilities: 10,000-vol. library of books on canals of the Towpath Era & canal related industries available for research on premises only; archives; reading room; research center; 50-seat auditorium. Transportation-related books & items for sale.
Activities: tours; lectures; films; formally organized education programs; loan & temporary exhibitions; hiking trails; boat rentals; mule-drawn canal boat rides.
Publications: quarterly, biannual books; Canal Currents; quarterly, Locktender; annual, Proceedings of Canal History and Technology Proceedings.
Hours & Admission Prices: Museum: Memorial Day to Labor Day Wed.-Sun. 12-5; Sept.-May see website for hours. Boat: May-Sept. Museum & Boat Ride: adults $11.75, seniors & military $10.50, children $9. &
Attendance: 20,000 (estimated)
Membership: Associate $25; Contributing $40; Supporting $60; Sustaining $100; Sponsoring $500; Patron $1,000; Benefactor $2,500. Corporate: Supporting $100-$499; Sponsoring $500-$999; Patron $1,000-$2,500; Benefactor $2,500 & up.

SIGAL MUSEUM - NORTHAMPTON COUNTY HISTORICAL AND GENEALOGICAL SOCIETY, 342 Northampton St., Easton, PA 18042. Tel.: 610-253-1222. Fax. 610-253-4701.
E-mail: director@northamptonctymuseum.org
Web Site: northamptonctymuseum.org
Founded: 1906.
Congressional District: 15
Key Personnel: Exec. Dir., Barbara Kowitz; Pres. (V), L. Anderson Daub; Librarian (V), Mrs. Roland S. Moyer; Cur., Andria Zaia.

Personnel Profile: Full-Time Paid 3; Full-Time Volunteers 2; Part-Time Paid 4; Part-Time Volunteers 90.
Governing Authority: society; nonprofit. Subsidiary Institutions: History Learning Center; Jane S. Moyer Library; Nicholas Children's Museum; Kressler Memorial Gardens; Northampton County Museum. Tax-exempt: 501(c)(3).
Institution Type/Description: Historical Society Museum.
Collections: 19th-century portraits; Native American artifacts; period dolls; military equipment & uniforms; 19th-century costumes; folk art items; manuscripts; early craft & trade implements; furniture, food prep implements; agricultural equipment & tools; musical instruments; cultural arts; maps; photographs.
Research Fields: genealogy; 18th- to 20th-century local history.
Facilities: 5,000-vol. library of local history & genealogical books available for use by request.
Activities: guided tours; lectures; permanent & changing exhibitions.
Publications: approx. 20 books & pamphlets on subjects related to local history; newsletter, Tales From The Grapevine.
Hours & Admission Prices: Museum: Tues.-Sat. 9:30-3, Sun. 12-4. Library: Tues.-Sat. 9:30-3. Adults $7; discounts to members. &
Attendance: 3,000 (estimated)
Membership: Historian (under 18) $10; Historian (over 18) $25; Individual $50; Family $75; Friend $100; Patron $250; Benefactor $500; Director's Circle $1,000. Corporate: $250; $500; $1,500; $2,500; $5,000.

Ebensburg

CAMBRIA COUNTY HISTORICAL SOCIETY, 615 N. Center, Ebensburg, PA 15931-1122. Mailing Address: P.O. Box 278, Ebensburg, PA 15931-0278. Tel.: 814-472-6674.
E-mail: awbuck@verizon.net
Web Site: www.cambriacountyhistorical.com
Founded: 1925.
Congressional District: 12
Key Personnel: C.E.O. & Pres. (V), Dave Huber; Cur., Kathy Jones.
Personnel Profile: Part-Time Paid 1; Part-Time Volunteers 25.
Governing Authority: private; nonprofit organization. Tax-exempt: 501(c)(3).
Institution Type/Description: General Museum: housed in the 1889 A.W. Buck house.
Collections: 19th-20th century artifacts; history; agriculture; archives; Indian artifacts; industry; numismatic; toys; military; county census; newspapers from 1831.
Research Fields: genealogy; county history.
Facilities: approx. 3,000-vol. library of archives, State's history, biography, genealogy, Indian history, county censuses & newspapers available on premises; reading room.
Activities: guided tours; lectures; films.
Publications: reprint of 1890 Caldwell Atlas of Cambria County; quarterly newsletter, The Cambria County Heritage.
Hours & Admission Prices: Tues.-Fri. 10-4, Sat. 9-1. No charge; donations accepted; Research Library $3 non members. &
Attendance: 2,000 (estimated)
Membership: Annual $20; Family $25; Donor $30; Patron $50; Life $250; Benefactor $500.

Eckley

ECKLEY MINERS' VILLAGE, (M), Main St., Eckley, PA 18255. Mailing Address: 2 Eckley Main St., Weatherly, PA 18255-5030. Tel.: 570-636-2070. Fax: 570-636-2938. TDD: 800-654-5988.
Web Site: www.eckleyminersvillagemuseum.com
Founded: 1969.
Congressional District: 11
Key Personnel: Dir., Bode J. Morin, Ph.D.; Pres. (V), Margie Bogash; Museum Shop Mgr., Melissa Stachaz.
Personnel Profile: Full-Time Paid 5; Part-Time Paid 4; Part-Time Volunteers 55; Interns 3.
Governing Authority: state. Pennsylvania Historical & Museum Commission, P.O. Box 1026, Harrisburg, 17120. Tax-exempt: 170(c)(1).
Institution Type/Description: Historic Village: 54 houses built in the 1850s as coal patch town including Roman Catholic Church, Episcopal Church, doctor's office; coal breaker, visitor's center & mule barn.
Collections: miner's houses; mine owner's homes; churches; artifacts relating to mining & life of the anthracite families.
Major Exhibits: Breaker Boys - Photographs, 11/13-2/14; Scott Herring - Photography, 2/14-6/14.
Research Fields: anthracite coal mining & social & cultural history of anthracite region.
Facilities: library of research material; visitor's center; theater. Museum-related items for sale.

Activities: guided tours; lectures; films; plays; education programs for primary, secondary & college students; permanent exhibitions. Special Events: Halloween; Patch Town Days; Civil War Encampment; Christmas Program.
Publications: quarterly newsletter; pamphlets.
Hours & Admission Prices: Mon.-Sat. 9-5, Sun. 12-5. Adults $8, senior citizens 60 & over $7, children 6-12 $6; discounts to AAM, ICOM members, PHML Trails of History & all professional organizations; members no charge. Walking Tours: $2 per person. Closed Martin Luther King Jr. Day; Veterans Day; Columbus Day; Thanksgiving, Christmas. &
Attendance: 20,000 (estimated)
Membership: Individual $15; Family $20; Contributor $30; Sponsor $50; Benefactor $100; Patron $500.

Eldred

ELDRED WORLD WAR II MUSEUM, 201 Main St., Eldred, PA 16731. Mailing Address: P.O. Box 273, Eldred, PA 16731-0273. Tel.: 814-225-2220; 866-686-9944 (Toll Free). Fax: 814-225-4407.
Founded: 1996.
Key Personnel: Dir., Jay P. Tennies
Institution Type/Description: Military History Museum.
Collections: WWII history; military equipment, uniforms, & personal stories; photographs; personal artifacts.
Activities: special events.
Hours & Admission Prices: Sun. 1-4, Tues.-Sat. 10-4. Adults $5; discounts to groups; children 18 & under no charge. Closed New Year's Eve & Day; Independence Day; Thanksgiving & day after; Christmas Eve, Day & day after.

Elizabethtown

WINTERS HERITAGE HOUSE MUSEUM, 41-47 E. High St., Elizabethtown, PA 17022. Mailing Address: P.O. Box 14, Elizabethtown, PA 17022-0014. Tel.: 717-367-4672. Fax: 717-367-9991.
Web Site: www.elizabethtownhistory.org
Founded: 1988.
Congressional District: 16
Key Personnel: Exec. Dir., Lori B. Donofrio-Galley; Pres. (V), Dr. Michael A. Worman.
Personnel Profile: Full-Time Volunteers 100; Part-Time Paid 2; Part-Time Volunteers 3; Interns 1.
Governing Authority: Parent Institution: Elizabethtown Preservation Assoc. Tax-exempt.
Institution Type/Description: Historic House Museum: c.1750.
Collections: family histories; maps; deeds; obituaries; local history & memorabilia; oral histories; yearbooks; photographs; architecture.
Research Fields: genealogy; local history.
Facilities: genealogy library.
Activities: hands-on demonstrations; classes; workshops; research; walking tours; mural interpretation. Annual Events: Appraisal Fair; Holiday Craft Show.
Publications: member newsletter.
Hours & Admission Prices: March-May & Sept.-Dec. Thurs.-Fri. 9:30-3:30, Sat. 9:30-12; June-Aug. Thurs.-Fri. 9:30-3:30. Museum: adults $2; members no charge. Research: non-members $5 daily. Closed holidays.
Attendance: 1,000 (estimated)
Membership: Individual $20; Family $40; Sustaining $60; Patron $100; Sponsor $250; Benefactor $500; Associate & Life $1,000.

Elkins Park

RICHARD WALL HOUSE MUSEUM - THE IVY, 1 Wall Park Dr., Elkins Park, PA 19027. Mailing Address: 8230 Old York Rd., Elkins Park, PA 19027-1589. Tel.: 215-887-9159 & 6200, ext. 114. Fax: 215-887-1561.
Web Site: www.cheltenhamtownship.org
Founded: 1980.
Congressional District: 13
Key Personnel: Chm., Jack Washington; Vice Chm., James McCann; Cur., Dorothy Spruill; Township Mgr., Bryan T. Havir.
Personnel Profile: Part-Time Volunteers 27; Interns 7.
Volunteer Hours: 802
Governing Authority: Parent Institution: Cheltenham Township Board of Commissions.
Institution Type/Description: Historic House Museum: listed on the National Register of Historic Places.
Collections: local history & culture from late 1700s to 1950s; period furnishings; personal artifacts; springhouse; carriage house.
Publications: semi-annual newsletter, Wallpaper.
Hours & Admission Prices: Sun. 1-4; other times by appointment. No charge; donations accepted.

Attendance: 315 (estimated)

THE TEMPLE JUDEA MUSEUM, (M), 8339 Old York Rd., Elkins
　Park, PA 19027-1515. Tel.: 215-887-2027. Fax: 215-887-1070.
E-mail: tjmuseum@aol.com
Web Site: www.kenesethisrael.org/museum.html
Founded: 1984.
Congressional District: 2
Key Personnel: Dir. & Cur., Rita Rosen Poley; Chm. (V), Karen Shain Schloss.
Personnel Profile: Part-Time Paid 1; Part-Time Volunteers 20.
Governing Authority: denominational group; nonprofit. Tax-exempt: 501(c)(3).
Institution Type/Description: Religious Museum.
Collections: over 1,000 Jewish artifacts from around the world; fabrics crafted
　for religious ceremonial, folk & cultural use; paintings; prints; lithographs;
　photographs; silver ceremonial objects; embroidered Torah binder, 1695;
　second oldest American Ketubah (marriage contract) PA, 1778; Jewish
　music & performing arts; ephemera.
Facilities: 500-vol. library; 197 sq. ft. exhibit space.
Activities: guided tours; lectures; loan & temporary exhibitions. Annual Event:
　Artisans' Festival.
Publications: newsletter, Friends of the Museum.
Hours & Admission Prices: Mon.-Fri. 1-4; other times by appointment. No
　charge. Closed Jewish holidays; legal holidays. &
Attendance: 2,000 (estimated)
Membership: Friend $36; Patron $90; Benefactor $180.

Elverson

HOPEWELL FURNACE NATIONAL HISTORIC SITE, 2 Mark
　Bird Lane, Elverson, PA 19520-9535. Tel.: 610-582-8773. Fax:
　610-582-2768.
E-mail: hofu_superintendent@nps.gov
Web Site: www.nps.gov/hofu
Formerly: Hopewell Village National Historic Site
Founded: 1938.
Congressional District: 7
Key Personnel: Supt., Edie Shean-Hammond; Facility Mgr., George Martin;
　Pres. Friends of Hopewell Furnace NHS, Jim Thorne; Cultural Resource
　Mgr., Rebecca Ross; Chief Interpretation, Frances Delmar; Museum Shop
　Mgr., Christine Hawthorne; Volunteer Coord., Frank Hebblethwaite.
Personnel Profile: Full-Time Paid 12; Part-Time Paid 8; Part-Time Volunteers
　543.
Governing Authority: federal. Parent Institution: National Park Service, Dept.
　of Interior, Washington, DC. Tax-exempt.
Institution Type/Description: Industrial Museum.
Collections: 1770-1880 structures with period furniture; tenant houses; oper-
　ating water-wheel; restored furnace structures; office; store; barn; archives
　with microfilm of original records & documents; history of iron business;
　manuscripts; CCC records; WPA & WWII records; CCC structures &
　artifacts; living history farm; apple orchard; ethnographic database 1771-
　1930s.
Research Fields: late 18th- & 19th-century history of iron production & village
　life in Pennsylvania; CCC & WPA history; African American history
　including underground railroad.
Facilities: 700-vol. library of books, microfilm & dissertations pertaining to
　the history of iron business; auditorium; nature trails; 848-acre site; apple
　orchard. Postcards, furnace collectibles, books, arts & crafts for sale.
Activities: pre-arranged guided tours; permanent exhibitions; film & slide
　presentation; school loan service of films; curriculum-based interactive
　school programs; research; sheep shearing; living history costumed inter-
　pretation; charcoal-making demonstrations; apple picking; arts & crafts
　demonstrations.
Publications: handbook, Hopewell Furnace; American Charcoal Making;
　newsletter, Hopewell Volunteer Reporter; video, Hopewell Furnace Dem-
　onstration; A Summer Day at a Charcoal Iron Furnace; Junior History Guide
　to Hopewell Furnace NHS; Hopewell Village - The Dynamics of a 19th
　Century Iron Making Community; Hopewell Furnace Educational Jigsaw
　Puzzle.
Hours & Admission Prices: Grounds: daily 9-5. Visitor Center & Village:
　Wed.-Sun. 9-5. No charge; donations accepted. Closed New Year's Day;
　Martin Luther King Jr. Day; Presidents' Day; Thanksgiving; Christmas. &
Attendance: 60,000 (estimated)

Emlenton

PUMPING JACK MUSEUM, Crawford Center on Hill St., Hill St.,
　Emlenton, PA 16373. Mailing Address: P.O. Box 25, Emlenton, PA
　16373-0025. Tel.: 724-867-0030.
E-mail: pumpingjackmuseum@csonline.com
Web Site: www.pumpingjack.com

Key Personnel: Pres. (V), Joyce Beikert.
Personnel Profile: Full-Time Volunteers 8.
Institution Type/Description: History Museum.
Collections: oil history; local heritage; equipment; personal artifacts; photo-
　graphs.
Research Fields: genealogy.
Publications: book, Emlenton Walking Tour; Allegheny River guide, Oil On
　The Brain.
Hours & Admission Prices: May-Oct. Sat.-Sun. 12-3; other times by appoint-
　ment. No charge; donations accepted. &
Attendance: 100 (estimated)
Membership: Student $3; Individual $10; Family $15.

Emmaus

SHELTER HOUSE SOCIETY, 601 S. 4th St., Emmaus, PA
　18049-3934. Mailing Address: Box 254, Emmaus, PA 18049-0254.
　Tel.: 610-965-9258.
E-mail: jwmaulfair@verizon.net
Founded: 1951.
Congressional District: 15
Key Personnel: Chm., Noreen Yamamoto; Pres., Jane Maulfair; Museum Shop
　Mgr., Dean Bortz; Asst. Sec., Ruth L. Grohol.
Governing Authority: board of directors. Tax-exempt.
Institution Type/Description: Historic House: 1734-1741 Shelter House.
Collections: history.
Research Fields: history.
Facilities: 500-vol. library of books & material pertaining to local & family
　history available for use on premises; reading room; pavilion for cultural
　group meetings; outdoor fireplaces & recreation areas.
Activities: cultural gatherings.
Publications: annual magazine, The Hearthstone.
Hours & Admission Prices: Daily by appointment only; closed Christmas. No
　charge; donations accepted.
Attendance: 400 (estimated)
Membership: Individual $15; Family $25; Patriots $50; Life (Founders) $100.

Ephrata

EICHER ARTS CENTER AND INDIAN MUSEUM, 407 Cocalico
　St., Ephrata, PA 17522. Mailing Address: P.O. Box 601, Ephrata,
　PA 17522. Tel.: 717-738-3084.
Web Site: www.virtualephrata.org
Key Personnel: Coord., James DeFilippis
Institution Type/Description: History Museum: built in 1733 by members of
　the Ephrata Cloister Brotherhood.
Collections: local history & culture; Native American artifacts; period furnish-
　ings; photographs.
Facilities: Museum-related items for sale.
Activities: rental facilities.
Hours & Admission Prices: By appointment. No charge.

✻　**EPHRATA CLOISTER, (M),** 632 W. Main St., Ephrata, PA
　17522-1717. Tel.: 717-733-6600.
E-mail: ephrata1732@gmail.com
Web Site: www.ephratacloister.org
Founded: 1732.
Congressional District: 16
Key Personnel: Dir., Elizabeth Bertheaud; Pres., Drew Myers; Museum Shop
　Mgr., Susan Shober.
Personnel Profile: Full-Time Paid 7; Part-Time Paid 5; Part-Time Volunteers
　120.
Volunteer Hours: 2,558
Operating Expenses: 263,223
Operating Income: 258,716
Governing Authority: state. Parent Institution: Pennsylvania Historical and
　Museum Commission, Keystone Bldg., Plaza Level, 400 North St., Harris-
　burg, PA 17120-0053. Subsidiary Institution: Ephrata Cloister Associates.
　Tax-exempt.
Institution Type/Description: Historic Site: comprising of 9 mid-18th century
　buildings of Germanic architectural style, located on original site of a
　celibate religious community.
Collections: books printed here in 18th & early 19th centuries; furniture
　produced by 18th century society; buildings representative of unique
　architectural significance.
Research Fields: archaeological; architectural; calligraphy & music; religion.
Facilities: 600-vol. library; 100-seat auditorium; 450-seat outdoor amphithe-
　atre. Cloister inspired reproductions & folk art items for sale.

Activities: guided tours; lectures; concerts; formally organized educational programs; docent program or council; history classes; visitor orientation video.

Publications: books, booklets on Pennsylvania history.

Hours & Admission Prices: Jan.-Feb. Wed.-Sat. 9-5, Sun. 12-5; March & Nov.-Dec. Tues.-Sat. 9-5; April-Oct. Mon.-Sat. 9-5, Sun. 12-5. Adults $10, senior citizens 60 & over $9, youth 6-17 $6; discounts to groups, all motor clubs, National Historical Society, AAM & ICOM members; members no charge. Closed New Year's Day; Easter; Columbus Day; Veterans Day; Thanksgiving & day after; Christmas.

Attendance: 16,000 (accurate)

Membership: Senior Individual 60 & over $20; Individual $25; Contributing (2 person) $40; Family $50; Business, Civic & Professional $100; Benefactor $500.

HISTORICAL SOCIETY OF THE COCALICO VALLEY,
237/249 W. Main St., Ephrata, PA 17522. Mailing Address: P.O. Box 193, Ephrata, PA 17522-0193. Tel.: 717-733-1616.

E-mail: cjmarquet@gmail.com

Web Site: www.cocalicovalleyhs.org

Founded: 1957.

Congressional District: 16

Key Personnel: Pres. (V), Lowell Haws; Librarian, Cynthia J. Marquet.

Personnel Profile: Part-Time Paid 1.

Governing Authority: society. Tax-exempt.

Institution Type/Description: History Museum.

Collections: history of Cocalico Valley; reference library; manuscripts; papers of Col. George S. Howard, 1st Dir. of U.S.A.F. Band; Harry F. Stauffer photographs; Pennsylvania German art & culture.

Research Fields: genealogy; local history; U.S.A.F. Band history; history & culture of Pennsylvania Germans.

Facilities: library of books on genealogy & history available for use on premises.

Activities: lectures; films; permanent & temporary exhibitions.

Publications: Journal of the Historical Society of the Cocalico Valley; occasional books of historical photographs.

Hours & Admission Prices: Museum: Sat. 10-4. No charge; donations accepted. Library & Research Center: Mon. & Wed.-Thurs. 9:30-6, Sat. 8:30-5. Research: $3.

Attendance: 900 (estimated)

Membership: Regular $25; Family $35; Life $500.

Erie

ERIE ART MUSEUM, (M),
411 State St., Erie, PA 16501-1106. Tel.: 814-459-5477. Fax: 814-452-1744. Facebook: Erie Art Museum.

Web Site: www.erieartmuseum.org

Formerly: Erie Art Center, Art Club of Erie

Founded: 1962.

Congressional District: 21

Key Personnel: Exec. Dir., John L. Vanco; Pres. (V)., Barb Haggerty; Dir. Mktg., Carolyn Eller; Dir. Administration & Finance, Mary Pruchniewski; Dir. Devel., Jennifer Dobbs; Dir. Education & Folk Art Coord., Kelly Armor; Frame Shop Mgr., Lin-Lang Su; Coord. Publications, Andrea Krivak; Education Coord., Ally Thomas; Building & Events Operations, Charles Cehovin; Registrar, Vance Lupher; Asst. Cur., Amanda Steadman.

Personnel Profile: Full-Time Paid 11; Part-Time Paid 7; Part-Time Volunteers 150; Interns 20.

Governing Authority: nonprofit organization. Tax-exempt: 501(c)(3).

Institution Type/Description: Art Museum: housed in the former Erie branch of the US Bank of Pennsylvania; built in 1839.

Collections: paintings; sculpture; graphics; decorative arts; photography; American ceramics; Indian bronze & stone sculpture; oriental porcelains & jade; contemporary baskets; Tibetan paintings.

Major Exhibits: Dave Bennett, 11/29/13-3/1/14; Robert Polidori, 1/18/14-4/13/14; Kids as Curators, 1/25/14-3/23/14; Anna Weaver, Amish Painter, 3/7/14-6/14/14; 91st Annual Spring Show, 4/26/14-7/20/14; Pat Kerney, 6/20/14-9/13/14; Environmental Impact (T), 8/14-9/14; Focus Fiber (T), 9/26/14-1/18/15; Wilbur Adams, 10/4/14-1/4/15.

Research Fields: American ceramic art.

Facilities: studios; classrooms; auditorium; multi-purpose room; cafe.

Activities: guided tours; lectures; films; gallery talks; formally organized educational programs; temporary & traveling exhibitions; circulating exhibitions; jazz & new music concerts. Museum Sponsors: Annual Spring Show Competition; Erie Art Museum Blues & Jazz Festival.

Publications: monthly e-newsletter; members' newsletter; exhibition announcements; exhibition catalogs.

Hours & Admission Prices: Tues.-Sat. 11-5, Sun. 1-5. Adults $7, students, senior citizens & children under 12 $5; Wed. and AAM, ICOM & museum

members no charge. Closed New Year's Day; Easter; Independence Day; Thanksgiving; Christmas.

Attendance: 36,000 (estimated)

Membership: Student $20; Senior Citizen $25; Individual $45; Family $75; Patron $100; Friend $250; Advocate $500; Contributor $1,000; Benefactor $2,500; Fellow $5,000. Green (add $10 to any level).

ERIE COUNTY HISTORY CENTER & CASHIER'S HOUSE,
417-419 State St., Erie, PA 16501. Mailing Address: 419 State St., Erie, PA 16501-1106. Tel.: 814-454-1813. Fax: 814-454-6890.

E-mail: echs@eriecountyhistory.org

Web Site: www.eriecountyhistory.org

Founded: 1903.

Congressional District: 3

Key Personnel: Exec. Dir., Caleb Pifer; Operations Dir., Melanie Kuebel-Stankey.

Personnel Profile: Full-Time Paid 4; Part-Time Paid 2; Part-Time Volunteers 10; Interns 8.

Governing Authority: society. Parent Institution: Erie County Historical Society. Tax-exempt: 501(c)(3).

Institution Type/Description: Historical Society Museum.

Collections: 1735-present, northwestern Pennsylvania & Great Lakes history.

Research Fields: history of Erie County area, industry & architecture.

Facilities: 3,500-vol. library containing books, documents, manuscripts, dealing with Erie County & northwest Pennsylvania history.

Activities: lectures; formally organized educational programs; temporary exhibitions.

Publications: Journal of Erie Studies; books; periodic newsletter.

Hours & Admission Prices: Tues.-Sat. 11-4. Family $10, adults $5, seniors & students $4, children $3; discounts to AAM & AASLH members; members no charge. Closed holidays.

Attendance: 5,000 (estimated)

Membership: Student $15; Individual $45; Family & Grandparent $60; Patron $75; Sustaining $100; Sponsor $250; Corporate $500.

ERIE MARITIME MUSEUM & FLAGSHIP NIAGARA, (M),
Bayview Commons, 150 E. Front St., Ste. 100, Erie, PA 16507-1594. Tel.: 814-452-2744. Fax: 814-455-6760.

E-mail: sail@flagshipniagara.org

Web Site: www.eriemaritimemuseum.org

Formerly: Erie Maritime Museum, Homeport U.S. Brig Niagara

Founded: 1998.

Congressional District: 3

Key Personnel: Chm. PHMC, Andrew E. Masich; Exec. Dir. PHMC, James M. Vaughan; Pres. Flagship Niagara League, Tim Goldsmith; Site Admin. & Sr. Captain, Walter Rybka; Museum Shop Mgr., Barbara Corbett; Museum Shop Mgr., Judy Tucci.

Personnel Profile: Full-Time Paid 2; Full-Time Volunteers 1; Part-Time Paid 2; Part-Time Volunteers 160; Interns 2.

Governing Authority: state. Administered by Pennsylvania Historical & Museum Commission, Box 1026, Harrisburg, PA 17108. Subsidiary Institution: Flagship Niagara League. Tax-exempt.

Institution Type/Description: Historic Ship & Maritime History Museum.

Collections: Battle of Lake Erie; War of 1812; 19th-century sailing & regional maritime history; homeport to the 1813 U.S. Brig Niagara; live fire exhibit of the U.S. Brig Lawrence; U.S.S. Michigan/U.S.S. Wolverine; photographs; video.

Research Fields: War of 1812; Battle of Lake Erie; naval architecture; Lake Erie maritime history; U.S.S. Michigan/U.S.S. Wolverine; US Lifesaving Service; Erie Lighthouses.

Facilities: library on Lake Erie maritime history available to scholars & members; visitor center; theater; 299-seat auditorium. Museum-related items for sale.

Activities: guided tours; special programs; on-board ship demonstrations; hands-on & interactive exhibits; concerts; films; lectures; one day sailing school vessel; 2-4 week live-aboard sail training; educational programs.

Publications: quarterly newsletter, Niagara League News; weekly e-newsletter, Niagara.

Hours & Admission Prices: April-Oct. Mon.-Sat. 9-5, Sun. 12-5; Nov.-March Thurs.-Sat. 9-5, Sun. 12-5; call for ships schedule. Adults $8, senior citizens $7, youth 6-17 $5; discounts to AAA, AAM & ICOM members; active military & their family, children under 3 & league members no charge. Closed New Year's Day; Martin Luther King Jr. Day; Presidents' Day; Columbus Day; Veterans Day; Thanksgiving; Christmas.

Attendance: 24,869 (accurate)

Membership: Senior Individual & Senior Couple $30; Individual & Couple $35; Family $50; Plank $100; Deck $250; Mast $500; Ship $1,000. Business: Small $250; Large & Corporate $500.

ERIE PLANETARIUM, 356 W. Sixth St., Erie, PA 16507-1245. Mailing Address: 419 State St., Erie, PA 16501-1106. Tel.: 814-871-5790 & 454-1813. Fax: 814-454-6890.
E-mail: echs@eriecountyhistory.org
Web Site: www.eriecountyhistory.org/erie-planetarium
Founded: 1959.
Congressional District: 3
Key Personnel: Dir., Caleb Pifer; Dir. Operations & Visitor Member Svcs., Melanie Kuebel-Stankey; Planetarium Coord., Jim Gavio.
Personnel Profile: Part-Time Paid 1; Part-Time Volunteers 6.
Governing Authority: private; nonprofit organization. Parent Institution: Erie County Historical Society. Tax-exempt: 501(c)(3).
Institution Type/Description: Planetarium.
Collections: geology; astronomy.
Facilities: Spitz A3P projector; 20 ft. dome.
Activities: planetarium shows; scout programs; birthday parties; special event programs; hands-on activities.
Hours & Admission Prices: Call for hours. Family $10, adults $5, senior $4, children $3; discounts to AAM members & groups; members no charge.
Attendance: 5,500 (estimated)
Membership: Student $15; Individual $45; Family $60; Patron $75; Sustaining $100; Sponsor $250; Corporate $500.

ERIE ZOOLOGICAL PARK & BOTANICAL GARDENS OF NORTHWESTERN PENNSYLVANIA, 423 W. 38th St., Erie, PA 16508-2701. Mailing Address: P.O. Box 3268, Erie, PA 16508-0268. Tel.: 814-864-4091. Fax: 814-864-1140.
E-mail: info@eriezoo.org
Web Site: www.eriezoo.org
Founded: 1962.
Key Personnel: Pres. & C.E.O., Scott Mitchell; Coord. Mktg. & Events, Kelly Miller.
Personnel Profile: Full-Time Paid 27; Part-Time Paid 70; Part-Time Volunteers 412.
Governing Authority: society. Parent Institution: Erie Zoological Society. Tax-exempt.
Institution Type/Description: Zoo & Botanical Garden.
Collections: 2,500 specimens representing 600 species of plants; over 500 animals from 118 species.
Publications: member newsletter, ZooNews.
Hours & Admission Prices: Daily 10-5. Adults $7, senior citizens $6, children 2-11 $4; children under 2, members and physically & mentally challenged no charge.
Attendance: 462,100 (estimated)
Membership: Individual, Single Parent, Grandparent & Family $50; Family Plus & Grandparent Plus $60.

EXPERIENCE CHILDREN'S MUSEUM, 420 French St., Erie, PA 16507-1541. Tel.: 814-453-3743. Fax: 814-459-9735.
E-mail: executivedirector@eriechildrensmuseum.org
Web Site: www.eriechildrensmuseum.org
Founded: 1992.
Congressional District: 21
Key Personnel: Exec. Dir., Ainslie Brosig.
Personnel Profile: Full-Time Paid 3; Part-Time Paid 4; Part-Time Volunteers 220.
Governing Authority: nonprofit; board of directors. Tax-exempt.
Institution Type/Description: Children's/Youth Museum.
Collections: 55 hands-on exhibits designed for children ages 2-10.
Facilities: theater. Gift items for sale.
Publications: newsletter, Kaleidoscope.
Hours & Admission Prices: Tues.-Sat. 10-4, Sun. 1-4. Admission $6; discount to ACM members; members & children under 2 no charge. Closed New Year's Day, Easter, Thanksgiving, Christmas Eve & Day.
Attendance: 37,200 (accurate)
Membership: Single Parent $55; Family $65; Single Grandparent $50; Grandparent Couple $60.

FIREFIGHTERS HISTORICAL MUSEUM, 428 Chestnut St., Erie, PA 16507-1224. Tel.: 814-456-5969.
Web Site: firefightershistoricalmuseum.com
Founded: 1976.
Institution Type/Description: Firefighting History Museum.
Collections: firefighting history & equipment; fire trucks; hose carts; photographs.
Hours & Admission Prices: June-July Sat. 11-4, Sun. 1-4; Sept.-Oct. Sat.-Sun. 1-4. Adults $4, senior citizens & firefighters $2.50, children 6-12 $1.

WATSON-CURTZE MANSION, 356 W. Sixth St., Erie, PA 16507-1245. Mailing Address: 419 State St., Erie, PA 16501-1106. Tel.: 814-871-5790 & 454-1813. Fax: 814-454-6890.
E-mail: echs@eriecountyhistory.org
Web Site: www.eriecountyhistory.org
Formerly: Erie Historical Museum and Planetarium
Founded: 1999.
Congressional District: 3
Key Personnel: Dir., Caleb Pifer; Operating Dir., Melanie Kuebel-Stankey.
Personnel Profile: Full-Time Paid 3; Part-Time Volunteers 15.
Governing Authority: private; nonprofit organization. Parent Institution: Erie County Historical Society. Tax-exempt: 501(c)(3).
Institution Type/Description: Historic House.
Collections: 19th- to 20th-century Victorian & American Renaissance decorative arts, Eugene Iverd paintings & prints; textiles & costumes; civil war history; Moses Billings paintings; planetarium. Historic House: 1891 Watson-Curtze Mansion.
Research Fields: 19th-century decorative arts.
Activities: guided tours; lectures; educational programs; permanent & temporary exhibitions; special events. Annual Events: Victorian Holiday exhibit mid-November to mid-January.
Publications: brochures; flyers, pamphlets on events that occur at the museum & planetarium; Guide to the Mansion.
Hours & Admission Prices: Call for hours. Family $10, Adults $5, seniors $4, children 2-12 $3; discounts to AAM members & Wed.; members & children under 2 no charge.
Attendance: 5,075 (accurate)
Membership: Student $15; Individual $45; Family $60; Patron $75; Sustaining $100; Sponsor $250; Corporate $500.

Fallsington

HISTORIC FALLSINGTON, INC., 4 Yardley Ave., Fallsington, PA 19054-1117. Tel.: 215-295-6567. Fax: 215-295-6567.
E-mail: info@historicfallsington.org
Web Site: www.historicfallsington.org
Founded: 1953.
Congressional District: 8
Key Personnel: Exec. Dir., Erica Armour; Pres. (V), Robert L.B. Harman.
Personnel Profile: Full-Time Paid 1; Part-Time Paid 3; Part-Time Volunteers 90; Interns 2.
Governing Authority: board of trustees. Tax-exempt: 501(c)(3).
Institution Type/Description: 17th to 20th-Century Quaker Village.
Collections: furnishings & documents; local history. Historic Houses: 1809 Burges-Lippincott House; 1700s Stage-Coach Tavern; 1760s Moon-Williamson Log House; 1758 schoolmaster's house; 1728 former Friends' Meeting House.
Research Fields: local village & Quaker history & culture, 1685-present.
Facilities: picnic area. Museum-related items for sale.
Activities: guided tours; programs & special events; lectures; crafts groups.
Publications: booklet, Historic Fallsington.
Hours & Admission Prices: mid-May to mid-Oct. Tues.-Sat. 10:30-3:30; mid-Oct. to mid-May Tues.-Fri. by appointment. Adults $6, senior citizens $5, children 6-18 $3; discounts to AAA members; members & children 5 and under no charge. Closed New Year's Day; Presidents' Day; Easter; Memorial Day; Independence Day; Labor Day; Christmas.
Attendance: 7,500 (estimated)
Membership: Individual $25; Family $45; Donor $100; Benefactor $250; Patron $1,000; Preservationist $2,500.

Farmington

FORT NECESSITY NATIONAL BATTLEFIELD, One Washington Pkwy., Farmington, PA 15437-9501. Tel.: 724-329-5819. Fax: 724-329-8682.
E-mail: lawren_dunn@nps.gov
Web Site: www.nps.gov/fone
Founded: 1931.
Congressional District: 12
Key Personnel: Deputy Supt., Keith Newlin; Supt., Jeffrey Reinbold; Unit Mgr., Chief Law Enforcement Ranger, Norman W. Nelson; Cultural Resource Mgr. & Cur., Lawren Dunn; Museum Shop Mgr., James Tomasek.
Personnel Profile: Full-Time Paid 15; Part-Time Paid 2; Part-Time Volunteers 390; Interns 3.
Governing Authority: federal. Parent Institution: Dept. of the Interior. Subsidiary Institution: National Park Service. Tax-exempt.
Institution Type/Description: History Museum.
Collections: history; 1820s stage tavern; 1827 Mount Washington Tavern; 1754 French & Indian war site; George Washington's first battle; reconstructed Fort Necessity Stockade.

Major Exhibits: Seneca, 4/14-4/15.
Research Fields: history.
Facilities: 1200-vol. library of historic source material pertaining to the Great Meadows & Braddock's Campaigns along with material on the National Road & early transportation and the French & Indian War; visitor center.
Activities: permanent exhibits; guided tours & living history programs during summer months; audiovisual program.
Publications: Park Handbook, A Charming Field; park folders & inserts.
Hours & Admission Prices: Grounds: daily dawn to dusk. Center: daily 9-5. Adults 16 & over $5. Closed New Year's Day; Martin Luther King Jr. Day; Thanksgiving; Christmas. ♿
Attendance: 125,000 (estimated)

Fayetteville

PENNSYLVANIA FOREST FIRE MUSEUM, 3050 Lincoln Way E., Fayetteville, PA 17222-9556. Mailing Address: P.O. Box 176, Fayetteville, PA 17222-0176. Tel.: 717-352-2815.
Institution Type/Description: Forest Fire History Museum.
Collections: forest fire protection history; forest resources.
Facilities: library; classrooms; auditorium; outreach programs. Museum-related items for sale.
Activities: educational programs.
Hours & Admission Prices: Call for hours.

Fort Loudoun

STATE HISTORIC SITE OF FORT LOUDOUN, 1720 Brooklyn Rd., Fort Loudoun, PA 17724. Mailing Address: P.O. Box 181, Fort Loudoun, PA 17224-0181. Tel.: 717-369-3318. Fax: 717-783-1073.
E-mail: secretary@fortloudoun-pa.com
Web Site: www.fortloudoun-pa.com/
Founded: 1968.
Congressional District: 9
Key Personnel: Property Placement Officer, Robert N. Sieber.
Governing Authority: state. Peters Township Board of Supervisors. Operated by the Fort Loudon Historical Society in agreement with the Pennsylvania Historical and Museum Commission, Box 1026, Harrisburg, PA 17120.
Institution Type/Description: Historic Site: site of Fort Loudoun, erected in 1756 by the British.
Collections: historical displays of the fort occupation.
Facilities: picnic facilities.
Activities: reenactments.
Hours & Admission Prices: Memorial Day to Labor Day Sat.-Sun. 12-5; other times by appointment. No charge; donations accepted.
Membership: Individual $5; Family $10; Life $100.

Fort Washington

THE HIGHLANDS, (M), 7001 Sheaff Lane, Fort Washington, PA 19034-2005. Tel.: 215-641-2687. Fax: 215-641-2556.
E-mail: mbb@highlandshistorical.org
Web Site: www.highlandshistorical.org
Founded: 1975.
Congressional District: 13
Key Personnel: Pres., Charles L. Sheppard, II; Dir., Margaret Bleecker Blades; Cur. Education, Elizabeth Gavrys.
Personnel Profile: Full-Time Paid 2; Part-Time Paid 3; Part-Time Volunteers 60.
Governing Authority: state. Administered by the Highlands Historical Society in cooperation with the Pennsylvania Historical & Museum Commission, Box 1026, Harrisburg, PA 17120. Tax-exempt.
Institution Type/Description: Historic House: 1796 Georgian country house built by Anthony Morris, Speaker of the Pennsylvania Senate.
Collections: nine outbuildings; 44 acre estate; 2 acre formal gardens.
Facilities: formal gardens.
Activities: restoration & public programs.
Publications: quarterly newsletter.
Hours & Admission Prices: Tours: Mon.-Fri. 1:30 & 3, Sat.-Sun. by appointment. Adults $5; discounts to AAM & ICOM members; AABGA members no charge.
Attendance: 12,000 (accurate)
Membership: Individual $40; Family $60; Friend $100; Sheaff Society $250; Sinkler/Roosevelt Society $500; Anthony Morris Circle $1,000.

HISTORICAL SOCIETY OF FORT WASHINGTON, 473 Bethlehem Pike, Fort Washington, PA 19034-2313. Tel.: 215-646-6065.
Web Site: www.amblerhistory.com
Founded: 1935.

Congressional District: 13
Key Personnel: Pres. (V), H. Roy Thompson; Treas., Ingrid Rivel; Librarian, William Amey.
Personnel Profile: Part-Time Volunteers 25.
Governing Authority: executive board. Tax-exempt.
Institution Type/Description: Historical Society & Genealogical Library: housed in 1801 Clifton House.
Collections: 1,800-vol. library.
Research Fields: county history; genealogy; biography; Revolutionary War, Civil War, local, state & U.S. history.
Facilities: research library & reading room.
Activities: lectures; formally organized educational programs; permanent exhibitions.
Publications: newsletter; brochure.
Hours & Admission Prices: Sept.-June Wed. 2-4; other times by appointment. No charge; donations accepted. Closed holidays.
Attendance: 800 (estimated)
Membership: Students $10; Single $25; Couple $40; Business $100; Business Sponsor & Life Individual $200; Life Couple $300.

HOPE LODGE AND MATHER MILL, (M), 553 S. Bethlehem Pike, Fort Washington, PA 19034. Tel.: 215-646-1595. Fax: 215-628-9471.
E-mail: jhauger@state.pa.us
Web Site: www.ushistory.org/hope
Founded: 1957.
Congressional District: 13
Key Personnel: Site Admin., Joan Hauger; Pres. (V) Friends of Hope Lodge, Jack Gumbrecht; Museum Shop Mgr., Wanda Rauch.
Personnel Profile: Full-Time Paid 2; Part-Time Volunteers 60.
Governing Authority: state. Parent Institution: Pennsylvania Historical & Museum Commission, Commonwealth Keystone Building, Plaza Level, 400 North St., Harrisburg, PA 17120-0053. Tax-exempt.
Institution Type/Description: Historic Buildings: 1743, Hope Lodge, built for Samuel Morris; c.1820 Mather mill, stone grist mill built on the site of former 17th-century mill by Edward Farmar.
Collections: 18th & 19th century furnishing & decorative arts, paintings & prints.
Research Fields: social & labor history; architecture; agriculture.
Facilities: landscape gardens. Museum-related items for sale.
Activities: guided tours; lectures; craft demonstrations; school tours. Museum Sponsors: traditional music events; reenactment of 1777 Whitemarsh Encampment; British Car Show.
Publications: cookbook, Hope Lodge Favorite Recipes; History of Mather Mill; biography of Samuel Morris. Hope Lodge and Mather Mill: a Pennsylvania Trail of History Guide; History of William & Alice Degin (Preservers of a National Treasure); Lime Industry at Hope Lodge.
Hours & Admission Prices: Fri.-Sat. 10-5, Sun. 12-5. Adults $6, senior citizens $5, youth 3-11 $3; discounts to AAM members, groups, PA Trail of History; Travel & Motor Club members; active military & their families, children under 3, museum professionals, PA Heritage Society members & members no charge. Closed New Year's Day; Martin Luther King Jr. Day; Presidents' Day; Columbus Day; Veterans Day; Thanksgiving & day after; Christmas. ♿
Attendance: 8,300 (accurate)
Membership: Individual $20; Family $35; Patron $50; Sponsor $100.

Forty Fort

NATHAN DENISON ADVOCATES, 35 Denison St., Forty Fort, PA 18704-4311. Mailing Address: 23 Ivy Ln., Dupont, PA 18641. Tel.: 570-288-5531 & 5623.
Founded: 1970.
Congressional District: 11
Key Personnel: Pres. (V), Mrs. Louise Robinson; Property Placement Officer PA Historic & Museum Commission, Robert Sieber.
Personnel Profile: Part-Time Volunteers 22.
Governing Authority: state; nonprofit. Managed by the Denison Advocates. Supported by the Pennsylvania Historical & Museum Commission, Box 1026, Harrisburg, PA 17108-1026. Tax-exempt.
Institution Type/Description: Historic House: 1790 Denison House, Connecticut style architecture, built in the Wyoming Valley; home of Col. Nathan Denison, 1790-1809.
Collections: 18th-century furnishings of house, authentic & reproductions; quill pen; ink well; sealing wax & stamp molds; red ware sand blotter shaker; pictures; paintings; 1684 land sale document.
Facilities: Gift items for sale.
Activities: guided tours. Special Events: May-October.
Publications: booklet, Nathan Denison In The Wyoming Valley, The Nathan Denison Story: A Coloring Booklet; video, The Battle of Wyoming.

Hours & Admission Prices: Tours by appointment. Adults $3, children $2.
Attendance: 1,000 (accurate)

Franklin

DEBENCE ANTIQUE MUSIC WORLD, (M), 1261 Liberty St., Franklin, PA 16323-1361. Tel.: 814-432-8350. Fax: 814-437-7193.
E-mail: debencemuseum@verizon.net
Web Site: www.debencemusicworld.com
Founded: 1994.
Key Personnel: Dir. & Chm. (V), Prescott Greene; Museum Shop Mgr., Mary C. Nicklin.
Personnel Profile: Full-Time Volunteers 1; Part-Time Paid 1; Part-Time Volunteers 37.
Volunteer Hours: 5,000
Governing Authority: Tax-exempt.
Institution Type/Description: Musical Instruments Museum.
Collections: over 200 period automatic instruments including calliopes, carousel band organs, music boxes, nickelodeons, & orchestrions.
Activities: monthly music-related program; guided tours for area 4th graders.
Publications: biannual newsletter.
Hours & Admission Prices: April-Oct. Tues.-Sat. 11-4, Sun. 12:30-4; Nov.-March by appointment. Adults $8, senior citizens $7, students $5, youth 3-14 $3; discounts to AAA & WDUQ members and groups of 10 or more; children under 3, area 4th graders & basement exhibits no charge. &

Attendance: 3,000 (accurate)
Membership: Individual $35; Family $50; Edison Club $100; Music Box Club $250; Nickelodeon Club $500; Calliope Club $1,000.

VENANGO COUNTY HISTORICAL SOCIETY, 301 S. Park St., Franklin, PA 16323-1238. Tel.: 814-437-2275.
E-mail: vchistory@csonline.net
Key Personnel: Pres., Rainy Linn; Archivist, Marianne Battista
Institution Type/Description: Historical Society: housed in the Hoge-Osmer House, built c.1865.
Collections: local history & culture; photographs; personal artifacts.
Hours & Admission Prices: Jan.-April Sat. 10-2; May-Dec. Tues.-Thurs. & Sat. 10-2.

Galeton

PENNSYLVANIA LUMBER MUSEUM, (M), 5660 U.S. 6 W., Galeton, PA 16922. Mailing Address: P.O. Box 239, Galeton, PA 16922-0239. Tel.: 814-435-2652. Fax: 814-435-6361.
E-mail: pberberich@pa.gov
Web Site: www.lumbermuseum.org
Founded: 1970.
Congressional District: 5
Key Personnel: Site Admin., Jeff Bliemeister; Pres., Robert Miller; Museum Shop Mgr., Gloria Harris; Records Mgr., Patricia Berberich.
Personnel Profile: Full-Time Paid 3; Part-Time Volunteers 14.
Governing Authority: state. Pennsylvania Historical and Museum Commission, 3rd & North Sts., P.O. Box 1026, Harrisburg, PA. 17120. Tax-exempt: 170(c)1.
Institution Type/Description: History and Lumber Museum.
Collections: logging tools & equipment; sawmill; logging camp; outdoor museum; two locomotive engines; log loader; railroad log cars; 1917 Model-T Ford Runabout.
Research Fields: pertaining to collections.
Facilities: 2,000-vol. library of books on forestry, natural history & lumbering history available for use on premises; auditorium. Gift items for sale.
Activities: guided tours; lectures; films; festivals; formally organized educational programs; permanent & temporary exhibitions.
Publications: walking guide to the museum.
Hours & Admission Prices: April-Nov. Wed.-Sun. 9-5. Adults $6, senior citizens $5, children 3-11 $3; discounts to AAM members; members no charge. Closed winter holidays. &
Attendance: 9,000 (estimated)
Membership: Single $10; Family $15; Patron $30; Supporting $50; Corporate $75; Benefactor $100.

Gallitzin

ALLEGHENY PORTAGE RAILROAD NATIONAL HISTORIC SITE AND JOHNSTOWN FLOOD NATIONAL MEMORIAL, 110 Federal Park Rd., Gallitzin, PA 16641-2000. Tel.: 814-886-6116. Fax: 814-884-0206.
E-mail: nancy_smith@nps.gov
Web Site: www.nps.gov/alpo

Founded: 1964.
Congressional District: 12
Key Personnel: Supt., Keith Newlin; Cur., Nancy Smith; Museum Shop Mgr., Doug Bosley.
Personnel Profile: Full-Time Paid 10; Part-Time Paid 5.
Governing Authority: federal. Parent Institution: U.S. Dept. of the Interior, National Park Service, Washington, DC 20240. Tax-exempt: 501(c)(3).
Institution Type/Description: Visitor Center:
Collections: books, articles, books; photographs; 1889 Johnstown flood artifacts; photographs & other artifacts relating to Allegheny Portage Railroad, Pennsylvania Mainline Canal, transportation; Lemon House: transportation, social & economic story.
Research Fields: Allegheny Portage Railroad; Pennsylvania Mainline Canal; Johnstown flood of 1889; 19th century relief efforts; immigrants.
Facilities: 35-vol. library of secondary source material on Pennsylvania Main Line Canal, Allegheny Portage Railroad, & 1889 Johnstown Flood available for use on premises; 50-seat auditorium; picnic area; theater. Leaflets & books for sale.
Activities: guided walks; permanent & temporary exhibits; audiovisual programs; period costume demonstrations; hiking.
Publications: park folders; park inserts; park handbook, The Lemon House, A Place in History; The Sylvester Welch's Report on the Allegheny Portage Railroad; Allegheny, Old Portage Railroad 1834-1854; books.
Hours & Admission Prices: Daily 9-5. Adults $4; children under 16 no charge. Closed holidays in winter. &
Attendance: 140,000

Gettysburg

ADAMS COUNTY HISTORICAL SOCIETY, (M), 368 Springs Ave., Gettysburg, PA 17325-1718. Mailing Address: P.O. Box 4325, Gettysburg, PA 17325-4325. Tel.: 717-334-4723, ext. 201. Fax: 717-334-0722.
E-mail: info@achs-pa.org
Web Site: www.achs-pa.org
Founded: 1940.
Congressional District: 19
Key Personnel: Dir., Benjamin Neely.
Personnel Profile: Full-Time Paid 1; Part-Time Paid 4; Part-Time Volunteers 40; Interns 2.
Governing Authority: nonprofit organization. Tax-exempt: 501(c)(3).
Institution Type/Description: Historical Society Museum.
Collections: Civil War archives.
Facilities: 5,000-vol. library of local history, genealogy, newspapers, tax records, estate papers; Civil War books available for research on premises during public hours; research center; reading room.
Activities: lectures; research; workshops; classes.
Publications: quarterly newsletter; annual volume.
Hours & Admission Prices: Wed. & Fri.-Sat. 9-12 & 1-4, Thurs. 9-12, 1-4 & 6-9. Research: adults $5; members no charge. Closed holidays.
Attendance: 2,100 (estimated)
Membership: Supporting $35; Family $50; Bermudian Settlement $100; Carroll's Delight $500; Digges Choice $1,000; Corporate $100-$10,000.

THE DAVID WILLS HOUSE, 8 Lincoln Sq., Gettysburg, PA 17325-2205. Mailing Address: 1195 Baltimore St., Gettysburg, PA 17325. Tel.: 877-874-2478. Fax: 717-334-5796.
E-mail: info@gettysburgfoundation.org
Web Site: www.gettysburgfoundation.org
Founded: 2009.
Congressional District: 19
Key Personnel: Mgr., Cheryl Cline.
Governing Authority: Parent Institution: National Park Service & Gettysburg Foundation. Tax-exempt.
Institution Type/Description: History Museum.
Collections: local history & culture; Abraham Lincoln's Gettysburg Address; replicas of Lincoln's bedroom & David Wills' law office.
Hours & Admission Prices: Call for hours. &
Membership: Individual $35; Family $63.

EISENHOWER NATIONAL HISTORIC SITE, 1195 Baltimore Pike, Ste. 100, Gettysburg, PA 17325-7034. Tel.: 877-874-2478. Fax: 717-338-0821. TDD: 717-334-1382.
Web Site: www.nps.gov/eise
Founded: 1967.
Congressional District: 4
Key Personnel: Supt., Bob Kirby; Museum Cur., Michael R. Florer; Supervisory Historian, Carol A. Hegeman; Museum Shop Mgr., Lisa Kamps.

Personnel Profile: Full-Time Paid 6; Part-Time Paid 6; Part-Time Volunteers 707; Interns 5.
Governing Authority: federal. Parent Institution: National Park Service, U.S. Dept. of Interior, Washington, DC 20240. Tax-exempt: 501(c)(3).
Institution Type/Description: Historic Site: presidential & retirement home of Dwight D. Eisenhower, 34th President of the U.S. & Supreme Commander of Allied forces in Europe during World War II.
Collections: historic structures; furnishings; farm equipment & other vehicles; Eisenhower personal possessions; Eisenhower presidential memorabilia; World War II memorabilia; archaeology.
Research Fields: Eisenhower presidency; 20th-century presidential history.
Facilities: information & visitor center. Museum-related items for sale.
Activities: self-guided tours of home & grounds; farm walking tour; Junior Secret Service Agent program; audiovisual program; conducted tours and programs.
Publications: brochure.
Hours & Admission Prices: Visitor Center: daily 8-5. House: daily 9-4. Tours via shuttle bus only. Adults $7.50, children 6-12 $5; discounts to groups of 16 or more. Closed New Year's Day; Thanksgiving; Christmas. &
Attendance: 70,000 (accurate)

GETTYSBURG FIRE DEPARTMENT MUSEUM, 35 N. Stratton St., Gettysburg, PA 17325. Tel.: 717-334-8300.
Institution Type/Description: Firefighting History Museum.
Collections: firefighting history & equipment; fire trucks; helmets; uniforms; speaking trumpets; buckets; paintings; medals.
Hours & Admission Prices: Call for hours.

GETTYSBURG NATIONAL MILITARY PARK, 1195 Baltimore Pike, Ste. 100, Gettysburg, PA 17325-7034. Tel.: 877-874-2478. Fax: 717-334-1891. TDD: 717-334-1382.
Web Site: www.gettysburgfoundation.org
Founded: 1895.
Congressional District: 4
Key Personnel: Chief Cur. Museum, Greg Goodell.
Personnel Profile: Full-Time Paid 90; Part-Time Volunteers 1,929.
Governing Authority: federal. Parent Institution: National Park Service, Dept. of the Interior. Tax-exempt.
Institution Type/Description: Military Museum & Battlefield.
Collections: Gettysburg Museum of the Civil War; Paul Philippoteaux's Gettysburg Cyclorama. Historic Site: Gettysburg National Cemetery.
Research Fields: Civil War & related topics.
Facilities: 2,700-vol. library primarily on the Battle of Gettysburg, Civil War & environmental books, available for use on premises by appointment. Books, postcards & slides for sale.
Activities: guided tours; films; permanent exhibits; site talks; campfire programs; Electric Map orientation program of the Battle of Gettysburg.
Hours & Admission Prices: April-May & Sept.-Oct. daily 8-6; June-Aug. daily 8-7; Nov.-March daily 8-5. Closed New Year's Day; Thanksgiving; Christmas. &
Attendance: 1,900,000 (accurate)

LINCOLN TRAIN MUSEUM, 425 Steinwehr Ave., Gettysburg, PA 17325-2930. Tel.: 717-334-5678. Fax: 717-334-0224. Facebook: Lincoln Train Museum.
E-mail: gettsburglincolntrain@gmail.com
Web Site: www.lincolntrain.com
Congressional District: 19
Key Personnel: Museum Shop Mgr., Karen Saylor.
Personnel Profile: Full-Time Paid 1; Part-Time Paid 6.
Operating Expenses: 150,000
Operating Income: 175,000
Governing Authority: private; profit organization. Parent Institution: Lincoln Holding Co. LLC.
Institution Type/Description: History Museum.
Collections: Lincoln-related artifacts; train collection; railroad dioramas; hands-on exhibits; American history & artifacts of America.
Facilities: Gift items for sale.
Activities: simulated ride; audio/visual dioramas; interactive operating train layout
Hours & Admission Prices: Spring & Fall Sun.-Thurs. 9-7, Fri.-Sat. 9-9; June & July daily 9-9; Dec.-March Fri.-Sat. & holidays 9-6. Adults $7, adult members $6, children 6-12 $4, discounts to AAA members, seniors, police, fire, EMS & military & reenactors in period dress; children 5 & under no charge. Prices subject to change. Closed New Year's Day, Thanksgiving Day & Christmas Day. &
Attendance: 30,000 (estimated)

SCHMUCKER ART GALLERY, (M), Gettysburg College, 300 N. Washington St., Gettysburg, PA 17325-1483. Mailing Address: Box 2452, Gettysburg, PA 17325. Tel.: 717-337-6080. Fax: 717-337-6099. Facebook: Schmucker Art Gallery.
E-mail: segan@gettysburg.edu
Web Site: www.gettysburg.edu/gallery
Congressional District: 91
Key Personnel: Dir., Shannon Egan, Ph.D.
Personnel Profile: Full-Time Paid 1; Part-Time Paid 1; Part-Time Volunteers 1; Interns 8.
Governing Authority: Parent Institution: Gettysburg College. Tax-exempt.
Institution Type/Description: Art Gallery.
Collections: works by local, national & international contemporary artists.
Major Exhibits: Out of Rubble, 1/23/14-3/8/14; Glenn Ligon: Narratives (Disembark) Suite, 1/24/14-3/8/14; Juried Student Art Exhibition, 3/28/14-4/19/14; Juror's Exhibition, 3/28/14-4/19/14; Capstone 2014, 4/30/14-5/18/14; Adams County Arts Council Juried Art Exhibition, 5/30/14-6/22/14; Michael Scoggins, 9/10/14-12/6/14.
Research Fields: nineteenth & twentieth century art.
Facilities: 1,600 sq. ft. project space.
Activities: temporary & traveling exhibitions.
Publications: exhibitions catalogs.
Hours & Admission Prices: Tues.-Sat. 10-4. No charge; donations accepted. Closed college holidays. &
Attendance: 4,000 (estimated)

SHRIVER HOUSE MUSEUM, 309 Baltimore Ave., Gettysburg, PA 17325-2602. Tel.: 717-337-2800. Facebook: Shriver House Museum.
E-mail: mail@shriverhouse.org
Web Site: www.shriverhouse.org
Founded: 1996.
Key Personnel: Dir. & Chm. (V), Nancie W. Gudmestad; Pres. (V), Del Gudmestad.
Personnel Profile: Full-Time Paid 1; Full-Time Volunteers 1; Part-Time Paid 9; Part-Time Volunteers 1.
Institution Type/Description: Historic House Museum: housed in the former home of George & Hettie Shriver; built in 1860.
Collections: family & local history; period furnishings; photographs.
Facilities: Books, videos, prints, period games & puzzles for sale.
Activities: rental facilities.
Publications: The Shrivers of Gettysburg - Eyewitness to The Battle.
Hours & Admission Prices: March Sat.-Sun. 10-5; April-Oct. Sun.-Thurs. 10-5, Fri. -Sat. 10-6; Nov. 12-5; Thanksgiving 5-10; Dec. Sat. 12-10; other times by appointment. Adults $8.95, children 7-12 $6.95; discounts to groups (10+). Closed Jan. & Feb., Christmas Eve, Day & week.
Attendance: 30,000 (accurate)

SOLDIERS NATIONAL MUSEUM, 777 Baltimore St., Gettysburg, PA 17325-2600. Tel.: 717-334-4890. Fax: 717-334-9100.
E-mail: info@gburgtours.com
Web Site: www.gettysburgbattlefieldtours.com
Founded: 1950.
Congressional District: 19
Key Personnel: Administrative Asst., Bonnie Jacoby.
Personnel Profile: Full-Time Paid 3.
Governing Authority: private. Parent Institution: Gettysburg Heritage Enterprises, Inc., Gettysburg, PA.
Institution Type/Description: Military Museum.
Collections: artifacts; military memorabilia; confederate encampment; miniature dioramas.
Activities: self guided tours; loan exhibitions.
Hours & Admission Prices: March-June & Sept. to late Nov. daily 9-5; July to mid-Aug. daily 9-7. Adults $7.50, senior citizens $6.75, children 6-12 $3.50. Closed Thanksgiving & Sun. after.
Attendance: 5,000 (estimated)

Girard

THE BATTLES MUSEUMS OF RURAL LIFE, 436 Walnut St., Girard, PA 16417-1650. Mailing Address: 419 State St., Erie, PA 16501-1106. Tel.: 814-454-1813. Fax: 814-454-6890.
E-mail: echs@eriecountyhistory.org
Web Site: www.eriecountyhistory.org
Founded: 1989.
Congressional District: 6
Key Personnel: Dir., Caleb Pifer; Operations Dir., Melanie Kuebel-Stankey; Pres. (V), Jack Watts.

Personnel Profile: Full-Time Paid 1; Part-Time Paid 1; Part-Time Volunteers 14.

Governing Authority: private; nonprofit organization. Parent Institution: Erie County Historical Society, 417 State St., Erie, PA 16501. Tax-exempt: 501(c)(3).

Institution Type/Description: Historic & Agriculture Museum: housed on 130 acres of farmland & woodland.

Collections: 2 & 3 dimensional artifacts from the Battles family of Girard, PA & from the region 1840-1952. Historic Buildings: 1858 & 1861 farmhouses.

Research Fields: 1920's costumes; 1860 Lifestyles.

Facilities: ecology trail.

Activities: docent program; formal education programs for adults, children & under-graduate or graduate archaeology students with Edinboro University; guided tours; lectures; temporary exhibitions.

Hours & Admission Prices: By appointment. Family $10, Adults $5, senior citizens $4, children $3; discounts to AAM members; members & children under 6 no charge. &

Attendance: 3,562 (accurate)

Membership: Student $15; Individual $45; Family $60; Patron $75; Sustaining $100; Sponsor $250; Corporate $500.

HAZEL KIBLER MEMORIAL MUSEUM, 522 Main St., Girard, PA 16417-1713. Tel.: 814-774-4168.

E-mail: swinick@aol.com

Web Site: westcountyhistorical.com

Key Personnel: Pres., Stephanie Wincik.

Governing Authority: Parent Institution: West County Historical Society.

Institution Type/Description: History Museum.

Collections: Marx toys; clown & circus owner, Dan Rice memorabilia; period furnishings; photographs; newspapers.

Hours & Admission Prices: mid-May to Sept. Sun. 2-5; groups by appointment. Adults $2, children $1.

Gladwyne

HENRY FOUNDATION FOR BOTANICAL RESEARCH, 801 Stony Lane, Gladwyne, PA 19035-1460. Mailing Address: P.O. Box 7, Gladwyne, PA 19035-0007. Tel.: 610-525-2037. Fax: 610-525-4024.

Founded: 1949.

Key Personnel: Pres. & Exec. Dir., Susan P. Treadway; Cur., Betsey W. Davis.

Personnel Profile: Full-Time Paid 2; Part-Time Paid 1; Part-Time Volunteers 8; Interns 3.

Governing Authority: nonprofit. Tax-exempt.

Institution Type/Description: Botanical Garden.

Collections: Native American Flora; Mary G. Henry Collection.

Research Fields: Native American flora.

Facilities: library pertaining to botany; botanical garden.

Activities: guided tours; lectures; organized educational programs; special events; classes.

Publications: newsletter.

Hours & Admission Prices: Tues.-Thurs. 10-3 by appointment. Admission $5.

Attendance: 8,000 (estimated)

Membership: Individual $25; Dual $40; Contributor $100; Sustainer $250; Patron $500; Benefactor $1,000; Guardian $2,500.

RIVERBEND ENVIRONMENTAL EDUCATION CENTER, 1950 Spring Mill Rd., Gladwyne, PA 19035-1000. Tel.: 610-527-5234. Fax: 610-527-1161.

E-mail: info@riverbendeec.org

Web Site: www.riverbendeec.org

Founded: 1974.

Congressional District: 13

Key Personnel: Exec. Dir., Laurie Bachman; Chm. (V) & Pres. (V), Beverly Galloway.

Personnel Profile: Full-Time Paid 6; Part-Time Paid 12; Part-Time Volunteers 10; Interns 5.

Governing Authority: nonprofit organization. Tax-exempt: 501(c)(3), 170(b)(1)(A).

Institution Type/Description: Nature Center.

Collections: taxidermy; live animals in tanks.

Facilities: 1,000-vol. library; 30 acres of open space with ponds, streams, meadow & wooded areas; nature center with teaching & meeting space.

Activities: school & camp educational programs; public programs; special events; organized educational programs; summer day camp; internship program; Native American program; birthday parties; barn rentals.

Publications: newsletter, Round the Bend; camp brochure; school program brochure; general brochure on mission.

Hours & Admission Prices: Grounds: daily dawn to dusk. Center: Mon.-Fri.

9-5. Fee for camp & school programs. Walking Trails: no charge. Closed national holidays. &

Attendance: 12,000 (estimated)

Membership: Basic Family & Individual $50; Family & Individual Plus $100; Birdwatcher $250; Nature Lover $500; Riverkeeper $1,000.

Glenside

ARCADIA UNIVERSITY ART GALLERY, (M), Church & Easton Rds., Glenside, PA 19038. Mailing Address: 450 S. Easton Rd., Glenside, PA 19038-3295. Tel.: 215-572-2133, 2131 & 2900. Fax: 215-881-8774.

E-mail: torchiar@arcadia.edu

Web Site: www.arcadia.edu

Formerly: Beaver College Art Gallery

Founded: 1853.

Congressional District: 135

Key Personnel: Pres., Jerry Griener; Dir., Richard Torchia; Vice Pres. Institutional Advancement, Nick Costa; Public Rels., Lori Bauer.

Personnel Profile: Full-Time Paid 1; Part-Time Paid 9; Part-Time Volunteers 13.

Governing Authority: college; nonprofit. Parent Institution: Arcadia University. Tax-exempt: 501(c)(3).

Institution Type/Description: Art Gallery: housed in c.1893 historic building.

Collections: paintings; sculpture; prints; works by regional artists; works by Benton Spruance, long time chair of the Fine Arts Dept.

Research Fields: contemporary visual art.

Facilities: 1,200 sq. ft. exhibit space; 150-seat theatre. Exhibition catalogues for sale.

Activities: guided tours; lectures. Annual Events: Exhibitions & Symposia.

Publications: exhibition catalogues.

Hours & Admission Prices: Sept.-May Tues.-Wed. & Fri. 10-3, Thurs. 10-8, Sat.-Sun. 12-4. No charge; donations accepted. Closed Thanksgiving; Christmas.

Attendance: 6,500 (estimated)

Membership: Artist & Senior Citizen $25; Friend $50; Family $100; Patron $200; Benefactor $500.

Grantham

M. LOUISE AUGHINBAUGH GALLERY AT MESSIAH COLLEGE, Climenhaga Fine Arts Ctr., One College Ave., Ste. 3400, Grantham, PA 17055. Tel.: 717-766-2511, ext. 2486.

E-mail: sbiddle@messiah.edu

Web Site: messiah.edu

Founded: 1979.

Key Personnel: Dir., Christine A. Forsythe.

Governing Authority: private college. Tax-exempt.

Institution Type/Description: Art Museum.

Collections: paintings.

Activities: lectures; education programs.

Hours & Admission Prices: Mon.-Thurs. 9-4, Fri. 9-9, Sun. 2-5. No charge. &

THE OAKES MUSEUM OF NATURAL HISTORY, (M), Messiah College, One College Ave., Ste. 3029, Grantham, PA 17055. Tel.: 717-691-6082. Fax: 717-691-6046.

E-mail: oakesmuseum@messiah.edu

Web Site: www.messiah.edu/Oakes

Founded: 2002.

Key Personnel: Dir., Kenneth A. Mark; Education Coord., Helena Cicero; Education Coord., Beth Erikson; Treas., Lois Voigt; Cur. Herpetology & Ornithology, Dr. Erik Lindquist; Cur. Botany & Entomology, Dr. David Foster; Cur. Geology, Mr. Edwin Charles; Cur. Concology, Mrs. Ruth Bierbower; Cur. Mycology, Dr. Gary Emberger; Cur. Archaeology, Dr. David Pettegrew.

Personnel Profile: Full-Time Paid 2; Part-Time Paid 6; Part-Time Volunteers 18; Interns 1.

Governing Authority: private college. Parent Institution: Messiah College. Tax-exempt: 501(c)(3).

Institution Type/Description: Natural History Museum.

Collections: African & North American mammals, including a full body elephant skeleton; bird eggs & nests, seashells, minerals, insects, fungi, snakes & frogs; Native American artifacts.

Research Fields: vernal pool study; Project Golden Frog; Yellow Breeches Restoration Project.

Facilities: 10,000 sq. ft. exhibit space; 75 acre outdoor program area; classrooms; labs. Museum-related items for sale.

Activities: formal education programs for college students, adults & children; guided tours; lectures; participatory & temporary exhibits; scout programs;

senior citizen customized programming to meet education requirements for Pennsylvania state standards.

Hours & Admission Prices: Sat. 1-5. Adults $6, senior citizens, students and children 3-12 $3.50; discounts to groups & AAM members. Closed major holidays. &

Attendance: 16,674 (accurate)

Green Lane

GOSCHENHOPPEN FOLKLIFE LIBRARY AND MUSEUM, 116 Gravel Pike, Red Men's Hall, Green Lane, PA 18054-0476. Mailing Address: P.O. Box 476, Green Lane, PA 18054-0476. Tel.: 215-234-8953.

E-mail: redmens_hall@goschenhoppen.org

Web Site: www.goschenhoppen.org

Founded: 1965.

Congressional District: 15

Key Personnel: C.E.O. & Chm., George Spotts; Pres. (V), Edward C. Johnson; Chm. Museum & Library, D.F. Abe Roan; Sec., Susan Cook; Office Mgr., Jodie Ritto.

Governing Authority: society. Parent Institution: Goschenhoppen Historians, Inc. Tax-exempt: 501(c)(3).

Institution Type/Description: Folklife Museum: housed in c.1900 three-story brick Red Men's Hall; Country Store Museum: a recreation of a country store c.1870-1930; Henry Antes House: a restored 1736 Germanic house containing two floors in the attic & located on the site of the Pottsgrove encampment of the Continental Army, September, 1777.

Collections: Pennsylvania German folklore & folk culture; local history, & archaeology; archives; agriculture; costumes; 1864 & 1820 pipe organ; decorative arts; textiles; weaver's shop; hearth kitchen; PA German Parlor; blacksmith shop; country store museum.

Research Fields: folklore; PA German folk culture; regional history from 1683 on.

Facilities: 700-vol. library of Pennsylvania history & folk culture available for use on premises; 250-seat auditorium; meeting & conference rooms.

Activities: guided tours; concerts; lectures; formally organized educational programs; permanent & temporary exhibitions. Annual Event: Goschenhoppen Folk Festival in August.

Publications: monthly newsletter, Goschenhoppen Newsletter; cookbook, Goschenhoppen Recipes; annual festival program, The Goschenhoppen Intelligencer, a reproduction of the German newspaper, Bauren-Freund; books, Just a Quilt, local oral history & local designs; Lest I Shall Be Forgotten-color photos, anecdotes & traditions of quilts from the Montgomery County Quilt Documentation; James E. Frill's Music Book, 1830s; Berks Co. Fiddle Tunes.

Hours & Admission Prices: Museum: April-Oct. Sun. 1:30-4; other times by appointment. No charge; donations accepted. Library: by appointment only. &

Attendance: 7,850 (estimated)

Membership: SIngle $20; Family $30; Contributing $50; Patron $80; Individual Life $250; Couple Life $350.

Greencastle

ALLISON-ANTRIM MUSEUM, INC., 365 S. Ridge Ave., Greencastle, PA 17225-1157. Tel.: 717-597-9010.

E-mail: aamuseum@greencastlemuseum.org

Web Site: www.greencastlemuseum.org

Founded: 1995.

Congressional District: 9

Key Personnel: Pres., Bonnie A. Shockey; Treas., David McCarney.

Personnel Profile: Full-Time Volunteers 1; Part-Time Volunteers 45.

Governing Authority: public; nonprofit organization. Tax-exempt: 501(c)(3).

Institution Type/Description: General Museum.

Collections: 20th-century paintings by African American artist, Walter Washington Smith; memorabilia of Henry P. Fletcher, a U.S. ambassador and diplomat to six countries for 51 years under eight presidents; Pennsylvania governors' signatures on primary documents dating back to 1715; Civil War collection, including primary documents, letters and uniforms; the Carl's Drugstore collection of medical and drug store items; early agricultural equipment and implements.

Facilities: 1,500-vol. library Greencastle & Antrim Township local history & genealogy; 2,200 sq. ft. exhibit space.

Activities: docent program; guided tours; lectures; temporary exhibitions; monthly speaker series; monthly special exhibits; open houses.

Publications: bimonthly newsletter, Allison-Antrim Annals.

Hours & Admission Prices: Mon.-Fri. 12-4; other times by appointment. No charge; donations accepted. Closed major holidays. &

Attendance: 1,000 (accurate)

Membership: Student $5; Individual $10; Family $25; Patron $26-$99;

Supporting & Business $100-$199; Sustaining $200-$799; Lifetime $800; Lifetime Couple $1,500.

Greensburg

BALTZER MEYER HISTORICAL SOCIETY, 642 Baltzer Meyer Pike, Greensburg, PA 15601-9711. Tel.: 724-836-6915.

E-mail: baltzermeyer@juno.com

Web Site: baltzermeyer.pa-roots.com/Pages/home.html

Institution Type/Description: Historical Society Museum.

Collections: family, church, & cemetery records. Historic Buildings: 1881 one-room schoolhouse; 1884 Old Zion Lutheran church.

Activities: research.

Hours & Admission Prices: mid-April to mid-Oct. Mon. & Wed. 10-2; mid-Oct. to mid-April Wed. 10-2, call to confirm. Closed New Year's Eve; Christmas Eve.

WESTMORELAND COUNTY HISTORICAL SOCIETY, 362 Sand Hill Rd., Ste. 1, Greensburg, PA 15601. Tel.: 724-532-1935. Fax: 724-532-1938.

E-mail: history@westmorelandhistory.org

Web Site: www.westmorelandhistory.org

Founded: 1908.

Congressional District: 18

Key Personnel: Dir., Lisa C. Hays.

Personnel Profile: Full-Time Paid 4; Part-Time Paid 3; Part-Time Volunteers 175.

Governing Authority: Tax-exempt.

Institution Type/Description: Historical Society Museum.

Collections: local history & culture; period furnishings; personal artifacts; photographs.

Facilities: library; archives.

Activities: educational programs; research.

Publications: triannual magazine, Westmoreland History.

Hours & Admission Prices: Tues.-Fri. 9-5; other times by appointment. Library: $3; Historic Site: May-Oct. $5; discounts to AAA members. &

Membership: Student $18; Individual $38; Family $48; Sustaining $55; Contributor & Business $105; Benefactor $250.

❋ **WESTMORELAND MUSEUM OF AMERICAN ART, (M),** 221 N. Main St., Greensburg, PA 15601-1898. Tel.: 724-837-1500. Fax: 724-837-2921.

E-mail: info@wmuseumaa.org

Web Site: www.wmuseumaa.org

Founded: 1949.

Congressional District: 20

Key Personnel: Dir. & C.E.O., Judith H. O'Toole; Pres. (V), Bruce M. Wolf; Chief Cur., Barbara L. Jones; Asst. Dir. Advancement, Amy B. Baldonieri; Asst. Public & Financial Devel., Pat Erdelsky; Dir. Education & Visitor Svcs., Katie Barnard; Dir. Mktg. & Information Technology, Judy Linz Ross; Preparator, P.J. Zimmerlink; Mgr. Collections, Douglas W. Evans; Museum Shop Mgr., Ginnie Leiner.

Personnel Profile: Full-Time Paid 14; Part-Time Paid 16; Part-Time Volunteers 185; Interns 4.

Governing Authority: nonprofit organization. Tax-exempt: 501(c)(3) & 4942(j)(3).

Institution Type/Description: American Art Museum.

Collections: 18th- to 20th-century American paintings; sculpture; drawings; prints; furniture & decorative arts; 19th & 20th-century toy collection.

Research Fields: paintings by Southwestern Pennsylvania & American artists.

Facilities: cafe. Museum-related items for sale.

Activities: docent guided tours; in-house lectures & gallery talks; educational outreach programs for grades K-12; children art classes & art camp; internships; inter-museum loans; ongoing educational activities; permanent & temporary exhibitions; symposia.

Publications: book, 250 Years of Art in Pennsylvania; catalogs, George Hetzel & the Scalp Level Tradition; The Permanent Collection of the Westmoreland Museum of Art; Southwestern Pennsylvania Painters, 1800-1945; Penn's Promise: Still Life Painting in Pennsylvania, 1795-1930; Southwestern Pennsylvania Painters From the Westmoreland Museum of Art Permanent Collection; exhibition catalogs; gallery guides; Born of Fire: The Valley of Work; American Scenery: Different Views in Hudson River School Painting; Samuel Rosenberg: Portrait of a Painter; Made in Pennsylvania: A Folk Art Tradition; Painting in the United States 2008.

Hours & Admission Prices: Wed. & Fri.-Sun. 11-5, Thurs. 11-9. Suggested Donation: adults $5; discounts to AAM members; members, students & children under 12 no charge. Closed New Year's Day; Easter; Thanksgiving; Christmas. &

Attendance: 23,300 (accurate)

Membership: Student & Senior $25; Individual $40; Dual & Family $60;

Donor $125; Patron's Circle $450; Director's Circle $1,000; President's Circle & Corporate $2,000. Small Business $125, $450 or $1,000.

Greenville

GREENVILLE RAILROAD PARK AND MUSEUM, 314 Main St., Greenville, PA 16125-2615. Tel.: 724-588-4009.
E-mail: greenvillerailroadpark@gmail.com
Web Site: www.greenvilletrainmuseum.org
Institution Type/Description: Railroad Museum.
Collections: railroad history; switch engines; coal tender; hopper car; 1913 Empire touring car; 1952 caboose.
Hours & Admission Prices: May & Sept.-Oct. daily 1-5; mid-June to Labor Day Tues.-Sun. 1-5.

Grove City

GROVE CITY HISTORICAL SOCIETY, 111 College Ave., Grove City, PA 16127. Mailing Address: P.O. Box 764, 111 College Ave., Grove City, PA 16127. Tel.: 724-458-1798. Facebook: Grove City Historical Society.
E-mail: gcahs@zoominternet.net
Web Site: grovecityhistoricalsociety.org
Formerly: Grove City Historical Society
Founded: 1998.
Congressional District: 3
Key Personnel: Pres. (V), Linda P. Bennett.
Personnel Profile: Part-Time Volunteers 60.
Governing Authority: bd. directors. Tax-exempt.
Institution Type/Description: Historical Society Museum: housed in the former Traveler's Hotel.
Collections: local history, heritage & culture; period furnishings; personal artifacts; photographs.
Activities: school tours; seminars.
Publications: The Way We Were; Grandma Left the Light On; Historical Happenings-3 times a year.
Hours & Admission Prices: Tues.-Sat. 12-3. No charge; donations accepted. &
Attendance: 1,400 (accurate)
Membership: Student and Senior 55 & over $10; Individual $20; Family $40; Corporate $50; Life $500; Founding $1,000.

Halifax

LAKE TOBIAS WILDLIFE PARK, 760 Tobias Rd., Halifax, PA 17032-9474. Tel.: 717-362-9126. Fax: 717-362-9993. Facebook: Lake Tobias Wildlife Park.
E-mail: info@laketobias.com
Web Site: laketobias.com
Founded: 1965.
Key Personnel: Park Mgr., Ern Tobias; Public Rels. & Mktg., Jan Tobias-Kieffer
Institution Type/Description: Zoo.
Collections: animals from around the world including African lions, bears, tigers, zebra, cougar, donkeys, kangaroo, & birds.
Activities: zoo exhibits; safari tours; reptiles & exotics; petting zoo; special events.
Hours & Admission Prices: May & Sept. Sat.-Sun. 10-7; Memorial Day to Labor Day Mon.-Fri. 10-6, Sat.-Sun. 10-7; Oct. Sat.-Sun. 10-6. Park: 3 & over $6; children under 3 no charge. Safari: 3 & over $6; children under 3 no charge. &
Attendance: 160,000 (estimated)

Hamburg

READING RAILROAD HERITAGE MUSEUM, 500 S. Third St., Hamburg, PA 19526. Mailing Address: P.O. Box 15143, Reading, PA 19612-5143.
Web Site: www.readingrailroad.org
Founded: 1976.
Key Personnel: Devel., John Brown; Pres. (V), Duane E. Engle; Treas., Jim Adams; Archivist, Richard Bates.
Personnel Profile: Part-Time Volunteers 100.
Governing Authority: private; nonprofit organization. Tax-exempt: 501(c)(3).
Institution Type/Description: Transportation Museum.
Collections: Reading Railroad history & equipment, artifacts & paperwork; operational equipment.
Facilities: library; 41,200 sq. ft. exhibit space. Museum-related items for sale.
Activities: guided tours; loan, temporary & traveling exhibitions; operating train trips; equipment restoration.

Publications: monthly newsletter, Crusader; quarterly magazine, Beeline.
Hours & Admission Prices: Sat. 10-4, Sun. 12-4. Adults $5; members no charge.
Attendance: 2,000 (estimated)
Membership: Student $15; Individual $35; Family $43; Contributing $55; Sustaining $100; Corporate $150.

Hanover

HANOVER FIRE MUSEUM, Wirt Park Fire Station, 201 N. Franklin St., Hanover, PA 17331. Tel.: 717-637-6674 & 6671.
Founded: 1980.
Institution Type/Description: Firefighting History Museum.
Collections: firefighting history & equipment; 1882 Silsby steamer 550 GRM; c.1830 hand pumper; c.1770 Nushem grinder hand engine; 1911 alarm board.
Hours & Admission Prices: Daily 10-8.

Harleysville

MENNONITE HERITAGE CENTER, 565 Yoder Rd., Harleysville, PA 19438-1020. Tel.: 215-256-3020. Fax: 215-256-3023. Facebook: Mennonite Heritage Center.
E-mail: info@mhep.org
Web Site: www.mhep.org/
Founded: 1974.
Congressional District: 5
Key Personnel: Dir., Sarah Wolfgang Heffner; Asst. Dir., Rose A. Moyer; Cur. & Librarian, Joel D. Alderfer; Archivist, Forrest Moyer; Volunteer Coord. & Museum Shop Mgr., Sara Kolb.
Personnel Profile: Full-Time Paid 3; Full-Time Volunteers 4; Part-Time Volunteers 50.
Governing Authority: nonprofit organization. Parent Institution: Mennonite Historians of Eastern Pennsylvania. Tax-exempt: 501(c)(3).
Institution Type/Description: Heritage Center.
Collections: Pennsylvania German Fraktur (illuminated writing); costumes; tools; archival materials; 16th to 19th-century books & Bibles; ethnology; photographs; needlework; quilts & coverlets; ceramics; pottery; glassware; local history & genealogy; early southeastern Pennsylvania German & Mennonite history.
Research Fields: local genealogy.
Facilities: library; archives; 3,000 sq. ft. exhibit space. Mennonite-Anabaptist history books, Pennsylvania German culture books, local cookbooks, Fraktur prints & local genealogy books for sale.
Activities: audiovisual presentations; educational, interpretive & slide lectures; educational programs for adults; religious programs; traditional arts workshop. Museum Sponsors: Apple Butter Frolic; Pennsylvania German Folk Art Sale; Whack & Roll Crocquet Tournament.
Publications: quarterly newsletter.
Hours & Admission Prices: Tues.-Fri. 10-5, Sat. 10-2. Donation requested. Closed New Year's Day; Good Friday; Easter; Independence Day; Thanksgiving; Christmas. &
Attendance: 7,000 (estimated)
Membership: Student & Senior $25; Individual $35; Family $50; Contributing $75; Supporting $100; Sustaining $250; Associate $1,000.

Harrisburg

ART ASSOCIATION OF HARRISBURG, 21 N. Front St., Harrisburg, PA 17101-1625. Tel.: 717-236-1432. Fax: 717-236-6631.
E-mail: carrie@artassocofhbg.com
Web Site: www.artassocofhbg.com
Founded: 1926.
Key Personnel: Dir., Pres. & Museum Shop Mgr., Carrie Wissler Thomas; Chm. (V), Rick LeBlanc.
Personnel Profile: Part-Time Paid 5.
Operating Expenses: 216,817
Operating Income: 219,042
Governing Authority: not-for-profit organization. Tax-exempt.
Institution Type/Description: Art: 1810 four-story Italianate-style home housing the Art Association building.
Collections: paintings; graphics.
Major Exhibits: Figuratively Speaking, 1/12/14-2/13/14; Invitational, 2/21/14-3/27/14; Invitational, 4/4/14-5/8/14; 86th Juried Show, 5/16/14-6/19/14; Art School Annual, 6/27/14-7/24/14; Invitational, 8/15/14-9/4/14; Fall Members, 9/12/14-10/16/14; Invitational, 10/24/14-11/26/14; Invitational, 12/5/14-1/8/15.
Facilities: 400-vol. art library; educational facilities. Museum-related items for sale.
Activities: formal educational programs; guided tours; concerts; community

exhibitions. Annual Events: Juried Exhibition; Meet the Artist in summer; Gallery Walk in September; Costume Ball.
Publications: quarterly newsletter; school catalogue; monthly show invitations.
Hours & Admission Prices: Mon.-Thurs. 9:30-9, Fri. 9:30-4, Sat. 10-4, Sun. 2-5. No charge. Closed New Year's Day; Easter; Thanksgiving; Christmas.
Attendance: 12,000 (estimated)
Membership: Student $25; Artist & Supporter $50; Promoter $75; Patron $125; Benefactor $250; Sponsor $500; Gold $750; Friend $1,000.

BENJAMIN OLEWINE III NATURE CENTER, 100 Wildwood Way, Harrisburg, PA 17110-2914. Tel.: 717-221-0292. Fax: 717-221-0318.
E-mail: friendsofww@wildwoodlake.org
Web Site: www.wildwoodlake.org
Institution Type/Description: Nature Center.
Collections: natural history; wildlife & their habitats; wetland exhibits; photographs.
Facilities: nature trails. Museum-related items for sale.
Activities: educational programs; bird observation area.
Hours & Admission Prices: Park: daily dawn to dusk. Center: Tues.-Sun. 10-4. Closed holidays. &
Attendance: 85,000 (estimated)

FORT HUNTER MANSION & PARK, 5300 N. Front St., Harrisburg, PA 17110-1718. Tel.: 717-599-5751. Fax: 717-599-5838.
Web Site: www.forthunter.org
Founded: 1933.
Key Personnel: Mansion Mgr., Mary Trost; Park Mgr., Julia Hair; Educator, Elizabeth Johnson.
Personnel Profile: Full-Time Paid 1; Part-Time Paid 7.
Governing Authority: county; board of trustees; nonprofit organization. Parent Institution: County of Dauphin. Subsidiary Institution: Board of Trustees for Fort Hunter. Tax-exempt: 501(c)(3).
Institution Type/Description: Historic House & Park: housed in 1814 Federal style stone mansion, on the site of French & Indian War fort.
Collections: carriages; photographs; clothing; china; correspondence; elliptical staircase; Early American, Empire and Victorian furnishings. Historic Structures: historic barns; icehouse; tavern; springhouse.
Facilities: 8,000 sq. ft. exhibit space; nature area; 19th-century boxwood garden; herb gardens; pavilion for picnics; trails. Museum-related items for sale.
Activities: guided tours; lectures; arts festivals; organized educational programs for children; docent program; temporary exhibitions; performing arts; nature walks; military reenactments. Museum Sponsors: Fort Hunter Day in September; Yuletide celebrations; seasonal fairs & craft shows; Christmas celebration.
Publications: quarterly newsletter, The Chronicler.
Hours & Admission Prices: May-Dec. Tues.-Sat. 10-5, Sun. 12-5. Adults $5, senior citizens $4, children $3; members no charge. &
Attendance: 9,000 (estimated)
Membership: Individual $25; Family & Organization $35; Supporting $50; Patron $100; Sponsor & Corporate $250.

THE HISTORICAL SOCIETY OF DAUPHIN COUNTY, The John Harris/Simon Cameron Mansion, 219 S. Front St., Harrisburg, PA 17104-1619. Tel.: 717-233-3462. Fax: 717-233-6059. Facebook: Dauphin County History.
E-mail: office@dauphincountyhistory.org
Web Site: www.dauphincountyhistory.org
Founded: 1869.
Congressional District: 17
Key Personnel: Exec. Dir., Nicole McMullen; Cur., Stephen Bachmann; Librarian, Ken Frew.
Personnel Profile: Full-Time Paid 1; Part-Time Paid 6; Part-Time Volunteers 65.
Governing Authority: board of trustees. Tax-exempt.
Institution Type/Description: Historic House Museum: 1766 built by John Harris, Jr., founder of Harrisburg; enlarged by Simon Cameron, Lincoln's first secretary of war.
Collections: decorative arts; household items; tools; toys; textiles; Indian artifacts; ceramics; glass; clocks; John Harris furniture; Simon Cameron furniture.
Research Fields: genealogy; county history; structure history.
Facilities: library; youth education building; reception & meeting rooms.
Activities: guided tours; research; family & school programs; workshops; bus tours; weddings; private & community events; adult programs.
Publications: newsletter, The Oracle.

Hours & Admission Prices: Tours: Mon.-Thurs. 1-4, last tour begins at 3. Tour: adults $8, senior citizens $7, children 6-12 $6; discounts to groups; children under 6 & members no charge. Library: Tues.-Fri. 1-4. Adults $10. Closed most holidays. &
Attendance: 2,500 (estimated)
Membership: Individual $35; Family $45; Heritage Circle $55; Haldeman Circle $100; Cameron Circle $250; Harris Circle $500; Penn Circle $1,000.

THE NATIONAL CIVIL WAR MUSEUM, (M), One Lincoln Center at Reservoir Park, Harrisburg, PA 17103. Tel.: 717-260-1861. Fax: 717-260-9599.
E-mail: info@nationalcivilwarmuseum.org
Web Site: www.nationalcivilwarmuseum.org
Founded: 2001.
Key Personnel: C.E.O., Wayne E. Motts; Chm. (V), Paul Wipple; Museum Shop Mgr., Kate McDermott.
Personnel Profile: Full-Time Paid 9; Full-Time Volunteers 10; Part-Time Volunteers 27; Interns 25.
Governing Authority: Tax-exempt.
Institution Type/Description: Military & History Museum.
Collections: personal & military artifacts; manuscripts; documents; photographs.
Facilities: research library; galleries; collection storage; offices; educational, retail & public space.
Activities: special events; book signings; lectures; encampments; music.
Hours & Admission Prices: Mon., Tues., Thurs. & Fri., Wed. 10-8, Sat. 10-5, Sun. 12-5. Family $35, adults $10, senior citizens $9, students $8, closed New Years, Christmas, Easter & Thanksgiving. &
Attendance: 40,000 (accurate)
Membership: Student $25; Military, Educator & Senior $35; One Adult Household $50; Two Adult Household $85.

PENNSYLVANIA FEDERATION OF MUSEUMS AND HISTORICAL ORGANIZATIONS, 234 N. 3rd St., Harrisburg, PA 17101-1516. Tel.: 717-909-4950. Fax: 717-909-3996.
E-mail: pamuseums@pamuseums.org
Web Site: www.pamuseums.org
Founded: 1907.
Congressional District: 17
Key Personnel: Exec. Dir., Deborah M. Filipi; Deputy Dir., Janet MacGregor.
Personnel Profile: Full-Time Paid 3; Part-Time Paid 1.
Governing Authority: nonprofit organization.
Institution Type/Description: Museum Service Organization.
Activities: training programs for volunteer & professional museum staff.
Publications: quarterly newsletter, Tapestry; Keystone Treasures: Guide to Museums and Historical Organizations in Pennsylvania.
Hours & Admission Prices: Mon.-Fri. 8:30-5. Closed federal & state holidays. &
Membership: Individual $35; Affiliate $95; Institutional $55-$225.

PENNSYLVANIA HISTORICAL & MUSEUM COMMISSION, (M), 300 North St., Harrisburg, PA 17120-0101. Tel.: 717-787-2891. Fax: 717-783-1073.
Web Site: www.phmc.state.pa.us
Founded: 1945.
Congressional District: 17
Key Personnel: Exec. Dir., James M. Vaughan; Chm. (V), Andrew E. Masich; Dir. Bureau of Historic Sites & Museums, Brenda Reigle; Dir. Bureau of Archives & History, David Haury; Dir. Bureau of The State Museum of Pennsylvania, David Dunn; Dir. Bureau of Management Svcs., Thomas Leonard; Div. Chief Historic Sites & Museums, Michael Bertheaud; Chief Div. Architecture & Preservation, Barry Loveland.
Governing Authority: state. Branch Museums & Historic Sites: State Museum of Pennsylvania, Harrisburg; Landis Valley Village & Farm Museum, Lancaster; Pennsylvania Military Museum, Boalsburg; Ft. Pitt Museum, Pittsburgh; Pennsylvania Lumber Museum, Galeton; Somerset Historical Center, Somerset; Old Economy Village, Ambridge; Daniel Boone Homestead, Birdsboro; Conrad Weiser Homestead, Womelsdorf; French Azilum, Wysox; Brandywine Battlefield Park, Chadds Ford; Pennsbury Manor, Morrisville; Historic Peace Church, Camp Hill; Morton Homestead, Prospect Park; United States Brig Niagara & Erie Maritime Museum, Erie; Ft. LeBoeuf, Waterford; Judson House, Waterford; Old Mill Village, New Milford; Railroad Museum of Pennsylvania, Strasburg; Scranton Iron Furnaces, Scranton; Old Chester Court House, Chester; Tuscarora Academy, Academia; Ephrata Cloister, Ephrata; Robert Fulton Birthplace, Quarryville; Cornwall Iron Furnace, Cornwall; McCoy House, Lewistown; Graeme Park, Horsham; Hope Lodge, Mather Mill, Ft. Washington; Joseph Priestley House, Northumberland; Historic Warrior Run Church, McEwensville; Drake Well Museum, Titusville; Historic Pithole City, Pithole; David

Bradford House, Washington; Bushy Run Battlefield, Jeannette; The Highlands, Whitemarsh Township; Washington Crossing Historic Park, Washington Crossing; Pennsylvania State Archives, Harrisburg; Pennsylvania Anthracite Heritage Museum, Scranton; Eckley Miners' Village, Eckley; Museum of Anthracite Mining, Ashland; Nathan Denison House, Forty Fort; Bowman's Hill Wildflower Preserve, Washington Crossing. Tax-exempt.
Institution Type/Description: State Agency; Conservation Center.
Collections: arts & crafts; china; glass; silver; folk arts; textile; military; preservation project; anthropology; ethnology; Indian artifacts; archaeology; entomology; insects; geology; mineralogy; paleontology; herbarium; medical; dental; health; natural history; natural science; science; agriculture; antiques; forestry; guns; industrial; logging & lumber; mining; religious; technology; transportation; rural life history; maritime; railroad; historic houses; ruins; industrial sites; religious & political history.
Research Fields: history; archaeology; geology; natural science; exhibits technology; science & technology; fine arts; decorative arts; museology; genealogy; military history; maritime.
Facilities: 99,000-vol. library of history, science, natural history, art, folk art, archaeology available for inter-library loan & on premises by written request; nature/conservation center; planetarium; reading room; 400-seat auditorium; theater; classrooms. Reproductions & items relating to collections for sale.
Activities: guided tours; lectures; films; concerts; dance recitals; arts festivals; study clubs; hobby workshops; TV & radio programs; formally organized education programs; inter-museum loan, permanent, temporary & traveling exhibitions.
Publications: books; leaflets; post cards; monographs; reproductions; calendars; posters; history magazine; rack cards.
Hours & Admission Prices: See individual listings for museums, historic sites, & houses; discounts for active military & their families and AAM & ICOM members; PA Heritage Foundation, State Legislature, media, State Museum of Pennsylvania members, teachers, tour leaders & children under 3 no charge. &
Attendance: 1,100,000 (estimated)
Membership: Pennsylvania Heritage Foundation: Individual $50; Family $70; Contributor $100; Patron $250; Benefactor $500.

THE PENNSYLVANIA NATIONAL FIRE MUSEUM, 1820 N. 4th St., Harrisburg, PA 17102. Tel.: 717-232-8915. Fax: 717-232-8916.

E-mail: info@pnfm.org
Web Site: www.pnfm.org
Founded: 1993.
Key Personnel: Pres. (V), Dave Warren, Jr.; Museum Shop Mgr., John Wagner.
Personnel Profile: Part-Time Volunteers 20.
Governing Authority: Parent Institution: Pennsylvania National Fire Museum, Inc. Tax-exempt.
Institution Type/Description: Fire Museum: housed in the 1899 Victorian firehouse Reily Hose Company No. 10.
Collections: firefighting history & equipment from hand-drawn fire apparatus to modern tools.
Hours & Admission Prices: Tues.-Sat. 10-4, Sun. 1-4. Adults $6, seniors $5, student $4; discounts to groups. Closed holidays.
Attendance: 5,000 (estimated)

ROSE LEHRMAN ART GALLERY, (M), Rose Lehrman Art Center, Harrisburg Area Community College, 1 HACC Dr., Harrisburg, PA 17110-2903. Tel.: 717-780-2435.

E-mail: kebanist@hacc.edu
Web Site: www.hacc.edu/RoseLehrmanArtsCenter/ArtGallery/index.cfm
Key Personnel: Cur., Kim Banister
Institution Type/Description: Art Gallery.
Collections: art exhibitions.
Hours & Admission Prices: Summer & Fall: Mon., Wed. & Fri. 11-3, Tues. & Thurs. 11-3 and 5-7 or by appointment.

∗ THE STATE MUSEUM OF PENNSYLVANIA, (M), 300 North St., Harrisburg, PA 17120-0101. Tel.: 717-787-4980. Fax: 717-783-4558.

E-mail: bhager@state.pa.us
Web Site: www.statemuseumpa.org
Founded: 1905.
Congressional District: 17
Key Personnel: Museum Dir., David W. Dunn; Exec. Dir., PA Historical & Museum Commission., James M. Vaughn; Chief, Education & Outreach, Beth A. Hager.

Governing Authority: state. Parent Institution: Pennsylvania Historical and Museum Commission. Tax-exempt: 170(c)(1).
Institution Type/Description: General Museum.
Collections: Pennsylvania collections of fine arts; Native American & historic archaeology; natural science including birds, fish, mammals, insects, plants; science; technology including vehicles, appliances, tools, machinery; military & political history; community & domestic life; geology & paleontology.
Research Fields: Pennsylvania art history, history, mammals & plants; archaeology; technology; decorative art; geology & paleontology.
Facilities: 25,000-vol. library of material on general & Pennsylvania archaeology, technology, military history, Pennsylvania biography & natural science available for use on premises or by arrangement; reading room; planetarium; 395-seat auditorium; classrooms. Museum-related items for sale.
Activities: tours; lectures; films; gallery talks; concerts; arts festivals; TV & radio programs; formally organized education programs; docent program; inter-museum loan & temporary exhibitions.
Publications: Susquehanna Indians; quarterly Calendar of Events; Natural History Notes; Ephrat Archaeology Series.
Hours & Admission Prices: Wed.-Sat. 9-5, Sun. 12-5. Adults $5; seniors 60 & up and children 1-12 $4; PA Heritage Society members no charge. &
Attendance: 300,000 (estimated)
Membership: PA Heritage Society: Individual $50; Family $70.

SUSQUEHANNA ART MUSEUM, (M), 15 N. 3rd St., Harrisburg, PA 17108. Mailing Address: P.O. Box 11818, Harrisburg, PA 17108-1818. Tel.: 717-233-8668. Fax: 717-233-8155.

E-mail: info@sqart.org
Web Site: www.sqart.org
Founded: 1989.
Congressional District: 17
Key Personnel: Exhibitions Mgr., Amy Hammond; Dir. Outreach Education, Wendy Sweigart; Museum Operations Coord., Susan Bennett.
Personnel Profile: Full-Time Paid 3; Part-Time Paid 3; Part-Time Volunteers 10; Interns 3.
Governing Authority: Tax-exempt.
Institution Type/Description: Art Museum.
Collections: works by regional & international artists.
Hours & Admission Prices: Call for hours & admission prices. &
Attendance: 10,000 (estimated)
Membership: Individual $30; Family $50; Supporter $100; Patron $250; Benefactor $500; Sustainer Circle $1,000.

WHITAKER CENTER FOR SCIENCE AND THE ARTS, 222 Market St., Harrisburg, PA 17101-2113. Mailing Address: 225 Market St., 2nd Fl., Harrisburg, PA 17101-2126. Tel.: 717-214-ARTS (2787). Fax: 717-221-8208. Facebook: Whitaker Center Hbg.

E-mail: info@whitakercenter.org
Web Site: www.whitakercenter.org
Formerly: Museum of Scientific Discovery
Founded: 1993.
Congressional District: 17
Key Personnel: Pres. & C.E.O., Michael L. Hanes; Vice Pres. Science & IMAX Programs, Steve Bishop; C.F.O., Jacqueline Wolpert; Bd. Chm., Gary St. Hilaire; Dir. Finance, Margaret Freedman; Vice Pres. Operations, Lisa Kreider; Museum Shop Mgr., Teresa Griffin; Dir. Education, Lori Lauver.
Personnel Profile: Full-Time Paid 27; Part-Time Paid 27; Part-Time Volunteers 145; Interns 2.
Governing Authority: private; not-for-profit organization. Tax-exempt: 501(c)(3).
Institution Type/Description: Science Center.
Collections: physical, natural & life science exhibits.
Facilities: 200-seat IMAX(R) theater; 664-seat performing arts theater; children's hall; classrooms. Museum-related items for sale.
Activities: interactive exhibits; live science theatre & demonstrations; outreach program; workshops for children, families & adults; teacher professional development workshops; programs for schools, scouts & groups; volunteer programs.
Publications: members newsletter, Passport; Educator's Planning Guide; theater program, Spotlight; monthly events guides.
Hours & Admission Prices: Science Center: Tues.-Sat. 9:30-5, Sun. 11:30-5. Adults $16, senior citizens (55+) & military with valid ID $14, junior (3-17) $12.50, discounts to ASTC members; members & children under 3 no charge. Hollywood IMAX Movies: adults $13.75, senior citizens (55+) & military with valid ID $12.75, juniors (3-17) $11.75. IMAX: adults $9.50, junior (3-17), senior citizens (55+) & military with valid ID $8; discounts

to members. Combo: adults $19.75, senior (55+) & military with valid ID $18, junior (3-17) $16.75. Holiday hours vary. &
Attendance: 400,000 (estimated)
Membership: Individual $89; Family & Grandparent $119; Friend $170.

Hartsville

WARWICK TOWNSHIP HISTORICAL SOCIETY & MOLAND HOUSE HISTORIC PARK, 1641 Old York Rd., Hartsville, PA 18974. Mailing Address: P.O. Box 107, Jamison, PA 18929-0107. Tel.: 215-822-2061 & 918-1754. Facebook: The Moland House.
E-mail: events@moland.org
Web Site: www.moland.org
Formerly: Moland House
Founded: 1997.
Key Personnel: Pres., Edward Price; Museum Shop Mgr., JoAnne Mullen
Governing Authority: Parent Institution: Warwick Township. Subsidiary Institution: Warwick Township Historical Society. Tax-exempt.
Institution Type/Description: Historical House Museum: housed in Moland family's farmhouse which was used as George Washington's headquarters from August 10, 1777 to August 23, 1777.
Collections: photographs; historical records.
Facilities: colonial history library.
Activities: Museum Sponsors: reenactment in August.
Publications: Moland Gazette.
Hours & Admission Prices: Call for hours. Adults $3.
Attendance: 1,000 (estimated)
Membership: Individual $10; Family $15; Corporate $50.

Hatboro

AMY B. YERKES MUSEUM - MILLBROOK SOCIETY, 32 N. York Rd., Hatboro, PA 19040. Mailing Address: P.O. Box 506, Hatboro, PA 19040. Tel.: 215-957-1877.
E-mail: museum@millbrooksociety.org
Web Site: www.millbrooksociety.org
Founded: 1984.
Key Personnel: Dir., Mary Porter; Pres. (V), L. Maguha.
Volunteer Hours: 1,300
Operating Expenses: 11,500
Operating Income: 10,000
Institution Type/Description: History Museum.
Collections: local history & culture; period furnishings; personal artifacts; photographs.
Facilities: archives.
Hours & Admission Prices: Wed. 7:30pm-9:30pm. No charge; donations accepted.

Haverford

CANTOR FITZGERALD GALLERY - HAVERFORD COLLEGE, Whitehead Campus Center, 370 Lancaster Ave., Haverford, PA 19041. Tel.: 610-896-1287.
E-mail: hcexhibits@gmail.com
Web Site: www.haverford.edu/HHC/exhibits
Key Personnel: Campus Exhibit Coord., Matthew Seamus Callinan
Institution Type/Description: Art Gallery.
Collections: paintings; sculpture.
Hours & Admission Prices: Mon.-Tues. & Thurs.-Fri. 11-5, Wed. 11-8, Sat.-Sun. 12-5. No charge.

HAVERFORD COLLEGE ARBORETUM, 370 Lancaster Ave., Haverford, PA 19041-1392. Tel.: 610-896-1101. Fax: 610-896-1095.
E-mail: arbor@haverford.edu
Web Site: www.haverford.edu/Arboretum
Founded: 1834.
Congressional District: 13
Key Personnel: Dir., William Astifan; Cur., Martha Van Artsdalen.
Personnel Profile: Full-Time Paid 5; Part-Time Volunteers 3.
Governing Authority: college. Parent Institution: Haverford College. Tax-exempt: 501(c)(3).
Institution Type/Description: Arboretum: Haverford College is a Quaker institution, the architecture reflects a simple style starting in 1833.
Collections: 216 acres of oaks, maples, elms, beech & conifers in 18 acre pinetum; peace garden; 2 Asian gardens.
Facilities: 216 acres; gardens; nature trail.
Activities: self-guided tours; lectures; seasonal guided tours; garden bus trips.

Publications: self-guided tree tours; newsletters; annual report.
Hours & Admission Prices: Daily dawn-dusk. No charge; donations accepted. &
Membership: Students $10; Individual $25; Family $40; Sustaining $60; Supporter $100; Benefactor $250; Patron $500; Life Member $1,000.

MAIN LINE ART CENTER, 746 Panmure Rd., Haverford, PA 19041-1218. Tel.: 610-525-0272. Fax: 610-525-5036.
Key Personnel: Exec. Dir., Judy Herman
Institution Type/Description: Art Gallery.
Collections: paintings; drawings; sculpture; ceramics; photographs; printmaking.
Activities: outreach programs; special events.
Hours & Admission Prices: Call for hours.

Havertown

HAVERFORD TOWNSHIP HISTORICAL SOCIETY, Karakung Dr., Powder Mill Valley Park, Havertown, PA 19083. Mailing Address: Box 825, Havertown, PA 19083-0825. Tel.: 610-446-7988.
E-mail: info@haverfordhistoricalsociety.org
Web Site: www.haverfordhistoricalsociety.org
Founded: 1939.
Congressional District: 7
Key Personnel: Cur., Carolyn Joseph; Pres., Amy Wolfe.
Personnel Profile: Part-Time Volunteers 12.
Governing Authority: society. Tax-exempt.
Institution Type/Description: Local History Museum: housed in c.1710 Lawrence Cabin, adjoining c.1810 Nitre Hall home of the Powder Master & c.1797 Federal School.
Collections: period furnishings; costumes; glass photographic plates of Philadelphia & Westchester Traction Co., 1903-1915; railroads; engines; paper archives of township historical sites, buildings, organizations & government.
Research Fields: history of Haverford Township, powder mills & manufacture of black powder.
Facilities: 75-vol. library of local & Pennsylvania history, architecture & preservation, available by appointment.
Activities: guided tours; lectures; permanent & temporary exhibitions; a full day of life, work &/or school in the 1700s & 1800s programs. Museum Sponsors: Craft Days; A Day in a One-Room School; Heritage Festival in June; Holiday House Tour in December.
Publications: brochure; newsletter.
Hours & Admission Prices: May-Sept. 1st Sat.-Sun. each month 1-4. Adults $2, children 6-18 $1; members no charge.
Attendance: 1,000 (estimated)
Membership: Student $5; Individual $10; Family $15; Contributing Organization $25.

Hazleton

GREATER HAZLETON HISTORICAL SOCIETY & MUSEUM, 55 N. Wyoming St., Hazleton, PA 18201-6069. Tel.: 570-455-8576. Facebook: Greater Hazleton Historical Society and Museum.
E-mail: hazeltonmuseum@gmail.com
Web Site: hazeltonmuseum.org
Founded: 1983.
Congressional District: 11
Key Personnel: Pres., Thomas Gabos.
Governing Authority: nonprofit organization. Tax-exempt.
Institution Type/Description: Historical Society Museum.
Collections: Indian artifacts; local historical & industrial items; photographs; mining industry items; anthropology.
Research Fields: Pennsylvania newspaper project; local history.
Facilities: local history library including newspapers & genealogical research materials; 100-seat auditorium; 1,850 sq. ft. exhibit space.
Activities: guided tours; films; temporary exhibitions.
Publications: quarterly newsletter.
Hours & Admission Prices: By appointment only. &
Attendance: 1,000 (estimated)

Hellertown

GILMAN MUSEUM, 726 Durham St., at the Cave, Hellertown, PA 18055. Mailing Address: P.O. Box M, Hellertown, PA 18055-0220. Tel.: 610-838-8767. Fax: 610-838-2961.
Web Site: www.lostcave.com

Founded: 1955.
Congressional District: 15
Key Personnel: Dir., C.E.O. & Chief Cur., Beverly L. Rozewicz; Chm. (V), Robert Gilman.
Governing Authority: nonprofit.
Institution Type/Description: Natural History & Antique Weapons: located on the site of an old Limestone Quarry, at the entrance of a natural underground series of caverns.
Collections: guns; mounted natural history specimens from around the world; gem stones; mineral specimens; early colonial items; fossils.
Research Fields: weaponry; gems; minerals; plants; ancient curios.
Facilities: botanical garden. Museum-related items for sale.
Activities: guided cave tours; permanent exhibitions.
Hours & Admission Prices: Memorial Day to Labor Day daily 9-6; Sept.-May daily 9-5. Cavern Tours: adults $12.50, children 3-12 $8. Closed New Year's Day; Easter morning; Thanksgiving; Christmas.
Attendance: 25,000 (estimated)

LOWER SAUCON TOWNSHIP HISTORICAL SOCIETY - LUTZ FRANKLIN SCHOOL, 4216 Countryside Lane, Hellertown, PA 18055. Mailing Address: P.O. Box 176, Hellertown, PA 18055. Tel.: 610-625-8771.
E-mail: lshistorical@yahoo.com
Web Site: www.lutzfranklin.com
Founded: 2003.
Congressional District: 131
Key Personnel: Pres. (V), Susan Horiszny
Institution Type/Description: History Museum: housed in a former schoolhouse, c.1826.
Collections: local history & culture; period furnishings; personal artifacts; photographs.
Hours & Admission Prices: Call for hours. No charge; donations accepted.
Attendance: 900 (estimated)
Membership: Student $5; Individual $15; Family $25; Lifetime $250; Corporation $350.

Hershey

AACA MUSEUM - ANTIQUE AUTO MUSEUM, (M), 161 Museum Dr., Hershey, PA 17033-2462. Tel.: 717-566-7100. Fax: 717-566-7300.
E-mail: info@aacamuseum.org
Web Site: aacamuseum.org
Founded: 1993.
Key Personnel: Exec. Dir., Mark Lizewskie; Pres. (V), William Edmunds; Human Resources & Volunteer Mgr., Rusty Sellers; Mktg. & Public Rels., Nancy Gates; Special Events, Michael Patterson; Retail Store, Rochelle Coslow-Robinson
Institution Type/Description: Automobile Museum.
Collections: automobile history; period automobiles & buses.
Facilities: Museum-related items for sale.
Activities: hands-on activities; license plate rubbing; village mat with cars. Annual Event: Model Railroad November to December.
Publications: email newsletters.
Hours & Admission Prices: Daily 9-5. Adults $10, seniors 61 & over $9, children 4-12 $7; members and children 3 & under no charge. Closed New Year's Day; Thanksgiving; Christmas Eve & Day.
Attendance: 60,000 (accurate)
Membership: Individual $35; Family $80; Supporting $120; Life $1,000.

HERSHEY-DERRY TOWNSHIP HISTORICAL SOCIETY, 40 Northeast Dr., Hershey, PA 17033-2732. Tel.: 717-520-0748.
E-mail: hdths@hersheyhistory.org
Web Site: hersheyhistory.org
Institution Type/Description: History Museum.
Collections: local history & artifacts; Milton Hershey's life & career; period furnishings; personal artifacts; genealogy; photographs; Native American artifacts; farm implements; household utensils; Hershey history & memorabilia.
Facilities: Museum-related items for sale.
Hours & Admission Prices: Mon., Wed. & Fri. 9-4:30, Sat. 9-1.

✳ **HERSHEY GARDENS, (M),** 170 Hotel Rd., Hershey, PA 17033-9507. Tel.: 717-534-3492. Fax: 717-533-5095.
E-mail: info@hersheygardens.org
Web Site: www.hersheygardens.org
Founded: 1937.
Congressional District: 17
Key Personnel: Dir., Marta Howell; Dir. Horticulture, Barbara Whitcraft; Dir.

Public Rels. & Mktg., Jill Manley; Ground Mgr., Jamie Shiffer; Coord. Special Events, Tammy Harris.
Personnel Profile: Full-Time Paid 9; Part-Time Paid 22; Part-Time Volunteers 350; Interns 4.
Governing Authority: private; nonprofit organization. Parent Institution: The M.S. Hershey Foundation, 1 W. Chocolate Ave., Ste. 200, Hershey, PA 17033. Tax-exempt: 501(c)(3).
Institution Type/Description: Arboretum/ Botanical Gardens & Horticultural Society: retains original 1937 landscape design & structures.
Collections: over 7,000 roses; more than 75,000 spring bulbs; rare trees & plants; summer flowering annuals; horticulturally themed gardens; mature specimen trees; seasonal butterfly house of North American species; children's garden; giant sequoia.
Facilities: 23-acres exhibit space; botanical gardens. Museum-related items for sale.
Activities: formal education programs; lectures; temporary exhibits; special tours; member's reception; volunteer reception. Annual Events: Hershey Community Gardenfest; Fall Family Fun Fest; A Day of Wines & Roses; Jack-O-Lantern Jamboree; Biergarten; Brick and Bench Dedication Ceremonies.
Publications: quarterly newsletter, Twigs & Gigs.
Hours & Admission Prices: Jan.-Feb. Fri.-Sun. 10-4; March & Nov.-Dec. daily 10-4; April-May & Sept.-Oct. daily 9-5; June-Aug. daily 9-8. Adults $10, seniors 62 & over $9, children 3-12 $6; discounts to AAA members & groups; children under 3 & members no charge. Closed Thanksgiving; Christmas.
Attendance: 120,800 (accurate)
Membership: Individual $35; Household $60; Crocus $150; Tulip $250; Lily $500; Rose $1,000.

THE HERSHEY STORY, 63 W. Chocolate Ave., Hershey, PA 17033-1558. Tel.: 717-534-3439. Fax: 717-534-8940.
E-mail: info@hersheymuseum.org
Web Site: www.hersheystory.org
Formerly: Hershey Museum
Founded: 1933.
Congressional District: 17
Key Personnel: Interim Dir., Don Papson; Assoc. Dir., Amy Bischof; Mgr. Public Programs, Lois Miklas; Dir. Education, Mariella Trosko; Mgr. Chocolate Lab, Kyle Nagurny; School Programs Supvr., Beth Hiner; Mgr. Collections, Valerie Seiber; Mgr. Visitor Experience, Lisa Morelli; Trips & Excursions Consultant, Janet Hester; Mgr. Systems, Information & Accounting, Sharon Smith; Coord. Membership, Barb Latz.
Personnel Profile: Full-Time Paid 10; Part-Time Paid 17; Part-Time Volunteers 86.
Governing Authority: foundation. Parent Institution: The M.S. Hershey Foundation. Tax-exempt: 501(c)(3).
Institution Type/Description: History Museum.
Collections: Milton Hershey's life & legacy.
Research Fields: Milton S. Hershey; Hershey chocolate history; Pennsylvania German life & arts; North American Indian ethnographic & archeological materials.
Facilities: 12,500 sq. ft. exhibit space.
Activities: chocolate lab; chocolate tasting; Museum EduQuests for school groups.
Publications: book, Built on Chocolate: The Story of the Hershey Chocolate Company.
Hours & Admission Prices: Daily 9-5:30. Adults $10, senior citizens 62 & over $9, youth 3-12 $7.50; discounts to groups, ICOM, AAM & AAA members; members no charge. Closed Thanksgiving; Christmas.
Attendance: 90,215 (accurate)
Membership: Senior 60 & over $40; Individual $50; Couple $75; Family $100.

THE MUSEUM OF BUS TRANSPORTATION, INC., 161 Museum Dr., Hershey, PA 17033-2462. Tel.: 717-566-7100, ext. 119. Fax: 717-566-7300.
Web Site: busmuseum.org
Founded: 1991.
Key Personnel: C.E.O., Pres. & Mgr., J. Thomas Collins; Vice Pres., Robert Smith; Treas., Edwin P. Wolf; Museum Shop Mgr., O.J. Ogden.
Personnel Profile: Part-Time Volunteers 12.
Governing Authority: private; nonprofit organization. Tax-exempt: 501(c)(3).
Institution Type/Description: Transportation Museum.
Collections: 12 buses from 1912-1987; photographs; scale models; bus stop signs; bus station signs.
Facilities: library; 6,000 sq. ft. exhibit space. Museum-related items for sale.
Activities: guided tours; lectures; special events. Annual Events: meeting and open house in October.
Publications: quarterly newsletter, Bus Musings.

Hours & Admission Prices: Summer: daily 9-5; Labor Day to Memorial Day Wed.-Sun. 9-5. Adults $7; discount to members. Closed New Year's Day; Thanksgiving; Christmas. &
Attendance: 50,000 (accurate)
Membership: Annual $30; Lifetime $500.

ZOOAMERICA NORTH AMERICAN WILDLIFE PARK, 100 W. Hershey Park Dr., Hershey, PA 17033. Mailing Address: P.O. Box 866, Hershey, PA 17033-2727. Tel.: 717-534-3900. Fax: 717-534-3151.
E-mail: zooamerica@hersheypa.com
Web Site: www.zooamerica.com
Founded: 1978.
Congressional District: 17
Key Personnel: Dir., Troy E. Stump; Cur., Dale Snyder; Supvr. Southern Swamps, Pat McCann; Supvr. Great Southwest, Katie Govern; Supvr. Education Programs, Elaine Gruin; Supvr. Exterior Exhibits, Tal Wenrich; Supvr. Exterior Exhibits, Tim Becker; Administrative Support, Dee Nixon.
Personnel Profile: Full-Time Paid 11; Part-Time Paid 25; Part-Time Volunteers 12.
Governing Authority: Parent Institution: Hershey Entertainment and Resorts.
Institution Type/Description: Zoological Park.
Collections: wild animals & plants of North America.
Research Fields: veterinary medicine.
Facilities: zoological park; aquarium; outdoor snack shop. Materials relating to natural history of North America for sale.
Activities: guided tours; lectures; films; gallery talks; formally organized education programs for children.
Publications: biannual, ZooAmerica Newsletter.
Hours & Admission Prices: Call for hours. Adults $11, senior citizens, children 3-8 $9; children under 2 no charge. Closed New Year's Day; Thanksgiving; Christmas. &
Membership: Individual $45; Family $125.

Honesdale

HONESDALE FIRE MUSEUM, Protection Engine Company No. 3, 1205 Main St., Honesdale, PA 18431.
E-mail: srpratt@verizon.net
Web Site: www.engine3.org/Museum.html
Institution Type/Description: Fire Museum.
Collections: early fire prints; photographs; helmets; speaking trumpets; early fire memorabilia; 1874 Silsby steamer no. 483; 1936 Seagrave fire engine.
Hours & Admission Prices: By appointment.

WAYNE COUNTY HISTORICAL SOCIETY, 810 Main, Honesdale, PA 18431-1847. Mailing Address: P.O. Box 446, Honesdale, PA 18431-0446. Tel.: 570-253-3240. Fax: 570-253-5204.
E-mail: wchs@ptd.net
Web Site: www.waynehistorypa.org
Founded: 1917.
Congressional District: 10
Key Personnel: Exec. Dir., Tammy Satter; Pres., Elaine Hillier; 1st Vice Pres., Lars Hanson; 2nd Vice Pres., Linda Lee; Sec., Dorothy Kieff; Treas., Thomas Colbert; Museum Shop Mgr., Kay Stephenson; Librarian, Alice Scott.
Personnel Profile: Full-Time Paid 1; Part-Time Paid 3; Part-Time Volunteers 100.
Governing Authority: society. Tax-exempt.
Institution Type/Description: History Museum.
Collections: archives; paintings; costumes; cut glass; history; 1829 The Stourbridge Lion steam locomotive replica; D&H Canal Co. exhibit; Native American artifacts including 4,600 pieces archaeology collection of Dr. Vernon Leslie.
Research Fields: genealogy & general Wayne County history.
Facilities: library of historical records & genealogy books available for use on premises.
Activities: permanent & rotating exhibitions.
Publications: quarterly newsletter; D & H Canal, A History by Edwin LeRoy; Canal Town-Honesdale 1850-1875 & Honesdale, the Early Years; Things Forgotten-Wayne Co. 1876-89; Honesdale & The Sturbridge Lions, History of Wayne Co. PA 1788-1998, Of Pulley, Ropes and Gear; The Pride and The Lion; Rural Schools of Wayne County; Murder, Mayhem & Sundry Misadventures in Wayne County, PA 1850-1910.
Hours & Admission Prices: Call for hours. Adults $5, children 12-18 $3; discounts to groups & AAA members; members & students no charge. Closed New Year's Day; Thanksgiving; Christmas. &
Attendance: 5,000 (estimated)
Membership: Junior $10; Individual $35; Family $50; Life Member $1,000.

Horsham

GRAEME PARK, (M), 859 County Line Rd., Horsham, PA 19044-1401. Tel.: 215-343-0965. Fax: 215-343-2223. Facebook: Graeme Park.
E-mail: ra-graemepark@state.pa.us
Web Site: www.graemepark.org
Founded: 1958.
Congressional District: 13
Key Personnel: Pres. (V), Beth McCausland; Dir., Office Mgr, Historic Site Admin. & Museum Educator, Carla Loughlin; Museum Shop Mgr., Toni Kistner.
Personnel Profile: Part-Time Paid 2; Part-Time Volunteers 50.
Governing Authority: state. Parent Institution: Pennsylvania Historical & Museum Commission, Keystone Bldg., 400 North St., Harrisburg, PA 17108. Subsidiary Institution: Friends of Graeme Park. Tax-exempt.
Institution Type/Description: Historic House: 1722, Keith Mansion, located within 42 acres at Graeme Park.
Collections: period furnishings; Georgian style wood paneling installed by Dr. Thomas Graeme, the son-in-law of Gov. Keith, in 1760; reproduction of an outdoor kitchen. Historic Buildings: 19th-century barn; The Keith House.
Research Fields: 18th-century architecture; social history; food preparation; herb gardens; labor history; gender history; African American history; slavery.
Facilities: visitor center; picnic area; rental facilities. Museum-related items for sale.
Activities: guided tours; organized education programs for adults & school groups; seasonal hearth cooking classes; paranormal investigators; rental facilities April-Nov. Museum Sponsors: Sweethearts Tour in February; Charter Day in March; WWII Weekend in April; Celtic Heritage Festival in July; Halloween Tours in October.
Publications: brochure, Graeme Park; books; The Penrose Family at Graeme Park; newsletter, Graeme park Gazette; Monograph on Sir William Keith; Trail of History Guide to Graeme Park.
Hours & Admission Prices: Fri.-Sat. 10-4, Sun. 12-4; last tours are one hour before closing. Adults $6, seniors 65 & over and groups $5, youth 3-11 $3; discounts to PHMC Trail of History, AAM, ICOM & AAA members; PA Heritage members, members & children under 3 no charge. Closed New Year's Day; Martin Luther King Jr. Day; Presidents' Day; Columbus Day; Veterans Day; Thanksgiving & day after; Christmas.
Attendance: 8,311 (estimated)
Membership: The Friends of Graeme Park: Individual $25; Family $40; Poet $75; Doctor $150; Governor $300.

THE HAROLD F. PITCAIRN WINGS OF FREEDOM AVIATION MUSEUM, 1155 Easton Rd., Horsham, PA 19044. Mailing Address: P.O. Box 747, Horsham, PA 19044. Tel.: 215-672-2277. Facebook: Wings of Freedom Aviation Museum.
E-mail: dvhaa@wingsoffreedommuseum.org
Web Site: www.wingsoffreedommuseum.org
Founded: 2001.
Key Personnel: Pres., John Rehfuss; Vice Pres., Mark Horwitz; Chm. (V), Gen. Ronald Nelson; Museum Shop Mgr., Shirley Luvender.
Personnel Profile: Part-Time Volunteers 50.
Governing Authority: Parent Institution: Delaware Valley Historical Aircraft Assoc. Tax-exempt.
Institution Type/Description: Aviation Museum.
Collections: 17 aircrafts; World War I; Pitcairn Era' World War II; Tuskegee Airmen; Cold War; women in aviation; Korea; Southeast Asia; contemporary aviation; space exploration; over 200 hand-crafted scale models; videos; films.
Facilities: archives.
Publications: quarterly members' magazine.
Hours & Admission Prices: Wed.-Fri. 10:30-3, Sat.-Sun. 10:30-4. No charge; donations accepted. &
Attendance: 12,000 (accurate)
Membership: Individual $35; Sponsor $150; Lifetime $400; Corporate $1,500.

Hummelstown

HUMMELSTOWN AREA HISTORICAL SOCIETY MUSEUM AND PARISH HOUSE, Rosanna St. & N. Alley, Hummelstown, PA 17036. Mailing Address: 32 W. Main St., Hummelstown, PA 17036-1515. Tel.: 717-566-6314.
E-mail: hahs@hummelstownhistoricalsociety.org
Web Site: www.hummelstownhistoricalsociety.org
Founded: 1971.
Congressional District: 17
Key Personnel: Pres., Chad Lister.
Personnel Profile: Part-Time Paid 1.

Governing Authority: Tax-exempt.
Institution Type/Description: Historical Society Museum: housed in a former Zion Lutheran Church; built in 1815.
Collections: local history & culture; period furnishings; personal artifacts; photographs; clothing; Susquehannock Indian artifacts.
Facilities: research library.
Activities: monthly membership meetings & speakers; annual member picnic; period bus trips to historic locations. Annual Event: Early American Christmas in December.
Publications: quarterly member's newsletter.
Hours & Admission Prices: Museum: by appointment. Library: Mon. & Wed. 10-4, Thurs. 10-2. Museum: no charge. Library: $5. &
Attendance: 1,000 (estimated)
Membership: Annual $15; Contributing $30; Business $50; Life $200.

Huntingdon

HUNTINGDON COUNTY HISTORICAL SOCIETY, 106 4th St., Huntingdon, PA 16652-1418. Mailing Address: P.O. Box 305, Huntingdon, PA 16652-0305. Tel.: 814-643-5449. Fax: 814-643-2711.
E-mail: mail@huntingdonhistory.org
Web Site: www.huntingdonhistory.org
Founded: 1936.
Congressional District: 9
Key Personnel: Exec. Dir., Kelley Kroecker; Pres., Fred Lang.
Personnel Profile: Part-Time Paid 1; Part-Time Volunteers 7.
Governing Authority: society. Tax-exempt.
Institution Type/Description: History Museum.
Collections: county historical materials; photographs; house furnishings; military artifacts; fine arts; costumes; manuscripts; historic house.
Research Fields: Local Sites Survey; county history; genealogy; industrial & Civil War history of county.
Facilities: research library.
Activities: programs; permanent & temporary exhibits; tours; research projects; workshops.
Publications: genealogy; quarterly newsletter; Glazier Stoneware; Atlas Blair & Huntingdon Counties; Second Century: A Huntingdon County Bicentennial Album; Two Centuries in Huntingdon; Along the Raystown; Huntingdon County Cemetery Guide.
Hours & Admission Prices: Victorian House Museum & Library: March to mid-Nov. Wed.-Fri. 9-4; other times by appointment. Office: Mon.-Fri. 9-4. House Tours: $5 per person. Library: $7 per person per day; members no charge.
Attendance: 1,800 (estimated)
Membership: Individual $25; Family $30; Supporter & Business $60; Benefactor $125; Corporate $250-$1,000; Life $500.

ISETT ACRES MUSEUM, Stone Creek Ridge Rd., Huntingdon, PA 16652. Mailing Address: P.O. Box 419, Huntingdon, PA 16652. Tel.: 814-643-9600.
Institution Type/Description: History Museum.
Collections: local history & culture; period furnishings; personal artifacts; photographs; farm equipment; toys; Native American artifacts; cameras.
Hours & Admission Prices: Mon.-Sat. 8-5, Sun. 12-5. Adults $6, students $4. &

JUNIATA COLLEGE MUSEUM OF ART, Moore & 17th St., Huntingdon, PA 16652. Tel.: 814-641-3505. Fax: 814-641-3607.
E-mail: maloney@juniata.edu
Web Site: www.juniata.edu/services/museum
Founded: 1998.
Congressional District: 9
Key Personnel: Dir., Judy Maloney; Cur., Jennifer Streb.
Personnel Profile: Full-Time Paid 2; Interns 4.
Governing Authority: private college. Parent Institution: Juniata College. Tax-exempt: 501(c)(3).
Institution Type/Description: Art Museum: housed in former Carnegie Library built in 1906; Beaux-Arts style, Greek cross with rotunda & stained glass windows.
Collections: American & European paintings, drawings & prints from the 17th-21st centuries; emphasis on the Hudson River School & American portrait miniatures; local history.
Major Exhibits: Kharakoffel Installation, 11/13-2/14.
Research Fields: Hudson River School; American portrait miniatures.
Facilities: 2,000 sq. ft. exhibit space.
Activities: loan, traveling & temporary exhibitions.
Publications: Minna Citron: The Uncharted Course from Realism to Abstraction.

Hours & Admission Prices: May-Aug. Wed.-Fri. 12-4; Sept.-April Mon.-Fri. 10-4, Sat. 12-4. No charge. Closed major holidays; college holidays. &
Attendance: 2,000 (estimated)
Membership: Student & Senior Citizen $15; Family $35; Contributor $100; Supporter $250; Patron $500.

SWIGART MUSEUM, (M), 12031 William Penn Hwy., Museum Park, Rte. 22 E., Huntingdon, PA 16652. Mailing Address: P.O. Box 214, Huntingdon, PA 16652-0214. Tel.: 814-643-0885. Fax: 814-643-2857.
E-mail: tours@swigartmuseum.com
Web Site: www.swigartmuseum.com
Founded: 1927.
Congressional District: 9
Key Personnel: Dir. & Museum Shop Mgr., Marjorie E. Cutright; Pres. (V), Patricia B. Swigart.
Governing Authority: individual operation.
Institution Type/Description: Automotive Museum.
Collections: 2 Tuckers - The Tin Goose & #1013; Herbie The Love Bug; 1920 Carroll; 1916 Scripps-Booth; 1908 Studebaker Electric; license plates; name plates; toys; photographs; paintings; automobiliana.
Research Fields: automobile history; development of automotive transportation.
Facilities: 20,000-vol. library of manuscripts, books, manuals, journals, catalogs, sales literature available for research upon application. Transportation & automobile-related books & items for sale.
Activities: guided tours; lectures; special programs for schools.
Hours & Admission Prices: Memorial Day-Oct. daily 10-5. Adults $7, senior citizens $6.50, children 6-12 $3.50; discounts to AAM members & groups of 20 or more; children under 6 no charge. &
Membership: Junior $15; Regular $30; Senior Couple $50; Family $100; Supporter $500; Patron $1,000.

Indiana

HISTORICAL AND GENEALOGICAL SOCIETY OF INDIANA COUNTY, 621 Wayne Ave., Indiana, PA 15701-3072. Tel.: 724-463-9600. Fax: 724-463-9899.
E-mail: ichistoricalsociety@gmail.com
Web Site: www.rootsweb.ancestry.com/~paicgs/
Founded: 1938.
Congressional District: 12
Key Personnel: Pres., Thomas Crumm; Chm., JoAnne McQuilkin.
Personnel Profile: Full-Time Paid 1; Part-Time Volunteers 30.
Volunteer Hours: 4,000
Governing Authority: private; nonprofit. Tax-exempt: 501(c)(3).
Institution Type/Description: Local History Museum.
Collections: genealogy; county history; archives relating to local history; artifacts of local manufacture & use; fine arts; tools; underground railroad in Indiana County, PA.
Research Fields: genealogy; county history; historic preservation; western Pennsylvania history.
Facilities: 10,000-vol. library of genealogy & local history books available for use on premises; reading room. Magazines, postcards, booklets, genealogical supplies for sale.
Activities: meetings; workshops; tours; seminars; historic preservation consultation.
Publications: annual magazine; monthly newsletter, The Clark House News.
Hours & Admission Prices: Tues.-Fri. 9-4, Sat. 10-3. Library: $3 per person; members & students no charge. &
Attendance: 2,300 (accurate)
Membership: Individual $25; Family $30; Sustaining & Small Business $60; Corporate $125; Life $500; Life (Husband & Wife) $750; Society Benefactor $1,000.

THE JIMMY STEWART MUSEUM, 835 Philadelphia St., Indiana, PA 15701-3907. Mailing Address: P.O. Box One, Indiana, PA 15701. Tel.: 724-349-6112; 800-83-JIMMY. Fax: 724-349-6140.
E-mail: tharley@jimmy.org
Web Site: www.jimmy.org
Founded: 1994.
Congressional District: 12
Key Personnel: C.E.O., Timothy F. Harley; Pres., Pauline Simms; Treas., David Finui; Museum Shop Mgr., Elizabeth Leeper.
Personnel Profile: Full-Time Paid 1; Part-Time Paid 1; Part-Time Volunteers 25.
Governing Authority: private; nonprofit. Parent Institution: The James M. Stewart Museum Foundation, Indiana, PA. Tax-exempt: 501(c)(3).
Institution Type/Description: History & Audio-Visual Film Museum.

Collections: artifacts documenting the history of the Stewart family in western Pennsylvania; original movie posters & stills; personal & professional memorabilia of Jimmy Stewart, including awards, movie props, military artifacts, photographs, correspondence from directors & actors.
Facilities: 40-vol. library of books on Stewart's career, military service, family history, correspondence, radio, theater, film and TV career; classrooms; 5,000 sq. ft. exhibit space; 50-seat theater. Videos, posters, books, lobby cards, half sheets, film memorabilia, postcards and T-shirts for sale.
Activities: docent program; films; formal education programs for children and Indiana (PA) University students; guided tours; lectures; temporary exhibitions; broadcast programs. Annual Events: Harvey Award; It's A Wonderful Life Celebration in November.
Publications: The Stewart Sentinel - The Newsletter of The Jimmy Stewart Museum.
Hours & Admission Prices: Mon.-Sat. 10-4, Sun. 12-4. Adults $7, senior citizens & students $6, children $5; discounts to AAM members; children under 7 & members no charge. Closed New Year's Day; Martin Luther King, Jr. Day; President's Day Easter; Independence Day; Labor Day; Thanksgiving; Christmas. &
Attendance: 5,600 (estimated)
Membership: Student & Senior $25; Adult $35; Senior Couple $45; Family $50.

KIPP GALLERY AT INDIANA UNIVERSITY OF PENNSYL-VANIA, College of Fine Arts, Sprowls Hall, 1st Fl., 470 S. 11th St., Indiana, PA 15705. Tel.: 724-357-2530. Fax: 724-357-7778.
Web Site: www.arts.iup.edu/kipp
Founded: 1971.
Congressional District: 12
Key Personnel: Dean College of Fine Arts, Michael Hood; Dir., Kyle Houser.
Personnel Profile: Full-Time Paid 1; Part-Time Paid 4; Interns 3.
Governing Authority: university; not for profit organization. Parent Institution: Indiana University of Pennsylvania. Subsidiary Institution: IUP Student Cooperative Association.
Institution Type/Description: University Art Gallery.
Collections: paintings; photographs; sculpture.
Publications: quarterly bulletin, Insight.
Hours & Admission Prices: Tues.-Fri. 11-4, Sat. 1-4. No charge. &

THE UNIVERSITY MUSEUM, (M), Sutton Hall, Indiana University of Pennsylvania, 1011 South Dr., Indiana, PA 15705. Mailing Address: Sutton Hall, Indiana Univ. of Penn., Room 111, Indiana, PA 15705. Tel.: 724-357-2397. Fax: 724-357-7778.
E-mail: mhood@iup.edu
Web Site: www.iup.edu/museum
Founded: 1981.
Congressional District: 12
Key Personnel: Pres., William Double; Dean, Michael Hood.
Personnel Profile: Part-Time Paid 1; Part-Time Volunteers 40; Interns 3.
Governing Authority: university. Parent Institution: The Foundation For IUP. Tax-exempt: 501(c)(3).
Institution Type/Description: Art & History Museum: National Historic Landmark Site.
Collections: art & cultural history; folk art; 19th & 20th-century American art works; American Indian artifacts; 185 paintings & drawings by Milton Bancroft; Inuit Indian sculptures.
Research Fields: Collections: Milton Bancroft; Wilbur Coffman Photographs; James & Mary Jack Inuit; Ros Purdy Memorial Graphics.
Activities: guided tours; docent programs; lectures; seminars; participatory, loan & temporary exhibitions.
Publications: Insight University Museum quarterly bulletin.
Hours & Admission Prices: Tues.-Wed. & Fri. 2-6:30, Thurs. 12-7:30, Sat. 12-4. No charge. Closed university holidays. &
Attendance: 8,000
Membership: Museum Friends: Student $5; Senior Citizen $15; Individual $25; Supporting $50; Active $100; Patron $250; Show Sponsorship $500; Benefactor $1,000 & up.

Jamestown

PYMATUNING DEER PARK, 804 E. Jamestown Rd., Jamestown, PA 16134. Mailing Address: 842 E. Jamestown Rd., Jamestown, PA 16134-9502. Tel.: 724-932-3200. Fax: 724-932-3198.
E-mail: info@pymatuningdeerpark.com
Web Site: www.pymatuningdeerpark.com
Founded: 1953.
Institution Type/Description: Zoo.
Collections: over 250 animals & birds from around the world including Siberian Tigers, African lions, camels, Black Bears, primates, & deer.

Facilities: Park-related items for sale.
Activities: petting zoo; train rides; pony rides.
Hours & Admission Prices: Memorial Day to Labor Day daily 10-6; Sept. Sat.-Sun. 10-6. Adults 13 & over $8, senior citizens $7, children 2-12 $6; children under 2 no charge. Train Ride: $2; children under 2 no charge.

Jeannette

BUSHY RUN BATTLEFIELD, (M), 1253 Bushy Run Rd., Jeannette, PA 15644. Mailing Address: P.O. Box 468, Harrison City, PA 15636-0468. Tel.: 724-527-5584. Fax: 724-527-5610.
E-mail: brbhs@winbeam.com
Web Site: www.bushyrunbattlefield.com
Founded: 1933.
Congressional District: 12
Key Personnel: Pres. (V), Bonnie Ramus; Museum Shop Mgr., Kelly Ruoff.
Personnel Profile: Full-Time Paid 1; Part-Time Paid 1; Part-Time Volunteers 75.
Governing Authority: state. Parent Institution: Pennsylvania Historical & Museum Commission, Box 1026, Harrisburg, PA 17108-1026. Subsidiary Institution: Bushy Run Battlefield Heritage Society. Tax-exempt.
Institution Type/Description: Military Museum: located on the site of Bushy Run Battlefield, used during Pontiac's Rebellion, 1763.
Collections: 18th-century military artifacts.
Research Fields: middle 18th-century military.
Facilities: amphitheatre; picnic area; visitor's center; trails. Military publications for sale.
Activities: military reenactments; education programs; lectures; special events; permanent exhibitions; discovery trail; environmental trails.
Publications: book, War for Empire in Western Pennsylvania; Bushy Run Battlefield: Pennsylvania Trail of History Guide, The Battle of Bushy Run.
Hours & Admission Prices: Visitor's Center: April-Oct. Wed.-Sun. 9-5. Park: Wed.-Sun. 9-5; groups by appointment. Adults $5, seniors & AAA members $4.50, children 6-17 $3; members no charge. &
Attendance: 33,000 (estimated)
Membership: Active Individual $15; Family $25; Sponsor $50; Corporate $100-$499; Endowment $500-$999; Lifetime $1,000.

Jenkintown

ABINGTON ART CENTER, 515 Meetinghouse Rd., Ste. 1, Jenkintown, PA 19046-2964. Tel.: 215-887-4882. Fax: 215-887-5789.
E-mail: info@abingtonartcenter.org
Web Site: www.abingtonartcenter.org
Founded: 1939.
Congressional District: 13
Key Personnel: Exec. Dir., Laura Burnham; Asst. Dir., Heather Rutledge; Chm. (V), Eric Weckel.
Personnel Profile: Full-Time Paid 4; Part-Time Paid 7; Part-Time Volunteers 32.
Governing Authority: private; nonprofit. Tax-exempt: 501(c)(3).
Institution Type/Description: Art Center & Sculpture Park.
Collections: regional contemporary art & sculpture park.
Research Fields: outdoor contemporary sculpture.
Facilities: studio school; sculpture park.
Activities: formal educational program; guided tours; lectures; rental facility. Museum Sponsors: Outreach programs to many organizations including 500 school children. Annual Event: concert series, Festival for young artists in August.
Publications: a combined quarterly newsletter and class listing.
Hours & Admission Prices: Wed. & Fri. 10-5, Thurs. 10-7, Sat.-Sun. 10-3. No charge; donations accepted. Closed New Year's Eve & Day; Independence Day; Thanksgiving; Christmas week. &
Attendance: 31,421 (accurate)
Membership: Individual $50; Dual & Family $65; Member Plus $75; Patron $125 & up.

OLD YORK ROAD HISTORICAL SOCIETY, 515 Meetinghouse Rd., Jenkintown, PA 19046-2964. Tel.: 215-886-8590.
E-mail: oldyorkroadhistory@gmail.com
Web Site: www.oyrhs.org
Founded: 1936.
Congressional District: 13
Key Personnel: Pres. (V), David B. Rowland; Photo Archivist, Leslie Bell; Archivist, Linda Stanley.
Personnel Profile: Part-Time Paid 1; Part-Time Volunteers 13.
Governing Authority: society. Tax-exempt.
Institution Type/Description: Local History Research Library.
Collections: local historical items; 165,000 photographic images, vertical files; property atlas collection; journals; books, newspapers, magazines, &

manuscripts pertaining to Old York Road area of southeastern Pennsylvania (Eastern Montgomery County).
Research Fields: genealogy; preservation of historical buildings.
Activities: lecture series (5 events); organized tours of local historic sites.
Publications: annual journal, Old York Road Historical Society Bulletin; semiannual newsletter, The Corridor.
Hours & Admission Prices: Mon. 7pm-9pm, Tues. 11-2, Wed. 11-3; other times by appointment. No charge; donations accepted.
Attendance: 350 (accurate)
Membership: Individual $35; Family $55; Patron $90; Patron Plus $150; Contributor $250; Sustainer $500; Benefactor $1,000.

Jersey Shore

JERSEY SHORE HISTORICAL SOCIETY, 200 S. Main St., Jersey Shore, PA 17740-1812. Tel.: 717-398-1973.
Founded: 1963.
Institution Type/Description: Historical Society Museum.
Collections: local history & heritage; photographs; personal artifacts; period furnishings.
Hours & Admission Prices: Call for hours.
Membership: Individual $15; Life $100.

Jim Thorpe

ASA PACKER MANSION, Packer Hill, Jim Thorpe, PA 18229-0108. Mailing Address: P.O Box 108, Jim Thorpe, PA 18229-0108. Tel.: 570-325-3229.
E-mail: abretzik@yahoo.com
Web Site: www.asapackermansion.com
Founded: 1913.
Congressional District: 11
Key Personnel: Dir. & Historian, Ava Bretzik.
Governing Authority: municipal; society. Parent Institution: Borough of Jim Thorpe. Subsidiary Institution: Jim Thorpe Lions Club.
Institution Type/Description: Historic Building: 1860 Victorian mansion.
Collections: furnishings.
Facilities: 500-vol. library of owner's personal library available for use by reference. Museum-related items for sale.
Activities: guided tours; lectures; permanent & temporary exhibitions.
Hours & Admission Prices: April-May Sat.-Sun. 11-4:15; Memorial Day-Nov. 1. daily 11-4:15. Adults $8, senior citizens 55 & over $7, students & children 6-18 $5; children under 5 no charge.
Attendance: 30,000

Johnstown

JOHNSTOWN AREA HERITAGE ASSOCIATION, 201 6th Ave., Johnstown, PA 15906-2500. Mailing Address: P.O. Box 1889, Johnstown, PA 15907-1889. Tel.: 814-539-1889. Fax: 814-535-1931.
E-mail: info@jaha.org
Web Site: www.jaha.org
Formerly: Johnstown Flood Museum
Founded: 1971.
Congressional District: 12
Key Personnel: Exec. Dir., Richard A. Burkert; Pres. (V), Mark Pasquerilla; Designer & Archivist, Marcia Kelly; Dir. Visitor Svcs., Kim Baxter; Cur., Kaytlin Sumner.
Personnel Profile: Full-Time Paid 12; Part-Time Paid 15; Part-Time Volunteers 30; Interns 3.
Governing Authority: nonprofit organization. Parent Institution: Johnstown Area Heritage Association. Tax-exempt: 501(c)(3).
Institution Type/Description: History Museum.
Collections: industry, transportation & flooding; manuscripts; photographs; Academy Award-winning film portray the legendary 1889 Johnstown Flood. Heritage Discovery Center: interactive media & exhibits on life in industrial Johnstown.
Research Fields: influence of flooding in the development of Johnstown; early transportation; steel industry in the area; coal industry in Western Pennsylvania; immigration & ethnicity.
Facilities: library of local history manuscripts & photographic collection available for research. Books & museum-related items for sale.
Activities: guided tours; lectures; films; loan, permanent & temporary exhibitions; film festival; music festival; walking tours.
Publications: 8 or more misc. local history publications; quarterly newsletter, Walking Tours.

Hours & Admission Prices: Jan. March Tues.-Sun. 10-5; April-Dec. daily 10-5. Two Sites: adults $8, senior citizens $7, children $6; discounts to AAM & ICOM members; members & children no charge.
Attendance: 68,000 (accurate)
Membership: Individual $50; Family $75.

SOUTHERN ALLEGHENIES MUSEUM OF ART AT JOHNSTOWN, Pasquerilla Performing Arts Center, University of Pittsburgh at Johnstown, 450 Schoolhouse Rd., Johnstown, PA 15904-2912. Tel.: 814-269-7234. Fax: 814-269-7240.
E-mail: johnstown@sama-art.org
Web Site: www.sama-art.org
Founded: 1982.
Congressional District: 72
Key Personnel: Exec. Dir., G. Gary Moyer; Coord. & Education Admin., Tina Lehman.
Personnel Profile: Full-Time Paid 1; Part-Time Paid 2; Part-Time Volunteers 2.
Governing Authority: nonprofit organization. Parent Institution: Southern Alleghenies Museum of Art, Saint Francis University Mall, Loretto, PA 15940. Tax-exempt: 501(c)(3).
Institution Type/Description: Art Museum.
Collections: paintings; photographs; sculpture.
Activities: changing exhibitions; gallery tours; lectures; artists in residence; classes & workshops; performing arts; summer children's events; art film festival.
Publications: exhibition catalogues.
Hours & Admission Prices: Mon.-Fri. 9:30-4:30. No charge; donations accepted. Closed major holidays.
Attendance: 50,428 (accurate)
Membership: See separate listing for Southern Alleghenies Museum of Art.

Kempton

ALBANY TOWNSHIP HISTORICAL SOCIETY, 404 Old Philly Pike, Kempton, PA 19529-9306. Mailing Address: P.O. Box 95, Kempton, PA 19529. Tel.: 610-756-6144.
E-mail: info@albanyths.org
Web Site: albanyths.org
Founded: 1997.
Institution Type/Description: Historical Society Museum: housed in a former grain & feed warehouse; built in 1917.
Collections: local history & culture; photographs; personal artifacts; period furnishings.
Activities: special events.
Publications: quarterly newsletter; book, Folk Art and Foodways of the Pennsylvania Dutch.
Hours & Admission Prices: 3rd Sun. of month 1-4. No charge.
Attendance: 450 (estimated)
Membership: Student $10; Individual $20; Family $25; Corporate $100; Lifetime $300; Lifetime Family $375.

HAWK MOUNTAIN SANCTUARY, 1700 Hawk Mountain Rd., Kempton, PA 19529-9379. Tel.: 610-756-6961. Fax: 610-756-4468.
E-mail: info@hawkmountain.org
Web Site: www.hawkmountain.org
Founded: 1934.
Congressional District: 6
Key Personnel: Pres., Jerry Regan; Museum Shop Mgr., Mary Therese Grob.
Personnel Profile: Full-Time Paid 18; Part-Time Paid 5; Part-Time Volunteers 250; Interns 10.
Governing Authority: Tax-exempt.
Institution Type/Description: Nature Center.
Collections: natural area; plant garden; observation point; visitor center.
Research Fields: ornithology & ecology.
Facilities: visitor center; 8 miles of trails & picnic lookouts; 2,600 acre natural area.
Activities: public programs & year-round special events; eye-level views of raptors during their autumn migration.
Publications: Raptor Watch: a global directory of raptor migration sites; Hawks Aloft by Maurice Brown; Hawk Mountain 1,000: A Species Checklist; Raptors of Hawk Mountain Coloring Book; Flight Guide.
Hours & Admission Prices: Sept.-Nov. daily 8-5; Dec.-Aug. daily 9-5. Adults $5 ($7 on autumn weekends), children 6-12 $3; members no charge.
Attendance: 70,000 (estimated)
Membership: Individual $35; Family $40; Family Plus $50; Sustaining $75; Broadwing Club $100.

Kennett Square

LONGWOOD GARDENS, 1001 Longwood Rd., Kennett Square, PA 19348. Mailing Address: P.O. Box 501, Kennett Square, PA 19348-0501. Tel.: 610-388-1000. Fax: 610-388-2294.
Web Site: www.longwoodgardens.org
Founded: 1906.
Congressional District: 5
Key Personnel: C.E.O., Nathan Hayward; Dir., Paul Redman; Head Horticulture Dept., Sharon Loving; Head Maintenance Dept., Mark Winnicki; Head Administration Dept., Dennis Fisher; Cur. Plants, Tomasz Anisko; Mgr. Mktg., Marnie Conley; Performing Arts Coord., Tom Warner.
Personnel Profile: Full-Time Paid 160; Part-Time Paid 170; Part-Time Volunteers 700; Interns 55.
Governing Authority: nonprofit. Parent Institution: Longwood Gardens, Inc., P.O. Box 501, Kennett Square, PA 19348. Cooperates with University of Delaware. Tax-exempt: 501(c)(3).
Institution Type/Description: Arboretum & Horticultural Display Garden.
Collections: 11,000 tropical & hardy woody & herbaceous plants. Historic House: Peirce-du Pont House containing the Longwood Heritage exhibit.
Research Fields: hybridization, selection of ornamental plants; studies of improved horticultural techniques.
Facilities: 21,500-vol. library of books on gardening, horticulture, floras, landscape design available for inter-library loan & for research to qualified persons; 1,077 acres of developed gardens; 4 acres under glass; 20 indoor gardens; open air theater; 400-seat restaurant. Books, videos, items for sale.
Activities: guided tours; lectures; concerts; formally organized programs for adults, undergraduate & graduate college students; professional gardener training program; Idea Garden for home gardeners; horticultural exhibitions; arts & crafts exhibits; evening holiday displays; performing arts events.
Publications: picture book, Longwood Gardens; video, Longwood Gardens/A Video Visit Throughout the Year; books, The Planning Vision; Tulip Trees & Quaker Gentlemen; booklets on various plant groups; compact discs of the Longwood Organ; video, Horticultural Career Training at Longwood Gardens.
Hours & Admission Prices: April-Aug. daily 9-6; Sept.-March daily 9-5. Illuminated evening fountain display: June-Aug. Thurs.-Sat. evenings at dusk. Adults $18, seniors 62 & over $15, youth 5-18 $8; children 4 & under no charge. &
Attendance: 1,024,000 (accurate)
Membership: Student $30; Individual $65; Dual $95; Family $120.

King of Prussia

VALLEY FORGE NATIONAL HISTORICAL PARK, 1400 N. Outer Line Dr., King of Prussia, PA 19406-1009. Tel.: 610-783-1077. Fax: 610-783-1060.
Web Site: www.nps.gov/vafo
Founded: 1893.
Congressional District: 7
Key Personnel: Supt., Michael A. Caldwell; Volunteer Coord., Ernestine White; Cur., Dona McDermott; Museum Shop Mgr., Daria Fink.
Personnel Profile: Full-Time Paid 72; Part-Time Volunteers 124.
Governing Authority: federal. Parent Institution: Dept. of Interior, National Park Service. Tax-exempt.
Institution Type/Description: Historic Site: c.1777-1778 site of Continental Army winter encampment.
Collections: period furnishings, 1770-1780; original headquarters of George Washington & other generals; replica huts of the Continental soldiers; archaeological artifacts from encampment; George C. Neumann collection of revolutionary arms & accoutrements; military manuscripts; 18th century military weaponry.
Research Fields: George Washington; encampments of Revolutionary army; military history of the period.
Facilities: research library; visitor center; tour road; picnic areas; bicycle, hiking & bridle trails. Publications & postcards for sale.
Activities: guided walks & tours; living history demonstrations; orientation film; special events.
Hours & Admission Prices: Daily 9-5. No charge. Closed New Year's Day; Thanksgiving; Christmas. &
Attendance: 1,200,000

Kinzers

ROUGH & TUMBLE ENGINEERS HISTORICAL ASSOCIATION, Rte. 30, Kinzers, PA 17535. Mailing Address: Box 9, Kinzers, PA 17535-0009. Tel.: 717-442-4249.
Web Site: www.roughandtumble.org
Founded: 1948.

Key Personnel: Pres., Harvey Bashore; Cur., David Adams.
Governing Authority: society. Tax-exempt: 501(c)(3).
Institution Type/Description: Agricultural & Mechanical Technology & History.
Collections: steam engines; tractors; generators; small models; covered wagon; gas engines; sawmill; trains; large industrial trailers, commercial vehicles (trucks, construction equipment) agricultural & household.
Research Fields: history of technology.
Facilities: 200-vol. library pertaining to steam power, traction engines & early gas power for public use. Museum-related items for sale.
Activities: guided tours; lectures; broadcast programs; participatory, loan & temporary exhibitions.
Publications: quarterly newsletter, Whistle; book, Rough and Tumble Engineering.
Hours & Admission Prices: Call for hours & admission. &
Membership: Individual $20; Family $25; Life $200.

Kittanning

ARMSTRONG COUNTY HISTORICAL MUSEUM AND GENEALOGICAL SOCIETY, INC., 300 N. McKean St., Kittanning, PA 16201-1373. Mailing Address: P.O. Box 735, Kittanning, PA 16201-0735. Tel.: 724-548-5707.
E-mail: achmgs@windstream.net
Web Site: WWW.achmgs.yolasite.com
Formerly: McCain House - Armstrong County Historical Society
Founded: 1918.
Congressional District: 3
Key Personnel: Pres. (V), Lee J. Calarie.
Personnel Profile: Part-Time Volunteers 15; Interns 1.
Volunteer Hours: 1,200
Operating Expenses: 34,276
Operating Income: 33,241
Governing Authority: Tax-exempt.
Institution Type/Description: Historic House Museum: built in 1842.
Collections: local history, culture, & heritage; period furnishings; personal artifacts; photographs.
Research Fields: genealogy.
Facilities: Mildred Lankerd-Thomas Genealogical library.
Activities: Genealogy club monthly meetings 1st Sun.
Publications: quarterly Genealogy & Museum News.
Hours & Admission Prices: April to mid-Nov. 1st Sun. each month & Wed. 1-4. No charge, donations accepted.
Attendance: 650 (estimated)
Membership: Single $25; Family $30.

Knoxville

KNOXVILLE PUBLIC LIBRARY, 112 Main St., Knoxville, PA 16928-0277. Mailing Address: P.O. Box 277, Knoxville, PA 16928-0277. Tel.: 814-326-4448. Fax: 814-326-4448.
E-mail: kplibrary@verizon.net
Web Site: www.knoxvillepubliclibrary.com
Founded: 1921.
Congressional District: 10
Key Personnel: Pres., Eugene A. Seelye; Librarian, Ellen Williams; Asst. Librarian, Elaine Van Sickle.
Personnel Profile: Part-Time Paid 2; Part-Time Volunteers 2.
Governing Authority: nonprofit organization. Tax-exempt: 501(c)(3).
Institution Type/Description: History Museum.
Collections: local history; ephemera; photographs.
Facilities: 200-vol. library of Tioga County, manuscripts, Pennsylvania & southern tier New York history.
Activities: permanent & temporary exhibitions.
Hours & Admission Prices: Mon. 9-8:30, Wed. 9-6:30, Fri.-Sat. 9-4. No charge.
Attendance: 100 (estimated)

Kutztown

PENNSYLVANIA GERMAN CULTURAL HERITAGE CENTER AT KUTZTOWN UNIVERSITY, (M), 22 Luckenbill Rd., Kutztown, PA 19530-9203. Tel.: 610-683-1589; 484-646-4165 (Library) & 4166 (office). Fax: 610-683-1330.
E-mail: heritage@kutztown.edu
Web Site: www.kutztown.edu/community/PGCHC
Formerly: Pennsylvania Dutch Folk Culture Society, Inc.
Founded: 1992.
Congressional District: 187
Key Personnel: Dir., Dr. Robert Reynolds; Registrar, Patty Frandsen; Bldg.

Conservator & Exhibit Specialist, Patrick Donmoyer; Coord. Public Rels., Amanda Richardson.
Personnel Profile: Full-Time Paid 1; Full-Time Volunteers 2; Part-Time Paid 3; Part-Time Volunteers 4; Interns 1.
Governing Authority: nonprofit organization. Parent Institution: Kutztown University Foundation. Tax-exempt: 501(c)(3).
Institution Type/Description: Folk Culture Museum.
Collections: folk items & art; taufscheins; agriculture; costumes; colonial period & turn of the century artifacts; third generation tinsmith shop; tools & items made by former tinsmith; 1800-1910 PA German culture; home butchering equipment; painted furniture & dower chest. Historic Buildings: Swiss bank barn c.1860's; summerhouse c.1810; washhouse & butchering shop; one room schoolhouse c.1850's; 19th century farmhouse.
Research Fields: folklore; genealogy & local history; PA German culture.
Facilities: library of books, tapes, pictures, slides pertaining to genealogy folklore. Handmade, local & regional history & culture & museum-related items for sale.
Activities: guided tours; lectures; formally organized education programs for children & adults; folklife demonstrations; children's cultural camp.
Publications: newsletter, Heritage Center News.
Hours & Admission Prices: Mon.-Fri. 10-4; tours by appointment. Adults $5; members no charge. Closed most holidays. &
Attendance: 12,000 (estimated)
Membership: Adult $25-$500.

Lackawaxen

ZANE GREY MUSEUM, 135 Scenic Dr., Lackawaxen, PA 18435. Mailing Address: 274 River Rd., Beach Lake, PA 18405-4046. Tel.: 570-685-4871. Fax: 570-685-4874.
E-mail: upde_interpretation@nps.gov
Web Site: www.nps.gov/upde
Founded: 1978.
Congressional District: 10
Key Personnel: Supt., Vidal Martinez; Chief Interpretation, Loren Goering; Museum Cur., Dorothy Moon; Museum Shop Mgr., Connie Lloyd.
Personnel Profile: Full-Time Paid 3; Part-Time Paid 3.
Governing Authority: federal government. Parent Institution: National Park Service, Dept. of the Interior, Washington, DC. Tax-exempt.
Institution Type/Description: History Museum.
Collections: books, archival; Zane Grey memorabilia.
Research Fields: life & writing of Zane Grey.
Facilities: archives.
Activities: guided tours.
Publications: brochures; Zane Grey novels; audiotapes; videotapes.
Hours & Admission Prices: Memorial Day-Labor Day Fri.-Sun. 10-5; Sept. to mid-Oct. Sat.-Sun. 10-5. No charge; donations accepted.
Attendance: 8,000 (accurate)

Lake Ariel

CLAWS 'N' PAWS WILD ANIMAL PARK, 1475 Ledgedale Rd., Lake Ariel, PA 18436-5589. Tel.: 570-698-6154.
Web Site: www.clawsnpaws.com
Founded: 1974.
Institution Type/Description: Zoo.
Collections: over 300 animals representing 120 species.
Facilities: Museum-related items for sale.
Activities: animal shows; special events.
Hours & Admission Prices: May to mid-Oct. daily 10-6. Adults 12 & over $14.95, seniors 65 & over $13.95, children 2-11 $10.95; children one & under no charge.

Lancaster

THE AMISH FARM & HOUSE, 2395 Lincoln Hwy. E., Lancaster, PA 17602-1174. Mailing Address: 2395 Covered Bridge Dr., Lancaster, PA 17602-1174. Tel.: 717-394-6185. Fax: 717-394-4857.
E-mail: info@amishfarmandhouse.com
Web Site: www.amishfarmandhouse.com
Founded: 1955.
Congressional District: 16
Key Personnel: Gen. Mgr., Mark Andrews; Dir. Mktg., Eric Conner
Institution Type/Description: Historic House Museum.
Collections: Amish life & culture; period furnishings; personal artifacts; farm equipment; tools; farm animals. Historic Buildings: 1805 farmhouse; barns; blacksmith shop; one-room schoolhouse; 1855 covered bridge.
Hours & Admission Prices: Jan.-March daily 10-4; April-May & Sept.-Oct. daily 9-5; June-Aug. daily 9-6; Nov.-Dec. daily 9-4. General Admission:

adults 12 & over $8.50, seniors 60 & over $7.75, children 5-11 $5.50; children 4 & under no charge. Closed New Year's Day; Thanksgiving; Christmas.

DEMUTH MUSEUM, 120 E. King St., Lancaster, PA 17602-2832. Tel.: 717-299-9940. Fax: 717-299-9749.
E-mail: info@demuth.org
Web Site: www.demuth.org
Founded: 1981.
Key Personnel: Dir., Anne M. Lampe.
Personnel Profile: Full-Time Paid 3; Part-Time Paid 1.
Governing Authority: Tax-exempt.
Institution Type/Description: Art Museum.
Collections: Charles Demuth's artwork; archive.
Facilities: library; archives.
Hours & Admission Prices: Feb.-Dec. Tues.-Sat. 10-4, Sun. 1-4. No charge; donations accepted.

HANDS-ON HOUSE, CHILDREN'S MUSEUM OF LANCASTER, 721 Landis Valley Rd., Lancaster, PA 17601-4888. Tel.: 717-569-KIDS. Fax: 717-581-9283. Facebook: Hands-on House, Children's Museum of Lancaster.
E-mail: info@handsonhouse.org
Web Site: handsonhouse.org
Founded: 1987.
Key Personnel: Exec. Dir., Lynne Morrison.
Personnel Profile: Full-Time Paid 5; Part-Time Paid 12.
Governing Authority: nonprofit. Tax-exempt: 501(c)(3).
Institution Type/Description: Children's Museum.
Collections: hands-on exhibits.
Facilities: Children's toys for sale.
Activities: organized education programs for children; participatory exhibits.
Hours & Admission Prices: Memorial Day to Labor Day Mon.-Thurs. & Sat. 10-5, Fri. 10-8, Sun. 12-5; Sept.-May Tues.-Thurs. 11-4, Fri. 11-8, Sat. 10-5, Sun. 12-5. Admission $8.50; discounts to groups & ACM members; members no charge. Closed New Year's Day; Easter; Memorial Day; Independence Day; Labor Day; Thanksgiving; Christmas Eve & Day. &
Attendance: 65,000 (estimated)
Membership: Individual $33; Grandparent $55; Family $89; Family Plus $99; Contributing $139.

HERITAGE CENTER OF LANCASTER COUNTY, INC., 13 W. King St., Lancaster, PA 17603-3813. Mailing Address. P.O. Box 417, Lancaster, PA 17608-0417. Tel.: 717-299-6440. Fax: 717-299-6916.
E-mail: info@lancasterheritage.com
Web Site: www.lancasterheritage.com
Founded: 1973.
Congressional District: 16
Key Personnel: Exec. Dir., Wendy Nagle; Cur., Wendell Zercher; Museum Shop Mgr., Sue Schumann.
Personnel Profile: Full-Time Paid 6; Part-Time Paid 14; Part-Time Volunteers 10; Interns 2.
Governing Authority: nonprofit organization. Tax-exempt: 501(c)(3).
Institution Type/Description: General Museum.
Collections: Lancaster County decorative arts; furniture; quilt & textiles; fraktur; paintings; ceramics; glass; iron; silver; pewter; regional history.
Research Fields: 18th, 19th & 20th-century decorative arts of Lancaster County; Pennsylvania & mid-Atlantic region.
Facilities: rental facilities. Museum-related items for sale.
Activities: lectures; loan, permanent & temporary exhibitions; special events.
Publications: newsletter; exhibit catalogs.
Hours & Admission Prices: Temporarily closed. Quilt Museum: adults $6; children no charge. Heritage Center: no charge. Closed New Year's Day; Easter; Memorial Day; Independence Day; Labor Day; Thanksgiving; Christmas. &
Attendance: 75,000 (accurate)
Membership: Individual $25; Dual $50; Family & Friends $75; Museum Circle $100; Curator Circle $500; Benefactor $1,000; Collector's Circle $2,500; Heritage Guild $5,000.

LANCASTER MUSEUM OF ART, 135 N. Lime St., Lancaster, PA 17602-2952. Tel.: 717-394-3497.
E-mail: info@Lmapa.org
Web Site: www.Lmapa.org
Founded: 1965.
Congressional District: 16
Key Personnel: Admin., Judy W. Smith; Bd. Pres., Laurin J. Bloom.

Personnel Profile: Full-Time Paid 1; Part-Time Paid 1; Part-Time Volunteers 30; Interns 2.
Governing Authority: nonprofit organization. Tax-exempt: 501(c)(3).
Institution Type/Description: Regional Visual Arts Museum.
Collections: various media of art.
Research Fields: The Grubb Family; Lancaster City Mansion.
Activities: changing exhibitions; guided tours; lectures; concerts; arts festivals; organized education programs.
Publications: newsletter; exhibitions catalogues; e-mail newsletters.
Hours & Admission Prices: Tues.-Sat. 10-4, Sun. 12-4. No charge; donations requested. Closed national holidays.
Attendance: 35,000 (estimated)
Membership: Individual $35; Family $65; Sponsor $125; Patron $250; Curator's Circle $500; Director's Circle $1,250.

LANCASTER QUILT & TEXTILE MUSEUM, 37 Market St., Lancaster, PA 17603. Mailing Address: c/o Heritage Center of Lancaster County, Inc., 13 W. King St., Lancaster, PA 17603. Tel.: 717-397-2970 & 299-6440. Fax: 717-299-6916.
E-mail: info@lancasterheritage.com
Web Site: www.quiltandtextilemuseum.com
Institution Type/Description: Textile Museum.
Collections: Amish quilts from 1870-1940s; Amish history & culture.
Facilities: rental facilities.
Activities: educational programs; demonstrations.
Hours & Admission Prices: Mon. & Wed.-Thurs. 10-5, Tues. & Fri.-Sat. 9-5, 1st Fri. of month 9-9. Adults $6, students $4; discounts to groups of 15 or more; children 17 & under no charge.

LANCASTER SCIENCE FACTORY, 454 New Holland Ave., Lancaster, PA 17602. Tel.: 717-509-6363. Fax: 717-509-6386.
E-mail: info@tlsf.org
Web Site: www.lancastersciencefactory.org
Founded: 2008.
Key Personnel: Exec. Dir., Bob Herbert.
Personnel Profile: Full-Time Paid 5; Part-Time Paid 2; Part-Time Volunteers 80.
Governing Authority: Tax-exempt.
Institution Type/Description: Science Museum.
Collections: hands-on exhibitions; physical sciences; technology; engineering; mathematics.
Activities: educational programs; special events; school groups.
Hours & Admission Prices: May-Aug. Mon.-Sat. 10-5, Sun. 12-5; Sept.-April Tues.-Sat. 10-5, Sun. 12-5. Adults $8, seniors 60 & over $7, children 3-15 $6.50; discounts to groups of 10 or more; ASTC members and children 2 & under no charge. Closed New Year's Day; Easter; Thanksgiving; Christmas. &
Membership: Family & Grandparent $80; Family Plus $105.

LANCASTERHISTORY.ORG - LANCASTER COUNTY'S HIS-TORICAL SOCIETY & PRESIDENT JAMES BUCHANAN'S WHEATLAND, (M), 230 N. President Ave., Lancaster, PA 17603-3125. Tel.: 717-392-4633. Facebook: Lancaster History.
E-mail: info@lancasterhistory.org
Web Site: www.lancasterhistory.org
Founded: 1886.
Congressional District: 16
Key Personnel: Pres. & C.E.O., Thomas R. Ryan; Vice Pres., Robin Sarratt; Dir. Wheatland, Patrick Clarke; Cur., Barry Rauhauser; Dir. Library Svcs., Marjorie Bardeem; Archivist, Heather Tennies; Genealogist, Kevin Shue.
Personnel Profile: Full-Time Paid 15; Part-Time Paid 10; Part-Time Volunteers 374; Interns 7.
Governing Authority: private; nonprofit corporation. Tax-exempt: 501(c)(3).
Institution Type/Description: Historical Society Archives, Library & Exhibitions.
Collections: Lancaster County history; glass; pewter; costumes; books; Judge Yeates Law library with 1,500 vols.; President James Buchanan's 1828 mansion, Wheatland; county archives 1729-1850; local newspapers 1787-1936; law office & home of Thaddeus Stevens; Esprit collection of Amish Quilts; manuscript material; Jacob Eichholtz portraits; Arthur Armstrong portraits; local history; historic house.
Major Exhibits: County, Commonwealth & Country, 1/14-12/14; Lancaster: The Small City That Changed America, 7/14-12/14.
Research Fields: economic, political, social, democracy & political process in 19th-century America, genealogical & technological history of Lancaster County.
Facilities: 15,000-vol. library & archives; orientation theater; auditorium; reading room; meeting room; research facilities; classroom. Bookstore and museum-related items for sale.

Activities: tours; lectures; living history events; temporary exhibits; arboretum walk.
Publications: Journal of the Lancaster County Historical Society; newsletter, The Historian.
Hours & Admission Prices: Mon., Wed. & Fri.-Sat. 9:30-5, Tues. & Thurs. 9:30-8. Library: adults $7; discounts to AAM members; members no charge. Exhibitions: adults $7; members no charge. Wheatland Tours: $10; members no charge. Combo pricing available. Closed national holidays. &
Attendance: 34,000 (accurate)
Membership: High School Students free upon application; College Students $5 (semester); Regular $50; Family $75.

LANDIS VALLEY VILLAGE AND FARM MUSEUM, (M), 2451 Kissel Hill Rd., Lancaster, PA 17601-4809. Tel.: 717-569-0401, ext. 216 & 208. Fax: 717-560-2147.
E-mail: jlewars@pa.gov
Web Site: landisvalleymuseum.org
Founded: 1925.
Congressional District: 16
Key Personnel: Bd. Pres., Steven Petersen; Dir., James A. Lewars; Events Coord., Cindy Kirby-Reedy; Maintenance & Security, William Morrow; Farm & Garden Mgr., Joseph Schott; Coord. Heirloom Seed Project, Beth Leensvaart; Cur., Bruce Bomberger; Interpretation Supvr., Karen Cunningham; Museum Educator, Tim Essig; Museum Educator, Mike Emery; Museum Preparator, Donna Horst; Dir. Group Sales, Joyce Perkinson; Scholar in Residence, Dr. Irwin Richman.
Personnel Profile: Full-Time Paid 25; Full-Time Volunteers 2; Part-Time Paid 44; Part-Time Volunteers 409; Interns 3.
Governing Authority: state. Parent Institution: Pennsylvania Historical and Museum Commission. Subsidiary Institution: Landis Valley Assoc. Tax-exempt: 501(c)(3).
Institution Type/Description: Rural Life & Culture Village Museum Complex.
Collections: agriculture; Pennsylvania German decorative arts; folk culture; textiles; transportation. Historic Buildings: Mennonite & Amish religion; tavern; gun shop; craft building; 1870s Isaac Landis farm complex; late 1700 Log Farm; 1830s-1840s Brick Farmstead; 1870s Pierce Landis complex; 1890 Schoolhouse; 1857 Landis Valley Hotel.
Research Fields: agriculture; Pennsylvania German history & material culture; folk culture; science & technology; textiles; transportation.
Facilities: 12,000-vol. library of agricultural journals & texts, folk art, farm & history books available for use on premises by written request. Publications, gift items, museum reproductions & ceramics for sale.
Activities: guided tours; lectures; films; formally organized education programs; permanent & temporary exhibitions; workshops; youth classes; holiday events. Museum Sponsors: Harvest Days; Pennsylvania Institute of Rural Life & Culture.
Publications: history & guidebook, Landis Valley; Heirloom Seed Catalog (annual); quarterly The Valley Gazette; PA German Food & Traditions; PA German Farms, Gardens & Seeds; 400 Years at Landis Valley; book, The Art of Seeds; Lancaster County postcards; the Landis family album.
Hours & Admission Prices: Mon.-Sat. 9-5, Sun. 12-5. Adults $12, senior citizens $10, youth 6-12 $8; discount to groups, AAA, AAM & ICOM members, student groups & for tickets from other state sites & local museums; children under 6 & members no charge. Closed New Year's Day; Thanksgiving; Christmas. &
Attendance: 65,737 (accurate)
Membership: Student $20; Senior Citizen $25; Individual $35; Family & Grand Family $55; Patron $100; Heirloom Club $250; Landis Club $500.

NORTH MUSEUM OF NATURAL HISTORY & SCIENCE, 400 College Ave., Lancaster, PA 17603-3393. Tel.: 717-291-3941.
E-mail: info@northmuseum.org
Web Site: www.northmuseum.org
Founded: 1953.
Congressional District: 16
Key Personnel: Dir., Margaret M. Marino; Pres. (V), Ned Wehler; Museum Shop Mgr., Kent Strickler.
Personnel Profile: Full-Time Paid 9; Full-Time Volunteers 1; Part-Time Paid 15; Part-Time Volunteers 150.
Governing Authority: Tax-exempt.
Institution Type/Description: Natural History & Science Museum.
Collections: hands-on natural history & science exhibitions.
Facilities: Museum-related items for sale.
Activities: hands-on exhibitions; educational programs; group & school tours; special events.
Hours & Admission Prices: Tues.-Sat. 10-5, Sun. 12-5. Museum: adults $7.50, students 3-17 and seniors 65 & over $6.50; discounts to AAM members; members and children 2 & under no charge. Museum & Planetarium: adults $10, students 3-17 and seniors 65 & over $9; discounts to AAM members;

members no charge. Closed New Year's Day; Easter; Independence Day; Labor Day; Thanksgiving; Christmas. &

Attendance: 30,000 (estimated)

THE PHILLIPS MUSEUM OF ART, FRANKLIN & MARSHALL COLLEGE, (M), 700 College Ave., Lancaster, PA 17604. Mailing Address: P.O. Box 3003, Lancaster, PA 17604-3003. Tel.: 717-291-3879. Fax: 717-358-4441.

E-mail: claire.giblin@fandm.edu
Web Site: www.fandm.edu/phillipsmuseum.xml
Founded: 2000.
Congressional District: 16
Key Personnel: Dir., Eliza J. Reilly; Cur. Exhibitions, Claire Giblin; Registrar & Collections Mgr., Maureen Lane; Exhibition Coord., Russell O'Connell.
Personnel Profile: Full-Time Paid 3; Part-Time Paid 1; Part-Time Volunteers 2; Interns 20.
Governing Authority: nonprofit; private college. Parent Institution: Franklin & Marshall College, Lancaster, PA. Tax-exempt: 501(c)(3).
Institution Type/Description: Art Museum.
Collections: American art & material culture; African American art; 20th century photography; works on paper.
Research Fields: African American abstract art.
Facilities: cafeteria; educational facilities.
Activities: lectures; loan, temporary & traveling exhibitions.
Publications: exhibition catalogues.
Hours & Admission Prices: Academic Year: Tues.-Fri. 11:30-4:30, Sat.-Sun. 12:30-4:30; Summer: call for hours. No charge. Closed legal holidays. &
Attendance: 5,000 (estimated)
Membership: General $35; Contributor $100; Franklin $500; Trustee Assoc. $1,000; Fellow $5,000.

ROCK FORD PLANTATION, 881 Rockford Rd., Lancaster, PA 17602-1225. Tel.: 717-392-7223. Fax: 717-392-7283 (call first).
Web Site: www.rockfordplantation.org
Formerly: Historic Rock Ford
Founded: 1958.
Congressional District: 97
Key Personnel: Exec. Dir., Samuel C. Slaymaker; Pres. (V), Raymond Bradley; Museum Shop Mgr., Lisa A. Haldy.
Personnel Profile: Full-Time Paid 1; Part-Time Paid 3; Part-Time Volunteers 50; Interns 3.
Governing Authority: nonprofit organization. Parent Institution: Rock Ford Foundation, Inc. Tax-exempt: 501(c)(3).
Institution Type/Description: Historic House: 1794 Georgian mansion of Gen. Edward Hand.
Collections: late 18th & early 19th century American furnishings; art belonging to General Edward Hand & his family.
Research Fields: Hand family.
Facilities: Gift items for sale.
Activities: guided tours; school programs; permanent exhibitions. Museum Sponsors: Yuletide Tours; Revolutionary War Encampment; Halloween program.
Publications: Rock Ford Guidebook; school packets, Gen. E. Hand biographies.
Hours & Admission Prices: April-Oct. Wed.-Sun. 11-3; last tour begins at 3; Nov.-March by appointment. Adults $6, seniors $5, students 6-12 $4; discounts to AAA members & groups; members & children under 6 no charge.
Attendance: 3,000 (estimated)
Membership: Individual $25; Family $50; Business $75; Patron $100.

SEHNER-ELLICOTT-VON HESS HOUSE, 123 N. Prince St., Lancaster, PA 17603-3525. Tel.: 717-291-5861. Fax: 717-291-2251.
E-mail: tas@hptrust.org
Web Site: www.hptrust.org
Founded: 1966.
Key Personnel: Exec. Dir., Timothy Smedick.
Governing Authority: Parent Institution: Historic Preservation Trust of Lancaster County.
Institution Type/Description: Historic House: c.1787.
Collections: county history & culture; period furnishings; photographs.
Facilities: library.
Hours & Admission Prices: Mon.-Fri. 9-3:30. No charge; donations accepted. Closed holidays.
Membership: Individual $35; Family $50; Business $100.

WOLF MUSEUM OF MUSIC AND ART, 423 W. Chestnut Ave., Lancaster, PA 17603-3405. Mailing Address: P.O. Box 701, Lancaster, PA 17608-0701. Tel.: 717-392-6382.
Web Site: wolfmuseum.org
Founded: 1974.
Institution Type/Description: Music & Art Museum: housed in the former home & studio of Dr. William A. Wolf & his wife Frances; built in 1886.
Collections: family history; period furnishings; personal artifacts; two 1915 Knabe concert grand pianos; paintings.
Activities: public recitals.
Hours & Admission Prices: By appointment & for public recitals.

Landisville

AMOS HERR HOUSE FOUNDATION AND HISTORIC SOCIETY, 1756 Nissley Rd., Landisville, PA 17538-1360. Mailing Address: P.O. Box 52, Landisville, PA 17538-0052. Tel.: 717-898-8822.
E-mail: info@herrhomestead.org
Web Site: www.herrhomestead.org
Founded: 1990.
Key Personnel: Pres. (V) & Museum Shop Mgr., John Houston; Treas., Nancy Smith; Cur. & Archivist, Eileen Johns; Sec., Mrs. Millie Brubaker.
Personnel Profile: Part-Time Volunteers 120.
Volunteer Hours: 1,325
Governing Authority: private; nonprofit organization. Tax-exempt.
Institution Type/Description: History Museum.
Collections: furnishings; 19th-20th century farm equipment.
Facilities: library; nature center. Museum-related items for sale.
Activities: docent program; guided tours; lectures; school loan service; participatory & temporary exhibitions. Annual Events: historic lectures; Amos Herr 5K Honey Run.
Publications: newsletter, three times a year, AHHF newsletter.
Hours & Admission Prices: April-Oct. Sat.-Sun. 1-4; private tours by appointment. No charge; donations accepted. Closed New Year's Day; Easter; Memorial Day; Independence Day; Labor Day; Thanksgiving; Christmas. &
Attendance: 390 (estimated)
Membership: Individual $15; Family $25; Supporting $50; Contributing $100; Sustaining $250; Guarantor $500; Benefactor $1,000.

Langhorne

HISTORIC LANGHORNE ASSOCIATION, 160 W. Maple Ave., Langhorne, PA 19047-2820. Tel.: 215-757-1888 & 6158. Fax: 215-741-5767.
E-mail: historiclanghorne1@verizon.net
Web Site: historiclanghorne.org
Founded: 1965.
Congressional District: 8
Key Personnel: Pres., James Maier; Vice Pres. & Archivist, Lawrence Langhans; Treas., Jack Fulton; Librarian, Jean Noble; Archivist, Museum Shop Mgr. & Recording Sec., Evelyn Aicher.
Personnel Profile: Part-Time Volunteers 15.
Governing Authority: nonprofit organization. Tax-exempt: 501(c)(3).
Institution Type/Description: Historical Association.
Collections: early Quaker heritage & Bucks County pre-Revolutionary War to present, history; artifacts; clothing; quilts; art; architecture.
Research Fields: homes & other buildings of historic value in Langhorne; genealogy.
Facilities: 275-vol. library on American, Pennsylvania & Bucks County history; colonial artifacts available to the public; 75-seat auditorium; 1,600 sq. ft. exhibit space. Museum-related items for sale.
Activities: guided tours; lectures; loan & temporary exhibitions; monthly general meeting with guest speaker. Annual Events: Walking tour of Langhorne; Memorial Day Open House; October Historic Ghost Tours of Langhorne; Christmas house tour of Langhorne; Santa's visit & carol sing-a-long.
Publications: quarterly newsletter.
Hours & Admission Prices: Wed. & Sat. 10-12 & 7-9; other times by appointment. No charge; donations accepted. &
Attendance: 1,200 (accurate)
Membership: Individual $25; Family $30; Patron $50; Century $100; Historic $250; Worthington $500; Williamson $1,000.

Lansdale

JENKINS HOMESTEAD AND LANSDALE HISTORICAL RE-SEARCH CENTER, 137 Jenkins Ave., Lansdale, PA 19446. Tel.: 215-855-1872. Fax: 215-393-8919.
E-mail: info@lansdalehistory.org
Institution Type/Description: History Museum.
Collections: local history & culture; period furnishings; personal artifacts; photographs. Historic Building: c.1770 farmstead.
Hours & Admission Prices: Wed.-Thurs. 11-4, Sat. 9-12.

Lansford

NO. 9 COAL MINE & MUSEUM, 9 Dock St., Lansford, PA 18232-1202. Mailing Address: P.O. Box 287, Lansford, PA 18232-1202. Tel.: 570-645-7074.
E-mail: humblebe@ptd.net
Web Site: www.no9mine.com
Founded: 1992.
Key Personnel: C.E.O. & Pres. (V), David Kuchta; Museum Shop Mgr., Jan Levan.
Personnel Profile: Part-Time Paid 8; Part-Time Volunteers 50.
Governing Authority: Tax-exempt.
Institution Type/Description: Mining Museum.
Collections: local mining history, equipment, & tools; personal artifacts; photographs; paintings; railroad artifacts; fossils; gems.
Facilities: Gift items for sale.
Activities: special events; train ride into mine. Museum Sponsors: Memorial & Labor Day picnics; Miners Heritage Festival; Haunted Mine Tours.
Publications: annual newsletter.
Hours & Admission Prices: Museum: Wed.-Sun. 12-4. Mine Tours: May Fri.-Sun. 11-3; July-Aug. Wed.-Sun. 11-3; Sept.-Oct. Sat.-Sun. 11-3. Museum: $3 per person. Mine Tour & Museum: $8 per person; discounts to AAA members & groups of 20 or more. &
Attendance: 10,000 (accurate)
Membership: Individual $10.

Latrobe

KLBE LATROBE AIR MUSEUM, Arnold Palmer Rgnl. Airport, 148 Aviation Lane, Latrobe, PA 15650. Tel.: 724-787-8396.
Key Personnel: Dir., Dave Austin; Dir., Sam Schrecengost
Institution Type/Description: Aviation History Museum.
Collections: aviation history; aircraft; OX-5 airplane engine.
Facilities: theater.
Hours & Admission Prices: Summer: Sat.-Sun. 10-2; other times by appointment. No charge.

Laughlintown

COMPASS INN MUSEUM, 1382 Rte. 30 E., Laughlintown, PA 15655. Mailing Address: P.O. Box 167, Laughlintown, PA 15655-0167. Tel.: 724-238-4983. Fax: 724-238-3968. Facebook: Compass Inn Museum.
E-mail: mistressinnkeeper@compassinn.com
Web Site: www.compassinn.com
Founded: 1972.
Congressional District: 18
Key Personnel: Pres. (V), Sean Murphy; Inn Keeper, Nicole Bosley; Dir. Operations, Tina Yandrick.
Personnel Profile: Full-Time Paid 2; Part-Time Paid 1; Part-Time Volunteers 40.
Governing Authority: Parent Institution: Ligonier Valley Historical Society. Tax-exempt.
Institution Type/Description: History Museum: housed in a restored stagecoach stop; built in 1799. Listed on the National Register of Historic Places.
Collections: early 1800s life; period furnishings; Conestoga wagon; stagecoach; blacksmith tools; pots & utensils; 1790 beehive oven; cookhouse; blacksmith shop; barn; carpenter shop.
Facilities: Museum-related items for sale.
Activities: school programs; tours. Museum Sponsors: Living History Weekends June to August; Halloween Storytelling Event in October; Candlelight Tours November to mid-December.
Publications: quarterly newsletter.
Hours & Admission Prices: May-Oct. Tues.-Sat. 11-4, Sun. 1-5; groups of 10 or more by appointment. Adults $9, students $6; discounts to senior citizens; members and children 5 & under no charge.
Attendance: 5,000 (estimated)
Membership: Senior $10; Individual $15; Household $25; Institution & Professional $50; Patron $70; Friend $100; Benefactor $250.

Lebanon

THE LEBANON COUNTY HISTORICAL SOCIETY, 924 Cumberland St., Lebanon, PA 17042-5186. Tel.: 717-272-1473. Fax: 717-272-7474.
E-mail: office@lchsociety.org
Web Site: lchsociety.org
Formerly: The Stoy Museum of the Lebanon County Historical Society
Founded: 1898.
Congressional District: 17
Key Personnel: Pres., Barbara Gaffney; Treas., Carol Christ; Librarian, Brian C. Kissler; Office Coord., Tina Valgenti.
Personnel Profile: Full-Time Paid 2; Part-Time Volunteers 82.
Governing Authority: nonprofit organization. Subsidiary Institution: Union Canal Tunnel Park, 25th St. & Union Canal Dr., Lebanon, PA. Tax-exempt.
Institution Type/Description: Historical Society Museum.
Collections: historical interest; period items. Historic Site: Union Canal Tunnel, oldest existing transportation tunnel in the U.S.
Research Fields: local history & genealogy.
Facilities: 3,000-vol. library of genealogy, Civil War and history books available for use on premises; microfilms of local newspapers from 1838-present; reading room.
Activities: guided tours; lectures; field trips; permanent & temporary exhibitions; historical information. Society Sponsors: Union Canal Tunnel Fair; Scare Affair in October.
Publications: annual booklet; bimonthly newsletter, Seeds of History.
Hours & Admission Prices: Museum: Mon. & Fri. 1-8, Thurs. 9-4:30, Sat. 9-3, Sun. 1-4:30; groups by appointment. Adults $5, senior citizens $4, children 9-18 $2; members & children under 9 no charge. Library: adults $5; students & members no charge. Union Canal Rides: July-Oct. Sun. 12:30-4:30. Adults $5, children 6-14 $3. Closed national holidays.
Attendance: 2,000 (accurate)
Membership: Individual $35; Family $60; Friend $75; 1898 Circle Individual $150; 1898 Circle Family $200; Patron $300; Historian $500; Benefactor $1,000 & up.

Leechburg

DAVID LEECH HOUSE - LEECHBURG AREA MUSEUM AND HISTORICAL SOCIETY, 118 First St., Leechburg, PA 15656-1304. Mailing Address: P.O. Box 156, Leechburg, PA 15656-0156. Tel.: 724-845-8914.
E-mail: lamahs@windstream.net
Web Site: www.leechburgmuseum.org
Institution Type/Description: Historic House Museum: housed in the home of Leechburg founder, David Leech; c.1830.
Collections: local history & culture; period furnishings; personal artifacts; photographs; paintings; newspapers; postcards.
Facilities: library. Museum-related items for sale.
Activities: research; special events.
Hours & Admission Prices: March-Dec. Wed. & Sat. 12-3.

Lewisburg

PACKWOOD HOUSE MUSEUM, 8 Market St., Lewisburg, PA 17837. Mailing Address: 15 N. Water St., Lewisburg, PA 17837-1569. Tel.: 570-524-0323. Fax: 570-524-0548.
E-mail: info@packwoodhousemuseum.com
Web Site: www.packwoodhousemuseum.com
Founded: 1976.
Congressional District: 10
Key Personnel: Admin., Jennifer Snyder; Chm. (V), Christine Sperling; Museum Shop Mgr., Sue Hornberger.
Personnel Profile: Full-Time Paid 1; Part-Time Paid 1; Part-Time Volunteers 50; Interns 4.
Governing Authority: nonprofit organization. Parent Institution: The Fetherston Foundation. Tax-exempt: 501(c)(3).
Institution Type/Description: Historic House Museum: housed in a former log tavern (1790-1830), hotel (1830-1886), apartments (1893-1936) and later residence of John & Edith Fetherston.
Collections: decorative arts, including American 18th- to 20th-century furniture, glassware, stoneware, chinaware, quilts, coverlets; primitive farm accessories; Oriental rugs & objects; central Pennsylvania pieces.
Research Fields: decorative arts from the central Susquehanna Valley.
Facilities: botanical garden; rental facilities available. Museum-related items for sale.
Activities: guided tours; lectures; bus trips; changing exhibits; special seasonal programs; craft classes.
Publications: three newsletters per year; annual report.
Hours & Admission Prices: Tues.-Sat. 10-5; other times by appointment.

Guided Tours: 11, 1 & 3:30. Adults $10, senior citizens $8, students $7; discounts to AAM, ICOM & AASLH members; members & children under 12 no charge. Closed holidays.
Attendance: 4,652 (accurate)
Membership: Student $5; Individual $30; Family $45; Business $100; Sponsor $125; Patron $250; Benefactor $500; Life $2,500.

SAMEK ART GALLERY, (M), Elaine Langone Center, Bucknell University, Lewisburg, PA 17837. Tel.: 570-577-3981. Fax: 570-577-3215.
E-mail: peltier@bucknell.edu
Web Site: www1.bucknell.edu/samek/
Formerly: Bucknell Art Gallery
Founded: 1979.
Congressional District: 5
Key Personnel: Operations Mgr., Cynthia Peltier.
Personnel Profile: Full-Time Paid 3; Part-Time Paid 12; Interns 2.
Governing Authority: university. Parent Institution: Bucknell University. Tax-exempt: 501(c)(3).
Institution Type/Description: University Art Museum.
Collections: Kress Collection: Renaissance Art, 19th- & 20th-century European & American Art; The Sordoni Collection of Japanese Art; Cook Collection of Musical Instruments.
Research Fields: contemporary & 20th-century American art; photography.
Facilities: 125-seat auditorium.
Activities: guided tours; lectures; films; concerts; dance recitals; arts festivals; organized education programs; docent program; loan, temporary & traveling exhibitions.
Publications: exhibition catalogues.
Hours & Admission Prices: Aug.-May Mon.-Wed. & Fri. 11-5, Thurs. 11-8, Sat.-Sun. 1-5; call to confirm show. No charge. Closed major holidays; university breaks. &
Attendance: 24,853 (accurate)

SLIFER HOUSE, 80 Magnolia Dr., Lewisburg, PA 17837-6312. Tel.: 570-524-2245. Fax: 570-524-2245. Facebook: Slifer House.
E-mail: jessica.owens@albrightcare.org
Web Site: www.sliferhouse.org
Founded: 1976.
Congressional District: 5
Key Personnel: Dir., Jessica Owens; Chm. (V), Paul Mauger; C.F.O., Jackie Dancho.
Personnel Profile: Part-Time Paid 1; Part-Time Volunteers 20.
Governing Authority: private; nonprofit organization. Parent Institution: Albright Care Services. Tax-exempt: 501(c)(3).
Institution Type/Description: Historic House: 1861 Tuscan villa, designed by architect Samuel Sloan for secretary of the Commonwealth of PA, Eli Slifer.
Collections: Victorian furnishings & decorative arts representative of life in a country estate in central Pennsylvania.
Research Fields: life of Colonel Eli Slifer; Victorian era; decorative arts; political history.
Facilities: library & archives; 400 sq. ft. exhibit space. Postcards, books, prints, other museum-related items for sale.
Activities: guided tours; lectures; concerts; docent program. Annual Events: Lincoln Program; summer concert series; May Day celebration; Holiday Soiree; Civil War Encampment; Children's History Camp.
Publications: Delta Place Newsletter, biannual.
Hours & Admission Prices: Jan.-April 15 by appointment; April 16-Dec. Wed.-Sun. 1-4 Adults $7, senior citizens 60 & over $6, college students with ID $5, youth 5-17 $3; members & children 4 & under no charge. Closed Easter; Thanksgiving; Christmas.
Attendance: 3,200 (accurate)
Membership: Senior Citizen $20; Patron $40; Family $75; Contributing $100; Business $250; Lifelong $1,000.

UNION COUNTY HISTORICAL SOCIETY, Union County Courthouse, 2nd and St. Louis Sts., Lewisburg, PA 17837-1903. Mailing Address: Union County Courthouse, 103 S. 2nd St., Lewisburg, PA 17837-1903. Tel.: 570-524-8666. Fax: 570-524-8743.
E-mail: hstoricl@ptd.net
Web Site: www.unioncountyhistoricalsociety.org
Founded: 1908.
Congressional District: 5
Key Personnel: Pres., M. Lois Huffines; Vice Pres., Jeannette Lasansky; Sec., Diane Meixell; Treas., David Milne.
Personnel Profile: Part-Time Paid 3; Part-Time Volunteers 10.

Governing Authority: society; nonprofit organization. Branch Museum: small museum at the Union County Court House, Lewisburg. Tax-exempt.
Institution Type/Description: Historical Society Museum.
Collections: local historical items; manuscripts; slides; local pictures, early county school records; genealogy; oral traditions.
Research Fields: oral traditions; genealogy; history.
Facilities: office.
Activities: house tours; lectures; films; permanent & temporary exhibitions; oral tradition project.
Publications: Union County Pennsylvania: A Celebration in History by Charles M. Synder; Study Guide for elementary students of Union County, PA; Union County Heritage paperback collections of research in county, published every two years since 1968; annual calendar with historic county scenes from 1976 to present; local postcards; Oral Traditions publications: Willow Oak and Rye, Buggy Town, Central Pennsylvania Redware, etc.
Hours & Admission Prices: Mon.-Fri. 8:30-12 & 1-4:30. Suggested Donation: adults $5. &
Attendance: 1,300 (estimated)
Membership: Individual $30; Family $45; Contributor $60; Patron $100; Sponsor $150; Life $400.

Lewistown

MCCOY HOUSE, (M), 17 N. Main St., Lewistown, PA 17044-1746. Mailing Address: One W. Market St., Lewistown, PA 17044. Tel.: 717-242-1022 & 248-4711. Fax: 717-242-3488.
E-mail: info@mifflincountyhistoricalsociety.org
Web Site: www.mccoyhouse.com
Founded: 1921.
Congressional District: 9
Key Personnel: Property Placement Officer, Michael A. Bertheaud; Pres., Harvey C. Eckert; Sec., Karen L. Aurand.
Personnel Profile: Part-Time Paid 1; Part-Time Volunteers 54.
Governing Authority: state. Operated by the Mifflin County Historical Society for the Pennsylvania Historical & Museum Commission, Box 1026, Harrisburg, PA 17120. Tax-exempt.
Institution Type/Description: Historic House: birthplace of Major General Frank McCoy, who served in the Army from the Spanish-American War through World War II.
Collections: General McCoy memorabilia; furniture & furnishings of the family; Mifflin County Historical Society artifacts.
Facilities: publications for sale.
Activities: bus trips; banquets.
Publications: local history & genealogy books; society newsletter, "Notes From Monument Square".
Hours & Admission Prices: Museum: mid-May to Sept. Sun. 1:30-4. No charge; donations accepted. Library: March to mid-Nov. Tues.-Wed. 10-4. Admission $5; members no charge.
Attendance: 1,500 (estimated)
Membership: Individual $20; Family $25; Supporting $40; Organizations $50; Contributing $75; Individual Life $200.

Ligonier

ANTIOCHIAN HERITAGE MUSEUM, Rte. 711, Ligonier, PA 15658. Mailing Address: 140 Church Camp Trail, Bolivar, PA 15923. Tel.: 724-238-3677. Fax: 724-238-2102.
E-mail: info@antiochianvillage.org
Web Site: www.antiochianvillage.org/center/heritage/museum.html
Key Personnel: Cur., Julia Ritter
Institution Type/Description: History Museum.
Collections: over 750 items including textiles, inlaid woodwork, metal crafts, jewelry & religious art of the Near East.
Hours & Admission Prices: Winter: Mon.-Fri. 10-4; Summer: call for extended hours; other times by appointment. No charge; donations accepted. &

FORT LIGONIER ASSOCIATION, 200 S. Market St., Ligonier, PA 15658-1242. Tel.: 724-238-9701. Fax: 724-238-9732.
E-mail: office@fortligonier.org
Web Site: www.fortligonier.org
Founded: 1946.
Congressional District: 12
Key Personnel: Dir., J. Martin West; Cur. Education, Penelope A. West.
Personnel Profile: Full-Time Paid 6; Part-Time Paid 18; Part-Time Volunteers 60; Interns 2.
Governing Authority: nonprofit foundation. Tax-exempt.
Institution Type/Description: History Museum.
Collections: archaeology; decorative arts; fine arts; military arms & equipment; reconstructed 18th-century fort.
Research Fields: archaeology; French & Indian War; 18th-century military &

fortifications; Native Americans; local history; original art by Reynolds, Ramsay & Morier.
Facilities: 150-seat auditorium. Museum-related items for sale.
Activities: military interpretation programs; living history interpretation programs; formal educational programming & classes; folkcrafts.
Publications: booklets & pamphlets.
Hours & Admission Prices: April 15-Nov. 15 Mon.-Sat. 10-4:30, Sun. 12-4:30. Adults $7, seniors $6, children 6-14 $4; discounts to groups, AAM & ICOM members; children under 5 & members no charge. &
Attendance: 37,112 (accurate)
Membership: Individual $35; Family $45; Sponsor $200.

SOUTHERN ALLEGHENIES MUSEUM OF ART AT LIGONIER VALLEY, One Boucher Lane, Rte. 711, Ligonier, PA 15658-2110. Tel.: 724-238-6015. Fax: 724-238-6281.
E-mail: ligonier@sama-art.org
Web Site: ww.sama-art.org
Founded: 1997.
Institution Type/Description: Art Museum.
Collections: paintings; paperweights.
Activities: permanent & temporary exhibitions; special events.
Hours & Admission Prices: Tues.-Fri. 10-5, Sat.-Sun. 1-5. No charge; donations accepted. Closed holidays.
Attendance: 6,000 (estimated)

Limerick

LIMERICK TOWNSHIP HISTORICAL SOCIETY - THE HUNSBERGER HOUSE, 545 W. Ridge Pike, Limerick, PA 19468-1417. Tel.: 610-495-5229.
Web Site: limerickpahistory.org
Founded: 1984.
Key Personnel: Chm. (V), William Miller; Pres. (V), Dorothy L. Jones.
Personnel Profile: Full-Time Volunteers 5; Part-Time Volunteers 5.
Governing Authority: Tax-exempt.
Institution Type/Description: Historical Society Museum: housed in the home of Isaac Tyson Hunsberger; built in 1827. Listed on the National Register of Historic Places.
Collections: local history & culture; photographs; early papers & documents; books; period artifacts. Historic Buildings: bake house/summer kitchen; cider press house.
Activities: temporary exhibitions.
Publications: Medinger Pottery; pamphlet, Rothermel; book, A Journey Through Time.
Hours & Admission Prices: Wed. 9-4, 2nd Sun. each month 1-4. No charge; donations accepted.
Attendance: 20 (estimated)
Membership: Student $1; Individual $10; Life $100.

Lititz

LITITZ HISTORICAL FOUNDATION MUSEUM & MUELLER HOUSE, 137-145 E. Main St., Lititz, PA 17543-2009. Mailing Address: P.O. Box 65, Lititz, PA 17543. Tel.: 717-627-4636.
E-mail: ihf@dejazzd.com
Web Site: www.lititzhistoricalfoundation.com
Founded: 1961.
Institution Type/Description: Historic House Museum: built in 1793.
Collections: local history & culture; period furnishings; personal artifacts.
Hours & Admission Prices: Memorial Day to Labor Day Mon.-Fri. 10-4. Museum: no charge; donations accepted. House Tours: adults $5, students $3; discount to AAA members; children under 10 & members no charge.

WILBUR CHOCOLATE CANDY AMERICANA ANTIQUE COLLECTIONS & CANDY STORE, 48 N. Broad St., Lititz, PA 17543-1005. Tel.: 888-294-5287 (Toll Free); 717-626-3249 (Store).
Web Site: www.wilburbuds.com
Key Personnel: Mgr., Amy Weik
Institution Type/Description: Candy Museum.
Collections: company history; period chocolate memorabilia; early candy machinery & molds; tins & boxes; marble slabs; starch trays; copper kettles; over 150 hand-painted European & Oriental porcelain chocolate pots; video.
Activities: video tour; see candy being made in the working kitchen.
Hours & Admission Prices: Mon.-Sat. 10-5. No charge. &

Lock Haven

HEISEY MUSEUM, CLINTON COUNTY HISTORICAL SOCIETY, 362 E. Water St., Lock Haven, PA 17745-1418. Tel.: 570-748-7254.
E-mail: heisey@clintoncountyhistory.com
Web Site: www.clintoncountyhistory.com
Founded: 1921.
Congressional District: 23
Key Personnel: Pres. (V), JoAnn Bowes.
Personnel Profile: Part-Time Volunteers 20.
Governing Authority: society; nonprofit organization. Parent Institution: Clinton County Historical Society. Subsidiary Institutions: Heisey Museum, Lock Haven, PA. Tax-exempt: 501(c)(3).
Institution Type/Description: Historical Society Museum.
Collections: Victorian period clothing; furniture; artifacts; logging & transportation history.
Major Exhibits: Boy Scout Memorabilia, 2/14; Lumber Heritage & Girl Scout Memorabilia, 3/14; 175 year Anniversary of Clinton County Leading Ladies, 4/14; 175 Anniversary Clinton County Highlighting Early Businesses, 5/14; 175 Anniversary Clinton County Early Male Leaders, 6/14; Display of Vintage Art, 7/14.
Research Fields: Victorian period; canal & lumber history; local history.
Facilities: 500-vol. non-circulating library on local & Clinton County history. Gift items for sale.
Activities: guided tours; lectures.
Publications: quarterly newsletter, News & Notes; monthly newspaper column.
Hours & Admission Prices: Wed.-Thurs. 10-3. Groups $3 per person, members no charge. Closed Thanksgiving; Christmas.
Attendance: 4,850 (estimated)
Membership: Annual: Student $5; Single $20; Member & Spouse $30; Life: $200-300.

PIPER AVIATION MUSEUM FOUNDATION, One Piper Way, Lock Haven, PA 17745-2266. Tel.: 570-748-8283. Fax: 570-893-8357.
E-mail: piper@kcnet.org
Web Site: www.pipermuseum.com
Founded: 1986.
Congressional District: 76
Key Personnel: Pres., John R. Bryerton; Treas., Kim Rockwell; Cur. & Registrar, Dr. Ira G. Masemore; Historian, Harry Mutter; Public Rels., Stacy Young.
Personnel Profile: Full-Time Paid 1; Full-Time Volunteers 4; Part-Time Paid 3; Part-Time Volunteers 18.
Governing Authority: private; nonprofit organization. Tax-exempt: 501(c)(3).
Institution Type/Description: Aviation Museum.
Collections: Piper Aircraft history from 1937 to modern times; early Piper Company including archives; aviation history.
Research Fields: history of Piper aircraft.
Facilities: library & archives of photographs, company records, slides, movies & scrapbooks available to the public by appointment; 27,700 sq. ft. exhibit space. Museum-related items for sale.
Activities: temporary exhibitions; Tomahawk flight simulator. Annual Events: Sentimental Journey to Cub Haven Fly-In in June; Tomahawk flight simulator; 3 day Wings & Things day camp in July; Veterans Day Open House; Wings Over Piper R/C Model Fly-In in August.
Publications: semi-annual newsletter, The Cub Reporter.
Hours & Admission Prices: Mon.-Fri. 9-4, Sat. 10-4, Sun. 12-4. Adults $6, seniors $5, children 7-15 $3; children under 6 & members no charge. Closed major holidays. &
Attendance: 3,500 (estimated)
Membership: Personal $20; Family $30; Brick $75; Personal with Brick $80; Family with Brick $100; Life $300.

Loretto

* **SOUTHERN ALLEGHENIES MUSEUM OF ART AT LORETTO,** Saint Francis University Mall, 110 Franciscan Way, Loretto, PA 15940-9709. Mailing Address: P.O. Box 9, Loretto, PA 15940-0009. Tel.: 814-472-3920. Fax: 814-472-4131.
E-mail: loretto@sama-art.org
Web Site: www.sama-art.org
Founded: 1975.
Congressional District: 12
Key Personnel: Exec. Dir., G. Gary Moyer; Pres. (V), John K. Duggan, Jr.; Interim Cur. & Collections Mgr., Bobby J. Moore.
Personnel Profile: Full-Time Paid 12; Part-Time Paid 13; Part-Time Volunteers 25; Interns 10.

Governing Authority: nonprofit organization. Branch Museums: Southern Alleghenies Museum of Art, 1210 11th Ave., Altoona, PA. Tel.: 814-946-4464. Southern Alleghenies Museum of Art, University of Pittsburgh at Johnstown, Johnstown, PA. Tel.: 814-269-7234. Southern Alleghenies Museum of Art at Ligonier Valley, One Boucher Lane, Rte. 711 S., Ligonier, PA. Tel.: 724-238-6015. Tax-exempt: 501(c)(3).
Institution Type/Description: Visual Art Museum: located on the Mall of St. Francis University.
Collections: 19th- & 20th-century American art.
Research Fields: American art.
Facilities: library; 250-seat auditorium; two classrooms; community arts center.
Activities: guided tours; lectures; concerts; formally organized education programs for children, adults & college students affiliated with St. Francis University; inter-museum loan; permanent & temporary exhibitions.
Publications: catalogues of special exhibitions & the permanent collection; Friends material; calendars; newsletter.
Hours & Admission Prices: Tues.-Fri. 10-5, Sat.-1-5. No charge. Closed holidays. &
Attendance: 75,000 (estimated)
Membership: Student & Senior Citizen $25; Individual Artist $35; Family $50-$99; Sponsor $100-$249; Sustaining $250-$499; Benefactor $500-$999; Connoisseur $1,000-$1,499; Exhibition Sponsor $1,500; Education Sponsor $2,500 & up.

Malvern

THE WHARTON ESHERICK MUSEUM, (M), 1520 Horseshoe Tr., Malvern, PA 19355. Mailing Address: P.O. Box 595, Paoli, PA 19301-0595. Tel.: 610-644-5822. Fax: 610-644-2244.
E-mail: information@whartonesherickmuseum.org
Web Site: www.whartonesherickmuseum.org
Founded: 1971.
Congressional District: 6
Key Personnel: Pres. (V), Laurence A. Liss; Exec. Dir. & Cur., Paul Eisenhauer; Dir. Business & Programs, Lauren Otero.
Personnel Profile: Full-Time Paid 2; Part-Time Paid 2; Part-Time Volunteers 24.
Governing Authority: nonprofit organization. Tax-exempt: 501(c)(3).
Institution Type/Description: Art & Woodworking Museum: housed in Wharton Esherick's handcrafted residence & studio, a national historic landmark for architecture.
Collections: 200 pieces of the artist's work including oil & watercolor paintings, woodcuts & prints, sculpture in wood, stone & ceramic, furniture, utensils & furnishings.
Facilities: library of catalogs, clippings & correspondence relative to Wharton Esherick, available for use on premises on request by scholars. Catalogs, postcards, slides, note cards, restrikes of woodcuts for sale.
Activities: guided tours; permanent & changing exhibitions; outside lectures.
Publications: quarterly newsletter; museum catalog, Wharton Esherick: Journey of a Creative Mind; UPenn exhibit catalog, Song of the Broad Axe.
Hours & Admission Prices: Guided Tours: March-Dec. Sat. 10-5, Sun. 1-5; groups weekdays 10-4; reservations required for all tours. Adults $12, children under 12 $6; discounts to AAM members; museum members no charge. Closed holidays. &
Attendance: 5,000 (estimated)
Membership: Individual $40; Family $75; Supporting $150; Contributing $250.

Manheim

MANHEIM FIRE COMPANY MUSEUM, 83 S. Main St., Manheim, PA 17545-1645. Tel.: 717-665-3661.
Institution Type/Description: Firefighting History Museum.
Collections: firefighting history, apparatus & equipment; helmets; parade uniforms; photographs.
Hours & Admission Prices: Call for hours.

McKeesport

MCKEESPORT REGIONAL HISTORY & HERITAGE CENTER, 1832 Arboretum Dr., McKeesport, PA 15132. Tel.: 412-678-1832. Fax: 412-678-7130.
E-mail: mckheritage@yahoo.com
Web Site: www.mckeesportheritage.org
Formerly: McKeesport Heritage Center
Founded: 1980.
Key Personnel: Dir., Michelle Wardle; Pres. (V), John Barna.
Personnel Profile: Part-Time Paid 2.
Institution Type/Description: History Museum.

Collections: local history & culture; period furnishings; personal artifacts; photographs.
Research Fields: genealogy; local history.
Publications: quarterly newsletter.
Hours & Admission Prices: Tues.-Thurs. 9-5, Sat. 9-3; other times by appointment. No charge; donations accepted. &
Membership: Individual $25; Family $35; Contributing $50; Benefactor $250; Life $500.

Meadville

ALLEGHENY COLLEGE ART GALLERIES (BOWMAN, PENELEC & MEGAHAN GALLERIES), (M), N. Main St., Meadville, PA 16335-3902. Mailing Address: Box 23, Meadville, PA 16335. Tel.: 814-332-4365.
E-mail: darren.miller@allegheny.edu
Web Site: www.allegheny.edu/artgalleries
Founded: 1970.
Congressional District: 21
Key Personnel: Dir., Darren Miller.
Personnel Profile: Full-Time Paid 1; Part-Time Paid 5.
Governing Authority: private college; nonprofit. Parent Institution: Allegheny College. Tax-exempt: 501(c)(3).
Institution Type/Description: Art Gallery.
Collections: 20th-century prints & photos; miscellaneous art works.
Research Fields: contemporary American art.
Facilities: classrooms; theater; 4,000 sq. ft. exhibit space.
Activities: guided tours; lectures; loan, temporary & traveling exhibitions.
Publications: brochures; catalogs.
Hours & Admission Prices: Tues.-Fri. 12:30-5, Sat. 1:30-5, Sun. 2-4. No charge. &
Attendance: 5,000 (estimated)

BALDWIN-REYNOLDS HOUSE MUSEUM, 639 Terrace St., Meadville, PA 16335-1733. Mailing Address: 411 Chestnut St., Meadville, PA 16335. Tel.: 814-724-6080.
E-mail: museum@baldwinreynolds.org
Web Site: www.baldwinreynolds.org
Founded: 1883.
Congressional District: 21
Key Personnel: Chm. (V), Bruce Barrett; Pres. (V), Beth Rekas; Cur., Joshua F. Sherretts.
Personnel Profile: Full-Time Volunteers 1; Part-Time Paid 1; Part-Time Volunteers 10; Interns 1.
Governing Authority: society. Parent Institution: Crawford County Historical Society. Tax-exempt: 501(c)(3).
Institution Type/Description: Historic House: 1841-43 Baldwin-Reynolds Mansion.
Collections: paintings; costumes; history; medicine; numismatic; genealogy books & artifacts; newspapers; maps; manuscripts. Historic House: Dr. J. R. Mosier Medical Office.
Research Fields: Holland Land Co., Pennsylvania Population Co., Reynolds Papers; Shadeland Papers; genealogy; Henry Baldwin; rural medicine.
Facilities: 3,000-vol. library of history books.
Activities: guided tours; lectures; films; permanent exhibitions; bus tours. Annual Event: Trees of Christmas in November & December.
Publications: books, In French Creek Valley; 1841 Diary of William Reynolds; Crawford County Cemetery Inscriptions, Vol. I, II, III, IV; Stories from French Creek Valley; The First 100 Years; Pioneers of Crawford County, 1788-1800; Atlas of the Oil Region of Pennsylvania, 1865; John Wilkes Booth; Atlas of Crawford County Penna. 1876; David Mead: Pennsylvania's Last Frontiersman; The Baldwin-Reynolds House; reprint, Tribune 1888 Centennial Edition; biannual newsletter, Crawford County History; The Civil War Diaries of Seth Waid III; The Lake As It Was (A History of Connezut Lake, Penns.); Pioneer Life In Crawford County, Pennsylvania; Images of America: Meadville.
Hours & Admission Prices: mid-May to Aug. Wed.-Sun. 12-4. Tours: June-Aug. 12, 1, 2, 3. Adults $5, children $3; members no charge.
Attendance: 4,000 (estimated)
Membership: Individual $15; Family $30; Patron $40; Benefactor $75; Sustaining $150; Lifetime $450.

Mechanicsburg

MECHANICSBURG MUSEUM ASSOCIATION, (M), 2 W. Strawberry Alley, Mechanicsburg, PA 17055-6213. Tel.: 717-697-6088. Fax: 717-697-6285.
E-mail: mechanicsburgmuseum@verizon.net
Web Site: www.mechanicsburgmuseum.org
Founded: 1975.

Congressional District: 19
Key Personnel: Dir. & Museum Shop Mgr., Steven B. Zimmerman; Pres. (V), Ruth N. Wrightstone, Ph.D.; Membership, Jean Souder; Membership, Fay GeeGee; Education, Nancy Cassel; Public Rels., Grace Rarick; Cur., Fern Oram; Security, Gerald Forry.
Personnel Profile: Part-Time Paid 1; Part-Time Volunteers 100.
Governing Authority: private; nonprofit organization. Tax-exempt: 501(c)(3).
Institution Type/Description: Local History Museum.
Collections: photographs; decorative arts; local industry & business; oral histories; manuscripts; advertising signs; railroad artifacts; musical instruments; clothing; Civil War; military; fine arts. Historic Buildings: Passenger Station; Museum at the Freight Station; Frankeberger Tavern; Stationmasters House; Washington Street Station.
Major Exhibits: All Aboard the Santa Express, 12/13-1/14; Mechanicsburg Artists, 2/14-6/14.
Research Fields: local history & chronology; building & property history; oral history of local people; community history; Cumberland Valley Railroad.
Facilities: 500-vol. library; educational facilities. Museum-related items for sale.
Activities: concerts; docent program; formal education programs; guided tours; lectures; participatory & temporary exhibits. Annual Events: auction; Garden Tour; Quilt Show; Golf Tournament; occasional papers.
Publications: newsletter, Mechanicsburg Museum Association Newsletter; annual book, A Walk Through Mechanicsburg's Past.
Hours & Admission Prices: Wed.-Sat. 12-3; other times by appointment. Frankeberger Tavern: May-Sept. Sat. 12-3. No charge; donations accepted. Closed holidays. &
Attendance: 3,500 (estimated)
Membership: Family $35; Sustaining $75; Benefactor $100; Corporate $250; Patron $1,000.

Media

DELAWARE COUNTY INSTITUTE OF SCIENCE, 11 Veterans Sq., Media, PA 19063-3201. Tel.: 610-566-5126.
Web Site: www.delcoscience.com
Founded: 1833.
Key Personnel: Pres. Pro-tem, Roger Mitchell
Institution Type/Description: Science Museum.
Collections: mounted birds & animals; herbarium; fossils; shells; corals; microscopes; minerals; ca. 5500 specimens.
Facilities: library.
Hours & Admission Prices: Mon., Thurs. & Sat. 9 to noon. No charge.
Attendance: 2,000 (estimated)

TYLER ARBORETUM, 515 Painter Rd., Media, PA 19063-4424. Tel.: 610-566-9134. Fax: 610-891-1490.
E-mail: info@tylerarboretum.org
Web Site: tylerarboretum.org
Founded: 1946.
Congressional District: 5
Key Personnel: Exec. Dir., Richard A. Colbert; Dir. Horticulture, Mike Karkowski.
Personnel Profile: Full-Time Paid 9; Part-Time Paid 15; Part-Time Volunteers 275; Interns 3.
Governing Authority: nonprofit educational institution. Tax-exempt.
Institution Type/Description: Arboretum.
Collections: plant collections from mid-1800s by Painter brothers and 1950s by Dr. Wister: rhododendron, cherry, crab apple, lilac, magnolia; 85-acre pinetum; contemporary specialty gardens: fragrant herb garden, bird habitat garden, butterfly garden, newly redesigned native woodland walk; meadow maze; 200 acres of special plant collections; 450 acres of forest & fields. Historic Buildings: c.1738 Lachford Hall, home of the Painter brothers; c.1863 Painter Library; c.1833 barn.
Facilities: visitors center; 20 miles of marked trails. Books & gift items for sale.
Activities: permanent exhibitions; educational lectures & workshops; informal walks.
Publications: quarterly newsletter; educational event schedule; various brochures; visitors map & guide; annual report.
Hours & Admission Prices: Seasonal hours, see website. Adults $7, seniors 65 & over $6, children 3-15 $4; discounts to AABGA members; children under 3 & members no charge; participates in AHS reciprocal admissions program.
Attendance: 52,000 (estimated)
Membership: Individual $50; Family $60; Contributing $130; Organizational $200; Circle of Friends $300-$3,000.

Mercer

MERCER COUNTY HISTORICAL SOCIETY, 119 S. Pitt St., Mercer, PA 16137-1211. Tel.: 724-662-3490.
E-mail: info@mchspa.org
Web Site: www.mchspa.org
Founded: 1946.
Congressional District: 21
Key Personnel: Exec. Dir., William C. Philson; Pres. & C.E.O., David M. Miller.
Personnel Profile: Full-Time Paid 1; Full-Time Volunteers 1; Part-Time Paid 5; Part-Time Volunteers 13.
Governing Authority: society; nonprofit organization. Tax-exempt.
Institution Type/Description: Historical Society Museum: housed in 1825 Magoffin House.
Collections: papers of Dr. John W. Goodsell, Griff Nichols, Junkin Estate; manuscripts; church & cemetery records; archives of local materials; Indian artifacts; restored church; restored print shop. Historic Buildings: Caldwell School; log cabin.
Facilities: 3,000-vol. library of history & genealogy available for use on premises or through photocopy; reading room; microfilm; archive material.
Activities: guided tours; films.
Publications: quarterly newsletter; manuscript, on Polar Trails.
Hours & Admission Prices: Tues.-Fri. 10-4:30, Sat. 10-3; group tours by appointment only. No charge; donations accepted. Closed national holidays. &
Attendance: 3,000 (estimated)
Membership: Student & Senior Citizen $6; Regular $10; Family $15; Contributing $25; Patron $50; Sustaining $100; Life $200.

Merion Station

SAINT JOSEPH'S UNIVERSITY - UNIVERSITY GALLERY, Merion Hall, Maguire Campus, 376 N. Latches Ln., Merion Station, PA 19066. Mailing Address: 5600 City Ave., Philadelphia, PA 19131. Tel.: 610-660-1840. Fax: 610-660-2278.
E-mail: jbracy@sju.edu
Web Site: www.sju.edu/gallery
Founded: 1976.
Key Personnel: Chm. Art Dept., Dennis McNally, S.J., Ph.D.; Assoc. Dir. Gallery, Jeanne Bracy.
Personnel Profile: Part-Time Paid 1.
Governing Authority: private university; nonprofit. Parent Institution: Saint Joseph's University. Tax-exempt: 501(c)(3).
Institution Type/Description: University Art Museum.
Collections: Ceramics
Activities: arts festivals; concerts; films; formal education programs for undergraduate & graduate college students; lectures; participatory exhibits; theater.
Hours & Admission Prices: Sept.-May Mon.-Fri. 9-7, Sat. 10-1. No charge. Closed national holidays. &
Attendance: 2,000 (estimated)

Middleburg

THE SNYDER COUNTY HISTORICAL SOCIETY, 30 E. Market St., Middleburg, PA 17842-1017. Mailing Address: P.O. Box 276, Middleburg, PA 17842-0276. Tel.: 570-837-6191. Fax: 570-837-4282.
E-mail: schs@snydercounty.org
Founded: 1898.
Congressional District: 10
Key Personnel: C.E.O. & Pres. (V), Teresa J. Berger; Editor, Ruth Roush; Sec., Lee E. Knepp.
Personnel Profile: Part-Time Volunteers 20; Interns 1.
Governing Authority: society. Tax-exempt: 501(c)(3).
Institution Type/Description: Historical Society Museum.
Collections: archives; genealogy; Indian artifacts; Civil War relics; history books & old newspapers available for use on premises.
Research Fields: history; genealogy; Pennsylvania German heritage.
Facilities: auditorium; library.
Activities: tours; meetings; presentations.
Publications: annual book, Snyder County Historical Bulletin; 5-vol. set of bulletins.
Hours & Admission Prices: Museum: May-Sept. Sun. 1:30-5; other times by appointment. Library: March-April & Oct.-Dec. Mon. & Thurs.-Fri. 10-3.30; May-Sept. Sun. 1:30-5, Mon. & Thurs.-Fri. 10-3:30. No charge; donations accepted. &
Attendance: 2,000 (estimated)
Membership: Junior $10; Individual $20; Life $200.

Mifflinburg

MIFFLINBURG BUGGY MUSEUM ASSOCIATION, INC., 598 Green St., Mifflinburg, PA 17844-1241. Tel.: 570-966-1355. Fax: 570-966-9231.
E-mail: buggymuseum@windstream.net
Web Site: www.buggymuseum.org
Founded: 1978.
Congressional District: 10
Key Personnel: Dir., Bronwen Sanders; Pres. (V), Gory Coddington.
Personnel Profile: Full-Time Paid 1; Part-Time Volunteers 65; Interns 1.
Governing Authority: private; nonprofit organization. Tax-exempt: 501(c)(3).
Institution Type/Description: Industrial Museum.
Collections: tools & equipment of the William A. Heiss Coachworks; Heiss family furnishings, documents & photographs; buggy industry in Mifflinburg; late 19th to early 20th century transportation.
Research Fields: history of Mifflinburg Buggy makers.
Facilities: library; auditorium. Museum-related items for sale.
Activities: guided tours; temporary exhibitions; trips for members; bingo. Annual Events: May Fest; Buggy Day (Heritage); Garden Tour; Cemetery Walk; temporary exhibits.
Publications: biannual newsletter, Coachwork Chronicle.
Hours & Admission Prices: Museum: April-Oct. Thurs.-Sat. 10-5, Sun. 1-5. Visitor Center: Nov.-Dec. Thurs.-Sat. 10-4. Adults $10, children $5; discounts to groups of 20 or more and AAA & AARP members; members no charge. Closed major holidays. &
Attendance: 4,000 (estimated)
Membership: Individual $20; Family $35; Friend $50; Sponsor $100; Patron $250; Guardian $500; Benefactor $1,000; Ambassador $2,500.

Milford

PIKE COUNTY HISTORICAL SOCIETY, 608 Broad St., Milford, PA 18337-1704. Mailing Address: P.O. Box 915, Milford, PA 18337-0915. Tel.: 570-296-8126.
E-mail: pikemuse@ptd.net
Web Site: www.pikecountyhistoricalsociety.org
Founded: 1930.
Congressional District: 12
Key Personnel: Dir., Lori Strelicki.
Governing Authority: society; executive committee. Tax-exempt.
Institution Type/Description: Local History Museum.
Collections: period farm & home implements; Lincoln assassination tableau; Lincoln Flag; Civil War exhibit; Indian relics; old photos; early schoolhouse memorabilia; restored Abbot & Downing Coucard Coach. Historic Building: restored one-room schoolhouse.
Facilities: 30-vol. collection of forestry materials available for use on premises by arrangement.
Activities: guided tours; lectures; permanent exhibitions.
Publications: cassette tapes, Ride Through Time.
Hours & Admission Prices: Columns Museum: July-Aug. Wed.-Sun. 1-4; Sept.-June Wed & Sat.-Sun. 1-4; other times by appointment. Adults $5, students $3; children & members no charge.
Attendance: 3,000 (estimated)
Membership: Junior Historian $5; Individual $25; Patron $50; Grand Patron $150; Museum Sponsor $500.

Mill Run

FALLINGWATER, (I), 1478 Mill Run Rd., Mill Run, PA 15464-1542. Mailing Address: P.O. Box R, Mill Run, PA 15464-0167. Tel.: 724-329-8501. Fax: 724-329-0881.
E-mail: fallingwater@paconserve.org
Web Site: www.fallingwater.org
Founded: 1964.
Congressional District: 12
Key Personnel: Pres., Thomas Saunders; Dir. & Vice Pres., Lynda S. Waggoner; Museum Shop Mgr., Betsy Poole.
Personnel Profile: Full-Time Paid 56; Part-Time Paid 52; Part-Time Volunteers 80; Interns 4.
Governing Authority: nonprofit organization. Parent Institution: Western Pennsylvania Conservancy. Tax-exempt: 501(c)(3).
Institution Type/Description: Historic House: designed by Frank Lloyd Wright in 1935 for the Edgar Kaufmann family, built over a waterfall.
Collections: decorative arts, including furniture by Frank Lloyd Wright; glass; ceramics; textiles; sculpture by Lipchitz, Arp & Voulkos; painting & graphic works by Picasso, Diego Rivera; 19th-century Japanese printmakers, J.J. Audubon.
Facilities: 500-vol. library of art & architecture; in-door restaurant; nature center; rental facilities. Books, contemporary & traditional crafts, slides & cards for sale.
Activities: general guided tours, in-depth tours, family & Land of Fallingwater tours; week-long residency programs for teachers & students; lectures; school programs; hiking; camping.
Publications: annual brochure; newsletter, Conserve.
Hours & Admission Prices: mid-March to Nov. Thurs.-Tues. 10-4; reservations essential. Adults $20. Grounds only: adults $8. In-depth tours by appointment only: adults $65; discounts to school groups, AAM members, Western Pennsylvania Conservancy & Frank Lloyd Wright Building Conservancy members. &
Attendance: 160,602 (accurate)
Membership: Individual $25; Dual $50; Contributing $100; Patron $250; Benefactor $500; Leadership $1,000.

Millersburg

HISTORICAL SOCIETY OF MILLERSBURG & UPPER PAXTON TOWNSHIP MUSEUM, 330 Center St., Millersburg, PA 17061. Mailing Address: P.O. Box 171, Millersburg, PA 17061-0171. Tel.: 717-692-4084.
Founded: 1980.
Congressional District: 17
Key Personnel: Dir., Don Smith; Dir., Leslie Smith.
Governing Authority: nonprofit organization. Tax-exempt: 501(c)(3).
Institution Type/Description: Local History Museum: housed in 1919 fire house & municipal building.
Collections: local life from prehistoric to modern times; pre-Indian & Indian artifacts; clothing; photographs; archives; paintings. Historic Building: 1898 railroad station.
Facilities: 100-vol. library of historical & genealogical material available to the public; 1,500 sq. ft. exhibit space. Books & other gift items for sale.
Activities: guided tours; temporary exhibitions. Museum Sponsors: Arts & Crafts Show.
Publications: quarterly newsletter, The Herald.
Hours & Admission Prices: May-Oct. Sat. 10-2, Sun. 2-4. No charge; donations accepted.
Attendance: 1,000 (accurate)
Membership: Student $5; Individual $10; Family $18; Life $200.

Montrose

SUSQUEHANNA COUNTY HISTORICAL SOCIETY & FREE LIBRARY ASSOCIATION, Two Monument Sq., Montrose, PA 18801-1115. Tel.: 570-278-1881. Fax: 570-278-9336.
E-mail: info@susqcohistsoc.org
Web Site: www.susqcohistsoc.org
Founded: 1890.
Congressional District: 10
Key Personnel: Administrator & Librarian, Susan Stone; Chm. Historical Committee, Sue Bennett Dyson; Chm. Bd., Cathy Chiarella; Cur., Elizabeth A. Smith.
Personnel Profile: Full-Time Paid 2; Part-Time Paid 2; Part-Time Volunteers 10.
Governing Authority: nonprofit organization. Tax-exempt.
Institution Type/Description: History Museum.
Collections: Susquehanna County history & culture; period artifacts.
Research Fields: Susquehanna County history & genealogy.
Facilities: 500-vol. library of genealogy, marriage, death, cemetery & local DAR records, complete genealogies, microfilms of county newspapers from 1816-present & census reports of Susquehanna County & neighboring counties from 1790-1930, available for use on premises with staff member present.
Activities: guided tours on request; permanent & rotating exhibitions. Museum Sponsors: special exhibit in December.
Publications: reprints, Centennial History of Susquehanna County, 1887; History of Susquehanna County, 1873; 1872 Beer's Atlas of Susquehanna County with complete index; Susquehanna County Historical Society Journal of Genealogy and Local History; Waiting for the Lord: Nineteenth Century Black Communities in Susquehanna County, PA.
Hours & Admission Prices: Museum & Genealogy Reference Room: May-Sept. Mon.-Fri. 9-5; Oct.-April Mon. & Thurs.-Fri. 9-5, Tues.-Wed. 12-5. Public Library: Mon.-Fri. 9-9, Sat. 9-4. No charge; donations accepted. Genealogy Reference Room: $5 research fee for visiting non-members; $65 for all mailed requests. Closed national holidays. &
Attendance: 2,700 (estimated)
Membership: Public Library: Contributing Individual $20; Family $30. Historical Society: Genealogy & Journal: Individual $20; Dual $25.

Morrisville

✳ PENNSBURY MANOR, (M), 400 Pennsbury Memorial Rd., Morrisville, PA 19067-6797. Tel.: 215-946-0400. Fax: 215-310-1011. TDD: 215-310-1016.
E-mail: willpenn17@aol.com
Web Site: www.pennsburymanor.org
Founded: 1939.
Congressional District: 8
Key Personnel: Dir., Douglas Miller; Museum Education, Mary Ellyn Kunz; Pres., Ron Schmid; Cur., Todd Galle; Volunteer Coord., Hannah Howard; Horticulturist, Mike Johnson; Mgr. & Public Rels., Tabitha Dardes; Supt. Bldgs. & Grounds, Joseph Cameli; Museum Shop Mgr., Cindy Praria.
Personnel Profile: Full-Time Paid 11; Part-Time Paid 6; Part-Time Volunteers 220; Interns 2.
Governing Authority: state; nonprofit organization. Parent Institution: Pennsylvania Historical and Museum Commission, P.O. Box 1026, Harrisburg, PA. 17108-1026. Subsidiary: The Pennsbury Society. Tax-exempt.
Institution Type/Description: Historic House Museum: 1683 Pennsbury Manor, residence of William Penn, reconstructed in 1939.
Collections: 17th century English furnishings; living collection of ornamental & useful plants; livestock.
Research Fields: William Penn; 17th-century Pennsylvania; Quaker social & cultural life; historic preservation movement.
Facilities: 1,100-vol. library of material pertaining to William Penn, decorative arts, & social history available on premises; 100-seat auditorium; farm; picnic pavilion; period gardens. Museum-related items for sale.
Activities: guided tours; programs for elementary, secondary schools, & families; formally organized education programs for adults; living history programs; craft workshops; temporary exhibitions. Museum Sponsors: Holly Nights.
Publications: A Pennsbury Manor Cookbook; Pennsbury Brewing Book; Chuse Thy Cloaths: Pennsbury Clothing; William Penn.
Hours & Admission Prices: Tues.-Sat. 9-5, Sun. 12-5. Adults $9, senior citizens 65 & over $7, children 3-11 $5; discounts to AAM & ICOM members. Grounds Pass: $3. Closed New Year's Day; Veterans Day; Columbus Day; Thanksgiving & day after; Christmas. ♿
Attendance: 27,500 (accurate)
Membership: Basic $25; Individual $35; Family $45; Individual Passport $55; Family Passport $65; Heritage $250; Penn's Partners $500 & up.

Mountainhome

CRESCO STATION MUSEUM, Rte. 390 & Sand Spring Rd., Mountainhome, PA 18342. Mailing Address: P.O. Box 358, Mountainhome, PA 18342-0358. Tel.: 570-595-2279.
E-mail: stationmuseum@verizon.net
Web Site: www.barretthistory.org
Founded: 1993.
Governing Authority: Parent Institution: Barrett Township Historical Society. Tax-exempt.
Institution Type/Description: History Museum.
Collections: local history & culture; photographs; personal artifacts; period furnishings.
Activities: temporary exhibitions; research. Museum Sponsors: monthly art shows; musical programs June to September.
Hours & Admission Prices: Memorial Day to June & Sept.-Oct. 12 Sun. 1-4; July-Aug. Wed. & Sat.-Sun. 1-4. No charge.
Attendance: 500 (estimated)

Muncy

MUNCY HISTORICAL SOCIETY, 40 N. Main St., Muncy, PA 17756. Mailing Address: P.O. Box 11, Muncy, PA 17756. Tel.: 570-546-5917.
E-mail: muncyhistorical@aol.com
Web Site: muncyhistoricalsociety.org
Institution Type/Description: Historical Society Museum.
Collections: local history & culture; period furnishings; personal artifacts; photographs.
Facilities: library.
Hours & Admission Prices: March-Nov. Mon.-Fri. 9-3; other times by appointment. Closed holidays.

Narberth

SWEET MABEL FOLK ART & FINE CRAFT GALLERY, 41 N. Narberth Ave., Narberth, PA 19072-2347. Tel.: 610-667-3041.
Institution Type/Description: Art Gallery.
Collections: works by regional & national artists; carvings; paintings; photographs.

Activities: workshops; classes; receptions; birthday parties.
Hours & Admission Prices: Tues.-Sat. 11-6, Sun. 12-5.

Nazareth

MARTIN GUITAR MUSEUM, 510 Sycamore St., Nazareth, PA 18064. Mailing Address: P.O. Box 329, Nazareth, PA 18064-0329. Tel.: 610-759-2837. Fax: 610-759-6360.
Institution Type/Description: Guitar Museum.
Collections: music history & culture; over 170 guitars; tools; guitarmaking.
Facilities: visitor center. Museum-related items for sale.
Activities: guided tours.
Hours & Admission Prices: Museum & Visitor Center: Mon.-Fri. 8-5. Factory: Mon.-Fri. 9-4. No charge; donations accepted. Closed national holidays. ♿
Attendance: 30,000 (estimated)

MORAVIAN HISTORICAL SOCIETY, 214 E. Center St., Nazareth, PA 18064-2209. Tel.: 610-759-5070. Fax: 610-759-2462.
E-mail: director@moravianhistoricalsociety.org
Web Site: www.moravianhistoricalsociety.org
Founded: 1857.
Congressional District: 15
Key Personnel: Exec. Dir., Megan van Ravenswaay; Pres., Rev. Dr. Craig Atwood; Vice Pres., Dr. Linda Shay Gardner.
Personnel Profile: Full-Time Paid 1; Part-Time Paid 6; Part-Time Volunteers 50; Interns 3.
Governing Authority: society. Tax-exempt.
Institution Type/Description: History Museum: housed in 1740 Whitefield House.
Collections: art & artifacts concerning the history of the Moravian Church; 18th century paintings by John Valentine Haidt; early musical instruments; manuscripts.
Research Fields: Moravian Church history.
Facilities: research library.
Activities: guided tours; permanent exhibitions.
Publications: Journal of Moravian History.
Hours & Admission Prices: Daily 1-4. Adults $5, children $3; children under 5, members, AAM & ICOM members no charge. Closed major holidays. ♿
Attendance: 5,000 (accurate)
Membership: Individual $50. Donor Clubs: Manor Club $100; Bicentennial Club $200; Whitefield Club $500; Spangenberg Club $1,000; Zinzendorf Club $5,000; Comenius Club $10,000 & up.

PENNSYLVANIA LONGRIFLE MUSEUM AND JOHN JOSEPH HENRY HOUSE MUSEUM, 402 Henry Rd., Nazareth, PA 18064. Mailing Address: P.O. Box 345, Nazareth, PA 18064-0345. Tel.: 610-759-9029. Fax: 610-759-9029. Facebook: Pennsylvania Longrifle Museum.
E-mail: jacobsburg@rcn.com
Web Site: www.jacobsburghistory.com
Founded: 1972.
Congressional District: 15
Key Personnel: Exec. Dir., Ira Hiberman.
Personnel Profile: Full-Time Paid 1; Part-Time Paid 1; Part-Time Volunteers 30.
Governing Authority: Parent Institution: Jacobsburg Historical Society. Tax-exempt.
Institution Type/Description: History Museum.
Collections: early American firearms history; Henry family history; early industrial America.
Activities: school programs; lectures; early American gunsmithing courses; Early American Craft Education (EACE).
Publications: The Jacobsburg Record.
Hours & Admission Prices: May-Oct. Sat.-Sun. 12-4. Adults $5; members & children under 12 no charge. Blue Star Museum.
Attendance: 5,000 (estimated)
Membership: Student $10; Individual $20; Family & Business $35; Life $500.

New Brighton

THE MERRICK ART GALLERY, 1100 Fifth Ave., New Brighton, PA 15066. Mailing Address: P.O. Box 312, New Brighton, PA 15066-0312. Tel.: 724-846-1130.
E-mail: merrickartgallery@verizon.net
Web Site: www.merrickartgallery.org
Founded: 1880.
Congressional District: 4
Key Personnel: Dir. & Education Dir., Cynthia A. Kundar; Trustee, Karen Capper.

Personnel Profile: Full-Time Paid 1; Part-Time Volunteers 35; Interns 1.
Governing Authority: nonprofit organization. Tax-exempt: 501(c)(3).
Institution Type/Description: Art Museum.
Collections: French, English, German & American 18th- & 19th-century paintings; 225 European & American 19th-century paintings; manuscript collections; works by Prud'hon Courbet, Thomas Sully, Thomas Hill, A.B. Durand; geological, zoological & entomological specimen collections.
Research Fields: 19th century.
Facilities: 1,200-vol. library, available for use under director's supervision; classrooms.
Activities: guided tours; arts festivals; lectures; concerts; hobby workshops; docent program; formally organized education programs; permanent & temporary exhibits monthly.
Publications: quarterly newsletter, Merrick Gallery Associates Newsletter; catalogue of 19th-century European & American paintings; brochure, The Merrick Art Gallery, a self-guiding tour of the permanent collection.
Hours & Admission Prices: Winter & Spring: Tues.-Sat. 10-4:30, Sun. 1-4; Summer & Fall: Wed.-Sat. 10-4, every other Sun. 1-4. No charge; donations accepted; docent guided tours $2. Closed New Year's Eve & Day; Memorial Day; Independence Day; Labor Day; Thanksgiving; Christmas to mid-Jan.; holiday weekends. &
Attendance: 4,200 (estimated)
Membership: Student $5; Senior $10; Individual $20; Family & Nonprofit $25; Contributing $40; Business Level I $50; Business Level II & Gustave Courbet Society $100; Lifetime & Merrick Circle $500.

New Castle

HOYT CENTER FOR THE ARTS, 124 E. Leasure Ave., New Castle, PA 16101-2398. Tel.: 724-652-2882. Fax: 724-657-8786.
E-mail: hoyt@hoytartcenter.org
Web Site: www.hoytartcenter.org
Formerly: The Hoyt Institute of Fine Arts
Founded: 1965.
Congressional District: 4
Key Personnel: C.E.O. & Exec. Dir., Kimberly B. Koller-Jones; Pres. (V), Richard E. Flannery; Exhibitions Coord., Patricia McLatchy; Program Dir., Robert Presnar; Mktg. Dir., Melissa Maiella.
Personnel Profile: Full-Time Paid 4; Part-Time Paid 7; Part-Time Volunteers 3; Interns 1.
Governing Authority: nonprofit organization. Tax-exempt: 501(c)(3).
Institution Type/Description: Art Museum: housed in two early 20th-century mansions, Greek Revival style & Tudor Revival style.
Collections: paintings; prints; ceramics; photographs; sculpture; furniture.
Major Exhibits: Lost & Found, 1/7-2/28/14; Edward Eberle, 3/4-4/25/14; Pittsburgh Pastel Society, 6/3-7/25/14; Black & White Collective, 7/29-9/19/14; Annual Hoyt Mid-Atlantic Juried Art Show, 9/23-10/31/14; Painting International, 11/4-12/23/14.
Facilities: 1,000-vol. library of fine arts books, available for use on premises only; classrooms. Art items for sale.
Activities: guided & self-guided tours; lectures; concerts; organized educational programs; temporary & traveling exhibitions. Institute Sponsors: Mid Atlantic Art Show; Regional Art Show.
Publications: quarterly newsletter.
Hours & Admission Prices: Tues. & Thurs. 11-8, Wed. & Fri.-Sat. 11-4, Sun. for special events; guided tours by appointment. East Mansion no charge. West Mansion Tours: Guided $5, Self-Guided $3; discounts to AAM members. Closed New Year's Day; Easter; Memorial Day; Independence Day; Labor Day; Thanksgiving; Christmas. &
Attendance: 22,500 (estimated)
Membership: Youth $15; Artist $25; Individual $30; Family $55; Associate $75; Century $100; Century II $200; Patron $300; Hoyt Donor $500; Director's Circle $1,000; Benefactor $2,500.

LAWRENCE COUNTY HISTORICAL SOCIETY, 408 N. Jefferson St., New Castle, PA 16101. Mailing Address: P.O. Box 1745, New Castle, PA 16103. Tel.: 724-658-4022. Fax: 724-658-8885.
E-mail: info@lawrencechs.com
Web Site: www.lawrencechs.com
Founded: 1938.
Key Personnel: Pres (V), Paul P. Kynch.
Personnel Profile: Full-Time Volunteers 1; Part-Time Volunteers 8.
Governing Authority: nonprofit organization. Tax-exempt: 501(c)(3).
Institution Type/Description: Historical Society Museum; 1904 18 room mansion.
Collections: local history & culture; period furnishings; personal artifacts; photographs; oral histories; China, fireworks; Civil War artifacts.
Hours & Admission Prices: Tues., Thurs. & Sat. 11-4; other times by appointment. &

Membership: Student/Senior $25; Individual $50; Family $75; Centennial $100; Bicentennial $200; Quin Centennial $500; Historical Donor $1,000.

New Hope

✳ **BOWMAN'S HILL WILDFLOWER PRESERVE, (M),** 1635 River Rd. (PA Rte. 32), New Hope, PA 18938. Mailing Address: P.O. Box 685, New Hope, PA 18938-0685. Tel.: 215-862-2924. Fax: 215-862-1846.
E-mail: bhwp@bhwp.org
Web Site: www.bhwp.org
Founded: 1934.
Congressional District: 8
Key Personnel: Dir., A. Miles Arnott; Chm. (V), Betsy Falconi; Coord. Education, Amy Hoffmann; Museum Shop Mgr., Nancy Apple.
Personnel Profile: Full-Time Paid 4; Part-Time Paid 8; Part-Time Volunteers 100; Interns 2.
Governing Authority: state. Parent Institution: Pennsylvania Historical & Museum Commission. Subsidiary Institution: Bowman's Hill Wildflower Preserve Assoc. Tax-exempt: 501(c)(3).
Institution Type/Description: Botanical Garden.
Collections: native plants of Pennsylvania.
Facilities: library; visitor & resource center; 2.5 miles of trails; 80-seat auditorium; indoor bird observatory; picnic pavilion. Museum-related items for sale.
Activities: guided tours; lectures; docent program; internship program; permanent & temporary exhibitions; symposia for professionals, classes, workshops, children's programs; Signature Plants self-guided tour.
Publications: newsletter; blooming guides; Ways With Wildflowers; seasonal calendar of events; Native Plant Information Sheets.
Hours & Admission Prices: Jan. 3-Dec. 23 daily 9-5. Adults $5, seniors 62 & over and students $3, children 4-14 $2; children under 4 & members no charge. Closed Thanksgiving. &
Attendance: 40,000 (estimated)
Membership: Student & Senior Citizen $20; Individual $25; Family $50; Benefactor $100; Patron $500; Sustaining $1,000.

THE PARRY MANSION MUSEUM, 45 S. Main St., New Hope, PA 18938. Mailing Address: P.O. Box 41, New Hope, PA 18938-0041. Tel.: 215-862-6729. Fax: 215-862-8227.
E-mail: newhopehs@verizon.net
Web Site: newhopehistoricalsociety.org
Founded: 1972.
Congressional District: 18
Key Personnel: Pres., Edwin Hild; Exec. Dir., Deborah Lang; Treas., Frank Policare; Archivist, Terry McNealy.
Personnel Profile: Full-Time Volunteers 6; Part-Time Paid 3; Part-Time Volunteers 3.
Governing Authority: nonprofit organization. Parent Institution: New Hope Historical Society. Tax-exempt.
Institution Type/Description: Decorative Arts Museum.
Collections: 1775-1900 furniture & decorative arts.
Activities: guided tours; docent program.
Publications: A Walking Tour of New Hope; Guide to the Parry Mansion Museum; The Parry Legacy; A Quaker Lady's Cookbook.
Hours & Admission Prices: May-Oct. Sat.-Sun. 1:30-3. No charge. Group tours of 10 or more available year round by reservation.
Attendance: 2,300 (estimated)
Membership: Individual $30; Family $60; Business $100; Business Partner $250; Life $500.

Newtown

HICKS ART CENTER GALLERY - BUCKS COUNTY COMMUNITY COLLEGE, (M), 275 Swamp Rd., Newtown, PA 18940-4106. Tel.: 215-504-8531.
E-mail: orlandof@bucks.edu
Web Site: www.bucks.edu/gallery
Key Personnel: Exec. Dir., Fran Orlando
Institution Type/Description: Art Gallery.
Collections: works by college students & county residents.
Activities: Annual Event: Bucks County High School Art Exhibition.
Hours & Admission Prices: Mon. & Fri. 9-4, Tues.-Thurs. 9-8, Sat. 9-12.

NEWTOWN HISTORIC ASSOCIATION, Court St. & Centre Ave., Newtown, PA 18940. Mailing Address: P.O. Box 303, Newtown, PA 18940-0303. Tel.: 215-968-4004. Fax: 215-968-8925.
E-mail: dcnhh@comcast.net

Web Site: www.newtownhistoric.org
Founded: 1964.
Congressional District: 8
Key Personnel: Pres. (V), Head Research Library & Security, David Callahan; Pres. (V), Head Research Library & Cur., Harriet Beckert; Treas., Marjorie Torongo; Librarian, Rick Booream; Membership, Mary J. Callahan; Museum Shop Mgr., Geno Peruzzi.
Personnel Profile: Part-Time Volunteers 16; Interns 2.
Governing Authority: nonprofit organization. Tax-exempt: 501(c)(3).
Institution Type/Description: Local History Museum: housed in early 1700s Half Moon Inn.
Collections: Edward Hicks artifacts; local artifacts of historical significance; genealogy.
Facilities: 1,000-vol. library of Newtown & area history books, available for use on premises.
Activities: guided tours; lectures. Annual Events: Market Day Craft Fair; Christmas Open House Tour.
Publications: newsletter, The Half Moon.
Hours & Admission Prices: June-Aug. Tues. 9-3, Thurs. 7-9, Sun. 2-4; Sept.-May Tues. 9-3, Thurs. 7-9. No charge; donations accepted. Closed holidays.
Attendance: 2,200 (estimated)
Membership: Individual $25; Family $35; Contributing $75; Corporate $150; Life $500.

Newtown Square

COLONIAL PENNSYLVANIA PLANTATION, (M), 3900 N. Sandy Flash Dr., Newtown Square, PA 19073. Mailing Address: Ridley Creek State Park, 1023 Sycamore Mill Rd., Media, PA 19063. Tel.: 610-566-1725. Fax: 610-566-4252.
E-mail: info@colonialplantation.org
Web Site: www.colonialplantation.org
Founded: 1973.
Congressional District: 7
Key Personnel: Pres., Patricia A. Theodore; Treas., Lynn Weber; Dir. Public Rels., James Adams; Office Mgr., Joy Woppert.
Personnel Profile: Full-Time Paid 3; Part-Time Paid 18; Part-Time Volunteers 50.
Volunteer Hours: 7,000
Operating Expenses: 185,000
Operating Income: 188,000
Governing Authority: nonprofit organization. Tax-exempt: 501(c)(3).
Institution Type/Description: Historical Museum: housed in c.18th-century Quaker farm.
Collections: 18th-century household furnishings; farm implements; farm animals.
Research Fields: food preservation; farm techniques; animal breeding.
Facilities: 2,000-vol. library pertaining to colonial history. Museum-related items for sale.
Activities: guided tours; hobby workshops; organized education programs for children; docent program.
Publications: Inside the Colonial Pennsylvania Plantation; An Activity Book of Colonial Times.
Hours & Admission Prices: Groups Tours: April-Nov. Mon.-Fri. Public Tours: Sat.-Sun. 11-5. Adults $8, children 4-12 $6; members & children under 4 no charge. Special weekend events may have different pricing. &
Attendance: 27,000 (accurate)
Membership: Student $25; Individual $40; Family $60; Contributing $125; Sustaining $250; Life $1,250.

Norristown

ELMWOOD PARK ZOO, 1661 Harding Blvd., Norristown, PA 19401. Tel.: 610-277-3825. Fax: 610-292-0332.
E-mail: admin@elmwoodparkzoo.org
Web Site: www.elmwoodparkzoo.org
Institution Type/Description: Zoo.
Collections: over 300 animals; mammals; birds; reptiles; fish; amphibians; butterflies.
Activities: zoo camp; scout programs; educational programs; Zoo-on-Wheels; birthday parties; special events; rental facilities. Annual Event: Beast of a Feast in June.
Hours & Admission Prices: Daily 10-5. Adults $12, seniors 65 & over and children 2-12 $9; discounts to groups, members & active military; children under 2 no charge. Closed New Year's Eve & Day; Thanksgiving; Christmas Eve & Day.

HISTORICAL SOCIETY OF MONTGOMERY COUNTY, 1654 DeKalb St., Norristown, PA 19401-5415. Tel.: 610-272-0297. Fax: 610-272-2609.
E-mail: contactus@hsmcpa.org
Web Site: www.hsmcpa.org
Key Personnel: Exec. Dir., Karen McCurdy Wolfe; Cur., Susan M. Pavlik; Archivist, Nancy Sullivan.
Governing Authority: Tax-exempt: 501(c)(3).
Institution Type/Description: Historical Society Museum.
Collections: local history & culture; period furnishings; photographs; personal artifacts.
Research Fields: genealogy; local history.
Activities: special events; guided tours.
Publications: bulletin.
Hours & Admission Prices: Mon. & Thurs. 9-5, Tues.-Wed. 1-9. &

North Huntingdon

BIG MAC MUSEUM, 9051 Rte. 30, North Huntingdon, PA 15642-2792. Tel.: 724-863-9837.
Web Site: www.bigmacmuseum.com
Founded: 2007.
Institution Type/Description: History Museum.
Collections: Big Mac history & advertising; photographs; memorabilia; 14 ft by 12 ft Big Mac.
Facilities: restaurant.
Hours & Admission Prices: Tours: by appointment.

North Wales

ROTH LIVING FARM MUSEUM OF DELAWARE VALLEY COLLEGE, 502 Dekalb Pike, North Wales, PA 19454-2236. Tel.: 215-699-3994.
E-mail: rothmuseum@delval.edu
Web Site: www.delval.edu/roth
Key Personnel: Mgr., George Gross.
Personnel Profile: Part-Time Paid 1; Part-Time Volunteers 4; Interns 4.
Institution Type/Description: Living Farm Museum.
Collections: 19th century American farming; period equipment; hands-on exhibits; horses; cattle; sheep; goats; chickens.
Activities: special events; demonstrations; school programs; seasonal activities; 4H & scout programs.
Hours & Admission Prices: Call for hours. Prices vary by program.
Attendance: 1,153

Northumberland

JOSEPH PRIESTLEY HOUSE, (M), 472 Priestley Ave., Northumberland, PA 17857-1226. Mailing Address: The Friends of Joseph Priestly House, P.O. Box 346, Lewisburg, PA 17837. Tel.: 570-473-9474.
E-mail: info@josephpriestleyhouse.org
Web Site: www.josephpriestleyhouse.org
Founded: 1960.
Congressional District: 10
Key Personnel: Pres. (V), Tom Bresenhan.
Personnel Profile: Part-Time Volunteers 20; Interns 1.
Governing Authority: state. Parent Institution: Pennsylvania Historical & Museum Commission, 300 North St., Harrisburg, PA 17120-0024. Tax-exempt.
Institution Type/Description: Historic House Museum: housed in the former home of Joseph Priestley; built in 1798.
Collections: period scientific equipment; decorative art & historic furniture of the period of Joseph Priestley.
Major Exhibits: Joseph Priestley Timeline, 11/12-11/15.
Research Fields: 18th-century chemistry; dissenting 18th-century English religion; social history; early US political history.
Facilities: visitor center gallery; meeting hall. Museum-related items for sale.
Activities: tours; seminars; education programs; videos; summer history camp; 18th century chemistry demonstrations (4 times a year); Commonwealth Charter Day (March); Oxygen Day (Aug.); Heritage Day (Nov.); Twelfth Day (Jan.).
Publications: Friends Newsletter.
Hours & Admission Prices: Visitor Center: mid-March to Nov. Sat.-Sun. Tours: 1, 2 & 3. Adults $6, senior citizens & groups $5.50, youth 3-11 $4; discounts to PHMC, AAM, ICOM & Federation members; members and active military & their family no charge. Closed Easter.
Attendance: 900 (accurate)
Membership: Individual $15; Family $25; Patron $100; Donor $500; Benefactor $1,000.

Oil City

VENANGO MUSEUM OF ART, SCIENCE & INDUSTRY, 270 Seneca St., Oil City, PA 16301-1304. Tel.: 814 676 2007. Fax: 814-678-6719.
E-mail: venangomuseum@verizon.net
Web Site: www.venangomuseum.org
Founded: 1961.
Congressional District: 2
Key Personnel: Pres. (V), Mary Balas; Financial Dir., Ola Cox; Exec. Dir., Security & Museum Shop Mgr., Betsy Kellner.
Personnel Profile: Full-Time Paid 1; Part-Time Paid 3; Part-Time Volunteers 6; Interns 2.
Governing Authority: nonprofit organization. Parent Institution: Venango Museum Corporation. Tax-exempt: 501(c)(3).
Institution Type/Description: Art, Science & Industry Museum: housed in c.1905 Beau Arts style Post Office Building.
Collections: archives; photographs; clocks; oil industry items & history; items relating to northwestern Pennsylvania history; 1928 Wurlitzer organ; 1937 Cord Automobile.
Research Fields: related to artifacts & individuals from Venango County.
Facilities: 750-vol. library of ledgers, stock certificates, periodicals & newspapers, for staff use only, includes records of early oil companies that developed in Venango County after Col. Edwin Drake discovered oil in Titusville; educational facilities. Museum-related items for sale.
Activities: guided tours; lectures; concerts; rental gallery; organized education programs for children; film documentaries.
Publications: quarterly newsletter, Venango Museum of Art, Science & Industry.
Hours & Admission Prices: April-Dec. Tues.-Fri. 10-4, Sat. 11-4, Sun. 2-5. Adults $4, senior citizens & students $2, children under 12 $1.50; members no charge. Closed New Year's Day; Easter; Memorial Day; Independence Day; Labor Day; Thanksgiving; Christmas. &
Attendance: 3,000 (estimated)
Membership: Senior Citizen $10; Individual $15; Family $25; Sustaining Friends Circle $100-$199; Director's Circle $250-$499; Trustees Circle $500-$999; Presidents' Circle $1,000-$2,999; Founders Circle $3,000 & up.

Orwell

HOME TEXTILE TOOL MUSEUM, 1819 Orwell Hill Rd., Orwell, PA 18837. Mailing Address: P.O. Box 153, Rome, PA 18837. Tel.: 570-247-7175.
E-mail: info@httm.org
Web Site: www.hometextiletoolmuseum.org
Founded: 1999.
Congressional District: 10
Governing Authority: Tax exempt: 501(c)(3).
Institution Type/Description: History Museum.
Collections: local history; spinning wheels; looms; early tools.
Facilities: 4 exhibit buildings.
Activities: workshops; demonstrations.
Hours & Admission Prices: Call for hours. Adults $5. &

Paradise

NATIONAL CHRISTMAS CENTER AND MUSEUM, 3427 Lincoln Hwy. E., Paradise, PA 17562-9621. Tel.: 717-442-7950. Fax: 717-442-9304.
E-mail: info@nationalchristmascenter.com
Web Site: www.nationalchristmascenter.com
Founded: 1998.
Key Personnel: Exec. Dir. & Museum Shop Mgr., Mandy Brown; Cur., Jim Morrison.
Personnel Profile: Full-Time Paid 2; Part-Time Paid 14.
Governing Authority: bd. of directors.
Institution Type/Description: Christmas History Museum.
Collections: Christmas history & traditions from around the world; homes hand-painted in folk art of each country; life-sized figures; period Christmas china; ornaments; Santa Claus images & stories; Santa's workshop & the North Pole; trains; cookie cutters; American Christmas collection; vintage merchandise, signage, fixtures; wooden & cardboard toys, v-mail, ration books & stamps, period greeting cards; 1897 letter, photos, family history of letter's author; hand-blown glass ornaments; contemporary & period nativities from around the world.
Research Fields: Christmas customs, celebrations, lore, traditions.
Facilities: 20,000 sq. ft. exhibit space. Gift items for sale.
Hours & Admission Prices: March-April Sat.-Sun. 10-6; May-Jan. 1 daily 10-6; other times by appointment. Call for extended holiday hours. Adults $12, children 3-12 $5. Closed New Year's Day; Easter; Thanksgiving; Christmas. &
Attendance: 30,000 (estimated)
Membership: Children (3-12) $12.50; Adults $29.

Pennsburg

SCHWENKFELDER LIBRARY & HERITAGE CENTER, (M), 105 Seminary St., Pennsburg, PA 18073-1898. Tel.: 215-679-3103. Fax: 215-679-8175.
E-mail: info@schwenkfelder.com
Web Site: schwenkfelder.com
Founded: 1913.
Congressional District: 15
Key Personnel: Administrative Asst., Michelle Pritt; Exec. Dir., David W. Luz; Pres., Jerry Heebner; Cur. Collections, Candace K. Perry; Archivist, Hunt Schenkel; Educator, Rebecca Lawrence; Assoc. Dir. Research, Dr. Allen Viehmeyer; Assoc. Dir. Theology, Dr. Peter C. Erb.
Personnel Profile: Full-Time Paid 5; Full-Time Volunteers 1; Part-Time Paid 3; Part-Time Volunteers 129.
Governing Authority: nonprofit organization. Tax-exempt: 501(c)(3).
Institution Type/Description: History Museum.
Collections: European & American culture of the Schwenkfelders, a German Protestant sect, 1720-present; local history of Perkiomen Valley, PA.
Research Fields: Schwenkfelder history & culture; Upper Perkiomen Valley history.
Activities: guided tours; permanent & changing exhibitions.
Publications: newsletter, Fraktor; book, Fraktu - Writings and Folk Art Drawings.
Hours & Admission Prices: Tues.-Wed. & Fri. 9-4, Thurs. 9-8, Sat. 10-3, Sun. 1-4. No charge, donations accepted. &
Attendance: 10,466 (accurate)
Membership: Friends of the Schwenkfelder Library and Heritage Center: Student $3; Individual $10; Family $15; Life $150; Corporate $250.

Perkasie

THE PEARL S. BUCK HOUSE, (M), 520 Dublin Rd., Perkasie, PA 18944-3000. Tel.: 215-249-0100; 800-220-BUCK. Fax: 215-249-9657.
E-mail: info@pearlsbuck.org
Web Site: www.psbi.org
Founded: 1964.
Congressional District: 8
Key Personnel: C.E.O., Janet C. Mintzer; Vice Pres. Operations, Teri Mandic; Cur., Donna C. Rhodes.
Personnel Profile: Full-Time Paid 1; Part-Time Paid 2; Part-Time Volunteers 150; Interns 2.
Governing Authority: nonprofit organization. Parent Institution: Pearl S. Buck International, Inc. Tax-exempt: 501(c)(3).
Institution Type/Description: Historic House Museum: housed in pre-1825 stone farmhouse.
Collections: Asian & American art, artifacts & furniture; books, awards & personal items of Pearl S. Buck.
Facilities: 8,200-vol. library of fiction, history, biography, juvenile & foreign books; first editions & other copies of Pearl S. Buck's works; 149,000 page archives; 1,600 sq. ft. exhibit space; 160-seat auditorium; meeting space; outdoor rental facility. Gift items for sale.
Activities: culture camps for children; writers' workshops; high school leadership program. Annual Events: holiday festival tours in December; Taste of the World signature event in May.
Publications: Connections; annual report; quarterly success stories.
Hours & Admission Prices: Mon.-Sat. 11, 1 & 2, Sun. 1 & 2. House Tours: call for hours; groups & foreign language tours by appointment. Adults $9, senior citizens $8, students $6; discounts to AAA members; members no charge. Holiday Tours: call for admission prices. Closed legal holiday & holiday weekends. &
Attendance: 22,000 (accurate)
Membership: Individual $30-$1,000.

Philadelphia

*** THE ACADEMY OF NATURAL SCIENCES OF DREXEL UNIVERSITY, (M),** 1900 Ben Franklin Pkwy., Philadelphia, PA 19103-1101. Tel.: 215-299-1000. Fax: 215-299-1028. Facebook: Academy of Natural Sciences.
E-mail: bclardo@ansp.org
Web Site: www.ansp.org
Formerly: The Academy of Natural Sciences

Founded: 1812.
Congressional District: 2
Key Personnel: Chm. Bd. Trustees, Cynthia Heckscher; Pres. & C.E.O., George W. Gephart, Jr.; Vice Pres. Collections, Edward Daeschler, Ph.D.; Vice Pres. Finance & Admin., David Rusenko; Vice Pres. Academy Science & Dept. Head Dept of Biodiversity, Earth & Environmental Science, David Velinsky, Ph.D.; Vice Pres. Public Experience & Strategic Initiatives, Sara Hertz; Vice Pres. Institutional Advancement, Amy Marvin; Dir. Center for Environmental Policy, Roland Wall; Mgr. Volunteer Svcs., Lois Kuter.
Personnel Profile: Full-Time Paid 110; Part-Time Paid 83; Part-Time Volunteers 325; Interns 25.
Volunteer Hours: 33,800
Operating Expenses: 12,331,360
Operating Income: 12,375,439
Governing Authority: Parent Institution: Drexel University; nonprofit organization. Tax-exempt: 501(c)(3).
Institution Type/Description: Natural Science Museum.
Collections: botany; diatoms; entomology; ichthyology; malacology; invertebrate paleontology; vertebrate paleontology; ornithology; mammalogy; herpetology; mineralogy; mounted specimens; art objects; portraits; ornithological photographic materials; live animal unit; dinosaurs; manuscripts; photographs; periodicals;
Major Exhibits: Dinosaurs Unearthed (T), 10/12/13-3/30/14; Unnatural History, 4/19/14-8/2/14; Birds of Paradise (T), 5/3/14-9/1/14; Chocolate, 10/11/14-1/24/15.
Facilities: library; dinosaur hall; live butterfly garden; Outside In (hands-on children's nature center); live animal center; auditorium; cafe.
Activities: lectures; films; gallery talks; nature workshops; educational programs; graduate & undergraduate teaching with various universities; temporary & traveling exhibitions; school loan service in school programs; natural science expeditions; global scientific research; environmental and sustainable free panel public programs.
Publications: scientific publications, Proceedings of the Academy of Natural Sciences; Monographs & Malacologia; members online newsletter; members magazine, Frontiers.
Hours & Admission Prices: Mon.-Fri. 10-4:30, Sat.-Sun. & holidays 10-5. Adults $15, senior citizens, military, college students & children 3-12 $13; discounts to groups & AAA members; members & children under 3 no charge. Closed New Year's Day; Thanksgiving; Christmas. ♿
Attendance: 200,000 (estimated)
Membership: Individual $50; Family $89; Family Plus $99; Supporting $150; Partners Club $250; Founders Club $500; Leadership Circles $1,000-$25,000.

THE AFRICAN AMERICAN MUSEUM IN PHILADELPHIA,
701 Arch St., Philadelphia, PA 19106-1504. Tel.: 215-574-0380, ext. 230. Fax: 215-574-3110.
E-mail: info@aampmuseum.org
Web Site: www.aampmuseum.org
Founded: 1976.
Congressional District: 1
Key Personnel: Pres. & C.E.O., Romona Riscoe Benson; Membership & Volunteer Coord., Cassandra Murray; Cur. Education & Public Programming, Leslie Willis-Lowry; Conservator & Cur. Collections, Leslie Guy; Cur. Exhibitions, Richard Watson; Museum Shop Mgr., Gladys M. Adams.
Personnel Profile: Full-Time Paid 25; Part-Time Paid 5; Part-Time Volunteers 80; Interns 6.
Governing Authority: municipal; nonprofit organization. Tax-exempt: 501(c)(3).
Institution Type/Description: Historical & Cultural Museum. This African American Museum in Philadelphia (AAMP) is dedicated to collecting, preserving and interpreting the material and intellectual culture of African Americans.
Collections: paintings & prints by African American artists; artifacts & memorabilia relating to 19th- & 20th-century African American history; Jack Franklin Photographic Collection 1940-present; performing arts memorabilia; Philadelphia Hilldale Baseball Club; sports; local politics & civil rights; beauticians; military history; public school education; local community organizations; African artifacts.
Research Fields: African American history & culture in Philadelphia, the Delaware Valley, Pennsylvania & the Americas.
Facilities: 1,500-vol. library of African American books; 200-seat auditorium. Museum-related items for sale.
Activities: guided tours; seminars; film & lecture series; jazz concerts; arts festivals; docent programs; workshops; theater; loan & traveling exhibitions; radio programs; African American artists exhibitions; book signings; collecting workshops; gallery talks; family programs.
Publications: exhibition catalog; brochures.
Hours & Admission Prices: Tues.-Sat. 10-5, Sun. 12-5. Adults $10, youth 4-12,

students with ID & senior citizens $8; discounts to groups, AAM & ICOM members; members no charge. Closed national holidays. ♿
Attendance: 60,000 (estimated)
Membership: Student & Senior Citizen $25; Individual $45; Family $60; other categories available.

AMERICAN CATHOLIC HISTORICAL SOCIETY, 263 S. 4th
St., Philadelphia, PA 19106-3819. Tel.: 215-925-5752.
E-mail: americancatholichistsoc@gmail.com
Web Site: www.amchs.org
Founded: 1884.
Key Personnel: Exec. Dir., Rev. Msgr. James P. McCoy; Pres. (V) & Co Editor, Rodger Van Allen, Ph.D.; Vice Pres., Michael Finnegan; Co-Editor, Kathleen Oxx; Sec., Nicholas Rademacher; Archivist, Joseph J. Casino.
Personnel Profile: Part-Time Volunteers 20.
Volunteer Hours: 2,000
Operating Expenses: 100,414
Operating Income: 112,842
Governing Authority: nonprofit organization; bd. of managers of the American Catholic Historical Society of Philadelphia. Subsidiary Institution: Philadelphia Archdiocesan Historical Research Center, 100 E. Wynnewood Rd., Wynnewood, PA 19096-3001. Tax-exempt: 501(c)(3).
Institution Type/Description: Religious & American History Museum.
Collections: archives; history.
Research Fields: American & American Catholic history.
Facilities: 35,000-vol. library of Americana, newspapers, pamphlets, artifacts & original manuscripts pertaining to the Catholic Church in the U.S. by appointment only.
Activities: guided tours; lectures; temporary exhibitions. Annual Event: Barry Award Dinner.
Publications: annual newsletter; quarterly, Journal: American Catholic Studies.
Hours & Admission Prices: by appointment. No charge; donations accepted.
Attendance: 1,000 (estimated)
Membership: Annual $50.

AMERICAN PHILOSOPHICAL SOCIETY MUSEUM (APS),
(M), Philosophical Hall, 104 S. Fifth St., Philadelphia, PA 19106. Mailing Address: Richardson Hall, 431 Chestnut St., Philadelphia, PA 19106-2426. Tel.: 215-440-3440. Fax: 267-386-3491.
E-mail: museum@amphilsoc.org
Web Site: www.apsmuseum.org
Founded: 2001.
Key Personnel: C.E.O., Mary Patterson McPherson; Dir. & Cur., Sue Ann Prince; Chm. (V), Henry A. Millon; Pres., Baruch S. Blumberg; Devel., Nanette Holben; Education, Jenni Drozdek; Treas., John Wolfe.
Personnel Profile: Full-Time Paid 5; Part-Time Paid 3; Part-Time Volunteers 12; Interns 3.
Governing Authority: private; nonprofit organization. Parent Institution: American Philosophical Society, Philadelphia, PA. Tax-exempt: 501(c)(3).
Institution Type/Description: History, Art & Science Museum.
Collections: 18th- & 19th-century history, art & science.
Research Fields: exploration, surveying & mapping (18th-20th centuries); natural history.
Facilities: auditorium; 2,000 sq. ft. exhibit space; nature center.
Activities: docent program; formal education program for University of Arts students; guided tours; workshops; lectures; artist residencies; participatory & temporary exhibitions.
Publications: exhibitions catalogues.
Hours & Admission Prices: Summer: Thurs.-Sun. 10-4. Winter: Fri.-Sun. 10-4; additional Wed. evening hours. No charge; donations requested. Closed holidays. ♿
Attendance: 68,212 (accurate)

AMERICAN SWEDISH HISTORICAL MUSEUM, (M), 1900
Pattison Ave., Philadelphia, PA 19145-5999. Tel.: 215-389-1776. Fax: 215-389-7701.
E-mail: info@americanswedish.org
Web Site: www.americanswedish.org
Founded: 1926.
Congressional District: 3
Key Personnel: Exec. Dir., Tracey Beck; Assoc. Dir., Birgitta W. Davis; Chm., Leonard Busby; Education & Visitor Services Mgr., Tricia Davis; Cur., Carrie Hogan; Coord. Membership & Mktg., Caroline Rossy; Maintenance, Frank Sanders.
Personnel Profile: Full-Time Paid 5; Part-Time Paid 6; Part-Time Volunteers 200; Interns 6.
Governing Authority: nonprofit organization. Parent Institution: American Swedish Historical Foundation. Tax-exempt: 501(c)(3).

Institution Type/Description: Swedish-American History & Art Museum: located on 17th-century Queen Christina land grant.
Collections: Swedish colonial & immigrant experience in the U.S. from 1638; manuscripts.
Research Fields: history of Swedes & American Swedes.
Facilities: 6,000-vol. reference library; meeting rooms; banquet facilities. Gift items for sale.
Activities: tours; lectures; films; concerts; reading room; inter-museum loan, permanent, temporary & traveling exhibitions.
Publications: newsletter; occasional & exhibition catalogs.
Hours & Admission Prices: Tues.-Fri. 10-4, Sat.-Sun. 12-4. Adults $8, senior citizens & students $6, children 5-11 $4; discounts to AAM members; members & children under 5 no charge. Closed holidays. &
Attendance: 11,000 (accurate)
Membership: Individual $50; Household $65; Friend $125; Sustaining $250; Patron $500; Key Contributor $1,000.

ARTHUR ROSS GALLERY, UNIVERSITY OF PENNSYLVA-NIA, (M), 220 S. 34th St., Philadelphia, PA 19104-3808. Tel.: 215-898-2083 & 1479. Fax: 215-573-2045.
E-mail: arg@pobox.upenn.edu
Web Site: www.upenn.edu/ARG
Founded: 1983.
Congressional District: 1
Key Personnel: Dir., Lynn Marsden-Atlass; Assoc. Dir., Dejay B. Duckett; Gallery Coord., Sara Stewart.
Personnel Profile: Full-Time Paid 3.
Governing Authority: private university. Parent Institution: University of Pennsylvania, Philadelphia. Tax-exempt: 501(c)(3).
Institution Type/Description: Art Gallery: housed in a building designed by Frank Furness in 1891 as the University of Pennsylvania main library. A National Historical Landmark.
Collections: fine arts; historic & contemporary art; archaeological artifacts; textiles; sculpture.
Facilities: 1,700 sq. ft. exhibit space.
Activities: children's programs; lectures; symposia.
Publications: catalogues; annual exhibition schedule.
Hours & Admission Prices: Tues.-Fri. 10-5, Sat.-Sun. 12-5. No charge. Closed New Year's Day; Easter; Independence Day; Thanksgiving; Christmas. &
Attendance: 12,000 (estimated)
Membership: Contributing Friend $100-$999; Sustaining Friend $1,000 and up.

THE ATHENAEUM OF PHILADELPHIA, (M), 219 S. 6th St., E. Washington Square, Philadelphia, PA 19106-3794. Tel.: 215-925-2688. Fax: 215-925-3755.
E-mail: sltatman@philaathenaeum.org
Web Site: www.philaathenaeum.org
Founded: 1814.
Congressional District: 1
Key Personnel: Exec. Dir., Dr. Sandra L. Tatman; Chm. (V), Lea C. Sherk; Cur. Architecture, Bruce Laverty; Circulation Librarian, Jill L. Lee; Asst. Dir. Programs, Eileen M. Magee; Digital Center Supvr., Michael Senaca.
Personnel Profile: Full-Time Paid 8; Part-Time Paid 6; Part-Time Volunteers 6; Interns 6.
Governing Authority: nonprofit corporation. Tax-exempt: 501(c)(3) & 509(a)(1).
Institution Type/Description: Library with Art Collections: housed in 1845-1847, Athenaeum of Philadelphia (national historic landmark), John Notman (1810-1865), architect.
Collections: architecture & design; 300,000 architectural photographs; 250,000 architectural drawings, c.1790-1945.
Research Fields: 19th- & early 20th-century American architecture, literature, decorative & book arts.
Facilities: 75,000-vol. reference library; manuscripts available for use by qualified students upon application; architectural archives; reading room; Dorothy W. & F. Otto Haas Gallery.
Activities: lectures; inter-museum permanent & temporary exhibitions; provides research grants in American architectural history & building technology prior to 1860.
Publications: newsletter; booklist; annual report.
Hours & Admission Prices: Mon.-Fri. 9-5, 1st & 3rd Sat. each month 11-3. Research: by appointment. No charge; donations accepted. Closed bank holidays. &
Attendance: 20,000 (estimated)
Membership: Annual $150; Patron $200; Sustaining $300; Supporting $500; Benefactor $1,000.

AUTOMOTIVE MUSEUM SIMEONE FOUNDATION, 6825-31 Norwitch Dr., Philadelphia, PA 19153-3412. Tel.: 215-365-7233. Fax: 215-365-8230.
E-mail: amanda@simeonemuseum.org
Web Site: www.simeonemuseum.org
Formerly: Simeone Automotive Museum
Key Personnel: Exec. Dir., Fred Simeone; Operations Admin., Volunteers, Amanda Bartley; Facilities Rental, Maureen Moroney; Cur., Kevin Kelly; Retail Operations, Darryl Northington; Communications & Public Rels., Harry Hurst.
Governing Authority: Tax-exempt.
Institution Type/Description: Sports Museum.
Collections: racing cars.
Hours & Admission Prices: Tues.-Fri. 10-6, Sat.-Sun. 10-4. Adults $12, seniors $10, students $8; members & children under 8 no charge.

AWBURY ARBORETUM, Francis Cope House, One Awbury Rd., Philadelphia, PA 19138-1505. Tel.: 215-849-2855. Fax: 215-849-0213.
E-mail: awbury@awbury.org
Web Site: www.awbury.org
Founded: 1916.
Congressional District: 2
Key Personnel: Gen. Mgr., Christopher R. van de Velde; Bd. Pres. (V), Mark Sellers.
Personnel Profile: Full-Time Paid 2; Full-Time Volunteers 1; Part-Time Paid 15; Part-Time Volunteers 12; Interns 15.
Governing Authority: two nonprofit organizations. City Parks Association, Trustees. Parent Institution: Awbury Arboretum Association; Subsidiary Institution: City Parks Association. Tax-exempt: 501(c)(3) & 509(a)(1).
Institution Type/Description: Arboretum & Cultural Landscape: 1860 Francis Cope House.
Collections: 55-acre English landscaped park; noteworthy trees include beeches, river birch, bigleaf linden, alder, ginkgo, oaks & ailanthus; champion archives with Cope family records; arboretum plans & records; historic furnishings; plants; euphusage 19th-century plants typical to a Victorian country estate, as well as native species & local plant communities. Archives include correspondence, books & diaries of Cope family; 1,000 historic photographs of Awbury and its inhabitants; plans for the landscape and historic structures; records of the city Parks Association established in 1880. Maps.
Research Fields: nature education; horticulture; local history.
Facilities: nature & conservation center.
Activities: guided tours; formally organized education programs; adult job training in horticulture.
Publications: newsletter, The Arbor; A Walk though Awbury with Sir Peter Shepheard; A History of Awbury; Awbury Memories; For Emancipation and Education: Black and Quakes,1680-1900.
Hours & Admission Prices: Grounds: daily dawn-dusk. House: Mon.-Fri. 9-5. No charge; donations accepted. &
Attendance: 10,000 (estimated)
Membership: Senior Citizen, Student & Individual $25; Family $50; Supporting $500; Sustaining $1,500.

BARNES FOUNDATION, (M), 2025 Benjamin Franklin Pkwy., Philadelphia, PA 19103. Tel.: 215-278-7000. Fax: 610-664-4026.
E-mail: info@barnesfoundation.org
Web Site: www.barnesfoundation.org
Founded: 1922.
Congressional District: 13
Key Personnel: Dir., Derek Gillman; Chm., Dr. Bernard Watson; Museum Shop Mgr., Julie Steiner.
Personnel Profile: Full-Time Paid 93; Full-Time Volunteers 73; Part-Time Paid 37; Part-Time Volunteers 73; Interns 2.
Governing Authority: Tax-exempt.
Institution Type/Description: Art Foundation: 18th-century historic site.
Collections: impressionist; post-impressionist; early modern; old masters; African sculpture; Native American art; American decorative art.
Major Exhibits: Yinka Shonibare: Magic Ladders, 1/25/14-4/28/14; The World is an Apple: The Still Lives of Paul Cezanne (T), 1/14/14-9/22/14; William Glackens (T), 11/8/14-2/2/15.
Facilities: art gallery; changing exhibition space; art library, restaurant; cafe; auditorium; classrooms.
Activities: ongoing classes in art appreciation & horticulture.
Publications: books, Great French Paintings; The Art of Renoir; The Art of Matisse; The Art of Painting; CD-ROM, A Passion for Art; The Barnes Foundation Masterworks; Renoir in the Barnes Foundation; Ellsworth Kelly: Sculpture on the Wall.

Hours & Admission Prices: Wed.-Mon. 10-6. Adults $22; children under 3 & members no charge. &

Attendance: 300,000 (accurate)

Membership: Student $45; Patron $90-$149; Contributor $150-$249; Supporter $250-$499; Sustainer $500-$999; Circle Member $1,000-$4,999; Circle Patron $5,000-$9,999; Directors Circle $10,000-$24,999.

BARTRAM'S GARDEN, 54th St. & Lindbergh Blvd., Philadelphia, PA 19143. Tel.: 215-729-5281. Fax: 215-729-1047.

E-mail: info@bartramsgarden.org

Web Site: www.bartramsgarden.org

Founded: 1893.

Congressional District: 1

Key Personnel: Pres., John Bartram Association, Elizabeth Bressi-Stoppe; Exec. Dir., Maitreyi Roy; Communications Mgr., Kim Massare.

Governing Authority: nonprofit organization. Affiliated with the John Bartram Assoc. Parent Institution: Fairmount Park. Tax-exempt: 501(c)(3); 170(b)(1)(A).

Institution Type/Description: Historic House & Botanical Garden: c.1700 Bartram House.

Collections: 18th & early 19th-century furnishings & decorative arts that reflect the Bartram tenancy; early Bartram related books; manuscripts; publications, horticultural & ornithological works. Historic Buildings: outbuildings; stable; sheds; barn; icehouse; seed house; stone cider mill.

Research Fields: 18th & 19th-century horticultural & botanical practices; the Bartram family.

Facilities: library of books, papers, articles that relate to the Bartrams & horticulture in the 18th & 19th centuries available for research upon application to the John Bartram Assoc.; botanical garden. Museum-related items for sale.

Activities: guided tours; lectures; formally organized educational programs; loan, permanent, temporary & traveling exhibitions.

Publications: Bartram Broadside; Bartram Leaf.

Hours & Admission Prices: April-Dec. Tues.-Sun. 10-4. Adults $12, senior citizens, students & children 16 & under $10; discounts to AAM members; children under 2 and members no charge. Closed holidays.

Attendance: 40,000 (estimated)

Membership: Gardener $40; Seed Saver $60; Family $75; Family & Friends $120; Botanist $275; Explorer $500; Collector $1,000; Sponsor $2,500.

BETSY ROSS HOUSE, 239 Arch St., Philadelphia, PA 19106-1999. Tel.: 215-686-1252. Fax: 215-686-1256.

E-mail: lisa@betsyrosshouse.org

Web Site: www.betsyrosshouse.org

Founded: 1898.

Congressional District: 1

Key Personnel: Dir., Lisa Moulder; Chm., Wayne Spilove; Mgr. Collections, Michelle Budeni; Facilities Mgr., Frank Fisher; Public Rels., Heather Kincaid; Museum Shop Mgr., Ashley Baney.

Personnel Profile: Full-Time Paid 6; Part-Time Paid 10; Part-Time Volunteers 3; Interns 2.

Governing Authority: private; nonprofit organization. Parent Institution: Historic Philadelphia, Inc. Tax-exempt: 501(c)(3).

Institution Type/Description: Historic House: 1773-1786 home of Betsy Ross, seamstress of the first American flag.

Collections: craftsman's tools; furnishings; personal artifacts; restored 18th-century upholstery shop.

Research Fields: 18th-19th centuries historic Philadelphia; history of the Betsy Ross house; Betsy Ross.

Facilities: courtyard. Museum-related items for sale.

Activities: concerts. Museum Sponsors: playlets & reenactments June-August; Flag Day celebration; Independence Day activities.

Hours & Admission Prices: April to mid-Oct. daily 10-5; mid-Oct. to March Tues.-Sun. 10-5. Adults & seniors $3, children & students $2; discounts to AAM & ICOM members; members no charge. Audio Guide, includes admission $4. Closed New Year's Day; Thanksgiving; Christmas. &

Attendance: 315,459 (accurate)

CAROLYN-FIEDLER-ALBER GALLERY AT ALLENS LANE ART CENTER, 601 W. Allens Lane, Philadelphia, PA 19119. Tel.: 215-248-0546. Fax: 215-248-0559.

Institution Type/Description: Art Gallery.

Collections: paintings; sculpture; photographs.

Hours & Admission Prices: Mon.-Fri. 10-5; other times by appointment.

CHESTNUT HILL HISTORICAL SOCIETY, 8708 Germantown Ave., Philadelphia, PA 19118-2717. Tel.: 215-247-0417. Fax: 215-247-9329.

E-mail: info@chhist.org

Web Site: www.chhist.org

Key Personnel: Dir., Jennifer S. Hawk.

Personnel Profile: Full-Time Paid 1; Part-Time Paid 4; Part-Time Volunteers 12; Interns 1.

Institution Type/Description: Historical Society Museum: housed in an 1870s Victorian house.

Collections: local history & culture; genealogy; architectural drawings & building records; photographs; maps; prints; drawings.

Activities: educational programs; tours.

Hours & Admission Prices: Tues.-Fri. 9:30-2:30, Sat. 11-4 by appointment. Research $15; members no charge.

Attendance: 800 (estimated)

Membership: General $50; Archivist $100; Sustainer $250.

THE CLAY STUDIO GALLERY, 139 N. 2nd St., Philadelphia, PA 19106. Tel.: 215-925-3453. Fax: 215-925-7774.

E-mail: info@theclaystudio.org

Web Site: www.theclaystudio.org

Institution Type/Description: Art Gallery.

Collections: ceramic arts; sculpture.

Facilities: Gallery-related items for sale.

Activities: educational programs; classes; workshops.

Hours & Admission Prices: Tues.-Sat. 11-7, Sun. 12-6.

CLIVEDEN, 6401 Germantown Ave., Philadelphia, PA 19144. Tel.: 215-848-1777. TDD: 215-848-1777.

E-mail: info@cliveden.org

Web Site: www.cliveden.org

Founded: 1972.

Congressional District: 201

Key Personnel: Exec. Dir., David W. Young; Bd. Chair, Jennifer Celata; Deputy Dir. Devel., Anne Roller.

Personnel Profile: Full-Time Paid 2; Part-Time Paid 15; Interns 2.

Governing Authority: nonprofit organization. Operated by Cliveden, Inc. Parent Institution: The National Trust for Historic Preservation, 1785 Massachusetts Ave., N.W., Washington, DC 20036. Tax-exempt: 501(c)(3).

Institution Type/Description: Historic House Museum: 1763-1767 residence built by Benjamin Chew & site of the Battle of Germantown, Oct. 4, 1777.

Collections: Philadelphia Chippendale furniture; Chew family memorabilia & papers; 18th to 19th-century decorative arts. Historic House: c.1798 Upsala home.

Research Fields: Chew Family; decorative arts; Germantown; historic preservation; Battle of Germantown; Philadelphia social history.

Activities: guided tours; community events. Museum Sponsors: Revolutionary Germantown Festival.

Publications: e-newsletter.

Hours & Admission Prices: April-Dec. Thurs.-Sun. 12-4; other times by appointment. Adults $10, students $6; discounts to AAA & Mobil travel members; members & Friends of Cliveden & National Trust no charge. Closed New Year's Day; Easter; Thanksgiving; Christmas. &

Attendance: 60,000 (estimated)

Membership: Individual $35; Family $60; Sponsor $250.

THE DESIGN CENTER AT PHILADELPHIA UNIVERSITY, 4200 Henry Ave., Philadelphia, PA 19144. Tel.: 215-951-5338. Fax: 215-951-2662.

E-mail: thedesigncenter@philau.edu

Web Site: www.philau.edu/designcenter

Formerly: The Design Center at Philadelphia University

Founded: 1978.

Congressional District: 2

Key Personnel: Dir., Hilary Jay; Asst. Dir., Carla Bednar; Cur. Collections, Nancy Packer.

Personnel Profile: Full-Time Paid 2; Part-Time Paid 6; Part-Time Volunteers 2; Interns 2.

Governing Authority: college; nonprofit organization. Parent Institution: Philadelphia University. Tax-exempt: 501(c)(3).

Institution Type/Description: Design Museum.

Collections: international textiles dating from 4th-century A.D. to present; 200,000 indexed swatches; fashion; costumes; clothing accessories; interior fabrics; historic textiles; quilts; lace 16th-century to present; fringe collection; color swatch cards; textile tools & implements; dye sample books.

Research Fields: history of textiles; textile industry; historic clothing; international textiles & costumes; lace; color science; design; Philadelphia history.

Facilities: 500-vol. library.

Activities: guided tours; lectures; loan & temporary exhibitions; outreach materials.

Publications: Florabunda, The Evolution of Floral Design on Fabric; The

Philadelphia System of Textile Manufacture, 1884-1984; Anything Goes: Textiles of the 20s & the 30s; The Art of the Textile Blockmaker; The Bauhaus Weaving Workshop: Source & Influence for American Textiles; Scalamandre: Preserving America's Textile Heritage, 1929 1989; The Art of African Textiles: Form and Function, 1995; Le Corbusier: Inside the Machine for Living 2001; Black & White 2002; What is Design Today? 2002.

Hours & Admission Prices: Galleries: Mon.-Fri. 10-4. Textile Collection: by appointment. No charge. Closed major & school holidays. &

Attendance: 6,000 (estimated)

EASTERN STATE PENITENTIARY HISTORIC SITE, (M), 22nd & Fairmount Ave., Philadelphia, PA 19130. Mailing Address: 2027 Fairmount Ave., Philadelphia, PA 19130. Tel.: 215-236-3300 & 5111. Fax: 215-236-5289.

E-mail: info@easternstate.org
Web Site: www.easternstate.org
Key Personnel: Exec. Dir., Sara Jane Elk; Program Dir., Sean Kelley; Asst. Program Dir. Operations & Special Events, Brett Bertolino
Institution Type/Description: Historic Site.
Collections: artifacts; inmate made crafts.
Hours & Admission Prices: Daily 10-5. Adults $12, senior citizens $10, students & children 8-17 $8; children under 7 not admitted. Closed New Year's Eve & Day; Easter; Thanksgiving; Christmas Eve & Day.
Attendance: 190,000 (estimated)

THE EBENEZER MAXWELL MANSION, INC., (M), 200 W. Tulpehocken St., Philadelphia, PA 19144-3210. Tel.: 215-438-1861. Fax: 215-438-0133.

E-mail: emaxwellmansion@yahoo.com
Web Site: www.ebenezermaxwellmansion.org
Founded: 1975.
Congressional District: 2
Key Personnel: Exec. Dir., Diane S. Richardson; Pres., Susan B. Harrison.
Personnel Profile: Full-Time Paid 1; Part-Time Paid 1; Part-Time Volunteers 80.
Governing Authority: nonprofit. Tax-exempt: 501(c)(3).
Institution Type/Description: Historic House: 1859 Victorian Villa.
Collections: period rooms, furnishings & gardens reflecting mid- & late Victorian life of 1850-1890; hands-on exhibits.
Research Fields: 19th-century architecture; decorative arts; lifestyles; technical restoration information.
Facilities: 250-vol. library pertaining to Victorian history, artifacts & restoration processes; educational facilities; period gardens.
Activities: guided tours; lectures; organized educational programs; docent program; temporary exhibitions; workshops; classes. Museum Sponsors: Old Fashioned Picnic in summer; Murder Mystery in October; Dickens Christmas Party in December.
Publications: brochure; semiannual newsletter; annual report; Max Fax; garden brochure; garden book.
Hours & Admission Prices: Thurs.-Sat. 12-4; groups by appointment. Adults $6; members no charge. Closed national holidays.
Attendance: 980 (accurate)
Membership: Mansion Friend $40; Mansion Sponsor $60; Business Sponsor $75; Mansion Conservator $125; Mansion Sustainer $200; Mansion Preservation Circle $500; Mansion Benefactor $1,000.

EDGAR ALLAN POE NATIONAL HISTORIC SITE, 532 N. Seventh St., Philadelphia, PA 19123. Mailing Address: 143 S. 3rd St., Philadelphia, PA 19106-2818. Tel.: 215-597-8780 & 8787. Fax: 215-861-4950. TDD: 215-597-8780.

Web Site: www.nps.gov/edal
Founded: 1978.
Congressional District: 3
Key Personnel: Mgr., Patricia Jones; Cur., Karie Diethorn.
Personnel Profile: Full-Time Paid 3; Part-Time Paid 4; Part-Time Volunteers 6.
Governing Authority: federal. Parent Institution: Independence National Historical Park, 143 S. 3rd St., Philadelphia, PA 19106. Tax-exempt.
Institution Type/Description: Historic Site: c.1843 Edgar Allan Poe brick house.
Collections: Poe's life & history; personal artifacts; period furnishings.
Research Fields: Poe's life; literary achievements.
Facilities: reference library of secondary materials on Edgar Allan Poe available for research on premises; reading room; theater. Books, tapes, cards on Edgar Allan Poe for sale.
Activities: Ranger guided tours offered with special theme events; lectures; films; formally organized education programs for children & college students; permanent exhibitions; Junior Ranger activities. Annual Events: Programs commemorating Poe's birth & death in January & October;

Poetry in April; Mystery in July; science fiction in August; Ghostly Grip Horror Tales in October.
Publications: park folder; teacher's handbook.
Hours & Admission Prices: Check website for hours.

ELFRETH'S ALLEY ASSOCIATION, 126 Elfreths Alley, Philadelphia, PA 19106-2006. Tel.: 215-574-0560. Fax: 215-922-7869.

E-mail: information@elfrethsalley.org
Web Site: www.elfrethsalley.org
Founded: 1934.
Congressional District: 1
Key Personnel: Pres., Erik William; Exec. Dir., Dena Ferrara.
Personnel Profile: Full-Time Paid 1; Part-Time Paid 3; Part-Time Volunteers 5; Interns 3.
Governing Authority: Tax-exempt.
Institution Type/Description: Historic House.
Collections: 18th-century & early 19th-century historic structures; historic street featuring 28 buildings interpreting urban life; archival documents; 18-century homewares & furniture.
Hours & Admission Prices: Tues.-Sat. 10-5, Sun. 12-5. Adults $5, children 6-18 & school groups $1; discount to AAM & ICOM members; Philadelphia schools and children under 6 no charge.
Attendance: 44,300 (accurate)

ESTHER M. KLEIN ART GALLERY AT UNIVERSITY CITY SCIENCE CENTER, 3600 Market St., Philadelphia, PA 19104-2641. Mailing Address: 3711 Market St., Ste. 800, Philadelphia, PA 19104-5504. Tel.: 215-966-6188. Fax: 215-966-6001.

E-mail: kleinart@sciencecenter.org
Web Site: www.kleinartgallery.org
Founded: 1976.
Congressional District: 2
Key Personnel: C.E.O. & Pres., Dr. Stephen Tang; Dir., Dan Schimmel; Cur., David Clayton.
Personnel Profile: Part-Time Paid 2; Part-Time Volunteers 10; Interns 2.
Governing Authority: private; nonprofit organization. Parent Institution: University City Science Center. Tax-exempt: 501(c)(3).
Institution Type/Description: Art Gallery.
Collections: paintings; sculpture; photography; styles range from realism to abstraction, folk art to traditional.
Research Fields: development of new & creative technologies.
Facilities: library of exhibition catalogues; 60 seat auditorium; 1,000 sq. ft. exhibit space.
Activities: art education programs for students & adults; arranged tours; continuing changing exhibitions by well-known & emerging artists; multicultural exhibitions; Art-in-Science & community service exhibitions; gallery talks; slide & panel discussions; education programs; studio workshops.
Publications: catalogues; brochures.
Hours & Admission Prices: Mon.-Sat. 9-5. No charge; donations accepted. &
Attendance: 100,000 (estimated)

THE FABRIC WORKSHOP AND MUSEUM, (M), 1214 Arch St., Philadelphia, PA 19107-2800. Tel.: 215-561-8888. Fax: 215-561-8887.

E-mail: info@fabricworkshopandmuseum.org
Web Site: www.fabricworkshopandmuseum.org
Founded: 1977.
Congressional District: 1
Key Personnel: Artistic Dir. & Founder, Marion Boulton Stroud; Pres. (V), Katherine Sokolnikoff; Project Technician/Printer, Jenn McTague; Asst. to Directors, Michele Bregande; Museum Shop Mgr., Tracey Blackman; Project Coord., Andrea Landau; Master Printer & Project Coord., Mary Anne Friel; Videographer, Tyler Henry; Master Printer, Virgil Marti; Project Technician/Printer, Tim Eads; Museum Education Coord., Sophie Sanders; Head Education, Apprentice Coord. & Master Printer, Christina Roberts; Devel. Assoc., Amy Cretaro; Coord. High School Apprentice, Lucy Lau Bigham.
Personnel Profile: Full-Time Paid 13; Part-Time Paid 7.
Governing Authority: nonprofit organization. Tax-exempt: 501(c)(3).
Institution Type/Description: Contemporary Art Museum & Studio focusing on work in new material & new media.
Collections: contemporary art produced in the Artist-in-Residence program, including print multiples, video, monoprints, sculptural objects, installations, performance costumes, furniture, functional objects, preliminary drawings, paintings, related ceramics & sculpture; photographs; sculpture.
Facilities: 4,500 sq. ft. exhibit space. Museum-related items for sale.

Activities: artist-in-residence program; lectures; guided tours; organized educational programs; loan, temporary & traveling exhibitions; apprentice program; video programs; slide library.

Publications: An Industrious Art: Innovation in Pattern & Print; A Decade of Pattern; Donald Lipski: Who's Afraid of Red, White & Blue; Mel Chin: Soil and Sky; Carrie Mae Weems; Jorge Pardo; Glen Ligon and Gary Simmons; Comfort Zone; New Material as New Media; Lee Bul: Live Forever; On the Wall; Swarm; Doug Aitken A-Z Book (Fractals); RN The Past, Present and Future of Nurses' Uniform; Experiments with the Truth; Will Stokes, Jr.; Ed Ruscha: Industrial Strength; Cai Guo-Qiang: Fallen Blossoms.

Hours & Admission Prices: Mon.-Fri. 10-6, Sat.-Sun. 12-5; groups by appointment. Adults $3; members no charge. Closed federal holidays. &

Attendance: 10,000 (estimated)

Membership: Artist $20; Individual $35; Family $45; Patron $100; Sustaining $250; Friend $500; Benefactor $1,000; Director's Circle $2,000.

FAIRMOUNT PARK HORTICULTURE CENTER, 100 N. Horticultural Dr., West Park, Philadelphia, PA 19131. Mailing Address: One Parkway, 10th Fl., 1515 Arch St., Philadelphia, PA 19102-1511. Tel.: 215-685-0096. Fax: 215-685-0103.

Web Site: www.fairmountpark.org/hortcenter.asp
Formerly: The Horticulture Center
Founded: 1979.
Congressional District: 3
Key Personnel: Exec. Dir., Mark A. Focht; Commissioner, Michael DiBerardinis; District Mgr., Lori Hayes.
Personnel Profile: Full-Time Paid 19; Part-Time Volunteers 15.
Governing Authority: municipal. Parent Institution: Fairmount Park Commission. Tax-exempt.
Institution Type/Description: Park Museum.
Collections: trees of Asian & North American origin.
Facilities: greenhouses; 32,000 sq. ft. exhibit space; 75-seat meeting room.
Activities: guided tours; lectures; hobby workshops.
Hours & Admission Prices: Display House: daily 9-3. Donation: adults $2. Grounds: daily 9-5. Closed national holidays. &
Attendance: 20,000 (estimated)

FIREMAN'S HALL MUSEUM, 147 N. 2nd St., Philadelphia, PA 19106-2097. Tel.: 215-923-1438. Fax: 215-923-0479.

E-mail: firemus@aol.com
Formerly: Philadelphia Fire Museum
Founded: 1967.
Congressional District: 3
Key Personnel: Pres., Charles M. Lillie; Museum Shop Mgr., Henry J. Magee.
Personnel Profile: Full-Time Paid 2; Part-Time Paid 4; Part-Time Volunteers 14; Interns 4.
Governing Authority: municipal government; nonprofit. Parent Institution: Philadelphia Fire Department; Subsidiary: Philadelphia Fire Department Historical Corp. Tax-exempt.
Institution Type/Description: Fire-Fighting Museum: housed in 1900 Philadelphia firehouse.
Collections: fire-fighting equipment & apparatus; tools; uniforms; prints; photographs; models; fire house furniture; fire marks.
Research Fields: history of fire fighting in Philadelphia & surrounding area.
Facilities: 1,000-vol. library of books, reports, log, ledgers & journals, available for use by public by appointment. Gift items for sale.
Activities: guided tours; lectures; films; organized educational programs.
Hours & Admission Prices: Tues.-Sat. 10-4:30. No charge; donations accepted. Closed city holidays. &
Attendance: 26,000
Membership: Individual $20; Family $30; Patron $100; Corporate $250-$5,000.

FORT MIFFLIN ON THE DELAWARE, Fort Mifflin & Hog Island Rds., Philadelphia, PA 19153-3990. Tel.: 215-685-4167. Fax: 215-685-4166.

Web Site: www.fortmifflin.us
Founded: 1984.
Congressional District: 1
Key Personnel: Exec. Dir., Elizabeth Beatty.
Personnel Profile: Full-Time Paid 3; Part-Time Paid 7; Part-Time Volunteers 200.
Governing Authority: municipal; nonprofit organization. Tax-exempt: 501(c)(3).
Institution Type/Description: Historic Site: 1777 fort.
Collections: 13 historic buildings; moat; cannon; maps.
Facilities: 3,000 sq. ft. exhibit space. Gift items for sale.
Activities: guided tours; lectures; weapons demonstrations; organized educational programs; docent program; Soldier Life; Revolutionary War & Civil War interpretation.
Publications: Ft. Mifflin: Valiant Defender; Fort Quarterly.
Hours & Admission Prices: April-Nov. Wed.-Sun. 10-4; other times by appointment. Adults $6, senior $5, children 6-12 $3; members and children 5 & under no charge. &
Attendance: 16,000 (accurate)
Membership: Student $30; Individual $40; Family $50.

＊ THE FRANKLIN INSTITUTE, (M), 222 N. 20th St., Philadelphia, PA 19103-1190. Tel.: 215-448-1200. Fax: 215-448-1109.

E-mail: dwint@fi.edu
Web Site: www.fi.edu
Founded: 1824.
Congressional District: 2
Key Personnel: Pres. & C.E.O., Dennis M. Wint, Ph.D.; Chm. (V), Marsha R. Perelman; Sr. Vice Pres. External Affairs, Larry Dubinski; Sr. Vice Pres. Science, Frederic Bertley, Ph.D.; Vice Pres. Operations & Capital Projects, Richard Rabena; Vice Pres. Human Resources, Reid Styles; Vice Pres. Devel., Marisa Wigglesworth; Sr. Vice Pres. Programs, Mktg. & Business Devel., Troy Collins.
Personnel Profile: Full-Time Paid 141; Part-Time Paid 163; Part-Time Volunteers 647; Interns 59.
Volunteer Hours: 72,888
Operating Expenses: 32,412,198
Operating Income: 33,959,880
Governing Authority: nonprofit organization. Tax-exempt: 501(c)(3).
Institution Type/Description: Science & Technology Museum, Planetarium & Omniverse Theater.
Collections: science; history; industry; technology; transportation; marine; naval; aeronautics; astronomy; space exploration; energy; mathematics; stamps & coins; Ben Franklin National Memorial.
Major Exhibits: One Day in Pompeii (T), 11/13-4/14; Circus: Science Under the Big Top (T), 6/14-9/14; 101 Inventions That Changed the World (T), 6/14-10/14; Body Worlds: Animals Inside Out, 11/14-4/15.
Research Fields: education; science; technology.
Facilities: 270-seat auditorium; 140-seat theatre for science demonstrations; 200-seat planetarium; 340-seat Omniverse/IMAX theater; 75-seat discovery theater; children's museum; classrooms; video conferencing center. Books, educational games & museum-related items for sale.
Activities: lectures; workshops; formally organized education programs for children; films; docent program or council; inter-museum loan, permanent, temporary & traveling exhibitions; traveling science shows; special educational programs with local school districts; museum-to-go; science activity kit & teacher training outreach programs; scout & group camp-in overnight programs.
Publications: bimonthly, Journal of The Franklin Institute; quarterly members' newsletter.
Hours & Admission Prices: Daily 9:30-5. IMAX Theater: Sun.-Thurs. 9:30-6, Fri.-Sat. 9:30-9. Adults $15.50, senior citizens, students & military $14.50, children $12; discounts to AAM & ASTC members; members no charge. Closed New Year's Day; Thanksgiving; Christmas Eve & Day. &
Attendance: 890,810 (accurate)
Membership: Individual $50; Family $99; Family Max $155; Premier $200.

FRED WOLF, JR. GALLERY/KLEIN BRANCH JEWISH COMMUNITY CENTER, 10100 Jamison Ave., Philadelphia, PA 19116-3832. Tel.: 215-698-7300. Fax: 215-673-7447.

E-mail: pactman@phillyjcc.com
Founded: 1975.
Congressional District: 3
Key Personnel: Dir. Klein Branch, Andre Krug; Co-Chair, Gary B. Freedman; Co-Chair, Max M. Berger; Dir. Art Gallery, Ruth Morley; Dir. Public Rels. & Mktg., Alison Polsky; Security, Ray Graham.
Personnel Profile: Full-Time Paid 1; Part-Time Volunteers 4.
Governing Authority: private; nonprofit Jewish community center. Parent Institution: Jewish Community Centers of Greater Philadelphia, Philadelphia, PA. Tax-exempt: 501(c)(3).
Institution Type/Description: Art and Arts & Crafts Museum.
Collections: Judaic artifacts from the Obermayer collection dating back to the 18th century in the US; prints, paintings, photographs, & sculptures from the early 1900s to the present; millennium exhibit of Judaica.
Facilities: 250-seat auditorium; 150 sq. ft. exhibit space; 400-seat theater; educational facilities; restaurant. Museum-related items for sale.
Activities: lectures; formal education programs for adults; opening receptions; gallery talks; temporary exhibitions.
Publications: newsletter of programs, Impact; catalogue of events, Directions.
Hours & Admission Prices: Mon.-Thurs. 9-9, Fri. & Sun. 9-4. No charge. Closed Jewish holidays. &
Attendance: 45,000 (estimated)

Membership: Jewish Community Center: call for fees. Fred Wolf, Jr. Gallery: no charge.

FREE LIBRARY OF PHILADELPHIA RARE BOOK DEPART-MENT, 1901 Vine St., 3rd Fl., Philadelphia, PA 19103-1189. Tel.: 215-686-5416. Fax: 215-563-3628.
E-mail: dewaltj@library.phila.gov
Web Site: www.freelibrary.org
Founded: 1891.
Key Personnel: Dept. Head, Jim DeWalt.
Personnel Profile: Full-Time Paid 5.
Governing Authority: municipal. Tax-exempt: 170(b)(1)(A).
Institution Type/Description: Public Library.
Collections: cuneiform tablets; European & Oriental manuscripts; incunabula; Pennsylvania German fraktur & imprints; early American children's books; Beatrix Potter, Kate Greenaway, Arthur Rackham, Charles Dickens and Palmer Cox collections; Carson collection on the Development of the Common Law; Gimbel collection of works of Edgar Allan Poe; Robert Lawson & Munro Leaf collections; Thornton Oakley collection of Howard Pyle and His Students; Biddle collection of Horace; Elkins collection of Americana and works of Oliver Goldsmith; Strouse collection of Letters of the Presidents; American Sunday School Union collection; Brewster collection of A.B. Frost.
Facilities: reading room; 400-seat auditorium.
Activities: lectures; temporary exhibitions.
Publications: books.
Hours & Admission Prices: Mon.-Fri. 9-5. &

THE GALLERIES AT MOORE, (M), 20th St. & The Parkway, Philadelphia, PA 19103. Tel.: 215-965-4027. Fax: 215-568-5921. Facebook: The Galleries at Moore.
E-mail: galleries@moore.edu
Web Site: www.thegalleriesatmoore.org
Founded: 1984.
Congressional District: 2
Key Personnel: Dir. & Chief Cur., Kaytie Johnson; Gallery Mgr., Gabrielle Lavin; Coord. Outreach & Programs, Elizabeth Gilly.
Personnel Profile: Full-Time Paid 3; Part-Time Paid 6.
Governing Authority: college. Parent Institution: Moore College of Art and Design. Tax-exempt: 501(c)(3).
Institution Type/Description: Art Gallery.
Collections: works by regional, national & international artists and designers.
Major Exhibits: Pretty Vacant: Punk/Post Punk Design, 1/25/14-3/15/14.
Research Fields: contemporary art & design.
Activities: guided tours; lectures; films; gallery talks; symposia; temporary & traveling exhibitions.
Publications: exhibition catalogues.
Hours & Admission Prices: Mon.-Thurs. & Sat. 11-5, Fri. 11-8. No charge; donations accepted. Closed all academic & legal holidays. &
Attendance: 49,500 (estimated)

THE GENEALOGICAL SOCIETY OF PENNSYLVANIA, 2207 Chestnut St., Philadelphia, PA 19103. Tel.: 215-545-0391. Fax: 215-545-0936.
E-mail: execdir@genpa.org
Web Site: www.genpa.org
Founded: 1892.
Congressional District: 1
Key Personnel: Pres., Claire Keenan Agthe; Administrative Mgr., Joyce Homan.
Personnel Profile: Full-Time Paid 1; Part-Time Paid 2; Part-Time Volunteers 25.
Governing Authority: bd. dirs. Tax-exempt: 501(c)(3).
Institution Type/Description: Genealogical Society Library.
Collections: genealogies; manuscripts; family, church & civil records.
Facilities: library.
Activities: lectures.
Publications: semiannual magazine, The Pennsylvania Genealogical Magazine; quarterly newsletter, Penn In Hand; 2-3 yearly monographs & special publications.
Hours & Admission Prices: Call for hours. &
Attendance: 900 (accurate)
Membership: Student, Teacher & Senior $50; Research $60; Patron $120; Philadelphia $500.

GRAND ARMY OF THE REPUBLIC MUSEUM & LIBRARY, 4278 Griscom St., Philadelphia, PA 19124-3954. Tel.: 215-289-6484 & 673-1688.
E-mail: garmuslib@verizon.net
Web Site: www.garmuslib.org
Founded: 1926.
Key Personnel: Dir., Elmer F. Atkinson; Sec., Michael E. Peter.
Personnel Profile: Part-Time Volunteers 20.
Governing Authority: nonprofit organization. Tax-exempt: 501(c)(3).
Institution Type/Description: Civil War Museum: housed in 1796 late Georgian-style house.
Collections: Civil War items, including guns, swords, drums, letters & pictures; war-related books.
Research Fields: Civil War historical & genealogical records.
Facilities: 3,000-vol. library pertaining to the Civil War for research on premises; 4,200 sq. ft. exhibit space.
Activities: guided tours; lectures; loan exhibitions; community outreach programs.
Publications: triannual newsletter.
Hours & Admission Prices: Tues. 12-4, first Sun. of month 12-5; other times by appointment. No charge; donations accepted. &
Attendance: 3,200
Membership: Individual $20; Family $30; Colonel's Guard $40; General's Staff $50.

HISTORIC GERMANTOWN, 5501 Germantown Ave., Philadelphia, PA 19144-2291. Tel.: 215-844-1683. Fax: 215-844-2831.
E-mail: programs@freedomsbackyard.com
Web Site: www.freedomsbackyard.com
Formerly: Germantown Historical Society
Founded: 1900.
Congressional District: 1
Key Personnel: Exec. Dir., Barbara Hogue; Pres. (V), Mark R. Sellers; Librarian & Archivist, Alex Bartlett.
Personnel Profile: Full-Time Paid 1; Part-Time Paid 3; Part-Time Volunteers 30; Interns 2.
Governing Authority: corporation; nonprofit. Tax-exempt.
Institution Type/Description: Local History Museum & Library.
Collections: 17th- to 20th-century artifacts relating to Germantown, Mt. Airy & Chestnut Hill (northwest Philadelphia); decorative arts; costumes; textiles; dolls; toys; household equipment; photographs spanning 300 years of Germantown's history.
Research Fields: genealogy; local railroads; industrial, commercial & architectural history; history of local organizations; gardening.
Facilities: 3,000-vol. library of local historical books; over 5,000 manuscripts, photographs & prints.
Activities: guided tours; lectures; seminars; school programs; facility rental.
Publications: magazine, Germantown Crier; pamphlets; small books on topics of local history; brochures; newsletter, Germantown Now and Then.
Hours & Admission Prices: Tues. 9-1, Thurs. 1-5, 1st & 3rd Sun. each month by appointment. Adults $10, students $5; discounts to AAM, AAA, ICOM & Philadelphia card members; members no charge. &
Attendance: 4,500 (estimated)
Membership: Individual $35; Household $50; History Hunter $100; Concord Club $250; Heritage Circle $500; Preservationist Guild $1,000.

HISTORICAL SOCIETY OF PENNSYLVANIA, 1300 Locust St., Philadelphia, PA 19107-5661. Tel.: 215-732-6200. Fax: 215-732-2680.
E-mail: library@hsp.org
Web Site: www.hsp.org
Founded: 1824.
Congressional District: 1
Key Personnel: Pres., Page Talbott; Dir. Library, Lee Arnold; Chm. (V), Bruce Fenton.
Personnel Profile: Full-Time Paid 27; Part-Time Paid 16; Part-Time Volunteers 9; Interns 8.
Governing Authority: society. Tax-exempt: 501(c)(3).
Institution Type/Description: Historical Research Library.
Collections: 21 million manuscripts, books, prints, drawings, maps & photographs; Balch Institute for Ethnic Studies and The Genealogical Society of Pennsylvania collections.
Research Fields: political, commercial & social history of the American colonies; mid-Atlantic & Southern states prior to the Civil War; Philadelphia, Commonwealth of Pennsylvania & Delaware Valley from colonial period to the present; immigration & ethnic studies.
Facilities: library of nongovernmental repositories of documentary materials including 560,000 books, 312,000 graphic works, and 20 million manuscripts renowned for its 17th-, 18th- & 20th-century holdings on Philadelphia & southeastern Pennsylvania collections are used for historical & genealogical research.
Activities: conferences; orientation to research; lectures.
Publications: Guide to the Manuscript Collections of the Historical Society of

Pennsylvania; quarterly journal, The Pennsylvania Magazine of History & Biography; magazine, Pennsylvania Legacies; video, All Aboard for Philadelphia; members' newsletter, Sidelights.
Hours & Admission Prices: Tues. & Thurs. 12:30-5:30, Wed. 12:30-8:30, Fri. 10-5:30. Library: adults $8, college students w/valid ID $5; high school students no charge. Closed national holidays. &
Attendance: 6,046 (accurate)
Membership: Research $75; Patron $130; Patron Plus $250; Contributor $500.

* **INDEPENDENCE NATIONAL HISTORICAL PARK, (M),** 143 S. 3rd St., Philadelphia, PA 19106-2818. Tel.: 215-597-8787. Fax: 215-861-4950.
Web Site: www.nps.gov/inde
Founded: 1948.
Congressional District: 3
Key Personnel: Supt., Cynthia MacLeod; Chief Cur., Karie Diethorn; Archivist & Library Mgr., Christian Higgins.
Governing Authority: federal. Parent Institution: National Park Service. Tax-exempt.
Institution Type/Description: General Museum.
Collections: decorative arts; history; archaeology; American portraits from 1770-1830 by Peale, Sharples, West, Sully, Pine; 18th century American period furnishings; objects and documents relating to Independence Hall, Congress Hall; personal memorabilia relating to Revolution and early Federal period; archives; military; graphics. Historic Buildings and Sites: 1732-1800 Independence Hall; Liberty Bell Pavilion; 1790-1800 Congress Hall; 1791-1800 Old City Hall; Second Bank Portrait Gallery; 1722-1790 Franklin Court; 1787-1836 Bishop White House; 1791-1793 Todd House; 1774-1800 City Tavern; 1775-1805 New Hall Military Museum; 1797-1798 Kosciuszko House; 1777 Graff House; 1772-1794 Deshler-Morris House.
Research Fields: decorative arts; history.
Facilities: 6,000-vol. library of books, manuscripts & microfilm relating to parks historic period available for inter-library loan and upon written request.
Activities: guided tours; off-site lectures; formally organized education programs for children; inter-museum loan, permanent & temporary exhibitions.
Publications: book, Treasures of Independence.
Hours & Admission Prices: Daily 9-5. Independence Visitor Center: 8:30-6; tour ticket required. &
Attendance: 5,500,000 (accurate)

* **INDEPENDENCE SEAPORT MUSEUM, (M),** Penn's Landing Waterfront, 211 S. Columbus Blvd. at Walnut St., Philadelphia, PA 19106-3101. Tel.: 215-413-8655. Fax: 215-925-6713.
E-mail: hcorse@phillyseaport.org
Web Site: www.phillyseaport.org
Founded: 1960.
Congressional District: 1
Key Personnel: Pres. & C.E.O., John Brady; Chm., William F. McLaughlin, Jr.; Devel., Lily Williams; Cur., Craig Bruns; Mktg. & Public Rels., Hope Koseff Corse; Librarian, Megan Good; Mgr. External Affairs, Janelle Carter; Museum Shop Mgr., Christopher Neapolitan.
Personnel Profile: Full-Time Paid 23; Part-Time Paid 19; Part-Time Volunteers 40; Interns 5.
Operating Expenses: 4,793,386
Operating Income: 1,709,237
Governing Authority: nonprofit organization. Tax-exempt: 501(c)(3).
Institution Type/Description: Maritime Museum.
Collections: USS Olympia; USS Becuna; ship models; historic small boats; figureheads; paintings; prints; drawings; weapons; memorabilia & artifacts related to ships & the sea; manuscripts; photographs; textiles.
Major Exhibits: Oh, Sugar!, 9/13-2/14; Saving the USS United States, 2/14-9/14; Marking Time - Voyage to Vietnam, 10/14-2/15.
Research Fields: maritime history with emphasis on Delaware Bay, River & tributaries.
Facilities: 10,000-vol. library available by appointment only; 530-seat concert hall. Museum-related items for sale.
Activities: lectures; films; permanent, temporary & traveling interactive exhibitions; boat conservation; boat building classes; inter-museum loans; community school & family.
Publications: newsletter, Masthead; Challenger Sketchbook; American Naval Broadsides; The Delaware Bay and River Defense of Philadelphia 1775-1777; George Robert Bonfield Marine Painter 1805-1898; The Titanic & Her Era; The Empress of China; Philadelphia on the River; The Tale of the Mermaid; Gone Fishing; Ironclad Intruder: U.S.S. MONITOR; Commerce & the Constitution; Marine Art & Antiques/Jack Tar/A Sailor's Life, 1750-1910; A Yachtsman's Eye/The Glen S. Foster Collection of Marine Paintings.
Hours & Admission Prices: Daily 10-5. Museum & Historic Ships: adults $13.50, seniors 65 & over, children, students & military $10; discounts to

AAM, AAA & WHYY members and groups of 10 or more; members & children under 2 no charge. Closed New Year's Day; Thanksgiving; Christmas. &
Attendance: 88,570 (accurate)
Membership: Individual $45; Young Friends of ISM $50; Family $75; Individual Plus $100; Navigator $150; First Mate $300; Pilot $500; Seaport Society $1,000-$5,000.

INSTITUTE OF CONTEMPORARY ART, UNIVERSITY OF PENNSYLVANIA, 118 S. 36th St., Philadelphia, PA 19104-3289. Tel.: 215-898-7108 & 5911. Fax: 215-898-5050.
E-mail: info@icaphila.org
Web Site: www.icaphila.org
Founded: 1963.
Congressional District: 1
Key Personnel: Dir., Amy Sadao; Chief Cur., Ingrid Schaffner; Asst. Cur., Kate Kraczon; Assoc. Cur., Anthony Elms; Program Cur., Alex Klein; Dir. Devel., Samantha Gibb Roff; Business Admin., Shannon Freitas; Dir. Mktg. & Communications, Jill Katz; Dir. Curatorial Affairs, Robert Chaney; Assoc. Registrar, Dana Hanmer; Head Preparator & Building Admin., Paul Swenbeck; Coord. Visitor Svcs. & Program Tech, William Hidalgo; Communications Assoc., Becky Huff Hunter; Assoc. Dir. Devel. Individual Gifts, Jeffrey Bussmann.
Personnel Profile: Full-Time Paid 18; Part-Time Paid 3; Part-Time Volunteers 20; Interns 8.
Operating Expenses: 2,800,000
Governing Authority: Parent Institution: University of Pennsylvania. Tax-exempt.
Institution Type/Description: Contemporary Art Gallery.
Collections: works by contemporary artists.
Major Exhibits: ICA@50, 2/12/14-8/17/14; Ruffneck Constructivists, 2/12/44-8/17/14.
Research Fields: contemporary art.
Facilities: archives of statements & notes by artists exhibited at ICA; Philadelphia artists registry; videotapes archives; auditorium.
Activities: public programs; readings; lectures; symposia; performances; films & videos.
Publications: exhibition catalogues; brochures; limited art editions made specifically for the ICA; monographs.
Hours & Admission Prices: Wed. 11-8, Thurs.-Fri. 11-6, Sat.-Sun. 11-5. No charge; donations accepted. Closed New Year's Day; Easter; Thanksgiving; Christmas. &
Attendance: 25,000 (accurate)
Membership: Individual $40; Contributor $100; Dual/Family $250; Participant $500; Leadership Circle $1,000; Art Council $2,500.

JOHNSON HOUSE UNDERGROUND RAILROAD MUSEUM, 6306 Germantown Ave., Philadelphia, PA 19144-1908. Tel.: 215-438-1768.
Institution Type/Description: Historic House: built c.1768. Listed on the National Register of Historic Places.
Collections: local history & culture; period furnishings; photographs; personal artifacts.
Activities: educational programs; workshops.
Hours & Admission Prices: Thurs.-Fri. 10-4 by appointment, Sat. 1-4.

LA SALLE UNIVERSITY ART MUSEUM, (M), 1900 W. Olney Ave., Philadelphia, PA 19141-1199. Tel.: 215-951-1221. Fax: 215-951-5096.
E-mail: vendelin@lasalle.edu
Web Site: www.lasalle.edu/museum
Founded: 1975.
Congressional District: 1
Key Personnel: Dir. Chief Cur., Klare Scarborough; Chm. (V), William E. Kelly, Jr.; Cur. Art, Carmen Vendelin; Cur. Education, Miranda Clark-Binder; Museum Asst., Rebecca Oviedo.
Personnel Profile: Full-Time Paid 4.
Governing Authority: university. A branch of La Salle University. Tax-exempt: 75243239.
Institution Type/Description: Art Museum.
Collections: European & American 15th- to 20th-century paintings, drawings, watercolors, prints; sculpture; Japanese prints; African sculpture; ancient Greek terra-cotta vessels & Tanagra figurines; Pre-Columbian terra-cotta vessels; Chinese, Vietnamese, & Japanese ceramics; Indian miniatures.
Major Exhibits: Printmakers of Baroque: 17th-Century Explorations of Space and Light, 12/13-2/14; American Scenes: WPA Era Prints of the 1930s and 1940s, 3/14-5/14; 7th Annual Art Faculty Exhibition, 8/14-9/14; Hung Lin, 9/14-12/14; Feminism & Domesticity, 12/14-2/15.
Activities: gallery talks; tours; temporary & special exhibitions; concerts; lectures; K-12 educational programming.

Publications: exhibit publications, LaSalle University Art Museum: Guide to the Collection (2002); survey.
Hours & Admission Prices: Mon.-Fri. 10-4. No charge; donations accepted. &
Attendance: 7,020 (accurate)
Membership: Hierarchy of Angels: Angels $25; Archangels $50; Principalities $75; Virtues $100; Powers $250; Dominations $500; Thrones $750; Cherubim $1,000; Seraphim $2,500 & up.

LEMON HILL MANSION, Lemon Hill & Sedgeley Dr., E., Fairmont Park, Philadelphia, PA 19130. Mailing Address: Colonial Dames of America, 303 W. Lancaster Ave., PMB 139, Wayne, PA 19087-3938. Tel.: 215-232-4337. Fax: 215-646-8472.
E-mail: epenniman149@gmail.com
Web Site: www.lemonhill.org
Founded: 1800.
Congressional District: 2
Key Personnel: Chm., Eleanor Penniman; Dir., Joyce Jones.
Personnel Profile: Part-Time Paid 1; Part-Time Volunteers 150.
Governing Authority: nonprofit organization. Affiliated with Colonial Dames of America, 421 E. 61st St., New York, NY 10021. Tax-exempt: 501(c)(3).
Institution Type/Description: Historic House & Site: c.1800 Lemon Hill Mansion built by Henry Pratt on land that had belonged to Robert Morris.
Collections: 1800-1836 decorative arts & furnishings.
Activities: guided tours; lectures; permanent & temporary exhibitions.
Publications: The Diary of Harriet Manigault, 1813-16; brochure; postcard.
Hours & Admission Prices: April-Dec. Wed.-Sun. 10-4; other times by appointment. Adults $5, students & senior citizens 60 and over $3.
Attendance: 10,000 (estimated)
Membership: Friends of Lemon Hill: $35-$1,000.

LEST WE FORGET BLACK HOLOCAUST MUSEUM OF SLAVERY, 3650 Richmond St., Philadelphia, PA 19134. Mailing Address: P.O. Box 26846, Philadelphia, PA 26846. Tel.: 215-205-4324 & 397-6060.
E-mail: lestweforgetmuseum@yahoo.com
Web Site: www.lestweforgetmuseumofslavery.com
Key Personnel: Owner, J. Justin Ragsdale; Owner, Gwen Ragsdale
Institution Type/Description: History Museum.
Collections: Black Holocaust history; slavery artifacts; shackles; branding irons; documents.
Hours & Admission Prices: Wed.-Sun. 10-6 by appointment. Adults $10.

LIBRARY COMPANY OF PHILADELPHIA, 1314 Locust St., Philadelphia, PA 19107-5698. Tel.: 215-546-3181. Fax: 215-546-5167.
E-mail: cking@librarycompany.org
Web Site: www.librarycompany.org
Founded: 1731.
Congressional District: 1
Key Personnel: Dir., John C. Van Horne; Pres. (V), B. Robert DeMento; Librarian, James N. Green; Cur. Prints, Sarah Weatherwax; Chief Conservation, Jennifer Woods Rosner; Chief Reference, Cornelia S. King; Chief Cataloguer, Holly Phelps; Cur. Printed Books, Rachel D'Agostino.
Personnel Profile: Full-Time Paid 18; Part-Time Paid 7; Part-Time Volunteers 3; Interns 4.
Governing Authority: nonprofit organization. Tax-exempt: 170(b)(1)(A) & 501(c)(3).
Institution Type/Description: Library of Rare Books.
Collections: libraries of James Logan, Benjamin Rush, William Byrd & Benjamin Franklin; books printed in or available in America before 1860; broadsides, manuscripts, newspapers, pamphlets & periodicals printed in & around Philadelphia; pamphlets of the American Revolution, Federal & Jacksonian Periods, and the Civil War; history of women, particularly of the 19th century; Afro-Americana; American Judaica; prints & photos of Philadelphia; 16th- & 17th-century European science, literature, history, classical language & religion books; paintings; sculpture; furniture.
Major Exhibits: That's So Gay: The Not-So-Hidden History of Gayness in Early American Culture, 2/14-8/14.
Research Fields: American & Anglo-American history to 1860; African American history; medical history; American architectural history; early American graphic arts; early American photography; American imprints; women's history, particularly 19th century.
Facilities: 500,000-vol. library of American history & literature, Afro-American history & other reference volumes for use by public for research on premises; print department with 75,000 graphic items; reading rooms.
Activities: lectures; guided tours by appointment; permanent & temporary exhibitions; public programs; fellowship program.
Publications: Annual Report of the Library Company; public exhibition catalogs & special publications; newsletter.

Hours & Admission Prices: Mon.-Fri. 9-4:45. No charge. Closed major holidays. &
Attendance: 6,106 (accurate)
Membership: Scholar $50; Friend $75; Shareholder $200 (one-time investment; dues $100 a year).

MARIAN ANDERSON HISTORICAL SOCIETY & MUSEUM, 762 Marian Anderson Way, Philadelphia, PA 19146-1822. Tel.: 215-732-9505.
Key Personnel: Cur. & Dir. Mktg., Phyllis Sims
Institution Type/Description: Historical Society Museum: housed in the former home of singer Marian Anderson. Listed on the National Register of Historic Places.
Collections: Marian Anderson's life & career; personal artifacts; period furnishings; photographs.
Hours & Admission Prices: Call for hours.

MARIO LANZA INSTITUTE/MUSEUM, 712 Montrose St., Philadelphia, PA 19147-3944. Mailing Address: P.O. Box 54624, Philadelphia, PA 19148-0624. Tel.: 215-238-9691. Fax: 215-238-9694.
E-mail: mariolanzamuseum@aol.com
Web Site: www.mariolanzainstitute.org
Founded: 1962.
Congressional District: 1
Key Personnel: Pres., William J. Ronayne; Publicity Dir. & Sec., Bill Ronayne; Treas. & Vice Pres., Jeanette Frese; Bd. Member, Ray Katz; Bd. Member, Dorothy Todaro; Bd. Member, Joan Burns; Bd. Member, Giovanna Cavaliere; Legal Counsel, Joseph Caruso, Esq.
Personnel Profile: Full-Time Volunteers 3; Part-Time Volunteers 1.
Governing Authority: private; nonprofit organization. Parent Institution: Mario Lanza Institute. Tax-exempt: 501(c)(3).
Institution Type/Description: Specialized Museum: one block from museum is the Mario Lanza Birthplace.
Collections: concentration on Mario Lanza; film-related memorabilia; private photographs, movie stills & posters; private audio recordings; video movie & TV programming; commercial records, compact discs & cassette tapes; paintings; biographies; newsletters; costumes.
Research Fields: biographical & recordings.
Facilities: 40-vol. library of ball fundraising events, clippings & photographs; 1,000 sq. ft. exhibit space. Museum-related items for sale.
Activities: concerts; films; guided tours; luncheons. Annual Events: Mario Lanza Ball; Spring Recital.
Publications: brochure; catalog; articles; newsletter.
Hours & Admission Prices: Mon.-Wed. & Fri.-Sat. 11-3. No charge; donations accepted.
Attendance: 2,500 (estimated)
Membership: Call for information.

MARVIN SAMSON CENTER FOR HISTORY OF PHARMACY/USP MUSEUM, 600 S. 43rd St., Philadelphia, PA 19104-4418. Tel.: 215-596-8721. Fax: 215-895-1113.
Web Site: www.usp.edu/museum
Founded: 1995.
Key Personnel: Dir. & Cur., Michael Brody
Institution Type/Description: Pharmacy Museum.
Collections: history of pharmacy.
Hours & Admission Prices: Call for hours. No charge.
Attendance: 5,000

THE MASONIC LIBRARY AND MUSEUM OF PENNSYLVANIA, Masonic Temple, One N. Broad St., Philadelphia, PA 19107-2520. Tel.: 215-988-1485 & 1900. Fax: 215-988-1953.
E-mail: dpbuttleman@pagrandlodge.org
Web Site: www.pagrandlodge.org
Founded: 1731.
Congressional District: 2
Key Personnel: Librarian, Glenys A. Waldman; Asst. Librarian, Catherine L. Giaimo; Cur., Dennis P. Buttleman, Jr.; Museum Shop Mgr., Carole Alpe.
Governing Authority: society. Parent Institution: Grand Lodge F. & A.M. of Pennsylvania. Tax-exempt.
Institution Type/Description: History Museum: housed in c.1873 Masonic Temple.
Collections: historic Masonic objects; jewels; textiles; ceramics; glass; documents; books; prints.
Research Fields: history of Freemasonry; Pennsylvania Freemasonry; Masonic & fraternal symbolism & mysticism; religion; philosophy.
Facilities: 70,000-vol. library available to the public; building available for rental. Masonic books & other museum-related items for sale.
Activities: guided tours; temporary & permanent exhibitions.

Publications: quarterly magazine, The Pennsylvania Freemason.
Hours & Admission Prices: Tues.-Fri. 9-5, Sat. 9-12; other times by appointment. Admission $8, members $5. Closed national holidays. &
Attendance: 30,000 (estimated)

*** MORRIS ARBORETUM OF THE UNIVERSITY OF PENNSYLVANIA,** 100 E. Northwestern Ave., Philadelphia, PA 19118-2697. Tel.: 215-247-5777. Fax: 215-248-4439.
E-mail: cranesj@upenn.edu
Web Site: www.morrisarboretum.org
Formerly: The Gardens at Morris Arboretum
Founded: 1933.
Congressional District: 2
Key Personnel: Dir., Paul W. Meyer; Dir. Facilities & Finance, Kevin Schrecengost; Dir. Botany, Dr. Timothy Block; Dir. Mktg., Susan Crane; Dir. Public Programs, Robert Gutowski; Museum Shop Mgr., Adele Waerig.
Personnel Profile: Full-Time Paid 34; Part-Time Paid 15; Part-Time Volunteers 401; Interns 9.
Volunteer Hours: 1,862
Operating Expenses: 7,043,654
Operating Income: 7,043,654
Governing Authority: college; nonprofit organization. Parent Institution: University of Pennsylvania.
Institution Type/Description: Arboretum.
Collections: mature specimens of native & introduced woody plants in landscape setting; herbarium; tropical fernery; modern sculpture.
Research Fields: adaptability of trees to urban sites; flora of Pennsylvania; plant introduction & evaluation; micro propagation research; bioassays of the living collection; integrated pest management.
Facilities: library; historic 166-acre arboretum; specialty gardens; herbarium; education center; plant pathology lab.
Activities: guided tours; lectures; educational programs; research programs; Center for Urban Forestry; permanent exhibits; special programs; seed exchange. Annual Events: Garden Railway late May to mid-October; Holiday Garden Railway late November to December.
Publications: quarterly newsletter; seed exchange list; interpretive brochures; annual report.
Hours & Admission Prices: April-Oct. Mon.-Fri. 10-4, Sat.-Sun. 10-5; Nov.-March 10-4 daily. Adults $16, senior citizens $14, students & youth 3-18 $7; discounts to PHS, AAA, WHYY, WXPN, PCVB members and Penn students, staff, & alumni; members no charge. Closed New Year's Day; Thanksgiving; Christmas Eve & Day. &
Attendance: 126,980 (accurate)
Membership: Far Away Friends $40; Individual $55; Dual $65; Regular Family $75; Beech $95; Chestnut $150; Holly $250; Oak $500; Laurel $1,000; Katsura $2,500.

THE MUSEUM OF NURSING HISTORY, INC., LaSalle Univ., St. Benilde Tower - 3rd Fl., 1900 W. Olney Ave., Philadelphia, PA 19141-1199. Tel.: 215-831-7819. Facebook: Museum Nursing History.
E-mail: sdavis10@earthlink.net
Web Site: www.nursinghistory.org
Founded: 1976.
Key Personnel: Pres. & Nurse Historian, Sandra Davis, Ed.D., R.N.; Treas., Jane Early.
Personnel Profile: Part-Time Volunteers 15.
Volunteer Hours: 200
Operating Expenses: 3,810
Operating Income: 3,000
Governing Authority: volunteer board of directors. Tax-exempt: 501(c)(3).
Institution Type/Description: Nursing History: originally located in Pennsylvania Hospital, the first hospital in the U.S. (1751).
Collections: nursing memorabilia from 1860-1960 with emphasis on nurses & nursing in Pennsylvania; military nurses uniform; archival text collection; original nursing organization papers.
Research Fields: nurses & nursing in Pennsylvania, 1860-1960; nursing texts, biographies, papers & reports.
Facilities: 200-vol. library on nursing; 250-seat theater; 800 sq. ft. exhibit space.
Activities: formal education programs for all area nursing schools. Annual Events: Fall Educational Program; Spring Historical Presentation.
Publications: biannual newsletter, The Museum Muse.
Hours & Admission Prices: Mon.-Fri. 9-5, special arrangements for group tours. No charge. &
Attendance: 120 (estimated)
Membership: Student & Retired $20; Individual $35; Sponsor $50; Joan Large Memorial Fund $100; Patron $150. Organization: Alice Fischer $250; Lillian Wald $500; Lavinia Dock $750; Adelaide Nutting $1,000.

MUSEUM OF THE AMERICAN REVOLUTION, (M), 123 Chestnut St., Ste. 401, Philadelphia, PA 19106-3059. Tel.: 215-253-6732.
E-mail: info@amrevmuseum.org
Web Site: www.amrevmuseum.org
Formerly: The American Revolution Center
Founded: 2000.
Key Personnel: Pres. & C.E.O., Bruce Cole; Chm. Bd., H.F. Lensest; Senior Vice Pres., ZeeAnn Mason.
Personnel Profile: Full-Time Paid 2; Part-Time Paid 5.
Governing Authority: society. Tax-exempt: 501(c)(3).
Institution Type/Description: History Museum: located on the site of Valley Forge encampment area.
Collections: Washington memorabilia & artifacts; 18th-century firearms & accoutrements; decorative arts; archival information including books, manuscripts & documents.
Research Fields: firearms of Revolutionary period.
Hours & Admission Prices: Museum under construction.

MUTTER MUSEUM OF THE COLLEGE OF PHYSICIANS OF PHILADELPHIA, (M), 19 S. 22nd St., Philadelphia, PA 19103-3097. Tel.: 215-563-3737. Fax: 215-561-6477.
Web Site: www.collegeofphysicians.org
Founded: 1863.
Congressional District: 2
Key Personnel: Dir., Robert Hicks, Ph.D.; C.E.O. The College of Physicians of Philadelphia, George M. Wohlreich, M.D.; Chm. (V), Bennett Lorber, M.D.; Cur., Anna N. Dhody; Dir. Communications, J. Nathan Bazzel; Museum Shop Mgr., Marie Fury.
Personnel Profile: Full-Time Paid 4; Part-Time Volunteers 20; Interns 2.
Governing Authority: nonprofit organization. Parent Institution: The College of Physicians of Philadelphia. Tax-exempt: 501(c)(3).
Institution Type/Description: Medical Museum.
Collections: human anatomy & pathology; medical history & biography; development of medical instrumentation; development of fetus & anomalies; folklore; quackery; military medicine; nursing & apothecary artifacts; memorabilia of physicians; period medical items; oil portraits; prints; photographs; sculpture.
Research Fields: medical history & biography; pathology of bone & tissue; paleopathology; physical anthropology; biology; anatomy; military medicine; folklore; genetic defects; teratology; anomalies; medical specialties & instrumentation; medical art illustration.
Facilities: reference library with medical journals, documentation & bibliographic services available for inter-library use; herb garden; meeting rooms; audiovisual equipment. Reprints, slides, photographs of collections for sale.
Activities: audio guide; guided tours; lectures by appointment; permanent & temporary exhibits.
Publications: quarterly journal; Transactions & Studies of the College of Physicians of Philadelphia.
Hours & Admission Prices: Library: by appointment. Museum: daily 10-5. Adults $14, senior citizens 65 & up, college students with I.D. & children 6-18 $10; discounts to groups, AAM & ICOM members; children under 6 no charge. Closed New Year's Day; Thanksgiving; Christmas Eve & Day. &
Attendance: 104,086 (accurate)
Membership: Student $35; Individual $55; Family $85.

NAOMI WOOD COLLECTION AT WOODFORD MANSION, (M), 33rd & Dauphin St., E. Fairmount Park, Philadelphia, PA 19132. Tel.: 215-229-6115.
Web Site: woodfordmansion.org
Founded: 1926.
Key Personnel: Co-Trustee, Lawrence H. Berger; Site Mgr., Martha Moffat.
Personnel Profile: Full-Time Paid 1; Part-Time Paid 6; Part-Time Volunteers 30.
Governing Authority: nonprofit organization. Tax-exempt.
Institution Type/Description: Historic House Museum: 1756 Woodford Mansion.
Collections: decorative arts; English Delftware; Colonial Philadelphia & American furnishings; pewter; silver; needlework; portraits.
Facilities: picnic area.
Activities: guided tours; orchard harvest festivals. Annual Event: Holiday Tours in December.
Publications: book, The Story of The Naomi Wood Collection and Woodford Mansion.
Hours & Admission Prices: Tues.-Sun. 10-4. Adults $5, senior citizens $3, children $2. Closed major holidays.
Attendance: 5,000 (accurate)

NATIONAL ARCHIVES AT PHILADELPHIA, (M), 900 Market St., Philadelphia, PA 19107-4292. Tel.: 215-606-0100. Fax: 215-606-0111. Facebook: NARA at Philadelphia.
E-mail: philadelphia.archives@nara.gov
Web Site: www.archives.gov/philadelphia/
Formerly: National Archives and Records Administration - Mid-Atlantic Region
Founded: 1934.
Congressional District: 1
Key Personnel: Field Support Officer, David Roland; Admin. Officer, Brenda Bernard; Dir., Federal Records Center, Aaron Swann; Public Programs, Beth Levitt; Teacher Workshops & School Tours, Andrea Reidell.
Governing Authority: federal government. Parent Institution: National Archives & Records Administration, Washington D.C. Tax-exempt.
Institution Type/Description: Archives: housed in 1938 Works Project Administration Construction building.
Collections: photographs, documents & artifacts related to cultural, social, political & economic history of the Federal Government; Federal Government activities in the Mid-Atlantic region; archives; 1790-1965, documents, drawings & photographs; East Coast lighthouse drawings; records of Federal agencies & courts in Delaware, Maryland, Pennsylvania, Virginia & West Virginia.
Research Fields: Federal records; Chinese immigration & exclusion laws; Civil War; Confederate blockade running; patent materials; 19th-century military & federal arsenals; U.S. census; immigration & naturalization; Federal Court records; U.S. mint records; U.S. Corp. of Engineers; African-American history; women's history.
Facilities: 500-vol. guide to the archives; archives available for use by public; research & reference rooms; classroom; conference room; field research station; 600 sq. ft. exhibit space. Gift items for sale.
Activities: guided tours; lectures; films; organized education programs for children, adults & undergraduate or graduate college students; participatory, loan, temporary & traveling exhibitions; National History Day local coordinator.
Publications: periodic guides to various collections; periodical, Guide to Records in the National Archives-Mid Atlantic Region; exhibition catalogs.
Hours & Admission Prices: Mon.-Fri. 8:30-4:45, 2nd Sat. of month 8-4. No charge; donations accepted. Closed federal holidays. &
Attendance: 12,000 (accurate)

✳ **NATIONAL CONSTITUTION CENTER, (M),** 525 Arch St., Independence Mall, Philadelphia, PA 19106-1595. Tel.: 215-409-6600.
Web Site: www.constitutioncenter.org
Key Personnel: Pres. & C.E.O., Jeffrey Rosen; C.O.O., Vince Stango; Vice Pres. Mktg. & Communications, Steve Rosenberg
Institution Type/Description: History Museum.
Collections: the Constitution & its history; photographs.
Activities: special events.
Hours & Admission Prices: Mon.-Fri. 9:30-5, Sat. 9:30-6, Sun. 12-5. Adults $14.50, senior citizens 65 & over and students with ID $13.50, children 4-12 $8; military & children under 4 no charge. Closed New Year's Day; Thanksgiving; Christmas.

NATIONAL LIBERTY MUSEUM, (M), 321 Chestnut St., Philadelphia, PA 19106-2707. Tel.: 215-925-2800. Fax: 215-925-3800.
E-mail: jgriesemer@libertymuseum.org
Web Site: www.libertymuseum.org
Founded: 2000.
Key Personnel: C.E.O. & Vice Pres., Gwen Borowsky; Founder & Chm. (V), Irvin J. Borowsky; Pres. (V), Douglas Tozour; Controller, Ronald Kroes; Sr. Vice Pres. Devel., Ken Boyden; Devel., Peggy Sweeney; Dir. Programs, Kevin O'Rangers; Mktg., Sherry Hawk; Public Rels., Jan Griesemer; Mktg., Bob Hawk; Museum Shop Mgr., Leroy Ford.
Personnel Profile: Full-Time Paid 24; Full-Time Volunteers 1; Part-Time Paid 7; Part-Time Volunteers 10.
Governing Authority: private; nonprofit organization. Tax-exempt: 501(c)(3).
Institution Type/Description: Art Museum.
Collections: glass sculptures; film; artwork.
Facilities: education center; 30,000 sq. ft. exhibit space. Museum-related items for sale.
Activities: docent program; films; guided tours; participatory exhibits. Annual Events: Glass Weekend - includes dinner, auction & view of private collections; Heroes of Liberty dinner; Police & Firefighters awards of valor; International Interfaith awards; Inspirations award; Young Heroes award.
Publications: newsletters, National Liberty Museum; Coalition of Collectors and Artists; book, Heroes of Liberty From Around the World; Facebook; Twitter.
Hours & Admission Prices: Summer: daily 10-5. Sept.-June Tues.-Sat. 10-5,

Sun. 12-6. Adults $7, seniors $6, students $5; scheduled tours & members no charge. Closed Christmas. &
Attendance: 62,000 (estimated)
Membership: Individual $35; Family $60; Committee to Difuse Violence $100; Supporters of Freedom $250; Liberty Circle $500; Founder's Committee $1,000; Wall of Honor Circle & Major Donor $10,000 & up.

NATIONAL MUSEUM OF AMERICAN JEWISH HISTORY, (M), 101 South Independence Mall East, Philadelphia, PA 19106-2517. Tel.: 215-923-3811. Fax: 215-923-0763.
E-mail: nmajh@nmajh.org
Web Site: www.nmajh.org
Founded: 1976.
Congressional District: 1
Key Personnel: Pres. & C.E.O., Michael Rosenzweig; Dir. Gwen Goodman Museum, Ivy L. Barsky; Co-Chm., Ronald Rubin; Dir. Finance, Don Maedche; Deputy Dir. Programming & Museum Historian, Josh Perelman; Dir. Public Rels., Jay Nachman; Dir. Education, Linda Steinberg; Retail Sales Mgr., Kristen Kreider.
Personnel Profile: Full-Time Paid 48; Full-Time Volunteers 35; Part-Time Paid 15; Part-Time Volunteers 125; Interns 4.
Governing Authority: nonprofit. Tax-exempt: 501(c)(3).
Institution Type/Description: Social & Ethnic History Museum: located on Independence Mall.
Collections: artifacts reflect the occupational, domestic, communal & religious aspects of Jewish life in America from Colonial times to the present.
Research Fields: History of Jews & Judaism in America.
Facilities: 1,500-vol. library of monographs & reference works on American Jewish history, books, pamphlets & other printed primary source material available to scholars & interested researchers with permission of the registrar; auditorium; classrooms; theater in gallery; film library focusing on American Jewish contributions to the cinema and American Jewish history. Books on American Jewish life & history, ceremonial objects, photos, graphics & reproductions of historical artifacts for sale.
Activities: guided tours; lectures; films; gallery talks; concerts; docent program or council; loan, permanent & temporary exhibitions.
Publications: newsletter; exhibition catalogs.
Hours & Admission Prices: Tues.-Sun. 10-5. Adults $12, seniors $11; members, active military & children no charge. Closed New Year's Day; Martin Luther King, Jr. Day; Memorial Day; Independence Day; Labor Day; Thanksgiving; major Jewish holidays & festivals. &
Attendance: 250,000 (estimated)
Membership: Student & Senior $20; Senior Couple $25; Individual $54; Dual $70; Family $90; Friends Circle Supporter $150; Family Plus $180; Friends Circle $300; Friends Sponsor $500; Friends Circle Benefactor $1,000.

PAINTED BRIDE ART CENTER, 230 Vine St., Philadelphia, PA 19106-1293. Tel.: 215-925-9914. Fax: 215-925-7402.
E-mail: info@paintedbride.org
Web Site: paintedbride.org
Key Personnel: Exec. Dir., Laurel Raczka; Assoc. Dir., Lisa Nelson-Haynes
Institution Type/Description: Art Center.
Collections: works by local, national & international artists.
Facilities: performing arts theater.
Activities: performances.
Hours & Admission Prices: Tues.-Sat. 12-6 & during performances.

THE PAUL ROBESON HOUSE, 4951 Walnut St., Philadelphia, PA 19139-4228. Mailing Address: West Philadelphia Cultural Alliance, 4949/4951 Walnut St., Philadelphia, PA 19139-4228. Tel.: 215-747-4675.
E-mail: wpca@wpcalliance.com
Web Site: www.paulrobesonhouse.org
Institution Type/Description: Historical House Museum: housed in the former home of Paul Robeson.
Collections: Robeson's life, family & career; photographs; personal artifacts; paintings; period furnishings.
Hours & Admission Prices: By appointment. Adults $5, children $4; discounts to groups of 10 or more.

✳ **PENNSYLVANIA ACADEMY OF THE FINE ARTS, (M),** 128 N. Broad St., Philadelphia, PA 19102-1424. Tel.: 215-972-7600 & 7642. Fax: 215-972-5564, 567-2429 & 569-0153.
E-mail: pafa@pafa.org
Web Site: www.pafa.org
Founded: 1805.
Congressional District: 2
Key Personnel: Chm. (V), Kevin F. Donohoe; Pres. & C.E.O., David R. Brigham; Edna S. Tuttleman Museum Dir., Harry Philbrick; Dir. Human

Resources, James Gaddy; Exec. Vice Pres. Devel., Melissa DeRuiter; Mgr. Rights & Reproductions, Judith Thomas; Sr. Registrar, Gale Rawson; Asst. Registrar, Jennifer Johns; Cur. Contemporary Art, Julien Robson; Cur. Modern Art, Robert Cozzolino; Dir. Retail Sales & Visitor Svcs., Mark De Lelys; Cur. Historical American Art, Anna Marley; Dir. Museum Education, Monica Zimmerman.
Personnel Profile: Full-Time Paid 76.
Governing Authority: nonprofit. Tax-exempt: 501(c)(3).
Institution Type/Description: Art Museum & School: housed in 1876 Centennial building designed by Furness & Hewitt.
Collections: 18th- to 21st-century American paintings, drawings, sculpture & prints; works by Washington Allston, Benjamin West, the Peale Family, Gilbert Stuart, Thomas Sully, William Rush, William Sidney Mount, Thomas Eakins, Winslow Homer, Edward Hopper, Childe Hassam, Arthur B. Carles, Robert Henri, Cecilia Beaux, Horace Pippin, William Bailey, Alex Katz, William Merritt Chase, Richard Diebenkorn, Jacob Lawrence, Robert Motherwell, Sidney Goodman & Mary Frank.
Research Fields: American painting; sculpture; works on paper; architecture; archives.
Facilities: 9,000-vol. library of visual arts & Academy's archives open to public for reference by appointment; 130-seat auditorium; classrooms; cafe. Catalogs, books, jewelry, postcards, posters & other museum-related items for sale.
Activities: guided tours; lectures; permanent collection; studio classes; temporary & inter-museum traveling exhibitions; symposia; loan program.
Publications: exhibition catalogues; newsletter; brochures; quarterly magazine; annual report.
Hours & Admission Prices: Tues.-Sat. 10-5, Sun. 11-5. Permanent Collection: adults $10, senior citizens & students with ID $8, youth 13-18 $6; discounts to AAM members; members & children 12 & under no charge. Combination: adults $15, seniors & students with ID $12, youth 13-18 $10; discounts to AAM members; children 12 & under no charge. Closed major holidays. &
Attendance: 132,532 (accurate)
Membership: Student $20; National $30; Individual $55; Young Friends $60; Family $80; Alumni Circle $85; Friend $150; Patron $300; Potamkin Society Young Collector's Under 40 $350; Benefactor $500; Potamkin Collector's Society $750; The Peale Circle $1,000 & up.

THE PHILADELPHIA ART ALLIANCE, 251 S. 18th St., Philadelphia, PA 19103-6168. Tel.: 215-545-4302. Fax: 215-545-0767. Facebook: Phil Art Alliance.
E-mail: info@philartalliance.org
Web Site: www.philartalliance.org
Founded: 1915.
Congressional District: 2
Key Personnel: Exec. Dir., Lesly Attarian; Pres., Diane Dalto; Dir. Exhibitions, Melissa Caldwell; Chief Cur., Sarah Archer; Dir. Programs, Mat Tomezsko; Coord. Membership, Joanna Grim; Receptionist, Amanda Childs; Office Mgr., Jamie DeAngelis.
Personnel Profile: Full-Time Paid 3; Part-Time Paid 3; Part-Time Volunteers 12; Interns 4.
Governing Authority: nonprofit organization. Tax-exempt: 501(c)(3).
Institution Type/Description: Multi-Disciplinary Arts Center: housed in c.1906 home, designed by Klauder of the architectural firm Day & Klauder; located on Rittenhouse Square.
Collections: contemporary craft & design.
Activities: contemporary visual arts exhibitions program; guided tours; lectures; gallery talks; concerts; readings of screen plays; poetry readings; book launchings & signings; temporary exhibitions.
Publications: newsletter; exhibition catalogs.
Hours & Admission Prices: Tues.-Sun. 11-5. Adults $5, seniors $3; members no charge. Closed national holidays.
Attendance: 20,000 (estimated)
Membership: Individual $75; Supporter $100; Patron $250; Benefactor $500.

PHILADELPHIA DOLL MUSEUM, 2253 N. Broad St., Philadelphia, PA 19132. Tel.: 215-787-0220. Fax: 215-787-0226.
Web Site: www.philadollmuseum.com
Founded: 1988.
Institution Type/Description: Doll Museum.
Collections: 500 black dolls from around the world including African, European, American Folk Art, Roberta Bell; doll history; photographs.
Facilities: library.
Activities: seminars; lectures; workshops; educational programs.
Hours & Admission Prices: Thurs.-Sat. 10-4, Sun. 12-4. Adults $4, seniors, students & children under 12 $3; discounts to groups.

PHILADELPHIA HISTORY MUSEUM, (M), 15 S. 7th St., Philadelphia, PA 19106-2313. Tel.: 215-685-4830. Fax: 215-685-4837. Facebook: Philadelphia History Museum.
E-mail: info@philadelphiahistory.org
Web Site: www.philadelphiahistory.org
Formerly: Atwater Kent Museum of Philadelphia dba Philadelphia History Museum
Founded: 1938.
Congressional District: 1
Key Personnel: C.E.O. & Exec. Dir., Charles Croce; Exec. Admin., Emily Cooper Moore; Senior Cur., Jeffrey R. Ray; Dir. Collection, Kristen Froehlich; Registrar, Susan Drinan; Historian, Cynthia Little, Ph.D.; Dir. Devel. & Mktg. Communications, Mary-Anne Smith-Harris; Communications Coord., Kelly Murphy.
Personnel Profile: Full-Time Paid 9; Part-Time Volunteers 31; Interns 7.
Volunteer Hours: 2,685
Operating Expenses: 1,500,000
Operating Income: 1,200,000
Governing Authority: bd. trustees. Philadelphia, PA. Tax-exempt.
Institution Type/Description: History Museum: housed in 1826 building designed by John Haviland.
Collections: 50,000 prints, photographs & two-dimensional ephemera; 43,000 three-dimensional objects, including 2,700 textiles, 2,000 toys & 2,000 craft & industry artifacts; 7,500 books, pamphlets & manuscripts reflecting Philadelphia history 1680 to the present; Historical Society of Pennsylvania Art and Artifact Collection (11,400 items); Insurance Company of North America (CIGNA) (350 items); The Balch Institute for Ethnic Studies (250 items)
Major Exhibits: Authentic Philly: Tony Autus Cartoons of Philadelphia, 10/13-2/14; Gifts That Gleam: Stories in Silver, 10/13-6/14; Behind the Scenes of the Nutcracker, 11/13-1/14.
Research Fields: social, cultural, industrial history of Philadelphia.
Facilities: collection study center for research; gallery space.
Activities: educational programs. Museum Sponsors: Community History Gallery; National History Day; Quest for Freedom; Conversations, Where in the World is Philadelphia.
Publications: books, Invisible Philadelphia: Community Through Voluntary Organizations; Rebels and Loyalists: The Revolutionary Soldier in Philadelphia; Magic Lantern of Dr. Thomas Story Kirkbridge.
Hours & Admission Prices: Tues.-Sat. 10:30-4:30. Adults $10, seniors $8, students & teens 13-18 $6; members, active military, children 12 & under no charge. &
Attendance: 17,176 (estimated)
Membership: Senior & Student $25; Individual $35; Young Friend $45; Household $55; Grandparents $65; Participating $75; Sustaining $125; Contributing $250; Sponsor $500; Atwater Kent Society $1,000.

PHILADELPHIA MUMMERS MUSEUM, 1100 South 2nd St., Philadelphia, PA 19147-5497. Tel.: 215-336-3050. Fax: 215-389-5630.
E-mail: mummersmus@aol.com
Web Site: www.mummersmuseum.org
Founded: 1976.
Congressional District: 1
Key Personnel: Exec. Dir., Palma B. Lucas; Pres., Rocco Galleli; Cur. & Library Coord., Jack Cohen; Museum Shop Mgr., Eileen Garbarino.
Personnel Profile: Full-Time Paid 2; Full-Time Volunteers 1; Part-Time Paid 7; Part-Time Volunteers 30.
Governing Authority: municipal; nonprofit organization. Parent Institution: New Year's Shooters & Mummers Museum, Inc. Tax-exempt: 501(c)(3).
Institution Type/Description: Audio-Visual Film & Costume Museum: located on 1600-1800s site of original route of Mummers.
Collections: memorabilia of Philadelphia Mummers including comic, fancy brigade & string band divisions; Mummer suits of present & past years; photographs of Mummers & past parades; ornamental organizational badges; video tapes & films of past and present parades.
Research Fields: Philadelphia Mummery.
Facilities: library of past parade program books, old photographs, films, slides, recordings & tape interviews with veteran Mummers & newspaper clippings, available for use by museum members only; reading room; theater. Gift items for sale.
Activities: guided tours; films; permanent exhibitions; band concerts; special events with Philadelphia Mummers as the main theme.
Publications: monthly newsletter, Mummers Museum News.
Hours & Admission Prices: May-Sept. Tues. & Thurs. 9:30-9:30, Wed. & Fri-Sat. 9:30-4:30; Oct.-April Wed.-Sat. 9:30-4:30. Adults $3.50, senior citizens & children $2.50; discounts to international students, groups, AAA & AARP members; members no charge. Closed holidays. &
Attendance: 12,000 (estimated)

Membership: Individual $15; Family $30; Contributing $100; Patron $500; Corporate $1,000.

✠　**PHILADELPHIA MUSEUM OF ART,** 26th St. & Benjamin Franklin Pkwy., Philadelphia, PA 19130. Mailing Address: P.O. Box 7646, Philadelphia, PA 19101-7646. Tel.: 215-763-8100 & 235-SHOW (ticketing line). Fax: 215-236-4465.
E-mail: visitorservices@philamuseum.org
Web Site: www.philamuseum.org
Founded: 1876.
Congressional District: 3
Key Personnel: Chm. Bd., Constance H. Williams; Dir. The George D. Widener & C.E.O., Timothy Rub; Pres. & C.O.O., Gail M. Harrity; Deputy Dir. Collections & Exhibitions, Alice Beamesderfer; Gen. Counsel & Sec., Lawrence Berger; C.F.O., Robert Rambo; Assoc. Cur. American Art & Mgr. Center of American Art, Mark Mitchell; The Nancy M. McNeil Assoc. Cur. American Modern & Contemporary Crafts & Decorative Arts, Elisabeth Agro; The H. Richard Dietrich, Jr. Cur. American Decorative Arts, David Barquist; The Robert L. McNeil, Jr. Sr. Cur. American Art & Dir. Center for American Art, Kathleen A. Foster; The Montgomery-Garvan Assoc. Cur. American Decorative Arts, Alexandra A. Kirtley; The Jack M. & Annette Y. Friedland Sr. Cur. Costume & Textiles, Dilys E. Blum; Assoc. Cur. Costume & Textiles and Supervising Cur. Study Room & Academic Rels., H. Kristina Haugland; The Luther W. Brady Cur. Japanese Art & Cur. East Asian Art, Felice Fischer; The Maxine & Howard Lewis Assoc. Cur. Korean Art, Hyunsoo Woo; Cur. European Decorative Arts after 1700, Kathryn Bloom Hiesinger; The J.J. Medveckis Assoc. Cur. Arms & Armor, Pierre Terjanian
The Gisela & Dennis Alter Sr. Cur. European Painting Before 1900, & Sr. Cur. John G. Johnson Collection & the Rodin Museum Joseph Rishel; The Gloria & Jack Drosdick Assoc. Cur. European Painting & Sculpture Before 1900 & the Rodin Museum, Jennifer Thompson; The Stella Kramrisch Cur. Indian & Himalayan Art, Darielle Mason; The Keith L. & Katherine Sachs Cur. Contemporary Art, Carlos Basualdo; The Muriel & Philip Berman Cur. Modern Art, Michael Taylor; The Brodsky Cur. Photographs, Alfred Stieglitz Ctr., Peter Barberie; The Kathy & Ted Fernberger Cur. Prints, John Ittmann; The Audrey & William H. Helfand Sr. Cur. Prints, Drawings, & Photographs, Innis Howe Shoemaker; Cur. Drawings, Ann Percy; Dir. Editorial & Graphic Design, Ruth Abrahams; The William T. Ranney Dir. Publishing, Sherry Babbitt; Dir. Rights & Reproductions, Conna Clark; Dir. Communications, Norman Keyes; Dir. External Affairs, Cheryl McClenney-Brooker; Exec. Dir. Devel., Kelly O'Brien; Dir. Human Resources, Robin Proctor; Dir. Engineering Facilities & Operations, Al Shaikoli; Dir. Visitor Svcs., Jessica Sharpe; Sr. Registrar, Irene Taurins; Dir. Information Svcs., William Weinstein; Dir. Special Exhibitions Planning, Suzanne F. Wells; Dir. Membership, Beth Yeagle; Arcadia Dir. Library & Archives, Danial Elliott; The Martha Hamilton Morris Archivist, Susan Anderson; The Kathleen C. Sherrerd Sr. Cur. Education, Marla K. Shoemaker; The Constance Williams Cur. Education, School & Teacher Programs, Barbara A. Bassett; Sr. Conservator Works of Art on Paper, Nancy Ash; The Elaine S. Harrington Sr. Conservator Furniture & Woodwork, David deMuzio; The Neubauer Family Chair Conservation, Sr. Conservator Decorative Arts & Sculpture, P. Andrew Lins; Sr. Scientist - Conservation, Beth A. Price; Vice Chm. Conservation & The Aronson Sr. Conservator Paintings, Mark S. Tucker.
Personnel Profile: Full-Time Paid 336; Full-Time Volunteers 525; Part-Time Paid 147; Part-Time Volunteers 632; Interns 20.
Governing Authority: municipal; nonprofit organization. Affiliated Museums: Rodin Museum, 22nd St. & The Parkway, 19130; Samuel S. Fleisher Art Memorial, 715-719 Catharine St., 19147; Mount Pleasant and Cedar Grove Fairmount Park Houses. Tax-exempt.
Institution Type/Description: Art Museum: located in Fairmount Park.
Collections: Indian and Himalayan, Far & Near Eastern art; European medieval & Renaissance art including Foulc & Barnard; European painting, sculpture & decorative arts, including John G. Johnson; Eastern & Western period rooms from 12th- to 19th-centuries; American painting, sculpture & decorative arts, including Pennsylvania German arts & the Lorimer glass; 20th century painting & sculpture, including Gallatin & Arensberg; contemporary art; prints & drawings; Alfred Stieglitz Center collection of photography; Kienbusch arms & armor. Historic Houses: 1721 Cedar Grove; 1761 Mount Pleasant.
Research Fields: art history; conservation.
Facilities: 130,000-vol. library of art reference books available for inter-library loan & for use on premises; 395-seat auditorium; classrooms; restaurant & cafeteria. Art books, reproductions, jewelry, postcards for sale.
Activities: guided tours; lectures; films; gallery talks; concerts; arts festivals; formally organized education programs for children, adults, families & the handicapped; guide programs; inter-museum loan, special & traveling exhibitions; audio tours; Friday evening programs.

Publications: Bulletin Philadelphia Museum of Art; exhibition & collection catalogs; monthly calendar.
Hours & Admission Prices: Tues.-Thurs. & Sat.-Sun. 10-5, Fri. 10-8:45. Adults $16, senior citizens 65 & over $14, students & children 13-18 $12; discounts to AAM & ICOM members; members and children 12 & under no charge. General admission does not include special ticketed exhibitions. Closed Independence Day; Thanksgiving; Christmas. ♿
Attendance: 780,000 (estimated)
Membership: Student $40; Individual $70; Dual Plus $115; Family & Friends $185; Ambassador $225; Sustainer $400; Sponsor $750; Patron $1,000; Associates $2,000 & up.

PHILADELPHIA　　MUSEUM　　OF　　JEWISH ART/CONGREGATION RODEPH SHALOM, (M), 615 N. Broad St., Philadelphia, PA 19123-2417. Tel.: 215-627-6747. Fax: 215-627-1313.
E-mail: pmja@rodephshalom.org
Web Site: www.rodephshalom.org
Founded: 1975.
Key Personnel: Dir., Wendi Furman; Chm., Gail Rosenberg; Cur., Matthew Singer.
Governing Authority: synagogue; nonprofit.
Institution Type/Description: Religious Museum: housed in the oldest German synagogue in the Western hemisphere.
Collections: Jewish art; religious ceremonials.
Facilities: 150 sq. ft. exhibit space.
Hours & Admission Prices: Mon.-Thurs. 10-4, Fri. 10-2. No charge; donations accepted. Closed Jewish & public holidays. ♿
Attendance: 6,000

PHILADELPHIA SOCIETY FOR THE PRESERVATION OF LANDMARKS, (M), 321 S. 4th St., Philadelphia, PA 19106-4218. Tel.: 215-925-2251. Fax: 215-925-7909.
E-mail: info@philalandmarks.org
Web Site: www.philalandmarks.org
Founded: 1931.
Congressional District: 1
Key Personnel: Exec. Dir., Brandi Levine; Chm. (V), George Haitsch; Admin., Jorja Fullerton.
Personnel Profile: Full-Time Paid 4; Full-Time Volunteers 8; Part-Time Paid 4; Part-Time Volunteers 80.
Governing Authority: nonprofit organization. Historic House Museums: 1744 Grumblethorpe, 5267 Germantown Ave.; 1765 Powel House, 244 S. 3rd St.; 1786 Physick House, 321 S. 4th St.; 1724 Historic Waynesborough, 2049 Waynesborough Rd., Paoli. Tax-exempt: 501(c)(3).
Institution Type/Description: Historic Building/Site.
Collections: 1700-1840 Philadelphia & American-made furniture & decorative arts.
Research Fields: culinary history; medical history pre-1840; gardens 1744-1855.
Activities: tours; lectures; special events; annual meeting; programs; workshops; Elderhostel program in Philadelphia.
Publications: newsletter.
Hours & Admission Prices: Powel & Physick House: Thurs.-Sat. 12-4, Sun. 1-4. Grumblethorpe: by appointment. Waynesborough: mid-March to Dec. Wed.-Sun. 1-3. Family $12, Adults $5, students & senior citizens $4, groups of 10 or more $3; children under 6 & Landmarks' members no charge.
Attendance: 21,000 (estimated)
Membership: Individual $25; Household $50.

PHILADELPHIA ZOO, 3400 W. Girard Ave., Philadelphia, PA 19104-1139. Tel.: 215-243-1100, ext. 0. Fax: 215-243-5385.
E-mail: lastname.firstname@phillyzoo.org
Web Site: www.philadelphiazoo.org
Formerly: The Philadelphia Zoo and Zoological Garden
Founded: 1859.
Congressional District: 2
Key Personnel: Chm. Bd., Jay H. Calvert, Jr.; C.E.O. & Pres., Vikram Dewan; C.O.O., Dr. Andrew Baker; C.F.O., Joseph Steuer; C.M.O., Amy Shearer; Vice Pres. Animal Health, Keith C. Hinshaw, DVM; Vice Pres. Conservation, Kim Lengel; Vice Pres. Facilities, Nina Bisbee; Vice Pres. Community Affairs & Rgnl. Initiatives, Kenneth Woodson.
Personnel Profile: Full-Time Paid 195; Part-Time Paid 178; Part-Time Volunteers 600; Interns 131.
Governing Authority: society; nonprofit corporation; board of directors. Parent Institution: Zoological Society of Philadelphia. Tax-exempt: 501(c)(3).
Institution Type/Description: Zoo.
Collections: 1,200 species of wild animals; 30,000 species of plants; birds; mammals; reptiles; amphibians; invertebrates; pachyderm; bear. Historic Buildings: Solitude, 1785; Furness Gates; Bird House.

Research Fields: animal behavior; animal nutrition; exotic animal pathology; visitor behavior; animal husbandry.

Facilities: 1,400-vol. library of zoology & natural history available for use on premises; 42 acre Victorian garden; zoo.

Activities: guided tours; travel program; lectures; formally organized education programs for children, adults & graduate students affiliated with the University of Pennsylvania; children & adult workshops; docent programs; participatory, permanent exhibitions; mobile vans; zoo camp; junior zoo intern program; children's zoo volunteer program; education internships; live animal presentations; youth programming. Annual Events: Zoobilee; Adopt Day; International Migratory Bird Days; Teach Fest; Boo At The Zoo; Party for the Planet, Rock 'N' Roar; Sunset Safari.

Publications: monthly e-newsletter; member news.

Hours & Admission Prices: March-Nov. daily 9:30-5; Dec.-Feb. daily 9:30-4. March-Nov. adult $18, child 2-11 $15; Dec.-Feb. adults $14; discounts to groups with advanced reservations; children under 2 & members no charge. Closed New Year's Eve & Day; Thanksgiving; Christmas Eve & Day. &

Attendance: 1,200,000 (accurate)

Membership: Individual $59; Family $99; Family Plus $129; Family Deluxe $229; Contributor $295; Supporter $400; Benefactor $800; Associate $1,300; Patron $2,500; Presidents Circle $5,000; Leadership Partner $10,000; Founder $15,000; Keeper of the Kingdom $20,000.

***　PLEASE TOUCH MUSEUM, (M),** Memorial Hall, Fairmount Park, 4231 Ave. of the Republic, Philadelphia, PA 19131-3719. Tel.: 215-581-3181. Fax: 215-581-3182.

E-mail: info@pleasetouchmuseum.org

Web Site: www.pleasetouchmuseum.org

Founded: 1976.

Congressional District: 3

Key Personnel: Interim Pres. & C.E.O., Lynn McMaster; Vice Pres. Operations, John McDevitt; Vice Pres. External Rels., Joe Costello; Chm. Bd., Sally Stetson; Vice Pres. Finance & C.F.O., Michael Armento; Vice Pres. Devel., Stephanie Capello; Vice Pres. Education, Trapeta Mayson; Vice Pres. Community Learning, Leslie Walker; Cur., Stacey Swigart; Dir. Retail Operations, Beth Kirk; Exec. Asst., Caitlyn Dion.

Personnel Profile: Full-Time Paid 72; Part-Time Paid 46; Part-Time Volunteers 4.

Governing Authority: nonprofit organization. Tax-exempt: 501(c)(3).

Institution Type/Description: Children's Museum.

Collections: contemporary American toys post-1945; period children's artifacts; archives; art & sculpture; photographs; 1876 Centennial; operational Denzel Carousel.

Research Fields: play; value of play; cognitive learning in informal environments; toy history; toy safety; play behavior; 1876 Centennial exhibition; history of childhood in the Delaware Valley.

Facilities: 35,886 sq. ft. exhibit space; 130-seat theatre; classrooms; meeting rooms; cafe. Retail items including children's books, toys, games and gift items for sale.

Activities: special events; group self-guided programs; guide tours; community outreach; theater performances & creative dramatics programs including dance & puppetry; science demonstrations; hands-on exhibits; literary events; book award; studio workshops.

Publications: monthly newsletter; annual report; PTH notebook; pamphlets; fact sheet on museum history & Centennial 1876.

Hours & Admission Prices: Mon.-Sat. 9-5, Sun. 11-5. Admission $16; discounts to AAM & ACM members; children under one no charge. Closed Thanksgiving; Christmas. &

Attendance: 592,000 (estimated)

Membership: Nanny/Caregiver $30; Basic $150; Explorer $180; Premier $220; Centennial Guild $350 & up.

PRESBYTERIAN HISTORICAL SOCIETY, 425 Lombard St., Philadelphia, PA 19147-1516. Tel.: 215-627-1852. Fax: 215-627-0509.

E-mail: refdesk@history.pcusa.org

Web Site: www.history.pcusa.org

Founded: 1852.

Congressional District: 1

Key Personnel: Dir., Frederick J. Heuser, Jr.

Governing Authority: church. Affiliated with the Dept. of History, office of General Assembly, Presbyterian Church. (U.S.A.). Tax-exempt: 501(c)(3).

Institution Type/Description: Religious Museum.

Collections: oil portraits; communion tokens; silver; pewter; pictures; maps; manuscript collection of 400,000 letters; early European imprints before 1800; early Bibles; archives; numismatic.

Facilities: 117,000-vol. library of books on Presbyterian & Reformed Church history available for use on premises; reading room.

Activities: guided tours; lectures; permanent & temporary exhibitions.

Publications: quarterly journal, American Presbyterians: Journal of Presbyterian History; books, Presbyterian Historical Society Publication Series.

Hours & Admission Prices: Mon.-Fri. 8:30-4:30. No charge. Closed national holidays. &

Attendance: 1,000

Membership: Lois Harkrider Stair Associates $1-$99; Hallie Paxson Winsborough Club $100-$249; William H. Sheppard Council $250-$499; Lucy Craft Laney Scholars $500-$999; John Wanamaker Society $1,000-$4,999; John Chavis Fellows $5,000-$9,999; John Witherspoon Circle $10,000-$14,999; Francis Makemie Society $15,000 & up.

THE PRINT CENTER, 1614 Latimer St., Philadelphia, PA 19103-6308. Tel.: 215-735-6090. Fax: 215-735-5511.

E-mail: info@printcenter.org

Web Site: www.printcenter.org

Founded: 1915.

Congressional District: 2

Key Personnel: Exec. Dir., Elizabeth F. Spungen; Asst. Dir., Ashley Peel Pinkham; Pres. (V), Hester Stinnett; Cur., John Caperton; Gallery Store Mgr., Eli VandenBerg.

Personnel Profile: Full-Time Paid 4; Part-Time Paid 3; Part-Time Volunteers 50; Interns 10.

Governing Authority: nonprofit organization. Tax-exempt: 501(c)(3).

Institution Type/Description: Arts Center.

Collections: permanent print collection at the Philadelphia Museum of Art; archive collection at the Historical Society of Pennsylvania.

Research Fields: prints & photographs.

Facilities: reference service. Museum-related items for sale.

Activities: lectures; workshops; temporary & traveling exhibitions; photography & print shows; consultations; referrals; consignment photo & print sales.

Publications: exhibition catalogs.

Hours & Admission Prices: Tues.-Sat. 11-6. No charge; donations accepted.

Attendance: 8,000 (estimated)

Membership: Student $30; Artist & Individual $40; Contributing (2 individuals) $75; Sustaining $150; Collector $300; Zinc $500; Copper $1,000; Silver $2,500; Corporate $3,500; Gold $5,000; Platinum $10,000.

RODIN MUSEUM, Benjamin Franklin Pkwy. at 22nd St., Philadelphia, PA 19130. Mailing Address: c/o Philadelphia Museum of Art, P.O. Box 7646, Philadelphia, PA 19101-7646. Tel.: 215-763-8100. Fax: 215-235-0050. TDD: 215-684-7600.

Web Site: www.rodinmuseum.org

Founded: 1926.

Congressional District: 3

Key Personnel: Sr. Cur. The Gisela & Dennis Alter European Paintings Before 1900 & The John G. Johnson Collection, Joseph J. Rishel; Museum Shop Mgr., Stuart Gerstein.

Governing Authority: municipal; nonprofit. Parent Institution: The Philadelphia Museum of Art, P.O. Box 7646, 19101. Tax-exempt.

Institution Type/Description: Art Museum.

Collections: Auguste Rodin sculpture & drawings.

Research Fields: art history; conservation.

Facilities: Books & replicas of Rodin sculptures for sale.

Activities: guided tours; lectures; gallery talks; inter-museum loan & permanent exhibitions; audio tours.

Publications: handbook; catalog, The Sculpture of Auguste Rodin.

Hours & Admission Prices: Tues.-Sun. 10-5. Suggested Donation: adults $3. Closed holidays. &

Attendance: 52,480 (accurate)

Membership: Included in Philadelphia Museum of Art membership. Student $35; Individual $50; Household $85; Contributing $110; Supporting $150; Sustaining $250; Sponsor $500; Patron $1,000; Museum Associates $1,500 & up.

ROSENBACH MUSEUM & LIBRARY, (M), 2008-2010 DeLancy Place, Philadelphia, PA 19103-6584. Tel.: 215-732-1600. Fax: 215-545-7529.

E-mail: info@rosenbach.org

Web Site: www.rosenbach.org/

Founded: 1954.

Congressional District: 2

Key Personnel: Chm., Susan B. Miller; Dir., Derick Dreher; Cur. & Dir. Collections, Judith M. Guston; Registrar, Karen Schoenewaldt; Librarian, Elizabeth E. Fuller; Facilities Mgr., Stacey Hendricks; Buyer & Museum Shop Mgr., Candace Wilkin; Visitor Svcs. Mgr. & IT Coord., Lauren Abshire; Exec. Asst., Cathleen Chandler; Publicist, Megan Wendell; Dir. Devel., Christina Deemer; Hirsig Family Dir. Education, Emilie Parker;

Asst. Dir. Education, Farrar Fitzgerald; Asst. Cur., Katherine Hurwich Haas; Devel. Assoc., Mary Duffy; Traveling Exhibitions Coord., Patrick Rodgers; Mktg. & Special Events Assoc., Elyse Poinsett.

Personnel Profile: Full-Time Paid 17; Full-Time Volunteers 1; Part-Time Paid 4; Part-Time Volunteers 42.

Governing Authority: nonprofit corporation. Tax-exempt.

Institution Type/Description: Rare Books & Library Museum.

Collections: rare books & manuscripts: British & American literature, Chaucer-present; Joyce's Ulysses manuscript; American history, 1504-1941; 18th- & 19th-century English, French & American furniture, paintings, silver & other decorative arts; incunabula; over 7,000 drawings & watercolors by Maurice Sendak; the Marianne Moore literary archives; prints, drawings, book illustrations & portrait miniatures.

Research Fields: bibliography; decorative arts; book illustrations; American & English literature; American history.

Facilities: banquet facility. Museum-related items for sale.

Activities: museum & gallery tours; public programs; children's programming; permanent & temporary exhibitions; museum membership program; volunteer docent program; rental space available. Museum Sponsors: Bloomsday celebration in June; special events in conjunction with exhibitions.

Publications: newsletter; brochure; catalog, A Selection From Our Shelves; publications of Maurice Sendak; Cook's Choice: A Selection of Recipes; Passing Through: Letters & Documents Written in Philadelphia by Famous Visitors; Rosenwald & Rosenbach: Two Philadelphia Bookmen; facsimile of James Joyce's autograph manuscript of Ulysses; facsimile of Poor Richard's Almanac by Benjamin Franklin; 18th-century French Book Illustration; Drawings of Jean-Baptiste LePrince; Marianne Moore: Vision Into Verse; Rosenbach Abroad; Blake to Beardsley; Rosenbach Redux: Further Book Adventures in England & Ireland; Sendak at the Rosenbach; The Poet's Progress: Robert Burns; Bram Stroker's Dracula: A Centennial Exhibition; The Advent of Alice: A Celebration of the Carroll Centenary; Nakahama Manjiro's Hyosen Kiryaku: A Companion Book; Ulysses in Hand: The Rosenbach Manuscript.

Hours & Admission Prices: Tues. & Fri. 12-5, Wed.-Thurs. 12-8, Sat.-Sun. 12-6. Guided Tours: adults $10, senior citizens 65 & over $8, children & students with ID $5; discounts to groups, AAM & ICOM members and museum professionals; children under 5 & members no charge. Closed national holidays. &

Attendance: 11,500 (accurate)

Membership: Student & Senior Citizen $25; Friend $45; Household $75; Connoisseur $150; Patron $300; Aficionado $500; Bibliophile $1,000.

ROXBOROUGH MANAYUNK ART CENTER, 419 Green Lane, Philadelphia, PA 19128-3325. Tel.: 215-482-3363.

Institution Type/Description: Art Gallery.

Collections: paintings; sculpture.

Activities: special events.

Hours & Admission Prices: Sat.-Sun. 10-4.

RYERSS MUSEUM & LIBRARY, Burholme Park, 7370 Central Ave., Philadelphia, PA 19111-3059. Tel.: 215-685-0544 & 0599.

E-mail: ryerssmuseum@hotmail.com

Web Site: ryerssmuseum.org

Founded: 1910.

Congressional District: 4

Key Personnel: Park Historian & Admin. Supvr., Theresa Stuhlman.

Personnel Profile: Full-Time Paid 1; Part-Time Paid 5; Part-Time Volunteers 2.

Governing Authority: municipal; nonprofit. Parent Institution: Philadelphia Dept. of Parks & Recreation.

Institution Type/Description: Decorative Art & General Museum: housed in 1859 Victorian House.

Collections: 18th- & 19th-century decorative arts; China Trade & Oriental collections; Victorian & period furnishings.

Research Fields: pertaining to the collections.

Facilities: 20,000-vol. library of general and Victorian era books; reading room.

Activities: children's crafts; adult craft workshops; stamp club meetings; used book sale.

Hours & Admission Prices: Fri.-Sun. 10-4. No charge; donations accepted. Closed New Year's Day, Easter, Thanksgiving, Christmas. &

Attendance: 6,171 (accurate)

Membership: Friends of Ryerss $10.

ST. GEORGE'S UNITED METHODIST CHURCH, 235 N. Fourth St., Philadelphia, PA 19106-1194. Tel.: 215-925-7788.

E-mail: office@historicstgeorges.org

Web Site: www.historicstgeorges.org

Founded: 1767.

Key Personnel: Admin., Donna Miller

Governing Authority: church. Tax-exempt.

Institution Type/Description: Religious Museum: housed in 1763 St. George's United Methodist Church, built of British brick & located on the original site of construction.

Collections: original John Wesley Chalice Cup sent from England to Francis Asbury in 1784; Francis Asbury's personal Bible brought with him from England in 1771; manuscripts; Asbury watch.

Research Fields: British & American church history; records of closed churches.

Facilities: 7,000-vol. library of European & American books on Methodism available for research on premises; reading room.

Activities: guided tours; temporary exhibitions. Museum Sponsors: Associate Membership Program.

Publications: quarterly newsletter to associate members.

Hours & Admission Prices: Mon.-Fri. 10-3, Sat.-Sun. by appointment. No charge; donations accepted.

Attendance: 2,000 (estimated)

SAMUEL S. FLEISHER ART MEMORIAL, 719 Catharine St., Philadelphia, PA 19147-2811. Tel.: 215-922-3456, ext. 318.

E-mail: info@fleisher.org

Web Site: www.fleisher.org

Key Personnel: Bd. Pres., Liz Price; Exec. Dir., Matthew Braun; Exhibitions Mgr., Warren Angle.

Governing Authority: Parent Institution: Philadelphia Museum of Art.

Institution Type/Description: Art Museum.

Collections: works by regional artists.

Activities: Annual Event: The Challenge (artists competition).

Hours & Admission Prices: Mon.-Fri. 11-5; call for additional hours.

SCHMIDT-DEAN GALLERY, 1719 Chestnut St., Philadelphia, PA 19103. Tel.: 215-569-9433. Fax: 215-569-9434.

E-mail: schmidtdean@netzero.net

Web Site: schmidtdean.com

Founded: 1988.

Institution Type/Description: Art Gallery.

Collections: works by regional, national & international artists.

Major Exhibits: Robert Straight, 12/13-1/14.

Hours & Admission Prices: Tues.-Sat. 10:30-6. No charge.

THE SCHUYLKILL CENTER FOR ENVIRONMENTAL EDUCATION, 8480 Hagy's Mill Rd., Philadelphia, PA 19128-1998. Tel.: 215-482-7300, ext. 110. Fax: 215-482-8158.

E-mail: scee@schuylkillcenter.org

Web Site: www.schuylkillcenter.org

Founded: 1965.

Congressional District: 2

Key Personnel: Exec. Dir., Dennis Burton; Pres. Bd. Trustees, John Howard; Bd. Vice Pres., Anne Bower; Bd. Vice Pres., Jeffery Hayes; Bd. Vice Pres., Lara Herzig Malatests; Bd. Vice Pres., Wendy Willard; Bd. Sec., Ron Varnum; Dir. Education, Virginia Ranly; Dir. Wildlife Rehabilitation, Rick Schubert; Dir. Art Programs, Mary Salvante; Dir. Land Restoration, Fran Lawn; Grant Coord., Emily Simmons.

Personnel Profile: Full-Time Paid 15; Part-Time Paid 15; Part-Time Volunteers 108.

Governing Authority: nonprofit organization. Parent Institution: SCEE, Inc. Tax-exempt: 501(c)(3).

Institution Type/Description: Nature Center.

Collections: Over 400-acres of the nature center; insect collection; taxidermied birds; mineral & herbarium; rare books; solar panel array & green roof.

Research Fields: educational process; ecological restoration; environmental education.

Facilities: nature center; field research station; educational facilities; 200-seat auditorium; K-8 charter school; interactive discovery center. Museum-related items for sale.

Activities: guided tours; lectures; arts festivals; organized education programs children, adults & undergraduate or graduate college students affiliated with Philadelphia, Arcadia & Temple Universities; school loan service.

Publications: quarterly newsletter, The Quill; seasonal brochures; annual report.

Hours & Admission Prices: Center: Mon.-Sat. 8:30-5. Trails: daily 8:30-4:30. No charge; donations accepted. Closed New Year's Day; Easter; Memorial Day; Independence Day; Labor Day; Thanksgiving; Christmas. &

Attendance: 55,500 (accurate)

Membership: Individual $40; Family $60; Wildlife $75; Contributing & Organizational $100; Fellows $250-$1,000.

THE STEPHEN GIRARD COLLECTION, (M), Girard College #116, 2101 S. College Ave., Philadelphia, PA 19121-4857. Tel.: 215-787-4434. Fax: 215-787-4404.
E-mail: elaurent@girardcollege.edu
Web Site: www.girardcollege.edu
Founded: 1831.
Congressional District: 2
Key Personnel: Dir. Historical Resources, Elizabeth M. Laurent; Pres. Girard College, Clay Armbrister.
Personnel Profile: Full-Time Paid 1; Part-Time Volunteers 4.
Governing Authority: Parent Institution: Board of Directors of City Trusts. Subsidiary Institution: Girard College. Tax-exempt.
Institution Type/Description: Period Furniture & Decorative Arts Museum: located in Founder's Hall, Greek Revival Building designed by Thomas U. Walter, on the grounds of Girard College.
Collections: personal artifacts of Stephen Girard (1750-1831) including furniture by Trotter, Connelly & Haines; American, French & English Silver; China trade items; paintings, books, porcelain; S.G. papers: 1780-1831 correspondence, including U.S. presidents, statesmen, naval officers & diplomats; ledgers & journals; charts; maps; broadsides; drawings; ships & shipping; 1832 architectural competition drawings; T.U. Walter construction drawings for Founder's Hall.
Activities: guided tours; Stephen Girard papers on microfilm are available to scholars by appointment at the American Philosophical Society in Philadelphia.
Publications: Stephen Girard Collection; A Catalogue of the Personal Library of Stephen Girard; Monument to Philanthropy: The Design and Building of Girard College, 1832-1848; Girard College - A Living History.
Hours & Admission Prices: Thurs. 9-2; group tours available by appointment at other times. Admission fee charged.
Attendance: 2,000 (accurate)

TALLER PUERTORRIQUENO, 2721 N. 5th St., Philadelphia, PA 19133. Tel.: 215-426-3311. Fax: 215-426-5682. Facebook: Taller Puertorriqueno.
Web Site: www.tallerpr.org
Formerly: Lorenzo Homar Gallery
Founded: 1974.
Key Personnel: Dir., Carmen Febo San Miguel; Museum Shop Mgr., Aida Devine
Institution Type/Description: Art Gallery.
Collections: works of Latin American & Caribbean art.
Major Exhibits: Modern Madonna's Counterpoint, 1/21/14-4/5/14; Iconography of Meaning in Contemporary Art, 4/25/14-7/26/14.
Activities: special events.
Hours & Admission Prices: Mon.-Fri.9-5, Sat. 10-6. No charge, donations accepted.
Membership: Student $20; Individual $30.

TEMPLE CONTEMPORARY, TYLER SCHOOL OF ART OF TEMPLE UNIVERSITY, 2001 N. 13th St., Philadelphia, PA 19122-6016. Tel.: 215-777-9144. Fax: 215-777-9143.
E-mail: exhibitions@temple.edu
Web Site: www.templecomtemporary.org
Founded: 1985.
Key Personnel: Dir., Robert Blackson; Asst. Dir., Sarah Biemiller.
Personnel Profile: Full-Time Paid 2; Part-Time Paid 3.
Governing Authority: public university; nonprofit. Parent Institution: Temple University. Subsidiary Institution: Tyler School of Art. Tax-exempt.
Institution Type/Description: Art Gallery.
Collections: works by contemporary artists.
Research Fields: contemporary art.
Activities: films; lectures; symposiums; traveling exhibitions.
Publications: exhibition catalogues.
Hours & Admission Prices: Wed.-Sat. 11-6; other times by appointment. No charge. ♿
Attendance: 20,000 (estimated)
Membership: Friends of the Gallery $25; $100; $1,000; $5,000.

TREASURY OF FAITH MUSEUM, 810 N. Franklin St., Philadelphia, PA 19123. Tel.: 215-627-3389. Fax: 215-627-4225.
E-mail: tofmuseum@catholic.org
Institution Type/Description: Religious Museum.
Collections: Ukrainian Catholic Church history & roots; religious artifacts; altar; paintings; signed antimensions; iconostas from the original Ukrainian Catholic Cathedral of the Immaculate Conception; bishop's throne; tabernacle; liturgical books; sacred vessels.

Facilities: classrooms.
Hours & Admission Prices: Call for hours.

✳ UNIVERSITY OF PENNSYLVANIA MUSEUM OF ARCHAEOLOGY AND ANTHROPOLOGY, (M), 3260 South St., Philadelphia, PA 19104-6324. Tel.: 215-898-4000. Fax: 215-898-0657.
E-mail: info@pennmuseum.org
Web Site: www.penn.museum
Founded: 1887.
Congressional District: 1
Key Personnel: Chm. (V), Bd. Overseers, Michael J. Kowalski; Dir., Richard Hodges, Ph.D.; Cur. Mediterranean Section, Brian Rose, Ph.D.; C.O.O., Melissa Smith; Assoc. Dir. Admin. & Finance, Alan Waldt; Chief of Staff, James Mathieu; Dir. Public Rels., Pam Kosty; Dir. Devel., Amanda Mitchell-Boyask; Dir. Community Engagement, Jean Byrne; Senior Archivist, Alex Pezzati; Head Librarian, John Weeks, Ph.D.; Dir. Exhibits & Lead Exhibit Designer, Kate Quinn; Asst. Dir. Special Events, Tena Thomason; Senior Registrar, Xiuqin Zhou; Head Conservator, Lynn Grant; Assoc. Cur. Babylonian Section, Stephen Tinney, Ph.D.; Cur. American Section, Robert Preucel, Ph.D.; Cur. Egyptian Section, David Silverman, Ph.D.; Assoc. Cur. Near East Section, Richard Zettler, Ph.D.; Assoc. Cur. Historical Archaeology Section, Robert L. Schuyler, Ph.D.; Chair Women's Committee, Barbara Rittenhouse; Volunteer & Staffing Mgr., Jane Nelson; Volunteer Docents Chair, Lynn Smith; Museum Shop Mgr., Kevin Freitag.
Personnel Profile: Full-Time Paid 102; Full-Time Volunteers 9; Part-Time Paid 54; Part-Time Volunteers 250; Interns 15.
Governing Authority: university. Parent Institution: The University of Pennsylvania. Tax-exempt: 501(c)(3).
Institution Type/Description: University Archaeology & Anthropology Museum.
Collections: Egyptian, Mediterranean, Near Eastern, African, South, Southeast & East Asian, Oceanic & Australian, American, Mesoamerican archaeological & ethnographic materials.
Research Fields: fields related to collections.
Facilities: 80,000-vol. library on archaeology, anthropology & ethnology available for use of University of PA students, visiting scholars & staff; archives containing 2,000 linear feet and 300,000 photographs of expedition & administrative records for use by qualified researchers; 60,375 sq. ft. of exhibition space; two auditoriums; classrooms; cafeteria.
Activities: lectures; films; gallery talks; concerts; docent program; travel program; community school programs; formally organized education programs for undergraduate & graduate students; inter-museum loan, permanent, temporary & traveling exhibitions; school loan service; family-oriented world culture days.
Publications: magazine 3 times a year, EXPEDITION; quarterly calendar of events; quarterly membership newsletter; annual report; monographs; guide books; catalogues; edited volumes.
Hours & Admission Prices: Tues. & Thurs.-Sun. 10-5, Wed. 10-8. Adults $10, senior citizens 65 & over, children 6-17 & students $7; discounts to AAM & ICOM members; children under 6, Penn Card holders & members no charge. Closed holidays. ♿
Attendance: 145,000 (estimated)
Membership: Student $40; Associate $50; Individual $55; Dual $80; Household $95; Friends & Family $150.

THE UNIVERSITY OF THE ARTS - ROSENWALD-WOLF GALLERY, 333 S. Broad St., Philadelphia, PA 19107-5839. Mailing Address: 320 S. Broad St., Philadelphia, PA 19102-4994. Tel.: 215-717-6480. Fax: 215-717-6468.
E-mail: ssachs@uarts.edu
Web Site: www.uarts.edu
Founded: 1876.
Congressional District: 1
Key Personnel: Pres. (V), Sean Buffington; Dir., Sid Sachs.
Personnel Profile: Full-Time Paid 2; Part-Time Paid 10.
Governing Authority: college; nonprofit organization. Parent Institution: University of the Arts. Tax-exempt: 501(c)(3).
Institution Type/Description: University Art Gallery.
Collections: paintings.
Research Fields: contemporary design, fine arts and crafts.
Activities: temporary exhibitions, lectures & gallery talks; conferences; performances open to the public.
Publications: exhibition catalogues & illustrated brochures.
Hours & Admission Prices: Mon.-Fri. 10-5, Sat. 12-5. No charge. Closed academic holidays. ♿
Attendance: 30,000 (estimated)

U.S. MINT-PHILADELPHIA, 5th and Arch Sts., Philadelphia, PA 19106. Mailing Address: 151 N. Independence Mall E., Philadelphia, PA 19106-1886. Tel.: 215-408-0114. Fax: 215-408-2780. TDD: 215-597-0077.
E-mail: tgrant@usmint.gov
Web Site: www.usmint.gov/
Founded: 1792.
Key Personnel: Dir., Edmund Moy; Public Rels. & Exhibits, Timothy Grant.
Governing Authority: federal; nonprofit. Dept. of Treasury. Tax-exempt.
Institution Type/Description: Numismatics Museum.
Collections: coin production equipment; mint artifacts; coins; medals.
Research Fields: coins & medals produced by the U.S. Mint.
Facilities: 500-vol. library of coins, medal and mint history material; 22,000 sq. ft. exhibit space. Current U.S. special coins and medals for sale.
Publications: tour brochure.
Hours & Admission Prices: Mon.-Fri. 9-4:30. No charge. Closed federal holidays. &
Attendance: 240,000 (accurate)

THE VICTORIAN SOCIETY IN AMERICA, 1636 Sansom St., Philadelphia, PA 19103-5404. Tel.: 215-636-9872. Fax: 215-636-9873.
E-mail: info@victoriansociety.org
Web Site: www.victoriansociety.org
Founded: 1966.
Key Personnel: Pres. (V), Bruce Davies; Business Mgr., Robert McCown.
Personnel Profile: Full-Time Paid 1; Part-Time Paid 1; Part-Time Volunteers 40.
Governing Authority: society. Tax-exempt: 501(c)(3).
Institution Type/Description: Historical Society: housed in an1899 Victorian church.
Collections: documentary resources.
Research Fields: 19th-century America & Great Britain.
Facilities: library of 19th-century American History available for research in the Athenaeum only.
Activities: symposia; lectures; British & American summer schools; international & study tours; festivals.
Publications: quarterly newsletter, The Victorian; semiannual scholarly magazine, 19th Century.
Hours & Admission Prices: Mon.-Fri. 9-5. No charge. Closed New Year's Day; Memorial Day; Independence Day; Labor Day; Thanksgiving; Christmas.
Membership: Student $30; Library/Historic House & Museums/University $45; Individual $50; Household $60; Sustaining $100; Contributing & Business $250; Life $1,500.

WAGNER FREE INSTITUTE OF SCIENCE, (M), 1700 W. Montgomery Ave., Philadelphia, PA 19121-3227. Tel.: 215-763-6529. Fax: 215-763-1299.
E-mail: info@wagnerfreeinstitute.org
Web Site: www.wagnerfreeinstitute.org
Founded: 1855.
Congressional District: 3
Key Personnel: Dir., Susan Glassman; Museum Admin., Pat Warner; Librarian & Archivist, Lynn Dorwaldt; Dir. Children's Education, Dana Semos; Children's Educator, Holly Clark.
Personnel Profile: Full-Time Paid 9; Part-Time Paid 3; Part-Time Volunteers 60; Interns 8.
Governing Authority: nonprofit organization. Tax-exempt: 501(c)(3).
Institution Type/Description: Natural History & Natural Science Museum.
Collections: mineralogy; paleontology; entomology; botany; shells; mammals; fish & reptiles; birds; 19th-century scientific instruments.
Research Fields: history of science; systematics.
Facilities: 42,000-vol. library; archives; reading room; 450-seat auditorium.
Activities: lectures; films; formally organized educational programs; permanent exhibitions; symposia; tours; special weekend programs.
Publications: bulletin, Transactions; course catalogue.
Hours & Admission Prices: Tues.-Fri. 9-4. Museum: $8 suggested donation. Guided Tours: adults $15, seniors $10, children $5; discounts to members & groups. Education Programs: no charge. Closed national holidays.
Attendance: 18,000 (estimated)
Membership: Individual Friend $25; Family Friends $35; Contributor $50; Donor $100; Sustainer $250; Patron $500; Benefactor $1,000 & up.

✻ **WOODMERE ART MUSEUM, (M),** 9201 Germantown Ave., Philadelphia, PA 19118-2618. Tel.: 215-247-0476. Fax: 215-247-2387.
Web Site: www.woodmereartmuseum.org
Founded: 1940.

Congressional District: 13
Key Personnel: Dir. & C.E.O., William Valerio, Ph.D.; Pres., John Affleck; Cur. Education, Pamela Birmingham; Registrar, Sally Larson; Museum Educator, Hildy Tow; Museum Shop Mgr., Stephen Kersner.
Personnel Profile: Full-Time Paid 14; Part-Time Paid 8; Part-Time Volunteers 78.
Governing Authority: nonprofit organization. Tax-exempt: 501(c)(3).
Institution Type/Description: Art Museum: housed in 1867 Victorian mansion with modern additions.
Collections: works of Philadelphia artists; European collection of founder Charles Knox Smith including Oriental carpets, Royal Sevres & Meissen ware.
Research Fields: 19th- & 20th-century American Art.
Facilities: 2,000-vol. library of basic reference materials, including a collection of monographic volumes dedicated to artists represented in the collections. Works by regional artists including paintings, prints, ceramics & textiles for sale.
Activities: guided tours; gallery talks; lectures; concerts; permanent & temporary exhibitions; children's gallery.
Publications: exhibition catalogues; calendar of events; newsletter; program guide.
Hours & Admission Prices: Tues.-Thurs. & Sun. 10-5, Fri. 10-8:45, Sat. 10-6. Special Exhibitions: adults $10, senior citizens $7; discounts to AAM & ICOM members & Cultural Pass holders; Sun., members, children & college students with ID no charge. Closed selected holidays. &
Attendance: 48,000 (estimated)
Membership: Student $20; Individual $40; Family $70; Family & Friends $100; Patron $400.

WYCK, 6026 Germantown Ave., Philadelphia, PA 19144-2191. Tel.: 215-848-1690. Fax: 215-848-1612.
E-mail: wyck@wyck.org
Web Site: www.wyck.org
Founded: 1973.
Key Personnel: Exec. Dir., Eileen Rojas; Chm. (V), Rob Fleming; Vice Chm., Emily Lind Baker; Cur., Laura Keim; Horticulturist, Nicole Juday.
Personnel Profile: Full-Time Paid 1; Part-Time Paid 5; Part-Time Volunteers 20.
Governing Authority: nonprofit organization; administered by The Wyck Association. Tax-exempt: 170(b)(1)(A).
Institution Type/Description: Historic Building Museum: housed in 18th-century building.
Collections: furniture; glass; ceramics; metals; textiles; archives with nine generations of family manuscripts; rose & vegetable garden.
Research Fields: Germantown gardens & houses; social history.
Facilities: 2,000-vol. library pertaining to history, religion & science. Museum-related items for sale.
Activities: guided tours; organized education programs for children & adults; intern program for gardening students. Museum Sponsors: Farmer's Market May to November.
Publications: pamphlet, Germantown & Its Founders; brochure, Wyck.
Hours & Admission Prices: April-Dec. 15 Tues., Thurs. & Sat. 1-4. Family $10, adults & children $5, senior citizens $4; discounts to AAM & ICOM members; members no charge. Closed New Year's Day; Independence Day; Christmas; Thanksgiving. &
Attendance: 3,681 (accurate)
Membership: Individual $50; Family $75; Donor $150; Contributor $250; Benefactor $500; Patron $1,000.

Phoenixville

HISTORICAL SOCIETY OF THE PHOENIXVILLE AREA, 204 Church St., Phoenixville, PA 19460-3414. Tel.: 610-933-7646.
E-mail: hspa@verizon.net
Web Site: www.hspa-pa.org
Founded: 1980.
Congressional District: 6th
Key Personnel: Pres., Susan C. Marshall; Treas., Duane Parker; Devel., Richard Lusch; Public Rels., Martha Parker; Archivist, John R. Ertell; Office Mgr., Robert Deger.
Personnel Profile: Part-Time Volunteers 100.
Governing Authority: private; nonprofit organization. Tax-exempt: 501(c)(3).
Institution Type/Description: Historical Society Museum.
Collections: local history; Native American; industrial history; Etruscan Majolica pottery; documents; photographs.
Research Fields: regional & community history; genealogy.
Facilities: 275-vol. library; archives; 1,700 sq. ft. exhibit space.
Activities: formal education programs for adults; guided tours; temporary exhibitions; broadcast programs. Annual Events: Strawberry Festival; Flea Market; town walk; banquet.

Publications: quarterly newsletter, Newsletter of the Historical Society of the Phoenix Area.
Hours & Admission Prices: Wed.-Fri. 9-3, 1st Sun. of month 1-4; other times by appointment. No charge; donations accepted. &
Attendance: 750 (accurate)
Membership: Student & Senior Citizen $18; Individual $25; Family $30; Business $50; Friend of HSPA $75; Contributor $150; Sponsor $250.

Pittsburgh

THE ANDY WARHOL MUSEUM, (M), 117 Sandusky St., Pittsburgh, PA 15212-5890. Tel.: 412-237-8300. Fax: 412-237-8340.
Web Site: www.warhol.org
Founded: 1994.
Congressional District: 18
Key Personnel: Chm. Bd. Overseers, Randall Dearth; Pres. & C.E.O., David M. Hillenbrand; Dir. AWM, Thomas Sokolowski; Deputy Dir., Colleen Russell Criste; Archivist, Matt Wrbican; Cur. Film & Video, Geralyn H. Huxley.
Governing Authority: nonprofit organization. Parent Institution: Carnegie Institute. Tax-exempt: 501(c)(3).
Institution Type/Description: Art Museum: industrial warehouse with ornate terra cotta clad facade, built 1911-22.
Collections: concentrations on the works, creative process & life of Andy Warhol including paintings, prints, drawings, films, videos, audio-tapes, sculpture, photographs, installations & archival study center.
Research Fields: collections & cultural studies relating to the second half of the 20th-century.
Facilities: archives study center; education resource center & studio; 110-seat theater; cafe. Museum-related items for sale.
Activities: lectures; art activities for adults & children; films; temporary special exhibitions; rotating exhibitions of museum collections; group visits; teacher education workshops; opening and other events.
Publications: inaugural publication, The Andy Warhol Museum; Andy Warhol 1956-86: Mirror of His Time; The Warhol Look/Glamour Style Fashion; Andy Warhol: 365 Takes; magazine, CARNEGIE.
Hours & Admission Prices: Tues.-Thurs. & Sat.-Sun. 10-5, Fri. 10-10. Adults $15, senior citizens 55 & over $9, students & children 3-18 $8; discounts Fri. 5-10 and AAM & ICOM members; members of Carnegie Museums of Pittsburgh no charge. Closed legal holidays. &
Attendance: 92,000
Membership: Senior 65 & over $50; Individual $75; Dual $100; Family $130; Premium $200.

ASSOCIATED AMERICAN JEWISH MUSEUMS, 4905 Fifth Ave., Pittsburgh, PA 15213-2941. Tel.: 412-621-6566. Fax: 412-621-5475.
E-mail: jacob@rodefshalom.org
Founded: 1972.
Key Personnel: C.E.O. & Pres. (V), Walter Jacob; Financial Dir., Jeff Herzog; Education, F. Pomerantz; Public Rels., Francine Rickenbach.
Personnel Profile: Part-Time Volunteers 10.
Governing Authority: private; nonprofit organization. Tax-exempt: 501(c)(3).
Institution Type/Description: Ethnic History Association.
Collections: historical & artistic Judaica; synagogue collections.
Research Fields: modern Ketubot; rural American synagogues; Samuel Rosenberg, the artist.
Activities: lectures; loan & traveling exhibitions. Annual Event: Symposia.
Publications: quarterly newsletter, Gallery.
Hours & Admission Prices: No charge; donations accepted.
Attendance: 5,000 (estimated)
Membership: Synagogues $25-$106.

AUGUST WILSON CENTER FOR AFRICAN AMERICAN CULTURE, 980 Liberty Ave., Pittsburgh, PA 15222-3736. Tel.: 412-258-2700 & 338-8725. Fax: 412-258-2701.
E-mail: swhite@augustwilsoncenter.org
Web Site: www.augustwilsoncenter.org
Founded: 2002.
Congressional District: 14
Key Personnel: Co-Exec. Dir., Oliver Byrd; Co-Exec. Dir., Sala Udin; Chm. (V), Aaron Walton.
Personnel Profile: Full-Time Paid 19; Part-Time Paid 1; Part-Time Volunteers 140; Interns 3.
Governing Authority: Tax-exempt.
Institution Type/Description: Cultural Center.
Collections: African American culture & history; photography; paintings; period artifacts; communication artifacts.
Facilities: 486-seat theater; cultivation center; dance studio; meeting room; cafe.

Activities: visual & performing arts programs; educational programs.
Publications: season brochure; gallery guides; marketing collateral; curriculum guides.
Hours & Admission Prices: Tues.-Sat. 11-6. Adults $8, teachers $6, seniors & students $4; discounts to groups; members & NARM members no charge. Closed major holidays. &
Attendance: 27,000 (estimated)
Membership: Basic Individual $25; Basic Family $75; Premium Patron $250; Premium Leader $500; Premium President $1,000; Premium August Wilson Circle $2,500.

✱ **CARNEGIE MUSEUM OF ART, (M), (I),** 4400 Forbes Ave., Pittsburgh, PA 15213-4080. Tel.: 412-622-3301. Fax: 412-622-5787. Facebook: Carnegie Museum of Art.
Web Site: www.cmoa.org
Founded: 1895.
Congressional District: 14
Key Personnel: Chm., Martin G. McGuinn; Dir. The Henry J. Heinz II, Lynn Zelevansky; Cur. Fine Arts, Louise Lippincott; Cur. Architecture, Tracy Myers; Cur. Architecture, Raymund Ryan; Cur. Education, Marilyn M. Russell; Cur. Dec. Arts, Rachel Delphia; Chief Preparator, Kurt Christian; Chief Conservator, Ellen Baxter; Dir. Publications, Katie Reilly; Registrar, Monika Tomko; Dir. Devel., Jamie McMahon; Dir. Mktg., Nancy Netchi; Dir. Admin. Operations & Exhibitions, Sarah Minnaert; Web & Digital Media Mgr., Jeff Inscho; Media Relations Mgr., Jonathan Gaugler.
Governing Authority: nonprofit organization. Parent Institution: Carnegie Institute. Tax-exempt: 501(c)(3).
Institution Type/Description: Art Museum: housed in 1896-1907, Alden and Harlow American Renaissance-style building & 1974 Edward Larrabee Barnes addition.
Collections: European & American paintings with emphasis on 19th- & 20th-century American, French Impressionist & Post-Impressionist; sculpture; prints; drawings; watercolors; photographs; international & American contemporary art, including film & video; Asian & African art; European & American decorative arts; architectural & sculptural casts; architectural drawings & models.
Major Exhibits: 2013 Carnegie International, 10/5/13-3/16/14; Architecture + Photography, 4/13/14-5/25/14; David Hartt: Stray Light (Forum Gallery) (T), 5/17/14-8/10/14; Old Masters Works on Paper (working title), 5/31/14-9/14/14; Pittsburgh Biennial, 7/14-9/14; Gallery CSI: Renaissance Paintings (working title), 7/28/14-9/14/14; Sebastian Errazuriz, 9/6/14-11/14; Storyteller: The Photographs of Duane Michals (T), 11/14-2/15/15; The Duane Michals Collection (working title), 11/14-3/15/15.
Research Fields: pertaining to collections.
Facilities: library; outdoor sculpture court; theater; conservation laboratory; cafe. Museum-related items for sale.
Activities: guided & audio tours; lectures; classes; gallery talks; inter-museum loans; traveling exhibition program; special exhibitions; docent program; visiting filmmaker & video artist programs. Museum Sponsors: Carnegie International exhibition; Annual Associated Artists of Pittsburgh exhibition; Three Rivers Arts Festival.
Publications: exhibition & collection catalogues; educational materials; brochures; film program notes; CARNEGIE Magazine.
Hours & Admission Prices: Wed.-Mon. Adults $17.95, seniors $14.95, students w/ID & children 3-18 $11.95; discounts to AAM & ICOM members; members & children under 3 no charge. Closed Tues., New Year's Day; Easter; Independence Day; Labor Day; Thanksgiving; Christmas. &
Attendance: 645,200 (accurate)
Membership: Senior 65 & over $50; Individual $75; Dual $100; Family $150; Premium $250.

✱ **CARNEGIE MUSEUM OF NATURAL HISTORY, (M),** 4400 Forbes Ave., Pittsburgh, PA 15213-4080. Tel.: 412-622-3131. Fax: 412-622-8837.
E-mail: cmnhweb@carnegiemnh.org
Web Site: www.carnegiemnh.org
Founded: 1896.
Congressional District: 14
Key Personnel: Bd. Chm., Joseph Guyaux; Pres. & C.E.O., David M. Hillenbrand; Co Dir., Ron Baillie; Co Dir., Ann Metzger; Dir. Finance, Robert Spoharski; Cur. Vertebrate Paleontology, Dr. K. Christopher Beard; Cur. Vertebrate Paleontology, Dr. Dave Berman; Section Head & Collection Mgr., Marc Wilson; Collection Mgr. Amphibians & Reptiles, Stephen Rogers; Assoc. Cur. Botany, Dr. Cynthia Morton; Assoc. Cur., Invertebrate Zoology, Dr. John E. Rawlins; Cur. Anthropology, Dr. James B. Richardson, III; Cur. Emeritus Anthropology, Dr. David R. Watters; Assoc. Cur. Mollusks, Timothy Pearce; Cur. Public Programs, MaryAnn Steiner; Dir. Exhibitions & Visitor Experiences, Sarah Cole; Dir. Powdermill Nature

Reserve, Dr. John Wenzel; Mgr. Library, Xianghua Sun, Dir. Mktg., Nancy-Rose Netchi; Sr. Dir. Devel., Foundation Rels. & Science, Danni Piccolo; Dir. Devel., Alyssa DeLuca; Vice Pres. CMP Devel., Dolores F. Ellenberg.
Volunteer Hours: 29,873
Operating Expenses: 34,711,000
Operating Income: 34,711,000
Governing Authority: nonprofit organization. Parent Institution: Carnegie Institute. Branch Station: Powdermill Nature Reserve, 1847, Rte. 381, Rector, PA 15677-9605. Tax-exempt: 501(c)(3).
Institution Type/Description: Natural History & Anthropology Museum: housed in c.1896 American Renaissance style building.
Collections: geology & minerals; invertebrate & vertebrate paleontology; paleobotany; botany; herbarium; entomology; mollusks; amphibians; reptiles; ornithology; mammalogy; anthropology; archaeology; historic; wildlife art.
Research Fields: geology; invertebrate & vertebrate paleontology; paleobotany; botany; entomology; arachnids; amphibians; reptiles; ornithology; malacology; mammalogy; anthropology; archaeology; conservation; ecology.
Facilities: 150,000-vol. library of reference works on natural history & anthropology; maps; pamphlets, available for inter-library loan & to qualified students & research personnel; molecular & conservation laboratories; classrooms; lecture hall; cafe; 2,200-acre biological field station in western Pennsylvania. Reproductions, original objects & museum-related items for sale.
Activities: guided tours; teen docent program; lectures; films; gallery talks; in-service programs for teachers; formally organized education programs for children & adults; outreach programs to schools, hospitals & institutions with special-needs populations; training programs for professional museum workers; pre- & postdoctoral fellowship program; permanent & temporary exhibitions; distance learning program; school loan service.
Publications: scientific journals, Annals of Carnegie Museum, Bulletin Series; Special Publications Series; CARNEGIE Magazine; exhibition catalogues; pamphlets; popular books for adults & children.
Hours & Admission Prices: Mon., Wed. & Fri.-Sat. 10-5, Thurs. 10-8, Sun. 12-5; call to confirm. Adults $17.95, senior citizens $14.95, students with ID & children 3-18 $11.95; discounts to AAM & ICOM members and Thurs. after 4pm; members no charge. Closed legal holidays. &
Attendance: 921,198 (accurate)
Membership: Senior 65 & over $50; Individual $75; Dual $100; Family $150; Premium $250.

CARNEGIE MUSEUMS OF PITTSBURGH (CARNEGIE INSTITUTE), (M), 4400 Forbes Ave., Pittsburgh, PA 15213-4080. Tel.: 412-622-3131.

Web Site: www.carnegiemuseums.org
Founded: 1896.
Congressional District: 14
Key Personnel: Pres. & C.E.O., David M. Hillenbrand; C.F.O., Kevin D. Hiles; Vice Pres. Devel., Dolores F. Ellenberg; Vice Pres. Facilities, Planning & Operations, Tony Young; Dir. Corporate Human Resources, Beverly McGrath; Co-Dir. Henry Buhl, Jr., Carnegie Science Center, Ronald Baillie; Vice Pres. Mktg. & Co-Dir. Henry Buhl, Jr., Carnegie Science Center, Ann Metzger; Dir. The Andy Warhol Museum, Eric Shiner; The Henry J. Heinz II Dir. Carnegie Museum of Art, Lynn Zelevansky.
Personnel Profile: Full-Time Paid 412; Part-Time Paid 696; Part-Time Volunteers 614; Interns 47.
Volunteer Hours: 29,873
Operating Expenses: 34,711,000
Operating Income: 34,711,000
Governing Authority: nonprofit organization. Administers: The Andy Warhol Museum; Carnegie Museum of Art; Carnegie Museum of Natural History; Carnegie Science Center. Tax-exempt: 501(c)(3).
Institution Type/Description: General Museum.
Collections: located at The Andy Warhol Museum; Carnegie Museum of Art; Carnegie Museum of Natural History; Carnegie Science Center.
Major Exhibits: RACE: Are We So Different? (T), 5/10/13-10/26/14; 2013 Carnegie International, 10/5/13-3/16/14; Storyteller: The Photographs of Duane Michals (T), 10/14-2/15/15.
Research Fields: pertaining to collections.
Facilities: see The Andy Warhol Museum, Carnegie Museum of Art, Carnegie Museum of Natural History and Carnegie Science Center for additional facilities.
Activities: permanent exhibitions; educational programs for adults & children; guided tours; lectures; concerts; public events; member activities; member trips. See The Andy Warhol Museum, Carnegie Museum of Art, Carnegie Museum of Natural History, and Carnegie Science Center for additional activities.
Publications: CARNEGIE Magazine. See The Andy Warhol Museum, Carnegie Museum of Art, Carnegie Museum of Natural History, and Carnegie Science Center for additional publications.
Hours & Admission Prices: See The Andy Warhol Museum, Carnegie Museum of Art, Carnegie Museum of Natural History and Carnegie Science Center for hours & admission prices. &
Attendance: 921,198 (accurate)
Membership: Senior 65 & up $50; Individual $75; Dual $100; Family $150; Premium $250. (All memberships subject to change).

* CARNEGIE SCIENCE CENTER, (M), One Allegheny Ave., Pittsburgh, PA 15212-5895. Tel.: 412-237-3326. Fax: 412-237-3375.

Web Site: www.carnegiesciencecenter.org
Founded: 1991.
Congressional District: 18
Key Personnel: Henry Buhl, Jr. Co Dir., Ann Metzger; Henry Buhl, Jr. Co Dir., Ronald J. Baillie; Dir. Visitor Experience, Jessica Lausch; C.F.O., Kevin D. Hiles; Dir. Finance, Nancy Sherron; Vice Pres. Devel., Dolores F. Ellenberg; Dir. Devel. & Science, Danni Piccolo; Museum Shop Mgr., Donna Riedford.
Personnel Profile: Full-Time Paid 80; Part-Time Volunteers 170; Interns 15.
Volunteer Hours: 27,649
Operating Expenses: 9,430,000
Operating Income: 9,430,000
Governing Authority: nonprofit organization. Parent Institution: Carnegie Institute. Tax-exempt: 501(c)(3).
Institution Type/Description: Science and Technology Center.
Collections: Cold War-era submarine; space artifacts; model railroad & village.
Facilities: 150-seat planetarium; 300-seat auditorium; 340-seat OMNIMAX theater; coral reef aquarium; classrooms; observatory; computer learning lab; cafe; concession stand. Museum-related items for sale.
Activities: interactive exhibits; demonstrations; sky shows; laser shows; workshops for children & adults; science & engineering fair; teacher education; outreach programs; volunteer program; lab programs; overnighters; birthday parties; science apprenticeships.
Publications: astronomical calendar; magazine, CARNEGIE; newsletter, Science Impact.
Hours & Admission Prices: Sun.-Thurs. 10-5, Fri.-Sat. 10-7; call to confirm. Adults $17.95, children $9.95; discounts to groups, AAM, ASTC & ICOM members; members no charge. Closed Thanksgiving; Christmas; occasional Steelers home game days. &
Attendance: 535,000 (accurate)
Membership: Senior 65 & over $50; Individual $75; Dual $100; Family $150; Premium $250.

CENTER FOR AMERICAN MUSIC, University of Pittsburgh, Stephen Foster Memorial, 4301 Forbes Ave., Pittsburgh, PA 15260. Tel.: 412-624-4100. Fax: 412-624-7447.

E-mail: dir@pitt.edu
Web Site: www.pitt.edu/~amerimus/CAM1.htm
Founded: 1937.
Congressional District: 14
Key Personnel: Dir., Deane L. Root; Assoc. Dir., Kathryn Miller Haines.
Personnel Profile: Full-Time Paid 2; Part-Time Paid 1; Part-Time Volunteers 20; Interns 2.
Governing Authority: university. Parent Institution: University of Pittsburgh. Tax-exempt: 170(b)(1)(A).
Institution Type/Description: Music History Museum.
Collections: material relating to music in American Life; composer Stephen Collins Foster, 1826-1864; manuscripts; musical instruments; photographs.
Research Fields: life & works of Stephen Collins Foster; American music history.
Facilities: 30,000-item research library relating to American music 1830s-1930s emphasis; Life & Works of Stephen Collins Foster; archives; reading room; 580-seat auditorium; classroom.
Activities: guided tours; permanent exhibitions; performances; radio broadcasts; concert series; educational programs; lectures; curriculum guides.
Publications: songbook, Songs of Stephen Foster; research pamphlets; scores; sound recordings; biographical sketch of Foster.
Hours & Admission Prices: By appointment only. Closed university holidays. &
Attendance: 3,497 (estimated)

CHILDREN'S MUSEUM OF PITTSBURGH, (M), 10 Children's Way, Pittsburgh, PA 15212-5250. Tel.: 412-322-5058. Fax: 412-322-4932.

E-mail: stuffee@pittsburghkids.org
Web Site: www.pittsburghkids.org

Formerly: Pittsburgh Children's Museum
Founded: 1980.
Congressional District: 14
Key Personnel: Exec. Dir., Jane Werner; Deputy Exec. Dir., Chris Siefert; Pres., Jennifer Broadhurst; Vice Pres., Michael Duckworth; Sec., Winston Simmonds; Treas., Robert Denove; Dir. Finance, Rebecca McNeil; Dir. Visitor Svcs., Admissions & Museum Shop Mgr., George Brzezinski; Dir. Mktg., Bill Schlageter; Dir. Devel., Gina Focareta Evans; Dir. Exhibits, Penny Lodge; Dir. New Media, Suzanne McCaffrey; Dir. Learning & Research, Lisa Brahms.
Personnel Profile: Full-Time Paid 37; Part-Time Paid 120; Part-Time Volunteers 30.
Governing Authority: nonprofit organization. Tax-exempt: 501(c)(3).
Institution Type/Description: Children's Museum.
Collections: hands-on exhibits; renewable & educational resources; Margo Lovelace puppet & mask collection; communication artifacts; Andy Warhol's Myths; 10 silkscreen prints; Jim Henson puppets & props; Fred Rogers puppets; kids' climber; shimmering wind sculpture. Historic Buildings: c.1897 Post Office; c.1939 Planetarium.
Research Fields: anthropological studies relevant to international artifacts contained in Margo Lovelace puppet & mask collection; printmaking in the contemporary U.S.; interactive artwork; artists.
Facilities: 125-seat theater; classrooms; party rooms; 31,000 sq. ft. exhibit space; cafe. Museum-related items for sale.
Activities: participatory exhibits including play with real stuff for children 2-14; guided tours; organized education programs for children, undergraduate & graduate college students; volunteer program; traveling, loan, permanent & temporary exhibitions; school rental programs; school outreach programs; performing arts; puppet shows; plays; storytelling; residencies; afterschool & overnight programs for adolescents; literacy program for families; school & outreach programs; birthday party program; brochures; kids' climbers; rental facilities; family learning studies.
Publications: bimonthly e-newsletter; brochures; posters; banners; eblasts.
Hours & Admission Prices: Mon.-Sun. 10-5. Adults $13, senior citizens & children over 2 $12; discounts to ASTC & ACM reciprocal memberships; members no charge. Closed New Year's Day; Memorial Day; Independence Day; Labor Day; Thanksgiving; Christmas. &
Attendance: 248,971 (accurate)
Membership: Family $99; Extended Family & Grandparents $130; Deluxe Family $199.

THE FORT PITT BLOCK HOUSE, Point State Park, Pittsburgh, PA 15222. Mailing Address: 101 Commonwealth Place, Point State Pk., Pittsburgh, PA 15222-1212. Tel.: 412-471-1764.
Founded: 1894.
Congressional District: 14
Key Personnel: Pres. (V), Fort Pitt Society, Joanne Ostergaard; Cur. & Museum Shop Mgr., Emily Hoover.
Governing Authority: private; nonprofit organization. Parent Institution: Fort Pitt Society; Subsidiary Institution: Pittsburgh Chapter, NSDAR. The Fort Pitt Society of the Daughters of the American Revolution of Allegheny County, Pennsylvania; Pittsburgh Chapter DAR.
Institution Type/Description: Historic Building: house in the Fort Pitt Block House; built in 1764.
Collections: 1750-1800 historic artifacts from Pittsburgh; military & frontier artifacts; tools; muskets; cannon; furniture; maps; documents.
Research Fields: French & Indian War; early colonial, western Pennsylvania, & Pittsburgh history.
Facilities: private library for use by curators; lectures; 100-seat auditorium; classrooms. Reproduction tinware, books & other museum-related items for sale.
Activities: guided tours; lectures; formally organized education programs for children; docent program; permanent & temporary exhibits. Museum Sponsors: 18th-century music performances June to August; Royal American Regiment drills.
Hours & Admission Prices: April-Oct. Wed.-Sun. 10:30-4:30; Nov.-March Fri.-Sun. 10:30-4:30. No charge; donations accepted. Closed most legal holidays. &
Attendance: 10,000 (estimated)

THE FRICK ART & HISTORICAL CENTER, (M), 7227 Reynolds St., Pittsburgh, PA 15208-2919. Tel.: 412-371-0600. Fax: 412-371-6104.
E-mail: info@thefrickpittsburgh.org
Web Site: www.thefrickpittsburgh.org
Founded: 1990.
Congressional District: 14
Key Personnel: Dir., William B. Bodine, Jr.; Chm. (V), David A. Brownlee; Dir. Finance & Administrative Svcs., Lori E. Presto; Dir. External Affairs,

Susan S. Neszpaul; Dir. Operations & Visitor Svcs., Bill Nichols; Dir. Education, Amanda D. Gillen; Dir. Curatorial Affairs, Sarah Hall; Museum Store Mgr., Kate Blumen.
Personnel Profile: Full-Time Paid 36; Part-Time Paid 135; Part-Time Volunteers 5.
Volunteer Hours: 425
Operating Expenses: 5,200,378
Operating Income: 5,238,155
Governing Authority: nonprofit organization. Subsidiary Institution: The Frick Art Museum; Clayton, The Henry Clay Frick Estate; Car & Carriage Museum. Tax-exempt: 501(c)(3).
Institution Type/Description: Art & History Museums.
Collections: Frick Art Museum: European paintings with emphasis on Italian paintings of the early Renaissance; 18th- century French paintings & sculptures; bronze & terracotta sculpture; tapestries; decorative arts; Chinese porcelain. Clayton, The Henry Clay Frick Estate: Frick family heirlooms; decorative arts; period furniture & furnishings; 19th-century art; porcelain; glass; costumes & textiles. Car & Carriage Museum: period carriages & vintage automobiles.
Major Exhibits: An American Odyssey: The Warner Collection of American Paintings (T), 3/14-5/25/14; Edgar Degas: The Private Impressionist, Works on Paper by the Artist and His Circle (T), 6/27/14-10/4/14; Charles Courtney Curran: Seeking the Beautiful (working title) (T), 11/14-2/1/15.
Research Fields: pertaining to collections & programming.
Facilities: 165-seat auditorium; sketching & workshop studio; education center; restaurant. Museum-related items for sale.
Activities: guided tours; educational programs; lectures; traveling & permanent exhibitions; member receptions; special events; concerts; public events; film & video programs; cell phone tours.
Publications: member's newsletter & calendar; books, The Frick Art & Historical Center; My Father, Henry Clay Frick; catalogues, Renaissance & Baroque Bronzes in the Frick Art Museum; Collecting in the Gilded Age: Art Patronage in Pittsburgh 1890-1910; Clayton Days - Picture Stories; Aaronel de Roy Gruber: The Frick Landscapes; The Early Works of Henry Koerner; Pittsburgh Collects: European Drawings, 1500 to 1800; Artistry & Innovation in Pittsburgh Glass, 1808-1882: From Bakewell & Ensell to Bakewell, Pears & Co.; Impressions of Interiors: Gilded Age Paintings by Walter Gay.
Hours & Admission Prices: Tues.-Sun. 10-5. Clayton Tour: adults $12, seniors & students $10; discounts to AAM & ICOM members; members no charge. Other Venues: no charge; donations accepted. Closed New Year's Day; Independence Day; Thanksgiving; Christmas Eve & Day. &
Attendance: 130,749 (accurate)
Membership: Student $15; Senior $30; Individual & Teacher $55; Family $75; Fellow $125; Patron $250; Benefactor $500; Founders' Circle $1,000.

HISTORIC HARTWOOD MANSION, 200 Hartwood Acres, Pittsburgh, PA 15238-1193. Tel.: 412-767-9200. Fax: 412-767-0171.
E-mail: patti.benaglio@alleghenycounty.us
Web Site: www.county.allegheny.pa.us/parks
Founded: 1978.
Congressional District: 4
Key Personnel: Manager, Patti Benaglio; Chm. (V), Kathy Nicklas.
Personnel Profile: Part-Time Paid 9; Part-Time Volunteers 25.
Governing Authority: Parent Institution: Allegheny County Parks. Tax-exempt.
Institution Type/Description: Historic House Museum: housed in the former estate of John and Mary Flinn Lawrence, daughter of state Sen. William Flinn; built in 1929.
Collections: 16th-18th century furniture; personal artifacts; family photographs.
Facilities: stable complex.
Activities: weddings. Annual Event: holiday decorations mid-November to December; Holiday Teas.
Hours & Admission Prices: Mon.-Sat. 10-3, Sun. 12-4. Adults $6, senior citizens 60 & over and children 13-17 $4, children 6-12 $2, children 5 & under $1. Public Holiday Teas: $33. Closed holidays.
Attendance: 3,000 (estimated)

HUNT INSTITUTE FOR BOTANICAL DOCUMENTATION, Carnegie Mellon University, 5000 Forbes Ave., Pittsburgh, PA 15213-3815. Tel.: 412-268-2434. Fax: 412-268-5677.
Web Site: huntbot.andrew.cmu.edu
Founded: 1961.
Congressional District: 21
Key Personnel: Dir., Dr. Robert W. Kiger; Asst. Dir., Dr. T. D. Jacobsen; Business Officer & Sales Mgr., Lana M. Vernacchio; Art Cur., Lugene B. Bruno; Librarian, Charlotte A. Tancin; Archivist, J. Dustin Williams.
Personnel Profile: Full-Time Paid 17.
Governing Authority: university. Parent Institution: Research Division of Carnegie Mellon University. Tax-exempt.

Institution Type/Description: Science Art Institute.
Collections: botanical art & illustrations; watercolors, drawings & original prints; autograph collection from 18th through the 20th century; private papers of botanists; iconographic collection; Strandell collection of Linnaeana; Michel Adanson's library; Torner collection of Sesse & Mocino biological illustrations.
Research Fields: history of botany.
Facilities: 25,000-vol. library of books & journals on botany & its history, especially systemic botany, emphasizing pre-1850, available for research on premises; gallery.
Activities: temporary & traveling exhibitions.
Publications: brochures; journal; books; exhibition catalogues; bulletin.
Hours & Admission Prices: Mon.-Fri. 8:30-12 & 1-5, Sun. 1-4. No charge. Closed holidays.
Membership: Associate $35; Patron $100 & up.

MATTRESS FACTORY, LTD., (M), 500 Sampsonia Way, Pittsburgh, PA 15212-4444. Tel.: 412-231-3169. Fax: 412-322-2231.
E-mail: info@mattress.org
Web Site: www.mattress.org
Founded: 1977.
Congressional District: 14
Key Personnel: C.E.O. & Pres. (V), Barbara Luderowski; Cur., Michael Olijnyk; Dir. Operations, Catena Bergevin.
Personnel Profile: Full-Time Paid 12; Part-Time Paid 12; Part-Time Volunteers 40; Interns 5.
Governing Authority: nonprofit organization. Tax-exempt: 501(c)(3).
Institution Type/Description: Art Museum.
Collections: works by James Turrell, Winifred Lutz, Rolf Julius, Jene Highstein, Bill Woodrow, Allan Wexler, William Anastasi; collector's prints by Julius, Jene Highstein, William Anastasi, & Jessica Stockholder; installation art.
Major Exhibits: Janine Antoni at the MF, 5/24/13-5/14; Artists in Residence: Detroit, 9/13/13-5/25/14.
Facilities: library, including videos, tapes, slides, & books, pertaining to installation & performance art; 7,820 sq. ft. exhibit space; education studio space; cafe. Museum-related items for sale.
Activities: guided tours; youth & adult education workshops; lectures; participatory exhibits; performances by performance artists.
Publications: biannual newsletter, Installation & Performance; retrospective catalogue.
Hours & Admission Prices: Tues.-Sat. 10-5, Sun. 1-5; other times by appointment. Adults $15, senior citizens & students $10; discounts to North American Reciprocal Membership Program & AAM members, members & children under 6 no charge. Closed New Year's Day; Easter; Memorial Day; Thanksgiving; Christmas. &
Attendance: 75,000 (accurate)
Membership: Student $25; Senior $30; International $40; Household $125; Associate $250; Factory 500 $500; Patron $1,000.

THE MILLER GALLERY AT CARNEGIE MELLON, Purnell Center for the Arts, Carnegie Mellon University, Pittsburgh, PA 15213-3890. Mailing Address: 5000 Forbes Ave., Pittsburgh, PA 15213-3890. Tel.: 412-268-3618. Fax: 412-268-4746.
Web Site: www.cmu.edu/millergallery
Formerly: Regina Gouger Miller Gallery
Founded: 2000.
Key Personnel: Dir., Astria Suparak; Graphics & Office Coord., Margaret Cox.
Personnel Profile: Full-Time Paid 3; Part-Time Paid 15; Interns 2.
Governing Authority: Parent Institution: Carnegie Mellon. Tax-exempt.
Institution Type/Description: Art Gallery.
Collections: paintings.
Facilities: 9,000 sq. ft. exhibit space.
Activities: special events.
Publications: exhibition catalogues & brochures.
Hours & Admission Prices: Tues.-Sun. 12-6. No charge; donations accepted. &
Attendance: 15,000 (estimated)
Membership: Student $10; Individual $40

NATIONAL AVIARY IN PITTSBURGH, INC., Allegheny Commons West, 700 Arch St., Pittsburgh, PA 15212-5248. Tel.: 412-323-7235. Fax: 412-321-4364. Facebook: National Aviary.
E-mail: info@aviary.org
Web Site: www.aviary.org
Founded: 1952.
Congressional District: 14
Key Personnel: Mng. Dir., Cheryl Tracy; Museum Shop Mgr., Lori Urbowitz.

Personnel Profile: Full-Time Paid 30; Part-Time Paid 5; Part-Time Volunteers 69; Interns 7.
Governing Authority: private. Parent Institution: National Aviary in Pittsburgh, Inc. Tax-exempt: 501(c)(3).
Institution Type/Description: Aviary.
Collections: 883 live birds; plants.
Research Fields: bird behavior; diets; avian ecology; breeding & reproduction.
Facilities: library of natural history books available on premises by permission; zoological park; botanical garden. Bird guide books, nature oriented gift items for sale.
Activities: guided tours on special request; permanent exhibitions; live exhibits; volunteer program. Center Sponsors: special instruction for talented school children by arrangement.
Publications: member newsletter, Bird Calls; conservation newsletter, Flight Paths.
Hours & Admission Prices: Daily 10-5. Adults $13, seniors 60 & up $12, children 2-12 $11; discounts to AZA Reciprocal members; members & children 2 & under no charge. Closed Thanksgiving; Christmas Eve & Day. &
Attendance: 140,000 (accurate)
Membership: Senior $40; Individual $50; Dual $65; Family $85; Family Premium $125; Joint Membership with Children's Museum $180.

PHIPPS CONSERVATORY AND BOTANICAL GARDENS, (M), 1 Schenley Park, Pittsburgh, PA 15213-3830. Mailing Address: 1059 Shady Ave., Pittsburgh, PA 15232-2912. Tel.: 412-622-6914. Fax: 412-622-7363.
E-mail: info@phipps.conservatory.org
Web Site: www.phipps.conservatory.org
Founded: 1892.
Congressional District: 14
Key Personnel: Dir., Richard Piacentini; Chm. (V), Susan Golomb
Governing Authority: municipal; nonprofit organization. Tax-exempt.
Institution Type/Description: Conservatory.
Collections: orchids; palms; cacti & succulents; tropical plants; ferns; bonsai; plants from all over the world.
Facilities: 200-vol. library on horticulture available for reference; botanical gardens; display gardens under glass; outdoor perennial, rose, Japanese, & aquatic gardens. Gardening, conservatory-related items for sale.
Activities: permanent exhibits; seasonal flower shows; educational classes; guided tours.
Publications: e-newsletter, Phipps Welcomes the World; schools & programs brochure.
Hours & Admission Prices: Fri. 9:30am-10pm, Sat.-Thurs. 9:30-5. Adult $12, seniors & students $11, children 2-18 $9; discounts to groups of 15 or more & AAA members; children under 2 & members no charge. Closed Thanksgiving & Christmas. &
Attendance: 210,000 (estimated)
Membership: Student & Senior Citizen $50; Individual $55; Dual $75; Grandparents & Family/Household $90; Family/Household Plus $125; Supporting $250; Sustaining $500; Benefactor $1,000; Henry Phipps Associate $1,250.

PITTSBURGH FILMMAKERS/PITTSBURGH CENTER FOR THE ARTS, 6300 Fifth Ave., Pittsburgh, PA 15232-2922. Tel.: 412-361-0873 & 0455. Fax: 412-361-8338.
E-mail: info@pittsburgharts.org
Web Site: www.pittsburgharts.org
Founded: 1945.
Congressional District: 14
Key Personnel: Exec. Dir., Charlie Humphrey; Dir., Laura Domencic; Dir. Devel., Loretta Stanish; Cur., Adam Welch; Museum Shop Mgr., Jen Carter.
Personnel Profile: Full-Time Paid 38; Part-Time Paid 40; Part-Time Volunteers 50; Interns 4.
Governing Authority: nonprofit organization. Tax-exempt: 501(c)(3).
Institution Type/Description: Contemporary Arts Center.
Collections: various forms of art media.
Facilities: artworks by more than 600 regional & national artists for sale.
Activities: guided tours; lectures; organized education programs for children & adults with special needs, high school program; regional, national, international & traveling exhibitions, Museum Sponsors: Holiday Sale; Art Camp.
Publications: class schedule; email announcements & updates.
Hours & Admission Prices: Tues.-Sat. 10-5, Sun. 12-5. Suggested Donation $5; members no charge. Closed Christmas. &
Attendance: 36,067 (accurate)
Membership: Associate: Individual $60; Family $90. Access: Individual $75; Family $150.

PITTSBURGH ZOO AND PPG AQUARIUM, One Wild Place, Pittsburgh, PA 15206. Tel.: 412-665-3639, ext. 0. Fax: 412-665-3661.
E-mail: djones@pittsburghzoo.org
Web Site: zoo.pgh.pa.us
Founded: 1898.
Key Personnel: C.E.O. & Pres., Dr. Barbara T. Baker; Chm. (V), Beverlynn Elliott; Zoo Veterinarian, Dr. Ginger Takle; Education Coord., Margie Marks; Gift Shop Mgr., Eli Grill.
Personnel Profile: Full-Time Paid 110; Part-Time Paid 35; Part-Time Volunteers 165; Interns 6.
Governing Authority: Parent Institution: Zoological Society of Pittsburgh. Tax-exempt.
Institution Type/Description: Zoo & Aquarium.
Collections: birds, reptiles fish & mammals from all over the world; 4,000 specimens; 434 species.
Research Fields: Inia management & biology; African elephant reproduction.
Facilities: 500-vol. library of reference books.
Activities: art classes; twilight tours; summer camp program; workshops. Zoo Sponsors: Members' Day; Radio Stations Day.
Publications: bimonthly, Zoo Insider; Zoo Explorer.
Hours & Admission Prices: Spring & Fall daily 9-5; Winter daily 9-4 (gates close at 3); Memorial Day to Labor Day daily 9-6 (gates close at 4:30). Adults $14, senior citizens $13, children 2-13 $12; discount to groups; members & AAZPA members no charge. Closed New Year's Day; Thanksgiving; Christmas. &
Attendance: 941,116 (accurate)
Membership: Individual & Senior Couple $60; Family & Grandparent $85; Family Plus $120; Contributing $170.

RODEF SHALOM BIBLICAL BOTANICAL GARDEN, 4905 5th Ave., Pittsburgh, PA 15213-2941. Tel.: 412-621-6566. Fax: 412-621-5475.
E-mail: jacob@rodefshalom.org
Web Site: www.biblicalgardenpittsburgh.org
Founded: 1987.
Key Personnel: C.E.O., Charles Deaktor; Dir., Irene Jacob.
Personnel Profile: Full-Time Volunteers 2; Part-Time Volunteers 45.
Governing Authority: nonprofit. Parent Institution: Rodef Shalom congregation. Tax-exempt: 501(c)(3).
Institution Type/Description: Biblical Botanical Garden.
Collections: more than 200 temperate & tropical species of plants.
Research Fields: biblical plants; near Eastern plants.
Facilities: gardens.
Activities: guided tours; lectures; organized education programs for adults; temporary exhibitions; annual events.
Publications: periodic newsletter, Papyrus; biannual papers of symposia; annual booklets; books.
Hours & Admission Prices: June to mid-Sept. Sun.-Tues. & Thurs. 10-2, Wed. 7-9, Sat. 12-1. No charge. &
Attendance: 3,500 (estimated)

✳ SENATOR JOHN HEINZ HISTORY CENTER, (M), 1212 Smallman St., Pittsburgh, PA 15222-4200. Tel.: 412-454-6000. Fax: 412-454-6031.
E-mail: nschano@heinzhistorycenter.org
Web Site: www.heinzhistorycenter.org
Formerly: Senator John Heinz Pittsburgh Regional History Center
Founded: 1879.
Congressional District: 14
Key Personnel: Pres. & C.E.O., Andrew E. Masich; Chm. (V), Robert J. Cindrich; Museum Div. Dir., Anne Madarasz; Senior Vice Pres., Betty Arenth; Dir. Mktg. & Communications, Ned Schano; Dir. Human Resources, Renee Falbo; Dir. Meadowcroft Rockshelter & Historic Village, David Scofield; Dir. Library & Archives, Alexis Macklin; Museum Shop Mgr., Beth Muth.
Personnel Profile: Full-Time Paid 77; Part-Time Paid 40; Part-Time Volunteers 700; Interns 20.
Governing Authority: board of trustees. Parent Institution: Historical Society of Western Pennsylvania. Subsidiary Institution: Meadowcroft Rockshelter & Historic Village, Avella, PA; Fort Pitt Museum; Western Pennsylvania Sports Museum. Tax-exempt: 501(c)(3).
Institution Type/Description: History Museum.
Collections: books; archives; costumes; tools; vehicles; industrial artifacts; glass & decorative arts; toys & other items related to Western Pennsylvania history.
Research Fields: western Pennsylvania & Pittsburgh history; American business & industry; ethnicity & folklife; urban & rural history; family history; women's history.

Facilities: 35,000-vol. library & archive of regional history books, manuscripts, maps, family history, music & newspapers, available for interlibrary loan & for public use; reading room; 80-seat restaurant; classrooms; 7,000 sq. ft. Children's Discovery Hall. Museum-related items for sale.
Activities: concerts; guided tours; dance recitals; docent program; formal education for children; lectures; hobby workshops; participatory, permanent, temporary & traveling exhibits; rental gallery; theater; library & archives research facilities. Annual Events: History Makers Dinner; History Uncorked; National History Day; 1879 Founders Circle Dinner; History Book Fair; Asian Heritage Festival; Glass Expo.
Publications: quarterly magazine, Western Pennsylvania History; quarterly newsletter, Making History; books, Boundless Lives: Italian Americans of Western Pennsylvania; Glass: Shattering Notions; Points in Time: Building a Life in Western Pennsylvania; Pittsburgh's Strip District: Around the World in a Neighborhood; Pittsburgh Born, Pittsburgh Bred; Soul Soldiers: African Americans and the Vietnam Era; local history curriculum kits.
Hours & Admission Prices: History Center: daily 10-5. Sports Museum: daily 10-5. Library & Archives: Thurs.-Sat. 10-5. Meadowcroft: Memorial Day-Labor Day Wed.-Sat. 12-5, Sun. 1-5. History Center, Sports Museum, Library & archives: adults $10, seniors $9, children 5-17 $5; discounts to AAM members; children under 5 & Time Travelers no charge. Meadowcroft: adults $10, seniors $9, children 5-17 $5; discounts to AAM members & groups; children under 5 & members no charge. Closed New Year's Day; Easter; Thanksgiving; Christmas. &
Attendance: 160,000 (estimated)
Membership: Student $25; Individual $57; Grandparent $80; Family $85; Contributor $125; Patron $250.

SILVER EYE CENTER FOR PHOTOGRAPHY, 1015 E. Carson St., Pittsburgh, PA 15203-1109. Tel.: 412-431-1810. Fax: 412-431-5777.
E-mail: info@silvereye.org
Web Site: www.silvereye.org
Founded: 1985.
Institution Type/Description: Art Gallery.
Collections: photographs.
Activities: temporary exhibitions.
Hours & Admission Prices: Tues.-Sat. 12-6. No charge; donations accepted. &
Membership: Seniors 60 & over $25; Individual & College Student $35; Household $50; Gallery $70; International $100.

SOCIETY FOR CONTEMPORARY CRAFT, 2100 Smallman St., Pittsburgh, PA 15222-4440. Tel.: 412-261-7003. Fax: 412-261-1941.
E-mail: info@contemporarycraft.org
Web Site: www.contemporarycraft.org
Founded: 1971.
Congressional District: 14
Key Personnel: Exec. Dir., Janet L. McCall; Pres. Bd., David Blair; Dir. Exhibitions, Kate Lydon; Studio Program Coord., Rachel Saul; Office Mgr. & Devel. Asst., Erin McKevitt; Dir. Devel., Pamela Quatchak; Financial Asst., Yu-San Cheng; Museum Store Mgr., Megan Crowell; Marketing Mgr., Norah Guignon.
Personnel Profile: Full-Time Paid 8; Part-Time Paid 1; Part-Time Volunteers 40; Interns 3.
Governing Authority: nonprofit organization. Tax-exempt.
Institution Type/Description: Contemporary Craft Museum.
Collections: contemporary works of art by nationally & internationally known artists including clay, wood, metal, glass, jewelry, fiber & mixed media.
Major Exhibits: Enough Violence: Artists Speak Out (T), 9/27/13-3/22/14; Transformation 9: Contemporary Works in Clay, 4/11/14-8/23/14.
Research Fields: contemporary art with a focus on crafts.
Facilities: 1,700-vol. library; 3,000 sq. ft. exhibit space. Satellite gallery located at One Mellon Center, Downtown Pittsburgh. Contemporary crafts for sale.
Activities: guided tours; artist lectures; adult studio classes; quarterly public opening receptions; drop-in studio for kids; artist demonstrations.
Publications: exhibition catalogs; monthly e-newsletter; studio catalogs.
Hours & Admission Prices: Mon.-Sat. 10-5. No charge; donations accepted. Closed major holidays. &
Attendance: 128,000 (estimated)

SOLDIERS & SAILORS MILITARY MUSEUM & MEMORIAL, 4141 Fifth Ave., Pittsburgh, PA 15213. Tel.: 412-621-4253. Fax: 412-683-9339.
E-mail: curator@soldiersandsailorshall.org
Founded: 1910.
Congressional District: 14
Key Personnel: C.E.O., John F. McCabe.

Personnel Profile: Full-Time Paid 13; Part-Time Paid 7; Interns 1.
Governing Authority: Tax-exempt.
Institution Type/Description: Military History Museum.
Collections: military history; photographs; personal artifacts; uniforms.
Research Fields: military history; regimental history; membership records.
Hours & Admission Prices: Mon.-Sat. 10-4. Adults $5-$8; members, military & veterans no charge. Closed New Year's Day; Independence Day; Labor Day; Thanksgiving; Christmas. &
Membership: $30-$1,000.

UNIVERSITY ART GALLERY, UNIVERSITY OF PITTS-BURGH, 650 Schenley Dr., Frick Fine Arts Bldg., University of Pittsburgh, Pittsburgh, PA 15260-7601. Tel.: 412-648-2423. Fax: 412-648-2792.
E-mail: uag@pitt.edu
Web Site: www.haa.pitt.edu/collections/university-art-gallery
Founded: 1966.
Key Personnel: Dept. Chm., Barbara McCloskey.
Personnel Profile: Full-Time Paid 1; Interns 1.
Governing Authority: private university. Parent Institution: University of Pittsburgh. Tax-exempt.
Institution Type/Description: Art Gallery: housed in 1965 building by Helen Clay Frick.
Collections: emphasis on western Pennsylvania; paintings; Callot prints; Asian decorative arts; western Pennsylvania paintings; works by Gertrude Quastler.
Facilities: library; 200-seat auditorium; educational facilities.
Activities: lectures; loan, temporary & traveling exhibitions.
Hours & Admission Prices: Mon.-Fri. 10-4 & by appointment. No charge. Closed holidays; when the University is not in session. &

WOOD STREET GALLERIES, 601 Wood St., Pittsburgh, PA 15222-2503. Tel.: 412-471-5605. Fax: 412-232-3262.
Web Site: www.woodstreetgalleries.org
Key Personnel: Cur., Murray Horne; Asst. Cur., Kate Little; Preparator, George Dun; Preparator, Chris Korch; Preparator, Chris Beauregard; Preparator, Ian Brill; Preparator, Marc Burgess; Preparator, Jonathan Chamberlain; Preparator, Ryan Emmett; Preparator, Curt Riegelnegg
Institution Type/Description: Contemporary Gallery.
Collections: works by contemporary artists.
Hours & Admission Prices: Tues.-Thurs. 11-6, Fri.-Sat. 11-8. No charge; donations accepted.
Attendance: 1,500 (accurate)

Pleasantville

PITHOLE VISITOR CENTER, 14118 Pithole Rd., Pleasantville, PA 16341. Mailing Address: 202 Museum Lane, Titusville, PA 16354-7658. Tel.: 814-827-2797. Fax: 814-827-4888 (Drake Well).
E-mail: drakewell@verizon.net
Founded: 1975.
Congressional District: 23
Key Personnel: C.E.O., James M. Vaughan; Chm., Stephen Miller; Site Mgr., Barbara Zolli.
Personnel Profile: Part-Time Volunteers 6.
Governing Authority: state. Parent Institution: Pennsylvania Historical & Museum Commission, Commonwealth Keystone Bldg., 400 North St., Harrisburg, PA 17120. Tel.: 717-787-3115. Tax-exempt.
Institution Type/Description: Oil Well Museum.
Collections: 1865-75 oil industry tools & equipment; model of former oil boom town.
Research Fields: 19th-century oil industry.
Facilities: visitor center; picnic area; archaeology site. Museum-related items for sale.
Activities: self-guided tours; multi-media programs.
Hours & Admission Prices: Call for hours.
Attendance: 2,000 (estimated)

Plymouth

PLYMOUTH HISTORICAL SOCIETY, INC., 115 Gaylord Ave., Plymouth, PA 18651-2200. Mailing Address: 157 Nottingham St., Plymouth, PA 18651. Tel.: 570 779 5840.
E-mail: georgettapotoski@aol.com
Web Site: www.locallivinghistorypa.com/phs/home.htm
Founded: 1986.
Key Personnel: Dir. & Pres. (V), Stephen Kondrad; Treas., Helen Yonells; Public Rels., Chris Pagoda; Museum Shop Mgr., Mary Langdon.
Personnel Profile: Part-Time Volunteers 6.

Governing Authority: private; nonprofit organization. Tax-exempt: 501(c)(3).
Institution Type/Description: Historical Society Museum.
Collections: local artifacts & memorabilia; Native American; coal mining; genealogy; early 1900s kitchen; utensils; dishes; sewing machine; carpet loom; period furnishings; Civil War artifacts; Gov. Arthur James memorabilia; period clothing; shoe making; early photography equipment; military artifacts; photographs; local yearbooks, census, church & cemetery records; maps; Pittston post office; anthracite coal mining artifacts.
Research Fields: church & cemetery histories; genealogy; coal mining; military; early local history
Facilities: 500-vol. library; 350-seat auditorium; 3,000 sq. ft. exhibit space; classrooms.
Activities: formal education programs for adults; guided tours; school group tours; temporary exhibitions; lectures; special events.
Publications: annual newsletter, Plymouth Historical Society, Inc.
Hours & Admission Prices: Thurs. & Sat. 12-4; other times by appointment. No charge; donations accepted.
Attendance: 500 (accurate)
Membership: Individual $20; Family $25.

Point Marion

FRIENDSHIP HILL NATIONAL HISTORIC SITE, 223 New Geneva Rd., Point Marion, PA 15474. Mailing Address: 1 Washington Pkwy., Farmington, PA 15437-9501. Tel.: 724-725-9190. Fax: 724-725-1999.
E-mail: lawren_dunn@nps.gov
Web Site: www.nps.gov/frhi
Founded: 1978.
Congressional District: 20
Key Personnel: Deputy Supt., Keith Newlin; Supt., Jeffrey P. Reinbold; Unit Mgr. & Chief Law Enforcement Ranger, Norman W. Nelson; Cultural Resource Mgr. & Cur., Lawren Dunn; Museum Shop Mgr., James Tomasek.
Personnel Profile: Full-Time Paid 3; Part-Time Paid 1; Part-Time Volunteers 390.
Governing Authority: federal. Parent Institution: Dept. of the Interior. Subsidiary Institution: National Park Service. Tax-exempt.
Institution Type/Description: History Museum.
Collections: Albert Gallatin; whiskey rebellion.
Research Fields: career of Albert Gallatin; history of southwest Pennsylvania.
Facilities: library; visitor center; hiking trails.
Activities: guided tours; self-guided tours; trail guides for hiking & cross-country skiing.
Publications: park folders & inserts only.
Hours & Admission Prices: Park: daily dawn to dusk. Building: Summer: daily 9-5; Winter: Sat.-Sun. 9-5. No charge; donations accepted. Closed Federal holidays; New Year's Day; Martin Luther King Jr. Day; George Washington's Birthday; Veterans Day; Thanksgiving; Christmas. &
Attendance: 50,000 (estimated)

Pottstown

POTTSGROVE MANOR, (M), 100 W. King St., Pottstown, PA 19464-6318. Tel.: 610-326-4014. Fax: 610-326-9618. Facebook: Pottsgrove Manor.
E-mail: pottsgrovemanor@montcopa.org
Web Site: www.montcopa.org/pottsgrovemanor
Founded: 1988.
Congressional District: 5
Key Personnel: Deputy Dir. Parks, Trails & Historic Sites Div., Ronald Ahlbrandt; Site Supvr., Laura Adie; Cur., Amy Reis.
Personnel Profile: Full-Time Paid 2; Part-Time Paid 3; Part-Time Volunteers 60.
Governing Authority: Parent Institution: Parks Trails & Historic Sites Div., Assets & Infrastructure Dept., Montgomery County. Tax-exempt.
Institution Type/Description: Historic House: 1752 Pottsgrove Manor.
Collections: 1752-1783 furnishings & interpretation; lifestyle.
Major Exhibits: To The Manor Worn: Clothing the 18th Century Household, 3/14-11/14.
Research Fields: early Pennsylvania iron industry; Potts family & 18th-century lifestyle; colonial Black history.
Facilities: colonial revival boxwood formal & flower gardens. Museum-related items, handcrafted colonial reproductions & books for sale.
Activities: guided tours; children's educational programs & special events.
Publications: volunteer newsletter.
Hours & Admission Prices: Tues.-Sat. 10-4, Sun. 1-4. No charge; donations accepted. Closed New Year's Day; Easter; Independence Day; Thanksgiving; Christmas. &
Attendance: 6,188 (accurate)

Pottsville

HISTORICAL SOCIETY OF SCHUYLKILL COUNTY, (M), 305 N. Centre St., Pottsville, PA 17901-2512. Mailing Address: P.O. Box 1356, Pottsville, PA 17901-7356. Tel.: 570-622-7540. Fax: 570-628-2012.
Key Personnel: Dir., Dr. Peter Yasenchak.
Institution Type/Description: Historical Society Museum.
Collections: county history & culture; photographs; personal artifacts.
Hours & Admission Prices: Wed. 1:30-7, Thurs.-Fri. 10-4, Sat. 9-1. Adults $3; members no charge.

JEWISH MUSEUM OF EASTERN PENNSYLVANIA, 2400 W. End Ave., Pottsville, PA 17901. Tel.: 570-622-5890.
Institution Type/Description: Jewish History Museum.
Collections: Jewish history & culture; personal artifacts; period furnishings; photographs.
Activities: special events.
Hours & Admission Prices: By appointment only. No charge. ♿

Prospect Park

MORTON HOMESTEAD, (M), 100 Lincoln Ave., Prospect Park, PA 19076. Mailing Address: 923 7th Ave., Prospect Park, PA 19076-2307. Facebook: Prospect Park Historical Society.
E-mail: PPHS19076@gmail.com
Founded: 1939.
Congressional District: 7
Key Personnel: Pres. & C.E.O., Glen J. Schwenke; Vice Pres., John R. Shemeluk; Sec. & Treas., Patricia J. Schwenke.
Governing Authority: state. Parent Institution: Pennsylvania Historical & Museum Commission. Tax-exempt.
Institution Type/Description: History Museum: housed in late 17th-century Morton Homestead.
Collections: outdoor exhibits on history of the site.
Facilities: outdoor exhibits; picnic area.
Activities: tours.
Hours & Admission Prices: Temporarily closed. ♿
Attendance: 2,000 (estimated)

Punxsutawney

PUNXSUTAWNEY AREA HISTORICAL & GENEALOGICAL SOCIETY, 400-401 W. Mahoning St., Punxsutawney, PA 15767. Mailing Address: P.O. Box 286, Punxsutawney, PA 15767-0286. Tel.: 814-938-2555.
Founded: 1978.
Key Personnel: Chm. (V), Elmer Reed; Pres. (V), Martha A. Armstrong; Museum Shop Mgr., Karen Curry.
Personnel Profile: Full-Time Volunteers 2; Part-Time Volunteers 25.
Governing Authority: Tax-exempt.
Institution Type/Description: Historical Society Museum.
Collections: local history & culture; Native American artifacts; period tools & utensils; regional lumbering; area coal mining & coke production; local railroading; clothing; quilts; radio & early televisions; photographs; Groundhog Day history; genealogy.
Facilities: Museum-related items for sale.
Activities: children's art workshops in the spring & fall; children's history camp in summer.
Publications: quarterly newsletter, Histo-Report.
Hours & Admission Prices: Bennis House & Lattimer House: Thurs.-Sun. 1-4. Genealogy: Lattimer House Thurs. & Sat. 10-1. No charge; donations accepted. ♿
Attendance: 1,000 (estimated)
Membership: Individual $15; Family $25; Life $300.

PUNXSUTAWNEY WEATHER DISCOVERY CENTER, 201 N. Findley St., Punxsutawney, PA 15767. Mailing Address: P.O. Box 72, Punxsutawney, PA 15767. Tel.: 814-938-1000.
E-mail: info@weatherdiscovery.org
Web Site: www.weatherdiscovery.org
Founded: 2001.
Institution Type/Description: Science Museum.
Collections: hands-on exhibits; meteorology; science; weather-related exhibits; photographs.
Facilities: theater; classrooms. Center-related items for sale.
Activities: educational programs & classes; interactive exhibits; videos.
Hours & Admission Prices: Mon.-Tues. & Thurs.-Sat. 10-4. Admission: $4 per person; children under 2 no charge.

Quarryville

SOUTHERN LANCASTER COUNTY HISTORICAL SOCIETY, INC., (M), 1932 Robert Fulton Hwy., Quarryville, PA 17566. Mailing Address: P.O. Box 33, Quarryville, PA 17566-0033. Tel.: 717-548-2679.
E-mail: solancohistorical@frontier.com
Web Site: www.rootsweb.ancestry.com/~paslchs
Formerly: Solanco Historical Society
Founded: 1970.
Congressional District: 16
Key Personnel: Dir., Kenneth Brown; Dir., Robert Highfield; Dir., Elizabeth Miller; Pres. (V), Stanley T. White; Museum Shop Mgr., Donna McCool.
Personnel Profile: Part-Time Volunteers 35.
Governing Authority: state. Parent Institution: Pennsylvania Historical & Museum Commission, Box 1026, Harrisburg, PA. 17120. Tax-exempt 501(c)(3).
Institution Type/Description: Historic House: c.1760 Robert Fulton birthplace; historical society archives.
Collections: local history; period furniture; genealogical material; 4 historic buildings; historical photographs.
Research Fields: agriculture & mining of the southern Lancaster County area; Lancaster, Oxford, Southern Railroad; genealogy; local history.
Facilities: garden; restrooms.
Activities: guided tours; permanent exhibitions.
Publications: annual calendar; quarterly newsletter.
Hours & Admission Prices: Birthplace: Memorial Day-Labor Day Sat. 11-4, Sun. 1-5; SLCHS Archives Warehouse Wed. 9-12. Adults $4, children 6-12 $2; children under 6 no charge. ♿
Attendance: 1,000 (estimated)
Membership: Students 23 & Under $5; Seniors 60 & Over $10; Annual $12; Life 55 & Over $120.

Reading

BERKS COUNTY HERITAGE CENTER, 1102 Red Bridge Rd., Reading, PA 19605. Mailing Address: 2201 Tulpehocken Rd., Wyomissing, PA 19610-1020. Tel.: 610-374-8839. Fax: 610-373-7049.
E-mail: cwegener@countyofberks.com
Web Site: www.co.berks.pa.us/parks/cwp
Founded: 1981.
Congressional District: 17
Key Personnel: Dir., Cathy L. Wegener.
Personnel Profile: Full-Time Paid 1; Part-Time Paid 7; Part-Time Volunteers 125.
Governing Authority: Parent Institution: Berks County Parks & Recreation Dept. Tax-exempt.
Institution Type/Description: History Museum.
Collections: local history & culture; wagon wheel manufacturing; period artifacts; canal transportation. Historic Building: Gruber Wagon Works.
Activities: lectures; workshops. Annual Event: Heritage Festival in October.
Publications: Gruber Wagon Works: The Place Where Time Stood Still.
Hours & Admission Prices: May-Oct. Tues.-Sat. 10-4, Sun. 12-5. Adults $5, senior citizens 60 & over $4, students 7-18 $3; children under 7 no charge. ♿
Attendance: 10,000 (estimated)

CENTRAL PENNSYLVANIA AFRICAN AMERICAN MUSEUM, Old Bethel African Methodist Episcopal Church, 119 N. Tenth St., Reading, PA 19601-3704. Tel.: 610-371-8713. Fax: 610-371-8739.
E-mail: CPAAM@verizon.net
Web Site: www.cpaammuseum.org
Founded: 1998.
Key Personnel: Dir., C.E.O. & Pres. (V), Frank L. Gilyard, Sr.
Personnel Profile: Part-Time Volunteers 3.
Governing Authority: Tax-exempt.
Institution Type/Description: History Museum.
Collections: local history & culture; period furnishings; personal artifacts; photographs; books; Underground Railroad history.
Activities: Museum Sponsors: Annual Awards Event; Black History Month; Art Festival; Juneteenth.
Publications: Students $5; Friend of Museum $15; Individual $25; Family $35.
Hours & Admission Prices: Wed. & Fri. 10:30-1:30, Sat. 1-4; other times by appointment. Adults $3, senior citizens $2, children 12 & under $1.50; discounts to groups.
Attendance: 300

FREEDMAN GALLERY-ALBRIGHT COLLEGE, (M), 13th St. & Bern St., Reading, PA 19604. Mailing Address: Box 15234, Reading, PA 19612-5234. Tel.: 610-921-7541. Fax: 610-921-7768.
E-mail: gallery@alb.org
Web Site: www.albright.edu/freedman
Founded: 1976.
Congressional District: 6
Key Personnel: Dir. Center for the Arts, David M. Tanner; Cur., Erin Riley-Lopez; Cur. Education, Beth Krumholz; Collections Mgr. & Preparator, Nancy Sarangoulis.
Personnel Profile: Full-Time Paid 2; Part-Time Paid 3; Interns 1.
Governing Authority: college. Parent Institution: Albright College. Tax-exempt: 501(c)(3).
Institution Type/Description: College Art Gallery.
Collections: paintings, drawings, photographs, sculpture & prints by contemporary American artists.
Research Fields: contemporary art in all mediums.
Facilities: 2,800 sq. ft. exhibit space; 250-seat auditorium; educational facilities.
Activities: guided tours; lectures; films; broadcast programs; organized education programs for undergraduate or graduate college students; participatory, loan, temporary & traveling exhibitions.
Publications: exhibition catalogs; brochures; discovery guides.
Hours & Admission Prices: Sept.-May Tues.-Fri. 10-5, Sat.-Sun. 1-4; Call for summer hours. No charge; donations accepted. Closed major holidays & college breaks. &
Attendance: 4,457 (accurate)
Membership: Individual $25; Family & Contributing $50; Supporting $100; Patron $500; Benefactor $1,000.

HISTORICAL SOCIETY OF BERKS COUNTY, (M), 940 Centre Ave., Reading, PA 19601-2198. Tel.: 610-375-4375. Fax: 610-375-4376.
E-mail: history@berkshistory.org
Web Site: www.berkshistory.org
Founded: 1869.
Congressional District: 6
Key Personnel: C.E.O., Beulah B. Fehr; Dir., Sime B. Bertolet; Museum Shop Mgr., Kimberly Longlott.
Personnel Profile: Full-Time Paid 5; Part-Time Paid 4; Part-Time Volunteers 100; Interns 10.
Governing Authority: society. Tax-exempt: 501(c)(3).
Institution Type/Description: History Museum.
Collections: agriculture items; archives; decorative arts; history; Indian artifacts; Pennsylvania German arts & crafts.
Major Exhibits: Berks County During the Civil War, 7/13-7/14.
Research Fields: county history.
Facilities: 10,000-vol. library of local history books available on premises; 120-seat auditorium. Gift items for sale.
Activities: lectures; concerts; formally organized education programs for children; docent program permanent & temporary exhibitions; school loan service; children's hands-on room.
Publications: quarterly magazine, Historical Review of Berks County; The Passing Scene.
Hours & Admission Prices: Tues.-Sat. 9-4. Adults $5, senior citizens $3, children $2; discount to AAM members; members no charge. Library: $5. &
Attendance: 15,000 (estimated)
Membership: Individual $55; Family $65; Contributing $100; Benefactor $150; Sponsor $250; Guarantor $500; Patron $1,000; Founder's League $2,500.

MID ATLANTIC AIR MUSEUM, 11 Museum Dr., Reading, PA 19605-9407. Tel.: 610-372-7333. Fax: 610-372-1702.
E-mail: maam@maam.org
Web Site: www.maam.org
Founded: 1980.
Congressional District: 14
Key Personnel: Chm. & Pres., Russell A. Strine; Cur. & Museum Shop Mgr., Linda T. Strine; Business Office Mgr., Brenda Saylor; Special Projects Coord., David Schott.
Personnel Profile: Full Time Paid 5; Part-Time Paid 5; Part-Time Volunteers 150; Interns 1.
Governing Authority: private; nonprofit organization. Tax-exempt: 501(c)(3).
Institution Type/Description: Aviation Museum.
Collections: 70 aircrafts; aviation with emphasis on Mid-Atlantic states history.
Research Fields: historic aviation.

Facilities: 10,000-vol. library available for use on the premises; field research station; classroom; 25-seat meeting room. Museum-related items for sale.
Activities: concerts; docent program; guided tours; hobby workshops; lectures; rental gallery; school loan service; loan, temporary & traveling exhibitions; training programs for professional museum workers; speakers bureau activities. Annual Events: World War II Weekend; Planes, Trains & Automobiles Weekend.
Publications: bimonthly newsletter, Museum Review.
Hours & Admission Prices: Daily 9:30-4. Adults $6, children $3; discounts to groups; members & children 5 and under no charge. Closed New Year's Day; Easter; Independence Day; Thanksgiving; Christmas. &
Attendance: 30,000 (accurate)
Membership: Individual $45; Sustaining $100; Corporate $500; Life $750.

PLANETARIUM AT THE READING PUBLIC MUSEUM, 1211 Parkside Dr. S., Reading, PA 19611-1441. Mailing Address: 500 Museum Rd., Reading, PA 19611-1425. Tel.: 610-371-5850, ext. 244. Fax: 610-371-5632.
E-mail: planetarium@readingpublicmuseum.org
Web Site: www.readingpublicmuseum.org/planetarium
Founded: 1967.
Congressional District: 6
Key Personnel: C.E.O. & Dir., Ronald C. Roth; Chm. (V), Rolf D. Schmidt; 1st Vice Chm. (V), Kathleen W. Kleppinger; 2nd Vice Chm. (V), Donald Bristol; Dir. Planetarium, Mark J. Mazurkiewicz.
Personnel Profile: Full-Time Paid 1; Part-Time Paid 3.
Governing Authority: Foundation. Parent Institution: Reading Public Museum. Tax-exempt: 170 (b)(1)(A).
Institution Type/Description: Planetarium.
Collections: astronomical artifacts; meteorites; serigraphs.
Research Fields: planetarium production techniques.
Facilities: 1,000-vol. library of astronomical materials, tapes, records, films & 7,000 photographic slides; satellite weather station; spacesphere.
Activities: foreign language programs; special education classes; special provisions for the physically handicapped; college classes.
Publications: M.A.P.S.; Planetary Society; I.P.S.
Hours & Admission Prices: Tues.-Thurs. & Sat. 11-5, Fri. 11-8, Sun. 12-5. Museum: adults $7, children $5. Star Shows: adults $6, seniors, children & students $4; discounts to AAM & ICOM members; children under 4 & members no charge. Closed New Year's Day; Martin Luther King Jr. Day; Labor Day; Thanksgiving; Christmas. &
Attendance: 17,000 (accurate)
Membership: Full-time Student $20; Senior $25; Individual $35; Family Dual $60; Contributor $125.

READING AREA FIREFIGHTERS MUSEUM, INC., 501 S. 5th St., Reading, PA 19602. Tel.: 610-655-6080.
Founded: 2002.
Key Personnel: Pres. (V), William H. Rehr, III
Personnel Profile: Part-Time Volunteers 20.
Governing Authority: Tax-exempt.
Institution Type/Description: Firefighting History Museum.
Collections: firefighting history & equipment; early furnishings; uniforms; photographs; fire dispatch equipment.
Hours & Admission Prices: Call for hours. No charge; donations accepted.
Membership: Seniors & Students 18 & under; $5; Individuals $10; Family $15; Fire Company $50; Business $100.

*** READING PUBLIC MUSEUM, (M),** 500 Museum Rd., Reading, PA 19611-1425. Tel.: 610-371-5850, ext. 222. Fax: 610-371-5632. Facebook: Reading Public Museum.
E-mail: info@readingpublicmuseum.org
Web Site: www.readingpublicmuseum.org
Founded: 1904.
Congressional District: 6
Key Personnel: Chm., Kathleen Kleppinger; Dir. & C.E.O., John Graydon Smith; 1st Vice Chm., Richard W. Zuidema; Sec., Judy Phelps; Asst. Sec., Charles Harenza, Esq.; Treas., Bruce Rhoads.
Personnel Profile: Full Time Paid 17; Part-Time Paid 6; Part-Time Volunteers 300; Interns 2.
Governing Authority: foundation. Parent Institution: Foundation for the Reading Public Museum. Tax-exempt: 501(c)(3).
Institution Type/Description: Art & Science Museum.
Collections: anthropology; arboretum; archaeology; paintings; sculpture; graphics; decorative arts; entomology; Pennsylvania folk art; American Indian artifacts; mineralogy; natural history; paleontology.
Major Exhibits: Painters of Berks County, 5/13-9/14; Norman Rockwell and Scouting, 10/13-1/14; Scouting Through the Years, 10/13-1/14.

Research Fields: paintings; graphics; entomology.
Facilities: 18,000-vol. library of art & natural history books available on premises by request; arboretum; 200-seat auditorium; planetarium. Fine art, local art, reproductions, crafts, museum-related items for sale.
Activities: guided tours; lectures; films; gallery talks; concerts; formally organized educational programs; gallery teachers program; outreach & collaborations with many local organizations; inter-museum loan, permanent, temporary & traveling exhibitions; school loan service; organized trips.
Publications: exhibit catalogues; Museum Road.
Hours & Admission Prices: Daily 11-5. Adults $10, children 4-17, seniors & students $6; discounts as part of ASTC reciprocal agreement; members no charge. Closed New Year's Day; Thanksgiving; Christmas. &
Attendance: 70,000 (estimated)
Membership: Student, Senior Citizen & Educator $30; Individual $40; Family & Grandparent $60; Contributor $120; Artist's Studio $250 & up.

Robertsdale

FRIENDS OF THE EAST BROAD TOP MUSEUM, 550 Main St., Robertsdale, PA 16674. Mailing Address: P.O. Box 68, Robertsdale, PA 16674. Tel.: 814-635-2388.
Institution Type/Description: History Museum.
Collections: local history & culture; period furnishings; personal artifacts; photographs. Historic Buildings: 1915 post office; 1917 railroad depot.
Hours & Admission Prices: Call for hours.

Rockhill Furnace

ROCKHILL TROLLEY MUSEUM, 430 Meadow St., Rockhill Furnace, PA 17249. Mailing Address: P.O. Box 203, Rockhill Furnace, PA 17249. Tel.: 814-447-9576 (weekends); 610-437-0448 (weekdays). Facebook: Rockhill Trolley.
E-mail: info@rockhilltrolley.org
Web Site: www.rockhilltrolley.org
Founded: 1962.
Congressional District: 9
Key Personnel: Pres., Joel Salomon; Archivist, Douglas Peters; Museum Shop Mgr., Charles T. Kumpas.
Personnel Profile: Part-Time Volunteers 35.
Governing Authority: private; nonprofit organization. Parent Institution: Railways To Yesterday Inc., P.O. Box 1601, Allentown, PA 18105. Tax-exempt: 501(c)(3).
Institution Type/Description: Transportation Museum.
Collections: 22 trolleys from 1899-1947 including cars from Pennsylvania; photographs; railway artifacts.
Facilities: library. Seven operating passenger trolleys; two operating snowsweeper trolleys. Museum-related items for sale.
Activities: demonstration trolley ride; guided tours; restoration shop viewing area; special events. Museum Sponsors: Fall Spectacular; Member's Day; Santa's Trolley.
Publications: quarterly newsletter, The Retriever.
Hours & Admission Prices: June-Oct. Sat.-Sun. 11-4. Adults $7, children 2-12 $4; discounts to groups, AAA & AARP members; members no charge.
Attendance: 5,000 (estimated)
Membership: Associate $25; Sustaining $40.

Rome

P. P. BLISS GOSPEL SONGWRITERS MUSEUM, Main St., Rome, PA 18837. Mailing Address: P.O. Box 84, Rome, PA 18837. Tel.: 570-247-2228.
E-mail: ppbmuseum@cableracer.com
Key Personnel: Dir., Deborah Barrett
Institution Type/Description: History Museum: housed in the former home of Mr. Bliss' parents.
Collections: history of P.P. Bliss & other gospel songwriters; musical instruments; personal artifacts; song books & sheet music; clothing; news articles.
Hours & Admission Prices: Memorial Day to Labor Day Wed. & Sat. 1-4; other times by appointment.

Russell

SHED MUSEUM, 7159 Scandia Rd., Russell, PA 16345-6941. Tel.: 814-757-4443.
E-mail: thehedscandia@verizon.net
Web Site: www.theshedscandia.com
Founded: 2005.

Congressional District: 65
Key Personnel: Dir., Debbie Fitzsimmons
Institution Type/Description: History Museum.
Collections: local history & culture; period artifacts; early cars; gas engines; photographs.
Facilities: Museum-related items for sale.
Hours & Admission Prices: May-Oct. Fri.-Sun. 11-5 & by appointment. No charge.
Attendance: 400 (estimated)

Saint Marys

HISTORICAL SOCIETY OF ST. MARYS AND BENZINGER TOWNSHIP, 99 Erie Ave., Saint Marys, PA 15857-1408. Tel.: 814-834-6525.
E-mail: stmaryshistoricalsociety@windstream.net
Web Site: smhistoricalsociety.com
Founded: 1960.
Congressional District: 5
Key Personnel: Pres., Jeanette Donachy; Sec., Donna M. Burgess; Cur., Alice Beimel.
Personnel Profile: Part-Time Volunteers 30.
Governing Authority: municipal; nonprofit organization. Tax-exempt.
Institution Type/Description: History Museum.
Collections: photographic representation of early history of St. Marys; documents; artifacts; manuscript collections.
Research Fields: local history; videotaped interviews of older residents; family histories; photos; artifacts; deeds, maps & other documents.
Facilities: library of photographs, artifacts, deeds, maps & documents available for loan and on premises; reading room.
Activities: guided tours; lectures; formally organized educational programs; permanent exhibitions.
Publications: weekly historical column; annual member newsletter.
Hours & Admission Prices: Tues. 10-4, Thurs. 1-4 & 6-8; other times by appointment. No charge. &
Attendance: 1,500 (estimated)
Membership: Student $1; Single $6; Family $10; Patron $25; Benefactor $50; Corporate $100.

Schaefferstown

HISTORIC SCHAEFFERSTOWN, INC., 106 N. Market St., Schaefferstown, PA 17088. Mailing Address: P.O. Box 307, Schaefferstown, PA 17088-0307. Tel.: 717-949-2444.
E-mail: info@hsimuseum.org
Web Site: www.hsimuseum.org
Founded: 1965.
Congressional District: 17
Key Personnel: Pres., Alice Oskam; Museum Shop Mgr., Betty Fromm.
Personnel Profile: Part-Time Paid 1; Part-Time Volunteers 205.
Governing Authority: nonprofit organization. Tax-exempt: 501(c)(3).
Institution Type/Description: Village Museum.
Collections: portrayal of 18th- & 19th-century Pennsylvania German life & culture; tools; implements; furniture; quilts; Indian artifacts. Historic Buildings: c.1740 Alexander Schaeffer Farm Museum; c.1740s Gemberling-Rex House; 1969 Thomas R. Brendle Memorial Library & Museum.
Research Fields: agricultural practices from 18th-19th century to present in Pennsylvania Dutch country; folklife of the Pennsylvania Germans.
Facilities: 700-vol. library of books on Pennsylvania history, the history of the Pennsylvania Germans, folklore & folklife of the Pennsylvania Germans available for use upon request; reading room.
Activities: guided tours; lectures; permanent & temporary exhibitions.
Publications: quarterly bulletin, Historic Schaefferstown Record; pamphlets; booklets, Volume I of the Brendle Collection of Pennsylvania German Folklore.
Hours & Admission Prices: Alexander Schaeffer Farm Museum: April-Oct. 4 1st Sat. of month 10-4. Festivals: 10-5. Adults $5; children under 12 & members no charge. Brendle Museum & Gemberling Rex House: April 4 to Oct. 4 Tues. 1-4, 1st Sat. of month 10-4; other times by appointment. Adults $3 per site; discounts to AAM & ICOM members; members no charge.
Attendance: 4,863 (accurate)
Membership: Individual $17; Family $27.

Schnecksville

LEHIGH VALLEY ZOO, 5150 Game Preserve Rd., Schnecksville, PA 18078-0519. Mailing Address: P.O. Box 519, Schnecksville, PA 18078-0519. Tel.: 610-799-4171. Fax: 610-799-4170.
Institution Type/Description: Zoo.

Collections: wildlife & their habitats.
Activities: special events; educational programs; rental facilities.
Hours & Admission Prices: April-Oct. daily 10-4; Nov.-March daily 10-3. April-Oct.: adults $9, seniors 65 & over $7.25, children 2-11 $6.50; Nov.-March: adults $6.50, seniors 65 & over and children 2-11 $6.50. Closed New Year's Day; Thanksgiving; Christmas.

Schwenksville

PENNYPACKER MILLS, 5 Haldeman Rd., Schwenksville, PA 19473-1844. Mailing Address: c/o Montgomery County Parks, Trails, & Historic Sites, P.O. Box 311, Norristown, PA 19404-0311. Tel.: 610-287-9349. Fax: 610-287-9657. Facebook Pennypacker Mills.
E-mail: pennypackermills@montcopa.org
Web Site: www.montcopa.org/pennypackermills
Founded: 1981.
Congressional District: 6
Key Personnel: Historic Site Administrator, Ella Aderman.
Governing Authority: county. Parent Institution: Montgomery Co. (PA) Dept. of Parks & Heritage Services, 1430 Dekalb St., Norristown, PA 19404. Tax-exempt.
Institution Type/Description: Historic House: c.1901-1916 colonial revival building.
Collections: Pennypacker family memorabilia; 18th- to 20th-century furnishings; letters; books; photos; historic landscape.
Research Fields: 17th to early 20th-century family history; 20th-century landscape history; colonial revival architectural history; late 19th to early 20th-century Pennsylvania political history.
Facilities: 3,000-vol. library pertaining to local history; 50,000 Pennypacker letters; exhibit gallery; informal landscape garden; mansion; barn; 170 acre rural property.
Activities: guided tours; organized educational programs; docent program; workshops; nature walks; loan & temporary exhibitions. Special Events: Summer Social; Children's Halloween; Civil War Reunion; Christmas Open House.
Publications: bimonthly newsletter.
Hours & Admission Prices: Tues.-Sat. 10-4, Sun. 1-4; last tour 3:30. Suggested Donation: $2. Closed New Year's Eve & Day; Easter; Independence Day; Thanksgiving; Christmas Eve & Day. &
Attendance: 25,000 (accurate)

Scottdale

WEST OVERTON MUSEUMS, (M), 109 West Overton Rd., Scottdale, PA 15683. Tel.: 724-887-7910.
E-mail: info@westovertonvillage.org
Web Site: www.westovertonvillage.org
Formerly: Westmoreland Fayette Historical Society
Founded: 1928.
Congressional District: 12
Key Personnel: Pres. Bd., Brian Corcoran; Dir., Kelly Linn; Asst. Dir., Jessica Kadie-Barclay.
Personnel Profile: Full-Time Paid 2; Part-Time Paid 1; Part-Time Volunteers 20.
Governing Authority: society. Tax-exempt: 501(c)(3).
Institution Type/Description: Museum Complex: c.1850 rural industrial village.
Collections: 1839-1939 household & farm implements; industrial implements; archives.
Facilities: auditorium; media center & reference library; archives; photographs. Museum-related items for sale.
Activities: guided tours; lectures; theater; organized education programs for children; participatory, permanent & temporary exhibitions.
Publications: Oberholtzer & Nash Family History; quarterly newsletter, The Village Progress; a collection of Civil War letters, Remember Your Friend Until Death.
Hours & Admission Prices: Call or see website for information. &
Attendance: 5,000 (estimated)
Membership: Individual $15; Family $30; Homestead $50; Distillery $100; Village $250.

Scranton

ELECTRIC CITY TROLLEY MUSEUM, 300 Cliff Ave., Scranton, PA 18503. Tel.: 570-963-6590. Fax: 570-963-6447.
Web Site: ectma.org
Institution Type/Description: Trolley Museum: housed in a late 19th century mill building.
Collections: trolley history & artifacts; hands-on exhibitions; restored trolley;

photographs; trolley models; stock certificates; map; mine railway equipment; signals; signal equipment & relays; early railroad yard lights.
Facilities: 50-seat theater.
Activities: trolley excursions. Museum Sponsors: Trolley Film Festival.
Hours & Admission Prices: Jan.-April Wed.-Sun. 9-4; May-Dec. daily 9-5. Trolley Excursions: Thurs.-Sun. 10:30, 12, 1:30 & 3. Closed New Year's Day; Thanksgiving; Christmas. &
Attendance: 27,600 (accurate)

EVERHART MUSEUM: NATURAL HISTORY, SCIENCE AND ART, (M), 1901 Mulberry St., Scranton, PA 18510-2390. Tel.: 570-346-7186. Fax: 570-346-0652.
E-mail: general.information@everhart-museum.org
Web Site: www.everhart-museum.org
Founded: 1908.
Congressional District: 10
Key Personnel: C.E.O. & Dir., Cara A. Sutherland; Chm. Bd., Joseph T. Wright; Dir. Interpretive Programs, Stefanie Colarusso; Cur., Nezka Pfeifer; Museum Shop Mgr., Deb Burke; Administrative Asst., Nancy Casey; Dir. Resources, Deborah L. Pann.
Personnel Profile: Full-Time Paid 6; Part-Time Paid 10; Part-Time Volunteers 60; Interns 10.
Governing Authority: nonprofit organization. Tax-exempt: 501(c)(3).
Institution Type/Description: Art, Science & Natural History Museum.
Collections: 19th- & 20th-century American art; European paintings & works on paper; Dorflinger Glass; American folk art; Asian Art; Ethnographic Arts from Oceanic, African, Native American & Pre-Columbian cultures; natural history collection - animals, birds, insects, fossils, rocks, shells, vertebrates, minerals.
Research Fields: John Willard Raught (1857-1931) American Impressionist Painter; American folk art; geology; regional art; Dorflinger Glass.
Facilities: Museum-related items for sale.
Activities: lectures; gallery talks; educational programs; permanent & temporary exhibitions.
Publications: newsletter; brochures; catalogs.
Hours & Admission Prices: Feb.-Dec. Mon. & Thurs.-Fri. 12-4, Sat. 10-5, Sun. 12-5. Adults $5, students & seniors $3, children $2; discounts to AAM & ICOM members; children under 6 & members no charge. &
Attendance: 30,000 (estimated)
Membership: Student/Academic & Senior 60 and over $25; Individual $30; Dual Senior $40; Family $45; Sustainer $100; Patron $250; Curator's Circle $500; Artist's Circle $1,000; Director's Circle $2,000 & up.

THE HOUDINI MUSEUM & THEATER, 1433 N. Main Ave., Scranton, PA 18508-1822. Mailing Address: 229 Willow Ave., Olyphant, PA 18447-1443. Tel.: 570-342-5555.
Web Site: www.houdini.org
Founded: 1992.
Key Personnel: C.E.O. & Pres. (V), John Bravo; Chm. (V), Joann Englehardt; Financial Dir., Dorothy Krugger; Cur. & Archivist, Freida Schmidt; Devel., Penny Wilkes; Public Rels., Dick Brooks; Security & Museum Shop Mgr., Ray Carter; Education, Dorothy Dietrich.
Personnel Profile: Full-Time Volunteers 2; Part-Time Volunteers 10.
Governing Authority: private; nonprofit organization. Tax-exempt: 501(c)(3).
Institution Type/Description: History Museum.
Collections: history room containing pictures & artifacts designed to map Houdini's life from birth to death; prop room displays artifacts & props belonging to Houdini; theater room displays posters; film & artifacts.
Research Fields: research is ongoing to uncover dates & significant appearances of Houdini as he traveled the world.
Facilities: 5,000-vol. library of books & documents on Houdini; educational facilities; 4,000 sq. ft. exhibit space; 150-seat large screen theater. Museum-related items for sale.
Activities: films; formal education programs for children; guided tours; lectures; theater; magic show. Museum Sponsors: Halloween Spooktacular Shows.
Hours & Admission Prices: Open Memorial Day to Labor Day daily for tours & magic shows, call for hours. Adults $14.95, children 11 & under $11.95; discounts to AAM & ICOM members and groups of 20 or more.
Attendance: 40,000 (estimated)
Membership: Individual $25; Patron $50; Silver $100; Gold $200.

THE LACKAWANNA HISTORICAL SOCIETY AT THE CATLIN HOUSE, 232 Monroe Ave., Scranton, PA 18510-2104. Tel.: 570-344-3841. Fax: 570-344-3815.
E-mail: lackawannahistory@gmail.com
Web Site: lackawannahistory.org
Founded: 1886.
Congressional District: 10

Key Personnel: Exec. Dir., Mary Ann Moran-Sarakinus; Pres., Michael Gilmartin; Treas., Douglas Forrer; Sec., William Conlogue; Museum Asst., Sarah Piccini.

Personnel Profile: Full-Time Paid 1; Part-Time Paid 2; Part-Time Volunteers 8.

Governing Authority: nonprofit organization. Tax-exempt: 501(c)(3).

Institution Type/Description: Historic Society Museum: housed in The Catlin House, a late Victorian mansion in English Tudor Revival architecture.

Collections: archives containing manuscripts, letters, documents, maps, books, & photographs pertaining to the settlement of Lackawanna County, Pennsylvania & the development of the Iron, Coal, Railroad & Lace industries from 1840-present; period furnishings; clothing; paintings; agricultural items; Civil War items; bicycles; typewriters; local memorabilia.

Research Fields: iron manufacturing; anthracite coal mining; Pennsylvania railroads & street car lines; 19th-century architecture; photographic history; lace manufacture & genealogy.

Facilities: 2,000-vol. library of books; 322 manuscripts; archives of coal companies, lace manufacturing company, railroads, genealogies, local histories, photographs, maps, letters, ledgers & reports available to the public.

Activities: guided tours by prior appointment; lectures; concerts; docent program; organized education programs for undergraduate or graduate college students affiliated with Marywood College & University of Scranton, Scranton, PA; loan, temporary & traveling exhibitions; traveling lectures to schools.

Publications: journal, The Lackawanna Historical Society Journal; brochure, Welcome to the Lackawanna Historical Society; book, Greetings from Scranton.

Hours & Admission Prices: Tues.-Fri. 10-5, Sat. 12-3. Guided Tours: Tues.-Fri. 1-3, Sat. 12-3 or by appointment. Museum: $2 per person; donations accepted. Library: $5 per person. Closed New Year's; Good Friday; Memorial Day; Independence Day; Labor Day; Thanksgiving; Christmas Eve & Day.

Attendance: 4,000 (estimated)

Membership: Student $10; Individual $25; Family $35; Contributing $75; Sustaining $150; Organization & Business available.

PENNSYLVANIA ANTHRACITE HERITAGE MUSEUM, (M),

22 Bald Mountain Rd., McDade Park, Scranton, PA 18504. Tel.: 570-963-4804 & 4845. Fax: 570-963-4194.

E-mail: ckulesa@state.pa.us

Web Site: www.anthracitemuseum.org

Founded: 1970.

Congressional District: 10

Key Personnel: Pres. (V), Mr. Francis Tartella; Admin., Chester J. Kulesa; Cur., Richard Stanislaus; Museum Shop Mgr., Margaret Reese.

Personnel Profile: Full-Time Paid 7; Part-Time Paid 3; Part-Time Volunteers 40; Interns 2.

Governing Authority: state. Pennsylvania Historical and Museum Commission, Bureau of Historic Sites and Museums Commonwealth Keystone Building, Plaza Level, 400 North St., Harrisburg, PA 17120-0053. Tax-exempt: 170(c)(1).

Institution Type/Description: History Museum.

Collections: religious, medical & education materials; mining tools & machinery; silk & lace machinery; photographs; maps; vehicles; prints; paintings.

Research Fields: history of northeast Pennsylvania including transportation, urban, mining, textile, social & cultural areas, ethnicity, immigration.

Facilities: library of research materials; 25,000 sq. ft. exhibit hall; auditorium. Museum-related items for sale.

Activities: tours; films; lectures; education programs for students; arts fair; drama & music performances.

Publications: quarterly newsletter; pamphlets; books.

Hours & Admission Prices: April-Nov. Mon.-Sat. 9-5, Sun. 12-5; Dec.-March Tues.-Sat. 9-5, Sun. 12-5. Adults 12-64 $6, senior citizens 65 & over $5.50, youth 3-11 $4; discounts to AAM & ICOM; active military, family, children under 6 & members no charge. Closed New Year's Day; Civil Rights Day; Presidents' Day; Columbus Day; Veterans Day; Thanksgiving; Christmas.

Attendance: 20,000 (accurate)

Membership: Senior Citizen & Student $15; Individual $20; Family $30.

STEAMTOWN NATIONAL HISTORIC SITE,

350 Cliff St. off Lackawanna Ave., Scranton, PA 18503. Mailing Address: 150 S. Washington Ave., Scranton, PA 18503-2079. Tel.: 570-340-5200 & 5201. Fax: 570-340-5235.

Web Site: www.nps.gov/stea

Founded: 1986.

Congressional District: 11

Key Personnel: Park Supt., Harold "Kip" Hagen.

Personnel Profile: Full-Time Paid 60; Part-Time Paid 1; Part-Time Volunteers 157.

Governing Authority: federal. Parent Institution: National Park Service. Tax-exempt: 501(c)(3).

Institution Type/Description: Transportation & Technology Museum; located in Scranton's DL&W railroad yard.

Collections: steam locomotives; passenger coaches; maintenance of way equipment; locomotive shop & turntable; small objects; restored roundhouse; operating railyard; signal tower.

Research Fields: 20th-century steam railroading.

Facilities: visitor center.

Activities: steam excursions; tours conducted by park rangers.

Hours & Admission Prices: Jan. 6-March 29 daily 10-4; March 30-Jan. 5 daily 9-5. Adults $7; children 15 & under no charge. Train Excursion Ride: call for information. Closed New Year's Day; Thanksgiving; Christmas.

Attendance: 120,000 (accurate)

SURACI GALLERY, MAHADY GALLERY, AND THE MASLOW COLLECTION, (M),

2300 Adams Ave., Scranton, PA 18509-1598. Tel.: 570-348-6278, ext. 2428. Fax: 570-340-6023.

E-mail: gallery@marywood.edu

Web Site: www.themaslowcollection.org

Formerly: Marywood University Art Galleries (Suraci Gallery and Contemporary Gallery)

Founded: 1915.

Congressional District: 11

Key Personnel: C.E.O. & Pres., Sister Anne Munley, I.H.M., Ph.D.; Dir., Sandra Ward Povse.

Personnel Profile: Full-Time Paid 2; Part-Time Paid 14; Interns 1.

Governing Authority: private college; nonprofit. Parent Institution: Marywood University. Tax-exempt: 501(c)(3).

Institution Type/Description: University Art Gallery.

Collections: Suraci Gallery: 19th-century Chinese trade furniture & Russian bronzes; European marble sculpture; Oriental decorative arts (ceramics & ivories); 19th-century European ceramics & American and European furniture. Mahady Gallery: contemporary art exhibits in all media. Maslow Collection: contemporary art including paintings, photographs, & major prints by Jasper Johns, Frank Stella, Robert Rauschenberg, & Andy Warhol; contemporary 2-dimensional.

Research Fields: 19th-century Chinese trade furniture; Oriental decorative arts; Russian bronzes & European marble sculpture; 19th-century European & American furniture; 19th-century European ceramics; contemporary paintings by emerging NY artists from late 1970s to early 1990s.

Facilities: 5,500 sq. ft. exhibit space.

Activities: guided tours; lectures; temporary exhibitions; field experience & independent studies for undergraduate & graduate college students. Biennial Events: Regional Art Exhibition, a juried exhibit open to artists in northeastern Pennsylvania.

Publications: exhibition catalogs, Women's Work; Soul Full: the personal & the political; Transcending the Surface; The Poetic Object: Paintings of Rebecca Purdum; Zeuxis: Still Life Painting; Anne Tabachnick: Painting about Paintings; Rayuela/Hopscotch: 15 Contemporary Latin American Artists; Potter's Apprentice.

Hours & Admission Prices: Winter: Mon. & Thurs.-Fri. 9-4, Tues.-Wed. 9-8, Sat.-Sun. 1-4. Summer: Mon.-Fri. 12-3. No charge. Closed major holidays; semester breaks.

Attendance: 17,525 (estimated)

TIMMY'S TOWN CENTER,

The Mall at Steamtown, 2nd Fl., Scranton, PA 18503. Mailing Address: 108 N. Washington Ave., Ste. 400, Scranton, PA 18503. Tel.: 570-341-1511. Fax: 540-504-3209.

E-mail: timmystowncenter@verizon.net

Web Site: www.timmystowncenter.org

Founded: 2005.

Congressional District: 10

Key Personnel: Dir., Megan Swann; C.E.O., Alexis Kelly

Institution Type/Description: Children's Museum.

Collections: hands-on exhibitions.

Activities: permanent & temporary exhibitions; workshops; programs; activities; special events.

Hours & Admission Prices: Thurs.-Sat. 10-4, Sun. 12-4. Admission $3; discounts to ACM members; members & children under 2 no charge.

Attendance: 10,000 (accurate)

Membership: Individual $25; Family $100.

Selinsgrove

LORE DEGENSTEIN GALLERY, (M), Susquehanna University,
514 University Ave., Selinsgrove, PA 17870-1001. Tel.: 570-372-
4059. Fax: 570-372-2775.
E-mail: olivetti@susqu.edu
Web Site: www.susqu.edu/art_gallery
Founded: 1993.
Congressional District: 17
Key Personnel: Dir., Daniel Olivetti.
Personnel Profile: Full-Time Paid 1; Part-Time Paid 15; Interns 2.
Governing Authority: private college; nonprofit. Parent Institution: Susque-
hanna University. Tax-exempt.
Institution Type/Description: Art Gallery.
Collections: American & Pennsylvania regional art including paintings,
sculpture, prints, drawings & photographs; French advertising posters from
1887-1990; French literary magazines from 1895-1905.
Research Fields: American art; French advertising posters.
Facilities: 450-seat auditorium; 2,500 sq. ft. exhibit space.
Activities: formal education programs for undergraduate students affiliated
with Susquehanna University; guided tours; lectures; loan, temporary &
traveling exhibitions.
Publications: exhibition catalog, Encountering the Narrative in the Recent
Work of Florence Putterman; Hans Moller, Purveyor of Color: The Essence
of Vision 1943-1995.
Hours & Admission Prices: Daily 12-4. No charge. Closed university holidays
& breaks. &
Attendance: 3,592 (accurate)

Sewickley

SEWICKLEY VALLEY HISTORICAL SOCIETY, 200 Broad St.,
Sewickley, PA 15143-1525. Tel.: 412-741-5315.
E-mail: sewickleyhistory@verizon.net
Web Site: www.sewickleyhistory.org
Key Personnel: Pres., L. John Kroeck
Institution Type/Description: Historical Society Museum.
Collections: local history & culture; period furnishings; personal artifacts;
photographs.
Activities: lectures; seminars.
Hours & Admission Prices: Tues.-Fri. 10-2; other times by appointment.

Shenandoah

SCHUYLKILL HISTORICAL FIRE SOCIETY, 105 S. Jardin St.,
Shenandoah, PA 17976. Tel.: 570-628-3691.
Key Personnel: Pres., Michael Kitsock
Institution Type/Description: Firefighting History Museum.
Collections: firefighting history & equipment; hand pumpers; hose carts; water
wagon; photographs; uniforms; personal artifacts.
Activities: group tours.
Hours & Admission Prices: By appointment. No charge; donations accepted.

Shippensburg

SHIPPENSBURG HISTORICAL SOCIETY MUSEUM, 52 W.
King St., Shippensburg, PA 17257-0539. Mailing Address: P.O.
Box 539, Shippensburg, PA 17257-0539. Tel.: 717-532-6727.
Facebook: Shippensburg Historical Society.
E-mail: shiphist@pa.net
Web Site: www.shippenburghistory.org
Founded: 1945.
Congressional District: 9
Key Personnel: Pres., Lizzie Bailey; Vice Pres., Bruce Hockersmith; Dir.,
Tiffany Weaver; Cur., John McCorriston.
Personnel Profile: Part-Time Paid 1; Part-Time Volunteers 50.
Governing Authority: society. Affiliated with Shippensburg Historical Society.
Tax-exempt: 170(b).
Institution Type/Description: General Museum.
Collections: area history; archives; material pertaining to local events &
history; general items; Native American artifacts; WPA projects; medical
related items; photographs; Laughlin postcards; clothing; goblets; toys;
dolls; inkwells; miniature carousels; fans & ivory artifacts.
Research Fields: genealogy & local history.
Facilities: library of early first editions, Pennsylvania literature & history and
historical society books available for research on premises.
Activities: guided tours; permanent & temporary exhibitions; programs &
special events.
Publications: books, Forbes Expedition - Cowan's Gap to Juniata Crossing;
1821 & 1858 maps of Shippensburg; Shippen's Survey 1737-1762; Yester-

day's Shippensburg 1994; Cemetery Lists: Records In Stone, Vols. I, II, III,
IV & V; pictorial history, Shippensburg area; The Elusive Fort Morris; Do
You Remember?; Clyde A. Laughlin, Postcard King of the Cumberland
Valley; Shippensburg area historical postcards; The Caledonia Story;
newsletter for members, The Shippensburg Historical Society: A Fifty Year
Retrospective 1945-1995; Indians, Indians; 1860 U.S. Census of Shippens-
burg Borough, Cumberland Co., PA; Shippensburg in the Civil War; Black
History of Shippensburg, Pennsylvania: 1860-1936; Samuel Davis Sturgis
1822-1889: Shippensburg's Forgotten Hero; Battle of the Bulge: One Small
Corner.
Hours & Admission Prices: Wed. & Fri.-Sat. 1-4 & by appointment. No
charge; donations accepted.
Attendance: 1,800 (estimated)
Membership: Single $20; Couple $30; Contributing $50; Business $100;
Corporate $250; Life $300.

**SHIPPENSBURG UNIVERSITY FASHION ARCHIVES AND
MUSEUM,** 1871 Old Main Dr., Harley Hall, Lower Level,
Shippensburg, PA 17257-2299. Tel.: 717-477-1239. Facebook;
Shippensburg University Fashion Archives & Museum.
Web Site: www.fashionarchives.org
Founded: 1980.
Key Personnel: Dir., Dr. Karin J. Bohleke.
Personnel Profile: Part-Time Paid 5; Part-Time Volunteers 6.
Governing Authority: public university. Parent Institution: Shippensburg Uni-
versity. Tax-exempt: 501(c)(3).
Institution Type/Description: University Costume Museum.
Collections: 19th-20th century clothing & accessories.
Facilities: 1,500-vol. library; 800 sq. ft. exhibit space.
Activities: lectures; temporary exhibitions.
Publications: annual newsletter, A Sense of Style; exhibit catalogs.
Hours & Admission Prices: Academic Year: Mon.-Thurs. 12-4; other times by
appointment. No charge; donations accepted. Closed holidays.
Attendance: 900 (estimated)

Somerset

**HISTORICAL AND GENEALOGICAL SOCIETY OF SOMER-
SET COUNTY,** 10649 Somerset Pike, Somerset, PA 15501-7357.
Tel.: 814-445-6077. Fax: 814-443-6621.
E-mail: bsarver@state.pa.us
Web Site: www.somersethistoricalcenter.org
Founded: 1959.
Congressional District: 12
Key Personnel: Pres., George Kaufman; Cur., Carrie Blough; Asst. Cur., Gail
Smith; Librarian, Susan Seese; Museum Shop Mgr., Mark Ware.
Personnel Profile: Full-Time Paid 2; Part-Time Paid 1; Part-Time Volunteers
35.
Governing Authority: nonprofit corporation. Tax-exempt: 501(c)(3).
Institution Type/Description: Local History Museum; SW Pennsylvania rural
history.
Collections: agriculture; archaeology; genealogy; archives; cider press; maple
sugar camp. Historic Houses: 1804 log house; 1773 first settlers log cabin.
Historic Site: covered bridge; agriculture
Research Fields: agriculture; archaeology; genealogy; manuscripts; archives.
Facilities: 1,000-vol. library of history books available for use on premises;
reading room.
Activities: lectures; tours; craft workshops; genealogical research. Museum
Sponsors: Mountain Craft Days in September.
Publications: quarterly magazine, Laurel Messenger.
Hours & Admission Prices: Tues.-Sat. 9-5, Sun. 12-5. Adults $6, seniors $3.50,
children 6-17 $2; discounts with tickets from other PHMC sites & AAM
members; members no charge. Closed most holidays. &
Attendance: 23,000 (accurate)
Membership: Student $5; Individual $25; Family $35; Sustaining $55; Life
$1,000.

LAUREL ARTS, 214 S. Harrison Ave., Somerset, PA 15501-1803.
Mailing Address: P.O. Box 414, Somerset, PA 15501. Tel.: 814-
443-2433. Fax: 814-443-3870.
E-mail: arts@laurelarts.org
Web Site: www.laurelarts.org
Founded: 1975.
Congressional District: 12
Key Personnel: Pres. (V), Hank Parke.
Personnel Profile: Full-Time Paid 3; Part-Time Paid 10; Part-Time Volunteers
500.
Governing Authority: private; nonprofit organization. Subsidiary Institution:
Guild of American Papercutters National Museum. Tax-exempt: 501(c)(3).

Institution Type/Description: Art Museum.

Collections: local & regional works of art including oils, watercolors, pastels, serigraphs, lithographs, weavings, mosaics, photographs, papercuttings, & bronzes.

Facilities: library; 1,600 sq. ft. exhibit space.

Activities: permanent & temporary exhibitions; art classes & workshops; performing arts programs; summer camps; arts festivals; dance recitals; formal education programs. Annual Event: Somerfest.

Publications: bimonthly newsletter, ArtLink.

Hours & Admission Prices: Tues.-Thurs. 10-6, Fri. 10-4, Sat. 12-4. No charge. Closed New Year's Day; Easter; Memorial Day; Independence Day; Labor Day; Veterans Day; Thanksgiving & day after; Christmas week.

Attendance: 51,015 (estimated)

Membership: Single $25; Couple $35; Family $40; Associate $100; Platinum $250; Sustaining $500 & up.

SOMERSET HISTORICAL CENTER, (M), 10649 Somerset Pike, Somerset, PA 15501-7357. Tel.: 814-445-6077. Fax: 814-443-6621.

E-mail: c-mware@pa.gov

Web Site: www.somersethistoricalcenter.org

Founded: 1969.

Congressional District: 12

Key Personnel: Dir., Mark Ware; Education Coord., Katie Cordek; Cur., Jacob Miller; Maintenance, Eric Sadler.

Personnel Profile: Full-Time Paid 2; Part-Time Paid 2; Part-Time Volunteers 25; Interns 1.

Governing Authority: state. Parent Institution: Pennsylvania Historical and Museum Commission, Commonwealth Keystone Bldg., 400 North St., Harrisburg, PA. 17120-0053. Subsidiary: Historical & Genealogical Society of Somerset Co. Tax-exempt: 501(c)(3).

Institution Type/Description: History Museum.

Collections: agriculture; rural history; 1880 cider press. Historic Buildings & Sites: 1798 two-story log house; 1860 maple sugar camp; 1850 covered bridge; 1773 reconstructed log cabin; comprehensive agricultural exhibits.

Research Fields: agriculture & rural life; southwest PA 1740-1940 genealogy; vernacular architecture.

Facilities: 300-vol. library of genealogical and history books available for use on premises only; 1,000 family history files; folklife center.

Activities: guided tours; lectures; films; formally organized educational programs; permanent & temporary exhibitions. Museum Sponsors: Mountain Craft Days.

Publications: quarterly newsletter, Laurel Messenger.

Hours & Admission Prices: April-Oct. Tues.-Sat. 9-5, Sun. 12-5; Nov.-March Tues.-Sat. 9-5. Adults 18-60 $6, seniors $5.50, youth $3; discounts to groups, tickets from other PHMC sites, AAA, ICOM & AAM members; members no charge. Closed major holidays. &

Attendance: 23,000 (accurate)

Membership: Single $25; Family $35.

South Fork

JOHNSTOWN FLOOD NATIONAL MEMORIAL, 733 Lake Rd., South Fork, PA 15956-3602. Tel.: 814-495-4643. Fax: 814-495-7463.

Web Site: www.nps.gov/jofl

Founded: 1964.

Congressional District: 12

Key Personnel: Supt., Keith Newlin; Chm. (V) & Museum Shop Mgr., Doug Richardson; Cur., Nancy Smith.

Personnel Profile: Full-Time Paid 4; Part-Time Volunteers 100.

Governing Authority: federal. U.S. Dept. of the Interior, National Park Service, Washington, DC 20240. Tax-exempt: 501(c)(3).

Institution Type/Description: Park Museum Memorial: located at the site of the dam which collapsed May 31, 1889, causing the Johnstown Flood.

Collections: books; articles; photos; artifacts relating to 1889 Johnstown Flood.

Research Fields: Pennsylvania Mainline Canal; Lake Conemaugh; South Fork Fishing & Hunting Club; 1889 Johnstown Flood.

Facilities: visitor center; picnic area. Books for sale.

Activities: guided tours; hiking; costumed interpretation

Hours & Admission Prices: Memorial Day-Labor Day daily 9-5. Adults 15 & over $5. Closed New Year's Day; Martin Luther King Jr. Day; Presidents' Day; Veterans Day; Thanksgiving; Christmas. &

Attendance: 190,000

South Williamsport

WORLD OF LITTLE LEAGUE: PETER J. MCGOVERN MUSEUM AND OFFICIAL STORE, (M), 525 U.S. Rte. 15, (Montgomery Pike), South Williamsport, PA 17702. Mailing Address: P.O. Box 3485, Williamsport, PA 17701-0485. Tel.: 570-326-3607. Fax: 570-326-2267. Facebook: Little League Museum.

E-mail: lvanauken@littleleague.org

Web Site: www.littleleague.org

Founded: 1982.

Congressional District: 5

Key Personnel: Pres. & C.E.O., Stephen D. Keener; Vice Pres. & Exec. Dir., Lance W. Van Auken; Dir. Public Programming & Outreach, Janice L. Ogurcak; C.F.O., David Houseknecht; Cur. & Asst. Dir., Adam Thompson.

Personnel Profile: Full-Time Paid 2; Part-Time Paid 6; Part-Time Volunteers 12; Interns 1.

Governing Authority: nonprofit organization. Parent Institution: Little League Baseball, Inc. Tax-exempt: 501(c)(3).

Institution Type/Description: Little League Baseball & Softball Museum.

Collections: growth & development of Little League Baseball, 1939-present; equipment; uniforms; photos; 1954-present team rosters; 1947-present Little League World Series programs; magazines featuring youth baseball; films.

Facilities: 110-vol. library pertaining to the sport of baseball & Little League, pamphlets, booklets, rule books, articles, archives tracing the growth & development of Little League, available for use by appointment. Museum-related items for sale.

Activities: films; guided tours; loan, temporary & participatory exhibitions.

Publications: newsletter, Pieces of the Past.

Hours & Admission Prices: Daily 9-5. Adults $5, seniors $3, children $2. &

Attendance: 18,000 (accurate)

Springdale

RACHEL CARSON HOMESTEAD, 613 Marion Ave., Springdale, PA 15144-1242. Mailing Address: P.O. Box 46, Springdale, PA 15144-0046. Tel.: 724-274-5459. Fax: 724-275-1259.

E-mail: info@rachelcarsonhomested.org

Web Site: www.rachelcarsonhomestead.org

Founded: 1975.

Congressional District: 4

Key Personnel: Pres. (V), William Schillinger; Exec. Dir., Patricia DeMarco, Ph.D.

Personnel Profile: Full-Time Paid 4; Part-Time Volunteers 10; Interns 1.

Governing Authority: private; nonprofit. Tax-exempt: 501(c)(3).

Institution Type/Description: Historic House: located at the birthplace & childhood home of ecologist & Silent Spring author Rachel Carson.

Collections: Carson family period artifacts; environmental history.

Facilities: 150-vol. library on Rachel Carson, environmental issues & history; educational facilities; 800 sq. ft. exhibit space; nature/conservation center. Museum-related items for sale.

Activities: formal education programs for children; summer camp grades K-8; guided tours; lectures; loan exhibitions; volunteer programs. Annual Event: Rachel Carson Day in May.

Publications: quarterly newsletter, The Spring.

Hours & Admission Prices: By appointment. Adults $10, senior citizens & children 5-12 $5; discounts to groups of ten or more & members. Closed major holidays. &

Attendance: 2,000 (estimated)

Membership: Kid's Club $10; Student, Senior & Low Income $15; Individual $30; Family $50; Supporting Member $125; Sustaining Member $275; Patron $500; Benefactor $1,000.

Springs

SPRINGS MUSEUM, 134 River Rd., Springs, PA 15562. Mailing Address: P.O. Box 62, Springs, PA 15562-0062. Tel.: 814-662-2625.

E-mail: jfb@qcol.net

Web Site: www.springspa.org

Founded: 1957.

Congressional District: 12

Key Personnel: Pres. (V), Joseph Bender.

Personnel Profile: Part-Time Paid 7; Part-Time Volunteers 6.

Volunteer Hours: 150

Operating Expenses: 151,005

Operating Income: 129,220

Governing Authority: society. Parent Institution: Spring Historical Society of Casselman Valley. Tax-exempt.

Institution Type/Description: General Museum.
Collections: fossils; pioneer tools; furniture.
Facilities: library of scrapbooks of historical value available on premises.
Activities: guided tours. Society Sponsors: Folk Festival in October.
Publications: annual magazine, Casselman Chronicle.
Hours & Admission Prices: Memorial Day to Labor Day Wed.-Fri. 1-5, Sat. 9-2. No Charge; donations accepted. &
Attendance: 5,000 (estimated)
Membership: Individual $20; Family $30.

State College

CENTRE COUNTY HISTORICAL SOCIETY, (M), Centre Furnace Mansion, 1001 E. College Ave., State College, PA 16801-6898. Tel.: 814-234-4779.
E-mail: info@centrecountyhistory.org
Web Site: www.centrehistory.org
Founded: 1904.
Congressional District: 5
Key Personnel: Pres. (V), Jacqueline J. Melander; Exec. Dir., Mary Sorensen; Program Coord., Megan Orient.
Personnel Profile: Full-Time Paid 2; Part-Time Volunteers 250; Interns 1.
Volunteer Hours: 5,000
Governing Authority: private; nonprofit organization. Tax-exempt: 501(c)(3).
Institution Type/Description: Historic House Museum.
Collections: Victorian furnishings; 19th-20th century Centre County clothing, tools, business memorabilia & personal artifacts; area business history; Thompson family papers; photographs, manuscripts, ledgers & ephemera of area residents; sculptor, George Gray Barnard papers.
Major Exhibits: Veiled Arts of Victorian Women, 3/16/14-8/31/14.
Research Fields: 19th-century Pennsylvania; historic gardens; Centre County social, cultural, economic & architectural history.
Facilities: 1,000-vol. library; 500 sq. ft. exhibit space; gardens; walking trail with interpretive signs. Museum-related items for sale.
Activities: docent guided tours; lectures; walking trails; group tours; meeting/special events facilities. Annual Events: Plant Celebration; Stocking Stuffer; Boogersburg School Open House.
Publications: quarterly newsletter, Mansion Notes; annual journal, Centre County Heritage.
Hours & Admission Prices: Tours: Sun., Wed. & Fri. 1-4. Library & Archives: by appointment. No charge; donations accepted. Closed New Year's Eve & Day; Easter; Memorial Day; Independence Day; Labor Day; Thanksgiving; Christmas Eve, Day & week. &
Attendance: 7,000 (estimated)
Membership: Individual $35; Family $50; Friend $100; Patron $250; Benefactor $500; Steward $1,000; Ironmaster $1,500.

DISCOVERY SPACE OF CENTRAL PENNSYLVANIA, (M), 112 W. Foster Ave., Ste. 1, State College, PA 16801. Tel.: 814-234-0200.
E-mail: info@mydiscoveryspace.org
Web Site: mydiscoveryspace.org
Founded: 2011.
Key Personnel: Dir., Allayn Beck; Asst. Dir., Michele Crowl
Institution Type/Description: Children's Museum.
Collections: hands-on exhibitions.
Facilities: 4,000 sq. ft. exhibit space.
Activities: educational programs; classes; birthday parties; special events; field trips; summer camps.
Hours & Admission Prices: Wed.-Sat. 10-5, Sun. 12-5. Adults $6; discounts to groups; children under two no charge; discounts to ASTC & AAA members. Closed New Year's Eve & Day; Easter; Memorial Day; Thanksgiving; Christmas Eve & Day.
Membership: Single Parent $65; Grandparent $65; Family $75.

Sterling Run

LITTLE MUSEUM - CAMERON COUNTY HISTORICAL SOCIETY, Rte. 120, Sterling Run, PA 15834. Mailing Address: P.O. Box 433, Emporium, PA 15834-0433. Tel.: 814-486-0213.
E-mail: info@thelittlemuseum.org
Web Site: thelittlemuseum.org
Founded: 1921.
Institution Type/Description: Historical Society Museum.
Collections: local history & culture; period furnishings; personal artifacts; photographs.
Facilities: Museum-related items for sale.
Hours & Admission Prices: Wed. & Sat.-Sun. 1-4; other times by appointment.

Stewartstown

STEWARTSTOWN HISTORICAL SOCIETY, 17 Mill St., Stewartstown, PA 17363. Mailing Address: P.O. Box 82, Stewartstown, PA 17363. Tel.: 717-993-5003.
E-mail: stewhist@yahoo.com
Web Site: www.stewhist.org
Founded: 1984.
Key Personnel: Pres. (V), Donald Linebaugh, Ph.D.
Personnel Profile: Part-Time Volunteers 15.
Volunteer Hours: 400
Operating Expenses: 13,587
Operating Income: 11,211
Governing Authority: Tax-exempt.
Institution Type/Description: Historical Society Museum.
Collections: local history & culture; photographs; postcards; local advertising; genealogy; family histories.
Hours & Admission Prices: Sun. 2-4. No charge; donations accepted. &
Attendance: 500 (accurate)
Membership: Single $10; Family $15; Corporate $25.

Strasburg

THE NATIONAL TOY TRAIN MUSEUM, (M), 300 Paradise Ln., Strasburg, PA 17579-0248. Mailing Address: P.O. Box 248, Strasburg, PA 17579-0248. Tel.: 717-687-8976 & 8623. Fax: 717-687-0742. Facebook: National Toy Train Museum.
E-mail: jluppino@traincollectors.org
Web Site: www.nttmuseum.org
Founded: 1954.
Congressional District: 16
Key Personnel: Operations Mgr., John V. Luppino; Pres., Carol R. McGinnis; Past Pres., Ronald A. Stowell; Quarterly Editor, Mark Boyd; Publications Editor, Timothy Stier.
Personnel Profile: Full-Time Paid 6; Part-Time Paid 7; Part-Time Volunteers 20.
Governing Authority: nonprofit organization. Parent Institution: Train Collectors Association, P.O. Box 248, Paradise Lane, Strasburg, PA 17579. Tax-exempt: 501(c)(3).
Institution Type/Description: Toy & Model Train Museum.
Collections: toy trains, primarily Tinplate dating from 1840's; model trains & related paraphernalia & accessories; railroadiana; manuscript collections.
Research Fields: toy trains; antique toys; railroadiana.
Facilities: library of books on toys, toy & model trains and railroads, hobby & train magazines, photos and prints, manufacturers' catalogues & archival material. Books, gifts & toys for sale.
Activities: five operating layouts; videos; changing exhibitions.
Publications: quarterly magazine, Train Collectors Quarterly; bimonthly newsletter, National Headquarters News; Train Collectors quarterly directory; online magazine, "e-train" at www.tcaetrain.org; book: Lionel, Standard of the World; pamphlets; monthly members' newsletter, On Track With TCA.
Hours & Admission Prices: April & Nov.-Dec. Sat.-Sun. & holidays 10-5; May & Sept.-Oct. Fri.-Mon. 10-5; June-Aug. daily 10-5. Adults 13-64 $6, seniors 64+ $5, children 6-12 $3, family $15, season pass $18,; discount to AAA members; AAM members, members & children under 6 no charge. Tours: adults $2.50, children $1. Closed New Year's Day; Thanksgiving; Christmas. &
Attendance: 53,000 (accurate)
Membership: Individual & Junior $35; Family: additional $1 per family member.

RAILROAD MUSEUM OF PENNSYLVANIA, (M), 300 Gap Rd., Strasburg, PA 17579. Mailing Address: P.O. Box 15, Strasburg, PA 17579-0015. Tel.: 717-687-8629 & 8628, ext. 3001. Fax: 717-687-0876.
E-mail: info@rrmuseumpa.org
Web Site: www.rrmuseumpa.org
Founded: 1963.
Congressional District: 16
Key Personnel: Dir., Charles Fox; Pres., Al Giannantonio; Education, Patrick Morrison; Librarian & Archivist, Nick Zmijewski; Restoration Mgr., Allan Martin; Security, Dennis Keperling; Gift Shop Mgr., Laura Martin.
Personnel Profile: Full-Time Paid 20; Part-Time Paid 5; Part-Time Volunteers 150; Interns 2.
Governing Authority: state. Parent Institution: Pennsylvania Historical and Museum Commission, The State Museum Bldg., 300 North St., Harrisburg, PA 17120-0090. Tax-exempt: 170(c)(1).
Institution Type/Description: History & Transportation Museum.

Collections: transportation; railroad history, locomotives; railroad cars; railroad models; tools; telegraph equipment; lanterns; photographs; artwork; paper railroadiana.
Research Fields: railroad history & technology; photographs & negatives.
Facilities: 6,000-vol. library of reports, books, timetables on railroad history; 105,000 sq. ft. exhibit space; 4-acre exhibit space apart from museum; 30-seat theater; outdoor railyard. Museum-related items for sale.
Activities: guided tours; railroad skills demonstrations; live steam operations; orientation programs; children's programs; concerts; docent program; films; loan, participatory & traveling exhibitions.
Publications: quarterly journal, The Milepost; Pennsylvania Railroad Cookbook; booklet, The Broadway Limited; guidebook, The Railroad Museum of Pennsylvania; booklet, The Lancaster Locomotive Works; Baldwin Locomotive Works Negative Catalogue; Benkline Collection of Negatives.
Hours & Admission Prices: April-Oct. Mon.-Sat. 9-5, Sun. 12-5; Nov.-March Tues.-Sat. 9-5, Sun. 12-5. Adults $8, senior citizens $7, children 6-17 $6; discounts to AAM & AAA members; children 5 and under & members no charge. Closed New Year's Day; Thanksgiving; Christmas. &
Attendance: 133,500 (accurate)
Membership: Student $20; Basic $35; Family $45; Supporting $60; Benefactor $70-$5,000.

Strongstown

DANE CASTLE, Frederick Rd., Strongstown, PA 15957. Mailing Address: P.O. Box 10, Route 403 North, Strongstown, PA 15957-0010. Tel.: 814-749-7341.
Web Site: www.danecastle.com
Founded: 2002.
Institution Type/Description: History Museum.
Collections: medieval art; weapons; suits of armor; shields; helmets; photographs.
Activities: rental facilities; group tours. Museum Sponsors: Castle Great Hall Tours; school group tours by appointment April, May, Sept. & Oct.
Hours & Admission Prices: April-Nov. 1st by appointment. Admission $5. &

Stroudsburg

MONROE COUNTY HISTORICAL ASSOCIATION/STROUD MANSION, 900 Main St., Stroudsburg, PA 18360-1604. Tel.: 570-421-7703. Fax: 570-421-9199.
E-mail: mcha@ptd.net
Web Site: www.monroehistorical.org
Formerly: Stroud Mansion/Monroe County Historical Association
Founded: 1921.
Congressional District: 10
Key Personnel: Pres., David Thomas; Exec. Dir., Amy Leiser.
Personnel Profile: Full-Time Paid 1; Part-Time Paid 2; Part-Time Volunteers 10.
Governing Authority: nonprofit organization. Parent Institution: Monroe County Historical Association. Tax-exempt: 501(c)(3).
Institution Type/Description: Local History Museum: housed in the 1795 Strond Mansion built by Jacob Stroud, founder of Stroudsburg.
Collections: decorative arts; toys; textiles; costumes; textile working tools; weapons; Indian artifacts; Victorian paintings; prints; photographs.
Research Fields: Monroe County history & genealogy.
Facilities: 8,000-vol. library; 2,800 sq. ft. exhibit space. Historical Society publications for sale.
Activities: guided tours; lectures; hobby workshops; school loan service; outreach educational programs for schools & community groups.
Publications: quarterly newsletter, The Fan Light; brochure about sites & services.
Hours & Admission Prices: Tues.-Fri. 11-4, 1st & 3rd Sat. 10-4. Adults $8; discounts to AAM members; members no charge. Closed major holidays. &
Attendance: 1,895 (accurate)
Membership: Student $10; Senior Individual $15; Individual $25; Family $35; Patron $50; Century $100; Small Business $150; Corporate $250.

QUIET VALLEY LIVING HISTORICAL FARM, (M), 347 Quiet Valley Rd., Stroudsburg, PA 18360-9455. Tel.: 570-992-6161. Fax: 570-992-9587.
E-mail: farm@quietvalley.org
Web Site: www.quietvalley.org
Founded: 1963.
Congressional District: 15
Key Personnel: Bd. Pres., David Durham; Dir., Janet F. Mishkin.
Personnel Profile: Full-Time Paid 3; Part-Time Paid 20; Part-Time Volunteers 400.
Governing Authority: nonprofit organization. Tax-exempt: 501(c)(3).

Institution Type/Description: Living Farm Museum: housed on the 1765 site of a Pennsylvania homestead.
Collections: furnishings; implements; Pennsylvania German agriculture. Historic Buildings: farm house & barn.
Research Fields: early farm crafts; 18th-19th century agriculture; rural life.
Facilities: 500-vol. library of material relating to early farm life available to the staff only. Handmade history-related items, commercial items & books for sale.
Activities: guided tours for school groups; tours for general public; workshops; docent program; day camp. Annual Events: Harvest Festival; Farm Animal Frolic; Old Time Christmas.
Publications: newsletter, five times a year.
Hours & Admission Prices: 3rd Sat. in June-Labor Day Tues.-Sat. 10-5, Sun. 12-5; school groups by appointment. Adults $10, children 3-12 $5; discounts to groups & AAA members; members no charge. Special Events: Farm Animal Frolic: adults $8, children 3-12 $5, Harvest Festival: adults $10, children 3-12 $5, Old Time Christmas: adults $10, children 3-12 $5. &
Attendance: 27,000 (estimated)
Membership: Student $15; Senior $20; Individual $25; Family $55; Contributing $100; Sustaining $150; Donor $500; Life $1,000.

Sunbury

THE NORTHUMBERLAND COUNTY HISTORICAL SOCIETY, (M), 1150 N. Front St., Sunbury, PA 17801-1126. Tel.: 570-286-4083.
E-mail: info@northumberlandcountyhistoricalsociety.org
Web Site: www.northumberlandcountyhistoricalsociety.org
Founded: 1925.
Congressional District: 17
Key Personnel: Pres. (V), Scott Heintzelman; Office Mgr., Charlotte Rhinehart.
Personnel Profile: Part-Time Volunteers 45.
Governing Authority: Tax-exempt.
Institution Type/Description: Historic Site: 1756-94, frontier outpost; 1852, Hunter Mansion.
Collections: military & Indian artifacts; documents; clothing.
Research Fields: genealogy; local history.
Facilities: library.
Activities: guided tours; special programs.
Publications: books & booklets on Pennsylvania history; Society proceedings & reprints.
Hours & Admission Prices: Mon., Wed. & Fri. 1-4. Museum: no charge; donations accepted. Library $5; members no charge. &
Attendance: 3,900 (estimated)
Membership: Individual $20; Family $35; Life $300.

Swarthmore

* **SCOTT ARBORETUM OF SWARTHMORE COLLEGE, (M),** 500 College Ave., Swarthmore, PA 19081-1306. Tel.: 610-328-8025. Fax: 610-328-7755. Facebook: Scott Arboretum.
E-mail: scott@swarthmore.edu
Web Site: www.scottarboretum.org
Formerly: The Scott Foundation
Founded: 1929.
Congressional District: 7
Key Personnel: Dir., Claire Sawyers; Pres. (V), Laura Fetterman; Education Coord., Julie Jenny; Horticultural Coord., Jeff Jabco; Public Rels. Coord. (V), Becky Robert; Cur., Andrew Bunting; Office Mgr., Jacqui West.
Personnel Profile: Full-Time Paid 22; Part-Time Paid 19; Part-Time Volunteers 200; Interns 1.
Operating Expenses: 1,245,992
Operating Income: 1,236,107
Governing Authority: college. Parent Institution: Swarthmore College. Tax-exempt: 501(c)(3).
Institution Type/Description: Arboretum.
Collections: horticulture; botany; arboretum; plants with significant ornamental value.
Research Fields: plant evaluation programs.
Facilities: 1,000-vol. library of reference & research books available on premises; classrooms.
Activities: guided tours; lectures; formally organized education programs; permanent & temporary exhibitions; Scott medal & award.
Publications: newsletter, Hybrid; class schedules, annual report, interpretive brochures on special collections; book, The Scott Arboretum of Swarthmore College: The First 75 Years.
Hours & Admission Prices: Arboretum: daily dawn-dusk. Main office: Mon.-Fri. 8:30-12 & 1-4:30. No charge; donations accepted. Office closed New Year's Eve & Day; Independence Day; Thanksgiving; Christmas Eve, Day & week. &

Attendance: 30,000 (estimated)
Membership: Student $10; Individual $40; Dual $55; Contributor $75; Organization $125; Sponsor $150; Benefactor $250; Patron $500; Director's Circle $1,000; Philanthropist $3,000.

Tarentum

ALLEGHENY-KISKI VALLEY HISTORICAL SOCIETY AND HERITAGE MUSEUM, 224 E. Seventh Ave., Tarentum, PA 15084-1513. Tel.: 724-224-7666. Fax: 724-224-7642. Facebook: Allegheny-Kiski Valley Historical Society and Heritage Museum.
E-mail: akvhs@salsgiver.com
Web Site: akvhs.org
Formerly: Allegheny-Kiski Valley Historical Society
Founded: 1966.
Key Personnel: Pres. (V), Dolly Mistrik.
Personnel Profile: Part-Time Paid 1.
Governing Authority: Tax-exempt.
Institution Type/Description: Heritage Museum: housed in the former home of American Legion Post 85; built in 1931.
Collections: local history & culture; period furnishings; mural; photographs; personal artifacts; industry; military, local artwork.
Major Exhibits: Eddie Adams Paris Collection, 6/14.
Activities: internships.
Publications: bimonthly, Chronicle.
Hours & Admission Prices: Wed. & Sat. 11-3. Adults $5; members & veterans no charge. &
Attendance: 4,000 (estimated)
Membership: Student $5; Individual $25; Family $35; Friend $75; Patron $150.

Tidioute

SIMPLER TIMES MUSEUM, 111 Simpler Times Lane, St. U.S. Rte. 62, Tidioute, PA 16351. Mailing Address: 111 Simpler Times Lane, Tidioute, PA 16351. Tel.: 814-484-3483.
Founded: 1992.
Key Personnel: Pres. (V), Bruce E. Ziegler.
Personnel Profile: Part-Time Volunteers 8.
Governing Authority: Tax-exempt.
Institution Type/Description: Local History Museum.
Collections: local oil, gas, timber, & agriculture history; 90 gasoline pumps; hit and miss engines; farm tractors & equipment; Ford cars from 1913 & up; gasoline signs, globes & containers.
Hours & Admission Prices: Call for hours. Adults $4; 4-H, Girl Scouts, Boy Scouts, & church groups no charge.
Attendance: 4,500 (estimated)

Tionesta

FOREST COUNTY HISTORICAL SOCIETY, 206 Elm St., Tionesta, PA 16353. Mailing Address: P.O. Box 546, Tionesta, PA 16353-0546. Tel.: 814-755-4422.
Institution Type/Description: Historical Society Museum.
Collections: local history & culture; period artifacts; photographs.
Hours & Admission Prices: late May to late Sept. Mon.-Sat. 10-4; other times by appointment.

Titusville

* **DRAKE WELL MUSEUM, (M),** 202 Museum Lane, Titusville, PA 16354-8902. Tel.: 814-827-2797. Fax: 814-827-4888. Facebook: Drake Well Museum.
E-mail: drakewell@verizon.net
Web Site: www.drakewell.org
Founded: 1934.
Congressional District: 23
Key Personnel: C.E.O., James M. Vaughan; Pres. (V), Harry Wilmoth; Museum Shop Mgr., Sheri Hamilton.
Personnel Profile: Full-Time Paid 7; Part-Time Paid 6; Part-Time Volunteers 10; Interns 2.
Governing Authority: state. Administered by the Pennsylvania Historical & Museum Commission, Commonwealth Keystone Bldg., 400 North St., Harrisburg, PA 17120. Tax-exempt: 170(b)(1)(A).
Institution Type/Description: Industrial Oil Museum: located on site of first commercially successful oil well.
Collections: early lighting devices; oil well drilling & production tools and equipment; collodion wet plate negatives of development of Pennsylvania oil industry; replica of Drake's derrick & engine house.

Research Fields: oil industry history.
Facilities: approx. 5,000-vol. library on oil production & its history available for research; 140-seat auditorium; theater; picnic area. Museum-related items for sale.
Activities: guided & self-guided tours; concerts; formally organized education programs for children; permanent & temporary exhibitions; film & video programs.
Publications: newsletter, The Barker.
Hours & Admission Prices: Jan.-March Wed.-Sat. 9-5, Sun. 12-5; April-Dec. Tues.-Sat. 9-5, Sun. 12-5. Adults 12-64 $10, senior citizens 65 & over $8, youth 3-11 $5; discounts to AAA members; children under 3 no charge. Closed New Year's Day; Martin Luther King Jr. Day; Presidents' Day; Columbus Day; Veterans Day; Thanksgiving & day after; Christmas. &
Attendance: 42,000 (estimated)
Membership: Individual $30; Corporate $%150-$250.

Towanda

BRADFORD COUNTY HISTORICAL SOCIETY, 109 Pine St., Towanda, PA 18848-1701. Tel.: 570-265-2240.
E-mail: info@bradfordhistory.com
Web Site: www.bradfordhistory.com
Founded: 1870.
Congressional District: 10
Key Personnel: Dir., Matthew Carl.
Personnel Profile: Full-Time Paid 1; Part-Time Paid 1; Part-Time Volunteers 25.
Governing Authority: nonprofit organization. Tax-exempt: 501(c)(3).
Institution Type/Description: Historical Society Museum: Bradford County jail building.
Collections: furniture & furnishings of the 19th century; Civil War memorabilia; Indian artifacts; boy & girl scouting items; toys; baby carriages; tools; handmade items. Replica: c.1905 log cabin.
Research Fields: genealogy & local history.
Facilities: 1,000-vol. library of books & microfilm pertaining to local history & genealogy available for research.
Activities: guided tours.
Publications: quarterly magazine, The Settler; 1995 History of Bradford County PA; book, Barclay MT - A History.
Hours & Admission Prices: Library: Wed.-Fri. 10-4, Sat. by appointment. Museum: June-Sept. Thurs.-Sat. 10-4; groups by appointment. No charge; donations accepted. Genealogy Research: $5 daily. Closed Federal holidays. &
Attendance: 1,500
Membership: Student $15; Basic $25; Sustaining $35; Family $45; Benefactor $100; Patron $500; Life $1,000.

FRENCH AZILUM, INC., 469 Queens Rd., Towanda, PA 18848-9107. Tel.: 570-265-3376.
E-mail: frenchazilum@epix.net
Web Site: www.frenchazilum.com
Founded: 1954.
Congressional District: 10
Key Personnel: Pres. Bd., Phillip Swank; Site Mgr., Danielle Lambert.
Personnel Profile: Full-Time Paid 1; Part-Time Paid 2; Part-Time Volunteers 12.
Governing Authority: state. Parent Institution: Pennsylvania Historical & Museum Commission, Box 1026, Harrisburg, PA 17120. Tax-exempt.
Institution Type/Description: Historic House & Site: 1793-1803 Refuge of the French Royalists; 1836 Laporte House.
Collections: heirlooms of the emigres; antique farm tools; blacksmith & carpenters tools; spinning & weaving implements; 18th- & 19th-century period furniture.
Research Fields: archaeological sites.
Facilities: library of material pertaining to French Revolution; Marie Antoinette; French Azilum & local history; nature trail. Museum-related items for sale.
Activities: site interpretation. Museum Sponsors: Heritage Day - Living History Stations, Stonewall Workshop.
Hours & Admission Prices: May 23-Sept. 7 Fri.-Mon. 11-5; 1st tour at 12pm, last tour at 4pm; Sept. 13-Oct. 5 Sat.-Sun. 11-5; other times by appointment. Adults $5, students $3; children under 12 & members no charge. &
Attendance: 3,750 (estimated)
Membership: Basic $15; Individual $25; Family $40; Supporting $75; Contributing $100.

Trappe

HENRY MUHLENBERG HOUSE, 201 W. Main St., Trappe, PA 19426. Mailing Address: P.O. Box 26708, Collegeville, PA 19426-0708. Tel.: 610-489-7560.
E-mail: info@trappehistoricalsociety.org
Governing Authority: Parent Institution: Trappe Historical Society.
Institution Type/Description: Historic House Museum: built in 1776. Listed on the National Register of Historic Places.
Collections: Muhlenberg family history; period furnishings; personal artifacts; photographs.
Activities: special programs & events.
Publications: quarterly newsletter.
Hours & Admission Prices: Call for hours.

THE HISTORICAL SOCIETY OF TRAPPE, COLLEGEVILLE, PERKIOMEN VALLEY, INC., 301 W. Main St., Trappe, PA 19426. Mailing Address: P.O. Box 26708, Collegeville, PA 19426-0708. Tel.: 610-489-7560. Fax: 610-489-7560.
E-mail: info@trappehistoricalsociety.org
Web Site: www.trappehistoricalsociety.org
Founded: 1964.
Congressional District: 13
Key Personnel: Pres., Rev. Robert A. Meschke; Treas., Patricia A. Quinty; Historian, Rev. Judith A. Meier.
Personnel Profile: Part-Time Paid 1; Part-Time Volunteers 20; Interns 1.
Governing Authority: society; nonprofit organization. Branch Museum: Muhlenberg House, 201 W. Main St., Trappe, PA. Tax-exempt: 501(c)(3).
Institution Type/Description: Local History Museum: housed in early 18th- & 19th-century Dewees Tavern.
Collections: archives; genealogy; history; preservation project; manuscript collections; historical markers.
Research Fields: genealogy; Muhlenberg & Revolutionary War period.
Facilities: 500-vol. library of research books & manuscripts on genealogy available for use on premises by appointment. Gifts & pamphlets for sale.
Activities: guided tours; lectures; permanent exhibitions; historical markers.
Publications: quarterly, The Chronicle; newsletter; map, Washington's Itinerary in Montgomery County Sept.-Dec. 1777; Registry of Historic Building in Trappe.
Hours & Admission Prices: House: June-Sept. Sun. 1:30-4. Museum: Tues. & Thurs. 9-12. Donations: $2.
Attendance: 1,100 (estimated)
Membership: Student $10; Individual $25; Family $50; Supporting $100-$249; Patron $250-$499; Sustaining $500-$999; Benefactor $1,000 & up.

University Park

EARTH & MINERAL SCIENCES MUSEUM AND ART GALLERY, (M), 19 Deike Burrows Rd., University Park, PA 16802-5000. Mailing Address: 207 Deike Bldg., Pennsylvania State University, University Park, PA 16802-5000. Tel.: 814-865-6336.
E-mail: rgraham@ems.psu.edu
Web Site: www.ems.psu.edu/museum
Key Personnel: C.E.O. & Dir., Russell W. Graham; Asst. Dir., Julianne Snider.
Personnel Profile: Full-Time Paid 2; Part-Time Paid 1.
Governing Authority: Parent Institution: Pennsylvania State University. Tax-exempt.
Institution Type/Description: Natural History, Science & Art Museum.
Collections: rocks; minerals; fossils; glasses; ceramics; metals; plastics; synthetic materials; mining & scientific equipment; archaeological artifacts; paintings; sculptures; drawings.
Research Fields: mineralogy & paleontology.
Facilities: two exhibit galleries; research & collection facility; curator office; exhibits & collection office.
Activities: school tours & special programs.
Publications: annual report.
Hours & Admission Prices: Mon.-Fri. 9:30-5. No charge; donations accepted. Closed legal holidays; university's recess. &
Attendance: 8,000 (estimated)
Membership: Regular $20; Pyrite $100; Gypsum $500; Quartz & Gold $1,000.

THE FROST ENTOMOLOGICAL MUSEUM, DEPT. OF ENTOMOLOGY, THE PENNSYLVANIA STATE UNIVERSITY, Headhouse #3, Curtin Rd., University Park, PA 16802-1009. Tel.: 814-863-2865. Fax: 814-865-3048.
E-mail: kck@psu.edu
Web Site: ento.psu.edu/facilities/frost
Founded: 1969.
Congressional District: 23

Key Personnel: Cur. Emeritus, Dr. Ke Chung Kim.
Personnel Profile: Part-Time Volunteers 1; Interns 2.
Governing Authority: university. Parent Institution: The Pennsylvania State University. Tax-exempt.
Institution Type/Description: Specialized Natural History Museum.
Collections: entomology; zoology; biodiversity.
Research Fields: insect systematics; ecology; faunistics; biodiversity survey.
Facilities: 400-vol. library of taxonomy, entomology, monographs & catalogs available for research by special arrangement; separate laboratory operation.
Activities: guided tours; formally organized education programs; permanent & temporary exhibitions; lectures.
Publications: research papers.
Hours & Admission Prices: Mon.-Fri. 9:30-4:30. No charge; donations accepted. &
Attendance: 5,000 (estimated)

MATSON MUSEUM OF ANTHROPOLOGY, 409 Carpenter Bldg., Pennsylvania State University, University Park, PA 16802-3401. Tel.: 814-865-3853.
Web Site: www.anthro.psu.edu/matson_museum/index.shtml
Key Personnel: Dir., Dr. Clair McHale Milner
Institution Type/Description: Anthropology Museum.
Collections: history of human cultures; cultural & biological differences; photographs.
Facilities: Museum-related items for sale.
Hours & Admission Prices: Fall & Spring: Mon.-Thurs. 9-4, Fri. 9-3; Summer: call for hours. No charge.

PALMER MUSEUM OF ART, THE PENNSYLVANIA STATE UNIVERSITY, (M), Curtin Rd., University Park, PA 16802-2507. Tel.: 814-865-7672. Fax: 814-863-8608.
E-mail: egw4@psu.edu
Web Site: www.palmermuseum.psu.edu
Founded: 1972.
Congressional District: 5
Key Personnel: Dir., Jan Keene Muhlert; Advisory Bd. Chm., John P. Driscoll, Ph.D.; Cur., Joyce Robinson, Ph.D.; Cur. Charles V. Hallman, Patrick J. McGrady, Ph.D.; Coord. Membership & Public Rels., Jennifer Cozad Feehan; Cur. Education, Dana Carlisle Kletchka, Ph.D.; Registrar, Beverly Balger Sutley; Administrative Asst., Elizabeth Warner; Sr. Exhibition Preparator, Richard Hall; Preparator, Craig Witter; Museum Security & Facility Mgr., Jeremy R. Warner; Museum Store Mgr., Lynne McCormack.
Personnel Profile: Full-Time Paid 12; Part-Time Paid 25; Part-Time Volunteers 84; Interns 5.
Governing Authority: university. Parent Institution: The Pennsylvania State University. Tax-exempt.
Institution Type/Description: Art Museum.
Collections: American & European paintings, drawings, photographs, prints & sculpture; Asian ceramics, painting & prints; limited material in ancient, African & Near Eastern areas; ancient Peruvian ceramics; contemporary European & Japanese studio ceramics.
Major Exhibits: British Watercolors from the Permanent Collection, 1/7/14-5/4/14; Forging Alliances, 1/7/14-5/11/14; Surveying Judy Chicago: Five Decades, 1/21/14-5/11/14; American Prints, 5/13/14-8/24/14; Seeing America: Photographs from the Permanent Collection, 5/20/14-8/17/14; Window on the West: Views from the American Frontier: The Phelan Collection (T), 6/10/14-8/31/14; Marcellus Shale Documentary Project (T), 9/14-12/14; Recent Acquisitions, 9/14-12/14; Henry Varnum Poor: Studies for the Land Grant Frescoes, 9/14-12/14.
Research Fields: Western art from antiquity to the present day; American art.
Facilities: print study room; 150-seat auditorium. Museum-related items sold in store.
Activities: guided tours; lectures; temporary & traveling exhibitions; symposia; teachers' workshops; family programs; print & drawing study club.
Publications: catalogs, pamphlets & brochures of exhibitions; triannual newsletter.
Hours & Admission Prices: Tues.-Sat. 10-4:30, Sun. 12-4. No charge; donations accepted. Closed national holidays; New Year's Eve & Day; Christmas Eve, Day & week. &
Attendance: 38,562 (accurate)
Membership: Student $10-$34; Individual $35-$59; Family & Household $60-$99; Sustaining $100-$299; Benefactor $300-$599; Sponsor $600-$999; Director's Circle $1,000 & up; Corporate Circle $2,500 & up. Senior Citizens 60 & over 10% discount.

PENN STATE ALL-SPORTS MUSEUM, (M), Beaver Stadium, University Park, PA 16802. Tel.: 814-865-0044.
Founded: 2002.

Key Personnel: Dir., Ken Hickman; Programming & Education, Aimee Brown.
Personnel Profile: Full-Time Paid 3; Part-Time Paid 10; Part-Time Volunteers 50.
Governing Authority: Parent-Institution: The Pennsylvania State University. Tax-exempt.
Institution Type/Description: Sports Museum.
Collections: Penn State's sports history; student-athletes & coaches; photographs; uniforms; trophies; Olympic memorabilia.
Facilities: 30-seat theater.
Hours & Admission Prices: Tues.-Sat. 10-4, Sun. 12-4; call for additional hours. Adults $5, seniors citizens 65 & over, students and children under 14 $3. Closed New Year's Eve & Day; Easter; Memorial Day; Thanksgiving; Christmas Eve & Day. ♿
Attendance: 22,000 (estimated)

Upland

THE FRIENDS OF THE CALEB PUSEY HOUSE, INC., (M), 15 Race St., Upland, PA 19015. Mailing Address: P.O. Box 1183, Upland, PA 19015-0183. Tel.: 610-874-5665.
E-mail: calebpuseyhouse@comcast.net
Founded: 1960.
Congressional District: 7
Key Personnel: Pres. (V), Harold R. Peden; Vice Pres., Arthur Hickey; Vice Pres., Ruth Moll.
Personnel Profile: Part-Time Volunteers 27.
Governing Authority: board of directors. Tax-exempt: 501(c)(3).
Institution Type/Description: General Museum: housed in 1683 Caleb Pusey House.
Collections: furniture; architecture; archaeology; herbarium; linens. Historic Houses: 1683 schoolhouse; c.1790 Pennock Log House.
Research Fields: Pusey genealogy.
Facilities: reception area.
Activities: tours; lectures.
Publications: newsletter, The Weathervane.
Hours & Admission Prices: May-Oct. Sat.-Sun. 1-4; groups by special arrangement. No charge; donations accepted. ♿
Attendance: 1,500 (estimated)

Vandergrift

VICTORIAN VANDERGRIFT MUSEUM & HISTORICAL SO-CIETY, 184 Sherman Ave., Vandergrift, PA 15690-1136. Tel.: 724-568-1990.
E-mail: vvmhs@comcast.net
Web Site: www.vvmhs.org
Founded: 1990.
Key Personnel: Dir., Elizabeth Caporali; Pres. (V), Anthony Ferrante.
Personnel Profile: Part-Time Paid 1; Part-Time Volunteers 10.
Governing Authority: Tax-exempt.
Institution Type/Description: Historical Society Museum: housed in the former Sherman School building.
Collections: local history & culture; photographs; personal artifacts; early furnishings.
Facilities: genealogy library.
Publications: monthly newsletter, Winding Streets.
Hours & Admission Prices: Mon.-Sat. 10-3. No charge; donations accepted.
Membership: Individual $20.

Wallingford

FAY FREEDMAN GALLERY, 414 Plush Mill Rd., Wallingford, PA 19086. Tel.: 610-566-1713. Fax: 610-566-0547.
E-mail: info@communityartscenter.org
Web Site: www.communityartscenter.org
Key Personnel: Exec. Dir., Deborah R. Yoder.
Institution Type/Description: Art Gallery.
Collections: paintings; sculpture.
Facilities: Gallery-related items for sale.
Activities: classes; workshops.
Hours & Admission Prices: Call for hours.

Warren

WARREN COUNTY HISTORICAL SOCIETY, 210 Fourth Ave., Warren, PA 16365-2318. Mailing Address: P.O. Box 427, Warren, PA 16365-0427. Tel.: 814-723-1795.
E-mail: warrenhistory@kinzua.net
Web Site: www.warrenhistory.org

Founded: 1900.
Congressional District: 5
Key Personnel: Mng. Dir., Michelle Gray.
Personnel Profile: Full-Time Paid 1; Part-Time Paid 2; Part-Time Volunteers 90, Interns 5.
Governing Authority: society. Subsidiary Institution: Wilder Museum. Tax-exempt: 501(c)(3).
Institution Type/Description: Local History Museum: housed in second empire mansion.
Collections: photographs; paintings; documents; manuscripts; letters; diaries; county records; genealogical records; books; pamphlets; tools & implements; utensils; maps; Victorian furniture.
Research Fields: area history; petroleum history.
Facilities: 2,000-vol. library of Pennsylvania history books available for use on premises; genealogy department; reading room. Books & pamphlets for sale.
Activities: guided tours; lectures; temporary exhibitions.
Publications: periodical, Stepping Stones; Index to Stepping Stones 1955-current; books, Historic Buildings in Warren County, Vols. 1,2,3,4,5; Of Prescriptions & Playbills, from the diaries of Michael V. Ball, M.D., 1884-86; History & Development of the Petroleum Industry in Warren County, PA; Cavalcade-Warren County's Second Century c.2000; Murder in the Courtroom; Pennsylvania Wilds; Vignettes of Yesteryear Vols. I & II; Images of America Series: Warren.
Hours & Admission Prices: Mon.-Fri. 8:30-4:30, May-Sept. Sat. 9 to noon; other times by appointment. Adults $1; members & children no charge. Closed national holidays. ♿
Attendance: 10,500 (estimated)
Membership: Student $15; Senior 65 & over $20; Individual 18-64 & Senior Family $30; Family (at one address) $40; Nonprofit Organization $50; Contributing Family (at one address) $75; Patron Family (at one address) $100; Contributing Business $100-$499; Life $500 & up; Patron Business $500-$1,000.

Washington

DAVID BRADFORD HOUSE, (M), 175 S. Main St., Washington, PA 15301-4948. Mailing Address: P.O. Box 537, Washington, PA 15301-0537. Tel.: 724-222-3604. Facebook: David Bradford House.
E-mail: bradfordhouse@verizon.net
Web Site: www.bradfordhouse.org
Founded: 1960.
Congressional District: 22
Key Personnel: Pres., William Price; Historical Dir., Clay Kilgore; Admin., Tracie Liberatore.
Personnel Profile: Part-Time Paid 2; Part-Time Volunteers 17.
Governing Authority: state. Operated by the Bradford House Historical Association for the Pennsylvania Historical & Museum Commission, P.O. Box 1026, Harrisburg, PA 17120. Tax-exempt.
Institution Type/Description: Historic House: housed in c.1788 restored home of a 1794 Whiskey Tax Rebellion leader.
Collections: historic period furnishings; 18th-century plant & herb garden.
Major Exhibits: Whiskey Rebellion, 7/14.
Facilities: Museum-related items for sale.
Activities: guided tours; lectures; formally organized education programs for children; demonstrations by guides in Colonial dress.
Publications: newsletter, The Bradford House.
Hours & Admission Prices: May-Dec. Wed.-Sat. 11-5. Adults $5, seniors $4, students 6-18 $3; discounts to AAM & AAA Motor Club members; children under 6 & members no charge.
Attendance: 1,500 (estimated)
Membership: Students $15; Adults $30; Family $50.

PENNSYLVANIA TROLLEY MUSEUM, (M), 1 Museum Rd., Washington, PA 15301-6133. Tel.: 724-228-9256. Fax: 724-228-9675.
E-mail: ptm@pa-trolley.org
Web Site: www.patrolley.org
Founded: 1953.
Congressional District: 20
Key Personnel: Chm. (V) & Pres. (V), Walt Pilof; Exec. Dir., Scott R. Becker; Sec., Ralph Ciccone; Treas., Joe Stolmack.
Personnel Profile: Full-Time Paid 4; Part-Time Paid 1; Part-Time Volunteers 150; Interns 2.
Governing Authority: nonprofit organization. Parent Institution: Pennsylvania Trolley Museum, Inc., 1 Museum Rd., Washington 15301. Tax-exempt: 501(c)(3).
Institution Type/Description: Railway Museum.

Collections: city & interurban trolley cars, diesel & electric locomotives; railroad cars.

Research Fields: Pennsylvania Electric Railway History.

Facilities: research library collection, periodicals, photo negatives, blueprints; Visitor Education Center; 2 mi. trolley line, trolley car barn; restoration shop; vintage trolley waiting shelters; 28,000 sq. ft. trolley display building; Museum-related items for sale.

Activities: operation of historic trolley cars; car restoration; meetings; lectures; films; tours; archives; education program; guided tours of collections.

Publications: bimonthly news magazine, Trolley Fare; annual report; guidebook of collections.

Hours & Admission Prices: April-May & Sept.-Dec. Mon. & Fri. 10-4, Sat.-Sun. 11-5; Memorial Day-Labor Day Mon.-Fri. 10-4, Sat.-Sun. 11-5. Adults $9, seniors 62 & over $8, children 3-15 $6; children 2 & under no charge. &

Attendance: 21,821 (accurate)

Membership: Associate $30; Voting $40; Couple $60; Family $75.

WASHINGTON COUNTY HISTORICAL SOCIETY - LEMOYNE HOUSE,

49 E. Maiden St., Washington, PA 15301-4941. Tel.: 724-225-6740. Fax: 724-225-8495.

E-mail: wchspa@verizon.net

Web Site: www.wchspa.org

Founded: 1900.

Congressional District: 20

Key Personnel: Dir., Clayton Kilgore; Pres. Bd. (V), David Budinger; Coord. Public Rels. & Education, Lynn Manning; Research Librarian, Charles Edgar; Administrative Asst., Charlotte Davidson.

Personnel Profile: Full-Time Paid 1; Part-Time Paid 4; Part-Time Volunteers 45; Interns 3.

Governing Authority: nonprofit corporation. Tax-exempt.

Institution Type/Description: Historic House: historic house & doctor's office used as part of the Underground Railroad. A National Historic Landmark.

Collections: archives; historical; Washington County artifacts & memorabilia; crematory.

Research Fields: local history, genealogy.

Facilities: library of county history.

Activities: guided tours; school programs; workshops; lectures; research service. Annual Events: Christmas Candlelight Tours; Art Show & Sale.

Publications: newsletter; books, Researching Washington County Kit; map, 1817 County Map; engraving, 1896 Bird's Eye View of Washington, PA.

Hours & Admission Prices: Tues.-Fri. 11-4. Adults $5, students $4; discounts to AAA & AAM members; members no charge. Closed holidays.

Attendance: 8,000 (estimated)

Membership: Single $25; Family $45; Sustaining $50; Life $300; Dr. Ausalom Baird Circle $250-$499; Captain David Acheson Circle $500-$999; William Holmes McGuffey Circle $1,000-$4,999; Dr. Francis Julius Lemoyne Circle $5,000 & up.

Washington Crossing

WASHINGTON CROSSING HISTORIC PARK, (M),

1112 River Rd., Washington Crossing, PA 18977-1202. Mailing Address: P.O. Box 103, Washington Crossing, PA 18977-0103. Tel.: 215-493-4076. Fax: 215-493-4820.

Web Site: www.ushistory.org/washingtoncrossing

Founded: 1917.

Congressional District: 8

Key Personnel: Historic Site Admin., Joan D. Hauger; Museum Cur., Kimberly McCarty; Maintenance Supvr., Loren Zuck.

Personnel Profile: Full-Time Paid 4; Part-Time Paid 6.

Governing Authority: state. Administered by Pennsylvania Historical & Museum Commission, Box 1026, Harrisburg, PA 17120. Subsidiary Institution: Friends of Washington Crossing Park. Tax-exempt.

Institution Type/Description: Historic Site & Recreational Park Land.

Collections: artifacts & paintings; manuscripts. Historic Buildings & Sites: 1702, 1757, 1788 Thompson-Neely; 1757 McKonkey's Ferry Inn; 1776 soldiers' graves; 1840 Thompson's Mill; 1816 Mahlon K. Taylor House.

Research Fields: American Revolution Colonial life & agriculture; Park related history.

Facilities: library of research material available for use by public with appointment; nature trails; picnic pavilions. Museum-related items for sale.

Activities: guided tours; lectures; films; gallery talks; drama; study clubs; formally organized education programs for children, adults & undergraduate college students; permanent & temporary exhibitions; school programs. Annual Event: sheep shearing; Harvest Festival; reenactment of Christmas crossing of the Delaware at 1pm Christmas Day.

Publications: books; park brochure; booklet; pamphlet; painting reproduction folder; guide book, Washington Crossing Historic Park.

Hours & Admission Prices: April-Dec. Thurs.-Sun. 10-4. Bowman's Hill

Tower: Tues.-Sun. 10-4, weather permitting. Grounds: 8am to dusk. Single Site: adults & children 4 & up $5; discounts to AAM members; children under 4 & members no charge. Three Sites: adults $9; children 4 & up $5; discounts to AAM members; children under 4 & members no charge. Grounds: no charge. Closed New Year's Day; Martin Luther King Jr. Day; Presidents' Day; Columbus Day; Thanksgiving. &

Attendance: 350,000 (estimated)

Membership: Patriot $25; Infantry $50; Oarsman $100; Lord Stirling $250; General Washington $500.

Waterford

FORT LEBOEUF MUSEUM,

123 S. High St., Waterford, PA 16441. Mailing Address: Pennsylvania Historical & Museum Commission, Box 622, Waterford, PA 16441. Tel.: 814-796-2542.

Web Site: www.fortleboeufhistoricalsociety.org

Founded: 1929.

Congressional District: 24

Key Personnel: Property Placement Officer, Robert N. Seiber; Chm. (V), Dan Kerr; Pres. (V), Judy Nelson; Museum Shop Mgr., Jim Edwards.

Governing Authority: state. Operated by the Fort LeBoeuf Historical Society under authority of PA. Historical & Museum Commission, Box 1026, Harrisburg, PA 17120. Tax-exempt.

Institution Type/Description: History Museum: located at the site of a succession of three forts, all named Fort LeBoeuf; built by the French in 1753; rebuilt by the British in 1760; rebuilt by the governor of Pennsylvania in 1795.

Collections: artifacts from French occupation; archives; history. Historic Sites: 1753 French Fort; 1760 British Fort; 1795 American Fort.

Research Fields: historic archaeology.

Facilities: 500-vol. library of historical reference books available for research on premises.

Activities: guided tours; lectures.

Publications: Books, booklets, postcards on Pennsylvania History; brochure, How to Arrange a Tour of the Fort LeBoeuf Museum.

Hours & Admission Prices: Closed.

Membership: Individual $10; Household Family $20; Civic Organization $50; Lifetime Individual $150; Lifetime Household Family $250.

Watsontown

HISTORIC WARRIOR RUN CHURCH, (M),

Intersection Susquehanna Trail & 8th St., Watsontown, PA 17777. Mailing Address: Fort Freeland Heritage Society, P.O. Box 26, Turbotville, PA 17772-0026. Tel.: 570-538-1308.

E-mail: info@freelandfarm.org

Web Site: www.freelandfarm.org/warrior_run_church.php

Founded: 1835.

Congressional District: 17

Key Personnel: Pres., E. Jane Koch.

Personnel Profile: Full-Time Volunteers 10; Part-Time Volunteers 20.

Governing Authority: state. Operated by The Warrior Run/Fort Freeland Heritage Society for the Pennsylvania Historical & Museum Commission, Bureau of Historic Sites & Museums, Commonwealth Keystone Bldg., 400 North St., Harrisburg, PA 17120-0053. Tax-exempt.

Institution Type/Description: Historic Building & Site: 1835 Country Greek Revival Style Church; Revolutionary War Graveyard.

Collections: church furniture & furnishings; books; records. Graveyard: burials include veterans of Revolutionary War, War 1812, Mexican War & Civil War, 1789-1940.

Research Fields: genealogy.

Facilities: picnic area.

Activities: educational programs. Annual Events: Strawberry Festival; Heritage Days; Christmas Candlelight Service.

Publications: Warrior Run Church Sampler; Reflexions & Images; Preservation of the Past; Tombstones of Historic Warrior Run Church; Fields of Honor.

Hours & Admission Prices: Tours by appointment. No charge; donations accepted. &

Attendance: 3,000 (estimated)

Membership: Student $1; Annual $20; Life $200.

Wayne

CHANTICLEER FOUNDATION,

786 Church Rd., Wayne, PA 19087-4713. Tel.: 610-687-4163. Fax: 610-293-0149.

E-mail: admin@chanticleergarden.org

Web Site: www.chanticleergarden.org

Founded: 1993.

Key Personnel: Exec. Dir., R. William Thomas.

Governing Authority: Tax-exempt.
Institution Type/Description: Garden.
Collections: plants; trees; flowers; vegetable gardens.
Activities: workshops; classes; educational programs.
Hours & Admission Prices: April-Oct. Wed.-Sun. 10-5; groups tours by appointment. Adults $10; discounts to museum & library members; garden professionals & children under 13 no charge. Season Pass: 1 Person $30, 2 Person $50, 3 Person $75. &
Attendance: 40,000 (accurate)

THE FINLEY HOUSE, 113 W. Beech Tree Lane, Wayne, PA 19087-3212. Tel.: 610-688-2668.
Web Site: radnorhistory.org
Founded: 1948.
Congressional District: 5
Key Personnel: Pres. (V), Ted Pollard.
Personnel Profile: Part-Time Volunteers 20; Interns 2.
Governing Authority: society. The Radnor Historical Society. Tax-exempt.
Institution Type/Description: Historical Society Museum: housed in c.1789 Finley House.
Collections: early vehicles; early local photographs; maps; manuscripts; decorative arts.
Research Fields: transportation archives; history.
Facilities: 400-vol. library of books on local history & genealogy available for use on premises or by appointment.
Activities: lectures; permanent & temporary exhibitions; field trips.
Publications: annual magazine, Radnor Historical Society Bulletin.
Hours & Admission Prices: Tues. & Sat. 2-4; other times by appointment. No charge; donations accepted.
Attendance: 400 (estimated)
Membership: Student $5; Individual $15; Family $25; Contributing $50; Patron $100; Benefactor $250.

Waynesboro

OLLER HOUSE, 138 W. Main St., Waynesboro, PA 17268-1564. Tel.: 717-762-1747.
E-mail: waynesborohistory@comcast.net
Web Site: www.waynesborohistory.com
Key Personnel: Mgr., Ken Beam.
Governing Authority: Parent Institution: Waynesboro Historical Society.
Institution Type/Description: Historic House: built in 1892.
Collections: local history & culture; period furnishings; personal artifacts; photographs.
Facilities: library.
Activities: special events.
Hours & Admission Prices: Wed. 1-5, Thurs. & Sat. 10-4, Fri. 10-1; other times by appointment. No charge; donations accepted.

RENFREW MUSEUM & PARK, 1010 E. Main St., Waynesboro, PA 17268-2338. Tel.: 717-762-4723. Fax: 717-762-6384.
E-mail: renfrew@innernet.net
Web Site: www.renfrewmuseum.org
Founded: 1973.
Congressional District: 12
Key Personnel: C.E.O., Douglas Tengler; Chm. (V), David Hykes; Admin., Bonnie Iseminger; Supvr. Grounds, John H. Frantz; Museum Shop Mgr., Cheryl Keyser.
Personnel Profile: Full-Time Paid 2; Part-Time Paid 3; Part-Time Volunteers 21.
Governing Authority: municipal. Subsidiary Institution: Renfrew Institute for Cultural & Environmental Studies. Tax-exempt: 170(b)(1)(A).
Institution Type/Description: Decorative Arts Museum: housed in 1812 Pennsylvania German farm house.
Collections: American decorative arts including furniture, furnishings & eight-legged sideboard signed by Aaron Colton of the American Federal period; Windsor chairs; Pennsylvania blanket chests; Chalkware; Canton; John Bell & Bell family pottery; early 19th-century Pennsylvania German farmstead with standing farmhouse, barn, miller's house, tannery & mill sites; Snow Hill archives - a religious off-branch of the Ephrata Cloister c.1770; farming tools & household items.
Research Fields: Pennsylvania German culture, historic preservation, colonization & settlement of the Cumberland Valley & Shenandoah regional pottery.
Facilities: library on the decorative arts, historical books, exhibition catalogues; museum studies on historical preservations & periodicals available for use on premises; 107-acre park with picnic area; nature trails; visitors center.
Activities: guided tours; lectures; films; gallery talks; concerts; changing

exhibits; sponsored trips; environmental & cultural education programs for school groups; traveling trunk program.
Publications: catalog, Highlights of the Renfrew Collection; quarterly newsletters; special exhibit catalogs; brochures for sites; John Bell catalog & brochure.
Hours & Admission Prices: April-Oct. Tues.-Fri. 12-4, Sat.-Sun. 1-4; other times by appointment. Adults $5, seniors $4.50, children 7-12 $3.50; discount to groups; children 6 & under and members no charge. Closed Mother's Day; Memorial Day; Father's Day; Independence Day; Labor Day.
Attendance: 14,500 (estimated)
Membership: Friends of Renfrew: Individual $25; Family $50; Supporting $100; Sustaining $150; Renfrew Society $250; Life $2,000.

Waynesburg

GREENE COUNTY HISTORICAL MUSEUM, (M), 918 Rolling Meadows Rd., Waynesburg, PA 15370-3470. Tel.: 724-627-3204. Fax: 724-627-3204.
E-mail: gchsmuseum@windstream.net
Web Site: www.greencountyhistory.com
Founded: 1925.
Congressional District: 50
Key Personnel: Pres., Gretchen Graham; Vice Pres., Linda Rush; Treas., Jim Weinschenker; Sec., Deborah Wilson; Admin., Eben Williams.
Personnel Profile: Full-Time Paid 1; Full-Time Volunteers 12; Part-Time Paid 3; Part-Time Volunteers 35; Interns 2.
Governing Authority: society. Parent Institution: Greene County Historical Society. Tax-exempt: 501(c)(3).
Institution Type/Description: Historical Society Museum: housed in 1861 former county home.
Collections: Victorian rooms; archives; archaeology; manuscript collections; Indian artifacts; glassware; pottery; military artifacts; Governor Martin.
Research Fields: local history.
Activities: guided tours; permanent & temporary exhibitions; guided tours for groups with reservations. Annual Event: Harvest Festival in October.
Publications: quarterly newsletter; books, Monongahela of Old; Waynesburg-Prosperous & Beautiful; Census of Green County 1800; The Tenmile Country and its Pioneer Families.
Hours & Admission Prices: Tues.-Sat. 10-3, Sun. 1-4. Adults $5, seniors 60 & over $3; discount to AAA members; children under 12 & members no charge. Closed Easter; Thanksgiving; Christmas. &
Attendance: 5,000 (estimated)
Membership: Senior $8; Single $10; Couple $15; Family $18; Friend $50; Donor $51-$99.

PAUL R. STEWART MUSEUM, Waynesburg University, 51 W. College St., Waynesburg, PA 15370-1258. Tel.: 724-627-8191 & 852-3214. Fax: 724-627-3225.
E-mail: waynesburgalumni@waynesburg.edu
Web Site: www.waynesburg.edu
Key Personnel: Cur., James "Fuzzy" Randolph
Institution Type/Description: History Museum.
Collections: local history; Native American artifacts; early Monongahela Valley glassware & pottery; fossils; rocks & minerals from around the world; county newspapers, letters & records from 1817 to present.
Hours & Admission Prices: Sept.-May Mon.-Fri. 9 to noon; other times by appointment. No charge. Closed university holidays & breaks.

West Chester

AMERICAN HELICOPTER MUSEUM & EDUCATION CENTER, (M), 1220 American Blvd., West Chester, PA 19380-4268. Tel.: 610-436-9600. Fax: 610-436-8642.
E-mail: info@helicoptermuseum.org
Web Site: www.americanhelicopter.museum
Founded: 1993.
Congressional District: 16
Key Personnel: Exec. Dir., G. Timlin Conaway, Jr.; Chm. Bd., Robert Beggs; Vice Pres., Chuck Schalch.
Personnel Profile: Full-Time Paid 3; Part-Time Paid 5; Part-Time Volunteers 125.
Governing Authority: private; nonprofit. Tax-exempt: 501(c)(3).
Institution Type/Description: Aeronautics Museum.
Collections: concentration on rotary-wing aircraft, autogiros, helicopters & convertaplanes manufactured or operated in the United States.
Facilities: library; archives; 72-seat auditorium; classroom; 14,000 sq. ft. exhibit space. Museum-related items for sale.
Activities: docent program; formal educational programs; guided tours; lectures; rental facilities; children's birthday parties; special events.

Publications: quarterly newsletter - Vertika.
Hours & Admission Prices: Wed.-Sat. 10-5, Sun. 12-5. Adults $10, seniors $8, students & children 3-18 $8; members, active military personnel with ID and children 2 & under no charge. Closed New Year's Day; Easter; Memorial Day; Independence Day; Labor Day; Thanksgiving; Christmas. &
Attendance: 30,000 (estimated)
Membership: Student & Military $25; Individual $50; Grandparents $60; Family $75; Friend $100; Sustaining $150; Patron $250; Benefactor $500; President's Circle $1,000.

CHESTER COUNTY HISTORICAL SOCIETY, 225 N. High St., West Chester, PA 19380-2658. Tel.: 610-692-4800. Fax: 610-692-4357.
E-mail: cchs@chestercohistorical.org
Web Site: www.chestercohistorical.org
Founded: 1893.
Congressional District: 16
Key Personnel: Pres., Robert Lukens; Chm. (V), Vincent Donahue; Photo Archivist, Pamela C. Powell; Mgr. Operations, Carol McLain; Dir. Collections & Cur., Ellen Endslow; Vice Pres. Devel., David Reinfeld; Librarian, Diane Rofini.
Personnel Profile: Full-Time Paid 8; Part-Time Paid 10; Part-Time Volunteers 300; Interns 8.
Volunteer Hours: 15,481
Operating Expenses: 1,500,000
Operating Income: 1,130,000
Governing Authority: society. Tax-exempt.
Institution Type/Description: History Museum.
Collections: local history & culture; manuscripts; paintings; photographs; decorative arts.
Major Exhibits: Profiles: Chester County Clothing of the 1800s, 11/13-8/14.
Research Fields: furniture; textiles; costumes; genealogy; local history; preservation.
Facilities: 25,000-vol. library of local history; 500,000 manuscripts; 80,000 photographs; 70,000 objects & decorative arts all available for study on premises. Collection-related items for sale.
Activities: guided tours; lectures; films; gallery talks; concerts; docent program seminars; field trips; inter-museum loan, permanent & temporary exhibitions; school tours; traveling trunks; hands-on exhibits; history lab; teacher inservice training. Annual Event: National History Day.
Publications: annual report; monthly calendar of activities.
Hours & Admission Prices: Museum & Library: Wed.-Sat. 9:30-4:30. Museum & Library: adults $6, seniors $5, students $3.50; discounts to AAA members; members no charge. Closed New Year's Day; Memorial Day; Independence Day; Labor Day; Thanksgiving; Christmas. &
Attendance: 32,000 (accurate)
Membership: Senior $40; Individual $45; Family Senior $50; Family $55; Contributor $125; Patron $250; Benefactor $500.

White Mills

DORFLINGER-SUYDAM WILDLIFE SANCTUARY & GLASS MUSEUM, Elizabeth St. & Long Ridge Rd., White Mills, PA 18473. Mailing Address: P.O. Box 356, White Mills, PA 18473-0356. Tel.: 570-253-1185. Fax: 570-253-5196.
E-mail: suydam@ptd.net
Web Site: www.dorflinger.org
Founded: 1980.
Congressional District: 10
Key Personnel: Exec. Dir., Joan G. Gillner.
Personnel Profile: Full-Time Paid 3; Part-Time Volunteers 100.
Governing Authority: Tax-exempt.
Institution Type/Description: Wildlife Sanctuary & Glass Museum.
Collections: local natural history & culture; Dorflinger glass including over 1,000 pieces of cut, engraved, etched, gilded & enameled crystal.
Facilities: nature trails.
Activities: educational programs; special events.
Publications: biannual newsletter, The Sanctuary.
Hours & Admission Prices: Sanctuary: daily dawn to dusk. Museum: May-Oct. Wed.-Sat. 10-4, Sun. 1-4. Adults $5, seniors 55 & over $4; discounts to members.
Attendance: 3,000 (estimated)
Membership: Individual $25; Family $50; Donor $100.

Wilkes-Barre

LUZERNE COUNTY HISTORICAL SOCIETY, (M), 69 S. Franklin St., Wilkes-Barre, PA 18701. Mailing Address: 49 S. Franklin St., Wilkes-Barre, PA 18701-1290. Tel.: 570-823-6244. Fax: 570-823-9011.
E-mail: mriccetti@luzernehistory.org
Web Site: www.luzernehistory.org
Formerly: Wyoming Historical and Geological Society
Founded: 1858.
Congressional District: 11
Key Personnel: Dir. Operations, Mark J. Riccetti; Pres., Janet Flach; Librarian & Archivist, Amanda Fontenova; Cur., Mary Ruth K. Burke.
Personnel Profile: Full-Time Paid 3; Part-Time Paid 2; Part-Time Volunteers 3; Interns 5.
Governing Authority: nonprofit organization. Branch Museums: 1790 Nathan Denison House, 35 Denison St., Forty Fort, PA; 1803 Swetland Homestead, 885 Wyoming Ave., Wyoming, Pa. Tax-exempt: 501(c)(3).
Institution Type/Description: History Museum.
Collections: archaeology; archives; costumes; geology; Indian artifacts; industry; military; mineralogy; textiles; manuscripts; historic houses.
Research Fields: anthracite coal mining; archaeology; geology; genealogy; history of Luzerne County, Pennsylvania.
Facilities: 6,000-vol. library of local history & history of coal industry; reading room.
Activities: formally organized educational programs; permanent & temporary exhibitions.
Publications: books, The Susquehanna Company Papers; Proceedings & Collections; Steamboats on the Susquehanna; Gone But Not Forgotten; Civil War Veterans of Northeastern Pennsylvania; A Story Runs Through It: The Wyoming Valley Levee System; Coloring the Pieces of Luzerne County.
Hours & Admission Prices: Library: Tues.-Fri. 12-4, Sat. 10-4. Museum: Tues.-Fri. 12-4, Sat. 12-4. Adults $5, children $2.
Attendance: 12,000 (estimated)
Membership: Student & Senior $20; Individual $40; Family $60; Patron $100; Business $250; Life $1,000.

SORDONI ART GALLERY, (M), Wilkes University, 150 S. River St., Wilkes-Barre, PA 18766. Tel.: 570-408-4325. Fax: 570-408-7733. Facebook: Sordoni Art Gallery.
E-mail: brittany.kramer@wilkes.edu
Web Site: wilkes.edu/sordoniartgallery
Founded: 1973.
Congressional District: 11
Key Personnel: Dir., Brittany Kramer DeBalko; Chm. (V), Joel Zitofsky.
Personnel Profile: Full-Time Paid 6; Part-Time Paid 8; Interns 1.
Volunteer Hours: 50
Governing Authority: Wilkes University. Tax-exempt.
Institution Type/Description: Art Gallery.
Collections: 19th- & 20th-century American Art.
Major Exhibits: Paintings by George Cabin, 1/1-5/14; Permanent Collection, 6/1-8/14; Faculty Exhibition, 2014, 9/1-12/14.
Research Fields: early 20th-century American & Contemporary Art.
Facilities: 1,600 sq. ft. exhibition space; storage; offices; workrooms.
Activities: guided tours; lectures; gallery talks; formally organized education programs for undergraduate students; loan, temporary & traveling exhibitions; artist-in-residence.
Publications: exhibition catalogs, American Art.
Hours & Admission Prices: Tues-Sun. 12-4:30. No charge. Closed major holidays. &
Attendance: 5,000 (accurate)
Membership: Student $10; Individual $50; Patron $100; Connoisseur $250; Old Master's Circle $500; Collector's Society $1,000 & up.

Williamsport

CHILDREN'S DISCOVERY WORKSHOP, Williamsport YMCA, 343 W. 4th St., Williamsport, PA 17701-6401. Tel.: 570-323-7134.
Web Site: www.williamsportymca.org/cdw/index.html
Institution Type/Description: Children's Museum.
Collections: hands-on exhibits.
Activities: workshops; programs; special events.
Hours & Admission Prices: Call for hours. &
Membership: Single Parent $42; Family $54; Corporate $111. (additional adults $12).

PETER HERDIC TRANSPORTATION MUSEUM, 810 Nichols Place, Williamsport, PA 17701. Mailing Address: 1500 W. 3rd St., Williamsport, PA 17701. Tel.: 570-601-3455.
Institution Type/Description: Transportation Museum.
Collections: period vehicles including cars, trains, & boats.
Hours & Admission Prices: Tues.-Sat. 10-3; other times by appointment.

*** THE THOMAS T. TABER MUSEUM OF THE LYCOMING COUNTY HISTORICAL SOCIETY, (M),** 858 W. 4th St., Williamsport, PA 17701-5824. Tel.: 570-326-3326. Fax: 570-326-3689.
E-mail: lchsmuseum@verizon.net
Web Site: www.tabermuseum.org
Formerly: Lycoming County Historical Society and Museum
Founded: 1907.
Congressional District: 10
Key Personnel: Dir., Gary W. Parks; Chm. (V), Martha Huddy; Cur., Scott Sagar; Museum Store Mgr., Anne Persun.
Personnel Profile: Full-Time Paid 3; Part-Time Paid 4; Part-Time Volunteers 90; Interns 3.
Governing Authority: society. Tax-exempt: 501(c)(3).
Institution Type/Description: Regional History Museum.
Collections: American Indian artifacts; archives; 1850-1900 lumber industry; Civil War items; Victorian era period room; blacksmith shop; general store; gristmill; industry; industry trades; military; textiles; farm & home utensils; agriculture; costumes; transportation; music; wildlife; toy train collection; local history & genealogy; fine arts.
Research Fields: regional archaeology; regional history; 1870-1900 lumber industry.
Facilities: Pamphlets, crafts & museum-related items for sale.
Activities: guided tours; lectures; gallery talks; arts festivals; formally organized education programs for children; family programs; inter-museum loan, permanent & temporary exhibitions.
Publications: bimonthly newsletter; annual Journal; reprints of OP books on local history.
Hours & Admission Prices: May-Oct. Tues.-Fri. 9:30-4, Sat. 11-4, Sun. 1-4; Nov.-April Tues.-Fri. 9:30-4, Sat. 11-4. Adults $7.50, senior citizens 65 & over $6, children 3-12 $5; discount to military & AAA members; members no charge. Closed national holidays; Sun. Nov.-April. &
Attendance: 13,000 (accurate)
Membership: Individual $35; Family $45; Patron $75; Contributor $100; Business $100-$249; Corporate $250-$500; Benefactor $500; Life $5,000.

Willow Street

1719 HANS HERR HOUSE & MUSEUM, 1849 Hans Herr Dr., Willow Street, PA 17584-9536. Tel.: 717-464-4438.
E-mail: info@hansherr.org
Web Site: www.hansherr.org
Founded: 1974.
Congressional District: 100
Key Personnel: Dir., Becky Gochnauer; Administrative Asst., Donnalee Mylin.
Personnel Profile: Part-Time Paid 2; Part-Time Volunteers 40.
Governing Authority: Parent Institution: Lancaster Mennonite Historical Society. Tax-exempt: 501(c)(3).
Institution Type/Description: Historic House: 1719 Herr House.
Collections: Mennonite rural life; pre-1750 furnishings; agricultural items; replica eastern woodland Native American longhouse.
Research Fields: 1700s Pennsylvania German life; Mennonite history; life in Lancaster County.
Facilities: 2,200 sq. ft. exhibit space; 18th-century garden & orchard.
Activities: guided tours; lectures; organized educational programs. Special Events: candlelight tours at Christmas; 18th-century crafts festival; Fall apple tasting festival.
Publications: quarterly newsletter, Herr House Foundation.
Hours & Admission Prices: April to 1st week Dec. Mon.-Sat. 9-4. Adults $8, children 7-12 $4; discounts to groups of 10 or more. Closed Good Friday; Thanksgiving.
Attendance: 8,200 (estimated)
Membership: Individual $25; Family $40; Donor $50; Partner $75; Supporter $100; Honorary $250.

Windber

WINDBER COAL HERITAGE CENTER, 501 15th St., Windber, PA 15963-0115. Tel.: 814-467-6680; 877-826-3933. Fax: 814-467-8715.
E-mail: scott.johnson@windbercoal.org
Web Site: www.windbercoal.org

Founded: 1997.
Institution Type/Description: Mining History Museum.
Collections: coal mining history, tools & equipment; coal miners & their families; photographs.
Facilities: visitor's center. Museum related items for sale.
Activities: interactive exhibits; temporary exhibitions; educational programs; special events. Annual Event: Miner's Memorial Day in June.
Hours & Admission Prices: Call for hours. Discount to members & group tours over 20. &

Womelsdorf

CONRAD WEISER HOMESTEAD AND MEMORIAL PARK, (M), 30 Weiser Lane, Womelsdorf, PA 19567-9768. Tel.: 610-589-2934. Fax: 610-589-9458.
E-mail: info@conradweiserhomestead.org
Web Site: www.conradweiserhomestead.org
Founded: 1928.
Congressional District: 6
Key Personnel: Pres. (V), David G. Sonnen; Groundskeeper, Arnel Greth; Museum Shop Mgr., Brian Beamesderfer.
Personnel Profile: Part-Time Paid 1; Part-Time Volunteers 30.
Governing Authority: state. Administered by Pennsylvania Historical & Museum Commission, Commonwealth Keystone Bldg., Plaza Level, 400 North St., Harrisburg, PA 17120. Tax-exempt.
Institution Type/Description: Historic House: 1729-1760 Conrad Weiser homestead.
Collections: period furnishings; artifacts.
Facilities: visitor center; picnic facilities; 26 acre park.
Activities: self-guided tours; craft demonstrations; special events. Annual Events: Charter Day in March; Candlelight Tour in November.
Publications: newsletter.
Hours & Admission Prices: Grounds: dawn to dusk. Homestead: April-Nov. 1st Sun. each month 12-5. No charge; donations accepted. &
Attendance: 25,000 (estimated)
Membership: Junior 16 & under $5; Individual $20; Family $30; Patron $75; Sponsor $100; Benefactor $500. (Special PA Heritage Society Membership package available for $25 extra).

Worcester

PETER WENTZ FARMSTEAD, Shearer Rd., Worcester, PA 19490. Mailing Address: P.O. Box 240, Worcester, PA 19490-0240. Tel.: 610-584-5104. Fax: 610-584-6860.
E-mail: peterwentzfarmstead@mail.montcopa.org
Web Site: www.montcopa.org
Founded: 1976.
Congressional District: 13
Key Personnel: Pres. (V), Anne Condon; Admin., Dianne M. Cram; Asst. Admin., John Schilling; Farm Mgr., James Nichols; Cur., Morgan McMillan; Educator, Kimberly Boice; Asst. Farm Mgr., Jay Ryan.
Personnel Profile: Full-Time Paid 6; Part-Time Volunteers 190.
Governing Authority: county. Subsidiary Institution: Peter Wentz Farmstead Society. Tax-exempt.
Institution Type/Description: Historic Building & Site: 1758 farmstead belonging to Peter Wentz; used twice by General Washington where he planned the Battle of Germantown & received the word of victory at Saratoga.
Collections: period furniture; decorative arts; costumes; tools; farm implements; working 18th-century farm.
Research Fields: Pennsylvania German history; Revolutionary War; Wentz Family; Schultz Family.
Facilities: research library; classroom. Handmade museum-related items for sale.
Activities: guided tours; lectures; organized education programs for children, adults & undergraduate or graduate college students; docent program; loan & temporary exhibitions; period craft demonstrations; seminars & workshops; intern program for college students; summer day camp (colonial crafts featured) for 3rd, 4th & 5th grades.
Publications: quarterly newsletter, Wentz Post; annual brochures; booklets.
Hours & Admission Prices: Tues.-Sat. 10-4, Sun. 1-4. No charge; donations accepted. Closed county holidays. &
Attendance: 10,000 (estimated)
Membership: Peter Wentz Farmstead Society, a support group for Farmstead: Student $10; Single Adult $20; Family $30.

Wrightsville

WRIGHTSVILLE HISTORICAL MUSEUM, 309 Locust St., Wrightsville, PA 17368-1221. Tel.: 717-252-1169.
Institution Type/Description: History Museum.

Collections: local history & culture; photographs; period artifacts.
Activities: temporary exhibits.
Hours & Admission Prices: Call for hours. No charge; donations accepted.

Wyalusing

WYALUSING VALLEY MUSEUM, 28 Homer Ln., Wyalusing, PA 18853. Mailing Address: P.O. Box 301, Wyalusing, PA 18853-0301. Tel.: 570-746-3979.
E-mail: wyalusingmuseum@frontiernet.net
Web Site: www.wyalusingmuseum.com
Founded: 1980.
Key Personnel: Pres. (V), Mary Skillings; Treas., Karl Peterson; Cur., Morgan Clinton.
Personnel Profile: Part-Time Paid 2; Part-Time Volunteers 15.
Governing Authority: Tax-exempt.
Institution Type/Description: History Museum.
Collections: Wyalusing area history; schools; Native Americans; military artifacts; furnishings; personal artifacts.
Facilities: Museum-related items for sale.
Activities: Annual Event: Wine Festival.
Publications: local history gleaned from articles found in the Wyalusing Rocket-Courier.
Hours & Admission Prices: May-Oct. Sat.-Sun. 12-4; other times by appointment. No charge; donations accepted.
Membership: Individual $15; Household $25; Bronze $50; Silver $75; Gold $100.

York

POLICE HERITAGE MUSEUM, 54 W. Market St., York, PA 17401-1228. Mailing Address: P.O. Box 1941, York, PA 17405-1941. Tel.: 717-845-2677.
Key Personnel: Chm., John Stine
Institution Type/Description: History Museum.
Collections: law enforcement history & artifacts; photographs; police bicycles & motorcycles; documents; police & prison equipment; badges; patches; personal artifacts; uniforms.
Hours & Admission Prices: April-Oct. Sat. 9-4; group tours by appointment. Adults $3, children 6-12 $1; children under 6 no charge.

USA WEIGHTLIFTING HALL OF FAME, 3300 Board Rd., York, PA 17406-8409. Tel.: 717-767-6481.
Key Personnel: Controller, Dave Kogut
Institution Type/Description: Sports Museum.
Collections: barbells; trophy cups; photographs; barbell mobile; Hall of Fame inductees.
Hours & Admission Prices: Mon.-Sat. 10-5.

YORK COLLEGE GALLERIES, Wolf Hall, 441 Country Club Rd., York, PA 17403-3643. Tel.: 717-815-1354 & 1528. Facebook: York College Galleries.
E-mail: mclayrob@ycp.edu
Web Site: www.ycp.edu
Key Personnel: Gallery Dir., Matthew Clay-Robison
Institution Type/Description: Art Gallery.
Collections: works by local & national artists.
Activities: lectures; workshops.
Hours & Admission Prices: Mon.-Tues. & Fri. 9-5, Wed.-Thurs. 9-9, Sat. 10-4. No charge. &

YORK COUNTY HERITAGE TRUST, AGRICULTURAL AND INDUSTRIAL MUSEUM, 217 W. Princess St., York, PA 17403-2013. Mailing Address: 250 E. Market St., York, PA 17403-2013. Tel.: 717-846-6452. Fax: 717-812-1204.
E-mail: info@yorkheritage.org
Web Site: www.yorkheritage.org
Founded: 1999.
Congressional District: 19
Key Personnel: Dir., Joan Mummert; Dir. Exhibits & Collections, Jennifer Royer; Dir. Education, Daniel Roe; Dir. Library & Archives, Lila Fourhman-Shaull; Museum Shop Buyer, Carl Preate.
Personnel Profile: Full-Time Paid 15; Part-Time Paid 6; Part-Time Volunteers 150; Interns 15.
Governing Authority: nonprofit organization. Parent Institution: York County Heritage Trust. Subsidiary Institution: Historical Society Museum, Library & Archives; Fire Museum of York County; Colonial Complex; Bonham House. Tax-exempt: 501(c)(3).
Institution Type/Description: Agriculture & Industrial Museum; 1874-1955 former industrial complex.
Collections: industrial artifacts dating from 19th century to present used in local industries such as wallpaper & wire cloth manufacturing, defense, dentifrices, physical fitness, refrigeration & air conditioning, & automobile manufacturing; agricultural artifacts including a three-story working grist mill.
Research Fields: agriculture & industry.
Facilities: 1,800-vol. library & archives; cafeteria. Museum-related items for sale.
Activities: rental facilities; self-guided & guided tours; permanent & temporary exhibits; education programs for children; docent program; volunteer opportunities.
Publications: quarterly newsletter, The Chronicle; membership newsletter, Trust Talk.
Hours & Admission Prices: Agricultural & Industrial Museum: Tues.-Sat. 10-4. Colonial Complex: Tours Tues.-Sat. 10-4. Historical Society Museum: Tues.-Sat. 10-4. Fire Museum: Sat. 10-4. York County Heritage Sites: adults $10, students $7, children 8-18 $5; discounts for military, AAA, ICOM & AAM members; members no charge. Library $6. Closed New Year's Day; Easter; Memorial Day; Independence Day; Labor Day; Thanksgiving; Christmas. &
Attendance: 48,250 (accurate)
Membership: York County Heritage Trust membership: Senior Citizen $35; Individual $40; Senior Couple $45; Family $55.

YORK COUNTY HERITAGE TRUST - BONHAM HOUSE, 250 E. Market St., York, PA 17403. Tel.: 717-846-6452.
Institution Type/Description: Historic House Museum: housed in the former home of Horace & Rebekah Bonham, built in 1875.
Collections: Bonham family history & personal artifacts; period furnishings; paintings.
Hours & Admission Prices: April-Dec. Sat. 10-4 by appointment.
Membership: Senior Citizen $40; Individual $50; Senior Citizen Couple $55; Family $60; Pioneer Circle $100; Bonham Circle $250; Confederation Circle $500; Golden Plough Circle $1,000; Lafayette Circle $2,500 & up.

YORK COUNTY HERITAGE TRUST - COLONIAL COMPLEX, 250 E. Market St., York, PA 17403. Tel.: 717-846-6452.
Institution Type/Description: History Museum.
Collections: local history & culture; period furnishings. Historic Buildings: General Gates house c.1751; Golden Plough Tavern 1741; Barnett Bobb log house; Colonial Court House.
Hours & Admission Prices: April-Dec. Tues.-Sat. 10-4; groups by appointment. Adults $10, children 8-18 $5. Closed New Year's Day; Easter; Memorial Day; Independence Day; Labor Day; Thanksgiving; Christmas.
Membership: Senior Citizen $40; Individual $50; Senior Citizen Couple $55; Family $60; Pioneer Circle $100; Bonham Circle $250; Confederation Circle $500; Golden Plough Circle $1,000; Lafayette Circle $2,500 & up.

YORK COUNTY HERITAGE TRUST, FIRE MUSEUM OF YORK COUNTY, 757 W. Market St., York, PA 17401-3650. Mailing Address: 250 E. Market St., York, PA 17403-2013. Tel.: 717-848-1587. Fax: 717-812-1204.
E-mail: info@yorkheritage.org
Web Site: www.yorkheritage.org
Founded: 1973.
Congressional District: 19
Key Personnel: Pres. & CEO, Joan Mummert; Dir. Exhibits & Collections, Jennifer Royer; Dir. Education, Daniel Roe; Dir. Library & Archives, Lila Fourhman-Shaull; Museum Shop Buyer, Carl Preate.
Personnel Profile: Full-Time Paid 15; Part-Time Paid 6; Part-Time Volunteers 250; Interns 15.
Governing Authority: nonprofit. Parent Institution: York County Heritage Trust. Subsidiary Institutions: Historical Society Museum; Agricultural and Industrial Museum; Colonial Complex; Bonham House; Murals of York. Tax-exempt: 501(c)(3).
Institution Type/Description: Fire-fighting Museum: housed in 1903-1904 Royal Fire Station #6.
Collections: 1700s-present fire-fighting history in York County.
Facilities: 1,800-vol. library & archives on fire service history; 12,500 sq. ft. exhibit space.
Activities: self-guided & guided tours; volunteer opportunities; permanent & temporary exhibits; education programs.
Publications: quarterly newsletter, The Chronicle.
Hours & Admission Prices: Fire Museum: Sat. 10-4; tours by appointment. Historical Society Museum & Agricultural and Industrial Museum: Tues.-Sat. 10-4. Colonial Complex: Guided Tours Tues.-Sat. 10-4. Bonham House: by appointment. York County Heritage Trust Sites: adults $10, students $7, children 8-18 $5; discounts to groups, seniors, military, AAA,

ICOM & AAM members; members no charge. Library $6. Closed New Year's Day; Easter; Memorial Day; Independence Day; Labor Day; Thanksgiving; Christmas. &

Attendance: 48,250 (accurate)

Membership: College Students w/ID $30; Senior Citizen 65 & over $35; Individual $40; Senior Couple $45; Family $55; Sustainer $100; Contributor $250; Patron $500; Patriot $1,000.

YORK COUNTY HERITAGE TRUST, HISTORICAL SOCIETY AND LIBRARY/ARCHIVES, (M), 250 E. Market St., York, PA 17403-2013. Tel.: 717-848-1587. Fax: 717-812-1204.

E-mail: info@yorkheritage.org
Web Site: yorkheritage.org
Formerly: The Historical Society of York County
Founded: 1895.
Congressional District: 19
Key Personnel: Pres. & C.E.O., Joan Mummert; Chm. Bd., Nancy Ahalt; Dir. Education, Dan Roe; Dir. Library & Archives, Lila Fourhman-Shaull; Museum Shop Buyer, Judy Bono.
Personnel Profile: Full-Time Paid 15; Part-Time Paid 5; Part-Time Volunteers 150; Interns 15.
Governing Authority: nonprofit. Parent Institution: York County Heritage Trust. Subsidiary Institutions: Agricultural & Industrial Museum, 217 W. Princess St., York, PA; Colonial Complex (Gates House, Plough Tavern, Bobb Log House & Colonial Court House), 157 W. Market St., York, PA; Bonham House, 152 E. Market St., York, PA; Fire Museum, 757 W. Market St., York, PA. Tax-exempt: 501(c)(3).
Institution Type/Description: History Museum.
Collections: costumes; decorative arts; industrial & agricultural machines & products; folk art; historical domestic & cultural artifacts of York Co.; library collections including published manuscripts & photographical materials; genealogy. Historic Houses: 1741 Golden Plough Tavern; 1751 General Gates House; 1812 Log House; 1870s Bonham House; Historical Society Museum & Library; Fire Museum; Agricultural & Industrial Museum; Colonial Court House.
Research Fields: genealogy; Pennsylvania decorative & folk art; history; York County history.
Facilities: 35,000-vol. library.
Activities: guided tours; lectures; field trips; gallery talks; formally organized education programs for children; volunteer program; rental facilities; sleepovers; permanent & temporary exhibitions. Annual Fund Raising Events: Oyster Festival; Celebrity Art & Leisure Auction, Brew Fest; Book Blast.
Publications: quarterly newsletter, The Chronicle; membership newsletter, Trust Talk.
Hours & Admission Prices: Historical Society Museum: Tues.-Sat. 10-4. Adult $10, children 8-18 $5. Family Pass: $30. Combination tickets available. Library & Archives: Tues.-Sat. 9-5. Library & Archives: adults $6. Closed New Year's Day; Easter Monday; Memorial Day; Independence Day; Labor Day; Thanksgiving; Christmas. &
Attendance: 40,244 (accurate)
Membership: Senior Citizen $40; Individual $50; Senior Citizen Couple $55; Family $60; Pioneer Circle $100; Bonham Circle $250; Confederation Circle $500; Golden Plough Circle $1,000; Lafayette Circle $2,500 & up.

York Springs

EASTERN MUSEUM OF MOTOR RACING, 100 Baltimore Rd., York Springs, PA 17372. Mailing Address: P.O. Box 688, Mechanicsburg, PA 17055-0688. Tel.: 717-528-8279. Facebook: Eastern Museum of Motor Racing Group.

E-mail: admin@emmr.org
Web Site: www.emmr.org
Founded: 1975.
Congressional District: 19
Key Personnel: Administrative Coord., Amanda J. Eshenour; Museum Shop Mgr., Larry Garland; Museum Shop Mgr., Kim Garland.
Personnel Profile: Part-Time Paid 1; Part-Time Volunteers 50.
Institution Type/Description: General Museum
Collections: period race cars; photographs; sprint cars; midgets; stock cars; motorcycles; Indy cars; NASCAR; drag racing.
Major Exhibits: Drag Car Exhibit, 2/14-12/14; Motocross Exhibit, 2/14-12/14; Road Race Car Exhibit (1 to 2 cars), 2/14-12/14; Sprint & Midget Cars, 2/14-12/14.
Facilities: research library. Museum-related items for sale.
Activities: special events; see website for schedule.
Publications: 4 yrly. newsletters.
Hours & Admission Prices: April-Oct. Fri.-Sun. 10-4. No charge; donations accepted.
Membership: Annual $20; Lifetime $200.

Zelienople

ZELIENOPLE HISTORICAL SOCIETY, 243 S. Main St., Zelienople, PA 16063-1151. Tel.: 724-452-9457.

E-mail: zhs@zelienoplehistoricalsociety.com
Web Site: www.zelienoplehistoricalsociety.com
Founded: 1975.
Congressional District: 4
Key Personnel: C.E.O., Elizabeth Kelleher; Pres. (V), Janet Rogan; Museum Shop Mgr., Mary Cameron.
Personnel Profile: Full-Time Paid 1; Part-Time Paid 1; Part-Time Volunteers 30.
Governing Authority: nonprofit. Tax-exempt: 501(c)(3).
Institution Type/Description: Historic Houses: Passavant House c.1808 Federal-Georgian brick & frame structure; c.1805 Buhl House.
Collections: early western Pennsylvania living customs; furniture; clothing; cooking; tools; documents; decorative arts; glass; archives; photographs; postal; textiles; rare books; dolls; over 1,000 family letters; genealogy; library of history.
Research Fields: genealogy; religious practices; local history.
Facilities: 1,300-vol. library pertaining to history, genealogy, Communal societies & religion.
Activities: guided tours; lectures; arts festivals.
Publications: quarterly newsletter, Zelienople Historical Society Newsletter; brochures; booklets.
Hours & Admission Prices: Tours: Mon.-Fri. 9:30am; other times by appointment. Adults $5, students $3; members no charge. Office: Mon.-Fri. 9-12. Library: by appointment. &
Attendance: 1,000 (estimated)
Membership: Student $5; Individual $20; Couple $30; Family $35; Individual Life & Corporate $200; Life Couple $250; Life Patron $350 & up.

RHODE ISLAND

(110 listings)

Adamsville

GRAY'S STORE, 4 Main St., Adamsville, RI 02801. Mailing Address: P.O. Box 53, Adamsville, RI 02801-0053. Tel.: 401-635-4566.

Congressional District: 1
Key Personnel: Owner, Grayton T. Waite
Institution Type/Description: Historic Building: housed in a general store built by Samuel Church in 1788; also includes first post office in Little Compton (1804).
Collections: local history; period furnishings & artifacts; soda fountain; post office; cigar & tobacco cases.
Hours & Admission Prices: Mon.-Sat. 9-5, Sun. & holidays 12-4. No charge; donations accepted.

Barrington

BARRINGTON PRESERVATION SOCIETY & MUSEUM, (M), Barrington Public Library, Lower Level, 281 County Rd., Barrington, RI 02806-2406. Tel.: 401-289-0802.

E-mail: museumdirector@barrpreservation.org
Web Site: www.barrpreservation.org
Key Personnel: Dir., Carole Villucci
Institution Type/Description: History Museum.
Collections: local history & culture; photographs; personal artifacts; government records; toys; local business; farm implements; posters; period furnishings & clothing.
Facilities: library.
Activities: research; permanent & temporary exhibits.
Hours & Admission Prices: Wed.-Thurs. & Sat. 11-3; other times by appointment. No charge.

Block Island

BLOCK ISLAND HISTORICAL SOCIETY MUSEUM, Old Town Rd., Block Island, RI 02807. Mailing Address: P.O. Box 79, Block Island, RI 02807-0079. Tel.: 401-864-4357.

E-mail: blockhistory@mc.com
Web Site: www.blockislandtimes.com/listings/2867912/the-block-island-historical-society
Key Personnel: Exec. Dir., Pamela Gasner; Pres., Dr. Gerald Abbott
Institution Type/Description: Historical Society Museum.
Collections: Block Island history; farming & maritime; period furniture. Historic House: c.1850 farmhouse.

Hours & Admission Prices: Summer: daily 11-4; Spring & Fall: Sat.-Sun. 11-4. Adults $5.50, seniors & students $3; members & children 16 & under no charge.

Bristol

AUDUBON SOCIETY OF RHODE ISLAND ENVIRONMENTAL EDUCATION CENTER, 1401 Hope St., Bristol, RI 02809-1153. Mailing Address: 12 Sanderson Rd., Smithfield, RI 02917-2606. Tel.: 401-245-7500. Fax: 401-245-9339.

E-mail: adimonti@asri.org
Web Site: www.asri.org
Founded: 1897.
Congressional District: 2
Key Personnel: Exec. Dir., Lawrence Taft; Pres., Candace Powell; Treas., Mark Carrison; Sr. Dir. Advancement, Jeffery Hall; Dir., Anne DiMonti; Sr. Dir. Education, Kristen Swanberg; Museum Shop Mgr., Jan Weyant.
Personnel Profile: Full-Time Paid 20; Part-Time Paid 12; Part-Time Volunteers 100; Interns 6.
Governing Authority: private; nonprofit organization. Parent Institution: Audubon Society of Rhode Island. Tax-exempt: 501(c)(3).
Institution Type/Description: Natural History Museum.
Collections: flora, fauna & natural history of Rhode Island; fresh & saltwater aquaria; life-sized models of right whale & harbor seal.
Research Fields: wildlife & habitat change.
Facilities: aquarium; botanical garden; 4,000 sq. ft. exhibit space; 28 acre refuge on Narragansett Bay; 2 classrooms; 100-seat auditorium; nature center; interpretive trails; boardwalk through Red Maple Swamp & Salt Marsh. Museum-related items for sale.
Activities: docent program; formal education programs for adults & children; guided tours; lectures; participatory exhibits; study clubs; touch tank of tidal pool.
Hours & Admission Prices: Memorial Day-Sept. daily 9-5; Oct.-May Mon.-Sat. 9-5, Sun. 12-5. Adults $6, children 4-12 $4; children under 4 & members no charge. &
Attendance: 33,000 (accurate)
Membership: Individual $35; Family $45; Family Plus $55; Steward $75; Defender $100; Business $125; Advocate $250; Conservator $500; Benefactor $1,000; Life $3,000.

BLITHEWOLD MANSION, GARDENS & ARBORETUM, 101 Ferry Rd., (Rt. 114), Bristol, RI 02809-2902. Tel.: 401-253-2707. Fax: 401-253-0412.

E-mail: info@blithewold.org
Web Site: www.blithewold.org
Founded: 1976.
Congressional District: 1
Key Personnel: Exec. Dir., Karen Binder; Chm. (V), Noreen Ackerman; Chm. Emeritus (V), Mike Szostak; Dir. Special Events, Karen Bellavance; Grounds Mgr., Fred Perry; Mgr. Education & Programs, Julie Murphy; Dir. Communications & Visitor Experience, Tree Callanan; Cur., Margaret Whitehead; Museum Shop Mgr., Sue Legault.
Personnel Profile: Full-Time Paid 8; Part-Time Paid 17; Part-Time Volunteers 280; Interns 5.
Governing Authority: nonprofit organization. Parent Institution: Blithewold, Inc. Tax-exempt.
Institution Type/Description: Historic Building & Arboretum: 1908 English manor house & summer residence of Augustus Van Wickle.
Collections: 1,500 collection of exotic trees & shrubs; 200 varieties of woody plants; herbaceous garden plants; Van Wickle family furniture & fine art.
Research Fields: plant hardiness testing.
Facilities: 200-vol. library of books on horticultural & garden reference available for research on premises; historic landscape; 33-acre arboretum & flower gardens. Museum-related items for sale.
Activities: guided tours; lectures; formally organized education programs for adults & undergraduate college students; plant sales; concerts; summer camp. Annual Events: 2nd Daffodil Days mid-April to early May; Christmas at Blithewold from late November to early January.
Publications: seasonal newsletter.
Hours & Admission Prices: Mansion: April to Columbus Day & day after Thanksgiving to New Year's Day Tues.-Sun. 10-4. Grounds: April to Columbus Day & day after Thanksgiving to New Year's Day daily 10-5. Family $24, adults $11, seniors, military & students $9, children 6-17 $3; discounts to AAA members; children 5 & under and members no charge. Closed New Year's Eve & Day; Christmas Eve & Day. &
Attendance: 32,500 (accurate)
Membership: Individual $37; Family $57; Contributing $100; Business $150; Sponsor $250; Patron $500.

COGGESHALL FARM MUSEUM INC., (M), 1 Coggeshall Farm Rd., Bristol, RI 02809-1019. Mailing Address: P.O. Box 562, Bristol, RI 02809-0562. Tel.: 401-253-9062.

E-mail: info@coggeshallfarm.org
Web Site: www.coggeshallfarm.org
Founded: 1968.
Key Personnel: Dir. Historic Interpretation, Justin Squizzero; Pres. (V), Andy Tyska; Interpreter, Mary Betts; Farm Mgr., Jonny Larason; Farm Mgr., Shelley Otis.
Personnel Profile: Full-Time Paid 3; Part-Time Paid 2.
Governing Authority: nonprofit organization. Tax-exempt: 501(c)(3).
Institution Type/Description: Living History Museum.
Collections: cooling house, spring house, 18th-century kitchen garden & farming tools, blacksmith shop, farm out buildings, minor breed animals, heirloom vegetable gardens, salt marsh, hayfields & period hand tools. Historic Building: 18th century tenant farmhouse.
Research Fields: agriculture & farm life in Rhode Island in the 1790s.
Facilities: 48 acre farm; picnic area. Gift items for sale.
Activities: costumed interpreters; formally organized education programs; docent program; historic foodways workshops; family programs; scheduled weekend events year round.
Publications: quarterly newsletter, The Coggeshall Farmer Newsletter.
Hours & Admission Prices: Tues.-Sun. 10-4. Adults $5, children $3; members no charge. Special Events: additional fee. Closed selected holidays. &
Attendance: 9,000 (estimated)
Membership: Student & Individual $20; Family $40; Supporting $50; Sustaining $100; Sponsor $250.

HERRESHOFF MARINE MUSEUM/AMERICA'S CUP HALL OF FAME, (M), One Burnside St., Bristol, RI 02809. Mailing Address: P.O. Box 450, Bristol, RI 02809-0420. Tel.: 401-253-5000. Fax: 401-253-6222. Facebook: Herreshoff Marine Museum.

E-mail: info@herreshoff.org
Web Site: www.herreshoff.org
Founded: 1971.
Congressional District: 2
Key Personnel: Chm. Bd. (V), David Ford; C.E.O., Wm. H. Dyer Jones; C.O.O., Lawrence D. Lavers.
Personnel Profile: Full-Time Paid 6; Part-Time Paid 4; Part-Time Volunteers 60; Interns 6.
Governing Authority: private; nonprofit organization. Subsidiary Institute: Herreshoff Institute & America's Cup Hall of Fame. Tax-exempt: 501(c)(3).
Institution Type/Description: Maritime Museum: located at the site of the former Herreshoff Manufacturing Company.
Collections: 60 sail & power yachts, all built by Herreshoff and many of which were built around 1900; yachts range in size from 8' to 75'; photographs; America's Cup memorabilia; 1/6 scale sailing vessel, Reliance.
Facilities: research library covering era of Herreshoff Yacht Building & America's cup.
Activities: N.G. Herreshoff model room.
Publications: e-newsletters, Current.
Hours & Admission Prices: April 28-Oct. 31 daily 10-5, Nov.-Dec. call for hours. Adults $12, senior citizens $10, military $8, students 10 & over $5; children 9 & under, CAMM & museum members no charge. Closed Independence Day. &
Attendance: 8,000 (accurate)
Membership: Individual $40; Family $60; Vigilant $125; Columbia $250; Defender $500; Resolute $1,000; Amaryllis $1,000 per year for 5 years; Reliance Society $2,000 per year for 5 years; Gloriana $5,000 per year for 5 years.

LINDEN PLACE, 500 Hope St., Bristol, RI 02809-1808. Tel.: 401-253-0390. Fax: 401-253-4106.

E-mail: info@lindenplace.org
Web Site: www.lindenplace.org
Key Personnel: Exec. Dir., James Burke Connell; Tour Dir., Joan Doyle Roth; Site Admin., Susan E. Battle
Institution Type/Description: Historic Buildings: housed in an 1810 Federal-style mansion built by General George De Wolf; the estate was featured in the film The Great Gatsby.
Collections: period furnishings; gardens; historic buildings.
Activities: guided tours; special events; summer camps.
Hours & Admission Prices: May to Columbus Day Tues.-Sat. 10-4; other times by appointment. Adults $8, seniors & students $6, children 6-12 $5; discounts to PBS, AAA, & NE Museum members.

MOUNT HOPE FARM, 250 Metacom Ave., Bristol, RI 02809-5180. Mailing Address: P.O. Box 66, Bristol, RI 02809-0066. Tel.: 401-254-1745. Fax: 401-254-1270.
Web Site: www.mounthopefarm.com
Founded: 1998.
Congressional District: 1
Key Personnel: Dir. & Pres. (V), James W. Farley.
Governing Authority: Parent Institution: The Mount Hope Trust in Bristol. Tax-exempt.
Institution Type/Description: Historic Site & Building: housed in Governor Bradford House, c.1745. Listed on the National Register of Historic Places.
Collections: period furnishings; personal artifacts.
Facilities: nature trails; garden.
Activities: rental facilities; retreats; lecture series; education programs.
Hours & Admission Prices: May-Oct. & Dec. Wed.-Sat. 12-4. Admission $6.

Chepachet

GLOCESTER HERITAGE SOCIETY, 1181 Putnam Pike, Chepachet, RI 02814. Mailing Address: P.O. Box 269, Chepachet, RI 02814-0269. Tel.: 401-568-8967.
E-mail: info@glocesterheritagesociety.org
Web Site: www.glocesterheritagesociety.org
Formerly: Job Armstrong Store
Founded: 1967.
Key Personnel: Pres., Roland Rivet; Vice Pres., Marie Sweet.
Governing Authority: Parent Institution: Glocester Heritage Society. Tax-exempt: 501(c)(3).
Institution Type/Description: Historic Building: housed in an early 1800s store.
Collections: local history; weaving; quilting; rug hooking; photographs.
Research Fields: local history; genealogy.
Facilities: archives; reading room; meeting room. Museum-related items for sale.
Activities: Museum Sponsors: Gala & Silent Auction in spring; Dorr Rebellion Days in June; Heritage Day in September; Peddlar's Faire in November; Candlelight Shopping in December.
Publications: newsletter, Glocester Heritage Society.
Hours & Admission Prices: Thurs. 11-2 by appointment. No charge; donations accepted.
Membership: Senior & Student $10; Single $15; Family $25; Lifetime $300.

Coventry

GENERAL NATHANAEL GREENE HOMESTEAD, 50 Taft St., Coventry, RI 02816-5314. Tel.: 401-821-8630.
Institution Type/Description: Historic House Museum: built in 1770. Listed on the National Register of Historic Places.
Collections: Greene family history & memorabilia; period furniture; photographs; personal artifacts.
Hours & Admission Prices: April-Oct. Wed. & Sat. 10-5, Sun. 1-5; other times by appointment.

PAINE HOUSE, 7 Station St., Coventry, RI 02816. Mailing Address: Western Rhode Island Civic Historical Society, P.O. Box 2, Coventry, RI 02816. Tel.: 401-615-2426.
Governing Authority: Parent Institution: The Western Rhode Island Civic Historical Society.
Institution Type/Description: Historic House Museum.
Collections: local history & culture; personal artifacts; clothing; photographs; quilts; tools; books.
Hours & Admission Prices: May-Sept. Sat. 10-3; other times by appointment. Adults $3, children 6-12 $1.

WESTERN RHODE ISLAND CIVIC HISTORICAL SOCIETY, 7 Station St., Coventry, RI 02816. Mailing Address: P.O. Box 2, Coventry, RI 02816. Tel.: 401-385-9997.
E-mail: info@westernrihistory.org
Web Site: westernrihistory.org
Founded: 1945.
Congressional District: 1
Key Personnel: Pres., Norma Smith; Vice Pres., Marilyn Nagy; Sec., Katie McDonald.
Personnel Profile: Part-Time Volunteers 1.
Governing Authority: nonprofit organization. Branch Museum: Paine House, 1 Station St., Washington, RI. Tax-exempt.
Institution Type/Description: Local History Museum.

Collections: costumes; tools; utensils; period furnishings.
Research Fields: history of Western Rhode Island.
Facilities: library of historical books; scrapbooks; old tax books.
Activities: guided tours; permanent exhibitions.
Publications: booklet, The Paine House.
Hours & Admission Prices: May-Dec. Sat. 1-4 by appointment. Adults $3, students & children $1; members no charge.
Attendance: 75 (estimated)
Membership: Individual $10.

Cranston

GOVERNOR SPRAGUE MANSION, 1351 Cranston St., Cranston, RI 02920. Tel.: 401-944-9226.
E-mail: smoyer3@verizon.net
Web Site: www.cranstonhistoricalsociety.org
Founded: 1966.
Congressional District: 2
Key Personnel: Pres. (V), Sandra Moyer.
Personnel Profile: Part-Time Volunteers 15.
Governing Authority: Parent Institution: Cranston Historical Society. Tax-exempt.
Institution Type/Description: Historic House Museum: housed in the former home of the Sprague family, built in 1790.
Collections: period furnishings; personal artifacts; photographs.
Hours & Admission Prices: Call for hours. Adults 12 & over $10, children under 12 $5. &
Attendance: 2,000 (estimated)
Membership: Single $25; Family $35; Life $200; Endowment $1,000.

JOY HOMESTEAD, 156 Scituate Ave., Cranston, RI 02921. Mailing Address: Cranston Historical Society, 1351 Cranston St., Cranston, RI 02920-6721. Tel.: 401-944-9226.
Founded: 1949.
Congressional District: 2
Key Personnel: Pres. (V), Sandra Moyer.
Personnel Profile: Part-Time Volunteers 10.
Governing Authority: Parent Institution: Cranston Historical Society. Tax-exempt.
Institution Type/Description: Historic House: housed in the former home of Job Joy; built in 1764. Listed on the National Register of Historic Places.
Collections: local history & culture; period furnishings; personal artifacts; photographs.
Activities: educational programs & activities.
Hours & Admission Prices: Call for hours.
Attendance: 750 (estimated)

STEAMSHIP HISTORICAL SOCIETY OF AMERICA, 30 Kenney Dr., Ste. 3, Cranston, RI 02920-4404. Tel.: 401-463-3570. Fax: 401-463-3572.
E-mail: info@sshsa.org
Web Site: www.sshsa.org
Founded: 1935.
Congressional District: 2
Key Personnel: Exec. Dir., Matthew S. Schulte; Pres. (V), John F. Hamma; Editor, Jim Pennypacker; Devel. Dir., Vera Harsh; Membership Dir., Diana Moraco; Research Asst., Astrid Drew; Part-time Staff, Alissa Cafferky; Part-time Staff, Karen Sylvia.
Personnel Profile: Full-Time Paid 3; Full-Time Volunteers 1; Part-Time Paid 5; Part-Time Volunteers 3; Interns 1.
Governing Authority: society. Tax-exempt: 501(c)(3).
Institution Type/Description: Research Library.
Collections: steamship & navigation-related items.
Research Fields: powered shipping & navigation.
Facilities: 10,000-vol. library of books & periodicals; 200,000 photographs in field of powered shipping & navigation; 25,000 postcards; pamphlet & brochure files; ship menus & plans; shipping ephemera.
Activities: marine research, in person & by mail.
Publications: magazine, Powerships; maritime newsletter, AHOY!; books, Canadian Coastal and Inland Steam Vessels; Merchant Steam Vessels of the United States; Steamboats on the Muskingum; Steamboats For Rondout; Paddlewheel Inboard; Photographic Portraits of American Ocean Steamships, 1850-1870; Hollywood to Honolulu; booklets, Steam Navigation on the Carolina Sounds and the Chesapeake in 1892; 40th Anniversary.
Hours & Admission Prices: Mon.-Fri. 8:30-4:30. No charge. &
Membership: Student $30; Annual $50; Family (per additional person) $2; Contributing $75; Sustaining $100; Benefactor $1,000; Life $2,500.

East Greenwich

JAMES MITCHELL VARNUM HOUSE AND MUSEUM, 57
Pierce St., East Greenwich, RI 02818. Mailing Address: 6 Main St.,
East Greenwich, RI 02818. Tel.: 401-884-1776.
E-mail: k8bcm@cox.net
Web Site: www.varnumcontinentals.org
Founded: 1939.
Congressional District: 2
Key Personnel: Dir., Col. Bruce C. MacGunnigle; Caretaker, Barlow B. Healy;
Cur., Skip Healy.
Personnel Profile: Full-Time Volunteers 2; Part-Time Volunteers 6.
Governing Authority: nonprofit organization. Parent Institution: Varnum Con-
tinentals, 6 Main St., East Greenwich, RI 02818. Tax-exempt: 501(c)(3).
Institution Type/Description: Historic House Museum: 1773 home of Major
General James Mitchell Varnum.
Collections: eight authentic 18th-century furnished rooms.
Facilities: Museum-related gifts for sale.
Activities: guided tours; permanent exhibitions.
Publications: brochure; book, James Mitchell Varnum 1748-1789: The Man
and His Mansion.
Hours & Admission Prices: June-Aug. Sat.-Sun. 10-4; call to confirm hours.
Suggested Donations: $5; discounts to AAM & ICOM members. &
Attendance: 600 (estimated)
Membership: Annual $35.

NEW ENGLAND WIRELESS & STEAM MUSEUM INC., (M),
1300 Frenchtown Rd., East Greenwich, RI 02818-1329. Tel.:
401-885-0545. Fax: 401-884-0683.
E-mail: newsm@newsm.org
Web Site: www.newsm.org
Founded: 1964.
Congressional District: 2
Key Personnel: Emeritus Pres., Robert W. Merriam; Pres., Frederick L. Jaggi.
Personnel Profile: Full-Time Volunteers 2; Part-Time Volunteers 73.
Governing Authority: nonprofit organization. Tax-exempt: 501(c)(3).
Institution Type/Description: History Museum: listed on the National Register
of Historic Places.
Collections: devices from the days of the American Institute of Electrical
Engineers (now IEEE) & the American Society of Mechanical Engineers
(ASME); electrical communications equipment; stationary steam engines;
radio & wireless equipment.
Research Fields: electrical & mechanical engineering history.
Facilities: 10,000-vol. non-circulating library on engineering available on
premises only; 110-seat auditorium & theater.
Activities: guided tours; lectures; films; hobby workshops; educational courses
in electrical & mechanical engineering. Museum Sponsors: Yankee
Tune-Up in Summer; Yankee Steam-Up in Fall.
Publications: biannual historical monographs; book, Wireless Communication
in the U.S. 1890-1920.
Hours & Admission Prices: Groups by appointment only. Adults $15; discount
to groups of 10 or more. &
Attendance: 3,000 (estimated)

VARNUM MEMORIAL ARMORY & MILITARY MUSEUM, 6
Main St., East Greenwich, RI 02818-3827. Tel.: 401-884-4110.
E-mail: armory@varnumcontinentals.org
Web Site: www.varnumcontinentals.org
Founded: 1907.
Congressional District: 2
Key Personnel: C.E.O., Col. John F. Cuddy; Cur., Maj. Donald Marcum.
Personnel Profile: Part-Time Paid 2; Part-Time Volunteers 8.
Governing Authority: private; nonprofit organization. Branch Museum: Var-
num House Museum, 57 Pierce St., East Greenwich. Tax-exempt:
501(c)(3).
Institution Type/Description: Military Museum: 1913 medieval-style armory;
headquarters of Varnum Continentals military command dating from 1775,
built 1913 in Medieval style & occupied by Varnum Continentals.
Collections: military artifacts; 76th Division artifacts; historical documents
from 17th-19th centuries; over 900 WWI & WWII posters; Revolutionary
War R.I. train artillery helmet.
Publications: newsletter.
Hours & Admission Prices: By appointment only. No charge; donations
accepted.
Attendance: 600 (estimated)
Membership: Annual $35.

East Providence

**HUNT HOUSE MUSEUM - EAST PROVIDENCE HISTORI-
CAL SOCIETY,** 65 Hunts Mills Rd., East Providence, RI 02916.
Mailing Address: P.O. Box 4774, East Providence, RI 02916. Tel.:
401-438-1750.
E-mail: info@ephist.org
Web Site: ephist.org
Founded: 1966.
Congressional District: 1
Key Personnel: Pres. (V), Margaret Dooley.
Personnel Profile: Part-Time Volunteers 30; Interns 1.
Governing Authority: Parent Institution: East Providence Historical Society.
Tax-exempt.
Institution Type/Description: Historic House Museum.
Collections: local history; period artifacts; Rumford Baking Powder Company
history & artifacts, 1855-1960; genealogy; photographs.
Research Fields: neighborhoods; early history of East Providence.
Facilities: library.
Activities: monthly meetings.
Publications: monthly, The East Providence Gazette; Images of America - East
Providence.
Hours & Admission Prices: March-June & Sept.-Dec. 2nd Sun. each month
1-3:30; other times by appointment. No charge; donations accepted.
Attendance: 100 (estimated)
Membership: Individual $15; Family & Corporate $25; Life $150.

Exeter

TOMAQUAG INDIAN MEMORIAL MUSEUM, Arcadia Village,
390 A Summit Rd., Exeter, RI 02822-1808. Tel.: 401-491-9063.
Fax: 401-491-9063. Facebook: Tomaquag Museum.
E-mail: lorenspears@tomaquagmuseum.com
Web Site: www.tomaquagmuseum.com
Founded: 1958.
Congressional District: 2
Key Personnel: Dir., Loren Spears; Chm. (V), Maria Lawrence, Ph.D.
Personnel Profile: Full-Time Paid 1; Part-Time Paid 2; Part-Time Volunteers
20; Interns 10.
Volunteer Hours: 560
Operating Expenses: 70,000
Operating Income: 70,000
Governing Authority: Tomaquag bd. dirs. Tax-exempt.
Institution Type/Description: Native American Museum.
Collections: Native American artifacts; Southern New England ash splint
baskets; traditional clothing; dolls; variety of exhibits depicting the history
& culture of Native American people from the past to contemporary times.
Facilities: conference space. Museum-related items for sale.
Activities: special & public events; tours.
Publications: educational materials; curriculum; book, Through Our Eyes: An
Indigenous View of Mashapaug Pond.
Hours & Admission Prices: Spring, Summer & Fall: Wed. 10-5, Sat. 10-2;
tours, groups & winter by appointment. Adults $5, students & seniors $4,
children $3; discounts to Blue Star Museum members; military families
Memorial Day-Labor Day no charge. &
Attendance: 2,000 (estimated)

Foster

FOSTER TOWN HOUSE, 181 Howard Hill Rd., Foster, RI 02825-
1226. Tel.: 401-392-9200. Fax: 401-702-5010.
Congressional District: 2
Key Personnel: Dir., Carol Sholly.
Governing Authority: Parent Institution: town of Foster.
Institution Type/Description: Historic House: built in 1781
Collections: local history; period furnishings; personal artifacts.
Hours & Admission Prices: Summer: Mon.-Thurs. 8:30-5:30. No charge. &
Attendance: 300 (estimated)

Jamestown

BEAVERTAIL LIGHTHOUSE MUSEUM, Beavertail State Park,
Jamestown, RI 02835. Mailing Address: Beavertail Lighthouse
Museum Association, P.O. Box 83, Jamestown, RI 02835-0083.
Tel.: 401-423-3270.
E-mail: info@beavertaillight.org
Web Site: www.beavertaillight.org
Founded: 1898.
Congressional District: 1

Key Personnel: Pres. (V), Guy Archambault; Museum Shop Mgr., Dorrie Lynn.
Personnel Profile: Part-Time Volunteers 50.
Governing Authority: Parent Institution: Rhode Island Dept. of Environmental Management. Tax-exempt.
Institution Type/Description: Lighthouse Museum: the third oldest lighthouse on the Atlantic seacoast.
Collections: lighthouse history; artifacts.
Facilities: Museum-related items for sale.
Activities: panoramic view of Narragansett Bay.
Publications: newsletter.
Hours & Admission Prices: May 24 to mid-June & Sept. to Columbus Day Sat.-Sun. 12-3; June 16 to Labor Day daily 10-4. No charge; donations accepted.
Attendance: 27,000 (estimated)

JAMESTOWN FIRE DEPARTMENT MEMORIAL MUSEUM, 50 Narragansett Ave., Jamestown, RI 02835-1167. Tel.: 401-423-0062. Fax: 401-423-7278.
E-mail: jamestownfd@msn.com
Web Site: www.jamestownfd.com/museum.htm
Key Personnel: Museum & Web, Kenneth H. Caswell
Institution Type/Description: Fire Fighting Museum.
Collections: period fire fighting equipment; photographs.
Hours & Admission Prices: Call for hours.

JAMESTOWN MUSEUM, 92 Narragansett Ave., Jamestown, RI 02835-1174. Mailing Address: P.O. Box 156, Jamestown, RI 02835-0156. Tel.: 401-423-0784.
E-mail: info@jamestownhistoricalsociety.org
Web Site: jamestownhistoricalsociety.org
Founded: 1972.
Congressional District: 1
Key Personnel: Pres. Historical Society, Linnea Petersen.
Personnel Profile: Part-Time Volunteers 100.
Governing Authority: society. Parent Institution: Jamestown Historical Society. Tax-exempt.
Institution Type/Description: Local History Museum: housed in 19th-century schoolhouse.
Collections: history of Jamestown & the old ferry system; Colonial Battery; 18th-century windmill; 18th-century friends meeting house.
Research Fields: local history; genealogy; architecture.
Activities: temporary exhibitions.
Publications: newsletter; The Jamestown Bridge, 1940-2007; Concept to Demolition, Keepers of the Dutch Island Light.
Hours & Admission Prices: June to early Sept. Wed.-Sun. 1-4; early Sept.-Oct. Sat.-Sun. 1-4. No charge; donations accepted. ᛔ
Attendance: 1,000 (estimated)
Membership: Single $25; Family $40; Patron $500; 1657 Member $1,000.

SYDNEY L. WRIGHT MUSEUM, Jamestown Philomenian Library, 26 North Rd., Jamestown, RI 02835-1434. Tel.: 401-423-7281.
Web Site: www.jamestownri.com/library/museum.htm
Founded: 1973.
Institution Type/Description: Archaeology Museum.
Collections: Native American artifacts relating to Conanicut Island.
Hours & Admission Prices: mid-May to mid-June Mon.-Tues. 10-9, Wed. 10-5 & 7pm-9pm, Thurs. 12-5 & 7pm-9pm, Fri.-Sat. 10-5; June 15-Sept. 15 Mon.-Tues. 10-9, Wed. 10-5 & 7pm-9pm, Thurs. 12-5 & 7pm-9pm, Fri. 10-5, Sat. 10-2; mid-Sept. to mid-May Mon.-Tues. 10-9, Wed. 10-5 & 7pm-9pm, Thurs. 12-5 & 7pm-9pm, Fri.-Sat. 10-5, Sun. 1-5. ᛔ

✳ **WATSON FARM, (M),** 455 North Rd., Jamestown, RI 02835-2238. Tel.: 401-423-0005. Fax: 401-423-2554.
Web Site: www.historicnewengland.org
Key Personnel: Pres., Carl Nold; Farm Mgr., Don Minto; Farm Mgr., Heather Minto.
Governing Authority: society; nonprofit organization. Parent Institution: Historic New England, 141 Cambridge St. Boston, MA 02114. Tax-exempt: 501(c)(3).
Institution Type/Description: Historic Farm: c.1796 Watson Farm, 285-acre working farm.
Collections: heritage breed cattle & sheep; hayfields; pastures.
Activities: guided tours; special events & programs; school programs.
Publications: Guide to Historic New England.
Hours & Admission Prices: June-Oct. 15 Tues., Thurs. & Sun. 1-5. Adults $4; discounts to seniors, WGBH, AAA, AAM & ICOM members; Historic New England members no charge. Call for further information.

Attendance: 3,059 (accurate)
Membership: National $35; Individual $45; Household $55; Garden & Landscape $75; Contributing and Library & School $100; Historic New England Affiliate $100-$350; Young Friends of Historic New England $100-$1,500; Friends of the Library and Archives $125; Historic Homeowner $200; Business & Ogden Codman Design Group $250; Appleton Circle $1,750-$3,500.

Kingston

FINE ARTS CENTER MAIN GALLERY, 105 Upper College Rd., Kingston, RI 02881. Tel.: 401-874-5821. Fax: 401-874-2729.
E-mail: artdept@etal.uri.edu
Web Site: www.uri.edu/artsci/art/galleries.html
Founded: 1968.
Key Personnel: Dir., Ronald Hutt.
Governing Authority: Parent Institution: Dept. of Art & Anthropology, University of Rhode Island. Tax-exempt.
Institution Type/Description: Art Gallery.
Major Exhibits: Robert Dilworth, 2/14-3/14; John Udvardy, 3/14-4/14; Annual Juried Exhibition, 4/14-5/14; Providence Collector Dr. Joseph Hazan, 6/14-8/14; Annual Faculty Exhibition, 9/14-10/14.
Facilities: main gallery; experimental gallery; corridor gallery; video corridor.
Activities: special events; temporary exhibitions.
Hours & Admission Prices: Mon.-Sat. 12-4. No charge; donations accepted.
Attendance: 10,262 (accurate)

HELME HOUSE GALLERY, 2587 Kingstown Rd., Kingston, RI 02881-1605. Tel.: 401-783-2195.
E-mail: socart@verizon.net
Web Site: www.southcountyart.org
Founded: 1929.
Key Personnel: Dir., Jason Fong
Institution Type/Description: Art Gallery.
Collections: paintings; photographs; sculpture.
Facilities: Museum-related items for sale.
Activities: rental facilities.
Hours & Admission Prices: Wed.-Sun. 1-5. Closed holidays.
Membership: Individual $65; Family $95.

PETTAQUAMSCUTT HISTORICAL SOCIETY, 2636 Kingstown Rd., Kingston, RI 02881-1624. Tel.: 401-783-1328. Facebook: Pettaquamscutt Historical.
E-mail: washingtoncountyhistory@gmail.com
Web Site: www.washingtoncountyhistory.org
Founded: 1958.
Congressional District: 2
Key Personnel: Dir., Lori Urso; Pres. (V), H. Graham Nye.
Personnel Profile: Full-Time Paid 1; Part-Time Paid 2; Part-Time Volunteers 40.
Governing Authority: nonprofit organization. Tax-exempt: 501(c)(3).
Institution Type/Description: Historical Society Museum: housed in 1858 & 1861 Old Washington County Jail.
Collections: Civil War artifacts; period furniture; house, shop & farm furnishings; artifacts from late 18th-century to present; local historic manuscripts & printed materials; mural by Ernest Hamlin Baker.
Research Fields: local history; genealogy.
Facilities: 1,000-vol. local history & genealogy library on site.
Activities: guided tours; lectures; permanent & temporary exhibitions; seasonal events.
Publications: quarterly newsletter; Lost South Kingstown; The Family of John Watson of the Narragansett Country; Crime and Punishment and the Old Washington County Jail.
Hours & Admission Prices: Tues., Thurs. & Sat. 1-4. Adults $5, students & seniors $3; discounts to AAM members; members and children 12 & under no charge.
Attendance: 2,000 (estimated)
Membership: Individual $20; Family $35; Friend $50; Sponsor $100; Patron $250; Benefactor $500; Life $1,000.

Kingstown

FAYERWEATHER HOUSE, 1895 Mooresfield Rd., Kingstown, RI 02881-1715. Mailing Address: P.O. Box 222, Wakefield, RI 02880. Tel.: 401-789-9072. Facebook: Fayerweather Craft Guild.
Web Site: www.fayerweatherhouse.8m.net
Institution Type/Description: Historic House Museum: built in 1820. Listed on the National Register of Historic Places.
Collections: family history; period furnishings; personal artifacts; photographs.

Hours & Admission Prices: May-Dec. Tues.-Sat. 10-4. No charge.

Lincoln

✱ **ARNOLD HOUSE, (M),** 487 Great Rd., Lincoln, RI 02865. Mailing Address: Historic New England, 141 Cambridge St., Boston, MA 02114-2702. Tel.: 401-728-9696.
Web Site: www.historicnewengland.org
Key Personnel: Site Mgr., Dan Santos.
Governing Authority: Parent Institution: Historic New England.
Institution Type/Description: Historic House: built in 1693.
Collections: historic house.
Activities: guided tours with focus on architecture.
Hours & Admission Prices: Sat.-Sun. 11-5. Adults $5; discounts to AAM & ICOM members; Historic New England members no charge.
Attendance: 2,672 (accurate)
Membership: National $35; Individual $45; Household $55; Garden & Landscape $75; Contributing and Library & School $100; Young Friends of Historic New England $100-$1,500; Friends of the Library and Archives $125; Historic Homeowner $200; Business & Ogden Codman Design Group $250; Appleton Circle $2,500 & up.

FRIENDS OF HEARTHSIDE, INC., 677 Great Rd., Lincoln, RI 02865-1401. Mailing Address: 757 Great Rd., Lincoln, RI 02865-3820. Tel.: 401-726-0597.
E-mail: kathy.hartley@hearthsidehouse.org
Web Site: www.hearthsidehouse.org
Founded: 2001.
Congressional District: 1
Key Personnel: Chm. (V), Kathryn A. Hartley.
Personnel Profile: Part-Time Volunteers 25; Interns 1.
Governing Authority: Tax-exempt.
Institution Type/Description: Historic House: housed in an 1810 mansion.
Collections: textiles; period furnishings; personal artifacts; blacksmith shop.
Activities: tours; special events; blacksmithing classes.
Publications: email newsletter.
Hours & Admission Prices: Tours: March-Dec. 2nd Sat. of month. No charge; donations accepted.
Attendance: 1,000 (estimated)
Membership: $15-$1,000 & up.

Little Compton

LITTLE COMPTON HISTORICAL SOCIETY, (M), 548 W. Main Rd., Little Compton, RI 02837-1123. Mailing Address: P.O. Box 577, Little Compton, RI 02837-0577. Tel.: 401-635-4035. Fax: 401-635-4035.
E-mail: lchistory@littlecompton.org
Web Site: www.littlecompton.org
Founded: 1937.
Congressional District: 2
Key Personnel: Mng. Dir., Marjory O'Toole; Pres. (V), Shelley Bowen; Admin., Nancy Carignan; Mgr. Collections, Fred Bridge.
Personnel Profile: Full-Time Paid 1; Part-Time Paid 1; Part-Time Volunteers 50; Interns 1.
Governing Authority: society. Tax-exempt: 501(c)(3)
Institution Type/Description: History Museum: housed in 1850 Wilbor Barn & 1690-1860 Wilbor House.
Collections: Museum: built c.1690 includes 17th, 18th & 19th century furnishings & artifacts. 19th century Friends Meeting House. Outbuildings: include a carriage house with restored sleighs & coaches; a one-room schoolhouse with early school books, slates, & maps; barn houses farm tools & equipment including dairying utensils, vehicles, and looms; Portuguese artifacts; photographs; documents of Little Compton over the last century; artist, Sidney Burleigh's studio built over the hull of a rescued 19th century catboat and a cookhouse where food for the Rhode Island Red Chickens was prepared; Native American artifacts including, stone tools, pottery and baskets.
Major Exhibits: Little Compton School Houses, 7/14-10/14.
Research Fields: 17th, 18th & 19th century American social history.
Facilities: research facility.
Activities: guided tours; permanent & temporary exhibitions; summer camps; educational programs.
Publications: Little Compton Families; Little Compton Remembers World War II; newsletter, Past Times; Compton Land Revisited; L Is For Little Compton; The History of Little Compton - First Light Sakonnet - 1660-1820; pamphlets, Portraits in Time; A Time To Play; Terra Nova, Vida Nova; Life & Works of Sydney Burleigh; books, Sakonnet Point Perspectives; The History of Little Compton - A Home By The Sea, 1820-1954; Remembering Adamsville - Oral Histories.

Hours & Admission Prices: late June to Labor Day Thurs.-Sun. 1-5; Sept.-Oct. Sat.-Sun. 1-5; other times by appointment. Adults $7.50, children $5; discounts to NEMA members; reciprocal membership; members no charge.
Attendance: 2,750 (estimated)
Membership: Individual $20; Family $30; Contributing $50; Corporate $75; Supporting $100; School Program Sponsor $150.

Middletown

NORMAN BIRD SANCTUARY, 583 Third Beach Rd., Middletown, RI 02842-5738. Tel.: 401-846-2577. Fax: 401-846-2772.
E-mail: nharrison@normanbirdsanctuary.org
Web Site: www.normanbirdsanctuary.org
Key Personnel: Exec. Dir., Natasha Harrison; Dir. Properties, Joseph McLaughlin; Dir. Education, Kim Botelho; Coord. Education, Rachel Holbert; Coord. Education, Nicole Lavoie; Administrative Dir., Lesley Muir; Dir. Devel., Suzanne Garvin; Devel. Asst., Erika Gibb
Institution Type/Description: Bird Sanctuary.
Collections: native birds & plants.
Hours & Admission Prices: Daily 9-5. Trail: adults $6, children 3-12 $3; children under 3 & members no charge. Closed major holidays.
Attendance: 40,000 (estimated)
Membership: Individual $40; Family $55; Extended Family 100; Business $125; Supporting $250; Patron $500; Benefactor $1,000; Life $5,000.

PRESCOTT FARM, 2009 W. Main Rd., Middletown, RI 02842-7963. Mailing Address: Newport Restoration Foundation, 51 Touro St., Newport, RI 02840. Tel.: 401-849-7300.
Web Site: www.newportrestoration.org/visit/prescott_farm/
Key Personnel: Exec. Dir., Pieter Roos
Institution Type/Description: History Museum.
Collections: farm history; early American architecture & landscape; period artifacts. Historic Structure: 1812 windmill.
Facilities: nature trails.
Activities: educational programs; rental facilities.
Hours & Admission Prices: June-Sept. Tues.-Sat. 10-3. Guided Tour: adults $5; children under 12 no charge. Grounds: no charge.

SACHUEST POINT NATIONAL WILDLIFE REFUGE, 769 Sachuest Point Rd., Middletown, RI 02842. Mailing Address: 50 Bend Rd., Charlestown, RI 02813-2503. Tel.: 401-847-5511.
Institution Type/Description: Wildlife Refuge.
Collections: over 200 bird species; hands-on exhibits.
Facilities: 242 acre site; visitor center; nature trails; classrooms.
Activities: viewing platforms; fishing; special events.
Hours & Admission Prices: Call for hours.
Attendance: 65,000 (estimated)

WHITEHALL MUSEUM HOUSE, 311 Berkeley Ave., Middletown, RI 02842-5392. Mailing Address: 29 Longfellow Rd., Jamestown, RI 02835. Tel.: 401-846-3116.
E-mail: ekcburgess@aol.com
Web Site: whitehallmuseumhouse.org
Founded: 1900.
Key Personnel: Chm. (V) & Pres. (V), Eleanor Burgess; Museum Shop Mgr., Maris Humphreys.
Personnel Profile: Full-Time Volunteers 10; Part-Time Volunteers 14.
Governing Authority: Parent Institution: The National Society of The Colonial Dames of America in the State of RI. Subsidiary Institution: Whitehall Committee. Tax-exempt.
Institution Type/Description: Historic House Museum: housed in the former home of Anglican Bishop, George Berkeley; built in 1729.
Collections: period furnishings; herb garden; orchard; books by & about George Berkeley; cooking tools.
Research Fields: philosophy.
Facilities: library; archives.
Activities: Museum Sponsors: Open House in May, September, October & December; guided tours by philosophers in July and August.
Publications: Welcome to Whitehall.
Hours & Admission Prices: July-Aug. Tues.-Sun. 10-4; other times by appointment. Adults $5; discounts to AAA & International Berkeley Society members and Anglican clergy; members no charge.
Attendance: 534 (accurate)
Membership: Friend of Whitehall $25.

Narragansett

SOUTH COUNTY MUSEUM INC., (M), 115 Strathmore St., Narragansett, RI 02882. Mailing Address: Box 709, Narragansett, RI 02882-0709. Tel.: 401-783-5400. Fax: 401-783-0506.
E-mail: info@southcountymuseum.org
Web Site: www.southcountymuseum.org
Founded: 1933.
Congressional District: 2
Key Personnel: Pres., Daryl Anne Anderson; Vice Pres., Doris Manganaro; Staff Dir., Jim Crothers; Business Mgr., Carolyn Shea.
Personnel Profile: Part-Time Paid 4; Interns 1.
Governing Authority: nonprofit organization. Tax-exempt.
Institution Type/Description: History Museum.
Collections: rural early America; agriculture; technology; transportation; print shop; blacksmith shop; carpentry shop; household items of Rhode Island rural life 1800-1930; living history farm; maritime; textiles; weaving & cottage industry.
Research Fields: 1800-1930 culture & social history.
Facilities: picnic area.
Activities: educational programs; temporary & permanent exhibits; special events; lectures.
Publications: newsletter, Canonchet Gazette.
Hours & Admission Prices: May-June & Sept. Fri.-Sat. 10-4; July-Aug. Wed.-Sat. 10-4. Adults $6, seniors $5, children 6-12 $2; discounts to AAM, AAA, NEMA, & AASLH members; children under 6 & members no charge. &
Attendance: 3,017 (accurate)
Membership: Individual $25; Family $35; Business $125; Benefactor $100; Life $1,000.

Newport

ARTILLERY COMPANY OF NEWPORT MILITARY MUSEUM, 23 Clarke St., Newport, RI 02840-3023. Mailing Address: P.O. Box 14, Newport, RI 02840. Tel.: 401-846-8488.
E-mail: info@newportartillery.org
Web Site: www.newportartillery.org
Founded: 1959.
Congressional District: 1
Key Personnel: Colonel Commanding, Col. Comm. Robert Edenbach; Exec. Officer, Michael Pine; Operations Officer, Major Steven Colonies; Adjutant & Communications Officer, Joanne Pike; Finance Officer, Pvt. Corinne Edenbach
Governing Authority: artillery company. Subsidiary Institution: Artillery Company of Newport Foundation. Tax-exempt: 501(c)(3) & 170(b)(1)(A).
Institution Type/Description: Military Museum: housed in c.1836 Armory.
Collections: uniforms & militaria from 125 countries; artillery artifacts include 4 original Paul Revere Cannons.
Research Fields: history of the artillery company & Rhode Island militia.
Activities: guided tours.
Hours & Admission Prices: May-Oct. Sat. 10-4. No charge; donations accepted.
Membership: Annual $45.

COLONY HOUSE, Washington Square, Newport, RI 02840. Mailing Address: c/o Newport Historical Society, 82 Touro St., Newport, RI 02840-2931. Tel.: 401-846-0813. Fax: 401-846-1853.
Web Site: www.newporthistorical.org
Founded: 1739.
Personnel Profile: Part-Time Paid 3; Interns 5.
Governing Authority: state. Parent Institution: State of Rhode Island. Managed by the Newport Historical Society. Tax-exempt: 501(c)(3).
Institution Type/Description: Historic State House Building: 1739 Colony House. Listed on the National Register of Historic Places.
Collections: period furnishings; paintings.
Hours & Admission Prices: Call for hours. &
Attendance: 2,750 (estimated)

EDWARD KING HOUSE, 35 King St., Newport, RI 02840-3595. Tel.: 401-846-7426.
Institution Type/Description: Historic House: former home of China Trade merchant, Edward King; built in c.1845.
Collections: historic house; gardens.
Facilities: banquet facilities.
Activities: tours; rental facilities; weddings; special events.
Hours & Admission Prices: Tours: Mon.-Fri. 9-4; other times by appointment.

FORT ADAMS STATE PARK, 90 Fort Adams Dr., Newport, RI 02840. Tel.: 401-841-0707. Fax: 401-841-0790.
E-mail: info@fortadams.org
Web Site: www.fortadams.org
Founded: 1994.
Key Personnel: Dir., Robert McCormack
Institution Type/Description: Park & History Museum.
Collections: local & military history; photographs; period furnishings; personal artifacts.
Facilities: rental facilities. Museum-related items for sale.
Activities: rental facilities; special events; military reenactments; music festivals; classic vehicle shows; guided tours.
Hours & Admission Prices: Memorial Day to Columbus Day 10-4. Adults $10, children 6-17 $5; discounts to groups; children 5 & under no charge.
Attendance: 20,000 (estimated)

FRIENDS MEETING HOUSE (QUAKER), 30 Marlborough St., Newport, RI 02840. Mailing Address: Newport Historical Society, 82 Touro St., Newport, RI 02840-2931. Tel.: 401-846-0831 (Historical Society).
Institution Type/Description: Historic House: built in 1704
Collections: period furnishings.
Hours & Admission Prices: Call for hours.

✻ INTERNATIONAL TENNIS HALL OF FAME & MUSEUM, (M), 194 Bellevue Ave., Newport, RI 02840-3586. Tel.: 401-849-3990; 800-457-1144. Fax: 401-849-8780 & 851-7920 (Research Center).
E-mail: newport@tennisfame.com
Web Site: www.tennisfame.com
Founded: 1954.
Congressional District: 1
Key Personnel: C.E.O., Mark Stenning; Dir., Douglas Stark; Chm., Christopher E. Clouser; Pres., Stan Smith; Vice Pres. & Dir. Administration, Nancy Cardoza; Dir. Retail, Leslie Thomas.
Personnel Profile: Full-Time Paid 24; Part-Time Paid 6; Part-Time Volunteers 8; Interns 8.
Governing Authority: nonprofit corporation. Parent Institution: International Tennis Hall of Fame, Inc. Tax-exempt: 501(c)(3).
Institution Type/Description: Sports Museum: housed in the 1880 Newport Casino, site of the first national tennis championships, held in 1881.
Collections: tennis racquets; equipment; trophies; photographs; prints; paintings; sculpture; tennis costume; tennis history; player memorabilia.
Research Fields: history of tennis & its players; social history of Newport, the Gilded Age; McKim, Mead & White architecture.
Facilities: 5,000-vol. Information Research Center by appointment only; 13 public access grass courts. Museum-related gifts for sale.
Activities: permanent & temporary exhibitions; concert series; men's professional grass court championship. Museum Sponsors: opening night jazz festival.
Publications: brochures; newsletters.
Hours & Admission Prices: Daily 9:30-5. Adults $12, seniors, military & student $10; discounts to USTA, AAA, NEMA, ISHA, ICOM & AAM members; members and children 16 & under no charge. Closed Thanksgiving; Christmas. &
Attendance: 25,000 (estimated)
Membership: Baseline $35; Classic $50; Ace $100; Court $250; Center Court $500; Chairman's Circle $1,000.

MUSEUM OF YACHTING, Fort Adams State Park, 449 Thames St., Newport, RI 02840-6720. Tel.: 401-847-1018. Fax: 401-847-8320.
E-mail: info@museumofyachting.org
Web Site: www.moy.org
Founded: 1983.
Congressional District: 1
Key Personnel: Pres., Terry Nathan; Vice Pres. Mktg., Susan Daly; Facilities Mgr., Per Peterson; Cur., Jay Picotte.
Personnel Profile: Full-Time Paid 3; Part-Time Paid 2; Part-Time Volunteers 100; Interns 1.
Governing Authority: nonprofit organization. Tax-exempt: 501(c)(3).
Institution Type/Description: Yachting Museum: located in Fort Adams State Park.
Collections: history of yachting; small boats; Sailors Hall of Fame; America's Cup history; photographs; trophies; in-water boats.
Research Fields: restoration of old wooden boats; restoration materials, techniques, etc.
Facilities: 3,000-vol. library of sailing history; America's Cup; sailing &

boating magazines, available for use by members; 12,000 sq. ft. exhibit space. Gift items for sale.

Activities: formal education programs for adults; experiment with lifting systems; guided tours; lectures; temporary & traveling exhibition. Annual Events: Labor Day weekend, Classic Yacht Regatta; Small Boat Regatta; Unlimited Regatta; Newport 12 Meter Regatta.

Publications: quarterly, Spinnaker.

Hours & Admission Prices: Tues.-Sat. 12-5. Adults $5; discounts to groups; members, students & children under 18 no charge.

Attendance: 8,000 (accurate)

Membership: Crew (Individual) $30; Starboard Watch (Family) $40; Helmsman $50; Watch Captain $100; Navigator $250; Captain $500; Founder $1,000.

NATIONAL MUSEUM OF AMERICAN ILLUSTRATION, (M),

Vernon Court, 492 Bellevue Ave., Newport, RI 02840-4127. Tel.: 401-851-8949. Fax: 401-851-8974.

E-mail: art@americanillustration.org

Web Site: www.americanillustration.org

Founded: 1998.

Congressional District: 1

Key Personnel: Dir. & Co-Founder, Judy Goffman Cutler; C.E.O. & Co-Founder, Laurence S. Cutler; Financial Dir., David Breznicky; Asst. to Chm., Chelsea McInnis; Cur., Jennifer P. Greenawalt; Museum Shop Mgr., Jill Perkins; Dir. External Affairs, Jonathan Trumbull Isham; Public Rels., Dallas Pell; Admin. Exterior, Anthony Bolarhino; Graphic Designer, Rachel Sowersby.

Personnel Profile: Full-Time Paid 9; Full-Time Volunteers 6; Part-Time Paid 2; Part-Time Volunteers 25; Interns 3.

Governing Authority: private; nonprofit organization. Parent Institution: American Civilization Foundation. Tax-exempt: 501(c)(3).

Institution Type/Description: Art Museum.

Collections: works by over 150 illustrators including Norman Rockwell, Maxfield Parrish, N.C. Wyeth, J.C. Leyendecker, Charles Dana Gibson, Jessie Willcox Smith, Howard Pyle, H.C. Christy.

Research Fields: life & works of J.C. Leyendecker, Edith Wharton & Maxfield Parrish; murals by Tiffany studios & James Wall Finn; Norman Rockwell.

Facilities: 4,300-vol. library; 30,000 sq. ft. exhibit space; gardens & park. Museum-related items for sale.

Activities: tours; lectures; travelling exhibitions loaned to other museums.

Publications: newsletter, From Time to Time; Guidebook: The Grand Tour; books, Maxfield Parrish & The American Imagists; J.C. Leyendecker; miniature maps; Norman Rockwell & His Mentor, J.C. Leyendecker; Norman Rockwell's America...in England; quarterly, MuseNews; Norman Rockwell's America; The American Muse; Tom Wolfe Illustrations.

Hours & Admission Prices: Mon.-Fri. 9-5 or by appointment. Adults $18, senior citizens 60 & over and military with ID $16, students with ID $12, children 6-12 $8; discounts to groups; AAM, ICOM, NEMA, & National Arts Club members no charge. Children 5 & under not admitted.

Attendance: 39,000 (estimated)

NAVAL WAR COLLEGE MUSEUM, (M), 686 Cushing Rd.,

Coasters Harbor Island, Newport, RI 02841-1207. Tel.: 401-841-4052 & 2101. Fax: 401-841-7074.

E-mail: museum@usnwc.edu

Web Site: www.usnwc.edu/museum

Founded: 1952.

Congressional District: 2

Key Personnel: Dir., Prof. John Hattendorf; Dir. Education, John W. Kennedy; Cur., Robert Cembrola; Mng. Dir., John Pentangelo; Sec., Kelly Folger; Museum Shop Mgr., Julia Koster.

Personnel Profile: Full-Time Paid 4; Part-Time Volunteers 8.

Governing Authority: federal-Navy Department. Parent Institution: Naval History and Heritage Command, Washington, DC. Subsidiary Institution: Naval War College. Tax-exempt.

Institution Type/Description: Naval Museum: housed in 1820 Founders Hall, a National Historic Landmark.

Collections: art, artifacts, imprints & prints on the history of naval warfare; naval history & heritage of the Narragansett Bay region.

Research Fields: military history; naval history; naval regional history; Naval War College institutional history.

Facilities: 200-vol. library of books & 10 file drawers of principally imprint (loose) items on Naval warfare, Naval regional history & museum studies.

Activities: guided tours; permanent, temporary & loan exhibitions; staff talks; guest lecturers.

Publications: leaflets; descriptive brochure; exhibit catalogs; National Historic Site (Founders Hall); President's House; museum outdoor displays.

Hours & Admission Prices: June-Sept. Mon.-Fri. 10-4:30, Sat.-Sun. 12-4:30; Oct.-May Mon.-Fri. 10-4:30; by appointment. Adults 18 & over require photo ID. No charge; donations accepted. Closed Federal holidays. ⅊

Attendance: 36,817 (accurate)

✱ NEWPORT ART MUSEUM & ART ASSOCIATION, (M), 76

Bellevue Ave., Newport, RI 02840-7411. Tel.: 401-848-8200. Fax: 401-848-8205. Facebook: Newport Art Museum.

E-mail: info@newportartmuseum.org

Web Site: www.newportartmuseum.org

Founded: 1912.

Congressional District: 1

Key Personnel: Exec. Dir., Elizabeth A. Goddard; Pres., Peter Englehart; Cur., Nancy Whipple Grinnell; Dir. Education, Maggie Anderson.

Personnel Profile: Full-Time Paid 9; Part-Time Paid 12; Part-Time Volunteers 90; Interns 11.

Governing Authority: nonprofit organization. Tax-exempt.

Institution Type/Description: Art Museum.

Collections: contemporary & historic Rhode Island & New England artists. Historic House: c.1864 designed by Morris Hunt.

Major Exhibits: Irving Barrett: Aquidneck Aviary, 1/18/14-5/11/14; Aquidneck Industries 2014, 2/8/14-5/4/14; Elizabeth Congdon: Flowers, 5/10/14-8/12/14; Mary Jameson: Marine Botanical, 5/17/14-9/1/14; Corinne Colaruss: Paintings, 5/17/14-9/7/14; Very Simple Charm: The Life and Work of Richard Morris Hunt, 5/31/14-9/14/14; Boston Printmakers: From Palette to Plate, 8/30/14-1/4/15; Claudia Flynn: Sculpture, 9/6/14-1/7/15; The Eternal Feminine: Selections, 9/27/14-1/15/15.

Research Fields: American Art.

Facilities: 7 galleries; art school.

Activities: art classes; summer camps; lecture program; films; concerts; guided tours; temporary exhibitions.

Publications: catalogs; annual report; calendar; newsletter.

Hours & Admission Prices: Tues.-Sat. 10-4, Sun. 12-4. Open holiday Mondays. Adults $10, seniors 65+ $8, active military with ID & students $6, children 5 & under & members no charge, discounts to AAA & NARM members; Sat. 11-12 no charge. ⅊

Attendance: 19,619 (accurate)

Membership: Student $30; Active Military $40; Senior $50; Senior Household $60; Military Household $60; Patron $175; Supporting $275; Council $550; Benefactor $1,000.

NEWPORT HISTORICAL SOCIETY & THE MUSEUM OF NEWPORT HISTORY, 127 Thames St., Newport, RI 02840-

6627. Mailing Address: 82 Touro St., Newport, RI 02840-2931. Tel.: 401-846-0813 & 841-8770. Fax: 401-846-1853.

E-mail: info@newporthistorical.org

Web Site: newporthistorical.org

Founded: 1854.

Congressional District: 1

Key Personnel: Exec. Dir., Ruth Taylor; Pres. (V), Thomas Goddard; Reference Librarian & Genealogist, Bertram Lippincott, III; Dir. of Education, Ingrid Peters; Administrative Mgr., Loraine Byrne; Dir. Retail Operations, Kathleen Vandeveer.

Personnel Profile: Full-Time Paid 4; Part-Time Paid 10; Part-Time Volunteers 50; Interns 5.

Governing Authority: nonprofit. Tax-exempt: 501(c)(3).

Institution Type/Description: History Museum.

Collections: Newport County & Rhode Island history; computerized survey of Newport Historical Landmark District; town records & documents from earliest days of Newport; 200,000 photographs; family & business records; three centuries worth of manuscripts & letters; paintings; Townsend-Goddard furniture; glass; silver; pewter; tall clocks; textiles; maps; atlases. Historic Houses: c.1697 Wanton-Lyman-Hazard House; c.1699 Great Friends Meeting House; c.1730 Seventh Day Baptist Meeting House; Museum of Newport History at the c.1762 Brick Market.

Research Fields: Newport County & Rhode Island history; architectural history; genealogy; archaeology.

Facilities: 12,000-vol. research library; 17th, 18th & 19th-century manuscripts department; audiovisual programs; tours; lecture series.

Activities: guided & self-guided walking tours; permanent & temporary exhibits; educational programs;

Publications: journal, Newport History; book, Newport: A Short History; book, African Americans in Newport.

Hours & Admission Prices: Newport Historical Society Library & Gallery: Tues.-Fri. 9:30-4:30, Sat. 9:30-12. No charge. Museum of Newport History: Winter: daily 10-5. Summer: Mon.-Sat. 10-6, Sun. 10-5. Suggested Donation: adults $4, children $2; discounts to AAM & ICOM members; members & children under 5 no charge.

Attendance: 30,000 (estimated)

Membership: Student $25; Library & Museum $35; Individual $50; Family $100; Sponsor $250; 1854 Society $1,000.

NEWPORT RESTORATION FOUNDATION, 51 Touro St., Newport, RI 02840-2932. Tel.: 401-849-7300. Fax: 401-849-0125. Facebook: Newport Restoration Foundation.
E-mail: info@newportrestoration.org
Web Site: newportrestoration.org
Founded: 1968.
Key Personnel: Exec. Dir., Pieter N. Roos; Dir. Collections, A. Bruce MacLeish; Dir. Finance, Amy Winsor; Dir. Preservation, Robert Foley; Dir. Education & Public Programs, Lisa Dady; Mgr. Mktg. & Public Rels., Jeanine Kober; Human Resources Admin., Maeve Sheehan.
Personnel Profile: Full-Time Paid 28; Part-Time Paid 26; Interns 1.
Governing Authority: private; nonprofit organization. Branch Museums: Rough Point, Newport, RI; Samuel Whitehorne House, Newport, RI; Prescott Farm, Middletown, RI. Tax-exempt.
Institution Type/Description: Historic House Museum.
Collections: 18th & early 19th century houses; European furniture & decorative arts; European fine art; Rhode Island furniture; oriental porcelain; architectural drawings; photographs.
Major Exhibits: No Rules: The Personal Style of Doris Duke at Rough Point, 4/14-11/14; Newport Style: High Chests by Newport Master Cabinetmakers at Whitehorne House, 5/14-10/14.
Research Fields: American architecture; vernacular architecture; preservation methods; American furniture; 20th century clothing.
Facilities: 2,000 sq. ft. exhibit space.
Activities: docent program; formal education programs for adults & students; school programs; guided tours; hobby workshops; temporary exhibitions; training programs for professional museum workers.
Hours & Admission Prices: Rough Point: mid-April to mid-May Thurs.-Sat. 10-1:45; mid-May to early Nov. Tues.-Sat. 10-3:45. Adults $25; discounts to AAM members; children 12 & under no charge. Whitehorne House: May-Oct. Thurs.-Mon. 11-3. Adults $6; discounts to AAM members; children 12 & under no charge. Guided Tours: 10:30 & 3. Adults $12; discounts to AAM members. Prescott Farm Grounds: daily dawn to dusk. No charge. &
Attendance: 21,455 (estimated)

*** THE PRESERVATION SOCIETY OF NEWPORT COUNTY/THE NEWPORT MANSIONS, (M),** 424 Bellevue Ave., Newport, RI 02840-6924. Tel.: 401-847-1000.
E-mail: info@newportmansions.org
Web Site: www.newportmansions.org
Founded: 1945.
Congressional District: 1
Key Personnel: C.E.O., Trudy Coxe; Chm., Donald O. Ross; Dir. Mktg. Svcs. & Museum Experience, John G. Rodman; Dir. Museum Affairs, John R. Tschirch; Properties Dir., Curt Genga; Chief Conservator, Charles J. Moore; Cur., Paul F. Miller; Dir. Special Events, Philip F. Pelletier; Chief Institutional Advancement, Mary B. Kozik; Dir. Finance, Jim Burress; Dir. Gardens & Grounds, Jeff Curtis; Dir. Retail Sales, Cynthia O'Malley; Grant Writer, Katherine Long; Special Projects, Terry Dickinson.
Personnel Profile: Full-Time Paid 103; Part-Time Paid 246.
Governing Authority: society; nonprofit organization. Tax-exempt: 501(c)(3).
Institution Type/Description: Preservation Project & Historical District: c.1748-1902, group of 11 mansions located within 1-1/2 miles of each other; topiary garden in Portsmouth.
Collections: furniture; furnishings; paintings. Historic Mansions & Museums: Green Animals, Cory's Lane; Portsmouth; 1748 Hunter House, 54 Washington St., a National Historic Landmark; 1839 Kingscote, Bellevue Ave., a National Historic Landmark; 1852 Chateau-sur-Mer, Bellevue Ave., a National Historic Landmark; 1892 Marble House, Bellevue Ave., a National Historic Landmark; 1901 The Elms, Bellevue Ave., a National Historic Landmark; 1895 The Breakers, Ochre Point Ave., a National Historic Landmark; 1902 Rosecliff, Bellevue Ave; 1895 The Breakers Stable, Coggeshall Ave; 1883 Isaac Bell House, Bellevue Ave., a National Historic Landmark; 1860 Chepstow, Narragansett Ave.
Research Fields: belle epoch social history; history of architecture & decorative arts.
Activities: guided tours. Museum Sponsors: tours for students of Newport schools; members' tours & lectures; behind the scene tours; walking tours; audio & garden tours; preservation in action tours. Annual Events: The Newport Symposium in Spring; Newport Flower Show; Newport Mansions Wine & Food Festival; The Newport Student Forum in November; Christmas at The Newport Mansions.
Publications: The Newport Gazette; annual report.
Hours & Admission Prices: Call for hours. $14-$25 per house; discounts to groups of 20 or more & AAM members; members no charge. Combination tickets available. &
Attendance: 779,769 (accurate)
Membership: Single $50; Household $90; Steward $250; Patron $500; Benefactor $1,000; President's Circle $5,000.

REDWOOD LIBRARY AND ATHENAEUM, 50 Bellevue Ave., Newport, RI 02840-3292. Tel.: 401-847-0292. Fax: 401-841-5680.
E-mail: redwood@redwoodlibrary.org
Web Site: www.redwoodlibrary.org
Founded: 1747.
Congressional District: 1
Key Personnel: Dir. Special Collections, Whitney Pape; Pres. (V), Edwin Fischer, M.D.
Personnel Profile: Full-Time Paid 10; Part-Time Paid 13; Part-Time Volunteers 8; Interns 1.
Governing Authority: society. Tax-exempt: 501(c)(3).
Institution Type/Description: Historic Building: c.1750 designed by architect, Peter Harrison.
Collections: paintings by Gilbert Stuart, Charles Bird King, Robert Feke, Rembrandt Peale, Thomas Sully; sculpture; manuscript collections; 18th- & 19th-century material culture collections.
Research Fields: 18th- & 19th-century Americas.
Facilities: 200,000-vol. library of general and special collection books, may be used by researchers.
Activities: guided tours; exhibitions.
Publications: annual report; quarterly booklist; exhibition checklists; occasional publications.
Hours & Admission Prices: Mon.-Wed. & Fri.-Sat. 9:30-5:30, Thurs. 9:30-8, Sun. 1-5; No charge; donations accepted for tours or research. Closed holidays. &
Attendance: 18,000 (accurate)
Membership: Youth $25; Individual $70; Household $100.

SAVE THE BAY EXPLORATION CENTER AND AQUARIUM, Rotunda at Easton's Beach, 175 Memorial Blvd., Newport, RI 02840-3659. Tel.: 401-324-6020. Fax: 401-324-6022.
E-mail: savebay@savebay.org
Web Site: www.savebay/aquarium
Founded: 2006.
Congressional District: 1
Personnel Profile: Full-Time Paid 1; Part-Time Paid 3; Part-Time Volunteers 23.
Governing Authority: Parent Institution: Save The Bay Narragansett Bay. Tax-exempt.
Institution Type/Description: Marine Science Learning Center.
Collections: hands-on interactive exhibits.
Hours & Admission Prices: Memorial Day to Labor Day daily 10-4; Oct.-April Sat.-Sun. 10-4. Adults $6; children 3 & under no charge. &
Attendance: 11,000 (accurate)

TOURO SYNAGOGUE, 85 Touro St.; Newport, RI 02840-2969. Tel.: 401-847-4794.
Web Site: www.tourosynagogue.org
Key Personnel: Admin., Meryle Cawley.
Governing Authority: Parent Institution: Touro Synagogue Foundation. Tax-exempt.
Institution Type/Description: Historic Building: built in 1763. A National Historic Site.
Collections: Jewish history & culture; religious artifacts; period furnishings; photographs; paintings.
Facilities: visitor center.
Activities: tours.
Hours & Admission Prices: See website for hours. Adults $12, seniors $10, students, military, National Trust & Park Service with ID $8; discounts to groups of 10 or more; children 13 & under no charge.
Attendance: 30,000 (estimated)
Membership: Individual $25; Family $50; Contributing $100; Sustaining $180; Founder $500.

North Kingstown

QUONSET AIR MUSEUM, 488 Eccleston Ave., North Kingstown, RI 02852-7406. Mailing Address: P.O. Box 1571, North Kingstown, RI 02852-0629. Tel.: 401-294-9540. Fax: 401-294-9887.
E-mail: support@quonsetairmuseum.com
Web Site: quonsetairmuseum.com
Founded: 1991.
Congressional District: 2
Key Personnel: Pres., David Stecker; Dir. & Museum Shop Mgr., David H. Payne.
Personnel Profile: Part-Time Volunteers 28; Interns 2.
Governing Authority: Tax-exempt.
Institution Type/Description: Air Museum.
Collections: Rhode Island's aviation heritage; military history.

Activities: educational programs.
Publications: Quonset Air Museum's Scout.
Hours & Admission Prices: June-Sept. daily 10-3; Oct.-May Sat.-Sun. 10-3; groups & other times by appointment. Adults $7, seniors 65 & over $6, children under 12 $3; discounts to groups; military no charge. Closed New Year's Day; Easter; Thanksgiving; Christmas. &
Attendance: 8,500 (estimated)
Membership: Individual $35; Family $50; Century Club $100; Corporate $500; Lifetime $1,000.

SEABEE MUSEUM AND MEMORIAL PARK, 21 Iafrate Way, North Kingstown, RI 02852-1792. Tel.: 401-294-7233. Fax: 401-294-9501.
E-mail: info@seabeemuseum.com
Web Site: www.seabeemuseum.com
Founded: 1995.
Congressional District: 2
Key Personnel: Pres. (V), Nicholas Fisch; Museum Shop Mgr., Robert Schwab
Governing Authority: Tax-exempt.
Institution Type/Description: Military Museum.
Collections: military artifacts & memorabilia relating to the Seabees.
Hours & Admission Prices: May-Oct. daily 9:30-2; Nov.-April Wed. & Sat.-Sun. 9:30-2; other times by appointment. No charge; donations accepted.
Attendance: 3,500 (estimated)
Membership: Regular $25; Silver $35; Gold $50; Life $100.

North Scituate

BORDERS FARM MUSEUM, 590 Danielson Pike, Apt. 4, North Scituate, RI 02825-1321. Tel.: 401-647-4984.
E-mail: farmer@bordersfarm.org
Web Site: www.bordersfarm.org
Key Personnel: Owner, Mary Thomas
Institution Type/Description: Farm Museum.
Collections: farm & agricultural history; farm tools & machinery. Historic Building: 1849 farmhouse.
Facilities: 200 acre farmland.
Activities: educational programs & activities.
Hours & Admission Prices: Daily 8-5.

Pawtucket

DAGGETT HOUSE, Slater Memorial Park, Next to Loof Carousel, Pawtucket, RI 02861. Mailing Address: 16 Second St., Pawtucket, RI 02861-3133. Tel.: 401-722-6931.
Founded: 1902.
Key Personnel: Chm. (V), Joslin A. Brooks.
Governing Authority: Parent Institution: DAR Pawtucket Chapter. Tax-exempt.
Institution Type/Description: Historic House: built 1685. Listed on the National Register of Historic Places.
Collections: period furnishings; family history & personal artifacts; needlework; Civil War; uniforms.
Activities: rental facilities.
Hours & Admission Prices: Call for hours & admission prices.
Attendance: 700 (estimated)

✳ **SLATER MILL, (M),** 67 Roosevelt Ave., Pawtucket, RI 02860-2127. Mailing Address: P.O. Box 696, Pawtucket, RI 02862-0696. Tel.: 401-725-8638. Fax: 401-722-3040.
E-mail: info@slatermill.org
Web Site: www.slatermill.org
Founded: 1921.
Congressional District: 1
Key Personnel: Exec. Dir., Susan A. Whitney; Pres. (V), James M. Sweeney; Cur., Andrian Paquette; Museum Shop Mgr., Matthew R. Davis.
Personnel Profile: Full-Time Paid 6; Part-Time Paid 21; Part-Time Volunteers 3.
Governing Authority: nonprofit organization. Parent Institution: Old Slater Mill Association. Tax-exempt.
Institution Type/Description: History Museum of Textile Industry & Arts; Historic site.
Collections: textile machinery & domestic textile implements; machine tools & shop equipment; labor & social history; Artisan's furniture & furnishings; mechanical power generation & transmission equipment Historic Buildings: 1793 Old Slater Mill; 1758 Sylvanus Brown House; 1810 Wilkinson Mill; 19th-century machine shop; reconstructed 1820s water wheel & raceway system.
Research Fields: industrial & labor history; technology; waterpower; textiles.

Facilities: 500 vol. library of history & textile books; photograph collection of local industrial sites & textile processes. Museum-related gifts for sale.
Activities: living history portrayals; flexible visiting hours; craft demonstrations; hands-on fibers program for students; fiber arts workshops; special events; membership program; vacation activities for children.
Publications: quarterly newsletter, The Millwright; annual research publications.
Hours & Admission Prices: March-April Sat.-Sun. 11-3; May-Oct. daily 10-4. Adults $12, senior citizens 65 & over $10, children 6-12 $8.50; discounts to groups with reservation, AAA, NEMA, AAM & ICOM members; members & children under 6 no charge. &
Attendance: 25,000 (estimated)
Membership: Individual $35; Family $50; Contributor $75; Keeper of the Wheel $100; Sponsor $200; Donor $500; Benefactor $1,000. Corporate Contributor $100; Corporate Sponsor $250; Corporate Donor $500; Corporate Patron $1,000.

Peace Dale

MUSEUM OF PRIMITIVE ART AND CULTURE, 1058 Kingstown Rd., Ste. 5, Peace Dale, RI 02879-2487. Tel.: 401-783-5711.
E-mail: mpaac@verizon.net
Web Site: www.primitiveartmuseum.org
Founded: 1892.
Congressional District: 2
Key Personnel: Pres. (V), Virginia Williams.
Personnel Profile: Part-Time Paid 2; Part-Time Volunteers 11.
Governing Authority: private. Tax-exempt.
Institution Type/Description: Anthropology Museum.
Collections: artifacts from New England, North America & other continents.
Research Fields: archaeology; ethnology.
Activities: public education program, including traveling exhibit & lectures at public schools; monthly membership programs; film series; thematic walking & driving tours of the area.
Publications: quarterly newsletter; catalogue, Rhode Island's Prehistoric Past; catalogue, The 19th Century American Collector: A Rhode Island Perspective.
Hours & Admission Prices: Wed. 10-2; other times by appointment. Suggested Donation $1; discounts to AAM, ICOM & NEMA members; members no charge. Closed holidays.
Attendance: 3,000 (estimated)
Membership: Student $5; Individual $20; Family $25; Corporate $35 & up; Contributing $35; Business $50 & up; Sustaining $60; Supporting $100.

Portsmouth

PORTSMOUTH HISTORICAL SOCIETY, 870 E. Main Rd. & Union St., Portsmouth, RI 02871. Mailing Address: P.O. Box 834, Portsmouth, RI 02871-0834. Tel.: 401-683-9178.
E-mail: portsmouthhistorical@yahoo.com
Web Site: www.portsmouthhistorical.com
Founded: 1938.
Congressional District: 1
Key Personnel: Pres., Herbert Hall; Cur. & Librarian, Marjorie P. Webster.
Personnel Profile: Part-Time Volunteers 12.
Governing Authority: society; nonprofit organization. Tax-exempt.
Institution Type/Description: Local History Museum: housed in 1865 former Portsmouth Christian Union Church. Listed on the National Historic Register.
Collections: farm and home tools; furniture; clothing; hymnals; old toys; horsedrawn vehicles. Historic Building: 1725 The Southernmost Schoolhouse.
Major Exhibits: Genealogy Reflected in Sherman Diaries, 5/26/14-10/13/14; Lost in Time, 5/26/14-10/13/14.
Facilities: 400-vol. library of history books available for use on premises.
Activities: lectures; permanent & temporary exhibits; Harvest Social.
Publications: member newsletters.
Hours & Admission Prices: Memorial Day-Columbus Day Sun. 2-4; other times by appointment. No charge; donations accepted.
Membership: Individual $10; Family $20.

Providence

ANNMARY BROWN MEMORIAL, 21 Brown St., Providence, RI 02912-9005. Mailing Address: Box A Brown Univ., Providence, RI 02912.
Web Site: dl.lib.brown.edu/libweb/about/amb
Founded: 1905.
Key Personnel: Head, Business & Facilities, Barbara Schulz; Dir. Special Collections, Dominique Coulombe.

Personnel Profile: Full-Time Paid 1.
Governing Authority: university. Parent Institution: Brown University Library. Tax-exempt: 103(6).
Institution Type/Description: Art Museum.
Collections: period master paintings; Brown family heirlooms & correspondence; British swords 1700-1900.
Research Fields: American Revolution & Civil War.
Activities: permanent, temporary & traveling exhibitions.
Publications: brochure, Annmary Brown Memorial.
Hours & Admission Prices: Labor Day to Memorial Day Mon.-Fri. 1-5. No charge. Closed national holidays.

BERT GALLERY, 540 S. Water St., Providence, RI 02903-4322. Tel.: 401-751-2628.
E-mail: info@bertgallery.com
Institution Type/Description: Art Gallery.
Collections: paintings; drawings; woodcuts; sculpture.
Facilities: 1,000 sq. ft. exhibit space; archives. Museum-related items for sale.
Activities: special events.
Hours & Admission Prices: Tues.-Fri. 11-5, Sat. 12-4; other times by appointment.

BETSEY WILLIAMS COTTAGE, Roger Williams Park, Providence, RI 02907. Tel.: 401-785-9457. Fax: 401-461-5146.
Founded: 1871.
Congressional District: 2
Key Personnel: Dir. Roger Williams Park Museum, Renee Gamba; Cur., Marilyn Massaro; Educator, Dawn Valentim.
Personnel Profile: Full-Time Paid 4.
Governing Authority: municipal. Parent Institution: City of Providence, RI. Subsidiary Institution: Parks Department. Tax-exempt.
Institution Type/Description: Historic House: c.1782 Betsey Williams Cottage.
Collections: furniture & furnishings of early 19th century.
Research Fields: history of Roger Williams Park; social history of Providence & Rhode Island; late 18th to early 19th-century rural life in Rhode Island.
Activities: permanent exhibitions; social history programs.
Hours & Admission Prices: House temporarily closed for restoration.

CHAZAN GALLERY @ WHEELER, 228 Angell St., Providence, RI 02906. Tel.: 401-421-9230. Fax: 401-751-7674. Facebook: Chazan Gallery.
E-mail: info@chazangallery.org
Web Site: www.chazangallery.org
Key Personnel: Dir., Liz Kilduff; Asst. Dir., Elena Lledo
Institution Type/Description: Art Gallery.
Collections: works by local contemporary artists.
Major Exhibits: Hannah Antalek, Harrison Bucy, Jenn Houle, Anne Rogers, Margaret Rogers, Jodi Stevens, 1/17/14-1/30/14; Jane Masters, Jacqueline Ott, 2/7/14-2/27/14; Sam Duket, Bradley Fesmire, Shawn Gilheeney, Buck Hastings, Neal Walsh, 3/7/14-3/27/14; Jeffrey Bertwell, Saberah Malik, Laurie Sverdlove, 9/14-10/14; Michelle Benoit, Susan Doyle, Joan Wyand, 10/14-11/14; Juan Jose Barboza-Gubo, 11/14-12/14.
Hours & Admission Prices: Tues.-Sat. 11-4, Sun. 2-4. No charge.
Attendance: 3,000 (estimated)

CULINARY ARTS MUSEUM, (M), 315 Harborside Blvd., Providence, RI 02905-5202. Tel.: 401-598-2805. Fax: 401-598-2807.
E-mail: museum@jwu.edu
Web Site: www.culinary.org
Formerly: Culinary Archives & Museum, Johnson & Wales University
Founded: 1989.
Congressional District: 2
Key Personnel: Cur., Richard J.S. Gutman; Mgr. Operations, Stephen Spencer; Mgr. Collections, Erin M. Williams; Events & Program Coord., Kristin Zosa Puleo; Museum Asst., Deborah Pinkham.
Personnel Profile: Full-Time Paid 5; Part-Time Paid 5; Part-Time Volunteers 2; Interns 4.
Governing Authority: private university. Parent Institution: Johnson & Wales University. Tax-exempt.
Institution Type/Description: Culinary & Gastronomy Museum.
Collections: 60,000 cookbooks; 20,000 menus; 80,000 culinary-related postcards; 10,000 culinary pamphlets; tens of thousands of culinary objects; fashion & food prints; silver.
Facilities: library available to public for research; educational facilities.
Activities: lectures & demonstrations; adult education programs; loan, temporary & traveling exhibitions.
Hours & Admission Prices: Tues.-Sun. 10-5; research by appointment only.

Adults $7, senior citizens $6, college students $4, children 5-18 $2; discounts to groups & AAA members; children under 5 no charge. Closed major holidays.
Attendance: 20,000 (accurate)

DAVID WINTON BELL GALLERY, (M), List Art Center, Brown University, 64 College St., Providence, RI 02912. Tel.: 401-863-2932 & 2929. Fax: 401-863-9323.
Web Site: www.brown.edu/bellgallery
Founded: 1971.
Congressional District: 1
Key Personnel: Dir., Jo-Ann Conklin; Cur., Maya Allison; Prep., Cameron Shaw; Admin., Terrance Abbott.
Personnel Profile: Full-Time Paid 1; Part-Time Paid 20; Interns 2.
Governing Authority: university. Parent Institution: Brown University. Tax-exempt: 501(c)(3).
Institution Type/Description: University Art Gallery.
Collections: 5,000 prints, drawings & photographs from 1500 to present; 200 paintings & sculptures, including works by Olitski, Stella, Motherwell, Bontecou & Caro.
Research Fields: 20th-century art, prints (Old Master to Contemporary).
Facilities: 250-seat auditorium; classrooms; gallery.
Activities: lectures; films; gallery talks; formally organized programs for undergraduate & graduate college students affiliated with Brown University; loan, temporary & traveling exhibitions.
Publications: exhibition catalogues; calendar of events.
Hours & Admission Prices: Sept.-July Mon.-Fri. 11-4, Sat.-Sun. 1-4. No charge; donations accepted. Closed New Year's Day; Memorial Day; Independence Day; Thanksgiving; Christmas.
Attendance: 10,000 (estimated)

EDWARD MITCHELL BANNISTER GALLERY, Rhode Island College - Roberts Hall, 124, 600 Mt. Pleasant Ave., Providence, RI 02908. Tel.: 401-456-9765.
E-mail: bannistergallery@ric.edu
Web Site: www.ric.edu/bannister/info.php
Key Personnel: Dir., James Montford
Institution Type/Description: Art Gallery.
Collections: works by local, regional, & contemporary artists.
Hours & Admission Prices: Tues.-Fri. 12-8. No charge. Closed holidays.

GOVERNOR HENRY LIPPITT HOUSE MUSEUM, 199 Hope St., Providence, RI 02906-2136. Mailing Address: 957 N. Main St., Providence, RI 02904-5715. Tel.: 401-453-0688. Fax: 401-453-8221.
E-mail: lippitthouse@preserveri.org
Web Site: www.preserveri.org
Founded: 1991.
Key Personnel: Chm. (V), Patrice Hagan; Exec. Dir., Valerie Talmage; Admin., David Wrenn.
Personnel Profile: Full-Time Paid 1; Part-Time Paid 3.
Governing Authority: private; nonprofit organization. Parent Institution: Preserve Rhode Island. Tax-exempt: 501(c)(3).
Institution Type/Description: Historic House: 1865 renaissance-revival style mansion; a National Historic Landmark.
Collections: period furnishings; decorative arts; personal family memorabilia; photographs; documents; woodcarvings; etched & stained glass windows; stenciling; faux marble & wood finishes.
Research Fields: architecture; decorative arts; antique furnishings; historic photographs; domestic services; fine art; 19th century Victorian culture.
Facilities: restrooms; museum & grounds available for rentals.
Activities: guided tours; concert series; holiday open house.
Publications: Progress in Preservation: Annual Report for Preserve Rhode Island and end of year; newsletter.
Hours & Admission Prices: May-Oct. Fri. 11-3; other times by appointment. Adults $10; discount to groups & AAM members; children 12 & under no charge. Closed all holidays.
Attendance: 1,350 (estimated)

HAFFENREFFER MUSEUM OF ANTHROPOLOGY, BROWN UNIVERSITY, (M), 21 Prospect St., Providence, RI 02912. Mailing Address: 300 Tower St., Bristol, RI 02809. Tel.: 401-253-8388. Fax: 401-253-1198.
E-mail: haffenreffermuseum@brown.edu
Web Site: www.brown.edu/Facilities/Haffenreffer/index.html
Founded: 1956.
Congressional District: 1
Key Personnel: Dir., William Simmons; Pres., Jeffrey Schreck; Deputy Dir. &

Chief Cur., Kevin P. Smith; Cur. Res., Douglas Anderson; Assoc. Cur., Thierry Gentis; Consultant Conservator, Alexandra Allardt; Conservation Asst., Storage Mgr. & Exhibit Preparator, Rip Gerry; Office Mgr., Carol Dutton; Post-Doctoral Fellow, Jennifer Stampe; Cur. Education, Geralyn Ducady.

Personnel Profile: Full-Time Paid 7; Part-Time Paid 2; Part-Time Volunteers 3; Interns 3.

Governing Authority: university. Parent Institution: Brown University, Providence, RI. Tax-exempt: 170(b)(1)(A).

Institution Type/Description: Anthropology Museum.

Collections: American Indian: North, Central & South America, Arctic; African, Pacific, Eastern European, Mid-Eastern, Asian collections; anthropology; archaeology; ethnology.

Research Fields: anthropology; archaeology; ethnology; ethnohistory.

Facilities: 4,000-vol. library of anthropology books available for research on premises.

Activities: lectures; films; gallery talks; formally organized education programs for children, adults, graduate & undergraduate students; permanent & temporary exhibitions; training programs in museology; Friends Association.

Publications: Friends' newsletters; The Cashinahua of Eastern Peru; Burr's Hill: A Seventeenth Century Wampanoag Burial Ground; Hau Kola: The Plains Indian Collection of the Haffenreffer Museum of Anthropology; Traditional Art of Africa; Female Costume of the Sarakatsani; What Cheer! Selections from A Key Into The Language of America; Out of the North: The Subarctic Collection of the Haffenreffer Museum of Anthropology; History on Birchbark: The Art of Tomah Joseph, Passamaquoddy; Passionate Hobby: Rudolf Frederick Haffenreffer & the King Philip Museum; Gifts of Pride & Love: Kiowa and Comanche Cradles; Kayak, Umiak, Canoe; Model Kayaks, Umiaks and Canoes from the North Pacific in Haffenreffer Museum of Anthropology Collections.

Hours & Admission Prices: Manning Hall Gallery: Tues.-Sun. 10-4. No charge; donations accepted. Closed federal holidays; university breaks. &

Attendance: 17,000 (estimated)

Membership: Student $15; Individual $25; Dual & Couple $30; Family $35; Contributing $50-$99; Saville Society $100-$249; Giddings Society $250-$499; Mount Hope Society $500-$999; Haffenreffer Society $1,000 & up.

HUNT-CAVANAGH GALLERY - PROVIDENCE COLLEGE, Hunt-Cavanagh Hall, 549 River Ave., Providence, RI 02918. Tel.: 401-865-2400.

E-mail: gwallace@providence.edu

Web Site: www.providence.edu/art/hunt+cavanagh

Institution Type/Description: Art Gallery.

Collections: photographs; paintings; sculpture; drawings.

Activities: temporary exhibitions.

Hours & Admission Prices: During Academic Year: Mon.-Fri. 9-4. No charge. Closed spring break.

JOHN HAY LIBRARY, Brown Univ., 20 Prospect St., Providence, RI 02912. Mailing Address: Box A, Providence, RI 02912-9039. Tel.: 401-863-3723.

Institution Type/Description: Library.

Collections: books; manuscripts; archives; poetry; plays; literature; art; science; history.

Facilities: library.

Activities: research.

Hours & Admission Prices: Academic Year: Mon.-Fri. 9-6, Sun. 1-5; Summer: call for hours.

* **JOHN NICHOLAS BROWN CENTER FOR PUBLIC HUMANITIES AND CULTURAL HERITAGE, (M),** 357 Benefit St., Providence, RI 02903-2929. Mailing Address: Box 1880, Brown University, Providence, RI 02912-1880. Tel.: 401-863-1177. Fax: 401-863-7777.

E-mail: publichumanities@brown.edu

Web Site: www.brown.edu/jnbc

Founded: 1983.

Congressional District: 1

Key Personnel: Dir., Steven Lubar; Asst. Dir. & Cur., Ron M. Potvin; Assoc. Dir. Programs, Anne Valk; Mgr., Jenna Legault.

Personnel Profile: Full-Time Paid 4.

Governing Authority: nonprofit organization. Parent Institution: Brown University. Tax-exempt: 501(c)(3).

Institution Type/Description: Brown University Public Humanities MA Program.

Collections: decorative arts & furniture of America & Europe; historic wallpaper.

Research Fields: public humanities; cultural heritage; oral history; public art; historic house museums.

Facilities: research center.

Activities: fellowship program; lectures; conferences; seminars; professional workshops; special events; permanent exhibitions.

Hours & Admission Prices: Call for information.

Attendance: 1,500 (estimated)

MEETING HOUSE OF THE FIRST BAPTIST CHURCH IN AMERICA, 75 N. Main St., Providence, RI 02903-1307. Tel.: 401-454-3418.

E-mail: fbc_inamerica@verizon.net

Web Site: www.fbcia.org

Founded: 1638.

Congressional District: 1

Key Personnel: Minister, Dr. Dan Ivins; Moderator, David Coon, Esq.

Personnel Profile: Full-Time Paid 3; Part-Time Paid 3.

Governing Authority: Tax-exempt.

Institution Type/Description: Historic Building: a National Historic Landmark.

Collections: local history; period furnishings; paintings.

Facilities: auditorium. Museum-related items for sale.

Activities: guided & self-guided tours; concerts, Music in the Meeting House.

Publications: FIRST: History of the First Baptist Church in America.

Hours & Admission Prices: Mon.-Fri. 10-12 & 1-3; groups by appointment. Guided Tours: Sun. after church services. $2 per person. Closed National holidays. &

Attendance: 9,000 (estimated)

* **MUSEUM OF ART, RHODE ISLAND SCHOOL OF DESIGN, (M),** 224 Benefit St., Providence, RI 02903-2723. Tel.: 401-709-8402. Fax: 401-454-6556. TDD: 800-745-5555.

E-mail: museum@risd.edu

Web Site: www.risd.edu/museum.cfm

Founded: 1877.

Congressional District: 1

Key Personnel: Dir., John W. Smith; Chm., William Tsiaras; Dir. Planning, Ann S. Woolsey; Financial Mgr., Glenn Stinson; Cur. Painting & Sculpture, Maureen O'Brien; Cur. Ancient Art, Gina Borromeo; Cur. Costume & Textiles, Kate Irvin; Cur. Prints, Drawings & Photographs, Jan Howard; Cur. Contemporary Art, Dominic Molon; Cur. Decorative Arts & Design, Elizabeth A. Williams; Coord. Special Events, Pam Kimel; Registrar, Deborah Diemente; Mgr. Safety & Security, Philip Lessard; Museum Shop Mgr., Charles Flora.

Personnel Profile: Full-Time Paid 60; Part-Time Paid 20; Part-Time Volunteers 60; Interns 8.

Operating Expenses: 8,600,000

Operating Income: 8,600,000

Governing Authority: college; nonprofit organization. Parent Institution: Rhode Island School of Design. Tax-exempt: 501(c)(3).

Institution Type/Description: Art Museum.

Collections: classical art; medieval art; 15th-20th century painting, sculpture & decorative arts; Albert Pilavin Collection of 20th-Century American Art; Nancy Sayles Day Collection of modern Latin American Art; Pendleton House Collection of American furniture, silver, china & decorative arts of the 18th-19th centuries; costume collection; Oriental textiles & art; Lucy Truman Aldrich Collection of 18th-century European porcelain; Abby Aldrich Rockefeller Collection of Japanese prints; contemporary art; Aaron Siskind Center for the Study of Photography.

Major Exhibits: Making It America, 10/11/13-2/9/14; Andy Warhol's Screen Tests, 11/15/13-5/11/14; Andy Warhol's Photographs, 1/3/14-6/29/14; Arlene Shechet: Meissen Recast, 1/17/14-6/8/14; Graphic Design: Now in Production (T), 3/28/14-8/3/14.

Facilities: educational facilities. Museum-related items for sale.

Activities: formal education programs for adults, children, undergraduate, & graduate college students; concerts; poetry readings; docent program; videos; guided tours; lectures; loan, temporary & traveling exhibitions; visits to senior centers & children's hospital.

Publications: monthly calendar of events; catalogs; semiannual journal, Manual: A journal about art and its making.

Hours & Admission Prices: Tues.-Wed., Fri.-Sun. 10-5, Thurs. 10-9. Adults $12, senior citizens 62 & over $10, college students $5, children 5-18 $3; discounts to AAA, AAM & ICOM members; children under 5 & members no charge. Closed New Year's Day; Independence Day; Thanksgiving; Christmas. &

Attendance: 100,000 (accurate)

Membership: Individual $50; Family $75; Friend $150; Contributing Friend $250; Supporting Friend $500; Radeke Patron $1,500; Radeke Leader $2,500; Radeke Benefactor $5,000.

MUSEUM OF NATURAL HISTORY AND PLANETARIUM,
(M), Roger Williams Park, Providence, RI 02907. Tel.: 401-785-9457. Fax: 401-461-5146. TDD: 401-751-0203.
E-mail: info@musnathist.com
Web Site: providenceri.com/museum
Formerly: Museum of Natural History, Roger Williams Park
Founded: 1896.
Congressional District: 2
Key Personnel: Dir., Renee Gamba; Cur., Marilyn Massaro; Education Cur., Lily Benedict; Asst. Cur., Michael Kieron; Educator, Dawn Temple; Office Mgr., Yamilet Reyes.
Personnel Profile: Full-Time Paid 5; Part-Time Paid 3; Part-Time Volunteers 4.
Governing Authority: municipal. Parent Institution: Providence Parks & Recreation Dept. Subsidiary Institution: Betsey Williams Cottage. Tax-exempt: 501(c)(3).
Institution Type/Description: Natural History Museum.
Collections: regional natural artifacts; Native American & Pacific island anthropology & archaeology.
Research Fields: anthropology; ethnology; natural sciences & education; Native American; Pacific Islands; Africa.
Facilities: 125-seat auditorium; planetarium; classroom.
Activities: planetarium shows; educational programs; semi-permanent exhibitions; film & video; lectures; performances; public forums.
Publications: exhibition catalogs; educational materials; member newsletters.
Hours & Admission Prices: Museum: daily 10-5. General admission $2; discounts to AAM & AAA members; children under 4 no charge. Planetarium Shows: July-Aug. daily 2 pm; Sept.-June Sat.-Sun. 2 pm; children under 4 not permitted. Admission $3; discounts to AAM & AAA members. ♿
Attendance: 100,000 (estimated)
Membership: Student $20; Individual $25; Senior Citizens $40; Family $50.

PROVIDENCE ATHENAEUM, 251 Benefit St., Providence, RI 02903-2799. Tel.: 401-421-6970. Fax: 401-421-2860.
E-mail: info@providenceathenaeum.org
Web Site: www.providenceathenaeum.org
Founded: 1836.
Key Personnel: Exec. Dir., Alison Maxell.
Personnel Profile: Full-Time Paid 7; Part-Time Paid 17.
Institution Type/Description: Historic Building: built in 1838.
Collections: books; works by local artists.
Facilities: library.
Activities: cultural programs; lectures; readings; theatrical presentations; musical performances.
Hours & Admission Prices: June-Aug. Mon.-Thurs. 9-7, Fri. 9-5, Sat. 9-1; Sept.-May Mon.-Thurs. 9-7, Fri.-Sat. 9-5, Sun. 1-5. No charge; donations accepted. Closed New Year's Day; Martin Luther King Jr. Day; Presidents' Day; Easter; Memorial Day; Independence Day; Victory Day; Labor Day; Columbus Day; Veteran's Day; Thanksgiving; Christmas Eve & Day.
Membership: Introductory $65; Individual $165; Household $200.

PROVIDENCE CHILDREN'S MUSEUM, 100 South St., Providence, RI 02903-4749. Tel.: 401-273-5437 & 5434, ext. 234. Fax: 401-273-1004.
E-mail: provcm@childrenmuseum.org
Web Site: www.childrenmuseum.org
Founded: 1976.
Congressional District: 1
Key Personnel: Exec. Dir., Janice O'Donnell; Business Mgr., Marvin Ronning; Dir. Devel., Jennifer Laurelli; Early Childhood, Mary Scott Hackman; Visitor Svcs. & Volunteer Svcs., Kelly Fenton; Dir. Education, Cathy Saunders; Program & Exhibit Developer, Carly Loeper; Special Events, Kristen Haffenreffer Moran; Museum Shop Mgr., Sua Xiong; Public Rels., Megan Fischer.
Personnel Profile: Full-Time Paid 17; Full-Time Volunteers 9; Part-Time Paid 25; Part-Time Volunteers 100; Interns 15.
Governing Authority: nonprofit organization. Tax-exempt: 501(c)(3).
Institution Type/Description: Children's Museum.
Collections: state & local history; mirrors & lens; water; childhood development; c.1930 marionettes; art for and by children; human skeletal system; bridges & roadways; geometry & building.
Research Fields: child development; subject areas pertinent to exhibits.
Facilities: conference room; classroom; birthday party room; children's garden. Educational items for sale.
Activities: docent & intern program; formally organized educational programs for children; family preservation program in conjunction with social service agencies; scout program; AmeriCorps National Service program; professional development workshops for teachers & childcare providers.

Publications: bimonthly newsletter, Dragon's Tale; annual report; Play With Your Kids! A How To, Why To Guide for Parents.
Hours & Admission Prices: April to Labor Day daily 9-6; Sept.-March Tues.-Sun. 9-6; call for additional hours. Adults $9; discounts to groups and AAM, ICOM & NEMA members; first Sun. of the month, members, ACM members, children under one, preschool & elementary school teachers no charge. Closed Thanksgiving; Christmas Eve & Day. ♿
Attendance: 144,684 (accurate)
Membership: Couple $75 (add $10 for each additional family member); Reciprocal for 4 people (add $10 for each additional family member) $125; Library for 4 people $250 half pass, $350 full pass; Corporate for 4 people $1,000.

PROVIDENCE CITY HALL, 25 Dorrance St., Providence, RI 02903-1738. Tel.: 401-421-7740.
Institution Type/Description: Historic Building: built in 1878. Listed on the National Register of Historic Places.
Collections: city & state history and government; photographs; paintings; personal artifacts.
Activities: tours.
Hours & Admission Prices: Tours: Summer: Mon.-Fri. 8:30-4; Winter: Mon.-Fri. 8:30-4:30 by appointment.

THE PROVIDENCE JEWELRY MUSEUM, 4 Edward St., Providence, RI 02904-2404. Tel.: 401-274-0999; 781-3100.
E-mail: info@providencejewelrymuseum.com
Web Site: www.providencejewelrymuseum.com
Institution Type/Description: Jewelry Museum.
Collections: fine & fashion jewelry from the 18th century to the present including match safes, pen & fruit knives, card cases, dresser items.
Hours & Admission Prices: By appointment.

PROVIDENCE PRESERVATION SOCIETY, 21 Meeting St., Providence, RI 02903-1214. Tel.: 401-831-7440. Fax: 401-831-8583.
E-mail: info@ppsri.org
Web Site: www.ppsri.org
Founded: 1956.
Congressional District: 1
Key Personnel: Exec. Dir., James Hall; Coord. Devel., Lauren Goldenberg; Coord. Advocacy & Education, Paul Wackrow; Office Admin., Angela Kondon.
Governing Authority: society. Tax-exempt: 501(c)(3).
Institution Type/Description: Preservation Society: housed in 1772 Shakespeare's Head House & Garden.
Collections: structures.
Research Fields: preservation projects; Providence history, architecture, & planning issues.
Facilities: Brick school house; meeting & lecture room.
Activities: guided tours; lectures; field trips; films; docent training program; consultant bureau services; workshops on use of historic buildings. Museum Sponsors: Festival of Historic Houses in June.
Publications: quarterly newsletter; PPS News; broadsides, A Mile of History: Benefit Street; Providence in 1776; Elmwood, A Victorian Neighborhood; Brown University: A Microcosm of American Architecture; Broadway, A Victorian Boulevard; Downtown Providence: Commerce in Architecture; The Armory District: Harmonious Streetscapes; brochures; The Providence Waterfront; Stained Glass in Providence; PPS/AIA RI Guide to Providence Architecture.
Hours & Admission Prices: Office: Mon.-Fri. 9-5. Closed all major holidays. ♿
Membership: Individual $50; Household $100; Benefactor $250; Leadership $500.

PROVIDENCE PUBLIC LIBRARY, 150 Empire St., Providence, RI 02903-3219. Tel.: 401-455-8000 & 8021.
Key Personnel: Special Collections Librarian, Richard J. Ring
Institution Type/Description: Library.
Collections: books; manuscripts; pamphlets; newspapers; maps; art; period artifacts.
Activities: special events.
Publications: Occasional Nuggets.
Hours & Admission Prices: Mon. 1-8, Tues. 9:30-5, Thurs.-Fri. 1-5. No charge. ♿

RHODE ISLAND BLACK HERITAGE SOCIETY, 101 Dyer St., Providence, RI 02903. Tel.: 401-421-0606. Facebook: Rhode Island Black Heritage Society.
E-mail: riblackheritagesociety@gmail.com
Web Site: ribhs.org
Founded: 1975.
Congressional District: 1
Key Personnel: Pres., Charles C. Newton; Dir., Joaquina Teixeira; Office Mgr., Lee Campinha-Schenck.
Personnel Profile: Full-Time Paid 1.
Governing Authority: nonprofit organization. Tax-exempt: 501(c)(3).
Institution Type/Description: Black Heritage Society Museum.
Collections: Rhode Island Historical documents & art; photographs; Afro-American historical collection; manuscripts.
Research Fields: Afro-Americans in Rhode Island.
Facilities: 2,500-vol. library of Afro-American history available for research by request to institutions; reading room.
Activities: guided tours; lectures; films; concerts; TV programs; formally organized education programs for children; permanent, temporary & traveling exhibitions.
Publications: quarterly newsletter, Rhode Island Black Heritage Society Bulletin; annual, A Selected Scholarly Document.
Hours & Admission Prices: Mon.-Fri. 10:30-3:30; other times by appointment. Special collections research by appointment only; fee charged for searches & archival assistance. No charge; donations accepted. Closed holidays. &
Membership: Student & Senior Citizen $15; Individual $25; Family $50.

RHODE ISLAND HISTORICAL PRESERVATION AND HERITAGE COMMISSION, Old State House, 150 Benefit St., Providence, RI 02903-1209. Tel.: 401-222-2678. Fax: 401-222-2968.
E-mail: info@preservation.ri.gov
Web Site: www.preservation.ri.gov/
Formerly: Rhode Island Historical Preservation Commission
Key Personnel: Exec. Dir., Edward F. Sanderson.
Governing Authority: state. Branch Museums: 1901 State House, Providence, RI; 1707-1740 Governor Steven Hopkins House, Providence; Fort Adams State Park, Newport, RI; 1739 Colony House, Newport, RI.
Institution Type/Description: Historical Preservation Commission: housed in 1762 Old State House.
Collections: portraits; furnishings & furniture from the 19th century and primarily from the early 20th century.
Research Fields: historical development; architectural history; archaeology.
Publications: series of historic site & survey reports.
Hours & Admission Prices: Office: Mon.-Fri. 8:30-4:30. No charge.

* **RHODE ISLAND HISTORICAL SOCIETY, (M),** 110 Benevolent St., Providence, RI 02906-3103. Tel.: 401-331-8575. Fax: 401-351-0127.
Web Site: www.rihs.org
Founded: 1822.
Congressional District: 1
Key Personnel: Dir., C. Morgan Grefe, Ph.D.; Pres., Barry G. Hittner; Vice Pres., William Simmons, Ph.D.; Controller, Charmyne Goodfellow; Dir. Collections, Kirsten Hammerstrom; Reference Librarian, Lee Teverow; Editor, Elizabeth Stevens, Ph.D.; Dir. Devel., Kathy Clarendon; Dir. Education, Elyssa Tardif; Museum Shop Mgr., Patty Reynolds; Museum Shop Mgr., Darline Amaral.
Personnel Profile: Full-Time Paid 14; Part-Time Paid 21; Part-Time Volunteers 179; Interns 5.
Governing Authority: society. Branch Museums: 1786 John Brown House; 1822 Aldrich House; 1874 library; Museum of Work and Culture, Woonsocket. Tax-exempt: 509(a)(1).
Institution Type/Description: Historical Society Museums.
Collections: Rhode Island furniture; works of local artists; historical objects; Chinese porcelain & furniture; dolls & glassware; textiles & clothing; wallpaper; military; Indian artifacts; musical instruments; folk art; silver, domestic & ethnic items; archives; photographs; manuscripts; film & video collections; genealogy. Historic Building: John Brown House, national landmark built in 1786; Aldrich House, national landmark built c.1820.
Research Fields: Rhode Island history; genealogy.
Facilities: 100,000-vol. library of books, 400,000 images, 7,000 manuscripts; archives; reading room.
Activities: guided tours; lectures; formally organized educational programs; docent program; permanent & temporary exhibitions; films; gallery talks; history exhibits.
Publications: peer-reviewed journal, Rhode Island History; newsletter; The Papers of General Nathanael Greene (13 vols.); The John Brown House Museum: A Passion for the Past (2010).
Hours & Admission Prices: Library: Wed.-Fri. 10-5, 2nd Sat. of the month

10-5. Adults $5. John Brown House: Tues.-Sat. 9:30-4:30. Family max. $15, adults $7, students & senior citizens $5.50, children 7-17 $4; members no charge. Museum of Work & Culture: Tues.-Fri. 9:30-4, Sat. 10-5, Sun. 1-5. Adults $6, senior citizens & students $4; discounts groups of 20 or more; children under 10 with adult & RIHS members no charge. Closed New Year's Day; Easter; Independence Day; Thanksgiving; Christmas. &
Attendance: 30,000 (accurate)
Membership: Student & Senior $30; Basic $40; Gaspee Group $50-$99; May 4th Circle $100-$249; Friends of Roger Williams $250 & up; Life $2,000.

RHODE ISLAND STATE ARCHIVES, 337 Westminster St., Providence, RI 02903-3302. Tel.: 401-222-2353. Fax: 401-222-3199.
E-mail: statearchives@sos.ri.gov
Web Site: sos.ri.gov/archives
Founded: 1647.
Congressional District: 1
Key Personnel: State Archivist, R. Gwenn Stearn.
Governing Authority: state. Affiliated with Secretary of State's Office. Tax-exempt.
Institution Type/Description: State Archives.
Collections: archives; history; legislature; military; manuscripts.
Research Fields: genealogy; military; legislative; state government.
Facilities: library of manuscripts available in office only; reading room.
Hours & Admission Prices: Mon.-Fri. 8:30-4:30. No charge. &

ROGER WILLIAMS NATIONAL MEMORIAL, 282 N. Main St., Providence, RI 02903-1240. Tel.: 401-521-7266. Fax: 401-521-7239.
Web Site: www.nps.gov/rowi
Founded: 1965.
Congressional District: 1
Key Personnel: Site Mgr., Jennifer Gonsalves; Supt., Jan Reitsma.
Personnel Profile: Full-Time Paid 5; Part-Time Paid 1; Part-Time Volunteers 3.
Governing Authority: federal. Parent Institution: National Park Service Affiliated with the U.S. Dept. of the Interior. Tax-exempt.
Institution Type/Description: Historic Site.
Collections: exhibits on Roger Williams, the founder of Rhode Island & one of the earliest proponents of religious freedom and democracy in North America; history of Rhode Island.
Research Fields: Roger Williams biography; colonial history.
Facilities: visitor center; 4 acre park.
Activities: talks; slide show; hands-on presentations.
Publications: educational programs brochure; brochure, Roger Williams.
Hours & Admission Prices: Summer: daily 9-5; Columbus Day to May daily 9-4:30. No charge; donations accepted. Closed New Year's Day; Thanksgiving; Christmas. &
Attendance: 121,960 (estimated)

ROGER WILLIAMS PARK ZOO, 1000 Elmwood Ave., Providence, RI 02907-3659. Tel.: 401-785-3510 & 941-3910. Fax: 401-941-3988.
E-mail: info@rwpzoo.org
Web Site: www.rwpzoo.org
Founded: 1872.
Congressional District: 2
Key Personnel: Exec. Dir. Zoo & RI Zoological Society, Jeremy Goodman, D.V.M.; Chm. (V), Maribeth Williamson; Cur. Education, Shareen Knowlton; Veterinarian, Dr. Michael McBride; Deputy Dir. Animal Care, Tim French.
Personnel Profile: Full-Time Paid 81; Part-Time Paid 54; Part-Time Volunteers 256; Interns 10.
Governing Authority: owned by the city of Providence, RI. Managed and supported by the nonprofit Rhode Island Zoological Society.
Institution Type/Description: Zoo.
Collections: over 100 species from around the world displayed in natural habitats including, The Fabric of Africa; Tropical America; Australasia; North America.
Research Fields: animal behavior; reproductive physiology; invasive plant eradication via beneficial native insects; field conservation.
Facilities: 2,400-vol. library of general animal information; nature conservation center; nature trails; concessions.
Activities: guided tours; informal learning education programs for children; docent program; traveling exhibitions; live animal demonstrations; overnight zoo tours for children; annual special events; conservation lectures; field research.
Publications: membership newsletter, Wild; docent e-newsletter, The Screech

Owl; education department newsletter, Wild Adventures; e-newsletter, ZooNews; membership e-newsletter, Peck-A-Zoo; membership publication, WILD magazine.
Hours & Admission Prices: Daily 9-4. Adults $14.95, senior citizens 62 & over $12.95, children 3-12 $9.95; children 2 & under no charge. Closed Thanksgiving; Christmas Eve & Day. &
Attendance: 615,664 (accurate)
Membership: Individual $49; Family & Grandparent $79; Family Plus $89; Zookeeper $125; Zoo Guardian $250 & up.

STEPHEN HOPKINS HOUSE, 15 Hopkins St., Providence, RI 02903. Tel.: 401-421-0694.
E-mail: shh1707@gmail.com
Web Site: www.stephenhopkins.org
Founded: 1707.
Key Personnel: Chair, Kim N. Clark; Past Chair, Mimi Freeman; Pres., Nancy Bredbeck.
Personnel Profile: Full-Time Volunteers 5.
Governing Authority: National Society of Colonial Dames of America in the State of Rhode Island. Tax-exempt.
Institution Type/Description: Historic Museum House: 1707 Governor Stephen Hopkins home.
Collections: furnishings; Parterre Garden; Stephen Hopkins furnishings.
Facilities: Postcards & brochures for sale.
Activities: guided tours.
Hours & Admission Prices: May-Nov. Sat. 10-4; other times by appointment. No charge; donations accepted.
Attendance: 1,200 (accurate)

Rose Island

ROSE ISLAND LIGHTHOUSE, Rose Island, RI 02840. Mailing Address: Rose Island Lighthouse Foundation, P.O. Box 1419, Newport, RI 02840-0014. Tel.: 401-847-4242. Fax: 401-847-7262.
E-mail: david@roseisland.org
Web Site: www.roseislandlighthouse.org
Key Personnel: Exec. Dir., David McCurdy.
Governing Authority: nonprofit organization. Tax-exempt: 501(c)(3).
Institution Type/Description: Lighthouse Museum: listed on the National Register of Historic Places.
Collections: lighthouse history.
Activities: group tours; scout campovers; meetings. Museum Sponsors: school & youth group programs April to October; seal tours November to April.
Hours & Admission Prices: July to Labor Day daily 10-4; other times by appointment. Island accessible by ferry. Museum: adults $5, seniors 65 & over and children 6-12 $4.

Saunderstown

* **CASEY FARM, (M),** Route 1A, 2325 Boston Neck Rd., Saunderstown, RI 02874. Mailing Address: 141 Cambridge St., Boston, MA 02114-2702. Tel.: 401-295-1030; 617-227-3956.
Web Site: www.historicnewengland.org
Founded: 1955.
Congressional District: 2
Key Personnel: Pres. & C.E.O., Carl Nold; Interim Farm Mgr., Ashley Coerdt.
Governing Authority: society; nonprofit organization. Parent Institution: Historic New England, 141 Cambridge St. Boston, MA 02114. Tel.: 617-227-3956. Tax-exempt: 501(c)(3).
Institution Type/Description: Historic Homestead: c.1750, homestead still functioning as a working farm.
Collections: barns; out-buildings.
Activities: guided tours; school programs; special events & tours; CSA program. Museum Sponsors: Farmers Market.
Publications: magazine, Historic New England.
Hours & Admission Prices: June-Oct. 15 Tues. & Thurs. 1-5, Sat. 9-2. Adult $4; discount to seniors, AAM, ICOM, AAA & WGBH members; Historic New England members no charge. Call for further information.
Attendance: 30,093 (accurate)
Membership: National $35; Individual $45; Household $55; Garden & Landscape $75; Contributing and Library & School $100; Young Friends of Historic New England $100-$1,500; Friends of the Library and Archives $125; Historic Homeowner $200; Business & Ogden Codman Design Group $250; Appleton Circle $2,500.

THE GILBERT STUART BIRTHPLACE AND MUSEUM, 815 Gilbert Stuart Rd., Saunderstown, RI 02874-2911. Tel.: 401-294-3001. Fax: 401-294-3869.
E-mail: info@gilbertstuartmuseum.org
Web Site: gilbertstuartmuseum.org
Founded: 1930.
Congressional District: 2
Key Personnel: Dir. & Museum Shop Mgr., Margaret O'Connor.
Personnel Profile: Full-Time Paid 2; Part-Time Paid 1; Part-Time Volunteers 65.
Governing Authority: nonprofit organization. Tax-exempt.
Institution Type/Description: Historic House Museum: 1755 birthplace of artist Gilbert Stuart.
Collections: period furnishings. Historic Buildings: 1750 Snuff Mill; 1662 Grist Mill.
Facilities: house museums.
Activities: guided tours; formally organized education programs for children; permanent exhibitions. Museum Sponsors: Herring Run in April.
Publications: Gilbert Stuart Museum Pre and Post Visits Lesson Plans for Teachers.
Hours & Admission Prices: early May to Sept. Mon. & Thurs.-Sat. 11-4, Sun. 12-4. Oct. call for hours. Adults $7, children 6-12 $4; members & children under 6 no charge.
Attendance: 5,400 (estimated)
Membership: Single $30; Family $40.

Smithfield

AUDUBON SOCIETY OF RHODE ISLAND, 12 Sanderson Rd., Smithfield, RI 02917-2600. Tel.: 401-949-5454. Fax: 401-949-5788.
E-mail: audubon@asri.org
Web Site: www.asri.org
Founded: 1897.
Congressional District: 2
Key Personnel: Exec. Dir., Lawrence Taft; Bd. Pres. (V), Candace E. Powell; Financial Dir., Susan Mansolillo; Cur., Eugenia Marks; Sr. Dir. Educational Programs, Kristen Swanberg; Publications Asst., Hope Foley; Museum Shop Mgr., Jan Weyant.
Personnel Profile: Full-Time Paid 22; Part-Time Paid 15; Part-Time Volunteers 150.
Governing Authority: society; nonprofit organization. Subsidiary Institution: Environmental Education Center, Bristol, RI. Branch Museums: Caratunk Wildlife Refuge, Seekonk, MA; Parker Woodland, Coventry, RI; Eppley Wildlife Sanctuary, Exeter, RI; Kimball Wildlife Refuge, Charlestown, RI; Environmental Education Center, Bristol, RI; Audubon Society, RI. Tax-exempt: 501(c)(3).
Institution Type/Description: Nature Conservation Center.
Collections: area natural history; 19th-century art; 19th-century natural history books; Rhode Island bird skins; Rhode Island shells; miscellaneous natural materials & objects.
Research Fields: ecology.
Facilities: 70-seat auditorium; nature & conservation center; educational facilities. Natural history books, optical equipment & other museum-related items for sale.
Activities: guided tours; lectures; films; study clubs; organized education programs; school loan service. Annual Event: Bird & Wildlife Carving in October.
Publications: quarterly newsletter, Audubon Society of Rhode Island Report.
Hours & Admission Prices: Mon.-Fri. 9-5. No charge; donations accepted for use of trails. Closed New Year's Day; Thanksgiving; Christmas. &
Attendance: 25,000 (estimated)
Membership: Student & Senior $35; Basic $45; Steward $75; Sustaining $100; Associate $250; Benefactor $500.

SMITH-APPLEBY HOUSE, 220 Stillwater Rd., Smithfield, RI 02917-1849. Tel.: 401-231-7363.
E-mail: contact@smithapplebyhouse.org
Web Site: www.smithapplebyhouse.org
Institution Type/Description: Historic House Museum: housed in an 18th century farm house.
Collections: period furnishings; personal artifacts.
Activities: seasonal events.
Hours & Admission Prices: By appointment.

South Kingstown

RHODE ISLAND RAILROAD MUSEUM AT KINGSTON STATION, 1 Railroad Ave., South Kingstown, RI 02982. Mailing Address: P.O. Box 191, West Kingston, RI 02892-0191. Tel.: 401-789-3327.
E-mail: birdman@uri.edu
Founded: 1997.
Congressional District: 2

Governing Authority: Parent Institution: Friends of the Kingston Station. Tax-exempt.
Institution Type/Description: Railroad Museum.
Collections: railroad history & artifacts; Narragansett Railroad; photographs.
Hours & Admission Prices: Sun. 2-5; other times by appointment. No charge; donations accepted. ♿
Attendance: 6,000 (estimated)
Membership: Annual $15.

Tiverton

BUTTERFLY ZOO, 409 Bulgarmarsh Rd., Tiverton, RI 02878. Mailing Address: 1151 Aquidneck Ave., Middletown, RI 02842. Tel.: 401-849-9519.
Key Personnel: Owner, Marc Schenck
Institution Type/Description: Butterfly Zoo.
Collections: over 30 species of butterflies.
Facilities: Zoo-related items for sale.
Activities: guided & self-guided tours.
Hours & Admission Prices: Memorial Day to Labor Day Mon.-Sat. 11-4, Sun. 12-4. Adults $6, children $4; children under 3 no charge.

CHASE-CORY HOUSE, 3908 Main Rd., Tiverton, RI 02878-4809. Mailing Address: P.O. Box 98, Adamsville, RI 02801. Tel.: 401-624-2096.
E-mail: info@tivertonfourcorners.com
Web Site: www.tivertonfourcorners.com
Institution Type/Description: Historic Houses.
Collections: historic buildings; period furnishings.
Activities: special events.
Hours & Admission Prices: By appointment. No charge.

EMILIE RUECKER WILDLIFE REFUGE, Seapowet Ave., Tiverton, RI 02878. Tel.: 401-949-5454.
Institution Type/Description: Wildlife Refuge.
Collections: birds including Great Egrets, Snowy Egrets, & Glossy Ibis.
Facilities: 50 acre site; nature trails.
Activities: hiking; bird watching.
Hours & Admission Prices: Call for hours.

FORT BARTON, Lawton & Highland Aves., Tiverton, RI 02878-4401. Mailing Address: Town Hall, 343 Highland Rd., Tiverton, RI 02878-4499. Tel.: 401-625-6700 & 624-2549.
Key Personnel: Co Chm., Garry Plunkett; Co Chm., Ginger Lacy.
Governing Authority: Parent Institution: Tiverton Open Space and Land Preservation Commission.
Institution Type/Description: History Museum: site was the staging area for the invasion of Aquidneck Island & named after Lt. Col. William Barton. Listed on the National Register of Historic Places.
Collections: Fort Barton history.
Facilities: observation tower.
Activities: nature walks.
Hours & Admission Prices: Call for hours.

Warren

CHARLES W. GREENE MUSEUM, George Hail Free Library, 530 Main St., Warren, RI 02885-4368. Tel.: 401-245-7686. Facebook: George Hail Free Library.
E-mail: epatricrn@yahoo.com
Web Site: www.georgehail.org/cwgMuseum.htm
Founded: 1871.
Congressional District: 1
Key Personnel: Dir., E. Patricia Redfearn; Chm. (V), John Chaney.
Personnel Profile: Full-Time Paid 1; Part-Time Volunteers 1.
Governing Authority: Parent Institution: George Hail Free Library. Tax-exempt.
Institution Type/Description: History Museum.
Collections: Native American artifacts; domestic artifacts; military artifacts; Warren-related artifacts; local authors.
Hours & Admission Prices: Museum: Wed. 2-4 or by appointment. Library: Mon.-Thurs. 10-8, Fri-Sat. 10-5; July-Aug. Mon.-Thurs. 10-8, Fri. 10-5, Sat. 10-3. No charge; donations accepted.
Attendance: 125 (estimated)

FIREMEN'S MUSEUM, 38 Baker St., Warren, RI 02885-3107. Mailing Address: 1 Joyce St., Warren, RI 02885-3238. Tel.: 401-245-7600.
Key Personnel: Dir., Chief Galinelli
Institution Type/Description: Firemen's Museum: housed in the former Narragansett Steam Fire Company Station Number 3.
Collections: fire department history; photographs; equipment; 1802 fire truck.
Hours & Admission Prices: By appointment only. No charge.

TOUISSET MARSH WILDLIFE REFUGE, 107 Touisset Rd., Warren, RI 02885. Mailing Address: 12 Sanderson Rd., Smithfield, RI 02917-2606. Tel.: 508-761-8230.
Web Site: www.asri.org/refuges/touisset-marsh-wildlife.html
Institution Type/Description: Wildlife Refuge.
Collections: wildlife & their habitats; plants; ecology.
Facilities: nature trails.
Activities: hiking; bird watching.
Hours & Admission Prices: Daily dawn-dusk. No charge; donations accepted.

Warwick

CLOUDS HILL VICTORIAN HOUSE MUSEUM, 4157 Post Rd., Warwick, RI 02886. Mailing Address: P.O. Box 522, East Greenwich, RI 02818-0522. Tel.: 401-884-9490.
E-mail: office@cloudshill.org
Web Site: www.cloudshill.org
Key Personnel: Dir., Wayne Cabral
Institution Type/Description: Historic House Museum: built in 1872 by William Smith Slater for his daughter, Elizabeth Ives Slater.
Collections: period furnishings; personal artifacts; photographs.
Activities: special events.
Hours & Admission Prices: By appointment. Adults $15, seniors $10; discounts to groups.

GREENWOOD VOLUNTEER FIRE COMPANY AND MUSEUM, 45 Kernick St., Warwick, RI 02886. Tel.: 401-667-0447.
E-mail: gvfd555@verizon.net
Web Site: greenwoodfirecompany.webs.com
Founded: 1980.
Key Personnel: Dir. & Fire Chief, Kenneth Smith; Pres., Howard Fleming.
Governing Authority: private; nonprofit organization. Tax-exempt: 501(c)(3).
Institution Type/Description: Historic Building: housed in a former fire station; built in 1924.
Collections: fire company history; fire apparatus; photographs; 1940 Mack pumper; 1954 Maxim pumper; 1825 Hunneman Tuscatucket.
Facilities: library of fire-related books; 2,000 sq. ft. exhibit space. Museum-related items for sale.
Activities: guided tours; participatory & traveling exhibitions. Annual Events: Steak Fry in May & October.
Hours & Admission Prices: By appointment. No charge; donations accepted.
Attendance: 400 (estimated)
Membership: Associate & Life $20.

JOHN WATERMAN ARNOLD HOUSE, 25 Roger Williams Circle, Warwick, RI 02888. Tel.: 401-467-7647. Facebook: John Waterman Arnold House.
E-mail: fag311@cox.net
Web Site: www.warwickhistoricalsocietyonline.org
Founded: 1932.
Key Personnel: Dir., Felicia Gardella.
Volunteer Hours: 600
Governing Authority: Parent Institution: Warwick Historical Society. Tax-exempt.
Institution Type/Description: Historic House Museum: built in 1786. Listed on the National Register of Historic Places.
Collections: Waterman family & Warwick history; period furnishings; personal artifacts; photographs; genealogy; early maps; Warwick historical documents.
Publications: Warwick history pamphlets.
Hours & Admission Prices: Wed. 11-3, 2nd Sat. each month 11-2. No charge; donations accepted.
Attendance: 600 (estimated)
Membership: Student $5; Individual $10; Family $15; Organizations $50.

WARWICK MUSEUM OF ART, Kentish Artillery Armory, 3259 Post Rd., Warwick, RI 02886-7145. Tel.: 401-737-0010. Fax: 401-737-1796.
E-mail: info@warwickmuseum.org

Web Site: www.warwickmuseum.org
Founded: 1976.
Congressional District: 2
Key Personnel: Pres. Bd., Deborah Mercer; Program Dir., Simone Spruce-Torres.
Personnel Profile: Part-Time Paid 1; Part-Time Volunteers 7; Interns 1.
Governing Authority: nonprofit. Tax-exempt: 501(c)(3).
Institution Type/Description: Art Gallery.
Collections: paintings; photographs; prints; sculpture.
Facilities: 1912 Armory.
Activities: lectures; performance art (10 in series); studio art and craft classes; inter-museum loan & traveling exhibitions. Museum Sponsors: monthly musical performances.
Publications: newsletters; exhibit catalogues.
Hours & Admission Prices: Mon.-Fri. 10-5, Sat. 10-3. No charge; donations accepted. Closed New Year's; Easter; Memorial Day; Independence Day; Labor Day; Rosh Hashanah; Thanksgiving; Christmas. &
Attendance: 9,000 (estimated)
Membership: Friend $5; Family $10; Patron $15.

Westerly

BABCOCK-SMITH HOUSE MUSEUM, 124 Granite St., Westerly, RI 02891-2435. Tel.: 401-596-5704.
Web Site: www.babcock-smithhouse.org
Founded: 1972.
Key Personnel: Chm. (V), John B. Coduri.
Personnel Profile: Part-Time Volunteers 60.
Governing Authority: nonprofit organization. Tax-exempt.
Institution Type/Description: Historic House: housed in an early Georgian-style mansion, built c.1734. Listed on the National Register of Historic Places.
Collections: period furnishings; personal artifacts; westerly granite industry history & records.
Facilities: Museum-related items for sale.
Publications: biannual newsletter; book, Built From Stone - The Westerly Granite Story.
Hours & Admission Prices: May-June & Sept.-Oct. Sat. 2-5; July-Aug. Thurs.-Sun. 2-5; other times by appointment. Adults $5, children $1; members no charge.
Membership: Individual $25; Family $50; Historian $100; Custodian $250; Curator $500; Conservator $1,000; Preservationist $2,500.

WESTERLY PUBLIC LIBRARY, 44 Broad St., Westerly, RI 02891-1856. Tel.: 401-596-2877. Fax: 401-596-5600.
E-mail: ktaylor@westerlylibrary.org
Web Site: www.westerlylibrary.org
Founded: 1894.
Congressional District: 51
Key Personnel: Interim Exec. Dir., Dan Snydacker; Pres., William S. Brown.
Personnel Profile: Full-Time Paid 17; Part-Time Paid 13; Part-Time Volunteers 40.
Governing Authority: board of trustees; nonprofit organization. Parent Institution: Memorial & Library Association of Westerly. Tax-exempt: 501(c)(3).
Institution Type/Description: Art & History Museum: housed in Civil War Memorial building.
Collections: Civil War artifacts; paintings; furniture; minerals; toys; dolls & doll furniture; original art objects; local history & genealogical material.
Facilities: 120,000-vol. library available on premises; archives; 150-seat auditorium; 18-acre Wilcox Park; gallery.
Activities: lectures; gallery talks; concerts; arts festivals; workshops; temporary & traveling exhibitions; school loan & art loan service.
Hours & Admission Prices: Mon.-Wed. 9-8, Thurs.-Fri. 9-6, Sat. 9-4. No charge; donations accepted. Closed major holidays. &
Attendance: 280,000 (estimated)

Woonsocket

MUSEUM OF WORK & CULTURE, 42 S. Main St., Woonsocket, RI 02895-4274. Tel.: 401-769-9675.
Web Site: www.rihs.org/Museums.html
Governing Authority: Parent Institution: Rhode Island Historical Society.
Institution Type/Description: History Museum.
Collections: interactive exhibits; French Canadians who settled in New England; cultural artifacts.
Hours & Admission Prices: Tues.-Fri. 9:30-4, Sat. 10-4, Sun. 1-4. Adults $8, seniors & students $6; discounts to groups; members & children under 10 no charge.

SOUTH CAROLINA

(233 listings)

Abbeville

BURT-STARK MANSION, 400 N. Main St., Abbeville, SC 29620-1706. Mailing Address: P.O. Box 164, Abbeville, SC 29620-0164. Tel.: 864-366-0166.
E-mail: info@burt-stark.com
Web Site: www.burt-stark.com
Institution Type/Description: Historic House: built in 1830s. A National Historic Landmark.
Collections: Stark family's personal artifacts; period furnishings.
Hours & Admission Prices: Feb.-Dec. Fri.-Sat. 1-5; other times by appointment. Adults $10. Closed major holidays.

DR. SAMUEL R. POLIAKOFF COLLECTION OF WESTERN ART - ABBEVILLE COUNTY LIBRARY, 201 S. Main St., Abbeville, SC 29620. Tel.: 864-459-4009. Fax: 864-459-4009.
Key Personnel: Dir., Mary Elizabeth Land
Institution Type/Description: Library.
Collections: contemporary southwestern Indian pottery & textiles; Western paintings & bronzes; books; photographs.
Hours & Admission Prices: Mon., Wed. & Fri. 9-5:30, Tues. & Thurs. 9-8, Sat. 9-3. No charge. Closed major holidays.

Aiken

AIKEN CENTER FOR THE ARTS, 122 Laurens St., S.W., Aiken, SC 29801-3888. Tel.: 803-641-9094. Fax: 803-641-2009. Facebook: Aiken Center for the arts.
E-mail: acaexecdir@bellsouth.net
Web Site: www.aikencenterforthearts.org
Founded: 1972.
Congressional District: 3
Key Personnel: Exec. Dir., Elizabeth Williamson; Pres., Skipper Perry; Treas., Janice Williams; Gallery Store Mgr., Michelle Petty.
Personnel Profile: Full-Time Paid 2; Part-Time Paid 7; Part-Time Volunteers 217.
Governing Authority: private; nonprofit organization. Tax-exempt: 501(c)(3).
Institution Type/Description: Art Museum.
Collections: paintings.
Facilities: 8,000 sq. ft. exhibit space; 180-seat theater; educational facilities. Museum-related items for sale.
Activities: rental facilities; formal education programs; Aiken youth orchestra; special events; concerts; films; guided tours; lectures; participatory & traveling exhibits; theater. Annual Events: Taste of Wine & Art; Antiques in the Heart of Aiken.
Publications: quarterly newsletter, Art Matters.
Hours & Admission Prices: Mon.-Sat. 10-5. No charge. Closed Memorial Day; Independence Day; Labor Day; Thanksgiving; Christmas. &
Attendance: 25,000 (estimated)
Membership: Young Contemporary $30; Individual $50; Family $75; Contributor $150; Patron $350; Benefactor $1,000; Saint & Corporate $2,500.

AIKEN COUNTY HISTORICAL MUSEUM, 433 Newberry St., S.W., Aiken, SC 29801-4844. Tel.: 803-642-2017 & 2015. Fax: 803-642-2016. Facebook: Friends of Aiken County Historical Museum.
E-mail: elevy@aikencountysc.gov
Web Site: www.aikencountyhistoricalmuseum.org
Founded: 1970.
Congressional District: 8
Key Personnel: Dir., Elliott Levy; Asst. Dir. & Educator, Mary White; C.E.O., Owen Clary; Asst. Dir. & Collections Mgr., Brenda Baratto; Museum Shop Mgr., Nancy Goelz.
Personnel Profile: Full-Time Paid 3; Part-Time Paid 1; Part-Time Volunteers 85.
Governing Authority: county; nonprofit. Parent Institution: Aiken County. Tax-exempt.
Institution Type/Description: History Museum: 1931 winter colony mansion.
Collections: Aiken County artifacts; archeology collection; industrial, kaolin, textile & nuclear exhibits; restored, mid 1800s buttery; Black history & agriculture. Historic Buildings: 1890 one-room schoolhouse; 1808 furnished log cabin; 1920s miniature circus; winter colony residence.
Research Fields: historical sites; local history.
Facilities: 200-vol. staff research library of historical books; files; photographs; maps; conference room. Museum-related items for sale.

Activities: guided tours; lectures, films, docent program; formally organized education programs for students; permanent & traveling exhibitions. Museum Sponsors: Garden Show in May.
Publications: quarterly newsletter.
Hours & Admission Prices: Tues.-Sat. 10-5, Sun. 2-5. No charge; donations accepted. Closed national holidays. &
Attendance: 20,778 (accurate)
Membership: Individual $10; Family $15; Sponsor $25; Benefactor $50; Patron $100.

AIKEN THOROUGHBRED RACING HALL OF FAME AND MUSEUM, 135 Dupree Place, Aiken, SC 29801. Mailing Address: P.O. Box 1177, Aiken, SC 29802-1177. Tel.: 803-642-7631 & 7650. Fax: 803-643-4780.
E-mail: halloffame@cityofaikensc.gov
Web Site: www.aikenracinghalloffame.com
Founded: 1977.
Congressional District: 81
Key Personnel: C.E.O., Lisa J. Hall.
Personnel Profile: Full-Time Paid 1; Part-Time Volunteers 25.
Governing Authority: nonprofit. Tax-exempt.
Institution Type/Description: Specialized Museum: housed in a restored carriage house.
Collections: Thoroughbred horses; trainers & owners that trained in Aiken.
Facilities: 1,500 sq. ft. exhibit space. Museum-related items for sale.
Activities: loan exhibitions. Museum Sponsors: Induction of Thoroughbreds of Fame into Hall of Fame.
Hours & Admission Prices: June-Aug. Sat.-Sun. 2-5; Sept.-May Tues.-Fri. & Sun. 2-5, Sat. 10-5. No charge; donations accepted. Closed New Year's Day; Martin Luther King Jr. Day; Good Friday; Easter; Memorial Day; Independence Day; Labor Day; Thanksgiving; Christmas. &
Attendance: 20,000 (estimated)

AIKEN VISITORS CENTER AND TRAIN MUSEUM, 406 Park Ave., S.E., Aiken, SC 29801. Tel.: 803-293-7846; 888-245-3672.
Institution Type/Description: Historic Building: housed in the former Aiken Railroad Depot.
Collections: local & railroad history; HO scale dioramas.
Facilities: rental facilities.
Activities: special events; rental facilities.
Hours & Admission Prices: Wed.-Fri. 10-5, Sat. 9-2.

Anderson

ANDERSON CITY FIRE DEPARTMENT MUSEUM, 400 S. McDuffie, Anderson, SC 29624. Tel.: 864-231-2256.
Institution Type/Description: Firefighting History Museum.
Collections: firefighting history, vehicles & equipment; hose cart; helmets; uniforms; photographs.
Hours & Admission Prices: Mon.-Fri. 9-5; other times by appointment. No charge.

ANDERSON COUNTY ARTS CENTER, 110 Federal St., Anderson, SC 29625. Tel.: 864-222-2787. Fax: 864-224-8864.
E-mail: info@andersonartscenter.org
Web Site: www.andersonartscenter.org
Founded: 1972.
Congressional District: 3
Key Personnel: Exec. Dir., Kimberly Spears; Administrative Dir., Annette Buchanan; Pres., Dr. Bob Austin; Dir. Communications, Stacey McAdams.
Personnel Profile: Full-Time Paid 4; Part-Time Paid 1; Part-Time Volunteers 5.
Governing Authority: nonprofit. Tax-exempt: 170(b)(1)(A).
Institution Type/Description: Arts Center: housed in 1908 Carnegie Library Building.
Collections: changing exhibits.
Facilities: classrooms; pottery studio; photography lab; kitchen & banquet facilities.
Activities: lectures; gallery talks; Summer Soiree' festival; rural arts program, changing monthly exhibits, featuring international, national & local artists; workshops; art school classes; musical performances; dance programs; formally organized education programs. Annual Event: Mother-Daughter Christmas Tea.
Publications: art school brochures; monthly invitations; gallery announcements.
Hours & Admission Prices: Tues.-Fri. 9:30-5:30, Sat. open to groups by special request only. No charge; donations accepted. Closed Memorial Day; Labor Day; Thanksgiving & day after; Independence Day week.
Membership: Individuals $30; Family & Patron $50; Donor $100; Sustaining Donor $250; Benefactor $500.

ANDERSON COUNTY MUSEUM, (M), 202 E. Greenville St, Anderson, SC 29621-5509. Tel.: 864-260-4737. Fax: 864-332-5320.
E-mail: bchilds@andersoncountysc.org
Web Site: www.andersoncountymuseum.org
Founded: 1983.
Congressional District: 3
Key Personnel: Exec. Dir., Beverly R. Childs; Cur. Collections, Alison Hinman.
Personnel Profile: Full-Time Paid 3; Part-Time Paid 1; Part-Time Volunteers 97; Interns 3.
Governing Authority: nonprofit. Parent Institution: Anderson County. Tax-exempt.
Institution Type/Description: History Museum.
Collections: local history photographs & artifacts. Historic Buildings: restored harness shop; restored one-room schoolhouse.
Major Exhibits: Desolate Pride - Civil War Exhibit, 5/12-6/15.
Research Fields: area & local history.
Facilities: 1,600-vol. library of local history material available to the public.
Activities: guided tours; temporary & permanent exhibitions.
Publications: membership newsletter, Musings; volunteer newsletter, Muse News.
Hours & Admission Prices: Museum: Tues. 10-7, Wed.-Sat. 10-4. The Anderson County Reading Room: Thurs. 1-4; other times by appointment. No charge; donations accepted. &
Attendance: 16,500 (accurate)
Membership: Individual Associate $35; Family Contributor's Circle $65; Sustainer's Society $125; Curator's Club $250; Director's Guild $500; Foundation Fellows $1,000; Corporate Partner $1,500.

Awendaw

CENTER FOR BIRDS OF PREY, 4872 Seewee Rd., Awendaw, SC 29429. Mailing Address: P.O. Box 1247, Charleston, SC 29402-1247. Tel.: 843-971-7474. Fax: 843-971-7029. Facebook: Center for Birds of Prey.
E-mail: info@thecenterforbirdsofprey.org
Web Site: www.thecenterforbirdsofprey.org
Founded: 1991.
Key Personnel: Dir., Jim Elliott, Jr.
Institution Type/Description: Nature Center.
Collections: over 50 species of birds of prey including eagles, hawks, owls, falcons, kites, & vultures.
Research Fields: hawk watch; swallow-failed kite citizen science project.
Activities: educational programs; flight demonstrations.
Hours & Admission Prices: Guided Tours: Thurs.-Sat. 10:30 & 2. Flight Demonstrations: Thurs.-Sat. 11:30 & 3. Adults $12; discounts or complimentary depending on membership level.
Attendance: 9,000 (estimated)

Barnwell

BARNWELL COUNTY MUSEUM, 9426 Marlboro Ave., Barnwell, SC 29812. Mailing Address: P.O. Box 422, Barnwell, SC 29812-0422. Tel.: 803-259-1916. Fax: 803-259-1916.
E-mail: barnwell.museum@att.net
Founded: 1978.
Congressional District: 2
Key Personnel: Chm., Anne W. Hagood; Dir., Jerry Morris; Museum Shop Mgr., Marie Peeples.
Personnel Profile: Part-Time Paid 1; Part-Time Volunteers 10.
Governing Authority: county; nonprofit organization. Affiliated with Museum Association & County Council. Tax-exempt.
Institution Type/Description: History Museum: housed in the Fuller House.
Collections: papers & books on Barnwell County dating from Revolutionary War times; Tarleton Brown Revolutionary soldier collection; costumes; dolls; implements of plantation farm life; silver; china; photographs; paintings & posters; maps; pottery; Civil War memorabilia; uniforms; medals; clippings; newspapers from World War I to Desert Storm; William Gilmore Simms.
Activities: guided tours; lectures; gallery talks; concerts; loan, permanent, temporary & traveling exhibitions.
Publications: Memoirs of Tarleton Brown (reprint).
Hours & Admission Prices: Tues.-Thurs. & Sun. 3-5:30. No charge; donations accepted. Closed holidays. &
Attendance: 4,500 (estimated)
Membership: Individual $15; Family $30; Friend $50-$249; Benefactor $250 & up.

Beaufort

THE BEAUFORT ARSENAL, 713 Craven St., Beaufort, SC 29902-5571. Mailing Address: P.O. Box 11, Beaufort, SC 29901-0011. Tel.: 843-379-3331. Fax: 843-379-3371.
E-mail: eryan@historicbeaufort.org
Web Site: www.historic-beaufort.org
Formerly: Beaufort Museum at the Arsenal
Founded: 1939.
Congressional District: 2
Key Personnel: Dir. Museums, Elizabeth G. Ryan.
Personnel Profile: Full-Time Paid 3; Part-Time Paid 6; Part-Time Volunteers 90.
Governing Authority: municipality. Parent Institution: Historic Beaufort Foundation. Tax-exempt.
Institution Type/Description: History Museum: built in 1798.
Collections: 1798 Arsenal & 1917 Carnegie Library; history of Beaufort & the Low Country; Native American artifacts; textiles; costumes; household; personal furnishings & accessories; Low Country artwork & photographs.
Research Fields: Beaufort & Low Country history.
Facilities: meeting room. Museum-related items for sale.
Activities: educational programs in galleries; speakers series; slide lecture programs; special events; Summer in the Courtyard.
Publications: newsletter, Museum Muse.
Hours & Admission Prices: Mon.-Tues. & Thurs.-Sat. 10-5. Adults $3; children 6 & under and members no charge. Closed holidays.
Attendance: 12,000 (estimated)
Membership: Student $15; Individual $25; Family $50; Supporter $100; Donor $250; Patron $500; Benefactor $1,000; Sustainer $2,500; Danner $5,000.

THE VERDIER HOUSE, 801 Bay St., Beaufort, SC 29902-5565. Mailing Address: Historic Beaufort Foundation, P.O. Box 11, Beaufort, SC 29901-0011. Tel.: 843-379-3331. Fax: 843-379-3371.
E-mail: director@historicbeaufort.org
Web Site: www.historicbeaufort.org
Founded: 1977.
Congressional District: 2
Key Personnel: Mgr., Sandy Patterson; Interim Exec. Dir., Maxine Lutz.
Personnel Profile: Full-Time Paid 1; Part-Time Paid 4; Part-Time Volunteers 60.
Governing Authority: nonprofit organization. Parent Institution: Historic Beaufort Foundation Inc. Tax-exempt: 501(c)(3).
Institution Type/Description: Historic House: housed in the former home of John Mark Verdier; built c.1805.
Collections: 1790-1825 Federal period furnishings & artifacts.
Research Fields: Beaufort style architecture.
Facilities: historic preservation library; meeting space. Museum-related items for sale.
Activities: guided tours; lectures; docent program; preservation of historical buildings & sites in Beaufort County. Museum Sponsors: Lafayette Soiree; Fall Tours in October; Christmas Open House.
Publications: quarterly newsletter, A Guide to Historic Beaufort.
Hours & Admission Prices: Mon.-Sat. 10-3:30; other times by appointment for groups of 8 or more. Verdier House: adults $10; students, children under 18, active military & foundation members no charge.
Attendance: 10,000
Membership: Single $25; Family $50; Supporting $100; Donor $250; Patron $500; Benefactor $1,000.

Beech Island

BEECH ISLAND HISTORICAL SOCIETY, 144 Old Jackson Hwy., Beech Island, SC 29842-4568. Tel.: 803-867-3600. Fax: 803-867-3600.
E-mail: bihs@comcast.net
Web Site: www.beechislandhistory.org
Founded: 1985.
Congressional District: 3
Key Personnel: Dir. & C.E.O., Jackie Bartley.
Personnel Profile: Part-Time Volunteers 1.
Governing Authority: Tax-exempt.
Institution Type/Description: History & Agricultural Museum.
Collections: Beech Island & agricultural history; personal artifacts.
Activities: special programs & events; monthly meetings; spring house tour.
Publications: newsletter, Four Centuries & More.
Hours & Admission Prices: Wed.-Thurs. 11-1; other times by appointment. No charge; donations accepted. &
Attendance: 350 (accurate)
Membership: Single $25; Family $30.

REDCLIFFE PLANTATION STATE HISTORIC SITE, 181 Redcliffe Rd., Beech Island, SC 29842-9535. Tel.: 803-827-1473.
E-mail: redcliffe@scprt.com
Web Site: www.southcarolinaparks.com/park-finder/state-park/2015.aspx
Personnel Profile: Full-Time Paid 3; Part-Time Volunteers 2.
Governing Authority: Parent Institution: South Carolina Dept. of Park, Recreation & Tourism.
Institution Type/Description: Historic House: housed in the home of South Carolina Sen. James Henry Hammond; built in 1859. Listed on the National Register of Historic Places.
Collections: local history & culture; period furnishings; personal artifacts; portraits; photographs.
Research Fields: African American history; genealogy; agriculture; reconstruction; Civil War; slavery; architecture; 19th century daily life.
Activities: educational programs.
Hours & Admission Prices: Tours: Thurs.-Mon. 1, 2, & 3. Adults 16 & over $4, children 6-15 $3, SC senior citizens $2.50.

Belton

BELTON CENTER FOR THE ARTS, 306 City Sq., Belton, SC 29627. Mailing Address: P.O. Box 368, Belton, SC 29627-0368. Tel.: 864-338-8556. Fax: 864-338-0280.
Institution Type/Description: Art Gallery.
Collections: paintings; sculpture.
Facilities: Gallery-related items for sale.
Activities: special events; classes.
Hours & Admission Prices: Tues.-Fri. 10-5:30, Sat. 10-2.

SOUTH CAROLINA TENNIS HALL OF FAME, 50 N. Main St., Belton, SC 29627. Mailing Address: P.O. Box 843, Belton, SC 29627-0843. Tel.: 864-338-7400. Fax: 864-338-4034.
E-mail: bama7400@aol.com
Web Site: www.beltonsc.com
Founded: 1984.
Congressional District: 6
Key Personnel: C.E.O., Rex Maynard; Chm. (V), Paul Pittman.
Personnel Profile: Part-Time Volunteers 14.
Governing Authority: nonprofit organization. Parent Institution: South Carolina Tennis Patrons Foundation. Tax-exempt: 501(c)(3).
Institution Type/Description: Tennis Museum.
Collections: portraits of inductees; trophies & tennis memorabilia; books; magazines; rackets; oil paintings; mementos.
Research Fields: tennis.
Facilities: 400 sq. ft. exhibit space.
Activities: tours. Museum Sponsors: annual banquet.
Hours & Admission Prices: Wed.-Fri. 10-4, Sat. 10-2. No charge. Closed holidays. &
Attendance: 2,500 (estimated)
Membership: Member $25; Friend $50; Patron $100; Sponsor $250; Benefactor $500; Champion $1,000.

Bennettsville

JENNINGS-BROWN HOUSE FEMALE ACADEMY, 121 S. Marlboro St., Bennettsville, SC 29512-4031. Mailing Address: P.O. Box 178, Bennettsville, SC 29512-0178. Tel.: 843-479-5624.
E-mail: marlborough@mecsc.net
Founded: 1967.
Congressional District: 6
Key Personnel: Bd. Pres. Historical Society (V), Marty Rankin; Exec. Dir. Museum, Lucille Carabo.
Personnel Profile: Full-Time Paid 2; Part-Time Volunteers 13.
Governing Authority: county; society. Parent Institution: Marlboro County Historic Preservation Commission. Affiliated with Marlboro County Historical Museum, 123 S. Marlboro St., Bennettsville, SC 29512. Tax-exempt: 501(c)(3).
Institution Type/Description: Historic House: c.1826-27 Jennings-Brown House.
Collections: vintage furnishings and furniture pre-dating 1850.
Facilities: County historical & museum-related items for sale.
Activities: guided tours; traveling exhibitions.
Hours & Admission Prices: Mon.-Thurs. 10-5, Fri. 10-1. Adults $2, students $2.
Attendance: 600 (estimated)
Membership: Adult $10; Couple $15; Family $20; Patron $50-$199; Angel $200 & up.

MARLBORO COUNTY HISTORICAL MUSEUM, 123 S. Marlboro St., Bennettsville, SC 29512-4031. Tel.: 843-479-5624.
E-mail: marlborough@mecsc.net
Founded: 1970.
Congressional District: 6
Key Personnel: Pres. (V), Marty Rankin; Exec. Dir., Susan Cloer.
Personnel Profile: Full-Time Paid 2; Part-Time Volunteers 40; Interns 2.
Governing Authority: society. Parent Institution: Marlborough Historical Society. Tax-exempt: 501(c)(3).
Institution Type/Description: Historical & Preservation Society: housed in the former home of Dr. & Mrs. John Frank Kinney; built in 1902.
Collections: pre-1850 southern antiques; local history items from Indians through World War II. Historic Houses: 1826 Jennings Brown House; 1833 Bennettsville Female Academy.
Facilities: Tour booklet, historic site tiles & other history-related items for sale.
Activities: guided tours; lectures; permanent & temporary exhibitions.
Hours & Admission Prices: Mon.-Fri. 10-4. House: adults $2, children $1. Closed major holidays.
Attendance: 2,547 (accurate)
Membership: Individual $10; Couple $15; Family $20; Patron $50-$199; Angel $200 & up.

Bishopville

FRYAR TOPIARY GARDENS, 145 Broad Acres Rd., Bishopville, SC 29010-2819. Tel.: 803-484-5581.
Institution Type/Description: Garden.
Collections: sculptured plants.
Hours & Admission Prices: Tues.-Sat. 10-4. No charge; donations accepted.

SOUTH CAROLINA COTTON MUSEUM, (M), 121 W. Cedar Ln., Bishopville, SC 29010-1454. Tel.: 803-484-4497. Fax: 803-484-5203.
E-mail: sccottonmus@ftc-i.net
Web Site: www.sccotton.org
Founded: 1993.
Congressional District: 5
Key Personnel: Exec. Dir., Janson L. Cox; Pres., Gail Player; Business Mgr., Melissa Brundage.
Personnel Profile: Full-Time Paid 1; Full-Time Volunteers 1; Part-Time Paid 5; Part-Time Volunteers 25.
Governing Authority: private; nonprofit organization. Tax-exempt: 501(c)(3).
Institution Type/Description: History Museum.
Collections: economic, social & political impact of cotton in the south.
Research Fields: veterans oral history project, partner with the Library of Congress.
Facilities: meeting room; conference room (both available for rent).
Activities: educational programs; audiovisual programming for groups; Veterans Oral History program as a partner with the Library of Congress.
Publications: periodic newsletter.
Hours & Admission Prices: Mon.-Fri. 10-4:30, Sat. 10-4. Adults $6, senior citizens $4, students $3; discounts to AAA & ICOM members; active duty military, members, children 5 & under no charge. &
Attendance: 7,842 (accurate)
Membership: Adult $25; Family $35; Business $100 & up; Corporate $1,000.

Blacksburg

KINGS MOUNTAIN NATIONAL MILITARY PARK, 2625 Park Rd., Blacksburg, SC 29702-7325. Tel.: 864-936-7921. Fax: 864-936-9897. TDD: 864-936-7921.
Web Site: www.nps.gov/kimo.htm
Founded: 1931.
Congressional District: 5
Key Personnel: Chief Interpretation & Resource Mgmt., Chris Revels; Bookstore Mgr., Wilma Scoggins.
Personnel Profile: Full-Time Paid 12; Part-Time Paid 5; Part-Time Volunteers 40; Interns 3.
Governing Authority: federal; nonprofit organization. Parent Institution: National Park Service. Tax-exempt.
Institution Type/Description: Military Park Museum.
Collections: military artifacts; 18th-century frontier life.
Research Fields: southern campaign of Revolution & battle of Kings Mountain.
Facilities: 300-vol. library of history of Revolutionary War available for use on premises. Publications for sale.
Activities: movies; permanent & temporary exhibitions.
Hours & Admission Prices: Daily 9-5. No charge; donations accepted. Closed New Year's Day; Thanksgiving; Christmas. &
Attendance: 115,000 (accurate)

Blackville

AGRICULTURAL HERITAGE CENTER, Clemson University's Edisto Research and Educ. Ctr., 64 Research Rd., Blackville, SC 29817. Tel.: 803-284-3343. Fax: 803-284-3684.
Institution Type/Description: History Museum.
Collections: local agricultural history; rural life; regional culture & economics; period equipment & artifacts.
Hours & Admission Prices: Mon.-Sat. 9-4. Tours: $2 per person; children K-12 no charge.

Branchville

BRANCHVILLE RAILROAD SHRINE AND MUSEUM, INC., 7504 Freedom Rd., Branchville, SC 29432-2310. Mailing Address: Town of Branchville, P.O. Box 85, Branchville, SC 29432-0085. Tel.: 803-274-8820. Fax: 803-274-8760.
Founded: 1969.
Congressional District: 2
Key Personnel: Pres., Johnny Norris; Vice Pres., Mike Norris.
Governing Authority: nonprofit. Tax-exempt: 501(c)(3).
Institution Type/Description: Transportation Museum: housed in 1877 Branchville Southern Railroad Depot at the site of the first railroad junction of the world.
Collections: railroad artifacts; model trains; historical items used when the town was a railroad center.
Research Fields: railroading.
Facilities: 50-seat restaurant. Gift items for sale.
Activities: guided tours; loan & permanent exhibitions. Museum Sponsors: R.R. Daze Festival in September.
Hours & Admission Prices: Fri.-Sat. 10-2, Sun. 2-5; other times by appointment. No charge; donations accepted. Closed major holidays. &
Membership: Individual $5.

Camden

CAMDEN ARCHIVES & MUSEUM, 1314 Broad St., Camden, SC 29020-3535. Tel.: 803-425-6050. Fax: 803-424-4053.
Web Site: www.camdenarchives.org
Founded: 1973.
Congressional District: 5
Key Personnel: Dir., Howard Branham; Deputy Dir., Katherine Richardson; Chm. (V), Frank Goodale; Assoc. Archivist, Peggy Brakefield; Administrative Asst., Barbara Rogers.
Personnel Profile: Full-Time Paid 2; Part-Time Paid 4; Part-Time Volunteers 1.
Governing Authority: municipal; nonprofit. Parent Institution: City of Camden. Tax-exempt.
Institution Type/Description: Archives & Local History Museum.
Collections: genealogy; archives; local artifacts; memorabilia; sections on American Revolution; Civil War; South Carolina history & biography; The South; local history; collections of South Carolina Society, D.A.R. & South Carolina Chapter, Colonial Dames, 17th-century; manuscripts.
Research Fields: genealogy; local history.
Facilities: 2,800-vol. library available for research on premises only; reading room.
Activities: guided tours; special exhibits.
Publications: brochures.
Hours & Admission Prices: Mon.-Fri. 8-5, Sat. 10-4, 1st Sun. each month 1-5. No charge; donations accepted. Closed city holidays. &
Attendance: 6,000 (estimated)
Membership: Friends: Individual $25; Family $45; Contributor $100-$249; Patron $250-$499; Benefactor $500 & up.

FINE ARTS CENTER OF KERSHAW COUNTY, INC., 810 Lyttleton St., Camden, SC 29020-4411. Mailing Address: P.O. Box 1498, Camden, SC 29021-8498. Tel.: 803-425-7676. Fax: 803-425-7679.
E-mail: kcobb@fineartscenter.org
Web Site: www.fineartscenter.org
Founded: 1976.
Congressional District: 5
Key Personnel: Exec. Dir., Kristin Cobb; Pres. Bd. Dir., Karen Eckford; Dir. Facility, Dianne Edwards; Dir. Finance, Daphne Cantey; Dir. Education, Steve LeVan; Dir. Mktg., Jane Peterson.
Personnel Profile: Full-Time Paid 6; Part-Time Paid 6; Part-Time Volunteers 30.
Governing Authority: nonprofit organization. Tax-exempt: 501(c)(3).
Institution Type/Description: Art Gallery, housed in Bassett Memorial Building.

Collections: Carroll Bassett & sporting art equestrian bronzes; art works; Carroll K. Bassett Memorial Building.
Facilities: 274-seat auditorium; educational facilities.
Activities: lectures; gallery talks; concerts; education programs for children & adults; offerings for all ages in the arts including theatre & visual arts; traveling & permanent exhibits; photography club programs; art association.
Publications: monthly newsletter; season & class brochures.
Hours & Admission Prices: Winter: Mon.-Fri. 12-6, Sat. by appointment. Summer: Mon.-Fri. 9-5, Sat. by appointment. Gallery: no charge; donations accepted. Performance Series: call for admission fee. Closed Easter; Memorial Day; Independence Day; Labor Day; Thanksgiving; Christmas Eve to New Year's Day. &
Attendance: 70,000 (accurate)
Membership: Contributor (individual) $35; Supporter $50; Sustainer $150; Advocate $500; Patron $1,000; Benefactor $2,500; Life $10,000.

HISTORIC CAMDEN REVOLUTIONARY WAR SITE, 222 Broad St., Camden, SC 29020. Mailing Address: P.O. Box 710, Camden, SC 29021-0710. Tel.: 803-432-9841. Fax: 803-432-3815.
E-mail: hiscamden@truvista.net
Web Site: www.historic-camden.net
Founded: 1967.
Congressional District: 5
Key Personnel: Chm., M. Traysee Dunaway; Vice Chm., Wm. Davie Beard; Dir. & Museum Shop Mgr., Joanna Craig.
Personnel Profile: Full-Time Paid 1; Part-Time Paid 4; Part-Time Volunteers 65.
Governing Authority: nonprofit. Parent Institution: Historic Camden Foundation. Tax-exempt: 501(c)(3).
Institution Type/Description: History Museum: housed in reconstructed 1777 Kershaw-Cornwallis House, located on 106 acre archeological park encompassing the colonial to late 18th-century village of Camden.
Collections: archaeological remains from pre-European, colonial & antebellum periods; tools & furnishings from late 18th to the early 19th century.
Research Fields: Indian, colonial & revolutionary periods.
Facilities: 400-vol. library pertaining to Indian, colonial & revolutionary periods; catering service available. Books, local crafts & museum-related items for sale.
Activities: guided tours; lectures; concerts; organized education programs for children & adults; living history demonstrations.
Publications: annual newsletter; calendar of events.
Hours & Admission Prices: Tues.-Sat. 10-5, Sun. 2-5. Guided Tours: Tues.-Fri. 10:30 & 3, Sat. 10:30-4, Sun. 2-4. Adults $5, senior citizens $4, students $3; discounts to Mobil, AAA, AAM & ICOM members; members & children under 6 no charge. Closed major holidays.
Attendance: 20,191 (estimated)
Membership: Senior Citizen $25; Individual $30; Family $40; Contributing $50; Sustaining $100; Leadership $250; Fellow $500; Patron $1,000.

HISTORIC ROBERT MILLS COURTHOUSE, 607 Broad St., Camden, SC 29020-4703. Mailing Address: P.O. Box 605, Camden, SC 29021. Tel.: 800-968-4037; 803-432-2525. Fax: 803-432-4181.
Key Personnel: Exec. Dir., Liz Horton.
Governing Authority: Parent Institution: Kershaw County Chamber of Commerce & Visitors Center.
Institution Type/Description: Historic Building: designed by South Carolina architect, Robert Mills; built in 1827.
Collections: local history & culture; architecture; period furnishings; personal artifacts; photographs.
Activities: guided tours.
Hours & Admission Prices: Mon.-Fri. 9-5, Sat. 11-4. No charge; donations accepted. &

KERSHAW COUNTY HISTORICAL SOCIETY, 811 Fair St., Camden, SC 29020-4404. Mailing Address: P.O. Box 501, Camden, SC 29021-0501. Tel.: 803 425 1123.
E-mail: kchistory@camden.net
Web Site: www.kershawcountyhistoricalsociety.org
Founded: 1954.
Congressional District: 5
Key Personnel: Pres., H.W. Funderburk, Jr.; Dir., Kathleen P. Stahl.
Personnel Profile: Part-Time Paid 1; Part-Time Volunteers 18.
Governing Authority: society; nonprofit organization. Tax-exempt: 501(c)(3).
Institution Type/Description: Historic House: housed in the Bonds Conway House; c.1812.
Collections: local history & culture; period furnishings.

Research Fields: listing of historic battle sites; historic houses in county.
Facilities: Local history books for sale.
Activities: guided tours; organized education programs for adults.
Publications: quarterly newsletter, Update; brochure; books, Historic Camden, Vols. I & II; Camden Homes & Heritage; Decorative Arts in Camden & Kershaw County; Kershaw County Cemetery Survey, Vols. I, II & III; Kershaw County Census, 1800, 1810, 1820, 1830; History and Homes of Liberty Hill, SC; A Guide to Selected Historical Sites in Kershaw County, SC; Kershaw County Legacy I & II; A Guide to Historic Sites in Camden, SC; series of reprints of old pamphlets on county history; Kershaw County State Census of White Population 1839; Kershaw County State Agricultural Census 1868; In the Sunny South: A Winter Colonist's View of Camden, SC & Vicinity in 1901; Battle of Camden, SC; Kershaw County District Business Directory, 1854-1900; Gold Rush Letter from William Lemond, a 49er from Carolina; Kershaw County Confederate Miscellany; Horatio Gates & Battle of Camden; Just Mud Kenshaw County SC Pottery; Wateree River Plantation Rosny from 1815; Genealogy of the McWillie and Cunningham Families; Index for the Genealogy of the McWillie and Cunningham Families; A History of Kershaw County, SC.
Hours & Admission Prices: Thurs. 1-5; other times by appointment. No charge; donations accepted. Closed holidays.
Attendance: 150 (estimated)
Membership: Student $10; Senior Citizen $20; Individual $25; Family $35; Business $50; Contributing $75; Sustaining $100; Life $1,000.

NATIONAL STEEPLECHASE MUSEUM, Springdale Race Course, 200 Knights Hill Rd., Camden, SC 29020-2154. Mailing Address: P.O. Box 2424, Camden, SC 29020-8008. Tel.: 803-432-6513; 800-780-8117. Fax: 803-432-4062.
E-mail: hope@steeplechasemuseum.org
Web Site: www.steeplechasemuseum.org
Founded: 1998.
Congressional District: 5
Key Personnel: Exec. Dir., Hope Cooper; Chm. (V), Beverly Steinman.
Personnel Profile: Full-Time Paid 1; Part-Time Paid 1; Part-Time Volunteers 1.
Governing Authority: Tax-exempt: 501(c)(3).
Institution Type/Description: Steeplechase History Museum.
Collections: history of American steeplechasing; photographs; memorabilia; art; trophies.
Major Exhibits: Americans at Aintree, 3/14-12/14.
Research Fields: steeplechase history.
Facilities: library; archives.
Activities: interactive exhibits; special events.
Publications: newsletter.
Hours & Admission Prices: Sept.-May Wed.-Sat. 10-4; other times by appointment. No charge; donations accepted. Closed major holidays. &
Attendance: 5,000 (estimated)
Membership: Starting Gate $50; Sweepstakes $100; Syndicate $200; Home Stretch $500; Winner's Circle $1,000; President's Council $2,500; Champion's Club $5,000.

Cayce

CAYCE HISTORICAL MUSEUM, City of Cayce Municipal Complex, 1800 12th St., Cayce, SC 29033-2935. Tel.: 803-739-5385. Fax: 803-796-9072.
E-mail: caycemuseum@historysc.com
Key Personnel: Dir., Leo Redmond
Institution Type/Description: History Museum.
Collections: Native American artifacts; period artifacts; furnishings; exhibits relating to the periods of Colonial trade, agricultural development & transportation from the 18th century to the present.
Hours & Admission Prices: Tues.-Fri. 9-4, Sat.-Sun. 2-5. Adults $2, senior citizens & students 13 and over $1, students 12 & under $.50; Sun. no charge. Closed city holidays.

Central

ASHTABULA HISTORIC HOUSE, 2725 Old Greenville Hwy., Central, SC 29630. Mailing Address: Pendleton Historic Foundation, P.O. Box 444, Pendleton, SC 29670-0444. Tel.: 864-646-7249.
E-mail: info@pendletonhistoricfoundation.org
Web Site: www.pendletonhistoricfoundation.org
Founded: 1960.
Congressional District: 3
Key Personnel: Chm. (V) & Museum Shop Mgr., Ellen Harrison; Pres. (V), Jackie Reynolds; Treas., Elizabeth Vogt; Sec., Tim Drake; Exec. Dir., Les McCall.
Personnel Profile: Full-Time Paid 1; Full-Time Volunteers 2; Part-Time Paid 3; Part-Time Volunteers 17; Interns 3.

Governing Authority: private; nonprofit organization. Parent Institution: Pendleton Historic Foundation. Tax-exempt: 501(c)(3).
Institution Type/Description: Historic House Museum.
Collections: period furnishings; period tools. Historic Houses: c.1825 2-story antebellum plantation house; c.1790 original brick 2-story house.
Research Fields: genealogy 1830-1908.
Activities: guided tours; permanent exhibitions; historical reenactments; first-person interpretive tours; teas & lecture series; docent training programs. Annual Events: Special Themed Monthly Tours; Special Christmas Event & Tours in December.
Publications: brochures; quarterly, Pendleton Historic Foundation Newsletter; history books.
Hours & Admission Prices: See website for hours & admission prices. &
Attendance: 4,383 (accurate)
Membership: Individual $35; Family $50; Patron $100 & up.

CENTRAL HERITAGE SOCIETY, 416 Church St., Central, SC 29630-9152. Mailing Address: P.O. Box 1162, Central, SC 29630-9152. Tel.: 864-639-2794 & 2156.
Institution Type/Description: History Museum.
Collections: local history & culture; photographs; period furnishings; personal artifacts.
Hours & Admission Prices: Sun. 2-4. Closed holidays.

Charleston

AIKEN-RHETT HOUSE, 48 Elizabeth St., Charleston, SC 29403-6250. Tel.: 843-723-1159.
E-mail: vperry@historiccharleston.org
Web Site: www.historiccharleston.org
Institution Type/Description: Historic House Museum: housed in the former home of Gov. William Aiken, Jr.; built in 1818. Listed on the National Register of Historic Places.
Collections: Aiken-Rhett family history; period furnishings; personal artifacts; paintings; sculpture.
Hours & Admission Prices: Mon.-Sat. 10-5, Sun. 2-5. Adults $10. Closed Thanksgiving; Christmas Eve & Day.

AMERICAN MILITARY MUSEUM, 2070 Sam Rittenberg Blvd., Ste. 216, Charleston, SC 29407. Tel.: 843-577-7000. Fax: 843-577-7008.
E-mail: info@americanmilitarymuseum.org
Web Site: www.americanmilitarymuseum.org
Founded: 1987.
Congressional District: 1
Key Personnel: Dir., George E. Meagher; Cur., Michael Lussier; Museum Shop Mgr., Sec. & Treas., Randi Meagher.
Personnel Profile: Full-Time Paid 2; Full-Time Volunteers 3; Part-Time Volunteers 10.
Governing Authority: individual operation. Tax-exempt.
Institution Type/Description: American Military History Museum.
Collections: U.S. military uniforms, insignias, & medals; weapons & personal artifacts; American military history from Revolutionary War to present; military miniatures.
Activities: guided tours. Special events: POW/MIA, Memorial Day, Armed Forces Week.
Hours & Admission Prices: Mon.-Sat. 10-6, Sun. 1-5; school groups by appointment. Adults $7, seniors 55 & over and retired military $5, students 13-18 $3, children 6-12 $2; discounts to groups; active military no charge. Closed Thanksgiving; Christmas. &

AVERY RESEARCH CENTER FOR AFRICAN AMERICAN HISTORY & CULTURE, 125 Bull St., College of Charleston, Charleston, SC 29424. Mailing Address: 66 George St., Charleston, SC 29424. Tel.: 843-953-7609; 843-953-7608. Fax: 843-953-7607.
E-mail: lessanepw@cofu.edu
Web Site: www.cofc.edu/avery
Founded: 1985.
Congressional District: 1
Key Personnel: Exec. Dir., Patricia Williams Lessane, Ph.D.; Processing Archivist, Georgette Mayo; Cur., Coord. Public Programs & Facilities Mgr., Curtis J. Franks; Assoc. Dir., Deborah Wright.
Governing Authority: college; State of South Carolina. Parent Institution: College of Charleston. Tax-exempt: 501(c)(3).
Institution Type/Description: History Museum & Research Center.
Collections: print materials; photographs; audio & videotape; manuscripts; organizational records related to the history & culture of South Carolina and Low Country African Americans.

Research Fields: African American history; Southern history; folklore; anthropology; linguistics.
Facilities: reference library; reading room.
Activities: reference & referral services to scholars, students & the community; public programs.
Publications: quarterly, The Avery Review; biannual newspaper, The Bulletin; quarterly calendar of events.
Hours & Admission Prices: Museum Tours: Mon.-Fri. 10:30, 11:30, 1:30, 2:30 & 3:30; Reading Room: Mon.-Fri. 10-12:30 & 1:30-5. No charge; donations accepted. Closed university holidays. &
Attendance: 8,500 (estimated)

CALHOUN MANSION, 14-16 Meeting St., Charleston, SC 29401-2706. Tel.: 843-722-8205. Fax: 843-723-1147.
E-mail: cmansion@tmo.blackberry.net
Institution Type/Description: Historic House: former home of George W. Williams, built in 1876.
Collections: local history & culture; period furnishings; decorative painting & lighting.
Facilities: garden. Museum-related items for sale.
Hours & Admission Prices: Tours: March-Nov. daily 11-5; Dec.-Feb. daily 11-4:30. Adults $15.

CHARLES TOWNE LANDING STATE HISTORIC SITE, 1500 Old Towne Rd., Charleston, SC 29407-6099. Tel.: 843-852-4200. Fax: 843-852-4205.
E-mail: charlestowne@scprt.com
Web Site: www.southcarolinaparks.com
Founded: 1970.
Key Personnel: Park Mgr., Rob Powell.
Governing Authority: state. Parent Institution: S.C. Dept. of Parks, Recreation & Tourism, 1205 Pendleton St., Columbia, SC 29201. Tax-exempt: 170(b)(1)(A).
Institution Type/Description: Historic Site: 1670 site of the first permanent European settlement in South Carolina.
Collections: prehistory to plantation era; native animals; crafts. Reproduction Historic Ship: trading ketch 1670 Adventure.
Research Fields: early South Carolina towne life; animal husbandry; agricultural relating to 1670.
Facilities: botanical garden; zoological park; vending area. Museum-related items for sale.
Activities: formally organized education programs for children, adults & undergraduate college students; docent program; training programs for professional museum workers; permanent exhibitions.
Publications: monthly newsletter.
Hours & Admission Prices: Daily 9-5. Adults $10, seniors $6.50, children 6-15 $6; children 5 & under no charge. Closed Christmas Eve & Day. &
Attendance: 185,000 (accurate)

CHARLESTON LIBRARY SOCIETY, 164 King St., Charleston, SC 29401-2269. Tel.: 843-723-9912. Fax: 843-723-3500.
E-mail: info@charlestonlibrarysociety.org
Web Site: www.charlestonlibrarysociety.org
Founded: 1748.
Congressional District: 1
Key Personnel: Exec. Dir., Anne W. Cleveland.
Personnel Profile: Full-Time Paid 5; Part-Time Paid 2; Interns 1.
Governing Authority: society. Tax-exempt: 501(c)(3).
Institution Type/Description: Historical & Cultural Library.
Collections: 100,000 books; Charleston newspapers dating from 1732; 18th-century books, pamphlets & periodicals; manuscripts; 1,686 microfilm reels.
Research Fields: South Carolina history; Charleston history; genealogy; military history; S.C. authors.
Facilities: 90,000-vol. library.
Activities: lectures; concerts; Life-Long Learning Series.
Hours & Admission Prices: Mon.-Fri. 9:30-5:30, Sat. 9:30-2. Research: $5; members no charge. Closed holidays. &
Attendance: 11,500
Membership: College Student and Senior 65 & over $50; Individual $75.

＊ THE CHARLESTON MUSEUM, (M), 360 Meeting St., Charleston, SC 29403-6297. Tel.: 843-722-2996. Fax: 803-722-1784.
E-mail: jbrumgardt@charlestonmuseum.org
Web Site: www.charlestonmuseum.org
Founded: 1773.
Congressional District: 1

Key Personnel: Dir. & C.E.O., Dr. John R. Brumgardt; Bd. Trustees Pres., Dr. John Rashford; Admin. Svcs., Vickie Styles; Asst. Dir., Carl P. Borick; Cur. Historical Archaeology, Martha Zierden; Cur. History, J. Grahame Long; Education Coord., Stephanie Thomas; Registrar, Jan Z. Hiester; Archivist, Jennifer E. Scheetz; Admin. Mgr., Susan McKellar; Bldg. Mgr., Robert Gabel; Public Rels. & Events Coord., Rachel Chesser.
Personnel Profile: Full-Time Paid 26; Part-Time Paid 36; Part-Time Volunteers 55.
Governing Authority: nonprofit organization. Branch Museums: Heyward-Washington House, 87 Church St.; Joseph Manigault House, 350 Meeting St; Dill Sanctuary. Tax-exempt: 501(c)(3).
Institution Type/Description: General Museum.
Collections: archaeology & ethnology; natural science; history; decorative arts; major collections' emphasis on Charleston & South Carolina coastal region with comparative material of worldwide scope; regional collections of birds, mammals, fish, reptiles, amphibians, invertebrates, plants, fossils; furniture; silver; textiles; historic archaeology.
Research Fields: whales; historic archaeology; furniture; textiles; silver; ornithology.
Facilities: 20,000-vol. library of books, periodicals, maps, photographs, sheet music, available on inter-library loan & on premises; 300-seat auditorium; 75-seat orientation room; classrooms. Museum-related items for sale.
Activities: tours; lectures; films; formally organized education programs for children; publications program; docent program; inter-museum loan, permanent & temporary exhibitions.
Publications: quarterly newsletter, Charleston Museum Newsletter; books, Status of South Carolina Birds; Story of Francis Simmons Holmes; Audubon: The Charleston Connection; Charleston Silver; The Story of Sea Island Cotton.
Hours & Admission Prices: Mon.-Sat. 9-5, Sun. 1-5. Adults $10, children $5; discounts to AAM members; museum members no charge. Closed New Year's Day; Easter; Thanksgiving; Christmas. &
Attendance: 116,528 (accurate)
Membership: Individual $50; Family $60; Milby Burton Society $125; Business Donor $250; Manigault Society $550; 1773 Society $1,000.

CHILDREN'S MUSEUM OF THE LOWCOUNTRY, 25 Ann St., Charleston, SC 29403-6213. Tel.: 843-853-8962. Fax: 843-853-1042.
E-mail: info@explorecml.org
Web Site: www.explorecml.org
Founded: 2003.
Congressional District: 1
Key Personnel: Exec. Dir., Denis R. Chirles.
Personnel Profile: Full-Time Paid 8; Full-Time Volunteers 21; Part-Time Paid 20; Part-Time Volunteers 75; Interns 5.
Volunteer Hours: 500
Governing Authority: Tax-exempt.
Institution Type/Description: Children's Museum.
Collections: hands-on exhibits.
Major Exhibits: The Big Adventure (T), 5/14-9/14.
Activities: special programs; classes.
Hours & Admission Prices: Tues.-Sat. 9-5, Sun. noon-5. Admission $10, SC residents $8; discounts to military & educators; members & children under one no charge. Closed Easter; Independence Day; Thanksgiving; Christmas Eve & Day. &
Attendance: 113,205 (accurate)
Membership: Grandparent $65; Family $80; Explorer $135; Play $395.

THE CITADEL ARCHIVES & MUSEUM, 171 Moultrie St., Charleston, SC 29409-6141. Tel.: 843-953-6846. Fax: 843-953-6956.
Web Site: www.citadel.edu/archivesandmuseum
Founded: 1842.
Congressional District: 1
Governing Authority: state. Affiliated with The Citadel, Military College of South Carolina. Tax-exempt: 170(b)(1)(A).
Institution Type/Description: College History Museum.
Collections: cadet uniforms & arms; photographs relating to The Citadel, The Military College of South Carolina.
Research Fields: The Citadel (Institutional).
Facilities: archives by appointment only.
Activities: tours.
Publications: brochure.
Hours & Admission Prices: Daily 12-5. No charge. Closed college, religious & national holidays. &
Attendance: 8,378 (accurate)

CITY HALL COUNCIL CHAMBER GALLERY, 80 Broad St., Charleston, SC 29401-2225. Tel.: 843-724-3729. Fax: 843-720-3959.
E-mail: maybankv@charleston-sc.gov
Web Site: www.charleston-sc.gov/index.aspx?NID=179
Founded: 1818.
Congressional District: 1
Key Personnel: Chief Tourism Official, Vanessa Turner Maybank.
Governing Authority: municipal. Parent Institution: City of Charleston. Tax-exempt: 170(b)(1)(A).
Institution Type/Description: Art Gallery: housed in c.1801 1st U.S. Bank Building.
Collections: oil portraits of U.S. Presidents; marble sculptures; period furnishings; Edison light bulbs; council chamber.
Research Fields: local political history.
Facilities: library of municipal government records.
Activities: guided tours; lectures; gallery talks; permanent exhibitions.
Publications: catalogue.
Hours & Admission Prices: Mon.-Fri. 9-5. No charge; donations accepted. Closed major holidays. &
Attendance: 20,000 (estimated)

CONFEDERATE MUSEUM, Market Hall, Corner of Meeting & Market Sts., Upstairs, 188 Meeting St., Charleston, SC 29401. Mailing Address: P.O. Box 20997, Charleston, SC 29413-0997. Tel.: 843-723-1541.
Founded: 1898.
Key Personnel: Dir., C.E.O. & Chm., June Murray Wells.
Personnel Profile: Part-Time Paid 3; Part-Time Volunteers 15.
Governing Authority: Parent Institution: Charleston Chapter #4, United Daughters of the Confederacy. Tax-exempt.
Institution Type/Description: History Museum.
Collections: Confederate military history; local history & culture; uniforms; flags; weapons; cannon; photographs; period artifacts. Historic Building: Market Hall built 1841.
Research Fields: Confederate history & genealogy.
Facilities: research library.
Activities: guided group tours.
Publications: museum brochure.
Hours & Admission Prices: Tues.-Sat. 11-3:30; group tours for 15 or more by appointment. Adults $5, children $3; members no charge. Closed major holidays. &
Attendance: 18,000 (estimated)

DRAYTON HALL, (M), 3380 Ashley River Rd., Hwy. 61, Charleston, SC 29414-7105. Tel.: 843-769-2600. Fax: 843-766-0878.
E-mail: info@draytonhall.org
Web Site: www.draytonhall.org
Founded: 1974.
Congressional District: 1
Key Personnel: Exec. Dir., George W. McDaniel; Exec. Asst., Dawn Brogan; Dir. Finance, Paula Marion; Property Council Chm. (V), Anthony Wood; Supt. Bldgs. & Grounds, John M. Kidder; Group Sales Coord., Debbi Zimmerman; Dir. Mktg., Vera Ford.
Personnel Profile: Full-Time Paid 24; Part-Time Paid 30; Part-Time Volunteers 10; Interns 5.
Governing Authority: nonprofit organization. Property of the National Trust for Historic Preservation 1785 Massachusetts Ave., N.W., Washington, DC 20036. Tax-exempt: 501(c)(3).
Institution Type/Description: Historic House Museum: housed in c.1738-42, Drayton Family Residence.
Collections: Georgian architecture; historic landscape.
Research Fields: Drayton genealogy; 18th & 19th century plantation system; African-American history; historic preservation; architectural history.
Facilities: walking history/nature trails. Museum-related books & gifts for sale.
Activities: guided tours; school programs & special events; lectures.
Publications: newsletter for members; written tour of Drayton Hall (for hearing impaired people; French and German readers).
Hours & Admission Prices: March-Oct. daily 8:30-5; Nov.-Feb. daily 9:30-5. Adults $14, youth 12-18 $8, children 6-11 $6; discounts to AAA members & military; children under 5 & National Trust members no charge. Closed New Year's Day; Thanksgiving; Christmas. &
Attendance: 53,000 (accurate)
Membership: Friends of Drayton Hall $40; Joint Membership in the Drayton Hall & National Trust for Historic Preservation $45; Associate $75; Partner $100; Contributor $250; Sustaining $500; Drayton Hall Society $1,000.

* **GIBBES MUSEUM OF ART, (M),** 135 Meeting St., Charleston, SC 29401-2297. Tel.: 843-722-2706. Fax: 843-720-1682.
E-mail: amercer@gibbesmuseum.org
Web Site: www.gibbesmuseum.org
Founded: 1858.
Congressional District: 1
Key Personnel: Exec. Dir., Angela D. Mack; Pres., Laura Gates; Dir. Programs & Events, Lasley Steaver; Dir. Collections Administration, Zinnia Willits; Museum Sales Mgr., Sara Meyers; Dir. Operations, Gregory Jenkins.
Personnel Profile: Full-Time Paid 12; Part-Time Paid 8; Part-Time Volunteers 30; Interns 2.
Governing Authority: nonprofit organization. Parent Institution: Carolina Art Association. Tax-exempt: 501(c)(3).
Institution Type/Description: Art Museum.
Collections: American portraits; prints & miniatures; contemporary art; sculpture; Japanese prints; photography.
Major Exhibits: The Great Wave: Japanese in Charleston, 1/17/14-3/23/14; Romantic Spirits: Nineteenth Century Paintings of the South, 1/17/14-3/23/14; Beyond the Darkroom: Photography in the 21st Century, 4/4/14-6/29/14; John Weshmark: Narratives, 4/4/14-6/29/14.
Research Fields: Charleston art & architecture; photography.
Facilities: library.
Activities: guided tours; lectures; films; gallery talks; arts festivals; studio classes for children & adults; temporary & permanent exhibitions.
Publications: books, The Miniature Portrait Collection of The Carolina Art Association; catalogue, Selections from Carolina Art Association Collection; tricentennial catalogue, Art in South Carolina 1670 & 1970; Charles Fraser of Charleston; Oystering: A Way of Life; This is Charleston; Alice Ravenel Huger Smith: An Artist, A Place & A Time; In Pursuit of Refinement: Charlestonians Abroad, 1740-1860; Henry Benbridge: Charleston Portrait Painter; Rhythms of Life: The Art of Jonathan Green; Landscape of Slavery: The Plantation in American Art; Landscape of Slavery: The Plantation in American Art; The Life and Art of Alfred Hutty (Dec. 2011).
Hours & Admission Prices: Tues.-Sat. 10-5, Sun. 1-5. Adults $9, seniors, military & students $7, children 6-12 $5; NARM reciprocal memberships; children under 6 & members no charge. Closed national holidays. &
Attendance: 60,000 (estimated)
Membership: Student $30; Individual $45; Family $100.

HALSEY INSTITUTE OF CONTEMPORARY ART AT THE COLLEGE OF CHARLESTON, (M), 161 Calhoun St., Charleston, SC 29424. Tel.: 843-953-4422. Fax: 843-953-7890.
E-mail: sloanm@cofc.edu
Web Site: www.halsey.cofc.edu
Formerly: Halsey Gallery, School of the Arts, College of Charleston
Founded: 1978.
Congressional District: 1
Key Personnel: Dir., Mark Sloan; Chm. (V), Susan Bass.
Personnel Profile: Full-Time Paid 5; Part-Time Paid 5; Part-Time Volunteers 9; Interns 4.
Operating Expenses: 700,000
Operating Income: 700,000
Governing Authority: college; nonprofit. Parent Institution: College of Charleston. Tax-exempt.
Institution Type/Description: Art Gallery.
Collections: works by contemporary artists.
Major Exhibits: Jody Zellen: Above the Fold, 1/14-3/14; Bob Trotman: Business as Usual, 1/14-3/14; An Art That Nature Makes: The Photography of Rosamond Purcell (T), 8/14-10/14; Jumaadi: Forgive Me Not to Miss You Not (T), 10/14-12/14; Picasso: Diurnes, 10/14-12/14.
Research Fields: Contemporary art.
Facilities: reference library; archives; video screening room; conference room; 275-seat auditorium; artist-in-residence studios.
Activities: lectures; films; symposia; temporary exhibits; visiting artist program; international artist-in-residence program.
Publications: catalogs, Effective Sight: The Paintings of Juan Logan; Self-Made Worlds: Visionary Folk Art Environments; Hung Liu: Washington Town Blues; The Right to Assemble; With Beauty Before Us: The Navajo of Chil Chen Beto; Pop-Luxe: The Language of the Garment; Cheryl Goldsleger: Improvisations; Black Boiled Coffee and the Cacophony of Frogs; Evon Streetman in Retrospect; Appropriate to the Moment: The Paintings of Michael Tyzack; Cheryl Peacock's Failure Journal; No Man's Land: Fragile Ecologies and Contemporary Photographs; Alive Inside: The Lure & Lore of the Sideshow; Force of Nature: Site Installations by Ten Japanese Artists; Aldwyth: work v./work n; Palmetto Portraits Project; Aggie Zed: Keeper's Keep; Return to the Sea: Saltworks by Motoi Yamamoto; The Paternal Suit: Heirlooms from the F. Scott Hess Family Foundation; The Pulse Dome Project: Art & Design; Tales of the Conjure Woman: Renee Stout.

Hours & Admission Prices: Sept.-June Mon.-Sat. 11-4. No charge; donations accepted. &
Attendance: 16,000 (estimated)
Membership: No Monet (Student) $15; Minimalist $30; Modernists $55; Futurist $125; Postmodernist $350; Conceptualist $2,500; Utopian $5,000; Sugar Dada/Mama $10,000.

HISTORIC CHARLESTON FOUNDATION, 40 E. Bay, Charleston, SC 29401-2547. Mailing Address: P.O. Box 1120, Charleston, SC 29402-1120. Tel.: 843-723-1623. Fax: 843-577-2067.
E-mail: krobinson@historiccharleston.org
Web Site: historiccharleston.org
Founded: 1947.
Congressional District: 1
Key Personnel: Exec. Dir., Katharine S. Robinson; Dir. Museums & Preservation, Winslow Hastie; Dir. Reproductions, Stephen Hanson; Dir. Communications, Leigh Handal; Dir. Finance, Cynthia Ellis; Dir. Devel., Kathryn Matthew; Dir. Operations, Betty Guerard; Cur., Brandy Culp; Museum Shop Mgr., Rich Gaskalla.
Personnel Profile: Full-Time Paid 24; Part-Time Paid 76; Part-Time Volunteers 500; Interns 6.
Governing Authority: nonprofit organization. Branch Museums: c.1808 Nathaniel Russell House, 51 Meeting St.; c.1818 Aiken-Rhett House, 48 Elizabeth St. Tax-exempt: 501(c)(3).
Institution Type/Description: Foundation Operated House Museums: housed in the Captain James Missroon House; built in 1789.
Collections: fine & decorative arts of the period 1750-1860 with emphasis on Charleston made objects.
Research Fields: preservation of historic houses & buildings; neighborhood revitalization; decorative & fine arts; Charleston social history, African-American history.
Activities: guided & audio tours; lectures; formally organized education programs for docents.
Publications: Grandeur Preserved: The House Museums of Historic Charleston Foundation (2008); occasional technical publications; newsletter, Historic Charleston Foundation; HCF Annual Report.
Hours & Admission Prices: Mon.-Sat. 10-5, Sun. 2-5. $10 per house or combination with Aiken-Rhett House $16, youth 6-16 $5. Private Home Tours: mid-March to mid-April. Adults $45. Closed Thanksgiving; Christmas Eve & Day. &
Attendance: 80,000 (accurate)

JOHN RIVERS COMMUNICATIONS MUSEUM, 58 George St., College of Charleston, Charleston, SC 29424. Tel.: 843-953-5810.
E-mail: zenderr@cofc.edu
Web Site: www.cofc.edu/~jrmuseum
Formerly: The Broadcast Museum
Founded: 1989.
Key Personnel: Cur., Rick Zender.
Personnel Profile: Full-Time Paid 1; Part-Time Paid 1; Part-Time Volunteers 3.
Governing Authority: state; public college; nonprofit. Parent Institution: College of Charleston.
Institution Type/Description: Communications Museum: housed in the c.1803 Barnard Elliott House.
Collections: historical artifacts which provide a forum for understanding the scientific & social implications of communications; authentic communications hardware including the pre-1950 phonograph, radio, telegraph and television.
Facilities: 30-seat auditorium.
Activities: documentary film & discussion series; formal education programs for adults & children; guided tours; lectures; school loan service.
Hours & Admission Prices: Mon.-Fri. 12-4 & by appointment. Summer hours vary. No charge; donations accepted. Closed federal, national, state & school holidays.
Attendance: 3,000 (estimated)

MACAULAY MUSEUM OF DENTAL HISTORY, Medical University of South Carolina, 175 Ashley Ave., Charleston, SC 29425. Mailing Address: 175 Ashley Ave., MSC 403, Charleston, SC 29425. Tel.: 843-792-2288. Fax: 843-792-8619.
E-mail: waringhl@musc.edu
Web Site: waring.library.musc.edu
Founded: 1975.
Congressional District: 1
Key Personnel: Cur., Susan Hoffius.
Personnel Profile: Full-Time Paid 3; Part-Time Paid 1.
Governing Authority: state. Parent Institution: Medical University of South Carolina. Tax-exempt.

Institution Type/Description: Dental Museum.
Collections: 6,000 items including early dental chairs, foot powered drills, wooden dental cabinets, dental lathe, old dental X-ray units, itinerant dentists' medicine cases, cases of molds for crowns, dental turn keys, & period dental instruments; manuscripts.
Research Fields: antique dental instruments and equipment of South Carolina dentists.
Facilities: 200-vol. library of medical and dental textbooks of the late 19th and early 20th centuries and books on the history of dentistry available for research on premises.
Activities: guided tours; lectures; permanent exhibitions.
Publications: brochure.
Hours & Admission Prices: Mon.-Fri. 8:30-5; by appointment. No charge; donations accepted. Closed holidays.
Attendance: 250
Membership: Student $5; Regular $20; Contributing $100; Life $1,000.

MAGNOLIA PLANTATION AND GARDENS, 3550 Ashley River Rd., Charleston, SC 29414-7127. Tel.: 843-571-1266. Fax: 843-571-5346. Facebook: Magnolia Plantation and Gardens.
E-mail: frazierh@aol.com
Web Site: magnoliaplantation.com
Founded: 1676.
Congressional District: 1
Key Personnel: Exec. Dir., Tom Johnson; Dir. Operations, Mary Ann Johnson; Mktg. & Public Rels. Mgr., Herb Frazier; Group Sales, Sharon Newton; Museum Shop Mgr., Debbie Kieter.
Personnel Profile: Full-Time Paid 40; Part-Time Paid 25; Part-Time Volunteers 18; Interns 5.
Volunteer Hours: 1,521
Operating Expenses: 3,708,764
Operating Income: 5,018,973
Governing Authority: Parent Institution: Magnolia Plantation Corp.
Institution Type/Description: Historic House and Gardens: c.1680s Drayton family home & gardens.
Collections: early natural history & wildlife art; early American furniture.
Research Fields: general horticulture & history; low country history; Drayton family.
Facilities: botanical garden; zoological park; 60 acres Audubon swamp garden; nature trails.
Activities: guided tours; holiday special events
Publications: brochure; historic book of plantation; art book.
Hours & Admission Prices: Daily 9-5. Plantation: adults $15, children 6-12 $10; discounts to educational groups & groups of 15 or more and AAM & ICOM members; children under 6 no charge. House Museum: admission $8; children under 6 no charge. Audubon Swamp Garden: admission $8; children under 6 no charge. Nature Train Tour, Nature Boat Tour & Slavery to Freedom Tour $8.
Attendance: 135,500 (estimated)
Membership: Individual $55; Family $100.

MIDDLETON PLACE, 4300 Ashley River Rd., Charleston, SC 29414-7206. Tel.: 843-556-6020. Fax: 843-766-4460.
E-mail: ttodd@middletonplace.org
Web Site: middletonplace.org
Formerly: Middleton Place House Museum
Founded: 1974.
Key Personnel: Pres., Charles H.P. Duell; Vice Pres. Accounting, Ileen Grange; Vice Pres. Museums, M. Tracey Todd; Archivist, Barbara Doyle; Mktg. Asst., Sabrina Fidler; Dir. Mktg. & Public Rels., Warren Cobb; Membership Coord., Sue Braund; Security, Jim Woodle; Museum Shop Mgr., Maria Keneally.
Personnel Profile: Part-Time Volunteers 300.
Governing Authority: public; nonprofit foundation. Parent Institution: Middleton Place Foundation. Subsidiary Institution: Edmondston-Alston House, Charleston, SC. Tax-exempt: 501(c)(3).
Institution Type/Description: Historic House: restored 1755 structure.
Collections: concentration on mid-18th century to present American history; special emphasis on agrarian life of 19th-century Carolina Low Country.
Research Fields: life & work of artist John Izard Middleton, 1785-1849; Henry Middleton, 1770-1846, Minister to Russia (1820-1830); diplomacy; family history; political, cultural & decorative arts influences of period; African-American history & culture; botanical history of gardens & landscape architecture; Arthur Middleton, signer of the Declaration of Independence (1742-1787).
Facilities: living history stableyards with heritage breed livestock, 65-acre garden; 250-seat restaurant. Craft items, folk art, historic reproductions of glassware, furniture & porcelain, books on gardening & estate jewelry and other museum-related items for sale.

Activities: guided tours; lectures; concerts; docent program; participatory exhibits.
Publications: quarterly newsletter, The Notebook.
Hours & Admission Prices: Garden & Plantation Stableyards: daily 9-5. House Museum: Mon. 12-4:30, Tues.-Sun. 10-4:30. House Guided Tour: $10. Garden & Planetarium: adult $25, children 7-15 $5. &
Attendance: 100,000 (estimated)
Membership: Couple $100; Multiple $200.

NATHANIEL RUSSELL HOUSE, 51 Meeting St., Charleston, SC 29401-2536. Tel.: 843-724-8481.
E-mail: vperry@historiccharleston.org
Web Site: www.historiccharleston.org
Institution Type/Description: Historic House Museum: built in 1808. Listed on the National Register of Historic Places.
Collections: Nathaniel Russell's life & career; period furnishings; personal artifacts; paintings.
Activities: guided tours.
Hours & Admission Prices: Tours: Mon.-Sat. 10-5, Sun. 2-5. Adults $10, children 6-16 $5; children 5 & under no charge. Closed Thanksgiving; Christmas Eve & Day.

OLD EXCHANGE & PROVOST DUNGEON, 122 E. Bay St., Charleston, SC 29401-2103. Tel.: 843-727-2165; 888-763-0448. Fax: 843-727-2163.
E-mail: oldexchange@infoave.net
Web Site: www.oldexchange.com
Founded: 1976.
Congressional District: 3
Key Personnel: Dir., Tony Youmans; Commission Chm., Mrs. Laura Kennedy LeGrand; Museum Shop Mgr., Suzanne Houser.
Personnel Profile: Full-Time Paid 9; Part-Time Paid 17; Part-Time Volunteers 1.
Governing Authority: state government. Parent Institution: Daughters of American Revolution. Subsidiary Institution: State of South Carolina. Management: City of Charleston, SC. Tax-exempt.
Institution Type/Description: Historic Building: c.1771 The Old Exchange Building, based on its architectural distinction & its role in the formation of the United States, it is historically one of America's significant structures; an exchange & customs house, the building has maintained service to the people politically, socially, economically & educationally.
Collections: colonial artifacts; postal exhibits; civil war art; photographs and history of pirates of the Carolinas.
Facilities: banquet facilities. Gift items for sale.
Activities: tours; lectures; organized education programs. Museum Sponsors: Colonial Christmas Program; Halloween Ghostly Guests story-telling presentation; Constitution Week in September.
Hours & Admission Prices: Daily 9-5. Adults $8, children 7-12 $4; discounts to AAA & AARP members, seniors, groups, military & students; Friends of the Old Exchange members and children 6 & under no charge. Closed major holidays. &
Attendance: 56,242 (accurate)
Membership: Friends of the Old Exchange: Students $10; Members $25; $50; $75; $100.

THE OLD SLAVE MART MUSEUM, 6 Chalmers St., Charleston, SC 29401-3005. Tel.: 843-724-3746. Fax: 843-724-3734.
E-mail: vaughn@ci.charleston.sc.us
Key Personnel: Dir., Nichole Green; Dir. Media Rels., Barbara Vaughn; Cur., Elaine Nichols; Archivist, Harlan Greene.
Governing Authority: city.
Institution Type/Description: History Museum: housed in a former slave auction gallery; built in 1859.
Collections: African American history, arts & crafts.
Hours & Admission Prices: Mon.-Sat. 9-5. Closed New Year's Day; Thanksgiving; Christmas.

THE POWDER MAGAZINE, 79 Cumberland St., Charleston, SC 29401-3112. Tel.: 843-722-9350. Fax: 843-722-3711.
E-mail: info@powdermag.org
Web Site: www.powdermag.org
Founded: 1713.
Congressional District: 1
Key Personnel: Dir., Alan Stello; Chm. (V), Mary Mundy.
Personnel Profile: Full-Time Paid 1; Part-Time Paid 2; Part-Time Volunteers 5.
Governing Authority: Parent Institution: National Society of the Colonial Dames of America in the State of South Carolina. Tax-exempt.

Institution Type/Description: History Museum: housed in c. 1713 Powder Magazine.
Collections: local & military history; period artifacts.
Research Fields: SC colonial military history.
Activities: colonial living history; demonstrations; permanent exhibitions.
Publications: The Powder Keg.
Hours & Admission Prices: Mon.-Sat. 10-4, Sun. 1-4. Family $6, adults $3, children $1.
Attendance: 15,000 (estimated)

SOUTH CAROLINA AQUARIUM, 100 Aquarium Wharf, Charleston, SC 29401-6300. Tel.: 843-720-1990. Fax: 843-720-3861.
Web Site: www.scaquarium.org
Founded: 2000.
Congressional District: 1
Key Personnel: Cur., Rachel Kalisperis; Education, Whit McMillan; Public Rels., Kate Dittloff; Dir. Human Resources, Keisha Legerton.
Personnel Profile: Full-Time Paid 70; Part-Time Volunteers 350; Interns 16.
Governing Authority: private; nonprofit organization. Tax-exempt: 501(c)(3).
Institution Type/Description: Aquarium.
Collections: South Carolina's marine & freshwater animals & plants; aquatic organisms from around the world; videos.
Research Fields: fisheries; shellfish; plankton; marketing & visitor studies; sea turtle rehabilitation.
Facilities: library; aquarium; educational facilities; 75,000 sq. ft. exhibit space. Aquarium-related items for sale.
Activities: docent program; formal education programs; participatory exhibits; rental gallery; teachers training program. Annual Events: Community Appreciation Days in January; members' events; Scuba-do fundraiser.
Publications: quarterly members' magazine, Tributaries.
Hours & Admission Prices: March-Aug. daily 9-5; Sept.-Feb. daily 9-4. Adults $17.95, senior citizens $16.95, children 2-11 $10.95; discounts to groups; children one & under and members no charge. Closed Thanksgiving; Christmas. &
Attendance: 427,101 (accurate)
Membership: Individual $50; Dual $65; Individual Plus $75; Grandparent $80; Family $90; Family Plus $135; Friend $250 & up; Associate $500 & up; Patron $1,000 & up; Benefactor $2,500 & up.

SOUTH CAROLINA HISTORICAL SOCIETY, 100 Meeting St., Charleston, SC 29401-2215. Tel.: 843-723-3225. Fax: 843-723-8584.
E-mail: faye.jensen@schsonline.org
Web Site: www.schsonline.org
Founded: 1855.
Congressional District: 1
Key Personnel: C.E.O. & Dir., Faye L. Jensen, Ph.D.; Chm. (V), J. Paul Trouche; Asst. Dir., John Tucker; Archivist, Reference & Research Consultant, Mary Jo Fairchild; Membership, Halley Cella; Publications, Matt Lockhart; Programs, Virginia Ellison.
Personnel Profile: Full-Time Paid 7; Part-Time Paid 4; Part-Time Volunteers 5; Interns 2.
Governing Authority: nonprofit organization. Tax-exempt: 501(c)(3).
Institution Type/Description: Historical Society Library & Archives: housed in 1822 Robert Mills Building.
Collections: manuscripts; photographs; art; architectural records; maps, plats & monuments; printed material; books; genealogical charts; maritime Civil War research materials & artifacts.
Research Fields: Civil War artifacts; 1700-1900 manuscripts.
Facilities: 40,000-vol. library of books & manuscripts, available to the public; reading room.
Activities: lectures; loan & temporary exhibitions. Society Sponsors: fall plantation tour; annual meeting & house tour; symposia.
Publications: quarterly magazines, South Carolina Historical Magazine; Carologue.
Hours & Admission Prices: Tues.-Fri. 9-4, Sat. 9-2. Adults $5; children & members no charge. Closed federal holidays; New Year's Eve & Day; Christmas Eve, Day & week.
Attendance: 3,500 (estimated)
Membership: Young Carolinians $35; Regular & Libraries $55; Sustainer $100 & up; Business & Corporation $1,000 & up; Sponsor $500; Patron $1,000; Benefactor $2,000 & up.

THOMAS ELFE HOUSE, 54 Queen St., Charleston, SC 29401-2806. Tel.: 843-722-9161.
E-mail: info@thomaselfehouse.com
Web Site: www.thomaselfehouse.com
Institution Type/Description: Historic House Museum.

Collections: local history & culture; period furnishings; personal artifacts; photographs.
Hours & Admission Prices: By appointment. Adults $8, children $4; discounts to groups.

WARING HISTORICAL LIBRARY, Medical University of South Carolina, 175 Ashley Ave., Charleston, SC 29425. Mailing Address: MSC 403, Charleston, SC 29425. Tel.: 843-792-2288. Fax: 843-792-8619.
Web Site: waring.library.musc.edu
Founded: 1966.
Congressional District: 1
Key Personnel: Dir., W. Curtis Worthington, Jr., M.D.; Cur., Susan Hoffius.
Personnel Profile: Full-Time Paid 3; Part-Time Paid 1.
Governing Authority: state. Parent Institution: Medical University of South Carolina. Subsidiary Institution: Health Sciences Library. Tax-exempt.
Institution Type/Description: University Medical Museum & Library: housed in 1894 building designed by architect John Snook.
Collections: medical objects including doctors' saddle bags, medicine chests, amputation kits, electro-therapeutic machines, bleeding instruments, obstetrical specula and forceps; pharmaceutical items; manuscripts; books authored by South Carolinians.
Research Fields: period instruments & equipment of physicians, nurses and pharmacists of South Carolina.
Facilities: 14,000-vol. library of books on the history of the health sciences, particularly medicine; old & new books by South Carolinians available for research.
Activities: guided tours; permanent & temporary exhibitions.
Publications: brochure; newsletter.
Hours & Admission Prices: Mon.-Fri. 8:30-5. No charge; donations accepted. Closed holidays. &
Membership: Student $5; Regular $20; Contributing $100; Life $1,000.

Cheraw

CHERAW LYCEUM MUSEUM, 200 Market St., Cheraw, SC 29520-2414. Mailing Address: P.O. Box 219, Cheraw, SC 29520-0219. Tel.: 843-537-8401 & 8425. Fax: 843-537-8407.
Web Site: www.cheraw.com
Founded: 1962.
Congressional District: 5
Key Personnel: C.E.O., J. William Taylor; Treas., Helen D. Funderburk; Cur., Sarah C. Spruill.
Personnel Profile: Part-Time Paid 1; Part-Time Volunteers 2.
Governing Authority: municipal; nonprofit. Parent Institution: Town of Cheraw. Tax-exempt: 501(c)(3).
Institution Type/Description: History Museum: housed in a former court building; built in 1825
Collections: Revolutionary War artifacts; clothing & personal items of early 19th century; Civil War artifacts; products of present industries; Cheraw Indian artifacts & displays; artifacts from steamboat era; 19th-century bank notes & commercial ledgers.
Research Fields: local history; life on Great Pee Dee River; Cheraw Indians; Confederate War; steamboat era; Dizzy Gillespie memorabilia.
Activities: guided tours; arts festivals.
Publications: brochures.
Hours & Admission Prices: Mon.-Fri. 9-5, Sat.-Sun. by appointment. No charge; donations accepted.
Attendance: 4,000 (estimated)

Chesnee

CAROLINA FOOTHILLS ARTISAN CENTER, 124 W. Cherokee, Chesnee, SC 29323. Tel.: 864-461-3050.
Institution Type/Description: Art Gallery.
Collections: works by local artists.
Facilities: 2,000 sq. ft. exhibit space.
Activities: classes; special events.
Hours & Admission Prices: Mon.-Sat. 10-5:30.

Chester

CHESTER COUNTY HISTORICAL SOCIETY MUSEUM, 107 McAliley St., Chester, SC 29706-1741. Mailing Address: P.O. Box 811, Chester, SC 29706-0811. Tel.: 803-385-2332. Facebook: Chester County, SC Historical Society.
E-mail: ccmuseum@truvista.net
Web Site: chesterscmuseum.org
Founded: 1959.

Congressional District: 6
Key Personnel: Museum Dir., Liz Anderson
Governing Authority: society; nonprofit organization. Tax-exempt.
Institution Type/Description: Historical Society Museum: housed in 1914 Chester County Jail.
Collections: Gatlin Catawba Indian collection of more than 30,000 artifacts; 1825 Mills' Atlas of South Carolina; portraits of the four former Chester County residents who were signers of the Ordinance of Secession; switchboard; phone booth.
Activities: permanent exhibitions.
Publications: quarterly newsletter.
Hours & Admission Prices: Wed. & Fri.-Sat. 10-3. Adults $3; members no charge.
Attendance: 1,293 (accurate)
Membership: Student & Senior $15; Single $20; Family $40.

CHESTER COUNTY TRANSPORTATION MUSEUM, 157 Wylie St., Chester, SC 29706. Tel.: 803-385-2330.
E-mail: ccmuseum@truvista.net
Institution Type/Description: Transportation Museum: housed in the former Seaboard Railroad freight depot.
Collections: local railroad history; period furnishings; railroad artifacts; early automobiles & bicycles; a horse-drawn buggy; model railroads.
Facilities: Museum-related items for sale.
Hours & Admission Prices: Tues.-Sat. 11-3. Adults $3, students & senior citizens $2.

Clemson

BOB CAMPBELL GEOLOGY MUSEUM, 140 Discovery Lane, Clemson University, Clemson, SC 29634. Tel.: 864-656-4600. Fax: 864-656-6230.
E-mail: bcgm@clemson.edu
Web Site: www.clemson.edu/geomuseum
Founded: 1989.
Key Personnel: Dir., Dr. Patrick McMillan; Cur. Education, Allison Jones.
Personnel Profile: Full-Time Paid 1; Part-Time Paid 3; Part-Time Volunteers 12.
Governing Authority: public university; nonprofit. Parent Institution: Clemson University. Tax-exempt.
Institution Type/Description: Geology Museum.
Collections: minerals; gems; fossils; rocks from around the world; meteorites; mining artifacts; fluorescent minerals; skeletal reproduction of saber-toothed cat.
Research Fields: vertebrate paleontology.
Facilities: 1,200-vol. library of geological books, journals & special reports; botanical garden; 2,800 sq. ft. exhibit space. Museum-related items for sale.
Activities: guided tours; scout programs by request; after school programs; monthly hikes. Museum Sponsors: Open House in October; Identification Day.
Publications: semiannual member newsletter, The Geological Record.
Hours & Admission Prices: Museum: Wed.-Sat. 10-5, Sun. 1-5. Adults $3, children $2; discounts to SCFM (South Carolina Federation of Museums); museum members, children under 2 & Clemson University students no charge. Reciprocal member agreement. Closed major holidays & home football games. &
Attendance: 8,000 (estimated)
Membership: Individual $45; Family $65; Patron $1,000.

FORT HILL (THE JOHN C. CALHOUN HOUSE), Fort Hill St., Clemson University, Clemson, SC 29634-5615. Mailing Address: Trustee House, Clemson University, Box 345615, Clemson, SC 29634. Tel.: 864-656-2475. Fax: 864-656-1026.
E-mail: hiottw@clemson.edu
Web Site: www.clemson.edu/about/history/properties/fort-hill.html
Founded: 1889.
Congressional District: 3
Key Personnel: Dir., William D. Hiott, Sr
Personnel Profile: Full-Time Paid 2; Part-Time Paid 6; Part-Time Volunteers 10; Interns 18.
Governing Authority: university; society. Parent Institution: Clemson University. Tax-exempt: 501(c)(3).
Institution Type/Description: Historic House Museum: c.1803 home of John C. Calhoun, 1825-1850.
Collections: original furnishings; Flemish & family portraits; personal artifacts; government documents.
Research Fields: American & Clemson history; architecture; decorative arts; horticulture.
Facilities: Museum-related items for sale.

Activities: guided tours; lectures.
Publications: brochure.
Hours & Admission Prices: Mon.-Sat. 10-12 & 1-4:30, Sun. 2-4:30. Suggested Donation: adults $5, senior & student $4, children $2; discounts to groups, AAM & ICOM members. Closed university holidays.
Attendance: 23,583 (accurate)

HANOVER HOUSE, South Carolina Botanical Garden, Perimeter Rd., Clemson University, Clemson, SC 29634. Mailing Address: Box 345615, Historic Properties, Clemson University, Clemson, SC 29634-5615. Tel.: 864-656-2241 & 2475. Fax: 864-656-1026.
E-mail: hiottw@clemson.edu
Web Site: www.clemson.edu/about/history/properties/hanover-house.html
Founded: 1941.
Congressional District: 3
Key Personnel: Dir. & Cur., William D. Hiott, Sr.
Governing Authority: university; society. Parent Institution: Clemson University. Tax-exempt: 501(c)(3).
Institution Type/Description: Historic House Museum: home built in 1716.
Collections: period furnishings; 18th-19th century artifacts.
Research Fields: early modern Europe; Colonial America; early American Colonial architecture; Huguenots.
Activities: guided tours; lectures.
Publications: Hanover House.
Hours & Admission Prices: Sat. 10-12 & 1-5, Sun. 2-4:30; groups & other times by appointment. Suggested Donations: adults $5, seniors $4, children $2; discounts for AAM & ICOM members; members no charge. Closed university holidays. &

RUDOLPH E. LEE GALLERY, G-50 Lee Hall, Clemson University, Clemson, SC 29634. Mailing Address: P.O. Box 340509, Clemson, SC 29634. Tel.: 864-656-3899. Fax: 864-656-7523.
E-mail: woodwaw@exchange.clemson.edu
Web Site: www.clemson.edu/caah/leegallery
Founded: 1956.
Key Personnel: Pres., James F. Barker; Dir., Denise Woodward-Detrich; Membership, Jennifer Staley; Volunteer Coord., Fleming Markel.
Governing Authority: university. Parent Institution: Clemson University. Tax-exempt.
Institution Type/Description: Art Gallery.
Collections: paintings; graphics; photography; architecture projects.
Facilities: 2,400 sq. ft. exhibit space.
Activities: guided tours; lectures; films; gallery talks; inter-museum loan, permanent, temporary & traveling exhibitions.
Publications: exhibition posters; Clemson National Print and Drawing catalog.
Hours & Admission Prices: Mon.-Thurs. 9-4:30, Sun. 2-5. No charge. Closed university, state & national holidays.

THE SOUTH CAROLINA BOTANICAL GARDEN, 150 Discovery Lane, Clemson University, Clemson, SC 29634-0174. Tel.: 864-656-3405 & 2458. Fax: 864-656-6230.
E-mail: scbg@clemson.edu
Web Site: www.clemson.edu/scbg
Founded: 1961.
Congressional District: 3
Key Personnel: Dir., Patrick McMillan; Dir. Education, Lisa Wagner; Garden Mgr., James Arnold; Sr. Horticulturist, John Bodiford; Horticulturist, Kathy Bridges; Facilities Mgr., Eric Soto; Garden Rentals & Visitor Svcs. Mgr., Darlene Evand.
Personnel Profile: Full-Time Paid 8; Part-Time Paid 8; Part-Time Volunteers 50; Interns 1.
Governing Authority: university. Parent Institution: Clemson University. Tax-exempt.
Institution Type/Description: Botanical Garden: located on part of the original John C. Calhoun Plantation Estate.
Collections: niche gardens; natural woodlands; geology; historic buildings.
Research Fields: plant introductions with USDA; turf grass research plots.
Facilities: botanical garden; classrooms; visitor center; cafe.
Activities: guided tours; lectures; formally organized education programs for undergraduate & graduate students; visual & performing arts programs; nature, science & gardening programs; nature walks; cultural festivals; educational series.
Publications: quarterly newsletter, The Gardens Gate.
Hours & Admission Prices: Daily dawn-dusk. No charge; donations accepted. &
Attendance: 100,000 (estimated)

Membership: Contributor $45-$99; Sustaining $100-$249; Associate $250-$499; Patron $500-$999; Benefactor $1,000-$4,999; Fellow $5,000-$9,999; Founder $10,000 & up.

STROM THURMOND INSTITUTE, Clemson University, Silas Pearman Blvd., Clemson, SC 29634. Tel.: 864-656-4700. Fax: 864-656-4780.
E-mail: sti-web@strom.clemson.edu
Web Site: www.strom.clemson.edu
Key Personnel: Dir., Dr. Robert H. Becker.
Governing Authority: nonprofit.
Institution Type/Description: Archive.
Collections: Senator Thurmond's personal papers & memorabilia.
Hours & Admission Prices: Mon.-Fri. 8-4:30. No charge. Closed New Year's Day; Spring break in March; Independence Day; Oct. break; Thanksgiving weekend; Christmas break.

Clinton

MUSGROVE MILL STATE HISTORIC SITE VISITOR CENTER, 398 State Park Rd., Clinton, SC 29325. Tel.: 864-938-0100.
Institution Type/Description: Visitor Center: housed on the historic site of the Battle of Musgrove Mill in 1780.
Collections: local history & the Battle of Musgrove Mill; photographs; period artifacts.
Facilities: 2.5 mile nature trail.
Activities: living history programs; interpretive trails; special events; encampments.
Hours & Admission Prices: Call for hours.

Columbia

COLUMBIA FIRE DEPARTMENT MUSEUM, 1800 Laurel St., Columbia, SC 29201-2627. Tel.: 803-733-8350. Fax: 803-733-8311.
E-mail: cfdjreich@columbiasc.net
Web Site: www.columbiasouthcarolina.com/fire-museum.html
Founded: 1996.
Key Personnel: Cur., John G. Reich.
Governing Authority: municipal government; nonprofit.
Institution Type/Description: Fire-Fighting Museum.
Collections: 1903 metropolitan horse-drawn steamer; history of fire service in Columbia, South Carolina.
Research Fields: black history; oral history.
Facilities: 100-vol. library of scrapbooks & yearbooks; classroom.
Publications: committee newsletter published three times annually.
Hours & Admission Prices: Mon.-Fri. 8:30-5, Sat.-Sun. by appointment. No charge; donations accepted. &
Attendance: 10,000 (estimated)

✳   **COLUMBIA MUSEUM OF ART,** 1515 Main St., Columbia, SC 29201. Mailing Address: P.O. Box 2068, Columbia, SC 29202-2068. Tel.: 803-799-2810. Fax: 803-343-2150.
E-mail: ahorne@columbiamuseum.org
Web Site: www.columbiamuseum.org
Founded: 1950.
Congressional District: 2
Key Personnel: Exec. Dir., Karen Brosius; Chief Cur., Will South; Dir. External Affairs, Joelle Ryan Cook; Dir. Education, Kerry Kuhlkin-Hornsby; Dir. Facility Operations, Michael Roh; Dir. Devel., Lowndes Macdonald; Mgr. Human Resources, Teri Keener Mewhorter; Museum Shop Mgr., Bohumila Augustinova.
Personnel Profile: Full-Time Paid 30; Part-Time Paid 21; Part-Time Volunteers 100; Interns 4.
Governing Authority: nonprofit organization. Tax-exempt: 501(c)(3).
Institution Type/Description: Art Museum.
Collections: European & American fine and decorative arts from the 14th century to present: Samuel H. Kress Collection of Renaissance and Baroque art; painting; prints; sculpture; furniture; ceramics; 19th & 20th century design; photography.
Research Fields: items pertaining to the collections: European & American fine design & decorative arts; general art.
Facilities: 14,000-vol. library of art history books, journals, and artist files available for on-site research; art studios; 154-seat auditorium; rental facilities. Museum-related items for sale.
Activities: guided tours; training programs; lectures; concerts; films; gallery talks; art classes; family programs.
Publications: collections; annual report; exhibition catalogs.
Hours & Admission Prices: Tues. Fri. 11-5, Sat. 10-5, 1st Fri. of month 11-8,

Sun. 12-5. Adults $10, seniors 65 & over and military $8, students $5; discounts to AAM & ICOM members; children 5 & under, members & Sun. no charge. Closed major holidays. &
Attendance: 135,000 (accurate)
Membership: Individual $45; Dual $65; Kids Plus $75; Patron $200; Premier Society $500 & up.

EDVENTURE, INC., (M), 211 Gervais St., Columbia, SC 29201-3067. Mailing Address: P.O. Box 1638, Columbia, SC 29202-1638. Tel.: 803-779-3100. Fax: 803-779-3144.
E-mail: info@edventure.org
Web Site: www.edventure.org
Founded: 1994.
Congressional District: 2
Key Personnel: C.E.O. & Pres., Catherine Wilson Horne; Chm., Daniel B. Amaker, Jr.; C.F.O., Julia Kennard; Vice Pres. Education, Nikki Williams; Dir. Mktg., Wayne Thornley.
Personnel Profile: Full-Time Paid 34; Part-Time Paid 71.
Operating Expenses: 4,124,259
Operating Income: 4,127,751
Governing Authority: private; nonprofit organization. Parent Institution: Ed-Venture Inc. Tax-exempt: 501(c)(3).
Institution Type/Description: Children's Museum.
Collections: hands-on learning & discovery; exhibits & programs on health & wellness, communities, communications, environment & international cultures.
Activities: formal education programs for children; traveling exhibitions; school outreach programs.
Publications: bimonthly newsletter.
Hours & Admission Prices: Summer: Mon.-Sat. 9-5, Sun. 12-5; Winter: Tues.-Sat. 9-5, Sun. 12-5. Adults & children $11.50; members no charge. Closed Thanksgiving; Christmas Eve & Day. &
Attendance: 233,000 (estimated)
Membership: Weekday $95; Anytime $150; Explorer Circle $200; Developer's Circle $500; Leader's Circle $1,000.

GOVERNOR'S MANSION, 800 Richland St., Columbia, SC 29201-2397. Tel.: 803-737-1710. Fax: 803-737-3860.
E-mail: nancybunch@gov.sc.gov
Web Site: www.scgovernorsmansion.org
Founded: 1855.
Key Personnel: C.E.O., Gov. Nikki Haley; Chm. (V), Mary Ross; Cur. & Tour Dir., Nancy B. Bunch.
Personnel Profile: Full-Time Paid 8; Full-Time Volunteers 40; Part-Time Paid 1; Part-Time Volunteers 50.
Governing Authority: state. Governor's Mansion Commission. Tax-exempt.
Institution Type/Description: Historic House: 1855 Governor's Residence.
Collections: furniture; paintings; objects of art which pertain to the history of this state; historical garden. Historic Houses: 1854 The Lace House; 1830 The Caldwell-Boylston House.
Facilities: botanical garden.
Activities: guided tours by appointment; docent program or council; loan, permanent & temporary exhibitions.
Hours & Admission Prices: Tours: summer: Tues.-Wed. 10, 10:30 & 11; winter, spring & fall: Tues.-Thurs. 10, 10:30 & 11. No charge; donations accepted. &
Attendance: 15,000 (accurate)

✳   **HISTORIC COLUMBIA FOUNDATION, (M),** 1601 Richland St., Columbia, SC 29201-2633. Tel.: 803-252-7742, ext. 14. Fax: 803-929-7695.
E-mail: rwaites@historiccolumbia.org
Web Site: www.historiccolumbia.org
Founded: 1961.
Congressional District: 6
Key Personnel: Exec. Dir., Robin Waites; Chmn. (V), Michael Edens; Dir. Cultural Resources, John Sherrer, III; Dir. Programs, Sarah Bladwell.
Personnel Profile: Full-Time Paid 15; Part-Time Paid 15; Part-Time Volunteers 200; Interns 6.
Governing Authority: nonprofit organization. Subsidiary Institutions: Mann-Simons Cottage; Woodrow Wilson Family Home; Robert Mills House; Hampton-Preston Mansion; Modjeska Simkins House. Tax-exempt: 501(c)(3).
Institution Type/Description: Historical Society Museum.
Collections: Empire period decorative arts; Wade Hampton family artifacts; Victorian items & decorative arts; Woodrow Wilson material. Historic Houses: Seibels House & Big Apple Club.
Research Fields: local & state history; 19th-century decorative arts; Wade

Hampton & Woodrow Wilson families; 19th- & 20th-century architectural history; African-American consumption patterns, 19th-20th century; reconstruction era.

Facilities: 11 acres of historical gardens.

Activities: historic house & walking tours; lectures; films; heritage tours; workshops; speaker's bureau; preservation advocacy; rental facility.

Publications: Historic Columbia Foundation News; Richland County's Rural African American School, 1895-1954; Experience Historic Columbia, Main Street Columbia, SC.

Hours & Admission Prices: Robert Mills, Hampton-Preston, Woodrow Wilson & Mann-Simions: Tues.-Sat. 10-4, Sun. 1-5. Houses: $6 each; discounts to AAM, AAA & SEMC members, senior citizens, teachers, students, military; members, Seibels House & Big Apple no charge.

Attendance: 35,000 (estimated)

Membership: Individual $35; Family $50; Friend $100; Woodrow Wilson Society $250; Mamn-Simons Society $500; Hampton-Preston Society $1,000; Seibels Society $2,500; Robert Mills Society $5,000.

THE MUSEUM OF EDUCATION, University of South Carolina, Wardlaw Hall, 820 Main St., Columbia, SC 29208. Tel.: 803-777-5741.

E-mail: museumofeducation@sc.edu

Web Site: www.ed.sc.edu/museum

Founded: 1977.

Key Personnel: Cur., Dr. Craig Kridel

Institution Type/Description: Education Museum.

Collections: exhibitions, publications & programs pertaining to the issues of education.

Hours & Admission Prices: Mon.-Fri. 9-4. No charge. ♿

PONDER FINE ARTS GALLERY - BENEDICT COLLEGE, 1600 Harden St., Columbia, SC 29204. Tel.: 803-253-5000.

Institution Type/Description: Art Gallery.

Collections: paintings; photographs; sculpture.

Activities: permanent & temporary exhibits.

Hours & Admission Prices: Mon.-Fri. 10-4. No charge.

RIVERBANKS ZOO & GARDEN, 500 Wildlife Pkwy., Columbia, SC 29210-8093. Mailing Address: P.O. Box 1060, Columbia, SC 29202-1060. Tel.: 803-779-8717. Fax: 803-253-6381.

Web Site: www.riverbanks.org

Founded: 1974.

Congressional District: 6

Key Personnel: Exec. Dir., Palmer E. Krantz, III; Chm., Lloyd S. Liles; Dir. Mktg., Tommy Stringfellow; Financial Dir., George R. Davis; Dir. Animal Collections, Ed Diebold; Retail Sales Mgr., Jason Painter.

Personnel Profile: Full-Time Paid 126; Part-Time Paid 41; Part-Time Volunteers 349; Interns 12.

Governing Authority: nonprofit organization. Parent Institution: Riverbanks Park Commission. Tax-exempt: 501(c)(3).

Institution Type/Description: Zoo & Botanical Garden.

Collections: Specimens: 265 birds; 106 mammals; 326 reptiles; 681 fish; 130 amphibians; Species: 54 birds; 43 mammals; 61 reptiles, 151 fish, 19 amphibians.

Research Fields: reproductive physiology.

Facilities: botanical garden; zoological park; cafeteria. Museum-related items for sale.

Activities: self-guided tours; lectures; concerts; radio programs; formally organized education programs for children & adults.

Publications: 6 issue magazine, Riverbanks.

Hours & Admission Prices: April 4-Oct. 4 Mon.-Fri. 9-5, Sat.-Sun. 9-6; Oct. 5-April 3 daily 9-5. Adults $9.75, military and seniors 62 & over $8.75, children 3-12 $7.25; discount to AZA members; AAZPA & AABGA members, reciprocal zoo societies, children under 3 & members no charge. Closed Thanksgiving; Christmas. ♿

Attendance: 872,000 (accurate)

Membership: Individual $34; Individual Plus $49; Family $59; Family Plus $74; Patron $125; Curators' Circle $250; Director's Circle $500.

✳ SOUTH CAROLINA CONFEDERATE RELIC ROOM AND MILITARY MUSEUM, (M), 301 Gervais St., Ste. 1, Columbia, SC 29201-3073. Tel.: 803-737-8095. Fax: 803-737-8099.

E-mail: jcassidy@crr.sc.gov

Web Site: www.crr.sc.gov.

Founded: 1896.

Congressional District: 2

Key Personnel: Dir., W. Allen Roberson; Admin. Coord. & Museum Shop Mgr., Shirley D. Schoonover; Cur. Education, William Joe Long; Cur.

History, Kristina Dunn Johnson; Registrar, Rachel H. Cockrell; Cur. Exhibits & Design, Jami Cassidy; Coord. Museum Advancement, Kaela Harmon.

Personnel Profile: Full-Time Paid 7; Part-Time Paid 3; Part-Time Volunteers 2; Interns 2.

Governing Authority: state. Tax-exempt.

Institution Type/Description: History Museum.

Collections: Civil War flags; South Carolina uniforms; weapons; archives; 19th-20th century textiles; South Carolina's military history from Revolutionary War to present.

Research Fields: South Carolina history.

Facilities: 5,550-vol. library of historical books, pamphlets & brochures available on the premises.

Activities: guided tours; lectures; permanent & temporary exhibitions.

Publications: e-newsletter, The Regimental Courier.

Hours & Admission Prices: Tues.-Sat. 10-5, 1st Sun. of month 1-5. Adults 18-61 $5, military & seniors $4, youth 13-17 $2; discounts 1st Sun. of month; members and children 12 & under no charge. Closed most state holidays. ♿

Attendance: 22,000 (accurate)

Membership: Individual $35; Cadet $40; Dual $45; Family $60; Colonel $150; General $250.

SOUTH CAROLINA DEPARTMENT OF ARCHIVES & HISTORY, 8301 Parklane Rd., Columbia, SC 29223-4905. Tel.: 803-896-6100. Fax: 803-896-6186.

Web Site: scdah.sc.gov/

Founded: 1905.

Congressional District: 2

Key Personnel: Dir., W. Eric Emerson, Ph.D.; Chm. (V), A.V. Huff, Jr., Ph.D.; Dir. Historical Svcs., Elizabeth M. Johnson; Dir. Archives, Steve Tuttle; Dir. Records Mgmt., H. Richard Harris; Museum Shop Mgr., Terry Mulholland.

Personnel Profile: Full-Time Paid 29; Part-Time Paid 1; Part-Time Volunteers 4; Interns 1.

Governing Authority: state.

Institution Type/Description: State Government Historical Agency.

Collections: 27,000 cubic ft. of archival records; 5,000 books and periodicals; 20,000 reels of microfilm; 3,000 item map collection.

Research Fields: history of the state of South Carolina.

Facilities: reading room; meeting rooms; auditorium; multi-purpose classrooms.

Activities: temporary exhibitions; curriculum resources development; records management for state & local government; historic markers; national register nominations; historic preservation program; symposium; fundraisers; heritage lecture tours. Annual Event: Statewide Historic Preservation Conference.

Publications: books, Biographical Directory of the House of Representatives; Colonial Records of South Carolina; State Records of South Carolina; brochures; documentary microfilm; popular historical booklets; electronic newsletter.

Hours & Admission Prices: Tues.-Sat. 8:30-5. No charge. Reference Room: closed state holidays. ♿

Attendance: 5,947 (accurate)

Membership: Foundation: Individual $35; Family $60; Advocate $100; Guardian $250; Ambassador $500; Friend $1,000.

SOUTH CAROLINA DEPARTMENT OF PARKS, RECREATION AND TOURISM, 1205 Pendleton St., Columbia, SC 29201-3756. Tel.: 803-734-0156. Fax: 803-734-1017.

Web Site: www.southcarolinaparks.com

Founded: 1967.

Congressional District: 2

Key Personnel: Exec. Dir., Chad Prosser; Dir. Parks, Phil Gaines; Chief Resource Mgmt., Irvin Pitts; Chief Education & Interpretation, Terry Hurley; Recreation Mgr. & Biology Cur., Stan Hutto.

Personnel Profile: Full-Time Paid 30; Part-Time Paid 15; Part-Time Volunteers 10; Interns 3.

Governing Authority: state. Tax-exempt.

Institution Type/Description: State Park & Museums.

Collections: Historic Houses and Sites: 1832 Rose Hill Plantation Home, Union County; 1756 settler's log dwelling, Richland County; c.1750 Hampton Plantation Home, Charleston County; 1858 Redcliffe Plantation Home, Aiken County. Atalaya, Huntington Beach State Park, Georgetown County: c.1933 Winter Home Archer & Anna Huntington. Historic Sites: Andrew Jackson State Park, Lancaster County: exhibits on country farm-life; Hunting Island Lighthouse Complex, Beaufort: maritime history; Old Dorchester State Park, Dorchester County: 18th century trading and religious settlement, exhibit; Rivers Bridge State Park, Bamberg County: Civil War Battle Site; 1823-50 Landsford Canal State Park, Chester County; 1670 Charles Towne Landing, Charleston County; Kings Mountain State

Park, York County; living history farm; Keowee-Toxaway State Park, Pickens County: Cherokee Indian exhibits; Caesar's Head State Park, Greenville County; Hunting Island State Park, Beaufort County; Myrtle Beach State Park, Horry County; Oconee State Park, Oconee County; Poinsett State Park, Midlands-Sumter County.

Research Fields: history; decorative arts; archaeology; folklore; Indian artifacts; industry; marine; transportation; architecture; historic preservation.

Activities: guided tours; lectures; permanent & temporary exhibitions; school programs; organized historical & nature programming.

Publications: brochures; visitors guide.

Hours & Admission Prices: Contact individual park or site for hours & admissions.

SOUTH CAROLINA INSTITUTE OF ARCHAEOLOGY & ANTHROPOLOGY, 1321 Pendleton St., University of South Carolina, Columbia, SC 29208-4103. Tel.: 803-576-6573.

E-mail: nrice@sc.edu

Web Site: www.cas.sc.edu/sciaa

Founded: 1963.

Congressional District: 2

Key Personnel: Interim Dir., Charles Cobb; State Archaeologist, Jonathan M. Leader; Business Mgr., Susan Lowe; Cur., Sharon L. Pekrul.

Personnel Profile: Full-Time Paid 50; Part-Time Paid 24; Part-Time Volunteers 17; Interns 4.

Governing Authority: state; public college; nonprofit. Parent Institution: State of South Carolina. Subsidiary Institution: The University of South Carolina. Tax-exempt: 170(b)(1)(A).

Institution Type/Description: State Agency; Archaeology & Anthropology Research Institute.

Collections: pre-historic & historic period Native American, African American and Euro American archaeological collections; ethnographic material; fossils & geological samples.

Research Fields: pre-history & history of South Carolina; French & Spanish presence in South Carolina; Civil War & plantation studies; Paleo-Indian studies; maritime studies; protohistoric Indian studies; private artifact collections survey.

Facilities: 29,000-vol. library of archaeology & anthropology reports & reference materials; archaeological site files, field & lab records and maps, photographs & slides; 3 laboratories; conference room; field research station.

Activities: guided tours; lectures; organized education programs for undergraduate & graduate college students affiliated with University of South Carolina. Institution Sponsors: Archaeology Conference; Fall Field Day.

Publications: periodic research reports.

Hours & Admission Prices: Mon.-Fri. 8:30-5. No charge. Closed state holidays. &

SOUTH CAROLINA LAW ENFORCEMENT OFFICERS HALL OF FAME, 5400 Broad River Rd., Columbia, SC 29212-3540. Tel.: 803-896-8199. Fax: 803-896-8067.

Web Site: www.scdps.org/hof

Formerly: South Carolina Criminal Justice Hall of Fame

Founded: 1979.

Key Personnel: Admin., Marsha T. Ardila; Administrative Specialist, Emily C. Harrison.

Personnel Profile: Full-Time Paid 2.

Governing Authority: state; nonprofit organization. Parent Institution: Dept. of Public Safety.

Institution Type/Description: Law Museum.

Collections: artifacts related to the history of law enforcement in South Carolina.

Research Fields: historical development of law enforcement in South Carolina.

Facilities: 75-seat auditorium.

Activities: guided tours; permanent & temporary exhibitions.

Publications: brochures on Hall of Fame programs.

Hours & Admission Prices: Mon.-Fri. 8:30-5. No charge. Closed state holidays. &

Attendance: 7,000 (accurate)

SOUTH CAROLINA MILITARY MUSEUM, (M), One National Guard Rd., Columbia, SC 29201-4752. Tel.: 803-806-4440. Fax: 803-806-2103.

Founded: 2007.

Congressional District: 5

Key Personnel: Dir. & Cur., Ewell G. Sturgis, Jr.; Chm. (V), Steven W. Vinson; Registrar, Joy Maples; Museum Shop Mgr., Edward Y. Hall.

Personnel Profile: Full-Time Paid 2; Part-Time Volunteers 24.

Governing Authority: state; nonprofit. Parent Institution: South Carolina National Guard. Subsidiary Institution: South Carolina Military History Foundation Inc. Tax-exempt.

Institution Type/Description: Military Museum.

Collections: South Carolina militia & National Guard memorabilia artillery.

Activities: guided tours.

Hours & Admission Prices: Tues.-Sat. 10-4, Sun. 1-4. No charge; donations accepted. &

Attendance: 6,000 (estimated)

* SOUTH CAROLINA STATE MUSEUM, (M), 301 Gervais St., Ste. 2, Columbia, SC 29201-3073. Mailing Address: 301 Gervais St., Loading Zone D, Columbia, SC 29201. Tel.: 803-898-4983. Fax: 803-898-4969.

E-mail: merritt.mchaffie@scmuseum.org

Web Site: www.scmuseum.org

Founded: 1973.

Congressional District: 2

Key Personnel: Dir., William Calloway; Chm. (V), Gray Culbreath; Mgr. Human Resources, Susan Worthy; Exhibit Dir., A. Michael Fey; Chief Registrar, Michelle Baker; Dir. Education, Tom Falvey; Cur. Art, Paul Matheny; Museum Shop Mgr., Scottie Ash; Dir. Mktg., Merritt McHaffie.

Personnel Profile: Full-Time Paid 26; Part-Time Paid 36; Part-Time Volunteers 159; Interns 3.

Governing Authority: state. Parent Institution: State of South Carolina. Subsidiary Institution: South Carolina Museum Foundation. Tax-exempt: 170(b)(1)(A).

Institution Type/Description: General Museum: housed in former textile mill building.

Collections: South Carolina history, natural history, art, science & technology; observatory; planetarium; 4D theater.

Major Exhibits: Tutankhamun, 1/14-5/14; Mama Let's Make a Moon, 3/21/14-6/29/14; Conrad Wise Chapman, 7/11/14-1/18/15.

Research Fields: South Carolina art; local natural history; South Carolina culture.

Facilities: 236-seat auditorium; multi-purpose meeting room; science demonstration theater; hands-on discovery center for children; two educational activity centers; conservation laboratory; 4D theater; planetarium; observatory. Museum-related items for sale.

Activities: long-term, temporary & traveling exhibits; tours; education study visits; inter-museum loan exhibits; science demonstrations; museum personnel workshops; teacher workshops; facility use program; films; lectures; advisory services to other museums; nature programs; 4D programming; observatory distance learning programs; planetarium shows.

Publications: e-newsletter; bulletins; brochures, Common Snakes of South Carolina; magazine, Images.

Hours & Admission Prices: Memorial Day-Labor Day Mon.-Sat. 10-5, Sun. 1-5; Sept.-May Tues.-Sat. 10-5, Sun. 1-5. Adults 13 & over $7, military $6, senior citizens 62 & over $6, children 3-12 $5; discounts to groups of 10 or more, Southeastern Museums Conference & South Carolina Federation of Museums members; members, children under 3 & South Carolina school groups with advance notice no charge. Closed Easter; Thanksgiving; Christmas Eve & Day. &

Attendance: 152,623 (accurate)

Membership: Individual $35; Dual $50; Family & Grandparent $70; Premier $120; Charter Collection $150; Director's Guild $250; Foundation Fellows $500; Trustee's Council $1,000.

* THE UNIVERSITY OF SOUTH CAROLINA MCKISSICK MUSEUM, (M), 816 Bull St., Columbia, SC 29208. Tel.: 803-777-7251. Fax: 803-777-2829.

E-mail: robertso@mailbox.sc.edu

Web Site: www.cas.sc.edu/mcks/

Founded: 1976.

Congressional District: 2

Key Personnel: Exec. Dir., Lynn Robertson; Chief Cur. Exhibitions, Jason Shaiman; Chief Cur. Research & Folklife, Saddler Taylor; Cur. Temporary Exhibitions, Nathan Stalvey; Cur. Collections, Jill Koverman; Business Mgr., Peggy Nunn; Cur. Faculty, Lana Burgess; Visitor Svcs. Mgr., Ja-Nae Epps.

Personnel Profile: Full-Time Paid 8; Part-Time Paid 10; Part-Time Volunteers 30; Interns 2.

Governing Authority: university. Parent Institution: University of South Carolina. Tax-exempt: 170(b)(1)(A).

Institution Type/Description: University Museum.

Collections: 19th- & 20th-century decorative arts; folk art & material culture of South Carolina and the South; Bernard Baruch silver; gems & minerals.

Research Fields: university history; material & folklife culture of the Southeast; interdisciplinary Southern studies.

Facilities: library; folklife resource center; meeting room; auditorium.

Activities: guided tours; lectures; permanent, temporary & traveling exhibitions; concerts; demonstrations; academic classes; educational programs; workshops; meetings.

Publications: magazine; educator curriculum materials; catalogs; brochures; guides.
Hours & Admission Prices: Mon.-Fri. 8:30-5, Sat. 11-3. No charge; donations accepted. Closed New Year's Eve & Day; Independence Day; Labor Day; Thanksgiving; Christmas Eve, Day & week; University Holidays. &
Attendance: 35,000 (estimated)
Membership: Student $10; Individual $25; Dual $35; Household $50; Patron $100-$249; Sponsor $250-$499; Benefactor $500-$999; Director's Circle $1,000.

U.S. ARMY FINANCE CORPS MUSEUM, 4392 Magruder Ave., Columbia, SC 29207. Mailing Address: Commandant, U.S. Army Finance School, 10,000 Hampton Pkwy., Fort Jackson, SC 29207-7050. Tel.: 803-751-3771. Fax: 803-751-1749.
E-mail: atsg-fsa@jackson.army.mil
Founded: 1954.
Key Personnel: Cur., Henry D. Howe, III
Personnel Profile: Full-Time Paid 1.
Governing Authority: federal. Parent Institution: U.S. Army Center of Military History. Tax-exempt.
Institution Type/Description: Military Museum.
Collections: history of the U.S. Army Finance Corps.; American currency from revolutionary war to present; U.S. military currency from Word War II through Vietnam.
Research Fields: military currency.
Activities: guided tours; lectures; traveling exhibitions.
Hours & Admission Prices: Tues.-Fri. 10-4. No charge. Closed federal holidays. &
Attendance: 12,000 (accurate)

Conway

HORRY COUNTY MUSEUM, (M), 428 Main St., Conway, SC 29526-4308. Tel.: 843-915-5320. Fax: 843-248-1854.
E-mail: hcgmuseum@horrycounty.org
Web Site: www.horrycountymuseum.org
Founded: 1979.
Congressional District: 6
Key Personnel: Dir., R. Walter Hill, IV; Chm. (V), Jody Nyers; Education Specialist, Marion Haynes; Museum Shop Mgr., Julie Pinckney.
Personnel Profile: Full-Time Paid 4; Part-Time Paid 2; Part-Time Volunteers 12; Interns 2.
Governing Authority: Parent Institution: Horry County Government.
Institution Type/Description: Anthropology, History & Archaeology Museum.
Collections: material relating to local history, prehistory & natural history of the county.
Research Fields: local history; archaeology; technology; folklore.
Facilities: photographic archives.
Activities: guided tours; lectures; hobby workshops; formally organized education programs for children; loan, permanent, temporary & traveling exhibitions; school loan service.
Hours & Admission Prices: Tues.-Sat. 9-5. No charge; donations accepted. Closed county government holidays. &
Attendance: 40,000 (estimated)

Darlington

DARLINGTON RACEWAY STOCK CAR MUSEUM, NMPA HALL OF FAME, 1301 Harry Byrd Hwy., Darlington, SC 29532-3517. Mailing Address: P.O. Box 500, Darlington, SC 29540-0500. Tel.: 843-395-8900. Fax: 803-393-3911.
E-mail: dworden@darlingtonraceway.com
Web Site: www.darlingtonraceway.com
Formerly: NMPA Stock Car Hall of Fame, Joe Weatherly Museum
Founded: 1965.
Congressional District: 6
Key Personnel: Museum Shop Mgr., Vicki Sanders.
Governing Authority: nonprofit organization. Parent Institution: Americrown.
Institution Type/Description: Stock Car Museum.
Collections: stock cars; photos; trophies; engines; stock car racers' Hall of Fame.
Facilities: Racing items for sale.
Activities: track tours.
Hours & Admission Prices: Mon.-Fri. 10-5, Sat. 10-4. Adults $5; children under 12 no charge. &
Attendance: 250,000 (estimated)

Dillon

JAMES W. DILLON HOUSE MUSEUM, 1302 W. Main St., Dillon, SC 29536. Mailing Address: Dillon County Historical Society, P.O. Box 1806, Dillon, SC 29536-1806. Tel.: 843-774-6122 & 441-5273. Fax: 843-774-5521.
E-mail: bbar830771@aol.com
Web Site: www.dillonmuseum.com
Founded: 1961.
Congressional District: 6
Key Personnel: Resource Mgr., Betty Barclay.
Personnel Profile: Part-Time Paid 1; Part-Time Volunteers 10.
Governing Authority: society; nonprofit. Parent Institution: Dillon County Historical Society. Tax-exempt: 501(c)(3).
Institution Type/Description: Historic House.
Collections: period furniture & clothing; archives; 1910 kitchen.
Activities: guided tours. Museum Sponsors: Christmas at the Dillon House in December.
Publications: brochures.
Hours & Admission Prices: Tues.-Thurs. 2-4; other times by appointment. No charge; donations accepted. Closed Easter; Independence Day; Thanksgiving; Christmas. &
Attendance: 364 (estimated)

Due West

BOWIE ARTS CENTER, Erskine College, Two Washington St., Due West, SC 29639. Mailing Address: P.O. Box 338, Due West, SC 26939-0338. Tel.: 864-379-8867. Fax: 864-379-2167.
E-mail: burton@erskine.edu
Founded: 1995.
Congressional District: 3
Key Personnel: Dir., Jan B. Walker; Acting Cur., Ruth M. Burton.
Personnel Profile: Part-Time Paid 3.
Governing Authority: Parent Institution: Erskine College.
Institution Type/Description: Art Gallery and Antiques Museum.
Collections: antique mechanical musical instruments; clocks; decorative arts; glass & porcelain; furnishings from the 19th & early 20th centuries; photographs.
Hours & Admission Prices: Call for hours. No charge. Closed major holidays; college breaks.

Eastover

KENSINGTON MANSION/SCARBOROUGH-HAMER FOUNDATION, 4101 McCords Ferry Rd./ US 601, Eastover, SC 29044. Mailing Address: P.O. Box 237, Eastover, SC 29044-0237. Tel.: 803-353-0456.
E-mail: staff@kensingtonmansion.org
Web Site: www.kensingtonmansion.org
Founded: 1996.
Key Personnel: Dir., Rickie Good; Chm. (V), Mary B. Waters.
Personnel Profile: Part-Time Paid 3; Part-Time Volunteers 15.
Governing Authority: nonprofit. Partner Institution: International Paper. Tax-exempt: 501(c)(3).
Institution Type/Description: Historic House Museum: 1854 Italianate mansion built by Matthew Richard Singleton.
Collections: Victorian textiles & decorative arts; tools & farm implements.
Activities: Museum Sponsors: Christmas at Kensington.
Publications: quarterly newsletter, Kensington Times.
Hours & Admission Prices: Tours: March-July & Sept.-Dec. Thurs.-Sat. 9:30, 11, 1 & 2:30. Adults 13-59 $7, seniors 60 & over and active military $6, children 6-12 $5; children under 6 no charge. Closed major holidays.
Attendance: 3,000 (accurate)
Membership: Senior $30; Single 35; Dual $50; Family $65; Patron $100; Executive $500; Benefactor $1,000.

Edgefield

FRESHWATER COAST DISCOVERY CENTER, 405 Main St., Edgefield, SC 29824-1301. Mailing Address: South Carolina National Heritage Corridor, P.O. Box 477, Belton, SC 29627. Tel.: 803-637-0877. Fax: 803-637-6237.
E-mail: edgefielddc@scprt.com
Web Site: www.scnhc.org
Formerly: The Joanne T. Rainsford Heritage Discovery Center
Key Personnel: Pres. & C.E.O., Michelle McCollum.
Governing Authority: Parent Institution: South Carolina National Heritage Corridor.

Institution Type/Description: History Museum.

Collections: local history & culture; cotton industry; period artifacts; photographs.

Hours & Admission Prices: Tues.-Sat. 10-5. No charge. Closed New Year's Day; Independence Day; Thanksgiving; Christmas Day.

OAKLEY PARK MUSEUM, 300 Columbia Rd., Edgefield, SC 29824-1224. Tel.: 803-637-4027.

E-mail: libby.ready1229@yahoo.com

Formerly: Oakley Park, UDC Shrine

Founded: 1941.

Congressional District: 3

Key Personnel: Pres. (V) & Dir. (V), Elizabeth Ready.

Personnel Profile: Full-Time Volunteers 2; Part-Time Volunteers 5.

Governing Authority: nonprofit. Parent Institution: South Carolina Div. of the UDC, Columbia, SC. Tax-exempt.

Institution Type/Description: Historic House: 1835 Oakley Park, where the Red Shirts of South Carolina were organized to ease reconstruction following the Civil War; ancestral home of Gen. Martin W. Gary & Gov. John Gary Evans.

Collections: Civil War memorabilia; period furniture & furnishings; photographs; manuscripts; 1850-1970 letters; quilts; kitchen building; covered well.

Research Fields: Civil War, Edgefield District; Revolutionary War, Old Ninety Six District.

Activities: tours; luncheons; living history program; meetings; weddings; receptions. Museum Sponsors: Civil War encampments.

Publications: newsletter.

Hours & Admission Prices: Thurs.-Sat. 10-4; other times by appointment. Adults $5, students $3; discount to senior, groups of 20 or more & AAM members with ID; children under 5 no charge. Closed New Year's Day; Thanksgiving; Christmas. &

Attendance: 4,000 (estimated)

Membership: Individual $5.

TERRY FERRELL'S ANTIQUES & MUSEUM, Turner's Corner Store, 101 Courthouse Square, Edgefield, SC 29824-1362. Tel.: 803-637-4618.

Key Personnel: Owner, Terry Ferrell

Institution Type/Description: History Museum.

Collections: local history & culture; pottery; quilts; glassware.

Hours & Admission Prices: Thurs.-Sat. 10-5; other times by appointment.

WILD TURKEY CENTER AND WINCHESTER MUSEUM, 770 Augusta Rd., Edgefield, SC 29824-1573. Mailing Address: P.O. Box 530, Edgefield, SC 29824-0530. Tel.: 800-THE-NWTF & 637-7626.

Web Site: www.nwtf.org

Founded: 1998.

Congressional District: 3

Institution Type/Description: Wild Turkey History Museum.

Collections: wild turkey history; turkey calls & hunting; habitat management.

Facilities: theater; nature trails. Museum-related items for sale.

Activities: video; tours; school field trips; educational programs. hiking trails; interactive exhibits.

Hours & Admission Prices: Mon.-Fri. 8:30-5, Sat. 10-2. No charge; donations accepted. Closed national holidays.

Attendance: 10,000 (estimated)

Edisto Island

EDISTO ISLAND HISTORIC PRESERVATION SOCIETY MUSEUM, 8123 Chisolm Plantation Rd., Edisto Island, SC 29438-6618. Tel.: 843-869-1954. Fax: 843-869-2754.

E-mail: gsmith@edistomuseum.org

Web Site: www.edistomuseum.org

Founded: 1986.

Congressional District: 6

Key Personnel: C.E.O. & Dir., Gretchen M. Smith; Pres. (V), Jane Darby.

Personnel Profile: Full-Time Paid 1; Part-Time Paid 3; Part-Time Volunteers 27.

Governing Authority: private; nonprofit. Tax-exempt: 501(c)(3).

Institution Type/Description: Historical & Preservation Society Museum.

Collections: prehistoric low country to 1920, concentrating on Edisto Island, S.C.

Facilities: 2,500 sq. ft. exhibit space. Museum-related items for sale.

Activities: guided tours; lectures; temporary exhibitions. Museum Sponsors: quarterly meetings with speakers.

Publications: quarterly newsletter, Edisto Echoes; book, Indigo, Gullah & Edisto in 1808.

Hours & Admission Prices: Jan.-Feb. Tues., Thurs. & Sat. 1-4; March-Oct. Tues.-Sat. 12-5; Nov.-Dec. Tues.-Sat. 11-4. Adult $4; discounts to AAM members; members and children 10 & under no charge. &

Attendance: 6,000 (estimated)

Membership: Individual $30; Family $50; Toogoodoo Society $100; Russell Creek Society $250; Dauhoo Society $500; Edisto Society $1,000.

EDISTO ISLAND SERPENTARIUM, 1374 Hwy. 174, Edisto Island, SC 29438-6816. Tel.: 843-869-1171. Fax: 843-869-1959.

E-mail: info@edistoserpentarium.com

Web Site: www.edistoserpentarium.com

Institution Type/Description: Herpetology Museum.

Collections: reptiles from around the world including alligators, turtles, & snakes.

Activities: special events.

Hours & Admission Prices: April 29-May 22 & Aug. 19-Sept. 4 Thurs.-Sat. 10-6; May 24-Aug. 14 Mon.-Sat. 10-6; call for additional hours. Adults 13 & over $12.95, seniors 65 & over $11.95, children 6-12 $9.95, children 4-5 $5.95; discounts to AAA members; children 3 & under no charge.

Ehrhardt

RIVERS BRIDGE STATE HISTORIC SITE, 325 State Park Rd., Ehrhardt, SC 29081-9157. Tel.: 803-267-3675. Fax: 803-267-3675.

E-mail: riversbridgesp@scprt.com

Web Site: scprt.com

Key Personnel: Park Mgr., John White.

Personnel Profile: Full-Time Paid 2; Part-Time Volunteers 1.

Governing Authority: Parent Institution: S.C. Park, Recreation and Tourism.

Institution Type/Description: Historic Site: battle between the Confederacy & the Union army, Feb. 2-3, 1865. Listed on the National Register of Historic Places.

Collections: Civil War history.

Activities: special programs; rental facilities.

Hours & Admission Prices: Daily 9-6. No charge; donations accepted. Closed holidays. &

Attendance: 6,000 (estimated)

Elloree

ELLOREE HERITAGE MUSEUM & CULTURAL CENTER, INC., 2714 Cleveland St., Elloree, SC 29047. Mailing Address: P.O. Box 54, Elloree, SC 29047-0054. Tel.: 803-897-2225. Fax: 803-897-2252. Facebook: Elloree Heritage Museum.

E-mail: elloreemuseum@ntinet.com

Web Site: www.elloreemuseum.org

Founded: 1998.

Key Personnel: Dir., Howard Shirer; Chm. (V), Jan Miller; Museum Shop Mgr., Jane Livingston

Institution Type/Description: History Museum.

Collections: plantation gin house with original gin, cotton press & mechanicals; horse-drawn plows, planters & cultivators; photographs.

Hours & Admission Prices: Wed.-Sat. 10-5. Adults $5, seniors 60 & over $4, children 6-18 $3; discounts to groups; children under 6 no charge. &

Membership: Individual $25; Family $50; Collector's Club $100; Director's Guild $250; Historian's Round Table $500; Trustee's Council $1,000.

J'S TEA-RIFIC TEAPOT MUSEUM, 2732 Cleveland St., Elloree, SC 29047. Tel.: 803-897-2133.

Institution Type/Description: Teapot Museum.

Collections: teapots from around the world.

Hours & Admission Prices: Wed.-Fri. 11-5; other times by appointment.

Florence

DOOLEY PLANETARIUM - FRANCIS MARION UNIVERSITY, Cauthen Educational Media Center, 2nd Fl., Florence, SC 29506. Mailing Address: P.O. Box 100547, Florence, SC 29502-0547. Tel.: 843-661-1381.

Institution Type/Description: Planetarium.

Collections: astronomy; space science.

Hours & Admission Prices: By appointment.

FLORENCE COUNTY MUSEUM, (M), 111 W. Cheves St., Florence, SC 29501. Tel.: 843-662-3351. Fax: 843-665-9527.

E-mail: florencemuseum@me.com

Web Site: www.flocomuseum.org
Formerly: Florence Museum of Art, Science & History and the Florence
   Railroad Museum
Founded: 1924.
Congressional District: 6
Key Personnel: Exec. Dir., Andrew Russel Stout; Cur., Interpretation &
   Collections, Stephen Motte; Cur., Education, Kim Groom.
Governing Authority: quasi-public nonprofit organization. Subsidiary Institu-
   tion: The Railroad Museum. Tax-exempt: 501(c)(3).
Institution Type/Description: General Museum.
Collections: paintings; Native American artifacts; Civil War memorabilia;
   South Carolina history; Asian, Greek & Roman artifacts; American Art;
   railroad cars.
Research Fields: art; history & science.
Activities: lectures; gallery talks; study clubs; inter-museum loans; research;
   art classes; juried competitions; temporary exhibitions.
Publications: quarterly newsletter, Florence Museum.
Hours & Admission Prices: Temporarily closed for relocation.
Attendance: 19,000 (estimated)

WAR BETWEEN THE STATES MUSEUM, 107 S. Guerry St.,
   Florence, SC 29501-4328. Tel.: 843-669-1266.
Web Site: peedeerifles.homestead.com/wbtsmuseum.html
Founded: 1988.
Key Personnel: Dir., Carl Hill, Jr.
Personnel Profile: Full-Time Volunteers 1; Part-Time Volunteers 1.
Governing Authority: bd. of directors. Parent Institution: Sons of Confederate
   Veterans. Subsidiary Institution: Pee Dee Rifles Camp #1419. Tax-exempt.
Institution Type/Description: History Museum.
Collections: Civil War history; photographs; military artifacts; manuscripts.
Facilities: Museum-related items for sale.
Hours & Admission Prices: Wed. & Sat. 10-5. Adults $2, children $1. Closed
   Christmas.
Attendance: 2,000 (estimated)

Fort Jackson

FORT JACKSON MUSEUM, (M), 4442 Jackson Blvd., Fort
   Jackson, SC 29207-5100. Mailing Address: 2179 Sumter St., Fort
   Jackson, SC 29207-6102. Tel.: 803-751-7419. Fax: 803-751-4434.
E-mail: williamsb1@jackson.army.mil
Web Site: www.jackson.army.mil/museum/index.htm
Founded: 1974.
Congressional District: 2
Key Personnel: Cur., Bessie Williams; Museum Specialty, Brandon Wiegand.
Governing Authority: federal. Tax-exempt.
Institution Type/Description: Military Museum.
Collections: military objects of the US Army & history of Fort Jackson,
   beginning with World War I to present; military objects of U.S. allies &
   enemies from W.W. I through Desert Storm; local military history including
   Indian Warriors, the Revolution, War of 1812 Mexican, Civil, & Spanish
   American War periods, training of individual soldier.
Research Fields: U.S. military history.
Facilities: 1,500-vol. library of military manuals & history sources available
   for research by special request.
Activities: guided tours; lectures; films; temporary exhibitions.
Hours & Admission Prices: Mon.-Fri. 9-4. No charge; donations accepted.
   Closed federal holidays. ♿
Attendance: 32,000

U.S. ARMY CHAPLAIN MUSEUM, (M), USACHCS, 10100 Lee
   Rd., Fort Jackson, SC 29207-7000. Tel.: 803-751-8827 & 8079.
   Fax: 803-751-8890.
E-mail: marcia.mcmanus@us.army.mil
Web Site: www.chapnet.army.mil/usachcs
Founded: 1957.
Congressional District: 2
Key Personnel: Dir. & Cur., Marcia McManus; Museum Technician, Tim
   Taylor.
Personnel Profile: Full-Time Paid 2; Part-Time Volunteers 6.
Governing Authority: federal. Parent Institution: U.S. Army Chaplain Center &
   School. Subsidiary Institution: U.S. Army Center of Military History.
   Tax-exempt
Institution Type/Description: Military History Museum.
Collections: artifacts; photos & documents depicting the U.S. Army Chap-
   laincy; manuscript collections.
Research Fields: U.S. Army Chaplaincy; religious writings; military uniforms.
Facilities: 4,000-vol. library of theological-military history; archives; photo
   collections; reading room.

Activities: guided tours; lectures.
Publications: newsletter.
Hours & Admission Prices: Mon.-Fri. 9-4. No charge. Closed federal holidays.
   ♿
Membership: Chaplain Regimental Museum Association: Annual $20.

Gaffney

CHEROKEE COUNTY HISTORY AND ARTS MUSEUM, (M),
   301 College Dr., Gaffney, SC 29340-3006. Mailing Address: P.O.
   Box 8113, Gaffney, SC 29340. Tel.: 864-489-3988. Fax: 864-489-
   3988.
E-mail: chaps@cherokeecountyhistory.org
Web Site: www.cherokeecountyhistory.org
Founded: 2000.
Key Personnel: Dir., Jane Waters; Pres. (V), Dinah Hamrick; Admin., Lisa
   Hissins.
Personnel Profile: Full-Time Paid 1; Part-Time Paid 3; Part-Time Volunteers
   43.
Volunteer Hours: 1,114
Operating Expenses: 109,062
Operating Income: 115,493
Governing Authority: Parent Institution: CHAPS. Tax-exempt.
Institution Type/Description: History Museum.
Collections: local history & culture; period furnishings; personal artifacts;
   photographs.
Activities: educational & cultural outreach programs.
Publications: journal.
Hours & Admission Prices: Wed.-Fri. 10-4, Sat. call for special events. Adults
   $5, children $3; members no charge. ♿
Attendance: 3,000 (estimated)

CHEROKEE COUNTY VETERANS MUSEUM, 200 S. Logan St.,
   Gaffney, SC 29340. Tel.: 864-812-4136.
Institution Type/Description: Military History Museum.
Collections: military history & artifacts; photographs; personal artifacts.
Hours & Admission Prices: Mon. 1-3, Sat. 9am to noon, Sun. 2-4; groups by
   appointment. No charge. Closed holidays.

COWPENS NATIONAL BATTLEFIELD, 4001 Chesnee Hwy.,
   Gaffney, SC 29341. Mailing Address: P.O. Box 308, Chesnee, SC
   29323-0308. Tel.: 864-461-2828. Fax: 864-461-7795.
E-mail: cowp_interpretation@nps.gov
Web Site: www.nps.gov/cowp/
Founded: 1933.
Congressional District: 5
Key Personnel: Supt., Tim Stone; Administrative Officer, Michelle Lester;
   Chief Ranger, Kathy McKay.
Personnel Profile: Full-Time Paid 1; Part-Time Paid 2.
Governing Authority: federal. Affiliated with the Department of Interior,
   National Park Service, Washington, DC. Tax-exempt.
Institution Type/Description: Military Museum: located on site of 1781 Battle
   of Cowpens.
Collections: weapons of artillery of Revolutionary period; records of men who
   fought in the battle. Historic House: 1830 Robert Scruggs House.
Research Fields: American Revolution; Southern Campaign.
Facilities: theater; picnic area with shelter. Books & pamphlets for sale.
Activities: guided tours on request, if staff is available; demonstrations at
   special events; introductory DVD; fiber optics map. Annual Event: Living
   History Encampment in January.
Publications: handbook; DVD presentation; books; postcards.
Hours & Admission Prices: Daily 9-5. No charge; donations accepted. Closed
   New Year's Day; Thanksgiving; Christmas. ♿
Attendance: 208,566 (accurate)

WINNIE DAVIS HALL OF HISTORY, Limestone College, 1115
   College Dr., Gaffney, SC 29340-3778. Tel.: 864-488-4575.
E-mail: phoskins@limestone.edu
Web Site: www.limestone.edu
Founded: 1976.
Congressional District: 5
Key Personnel: College Historian, Patricia Hoskins.
Governing Authority: college. Parent Institution: Limestone College. Tax-
   exempt.
Institution Type/Description: College Museum: housed in 1898-1901 Winnie
   Davis Hall of History, built to house Confederate records.
Collections: college memorabilia; local & South Carolina history; manu-
   scripts.
Research Fields: local history; Revolutionary War; Civil War.

Hours & Admission Prices: By appointment only. No charge.

Georgetown

GEORGETOWN COUNTY MUSEUM, (M), Georgetown County Historical Society, 120 Broad St., Georgetown, SC 29440-3630. Tel.: 843-545-7020. Fax: 843-545-7020.
E-mail: georgetownmuseum@gmail.com
Web Site: www.georgetowncountymuseum.com
Founded: 2005.
Key Personnel: Dir., Jill Santropietro.
Personnel Profile: Full-Time Paid 1; Part-Time Paid 1; Part-Time Volunteers 8.
Governing Authority: Tax-exempt.
Institution Type/Description: Historical Museum.
Collections: Native Americans; maps; Revolutionary & Civil War artifacts; presidential letters; Low country memorabilia; slavery 19-20th century.
Hours & Admission Prices: Tues.-Fri. 10-5, Sat. 10-3. Adults $4, seniors 64 & over $3; members no charge.
Membership: Individual $35; Family $70; Contributor $100; Silver Sponsor $250; Gold Sponsor $500; Partner $750; Patron $1,000; Benefactor $2,500.

HOBCAW BARONY DISCOVERY CENTER, 22 Hobcaw Rd., Georgetown, SC 29440-9500. Tel.: 843-546-4623. Fax: 843-545-7231.
E-mail: hobcaw@belle.baruch.sc.edu
Web Site: www.hobcawbarony.org
Formerly: Bellefield Nature Center
Governing Authority: Parent Institution: The Belle W. Baruch Foundation. Tax-exempt.
Institution Type/Description: History Museum.
Collections: local history & ecology; 1,200 gallon fish tank; pine & cypress ecosystem; photographs.
Facilities: classrooms.
Activities: lectures; educational programs; research.
Hours & Admission Prices: Center: Mon.-Fri. 9-5. No charge; donations accepted. Guided Tours: Tues.-Fri. by appointment. $20 per person.
Attendance: 12,000 (accurate)

HOPSEWEE PLANTATION, 494 Hopsewee Rd., Georgetown, SC 29440-5598. Tel.: 843-546-7891. Fax: 843-546-3957.
E-mail: fdbeatie@gmail.com
Web Site: www.hopsewee.com
Founded: 1970.
Congressional District: 6
Key Personnel: Owner, Frank Beattie; Mgr. River Oaks Tea Room, Raejean Beattie; Docent, Jean Efird; Docent, Sara Morrison; Docent, Patricia Candale.
Personnel Profile: Full-Time Paid 4; Part-Time Paid 7.
Governing Authority: individual operation.
Institution Type/Description: Historic House: c.1735-40 birthplace of Thomas Lynch Jr., signer of the Declaration of Independence.
Collections: 18th-19th century furniture & furnishings.
Activities: guided tours.
Hours & Admission Prices: House & Tea Room: Feb.-Nov. Tues.-Fri. 10-4, Sat. 12-4. Adults $17.50, children 5-17 $7.50; discounts to AAM members. Grounds: daily $5 per car. Closed Thanksgiving.
Attendance: 9,500 (estimated)

KAMINSKI HOUSE MUSEUM, (M), 1003 Front St., Georgetown, SC 29440-3521. Mailing Address: P.O. Drawer 939, Georgetown, SC 29442. Tel.: 843-546-7706. Fax: 843-527-4871.
E-mail: ckinder@cogsc.com
Web Site: www.kaminskihousemuseum.org
Founded: 1973.
Congressional District: 6
Key Personnel: Dir., Cindy Kinder; Pres. (V), Chris Miller; Museum Shop Mgr., Becky Berry.
Personnel Profile: Full-Time Paid 1; Part-Time Paid 3; Part-Time Volunteers 15.
Governing Authority: municipal government. Parent Institution: City of Georgetown, SC. Tax-exempt: 170(b)(1)(A).
Institution Type/Description: Historic House: 1769 Harold Kaminski House.
Collections: 18th, 19th & 20th-century furniture, glass, silver, china & fine period items.
Research Fields: Social History of House.
Facilities: Museum-related items for sale.
Activities: guided tours; decorative arts lecture series; music series; story telling.
Publications: newsletter, Kaminski House Museum Book.

Hours & Admission Prices: Mon.-Sat. 9-5, call for tour times. Adults $7, seniors, AAA & AARP members $5, children 6-12 $3; discounts to groups, AAM, SEMC & SCFM members; members no charge.
Attendance: 13,000 (accurate)
Membership: Individual $20; Family $35; Commanders Club $100.

THE RICE MUSEUM, Lafayette Park, Front and Screven Sts., 633 Front St., Georgetown, SC 29440. Mailing Address: P.O. Box 902, Georgetown, SC 29442-0902. Tel.: 843-546-7423. Facebook: The Rice Museum.
E-mail: thericemuseum@sc.rr.com
Web Site: www.ricemuseum.org
Founded: 1968.
Congressional District: 1
Key Personnel: Dir., James A. Fitch; Chm. (V), Frank Beatty.
Governing Authority: nonprofit. Parent Institution: Rice Museum, Inc. Tax-exempt: 502(c)(3).
Institution Type/Description: History Museum: housed in the 1842 Old Market Building located on the Old Market site.
Collections: rice culture in South Carolina low country. Historic Building: 1845 Clock & Bell Tower.
Facilities: 50-seat auditorium. Rice Cook Book, notepaper & postcards & museum-related items for sale.
Activities: guided tours; lectures; films; changing exhibits.
Publications: Guide to Historic Georgetown County, S.C.; historic Georgetown county leaflets; Pass the Pilau, Please.
Hours & Admission Prices: Mon.-Sat. 10-4:30. Adults $7, seniors $5, students 6-21 $3; discounts to groups; children under 6 no charge. Closed New Year's Day; Thanksgiving; Christmas. ♿
Attendance: 30,000 (estimated)
Membership: Individual $50; Family $100; Contributor $150; Subscriber $250; Patron $500; Benefactor $1,000 & up.

SOUTH CAROLINA MARITIME MUSEUM, 729 Front St., Georgetown, SC 29440. Mailing Address: P.O. Box 2228, Georgetown, SC 29440. Tel.: 843-520-0111. Facebook: SC Maritime Museum.
E-mail: info@sc-mm.org
Governing Authority: Tax-exempt: 501(c)(3).
Institution Type/Description: Maritime History Museum.
Collections: maritime history & artifacts; photographs; paintings.
Hours & Admission Prices: Mon.-Sat. 11-5.

Green Sea

COUNTRY FARM MUSEUM, 1991 Fair Bluff Hwy., Green Sea, SC 29545-4452. Tel.: 843-756-1682.
Institution Type/Description: Farm Museum.
Collections: local history; agriculture; farm machinery & implements; period artifacts & farm toys; hands-on exhibits.
Hours & Admission Prices: Call for hours.

Greenville

BOB JONES UNIVERSITY MUSEUM & GALLERY, INC., 1700 Wade Hampton Blvd., Greenville, SC 29614. Tel.: 864-770-1331. Fax: 864-770-1306.
E-mail: contact@bjumg.org
Web Site: www.bjumg.org
Founded: 1951.
Congressional District: 4
Key Personnel: Dir., Erin Jones; Cur., John Nolan; Dir. Educations, Donnalynn Hess; Registrar, Barbara Sicko; Events Coord., Amy Basinger; Dir. Devel., Frank Richards; Grant Specialist, Blaine Welgraven.
Personnel Profile: Full-Time Paid 12; Part-Time Paid 4; Part-Time Volunteers 1; Interns 7.
Governing Authority: bd. of trustees. Subsidiary Institution: Museum & Gallery at Heritage Green. Tax-exempt.
Institution Type/Description: Old Masters Museum.
Collections: 14th-19th century European old masters paintings; archaeological & illustrative material relating to the Holy Lands; decorative arts; sculpture; textiles.
Major Exhibits: Victorian Exhibition, 8/13-2/15; Charles Dickens: The Continuing Victorian Narrative, M&G at Heritage Green, 10/13-5/15.
Facilities: Museum-related items for sale.
Activities: guided tours for adults and children first grade & above; audio tours; formally organized education programs for educators (private, public & home school); inter-museum loan & permanent exhibitions; performances; interactive learning center.

Publications: catalogucs; Bob Jones University Collection of Religious Art; Italian Paintings; John the Baptist and the Baroque Vision; Selected Masterworks from BJU Museum & Gallery; Icon; quarterly newsletter, Gallery News; Discovering a Pre-Renaissance Master: Tommaso del Mazza.
Hours & Admission Prices: mid-Jan. to mid-Dec. Tues.-Sun. 2-5. Adults $5, senior citizens $4, students $3; members and children 12 & under no charge. Audio Tour: $5. North American and the Southeastern reciprocal membership programs. Closed Commencement Day; Independence Day; Thanksgiving weekend. M&G at Heritage Green: Tues.-Sat. 10-5, Sun. 2-5. Adults $5, senior citizens $4, students $3; members & children 12 & under no charge. Closed New Year's Day; Thanksgiving; Christmas Eve & Day. &
Attendance: 21,550 (accurate)
Membership: Individual $50; Family & Dual $150.

THE CHILDREN'S MUSEUM OF THE UPSTATE, 300 College St., Greenville, SC 29601. Tel.: 864-233-7755. Facebook: TCM Upstate.
E-mail: info@tcmgreenvillesc.org
Web Site: www.tcmupstate.org
Founded: 2009.
Congressional District: 4
Key Personnel: C.E.O. & Pres., Nancy Halverson; Pres. (V), Cary Weekes; Museum Shop Mgr., Charlie Bishop.
Governing Authority: Tax-exempt.
Institution Type/Description: Children's Museum.
Collections: hands-on exhibits.
Activities: educational programs; special events.
Hours & Admission Prices: Memorial Day to Labor Day Mon.-Wed. & Fri.-Sat. 9-5, Thurs. 9-7; Sept.-May Tues.-Sat. 9-5, Sun. 11-5; groups by appointment. Adults $10, seniors & military $9.50, children 2-12 $9; members no charge. Closed New Year's Day; Easter; Thanksgiving; Christmas. &
Attendance: 150,000 (estimated)
Membership: Family Basic & Grandparent Basic $150; Family Plus & Grandparent Plus $350; Family Patron & Grandparent Patron $650.

＊ GREENVILLE COUNTY MUSEUM OF ART, (M), 420 College St., Greenville, SC 29601-2099. Tel.: 864-271-7570. Fax: 864-271-7579.
E-mail: info@gcma.org
Web Site: www.gcma.org
Founded: 1963.
Congressional District: 4
Key Personnel: Dir., Thomas W. Styron; Comptroller, Jeanne Marsh; Devel. Officer, Anne Q. Barr; Head Collections Mgmt. & Security, Claudia Beckwith; Communications, Paula Angermeier.
Personnel Profile: Full-Time Paid 23; Part-Time Paid 8; Part-Time Volunteers 200.
Governing Authority: appointed commission; nonprofit organization. Tax-exempt: 501(c)(3) & 170(b)(1)(A).
Institution Type/Description: Art Museum. American art from the 18th century to present.
Collections: American painting, sculpture & works on paper.
Research Fields: American paintings & sculpture; emphasis on Southern-related art; Andrew Wyeth; Jasper Johns.
Facilities: 76,000 sq. ft. exhibit space; 150-seat film theatre; art studios.
Activities: guided tours; lectures; gallery talks; formally organized education programs for children, adults, undergraduate & graduate college students; docent program; permanent collection of American art; temporary & traveling exhibitions; studio courses for adults & children in painting, drawing, photography, ceramics, printmaking, metal sculpture.
Publications: membership newsletter; brochures; catalogs for exhibitions; Greenville County Museum of Art: The Southern Collection; Andrew Wyeth: America's Painter.
Hours & Admission Prices: Wed.-Sat. 10-6, Sun. 1-5. No charge; donations accepted. Closed major holidays. &
Attendance: 100,000 (estimated)
Membership: Friend $50; Young Collectors $75; Contributing $100; Sponsor $500; Grand Benefactor $2,500; Director's Circle $5,000; Chairman's Circle $10,000.

GREENVILLE ZOO, 150 Cleveland Park Dr., Greenville, SC 29601-3147. Tel.: 864-467-4300. Fax: 864-467-4314.
E-mail: zooinfo@greenvillesc.gov
Web Site: www.greenvillezoo.com
Founded: 1960.
Congressional District: 4

Key Personnel: Dir., Jeff Bullock; Mgr. Public Svcs., Crystal Rose.
Personnel Profile: Full-Time Paid 32; Part-Time Paid 2; Part-Time Volunteers 100; Interns 6.
Governing Authority: Parent Institution: City of Greenville, SC.
Institution Type/Description: Zoo.
Collections: 300 animals including large mammals & reptiles,
Facilities: picnic areas. Museum-related items for sale.
Hours & Admission Prices: mid-February to Jan. daily 10-4:30. Adults $7.75, children 3-15 $4.50; discounts to groups & AZA reciprocal zoo members; members & children under 3 no charge. Closed New Year's Day; Thanksgiving; Christmas. &
Attendance: 275,000 (estimated)
Membership: Individual $39; Grandparent $49; Family $57; Family Plus $77; Peacock Society $112.

MUSEUM AND LIBRARY OF CONFEDERATE HISTORY, 15 Boyce Ave., Greenville, SC 29601-3109. Tel.: 864-421-9039. Fax: 864-421-9039.
E-mail: confedmuseum@att.net
Web Site: www.confederatemuseum.org
Formerly: South Carolina SCV Confederate Museum
Key Personnel: Dir., V. Michael Couch, Sr.; Chm. (V), Terry Lee Rude; Museum Shop Mgr., Greg Harrison.
Personnel Profile: Part-Time Volunteers 45.
Governing Authority: bd. of directors. Parent Institution: 16th Regiment South Carolina Volunteers, Camp 36, Sons of Confederate Veterans, Greenville, SC 29601. Tax-exempt.
Institution Type/Description: History Museum.
Collections: Civil War history; personal artifacts; military equipment; War for Southern Independence artifacts & documents.
Research Fields: genealogy; historical.
Activities: Museum Sponsors: Christmas in Dixie in December.
Publications: quarterly newsletter.
Hours & Admission Prices: Mon. & Wed. 10-3, Fri. 1-9, Sat. 10-5, Sun. 1-5; other times by appointment. No charge; donations accepted. Closed New Year's Day; Christmas. &
Attendance: 12,000 (estimated)
Membership: Wade Hampton Legion of Honor $100; Stonewall Jackson's Honor Brigade $500; Robert E. Lee's Aide-de-Camp $1,000.

PARK CENTER - PARIS MOUNTAIN STATE PARK, 2401 State Park Rd., Greenville, SC 29609. Tel.: 864-244-5565.
Institution Type/Description: Park Museum: housed in a renovated bathhouse built by the Civilian Conservation Corps in the 1930s.
Collections: local history; period artifacts; 3-D park map.
Facilities: nature trails; classroom.
Hours & Admission Prices: Call for hours. Adults $2, senior citizens $1.25; children 15 & under no charge.

ROPER MOUNTAIN SCIENCE CENTER, 402 Roper Mountain Rd., Greenville, SC 29615-4298. Tel.: 864-355-8900. Fax: 864-355-8948.
E-mail: education@ropermountain.org
Web Site: www.ropermountain.org
Founded: 1978.
Key Personnel: Dir., Greg Cornwell; Cur. Space Science, Charles St. Lucas; Cur. Physical Science, Latongia Pepper; Plant Engineer, Charles Head; Sec./Bookkeeper, Kelli Cox; Cur. Health & Web Coord., Kathy Hutchins; Cur. Computer Technology & Publishing, Cynthia Childers; Cur. Marine & Earth Science, Brandis Hartsell; Cur. Life Science, Tim Taylor.
Governing Authority: public school district. Parent Institution: Greenville County School District. Tax-exempt: 170(b)(1)(A).
Institution Type/Description: Science Center.
Collections: hands-on exhibits; living history farm.
Facilities: arboretum; life science labs; marine lab; observatory; physical science lab; 300-seat auditorium; 170-seat planetarium; nature trails; tropical rainforest conservatory.
Activities: summer concerts; teacher courses; summer science camps; formal education lessons for students K-12; science equipment; public observatory viewings; second Saturday events. Annual Event: Starry Friday Nights Programs January to late November.
Publications: newsletter; teacher guides to services.
Hours & Admission Prices: Mon.-Fri. 8:30-5. No charge. &
Attendance: 252,626 (accurate)
Membership: Senior Citizen $20; Individual $25; Teachers' Family $35; Family $45; Sponsor $100; Benefactor $250.

SHOELESS JOE JACKSON MUSEUM AND BASEBALL LIBRARY, 356 Field St., Greenville, SC 29601-3541. Mailing Address: P.O. Box 4755, Greenville, SC 29608-4755. Tel.: 864-346-4867.
E-mail: info@shoelessjoejacksonmuseum.org
Web Site: www.shoelessjoejackson.org
Key Personnel: Pres., Cur. & Publicist, Arlene Marcley.
Governing Authority: Tax-exempt: 501(c)(3).
Institution Type/Description: History Museum: housed in the former home of Joe Jackson.
Collections: records, artifacts, photographs & film relating to the life & baseball career of Shoeless Joe Jackson.
Facilities: Museum-related items for sale.
Hours & Admission Prices: Sat. 10-2; other times by appointment. No charge; donations accepted.

THOMPSON GALLERY - FURMAN UNIVERSITY, Art Dept., 3300 Poinsett Hwy., Greenville, SC 29613. Tel.: 864-294-2074.
Web Site: www2.furman.edu/academics/art/pages/default.aspx
Institution Type/Description: Art Gallery.
Collections: paintings; sculpture; drawings.
Activities: temporary exhibitions.
Hours & Admission Prices: Academic Year: Mon.-Fri. 9-5.

UPCOUNTRY HISTORY MUSEUM - FURMAN UNIVERSITY, (M), 540 Buncombe St., Greenville, SC 29601-1906. Tel.: 864-467-3100. Fax: 864-467-3105. Facebook: Upcountry History Museum.
E-mail: info@upcountryhistory.org
Web Site: www.upcountryhistory.org
Founded: 1983.
Congressional District: 4
Key Personnel: Chm., Cy Burgess; Exec. Dir., Dana L. Thorpe; Dir. Devel., Tim Bishop; Dir. Education, Kathy Kime; Event Coord., Kimberly Adams; Mgr. IT & Facilities, Ellen Hawkings; Information & Data Mgr., Sarah Harrison; Cur., Heather Yenco; Visitor Svcs. Coord., Jean Evans.
Personnel Profile: Full-Time Paid 8; Part-Time Paid 12; Part-Time Volunteers 30; Interns 9.
Volunteer Hours: 300
Operating Expenses: 1,100,000
Operating Income: 1,100,000
Governing Authority: Parent Institution: Furman University. Tax-exempt.
Institution Type/Description: History Museum.
Collections: Upcountry South Carolina history; interactive exhibits; oral histories; photographs.
Major Exhibits: Protests, Prayers, and Progress: Greenville's Civil Rights Movement, 1/14-6/14; The World of Jan Brett, 2/14-4/14; The Amazing Castle (T), 6/14-9/14; What Lies Beneath: Underwater Archaeology in the Carolinas, 10/14-5/15.
Research Fields: textile history; WWII; civil rights.
Facilities: 65-seat theatre; classroom; archives. Museum-related items for sale.
Activities: presentations; permanent & temporary exhibits; lectures; living history tours; family programs; school programs; group tours.
Hours & Admission Prices: Tues.-Sat. 10-5, Sun. 1-5. Adults $5; discounts to NARM members; members no charge. Closed major holidays. &
Attendance: 45,000 (estimated)
Membership: Individual $35; Family $65; Supporting $125; $250; $500; $1,000; $5,000.

Greenwood

THE BENJAMIN E. MAYS HISTORIC SITE, 239 N. Hospital St., Greenwood, SC 29646. Mailing Address: P.O. Box 1326, Greenwood, SC 29648. Tel.: 864-229-8801 & 223-8434.
Web Site: www.mayshousemuseum.org
Institution Type/Description: Historic House Museum: housed in the former home of Morehouse College President, Dr. Benjamin Mays.
Collections: Mays life & career; personal artifacts; photographs; early 1900 furnishings; books; articles; memorabilia. Historic Building: 1800s one-room school.
Facilities: theater.
Hours & Admission Prices: By appointment.

THE MUSEUM, 106 Main St., Greenwood, SC 29646-2763. Mailing Address: c/o Greenwood Arts Council, P.O. Box 3366, Greenwood, SC 29646. Tel.: 864-229-7093. Fax: 864-229-9317.
E-mail: themuseum@greenwood.net
Web Site: www.emeraldtriangle.sc/museum

Formerly: The Greenwood Museum
Founded: 1967.
Congressional District: 3
Key Personnel: Exec. Dir., Stacey Thompson; Pres. (V), Sandra Johnson; Collections Tech, Clay Barton; Museum Shop Mgr., April Miller.
Personnel Profile: Full-Time Paid 1; Part-Time Paid 3; Part-Time Volunteers 25; Interns 6.
Governing Authority: nonprofit organization. Subsidiary Institution: Railroad Historical Center. Tax-exempt: 501(c)(3) & 170(b)(1)(A).
Institution Type/Description: General Museum.
Collections: geology; archaeology; fossils; Indian & African artifacts; bottles & glass; early kitchen; replicas of doctors office, parlor, country store, early century school room; 1900 drug store; Thomas A. Edison display including replica of first phonograph made; old photographs; mounted butterflies, birds & moths; shells; firearms; war relics; zoology; animals; animal heads; operational commercial loom; hand loom; c.1910 linotype machine; hand-operated paper cutter; foot-operated job press; book binder press; process from raw bale to finished cloth; full scale replica of Cinderella coach; African collection of Frank Delano.
Facilities: Museum-related items for sale.
Activities: guided tours; formally organized education programs for children; permanent & temporary exhibitions; outreach programs; internships; summer camp; birthday parties; facility rental.
Publications: quarterly bulletin, Museum Newsletter.
Hours & Admission Prices: Museum: Wed.-Sat. 10-5. Railroad History Center: April-Oct. Sat. 10-4. Adults $5, seniors & children $2; discounts to AAM members & groups of 10 or more; members no charge. Closed legal holidays. &
Attendance: 6,000 (accurate)
Membership: Student $10; Senior Individual $15; Senior Duo & Individual $25; Family & Supporter $50-$199; Pillar Bronze $200 & over; Pillar Silver $500 & over; Pillar Gold $1,000 & over; Pillar Platinum $1,500 & over.

Greer

GREER HERITAGE MUSEUM, 106 S. Main St., Greer, SC 29651-3430. Mailing Address: P.O. Box 995, Greer, SC 29652-0995. Tel.: 864-877-3377.
E-mail: greerheritagemuseum@yahoo.com
Founded: 1993.
Key Personnel: Dir., Cur. & Pres. (V), David V. Duncan; Founder, Carm Hudson.
Governing Authority: nonprofit organization. Tax-exempt: 501 (c)(3).
Institution Type/Description: History Museum housed in the former City Hall & 1935 post office (WPA Building).
Collections: local history & culture relating to the city of Greer.
Facilities: library.
Activities: school groups; special events.
Hours & Admission Prices: Fri.-Sat. 10-4; groups of 10 or more by appointment. No charge; donations accepted. &

ZENTRUM MUSEUM, 1400 Hwy. 101 S., Greer, SC 29651-6731. Tel.: 864-989-5528 & 6000.
E-mail: amber.scruggs@bmwmc.com
Web Site: www.bmwusfactory.com/#/zentrum/1248
Institution Type/Description: BMW History Museum.
Collections: BMW history; racing & touring vehicles; technology & engineering.
Hours & Admission Prices: Mon.-Fri. 9:30-5:30; see website for holidays hours. No charge.

Hampton

HAMPTON COUNTY MUSEUM AT THE OLD JAIL, 702 W. 1st St., Hampton, SC 29924. Mailing Address: P.O. Box 152, Hampton, SC 29924. Tel.: 803-943-5484.
Formerly: Hampton County Historical Society Museum
Founded: 1979.
Congressional District: 2
Key Personnel: Pres. (V) Hampton County Historical Society, LaClaire W. Laffitte; Museum Dir., Mary Ann Sowell; Vice Pres., Durena Marcum; Treas., Virginia Sinclair; Corresponding Sec., Marian Platts.
Personnel Profile: Part-Time Volunteers 12.
Governing Authority: society; nonprofit organization. Hampton County Historical Society. Tax-exempt.
Institution Type/Description: History Museum: housed in c.1878 Old Jail. Listed on the National Register of Historic Places.
Collections: local history; veterans' of all wars personal artifacts; Prisoners of War artifacts from WWII & Korea; photographs; book & manuscript by

Manny Lawton; cemetery records; swords, bayonets & Civil War memorabilia; children's toys & period artifacts; 1878-1930s country store artifacts; Civil War & WWII history; black history; early 1800s-1900s women's clothing; genealogy; Native American artifacts; Korean, Vietnam & Desert Storm War memorabilia; midwives; natural history room; farm-related artifacts; church records, medical items.

Facilities: library of cemetery, church & family records, books & pamphlets available for research on premises; reading & research room.

Activities: guided tours; permanent & temporary exhibitions. Annual Event: Hampton County Watermelon Festival; Museum Observes: Black History Month; Women's History Month; Veterans Day; Memorial Day.

Publications: brochures; books, Both Sides of the Swamp; From the Salkehatchic to the Savannah (A Visual Journey Through Hampton County); Founding Citizens of Our New Country (Late 1860s-Early 1870s).

Hours & Admission Prices: Thurs. & Sun. 2-5 & by appointment. No charge; donations accepted. &

Attendance: 500 (estimated)

Membership: Student $5; Adults $20; Couple $35.

Hartsville

CECELIA COKER BELL GALLERY, Coker College, Art Dept., 300 E. College Ave., Hartsville, SC 29550-3742. Tel.: 843-383-8156 & 8150. Fax: 843-383-8033.

E-mail: artgallery@coker.edu

Web Site: www.ceceliacokerbellgallery.com

Founded: 1983.

Congressional District: 5

Key Personnel: Dir., Larry Merriman.

Personnel Profile: Part-Time Paid 1; Interns 2.

Governing Authority: private college. Tax-exempt.

Institution Type/Description: Art Gallery & Teaching Gallery.

Collections: works by national & international artists.

Major Exhibits: 41st Annual Faculty Show, 1/16/14-1/31/14; Laura Carpenter Truitt: Thresholds, 2/3/14-2/28/14; Dawn Gettler: Maybe If I Called You Darling, 3/3/14-3/28/14; Senior Exhibitions, 3/31/14-5/9/14; 9th Annual Student Summer Show, 5/20/14-8/22/14; Solo Exhibition, 8/14-9/14; Solo Exhibition, 9/14-10/14; Solo Exhibition, 10/14-11/14; 42nd Annual Student Competition, 11/17/14-11/25/14.

Facilities: educational facilities; 750 sq. ft. exhibit space.

Activities: guided tours. Annual Events: exhibitions & lectures by visiting artists.

Publications: posters; catalogues.

Hours & Admission Prices: Academic Year Mon.-Fri. 10-4; Summer: by appointment only. No charge. &

Attendance: 4,000 (estimated)

HARTSVILLE MUSEUM, 222 N. Fifth St., Hartsville, SC 29550-4136. Mailing Address: P.O. Box 431, Hartsville, SC 29551-0431. Tel.: 843-383-3005. Fax: 843-383-2477.

E-mail: info@hartsvillemuseum.org

Web Site: www.hartsvillemuseum.org

Founded: 1980.

Congressional District: 6

Key Personnel: C.E.O. & Dir., Kathy M. Dunlap; Chm., Glenn J. Lawhon, Jr.; City Mgr., James Pennington; Museum Shop Mgr., Penny Anthony.

Personnel Profile: Full-Time Paid 2; Part-Time Paid 3; Part-Time Volunteers 21; Interns 4.

Governing Authority: municipal. Parent Institution: City of Hartsville. Tax-exempt.

Institution Type/Description: Local History Museum: housed in restored 1930 U.S. Post Office Building, changing gallery for Arts Programming.

Collections: Eastern Carolina Silver Co. holloware c.1900; iron & wood household objects; historical photos; costumes; documents; papers; ornaments; personal property pertinent to Hartsville & surrounding areas; 1899 locomobile steam car; first car in SC; sculpture courtyard.

Research Fields: local history.

Facilities: conference room.

Activities: guided tours; lectures; loan, permanent & temporary exhibitions.

Publications: booklet, Hartsville's Eastern Carolina Silver Company; Milestones, Hartsville's Centennial; book, Recollections of the Major.

Hours & Admission Prices: Mon.-Fri. 10-5, Sat. 10-2. No charge; donations accepted. Closed state holidays. &

Attendance: 12,500 (accurate)

Membership: Individual $25; Family $40; Patron $60; Sustainer $100; Business $150; Showcase $250; Heritage $500; "The Mayor" $1,000; Benefactor $3,000.

JACOB KELLEY HOUSE MUSEUM, 2585 Kellytown Rd., Hartsville, SC 29550. Mailing Address: 204 Hewitt St., Darlington, SC 29532-3214. Tel.: 843-332-6401 & 339-9093. Fax: 843-332-8017.

E-mail: dchc@darcosc.com

Congressional District: 5

Key Personnel: Pres. (V), Jo Ann K. Lee; C.E.O. & Dir., Doris Gandy.

Governing Authority: county; nonprofit. Parent Institution: Darlington County. Subsidiary Institution: Jacob Kelley House Guild. Tax-exempt.

Institution Type/Description: Historic House: 1820 Federal style plantation house used as General Sherman's headquarters during Civil War.

Collections: handcrafted furniture; decorative arts c.1820-1874. Historic Building: barn.

Activities: guided tours.

Publications: book, The Jacob Kelley House.

Hours & Admission Prices: Feb.-Nov. first Sun. of the month 3-5. No charge; donations accepted.

Membership: Guild $15.

KALMIA GARDENS, 1624 W. Carolina Ave., Hartsville, SC 29550-4906. Tel.: 843-383-8145. Fax: 843-383-8149.

Web Site: kalmiagardens.org

Key Personnel: Dir., Mary R. Ridgeway

Governing Authority: Parent Institution: Coker College.

Institution Type/Description: Botanical Garden & Historic House Museum: home of Thomas E. Hart, built in 1820. Listed on the National Register of Historic Places.

Collections: period furnishings; personal artifacts; rhododendrons; camellias; azaleas; wisteria; tea-olives; dogwood; Kalmia latifolia.

Facilities: 35 acre garden.

Hours & Admission Prices: Daily. No charge.

Hilton Head Island

COASTAL DISCOVERY MUSEUM AT HONEY HORN, 100 William Hilton Pkwy., Hilton Head Island, SC 29926-1216. Mailing Address: P.O. Box 23497, Hilton Head Island, SC 29925-3497. Tel.: 843-689-6767. Fax: 843-689-3035. Facebook: Coastal Discovery Museum.

E-mail: info@coastaldiscovery.org

Web Site: coastaldiscovery.org

Founded: 1985.

Congressional District: 2

Key Personnel: Pres. & C.E.O., Michael J. Marks; Vice Pres. Finance & Administration, Jennifer Stupica; Vice Pres. Programs, Natalie Hefter; Vice Pres. Mktg. & Devel., Robin Swift; Mgr. Natural History, Carlos Chacon; Cur. Education, Dawn Brut.

Governing Authority: bd. of trustees; nonprofit. Tax-exempt: 501(c)(3).

Institution Type/Description: General Museum.

Collections: local & North American natural history specimens; South Carolina archaeological materials; historical material relevant to history of the island, Colonial, Plantation, Civil War & current day.

Facilities: lecture hall & classroom; picnic facilities. Museum-related items for sale.

Activities: guided tours; lectures; films; gallery talks; classes; formally organized program for on- & off-site curriculum-based programming for schools; land, marsh & water programming.

Publications: newsletter; info-guides.

Hours & Admission Prices: Mon.-Sat. 9-4:30, Sun. 11-3. &

Attendance: 96,500 (accurate)

Membership: Individual $35; Family $50; Patron $100.

HARBOUR TOWN LIGHTHOUSE MUSEUM, 149 Lighthouse Rd., Hilton Head Island, SC 29928. Tel.: 866-305-9814 (Toll Free); 843-671-2810.

E-mail: info@harbourtownlighthouse.com

Web Site: www.harbourtownlighthouse.com

Institution Type/Description: Lighthouse Museum.

Collections: local history; Civil War artifacts; photographs; period artifacts.

Hours & Admission Prices: Call for hours.

THE SANDBOX, AN INTERACTIVE CHILDREN'S MUSEUM, 18A Pope Ave., Hilton Head Island, SC 29928-4708. Tel.: 843-842-7645.

E-mail: executivedirector@thesandbox.org

Web Site: www.thesandbox.org

Founded: 2005.

Key Personnel: Exec. Dir., Steve Maglione; Operations Mgr., Caroline Rinehart.

Personnel Profile: Full-Time Paid 2; Part-Time Paid 8; Part-Time Volunteers 1.
Governing Authority: Tax-exempt.
Institution Type/Description: Children's Museum.
Collections: hands-on exhibits.
Activities: special events; birthday parties; rental facilities. Museum Sponsors: Parent's Night Out.
Hours & Admission Prices: April & June-Aug. Mon.-Sat. 10-5; May & Sept.-March Tues.-Sat. 10-5. Admission $6; members & children under one no charge. Closed Thanksgiving; Christmas. &
Attendance: 28,465 (accurate)
Membership: Basic $75 (additional adult $20).

Hopkins

CONGAREE NATIONAL PARK, 100 National Park Rd., Hopkins, SC 29061-8320. Tel.: 803-776-4396. Fax: 803-783-4241.
Web Site: www.nps.gov/cong
Formerly: Congaree Swamp National Monument
Founded: 1976.
Congressional District: 6
Key Personnel: Mgr. Visitor Center, Fran Rametta; Supt., Tracy Swartout; Chief Resources, Bill Hulslander.
Personnel Profile: Full-Time Paid 12; Part-Time Paid 3; Part-Time Volunteers 100; Interns 2.
Governing Authority: federal. Parent Institution: National Park Service. Tax-exempt.
Institution Type/Description: National Park & Preservation Project: 26,800-acre National Park.
Collections: 5 national Champion trees; 20 state record trees; 86 tree species; loblolly pines; American elms, 17 feet in circumference; bald cypress; 26 feet in circumference; 779 plants; 83 mushrooms; 24 amphibians; 29 reptiles; 44 fish & 200 bird species.
Research Fields: biology; geology; hydrology; ecology.
Facilities: visitor center & headquarters; 20 miles of hiking trail; self-guided nature trail.
Activities: guided tours; hiking; picnicking; primitive camping with permit; 16 mile marked canoe trail.
Publications: Official Map & Guide, Official Congaree National Park Map & Guide; Birds of Congaree National Park.
Hours & Admission Prices: Visitor Center: Summer: daily 8:30-7; Winter: daily 8:30-5. No charge; donations accepted. Closed Christmas. &
Attendance: 134,000 (accurate)

Hunting Island

HUNTING ISLAND NATURE CENTER, 2555 Sea Island Pkwy., Hunting Island, SC 29920. Tel.: 843-838-7437.
E-mail: huntingisland@scprt.com
Web Site: www.huntingisland.com
Institution Type/Description: Nature Center.
Collections: live reptiles; natural resources of the park.
Hours & Admission Prices: Summer: daily 9-5; Winter: Tues.-Sat. 9-5.

Jackson

SILVER BLUFF AUDUBON CENTER, 4542 Silver Bluff Rd., Jackson, SC 29831. Tel.: 803-471-0291.
E-mail: pkoehler@audubon.org
Web Site: www.sc.audubon.org/Centers_SB.html
Institution Type/Description: Audubon Center.
Collections: local history, culture & heritage; wildlife & their habitats.
Facilities: nature trails. Center-related items for sale.
Activities: walking trails; special events; educational programs.
Hours & Admission Prices: Call for hours.

Johnston

EDGEFIELD COUNTY PEACH MUSEUM, 416 Calhoun St., Johnston, SC 29832-1317. Tel.: 803-275-0010. Fax: 803-275-3586.
Institution Type/Description: History Museum.
Collections: peach history, industry, & production; early pioneers.
Hours & Admission Prices: Mon.-Fri. 8:30-12:30; other times by appointment. Closed holidays.

Kingstree

THORNTREE PLANTATION, Nelson Blvd., Kingstree, SC 29556-3335. Mailing Address: Williamsburg Historical Society, 135 Hampton Ave., Kingstree, SC 29556-3423. Tel.: 843-355-3306.
Governing Authority: Parent Institution: Williamsburgh Historical Society.
Institution Type/Description: Historic House Museum: housed in the home of James Witherspoon; built in 1749.
Collections: local history & heritage; period furnishings; personal artifacts; portraits; photographs.
Hours & Admission Prices: By appointment.

WILLIAMSBURGH HISTORICAL MUSEUM, 135 Hampton Ave., Kingstree, SC 29556-3423. Tel.: 843-355-3306.
E-mail: history1@ftc-i.net
Key Personnel: Dir., Joanne B. Brown.
Governing Authority: Parent Institution: Williamsburgh Historical Society. Tax-exempt.
Institution Type/Description: History Museum.
Collections: historic archives & artifacts; antique china displays; early maps; dug-out canoe.
Hours & Admission Prices: Tues.-Thurs. 10-3. No charge; donations accepted. Closed major holidays.

Lake City

BROWNTOWN MUSEUM, Hwy. 341, Lake City, SC 29560. Mailing Address: 414 Main St., Hemingway, SC 29554-9190. Tel.: 843-558-2355.
Key Personnel: Dir. & Cur., Nell Morris; Pres., Kathy Loyd; Treas., Mona Prosser; Sec., Carol Cockfield.
Governing Authority: nonprofit organization. Parent Institution: Three Rivers Historical Society. Tax-exempt.
Institution Type/Description: Historic Site & Preservation Project.
Collections: cotton gin; corn crib; smokehouse; farm equipment. Historic House: 1845 Brown-Burrows House.
Research Fields: Brown family genealogy; census records; equity rolls; family Bibles; cemeteries; court house records.
Facilities: 44-vol. library.
Activities: guided tours; lectures; organized educational programs for children & adults; annual events.
Publications: quarterly newsletter, The Three Rivers Chronicle.
Hours & Admission Prices: Sat. 9-4, groups by appointment. Adults $5, children $2. &
Attendance: 350 (estimated)
Membership: Annual $20.

NATIONAL BEAN MARKET MUSEUM, 111 Henry St., Lake City, SC 29560. Mailing Address: P.O. Box 943, Lake City, SC 29560-0943. Tel.: 843-374-1500.
Institution Type/Description: Historic Building: built in 1936. Listed on the National Register of Historic Places.
Collections: early farm life; local history & culture; pole tobacco barn; agriculture.
Hours & Admission Prices: Mon.-Thurs 8-5, Fri. 8-4. No charge. Closed major holidays.

Lancaster

ANDREW JACKSON STATE PARK, 196 Andrew Jackson Park Rd., Lancaster, SC 29720-6404. Tel.: 803-285-3344.
E-mail: ajacksonsp@scprt.com
Web Site: southcarolinaparks.com/park-finder/state-park/1797.aspx
Founded: 1954.
Congressional District: 5
Key Personnel: Supt., Kirk Johnston.
Personnel Profile: Full-Time Paid 3; Part-Time Paid 2; Part-Time Volunteers 5.
Governing Authority: Parent Institution: South Carolina State Parks.
Institution Type/Description: Park & History Museum: dedicated to Andrew Jackson, the 7th president of the United States.
Collections: Andrew Jackson's life & history; replica 18th-century one-room schoolhouse; statue.
Facilities: amphitheatre; picnic area; nature trails.
Activities: living history programs; camping; fishing; picnicking; rental facilities; concerts; special events. Annual Events: Andrew Jackson Birthday in March; Lantern Tour in November.

Hours & Admission Prices: Park: Summer, daily 9-9; Winter, daily 8-6. Schoolhouse: mid-March to Nov. Sat. 1-5, Sun. 2-5. Museum: Sat.-Sun. 1-5; other times by appointment. Adults $2, seniors $1.25; children 15 & under no charge.
Attendance: 51,000 (estimated)

LANCASTER & CHESTER RAILWAY MUSEUM, 512 S. Main St., 2nd Fl., Lancaster, SC 29720-3622. Mailing Address: P.O. Box 1450, Lancaster, SC 29721-1450. Tel.: 803-286-2100 & 2102. Fax: 803-286-4158.
Institution Type/Description: Railway Museum.
Collections: railway history; scale model replica of the railway route; photographs; railway memorabilia.
Hours & Admission Prices: 1st & 3rd Sat. of the month 10-4. No charge.

Lando

LANDO MANETTA MILLS HISTORY CENTER, 3801 Lando Rd., Lando, SC 29724-9600. Tel.: 803-789-6361.
E-mail: paul@landomanettamillshistorycenter.com
Web Site: landomanettamillshistorycenter.com
Governing Authority: nonprofit organization.
Institution Type/Description: History Museum: housed in the former Manetta Mills Company office.
Collections: local history & culture; personal artifacts; period furnishings; photographs.
Activities: Annual Event: Lando Days in October.
Hours & Admission Prices: Mon. & Wed. 9 to noon, Sun. 2-5; other times by appointment.

Latta

DILLON COUNTY MUSEUM, 101 S. Marion St., Latta, SC 29565-1558. Mailing Address: Dillon County Historical Society, P.O. Box 1806, Dillon, SC 29536-1806. Tel.: 843-441-5273 & 774-6122.
E-mail: bbar830771@aol.com
Congressional District: 6
Key Personnel: Dir., Betty L. Barclay.
Personnel Profile: Part-Time Paid 1; Part-Time Volunteers 10.
Governing Authority: Tax-exempt.
Institution Type/Description: Historic Building: housed in the former office of Dr. Henry Edwards; built in 1925. Listed on the National Register of Historic Places.
Collections: family & local history; Dr. Edwards' dental office; documents; photographs; personal artifacts; period furnishings; cotton & tobacco industries; military artifacts; period clothing.
Hours & Admission Prices: Tues.-Thurs. 2-4; other times by appointment. No charge; donations accepted. &
Attendance: 500 (estimated)

Laurens

THE CHARLES H. DUCKETT HOUSE, 105 Downs St., Laurens, SC 29360. Tel.: 803-896-6196.
Institution Type/Description: Historic House Museum: housed in the former home of prominent black businessman, Charlie Duckett; built c.1892. Listed on the National Register of Historic Places.
Collections: Duckett life & career; period furnishings; photographs; local history.
Hours & Admission Prices: Call for hours.

THE JAMES DUNKLIN HOUSE, 544 W. Main St., Laurens, SC 29360. Mailing Address: 2009 Lakeview Dr., Laurens, SC 29360-5132. Tel.: 864-984-4735 & 683-2432.
Founded: 1972.
Congressional District: 5
Key Personnel: Pres., James C. Todd, III; Mgr., Shawn Brown.
Governing Authority: nonprofit organization. Affiliated with the Laurens County Landmarks Foundation, 544 W. Main St., Laurens, SC 29360. Tax-exempt: 170(b)(1)(A).
Institution Type/Description: Historic Foundation: housed in 1812 The James Dunklin House. Listed on the National Register of Historic Places.
Collections: furniture.
Activities: guided tours.
Hours & Admission Prices: 1st Sun. of month 2-5; other times by appointment. Adults $2, children 11 & under $1.
Attendance: 300 (accurate)

Lexington

LEXINGTON COUNTY MUSEUM, 231 Fox St., Lexington, SC 29072-2654. Mailing Address: P.O. Box 637, Lexington, SC 29071-0637. Tel.: 803-359-8369. Fax: 803-808-2160. Facebook: Lexington County Museum.
E-mail: museum@lex-co.com
Web Site: www.lex-co.sc.gov/museum
Founded: 1970.
Congressional District: 2
Key Personnel: Chm. (V), Bill Kiesling; Dir., J.R. Fennell.
Personnel Profile: Full-Time Paid 2; Part-Time Paid 6; Interns 2.
Governing Authority: county. Lexington County Government. Tax-exempt: 170(b)(1)(A).
Institution Type/Description: History Museum.
Collections: Native American artifacts; 18th & 19th century furniture made in Lexington County area; textiles; manuscripts; looms; spinning wheels; farm implements including cotton gin, cane press, hand implements. Historic Buildings: 1832 John Fox House; 1834 Ernest Hazelius House; 1771 Lawrence Corley Log Cabin; 1772 Henrich Senn Log Cabin; 18th century, Lorick Log Cabin; 1820 lawyer's office; 1850 gin house and barn; 1815, school; 1810 Leaphart Harman House.
Research Fields: local history with emphasis on Swiss & German settlers.
Activities: guided tours; lectures; permanent exhibitions; demonstrations of yarn spinning & cloth weaving.
Publications: brochure.
Hours & Admission Prices: Tues.-Sat. 10-4, Sun. 1-4; last tour 3. Adults $5; members no charge. Closed major holidays.
Attendance: 20,000 (accurate)
Membership: Student $10; Senior & Military $15; Individual $20; Family $25; Organizational $50; Patron $100; Corporate $1,000; Benefactor $2,500.

Manning

CLARENDON COUNTY ARCHIVES AND HISTORY CENTER, Old Manning Library, N. Brooks St., Manning, SC 29103-3209. Mailing Address: 211 N. Brooks St., Manning, SC 29102-3209. Tel.: 803-435-0328.
Institution Type/Description: History Museum.
Collections: county history & culture; photographs; personal artifacts; maps; manuscripts.
Hours & Admission Prices: Mon.-Fri. 9:30-5, Sat. by appointment.

Marion

MARION COUNTY MUSEUM, 101 Willcox Ave., Marion, SC 29571-2809. Tel.: 843-423-8299.
Key Personnel: Dir., Thomas Gett
Institution Type/Description: Historic Building: housed in a former schoolhouse, c.1886. Listed on the National Register for Historic Sites.
Collections: period furnishings; oriental rugs.
Hours & Admission Prices: Tues.-Fri. 9-12 & 1-5; other times by appointment. No charge.

McClellanville

HAMPTON PLANTATION STATE HISTORIC SITE, 1950 Rutledge Rd., McClellanville, SC 29458-9588. Tel.: 843-546-9361. Fax: 843-527-4995.
E-mail: hampton@scprt.com
Web Site: www.southcarolinaparks.com
Formerly: Hampton Plantation State Park
Founded: 1971.
Congressional District: 1
Key Personnel: Rgnl. Chief, Ray Stevens; Park Mgr., Dale Purvis.
Personnel Profile: Full-Time Paid 2; Part-Time Volunteers 1.
Governing Authority: state. Affiliated with South Carolina Dept. of Parks, Recreation & Tourism, Brown Bldg., 1205 Pendleton St., Columbia, SC. 29201. Tel.: 803-758-7507. Tax-exempt.
Institution Type/Description: Historic House: c.1750 Hampton House, former rice plantation.
Collections: local history & culture; photographs; period furnishings; personal artifacts.
Research Fields: rice plantation culture.
Facilities: gardens.
Activities: guided & self-guided tours; nature trail.
Publications: brochure, Hampton Plantation Visitors Guide.
Hours & Admission Prices: Park: daily 9-5. Mansion Tours: Sat.-Tues. 1pm, 2pm & 3pm. Adults $7.50, senior citizens $3.75, children $3.50. Closed Christmas. &

Attendance: 30,000 (accurate)

THE VILLAGE MUSEUM, 401 Pinckney St., McClellanville, SC 29458. Mailing Address: P.O. Box 595, McClellanville, SC 29458. Tel.: 843-887-3030.
E-mail: villagemuseum@tds.net
Institution Type/Description: History Museum.
Collections: local history & culture; early pioneers; Native American artifacts.
Hours & Admission Prices: Thurs.-Sat. 10-12 & 1-5. Adults $3; discounts to groups; members, children & students no charge.

McConnells

HISTORIC BRATTONSVILLE, 1444 Brattonsville Rd., McConnells, SC 29726-8768. Tel.: 803-684-2327. Fax: 803-684-0149.
E-mail: hbratton@chmuseums.org
Web Site: www.chmuseums.org
Founded: 1976.
Congressional District: 5
Key Personnel: Dir., Van Shields; Site Dir., Chuck LeCount; Museum Shop Mgr., Mark Cockerille.
Personnel Profile: Full-Time Paid 15; Part-Time Paid 1; Part-Time Volunteers 200.
Governing Authority: nonprofit organizations. Affiliated with York County Council. Parent Institution: York County Historical Commission. Tax-exempt.
Institution Type/Description: Preservation Project: housed in 1823 federal style mansion located on the site of Huck's Defeat in battle during the Revolutionary War.
Collections: 29 historic structures & programs chronicling Carolina and Piedmont development from the 1750s over 720 acres of a living history village & battlefield. Historic Houses: 1776 Col. William Bratton house; 1823 The Homestead; 1843 Brattonsville Female Seminary.
Research Fields: county records; genealogy; craftsman & occupational data in York County.
Facilities: picnic area.
Activities: guided tours; organized educational programs; festivals. Project Sponsors: Living History Days, Saturdays, March-November.
Publications: book, Historic Brattonsville.
Hours & Admission Prices: Mon.-Sat. 10-5, Sun. 1-5; groups & Christmas tours by appointment. Adults $6, seniors 60 & over $5, youth 4-17 $3; members and children 3 & under no charge. Closed New Year's Day; Thanksgiving; Christmas. &
Attendance: 21,920 (accurate)
Membership: Students $20; Senior $25; Individual & Jade Society Individual $35; Senior Family $40; Family & Jade Society Family $50; Grandparents $65; Family 2 Year $90.

McCormick

DORN MILL CENTER, 200 N. Main St., McCormick, SC 29835. Mailing Address: 300 Pineview, McCormick, SC 29835.
E-mail: mcparnell@wctel.net
Web Site: www.mccormickschistory.org
Founded: 1980.
Congressional District: 3
Key Personnel: Chm. (V), Marian Parnell.
Personnel Profile: Part-Time Volunteers 6.
Governing Authority: Parent Institution: McCormick Historical Commission. Subsidiary Institution: South Carolina's Governor's Office. Tax-exempt.
Institution Type/Description: Interpretive Center.
Collections: America's agricultural & industrial history; gristmill; cotton gin; weigh station; mill equipment; period artifacts.
Facilities: Museum-related items for sale.
Activities: craft demonstrations.
Hours & Admission Prices: By appointment. No charge; donations accepted. &

Moncks Corner

BERKELEY COUNTY MUSEUM & HERITAGE CENTER, 950 Stony Landing Rd., Moncks Corner, SC 29461-2944. Tel.: 843-899-5101. Fax: 843-899-5101.
E-mail: berkmuseum@homesc.com
Formerly: Berkeley Museum, Inc.
Founded: 1992.
Congressional District: 1
Key Personnel: Chm. & Pres. (V), Willard Strong; Mgr., Carolyn Pilgrim.
Personnel Profile: Part-Time Paid 2.

Governing Authority: Tax-exempt.
Institution Type/Description: History Museum. Home of the Little David; birthplace of Francis Marion.
Collections: works of local artists & artisans; replica of CSS David, a semi-submersible torpedo boat.
Facilities: research library.
Activities: Annual Event: Antique Tractor and Engine Show in November.
Publications: newsletter, Historical Society.
Hours & Admission Prices: Tues.-Sat. 9-4:30, Sun. 1-4:30. Adults $3, seniors $2; discounts to groups; children 6 & under no charge. Admission to Old Santee Canal Park includes entry to Berkeley Museum. Closed Easter; Thanksgiving; Christmas. &
Attendance: 10,000 (estimated)
Membership: Individual $25; Family $35; Donor $100; Contributor $250; Supporting $500; Benefactor $1,000; Sustaining $2,500; Patron $5,000; Founder $10,000; Chairman's Circle $10,000 & up.

CYPRESS GARDENS, 3030 Cypress Gardens Rd., Moncks Corner, SC 29461-6447. Tel.: 843-553-0515.
E-mail: tcook@berkeleycountysc.gov
Web Site: www.cypressgardens.info
Institution Type/Description: Gardens.
Collections: gardens; wildlife & their habitats; archaeology; 12 species of butterflies.
Facilities: nature trails.
Activities: swamp boat rides.
Hours & Admission Prices: Daily 9-5. Adults $10, seniors 65 & over $9, children 6-12 $5; children 5 & under no charge. Closed New Year's Day; Thanksgiving; Christmas Eve & Day.
Attendance: 45,000 (estimated)

OLD SANTEE CANAL PARK, 900 Stony Landing Rd., Moncks Corner, SC 29461-2944. Tel.: 843-899-5200. Fax: 843-761-7032.
E-mail: parkinfo@santeecooper.com
Web Site: www.oldsanteecanalpark.org
Formerly: Old Santee Canal State Historic Site
Founded: 1991.
Congressional District: 1
Key Personnel: C.E.O., Lonnie N. Carter; Museum Shop Mgr., Cindy Moyer.
Personnel Profile: Full-Time Paid 7; Part-Time Paid 2; Part-Time Volunteers 20; Interns 3.
Governing Authority: state; not-for-profit organization. Parent Institution: Santee Cooper.
Institution Type/Description: Park Museum: located on historic Stony Landing Plantation at southern terminus of the Santee Canal (1800-1850s), the first true canal constructed in the United States.
Collections: Santee Canal & activities of the Stony Landing Plantation; construction of the CSS Little David; natural history of a swamp/wetlands environment.
Activities: monthly events.
Hours & Admission Prices: Daily 9-5. Adults $3, senior citizens over 65 $2; discounts to AAA members; children under 5 no charge. Closed New Year's Day; Easter; Thanksgiving; Christmas Eve & Day. &
Attendance: 24,834 (accurate)
Membership: Individual Pass $25; Family Pass $35.

Mount Pleasant

BOONE HALL PLANTATION & GARDENS, 1235 Long Point Rd., Mount Pleasant, SC 29464-9020. Tel.: 843-884-4371. Fax: 843-884-0475.
E-mail: info@boonehallplantation.com
Web Site: www.boonehallplanation.com
Founded: 1955.
Institution Type/Description: History Museum: house built in 1936.
Collections: plantation history from 1681 to present; period furnishings; photographs. Historic Buildings: smokehouse; slave cabins.
Activities: educational programs; guided tours.
Hours & Admission Prices: late March to Labor Day Mon.-Sat. 8:30am-6:30pm, Sun. 1-5; Sept. to late March Mon.-Sat. 9-5, Sun. 1-4. Adults $19.50, children 6-12 $10; children 5 & under no charge.

CHARLES PINCKNEY NATIONAL HISTORIC SITE, 1254 Long Point Rd., Mount Pleasant, SC 29464. Mailing Address: 1214 Middle St., Sullivan's Island, SC 29482-9717. Tel.: 843-881-5516. Fax: 843-881-7070.
Web Site: www.nps.gov/chpi
Founded: 1988.
Congressional District: 1

Key Personnel: Supt., Tim Stone; Chief Interpretation, Dawn Davis; Cur., Catherine Fowler; Museum Shop Mgr., Kevin Bates.
Personnel Profile: Full-Time Paid 7; Part-Time Volunteers 6.
Governing Authority: federal government; nonprofit. Parent Institution: National Park Service. Tax-exempt.
Institution Type/Description: Historic Site: located in a c.1828 plantation house.
Collections: emphasis on the life & political career of Charles Pinckney, the U.S. Constitution & plantation life and slavery in early America; archaeological artifacts connected to the Pinckney family occupation, 1754-1817; artifacts connected to Snee Farm plantation during the 17th-20th centuries.
Research Fields: Revolutionary war; U.S. Constitution; slavery; U.S. history to 1825; Gullah.
Facilities: classroom. Museum-related items for sale.
Activities: self-guided tours; video presentation; formally organized education programs for children & adults by reservation; temporary exhibits.
Publications: site brochure; site archaeology bulletin; African Americans at Snee Farm; site bulletin.
Hours & Admission Prices: Daily 9-5. No charge; donations accepted. Closed New Years Day; Thanksgiving; Christmas. &
Attendance: 35,000 (accurate)

PATRIOTS POINT NAVAL AND MARITIME MUSEUM, 40 Patriots Point Rd., Mount Pleasant, SC 29464-4377. Tel.: 843-884-2727.
Web Site: www.patriotspoint.org
Founded: 1976.
Congressional District: 1
Key Personnel: Chm., John B. Hagerty; Exec. Dir., Mac Burdette; Dir. Operations, Bob Howard; Collections, Melissa Buchanan; Sr. Cur., David A. Clark; Museum Shop Mgr., Samuel Derrick.
Personnel Profile: Full-Time Paid 76; Part-Time Paid 36; Part-Time Volunteers 50.
Governing Authority: state; nonprofit. Affiliated with Patriots Point Development Authority. Tax-exempt: 501(c)(3).
Institution Type/Description: Maritime and Naval Museum: housed in the aircraft carrier USS Yorktown, destroyer USS Laffey, & submarine USS Clamagore located in Charleston Harbor; also houses Congressional Medal of Honor Museum & Carrier Aviation Hall of Fame.
Collections: Navy & Marine Corps aircraft; naval & maritime artifacts; manuscripts.
Research Fields: naval & maritime history.
Facilities: theater; 240-seat auditorium; catering; restaurant. Naval items depicting the five ships for sale.
Activities: films; loan, permanent & temporary exhibitions; concerts; radio programs; scout camping program.
Hours & Admission Prices: Daily 9am to various seasonal closing times. Adults $18, senior citizens & active military $13, children 6-11 $8; discounts to Historic Naval Ships Association; children under 6 no charge. Closed Christmas. &
Attendance: 263,000 (accurate)
Membership: Student $25; Individual $35; One Star Family $75; 2 Star Family $90; 3 Star Family $125; Corporate $500.

Mullins

S.C. TOBACCO MUSEUM, 104 NE Front St., Mullins, SC 29574-2810. Tel.: 800-207-7967; 843-464-8194.
E-mail: cityofmullins@mullinssc.us
Web Site: www.mullinssc.us/sctobaccomuseumindex.html
Founded: 1998.
Congressional District: 6
Key Personnel: Dir., Reginald McDaniel.
Personnel Profile: Full-Time Paid 2; Part-Time Volunteers 2.
Governing Authority: city of Mullins.
Institution Type/Description: History Museum: housed in an historic train depot.
Collections: history of tobacco growing; early farm life; tobacco-related equipment; blacksmith tools; documentary video; original oil paintings of tobacco farm life.
Major Exhibits: Farm Food Bounty, 2011-2015.
Facilities: research library. Museum-related items for sale.
Activities: research; video; tours.
Hours & Admission Prices: Mon.-Fri. 9-5; groups by appointment. Adults $2, children & seniors $1. Closed holidays. &
Attendance: 2,000 (estimated)

Murrells Inlet

* **BROOKGREEN GARDENS, (M),** 1931 Brookgreen Dr., Murrells Inlet, SC 29576. Mailing Address: P.O. Box 3368, Pawleys Island, SC 29585-3368. Tel.: 843-235-6000; 800-849-1931. Fax: 843-235-6039.
E-mail: info@brookgreen.org
Web Site: www.brookgreen.org
Founded: 1931.
Congressional District: 6
Key Personnel: C.E.O. & Pres., Robert Jewell; Mgr., Jay Rowe; Vice Pres. Finance, Kathleen Zeiss; Vice Pres. & Cur. of Sculpture, Robin R. Salmon; Vice Pres. Horticulture & Conservation, Sara Millar; Vice Pres. Devel., Phillip A. Tukey; Vice Pres. Mktg., Helen Benso; Museum Shop Mgr., Ashley Gray.
Personnel Profile: Full-Time Paid 69; Part-Time Paid 34; Part-Time Volunteers 400.
Governing Authority: nonprofit organization. Tax-exempt: 501(c)(3).
Institution Type/Description: Art Museum & Botanical Garden.
Collections: sculpture; botany; native fauna.
Research Fields: American figurative sculpture; maintenance of outdoor sculpture; longleaf pine forest ecosystem.
Facilities: American sculpture garden; nature center; cafe. Museum-related items for sale.
Activities: guided tours; temporary & permanent exhibitions; adult education program; day camp (Camp Brookgreen). Annual Events: Cool Summer Evenings in summer; Nights of A Thousand Candles in December.
Publications: biannual, Brookgreen Journal; Catalogue of Sculpture Collection; newsletter, 2 issues per year.
Hours & Admission Prices: June 17-Aug. 14 daily 9:30-7; Aug. 15-June 16 daily 9:30-5. Adults $14, seniors $12, children 4-12 $7; children 3 & under and members no charge. Admission valid for 7 days. Closed Christmas. &
Attendance: 270,350 (accurate)
Membership: Individual $60; Family $90; President's Council $250; Chairman's Council $1,000; Huntington Society $2,500; Atalaya Society $5,000.

Myrtle Beach

CHILDREN'S MUSEUM OF SOUTH CAROLINA, 2204 N. Oak St., Myrtle Beach, SC 29577-3054. Tel.: 843-946-9469. Fax: 843-946-7011.
E-mail: marketing@cmsckids.org
Web Site: www.cmsckids.org
Founded: 1993.
Congressional District: 1
Key Personnel: Gen. Mgr., Melanie McMurrain; Pres., Melody Breeder; Museum Shop Mgr., Rhonda Perry.
Personnel Profile: Full-Time Paid 3; Part-Time Paid 4.
Governing Authority: private; nonprofit organization. Tax-exempt: 501(c)(3).
Institution Type/Description: Children's Museum.
Collections: hands-on exhibits.
Activities: interactive learning experiences; demonstrations; activities; outreach programs.
Publications: monthly e-newsletter.
Hours & Admission Prices: June-Aug. Mon.-Sat. 9-4, Sun. 12-5; Sept.-May Tues.-Sat. 9-3, Sun. 12-5. Admission $8; discounts to groups of 10 or more; children under 2 & members no charge. Closed New Year's Day; Memorial Day; Independence Day; Labor Day; Thanksgiving; Christmas Eve & Day. &
Attendance: 35,000 (accurate)
Membership: Family $90; Family Plus $125.

FRANKLIN G. BURROUGHS-SIMEON B. CHAPIN ART MUSEUM, (M), 3100 S. Ocean Blvd., Myrtle Beach, SC 29577-4858. Tel.: 843-238-2510. Fax: 843-238-2910. Facebook: Franklin G. Burroughs-Simeon B. Chapin Art Museum.
E-mail: pgoodwin@myrtlebeachartmuseum.org
Web Site: myrtlebeachartmuseum.org
Founded: 1989.
Congressional District: 1
Key Personnel: Dir., Patricia Goodwin; Chm., John C. Stewart; Treas., Bill Pritchard; Museum Shop Mgr., Casey Church.
Personnel Profile: Full-Time Paid 4; Part-Time Paid 3; Part-Time Volunteers 10; Interns 1.
Governing Authority: private; nonprofit. Tax-exempt: 501(c)(3).
Institution Type/Description: Art Museum.
Collections: regional artists.
Major Exhibits: Julyan Davis: Appalachian Ballads, 1/7/14-3/16/14; Fiber Arts International, 1/19/14-4/24/14; Horry/Georgetown High Schools Juried Art Exhibition, 3/23/14-4/20/14; Track of the Rainbow Serpent: Australian

Aboriginal Paintings of the Wolfe Creek Crater, 4/24/14-9/14/14; Carved American Folk Canes: The Pat Connell Gift from the Mobile Museum of Art, 4/27/14-9/14/14; Waccamaw Arts and Crafts Guild 17th Annual Juried Exhibition, 5/8/14-5/22/14; Claire Farrell: A is for Art, 5/27/14-9/14/14; Classic Images: Photographs by Ansel Adams, 6/5/14-9/21/14; Hurricane Hugo Recalled: The Batiks of Leo Twiggs, 9/21/14-12/28/14; Dixie Dugan: A Retrospective, 9/21/14-12/28/14; Shadow of the Turning: The Art of Binh Pho, 10/5/14-1/4/15.
Facilities: library.
Activities: formal education programs for adults & children; guided tours; temporary exhibitions.
Publications: quarterly newsletter, Villa Voice.
Hours & Admission Prices: Tues.-Sat. 10-4, Sun. 1-4; call for holiday hours. No charge; donations accepted. ♿
Attendance: 26,000 (estimated)
Membership: Student $15; Individual $50; Family $75; Donor $100; Patron $250; Advocate $500; Leader $1,000.

FREEWOODS FARM, 9515 Freewoods Rd., Myrtle Beach, SC 29588. Tel.: 843-650-9139.
Web Site: www.freewoodsfarm.com
Founded: 1987.
Institution Type/Description: African American Historical Farm Museum.
Collections: African American farming history; farm equipment & tools; re-creation of main street; wetlands; period furnishings; personal artifacts; photographs.
Facilities: 40-acre farm.
Activities: special events; educational programs. Annual Events: Emancipation Day Celebration in January; Black History Month in February; VeggieFest in June & July.
Hours & Admission Prices: Call for hours. No charge; donations accepted. ♿

MYRTLE BEACH STATE PARK NATURE CENTER, 4401 S. Kings Hwy., Myrtle Beach, SC 29575-4936. Tel.: 843-238-5325 & 0874. Fax: 843-238-9483.
E-mail: awilson@scprt.com
Web Site: www.southcarolinaparks.com
Founded: 1970.
Congressional District: 6
Key Personnel: Naturalist, Ann Malys Wilson.
Personnel Profile: Full-Time Paid 1; Part-Time Paid 2; Part-Time Volunteers 803; Interns 3.
Governing Authority: state; nonprofit.
Institution Type/Description: Nature Center.
Collections: shells; animals, reptiles; natural history.
Research Fields: shells; coastal flora & fauna.
Facilities: 100-vol. library; aquarium; nature center; educational facilities.
Activities: guided tours; lectures; organized educational programs for children, adults & college students; participatory exhibits; hands-on natural history programs & walks; curriculum-related school programs for grades 1-5.
Hours & Admission Prices: June-July Tues.-Sun. 11:30-4:30. Varied hours in off season. Park: adult $5, children 6-15 $3; children under 5 no charge. Nature Center: no charge. South Carolina State Park Passport: $75 annual pass. ♿
Attendance: 28,034 (accurate)

RIPLEY'S AQUARIUM, 1110 Celebrity Circle, Myrtle Beach, SC 29577-7465. Tel.: 800-734-8888 (Toll Free); 843-916-0888.
E-mail: info@ripleysaquarium.com
Web Site: myrtlebeach.ripleyaquariums.com
Institution Type/Description: Aquarium.
Collections: marine animals including sharks, octopus, stingrays, & fish; coral reef.
Activities: touch tank; educational programs; camps.
Hours & Admission Prices: Daily 9am-10pm. Adults 12 & over $18.99, children 6-11 $9.99, children 2-5 $3.99.

SOUTH CAROLINA HALL OF FAME, Myrtle Beach Convention Center, 2101 N. Oak St., Myrtle Beach, SC 29577. Mailing Address: P.O. Box 2115, Myrtle Beach, SC 29578-2115. Tel.: 843-626-7444. Fax: 843-448-3007.
E-mail: info@palmettoeventproductions.com
Web Site: southcarolinahalloffame.com
Founded: 1973.
Congressional District: 7
Key Personnel: Dir., Brad Dean; Chm. (V), Leo Twiggs
Governing Authority: Tax-exempt.
Institution Type/Description: Hall of Fame.

Collections: South Carolina's history & heritage; Hall of Fame inductees.
Hours & Admission Prices: Daily 8:30-5. No charge.

WACCATEE ZOO, 8500 Enterprise Rd., Myrtle Beach, SC 29588-6626. Tel.: 843-650-8500.
E-mail: site_mgr@waccateezoo.com
Web Site: www.waccateezoo.com
Key Personnel: Owner, Kathleen Futrell; Owner, Archie Futrell
Institution Type/Description: Zoo.
Collections: over 100 species of animals.
Hours & Admission Prices: Daily 10-5. Adults 13 & over $8, children 1-12 $4; discounts to groups; children under one no charge.

Neeses

NEESES FARM MUSEUM, 6449 Savannah Hwy., Neeses, SC 29107. Mailing Address: P.O. Box 70, Neeses, SC 29107-0070. Tel.: 803-247-5811. Fax: 803-247-5811.
Founded: 1976.
Congressional District: 1
Key Personnel: Chm. (V) & Town Clerk, Sonja Gleaton.
Personnel Profile: Part-Time Volunteers 2.
Governing Authority: municipal. Parent Institution: Town of Neeses. Tax-exempt.
Institution Type/Description: Agriculture Museum.
Collections: period harvest equipment; World War I tools & relics; cotton gin; stone grinder; Edgefield pottery; wood burning cook stove; Farmall tractor; Native American cultural display.
Facilities: picnic shed.
Activities: permanent exhibitions; dancing demonstrations by Pee Dee Indian tour guide by appointment.
Hours & Admission Prices: By appointment. No charge; donations accepted. ♿
Attendance: 500 (estimated)

Newberry

NEWBERRY COUNTY HISTORICAL & MUSEUM SOCIETY, 1503 Nance St., Newberry, SC 29108-2740. Mailing Address: P.O. Box 186, Newberry, SC 29108-0186. Tel.: 803-276-8610.
Web Site: www.newberrycountyhistory.com/museum.html
Founded: 1964.
Key Personnel: Dir., Ernie Sheely; Pres. (V), Peggie West.
Governing Authority: Tax-exempt.
Institution Type/Description: History Museum.
Collections: artifacts & displays relating to the history of Newberry County. Historic House: Gauntt House, c.1808.
Publications: semi-annual bulletin.
Hours & Admission Prices: 1st & 3rd Sat. of each month 1-4; other times by appointment. No charge; donations accepted.

Ninety Six

NINETY SIX NATIONAL HISTORIC SITE, 1103 Hwy. 248 S., Ninety Six, SC 29666-8611. Tel.: 864-543-4068. Fax: 864-543-2058.
Web Site: www.NPS.gov/NISI
Founded: 1976.
Congressional District: 3
Key Personnel: Chief Ranger, Tim Cruze; Cur., Sarah Cunningham.
Personnel Profile: Full-Time Paid 2; Part-Time Paid 1; Part-Time Volunteers 40.
Governing Authority: federal. National Park Service, Dept. of the Interior. Tax-exempt.
Institution Type/Description: History Museum & Site: commemorates community & village of the settlement of Ninety Six during the 18th-century & the siege of Ninety Six during the American Revolution interpreting the role of slavery.
Collections: 18th-century Indian artifacts; trade items; tools related to cultivation of flax & building of log cabins; military artifacts; pioneers items; findings of archaeological research. Historic Structure: c.1787 two-story log cabin.
Research Fields: history; archaeology.
Facilities: restrooms.
Activities: lectures; videos; movie; special events; wayside exhibits. Museum Sponsors: Living History Events.
Publications: brochures.
Hours & Admission Prices: Visitor Center: daily 9-5. Cabin: open for special occasions & during Living History events. No charge; donations accepted. Closed New Year's Day; Thanksgiving; Christmas. ♿

Attendance: 50,000 (estimated)

North Charleston

HUNLEY SUBMARINE - WARREN LASCH CONSERVATION CENTER, 1250 Supply St., North Charleston, SC 29405-2219. Tel.: 843-743-4865, ext. 32. Fax: 843-744-1480.
E-mail: correia@hunley.org
Web Site: www.hunley.org
Institution Type/Description: History Museum.
Collections: Hunley submarine & artifacts; Hunley Captain's & crew history; life size Hunley model; photographs.
Facilities: Museum-related items for sale.
Activities: guided tours.
Publications: membership newsletter, The Blue Light.
Hours & Admission Prices: Sat. 10-5, Sun. 12-5. Adults $12; discounts to members, senior citizens & military.
Membership: Senior $35; Individual $40; Joint Senior $55; Joint $65; Family $80.

NORTH CHARLESTON AND AMERICAN LAFRANCE FIRE MUSEUM AND EDUCATIONAL CENTER, (M), 4975 Centre Pointe Dr., North Charleston, SC 29418-6945. Mailing Address: P.O. Box 190016, North Charleston, SC 29419-9016. Tel.: 843-740-5550. Fax: 843-740-5551.
E-mail: reneefryenc@yahoo.com
Web Site: www.legacyofheroes.org
Founded: 2007.
Key Personnel: Dir., Renee B. Frye.
Personnel Profile: Full-Time Paid 2; Part-Time Paid 10.
Governing Authority: city.
Institution Type/Description: Fire Museum.
Collections: fire fighting history; period fire fighting equipment & trucks; American LaFrance period vehicles; interactive fire & home safety exhibits.
Activities: educational programs.
Hours & Admission Prices: Mon.-Sat. 10-5, Sun. 1-5. Adults 13 & over $6; discounts to groups of 15 or more; children 12 & under no charge. Closed New Year's Day; Thanksgiving; Christmas Eve & Day. &
Attendance: 32,000 (estimated)

North Myrtle Beach

NORTH MYRTLE BEACH AREA HISTORICAL MUSEUM, 799 2nd Ave. N., North Myrtle Beach, SC 29582. Mailing Address: P.O. Box 4471, North Myrtle Beach, SC 29597. Tel.: 843-427-7668. Facebook: NMB Museum.
E-mail: beachhistory@nmbmuseum.com
Web Site: www.nmbmuseum.com
Founded: 2005.
Key Personnel: Dir., Jenean Neilsen Todd; Chm. (V), Dick Hester; Treas., Dee Myers.
Personnel Profile: Full-Time Paid 1; Part-Time Paid 3; Part-Time Volunteers 25; Interns 3.
Volunteer Hours: 1,573
Governing Authority: private; nonprofit organization. Tax-exempt: 501(c)(3).
Institution Type/Description: History Museum.
Collections: cultural history of Little River, North Myrtle Beach, Atlantic Beach, Longs, & surrounding areas in northeastern South Carolina; 20th-century artifacts & photographs; oral history recordings; Native American artifacts.
Research Fields: cultural history of Little River, Cherry Grove, Ocean Drive, Crescent Beach, Windy Hill, Atlantic Beach, Wampee, Brooksville, Barefoot Resort, & Tidewater Plantation.
Facilities: 75 seat multi-purpose area; 3,600 sq. ft. exhibition space. Museum-related items for sale.
Activities: arts festivals; concerts; docent program; formal education programs for children; guided tours; hobby workshops; lectures; participatory, temporary & traveling exhibitions; theater. Annual Event: Founder's Day.
Hours & Admission Prices: Wed.-Sat. 10-4. Adults $5, seniors 60 & over and students & military with ID $4, youth 5-17 $3; discounts to AAM & ICOM members; members no charge. Closed New Year's Day; Easter; Thanksgiving; Christmas. &
Attendance: 7,500 (estimated)
Membership: Individual $25; Family $60; Sandlapper League $1,000; Waccamaw Club $2,500; Palmetto Patron $5,000; Crescent Society $10,000.

Orangeburg

I.P. STANBACK MUSEUM & PLANETARIUM, (M), 300 College St., N.E., South Carolina State University, Orangeburg, SC 29117. Mailing Address: Stanback Planetarium & NASA ERC, P.O. Box 7636, South Carolina State Univ., Orangeburg, SC 29117. Tel.: 803-536-7174 & 8711. Fax: 803-536-8309.
E-mail: bmille26@scsu.edu
Web Site: www.scsucrash.blogspot.com
Founded: 1980.
Congressional District: 2
Key Personnel: Pres., Dr. George E. Cooper; Dir., Ellen Zisholtz.
Personnel Profile: Part-Time Paid 5; Interns 1.
Governing Authority: state; nonprofit organization. Parent Institution: South Carolina State University. Tax-exempt: 501(c)(3).
Institution Type/Description: Planetarium & Art Museum.
Collections: contemporary Afro-American art including William H. Johnson, Ellis Wilson, Elton Fax, Romare Bearden; Jacob Lawrence; Lois Jones; bronze statuary from Benin; over 200 pieces from Cameroons & parts of West Africa; 300-400 photographs, Harlem on My Mind Exhibit, including Van Der Zee, Gordon Parks & Susskind.
Research Fields: contemporary Afro-American art; African art.
Facilities: planetarium; education resource center.
Activities: gallery guided tours; formally organized education programs for K-12 students & undergraduate college students; permanent, temporary & traveling exhibitions.
Publications: annual brochure; occasional catalogs for special exhibits.
Hours & Admission Prices: Museum: Mon.-Fri. 9-5. No charge. Planetarium Shows: Tues.-Fri. 4pm. No charge. &
Attendance: 30,000 (estimated)
Membership: Students $15; Individual $25; Family $75; Founding $100-$249; Supporting $250-$499; Sustaining $500-$999; Patron's Circle $1,000 & up; Corporate $5,000.

Parris Island

PARRIS ISLAND MUSEUM, (M), Bldg. 111, Panama St., MCRD, Parris Island, SC 29905. Mailing Address: Commanding Gen., ATTN MUS, MCRD ERR, Box 19320, Parris Island, SC 29905-9001. Tel.: 843-228-2951. Fax: 843-228-3065.
E-mail: stephen.wise@usmc.mil
Founded: 1976.
Congressional District: 2
Key Personnel: Dir., Dr. Stephen R. Wise; Archaeologist, Kimberly Zawacki; Museum Shop Mgr., Rebecca Smith.
Personnel Profile: Full-Time Paid 6; Part-Time Volunteers 3.
Governing Authority: federal. Parent Institution: Marine Corps History Division. Tax-exempt.
Institution Type/Description: Historic Site & Military Museum: located on the site of c.1566-1587, Spanish settlement, Santa Elena; later the 19th century Port Royal Navy Yard & Marine Corps Recruit Depot.
Collections: ceramics; metal work; uniforms; firearms; archives; photograph collection.
Research Fields: Spanish & French settlements of the Southeast; history of the Port Royal area, Marine Corps history; current recruit training.
Facilities: 200-vol. library pertaining to history of Port Royal & U.S. Marine Corps available for use on the premises.
Activities: lectures.
Publications: quarterly newsletter; brochure, Driving Tour.
Hours & Admission Prices: Daily 10-4:30. No charge; donations accepted. Closed New Year's Day; Easter; Thanksgiving; Christmas. &
Attendance: 108,314 (accurate)
Membership: Student $10; Individual $20; Family $30; Contributor $100; Business Sponsor $100 & up; Sustaining $1,000 or $100 for 10 yrs.

Pendleton

BART GARRISON AGRICULTURAL MUSEUM OF SOUTH CAROLINA, 120 History Lane, Pendleton, SC 29670. Mailing Address: P.O. Box 565, Pendleton, SC 29670-0565. Tel.: 864-646-3782; 800-862-1795. Fax: 864-646-7768.
E-mail: history@pendletondistrict.org
Web Site: www.pendletondistrict.org
Formerly: Pendleton Districk Agricultural Museum
Founded: 1976.
Congressional District: 3
Key Personnel: Dir., Vicki Fletcher; Chm., Dr. William Steirer; Cur., Les McCall.
Personnel Profile: Full-Time Paid 3; Part-Time Volunteers 2.

Governing Authority: nonprofit. Parent Institution: Pendleton District Historical, Recreational & Tourism Commission. Tax-exempt: 501(c)(3).
Institution Type/Description: Agricultural Museum.
Collections: statewide collection of agricultural tools & equipment.
Major Exhibits: The Future of Agriculture, 7/13-7/14; Cotton in South Carolina, 7/13-7/14.
Activities: guided tours.
Hours & Admission Prices: By appointment only. No charge; donations accepted. &
Attendance: 450 (estimated)

PENDLETON DISTRICT HISTORICAL, RECREATIONAL AND TOURISM COMMISSION, (M), 125 E. Queen St., Pendleton, SC 29670-1309. Mailing Address: P.O. Box 565, Pendleton, SC 29670-0565. Tel.: 864-646-3782 & 800-862-1795. Fax: 864-646-7768.
E-mail: les@pendletondistrick.org
Web Site: www.pendletondistrict.org
Founded: 1966.
Congressional District: 3
Key Personnel: Chm., Dr. Bill Steiser; Dir., Vicki B. Fletcher; Cur. Collections, Les McCall; Events & Tours Coord., Brook Havice.
Personnel Profile: Full-Time Paid 3; Interns 1.
Governing Authority: county; state. Branch Museum: Agricultural Museum, U.S. 76, Pendleton, SC. Tax-exempt: 170(b)(1)(A).
Institution Type/Description: History Museum.
Collections: 1850 Hunter's Store artifacts; manuscripts; local history library of Anderson, Oconee, Pickens Counties; family genealogies; regional photo collection, 3,000 prints; early farm tools; Civil War military; textile mills records.
Research Fields: general history of Anderson, Oconee, Pickens Counties; family genealogies.
Facilities: 2,000-vol. library of history books available on premises; photostats on request; reading room. Books, arts & crafts, postcards for sale.
Activities: guided tours; lectures; films; arts festivals; formally organized education programs for children and adults; permanent, temporary & traveling exhibitions.
Publications: monthly newsletter, Friends of the Pendleton District.
Hours & Admission Prices: April-Oct. Mon.-Fri. 9-4:30, Sat. 10-3; Nov.-March Mon.-Fri. 9-4:30. No charge; donations accepted. Closed state holidays. &
Attendance: 5,000 (estimated)
Membership: Friends of the Pendleton District $25, $50, $100 & $250.

WOODBURN HISTORIC HOUSE, 130 History Lane, Pendleton, SC 29670-8700. Mailing Address: Pendleton Historic Foundation, P.O. Box 444, Pendleton, SC 29670. Tel.: 864-646-7249.
E-mail: info@pendletonhistoricfoundation.org
Web Site: www.pendletonhistoricfoundation.org
Formerly: Woodburn Plantation
Founded: 1960.
Congressional District: 3
Key Personnel: Pres. (V), Jackie Reynolds; Treas., Elizabeth Vogt; Sec., Tim Drake; Exec. Dir., Les McCall; Museum Shop Mgr., Ellen Harrison.
Personnel Profile: Full-Time Paid 1; Full-Time Volunteers 2; Part-Time Paid 3; Part-Time Volunteers 17; Interns 3.
Governing Authority: foundation; nonprofit organization. Parent Institution: Pendleton Historic Foundation. Tax-exempt: 501(c)(3).
Institution Type/Description: Historic House: c.1830 four-story Greek Revival Plantation House.
Collections: period furnishings; old tools. Historic Houses: c.1800 log house; carriage house; c. 1810 log cook house; slave cabin.
Research Fields: genealogy 1830-1908.
Facilities: walking trail.
Activities: guided tours; permanent exhibitions; historical reenactments; first-person interpretive tours; tea & lecture series; children's educational programs & tours; docent training program. Annual Event: Special Themed Tours.
Publications: brochures; quarterly newsletters; history books; e-newsletter.
Hours & Admission Prices: See website for hours & admission prices.
Attendance: 4,384 (accurate)
Membership: Individual $35; Family $50; Patron $100 & up.

Pickens

HAGOOD-MAULDIN HOUSE AND IRMA MORRIS MUSEUM OF FINE ARTS, 104 N. Lewis St., Pickens, SC 29671-2311. Mailing Address: P.O. Box 775, Pickens, SC 29671-0775. Tel.: 864-898-5963.
E-mail: pickenscohistory@gmail.com
Web Site: pickenscountyhistoricalsociety.com
Founded: 1958.
Congressional District: 3
Key Personnel: Pres., Ken Nabors; Vice Pres., Wayne Kelley.
Personnel Profile: Part-Time Volunteers 50.
Governing Authority: Parent Institution: Pickens County Historical Society. Tax-exempt.
Institution Type/Description: Art Museum: housed in the former home of attorney James Hagood; built c.1856.
Collections: 17th & 18th century art & furnishings.
Activities: Museum Sponsors: Art in the Garden Event; Azalea Festival Event; July 4th Event; Special Christmas Event in December.
Publications: newsletters.
Hours & Admission Prices: April-Nov. 1st & 3rd Sat. 11-4; other times by appointment. Adults $3, students $1. Closed Independence Day; Labor Day; Thanksgiving; Christmas.
Attendance: 350 (estimated)
Membership: Student $5; Individual $10; Couple $15; Lifetime $250; Miss Queen $1,000; John C. Calhoun $1,200; Gen. Andrew Pickens $1,500 & up.

PICKENS COUNTY MUSEUM OF ART & HISTORY, (M), 307 Johnson St., Pickens, SC 29671-2463. Tel.: 864-898-5963. Fax: 864-898-5580.
E-mail: picmus@co.pickens.sc.us
Web Site: www.pickenscountymuseum.org, www.hagoodmill.org
Founded: 1976.
Congressional District: 3
Key Personnel: Exec. Dir., C. Allen Coleman; Chm., Wayne Kelley; Cur., Helen Hockwelt; Mill Site Mgr., Ed L. Bolt; Chief Preparator, Dan M. Brennan; Museum Shop Mgr., Den Keys.
Personnel Profile: Full-Time Paid 5; Part-Time Volunteers 70.
Governing Authority: county government; nonprofit. Parent Institution: Pickens County, SC. Subsidiary Institution: Pickens County Cultural Commission. Tax-exempt.
Institution Type/Description: Art & History Museum.
Collections: regional 20th & 21st century art; art, artifacts & antiquities pertaining to the regional history including Native Americans & settlers; period furnishings; pictures; native plant gardens. Off-site: Hagood Mill Historic Site & Folklife Center; 1845 gristmill & historic park.
Research Fields: regional folklife & history.
Activities: guided tours; lectures; gallery talks; concerts; drama; formally organized education programs; permanent & temporary exhibitions; workshops; classes. Museum Sponsors: Autumn music series; Folklife Festival annual in September; Storytelling Festival in October; Statewide Juried Art Competition in Spring; Youth arts programs; Selugadu: A Native American Celebration in November.
Publications: newsletter, Old Gaol Gazette.
Hours & Admission Prices: Tues.-Wed. & Fri. 9-5, Thurs. 9-7:30, Sat. 9-4:30. No charge; donations accepted. Closed major holidays. &
Attendance: 36,000 (estimated)
Membership: Student $10; Senior $20; Individual $25; Family $50; Contributor $100; Director $250; Patron $1,000; Benefactor $2,500.

Ravenel

RAVENEL CAW CAW INTERPRETIVE CENTER, 5200 Savannah Hwy., Ravenel, SC 29470-5542. Tel.: 843-889-8898 & 795-4386.
Institution Type/Description: Environmental Center.
Collections: wildlife & their habitats; photographs; natural, cultural & historical resources.
Activities: educational programs; nature trails.
Hours & Admission Prices: Wed.-Sun. 9-5. Admission $1 per person; children under 2 no charge.

Ridgeland

BLUE HERON NATURE CENTER, 321 Bailey Lane, Ridgeland, SC 29936-8597. Tel.: 843-726-7611. Fax: 843-726-3263.
Institution Type/Description: Nature Center.
Collections: local natural heritage; ecology; wildlife & their habitats; photographs.
Facilities: nature trail; classroom.
Activities: educational programs; teacher workshops.
Hours & Admission Prices: Call for hours.

PRATT MEMORIAL LIBRARY & WEBEL MUSEUM, 451 E. Wilson St., Ridgeland, SC 29936. Mailing Address: P.O. Box 1540, Ridgeland, SC 29936-2626. Tel.: 843-726-7744. Fax: 843-726-7813.
Key Personnel: Branch Mgr., Marcia Cleland
Institution Type/Description: History Museum & Library.
Collections: 250 rare books on the history of Lowcountry; Indian artifacts from Jasper County; 200 portraits & maps; rice culture dioramas; historical materials on the Revolution & Civil Wars.
Facilities: library.
Hours & Admission Prices: Library: Mon.-Thurs. 10:30-5:30, Fri. 10:30-4:30, Sat. 11-2. Museum: Mon. 10:30-6, Tues.-Thurs. 10:30-5:30, Fri. 10:30-4:30, Sat. 10-1. No charge. Closed holidays.

Rock Hill

CATAWBA CULTURAL CENTER, 1536 Tom Stevens Rd., Rock Hill, SC 29730. Tel.: 803-328-2427. Fax: 803-328-5791.
Institution Type/Description: Native American History Museum.
Collections: Catawba Indian culture, history & artifacts; pottery; photographs.
Facilities: nature trails. Center-related items for sale.
Activities: nature trails; videos; guided tours.
Hours & Admission Prices: Mon.-Sat. 9-5. No charge; donations accepted. Closed New Year's Day; Indepedence Day; Thanksgiving; Christmas.

CENTER FOR THE ARTS - DALTON GALLERY, 121 E. Main St., Rock Hill, SC 29730-4539. Mailing Address: P.O. Box 2797, Rock Hill, SC 29732-4797. Tel.: 803-328-2787. Fax: 803-328-2165. Facebook: York County Arts.
E-mail: arts@yorkcountyarts.org
Web Site: www.yorkcountyarts.org
Founded: 1978.
Key Personnel: Exec. Dir., Debra Heintz
Institution Type/Description: Art Gallery.
Collections: paintings.
Major Exhibits: Tony DiGiorgio, 1/10/14-2/16/14; Rock Hill Teachers' Choice, 2/21/14-3/16/14; Earth Wind & Water, 3/21/14-4/27/14; Zeigler, Farfan, Bailey, 5/2/14-6/15/14; 10th Annual Phography Competition, 6/19/14-7/27/14; 25th Annual Juried Exhibition, 8/1/14-9/14/14; Christie Blizard, 9/19/14-11/2/14; Christmas Ville - Vernon Grant's Santas, 11/14/14-12/28/14.
Activities: performing arts; classes.
Hours & Admission Prices: 2nd & 4th Sat. 10-2, Sun. 2-4 each month. No charge.
Membership: Student $10; Individual $35; Family & Artist $50; Friend $75; Contributor $100; Civitas Society $300; Bronze Civitas Society $500; Silver Civitas Society $1,000; Gold Civitas Society $2,500; Platinum Civitas Society $5,000. Corporate: Friend $75; Contributor $100; Partner $300; Patron $500; Sustainer $1,000; Sponsor $2,500; Investor $5,000; Benefactor $7,500; Guarantor $10,000.

COMPORIUM TELEPHONE MUSEUM, 117 Elk Ave., Rock Hill, SC 29730. Mailing Address: P.O. Box 470, Rock Hill, SC 29731. Tel.: 803-324-4030.
Formerly: Rock Hill Telephone Company Museum
Governing Authority: Parent Institution: Rock Hill Telephone Company.
Institution Type/Description: Company History Museum.
Collections: telephone company & communications history; period telephone equipment; 1917 company truck; hands-on exhibits; early switchboard; directories.
Activities: video.
Hours & Admission Prices: Mon., Wed. & Fri.-Sat. 10-2; groups by appointment.

MAIN STREET CHILDREN'S MUSEUM, 133 E. Main St., Rock Hill, SC 29730. Tel.: 803-684-2327.
Institution Type/Description: Children's Museum.

Collections: hands-on exhibitions.
Activities: educational programs; special events; birthday parties.
Hours & Admission Prices: Tues.-Sat. 10-5, Sun. 12-5. Admission $5 per person; discounts to members & groups of 15 or more; children under one no charge. &

*   **MUSEUM OF YORK COUNTY, (M),** 4621 Mount Gallant Rd., Rock Hill, SC 29732-9637. Tel.: 803-329-2121. Fax: 803-329-5249.
E-mail: info@chmuseums.org
Web Site: www.chmuseums.org
Founded: 1950.
Congressional District: 5
Key Personnel: Dir., Van W. Shields; Deputy Dir. Mktg. & Visitor Svcs., Jeannie Marion.
Personnel Profile: Full-Time Paid 56; Part-Time Paid 20; Part-Time Volunteers 489; Interns 16.
Governing Authority: county; bd. of trustees. Parent Institution: Culture & Heritage Museums. Tax-exempt: 501(c)(3).
Institution Type/Description: General Museum.
Collections: mounted African hoofed mammals; local natural history specimens; mounted North American animals; cultural materials; ethnological materials from Africa; art by Vernon Grant; regional art; York County history.
Research Fields: African animals, astronomy; local natural history; regional historical material culture.
Facilities: picnic facilities; planetarium; auditorium; nature trail; outdoor amphitheatre.
Activities: guided tours; lectures; films; gallery talks; classes; workshops; formally organized education programs for children, adults & undergraduate college students; inter-museum loan, permanent, temporary & traveling exhibitions.
Publications: bimonthly newsletter; posters; brochures; gallery guides; members magazine.
Hours & Admission Prices: Museum: Mon.-Sat. 10-5, Sun. 1-5. Office: Mon.-Fri. 8:30-5:30. Adult $5, senior citizens 60 & over $4, children 4-17 $3; discount to AAM members; members & children 3 and under no charge. Closed New Year's Day; Thanksgiving; Christmas Eve & Day. &
Attendance: 60,000 (estimated)
Membership: Senior Citizen $30; Adult $40; Dual $50; Household $65; Sustainer $150; Partner $300; Sponsor $500; Patron $1,000; Benefactor $2,500.

WINTHROP UNIVERSITY GALLERIES, (M), 126 McLaurin Hall, Rock Hill, SC 29733. Tel.: 803-323-2493.
E-mail: derksenk@winthrop.edu
Web Site: www.winthrop.edu/vpa/galleries
Key Personnel: Dir., Karen Derksen.
Personnel Profile: Full-Time Paid 1; Interns 5.
Governing Authority: Parent Institution: Winthrop University.
Institution Type/Description: Art Galleries.
Collections: art exhibitions.
Hours & Admission Prices: Mon.-Fri. 9-5. No charge.

Roebuck

WALNUT GROVE PLANTATION, 1200 Otts Shoals Rd., Roebuck, SC 29376-3518. Tel.: 864-576-6546. Fax: 864-576-4058. Facebook: Walnut Grove Plantation.
E-mail: walnutgrove@spartanburghistory.org
Web Site: www.spartanburghistory.org
Founded: 1961.
Congressional District: 4
Key Personnel: C.E.O. & History Assoc., Jennifer Furrow; Exec. Dir., Becky Slayton; Dir., Zac Cunningham.
Personnel Profile: Full-Time Paid 1; Part-Time Paid 16; Part-Time Volunteers 2.
Governing Authority: society; nonprofit. Parent Institution: Spartanburg County Historical Association. Tax-exempt: 501(c)(3).
Institution Type/Description: Historic House Museum: c.1765 Walnut Grove Plantation, pre-Revolutionary manor house built on land grant from George III to Charles Moore.
Collections: 1760-1805 furnishings; The Manor House and kitchen; Rocky Springs academy; doctor's office; outbuildings.
Facilities: visitor's center; picnic pavilion. Museum-related items for sale.
Activities: guided tours; permanent exhibitions; living history demonstration with reservation. Annual Event: FestiFall, Living History Festival in October.

Publications: The Drover's Post; Sal & Amanda Visit Walnut Grove Plantation.
Hours & Admission Prices: April-Oct. Tues.-Sat. 11-5, Sun. 2-5; Nov.-March Sat. 11-5; other times by appointment. Adults $6, seniors $5.50, children under 18 $3; discounts to AAA members; members no charge. Closed holidays. &
Attendance: 13,000 (accurate)
Membership: Student 18 & under $10; Individual $40; Dual $60; Family $75; Spartan Regiment Society $100 & up; Kate Moore Barry Society $250 & up; Lewis P. Jones Society $500 & up.

Saint Helena Island

PENN CENTER, INC. NATIONAL HISTORIC LANDMARK/YORK W. BAILEY MUSEUM, 16 Penn Center Cir., W., Saint Helena Island, SC 29920. Mailing Address: P.O. Box 126, Saint Helena Island, SC 29920-0126. Tel.: 843-838-2432. Fax: 843-838-8545.
E-mail: info@penncenter.com
Web Site: www.penncenter.com
Congressional District: 2
Key Personnel: C.E.O. & Dir., Walter Mack; Chm. (V), John Smalls; Dir. History & Culture, Rosalyn Browne; Museum Shop Mgr., Karen Ward.
Personnel Profile: Full-Time Paid 2; Part-Time Volunteers 7.
Governing Authority: Parent Institution: Penn Center, Inc. Tax-exempt.
Institution Type/Description: History Museum.
Collections: Sea Island & Gullah African American history; archives including historic photographs & manuscripts for Penn School (1862-1960).
Research Fields: Reconstruction Era; education of freed slaves.
Facilities: Museum-related items for sale.
Activities: cultural lessons; demonstrations; educational programs. Museum Sponsors: Annual Penn Center Heritage Days Celebration in November.
Publications: membership newsletter.
Hours & Admission Prices: Mon.-Sat. 11-4; other times by appointment. Adults $5, children $3. Closed New Year's Day; Independence Day; Labor Day; Thanksgiving; Christmas. &
Attendance: 30,000 (estimated)
Membership: Supporting Friends $100-$499; Benefactors $500-$1,999; Leading Friends $2,000-$4,999; Friends for Life $5,000-$10,000.

Saint Matthews

CALHOUN COUNTY MUSEUM & CULTURAL CENTER, (M), 313 Butler St., Saint Matthews, SC 29135-1409. Tel.: 803-874-3964. Fax: 803-874-4790. Facebook: Calhoun County Museum.
E-mail: calmus@oburg.net
Web Site: www.calhouncountymuseumandculturalcenter.org
Founded: 1954.
Congressional District: 93
Key Personnel: Dir., Debbie U. Roland; Chm. & Pres. (V), Ann Haigler.
Personnel Profile: Full-Time Paid 3.
Governing Authority: county. Tax-exempt.
Institution Type/Description: General Museum.
Collections: archives; costumes; agriculture; archaeology; history; medical; preservation project.
Research Fields: history; genealogy; archaeology; natural history.
Facilities: 2,000-vol. library of rare books available on premises.
Activities: guided tours; lectures; film series; art exhibits; drama.
Publications: booklet, Brief History of Calhoun County.
Hours & Admission Prices: Tues.-Fri. 9-4; groups of 10 or more & researchers by appointment. No charge; donations accepted. &
Attendance: 7,000 (estimated)
Membership: Friend of Museum: Single $35; Family $55; Patron $100; Corporate $500; Benefactor $1,000 & up.

Saluda

SALUDA COUNTY HISTORICAL SOCIETY MUSEUM AND THEATER, 105 Law Range, Saluda, SC 29138-1701. Mailing Address: P.O. Box 22, Saluda, NC 29138. Tel.: 864-445-8550.
E-mail: info@saludacountyhistoricalsociety.org
Web Site: www.saludacountyhistoricalsociety.org
Founded: 1987.
Key Personnel: Dir., Meade Hendrix; Chm. (V), Bela Herlong; Pres. (V), Tommy Willis.
Personnel Profile: Part-Time Paid 1.
Governing Authority: Tax-exempt.
Institution Type/Description: Historical Society Museum: listed on the National Register of Historical Places.
Collections: local history & culture; farm tools; liquor still; period scrapbooks; arts & crafts; photographs; personal artifacts. Historic Building: theater, c.1936.

Facilities: theater.
Hours & Admission Prices: Mon.-Fri. 10-4, Sat. 10-1. Closed major holidays.

Seneca

LUNNEY MUSEUM, 211 W. South 1st St., Seneca, SC 29678-3307. Tel.: 864-882-4811.
Key Personnel: Dir., Judy Havice.
Governing Authority: Operated by: Oconee County Museum.
Institution Type/Description: Historic House: housed in a California-style bungalow, built in 1909 by Dr. & Mrs. W.J. Lunney. Listed on the National Register of Historic Places.
Collections: Victorian furniture; personal artifacts; Oconee County history.
Hours & Admission Prices: Thurs.-Sun. 1-5. No charge. Closed major holidays.

THE WORLD OF ENERGY, 7812 Rochester Hwy., Seneca, SC 29672-0752. Tel.: 800-777-1004, ext. 1. Fax: 864-885-4605.
E-mail: worldofenergy@duke-energy.com
Web Site: www.duke-energy.com/worldofenergy
Formerly: World of Energy at Keowee-Toxaway
Founded: 1969.
Congressional District: 3
Key Personnel: Mgr. Community Rels., BJ Gatten.
Personnel Profile: Full-Time Paid 3; Part-Time Paid 2.
Governing Authority: company. Duke Energy Company.
Institution Type/Description: Science & Technology Center (Energy & Electricity emphasis): overlooking the Oconee Nuclear Station & the nearby lakes Keowee & Jocassee.
Collections: hands-on energy exhibits; hydro chamber with waterwheel; leaf chamber; coal chamber; fission chamber; model of nuclear reactor; fiber optics display; electronic energy conservation quiz; home energy conservation & recycling exhibit.
Facilities: 120-seat auditorium; classroom; nature trail; picnic shelter; seasonal butterfly garden.
Activities: lectures; films; special events; formally organized education programs; permanent exhibitions; self-guided tours.
Publications: pamphlets, Explore Energy; The Forests & Flowers of Keowee-Toxaway; Catalog of Educational Services; films.
Hours & Admission Prices: Mon.-Fri. 9-5, Sat. 12-5. No charge. Closed holidays. &
Attendance: 20,000 (estimated)

Spartanburg

HATCHER GARDEN AND WOODLAND PRESERVE, 820 John B. White Blvd., Spartanburg, SC 29306-4043. Mailing Address: P.O. Box 2337, Spartanburg, SC 29304. Tel.: 864-574-7724.
Institution Type/Description: Garden.
Collections: plants; trees; flowers; wildlife & their habitats.
Facilities: pavilion.
Activities: classes.
Hours & Admission Prices: Daily dawn to dusk. No charge. &
Attendance: 35,000 (accurate)

MILLIKEN GALLERY, CONVERSE COLLEGE, 580 E. Main St., Spartanburg, SC 29302-0006. Tel.: 864-596-9214. Fax: 864-596-9606.
E-mail: kathryn.boucher@converse.edu
Web Site: www.converse.edu
Founded: 1971.
Congressional District: 3
Key Personnel: Dir., Kathryn Boucher.
Personnel Profile: Part-Time Paid 6; Interns 4.
Governing Authority: college.
Institution Type/Description: College Art Gallery.
Collections: changing exhibitions.
Activities: visiting artists; workshops; community outreach.
Hours & Admission Prices: Mon.-Fri. 9-5, Sun. 2-5. No charge. Closed school holidays & breaks. &
Attendance: 1,700

THE SANDOR TESZLER LIBRARY, Wofford College, 429 N. Church St., Spartanburg, SC 29303-3663. Tel.: 864-597-4300. Fax: 864-597-4329.
E-mail: coburnoh@wofford.edu
Web Site: www.wofford.edu/library/
Congressional District: 4
Key Personnel: Dean Library, Oakley H. Coburn.

Governing Authority: college. Wofford College.
Institution Type/Description: Art Gallery & College Museum: located on 1854 campus.
Collections: books; prints; posters; framed paintings & graphics; original & reproduction sculpture.
Facilities: 255,000-vol. library available for inter-library loan; planetarium; 1,000-seat auditorium; classrooms; 700-seat cafeteria.
Activities: films; arts festivals; formally organized education programs for adults & undergraduate college students; temporary & traveling exhibitions.
Publications: exhibition brochures; catalogs.
Hours & Admission Prices: Fall, Interim & Spring Semesters: Mon.-Thurs. 8am to midnight, Fri. 8-7, Sat. 10-5, Sun. 1pm to midnight. No charge. &

THE SEAY HOUSE, 106 Darby Rd., Spartanburg, SC 29306. Mailing Address: P.O. Box 887, Spartanburg, SC 29304-0887. Tel.: 864-596-3501. Fax: 864-596-2399.
E-mail: seayhouse@spartanburghistory.org
Web Site: www.spartanburghistory.org
Founded: 1974.
Congressional District: 4
Key Personnel: Dir., Caroline Sexton.
Personnel Profile: Full-Time Paid 1; Part-Time Paid 1.
Governing Authority: society; nonprofit organization. Parent Institution: Spartanburg County Historical Association. Tax-exempt: 501(c)(3).
Institution Type/Description: Historic House: c.1890 home.
Collections: c.1890 decorative arts, focus on women's history; Seay family belongings.
Activities: guided tours.
Publications: The Drover's Post.
Hours & Admission Prices: April-Oct. 3rd Sat. of month 11-5; other times by appointment. No charge; donations accepted. Closed holidays. &
Attendance: 100 (estimated)
Membership: Student $10; Individual $40; Family $75; Spartan Regiment Society $100; Kate Barry Society $250; Lewis P. Jones Society $500.

SPARTANBURG ART MUSEUM, (M), 200 E. Saint John St., Spartanburg, SC 29306-5124. Tel.: 864-582-7616. Fax: 864-948-5353. TDD: 864-583-2776.
E-mail: museuminfo@spartanburgartmuseum.org
Web Site: www.spartanburgartmuseum.org
Formerly: Spartanburg County Museum of Art
Founded: 1969.
Congressional District: 4
Key Personnel: Dir., Scott Cunningham; Dir. Art School, Kathleen Moore; Outreach Coord., Angie Shuman.
Personnel Profile: Full-Time Paid 1; Part-Time Paid 6; Part-Time Volunteers 5; Interns 4.
Governing Authority: nonprofit organization. Tax-exempt: 501(c)(3).
Institution Type/Description: Art Museum.
Collections: paintings, graphics & decorative arts by regional, nationally & internationally known artists.
Facilities: 3,000 sq. ft. exhibit space.
Activities: guided tours; lectures; films; traveling exhibitions; Converse College intern program; temporary exhibitions; Art School; C.O.L.O.R.S., a free studio class for inner-city youth; docent programs; art trips; exhibition series. Museum Sponsors: annual meeting; 12-24 art exhibitions annually.
Publications: quarterly art events listing, Art Calendar; membership campaign; special events & exhibits brochures.
Hours & Admission Prices: Museum: Tues.-Sat. 10-5. Adults $4, seniors & active military $3, students with college ID $2.50, children 6-18 $2; children 5 & under no charge. Artwalk: 3rd Thurs. each month 6pm-9pm. Closed New Year's Day; Veterans Day; Martin Luther King Jr. Day; Easter Monday; Memorial Day; Independence Day; Labor Day; Columbus Day; Presidents' Day; Thanksgiving; Christmas Eve & Day. &
Attendance: 15,000 (estimated)
Membership: Senior 65 & over $25; Basic $35; Friend $65; Supporter $150; Patron $300.

SPARTANBURG REGIONAL MUSEUM OF HISTORY, 200 E. St. John St., Spartanburg, SC 29306-5124. Mailing Address: P.O. Box 887, Spartanburg, SC 29304-0887. Tel.: 864-596-3501. Fax: 864-596-2399.
E-mail: regionalmuseum@spartanburghistory.org
Web Site: www.spartanburghistory.org
Formerly: Spartanburg County Regional Museum of History
Founded: 1961.
Congressional District: 4
Key Personnel: Dir., Caroline Sexton.

Personnel Profile: Full-Time Paid 1; Part-Time Volunteers 6; Interns 2.
Governing Authority: society. Parent Institution. Spartanburg County Historical Association. Tax-exempt: 501(c)(3).
Institution Type/Description: Local History Museum.
Collections: Sloan doll collection; photographs throughout Spartanburg in many time periods; quilts; decorative arts.
Facilities: 4,500 sq. ft. exhibit space.
Activities: permanent & temporary exhibitions. Annual Event: Christmas Toys.
Publications: quarterly newsletter, Drover's Post.
Hours & Admission Prices: Tues.-Sat. 10-5. No charge; donations accepted. Closed major holidays. &
Attendance: 7,000 (accurate)
Membership: Student $10; Individual $40; Family $75; Spartan Regiment Society $100; Kate Barry Society $250; Lewis P. Jones Society $500.

SPARTANBURG SCIENCE CENTER, 200 E. Saint John St., Spartanburg, SC 29306-5124. Tel.: 864-583-2777. Fax: 864-948-5353.
E-mail: science@spartanarts.org
Web Site: spartanburgsciencecenter.org
Founded: 1978.
Congressional District: 4
Key Personnel: Exec. Dir., John F. Green.
Personnel Profile: Full-Time Paid 1; Part-Time Paid 1; Part-Time Volunteers 5; Interns 2.
Governing Authority: nonprofit organization. Tax-exempt: 501(c)(3).
Institution Type/Description: Science Museum & Center.
Collections: skulls; skeletons; fossils; minerals; Indian artifacts; living cold-blooded vertebrates; aquaria; old texts on natural history; 35 mm slide collection of regional flora & fauna and local rocks & minerals; South Carolina wildlife; Thomas Edison; Furman T. Wallace wildlife exhibit; science including astronomy, the human body, insects, plants & pure science.
Research Fields: regional natural history including birds, reptiles & plants.
Facilities: 100-vol. library of books on the sciences & natural history, available for inter-library loan; microscopes; nature center; aquarium; science camp location.
Activities: lectures; films; science summer camps; birthday parties; formally organized education programs for children & adults; permanent & traveling exhibitions; school loan service; Health & Wellness program (preschool); community out reach program; portable planetarium, Starlab, used in local schools; science club.
Publications: biannual newsletter.
Hours & Admission Prices: Thurs.-Sat. 10-5, Sun. 1-5. Adults $4, seniors $3, college students $2.50, children 6-18 $2; children 5 & under and members no charge. &
Attendance: 25,000 (accurate)
Membership: Student $30; Individual $50; Grandparent $60; Family $75; Family Plus $125; Supporter $300.

Sullivan's Island

*** FORT SUMTER NATIONAL MONUMENT,** 1214 Middle St., Sullivan's Island, SC 29482-9748. Tel.: 843-883-3123, ext. 23 (chief ranger); ext. 22 (park historian). Fax: 843-883-3910.
E-mail: bill_martin@nps.gov
Web Site: www.nps.gov/fosu
Founded: 1948.
Congressional District: 1
Key Personnel: Supt., Tim Stone; Chief Interpretation, Dawn Davis; Historian, Richard Hatcher; Cur., Catherine Fowler; Museum Shop Mgr., Kevin Bates.
Personnel Profile: Full-Time Paid 25; Part-Time Paid 5; Part-Time Volunteers 17; Interns 2.
Governing Authority: federal government. Parent Institution: National Park Service. Tax-exempt: 101(6).
Institution Type/Description: Military Museums: housed in Fort Moultrie Visitor Center & 1829 Fort Sumter.
Collections: artifacts & manuscripts, connected with Fort Sumter & Fort Moultrie; military uniforms; military equipment.
Major Exhibits: Honorable in Defeat, Humble in Victory: Robert Anderson at Fort Sumter, 10/26/13-5/15.
Research Fields: Seacoast Defense; Civil War; Revolutionary War in Charleston; Spanish American War; World War I; World War II.
Facilities: 2,000-vol. library dealing with American Revolution, Civil War & American seacoast defense, available for use under supervision of park employee; reading room; 120-seat auditorium; theater. Museum-related items for sale.
Activities: guided tours; lectures; films; formally organized education programs for children; temporary exhibitions; living history programs.
Publications: park brochures; site bulletins; teacher's guides.

Hours & Admission Prices: Fort Sumter: call for hours. Accessible only by boat. Tour Boat: adults $17, seniors $15, children 6-11 $10; children 5 & under no charge. Fort Moultrie: daily 9-5. Families $5, adults $3, seniors 62 & over $1; children 16 & under no charge. Closed New Year's Day; Thanksgiving, Christmas. &

Attendance: 337,000 (estimated)

Summerton

SANTEE NATIONAL WILDLIFE REFUGE, 2125 Fort Watson Rd., Summerton, SC 29148-8638. Tel.: 803-478-2217. Fax: 803-478-2314.

E-mail: santee@fws.gov
Web Site: www.fws.gov/santee
Founded: 1941.
Institution Type/Description: Wildlife Refuge: housed on Santee Indian Mound/Fort Watson historic site.
Collections: wildlife & their habitats; photographs.
Facilities: nature trails.
Activities: educational programs; walking trails; wildlife observation.
Hours & Admission Prices: Center: Tues.-Sat. 8-4. Grounds: daily sunrise to sunset. No charge.
Attendance: 181,000 (estimated)

Summerville

OLD DORCHESTER STATE HISTORIC SITE, 300 State Park Rd., Summerville, SC 29485-8431. Tel.: 843-873-1740. Fax: 843-873-1740.

E-mail: colonialdorchester@scprt.com
Web Site: www.southcarolinaparks.com
Founded: 1960.
Congressional District: 1
Key Personnel: Park Mgr. & Archaeologist, Ashley Chapman.
Governing Authority: state. Affiliated with the South Carolina Dept. of Parks, Recreation & Tourism, Brown Bldg., 1205 Pendleton St., Columbia, SC 29201. Tel.: 803-734-0168. Tax-exempt.
Institution Type/Description: Preservation Project: archaeological site of the colonial Village of Dorchester founded in 1697 by Congregationalists from Massachusetts.
Collections: archaeological remnants of 18th-century village. Historical Buildings: tabby fort 1757-60; Church Tower 1751.
Research Fields: 18th-century trading town of Dorchester.
Activities: educational programs; public excavations 6 days a month & special events offered periodically.
Publications: brochure, Historical Visitors Guide.
Hours & Admission Prices: Daily 9-6. Adult $2, seniors $1.25; children 15 & under no charge. &
Attendance: 118,300

SUMMERVILLE DORCHESTER MUSEUM, 100 E. Doty Ave., Summerville, SC 29483. Mailing Address: P.O. Box 1873, Summerville, SC 29484. Tel.: 843-875-9666.

Institution Type/Description: History Museum: housed in the former Summerville Police Station.
Collections: local history & culture; period furnishings; personal artifacts; photographs.
Activities: lectures.
Hours & Admission Prices: Mon.-Sat. 9-2. Closed Thanksgiving; Christmas.

Sumter

SUMTER COUNTY GALLERY OF ART, 200 Hasel St., Sumter, SC 29150-4506. Mailing Address: Box 1316, Sumter, SC 29151-1316. Tel.: 803-775-0543. Fax: 803-778-2787.

E-mail: director@sumtergallery.com
Web Site: www.sumtergallery.org
Founded: 1970.
Congressional District: 5
Key Personnel: Exec. Dir., Karen Watson; Dir. Art Education, Amanda Cox; Asst. Dir. & Cur., Frank McCauley.
Personnel Profile: Full-Time Paid 3; Part-Time Paid 2; Part-Time Volunteers 10.
Governing Authority: nonprofit organization. Tax-exempt: 501(c)(3).
Institution Type/Description: Art Museum.
Collections: Elizabeth White collection; regional contemporary & traditional art.
Facilities: classrooms. Museum-related items for sale.

Activities: opening receptions; artist lectures; panel discussions; artists-in-residence; summer art camp & scholarship fund; children & adult art classes.
Publications: quarterly newsletters; brochure, Elizabeth White House; general information & membership brochure; monthly gallery guides; art school flyers; exhibition catalogs.
Hours & Admission Prices: Tues.-Sat. 11-5, Sun. 1:30-5. No charge; donations accepted. Closed holidays. &
Attendance: 11,000 (estimated)
Membership: Individual & Military $50; Family $75; Patron $100; Bronze $250; Silver $500; Gold $1,000.

THE SUMTER COUNTY MUSEUM, (M), 122 N. Washington St., Sumter, SC 29150-4920. Mailing Address: P.O. Box 1456, Sumter, SC 29151-1456. Tel.: 803-775-0908.

E-mail: rgood@sumtercountymuseum.org
Web Site: www.sumtercountymuseum.org
Founded: 1976.
Congressional District: 5
Key Personnel: Exec. Dir, Annie Rivers.
Personnel Profile: Full-Time Paid 1; Part-Time Paid 2; Part-Time Volunteers 40; Interns 1.
Governing Authority: bd. of trustees. Subsidiary Institution: Genealogical & Historical Research Center, 219 W. Liberty St., Sumter. Tel. 803-773-9144. Tax-exempt.
Institution Type/Description: Historic House: Williams-Brice House, a 1916 three story brick house; reconstructed homestead representing life in 1800.
Collections: area history; early farm & household equipment; textiles; archives, manuscripts, documents, photos.
Major Exhibits: Mary McLeod Bethune: From Maysville to National Hero, 11/13-3/14; 100 Years of Healthcare, 4/14-8/14.
Research Fields: genealogy; history.
Facilities: meeting rooms; genealogical & research center; reception hall. Museum-related items for sale.
Activities: guided tours; lectures; temporary & permanent exhibits.
Publications: museum newsletter; historical documents transcribed.
Hours & Admission Prices: Museum: Thurs.-Sat. 10-5. Research Center: Tues.-Sat. 10-1 & 2-5. Museum: adults $3, children 6-17 $1. Research Center: $5 per day for non-members. &
Attendance: 9,000 (accurate)
Membership: Individual $30; Family $50; Patron $100; Benefactor $250; Gamecock $500; Heritage $1,000.

SWAN LAKE IRIS GARDENS, 822 W. Liberty St., Sumter, SC 29150. Mailing Address: P.O. Box 1449, 21 N. Main St., Sumter, SC 29150. Tel.: 803-436-2640. Fax: 803-436-2615.

E-mail: tourism@sumter.sc.com
Web Site: sumtertourism.com
Institution Type/Description: Garden.
Collections: Japanese iris; camellias; azaleas; day lilies; Japanese magnolias; swans from around the world; Grainge McKoy recovery sculpture.
Facilities: nature trails; visitor's center.
Activities: docent guided tours; rental facilities; nature trails. Museum Sponsors: Sumter Iris Festival in May.
Hours & Admission Prices: Daily 7:30 am to dusk. No charge.

Sunset

JOCASSEE GORGES VISITOR CENTER, Keowee-Toxaway State Natural Area, 108 Residence Dr., Sunset, SC 29685-2128. Tel.: 864-868-2605.

E-mail: keoweetoxaway@scprt
Institution Type/Description: Natural History Museum: housed in the former Holly Springs Baptist Church.
Collections: natural & cultural history; raised topographic map; photographs; Native American artifacts; early settlers; wildlife & their habitats.
Facilities: nature trails.
Hours & Admission Prices: Daily 11am to noon & 4-5.

Union

ROSE HILL PLANTATION STATE HISTORIC SITE, 2677 Sardis Rd., Union, SC 29379-7904. Tel.: 864-427-5966. Fax: 864-427-5966.

E-mail: rosehill@scprt.com
Web Site: www.southcarolinaparks.com
Founded: 1960.
Key Personnel: Park Mgr., Trampas Alderman.

Governing Authority: state; not-for-profit organization. Parent Institution: South Carolina Parks; Recreation & Tourism. Subsidiary Institution: State Parks.
Institution Type/Description: State Park: housed in the former home of South Carolina Governor William Henry Gist; built in 1828.
Collections: 19th-century Southern household & personal items; artifacts owned by Governor Gist. Historic Building: kitchen house.
Activities: exhibits; picnicking; hiking; special programs.
Hours & Admission Prices: Mansion Tours: March-Oct. daily 1, 2 & 3; Nov.-Feb. Thurs.-Mon. 1, 2 & 3. Adults $5, students 6-16 $4, seniors $3. Park Grounds: daily 9-6. No charge. Closed Thanksgiving, Christmas Eve & Day.
Attendance: 3,500 (estimated)

UNION COUNTY MUSEUM, 127 W. Main St., Union, SC 29379. Mailing Address: P.O. Box 220, Union, SC 29379-0220. Tel.: 864-429-5081. Facebook: Union County Historical Society.
E-mail: uncomus@bellsouth.net
Web Site: unioncountymuseum.com
Founded: 1976.
Congressional District: 4
Key Personnel: Exec. Dir., Ola Jean Kelly; Pres. (V), Charles E. Smith.
Personnel Profile: Part-Time Paid 3; Part-Time Volunteers 15.
Operating Expenses: 87,575
Operating Income: 141,844
Governing Authority: Parent Institution: Union County Historical Society. Tax-exempt.
Institution Type/Description: History Museum.
Collections: local history & culture; photographs; period furnishings; personal artifacts.
Major Exhibits: Living History at Cross Keys Plantation, 4/14; Wedding Attire, 6/14-7/14.
Research Fields: genealogy.
Facilities: Museum-related items for sale.
Activities: special events. Museum sponsors: living history event last weekend in April.
Publications: quarterly newsletter.
Hours & Admission Prices: Tues. & Thurs.-Fri. 9-4, Sat. 1-4; other times by appointment. No charge; donations accepted. Closed major holidays.
Attendance: 4,019 (accurate)
Membership: Student $5; Adult $15; Family $20; Life $250; Sponsor $500; Corporate/Benefactor $1,000.

Wagener

WAGENER MUSEUM, 12 Short St., Wagener, SC 29164. Mailing Address: P.O. Box 1004, Wagener, SC 29164-1004. Tel.: 803-564-3412 & 3507. Facebook: Wagener Museum.
E-mail: wagenermuseum@yahoo.com
Web Site: www.wagenersc.com
Founded: 1989.
Congressional District: 2
Key Personnel: Chm. (V), Cynthia R. Hardy.
Personnel Profile: Part-Time Volunteers 5.
Governing Authority: Parent Institution: Town of Wagener.
Institution Type/Description: History Museum.
Collections: local history & culture; photographs.
Publications: Local History Vol. I 1887-1990; Vol. II 1990-2005.
Hours & Admission Prices: Mon.-Tues. & Thurs.-Fri. 9-4, Sat. by appointment (through Town Hall entrance). No charge; donations accepted. Closed national holidays. &
Attendance: 90 (estimated)

Walhalla

OCONEE HERITAGE CENTER, (M), 123 Brown Square Dr., Walhalla, SC 29691. Mailing Address: P.O. Box 395, Walhalla, SC 29691. Tel.: 864-638-2224.
E-mail: info@oconeeheritagecenter.org
Web Site: www.oconeeheritagecenter.org
Founded: 1999.
Key Personnel: Dir., Leslie White; Asst. Cur., Jennifer Moss.
Governing Authority: Tax-exempt.
Institution Type/Description: History Museum.
Collections: county history & culture; personal artifacts; photographs; period furnishings.
Research Fields: Oconee County history & genealogy.
Facilities: classroom.
Activities: Museum Sponsors: Oconee Appalachian Kids; Oconee Appalachian Kids Harvest Moon Gala.

Hours & Admission Prices: Thurs.-Fri. 12-6, Sat. 10-3; other times by appointment. No charge; donations requested. &
Attendance: 4,000 (accurate)
Membership: Individual $15; Family $25.

OCONEE STATION STATE HISTORIC SITE, 500 Oconee Station Rd., Walhalla, SC 29691-3126. Tel.: 864-638-0079.
Institution Type/Description: Historic Site: the area served as a military compound against attack from the Cherokee Indians.
Collections: local history, heritage, & culture; period furnishings; personal artifacts; photographs. Historic Buildings: Oconee Station, a stone blockhouse built in 1792; William Richards House built in 1805.
Activities: educational programs.
Hours & Admission Prices: Sat.-Sun. 1-5; other times by appointment.

PATRIOT'S HALL-OCONEE VETERAN'S MUSEUM, 13 Short St., Walhalla, SC 29691-2229. Mailing Address: P.O. Box 591, Walhalla, SC 29691-0591. Tel.: 864-638-5455 & 972-8173.
Web Site: www.oconeeveteransmuseum.org
Founded: 2004.
Key Personnel: Dir. & Chm. (V), A.J. Smith.
Governing Authority: Tax-exempt.
Institution Type/Description: Veterans Museum: housed in the Old Rock Building; built in 1933.
Collections: artifacts & memorabilia dedicated to Oconee veterans.
Hours & Admission Prices: Sat. 10-3; other times by appointment. No charge; donations accepted. &
Attendance: 1,900 (estimated)
Membership: Student & Veteran $10; Individual $15; Family $25.

Walterboro

COLLETON MUSEUM, 506 E. Washington St., Walterboro, SC 29488-4028. Tel.: 843-549-2303. Fax: 843-549-7215.
E-mail: museum@colletoncounty.org
Web Site: www.colletonmuseum.com
Founded: 1985.
Congressional District: 6
Key Personnel: C.E.O., Dir. & Archivist, Gary Brightwell; Chm. (V) & Education Asst., Elaine Inabinet.
Personnel Profile: Full-Time Paid 2; Part-Time Paid 1; Part-Time Volunteers 20; Interns 2.
Governing Authority: county; nonprofit.
Institution Type/Description: History Museum: housed in c.1855 old jail, a two-story neo-Gothic structure resembling a castle, designed by Edward C. James & Francis P. Lee.
Collections: 19th-century furnishings, personal artifacts & tools; natural history exhibit Animals of the ACE Basin.
Research Fields: history & culture of Colleton County; natural history.
Facilities: 1,200 sq. ft. exhibit space. Books, posters, prints, postcards, jewelry, pewter, educational children's toys, ceramic gifts & other museum-related items for sale.
Activities: guided tours; lectures; docent program; receptions; teachers environmental network; traveling trunks.
Publications: newsletter, Notes From the Old Jail.
Hours & Admission Prices: Tues.-Fri. 10-1 & 2-5, Sat. 12-4. No charge; donations accepted. &
Attendance: 7,980 (accurate)
Membership: Student & Senior $20; Individual $30; Associate $50; Beaulah Glover Society $100; Walter Family Society $250; Colonel Isaac Hayne Society $500; Sir John Colleton Society $1,000 & up.

SLAVE RELICS MUSEUM, 208 Carn St., Walterboro, SC 29488-3965. Tel.: 843-549-9130.
Web Site: slaverelics.org
Institution Type/Description: History Museum.
Collections: slave-era history & culture; documents; personal artifacts; photographs; period furnishings.
Hours & Admission Prices: Mon.-Thurs. 9:30-5, Sat. 10-3. Adults $6, children $5; discounts to groups.

SOUTH CAROLINA ARTISANS CENTER, 318 Wichman St., Walterboro, SC 29488-2921. Tel.: 843-549-0011. Fax: 843-549-7433.
E-mail: info@scartisanscenter.com
Web Site: scartisanscenter.com
Founded: 1994.
Congressional District: 6

Key Personnel: Pres., Chris Bickley; Exec. Dir., Gale M. Doggette; Treas., Dolly Droze.
Personnel Profile: Full-Time Paid 2; Part-Time Paid 7; Part-Time Volunteers 10.
Governing Authority: private; nonprofit organization. Tax-exempt: 501(c)(3).
Institution Type/Description: Arts & Crafts Museum: housed in a 1910 Victorian cottage in the Hickory Valley Historic District.
Collections: South Carolina fine craft & folk artists; cultural heritage of South Carolina.
Facilities: educational facilities; 4,800 sq. ft. exhibit space. Fine crafts & folk art created by South Carolina residents for sale.
Activities: arts festivals; educational programs for adults & children which include workshops & artist demonstrations; guided tours.
Publications: monthly newsletter.
Hours & Admission Prices: Mon.-Sat. 9-5, Sun. 1-5. No charge. Closed New Year's Day; Easter; Independence Day; Thanksgiving; Christmas. &
Attendance: 125,000 (estimated)
Membership: Artist & Student $15; Friend $25; Family $50; Contributor $100; Supporter $250; Patron $500; Corporations & Organizations: Non-Profit $50; Contributor $100; Supporter $250; Patron $500; Fellow $501 & up.

Wellford

HOLLYWILD ANIMAL PARK, 2325 Hampton Rd., Wellford, SC 29385-9010. Mailing Address: P.O. Box 683, Inman, SC 29349-0683. Tel.: 864-472-2038.
E-mail: hollywildanimalpark@gmail.com
Web Site: www.hollywild.com
Institution Type/Description: Zoo.
Collections: wildlife from around the world.
Activities: safari rides; videos; birthday parties; educational programs.
Hours & Admission Prices: Call for hours and admissions.

Winnsboro

FAIRFIELD COUNTY MUSEUM, 231 S. Congress St., Winnsboro, SC 29180-1105. Mailing Address: P.O. Box 6, Winnsboro, SC 29180-0006. Tel.: 803-635-9811. Fax: 803-815-9811.
E-mail: fairfieldmus@truvista.net
Web Site: www.fairfieldsc.com/secondary.aspx?pageID=125
Founded: 1963.
Congressional District: 5
Key Personnel: Dir. & Museum Shop Mgr., Pelham Lyles.
Personnel Profile: Full-Time Paid 1; Part-Time Volunteers 2.
Governing Authority: society. Parent Institution: Fairfield County Council. Subsidiary Institution: Friends of the Museum. Tax-exempt.
Institution Type/Description: Historic Building & Site: housed in 1830 Cathcart-Ketchin, 3-story Federal brick structure.
Collections: local living utensils; guns; china; kitchenware; letters; quilts; genealogical data; Indian artifacts; toys; period clothing.
Research Fields: genealogy; local history.
Activities: guided tours; lectures; arts festivals; oral history film project; archaeological field studies; preservation projects; battle reenactments.
Publications: FOTM newsletter.
Hours & Admission Prices: Tues.-Fri. 10-5, Sat. 10-3. No charge.
Attendance: 3,500 (estimated)
Membership: Student $8; Individual $10; Couple $15; Family $20.

THE SOUTH CAROLINA RAILROAD MUSEUM, 110 Industrial Park Rd., Winnsboro, SC 29180-9113. Mailing Address: P.O. Box 7246, Columbia, SC 29202-7246. Tel.: 803-635-4242 & 9893.
E-mail: info@scrm.org
Web Site: www.scrm.org
Founded: 1973.
Congressional District: 3
Key Personnel: Chm. (V), Kelvin Woods; Pres., Rufus Timms; Museum Shop Mgr., Joe Palma.
Personnel Profile: Part-Time Volunteers 30; Interns 1.
Governing Authority: Tax-exempt.
Institution Type/Description: Railroad Museum.
Collections: historical artifacts & photographs; South Carolina railroad heritage; locomotives; passenger and freight cars; cabooses.
Facilities: Museum-related items for sale.
Activities: train rides; special events.
Hours & Admission Prices: June-Aug. Sat.; see website for special events. &
Attendance: 10,000 (accurate)
Membership: Youth $17; Senior $18; Individual $25; Family $35.

Woodruff

HISTORIC PRICE HOUSE, 1200 Oak View Farms Rd., Woodruff, SC 29388-8313. Mailing Address: 1200 Otts Shoals Rd., Roebuck, SC 29376-3518. Tel.: 864-576-6546. Fax: 864-576-4058. Facebook: Historic Price House.
E-mail: pricehouse@spartanburghistory.org
Web Site: www.spartanburghistory.org
Founded: 1972.
Congressional District: 4
Key Personnel: Dir., Zac Cunningham.
Personnel Profile: Full-Time Paid 1; Part-Time Paid 6; Part-Time Volunteers 2.
Governing Authority: society; nonprofit. Parent Museum: Spartanburg County Historical Association. Tax-exempt: 501(c)(3).
Institution Type/Description: History Museum: housed in c.1795, three story brick structure, once located on 2,000 acres.
Collections: 1790-1820 decorative arts & federal pieces; separate kitchen; double-pen slave cabin.
Activities: guided tours. Annual Event: Taste of the Backcountry Festival in April; Taste of the Backcountry in May.
Publications: The Drover's Post.
Hours & Admission Prices: May-Oct. Sat. 11-5, Sun. 2-5. Adults $5, children 6-17 $3; discount to groups; children under 6 no charge. Closed holidays.
Attendance: 400 (accurate)
Membership: Student 18 & under $10; Individual $40; Dual $60; Family $75; Spartan Regiment Society $100; Kate Moore Barry Society $250; Lewis P. Jones Society $500.

Yemassee

THE LOWCOUNTRY VISITORS CENTER & MUSEUM, 1 Lowcountry Lane, Yemassee, SC 29945. Mailing Address: P.O. Box 615, Yemassee, SC 29945-0615. Tel.: 843-717-3090; 800-528-6870. Fax: 843-717-2888. Facebook: SC Lowcountry Tourism Commission.
E-mail: peach@southcarolinalowcountry.com
Web Site: www.southcarolinalowcountry.com
Institution Type/Description: Historic House Museum: housed in the former home of the Frampton family; built in 1868.
Collections: family history; period furnishings & artifacts; photographs.
Facilities: Museum-related items for sale.
Hours & Admission Prices: Daily 9-5:30. No charge. Closed federal & state holidays.

SOUTH DAKOTA

(141 listings)

Aberdeen

DACOTAH PRAIRIE MUSEUM, (M), 21 S. Main St., Aberdeen, SD 57401-4218. Tel.: 605-626-7117. Fax: 605-626-4026.
E-mail: dacotahprairiemuseum@gmail.com
Web Site: www.dacotahprairiemuseum.com
Founded: 1964.
Congressional District: 1
Key Personnel: Dir., Sue Gates; Pres. (V), Jacie Schley; Cur. Exhibits, Lora Schaunaman; Cur. Education, Sherri Rawstern.
Personnel Profile: Full-Time Paid 5; Part-Time Paid 2; Part-Time Volunteers 65.
Governing Authority: county. Tax-exempt: 501(c)(3).
Institution Type/Description: General Museum: housed in 1889 bank.
Collections: pioneer artifacts; period rooms; furniture; photographs; natural history; history; Native American artifacts & crafts; archives; art gallery; manuscripts; clothing.
Research Fields: Brown County, SD; railroad history in SD.
Facilities: 500-vol. library of history books available on premises; reading room. Handicrafts, books, notecards & local art for sale.
Activities: guided tours; lectures; gallery talks; formally organized education programs for children; temporary exhibitions; school loan service.
Publications: Architectural Records in Brown County Collections.
Hours & Admission Prices: Tues.-Fri. 9-5, Sat.-Sun. 1-4. No charge; donations accepted. Closed national holidays. &
Attendance: 76,024 (accurate)
Membership: Student & Senior Citizen $20; Individual $25; Family & Pioneer $50; Collector $125; Curator $250; Benefactor $500; Patron $1,000 & up.

NORTHERN GALLERIES, Northern State University, 1200 S. Jay St., Aberdeen, SD 57401-7198. Tel.: 605-626-7766 & 7762. Fax: 605-626-2263.
E-mail: kilianp@northern.edu
Founded: 1975.
Key Personnel: Dir., Greg Blair.
Personnel Profile: Part-Time Paid 1.
Governing Authority: Parent Institution: Northern State University. Tax-exempt.
Institution Type/Description: University Art Gallery.
Collections: contemporary art.
Hours & Admission Prices: Sept.-May Mon.-Fri. 8-4:30. No charge.
Attendance: 4,700 (accurate)

STORYBOOK LAND, Wylie Park, N. Hwy 281, Aberdeen, SD 57401. Mailing Address: Aberdeen Parks, Recreation and Forestry Department, 225 SE 3rd Ave., Aberdeen, SD 57401-4245. Tel.: 605-626-7015. Fax: 605-626-7989.
E-mail: prf@aberdeen.sd.us
Web Site: www.aberdeen.sd.us/storybookland
Institution Type/Description: Park.
Collections: story book & nursery rhyme sculptures & scenes; railroad history. Historic Building: 1881 train depot.
Facilities: visitor center. Gift items for sale.
Activities: train rides; carousel & Zamperla balloon rides. Annual Event: Storybook Land Festival in July.
Hours & Admission Prices: mid-ADepot Museum: Memorial Day to Labor Day daily 10-9. No charge. &
Attendance: 250,000 (estimated)

WEIN GALLERY, Presentation College, 1500 N. Main St., Aberdeen, SD 57401-1280. Tel.: 605-229-8585.
E-mail: elaine.kling@presentation.edu
Web Site: www.presentation.edu/weingallery
Key Personnel: Dir., Elaine Kling
Institution Type/Description: Art Gallery.
Collections: artwork by local artists.
Hours & Admission Prices: Mon.-Fri. 8-8, Sat. 1-7. No charge.

Armour

DOUGLAS COUNTY MUSEUM COMPLEX, Courthouse Grounds, Armour, SD 57313. Mailing Address: P.O. Box 638, Armour, SD 57313-0638. Tel.: 605-724-2129.
Founded: 1958.
Congressional District: 2
Key Personnel: Pres. & Cur., Sharon A. Wiese; Dir., Laverne Vanderwerff; Asst. Dir., Dot Hoveng.
Operating Income: 3,200
Governing Authority: county; society; nonprofit. Parent Institution: Douglas County Historical Society. Tax-exempt.
Institution Type/Description: Historical Society Museum: housed in 1904 County Office Bldg.
Collections: historical items from people of the area: clothing, dishes, farm implements; rock collection; Indian artifacts; photos; records of early organizations. Historic Buildings: c.1904 County Office Bldg.; c.1886 restored railroad house; one-room country school.
Research Fields: history of Douglas County.
Activities: guided tours; temporary exhibitions. Museum Sponsors: school days in May.
Publications: book, Douglas County-The Little Giant.
Hours & Admission Prices: Tues., Fri. & holidays 1-5; other times by appointment. No charge; donations accepted.
Attendance: 600 (estimated)
Membership: Annual $10.

Belle Fourche

TRI-STATE MUSEUM, (M), 415 5th Ave., Belle Fourche, SD 57717-1435. Tel.: 605-723-1200.
E-mail: tristatemuseum@rushmore.com
Web Site: www.tristatemuseum.com
Founded: 1955.
Key Personnel: Dir., Rochelle Silva; Chm. (V), Charlie Johnson; Exhibits Coord., DeEtte Gross; Admin. Asst., Morgan Callan Rogers.
Personnel Profile: Full-Time Paid 1; Part-Time Paid 4; Part-Time Volunteers 24.
Governing Authority: Parent Institution: City of Belle Fourche. Tax-exempt.
Institution Type/Description: History Museum.
Collections: over 5,000 artifacts; rodeo memorabilia; historical records; antiques; collectibles; dolls; military; fossils. Historic Building: 1876 cabin.
Hours & Admission Prices: Summer: Mon.-Sat. 9-6, Sun. 12-5; Winter: Tues.-Sat. 10-4. Suggested donations: adult $5, children under 12 $2.
Attendance: 14,000 (estimated)
Membership: Individual $20; Family $35; Business $100.

Brookings

BROOKINGS ARTS COUNCIL, 524 Fourth St., Brookings, SD 57006-2045. Tel.: 605-692-4177. Fax: 605-692-8298.
E-mail: directorbac@swiftel.net
Web Site: www.brookingsartscouncil.org
Founded: 1977.
Congressional District: 1
Key Personnel: Pres., Jean Jostad; Dir., Heather Kallhoff; Gallery Asst., Ladan Bahmani; Gallery Asst., Julie Luke.
Governing Authority: nonprofit organization. Tax-exempt.
Institution Type/Description: Cultural Arts Center: housed in 1914 Carnegie Library Building.
Collections: paintings; photographs.
Facilities: meeting space. Museum-related items for sale.
Activities: theatre performances; dance, poetry & music recitals; visual arts; humanities; formally organized education programs; special public events; lectures; arts workshops; temporary exhibits.
Hours & Admission Prices: Tues.-Sat. 12-5. No charge; donations accepted. Closed national holidays. &
Attendance: 12,000 (estimated)
Membership: Student & Senior $10; Individual $25-$49; Friend & Family $50-$99; Supporter $100-$199; Patron $200-$499; Benefactor $500 & up.

CHILDREN'S MUSEUM OF SOUTH DAKOTA, 521 4th St., Brookings, SD 57006. Tel.: 605-692-6700. Facebook: Children's Museum of South Dakota.
E-mail: info@prairieplay.org
Web Site: www.prairieplay.org
Founded: 2010.
Key Personnel: Exec. Dir., Suzanne Hegg
Institution Type/Description: Children's Museum.
Collections: hands-on exhibitions.
Facilities: cafe.
Activities: educational programs; special events; birthday parties.
Hours & Admission Prices: Tues.-Sat. 10-5, Sun. 12-5. Admission $6 per person; children under one & members no charge. Closed Independence Day; Thanksgiving; Christmas Eve & Day. &
Attendance: 120,000
Membership: Family & Grandparent $95; Reciprocal $125.

MCCRORY GARDENS AT SOUTH DAKOTA STATE UNIVERSITY, SMVC 106, 22nd Ave., Brookings, SD 57007. Mailing Address: Box 2140A, SMVC 106, Brookings, SD 57007. Tel.: 605-688-6707. Fax: 605-688-4452. Facebook: McCrory Gardens.
E-mail: david.graper@sdstate.edu
Web Site: www.mccrorygardens.com
Founded: 1964.
Key Personnel: Dir., David F. Graper, Ph.D.; Assoc. Dir., Martin Maca.
Personnel Profile: Full-Time Paid 3; Part-Time Paid 10; Part-Time Volunteers 24; Interns 4.
Governing Authority: Parent Institution: South Dakota State University. Tax-exempt.
Institution Type/Description: Arboretums.
Collections: ornamental plants; trees; shrubs; grasses; flowers; rock garden; prairie garden.
Research Fields: ornamental horticulture.
Activities: children's maze; rock garden.
Hours & Admission Prices: Dawn to dusk. Adults $6; discounts to American Horticulture Society Reciprocal Program members; members no charge. &
Attendance: 6,000 (estimated)
Membership: Individual $25; Family $50; Business $250; Orchid $500; All America Supporter $1,000.

*** SOUTH DAKOTA ART MUSEUM, (M),** Medary Ave. & Harvey Dunn, Brookings, SD 57007. Mailing Address: P.O. Box 2250, Brookings, SD 57007. Tel.: 605-688-5423; 866-805-7590. Fax: 605-688-4445.
E-mail: sdsu.sdam@sdstate.edu
Web Site: www.southdakotaartmuseum.com
Founded: 1969.
Congressional District: 1

Key Personnel: Dir., Lynn Verschoor; Cur. Collections, Lisa Scholten; Cur. Exhibits, Jodi Lundgren; Mktg. & Devel. Coord., Stacy Buehner; Museum Shop Mgr., Pam Adler.
Personnel Profile: Full-Time Paid 5; Part-Time Paid 8; Part-Time Volunteers 20.
Governing Authority: state; university. Parent Institution: South Dakota State University. Tax-exempt: 170(b)(1)(A)(iv).
Institution Type/Description: Art Museum.
Collections: state, regional & national art of 19th-20th centuries; Harvey Dunn & Oscar Howe paintings; Native American art; Marghab embroidered linen from Madeira.
Major Exhibits: Stephen Braun: Oblique Legacies, 1/14/14-5/11/14; Bob H. Miller: Glow Pop Art (T), 2/15/14-6/1/14; Nathan Holman, 3/25/14-7/27/14; Howard Pyle and Harvey Dunn: Teacher and Student (T), 4/8/14-8/3/14; Shannon Sargent: Objects Found for the Purpose of Understanding, 5/20/14-10/12/14; Robert Jackson, 6/10/14-9/21/14; Gerald Cournoyer, 6/10/14-9/21/14; SD Governor's 6th Biennial Art Exhibition (T), 9/30/14-1/4/15.
Research Fields: South Dakota & Native American art.
Facilities: 2,000-vol. art library available by reservation only; reading room; 150-seat auditorium; multi-purpose room; classroom. Museum-related items for sale.
Activities: guided tours; films; gallery talks; inter-museum loan, permanent, special & touring exhibitions; art in the schools program; kids sensation station for families.
Publications: exhibition catalogues; fall & spring mailers, brochures, postcards.
Hours & Admission Prices: Jan.-March Mon.-Fri. 10-5, Sat. 10-4; April-Dec. Mon.-Fri. 10-5, Sat. 10-4, Sun. 12-4. No charge; donations accepted. Closed New Year's Day; Thanksgiving; Christmas; state holidays. &
Attendance: 150,000 (estimated)
Membership: Individual $30; Family $45; Business $250.

STATE AGRICULTURAL HERITAGE MUSEUM, (M), South Dakota State University, 925 11th St., Brookings, SD 57007. Mailing Address: SDSU Box 601, Brookings, SD 57007. Tel.: 605-688-6226. Fax: 605-688-6303.
E-mail: sdsu.agmuseum@sdstate.edu
Web Site: www.sdsu.edu/agmuseum
Founded: 1967.
Congressional District: 1
Key Personnel: Interim Dir., Barry H. Dunne; Cur., Carrie Van Buren; Cur. & Museum Shop Mgr., Michelle Glanzer; Cur., Dawn Stephens.
Personnel Profile: Full-Time Paid 5; Full-Time Volunteers 8; Part-Time Volunteers 4.
Governing Authority: state. Parent Institution: South Dakota State University. Tax-exempt: 170(c).
Institution Type/Description: Agricultural Museum.
Collections: South Dakota history 1860-1950; agriculture & technology publications; photographic archives; ethnography; homesteading; farm machinery; claim shanty; major farm machinery archive.
Research Fields: agricultural history.
Facilities: 2,500-vol. library; archives; 6,500 sq. ft. exhibit space. Gift items for sale.
Activities: docent program; guided tours; lectures; temporary exhibitions; study & work programs for University students.
Hours & Admission Prices: Jan.-March Mon.-Sat. 10-5; April-Dec. Mon.-Sat. 10-5, Sun. 1-5. No charge; donations accepted. &
Attendance: 15,011 (accurate)
Membership: Individual $30; Family $40; Business $50; Patron $100; Program Sponsor $500-$5,000; Exhibit Sponsor $1,000-$5,000.

Buffalo

BUFFALO HISTORICAL MUSEUM AND ONE-ROOM SCHOOLHOUSE, Hwy. 85, Buffalo, SD 57720. Mailing Address: P.O. Box 391, Buffalo, SD 57720. Tel.: 605-375-3800 & 3787.
Founded: 1967.
Key Personnel: Pres. (V), Ray Anderson; Cur. & Museum Shop Mgr., Nora E. Boyer
Governing Authority: Tax-exempt.
Institution Type/Description: History Museum.
Collections: local history from the cattle-ranch & homesteading eras. Historic Building: 1914 one-room schoolhouse.
Hours & Admission Prices: Memorial Day to Labor Day Mon.-Fri. 10:30-2:30; other times by appointment. No charge. &

Carthage

CAMPBELL ORIGINAL STRAW BALE BUILT MUSEUM, 206 Main St., E., Carthage, SD 57323. Mailing Address: P.O. Box 3, Carthage, SD 57323-0003. Tel.: 605-772-4166 & 5778.
E-mail: madfarms@alliancecom.net
Web Site: strawbalemus.com
Institution Type/Description: History Museum: housed in a museum insulated with home-grown straw.
Collections: local history & culture; photographs.
Hours & Admission Prices: Call for hours.

Chamberlain

AKTA LAKOTA MUSEUM AND CULTURAL CENTER, (M), 1301 N. Main St., Chamberlain, SD 57325-1656. Mailing Address: P.O. Box 89, Chamberlain, SD 57325-0089. Tel.: 605-234-3452 & 3300. Fax: 605-234-3388.
E-mail: aktalakota@stjo.org
Web Site: www.aktalakota.org
Founded: 1991.
Congressional District: 20
Key Personnel: C.E.O., Father Stephen Huffstetter, S.C.J.; Dir., Dixie Thompson; Cur., Sara A. Caspi; Museum Shop Mgr., Vickie Brennan.
Personnel Profile: Full-Time Paid 2; Part-Time Paid 1; Part-Time Volunteers 2.
Governing Authority: Parent Institution: Congregation of the Priests of the Sacred Heart, Inc. of South Dakota. Subsidiary Institution: St. Joseph's Indian School. Tax-exempt.
Institution Type/Description: History Museum: Lakota culture & heritage.
Collections: Native American artifacts; quillwork; beadwork; art collection.
Research Fields: Plains Indians ethnology.
Facilities: library of Native American material & culture; Native American art available for research on premises. Gift items for sale.
Activities: guided tours; permanent & temporary exhibits; gallery talks; workshops; inter-museum loan.
Publications: Children of the Earth.
Hours & Admission Prices: May-Oct. Mon.-Sat. 8-6, Sun. 9-5; Nov.-April Mon.-Fri. 8-5. No charge; donations accepted. Closed legal holidays. &
Attendance: 30,000 (estimated)

SOUTH DAKOTA HALL OF FAME, 1480 S. Main, Chamberlain, SD 57325. Mailing Address: 1485 Main St., Chamberlain, SD 57325. Tel.: 605-734-4216. Fax: 605-734-4216.
E-mail: info@sdhalloffame.com
Web Site: www.sdhalloffame.com
Founded: 1974.
Congressional District: 2
Key Personnel: Chm., Richard Ekstrum; Vice Chm., Glenn Jergenson; Sec. & Treas., Lynne Duling.
Personnel Profile: Full-Time Paid 2; Part-Time Paid 2; Part-Time Volunteers 15.
Governing Authority: nonprofit organization. Tax-exempt: 501(c)(3).
Institution Type/Description: Historic Research Institute & Heritage Center.
Collections: biographical information on persons who were instrumental in building the state of South Dakota; county history books; artwork; artifacts from pioneer & Indian days; South Dakota culture.
Research Fields: personal information on inductees to the hall of fame; biographical memorials & life members.
Facilities: Museum-related items for sale.
Activities: permanent, temporary & traveling exhibitions; scavenger hunt for elementary school children.
Publications: annual, South Dakota Hall of Fame Magazine.
Hours & Admission Prices: Memorial Day to Labor Day Mon.-Fri. 10-5, Sat. 10-4, Sun. 1-4. Winter: Mon.-Fri. 10-5. No charge; donations accepted. Closed state holidays. &
Attendance: 4,000 (estimated)
Membership: Family $60; Business $100; Friend; $100-$249; Helper $250-499; Backer $500-$999.

Clark

BEAUVAIS HERITAGE COMPLEX, Hwy. 212, Clark, SD 57225. Mailing Address: c/o Clark Chamber of Commerce, P.O. Box 163, Clark, SD 57225. Tel.: 605-532-5772.
Web Site: www.clarksd.com/museum/museum.htm
Founded: 1975.
Congressional District: 7
Key Personnel: Pres., Clark County Historical Society, Greg Furness.
Governing Authority: nonprofit organization. Parent Institution: Clark County Historical Society. Tax-exempt: 170(b)(1)(A).

Institution Type/Description: Cultural & Historical Center.
Collections: memorabilia of 1870-1950; personal artifacts of Gov. Elrod of South Dakota; cup & saucer collection; toys; hand-crafted embroidery crochet; quilts; machinery; area business, family & county records; family & local town histories; military display. Historic Buildings: c.1880 claim shanty; Clark 1910 depot; 1911 church; Heritage school; c.1880 house; machinery building.
Research Fields: local & county history.
Facilities: 100-vol. library of county records, township & city records, private business & personal ledgers, old magazines & newspapers, local history videotapes; 900 family history files; county & township history files; available for research with permission from President of Historical Society.
Activities: guided tours; permanent & temporary exhibitions; tours; entertainment. Center Sponsor: Heritage Day in June; Threshing Bee & Tractor displays in August; competitions in August.
Publications: annual newsletter.
Hours & Admission Prices: Tours by appointment. No charge; donations accepted. &
Attendance: 500 (estimated)
Membership: Individual $10; Business $50; Life $200.

Crazy Horse

INDIAN MUSEUM OF NORTH AMERICA AT CRAZY HORSE MEMORIAL, (M), 12151 Avenue of the Chiefs, Crazy Horse, SD 57730-8900. Tel.: 605-673-4681. Fax: 605-673-2185.
E-mail: memorial@crazyhorse.org
Web Site: www.crazyhorsememorial.org
Founded: 1972.
Key Personnel: C.E.O., Ruth Ziolkowski; Registrar, Janeen Melmer; Librarian, Marguerite Cullum.
Personnel Profile: Full-Time Paid 3; Part-Time Paid 1; Part-Time Volunteers 2.
Governing Authority: nonprofit. Parent Institution: Crazy Horse Memorial, 12151 Avenue of the Chiefs, Crazy Horse, SD 57730-8900. Tax-exempt: 501(c)(3).
Institution Type/Description: American Indian Museum.
Collections: American Indian art; artifacts; mountain carving of Crazy Horse in progress, when completed it will be the world's largest mountain carving at 563' high & 641' long.
Facilities: 14,000-vol. library of books on Indian history; 2 theatres. American Indian made items for sale.
Publications: book, Indian Museum of North America; Crazy Horse Coloring Book; Carving a Dream (History of Crazy Horse Memorial & Korczak).
Hours & Admission Prices: Daily dawn dusk. Car $27, adults $10; discounts to senior citizens, Custer County residents, scouts & servicemen in uniform; members, children under 6 & American Indians no charge. &
Attendance: 1,000,000 (estimated)
Membership: Grass Roots $44; Single Jack $125; Korczak's Dream $174; Driller $250; Blaster $500; Ruth's Friend $1,000; Crazy Horse Bronze $1,500.

Custer

CUSTER COUNTY 1881 COURTHOUSE MUSEUM, 411 Mt. Rushmore Rd., Custer, SD 57730. Mailing Address: P.O. Box 826, Custer, SD 57730-0826. Tel.: 605-673-2443. Fax: 605-673-2443.
E-mail: cchstsoc@gwtc.net
Web Site: www.1881courthousemuseum.com
Founded: 1974.
Key Personnel: Chm. (V), Leon Nepper; Dir. & Museum Shop Mgr., Sandy Ackman.
Personnel Profile: Part-Time Paid 2; Part-Time Volunteers 75.
Governing Authority: society. Parent Institution: Custer County Historical Society, P.O. Box 826, Custer, SD. 57730. Tax-exempt.
Institution Type/Description: Historical Society Museum: housed in 1881 County Court House. Western, pioneer, mining & lumbering.
Collections: rocks; minerals; American Indian; period clothing; mining, ranching & lumbering; Custer 1874 expedition, Illingsworth pictures; mounted animals/birds of area, military uniforms from The Civil War to the present.
Research Fields: county & Black Hills history.
Facilities: library of books of historical value available for research on premises. Books of local history & pioneer information for sale.
Activities: permanent exhibitions.
Publications: Visitors' Brochure.
Hours & Admission Prices: May & Sept. daily 10-4; Memorial Day-Labor Day Mon.-Sat. 10-7, Sun. 1-7. Adults $6, seniors $5, youth 12-18 yrs $2. &
Attendance: 4,500 (accurate)
Membership: Custer County Historical Society $10 annual; Lifetime Society $40.

CUSTER STATE PARK, 13329 US Hwy. 16A, Custer, SD 57730-8351. Tel.: 605-255-4464. Fax: 605-255-4460.
E-mail: custerstatepark@state.sd.us
Web Site: www.custerstatepark.com
Founded: 1912.
Congressional District: 32
Key Personnel: Mgr. Black Hills Region, Matt Snyder; Naturalist, Julie Brazell; Visitor Svcs. Coord., Craig Pugsley.
Personnel Profile: Full-Time Paid 33; Part-Time Paid 4; Part-Time Volunteers 17; Interns 5.
Governing Authority: state. Parent Institution: South Dakota Department of Game, Fish & Parks, Foss Bldg., 523 E. Capitol, Pierre 57501.
Institution Type/Description: Park & Visitor Center.
Collections: historical & natural history exhibits.
Research Fields: history of programs, park development, culture & nature.
Facilities: visitor's center; hiking trails.
Activities: guided nature hikes; lectures; junior naturalist programs for children 7-12; gold panning & living history demonstrations; outdoor recreational activities; on-site school field trips; environmental education programs; teacher workshops; special events; permanent exhibits; camping.
Publications: magazine, Tatanka; Wildlife; Fisheries and Aquatic Resources; Charles Badger Clark; Peter Norbeck; Bird Checklist; Wildflower Checklist; brochure, Geocaching.
Hours & Admission Prices: Park: daily. Peter Norbeck Visitor Center: April-Nov. 9-5. Wildlife Station Visitor Center: May-Oct. 10-4. Badger Hole Historical Site: May-Sept. 10-5. Entrance License: $15 vehicles &
Attendance: 1,800,000 (estimated)

FOUR MILE OLD WEST TOWN, 11921 W. Hwy. 16, Custer, SD 57730-7114. Tel.: 605-673-3905.
Web Site: www.fourmilesd.com
Formerly: Four Mile Ghost Town
Founded: 1995.
Key Personnel: Dir., Mary Krogman.
Personnel Profile: Full-Time Volunteers 2; Part-Time Volunteers 3.
Institution Type/Description: History Museum.
Collections: local history; period furnishings; photographs; postal records. Historic Buildings: town hall; saloon; general store; church; bank; sheriff's office; Dakota Territory Jails; Slat Iron jail cell.
Activities: presentations; melodramas.
Hours & Admission Prices: mid-May to early Oct. daily 8:30-8. Admission $5; discounts to groups, home schoolers, AAM & museum members; children 6 & under no charge. &
Attendance: 18,000 (estimated)

JEWEL CAVE NATIONAL MONUMENT, 11149 US Hwy. 16, Bldg. B12, Custer, SD 57730-8166. Tel.: 605-673-8300. Fax: 605-673-8397.
E-mail: jeca_interpretation@nps.gov
Web Site: www.nps.gov/jeca
Founded: 1908.
Key Personnel: Supt., Larry Johnson; Museum Shop Mgr., Patty Ressler
Governing Authority: Parent Institution: Dept. of the Interior. Subsidiary Institution: National Park Service. Tax-exempt.
Institution Type/Description: Visitor Center and Cave.
Collections: local history; geology.
Activities: guided nature hikes; interpretive patio talks; junior ranger programs; cave tours; special events.
Publications: newspaper; park brochure; site bulletins; rack cards.
Hours & Admission Prices: Visitor Center: late May to mid-Sept. daily 9-5:30; late Sept. to mid-May daily 8:30-4:30. Scenic Tour: adults 17 & over $8, youth 6-16 & senior citizens $4; children under 5 no charge. Jewel Cave Discovery: adults 16 & over $4; senior citizens & children 15 and under no charge. &
Attendance: 109,279 (accurate)

NATIONAL MUSEUM OF WOODCARVING, 12111 W. Hwy. 16, Custer, SD 57730. Mailing Address: P.O. Box 747, Custer, SD 57730-0747. Tel.: 605-673-4404. Fax: 605-673-3843.
E-mail: woodcarv@gwtc.net
Web Site: www.blackhills.com/woodcarving
Founded: 1972.
Key Personnel: C.E.O. & Owner, Dale Schaffer; Museum Shop Mgr., Lois Massa.
Governing Authority: individual operation.
Institution Type/Description: National Woodcarving Museum: located in the area of the Black Hills where Custer & party found gold in 1874.

Collections: Dr. Niblack (original Disneyland animator) life-work of wood-carvings; over 75 woodcarvers' carvings for sale.
Facilities: snack shop; theater. Gift items for sale.
Activities: self-guided tours; gallery talks; permanent & temporary exhibitions; woodcarving classes thru the summer.
Hours & Admission Prices: May-Oct. 25 daily 9-5. Adults $9.79; discounts to senior citizens & groups. &
Attendance: 100,000 (estimated)

De Smet

DE SMET DEPOT MUSEUM, 104 Calumet Ave., N.E., De Smet, SD 57231. Mailing Address: P.O. Drawer 70, De Smet, SD 57231-0007. Tel.: 605-854-3991 & 3731. Fax: 605-854-3731.
Founded: 1965.
Congressional District: 1
Key Personnel: C.E.O. & Mayor, Gary Wolkow; Finance Officer, Eileen Wolkow.
Personnel Profile: Part-Time Paid 2; Part-Time Volunteers 30.
Governing Authority: municipal. Parent Institution: City of De Smet. Tax-exempt.
Institution Type/Description: History Museum: housed in Old Chicago North-western Depot & City Bldg.
Collections: pioneer collections; CNW railroad depot.
Activities: tours; pageants.
Publications: brochure
Hours & Admission Prices: June-Aug. Mon.-Sat. 10-5. No charge; donations accepted. &
Attendance: 2,500 (estimated)

LAURA INGALLS WILDER MEMORIAL SOCIETY, INC., 105 Olivet Ave., S.E., De Smet, SD 57231. Mailing Address: P.O. Box 426, De Smet, SD 57231-0426. Tel.: 800-880-3383.
E-mail: laura@discoverlaura.org
Web Site: www.discoverlaura.org
Institution Type/Description: General Museum: housed in the original Survey-or's House from Laura's book By the Shores of Silver Lake.
Collections: life & history of Laura Ingalls Wilder.
Facilities: Museum-related items for sale.
Hours & Admission Prices: May & Sept. Mon.-Sat. 9-4; June-Aug. daily 9-5:50, Sun. 10-5:30; Oct.-April Mon.-Fri. 9-4. Adults $8, children 6-12 $4; children 5 & under no charge. &

Deadwood

DAYS OF '76 MUSEUM, 18 Seventy Six Dr., Deadwood, SD 57732-1527. Mailing Address: P.O. Box 252, Deadwood, SD 57732-0391. Tel.: 605-578-1657. Fax: 605-717-0052. Facebook: Deadwood History.
E-mail: Mary@deadwoodhistory.com
Web Site: www.deadwoodhistory.com
Founded: 1924.
Key Personnel: Exec. Dir., Mary Kopco; Museum Shop Mgr., Michele Schulz.
Operating Expenses: 299,885
Operating Income: 315,000
Governing Authority: Parent Institution: Deadwood History, Inc. Tax-exempt.
Institution Type/Description: History Museum.
Collections: over 50 horse-drawn vehicles; costumes; archives; photographs; rodeo memorabilia; Old West & Native American art & artifacts; early pioneer history; American Indian artifacts; firearms; carriages.
Publications: quarterly newsletter, Deadwood History Banner.
Hours & Admission Prices: Call for hours. Adults $5.50, children 7-13 $2.50; discounts to AAM members; children 6 & under and members no charge. &
Attendance: 15,000 (estimated)
Membership: Senior $25; Individual $30; Senior Family $35; Family $45; Potato Creek Johnny $50; Deadwood Dick $100; Calamity Jane $250; Wild Bill $500.

DEADWOOD HISTORY, INC., (M), 150 Sherman, Deadwood, SD 57732. Mailing Address: P.O. Box 252, Deadwood, SD 57732-0252. Tel.: 605-722-4800 & 578-1714. Fax: 605-717-0052. Face-book: Deadwood History.
E-mail: mary@deadwoodhistory.com
Web Site: www.deadwoodhistory.com
Formerly: Adams Museum & House
Founded: 1930.
Congressional District: 31
Key Personnel: Chm. (V), David Wolff; Exec. Dir., Mary Kopco; Bookkeeper,

April Hoover; Archivist, Carolyn Weber; Cur. Exhibits, Darrel Nelson; Dir. Communications, Rose Speirs; Coord. Facilities, Brian Ledbetter; Cur. Interpretation, Ellyn Van Evra; Educator, Chelsie Bauer; Museum Shop Mgr., Michele Schulz; Dir. Museum Svcs., Karin Savoie.
Personnel Profile: Full-Time Paid 20; Part-Time Paid 15; Part-Time Volunteers 30; Interns 8.
Governing Authority: bd. of dirs. Subsidiary Institution: Adams Museum; Days of '76 Museum; Adams Research and Cultural Center; Historic Adams House, 22 Van Buren Ave., Deadwood, SD 57732. Tel.: 605-722-4800, Fax: 605-717-0052. Tax-exempt.
Institution Type/Description: History Museums.
Collections: costumes; geology; Native American artifacts; mineralogy; mili-tary; paleontology; pioneer room with artifacts; photographs of Deadwood & western personalities; guns; 1879 locomotive; 1834 The Thoen Stone; Russian artifacts; Victorian & American furniture; mining history; Days of '76 celebration & rodeo; Don Clowser collection.
Research Fields: Deadwood history; Black Hills mining history.
Facilities: Museum-related items for sale.
Activities: permanent & temporary exhibitions; public programs; lectures.
Publications: calendar of historic photographs; Historical Guide to Early Deadwood; quarterly, Deadwood History Banner; The Adams House Revealed; Raiding Deadwood's Bad Lands; Adams A to Z; Adams Art Gallery Guide.
Hours & Admission Prices: Adams Museum & Days of '76 Museum: April & Oct. Mon.-Sat. 9-5; May-Sept. daily 9-5; Nov.-March Tues.-Sat. 10-4; other times by appointment. Museum: no charge; donations accepted. Historic Adams House: April-Oct. Mon.-Sat. 9-5; May-Sept. daily 9-5; other times by appointment. Days of '76 Museum: adults $5.50, children 7-13 $2.50; discounts to AAM & AAA members; children 6 & under and members no charge. &
Attendance: 100,000 (estimated)
Membership: Senior $25; Individual $30; Senior Family $35; Family $45; Potato Creek Johnny $50; Deadwood Dick $100; Calamity Jane $250; Wild Bill $500.

Dell Rapids

DELL RAPIDS SOCIETY FOR HISTORICAL PRESERVA-TION, 407 E. 4th St., Dell Rapids, SD 57022-1927. Mailing Address: P.O. Box 143, Dell Rapids, SD 57022-0143. Tel.: 605-428-4821.
Founded: 1998.
Key Personnel: Pres. (V) & Museum Shop Mgr., Alice Chamley; Vice Pres., Jean Rave.
Volunteer Hours: 400
Governing Authority: Tax-exempt.
Institution Type/Description: History Museum.
Collections: local history & culture; documents; newspapers; photographs; period artifacts.
Activities: group tours.
Hours & Admission Prices: Memorial Day to Labor Day Tues.-Sat. 1-4. No charge.
Membership: Single $10; Family $15.

LITTLE VILLAGE FARM MUSEUM, 47582 240th St., Dell Rapids, SD 57022-6113. Tel.: 605-428-5979. Fax: 605-428-4999.
Founded: 1995.
Key Personnel: Chm. (V) & Museum Shop Mgr., Joan Redder-Lacey; Pres. (V), James Lacey
Institution Type/Description: Pioneer History Museum.
Collections: pioneer history; historic buildings.
Hours & Admission Prices: April-Oct. daily 8am to early evening by appointment. Adults $5, children $2. &
Attendance: 750 (estimated)

Elk Point

UNION COUNTY HISTORICAL SOCIETY, 124 E. Main, Elk Point, SD 57025. Mailing Address: P.O. Box 552, Elk Point, SD 57025-0552. Tel.: 605-356-3273.
Founded: 1988.
Key Personnel: Pres. (V), Mark Turner; Sec., Sondra Stickney
Governing Authority: Tax-exempt.
Institution Type/Description: History Museum.
Collections: Union County history; photographs; personal artifacts; genealogy; Lewis & Clark.
Publications: quarterly newsletter, The Sands of Time.
Hours & Admission Prices: Mon.-Tues. & Thurs.-Fri. 11:30-3:30; other times by appointment. No charge; donations accepted. Closed holidays. &
Attendance: 290 (accurate)

Membership: Single $20; Family $30; Life $300; Couple Life $500.

Elkton

ELKTON COMMUNITY MUSEUM AND HISTORICAL SOCI-
ETY, 206 Elk St., Community Center, Elkton, SD 57026. Mailing
Address: 21539 485th Ave., Elkton, SD 57026-8821. Tel.: 605-542-
2451 & 3991; 605-542-5411.
E-mail: dikamp@itctel.com
Founded: 1989.
Key Personnel: Pres. (V), Beverly Schwing.
Personnel Profile: Part-Time Volunteers 4.
Governing Authority: Tax-exempt.
Institution Type/Description: History Museum.
Collections: local history & culture; period artifacts; newspapers; photographs;
miniature shoes; church history; clothing; genealogy.
Hours & Admission Prices: By appointment. No charge; donations accepted.
Closed holidays. &
Attendance: 100
Membership: Individual $5.

Ellsworth AFB

SOUTH DAKOTA AIR AND SPACE MUSEUM, 2890 Davis Dr.,
Ellsworth AFB, SD 57706. Mailing Address: P.O. Box 871,
Ellsworth Heritage Foundation, Inc., Box Elder, SD 57719-0871.
Tel.: 605-385-5189. Fax: 605-385-6295.
E-mail: sdasm@midconetwork.com
Web Site: www.sdairandspacemuseum.com
Founded: 1982.
Key Personnel: Dir., Carl Engwall; Museum Shop Mgr., Beverly A. LeCates.
Personnel Profile: Full-Time Paid 2; Part-Time Volunteers 5.
Governing Authority: Tax-exempt.
Institution Type/Description: Military Museum.
Collections: over 25 aircraft including historic bombers, fighters, & utility
aircraft; missiles; aviation memorabilia; Aviation Hall of Fame.
Activities: interactive aircraft cockpit; simulators.
Hours & Admission Prices: June to Labor Day daily 8:30-6; Sept.-May daily
8:30-4:30. No charge. Closed New Year's Day; Easter; Thanksgiving;
Christmas. &
Attendance: 103,000
Membership: Adult $10; Family $25; Nonprofit Organization $50; Sustaining
$50-$999; Corporate $100-$500; Life $1,000 & up.

Eureka

EUREKA PIONEER MUSEUM OF MCPHERSON COUNTY,
INC., 1708 J Ave., Eureka, SD 57437. Mailing Address: P.O. Box
902, Eureka, SD 57437-0902. Tel.: 605-284-2987. eurekasd.com.
Web Site: www.glpta.org/eurekamuseum.htm
Founded: 1978.
Congressional District: 16
Key Personnel: Chm. (V), Asst. Dir. & Cur., Hulda Opp; Pres., Sonja
Anderson; Vice Pres., Arlo Mehlhaff; Dir., Cur. & Museum Shop Mgr.,
Edmund Opp; Sec., Claudia Merthan; Treas., Jean Bertsch; Dir. Public
Rels., Linda Bergman.
Personnel Profile: Full-Time Paid 1; Full-Time Volunteers 1; Part-Time
Volunteers 11.
Volunteer Hours: 500
Operating Expenses: 13,832
Operating Income: 23,762
Governing Authority: county; nonprofit organization. Tax-exempt.
Institution Type/Description: History Museum with focus on the lives and
work of pioneers.
Collections: local memorabilia; early machinery; manuscripts. Historic Build-
ings: schoolhouse; sod house; church.
Research Fields: local history.
Facilities: archives of documents & photographs of local nature; reading room.
Activities: guided tours; lectures; arts festivals; study clubs.
Publications: brochure.
Hours & Admission Prices: April 15-Nov. 1 Wed.-Fri. 1-5, Sat.-Sun. 2-5. No
charge; donations accepted. &
Attendance: 2,500 (estimated)
Membership: Annual $10; Family $20; Contributing $40; Business $50; Patron
$100.

Faulkton

FAULK COUNTY MUSEUM, 814 Court St., Faulkton, SD 57438-
2208. Mailing Address: Faulkton Historical Society, P.O. Box 584,
Faulkton, SD 57438-0584. Tel.: 605-598-4285.
E-mail: jody.moritz@k12.sd.us
Web Site: www.faulktoncity.org/historic_sites.htm
Founded: 1971.
Key Personnel: Pres. (V), Jody Moritz; Treas., Joyce Arnold; Cur., Judy Dixon.
Personnel Profile: Part-Time Volunteers 12.
Governing Authority: nonprofit. Parent Institution: Faulk County Historical
Society, Inc. Tax-exempt: 501(c)(3).
Institution Type/Description: History Museum.
Collections: Faulk County history & culture; personal artifacts; paintings;
photographs.
Activities: formal education programs for children.
Hours & Admission Prices: Summer: Wed. & Fri. 1-4. No charge; donations
accepted.
Attendance: 50
Membership: Adult $5.

PICKLER MANSION, 900 8th Ave., Faulkton, SD 57438. Mailing
Address: P.O. Box 584, Faulkton, SD 57438-0584. Tel.: 605-598-
4285.
E-mail: jody.moritz@k12.sd.us
Web Site: www.faulktoncity.org/historic-sites.htm
Founded: 1971.
Key Personnel: Pres. & Chm. (V), Jody Moritz; Cur., Judy Dixon; Dir., Janet
G. Reed.
Personnel Profile: Part-Time Paid 1; Part-Time Volunteers 6.
Volunteer Hours: 100
Operating Expenses: 1,600
Operating Income: 1,700
Governing Authority: Parent Institution: Faulk County Historical Society, Inc.
Tax-exempt.
Institution Type/Description: Historic House Museum: home of South Dako-
ta's first U.S. Congressman & his wife, John & Alice Pickler.
Collections: furnishings; personal artifacts.
Hours & Admission Prices: Memorial Day to Labor Day daily 1-4. Adults
$5.25.
Attendance: 300 (estimated)
Membership: $5 per year.

Flandreau

MOODY COUNTY HISTORICAL SOCIETY MUSEUM & RE-
SEARCH CENTER, 706 E. Pipestone Ave., Flandreau, SD 57028.
Mailing Address: P.O. Box 25, Flandreau, SD 57028-0025. Tel.:
605-997-3191.
E-mail: mchs1@knology.net
Web Site: www.moodycountymuseum.com
Founded: 1964.
Key Personnel: Dir., Dale A. Johnson; Pres. (V), Carole Hurley; Vice Pres.,
Warren Jackson.
Personnel Profile: Part-Time Paid 1; Part-Time Volunteers 10.
Governing Authority: society. Parent Institution: Moody County Historical
Society. Tax-exempt.
Institution Type/Description: History Museum.
Collections: carriage; early dental office; medical exam room; living room;
church pulpit; organ; barber shop; pioneer piano; clothing; dolls; bedroom;
post office; pictures; tools; rural schoolhouse; Indian artifacts; old photo-
graphs; family histories. Historic Buildings: 1881 Milwaukee railroad
depot; 1871 Indian River Bend meeting house/church; one-room country
schoolhouse.
Research Fields: genealogy; county history.
Activities: guided tours. Museum Sponsors: Fourth of July Festival; monthly
brown bag lunch programs Sept.-April; Christmas Tree Festival in December
Publications: quarterly newsletter, Moody County Pioneer.
Hours & Admission Prices: Tues.-Fri. 11-4, Sat. 10-3; other times by
appointment. No charge; donations accepted. Closed New Year's Eve &
Day; Memorial Day; Thanksgiving weekend; Christmas Eve & Day. &
Attendance: 1,500 (estimated)
Membership: Annual $20.

Fort Meade

OLD FORT MEADE MUSEUM AND HISTORIC RESEARCH ASSOCIATION, Bldg. 55 Sheridan St., Fort Meade, SD 57741. Mailing Address: P.O. Box 164, Fort Meade, SD 57741-0164. Tel.: 605-347-9822.
E-mail: support@fortmeademuseum.org
Web Site: www.fortmeademuseum.org
Founded: 1964.
Key Personnel: Dir. & Museum Shop Mgr., Bob Kusser; Pres. (V), Peg Aplan; Sec., Marshall Williams.
Personnel Profile: Full-Time Paid 1; Full-Time Volunteers 16; Part-Time Paid 3; Part-Time Volunteers 16; Interns 1.
Governing Authority: private; nonprofit. Tax-exempt.
Institution Type/Description: Military Museum: U.S. Cavalry fort built in 1878 by Seventh Cavalry.
Collections: memorabilia of the units and troopers; chronological history of Old Fort Meade as a military installation from 1878-1944; Custer's 1874 Black Hills expedition; Black Hills Gold Rush; 25th Infantry (Buffalo soldiers); 1935 National Geographic stratosphere balloon flight; C.C.C. camps; German P.O.W.'s; 4th Cavalry horseback maneuvers.
Research Fields: history relating to 1878-1944 military post.
Facilities: Gift items & books for sale.
Activities: tours; self-guided tours; organized education programs for children & adults; Ft. Meade story slide/tape film program; video program; Cavalry Days in June; Re-enactments; living history programs.
Publications: book, The Peace Keeper Post on the Dakota Frontier-1878-1944; videotape; Fort Meade and the Black Hills.
Hours & Admission Prices: May 15th & Sept. 15th 9-5; Memorial Day to Labor Day daily 9-5. Adults $5, groups $3; discounts to AAM members; children under 12, school groups & members no charge.
Attendance: 8,000 (estimated)
Membership: Individual $25; Family $35; Life $100.

Fort Pierre

VERENDRYE MUSEUM, 115 Deadwood St., Fort Pierre, SD 57532. Mailing Address: PO Box 665, Fort Pierre, SD 57532-0665. Tel.: 605-223-7697.
Web Site: www.fortpierce.com
Founded: 1968.
Key Personnel: Pres. (V), Darby Nutter.
Governing Authority: Tax -exempt.
Institution Type/Description: History Museum.
Collections: local history; photographs; personal artifacts.
Hours & Admission Prices: Memorial Day to Labor Day daily 9-5; other times by appointment. No charge; donations accepted.
Attendance: 600 (estimated)
Membership: Single $25; Family $35.

Frankfort

FISHER GROVE COUNTRY SCHOOL/FISHER GROVE STATE PARK, 17290 Fishers Lane, Frankfort, SD 57440-6700. Tel.: 605-472-1212 & 1336.
E-mail: lake.louise@state.sd.us
Web Site: gfp.sd.gov/state-parks/directory/fisher-grove/
Founded: 1884.
Key Personnel: Park Mgr., Charles Jones.
Governing Authority: state. Tax-exempt.
Institution Type/Description: Historic Building: 1884 Deiter School.
Collections: articles from schoolhouse; interpretive display.
Activities: recorded lecture.
Hours & Admission Prices: Call for confirmation of hours.

Freeman

HERITAGE HALL MUSEUM & ARCHIVES, 748 S. Main St., Freeman, SD 57029-2317. Mailing Address: P.O. Box 693, Freeman, SD 57029-1000. Tel.: 605-925-7545. Fax: 605-925-4271.
E-mail: museum@freemaninfo.com
Web Site: www.freemanmuseum.org
Founded: 1930.
Key Personnel: Dir., Stacey Waltner; Pres. (V), Kevin Albrecht; C.E.O., Pam Tieszen; Vice Pres., Jeremy Waltner; Archivist, Duane Schrag; Cur., Roy Kaufman.
Personnel Profile: Part-Time Paid 2; Part-Time Volunteers 18.
Governing Authority: 10 community churches. Parent Institution: Freeman Academy. Tax-exempt.
Institution Type/Description: Ethnic Museum.
Collections: German-Russian cultural artifacts; over 450 19th century Plains Indian artifacts; history of pioneer settlers; early life in South Dakota; autos; motorcycles; farm implements; airplane.
Facilities: 1,808-vol. library of South Dakota history, Mennonite history, German Mennonite history, & religious history available for limited inter-library loan & use by the public.
Activities: guided tours; lectures.
Hours & Admission Prices: Memorial Day to Labor Day Tues.-Sat. 11-4:30, Sun. 1-4:30; other times by appointment. Adults $5, students grades 1-12 $3; pre-school no charge. &
Attendance: 4,000 (estimated)
Membership: Personnel $100; Business $200; Founding $500.

Garretson

GARRETSON HISTORICAL SOCIETY MUSEUM, 609 Main St., Garretson, SD 57030. Mailing Address: P.O. Box 614, Garretson, SD 57030-0614. Tel.: 605-594-6694.
Founded: 1987.
Key Personnel: Pres. (V), Oran Sorenson
Institution Type/Description: Historical Society Museum.
Collections: veterans' display; obituary scrapbook; Garretson school & local history; period artifacts; furnishings; photographs.
Hours & Admission Prices: Mon.-Fri. 11-5; other times by appointment. &
Membership: Single $15; Family $25.

Geddes

GEDDES HISTORIC DISTRICT VILLAGE, 311 Main St., Geddes, SD 57342-0097. Mailing Address: P.O. Box 97, Geddes, SD 57342-0097. Tel.: 605-337-2501. Fax: 605-337-3535.
E-mail: dufsdfek@midstatesd.net
Web Site: www.geddes.org
Formerly: Charles Mix County Historical Restoration Society
Founded: 1969.
Congressional District: 1
Key Personnel: C.E.O., Chm. (V) & Sec., Ronald D. Dufek; Chm. (V) & Museum Shop Mgr., Irene Merkwan.
Personnel Profile: Full-Time Volunteers 5; Part-Time Volunteers 15.
Volunteer Hours: 225
Operating Expenses: 16,000
Operating Income: 15,000
Governing Authority: nonprofit organization. Parent Institution: Charles Mix County Historical Restoration Society. Tax-exempt: 501(c)(3).
Institution Type/Description: History Museum.
Collections: 1850-1910 period artifacts; desks; maps; photographs; tools; furs; 1910 & 1958 era furniture; partial L&C keel boat replica; pioneer cultural & heritage artifacts; Peter Norbeck documents. Historical Buildings: 1900 Padley Hotel; The Red Oak Bar; 1895 Red, White & Blue Rural Schoolhouse; 1857 Papineau (French Canadian) Trading Post (log cabin); former Gov. Norbeck Boyhood Home; 1804 Lewis & Clark Fur Trader Museum; 1895 Chaims Shanty.
Major Exhibits: Geddes Museum, 1/14-12/14; Peter Worbeck Boyhood Home, 5/14-10/1/14; Red White Blue Rural Schoolhouse, 5/14-10/1/14; Papineau Trading Post 1857, 5/14-10/1/14.
Research Fields: regional & state history.
Facilities: library.
Activities: guided tours. Annual Events: Spring Flea Market & Antique Auction; Tour of Homes in June; Peter Norbeck Day in June; Fur Trader Day in August; Fall Flea Market & Antique Auction; Fall Demolition Derby.
Publications: brochures; City of Geddes Profile; book, 100 Year Memories 1900-2000.
Hours & Admission Prices: mid-May to mid-Sept. daily 9-7. No charge; donations requested. &
Attendance: 400 (estimated)
Membership: Annual $10; Business $25; Lifetime $50.

Gettysburg

DAKOTA SUNSET MUSEUM, 205 W. Commercial Ave., Ste. 104, Gettysburg, SD 57442-1103. Tel.: 605-765-9480.
E-mail: dakotasunset@venturecomm.net
Web Site: www.dakotasunsetmuseum.com
Founded: 1984.
Key Personnel: Pres. (V), Bob Potts; Dir., Cur. & Museum Shop Mgr., Kathleen Nagel; Archivist, Eileen Jost; Archivist, Mary Carol Potts.
Personnel Profile: Part-Time Paid 4; Part-Time Volunteers 12.
Governing Authority: private; nonprofit organization. Tax-exempt: 501(c)(3).

Institution Type/Description: History Museum.
Collections: Potter County history from 1880s to 1960s; Medicine Rock, a 40-ton boulder with human footprints & a handprint, sacred to Native Americans; Civil War artifacts; big game animals; home furnishings; period barber shop; barn; country school; blacksmith shop; military artifacts; banking display; medical & dental artifacts; Native American artifacts including a Dentalium shell trade wood dress; general store; family histories & obituaries; school & church records; newspapers; homestead maps.
Facilities: library; 6,000 sq. ft. exhibit space. Museum-related items for sale.
Activities: guided tours; temporary & traveling exhibitions; monthly historical meetings; research. Annual Event: Christmas Tree Extravaganza.
Publications: newsletter, Dakota Sunset Newsletter.
Hours & Admission Prices: Summer: Mon.-Sat. daily 1-5; Winter: Tues.-Sat. 1-5. No charge; donations accepted. Closed New Year's Day; Easter; Memorial Day; Independence Day; Labor Day; Thanksgiving; Christmas.
Attendance: 1,200 (estimated)
Membership: Adult $5.

Groton

GRANARY RURAL CULTURAL CENTER, 40161 128th St., Groton, SD 57445-5405. Mailing Address: c/o Dacotah Prairie Museum, 21 S. Main St., Aberdeen, SD 57401. Tel.: 605-626-7117. Fax: 605-626-4026.
E-mail: sue.gates@browncounty.sd.gov
Web Site: www.granaryfinearts.org
Founded: 1996.
Key Personnel: Chm. & Pres. (V), Sue Gates; Dir., Lora Schaunanman.
Personnel Profile: Full-Time Volunteers 1; Part-Time Paid 1.
Governing Authority: Parent Institution: Dacotah Prairie Museum, 21 S. Main St., Aberdeen, SD 57401. Tax-exempt.
Institution Type/Description: Art Gallery.
Collections: works by local artists.
Activities: Museum Sponsors: All-Dakota High School Exhibition in April.
Hours & Admission Prices: Call for hours. No charge; donations accepted. &
Attendance: 3,500 (estimated)

Hermosa

HERMOSA ARTS AND HISTORY MUSEUM, 25 N. Second St., Hermosa, SD 57744. Mailing Address: P.O. Box 175, Hermosa, SD 57744-0175. Tel.: 605-431-0708.
E-mail: hermosamuseum@yahoo.com
Founded: 1999.
Congressional District: 1
Key Personnel: Dir., Doug Hesnard.
Personnel Profile: Part-Time Paid 1; Part-Time Volunteers 8.
Governing Authority: Parent Institution: Hermosa Arts and History Association, Hermosa, SD. Tax-exempt.
Institution Type/Description: History Museum: housed in the 1889 Hermosa school.
Collections: local history & culture; photographs; art; Calvin Coolidge; railroad & agricultural memorabilia.
Activities: speakers; book signings; readings. Museum Sponsors: Open House in July & August.
Hours & Admission Prices: By appointment. No charge. &

Hill City

BLACK HILLS MUSEUM OF NATURAL HISTORY, 117 Main St., Hill City, SD 57745. Mailing Address: P.O. Box 614, Hill City, SD 57745-0614. Tel.: 605-574-4505. Fax: 605-574-2518.
E-mail: neal@bhmnh.org
Web Site: www.bhmnh.org
Founded: 1990.
Key Personnel: Exec. Dir., Pres. (V) & Cur., Neal L. Larson; Chm. (V), Joe Harris; Treas. & Cur., Robert A. Farrar; Mktg., Deb Casey; Membership & Public Rels., June Zeitner; Museum Shop Mgr., V. Brenda Larson.
Personnel Profile: Full-Time Volunteers 15; Part-Time Paid 1; Part-Time Volunteers 18; Interns 3.
Governing Authority: private; not-for-profit organization. Tax-exempt: 501(c)(3).
Institution Type/Description: Natural History Museum.
Collections: worldwide paleontological specimens; cretaceous ammonites; end of the cretaceous dinosaurs; T-Rex artifacts; invertebrate & vertebrate dinosaur fossils; minerals; meteorites; mining photographs; deeds; Black Hills history.
Research Fields: ammonite locality & stratigraphy; fossil; dinosaur; fishes; mammal & minerals.

Facilities: library of paleontology books; 4,000 sq. ft. exhibit space; field research station.
Activities: talks & lectures; loan, traveling & temporary exhibitions; school loan service; bi-annual symposium. Annual Events: Natural History Festival; auction.
Hours & Admission Prices: Summer: Mon.-Sat. 9-7, Sun. 10-6; Winter: call for hours; Spring: Mon.-Sat. 9:30-5, Sat.-Sun. 12-4. Adults $7.50, senior citizens, active military & veterans $6, children 6-15 $4; children 5 & under no charge. &
Attendance: 76,465 (accurate)

Hot Springs

BLACK HILLS WILD HORSE SANCTUARY & VISITORS CENTER, Highland Rd., Hot Springs, SD 57747. Mailing Address: P.O. Box 998, Hot Springs, SD 57747-0998. Tel.: 800-252-6652; 605-745-5955. Fax: 605-745-4339.
E-mail: iram@gwtc.net
Web Site: wildmustangs.com
Key Personnel: Program Mgr., Susan Watt.
Governing Authority: nonprofit organization. Tax-exempt: 501(c)(3).
Institution Type/Description: Wild Horse Sanctuary.
Collections: wild horses; wild horse, pioneer, & Native American history; Crazy Horse movie set; Native American ceremonial sites; pioneer homesteads; one-room school house.
Facilities: Museum-related items for sale.
Activities: guided tours.
Hours & Admission Prices: Guided Tours: by appointment. 2 Hour Guided Tour: adults 19 & over $50, seniors 55 & over $45, youth 13-18 $15; children 5-12 $7.50; children 4 & under no charge. Additional tours available.

* **MAMMOTH SITE OF HOT SPRINGS, INC., (M),** 1800 US 18 Bypass, Hot Springs, SD 57747-9604. Mailing Address: P.O. Box 692, Hot Springs, SD 57747-0692. Tel.: 605-745-6017. Fax: 605-745-3038.
E-mail: joem@mammothsite.org
Web Site: www.mammothsite.org
Founded: 1975.
Congressional District: 2
Key Personnel: C.O.O., Joe Muller; Pres., Linda Stoll; Vice Pres., Julie Mossman; Sec., Anna Merrill.
Personnel Profile: Full-Time Paid 12; Part-Time Paid 41; Part-Time Volunteers 50; Interns 9.
Governing Authority: nonprofit organization. Tax-exempt: 501(c)(3).
Institution Type/Description: Paleontology Museum.
Collections: mammoth tusks; skulls; bones of mammoth; pleistocene fauna.
Research Fields: paleontology; sedimentology; paleobotany.
Facilities: library of books on paleontology & sedimentology available for research; video presentation. Mammoth site booklets for sale.
Activities: guided tours; lectures; permanent exhibitions; junior paleontologist, advanced paleontologist & Atlatl programs.
Publications: booklet, Mammoth Graveyard; book, Mammoth Site of Hot Springs, South Dakota; brochures; Megafauna & Man (1989 Symposium volume); Annotated Bibliography of North American Mammoths 1940-1990.
Hours & Admission Prices: April to mid-May & Sept.-Oct. daily 9-5; mid-May to Aug. daily 8-8; Nov.-March daily 9-3:30. Adults $9, seniors 60 & over $8, children 4-12 $7; discounts to AAM members; children 3 & under and members no charge. &
Attendance: 97,095 (accurate)
Membership: Single $20; Family $49; Donor $85; Contributing $150; Supporting $250; Sustaining $500 & up.

PIONEER MUSEUM IN HOT SPRINGS, 300 N. Chicago St., Hot Springs, SD 57747-1657. Mailing Address: P.O. Box 361, Hot Springs, SD 57747-0361. Tel.: 605-745-5147.
E-mail: pioneer@pioneer-museum.com
Formerly: Fall River County Historical Museum
Founded: 1961.
Congressional District: 2
Key Personnel: Pres., David Purtill.
Personnel Profile: Full-Time Paid 1; Part-Time Paid 2; Part-Time Volunteers 40.
Governing Authority: society; nonprofit. Parent Institution: Fall River Co. Historical Society. Tax-exempt: 501(c)(3).
Institution Type/Description: Historical Society Museum: housed in 1893 schoolhouse.

Collections. artifacts; furniture; tools; dishes; books & other items showing how pioneers lived; photos; quilts; toys; paintings; photos; medical artifacts; cameras.
Research Fields: local history.
Activities: guided tours; formally organized education programs for children; permanent & temporary exhibitions. Museum Sponsors: Pioneer Days.
Publications: books, Fall River County Pioneers Histories; IGLOO: A History of the Black Hills Ordinance Depot.
Hours & Admission Prices: May 15-Oct. 15 Mon.-Sat. 9-5. Adults $5, seniors $4; discounts to families, school groups & bus tours; children 12 & under no charge. &
Attendance: 10,000 (estimated)
Membership: Senior $8; Individual $10; Family $15; Business $25; Lifetime $150.

WIND CAVE NATIONAL PARK, Hot Springs, SD 57747. Mailing Address: 26611 U.S. Hwy. 385, Hot Springs, SD 57747-6027. Tel.: 605-745-4600. Fax: 605-745-4207.
E-mail: tom_farrell@nps.gov
Web Site: www.nps.gov/wica/
Founded: 1903.
Congressional District: 2
Key Personnel: Supt., Videl Davila; Cur., Tom Farrell.
Personnel Profile: Full-Time Paid 24; Part-Time Paid 40; Part-Time Volunteers 2; Interns 2.
Governing Authority: federal. Tax-exempt.
Institution Type/Description: Natural History Museum.
Collections: geology; natural history; American Indian artifacts; area history.
Facilities: 1,000-vol. library of natural history books available for use on the premises.
Activities: guided cave tours; lectures; campfire programs.
Hours & Admission Prices: Daily 8-4:30. Cave Tours: Natural Entrance Tour & Fair Grounds Tour: adults $9, children 6-16 & senior pass holders $4.50. Garden of Eden Tour: adults $7, children 6-16 $3.50; children under 5 no charge. Museum: no charge. Closed New Year's Day; Thanksgiving; Christmas. &
Attendance: 102,000 (estimated)

Howard

MINER COUNTY RURAL LIFE MUSEUM, 127 S. Main St., Howard, SD 57349. Mailing Address: P.O. Box 245, Howard, SD 57349-0245. Tel.: 605-772-5677.
E-mail: mchs@alliancecom.net
Key Personnel: Pres. (V), Mary E. Leary.
Personnel Profile: Part-Time Volunteers 12.
Governing Authority: Parent Institution: Miner County Historical Society. Tax-exempt.
Institution Type/Description: History Museum.
Collections: local history & culture; photographs; personal artifacts.
Activities: group tours; lessons in traditional arts; hand quilting.
Publications: Howard, SD - The First Four Years, 1881-1884.
Hours & Admission Prices: Mon. 11-2, Fri. 1-4. No charge; donations accepted. &
Membership: Miner County Historical Society: Individual $35 ($10 thereafter).

Huron

DAKOTALAND MUSEUM, State Fair Grounds, 3rd St., Huron, SD 57350-1254. Mailing Address: P.O. Box 1254, Huron, SD 57350-1254. Tel.: 605-352-4626.
E-mail: rachelclenden@yahoo.com
Web Site: www.dakotalandmuseum.org
Founded: 1960.
Congressional District: 32
Key Personnel: Chm. (V), Peggy Gibson; Dir., Rachel Farrell.
Personnel Profile: Full-Time Paid 1; Part-Time Paid 5; Part-Time Volunteers 12.
Governing Authority: Parent Institution: Dakotaland Museum Board. Subsidiary Institution: Huron City, Beadle County. Tax-exempt.
Institution Type/Description: History Museum.
Collections: pioneer exhibits; American Indian artifacts, including 1,500 arrowheads, beadwork, peace pipes, moccasins, purses, baskets; history; Kouf Natural History including 370 mounted mammals & birds; 1,600 items of clothing dating back to 1860; toys; dolls; furnished bedroom, parlor, physician's office & kitchen including a burled walnut bed & antique organ; hand tools; 1881 corn planter, walking plows & other vintage farm implements; clocks; surrey; radios; Model T car; 1914 Monroe automobile; fire engine; glassware; cameras & projectors; baby buggies & cradles;

musical instruments; Civil War memorabilia; compressed air locomotive. Historic Buildings: 1894 Pyle House; 1870s log cabin; 1887 Centennial Stone Church Center.
Research Fields: history books; historic books; regional photographs.
Activities: guided tours; permanent & temporary exhibitions.
Publications: brochure.
Hours & Admission Prices: Memorial Day-Labor Day Mon.-Fri. 10-6, Sat. 2-6, Sun. 1-3. Dakotaland Museum: adults $3, children 6-12 $1. &
Attendance: 1,300 (estimated)

PYLE HOUSE MUSEUM, 376 Idaho Ave., S.E., Huron, SD 57350-2527. Mailing Address: P.O. Box 1254, Huron, SD 57350-1254. Tel.: 605-352-2528.
E-mail: rachelclenden@yahoo.com
Web Site: www.dakotalandmuseum.org/pylehouse
Founded: 1985.
Congressional District: 32
Key Personnel: Dir., Rachel Farrell; Chm. (V), Peggy Gibson.
Personnel Profile: Full-Time Paid 1; Part-Time Volunteers 5.
Governing Authority: Parent Institution: Dakotaland Museum/Pyle. Subsidiary Institution: Huron City/Beadle Co. Tax-exempt.
Institution Type/Description: Historic House Museum: housed in the home of Gladys Pyle, the first elected woman U.S. senator, built in 1894.
Collections: period artifacts; furnishings; personal artifacts; original blueprint, architect Dracon 1893-1894.
Hours & Admission Prices: Summer: Mon.-Fri. 1-3:30; other times by appointment. Adults $3; children 12 & under no charge. Closed major holidays.
Attendance: 500 (estimated)

Interior

BADLANDS NATIONAL PARK, 25216 Ben Reifel Rd., Interior, SD 57750. Mailing Address: P.O. Box 6, Interior, SD 57750. Tel.: 605-433-5361. Fax: 605-433-5404. TDD: 605-433-5361.
Web Site: www.nps.gov/badl
Founded: 1939.
Key Personnel: Supt., Eric Brunnemann.
Governing Authority: federal. Affiliated with the U.S. Dept. of the Interior, National Park Service, Washington, DC 20240. Tax-exempt.
Institution Type/Description: Natural History Museum.
Collections: Ben Reifel Visitor Center: life during the Eocene, Oligocene, & Pleistocene periods; films. White River Visitor Center: Sioux Indian artifacts; films.
Research Fields: wildlife: buffalo; prairie dogs; mule deer; Rocky Mountain bighorn; Paleontology; woody draws ecology; grassland ecology problems.
Facilities: visitor centers. Publications for sale.
Activities: Ben Reifel Visitor Center: slide talks; guided nature walks; hikes; night prowls & night sky observation; nature trails; fossil preparation demonstrations; films; touch screens; introductory program.
Publications: booklets, Where Buffalo Roam; Badlands: Its Life and Landscape; Wildflowers of the Northern Plains and Black Hills; This Curious Country; Video: Land of Stone and Light.
Hours & Admission Prices: Ben Reifel Visitor Center: mid-April to mid-May and early Sept. to late Oct. daily 8-5; late Oct. to mid-April daily 8-4; mid-May to early Sept. 8-7. Car $15, motorcycle $10, individual (hike, bicycle) $7. Closed New Year's Day; Thanksgiving; Christmas. &
Attendance: 906,868 (accurate)

Ipswich

J.W. PARMLEY HISTORICAL HOME SOCIETY, 115 Main St., Ipswich, SD 57451. Mailing Address: P.O. Box 111, Ipswich, SD 57451-0111. Tel.: 605-426-6024 & 6949.
Web Site: www.ipswich-sd.com
Formerly: J.W. Parmley Historical Home Museum
Founded: 1981.
Key Personnel: Cur., Ray Kub.
Governing Authority: bd. Branch Museum: 319 4th St., Ipswich, SD 57451. Tax-exempt.
Institution Type/Description: Historic Building.
Collections: local history; pioneers; early businesses; animal exhibits complied by pioneers; clothing; military; photographs; archives; 1900 business office; 1921 home
Activities: permanent exhibitions; special tours. Museum Sponsors: Barbeque at the Land Office in June; Ice Cream Social at the Home in August; Fundraiser in October.
Hours & Admission Prices: Memorial Day to Labor Day Fri. & Sun. 2-5; other times by appointment. No charge; donations accepted.
Attendance: 500 (accurate)

Kadoka

BADLANDS PETRIFIED GARDENS, Interstate 90, Exit 152, Kadoka, SD 57543. Mailing Address: P.O. Box 27, Kadoka, SD 57543-0027. Tel.: 605-837-2448.
Web Site: www.badlandspetrifiedgardens.com/default.html
Founded: 1956.
Key Personnel: Pres. & Deputy Dir., Robert Fugate; Vice Pres., Cathy Fugate; Asst. Dir., Patty Ulman; Business Officer, Floy Fugate; Museum Shop Mgr., Bill Fugate.
Personnel Profile: Full-Time Paid 1; Part-Time Paid 4.
Governing Authority: company organized for profit.
Institution Type/Description: Natural History Museum.
Collections: Badlands petrified wood, logs & stumps; fossils; agate & minerals; fluorescent minerals.
Facilities: Minerals, gemstones, jewelry & gift items for sale.
Activities: tours & permanent exhibitions.
Hours & Admission Prices: mid-April to Oct. daily 7-7. Adults $5, children 6-16 $2.50; discounts to groups. &

Attendance: 15,000 (estimated)

KADOKA DEPOT MUSEUM, S. Main St., Kadoka, SD 57543. Mailing Address: City of Kadoka, Box 58, Kadoka, SD 57543-0058. Tel.: 605-837-2229. Fax: 605-837-1262.
E-mail: kadokacity@wcenet.com
Web Site: www.kadokasd.com
Formerly: Jackson Washabaugh Historical Museum
Founded: 1980.
Key Personnel: Financial Officer, Patty Ulmen.
Governing Authority: nonprofit; municipal. Kadoka City Council.
Institution Type/Description: Historical Society Museum: housed in 1906 Chicago, Milwaukee, St. Paul Railroad Depot.
Collections: railroad; homestead; bank; school; army & navy photos; local memorabilia.
Research Fields: local memorabilia.
Activities: loan, permanent & temporary exhibitions.
Hours & Admission Prices: Call for confirmation of hours.
Attendance: 500 (estimated)

Keystone

BIG THUNDER GOLD MINE, 604 Blair, Keystone, SD 57751. Mailing Address: Box 459, Keystone, SD 57751-0459. Tel.: 605-666-4847. Fax: 605-666-4566.
E-mail: mclainsandra@aol.com
Web Site: www.bigthundermine.com
Founded: 1957.
Congressional District: 1
Key Personnel: Dir. & C.E.O., Sandi McLain; Museum Shop Mgr., Adam McLain.
Personnel Profile: Full-Time Paid 2; Part-Time Paid 22; Part-Time Volunteers 1.
Operating Expenses: 450,000
Operating Income: 529,000
Governing Authority: corp. Parent Institution: Big Thunder, Inc.
Institution Type/Description: Historic Site: 1890s gold mine.
Collections: mining equipment; photographs; 1880's historical equipment.
Facilities: 4,500 sq. ft. gold mine. Jewelry and gold panning supplies for sale.
Activities: guided mine tour; gold panning; lectures; prospecting.
Publications: Black Hills Mining History.
Hours & Admission Prices: May-Sept. daily 8-8. Adults $9.50, children 6-12 $6.50; discounts to members, AAM & ICOM members. Gold Panning: $9.50 without tour, $7.50 with tour. &

Attendance: 85,000 (estimated)

KEYSTONE AREA HISTORICAL SOCIETY, 410 3rd St., Keystone, SD 57751. Mailing Address: P.O. Box 177, Keystone, SD 57751-0177. Tel.: 605-666-4494. Fax: 605-666-4566.
E-mail: mclainsandra@aol.com
Web Site: www.keystonehistory.com
Founded: 1983
Congressional District: 1
Key Personnel: Pres. (V), Sandra McLain; Dir., Arlene Robinson; Treas., Wally Hunsacker.
Personnel Profile: Part-Time Paid 1; Part-Time Volunteers 10.
Operating Expenses: 10,000
Operating Income: 16,900
Governing Authority: nonprofit. Tax-exempt: 501(c)(4).

Institution Type/Description: General Museum: housed in c.1900 three-story Victorian frame school house.
Collections: early settlement of the local area & mining camps around Keystone; mining tools & equipment; Victorian period artifacts; Carrie Ingalls & family person artifacts, former residents; Bower Family Band collection; photographs; books; records; equipment from Mount Rushmore memorial; rock & gem collection.
Research Fields: mining camps; mineralogy; local genealogy; historic sites.
Facilities: one-room school. Booklets & prints for sale.
Activities: self-guided walking tours; lectures; films; loan exhibitions; living history one-room school for grades 1-7. Museum Sponsors: Carrie Ingalls Day in August.
Publications: newsletter, Holy Terror Tattler; book, Keystone & Its Colorful Characters.
Hours & Admission Prices: May 15-Sept. 15 Mon.-Sat. 11-4. No charge; donations accepted. &

Attendance: 3,674 (accurate)
Membership: Child $1; Adult $5; Charter Single $10; Charter Couple $15.

MOUNT RUSHMORE NATIONAL MEMORIAL, 13000 Hwy. 244, Keystone, SD 57751-0268. Mailing Address: 13000 Hwy. 244, Bldg. 31, Ste. 1, Keystone, SD 57751-0268. Tel.: 605-574-3163. Fax: 605-574-2307.
E-mail: zane_martin@nps.gov
Web Site: www.nps.gov/moru/
Founded: 1925.
Key Personnel: Supt., Gerard Baker; Chief Ranger, Mike Pflaum; Chief Interpretation, Judy Olson; Park Cur., Bruce Weisman.
Governing Authority: federal government. Parent Institution: National Park Service, Washington, DC. Tax-exempt.
Institution Type/Description: Park Museum: massive granite sculpture, carved into a mountainside, memorializing the likenesses of four American Presidents; Washington, Jefferson, Theodore Roosevelt & Lincoln.
Collections: 3,500 tools & equipment; 2,000 historic photographs & negatives; 40,000 pieces of archival materials.
Research Fields: background & construction of Mount Rushmore.
Facilities: library of historic papers & journals available for use on premises; 1,300-seat outdoor amphitheater; 5,200 sq. ft. exhibit space; restaurant. Museum-related items & literature for sale.
Activities: lectures; films; summer evening amphitheater program; ranger conducted talks in studio; permanent & temporary exhibits; film loan service.
Publications: brochures in five languages, English, Spanish, French, German & Japanese and Braille.
Hours & Admission Prices: Information Center: May 29-Aug. 14 daily 8am-10pm; Aug. 15-Sept. daily 8am-9pm; Oct.-May daily 8-5. Office: Mon.-Fri. 8-4:30. Sculptor's Studio: May 15-May 22 daily 9-4; May 23-Aug. 14 daily 8-7; Aug. 15-Oct. 5 daily 8-5. Donations accepted. Parking fee: $10 annual pass; $50 per bus per entry. &

Attendance: 2,754,261 (accurate)

NATIONAL PRESIDENTIAL WAX MUSEUM, Hwy. 609 16-A, Keystone, SD 57751. Mailing Address: P.O. Box 238, Keystone, SD 57751-0238. Tel.: 605-666-4455. Fax: 605-666-4455.
Founded: 1970.
Key Personnel: Museum Shop Mgr., Michelle Anderson.
Governing Authority: company organized for profit.
Institution Type/Description: Wax Museum.
Collections: over 100 life-size wax figures, including all of the U.S. presidents & other famous Americans.
Facilities: Gift items for sale.
Activities: lectures; taped tour.
Publications: brochures.
Hours & Admission Prices: April-May & Sept.-Oct. 9-5; Memorial Day to Labor Day 9-8. Adults $10, senior citizens $8, children 6-12 $7; children under 6 no charge.
Attendance: 28,000

RUSHMORE BORGLUM STORY, 342 Winter St., Keystone, SD 57751-2036. Tel.: 605-666-4448. Fax: 605-666-4482.
E-mail: borglum@gwtc.net
Web Site: www.rushmoreborglum.com
Institution Type/Description: History Museum.
Collections: life of Gutzon Borglum's, who at age 60 began carving Mt. Rushmore; paintings, sculptures & personal artifacts.
Facilities: Museum-related items for sale.
Activities: video.

Hours & Admission Prices: May & Sept. call for hours; June-Aug. daily 8:30-4:30. Adults $10. &

Kimball

SOUTH DAKOTA TRACTOR MUSEUM, 501 S. West St., Kimball, SD 57355. Mailing Address: P.O. Box 418, Kimball, SD 57355-0418. Tel.: 605-778-6513.
E-mail: rbickner@midstatesd.net
Web Site: sdtractormuseum.home.comcast.net
Founded: 2000.
Key Personnel: Pres. (V), Maynard Konechne; Sec., Maxine Bickner; Museum Shop Mgr., Dale Stanek.
Personnel Profile: Full-Time Volunteers 31; Part-Time Volunteers 37.
Governing Authority: private; nonprofit organization. Tax-exempt: 501(c)(3).
Institution Type/Description: Agriculture & Antiques Museum.
Collections: South Dakota agriculture; farm machinery; household items. Historic Buildings: 1930s one-room country school; blacksmith shop.
Facilities: library. Craft items, books & museum-related items for sale.
Activities: guided tours; loan, participatory, traveling & temporary exhibitions; broadcast programs; bus tour groups. Annual Events: Independence Day Parade; Desperado Days with Kimball Chamber in July; Homecoming Parade; guided tours given by volunteers.
Hours & Admission Prices: Memorial Day to Oct. 1 Mon.-Sat. 9-5, Sun. 1-5; other times by appointment. No charge; donations accepted. &
Attendance: 2,400 (accurate)
Membership: Adult $5.

Kyle

OGLALA LAKOTA COLLEGE HISTORICAL CENTER, 3 Mile Creek Rd., Kyle, SD 57752-0310. Mailing Address: P.O. Box 490, Kyle, SD 57752-0490. Tel.: 605-455-6000.
E-mail: mpourier@olc.edu
Web Site: www.olc.edu/about/historical_center
Founded: 1971.
Governing Authority: Parent Institution: Oglala Lakota College. Tax-exempt.
Institution Type/Description: Historical Center.
Collections: local history & culture; photographs & artwork from the early 1800s to the Wounded Knee Massacre in 1890.
Hours & Admission Prices: June-Sept. Mon.-Sat. 9-5. No charge; donations accepted. &

Lake City

FORT SISSETON HISTORIC STATE PARK, 11907 434th Ave., Lake City, SD 57247-6153. Mailing Address: South Dakota Game Fish and Parks, 523 E. Capitol Ave., Pierre, SD 57501. Tel.: 605-448-5474. Fax: 605-448-5572. Facebook: Ft. Sisseton.
E-mail: fortsisseton@state.sd.us
Web Site: gfp.sd.gov/state-parks/directory/fort-sisseton/
Founded: 1972.
Congressional District: 1
Key Personnel: Park Mgr., Katie Ceroll.
Governing Authority: state. Parent Institution: South Dakota Dept. Game, Fish, & Parks. Subsidiary Institution: Div. of Parks & Recreation. Tax-exempt.
Institution Type/Description: Historical Museum: located on site of c.1864 Ft. Wadsworth (Ft. Sisseton).
Collections: Indian artifacts; Civil War weapons & military uniforms; Sam Brown family collection.
Research Fields: Fort Sisseton history.
Facilities: audio-visual room; campground.
Activities: guided tours; films & slides; formally organized education programs for children. Park Sponsors: annual Fort Sisseton Historical Festival in June.
Publications: Fort Sisseton; Chilson's History of Fort Sisseton.
Hours & Admission Prices: Memorial Day-Labor Day daily 10-6; Sept. Mon.-Fri. 8-4:30; special group tours on request. Call for fees. &
Attendance: 65,000 (estimated)

Lake Norden

SOUTH DAKOTA AMATEUR BASEBALL HALL OF FAME, 519 Main Ave., Lake Norden, SD 57248. Mailing Address: P.O. Box 80, Lake Norden, SD 57248-0080. Tel.: 605-785-3553. Fax: 605-785-3315.
Founded: 1976.
Key Personnel: Pres., Scott Fiedler; Vice Pres., Jerry Des Lauriers; Exec. Sec. & Cur., Rusty Antonen.

Governing Authority: nonprofit organization. Tax-exempt: 501(c)(3).
Institution Type/Description: Sports Museum.
Collections: pictorial history of amateur baseball in South Dakota; artifacts; equipment; historical documents; Hall of Fame inductees displayed.
Research Fields: amateur baseball in South Dakota.
Activities: concerts; permanent exhibitions.
Hours & Admission Prices: May to Sept. daily 9-7; other times by appointment. No charge; donations accepted. &

Lead

BLACK HILLS MINING MUSEUM, 323 W. Main, Lead, SD 57754-1604. Tel.: 605-584-1605.
E-mail: bhminingmuseum@rushmore.com
Web Site: www.mining-museum.blackhills.com
Founded: 1986.
Congressional District: 1
Key Personnel: Pres., Todd Duex; Dir., Mallory Everett.
Personnel Profile: Full-Time Paid 1; Part-Time Paid 1; Part-Time Volunteers 12.
Governing Authority: nonprofit organization. Tax-exempt: 501(c)(3).
Institution Type/Description: Mining History Museum.
Collections: photographs; historic records from Black Hills mines, 1876-1940s.
Research Fields: Black Hills mining history, 1875-modern times.
Facilities: library available to public for research in facility only; archives; 12,000 sq. ft. exhibit space; 30-seat theater. Museum-related items for sale.
Activities: guided tours; gold panning.
Publications: biannual newsletter, Black Hills Mining Museum; newspaper, The Gold Belt Miner.
Hours & Admission Prices: May-Sept. daily 9-5; Oct.-April call for hours. Call for admission prices. &
Attendance: 20,000 (estimated)
Membership: Individual & Couple $5; Friend $100; Contributor $250; Donor $500; Sponsor $1,000; Benefactor $2,500; Major Contributor $5,000; Gold Star Contributor $10,000.

HOMESTAKE GOLD MINE VISITOR CENTER, 160 W. Main St., Lead, SD 57754-1362. Tel.: 605-584-3110.
E-mail: hvc@rushmore.com
Web Site: www.homestakevisitorcenter.com
Key Personnel: Exec. Dir., Melissa Johnson
Institution Type/Description: Mining History Museum.
Collections: history of mining including hoisting, crushing & milling; open mine pit.
Facilities: Museum-related items for sale.
Activities: guided tour; pan for gold; video.
Hours & Admission Prices: May-Sept. daily 8-6; groups by appointment. Tours: adults $7.50, seniors $6.75, students $6.50; discounts to groups of 10 or more and families with 4-6 members.

PRESIDENTS PARK SCULPTURE GARDEN, 11249 Presidents Park Loop, Lead, SD 57754-3846. Mailing Address: 104 S. Galena St., Lead, SD 57754-1674. Tel.: 605-584-9925.
Key Personnel: Mgr., Dave Olmstead
Institution Type/Description: History Museum.
Collections: 20 ft. tall sculptures of all 42 Presidents.
Facilities: visitors center; cafe. Museum-related items for sale.
Hours & Admission Prices: Daily 9 to sunset; weather permitting. Adults $8, seniors 60 & over $6.50, children 5-15 $6; discounts to military & groups; children 5 & under no charge. &

Lemmon

GRAND RIVER MUSEUM, 114 10th St. W., Lemmon, SD 57638-2202. Tel.: 605-374-3911 & 7574.
E-mail: grmuseum@sdplains.com
Web Site: www.grandrivermuseum.org/index.htm
Founded: 1998.
Key Personnel: Pres. (V), Stuart T. Schmidt; Dir. & Museum Shop Mgr., Phyllis Schmidt.
Personnel Profile: Part-Time Paid 1; Part-Time Volunteers 5.
Governing Authority: Tax-exempt.
Institution Type/Description: History Museum.
Collections: Grand River area history; animal & plant fossils; Native American people & culture; area ranching.
Facilities: theatre. Museum-related items for sale.
Activities: special programs; videos.
Publications: quarterly newsletter.

Hours & Admission Prices: May-Sept. Mon.-Sat. 9-6, Sun. 12-5. No charge; donations accepted.
Attendance: 5,500 (estimated)

PETRIFIED WOOD PARK & MUSEUM, 500 Main St., Lemmon, SD 57638-1523. Tel.: 605-374-3964.
Web Site: www.lemmonsd.com/petrified.html
Key Personnel: Dir., Carolyn Penfield
Institution Type/Description: Geological & Historical Museum.
Collections: artifacts made from petrified wood; sculptures; personal artifacts.
Hours & Admission Prices: Museum: Memorial Day-Labor Day daily 9-5. No charge; donations accepted.

Madison

KARL E. MUNDT HISTORICAL & EDUCATIONAL FOUN-DATION, Karl Mundt Library, Dakota State University, Madison, SD 57042. Mailing Address: 820 N. Washington Ave., Madison, SD 57042-0483. Tel.: 605-256-5211.
Web Site: www.departments.dsu.edu/library/archive/archives.htm
Founded: 1963.
Personnel Profile: Full-Time Paid 1.
Institution Type/Description: History Museum.
Collections: Senator Mundt's personal & political life; documents; films; photographs; tapes; scrapbooks.
Hours & Admission Prices: Research: Mon.-Fri. 8:30-4. No charge. Closed legal holidays.

PRAIRIE VILLAGE, W. Hwy. 34, Madison, SD 57042. Mailing Address: P.O. Box 256, Madison, SD 57042-0256. Tel.: 605-256-3644; 800-693-3644.
E-mail: prairiev@rapidnet.com
Web Site: www.prairievillage.org
Founded: 1966.
Congressional District: 1
Key Personnel: Pres., George Lee; Vice Pres., Kevin Bowman; Museum Mgr., Stan Rauch.
Personnel Profile: Part-Time Paid 6; Part-Time Volunteers 100.
Governing Authority: nonprofit organization. Prairie Historical Society, Inc. Tax-exempt: 501(c)(3).
Institution Type/Description: Village Museum.
Collections: furniture & furnishings; steam tractors & other steam equipment; threshing machines, locomotive; rocks; 3 steam trains. Historic Structures: 1906 church; 1912 Socialist Hall; 1877 claim shanty; 1878 Old Madison Hotel; 1880s country school; jail; print shop; barbershop; dentist office; 1893 steam carousel.
Facilities: library; 350-seat auditorium; classrooms; gift shop; large picnic shelter; 234 camp sites; 2 shower houses.
Activities: elementary school tours, train & carousel rides. Museum Sponsors: steam & horse threshing at jamboree in August.
Publications: weekly newspaper column, Madison Daily Leader.
Hours & Admission Prices: Village: Mother's Day weekend to Labor Day Mon.-Sat. 10-5, Sun. 11-5. Train Rides: Mother's Day to Sept. 1 Sat. Adults $5; discounts for senior citizens, group tours. Season Pass: $25.
Attendance: 40,000 (estimated)
Membership: 100 hours of volunteer work or $500 contribution.

SMITH-ZIMMERMANN HERITAGE MUSEUM, 221 N.E. 8th St., Madison, SD 57042-1639. Tel.: 605-256-5308.
E-mail: smith.zimmermann@dsu.edu
Web Site: www.smith-zimmermann.dsu.edu
Founded: 1952.
Congressional District: 1
Key Personnel: Pres., Susan Larsen; Coord., Cynthia Mallery.
Personnel Profile: Part-Time Paid 1; Part-Time Volunteers 25.
Governing Authority: Parent Institution: Lake County Historical Society. Tax-exempt.
Institution Type/Description: History Museum.
Collections: Lake County, SD history & pioneer artifacts from 1870-1970s; murals depicting pioneer life.
Research Fields: cultural history of eastern South Dakota; genealogical records; pictorial images
Facilities: period rooms. Books for sale.
Activities: guided tours; changing exhibits; brown-bag programs; educational programs; ethnic cooking programs.
Publications: quarterly newsletter; research reports.
Hours & Admission Prices: Tues.-Fri. 1-4:30; tours by appointment. No charge; donations accepted.
Attendance: 5,598 (accurate)

Membership: Family $15.

McLaughlin

MAJOR JAMES MCLAUGHLIN HERITAGE CENTER, Main St., McLaughlin, SD 57642. Mailing Address: P.O. Box 228, McLaughlin, SD 57642-0642. Tel.: 605-823-4590.
Founded: 1989.
Key Personnel: Pres. (V) & Museum Shop Mgr., Sharon Walker.
Governing Authority: Tax-exempt.
Institution Type/Description: History Museum.
Collections: local history; agriculture; Native American; Mahto Post Office; country school; James McLaughlin family history; documents.
Hours & Admission Prices: May-Sept. Mon.-Fri. 12-5. No charge; donations accepted.
Attendance: 75 (estimated)

Midland

MIDLAND PIONEER MUSEUM, Main St., Midland, SD 57552. Mailing Address: 25135 Capa Rd., Midland, SD 57552-3201. Tel.: 605-843-2150.
E-mail: kry@gwtc.net
Web Site: www.gwtc.net/~kry
Founded: 1974.
Key Personnel: C.E.O. & Dir., Janice D. Bierle.
Personnel Profile: Part-Time Paid 1.
Governing Authority: Tax-exempt.
Institution Type/Description: History Museum.
Collections: pioneer machinery; period artifacts; photographs.
Hours & Admission Prices: June-Aug. Mon.-Tues. & Fri. 1:30-4; other times by appointment. No charge; donations accepted.

Milbank

GRANT COUNTY HISTORICAL MUSEUM, Third Ave. & Third St., Milbank, SD 57252. Mailing Address: P.O. Box 201, Milbank, SD 57252-0201. Tel.: 605-432-9332.
E-mail: ssteeg65@itctel.com
Web Site: www.grantcountysdhistory.com
Founded: 1970.
Key Personnel: Dir. & Cur., Sharon Steege; Pres. (V), Arlo Levisen.
Personnel Profile: Part-Time Paid 1; Part-Time Volunteers 5.
Governing Authority: Tax-exempt.
Institution Type/Description: Historical Museum. Listed on the National Register of Historic Buildings.
Collections: artifacts from Grant County; photographs; plat maps; books.
Research Fields: genealogy.
Publications: newsletter.
Hours & Admission Prices: Memorial Day-Labor Day Sun. 2-5; other times by appointment. No charge; donations accepted.
Attendance: 250 (accurate)
Membership: Student $5; Adult $10; Group & Organization $20; Lifetime $100.

Mission

SICANGU HERITAGE CENTER, Sinte Gleska University, Antelope Lake Campus, Mission, SD 57555. Mailing Address: P.O. Box 675, Mission, SD 57555-0675. Tel.: 605-856-8211.
E-mail: heritagecenter@sintegleska.edu
Web Site: www.sintegleska.edu/heritage-center.html
Key Personnel: Dir., Marcella Cash; Museum Collections Mgr., Keli Herman; Gen. Asst., Terry Gray.
Governing Authority: bd. of regents. Parent Institution: Sinte Gleska University.
Institution Type/Description: Heritage Center.
Collections: Sicangu culture & history; personal artifacts.
Activities: group tours.
Hours & Admission Prices: Mon.-Fri. 9-5. No charge; donations accepted. Closed holidays.

Mitchell

CARNEGIE RESOURCE CENTER, 119 W. 3rd, Mitchell, SD 57301-3410. Mailing Address: P.O. Box 263, Mitchell, SD 57301-0263. Tel.: 605-996-3209.
Web Site: mitchellcarnegie.com
Formerly: Oscar Howe Art Center & YWCA

Founded: 2006.
Key Personnel: Pres. (V), Lyle W. Swenson.
Personnel Profile: Part-Time Volunteers 15.
Governing Authority: Parent Institution: Mitchell Area Historical Society. Subsidiary Institution: Mitchell Area Genealogical Society. Tax-exempt.
Institution Type/Description: Art Center.
Collections: Corn Palace historical items from 1892-present; dome with Oscar Howe mural; local historical & genealogical materials.
Hours & Admission Prices: Mon.-Sat. 1-5. No charge; donations accepted. Closed holidays.
Membership: Single $10; Family $15.

DAKOTA DISCOVERY MUSEUM, 1300 McGovern Ave., Mitchell, SD 57301-7901. Mailing Address: P.O. Box 1071, Mitchell, SD 57301-7071. Tel.: 605-996-2122. Fax: 605-996-0323. Facebook: Dakota Discovery Museum.
E-mail: history@dakotadiscovery.com
Web Site: www.dakotadiscovery.com
Formerly: Middle Border Museum & Oscar Howe Art Center
Founded: 1939.
Congressional District: 17
Key Personnel: Exec. Dir., Lori Holmberg; Pres., Dianne Carr.
Personnel Profile: Full-Time Paid 1; Part-Time Paid 1; Part-Time Volunteers 48.
Governing Authority: society. Parent Institution: Friends of the Middle Border, Inc. Tax-exempt: 501(c)(3).
Institution Type/Description: Historic Village Museum and Art Center.
Collections: items pertaining to the Middle Border (ND, SD & portions of adjoining states) settlement history exhibits; 1600-1939 era; American Indian artifacts; Charles Hargens Studio & Gallery; LeLand Case Office & Gallery; Oscar Howe, Harvey Dunn, Charles Greener, James Earle Fraser art collections; 4 historical buildings.
Research Fields: early history of ND, SD, NE & surrounding region.
Activities: guided tours; lectures; gallery talks; education programs for children & adults; permanent exhibitions; children's hands on activities with changing themes. Annual Events: Old Fashion Independence Day in July; Haunted Village in October; Victorian Christmas in December.
Publications: quarterly newsletter; tour guide.
Hours & Admission Prices: May & Sept. Mon.-Tues. & Thurs.-Sat. 9-5; June-Aug. Mon.-Tues. & Thurs.-Sat. 9-7; Oct.-April Tues.-Fri. 10-4, Sat. 1-4. Adults $7, senior citizens $6, children 6-17 $3; discounts to groups, and AAM, ICOM & AAA members; members no charge. Closed New Year's Eve & Day; Easter Mon.; Boxing Day; Thanksgiving; Christmas. &
Attendance: 7,000 (accurate)
Membership: Individual $25; Individual Premier $35; Family & Grandparent $40; Patron $100.

MITCHELL PREHISTORIC INDIAN VILLAGE & MUSEUM, (M), 3200 Indian Village Rd., Mitchell, SD 57301. Tel.: 605-996-5473.
E-mail: info@mitchellindianvillage.org
Web Site: www.mitchellindianvillage.org
Institution Type/Description: Archaeological Site.
Collections: reconstructed earth lodge; exhibits focusing on trade networks, pottery, tools & spear points.
Hours & Admission Prices: April & Oct. Mon.-Fri. 9-4; May & Sept. daily 9-4; Memorial Day-Labor Day daily 8-6. Adults $6, seniors 60 & over $5, children 6-18 $4; children 5 & under no charge.

Mobridge

KLEIN MUSEUM, 1820 W. Grand Crossing, W. Hwy. 12, Mobridge, SD 57601-1114. Tel.: 605-845-7243.
E-mail: kleinmuseum@westriv.com
Web Site: mobridgekleinmuseum.com
Founded: 1976.
Congressional District: 50
Key Personnel: Chm. (V), Judy Curran; Pres. (V), Sally Perman; Museum Shop Mgr., Diane Kindt.
Personnel Profile: Full-Time Paid 1; Part-Time Paid 1; Part-Time Volunteers 40.
Governing Authority: nonprofit organization; bd. of directors. Tax-exempt: 501(c)(3).
Institution Type/Description: General Museum.
Collections: pioneer & Native American artifacts. Historic Buildings: 1920s house; post office; 1920 school; tool shed.
Research Fields: local histories; photographs.
Facilities: Indian pottery & jewelry, books & museum-related items for sale.
Activities: guided tours; permanent & temporary exhibitions.

Publications: Klein Museum: More Than A Museum - Taking South Dakota Home With You.
Hours & Admission Prices: April-Oct. Mon. & Wed.-Fri. 9-5, Sat.-Sun. 1-5. Adults $3, students $2; members no charge. &
Attendance: 4,000 (estimated)
Membership: Single $17; Family $30; Business $45.

Murdo

1880 TOWN, I-90 Exit 170, Murdo, SD 57559. Mailing Address: P.O. Box 507, Murdo, SD 57559-0507. Tel.: 605-344-2236. Fax: 605-344-2236.
E-mail: info@1880town.com
Web Site: www.1880town.com
Founded: 1972.
Key Personnel: Mgr., Richard Hollinger.
Personnel Profile: Part-Time Paid 30.
Institution Type/Description: General Museum: built as a movie set but never used for filming.
Collections: period artifacts; Dances with Wolves movie props; 30 buildings.
Hours & Admission Prices: May & Oct. daily 8am to sunset; June-Aug. 6am-9pm; Sept. 7am to sunset. Adults $12, senior citizens $10, teens 13-18 $7, children 6-12 $5; discounts to groups; handicapped & children under 5 no charge. &

PIONEER AUTO MUSEUM, 503 E. 5th St., Murdo, SD 57559. Mailing Address: P.O. Box 76, Murdo, SD 57559-0076. Tel.: 605-669-2691. Fax: 605-669-3217.
E-mail: pas@pioneerautoshow.com
Web Site: www.pioneerautoshow.com
Founded: 1953.
Key Personnel: C.E.O., Dir. & Cur., Dave Geisler; Deputy Dir. & Cur., David M. Geisler; Museum Shop Mgr., Tennille Edwards.
Personnel Profile: Full-Time Paid 1; Part-Time Paid 14.
Governing Authority: company organized for profit.
Institution Type/Description: Transportation Museum & Antique Town: located at the northern head of the Texas Cattle Trail.
Collections: over 200 antique & classic cars; early motorcycles; Elvis Presley's motorcycle; bicycles; tractors; farm equipment; buggies; horse-drawn vehicles; musical instruments; 1906 Case steam engine; costumes; toys & dolls; Zeitner rocks, gems & fossils; circus artifacts; furniture; collectibles; early guns. Town Buildings: Jack's Jewelry; Homesteader Claim Shack; Murdo Bank; Blacksmith Shop; Barber Shop; Jail. Historic Buildings: 1906 Milwaukee Railroad Depot & Caboose; 1906 pioneer church; 1806 one-room country school; 1906 general store; butcher's shop.
Research Fields: automobiles.
Facilities: 50-seat cafeteria. Gifts & museum-related items for sale.
Activities: auctions; sales; Swap Meet.
Hours & Admission Prices: Memorial Day-Labor Day 7-10; Winter: 9-6. Adults $10, members $9, children 5-13 $5; discounts to groups, AAM, AAA & ICOM members. Closed New Year's Day; Easter; Thanksgiving; Christmas. &
Attendance: 126,000 (estimated)

Newell

NEWELL MUSEUM, 108 3rd St., Newell, SD 57760. Mailing Address: P.O. Box 433, Newell, SD 57760-0433. Tel.: 605-456-1310. Fax: 605-456-9820.
E-mail: newellmuseum@yahoo.com
Web Site: www.cityofnewell.com
Founded: 1983.
Congressional District: 1
Key Personnel: Chm., Donald Ericson; Vice Chm., David Morrel; Treas., Annitta Stolnack; Cur., Linda Velder; Archivist, Lauren Babb; Sec., Sharyl Scott.
Personnel Profile: Full-Time Paid 1; Part-Time Volunteers 3; Interns 1.
Governing Authority: municipal government; nonprofit. Parent Institution: Town of Newell. Tax-exempt.
Institution Type/Description: History Museum.
Collections: household items; clothing; Native American artifacts & culture; fossils; antique toys & dolls; musical instruments; restored one-room schoolhouse; barn; buggy. Historic Buildings: 1890 one-room log cabin; 1911 church.
Research Fields: 1874 Custer expedition; travels of Father DeSmet in Dakota Territory; local pioneer family histories; fur trade history; Orman Dam construction (1903-1912); Gen. George Crook's Horse Meat March after the Battle of Slim Buttes (1876); cattle drives & cowboys; rural schools in Butte County; Lewis & Clark Corp. of Discovery; Murder of Father Arthur Belknap, Lead, SD 1921; Congregational Church History 1907-2006;

Newell Methodist Church History 1910-1937; All Trails Lead to Deadwood, Prohibition; Homesteading in Butte County 1879-1917; area cemeteries.
Facilities: library.
Activities: guided tours; participatory & temporary exhibitions. Annual Event: Labor Day Weekend Open House.
Hours & Admission Prices: May 30-Sept. Tues.-Sat. 1-5; other times by appointment. No charge; donations accepted. &
Attendance: 670 (accurate)

Oldham

LORIKS PETERSON HERITAGE HOUSE, 108 E. Williams St., Oldham, SD 57051-7216. Mailing Address: 21617 443rd Ave., Oldham, SD 57051-7216. Tel.: 605-482-8640.
Web Site: www.sdmuseums.org
Founded: 1975.
Key Personnel: Vice Pres., Patricia Folsland
Governing Authority: Tax-exempt.
Institution Type/Description: Historic House Museum.
Collections: period farm machinery; pioneer artifacts; blacksmith equipment.
Hours & Admission Prices: Memorial Day to Labor Day Sun. 1-4; other times by appointment. No charge; donations accepted. &
Attendance: 310 (estimated)

Philip

MINUTEMAN MISSILE NATIONAL HISTORIC SITE, 21280 SD Hwy. 240, Philip, SD 57567-7102. Tel.: 605-433-5552. Fax: 605-433-5558.
Web Site: www.nps.gov/mimi
Institution Type/Description: National Historic Site.
Collections: Minuteman history & artifacts.
Hours & Admission Prices: Visitor Contact Station: Memorial Day to Labor Day Mon.-Sat. 8-4:30; Sept.-May Mon.-Fri. 8-4:30.

PRAIRIE HOMESTEAD, Exit 131 off Interstate 90 Hwy. 240, Philip, SD 57567-7007. Mailing Address: 21070 SD Hwy. 240, Philip, SD 57567-7007. Tel.: 605-433-5400.
Web Site: www.prairiehomestead.com
Founded: 1962.
Key Personnel: Owner Operator, Grady Crew; Owner Operator, Bernice Crew; Museum Shop Mgr., Heidi Porch.
Personnel Profile: Part-Time Paid 13.
Governing Authority: individual operation.
Institution Type/Description: Historic House: 1909 sod dugout, original home of Mr. & Mrs. Ed Brown.
Collections: portion of original furnishings & other furnishings typical of the Sodbusters in the area.
Activities: guided tours.
Publications: books, Homesteading Era Lesson Plan; Prairie Homestead.
Hours & Admission Prices: May-Oct. dawn to dusk. Adults $7, seniors $6.30, children 10-17 $6; discount to AAM & ICOM members; children under 10 no charge when accompanied by an adult. &
Attendance: 15,000 (estimated)

Piedmont

PETRIFIED FOREST OF THE BLACK HILLS, 8228 Elk Creek Rd., Piedmont, SD 57769-7208. Tel.: 605-787-4560; 800-846-2267. Fax: 605-787-6477.
E-mail: info@elkcreekresort.net
Web Site: www.elkcreekresort.net
Founded: 1929.
Key Personnel: Dir. & Museum Shop Mgr., Arvid Scott; Museum Shop Mgr., Tim Scott.
Personnel Profile: Full-Time Paid 5; Part-Time Paid 1.
Governing Authority: private.
Institution Type/Description: Natural History Museum.
Collections: cut & polished petrified wood; oligocene animal fossils; minerals; excavation site of dinosaur & barosaurus was found & taken to Yale Univ. by Dr. Marsh in 1889.
Facilities: Museum-related items for sale.
Activities: self guided tours with audio stations; trail maps for walking tour; lectures.
Publications: video, The Black Hills of South Dakota: Geological Gem of the West.
Hours & Admission Prices: May & Oct. 15 daily 9-5; June-Aug. daily 8:30-6. Adults $6.50, senior citizens & youths 13-18 $5, children 6-12 $4;

discounts to tour & group buses, AAM, AARP & ICOM members; children under 5 no charge. (prices include state sales tax) &
Attendance: 10,000 (accurate)

Pierre

SOUTH DAKOTA DISCOVERY CENTER, 805 W. Sioux Ave., Pierre, SD 57501-1858. Tel.: 605-224-8295. Fax: 605-224-2865. Facebook: South Dakota Discovery Center.
E-mail: kristiemaher@sd-discovery.com
Web Site: www.sd-discovery.com
Founded: 1989.
Congressional District: 1
Key Personnel: C.E.O., Kristie Maher; Pres. (V), Lynn Beck; Educational Dir., Sue Douglas; Special Programs, Anne Lewis; Museum Shop Mgr., Uncle Matt; Community Wellness Coord., Danette Jarzab.
Personnel Profile: Full-Time Paid 2; Part-Time Paid 10; Part-Time Volunteers 30; Interns 3.
Operating Income: 400,000
Governing Authority: private; nonprofit organization. Tax-exempt: 501(c)(3).
Institution Type/Description: Science Museum.
Collections: hands-on science exhibits.
Major Exhibits: Peanuts Naturally, Summer 2014 (T).
Facilities: classroom; 9,000 sq. ft. exhibit space; planetarium; meeting room; classrooms. Museum-related items for sale.
Activities: participatory exhibits; in-house workshops; teacher's workshops; math & science programs; traveling exhibits.
Publications: quarterly program guide; e-news.
Hours & Admission Prices: Memorial Day-Labor Day Mon.-Sat. 10-5, Sun. 1-5. Winter Sun.-Fri. 1-5, Sat. 10-5. Adults $4, children $3; discounts to AAA, ASTC members, museum members & groups. Closed New Year's Day; Good Friday; Easter; Thanksgiving; Christmas Eve & Day.
Attendance: 20,000 (accurate)
Membership: Individual $20; Family or Grandparent & Grandchildren $40; Benefactor $150; Friends of Discovery Center $250 & up.

SOUTH DAKOTA NATIONAL GUARD MUSEUM, 301 E. Dakota Ave., Pierre, SD 57501-3225. Tel.: 605-224-9991.
E-mail: sonja.johnson@state.sd.us
Web Site: ngmuseum.sd.gov
Founded: 1980.
Key Personnel: Dir., Sonja Johnson; Cur., Seb Axtman.
Governing Authority: nonprofit organization.
Institution Type/Description: Military Museum.
Collections: Civil War, Spanish American War, WWI & WWII, Korean War, Desert Storm, & Bosnian Peace Keeping Mission memorabilia; historical documents; military equipment, records, & relics; A-7-D jet; Sherman tank; armored personnel carrier; 75mm cannon; 105mm Howitzer; anti-aircraft guns; Iraqi Freedom and Enduring Freedom.
Hours & Admission Prices: Mon.-Fri. 9-4; other times by appointment. No charge; donations accepted. &
Attendance: 3,000 (accurate)

SOUTH DAKOTA STATE ARCHIVES, 900 Governors Dr., Pierre, SD 57501-2200. Tel.: 605-773-3458 & 3804. Fax: 605-773-6041.
E-mail: Archref@state.sd.us
Web Site: www.sdhistory.org
Founded: 1974.
Congressional District: 1
Key Personnel: State Archivist, Chelle Somsen; Archivist, Virginia Hanson; Archivist, Sara Casper; Archivist, Matthew Reitzel; Microfilm Supvr., Chris Harmon; Microfilm Asst., Marcia Wickett; Research Room Admin., Ken Stewart.
Personnel Profile: Full-Time Paid 8; Part-Time Volunteers 8.
Governing Authority: state. Parent Institution: South Dakota State Historical Society, Pierre, SD 57501. Tax-exempt.
Institution Type/Description: State Archives.
Collections: state agency records; South Dakota local government records; inter-state & federal government records directly related to South Dakota history & culture; photographs; maps; manuscripts.
Research Fields: South Dakota; Upper Midwest; North Plains; Prairie Culture; High Plains.
Facilities: 15,000-vol. library of South Dakota published documents & reference books, available for research; reading room.
Activities: internships for undergraduate students.
Hours & Admission Prices: Mon.-Fri. & first Sat. of month 9-4:30. No charge; donations accepted. Closed national & state holidays. &
Attendance: 5,200 (accurate)

SOUTH DAKOTA STATE HISTORICAL SOCIETY, (M), 900 Governors Dr., Pierre, SD 57501-2200. Mailing Address: Cultural Heritage Center, 900 Governors Dr., Pierre, SD 57501-2217. Tel.: 605-773-3458. Fax: 605-773-6041.
E-mail: jay.vogt@state.sd.us
Web Site: www.sdhistory.org
Founded: 1901.
Congressional District: 1
Key Personnel: Exec. Dir., Jay D. Vogt; Pres. (V) & Chm. (V), Brad Tennant; Dir. State Archaeological Research Center, Jim Haug; Dir. Research & Publications, Nancy Tystad Koupal.
Personnel Profile: Full-Time Paid 43; Part-Time Paid 6; Part-Time Volunteers 65.
Governing Authority: state. Parent Institution: State of South Dakota. Programs: State Archives, Pierre; Museum of the South Dakota Historical Society, Pierre; State Historical Preservation Center, Pierre; State Archaeological Research Center, Rapid City; Research & Publishing, Pierre. Tax-exempt.
Institution Type/Description: State Historical Society.
Collections: Native American artifacts; South Dakota cultural history; archives.
Research Fields: Native American history; South Dakota history.
Facilities: library of rare books; state archives; research room; observation gallery. Gift items for sale.
Activities: guided tours; permanent, temporary & traveling exhibits; workshops; educational programs; suitcase education kits; history conference.
Publications: quarterly, South Dakota History; newsletter, South Dakota History Notes.
Hours & Admission Prices: Memorial Day to Labor Day Mon.-Sat. 9-6:30, Sun. 1-4:30; Sept.-May Mon.-Sat. 9-4:30, Sun. 1-4:30. Adults $4, senior citizens 60 & over $3; discounts to AAM members; members no charge. Closed New Year's Day; Easter; Thanksgiving; Christmas. &
Attendance: 19,899 (accurate)
Membership: Student $30; Individual $35; Family & History-Related Group $45; Foreign $60; Heritage Circle $100 & up.

Pine Ridge

THE HERITAGE CENTER, 100 Mission Dr., Pine Ridge, SD 57770-2100. Tel.: 605-867-5491. Fax: 605-867-1291.
E-mail: heritagecenter@redcloudschool.org
Web Site: www.redcloudschool.org
Founded: 1968.
Congressional District: 1
Key Personnel: Dir., Peter Strong; Pres. (V), Rev. Peter Klink, S.J.; Chm. (V), Norma Tibbitts; Museum Shop Mgr., Myrtle Cedar Face.
Personnel Profile: Full-Time Paid 4; Part-Time Paid 2; Interns 3.
Governing Authority: nonprofit organization. Parent Institution: Red Cloud Indian School, Inc. Tax-exempt: 501(c)(3).
Institution Type/Description: Native American Art Museum: housed in c.1888 Holy Rosary Mission, scene of battle the day after the Wounded Knee Massacre.
Collections: paintings by Native American artists; star quilt collection; beadwork; quill & pottery collection; Native American print collection; Northwest Coast prints.
Research Fields: Native American art; Lakota artifacts; Lakota culture.
Facilities: 900-vol. library of Native American material & culture, Native American & Western art, available for research on premises only; reading room. Oglala Lakota beadwork & quillwork for sale.
Activities: guided tours; arts festivals; loan, temporary & traveling exhibitions.
Hours & Admission Prices: Mon.-Fri. 9-5. No charge; donations requested. &
Attendance: 12,000 (estimated)

Pollock

POLLOCK VISITORS/INTERPRETIVE CENTER, 110 Main St., Pollock, SD 57648. Mailing Address: P.O. Box 142, Pollock, SD 57648-0057. Tel.: 605-889-2450.
Founded: 2002.
Key Personnel: Chm. (V), Delores Kluckman; Museum Shop Mgr., Vina LaFave.
Personnel Profile: Part-Time Paid 1.
Institution Type/Description: Interpretive Center.
Collections: Native American artifacts; historical tools; handmade clocks; Lewis & Clark pictures & paintings; local history scrapbooks, 1888 to present.
Hours & Admission Prices: Mon.-Tues. & Fri. 9-4, Sat. 12:30-4. No charge; donations accepted. &
Attendance: 250 (estimated)

Rapid City

BEAR COUNTRY U.S.A., 13820 South Hwy. 16, Rapid City, SD 57702-6581. Tel.: 605-343-2290. Fax: 605-341-3206.
E-mail: pabear@bearcountryusa.com
Web Site: www.bearcountryusa.com
Founded: 1972.
Key Personnel: Owner/Operator, Pauline Casey.
Governing Authority: company organized for profit.
Institution Type/Description: Zoo. Drive-through Wildlife Park.
Collections: North American wildlife on display.
Research Fields: reproductive physiology of the American black bear.
Facilities: zoological park. Wildlife-related items for sale.
Activities: guided tours; lectures.
Publications: brochure.
Hours & Admission Prices: May & Sept.-Oct. daily 9-4; June-Aug. daily 8-6; Nov. daily 9-3; groups of 15 or more by appointment. Adults $16, seniors 62 & over and military adult $13, children 5-12 $10, military child $7; discount to groups; children 4 & under no charge. &

BLACK HILLS REPTILE GARDENS, INC., 8955 S. Hwy. 16, Rapid City, SD 57702. Mailing Address: P.O. Box 620, Rapid City, SD 57709-0620. Tel.: 605-342-5873.
E-mail: joe-m@reptilegardens.com
Web Site: www.reptilegardens.com
Founded: 1937.
Key Personnel: C.E.O., Pres. & Dir., Joe Maierhauser; Gen. Mgr. & Vice Pres., Tom Lang; Cur. & Reptile Project Dir., Ken Earnest; Gift Store Mgr., Jeff Oldham; Public Rels., John Brockelsby.
Personnel Profile: Full-Time Paid 20; Part-Time Paid 75.
Governing Authority: individual operation; S corporation.
Institution Type/Description: Reptile Museum.
Collections: 1,000 specimens of over 200 species of reptiles; herpetology.
Research Fields: breeding rare & endangered reptiles; growth in relation to food on snakes.
Facilities: cafe; arboretum; aviary. Gift items for sale.
Activities: lectures; formally organized education programs for children; permanent exhibitions; lectures; animal shows.
Hours & Admission Prices: April-May & Sept.-Oct. daily 9-4; Memorial Day-Labor Day daily 8-6. Adults $16, senior citizens $14.50, children $11; discount to groups; children 4 & under no charge. &
Attendance: 500,000

DAHL ARTS CENTER, 713 Seventh St., Rapid City, SD 57701-3695. Tel.: 605-394-4101. Fax: 605-394-6121.
E-mail: contact@thedahl.org
Web Site: www.thedahl.org
Founded: 1974.
Congressional District: 1
Key Personnel: Exec. Dir., Linda Anderson; Pres. (V), Tim Trithart.
Personnel Profile: Full-Time Paid 8; Part-Time Paid 5; Part-Time Volunteers 50; Interns 2.
Governing Authority: nonprofit organization. Parent Institution: Rapid City Arts Council.
Institution Type/Description: Community Arts Center.
Collections: 200 foot cycloramic oil on canvas mural depicting history of the United States; 140 paintings historically important to Black Hills region; national prints; early Oscar Howe tempera; contemporary artwork by SD area artists.
Facilities: three classrooms; 240-seat theater; children's gallery; two conference rooms.
Activities: temporary exhibition; gallery talks; guided tours; theater productions; concerts; classes; workshops; films; literary events; community events.
Publications: bimonthly newsletter; exhibition brochures; catalogues.
Hours & Admission Prices: Summer: Tues.-Fri. 12-8, Sat.-Sun. 1-5. Adults $2.50; members no charge. Closed major holidays. &
Attendance: 63,000 (estimated)
Membership: Student $15; Senior $20; Individual $35; Household $50; Master $150; Michelangelo $250; Medici $500.

THE JOURNEY MUSEUM, 222 New York St., Rapid City, SD 57701-1199. Tel.: 605-394-6923 & 2249. Fax: 605-394-6940.
E-mail: pchristie@journeymuseum.org
Web Site: www.journeymuseum.org
Founded: 1997.
Key Personnel: Facilities Mgr., Mark Slocum.
Personnel Profile: Full-Time Paid 12; Part-Time Paid 5; Part-Time Volunteers 200; Interns 2.

Governing Authority: Parent Institution: The Museum Alliance of Rapid City. Tax-exempt.
Institution Type/Description: Regional history museum with emphasis on Native Americans & pioneers.
Collections: participating museums and collections include: Museum of Geology, South Dakota School of Mines and Technology; State Archaeological Research Center; Sioux Indian Museum; Minnilusa Pioneer Museum; Duhamel Plains Indians Artifact Collection. Collections tell 2.5 billion years of Black Hills history.
Research Fields: paleontology, geology, archaeology, history, Native American history and culture.
Facilities: planetarium. Museum-related items for sale.
Activities: interactive exhibits; special shows & events.
Publications: exhibition brochures; quarterly newsletter.
Hours & Admission Prices: Summer: daily 9-5; Winter: Mon.-Sat. 10-5, Sun. 1-5. Adults $8.25, seniors 62 & over $7, students 11-17 $6; discounts to military, AAA, AAM & ICOM members; children 10 & under and members no charge. &
Attendance: 31,896 (estimated)
Membership: Student $15; Individual $25; Family $50; Grand Family $100; Lifetime $1,000.

MOTION UNLIMITED MUSEUM & CLASSIC CAR LOT, 6180 S. Hwy. 79, Rapid City, SD 57702-8467. Tel.: 605-348-7373.
E-mail: happymotoring@bluebottle.com
Web Site: www.motionunlimitedmuseum.com
Founded: 1972.
Key Personnel: Owner, Bill Napoli; Owner, Peggy Napoli
Institution Type/Description: Automobile, Motorcycle, & Toy Museum.
Collections: classic cars, pickups, sedans, coups, convertibles & motorcycles; period gas pumps; porcelain, tin & neon signs; posters; photographs; toys; period clothing; over 100 pedal vehicles.
Facilities: Cars & museum-related items for sale.
Hours & Admission Prices: May-Oct. Mon.-Fri. 9-6, Sat. 9-4, Sun. by appointment. Adults $5; discounts to members and AAM & ICOM members; active military and children 12 & under no charge. Closed New Year's Day; Easter; Independence Day; Thanksgiving, Christmas. &
Attendance: 1,500 (accurate)

MUSEUM OF GEOLOGY, SOUTH DAKOTA SCHOOL OF MINES AND TECHNOLOGY, (M), 501 E. St. Joseph, Rapid City, SD 57701-3901. Tel.: 605-394-2467. Fax: 605-394-6131.
E-mail: museum@sdsmt.edu
Web Site: museum.sdsmt.edu
Founded: 1885.
Key Personnel: Pres., Dr. Heather Wilson; Provost, Dr. Duane Hrncir; Dir., Dr. Laurie C. Anderson; Assoc. Dir., Sally Shelton; Cur., Dr. Darrin Pagnac; Cur., Christina Belanger; Cur., Maribeth Price; Cur., James Fox; Program Asst., Samantha Hustoft.
Personnel Profile: Full-Time Paid 4; Full-Time Volunteers 1; Part-Time Paid 5; Part-Time Volunteers 7; Interns 1.
Governing Authority: state. Parent Institution: South Dakota School of Mines & Technology. Tax-exempt: 501(c)(3).
Institution Type/Description: Science Museum.
Collections: rocks; minerals; fossil vertebrates; invertebrates; biology; maps; plants; books; archives.
Research Fields: paleontology; mineralogy; history of mining & exploration; biostratigraphy; geological conservation; paleontology rescue management.
Facilities: library; archives.
Activities: student & family summer field camps; permanent & temporary exhibitions; research; public outreach.
Publications: Dakoterra.
Hours & Admission Prices: Summer: Mon.-Fri. 9-5, Sat. 9-6, Sun. 12-5; Winter: Mon.-Fri. 9-4, Sat. 10-4. No charge; donations accepted. Closed holidays. &
Attendance: 26,000 (accurate)

SIOUX INDIAN MUSEUM, 222 New York St., Rapid City, SD 57701-1199. Tel.: 605-394-2381. Fax: 605-348-6182.
E-mail: sim@journeymuseum.org
Web Site: www.journeymuseum.org
Founded: 1939.
Congressional District: 2
Personnel Profile: Full-Time Paid 2.
Governing Authority: federal. Parent Institution: Indian Arts & Crafts Board, MS 2528-MIB, U.S. Department of the Interior, Washington, DC 20240. Tax-exempt.
Institution Type/Description: Indian Art Museum.

Collections: historic & contemporary arts of the Sioux; Native American arts & crafts; beadwork; quillwork.
Research Fields: contemporary Native American art.
Facilities: Beadwork, jewelry, moccasins, tobacco pouches, quillwork, dance costumes, accessories, headdresses, dolls & paintings.
Activities: guided tours; lectures; gallery talks; permanent & traveling exhibitions; demonstrations; events honoring Native Americans who have achieved fame in literature & performing arts; annual one-person exhibitions.
Publications: exhibition brochures.
Hours & Admission Prices: mid-Jan. to May & Sept.-Dec. Mon.-Sat. 10-5, Sun. 1-5; Memorial Day-Labor Day daily 9-6. Museum: adults 18-61 $8, seniors 62 & over $6.90, students 11-17 & college students $5.75; discounts to groups. Planetarium: adults 18-61 $5.75, seniors 62 & over $4.60, students 11-17 $3.45, children 10 & under $2; discounts to school groups. Closed New Year's Day; Easter; Thanksgiving; Christmas. &
Attendance: 45,000 (estimated)
Membership: Student $15; Individual $25; Family $50; Grand Family $100; Lifetime $1,000.

Redfield

REDFIELD'S HISTORIC CHICAGO AND NORTHWESTERN RR DEPOT, 715 3rd St. W., Redfield, SD 57469-1173. Mailing Address: 626 Main St., Redfield, SD 57469-1127. Tel.: 605-472-4566.
E-mail: cnwhistoricrrdepot@redfield.com
Founded: 1914.
Key Personnel: Chm. (V) & Museum Shop Mgr., Kathy Maddox
Institution Type/Description: History Museum: housed in the restored C&NW Depot built in 1914.
Collections: local history; telegraph equipment; railroad memorabilia.
Activities: group tours.
Hours & Admission Prices: May 15-Nov. 1 Thurs.-Sun. 1-5; other times by appointment. No charge; donations accepted. &

SPINK COUNTY HISTORICAL SOCIETY, Courthouse Square, 225 E. 8th Ave., Redfield, SD 57469. Mailing Address: 213 E. 2nd St., Redfield, SD 57469. Tel.: 605-472-0758.
Key Personnel: Treas., Jerry Hansen; Pres. (V), Al Evans.
Personnel Profile: Part-Time Volunteers 5.
Institution Type/Description: History Museum.
Collections: early pioneer life & history; birds; butterflies; Native American artifacts; farm machinery; military equipment; newspapers photographs.
Hours & Admission Prices: June-Aug. Thurs.-Sun. 1-5. No charge; donations accepted. &
Membership: Annual $5; Lifetime $50.

Roslyn

INTERNATIONAL VINEGAR MUSEUM, 502 Main St., Roslyn, SD 57261. Mailing Address: P.O. Box 201, Roslyn, SD 57261-0201. Tel.: 605-486-0075. Facebook: International Vinegar Museum.
E-mail: museum@internationalvinegarmuseum.com, marydwagner@gmail.com
Web Site: internationalvinegarmuseum.com
Founded: 1999.
Key Personnel: Chm. (V), Richard Snaza; Museum Shop Mgr., Mary Wagner.
Personnel Profile: Part-Time Paid 3; Part-Time Volunteers 3.
Governing Authority: Parent Institution: CARE. Tax-exempt.
Institution Type/Description: Vinegar Museum.
Collections: vinegar from around the world; paper made from vinegar; pottery.
Activities: vinegar tasting bar; guided tours. Annual Event: International Vinegar Festival in June; Dinner of Elegance.
Hours & Admission Prices: June to Labor Day Thurs.-Sat. 10-6. Adults $2.
Attendance: 1,500 (estimated)

Saint Francis

BUECHEL MEMORIAL LAKOTA MUSEUM, St. Francis Mission, 350 S. Oak St. on Rosebud Reservation, Saint Francis, SD 57572. Mailing Address: P.O. Box 499, Saint Francis, SD 57572-0499. Tel.: 605-747-2745. Fax: 605-747-2361.
E-mail: museum@gwtc.net
Web Site: www.sfmission.org/museum
Founded: 1915.
Congressional District: 23
Key Personnel: Dir., Fr. John Hatcher, S.J.; Museum Shop Mgr., Marie Kills In Sight.

Personnel Profile. Full-Time Paid 2; Part-Time Paid 2; Interns 2.
Governing Authority: church. Parent Institution: St. Francis Mission. Tax-exempt.
Institution Type/Description: Lakota Indian Museum.
Collections: Rosebud & Pine Ridge Sioux history.
Research Fields: history; ethnology.
Facilities: 190-vol. library of photographs available by request. Indian arts & crafts for sale.
Activities: guided tours.
Publications: books, A Grammar of Lakota; Lakota Dictionary; Sursoum Corda: Photo Album; Crying for a Vision: A Rosebud Sioux Trilogy; A Grammar of Lakota.
Hours & Admission Prices: Memorial Day to Labor Day Mon.-Thurs. & Sat. 8-5, Fri. 8-2:30, Sun. 10-4. No charge; donations accepted. Closed holidays.
Attendance: 3,600 (accurate)

Scotland

SCOTLAND HERITAGE CHAPEL & MUSEUM, 811 6th St., Scotland, SD 57059. Mailing Address: 351 4th St., Scotland, SD 57059-2112. Tel.: 605-583-4568, 2507 & 4144.
Founded: 1976.
Congressional District: 1
Key Personnel: Dir., Don Schmidt; Dir., Marvin Thum; Pres. (V), Linda Kluthe; Vice Pres., Betty Woehl; Treas., Lori Schmidt; Sec., Carolyn Thaler.
Personnel Profile: Part-Time Volunteers 8.
Governing Authority: nonprofit. Parent Institution: Scotland Historical Society. Tax-exempt: 501(c)(3).
Institution Type/Description: Local History Museum: housed in c.1874 Methodist & Siemantal Churches.
Collections: historical artifacts; books; furniture; clothes; household articles; farm tools; implementary & medical supplies; tools; farm machinery; railroad handcar; pencils & pens; local history; family records.
Activities: guided tours.
Publications: brochure.
Hours & Admission Prices: Summer: holidays & special occasions; other times by appointment. No charge; donations accepted. &
Attendance: 75 (estimated)
Membership: Annual $2; Life $25.

Sioux Falls

BATTLESHIP SOUTH DAKOTA MEMORIAL, Sherman Park, 12th St. & Kiwanis Ave., Sioux Falls, SD 57101. Mailing Address: Sioux Falls Parks & Recreation Dept., 100 E. 6th St., Sioux Falls, SD 57104. Tel.: 605-367-7141. Fax: 605-367-8234.
E-mail: ussdakota@aol.com
Web Site: www.usssouthdakota.com
Founded: 1968.
Congressional District: 1
Key Personnel: Chm. (V), David Witte; Museum Shop Mgr., Opal Engbom.
Personnel Profile: Full-Time Paid 2; Part-Time Volunteers 20.
Volunteer Hours: 3,850
Operating Expenses: 16,159
Operating Income: 28,800
Governing Authority: municipal. Parent Institution: City of Sioux Falls. Tax-exempt.
Institution Type/Description: Military & Nautical Museum.
Collections: memorabilia of the Battleship U.S.S. South Dakota; silver services; gun barrels; 1/4 scale model of ship; ship's log; bell; books; photographs; mast; anchors; flags.
Research Fields: Naval history.
Facilities: Commemorative coins for sale.
Activities: films; permanent exhibitions. Museum Sponsors: biennial crew member reunion.
Publications: brochure.
Hours & Admission Prices: Memorial Day-Labor Day daily 10-6. No charge; donations accepted.
Attendance: 10,000 (estimated)

THE CENTER FOR WESTERN STUDIES, (M), Augustana College, 2121 S. Summit, Sioux Falls, SD 57197. Mailing Address: 2001 S. Summit Ave., Sioux Falls, SD 57197. Tel.: 605-274-4007. Fax: 605-274-4999.
E-mail: cws@augie.edu
Web Site: www.augie.edu/cws/
Founded: 1970.
Congressional District: 1
Key Personnel: Exec. Dir., Dr. Harry F. Thompson; Office Coord., Amy

Nelson; Chm. (V), Tony Haga; Education Asst., Kristi Thomas; Collections Asst., Elizabeth Thrond.
Personnel Profile: Full-Time Paid 4; Part-Time Volunteers 6; Interns 4.
Volunteer Hours: 1,500
Operating Expenses: 475,000
Operating Income: 508,000
Governing Authority: college. Parent Institution: Augustana College. Tax-exempt.
Institution Type/Description: Cultural, Research & Archival Agency.
Collections: regional artifacts; Norwegian rosemaled furniture; paintings & prints by regional artists; Episcopal Diocese of South Dakota archives; United Church of Christ-South Dakota Conference archives; Stephen Riggs papers; Herbert Krause papers; Augustana college archives; John R. Milton papers; Harold Shunk papers; manuscripts; Blue Cloud Abbey American Indian culture.
Major Exhibits: Art of Cathleen Benberg & Steve Beaubien, 1/14-5/14; South Dakota 2014: Invitational Art Show & Sale in Observance of Statehood, 6/14-9/14.
Research Fields: Dakota (Sioux); South Dakota & Northern Plains history & cultures; Western American literature.
Facilities: 40,000-vol. library of material dealing with the Trans-Mississippi West & the Upper Great Plains region available for use on premises; archives; research center.
Activities: conferences on regional themes; annual art show; rotating exhibits; courses & workshops on regional history & historical document editing; forum on public affairs with national & international leaders, Gen. Colin Powell & George Bush; Mikhail Gorbachev; John Major; Barbara Bush; Queen Noor of Jordan; Al Gore; Vicente Fox; Sandra Day O'Connor; Perrez Musharraf; Madeleine Albright; Mary Robinson; Jared Cohen; Jon Huntsman.
Publications: books, Reveille for Sioux Falls: How a World War II Army Air Forces technical school changed a midwestern city; A Harvest of Words: Contemporary South Dakota Poetry; Birding in the Northern Plains; A New South Dakota History; The Northern Pacific Railroad and the Selling of the West; Yanktonai Sioux Water Colors; Sundancing At Rosebud And Pine Ridge; Where The West Begins; The Wind Blows Free; Frederick Manfred: A Bibliography & Publication History; Boy off the Farm; Next Year Will Be Better; What the Tallgrass Says; Prairie Architect; Over a Century of Leadership; South Dakota Territorial and State Governors; The Quartizite Border: Surveying and Marking the North Dakota-South Dakota Boundary, 1891-1892; The Last Contrary; Tomahawk and Cross; An Illustrated History of the Arts in South Dakota; A Noble Calling; Poems and Essays of Herbert Krause; Natural History of the Black Hills and Badlands; Guide to Collections Relating to South Dakota Norwegian-Americans; The Geography of South Dakota; Fort Sisseton; The Lizard Speaks; Essays on the Writings of Frederick Manfred; What It Took: A History of the USGS EROS Data Center; The Family Farmer's Advocate: South Dakota Farmers Union 1914-2000; Soldier, Settler, and Sioux: Fort Ridgely and the Minnesota River Valley, 1853-1967; The Lewis and Clark Expedition: Food, Nutrition and Health; The Lewis and Clark Expedition: Then and Now; Joseph Nicollet and His Map: Exploring the Upper Mississippi River; Sandra Day O'Connor.
Hours & Admission Prices: Mon.-Fri. 8-5, Sat. 10-2. No charge; donations accepted. Closed Independence Day; Christmas. &
Attendance: 7,000 (estimated)
Membership: Contributor $50; Explorer $100; Partner $250; Scout $500; Pioneer $1,000; Ranger $2,500; Westerner $5,000.

EIDE-DALRYMPLE ART GALLERY, Augustana College, 30th St., & Grange Ave., Sioux Falls, SD 57197. Mailing Address: Augustana College, 2001 S. Summit Ave., Sioux Falls, SD 57197. Tel.: 605-274-4609. Fax: 605-274-5323.
Web Site: www.augie.edu/gallery
Key Personnel: Dir., Lindsay Twa.
Governing Authority: Parent Institution: Augustana College. Tax-exempt.
Institution Type/Description: Art Gallery.
Collections: works by local & regional artists; European & American prints.
Activities: temporary & permanent exhibits; special events.
Hours & Admission Prices: Mon.-Fri. 10-5, Sat. 12-5. No charge. Closed major holidays.
Attendance: 3,000 (estimated)

GREAT PLAINS ZOO & DELBRIDGE MUSEUM OF NATURAL HISTORY, (M), 805 S. Kiwanis Ave., Sioux Falls, SD 57104-3798. Tel.: 605-367-7003. Fax: 605-367-8340.
E-mail: dsimon@gpzoo.org
Web Site: greatzoo.org
Founded: 1957.
Congressional District: 1

Key Personnel: C.E.O. & Pres., Elizabeth Whealy; Vice Pres. Operations, Dan Simon; Mgr. Guest Svcs., Maggie Erstad; Dir. Animal Programs, Lisa Smith.
Personnel Profile: Full-Time Paid 40; Part-Time Paid 40; Part-Time Volunteers 90.
Governing Authority: nonprofit organization. Zoological Society of Sioux Falls. Tax-exempt.
Institution Type/Description: Zoo & Natural History Museum.
Collections: Henry Brockhouse collection of mounted animals from 1950-1970; mammals; birds; reptiles; insects.
Facilities: 45-acre grounds; concessions. Gift items for sale.
Activities: outreach; summer camps; special events; on-site programs; fund-raisers; children's zoo; carousel; train rides.
Publications: quarterly membership newsletter; education materials; brochure; map.
Hours & Admission Prices: April-Sept. daily 9-6; Oct.-March daily 10-4. Adults $8.50, seniors $7.75, child $5.50; children under 2, zoo & museum members no charge. Closed New Year's Day; Thanksgiving; Christmas.
Attendance: 251,000 (accurate)
Membership: Individual $35; Family $72; Patron $108.58.

KIRBY SCIENCE DISCOVERY CENTER AT THE WASHINGTON PAVILION, 301 S. Main Ave., Sioux Falls, SD 57104-6311. Tel.: 605-367-7397, ext. 2307. Fax: 605-367-7399.
E-mail: elacey@washingtonpavilion.org
Web Site: washingtonpavilion.org
Founded: 1999.
Key Personnel: Pres. & C.E.O., Larry Toll; Dir., Erica Lacey.
Personnel Profile: Full-Time Paid 50; Part-Time Paid 100; Part-Time Volunteers 50.
Governing Authority: private; nonprofit organization. Parent Institution: Washington Pavilion Management, Inc. Tax-exempt.
Institution Type/Description: Science Technology Center.
Collections: interactive exhibits.
Facilities: large format movie theater.
Activities: interactive exhibits; programs; films.
Publications: bimonthly newsletter, Arts & Science Adventures.
Hours & Admission Prices: Center: late April to Labor Day Mon.-Thurs. & Sat. 10-5, Fri. 10-8, Sun. 12-5; Sept. to Easter Tues.-Thurs. & Sat. 10-5, Fri. 10-8, Sun. 12-5. CineDome: late April to Labor Day Mon.-Thurs. & Sat. 11-4, Fri. 11-8, Sun. 1-4; Sept. to Easter Tues.-Thurs. & Sat. 10-5, Fri. 10-8, Sun. 12-5. Combination Tickets: adults $9.50, children & seniors $7.50; discounts to groups; ASTC Travel Passport members & members no charge. Closed New Year's Day; Thanksgiving; Christmas Eve & Day.
Attendance: 100,000 (estimated)
Membership: Individual $35; Membership for Two $50; Grandparent & Household $80; Household Plus $125.

MUSEUM OF VISUAL MATERIALS, 500 N. Main Ave., Sioux Falls, SD 57104-5902. Tel.: 605-271-9500. Fax: 605-271-4793.
E-mail: jeremy@sfmvm.com
Web Site: www.sfmvm.com
Founded: 2007.
Key Personnel: Dir., Jeremy Brech
Institution Type/Description: Arts and Crafts Museum.
Collections: over 4,500 books; vinyl records & 8-track tapes; over 80,000 sewing buttons; fabrics; arts & crafts.
Hours & Admission Prices: Mon.-Wed. & Fri.-Sat. 10-4, Thurs. 10-7, Sun. 1-4.

PETTIGREW HOME AND MUSEUM, 131 N. Duluth, Sioux Falls, SD 57104. Mailing Address: 200 W. 6th St., Sioux Falls, SD 57104-6001. Tel.: 605-367-7097. Facebook: Pettigrew Home and Museum.
E-mail: museum@minnehahacounty.org
Web Site: www.siouxlandmuseums.com
Key Personnel: Dir., Bill Hoskins; Chm. (V), Julie Brue; Museum Shop Mgr., Shelly Sjovold.
Personnel Profile: Full-Time Paid 17; Part-Time Paid 10; Part-Time Volunteers 25; Interns 1.
Governing Authority: Parent Institution: Siouxland Heritage Museums. Tax-exempt.
Institution Type/Description: Historic House Museum: c.1889 home of South Dakota's first senator, Richard Pettigrew.
Collections: photographs; personal artifacts; furnishings.
Activities: video; interactive computer stations.
Publications: newsletter, Siouxland Heritage Museums Report.
Hours & Admission Prices: May-Sept. Mon.-Sat. 9-5, Sun. 12-5; Oct.-April daily 12-5. No charge; donations accepted. Closed major holidays.
Attendance: 9,132 (accurate)

SERTOMA BUTTERFLY HOUSE & MARINE COVE, 4320 Oxbow Ave., Sioux Falls, SD 57106-4110. Tel.: 605-334-9466. Fax: 605-334-9662.
E-mail: director@sertomabutterfly.org
Web Site: www.sertomabutterflyhouse.org
Key Personnel: Exec. Dir., Audrey Willard
Institution Type/Description: Butterfly House & Aquarium.
Collections: butterflies & fish from around the world.
Hours & Admission Prices: Memorial Day to Labor Day Mon.-Sat. 10-6, Sun. 11-5; Sept.-May Mon.-Sat. 10-4, Sun. 1-4. Adults 13-59 $8.50, senior citizens 60 & over $7, youth 5-12 $5.50, children 3-4 $3; children 2 & under no charge. Closed Easter; Thanksgiving; Christmas.
Membership: $30 & up.

19382.logo.tif

SIOUX COUNCIL SCOUT MUSEUM, 800 N. West Ave., Sioux Falls, SD 57104-5720. Tel.: 605-361-2697. Fax: 605-361-2381.
E-mail: sioux.council@scouting.org
Web Site: www.siouxbsa.org
Founded: 2007.
Key Personnel: Chm. (V), Reid Christopherson.
Governing Authority: Parent Institution: Sioux Council BSA. Tax-exempt.
Institution Type/Description: History Museum.
Collections: history of Boy Scouts of America since its founding in 1910; scout uniforms, equipment, & manuals; photographs.
Hours & Admission Prices: Mon. & Wed.-Fri. 9-6, Tues. 9-8, Sat. 9-3. No charge.

SIOUX EMPIRE MEDICAL MUSEUM, 1305 W. 18th St., Sioux Falls, SD 57117-5039. Mailing Address: Box 5039, 1305 W. 18th St., Sioux Falls, SD 57117-5039. Tel.: 605-333-6397.
Founded: 1975.
Congressional District: 1
Key Personnel: Chm. History Museum Committee, Thenetta Nield; Co-Chm. (V), Carol Turgeon.
Personnel Profile: Full-Time Volunteers 40.
Governing Authority: nonprofit organization. Sponsored by the Alumni Association of Sioux Valley Hospital School of Nursing, 1305 W. 18th, Sioux Falls, SD 57117. Tax-exempt: 501(c)(3).
Institution Type/Description: Medical Museum.
Collections: 1930 patient's room; orthopaedics; pediatrics; surgery; nursery; X-ray; two dental units; uniformed dolls; photographs of medical staff members; patent medicines.
Facilities: 100-vol. medical library available for use to professionals only.
Activities: guided tours; lectures; permanent & temporary exhibitions.
Hours & Admission Prices: Mon.-Fri. 11-4. No charge. Closed major holidays.
Attendance: 3,500 (accurate)

SIOUXLAND HERITAGE MUSEUMS, (M), 200 W. 6th St., Sioux Falls, SD 57104-6001. Tel.: 605-367-4210, ext. 0. Fax: 605-367-6004. Facebook: Siouxland Heritage Museums.
E-mail: bhoskins@minnehahacounty.org
Web Site: www.siouxlandmuseums.com
Founded: 1974.
Congressional District: 1
Key Personnel: Dir., William J. Hoskins; Mktg. Coord., Adam Nelson; Cur. Education, Kevin Gansz; Cur. Exhibits, Molly Engquist; Cur. Collections, Julie Breu; Preparator, William Booker; Events Coord., Accountant & Museum Shop Mgr., Staci Limoges.
Personnel Profile: Full-Time Paid 16; Part-Time Paid 9; Part-Time Volunteers 71.
Governing Authority: municipal; nonprofit. Parent Institution: Minnehaha County. Branch Museums: 1889 The Pettigrew Home & Museum, 131 N. Duluth, Sioux Falls, SD 57104.; 1889 The Old Courthouse Museum, 200 W. 6th St., Sioux Falls, SD 57104. Tax-exempt.
Institution Type/Description: General Museums: housed in 1889 The Pettigrew Home & Museum; housed in 1890 The Old Courthouse Museum.
Collections: local & regional natural history; Dakota & plains ethnology; local folk arts.
Major Exhibits: Minnehaha County Towns, 8/23/12-1/14; Chairs of Our Lives, 2/5/13-6/14.
Research Fields: local history; archaeology; folklife.
Facilities: reference library; archives & photographic archives; auditorium; classroom; inflatable planetarium.
Activities: tours; gallery talks; workshops; conferences; loan kits; formally

organized education programs for children & adults; audio-visual shows; permanent & changing exhibitions.

Publications: newsletter, Siouxland Heritage Museums Report.

Hours & Admission Prices: Old Courthouse Museum: Mon.-Wed. & Fri.-Sat. 8-5, Thurs. 8-9, Sun. 12-5. Pettigrew Home and Museum: May-Sept. Mon.-Wed. & Fri.-Sat. 9-5, Thurs, 9-9, Sun. 12-5; Oct.-April daily 12-5. No charge; donations accepted. Closed for major holidays. &

Attendance: 44,007 (accurate)

Membership: Individual $15; Family & Couple $25; Patron $50; Corporate $500.

WASHINGTON PAVILION OF ARTS AND SCIENCE, 301 S. Main Ave., Sioux Falls, SD 57104-6311. Tel.: 605-367-7397, ext. 2306. Fax: 605-367-7399.

E-mail: info@washingtonpavilion.org

Web Site: www.washingtonpavilion.org

Formerly: Visual Arts Center at the Washington Pavilion

Founded: 1999.

Key Personnel: Co-Pres., Larry Toll; Co-Pres., Scott Petersen; Dir., Visual Arts Center, David J. Merhib; Dir., Community Learning Center, Rose Ann Hofland; Dir., Kirby Science Discovery Center, Erica Lacey; Dir., Mktg. & Communications, Michele Wellman.

Governing Authority: private; nonprofit organization. Parent Institution: Washington Pavilion Management, Inc. Tax-exempt.

Institution Type/Description: Children's Museum.

Collections: paintings; graphics; photographs; sculpture.

Research Fields: regional, national, international art.

Facilities: art library; children's studio; classrooms; CineDome theater. Museum-related items for sale.

Activities: guided tours; gallery talks; hands-on exhibits. Annual Events: Wellness Festival; Outdoor Arts Festival.

Publications: members newsletter, Columns; catalogs.

Hours & Admission Prices: April-Sept. Mon.-Thurs. & Sat.-Sun. 10-5, Fri. 10-8; Sept.-Jan. Tues.-Thurs. & Sat. 10-5, Fri. 10-8, Sun. 12-5; Jan.-April Tues.-Sat. 10-5, Sun. 12-5. Kirby Science Discovery Center: adults $12, seniors & military $9, youth & students $6; Visual Arts Center: adults $7, seniors & military $5, youth & students $3.50; Combo Admission: adults $13, seniors & military $10, youth & students $7; members no charge. &

Attendance: 41,392 (accurate)

Membership: Dual $60; Family $100.

Spearfish

D.C. BOOTH HISTORIC NATIONAL FISH HATCHERY AND ARCHIVE, (M), 423 Hatchery Circle, Spearfish, SD 57783-2643. Tel.: 605-642-7730, ext. 0. Fax: 605-642-2336.

E-mail: dcbooth@fws.gov

Web Site: dcbooth.fws.gov

Formerly: Spearfish Station, Spearfish National Fish Hatchery

Founded: 1896.

Congressional District: 31

Key Personnel: Dir., Carlos Martinez; Cur., Randi Smith; Museum Shop Mgr., April Gregory.

Personnel Profile: Full-Time Paid 6; Part-Time Volunteers 30; Interns 2.

Governing Authority: federal; nonprofit. Parent Institution: U.S. Fish & Wildlife Service. Subsidiary Institution: Booth Society. Tax-exempt.

Institution Type/Description: Historic Fisheries Museum: housed in a fish hatchery building; built in 1899. Listed on the National Register of Historic Places.

Collections: historic fishery & fish culture artifacts; records; photographs; fish car material; U.S. Fish & Wildlife Service history; furnishings; Fish Culture Hall of Fame.

Research Fields: fish culture; fisheries; U.S. Fish & Wildlife Service history; fish stocking; U.S. Fish Commission, Bureau of Fisheries.

Facilities: nature trail; 10 acre site.

Activities: tours; wildlife observation; live trout feeding; underwater trout viewing windows; research.

Publications: semiannual newsletter.

Hours & Admission Prices: Grounds: daily. Fish Car Museum & Booth House & Fish Car: mid-May to mid-Sept. No charge; donations accepted. Group tour charge. &

Attendance: 150,000 (estimated)

Membership: Individual $35; Family $50; Business $100 & up.

DOLLS AT HOME MUSEUM, 435 Meier Ave., Spearfish, SD 57783-1977. Mailing Address: P.O. Box 489, Spearfish, SD 57783-0489. Tel.: 605-645-2192.

E-mail: bhpp@blackhills.com

Founded: 1999.

Key Personnel: Dir. & Pres. (V), Johanna Della Vecchia; Cur. & Museum Shop Mgr., Sonya Albers.

Personnel Profile: Part-Time Volunteers 4.

Volunteer Hours: 50

Operating Expenses: 5,499

Operating Income: 150,000

Governing Authority: private. Parent Institution: Aamphitheatre Affiliates. Tax-exempt.

Institution Type/Description:

Collections: dollhouses; modern & early one-of-a kind items by recognized miniature artists; furniture; period artifacts & collectibles; over 75 dollhouses & rooms, many with gardens & dioramas; historical Family Tree exhibit detailing history of immigrants to this Black Hills Area.

Facilities: Dolls & dollhouse furniture for sale.

Activities: seasonal puppet shows; special birthday parties. Museum Sponsors: Christmas holiday activities.

Hours & Admission Prices: Memorial Day to Labor Day by appointment. Season pass $10, adults $5, children $2.50; members no charge. &

Attendance: 500 (estimated)

HIGH PLAINS WESTERN HERITAGE CENTER, 825 Heritage Dr., Spearfish, SD 57783. Mailing Address: P.O. Box 524, Spearfish, SD 57783-0524. Tel.: 605-642-9378.

E-mail: info@westernheritagecenter.com

Web Site: www.westernheritagecenter.com

Founded: 1974.

Key Personnel: Exec. Dir., Peggy Ables

Institution Type/Description: History Museum.

Collections: Western art & artifacts; stagecoach; period kitchen; saddle shop; blacksmith shop; mining; ranching; rodeo; furnished log cabin; rural schoolhouse; period farm equipment.

Facilities: 200-seat theater. Museum-related items for sale.

Activities: special art events; performances.

Hours & Admission Prices: Daily 9-5. Adults 17-61 $7, seniors 62 & over $5, youth 6-16 $3.

Sturgis

BEAR BUTTE STATE PARK VISITORS CENTER, 20250 Hwy. 79, Sturgis, SD 57785. Mailing Address: P.O. Box 688, Sturgis, SD 57785-0688. Tel.: 605-347-5240. Fax: 605-347-7627.

E-mail: BearButte@state.sd.us

Web Site: gfp.sd.gov/state-parks/directory.bear-butte/

Founded: 1961.

Congressional District: 2

Key Personnel: Park Mgr., Jim Jandreau.

Personnel Profile: Full-Time Paid 1; Part-Time Paid 5; Part-Time Volunteers 2.

Governing Authority: state. Parent Institution: South Dakota Dept. of Game, Fish & Parks, Foss Bldg., 523 E. Capitol, Pierre, SD 57501-3182. Tel. 605-773-3391. Subsidiary Institution: Dept. of Parks. Tax-exempt.

Institution Type/Description: State Park Visitors Center Museum: located on a Native American traditional religious site.

Collections: Native American clothing & religious artifacts; archaeological site materials; plant displays; geological displays.

Research Fields: Native American Indian religion, anthropology, archaeology & geology.

Activities: guided tours; lectures; gallery talks; formally organized education programs for children & adults; loan, permanent & traveling exhibitions.

Publications: trail guide, Bear Butte Trail Guide; brochure, Bear Butte State Park brochure.

Hours & Admission Prices: Park: 8:30-5:30. Visitors Center: May to Sept. daily 9-6. Car $6, annual sticker $30. &

Attendance: 15,000

STURGIS MOTORCYCLE MUSEUM AND HALL OF FAME, (M), 999 Main St., Sturgis, SD 57785-1620. Mailing Address: P.O. Box 602, Sturgis, SD 57785-0602. Tel.: 605-347-2001. Fax: 605-720-0632.

E-mail: christine@sturgismuseum.com

Web Site: www.sturgismuseum.com

Founded: 2001.

Key Personnel: Exec. Dir., Christine Paige Diers; Pres. (V), Dave Davis; Museum Shop Mgr., Arlene Colaiacovo.

Personnel Profile: Full-Time Paid 2; Part-Time Paid 5.

Institution Type/Description: Motorcycle Museum.

Collections: motorcycles; motorcycle memorabilia; racing history; women in motorcycling; photographs; Hall of Fame inductees.

Activities: group tours.

Hours & Admission Prices: Daily call for hours. One adult $10, 2 adults $15,

discounts for seniors; members & children 12 & under no charge. Closed New Year's Day; Easter; Thanksgiving; Christmas. &
Attendance: 32,000 (accurate)
Membership: Individual $35; Life $500.

Timber Lake

TIMBER LAKE AND AREA HISTORICAL SOCIETY, 800 Main St., Timber Lake, SD 57656. Mailing Address: P.O. Box 181, Timber Lake, SD 57656-0181. Tel.: 605-865-3553 & 3546.
E-mail: timberlakemuseum@yahoo.com
Web Site: www.timberlakehistory.org
Institution Type/Description: Historical Society Museum.
Collections: fossils; traditional Lakota clothing; early reservation Native American items; photographs.
Hours & Admission Prices: By appointment.
Membership: Student $10; Individual $20; Sustaining $50; Patron $100; Benefactor $250.

Vermillion

AUSTIN-WHITTEMORE HOUSE MUSEUM, 15 Austin Ave., Vermillion, SD 57069-3055. Tel.: 605-624-8266.
E-mail: claycohistory@yahoo.com
Web Site: www.cchssd.org
Founded: 1968.
Personnel Profile: Full-Time Volunteers 1.
Governing Authority: Tax--exempt.
Institution Type/Description: Historic House.
Collections: Victorian furnishings; period artifacts.
Publications: From the River Valleys to the Rising Bluff; History of Dakota Territory.
Hours & Admission Prices: Memorial Day to Labor Day Mon.-Fri. 10-4:30, Sat.-Sun. by appointment. No charge; donations accepted.
Membership: Individual $15; Family $25; Institutional $50; Life $200.

✳ **NATIONAL MUSIC MUSEUM, (M),** Clark & Yale Sts., Vermillion, SD 57069-2390. Mailing Address: 414 E. Clark St., Vermillion, SD 57069-2390. Tel.: 605-677-5306. Fax: 605-677-6995.
E-mail: nmm@usd.edu
Web Site: www.nmmusd.org
Formerly: Shrine To Music Museum
Founded: 1973.
Congressional District: 1
Key Personnel: Dir., Dr. Cleveland Johnson; Chm. (V), Tom Lillibridge; Conservator, John Koster; Sr. Cur., Dr. Margaret Downie Banks; Cur., Dr. Sabine Klaus; Cur., Arian Sheets; Cur. Education, Dr. Deborah Check Reeves; Cur. Asst., Julie Boston; Cur. Asst., Micky Rasmussen; Collections Mgr., Rodger Kelly; Dir. Visitor Svcs., Vicky Lenz; Program Asst., Debra Schultz.
Personnel Profile: Full-Time Paid 12; Part-Time Paid 1; Part-Time Volunteers 5; Interns 9.
Governing Authority: nonprofit. Affiliated with University of South Dakota. Tax-exempt: 501(c)(3).
Institution Type/Description: Musical Instrument Museum.
Collections: John Powers saxophones; Paul & Jean Christian collection & archive; Holton & Leblanc Company archives; Arnold Ruskin brass collection; Arne B. Larson American, European & non-Western musical instruments; Wayne Sorenson 19th-century woodwinds; Witten-Rawlins early Italian stringed instruments, bows, tools, documentary source materials, & manuscripts; Higbee-Abbott-Zylstra early flutes & recorders; Cecil Leeson saxophones & archive; Rosario Mazzeo clarinets; Joe & Joella Utley brass instruments; Alan G. Bates harmonica & archive; William F. Ludwig II percussion collection.
Research Fields: history of musical instruments.
Facilities: library of music literature available for research on premises by appointment; conservation laboratory; concert hall. Postcards, catalogs, posters, CDs, & technical drawings for sale.
Activities: guided tours; lectures; gallery talks; concerts; formally organized education programs for undergraduate & graduate students; repairs & restorations; permanent exhibitions; workshops; conferences; self guided multi-media tours.
Publications: electronic newsletter; catalogs; recordings; scholarly books.
Hours & Admission Prices: Mon.-Sat. 9-5, Sun. 2-5. Adults $10; discounts to AAM & ICOM members; members no charge. Closed New Year's Day; Easter; Thanksgiving; Christmas. &
Attendance: 12,803 (accurate)
Membership: Member $40; Family $60; Contributing $100; Sustaining $250; Supporting $500; Life $1,500; Adolphe Sax Society $1,500-$2,499; Charles

W. Wheatstone Society $2,500-$4,999; John Franklin Stratton Society $5,000-$9,999; Franz Schwarzer Society $10,000-$24,999; Johann Wilhelm Haas Society $25,000-$49,999; August Grenser Society $50,000-$99,999; Christian Dieffenbach Society $100,000-$249,999; Andrea Guarneri Society $250,000-$499,999, Nanette Stein Streicher Society $500,000-$999,999; Jakob Stainer Society $1,000,000 -$2,499,999; Andreas Ruckers Society $2,500,000-$4,999,999; Antonio Stradivari Society $5,000,000 & up.

UNIVERSITY ART GALLERIES, Warren M. Lee Center, University of South Dakota, 414 E. Clark, Vermillion, SD 57069-2307. Tel.: 605-677-3177. Fax: 605-677-5988.
E-mail: alison.erazmus@usd.edu
Web Site: www.usd.edu/uag
Founded: 1976.
Key Personnel: Dir., Alison Erazmus.
Personnel Profile: Part-Time Paid 3; Part-Time Volunteers 3.
Governing Authority: state; university. Parent Institution: University of South Dakota. Tax-exempt.
Institution Type/Description: Art Gallery.
Collections: art work by contemporary Sioux artist, Oscar Howe; study collection with emphasis on modern & contemporary art on paper; historic art of South Dakota; Asian, African & Native American Art.
Research Fields: South Dakota artists; Northern Plains American Indian art.
Facilities: library.
Activities: lectures; films; gallery talks; loan, permanent, temporary & traveling exhibitions; school loan service.
Publications: exhibition catalogs.
Hours & Admission Prices: Main Gallery: Mon.-Fri. 8-5, Sat.-Sun. 1-5. Oscar Howe Gallery: Mon.-Sat. 1-5. No charge. Closed major holidays. &
Attendance: 20,000 (estimated)

W.H. OVER MUSEUM, (M), 1110 University, Vermillion, SD 57069. Mailing Address: 414 E. Clark, Vermillion, SD 57069-2307. Tel.: 605-677-5228.
E-mail: whover@usd.edu
Web Site: www.usd.edu/whover
Founded: 1883.
Congressional District: 1
Key Personnel: Pres. Bd., Larry Bradley.
Personnel Profile: Part-Time Paid 4; Part-Time Volunteers 20; Interns 2.
Governing Authority: Parent Institution: Friends of the Museum. Tax-exempt: 501(c)(3).
Institution Type/Description: Natural & Cultural History.
Collections: archeology; geology; natural history; regional history; Plains Indian ethnology; photographs; Stanley J. Morrow collection; David & Elizabeth Clark Memorial collection; contemporary Sioux paintings; Robert Penn.
Research Fields: Dakota & Sioux ethnology; regional & natural history.
Facilities: 4,000-vol. library; classroom. Museum-related items for sale.
Activities: craft classes & craftsmen in residence; tours; traveling exhibits; lecture & film series; school loan exhibits.
Publications: quarterly museum newsletter.
Hours & Admission Prices: Mon.-Sat. 10-4. No charge; donations accepted. Closed major holidays. &
Attendance: 12,416 (accurate)
Membership: Student $10; Senior Citizen $15; Individual $25; Family & Business $35; Sustaining $50; Patron $100; Benefactor $250; Life $1,000.

Volga

BROOKINGS COUNTY HISTORICAL SOCIETY MUSEUM, 207 Samara Ave., Volga, SD 57071. Mailing Address: P.O. Box 608, Volga, SD 57071-0608. Tel.: 605-627-4025.
E-mail: wagner3540@brookings.net
Founded: 1939.
Congressional District: 5
Key Personnel: C.E.O., Chm. & Pres. (V), Lawrence Barnett; Vice Chm., Harold Christianson; Bd. Member, Barbara Behrend; Sec., Joanne Murphy; Treas., Phillip Wagner; Bd. Member, Dorothy Husher; Bd. Member, Chuck Cecil; Bd. Member, Donald Kleinjan; Bd. Member, Lyle Strande; Bd. Member, Jerry Leslie; Bd. Member, Deanna Rude; Bd. Member, Grace Linn; Bd. Member, Dick Berreth; Bd. Member, Marvin Steinbeck; Bd. Member, Cynthia Jacobson, Bd. Member, Leland Schlimmer; Bd. Member, Ronald Ladegaard; Bd. Member, Robert Buchheim; Bd. Member, Darla Strande.
Personnel Profile: Part-Time Volunteers 60.
Governing Authority: nonprofit organization. Parent Institution: Brookings County Historical Society, Volga, SD. Tax-exempt.
Institution Type/Description: Local History Museum.

Collections: county history & artifacts. Historic Buildings: 1880 Old Medary School; 1872 Sundet log cabin; early farm equipment building; historic house.
Facilities: library of documents, books & periodicals relating to Brookings County history available for use on request.
Activities: guided tours.
Publications: quarterly newsletter, The Window.
Hours & Admission Prices: Memorial Day to Labor Day daily 1-4 by appointment. No charge; donations accepted. &
Attendance: 600 (estimated)
Membership: Individual $10; Family $15; Pathfinder $25-$49; Pioneer $50-$99; Homesteader $100 & up.

Wall

WOUNDED KNEE MUSEUM, 207 10th Ave., Wall, SD 57790. Mailing Address: PMB 348, Wall, SD 57790. Tel.: 605-279-2573.
E-mail: info@woundedkneemuseum.org
Web Site: www.woundedkneemuseum.org
Founded: 2001.
Institution Type/Description: History Museum.
Collections: history of the Lakota Indian tribe & the events surrounding the Wounded Knee Massacre; photographs; Native American artifacts.
Facilities: Museum-related items for sale.
Activities: tours.
Hours & Admission Prices: May-Oct. 12 daily 9-5; groups by appointment. Adults $6; children under 12 no charge. &

Watertown

BRAMBLE PARK ZOO, (M), 800 10th St., N.W., Watertown, SD 57201. Mailing Address: P.O. Box 910, Watertown, SD 57201-0910. Tel.: 605-882-6269. Fax: 605-882-5232.
E-mail: dmiller@watertownsd.us
Web Site: brambleparkzoo.com
Founded: 1912.
Key Personnel: C.E.O. & Dir., Dan D. Miller; Pres., Donna Schoenbeck; Treas. City of Watertown (V), Greg Dievsen; Cur., Jim Lloyd; Education, Jaime Stricker; Museum Shop Mgr., Kim Konard.
Personnel Profile: Full-Time Paid 9; Part-Time Paid 12; Part-Time Volunteers 25; Interns 2.
Governing Authority: city; nonprofit. Subsidiary Institution: Lake Area Zoological Society. Tax-exempt: 501(c)(3).
Institution Type/Description: Zoo.
Collections: native South Dakota wildlife, exotic species & selected endangered wildlife; living collection contains 500 specimens representing 130 species & 17 endangered species; 116 specimens of mammals representing 37 species; 215 specimens of birds representing 69 species; 16 reptiles of 10 species & 17 endangered species.
Research Fields: the social interactions of a mixed species exhibit of black lemurs and black & white ruffed lemurs; vaginal swelling and estrus cycling of female black & white lemurs; captive wildlife management; exotic zoo medicine; zoo education; South Dakota wildlife management.
Facilities: 450-vol. library of books on natural history, captive management, wildlife conservation & animal research; cafeteria. Zoo-related items for sale.
Activities: guided tours; broadcast programs; formal education programs for college students. Annual Events: Water Fest; Kids Day; Animal Enrichment Day; Roots & Shoots Day; Zoo Boo; chili cook-off.
Publications: bimonthly for members, LAZS Newsletter.
Hours & Admission Prices: Summer: daily 9-8. Winter: daily 10-4 (weather permitting). Adults $7, children 3-12 $4; discounts to AZA members; members & children under 3 no charge. Closed New Year's Day; Thanksgiving; Christmas. &
Attendance: 57,211 (accurate)
Membership: Single $40; Family & Grandparent $60; Donor $100; Sustaining $150; Life $500; Patron $1,000; Benefactor $5,000.

CODINGTON COUNTY HERITAGE MUSEUM, 27 First Ave., S.E., Watertown, SD 57201-3612. Tel.: 605-886-7335.
Web Site: www.cchsmuseum.org
Founded: 1975.
Congressional District: 1
Key Personnel: Exec. Dir., Christy Lickei; Pres., Donus Roberts.
Personnel Profile: Full-Time Paid 1; Part-Time Paid 1; Part-Time Volunteers 10.
Governing Authority: society; nonprofit organization. Parent Institution: Codington County Historical Society, Inc. Tax-exempt 501(c)(3).
Institution Type/Description: Local County History Museum: housed in c.1905-06 Carnegie Library Building.

Collections: local history.
Research Fields: local history & prehistory; local Civil War & Spanish American War diaries; local historical cookbooks & recipes; local railroad history; local architectural history; community bands; area homesteaders.
Facilities: meeting room.
Activities: guided tours; lectures; traveling exhibits; research assistance for scholars & family history buffs. Annual Event: Historic Homes Tour.
Publications: books. The Civil War Diary of Arthur Calvin Mellette; Maggie: The Civil War Diary of Margaret Wylie Mellette; bimonthly newsletter, The Codington County Courier; Pictorial History of Codington County: 1875-1987; walking & driving tour brochures; quarterly newsletter, The Codington County Courier.
Hours & Admission Prices: June-Aug. Mon.-Fri. 10-5, Sat. 10-2; Sept.-May Mon.-Fri. 1-5. No charge; donations accepted. Closed holidays.
Attendance: 3,000 (estimated)
Membership: Individual $25; Family $40; Bronze $75; Silver $100; Gold $150; Diamond $300. Business: Trailblazer $75; Pioneer $100; Homesteader $150; Community Builder $300.

MELLETTE HOUSE, 421 Fifth Ave., N.W., Watertown, SD 57201. Mailing Address: 900 S. Lake Dr., Watertown, SD 57201-5460. Tel.: 605-886-4730.
Web Site: www.mellettehouse.org
Founded: 1943.
Congressional District: 1
Key Personnel: Pres. (V), Prudence K. Calvin; Treas. & Sec., Ann Edelman.
Personnel Profile: Part-Time Paid 5; Part-Time Volunteers 15.
Governing Authority: society. Tax-exempt.
Institution Type/Description: Historic House: 1885 Mellette House, home of Arthur Calvin Mellette, first Governor of South Dakota.
Collections: original furnishings; family portraits; heirlooms; Indian collection.
Research Fields: Mellette family & early Watertown history.
Activities: guided tours.
Publications: Arthur's Civil War Diary; Maggie's Civil War Diary; The Mellette's: A Family and a Home.
Hours & Admission Prices: May-Oct. Tues.-Sun. 1-5. No charge; donations accepted.
Attendance: 2,750 (estimated)

REDLIN ART CENTER, 1200 Mickelson Dr., Ste. 314, Watertown, SD 57201-7261. Tel.: 605-882-3877. Fax: 605-882-3922.
E-mail: redlinac@redlinart.com
Web Site: www.redlinart.com
Founded: 1997.
Key Personnel: Exec. Dir., Julie Ranum
Institution Type/Description: Art Center.
Collections: over 150 oil paintings by Terry Redlin.
Activities: group tours.
Hours & Admission Prices: Mon.-Fri. 9-5, Sat. 10-4, Sun. 12-4. No charge. &

Webster

MUSEUM OF WILDLIFE, SCIENCE & INDUSTRY, 760 W. Hwy. 12, Webster, SD 57274-2212. Mailing Address: P.O. Box 235, Webster, SD 57274-0235. Tel.: 605-345-4751.
E-mail: info@sdmuseum.org
Web Site: www.sdmuseum.org
Founded: 1985.
Key Personnel: Pres., Jim Rood; Cur., Elaine Gilbertson.
Governing Authority: Tax-exempt: 501(c)(3).
Institution Type/Description: General Museum.
Collections: culture & heritage of Northeastern South Dakota; period furniture; cars; farm equipment; African big game hunting trophies; 23 buildings.
Hours & Admission Prices: May 15-Labor Day Mon.-Sat. 10-5, Sun. 1-5; Labor Day-Oct. 15 daily 1-5; other times by appointment. Suggested donation $5.
Attendance: 1,600 (accurate)

Wessington Springs

JERAULD COUNTY PIONEER MUSEUM, 105 Main St., Wessington Springs, SD 57382. Mailing Address: P.O. Box 363, Wessington Springs, SD 57382-0132. Tel.: 605-539-9211.
Key Personnel: Pres. (V), Judy Winegar.
Governing Authority: Tax-exempt.
Institution Type/Description: History Museum.
Collections: memorabilia & print information relating to Jerauld County history; artifacts & archives; photographs.

Hours & Admission Prices: April-Dec. Thurs. 9:30-12 & 1-4:30, Fri. 1-3:30; other times by appointment. No charge; donations accepted.
Attendance: 350 (estimated)

Winner

TRIPP COUNTY HISTORICAL SOCIETY MUSEUM, E. Hwy. 18, Winner, SD 57580. Mailing Address: P.O. Box 287, Winner, SD 57580-0287. Tel.: 605-842-0704 & 0647.
E-mail: gjbowar@gwtc.net
Founded: 1970.
Key Personnel: Dir., Steve Davis.
Personnel Profile: Part-Time Paid 2.
Governing Authority: Tax-exempt.
Institution Type/Description: Historical Society Museum.
Collections: local history & culture; furnishings; photographs; period artifacts; Native American artifacts.
Hours & Admission Prices: Memorial Day to Labor Day Wed.-Sun. 2-5. No charge; donations accepted.
Attendance: 500 (estimated)
Membership: Children $3; Single $10; Couple $15; Organization $25; Life Single $100; Life Couple $150.

Yankton

BEDE ART GALLERY, Mount Mary College, Art Office, 1105 W. 8th St., Yankton, SD 57078-3725. Tel.: 605-668-1574.
E-mail: dkahle@mtmc.edu
Web Site: www.mtmc.edu
Key Personnel: Dir., David Kahle
Institution Type/Description: Art Gallery.
Collections: student artwork.
Hours & Admission Prices: Mon.-Fri. 8-8. No charge. &
Attendance: 10,000 (estimated)

CRAMER-KENYON HERITAGE HOME, 509 Pine St., Yankton, SD 57078-4036. Tel.: 605-665-7470.
E-mail: aswif52@gmail.com
Web Site: cramer-kenyon.webs.com
Founded: 1976.
Congressional District: 1
Key Personnel: Pres., Anne Swift; Vice Pres., Doug Sall
Governing Authority: Parent Institution: Heritage Homes. Tax-exempt.
Institution Type/Description: Historic House Museum: built in 1882 by the Secretary of Dakota Territory.
Collections: 1880s decor; gas & electric chandeliers; oil paintings; period furnishings; art work.
Hours & Admission Prices: Memorial Day to Labor Day Wed.-Sun. 1-5; other times by appointment. Adults $5, students $2.
Attendance: 600 (estimated)

DAKOTA TERRITORIAL MUSEUM, 610 Summit St., Yankton, SD 57078-3858. Tel.: 605-665-3898.
E-mail: director@dakotaterritorialmuseum.org
Web Site: www.dakotaterritorialmuseum.org
Founded: 1961.
Congressional District: 1
Key Personnel: Dir., Crystal Nelson; Museum Shop Mgr., Heidi Henson.
Personnel Profile: Full-Time Paid 2; Part-Time Volunteers 11.
Governing Authority: society. Parent Institution: Yankton County Historical Society. Tax-exempt: 501(c)(3).
Institution Type/Description: Historical Society Museum.
Collections: Missouri River transportation material; manuscripts; books, photographs & objects pertaining to Dakota territory history; Yankton's role as territorial capital; Plains Indian ethnographic material; 19th-20th century Yankton county history; Max Copper fishing collection. Historic Buildings: c.1900 Gunderson rural schoolhouse; Dakota Territorial council building; c.1890 Great Northern RR depot; Cook blacksmith shop c.1880; 1860 Hovden pioneer cabin.
Research Fields: railroad history; Plains Indians; Dakota territory.
Facilities: Art prints & photo prints for sale.
Activities: guided tours, docent program, periodic lecture series.
Hours & Admission Prices: May-Sept. Mon.-Fri. 10-5, Sat.-Sun. 12-4; Oct.-April daily 12-4. No charge; donations accepted. Closed New Year's Eve & Day; Good Friday; Easter; Thanksgiving; Christmas Eve & Day. &
Attendance: 5,780 (estimated)
Membership: Individual $35; Family $50; Pioneer $100; Explorer $250; Riverboat Captain $1,000.

GAVINS POINT NATIONAL FISH HATCHERY AND AQUARIUM, 31227 436th Ave., Yankton, SD 57078-6364. Tel.: 605-665-3352. Fax: 605-665-3360.
E-mail: r6ffa_gav@fws.gov
Web Site: www.fws.gov/gavinspoint
Founded: 1960.
Congressional District: 1
Key Personnel: Project Leader, Marc Jackson.
Personnel Profile: Full-Time Paid 7; Part-Time Paid 1; Part-Time Volunteers 5.
Governing Authority: federal. Parent Institution: U.S. Department of the Interior, Washington, DC. Subsidiary Institution: U.S. Fish & Wildlife Service. Tax-exempt.
Institution Type/Description: Aquarium.
Collections: aquatic displays consisting of 40-50 Missouri River Basin fish, reptiles & amphibians.
Activities: guided tours.
Publications: brochure, Gavins Point National Fish Hatchery.
Hours & Admission Prices: Aquarium: April-Oct. daily 10-4. Hatchery: daily 8-4; guided group tours by appointment. No charge; donations accepted. &
Attendance: 110,000 (accurate)

TENNESSEE

(312 listings)

Adamsville

BUFORD PUSSER HOME & MUSEUM, 342 Pusser St., Adamsville, TN 38310-2338. Mailing Address: P.O. Box 301, Adamsville, TN 38310. Tel.: 731-632-4080. Fax: 731-632-4085.
E-mail: info@bufordpussermuseum.com
Web Site: www.bufordpussermuseum.com
Founded: 1988.
Congressional District: 7
Key Personnel: Cur., Renee Moss.
Personnel Profile: Part-Time Paid 3; Part-Time Volunteers 2.
Governing Authority: city. Tax-exempt.
Institution Type/Description: Historic House Museum.
Collections: Pusser's life & career; period furnishings; personal artifacts; photographs; cars; guns.
Facilities: Museum-related items for sale.
Hours & Admission Prices: May-Oct. Mon.-Fri. 10-5, Sat. 9-5, Sun. 1-5; Nov.-April Mon.-Fri. 11-4, Sat. 9-4, Sun. 1-4.
Attendance: 4,680 (accurate)

Alamo

TENNESSEE SAFARI PARK, 637 Conley Rd., Alamo, TN 38001-4106. Tel.: 731-696-4423; 901-734-6005. Fax: 731-696-2400.
E-mail: information@tennesseesafaripark.com
Web Site: www.tennesseesafaripark.com
Institution Type/Description: Zoo.
Collections: wildlife & their habitats including over 60 species.
Activities: special events; educational programs; birthday parties.
Hours & Admission Prices: early Spring to late Fall Mon.-Sat. 10-4:30 (last car going through), Sun. 12-4:30 (last car going through). Adults $12, children 2-12 $8; children one & under no charge.

Athens

MCMINN COUNTY LIVING HERITAGE MUSEUM, 522 W. Madison Ave., Athens, TN 37303. Mailing Address: P.O. Box 889, Athens, TN 37371-0889. Tel.: 423-745-0329 & 744-8100. Fax: 423-745-0329.
E-mail: livingheritagemuseum@livingheritagemuseum.com
Web Site: www.livingheritagemuseum.com
Founded: 1982.
Congressional District: 2
Key Personnel: Pres., Gerald Martin; Vice Pres., Laura Brown LeNoir; Treas., Tom Biddle; Trustee, Muriel Mayfield; Exec. Dir., Diane Hutsell, Museum Shop Mgr., Robin Muller.
Personnel Profile: Full-Time Paid 3; Full-Time Volunteers 2; Part-Time Paid 3; Part-Time Volunteers 175; Interns 2.
Governing Authority: nonprofit organization. Tax-exempt: 501(c)(3).
Institution Type/Description: History Museum.
Collections: pioneer & Victorian furnishings; decorative arts; needlework; fine art; period costumes; 19th- & early 20th-century quilts; 1880-1930 Burn Industrial Room General Store; 20th-century industrial items; 19th- & 20th-century bisque & china head dolls, toys & clothing; 1850-1935 desks, books

& maps; glass ceramics; farming equipment, Indian artifacts, 1820-1900 religious artifacts; 1850-1930 Tennessee Wesleyan College artifacts; 1860-1930 medical, legal & legislative artifacts; war artifacts.
Facilities: 65-vol. library for inter-library loan. Books & related items for sale.
Activities: guided tours; lectures; hobby workshops; organized education programs for children; participatory, loan, temporary & traveling exhibitions. Museum Sponsors: Annual Quilt Show.
Publications: brochure, In Preparation for Life 1840-1940; William E. Nash-His Life; The Heritage Cookbook (2nd printing); Living Legacies-A Guide to Museum Areas & Artifacts.
Hours & Admission Prices: Mon.-Fri. 10-5. Adults $5, senior citizens & students $3; discounts to AAA & AAM members; members no charge. Closed New Year's Day; Easter; Memorial Day; Independence Day; Labor Day; Thanksgiving; Christmas. &
Attendance: 22,000 (accurate)
Membership: Student & Senior Citizen $10; Individual $15; Family $35; Life $300; Patron $1,000.

SWIFT MUSEUM, McMinn County Airport, 217 County Rd. 552, Athens, TN 37371. Mailing Address: P.O. Box 644, Athens, TN 37371. Tel.: 423-745-9547. Fax: 423-745-9869.
E-mail: swiftlypam@aol.com
Web Site: www.swiftmuseumfoundation.org
Key Personnel: Exec. Dir., Pam Nunley
Institution Type/Description: History Museum.
Collections: Globe & Temco Swift aircraft from 1946-1951; T-35 Buckaroo USAF trainers; photographs; memorabilia.
Hours & Admission Prices: Mon.-Sat. 8-4 by appointment.

Belvidere

MUSEUM OF POWER & INDUSTRY AT FALLS MILL, 134 Falls Mill Rd., Belvidere, TN 37306-2225. Tel.: 931-469-7161. Fax: 931-469-7161.
E-mail: fallsmill@tnco.net
Web Site: www.fallsmill.com
Key Personnel: Owner, John Lovett; Owner, Jane Lovett
Institution Type/Description: Historic Building: an operating water-powered grain mill; built in 1873. Listed on the National Register of Historic Places.
Collections: mill history; period furnishings & machinery; printing press; dog-powered butter churn; hand looms; spinning wheels; 19th century power looms; wool carding machines.
Facilities: Mill-related items for sale.
Hours & Admission Prices: mid-Jan. to Dec. 23 Mon.-Tues. & Thurs.-Sat. 9-4, Sun. 12:30-4. Adults $4, senior citizens $3, children under 14 $2. Closed Thanksgiving.

Benton

NANCY WARD MUSEUM - POLK COUNTY HISTORICAL AND GENEALOGICAL SOCIETY, Commerce & Poplar Sts., Benton, TN 37307. Mailing Address: P.O. Box 636, Benton, TN 37307. Tel.: 423-338-1005.
Institution Type/Description: Historical Society Museum.
Collections: local history, culture & heritage; life of Nancy Ward; Cherokee Indian history; photographs; period artifacts; genealogical records.
Hours & Admission Prices: Call for hours.

Blountville

OLD DEERY INN AND MUSEUM, 3411 Hwy. 126, Blountville, TN 37617. Mailing Address: P.O. Box 3179, Blountville, TN 37617. Tel.: 423-323-4660.
E-mail: info@sullivancountytn.gov
Web Site: www.historicsullivan.com
Founded: 2000.
Key Personnel: Dir., Shelia Hunt; Pres., Dennis Houser.
Personnel Profile: Full-Time Volunteers 1; Part-Time Volunteers 5.
Governing Authority: county government. Tax-exempt: 501(c)(3).
Institution Type/Description: Historic Building: built in c.1801 by William Deery.
Collections: Tennessee history; early artifacts & furnishings; paintings; personal artifacts.
Activities: guided tours; special events.
Hours & Admission Prices: By appointment. Adults $5, children $3.
Attendance: 750 (estimated)

Bristol

ERNIE FORD HOME & MUSEUM, 1223 Anderson St., Bristol, TN 37620. Mailing Address: P.O. Box 204, Bristol, TN 37621. Tel.: 423-989-4850.
E-mail: bha@bristolhistoricalassociation.com
Web Site: www.bristolhistoricalassociation
Founded: 1979.
Key Personnel: Chm. (V), Brenda Otis; Pres. (V), Tim Buchanan; Museum Shop Mgr., Brenda Otis.
Personnel Profile: Full-Time Paid 1; Part-Time Volunteers 4.
Volunteer Hours: 60
Operating Expenses: 1,500
Operating Income: 100
Governing Authority: Parent Institution: Bristol Historical Assn. Tax-exempt.
Institution Type/Description: Historic House Museum: housed in the birthplace of Ernie Ford; c.1900.
Collections: Ernie Ford's life, family & career; local history & culture; photographs; early musicians; personal artifacts; biographical; recordings; publicity.
Publications: Newsletter; month email.
Hours & Admission Prices: By appointment. Adults $3. &
Attendance: 100 (estimated)
Membership: Student $5; Single $35; Family $45.

STEELE CREEK PARK NATURE CENTER, 80 Lake Shore Dr., Bristol, TN 37620. Tel.: 423-989-5616.
E-mail: jstout@bristoltn.org
Web Site: www.bristoltn.org/naturecenter.cfm
Institution Type/Description: Nature Center.
Collections: local natural history; live animals.
Facilities: nature trails.
Activities: educational programs; guided tours; summer nature camps.
Hours & Admission Prices: Nature Center: Mon.-Fri. 10-4, Sat. 10-6, Sun. 1-6. Park: daily 9-9. Park Entrance: $1 per car.

Brownsville

JOHN ESTES HOUSE, 121 Sunny Hill Cove, Brownsville, TN 38012-8367. Tel.: 731-779-9000.
E-mail: info@westtnheritage.com
Web Site: www.westtnheritage.com
Institution Type/Description: Historic House Museum: housed in the former home of blues legend, John Adam Estes.
Collections: John Adam Estes' life & career; photographs; biographies.
Hours & Admission Prices: Call for hours. No charge; donations accepted.
Attendance: 20,000 (accurate)

Bulls Gap

ARCHIE CAMPBELL TOURISM COMPLEX AND MUSEUM, 139 S. Main St., Bulls Gap, TN 37711. Mailing Address: P.O. Box 181, Bulls Gap, TN 37711. Tel.: 423-235-5216.
Institution Type/Description: History Museum: housed in the childhood home of Hee-Haw star, Archie Campbell.
Collections: Archie Campbell's life & career; personal artifacts; photographs; memorabilia.
Hours & Admission Prices: By appointment.

Burns

CUMBERLAND PRESBYTERIAN CHURCH BIRTHPLACE SHRINE, Montgomery Bell State Park, Burns, TN 37029. Mailing Address: 8207 Traditional Place, Cordova, TN 38016-7414. Tel.: 901-276-8602. Fax: 901-272-3913.
E-mail: archives@cumberland.org
Web Site: www.cumberland.org/hfcpc/
Founded: 1810.
Congressional District: 6
Key Personnel: Dir. & C.E.O., Susan Knight Gore; Pres. (V), Gwen McReynolds.
Personnel Profile: Full-Time Paid 1; Part-Time Volunteers 2.
Governing Authority: church. Parent Institution: Cumberland Presbyterian Church, 8207 Traditional Place, Cordova, TN. Tax-exempt: 501(c)(3).
Institution Type/Description: Historic Shrine: located in Montgomery Bell State Park, Dickson, Tennessee.
Collections: c.1800 replica Rev. Samuel McAdow home.
Facilities: 100-seat chapel.
Activities: Sunday morning worship service from June-Sept.; weddings.

Publications: booklets.
Hours & Admission Prices: Daily 8-6. No charge.

Butler

BUTLER MUSEUM, 123 Selma Curtis Rd., Butler, TN 37640. Tel.: 423-768-3880.
Web Site: www.thebutlermuseum.com
Institution Type/Description: History Museum.
Collections: local history, culture & heritage; personal artifacts; period furnishings; photographs; farming, manufacturing, & lumbering equipment; clothing. Historic Building: W.S. Stout's General Store.
Facilities: Museum-related items for sale.
Activities: group tours.
Hours & Admission Prices: Memorial Day to Labor Day Sat.-Sun. 1-4; other times by appointment. Adults $3, children $2.

Byrdstown

BORDERLANDS EXHIBIT, 109 W. Main St., Byrdstown, TN 38549. Tel.: 931-864-7195; 888-406-4704.
E-mail: info@theborderlands.org
Institution Type/Description: History Museum.
Collections: local history & culture; Civil War artifacts; period artifacts; photographs.
Hours & Admission Prices: Call for hours.

CORDELL HULL BIRTHPLACE & MUSEUM STATE HISTORIC PARK, 1300 Cordell Hill Memorial Dr., Byrdstown, TN 38549-4627. Tel.: 931-864-3247.
E-mail: david.delk@tn.gov
Web Site: www.cordellhullmuseum.com
Founded: 1997.
Institution Type/Description: History Museum: birthplace of Cordell Hull, appointed Secretary of State by President Franklin D. Roosevelt.
Collections: life of Cordell Hull; personal artifacts; photographs;
Facilities: Museum-related items for sale.
Activities: special events & programs. Museum Sponsors: Cordell Hull Folk Festival in September.
Hours & Admission Prices: April-Oct. daily 9-5; Nov.-March daily 9-4. No charge. Call the park for closings. &
Attendance: 86,400 (accurate)

Camden

TENNESSEE RIVER FRESHWATER PEARL MUSEUM FARM TOUR, Birdsong Resort, Marina & Campground, 255 Marina Rd., Camden, TN 38320-7832. Tel.: 731-584-7880, ext. 1 (Bookings). Fax: 731-584-3625.
E-mail: bob@birdsongresort.com
Web Site: www.tennesseeriverpearls.com
Formerly: A Pearl of a Tour
Founded: 1961.
Key Personnel: Owner & Tour Guide, Bob Keast.
Personnel Profile: Full-Time Paid 12.
Institution Type/Description: General Museum.
Collections: freshwater pearls & musseling industry.
Hours & Admission Prices: Mon.-Sat. 8-5, Sun. 1-4. Full Tour: adults $49.50. Mini Tour: adults $29.50. &
Attendance: 250,000 (accurate)

Carthage

SMITH COUNTY HERITAGE MUSEUM, 107 3rd St., Carthage, TN 37030-1472. Mailing Address: P.O. Box 73, Carthage, TN 37030-0073. Tel.: 615-735-1104.
Institution Type/Description: History Museum.
Collections: farming; agriculture; military; education; local churches; architecture; photographs.
Facilities: library. Museum-related items for sale.
Hours & Admission Prices: Wed. & Fri.-Sat. 10-2. No charge.

Castalian Springs

CRAGFONT-HISTORIC CRAGFONT, INC., 200 Cragfont Rd., Castalian Springs, TN 37031-4743. Mailing Address: 1011 Durham Dr., Gallatin, TN 37066-3411. Tel.: 615-452-7070.
Web Site: www.cragfont.com
Founded: 1958.
Congressional District: 4
Key Personnel: C.E.O., Mrs. Judith Wherry; Chm. Restoration, Mrs. James Gourley; Cur., Mrs. Lowell Fayna; Museum Shop Mgr., Margaret Fayna.
Personnel Profile: Full Time Paid 2; Part Time Volunteers 30.
Governing Authority: Tax-exempt: 501(c)(3)
Institution Type/Description: Historic House Museum: c.1798-1802 home of General James Winchester.
Collections: American Federal furnishings; stenciled walls in parlor; period farming tools; archives.
Research Fields: archives.
Facilities: 90-vol. library of original Cragfont books dating from 18th-19th century; restored original flower garden.
Activities: guided tours; lectures; permanent exhibitions.
Publications: descriptive brochure; cookbook; booklets, Cragfont; Haunts, Hauntings & Legends.
Hours & Admission Prices: mid-April to Oct. Tues.-Sat. 10-5, Sun. 1-5; Nov. to mid-April, by appointment. Adults $5, children 6-12 $3; discounts to seniors & groups; members & children under 6 no charge. &
Attendance: 4,230 (accurate)
Membership: Individual $15; Family $25.

Celina

CLAY COUNTY MUSEUM, INC., 805 Brown St., Celina, TN 38551. Mailing Address: P.O. Box 684, Celina, TN 38551. Tel.: 931-243-4220.
Founded: 1986.
Congressional District: 4
Key Personnel: Pres., Mary Loyd Reneau
Institution Type/Description: History Museum.
Collections: county history & culture; period furnishings; personal artifacts; photographs. Historic Building: law office.
Hours & Admission Prices: Call for hours. No charge; donations accepted. &

Centerville

GRINDER'S SWITCH CENTER, 405 W. Public Sq., Centerville, TN 37033-1606. Mailing Address: c/o Hickman County Chamber of Commerce, P.O. Box 126, Centerville, TN 37033. Tel.: 931-729-5774.
Institution Type/Description: History Museum.
Collections: local history; period furnishings & clothing; personal artifacts; photographs.
Hours & Admission Prices: Call for hours.

Chattanooga

BESSIE SMITH CULTURAL CENTER, 200 E. Martin Luther King Blvd., Chattanooga, TN 37403. Mailing Address: P.O. Box 11493, Chattanooga, TN 37401-2493. Tel.: 423-266-8658 & 267-1628. Fax: 423-267-1076.
E-mail: info@bessiesmithcc.org
Web Site: www.bessiesmithcc.org
Formerly: Chattanooga African Museum/Bessie Smith Hall
Founded: 1983.
Congressional District: 3
Key Personnel: Exec. Dir., Rose M. Martin; Pres. (V), Irvin Overton; Membership, Pat Lott; Education, Carman Davis; Facilities, Danny Toney.
Personnel Profile: Full-Time Paid 4; Part-Time Paid 2; Part-Time Volunteers 19; Interns 2.
Governing Authority: nonprofit organization. Tax-exempt: 501(c)(3).
Institution Type/Description: African-American History Museum.
Collections: African artifacts; recordings; oral history tapes; local artists; VHS tapes of national personalities; photographs of early local prominent citizens; prints by national artists; original clothing from local donors; Bessie Smith collection of local African-American history; oral history tapes (VHS); local history.
Facilities: 10,000-vol. library of fiction & reference books available to the public for research on premises; auditorium. Museum-related items for sale.
Activities: arts festivals; guided tours; lectures; formally organized education programs for children, undergraduate & graduate students; summer art programs featuring crafts, African drumming, dance, music history & museum intern career programs; art fun factory after school program; community outreach program.
Publications: quarterly newsletter, The Heritage; guide to the Chattanooga permanent exhibit; Life Series - Roland Carter, Russell Goode, Black America Series, Chattanooga; workbooks, Chattanooga African American Legacy; Biography & Achievements of the Colored Citizens of Chattanooga in 1904.

Hours & Admission Prices: Mon.-Fri. 10-5, Sat. 12-4. Adults $7, seniors & students with I.D. $5, children 6-12 $3; discounts to groups, and AAM & NAAM members; children 5 & under no charge. Closed national & legal holidays. &

Attendance: 156,000 (accurate)

Membership: Senior & Student $20; Individual $40; Family $70; Contributing $100; Organization $250; Corporate $600; Thousandaire $1,000.

CHATTANOOGA AFRICAN AMERICAN MUSEUM, 200 E. Martin Luther King Jr. Blvd., Chattanooga, TN 37403. Mailing Address: P.O. Box 11493, Chattanooga, TN 37401. Tel.: 423-266-8658.

Institution Type/Description: History Museum.

Collections: African American history, culture & industry; Bushman's hut replica; Ethiopia church model; mural; photographs; clothing; personal artifacts.

Hours & Admission Prices: Mon.-Fri. 10-5, Sat. 12-4.

CHATTANOOGA ARBORETUM & NATURE CENTER, 400 Garden Rd., Chattanooga, TN 37419-1807. Tel.: 423-821-1160, ext. 108. Fax: 423-821-1702.

E-mail: dmorgan@chattanature.org

Web Site: chattanoogaanc.org

Formerly: Tennessee Wildlife Center

Founded: 1978.

Congressional District: 3

Key Personnel: Exec. Dir., Dr. Jean Lomino; Pres. (V), John Mitchum; Dir. Education, Corey Hagen; Naturalist, Public Program Coord., Susan Russell; Member Svcs. Mgr. & Museum Shop Mgr., Diane Morgan; Cur. Wildlife, Tish Gailmard; Dir. Devel. & Mktg., Tina Harvey Crawford.

Personnel Profile: Full-Time Paid 5; Part-Time Paid 7; Part-Time Volunteers 58.

Governing Authority: nonprofit organization. Tax-exempt: 501(c)(3).

Institution Type/Description: Nature Center: located on Lookout Creek on the western side of Lookout Mountain, where the Battle of Lookout Mountain began; now part of species survival program for endangered Red wolves & other animals indigenous to the southeast region.

Collections: wolves; bald eagle; bobcats; raptors; other mammals.

Research Fields: herpetology; botany; wildflower ecology; endangered red wolves.

Facilities: botanical garden; nature & conservation center; reading room; 60-seat auditorium; 1,200 ft. wetland walkway; classrooms. Museum-related items for sale.

Activities: guided tours; lectures; formally organized education programs; permanent, temporary & traveling exhibitions.

Publications: monthly newsletter, Native Ground; wildlife booklet, First Aid & Home Care For Wildlife.

Hours & Admission Prices: April-Oct. Mon.-Sat. 9-5, Sun. 1-5; Nov.-March Mon.-Sat. 9-5. Adults $8, children & senior citizens $5; members no charge. Closed New Year's Day; Easter; Memorial Day; Independence Day; Labor Day; Thanksgiving; Christmas. &

Attendance: 61,000 (estimated)

Membership: Guest Option $25; Individual Plus $45; Family $60; Friend $150-$499; Patron $500-$999; Curator $1,000-$4,999; Conservator $5,000.

CHATTANOOGA HISTORY CENTER, (M), 2 W. Aquarium Way, Ste. 200, Chattanooga, TN 37402. Tel.: 423-265-3247. Fax: 423-266-9280.

E-mail: dblack@chattanoogahistory.org

Web Site: www.chattanoogahistory.org

Founded: 1978.

Congressional District: 3

Key Personnel: Exec. Dir. & Cur., Dr. Daryl Black; Deputy Dir., Marlene Payne; Pres. (V), Jo Coke.

Personnel Profile: Full-Time Paid 3; Part-Time Paid 2; Part-Time Volunteers 100.

Governing Authority: nonprofit organization. Tax-exempt: 501(c)(3).

Institution Type/Description: Regional History Museum.

Collections: objects pertaining to 10,000 years of human history in southeastern Tennessee, northern Georgia & northeastern Alabama from 1800-present.

Research Fields: Cherokee, Civil War, mid 20th century business & industry, local environmental changes.

Activities: guided tours; lectures; workshops for children & adults; special events.

Publications: membership newsletter, The Connection.

Hours & Admission Prices: Office: Mon.-Fri. 9-5. &

Membership: Individual $50; Family $75.

CHATTANOOGA ZOO AT WARNER PARK, 301 N. Holtzclaw Ave., Chattanooga, TN 37404-2823. Tel.: 423-697-1319. Fax: 423-697-1329.

E-mail: long_d@chattzoo.org

Web Site: www.chattzoo.org

Formerly: Warner Park Zoo

Founded: 1937.

Congressional District: 3

Key Personnel: Dir. & Museum Shop Mgr., Dardenelle Long; Pres. (V), Gary Chazen.

Personnel Profile: Full-Time Paid 25; Part-Time Paid 24; Part-Time Volunteers 85; Interns 2.

Governing Authority: nonprofit. Parent Institution: City of Chattanooga. Subsidiary Institution: Friends of the Zoo. Tax-exempt.

Institution Type/Description: Zoo.

Collections: animals from around the world.

Activities: summer camp; outreach programs.

Publications: magazine, ChattaZooga.

Hours & Admission Prices: Daily 9-5. Adults $8.95; discounts to AZA members; members no charge. Closed New Year's Day; Thanksgiving; Christmas. &

Attendance: 207,000 (accurate)

Membership: Individual $35.95; Family & Grandparent $55.95; Family Plus $75; Family Deluxe $139.95.

CREATIVE DISCOVERY MUSEUM, (M), 321 Chestnut St., Chattanooga, TN 37402-4902. Mailing Address: P.O. Box 6339, Chattanooga, TN 37401-6339. Tel.: 423-756-2738. Fax: 423-267-9344.

E-mail: info@cdmfun.org

Web Site: www.cdmfun.org

Founded: 1995.

Congressional District: 3

Key Personnel: C.E.O., Henry H. Schulson; Chm. (V), Kristina Montague; Devel. & Public Rels., Lynda LeVar; Education, Jayne Griffin; Bldg. & Exhibit Maintenance, Ken Crider; Visitor Svcs., Carla Parrish; Museum Shop Mgr., Brenda Baskette.

Personnel Profile: Full-Time Paid 21; Part-Time Paid 41; Part-Time Volunteers 144.

Governing Authority: private; nonprofit organization. Tax-exempt: 501(c)(3).

Institution Type/Description: Children's Museum.

Collections: hands-on exhibits.

Facilities: 26,000 sq. ft. exhibit space; 120-seat auditorium; 25-seat cafe; classroom; art workshop; conservatory. Museum-related items for sale.

Activities: arts festivals; concerts; dance recitals; docent program; films; guided tours; hobby workshops; lectures; rental gallery; theater; participatory & traveling exhibitions; homeschool workshops; birthday program; overnight program; science demonstrations; walk-up art & music lessons.

Publications: quarterly newsletter, GO!; educator's newsletter, Discoveries.

Hours & Admission Prices: March-June 20 Mon.-Sat. 10-5; June 21-Aug. 10 daily 9:30-5:30; Aug. 11-Sept. 1 daily 10-5; Sept. 2-Nov. Mon.-Tues. & Thurs.-Sat. 10-5, Sun. 12-5; Dec.-Feb. Mon.-Tues. & Thurs.-Sat. 10-5. Adults $11.95; members no charge. Closed Thanksgiving; Christmas Eve & Day. &

Attendance: 206,323 (accurate)

Membership: My Child & Me $75; My Family & Grandparents $95; Passport $150.

CRESS GALLERY OF ART, 752 Vine St., Chattanooga, TN 37403. Mailing Address: University of Tennessee at Chattanooga, 615 McCallie Ave., #1305, Chattanooga, TN 37403-2504. Tel.: 423-304-9789. Fax: 423-425-2101.

E-mail: ruth-grover@utc.edu

Web Site: www.cressgallery.org

Founded: 1980.

Congressional District: 3

Key Personnel: Dir. & Cur., Ruth Grover.

Personnel Profile: Full-Time Paid 1; Part-Time Paid 3.

Governing Authority: university. Affiliated with University of Tennessee at Chattanooga. Tax-exempt: 501(c)(3).

Institution Type/Description: Art Institute.

Collections: graphics; paintings; sculpture; photographs.

Major Exhibits: UTC Dept. of Art Juried Student Exhibition, 1/14; Michelle Segre: Sculpture and Drawing, 2/14-3/14; UTC BFA Senior Thesis Exhibitions, 4/14.

Activities: gallery talks; inter-museum loan, temporary & traveling exhibitions; visiting artist series; annual juried student & senior thesis exhibitions; biennial art faculty exhibition.

Hours & Admission Prices: Sept.-May Mon.-Fri. 9:30-7:30, Sat.-Sun. 1-4. No charge. Closed university holidays. &

Attendance: 6,500 (estimated)

Membership: Friends of the Gallery $50 & up.

DRAGON DREAMS MUSEUM & GIFT SHOP, 6722 E. Brainerd Rd., Chattanooga, TN 37421. Tel.: 423-892-2384.

Web Site: www.dragonvet.com/

Founded: 1990.

Key Personnel: Dir. & Owner, Barbara Newton, DVM

Personnel Profile: Full-Time Paid 1; Full-Time Volunteers 1.

Institution Type/Description: Dragon Art Museum.

Collections: hand-crafted figurines & furniture; wood, silver, jade, ivory, pewter, porcelain & fabric dragons.

Facilities: Museum-related items for sale.

Activities: special events.

Hours & Admission Prices: Wed.-Sat. 10-6, Sun. 1-6. Adults $10, children $4.50. Closed New Year's Eve & Day; Independence Day; Thanksgiving; Christmas Eve & Day. &

HOUSTON MUSEUM OF DECORATIVE ARTS, 201 High St., Chattanooga, TN 37403-1185. Tel.: 423-267-7176.

E-mail: thehoustonmuseum@gmail.com

Web Site: www.thehoustonmuseum.com

Founded: 1949.

Congressional District: 3

Key Personnel: Pres. (V), Barbara Wofford.

Personnel Profile: Part-Time Paid 1; Part-Time Volunteers 100.

Governing Authority: nonprofit organization. Tax-exempt: 501(c)(3).

Institution Type/Description: Antique Museum: housed in 1890 building. Decorative Arts Museum.

Collections: early American, French & English glass; Victorian art glass; pressed glass; Tiffany glass; 19th-century decorative arts; colonial through 19th-century furniture; pewter & Mettlach steins; porcelains; dolls; pitcher collection.

Research Fields: glass; pottery; porcelain; furniture; dolls; textiles.

Facilities: library.

Activities: guided tours; lectures; films; gallery talks; TV programs; docent program; permanent exhibitions.

Publications: newssheet.

Hours & Admission Prices: Thurs.-Sun. 12-4. Adults $9; discount to AAM members; children under 3 & members no charge. Closed major holidays.

Attendance: 10,000 (estimated)

Membership: Student & Senior $25; Individual $35; Family $45; Sponsor $50; Patron $100; Benefactor $250; Grand Benefactor $500; Anna Safley Houston Club $1,000, $2,000 & $5,000.

✱ **HUNTER MUSEUM OF AMERICAN ART, (M),** 10 Bluff View, Chattanooga, TN 37403-1197. Tel.: 423-267-0968. Fax: 423-267-9844.

Web Site: www.huntermuseum.org

Founded: 1952.

Congressional District: 3

Key Personnel: Chm. (V), Norma Mills; Chief Cur., Ellen Simak; Cur. Contemporary, Nandini Makrandi; Registrar, Theresa Slowikowski; Exhibits Designer, John Hare; Museum Shop Mgr., Mary Phlaum.

Personnel Profile: Full-Time Paid 20; Part-Time Paid 36; Part-Time Volunteers 99; Interns 2.

Governing Authority: nonprofit. Tax-exempt: 501(c)(3).

Institution Type/Description: Art Museum: housed in a restored mansion.

Collections: American painting, graphics & sculpture of 18th-21st centuries.

Research Fields: American art.

Facilities: 250-vol. non-circulating library of art books & periodicals; 150-seat auditorium; cafe; studios. Gift items for sale.

Activities: guided tours; lectures; films; gallery talks; permanent, temporary & traveling exhibitions; docent program; workshops; studio classes. Special Events: Spectrum, silent & live auction gala.

Publications: quarterly members magazine; exhibitions catalogs; A Catalogue of the American Collection, Hunter Museum of American Art, Volume I & II.

Hours & Admission Prices: Mon.-Tues. & Fri.-Sat. 10-5, Wed. & Sun. 12-5, Thurs. 10-8. Adults $9.95, seniors $8.95, youth 3-17 $4.95; discounts to AAM & ICOM members; children under 3 & members no charge. North American reciprocal admission. &

Attendance: 57,182 (accurate)

Membership: Individual $45; Family $65; Donor $100; Sponsor $350; Chairman's Circle Member $1,000; Chairman's Circle Benefactor $2,500; Chairman's Circle Grand Benefactor $5,000.

INTERNATIONAL TOWING & RECOVERY HALL OF FAME & MUSEUM, 3315 S. Broad St., Chattanooga, TN 37408-3052. Tel.: 423-267-3132.

E-mail: internationaltowingmuseum@comcast.net

Web Site: www.internationaltowingmuseum.org

Key Personnel: Exec. Dir., Cheryl H. Mish

Institution Type/Description: History Museum.

Collections: early towing & recovery history; Hall of Fame; photographs period wreckers & equipment; toys; tools; industry.

Activities: special events.

Hours & Admission Prices: March-Oct. Mon.-Sat. 9-5, sun. 11-5; Mon.-Sat. 10-4:30, Sun. 11-5. Adults $8, seniors 55 & over $7, children 6-18 $4; discounts to AAA members; children under 5 no charge. Closed major holidays.

NATIONAL MEDAL OF HONOR MUSEUM OF MILITARY HISTORY, Northgate Mall, Hwy. 153, Chattanooga, TN 37403. Mailing Address: P.O. Box 11467, Chattanooga, TN 37401-2467. Tel.: 423-877-2525 (museum) & 698-4511 (archives).

E-mail: info@mohm.org

Web Site: www.mohm.org

Formerly: Medal of Honor Museum

Founded: 1987.

Congressional District: 3

Key Personnel: Dir., Jim Wade; Chm. (V), Dan Saieed.

Personnel Profile: Part-Time Volunteers 17.

Governing Authority: nonprofit. Tax-exempt: 501(c)(3).

Institution Type/Description: History & Military Museum.

Collections: uniforms, pre-Civil War to present; weapons, 1776-present; military prints & paintings; medals; badges; Nuremburg collection; Confederate collection; currency collection, 1774-present; helmets; military history library; international marching band recording collection; Medal of Honor citations; photographic archives.

Publications: museum brochure; newsletter.

Hours & Admission Prices: Tues.-Fri. 11-5, Sat.-Sun. 1-5. No charge; donations accepted. Closed New Year's Day; Easter; Thanksgiving; Christmas. &

Attendance: 6,500 (estimated)

Membership: Friends: Squad $20; Corporate $1,000.

REFLECTION RIDING ARBORETUM & BOTANICAL GARDEN, 400 Garden Rd., Chattanooga, TN 37419. Tel.: 423-821-9582.

Institution Type/Description: Arboretum & Botanical Garden

Collections: plants; trees; flowers.

Facilities: 300-acre site.

Activities: educational programs; rental facilities.

Hours & Admission Prices: April-Oct. Mon.-Sat. 9-5, Sun. 1-5; Nov.-March Mon.-Sat. 9-5.

SISKIN MUSEUM OF RELIGIOUS AND CEREMONIAL ART, 1101 Carter St., Chattanooga, TN 37402-5017. Tel.: 423-648-1700. Fax: 423-648-1749.

E-mail: jan.hollingsworth@siskin.org

Web Site: www.siskin.org/museum

Founded: 1950.

Congressional District: 3

Key Personnel: Pres., Gerald Jensen; Dir. Mktg., Jan Hollingsworth.

Governing Authority: individual operation. Parent Institution: Siskin Memorial Foundation. Tax-exempt: 501(c)(3).

Institution Type/Description: Religious Museum.

Collections: 453 objects of art in silver, ivory, wood, brass & pewter; religious & ceremonial arts; objects pertaining to Judaica & Christian religions, Buddhism, Hinduism, Confucionism and others; rare books.

Research Fields: religious art & artifacts from Europe & Asia.

Activities: guided tours; lectures; permanent exhibitions.

Hours & Admission Prices: Mon.-Fri. 9-4; other times by appointment. No charge; donations accepted. Closed New Year's Eve & Day; Christmas Eve, Day & week; national holidays. &

TANNER HILL GALLERY, 3069 S. Broad St., Ste. 3, Chattanooga, TN 37408. Tel.: 423-280-7182.

Key Personnel: Dir., Angela Usrey; Dir., Annie Harpe

Institution Type/Description: Art Gallery.

Collections: paintings; sculpture.

Hours & Admission Prices: Wed.-Fri. 12-5.

TENNESSEE AQUARIUM, One Broad St., Chattanooga, TN 37402-1023. Mailing Address: P.O. Box 11048, Chattanooga, TN 37401-2048. Tel.: 800-262-0695; 423-265-0695. Fax: 423-267-3561.
Web Site: www.tennesseeaquarium.org
Founded: 1992.
Congressional District: 3
Key Personnel: Pres., Charles Arant; Retail Sales Mgr., Judy Powell; Public Rels., Thom Benson; Museum Shop Mgr., Laura Kroeger.
Personnel Profile: Full-Time Paid 156; Part-Time Paid 101; Part-Time Volunteers 575; Interns 15.
Governing Authority: Tax-exempt.
Institution Type/Description: Aquarium.
Collections: freshwater & saltwater aquariums.
Facilities: IMAX theater.
Hours & Admission Prices: Daily 10-6. Adults $24.95, children 3-12 $14.95; children under 3 no charge. IMAX Theater: adults $8.50, children 3-12 $6; children under 3 no charge. Combination Ticket: adults $29.95, children 3-12 $19.95; children under 3 no charge. &
Attendance: 700,000 (estimated)
Membership: Individual Plus $85; Family $115; Family Plus $155.

TENNESSEE VALLEY RAILROAD MUSEUM INC., 4119 Cromwell Rd., Chattanooga, TN 37421-2164. Tel.: 423-894-8028, ext. 12. Fax: 423-894-8029.
E-mail: info@tvrail.com
Web Site: www.tvrail.com
Founded: 1961.
Congressional District: 3
Key Personnel: Pres., Timothy Andrews; Vice Pres. Mechanical, L. James Miller, III; Sec., Randall Freer; Shop Foreman, George Walker; Museum Shop Mgr., Joyce Soule.
Personnel Profile: Full-Time Paid 7; Full-Time Volunteers 4; Part-Time Paid 10; Part-Time Volunteers 10.
Governing Authority: nonprofit organization. Tax-exempt: 501(c)(3).
Institution Type/Description: Railroad Transportation Museum.
Collections: operating Steam Locomotive 6910, 630, 4501, 610, 509; Pullman car; dining car 3158; c.1929 NC & ST.L. Drovers Caboose; ex-Air Force RSD1 diesel locomotives; ACT 1 experimental electric trains; original tracings from Baldwin Locomotive works; technical books; replica of Union Station; telegraph keys & sounders; artifacts indicative of daily train operations; c.1920-1950 coaches. Historic Buildings: c.1920 East Chattanooga Depot; Grand Junction Depot.
Research Fields: railroad history.
Facilities: cafeteria. Gift items for sale.
Activities: guided tours; daily 6 mile roundtrip ride over historic 1856 railroad through Missionary Ridge tunnel using vintage, restored equipment from the Golden Age of Railroad Travel; lectures; formally organized education programs. Museum Sponsors: steam-pulled excursions 6 times a year.
Publications: quarterly, Smoke and Cinders.
Hours & Admission Prices: mid-March to mid-June & mid-Aug. to Oct. Mon.-Fri. 10-1, Sat.-Sun. 10-5; mid-June to mid-Aug. daily 10-5; Nov. Sat.-Sun. 10-5. Adults $14, children 3-12 $8; discounts to AAA & NRHS members; members no charge. &
Attendance: 80,000 (estimated)
Membership: Individual $35, Family $55.

Clarksville

CLARKSVILLE-MONTGOMERY COUNTY MUSEUM DBA CUSTOMS HOUSE MUSEUM & CULTURAL CENTER, (M), 200 S. Second St., Clarksville, TN 37040-3400. Mailing Address: P.O. Box 383, Clarksville, TN 37041-0383. Tel.: 931-648-5780. Fax: 931-553-5179.
E-mail: info@customshousemuseumorg
Web Site: www.customshousemuseum.org
Founded: 1984.
Congressional District: 7
Key Personnel: Chm. Bd., Bill Hoy; Dir., Alan Robison; Chm. (V), Dan Hanley; Asst. Dir., Linda Maki; Pres. Guild, Mary Luther; Cur. Collections, Amy Andersen; Registrar, Amy Lewellen; Exhibit Technician, Randall Spurgeon; Cur. Education, Sue Lewis; Dir. Membership, Harriett Silvey; Public Rels. & Rentals Mgr., Terri Jordan; Museum Shop Mgr., Janie McGregor.
Personnel Profile: Full-Time Paid 8; Part-Time Paid 3; Part-Time Volunteers 2; Interns 1.
Governing Authority: nonprofit organization. Tax-exempt: 501(c)(3).
Institution Type/Description: General Museum: housed in c.1898 Post Office and Customs House.

Collections: photographs; manuscripts; late 19th century horse-drawn vehicles; toys; clothing; art history of tobacco. Memory Lane: fire department, print shop, Mrs. Clarks drapers shop, 1842 log house.
Research Fields: history; genealogy.
Facilities: 400-vol. reference library; rental facilities; 200-seat fully accessible auditorium. Museum-related items for sale.
Activities: guided tours; lectures; theatrical performances; organized education programs; docent program; loan exhibitions.
Publications: monthly newsletter, Art-N-Facts; books, Edibles from the Archives; Clarksville In The Civil War, A Chronology; Nineteenth Century Heritage; Ordeal by Fire.
Hours & Admission Prices: Tues.-Sat. 10-5, Sun. 1-5. Adults $5, senior citizens $3, college students $2, children 6-18 $1; discounts to AAM, ICOM, TAM, SEMC, AAA & AARP members; members, children 5 and under & Sun. no charge. Closed major holidays. &
Attendance: 35,000 (accurate)
Membership: Senior Citizen $15; Senior Citizen Couple $20; Active Military Family $25; Family $30; Patron $50; Contributor $100; Founder $500.

MARGARET FORT TRAHERN GALLERY, College & Eighth Sts., Clarksville, TN 37044. Mailing Address: P.O. Box 4677, Clarksville, TN 37044. Tel.: 931-221-7333. Fax: 931-221-7432.
Web Site: www.apsu.edu/art/gallery/index.html
Founded: 1974.
Congressional District: 7
Key Personnel: Dept. Art, Dixie Webb.
Personnel Profile: Full-Time Paid 1.
Governing Authority: Austin Peay State University. Branch Gallery: Larson Gallery, Harned Hall Building. Tax-exempt: SPA6-B.
Institution Type/Description: University Art Gallery.
Collections: contemporary prints; watercolors; Larson drawing collection; photography.
Activities: temporary & traveling exhibitions; formally organized education programs for adults; biennial & student annual exhibitions; biennial national drawing competition.
Publications: biennial exhibit catalog: Border to Border; exhibit announcement cards.
Hours & Admission Prices: Mon.-Fri. 9-4, Sat. 10-2, Sun. 1-4. No charge. Closed academic holidays. &
Attendance: 7,000

SMITH TRAHERN MANSION HOME OF FAMILY AND COMMUNITY EDUCATION, 101 McClure St., Clarksville, TN 37040. Mailing Address: P.O. Box 852, Clarksville, TN 37041-0852. Tel.: 931-648-9998.
E-mail: smithtrahern@hotmail.com
Founded: 1984.
Institution Type/Description: Historic House Museum; housed in the home of tobacconist, Christopher Smith; built in 1858. Listed on the National Register of Historic Places.
Collections: local history & culture; period furnishings; photographs; personal artifacts.
Activities: Museum Sponsors: Open House in December; Trees of Christmas in December.
Hours & Admission Prices: Mon.-Fri. 9:30-2:30; other times by appointment. Tours: $2.

Cleveland

MUSEUM CENTER AT 5IVE POINTS, (M), 200 Inman St. E., Cleveland, TN 37311-6039. Tel.: 423-339-5745. Fax: 423-476-7922.
E-mail: info@museumcenter.org
Web Site: www.museumcenter.org
Founded: 1999.
Key Personnel: Operations Mgr., Ken Cagle; Museum Shop Mgr., Tracy O'Connell.
Personnel Profile: Full-Time Paid 2; Part-Time Paid 4; Part-Time Volunteers 20.
Governing Authority: Tax-exempt.
Institution Type/Description: History Museum.
Collections: local history & culture; photographs; personal artifacts.
Activities: special events; rental facilities; birthday parties.
Hours & Admission Prices: Tues.-Fri. 10-5, Sat. 10-3. Adults $5; discounts to AAM members; members no charge. &
Attendance: 25,000 (accurate)
Membership: Individual $25; Family $50; Sponsor $100; Friend $250; Supporter $500; Sustainer $1,000.

RED CLAY STATE HISTORIC PARK, 1140 Red Clay Park Rd.,
   Cleveland, TN 37311. Tel.: 423-478-0339. Facebook: Red Clay
   State Historic Park.
E-mail: erin.medley@tn.gov
Web Site: tnstateparks.com/parks/about/red-clay
Founded: 1979.
Congressional District: 3
Key Personnel: Park Mgr., Erin Medley.
Personnel Profile: Full-Time Paid 5; Part-Time Paid 1; Part-Time Volunteers 1.
Governing Authority: state. Parent Institution: TN State Dept. of Environment
   & Conservation. Tax-exempt.
Institution Type/Description: Historic Site Park: c.1832-38 seat of the Chero-
   kee government & site of eleven general councils national affairs.
Collections: Paleo, Archaic, Mississippian, Woodland & historic period
   artifacts; 19th-century Cherokee artifacts.
Research Fields: Cherokee removal story c.1832-1838.
Facilities: library of material dealing with Cherokee & American Indian
   culture, history & natural history available for research on premises;
   reading room; 54-seat auditorium.
Activities: guided tours; lectures; arts festivals; permanent exhibitions.
Hours & Admission Prices: Daily 8-4:30. No charge; donations accepted. &
Attendance: 188,400 (accurate)

Clifton

T.S. STRIBLING MUSEUM, Clifton Library, 300 E. Water St.,
   Clifton, TN 38425. Tel.: 931-676-3188.
E-mail: johnstonw@k12tn.net
Institution Type/Description: Historic Site: housed in the former home of
   Pulitzer Prize-winning author, T.S. Stribling.
Collections: artifacts & memorabilia of the Stribling family.
Hours & Admission Prices: Tues.-Fri. 11:30-6.

Clinton

GREEN MCADOO CULTURAL CENTER & MUSEUM, 101
   School St., Clinton, TN 37717. Mailing Address: P.O. Box 1214,
   Clinton, TN 37717-1214. Tel.: 865-463-6500.
Institution Type/Description: History Museum.
Collections: local history & culture; desegregation in Clinton; 1950s period
   classroom; CBS broadcast news film; personal artifacts; photographs;
   biographies.
Activities: interactive exhibitions.
Hours & Admission Prices: Tues.-Sat. 10-5; groups by appointment.

MUSEUM OF APPALACHIA, 2819 Andersonville Hwy., Clinton,
   TN 37716-6756. Mailing Address: P.O. Box 1189, Norris, TN
   37828-1189. Tel.: 865-494-7680. Fax: 865-494-8957. Facebook:
   Museum of Appalachia.
E-mail: museum@museumofappalachia.org
Web Site: www.museumofappalachia.org
Founded: 1969.
Congressional District: 3
Key Personnel: Pres., Elaine I. Meyer; Museum Shop Mgr., JoEllen Emert.
Personnel Profile: Full-Time Paid 7; Part-Time Paid 18; Part-Time Volunteers
   56.
Governing Authority: Tax-exempt: 501(c)(3).
Institution Type/Description: Appalachian theme; Historic Village and Folk Art
   Museum: housed in thirty restored pioneer log structures.
Collections: over 200,000 early American & pioneer artifacts. Historic
   Buildings: Mark Twain family cabin; Cantelever barn; school house;
   chapel.
Research Fields: culture, heritage, lifestyle of frontier people through study of
   artifacts; Southern Appalachia mountain region; genealogy.
Facilities: cafe; rental facilities. Crafts, antiques & other museum-related items
   for sale.
Activities: permanent exhibitions, seasonal Appalachian music; rental facili-
   ties. Museum Sponsors: 4th of July Celebration; Days of the Pioneer
   Antique Show; Student Heritage Day; Tennessee Fall Homecoming; Christ-
   mas in Old Appalachia.
Publications: books, Music & Musical Instruments of the Southern Appala-
   chian Mountains; Guns and Gun Making Tools of the Southern Appala-
   chians; Baskets & Basketmakers of Southern Appalachia; A People and
   Their Quilts; Alex Stewart: Portrait of A Pioneer; The Museum of
   Appalachia Story. A People and Their Music.
Hours & Admission Prices: Seasonal hours. Call for admission prices. &
Attendance: 100,000 (estimated)
Membership: Please call museum for details.

Collierville

BIBLICAL RESOURCE CENTER & MUSEUM, (M), 140 E.
   Mulberry, Collierville, TN 38017-2675. Tel.: 901-854-9578. Fax:
   901-854-9883.
E-mail: info@biblical-museum.org
Web Site: www.biblical-museum.org
Founded: 1995.
Key Personnel: Chm. & C.E.O., Donald E. Bassett; Pres. (V), Larry Papasan;
   Museum Dir., Nancy W. Bassett; Collections Cur., L. Jacob Shock; Program
   Dir., Sara Hansen.
Personnel Profile: Full-Time Paid 2; Full-Time Volunteers 1; Part-Time Paid 2;
   Part-Time Volunteers 4; Interns 1.
Governing Authority: private; nonprofit organization. Tax-exempt: 501(c)(3).
Institution Type/Description: Religious & History Museum.
Collections: biblical, Egyptian & Ancient Near Eastern archaeological artifacts
   & replicas.
Facilities: 820-vol. library of Biblical reference, ecclesiastical history, educa-
   tional videos of Bible/Archaeology & Archaeological periodical; educa-
   tional facilities; 300 sq. ft. exhibit space. Museum-related items for sale.
Activities: formal educational programs for children; guided tours; lectures;
   school loan service; traveling exhibitions; summer archeological workshop
   for students; travel programs.
Publications: quarterly newsletter.
Hours & Admission Prices: Tues.-Sat. 10-5. Groups $3 per person; discounts
   to AAM & ICOM members. Closed major holidays. &
Attendance: 10,500 (accurate)
Membership: Individual $25; Family $50.

Columbia

AMUSE'UM COLUMBIA CHILDREN'S MUSEUM, 123 W. 7th
   St., Columbia, TN 38401. Tel.: 931-223-6337.
E-mail: columbiacm@hotmail.com
Web Site: amuseumcolumbia.org
Institution Type/Description: Children's Museum.
Collections: hands-on exhibitions.
Activities: educational programs; special events; birthday parties.
Hours & Admission Prices: Mon.-Tues. & Thurs. 10-4, Fri.-Sat. 10-5.
   Admission $5; children under one no charge.

JAMES K. POLK ANCESTRAL HOME, (M), 301 W. 7th St.,
   Columbia, TN 38401-3132. Mailing Address: P.O. Box 741,
   Columbia, TN 38402-0741. Tel.: 931-388-2354. Fax: 931-388-
   5971.
E-mail: jameskpolk@bellsouth.net
Web Site: www.jameskpolk.com
Founded: 1924.
Congressional District: 7
Key Personnel: Dir., John C. Holtzapple; Pres. (V), Lisa Butler; Cur.
   Collections, Thomas E. Price.
Personnel Profile: Full-Time Paid 2; Part-Time Paid 8; Part-Time Volunteers
   32.
Volunteer Hours: 2,200
Operating Expenses: 268,353
Operating Income: 275,054
Governing Authority: society. Parent Institution: James K. Polk Memorial
   Association, Columbia, TN 38402. Tax-exempt: 501(c)(3).
Institution Type/Description: Historic House Museum: 1816 James K. Polk
   Ancestral Home.
Collections: artifacts & documents that belonged to James K. Polk or relate to
   his presidency.
Research Fields: James K. Polk's political career; collections.
Facilities: 175-vol. library of books on James K. Polk available for use on
   premises. Postcards, & books for sale.
Activities: guided tours; permanent & temporary exhibitions; orientation video;
   outreach programs for area schools & nursing homes; lectures; discussion
   group; summer history camp.
Publications: brochure, Ancestral Home of James Knox Polk; booklet, A
   Special House, cookbook, Provisions & Politics.
Hours & Admission Prices: April-Oct. Mon.-Sat. 9-5, Sun. 1-5; Nov.-March
   Mon.-Sat. 9-4, Sun. 1-5. Adults $10, seniors 60 & up $8, youth 13-18 $7,
   childre 6-12 $5; discounts to groups & AAM members; members no charge.
   Closed New Year's Day; Thanksgiving; Christmas Eve & Day. &
Attendance: 10,266 (accurate)
Membership: Senior Citizen $40; Individual & Family $50.

Cookeville

COOKEVILLE CHILDREN'S MUSEUM, 36 W. 2nd St., Cookeville, TN 38503. Mailing Address: P.O. Box 204, Cookeville, TN 38503. Tel.: 931-520-3866.
Institution Type/Description: Children's Museum.
Collections: hands-on exhibitions.
Activities: educational programs; special events; birthday parties.
Hours & Admission Prices: Mon.-Wed. & Fri.-Sat. 10-4, Thurs. 12-5:30. Adults $4, children 2 & over $3; children under 2 no charge.

COOKEVILLE DEPOT MUSEUM, 116 N. Cedar Ave., Cookeville, TN 38501. Mailing Address: Dept. of Leisure Svcs., P.O. Box 998, Cookeville, TN 38503-0998. Tel.: 931-528-8570. Fax: 931-526-1167.
E-mail: depot@cookeville-tn.org
Web Site: www.cookevilledepot.com
Founded: 1984.
Congressional District: 6
Key Personnel: C.E.O. & Museums Admin., Judy Duke; Pres., John Buck.
Personnel Profile: Full-Time Paid 3; Full-Time Volunteers 1; Part-Time Paid 1; Part-Time Volunteers 10.
Governing Authority: municipal government; nonprofit. Parent Institution: City of Cookeville. Subsidiary Institution: Department of Leisure Services. Tax-exempt.
Institution Type/Description: Transportation Museum: built in 1909 by the Tennessee Central Railroad, the Cookeville depot is highlighted by its unique pagoda roof line.
Collections: photographs; memorabilia; artifacts of the Tennessee Central Railroad, particularly in Putnam County; two cabooses; 1913 Baldwin Ten Wheeler steam locomotive, static display.
Publications: biannual, The Highballer.
Hours & Admission Prices: Tues.-Sat. 10-4. No charge; donations accepted. Closed Good Friday; Independence Day; Labor Day; Thanksgiving; Christmas Eve & Day. &
Attendance: 10,500 (estimated)
Membership: Family $15; Contributor $30; Sustainer $50; Benefactor $100; Preserver $200; Conductor $500.

COOKEVILLE HISTORY MUSEUM, 40 E. Broad St., Cookeville, TN 38501-3210. Tel.: 931-520-5455.
E-mail: historymuseum@cookeville-tn.org
Web Site: www.cookevillehistorymuseum.com
Founded: 2003.
Congressional District: 6
Personnel Profile: Full-Time Paid 2; Part-Time Paid 1.
Governing Authority: city. Tax-exempt.
Institution Type/Description: History Museum.
Collections: local history & culture; period furnishings; personal artifacts; photographs.
Activities: special events; temporary exhibitions. Museum Sponsors: Children's History Hour.
Publications: The Exhibitor.
Hours & Admission Prices: Wed.-Sat. 10-4. No charge; donations accepted. &
Attendance: 10,000 (estimated)

Cornersville

LAIRDLAND FARM HOUSE, 3238 Blackburn Hollow Rd., Cornersville, TN 37047-7001. Tel.: 931-363-2205.
E-mail: info@lairdlandfarmhouse.com
Web Site: www.lairdlandfarmhouse.com
Founded: 2004.
Key Personnel: Dir., Bennita B. Rouleao; Chm. (V), Donald L. Rouleao.
Personnel Profile: Full-Time Volunteers 2.
Institution Type/Description: Historic House Museum: housed in the home of John Laird who received a 5,000 acre land grant from NC for his service during the Revolutionary War. Listed on the National Register of Historic Places.
Collections: Laird family history; period furnishings; personal artifacts; military equipment, uniforms & weapons; photographs; gardens; Civil War artifacts; edged weapons; arms; 1st ed. books.
Facilities: rental facilities.
Activities: rental facilities. Museum Sponsors: Living History Weekend in September.
Hours & Admission Prices: By appointment. Adults $12, students $5; children 8 & under no charge.
Attendance: 2,500 (estimated)

Covington

TIPTON COUNTY MUSEUM, VETERAN'S MEMORIAL & NATURE CENTER, 751 Bert Johnston Ave., Covington, TN 38019-2414. Mailing Address: P.O. Box 768, Covington, TN 38019-0768. Tel.: 901-476-0242. Fax: 901-476-0261.
E-mail: afisher@covingtontn.com
Web Site: www.tiptonco.com/museum.htm
Founded: 1998.
Congressional District: 8
Key Personnel: Dir., Alice Fisher; Program Coord., Elizabeth Newman.
Personnel Profile: Full-Time Paid 2.
Governing Authority: private; nonprofit organization. Tax-exempt.
Institution Type/Description: History Museum & Nature Center.
Collections: military history; environmental education; Tipton County residents personal artifacts; ecosystem of West Tennessee.
Facilities: 20 acre wildlife area; nature trail; wetlands.
Activities: history exhibits; programs & events for adults & children.
Publications: newsletter.
Hours & Admission Prices: Tues.-Fri. 9-5, Sat. 9-3. No charge. Closed major holidays. &
Attendance: 7,000 (accurate)
Membership: Individual, Senior & Military $25; Family $30; Contributor $50; Philanthropist $100; Benefactor $200; Corporate $500 & up.

Cowan

COWAN RAILROAD MUSEUM, 108 Front St., Cowan, TN 37318-0053. Mailing Address: P.O. Box 53, Cowan, TN 37318-0053. Tel.: 931-967-3078.
E-mail: secretary@cowanrailroadmuseum.org
Web Site: cowanrailroadmuseum.org
Institution Type/Description: Railroad Museum: housed in a former railroad depot, c.1904.
Collections: railroad history & artifacts; photographs; steam locomotive; flat car; caboose; Fairmont motor cars; HO scale model railroads.
Activities: Museum Sponsors: Railroad Days in June.
Hours & Admission Prices: May-Oct. Mon. & Thurs.-Sat. 10-4, Sun. 1-4; other times by appointment. No charge; donations accepted.
Membership: Annual $10.

Cross Plains

CROSS PLAINS HERITAGE MUSEUM, LIBRARY & ARCHIVES, 7821 Hwy. 25 E., Cross Plains, TN 37049-4851. Tel.: 615-654-2992.
Institution Type/Description: History Museum.
Collections: local history & culture; photographs; personal artifacts.
Hours & Admission Prices: Tues. & Thurs. 10-4. No charge.

Crossville

HOMESTEADS TOWER MUSEUM, 2611 Pigeon Ridge Rd., Crossville, TN 38555. Mailing Address: Cumberland Homesteads Tower Assoc., 96 Hwy. 68, Crossville, TN 38555. Tel.: 931-456-9663.
Web Site: www.cumberlandhomesteads.org
Institution Type/Description: Historic Building Museum: built in 1938.
Collections: local history; photographs; documents; personal artifacts; period furnishings.
Activities: lookout platform.
Hours & Admission Prices: April-Oct. Mon.-Sat. 10-4. Adults $4, youth $2; children under 6 & members no charge.

MILITARY MEMORIAL MUSEUM, 20 S. Main St., Crossville, TN 38555-4518. Tel.: 931-456-5520.
E-mail: nitaborinc@yahoo.com
Founded: 2002.
Key Personnel: Dir., Nita M. Boring; Pres. (V), Sean A. Boring.
Personnel Profile: Full-Time Volunteers 2; Part-Time Volunteers 12.
Governing Authority: Tax-exempt: 501(c)(3).
Institution Type/Description: Military History Museum.
Collections: military history including Civil War, WWI, WWII, Korean, Vietnam, Desert Storm & Iraq; personal artifacts; photographs; uniforms.
Research Fields: military history.
Hours & Admission Prices: Mon.-Fri. 9-4; other times by appointment. No charge; donations accepted.
Attendance: 2,473 (accurate)

Dandridge

JEFFERSON COUNTY MUSEUM & ARCHIVES, 202 W. Main St., Dandridge, TN 37725. Mailing Address: P.O. Box 1193, Dandrige, TN 37725. Tel.: 865-397-4904.
E-mail: archives@jeffersoncountytn.gov
Founded: 1958.
Congressional District: 2
Key Personnel: Dir., Chm. (V) & Museum Shop Mgr., Lura B. Hinchey
Institution Type/Description: History Museum: housed in a former courthouse; built in 1845.
Collections: local history & culture; county documents back to 1792; period furnishings; personal artifacts; photographs.
Hours & Admission Prices: Call for hours.
Attendance: 1,000 (estimated)

Dickson

CLEMENT RAILROAD HOTEL MUSEUM, 100 Frank Clement Place, Dickson, TN 37055. Mailing Address: P.O. Box 306, Dickson, TN 37056-0306. Tel.: 615-446-0500.
E-mail: execdirector@clementrailroadmuseum.org
Web Site: clementrailroadmuseum.org
Founded: 2009.
Congressional District: 5
Key Personnel: Dir., Terry Vaughan; Pres. (V), Jerry V. Smith
Governing Authority: Tax-exempt.
Institution Type/Description: Historic Building Museum: built in 1913. Listed on the National Register of Historic Places.
Collections: local history & culture; railroading; Civil War; Tennessee Governor Frank G. Clement & his family; personal artifacts; period furnishings; photographs.
Major Exhibits: Vietnam War Exhibit, 11/13-1/14.
Hours & Admission Prices: Mon.-Fri. 10-5, Sat. 10-4. Adults $4, seniors 50 & over and students 15 & over $2; children 14 & under no charge. &
Attendance: 6,000 (estimated)
Membership: Drummers - 18 & under $15; Seniors $25; Lineman $35; Conductor - Basic Family $65; Engineer - Standard Family $95; Station Agent - Premium Family $125-$250; Volunteer Train - Corporate $250; Railraod President - Patronage $250-$500.

Dover

FORT DONELSON NATIONAL BATTLEFIELD, 120 Fort Donelson Rd., Dover, TN 37058. Mailing Address: P.O. Box 434, Dover, TN 37058-0434. Tel.: 931-232-5706 (Visitor Center) & 5348 (office). Fax: 931-232-6331.
E-mail: fodo_ranger_activities@nps.gob
Web Site: www.nps.gov/fodo
Founded: 1928.
Congressional District: 8
Key Personnel: Supt., Brian McCutchen; Chief Interpreter & Resource Mgr., Douglas J. Richardson
Governing Authority: federal. A unit of the National Park Service, U.S. Department of the Interior, Washington, DC 20240. Tax-exempt.
Institution Type/Description: National Battlefield Museum.
Collections: Civil War artifacts; archival materials.
Research Fields: Civil War history.
Facilities: library of Civil War books available for use on premises.
Activities: self-guided tours; lectures; films; formally organized education programs for children, adults & undergraduate college students; permanent exhibitions; interpretive demonstrations.
Publications: booklet, Eastern National Parks & Monument Association.
Hours & Admission Prices: Office: Mon.-Fri. 8-4:30. Visitor Center: daily 8-4:30. No charge; donations accepted. Closed New Year's Day; Thanksgiving; Christmas. &
Attendance: 483,614 (accurate)

Ducktown

DUCKTOWN BASIN MUSEUM, 212 Burra Burra St., Ducktown, TN 37326. Mailing Address: P.O. Box 458, Ducktown, TN 37326-0458. Tel.: 423-496-5778.
Key Personnel: Dir., Ken Rush
Institution Type/Description: History Museum: housed on the Burra Burra mine site. Listed on the National Register of Historic Places.
Collections: history of mining & processing operations.
Facilities: Museum-related items for sale.
Hours & Admission Prices: May-Oct. Mon.-Sat. 10-4:30; Nov.-April Mon.-Sat. 9:30-4. Adults $4, seniors $3, children 13-18 $1, children 12 & under $.50. Closed Thanksgiving; Christmas. &

Dunlap

DUNLAP COKE OVENS PARK AND MUSEUM, 114 Walnut St., Dunlap, TN 37327. Tel.: 423-949-3483.
Institution Type/Description: History Museum.
Collections: local history & culture; photographs; period furnishings; personal artifacts.
Activities: special events.
Hours & Admission Prices: By appointment.

Dyersburg

DYERSBURG STATE COMMUNITY COLLEGE - WALTER E. DAVID WILDLIFE MUSEUM, 1501 Lake Rd., Dyersburg, TN 38024. Tel.: 731-286-3200.
E-mail: kjones@dscc.edu
Founded: 1986.
Key Personnel: Dir., Dr. Karen Bowyer; Cur., Ken Jones.
Governing Authority: public college. Tax-exempt.
Institution Type/Description: Wildlife Museum.
Collections: mounted native & African wild game specimens.
Facilities: classrooms.
Hours & Admission Prices: Mon.-Fri. 8 am-9 pm. No charge.
Attendance: 350 (estimated)

Elizabethton

JOHN & LANDON CARTER MANSION, 1013 Broad St., Elizabethton, TN 37643. Mailing Address: Sycamore Shoals Parks, 1651 W. Elk Ave., Elizabethton, TN 37643. Tel.: 423-543-5808. www.tnstateparks.com/sycamoreshoals.
E-mail: jennifer.bauer@tn.gov
Web Site: www.sycamoreshoalstn.org
Founded: 1780.
Key Personnel: Park Manager, Jennifer Bauer.
Personnel Profile: Full-Time Paid 6; Part-Time Paid 4.
Governing Authority: Parent Institution: Tennessee State Parks.
Institution Type/Description: Historic House Museum: built c.1780.
Collections: local history & culture; period furnishings; paintings; historic garden.
Hours & Admission Prices: Tours: Memorial Day to mid-Aug. daily 2 pm. No charge; donations accepted.
Attendance: 13,000 (estimated)

SYCAMORE SHOALS STATE HISTORIC AREA, 1651 W. Elk Ave., Elizabethton, TN 37643. Tel.: 423-543-5808. Fax: 423-543-0078.
Web Site: www.sycamoreshoalstn.org
Founded: 1976.
Key Personnel: Park Mgr., Jennifer Bauer.
Personnel Profile: Full-Time Paid 6; Part-Time Paid 4.
Governing Authority: Parent Institution: Tennessee State Parks.
Institution Type/Description: Park & Visitor Center.
Collections: local history & heritage; period furnishings; personal artifacts; photographs; 18th century history.
Facilities: nature trail; theater.
Activities: educational programs; special events.
Hours & Admission Prices: Center: Mon.-Sat. 8-4:30, Sun. 1-4:30. Park: sunrise to sunset. No charge; donations accepted. &
Attendance: 305,000 (accurate)

Elkton

MATT GARDNER HOMESTEAD MUSEUM, 110 Dixon Town Rd., Elkton, TN 38455. Mailing Address: P.O. Box 269356, Indianapolis, IN 46226. Tel.: 931-309-9695.
E-mail: info@mattgardnerhomestead.org
Web Site: www.mattgardnerhomestead.org
Institution Type/Description: History Museum.
Collections: US, Tennessee & African American history & culture; quilts; period furnishings; personal artifacts; photographs; agriculture; slavery; religion.
Hours & Admission Prices: Call for hours.

Englewood

ENGLEWOOD TEXTILE MUSEUM, 17 N. Niota St., Englewood, TN 37329-3245. Mailing Address: P.O. Box 253, Englewood, TN 37329. Tel.: 423-887-5455.
Key Personnel: Dir. & Chm. (V), Mark Cochran; Museum Shop Mgr., Gail Anderson.
Governing Authority: Tax-exempt.
Institution Type/Description: History Museum.
Collections: textile industry history; life of women in the mills.
Hours & Admission Prices: Mon.-Sat. 10-5. No charge; donations accepted. Closed major holidays. &
Attendance: 1,000 (estimated)

Erwin

UNICOI COUNTY HERITAGE MUSEUM, 529 Federal Hatchery Rd., Erwin, TN 37650. Mailing Address: Unicoi County Chamber of Commerce, P.O. Box 713, Erwin, TN 37650. Tel.: 423-743-9449 & 8923.
Institution Type/Description: History Museum: housed in an early 1900s home.
Collections: local history & culture; Blue Ridge pottery; Clinchfield Railroad memorabilia; personal artifacts; period furnishings; photographs.
Facilities: nature trails.
Hours & Admission Prices: May-Oct. & Dec. daily 1-5; other times by appointment.

Eva

TENNESSEE RIVER FOLKLIFE INTERPRETIVE CENTER AND MUSEUM, Nathan Bedford Forrest State Park, 1825 Pilot Knob Rd., Eva, TN 38333. Tel.: 731-584-6356. Fax: 731-584-1841.
Institution Type/Description: History Museum.
Collections: local history, heritage & culture; early pioneer life & customs; musseling; commercial fishing.
Facilities: Gift items for sale.
Activities: special events.
Hours & Admission Prices: Daily 8-11 & 12-4:30.

Fairview

BOWIE PARK & NATURE CENTER, 7211 Bowie Lake Rd., Fairview, TN 37062. Tel.: 615-799-5544.
E-mail: bowiepark@fairview-tn.org
Web Site: www.fairview-tn.org/bowie-park
Institution Type/Description: Nature Center.
Collections: native wildlife & their habitats; plants; trees; birds.
Facilities: nature trails.
Activities: rental facilities; outdoor classroom.
Hours & Admission Prices: Mon.-Sat. 9-4, Sun. 11-4.

Farragut

FARRAGUT FOLKLIFE MUSEUM, 11408 Municipal Center Dr., Farragut, TN 37934-2830. Tel.: 865-966-7057. Fax: 865-675-2096.
E-mail: julia.barham@townoffarragut.org
Web Site: www.townoffarragut.org/index.aspx?nid=186
Founded: 1986.
Congressional District: 2
Key Personnel: Dir., Julia Barham; Museum Coord., Sarah Julia Jones; Public Rels., Chelsea Riemann.
Personnel Profile: Full-Time Paid 1; Part-Time Volunteers 73.
Governing Authority: municipal. Tax-exempt: 170(b)(1)(A).
Institution Type/Description: History Museum.
Collections: East Tennessee area artifacts; personal artifacts of Admiral David Glasgow Farragut, first Admiral of the U.S. Navy & Civil War hero.
Facilities: 2,000 sq. ft. exhibit space. Museum-related items for sale.
Activities: docent program; guided tours; lectures; special events; open houses. Annual Event: Membership Buffet.
Publications: newsletter, Farragut Folklife Museum Newsletter.
Hours & Admission Prices: Mon.-Fri. 10-4:30. No charge; donations accepted. Closed New Year's week; Martin Luther King Jr. Day; Good Friday; Memorial Day; Independence Day; Labor Day; Thanksgiving & day after; Christmas week. &
Attendance: 4,590 (accurate)
Membership: Student $1; Individual $24; Family $36; Business $50; Lifetime $500.

Fayetteville

FAYETTEVILLE LINCOLN COUNTY MUSEUM AND CIVIC CENTER, 521 Main St., S., Fayetteville, TN 37334-3447. Mailing Address: 2270 Lewisburg Hwy., Fayetteville, TN 37334-6449. Tel.: 931-433-2921.
Web Site: www.flcmuseum.com
Key Personnel: Pres., Marie Caldwell; Vice Pres., Danny Bryant; Sec., Jim Harwell, DVM; Treas., Mark Mitchell; Dir., Farris Beasley, DVM; Dir., Eugene Ham; Dir., Ley Jean; Dir., Patty Patrick; Dir., Dusten Stewart; Dir., Kay Ward-Woods; Dir., Delbert Wicks.
Institution Type/Description: History Museum.
Collections: James Buchanan Memorial Courtyard; Admiral Frank Kelso; military artifacts; one-room schoolhouse; country store & post office; agricultural tools; Native American; fossils; medical equipment; space & missile command.
Hours & Admission Prices: May-Nov. Thurs.-Sat. 12:30-4:30.

Franklin

CARNTON PLANTATION, 1345 Eastern Flank Cir., Franklin, TN 37064-3259. Tel.: 615-794-0903. Fax: 615-794-6563.
E-mail: info@battleoffranklintrust.org
Web Site: www.carnton.org
Founded: 1977.
Congressional District: 6
Key Personnel: C.O.O. & Historian, Eric Jacobson; Special Events Coord., Leigh Bawcom; Office & Membership Coord., Angell Wallace; Cur., Joanna Stephens.
Governing Authority: private; nonprofit organization. Tax-exempt: 501(c)(3).
Institution Type/Description: Historic Site: 1826 Federal House and Outbuildings.
Collections: furniture, decorative arts & artifacts connected to the McGavock family. Historic Structures: 1826 house including gardens & grounds, slave house, spring house and smokehouse; restored garden.
Research Fields: Civil War; gardens & landscaping; plantation life; slaves; the McGavock family.
Facilities: Civil War-related books & other museum-related items for sale.
Activities: guided tours; lectures; arts festivals; concerts; docent program; education programs for children; volunteer program. Annual Events: Southern Folklife Festival; Anniversary of Battle of Franklin in November; Christmas at Carnton.
Publications: newsletter, Columns.
Hours & Admission Prices: Mon.-Sat. 9-5, Sun. 12-5; last tour at 4. Adults $15, seniors 65 & up $12, children 6-12 $8; discount to groups of 10 or more; children under 5 no charge.
Attendance: 45,000 (accurate)
Membership: The Battle of Franklin Trust: Basic $40; Associate $65; Trust Donor $125; Fountain Branch Carter Society $500; Randal McGavock Society $1,000; Carrie McGavock Society $5,000; Tod Carter Society $10,000.

THE CARTER HOUSE, 1140 Columbia Ave., Franklin, TN 37064-3617. Tel.: 615-791-1861. Fax: 615-794-6563. Facebook: Carter House.
E-mail: info@battleoffranklintrust.org
Web Site: www.battleoffranklintrust.org
Founded: 1951.
Congressional District: 6
Personnel Profile: Full-Time Paid 20; Part-Time Paid 15; Part-Time Volunteers 3; Interns 1.
Governing Authority: Parent institution: The Battle of Franklin Trust. Tax-exempt.
Institution Type/Description: Historic house; battlefield & museum. Listed on the National Register of Historic Places.
Collections: 19th century artifacts including Civil War; decorative arts; archival collections; communication artifacts; recreational artifacts.
Research Fields: Civil War.
Activities: guided tours; exhibitions.
Hours & Admission Prices: Mon.-Sat. 9-5, Sun. 12-5. Adults $15, seniors $12, children 6-12 $8. Closed New Year's Day; Easter; Thanksgiving; Christmas Eve & Day.
Attendance: 45,000 (accurate)

LOTZ HOUSE MUSEUM, 1111 Columbia Ave., Franklin, TN 37064. Tel.: 615-790-7190. Fax: 615-790-7197.
E-mail: info@lotzhouse.com
Web Site: www.lotzhouse.com
Founded: 2008.

Key Personnel: Dir., J.T. Thompson.
Governing Authority: Tax-exempt.
Institution Type/Description: Historic House Museum.
Collections: Lotz family history; Civil War; personal artifacts; period furnishings.
Research Fields: Civil War; Lotz family.
Facilities: Museum-related items for sale.
Activities: antique fine art & decorative art appraisals.
Publications: The Lotz Family - Survivors of the Battle of Franklin.
Hours & Admission Prices: Mon.-Sat. 9-5, Sun. 1-4; other times by appointment. Adults $10, seniors 65 & over $9, children 7-13 $5; discounts to groups and AAM & ICOM members; children 6 & under no charge.

Gainesboro

JACKSON COUNTY HISTORICAL MUSEUM, Fred Lucas Haile Bldg., 105 Montpelier St., Gainesboro, TN 38562. Mailing Address: c/o Jackson County Historical Society, P.O. Box 647, Gainesboro, TN 38562.
Institution Type/Description: History Museum.
Collections: local history & culture; period furnishings; personal artifacts; photographs.
Hours & Admission Prices: Call for hours. No charge.

Gallatin

HISTORIC ROSE MONT, 810 S. Water Ave., Gallatin, TN 37066-3735. Tel.: 615-451-2331; 888-451-2331.
E-mail: historicrosemont@att.net
Web Site: www.historicrosemont.com
Founded: 1993.
Key Personnel: Pres. (V), John L. Glover; Museum Shop Mgr., Erleen Blurton.
Personnel Profile: Full-Time Paid 1; Part-Time Volunteers 25.
Governing Authority: Tax-exempt.
Institution Type/Description: Historic House Museum: housed in the former home of Judge Walter Guild and his wife Bettie Alexander Guild; built in 1842.
Collections: local history & culture; period furnishings; personal artifacts; photographs.
Activities: rental facilities; special events. Museum Sponsors: Rose Mont Festival in June.
Hours & Admission Prices: April 15 to Oct. Tues.-Sat. 10-4, Sun. 1-4. Adults $5, seniors over 55 $4, children 6-12 $3; discounts to groups; children under 6 no charge. Closed holidays. &
Attendance: 5,000 (estimated)
Membership: Individual $10; Family $15.

SUMNER COUNTY MUSEUM, 183 W. Main St., Gallatin, TN 37066-3252. Mailing Address: P.O. Box 1163, Gallatin, TN 37066. Tel.: 615-451-3738. Fax: 651-451-0878.
E-mail: contact@sumnercountymuseum.org
Web Site: www.sumnercountymuseum.org
Founded: 1972.
Congressional District: 6
Key Personnel: Dir., Juanita Frazor; Chm. (V), Danny Sullivan; Cur., Allen Haynes.
Personnel Profile: Part-Time Paid 2; Part-Time Volunteers 2.
Volunteer Hours: 900
Operating Expenses: 19,000
Operating Income: 24,000
Governing Authority: Tax-exempt.
Institution Type/Description: History Museum.
Collections: local history & culture; period artifacts; early pioneer life; Civil War; World War I & II.
Research Fields: local history & education.
Activities: guided tours.
Hours & Admission Prices: April-Oct. Wed.-Sat. 9-4:30, Sun. 1-4:30. Adults $3, children 6-12 $1; members & children under 6 no charge donations accepted.
Attendance: 600 (estimated)
Membership: Individual $20; Family $35; Colleague $60; Guardian $200.

TENNESSEE AVIATION HALL OF FAME, 395 Devon Chase Hill, #101, Gallatin, TN 37066. Tel.: 615-452-3696.
E-mail: info@tnaviationhof.org
Web Site: www.tnaviationhof.org
Governing Authority: nonprofit organization. Tax-exempt: 501(c)(3).
Institution Type/Description: History Museum.
Collections: aviation history; Hall of Fame inductees.

Activities: Annual Event: Induction Ceremony.
Hours & Admission Prices: Call for hours.

Gatlinburg

ARROWMONT SCHOOL OF ARTS & CRAFTS, 556 Parkway, Gatlinburg, TN 37738-3202. Mailing Address: P.O. Box 567, Gatlinburg, TN 37738-0567. Tel.: 865-436-5860. Fax: 865-430-4101.
E-mail: info@arrowmont.org
Web Site: www.arrowmont.org
Founded: 1945.
Congressional District: 1
Key Personnel: Exec. Dir., Bill May; Gallery Mgr., Stefanie Garber Darr; Dir., Finance, Julia Clinton; Supply Store Mgr., Heather Ashworth; Dir., Programs, Bill Griffith; Communications Mgr., Chris Barrett; Devel. Mgr., Jennifer Brown; Dir., Facilities & Opers., Steve Reilly.
Governing Authority: private; nonprofit organization. Parent Institution: Pi Beta Phi fraternity. Tax-exempt: 501(c)(3).
Institution Type/Description: Art Gallery.
Collections: arts & crafts from 1950 to present.
Facilities: 5,000-vol. library of art & art making materials; 175-seat auditorium; 175-seat cafeteria; 10 studios; 3,700 sq. ft. exhibit space. Books, supplies & art-related items for sale.
Activities: concerts; formal education programs for adults, children & Univ. of Tennessee students; guided tours; lectures; loan, temporary & traveling exhibitions; mobile vans. Museum Sponsors: resident artist, studio assistant, work study & gallery internship programs.
Publications: quarterly, Arrowmont School of Arts & Crafts Newsletter; quarterly newsletter magazine, The Arrow; Spring & Summer: biannually, Workshop Schedule; Descriptive Poster; electronic newsletter, E-Visions; annual workshop schedule.
Hours & Admission Prices: Mon.-Fri. 8-5; call for holiday & weekend hours. No charge; donations accepted. &
Attendance: 4,150 (estimated)

COOTER'S PLACE GATLINBURG, 542 Pkwy., Gatlinburg, TN 37738-3202. Tel.: 865-430-9909.
E-mail: info@cootersplace.com
Web Site: www.cootersplace.com
Institution Type/Description: History Museum.
Collections: Dukes of Hazzard memorabilia; photographs; props.
Hours & Admission Prices: Call for hours. No charge.

GREAT SMOKY MOUNTAINS, SUGARLANDS VISITOR CENTER, Great Smoky Mountains National Park, 107 Park Headquarters Rd., Gatlinburg, TN 37738-4102. Tel.: 865-436-1200. Fax: 865-436-1220.
Web Site: www.nps.gov/grsm
Founded: 1961.
Congressional District: 1
Key Personnel: Dep. Supt., Patty Wissinger.
Governing Authority: federal. Parent Institution: National Park Service. Tax-exempt.
Institution Type/Description: Park Museum.
Collections: herbarium & insect specimens; bird & mammal skins; wet specimens (amphibians & reptiles); art collection; photographs of park habitats; freeze-dried specimens of birds & smaller mammals, mounted larger specimens; plant specimens & plastic reproductions of reptiles, amphibians & fungi & plants; herbarium.
Research Fields: natural & cultural history.
Facilities: 5,000-vol. library of general reference books available for use by permission on premises only; 150-seat auditorium. Postcards, slides, film & natural history books for sale.
Activities: permanent exhibitions; general orientation film about the park.
Publications: varied scientific reports.
Hours & Admission Prices: March & Nov. daily 8-5; April-May & Sept.-Oct. daily 8-6; June-Aug. daily 8-7; Dec.-Feb. daily 8 4:30. No charge; donations accepted. Closed Christmas. &
Attendance: 874,393 (accurate)

GUINNESS WORLD RECORDS MUSEUM, Baskins Square Mall, 631 Pkwy., Ste. B-11, Gatlinburg, TN 37738-3258. Tel.: 865-436-9100 & 430-7800.
Institution Type/Description: World Record Museum.
Collections: world record feats, facts & record holders; photographs; personal artifacts; movie memorabilia.
Activities: special events.

Hours & Admission Prices: Daily 9am. Adults 12 & over $9.99, children 6-11 $5.99; children 5 & under no charge.

HOLLYWOOD STAR CARS MUSEUM, 914 Pkwy., Gatlinburg, TN 37738-3104. Tel.: 865-430-2200.
Web Site: www.starcarstn.com
Formerly: Star Cars Museum
Key Personnel: Owner, Charles Moore
Institution Type/Description: Car Museum.
Collections: cars from Hollywood movies & TV shows of the last 50 years.
Hours & Admission Prices: Daily 9am to 10pm. Adult $11.99, children 6-12 $6.99 children under 6 no charge.

RIPLEY'S BELIEVE IT OR NOT! ODDITORIUM, 800 Parkway, Gatlinburg, TN 37738. Mailing Address: 88 River Rd., Gatlinburg, TN 37738. Tel.: 865-436-5096. Fax: 865-436-4145.
E-mail: gatlinburg@ripleys.com
Web Site: www.ripleysgatlinburg.com
Institution Type/Description: General Museum.
Collections: over 500 artifacts from around the world; hands-on exhibitions.
Activities: birthday parties; sleepovers.
Hours & Admission Prices: Daily 10-9. Adults 12 & over $14.99, children 6-11 $7.99; discounts to groups; children 5 & under no charge. &

SALT AND PEPPER SHAKER MUSEUM, 527 Cherry St., Gatlinburg, TN 37738-4706. Tel.: 888-778-1802; 865-430-5515.
Web Site: www.thesaltandpeppershakermuseum.com
Key Personnel: Owner, Andrea Ludden; Owner, Rolf Ludden
Institution Type/Description: General Museum.
Collections: over 20,000 salt & pepper shakers from around the world; pepper mills.
Facilities: Museum-related items for sale.
Hours & Admission Prices: Daily 10-4. Adults $3; children 12 & under no charge.

Germantown

GERMANTOWN REGIONAL HISTORY AND GENEALOGY CENTER (GRHGC), 7779 Poplar Pike, Germantown, TN 38138-5952. Tel.: 901-757-8480.
E-mail: jbaker@germantown-tn.gov
Web Site: germantown-library.org
Founded: 2006.
Key Personnel: Dir., Melody Pittman; Chm. (V), Byron Crain.
Personnel Profile: Part-Time Paid 1; Part-Time Volunteers 30.
Governing Authority: Parent Institution: Germantown Community Library. Subsidiary Institution: Tennessee Genealogical Society. Tax-exempt.
Institution Type/Description: Library.
Collections: cultural, architectural, genealogical & historical heritage of the South; books; microfilm; periodicals; photographs; maps; ephemera.
Research Fields: genealogy.
Facilities: library.
Activities: genealogy & history classes. Museum Sponsors: Annual Show and Tell History Fair.
Publications: quarterly periodical, Ansearchin' News.
Hours & Admission Prices: Mon. 10-2, Tues. & Thurs. 10-4, Sat. 9-5. No charge; donations accepted. Closed New Year's Day; Martin Luther King Jr. Day; Presidents' Day; Easter; Memorial Day; Independence Day; Labor Day; Thanksgiving & day after; Christmas Eve & Day.
Attendance: 2,400 (estimated)
Membership: Tennessee Genealogy Society: Single $25; Couple $35.

PT BOATS MUSEUM & ARCHIVES, 1384 Cordova Rd., Ste. 2, Germantown, TN 38138-2219. Mailing Address: P.O. Box 38070, Germantown, TN 38183-0070. Tel.: 901-755-8440. Fax: 901-751-0522.
E-mail: ptboats@ptboats.org
Web Site: www.ptboats.org
Formerly: The P.T. Boat Museum & Library
Founded: 1946.
Congressional District: 10
Key Personnel: C.E.O., Dir. & Treas., Alyce N. Guthrie; Pres. (V), Charles B. Jones; P.T. Boat Cur., Don Shannon; Administrative Asst., Allyson Bethune; Museum Shop Mgr., Rick Bethune.
Personnel Profile: Full-Time Paid 2; Part-Time Paid 1; Part-Time Volunteers 1.
Governing Authority: private; nonprofit organization. P.T. Boats, Inc. Tax-exempt: 501(c)(3).

Institution Type/Description: Maritime, Naval Museum: located at Battleship Cove.
Collections: World War II P.T. boats; one-man Japanese suicide demolition boat; books, diaries, insignias, memorabilia of 43 operating squadrons of World War II P.T. boats; U.S. mosquito fleet; tenders; bases; films; archives; 10,000 photographs; plans.
Research Fields: World War II P.T. boat operations; bases; tender ships.
Facilities: 250-vol. library. Books & history-related items for sale.
Activities: lectures; films; restoration; research; photographic services.
Publications: semiannual newsmagazine; books, The U.S. Mosquito Fleet; Early Elco P.T. Boats; P.T. Squadrons, Bases, Tenders ALL HANDS.
Hours & Admission Prices: Battle Ship Cove: Spring to May 23 daily 9-4:30; Summer daily 9-5. Adults $15, seniors $13, children 6-12 $9; discounts to AAA & HNSA members; children under 6 & military in uniform no charge. Germantown Headquarters: Mon.-Fri. 8-4. Closed New Year's Eve & Day; Independence Day; Thanksgiving; Christmas week.
Attendance: 100,000 (estimated)
Membership: Veteran of P.T. Any Amount; Non-Veteran of P.T. $25; Overseas $50; Citation $500; Bronze Medal of Honor $1,000; Silver Medal of Honor $5,000; Gold Medal of Honor $10,000.

Goodlettsville

MANSKER'S STATION & BOWEN PLANTATION HOUSE, 705 Caldwell Dr., Moss Wright Park, Goodlettsville, TN 37072. Tel.: 615-851-2241. Fax: 615-859-4563.
Web Site: www.manskersstation.org
Key Personnel: Dir., Allison Baker.
Personnel Profile: Full-Time Paid 2; Part-Time Paid 2; Part-Time Volunteers 25.
Governing Authority: Parent Institution: City of Goodlettsville, TN. Tax-exempt.
Institution Type/Description: Historic Sites & Preservation Societies.
Collections: House: period furnishings. Fort: 18th century pioneer life; Mansker's Station, reconstructed 1780 fort. Historic Building: Bowen Plantation House, c.1787.
Facilities: life center.
Activities: demonstrations; living history events.
Publications: brochures.
Hours & Admission Prices: Tours: Mon.-Fri. 9-12 & 1-4:30, last tour at 3:30. Adults $8; discounts to seniors, children under 12, AAM members & groups of 15 or more. Closed New Year's Day; Thanksgiving & day after; Christmas Eve & Day. &
Attendance: 8,000 (estimated)

Grand Junction

NATIONAL BIRD DOG MUSEUM, 505 Hwy. 57, Grand Junction, TN 38039-6059. Mailing Address: P.O. Box 774, Grand Junction, TN 38039-0774. Tel.: 731-764-2058. Fax: 731-764-3004.
E-mail: sportdog@bellsouth.net
Web Site: www.birddogfoundation.com
Founded: 1989.
Key Personnel: Exec. Dir., David Smith; Sec., Barbara Sweeney; Education Coord. & Librarian, Lucy Cogbill; Pres. (V), Don Driggers; Museum Shop Mgr., Patricia Belt.
Personnel Profile: Full-Time Paid 1; Part-Time Paid 3.
Governing Authority: Tax-exempt.
Institution Type/Description: General Museum.
Collections: pointing dog & retriever breeds history; art; paintings; photography; memorabilia; hunting; field trial activities; shooting sports; early history of the National Field Trials.
Facilities: library. Museum-related items for sale.
Activities: special events; meetings.
Hours & Admission Prices: Tues.-Fri. 9-2, Sat. 10-4, Sun. 1-4. No charge; donations accepted. &
Attendance: 6,000 (estimated)
Membership: Youth Supporter $10; Benefactor $50; Life Benefactor & Corporate $500; Life Patron & Patron Memorial $1,200; Founder $5,000.

Granville

GRANVILLE MUSEUM, INC., Clover St., Granville, TN 38564. Mailing Address: P.O. Box 26, Granville, TN 38564-0026. Tel.: 931-653-4511. Fax: 615-443-7117.
E-mail: rclemons@wilsonbank.com
Web Site: www.granvilletn.com
Founded: 1999.
Key Personnel: Dir. & Pres. (V), Randall Clemons; Treas., Suzanne Stafford; Cur., Patsy Yates; Archivist, Liz Bennett; Museum Shop Mgr., Peggy Sanders.

Personnel Profile: Part-Time Volunteers 135.
Governing Authority: private; nonprofit organization. Tax-exempt: 501(c)(3).
Institution Type/Description: History Museum.
Collections: business; military; schools; sports; music; church; family; river; clothing; period furnishings; early cars & tractors; agriculture; grist mill; blacksmith & weaving shops. Historic Structures: 1880 general store; 1880 Sutton homestead; 1950s service station; 1820 log homestead; late 1800's Saloon Museum; Early 1900's Ice Cream Parlor Museum.
Major Exhibits: A Woman's Work is Never Done, 3/6-6/10/14; Blue Ribbon Days: Getting Ready for the Fair, 6/11-9/2/14; If These Walls Could Talk, 9/3-11/4/14; Scarecrow Walk of Our Past, 10/4-10/31/14; Deck the Walls, 11/5-12/31/14.
Facilities: restaurant. Museum-related items for sale.
Activities: Annual Events: Granville Heritage Day; Genealogy Festival; Fall Celebration; Country Christmas.
Publications: biannual newsletter.
Hours & Admission Prices: Museum & Homestead: Wed.-Sat. 12-3; other times by appointment. Museum: no charge. Homestead: adults $5, seniors 62 & over $4, children 6-12 $3; children under 6 no charge. &
Attendance: 9,000 (estimated)
Membership: Friends $25.

Gray

EAST TENNESSEE STATE UNIVERSITY AND GENERAL SHALE BRICK NATURAL HISTORY MUSEUM AND GRAY FOSSIL SITE, (M), 1212 Suncrest Dr., Gray, TN 37615-4114. Mailing Address: P.O. Box 70300, Gray, TN 37615-9221. Tel.: 423-439-3659; 866-202-6223. Fax: 423-439-3658.
E-mail: info@grayfossilmuseum.com
Web Site: www.grayfossilmuseum.com
Founded: 2007.
Congressional District: 1
Key Personnel: Dir., Dr. Blaine Schubert; Museum Operations Mgr., April Season Nye.
Personnel Profile: Full-Time Paid 8; Part-Time Volunteers 100.
Governing Authority: Parent Institution: East Tennessee State University. Tax-exempt.
Institution Type/Description: Natural History Museum.
Collections: fossils from alligators, camels, sloth, elephant, rhino, tapirs & peccary.
Research Fields: paleontology; sedimentology; geomatics; geomorphology; geology; paleoecology.
Hours & Admission Prices: Daily 8:30-5. Gray Fossil Site Walk-in Tour: adults 5, seniors 65 & over $4, children 5-12 $3. All access Pass: adults $10, seniors 65 & over $9, children 5-12 $7. Closed New Year's Day; Thanksgiving; Christmas. &
Attendance: 53,798 (accurate)

Greeneville

ANDREW JOHNSON NATIONAL HISTORIC SITE, 101 N. College St., Greeneville, TN 37743-5607. Mailing Address: 121 Monument Ave., Greeneville, TN 37743-5552. Tel.: 423-638-3551 (Visitor Center) & 639-3711 (Admin.). Fax: 423-638-9194 (Visitor Center) & 798-0754 (Admin.).
Web Site: www.nps.gov/anjo/
Founded: 1942.
Congressional District: 1
Key Personnel: Supt., Lizzie Watts.
Governing Authority: federal. Parent Institution: U.S. Dept. of the Interior, National Park Service, Southeast Region, Appalachian Cluster, Atlanta Federal Center, 1924 Building, 100 Alabama St., S.W., Atlanta, GA 30303. Tax-exempt.
Institution Type/Description: History Museum and Historic Houses.
Collections: artifacts of the Johnson period, c.1820s-1875; Andrew Johnson National Cemetery. Historic Houses: 1830s-1851 Johnson House; 1830s-1840s Andrew Johnson Tailor Shop; 1851-1875 Andrew Johnson Homestead.
Research Fields: history.
Facilities: 550-vol. park library available on premises; visitor center Books & other museum-related items for sale.
Activities: ranger guided tours; self-guided tours; Video interpretive film at visitor center; film: Andrew Johnson Defender of the Constitution; film loan service; off-site programs upon request. Special Events: Wreath Laying at AJ Cemetery; Memorial Service at AJ Cemetery; Junior Ranger program.
Publications: park folder, Andrew Johnson National Historic Site; Impeachment Folder; Cemetery Folder; Andrew Johnson and His Slaves Folder; Junior Ranger Program.

Hours & Admission Prices: Daily 9-5. No charge; donations accepted. Closed New Year's Day; Thanksgiving; Christmas. &
Attendance: 48,553 (accurate)

DICKSON-WILLIAMS MANSION, 108 N. Irish St., Greeneville, TN 37743. Mailing Address: Dickson-Williams Historical Association, 1241 Tanglewood Dr., Greeneville, TN 37743. Tel.: 423-787-0500.
E-mail: director@mainstreetgreeneville.com
Web Site: www.mainstreetgreeneville.com
Governing Authority: Tax-exempt.
Institution Type/Description: Historic House Museum: housed in the home built by Irish immigrant William Dickson for his daughter Catherine and her husband Dr. Alexander Williams; built in 1821. The mansion served as headquarters for both Union and Confederate armies during the Civil War.
Collections: Williams family history; photographs; Civil War history & artifacts; period furnishings.
Hours & Admission Prices: Tours: daily 1pm; groups of 12 or more by appointment. Adults $10. Closed holidays.

DOAK HOUSE MUSEUM, (M), Tusculum College, Greeneville, TN 37743. Mailing Address: Tusculum College, P.O. Box 5026, Greeneville, TN 37743. Tel.: 423-636-8554 & 7348. Fax: 423-638-7166.
E-mail: clucas@tusculum.edu
Web Site: www.doakhouse.tusculum.edu
Founded: 1980.
Congressional District: 1
Key Personnel: Dir., George Collins; Assoc. Dir. & Cur., Cindy L. Lucas.
Personnel Profile: Full-Time Paid 2; Part-Time Paid 1; Part-Time Volunteers 24.
Governing Authority: college. Tax-exempt: 501(c)(3).
Institution Type/Description: College Museum; located in home of Samuel W. Doak, founder of Tusculum College. Listed on the National Register of Historic Places.
Collections: college-related artifacts; East Tennessee education artifacts & documents; period furniture.
Hours & Admission Prices: Mon.-Fri. 9-5; other times by appointment. Adults $2, children $1; discounts to AAM members. Closed New Year's Eve & Day; Good Friday; Memorial Day; Independence Day; Labor Day; Thanksgiving Day & day after; Christmas Eve, Day & week; college breaks.
Attendance: 6,509 (accurate)

NATHANAEL GREENE MUSEUM, 101 W. McKee St., Greeneville, TN 37743-4813. Tel.: 423-636-1558.
E-mail: director@nathanaelgreenemuseum.com
Web Site: nathanaelgreenemuseum.com
Founded: 1983.
Congressional District: 1
Governing Authority: Tax-exempt.
Institution Type/Description: Heritage Museum.
Collections: local history & heritage; photographs; personal artifacts; early furnishings.
Major Exhibits: Quilt Show, 10/1/14-10/18/14.
Hours & Admission Prices: Feb.-May & Sept.-Dec. Tues.-Sat. 11-5; June-Aug. Mon.-Sat. 11-5; other times by appointment. No charge; donations accepted. Closed holidays.
Attendance: 9,000 (accurate)
Membership: $25-$1,000.

PRESIDENT ANDREW JOHNSON MUSEUM AND LIBRARY, (M), 57 Gilland St., Greeneville, TN 37743. Mailing Address: Tusculum College, P.O. Box 5026, Greeneville, TN 37743. Tel.: 423-636-7348. Fax: 423-638-7166. Facebook: President Andrew Johnson Museum and Library.
E-mail: dboyd@tusculum.edu
Web Site: ajmuseum.tusculum.edu
Founded: 1993.
Congressional District: 1
Key Personnel: Dir., Dollie Boyd; Site & Events Mgr., Leah Walker; Archivist, Kathy Cuff.
Governing Authority: private college; nonprofit. Tax-exempt. 501(c)(3)
Institution Type/Description: Presidential Library: housed in 1841 college building. Listed on the National Register of Historic Places.
Collections: political memorabilia; music sheets; manuscripts; Johnson family possessions; books; pamphlets; maps; photographs; newspapers; printed & written media related to the growth, development & history of Tusculum College; college artifacts.

Hours & Admission Prices: Mon.-Fri. 9-5. No charge. Closed all Tusculum College holidays. &
Attendance: 2,800 (accurate)

Halls

THE VETERANS' MUSEUM, 100 Veterans' Dr., Halls, TN 38040-1342. Tel.: 731-836-7400. Fax: 731-836-7400. Facebook: The Veterans Museum.
E-mail: vetmuseumhalls@bellsouth.net
Web Site: www.dyaab.us
Founded: 1997.
Congressional District: 8
Key Personnel: Dir., Chm. (V) & Pres. (V), Patricia M. Higdon; Administrative Asst. & Museum Shop Mgr., Nancy Holman.
Personnel Profile: Full-Time Paid 1; Full-Time Volunteers 1; Part-Time Volunteers 26.
Governing Authority: Parent Institution: The Dyersburg Army Air Base Memorial Assoc., Inc. Tax-exempt.
Institution Type/Description: Military History Museum.
Collections: WWI & II, Korea, Vietnam, & Desert Storm memorabilia; military vehicles; A-7 airplane; CH-46 E Sea Knight helicopter; photographs; documents; diaries; personal & official letters; murals; personal artifacts; microfilm.
Research Fields: WWII history.
Facilities: library.
Activities: tours; special programs. Museum Sponsors: Armed Forces Day in May; Air Show in August; Veterans Day Celebration in November; Christmas Music in December.
Publications: members' newsletter.
Hours & Admission Prices: Sat.-Tues. 2-5. No charge; donations accepted. &
Attendance: 4,000 (estimated)
Membership: Individual $50.

Harrogate

ABRAHAM LINCOLN LIBRARY & MUSEUM, LINCOLN MEMORIAL UNIVERSITY, (M), 6965 Cumberland Gap Pkwy., Harrogate, TN 37752. Mailing Address: P.O. Box 2006, Harrogate, TN 37752. Tel.: 423-869-6235. Fax: 423-869-6350.
E-mail: thomas.mackie@lmunet.edu
Web Site: lmunet.edu/museum
Founded: 1897.
Congressional District: 4
Key Personnel: Dir., Thomas Mackie; Cur. & Asst. Dir., Steven Wilson; Education Coord., Program & Tourism Dir., Carol Campbell; Museum Archivist, Michelle Ganz; Guest Svcs., Jonathan Smallwood; Admin. Asst., Barbara Garman.
Personnel Profile: Full-Time Paid 5; Part-Time Paid 1; Part-Time Volunteers 5.
Governing Authority: bd. of advisors. Parent Institution: Lincoln Memorial University. Tax-exempt: 501(c)(3).
Institution Type/Description: History Museum.
Collections: Abraham Lincoln & Civil War materials: 30,000 artifacts; books; manuscripts; statuary; engravings; sheet music; pamphlets.
Major Exhibits: Cities of Comfort and Care, 10/13-12/14.
Research Fields: Abraham Lincoln; the field of Lincolniana; Civil War music, politics & military aspects of war.
Facilities: library; reading room; 140-seat auditorium & theater.
Activities: guided tours; permanent exhibitions; student training programs; workshops.
Publications: The Lincoln Herald; Lincoln Letters.
Hours & Admission Prices: March to late Nov. Mon.-Fri. 10-5, Sat. 12-5, Sun. 1-5; late Nov.-Feb. Mon.-Fri. 10-5, Sat. 12-5. Adults $5, senior citizens $3.50, children 6-12 $3; discounts to groups & AAA members; Lincoln Memorial University staff, students and faculty & children under 6 no charge. Closed New Year's Day; Easter; Thanksgiving; Christmas.
Attendance: 13,500 (accurate)

Hartsville

LIVING HISTORY MUSEUM, 101 White Oak St., Hartsville, TN 37074. Mailing Address: Chamber of Commerce, 240 Broadway, Hartsville, TN 37074-1336. Tel.: 615-374-9243. Fax: 615-374-0068.
E-mail: eford@hartsvilletrousdale.com
Web Site: www.hartsvilletrousdale.com
Founded: 1995.
Key Personnel: Dir. Chamber of Commece, Natalie Knatson; Chm. (V), John Oliver.
Personnel Profile: Part-Time Volunteers 2.

Volunteer Hours: 50
Operating Expenses: 1,000
Operating Income: 1,000
Governing Authority: Parent Institution: County Government. Subsidiary Institution: Chamber of Commerce. Tax-exempt.
Institution Type/Description: History Museum.
Collections: recreated 1930s farm; period furnishings.
Hours & Admission Prices: By appointment. No charge; donations accepted.
Attendance: 250 (estimated)

Hendersonville

HENDERSONVILLE ARTS COUNCIL, 1154 W. Main St., Hendersonville, TN 37077-0064. Mailing Address: P.O. Box 64, Hendersonville, TN 37077-0764. Tel.: 615-822-0789.
E-mail: artscouncil@monthaven.org
Web Site: www.hendersonvillearts.org
Founded: 1975.
Key Personnel: Exec. Dir., Alexander Brindley
Institution Type/Description: Art Gallery.
Collections: paintings; sculpture; photographs.
Hours & Admission Prices: Mon.-Fri. 9-3. No charge; donations accepted.

HISTORIC ROCK CASTLE, 139 Rock Castle Lane, Hendersonville, TN 37075-4522. Tel.: 615-824-5081. Facebook: Historic Rock Castle.
E-mail: info@historicrockcastle.com
Web Site: www.historicrockcastle.com
Founded: 1971.
Congressional District: 6
Key Personnel: Dir., Sara Beth Gideon.
Personnel Profile: Full-Time Paid 2; Part-Time Paid 2; Part-Time Volunteers 30.
Governing Authority: Tax-exempt.
Institution Type/Description: Historic House Museum.
Collections: local history & culture; period furnishings; photographs.
Facilities: Historic home; smokehouse; visitor's center; pavilion; family cemetery.
Activities: fall colonial fair; Daniel Smith days; October storytelling night; music series; classes & workshops; educational programs; weddings; special events.
Hours & Admission Prices: Tues.-Sat. 10-5, Sun. 1-5 (April - Nov.)', adults $7, seniors $6, children $5, mems. no charge. Closed Mon., Jan., all TN state holidays. &
Attendance: 12,000 (estimated)
Membership: Individual $30; Couple $50; Family 75.

OLD HICKORY LAKE VISITOR CENTER, No. 5 Power Plant Rd., Hendersonville, TN 37075-3467. Tel.: 615-822-4846 & 847-2395. Fax: 615-822-2743.
Web Site: www.lrn.usace.army.mil/op/old/rec/
Institution Type/Description: History Museum & Visitor Center.
Collections: local history; navigation locks; water safety; water resources development; U.S. Army Corps of Engineers; photographs; hands-on exhibits.
Activities: educational programs.
Hours & Admission Prices: Mon.-Fri. 7:30-4:30. No charge.
Attendance: 8,729,003 (accurate)

Henning

ALEX HALEY MUSEUM AND INTERPRETIVE CENTER, 200 S. Church St., Henning, TN 38041-7201. Mailing Address: P.O. Box 500, Henning, TN 38041-7201. Tel.: 731-738-2240. Fax: 731-738-2585.
E-mail: alexhaleymuseum@bellsouth.net
Web Site: www.alexhaleymuseum.com
Founded: 1986.
Key Personnel: Dir., Paula L. Boger; Chm. (V), Phillis Barlow.
Personnel Profile: Full-Time Paid 2; Part-Time Paid 2.
Governing Authority: Parent Institution: TN Historical Commission. Tax-exempt.
Institution Type/Description: Historic House Museum: housed in the boyhood home of renowned author & Pulitzer Prize winner, Alex Haley. Listed on the National Register of Historic Places.
Collections: period furnishings; personal artifacts.
Research Fields: genealogy.
Facilities: 6,500 sq. ft interpretive center; theatre; genealogy center. Museum-related items for sale.

Activities: Museum Sponsors: Black History Celebration in February; Alex Haley Birthday Celebration in August; Veterans Day Ceremony in November.
Publications: newsletter, The Front Porch.
Hours & Admission Prices: Tues.-Sat. 10-5, Sun. by appointment. Adults $6; members no charge. &
Attendance: 4,000 (estimated)

FORT PILLOW STATE HISTORIC PARK, 3122 Park Rd., Henning, TN 38041-5210. Tel.: 731-738-5581 & 5731. Fax: 731-738-9117.
Web Site: www.state.tn.us/environment/parks/FortPillow/index.shtml
Institution Type/Description: History Museum.
Collections: Civil War artifacts; local history & culture; photographs.
Facilities: nature trails.
Activities: video; guided tours; educational programs.
Hours & Admission Prices: Park: daily 8am to sunset. Museum: daily 8-11:30 & 12:30-4. No charge. Closed Thanksgiving; Christmas Eve & Day.

Hermitage

THE HERMITAGE: HOME OF PRESIDENT ANDREW JACKSON, 4580 Rachel's Lane, Hermitage, TN 37076-1331. Tel.: 615-889-2941. Fax: 615-889-9909.
E-mail: info@thehermitage.com
Web Site: www.thehermitage.com
Founded: 1889.
Congressional District: 5
Key Personnel: Dir. & C.E.O., Howard Kittell; Regent, Martha Cooper; Chief Cur. & Dir. Museum Svcs., Marsha A. Mullin; Dir. Finance, Kathy McCall; Dir. Mktg., Jason Nelson; Guest Svcs., Debbie Bourne; Exec. Asst. & Membership, Jane Maggard; Museum Shop Mgr., Amanda Millslagle.
Personnel Profile: Full-Time Paid 40; Part-Time Paid 40; Interns 9.
Governing Authority: nonprofit organization. Parent Institution: Ladies' Hermitage Assn.; trust deed from state of TN. Tax-exempt: 501(c)(3).
Institution Type/Description: Historic House.
Collections: archaeological; architectural; horticultural; gardens & grounds; furnishings; costumes; manuscripts; family possessions; carriages. Historic Buildings: 1804 log cabin; 1821-1836 Hermitage mansion & outbuildings; 1836 Tulip Grove mansion; 1823; Old Hermitage church; 1833 President's Tomb.
Research Fields: Andrew Jackson & his family; 19th-century horticulture & gardens; slave life.
Facilities: library; 120-seat restaurant; meeting & party facilities; visitor center. Books & museum-related items for sale.
Activities: Hermitage Mansion restoration; orientation film; temporary exhibitions; tour; school & adult education programs; archaeology programs.
Publications: guidebook.
Hours & Admission Prices: April-Oct. 15 daily 8:30-5; Oct. 16-March daily 9-4:30. Adults $17, senior citizens $14, students 13-18 $11, children 6-12 $7; discounts to groups, AAA, AAM & ICOM members; children 5 & under and members no charge. Closed Thanksgiving; Christmas. &
Attendance: 235,603 (accurate)
Membership: Student & Teacher $20; Senior $40; Individual $45; Dual $50; Family $60; Congressman's Circle $125; Senator's Circle $500; General's Circle $1,000.

Hixson

HIXSON FLIGHT MUSEUM, 1824 E. Crabtree Rd., Hixson, TN 37343. Tel.: 423-228-2359.
E-mail: curator@hixsonflightmuseum.org
Web Site: www.hixsonflightmuseum.org
Institution Type/Description: Flight Museum.
Collections: military & commercial aircraft.
Facilities: Museum-related items for sale.
Activities: outreach program; special eventsl; air shows; plane ride.
Hours & Admission Prices: Wed.-Thurs. 9-3, Fri.-Sat. 9-4. Guided Tour: $5-$10.

Hohenwald

LEWIS COUNTY MUSEUM OF NATURAL HISTORY & HOHENWALD DISCOVERY CENTER, 108 E. Main St., Hohenwald, TN 38462. Tel.: 931-796-1550.
E-mail: questions@lewiscountymuseum.com
Web Site: www.lewiscountymuseum.com
Institution Type/Description: History Museum.
Collections: local history & culture; early pottery; Lewis & Clark expedition; Civil War artifacts; Gordonsburg mining; wildlife & their habitats.

Activities: educational programs.
Hours & Admission Prices: Tues.-Sat. 10-4, Sun. 1-4. Adults $5, seniors $4, students $2. Closed major holidays.

MERIWETHER LEWIS NATIONAL MONUMENT, 189 Meriwether Lewis Park, Hohenwald, TN 38462-5591. Mailing Address: National Park Service, Natchez Trace Pkwy., 2680 Natchez Trace Pkwy., Tupelo, MS 38804. Tel.: 800-305-7417. Fax: 662-680-4034.
Web Site: www.nps.gov/natr
Founded: 1936.
Congressional District: 3
Key Personnel: U.S. Ranger, Terry Kelly; District Ranger, Dave Hajdik.
Personnel Profile: Full-Time Paid 5; Part-Time Volunteers 2.
Governing Authority: federal. Affiliated with National Park Service, Natchez Trace Pkwy., 2680 Natchez Trace Pkwy., Tupelo, MS 38804. Tax-exempt.
Institution Type/Description: Historic Site Museum: 1809 death & burial site of Meriwether Lewis.
Collections: pertaining to Lewis' life & death, 1774-1809, Lewis' grave & monument.
Facilities: 34 campground & picnic areas.
Activities: Annual Event: Craft Fair in October.
Hours & Admission Prices: Daily 8-5. No charge.
Attendance: 9,000 (estimated)

Humboldt

WEST TENNESSEE REGIONAL ART CENTER, 1200 Main St., Humboldt, TN 38343-3339. Mailing Address: P.O. Box 951, Humboldt, TN 38343-0951. Tel.: 731-784-1787. Fax: 901-784-1573.
E-mail: wtrac@aeneas.net
Web Site: www.wtrac.tn.org
Founded: 1994.
Congressional District: 8
Key Personnel: Chm. Bd., Charles Guy; Treas, Carolyn Barnett; Cur., Bill Hickerson.
Personnel Profile: Full-Time Paid 1; Part-Time Volunteers 12.
Governing Authority: Tax-exempt.
Institution Type/Description: Art Center.
Collections: fine arts.
Hours & Admission Prices: Mon.-Fri. 9-4:30, Sat.-Sun. group tours by appointment. Upstairs Gallery: $2. Downstairs Gallery: no charge; donations accepted. Closed holidays. &
Attendance: 10,000 (estimated)

Huntsville

WORLD WAR II REMEMBRANCE MUSEUM, 400 Scott High Dr., Huntsville, TN 37756. Tel.: 423-663-2805.
E-mail: tennesseeman43@comcast.net
Institution Type/Description: Military History Museum.
Collections: World War II history; Battleship USS Tennessee; photographs; personal artifacts.
Hours & Admission Prices: Call for hours.

Hurricane Mills

COAL MINER'S DAUGHTER MUSEUM, 1877 Hurricane Mills Rd., Hurricane Mills, TN 37078. Tel.: 931-296-1840. Fax: 931-296-1839.
Web Site: www.lorettalynn.com
Governing Authority: Parent Institution: Loretta Lynn Foundation.
Institution Type/Description: History Museum.
Collections: Loretta Lynn's career including memorabilia & awards; photographs; personal artifacts.
Hours & Admission Prices: Daily 9-5. Museum: adult $12.50, children under 10 no charge. Home Tour Package: $25, children under 10 no charge. &

Jackson

CASEY JONES HOME AND RAILROAD MUSEUM, Casey Jones Village, 30 Casey Jones Ln., Jackson, TN 38305. Mailing Address: 56 Casey Jones Ln, Jackson, TN 38305. Tel.: 731-668-1222. Fax: 731-664-7782.
E-mail: caseyjonesmuseum@gmail.com
Web Site: www.caseyjonesvillage.com
Founded: 1956.
Congressional District: 7

Key Personnel: C.E.O., T. Clark Shaw.
Personnel Profile: Full-Time Paid 2; Part-Time Paid 2; Part-Time Volunteers 1; Interns 1.
Governing Authority: municipal. Parent Institution: Old Country Store Inc.
Institution Type/Description: Railroad Museum: housed in c.1900 home of Casey Jones.
Collections: house & furnishings; steam locomotive engine; old railroad passes; timetables; telegraph instruments; lanterns; steam whistles; railroad art.
Research Fields: life & home of Casey Jones; items pertaining to the steam era.
Facilities: 8,000 sq. ft. train station. Railroad items & stamps for sale.
Hours & Admission Prices: Daily 9-5. Adults $6.50, seniors $5.50, children 6-12 $4; discounts to groups, military, AAA & TAM members; children under 6 no charge. Closed Easter; Thanksgiving; Christmas. &

Attendance: 50,000 (estimated)

CYPRESS GROVE NATURE PARK - AERIE TRAIL RAPTOR CENTER, W. Airways Blvd., (Hwy. 70), Jackson, TN 38301. Mailing Address: 3 Westwood Gardens Dr., Jackson, TN 38301-4218. Tel.: 731-425-8316 & 8384.
E-mail: smacdiarmid@cityofjackson.net
Institution Type/Description: Nature Center.
Collections: native wildlife & their habitats; plants; trees; birds of prey including hawks, eagles, & owls.
Facilities: nature trails.
Activities: observation tower; educational programs; hiking; classes.
Hours & Admission Prices: Park: daily 7:30 to dusk. Center: call for hours.

DISCOVERY MUSEUM OF WEST TENNESSEE, 305 E. College, Jackson, TN 38301-6215. Tel.: 731-410-8621. Fax: 731-410-8622.
E-mail: info@wtndiscovery.org
Web Site: wtndiscovery.org
Institution Type/Description: History Museum.
Collections: local history & culture; period furnishings; personal artifacts; photographs.
Activities: special events; educational programs.
Hours & Admission Prices: Tues.-Sat. 9-4. &

HISTORIC DOWNTOWN N.C. & ST. L. DEPOT AND RAILROAD MUSEUM, (M), 582 S. Royal St., Jackson, TN 38301. Tel.: 731-425-8223. Fax: 731-425-8682.
E-mail: thedepot@cityofjackson.net
Web Site: www.cityofjackson.net
Founded: 1996.
Congressional District: 8
Key Personnel: Dir., David Falk.
Personnel Profile: Full-Time Paid 1; Part-Time Paid 3; Part-Time Volunteers 14.
Governing Authority: city. Tax-exempt.
Institution Type/Description: Historic Building: built in 1907. Listed on the National Register of Historic Places.
Collections: railroad & depot history; railroad artifacts & memorabilia; photographs; model replica; paintings.
Activities: special events; permanent & temporary exhibits; birthday parties.
Hours & Admission Prices: Mon.-Sat. 10-3. No charge. Closed city holidays. &

Attendance: 11,463 (accurate)

INTERNATIONAL ROCK-A-BILLY HALL OF FAME MUSEUM, 105 N. Church St., Jackson, TN 38301-6213. Mailing Address: 314 Edenwood Dr., Jackson, TN 38301-3433. Tel.: 731-427-6262.
E-mail: rock@rockabillyhall.org
Web Site: www.rockabillyhall.org
Key Personnel: Pres., Henry Harrison
Institution Type/Description: History Museum.
Collections: artifacts & memorabilia relating to early rock & roll hillbilly music.
Hours & Admission Prices: Mon.-Thurs. 10-5, Fri.-Sat. 10-2.

Jamestown

YE OLE JAIL MUSEUM, 114 Central Ave. W., Jamestown, TN 38556. Mailing Address: P.O. Box 1294, Jamestown, TN 38556. Tel.: 931-879-9948.
E-mail: leanns@jamestowntn.org

Web Site: jamestowntn.org
Founded: 1995.
Congressional District: 14
Governing Authority: Tax exempt.
Institution Type/Description: Historic Building: housed in a former jail, used from 1900-1979.
Collections: local history; jail cells; period furnishings & artifacts.
Hours & Admission Prices: Call for hours. No charge; donations accepted. Closed federal holidays.
Attendance: 5,000 (estimated)

Jefferson City

GLENMORE MANSION, 1280 N. Chucky Pike, Jefferson City, TN 37760-4926. Mailing Address: P.O. Box 403, Jefferson City, TN 37760-0403.
Formerly: The Oaks
Founded: 1971.
Congressional District: 2
Key Personnel: Pres. (V), Helen T. Gray; Museum Shop Mgr., Delene Wilson.
Personnel Profile: Full-Time Volunteers 2; Part-Time Volunteers 1.
Governing Authority: nonprofit organization. Parent Institution: Association for the Preservation of TN Antiquities. Tax-exempt.
Institution Type/Description: Historic House: built in 1868. Listed on the National Register of Historic Places.
Collections: period furnishings; personal artifacts; photographs.
Activities: rental facilities; educational programs.
Publications: biannual newsletter.
Hours & Admission Prices: May-Oct. Sat.-Sun. 1-5; other times by appointment. Adults $5, children under 12 $2.50; members no charge.
Attendance: 1,500 (estimated)
Membership: Junior $10; Individual $25.

Johnson City

GENERAL SHALE MUSEUM OF ANCIENT BRICK, 3015 Bristol Hwy., Johnson City, TN 37602. Mailing Address: P.O. Box 3547, Johnson City, TN 37602. Tel.: 423-282-4661; 800-414-4661. Fax: 423-952-4103.
Web Site: www.generalshale.com
Institution Type/Description: History Museum.
Collections: history of brick & brick construction; bricks from biblical & pre-biblical times including ancient Egypt, Jerusalem, Nimrud, Ur, & Jericho.
Hours & Admission Prices: Mon.-Fri. 8-5.

HANDS ON! REGIONAL MUSEUM, 315 E. Main St., Johnson City, TN 37601-5700. Tel.: 423-928-6508 & 6509. Fax: 423-928-6915.
E-mail: handson@handsonmuseum.org
Web Site: www.handsonmuseum.org
Founded: 1986.
Congressional District: 1
Key Personnel: Exec. Dir., Trish Patterson; Mktg. & Membership, Kristine Amerine Carter; Mgr. Finance, Kay Hobbs; Mgr. Education, Programs & Science Lab, April Bunch; Coord. Exhibits & Outreach, Franci Sloan; Reservations, Karen Deckard.
Personnel Profile: Full-Time Paid 7; Part-Time Paid 15; Part-Time Volunteers 100; Interns 1.
Governing Authority: nonprofit organization. Tax-exempt: 501(c)(3).
Institution Type/Description: Children's Museum.
Collections: ark animals; hands-on exhibits.
Facilities: aquarium; 27,000 sq. ft. exhibit space. Museum-related items for sale.
Activities: organized education programs for children; participatory & traveling exhibitions; Tennessee curriculum school programs; night rentals; birthday parties; sleepovers. Museum Sponsors: Festival of Trees; Museum-in-a-Box outreach program; Summer Learning vacation program.
Publications: quarterly newsletter, Handprints; annual report.
Hours & Admission Prices: June-Aug. Mon.-Fri. 9-5, Sat. 10-5, Sun. 1-5; Sept.-May Tues.-Fri. 9-5, Sat. 10-5, Sun. 1-5. Adults $8; discounts to groups & ASTC members; members & children under 3 no charge. Closed New Year's Day; Martin Luther King Jr. Day; Easter; Memorial Day; Independence Day; Labor Day; Thanksgiving; Christmas Eve & Day.
Attendance: 70,013 (accurate)
Membership: Individual $30; Grandparent $50; Family $60; Joint Parent/Grandparent $95; Friend $100; Associate $250; Sponsor $500; Principal $1,000; Benefactor $2,500; Sustainer $5,000; Guarantor $10,000; Presidential $20,000 and up.

MUSEUM AT MOUNTAIN HOME - EAST TENNESSEE STATE UNIVERSITY, Dept. Learning Resources, Johnson City, TN 37614-1710. Mailing Address: P.O. Box 70693, Johnson City, TN 37614-1710. Tel.: 423-439-8069. Fax: 423-439-7025.
Founded: 1994.
Congressional District: 1
Key Personnel: Pres. (V), Janet Fisher.
Personnel Profile: Part-Time Paid 1; Part-Time Volunteers 3.
Governing Authority: Tax-exempt.
Institution Type/Description: Medical & Military Museum.
Collections: medical & military artifacts.
Hours & Admission Prices: Tues. & Thurs. 9-11, Wed. 1:30-3:30. No charge; donations accepted. Closed holidays. &

✳ **THE REECE MUSEUM, (M),** Gilbreath Dr., East Tennessee State University, Johnson City, TN 37614. Mailing Address: P.O. Box 70660, ETSU, Johnson City, TN 37614-1701. Tel.: 423-439-4392. Fax: 423-439-4283.
E-mail: burchete@etsu.edu
Web Site: www.etsu.edu/reece
Founded: 1965.
Congressional District: 1
Key Personnel: Dir., Theresa Burchett; Pres. ETSU, Dr. Paul Stanton.
Personnel Profile: Full-Time Paid 3; Part-Time Volunteers 2; Interns 1.
Governing Authority: university. Parent Institution: East Tennessee State University. Tax-exempt: 501(c)(3).
Institution Type/Description: History & Art Museum.
Collections: 18th-19th century East Tennessee history & contemporary regional art; paintings; graphics; history; textiles; Tennessee crafts; costumes; folklore; B. Carroll Reece Memorial collection; printing.
Research Fields: paintings; graphics; history; textiles; Tennessee crafts; costumes; printing; musical instruments.
Activities: self guided tours; lectures; gallery talks; concerts; formally organized education programs for children & adults; inter-museum loan, permanent, temporary & traveling exhibitions.
Publications: catalogs; quarterly newsletter; exhibit brochures; calendar.
Hours & Admission Prices: Tues.-Wed. & Fri. 9-4, Thurs. 9-7. Suggested Donation: $3. Closed major holidays. &
Attendance: 10,000 (estimated)

SLOCUMB GALLERIES, Ball Hall, Dept. of Art & Design, ETSU, Johnson City, TN 37614. Mailing Address: Box 70708 ETSU, Johnson City, TN 37614-1710. Tel.: 423-483-3179. Fax: 423-439-4393.
E-mail: contrera@etsu.edu
Web Site: art.etsu.edu/slocumb
Founded: 1965.
Congressional District: 1
Key Personnel: Dir., Karlota I. Contreras-Koterbay.
Personnel Profile: Full-Time Paid 1; Interns 5.
Governing Authority: university. Parent Institution: East Tennessee State University. Subsidiary Institution: Dept. of Art & Design. Tax-exempt.
Institution Type/Description: Art Gallery.
Collections: diverse art media.
Activities: artists' talks; lectures; traveling show; invited artists' exhibit; loan & temporary exhibition. Museum Sponsors: BFA/MFA Graduate Shows; National Art Competition.
Publications: catalogs; e-newsletter; posters; art essays.
Hours & Admission Prices: Mon.-Fri. 8:30-4. No charge. Closed university holidays. &
Attendance: 5,000 (estimated)

TIPTON-HAYNES STATE HISTORIC SITE, 2620 S. Roan St., Johnson City, TN 37601-7585. Mailing Address: P.O. Box 225, Johnson City, TN 37605-0225. Tel.: 423-926-3631.
E-mail: tiptonhaynes@cmbarqmail.com
Web Site: tipton-haynes.org
Founded: 1965.
Congressional District: 1
Key Personnel: C.E.O., Penny McLaughlin; Chm. (V), Mark Edmonds.
Personnel Profile: Full-Time Paid 1; Full-Time Volunteers 2; Part-Time Paid 1; Part-Time Volunteers 202.
Governing Authority: state. Affiliated with Tennessee Historical Commission, 2941 Lebanon Rd., Nashville, TN 37243-0442. Tax-exempt: 501(c)(3).
Institution Type/Description: Historic Home: c.1850 Tipton & Haynes House.
Collections: agriculture; history; outdoor museum; folklore; 15 millstones.

Historic Buildings: c.1850 Haynes home & law office; George Hayne home (slave); c.1784 barn, corn crib, spring house, smokehouse, still house & pig pen.
Research Fields: families of Col. John Tipton & Landon Carter Haynes; State of Franklin; local history.
Facilities: Museum-related items for sale.
Activities: guided tours; permanent exhibitions; summer history enrichment program, grades 1-6. Annual Events: Andre Michaux Day in Spring; Sorghum Festival in September; Stories from the Pumpkin Patch in October; Visions of Christmas re-enactment in December.
Publications: quarterly newsletter, The Link; handbook for tour guides; brochures of site.
Hours & Admission Prices: April-Nov. Tues.-Sat. 9-4; Dec.-March call for hours. Adults $5, students 12 & under $2.50; discount AAA members, school & scout groups; members no charge. Closed New Year's Eve & Day; Thanksgiving & day after; Christmas week. &
Attendance: 9,000 (estimated)
Membership: Student $20; Individual $25; Family $35; Supporting $50; Sustaining $100; Benefactor $250; Grand Benefactor $500; Philanthropist $1,000.

Jonesborough

JONESBOROUGH-WASHINGTON COUNTY HISTORY MUSEUM, (M), 117 Boone St., Historic Jonesborough Visitors Center, Jonesborough, TN 37659. Mailing Address: The Heritage Alliance, 212 E. Sabin Dr., Jonesborough, TN 37659-1306. Tel.: 423-753-9580. Fax: 423-753-5281.
E-mail: info@heritageall.org
Web Site: www.heritageall.org
Founded: 1982.
Congressional District: 1
Key Personnel: Museum Dir., Deborah Montanti.
Governing Authority: nonprofit. Parent Institution: Heritage Alliance. Tax-exempt: 501(c)(3).
Institution Type/Description: History Museum.
Collections: local artifacts from 1770-present; prehistoric artifacts; history of Jonesborough-Washington County, Tennessee's first town & county; restored 1886 one-room school building.
Research Fields: Jonesborough & Washington County; history; historic preservation; architectural; An 1886 One-Room Schoolhouse Teacher's Resource and Curriculum Guide.
Facilities: 500-vol. library of local history, museum methodology, historic preservation available for use on premises; reading room; community room & auditorium available for use in visitor's center.
Activities: permanent & temporary exhibitions; guided tours of historic district; school tours; suitcase exhibits which travel to schools; scout programs; craft programs in summer; teacher in-service programs; films; lectures; heritage education program in one-room school.
Hours & Admission Prices: Mon.-Fri. 9-5, Sat.-Sun. No charge; donations accepted. &
Attendance: 5,735 (accurate)
Membership: The Heritage Alliance of Northeast Tennessee & Southeast Virginia: General $50; Business & Pioneer Circle $100; Franklin Circle $250; Heritage Circle $500; Foounders Circle $1,000.

Kingsport

BAYS MOUNTAIN PARK & PLANETARIUM, 853 Bays Mountain Park Rd., Kingsport, TN 37660. Tel.: 423-229-9447. Fax: 423-224-2589.
E-mail: willey@ci.kingsport.tn.us
Institution Type/Description: Nature Preserve & Planetarium.
Collections: astronomy; space science; native wildlife & their habitats including wolves, bobcats, raccoons, raptors & reptiles.
Facilities: theater; nature center; 3,500 acre nature preserve. Museum-related items for sale.
Activities: educational programs; special events.
Hours & Admission Prices: Nature Center: Mon.-Fri. 8:30-5, Sat.-Sun. 12-7. Park: March Mon.-Fri. 8:30-5, Sat. 8:30-8, Sun. 12-8, April-May & Sept.-Oct. Mon.-Tues. & Thurs.-Fri. 8.30-5, Wed. 8:30-7, Sat. 8:30-8, Sun. 12-8. Park: $4 per car. Planetarium Shows: adults & children 6 and over $4; children 5 & under no charge. Nature Programs: $2 per person.
Membership: Individual $25; Family $35; Supporting $75; Life $1,000.

EXCHANGE PLACE LIVING HISTORY FARM, 4812 Orebank Rd., Kingsport, TN 37664. Tel.: 423-288-6071.
E-mail: email@exchangeplace.info
Key Personnel: Chm. (V), Marshall Adesman; Museum Shop Mgr., Billee Moore

Institution Type/Description: History Museum.
Collections: local history, culture & heritage; period furnishings, equipment & tools; livestock; historic buildings.
Activities: demonstrations; day camps; school programs; special events; group tours. Annual Events: Spring Garden Fair; Kirsten, An American Girl Day of Discovery in May; Hamlett-Dobson Farm Fest in July; Fall Folk Arts Festival in September; Witches Wynd in October; Christmas in the Country in December.
Hours & Admission Prices: May-Oct. Sat.-Sun. 2-4:30; groups by appointment.

NETHERLAND INN HOUSE MUSEUM & BOATYARD COMPLEX, 2144 Netherland Inn Rd., Kingsport, TN 37660-3052. Mailing Address: P.O. Box 293, Kingsport, TN 37660. Tel.: 423-335-5552.
E-mail: jgibson@naxs.net
Web Site: netherlandinn.com
Founded: 1966.
Congressional District: 1
Key Personnel: C.E.O. & Steering Committee Chm., Mrs. Dennis Phillips; Museum Dir., Furnishings Maintenance Chm. & Catalog Dept. Chm., Mrs. Jane Gibson; Guide Chm., Annette Pannell; Museum Shop Mgr., Mrs. Lib Findley.
Personnel Profile: Part-Time Paid 2; Part-Time Volunteers 110.
Governing Authority: private; nonprofit association. Parent Institution: Netherland Inn/Exchange Place Association, Inc., Box 293, Kingsport. Tax-exempt: 501(c)(3).
Institution Type/Description: Historic House Museum: built in 1802.
Collections: late 18th- & early 19th-century period furnishings; costumes; documents; manuscripts; guns; musical instruments; toys; farm equipment; tools; cooking utensils; flatboat replica; stagecoach; railroading items; forts; period dolls; wagons; period kitchen; well house; historical monuments. Historic Structures: 1773-1775 two-story log cabin children's museum; 1795 Ross Ordinary; 1830s two story log visitor's center.
Research Fields: families, land transactions, history of area; furnishings, uses in connection with restorations.
Facilities: gardens.
Activities: school tours; lectures; slide tape programs; temporary & permanent exhibits; crafts; children's museum teaching activities; guest lecturers; walking & windshield tours of historic districts. Special Events: English High Tea in May; Wine & Cheese Party in June; Fun Inn the Sun in July; Old Time Fiddlers & Blue Grass Festival in August; East Tenn. Country Scottish Dancers in October; Concert on the Green; hearth-side cooking demonstrations; annual 1818 Netherland Inn Assoc. Christmas Party in December; banquet.
Publications: quarterly newsletter, Newsletter of Netherland Inn Assn.; brochures; Netherland Family Genealogy; Netherland Inn & Boat Yard Booklet; Netherland Inn Cookbook; historical map of the Long Island of the Holston; History of Early Kingsport, Prehistory to 1900; Kingsport Heritage: The Early Years 1700-1900; Lithograph, Birth of Kingsport; The Netherland Inn Chronicles.
Hours & Admission Prices: Guided Tours: May-Oct. Sat.-Sun. 2-4; groups & other times by appointment. Adults $4, children 6 & under $1; discount to groups; Tourism Bureau & museum members no charge. Additional fee for some special events. &
Attendance: 9,000 (estimated)
Membership: Individuals: Student $5; Individual $25; Family $40; Patron $100; Benefactor $500; Life $5,000. Corporate & Professional: Club $50; Patron $100; Benefactor $500.

Kingston

ROANE COUNTY MUSEUM OF HISTORY & ART, 119 Court St., Kingston, TN 37763-2810. Mailing Address: P.O. Box 738, Kingston, TN 37763-0738. Tel.: 865-376-9211.
Web Site: www.roanetnheritage.com
Key Personnel: Dir., Darlene Trent
Institution Type/Description: Art & History Museum: housed in former antebellum courthouse.
Collections: Roane County & Tennessee history.
Facilities: library.
Hours & Admission Prices: Mon.-Fri. 9-4. No charge.

Knoxville

BECK CULTURAL EXCHANGE CENTER, INC., 1927 Dandridge Ave., Knoxville, TN 37915-1909. Tel.: 865-524-8461. Fax: 865-524-8462.
E-mail: beckcenter@beckcenter.net
Web Site: www.beckcenter.net
Founded: 1975.
Congressional District: 2
Key Personnel: Exec. Dir. & C.E.O., Robert J. Booker, Sr.; 1st Vice Pres., Annazette Houston; 2nd Vice Pres., Arnold G. Cohen; Pres., Samuel P. Anderson; Archivist, Timothy Vasser; Administrative Asst., Andre Canty.
Personnel Profile: Full-Time Paid 1; Part-Time Paid 3; Part-Time Volunteers 40.
Governing Authority: nonprofit. Tax-exempt: 501(c)(3).
Institution Type/Description: History Museum, Cultural Center & Local Black History.
Collections: books written by local Black authors; Black weekly newspapers; works of art by local artists; oral histories; biographies; manuscripts; phonograph records & photographs; Federal Judge William A. Hastle collection; Black classical phonograph record & movies.
Major Exhibits: There's Music in the Air, 1/14-3/14; Our Professional Athletes, 4/14-6/14; Austin High School 135th Anniversary, 7/14-9/14; Men and Women in the Service Industry, 10/14-12/14; Civil and Social Clubs, 1/15-3/15.
Research Fields: local Black history; state & national history.
Facilities: library of books on Black history available for research on premises; reading room; video room; classrooms.
Activities: guided tours; lectures; permanent & temporary exhibitions; in-house movie screenings.
Publications: quarterly newsletter, Historically Speaking; books, Tales From Back Then; Down Memory Lane; Blacks in Knoxville-The First 100 Years, 1791-1891; Two Hundred Years of Black Culture In Knoxville, Tennessee 1791-1991; And There Was Light! 120 Year History of Knoxville College; History of Austin High School; The Story of East Knoxville; The Story of Mechanicsville; historic calendars.
Hours & Admission Prices: Tues.-Sat. 10-6. No charge; donations accepted. Closed holidays. &
Attendance: 30,000 (accurate)
Membership: Student $20; Individual $25; Family $50; Organization $200; Sustaining $250; Sponsor $100; Patron $500; Corporate $1,000; Corporate Gold $5,000; Corporate Platinum $10,000.

BLOUNT MANSION, 200 W. Hill Ave., Knoxville, TN 37902-1812. Mailing Address: P.O. Box 1703, Knoxville, TN 37901-1703. Tel.: 865-525-2375; 888-654-0016 (toll free). Fax: 865-546-5315. Facebook: Blount Mansion.
E-mail: info@blountmansion.org
Web Site: www.blountmansion.org
Founded: 1926.
Congressional District: 2
Key Personnel: C.E.O., Katie Stringer, PhD.
Personnel Profile: Full-Time Paid 2; Part-Time Paid 1; Part-Time Volunteers 10; Interns 2.
Governing Authority: nonprofit organization. Parent Institution: Blount Mansion Association. Branch Museum: 1818, Craighead-Jackson House, 1000 State St. Tax-exempt: 501(c)(3).
Institution Type/Description: Historic House Museum: 1792 home & office of William Blount, governor of the Territory of the United States South of the River Ohio (Southwest Territory).
Collections: late 18th-century decorative arts.
Major Exhibits: Richard LeFevre: The Civil War History, 2/1-4/1/14.
Facilities: 18th-century garden; visitors center.
Activities: guided tours; docent program; permanent & temporary exhibitions; monthly lecture series; educational programs for school groups; audio visual program.
Hours & Admission Prices: Tues.-Sat. 9:30-5. Adults $7, senior citizens and CAA & AAA members $6, children 6-17 $5; discounts to groups; children under 5 and AASLH members no charge. Closed all major holidays.
Attendance: 10,100 (accurate)
Membership: Student, Teacher or Senior Citizen $25; Individual $40; Family $60; Patron $100; Delegate $250; Federalist $500; Governor's Cabinet $1,000.

CONFEDERATE MEMORIAL HALL-BLEAK HOUSE, 3148 Kingston Pike, S.W., Knoxville, TN 37919-4627. Tel.: 865-522-2371.
E-mail: bleakhouseevents@yahoo.com
Web Site: www.knoxvillecmh.org
Founded: 1959.
Congressional District: 2
Key Personnel: Pres., Namuni Young.
Personnel Profile: Part-Time Volunteers 25.
Governing Authority: nonprofit organization. Affiliated with United Daughters of the Confederacy, Tennessee Division, Nashville, TN 37221. Tax-exempt: 501(c)(3).

Institution Type/Description: General Museum: housed in 1858 Bleak House.
Collections: Confederate history; furniture & relics of Civil War period; archives; botany; costumes.
Research Fields: Civil War; reference: war records of confederate soldiers; records of confederate cemeteries; battles of Civil War; Generals.
Facilities: 1,000-vol. library of books available for use by permission of Executive Board; reading room. Confederacy memorabilia & postcards for sale.
Activities: guided tours; programs; civic & historical meetings, weddings, receptions, teas, political rallies, documentaries & videos.
Publications: brochure.
Hours & Admission Prices: March-Dec. Wed.-Fri. 1-4; other times by appointment, school & bus tours by appointment. Adults $5, senior citizens $4, students $3, children 7-12 $1.50.
Attendance: 1,300 (estimated)

CRESCENT BEND/THE ARMSTRONG-LOCKETT HOUSE AND THE WILLIAM P. TOMS MEMORIAL GARDENS, 2728 Kingston Pike, Knoxville, TN 37919-4600. Tel.: 865-637-3163 & 544-3000. Fax: 865-637-1709.
E-mail: rbrett@tds.net
Web Site: www.korrnet.org/cresbend
Founded: 1975.
Key Personnel: Chm., Ron Grimm.
Governing Authority: foundation. The Toms Foundation. Tax-exempt.
Institution Type/Description: Decorative Arts Museum: housed in 1834 Armstrong-Lockett House.
Collections: c.1720-1820 American and English furniture; paintings; mirrors; 1610-1820 English silver collection.
Facilities: library of books on English silver, American furniture & decorative arts.
Activities: guided tours; lectures.
Hours & Admission Prices: March-Dec. Tues.-Sat. 10-4, Sun. 1-4. Adults $7, students $5; discounts to AAA members & groups of 20 or more; children 12 & under no charge. Closed major holidays.
Attendance: 9,250 (accurate)

DISCOVERY CENTER (EAST TENNESSEE DISCOVERY CENTER), 516 N. Beaman St., Chilhowee Park, Knoxville, TN 37914-4410. Mailing Address: P.O. Box 6204, Knoxville, TN 37914-0204. Tel.: 865-594-1494. Fax: 865-594-1469.
E-mail: etdc@comcast.net
Web Site: www.etdiscovery.org
Founded: 1960.
Congressional District: 2
Key Personnel: Chm. (V), Stacy Roettger; Exec. Dir. & Museum Shop Mgr., Margaret Maddox; Education Facilitator, Ashley Whitmire; Administrative Asst., Karen Sullivan; Planetarium Facilitator, Charles Ferguson.
Personnel Profile: Full-Time Paid 4; Part-Time Paid 11; Part-Time Volunteers 10.
Governing Authority: nonprofit organization. Tax-exempt: 501(c)(3).
Institution Type/Description: Science Museum & Planetarium.
Collections: physical, life & earth sciences; astronomy; insects; saltwater & freshwater aquaria; live arthropods; amphibians; reptiles.
Facilities: 30 ft. hyper-hemisphere dome in the planetarium; classrooms. Museum-related items for sale.
Activities: guided tours; lectures; films; gallery talks; hobby workshops; formally organized education programs for children; permanent, temporary & traveling exhibitions.
Publications: pamphlets; newsletters; program guides.
Hours & Admission Prices: Mon.-Fri. 9-5, Sat. 10-5. Adults $4, senior citizens & children 5 and over $3, children 3-4 $2; discounts to AAM & AAA members; museum & ASTC members and children under 3 no charge.
Attendance: 57,000 (accurate)
Membership: Student $15; Individual $25; Family $50; Supporter $100; Friend $250; Patron $500.

EAST TENNESSEE HISTORICAL SOCIETY, 601 S. Gay St., Knoxville, TN 37902-1604. Mailing Address: P.O. Box 1629, Knoxville, TN 37901-1629. Tel.: 865-215-8824. Fax: 865-215-8819.
E-mail: eths@eastTNhistory.org
Web Site: www.easttnhistory.org
Founded: 1834.
Congressional District: 2
Key Personnel: Dir., Cherel Henderson; Pres. (V), Susan Williams; Cur. Education, Lisa Oakley; Cur. Collections, Michele MacDonald; Cur.

Exhibits, Adam Alfrey; Dir. Devel., Lisa Belleman; TAH Grant Mgr., William Hardy; Museum Shop Mgr., Diane Bohannon; Exec. Asst., Stephanie Henry.
Personnel Profile: Full-Time Paid 5; Part-Time Paid 5; Part-Time Volunteers 50; Interns 1.
Governing Authority: private; nonprofit organization. Tax-exempt: 501(c)(3).
Institution Type/Description: Historical Society Museum: housed in the Old Custom House, which was built between 1870-1874.
Collections: concentration on East Tennessee history from mid-1700s to late 20th century.
Activities: lecture series; school programming; special events; permanent exhibits.
Publications: annual scholarly journal, Journal of East Tennessee History; triannual genealogy publication, Tennessee Ancestors; semi-annual newsletter, Newsline.
Hours & Admission Prices: Mon.-Fri. 9-4, Sat. 10-4, Sun. 1-5. Adults $5, senior citizens 55 & over $4; members and children 16 & under no charge. Closed New Year's Day; Easter; Independence Day; Thanksgiving; Christmas Eve & Day.
Attendance: 53,752 (accurate)
Membership: Student/Teacher & Affiliate $25; Nonprofit Institutional $35; Individual $35; Family $45; Contributing $75; Sustaining $125; Patron $250; Benefactor $500; Grand Benefactor $1,000; Founder's Circle $2,500.

EWING GALLERY - UNIVERSITY OF TENNESSEE, Art & Architecture Bldg., 1715 Volunteer Blvd., Knoxville, TN 37996-2410. Tel.: 865-974-3199.
Web Site: www.ewing-gallery.utk.edu
Key Personnel: Dir., Sam Yates
Institution Type/Description: Art & Architecture Museum.
Collections: art & architecture history & current trends; drawings; photographs.
Activities: workshops; lectures; films; research.
Hours & Admission Prices: Mon. & Thurs. 10-8, Tues.-Wed. & Fri. 10-5, Sun. 1-4. Closed national holidays.

HISTORIC RAMSEY HOUSE, 2614 Thorngrove Pike, Knoxville, TN 37914-9704. Tel.: 865-546-0745. Fax: 865-546-1851. TDD: 856-546-0745.
E-mail: info@ramseyhouse.org
Web Site: www.ramseyhouse.org
Formerly: Ramsey House Plantation
Founded: 1952.
Congressional District: 2
Key Personnel: Pres. Bd. Dir., Wayne Decker; Exec. Dir., Judy LaRose.
Personnel Profile: Full-Time Paid 1; Part-Time Volunteers 25.
Governing Authority: nonprofit organization. Parent Institution: Association for the Preservation of Tennessee Antiquities, Belle Meade Mansion, Leake Ave., Nashville, TN 37205. Subsidiary Institution: Knoxville Chapter, Association for the Preservation of Tennessee Antiquities. Tax-exempt: 501(c)(3).
Institution Type/Description: Historical House Museum: c.1796-97 Ramsey House, formerly Swan Pond, restored; Heirloom Gardens.
Collections: furnishings of the period c.1797-1820; 18th-century textiles.
Research Fields: authentic restoration of house and furnishings; Ramsey family; reconstruction of farm buildings; restoration of gardens; costumes of period & history of textiles.
Facilities: 50-vol. library of historical, religious, legal and medical books. Booklets, postcards, books & magazines for sale.
Activities: guided tours; special events; lectures; children's education program; permanent exhibitions.
Publications: brochures, Ramsey House; Lebanon In-the-Fork; Research Resume for Ramsey Period Dress; Presbyterian Church; quarterly newsletter; education guide.
Hours & Admission Prices: Wed.-Sat. 10-4. Adults $7, children 6-12 $5; discounts to seniors, AAA & National Trust members; Museums of Knoxville & Tennessee, Association of Museums members & children under 6 no charge. Closed major holidays.
Attendance: 11,382 (accurate)
Membership: Individual $35; Family $50; Contributing $100; Sustaining $250; Patron $500; Swan Pond $1,000; Col. Ramsey's Circle $1500.

IJAMS NATURE CENTER, 2915 Island Home Ave., Knoxville, TN 37920. Tel.: 865-577-4717.
E-mail: lbales@ijams.org
Institution Type/Description: Nature Center.
Collections: wildlife & their habitats; raptors; snapping turtles; conservation; hands-on exhibitions.

Facilities: amphitheater; visitor center; 165 acres; nature trails; classrooms. Museum-related items for sale.
Activities: educational programs.
Hours & Admission Prices: Visitor Center: Tues.-Sat. 9-5, Sun. 1-5; other times by appointment. Trails: daily 8-6.

JOHN C. HODGES LIBRARY, 1015 Volunteer Blvd., Knoxville, TN 37996-1000. Tel.: 865-974-4351.
Institution Type/Description: Library.
Collections: books; journals; periodicals; microfilm & fiche; audio; video; multimedia.
Facilities: 2.3 million vol. library; 150-seat auditorium.
Activities: research.
Hours & Admission Prices: Academic Year: call for hours; Summer: Mon.-Thurs. 7:30am-12am, Fri. 7:30-6, Sat. 10-6, Sun. 12pm-12am.

THE KNOXVILLE BOTANICAL GARDENS AND ARBORE-TUM, 2743 Wimpole Ave., Knoxville, TN 37914-5958. Tel.: 865-862-8717. Fax: 865-862-8721.
E-mail: info@knoxgarden.org
Web Site: knoxgarden.org
Key Personnel: Dir., Steve Seifreid; Admin. & Special Events Coord., Danielle Velez; Garden Mgr., Brian Campbell
Institution Type/Description: Botanical Garden & Arboretum.
Collections: Knoxville's cultural & horticultural history; gardens; arboretum; bird sanctuary; Civil War; Native American.
Facilities: 44-acre site; amphitheater.
Activities: plays; concerts; festivals; special events.
Hours & Admission Prices: Daily sunrise-sunset. No charge.

*** KNOXVILLE MUSEUM OF ART, (M),** 1050 World's Fair Park Dr., Knoxville, TN 37916-1653. Tel.: 865-525-6101, ext. 0 & ext. 243. Fax: 865-546-3635.
E-mail: info@knoxart.org
Web Site: www.knoxart.org
Founded: 1961.
Congressional District: 2
Key Personnel: Chm. (V), Jay McBride; Chm., Bernie Rosenblatt; Exec. Dir., David L. Butler; Cur., Stephen C. Wicks; Asst. Cur. Public Programs, Krishna Adams; Assoc. Cur. Education K-12, Rosalind Martin; Dir. Finance, Joyce Jones; Alive After Five Coord., Michael Gill; Dir. Devel., Susan Hyde; Dir. Mktg., Angela Thomas; Assoc. Dir. Devel., Margo Clark; Dir. Administration, Denise DuBose; Coord. Fundraising Events, Carla Pare; Museum Shop Mgr., Susan Creswell; Curatorial Asst., Clark Gillespie.
Personnel Profile: Full-Time Paid 14; Part-Time Paid 15; Part-Time Volunteers 290; Interns 4.
Governing Authority: nonprofit organization. Tax-exempt: 501(c)(3).
Institution Type/Description: Art Museum: housed in an Edward Larrabee Barnes-designed facility.
Collections: 20th & 21st century art; regional art & fine craft.
Research Fields: east Tennessee region - contemporary art.
Facilities: 2,497-vol. library of art books & slides available by request.
Activities: docent & audio guided tours; lectures; concerts; arts festivals; school, youth, family programs; outreach programs; docent program; visitor service representatives; inter-museum loan, permanent, temporary & traveling exhibitions; exploratory gallery; teacher professional development programs; youth & adult art classes.
Publications: biannual, Canvas; exhibition catalogues with scholarly essays for emerging artist series; educator's learning guides.
Hours & Admission Prices: Tues.-Thurs. & Sat. 10-5, Fri. 10-8, Sun. 1-5. No charge; donations accepted. Closed New Year's Day; Martin Luther King Jr. Day; Easter; Memorial Day; Independence Day; Labor Day; Thanksgiving; Christmas. ♿
Attendance: 49,352 (accurate)
Membership: Student $20; Senior $35; KMA Guild $35 & $40; Individual $40; Senior Couple $55; Family $60; Associate $125; Fellow $250; Curator's Circle $500; Corporate $500 & up; Director's Circle $1,000; Benefactor $2,500-$4,999; Sustaining $5,000-$9,999; Masters $10,000-$24,999; Grand Masters $25,000-$49,999; Chairman's Club $50,000 & up.

KNOXVILLE POLICE DEPARTMENT MUSEUM, 800 Howard Baker Jr. Ave., Knoxville, TN 37915. Tel.: 865-215-7000.
Institution Type/Description: History Museum.
Collections: Knoxville police department history; photographs; uniforms; badges; weapons; literature.
Hours & Admission Prices: By appointment. No charges.

KNOXVILLE ZOOLOGICAL GARDENS, 3500 Knoxville Zoo Dr., Knoxville, TN 37914. Mailing Address: P.O. Box 6040, Knoxville, TN 37914-0040. Tel.: 865-637-5331, ext. 300. Fax: 865-637-1943.
Web Site: www.knoxville-zoo.org
Founded: 1948.
Congressional District: 2
Key Personnel: Exec. Dir., Jim Vlna; Bd. Chm., Tim Williams; Dir. Animal Collection & Dir. Herpetology, Lisa New; Dir. Education & Dir. Mktg., Alison Swank; Dir. Guest Svcs., Josh Hurley; Dir. Operations, Keith Montgomery.
Personnel Profile: Full-Time Paid 108; Part-Time Paid 103; Part-Time Volunteers 200; Interns 53.
Governing Authority: nonprofit organization. Tax-exempt: 501(c)(3).
Institution Type/Description: Zoo.
Collections: large mammals; reptiles & birds; red pandas; white rhinos; African elephants; Chinese alligators; American black bears; chimpanzees; giraffe; gorillas.
Research Fields: biology; zoology.
Facilities: 1,700-vol. library of animal reference used by employees on premises only; 53 acres zoological park. Zoo related items for sale.
Activities: docent program; permanent & traveling exhibitions; SSP programs; tram ride; elephant demonstration; camel ride; special events; keeper chats; animal encounters.
Publications: zoo map; 4 times a year, membership news magazine.
Hours & Admission Prices: Winter: daily 10-4:30; Summer: daily 9:30-6. Adults $19.95, senior citizens 65 & up and children 2-12 $15.95; discount to groups & AAM members; children under 2, members & AZA members no charge. Closed Christmas. ♿
Attendance: 395,000 (accurate)
Membership: Individual $60; Grandparent & Dual $70; One Parent $85; Two Parent $95; Family Fun Pack $160.

MABRY-HAZEN HOUSE, 1711 Dandridge Ave., Knoxville, TN 37915-1905. Tel.: 865-522-8661. Fax: 865-522-8471.
E-mail: mabryhazenhouse@gmail.com
Web Site: mabryhazen.com
Founded: 1992.
Congressional District: 2
Key Personnel: Exec. Dir., Calvin Chappelle; Pres. (V), Ken Knight.
Personnel Profile: Full-Time Paid 1.
Governing Authority: private; nonprofit organization. Parent Institution: The Hazen Historical Museum Foundation, Inc. Tax-exempt.
Institution Type/Description: Historic House: an 1858 Italianate four over four home, built by Joseph Alexander Mabry; used by the South and then the North during the Civil War.
Collections: 3,000 original artifacts collected by the Mabry-Hazen families over 130 years as occupants of the house; glass; ceramics; painting; furniture; books; papers; photographs; letters.
Research Fields: genealogy of Mabry-Hazen families; history of Knoxville 1858-1987.
Facilities: 1,600-vol. library of textbooks (1858-1987), novels, letters, manuscripts & resource books on decorative arts and museology; botanical garden; 4,500 sq. ft. exhibit space; 60-seat theater. Publications for sale.
Activities: concerts; docent program; films; formal education programs for children & adults; guided tours; lectures. Annual Events: Victorian Christmas; Open Gardens-Dogwood Arts Festival; Teas in spring & fall.
Publications: The Seduction of Miss Evelyn Hazen.
Hours & Admission Prices: March-Dec. Wed.-Fri. 11-5, Sat. 10-3; Jan.-Feb. by appointment only. Adults $5, students K-12 $2.50; children 4 & under and members no charge. Closed New Year's Day; Independence Day; Thanksgiving; Christmas. ♿
Attendance: 2,000 (estimated)
Membership: Student $15; Individual $35; Family $50; Friend $100; Maime Winstead $250; Evelyn Hazen $500; Joe Mabry $1,000.

MARBLE SPRINGS STATE HISTORIC SITE - GOVERNOR JOHN SEVIER MEMORIAL ASSOCIATION, 1220 W. Gov. John Sevier Hwy., Knoxville, TN 37920. Mailing Address: P.O. Box 20195, Knoxville, TN 37940-1195. Tel.: 865-573-5508.
E-mail: marblesprings@gmail.com
Web Site: www.marblesprings.net
Founded: 1963.
Congressional District: 2
Key Personnel: C.E.O. & Dir., Anna Chappelle; Pres. (V), Ethiel Garlington; Museum Shop Mgr., Rebecca Sardella.
Personnel Profile: Full-Time Paid 1; Part-Time Paid 2; Part-Time Volunteers 10.
Governing Authority: state. Parent Institution: State of Tennessee. Affiliated

with Tennessee Historical Commission, State Library & Archives Bldg., Nashville, TN 37219. Tax-exempt.
Institution Type/Description: Historic House: 1783-1815 Marble Springs plantation home of Tennessee's first governor, John Sevier.
Collections: agriculture; pioneer furniture & artifacts; focus on frontier history and pioneer way of life, 1780-1815; influence of Sevier on early history of the settlements in North Carolina, State of Franklin, territory south of the River Ohio & Tennessee. Historic Buildings: loomhouse; springhouse; Sevier Main house, kitchen; Trading Post & Tavern.
Research Fields: Gov. John Sevier & his family records; pioneer life.
Facilities: rental facilities. Gift items for sale.
Activities: living history tours. Museum Sponsors: Storytelling Festival in April; Statehood Celebration in June; John Sevier Days in September; Fall Festival in October; Candlelight Tours in December.
Publications: books; quarterly newsletter.
Hours & Admission Prices: Jan.-Feb. Sat. 11-5, Sun. 1-5; March-Dec. Wed.-Sat. 10-5, Sun. 12-5; other times by appointment. Site: no charge. Guided Tours: adults $4; children 10 & under no charge. Closed New Year's Day; Easter; Thanksgiving; Christmas Eve & Day.
Attendance: 7,242 (accurate)
Membership: Individual $25; Family $45; Contributing $75; Sustaining $100; Governor's Cabinet $500; Marble Springs Benefactor $1,000.

✳ **MCCLUNG MUSEUM OF NATURAL HISTORY & CULTURE, (M),** University of Tennessee, 1327 Circle Park Dr., Knoxville, TN 37996-3200. Tel.: 865-974-2144. Fax: 865-974-3827.
E-mail: museum@utk.edu
Web Site: mcclungmuseum.utk.edu
Formerly: Frank H. McClung Museum
Founded: 1961.
Congressional District: 2
Key Personnel: Dir., Dr. Jefferson Chapman; Asst. Cur. & Social Media, Catherine Shteynberg; Cur. Paleoethnobotany, Dr. Gary Crites; Cur. Archaeology, Dr. Timothy Baumann; Cur. Malacology, Gerald Dinkins; Registrar, Robert Pennington; Photographer, Lindsay Kromer; Exhibits Coord., Steve Long; Exhibit Preparator, Chris Weddig; Museum Educator, Deborah Woodiel; Museum Shop Mgr., Katie Willocks.
Personnel Profile: Full-Time Paid 11; Part-Time Paid 8; Part-Time Volunteers 52; Interns 3.
Volunteer Hours: 2,070
Operating Expenses: 1,424,419
Operating Income: 1,424,419
Governing Authority: university. Parent Institution: The University of Tennessee. Tax-exempt: 170(b)(1)(A).
Institution Type/Description: General Museum.
Collections: Lewis-Kneberg collection of Tennessee archaeology; archaeological collections from Tennessee reservoirs; Eleanor Deane Audigier art collection; historical & natural science materials; ancient Egyptian; ethnological collections; paleoethnobotanical collection; fresh water mussel collection; ornithological lithographs; vertebrate paleontology.
Major Exhibits: Glass of the Ancient Mediterranean, 1/14-5/14; Brightly Beaded, North American Indian Beadwork, 1/14-5/14; Selections of American & European Art from the Collections, 6/14-8/14; Natural History Illustrations, 1600-1900, 9/14-1/15.
Research Fields: Tennessee archaeology; zooarchaeology; malacology; paleoethnobotany; vertebrate paleontology.
Facilities: 6,000-vol. library of archaeological journals, newsletters, museum catalogs & reference material available for use on premises; reading room; 260-seat auditorium.
Activities: guided tours; inter-museum loan, permanent, temporary & traveling exhibitions; lectures.
Publications: exhibit catalogues; books on regional archaeology; monthly bulletin; research notes.
Hours & Admission Prices: Mon.-Sat. 9-5, Sun. 1-5. No charge; donations accepted. Closed New Year's Day; Easter; Memorial Day; Independence Day; Labor Day; Thanksgiving; Christmas Eve & Day. &
Attendance: 37,950 (accurate)
Membership: University $15; Individual & Family $30; Contributing $50; Sustaining $100; Supporting $250; Patron $500; Benefactor $1,000; Grand Benefactor $2,500.

UNIVERSITY OF TENNESSEE GARDENS, 2431 Joe Johnson Dr., Knoxville, TN 37996. Tel.: 865-974-8265. Fax: 865-974-1947.
E-mail: utgardens@utk.edu
Web Site: utgardens.tennessee.edu
Key Personnel: Dir., Dr. Susan Hamilton
Institution Type/Description: Gardens.
Collections: annuals; perennials; herbs; tropicals; trees; shrubs; vegetables.

Activities: educational programs; seminars; guided tours; special events; rental facilities.
Hours & Admission Prices: Daily. No charge.

WOMEN'S BASKETBALL HALL OF FAME, (M), 700 Hall of Fame Dr., Knoxville, TN 37915. Tel.: 865-633-9000. Fax: 865-633-9294.
E-mail: dhart@wbhof.com
Web Site: wbhof.com
Founded: 1999.
Key Personnel: Chm. (V), Joan Cronan; Pres., Dana Hart; Dir. Basketball Operations, Josh Sullivan; Museum Shop Mgr., Scott Wilson.
Personnel Profile: Full-Time Paid 4; Part-Time Paid 13; Interns 3.
Governing Authority: Tax-exempt.
Institution Type/Description: Sports Museum.
Collections: basketball artifacts & history; photographs; scrapbooks; medals; trophies; uniforms; personal artifacts; early rule books; video & audio recordings; over 100 jerseys; Hall of Honor.
Major Exhibits: TN Sports Hall of Fame - Nashville, TN, 4/14-10/14.
Facilities: rental facilities.
Activities: basketball courts; video; rental facilities; special events.
Hours & Admission Prices: May-Sept. Mon.-Sat. 10-5; Labor Day to April Tues.-Fri. 11-5, Sat. 10-5. Adults $7.95, seniors 62 & over and children 6-15 $5.95; discounts to AAA & military members; members and children 5 & under no charge. &
Attendance: 14,419 (estimated)
Membership: Youth 6-15 $19.95; Senior 62 & over $29.95; Adult 16-61 $39.95; Family $49.95.

Lafayette

GALEN SCHOOL MUSEUM, 6435 Galen Rd., Lafayette, TN 37083. Tel.: 615-666-2470.
Institution Type/Description: History Museum: built in 1929.
Collections: local history & culture; period furnishings; photographs; early farm tools.
Hours & Admission Prices: Call for hours. No charge.

JOHNSTON CABIN AT KEY PARK, 208 Church St., Lafayette, TN 37083. Mailing Address: c/o Macon County Historical Society, P.O. Box 231, Lafayette, TN 37083. Tel.: 615-688-4247.
Key Personnel: Pres. (V), Teresa Whittemore; Museum Shop Mgr., Mickey Meador.
Governing Authority: Parent Institution: Macon County Historical Society.
Institution Type/Description: Historic House Museum.
Collections: local history & culture; photographs; books; period furnishings & clothing; personal artifacts.
Hours & Admission Prices: Call for hours. No charge, donations accepted. &

Lake City

COAL MINER'S MUSEUM, 216 N. Main St., Lake City, TN 37769. Mailing Address: Coal Creek Watershed Foundation, Inc., 3502 Overlook Cir., Knoxville, TN 37909. Tel.: 865-426-7914.
E-mail: bthacker2@coalcreekaml.com
Institution Type/Description: History Museum.
Collections: mining history, tools & artifacts; photographs.
Hours & Admission Prices: Call for hours. No charge.

Lawrenceburg

CHEROKEE MUSEUM AND CULTURAL CENTER, #1 Public Square, Lawrenceburg, TN 38464-3331. Mailing Address: P.O. Box 331, Lawrenceberg, TN 38464. Tel.: 931-762-3733.
E-mail: centralbandofcherokee@gmail.com
Web Site: www.centralbandofcherokee.org
Institution Type/Description: Native American History Museum.
Collections: Native American history, culture & heritage; personal artifacts; hands-on exhibits.
Facilities: Center-related items for sale.
Hours & Admission Prices: Mon.-Sat. 9:30-4.

JAMES D. VAUGHN GOSPEL MUSIC MUSEUM, 31 Public Square, Sun Trust Bank Bldg., 3rd Fl., Lawrenceburg, TN 38464-3351. Mailing Address: c/o Mainstreet Lawrenceburg, P.O. Box 607, Lawrenceburg, TN 38466-0607. Tel.: 931-762-8991.
Key Personnel: Cur., Tom Crews
Institution Type/Description: History Museum.

Collections: James D. Vaughan's life & career; southern gospel music history; photographs.
Activities: special events; festivals.
Hours & Admission Prices: Call for hours.

LAWRENCE COUNTY HISTORICAL SOCIETY, Waterloo St., Lawrenceburg, TN 38468. Mailing Address: 1106 Hickory, Lawrenceburg, TN 38464-3921. Tel.: 931-762-4397.
Institution Type/Description: Historical Society Museum: built in 1893.
Collections: local history & culture; period furnishings; personal artifacts; photographs.
Hours & Admission Prices: Tues.-Thurs. 10-2; other times by appointment.

Lebanon

FIDDLERS GROVE HISTORICAL VILLAGE, 945 E. Baddour Pkwy., Lebanon, TN 37087-4338. Tel.: 615-443-2626.
E-mail: info@fiddlersgrove.org
Web Site: www.fiddlersgrove.org
Institution Type/Description: Historic Village.
Collections: local history & heritage; period furnishings & artifacts; early automobiles.
Activities: guided tours.
Hours & Admission Prices: By appointment. Tours: $4 per person; discounts to school groups.

LEBANON MUSEUM AND HISTORY CENTER, 200 N. Castle Heights Ave., Lebanon, TN 37087-2740. Tel.: 615-443-2839. Fax: 615-443-2851.
Web Site: www.lebanontn.org
Founded: 1997.
Personnel Profile: Part-Time Volunteers 1.
Governing Authority: Parent Institution: City of Lebanon.
Institution Type/Description: History Museum.
Collections: local history & culture; photographs; period furnishings; personal artifacts.
Activities: guided tours.
Hours & Admission Prices: Mon.-Fri. 8-4; groups of 10 or more by appointment. No charge.
Attendance: 200

WILSON COUNTY MUSEUM AKA THE FESSENDEN HOUSE, 236 W. Main St., Lebanon, TN 37087. Mailing Address: 111 S. Greenwood St., Lebanon, TN 37087. Tel.: 615-444-9127.
Institution Type/Description: Historic House Museum.
Collections: local history & heritage; period furnishings; personal artifacts.
Hours & Admission Prices: By appointment. No charge.

Lexington

BEECH RIVER CULTURAL CENTER & MUSEUM, 26 S. Broad St., Lexington, TN 38351-2002. Tel.: 731-967-0306.
Institution Type/Description: History Museum.
Collections: Henderson County history & culture; geology; early settlers; wars; personal artifacts.
Hours & Admission Prices: March 2-Dec. Wed. & Fri.-Sat. 10-3. No charge; donations accepted.

Limestone

DAVY CROCKETT BIRTHPLACE STATE PARK, 1245 Davy Crockett Park Rd., Limestone, TN 37681. Tel.: 423-257-2167.
Institution Type/Description: History Museum.
Collections: Davy Crockett's life & career; photographs; period furnishings; replica cabin.
Activities: video.
Hours & Admission Prices: Call for hours. No charge.

Livingston

OVERTON COUNTY HERITAGE MUSEUM, 318 W. Broad St., Livingston, TN 38570-1804. Tel.: 931-403-0909.
E-mail: pstover@twlakes.com
Web Site: www.overtonmuseum.com
Institution Type/Description: History Museum.
Collections: local history & culture; period furnishings; personal artifacts; photographs.
Hours & Admission Prices: Thurs.-Sat. 10-2; groups by appointment.

Lookout Mountain

BATTLES FOR CHATTANOOGA MUSEUM, 1110 E. Brow Rd., Lookout Mountain, TN 37350-1016.
E-mail: battle@seerockcity.com
Web Site: www.battlesforchattanooga.com
Institution Type/Description: History Museum.
Collections: Civil War history; weapons; miniature battle presentations.
Facilities: Museum-related items for sale.
Hours & Admission Prices: Summer: 9-6; Winter: daily 10-5. Adults $8, children 5-12 $6; children 4 & under no charge. &

CRAVENS HOUSE, Point Park Visitor Center, IN 148 Scenic Hwy., Lookout Mountain, TN 37350. Tel.: 423-821-7786. Fax: 423-825-5129.
Congressional District: 3
Key Personnel: Supt., Cathleen Cook; Chief Interpretation, Kim Coons; District Interpreter, Anton J. Heinlein.
Personnel Profile: Full-Time Paid 6; Part-Time Paid 5; Part-Time Volunteers 3.
Governing Authority: federal. Dept. of the Interior, National Park Service. Parent Institution: Chickamauga & Chattanooga National Military Park. Tax-exempt.
Institution Type/Description: Historic House: 1866 Cravens House, Post-Civil War home.
Collections: furniture dating from 1830 through 1890s.
Activities: guided tours.
Publications: brochures, Park and Cravens House.
Hours & Admission Prices: Memorial Day to Labor Day Sat.-Sun. 1-5. No charge; donations accepted. Closed Christmas.
Attendance: 2,000 (accurate)

Loretto

RALPH J. PASSARELLA MEMORIAL MUSEUM, 134 S. Main St., Loretto, TN 38469. Tel.: 931-853-4351. Fax: 931-853-4329.
Key Personnel: Cur., Jean Willis; Asst. Cur., Patty Brown.
Personnel Profile: Full-Time Volunteers 2.
Volunteer Hours: 28
Governing Authority: Parent Institution: Loretto Telecom.
Institution Type/Description: History Museum.
Collections: local history & culture; period furnishings; photographs; newspapers; magazines; books; toys; tools; quilts; medical equipment; telephones; switchboards; soda fountain; wood burning stove; early clothing.
Hours & Admission Prices: Tues.-Wed. 9-4. Adults $6; discounts to groups of 20 or more; children 2 & under no charge.

Lynchburg

MOORE COUNTY JAIL MUSEUM, 231 Main St., Lynchburg, TN 37352-8300. Mailing Address: P.O. Box 421, Lynchburg, TN 37352-0421. Tel.: 931-759-4111.
E-mail: info@lynchburgtn.com
Web Site: www.lynchburgtn.com
Governing Authority: Parent Institution: Moore County Historical Society.
Institution Type/Description: Historic Building: housed in the county's first jail; built c.1893.
Collections: local history; period furnishings; photographs; early clothing.
Hours & Admission Prices: Thurs.-Sat. 12-3. Adults $1; discounts to groups; children under 16 no charge. Closed holidays. &

TENNESSEE WALKING HORSE MUSEUM, 183 Main St., Lynchburg, TN 37352-8300. Mailing Address: P.O. Box 1010, Shelbyville, TN 37160. Tel.: 931-759-5747.
Institution Type/Description: History Museum.
Collections: walking horse history; hands-on exhibits; photographs; breed registry; World Grand Champions; care & training; saddles; videos.
Activities: Annual Event: Tennessee Walking Horse National Celebration in August.
Hours & Admission Prices: Tues.-Sat. 9-5. No charge. Closed major holidays.

Lynnville

LYNNVILLE RAILROAD MUSEUM, 162 Mill St., Lynnville, TN 38472. Mailing Address: P.O. Box 156, Lynnville, TN 38472. Tel.: 931-478-0880.
Institution Type/Description: Railroad Museum.
Collections: railroad history & memorabilia; 1927 Baldwin steam locomotive; replica depot; period furnishings; personal artifacts.

Hours & Admission Prices: Mon.-Sat. 10-4, Sun. 12-4.

Madison

AMQUI STATION AND VISITORS CENTER - MUSIC AND THE RAILROAD MUSEUM, (M), 301 B Madison St., Madison, TN 37115. Tel.: 615-891-1154.
E-mail: execdirector@amquistation.org
Web Site: www.amquistation.org
Congressional District: 5
Key Personnel: Dir., Cate Hamilton; Chm. (V), Bill Beck; Pres. (V), Nathan Massey.
Personnel Profile: Full-Time Paid 1.
Governing Authority: Parent Institution: Discover Madison, Inc. Tax-exempt.
Institution Type/Description: History Museum & Visitors Center: housed in a former railroad switching and passenger depot; built in 1910.
Collections: railroad history, artifacts & music; photographs; local & regional history.
Hours & Admission Prices: Mon.-Fri. 9-5.

Manchester

ARROWHEADS TO AEROSPACE MUSEUM, 24 Campground Rd., Manchester, TN 37355-6541. Tel.: 931-723-1323. Facebook; Coffee County/Manchester;/Tullahoma Museum.
E-mail: jfworthington@bellsouth.net
Founded: 1987.
Congressional District: 4
Key Personnel: Dir., Judy Worthington; Chm. (V), Raleigh Jernigan.
Personnel Profile: Full-Time Volunteers 1; Part-Time Volunteers 15; Interns 2.
Governing Authority: Tax-exempt.
Institution Type/Description: History Museum.
Collections: local history; railroad artifacts; period furnishings; Native American; military; quilts; teapots; toys; model trains, buttons; Nativities.
Facilities: Museum-related items for sale.
Hours & Admission Prices: Daily 10-2; other times by appointment. Adults $6, students 7-17 and seniors 65 & over $5, children 2-5 $4; discounts to groups, military, AAM, AAA & AARP members; members no charge. Closed New Year's Day; Christmas Eve & Day. &
Attendance: 5,000 (estimated)
Membership: Individual $25; 2 Individuals $50; Friends $100; Patrons $500; Partners $1,000.

Martin

J. HOUSTON GORDON MUSEUM, 10 Wayne Fisher Dr., The University of Tennessee at Martin, Martin, TN 38238. Tel.: 731-881-7094. Fax: 731-881-7074.
E-mail: museum@utm.edu
Formerly: University Museum
Founded: 1981.
Congressional District: 8
Key Personnel: Dir., Samuel S. Richardson.
Personnel Profile: Part-Time Paid 1.
Governing Authority: university; nonprofit. Parent Institution: The University of Tennessee. Tax-exempt.
Institution Type/Description: General University Museum.
Collections: works by local & national artists.
Facilities: 600 sq. ft. exhibit space.
Activities: rotating & traveling exhibits.
Publications: brochure.
Hours & Admission Prices: Mon.-Fri. 8-4:30. No charge. Closed university holidays. &
Attendance: 1,000 (estimated)

Maryville

SAM HOUSTON MEMORIAL ASSOCIATION, 3650 Old Sam Houston School Rd., Maryville, TN 37804-5644. Tel.: 865-983-1550.
E-mail: samhoustonsch@aol.com
Web Site: samhoustonhistoricschoolhouse.org
Formerly: Sam Houston Historical Schoolhouse
Founded: 1965.
Congressional District: 2
Key Personnel: Pres., Enoch B. Simerly; Vice Pres., Clara Peals; Treas., Charles England; Museum Shop Mgr., Mary Lynne Bell.
Personnel Profile: Full-Time Paid 1; Part-Time Volunteers 10.
Governing Authority: state. Parent Institution: Sam Houston Memorial Association. Tax-exempt.
Institution Type/Description: Park Museum.
Collections: early school and pioneer artifacts; genealogy of Houston family. Historic Building: 1794 Log School.
Research Fields: Sam Houston's youth.
Facilities: picnic area; enclosed pavilion with kitchen. Museum-related items for sale.
Activities: Annual Events: Sam's Birthday Celebration in March; Pioneer Encampment in June; Fall Festival; Colonial Encampment in October; Christmas Open House in December.
Publications: book, Sam Houston, The Man; Sam Houston, Man of Destiny; Sam Houston, Man of Mystery; Sam Houston Schoolhouse Cookbook; local history books; Cherokee culture and characters; local cookbooks.
Hours & Admission Prices: Feb.-Dec. Tues.-Sat. 10-5, Sun. 1-5. Adults $3, children under 8 $1; members no charge. Closed New Year's Day; Easter; Thanksgiving; Christmas week. &
Attendance: 8,000 (estimated)
Membership: Individual $10; Family $15; Contributing $25; Life $200.

Maynardville

UNION COUNTY HERITAGE MUSEUM AND LIBRARY, 3824 Maynardville Hwy., Maynardville, TN 37807. Mailing Address: P.O. Box 95, Maynardville, TN 37807. Tel.: 865-992-2136.
Founded: 1980.
Key Personnel: Chm. (V), J.V. Waller; Pres. (V), Martha Carter.
Personnel Profile: Part-Time Volunteers 35.
Governing Authority: Parent Institution: Union County Historical Society, Inc. Tax-exempt.
Institution Type/Description: History Museum & Library.
Collections: local history & culture; country music stars; period furnishings; photographs; personal artifacts; genealogical records; articles from Sharps Fort dig.
Research Fields: genealogical.
Activities: lectures; genealogy classes for young children.
Publications: quarterly journal, Pathways.
Hours & Admission Prices: Sun. 1-5, Mon.-Tues. 10-4. No charge; donations accepted. Closed Christmas Day. &
Attendance: 3,000 (estimated)
Membership: Annual $20.

McKenzie

GORDON BROWNING MUSEUM & GENEALOGICAL LIBRARY, 640 Main St. N., McKenzie, TN 38201-1720. Tel.: 731-352-3510.
E-mail: gbmuseum640@gmail.com
Web Site: tn-roots.com/GordonBrowning
Founded: 1971.
Congressional District: 7
Key Personnel: Dir., Jere R. Cox; Pres., James E. Choate.
Personnel Profile: Full-Time Volunteers 1; Part-Time Volunteers 2.
Governing Authority: Parent Institution: Carroll County Historical Society. Tax-exempt: 501(c)(3).
Institution Type/Description: History Museum.
Collections: local history & culture; photographs; documents; WWI, WWII, Korean uniforms, flags & artifacts; Civil War artifacts.
Hours & Admission Prices: Mon.-Tues. & Thurs.-Fri. 9-4. No charge; donations accepted. Closed New Year's Day; Independence Day; Labor Day; Thanksgiving; Christmas. &
Attendance: 800 (estimated)

McMinnville

SOUTHERN MUSEUM & GALLERIES OF PHOTOGRAPHY, 210 E. Main St., McMinnville, TN 37110-2508. Tel.: 931-507-8102.
E-mail: artgallery@multipro.com
Institution Type/Description: Art Gallery.
Collections: photographs.
Hours & Admission Prices: Wed. & Fri.-Sat. 10-4.

Memphis

ART MUSEUM OF THE UNIVERSITY OF MEMPHIS, (M), 3750 Norriswood Ave., CFA Bldg., Memphis, TN 38152. Mailing Address: 142 Communication/Fine Arts Bldg., Memphis, TN 38152. Tel.: 901-678-2224. Fax: 901-678-5118.
E-mail: artmuseum@memphis.edu
Web Site: www.memphis.edu/amum

Founded: 1981.
Congressional District: 9
Key Personnel: Dir., Leslie Luebbers, PhD; Asst. Dir. & Registrar, Lisa F. Abitz; Cur. Asst., Anita Huggins; Museum Media Specialist, Jason N. Miller; Exhibit Specialist & Preparator, Eric T. Bork.
Personnel Profile: Full-Time Paid 5; Part-Time Paid 1; Part-Time Volunteers 1; Interns 3.
Governing Authority: university. Parent Institution: The University of Memphis. Subsidiary Institution: Friends of the Art Museum. Parallel Institution: Institute of Egyptian Art & Archaeology. Tax-exempt.
Institution Type/Description: Fine Arts Museum.
Collections: Egyptian collection: antiquities 3500 BC-700 AD: mummies, religious & funerary items; jewelry.
Major Exhibits: 31st Annual Juried Student Exhibition, 2/8/14-3/19/14; Relic: The Mystic Potency of Stuff. John Salvest, 3/28/14-6/28/14; Juvenile-in-Residence: Photographs by Richard Ross, 9/14-11/14.
Research Fields: Egyptian art & archaeology, history of art.
Facilities: 6,000-vol. Egyptian library.
Activities: temporary & traveling exhibitions; guided tours; lectures; museum study classes; internship; Institute of Egyptian Art & Archaeology symposia; I.E.A.A. Arts in School program; interdisciplinary colloquia & proceedings.
Publications: triannual newsletter, AM Edition; brochures; symposia papers; exhibition catalogs; collection catalogs; proceedings; educators guide to collections.
Hours & Admission Prices: Mon.-Sat. 9-5. No charge; suggested Donation: $2; discounts to AAM members. Closed New Year's Day; Martin Luther King Jr. Day; Memorial Day; Independence Day; Labor Day; Thanksgiving; Christmas. ♿
Attendance: 9,807 (accurate)
Membership: Student $20 (single & dual); Faculty $40 (single & dual); Good Friend $50 (single & dual); Close Friend $150 (single & dual); Best Friend $500 (single & dual); Corporate Friend $1,000-$5,000.

BELZ MUSEUM OF ASIAN & JUDAIC ART, 119 S. Main St., Memphis, TN 38103-3647. Tel.: 901-523-ARTS.
E-mail: info@belzmuseum.org
Web Site: www.belzmuseum.org
Key Personnel: Dir., Belinda Fish; Guest Svcs. Admin., Jade Powell; Guest Svcs. Asst., Amber Curtis; Research Assoc., Daniel Graubman; Research Asst., Michelle D. Williams
Institution Type/Description: Art Museum.
Collections: Asian art pertaining to China's Quing Dynasty, including works in jade & ivory; Chinese puppets; historic & literal pieces relating to Judaism.
Hours & Admission Prices: Tues.-Fri. 10-5:30, Sat.-Sun. 12-5. Adults $6, seniors $5, students $4; children 5 & under no charge. Closed New Year's Day; Easter; Independence Day; Thanksgiving; Christmas.

C.H. NASH MUSEUM AT CHUCALISSA, (M), 1987 Indian Village Dr., Memphis, TN 38109-3005. Tel.: 901-785-3160. Fax: 901-785-0519.
E-mail: chucalissa@memphis.edu
Web Site: www.chucalissa.memphis.edu
Founded: 1958.
Congressional District: 9
Key Personnel: Dir., Dr. Robert Connolly; Administrative Assoc., Alex Hutson; Museum Shop Mgr., Sonny Bell.
Personnel Profile: Full-Time Paid 4; Part-Time Paid 2; Interns 3.
Governing Authority: Parent Institution: University of Memphis, 38152. Tax-exempt.
Institution Type/Description: Archaeology Museum: S.E. Native American, Choctaw & Pre-historic Native Mississippian culture.
Collections: archaeological research collections from sites in western Tennessee & adjacent areas; Chert reference collection from various geological formations in Midsouth; ceramic type collection from region & projective point type collection from region.
Research Fields: archaeology of Midsouth.
Facilities: 2,000-vol. library of archaeology & geology books; laboratory; 100-seat auditorium.
Activities: guided tours; formally organized education programs for all ages; permanent & temporary exhibitions; Native American arts; special events with Native American games & dances; hands-on archaeology lab exhibit.
Publications: tour guide; occasional papers.
Hours & Admission Prices: Tues.-Sat. 9-5, Sun. 1-5. Adults $5, children & senior citizens $3; discounts to AAA, groups of 10 or more, Get Away from Memphis members, University of Memphis faculty & students; members no charge. No admittance after 4:30. ♿
Attendance: 10,000 (accurate)

Membership: Students & Seniors $15; Individual $25; Family $40; Patron $100; Donor $200; President's Circle $500; Sinti Circle $1,000.

CENTER FOR SOUTHERN FOLKLORE, 119 S. Main St., Memphis, TN 38103-3681. Tel.: 901-525-3655. Fax: 901-544-9965.
Institution Type/Description: Southern Performers History Museum.
Collections: paintings; photographs; southern music, arts, culture & traditions; southern history.
Facilities: Museum-related items for sale.
Activities: educational programs; videos; special events; music programs.
Hours & Admission Prices: Mon.-Fri. 10-5. ♿

THE CHILDREN'S MUSEUM OF MEMPHIS, (M), 2525 Central Ave., Memphis, TN 38104-5926. Tel.: 901-458-2678. Fax: 901-458-4033.
E-mail: children@cmom.com
Web Site: www.cmom.com
Founded: 1987.
Congressional District: 9
Key Personnel: C.E.O., Richard Hackett; Pres. (V), Ron Kastner; C.F.O. & Controller, Randy McKeel; Dir. Exhibits, Jim Hyde; Dir. Public Rels. & Mktg., Carrie Roberts; Group Programs, Keosha Williams; Visitor Svcs. Mgr., Brad Laney; C.O.O., Cliff Drake; Program Specialist, Tamara Williamson; Dir. Education, Felicia Peat; Coord. Member Svcs., Sharon Rogness; Coord. Visitor Svcs., Jana Smith; Program Specialist, Chara Mitchell; Program Specialist, Monica Sanchez; Program Specialist, Katrina Skefos.
Personnel Profile: Full-Time Paid 13; Part-Time Paid 10; Part-Time Volunteers 150; Interns 3.
Governing Authority: nonprofit organization. Tax-exempt: 501(c)(3).
Institution Type/Description: Children's Museum.
Collections: hands-on exhibits including kid-size city, flight & distribution; the Mississippi River and hydrology; visual & performing arts.
Facilities: 20,000 sq. ft. exhibit space; classrooms; 200-seat auditorium; cafe; 2 birthday party rooms. Exhibition-related items for sale.
Activities: organized education programs for children; participatory & traveling exhibitions; annual special events; outdoor splash park.
Publications: bimonthly newsletter, Sparks!; biannual newsletter, Learning Journeys; annual report.
Hours & Admission Prices: Daily 9-5. Admission $12; discounts to ACM members; children under one & members no charge. Closed Easter; Thanksgiving; Christmas. ♿
Attendance: 177,931 (accurate)
Membership: CMOM (2 people) $95; CMOM + I $110; Grandparent (4 people) $115; CMOM + 2 $125; CMOM + 3 $130; CMOM +4 $140; Supporter $175; Sponsor $350; Benefactor $500; Patron $1,000.

CLOUGH-HANSON GALLERY, RHODES COLLEGE, 2000 N. Parkway, Memphis, TN 38112-1690. Tel.: 901-843-3442. Fax: 901-843-3727.
Web Site: www.rhodes.edu/academics/5264.asp
Founded: 1970.
Congressional District: 9
Key Personnel: Dir., Hamlett Dobbins.
Personnel Profile: Part-Time Paid 1; Interns 4.
Governing Authority: private; college. Tax-exempt: 501(c)(3).
Institution Type/Description: College Art Gallery.
Collections: Edward Curtis photographs; Asian woodcut prints; porcelain; fabrics & other objects; regional painting & sculpture; 20th-century prints.
Activities: lectures; temporary & traveling exhibitions. Annual Event: World AIDS Day Event.
Publications: annual exhibition catalog.
Hours & Admission Prices: Tues.-Sat. 11-5. No charge. Closed Martin Luther King Jr. Day; spring break; Good Friday; Easter; Memorial Day; Labor Day; fall break; Thanksgiving; Christmas. ♿
Attendance: 2,300 (estimated)

THE COTTON MUSEUM AT THE MEMPHIS COTTON EXCHANGE, 65 Union Ave., Memphis, TN 38103-5157. Tel.: 901-531-7826. Fax: 901-531-7827.
E-mail: info@memphiscottonmuseum.org
Web Site: www.memphiscottonmuseum.org
Founded: 2006.
Key Personnel: Exec. Dir., Anna Mullins; Exhibits Coord., Melissa Farnis; Events Coord., Vanessa Clemmensen
Institution Type/Description: History Museum.

Collections: artifacts & memorabilia relating to the cotton industry; exhibits & films detailing the cotton industry.

Hours & Admission Prices: Mon.-Sat. 10-5, Sun. 12-5. Adults $10, seniors & students $9, children 6-12 $8; discounts to AAM & ICOM members. children under 6 no charge. &

Membership: Individual $45; Family $70.

DAVIES MANOR PLANTATION, (M), 3570 Davieshire Dr., Memphis, TN 38133. Mailing Address: P.O. Box 56, Brunswick, TN 38014-0056. Tel.: 901-386-0715. Fax: 901-388-4677.

E-mail: daviesmnassoc@bellsouth.net

Web Site: daviesmanorplantation.org

Founded: 1975.

Key Personnel: Dir., Nancy J. McDonough; Pres. (V), Henry Boyd.

Personnel Profile: Full-Time Paid 3; Part-Time Paid 2; Part-Time Volunteers 30; Interns 4.

Governing Authority: Parent Institution: Davies Manor Association. Tax-exempt.

Institution Type/Description: Historic House: log cabin built c.1830. Listed on the National Register of Historic Places. Designation on TN Civil War Trail.

Collections: local history, heritage, & culture; period furnishings; textiles; tools; 3 log buildings; 4 outbuildings.

Activities: video; tours.

Hours & Admission Prices: late March to mid-Nov. Tues.-Sat. 12-4. Adults $5, students $4; members no charge.

Attendance: 1,000 (estimated)

Membership: Senior $25; Individual $30; Family $50; Corporate $1,000.

*** DIXON GALLERY AND GARDENS, (M),** 4339 Park Ave., Memphis, TN 38117-4698. Tel.: 901-761-5250. Fax: 901-682-0943.

E-mail: ntrenthem@dixon.org

Web Site: www.dixon.org

Founded: 1976.

Congressional District: 9

Key Personnel: Chm., D. Stephen Morrow; Dir., Kevin Sharp; Dir. Horticulture, Dale Skaggs; Registrar, Neil O'Brien; Dir. Devel., Susan Johnson; Cur. Education, Margarita Sandino; Assoc. Cur., Julie Pierotti; Controller, Gail Hopper; Museum Shop Mgr., Nancy Robertson.

Personnel Profile: Full-Time Paid 30; Part-Time Paid 10; Part-Time Volunteers 175; Interns 6.

Governing Authority: nonprofit organization. Tax-exempt: 501(c)(3).

Institution Type/Description: Art Museum.

Collections: 18th, 19th & early 20th-century paintings, prints & sculpture with emphasis on French & American Impressionists & Post-Impressionists, includes works of Blanche, Bonnard, Boudin, Braque, Butler, Cals, Carpeaux, Carrier-Beleuse, Cassatt, Cezanne, Chagall, Corot, Cox, Cross, Daubigny, Degas, Dufy, Fantin-Latour, Forain, Gauguin, Grandjean, Guigou, Guillaumin, Harpignies, Jacques, Jongkind, LaTouche, Legros, Lepine, Luce, Marquet, Mathey, Matisse, Monet, Morisot, Munch, Noufflard, Oberteuffer, Piette, Pissarro, Prendergast, Raeburn, Raffaelli, Renoir, Reynolds, Rouart, Sargent, Signac, Sisley, Toulouse-Lautrec, Tucker, Utrillo, Walker & Vuillard; European porcelain; Stout collection of 18th-century German porcelain; Adler collection of European & American Pewter; Hooker collection of 18-19th century English porcelain; 17 acres of gardens.

Major Exhibits: Wait Watchers: Photography by Haley Morris-Cafiero (T), 1/19/14-3/14; Color! American Photography Transformed (T), 1/19/14-3/23/14; African American Art from the John and Susan Horseman Collection, 1/19/14-3/23/14; Memphis Designs, 4/13/14-7/6/14; Memphis - Milano: 1980s Italian Design, 4/13/14-7/13/14; Nick Pena, 7/13/14-10/5/14; Charles Courtney Curran: Seeking the Ideal (T), 7/27/14-10/5/14; Connecting the World: The Panama Canal at 100 (T), 7/27/14-10/5/14; Portraits and Figures: Joyce Gingold, 10/12/14-1/4/15; Rodin and Figures from the Cantor Collection, 10/19/14-1/4/15.

Research Fields: Impressionist period; 18th-century European porcelain, particularly German & English.

Facilities: 3,000-vol. library by appointment; 6,800 sq. ft. exhibit space; 17 acres of gardens; 250-seat auditorium.

Activities: guided tours; random access audio tour; lectures; gallery talks; films; performing arts programs; symposiums; docent program; museum sponsored foreign tours; art history, museology & horticulture internships; inter-museum, permanent, & temporary exhibitions; rental facilities.

Publications: quarterly newsletter; annual report; catalogues: Mary Cassatt and The American Impressionists; Impressionists in 1877; Henri-Joseph Harpignies; Henriette Amiard Oberteuffer and George Oberteuffer; Joseph Mallord William Turner; Homage to Camille Pissarro; The Last Years 1890-1903; The Genius of Van Gogh; An International Episode: Millet, Monet and their North American Counterparts; An Assemblage of Decora-

tive Arts: A View of Tennessee Silversmiths; Marc Chagall-Selected Works, 1911-1981; Sources of the Modern Vision: A Selection of French Paintings from the Phillips Collection; Milton Avery's Mexico; ceramics, The Chinese Legacy; Vital Diversity: Japanese Paintings and Ceramics of the Edo Period; The Impressionist Vision; The Passion of Rodin: Sculpture from the B. Gerald Cantor collection; From Arcadia to Barbizon: A Journey in French Landscape Painting; Toulouse-Lautrec: The Guardsmark Collection; Odilon Redon; The Ian Woodner Family Collection; Louis XV and Madame de Pompadour: A Love Affair with Style; The Lamps of Tiffany; Highlights from the Egon and Hildegard Neustadt Collection; Jean-Louis Forain: The Impressionist Years; The Dixon Gallery & Gardens Permanent Collection Catalogue; Celebrate America: 19th Century Paintings from the Manoogian Collection; Visualizing the Blues: Images of the American South; The Call of the Wild: Sporting Art in the Mississippi Flyway; Regional Dialect: American Scene Paintings from the John and Susan Horseman Collection; Bold, Cautious, True: Walt Whitman and American Art of the Civil War; Helen M. Turner: The Woman's Point of View; Modern Dialect: American Paintings from the John and Susan Horseman Collection; Double Vision: Brin and Dale Baucum, A Retrospective in Clay; The History of Eighteenth Century German Porcelain: The Warda Stevens Stout Collection.

Hours & Admission Prices: Tues.-Sat. 10-5, 3rd Thurs. 10-8, Sun. 1-5. Adults $7, seniors 65 & over and students 18 & over $5, children 7-17 $3; discounts to AAM members & groups of 10 or more.; Sat. 10am-12pm, members and children 6 & under no charge. Closed New Year's Day; Independence Day; Thanksgiving; Christmas. &

Attendance: 57,630 (accurate)

Membership: Individual $45; Dual & Family $60-$124; Sponsor $125-$249; Young At Art $150-$249; Donor $250-$499; Cosmopolitans $300 per couple; Patron $500-$999; Sustainer $1,000-$1,499; Contributor $1,500-$2,499; Supporter $2,500-$4,999; Benefactor $5,000 & up; Life $25,000.

FIRE MUSEUM OF MEMPHIS, 118 Adams Ave., Memphis, TN 38103-2012. Tel.: 901-320-5650. Fax: 901-529-8422.

E-mail: penny@firemuseum.com

Web Site: www.firemuseum.com

Founded: 1993.

Congressional District: 9

Key Personnel: Gen. Mgr., Penny McKinney Smith; Chm. (V), Susan Helms.

Governing Authority: private; nonprofit. Tax-exempt: 501(c)(3).

Institution Type/Description: Fire-Fighting Museum: housed in 1910 Fire Engine House No. 1, which served as a fire station until 1973. Listed on the National Register of Historic Places.

Collections: concentration on historic fire apparatus, fire-fighting tools, equipment & apparel; photographs; film; books; examples of modern technology & fire prevention materials.

Research Fields: history of fire fighting in Memphis & surrounding area; fire prevention and safety education.

Facilities: library on fire history & service; educational facilities; 30-seat theater; facilities available for private parties with advanced arrangements. Fire memorabilia & home safety items for sale.

Activities: films; formal education programs for children; participatory & temporary exhibitions.

Publications: quarterly newsletter.

Hours & Admission Prices: Mon.-Sat. 9-4:30. Adults $6, children 3-12 $5, seniors over 60 & military $4; discount to groups; members, children 2 & under no charge. &

Attendance: 33,000 (accurate)

Membership: Basic $40; Family $50; Family Plus $60; White Helmet $100; Black Helmet $250; Red Helmet $500; Gold Helmet $1,000.

GIBSON GUITAR FACTORY, 145 Lt. George W. Lee Ave., Memphis, TN 38103. Tel.: 901-544-7998, ext. 4080.

E-mail: groupsales.memphis@gibson.com

Web Site: www.gibson.com

Institution Type/Description: Guitar Factory.

Collections: guitar-making process.

Hours & Admission Prices: Tours: Mon.-Sat. 11, 12, 1, 2, 3 & 4, Sun. 12, 1, 2, 3 & 4. Adults $10.

GRACELAND, 3764 Elvis Presley Blvd., Memphis, TN 38116-4198. Mailing Address: P.O. Box 16508, Memphis, TN 38186-0508. Tel.: 901 332 3322. Fax: 901 344 3116. Facebook: Graceland; TDD: 901-344-3146.

E-mail: graceland@elvis.com

Web Site: www.elvis.com

Founded: 1982.

Congressional District: 8

Key Personnel: C.E.O. & Pres., Jack Soden; Gen. Mgr. & Vice Pres., Regina

Gambill; Vice Pres. Entertainment & Music Publishing, Gary Hovey; Vice Pres. Worldwide Licensing, Susan Meek; Dir. Merchandising, Danny Hiltenbrand; Dir. Archives, Angela Marchese.
Personnel Profile: Full-Time Paid 250; Part-Time Paid 250.
Governing Authority: corporation. Parent Institution: Core Media Group, 900 Third Ave., 23rd Fl., New York, NY 10022. Subsidiary Institution: Elvis Presley Enterprises, Inc.
Institution Type/Description: Historic Home & Music Museum: housed in 1939 mansion occupied by singer/entertainer Elvis Presley from 1957 until his death in 1977.
Collections: clothing; costumes; jewelry; guitars; gold records & awards; photographs; personal mementoes; furnishings; automobiles; personal books & papers; pianos; twentieth century music & popular culture; Lisa Marie/Jetstar Airplanes.
Research Fields: Elvis Presley.
Facilities: 14-acre estate. Visitor Center Complex: 100-seat theater; two restaurants; rental facility. Gift items for sale.
Activities: audio tours; weddings; special events. Annual Events: Elvis Presley Birthday Celebration in January; Elvis Week in August; Scout Day Lighting Ceremony in November; Christmas at Graceland in December.
Publications: brochures; pamphlets.
Hours & Admission Prices: March-May Mon.-Sat. 9-5, Sun. 10-4; Memorial Day to Labor Day Mon.-Sat. 9-5, Sun. 9-4; Sept.-Oct. Mon.-Sat. 9-5, Sun. 10-4; Nov. daily 10-4; Dec.-Feb. Wed.-Mon. 10-4; call for special event hours. Mansion: adults $34, students and seniors 62 & over $25.20, children 7-12 $12; discounts to students, military & AAA members; teachers and children 6 & under no charge. Platinum Tour: adults $37, students and seniors 62 & over $29.70, children 7-12 $15; discounts to students, military & AAA members; teachers and children 6 & under no charge. VIP Tour: adults $72. Closed Thanksgiving; Christmas. &
Attendance: 650,000 (estimated)

* **LICHTERMAN NATURE CENTER, (M),** 5992 Quince Rd., Memphis, TN 38119-7257. Tel.: 901-767-7322. Fax: 901-682-3050.
E-mail: andy.williams@memphistn.gov
Web Site: www.memphismuseums.org
Founded: 1983.
Key Personnel: Mgr., Andy Williams; Chief Teacher & Naturalist, Dory Lerner.
Personnel Profile: Full-Time Paid 6; Part-Time Paid 10; Part-Time Volunteers 70.
Governing Authority: municipal. Parent Institution: City of Memphis. Subsidiary Institution: Memphis Museums, Inc. Tax-exempt: 501(c)(3).
Institution Type/Description: Environmental Education Center.
Collections: 65-acres of forest, meadow and lake habitats; natural history materials & artifacts.
Facilities: visitors center; special events pavilion & lawn; demonstration gardens; trails.
Activities: guided tours; lectures; organized education programs for children, college students & adults; participatory exhibits; demonstrations of raising wild flowers. Museum Sponsors: family field trips.
Publications: Museumscope.
Hours & Admission Prices: Tues.-Thurs. 10-3, Fri.-Sat. 10-5. Adults $6, senior citizen $5.50, children 3-12 $4.50; discounts to ASTC members; members & children under 3 no charge. Closed New Year's Day; Thanksgiving; Christmas Eve & Day. &
Attendance: 40,000 (estimated)
Membership: Individual $50; Family $75; IMAX Club $100.

* **MAGEVNEY HOUSE, (M),** 198 Adams Ave., Memphis, TN 38103. Mailing Address: 3050 Central Ave, Memphis, TN 38111-3316. Tel.: 901-320-6326. Fax: 901-320-6391.
Web Site: www.memphismuseums.org
Founded: 1941.
Congressional District: 8
Key Personnel: Dir., Steve Pike; Pres. (V), Jeffrey R. Wills.
Personnel Profile: Part-Time Paid 1.
Governing Authority: municipal. Parent Institution: Pink Palace Family of Museums. Subsidiary Institution: Memphis Museum, Inc. Tax-exempt: 501(c)(3).
Institution Type/Description: Historic House: c.1836 Magevney House.
Collections: period furniture.
Research Fields: historical investigation of Memphis pertaining to the home & its occupants.
Activities: guided tours; organized education programs for children; docent program; kitchen garden tours.
Hours & Admission Prices: Temporarily closed.

* **MALLORY-NEELY HOUSE, (M),** 652 Adams Ave., Memphis, TN 38105. Mailing Address: 3050 Central Ave., Memphis, TN 38111-3316. Tel.: 901-636-6370. Fax: 901-636-6391.
Web Site: www.memphismuseums.org
Founded: 1973.
Congressional District: 9
Key Personnel: Dir., Steve Pike; Pres. (V), Jeffrey R. Wills.
Personnel Profile: Full-Time Paid 1; Part-Time Paid 4; Part-Time Volunteers 50.
Governing Authority: municipal. Parent Institution: Pink Palace Family of Museums. Tax-exempt: 501(c)(3).
Institution Type/Description: Historic House.
Collections: original furnishings of the Victorian period through the 1960s.
Research Fields: extensive research into structure of home for preservation purposes; historical investigation of Memphis pertaining to home & occupants.
Facilities: classroom. Gift items for sale.
Activities: docent programs; formal education programs; guided tours. Museum Sponsors: seasonal exhibits.
Hours & Admission Prices: Fri.-Sat. 10-4; last tour 3pm. Adults $7, seniors 60 & over $6, youth 3-12 $5; discounts to groups; children under 3 no charge. &

MEMPHIS BELLE MEMORIAL ASSOCIATION, 5118 Park Ave., Ste. 110, Memphis, TN 38117-5710. Mailing Address: P.O. Box 1942, Memphis, TN 38101-1942. Fax: 901-767-4612.
E-mail: doctorhar@aol.com
Web Site: www.memphisbelle.com
Formerly: Memphis Bell B17 Flying Fortress
Founded: 1967.
Key Personnel: Pres., George Burns.
Personnel Profile: Part-Time Volunteers 38.
Governing Authority: Memphis Belle Memorial Association & U.S. Air Force Museum.
Institution Type/Description: Military Museum.
Collections: B-17 aircraft.
Hours & Admission Prices: Summer daily. Island admission prices vary.

MEMPHIS BOTANIC GARDEN, GOLDSMITH CIVIC GARDEN CENTER, 750 Cherry Rd., Memphis, TN 38117-4699. Tel.: 901-576-4100. Fax: 901-682-1561.
E-mail: info@memphisbotanicgarden.com
Web Site: www.memphisbotanicgarden.com
Founded: 1964.
Congressional District: 9
Key Personnel: Exec. Dir., Jim Duncan; Bd. Pres., Bill Ferrell; Dir., Administration, Mary Helen Butler; Dir., Finance, Walton Griffin; Dir., Facilities, Stan Myers; Dir., Mktg. & Public Rels., Jana Wilson.
Governing Authority: City of Memphis. Parent Institution: Memphis Park Commission. Subsidiary Institution: Memphis Botanic Garden Foundation. Tax-exempt: 501(c)(3).
Institution Type/Description: Botanic Garden: home of the Goldsmith Civic Garden Center.
Collections: Goldsmith Civic Garden Center; Rose Garden; Tennessee Bicentennial Iris Garden; Fonville Four Seasons Garden; Wild Flower Woodland; Michie Magnolia Trail; Orchid House Azalea & Dogwood Trail; Charlotte Sawyer Memorial Daffodil Trail; Japanese Garden of Tranquility; W.C. Paul Arboretum; Conifer Collection; Little Garden Club Sensory Garden; Perennial Trial Garden; herb collection; open meadow; cactus collection; organic garden; sculpture garden; Audubon Lake & Pavilion; water garden; Daylily collection; Little Garden Club Orientation Theater; Hardin Hall.
Research Fields: plant material.
Facilities: 2,000-vol. library of books on plants available for research on premises only; botanic garden; reading room; 250-seat auditorium; classroom; visitor center. Gift items for sale.
Activities: guided tours; lectures; films; study clubs; hobby workshops; organized education programs for children, adults & undergraduate college students affiliated with University of Memphis; docent program; participatory & temporary exhibitions. Center Sponsors: art exhibits; plant shows; outdoor concerts.
Publications: magazine, The Dirt; brochures; curriculum guide; educational catalog.
Hours & Admission Prices: Daily 9-6; Winter: 9-4:30. Adults $8, seniors 62 & over $6.50, children 2-12 $5; discounts to groups, and AHS & AABGA Reciprocal Gardens members; children under 2 & members no charge. Closed New Year's Day; Thanksgiving; Christmas. &
Attendance: 130,000 (accurate)
Membership: Garden-Senior $60; Garden $75; Green Family & Garden Club Partner $100; Garden Supporter $150; Photography Member &175; Avant

Gardener $300; Seasoned Gardener $500; Director's Council $1,000; Woodland Society $5,000.

∗ MEMPHIS BROOKS MUSEUM OF ART, (M), 1934 Poplar Ave., Overton Park, Memphis, TN 38104-2765. Tel.: 901-544-6200. Fax: 901-725-4071.
E-mail: brooks@brooksmuseum.org
Web Site: www.brooksmuseum.org
Founded: 1916.
Congressional District: 9
Key Personnel: Dir., Cameron Kitchin; Pres. (V), Nathan G. Bicks; Chief Cur., Marina Pacini; Cur., Stanton Thomas; Dir. Education, Kathy Dumlao; Assoc. Dir. Education, Elesha Newberry; Assoc. Dir. Education, Jenny Hornby; Collections Mgr., Kip Peterson; Assoc. Registrar, Marilyn Masler; Devel. Assoc., Genevieve Hill-Thomas; Dir. Mktg., Claudia Towell; Art Dir., JoAnn Moss; Museum Shop Mgr., Eva Langsdon; Public Rels. & Public Events Mgr., Andria Lisle.
Personnel Profile: Full-Time Paid 46; Part-Time Paid 51; Part-Time Volunteers 300; Interns 5.
Volunteer Hours: 600
Operating Expenses: 4,848,735
Operating Income: 5,125,570
Governing Authority: nonprofit corp. Parent Institution: Memphis Brooks Museum of Art, Inc. Tax-exempt: 501(c)(3) & 170(b)(1)(A).
Institution Type/Description: Fine Arts Museum.
Collections: Kress collection of Italian, Medieval, Renaissance & Baroque paintings & sculpture; 16th-19th century Northern European paintings & sculpture; 17th & 18th century English portraits & landscapes; 19th & 20th-century American paintings & sculpture; French Impressionists; 5,000 prints; porcelain; glass; furniture; artists' books.
Major Exhibits: Dali & Film, 2/14-5/11/14; Marisol: Sculptures and Works on Paper, 6/14-9/8/14.
Research Fields: pertaining to permanent collection.
Facilities: 8,000-vol. art reference library; 273-seat auditorium; orientation theatre; restaurant. Museum-related items for sale.
Activities: guided tours; lectures; films; demonstrations; student art activity area; musical events; gallery talks; docent program inter-museum loan, permanent, temporary & traveling exhibitions; outreach school program.
Publications: members' magazine; exhibition catalogues; interpretative exhibition; general museum brochures; calendar of events.
Hours & Admission Prices: Wed. & Fri. 10-4, Thurs. 10-8, Sat. 10-5, Sun. 11-5. Adults $7, seniors $6, students $3; discounts to AAM and ICOM members & scheduled groups; children under 6 & members no charge. Closed New Year's Day; Independence Day; Thanksgiving; Christmas. &
Attendance: 65,000 (accurate)
Membership: Individual $50; Dual $65; Family $75; Advocate $150; Fellow $300; Benefactor $500; Patron $1,000; Masterpiece Circle $2,500; Moss Society $5,000.

MEMPHIS COLLEGE OF ART, 1930 Poplar Ave., Overton Park, Memphis, TN 38104-2756. Tel.: 901-272-5100; 800-727-1088. Fax: 901-272-5104. Facebook: Memphis College of Art.
E-mail: info@mca.edu
Web Site: www.mca.edu
Founded: 1936.
Congressional District: 9
Key Personnel: Pres., Ron Jones; Vice Pres. Finance & Administration, Sherry Yelvington; Vice Pres. College Advancement, Shawna Engel; Vice Pres. Enrollment & Student Affairs, Susan Miller; Registrar, Sean Scott; Coord. Exhibitions, Catherine Blackwell Pena.
Personnel Profile: Full-Time Paid 49; Part-Time Paid 24.
Governing Authority: college. Tax-exempt: 501(c)(3).
Institution Type/Description: Art College.
Collections: works by college graduates, current students, faculty & outside artists.
Facilities: 16,000-vol. library of visual art books available for inter-library loan for teachers & graduate students on premises; 342-seat auditorium; classrooms. Art material for sale.
Activities: guided tours; lectures; films; arts festivals; organized education programs for children, adults & undergraduate college students; temporary exhibitions.
Publications: occasional catalogs; newsletter.
Hours & Admission Prices: Mon.-Fri. 8-5, Sat. 9-4, Sun. 12-4. No charge. Closed holidays. &
Attendance: 15,000 (estimated)
Membership: Individual $40; Family $50; Associate $100; Benefactor $250; President's Circle $500; Chairman's Circle $1,000; Trustees Circle $2,500; The Academy Society $5,000.

MEMPHIS ROCK 'N' SOUL MUSEUM, 191 Beale St., Ste. 100, Memphis, TN 38103-3715. Tel.: 901-205-2533. Fax: 901-205-2534.
E-mail: info@memphisrocknsoul.org
Web Site: www.memphisrocknsoul.org
Founded: 2000.
Key Personnel: Exec. Dir., John Doyle; Chm. (V), Kevin Kane; Museum Shop Mgr., Pam Hetsel.
Personnel Profile: Full-Time Paid 4; Part-Time Paid 8; Part-Time Volunteers 3.
Institution Type/Description: Music History Museum.
Collections: rock & soul music history; musical instruments; musicians.
Activities: rental facilities.
Hours & Admission Prices: Daily 10-7. Adults $10, youth 5-17 $7; discounts to AAA, AARP, & military. Closed New Year's Day; Thanksgiving; Christmas Eve & Day. &

MEMPHIS ZOO, 2000 Prentiss Place, Memphis, TN 38112-5033. Tel.: 901-276-(WILD). Fax: 901-725-9305.
E-mail: zooinfo@memphiszoo.org
Web Site: www.memphiszoo.org
Founded: 1906.
Congressional District: 9
Key Personnel: Pres. & C.E.O., Dr. Charles A. Brady; Dir. Operations, Wayne Carlisle; Cur. Mammals, Matt Thompson; Cur. Birds, Herb Roberts; Museum Shop Mgr., Conne Bellett; Cur. Reptiles, Steve Reichling.
Personnel Profile: Full-Time Paid 159; Part-Time Paid 50; Part-Time Volunteers 25; Interns 5.
Governing Authority: managed by society; city-owned. Parent Institution: Memphis Park Commission. Tax-exempt.
Institution Type/Description: Zoo.
Collections: live animals.
Research Fields: animal behavior; captive behavior.
Facilities: library; aquarium; education center; classroom.
Activities: special events; children & family programs; field trips; school programs.
Publications: bimonthly magazine, Exzooberance.
Hours & Admission Prices: March-Oct. daily 9-5; Nov.-Feb. daily 9-4; last admission 1 hour before closing. Adults 12-59 $13, senior citizens 60 & over $12, children 2-11 $8; children under 2 & zoo members no charge. Closed Thanksgiving; Christmas Eve & Day. &
Attendance: 1,066,000 (estimated)
Membership: Individual $60; Dual $70; Family & Grandparent $79; Contributing $165; Associate $300; Patron $500.

METAL MUSEUM, (M), 374 Metal Museum Dr., Memphis, TN 38106-1514. Tel.: 901-774-6380. Fax: 901-774-6382.
E-mail: info@metalmuseum.org
Web Site: www.metalmuseum.org
Formerly: National Ornamental Metal Museum
Founded: 1976.
Congressional District: 96
Key Personnel: Dir., Carissa Hussong; Pres. (V), Rob Keeler; Registrar, Rosie Meindl; Studio Mgr., James Masterson; Museum Shop Mgr., Virginia Nuckolls.
Personnel Profile: Full-Time Paid 12; Part-Time Paid 2; Part-Time Volunteers 200; Interns 2.
Governing Authority: nonprofit organization. Tax-exempt: 501(c)(3).
Institution Type/Description: Metal Museum: historic & contemporary decorative & fine art metalwork.
Collections: decorative metalwork; sculpture; jewelry; tools; hollowware; architectural forged & cast iron; drawings; videos; slides.
Research Fields: ironwork; decorative ironwork; restoration of ferrous & non-ferrous objects; historic & contemporary ferrous & non-ferrous metalwork.
Facilities: 20,000-vol. library & slide collection; teaching & demonstration building; 3.2 acres of grounds & sculpture garden; 2,800 sq. ft. exhibit space. Handcrafted jewelry, ironwork, books & other items for sale.
Activities: guided tours; lectures; organized programs for children & adults; training programs for professional museum workers; loan, temporary & traveling exhibitions; classes & workshops for beginners & professional metalsmiths; metal conservation & restoration services available to public & private collectors. Museum Sponsors: Repair Days.
Publications: quarterly newsletter; exhibition catalogues.
Hours & Admission Prices: Tues.-Sat. 10-5, Sun. 12-5. Adults $6, senior citizens $5, students $4; members & children under 5 no charge. Closed New Year's Eve & Day; Easter; Independence Day; Thanksgiving; Christmas Eve, Day & day after; during exhibit changes. &
Attendance: 38,000 (estimated)
Membership: Student & Senior $35; Individual $40; Dual & Family $55;

Supporting Friend $100; Donor $250; Silver Corporate/Individual $500; Gold Corporate/Individual $1,000.

MISSISSIPPI RIVER MUSEUM AT MUD ISLAND RIVER PARK, 125 N. Front St., Memphis, TN 38103-1713. Mailing Address: 101 N. Island Dr., Memphis, TN 38103. Tel.: 901-576-7241; 800-507-6507. Fax: 901-576-6666.
E-mail: trey@mudisland.com
Web Site: www.mudisland.com
Founded: 1978.
Congressional District: 9
Key Personnel: General Mgr., Trey Giuntini; Museum Mgr., Alisa Bradley.
Personnel Profile: Full-Time Paid 2; Part-Time Paid 25.
Governing Authority: Parent Institution: Riverfront Development Corp. Tax-exempt.
Institution Type/Description: History Museum Complex.
Collections: full size reconstruction of riverboat & Civil War gunboat with period furnishings; artifacts & archival relating to prehistoric & historic Indians, settlement & boat development; Civil War on the Mississippi River; Delta Music from Blues to Rock-n-Roll; boat models & engines; river engineering; natural history & sciences; related art collection; outdoor exhibition: five-block-long scale model of river focusing on cultural & natural history of the lower Mississippi River.
Facilities: aquarium; 5,000-seat outdoor theater; 4 eating facilities; 105-seat theater. Museum-related items for sale.
Activities: guided tours; lectures; films; gallery talks; concerts; formally organized education programs for children & adults; permanent & temporary exhibitions; rental facilities.
Hours & Admission Prices: mid-April to Oct. Tues.-Sun. 10-5. Adults $10, senior $9, children 5-11 $7; discounts to groups; children 4 & under no charge. &

Attendance: 97,000 (estimated)

* **NATIONAL CIVIL RIGHTS MUSEUM AT THE LORRAINE MOTEL,** 450 Mulberry St., Memphis, TN 38103-4214. Tel.: 901-521-9699. Fax: 901-521-9740.
E-mail: cdyson@civilrightsmuseum.org
Web Site: www.civilrightsmuseum.org
Founded: 1991.
Congressional District: 9
Key Personnel: Chm. (V), Herb Hilliard; Pres. & Exec. Dir., Beverly C. Robertson; Chm. Exec. Committee, J.R. Hyde.
Personnel Profile: Full-Time Paid 32; Part-Time Paid 8.
Governing Authority: private; nonprofit organization. Tax-exempt: 501(c)(3).
Institution Type/Description: History Museum: located at the Lorraine Motel, site of the assassination of Dr. Martin Luther King Jr.
Collections: artifacts that document civil rights activities & associated segregationist activity chiefly from 1619-2000; books; organizational pamphlets; photographs; original correspondence; paintings; clothing; art works; film; video.
Facilities: 40,000 sq. ft. exhibit space.
Activities: Museum Sponsors: Freedom Award; King Holiday; April 4th Commemoration.
Publications: Movement Newsletter; annual report.
Hours & Admission Prices: June-Aug. Mon. & Wed.-Sat. 9-6, Sun. 1-6; Sept.-May Mon. & Wed.-Sat. 9-5, Sun. 1-5. Adults $13, senior citizens & students $11, children 4-17 $9.50; discounts to military, AAA, AAM & AARP members; members and children 3 & under no charge. Closed New Year's Day; Thanksgiving & Christmas Eve & Day. &
Attendance: 200,000 (accurate)
Membership: Individual $40; Family $65; Advocate $150.

* **PINK PALACE FAMILY OF MUSEUMS, (M),** 3050 Central Ave., Memphis, TN 38111-3399. Tel.: 901-636-6320 & 6398. Fax: 901-636-2391.
E-mail: steve.pike@memphistn.gov
Web Site: www.memphismuseums.org
Formerly: Memphis Pink Palace Museum & Sharpe Planetarium and IMAX Theater
Founded: 1928.
Congressional District: 9
Key Personnel: Dir., Stephen J. Pike; Pres. (V), Jeffrey R. Wills; Admin. Program Affairs, Wesley S. Creel; Mgr. Collections, Louella Weaver; Mgr. Exhibits & Graphic Svcs., Steve Masler; Mgr. Education, Alice A. "Alex" Eilers; Mgr. Historic Properties, Jennifer Tucker; Crew Training Intl. IMAX Theater, Tony Hardy; Admin. Public Affairs, Richard Pugh; Mgr. Lichterman Nature Center, Andy Williams; Museum Shop Mgr., Debbie Jordan.
Personnel Profile: Full-Time Paid 50; Part-Time Paid 75; Part-Time Volunteers 1,213; Interns 4.

Governing Authority: municipal. Headquarters. Pink Palace Family of Museums. Parent Institution: City of Memphis. Subsidiary Institution: Memphis Museums, Inc.; Lichterman Nature Center; Mallory/Neely House; Magevney House; Coon Creek Science Center. Tax-exempt: 501(c)(3).
Institution Type/Description: General, Natural History, Science & Cultural History Museum.
Collections: artifacts, specimens & documents relating to the cultural & natural history of Memphis & the mid southern region; African American artifacts.
Facilities: 2,500-vol. library of history & natural science available for inter-library loan & on premises; planetarium; snack area; classrooms; teaching laboratories; 240-seat IMAX theater; 200-seat mansion theater. Museum-related items for sale.
Activities: radio programs; formally organized education programs for children & adults; docent program or council; permanent, temporary & traveling exhibitions; school loan service; planetarium shows; IMAX shows.
Publications: bimonthly newsletter, Museumscope; annual, Educator's Guide.
Hours & Admission Prices: Mon.-Sat. 9-5, Sun. 12-5. Adults $11.75, senior citizens $11.25, children $6.25; discounts to ASTC & AAM members; museum on Tues. 1-5 no charge. Planetarium: adults $4.50, senior citizens & children $4. IMAX: adults $8.25, senior citizens $7.50, children $6.50; discounts for exhibit & planetarium packages. Closed New Year's Day; Thanksgiving; Christmas. &
Attendance: 520,113 (accurate)
Membership: E-Member $30; Individual $50; Family $75; Family Plus One $85; Family Plus Two $95; IMAX Club $100; IMAX Club One $110; IMAX Club Two $120; Palace Guard $175; Benefactor $250; Advocate $500; Director's Circle $1,000.

SLAVE HAVEN UNDERGROUND RAILROAD MUSEUM (BURKLE ESTATE), 826 N. Second St., Memphis, TN 38107-2302. Mailing Address: P.O. Box 3142, Memphis, TN 38173-0142. Tel.: 901-527-3427.
Institution Type/Description: Historic House: former home of Jacob Burkle, built in 1849.
Collections: memorabilia from the Underground Railroad.
Hours & Admission Prices: Summer: Mon.-Sat. 10-4; Winter: Wed.-Sat. 10-4. Tours: adults $6, students 4-17 $4.

STAX MUSEUM OF AMERICAN SOUL MUSIC, (M), 926 E. McLemore Ave., Memphis, TN 38106-3338. Tel.: 901-946-2535. Fax: 901-507-1463.
Web Site: www.staxmuseum.com
Founded: 2003.
Key Personnel: Interim C.E.O., Deanie Parker; Pres. (V), Howard Robertson; Senior Mgr. Operations, Susan Green; Vice Pres. Finance, Duke Herenton; Public Rels., Tim Sampson; Museum Shop Mgr., Steve Walker.
Personnel Profile: Full-Time Paid 3; Part-Time Paid 7; Part-Time Volunteers 10; Interns 3.
Governing Authority: private; nonprofit organization. Parent Institution: Soulsville, 870 McLemore Ave., Memphis, TN 38106. Tax-exempt: 501(c)(3).
Institution Type/Description: Soul Music Museum.
Collections: over 2,000 cultural artifacts; photographs; sound recordings; musical & recording equipment; stage costumes; advertising; albums.
Research Fields: oral histories; local artists & community members.
Facilities: 60-seat theater. Museum-related items for sale.
Activities: concerts; formal education programs for children; lectures; loan & traveling exhibitions; theater; rental facility. Annual Events: fundraisers; Black History Month; Black Music History Month.
Hours & Admission Prices: April-Oct. Mon.-Sat. 10-5, Sun. 1-5; Nov.-March Tues.-Sat. 10-5, Sun. 1-5. Adults $12, senior citizens, students & military $11, children $9; discounts to groups & AAA members; members no charge. Closed New Year's Day; Easter; Thanksgiving; Christmas. &
Attendance: 42,422 (accurate)
Membership: $50, $100, $250, $500.

W.C. HANDY HOUSE & MUSEUM, 352 Beale St., Memphis, TN 38103. Tel.: 901-527-3427. Fax: 901-527-8784.
Institution Type/Description: Historic House: housed in the former home of William Christopher Handy.
Collections: W.C. Handy's life & career; personal artifacts; period furnishings; photographs.
Hours & Admission Prices: Summer: Tues.-Sat. 10-5; Winter: Tues.-Sat. 11-4. Adults $3, children $2.

WOODRUFF-FONTAINE HOUSE MUSEUM, 680 Adams Ave., Memphis, TN 38105-4902. Tel.: 901-526-1469. Fax: 901-526-4531.
E-mail: wfhouse@bellsouth.net

Web Site: www.woodruff-fontaine.com
Governing Authority: Parent Institution: Association for the Preservation of Tennessee Antiquities (APTA).
Institution Type/Description: Historic House: French Victorian mansion built in 1870 along "Millionaires Row."
Collections: period artifacts.
Hours & Admission Prices: Wed.-Sun. 12-4. Closed holidays.

Milan

WEST TENNESSEE AGRICULTURAL MUSEUM, 3 Ledbetter Gate Rd., Milan, TN 38358-6543. Tel.: 731-686-8067.
E-mail: utagmuseum@utk.edu
Web Site: http://milan-tennessee.edu
Founded: 1987.
Congressional District: 8
Institution Type/Description: Agricultural Museum.
Collections: local history; pioneer life & culture; period furnishings & equipment; personal artifacts; photographs.
Activities: guided tours; special events.
Hours & Admission Prices: Mon.-Fri. 8-4; Sat. & groups by appointment. No charge. Closed holidays. &

Morristown

CROCKETT TAVERN MUSEUM, 2002 Morningside Dr., Morristown, TN 37814-5459. Tel.: 423-587-9900.
E-mail: info@crocketttavernmuseum.org
Web Site: www.crocketttavernmuseum.org
Founded: 1958.
Congressional District: 4
Key Personnel: Pres. (V) Hamblen County Chapter A.P.T.A. & Museum Shop Mgr., Sally A. Baker.
Personnel Profile: Part-Time Paid 1; Part-Time Volunteers 1.
Governing Authority: nonprofit organization. Affiliated with the Assoc. for the Preservation of Tennessee Antiquities, Hamblen County Chapter-APTA, Belle Meade Mansion, Harding Rd. at Leake Ave., Nashville, TN 37205. Tax-exempt: 170(b)(1)(A).
Institution Type/Description: History Museum: housed in 1794 replica of boyhood home of David Crockett.
Collections: furniture; pioneer utensils; historical articles used by early pioneers.
Facilities: Crockett books, prints, buttons, pioneer-related & other museum-related items for sale.
Activities: guided tours; permanent exhibitions. Museum Sponsors: Davy Crockett Birthday Party in August.
Hours & Admission Prices: May-Oct. Tues.-Sat. 11-5. Adults $5, students 5-18 $1; discount to groups; members, children under 5, APTA members, and school & group trip chaperones no charge.
Attendance: 1,500 (estimated)
Membership: Sustaining $20.

ROSE CENTER, 442 W. 2nd N. St., Morristown, TN 37814-4026. Mailing Address: P.O. Box 1976, Morristown, TN 37816-1976. Tel.: 423-581-4330. Fax: 423-581-4307.
E-mail: postmaster@rosecenter.org
Web Site: www.rosecenter.org
Founded: 1976.
Congressional District: 1
Key Personnel: Dir., Robert Lydick; Chm. (V), Pete Barile; Operations Coord., Patty Gracey; Coord. Education & Special Events, Beccy Hamm; Bldg. Mgr., Ray James.
Personnel Profile: Full-Time Paid 3; Part-Time Paid 1; Part-Time Volunteers 800.
Governing Authority: nonprofit organization. Tax-exempt.
Institution Type/Description: Civic Art & Cultural Center: housed in 1892 former high school.
Collections: Hamblen County TN history, culture & industry; early Morristown photographs; Murrell flying machine & its history; local Civil War era artifacts & photos; documents; Medal of Honor winners.
Facilities: auditorium; classrooms.
Activities: dance recitals; arts festivals; hobby workshops; formally organized education programs for children & adults; temporary exhibits.
Publications: monthly newsletter; quarterly arts calendar.
Hours & Admission Prices: Mon.-Fri. 9:30-5, Sat. 9-1. No charge. Closed New Year's Day; Martin Luther King Day; Memorial Day; Independence Day; Labor Day; Thanksgiving; Christmas. &
Attendance: 65,000 (estimated)
Membership: Student $10; Senior $35; Individual $50; Family $75; Contributor $125; Sustaining $250; Preservation $500; Directors $1,000.

Mount Pleasant

MT. PLEASANT - MAURY MUSEUM OF LOCAL HISTORY, 108 Public Sq., Mount Pleasant, TN 38474. Tel.: 931-379-9511.
Institution Type/Description: History Museum.
Collections: local history & culture; period furnishings; personal artifacts; photographs.
Hours & Admission Prices: Mon.-Sat. 9:30-4:30; other times by appointment.

Mountain City

JOHNSON COUNTY WELCOME CENTER & MUSEUM, 716 S. Shady St., Mountain City, TN 37683. Mailing Address: P.O. Box 1, Mountain City, TN 37683. Tel.: 423-727-5800. Fax: 423-727-4943. Facebook: Johnson County Welcome.
E-mail: johnsoncountywelcomecenter@centurylink.net
Institution Type/Description: History Museum.
Collections: local history & culture; period furnishings; photographs; personal artifacts.
Hours & Admission Prices: Daily. No charge.

Murfreesboro

BALDWIN PHOTOGRAPHIC GALLERY, Learning Resources Center, Murfreesboro, TN 37132. Mailing Address: MTSU, Box 305, Murfreesboro, TN 37132. Tel.: 615-898-2085 & 5628 (Off. of Secy.). Fax: 615-898-5682.
E-mail: tjimison@mtsu.edu
Founded: 1961.
Congressional District: 6
Key Personnel: Cur., Tom Jimison; Archivist, Valerie Menard.
Personnel Profile: Full-Time Paid 1; Part-Time Paid 1; Part-Time Volunteers 10; Interns 1.
Governing Authority: public university; nonprofit organization. Tax-exempt.
Institution Type/Description: Photography Art Museum.
Collections: contemporary photography of past 20 years.
Facilities: photograph archive.
Activities: lectures; traveling exhibitions.
Publications: monthly, show announcements; posters.
Hours & Admission Prices: mid-Jan. to June & Sept. to mid-Dec. Mon.-Fri. 8-4:30, Sat. 9-11:45, Sun. 6pm-10pm; July-Aug. by appointment. No charge. Closed Labor Day. &
Attendance: 30,000 (estimated)

DISCOVERY CENTER AT MURFREE SPRING, 502 S.E. Broad St., Murfreesboro, TN 37130-4237. Tel.: 615-890-2300.
E-mail: info@discoverycenteronline.org
Web Site: www.discoverycenteronline.org
Founded: 1986.
Key Personnel: C.E.O., Tara MacDougall; C.F.O. & Dir. Operations, Veronica King; Dir. Mktg. & Public Rels., Ann Mapp; Museum Shop Mgr., Chantel Preston.
Personnel Profile: Full-Time Paid 10; Part-Time Paid 30; Part-Time Volunteers 200; Interns 2.
Institution Type/Description: Children's Museum.
Collections: hands-on exhibits.
Facilities: restaurant. Museum-related items for sale.
Activities: special events; educational programs; school groups; community outreach; school break camps.
Hours & Admission Prices: Mon.-Sat. 10-5, Sun. 1-5. Admission $6; discounts to ACM members; children under 2 & members no charge. Closed major holidays. &
Attendance: 118,984
Membership: Teacher $50; Family & Grandparent $75; Passport $125; Explorer $300.

OAKLANDS HISTORIC HOUSE MUSEUM, 900 N. Maney Ave., Murfreesboro, TN 37130-2955. Mailing Address: P.O. Box 432, Murfreesboro, TN 37133-0432. Tel.: 615-893-0022. Fax: 615-893-0513. Facebook: Oaklands Historic House Museum.
E-mail: info@oaklandsmuseum.org
Web Site: www.oaklandsmuseum.org
Founded: 1959.
Congressional District: 6
Key Personnel: Exec. Dir., James W. Manning, Jr.; Pres. (V), Betty Hord; Dir. Education, Mary Beth Nevills; Museum Cur., Nila Gober; Special Events Coord., Raina van Setter.
Personnel Profile: Full-Time Paid 2; Part-Time Paid 5; Part-Time Volunteers 15.

Governing Authority: nonprofit organization. Parent Institution: Oaklands Association Inc. Tax-exempt: 501(c)(3).
Institution Type/Description: Historic House: 1818-1865 cotton plantation.
Collections: period furnishings; military; textiles; medical; Tennessee Native Tree Arboretum.
Major Exhibits: Wedding Dresses Through the Decades, 1/5/14-3/2/14; Days of Washing, Churning & Learning, 5/8/14-5/9/14; Autumn in the Oaks, 10/2/14-10/3/14; Widows, Weapons & Wakes, 10/13/14-10/31/14.
Research Fields: history of Maney family & Oaklands; Victorian history of life & times in Tennessee.
Facilities: nature trail; picnic areas; visitor center. Museum-related items for sale.
Activities: guided tours; lectures; arts festivals; docent program; temporary exhibitions; educational program; wetlands program. Annual Event: Christmas Tour of Homes in December.
Publications: cookbook, Dining With Oaklands; Hearthstones; The Story of Rutherford County Homes; DVD, The History of Oaklands; Caught in the Crossfire of the Civil War; newsletter, Oaklands Historic House Museum.
Hours & Admission Prices: Tues.-Sat. 10-4, Sun. 1-4; last tour starts at 3. Adults $10, senior citizens, AAA & military $7, college students & children 6-17 $5; discounts to groups of 10 or more; members & children under 5 no charge. Closed New Year's Day; Easter; Thanksgiving; Christmas. &
Attendance: 6,000 (estimated)
Membership: Student $20; Contributor $30-$49; Sponsor $50-$99; Patron $100-$499; Grand Patron $500-$999; Benefactor $1,000 & up.

STONES RIVER NATIONAL BATTLEFIELD, 3501 Old Nashville Hwy., Murfreesboro, TN 37129-8621. Tel.: 615-893-9501. Fax: 615-893-9508. TDD: 615-893-9501.
E-mail: stri_information@nps.gov
Web Site: www.nps.gov/stri
Founded: 1927.
Congressional District: 6
Key Personnel: C.E.O. & Supt., Stuart Johnson; Park Ranger & Museum Mgr., James B. Lewis; Museum Shop Mgr., Amber Brosbol.
Personnel Profile: Full-Time Paid 11; Part-Time Paid 6; Part-Time Volunteers 280; Interns 4.
Governing Authority: federal. Parent Institution: National Park Service, U.S. Department of Interior, Interior Building, Washington, DC 20240. Tax-exempt.
Institution Type/Description: Park & Military Museum: adjacent to Stones River National Cemetery.
Collections: Civil War items pertaining to the battle of Stones River; manuscripts.
Research Fields: Battle of Stones River.
Facilities: 1,000-vol. library, including 128 vols. of official records & state troop rosters, available for use on premises; no reading room. Museum-related items for sale.
Activities: lectures; films; formally organized education programs for children & adults; permanent exhibitions. Museum Sponsors: Living History demonstrations during summer months.
Publications: brochure, Stones River; area map.
Hours & Admission Prices: Daily 8-5. No charge; donations accepted. Closed Thanksgiving; Christmas. &
Attendance: 192,355 (accurate)
Membership: Friends of Stones River National Battlefield $5.

TODD ART GALLERY - MIDDLE TENNESSEE STATE UNIVERSITY, 1301 E. Main St., Murfreesboro, TN 37132. Mailing Address: P.O. Box 25, Murfreesboro, TN 37130. Tel.: 615-898-5653.
Web Site: www.mtsu.edu/art/
Congressional District: 4
Governing Authority: Parent Institution: Middle Tennessee State University. Tax-exempt.
Institution Type/Description: Art Gallery.
Collections: paintings; sculpture.
Hours & Admission Prices: Call for hours. No charge; donations accepted. &

Nashville

ADVENTURE SCIENCE CENTER, 800 Fort Negley Blvd., Nashville, TN 37203-4833. Tel.: 615-862-5160. Fax: 615-862-5178.
Web Site: www.adventuresci.com
Formerly: Cumberland Science Museum
Founded: 1944.
Congressional District: 5
Key Personnel: C.E.O. & Pres., Susan B. Duvenhage; Chm. (V), Edward Lang; Exhibits & Collections Mgr., Herschell Parker; Dir. Operations, Tina

Brown; Dir. Education, Jeri Hasselbring; Bldg. Supt., Dan Slayden; Dir. Planetarium, Kris McCall; Museum Shop Mgr., Polly DuBose; Dir. Devel., Rae Hummell.
Personnel Profile: Full-Time Paid 36; Part-Time Paid 90; Part-Time Volunteers 150.
Governing Authority: Tax-exempt.
Institution Type/Description: Science Center.
Collections: geology; science; technology; human health; hands-on exhibits.
Research Fields: informal education in contemporary science.
Facilities: 44,000 sq. ft. exhibit space; 166-seat planetarium; classrooms; theater.
Activities: lectures; demonstrations; science camps; planetarium programs; field trips; camp-ins; educational programs; public programs.
Publications: newsletters.
Hours & Admission Prices: Mon.-Sat. 10-5, Sun. 12:30-5:30. Adults $11, seniors, military, college students & youth 3-12 $9; discounts to ASTC members; members no charge. Planetarium Shows: $6 non-members, $4 members. Closed Thanksgiving; Christmas. &
Attendance: 264,576 (accurate)
Membership: Individual $55; Single Parent Family $65; Grandparent $85; Family $90; Family Deluxe $145.

ASSOCIATION FOR THE PRESERVATION OF TENNESSEE ANTIQUITIES, 110 Leake Ave., Nashville, TN 37205-3706. Tel.: 615-352-8247.
E-mail: apta1951@bellsouth.net
Web Site: www.theapta.org
Founded: 1951.
Congressional District: 5
Key Personnel: Exec. Dir., Elliott W. McNiel, CAE; Pres., Frank McMeen.
Personnel Profile: Part-Time Paid 1; Part-Time Volunteers 40.
Governing Authority: nonprofit organization. Parent Institution: APTA. Branch Museums: The Athenaeum, 808 Athenaeum Place, Columbia, TN 38401; Belle Meade Plantation, 5025 Harding Rd., Nashville, TN 37205; The Buchanan Log House, 2910 Elm Hill Pike, Nashville, TN 37214; Rachel H.K. Burrow Museum & Historic Post Office, Arlington, TN 38002; Crockett Tavern-Museum, 2002 Morningside Dr., Morristown, TN 37814; Glenmore, 1280 N. Chuckey Pike, Jefferson City, TN 37760; The Little Courthouse, E. Market St., Bolivar, TN 38008; The Pillars, Washington St., Bolivar, TN 38008; Ramsey House, 2614 Thorngrove Pike, Knoxville, TN 37914; Woodruff-Fontaine House & Lee Memorial House, 680-690 Adams, Memphis, TN 38102. Tax-exempt.
Institution Type/Description: Historic Preservation Society: housed in c.1853 Belle Meade Plantation garden house.
Collections: costumes; decorative items; Civil War memorabilia; ante-bellum, Victorian & early American furniture, books, textiles & fans.
Research Fields: historic preservation; Tennessee history.
Activities: guided tours; lectures; films; formally organized education programs for children & adults; permanent & temporary exhibitions.
Publications: biannual newsletter; advertising brochure.
Hours & Admission Prices: Call or write for information & details on chapter house museums. Closed New Year's Day; Thanksgiving; Christmas. &
Membership: Individual $25; Family $40. Each chapter has different membership dues.

BELLE MEADE PLANTATION, 5025 Harding Pike, Nashville, TN 37205-2810. Tel.: 615-356-0501 & 800-270-3991. Fax: 615-356-2336.
E-mail: info@bellemeadeplantation.com
Web Site: www.bellemeadeplantation.com
Founded: 1953.
Congressional District: 5
Key Personnel: Pres., Alton W. Kelley; Museum Shop Mgr., Joanne Hostettler-Floyd.
Personnel Profile: Full-Time Paid 20; Part-Time Paid 10; Part-Time Volunteers 30; Interns 1.
Governing Authority: nonprofit organization. Parent Institution: Association For The Preservation of Tennessee Antiquities, 110 Leake Ave., Nashville, TN 37205. Tel.: 615-352-8247. Tax-exempt.
Institution Type/Description: Historic House: 1820 Belle Meade Mansion.
Collections: costumes; period furniture; Civil War memorabilia; antebellum, Victorian & Early American furniture & books; carriages. Historic Structures: c.1790 log cabin; 1840 garden house; 1840 smoke house; 1840 mausoleum; 1853 Greek Revival mansion; 1884 creamery; c.1890 child's playhouse; c.1893 carriage house & stables; 1830s slave cabin; chicken coop.
Research Fields: lineage of thoroughbred studs & mares.
Facilities: visitor center; restaurant; education building; theatre; 30 acre historic site. Museum-related items for sale.

Activities: guided tours; lectures; films; organized education programs for children & adults; permanent & temporary exhibitions.
Publications: brochure; Widows, Weepers & Wakes - Mourning in Middle Tennessee; Purging Pestilence - Dealing with Disease in the 19th Century; Belle Meade Bloodlines; Belle Meade Plantation.
Hours & Admission Prices: Mon.-Sat. 9-5, Sun. 11-5. Adults 13-64 $16, senior citizens 65 & over $14, youth 13-18 $10, children 6-12 $8; discount to groups; children 5 & under and members no charge. Closed New Year's Day; Easter; Thanksgiving; Christmas Eve & Day. &
Attendance: 135,000 (accurate)
Membership: Student $25; Senior Individual 65 & over $40; Individual $50; Senior Family $65; Family $75; Gamma $125; Enquirer $250; Iroquois $500.

BELMONT MANSION ASSOCIATION, 1700 Adden Ave., Nashville, TN 37212. Mailing Address: 1900 Belmont Blvd., Nashville, TN 37212. Tel.: 615-460-5459. Fax: 615-460-5688.
E-mail: kate.wilson@belmont.edu
Web Site: www.belmontmansion.com
Founded: 1972.
Key Personnel: Dir., Mark Brown; Pres. (V), Angie Adams; Museum Shop Mgr., Dianne Berry.
Personnel Profile: Full-Time Paid 2; Part-Time Paid 30; Part-Time Volunteers 34; Interns 2.
Governing Authority: Tax-exempt.
Institution Type/Description: Historic House Museum.
Collections: period furnishings; personal artifacts.
Major Exhibits: Preparing for Battle: December 1864, 12/12/14-12/13/14.
Activities: Behind-the-Scenes tours; lectures; seminars. Annual Events: Decorative Arts Symposium; Adelicia's Birthday; Candlelight Tour; Christmas with the Acklens; Christmas Dinner & Lunch; .
Publications: quarterly newsletter, A View from the Monte.
Hours & Admission Prices: Mon.-Sat. 10-4, Sun. 1-4; groups by appointment. Adults $12, seniors 65 & over and military $11, children 6-12 $3; discounts to AAM, ICOM and AAA members & groups of 15 or more; members and children 5 & under no charge. Closed New Year's Day; Independence Day; Thanksgiving; Christmas.
Attendance: 20,000 (estimated)
Membership: Senior $25; Individual $30; Family $35; Patron $50; Grand Patron $100; Benefactor $500; Adelicia's Circle $1,000.

*** CHEEKWOOD BOTANICAL GARDEN & MUSEUM OF ART, (M),** 1200 Forrest Park Dr., Nashville, TN 37205-4242. Tel.: 615-353-6964. Fax: 615-353-0919.
Web Site: www.cheekwood.org
Founded: 1960.
Congressional District: 6
Key Personnel: Pres. & C.E.O., Jane Offenbach MacLeod; Chm. Bd., George B. Stadler; Cur. Art, Jochen Wierich, Ph.D.; Registrar, Sarah K. Ritter.
Personnel Profile: Full-Time Paid 49; Part-Time Paid 81; Part-Time Volunteers 401.
Governing Authority: nonprofit organization. Tax-exempt.
Institution Type/Description: Art Museum & Botanical Garden.
Collections: Museum: 19th & 20th-century American art featuring works by The Eight; sculpture by William Edmondson; contemporary outdoor sculpture; Worcester porcelain; American & European silver. Botanical Garden: Dogwood; native & historic collections.
Major Exhibits: More Love: Art, Politics and Sharing Since the 1990s, 9/13-1/14.
Research Fields: William Edmondson; Worchester porcelain Torreya taxifolia.
Facilities: 9,000-vol. library of botany & fine arts available for inter-library loan & for use on premises; nature center; learning center; 1mile sculpture trail; auditorium; classrooms; tea room. Museum-related items for sale.
Activities: guided tours; lectures; gallery talks; formally organized education programs; docent program; permanent, temporary & traveling exhibitions; contemporary series.
Publications: quarterly calendar; newsletter; catalogues; brochures; posters; exhibition catalogues.
Hours & Admission Prices: Tues.-Sat. 9:30-4:30, Sun. 11-4:30. Adults $12, senior citizens 65 & over $10, youth 3-17 $5; discounts to groups and AABGA & AAA members; children 2 & under no charge. Closed New Year's Day; 2nd Sat. June; Thanksgiving; Christmas. &
Attendance: 250,000 (estimated)
Membership: Student $30; Educator $35; Individual (Day Pass) $40; Senior Family $50; Individual (Full Benefits) $55; Educator Family $60; Family $75; Supporting $135; Pineapple Society $300; Boxwood $600; Benefactor $1,250; Founder $2,500; Cheekwood Round Table $5,000; Cheekwood Cornerstone $10,000; Bryant Fleming Circle $25,000.

COOTER'S PLACE NASHVILLE, 2613B McGavock Pike, Nashville, TN 37214-1215. Tel.: 615-872-8358. Fax: 615-872-8359. Facebook; Cooter's Place Nashville.
E-mail: info@cootersplace.com
Web Site: cootersplace.ocm
Founded: 1999.
Key Personnel: Museum Shop Mgr., Jaclyn Smith
Institution Type/Description: History Museum.
Collections: Dukes of Hazzard memorabilia, pictures, props, & costumes; Cooter's tow truck; Daisy's jeep; Rosco's patrol car; 1969 Dodge Charger, General Lee.
Hours & Admission Prices: Daily 9-8. No charge. Closed Thanksgiving; Christmas Day. &

*** COUNTRY MUSIC HALL OF FAME AND MUSEUM, COUNTRY MUSIC FOUNDATION,** 222 5th Ave., S., Nashville, TN 37203-4206. Tel.: 615-416-2001; 800-852-6437 (group reservations & special events). Fax: 615-255-2245.
E-mail: info@countrymusichalloffame.org
Web Site: www.countrymusichalloffame.org
Founded: 1964.
Congressional District: 5
Key Personnel: C.E.O., Kyle Young; Chm. (V), Steve Turner; Pres. (V), Vince Gill; Asst. to Dir., Rachel Neerman; Vice Pres. Finance & Operations, Nina Hammontree; Vice Pres. Museum Svcs., Carolyn Tate; Vice Pres. Devel., Pamela Johnson; Vice Pres. Public Rels., Liz Thiels, Sr.; Vice Pres. Sales & Mktg., Sharon Burns; Vice Pres. Museum Progs., Jay Orr; Museum Shop Mgr., Lisa Lane.
Personnel Profile: Full-Time Paid 50; Part-Time Paid 17; Part-Time Volunteers 170; Interns 12.
Governing Authority: nonprofit organization. Owned & operated by Country Music Foundation, Inc. Branch Museum: Studio B, Roy Acuff Place & Music Sq. W., Nashville, TN 37203; Hatch Show Print, 316 Broadway, Nashville, TN. Tax-exempt: 501(c)(3).
Institution Type/Description: Country Music Museum & Historic Recording Studio and Historic Woodblock Print Shop.
Collections: musical instruments & memorabilia associated with the development of country music; sight & sound exhibits; costumes; more than 200,000 sound recordings; videos, films & photographs; print collection of books, periodicals, songbooks, sheet music, manuscripts, newspaper & magazine clippings, artist biographies & publicity materials; biographical exhibits.
Research Fields: history of country, traditional, folk & American music.
Facilities: research library; 40,000 sq. ft. exhibit space; reading room; laboratory; 4 theaters. Studio B prints, books, tapes, crafts & country music related items for sale.
Activities: films; permanent & temporary exhibitions; oral history program; education department; outreach school programs & kits; rental facilities.
Publications: The Official Country Calendar; Bill Monroe & His Blue Grass Boys; Truth is Stranger than Publicity; Country Music Legends in the Hall of Fame; Sing Your Heart Out, Country Boy; Country: The Music and the Musicians; The Explosion of American Music: BMI's 50th Anniversary; Bob Wills: Hubbin' It.; Ramblin' Rose: The Life and Career of Rose Maddox; A Good Natured Riot: The Birth of the Grand Ole Opry by Charles Wolfe; True Adventures with the King of Bluegrass-Jimmy Martin by Tom Piazza. Various anthologies & television specials: Hank Williams: Just Me & My Guitar, The First Recordings, Rare Demos; Jim Reeves: Live at the Opry; Mark O'Connor: The Championship Years; Jerry & Tammy Sullivan: A Joyful Noise; Raise Your Window: A Cajun Music Anthology; Webb Pierce: King of the Honky-Tonk; Jean Shepard: Honky-Tonk Heroine; Johnny Paycheck: The Real Mr. Heartache, The Little Darlin' Years. Major Label Collaborations: Gene Autry: Columbia Historic Edition; Roy Rogers: Columbia Historic Edition; Elvis Presley: Elvis in Nashville, Known Only To Him; Jim Reeves: Welcome to My World; Patsy Cline: The Patsy Cline Collection, Live at the Opry, Live, Volume II; Heartaches by the Number; Singing in the Saddle: The History of the Singing Cowboy; Finding Her Voice: Women In Country Music.
Hours & Admission Prices: Jan.-Feb. Wed.-Mon. 9-5; March-Dec. daily 9-5. Adults $19.99, seniors 60 & over $17.99, children 6-12 $11.99; discounts to AAA, AAM, Military & Students; children 5 & under and members no charge. Closed New Year's Day; Thanksgiving; Christmas. &
Attendance: 421,151 (accurate)
Membership: National/International $30; Individual $40; Friends & Family $100.

DISCIPLES OF CHRIST HISTORICAL SOCIETY, 1101 19th Ave. S., Nashville, TN 37212-2196. Tel.: 615-327-1444; 866-834-7563 (Toll Free). Fax: 615-327-1445.
E-mail: mail@discipleshistory.org

Web Site: www.discipleshistory.org
Founded: 1941.
Congressional District: 5
Key Personnel: Pres., Glenn T. Carson; Vice Pres. & Chief Archivist, Sara Harwell; Assoc. Archivist, Elaine Philpott.
Personnel Profile: Full-Time Paid 2; Part-Time Paid 6; Part-Time Volunteers 3; Interns 1.
Governing Authority: society; bd. of trustees. Parent Institution: Christian Church (Disciples of Christ). Tax-exempt: 170(b)(1)(A).
Institution Type/Description: Religious Museum.
Collections: historical artifacts & art objects pertaining to the religious heritage, backgrounds, origins, development & general history of the Disciples of Christ, Christian Churches, Churches of Christ & related groups.
Research Fields: history of the Disciples of Christ, Christian Churches, Churches of Christ & related groups.
Facilities: 34,000-vol. library & archives of resources related to three church bodies, available for member use & inter-library loan.
Activities: guided tours; lectures; permanent & temporary exhibitions.
Publications: quarterly magazine, Discipliana; books.
Hours & Admission Prices: Tues.-Fri. 9-4. No charge. Closed national holidays. &

Attendance: 1,000 (estimated)
Membership: Individual $100; Friends of President $250; The Millennial Group $1,000; The Unity Circle $2,500.

THE FISK UNIVERSITY GALLERIES, Fisk University, 1000 17th Ave., N., Nashville, TN 37208-3051. Tel.: 615-329-8720.
E-mail: galleries@fisk.edu
Web Site: www.fisk.edu/fiskuniversitygalleries
Founded: 1949.
Congressional District: 5
Key Personnel: C.E.O., Dr. Hazel R. O'Leary; Dir., Victor Simmons; Chm. (V), Mr. Virgis W. Colbert; Chm. Art Committee, Aaronetta Pierce; Archivist, Beth Howse; Public Rels., Crystal Ghassemi; Administrative Asst., Tabitha Williams; Gallery Coord., Sara Estes.
Personnel Profile: Full-Time Paid 2; Part-Time Volunteers 20.
Governing Authority: university. Parent Institution: Fisk University. Branch Museums: Carl Van Vechten Gallery; Aaron Douglas Gallery. Tax-exempt: 501(c)(3).
Institution Type/Description: Art Gallery & Museum: art galleries housed in 1888 Neo-Romanesque building & university library building.
Collections: Afro-American paintings, sculpture & graphics; photography; the contemporary & classical African collection; the Cyrus Baldridge collection of drawings; the Alfred Stieglitz collection; the Winold Reiss collection; the Carl Van Vechten collection of photographs; the James collection of African-American folk art.
Research Fields: African and Afro-American art.
Facilities: 4,500-vol. library of materials pertaining to art available for use on premises; classroom. Art & museum-related items for sale.
Activities: self-guided tours; lectures; films; gallery talks; annual arts festival; inter-museum loan; traveling & temporary exhibitions; docent program.
Publications: catalogs; exhibition brochures; calendar of events; newsletter, Fisk Art Report.
Hours & Admission Prices: Summer: Mon.-Fri. 10-5; Sept.-May Tues.-Sat. 10-5. Adults $10, seniors $6, college students with ID $5; youth 18 & under no charge. &

Attendance: 42,000 (accurate)

FORT NASHBOROUGH, 170 1st Ave., N., Nashville, TN 37201-1924. Mailing Address: Centennial Park Office, Nashville, TN 37201. Tel.: 615-862-8400. Fax: 615-862-5493.
E-mail: jackie.jones@nashville.gov
Web Site: www.nashville.gov/parks
Founded: 1780.
Congressional District: 5
Key Personnel: Dir., Roy E. Wilson.
Governing Authority: city. Parent Institution: Metro Board of Parks & Recreation. Tax-exempt.
Institution Type/Description: Historic Building/Site Museum: 1780 site of Fort Nashborough, first settlement in Nashville.
Collections: five log cabin reproductions.
Research Fields: historical.
Activities: self-guided with directional signage.
Hours & Admission Prices: Daily 9-5. No charge. Closed legal holidays. &

✳ **FRIST CENTER FOR THE VISUAL ARTS, (M),** 919 Broadway, Nashville, TN 37203-3822. Tel.: 615-244-3340. Fax: 615-244-3339. Facebook: Frist Center.
E-mail: mail@fristcenter.org
Web Site: www.fristcenter.org
Founded: 2001.
Congressional District: 5
Key Personnel: C.E.O., Susan Edwards; Deputy Dir., Ashley Howell; Chm. & Pres., William R. Frist; Education, Anne Henderson; Public Rels., Ellen Pryor; Controller, Carol Vollbrecht; Registrar, Amie Geremia; Cur., Mark Scala; Security, Paul Cotter; Retail Mgr., Jon Emmitt.
Personnel Profile: Full-Time Paid 70; Part-Time Paid 10; Part-Time Volunteers 307; Interns 10.
Volunteer Hours: 22,663
Operating Expenses: 11,109,600
Operating Income: 11,157,655
Governing Authority: private; nonprofit organization. Tax-exempt: 501(c)(3).
Institution Type/Description: Art Museum.
Collections: changing exhibitions by local, regional, national & international artists.
Major Exhibits: American Chronicles: the Art of Norman Rockwell (T), 11/13-2/9/14; Lain York Selections from the National Gallery, 1/31/14-5/11/14; Looking East: Western Artists and the Allure of Japan (T), 1/31/14-5/11/14; Goya: The Disasters of War (T), 2/28/14-6/8/14; Steve Mumford's War Journals, 2/28/14-6/8/14; Marty Stuart: Photographs from the Road, 5/9/14-10/19/14; Maira Kalman: The Elements of Style, 6/6/14-9/1/14; Watch Me Move: The Animation Show (T), 6/6/14-9/1/14; Real/Surreal (T), 6/27/14-10/13/14; Helen Pashgian: Columns (T), 9/26/14-1/4/15; Kandinsky: A Retrospective (T), 9/26/14-1/4/15; Sanctity Pictured: The Art of the Dominican and Franciscan Orders in Renaissance Italy, 10/31/14-1/25/15.
Facilities: 406-vol. library; 225-seat auditorium; 120-seat restaurant; 20,000 sq. ft. exhibit space. Museum-related items for sale.
Activities: formal education programs; guided tours; lectures; traveling exhibitions; family day programs; films; music.
Publications: quarterly newsletter, Members Gallery.
Hours & Admission Prices: Mon.-Wed. & Sat. 10-5:30, Thurs.-Fri. 10-9, Sun. 1-5:30. Adults $10, senior citizens, military & college students $7; discounts to groups; members, children 18 & under & AAM & ICOM members no charge. Additional pricing for special exhibitions. Closed New Years Day; Thanksgiving; Christmas. &
Attendance: 171,696 (accurate)
Membership: Teacher & College Student $25; Senior $35; Individual & Dual Senior $45; Dual $60; Family $65; Friend $100; Patron $250; Benefactor $500; Director's Circle $1,000; President's Circle $2,500; Rembrandt's Circle $5,000; Picasso Circle $10,000.

HISTORIC TRAVELLERS REST PLANTATION & MUSEUM, 636 Farrell Pkwy., Nashville, TN 37220-1218. Tel.: 615-832-8197. Fax: 615-832-8169. Facebook: Travellers Rest.
E-mail: director@travellersrestplantation.org
Web Site: www.travellersrestplantation.org
Formerly: Travellers Rest Plantation and Museum
Founded: 1954.
Congressional District: 5
Key Personnel: Exec. Dir., C.E.O. & Pres., Mary Kerr; Dir. Education, Tonya Staggs; Cur., Brian Allison.
Personnel Profile: Full-Time Paid 6; Part-Time Paid 2; Part-Time Volunteers 10.
Governing Authority: society. Parent Institution: National Society of the Colonial Dames of America in TN. Tax-exempt.
Institution Type/Description: Historic House Museum: housed in 1799 home of Judge John Overton.
Collections: federal & early empire decorative arts; Civil War Nashville artifacts; Travelers Rest Arabian horse artifacts & archival materials; historic Nashville; Tennessee Society of Colonial Dame; historic houses.
Research Fields: Historic Nashville; Civil War; Arabian horses; early Tennessee decorative arts; African-American history.
Facilities: 800-vol. library that includes books, photographs, & archival materials related to historic Nashville; education barn available for special events & rentals; 11-acre grounds. Gift items for sale.
Activities: guided tours; permanent & temporary exhibitions. Special Events: 19th Century Trades Day (Sept.); Halloween Event (Oct.); Holiday Event (Dec.); Living History Series (all-year).
Publications: site book, Civil War DVD; Arabian Horse DVD; book, Judge John Overton.
Hours & Admission Prices: Mon.-Sat. 10-4:30, Sun. 1-4:30. Adults $10, seniors 65 & over $9, children 7-11 $5, children 6 & under, Blue Star Museum Program & active military w/ID no charge. Closed New Year's Day, Thanksgiving; Christmas. &

Attendance: 15,000 (estimated)

LANE MOTOR MUSEUM, 702 Murfreesboro Pike, Nashville, TN 37210-4522. Tel.: 615-742-7445.
E-mail: info@lanemotormuseum.org
Web Site: www.lanemotormuseum.org
Founded: 2002.
Key Personnel: Dir., Jeff Lane; Museum Shop Mgr., David Yando.
Personnel Profile: Full-Time Paid 8; Full-Time Volunteers 1; Part-Time Paid 2; Part-Time Volunteers 20.
Governing Authority: Tax-exempt.
Institution Type/Description: Transportation Museum.
Collections: 150 unique cars & motorcycles dating from the 1920s to the present.
Major Exhibits: Staff Favorites, 1/12-6/14.
Publications: quarterly, Breaking News.
Hours & Admission Prices: Thurs.-Mon. 10-5. Adults 18-64 $7, seniors 65 & over $5, youth 6-17 $2; members & children under 5 no charge. Closed New Year's Day; Thanksgiving; Christmas. &

Attendance: 22,000
Membership: Annual $50.

NASHVILLE ZOO AT GRASSMERE, 3777 Nolensville Pike, Nashville, TN 37211-3324. Tel.: 615-833-1534. Fax: 615-333-0728.
E-mail: pr@nashvillezoo.org
Web Site: www.nashvillezoo.org
Founded: 1990.
Congressional District: 5
Key Personnel: Pres., Rick Schwartz; Chm. (V), Jim Hunt; Admin., Beth Murdock.
Personnel Profile: Full-Time Paid 110; Full-Time Volunteers 125; Part-Time Paid 17.
Governing Authority: Parent Institution: Nashville Zoo. Tax-exempt.
Institution Type/Description: Zoological Park.
Collections: live animals. Historic Buildings: Croft home built in 1810.
Research Fields: animal care & behavior; conservation studies.
Facilities: 188-acre park; cafe. Museum-related items for sale.
Publications: Zoo View.
Hours & Admission Prices: March 15-Oct. 14 daily 9-6; Oct. 15-March 14 daily 9-4. Adults $15, seniors $13, children 2-12 $10; discounts to groups; members & children under 2 no charge. &
Attendance: 648,034 (accurate)
Membership: Individual $35; Grandparent & Family $85; Safari Set $175; Keepers Circle $300; Curator's Club $575; Claws, Paws & Jaws Society $1,000.

THE PARTHENON, (M), Centennial Park, 2600 West End Ave., Nashville, TN 37203. Mailing Address: P.O. Box 196340, Nashville, TN 37219-6340. Tel.: 615-862-8431. Fax: 615-880-2265. Facebook: The Parthenon.
E-mail: info@parthenon.org
Web Site: www.parthenon.org
Founded: 1897.
Congressional District: 5
Key Personnel: Dir., Wesley M. Paine; Pres. Conservancy for the Parthenon & Centennial Park, Sylvia Rapaport; Facilities Mgr., Lauren Buffer; Museum Shop Mgr., Timothy Cartmell.
Personnel Profile: Full-Time Paid 11; Part-Time Paid 3; Part-Time Volunteers 40; Interns 1.
Governing Authority: municipal. Parent Institution: Metropolitan Board of Parks & Recreation. Supported by The Conservancy for the Parthenon and Centennial Park. Tax-exempt: 501(c)(3).
Institution Type/Description: Art Museum: housed in exact reproduction of Athenian Parthenon.
Collections: 1850-1923, James M. Cowan collection of American paintings; pre-Columbian art from Mexico; contemporary art; 42-ft. replica of Athena Parthenos statue.
Major Exhibits: Kristen Llamas: The Socratic Dialogues, 10/13-2/14; James Cowan Rejoin, 2/14-8/14; Flex It! My Body My Temple, 9/14-12/14.
Facilities: Museum-related items for sale.
Activities: guided tours by reservation only; rotating art shows; permanent, temporary & traveling exhibitions.
Hours & Admission Prices: June-Aug. Tues.-Sat. 9-4:30; Sept.-May Tues.-Sat. 9-4:30; Sun. open year roound 12:30-4:30; groups by appointment. Adults $6, seniors & children 4-17 $4; discounts to AAM members; members & children under 4 no charge. Closed New Year's Day; Thanksgiving Day & Fri.; Christmas. &

Attendance: 150,000 (accurate)
Membership: Senior, Student & Individual $25; Family $50; Hero $125-$749; Champion $750-$1,499; Nike $1,500-$3,499; Olympian $3,500 & up.

THE PUBLIC LIBRARY OF NASHVILLE AND DAVIDSON COUNTY, Nashville Room, Special Collections Div., 615 Church St., Nashville, TN 37219-2314. Tel.: 615-862-5782. Fax: 615-862-5838.
E-mail: andrea.blackman@nashville.gov
Web Site: www.library.nashville.org
Founded: 1887.
Congressional District: 5
Key Personnel: Dir., Donna Nicely; Mgr. Nashville Rm., Andrea Blackman.
Personnel Profile: Full-Time Paid 12; Part-Time Volunteers 35; Interns 1.
Governing Authority: municipal. Parent Institution: Nashville Public Library. Tax-exempt: 170(b)(1)(A).
Institution Type/Description: Local History Library.
Collections: community culture & history; over 28,000 books; ephemera; photographs; maps; architectural drawings; manuscripts; oral histories for the Civil Rights Oral History Project & the Veterans History Project; The Nashville Banner newspaper archives.
Research Fields: history of Nashville & Davidson County; collection of Nashville authors; genealogy; image collection; Banner newspaper collection; civil rights.
Facilities: library pertaining to Nashville & Tennessee history.
Activities: guided tours; permanent & temporary exhibitions.
Hours & Admission Prices: Tues.-Fri. 9-6, Sat. 9-5, Sun. 2-5. No charge. Closed Mondays & national holidays. &
Attendance: 48,162 (accurate)

RYMAN AUDITORIUM, 116 Fifth Ave. N., Nashville, TN 37219. Tel.: 615-458-8700. Fax: 615-458-8701.
E-mail: glevy@ryman.com
Web Site: www.ryman.com
Founded: 1892.
Institution Type/Description: History Museum: housed in the home of the Grand Ole Opry from 1943-1974. A National Historic Landmark.
Collections: Opry history; personal artifacts; period furnishings; photographs.
Facilities: auditorium.
Activities: special events; concerts; performances.
Hours & Admission Prices: Tours: daily 9-4. Adults $13, children 4-11 $6.50. Closed New Year's Day; Thanksgiving; Christmas. &
Attendance: 150,000 (accurate)

SARRATT GALLERY AT VANDERBILT, 207 Sarratt Student Center, Nashville, TN 37240. Tel.: 615-322-2471. Fax: 615-343-8081.
E-mail: bridgette.kohnhorst@vanderbilt.edu
Web Site: www.vanderbilt.edu/sarrattgallery
Key Personnel: Dir., Bridgette Kohnhorst
Institution Type/Description: Art Gallery.
Collections: paintings; sculpture; prints.
Hours & Admission Prices: Academic Year: Mon.-Fri. 9-9, Sat.-Sun. 10-10; Summer: Mon.-Fri. 9-4:30. No charge.

SOUTHERN BAPTIST HISTORICAL LIBRARY AND ARCHIVES, 901 Commerce St., Ste. 400, Nashville, TN 37203-3628. Tel.: 615-244-0344. Fax: 615-782-4821.
E-mail: bill@sbhla.org
Web Site: www.sbhla.org
Founded: 1951.
Congressional District: 5
Key Personnel: Dir. Library & Archives, Bill Sumners; Accountant, Debbie Keen; Librarian, Joy DuBose; Archivist, Taffey Hall; Asst. Librarian, Jean Forbis.
Personnel Profile: Full-Time Paid 5; Full-Time Volunteers 1; Part-Time Paid 3.
Governing Authority: denominational group. Affiliated with Southern Baptist Convention, 901 Commerce St., Nashville, TN 37203 Tax-exempt.
Institution Type/Description: Religious Museum.
Collections: artifacts & archival materials related to Baptist history; manuscripts.
Research Fields: Baptist history.
Facilities: 32,000-vol. library of books, periodicals, denominational annuals, audio-visuals; microfilm collection of Baptist materials from Eastern Europe, Soviet Union, Baptist Missionary Society Archives, London, England, 1792-1914, available for use on premises; reading room.
Activities: guided tours; temporary exhibitions.
Hours & Admission Prices: Mon.-Fri. 9-4; other times by appointment. No charge. Closed national holidays. &

Attendance: 1,100 (estimated)

TENNESSEE AGRICULTURAL MUSEUM, Ellington Agricultural Center, Nashville, TN 37204. Tel.: 615-837-5197.
E-mail: tennessee.agricultural.museum@tn.gov
Web Site: tnagmuseum.org
Personnel Profile: Full-Time Paid 2; Part-Time Volunteers 110.
Governing Authority: Parent Institution: State of Tennessee, Tennessee Dept. of Agriculture.
Institution Type/Description: Agriculture Museum: housed in a former horse barn which was once part of the Brentwood Hall estate of financier Rogers Caldwell.
Collections: home & farm artifacts; prints; textiles; woodworking artifacts; wagons; garden.
Facilities: garden; nature trail.
Activities: summer Saturdays; Pioneer Journey traveling trunk; adult classes; demonstrations; educational programs. Annual Events: Historic Rural Life Festival in May; Music & Molasses Arts & Crafts Festival in October.
Publications: children's activity book, Through the Garden Gate.
Hours & Admission Prices: Mon.-Fri. 9-4; groups by appointment. Museum: no charge; donations accepted. Special Events: call for admission prices. Closed state holidays. &
Attendance: 22,203 (accurate)
Membership: Individual $15; Family $30; Life $200.

TENNESSEE CENTRAL RAILWAY MUSEUM, 220 Willow St., Nashville, TN 37210-2159. Tel.: 615-244-9001. Fax: 615-244-2120.
E-mail: hultman@bellsouth.net
Web Site: www.tcry.org
Founded: 1990.
Congressional District: 5
Key Personnel: C.E.O., Chm. (V) & Pres. (V), Terry L. Bebout; Treas., J. Allen Hicks; Cur. & Archivist, Don Strong; Devel., George Gilbert; Public Rels., Robert E. Hultman; Museum Shop Mgr., Charles Owens.
Personnel Profile: Full-Time Paid 1; Full-Time Volunteers 60; Part-Time Paid 1; Part-Time Volunteers 50.
Governing Authority: private; nonprofit organization. Parent Institution: Cumberland Division SER-NMRA, Inc. Tax-exempt: 501(c)(3).
Institution Type/Description: Railroad Museum.
Collections: rolling stock, passenger rolling stock; historic pieces relating to the TC RY, Nashville, Chattanooga & St. Louis Ry, and L&N RR; RR paper & hardware concentration in southeastern US railways.
Research Fields: various freight cars bought & operated by L&N RR & the NC&StL RY; TC Ry.
Facilities: 500-vol. library of books on railroad history; 400 DVDs; 700 sq. ft. exhibit space. Museum-related items for sale.
Activities: formal education programs for children; guided tours; hobby workshops; lectures; loan exhibitions; excursion trains. Museum Sponsors: Model Train Shows in the Middle Tennessee; Day Out with Thomas the Tank Engine.
Publications: newsletter published 12 times annually, The Order Board.
Hours & Admission Prices: Mon.-Fri. 8-4, Sat. 9-3. Museum: no charge; donations accepted. Model Train Shows: adults $4; discounts to active military & NRHS members.
Attendance: 50,000 (estimated)
Membership: Regular $30; Household & Family $35.

TENNESSEE HISTORICAL COMMISSION, 2941 Lebanon Rd., Nashville, TN 37243-0442. Tel.: 615-532-1550. Fax: 615-532-1549.
E-mail: patrick.mcIntyre@tn.gov
Web Site: www.state.tn.us/environment/hist
Founded: 1919.
Congressional District: 5
Key Personnel: Exec. Dir., Patrick McIntyre; Chm. (V), Norman J. Hill; Asst. Dir. State Programs, Linda T. Wynn; Fiscal Officer, Doyal Vaughan; Deputy State Historic Preservation Officer, Richard G. Tune; Historic Preservation Specialist, James Jones; Historic Preservation Supvr., Steve Rogers; Historic Preservation Specialist, Joe Garrison; Historic Preservation Specialist, Claudette Stager; Historic Preservation Specialist, Louis Jackson; Military Sites Preservation Specialist, Fred Prouty; Historic Preservation Specialist, Brian Beadles; Sec., Angela Staggs.
Personnel Profile: Full-Time Paid 15.
Governing Authority: state. Tax-exempt.
Institution Type/Description: State Historic Preservation Agency.
Collections: period furnishings; personal artifacts; photographs.
Facilities: 200-300-vol. library of history, architecture, archaeology, historic preservation & folklore available for research.

Activities: lectures; formally organized education programs for undergraduate & graduate college students affiliated with historic preservation internship; operate all state-owned historic sites under an agreement with local clubs & organizations; administer programs of The National Historic Preservation Act; operates historical markers program.
Publications: periodical, The Courier; catalog, Tennessee Historical Markers; magazine, Guide to Civil War in Tennessee; Houston & Crocket: Heroes of Tennessee & Texas; Biographical Directory of the Tennessee General Assembly Vols. 1-6; Messages of the Governors of Tennessee Vols. 1-10.
Hours & Admission Prices: Mon.-Fri. 8-4:30 by appointment. No charge. Closed holidays. &

TENNESSEE SPORTS HALL OF FAME MUSEUM, 501 Broadway, Nashville, TN 37203. Tel.: 615-242-4750. Fax: 615-242-4752. Facebook: Tennessee Sports Hall of Fame.
E-mail: tnsports@bellsouth.net
Web Site: www.tshf.net
Founded: 2000.
Congressional District: 5
Key Personnel: Exec. Dir., Dr. Bill Emendorfer.
Governing Authority: Tax-exempt.
Institution Type/Description: Sports Museum.
Collections: sports memorabilia including Olympics, college football & basketball; photographs; personal artifacts; Hall of Fame inductees.
Facilities: theater.
Activities: rental facilities; interactive & video games.
Publications: quarterly newsletter.
Hours & Admission Prices: Tues.-Sat. 10-5. Adults $3, seniors over 55 & children $2; discounts to groups of 10 or more. &
Membership: Individual $25; Corporation $150.

TENNESSEE STATE CAPITOL, 600 Charlotte Ave., Nashville, TN 37243-9034. Tel.: 615-741-2692 & 1621. Fax: 615-741-7231.
E-mail: jim.hoobler@tn.gov
Congressional District: 5
Key Personnel: Cur., James A. Hoobler.
Personnel Profile: Full-Time Paid 1.
Governing Authority: Parent Institution: Tennessee Capitol Commission.
Institution Type/Description: Historic Building: built in 1859. A National Historic Landmark.
Collections: state government; period murals & frescos; portraits; House and Senate chambers; Governor's Office; period artifacts.
Activities: guided tours.
Hours & Admission Prices: Mon.-Fri. 9-4. Guided Tours: Mon.-Fri. 9-11 & 1-3; groups of 10 or more by appointment. No charge.
Attendance: 30,000 (estimated)

TENNESSEE STATE LIBRARY AND ARCHIVES, 403 7th Ave. N., Nashville, TN 37243. Tel.: 615-741-2764. Facebook: Tennessee State Library and Archives.
Web Site: www.tn.gov/tsla
Key Personnel: State Librarian & Archivist, Charles A. Sherrill
Institution Type/Description: Library & Archives.
Collections: books; manuscripts; documents; photographs.
Facilities: library; archives.
Activities: research; educational outreach; workshops; internships. Annual Events: Book Festival in October; Tennessee History Day.
Hours & Admission Prices: Library Reading Room: Tues.-Sat. 8-4:30. Legislative History: Tues.-Fri. 8-4:30. Closed New Year's Day; Martin Luther King Jr. Day; Presidents' Day; Good Friday; Memorial Day; Independence Day; Labor Day; Columbus Day; Veterans Day; Thanksgiving; Christmas. &

* **TENNESSEE STATE MUSEUM, (M),** 505 Deaderick St., Nashville, TN 37243-1402. Tel.: 615-741-2692. Fax: 615-741-7231. Facebook: Tennessee State Museum.
E-mail: museuminfo@tnmuseum.org
Web Site: tnmuseum.org
Founded: 1937.
Congressional District: 5
Key Personnel: Exec. Dir., Lois S. Riggins-Ezzell; Dir. Collections, Dan E. Pomeroy; Dir. Administration, Mary Jane Crockett-Green; Dir. External Affairs, Leigh Hendry; Dir. Public Programs, Paulette Fox; Dir. Capitol Projects, Patricia Rasbury; Registrar, Bob White; Media Contact, Mary Skinner; Museum Shop Mgr., Sunshine Thompson.
Personnel Profile: Full-Time Paid 50; Part-Time Volunteers 20.
Governing Authority: state. Parent Institution: Douglas Henry State Museum

Commission. Branch Museum: Tennessee State Museum Military Branch, 7th & Union, Nashville 37243-1120. Tax-exempt: 170(b)(1)(A).
Institution Type/Description: History Museum.
Collections: military memorabilia from 1780-present; historic paintings of prominent Tennesseans; fine arts collections of contemporary artists; historic materials reflecting the history and culture of the state; objects related to famous personalities; early 19th-century Tennessee-made silver, firearms, furniture, quilts.
Major Exhibits: Slave and Slaveholders of Wessyngton Plantation (T), 2/14-8/14; A Creative Legacy: African American Art in Tennessee, 2/14-8/14.
Research Fields: History Art Museum.
Facilities: 5,000-vol. library of book on general history available for inter-library loan on request. Museum-related items for sale.
Activities: lectures; gallery talks; docent program or council; loan, temporary & traveling exhibitions; school loan service.
Publications: quarterly newsletter.
Hours & Admission Prices: Tues.-Sat. 10-5, Sun. 1-5. No charge; donations accepted. Closed New Year's Day; Easter; Thanksgiving; Christmas. ♿
Attendance: 180,000 (estimated)
Membership: Teacher $25; Individual $30; Dual $40; Family $50; Contributing $100; Sustaining $500; John Sevier Member $1,000.

UPPER ROOM CHAPEL MUSEUM, 1908 Grand Ave., Nashville, TN 37212-2188. Mailing Address: P.O. Box 340004, Nashville, TN 37203-0004. Tel.: 615-340-7206. Fax: 615-340-7293. Facebook: Upper Room Chapel Museum.
E-mail: kkimball@upperroom.org
Web Site: www.upperroom.org
Founded: 1953.
Congressional District: 5
Key Personnel: Editor & Publisher, Sarah Wilke; Dir., Cur. & Museum Shop Mgr., Kathryn A. Kimball.
Personnel Profile: Full-Time Paid 4.
Governing Authority: Parent Institution: United Methodist Church. Tax-exempt.
Institution Type/Description: Christian Art Museum.
Collections: art objects having religious significance.
Research Fields: church history & music to selected groups; John Wesley manuscripts & books.
Facilities: 13,000-vol. library of books & manuscripts on family devotion, worship, meditation, prayers & Methodist history available for inter-library loan; 200-seat chapel; reading room. Museum-related items for sale.
Activities: guided tours; permanent & temporary exhibitions.
Publications: daily devotional guide, The Upper Room; Weavings; Alive Now; Pockets; DevoZine.
Hours & Admission Prices: Mon.-Fri. 8-4:30. No charge; donations accepted. Closed major holidays. ♿
Attendance: 8,500 (estimated)

VANDERBILT UNIVERSITY FINE ARTS GALLERY, 1220 21st Ave. S., Nashville, TN 37203. Mailing Address: PMB 273, 230 Appleton Place, Nashville, TN 37203-5721. Tel.: 615-322-0605 & 343-1704. Fax: 615-343-1382.
Web Site: www.vanderbilt.edu/gallery
Founded: 1961.
Congressional District: 5
Key Personnel: Dir., Joseph S. Mella.
Personnel Profile: Full-Time Paid 2; Part-Time Paid 5; Part-Time Volunteers 2; Interns 6.
Governing Authority: university. Affiliated with Vanderbilt University. Tax-exempt.
Institution Type/Description: Art Gallery: housed in renovated 1928 building designed by McKim, Mead & White.
Collections: Vanderbilt art collection of paintings, sculpture & graphics; Harold P. Stern collection of Asian art; Samuel H. Kress study collection of Renaissance paintings; Anna C. Hoyt collection of old masters & modern prints; ceramics; contemporary works on paper & multiples.
Activities: lectures; inter-museum loan; temporary & traveling exhibitions.
Publications: exhibition catalogues.
Hours & Admission Prices: School Year: Mon.-Fri. 11-4, Sat.-Sun. 1-5; Summer: Tues.-Fri. 12-4, Sat. 1-5. No charge. Closed school holidays. ♿
Attendance: 5,000 (estimated)

WARNER PARK NATURE CENTER, 7311 Hwy. 100, Nashville, TN 37221. Tel.: 615-352-6299. Fax: 615-880-2282.
Institution Type/Description: Nature Center.
Collections: environment; natural history; gardens; geology.

Activities: educational programs; workshops; outdoor recreation programs; scout programs.
Hours & Admission Prices: Park: sunrise to sunset. Center: Tues.-Thurs. & Sat. 8:30-4:30, Fri. 8:30-6, Sun. 12:30-4:30. Closed major holidays.

WATKINS INSTITUTE - BROWNLEE O. CURRY JR. GALLERY, 2298 Rosa L. Parks Blvd., Nashville, TN 37228-1306. Tel.: 615-383-4848. Fax: 615-383-4849.
Web Site: watkins.edu
Founded: 1885.
Congressional District: 5
Key Personnel: Pres., Ellen Meyer; Dir. Art School, Terry Thacker.
Governing Authority: bd. of trustees; nonprofit organization. Tax-exempt: 501(c)(3).
Institution Type/Description: Art Gallery with Art School.
Collections: Tennessee All-State Art Collection of paintings, graphics & sculpture; 50 etchings by Nahum Tschacbasov; paintings from Childe Hassam Fund; seven Elihu Vedder paintings; nine period hand colored prints of classical interiors & artifacts.
Facilities: studio classrooms for commercial & fine arts.
Activities: lectures; formally organized education programs for children & adults; inter-museum loan, permanent & traveling exhibitions. Museum Sponsors: Annual Arts Festival; Student Art Show; Tennessee All-State purchase award show; Faculty Art Show.
Publications: school catalogs & brochures; annual all-state exhibition catalog.
Hours & Admission Prices: Mon.-Fri. 9-8, Sat. 10-4, Sun. 2-4. No charge. Closed New Year's Day; Independence Day; Labor Day; Thanksgiving; Christmas. ♿

WILLIE NELSON & FRIENDS GENERAL STORE & MUSEUM, 2613 McGavock Pike, Nashville, TN 37214-1215. Tel.: 615-885-1515. Fax: 888-517-6579. Facebook: Willie Nelson General Store.
E-mail: kay@willienelsongeneralstore.com
Web Site: www.willienelsonmuseum.com
Founded: 1979.
Key Personnel: C.E.O., Mark Hughes.
Personnel Profile: Full-Time Paid 7.
Institution Type/Description: Country Music Stars Museum.
Collections: Willie Nelson & other country music stars' memorabilia; personal artifacts; photographs.
Research Fields: country music; Willie Nelson; music; Nashville history.
Facilities: Museum-related items for sale; Nashville souvenirs.
Activities: special events; CMA fest 2nd weekend June.
Hours & Admission Prices: Sun.-Sat. 8:30-9. Adults $8, 10% discount to military. ♿
Attendance: 100,000 (estimated)

Newbern

NEWBERN DEPOT & RAILROAD MUSEUM, 108 Jefferson St., Newbern, TN 38059. Tel.: 731-627-3221.
Institution Type/Description: Railroad Museum.
Collections: local history; tools; uniforms; schedules; photographs; model railroads; art work.
Hours & Admission Prices: Call for hours.

Newport

NEWPORT - COCKE COUNTY MUSEUM, 433 Prospect Ave., Newport, TN 37821. Tel.: 423-623-7201.
Institution Type/Description: History Museum.
Collections: local history & heritage; period furnishings; personal artifacts; tools; Gov. Ben W. Hooper; early sewing machine; household artifacts; quilts.
Hours & Admission Prices: Wed. 1-5. No charge.

Niota

NIOTA RAILROAD DEPOT, 201 E. Main St., Niota, TN 37826. Tel.: 423-568-2584.
Institution Type/Description: Historic Building: housed in a former depot now serving as Niota City Hall; built in 1854. Listed on the National Register of Historic Places.
Collections: local history & culture; period furnishings; photographs.
Hours & Admission Prices: By appointment.

Norris

NORRIS MUSEUM, Town Center, One Norris Sq., Norris, TN 37828. Mailing Address: P.O. Box 1110, Norris, TN 37828-1110. Tel.: 865-494-6800.
Founded: 1999.
Key Personnel: Chm. (V), Ellalyn Crossno; Pres. (V), Jerry L. Crossno.
Personnel Profile: Part-Time Volunteers 8.
Governing Authority: Parent Institution: City of Norris. Tax-exempt.
Institution Type/Description: History Museum.
Collections: local history & culture; personal artifacts; photographs.
Hours & Admission Prices: March-Dec. Tues. & Sun. 2-4; other times by appointment. No charge; donations accepted.

WILL G. AND HELEN H. LENOIR MUSEUM, 2121 Norris Frwy. (Hwy. 441), Norris, TN 37828. Mailing Address: P.O. Box 1090, Norris, TN 37828. Tel.: 865-494-9688.
Institution Type/Description: History Museum.
Collections: local history; period furnishings; personal artifacts; photographs; early barrel organ. Historic Buildings: 18th century grist mill; threshing barn.
Hours & Admission Prices: Daily. No charge.

Oak Ridge

AMERICAN MUSEUM OF SCIENCE & ENERGY, (M), 300 S. Tulane Ave., Oak Ridge, TN 37830-6700. Tel.: 865-576-3200. Fax: 865-576-6024.
E-mail: jcomish@amse.org
Web Site: www.amse.org
Founded: 1949.
Congressional District: 3
Key Personnel: Exec. Dir., James R. Comish; Deputy Dir., Kenneth Mayes; Information Officer, Lissa Clarke; Chm. (V) & Volunteer Coord., Glenda Bingham; Exhibits Design, Jerry King; Facilities Mgr., Rex Haun; Museum Coord., Ann Armstrong; Museum Discovery Shop Mgr., Caroline Baker.
Personnel Profile: Full-Time Paid 12; Part-Time Paid 10; Part-Time Volunteers 36; Interns 1.
Governing Authority: federal. Operated by Enterprise Advisory Services, Inc. Parent Institution: Oak Ridge National Laboratory-operated by U.T. Battelle. Subsidiary Institution: U.T. Battelle. Tax-exempt: 501(c)(3).
Institution Type/Description: Science & Technology Museum.
Collections: World of the Atom; Y-12 & National Defense; Earth's Energy Resources; The Oak Ridge Story; AMSE Lab-Hands-On Science; history of the Manhattan Project.
Facilities: 312-seat auditorium; two 50-seat demonstration areas; laboratory classroom; picnic area. Educational & gift items for sale.
Activities: self-guided tours; lectures; demonstrations; films; traveling & permanent exhibitions; tours of Oak Ridge Manhattan Project sites; nature walks; internet experiences; outreach to schools.
Hours & Admission Prices: Mon.-Sat. 9-5, Sun. 1-5. Adults 18-64 $5, seniors 65 & over $4, children 6-17 $3; ASTC Passport members, members & children under 6 no charge. Closed New Year's Day; Thanksgiving; Christmas. &
Attendance: 90,000 (accurate)
Membership: Individual $25; Grandparent $30; Family $35; Family & Friends $75; The Sunday Punch Club $150; The U-235 Club $235.

CHILDREN'S MUSEUM OF OAK RIDGE, INC., 461 W. Outer Dr., Oak Ridge, TN 37830-3700. Tel.: 865-482-1074. Fax: 865-481-4889.
E-mail: chmor@bellsouth.net
Web Site: www.childrensmuseumofoakridge.org
Founded: 1973.
Congressional District: 3
Key Personnel: Exec. Dir., Mary Ann Damos; Pres. Bd. Trustees, Larry Burkholder; Designer, Peg Heddleson; Business Mgr., Dawn Van Eek; Dir. Education, Joyce Gralak; Deputy Dir., Carroll Welch; Collections Mgr., Kay Palmateer.
Personnel Profile: Full-Time Paid 5; Full-Time Volunteers 2; Part-Time Paid 4; Part-Time Volunteers 40.
Governing Authority: nonprofit organization. Tax-exempt: 501(c)(3).
Institution Type/Description: Children's Museum.
Collections: Appalachian primitives; Japanese Kokeski Dolls; foreign dolls; Liberian artifacts; Appalachian folk art; coal mining artifacts; Cherokee Indian artifacts; costumes; natural history; early Oak Ridge artifacts; doll house; life-size bird room; Brazilian Rainforest; model trains; history of Oak Ridge & the Manhattan Project.
Research Fields: Appalachian life.

Facilities: 1,000-vol. library of books; 500 tapes, video cassettes & video center, slides & photographs on Appalachian history; reading room; 1,000-seat auditorium; theater; classrooms. Local handcrafts, exhibit-related items for sale.
Activities: guided tours; lectures; films; gallery talks; concerts; arts festivals; drama; hobby workshops; TV & radio programs; formally organized education programs for children, adults & undergraduate college students; docent program; loan, permanent, temporary & traveling exhibitions.
Publications: books, Encyclopedia of East Tennessee; These Are Our Voices-The Story of Oak Ridge 1942-1970; quarterly newsletter, Children's Museum Newsletter; teachers manual, When Grandma Was a Girl; Ridges & Valleys, Vols. I & II; Anderson County: A Pictorial History; Oak Ridge & Me - From Youth to Maturity.
Hours & Admission Prices: June-Aug. Mon.-Fri. 9-5, Sat. 10-4, Sun. 1-4; Sept.-May Tues.-Fri. 9-5, Sat. 10-4, Sun. 1-4. Adults $7, senior citizens $6, children 3 & over $5; discounts to groups; children under 3 & members no charge. Closed Independence Day; Thanksgiving; Christmas.
Attendance: 150,000 (estimated)

OAK RIDGE ART CENTER, 201 Badger Ave., Oak Ridge, TN 37830-6216. Mailing Address: P.O. Box 7005, Oak Ridge, TN 37831-3305. Tel.: 865-482-1441. Fax: 865-482-1441 (call first).
E-mail: oakridgeartcenter@comcast.net
Web Site: oakridgeartcenter.org
Founded: 1952.
Congressional District: 3
Key Personnel: Dir., Leah Marcum-Estes.
Personnel Profile: Full-Time Paid 1; Part-Time Paid 3.
Governing Authority: nonprofit organization. Tax-exempt: 501(c)(3).
Institution Type/Description: Art Center.
Collections: Gomez Collection of post-World War II art; primitive to contemporary art.
Facilities: library; painting & drawing studios; ceramics studio.
Activities: guided tours; lectures; films; concerts; arts festivals; broadcast programs; organized education programs for children, adults & undergraduate college students affiliated with Roane State Community College; temporary & loan exhibitions; docent program. Center Sponsors: Hot Pots & Cool Art in summer; Spring Tea in April.
Publications: monthly newsletter; monthly arts calendar.
Hours & Admission Prices: Tues.-Fri. 9-5, Sat.-Mon. 1-4. No charge; donations accepted. Closed major holidays. &
Attendance: 50,000 (estimated)
Membership: Retired Individual $25; Individual & Retired Family $35; Family $45; Friend $50; Patron $100; Sponsor $250; Benefactor $350; Corporate $500; Life $1,000.

UNIVERSITY OF TENNESSEE ARBORETUM, 901 S. Illinois Ave., Oak Ridge, TN 37830-8032. Tel.: 865-483-3571. Fax: 865-483-3572.
E-mail: utforest@utk.edu
Web Site: forestry.tennessee.edu/arboretum
Founded: 1964.
Congressional District: 3
Key Personnel: Dir., Kevin P. Hoyt; Sec., Lynne Lucas.
Personnel Profile: Part-Time Volunteers 5; Interns 4.
Governing Authority: university. Parent Institution: University of Tennessee. Tax-exempt.
Institution Type/Description: Arboretum.
Collections: forestry; horticulture.
Research Fields: forest tree breeding & adaptation; shade tree & visual screening research; urban forestry; tree physiology.
Facilities: visitor center; greenhouse.
Activities: self-guided tours.
Publications: biannual, University of Tennessee Arboretum Society Bulletin; trail guides; nature pamphlets.
Hours & Admission Prices: Daily 8am-Sunset. Office: Mon.-Fri. 8-4:30. No charge; donations accepted. Closed national holidays.
Attendance: 35,000 (estimated)
Membership: Student $15; Individual $30; Family $45; Society Friend $100; Patron $500; Corporate $1,000.

Paris

PARIS-HENRY COUNTY HERITAGE CENTER, 614 N. Poplar St., Paris, TN 38242-3440. Mailing Address: P.O. Box 822, Paris, TN 38242-0822. Tel.: 731-642-1030.
E-mail: director@phchc.com
Web Site: www.phchc.com
Founded: 1989.

Congressional District: 8
Key Personnel: Dir., Norma B. Steele; Chm. (V), Gerry Scholes; Treas., Vicki Muzzall.
Personnel Profile: Full-Time Paid 1; Part-Time Volunteers 30.
Governing Authority: private; nonprofit organization. Tax-exempt: 501(c)(3).
Institution Type/Description: History Museum.
Collections: history of Henry County & the city of Paris; Camp Tyson, WWII.
Research Fields: Civil War; E.W. Grove; Gilded Age, Paris, TN.
Facilities: library; 832 sq. ft. exhibit space. Museum-related items for sale.
Activities: docent program; guided tours; lectures; temporary exhibitions. Annual Events: Fish Fry in April; Grapes & Gourmet in June; Mardi Gras Ball.
Publications: biannual newsletter, Inkwell.
Hours & Admission Prices: Tues.-Fri. 10-4, Sat. 10-2. No charge; donations accepted. Closed New Year's Eve & Day; Thanksgiving weekend; Christmas Eve, Day & week. &
Attendance: 2,500 (estimated)
Membership: Cavitt Place Friend $25; Legacy $50; Century $100; Historian $250; Sponsor $500; Heritage $1,000.

Parsons

PARSONS AND GREATER AREA HISTORICAL MUSEUM, 535 Tennessee Ave., Municipal Bldg., Parsons, TN 38363. Mailing Address: P.O. Box 128, Parsons, TN 38363-0128. Tel.: 731-847-6358. Fax: 731-847-9272.
E-mail: coordinator@cityofparsons.com
Web Site: www.cityofparsons.com/museum.htm
Founded: 2007.
Congressional District: 7
Key Personnel: Chm. (V), Branson Townsend; Treas., Judy Daugherty.
Personnel Profile: Full-Time Paid 1; Part-Time Paid 1; Part-Time Volunteers 12.
Governing Authority: Parent Institution: City of Parson, TN. Tax-exempt.
Institution Type/Description: History Museum.
Collections: area history & culture; Native American; railroad artifacts; minerals & fossils; farming; schools; churches.
Facilities: library.
Activities: school programs.
Hours & Admission Prices: Mon.-Tues. & Thurs.-Fri. 9-5, Wed. 9-2. Admission: $2; discounts to volunteers. &
Attendance: 1,000 (estimated)

Pigeon Forge

ELVIS PRESLEY MUSEUM, 2638 Parkway, Pigeon Forge, TN 37863-3246. Tel.: 865-428-2001.
Web Site: www.elvispresleymuseum.com
Key Personnel: Contact, Grace Savern
Institution Type/Description: History Museum.
Collections: Elvis' life, family history, & memorabilia; personal artifacts; 1973 Lincoln Continental limousine; 1967 honeymoon Cadillac; jewelry.
Facilities: theatre. Museum-related items for sale.
Activities: live performances by tribute performers; videos.
Hours & Admission Prices: Call for hours. Adults $17, seniors 65 & over and students $15, children 6-11 $12. Closed Thanksgiving; Christmas.

PARROT MOUNTAIN & TROPICAL BIRD SANCTUARY, 1471 McCarter Hollow Rd., Pigeon Forge, TN 37862. Tel.: 417-823-0981; 800-987-9852. Fax: 865-428-7798.
Web Site: www.parrotmountainandgardens.com
Founded: 2002.
Congressional District: 4
Institution Type/Description: Bird Sanctuary.
Collections: birds including toucans, parrots, macaws, cockatoos; trees; flowers; plants.
Facilities: Gift items for sale.
Activities: bird feeding & petting area.
Hours & Admission Prices: March-Nov. Mon.-Sat. 10-5. Adults 12 & over $14.95, seniors 65 & over $13.95, children 2-11 $7.95; discounts to groups of 10 or more. &

SMOKY MOUNTAIN CAR MUSEUM, 2970 Parkway, Pigeon Forge, TN 37863-3314. Mailing Address: P.O. Box 385, Pigeon Forge, TN 37868-0385. Tel.: 865-453-3433.
Founded: 1956.
Congressional District: 1
Institution Type/Description: Car Museum.
Collections: famous, classic, early cars; dolls; books; posters.

Hours & Admission Prices: April-June daily 10-6; July to Labor Day daily 10-7. Adults $7.50, children 3-10 $2.

SOUTHERN GOSPEL MUSIC HALL OF FAME & MUSEUM, Dollywood, 2700 Dollywood Park Blvd., Pigeon Forge, TN 37863. Mailing Address: Southern Gospel Music Assoc., P.O. Box 6729, Sevierville, TN 37864. Tel.: 865-908-4040.
E-mail: kim@sgma.org
Web Site: www.sgma.org/default.htm
Key Personnel: Exec. Dir., Charlie Waller
Institution Type/Description: Music Museum.
Collections: Gospel music's artists & legends; photographs; recordings; Hall of Fame inductees.
Activities: concerts; special events. Annual Event: Induction Ceremony.
Hours & Admission Prices: See Dollywood website for hours.

Piney Flats

* **ROCKY MOUNT MUSEUM, (M),** 200 Hyder Hill Rd., Piney Flats, TN 37686-4630. Mailing Address: P.O. Box 160, Piney Flats, TN 37686-0160. Tel.: 423-538-7396. Fax: 423-538-1086.
E-mail: info@rockymountmuseum.com
Web Site: www.rockymountmuseum.com
Founded: 1958.
Congressional District: 1
Key Personnel: C.E.O. & Exec. Dir., Gary Walrath; Pres. (V), James Hager; Exec. Asst., Delores Miller; Museum Shop Mgr., Evelyn Hunt.
Personnel Profile: Full-Time Paid 3; Part-Time Paid 14; Part-Time Volunteers 60.
Governing Authority: state; nonprofit organization. Affiliated with State of Tennessee & Rocky Mount Historical Association. Tax-exempt: 501(c)(3).
Institution Type/Description: Historic Site: 1790-92 original U.S. Territorial Capitol of Southwest Territory.
Collections: period furniture & furnishings of log house including kitchen & servants' quarters; barn; period artifacts; textile collection; manuscripts; farm area, including garden & crops of 1791 period & animals. Historic Buildings: c.1770-1772 log house; separate kitchen; weaving cabin; smokehouse; blacksmith's shop; slave cabin.
Research Fields: early history of Southwest Territory; early agricultural history.
Facilities: historical library available for use on premises; conference center; rental facilities.
Activities: guided tours; first person living history interpretation; changing & permanent exhibitions; public & educational programming; adult education series; rental facilities.
Publications: occasional publications; bimonthly newsletter.
Hours & Admission Prices: Living History Tours: March to mid-Dec. Tues.-Sat. 11-5; other times by appointment. Adults $8, seniors $7, children 6-17 $5; discounts to groups, AAM, AAA & AARP members; members no charge. Office: Mon.-Fri. 9-5. Closed Thanksgiving. &
Attendance: 22,400 (accurate)
Membership: Individual $35; Family $45; Contributing $75; Small Business & Patron $100; Corporate $200; Supporting Donor $250; Corporate Sponsor, Fellow Donor & Foundation Member $500; Corporate Patron, Sustaining Donor & Foundation Sponsor $1,000; Corporate Partner $2,500; Foundation Partner $5,000.

Pinson

PINSON MOUNDS STATE ARCHAEOLOGICAL AREA, 460 Ozier Rd., Pinson, TN 38366-9626. Tel.: 731-988-5614. Facebook: Pinson Mounds State Archaeological Park.
E-mail: wesley.williams@tn.gov
Web Site: tnstateparks.com/parks/about/pinson-mounds
Founded: 1980.
Congressional District: 7
Key Personnel: Park Mgr., Tim Poole; Park Ranger, Wes Williams.
Governing Authority: state. Managed by the Tennessee Dept. of Environment and Conservation, Div. of Parks and Recreation, 401 Church St., Nashville, TN 37243. Tel. 615-532-0001.
Institution Type/Description: Archaeological Site: Middle Woodland Period ceremonial site, with mounds & earthworks.
Collections: artifacts from on-site fieldwork; regional & site specific prehistory.
Research Fields: Native American cultures; archaeology, with emphasis on the mid-south.
Facilities: 617-vol. library of archaeology research material, site reports, ethnographic, popular, museological, available for research on premises only; field research station; separate lab operation; 80-seat theater. Museum-related items for sale.

Activities: guided tours; lectures; films; video programs; research.
Publications: booklet, Dept. of Conservation, Div. of Archaeology, Research Series.
Hours & Admission Prices: Daily 8-4:30. No charge. Closed state winter holidays. &
Attendance: 110,000 (estimated)

Pleasant Hill

PIONEER HALL MUSEUM, 459 E. Main St., Pleasant Hill, TN 38578. Mailing Address: P.O. Box 264, Pleasant Hill, TN 38578. Tel.: 931-277-5313.
Web Site: www.pioneerhall.com
Founded: 1985.
Key Personnel: Pres. (V), Jeanne C. Kingsbury; Cur. & Museum Shop Mgr., Sharon Weible.
Operating Expenses: 6,066
Operating Income: 11,456
Governing Authority: Parent Institution: Pleasant Hill Historical Society. Tax-exempt.
Institution Type/Description: History Museum: built in 1889. Listed on the National Register of Historic Places.
Collections: local history & culture; period furnishings; personal artifacts; photographs; paintings.
Publications: biannual newsletter, Echoes From The Past.
Hours & Admission Prices: Sun. 2-5, Wed. 10-4. No charge; donations accepted.
Attendance: 220 (estimated)
Membership: Ox Cart Driver $25 & up; Founder $50 & up; Settler $75 & up; Pioneer $100 & up; Historian $500 & up; Genealogist $1,000 & up.

Portland

COLD SPRINGS SCHOOL AND MUSEUM, 303 Portland Blvd., Portland, TN 37148. Tel.: 615-325-2279.
Institution Type/Description: History Museum: housed in a former one-room schoolhouse.
Collections: local history & culture; period furnishings; photographs; personal artifacts; early medicine.
Hours & Admission Prices: June-Sept. Sun. 1-4.

Pulaski

GILES COUNTY HISTORICAL SOCIETY & MUSEUM, 122 S. Second St., Pulaski, TN 38478-3219. Mailing Address: P.O. Box 693, Pulaski, TN 38478-0693. Tel.: 931-363-2720.
E-mail: newmangeorge@bellsouth.net
Key Personnel: Dir., George W. Newman
Institution Type/Description: History Museum.
Collections: local history & culture; period furnishings; Civil War artifacts; photographs; personal artifacts.
Activities: guided tours.
Hours & Admission Prices: Mon.-Wed. & Fri.-Sat. 10-4, Sun. 1-5. No charge; donations accepted. Closed holidays.
Attendance: 300 (estimated)

Red Boiling Springs

CYCLEMOS MOTORCYCLE MUSEUM, 319 E. Main St., Red Boiling Springs, TN 37150-2322. Tel.: 615-699-5049.
Web Site: www.cyclemos.com
Institution Type/Description: Motorcycle Museum.
Collections: motorcycles; paintings; memorabilia; period clothing.
Hours & Admission Prices: Thurs.-Sun. 10-5.

HISTORIC RED BOILING SPRINGS - THE THOMAS HOUSE, 520 E. Main St., Red Boiling Springs, TN 37150. Mailing Address: P.O. Box 408, Red Boiling Springs, TN 37150. Tel.: 615-699-3006.
Web Site: www.thethomashouse.com
Institution Type/Description: Historic House Museum.
Collections: period furnishings; Christmas decorations; toys; dolls; clothing; Gone With The Wind memorabilia.
Facilities: Museum-related items for sale.
Activities: tours.
Hours & Admission Prices: By appointment.

Ripley

LAUDERDALE COUNTY MUSEUM, 123 S. Jefferson St., Ripley, TN 38063-1553. Tel.: 731-635-9541. Fax: 731-635-9064.
E-mail: stodd@lauderdalecountytn.org
Web Site: www.lauderdalecountytn.org
Founded: 2000.
Congressional District: 8
Key Personnel: Dir., Susan Todd; Chm. (V), Keith Davidson; Treas., Maurice Gaines.
Personnel Profile: Part-Time Volunteers 3.
Governing Authority: private; nonprofit organization. Parent Institution: Lauderdale Chamber/ECD.
Institution Type/Description: Historic Building: housed in Sugar Hill Mansion; built in 1843.
Collections: local history & culture; period furnishings; photographs; personal artifacts; medical books; bicycles; early clothing; cotton scales; school books.
Facilities: 1,000 sq. ft. exhibit space. Museum-related items for sale.
Activities: guided tours; temporary & traveling exhibitions.
Publications: magazine, Lauderdale County.
Hours & Admission Prices: Mon.-Fri. 8-4:30; other times by appointment. No charge; donations accepted. Closed Independence Day; Thanksgiving; Christmas Eve & Day.
Attendance: 350 (estimated)

Rogersville

TENNESSEE NEWSPAPER & PRINTING MUSEUM, 415 S. Depot St., Rogersville, TN 37857-3331. Tel.: 423-272-1961. Fax: 423-272-1961.
Institution Type/Description: History Museum: housed in a restored 1890 Southern Railway depot. Listed on the National Register of Historic Places.
Collections: newspaper & printing industries; local history; period machinery & equipment; newspapers.
Hours & Admission Prices: Daily 10-4 by appointment. No charge; donations accepted.

Rugby

HISTORIC RUGBY, (M), 5517 State Hwy. 52, Rugby, TN 37733. Mailing Address: P.O. Box 8, Rugby, TN 37733-0008. Tel.: 423-628-2441. Fax: 423-628-2266.
E-mail: historicrugby@highland.net
Web Site: www.historicrugby.org
Founded: 1966.
Congressional District: 12
Key Personnel: Exec. Dir., Cheryl Cribbet; Pres., Jane Logan; Museum Shop Mgr., Jesse Gully.
Personnel Profile: Full-Time Paid 3; Part-Time Paid 16; Part-Time Volunteers 25.
Governing Authority: society; nonprofit organization. Tax-exempt: 501(c)(3); 170(b)(1)(A).
Institution Type/Description: Historic Site.
Collections: preservation project; archives; manuscripts. Historic Buildings: 1884 Kingstone Lisle; 1882 Thomas Hughes Library; 1884 Percy Cottage reconstruction; 1887 Christ Church, Episcopal; 1907 Rugby Public School; 1880 Pioneer Cottage; 1880 Newbury House Inn; 1881 Rugby Commissary Reconstruction; 1880 Board of Aid reconstruction.
Research Fields: regional culture, Victorian arts, crafts & music, 19th century British culture.
Facilities: 7,000-vol. library of books on Victorian literature available for use on premises by appointment; restaurant. Books for sale.
Activities: guided tours; lectures; concerts. Museum Sponsors: annual pilgrimage, spring music festival & craft festival, Christmas at Rugby, year round workshops, stage presentations.
Publications: biannual newsletter, The Rugbeian; books, Rugby Recipes; Distant Eden-A Rugby History; Images of America Historic Rugby.
Hours & Admission Prices: Jan. by appointment; Winter: Mon.-Sat. 9:30-4:30; Sun. 12-5:30; Summer call for extended hours. Adults $7, senior citizens $6, students $3; discount to groups; members no charge. Closed New Year's Day; Thanksgiving; Christmas Eve & Day. &
Attendance: 15,000 (estimated)
Membership: Explorer $15; Individual $35; Family $50; Supporter $85; Patron $150; Colonist $300; Benefactor $750; Rugbeian $1,250.

Rutherford

DAVID CROCKETT CABIN, 219 N. Trenton St., Rutherford, TN 38369. Mailing Address: 945 S. Trenton, Rutherford, TN 38369-9670. Tel.: 731-665-7253.
E-mail: jobne@msn.com
Web Site: www.davycrockettcabin.org
Founded: 1954.
Congressional District: 8
Key Personnel: Mgr. & Treas., Joe Bone; Chm. (V), Hobert Walker.
Personnel Profile: Part-Time Paid 2; Part-Time Volunteers 6.
Governing Authority: bd. of directors. Parent Institution: town of Rutherford. Subsidiary Institution: Rutherford Lions Club. Tax-exempt.
Institution Type/Description: Historic Building: 1800s David Crockett Cabin.
Collections: period furniture; grave of David Crockett's mother; Davy's letters to his family.
Activities: guided tours; permanent exhibitions. Museum Sponsors: Davy Crockett Parade in October.
Publications: postcards; brochures; book, The Fabulous Davy Crockett.
Hours & Admission Prices: Memorial Day to Labor Day Tues.-Sat. 9:30-4:30, Sun. 1:30-4:30; other times by appointment. Requested Donations: family $7, adults $3, children 6 & over $1.50; discounts to students & seniors groups; children under 6 no charge. &
Attendance: 1,500 (estimated)

Savannah

CHERRY MANSION, 265 Main St., Savannah, TN 38372. Tel.: 731-607-1208.
Institution Type/Description: Historic House: built in 1830, former home of W.H. Cherry & headquarters for General U.S. Grant in 1862.
Collections: period furnishings.
Hours & Admission Prices: By appointment. Tours: adults $10, students $5.

TENNESSEE RIVER MUSEUM, (M), 495 Main St., Savannah, TN 38372-2062. Tel.: 800-552-3866; 731-925-8181. Fax: 731-925-6987.
E-mail: rachel@tourhardincounty.org
Web Site: www.tennesseerivermuseum.org
Founded: 1992.
Key Personnel: Tourism Dir., Rachel Baker.
Governing Authority: Tax-exempt.
Institution Type/Description: History Museum.
Collections: local history & culture; photographs; personal artifacts; period furnishings.
Hours & Admission Prices: Mon.-Sat. 9-5, Sun. 1-5. Adults $3; children 18 & under no charge. &
Attendance: 7,500 (accurate)

Selmer

MCNAIRY COUNTY HISTORICAL MUSEUM, 114 N. Third St., Selmer, TN 38375-2112. Tel.: 731-646-0018.
Founded: 1997.
Congressional District: 7
Key Personnel: Dir., Judy Hammons
Governing Authority: Parent Institution: McNairy County Historical Society. Tax-exempt.
Institution Type/Description: History Museum.
Collections: local history, culture, & heritage; photographs; personal artifacts; period furnishings; agriculture; Civil War; WWI & WWII; medical.
Activities: quilting lessons.
Publications: newsletter 3 times a year.
Hours & Admission Prices: Mon., Wed. & Fri. 12-3. No charge; donations accepted. &
Attendance: 250 (estimated)
Membership: Individual $10; Family $20; Lifetime $100.

Sevierville

FLOYD GARRETT'S MUSCLE CAR MUSEUM, 320 Winfield Dunn Pkwy., Sevierville, TN 37876. Tel.: 865-908-0882.
E-mail: floydava@musclecarmuseum.com
Institution Type/Description: Car Museum.
Collections: over 90 muscle cars.
Facilities: Museum-related items for sale.
Hours & Admission Prices: Jan.-March daily 9-5; April-Dec. daily 9-6. Adults $9.75, children 8-12 $4; children under 8 no charge. Closed Thanksgiving; Christmas. &

NATIONAL KNIFE MUSEUM, INC., 2320 Winfield Dunn Pkwy., Sevierville, TN 37876-0557. Mailing Address: P.O. Box 217, Kodak, TN 37764. Tel.: 865-453-5871, ext. 259.
E-mail: nkmcurator@gmail.com
Web Site: www.nkcaknife.org
Founded: 1981.
Key Personnel: On Site Mgr., Michael Zavasky; Staff, Ronald Ward.
Personnel Profile: Full-Time Volunteers 1; Part-Time Paid 2; Part-Time Volunteers 3.
Governing Authority: private; nonprofit organization. Tax-exempt: 501(c)(3).
Institution Type/Description: Knife and Cutlery Museum.
Collections: knives, knife & sword memorabilia: fruit, glass cake, period, commemorative & Rambo knife collections; pocket cutlery 1,500 CE to present; daggers; stone artifacts; scalpels; exhibition knives 1860s-1920s; custom knives 1930s-present.
Research Fields: cutlery history; metallurgy; human culture & the knife's importance.
Facilities: 650-vol. library; 4,400 sq. ft. exhibit space.
Activities: docent program; formal education programs; self-guided tours; special events celebrating knives; youth programs; training programs for professional workers; demonstrations.
Hours & Admission Prices: Daily 10-9. No charge; donations requested. &
Attendance: 40,000 (accurate)
Membership: Friends of National Knife Museum: $30, $100; $500 & up.

RAINFOREST ADVENTURES DISCOVERY ZOO, 109 NASCAR Dr., Sevierville, TN 37862. Tel.: 865-428-4091. Fax: 865-908-5076.
E-mail: rainforest24@juno.com
Web Site: www.rfadventures.com
Founded: 2001.
Institution Type/Description: Zoo.
Collections: over 600 animals from around the world.
Hours & Admission Prices: Daily 9-5. Adults $11.99, seniors 55 & over $9.99, youth 3-12 $6.99; children under 3 no charge. Closed Christmas.

SEVIER COUNTY HERITAGE MUSEUM, 167 E. Bruce St., Sevierville, TN 37862-3501. Tel.: 865-453-4058.
Institution Type/Description: History Museum: housed in the former post office. Listed on the National Register of Historic Places.
Collections: local history & culture; period artifacts; early furnishings; photographs.
Activities: lectures; classes.
Hours & Admission Prices: Mon.-Fri. 12-5, Sat. 12-3. No charge.

SMOKY MOUNTAIN DEER FARM AND EXOTIC PETTING ZOO, 478 Happy Hollow Ln., Sevierville, TN 37876. Tel.: 865-428-3337. Fax: 865-429-2218.
E-mail: warden89@hotmail.com
Web Site: www.deerfarmzoo.com
Founded: 1988.
Institution Type/Description: Zoo.
Collections: animals from around the world including deer, goats, camels, reindeer, cattle, zebra, zonkeys, kangaroos, elk, emu, Norwegian fjord, Sicilian donkeys.
Facilities: Zoo-related items for sale.
Activities: petting zoo; horse & pony rides.
Hours & Admission Prices: Daily 10-5:30. Adults 13 & over $10.95, children 3-12 $6.99, children 1-2 $.99; children under one $.09; discounts to seniors, military & AAA members. Pony Ride: $6.99. Closed Thanksgiving; Christmas.

TENNESSEE MUSEUM OF AVIATION, (M), 135 Air Museum Way, Sevierville, TN 37862-8703. Mailing Address: P.O. Box 5587, 37864, TN Tel.: 865-908-0171 & 0760. Fax: 865-908-8421.
E-mail: rmelton1@earthlink.net
Web Site: www.tnairmuseum.com
Founded: 2000.
Congressional District: 1
Key Personnel: Pres., C.E.O. & Chm. (V), R. Neal Melton; Membership Coord. & Education, Sandra Layman; Cur. & Archivist, Tom Walker; Museum Shop Mgr., Lana Johnson; Volunteer, Rhonda Melton.
Personnel Profile: Full-Time Paid 4; Full-Time Volunteers 2; Part-Time Paid 2; Part-Time Volunteers 20.
Governing Authority: private; nonprofit organization. Tax-exempt: 501(c)(3).
Institution Type/Description: Aviation Museum: state's official Aviation Hall of Fame.

Collections: aviation history from early flight to modern times; Tennessee aviation history; WWII combat aircraft; uniforms.

Research Fields: WWII aviation; WWI aviation; women's service in WWII military; WWII home front efforts; uniform identification.

Facilities: library; 15,000 sq. ft. exhibit space; hangar area available for dinners or receptions. Museum-related items for sale.

Activities: docent program; guided tours; temporary & traveling exhibitions; theater; lecture series; story telling. Annual Events: Fly-Ins; Tennessee Aviation Hall of Fame Induction.

Hours & Admission Prices: Mon.-Sat. 10-6, Sun. 1-6. Adults $12.75, senior citizens $9.75, children 6-12 $6.75; discounts to AAM & ICOM members; members and children 5 & under no charge. &

Attendance: 50,000 (estimated)

Membership: Military & Veteran Silver $20; Student & Senior $25; Individual $35; Military & Veteran Gold $40; Couple $50; Family $65; Echelon $250-$499; Wings $500-$999; TN Aviation Historical Society $1,000-$9,999; Life $10,000 & up.

Sewanee

UNIVERSITY ART GALLERY, UNIVERSITY OF THE SOUTH, Guerry Hall, Georgia Ave., Sewanee, TN 37383. Mailing Address: 735 University Ave., Sewanee, TN 37383-1000. Tel.: 931-598-1223. Fax: 931-598-3335.

E-mail: sjmaclar@sewanee.edu

Web Site: www.sewanee.edu/gallery

Founded: 1965.

Congressional District: 4

Key Personnel: Gallery Dir., Shelley MacLaren.

Personnel Profile: Full-Time Paid 1; Part-Time Paid 10.

Governing Authority: university. Parent Institution: The University of the South. Tax-exempt: 501(c)(3).

Institution Type/Description: University Art Gallery.

Collections: works of contemporary art.

Activities: guided tours; lectures; films; gallery talks; formally organized education programs for undergraduate college & graduate students; loan, permanent, temporary exhibitions.

Publications: bulletin; two annual catalogues.

Hours & Admission Prices: Tues.-Fri. 10-5, Sat.-Sun. 12-4. No charge; donations accepted. Closed university holidays. &

Attendance: 6,000 (estimated)

Shiloh

SHILOH NATIONAL MILITARY PARK & CEMETERY, Shiloh Battlefield Unit, 1055 Pittsburg Landing Rd., Shiloh, TN 38376-4331. Tel.: 731-689-5696. Fax: 731-689-5450.

E-mail: shil_administration@nps.gov

Web Site: www.nps.gov/shil

Founded: 1894.

Congressional District: 6

Key Personnel: Chief Interpretation & Resource Mgmt., Stacy D. Allen; Supt., Haywood S. Harrell.

Personnel Profile: Full-Time Paid 26; Part-Time Paid 6; Part-Time Volunteers 1.

Governing Authority: federal. Affiliated with U.S. National Park Services, U.S. Department of the Interior, Washington, DC 20240. Branch Museum: Corinth Battlefield Unit, Corinth Civil War Interpretive Center, 501 W. Linden St., Corinth, MS 38834. Tel. 662-287-9273.

Institution Type/Description: Military Museum.

Collections: history; military; manuscripts; Indian artifacts; historic battlefield. Historic Building: pre-Civil War war cabin.

Research Fields: Civil War history.

Facilities: 1,500-vol. library of Civil War source books & documents available for use on premises. Historical publications for sale.

Activities: guided tours; lectures; films; permanent exhibitions; demonstrations.

Publications: Park folder & driving tour guide.

Hours & Admission Prices: Visitor Center: daily 8-5. Family $5, adults $3. Closed Christmas. &

Attendance: 470,000 (estimated)

Smithville

APPALACHIAN CENTER FOR CRAFT, 1560 Craft Center Dr., Smithville, TN 37166-7352. Tel.: 615-597-6801. Fax: 615-597-6803.

E-mail: craftcenter@tntech.edu

Web Site: www.tntech.edu/craftcenter

Founded: 1979.

Key Personnel: Dir., Ward Doubet; Gallery Mgr., Gail S. Looper.

Personnel Profile: Full-Time Paid 1; Part-Time Paid 4; Interns 1.

Governing Authority: state government & public university. Parent Institution: Tennessee Tech University. Tax-exempt.

Institution Type/Description: General Museum.

Collections: over 25 exhibitions annually of contemporary & traditional fine craft.

Facilities: Museum-related items for sale.

Activities: Bachelor of Fine Arts degree & craft certificate programs; craft workshops. Annual Events: Annual Celebration of Craft Silent Auction in April; Holiday Festival in November.

Publications: annual exhibition calendar; workshop catalog; academic brochures.

Hours & Admission Prices: Daily 9-5. No charge. Closed New Year's Eve & Day; Easter; Thanksgiving; Christmas Eve, Day & week. &

Attendance: 150,000 (estimated)

Smyrna

SAM DAVIS HOME, 1399 Sam Davis Rd., Smyrna, TN 37167-2744. Tel.: 615-459-2341. Fax: 615-220-6053.

E-mail: director@samdavishome.org

Founded: 1930.

Key Personnel: Exec. Dir., Madelyn Rush; Bd. Pres., Barbara Vincion.

Personnel Profile: Full-Time Paid 2; Part-Time Paid 8; Part-Time Volunteers 1; Interns 2.

Governing Authority: Parent Institution: Sam Davis Memorial Association. Tax-exempt.

Institution Type/Description: Historic House Museum.

Collections: period furnishings; personal artifacts.

Major Exhibits: Lessons for Ladies, 3/1/14-3/31/14; A House in Mourning, 10/1/14-10/31/14.

Facilities: wedding & party facilities.

Activities: Annual Events: Family Day; Teddy Bear Teas; Days on the Farm; Summer Camps; Heritage Days; Ghost Tours; Easter "Eggs"travaganza; It Was An Antebellum Christmas; Movie Nights.

Publications: quarterly newsletter, The Courier.

Hours & Admission Prices: June-Aug. Mon.-Sat. 9-5, Sun. by 72-hour advance appointment; Sept.-May Mon.-Sat. 10-4, Sun. by 72-hour advance appointment; Jan. by 72-hour advance appointment. House & Museum: adults $10, seniors/students $8, children 6-12 $6; children under 6 & members no charge; House or Museum only: adults $8, seniors/students $6, children 6-12 $4; children under 6 & members no charge.

Membership: Stewarts Creek Society $25; Honor Guard $50; Charles Lewis Davis Society $100; Jane Simmons Davis Society $250; Coleman Scouts $500; Sam Davis Society $1,000.

Sneedville

VARDY COMMUNITY HISTORICAL CHURCH MUSEUM, Vardy Blackwater Rd., Sneedville, TN 37869. Mailing Address: Vardy Community Historical Society, P.O. Box 554, Sneedville, TN 37869. Tel.: 423-733-2305.

E-mail: claudefreeda@aol.com

Web Site: vardyhistoricalsociety.org

Founded: 1998.

Congressional District: 1

Governing Authority: Parent Institution: Vardy Community Historical Society. Tax-exempt.

Institution Type/Description: Historical Society Museum: housed in a former Presbyterian Church built in 1889.

Collections: local history, culture & heritage; personal artifacts; period furnishings; photographs. Historic Building: Mahala log cabin.

Hours & Admission Prices: By appointment. No charge; donations accepted. &

Attendance: 1,000

Southside

HISTORIC COLLINSVILLE, 4711 Weakley Rd., Southside, TN 37171. Tel.: 931-648-9141. Fax: 931-648-9831.

E-mail: jintown@aol.com

Web Site: www.historiccollinsville.com

Founded: 1997.

Key Personnel: Dir., Chm. (V) & Pres. (V), JoAnn B. Weakley; C.E.O., Glenn N. Weakley; Museum Shop Mgr., Carolyn Gannaway.

Volunteer Hours: 600

Operating Expenses: 11,836

Operating Income: 11,960

Governing Authority: Tax-exempt: 501(c)(3).

Institution Type/Description: Historic Village.
Collections: local history; restored log houses & outbuildings; Native American artifacts; mounted wildlife.
Activities: Civil War reenactment in May; annual Civil War events; pioneer activities specializing in fiber arts.
Hours & Admission Prices: mid-May to mid-Oct. Thurs.-Sun. 1-5. Adults $5; school tours $7; guided tours $15. children under 5 no charge. &
Attendance: 4,500 (estimated)

Sparta

WHITE COUNTY HERITAGE MUSEUM, 144 S. Main St., Sparta, TN 38583-2215. Tel.: 931-837-3900.
E-mail: wcmuseum@blomand.net
Web Site: www.whitecountyheritagemuseum.org
Founded: 2008.
Congressional District: 4
Key Personnel: Dir. & Museum Shop Mgr., Peggie J. Hurteau; Asst. Dir., Brenda Templeton; C.E.O., Herd Sullivan.
Personnel Profile: Part-Time Paid 2; Part-Time Volunteers 6; Interns 1.
Operating Income: 12,000
Governing Authority: county. Tax-exempt.
Institution Type/Description: Heritage Museum.
Collections: local history, heritage & culture; personal artifacts; photographs; model railroad; Native American artifacts; local musician heritage; farming; military; one room school house.
Activities: permanent & temporary exhibitions.
Hours & Admission Prices: Thurs. 10-4, Fri. 9-4, Sat. 9-2. No charge; donations accepted.
Attendance: 2,000 (estimated)

Spring Hill

RIPPAVILLA PLANTATION, 5700 Main St., Spring Hill, TN 37174-2408. Mailing Address: P.O. Box 1169, Spring Hill, TN 37174-1169. Tel.: 931-486-9037. Fax: 931-486-3175. Facebook: Rippavilla Plantation.
E-mail: rippavilla@bellsouth.net
Web Site: www.rippavilla.org
Founded: 1998.
Congressional District: 4
Key Personnel: Exec. Dir., Pam Perdue.
Governing Authority: Tax-exempt: 509(a)(1).
Institution Type/Description: Historic House Museum: c.1860.
Collections: local history & culture; period furnishings; personal artifacts; photographs; Civil War artifacts.
Hours & Admission Prices: Tours: Mon.-Sat. 9:30-4:30, Sun. 1-4:30. House Tours: adults $10, senior citizens 62 & over $8, children 6-12 $5; discounts to groups; children 5 & under no charge. &
Attendance: 15,000 (accurate)
Membership: Individual $25; Young or Young at Heart Family (21-35 or 62 & over) $30; Family $40; Benefactor $100; Patron $250; Cheairs $500-$1,000.

Springfield

ROBERTSON COUNTY HISTORY MUSEUM, 124 Sixth Ave. W., Springfield, TN 37172-2405. Mailing Address: P.O. Box 1022, Springfield, TN 37172-1022. Tel.: 615-382-7173.
E-mail: rchs@bellsouth.net
Web Site: rchsonline.com
Founded: 1993.
Congressional District: 6
Key Personnel: Dir., Linda D. Dean; Pres. (V), David C. Allen; Museum Shop Mgr., Charlotte E. Reedy.
Personnel Profile: Full-Time Paid 1; Part-Time Volunteers 24.
Governing Authority: Parent Institution: Robertson County Historical Society. Tax-exempt.
Institution Type/Description: History Museum: housed in the former U.S. Post Office.
Collections: county history including tobacco & whiskey trades; photographs.
Publications: quarterly newsletter.
Hours & Admission Prices: Wed.-Fri. 10-4, Sat. by appointment. Adults $4, seniors $2, students $1; discounts to groups; members no charge. Closed most holidays. &
Attendance: 800 (estimated)
Membership: Individual $15; Household $25.

Stanton

HATCHIE NATIONAL WILDLIFE REFUGE, 6772 Hwy. 76 S., Stanton, TN 38069-3648. Tel.: 731-772-0501. Fax: 731-772-7839.
E-mail: hatchie@fws.gov
Web Site: www.fws.gov/hatchie
Key Personnel: Refuge Mgr., Michael Chouinard
Institution Type/Description: Wildlife Refuge.
Collections: waterfowl.
Hours & Admission Prices: Call for hours.

Sweetwater

THE LOST SEA, 140 Lost Sea Rd., Sweetwater, TN 37874-6724. Tel.: 423-337-6616.
Institution Type/Description: Natural History Museum: a Registered National Landmark.
Collections: underground lake; caverns; 18th century village; local history; Native American artifacts; photographs.
Activities: cave tours.
Hours & Admission Prices: March-April & Sept.-Oct. daily 9-6; May-June & Aug. daily 9-7; July daily 9-8; Nov.-Feb. daily 9-5. Adults $15.95, children 5-12 $7.45; discounts to groups or 20 or more by appointment.

SWEETWATER HERITAGE MUSEUM, North & High Sts., Sweetwater, TN 37874. Mailing Address: P.O. Box 143, Etowah, TN 37331.
Institution Type/Description: History Museum.
Collections: local history & culture; photographs; period furnishings; personal artifacts.
Hours & Admission Prices: March-Oct. Wed. & Sat.-Sun. 2-4.

Tellico Plains

CHARLES HALL MUSEUM, 229 Cherohala Skyway, Tellico Plains, TN 37385-5500. Tel.: 423-253-4369. Fax: 423-253-8000.
E-mail: charleshallmuseum@hotmail.com
Web Site: www.charleshallmuseum.com
Founded: 2003.
Congressional District: 3
Key Personnel: Dir., C.E.O. & Chm. (V), Charles Hall.
Personnel Profile: Full-Time Volunteers 1; Part-Time Volunteers 1.
Volunteer Hours: 3,285
Operating Expenses: 22,018
Operating Income: 22,018
Governing Authority: Tax-exempt.
Institution Type/Description: History Museum.
Collections: local history & culture; photographs; personal artifacts; early furnishings; over 300 guns; antique vehicles; telephones; coins.
Facilities: Museum-related items for sale.
Hours & Admission Prices: Daily. No charge; donations accepted. &
Attendance: 30,000 (estimated)

Tiptonville

CARL PERKINS VISITOR CENTER, 230 Crl Perkins Pkwy., Tiptonville, TN 38079. Tel.: 731-253-9922.
Institution Type/Description: Visitor Center & Historic House Museum: housed in the boyhood home of singer, Carl Perkins.
Collections: Carl Perkins' life & career; personal artifacts; photographs; recordings.
Hours & Admission Prices: Call for hours.

Townsend

CADES COVE VISITOR CENTER AND OPEN-AIR MUSEUM, Great Smoky Mountains National Park, 10042 Campgrounds Dr., Townsend, TN 37882-5004. Mailing Address: Great Smoky Mountains National Park, 107 Park Headquarters Rd., Gatlinburg, TN 37738-4102. Tel.: 877-444-6777. Fax: 865-436-1220.
E-mail: grsm_smokies_information@nps.gov
Web Site: www.nps.gov/grsm
Founded: 1951.
Key Personnel: Chief Resource Education, Cathleen Cook.
Governing Authority: federal. Tax-exempt.
Institution Type/Description: Preservation Project: Cable Mill area & other historic structures typical of Southern Appalachia at turn of the 20th century.
Collections: water powered grist mill; historic house.

Facilities: visitor center; nature trail. Postcards, slides & books for sale.
Activities: permanent exhibitions; self-guided auto tour & nature trail; demonstrations of pioneer crafts.
Publications: booklet, Cades Cove Auto Tour.
Hours & Admission Prices: Museum: daily dawn to dusk. Visitor Center: Feb. & Nov. daily 9-5; March & Sept.-Oct. daily 9-6; April-Aug. daily 9-7; Dec.-Jan. daily 9-4:30. No charge. &
Attendance: 1,000,000

GREAT SMOKY MOUNTAINS HERITAGE CENTER, (M), 123 Cromwell Dr., Townsend, TN 37882-4323. Mailing Address: P.O. Box 268, Townsend, TN 37882-0268. Tel.: 865-448-0044. Fax: 865-448-6975.
E-mail: gsmhcevents@yahoo.com
Web Site: www.gsmheritagecenter.org
Founded: 2006.
Congressional District: 2
Key Personnel: Dir., Robert Patterson; Pres., James K. Leach; Treas., Richard Maples; Admin. Asst., Pam Haaby; Exhibit Technician, Bob Hood; Dir. Mktg., Nancy Williams; Museum Shop Mgr., Don Alexander.
Personnel Profile: Full-Time Paid 4; Part-Time Paid 7; Part-Time Volunteers 170; Interns 1.
Governing Authority: private; nonprofit organization. Tax-exempt: 501(c)(3).
Institution Type/Description: History Museum.
Collections: East Tennessee mountain history & culture; transportation; Native American artifacts; horse-drawn vehicles; 8 historic buildings & furnishings; textiles; tools; archaeological artifacts.
Facilities: 100-vol. library; 100-seat auditorium; educational facilities; 500-seat amphitheater. Museum-related items for sale.
Activities: concerts; docent program; films; guided tours; hobby workshops; lectures; participatory exhibits; theater; broadcast programs. Annual Events: Heritage Happenings Fund-raiser; Winter Heritage Festival; Quilter's Road Show; Knoxville Symphony Orchestra Concerts; Rotary Storytelling Festival; Smoky Mountain Woodcarvers Festival; Sunset Music Series; Blue Ribbon Country Fair; Christmas Memories.
Publications: quarterly newsletter, Mountain Echoes; book, Flavors of Our Heritage Cookbook; video, Peace of Ground.
Hours & Admission Prices: Mon.-Sat. 10-5, Sun. 12-5. Adults $6, senior citizens 60 & over and students $4; discounts to groups of 8 or more; members & children under 6 no charge. Closed New Year's Day; Easter; Thanksgiving; Christmas Eve & Day. &
Attendance: 28,600 (accurate)
Membership: Individual $15; Family $25; Patron $50; Contributor $100.

LITTLE RIVER RAILROAD AND LUMBER COMPANY MUSEUM, 7747 E. Lamar Alexander Pkwy., Townsend, TN 37882. Mailing Address: P.O. Box 211, Townsend, TN 37882. Tel.: 865-448-2211. Fax: 865-448-2312.
E-mail: sandy@littleriverrailroad.org
Web Site: www.littleriverrailroad.org
Founded: 1982.
Congressional District: 8
Key Personnel: Pres. (V), Don Niday; Museum Shop Mgr., Sandy Headrick.
Personnel Profile: Part-Time Paid 3.
Volunteer Hours: 450
Operating Expenses: 24,732
Operating Income: 22,900
Governing Authority: Tax-exempt.
Institution Type/Description: History Museum.
Collections: railroad & lumber company history; period furnishings; personal artifacts; photographs; papers; tools; caboose; flatcars; wooden water tank; log loader.
Facilities: Museum-related items for sale.
Activities: Museum Sponsors: Railroad Days in May (1st weekend) & September (last weekend).
Hours & Admission Prices: April-May, Sept. & Nov. Sat. 10-5, Sun. 1-5; June-Aug. & Oct. Mon.-Sat. 10-5, Sun. 1-5; other times by appointment. No charge; donations accepted.
Attendance: 19,000 (estimated)
Membership: Annual $21.47; Lifetime $200.

Trenton

TRENTON NIGHT LIGHT TEAPOT MUSEUM, Trenton City Hall, 309 College St., Trenton, TN 38382. Tel.: 731-855-2013. Fax: 731-855-1091.
Web Site: www.teapotcollection.com/285621.ihtml
Formerly: Porcelain Veilleuses-Theieres Museum
Founded: 1961.

Congressional District: 79
Governing Authority: city. Tax-exempt.
Institution Type/Description: Teapot Museum.
Collections: over 500 teapots from around the world dated from 1750 to 1860.
Activities: Annual Event: Teapot Festival in April.
Hours & Admission Prices: Mon.-Fri. 9-5; groups by appointment. No charge; donations accepted.
Attendance: 5,000 (estimated)

Tullahoma

BEECHCRAFT HERITAGE MUSEUM, 570 Old Shelbyville Hwy., Tullahoma, TN 37388-4703. Mailing Address: P.O. Box 550, Tullahoma, TN 37388-0550. Tel.: 931-455-1974. Fax: 931-455-1994.
E-mail: info@beechcraftheritagemuseum.org
Web Site: beechcraftheritagemuseum.org
Key Personnel: Dir. & C.E.O., Wade D. McNabb; Museum Shop Mgr., Jo Chouinard
Institution Type/Description: Aviation Museum.
Collections: period Beechcraft airplanes & memorabilia.
Hours & Admission Prices: Tues.-Sat. 8:30-4:30. Adults $10, children 12-17 $5; members and children 11 & under no charge.
Attendance: 6,000 (estimated)

FLOYD AND MARGARET MITCHELL MUSEUM, South Jackson Civic Center, 404 S. Jackson St., Tullahoma, TN 37388. Mailing Address: P.O. Box 326, Tullahoma, TN 37388-0326. Tel.: 931-455-5321.
E-mail: sojack@lighttube.net
Web Site: southjackson.org/museum.html
Key Personnel: Chm., Blossom Merryman
Institution Type/Description: History Museum.
Collections: local history, business & culture; photographs; personal artifacts; period furnishings; Native American; military memorabilia.
Hours & Admission Prices: 1st Sun. of month 2-4; other times by appointment.

HANDS-ON SCIENCE CENTER, 101 Mitchell Blvd., Tullahoma, TN 37388. Tel.: 931-455-8387. Facebook: Hands-on Science Center.
E-mail: hosc@lighttube.net
Web Site: hosc.org
Founded: 1990.
Key Personnel: Dir., Misty Marshall; Pres. (V), Sean Smith; Museum Shop Mgr., Jan Griffin.
Personnel Profile: Full-Time Paid 2; Full-Time Volunteers 1; Part-Time Paid 5; Part-Time Volunteers 35.
Governing Authority: Tax-exempt.
Institution Type/Description: Children's Science Museum.
Collections: over 100 hands-on exhibits.
Facilities: Museum-related items for sale.
Activities: rental facilities; educational programs.
Hours & Admission Prices: Tues.-Sat. 10-5, Sun. 1-5. Admission $5; children 2 & under no charge. &
Attendance: 20,000 (estimated)
Membership: Family $50.

TULLAHOMA FINE ARTS CENTER REGIONAL MUSEUM OF ART, 401 S. Jackson St., Tullahoma, TN 37388-3469. Tel.: 931-455-1234. Fax: 931-455-1234.
E-mail: lucy@tullahomafinearts.org
Web Site: www.tullahomafinearts.org
Institution Type/Description: Art Museum.
Collections: Baillet sisters' original art; paintings; sculpture.
Facilities: classrooms.
Activities: classes.
Hours & Admission Prices: Call for hours.

Union City

DISCOVERY PARK OF AMERICA, 830 Everett Blvd., Union City, TN 38261. Mailing Address: P.O. Box 927, Union City, TN 38261. Tel.: 877-885-5455; 731-885-5455.
E-mail: info@discoveryparkofamerica.com
Web Site: www.discoveryparkofamerica.com
Institution Type/Description: General Museum.
Collections: nature; science; technology; history; art; energy; military; Native American; nature history; transportation; fossils; plants & flowers from

around the world; 1800s farm equipment & family artifacts; locomotive. Historic Buildings; log cabins; farm buildings; early 1900s church; train station.
Facilities: theater; aquarium; 70,000 sq. ft. exhibit space; 50-acre complex; rental facilities; gardens.
Activities: educational programs; special events; rental facilities; school groups.
Hours & Admission Prices: By appointment.

DIXIE GUN WORKS' OLD CAR MUSEUM, 1412 W. Reelfoot Ave., Union City, TN 38261-5508. Tel.: 731-885-0561. Fax: 731-885-0440.
E-mail: dixiegun@earthlink.net
Web Site: www.dixiegunworks.com
Founded: 1954.
Congressional District: 8
Key Personnel: Vice Pres., Hunter M.F. Kirkland.
Personnel Profile: Full-Time Paid 26.
Governing Authority: private; nonprofit. Parent Institution: Dixie Gun Works.
Institution Type/Description: Automobile Museum.
Collections: American period automotives; 16th to 19th-century firearms.
Facilities: 15,000 sq. ft. exhibit space. Museum-related items for sale.
Hours & Admission Prices: Mon.-Fri. 8-5, Sat. 8-12. Adults $2, senior citizens & children $1. Closed New Year's Day; Memorial Day; Independence Day; Labor Day; Thanksgiving; Christmas.
Attendance: 3,000 (estimated)

OBION COUNTY MUSEUM, (M), 1004 Edwards St., Union City, TN 38261-5316. Mailing Address: P.O. Box 323, Union City, TN 38281-0323. Tel.: 731-885-6774.
E-mail: adelledoss@gmail.com
Web Site: www.ocmuseum.com
Founded: 1970.
Congressional District: 8
Key Personnel: Chm. (V) & Treas., Larry Mink; Cur., Polly Brasher; Registrar, Jennifer Wildes.
Personnel Profile: Full-Time Volunteers 4; Part-Time Volunteers 8; Interns 5.
Governing Authority: private; nonprofit organization. Tax-exempt: 501(c)(3).
Institution Type/Description: History Museum.
Collections: local & regional history & cultural heritage; American military artifacts from Civil War to present; natural history; photographs.
Research Fields: regional history.
Facilities: library of regional history books. Museum-related items for sale.
Activities: guided tours; permanent exhibitions; community events. Annual Event: Day At The Museum.
Hours & Admission Prices: Mon.-Fri. 10-4. Adults $2, children $1; discounts to groups and AAM & ICOM members; members no charge. Closed New Year's Day; Labor Day; Thanksgiving; Christmas.
Attendance: 650 (estimated)
Membership: Individual $30; Family $50; Business $100; Heritage Circle $500; Patron $1,000.

Vonore

FORT LOUDOUN STATE HISTORIC PARK, 338 Fort Loudoun Rd., Vonore, TN 37885-2704. Tel.: 423-884-6217. Fax: 423-884-2287.
Institution Type/Description: Park Museum.
Collections: local history; 18th century British fort replica; Tellico Blockhouse ruins.
Facilities: nature trails.
Activities: reenactments; hiking. Annual Event: 18th Century Trade Faire in September.
Hours & Admission Prices: Call for hours.

THE SEQUOYAH BIRTHPLACE MUSEUM, 576 Hwy. 360, Vonore, TN 37885-2816. Mailing Address: P.O. Box 69, Vonore, TN 37885-0069. Tel.: 423-884-6246. Fax: 423-884-2102.
E-mail: seqmus@tds.net
Web Site: www.sequoyahmuseum.org
Founded: 1986.
Key Personnel: Dir., Charlie Rhodarmer; Chm. (V), Maxwell Ramsey; Museum Shop Mgr., Linda Bosket.
Personnel Profile: Full-Time Paid 2; Part-Time Paid 3.
Governing Authority: Parent Institution: EBCI. Tax-exempt.
Institution Type/Description: Cultural History Museum.
Collections: Cherokee Indian writing system; culture & history.
Facilities: Museum-related items for sale.
Activities: special events; workshops; school programs.

Hours & Admission Prices: Mon.-Sat. 9-5, Sun. 12-5. Closed New Year's Day; Thanksgiving; Christmas.
Attendance: 17,000 (accurate)

VONORE HERITAGE MUSEUM, 619 Church St., Vonore, TN 37885-2324. Tel.: 423-884-2989.
Founded: 1996.
Key Personnel: Chm. (V), Violet K. Wolfe; Pres. (V), Shirley Brown; Museum Shop Mgr., Tony White
Institution Type/Description: History Museum.
Collections: local history & culture; farm tools & equipment; household artifacts; photographs.
Publications: quarterly newsletter, Vonore Heritage Museum.
Hours & Admission Prices: Tues.-Wed. & Fri. 10-4; other times by appointment. No charge; donations accepted. Closed New Year's Day; Thanksgiving & day after; Christmas.
Attendance: 900 (estimated)
Membership: Seniors over 62 $10; Adults $15.

Waverly

HUMPHREYS COUNTY MUSEUM & CIVIL WAR FORT, 201 Fort Hill Dr., Waverly, TN 37185-2127. Tel.: 931-296-1099.
Institution Type/Description: History Museum: housed in a 1922 mansion.
Collections: local history; Civil War, WWI & WWII artifacts; period furnishings; personal artifacts; Jesse James memorabilia; Captain Anderson; photographs. Historic Building: Civil War Fort.
Activities: rental facilities.
Hours & Admission Prices: Fri.-Sun. -4. No charge; donations accepted. Closed holidays.

1978 WAVERLY PROPANE EXPLOSION MUSEUM, W. Railroad St. & Richland Ave., Waverly, TN 37185. Mailing Address: City of Waverly, 101 E. Main St., Waverly, TN 37185-2143. Tel.: 931-296-2101. Fax: 931-296-1434.
E-mail: bfrazier@waverlytn.org
Web Site: waverlytn.org
Founded: 2002.
Congressional District: 8
Key Personnel: Dir., W.B. (Buddy) Frazier.
Governing Authority: Parent Institution: City of Waverly, TN. Tax-exempt.
Institution Type/Description: History Museum: housed in a caboose on the site of the 1978 train derailment disaster.
Collections: disaster history; photographs; video; newspaper articles; personal accounts.
Activities: self-guided tour.
Hours & Admission Prices: Daily 7:30 am-8:30 pm. No charge.
Attendance: 1,000 (estimated)

White House

WHITE HOUSE INN LIBRARY & MUSEUM, 412 Hwy. 76, White House, TN 37188-9201. Tel.: 615-672-0239. Fax: 615-672-9733.
Institution Type/Description: Library.
Collections: books; magazines; reference materials.
Hours & Admission Prices: Mon.-Wed. & Fri. 10-5:30, Thurs. 12-8, Sat. 10-4.

Winchester

FRANKLIN COUNTY OLD JAIL MUSEUM, 400 Dinah Shore Blvd., Winchester, TN 37398-1421. Mailing Address: 7895 Sewanee Hwy., Cowan, TN 37318-3706. Tel.: 931-967-0524.
Founded: 1973.
Congressional District: 4
Key Personnel: C.E.O., Pres. (V) Chm. Old Jail Museum Commission, Mrs. Nancy Hall; Hostess, Mrs. Ruth McNutt; Treas., Harry Fanning; Sec., Kathy Howse.
Personnel Profile: Full-Time Volunteers 8; Part-Time Volunteers 15.
Governing Authority: society; nonprofit. Parent Institution: Old Jail Museum Commission. Tax-exempt.
Institution Type/Description: History of Franklin County.
Collections: history of Franklin County; tools; pictures; household articles; records.
Research Fields: local history; Normandy Invasion during World War II.
Facilities: Books, prints by local artists, and museum-related items for sale.
Activities: guided tours; slide program for schools.
Publications: newspaper.

Hours & Admission Prices: April-Nov. 14 Tues.-Sat. 10-4. Adults $1, youth & children under 12 $.50; discounts for senior citizens; special rates for groups.
Attendance: 450 (estimated)

Woodbury

THE CANNON CULTURAL MUSEUM AND THE MARLEY BERGER GALLERY, Arts Center of Cannon County, 1424 John Bragg Hwy., Woodbury, TN 37190-6173. Mailing Address: P.O. Box 111, Woodbury, TN 37190. Tel.: 800-235-9073.
Web Site: www.artscenterofll.com
Founded: 2008.
Key Personnel: Exec. Dir., Donald Fann
Institution Type/Description: Art & History Museum.
Collections: paintings; sculpture; photographs; baskets; pottery.
Activities: permanent & temporary exhibitions; educational programs.
Hours & Admission Prices: Call for hours. No charge; donations accepted. &

TEXAS

(672 listings)

Abilene

ABILENE ZOOLOGICAL GARDENS, 2070 Zoo Lane, Abilene, TX 79602-1996. Tel.: 325-676-6085. Fax: 325-676-6084.
E-mail: abilene.zoo@abilenetx.com
Web Site: www.abilenezoo.org
Founded: 1965.
Congressional District: 17
Key Personnel: Exec. Dir., William R. Gersonde; Pres. (V), Sara Core; Cur. Education, Joy Harsh; Gen. Cur., Tim Singiser; Museum Shop Mgr. & Admissions Mgr., John Black; Concessions Mgr., Emily Thompson; Admin. Coord., Jere Drake; Mktg. & Devel. Coord., Jo Ann Wilson.
Personnel Profile: Full-Time Paid 25; Part-Time Paid 10; Part-Time Volunteers 2.
Governing Authority: municipal; nonprofit organization. Affiliated with Abilene Zoological Society. Tax-exempt: 501(c)(3).
Institution Type/Description: Zoo.
Collections: mammals; birds; reptiles; amphibians; invertebrates; native plants.
Research Fields: captive reproduction; artificial incubation behavior; maps; Attwater's prairie chicken introduction project.
Facilities: 1,000-vol. library of books available for use in zoo society office only. Animal-related items for sale.
Activities: guided tours; lectures; TV & radio programs; educational programs.
Publications: quarterly newsletter, Pronghorn Press; video, Habitat Restoration.
Hours & Admission Prices: Memorial Day to Labor Day Thurs. 9-9, Fri.-Wed. 9-5; Sept.-May daily 9-5. Adults $5, seniors 60 & over $4, children 3-12 $2.50; discounts to groups, AZS & AZA members; members & children under 3 no charge. Closed New Year's Day; Thanksgiving; Christmas. &
Attendance: 217,038 (accurate)
Membership: Individual $30; Family $45; Circle of Life $75; Species Protector $100; Caregiver $15 additional to any level.

CENTER FOR CONTEMPORARY ARTS, 220 Cypress St., Abilene, TX 79601. Tel.: 325-677-8389. Fax: 325-677-1171.
E-mail: darla@center-arts.com
Web Site: www.center-arts.com
Founded: 1989.
Key Personnel: Exec. Dir., Darla Harmon.
Personnel Profile: Full-Time Paid 3; Part-Time Paid 5; Part-Time Volunteers 30; Interns 5.
Governing Authority: Tax-exempt.
Institution Type/Description: Art Gallery.
Collections: works by contemporary artists; paintings; sculpture; photographs.
Activities: educational programs.
Hours & Admission Prices: Tues.-Sat. 11-5. No charge; donations accepted. &
Attendance: 50,000 (estimated)
Membership: Individual $50; Family $75; Supporter $125; Patron $250; Sustainer $500.

FRONTIER TEXAS MUSEUM AND VISITOR CENTER, (M), 625 N. First St., Abilene, TX 79601. Tel.: 325-437-2800. Fax: 325-437-2804.
Web Site: www.frontiertexas.com
Key Personnel: Exec. Dir., Jeff Salmon

Institution Type/Description: History Museum & Visitor Center.
Collections: hands-on exhibits; frontier life & artifacts; Native American artifacts; photographs; video.
Facilities: Museum-related items for sale.
Activities: special events; educational programs.
Hours & Admission Prices: Mon.-Sat. 9-6, Sun. 1-5. Adults $8, seniors 60 & over $6, students & teachers $5, children 3-12 $4; children under 3 no charge. Closed New Year's Day; Thanksgiving; Christmas. &

✳ **THE GRACE MUSEUM, (M),** 102 Cypress St., Abilene, TX 79601-5817. Mailing Address: P.O. Box 33, Abilene, TX 79604-0033. Tel.: 325-673-4587. Fax: 325-675-5993. Facebook: The Grace Museum.
E-mail: info@thegracemuseum.org
Web Site: thegracemuseum.org
Founded: 1937.
Congressional District: 17
Personnel Profile: Full-Time Paid 10; Part-Time Paid 6; Part-Time Volunteers 123; Interns 2.
Governing Authority: nonprofit organization. Tax-exempt: 501(c)(3).
Institution Type/Description: Art, History & Children's Participatory Museums.
Collections: 20th-century American art with emphasis on printmaking from 1930s & 40s; modern graphics; Texas regional art; Abilene history 1900-45; Texas & Pacific Railway Co.
Major Exhibits: Drawn In - Drawn Out: Contemporary Drawing, 2/14-4/14; Art in the Dark: Black & White Photography, 2/14-4/14; Youth Art Month, 3/14; Young Masters, 4/14; Home on the Range: Where the Prairie Meets the Plains in Central West Texas, 5/14/8/14.
Research Fields: early Texas artists; Abilene history; Texas & Pacific Railway.
Facilities: restored 1909 hotel; classrooms; rental space; staff & volunteer offices; restored lobby & ballroom; galleries; courtyard; rooftop terrace.
Activities: tours; lectures; films; gallery talks; annual state competition; volunteer & docent program; temporary & traveling exhibitions; school loan service.
Publications: newsletter; annual report; brochures; catalogs.
Hours & Admission Prices: Tues.-Wed. & Fri.-Sat. 10-5, Thurs. 10-8. Adults $8, seniors, students & military $4, children 4-12 $3; children 3 & under, members & Thurs. 5-8 no charge. &
Attendance: 70,678 (accurate)
Membership: Friend $45; Family $65; Supporter $100; Advocate $300; Sustainer $600; Benefactor $1,200; Champion $2,400.

NATIONAL CENTER FOR CHILDREN'S ILLUSTRATED LITERATURE, (M), 102 Cedar, Abilene, TX 79601-5718. Tel.: 325-673-4586. Fax: 325-673-0085. Facebook: National Center for Children's Illustrated Literature.
E-mail: info@nccil.org
Web Site: www.nccil.org
Founded: 1997.
Congressional District: 19
Key Personnel: Exec. Dir., Debbie Lillick; Chm. (V), Lee Hamilton.
Personnel Profile: Full-Time Paid 1; Part-Time Paid 3; Part-Time Volunteers 8; Interns 2.
Governing Authority: Tax-exempt.
Institution Type/Description: Art Museum.
Collections: original illustrations in children's literature.
Hours & Admission Prices: Center: Tues.-Sat. 10-4. Artwalk: 2nd Thurs. each month 5:30-8. No charge; donations accepted. &
Attendance: 18,000 (estimated)
Membership: Gallery Circle $50; Author's Circle $100; Collector's Circle $250; Illustrator's Circle $500; Artists Circle $1,000.

12TH ARMORED DIVISION MEMORIAL MUSEUM, 1289 N. 2nd St., Abilene, TX 79601. Tel.: 325-677-6515.
E-mail: hellcatsmuseum@yahoo.com
Web Site: www.12tharmoredmuseum.com
Founded: 1999.
Congressional District: 17
Key Personnel: Dir. & Chm., Dale Cartee; Education & Cur., William Lenches; Public Rels. & Museum Shop Mgr., Jennifer King; Treas., Robert Hoeweler; Archivist, Harry Dhans.
Personnel Profile: Full-Time Paid 3; Full-Time Volunteers 3; Part-Time Volunteers 10.
Governing Authority: private; nonprofit organization. Tax-exempt: 501(c)(3).
Institution Type/Description: Military History Museum.
Collections: 12th Armored Division artifacts; personal artifacts; over 2,700 photographs; Japanese & German soldier artifacts; POW & concentration camp artifacts; military vehicles.

Research Fields: Holocaust & Liberation of concentration camps.
Facilities: 1,000-vol. library; field research station; 30-seat theater; 20,000 sq. ft. exhibition space. Museum-related items for sale.
Activities: docent program; films; formal education programs for children; guided tours; lectures; temporary exhibitions; theater; living history program.
Hours & Admission Prices: Tues.-Sat. 10-5. Adults $4, senior citizens & students $3, children $1. Closed federal holidays. &

Attendance: 3,000 (estimated)
Membership: Sponsor $30.

Addison

CAVANAUGH FLIGHT MUSEUM, 4572 Claire Chennault, (at Addison Airport), Addison, TX 75001. Tel.: 972-380-8800. Fax: 972-248-0907. Facebook: Cavanaugh Flight Museum.
Web Site: www.cavanaughflightmuseum.com
Key Personnel: Dir., Doug Jeanes; Asst. Dir., Kevin Raulie; Finance Mgr., Denise Wilson; Maintenance Dir., Russell Martin
Institution Type/Description: History Museum.
Collections: period aircraft.
Hours & Admission Prices: Mon.-Sat. 9-5, Sun. 11-5. Adults $10, seniors & military $8, children 4-12 $5; children 3 & under no charge. Closed New Year's Day; Thanksgiving; Christmas.

MARY KAY MUSEUM, 16251 Dallas Pkwy., Addison, TX 75001. Tel.: 972-687-5720.
Founded: 1993.
Key Personnel: Dir., Jennifer Cook.
Personnel Profile: Full-Time Paid 2.
Governing Authority: Parent Institution: Mary Kay Inc.
Institution Type/Description: Company Museum.
Collections: life & history of Mary Kay Ash and her company; awards; portraits; company product line; personal artifacts; photographs; clothing; sculpture; news articles.
Facilities: restaurant; theater.
Hours & Admission Prices: Self-Guided Museum Tours: Mon.-Fri. 9-4:30. No charge. Closed major holidays. &

Albany

* **THE OLD JAIL ART CENTER, (M),** 201 S. Second St., Albany, TX 76430-2503. Tel.: 325-762-2269. Fax: 325-762-2260. Facebook: The Old Jail Art Center.
E-mail: info@theoldjailartcenter.org
Web Site: www.theoldjailartcenter.org
Founded: 1977.
Congressional District: 17
Key Personnel: Chm. (V), Steve Waller; Interim Dir., Margaret Blagg; Preparator & Cur. Exhibits, Patrick Kelly; Treas., Dan Neff; Dir. Education, Erin Whitmore; Archivist & Librarian, Molly Sauder; Office Mgr., Dorothy Walker.
Personnel Profile: Full-Time Paid 7; Part-Time Paid 1; Part-Time Volunteers 80.
Governing Authority: nonprofit organization. Tax-exempt: 501(c)(3).
Institution Type/Description: Art Museum: housed in c.1877 stone two-storied, Victorian classic-style jail building.
Collections: 19th, 20th & 21st century American & European paintings, drawings, prints, & sculpture; Asian collections featuring Chinese terra cotta tomb figures; pre-Columbian collection; local history archives, local heritage room & memorabilia collection.
Major Exhibits: Acquisitions 2002-2012, 10/13-1/14; Cell Series: Anthony Sonnenberg, 10/13-1/14; West Texas Triangle: Danville Chadbourne, 10/13-1/14; Drawn In/Drawn Out, 2/14-5/14; Cell Series: Chris Sauter, 2/14-5/14; Allison V. Smith: Going West, 6/14-9/14; BosworthCell Series: Campbell, 6/14-9/14; James Mager: Kites, 6/14-8/14; Cell Services: Rachel Hecker, 9/14-1/15; Ron Cohen, 9/14-1/15; George Grammer, 9/14-1/15.
Research Fields: contemporary artists; local history.
Facilities: 2,500-vol. library pertaining to contemporary art, art history, world art & regional history available for inter-library loan & public use; 7,328 sq. ft. exhibition space; The Stasney Center for Education; 3,500 sq. ft. enclosed sculpture yard. Books on art & local history for sale.
Activities: guided tours; lectures; concerts; theatre; facilities rental Gallery; organized educational programs; docent program; loan, temporary & traveling exhibitions.
Publications: newsletter (published 3 times per year), The Old Jail Art Center.
Hours & Admission Prices: Tues.-Sat. 10-5, Sun. 2-5. No charge; donations accepted. Closed holidays. &
Attendance: 10,000 (estimated)

Membership: Associate Member $12.50; Member $25; Contributor $50; Friend $100; Patron $250; Sustainer $500; Founders Circle $1,000 & up.

Alice

SOUTH TEXAS MUSEUM, 66 S. Wright St., Alice, TX 78332-4904. Mailing Address: P.O. Box 3232, Alice, TX 78333-3232. Tel.: 512-668-8891.
E-mail: stmuseum@sbcglobal.net
Founded: 1975.
Congressional District: 20
Key Personnel: Chm., Bren Ball; Vice Chm., Marin Perez; Sec., Virginia Bedgood.
Personnel Profile: Full-Time Paid 1.
Governing Authority: nonprofit organization. Tax-exempt: 501(c)(3).
Institution Type/Description: History Museum: housed in c.1940 headquarters for the McGill Brothers ranching operations; Texas historic landmark.
Collections: firearms; Civil War items; farm & ranch tools; saddles; period furniture; clothing; barbed wire collection; photographs; family histories; typewriters.
Research Fields: local history.
Facilities: 186-vol. Civil War library.
Activities: guided tours; lectures; study clubs; organized educational programs.
Publications: brochure; STM newsletter.
Hours & Admission Prices: Mon.-Fri. 10-12 & 1-5, Sat. 9-1; other times by appointment. No charge; donations accepted. Closed Easter; Memorial Day; Independence Day; Labor Day; Thanksgiving; Christmas.
Attendance: 1,200 (estimated)
Membership: Student $5; Individual $20; Family $35; Business $50; Contributor $100; Sponsor $250; Patron $500.

THIRD COAST SQUADRON CAF MUSEUM, 1309 S. Airport Rd., Alice, TX 78332. Mailing Address: P.O. Box 8192, Corpus Christi, TX 78468-8192. Tel.: 361-563-0063.
Institution Type/Description: Military History Museum.
Collections: military history & aircraft.
Hours & Admission Prices: Tues.-Thurs. 9-4. Adults $3, students $2; discounts to groups.

Alpine

HALLIE'S HALL OF FAME MUSEUM, 430 HC 65, Alpine, TX 79830-9752. Tel.: 432-376-2211.
E-mail: info@stillwellstore.com
Web Site: stillwellstore.com/hall-of-fame
Institution Type/Description: History Museum.
Collections: mementos & relics from the late Hallie Stillwell's ranch.
Hours & Admission Prices: Call for hours.

LAST FRONTIER MUSEUM, Antelope Lodge, 2310 W. Hwy. 90, Alpine, TX 79830-4106. Tel.: 432-837-2451.
Institution Type/Description: Mineralogy Museum.
Collections: rocks, gems & minerals of the region.
Hours & Admission Prices: Daily 9-9.

MUSEUM OF THE BIG BEND, (M), Sul Ross State University, Alpine, TX 79832. Mailing Address: C-101, Alpine, TX 79832. Tel.: 432-837-8143. Fax: 432-837-8901.
E-mail: ejackson@sulross.edu
Web Site: www.sulross.edu/~museum/
Founded: 1926.
Congressional District: 21
Key Personnel: Dir. & C.E.O., Larry Francell; Cur., Matt Walter; Cur. & Collections Mgr., Mary Bridges; Asst. to Dir., Liz Jackson.
Personnel Profile: Full-Time Paid 4; Part-Time Paid 5; Part-Time Volunteers 25.
Governing Authority: Sul Ross State University. Tax-exempt: 501(c)(3).
Institution Type/Description: Regional History Museum.
Collections: late 19th- & early 20th-century historical materials; archaeological artifacts from paleo-man to modern ethnology; regional art; archival collections in ranching, mining & business; 19th-century photographs of west Texas, rare books; Texana & Big Bend region special books.
Research Fields: regional photographic history of Anglo- & Mexican-American settlement; historical & archaeological collections.
Facilities: 2,000-vol. library pertaining to American material culture & museum science, available for research use; primary materials, such as archaeological field notes & historical documents, available for use to professional researchers. Museum-related items & books for sale.

Activities: guided tours; university class tours; education programs for children, undergraduate & graduate college students; temporary exhibitions.
Publications: booklets; newsletter.
Hours & Admission Prices: Tues.-Sat. 9-5, Sun. 1-5; guided tours by appointment. No charge; donations accepted. &

Attendance: 18,000 (estimated)

Alto

CADDO MOUNDS STATE HISTORIC SITE, 1649 State Hwy. 21 W., Alto, TX 75925-5739. Tel.: 936-858-3218. Fax: 936-858-3227.
E-mail: anthony.souther@thc.state.tx.us
Web Site: www.thc.state.tx.us
Formerly: Caddoan Mounds State Historic Site
Founded: 1982.
Congressional District: 1
Key Personnel: Site Mgr., Anthony Souther; Museum Shop Mgr., Michelle Ivie.
Personnel Profile: Full-Time Paid 5; Interns 1.
Governing Authority: Parent Institution: Texas Historical Commission, P.O. Box 12276, Capitol Station, Austin, TX, 78711. Tax-exempt.
Institution Type/Description: State Historic Site: located on a former Caddo Village and Ceremonial Grounds, 750-1400 A.D.
Collections: artifacts & archaeological findings associated with early Caddo culture excavated at the site.
Research Fields: Caddo archaeology.
Facilities: library of archaeological reports, theses, dissertations and related anthropological material available for research on premises only; materials for teachers & youth group leaders; 3/4 mile interpretive trail with 3 prehistoric mounds.
Activities: guided tours; audiovisual shows; permanent exhibitions; scout & youth workshops; mock dig; cooking on a hot rock; pottery making; atlatl throw. Museum Sponsors: Caddo Culture Day; Texas Historical Commission; Caddo Nation.
Publications: pamphlet; book, Caddo Mounds: Temples and Tombs of an Ancient People.
Hours & Admission Prices: Tues.-Sun. 8:30-4:30. Adults & senior citizens $4, children 6-18 & college students $3; children 5 & under no charge. Closed New Year's Eve & Day; Thanksgiving; Christmas Eve & Day. &

Attendance: 5,000 (estimated)

Alvin

ALVIN HISTORICAL MUSEUM, 300 W. Sealy, Alvin, TX 77511. Mailing Address: P.O. Box 1902, Alvin, TX 77512. Tel.: 281-331-4469.
Founded: 1979.
Key Personnel: Chm. (V), Tom Stansel; Pres. (V), Mona M. Travis; Museum Shop Mgr., Ivan Self.
Personnel Profile: Part-Time Volunteers 65.
Governing Authority: Parent Institution: Alvin Museum Society. Tax-exempt.
Institution Type/Description: History Museum.
Collections: local history & culture; photographs; period artifacts.
Facilities: Museum-related items for sale.
Hours & Admission Prices: Thurs.-Fri. & 1st Sat. of month 11-3. Adults $3; children under 12 no charge. Closed holidays.
Attendance: 300 (estimated)
Membership: Friend $10; Organization $20; Patron $25; Business $30; Benefactor $50.

MARGUERITE ROGERS HOUSE MUSEUM, 113 E. Dumble, Alvin, TX 77511. Mailing Address: Alvin Museum Society, P.O. Box 1902, Alvin, TX 77512-1902. Tel.: 281-585-2803.
Founded: 1976.
Key Personnel: Dir., Bobbie Case
Governing Authority: Parent Institution: Alvin Museum Society. Tax-exempt.
Institution Type/Description: Historic House: built early 1900s.
Collections: local history & culture; period furnishings; photographs.
Hours & Admission Prices: Tours: Thurs.-Fri. & 1st Sat. of month 11-3. Adults $3; children under 12 no charge. Closed holidays.
Attendance: 180 (estimated)
Membership: Friend $10; Organization $20; Patron $25; Business $30; Benefactor $50.

NOLAN RYAN CENTER, 2925 S. Bypass 35, Alvin, TX 77511. Tel.: 281-388-1134. Fax: 281-388-1135.
Institution Type/Description: Sports History Museum.
Collections: Nolan Ryan's life & career; photographs; personal artifacts; baseball memorabilia; broadcast audio.

Facilities: theater.
Activities: interactive pitch-catch exhibit; videos.
Hours & Admission Prices: Mon.-Sat. 9-4. Adults $5, students & seniors $2.50; discounts to groups; children under 6 no charge.

Amarillo

AMARILLO BOTANICAL GARDENS, 1400 Streit Dr., Amarillo, TX 79106. Tel.: 806-352-6513.
Web Site: www.amarillobotanicalgardens.org
Institution Type/Description: Botanical Gardens.
Collections: flowers; plants; trees.
Facilities: Garden-related items for sale.
Activities: special events; guided tours; educational programs. Museum Sponsors: Annual Hunt Dinner and Auction.
Hours & Admission Prices: Summer: Tues.-Sat. 9-5. Winter: Tues.-Fri. 9-5, Sat. 1-5. Closed holidays.
Membership: Individual $30; Family $65.

*** AMARILLO MUSEUM OF ART, (M),** 2200 S. Van Buren, Amarillo, TX 79109-2407. Mailing Address: P.O. Box 447, Amarillo, TX 79178. Tel.: 806-371-5050. Fax: 806-345-5682.
E-mail: amoa@actx.edu
Web Site: www.amarilloart.org
Founded: 1972.
Congressional District: 67
Key Personnel: Business Dir., Kim Mahan; Pres. Bd., Sheri Brosier; Curatorial Dir., Arlette Klaric, Ph.D.; Cur. Education, Julie Talley; Dir. Devel., Kay Kennedy.
Personnel Profile: Full-Time Paid 5; Part-Time Paid 7; Interns 1.
Governing Authority: nonprofit organization. Tax-exempt: 501(c)(3).
Institution Type/Description: Art Museum.
Collections: 20th-century American paintings; prints; drawings; photographs; sculpture; Asian art; Southeast Asian art & sculpture; Japanese prints; Middle Eastern textiles.
Major Exhibits: Hydrologic: Recent Works by Judy Youngblood (T), 12/13-3/14; The Fred and Laura Bidwell Collection of Photography (T), 1/14-3/14; Side by Side Series: Romy Owens and Christopher Pekoc (T), 5/14-8/14.
Research Fields: American art history; contemporary art; photography; Asian & Southeast Asian art.
Facilities: library; sculpture courtyard; classrooms.
Activities: guided tours; lectures; films; gallery talks; formally organized educational programs; docent program; inter-museum loan, temporary & traveling exhibitions; scholarships for children who cannot afford to attend.
Publications: quarterly newsletter; exhibit catalogues; gallery guides for adults & children; posters.
Hours & Admission Prices: Tues.-Fri. 10-5, Sat.-Sun. 1-5. No charge; donations accepted. Closed major national holidays. &

Attendance: 40,000
Membership: Family $50; Contributor $150; Patron $250; Sponsor $500; Benefactor $1,000.

AMARILLO ZOO, N.E. 24th Ave. & Dumas Hwy., Amarillo, TX 79105. Mailing Address: P.O. Box 1971, Amarillo, TX 79105. Tel.: 806-381-7911. Fax: 806-381-7901.
E-mail: zoo@amarilloparks.org
Web Site: www.amarillozoo.org
Founded: 1955.
Institution Type/Description: Zoo.
Collections: over 60 species of animals; reptiles; horses; donkeys; bison; elk; sheep; goats; black bears; monkeys; lions; bobcats; servals.
Facilities: 15 acre site; picnic area; nature trails.
Hours & Admission Prices: Daily 9:30-5. Adults 13-61 $3, seniors 62 & over $2, children 3-12 $1; children under 3 & Mon. no charge. Closed New Year's Day; Martin Luther King Jr. Day; Thanksgiving; Christmas.

AMERICAN QUARTER HORSE HALL OF FAME & MUSEUM, (M), 2601 I-40 E, Amarillo, TX 79104-3405. Tel.: 806-376-5181. Fax: 806-376-1005.
E-mail: museum@aqha.org
Web Site: www.quarterhorsemuseum.com
Formerly: American Quarter Horse Heritage Center & Museum
Founded: 1991.
Congressional District: 13
Key Personnel: C.E.O., Don Treadway; Treas., Trent Taylor; Sr. Dir., Chris Sitz; Dir. Retail Sales, Dana Skipper; Cur., Crystal Phares.
Personnel Profile: Full-Time Paid 15; Part-Time Paid 2.

Governing Authority: Parent Institution: American Quarter Horse Assoc. Tax-exempt: 501(c)(3).
Institution Type/Description: Equine Hall of Fame & Museum.
Collections: equine items relating to all aspects of the American Quarter Horse; ranching; racing; rodeo; American Quarter Horse Association registry.
Major Exhibits: The Buffalo Soldier: An American Horseman, 2/14-8/14; Remount Horses, 5/14-7/14; America's Horse in Art Show & Sale, 8/14-11/14; Youth Art Show, 12/14-1/15.
Facilities: 5,000-vol. library of equine & historical works; educational facilities; 18,000 sq. ft. exhibit space; 85-seat theatre. Western-style gift items for sale.
Activities: films; formal education programs; lectures; participatory & temporary exhibitions.
Hours & Admission Prices: Mon.-Sat. 9-5. Adults $6, senior citizens $5, children 6-18 $2; discounts for groups of 15 or more, AAM & ICOM members; AQHA, AQHYA members & children 5 & under no charge. Closed New Year's Day; Thanksgiving Day; Christmas Eve & Day. &
Attendance: 30,000 (accurate)
Membership: Annual $35.

DON HARRINGTON DISCOVERY CENTER, 1200 Streit Dr., Amarillo, TX 79106-1759. Tel.: 806-355-9547, ext. 101. Fax: 806-355-5703.
E-mail: apan@dhdc.org
Web Site: www.dhdc.org
Founded: 1975.
Congressional District: 13
Key Personnel: Bd. Pres. (V), Jason Herrick; Exec. Dir., Aaron Pan; Dir. School Svcs., Mindy Morgan; Dir. Devel., Liz Bentley; Dir. Visitor Experience, Mandi Ried; Museum Shop Mgr., Hillary Patak.
Personnel Profile: Full-Time Paid 12; Part-Time Paid 16; Part-Time Volunteers 2.
Governing Authority: nonprofit organization. Parent Institution: Foundation For Health & Science Education, Inc. Tax-exempt: 501(c)(3); 170(b)(1)(A).
Institution Type/Description: Science Museum & Planetarium.
Collections: hands-on exhibits; helium monument.
Major Exhibits: Megalodon - Largest Shark that Ever Lived, 9/13-1/14.
Facilities: 95-seat Digistar 3 digital theater; 45,000 sq. ft. exhibit space; 24 projector multi-image system.
Activities: public & curriculum coordinated presentations in health, astronomy, physical & earth science.
Publications: Discoveries Newsletter of Science Education & Events.
Hours & Admission Prices: Summer: Mon.-Sat. 9:30-4:30, Sun. 12-4:30; Winter: Tues.-Sat. 9:30-4:30, Sun. 12-4:30. Adults $10, youth 3-22 and seniors 60 & up $7; discounts to ASTC members; members & children under 3 no charge. &
Attendance: 143,000 (accurate)
Membership: Individual Plus Guest $40; Family & Grandparents $70; Family Plus $100.

HARRINGTON HOUSE, 1600 S. Polk St., Amarillo, TX 79102. Tel.: 806-374-5490.
E-mail: harringtonhouse@att.net
Web Site: harringtonhousehistorichome.org
Institution Type/Description: Historic House Museum: housed in the former home of Don & Sybil Harrington. Listed on the National Register of Historic Places.
Collections: Harrington family history; period furnishings; personal artifacts; clothing; household artifacts; decorative & fine arts.
Activities: permanent & temporary exhibitions; lectures.
Hours & Admission Prices: Tues. & Thurs. 10-12:30. No charge.

JULIAN BIVINS MUSEUM, Main & Hwy. 385, Amarillo, TX 79174. Mailing Address: P.O. Box 1890, Amarillo, TX 79174. Tel.: 806-372-2341.
Institution Type/Description: History Museum.
Collections: local history & culture; Native American artifacts; cowboy & pioneer artifacts; photographs; Boys Ranch history & documents.
Activities: Museum Sponsors: Boys Ranch Youth Cowboy Poetry Gathering in June; Labor Day Rodeo in September.
Hours & Admission Prices: Daily 8-5. No charge. Closed major holidays.

KWAHADI MUSEUM OF THE AMERICAN INDIAN, 9151 I-40 E., Amarillo, TX 79120. Mailing Address: P.O. Box 32125, Amarillo, TX 79120-2125. Tel.: 806-335-3175.
Institution Type/Description: Native American History Museum.

Collections: Native American history & culture; paintings; beadwork; personal artifacts.
Hours & Admission Prices: June-Aug. Wed.-Sun. 1-5; Sept.-May Sat.-Sun. 1-5. Adults $5, youth $3.

TEXAS AIR & SPACE MUSEUM, 10001 American Dr., Amarillo, TX 79111-1213. Mailing Address: P.O. Box 31535, Amarillo, TX 79120-1535. Tel.: 806-335-9159. Fax: 806-665-4627.
E-mail: info@texasairandspacemuseum.org
Web Site: www.texasairandspacemuseum.org
Governing Authority: Tax-exempt: 501(c)(3).
Institution Type/Description: History Museum.
Collections: aviation & space history & artifacts; NASA shuttle trainer; C-7 Caribou; P-51 Mustang; Reno Racer; early aircraft.
Hours & Admission Prices: Mon.-Fri. 10-4; other times by appointment.

TEXAS PHARMACY MUSEUM, Texas Tech School of Pharmacy, 1300 S Coulter St., Amarillo, TX 79106-1712. Tel.: 806-356-4000, ext. 268. Fax: 806-356-4669.
E-mail: paul.katz@ttuhsc.edu
Web Site: www.ttuhsc.edu/sop/prospective/visitors/museum.aspx
Founded: 1998.
Congressional District: 13
Key Personnel: Dir. (V), Dean Arthur Nelson, R.Ph., Ph.D.; Cur., Paul Katz, Ph.D.
Personnel Profile: Part-Time Paid 2; Part-Time Volunteers 1.
Governing Authority: nonprofit; public college. Parent Institution: Texas Tech University, Health Sciences Center. Tax-exempt: 501(c)(3).
Institution Type/Description: Pharmacy Museum.
Collections: history of pharmacy in Texas, the U.S. & Western Europe; early 20th century pharmacy; tools; pharmacy products & delivery systems; pharmacy art; Texas practitioners; books; containers; furniture; laboratory glassware; medical items; show globes; mortars & pestles.
Facilities: 500-vol. library; 3,000 sq. ft. exhibit space; 250-seat auditorium.
Activities: guided tours; lectures; graduate course; temporary exhibitions.
Hours & Admission Prices: Mon.-Fri. 1-5. No charge. Closed New Year's Eve & Day; Martin Luther King Jr. Day; Memorial Day; Independence Day; Labor Day; Thanksgiving; Christmas Eve & Day. &
Attendance: 431 (accurate)

WILDCAT BLUFF NATURE CENTER, 2301 N. Soncy Rd., Amarillo, TX 79124-5766. Mailing Address: P.O. Box 52132, Amarillo, TX 79159-2132. Tel.: 806-352-6007. Fax: 806-352-2274.
Founded: 1992.
Institution Type/Description: Nature Center.
Collections: reptiles; birds; mammals; invertebrates.
Facilities: nature trails.
Activities: special events.
Hours & Admission Prices: Tues.-Sat. 9-5. Adults $3, children & seniors $2; children under 3 no charge. Closed holidays.
Membership: Student $20; Teacher & Senior $25; Individual $35; Grandparent $40; Family $50; Friend $100; Supporter $250; Sustaining $500; Benefactor Lifetime $1,000.

Angleton

BRAZORIA COUNTY HISTORICAL MUSEUM, (M), 100 E. Cedar, Angleton, TX 77515-4602. Tel.: 979-864-1208. Fax: 979-864-1217.
E-mail: bchm@bchm.org
Web Site: www.bchm.org
Founded: 1983.
Congressional District: 22
Key Personnel: Dir., Jackie Haynes; Pres., Morris Paschall.
Personnel Profile: Full-Time Paid 8.
Governing Authority: county. Tax-exempt: 501(c)(3).
Institution Type/Description: History Museum: housed in 1897 Brazoria County Courthouse.
Collections: Brazoria County history from prehistoric to modern times; historical & archaeological materials; archives; photographs.
Facilities: library; 4,437 sq. ft. exhibit space.
Activities: guided tours; lectures; films; organized education programs for children; docent program; temporary exhibition; summer concerts featuring Zydeco & other music.
Publications: monthly newsletter.
Hours & Admission Prices: Mon.-Fri. 9-5, Sat. 9-3. No charge; donations accepted. Closed major holidays. &

Attendance: 20,000 (estimated)
Membership: Student & Senior $30; Family $50; General $100; Commander $250; Presidential $500; Corporate $1,000.

Anson

ANSON JONES MUSEUM, 1302 Ave. K, Anson, TX 79501. Mailing Address: P.O. Box 613, Anson, TX 79501. Tel.: 325-823-3683.
Founded: 1983.
Key Personnel: Dir. & Pres. (V), Bill Carman.
Personnel Profile: Part-Time Volunteers 102.
Governing Authority: Tax-exempt.
Institution Type/Description: History Museum.
Collections: local history & culture; Dr. Anson Jones' personal artifacts; photographs; period furnishings.
Research Fields: genealogy.
Activities: local area trips (stage stops).
Hours & Admission Prices: Wed.-Sun. 2-4:30. No charge; donations accepted. Closed holidays.
Attendance: 377 (accurate)
Membership: Single $10; Couples $15.

Archer City

ARCHER COUNTY MUSEUM, N. Sycamore St., Archer City, TX 76351. Mailing Address: 2627 Loftin Rd., Windthorst, TX 76389-4641. Tel.: 940-423-6426 & 257-4048.
Founded: 1974.
Key Personnel: Cur., Jack Loftin.
Governing Authority: county. Tax-exempt.
Institution Type/Description: General Museum: housed in 1910 Old County Jail.
Collections: barbed wire; bridle bits; spurs; blacksmith tools; prehistoric lizard fossils/fossil bones; art & western paintings; old clothing; farm machinery; vehicles; grinding implements; Indian artifacts; early pictures & newspapers; surrey; hack oil field equipment & trucks; Archer County artifacts.
Activities: permanent exhibitions.
Hours & Admission Prices: May-Nov. Sat. 9-5, Sun. 1-5; other times by appointment. No charge; donations accepted.
Attendance: 500 (estimated)

Arlington

ARLINGTON MUSEUM OF ART, 201 W. Main St., Arlington, TX 76010. Tel.: 817-275-4600. Fax: 817-394-2030.
E-mail: ama@arlingtonmuseum.org
Web Site: www.arlingtonmuseum.org
Founded: 1987.
Congressional District: 24
Key Personnel: Pres. & Chm. (V), Walter Virden; Sec., Lynda Freeman.
Personnel Profile: Part-Time Paid 1; Part-Time Volunteers 25; Interns 5.
Governing Authority: private; nonprofit organization. Tax-exempt.
Institution Type/Description: Art Museum & Gallery: located in a former 1950s storefront building on Main St. in downtown Arlington.
Collections: works by emerging and established artists.
Facilities: library; 20,000 sq. ft. exhibit space.
Activities: guided tours; lectures; workshops; educational programs. Annual Events: Art Auction; Youth Encounters; summer art camp.
Publications: biannual newsletter, AMA news; Artivities, workbook for children; exhibition catalogues.
Hours & Admission Prices: Tues.-Sat. 10-5, Sun. 1-5. Minor Exhibitions: $1-$2. Major Exhibitions: $5-$8; donations accepted. Closed New Year's Day; Independence Day; Thanksgiving; Christmas. &
Attendance: 18,000 (estimated)
Membership: Individual $25; Household $50; Sustaining $100; Advocate $500; Director's Circle $1,000; Twenty-First Century $2,100; Patron $5,000.

THE GALLERY AT UTA, 502 S. Cooper St., Fine Arts Bldg., Arlington, TX 76019. Mailing Address: Box 19089, Arlington, TX 76019. Tel.: 817-272-3143 & 5658. Fax: 817-272-2805.
E-mail: bhuerta@uta.edu
Web Site: www.uta.edu/gallery
Formerly: CRCA: The Gallery at UTA
Founded: 1976.
Congressional District: 24
Key Personnel: Dir., Benito Huerta; Asst. Dir., Patricia Healy.
Governing Authority: university. Parent Institution: University of Texas at Arlington. Tax-exempt.

Institution Type/Description: Art Museum
Collections: monthly exhibitions.
Research Fields: contemporary art.
Facilities: 4,000 sq. ft. exhibit gallery.
Activities: traveling & loan exhibitions; video & film series; artist residencies & lectures.
Publications: exhibition announcements & brochures; posters.
Hours & Admission Prices: Mon.-Fri. 10-5, Sat. 12-5. No charge. Closed school holidays. &
Attendance: 7,400 (accurate)

INTERNATIONAL BOWLING MUSEUM AND HALL OF FAME, 621 Six Flags Dr., Arlington, TX 76011-6305. Tel.: 817-385-8210. Fax: 817-385-8268.
E-mail: info@bowlingmuseum.com
Web Site: www.bowlingmuseum.com
Founded: 1988.
Key Personnel: Dir., Eric Kearney; Chm. (V), Keith Hamilton; Cur., Kelli Thomerson; Museum Shop Mgr., Anna Murphy.
Personnel Profile: Full-Time Paid 3; Part-Time Paid 3.
Governing Authority: nonprofit. Tax-exempt: 501(c)(3).
Institution Type/Description: Sports Museum.
Collections: 18th to 20th-century historical artifacts; equipment; uniforms; graphics; trophies.
Research Fields: history of bowling, women's sports, international sports & American social history.
Facilities: research library; theater.
Activities: films; permanent exhibitions; old time bowling alley re-creation; math, science & social studies programs for school children; guided instructional tours.
Publications: brochures; periodic news releases and news photos.
Hours & Admission Prices: Tues.-Sat. 9:30-5; other times by appointment. Adults $9.50; discounts to military, AAA, AARP, USBC, BPAA, & IBPSIA members; members no charge. &
Attendance: 5,300 (accurate)
Membership: Heritage $500-$10,000 & up; Sustaining $1,000-$60,000 & up.

RIVER LEGACY LIVING SCIENCE CENTER, 703 N.W. Green Oaks Blvd., Arlington, TX 76006-2404. Tel.: 817-860-6752. Fax: 817-860-1595.
E-mail: jill@riverlegacy.org
Web Site: www.riverlegacy.org
Founded: 1996.
Congressional District: 6
Key Personnel: Exec. Dir., Jill Hill; Pres. Bd. (V), Larry Fowler; Financial Dir., Bob Murday; Museum Shop Mgr., Sylvia Greene.
Personnel Profile: Full-Time Paid 8; Part-Time Paid 6; Part-Time Volunteers 140.
Governing Authority: private; nonprofit organization. Parent Institution: River Legacy Foundation. Tax-exempt.
Institution Type/Description: Science & Nature Center.
Collections: live animals; preserved animals; botany; archaeology; geology.
Facilities: classrooms; 3,000 sq. ft. exhibit space; 1,300 acres of natural habitat; nature center; rental facilities. Museum-related items for sale.
Activities: formal education programs for children; self-guided tours; professional development workshops; lectures; study clubs; participatory, temporary & traveling exhibitions; nature study programs for preschool age through adult.
Publications: annual newsletter, River Legacy Parks; twice monthly e-newsletter.
Hours & Admission Prices: Mon.-Sat. 9-5. No charge; donations accepted. Closed New Year's Day; Easter; Independence Day; Labor Day; Thanksgiving; Christmas Eve & Day. &
Attendance: 75,000 (estimated)

Athens

EAST TEXAS ARBORETUM & BOTANICAL SOCIETY, 1601 Patterson Rd., Athens, TX 75751. Mailing Address: P.O. Box 2231, Athens, TX 75751-7231. Tel.: 903-675-5630. Fax: 903-675-1618. Facebook: East Texas Arboretum and Botanical Society.
E-mail: etabs@cvc.net
Web Site: www.easttexasarboretum.org
Key Personnel: Exec. Dir., Teresa Glasgow.
Governing Authority: nonprofit organization. Tax-exempt: 501(c)(3).
Institution Type/Description: History Museum.
Collections: Gardens: plants; flowers; trees. House: period furnishings; personal artifacts. Historic House: Wofford House, c.1800.
Facilities: 100 acre site; nature trails.

Activities: rental facilities; special events; hiking.
Publications: newsletter.
Hours & Admission Prices: Spring & Summer: daily 7:30-7:30; Fall & Winter: daily 8-6. Suggested Donations: adults 12 & over $2; members no charge.
Membership: Individual $25; Contributing $35; Associate $60; Sustaining $100; Benefactor $250; Patron $500; Friend $1,000.

HENDERSON COUNTY HISTORICAL SOCIETY MUSEUM, 217 N. Prairieville, Athens, TX 75751-2042. Mailing Address: P.O. Box 943, Athens, TX 75751-0943. Tel.: 903-677-3611.
Institution Type/Description: Historical Society Museum.
Collections: local history & culture; Native Indian artifacts; dinosaur bones; military; scout memorabilia; period furnishings.
Facilities: library.
Activities: guided tours; docent program; lectures.
Publications: quarterly newsletter.
Hours & Admission Prices: Fri.-Sat. 10-3. No charge; donations accepted.

Austin

ARTHOUSE, The Jones Center, 700 Congress Ave., Austin, TX 78701-3217. Tel.: 512-453-5312. Fax: 512-459-4830.
E-mail: info@arthousetexas.org
Web Site: www.arthousetexas.org
Formerly: Texas Fine Arts Association
Founded: 1911.
Congressional District: 10
Key Personnel: Exec. Dir., Sue Graze; Pres. (V), Julie Thornton; Devel., Melissa Berry; Devel., Jennifer Gardner; Education, Erin Gentry; Public Rels., Virginia Jones; Cur., Elizabeth Dunbar; Coord. Membership, Ben Slade; Exhibition Coord., Kirk Nickel; Coord. Capital Campaign, Caitlin Sweeney; Preparator, Nathan Green.
Personnel Profile: Full-Time Paid 6; Part-Time Paid 6; Part-Time Volunteers 2; Interns 3.
Governing Authority: private; nonprofit organization. Tax-exempt: 501(c)(3).
Institution Type/Description: Art Museum.
Collections: works by contemporary artists.
Activities: contemporary art exhibitions; educational programs for teens; guided tours; lectures; screenings; commission new art work; annual fundraisers; off-site programming. Museum Sponsors: Arthouse Texas Prize.
Publications: exhibition catalogs; member e-newsletter.
Hours & Admission Prices: Wed. 12-11, Thurs.-Sat. 12-9, Sun. 12-5; groups by appointment. No charge; donations accepted. Closed New Year's Day; Independence Day; Thanksgiving; Christmas Eve & Day. &
Attendance: 25,000 (estimated)
Membership: Student $15; Supporter $40; Patron $100; Benefactor $250; Sustainer $500; Collectors Circle $1,500; Center Circle $5,000.

AUSTIN HISTORY CENTER, 810 Guadalupe St., Austin, TX 78701. Mailing Address: P.O. Box 2287, Austin, TX 78768-2287. Tel.: 512-974-7480. Fax: 512-974-7483. Facebook: Austin History Center.
E-mail: ahc_reference@austintexas.gov
Web Site: www.austinhistorycenter.org
Founded: 1955.
Congressional District: 10
Key Personnel: Mgr., Mike Miller.
Personnel Profile: Full-Time Paid 14; Part-Time Paid 1; Part-Time Volunteers 35; Interns 10.
Governing Authority: municipal government. Parent Institution: Austin Public Library. Tax-exempt.
Institution Type/Description: Archives & Library: housed in 1933 former Austin Public Library building. A Texas Historic Landmark.
Collections: local history books; photographs; maps; ephemera; newspapers; archives; drawings; Austin & Travis County history.
Major Exhibits: Women in Politics, 2/14-4/14; In the Shadow of the Live Music Capital, 4/14-6/14.
Research Fields: Austin & Travis County.
Facilities: 60,000-vol. library of local history books; vertical files; photographs; archives, available for use by public; meeting rooms; photography lab; 1,000 sq. ft. exhibit space.
Activities: temporary exhibitions; author events; public programs. Annual Event: Archives Clinic in October.
Publications: The Republic of Austin (2010).
Hours & Admission Prices: Tues.-Sat. 10-6, Sun. 12-6. No charge; donations accepted. &
Attendance: 50,000 (accurate)

AUSTIN MUSEUM OF ART - DOWNTOWN, 823 Congress Ave., Austin, TX 78701-2405. Mailing Address: 3809 W. 35th St., Austin, TX 78703-1001. Tel.: 512-495-9224. Fax: 512-495-9029.
E-mail: info@amoa.org
Web Site: www.amoa.org
Founded: 1961.
Congressional District: 10
Key Personnel: Pres. & Chm. (V), Bettye H. Nowlin; Treas., Clay Cary; Sr. Dir. Education, Judith Sims; Registrar, Cassandra Smith; Sr. Dir. Devel., Tom Jackson; Dir. Facilities & Operations, Bill Nichols; Dir. Finance & Operations, Curt Shinaberry; Museum Shop Mgr., Justin Hearne.
Personnel Profile: Full-Time Paid 17; Part-Time Paid 14; Part-Time Volunteers 1,470; Interns 15.
Governing Authority: nonprofit organization. Branch Museum: Austin Museum of Art - Laguna Gloria, 3809 W. 35th St., Austin, TX 78703. Tel.: 512-458-8191. Tax-exempt: 501(c)(3).
Institution Type/Description: Art Museum.
Collections: Downtown: modern & contemporary art; paintings; sculpture; photography; prints; drawings; video; multimedia installations. Laguna Gloria: historic home & gardens.
Research Fields: 20th century & contemporary art.
Facilities: Downtown: Museum-related items for sale. The Art School: classrooms; ceramic studio.
Activities: changing exhibitions; art classes for children & adults; docent tours; lectures; performances; films; concerts; hands-on interactive space for schools & families. Fund Raisers: Art Ball (annual gala & art auction); La Dolce Vita (annual wine tasting & art auction); Austin Fine Arts Festival.
Publications: e-newsletter.
Hours & Admission Prices: Downtown: Tues.-Wed. & Fri. 10-5, Thurs. 10-8, Sat. 10-6, Sun. 12-5. Adults $5, students & seniors $4; discounts to AAM, ICOM, AICA, TAM & AMP members and Tues.; children under 12 & members no charge. 1st Sat. of month pay what you wish. Laguna Gloria: Grounds Mon.-Sat. 9-5, Sun. 11-5. Villa daily 11-4. Suggested donation $3. Closed holidays. &
Attendance: 275,510 (accurate)
Membership: Individual $30; Household $60; Associate $150; Advocate $250; Collector's Forum $500; Director's Circle of Fellows $1,000; Director's Circle of Patrons $2,500; Director's Circle of Benefactor's $5,000.

AUSTIN MUSEUM OF ART - LAGUNA GLORIA, 3809 W. 35th St., Austin, TX 78703-1001. Tel.: 512-458-8191 & 495-9224, ext. 313.
Institution Type/Description: Historic House & Art Museum: housed in the former home of author, playwright & politician Clara Driscoll and her husband Hal Sevier; built in 1916. Listed on the National Register of Historic Places.
Collections: paintings; drawings; photographs; sculpture.
Facilities: 12 acre grounds.
Activities: rental facilities; classes.
Hours & Admission Prices: Grounds: Mon.-Sat. 9-5, Sun. 10-5. House: Tues.-Wed. 12-4, Thurs.-Sun. 10-4. Suggested Donation: $3 per person.

AUSTIN NATURE & SCIENCE CENTER, 301 Nature Center Dr., Austin, TX 78746-5775. Tel.: 512-974-3888. Fax: 512-974-3885.
E-mail: kathy.maddox@austintexas.gov
Web Site: www.austintexas.gov/ansc
Founded: 1960.
Congressional District: 10
Key Personnel: Site Supvr., Kathy Maddox.
Personnel Profile: Full-Time Paid 15; Part-Time Paid 40; Part-Time Volunteers 15; Interns 2.
Governing Authority: municipal. Parent Institution: City of Austin, Parks & Recreation Dept.
Institution Type/Description: Nature Center.
Collections: native Texas mammals, fish, birds, reptiles & insects; microscope tables & exhibits, touch-table; participatory exhibits; animal adaptations; hands on trail; local natural artifacts.
Major Exhibits: Nano Technology (T), 1/14-5/14.
Research Fields: wildlife rehabilitation; ornithology; native plants; butterflies.
Facilities: 80-acres of trails; visitor center; gardens. Museum-related items for sale.
Activities: preschool classes; family nature series; environmental education programs; bird walks; plant hikes; wildlife volunteers; caving; canoeing; wildflower forays; night hikes; school tours & classroom programs; rocks & fossils hikes; field trips; Deep Eddy Community Gardens; organic gardening classes & workshops; school & teacher workshops.
Publications: Teachers Resource Guide; Natural Selections.
Hours & Admission Prices: Austin Nature Center: Mon.-Sat. 9-5, Sun. 12-5.

No charge; donations accepted. Closed New Year's Day; Independence Day; Thanksgiving & day after; Christmas Eve & Day.
Attendance: 250,000 (accurate)
Membership: Kids' and Critters' Club: Individual $35; Family $45.

AUSTIN ZOO & ANIMAL SANCTUARY, 10808 Rawhide Tr., Austin, TX 78736. Mailing Address: P.O. Box 91808, Austin, TX 78709. Tel.: 512-288-1490. Fax: 512-288-3972.
E-mail: info@austinzoo.org
Web Site: www.austinzoo.org
Founded: 1994.
Key Personnel: Dir. & Pres. (V), Patti R. Clark.
Personnel Profile: Full-Time Paid 20; Full-Time Volunteers 1; Part-Time Paid 6; Part-Time Volunteers 625; Interns 4.
Governing Authority: Tax-exempt.
Institution Type/Description: Zoo.
Collections: over 300 animals representing more than 100 species; tigers; lions; monkeys; birds; reptiles.
Facilities: Zoo-related items for sale.
Activities: petting zoo; birthday parties; corporate events; train ride.
Hours & Admission Prices: Feb.-Oct. daily 9:30-6; Nov.-Jan. daily 9:30-5:30. Adults $9, students, military & seniors $8, children 2-12 $6; discounts to members. Closed Thanksgiving; Christmas.
Attendance: 181,000 (accurate)
Membership: Individual $50; Family of Four $100 & $30 for each additional member.

BG JOHN C.L. SCRIBNER TEXAS MILITARY FORCES MUSEUM, 2200 W. 35th St., Austin, TX 78703-1222. Mailing Address: P.O. Box 5218, Austin, TX 78763-5218. Tel.: 512-782-5659. Fax: 512-782-6750.
E-mail: museum@tx.ngb.army.mil
Web Site: www.texasmilitaryforcesmuseum.org
Founded: 1992.
Congressional District: 10
Key Personnel: Dir., Jeff Hunt; Deputy Dir. & Registrar, Lisa Sharik; Pres., Tim Weitz; Vice Chm., Richard Gruetzner; Cur. Exhibits, Edward Zapedea.
Personnel Profile: Full-Time Paid 3; Part-Time Volunteers 95; Interns 2.
Governing Authority: Parent Institution: Texas Adjutant General's Dept., P.O. Box 5218, Austin, TX 78763-5218. Tax-exempt: 501(c)(3).
Institution Type/Description: Military Museum: housed at historic Camp Mabry.
Collections: personal artifacts; photographs; paintings; military history of Texas from 1823 to present; vehicles; equipment; uniforms; weapons.
Research Fields: history of Texas military forces from 1823 to present including Texas Revolution, Republic of Texas, Civil War, Indian Wars, Texas Rangers, Spanish American War, WWI, WWII, Cold War, Korean War, State Missions, Global War on Terror.
Facilities: 20,000-vol. library of military history books; classroom; field research station. Museum-related items for sale.
Activities: docent program; guided tours; temporary exhibitions; training programs for professional museum workers; symposia series. Annual Events: American Heroes Celebration - Army & Air Guard displays & reenactments; Close Assault 1944 Living History Demonstrations.
Publications: quarterly newsletter, Texas Military Forces Museum; outdoor display guide; Historical Camp Mabry Walking Tour.
Hours & Admission Prices: Tues.-Sun. 10-4. Photo ID required for entry. No charge; donations accepted. Closed New Year's Day; Thanksgiving; Christmas Eve, Day & day after.
Attendance: 33,000 (accurate)
Membership: Annual $20; Life $250.

✳ **BLANTON MUSEUM OF ART, (M),** 200 E. Martin Luther King Jr. Blvd., Austin, TX 78701. Mailing Address: 200 E. Martin Luther King Jr. Blvd., Stop D1303, Austin, TX 78712-1609. Tel.: 512-471-7324. Fax: 512-471-7023.
E-mail: info@blantonmuseum.org
Web Site: www.blantonmuseum.org
Founded: 1963.
Congressional District: 10
Key Personnel: Dir., Simone Wicha; Dir. Public Rels. & Mktg., Kathleen Brady Stimpert; Dir. Devel., Karen Sumner; Dir. Membership & Museum Svcs., Kim Theel; Dir. Collections & Exhibitions, Gabriela Truly; Mgr. Collections, Sue Ellen Jeffers; Cur. At-Large, Annette DiMeo Carlozzi; Sr. Cur. Prints, Drawings, & European Paintings, Francesca Consagra; Dir. Finance & Admin., Stacey Cilek; Dir. Education & Academic Affairs, Ray Williams; Dir. Facility Operation & Security, Chris Seebach.
Personnel Profile: Full-Time Paid 60; Part-Time Paid 62; Part-Time Volunteers 120.

Governing Authority: state. Parent Institution: The University of Texas at Austin. Tax-exempt: 501(c)(3).
Institution Type/Description: Art Museum.
Collections: antiquities; Renaissance & Baroque art (Suida-Manning Collection); 19th century American art (Michener Collection); contemporary Latin American art; prints & drawings from the 15th-21st centuries.
Research Fields: Renaissance & Baroque art; 20th-century American art; contemporary Latin American art; prints & drawings.
Facilities: 300-seat auditorium; cafe; print study room; classrooms; e-lounge. Gift items for sale.
Activities: concerts; docent program; guided tours; lectures; films; gallery talks; formally organized education programs for K-12 & college students; loan, temporary & traveling exhibitions; training programs for professional museum workers.
Publications: calendar of exhibitions; catalogues; newsletter.
Hours & Admission Prices: Tues.-Fri. 10-5, 3rd Thurs. of month 10-9, Sat. 11-5, Sun. 1-5. Adults $9, seniors $7, youth & college students $5; discounts to AAM members; members, current UT students, faculty & staff, and children 12 & under no charge.
Attendance: 158,000 (accurate)
Membership: Individual $50; Individual Plus & Dual $65; Family $80; Sustaining $125; Founding $350; Contributor $600; Patron $1,200.

✳ **BULLOCK TEXAS STATE HISTORY MUSEUM, (M),** 1800 N. Congress Ave., Austin, TX 78701-1342. Mailing Address: P.O. Box 12874, Austin, TX 78711-2874. Tel.: 512-936-8746. Fax: 512-936-4699.
E-mail: contactus@thestoryoftexas.com
Web Site: www.thestoryoftexas.com
Founded: 2001.
Key Personnel: Dir., Joan Marshall.
Personnel Profile: Full-Time Paid 60; Part-Time Paid 20; Interns 4.
Governing Authority: Parent Institution: State Preservation Board of Texas. Tax-exempt.
Institution Type/Description: History Museum.
Collections: Texas history & culture.
Major Exhibits: La Belle the Ship that Changed Texas History (T), 11/16/13-4/13/14.
Facilities: multi-media, special effects theater; IMAX theater; cafe; bookstore. Museum-related items for sale.
Activities: educational programs.
Hours & Admission Prices: Mon.-Sat. 9-6, Sun. 12-6. Adults $9, students $8, seniors & military $7, youth 4-17 $6; discounts to AAM members; members and children 3 & under no charge. Closed New Year's Day; Easter; Thanksgiving; Christmas Eve & Day.
Attendance: 500,000 (accurate)
Membership: Basic $55.

CAPITOL VISITORS CENTER, A DIVISION OF THE STATE PRESERVATION BOARD, 112 E. Eleventh St., Austin, TX 78701-2403. Mailing Address: P.O. Box 13286, Austin, TX 78711-3286. Tel.: 512-305-8400. Fax: 512-305-8401.
E-mail: cvc.cvc@tspb.state.tx.us
Web Site: www.texascapitolvisitorscenter.com
Founded: 1994.
Congressional District: 21
Key Personnel: Capitol Cur., Ali James; Program Supvr., Kyle Schlafer; Tour Coord., Elizabeth Garzore; Museum Shop Mgr., Shawn Goodnight.
Personnel Profile: Full-Time Paid 6; Part-Time Paid 1.
Governing Authority: state government. Parent Institution: Texas State Preservation Board. Tax-exempt.
Institution Type/Description: Visitors Center: housed in the c.1856 Old General Land Office Building, designed by German architect C.C. Stremme. Oldest state office building in Texas.
Collections: multimedia presentations on the Texas Capitol & Texas history.
Research Fields: land history; Texas State Capitol.
Activities: guided tours; interactive exhibits.
Hours & Admission Prices: Mon.-Sat. 9-5, Sun. 12-5. No charge; donations accepted. Closed New Year's Day; Easter; Thanksgiving; Christmas Eve & Day.
Attendance: 144,308 (accurate)

DOLPH BRISCOE CENTER FOR AMERICAN HISTORY, Sid Richardson Hall, Unit 2, Univ. of Texas at Austin, Austin, TX 78712-0335. Mailing Address: Sid Richardson Hall 2.101, 1 University Station D1100, Austin, TX 78712-0335. Tel.: 512-495-4515. Fax: 512-495-4542.
E-mail: cah.reference@austin.utexas.edu
Web Site: www.cah.utexas.edu

Formerly: Center for American History
Congressional District: 10
Key Personnel: Exec. Dir., Dr. Don Carleton; Assoc. Dir. Media, Alison Beck; Asst. Dir. Exhibits & Material Culture, Lynn Bell; Assoc. Dir. Research & Collections, Brenda Gunn; Assoc. Dir. Devel., Ramona Kelly; Assoc. Dir. Communications, Erin Purdy; Asst. Dir. Facilities & Construction Projects, Denise Mayorga; Asst. Dir. Administration, Echo Uribe.
Personnel Profile: Full-Time Paid 50; Part-Time Paid 31; Part-Time Volunteers 10; Interns 3.
Governing Authority: public university; nonprofit. Parent Institution: University of Texas at Austin. Subsidiary Institutions: Sam Rayburn Museum, P.O. Box 309, Bonham, TX 75418; Winedale, P.O. Box 11, Round Top, TX 78954; Briscoe-Garner Museum, 333 N. Park, Uvalde, TX 78801. Tax-exempt.
Institution Type/Description: History Museum.
Collections: special collections & programs relating to the history of Texas, the South, Rocky Mountain West and select national areas.
Publications: e-newsletter, Dolph Briscoe Center for American History.
Hours & Admission Prices: Research & Collections Division: Academic Year: Mon.-Fri. 10-5, Sat. 9-2. No charge. Closed university holidays; New Year's Day; Independence Day; Labor Day; Thanksgiving; Christmas. &
Attendance: 39,000 (estimated)

ELISABET NEY MUSEUM, (M), 304 E. 44th St., Austin, TX 78751-3813. Tel.: 512-458-2255. Fax: 512-453-0638.
E-mail: enm@austintexas.gov
Web Site: www.elisabetneymuseum.org
Founded: 1911.
Congressional District: 10
Personnel Profile: Full-Time Paid 2; Part-Time Paid 2; Part-Time Volunteers 15.
Governing Authority: City of Austin, Texas. Tax-exempt: 501(c)(1).
Institution Type/Description: Historic Site & Interpretive Center.
Collections: Elisabet Ney's portraits in marble & plaster & personal artifacts, letters & memorabilia, housed & displayed in her former Austin, Texas studio.
Research Fields: portrait sculpture; 19th-century art history; Texas cultural history; women's studies; women artists.
Facilities: restored studio/museum; classroom annex.
Activities: guided tours; lectures; concerts; educational & public programming.
Hours & Admission Prices: Wed.-Sun. 12-5. No charge; donations accepted.
Attendance: 13,000 (accurate)

THE FRENCH LEGATION MUSEUM, (M), 802 San Marcos St., Austin, TX 78702-2647. Tel.: 512-472-8180. Fax: 512-472-9547.
E-mail: director@frenchlegationmuseum.org
Web Site: www.frenchlegationmuseum.org
Founded: 1956.
Congressional District: 10
Key Personnel: Chm., Gayla Lawson; Dir., Lise Ragbir; Assoc. Cur., Noel Harris Freeze; Programs Coord., Franke L. Smith.
Personnel Profile: Full-Time Paid 2; Part-Time Paid 4; Part-Time Volunteers 15.
Governing Authority: Daughters of the Republic of Texas, Inc. Tax-exempt: 501(c)(3).
Institution Type/Description: Historic Building Museum: housed in the former home of the French Charge d'Affaires to the Republic of Texas; built in 1841.
Collections: period French & early American furnishings; artifacts reflecting life in early days of Republic of Texas, c.1836-1846.
Research Fields: early Austin, Texas history; westward migration; Texas in relation to the Atlantic world; the enslaved in Texas; early Texas medical practices.
Facilities: meeting room. Museum-related items for sale.
Activities: guided tours; concerts; weddings; seasonal events; children's summer camp.
Publications: books, The French Legation in Austin, Volumes I & II; A History of the French Legation in Texas.
Hours & Admission Prices: Tues.-Sun. 1-5. Adults $5, senior citizens $3, students & teachers $2; discounts to groups, Texas Assoc. of Museums, Daughters of the Republic of Texas, AAA & AAM members; members & children under 5 no charge. Closed holidays. &
Attendance: 14,000 (estimated)
Membership: Student, Teacher & Senior Friend $20; General $30; Family $50; Charge d'Affaires $100; Diplomat $250; Ambassador $500.

GEORGE WASHINGTON CARVER MUSEUM & CULTURAL CENTER, 1165 Angelina St., Austin, TX 78702-2034. Tel.: 512-974-4926. Fax: 512-974-3699.
E-mail: carver.museum@austintexas.gov
Web Site: www.carvermuseum.org
Key Personnel: Dir., Bernadette M. Phifer.
Governing Authority: city. Parent Institution: Austin Parks and Recreation Dept. Tax-exempt.
Institution Type/Description: African American History & Culture Museum.
Collections: African-American history, culture, & art.
Activities: special events; school programs.
Hours & Admission Prices: Mon. & Fri. 9:30-6, Tues.-Thurs. 9:30-8, Sat. 1-5. No charge. &
Attendance: 100,500 (accurate)

GOODWILL COMPUTER MUSEUM, (M), 1015 Norwood Park Blvd., Austin, TX 78753-6608. Tel.: 512-637-7539. Facebook: Goodwill Computer Museum.
E-mail: museum@austingoodwill.org
Web Site: www.goodwillcomputermuseum.org
Founded: 2005.
Personnel Profile: Full-Time Paid 1; Part-Time Volunteers 5; Interns 1.
Governing Authority: Parent Institution: Goodwill Industries of Central Texas. Tax-exempt.
Institution Type/Description: Computer Museum.
Collections: computing history; period artifacts; documentation; marketing materials; restored & functional early computers; hands-on exhibits; recycled computer art; early computer hardware & software.
Research Fields: legacy data recovery.
Facilities: 1,200 sq. ft. exhibit space; laboratories.
Activities: hands-on interactive displays; functional legacy computer hardware; early computer games.
Hours & Admission Prices: Mon.-Sat. 10-8, Sun. 11-6. No charge. &
Attendance: 3,000 (estimated)

HARRY RANSOM CENTER AT THE UNIVERSITY OF TEXAS AT AUSTIN, 21st & Guadalupe, Austin, TX 78712. Mailing Address: P.O. Box 7219, Austin, TX 78713-7219. Tel.: 512-471-8944. Fax: 512-471-9646.
E-mail: webmail@hrc.utexas.edu
Web Site: www.hrc.utexas.edu
Founded: 1957.
Congressional District: 10
Key Personnel: Dir., Dr. Stephen Enniss; Assoc. Dir. Exhibitions, Cathy Henderson; Librarian, Dr. Richard W. Oram; Cur. Art, Peter Mears; Cur. Photography, Jessica S. McDonald.
Personnel Profile: Full-Time Paid 110; Part-Time Volunteers 25.
Governing Authority: university; nonprofit. Parent Institution: University of Texas at Austin. Tax-exempt: 501(c)(3).
Institution Type/Description: Humanities Research Library.
Collections: Gutenberg bible c.1455; first photograph c.1826; film archives of David O. Selznick & Robert DeNiro; paintings by Frida Kahlo & Diego River; manuscripts of James Joyce, Ernest Hemingway, T.S. Eliot, D.H. Lawrence, Tennessee Williams, & Norman Mailer.
Major Exhibits: The World at War, 1914-1918, 2/11/14-8/3/14; The Making of Gone With the Wind, 9/9/14-1/4/15.
Research Fields: 20th-21st century American, British & anglophone literature, manuscripts & books; 20th-21st century photography; 20th-21st century French literature & culture; art; performing arts; film.
Facilities: library; archives; educational facilities.
Activities: guided tours; lectures; temporary exhibits; docent program; graduate student internship program; post-doctoral research fellowship award program.
Publications: newsletter & books for an imprint series.
Hours & Admission Prices: Library Reading Room: Mon.-Fri. 9-5, Sat. 9-12. Exhibits: Tues.-Wed. & Fri. 10-5, Thurs. 10-7, Sat.-Sun. 12-5. No charge; donations accepted. Closed university holidays. &
Attendance: 80,000 (accurate)
Membership: Individual $50; Dual $90; Sustainer $125; Alliance $250; Guild $500; Director's Circle $1,000.

LADY BIRD JOHNSON WILDFLOWER CENTER, 4801 La Crosse Ave., Austin, TX 78739-1702. Tel.: 512-232-0100. Fax: 512-232-0156.
Web Site: www.wildflower.org
Founded: 1982.
Key Personnel: Pres., Deacon Turner; Founder, Lady Bird Johnson; Founder, Helen Hayes; Exec. Dir., Susan K. Rieff; Dir. Communications, Saralee Tiede; Dir. Plant Conservation, Flo Oxley; Dir. Finance & Operations,

Richard Tomhave; Dir. Product Mktg. & Museum Shop Mgr., Joseph Hammer; Dir. Landscape Restoration, Steve Windhager, Ph.D.; Dir. Devel., Justin Michalka; Dir. Gardens, Andrea DeLong-Amaya.

Personnel Profile: Full-Time Paid 50; Part-Time Paid 13; Part-Time Volunteers 400; Interns 5.

Governing Authority: not for profit organization. Parent Institution: University of Texas at Austin. Tax-exempt: 501(c)(3).

Institution Type/Description: Botanical Garden & Nature Center.

Collections: plants from the Southwest U.S. & northern Mexico; 25,000 images of North American native plants.

Research Fields: urban ecological engineering; restoration; ecology; native plant propagation.

Facilities: 1,800-vol. library on native plants of temperate North America; 232-seat auditorium; cafe; 23 theme gardens; nature center. Museum-related items for sale.

Activities: guided tours; lectures; workshops; traveling exhibitions; education programs for children & adults; docent program; internship programs for college students. Annual Events: Wildflower Days; spring & fall plant sales; Luminations Winter Festival; Goblins in the Garden; Artisans Festival; Holiday Shopping Event.

Publications: monthly newsletter; quarterly magazine, Wildflower; quarterly calendar of events.

Hours & Admission Prices: Center: April-May daily 9-5:30; June-March Tues.-Sun. 9-5:30. Cafe: Tues.-Sat. 9-4, Sun. 11-4. Adults $7, students & senior citizens over 60 $6, children 5 & over $3; discounts to AAM members; children 4 & under and members no charge. Closed New Year's Day; Independence Day; Thanksgiving; Christmas Eve, Day & week. &

Attendance: 100,000 (accurate)

Membership: Students & Seniors $30; Students & Seniors Dual $38; Individual $40; Dual $50; Family $65; Supporting $100; Contributor $250; Sustaining $500; Champion $1,000; Sunflower Society $1,500.

LYNDON BAINES JOHNSON LIBRARY AND MUSEUM, 2313 Red River St., Austin, TX 78705-5737. Tel.: 512-721-0200. Fax: 512-721-0171 & 0170. Facebook: LBJ Library Friends.

E-mail: johnson.library@nara.gov

Web Site: www.lbjlibrary.org

Founded: 1971.

Congressional District: 10

Key Personnel: Dir., Mark Updegrove; Deputy Dir., Tina Houston; Registrar, Michael MacDonald; Asst. Registrar, Renee Bair; Volunteer Coord., Laura Eggert; Museum Shop Mgr., Blair Newberry.

Personnel Profile: Full-Time Paid 50; Part-Time Paid 14; Part-Time Volunteers 90.

Volunteer Hours: 9,591

Governing Authority: federal. Parent Institution: National Archives & Records Admin., Washington, DC 20408. Tax-exempt: 170(b)(1)(A).

Institution Type/Description: Presidential Library.

Collections: personal papers; government records; still photographs; motion picture films; audio & video tapes; sound recordings; head of state gifts; gifts from private citizens; political campaign items; personal & family memorabilia.

Major Exhibits: American Soldier (T), 11/13-2/14; White House Christmas, 11/13-1/14; 60 From The 60's, 3/14-1/15; Cornerstones of Civil Rights, 4/14; Sculpture of the American West, 5/14-11/14; World Leaders on Baseballs, 6/14-12/14; Christmas at the White House, 11/14-1/15.

Research Fields: 1960s; life, times, career & administration of the 36th President of the United States.

Facilities: 15,000-vol. library; 3,900 serials; auditoriums & conference rooms; manuscript reading room; photography & video laboratory. Museum-related items for sale.

Activities: guided tours; lectures; films; permanent, temporary & traveling exhibitions; organized educational programs for children, adults, undergraduate & graduate college students; docent program; symposiums.

Publications: exhibition catalogues; list of holdings; bibliography; books; general information brochures.

Hours & Admission Prices: Daily 9-5. Adults $8, seniors $5, college students with ID & youth 13-17 $3; children 12 & under, members, active-duty military, school groups with reservations, University of Texas faculty, staff & students, Martin Luther King Day, Presidents' Day, Memorial Day, Independence Day, LBJ's Birthday (Aug. 27th), Veterans Day, Austin Museum Day & Explore UT no charge. Closed Christmas. &

Attendance: 126,820 (accurate)

Membership: Senior Citizen $65; Individual $80; Senior Dual $120; Dual $150; Sustaining $250; Fellow $500; Legacy $1,000.

MEXIC-ARTE MUSEUM, (M), 419 Congress, Austin, TX 78701-3619. Mailing Address: P.O. Box 2273, Austin, TX 78768-2273. Tel.: 512-480-9373. Fax: 512-480-8626.

E-mail: info@mexic-artemuseum.org

Web Site: www.mexic-artemuseum.org

Founded: 1984.

Congressional District: 10

Key Personnel: C.E.O., Founding & Artistic Dir., Sylvia Orozco; Managing Dir., Frank M. Rodriguez; Pres. Bd., Dr. John Hogg; Pres. (V), Lulu Flores; Dir. Public Rels., Alexandra M. Landeros; Dir. Communications & Public Rels., Joy Diaz; Museum Shop Mgr., Joan Savodelli.

Personnel Profile: Full-Time Paid 6; Full-Time Volunteers 1; Part-Time Paid 3; Part-Time Volunteers 50; Interns 5.

Governing Authority: private; nonprofit organization. Tax-exempt: 501(c)(3).

Institution Type/Description: Art Museum; focus on Mexican & Latin American culture.

Collections: concentration on 20th-century Mexican art; prints from the workshop of Popular Graphics; Jose Guadalupe Posada prints; photographs by Agustin Casasola of the Mexican Revolution; masks from Guerrero; photographs documenting Mexican-American history in Austin.

Facilities: 5,000-vol. library.

Publications: newsletter; major exhibits catalogues.

Hours & Admission Prices: Mon.-Thurs. 10-6, Fri.-Sat. 10-5, Sun. 12-5. Adults $5; discounts to AAM members; children, members & Sun. no charge. Closed major holidays. &

Attendance: 50,000 (estimated)

Membership: $40 & up.

NEILL-COCHRAN HOUSE MUSEUM, (M), 2310 San Gabriel St., Austin, TX 78705-5014. Tel.: 512-478-2335. Fax: 512-478-1865.

E-mail: info@nchmuseum.org

Web Site: www.nchmuseum.org

Founded: 1962.

Congressional District: 10

Key Personnel: Dir., Kristen Hennessey; Project Coord., Bethany Boucher.

Personnel Profile: Full-Time Paid 2; Part-Time Paid 1; Part-Time Volunteers 10; Interns 2.

Governing Authority: NSCDA in Texas. Tax-exempt.

Institution Type/Description: Historic House: house in the Neill-Cochran House; built in 1855.

Collections: furniture & furnishings of the period 1855-1915.

Major Exhibits: Abner Cook, Master Builder, 1/14-12/14.

Facilities: 200-vol. library of genealogical books & material available for research by permission.

Activities: guided tours; speaker series; docent course; temporary exhibitions.

Publications: member newsletter, Among Friends.

Hours & Admission Prices: Tues.-Sat. 1-4. Adults $5; discounts to AAA members, Time Travelers & school groups; members no charge. Closed New Year's Eve & Day; Easter; Thanksgiving; Christmas Eve & Day. &

Attendance: 2,500 (accurate)

Membership: Student $20; House Society $45; Museum Society $60; Hill Society $120; Neill Society $240; Cochran Society $600; Abner Cook Society $1,000; Bluebonnet Circle $2,000.

O. HENRY MUSEUM, 409 E. 5th St., Austin, TX 78701-3705. Tel.: 512-472-1903. Fax: 512-472-7102.

E-mail: ohenrymuseum@ci.austin.tx.us

Web Site: www.ohenrymuseum.org

Founded: 1934.

Congressional District: 10

Key Personnel: Cur., Valerie Bennett.

Personnel Profile: Full-Time Paid 1; Part-Time Paid 1; Part-Time Volunteers 10.

Governing Authority: municipal. Parent Institution: Austin Parks and Recreation Dept. Tax-exempt.

Institution Type/Description: Historic House Museum: 1891 home of O. Henry.

Collections: furnishings; memorabilia; documents; letters; photographs.

Activities: guided tours; permanent exhibitions; special events; writing classes; historical & literary outreach. Museum Sponsors: Pun-Off.

Publications: newsletter, Pun-Intended.

Hours & Admission Prices: Wed.-Sun. 12-5. No charge; donations accepted. Closed New Year's Day; Independence Day; Labor Day; Thanksgiving; Christmas.

Attendance: 15,000 (estimated)

Membership: Student & Seniors $15; Individual $20; Family $25; Benefactor & Sponsor $100, $500, $1,000.

REPUBLIC OF TEXAS MUSEUM, 510 E. Anderson Lane, Austin, TX 78752-1218. Tel.: 512-339-1997. Fax: 512-339-1998.
E-mail: headquarters@drtinfo.org
Web Site: www.drtinfo.org
Founded: 1895.
Governing Authority: nonprofit organization. Parent Institution: The Daughters of the Republic of Texas. Tax-exempt.
Institution Type/Description: History Museum.
Collections: Republic of Texas history & culture; photographs.
Facilities: library. Museum-related items for sale.
Activities: lectures; special events; educational programs.
Hours & Admission Prices: Mon.-Fri. 10-4. Adults $5, seniors $3, children & students $2. &

TEXAS GOVERNOR'S MANSION, 1010 Colorado St., Austin, TX 78701-2334. Mailing Address: P.O. Box 12428, Austin, TX 78711. Tel.: 512-305-8524.
E-mail: mansion.tours@tspb.state.tx.us
Web Site: governor.state.tx.us/mansion
Founded: 1856.
Key Personnel: Admin., Liz Geise.
Governing Authority: state government; nonprofit.
Institution Type/Description: Historic House & Site: designated as a National Historic Landmark in 1975.
Collections: furnishings; American Federal & Empire periods; furniture & decorative arts.
Publications: video, Texas Governor's Mansion.
Hours & Admission Prices: Tours: Tues.-Thurs. 10 to noon. Reservation required. &

TEXAS HISTORICAL COMMISSION, 1511 Colorado, Austin, TX 78701-1664. Mailing Address: P.O. Box 12276, Austin, TX 78711-2276. Tel.: 512-463-7948. Fax: 512-463-7002. Facebook: Texas Historical Commission.
E-mail: thc@thc.state.tx.us
Web Site: www.thc.state.tx.us
Founded: 1953.
Congressional District: 5
Key Personnel: Chm. Commission, Matt Kreisle; Exec. Dir., Mark Wolfe; Archaeology Div., Pat Mercado-Allinger; Architecture Div., Sharon Fleming; Staff Svcs. Div., Lynn Ward; Public Information & Education, Heather McBride; History Programs Div., Bratten Thomason; Historic Sites Div., Donna Williams; Community Heritage Devel. Div., Brad Patterson.
Personnel Profile: Full-Time Paid 175.
Governing Authority: state. 20 Branch Museums: Acton State Historic Site; Caddo Mounds State Historic Site; Casa Navarro State Historic Site; Confederate Reunion Grounds State Historic Site; Eisenhower Birthplace State Historic Site; Fannin Battleground State Historic Site; Fort Griffin State Historic Site; Fort Lancaster State Historic Site; Fort McKavett State Historic Site; Fulton Mansion State Historic Site; Landmark Inn State Historic Site; Levi Jordan Plantation State Historic Site; Magoffin Home State Historic Site; National Museum of the Pacific War; Sabine Pass Battleground State Historic Site; Sam Bell Maxey House State Historic Site; Sam Rayburn House Museum; San Felipe de Austin State Historic Site; Starr Family Home State Historic Site; Varner-Hogg Plantation State Historic Site.
Institution Type/Description: State Agency.
Collections: historic sites; archeological materials; period furnishings; costumes; textiles; household artifacts; ephemera; early photographs; fine art; period weapons.
Research Fields: preservation of historic landmarks; architecture; archaeology; historic sites; cultural resource management; museum management & services; field work data processing.
Facilities: 1,500-vol. library on Texas history, museology, management, preservation, archaeology, conservation, for use on premises; curatorial repository & lab.
Activities: lectures; administration of historic preservation programs for state; workshops; conferences; public programs.
Publications: bimonthly newsletter, Medallion; biennial reports; heritage tourism brochures.
Hours & Admission Prices: See individual listings. &

TEXAS MUSIC MUSEUM, 1109 E. 11th St., Austin, TX 78761. Mailing Address: P.O. Box 16467, Austin, TX 78761-6467. Tel.: 512-472-8891 & 471-0520. Fax: 512-471-9600.
E-mail: cshorkey@mail.utexas.edu
Web Site: www.texasmusicmuseum.org
Founded: 1984.

Congressional District: 10
Key Personnel: Pres. (V), Dr. Clayton Shorkey; Vice Pres., Rudy Martinez; Sec., Lawrence Egle; Treas., Joyce Davids Christianson.
Personnel Profile: Part-Time Volunteers 20.
Governing Authority: private; nonprofit organization. Tax-exempt.
Institution Type/Description: Music Museum.
Collections: music instrument-makers; recording & reproduction devices; music art, education, publishing & mechanical devices; artifacts; achievements & memorabilia pertaining to the composition, performance, reproduction & promotion of musical arts in Texas.
Major Exhibits: Austin African American Musicians, 1/14-12/14; Texas Music Historical Journey (1930-2014), 2/14-9/14.
Research Fields: musical heritage of Texas.
Facilities: 100-vol. library of Texas music books.
Activities: educational & recreational programs; meetings; performances; rental gallery; traveling exhibitions.
Hours & Admission Prices: Mon.-Fri. 8-5. No charge; donations accepted. &
Attendance: 10,000 (estimated)

TEXAS NATURAL SCIENCE CENTER, (M), 2400 Trinity, Austin, TX 78705-5730. Tel.: 512-471-1604. Fax: 512-471-4794.
E-mail: tmmweb@uts.cc.utexas.edu
Web Site: www.texasmemorialmuseum.org
Formerly: Texas Memorial Museum
Founded: 1936.
Congressional District: 10
Key Personnel: Dir., Edward C. Theriot; Dir. Museum Operations, Margaret Fischer; Dir. External Affairs, Susan Romberg; Accountant, Sara Gray; Museum Educator, Christina Ramsey Cid; Webmaster, Sharon Ruether; Artist, John Maisano; Museum Shop Mgr., Louise Meeks; LAN Mgr., Melissa Winans; Cur. Ichthyology, Dean A. Hendrickson; Cur. Entomology, John Abbott; Cur. Arthropods, James R. Reddell; Cur. Herpetology, David C. Cannatella; Asst. Cur. Herpetology, Travis LaDuc; Dir. Vertebrate Paleontology Lab, Timothy Rowe; Professor Emeritus, Vertebrate Paleontology Lab, Wann Langston; Professor Emeritus, Vertebrate Paleontology, Ernest Lundelius; Preparator, Vertebrate Paleontology Lab, Matt Brown; Collection Coord., Non-Vertebrate Paleontology Collection, Ann Molineux; Collections Mgr. Ichthyology, Jessica Rosales; Sr. Paleontology Educator, Pamela Owen; Paleontology Educator, Laura Naski; Collections Mgr., Vertebrate Paleontology, Chris Sagebiel; Security, Mike Fallon.
Personnel Profile: Full-Time Paid 18; Part-Time Paid 4; Part-Time Volunteers 23; Interns 2.
Governing Authority: state. Parent Institution: University of Texas at Austin. Tax-exempt: 170(b)(1)(A).
Institution Type/Description: Natural Science Museum.
Collections: geology; mineralogy; meteorites; tektites; vertebrate & invertebrate paleontology; paleobotany; vertebrate zoology; marine invertebrate zoology; entomology; arachnology.
Research Fields: paleontology; geology; vertebrate & invertebrate zoology; entomology; ichthyology; herpetology.
Facilities: reference library of scientific materials; archives; laboratories; research space. Nature-related items for sale.
Activities: permanent & temporary exhibitions; offsite exhibits; guided tours; collections care and maintenance; inter-museum loans; teacher training; museum studies courses in collaboration with the Univ. of Texas at Austin; research & publications; self-guided tours; weekend special events & activities.
Publications: quarterly e-newsletter, Smart Bytes.
Hours & Admission Prices: Mon.-Fri. 9-5, Sat. 10-5, Sun. 1-5. No charge; donations accepted. Closed New Year's Day; Easter; Independence Day; Thanksgiving; Christmas. &
Attendance: 80,952 (accurate)
Membership: Teacher K-12 $15; Individual & Family $40; Naturalist $125; Explorer $250; Visionary $500; Director's Circle $1,000 & up.

TEXAS PARKS AND WILDLIFE DEPARTMENT, 4200 Smith School Rd., Austin, TX 78744-3292. Tel.: 512-389-4800 & 800-792-1112.
Web Site: www.tpwd.state.tx.us
Founded: 1963.
Congressional District: 10
Key Personnel: Exec. Dir., Carter Smith; Dir. State Parks, Brent Leisure; Dir. Cultural Resources Mgmt., Michael Strutt; Dir. Interpretive Svcs., Chris Holmes; Chief Cur., Joanne Avant; Dir. Natural Resources Management, David Riskind.
Governing Authority: state. Tax-exempt.
Institution Type/Description: Texas Parks & Wildlife Agency: administers over 130 state parks, recreation areas, historical parks, historic sites, historic structures & natural areas.

Collections: decorative arts; period furnishings; personal articles of early Texas settlers & statesmen; World War I & II military materials as they relate to the Battleship Texas; archaeological material culture; pictographs; petroglyphs; paleontological; natural history; living history reenactments; landscaping; mechanical systems; architecture; farming; ranching; agriculture; 19th-century transportation, railroad & stagecoach; 19th-century frontier military history; clothing & textile; photographs; department division collection.

Research Fields: natural history; threatened & endangered species; archaeology; decorative arts; military history; Texas & Civil War history; Texas Revolution; World Wars I & II; 19th-century costume & textile history; state, regional, county & local history; Civilian Conservation Corps in Texas.

Facilities: picnic areas; camping; hiking trails; visitor centers; interpretive centers; amphitheaters.

Activities: guided tours; self-guided tours; lectures; films; audiovisual programs; permanent, temporary & traveling exhibitions; education programs; volunteer & docent training.

Publications: brochures; maps; trail guides; pamphlets; books; newsletters; department monthly magazine; six hunting, fishing & non-game stamps issued annually.

Hours & Admission Prices: Headquarters: Mon.-Fri. 8-5, call sites for hours. Fees vary. Annual Pass: $60. &

Membership: Various friends, volunteers and other support groups attached to each particular park.

TEXAS STATE LIBRARY & ARCHIVES COMMISSION, 1201 Brazos St., Austin, TX 78701-1938. Mailing Address: Box 12927, Austin, TX 78711-2927. Tel.: 512-463-5460. Fax: 512-463-5436.
E-mail: info@tsl.state.tx.us
Web Site: www.tsl.state.tx.us
Founded: 1839.
Key Personnel: Dir. & Librarian, Peggy D. Rudd; Asst. State Librarian, Edward Seidenberg.
Governing Authority: state. Tax-exempt: 501(c)(3).
Institution Type/Description: History Museum.
Collections: documents; artifacts; firearms; archives; manuscripts; photographs.
Research Fields: Texas history.
Facilities: 35,000-vol. library of Texana & general reference books available for research on premises; reading room.
Activities: permanent exhibitions.
Hours & Admission Prices: Library: Mon.-Fri. 8-5. Genealogy Section: Mon.-Fri. 8-5. Sam Houston Center: Mon.-Fri. 8-5, Sat. 9-4. No charge. Closed major holidays. &
Attendance: 5,000

THINKERY, 1830 Simond Ave., Austin, TX 78723. Tel.: 512-472-2499. Fax: 512-472-2495. Facebook: The Thinkery.
E-mail: info@thinkeryaustin.org
Web Site: www.thinkeryaustin.org
Formerly: Austin Children's Museum
Founded: 1983.
Congressional District: 10
Key Personnel: Exec. Dir., Mike Nellis.
Personnel Profile: Full-Time Paid 38; Part-Time Paid 100; Part-Time Volunteers 1,500; Interns 12.
Governing Authority: nonprofit organization. Tax-exempt: 501(c)(3).
Institution Type/Description: Children's Museum.
Collections: hands on exhibits.
Research Fields: parent & child relations; cognition & learning; science literacy.
Facilities: 40,000 sq. ft. exhibit space. Educational & cultural items for sale.
Activities: storytime; group tours; gingerbread workshops; birthday parties; facility rental. Museum Sponsors: Engineering Saturday; Science Sunday; Community Night; Spring Break Camps; Summer Camps; Sleepovers; Baby Bloomers; Cub Club.
Publications: quarterly newsletter; monthly e-newsletter.
Hours & Admission Prices: Tues. & Thurs.-Sat. 10-5, Wed. 10-8, Sun. 12-5. Adults & children over 2 $9; Wed. 5pm-8pm by donation; children under 2 & Sun. 4-5 no charge. ASTC & ACM reciprocal memberships. Closed all major holidays. &
Attendance: 220,000 (estimated)
Membership: Family (Basic) $95; Premier $150. Club House $350; Clubhouse Visionary $1,000.

UMLAUF SCULPTURE GARDEN & MUSEUM, (M), 605 Robert E. Lee Rd., Austin, TX 78704-1453. Tel.: 512-445-5582. Fax: 512-445-5583.
E-mail: director@umlaufsculpture.org
Web Site: umlaufsculpture.org
Founded: 1991.
Congressional District: 10
Key Personnel: Exec. Dir., Nina Seely; Museum Cur., Katie Robinson Edwards.
Governing Authority: private; nonprofit organization. Tax-exempt: 501(c)(3).
Institution Type/Description: Sculpture Museum.
Collections: over 130 sculptures by Charles Umlauf; bronze castings; marble carvings; drawings; paintings; mediums used by Umlauf include exotic woods, stone, aluminum, pewter, cast stone, stoneware, terra cotta.
Research Fields: sculptor Charles Umlauf; locating Umlauf sculptures in public placements across the U.S. (in conjunction with Save Outdoor Sculpture); developing museum programs for special needs visitors; developing interactive workbooks about sculpture for school children visiting the museum & self-guided tour kits for families.
Facilities: library available for scholarly research; archives; xeriscape garden; 3,500 sq. ft. exhibit space. Museum-related items for sale.
Activities: guided tours; lectures; sculpture demonstrations; poetry readings; dance recitals; formal education programs for children; docent program; training programs for professional museum workers; arts festivals; temporary exhibits; museum video captioned for hearing-impaired visitors; touch tours available for vision-impaired visitors. Annual Event: Museum's Birthday Celebration in June.
Publications: biannual museum newsletter, Garden Grapevine; monthly volunteers newsletter, Volunteer View; interactive workbooks about sculpture for school children; self-guided tour kits for families.
Hours & Admission Prices: Wed.-Fri. 10-4, Sat.-Sun.12-4. Adults $5, senior citizens 60 & up $3, students $1; discounts for AAM, American Horticulture Society & Texas Association Museum members; members, active military personnel & family and children 12 & under no charge. Closed New Year's Eve & Day; Independence Day; Thanksgiving; Christmas Eve & Day. &
Attendance: 27,000 (accurate)
Membership: The Muse $25; Lotus $45; The Family $75; The Kiss $150; The Skater $250; The Diver $500; Spirit of Flight $1,000.

WOMEN & THEIR WORK ART GALLERY, 1710 Lavaca St., Austin, TX 78701. Tel.: 512-477-1064. Fax: 512-477-1090.
E-mail: info@womenandtheirwork.org
Web Site: www.womenandtheirwork.org
Key Personnel: Exec. Dir., Chris Cowden
Institution Type/Description: Art Gallery.
Collections: works by contemporary female artists.
Activities: workshops; educational programs.
Hours & Admission Prices: Mon.-Fri. 10-6, Sat. 12-5. No charge; donations accepted.

Austwell

ARANSAS NATIONAL WILDLIFE REFUGE & CLAUDE F. LARD VISITOR CENTER, 1 Wildlife Cir., Austwell, TX 77950. Mailing Address: P.O. Box 100, Austwell, TX 77950. Tel.: 361-286-3559. Fax: 361-286-3722.
Institution Type/Description: Wildlife Refuge.
Collections: wildlife & their habitats; plants; photographs.
Facilities: nature trails; auditorium. Center-related items for sale.
Activities: educational programs; observation tower; special events.
Publications: pamphlets.
Hours & Admission Prices: Refuge: daily sunrise to sunset. Visitor Center: daily 8:30-4:30. Closed Thanksgiving; Christmas.

Baird

CALLAHAN COUNTY PIONEER MUSEUM, 100 W. 4th, B-1, Baird, TX 79504-5305. Tel.: 325-854-5875. Fax: 325-854-5841.
E-mail: sonia.walker@callahancounty.org
Founded: 1940.
Congressional District: 87
Key Personnel: Cur., Sonia Walker.
Personnel Profile: Part-Time Paid 1.
Governing Authority: nonprofit organization. Parent Institution: Callahan County.
Institution Type/Description: History Museum.
Collections: local pioneer articles.
Hours & Admission Prices: Mon.-Fri. 1-5. No charge. Closed major holidays. &

Attendance: 372 (accurate)

Bandera

FRONTIER TIMES MUSEUM, (M), 510 13th St., Bandera, TX 78003. Mailing Address: P.O. Box 1918, Bandera, TX 78003-1918. Tel.: 830-796-3864.
E-mail: information@frontiertimesmuseum.org
Web Site: www.frontiertimesmuseum.org
Founded: 1933.
Congressional District: 21
Key Personnel: Pres., George Sharman; Vice Pres., Harry Harris; Treas., Carol Anne Boyle; Sec., Clare Jo Anderson; Dir., Rebecca Norton.
Personnel Profile: Full-Time Paid 1; Part-Time Paid 3; Part-Time Volunteers 17.
Governing Authority: nonprofit organization. Tax-exempt: 501(c)(3).
Institution Type/Description: Historic Building.
Collections: emphasizing the Old West & Texas's early pioneer & ranching days; Western art; historical photographs; local rodeo champions & the Texas Heroes Hall of Honor.
Major Exhibits: Medina Lake, 100 Years Lakeside (T).
Facilities: Museum-related items for sale.
Activities: Annual Events: National Day of the Cowboy.
Publications: annual membership letter.
Hours & Admission Prices: Mon.-Sat. 10-4:30. Adults $5, seniors $3, students 6-18 $2; discounts to groups upon approval; children under 6 & members no charge. Closed Easter; Thanksgiving; Christmas. &
Attendance: 10,000 (accurate)
Membership: Individual $25; Family $50; Business $100.

Bay City

MATAGORDA COUNTY MUSEUM, (M), 2100 Ave. F, Bay City, TX 77414-0851. Tel.: 979-245-7502. Fax: 979-245-1233.
E-mail: mcma@matagordamuseum.com
Web Site: www.matagordamuseum.com
Founded: 1965.
Congressional District: 14
Key Personnel: Dir., Barbara Smith.
Personnel Profile: Part-Time Paid 5; Part-Time Volunteers 40.
Governing Authority: nonprofit organization. Parent Institution: Matagorda County Museum Assn. Tax-exempt: 501(c)(3).
Institution Type/Description: Historical Society Museum: housed in c.1917 federal post office building.
Collections: Matagorda County history from the time of the Karankawa Indians; Stephen F. Austin colonization; Republic of Texas; Civil War; cattle ranching into the 20th century; LaBelle shipwreck.
Facilities: 150-vol. library on Texas History; 87 annuals dating back to 1910; probate records & marriage records (1837-1900), available for public use; 8,000 sq. ft. exhibit space. Note cards & postcards for sale.
Activities: guided tours; lectures; docent program; formal education programs for children; rental gallery; temporary & traveling exhibitions; dinner theatre. Museum Sponsors: Annual Christmas Tour; Historical Summer Camp - children's museum.
Publications: quarterly newsletter, Matagorda County Musings.
Hours & Admission Prices: Wed.-Sun. 1-5. Adults $4, seniors $3, children $2; discounts to AAM, ICOM, TAM & SETMA members; members & children under 2 no charge. Closed New Year's Day; Memorial Day; Good Friday; Labor Day; Thanksgiving; Christmas. &
Attendance: 10,500 (accurate)
Membership: Individual $20; Family $30; Friend $50; Donor $100; Sponsor $250; Patron $500; Benefactor $1,000.

Baytown

BAYTOWN HISTORICAL MUSEUM, 220 W. Defee St., Baytown, TX 77520-4010. Tel.: 281-427-8768.
Institution Type/Description: History Museum.
Collections: local history & culture; early pioneer life; period artifacts; photographs.
Hours & Admission Prices: Tues.-Sat. 10-2. No charge.

BAYTOWN NATURE CENTER, 6213 Bayway Dr., Baytown, TX 77520. Mailing Address: c/o Eddie V Gray Wetlands Center, 1724 Market, Baytown, TX 77520. Tel.: 281-932-1972.
Governing Authority: nonprofit organization.
Institution Type/Description: Nature Center.
Collections: wildlife including over 300 species of birds; butterfly garden.
Facilities: visitor center; children's discovery center.

Hours & Admission Prices: Daily sunrise to sunset. Adults $3; children 12 & under no charge.

EDDIE V. GRAY WETLANDS CENTER, 1724 Market St., Baytown, TX 77520. Tel.: 281-420-7128.
Governing Authority: city.
Institution Type/Description: Nature Center.
Collections: local history; alligators; turtles; fish; butterflies; mounted animals; photographs; aquariums; terrariums.
Facilities: 9,000 sq. ft. exhibit space.
Activities: hands-on exhibits; educational programs.
Hours & Admission Prices: Mon.-Fri. 9-4, Sat. 10-4; groups by appointment. Closed holidays.

Beaumont

* **ART MUSEUM OF SOUTHEAST TEXAS, (M),** 500 Main St., Beaumont, TX 77701-3213. Mailing Address: P.O. Box 3703, Beaumont, TX 77704-3703. Tel.: 409-832-3432. Fax: 409-832-8508.
E-mail: info@amset.org
Web Site: www.amset.org
Founded: 1950.
Congressional District: 5
Key Personnel: Exec. Dir., Lynn P. Castle; Cur. Exhibitions & Collections, Sarah Hamilton; Pres., Melanie Dishman; Administrator Finance & Personnel, Patricia Siebert; Registrar, Paula Rodriguez; Public Rels., Monique Sennet; Asst. to Dir., Elizabeth Gorris; Cur. Education, Sandra Laurette; Museum Shop Mgr., Kathy Boudreaux.
Personnel Profile: Full-Time Paid 7; Part-Time Paid 9; Part-Time Volunteers 165.
Governing Authority: nonprofit corporation. Tax-exempt: 501(c)(3).
Institution Type/Description: Art Museum.
Collections: 19th-21st century American painting, sculpture, photography & folk art.
Research Fields: 20th-21st century American art; 20th-21st century Texas & American self-taught art.
Facilities: 1,500-vol. library of books on art & 6,000 slides available for research; 120-seat auditorium; restaurant; classrooms. Museum-related items for sale.
Activities: guided tours; travel tours; lectures; films; gallery talks; mobile outreach program; public children's computer lab; docent program for teens & adults; summer art camp; after school programs; Head Start; high school art contest. Museum Sponsors: Family Arts Day.
Publications: exhibition catalogues; gallery handouts; quarterly newsletter; posters.
Hours & Admission Prices: Mon.-Fri. 9-5, Sat. 10-5, Sun. 12-5. No charge; donations accepted. Closed New Year's Eve & Day; Easter; Independence Day; Thanksgiving; Christmas Eve & Day. &
Attendance: 36,000 (accurate)
Membership: Individual $30; Family $35; Friend $100; Fellow $200; Patron $500; President's Club $1,000; Sustainer $2,500; Business $250-$10,000.

THE ART STUDIO, INC., 720 Franklin St., Beaumont, TX 77701-4424. Tel.: 409-838-5393. Fax: 409-838-4695. Facebook: The Art Studio, Inc.
E-mail: artstudio@artstudio.org
Web Site: www.artstudio.org
Founded: 1983.
Congressional District: 9
Key Personnel: C.E.O., Dir. & Public Rels., Greg Busceme; Devel. & Membership and Administrative Asst., Elizabeth French.
Personnel Profile: Full-Time Paid 1; Part-Time Paid 1; Part-Time Volunteers 16; Interns 5.
Governing Authority: nonprofit organization. Tax-exempt: 501(c)(3).
Institution Type/Description: Art Gallery with Exhibit Area.
Collections: Bob Willis Memorial Ceramic Collection & Memorial Art Library.
Major Exhibits: Elizabeth Fontenot, 2/1/14-2/21/14; Acquae Obscurae, 3/1/14-3/21/14; Tasimjae - Membership Show, 4/5/14-4/25/14; Lisa Reinauer & Ken Baskin, 5/3/14-5/23/14; The Alternative Show, 6/7/14-6/27/14; The Tenant Show, 9/6/14-9/26/14; Mark Nesmith, 10/4/14-10/24/14; Holiday Shop O Rama, 12/6/14-12/19/14.
Facilities: classrooms; 14,000 sq. ft. exhibit space; art studios; performance stage; darkroom.
Activities: guided tours; lectures; hobby workshops; organized educational programs; at risk youth outreach; participatory exhibits.
Publications: monthly alternative press open to artists, writers & musicians, Issue.

Hours & Admission Prices: Tues.-Fri. 2-5, Sat. 1-4; other times by appointment. No charge; donations accepted. Closed holidays. &

Attendance: 5,000 (estimated)

Membership: Individual $35; Family/Group $50; Friend/Business $100; Sustaining $250; Patron $500; Angel $1,000; Benefactor $2,000; Life $10,000.

BABE DIDRIKSON ZAHARIAS MUSEUM, 1750 E IH-10, Beaumont, TX 77701. Mailing Address: P.O. Box 3827, Beaumont, TX 77704-3827. Tel.: 409-833-4622. Fax: 409-880-3750. Facebook: Babe Didrikson Zaharias Museum.

E-mail: klewis@ci.beaumont.tx.us

Web Site: www.babedidriksonzaharias.org

Founded: 1976.

Key Personnel: Cur., Karen Lewis.

Personnel Profile: Part-Time Paid 4.

Governing Authority: municipal government; nonprofit. Tax-exempt.

Institution Type/Description: Sports Museum.

Collections: sports trophies; golf clubs & bags; newspaper clippings; photographs; Olympic medals & videotape relating to Babe Didrikson Zaharias.

Hours & Admission Prices: Daily 9-5. No charge; donations accepted. Closed Thanksgiving & Christmas. &

Attendance: 3,000 (estimated)

BEAUMONT ART LEAGUE, 2675 Gulf St., Beaumont, TX 77703-4417. Tel.: 409-833-4179. Facebook: Beaumont Art League.

E-mail: info@beaumontartleague

Web Site: beaumontartleague.org

Founded: 1943.

Congressional District: 9

Key Personnel: Pres./Chm., Richard Tallent; Gallery Dir., Sarah Hamilton.

Governing Authority: nonprofit organization. Tax-exempt: 501(c)(3).

Institution Type/Description: Art Museum & Gallery.

Collections: works by local & national artists; paintings; drawings; prints; photographs; sculpture.

Facilities: classroom; 4,757 sq. ft. exhibit space. Art work for sale.

Activities: educational programs; participatory exhibits; workshops.

Publications: monthly newsletter, Beaumont Art League; class schedule; exhibit invitations.

Hours & Admission Prices: Wed.-Sat. 11-3. No charge; donations accepted. Closed New Year's Day; Independence Day; Thanksgiving; Christmas. &

Attendance: 2,000 (estimated)

Membership: Student $25; Individual $40; Family $50; Patron $100; Lifetime $1,000.

BEAUMONT ART LEAGUE, BROWN-SCURLOCK GALLERIES, 2675 Gulf St., Beaumont, TX 77703-4417. Tel.: 409-833-4179. Facebook: Beaumont Art League.

E-mail: info@beaumontartleague.org

Web Site: www.beaumontartleague.org

Founded: 1943.

Key Personnel: Dir., Sarah Hamilton; Pres. (V), Richard Tallent.

Personnel Profile: Part-Time Paid 1; Part-Time Volunteers 15.

Governing Authority: Operated by Beaumont Art League.

Institution Type/Description: Art Gallery.

Collections: paintings; photographs; sculpture.

Activities: art classes.

Publications: bimonthly newsletter, Collage.

Hours & Admission Prices: Wed.-Sat. 11-3. No charge; donations accepted. Closed New Year's; Thanksgiving; Christmas. &

Attendance: 2,000 (estimated)

Membership: Student $20; Individual $35; Family $45; Patron $100; Lifetime $1,000.

BEAUMONT COUNCIL OF GARDEN CLUBS, 6088 Babe Zaharias Dr., Beaumont, TX 77705-6747. Tel.: 409-842-3135. Fax: 409-840-6456.

E-mail: bcgc@beaumontbotanicalgardens.org

Founded: 1971.

Congressional District: 9

Key Personnel: Treas. & Education, Randy Hammerling; Devel. & Membership, Frankie Pletzer.

Personnel Profile: Full-Time Paid 4; Full-Time Volunteers 1; Part-Time Paid 3; Part-Time Volunteers 12.

Governing Authority: nonprofit organization. Tax-exempt: 501(c)(3).

Institution Type/Description: Botanical Garden.

Collections: horticultural specimens of ferns, shrubs, annuals, perennials, & bulbs; rose garden with 260 bushes; Japanese Garden. Warren Loose Conservatory: tropical plants including bromeliads, ferns, palms, citrus.

Facilities: 500-vol. library of material on horticulture, native plants, artistic design of flowers & landscape design available to the public; 300-seat garden center; 14-acre botanical garden; nature & conservation center; horticultural center.

Activities: guided tours; lectures; films; study clubs; organized educational programs; flower shows; seminars. Annual Event: Herb Fest.

Publications: quarterly newsletter, Take Time to Smell the Flowers; brochures.

Hours & Admission Prices: Gardens: daily 7:30 a.m.-dusk. No charge; donations accepted. Warren Loose Conservatory: Mon.-Tues. & Thurs.-Fri. 9-4, Sat.-Sun. 1-5. No charge; donations accepted. &

Attendance: 12,300 (estimated)

Membership: $10-$20 per club according to number of members.

DISHMAN ART MUSEUM, (M), 1030 E. Lavaca, Beaumont, TX 77705. Mailing Address: P.O. Box 10027, Beaumont, TX 77710-0027. Tel.: 409-880-8959. Fax: 409-880-1799.

E-mail: alicia.hargreaves@lamar.edu

Web Site: dept.lamar.edu/cofac/deptart/dishman.asp

Founded: 1983.

Congressional District: 9

Key Personnel: Dir., Jessica M. Dandona; Museum Asst., Alicia Hargreaves.

Personnel Profile: Full-Time Paid 2; Part-Time Paid 1; Part-Time Volunteers 2; Interns 2.

Governing Authority: public university. Parent Institution: Lamar University. Tax-exempt.

Institution Type/Description: Art Museum.

Collections: African & New Guinea masks & shields; 140 19th-century paintings; 250 19th-century porcelains; contemporary paintings & prints.

Research Fields: 19th-century academic European & American paintings.

Facilities: 500-vol. library of 19th-century & modern books; 6,000 sq. ft. exhibit space.

Activities: arts festivals; lectures. Annual Events: fundraiser ball; artist in residence.

Publications: catalogues for selected exhibitions.

Hours & Admission Prices: Mon.-Fri. 8-5. No charge; donations accepted. Closed New Year's Day; Good Friday; Memorial Day; Labor Day; Thanksgiving & day after; Christmas. &

Attendance: 4,000 (estimated)

EDISON MUSEUM, 350 Pine St., Beaumont, TX 77701-2437. Tel.: 409-981-3089. Fax: 409-838-2361.

E-mail: info@edisonmuseum.org

Web Site: www.edisonmuseum.org

Formerly: Edison Plaza Museum

Founded: 1980.

Congressional District: 9

Key Personnel: C.E.O. & Dir., Mellissa BiJeaux; Chm. (V), Vernon Pierce.

Personnel Profile: Full-Time Paid 1; Part-Time Paid 1; Part-Time Volunteers 4.

Governing Authority: private; not-for-profit organization. Parent Institution: Entergy/Texas. Tax-exempt.

Institution Type/Description: History Museum: housed in the Travis Street substation; the first substation to distribute electric power in Southeast Texas.

Collections: focus is on inventions & innovations of Thomas A. Edison.

Activities: Edison Junior Inventor Program.

Publications: A Teacher's Guide To The Edison Plaza Museum; The Inventor Bulletin Series.

Hours & Admission Prices: Mon.-Fri. 9-5. No charge; donations accepted. Closed major holidays. &

Attendance: 3,380 (accurate)

FIRE MUSEUM OF TEXAS, (M), 400 Walnut at Mulberry, Beaumont, TX 77701. Mailing Address: P.O. Box 3827, Beaumont, TX 77704-3827. Tel.: 409-880-3927. Fax: 409-880-3914.

E-mail: firemuseum@ci.beaumont.tx.us

Web Site: www.firemuseumoftexas.org

Founded: 1986.

Congressional District: 2

Key Personnel: Museum Mgr., Ami Kamara; Fire Chief, Anne Huff; Pres. Association (V), Ted Hillin.

Personnel Profile: Full-Time Paid 1; Part-Time Volunteers 24; Interns 1.

Governing Authority: municipal; City of Beaumont Fire Dept. Tax-exempt: 501(c)(3).

Institution Type/Description: Firefighting Museum: 1927 Beaumont Fire Department Firehouse; Texas Historical Landmark; Spanish Renaissance Revival.

Collections: firefighting services in Texas; hands-on fire safety; photographs & patches; fire alarm systems; international firefighting memorabilia; fire pumpers: 1856 Hand Drawn Tub pumper, 1917 American LaFrance piston pumper, 1926 American LaFrance rotary pumper; 1909 American LaFrance aerial ladder; 911 Memorial; State of Texas Firefighter Memorial.

Major Exhibits: Texas City Burning: The 1947 Texas City Disaster, 4/6/14-6/6/14.

Research Fields: history of fire service in Beaumont & in Texas; fire alarm systems; fire prevention education.

Facilities: 100-vol. library available for research by permission only; theater; kitchen facilities; fire safety activity center. Fire service related items for sale.

Activities: guided tours; fire prevention education programs; docent program; children's hands-on fire safety activity center; fire safety summer camps; annual car show; fire safety special events.

Publications: brochure; annual report; quarterly newsletter.

Hours & Admission Prices: Mon.-Fri. 8-4:30, Sat.-Sun. by appointment. No charge; donations accepted. Closed New Year's Day; Martin Luther King Jr. Day; Good Friday; Memorial Day; Independence Day; Labor Day; Thanksgiving & day after; Christmas. &

Attendance: 22,837 (accurate)

Membership: Junior Firefighter 12 & under $5; Firefighter $15; Individual $20; Family $35; Driver $50; Captain $100; District Chief $250; Asst. Fire Chief $500; Fire Chief $1,000.

JOHN JAY FRENCH HOUSE, BEAUMONT HERITAGE SOCIETY, (M), 3025 French Rd., Beaumont, TX 77706-7920. Mailing Address: Beaumont Heritage Society, 2240 Calder Ave., Beaumont, TX 77701. Tel.: 409-832-4010. Fax: 409-832-4012. Facebook: Beaumont Heritage Society.

E-mail: chambershse@sbcglobal.net

Web Site: beaumontheritage.org

Founded: 1968.

Congressional District: 9

Key Personnel: Exec. Dir., Darlene Chodzinski.

Governing Authority: nonprofit organization. Parent Institution: Beaumont Heritage Society. Tax-exempt: 501(c)(3).

Institution Type/Description: Historic House Museum: c.1845 Greek Revival house, John Jay French home.

Collections: mid-1840s decorative arts; pre-Civil War era furnishing & accessories; blacksmith shop; semi-detached kitchen.

Research Fields: Beaumont history; decorative arts & crafts.

Facilities: 425-vol. research library; 680 sq. ft. exhibit space. Museum-related items for sale.

Activities: guided tours; lectures; docent training program; formal education program for children & undergraduate or graduate college students; school tours mornings Feb.-April. Museum Sponsors: Christmas Candlelight Tour; Tex Fest; ethnic food fair; heritage dinner.

Publications: quarterly newsletter.

Hours & Admission Prices: Tues.-Sat. 10-4; group tours by appointment. Adults $3, senior citizens $2, students $1. Closed major holidays. &

Attendance: 12,000 (estimated)

Membership: Student $25; Individual $35; Family $50; Patron $100; Benefactor $200; Life $1,500.

✳ MCFADDIN-WARD HOUSE, (M), Visitor Center, 1906 Calder Ave., Beaumont, TX 77701-1517. Mailing Address: 725 Third St., Beaumont, TX 77701-1629. Tel.: 409-832-1906 & 2134. Fax: 409-832-3483.

E-mail: arlene@mcfaddin-ward.org

Web Site: www.mcfaddin-ward.org

Founded: 1983.

Congressional District: 9

Key Personnel: Dir., Allen Lea; Cur. Interpretation & Education, Judy Linsley; Volunteer Coord., Becky Fertitta; Mgr. Facilities, Felix McFarland; Asst. Dir. & Admin., Arlene Christiansen.

Personnel Profile: Full-Time Paid 13; Part-Time Paid 6; Part-Time Volunteers 130; Interns 1.

Governing Authority: nonprofit. Tax-exempt: 501(c)(3).

Institution Type/Description: Historic House: 1906 Beaux Arts Colonial house built by early Texas oil & ranching family.

Collections: American factory-made furniture; arts & crafts; furniture & decorative arts; original 19th & 20th-century family furnishings including English & American silver, ceramics, glass, oriental rugs, paintings. Historic Building: carriage house.

Research Fields: local & southeast Texas history; 20th-century furniture; English & American silver; ceramics; glass; oriental rugs.

Facilities: visitor center.

Activities: guided tours; educational programs; special events; seasonal interpretations.

Publications: quarterly newsletter; annual report; exhibit catalog; historical conference proceedings.

Hours & Admission Prices: Tues.-Sat. 10, 11, 1:30 & 2:30, Sun. 1-3. McFaddin-Ward House: children under 8 not admitted. Carriage House: Tues.-Sat. 10-4, Sun. 1-3. Reservations recommended. Adults $3; discount to AAM & TAM members; members no charge. Closed major holidays. &

Attendance: 10,387 (accurate)

SPINDLETOP/GLADYS CITY BOOMTOWN MUSEUM, 5550 University Dr., Beaumont, TX 77705. Mailing Address: P.O. Box 10070, Beaumont, TX 77710-0070. Tel.: 409-835-0823. Fax: 409-832-1782.

E-mail: christy.marino@lamar.edu

Web Site: www.spindletop.org

Founded: 1975.

Congressional District: 9

Key Personnel: Cur., Christy Marino.

Personnel Profile: Full-Time Paid 1; Part-Time Paid 4; Part-Time Volunteers 10.

Governing Authority: state; nonprofit. Parent Institution: Lamar University. Tax-exempt: 501(c)(3).

Institution Type/Description: Recreation of Oil Boomtown: located at the site of the Spindletop oilfield, the first major oilfield in the U.S., discovered in 1901.

Collections: oilfield machinery & tools; paper goods; household items; furniture; textiles; glass items; photographs; books.

Research Fields: Spindletop oil field history; oral history collection.

Facilities: 15-building complex.

Activities: guided tours; educational programs; research.

Hours & Admission Prices: Tues.-Sat. 10-5, Sun. 1-5; school groups by appointment. Adults $5, senior citizens 60 & over $3, children 6-12 $2; children 5 & under no charge. Closed all major holidays. &

Attendance: 25,000 (estimated)

TEXAS ENERGY MUSEUM, 600 Main St., Beaumont, TX 77701-3305. Tel.: 409-833-5100. Fax: 409-833-4282.

E-mail: ryan@texasenergymuseum.org

Web Site: www.texasenergymuseum.org

Founded: 1987.

Congressional District: 9

Key Personnel: Dir. & C.E.O., D. Ryan Smith; Chm. (V), Vernon Pierce; Museum Shop Mgr., Raphaella Cortello.

Personnel Profile: Full-Time Paid 3; Part-Time Paid 2; Part-Time Volunteers 12.

Governing Authority: not-for-profit organization. Tax-exempt: 501(c)(3).

Institution Type/Description: Industry History Museum.

Collections: petroleum geology & technology; history of the petroleum industry especially relating to southeast Texas.

Research Fields: history, economics & technology of the world wide oil & gas industry, with emphasis on the U.S., Mexico & Canada; astronomy; chemistry; physics; geology; paleontology; oceanography.

Facilities: aquarium; educational facilities; 16,000 sq. ft. exhibit space. Museum-related items for sale.

Activities: docent program; guided tours; lectures; temporary exhibitions; children's programs.

Publications: brochures.

Hours & Admission Prices: Tues.-Sat. 9-5, Sun. 1-5. Adults $2, senior citizens & children 6-12 $1; discounts to AAM members; Lamar University students with valid I.D. & Texas Assoc. of Museums members no charge. &

Attendance: 16,000 (accurate)

TYRRELL HISTORICAL LIBRARY, 695 Pearl St., Beaumont, TX 77701. Mailing Address: P.O. Box 3827, 695 Pearl St., Beaumont, TX 77704-3827. Tel.: 409-833-2759. Fax: 409-833-5828.

E-mail: wgrace@ci.beaumont.tx.us

Web Site: http://www.beaumontlibrary.org

Personnel Profile: Full-Time Paid 4; Part-Time Paid 1; Part-Time Volunteers 4; Interns 1.

Governing Authority: Parent Institution: City of Beaumont.

Institution Type/Description: Historical Library: housed in a former Baptist Church; built in 1903. Listed on the National Register of Historic Places.

Collections: books; genealogy; paintings.

Research Fields: history; genealogy.

Facilities: library of Texas history books.

Activities: research.

Publications: quarterly, The Journal.

Hours & Admission Prices: Mon.-Sat. 9-6. No charge. Closed city holidays. &

Attendance: 5,257 (accurate)

Beeville

BEEVILLE ART MUSEUM, (M), 401 E. Fannin, Beeville, TX 78102-3515. Tel.: 361-358-8615. Fax: 361-358-0413.
Key Personnel: Dir., Tracy Bell Saucier.
Governing Authority: Tax-exempt: 501(c)(3).
Institution Type/Description: Art Museum.
Collections: works by local artists.
Facilities: library.
Activities: educational programs; summer art camps; workshops.
Hours & Admission Prices: Mon.-Fri. 9-5, Sat. 10-2. No charge.

Belton

BELL COUNTY MUSEUM, (M), 201 N. Main St., Belton, TX 76513-3160. Mailing Address: P.O. Box 1381, Belton, TX 76513-5381. Tel.: 254-933-5243. Fax: 254-933-5756.
E-mail: museum@co.bell.tx.us
Web Site: www.bellcountymuseum.org
Founded: 1975.
Congressional District: 11
Key Personnel: Dir., Stephanie Turnham; Chm., Allen Cloud; Cur., Emily Dossman.
Personnel Profile: Full-Time Paid 4; Full-Time Volunteers 1; Part-Time Volunteers 30; Interns 1.
Volunteer Hours: 2,000
Operating Income: 274,000
Governing Authority: county. Parent Institution: Bell County. Subsidiary Institution: Bell County Museum Association, Inc. Tax-exempt: 170(b)(1)(A).
Institution Type/Description: History Museum: housed in c.1904 Carnegie Library Building.
Collections: local history items; Governor Miriam A. (Ma) Ferguson collection; mammoth tusk.
Research Fields: The Governors Ferguson.
Facilities: 16,000 sq. ft. exhibit space.
Activities: theater; organized educational programs; tours; downtown Belton walking tour.
Publications: bimonthly newsletter.
Hours & Admission Prices: Tues.-Sat. 12-5. No charge; donations accepted. &
Attendance: 12,500 (estimated)
Membership: Friend $25; Family $50; Patron $100; Partner $250; Benefactor $500; Associate $1,000 & up.

Benjamin

WICHITA - BRAZOS MUSEUM & CULTURAL CENTER, 200 E. Hayes St., Benjamin, TX 79505. Mailing Address: Box 104, Benjamin, TX 79505-0104. Tel.: 940-459-2229. Fax: 940-459-2229.
E-mail: kchc@srcaccess.net
Web Site: www.knoxcountytexas.com
Formerly: Knox County Museum
Founded: 1966.
Congressional District: 13
Key Personnel: C.E.O. & Chm. (V), Mary Jane Young.
Personnel Profile: Part-Time Volunteers 26.
Governing Authority: county; nonprofit. Parent Institution: Friends of Knox County Historical Commission. Tax-exempt.
Institution Type/Description: History Museum.
Collections: photographs & books relating to life in Knox County from prehistoric period to early 20th-century; pioneer artifacts; Civil War items; family histories; Knox County Veterans Memorial listing over 2,000 area veterans.
Research Fields: Knox County & region prehistoric to present, emphasis on 1860-1945; family history file; the great southern buffalo herd hunt of 1870s.
Publications: bi-annual newsletter, Wichita Brazos Museum; survey of cemetery; compiling genealogical history of Knox County families 1800-present; book, Knox County Historical Cookbook; book, Knox County Veterans Book; Knox County annual calendar; brochures.
Hours & Admission Prices: Tues.-Fri. 1-5. No charge; donations accepted. Closed legal holidays. &
Attendance: 700 (accurate)

Big Bend National Park

BIG BEND NATIONAL PARK, Science and Resource Management Center, 266 Tecolote Dr., Big Bend National Park, TX 79834. Mailing Address: Science and Resource Management Center, P.O. Box 129, Big Bend National Park, TX 79834-0129. Tel.: 1944. Fax: 432-477-1153.
E-mail: kate_hogue@nps.gov
Web Site: www.nps.gov/bibe
Founded: 1944.
Congressional District: 23
Key Personnel: Cur., Kate Hogue; Chief, Science & Resource Management, Phil Wilson.
Personnel Profile: Part-Time Paid 1.
Governing Authority: federal. Parent Institution: U.S. Dept. of the Interior, National Park Service. Tax-exempt.
Institution Type/Description: National Park Visitor Centers.
Collections: herbarium; insects; herpetology; archaeological; historical; paleontological; geological; period photographs; archives including history of BBNP establishment & development.
Research Fields: biological; archaeological; archives.
Facilities: 1,500-vol. library of books about local, human & natural history available for use on premises; archives. Books, slides & posters for sale.
Activities: guided tours; formally organized education programs; illustrated programs.
Publications: orientation brochures; park brochures; site bulletins; park newspaper.
Hours & Admission Prices: Daily 8-5. Annual Pass $40; autos $20 per week; bicycles, motorcycles & bus passengers $10 per week; Golden Age Passports: U.S. citizens 62 & over $10. &
Attendance: 360,000 (accurate)
Membership: Big Bend Natural History Association: Regular $25; Associate $50; Life $250; Corporate $500; Benefactor $1,000.

Big Lake

HICKMAN MUSEUM, 609 Main St., Big Lake, TX 76932. Mailing Address: 1005 N. Plaza Ave., Big Lake, TX 76932. Tel.: 325-884-2082.
Founded: 1984.
Governing Authority: Tax-exempt.
Institution Type/Description: Historic House Museum: housed in the former home of Gracie Hickman.
Collections: local history & culture; period furnishings; personal artifacts; early photographs; school, business & industry artifacts; early clothing; Native American artifacts; quilts; telephone switchboard; farming & ranching tools.
Activities: guided tours.
Hours & Admission Prices: Closed for renovations.
Attendance: 260 (estimated)

Big Spring

HANGAR 25 AIR MUSEUM, 1911 Apron Dr., Big Spring, TX 79720-7807. Mailing Address: P.O. Box 2925, Big Spring, TX 79721-2925. Tel.: 432-264-1999. Fax: 432-466-0316.
E-mail: hangar25@crcom.net
Web Site: www.hangar25airmuseum.com
Founded: 1999.
Key Personnel: Pres. Bd., Bruce Schooler; Admin., Genevieve Stockburger; Museum Shop Mgr., Amber Stokes.
Personnel Profile: Full-Time Paid 1; Part-Time Paid 1; Part-Time Volunteers 5; Interns 1.
Governing Authority: Tax-exempt: 501(c)(3).
Institution Type/Description: Military History Museum.
Collections: military aviation history; aircraft; uniforms; photographs; newspapers.
Facilities: 14,000 sq. ft. exhibit space.
Activities: research; educational programs.
Publications: newsletter, The Briefing Room, three times year.
Hours & Admission Prices: Tues.-Fri. 10-4, Sat. 10-2. No charge; donations accepted. &
Attendance: 5,000 (estimated)
Membership: Basic $30; Veteran $45; Contributing $50; Sponsor $100; Supporting $250; Patron $500; Benefactor $1,000; Ambassador $5,000.

HERITAGE MUSEUM & POTTON HOUSE, 510 Scurry, Big Spring, TX 79720-2736. Tel.: 432-267-8255. Fax: 432-267-9998 (call first).
E-mail: heritagemus@gmail.com
Web Site: www.bigspringmuseum.com
Founded: 1971.
Congressional District: 17
Key Personnel: Pres., Katie Cathey; Dir., Nancy Raney; Cur., Tammy Schrecengost.
Personnel Profile: Full-Time Paid 3; Full-Time Volunteers 23; Part-Time Paid 3; Part-Time Volunteers 4.
Governing Authority: nonprofit. Subsidiary Institution: Potton House. Tax-exempt.
Institution Type/Description: Historic Museum.
Collections: paintings by Harvey W. Caylor; early artifacts of the area; pioneer settlement; cattle industry railroad; ranching, farm & industry items; photographs; clothing; prehistoric cultures; architecture.
Research Fields: local & county history.
Facilities: 150-vol. library of local & ranching history available to the public.
Activities: guided tours; films; docent program; school loan service; lectures; historical & modern slideshow with sound tape narration; gallery talks; formally organized educational programs; temporary & traveling exhibitions.
Publications: quarterly newsletter; books, Howard County Historian; Gettin' Started: Howard County's First 25 Years; H.W. Caylor: Frontier Artist; Howard County In The Making.
Hours & Admission Prices: Tues.-Fri. 8:30-4, Sat. 10-4 by appointment. Adults $2, students & seniors $1; members no charge. Closed national holidays. &
Attendance: 15,000 (accurate)
Membership: Individual $15; Sustaining $20; Sponsor $30; Patron $50; Endowment $100; Benefactor $500.

Blanket

BLANKET HISTORICAL MUSEUM, 1200 S. Broadway, Blanket, TX 76432. Tel.: 325-748-2491.
Institution Type/Description: History Museum: housed in the former Blanket State Bank; built in 1901.
Collections: local history & culture; period furnishings; personal artifacts; photographs; family histories; books; early tools & equipment; Model T Ford; bank safes; surgical equipment.
Hours & Admission Prices: By appointment.
Attendance: 50

Boerne

AGRICULTURAL HERITAGE MUSEUM, 102 City Park Rd., Boerne, TX 78006. Mailing Address: P.O. Box 1076, Boerne, TX 78006-1076.
E-mail: info@agmuseum.org
Web Site: www.agmuseum.us
Founded: 1986.
Congressional District: 21
Key Personnel: Dir. & Pres., Gary Luxon; Treas., John Barteau.
Personnel Profile: Full-Time Volunteers 6; Part-Time Volunteers 10.
Operating Expenses: 3,543
Operating Income: 3,315
Governing Authority: nonprofit. Tax-exempt: 501(c)(3).
Institution Type/Description: Agriculture Museum.
Collections: articles relating to outdoor ranching and farming life in Texas Hill country & south Texas; blacksmith shop; woodworking shop.
Facilities: 20-vol. library of agricultural material available to the public; 5 acres of exhibit space.
Activities: guided tours; organized education programs for children; traveling exhibitions.
Publications: biannual newsletter.
Hours & Admission Prices: late Jan. to late Nov. Sat. 10-4; other times by appointment. Adults $5; children 12 & under no charge. Closed major holidays. &
Attendance: 4,000 (estimated)

KUHLMANN KING HISTORICAL HOUSE AND MUSEUM, 402 E. Blanco, Boerne, TX 78006-2008. Mailing Address: P.O. Box 178, Boerne, TX 78006-0178. Tel.: 830-249-7277.
Web Site: www.nootsweb.com/~txkendal
Formerly: Boerne Area Historical Preservation Society
Founded: 1970.
Congressional District: 21
Key Personnel: Pres., Carolyn Harz Goodall; Vice Pres., Barbara Phillip; Treas., Louise Davis.

Personnel Profile: Part-Time Volunteers 20.
Governing Authority: nonprofit organization. Parent Institution: City of Boerne. Tax-exempt.
Institution Type/Description: Historic House: c.1880, Kuhlmann-King Family Home. Historical Landmark; historical business bldg. c.1850's.
Collections: telephone operator's switchboard; period telephones; Ad Toepperwein sharp shooter; early Ebensberger Mortuary tools; Boerne Village Band (1860-2007); family photographs & histories; pioneer kitchen; 1614 Low German bible; historic structures.
Research Fields: family history; Boerne area & Kendall County History.
Facilities: Museum-related items for sale.
Activities: tours; Dickens on Main; permanent & temporary exhibits; Kendall Co. Historical Commission Passport Events.
Publications: books, Gone But Not Forgotten; Vols. I & II; Cemetery Surveys; Kuhlman-King House Museum & Archives; self-guided tour of Boerne Cemetery; Journey to Boerne;.
Hours & Admission Prices: Appointment only. No charge; donations accepted. &
Attendance: 1,500 (estimated)
Membership: Individual $15; Couple $20; Booster $50; Support $100; Business $250.

Bonham

FANNIN COUNTY MUSEUM OF HISTORY, One Main St., Bonham, TX 75418-4345. Tel.: 903-583-8042.
E-mail: fcmuseum@hotmail.com
Founded: 1987.
Key Personnel: Dir., Jean Dodson; Pres. (V), Glenn Taylor.
Personnel Profile: Part-Time Paid 1.
Governing Authority: Parent Institution: Fannin County Museum Society. Tax-exempt.
Institution Type/Description: History Museum: housed in the former Texas and Pacific Railway Depot; c.1900. Listed on the National Register of Historic Places.
Collections: local history & culture; railroad memorabilia & artifacts; fossils; Native American artifacts; period clothing; telephone switchboard; personal artifacts; photographs; 1936 Ford convertible; early fire truck.
Hours & Admission Prices: April-Sept. 1 Tues.-Sat. 10-4; Sept. 2-March Tues.-Sat. 12-4. No charge; donations accepted.
Attendance: 3,500 (accurate)
Membership: Individual $15; Family $20; Sustaining $30; Docent $100; Curator $250; Life $500.

FORT INGLISH, Hwy. 56 & Chinner St., Bonham, TX 75418. Mailing Address: P.O. Box 395, Bonham, TX 75418-0395.
Founded: 1976.
Congressional District: 4
Key Personnel: Pres. (V), Tom Thornton; Chm. (V), Glenn Taylor; Museum Shop Mgr., Mildred Welch.
Personnel Profile: Full-Time Volunteers 1; Part-Time Volunteers 20.
Governing Authority: private; nonprofit. Tax-exempt: 501(c)(3).
Institution Type/Description: History Museum: housed in a replica of log fort used as protection against Indians, 1837-1843.
Collections: concentration on northeast Texas history, 1837-1850; period artifacts & furniture. Historic Buildings: blacksmith shop; general store; one-room cabin; school & church; 1830s log cabin.
Facilities: educational facilities; 1,800 sq. ft. exhibit space. Museum-related items for sale.
Activities: guided-tours by volunteers dressed in period attire; informal education programs for children; self-guided tours; lye soap made on premises.
Publications: annual newsletter, Fort Inglish Courier.
Hours & Admission Prices: April-Sept. 1 Tues.-Sat. 10-4; guided tours & groups by appointment. No charge; donations accepted. &
Attendance: 3,500 (accurate)
Membership: Single $20; Family $25; Benefactor $26-$249; Life $250.

SAM RAYBURN HOUSE MUSEUM, 890 W. Hwy. 56, Bonham, TX 75418. Mailing Address: P.O. Box 308, Bonham, TX 75418-0308. Tel.: 903-583-5558. Fax: 903-640-0800. Facebook: Visit Sam Rayburn House.
E-mail: srhm@thc.state.tx.us
Web Site: www.visitsamrayhouse.com
Founded: 1975.
Congressional District: 4
Key Personnel: Site Mgr., Carole Stanton; Cur., Anne Ruppert.
Personnel Profile: Full-Time Paid 4; Part-Time Volunteers 2.

Governing Authority: state. Operated by the Texas Historical Commission, P.O. Box 12276, Austin, TX. 78711. Tax-exempt.
Institution Type/Description: Historic House: 1916 two-story frame farmhouse with Colonial Revival facade, built by Sam Rayburn.
Collections: personal belongings & household furnishings of Sam Rayburn; vehicles owned by Sam Rayburn.
Research Fields: U.S. Government, particularly the office of the Speaker of the House of Representatives; local history; the functioning of the U.S. Congress & the electoral process; history of Sam Rayburn.
Facilities: Books & gifts for sale.
Activities: guided tours; lectures; community workshops; docent program.
Publications: quarterly newsletter.
Hours & Admission Prices: Tues.-Sun. 9-5. Adults $4, students 6-18 & adults in tour groups $3, students in school groups $1; discounts to groups of 8 or more. Closed New Year's Eve & Day; Thanksgiving; Christmas Eve & Day. &
Attendance: 7,500 (accurate)
Membership: Adult $10; Family $20; Corporate $50; Contributor $100; Patron $500; Founder $1,000.

THE SAM RAYBURN MUSEUM, A DIVISION OF THE DOLPH BRISCOE CENTER FOR AMERICAN HISTORY, 800 W. Sam Rayburn Dr., Bonham, TX 75418-4103. Mailing Address: P.O. Box 309, Bonham, TX 75418-0309. Tel.: 903-583-2455. Fax: 903-583-7394.
Web Site: www.cah.utexas.edu
Founded: 1957.
Congressional District: 4
Key Personnel: Dir., Dr. Don Carleton.
Personnel Profile: Full-Time Paid 2; Part-Time Paid 1.
Governing Authority: bd. of trustees. Affiliated with the University of Texas at Austin. Tax-exempt: 501(c)(3).
Institution Type/Description: Biographical & Historical Museum.
Collections: paintings; memorabilia; Sam Rayburn's personal papers on microfilm; American history & government books; reproduction of seals; replica of Speaker's official office in the Capitol Building with original furniture.
Research Fields: history & government.
Facilities: library; reading room; film room.
Activities: guided tours; films.
Hours & Admission Prices: Mon.-Fri. 9-4:30, Sat. 10-2. No charge; donations accepted. Closed all major holidays. &
Attendance: 7,500 (estimated)

Borger

HUTCHINSON COUNTY MUSEUM, 618 N. Main, Borger, TX 79007-3529. Tel.: 806-273-0130. Fax: 806-273-0128.
E-mail: hcmuseum@cableone.net
Web Site: www.hutchinsoncountymuseum.org
Founded: 1977.
Congressional District: 18
Key Personnel: Dir., Wesley Phillips; Admin., Lynn Hopkins; Museum Shop Mgr., Judy Mihm.
Personnel Profile: Full-Time Paid 2; Part-Time Paid 1; Part-Time Volunteers 20.
Governing Authority: county. Parent Institution: Hutchinson County Historical Commission. Tax-exempt.
Institution Type/Description: History Museum: housed in 1927 Grand Hardware, early store, located in 1926 Oil Boom Town.
Collections: local history of the county from prehistoric times to the present, emphasizing the Oil Boom of 1926 & adobe walls, 1874; oil industry artifacts; 95' cable tool drilling rig.
Research Fields: early history of the county.
Facilities: library; archives; reading room.
Activities: guided tours; lectures; films; formally organized educational programs; permanent & special exhibitions.
Publications: book, Hutchinson County History; Hutchinson County Museum: 1977-1997; Hutchinson County, A Pictorial Legacy, 2003.
Hours & Admission Prices: Mon.-Fri. 9-5, Sat. 11-4:30. No charge; donations accepted. Closed legal holidays. &
Attendance: 6,189 (accurate)

Brady

HEART OF TEXAS COUNTRY MUSIC MUSEUM, 1701 S. Bridge, Brady, TX 76825. Tel.: 325-597-1895. Fax: 325-597-0515.
E-mail: tracy@hillbillyhits.com
Web Site: www.heartoftexascountry.com
Founded: 2000.

Key Personnel: Dir., Mr. Tracy Pitcox; Chm. (V), Maxine Bradford; Museum Shop Mgr., Sharon Jackson.
Personnel Profile: Full-Time Paid 1.
Institution Type/Description: Music History Museum.
Collections: country music history & memorabilia; personal artifacts; costumes; musical instruments; autographs; posters; photographs.
Activities: group tours.
Publications: monthly newsletter, Heart of Texas Country Music Association.
Hours & Admission Prices: Fri. 12-4, Sat. 10-4, Sun. 12-5; other times by appointment. No charge. &
Attendance: 2,500 (estimated)
Membership: Individual $8; Couple $10.

HEART OF TEXAS HISTORICAL MUSEUM, 117 N. High St., Brady, TX 76825. Mailing Address: P.O. Box 48, Brady, TX 76825. Tel.: 325-597-0526.
E-mail: hothistoricalmuseum@yahoo.com
Web Site: www.heartoftexashistoricalmuseum.com
Founded: 1974.
Congressional District: 11
Institution Type/Description: History Museum: housed in the former McCulloch County Jail; built in 1910. Listed on the National Register of Historic Places.
Collections: local history; period furnishings; personal artifacts; photographs; jail cells.
Publications: book, Handbook of McCulloch County History, Vol. III.
Hours & Admission Prices: Fri.-Sat. 1-5, Sun. 1-4; other times by appointment.

Breckenridge

BRECKENRIDGE AVIATION MUSEUM, Stephens County Airport, Breckenridge, TX 76424. Mailing Address: P.O. Box 388, Breckenridge, TX 76424. Tel.: 254-559-2515.
Web Site: www.breckenridgetexas.com
Institution Type/Description: Aviation Museum.
Collections: aviation history; aircraft.
Hours & Admission Prices: By appointment.

BRECKENRIDGE FINE ARTS CENTER, 207 N. Breckenridge Ave., Breckenridge, TX 76424-3503. Tel.: 254-559-6602.
E-mail: info@breckenridgefineart.org
Web Site: www.breckenridgefineart.org
Founded: 1985.
Key Personnel: Dir., Victoria MacFarlane; Pres. (V), Jesse Geron.
Personnel Profile: Full-Time Paid 1; Part-Time Volunteers 18.
Governing Authority: Parent Institution: Breckenridge Library & Fine Arts Foundation. Tax-exempt.
Institution Type/Description: Fine Art Gallery.
Collections: paintings; dolls; sculpture.
Activities: special events; painting & sculpture workshops; classes; temporary & permanent exhibitions.
Hours & Admission Prices: Summer: Tues.-Fri. 10-5; Academic Year: Tues.-Fri. 10-5, Sat. 10-3. No charge; donations accepted. &
Attendance: 2,300 (estimated)

SWENSON MEMORIAL MUSEUM OF STEPHENS COUNTY, 116 W. Walker, Breckenridge, TX 76424-3530. Mailing Address: P.O. Box 350, Breckenridge, TX 76424-0350. Tel.: 254-559-8471.
Founded: 1970.
Congressional District: 17
Key Personnel: Dir. & Exhibits, Freda Mitchell; Business Officer, David Duggan; Sec., Margaret Ables; Chm. Bd., Burrell McKelvain.
Personnel Profile: Full-Time Paid 1; Part-Time Paid 2; Part-Time Volunteers 2.
Governing Authority: nonprofit organization. Subsidiary Institution: J.D. Sandefer Oil Annex. Tax-exempt: 501(c)(3).
Institution Type/Description: History Museum: housed in 1920 bank building.
Collections: farming; business; ranching; furniture & furnishings; oil equipment; photographs & memorabilia of oil boom times in county; J.D. Sandefer Oil Annex; school, doctor & lawyers offices.
Research Fields: petroleum history of North Central Texas.
Facilities: 300-vol. library of history & museum-related material available for research; reading room.
Activities: guided tours; lectures; gallery talks; school loan service; temporary & permanent exhibitions.
Publications: brochures.
Hours & Admission Prices: Tues.-Fri. 10-12 & 1-5. No charge; donations accepted. Closed Independence Day; Thanksgiving; Christmas. &
Attendance: 1,200 (accurate)
Membership: $25-$1,000.

Brenham

TEXAS BAPTIST HISTORICAL MUSEUM, 10405 FM 50, Brenham, TX 77833-6424. Tel.: 979-836-5117.
E-mail: phillip.hassell@texasbaptists.org
Web Site: texasbaptisthistoricalmuseum.weebly.com
Founded: 1965.
Congressional District: 14
Key Personnel: Dir. & Cur., Phil Hassell.
Personnel Profile: Full-Time Paid 1; Part-Time Volunteers 15.
Governing Authority: church. Parent Institution: Baptist General Convention of Texas, 333 N. Washington, Dallas, TX 75246-1798. Tax-exempt: 170(b)(1)(A).
Institution Type/Description: History Museum: housed in 1839 church, the original home of Baylor University.
Collections: artifacts of Sam Houston & family; items from Baylor University; furniture; church artifacts; manuscripts.
Research Fields: local early Texas history.
Facilities: 200-vol. library of books on early Texas & Texas Baptist history available for loan on premises; reading room.
Activities: guided tours; lectures; films; temporary exhibits.
Publications: brochures.
Hours & Admission Prices: Tues.-Sat. 9-4. No charge; donations accepted. Closed New Year's Day; Easter; Independence Day; Christmas. &
Attendance: 10,000 (estimated)

Bronte

FORT CHADBOURNE VISITORS CENTER & MUSEUM, Roberta Cole Johnson Bldg., 651 Fort Chadbourne, Bronte, TX 76933. Tel.: 325-473-5311.
Institution Type/Description: Visitor Center & History Museum.
Collections: local history; American Indian & fort artifacts; military & Western memorabilia; cannons; Zappe Saloon's bar; saddles; Butterfield stagecoach; cowboy & ranching heritage; early firearms; Medal of Honor medals & memorabilia.
Facilities: library. Museum-related items for sale.
Hours & Admission Prices: Tues.-Sat. 9-5. Donations requested.

Brookshire

WALLER COUNTY HISTORICAL MUSEUM, 906 Cooper St., Brookshire, TX 77423. Tel.: 281-934-2826.
Institution Type/Description: History Museum: housed in the former home of Dr. Paul Donigan; c.1910.
Collections: local history & culture; period furnishings; photographs; personal artifacts; documents; paintings.
Hours & Admission Prices: Wed. & Fri. 10-4, Sat. 9 to noon.

Brownfield

TERRY COUNTY HERITAGE MUSEUM, 600 E. Cardwell, Brownfield, TX 79316. Mailing Address: P.O. Box 152, Brownfield, TX 79316-0152. Tel.: 806-637-2467.
Web Site: www.brownfieldchamber.com
Institution Type/Description: History Museum.
Collections: local history & culture; photographs; personal artifacts.
Hours & Admission Prices: Tues.-Sat. 10-12 & 1-3. Adults $3, children $1.

Brownsville

BROWNSVILLE MUSEUM OF FINE ART, 660 E. Ringgold St., Brownsville, TX 78520. Tel.: 956-542-0941. Fax: 956-542-6931.
Web Site: www.brownsvillemfa.org
Formerly: Brownsville Art League
Founded: 1935.
Congressional District: 27
Key Personnel: Exec. Dir., Barry T. Horn; Pres. (V), Eddie Knebel; Administrative Dir. & Show Dir., Tencha Sloss; Cur., Jennifer Cahn, Ph.D.; Accountant, Deyanira Ramirez; Education Coord., Linda W. Marin.
Personnel Profile: Full-Time Paid 5; Part-Time Paid 2; Part-Time Volunteers 30; Interns 2.
Governing Authority: nonprofit organization. Tax-exempt: 501(c)(3).
Institution Type/Description: Art Museum.
Collections: fine art including oil paintings; watercolors; collages; prints; sculpture; acrylics; pastels; pencil.
Research Fields: late 19th-20th Century south Texas & north Mexico artists.
Facilities: library; reading room; classrooms; kitchen; rental facility.
Activities: lectures; gallery talks; arts festivals; workshops; formally organized education programs for adults; inter-museum, permanent, temporary & traveling exhibitions.
Hours & Admission Prices: Tues. & Thurs.-Sat. 10-4, Wed. 10-8. Adults $5, children 6-12, students, & seniors over 65 $3; discounts to groups; members, Wed. after 5pm and children under 6 no charge. Closed New Year's Day; Good Friday; Easter; Memorial Day; Independence Day; Veterans Day; Thanksgiving; Christmas. &
Attendance: 16,000 (accurate)
Membership: Seasonal, Student, & Friends $60; Family $75; Active $120; Family Active $150.

CHILDREN'S MUSEUM OF BROWNSVILLE, 501 Ringgold St., #5, Dean Porter Park, Brownsville, TX 78520. Tel.: 956-548-9300. Fax: 956-504-1348. Facebook: Brite Sun 5.
E-mail: guestrelations@cmofbrownsville.com
Web Site: www.cmofbrownsville.com
Founded: 2005.
Institution Type/Description: Children's Museum.
Collections: hands-on exhibitions.
Activities: educational programming; special events; birthday parties; workshops.
Hours & Admission Prices: Tues.-Sat. 10-5, Sun. 12-4. Admission $5 per person; ACM members, members & children under one no charge. &

COSTUMES OF THE AMERICAS MUSEUM, #5 Dean Porter Park, 501 Ringgold St., Brownsville, TX 78520. Mailing Address: P.O. Box 3790, Brownsville, TX 78523. Tel.: 956-547-6890.
E-mail: admin@costumesoftheamericasmuseum.net
Web Site: www.costumesoftheamericasmuseum.net
Institution Type/Description: Costume Museum.
Collections: costumes from around the world.
Activities: temporary & traveling exhibitions.
Hours & Admission Prices: Tues.-Sat. 9-5, Sun. 12-4. Admission $2 per person; children 10 & under no charge.

GLADYS PORTER ZOO, 500 Ringgold St., Brownsville, TX 78520-7998. Tel.: 956-546-7187 & 2177. Fax: 956-541-4940.
E-mail: admin@gpz.org
Web Site: www.gpz.org
Founded: 1971.
Congressional District: 27
Key Personnel: Dir., Dr. Patrick M. Burchfield; Facilities Dir., Greeley A. Stones; C.F.O., Marilyn White; Veterinarian, Thomas W. DeMaar, DVM; Gen. Cur., Colette Hairston Adams; Dir. Mktg., Cynthia Garza Galvan; Concessions Mgr., Mario Calderon; Comptroller, Oralia Berlanga.
Personnel Profile: Full-Time Paid 84; Part-Time Paid 45; Part-Time Volunteers 70; Interns 4.
Governing Authority: nonprofit organization. Parent Institution: Valley Zoological Society. Tax-exempt: 501(c)(3).
Institution Type/Description: Botanical Gardens & Zoo with Aquarium.
Collections: 345 animal species & 1,557 animal specimens; 150 plant species & over 2,000 specimens.
Research Fields: animal health; reproduction & behavior.
Facilities: 900-vol. library of zoology, ecology, botany, veterinary medicine, available for research on premises; botanical garden; aquarium; classrooms; snack shops. Museum-related items for sale.
Activities: guided tours; lectures; TV & radio programs; formally organized educational program; docent program or council.
Publications: newsletter, Gladys Porter Zoo News.
Hours & Admission Prices: Mon.-Fri. 9-5, Sat.-Sun. 9-5:30. Adults $10, seniors 65 & over $8.50, children 2-13 $7; members no charge. &
Attendance: 344,154 (accurate)
Membership: Senior 65 & up $45; Single $50; Senior Plus $60; Single Plus $65; Family & Grandparent $70; Grandparent Plus $80; Family Plus $85; Supporting $200; Corporate $250; Sustaining $500; Patron $1,000; Life $5,000.

HISTORIC BROWNSVILLE MUSEUM, 641 E. Madison St., Brownsville, TX 78520. Tel.: 956-548-1313.
Institution Type/Description: History Museum: housed in the old Southern Pacific Railroad Depot; built in 1928. Listed on the National Register of Historic Places.
Collections: local history & culture; period furnishings; photographs; personal artifacts.
Facilities: Museum-related items for sale.
Hours & Admission Prices: Call for hours.

RIO GRANDE VALLEY WING C.A.F. MUSEUM, 955 S. Minnesota Ave., Brownsville, TX 78521. Tel.: 956-541-8585.
Institution Type/Description: Air Museum.
Collections: military history; early military aircraft, vehicles & equipment; uniforms; memorabilia; WWII & Korean War artifacts & memorial.
Facilities: Museum-related items for sale.
Activities: video. Museum Sponsors: Annual Air Show in March.
Hours & Admission Prices: Wed.-Sat. 9:30-3:30. Adults $6, seniors 55 & over $5, youth 12-18 $3; children 11 & under no charge.

SOUTHMOST HERITAGE CENTER, 34 Tony Gonzalez Dr., Brownsville, TX 78521-4715. Mailing Address: 1325 E. Washington St., Brownsville, TX 78520. Tel.: 956-541-5560.
Institution Type/Description: History Museum.
Collections: local history & culture; photographs; personal artifacts.
Facilities: Museum-related items for sale.
Activities: presentations; workshops; temporary exhibitions; educational programs.
Hours & Admission Prices: Call for hours.

STILLMAN HOUSE MUSEUM & BROWNSVILLE HERITAGE MUSEUM, 1325 E. Washington St., Brownsville, TX 78520-5705. Tel.: 956-541-5560. Fax: 956-541-5524.
E-mail: info@brownsvillehistory.org
Web Site: www.brownsvillehistory.org
Institution Type/Description: Historic House Museum: built in 1850.
Collections: local history & culture; period furnishings; personal artifacts; photographs.
Hours & Admission Prices: Tues.-Sat. 10-4, Sun. 1-5. Adults $5, seniors $4, students $2; children under 6 no charge.

Brownwood

BROWN COUNTY MUSEUM OF HISTORY, INC., 212 N. Broadway, Brownwood, TX 76804. Mailing Address: P.O. Box 2006, Brownwood, TX 76804-2006. Tel.: 325-641-1926.
E-mail: steveb@harrisbb.com
Web Site: www.browncountyhistory.org
Institution Type/Description: History Museum: housed in the former Brown County Jail; built in 1902.
Collections: local history & culture; period furnishings; photographs; personal artifacts.
Hours & Admission Prices: Sat. 10-4; other times by appointment. No charge; donations accepted.

MARTIN & FRANCES LEHNIS RAILROAD MUSEUM, 700 E. Adams, Brownwood, TX 76801-7002. Mailing Address: P.O. Box 1389, Brownwood, TX 76804-1389. Tel.: 325-643-6376.
E-mail: kpeterson@ci.brownwood.tx.us
Web Site: www.ci.brownwood.tx.us/lrm
Founded: 2007.
Key Personnel: Dir., Kim Peterson.
Personnel Profile: Full-Time Paid 1; Full-Time Volunteers 1; Part-Time Paid 2; Part-Time Volunteers 2.
Governing Authority: city. Tax-exempt.
Institution Type/Description: Transportation Museum.
Collections: railroad history; railcars; railway china; lanterns; photographs.
Hours & Admission Prices: Tues.-Sat. 10-4. Adults $3, seniors $2.50, children 5 & over $2; discounts to members, AAM & ICOM members, TAM, & active military. Closed holidays. ♿
Attendance: 3,500 (accurate)
Membership: $25; $50; $100; $250; $500; $1,000.

Bryan

BRAZOS VALLEY MUSEUM OF NATURAL HISTORY, 3232 Briarcrest Dr., Bryan, TX 77802-3015. Tel.: 979-776-2195. Fax: 979-774-0252.
E-mail: dcowman@brazosvalleymuseum.org
Web Site: www.brazosvalleymuseum.org
Founded: 1961.
Congressional District: 8
Key Personnel: Exec. Dir., Dr. Deborah F. Cowman; Pres. Bd. Trustees, Jacque Flagg; Education Coord., Assoc. Dir. & Museum Shop Mgr., Maria Lazo; Cur. Collections & Exhibits, Elisabeth Manning.
Personnel Profile: Full-Time Paid 3; Part-Time Paid 3; Part-Time Volunteers 40; Interns 6.
Governing Authority: nonprofit organization. Tax-exempt: 501(c)(3).
Institution Type/Description: Natural History Museum.
Collections: paleontology, vertebrate & invertebrate; anthropology; archaeology; geology & mineralogy; mammalogy; ornithology; paleobotany; botany; entomology; Gulf Coast & Caribbean malacology; local history; cultural history; gems & minerals; prehistoric lithic industries of Texas.
Research Fields: archaeology; history.
Facilities: library; discovery room; nature lab; classrooms; nature trail; live animal & insect observatory; wildflower garden. Museum-related items for sale.
Activities: guided tours of current major exhibits & discovery room; lectures; formally organized education programs; permanent & temporary exhibitions; school loan service; discovery kits; summer nature camp; cultural & environmental tours; cooperative agreements with Department of Anthropology, History Department, and Texas Cooperative Wildlife Collections (Wildlife & Fisheries Science, Texas A & M University); demonstration wetlands, yaupon thicket, prairie & post oak Savanna.
Publications: biannual newsletter.
Hours & Admission Prices: Mon.-Sat. 10-5. Adults $5, seniors & students $4; members and children 3 & under accompanied by parent no charge. Prices and hours change with major exhibits. Closed New Year's Day; Easter; Memorial Day; Thanksgiving; Christmas. ♿
Attendance: 30,000 (estimated)
Membership: College Student & Seniors Citizen $30; Individual $50; Dual $75; Family $100; Sponsor $150; Contributor $250; Benefactor $500; President's Circle $1,000.

THE CHILDREN'S MUSEUM OF THE BRAZOS VALLEY, 111 E. 27th St., Bryan, TX 77803-6947. Tel.: 979-779-5437. Fax: 979-775-4908.
Web Site: www.mymuseum.com
Key Personnel: Pres., Dave Stevenson; Dir. Devel., Rebecca Christopher
Institution Type/Description: Children's Museum.
Collections: hands-on interactive exhibits.
Facilities: Museum-related items for sale.
Activities: parties; school programs; festivals.
Hours & Admission Prices: Tues.-Fri. 9-2, Sat. 10-5. Adults & children $5, senior citizens $4; children under one no charge. Parking: no charge.

Buffalo Gap

BUFFALO GAP HISTORIC VILLAGE, 133 N. William, Buffalo Gap, TX 79508. Mailing Address: P.O. Box 818, Buffalo Gap, TX 79508-0818. Tel.: 325-572-3365. Fax: 325-572-3991. Facebook: Buffalo Gap Historic Village.
E-mail: info@tfhcc.com
Web Site: www.tfhcc.com
Founded: 1956.
Institution Type/Description: Living History Museum.
Collections: history & heritage of the Texas frontier; period artifacts; personal artifacts.
Facilities: rental facilities.
Activities: special events; demonstrations; rental facilities. Museum Sponsors: July 4th Jeep Jamboree; Comanche Moon Social in October; Pillage-the-Village in October; Ghost Tours in October.
Hours & Admission Prices: Memorial Day to Labor Day Mon.-Sat. 10-6, Sun. 12-6. Adults $7, seniors & military $6, students $4; members & children under 5 no charge. Closed New Year's Day; Thanksgiving; Christmas Eve & Day.
Attendance: 15,000 (accurate)
Membership: Scout $45; Pioneer Family $75; Homestead $125; Settler $250; Trail Blazer $500; Founder $1,000.

Burnet

FORT CROGHAN MUSEUM, 703 Buchanan Dr., Burnet, TX 78611. Mailing Address: P.O. Box 74, Burnet, TX 78611-0074. Tel.: 512-756-8281.
E-mail: info@fortcroghan.org
Web Site: www.fortcroghan.org
Founded: 1957.
Key Personnel: Chm. (V), Paul Shell; Museum Dir., Mildred Williams.
Personnel Profile: Part-Time Paid 1; Part-Time Volunteers 20.
Governing Authority: Parent Institution: Burnet County Heritage Society. Tax-exempt.
Institution Type/Description: History Museum: located on the site of 1849 Fort Croghan.
Collections: stone & log cabins from the 1850s & 1860s; Old Fort building; blacksmith's shop; 5 pioneer cabins; farm implements; historic displays.
Research Fields: county & regional history.

Activities: Museum Sponsors: Annual Fort Croghan Day 2nd Saturday in October; Christmas at Fort Croghan 2nd Saturday in December.
Publications: leaflets, History of Fort Croghan & Museum; Burnet County Cemetery Records; Burnet County History, vols. I & II.
Hours & Admission Prices: April-Aug. Thurs.-Sat. 10-5. No charge; donations requested. &
Attendance: 2,000 (estimated)
Membership: Annual $5.

HIGHLAND LAKES SQUADRON COMMEMORATIVE AIR FORCE MUSEUM, Burnet Municipal Airport, Kate Craddock Field, U.S. Hwy. 281, Burnet, TX 78611. Mailing Address: P.O. Box 866, Burnet, TX 78611. Tel.: 512-756-2226.
E-mail: caf@tstar.net
Personnel Profile: Part-Time Volunteers 30.
Governing Authority: Tax-exempt.
Institution Type/Description: Air Force Museum.
Collections: military aircraft & memorabilia; personal artifacts; photographs.
Activities: special events.
Hours & Admission Prices: Wed. & Sun. 1-4, Sat. 10-4; other times by appointment. Discounts to active military & seniors. Closed New Year's; Christmas.

Burton

TEXAS COTTON GIN MUSEUM, (M), 307 N. Main, Burton, TX 77835. Mailing Address: P.O. Box 98, Burton, TX 77835-0098. Tel.: 979-289-3378. Fax: 979-289-5210. Facebook: Texas Cotton Gin Museum.
E-mail: burtoncottongin@earthlink.net
Web Site: www.cottonginmuseum.org
Formerly: Burton Cotton Gin & Museum
Founded: 1989.
Key Personnel: Dir., Linda Russell; Chm. (V), Tony Williams; Cur., Jerry Moore.
Personnel Profile: Full-Time Paid 2; Part-Time Volunteers 15.
Institution Type/Description: History Museum.
Collections: operational cotton gin; local history & culture; machines; tools; engines; vehicles; Burton farmers gin archives & records.
Facilities: archives.
Activities: guided tours.
Publications: quarterly newsletter, Cotton Tales.
Hours & Admission Prices: Museum: Tues.-Sat. 10-4. Gin Tours: Tues.-Sat. 10 & 2. Museum: no charge. Gin Tours: adults $6, students $4; discounts to groups of 10 or more & AAM members; members no charge. Closed major holidays. &
Attendance: 4,200 (estimated)
Membership: Individual $50; Family $100; Patron $150; Supporter $250; Benefactor $500; Bronze $1,000; Silver $2,500; Gold $5,000; Platinum $10,000.

Caldwell

BURLESON COUNTY HISTORICAL MUSEUM, Burleson County Courthouse, Caldwell, TX 77836. Mailing Address: P.O. Box 127, Caldwell, TX 77836-0127. Tel.: 979-567-7196.
E-mail: burlesoncohissoc@aol.com
Founded: 1968.
Congressional District: 10
Key Personnel: Chm. Burleson County Historical Commission, Tammy Kubecka.
Governing Authority: county. Parent Institution: Burleson County. Tax-exempt.
Institution Type/Description: Local History Museum.
Collections: Fort Tenoxtitlan artifacts; photographs; manuscript collections; country store exhibit; early settler's artifacts; Indian artifacts; exhibits on plantation days & ranching in Burleson County; antique toys & children's furniture; c.1860-1890s.
Research Fields: history of Burleson County & its relation to Texas history.
Activities: guided tours; permanent & temporary exhibitions. Annual Events: Heritage Week in March; Archaeology Awareness Week in October.
Publications: facsimile reprint, Reminiscences of Burleson County, Texas; book, Treasured Recipes of Burleson County; book, Astride the Old San Antonio Road, A History of Burleson county.
Hours & Admission Prices: Fri. 2-4:30; other times by appointment. No charge; donations accepted. &

Cameron

MILAM COUNTY HISTORICAL MUSEUM, MILAM COUNTY JAIL MUSEUM, & MILAM COUNTY MUSEUM ANNEX, 112 W. First St., Cameron, TX 76520-4275. Mailing Address: P.O. Box 966, Cameron, TX 76520-0966. Tel.: 254-697-4770 & 8963. Fax: 254-697-4770.
E-mail: cking@milamcounty.net
Web Site: cameron-tx.com
Founded: 1977.
Congressional District: 11
Key Personnel: Pres., Sandra O'Donnell; Chm. (V) & Museum Shop Mgr., Margia Barkemeyer; Dir., Charles King.
Personnel Profile: Full-Time Paid 1.
Governing Authority: nonprofit organization. Jail: 201 E. Main St., Cameron, TX 76520. Annex: 102 W. Main, Cameron, TX 76520. Tax-exempt. 501(c)(3).
Institution Type/Description: History Museum.
Collections: original Sam Houston & Benjamin Bryant documents; Civil War letters & documents; Spanish mission artifacts; WWI & II artifacts; clothing; household items; guns; dolls; farm tools; furniture; photographs; handcrafts; scale model of 1940s Old Town Cameron. Historic Building: 1895 Milam County jail.
Activities: guided tours; permanent & temporary loan exhibits.
Publications: publications, The Milam County Courthouse; And then Came the People; Milam County - Birthplace of a Region; Spanish Missions of Milan County; Matchless Milam; Tales from the Museum, Vol. III; CD, Santa Fe Town.
Hours & Admission Prices: Museum: Tues.-Sat. 8-5. Jail: Tues. & Thurs. 1-4; other times by appointment. Annex & Old Town Cameron: by appointment. No charge; donations accepted. &
Attendance: 4,500 (estimated)
Membership: Basic $35; Supporter $40; Active & Organization $50; Participating $100; Sustaining & Corporate $200.

Canadian

RIVER VALLEY PIONEER MUSEUM, 118 N. 2nd St., Canadian, TX 79014-2202. Mailing Address: P.O. Box 1201, Canadian, TX 79014-1201. Tel.: 806-323-6548. Fax: 806-232-8993.
E-mail: rvmuseum@cebridge.net
Web Site: rivervalleymuseum.org
Founded: 1986.
Congressional District: 13
Personnel Profile: Full-Time Paid 1; Part-Time Paid 1; Part-Time Volunteers 1.
Governing Authority: Tax-exempt.
Institution Type/Description: History Museum.
Collections: local history & culture; photographs; Native American artifacts; archaeological; ranching.
Facilities: Museum-related items for sale.
Activities: rental facilities; educational programs.
Hours & Admission Prices: Tues.-Fri. 9-12 & 1-4, Sat. 1-3. No charge; donations accepted. &
Attendance: 3,000 (accurate)
Membership: Individual $20; Family $30; Business $50; Friend $250; Patron $500; Lifetime $1,000.

Canyon

* **PANHANDLE-PLAINS HISTORICAL MUSEUM, (M),** 2503 Fourth Ave., Canyon, TX 79015-4183. Mailing Address: WTAMU, Box 60967, Canyon, TX 79016. Tel.: 806-651-2244. Fax: 806-651-2250.
E-mail: museum@pphm.wtamu.edu
Web Site: www.panhandleplains.org
Founded: 1921.
Congressional District: 13
Key Personnel: Dir., Guy C. Vanderpool; Auxiliary Chm., Jane Stephens; Pres. (V), Alice Hyde; Conservation Center Dir., Richard Trela; Dir. Mktg., Linda Moreland; Cur. Art, Michael R. Grauer; Dir. Education, Mary Ann Ruelas; Asst. Archaeologist, Rolla Shaller; Cur. History, Dr. William E. Green; Asst. History Cur., Susan Denney; Cur. Archaeology, Dr. Jeff Indeck; Asst. Archivist & Librarian, Betty Bustos; Exhibits Supervisor, Kenny Schneider; Registrar, Mary Moore; Programs Coord., Amy David; Museum Shop Mgr., Tammy St. Pierre.
Personnel Profile: Full-Time Paid 27; Part-Time Paid 24; Part-Time Volunteers 10.
Governing Authority: state. Parent Institution: West Texas A&M University. Subsidiary Institution: Panhandle Plains Historical Society. Tax-exempt: 501(c)(3) & 170(b)(1)(A).

Institution Type/Description: History Museum.
Collections: geology; Plains Indian ethnology; paleontology; manuscripts; 150,000 photographs; transportation; anthropology; archaeology; cattle industry; pioneer life; agriculture; art; clothing and textiles; firearms; science & technology; furniture; restored T-Anchor Ranch.
Research Fields: general Texas & Southwestern History; ranching history, westward expansion; history of windmills; Southwest Indians.
Facilities: 10,000-vol. library & archive on history of Texas & the Southwest. Museum-related items for sale.
Activities: guided tours; lectures; educational programs for schools; regional museum clinics; permanent & temporary exhibits.
Publications: journal, Panhandle-Plains Historical Review; newsletter, PPHM News.
Hours & Admission Prices: June-Aug. Mon.-Sat. 9-6; Sept.-May Tues.-Sat. 9-5. Adults $10, senior citizens 65 & over $9, children 4-12 $5; discounts to groups; members & children under 4 no charge. Closed New Year's Day; Thanksgiving; Christmas Eve & Day. &
Attendance: 75,000 (estimated)
Membership: Friend $50; Family $75; Contributor $100; Supporter $250; Patron $500; Goodnight Circle $1,000.

Canyon Lake

THE HERITAGE MUSEUM OF THE TEXAS HILL COUNTRY, 4831 FM 2673, Canyon Lake, TX 78133-0004. Mailing Address: P.O. Box 1598, Canyon Lake, TX 78133-0004. Tel.: 830-899-4542.
E-mail: museum@gvtc.com
Web Site: www.theheritagemuseum.com
Founded: 2000.
Governing Authority: nonprofit organization. Tax-exempt: 501(c)(3).
Institution Type/Description: History Museum.
Collections: local history & culture; over 350 dinosaur tracks; farm equipment; pioneer & Native American artifacts; Canyon Dam construction & operation; photographs; personal artifacts; period furnishings.
Facilities: Museum-related items for sale.
Activities: educational programs; guided tours.
Publications: newsletter.
Hours & Admission Prices: Daily 1-5. Adults $4, children 5-12 $3. Closed New Year's Day; Easter; Thanksgiving; Christmas.

Carmine

TEXAS BASKETBALL MUSEUM, 107 Augsberg Ave., Carmine, TX 78932. Mailing Address: P.O. Box 401, Carmine, TX 78932. Tel.: 979-278-4222; 800-364-4667. Fax: 979-278-4222.
E-mail: texasbkb@swbell.net
Web Site: www.texasbasketball.com
Founded: 2009.
Key Personnel: Dir., Bob Springer
Institution Type/Description: Sports Museum.
Collections: basketball history & uniforms; photographs; uniforms; basketballs; plaques; Texas High School Basketball Hall of Fame.
Publications: magazine, Texas Basketball.
Hours & Admission Prices: By appointment. No charge; donations accepted.
Attendance: 350 (accurate)

Carrollton

A. W. PERRY HOMESTEAD MUSEUM, 1509 N. Perry Rd., Carrollton, TX 75006-6122. Mailing Address: P.O. Box 110535, Carrollton, TX 75011-0535. Tel.: 972-466-6380.
Web Site: cityofcarrollton.com/museum
Founded: 1976.
Congressional District: 24
Key Personnel: Dir., Toyia Pointer.
Personnel Profile: Full-Time Paid 1; Part-Time Paid 1; Part-Time Volunteers 6; Interns 1.
Governing Authority: Parent Institution: city of Carrollton, TX. Tax-exempt.
Institution Type/Description: Historic House: housed in the former home of A.W. Perry; built in 1857, rebuilt in 1909 by his son using some of the lumber from the original house.
Collections: Perry family history & culture; early pioneer life; period artifacts & furnishings; photographs.
Activities: group tours; rental facilities; special events; research; birthday parties; classes. Annual Events: Silent Film Series; Mother's Day Concert; Old Fashioned Christmas.
Hours & Admission Prices: Wed.-Sat. 10-12 & 1-5. No charge. &
Attendance: 2,900 (accurate)

Castroville

STEINBACH HOUSE - CASTROVILLE AREA CHAMBER OF COMMERCE, 100 Karm St., Castroville, TX 78009. Mailing Address: P.O. Box 572, Castroville, TX 78009. Tel.: 830-538-3142; 800-778-6775. Fax: 830-538-3295.
E-mail: chamber@castroville.com
Web Site: www.castroville.com/steinbach_house.html
Institution Type/Description: Historic House Museum: housed in a home originally built c.1618 in Wahlbach, France and relocated to Castroville in 1998.
Collections: local history & culture; Alsatian furniture; personal artifacts; photographs.
Hours & Admission Prices: Call for hours.

Center

SHELBY COUNTY MUSEUM, 230 Pecan St., Center, TX 75935-3649. Mailing Address: P.O. Box 1542, Center, TX 75935-1542. Tel.: 936-598-3613.
E-mail: shelbymuseum@sbcglobal.net
Web Site: www.shelbycountytexashistory.org
Governing Authority: Parent Institution: Shelby County Historical Society. Tax-exempt: 501(c)(3).
Institution Type/Description: History Museum: housed in c.1900 Weaver-Oates House built by E.H. Barron.
Collections: personal artifacts; period furnishings; photographs.
Facilities: library.
Activities: research.
Hours & Admission Prices: Mon.-Fri. 12-4; other times by appointment. No charge; donations accepted.

Chappell Hill

CHAPPELL HILL HISTORICAL SOCIETY MUSEUM, (M), 9220 Poplar St., Chappell Hill, TX 77426-6312. Tel.: 979-836-6033. Fax: 979-836-7438.
E-mail: chmuseum@chappellhillmuseum.org
Web Site: www.chappellhillmuseum.org
Founded: 1964.
Congressional District: 14
Key Personnel: Administrative Dir., Ladonna Vest.
Personnel Profile: Part-Time Paid 1; Part-Time Volunteers 35.
Governing Authority: society; nonprofit. Parent Institution: Chappell Hill Historical Society. Tax-exempt.
Institution Type/Description: General Museum.
Collections: confederate, early Texas history of Chappell Hill; archives; farm implements; documents of early families; early carpenters tools; Johnnie Swearingen folk art; photographs.
Research Fields: local history; genealogy.
Facilities: 150-seat auditorium; 3,200 sq. ft. exhibit space.
Activities: guided tours; permanent exhibitions; special exhibits. Annual Events: Bluebonnet Festival in Spring; Scarecrow Festival in Fall.
Publications: monthly newsletter; annual, Chappell Hill Historical Review.
Hours & Admission Prices: Wed.-Fri. 10-4, Sat. 11-3, Sun. 1-4; groups by appointment. Tours: $40-$80. Closed New Year's Eve & Day; Independence Day; Thanksgiving; Christmas Eve & Day. &
Attendance: 3,500 (accurate)
Membership: Society: Individual $20; Family $30; Patron & Patron Business $50; Sustaining & Sustaining Business $100; Life $500.

Childress

CHILDRESS COUNTY HERITAGE MUSEUM, 210 3rd St., N.W., Childress, TX 79201-4540. Tel.: 940-937-2261.
E-mail: childressmuseum@sbcglobal.net
Web Site: www.biz.childresstexas.net/childressmuseum.com
Founded: 1976.
Congressional District: 13
Key Personnel: Exec. Dir., Ramona K. Garcia; Pres., John Preston.
Personnel Profile: Full-Time Paid 1; Part-Time Volunteers 3.
Governing Authority: county; nonprofit organization. Branch Museum: Antique Transportation Museum. Tax-exempt: 501(c)(3).
Institution Type/Description: History Museum.
Collections: history of the area from Paleolithic era to present; Republic of Texas; Texas under Six Flags; ranching; farming; pioneering; rocks & minerals; autos; scale model train; linotype. Historic Buildings: late 1800s one room house; one room school.
Major Exhibits: Alamo Images (T), 3/14.
Activities: slide programs to school & other organizations; local history

program for 2nd, 3rd & 7th grades. Museum Sponsors: District History Fair for 6-12th graders; School Tours K-8th grade; Youth Explorer Days.
Hours & Admission Prices: Tues.-Sat. 10-5. No charge; donations accepted. Closed national holidays. &
Attendance: 7,000 (estimated)
Membership: Single $35; Family $50; Business $100; Group $200; Life $1,500.

Cisco

CONRAD HILTON CENTER & MUSEUM, 309 Conrad Hilton Ave., Cisco, TX 76437. Mailing Address: P.O. Box 350, Cisco, TX 76437. Tel.: 254-442-2537. Fax: 254-442-2553.
E-mail: ciscoinfo@ciscotx.com
Web Site: ciscochamber.com
Founded: 1986.
Congressional District: 19
Key Personnel: Dir. & Pres. Hilton Bd. (V), Eris Ritchie; Cur., John Waggoner.
Personnel Profile: Part-Time Paid 1.
Governing Authority: private; nonprofit organization.
Institution Type/Description: History Museum.
Collections: The Innkeeper Gallery: Conrad Hilton's life, family & career; photographs; videos. Cisco Museum: Cisco memorabilia; tools; medical instruments; clothing; uniforms; household artifacts; military uniforms from WWI to present day; photographs; personal artifacts. The Walls of Fame: photographs & artifacts related to famous Cisco residents. The Cisco Pictorial: photographs; newspapers; personal artifacts. The Santa Claus Bank Robbery: photographs; news articles.
Facilities: library.
Activities: guided tours; theater.
Hours & Admission Prices: Mon.-Fri. 9-12 & 1-5. No charge, donations accepted. Closed New Year's Day; Memorial Day; Independence Day; Labor Day; Thanksgiving; Christmas Eve, Day & week. &
Attendance: 300 (accurate)

LELA LATCH LLOYD MUSEUM, 116 W. 7th St., Cisco, TX 76437. Mailing Address: P.O. Box 62, Cisco, TX 76437. Tel.: 325-794-4400, ext. 4428; 254-442-2374.
E-mail: dhale@rcgates.com
Institution Type/Description: History Museum.
Collections: local history & culture; period furnishings; personal artifacts; photographs; Civil War artifacts; early settlers; ranching; farming.
Hours & Admission Prices: Fri.-Sat. 1-5. No charge; donations accepted.

Clarendon

SAINTS' ROOST MUSEUM, 610 E. Harrington St., Clarendon, TX 79226. Mailing Address: P.O. Box 781, Clarendon, TX 79226. Tel.: 806-874-2746.
E-mail: sandyskelton@yahoo.com
Institution Type/Description: History Museum: housed in the former Adair Hospital; built in 1910.
Collections: local history & culture; JA ranch co-founder, Cornelia Adair; cattleman, Col. Charles Goodnight; artist, Harold D. Bugbee; photographs; personal artifacts; period furnishings; Native American artifacts; military uniforms; paintings.
Hours & Admission Prices: Summer: Mon.-Fri. 2-5, Sun. 1-5.

Claude

ARMSTRONG COUNTY MUSEUM, CHARLES GOODNIGHT HISTORICAL CENTER AND GEM THEATRE, 120 N. Trice, Claude, TX 79019. Mailing Address: P.O. Box 450, Claude, TX 79019-0450. Tel.: 806-226-2187.
Web Site: www.armstrongcountymuseum.com
Formerly: Armstrong County Museum and Gem Theatre
Founded: 1990.
Key Personnel: Dir., Tom Novak.
Governing Authority: Tax-exempt.
Institution Type/Description: History Museum.
Collections: local history; period furnishings; photographs; paintings; Native American; military artifacts.
Activities: special events.
Hours & Admission Prices: Tues.-Sat. 12-4; other times by appointment. No charge. &
Membership: Sidekick $35; Wrangler $50; Nester $100; Settler $250; Pioneer $500; Founder $1,000; Cattalo $$1,500; Molly Goodnight $2,500; Col. Goodnight $5,000.

Cleburne

LAYLAND MUSEUM, (M), 201 N. Caddo, Cleburne, TX 76031-4903. Tel.: 817-645-0940. Fax: 817-641-4161.
E-mail: museum@cleburne.net
Founded: 1964.
Congressional District: 6
Key Personnel: Dir., Julie P. Baker; Administrative Asst. & Museum Shop Mgr., Christy Morton; Pres. (V), Phyllis McPherson.
Personnel Profile: Full-Time Paid 2; Part-Time Paid 1; Part-Time Volunteers 12.
Governing Authority: municipal government. Parent Institution: City of Cleburne, 10 N. Robinson, Cleburne 76031; Subsidiary Institution: 1914 Smith History Center, 200 N. Main St., Cleburne 76031. Tax-exempt.
Institution Type/Description: History Museum: housed in 1905 Carnegie Library building.
Collections: Native American & American domestic artifacts; Marchbanks confederate uniform; Kit Carson Saddle; Gen. Cleburne pistol & saddle; household furnishings; clothing; firearms; manuscripts; documents; photographs.
Research Fields: American home life; shelter; foodways; crafts; leisure activities; county history.
Facilities: research library; 130-seat theatre; curatorial center; teaching kitchen. Gift items for sale.
Activities: formally organized educational programs; performance & lecture series; gallery talks; craft demonstration; school loan trunks; heritage tours; festival events; traveling exhibitions.
Publications: books, Civil War Veterans of Johnson County; Johnson County Marriages 1892-1919; The History of Cleburne & Johnson County; Christopher Columbus; What We Eat.
Hours & Admission Prices: Tues.-Fri. 10-5, Sat. 10-4. No charge; donations accepted. Closed national holidays. &
Attendance: 8,000 (accurate)
Membership: $15; $25; $50; $100; $500; $1,000; $2,500.

LOWELL SMITH SR. HISTORY CENTER, 200 N. Main St., Cleburne, TX 76031. Mailing Address: 201 N. Caddo, Cleburne, TX 76031.
Institution Type/Description: History Center: housed in a former grocery store; built in 1914.
Collections: local history & culture; period vehicles & artifacts; neon lighting.
Facilities: library; classroom.
Activities: special events; demonstrations.
Hours & Admission Prices: Call for hours.

Clifton

BOSQUE MUSEUM, (M), 301 S. Ave. O, Clifton, TX 76634. Mailing Address: P.O. Box 345, Clifton, TX 76634-0345. Tel.: 254-675-3845. Fax: 254-675-8801.
E-mail: info@bosquemuseum.org
Web Site: www.bosquemuseum.com
Formerly: Bosque Memorial Museum
Founded: 1924.
Congressional District: 17
Key Personnel: Dir., George W. Larson, Ph.D.; Chm., Kaye Johnson; Vice Pres., Steven Harr; Sec., Kathy Kruse; Treas., Mechelle Slaughter; Cur., Bill Calhoon.
Personnel Profile: Full-Time Paid 2; Part-Time Paid 2; Part-Time Volunteers 2.
Volunteer Hours: 2,599
Operating Expenses: 129,000
Operating Income: 130,000
Governing Authority: private; nonprofit organization. Tax-exempt: 501(c)(3).
Institution Type/Description: History Museum.
Collections: Bosque County history; Texas & Norwegian immigration; lithic; paleoamerican.
Major Exhibits: Bosque County Goes to War, 1/14-3/14; Immigrant Books from Home, 2/14-5/14; 4th Annual Bosque Wildflower Show, 3/15/14-5/15/14; Bosque County in WWII, 5/14-9/14; Southwestern Indian Pots, 9/14-11/14.
Research Fields: Lithic Indian occupation in Bosque County; Texas-Norwegian culture.
Facilities: education & research center.
Activities: guided tours; lectures; school loan service; temporary & traveling exhibitions. Annual Events: Community Festivals: Octoberfest; Norwegian Christmas; Archaeology Lecture & Program.
Publications: quarterly newsletter, Museum Musings; The Painted Hills - History of Art in Bosque County; The Lady with the Pen; Juana.
Hours & Admission Prices: Tues.-Sat. 10-5. Adults $5; members no charge. Closed New Year's Day; Independence Day; Thanksgiving; Christmas. &

Attendance: 4,700 (accurate)
Membership: Individual $40; Dual & Family $50; Patron $150; Sponsor $300; Sustaining $500; Benefactor $1,000.

Clute

BRAZOSPORT MUSEUM OF NATURAL SCIENCE, (M), 400 College Blvd., Clute, TX 77531-4778. Tel.: 979-265-7831. Fax: 979-265-6022.
E-mail: bmns@bcfas.org
Web Site: bmns.org
Founded: 1962.
Congressional District: 6
Key Personnel: C.E.O. & Pres. (V), Wayne Humbird.
Personnel Profile: Full-Time Volunteers 1; Part-Time Volunteers 43.
Governing Authority: bd. of trustees; nonprofit organization. Tax-exempt.
Institution Type/Description: Natural Science Museum.
Collections: archaeology; botany; geology; malacology; marine; mineralogy; paleontology; zoology. A.P. Beutel Hall of Minerals; Bryan Cooney Hall of Fossils; Mildred Tate Hall of Malacology; Raymond Walley Hall of Archaeology: Hall of Wildlife; Children's Hall.
Research Fields: malacology; archaeology.
Facilities: 2,000-vol. library of natural science; aquarium; nature center; classrooms. Shells, rocks & fossils for sale.
Activities: guided tours; lectures; films; study clubs; workshops; formally organized educational programs; inter-museum loan, permanent, & temporary exhibitions. Affiliated Groups: Brazosport Archaeological Society; Brazosport Birders; Brazosport Birders & Naturalists; Sea Shell Searchers of Brazoria County. Special Events: Discovery Week; Annual Open House.
Publications: booklet; annual report.
Hours & Admission Prices: Tues.-Sat. 10-4, Sun. 2-5. No charge; donations accepted. &
Attendance: 14,021 (accurate)
Membership: Individual $10; Family $20; Patron $50; Benefactor $100; Sustaining $250; Life $1,000.

Coleman

HERITAGE HALL COLEMAN MUSEUM, 400 W. College Ave., Coleman, TX 76834. Mailing Address: 114 E. Pecan St., Coleman, TX 76834. Tel.: 325-625-2000.
Institution Type/Description: History Museum.
Collections: local history & culture; period furnishings; personal artifacts; photographs; dolls.
Facilities: rental facilities.
Activities: temporary & permanent exhibitions.
Hours & Admission Prices: June-Aug. Fri.-Sat. 10-4, Sept.-May Sat. 10-4. No charge; donations accepted.
Membership: Individual $30; Family $40.

College Station

GEORGE BUSH PRESIDENTIAL LIBRARY AND MUSEUM, 1000 George Bush Dr. W., College Station, TX 77845. Tel.: 979-691-4000. Fax: 979-691-4050. TTY: 979-691-4091.
Web Site: bushlibrary.tamu.edu
Founded: 1997.
Congressional District: 31
Key Personnel: Dir., Warren L. Finch; Asst. Dir., Patricia Burchfield; Cur., Susie Cox; Administrative Officer, Karen Gonzalez; Education, Dr. Shirley Hammond; Supervisory Archivist, Dr. Robert Holzweiss; Public Rels., Will King; Registrar, Jay Patton; Volunteer Coord., Sharon Merrill; Museum Shop Mgr., Joyce Cain; Security, James Mullins.
Personnel Profile: Full-Time Paid 50; Part-Time Volunteers 200; Interns 8.
Governing Authority: federal government. Parent Institution: National Archives and Records Admin., Washington, D.C. 20408. Tax-exempt: 170(b)(1)(A).
Institution Type/Description: Presidential Library.
Collections: personal papers; government records; photographs; film; video tapes; head of state gifts; gifts from private citizens; political campaign items; personal & family memorabilia; 100,000 artifacts.
Research Fields: life, times, career & admin. of the 41st President of the U.S.
Facilities: library; archives; 600-seat auditorium; educational facilities; 17,000 sq. ft. exhibit space; 146-seat large screen theater; manuscript reference room. Museum-related items for sale.
Activities: concerts; docent program; films; formal education programs for children, adults & Texas A&M University students; guided tours; loan, traveling & temporary exhibitions; lectures. Annual Events: Easter Egg Roll; Holidays in the Rotunda; Independence Day Celebration.
Hours & Admission Prices: Mon.-Sat. 9:30-5, Sun. 12-5. Adults $9, seniors $7,

children 6-17 $3; discounts to groups & military; members and children 5 & under no charge. Closed New Year's Day; Thanksgiving; Christmas. &
Attendance: 138,252 (accurate)
Membership: Associates $50; Associate Patron $100; Associates Sponsor $500; Senior Associates $1,000; Presidents Club $10,000.

J. WAYNE STARK GALLERIES, (M), 1120 Memorial Student Ctr., Joe Routt Blvd., College Station, TX 77843-4229. Mailing Address: Texas A&M University Art Galleries Dept., 4229 TAMU, College Station, TX 77843-4229. Tel.: 979-845-6081 & 8501. Fax: 979-862-3381.
E-mail: uart@uart.tamu.edu
Web Site: uart.tamu.edu
Founded: 1992.
Congressional District: 8
Key Personnel: Dir., Catherine A. Hastedt; Mgr. Collections, Amanda Cagle; Cur. Education, Gregory Phillipy.
Personnel Profile: Full-Time Paid 3; Part-Time Paid 8; Part-Time Volunteers 40.
Governing Authority: state; university. Parent Institution: Texas A&M University. Tax-exempt: 170(b)(1)(A).
Institution Type/Description: University Art Gallery.
Collections: works by Texas artists of the 20th century including E.M. (Buck) Schiwetz, Michael Frary, Russell Waterhouse, Dorothy Hood, Charles Schorre, John Alexander, David Caton.
Activities: guided tours; lectures; films; gallery talks; docent program; permanent & traveling exhibitions.
Publications: exhibit catalogues.
Hours & Admission Prices: Tues.-Fri. 9-8, Sat.-Sun. 12-6. No charge. Closed university holidays. &
Attendance: 51,549 (accurate)

MSC FORSYTH CENTER GALLERIES, TEXAS A&M UNIVERSITY, (M), 275 Joe Routt Blvd., Ste. 2440, College Station, TX 77843-2440. Mailing Address: Memorial Student Center, Ste. 2440, 4229-TAMU, College Station, TX 77843-4229. Tel.: 979-845-9251. Fax: 979-845-5117.
E-mail: uart@uart.tamu.edu
Web Site: forsyth.tamu.edu
Founded: 1989.
Congressional District: 8
Key Personnel: Asst. Dir., Amanda Dyer; Collections Mgr., Heather Ann Bennett; Communications Specialist, Lynn McDaniel.
Governing Authority: Texas A&M University. Tax-exempt: 501(c)(3).
Institution Type/Description: Art Museum
Collections: 19th-century English Cameo glass; 19th- & early 20th-century American art glass; 1850-1950 American paintings; 20th century quilts; Guatemalan textiles; early 20th-century French Cameo glass; American rich cut glass from the American Cut Glass Association; late 19th-century American Wooten cabinet secretary.
Research Fields: American painting & art glass; English cameo glass.
Facilities: 2,100 sq. ft. exhibit space.
Activities: guided tours; lectures; films; docent program; international glass tours; student art committee; visual arts gallery; loan & traveling exhibitions.
Publications: exhibit catalogues, A Texas Paperweight Celebration; A Rich and Lasting Beauty: Masterpieces from the American Cut Glass Association National Collection.
Hours & Admission Prices: Tues.-Fri. 9-8, Sat.-Sun. 12-6; call to confirm hours. No charge; donations accepted. Closed university holidays. &
Attendance: 25,000 (estimated)
Membership: Friends of the Forsyth: Students $25; Friend $50; Friend Dual $75; Associate $100; Sponsor $250; Donor $500; Patron $1,000; Corporate Dual $2,000.

MUSEUM OF THE AMERICAN G.I., (M), 1300 Cherokee, College Station, TX 77845. Mailing Address: P.O. Box 9599, College Station, TX 77845. Tel.: 979-777-2820.
Institution Type/Description: Military History Museum.
Collections: military history & artifacts; transport & tracked armored vehicles; artillery pieces; weapons; uniforms.
Activities: Annual Event: Open House.
Hours & Admission Prices: Mon.-Fri. 10-5.

Colorado City

HEART OF WEST TEXAS MUSEUM, 340 E. 3rd St., Colorado City, TX 79512-6408. Tel.: 325-728-8285. Fax: 325-728-8944.
E-mail: museum@cityofcoloradocity.org

Web Site: www.coloradocitytexas.org/museum
Founded: 1960.
Congressional District: 17
Key Personnel: Pres. Bd., Gay Houston; Cur., Patty Pharis.
Personnel Profile: Part-Time Paid 1; Part-Time Volunteers 5.
Governing Authority: municipal. Parent Institution: City of Colorado City, Texas. Tax-exempt.
Institution Type/Description: History Museum.
Collections: paleontology; pioneer memorabilia; photographs of early settlers and scenes; old coaches; period artifacts; china; bison antiquities replica; Columbian mammoth exhibit; Chief Lone Wolf & Native American artifacts; war memorabilia; caboose; guns.
Activities: guided tours.
Publications: guide book showing historical markers.
Hours & Admission Prices: Tues.-Fri. 12-5. No charge; donations accepted. &
Attendance: 400 (accurate)
Membership: Student $5; Single $15; Family $25.

Comanche

COMANCHE COUNTY HISTORICAL MUSEUM, 402 Moorman Rd., Comanche, TX 76442.
Institution Type/Description: History Museum.
Collections: local history & culture; period furnishings; personal artifacts; photographs.
Hours & Admission Prices: Thurs. & Sat. 2-4.

Comstock

SEMINOLE CANYON STATE PARK AND HISTORIC SITE, U.S. Hwy. 90 W., Comstock, TX 78837. Mailing Address: P.O. Box 820, U.S. Hwy. 90 West, Park Rd. 67, Comstock, TX 78837-0820. Tel.: 432-292-4464. Fax: 432-292-4596.
E-mail: randy.rosales@tpwd.state.tx.us
Web Site: www.tpwd.state.tx.us
Formerly: Seminole Canyon State Historical Park
Founded: 1980.
Congressional District: 23
Key Personnel: C.E.O., Carter Smith; Supt., Randy Rosales; Chm. Commission, Peter M. Holt; Chief Cur., Tanya Petruney.
Personnel Profile: Full-Time Paid 8; Part-Time Volunteers 30.
Governing Authority: state. Parent Institution: Texas Parks & Wildlife Dept., 4200 Smith School Rd., Austin, TX 78744. Tel. 512-479-4882. Tax-exempt.
Institution Type/Description: Park & Archaeology Site.
Collections: prehistoric artifacts; 19th-century railroad artifacts; items relating to modern ranching; Native American Indian paintings; cave dwellings; Fate Bell Shelter containing 4,000-year old pictographs & some of the oldest cave dwellings in North America.
Facilities: picnic area; camping facilities; hiking trails; visitor center.
Activities: guided rock art tours; lectures; permanent exhibitions.
Publications: brochures.
Hours & Admission Prices: Daily 8-4:45. Adults over 12 $3; children under 12 no charge. Rock Art Tour: June-Aug. Wed.-Sun. 10 a.m.; Sept.-May Wed.-Sun. 10 & 3. Admission over 8 $5; children under 8 no charge. &
Attendance: 14,007 (accurate)
Membership: Texas State Parks Pass $70.

Corpus Christi

✳ **ART MUSEUM OF SOUTH TEXAS, (M),** 1902 N. Shoreline, Corpus Christi, TX 78401-1164. Tel.: 361-825-3500. Fax: 361-825-3520.
E-mail: artmuseum@tamucc.edu
Web Site: artmuseumofsouthtexas.org
Founded: 1943.
Key Personnel: Dir., Joseph B. Schenk; Asst. Dir., Sara Morgan; Cur. Education, Linda Rodriguez; Cur., Deborah Fullerton; Mktg., Cindy Anderson.
Governing Authority: nonprofit organization. Affiliated with Corpus Christi Art Foundation. Tax-exempt: 501(c)(3).
Institution Type/Description: Art Museum.
Collections: paintings; drawings; sculpture; graphics; photographs.
Major Exhibits: Ansel Adams: Western Exposure, 1/24/14-5/14/14; Metal in Motion II: Mark "Scrapdaddy" Bradford, 4/4/14-6/22/14; Target Texas: Contemporary Texas Studio Glass, 5/16/14-6/29/14; Conversations with Wood: Selections from the Waterbury Collection, 7/3/14-9/7/14; Shelia Rogers: Oceans of Plastic, 7/3/14-10/27/14; Deept in Art of Texas: 100 Years of Early Texas Painting, 9/18/14-1/4/15; Rock and Roll Photographs from the Will Vogt Collection, 11/14-12/14.
Facilities: 2,500-vol. library of art books available by request; 231-seat auditorium; classrooms. Gift items for sale.

Activities: temporary & traveling exhibitions; guided tours; lectures; films; gallery talks; concerts; formally organized education programs; docent program or council.
Publications: exhibition catalogs.
Hours & Admission Prices: Tues.-Sat. 10-5, Sun. 1-5. Adults $8; discounts to AAM members; children 12 & under and members no charge. Closed New Year's Day; Thanksgiving; Christmas. &
Attendance: 100,000 (estimated)
Membership: Student $20; Individual $40; Couple $50; Family $60; Supporting $100; Associate Patron $250; Patron $500; Donor Benefactor $1,000; Foundation Circle $3,000.

✳ **CORPUS CHRISTI MUSEUM OF SCIENCE AND HISTORY, (M),** 1900 N. Chaparral, Corpus Christi, TX 78401-1114. Tel.: 361-826-4667. Fax: 361-884-7392. Facebook: Corpus Christi Museum.
Web Site: www.ccmuseum.com
Founded: 1957.
Congressional District: 14
Key Personnel: Dir., Carol Rehtmeyer.
Personnel Profile: Part-Time Volunteers 10.
Governing Authority: public-private partnership. Parent Institution: Corpus Christi Museum Joint Venture and City of Corpus Christi. Tax-exempt: 170(b)(1)(A)(v).
Institution Type/Description: General Museum.
Collections: history; marine science; earth science; marine & terrestrial archaeology; Texas mollusks; South Texas plants.
Research Fields: local history; natural history; South Texas Native American cultures; marine archaeology.
Activities: permanent exhibits; educational programs; lectures.
Hours & Admission Prices: Tues.-Sat. 10-5, Sun. 12-5. Adults $9, seniors & children 3-12 $7, active military $4.50; members and children 2 & under no charge. Closed New Year's Day; Thanksgiving; Christmas. &
Attendance: 65,938 (accurate)
Membership: Single & Guest $75; Family (up to 5 people) $120 (additional person $24).

PADRE ISLAND NATIONAL SEASHORE, 20420 Park Rd. 22, Corpus Christi, TX 78418. Tel.: 361-949-8173, ext. 223. Fax: 361-949-8023.
E-mail: james_lindsay@nps.gov
Web Site: www.nps.gov/pais
Founded: 1962.
Congressional District: 14 & 15
Key Personnel: Supt., Joe Escoto.
Personnel Profile: Full-Time Paid 1.
Governing Authority: federal. Parent Institution: Dept. of the Interior, National Park Service, Washington, DC. Tax-exempt.
Institution Type/Description: National Park Visitor Center & Museum.
Collections: shells; fulgurite; insect; seashells; herbarium; archaeology; archives, photographs; oral history; artifacts from 1554 Spanish shipwrecks; Spanish coins.
Research Fields: natural history; archaeology.
Facilities: library of material dealing with the natural & cultural resources of the island available for use on premises with advance notice required.
Activities: scheduled programs year round.
Publications: orientation brochures; natural history publications; SPMA publication on the park; site bulletins on Flora, Fauna and cultural history.
Hours & Admission Prices: Headquarters: Mon.-Fri. 8-4:30. Museum: June-Aug. daily 8:30-6; Sept.-May daily 8:30-4:30. Visitors Center & Museum no charge. Parking fee $10, Annual Pass $20. &
Attendance: 60,888 (accurate)

THE SELENA MUSEUM, 5410 Leopard St., Corpus Christi, TX 78408.
Web Site: www.selenaetc.com/museum
Institution Type/Description: History Museum.
Collections: Selena's life & career; personal artifacts; photographs; costumes; Selena's red Porsche; awards; memorabilia.
Hours & Admission Prices: Mon.-Fri. 10-4. Adults 12 & over $2; children under 12 $1; children under one no charge.

SOUTH TEXAS BOTANICAL GARDENS & NATURE CENTER, 8545 S. Staples, Corpus Christi, TX 78413. Tel.: 361-852-2100. Fax: 361-852-7875.
E-mail: wmwomack@stxbot.org
Founded: 1983.
Congressional District: 27

Key Personnel: Exec. Dir., Michael Womack, Ed.D.; Dir. Mktg., Mary Jane Crull
Governing Authority: Tax-exempt.
Institution Type/Description: Botanical Gardens & Nature Center.
Collections: plants; trees; flowers; birds; butterfly garden.
Activities: rental facilities; demonstrations; special events.
Hours & Admission Prices: June to Labor Day daily 8-7; Sept.-May daily 8-6. Adults 13-59 $6, seniors 60 & over, military, and students $5, children 5-12 $3; discounts to American Horticulture Society; children under 5 & members no charge.

TEXAS STATE AQUARIUM, 2710 N. Shoreline Blvd., Corpus Christi, TX 78402. Tel.: 361-881-1200; 800-477-4853. Fax: 361-881-1257.
Web Site: www.texasstateaquarium.org
Institution Type/Description: Aquarium.
Collections: fish; turtles; dolphins; alligators; birds; flowers.
Activities: sea camp; presentations.
Hours & Admission Prices: Labor Day to Feb. daily 9-5; March to Sept. daily 9-6. Adults 13 & over $15.95, seniors & military $14.95, children 3-12 $10.95; children 2 & under no charge. Closed Thanksgiving; Christmas.

TEXAS STATE MUSEUM OF ASIAN CULTURES & EDUCA-TIONAL CENTER, 1809 N. Champarral St., Corpus Christi, TX 78401-1111. Tel.: 361-881-8827.
E-mail: texasasianculturesmuseumcc@gmail.com
Web Site: www.asianculturesmuseum.org
Founded: 1973.
Congressional District: 27
Key Personnel: Dir., Catherine Lacroix; Pres. Bd., Richard L. Bowers.
Personnel Profile: Full-Time Paid 2; Part-Time Paid 2; Part-Time Volunteers 7.
Governing Authority: nonprofit organization. Parent Institution: Billie Trimble Chandler Art Foundation, Inc. Tax-exempt: 501(c)(3).
Institution Type/Description: Cultural Museum.
Collections: decorative arts of Japan, China, India & Korea, including Hakata dolls; porcelains; metalware; cloisonne; lacquerware; Buddhist & Hindu images; clothing; costumes; oriental fan.
Research Fields: Japanese, Chinese, Indian, Korean & Philippine culture, & religion, art.
Facilities: 600-vol. library of books plus magazines, pamphlets available for research on premises only; reading room; classrooms. Oriental curios for sale.
Activities: lectures; art & culture festivals; hobby workshops; formally organized education programs; guided tours by appointment. Museum Sponsors: Ladies Auxiliary.
Publications: quarterly newsletter.
Hours & Admission Prices: Tues.-Sat. 11-5. Adults $6, seniors $5, students $4, children 12 & under $3; discounts to AAM members; members no charge. Closed New Year's Day; Easter; Thanksgiving; Christmas.
Attendance: 15,000 (estimated)
Membership: Student & Senior Citizen $15; Individual $25; Family $50; Supporting $100; Sustaining $500; Patron $1,000; Benefactor $5,000.

TEXAS SURF MUSEUM, 309 N. Water St., Corpus Christi, TX 78401. Tel.: 361-882-2364.
E-mail: info@texassurfmuseum.com
Web Site: www.texassurfmuseum.com
Institution Type/Description: Sports Museum.
Collections: surfing history & culture; surf boards; photographs; surfing memorabilia; personal artifacts.
Facilities: theater. Museum-related items for sale.
Activities: surf movies.
Hours & Admission Prices: Mon.-Thurs. 10-7, Fri.-Sat. 10-10, Sun. 11-5. No charge.

USS LEXINGTON MUSEUM ON THE BAY, 2914 N. Shoreline Blvd., Corpus Christi, TX 78402-1116. Mailing Address: P.O. Box 23076, Corpus Christi, TX 78403-3076. Tel.: 800-ladylex. Fax: 361-883-8361.
E-mail: debbie@usslexington.com
Web Site: www.usslexington.com
Founded: 1991.
Congressional District: 27
Key Personnel: Exec. Dir., Frank Montesano; Operations, Security & Exhibits Dir., M. Charles Reustle; Historian Curatorial Research & Registrar, Cecil Johnson; Public Rels., Deborah Crites; Museum Shop Mgr., Maria Robles.
Personnel Profile: Full-Time Paid 47; Part-Time Paid 11; Part-Time Volunteers 75.

Governing Authority: private; nonprofit organization. Tax-exempt: 501(c)(3).
Institution Type/Description: Naval Military Museum: housed in the USS Lexington Aircraft Carrier.
Collections: armed services (all branches with specialty toward U.S. Navy); aircraft; machinery; guns.
Research Fields: World War II, all areas of Navy & other military; exhibits on all phases of military; crew; ship.
Facilities: mess deck; ships store; event space for up to 600 guests; educational facilities; 3D Mega theater.
Activities: arts festivals; concerts; dance recitals; docent program; films; formal education programs for children; guided tours; participatory & temporary exhibitions.
Hours & Admission Prices: Daily 9-5. Adults $13.95, military & senior citizens $11.95, children 4-12 $8.95; members no charge. Closed Thanksgiving; Christmas.
Attendance: 340,000 (estimated)
Membership: Student $20; Individual $40; Couple $55; Family $75; Commander $125; Captain $250; Admiral $500. ($5 discount to Seniors & Military).

Corsicana

CAPEHART COMMUNICATIONS COLLECTION, 409 S. 9th St., Corsicana, TX 75110. Tel.: 903-872-0440. Fax: 903-872-0441.
E-mail: ritaanddoncapehart@sbcglobal.net
Web Site: www.telcomhistory.org/vm/museumsCapehart.shtml
Key Personnel: Owner, Don Capehart; Owner, Rita Capehart
Institution Type/Description: Communications Museum.
Collections: communications history & equipment; telephones; magazines; posters; records; Western Electric company history & products including sewing machines, switches, telephone booths & toilets; Bell Labs & AT&T documents, memorabilia & artifacts.
Activities: guided tours.
Hours & Admission Prices: By appointment.

CORSICANA FIELD AVIATION HERITAGE FOUNDATION, 9000 Navarro Rd., Corsicana, TX 75109. Tel.: 903-654-4847. Fax: 903-872-9911. cfahf.org.
E-mail: canifly@wifi45.com
Web Site: www.corsicanafield.org/index.html
Founded: 1998.
Congressional District: 6
Key Personnel: Dir., Gary Farley.
Personnel Profile: Part-Time Volunteers 8.
Governing Authority: nonprofit organization.
Institution Type/Description: Military History Museum.
Collections: airfield history; memorial monument; military artifacts; photographs; uniforms; memorabilia; books; newspapers; documents; maps.
Facilities: library.
Hours & Admission Prices: Call for hours. No charge, donations accepted.
Attendance: 1,250 (estimated)

NAVARRO COUNTY HISTORICAL SOCIETY, PIONEER VILLAGE, 912 W. Park Ave., Corsicana, TX 75110-2931. Tel.: 903-654-4846. Fax: 903-874-4441.
E-mail: byoung@ci.corsicana.tx.us
Founded: 1958.
Congressional District: 6
Key Personnel: Exec. Dir. Navarro County Historical Society, Dir. Pioneer Village Cur., Chm. (V) & Museum Shop Mgr., Bobbie Young; Pres. (V), Stephen Farris.
Personnel Profile: Full-Time Paid 1; Part-Time Paid 1.
Governing Authority: municipal; county; society. Parent Institution: Navarro County Historical Society; Subsidiary Institution: City of Corsicana, TX. Tax-exempt.
Institution Type/Description: Village Museum: eight log buildings constructed in Navarro County during 1838-1865, moved to City Park & restored.
Collections: period furnishings; newspapers; tools; cooking utensils; early country music from Texas; agriculture; archaeology; archives; period county artifacts; manuscripts; costumes; ethnology; folklore; general history; Native American artifacts; basketry; quilts; Hall of Fame. Historic Buildings: blacksmith shop; 1838 Indian trading post; 1846 pioneer kitchen; 1851 general store; 1854 frontier home; 1860 slave quarters; 1865 the old barn; 1890 blacksmith shop; tack shed; Redden House; Cotton Gin Scale House; McKie Playhouse; Peace Officers Museum; Sam Roberts Museum; Lefty Frizzell Museum; 1920s era gas station; Carriage House, houses a surrey with fringed top, 4-wheeled buckboards, 2-wheeled cart & 1930 tractor.
Research Fields: general history; genealogy.
Facilities: archives of research books & documents available for use by public.

Activities: guided group tours; lectures; permanent exhibitions.
Publications: annual book, Scroll; books, Navarro County Histories Vols. 1-7; The Women of Navarro County, Navarro County History Vol. 6; 1996 Sesquicentennial Year, Moments in Time Vol. 7.
Hours & Admission Prices: Mon.-Fri. 8-5, Sat. 9-5, Sun. 1-5; call for holiday hours. Adults $5, children 3-17 $3; discount to TX Association of Museums. &
Attendance: 2,620 (accurate)
Membership: Navarro County Historical Society: Individual $10; Family $15; Sustaining $120.

PEARCE MUSEUM, (M), 3100 W. Collin St., Corsicana, TX 75110-3904. Tel.: 903-875-7642; 800-988-5317. Fax: 903-875-7593.
E-mail: archives@navarrocollege.edu
Web Site: www.pearcecollections.us
Formerly: Pearce Collections Museum
Founded: 1996.
Congressional District: 6
Key Personnel: Dir., Holly B. Wait; Chm. & Dir., Navarro College Foundation, Inc., Dr. Tommy Stringer; Museum Shop Mgr., Pat Granger.
Personnel Profile: Full-Time Paid 3; Part-Time Paid 3; Part-Time Volunteers 30.
Governing Authority: public college. Parent Institution: Navarro College. Subsidiary Institutions: Navarro College Foundation, Inc., Corsicana, TX; Navarro College, Corsicana, TX. Tax-exempt: 501(c)(3).
Institution Type/Description: Civil War & Western Art Museum.
Collections: artifacts; photographs; memorabilia; ephemera; graphic materials; American Western art; history of Navarro College, Navarro County & U.S. Civil War.
Facilities: 500-vol. library of Civil War & American Western Art books; 200-seat auditorium; educational facilities; 15,000 sq. ft. exhibit space; planetarium; 42-seat theater. Museum-related items for sale.
Activities: arts festival; docent program; films; formal education programs; guided tours; lectures. Annual Events: art lectures; Civil War lectures; artist-in-residence program.
Hours & Admission Prices: Mon.-Fri. 10-4, Sat. 12-4. Adults $8, seniors $6, children 3-18 & students $4; discounts to groups. Closed New Year's Eve & Day; Thanksgiving; Christmas Eve & Day. &
Attendance: 8,000 (estimated)
Membership: Individual $30; Family $50; Contributor $100; Supporter $250; Patron $1,000.

WATKINS WILDLIFE EXHIBIT, Watkins Construction Company, 3229 S. 15th St. (under-pass under 1-45), Corsicana, TX 75110. Mailing Address: c/o Watkins Construction Company, 2029 Glynwood Cir., Corsicana, TX 75151. Tel.: 903-874-6587.
Founded: 1945.
Institution Type/Description: Wildlife Exhibit.
Collections: over 400 mounted animals from around the world.
Hours & Admission Prices: Mon.-Fri. 8-5. No charge.

Cotulla

BRUSH COUNTRY MUSEUM, 201 S. Stewart, Cotulla, TX 78014-3070. Mailing Address: P.O. Box 369, Cotulla, TX 78014-0369. Tel.: 830-879-2429.
Web Site: www.historicdistrict.com/museum
Founded: 1982.
Congressional District: 28
Key Personnel: Chm. (V), James C. Barbour; Treas., Nora Mae Tyler; Cur., Nita Gierisch; Cur., Elizabeth Seidel.
Personnel Profile: Part-Time Paid 2; Part-Time Volunteers 3.
Governing Authority: county; nonprofit. Tax-exempt.
Institution Type/Description: History Museum.
Collections: photographs; artifacts; history of La Salle County & the Brush Country of South Texas.
Activities: guided tours
Hours & Admission Prices: Tues. & Thurs. 10-12 & 2-4, Wed. & Fri.-Sat. 1-4. No charge; donations accepted. Closed New Year's Day; Independence Day; Thanksgiving; Christmas. &
Attendance: 600 (estimated)
Membership: Individual $15; Family $25; Pioneer Family $50.

Crane

MUSEUM OF THE DESERT SOUTHWEST, 409 S. Gaston, Crane, TX 79731-2621. Mailing Address: P.O. Box 398, Crane, TX 79731. Tel.: 432-558-2311.
E-mail: mtdsw@sbcglobal.net
Formerly: Crane County Museum
Founded: 1972.
Institution Type/Description: History Museum.
Collections: Native American, cowboy & Horsehead Crossing artifacts; Castle Gap fossils; petroleum industry.
Hours & Admission Prices: May-Sept. Sat. 9-12 & 1-5, Sun.-Fri. 1-4; Oct.-April Mon.-Fri. 1-4. No charge; donations accepted.

Crosbyton

CROSBY COUNTY PIONEER MEMORIAL MUSEUM, 101 W. Main, (intersection U.S. 82 & F.M. 651), Crosbyton, TX 79322-2252. Tel.: 806-675-2331.
E-mail: ccpmm@door.net
Web Site: www.crosbycountymuseum.com
Founded: 1958.
Congressional District: 17
Key Personnel: Admin., Gary Mitchell; Dir., Verna Anne Wheeler; Administrative Asst., Lynn Cruz.
Personnel Profile: Full-Time Paid 3; Full-Time Volunteers 1; Part-Time Paid 2; Part-Time Volunteers 18; Interns 2.
Governing Authority: nonprofit; municipal. Parent Institution: City of Crosbyton, TX. Tax-exempt.
Institution Type/Description: Local History Museum: partially housed in replica of Hank Smith rock house, original structure in 1876-77.
Collections: agriculture; home & family artifacts; Texas plains area artifacts; Native American artifacts; cowboy memorabilia; archives; cultural materials from prehistoric Plains Indians to 20th-century pioneers.
Research Fields: pioneer & oral histories; educational history of West Texas & Crosby County; community history of Crosby County; archaeological research; prehistoric Plains Indians to 20th-century pioneers; village & city development; agrarian society & technological advancements.
Facilities: library of early editions on education & history; archival research on people & places of Crosby County & West Texas available on premises; 500-seat auditorium; meeting room. Historical publications, maps & gifts for sale.
Activities: guided tours; lectures; films; gallery talks; arts festivals; drama; study clubs; hobby workshops; formally organized educational programs; permanent & temporary exhibitions; school loan service; cooperative education program.
Publications: bulletins; brochures; cards; historical maps; books, A History of Crosby County 1876-1977, Sun Rising on the West, Rock House Kitchen, Crosby County Cemetery Survey; A History of Black Families, Crosby County 1921-2001; Estacado Cradle of Culture & Civilization on the Staked Plains of Texas; Teachers Hand Book, Grades 2-5.
Hours & Admission Prices: Jan. to mid-Dec. Tues.-Sat. 9-12 & 1-5. No charge; donations accepted. Closed national holidays. &
Attendance: 8,136 (accurate)

MT. BLANCO FOSSIL MUSEUM, 124 W. Main, Crosbyton, TX 79322-2253. Mailing Address: P.O. Box 550, Crosbyton, TX 79322-0550. Tel.: 806-675-7777. Fax: 806-675-2421.
E-mail: mtblanco1@aol.com
Web Site: www.mtblanco.com
Founded: 1998.
Congressional District: 19
Key Personnel: Dir., Joe Taylor.
Personnel Profile: Part-Time Paid 1.
Institution Type/Description: Paleontology Museum.
Collections: fossils; earth history; sculpture; 13 ft. stegosaurus.
Research Fields: fossils.
Facilities: Museum-related items for sale.
Activities: annual ICA meeting.
Publications: Mt. Blanco Fossil News.
Hours & Admission Prices: By appointment. Adults $4, children $1; discounts to groups.
Attendance: 1,000 (estimated)
Membership: Individual $10.

Cross Plains

ROBERT E. HOWARD MUSEUM, 625 Southwest 5th St., Cross Plains, TX 76443. Tel.: 254-725-6114.
E-mail: jehanke@aol.com
Web Site: crossplainstx.com/howard-museum
Institution Type/Description: Historic House Museum: housed in the former home of Robert E. Howard, author of Conan the Barbarian.
Collections: Howard's life & career; personal artifacts; period furnishings; photographs.
Hours & Admission Prices: By appointment.

Crowell

FIREHALL MUSEUM, 116 N. Main St., Crowell, TX 79227. Mailing Address: P.O. Box 323, Crowell, TX 79227. Tel.: 940-684-1160.
E-mail: donna_baize@yahoo.com
Founded: 1909.
Congressional District: 28 & 13
Key Personnel: Dir. & Chm. (V), Donna Baize; Pres. (V), Duane Johnson; Museum Shop Mgr., George Allen.
Volunteer Hours: 20
Governing Authority: Parent Institution: Foard County Historical Society. Subsidiary Institution: Fire Hall Museum. Tax-exempt.
Institution Type/Description: History Museum.
Collections: local history & culture; household items; period furnishings; personal artifacts; photographs; farm & ranch implements; Gen. George B. McClellan's 1877 copper mine & campsite; Native American artifacts; veteran artifacts.
Hours & Admission Prices: Mon.-Fri. 2:30-5. No charge; donations accepted. &
Attendance: 300 (estimated)

SANTA FE DEPOT MUSEUM & LIBRARY, 203 N. Main St., Crowell, TX 79227. Mailing Address: P.O. Box 317, Crowell, TX 79227.
Institution Type/Description: Library & History Museum: depot built c.1908.
Collections: local history & culture; period furnishings; Cynthia Ann Parker artifacts; Foard County Courthouse; personal artifacts; books; photographs.
Facilities: library.
Hours & Admission Prices: Mon.-Fri. 1:30-5.

Cuero

DEWITT COUNTY HISTORICAL MUSEUM, 312 E. Broadway, Cuero, TX 77954-2806. Tel.: 361-275-6322.
Web Site: www.rootsweb.ancestry.com/~txdewitt/dewittcountyhistmuseum.htm
Founded: 1973.
Congressional District: 14
Key Personnel: Dir., Verna Smith.
Governing Authority: society. Affiliated with DeWitt County Historical Commission. Tax-exempt: 509(A).
Institution Type/Description: Local History Museum: housed in 1886 Bates-Sheppard home.
Collections: furnishings.
Research Fields: DeWitt County history.
Activities: tours; special group tours; preview parties; research; traveling exhibits. Annual Event: Wildflowers of Texas in April.
Hours & Admission Prices: Mon. & Thurs.-Fri. 8-5, Tues 10-3, Sun. 1-5. No charge; donations accepted. &
Attendance: 3,000 (estimated)
Membership: Student $2; Individual $10; Family $15; Sustaining $35; Corporate $100; Life $150.

Dalhart

XIT MUSEUM, (M), 108 E. 5th St., Dalhart, TX 79022. Mailing Address: P.O. Box 730, Dalhart, TX 79022-0730. Tel.: 806-244-5390. Fax: 806-244-3031.
E-mail: curator@xitmuseum
Web Site: www.xitmuseum.com
Founded: 1975.
Congressional District: 13
Key Personnel: Dir. & Cur., Nicky Olson.
Governing Authority: nonprofit organization. Parent Institution: Dallam-Hartley Counties Historical Assn., Inc. Tax-exempt.
Institution Type/Description: Historical Society Museum: housed in a terra-cotta brick building.
Collections: memorabilia from Dallam & Hartley Counties & the XIT Ranch history; paintings; photographs.
Research Fields: Dallam, Hartley Counties, & Dalhart history.
Activities: guided tours; concerts; arts festivals; study clubs; permanent & temporary exhibitions. Museum Sponsors: Open House during the XIT Reunion & Rodeo held yearly in August; four artist receptions per year.
Hours & Admission Prices: Tues.-Sat. 9-5. No charge; donations accepted. &
Attendance: 5,000 (accurate)
Membership: Annual Individual $35; Contributor $100; Supporter $250; Patron $500; Benefactor $1,000; Founder $5,000.

Dallas

AFRICAN AMERICAN MUSEUM, (M), 3536 Grand Ave., Dallas, TX 75210-1005. Mailing Address: P.O. Box 150157, Dallas, TX 75315-0157. Tel.: 214-565-9026; 877-852-3292 (toll free). Fax: 214-421-8204.
E-mail: info@aamdallas.org
Web Site: www.aamdallas.org
Founded: 1974.
Congressional District: 5
Key Personnel: Pres. & C.E.O., Dr. Harry Robinson, Jr.; Vice Pres. Institutional Advancement, Jane Jones; Cur., Collections & Exhibitions, John Spriggins; Mgr. Retail Operations, Patrick Finnell.
Governing Authority: nonprofit organization. Parent Institution: Foundation for African-American Art. Tax-exempt: 501(c)(3).
Institution Type/Description: African American Culture Museum.
Collections: African American fine art & folk art; Texas Black History; Texas Black Women's Archives.
Research Fields: Black history in Texas; folk art.
Facilities: 1,500-vol. library of books; 100-seat auditorium; classrooms; courtyard; cafeteria.
Activities: lectures; organized educational program; docent program; loan, temporary & traveling exhibitions. Museum Sponsors: history conference; literary conference; history fair.
Publications: quarterly newsletter.
Hours & Admission Prices: Tues.-Fri. 11-5, Sat. 10-5, Sun. 1-5. No charge; donations accepted. Docent Guided Tours: adults $5, students $3. Closed New Year's Day; Independence Day; Thanksgiving Day; Christmas Eve & Day. &
Attendance: 201,000 (estimated)
Membership: Student & Senior Citizen $15; Individual $25; Family $35; Friend $50; Patron $100; Associate $500; Benefactor $1,000 & up.

CASETA: CENTER FOR THE ADVANCEMENT OF EARLY TEXAS ART, 14070 Proton Rd., Ste. 100, Dallas, TX 75244. Mailing Address: c/o Bill Reaves, P.O. Box 667487, Houston, TX 77266. Tel.: 972-233-9107, ext. 215.
Web Site: www.caseta.org
Key Personnel: Exec. Dir., Olivia Thompson; Chm., Robert Summers
Institution Type/Description: Art Museum.
Collections: works by Texas artists including paintings; sculpture & watercolor; art history.
Hours & Admission Prices: Call for hours.

CHILDREN'S AQUARIUM AT FAIR PARK, 1462 1st Ave., Dallas, TX 75210-1010. Mailing Address: P.O. Box 150113, Dallas, TX 75315-0113. Tel.: 214-554-7340. Fax: 214-421-3239.
E-mail: stephen.walker@dallaszoo.com
Web Site: www.childrensaquariumfairpark.com
Formerly: The Dallas Aquarium at Fair Park
Founded: 1936.
Congressional District: 5
Key Personnel: C.E.O. & Dir., Greggory Hudson; Gen. Mgr., Stephen Walker; Supvr., Barrett Christie; Senior Aquarist, Charles Yancey; Aquarist, Martin Conricote; Aquarist, Eric Julius; Deputy Dir. Operations, Doug Dykman.
Personnel Profile: Full-Time Paid 11; Part-Time Paid 2; Part-Time Volunteers 26; Interns 1.
Operating Expenses: 820,000
Operating Income: 796,000
Governing Authority: municipal. Parent Institution: Dallas Zoo. Tax-exempt.
Institution Type/Description: Aquarium.
Collections: freshwater & marine fish; reptiles; amphibians; invertebrates.
Research Fields: desert fish; Edwards Aquifer endemics; reproductive studies; husbandry research; seahorse & elasmobranch conservation; unionid biology & conservation.
Facilities: classrooms.
Activities: guided tours; lectures; formally organized education programs for children; daily keeper presentations to the general public; touch tank.

Publications: Paradigm; Zookeeper; Tracks & Facts; The Wildlife Saver.
Hours & Admission Prices: Daily 9-4. Adults $8, seniors 65 & up and children 3-12 $6; members and children 3 & under no charge. Closed Thanksgiving Day & Christmas Day. &
Attendance: 150,000 (estimated)
Membership: Individual $40; Individual Plus $50; Family $55; Family Plus $75; Sustaining $125; Patron $250; Sponsor $500.

DALLAS ARBORETUM & BOTANICAL GARDEN, 8525 Garland Rd., Dallas, TX 75218-4335. Mailing Address: 8617 Garland Rd., Dallas, TX 75218-3914. Tel.: 214-515-6615 & 6501. Fax: 214-324-9801.
E-mail: customerservice@dallasarboretum.org
Web Site: www.dallasarboretum.org
Founded: 1974.
Key Personnel: Pres. & C.E.O., Mary Brinegar; Chm., Brian Shivers; Vice Pres. Education & Research, Maria Conroy; Vice Pres. Adv. & Promotions, Terry Lendecker.
Governing Authority: bd. of directors; nonprofit organization. Parent Institution: Dallas Arboretum & Botanical Garden Society. Tax-exempt.
Institution Type/Description: Arboretum & Botanical Garden.
Collections: plant materials.
Research Fields: horticulture; botany; ecology.
Facilities: botanical garden; reading room. Museum-related items for sale.
Activities: guided tours; lectures; concerts; festivals; study clubs; hobby workshops; TV & radio programs; formally organized education programs for children, adults, undergraduate & graduate students; docent program or council; permanent & temporary exhibitions; volunteer groups.
Publications: bimonthly newsletter, The Dallas Arboretum; plant information handouts.
Hours & Admission Prices: Daily 9-5. Adults $10, seniors 65 & over $9, children 3-12 $7; members no charge. Parking $7. Closed New Year's Day; Thanksgiving; Christmas. &
Attendance: 300,000 (accurate)
Membership: Individual $82; Family $125; Family Adventure $175; Sustaining $300; Patron $550; Bronze Leaf $2,500; Silver Leaf $3,500; Gold Leaf $5,000; Caroline Rose Hunt Society $10,000.

DALLAS CONTEMPORARY, 161 Glass St., Dallas, TX 75207-6903. Tel.: 214-821-2522. Fax: 214-821-9103. Facebook: Dallas Contemporary.
E-mail: info@dallascontemporary.org
Web Site: www.dallascontemporary.org
Founded: 1978.
Congressional District: 30
Key Personnel: Dir., Peter Doroshenko; Assoc. Dir. Devel., Genniva Bruce; Assoc. Mgr. Devel., Clare Chadwick; Assoc. Dir. Exhibitions, Erin Cluley; Learning Mgr., Autumn Hill; Adjunct Cur., Lilia Kudelia; Adjunct Cur., Pedro Alonzo; Adjunct Cur., Florence Ostende.
Personnel Profile: Full-Time Paid 8; Part-Time Paid 1; Part-Time Volunteers 100; Interns 12.
Governing Authority: nonprofit. Tax-exempt: 501(c)(3).
Institution Type/Description: Contemporary Art Museum.
Collections: contemporary art.
Major Exhibits: JR, 1/14-4/14; Arthur Pena: Slight Shift, Steady Hand, 1/14-4/14; Paula Crown: Inside My Head, 1/14-4/14; Les Miserables: Post Conceptual Art, 1/14-4/14; Lavish Design in Dallas, 1/14-4/14; Richard Phillips, 4/14; Julian Schnabel, 4/14.
Research Fields: contemporary Texas art in a global context; Texas artists; opportunities for artists.
Facilities: library & information center; sculpture garden.
Activities: art talks; studio classes; workshops on collecting art, resource & information center for artists. Annual Events: Membership Show; Mix! Series of artists with diverse ethnic backgrounds; Auction, Legends, Luncheon in May.
Publications: biannual newsletter; exhibition brochures.
Hours & Admission Prices: Tues.-Sat. 10-5. No charge; donations accepted. Closed major holidays. &
Attendance: 45,000 (estimated)
Membership: Glass: Student & Artist $35; Individual $60; Couple $100; Y@161 $161; Steel $400; Innovator $1,000; Industrialist $3,500.

DALLAS FIREFIGHTERS MUSEUM, INC., 3801 Parry Ave., Dallas, TX 75226-1753. Tel.: 214-821-1500. Facebook: Dallas Firefighters Museum.
E-mail: dallasfirefightermuseum@yahoo.com
Web Site: www.dallasfiremuseum.com
Founded: 1972.

Key Personnel: Pres. (V), Stuart Grant; Vice Pres., Rett Blankenship.
Personnel Profile: Part-Time Paid 3.
Governing Authority: nonprofit organization. Tax-exempt.
Institution Type/Description: Firefighter's Museum.
Collections: period fire trucks including an 1884 horse-drawn steam pumper & 1936 ladder truck; firefighting history; extinguishers; helmets; suits; uniforms; photographs.
Facilities: Museum-related items for sale.
Activities: fire & life safety.
Publications: 2008 calendar.
Hours & Admission Prices: Wed.-Sat. 9-4. Adults $4, children $2. Closed New Year's Day; Thanksgiving; Christmas.
Attendance: 3,500 (estimated)

* **DALLAS HERITAGE VILLAGE AT OLD CITY PARK, (M),** 1515 S. Harwood St., Dallas, TX 75215. Tel.: 214-421-5141. Fax: 214-428-6351.
E-mail: info@dallasheritagevillage.org
Web Site: www.dallasheritagevillage.org
Formerly: Old City Park: The Historical Village of Dallas
Founded: 1966.
Congressional District: 5
Key Personnel: Exec. Dir. & Pres., Gary N. Smith.
Personnel Profile: Full-Time Paid 3; Part-Time Paid 20; Part-Time Volunteers 300.
Governing Authority: private. Parent Institution: Dallas County Heritage Society. Tax-exempt: 501(c)(3).
Institution Type/Description: Village Museum.
Collections: 38 historic structures from the period c.1840-1910; 19th-century material culture & decorative arts with emphasis on North Central Texas.
Research Fields: north central Texas; Dallas; American decorative arts; crafts.
Facilities: library. Museum-related items for sale.
Activities: interpretive programs; outreach programs; lectures; films; docent program; orientation program; craft demonstrations; permanent & temporary exhibitions; weddings in the church & bandstand. Museum Sponsors: 3 major special events.
Publications: membership newsletter, Heritage News; newsletter for volunteers, Gazette.
Hours & Admission Prices: Tues.-Sat. 10-4, Sun. 12-4. Adults $9, seniors $7, children 4-12 $5; discounts to AAM & ICOM members; museum professionals & members no charge. Closed New Year's Day; Memorial Day; Thanksgiving; Christmas. &
Attendance: 70,000 (estimated)
Membership: Scout $60; Family $75; Settler $125; Pioneer $300; Jr. Curator's Circle $550; Trailblazer $600; Curator's Circle $1,250; Corporate levels also available.

DALLAS HISTORICAL SOCIETY, (M), Hall of State, Fair Park, 3939 Grand Ave., Dallas, TX 75210. Mailing Address: P.O. Box 150038, Dallas, TX 75315-0038. Tel.: 214-421-4500. Fax: 214-421-7500.
E-mail: nora@dallashistory.org
Web Site: www.dallashistory.org
Founded: 1922.
Congressional District: 5
Key Personnel: Exec. Dir., Jack Bunning; Chm. (V), Max Wells; Pres. (V), Margaret Keliher.
Personnel Profile: Full-Time Paid 6; Part-Time Paid 2; Part-Time Volunteers 60; Interns 4.
Governing Authority: society; nonprofit. Tax-exempt: 501(c)(3).
Institution Type/Description: History Museum.
Collections: period artifacts, books, archive materials relating to southwestern & U.S. history; manuscripts; costumes.
Research Fields: American, Texas & Dallas history.
Facilities: 14,000-vol. library & 3,000,000 pages of archives on Southwestern U.S. available for use on premises; reading room; 400-seat auditorium; rental facilities.
Activities: guided tours; formally organized education programs for kindergarten through adult; docent program; lectures; gallery talks; hobby workshops; training programs; loan, permanent & temporary exhibitions; Hall of State available for special events.
Publications: monthly newsletter, Dallas Historical Society Register; books, Dallas Rediscovered; When Dallas Becomes a City; semi-annual historical journal, Legacies.
Hours & Admission Prices: Tues.-Sat. 10-5, Sun. 1-5. No charge; donations accepted. Closed New Year's Day; Thanksgiving; Christmas. &
Attendance: 209,165 (estimated)
Membership: Individual & Family $50; Biographer $100; Collector $250; Scholar $500; Fellow $1,250.

DALLAS HOLOCAUST MUSEUM/CENTER FOR EDUCATION & TOLERANCE, 211 N. Record St., Ste. 100, Dallas, TX 75202-3361. Tel.: 214-741-7500. Fax: 214-747-2270.
E-mail: info@dallasholocaustmuseum.org
Web Site: www.dallasholocaustmuseum.org
Formerly: Dallas Holocaust Memorial Center
Founded: 1984.
Congressional District: 3
Key Personnel: Pres. & C.E.O., Mary Pat Higgins; Chm., Hylton Jonas; Dir. Mktg., Paula Nourse; Sr. Dir. Education, Sara Abosch; Dir. Finance & Administration, Belinda Griffin; Archivist, Anthony Cimino; Devel. Coord., Kim Overs; Front Desk Assoc., Adilene Hernandes; Front Desk Assoc., Anthony Frey.
Personnel Profile: Full-Time Paid 9; Part-Time Paid 2; Part-Time Volunteers 75; Interns 3.
Governing Authority: not-for-profit organization. Tax-exempt: 501(c)(3).
Institution Type/Description: History Museum.
Collections: photographs & artifacts from the Holocaust & Jewish life in Europe before the Holocaust.
Facilities: 3,500-vol. library on the Holocaust & anti-Semitism; 135-seat auditorium; 2,500 sq. ft. exhibit space.
Activities: docent program; films; guided tours; lectures; school loan service; teachers training seminar.
Publications: triannual, Dallas Holocaust Museum Newsletter.
Hours & Admission Prices: Mon.-Fri. 9:30-5, Sat.-Sun. 11-5. Adults $8, students under 18, seniors 55 & over, & active military $6; discounts to groups of 15 or more. Closed New Year's Day; Easter; Rosh Hashanah; Independence Day; Yom Kippur; Thanksgiving; Christmas Eve & Day. &
Attendance: 60,000 (estimated)
Membership: Student $5; Teacher $18; Family $36; Mentor $54; Supporter $108; Patron $250; Outreach Partner $500; Guardians of Memory $1,000; Teachers of Tolerance $2,500; Defenders of Honor $5,000; Liberators from Indifference $7,500; Keepers of the Eternal Flame $10,000 & up.

*** DALLAS MUSEUM OF ART, (M),** 1717 N. Harwood St., Dallas, TX 75201-2398. Tel.: 214-922-1200. Fax: 214-922-1350.
E-mail: jbernstein@dma.org
Web Site: www.DallasMuseumofArt.org
Founded: 1903.
Congressional District: 5
Key Personnel: Pres. Bd. Trustees, John Eagle; Chm. Bd. Trustees, Margot B. Perot; Dir., Maxwell L. Anderson; Deputy Dir., Robert Stein; Assoc. Dir. External Affairs, Anne Bergeron; Sr. Cur. Arts of Africa, the Pacific & the Americas, Dr. Roslyn A. Walker; Asst. Cur. of Arts of the Americas, Kimberly L. Jones; Cur. Decorative Arts & Design, Kevin W. Tucker; Sr. Cur. Contemporary Art, Jeffrey Grove; Sr. Cur. European & American Art & Assoc. Dir Cur. Affairs, Olivier Meslay; Assoc. Cur. American Art, Sue Cantebury; Dir. Exhibitions & Collections, Tamara Wooton-Bonner; Dir. Education, Nicole Stutzman; Dir. Libraries & Imaging Svcs., Jacqueline Allen; Dir. Visitor Svcs., Barbee Barber; Asst. Cur. Contemporary Art, Gabriel Ritter; Dir. Communications, Jill Bernstein; Dir. Human Resources, Pamela Autry.
Personnel Profile: Full-Time Paid 205; Part-Time Paid 23; Interns 8.
Governing Authority: nonprofit organization. Tax-exempt: 501(c)(3).
Institution Type/Description: Art Museum.
Collections: European, American & Latin American painting, sculpture & decorative arts and design; contemporary international art; ancient Mediterranean art; Asian art; Indonesian & Oceanic art; African art; pre-Columbian art.
Facilities: 50,000-vol. library of art books & reference material available on premises; reading room; 350-seat auditorium; 2 restaurants. Museum-related items for sale.
Activities: guided tours; lectures; films; family programs; gallery talks; concerts; late night events; formally organized educational programs; docent program; training programs for professional museum workers; inter-museum loan, permanent, temporary, traveling & participatory exhibitions, live literary program. Museum Sponsors: annual Awards to Artists program.
Publications: member magazine; exhibition catalogues & brochures; collection catalogues; annual report.
Hours & Admission Prices: Tues.-Wed. & Fri.-Sun. 11-5, Thurs. 11-9. Adults $10, senior citizens $7, students $5; discount to AAM members; 1st Tues. of month, members and children under 12 no charge. Special exhibition admission varies. Closed Thanksgiving; Christmas. &
Attendance: 542,000 (accurate)
Membership: Member $75; Sustainer $125; Friend $250; Advocate $500; Junior Associates Circle $625; Contributor $1,000; Associates Circle $2,000; Patrons Circle $5,000; Fellows Circle $10,000; Leaders Circle $15,000; Benefactors Circle $25,000; Director's Circle $50,000; President's Circle $100,000; Chairman's Circle $250,000.

THE DALLAS WORLD AQUARIUM, 1801 N. Griffin St., Dallas, TX 75202-1503. Tel.: 214-720-2224.
E-mail: info@dwazoo.com
Web Site: www.dwazoo.com/aquarian.html
Institution Type/Description: Aquarium.
Collections: Aquarium: marine life from around the world. Rainforest: flora & fauna.
Facilities: restaurants. Museum-related items for sale.
Hours & Admission Prices: Daily 10-5. Adults $18.95, seniors 60 & over $14.95, children 3-12 $10.95; children 2 & under no charge. Closed Thanksgiving; Christmas.

DALLAS ZOO, 650 S. R.L. Thornton Freeway, Dallas, TX 75203-3013. Tel.: 214-670-5656. Fax: 214-670-7450.
E-mail: info@dalzoo.org
Web Site: www.dallaszoo.com
Founded: 1888.
Congressional District: 24
Key Personnel: Dir. Zoo, Richard W. Buickerood; Pres. Dallas Zoological Society, Michael Meadows; Deputy Dir. Animal Management, Chuck Siegel; Cur. Birds, Chris Brown; Business Mgr., Kerry Rhines; Cur. Mammals, Ken Kaemmerer; Cur. Mammals, Todd Bowsher; Staff Veterinarian, Dr. Tom Alvarado; Volunteer Coord., Sherri Reneau; Deputy Dir. Operations, Doug Dykman; Cur. Research & Education, Dr. Cynthia Bennett; Facility Mgr., Jeff Rash.
Personnel Profile: Full-Time Paid 236; Part-Time Paid 16; Part-Time Volunteers 500; Interns 2.
Governing Authority: municipal. Parent Institution: City of Dallas/Parks Department. Subsidiary Institution: Dallas Zoological Society. Tax-exempt.
Institution Type/Description: Zoo.
Collections: mammals; birds; reptiles; amphibians.
Research Fields: reproductive physiology; animal behavior.
Facilities: 2,000-vol. library of zoology & natural history available for reference & research on premises; 100 seat auditorium.
Activities: guided & self-guided tours; lectures; formally organized education programs for children; docent program or council; zooniversity classes; permanent exhibitions.
Publications: Adopt-an-Animal newsletter; members' newsletter; annual report.
Hours & Admission Prices: Daily 9-5. Adults $12, senior citizens 65 & over and children 3-11 $9; members & children under 3 no charge. Parking: $5. Closed Christmas. &
Attendance: 576,300 (accurate)
Membership: Individual $30; Family $40; Family Plus $60; Company Member $250; Corporate Council Leopard $2,500; Corporate Council Lion $5,000; Corporate Council Elephant $10,000.

THE EIGHT TRACK MUSEUM, 2630 E. Commerce, Dallas, TX 75226-1402. Mailing Address: 3100 Main St. #414, Dallas, TX 75226. Tel.: 469-867-4074.
Founded: 2010.
Institution Type/Description: General Museum.
Collections: eight track tapes.
Publications: e-newsletter.
Hours & Admission Prices: Call for hours.
Membership: Individual $20; Life $80.

FRONTIERS OF FLIGHT MUSEUM, 6911 Lemmon Ave., Dallas, TX 75209-3603. Tel.: 214-350-3600 & 1651. Fax: 214-351-0101.
E-mail: info@flightmuseum.com
Web Site: www.flightmuseum.com
Founded: 1988.
Congressional District: 5
Key Personnel: C.E.O. & Pres. (V), J. Jan Collmer; Chm., Senator Kay Bailey Hutchison; Pres. (V), Ray Woodland; Exec. Dir., Dan Hamilton; Cur., Chris Woodul; Financial Dir., Sam Montgomery; Devel. & Education, Dr. Sharon Spalding; Vice Pres. Advancement, Ilene Stern; Vice Pres. Programs, Bruce Bleakley; Gift Shop Mgr., Brenda Magee; Asst. to Exec. Dir., Anne Marie Evans.
Personnel Profile: Full-Time Paid 10; Full-Time Volunteers 5; Part-Time Paid 4; Part-Time Volunteers 150.
Governing Authority: nonprofit organization. Affiliated with the History of Aviation Collection/University of Texas at Dallas. Tax-exempt: 501(c)(3).
Institution Type/Description: Aeronautical History Museum.
Collections: aviation history from pre-Wright brothers through modern space

age; photos; models; Admiral Charles E. Rosendahl's lighter than air collection; World War I & II artifacts & exhibits; Admiral Richard E. Byrd's Antarctic expedition collection; business & commercial aviation exhibits; one-of-a-kind artifacts. Historic Aircraft: Sopwith Pup; Temple Monoplane, Glasflugel Sailplane; F16B No. 2; Gossamer Penguin; F-105F; T-33A; F4 WST.
Research Fields: lighter than air; commercial aviation; aircraft production & aviation training in Texas.
Facilities: 90-seat auditorium & theater; 3,600 sq. ft. exhibit space. Gift items for sale.
Activities: guided tours; lectures; organized educational programs; docent program. Annual Event: George E. Haddaway Award Gala in May.
Publications: quarterly newsletter; tour guide; education brochures; volunteer newsletter; joint North Texas Aviation Museum brochure.
Hours & Admission Prices: Mon.-Sat. 10-5, Sun. 1-5. Adults $8, seniors 65 & over $6, children 3-17 $5; discounts to groups, AAM, ICOM & TAM members; students & members no charge. Closed New Year's Eve & Day; Easter; Memorial Day; Independence Day; Labor Day; Thanksgiving; Christmas. &
Attendance: 100,000 (accurate)
Membership: Individual $50; Family $75; Patron $100; Sustaining $500. Corporate: $1,000; $2,500; $5,000.

JUANITA J. CRAFT CIVIL RIGHTS HOUSE, 2618 Warren Ave., Dallas, TX 75215-2911. Tel.: 214-670-8637.
Institution Type/Description: Historic House Museum: housed in the former home of civil rights organizer, Juanita J. Craft; visited here by President Lyndon Johnson & Martin Luther King Jr. to discuss the future of the civil rights movement.
Collections: local history & culture; period furnishings; personal artifacts; photographs.
Hours & Admission Prices: Mon.-Fri. 9-5:30 by appointment.

THE MCKINNEY AVENUE CONTEMPORARY (THE MAC), 3120 McKinney Ave., Dallas, TX 75204. Tel.: 214-953-1212.
E-mail: macmembership@the-mac.org
Web Site: www.the-mac.org
Founded: 1994.
Key Personnel: Dir., Lisa Hees; Museum Shop Mgr., Claire Roseland.
Personnel Profile: Full-Time Paid 3.
Governing Authority: nonprofit organization. Tax-exempt: 501(c)(3).
Institution Type/Description: Art Gallery.
Collections: works by contemporary artists.
Major Exhibits: Suzanne Anker: While Darkness Sleeps, 1/11/14-3/1/14; Aqua-culture curated by Henry Sanchez, 1/11/14-3/1/14; Paul Bryan - Believe It Anyway, 1/11/14-3/1/14; 20th Anniversary Exhibition 1994-2003, 5/14-6/14; core Residency Program, 5/14-6/14; Annual Membership Exhibition, 7/14-8/14; 20th Anniversary Exhibition 2004-2013, 11/14-12/14; MAC PAC Sponsored Exhibition, 11/14-12/14.
Activities: art talks; panel discussions.
Hours & Admission Prices: Wed.-Sat. 11-9; groups by appointment. Suggested Donations: adults $5, seniors & students $4. Closed major holidays. &
Attendance: 12,000 (estimated)
Membership: Artist & Student $30; True Blue $50; Writer's Garret Combo $60; MAC PAC $85-$120; Amigo $100; Lover $250; Champion $500; Benevolent One $1,000; Art God $2,500.

MEADOWS MUSEUM, (M), Southern Methodist University, 5900 Bishop Blvd., Dallas, TX 75205. Tel.: 214-768-2516. Fax: 214-768-1688. Facebook: Meadows Museum Dallas.
E-mail: meadowsmuseminfo@smu.edu
Web Site: www.meadowsmuseumdallas.org
Founded: 1965.
Congressional District: 3
Key Personnel: Dir., Mark A. Roglan; Mktg. & Public Rels. Mgr., Carrie Hunnicutt; Dir. Education, M. Carmen Smith; Financial Officer, Roni Arifin; Assoc. Dir. & Cur. Exhibitions, Bridget Marx; Mgr. Membership, Sheri Anne MacNeil; Mgr. Operations, Charles Guijarro; Security Mgr., Brenda Laury, Museum Shop Mgr., Barbara West.
Personnel Profile: Full-Time Paid 20; Part-Time Paid 10; Part-Time Volunteers 75; Interns 5.
Governing Authority: university. Parent Institution: Southern Methodist University. Tax-exempt.
Institution Type/Description: Art Museum.
Collections: Algur H. Meadows collection of Spanish art; Elizabeth Meadows sculpture collection; University art; works by Texas artists.
Major Exhibits: Sorolla and America (T), 12/13-4/14.
Research Fields: Spanish art history.
Facilities: 1,000-vol. library of books & periodicals on history of Spanish art

available on request; reading room; works on paper research room; special event halls; auditorium. Museum-related items for sale.
Activities: guided tours; gallery talks; concerts; lecture series; film programs; formally organized education programs for undergraduate & graduate students affiliated with Southern Methodist University; members' exhibition previews.
Publications: newsletter; exhibition catalogues.
Hours & Admission Prices: Tues.-Wed. & Fri.-Sat. 10-5, Thurs. 10-9, Sun. 1-5. Adults $10; members, children under 12, SMU faculty, staff, students & staff members of other museums no charge. Closed New Year's Day; Easter; Thanksgiving; Christmas Eve & Day. &
Attendance: 69,000 (accurate)
Membership: Goya Friend $60; El Greco Circle $150; Velazquez Court $300; Ribera Patron $500; Murillo Benefactor $1,000; Fortuny Angel $2,500. Corporate memberships vary.

MUSEUM OF BIBLICAL ART, 7500 Park Lane, Dallas, TX 75225-2025. Mailing Address: P.O. Box 12727, Dallas, TX 75225-0727. Tel.: 214-368-4622. Fax: 214-361-1365.
E-mail: frontdesk@biblicalarts.org
Web Site: www.biblicalarts.org
Formerly: Biblical Arts Center
Founded: 1966.
Congressional District: 5
Key Personnel: Co Dir., R.J. Machacek; Co Dir., Scott Peck; Asst. Dir., Dr. Val Robinson.
Personnel Profile: Full-Time Paid 5; Part-Time Paid 15; Part-Time Volunteers 5; Interns 10.
Governing Authority: nonprofit organization. Parent Institution: Miracle at Pentecost Foundation. Tax-exempt: 501(c)(3).
Institution Type/Description: Religious Art Museum.
Collections: 16th century to present art in mediums including performing arts pertaining to the Bible.
Facilities: multi-purpose auditorium; theater.
Activities: docent program; lectures; permanent & traveling exhibitions; formally organized education courses; performing arts programs.
Publications: books, Creation of a Masterpiece.
Hours & Admission Prices: Mon.-Sat. 10-5, Sun. 1-5. Galleries: no charge; donations accepted. Program: adults $12; senior citizens & children 6-18 $10. Closed New Year's Day; Thanksgiving; Christmas Eve & Day. &
Attendance: 70,000 (accurate)
Membership. Individual & Basic $35; Family $75.

MUSEUM OF GEOMETRIC & MADI ART, (M), 3109 Carlisle St., Dallas, TX 75204-1194. Tel.: 214-855-7802. Fax: 214-855-5479.
E-mail: dorothy@geometricmadimuseum.org
Web Site: www.geometricmadimuseum.org
Formerly: MADI Museum
Founded: 2003.
Congressional District: 30
Key Personnel: C.E.O. & Cur., Dorothy Masterson; Chm. & Pres. (V), Laura Moffat; Dir. Education & Museum Shop Mgr., Patricia Canning; Dir. Membership & Registrar, Grace Wyatt; Public Rels., Rebecah Beauchamp.
Personnel Profile: Full-Time Volunteers 1; Part-Time Paid 5; Part-Time Volunteers 125; Interns 1.
Governing Authority: private; nonprofit organization. Parent Institution: MADI World Art Foundation Museum of Geometric & MADI Art. Tax-exempt: 501(c)(3).
Institution Type/Description: Art Museum.
Collections: works by founder Carmelo Arden Quin, Rothfuss, Kosice, Herbin, Blaszko, Vardanega, Carreno, Guevara; artists from South & North America, Japan, France, Italy, Hungary, & other European countries; New Zealand, Africa.
Major Exhibits: New Digital Art, 1/10/14-3/30/14; Quilts: The New Geometry, 4/4/14-7/6/14.
Facilities: Museum-related items for sale.
Activities: docent program; guided tours; lectures; participatory exhibits; printmaking & collage classes.
Publications: Biography of Carmelo Arden Quin; show catalogues.
Hours & Admission Prices: Tues.-Wed. & Fri.-Sat. 11-5, Thurs. 11-7, Sun. 1-5. No charge; donations accepted. Closed New Year's Day; Thanksgiving; Christmas. &
Attendance: 5,103 (accurate)
Membership: Student $25; Basic $50; Family $100; Sponsor $250; Patron $500; Benefactor $1,000; Angel $2,500.

NASHER SCULPTURE CENTER, (M), 2001 Flora St., Dallas, TX 75201-2336. Tel.: 214-242-5100. Fax: 214-242-5155.
Web Site: www.nashersculpturecenter.org
Founded: 2003.
Congressional District: 30
Key Personnel: Dir., Jeremy Strick; Devel., Jill Magnuson; Devel., Martha Hess; Education, Anna Smith; Public Rels., Kristen Gibbins; C.F.O., John McBride; Registrar, Melissa Durkee; Cur., Jed Morse; Museum Shop Mgr., Carolyn Spinelli; Security, Mike Jensen.
Personnel Profile: Full-Time Paid 23; Part-Time Paid 21; Part-Time Volunteers 40; Interns 3.
Governing Authority: private; nonprofit organization. Tax-exempt: 501(c)(3).
Institution Type/Description: Art Museum & Sculpture Garden.
Collections: Raymond and Patsy Nasher collection of modern & contemporary sculpture from 19th-century to present.
Facilities: 3,000-vol. library of sculpture books; 10,000 sq. ft. exhibit space; 100-seat restaurant; classrooms; 200-seat theater. Museum-related items for sale.
Activities: arts festivals; concerts; dance recitals; docent program; films; formal education programs; guided tours; lectures; loan, temporary & traveling exhibitions.
Publications: quarterly members newsletter.
Hours & Admission Prices: Tues.-Sun. 11-5. Adults $10, senior citizens $7, students $5; children under 12 & members no charge. Closed Independence Day; Thanksgiving; Christmas.
Attendance: 250,000
Membership: Giacometti $75; Moore $125; Calder $500; Miro $1,000; Brancusi $2,500; Matisse $5,000; Rodin $10,000.

PEROT MUSEUM OF NATURE AND SCIENCE, 2201 N. Field St., Dallas, TX 75201. Tel.: 214-428-5555. Facebook: Perot Museum.
E-mail: info@perotmuseum.org
Web Site: www.perotmuseum.org/
Formerly: Museum of Nature & Science, Southwest Museum of Science and Technology, The Science Place & TI Founders IMAX Theater, the Dallas Children's Museum
Founded: 2012.
Key Personnel: C.E.O., Nicole Small; Chm., Carolyn Perot Rathjen; C.F.O., Sally Pietsch; Dir. Finance, Nadine Chaffin; Dir. Mktg., Beth Hook; Dir. Exhibits, Mike Spiewak; Dir. Guest Svcs., Mark Boyer; Dir. Volunteers, Fyve Hilton; Vice Pres. Programs, Steve Hinkley; Cur. Earth Science, Dr. Anthony Fiorillo; Dir. Major Gifts, Rhealyn Carter; Dir. Sales, Linda Murdock; Membership Mgr., Angel Wolfe; Vice Pres. Devel., Elisabeth Galley.
Personnel Profile: Full-Time Paid 97; Part-Time Paid 195; Part-Time Volunteers 1,176.
Governing Authority: nonprofit organization. Tax-exempt: 501(c)(3) & 170(b)(1)(A).
Institution Type/Description: Natural History, Science & Technology Museum.
Collections: over 5,800 scientific bird specimens; 14,000 malacology specimens; 3,300 rocks & minerals; 3,500 mammal specimens; 37,000 paleontology specimens.
Research Fields: vertebrate paleontology.
Facilities: library; auditorium; 3-D digital cinema; classrooms; cafe. Gift items for sale.
Activities: lectures; films; gallery talks; demonstrations; formally-organized educational programs; training programs for museum workers; teacher training; seasonal workshops; electric theater; school programs (onsite & outreach); after school programs; day camps; family festival events; birthday parties; sleepovers; scouting adventures; parent-and-child drop-in programs.
Publications: triannual newsletter, Explore; teacher's guide; brochures; fliers.
Hours & Admission Prices: Mon.-Sat. 10-5, Sun. 12-5. General admission: adults $15, senior citizens 65 & over and students 12-17 $12, child 2-11 $10; discounts to groups & ASTC members; members & children under 2 no charge. Films: adults, seniors & children 2-17 $5-$8; discounts to ASTC members. Special exhibitions may have additional cost. Closed Thanksgiving; Christmas. &
Membership: Student $45; Individual $65; Dual $80; Family $100; Family Plus $150; Family Plus Platinum $200; Inventor $250; Naturalist $500; Explorer $1,000.

✱ **THE SIXTH FLOOR MUSEUM AT DEALEY PLAZA, (M), (I),** 411 Elm St., Dallas, TX 75202-3301. Tel.: 214-747-6660; 888-485-4854. Fax: 214-747-6662. Facebook: Sixth Floor Museum.
E-mail: jfk@jfk.org
Web Site: www.jfk.org
Founded: 1989.
Congressional District: 30
Key Personnel: Exec. Dir., Nicola Longford; Chm. Bd., Ken Menges; Dir. Finance, Liz Shipp; Dir. Operations & Special Initiatives, Tim Case; Dir. Collections & Intellectual Property, Megan Bryant; Cur., Gary Mack; Dir. Education & Public Programming, Sharron Conrad; Assoc. Dir. Public Rels. & Mktg., Carol Murray; Museum Shop Mgr., Janet Stieve.
Personnel Profile: Full-Time Paid 44; Part-Time Paid 9; Part-Time Volunteers 66; Interns 5.
Volunteer Hours: 1,521
Operating Expenses: 5,090,500
Operating Income: 5,798,500
Governing Authority: private; nonprofit organization. Parent Institution: Dallas County Historical Foundation. Tax-exempt.
Institution Type/Description: History Museum & Historical Site: the former Texas School Book Depository.
Collections: photographs; film & video footage; documents; period artifacts; oral histories related to the assassination & legacy of President John F. Kennedy.
Research Fields: Kennedy assassination & investigations; 1960s Dallas history & culture; presidential history.
Facilities: visitor center; rental facilities; reading room; cafe. Museum-related items for sale.
Activities: audio tour guide in six different languages; cell phone tour; education programs for children & adults; public programs; permanent & temporary exhibitions; special events.
Publications: guide book, Dealey Plaza National Historic Landmark; DVD, Films from The Sixth Floor; brochures; student gallery guides; souvenir book; book on history of museum.
Hours & Admission Prices: Mon. 12-6, Tues.-Sun. 10-6. Adults $16, seniors 65 & up $14, youth 6-18 $13, children 5 & under with audio guide $4; discounts to groups and AAM, TAM & ICOM members; children 5 & under without audio guide no charge. Audio Tour: available in six languages included with admission. Closed Thanksgiving; Christmas. &
Attendance: 342,000 (accurate)

TEXAS DISCOVERY GARDENS, 3601 Martin Luther King Blvd., Gate 6 at Fair Park, Dallas, TX 75210. Mailing Address: P.O. Box 152537, Dallas, TX 75315-2537. Tel.: 214-428-7476. Fax: 214-428-5338.
E-mail: tdg@texasdiscoverygardens.org
Web Site: www.texasdiscoverygardens.org
Formerly: Dallas Horticulture Center
Founded: 1941.
Congressional District: 5
Key Personnel: Exec. Dir., Dick Davis; Chm. (V), Michael Bosco; Public Rels. & Mktg. Mgr., Sarah Gardner; Dir., Horticulture, Roger Sanderson; Gift Shop Coord., Kerry Ragsdill
Governing Authority: nonprofit organization. Tax-exempt: 501(c)(3).
Institution Type/Description: Arboretum & Botanical Garden Museum, Conservatory & Horticultural Resource Center: c.1936. A National Historic Landmark.
Collections: gardens including native plants; water features & outdoor sculpture; live butterflies; insectarium.
Facilities: 7 1/2-acre grounds; botanical garden; visitor center; butterfly house.
Activities: seminars; lectures & workshops; education programs for children; permanent & temporary exhibitions; seasonal events & festivals; community resource for public & private events, performing arts groups, weddings & conferences; garden tours.
Publications: quarterly newsletter.
Hours & Admission Prices: Daily 10-5. Adults $8, seniors 60 & up $6, children 3-11 $4; discounts to groups; members, children under 3 & Tues. no charge. Closed New Year's Day, Thanksgiving; Christmas Eve & Day. &
Attendance: 180,000 (estimated)
Membership: Malachite $40; Monarch $65; Blue Morpho $150; Paper Kite $300; Glasswinged $600; Birdwing $1,200.

TEXAS FIRE MUSEUM, City of Dallas Fire Dept. Maintenance Facility, 2600 Chalk Hill Rd., Dallas, TX 75212-4506. Mailing Address: P.O. Box 560724, Dallas, TX 75356-0724. Tel.: 214-267-1867.
Key Personnel: Dir., Bill Carroll.
Personnel Profile: Part-Time Paid 2; Part-Time Volunteers 10.
Governing Authority: nonprofit organization. Tax-exempt: 501(c)(3).
Institution Type/Description: Fire Museum.
Collections: fire service history; period fire equipment & trucks; personal artifacts; photographs.
Facilities: restaurant. Museum-related items for sale.
Publications: monthly newsletter.

Hours & Admission Prices: Thurs.-Sat. 10-2. No charge; donations accepted.
&

Attendance: 350 (estimated)

THE TRAMMELL & MARGARET CROW COLLECTION OF ASIAN ART, (M), 2010 Flora St., Dallas, TX 75201-2335. Tel.: 214-979-6430. Fax: 214-979-6439.

E-mail: sadams@crowcollection.org
Web Site: www.crowcollection.org
Founded: 1998.
Key Personnel: Dir., Amy L. Hofland; Pres., Trammell S. Crow.
Personnel Profile: Full-Time Paid 11; Part-Time Paid 19; Part-Time Volunteers 45; Interns 6.
Governing Authority: private; nonprofit organization. Tax-exempt.
Institution Type/Description: Art Museum.
Collections: art from Japan, China, Cambodia, India, Thailand, Myanmar & Tibet.
Facilities: 100-vol. library of Asian art & culture; 9,000 sq. ft. exhibit space. Museum-related items for sale.
Activities: arts festivals; concerts; dance recitals; docent program; guided tours; lectures; temporary exhibitions; theater. Museum Sponsors: After Dark monthly; Family Days at the Crow Collection monthly.
Hours & Admission Prices: Tues.-Sun. 10-6. No charge; donations accepted. Closed New Year's Day, Independence Day; Thanksgiving; Christmas. &
Attendance: 80,000 (estimated)
Membership: Teacher & Student $30; Magnolia Circle $65; Pearl Circle $125; Bamboo Circle $250; Lotus Circle $500; Peony Circle $1,000; Bronze Circle $2,500; Ivory Circle $5,000; Crystal Circle $7,500; Imperial Circle $10,000.

Decatur

WISE COUNTY HERITAGE MUSEUM, 1602 S. Trinity, Decatur, TX 76234-2717. Mailing Address: P.O. Box 427, Decatur, TX 76234-0427. Tel.: 940-627-5586.

E-mail: wisemuseum@embarqmail.com
Web Site: www.wisehistory.com
Founded: 1967.
Congressional District: 13
Key Personnel: C.E.O., Dir. & Chm. Wise County Historical Commission, Kerry Clower; Dir., Rosalie Gregg; Vice Pres., Exhibits Chm. & Museum Shop Mgr., Bettye Jane Dodds; Sec. & Trustee, Patti Gillispie; Trustee, CeCe Peguese; Trustee, Sue Tackel; Trustee, Franklin Blank.
Personnel Profile: Part-Time Paid 1; Part-Time Volunteers 4.
Operating Expenses: 79,370
Operating Income: 97,594
Governing Authority: society. Parent Institution: Wise County Historical Society, Inc. Tax-exempt: 501(c)(3).
Institution Type/Description: Local History Museum.
Collections: historical artifacts; manuscripts; genealogy; members of the Lost Battalion (Prisoners of War of the Japanese during WWII); dedication to the survivors of the USS Houston CA-30 & members of armed forces of England, Australia & Holland who were also POW's; members of the 131st Field Artillery & the USS Houston CA-30.
Research Fields: genealogy; family histories.
Facilities: over 2,000-vol. library of material on county history & newspaper abstracts available for use on premises; reading room; 300 seat auditorium; newspapers on microfilm; Wise County census records; cemetery, census, birth, marriage records; material on Tennessee, Virginia & other states. Gift items, plaques, cards, pictures & books pertaining to Wise County for sale.
Activities: guided tours; films; drama; hobby workshops; permanent & temporary exhibitions; weddings. Museum Sponsors: Fundraising Events; Wise County Fine Arts Festival October to December.
Publications: monthly newsletters; History of Wise County, A Link With The Past Vol. I; Vol. II; Vol. III; Pioneer History reprint; History of Rhome, TX; memoirs by former World War II POWs of the Lost Battalion who spent 3-1/2 years building the railroad that led to the bridge over the River Kwai.
Hours & Admission Prices: Mon.-Sat. 10-3. Adults $2, children $1; members no charge. Closed major holidays. &
Attendance: 7,000 (estimated)
Membership: Individual $10; Husband & Wife $15; Life Time $150.

Del Rio

WHITEHEAD MEMORIAL MUSEUM, 1308 S. Main St., Del Rio, TX 78840-5998. Tel.: 830-774-7568.

E-mail: lee.lincoln@rocketmail.com
Web Site: www.whiteheadmuseum.org
Founded: 1962.
Congressional District: 23

Key Personnel: C.E.O., Dir. & Museum Shop Mgr, Lee Lincoln; Pres. (V), Mike Parker.
Personnel Profile: Full-Time Paid 1; Part-Time Paid 3; Part-Time Volunteers 8.
Governing Authority: municipal. Tax-exempt.
Institution Type/Description: Historical Museum.
Collections: 2-1/2 acre site with over 14 historical & replica buildings including the 1870 Perry Store; Black Seminole Indian Scouts exhibit; graves of Judge Roy Bean & son Sam; Jersey Lily Saloon replica; pioneer log cabin; a folk art exhibit, the Cadena Nativity.
Research Fields: Dr. John Brinkley; Judge Roy Bean; Seminole Negro Indian Scouts; U-2 Spy Plane (USAF); genealogy, railroad; Val Verde County; Buffalo Soldier; Texas Rock Art; Lower Percos River.
Facilities: visitor center. Museum-related gifts for sale.
Activities: guided tours.
Publications: brochures
Hours & Admission Prices: Tues.-Sat. 9-4:30, Sun. 1-5, guided tours by appointment. Adults $5, seniors $4, youths 13-18 $3, children 6-12 $2; discount to AAM, ICOM & TAM members; members no charge. &
Attendance: 11,000 (estimated)
Membership: Single $15; Family $25; Corporate $50; Contributing $75; Sustaining $100 & up.

Denison

EISENHOWER BIRTHPLACE STATE HISTORIC SITE, 609 S. Lamar Ave., Denison, TX 75021-4821. Tel.: 903-465-8908. Fax: 903-465-8988.

Web Site: www.visiteisenhowerbirthplace.com
Founded: 1946.
Congressional District: 4
Key Personnel: Site Mgr., Robin Gilliam.
Personnel Profile: Full-Time Paid 3; Part-Time Paid 3; Part-Time Volunteers 12.
Governing Authority: state. Parent Institution: Texas Historical Commission. Tax-exempt.
Institution Type/Description: Historic House: 1881 house where Dwight D. Eisenhower was born.
Collections: 1890 period furnishings; political campaign memorabilia.
Facilities: visitor center; covered pavilion; education center; meeting & community room. Museum-related items for sale.
Activities: guided tours; permanent exhibitions.
Publications: brochure.
Hours & Admission Prices: Tues.-Sat. 9-5, Sun. 1-5; last tour 4pm. Adults $3, children 12-18 $2; discounts for AAM & ICOM members; children under 12 & members no charge. Closed New Year's Day; Thanksgiving; Christmas Eve & Day. &
Attendance: 15,000 (estimated)

RED RIVER RAILROAD MUSEUM, 101 E. Main St., Ste. 145, Denison, TX 75021-3001. Tel.: 903-463-5289.

E-mail: rrrmuseum@yahoo.com
Web Site: www.redriverrailmuseum.org
Key Personnel: Dir., Tina DiToma; Chm. (V), Doug Hoover.
Personnel Profile: Part-Time Paid 1; Part-Time Volunteers 16.
Governing Authority: Tax-exempt.
Institution Type/Description: Railroad History Museum: housed in the former Katy Depot.
Collections: railroad history & artifacts; photographs; books; maps; personal artifacts; period furnishings.
Facilities: Museum-related items for sale.
Activities: train simulator.
Hours & Admission Prices: Thurs.-Sat. 11-4, Sun. 1-4. No charge; donations accepted.
Membership: Caller $10; Switchman $25; Brakeman $50; Engineer $100; Conductor $250; Yardmaster $500.

Denton

BAYLESS-SELBY HOUSE MUSEUM, 317 W. Mulberry St., Denton, TX 76201-6062. Tel.: 940-349-2865. Fax: 940-349-2851.

E-mail: peggy.riddle@dentoncounty.com
Web Site: www.dentoncounty.com/bsh
Founded: 2001.
Key Personnel: Mgr., Peggy Riddle.
Governing Authority: Parent Institution: Denton County.
Institution Type/Description: Historic House Museum.
Collections: Victorian period artifacts & furnishings.
Activities: lectures.
Hours & Admission Prices: Tues.-Fri. 10-4:30, Sat. 11-3; groups by appointment. Group Tours: $1 per person. &

DAR MUSEUM FIRST LADIES OF TEXAS HISTORIC COSTUMES COLLECTION, Texas Woman's University, Administration Conference Tower, 2nd Fl., Denton, TX 76204. Mailing Address: P.O. Box 425379, Denton, TX 76204-5379. Tel.: 940-898-3644. Fax: 940-898-3556.
Web Site: www.twu.edu
Founded: 1940.
Congressional District: 26
Key Personnel: Dir. Conference Svcs., David Sweeten.
Governing Authority: Parent Institution: Texas Woman's University. Subsidiary Institution: Dept. of Family Sciences.
Institution Type/Description: Costume Museum.
Collections: Inaugural Ball gowns of Texas first ladies, including wives of the presidents of the Republic of Texas & the governors of the State; gowns worn by the wives of Vice President John Nance Garner, President Dwight D. Eisenhower & President Lyndon B. Johnson.
Activities: tours by appointment.
Publications: descriptive booklet of the collection; Texas' First Ladies Historical Costume Collection.
Hours & Admission Prices: Mon.-Fri. 8-5; groups by appointment only. No charge. Closed national holidays; university holidays. &

DENTON COUNTY AFRICAN AMERICAN MUSEUM, 317 W. Mulberry St., Denton, TX 76201-6062. Tel.: 940-349-2865. Fax: 940-349-2851.
E-mail: peggy.riddle@dentoncounty.com
Web Site: www.dentoncounty.com/dcaam
Founded: 2008.
Key Personnel: Mgr., Peggy Riddle.
Governing Authority: Parent Institution: Denton County.
Institution Type/Description: African American History Museum.
Collections: photographs; texts; personal artifacts; period furnishings.
Activities: group tours.
Hours & Admission Prices: Tues.-Fri. 10-4:30, Sat. 11-3; groups by appointment. Groups: $1 per person. Closed holidays. &

DENTON COUNTY MUSEUMS - COURTHOUSE-ON-THE-SQUARE MUSEUM, (M), 110 W. Hickory, Denton, TX 76201-4116. Tel.: 940-349-2850. Fax: 940-349-2851.
Web Site: www.dentoncounty.com/chos
Founded: 1979.
Key Personnel: Dir., Peggy Riddle; Cur. Collections, Kim McCoig Cupit; Coord. Tourism, Gretel L'Heureux.
Personnel Profile: Full-Time Paid 4; Part-Time Paid 3; Part-Time Volunteers 30.
Governing Authority: county. Branch Museums: Historical Park of Denton County: Bayless-Selby House Museum & Denton County African American Museum, 317 W. Mulberry St., Denton, TX. Tax-exempt.
Institution Type/Description: History Museum: housed in the former county courthouse; built in 1896. Listed on the National Register of Historic Places.
Collections: local history & culture; period furnishings; personal artifacts.
Activities: lectures; special events.
Hours & Admission Prices: Mon.-Fri. 10-4:30, Sat. 11-3. Museum: no charge; donations accepted. Group Tours: $1 per person. Closed holidays & holiday weekends. &
Attendance: 86,000 (accurate)

HANGAR 10 FLYING MUSEUM, 1945 Matt Wright Lane, Denton Municipal Airport, Denton, TX 76207-4537. Tel.: 940-565-1945.
Governing Authority: nonprofit organization. Tax-exempt: 501(c)(3).
Institution Type/Description: Aviation History Museum.
Collections: civil & military aircraft; aviation history; military memorabilia; personal artifacts.
Hours & Admission Prices: Mon.-Sat. 8:30-3; other times by appointment. No charge; donations accepted.

TEXAS WOMAN'S UNIVERSITY ART GALLERIES, 1200 Frame St., Visual Arts Bldg., Denton, TX 76204. Mailing Address: P.O. Box 425469, TWU Station, Denton, TX 76204-5469. Tel.: 940-898-2530 & 2533. Fax: 940-898-2496.
E-mail: visualarts@twu.edu
Web Site: www.twu.edu
Founded: 1901.
Congressional District: 26
Key Personnel: Chm. Visual Art Dept., John Weinkein.
Governing Authority: state; university. Affiliated with Texas Woman's University.

Institution Type/Description: Art Museum.
Collections: paintings; photographs; sculpture.
Activities: guided tours; lectures; films; gallery talks; formally organized education programs for children, adults, undergraduate and graduate students; temporary exhibitions; student thesis exhibition show.
Hours & Admission Prices: Mon.-Fri. 8-5. No charge. Closed national holidays. &

UNIVERSITY OF NORTH TEXAS ART GALLERY, (M), College of Visual Arts & Design, 1201 W. Mulberry, Denton, TX 76203. Mailing Address: 1155 Union Cir. #305100, Denton, TX 76203-5107. Tel.: 940-565-4005 & 4001. Fax: 940-565-4717.
E-mail: gallery@unt.edu
Web Site: gallery.unt.edu
Founded: 1972.
Congressional District: 26
Key Personnel: Dir., Tracee W. Robertson; Mgr. Programming, Katy Stewart; Asst. Dir. Exhibitions & Collections, Victoria Estrada Berg DeCuir.
Governing Authority: university; nonprofit. Parent Institution: Univ. of North Texas. Subsidiary Institutions: Cora Stafford Gallery, 1201 W. Mulberry, Denton, TX; Lightwell Gallery, 1201 W. Mulberry, Denton, TX; UNT Artspace FW; Design Gallery. Tax-exempt.
Institution Type/Description: Art Gallery.
Collections: 20th century contemporary art most of which is by former students & faculty. Visual art: painting; ceramics; prints; graphics; sculpture. A significant portion of the collection was donated by patrons & artists (such as DeKooning & Motherwell).
Research Fields: internships available in Museum/Gallery studies.
Facilities: 3,150 sq. ft. exhibit space.
Activities: formally organized education programs for undergraduates & graduates; over 700 works in the permanent collection available for loan to faculty & staff; temporary, loan & traveling exhibitions. Museum Sponsors: annual juried competition; MFA Exhibition for graduates.
Publications: exhibition catalogues.
Hours & Admission Prices: Tues. 12-5, Wed.-Thurs. 9:30-8, Fri.-Sat. 12-5. Summer: Tues.-Sat. 12-5. No charge; donations accepted. &
Attendance: 8,000 (accurate)

Dickens

DICKENS HISTORICAL MUSEUM, 609 Montgomery St., Dickens, TX 79229. Mailing Address: P.O. Box 311, Dickens, TX 79229. Tel.: 806-623-5566.
Key Personnel: Dir., Leanora Houwen.
Personnel Profile: Full-Time Volunteers 1.
Governing Authority: Tax-exempt.
Institution Type/Description: History Museum: housed in a former hardware store.
Collections: local history & culture; period furnishings; personal artifacts; household items; school records; military artifacts; clothing; ranching; books; maps.
Facilities: library.
Hours & Admission Prices: By appointment. No charge; donations accepted.
Attendance: 6 (estimated)

Dripping Springs

DR. POUND HISTORICAL FARMSTEAD MUSEUM, Founder's Park, Dripping Springs, TX 78620. Mailing Address: Friends of the Pound House Foundation, P.O. Box 1150, Dripping Springs, TX 78620. Tel.: 512-858-2030.
E-mail: poundhouse@verizon.net
Web Site: drpoundhistoricalfarmstead.org
Formerly: Dr. Pound Pioneer Farmstead Museum
Key Personnel: Exec. Dir., Jeanette Ramirez
Institution Type/Description: History Museum: housed on the farmstead built by one of the city's founding families; built in 1854.
Collections: Pound family & local history; period furnishings; personal artifacts; photographs; blacksmith collection; quilts. Historic Buildings: barn; smokehouse; windmill.
Activities: special events. Museum Sponsors: Heritage Gala in April; Summer Music Series in June; Pioneer Kids Days in July; Pioneer Days Fall Fest in September.
Hours & Admission Prices: Wed.-Sat. 12-3; other times by appointment. Suggested donation $5.

HARBOR RANCH, U.S. 290 W., Dripping Springs, TX 78620. Mailing Address: 1200 Belmont Pkwy., Austin, TX 78703-1414. Tel.: 512-469-0716. Fax: 512-494-8574.
Formerly: Galloping Road Compound & Poker Alley; Harbor Ranch
Founded: 1950.
Congressional District: 14
Key Personnel: Owner & C.E.O., Harriet Rutland.
Governing Authority: individual operation. Branch Museum: Harbor Ranch, US 290, Dripping Springs, TX 78620.
Institution Type/Description: Historic House Museum.
Collections: Texas-made furniture; farm tools; period artifacts. Historic Buildings: bunkhouse; barns.
Activities: guided tours; study clubs.
Hours & Admission Prices: By appointment only. No charge.
Attendance: 100 (estimated)

Dublin

DUBLIN BOTTLING WORKS MUSEUM & WP KLOSTER MUSEUM ANNEX, 105 E. Elm, Dublin, TX 76446-2309. Mailing Address: 221 S. Patrick St., Dublin, TX 76446-2347. Tel.: 254-445-4210. Fax: 254-445-4677.
E-mail: lori@dublinbottlingworks.com
Web Site: www.dublinbottlingworks.com
Formerly: Dublin Dr Pepper Bottling Company Museum
Founded: 1995.
Key Personnel: Collections Mgr., Lori Dodd.
Personnel Profile: Full-Time Paid 2.
Governing Authority: Parent Institution: Dublin Dr Pepper.
Institution Type/Description: Company Museum.
Collections: company history, bottling, marketing & advertising; photographs; catalogs.
Activities: guided tours.
Hours & Admission Prices: Daily 10-5. Adults $5, senior citizens & children 5-12 $4; children 4 & under no charge. Closed New Year's Day; Easter; Thanksgiving; Christmas. &
Attendance: 70,000 (estimated)

Dumas

MOORE COUNTY ART ASSOCIATION, The Art Center, 1810 S. Dumas Ave., Dumas, TX 79029-6002. Tel.: 806-935-5312. Fax: 806-934-4447.
E-mail: fineart54@valornet.com
Web Site: www.dumasmuseumandartcenter.org
Founded: 1954.
Key Personnel: C.E.O. Procurement & Pres., Carolyn Stallwitz; Dir. & Museum Shop Mgr., Marti Christman; Cur. Finance, Business Officer & Memorials, Glynda Pflug; Vice Pres., Mary Ferris; Cur. Catalogue, Risa Franco.
Personnel Profile: Full-Time Paid 1; Part-Time Paid 2; Part-Time Volunteers 2.
Governing Authority: nonprofit organization. Subsidiary Institution: Moore County Texas. Tax-exempt: 170(b)(1)(A).
Institution Type/Description: Art Center.
Collections: paintings; photographs; prints; sculpture.
Research Fields: local history.
Activities: guided tours; loan, permanent & temporary exhibitions; school tours; Jon Birdsong watercolor workshop, April 5, 6, 7, 2013; Jerry Yarnell Acrylic Workshop, Aug. (exact dates TBD).
Publications: brochure.
Hours & Admission Prices: Mon.-Sat. 10-5. No charge; donations accepted. Closed New Year's Day; Thanksgiving; Christmas. &
Attendance: 6,600 (accurate)
Membership: Individual $15; Family $25; Patron $50; Friend $75 & up; Benefactor $500 & up.

MOORE COUNTY HISTORICAL MUSEUM DBA WINDOW ON THE PLAINS, Window on the Plains, 1820 S. Dumas Ave., Dumas, TX 79029-6002. Tel.: 806-935-3113. Fax: 806-934-3621
E-mail: dumasmuseum@windstream.net
Web Site: www.dumasmuseumandartcenter.org
Founded: 1976.
Congressional District: 31
Key Personnel: Dir. & Museum Shop Mgr., Terri George; Pres., William F. Watson; Cur. Finance, Business Officer & Memorials, Glynda Pflug.
Personnel Profile: Full-Time Paid 1; Part-Time Paid 1; Part-Time Volunteers 35.
Volunteer Hours: 42
Operating Expenses: 103,839

Operating Income: 104,903
Governing Authority: nonprofit organization. Subsidiary Institution: Moore County Texas. Tax-exempt: 170(b)(1)(A).
Institution Type/Description: History and Wildlife Museum.
Collections: local wild life; historical items; Indian artifacts; local photographs; oral tapes; newspapers; archives.
Research Fields: local history.
Activities: guided tours; loan, permanent & temporary exhibitions; school tours.
Publications: brochures concentrating on Texas & Dumas.
Hours & Admission Prices: Mon.-Sat. 10-5. No charge; donations accepted. Closed Thanksgiving; Christmas. &
Attendance: 5,136 (accurate)
Membership: Individual $15; Family $25; Patron $50; Friend $100; Benefactor $500.

Duncanville

INTERNATIONAL MUSEUM OF CULTURES, (M), 411 E. Hwy. 67, Duncanville, TX 75137. Mailing Address: P.O. Box 381608, Duncanville, TX 75138. Facebook: International Museum of Cultures.
E-mail: mfkamm@internationalmuseumofcultures.org
Web Site: www.internationalmuseumofcultures.org
Founded: 1974.
Congressional District: 24
Key Personnel: C.E.O., Mary Fae Kamm; Chm. (V), Florentine Ramirez.
Personnel Profile: Full-Time Paid 2; Full-Time Volunteers 2; Part-Time Paid 3; Part-Time Volunteers 24; Interns 3.
Governing Authority: nonprofit organization. Tax-exempt: 501(c)(3).
Institution Type/Description: Anthropology Museum.
Collections: ethnographic collections from South America, Papua New Guinea, Africa, Southeast Asia, East Asia, Mexico, Native American.
Research Fields: Cultural & Social Anthropology; Theoretical & Applied Linguistics.
Facilities: 15,000-vol. library of books on anthropology, linguistics & literacy available for use on premises; field research stations.
Activities: Film and lecture series on cultures of the world.
Publications: Museum of Cultures Publications Series (23 volumes).
Hours & Admission Prices: Mon.-Sat. 10-4. Scheduled Tours: adults $5, seniors & children $4; discounts to AAM, AAA & KERA members; members no charge. &
Attendance: 13,500 (accurate)
Membership: Student $25; Senior $30; Individual $45; Family $60; Patron $100; Legacy $500 & up.

Eagle Lake

PRAIRIE EDGE MUSEUM, 408 E. Main St., Eagle Lake, TX 77434-2534. Tel.: 979-234-7442.
E-mail: prairieedgemuseum@yahoo.com
Web Site: www.prairieedgemuseum.com
Key Personnel: Dir., Christine Owen
Institution Type/Description: History Museum.
Collections: area natural & cultural history; plant & animal life; Native American artifacts; early pioneers; military.
Activities: educational programs; presentations; celebrations.
Hours & Admission Prices: Mon.-Fri. 9-1, Sat.-Sun. 2-5. No charge; donations accepted. &

Eden

DON FREEMAN MEMORIAL MUSEUM, 120 Paint Rock St., Eden, TX 76837. Mailing Address: P.O. Box 915, Eden, TX 76837. Tel.: 325-869-2211 & 5074.
Web Site: www.edentexas.com
Founded: 2003.
Congressional District: 11
Key Personnel: Dir., Carolyn Moody.
Personnel Profile: Part-Time Volunteers 4.
Institution Type/Description: History Museum.
Collections: local history & culture; period furnishings; personal artifacts; photographs; military artifacts.
Hours & Admission Prices: Sat. 10-5, Sun. 1-4. No charge; donations accepted.
Attendance: 267 (accurate)

Edinburg

EDINBURG FIRE DEPARTMENT FIREFIGHTERS MU-SEUM, 212 W. McIntyre, Edinburg, TX 78541. Mailing Address: P.O. Box 1079, Edinburg, TX 78541. Tel.: 956-292-2101. Fax: 956-289-1853.
E-mail: ssnider@cityofedinburg.com
Founded: 2004.
Key Personnel: Dir., Shawn M. Snider.
Personnel Profile: Full-Time Paid 27; Full-Time Volunteers 85.
Governing Authority: private; nonprofit organization. Tax-exempt: 501(c)(3).
Institution Type/Description: Firefighting History Museum.
Collections: Edinburg Fire Department history & equipment; photographs; documents; personal artifacts.
Facilities: library; 800 sq. ft. exhibition space.
Activities: guided tours; lectures.
Hours & Admission Prices: Mon.-Fri. 8-5. No charge; donations accepted. Closed New Year's Day; Easter; Independence Day; Labor Day; Christmas. &
Attendance: 750

❋ **MUSEUM OF SOUTH TEXAS HISTORY, (M),** 200 N. Closner Blvd., Edinburg, TX 78541-3554. Tel.: 956-383-6911. Fax: 956-381-8518.
E-mail: mpena@mosthistory.org
Web Site: www.mosthistory.org
Formerly: Hidalgo County Historical Museum
Founded: 1967.
Congressional District: 15
Key Personnel: Exec. Dir., Shan Rankin; Chm. Bd. Trustees, Barbara Guerra; Devel., Lynne Beeching; Registrar, Lisa Adam; Coord. Education, Judy McClelland; Programming Officer, Melissa Tijerina; Public Rels., Martha Pena; Cur. Archives & Collections, Barbara Stokes; Receptionist, Sandra Luna; Sr. Cur., Tom A. Fort; IT Specialist & Archives Asst., Steve Lomas; Maintenance, Nazario Reyna; Bldg. Supvr., Joe Hernandez.
Personnel Profile: Full-Time Paid 13; Part-Time Volunteers 25; Interns 40.
Governing Authority: nonprofit organization. Tax-exempt: 501(c)(3).
Institution Type/Description: History Museum.
Collections: history & culture of the Rio Grande Valley, South Texas (south of the Nueces River) & northeastern Mexico region; clothing, books, photographs, furniture, machinery, weapons from prehistoric Indian era to 1900's; folklife; Spanish colonial settlement era; ranching; law enforcement; early agriculture; archival documentation. Historic Building: 1910 County jail including hanging room & trap door.
Research Fields: architecture; customs & traditions of Rio Grande Valley; Spanish settlement, colonization & early ranch life, genealogy; early agriculture, ranching, transportation, settlement.
Facilities: library; archives relating to the lower Rio Grande Valley, available for research or sale; 10,000 sq. ft. exhibit areas. Museum-related gift items & books for sale.
Activities: guided tours with advance notice (bilingual available); video presentations; docent & volunteer programs; permanent & temporary exhibitions; educational & community service programs.
Publications: books, Folk Life & Folklore of the Mexican Border; Wild Horse Desert: The Heritage of South Texas; Rio Grande Heritage: A Pictorial History; Heritage Cookbook: A Round-up of Wild & Regional Foods; Mesquite Country: Taste & Traditions from the Tip of Texas (regional recipes & history); quarterly newsletter; annual report; Jewel of the Rio Grande (a history of the Museum); The Heritage Sampler: selections from the rich & colorful history of the Rio Grande Valley; Borderlands: The Heritage of the Lower Rio Grande through the Art of Jose Cisneros; Museum of South Texas History.
Hours & Admission Prices: Tues.-Sat. 10-5. Adults $5.50, senior citizens 62 & over and military $4.50, students $4, children 4-12 $3; children 3 & under and members no charge. Closed New Year's Eve & Day; Easter; Memorial Day; Independence Day; Labor Day; Thanksgiving; Christmas Eve & Day. &
Attendance: 50,000 (accurate)
Membership: Student $10; Individual $35; Family $50; Sustainer $75; Contributor $100 & up; Benefactor $250; Patron $500; Heritage Associate Contributor $1,000; Heritage Assoc. Patron $2,500; Heritage Council $5,000; Chairman's Circle $10,000.

Edna

TEXANA MUSEUM AND LIBRARY ASSOCIATION, 403 N. Wells, Edna, TX 77957-2730. Tel.: 361-782-5431.
Founded: 1967.
Congressional District: 14
Key Personnel: C.E.O. & Pres. (V), Harrison Stafford, II

Personnel Profile: Part-Time Paid 1; Part-Time Volunteers 3.
Governing Authority: nonprofit organization. Tax-exempt: 501(c)(3).
Institution Type/Description: History Museum.
Collections: historical artifacts; documents.
Research Fields: local history.
Facilities: 400-vol. library of historical subject matter available for research on site; reading room.
Activities: guided tours.
Hours & Admission Prices: Wed.-Fri. 1-5. No charge; donations accepted. Closed holidays. &
Attendance: 1,200 (estimated)
Membership: Individual $5; Family $10; Friend $25; Patron $100; Corporate $250; Benefactor $500.

El Campo

EL CAMPO MUSEUM OF NATURAL HISTORY, 2350 N. Mechanic, El Campo, TX 77437-2343. Mailing Address: P.O. Box 23, El Campo, TX 77437-0023. Tel.: 979-543-6885. Fax: 979-543-5788.
Web Site: www.elcampomuseum.com
Formerly: El Campo Museum of Art, History and Natural Science
Founded: 1978.
Congressional District: 14
Key Personnel: Dir. & Museum Shop Mgr., Cheri McGuirk.
Personnel Profile: Full-Time Paid 1; Part-Time Paid 1; Part-Time Volunteers 30.
Governing Authority: nonprofit organization. Tax-exempt: 501(c)(3).
Institution Type/Description: Natural History Museum.
Collections: wildlife dioramas; seashells; clowns; children's hands-on exhibits.
Research Fields: animal specimens; local history.
Activities: guided tours; permanent, temporary & traveling exhibitions; interactive computers.
Publications: booklets & brochures, El Campo Museum.
Hours & Admission Prices: Tues.-Fri. 10-12 & 1-5, Sat. 10-3. No charge; donations accepted. Closed major holidays. &
Attendance: 7,000 (estimated)
Membership: Individual $20; Family $30; Nonprofit & Business $75; Patron $125; Explorer $250; Collector $500.

El Paso

CENTENNIAL MUSEUM AND CHIHUAHUAN DESERT GARDENS, Corner of University and Wiggins Rd., El Paso, TX 79968. Mailing Address: 500 W. University Ave., El Paso, TX 79968-8900. Tel.: 915-747-5565. Fax: 915-747-5411.
E-mail: museum@utep.edu
Web Site: www.utep.edu/museum/
Founded: 1936.
Congressional District: 16
Key Personnel: Dir., Dr. W. Warner Wood; Cur. Collections & Exhibits, Scott Cutler; Dir. Laboratory of Environmental Biology, Dr. Art Harris; Administrative Asst. & Museum Shop Mgr., Kaye Mullins; Cur. Chihuahuan Gardens, John White; Administrative Sec., Eurydice Saucedo; Grounds Keeper, Barb Bailey.
Personnel Profile: Full-Time Paid 5; Full-Time Volunteers 1; Part-Time Paid 1; Part-Time Volunteers 16; Interns 4.
Governing Authority: state. Parent Institution: University of Texas at El Paso, West University Ave. Tax-exempt: 170(b)(1)(A).
Institution Type/Description: Natural History Museum.
Collections: natural & human history; archaeology; botany; geology; paleontology; mineralogy; ethnology; anthropology; fauna of the El Paso area; desert gardens.
Research Fields: archaeology; ethnology; geology; history; paleontology; biology.
Facilities: Museum-related items & books for sale.
Activities: guided tours; lectures; films; permanent & temporary exhibitions; youth classes; adult workshops.
Hours & Admission Prices: Tues.-Sat. 10-4:30. No charge; donations accepted. Closed New Year's Day; Easter; Thanksgiving; Christmas; university holidays. &
Attendance: 15,000 (estimated)

CHAMIZAL NATIONAL MEMORIAL, 800 S. San Marcial, El Paso, TX 79905-4123. Tel.: 915-532-7273. Fax: 915-532-7240.
Web Site: www.nps.gov/cham
Founded: 1967.
Congressional District: 16
Key Personnel: Supt., Fernando "Gus" Sanchez.

Governing Authority: federal. Parent Institution: National Park Service, Dept. of Interior, Washington, DC 20008. Tax-exempt.
Institution Type/Description: Park Museum: located on land acquired from Mexico through 1963 Treaty.
Collections: history; art; ethnology.
Research Fields: modern & historic cultures of Western Hemisphere; border history.
Facilities: 1,000-vol. library of history & cultural management; 7,500-seat informal amphitheater; 503-seat theater; 55-acre urban park.
Activities: lectures; films; gallery talks; concerts; dance recitals; arts festivals; folk festivals; drama; inter-museum loan & permanent exhibitions. Museum Sponsors: Classic Spanish Drama Festival in March; Border Folk Festival in September; urban programs in Spanish; programs imported from Mexico, Latin American & Europe.
Publications: orientation brochures.
Hours & Admission Prices: Grounds: daily 5am-10pm. Offices: Mon.-Fri. 8-4:30. No charge. Call for theater information & pricing. Closed New Year's Day, Thanksgiving, Christmas &
Attendance: 226,353

EL PASO HOLOCAUST MUSEUM & STUDY CENTER, (M), 715 N. Oregon, El Paso, TX 79902-3911. Tel.: 915-351-0048. Fax: 915-351-0908.
E-mail: info@elpasoholocaustmuseum.org
Web Site: elpasoholocaustmuseum.org
Founded: 1992.
Congressional District: 16
Key Personnel: Pres. Bd., Debra P. Kanof; Treas., Dona Scurry; Exec. Dir., Lori P. Shepherd; Dir. Education, Jamie Williams; Business Mgr., Vanessa Nevarez.
Personnel Profile: Full-Time Paid 3; Part-Time Volunteers 30.
Governing Authority: private; nonprofit organization. Tax-exempt: 501(c)(3).
Institution Type/Description: History Museum.
Collections: period artifacts; dramatic displays; pictures; posters; authentic history of Europe during the Nazi era.
Facilities: library of children & adult books on WWII & Holocaust.
Activities: docent program; films; guided tours; lectures; traveling exhibitions. Annual Events: Fundraising Dinner Commemoration of the Holocaust; Tour De Tolerance Cycling/Walk/Run Event.
Publications: quarterly in house, El Paso Holocaust Museum & Study Center.
Hours & Admission Prices: Tues.-Fri. 9-4, Sat.-Sun. 1-5. No charge; donations accepted. Closed New Year's Day; Easter; Labor Day; Rosh Hashanah; Yom Kippur; Christmas. &
Attendance: 20,000 (estimated)
Membership: Student $18; Individual $35; Family $50; Peacemaker $100; Liberator $250; Righteous $500; Survivor $1,000.

EL PASO MUSEUM OF ARCHAEOLOGY, 4301 Transmountain Rd., El Paso, TX 79924-3753. Tel.: 915-755-4332. Fax: 915-759-6824. Facebook: Friends of the El Paso Museum of Archaeology.
E-mail: archaeologymuseum@elpasotexas.gov
Web Site: elpasotexas.gov/arch_museum
Founded: 1977.
Congressional District: 16
Key Personnel: Dir., Julia Bussinger; Cur. Education, Marilyn Guida; Sec. & Museum Shop Mgr., Rosie Enriquez.
Personnel Profile: Full-Time Paid 3; Part-Time Paid 1; Part-Time Volunteers 29.
Governing Authority: municipal. Parent Institution: Museum and Cultural Affairs Department, One Arts Festival Plaza, El Paso, TX 79901. Tax-exempt: 170(b)(1)(A).
Institution Type/Description: Anthropology Museum & Nature Center.
Collections: prehistoric artifacts of the American southwest & northern Mexico; ceramics; textiles; stone tools; American Indian southwest arts & crafts; pottery; basketry.
Research Fields: archaeology & ethnology of the American southwest & northern Mexico.
Facilities: 400-vol. library on southwest anthropology & archaeology available for research only; 100-vols. educational materials; 60 seat auditorium. Museum-related items for sale.
Activities: guided tours; traveling & temporary exhibitions; docent program; field trips; workshops. Annual Event: Texas Archaeology Awareness Week.
Publications: Jornada Mogollon Conference publication; Mimbres Twins: Icons of Pueblo Ideology; Prehistoric Indians of the El Paso Area.
Hours & Admission Prices: Tues.-Sat. 9-5, Sun. 12-5. No charge; donations accepted. Closed New Year's Day; Martin Luther King Jr. Day; Memorial Day; Independence Day; Labor Day; Thanksgiving & day after; Christmas. &
Attendance: 22,000 (accurate)
Membership: Student $15; Senior & Military Individual $20; Individual $25;

Military Family $35; Family $40; Supporter's Circle $100-$500; Donor's Circle & Corporate Circle $1,000 & up.

*** EL PASO MUSEUM OF ART, (M),** One Arts Festival Plaza, El Paso, TX 79901-1135. Tel.: 915-532-1707. Fax: 915-532-1010.
E-mail: arts@elpasotexas.gov
Web Site: www.elpasoartmuseum.org
Founded: 1930.
Congressional District: 16
Key Personnel: Dir., Michael Tomor, Ph.D.; Sr. Cur., Patrick Shaw Cable, Ph.D.; Cur., Christian Gerstheimer; Community Engagement Mgr., Laura Zamarripa; Head of Devel., Jeffrey Romney; Preparator, Nick Munoz; Registrar, Michelle Villa; Museum Shop Mgr., Norma Geller.
Governing Authority: municipal; nonprofit organization. Parent Institution: City of El Paso. Subsidiary Institution: El Paso Museum of Archaeology at Wilderness Park; El Paso Museum of History. Tax-exempt: 501(c)(3).
Institution Type/Description: Art Museum.
Collections: Samuel H. Kress collection of 13th- to 18th-century European paintings & sculpture; 18th- to 21st centuries American art; 20th-century paintings of the Taos & Santa Fe schools; 18th- to 19th-centuries Mexican colonial art; 17th- to 20th-centuries Mexican Retablo paintings; contemporary art from Texas, New Mexico & the U.S-Mexican border.
Research Fields: pertaining to collections.
Facilities: 2,000-vol. library of art reference books, an affiliate site of National Gallery Extension Programs, available for use by request; 200-seat auditorium; classrooms. Museum-related items for sale.
Activities: guided tours; lectures; films; gallery talks; arts festivals; formally organized educational programs; adult & teen docent program; permanent, temporary & traveling exhibitions.
Publications: quarterly newsletter; exhibition catalogs.
Hours & Admission Prices: Tues.-Wed. & Fri.-Sat. 9-5, Thurs. 9-9, Sun. 12-5. No charge; donations accepted. Closed New Year's Day; Martin Luther King Jr. Day; Memorial Day; Independence Day; Labor Day; Thanksgiving; Christmas. &
Attendance: 100,000 (estimated)
Membership: Student & Senior Citizen $20; Artist/Teacher & Military $25; Individual $30; Military Family $55; Family $60; Contributor $100; Supporters Circle $250; Collectors Circle $500.

EL PASO MUSEUM OF HISTORY, (M), 510 N. Santa Fe St., El Paso, TX 79901-1145.
E-mail: cityhistorymuseum@elpasotexas.gov
Web Site: www.elpasotexas.gov/history
Founded: 1974.
Congressional District: 16
Key Personnel: Dir., Julia H. Bussinger; Sr. Cur., Barbara J. Angus; Dir. Devel., Jim Murphy.
Personnel Profile: Full-Time Paid 12; Part-Time Volunteers 10.
Governing Authority: municipal. Parent Institution: History Museums Dept., City of El Paso, El Paso, TX 79901. Tax-exempt.
Institution Type/Description: History Museum.
Collections: guns; horse gear; historical clothing; photographs; cavalry gear; items pertaining to history of El Paso; leatherworking & bottle collections.
Research Fields: El Paso regional history.
Activities: guided tours; docent program; permanent & traveling exhibitions; organized education programs.
Hours & Admission Prices: Tues.-Wed. & Fri.-Sat. 10-5, Thurs. 9-9, Sun. 12-5. No charge; donations accepted. Closed city holidays. &
Attendance: 31,542 (accurate)
Membership: Student, Senior & Military $20; Individual $25; Family $50; Advocate $250; Corporate & Provider $1,000; Supporter $2,500; Champion $5,000; Patron $10,000; Benefactor $20,000.

EL PASO ZOO, 4001 E. Paisano, El Paso, TX 79905-4223. Tel.: 915-521-1850. Fax: 915-521-1857. Facebook: El Paso Zoo.
E-mail: elpasozoo@elpasotexas.gov
Web Site: www.elpasozoo.org
Founded: 1941.
Congressional District: 16
Key Personnel: Dir., Steve Marshall; Cur., John Kiseda; Mktg. & Public Rels., Karla Martinez; Volunteer Coord., Toni Marie Lopez; Exec. Dir. El Paso Zoological Society, Renee Neuert.
Governing Authority: municipal; nonprofit organization. Parent Institution: City of El Paso. Tax-exempt: 501(c)(3).
Institution Type/Description: Zoo.
Collections: zoological exhibits; animal skull collection; conservation biofacts; original art; cultural artifacts.
Research Fields: animal behavior, physiology & medical.

Facilities: library of animal-oriented, biology, zoology, behavior & captive management books available for research on premises only. Animal-related items for sale.
Activities: daily programs & presentations; formal & informal education programs; special events.
Hours & Admission Prices: Winter: daily 9:30-4; Summer: Mon.-Fri. 9:30-4, Sat.-Sun. 9:30-5. Ages 13-59 $10, ages 60 & over and active duty military & spouse with ID $7.50, ages 3-12 $6; discounts to AZA & reciprocating zoo members; children 2 & under and members no charge. Closed New Year's Day; Thanksgiving; Christmas. &
Attendance: 285,971 (accurate)
Membership: Spider Monkey $50; SunBear $65; Paraje Pals $70; Wolf Pack $75; Tiger Team $95; Critter Club $100; Safari Society $1,500.

INSIGHTS-EL PASO SCIENCE MUSEUM, 505 N. Santa Fe, El Paso, TX 79901-1144. Mailing Address: P.O. Box 9248, El Paso, TX 79995-9248. Tel.: 915-534-0000, ext. 0. Fax: 915-532-7416.
E-mail: insightsepmuseum@elp.rr.com
Web Site: www.insightselpaso.org
Founded: 1979.
Congressional District: 16
Key Personnel: C.E.O., Jim Stegall; Exec. Dir, Mandy Chew; Pres. (V), Aaron Velasco; Museum Shop Mgr., Lourdes Ramirez.
Personnel Profile: Full-Time Paid 1; Part-Time Paid 4; Part-Time Volunteers 5.
Governing Authority: nonprofit organization. Tax-exempt: 501(c)(3).
Institution Type/Description: Science Center.
Collections: participatory science exhibits on perception & energy; microcomputers; human anatomy.
Facilities: observatory; classrooms. Science-related gifts for sale.
Activities: formally organized education programs for children; loan & permanent exhibitions; sight, smell, touch & hearing exhibits; mobile unit programming in community; formally organized community education program; rooftop observatory for Friday evening observing 8-10pm, weather permitting; science classes; adult computer classes. Museum Sponsors: science camps in summer.
Publications: membership newsletter, The Insights Insider.
Hours & Admission Prices: Tues.-Sat. 10-5, Sun. 12-5. Adults $8, seniors 62 & over, military & students $6, children 4-11 $4; discounts to groups, ASTC, AAM & ICOM members; children under 3 & members no charge. Closed major holidays. &
Attendance: 35,000 (accurate)

INTERNATIONAL MUSEUM OF ART, 1211 Montana Ave., El Paso, TX 79902-5511. Tel.: 915-543-6747. Fax: 915-543-9222.
E-mail: iavatx@aol.com
Web Site: www.internationalmuseumofart.net
Formerly: International Association for the Visual Arts
Founded: 1947.
Key Personnel: Dir. & Museum Shop Mgr., Emma Castillo; Pres. (V), Deane Miller.
Personnel Profile: Part-Time Paid 1.
Governing Authority: Tax-exempt.
Institution Type/Description: Art Museum: housed in the Turney Home.
Collections: Asian & African art; Mexican Revolution collection including replicas of Pancho Villa's death mask & a Mexican casita.
Activities: permanent & temporary exhibits.
Hours & Admission Prices: Thurs.-Sun. 1-5.
Attendance: 18,960 (estimated)

MAGOFFIN HOME STATE HISTORIC SITE, 1120 Magoffin Ave., El Paso, TX 79901. Tel.: 915-533-5147. Fax: 915-544-4398.
E-mail: leslie.bergloff@thc.state.tx.us
Web Site: www.visitmagoffinhome.com
Founded: 1976.
Congressional District: 16
Key Personnel: Site Dir., Leslie Bergloff; Dir. THC Historic Sites, Donna Williams.
Personnel Profile: Full-Time Paid 5; Part-Time Volunteers 40.
Governing Authority: state. Parent Institution: Texas Historical Commission, P.O. Box 12276, Austin, TX 78711-2276. Tel. 512-463-6100. Tax-exempt.
Institution Type/Description: Historic House Museum: 1875 territorial adobe house built by pioneer, businessman, and civic leader Joseph Magoffin. Listed on the National Register of Historic Places.
Collections: household furnishings; memorabilia & documents of the Magoffin family; Southwestern decorative arts.
Research Fields: historical architecture; Southwest borderland history; El Paso & Texas history; military history; 19th & 20th century social & cultural history; family history; decorative arts.

Activities: guided tour; special events; outreach programs.
Publications: THC Interpretive Guide.
Hours & Admission Prices: Tues.-Sun. 9-5; tours on the hour; last tour begins at 4pm; school tours by appointment. Adults $4, children 6-18 & students $3; discounts to groups of 10 or more; children 5 & under no charge. Closed New Year's Eve & Day; Thanksgiving; Christmas Eve & Day. &
Attendance: 18,000 (accurate)

THE NATIONAL BORDER PATROL MUSEUM AND MEMORIAL LIBRARY, (M), 4315 Transmountain Rd., El Paso, TX 79924-3753. Tel.: 915-759-6060. Fax: 915-759-0992.
E-mail: nbpm@borderpatrolmuseum.com
Web Site: www.borderpatrolmuseum.com
Founded: 1985.
Key Personnel: Pres. (V), David Ham; Administrator, Brenda Tisdale; Museum Shop Mgr., San Juanita Gallegos.
Personnel Profile: Full-Time Paid 5; Part-Time Volunteers 6.
Governing Authority: nonprofit. Tax-exempt.
Institution Type/Description: U.S. Border Patrol History Museum.
Collections: uniforms; weapons; vehicles; USBP Line of Duty memorial; Newton/Azrak memorial; Anthony L. Oneto memorial; sculpture; NASA & border patrol NASCAR; border patrol bull riders uniform; ultra-light & homemade motorcycles; smugglers' vehicles captured by border patrol; hands-on exhibitions.
Facilities: Gift items for sale.
Activities: guided tours; hands-on exhibits including helicopter, Willy's Jeep, snowmobile, ATV's.
Publications: brochures; flyers; gift shop catalogues; historical photos; books, Tales of the Rio Grande; Recuerdos; No Flag for My Coffin; post cards; bookmarks; prints; posters; annual calendar.
Hours & Admission Prices: Tues.-Sat. 9-5; guided tours by appointment. No charge; donations accepted. Parking: no charge. Closed major holidays. &
Attendance: 25,000 (accurate)
Membership: Individual $35.

RAILROAD & TRANSPORTATION MUSEUM OF EL PASO, 400 W. San Antonio Ave., El Paso, TX 79901. Mailing Address: c/o Friends of the Railroad and Transportation Museum of El Paso, P.O. Box 5722, El Paso, TX 79955. Tel.: 915-422-3420.
Web Site: www.elpasorails.org
Institution Type/Description: Railroad Museum.
Collections: railroad artifacts & memorabilia including a restored 4-4-0 classic American 1857 locomotive.
Hours & Admission Prices: Tues.-Sat. 11-5, Sun. 1-5. No charge; donations accepted.

STANLEE & GERALD RUBIN CENTER FOR THE VISUAL ARTS, UNIVERSITY OF TEXAS, EL PASO, Dawson Dr. at Sun Bowl Dr., El Paso, TX 79902. Mailing Address: 500 W. University Ave., El Paso, TX 79968-8900. Tel.: 915-747-6151. Fax: 915-747-6067.
E-mail: rubincenter@utep.edu
Web Site: www.rubincenter.utep.edu
Founded: 2004.
Key Personnel: Dir., Kate Bonansinga; Asst. Dir., Kerry Doyle; Registrar & Preparator, Daniel Szwaczkowski.
Personnel Profile: Full-Time Paid 4; Interns 4.
Governing Authority: private; university. Parent Institution: University of Texas at El Paso. Tax-exempt.
Institution Type/Description: Art Museum.
Collections: contemporary fine art & design.
Research Fields: contemporary art.
Facilities: 3,500 sq. ft. exhibit space; auditorium.
Activities: films; formal education programs for undergraduate or graduate University of Texas at El Paso students; guided tours; lectures; loans, participatory & traveling exhibitions; young curators' program; "Talk Back" program for high school students; family programs; art making workshops; symposium.
Publications: exhibition catalogues.
Hours & Admission Prices: Tues.-Wed. & Fri. 10-5, Thurs. 10-7; weekend hours by appointment. No charge. Closed New Year's Eve & Day; Thanksgiving; Christmas Eve & Day. &
Attendance: 11,000 (estimated)
Membership: Student $25; Individual $50; Family $60.

Eldorado

SCHLEICHER COUNTY HISTORICAL SOCIETY MUSEUM,
100 E. Murchison St., Eldorado, TX 76936. Mailing Address: P.O.
Box 1053, Eldorado, TX 76936. Tel.: 915-853-2411.
Institution Type/Description: Historical Society Museum.
Collections: local history & culture; period furnishings; personal artifacts;
photographs.
Hours & Admission Prices: Call for hours.

Emory

A.C. MCMILLAN AFRICAN AMERICAN MUSEUM, 149 Texas
St., Emory, TX 75440. Mailing Address: P.O. Box 1046, Emory,
TX 75440-1046. Tel.: 903-473-6315. Fax: 214-298-6942. Face-
book: AC McMillan African American Museum.
E-mail: acmaam@aol.com
Web Site: www.acmcmillanafricanamericanmuseum.org
Founded: 2000.
Key Personnel: Dir., Gwendolyn McMillan Lawe; Chm. (V), Lafayshia
Meador.
Personnel Profile: Part-Time Volunteers 5.
Institution Type/Description: History Museum.
Collections: African American history & art; slavery; photographs; personal
artifacts; dolls of color; African American postage stamps; Buffalo Soldiers;
Jim Crow images; Negro baseball leagues.
Facilities: library. Museum-related items for sale.
Activities: summer day camp; classes; group tours.
Hours & Admission Prices: Thurs.-Sat. 10-4. No charge.

Fairfield

FREESTONE COUNTY HISTORICAL MUSEUM, (M), 302 E.
Main St., Fairfield, TX 75840-1530. Mailing Address: P.O. Box
524, Fairfield, TX 75840-0009. Tel.: 903-389-3738.
E-mail: freestonecomuseum@windstream.net
Web Site: www.freestonecomuseum.com
Founded: 1967.
Congressional District: 6
Key Personnel: Pres. (V), Leslie Tate; Cur., Nancy Taylor.
Personnel Profile: Part-Time Paid 1; Part-Time Volunteers 6.
Governing Authority: county. Parent Institution: Freestone County. Tax-
exempt: 501(c)(3).
Institution Type/Description: Local History & Telephone Museum.
Collections: historical items pertaining to Freestone County; manuscripts.
Historic Houses: 1845 Carter Log House; 1852 Watson Log House; 1930s
Assembly of God Church; 1880s jail, telephone museum.
Research Fields: Genealogy.
Activities: tours.
Publications: Orbit 1900-1905.
Hours & Admission Prices: Wed. & Fri.-Sat. 10-5; other times by appointment.
Adults $3; children under 5 no charge. &
Attendance: 4,000 (accurate)
Membership: Individual $15; Family $25.

Falfurrias

THE HERITAGE MUSEUM AT FALFURRIAS, INC., 512 N. St.
Mary's, Falfurrias, TX 78355. Mailing Address: P.O. Box 86,
Falfurrias, TX 78355-0086. Tel.: 361-325-2907.
Web Site: www.heritagemuseum-falfurrias.com
Founded: 1965.
Congressional District: 15
Key Personnel: Pres. Bd. (V), Alberto Huerta; Museum Shop Mgr., Ramiro
Rodriguez.
Personnel Profile: Part-Time Paid 2; Part-Time Volunteers 1.
Volunteer Hours: 156
Operating Expenses: 25,228
Operating Income: 20,697
Governing Authority: nonprofit organization: Brooks County Historical Com-
mission. Tax-exempt.
Institution Type/Description: History Museum
Collections: Texas Rangers; Dryden photo negatives; farm equipment; Falfur-
rias creamery; local history photographs; arrowheads & Native American
bow; Veterans Memorial.
Research Fields: area history; archaeology; genealogy; Texas Ranger History;
50-year span photo negative collection of area people & events, reprinting
by museum for small fee.
Facilities: archives.

Activities: lectures; demonstrations; tours by appointment. Annual Events:
Tribute to Trejano Music Legends in March; Native American Powwow in
November.
Publications: books, Don Pedrito Jarmillio, Faith Healer of Los Almos;
Falfurrias; Political Brooks County Elected Officials; Brooks Co. Diamond
Jubilee.
Hours & Admission Prices: Tues.-Sat. 10-4. No charge; donations accepted. &
Attendance: 1,000 (estimated)
Membership: Active $10; Sustaining $25; Life & Memorials $50; Gold Star
$100.

Farmers Branch

FARMERS BRANCH HISTORICAL PARK, 2540 Farmers
Branch Lane, Farmers Branch, TX 75234-6214. Mailing Address:
P.O. Box 819010, Farmers Branch, TX 75381-9010. Tel.: 972-406-
0184. Fax: 972-247-3939.
E-mail: historicalpark@farmersbranch.info
Web Site: www.farmersbranch.info
Founded: 1986.
Congressional District: 26
Key Personnel: Dir. Parks & Recreation, Jeff Harting; Park Supt., Derrick
Birdsall; Cur., Jamie Rigsby; Museum Educator, Barbara Judkins; Museum
Shop Mgr., Kim Chapman.
Personnel Profile: Full-Time Paid 5; Part-Time Paid 2; Part-Time Volunteers
50.
Governing Authority: municipal; nonprofit. Parent Institution: City of Farmers
Branch. Tax-exempt.
Institution Type/Description: Historical Park & Archives.
Collections: books, photographs, artifacts & archival material documenting the
history of Texas with an emphasis on Peters Colony & Farmers Branch;
Paleo Indian artifacts. Historic Houses: 1937 Dodson House (home of
Farmers Branch's first mayor; Texas Historic Landmark); 1900 one-room
school; 1891 church; 1885 Queen Anne Victorian cottage; 1876 depot; 1856
Gilbert house (oldest structure on original foundation in Dallas County &
National Register of Historic Places & Landmarks); 1840's log homestead
& barns; 1930's Marathon gas station replica; 1936 cab-over-engine Ford
truck; 1840s replica of The Peters Colony Land Grant Office.
Facilities: 500-vol. library of research books; archives housing diaries,
photographs & documents of local history material.
Activities: guided tours; lectures; concerts; festivals; archeology fair; organized
education programs for children; docent program; school loan service;
archival research. Museum Sponsors: Star Parties.
Publications: quarterly newsletter; book: Once Upon A Time in Farmers
Branch, in conjunction with 3rd grade educational curriculum; book,
Farmers Branch, Texas: A Pictorial History, 1842-1996.
Hours & Admission Prices: Mon.-Fri. 8-6, Sat.-Sun. 12-6. Tours: by reserva-
tion. No charge; donations accepted. Closed New Year's Day, Easter,
Thanksgiving & Christmas. &
Attendance: 50,000 (estimated)

Floydada

FLOYD COUNTY HISTORICAL MUSEUM, 105 E. Missouri St.,
Floydada, TX 79235. Mailing Address: P.O. Box 304, Floydada,
TX 79235-0304. Tel.: 806-983-2415.
E-mail: fchmuseum@sbcglobal.net
Founded: 1971.
Institution Type/Description: History Museum.
Collections: local history & genealogy; early pioneers; Thomas Montgomery
ranch house replica; photographs.
Facilities: library; history & genealogy center.
Publications: annual newsletter, Muse Briefs.
Hours & Admission Prices: Mon.-Fri. 1-5; other times by appointment. No
charge; donations accepted. &
Attendance: 3,000 (accurate)
Membership: Adult $5; Family $10; Contributing $25; Life $500.

Fort Bliss

FORT BLISS AND OLD IRONSIDE MUSEUMS, Marshall Rd.,
Bldg. 1735, Fort Bliss, TX 79916. Tel.: 915-568-5412. Fax:
915-568-7166.
Web Site: www.bliss.army.mil/museum/fort_bliss_museum.htm
Formerly: U.S. Army Air Defense Artillery Museum & Fort Bliss Museum
Founded: 1975.
Congressional District: 16
Key Personnel: Dir., Peter Poessiger; Cur., Jennifer Nielsen.
Governing Authority: federal. Parent Institution: U.S. Army Museums Div.
Subsidiary Institution: U.S. Army Air Defense Artillery Association. Tax-
exempt: 501(c)(3).

Institution Type/Description: Military Museum.
Collections: weaponry, uniforms, insignia, vehicles, radars, search lights, archival holdings, books, documents, photographs, ephemera, associated with the history of the Air Defense Artillery.
Research Fields: military history.
Facilities: 1,000-vol. library of books on military history, primarily related to the Antiaircraft & Air Defense branch of the army, available for research on premises or by prearranged inter-museum loan; reading room; theater; classrooms. Museum-related items for sale.
Activities: formally organized education programs for adults & undergraduate college students affiliated with University of Texas at El Paso; permanent & traveling exhibitions.
Publications: book, Reasons Why II; book, Pocket History of Air Defense Artillery.
Hours & Admission Prices: Mon.-Fri. 9-4. Sat. 10-3. No charge. Closed federal holidays. &
Attendance: 60,000 (accurate)

Fort Davis

CHIHUAHUAN DESERT RESEARCH INSTITUTE, 43869 SH 118, Fort Davis, TX 79734. Mailing Address: P.O. Box 905, Fort Davis, TX 79734-0010. Tel.: 432-364-2499. Fax: 432-364-2686.
Web Site: www.cdri.org
Founded: 1973.
Key Personnel: Exec. Dir., Cynthia Griffin; Pres. (V), Suzette Ashworth.
Governing Authority: nonprofit organization. Tax-exempt: 501(c)(3).
Institution Type/Description: Arboretum & Visitor Center.
Collections: trees, shrubs, cacti & succulents native to Chihuahuan Desert Region of the U.S. & Mexico.
Research Fields: natural science research in the Chihuahuan Desert Region.
Facilities: 22,000-vol. library of books, reprints & journals; botanical garden; outdoor classroom pavilion; nature & conservation center; nature trails. Native plants, books & gift items for sale.
Activities: guided tours; lectures; films; organized education programs.
Publications: semi-annual magazine, The Chihuahuan Desert Discovery.
Hours & Admission Prices: Mon.-Sat. 9-5. Adults $5, seniors 65 & over $4; children under 12 & members no charge. &
Attendance: 8,000 (estimated)
Membership: Living Lightly $35; Regular $50; Friend $100; Supporter $250; Patron $500; Benefactor $750; Life $1,000.

FORT DAVIS NATIONAL HISTORIC SITE, Lt. Flipper Dr., Fort Davis, TX 79734-0015. Mailing Address: P.O. Box 1379, Fort Davis, TX 79734-0015. Tel.: 432-426-3224, ext. 225 (Historian Office). Fax: 432-426-3122.
E-mail: FODA_Superintendent@nps.gov
Web Site: www.nps.gov/foda
Founded: 1961.
Congressional District: 23
Key Personnel: Park Ranger & Coord. Volunteers, Bill Manhart; Chief Interpretation, John Heiner; Museum Shop Mgr., Patricia Hartnett.
Personnel Profile: Full-Time Paid 13; Part-Time Paid 5; Part-Time Volunteers 200.
Governing Authority: federal. Parent Institution: National Park Service; Dept. of Interior. Tax-exempt.
Institution Type/Description: Military Museum: located on the site of 1854-1891 Fort Davis.
Collections: late 19th-century military & civilian artifacts including weapons, photographs, & regimental records; manuscripts; restored commanding officer's quarters; lieutenant's quarters; enlisted men's barracks; officer's kitchen; servant's quarters; commissary; late 19th century medical instruments.
Research Fields: American frontier military history; frontier Indian wars; Buffalo Soldier regiments.
Facilities: 2,500-vol. library including 150 reels of microfilm, 155 rare books & manuscript material on military history available for use on premises; reading room; 48-seat auditorium. Books, postcards, historic maps, records & tapes of military music for sale.
Activities: 14-minute video program; 18 minute recording of a Dress Retreat Parade; Bugle Calls; self guiding tours of restored & refurnished buildings and grounds; education programs for grades K-12. Special Events: costumed interpreters in summer.
Hours & Admission Prices: Daily 8-5; research by appointment. Adults 16 & over $3; children 15 & under, Golden Age, Golden Eagle, Golden Access Passport, & National Parks Pass holders no charge. Closed New Year's Day; Martin Luther King Jr. Day; Presidents' Day; Thanksgiving; Christmas. &
Attendance: 50,000 (accurate)

Fort Hood

1ST CAVALRY DIVISION MUSEUM, (M), Bldg. 2218, 56th & 761 Tank Bn. Ave., Fort Hood, TX 76544-0187. Mailing Address: P.O. Box 5187, Fort Hood, TX 76544-0187. Tel.: 254-287-3626 & 532-2075 (Gift Shop). Fax: 254-287-6423 & 532-6490 (Gift Shop).
E-mail: steven.c.draper@us.army.mil
Web Site: www.hood.army.mil/1stcavdiv/1cdmuseum/index
Founded: 1971.
Congressional District: 11
Key Personnel: Dir., Cur. & Chm. (V), Steven C. Draper; Museum Foundation, Terry Maddox; Deputy & Exhibit Specialist, Jack Dugan; Collections Specialist, Amber Hills; Museum Shop Mgr., Michelle Wolf.
Personnel Profile: Full-Time Paid 5; Part-Time Volunteers 15.
Governing Authority: federal. Parent Institution: 1st Cavalry Division, U.S. Army and Army Center of Military History. Tax-exempt: 501(c)(3).
Institution Type/Description: Military Museum.
Collections: military artifacts of 1st Cavalry Division & its regiments from 1855 to present.
Research Fields: history of 1st Cavalry Division; regimental & unit histories of the division; U.S. Cavalry.
Facilities: research library; archives. Museum-related items for sale.
Activities: permanent & temporary exhibits; guided tours; films; off site programs; living history demonstrations; children's programs.
Publications: History of the 1st Cavalry Division; Vehicles Park Guide; Children's Treasure Hunt; Museum Fact Sheets; Reference Guide.
Hours & Admission Prices: Mon.-Fri. 9-4, Sat. 10-4, Sun. 12-4. No charge; donations accepted. Closed New Year's Day; Easter; Thanksgiving; Christmas. &
Attendance: 50,000 (accurate)

THIRD CAVALRY MUSEUM, Bldg. 409, 761st Tank BN Ave., Fort Hood, TX 76544. Mailing Address: P.O. Box 5917, Fort Hood, TX 76544-0917. Tel.: 254-288-3590 & 287-8811. Fax: 254-287-3833.
E-mail: ellis.s.hamric.civ@mail.mil
Founded: 1930.
Congressional District: 11
Key Personnel: Dir. & Cur., Scott Hamric; Museum Shop Mgr., Harvey Reed.
Personnel Profile: Full-Time Paid 2.
Governing Authority: federal. Admin. by Dept. of Army (111 Corps & Fort Hood). Tax-exempt: 501(c)(3) & 170(b)(1)(A).
Institution Type/Description: Military Museum.
Collections: military & related material of the 3rd Cavalry Regiment & its predecessor, the Regiment of Mounted Riflemen & opposing forces from 1846 to present; 50 military vehicles.
Research Fields: history of the Regiment of Mounted Riflemen and the 3rd Cavalry Regiment from 1846 to present.
Facilities: reference library, archives & study collections available to researchers, advance notice recommended; audio-visuals. Museum-related items for sale.
Activities: self-guided tours; special guided tours upon request; audiovisual programs; lectures; classes.
Publications: brochure, Fort Hood Heritage.
Hours & Admission Prices: Mon.-Fri. 9-4, Sat. 10-4, Sun. 12-4. No charge. Closed New Year's Day; Thanksgiving; Christmas. &
Attendance: 17,000 (accurate)

Fort McKavett

FORT MCKAVETT STATE HISTORICAL SITE, 7066 FM 864, Fort McKavett, TX 76841. Mailing Address: P.O. Box 68, Fort McKavett, TX 76841. Tel.: 325-396-2358. Fax: 325-396-2818.
E-mail: ft-mckavett@thc.state.tx.us
Web Site: www.visitfortmckavett.com
Founded: 1968.
Congressional District: 21
Key Personnel: Site Mgr., Michael A. Garza; Office Mgr., Nancy Jacoby; Daily Operations Specialist, Ken Lester; Cur. & Interpreter, Cody Mobley; Maintenance Supvr., Russell Tipton; Maintenance Technician IV, Joy Wright; Clerk, Wanda Martinez.
Personnel Profile: Full-Time Paid 7.
Governing Authority: Parent Institution: Texas Historical Commission.
Institution Type/Description: Historic Site & Military Museum: 1852-1883 Fort McKavett, home to four Buffalo solder regiments.
Collections: 1852-59 & 1868-83 artifacts & interpretive materials associated with military occupation; late 19th-century civilian occupation; 18 restored structures & 9 ruins surrounding two parade areas.
Research Fields: late 19th century frontier military history; minority groups.

Facilities: research library; picnic area. Museum-related items for sale.
Activities: research; in-house living history interpretation: uniforms and equipment demonstrations of 1870s-1880s infantry & cavalry; permanent exhibits; self-guided or guided tours; audiovisual room.
Publications: brochure; newsletter; booklet, walking guide.
Hours & Admission Prices: Daily 8-5. Adults $4, college students & children 6-18 $3; discounts to groups; children 5 & under no charge. Closed New Year's Eve & Day; Thanksgiving; Christmas Eve & Day. &
Attendance: 5,500 (estimated)
Membership: Single $15; Family $25; Corporate $100.

Fort Sam Houston

FORT SAM HOUSTON MUSEUM, 2340 Stanley Rd., Fort Sam Houston, TX 78234-7501. Mailing Address: 1400 E. Grayson St., Ste. 152, San Antonio, TX 78234-7000. Tel.: 210-221-1886 & 0019. Fax: 210-221-1311.
Formerly: Fort Sam Houston Military Museum
Founded: 1967.
Congressional District: 20
Key Personnel: Cur., Jacqueline B. Davis; Museum Specialist, Martin L. Callahan.
Personnel Profile: Full-Time Paid 3; Part-Time Volunteers 5.
Governing Authority: federal. Parent Institution: U.S. Army North. Tax-exempt.
Institution Type/Description: Military Historical Museum: located on Fort Sam Houston National Historic Landmark.
Collections: army uniforms; equipment; accoutrements; weapons from 1835 to present; military vehicles & artillery; photographs; personal papers; memorabilia; microfilmed historical records of Ft. Sam Houston.
Research Fields: history of Fort Sam Houston; history of US Army in San Antonio; army operations in Texas since 1845; Army life of soldiers & families.
Facilities: 5,000-vol. library of military history, technical publications on types of military equipment & uniforms, available for research with approval of curator on premises; 30 seat auditorium.
Activities: guided tours; permanent, temporary & traveling exhibits; audiovisual programs; formal & informal educational programs. Museum Sponsors: Historic Neighborhood Awareness Program.
Publications: Pocket Guide to Historic Sam Houston; Surrounded by History; Pocket Guide to the New Post; Pocket Guide to the Cavalry & Light Artillery Post; Fort Sam Houston and the Korean War; Pocket Guide to the Staff Post; Hospitals at Fort Sam Houston; Fort Sam Houston: National Historic Landmark (tour map); Quadrangle, Hub of Military Activity in Texas; The Post at San Antonio 1845-1879; Commodious Homes For The Troops, A Centennial History of the Cavalry and Light Artillery Post, Fort Sam Houston, TX 1905-2005; Maneuver Camp, 1911: Transformation of the army at Fort Sam Houston.
Hours & Admission Prices: Wed.-Sun. 10-4. No charge; donations accepted. Closed New Year's Day; Thanksgiving; Christmas; Sun. before Mon. holidays. &
Attendance: 30,000 (estimated)

✱ **U.S. ARMY MEDICAL DEPARTMENT MUSEUM, (M),** 2310 Stanley Rd., Bldg. 1046, Fort Sam Houston, TX 78234-2636. Mailing Address: P.O. Box 340 244, Fort Sam Houston, TX 78234. Tel.: 210-221-6358. Fax: 210-221-6781. Facebook: U.S. Army Medical Department Museum.
E-mail: scott.schoner@amedd.army.mil
Web Site: www.ameddmuseumfoundation.com
Founded: 1955.
Congressional District: 20
Key Personnel: Dir., Scott Schoner; C.E.O., B.G. (Ret.) Daniel F. Perugini; Pres. (V), Beverly Greenlee-Davis; Registrar, Chuck Franson; Gift Shop Mgr., Cindy Sifuentes; Museum Specialist, Paula Ussery.
Personnel Profile: Full-Time Paid 5; Part-Time Volunteers 3.
Governing Authority: federal; nonprofit. Parent Institution: U.S. Army Medical Command. Subsidiary Institution: AMEDD Museum Foundation, Inc. Tax-exempt: 501(c)(3).
Institution Type/Description: Military Medical Museum.
Collections: Army medical equipment; uniforms; Army medical insignia; military ambulances; POW material; unit colors; artwork; archives; hospital train ambulance car; combat medic memorial.
Research Fields: 1775 to present, Army medical department.
Facilities: 50-seat auditorium & theater; 11,000 sq. ft. exhibit space; memorial garden & plaza; activity room. Books, prints, jewelry & other gift items for sale.
Activities: guided tours; lectures; films.
Hours & Admission Prices: Tues.-Sat. 10-4. No charge; donations accepted. Closed federal holidays. &

Attendance: 34,876 (accurate)

Fort Stockton

ANNIE RIGGS MEMORIAL MUSEUM, 301 S. Main St., Fort Stockton, TX 79735-7209. Tel.: 432-336-2167. Fax: 432-336-7529.
E-mail: annieriggs@sbcglobal.net
Web Site: annieriggsmuseum.com
Founded: 1955.
Congressional District: 21
Key Personnel: Dir., Lacey C. Johnson.
Personnel Profile: Full-Time Paid 1; Part-Time Paid 4; Part-Time Volunteers 3.
Governing Authority: nonprofit organization. Parent Institution: Fort Stockton Historical Society. Subsidiary Institution: Historic Fort Stockton. Tax-exempt.
Institution Type/Description: General Museum: housed in 1899 Riggs Hotel.
Collections: historic textiles; Native American artifacts; cowboy & ranch implements; geologic specimens; religious items; kitchen utensils; area archaeology.
Research Fields: history of Pecos County.
Facilities: historical library.
Activities: tours; special events; summer concert series.
Publications: newsletter.
Hours & Admission Prices: June-Aug. Mon.-Sat. 9-6; Sept.-May Mon.-Sat. 9-5. Adults $3, seniors 65 & over $2.50, children 2-12 $2. Closed New Year's Day; Easter; Thanksgiving; Christmas Eve, Day & day after. &
Attendance: 4,000 (accurate)
Membership: Individual $15; Family $20; Business $60; Lifetime $1,000.

HISTORIC FORT STOCKTON, 301 E. Third, Fort Stockton, TX 79735-5702. Mailing Address: 301 S. Main, Fort Stockton, TX 79735. Tel.: 432-336-2400 & 2167. Fax: 432-336-0575.
E-mail: annieriggs@sbcglobal.net
Web Site: historicfortstockton.com
Founded: 1990.
Congressional District: 21
Key Personnel: Dir. & Cur., Jennifer Justison.
Personnel Profile: Full-Time Paid 1; Part-Time Paid 5; Part-Time Volunteers 10.
Governing Authority: society; nonprofit. Parent Institution: City of Fort Stockton. Tax-exempt.
Institution Type/Description: Historic Site & Military Museum: 1867-1886 frontier fort during Indian Wars. Listed on National Register of Historic Places.
Collections: military & supporting contractors directly relating to Fort Stockton, 1867-1886; uniforms; weapons; archaeological artifacts; archives. Historic Structures: original guardhouse; 3 officers' quarters; reconstructed kitchens; two enlisted barracks.
Research Fields: Fort Stockton, TX (military post); Indian Wars; frontier forts; Indians of West Texas, 1850-1900; trails & routes through West Texas.
Facilities: 65-vol. library on Fort Stockton, Indian Wars & local history; 1,600 sq. ft. exhibit space. Museum-related items for sale.
Activities: guided tours; films; lectures; educational programs; docent program; living history interpreters. Annual Events: Living History Days.
Publications: Fort Stockton Historical Society Newsletter.
Hours & Admission Prices: June-Aug. Mon.-Sat. 9-6; Sept.-May Mon.-Sat. 9-5. Adults $3, senior citizens $2.50, children 6-12 $2; discounts to groups; TAM & AAM members and children under 6 no charge. Closed New Year's Day; Easter; Thanksgiving; Christmas. &
Attendance: 4,000 (estimated)
Membership: Individual $15; Family $20; Business $60; Lifetime $1,000.

Fort Worth

AMERICAN AIRLINES C.R. SMITH MUSEUM, (M), 4601 Texas Hwy. 360 at FAA Road, Fort Worth, TX 76155. Mailing Address: P.O. Box 619617, Dallas-Fort Worth Airport, TX 75261. Tel.: 817 967 1560. Fax: 817-967-5737.
E-mail: info.crsmithmuseum@aa.com
Web Site: www.crsmithmuseum.org
Founded: 1993.
Key Personnel: Exec. Dir., Jay Luippold; Cur. & Head Interpretation & Collections, Tim McElroy.
Personnel Profile: Full-Time Paid 7; Part-Time Paid 2; Part-Time Volunteers 64.
Governing Authority: private; nonprofit organization. Tax-exempt.
Institution Type/Description: Company Museum.
Collections: American Airlines history; corporation acquired airlines; general aviation history; life of C.R. Smith, early C.E.O. of American Airlines.

Research Fields: early airlines in the United States; individuals who help make up the American Airline's family.
Facilities: 500-vol. library on aviation; theater. Gift items for sale.
Activities: educational programs; Eagle Aviation Academy; films; guided & self- guided tours.
Publications: biannual newsletter.
Hours & Admission Prices: Tues.-Sat. 9-5; see Web site for extended holiday hours. Adults $7, military, seniors 65 & up, students w/ID and children 2-18 $4; infants 0-23 months & members no charge. &

Attendance: 50,000 (accurate)
Membership: Individual Annual $50; Family Annual $75; Supporter $100; Patron $300; Sustaining $500; Platinum Lifetime $1,000.

＊ AMON CARTER MUSEUM OF AMERICAN ART, (M), 3501 Camp Bowie Blvd., Fort Worth, TX 76107-2695. Tel.: 817-738-1933. Fax: 817-989-5099. Facebook: Amon Carter Museum of American Art.
E-mail: tracy.greene@cartermuseum.org
Web Site: www.cartermuseum.org
Founded: 1961.
Congressional District: 12
Key Personnel: Bd. Pres., Karen Hixon; Dir., Andrew Walker; Sr. Deputy Dir., Lori Eklund; C.F.O., Randy Ray; Exec. Asst. to Dir., Pamela Graham; Sr. Cur. Photographs, John Rohrbach; Cur. Paintings & Sculpture, Rebecca Lawton; Asst. Cur. Paintings & Sculpture, Shirley Reece-Hughes; Dir. Library, Sam Duncan; Dir. Devel., Carol Noel; Dir. Publications, Will Gillham; Public Information Officer, Tracy Greene; Dir. Education, Stacy Fuller; Deputy Dir. Art & Research, Margi Conrads; Registrar, Marci Caslin; Installation Mgr., Jim Belknap; Human Resources Mgr., Kathy Goodale; Head IT, Janice Craddock; Museum Shop & Cafe Mgr., John Reilly; Events Mgr., Lauren Tevis; Facilities Dir., Alfred Walker; Security Dir., Shannon Locke.
Personnel Profile: Full-Time Paid 74; Part-Time Paid 40; Part-Time Volunteers 15; Interns 2.
Governing Authority: nonprofit organization. Tax-exempt: 501(c)(3).
Institution Type/Description: Art Museum.
Collections: 19th & 20th century American painting; sculpture; works on paper; Remington & Russell works; over 40,000 photographs.
Major Exhibits: James McNeill Whistler: Lithographs from the Steven L. Block Collection at the Speed Art Museum, 1/25/14-4/27/14; Art & Appetite: American Art, Culture & Cuisine (T), 2/22/14-5/18/14; Underground Encounters: Photographs by Kathy Sherman Suder, 3/15/14-8/17/14; Archibald Motley: Jazz Age Modernist (T), 6/14/14-9/7/14; Navigating the West: George Caleb Bingham & the River (T), 10/2/14-1/18/15.
Research Fields: American art & history.
Facilities: 40,000-vol. library of North American history, art & photography available for inter-library loan to institutions & for individual researchers on premises by appointment; Regional Reference Center for Archives of American Art, Smithsonian Institution with 7,500 microfilm reels; 160-seat theatre; microfilm archive of 19th-century American newspapers & illustrated periodicals; reading room. Books, art reproductions, cards & slides for sale.
Activities: guided tours; docent program; inter-museum loan, temporary & traveling exhibitions; lectures; films; summer storytime; crafting; gallery talks; family fun days; book club.
Publications: books on American art & cultural history; biannual program of events & exhibitions.
Hours & Admission Prices: Tues.-Wed. & Fri.-Sat. 10-5, Thurs. 10-8, Sun. 12-5. No charge, donations accepted. Closed major holidays. &
Attendance: 106,984 (accurate)
Membership: Associate $65; Friend $100; Sustainer $250; Patron $500; Benefactors Circle $1,000; Directors Circle $3,000; Presidents Circle $5,000; Trustees Circle $10,000; Founders Circle $25,000.

THE ART GALLERIES AT TCU, 2805 S. University Dr., 245N N. Moudy Bldg., Fort Worth, TX 76129. Mailing Address: School of Art, P.O. Box 29800, Fort Worth, TX 76129. Tel.: 817-257-7643. Fax: 817-257-7399.
E-mail: s.packard@tcu.edu
Web Site: www.theartgalleries.tcu.edu
Founded: 1874.
Congressional District: 12
Key Personnel: Dir., Sally Packard; Cur., Christina Rees; Gallery Asst., Devon Nowlin.
Personnel Profile: Full-Time Paid 2.
Governing Authority: private university; nonprofit organization. Tax-exempt: 501(c)(3).
Institution Type/Description: University Art Gallery.

Collections: works on paper by traditional as well as contemporary artists.
Activities: lectures; education programs for adults, undergraduate & graduate students.
Hours & Admission Prices: Academic Year: Mon. 11-6, Tues.-Fri. 11-4, Sat.-Sun. 1-4. No charge; donations accepted. Closed Easter; Christmas. &
Attendance: 4,100 (accurate)

B-36 PEACEMAKER MUSEUM, 3300 Ross Ave., Fort Worth, TX 76106. Mailing Address: P.O. Box 121096, Fort Worth, TX 76121. Tel.: 800-575-0535.
E-mail: ceo@b-36peacemakermuseum.org
Web Site: www.b-36peacemakermuseum.org
Founded: 2003.
Institution Type/Description: Military History Museum.
Collections: aviation history, aircraft & artifacts.
Hours & Admission Prices: Wed. 9-4, Sat. 9-5, Sun. 11-5. Adults $5; active military no charge. &
Membership: Individual $40.

CATTLE RAISERS MUSEUM, 1600 Gendy St., Fort Worth, TX 76107-4062. Tel.: 817-332-8551. Fax: 817-336-8749.
Web Site: www.cattleraisersmuseum.org
Founded: 1981.
Congressional District: 12
Key Personnel: Pres. (V), Kit Moncrief; Dir., Pat Riley; Assoc. Dir., Kim Smith.
Personnel Profile: Full-Time Paid 2; Part-Time Volunteers 15.
Governing Authority: nonprofit organization. Parent Institution: Texas & Southwestern Cattle Raisers Foundation. Tax-exempt.
Institution Type/Description: History Museum.
Collections: history of the cattle industry in Texas; cowboy & ranching artifacts; photographs; murals; life-size longhorns; breed wall; audio-visual hands-on exhibit.
Research Fields: cattle & ranching industry.
Facilities: library by appointment.
Activities: educational & outreach programs; special exhibits.
Publications: magazine, The Cattleman.
Hours & Admission Prices: Daily 10-5. Admission to Fort Worth Museum of Science & History includes Cattle Raisers Museum. Adult $14, senior & children $10; discounts to groups. &
Attendance: 450,000 (estimated)
Membership: Adult $75; 2 Adults $125; Family $250; 6 Plus Guests $500.

FORT WORTH BOTANIC GARDEN, 3220 Botanic Garden Blvd., Fort Worth, TX 76107-3420. Tel.: 817-871-7686 & 7680. Fax: 817-871-7638.
E-mail: steve.huddleston@fortworthtexas.gov
Web Site: www.fwbg.org
Founded: 1934.
Congressional District: 12
Key Personnel: Dir., Henry Painter; Receptionist, Dolores Santos.
Personnel Profile: Full-Time Paid 40; Part-Time Paid 3; Part-Time Volunteers 821.
Governing Authority: municipal. Parent Institution: City of Fort Worth, TX. Tax-exempt.
Institution Type/Description: Botanical Garden.
Collections: roses; miniature roses; iris; orchids; begonias; native plants; tropical plants under glass; ficus; perennials; cactus; succulents; shrubs; trees; Japanese gardens.
Facilities: 5,000-vol. library of botany books available for use by holders of library cards; 241-seat lecture hall; rental facilities; conservatory; gardens; restaurant. Gift-related items for sale.
Activities: education programs for children & adults; annual fall & spring festival in Japanese garden; workshops for teachers; summer garden clubs for children; plant sales; summer concert series; juried art show.
Publications: booklet, Japanese Garden; members bimonthly newsletter, The Redbud; brochures.
Hours & Admission Prices: Garden: daily 8 am-dusk. No charge. Japanese Garden: April-Oct. daily 9-7; Nov.-March daily 10-5. Adults $4 weekdays & $4.50 weekends, children 4-12 $3; discounts to seniors & reciprocal gardens; members & children under 4 no charge. Conservatory: April-Oct. Mon.-Sat. 10-6, Sun. 1-6; Nov.-March Mon.-Sat. 10-4, Sun. 1-4. Adults $1, seniors & children 4-12 $.50; members no charge. Garden Center & Japanese Garden closed Christmas. &
Attendance: 700,000 (estimated)
Membership: Individual $40; Couple $60; Family $75.

*** FORT WORTH MUSEUM OF SCIENCE AND HISTORY, (M),** 1600 Gendy St., Fort Worth, TX 76107-4062. Tel.: 817-255-9300, ext. 0. Fax: 817-732-7635.
E-mail: webmaster@fwmsh.org
Web Site: www.fortworthmuseum.org
Founded: 1941.
Congressional District: 12
Key Personnel: Pres., Van A. Romans; Chm. (V), Alston Roberts; Exec. Vice Pres. Interpretation, Kit Goolsby; Exec. Vice Pres. Innovation, Colleen Blair; Chief Administrative Officer, William A. Bleibdrey; Exec. Vice Pres. Operations, Amy M. Duncan; Cur. History, Dr. Gene Allen Smith; Dir. Bldg. Svcs., Rick Aguirre; Asst. Cur. Science, Leishawn Spotted Bear; Asst. Cur. History, Renee Tucker; Archivist & Research Librarian, Chanin Scanlon.
Personnel Profile: Full-Time Paid 80; Part-Time Paid 108; Part-Time Volunteers 30; Interns 2.
Governing Authority: nonprofit organization. Tax-exempt: 501(c)(3).
Institution Type/Description: General Museum.
Collections: botany; entomology; malacology; ethnology; herpetology; zoology; anthropology; mammalogy; mineralogy; meteoritics; paleontology; local history; western history; Fort Worth Stock Show collection.
Major Exhibits: Curious George, 10/13-1/14; Penguins, 11/13-2/14; Wizard of Oz, 2/14-5/14; Indiana Jones, 3/14-8/14.
Facilities: 6,000-vol. library; planetarium; Omni IMAX theater; educational facilities; laboratories. Museum-related items for sale.
Activities: interactive exhibitions; DinoDig; children's museum; Omni IMAX Dome Theater; digital planetarium; museum school for pre-school learners; lectures; films; workshops; distance learning; educator training programs; permanent & temporary exhibitions.
Publications: newsletter, Spark!
Hours & Admission Prices: Mon.-Sat.10-5, Sun. noon-5. Exhibits: adults $15, seniors 65 & over $13, children 2-12 $11; discounts to ASTC & ACM members, schools and groups; children under 2 & members no charge. Omni IMAX Theater: adults $7-$12, seniors 65 & over and children 2-12 $6-$10; discounts to museum members. Some special engagement exhibitions may require a fee from members; combination tickets available with discounts. Closed Thanksgiving; Christmas Eve & Day. &
Attendance: 792,581 (accurate)
Membership: Basic 2 $65; Basic 5 $100; Basic 8 $165. MAX 2 $120; MAX 5 $190; MAX 8 $250; Friend of Science & Friend of History $300.

FORT WORTH NATURE CENTER AND REFUGE, 9601 Fossil Ridge Rd., Fort Worth, TX 76135-9148. Tel.: 817-392-7410. Fax: 817-392-7415.
Web Site: www.fwnaturecenter.org
Founded: 1964.
Congressional District: 12
Governing Authority: city. Tax-exempt.
Institution Type/Description: Nature Center & Refuge.
Collections: wildlife & their habitats; native flora & fauna.
Facilities: interpretive center.
Hours & Admission Prices: Summer: Mon.-Fri. 8-7, Sat.-Sun. 7-7; Winter: daily 8-5. Adult $4, seniors 65 & over $3, children 3-17 $2; children under 3 no charge. Closed Thanksgiving; Christmas. &
Attendance: 37,730 (accurate)
Membership: Individual $45; Family $75.

FORT WORTH ZOOLOGICAL ASSOCIATION, INC., (M), 1989 Colonial Pkwy., Fort Worth, TX 76110-6640. Tel.: 817-759-7500 & 7555. Fax: 817-759-7501.
E-mail: awilson@fortworthzoo.com
Web Site: www.fortworthzoo.org
Formerly: Fort Worth Zoological Park
Founded: 1909.
Congressional District: 12
Key Personnel: C.E.O., Michael Fouraker; C.F.O., Scott Wilcox; Co-Chm., Mrs. Lee M. Bass; Co-Chm., Mrs. Charles Moncrief; Pres. (V), Ardon Moore; Communications Dir., Alexis Wilson; Dir. Operations, Kelley Allred; Conservation & Animal Programs Dir., Tarren Wagener; Veterinarian, Nancy Lung; Bird Cur., Katie Unger; Mammal Cur., Ron Surratt; Dir. Education, Kathy Dorris.
Personnel Profile: Full-Time Paid 100; Part-Time Paid 200; Part-Time Volunteers 50; Interns 2.
Governing Authority: municipal. Parent Institution: city of Fort Worth. Management contracted to Fort Worth Zoological Assoc. Tax-exempt.
Institution Type/Description: Zoo.
Collections: aviary; herpetarium; African diorama; zoo geographic exhibits; Asian Falls; World of Primates; Raptor Canyon; Texas-related artifacts; wildlife art gallery; African savannah; penguins.
Research Fields: animal behavior; field conservation; International Rhino Foundation.
Facilities: Zoo-related gifts for sale.
Activities: formally organized education programs for children & families.
Publications: quarterly magazine, Roar!
Hours & Admission Prices: Daily 10-5; extended hours seasonally. Adults $10.50, children 3-12 $8, senior citizens 65 & over $7; discounts to groups & Wed.; children under 3 & members no charge. Parking $5. &
Attendance: 1,000,000 (estimated)
Membership: First Child $25; Second Child $20; Additional Child $15; First Adult $55; Second Adult $40; Additional Adult $50.

KIMBELL ART MUSEUM, (M), 3333 Camp Bowie Blvd., Fort Worth, TX 76107-2792. Tel.: 817-332-8451. Fax: 817-877-1264.
E-mail: info@kimbellmuseum.org
Web Site: www.kimbellart.org
Founded: 1972.
Congressional District: 12
Key Personnel: Dir., Eric M. Lee; Pres., Mrs. Ben J. Fortson; Deputy Dir., George T.M. Shackelford; Cur. European Art & Head of Academic Svcs., Nancy E. Edwards; Cur. Asian & Non-Western Art, Jennifer Casler Price; Cur. European Art, C. D. Dickerson; Dir. Conservation, Claire M. Barry; Mgr. Publications & Public Access, Megan Smyth; Dir. Finance & Admin., Susan R. Drake; Librarian, Chia-Chun Shih; Registrar, Patricia Decoster; Mgr. Membership, Robert McAn; Head Mktg. & Public Rels., Jessica Brandrup; Head Corporate Partnerships, Angie Bulaich; Buffet Mgr., Shelby Schafer; Mgr. Operations, Larry Eubank; Mgr. Security, David McMillan.
Personnel Profile: Full-Time Paid 111; Part-Time Paid 98; Part-Time Volunteers 73; Interns 6.
Governing Authority: nonprofit organization. Parent Institution: Kimbell Art Foundation. Tax-exempt: 501(c)(3).
Institution Type/Description: Art Museum.
Collections: artifacts from antiquity to 20th-century including western European paintings & sculptures; Egyptian, Near Eastern, Greek, & Roman antiquities; Asian, Precolumbian, & African art displayed in two museum buildings, the 1972 original building designed by Louis I. Kahn and the 2013 building designed by Renzo Piano.
Major Exhibits: The Age of Picasso and Matisse: Modern Masters from the Art Institute of Chicago (T), 10/6/13-2/16/14; Samurai: Armor from the Ann and Gabriel Barbier-Meuller Collection (T), 2/16/14-8/17/14; Faces of Impressionism: Portraits from the Musee d'Orsay (T), 10/14.
Research Fields: pertaining to collection.
Facilities: library with 45,000-vol. books; 5,200 bound periodicals; 14,000 auction catalogues; nearly 23,000 microfiche & microfilms; on-line research databases. Library materials are available for interlibrary loan via OCLC/RLP Shared Resources Program; access to the library is by appointment to professional art historians & graduate students. 180-seat auditorium; 298-seat auditorium; restaurant. Education Dept. with education greeters' room; education studios with projectors for presentations and videos; retractable walls to divide space if needed; classroom facilities with art supply storage; resource library for volunteer docents and teachers who request permission. Shop with scholarly & popular art books, catalogs, & children's books for sale.
Activities: guided tours; gallery talks; lectures; symposia; formally organized education programs for children, adults, undergraduate & graduate college students; family programs; studio workshops; teacher training; docent program; inter-museum loan; permanent & temporary exhibitions; online research databases.
Publications: exhibition catalogues; Kimbell Art Museum Guide; monographs, Kimbell Masterpiece Series; Light is the Theme: Louis I. Kahn and the Kimbell Art Museum; The Kimbell Cookbook; Kimbell Art Museum Address Book; biannual magazine, Calendar; brochures; posters; art prints.
Hours & Admission Prices: Tues.-Thurs. & Sat. 10-5, Fri. 12-8, Sun. 12-5. Permanent Collection: No charge. Special Exhibitions: prices vary; members & other museum employees no charge. Closed Monday; New Year's Day; Independence Day; Thanksgiving; Christmas. &
Attendance: 240,000 (estimated)
Membership: Patron $75; Curator's Council & Family $120; Sustaining $300; Director's Circle $600; Donor's Circle $1,000; Connoisseur's Circle $2,500; Collector's Circle & Corporate Supporter $5,000; Benefactor's Circle $10,000; Corporate Partner $15,000; Corporate Leader $25,000.

LOG CABIN VILLAGE, (M), 2100 Log Cabin Village Lane, Fort Worth, TX 76109-1000. Tel.: 817-392-5881.
E-mail: logcabinvillage@fortworthtexas.gov
Web Site: logcabinvillage.org
Founded: 1966.
Congressional District: 12

Key Personnel: Museum Dir., Kelli L. Pickard; Educator & Collections Mgr., Rena Lawrence; Clerk, Fred Gersch; Clerk, Mike Garrett; Clerk, Marilyn Tonn; Maintenance, Steven Suarez.
Personnel Profile: Full-Time Paid 3; Part-Time Paid 13; Part-Time Volunteers 110.
Volunteer Hours: 14,735
Operating Expenses: 427,589
Operating Income: 427,589
Governing Authority: municipal. Parent Institution: Parks & Community Services Dept., City of Fort Worth, 1000 Throckmorton, 76102. Tax-exempt: 501(c)(3).
Institution Type/Description: Village Museum.
Collections: 19th-century log structures, furniture, quilts & other artifacts & reproductions related to Texas pioneers; original & copies of documents relating to pioneer families; original & copies of photographs. Historic Buildings: 1848 Isaac Parker Cabin; 1853 John Baptist Tompkins Cabin; 1855 Isaac Seela Cabin; 1853 Harry A. Foster Log House; 1850 Hartsford Howard Cabin; 1853 Thomas J. Shaw Cabin converted into operating grist mill; a reproduction blacksmith shop; 1870's Marine School; 1860s Reynolds Smokehouse.
Research Fields: history; genealogical.
Activities: formally organized education programs for children; pioneer craft & grist mill demonstrations; permanent & temporary exhibitions.
Publications: booklets, Log Cabin Village: A History & Guide.
Hours & Admission Prices: Tues.-Fri. 9-4, Sat.-Sun. 1-5. Adult $5, senior citizens & youth 4-17 $4.50; discounts to groups, AAM & Texas Association of Museums members; members no charge. History Programs $2-$15. &
Attendance: 30,885 (accurate)
Membership: Student $10; Individual $25; Family $45; Pioneer $100; Settler $250; Merchant $500; Lifetime $1,000.

*** MODERN ART MUSEUM OF FORT WORTH, (M),** 3200 Darnell St., Fort Worth, TX 76107-2872. Tel.: 817-738-9215. Fax: 817-735-1161.
E-mail: info@themodern.org
Web Site: www.themodern.org
Founded: 1892.
Congressional District: 12
Key Personnel: Dir., Dr. Marla Price; Chief Cur., Michael Auping; Controller, Jo Garwood; Dir. Membership & Special Events, Suzanne Woo; Cur. Education, Terri Thornton; Bookstore Mgr., Lorri Wright; Head Design & Installation, Tony Wright; Cur., Andrea Karnes; Registrar, Brent Mitchell; Public Rels. Mgr., Kendal Smith Lake; Editor, Leslie Murrell; Supvr. Security, Mark Evans; Computer Systems Mgr., Ashley Anderson; Curatorial Administrative Asst., Susan Colegrove; Accounting, Karen Seidler; Human Resources Mgr., Sally McCracken; Tour & Docent Coord., Erin Starr White; Asst. Special Events, Tina Gorski.
Personnel Profile: Full-Time Paid 57; Part-Time Paid 35; Part-Time Volunteers 110; Interns 3.
Governing Authority: nonprofit corporation. Operated by Fort Worth Art Association. Tax-exempt: 501(c)(3).
Institution Type/Description: Art Museum.
Collections: Post World War II international art in all media.
Major Exhibits: David Bates, 2/14-5/14.
Research Fields: post 1940 modern & contemporary art.
Facilities: print study.
Activities: special exhibitions; lectures; films; gallery talks; guided tours; inter-museum loan; education programs; art camp.
Publications: biannual illustrated calendar for members; exhibition catalogs; posters; special mailings; exhibition brochures & cards.
Hours & Admission Prices: Adults $10; members no charge. Closed holidays. &
Attendance: 180,000 (estimated)
Membership: Basic $65; Associate $125; Sustainer $200; Contributor $500; Patron $1,200; President's Circle $5,000.

NATIONAL COWGIRL MUSEUM AND HALL OF FAME, 1720 Gendy St., Fort Worth, TX 76107-4064. Tel.: 817-336-4475. Fax: 817-336-2470.
E-mail: cschaefer@cowgirl.net
Web Site: www.cowgirl.net
Founded: 1975.
Key Personnel: Exec. Dir., Pat Riley; Pres. (V), Kit Moncrief; Assoc. Dir., Kim Smith; Dir. Finance, Barbara Mounts; Cur. & Collections Mgr., Heather McMaster; Assoc. Exec. Dir. Education & Exhibits, Diana Vela, Ph.D.; Devel., Emmy Lou Prescott; Assoc. Exec. Dir. Devel., Membership & Mktg., Craig Schaefer; Museum Shop Mgr., Sarah Garrett.
Personnel Profile: Full-Time Paid 13; Part-Time Paid 6; Part-Time Volunteers 62; Interns 3.

Governing Authority: nonprofit organization. Tax-exempt: 501(c)(3).
Institution Type/Description: Research Center: emphasis on all aspects of Western American women.
Collections: costumes; western attire; cowgirl & western photos; saddles; ropes; trophies; western art; pop culture.
Research Fields: western, pioneer & rodeo women; contribution of women to Western U.S. history.
Facilities: library; research center; auditorium. Museum-related items for sale.
Activities: temporary & traveling exhibits; research; rental gallery. Annual Event: Honoree Induction Luncheon; Desert Rose Society Gala.
Publications: brochures; newsletter.
Hours & Admission Prices: Memorial Day to Labor Day Mon.-sat. 10-5, Sun. 12-5; Sept.-May Tues.-Sat. 10-5, Sun. 12-5. Adults $10, seniors 60 & over & children 3-12 $8; children 2 & under and members no charge. Closed New Year's Day; Thanksgiving; Christmas Eve & Day. &
Attendance: 35,089 (accurate)
Membership: 2 People $60; Family $95; Rodeo Cowgirl $155; Cowgirl Legend $500; Hall of Fame $1,000; Silver Spur $2,500; Golden Spur $5,000; Platinum $10,000.

NATIONAL MULTICULTURAL WESTERN HERITAGE MUSEUM, 3400 Mount Vernon Ave., Fort Worth, TX 76103-2525. Mailing Address: 2401 Scott Ave., Fort Worth, TX 76103-2228. Tel.: 817-534-8801. Fax: 817-534-6277.
E-mail: info@cowboysofcolor.org
Web Site: www.cowboysofcolor.org
Formerly: National Cowboys of Color Museum and Hall of Fame
Founded: 2001.
Congressional District: 26
Key Personnel: Exec. Dir., Gloria R. Austin; Pres. Bd., James N. Austin, Jr.; Chm. (V), Tre Garrett; Vice Pres. Bd., Heather Johnson.
Personnel Profile: Full-Time Paid 1; Part-Time Paid 2; Part-Time Volunteers 20; Interns 1.
Governing Authority: Tax-exempt.
Institution Type/Description: History Museum.
Collections: African American, Hispanic American, Native American, & European American pioneers; western culture; photographs; Tuskegee Airmen; Buffalo Soldiers; Hall of Fame inductees.
Research Fields: Western heritage.
Activities: programs; workshops; storytelling. Museum Sponsors: Western Heritage Symposium; Hall of Fame Induction Ceremony.
Publications: e-newsletter, Western Round-Up.
Hours & Admission Prices: Wed.-Fri. 12-5 by appointment, Sat. 11-5. Adults $6, senior citizens $4, students $3; discounts to Blue Star Museum, AAM & ICOM members; members and children 5 & under no charge. &
Attendance: 10,000 (estimated)
Membership: Student $20; Senior Individual $25; Educator $40; Individual & Senior Family $50; Family $75; Rancher $250; Buckaroo $500; Lifetime Individual $750; Wrangler $1,000; Foreman $2,500; Partner $5,000.

OSCAR E. MONNIG METEORITE GALLERY, Texas Christian University, Sid Richardson Science Bldg., 2950 W. Bowie, Fort Worth, TX 76109. Mailing Address: Box 298830, Fort Worth, TX 76129. Tel.: 817-257-6277. Fax: 817-257-7789.
E-mail: t.moss@tcu.edu
Web Site: www.monnigmuseum.tcu.edu
Key Personnel: Dir., Teresa Moss; Cur., Dr. Arthur Ehlmann
Institution Type/Description: Meteorite Gallery.
Collections: hands-on exhibits pertaining to meteorites.
Hours & Admission Prices: Tues.-Fri. 1-4, Sat. 9-4. No charge.

SID RICHARDSON MUSEUM, (M), 309 Main St., Fort Worth, TX 76102-4006. Tel.: 817-332-6554. Fax: 817-332-8671.
E-mail: info@sidrichardsonmuseum.org
Web Site: www.sidrichardsonmuseum.org
Formerly: Sid Richardson Collection of Western Art
Founded: 1982.
Congressional District: 12
Key Personnel: Dir., Mary Burke; Dir. Studio Programs, Katherine Yount; Adult Audiences Mgr., Leslie Thompson; Dir. Education Resources, Betsy Thomas; Museum Shop Mgr., Chris Gensheimer.
Personnel Profile: Full-Time Paid 7; Part-Time Volunteers 16.
Governing Authority: nonprofit organization. Parent Institution: Sid W. Richardson Foundation. Tax-exempt: 501(c)(3).
Institution Type/Description: Art Museum.
Collections: works by Western artists including paintings by Frederic Remington & Charles M. Russell & their contemporaries.
Major Exhibits: Western Treasures, 10/13-9/14/14.
Facilities: Museum-related items for sale.

Activities: educational program & guided tours by appointment.
Publications: brochure; gallery guides.
Hours & Admission Prices: Mon.-Thurs. 9-5, Fri.-Sat. 9-8, Sun. 12-5. No charge. Closed holidays. &
Attendance: 36,000 (accurate)

STOCKYARDS MUSEUM, 131 E. Exchange Ave., Ste. 113, Fort Worth, TX 76164-8213. Tel.: 817-625-5087. Fax: 817-625-5083. Facebook: Stockyard, Fort Worth, TX.
E-mail: nfwhs@sbcglobal.net
Founded: 1985.
Key Personnel: Dir., Teresa Burleson; Pres. (V), Tom Wiederhold
Governing Authority: Parent Institution: North Fort Worth Historical Society. Tax-exempt: 501(c)(3).
Institution Type/Description: History Museum.
Collections: local history; photographs; railroading; cattle industry; wagon trail.
Research Fields: books; photos.
Facilities: 2,300 sq. ft. exhibit space. Museum-related items for sale.
Hours & Admission Prices: Mon.-Sat. 10-5. No charge; donations accepted. &
Attendance: 15,000 (estimated)

TEXAS ASSOCIATION OF MUSEUMS, 101 Summit Ave., Ste. 802, Fort Worth, TX 76102-2615. Tel.: 817-332-1177. Fax: 817-332-1179. Facebook: Texas Association of Museums.
E-mail: admin@texasmuseums.org
Web Site: www.texasmuseums.org
Founded: 1960.
Congressional District: 12
Key Personnel: Exec. Dir., Ruth Ann Rugg; Program Assoc., Sandy Sage.
Personnel Profile: Full-Time Paid 1; Part-Time Paid 1; Part-Time Volunteers 65.
Governing Authority: nonprofit organization. Tax-exempt: 501(c)(3).
Institution Type/Description: State Museum Association.
Collections: association of museums.
Activities: training programs for professional museum workers; workshops; annual conference.
Publications: Museum Forms Book; monthly e-newsletter, Museline.
Hours & Admission Prices: Call for hours.
Membership: Student $25; Retired Professional $35; Individual $50; Business $200; Institutional $100-$1,000.

TEXAS CHRISTIAN UNIVERSITY - MOUDY GALLERY, 2805 S. University Dr., 245 Moudy Bldg. N., Fort Worth, TX 76129. Mailing Address: P.O. Box 298000, Fort Worth, TX 76129. Tel.: 817-257-2588.
E-mail: theartgalleries@tcu.edu
Web Site: www.theartgalleries.tcu.edu
Key Personnel: Dir., Anne Bothwell
Institution Type/Description: Art Gallery.
Collections: works by students & professional artists.
Activities: temporary exhibitions; lectures; panel discussions; public programs.
Hours & Admission Prices: Academic Year: Mon. 11-6, Tues.-Fri. 10-4, Sat. 1-4. No charge.

TEXAS CIVIL WAR MUSEUM, 760 Jim Wright Freeway N., Fort Worth, TX 76108-1222. Tel.: 817-246-2323. Fax: 817-246-3951.
E-mail: questions@texascivilwarmuseum.com
Web Site: texascivilwarmuseum.com
Founded: 2006.
Key Personnel: Pres. (V) & Cur., Ray Richey; Exec. Dir., Cynthia L. Harriman; Cur., Judy Richey; Museum Shop Mgr., John Bell; Reservations, Nancy Gentry.
Personnel Profile: Full-Time Paid 2; Full-Time Volunteers 1; Part-Time Paid 4.
Governing Authority: nonprofit organization. Tax-exempt.
Institution Type/Description: History Museum.
Collections: Civil War artifacts; over 80 flags; firearms; uniforms; period clothing; Victorian era dresses.
Hours & Admission Prices: Tues.-Sat. 9-5. Adults $6, students 7-12 $3; discounts to AAM members, groups & active military; children under 6 no charge. Closed New Year's Day; Independence Day; Thanksgiving; Christmas Eve & Day. &
Attendance: 20,000 (estimated)

TEXAS COWBOY HALL OF FAME, 128 E. Exchange Ave., Fort Worth, TX 76164-8210. Tel.: 817-626-7131. Fax: 817-626-7171.
E-mail: info@texascowboyhalloffame.org

Web Site: www.texascowboyhalloffame.org
Founded: 2001.
Key Personnel: Exec. Asst., Julia Buswold; Museum Shop Mgr., DeeDee Ramirez.
Governing Authority: Tax-exempt.
Institution Type/Description: History Museum.
Collections: cowboy history; Texas men & women of rodeo and cutting; photographs; videos; Sterquell wagons.
Facilities: Museum-related items for sale.
Activities: rental facilities.
Hours & Admission Prices: Mon.-Thurs. 9-5, Fri.-Sat. 10-7, Sun. 11-5. Family Package (2 adults, 4 children 5-12): $15, adults 18-59 $5, seniors 60 & over & students 13-17 with valid college ID $4, children 5-12 $3; military with ID & children 4 & under no charge; discounts to groups. Closed Easter, Christmas. &
Attendance: 80,000
Membership: Student & Senior $30; Individual $40; Family $100. Charter Memberships: Bronze $500; Copper $1,000; Silver $2,500; Gold $5,000; Platinum $10,000.

VETERANS MEMORIAL AIR PARK, 3300 Ross Ave., Fort Worth, TX 76106. Mailing Address: P.O. Box 161966, Fort Worth, TX 76161. Tel.: 800-575-0535. Fax: 817-488-8170.
E-mail: obainfo@verizon.net
Web Site: www.veteransmemorialairpark.com
Founded: 1998.
Key Personnel: Chm. (V), Chuck Burin; Pres. (V), Jim Hodgson; Museum Shop Mgr., Donna Hodgson.
Personnel Profile: Part-Time Volunteers 30.
Governing Authority: Parent Institution: OV-10 Bronco Association, Inc. Tax-exempt.
Institution Type/Description: Aviation History Museum.
Collections: aviation history; aircraft from 1943 to present; Forward Air Control; North Texas aviation history.
Facilities: 3,500 sq. ft. exhibition space; 5 acre outdoor area.
Activities: special events. Museum Sponsors: Reunions; Fly-Ins; Movies Every Wednesday; Founders Day Family Picnic in April.
Publications: monthly newsletter.
Hours & Admission Prices: Wed. 9-4, Sat. 9-5, Sun. 11-5. Family $10, adults $5, youth 6-16 $1; Blue Star Museum, active duty military & children under 6 no charge. &
Attendance: 1,200 (estimated)
Membership: Annual $40; Life over 75 $250; Life 65-75 $500; Life under 65 $1,000

VINTAGE FLYING MUSEUM, Meacham International Airport, 505 N.W. 38th St., Hangar 33 S., Fort Worth, TX 76106-4386. Mailing Address: P.O. Box 820099, Fort Worth, TX 76182-0099. Tel.: 817-624-1935. Fax: 817-624-2840.
E-mail: vfm@vintageflyingmuseum.org
Web Site: www.vintageflyingmuseum.org
Formerly: B.C. Vintage Flying Machines
Founded: 1980.
Key Personnel: C.E.O./Dir., Chuckie Hospers.
Personnel Profile: Full-Time Volunteers 2; Part-Time Volunteers 20.
Governing Authority: Tax-exempt: 501(c)(3).
Institution Type/Description: Military Aircraft Museum.
Collections: WWII memorabilia & artifacts; aircraft models; 1930s & 1940s aircraft.
Activities: workshops; summer camps; educational programs; museum tours.
Publications: newsletter, Wingtips.
Hours & Admission Prices: Sat. 10-5, Sun. 12-5, Mon.-Fri. by appointment. Adults $8, children 13-17 & seniors $5, children 6-12 $3; children under 6 no charge. &
Attendance: 5,000 (estimated)
Membership: Individual $36; Family $45; Life$1,000.

Fredericksburg

GILLESPIE COUNTY HISTORICAL SOCIETY, 312 W. San Antonio St., Fredericksburg, TX 78624-3761. Tel.: 830-990-8441. Fax: 830-990-2906.
E-mail: gailclifton@austln.rr.com
Web Site: www.pioneermuseum.com
Founded: 1935.
Key Personnel: Exec. Dir., Stephen Vollmer; Pres., Edward Stroeher; Administrative Sec. & Museum Shop Mgr., Gail Clifton.
Personnel Profile: Full-Time Paid 2; Part-Time Paid 10; Part-Time Volunteers 300.

Governing Authority: private; nonprofit organization. Subsidiary Institutions: Pioneer Museum, Fredericksburg, TX; Vereins Kirche Museum, Fredericksburg, TX. Tax-exempt: 501(c)(3).
Institution Type/Description: Historical Society.
Collections: Pioneer Museum: 9 historic period furnished buildings; native plant gardens. Vereins Kirche Museum: artifacts pertaining to the immigration & settlement of Germans from 1846 to early 1900s.
Research Fields: agricultural history; German immigrant history; social history of Texas Hill Country 1846 to present; frontier fort history.
Facilities: 100-vol. library of various historical books; botanical garden; 10,000 sq. ft. exhibit space; 228-seat auditorium; field research station. Museum-related items for sale.
Activities: guided tours; lectures. Annual Events: Founders Day Festival in May; Home Tour in December; Tannenbaum Ball in December.
Publications: quarterly newsletter, Der Trompeter.
Hours & Admission Prices: Pioneer Museum: Memorial Day to Labor Day Mon.-Sat. 10-5; Sept.-May Tues.-Sat. 10-5. Vereins Kirche Museum: Memorial Day to Labor Day Mon.-Sat. 10-5; Sept.-May Tues.-Sat. 10-4:30. Adults $5, students 6-17 $3; children 5 & under no charge. Closed major holidays.
Attendance: 22,000 (estimated)
Membership: Individual $30; Dual $50; Family $75; Company $250.

NATIONAL MUSEUM OF THE PACIFIC WAR, 340 E. Main St., Fredericksburg, TX 78624-4612. Tel.: 830-997-8600, ext. 200. Fax: 830-997-8220.
Web Site: www.pacificwarmuseum.org
Founded: 1967.
Congressional District: 21
Key Personnel: Museum Dir., Joe Cavanaugh; Pres. & C.E.O., Gen. Michael W. Hagee; Exec. Asst., Laura Nelson; Program Dir., Helen McDonald; Dir. Retail, Krista Gratigni.
Personnel Profile: Full-Time Paid 27; Part-Time Paid 15; Part-Time Volunteers 275.
Governing Authority: state. Parent Institution: Texas Historical Commission, Austin, TX. Tax-exempt.
Institution Type/Description: History Museum: housed in 1850 Nimitz Hotel, a famous hostelry until about the turn of the century.
Collections: Pacific World War II related artifacts including uniforms, personal equipment, weapons, flags & souvenirs; George Bush related artifacts; Admiral Nimitz, Nimitz Hotel & Fredericksburg-related artifacts; archives including manuscripts, war diaries, correspondences, photos, scrapbooks, ephemera; macro collection along history walk of the Pacific War, including two American & two Japanese aircraft, Japanese & American tanks, Half-Track, DUKW, naval & land based artillery of both sides; prints & posters developed for Symposia Series.
Major Exhibits: Ring of Fire (T), 2/7/14-3/14.
Research Fields: historic Fredericksburg; career of Admiral Nimitz; World War II in the Pacific.
Facilities: research library on Admiral Nimitz & the war in the Pacific open by appointment; 100-seat auditorium; 23,000 sq. ft. exhibit space, 3.5 acre outdoor exhibit.
Activities: living history programs; annual Pacific War Symposium; guided tours; lectures; illustrated talks by advance arrangements; loan, temporary & permanent exhibitions; educational programs.
Publications: book; Tarawa, The Story of a Battle; books & booklets relating to life & career of Admiral Nimitz & World War II in the Pacific; Attack on Pearl Harbor by Two Who Were There; Day of Infamy; The Cactus Air Force; The Cruise of the Lanikai; A Guide to the National Museum of the Pacific War.
Hours & Admission Prices: Daily 9-5. Adults $14, seniors $12, active duty & retired military $10, students $7; discounts to TAM & AAM members; World War II veterans, children under 5 and members no charge. Closed Thanksgiving; Christmas. &
Attendance: 120,000 (estimated)
Membership: Patriot $35; Supporter $125; Patron $250; Chairman's Circle $500; Benefactor Circle $1,000; 5 Star Circle $3,000.

PIONEER MUSEUM AND VEREINS KIRCHE, 325 W. Main St., Fredericksburg, TX 78624-3711. Mailing Address: 312 W. San Antonio St., Fredericksburg, TX 78624-3760. Tel.: 830-990-8441. Fax: 830-990-2906.
E-mail: gailclifton@austin.rr.com
Web Site: www.pioneermuseum.com
Founded: 1936.
Congressional District: 21
Key Personnel: Pres. Gillespie County Historical Society (V), Edward Stroeher; Pres. (V), Janet Lindemann; Dir., Stephen Vollmer; Administrative

Asst. & Museum Shop Mgr., Gail Clifton; Lead Interpreter Pioneer Museum, Evelyn Stork; Lead Interpreter Vereins Kirche, Clinton Stork.
Personnel Profile: Full-Time Paid 3; Part-Time Paid 13; Part-Time Volunteers 2.
Governing Authority: society. Parent Institution: Gillespie County Historical Society. Vereins Kirche Museum, Center of Marktplatz, Fredericksburg, TX. Tax-exempt: 501(c)(3).
Institution Type/Description: History Museum, House & Sites.
Collections: over 1,000 objects relating to the domestic life, commerce, ranching & farming of Gillespie county; Ty Cox tools including over 850 19th & early 20th-century tools; archives including manuscripts, photographs & oral history; early settlement of Fredericksburg & Gillespie county & subsequent chronicle of daily life in the county; Vereins Kirche Museum (1935) replica of original built in 1847. Historic Houses: c.1935 Vereins Kirche; c.1849-87 Kammlah Family Home & outbuildings; c.1878-1959 Fassel-Roeder House; c.1904 Weber Sunday House; c.1880 Walton Smith Cabin; c.1880 Schandua House; c. late 1920s White Oak School House.
Research Fields: German immigration to Texas Hill Country, daily life in Gillespie county from 1846 onward.
Facilities: Museum-related items for sale.
Activities: daily guided tours; special German meal & tour for groups; school tours. Annual Events: Christmas candlelight tour of homes; founders day celebration & lectures.
Publications: quarterly newsletter; brochures; books: Gillespie County A View of Its Past, Pioneers in God's Hills, Vol. II.
Hours & Admission Prices: Memorial Day to Labor Day Mon.-Sat. 10-5; Sept.-May Tues.-Sat. 10-5. Adults $5, children 6-17 $3; children under 6 no charge. &
Attendance: 30,000 (estimated)
Membership: Individual $30; Dual $50; Family $75; Business $250.

Frisco

THE FRISCO HERITAGE MUSEUM, 6455 Page St., Frisco, TX 75034-3486. Tel.: 972-292-5111.
E-mail: phosp@friscotexas.gov
Founded: 2008.
Key Personnel: Dir., Pete Hosp
Governing Authority: Parent Institution: City of Frisco, TX.
Institution Type/Description: History Museum.
Collections: rare artifacts & photographs depicting Frisco's past.
Hours & Admission Prices: Wed.-Sat. 10-5, Sun. 1-5. Family $8, adults $4, children 5-11 $2; children 4 & under no charge. &
Attendance: 12,328 (accurate)

MUSEUM OF THE AMERICAN RAILROAD, (M), 6299 Cotton Gin Rd., Frisco, TX 75034. Mailing Address: 6455 Page St., Frisco, TX 75034. Tel.: 214-428-0101. Fax: 214-426-1937.
E-mail: info@historictrains.org
Web Site: www.historictrains.org
Formerly: Age of Steam Railroad Museum
Founded: 1963.
Congressional District: 30
Key Personnel: C.E.O., Robert LaPrelle; Chm., John D. Maxson, Ph.D.; Vice Chair, Robert Willis; C.O.O., Kellie Murphy; Dir. Programs & Svcs., Florentina Duquin; Dir. Education, Robin Jett.
Personnel Profile: Full-Time Paid 3; Part-Time Paid 1; Part-Time Volunteers 55; Interns 1.
Governing Authority: Tax-exempt: 501(c)(3).
Institution Type/Description: Railroad Museum.
Collections: 38 pieces of rolling stock including 4 steam locomotives, 8 diesel locomotives, 1 electric locomotive, 17 passenger cars, & 7 freight cars; books; archives; photographs. Historic Railroad Buildings: 1901 Dallas Depot; 1905 Santa Fe Railway Interlocking Tower.
Research Fields: history of U.S. railroads.
Facilities: Museum-related items for sale.
Activities: lectures; films; permanent & traveling exhibitions.
Publications: bimonthly newsletter, The Clearance Card; brochure, The Age of Steam Exhibit; book, Iron Horses of the Santa Fe Trail.
Hours & Admission Prices: Museum: Wed.-Sun. 10-5. State Fair: Oct. daily 10-6. Adults $7, children under 13 $3; discounts to AAM members; members no charge.
Attendance: 46,000 (accurate)
Membership: Trainman $30; Family $40; Station Master $60; Engineer $100; Conductor $500; Railroad Magnate $1,000.

Fritch

LAKE MEREDITH AQUATIC AND WILDLIFE MUSEUM, 104 N. Robey, Fritch, TX 79036. Mailing Address: P.O. Box 758, Fritch, TX 79036-0758. Tel.: 806-857-2458. Fax: 806-857-3229.
Web Site: www.geocities.com/lakemeredithmuseum
Founded: 1976.
Congressional District: 13
Key Personnel: Dir. & Museum Shop Mgr., Renee Laney; Chm. (V), Walt Poling; Pres. (V), Wes Phillips.
Personnel Profile: Full-Time Paid 1; Part-Time Paid 2; Part-Time Volunteers 3.
Governing Authority: municipal; nonprofit organization. Parent Institution: City of Fritch. Tax-exempt: 501(c)(3).
Institution Type/Description: Natural History Museum.
Collections: vegetation & wildlife by LaNelle Poling; Playa Lake.
Facilities: 2 aquariums. Museum-related items for sale.
Activities: guided tours; lectures; films; traveling exhibitions; school loan service; special programs for children.
Publications: brochure; monthly newsletter.
Hours & Admission Prices: Mon.-Sat. 10-5. No charge; donations accepted. Closed New Year's Day; Thanksgiving; Christmas. &
Attendance: 6,500 (estimated)
Membership: Membership Fee $5 & up.

Gainesville

MORTON MUSEUM OF COOKE COUNTY, (M), 210 S. Dixon St., Gainesville, TX 76240-4719. Mailing Address: P.O. Box 150, Gainesville, TX 76241-0150. Tel.: 940-668-8900. Fax: 940-668-0533.
E-mail: mortonmuseum@att.net
Web Site: www.mortonmuseum.org
Founded: 1968.
Congressional District: 17
Key Personnel: Museum Coord., Jayleane Smith; CCHS Pres., Robin Rose; Museum Shop Mgr., Misty Landers.
Personnel Profile: Full-Time Paid 2; Part-Time Paid 1; Part-Time Volunteers 1.
Governing Authority: society. Affiliated with Cooke County Heritage Society, Inc. Tax-exempt: 501(c)(3).
Institution Type/Description: Historic Building, Historic Site & History Museum: housed in 1884 fire station, city hall & jail.
Collections: preservation project; Cooke County, Texas artifacts & manuscripts.
Research Fields: county history.
Facilities: 100-vol. library of books on history of Cooke County available for use on premises. Books & museum-related items for sale.
Activities: guided tours; formally organized educational programs; temporary & traveling exhibitions.
Publications: booklet, Texas: Cooke County, Its People, Productions & Resources, 1888; newsletter, Heritage Highlights; books, Early Days in Cooke County, 1848-1873; Cooke County Texas-Where the South & the West Meet; History of Cooke County-A Pictorial Essay; Gainesville and Cooke County-A Pictorial History (in Images of American Series).
Hours & Admission Prices: Tues.-Fri. 10-5, Sat. 12-3. No charge; donations accepted. Closed New Year's Day; Thanksgiving; Christmas. &
Attendance: 5,000 (estimated)
Membership: Individual $15; Family $25; Contributing $50-$99; Business $100; Lifetime $500; Lifetime Benefactor %501-$999; Lifetime Patron $1,000 & up.

Galveston

THE BISHOP'S PALACE, 1402 Broadway, Galveston, TX 77550-2014. Mailing Address: 502 20th St., Galveston, TX 77550-2014. Tel.: 409-762-2475. Fax: 409-762-1801.
Founded: 1886.
Congressional District: 9
Key Personnel: Cur., Archbishop Daniel diNardo; Exec. Dir., Rev. William D. Bartniski.
Personnel Profile: Part-Time Paid 10, Part-Time Volunteers 2.
Governing Authority: Catholic Diocese of Galveston-Houston. Tax-exempt.
Institution Type/Description: Historic Building: 1886 Walter Gresham Home/Bishop's Palace.
Collections: Victorian furniture.
Facilities: Postcards, brochures & gift items for sale.
Activities: guided tours.
Publications: flyers; newspapers; magazines.
Hours & Admission Prices: Call for hours & admission prices.
Attendance: 40,124 (accurate)

1859 ASHTON VILLA & THE HERITAGE VISITOR CENTER, 2328 Broadway St., Galveston, TX 77550-4642. Mailing Address: Galveston Historical Foundation, 1861 Custom House, 502 20th St., Galveston, TX 77550. Tel.: 409-762-3933. Fax: 409-762-1904.
E-mail: foundation@galvestonhistory.org
Web Site: www.galvestonhistory.org
Founded: 1974.
Congressional District: 9
Key Personnel: C.E.O., Dwayne Jones; Chm. (V), Merri Edwards; Pres. (V), Raymond Lewis; Museum Shop Mgr., Brandon Ragen.
Personnel Profile: Full-Time Paid 3; Part-Time Paid 8.
Governing Authority: municipal; historic foundation. Parent Institution: Galveston Historical Foundation, 502 20 St., Galveston, TX 77550. Tax-exempt.
Institution Type/Description: Historic House: 1859 Ashton Villa, home of James Moreau Brown, an early Galveston community leader.
Collections: furnishings; paintings; art objects; textiles.
Research Fields: decorative arts; textiles; photographs; postcards.
Facilities: ballroom available. Museum-related items for sale.
Activities: guided tours; docent programs & exhibits. Foundation Sponsors: Christmas events.
Publications: brochure, Ashton; book, The History of Ashton Villa.
Hours & Admission Prices: Mon.-Sat. 10-4, Sun. 12-4. Adults $7, students 6-18 $6; children under 5 & members no charge. Closed Thanksgiving; Christmas Eve & Day. &
Attendance: 15,000 (estimated)
Membership: Senior Citizen 65 & over and Students $20; Individual $35; Family (1 household & its children under 19) $45.

GALVESTON ARTS CENTER, 2501 Market St., Galveston, TX 77550. Tel.: 409-763-2403.
E-mail: information@galvestonartscenter.org
Web Site: www.contemporaryartgalveston.org
Founded: 1987.
Congressional District: 14
Key Personnel: Exec. Dir., Alexandra L. Irvine; Pres. (V), Robert Clepper; Cur., Clint Willour; Administrative Dir., Robin Cushman.
Personnel Profile: Full-Time Paid 2; Part-Time Paid 3; Part-Time Volunteers 40; Interns 1.
Governing Authority: private; nonprofit organization. Tax-exempt: 501(c)(3).
Institution Type/Description: Contemporary Art Museum: housed in the 1878 First National Bank building located in the Strand Historic District.
Collections: works by artists from around the state.
Publications: Catalogue for Al Souza: Addenda, 2006; The Art Guys Go Public, 2007; Remembering Robin Utterback, 2009.
Hours & Admission Prices: Tues.-Sat. 11-5, Sun. 12-5. No charge. &
Attendance: 12,000 (estimated)
Membership: Student & Senior Citizen $25; Individual $40; Household $75; Patron $100; Arts Sponsor $250.

GALVESTON COUNTY MUSEUM, (M), Shearn Moody Plaza Bldg., Ste. 4157, 123 Rosenberg, Galveston, TX 77550-1454. Tel.: 409-766-2340. Fax: 409-766-4545.
E-mail: helen.mooty@co.galveston.tx.us
Web Site: www.co.galveston.tx.us/museum
Founded: 1972.
Congressional District: 9
Key Personnel: Dir., Helen D. Mooty.
Personnel Profile: Full-Time Paid 1; Part-Time Paid 1; Part-Time Volunteers 3.
Governing Authority: county. Parent Institution: Galveston County Historical Commission. Subsidiary Institution: Galveston County History, Inc.
Institution Type/Description: History Museum.
Collections: photographs; clothing; Galveston county history; lighthouse lens; Edison film of 1900 storm; archival materials.
Research Fields: county history; 1900 Hurricane; Texas City Disaster; State Historical marker records; local & regional architecture; local genealogy; cemetery listings.
Facilities: research library; 2,400 sq. ft. exhibit space.
Activities: lectures; films; interpretive events; formal education programs for children; temporary exhibitions.
Publications: book, Galveston: A City on Stilts.
Hours & Admission Prices: Temporarily closed. &

GALVESTON HISTORICAL FOUNDATION, INC., 502 20th St., Galveston, TX 77550-1661. Tel.: 409-765-7834. Fax: 409-765-7851.
Web Site: www.galvestonhistory.org
Founded: 1871.

Congressional District: 9
Key Personnel: C.E.O., Dwayne Jones; Chm., Cheryl Vaiani.
Personnel Profile: Full-Time Paid 30; Part-Time Paid 45; Part-Time Volunteers 2,000.
Governing Authority: foundation. Subsidiary Institutions: 1859 Ashton Villa; 1859 St. Joseph Church; 1839 Samuel May Williams House; Texas Seaport Museum, home of the 1877 barque Elissa; Great Storm Theater; 1838 Michele B. Menard House. Tax-exempt.
Institution Type/Description: Preservation Project: housed in the U.S. Custom House; built in 1861.
Collections: local history & culture; period furnishings; memorabilia; maritime artifacts.
Research Fields: immigration through port of Galveston. 19th century social & religious history.
Activities: guided tours; formally organized education programs for children; docent program; permanent & temporary exhibitions; special events; historic homes tour. Museum Sponsors: Victorian street festival.
Publications: quarterly newsletter, attractions & special events brochures.
Hours & Admission Prices: Call for hours. No charge; donations accepted. &
Attendance: 600,000 (accurate)
Membership: Student $30; Senior $35; Individual $50; Senior Couple $60; Dual $65; Family $75; Patron $125; Benefactor $250; Memorial Society $1,000.

GALVESTON RAILROAD MUSEUM, 25th at Strand, Galveston, TX 77550. Mailing Address: 2602 Santa Fe Place, Galveston, TX 77550-1493. Tel.: 409-765-5700. Fax: 409-765-5744.
E-mail: galvrrmuseum@sbcglobal.net
Web Site: www.galvestonrrmuseum.com
Founded: 1983.
Congressional District: 14
Key Personnel: Exec. Dir., Morris S. Gould; Chm. (V), Dr. John E. Bertini; Dir. Mktg. & Coord. Events, Sandi Cobb; Museum Shop Mgr., Betty Morris.
Governing Authority: Tax-exempt.
Institution Type/Description: Railroad Museum.
Collections: steam & diesel engines; HO scale trains; passenger cars; O scale layout; HO scale layout.
Facilities: Museum-related items for sale.
Activities: birthday parties; rental facilities; special events; locomotive & caboose rides. Museum Sponsors: Easter Egg Hunt & Brunch in April; Model Train Show in May; Independence Day Parade; Hobo Night in October; Santa Train in December.
Publications: newsletter.
Hours & Admission Prices: May-Sept. 10-5; Oct.-April 10-4. Adults $7, senior citizens 65 & over $6, children 4-12 $5; discounts to groups. Train Rides: Sat. 11-1. Closed New Year's Day; Thanksgiving; Christmas Eve & Day. &
Membership: Switchman $35; Train Crew $50; Brakeman $100; Conductor $250; Engineer $500; Trainmaster $1,000.

JOHN SYDNOR'S 1847 POWHATAN HOUSE, 3427 Avenue O, Galveston, TX 77550-6734. Tel.: 409-763-0077.
E-mail: evangelinewhorton@yahoo.com
Web Site: www.galvestongardenclub.org
Founded: 1938.
Congressional District: 17
Key Personnel: Pres. (V), Dwayne Johnson; Chm. (V), Evangeline Whorton.
Personnel Profile: Full-Time Volunteers 2; Part-Time Volunteers 45.
Governing Authority: nonprofit organization. Parent Institution: Galveston Garden Club, Inc. Subsidiary Institution: Powhatan House Headquarters. Tax-exempt: 501(c)(3).
Institution Type/Description: Historic Building & Site: c.1847 home of pioneer businessman & the first Mayor of Galveston, John Seabrook Sydnor; originally built for use as 24-room family dwelling and guest house; relocated in 1895. Listed on the National Register of Historic Places.
Collections: furnishings of antebellum & Victorian period; artifacts; personal papers; letters; Civil War battle account; documents, photographs, primary research journals, & research from John Seabrook Sydnor & family; architectural documents of Charles Ladd & Caroline Willis Ladd, who relocated structure to its present location.
Research Fields: gardening & horticulture; life and times of the Sydnor Family; the William Charles & Caroline Willis Ladd family.
Facilities: 350-vol. library of books on gardening & flower arranging; documents & published material about John S. Sydnor, available for study & reference on the premises.
Activities: self-guided tours; lectures; formally organized education programs; docent program; permanent exhibitions. Annual Events: Caladium Bulb Sales in March; Annual Powhatan Pansy Potpourri in November.
Publications: monthly newsletter, Broadside.

Attendance: 2,000 (estimated)
Membership: Individual $30.

LONE STAR FLIGHT MUSEUM/TEXAS AVIATION HALL OF FAME, 2002 Terminal Dr., Galveston, TX 77554-9279. Mailing Address: P.O. Box 3099, Galveston, TX 77552-0099. Tel.: 409-740-7722. Fax: 409-740-7612.
E-mail: flight@lonestarflight.org
Web Site: www.lsfm.org
Founded: 1986.
Congressional District: 9
Key Personnel: Dir., Larry Gregory.
Governing Authority: nonprofit organization. Tax-exempt: 501(c)(3).
Institution Type/Description: Aeronautics Museum.
Collections: history of aviation; 40 vintage military & civilian aircraft from 1930s-1970s, most in flying condition or undergoing restoration; history of aviation in Texas.
Research Fields: military & civilian Texas aviation.
Facilities: 102,000 sq. ft. exhibit hangars.
Activities: Warbird rides: B-17, B-25, T6, Stearman. Annual Events: Two-Day Airshow in April; Texas Aviation Hall of Fame Induction Gala & Flyday in November.
Publications: quarterly newsletter.
Hours & Admission Prices: Daily 9-5. Adults $8, senior citizens 65 & up & children 5-17 $5; discounts to groups & AARP members; Texas Assoc. of Museum members, children 4 & under and members no charge. Closed Thanksgiving; Christmas. &
Attendance: 75,000 (estimated)
Membership: Student $48; Crewmember $100; Participating $150; Contributing $250; Supporting $500; Sustaining $1,000; Life $3,500.

MICHEL B. MENARD HOUSE, 1604 33rd St., Galveston, TX 77550. Mailing Address: Galveston Historical Foundation, 1861 Custom House, 502 20th St., Galveston, TX 77550. Tel.: 409-762-3933.
E-mail: foundation@galvestonhistory.org
Web Site: www.galvestonhistory.org
Institution Type/Description: Historic House Museum: housed in the former home of Michel B. Menard, one of the founders of Galveston; built in 1838. Listed on the National Register of Historic Places.
Collections: Menard family history; period furnishings; personal artifacts; photographs.
Activities: rental facilities.
Hours & Admission Prices: By appointment.

THE MOODY MANSION MUSEUM, 2618 Broadway, Galveston, TX 77550-4427. Tel.: 409-762-7668. Fax: 409-762-7055.
E-mail: k.guernsey@northenendowment.org
Web Site: www.moodymansion.org
Founded: 1991.
Congressional District: 9
Key Personnel: Pres., Edward L. Protz; Dir. Endowment, Betty Massey; Security & Maintenance, Mary Hoehne; Visitor Operations & Museum Shop Mgr., Karen Guernsey.
Personnel Profile: Full-Time Paid 3; Part-Time Paid 10.
Governing Authority: Parent Institution: Center for Twentieth Century Texas Studies. Tax-exempt.
Institution Type/Description: Historic House Museum: built between 1893 & 1895 31-room mansion.
Collections: late 19th & early 20th-century architecture, decorative arts, social & local history; furniture; ceramics; glass; costume & textiles; photographs; Moody family possessions.
Research Fields: early 20th-century decorative arts & material culture; social, cultural, political & business history of Galveston, Texas & U.S. from 1820-1980.
Facilities: 1,500-vol. archive of books on decorative arts, material culture and U.S. & Texas history.
Activities: guided tours; seasonal special events; evenings & lectures.
Publications: newsletter; brochure; booklet.
Hours & Admission Prices: Labor Day to Memorial Day Mon.-Fri. 11-3, tours 11, 1 & 3, Sat.-Sun. 11-3, tours on the hour; Summer: daily 11-4, tours on the hour. Adults $10, senior citizens $8, students $5; discount to AAM & TAM members; children under 6 no charge. Closed New Year's Day; Easter; Thanksgiving; Christmas. &

OCEAN STAR OFFSHORE DRILLING RIG & MUSEUM, (M),
20th St. at Harborside Dr., 2002 Wharf C, Galveston, TX 77553-2040. Mailing Address: 200 N. Dairy Ashford Rd., #4119, Houston, TX 77079-1101. Tel.: 281-679-8040; 409-766-7827. Fax: 281-544-2441.
E-mail: oec@oceanstaroec.com
Web Site: www.oceanstaroec.com
Founded: 1997.
Key Personnel: Exec. Dir., Sandra Mourton; Account Mgr., Don Staples; Dir. Operations, Lisa Lisinicchia; Museum Shop Mgr., Margi Peterson.
Personnel Profile: Full-Time Paid 6; Part-Time Paid 11; Part-Time Volunteers 4.
Governing Authority: private; nonprofit organization. Tax-exempt: 501(c)(3).
Institution Type/Description: Science & Technology Museum.
Collections: offshore energy: drilling equipment, geology, seismic, production, well servicing & completions; science technology; safety, construction & transportation relating to offshore drilling; videos; rigs; platforms; vessels; environmental exhibit.
Research Fields: pioneers & history of industry.
Facilities: aquarium; 32-seat classroom; 36-seat theater. Museum-related items for sale.
Activities: formal education programs; guided tours; school loan service; scout overnight programs; summer camps; family days.
Publications: quarterly newsletter, The Star.
Hours & Admission Prices: Call for hours. Adults $8. Closed Thanksgiving; Christmas Eve & Day. &
Attendance: 36,000 (accurate)
Membership: Individual $35; Family $50-$100; Corporate $1,000.

ROSENBERG LIBRARY, (M), 2310 Sealy Ave., Galveston, TX 77550-2296. Tel.: 409-763-8854, ext. 125. Fax: 409-763-0275. TDD: 409-763-8854.
E-mail: ebarton@rosenberg-library.org
Web Site: www.rosenberg-library-museum.org
Founded: 1904.
Congressional District: 9
Key Personnel: Pres. (V), Mrs. Jan Coggeshall; C.E.O. & Exec. Dir., John Augelli; Head Special Collections, Casey Edward Greene; Cur., Eleanor Barton; Museum Asst., Travis Bible.
Personnel Profile: Full-Time Paid 20; Part-Time Volunteers 1.
Governing Authority: nonprofit organization. Tax-exempt: 401(6)(c).
Institution Type/Description: General Museum & Library.
Collections: Galveston & Texas history; personal artifacts; maritime history; American Indian artifacts; clothing & textiles; regional artwork: Grace Spaulding John, Paul R. Schumann; Boyer Gonzales, Sr., Julius Stockfleth; Jean Scrimgeour Morgan.
Research Fields: Galveston & Texas history, archives; regional art.
Facilities: 1,864,000-vol. library of books, documents, manuscripts, maps & special collections related to history of Texas & Southwest available for use on premises; reading room.
Activities: self-guided tours; lectures; loan program; rotating exhibits.
Publications: books, Henry Rosenberg, 1824-1893; Samuel May Williams, 1795-1858 Biography & Calendar to Samuel May Williams Papers; Mier Expedition Diary; The Diary of Millie Gray, 1832-1840; Julius Stockfleth, Gulf Coast Marine & Landscape Painter; manuscript sources in the Rosenberg Library, a selective guide; Cartographic Sources from the Rosenberg Library; Through A Night of Horrors: Voices from the 1900 Storm; With Bold Strokes: Boyer Gonzales, 1864-1934.
Hours & Admission Prices: Mon.-Sat. 9-6. No charge; donations accepted. Closed national holidays. &
Attendance: 351,755
Membership: Friends of Rosenberg Library: Student $10; Individual $15; Family $25; Sustainer $50 & up; Business & Organization $100.

TEXAS SEAPORT MUSEUM, Pier 21, No. 8, Galveston, TX 77550. Tel.: 409-763-1877. Fax: 409-763-3037.
E-mail: elissa@galvestonhistory.org
Web Site: www.tsm-elissa.org
Founded: 1982.
Congressional District: 9
Key Personnel: Dir., James L. White; C.E.O. & Exec. Dir. Galveston Historical Foundation, Dwayne Jones; Chm. (V), Greg Brannon; Coord. Education, Christine Hayes; Museum Shop Mgr., Tery Alexander.
Personnel Profile: Full-Time Paid 8; Part-Time Paid 10; Part-Time Volunteers 175.
Governing Authority: private; not-for-profit organization. Parent Institution: Galveston Historical Foundation. Tax-exempt: 501(c)(3).
Institution Type/Description: Maritime Museum.
Collections: artifacts, cultural manifestations and exhibits of skills which demonstrate the nature of maritime commerce of Texas and the Gulf Coast. Historic Ship: 1877 iron barque Elissa, a National Historic Landmark.
Facilities: 8,000 sq. ft. museum space.
Activities: sail training program.
Publications: newsletter, Elissa Log.
Hours & Admission Prices: Daily 10-5. Adults $8, students & children 6-18 $6; discounts to AAM members; members, children 5 & under and GHF members no charge. Closed Thanksgiving; Christmas. &
Attendance: 55,000 (estimated)
Membership: Senior Citizen (65 & over) & Student $20; Family (1 household & its children 19 & under) $45; Sustaining $65; Patron $100; all fees are annual.

George West

GRACE ARMANTROUT MUSEUM, Hwy. 281 S., George West, TX 78022. Mailing Address: P.O. Box 248, George West, TX 78022-0248. Tel.: 361-449-3325. Fax: 361-449-3295.
E-mail: armant@the-i.net
Web Site: www.the-i.net/~armant
Key Personnel: Dir., Mary R. Johnson
Institution Type/Description: History Museum.
Collections: local history; personal artifacts; 1860-1940 furniture; period firearms; china & glassware; paintings; period farm & ranch equipment; 850 paving bricks engraved with the names of those who have lived in Live Oak County; train caboose.
Facilities: Museum-related items for sale.
Activities: community field trips; student programs; classroom trunk shows; summer activities; arts & crafts; calligraphy. Museum Sponsors: Winter Bridge Tournament; Spring Barbecue.
Hours & Admission Prices: Wed.-Sun. 1-5. No charge; donations accepted.

Georgetown

WILLIAMSON MUSEUM, 716 S. Austin Ave., Georgetown, TX 78626. Tel.: 512-943-1670. Fax: 512-943-1672.
E-mail: mross@williamsonmuseum.org
Web Site: www.williamsonmuseum.org
Founded: 1997.
Key Personnel: Dir., Mickie Ross; Pres. (V), Tommy Coleman
Institution Type/Description: History Museum.
Collections: local history & culture; period furnishings; personal artifacts; photographs; hands-on exhibitions.
Activities: educational programs. Annual Events: Up the Chisholm Trail Cattle Drive & Chuckwagon Cook-off; Pioneer Day; Archaeology Day; Chisholm Trail Days.
Hours & Admission Prices: Wed.-Fri. 12-5, Sat. 10-5. Closed major holidays.
Attendance: 12,000
Membership: Student & Teacher $10; Individual $25; Family $50; Protector $100; Preserver $250; Contributor $500; Perennial $1,000.

Gilmer

FLIGHT OF THE PHOENIX AVIATION MUSEUM, Hangar One, Gilmer, TX 75644. Mailing Address: P.O. Box 610, Gilmer, TX 75644-0610. Tel.: 903-843-2457. Fax: 903-843-3123.
Institution Type/Description: Aviation Museum.
Collections: military aircraft & equipment.
Hours & Admission Prices: Mon.-Sat. 9-4.

HISTORIC UPSHUR MUSEUM, 119 Simpson St., Gilmer, TX 75644-2231. Tel.: 903-843-5483. Fax: 903-843-5483.
Institution Type/Description: History Museum.
Collections: local history; photographs; period furniture; military uniforms; costumes; pottery.
Facilities: Museum-related items for sale.
Activities: docent programs; lectures; workshops.
Hours & Admission Prices: Tues.-Sat. 10:30-4. No charge; donations accepted.

LITERARY MUSEUM, 917 Madelaine St., Gilmer, TX 75644-3047. Tel.: 903-843-2282. Fax: 903-845-2155.
Institution Type/Description: History Museum.
Collections: memorabilia pertaining to Presidents, British Royalty, American & British authors and characters; photographs; paintings; manuscripts; busts of American & British writers; dolls.
Facilities: library.
Hours & Admission Prices: By appointment.

Glen Rose

BARNARD'S MILL & ART MUSEUM, 307 S.W. Barnard St., Glen Rose, TX 76043. Mailing Address: P.O. Box 2537, Glen Rose, TX 76043-2537. Tel.: 254-897-7494.
Web Site: www.barnardsmill.org
Founded: 1989.
Key Personnel: Dir., Richard H. Moore; Chm. (V), S.C. Coconaur; Treas., David B. Morrow; Cur., Hollis Taylor.
Personnel Profile: Part-Time Paid 2.
Governing Authority: nonprofit. Parent Institution: Somervell History Foundation. Tax-exempt: 501(c)(3).
Institution Type/Description: Art & History Museums: housed in historic Barnard's Mill built in 1860. Listed on the National Register of Historic Places.
Collections: 200 paintings, bronzes & etchings, works of artist Amy Miears Jackson, Lester Hughes, Jack Bryant, Robert Summers, Adolph Dehn, Bryon Fullerton, Morris Henry Hobbs, Frederik Taubes, Harding Black, Sherry Jo Horton, Bill Chappell, Jim Powell & Robert Wood; Marchman collection; period furnishings; local history & culture.
Facilities: library of instructional art & antique books.
Activities: guided tours; formal education program for college students affiliated with Tarleton State University.
Hours & Admission Prices: Sat. 10-5, Sun. 1-5; guided tours by appointment. No charge; donations accepted. Closed Christmas Eve & Day. &
Attendance: 3,000 (estimated)
Membership: Bronze $40; Silver $75; Gold $150; Platinum $500.

CREATION EVIDENCE MUSEUM OF TEXAS, 3102 FM 205, Glen Rose, TX 76043-0309. Mailing Address: P.O. Box 309, Glen Rose, TX 76043-0309. Tel.: 254-897-3200. Fax: 254-897-3100.
E-mail: creation@creationevidence.org
Web Site: www.creationevidence.org
Key Personnel: Founder & Dir., Carl Baugh
Institution Type/Description: Science Museum.
Collections: archaeology; geology; paleontology.
Hours & Admission Prices: Thurs.-Sat. 10-4. Admission $5; children 5 & under no charge.

DINOSAUR VALLEY STATE PARK, off FM 205 on Park Rd. 59, Glen Rose, TX 76043. Mailing Address: P.O. Box 396, Glen Rose, TX 76043-0396. Tel.: 254-897-4588; 800-792-1112.
Web Site: www.tpwd.state.tx.us
Founded: 1969.
Key Personnel: Supt., Billy P. Baker.
Personnel Profile: Full-Time Paid 6; Part-Time Paid 3; Part-Time Volunteers 3; Interns 1.
Governing Authority: state. Parent Institution: Texas Parks & Wildlife Dept., 4200 Smith School Rd., Austin. Tax-exempt.
Institution Type/Description: Paleontological Site.
Collections: dinosaur tracks from the Cretaceous period; life-size dinosaur statues; dinosaur models.
Facilities: 1,523-acre park.
Hours & Admission Prices: Daily 8am-10pm. Adults $5, Texas resident seniors 65 & over $3; children under 13 no charge. &
Attendance: 400,000 (estimated)

DINOSAUR WORLD, 1058 Park Rd. 59, Glen Rose, TX 76043. Tel.: 254-898-1526. Fax: 254-898-1782. Facebook: Dinosaur World Texas.
E-mail: texas@dinosaurworld.com
Web Site: www.dinosaurworld.com/dinosaur_world_glen_rose_texas/
Key Personnel: Owner, Christer Svensson
Institution Type/Description: Natural History Museum.
Collections: over 150 life size dinosaur; dinosaur eggs; raptor claws.
Facilities: theatre; picnic area. Museum-related items for sale.
Activities: classes; educational programs; fossil dig; outreach programs; birthday parties.
Hours & Admission Prices: Daily 9-6. Adults $12.75, seniors over 60 $10.75, children 3-12 $9.75; discounts to groups & military dependents; active military no charge. Closed Thanksgiving; Christmas.

FOSSIL RIM WILDLIFE CENTER, 2299 County Rd. 2008, Glen Rose, TX 76043. Mailing Address: 2155 County Rd. 2008, Glen Rose, TX 76043. Tel.: 254-897-2960. Fax: 254-897-3785.
Web Site: www.fossilrim.org
Founded: 1974.
Congressional District: 17

Key Personnel: C.E.O. & Exec. Dir., Dr. Patrick Condy; Chm (V), Robert Baker; Dir. Finance, Pam Adams; Dir. Mktg. & Membership, Warren Lewis.
Governing Authority: nonprofit. Parent Institution: Earth Promise dba Fossil Rim Wildlife Center. Tax-exempt: 501(c)(3).
Institution Type/Description: Wildlife Conservation Center.
Collections: ecological facts; lifestyle conservation issues; local, national and international endangered species issues; nature & wildlife awareness.
Research Fields: social behavior, metapopulation structure, genetic variability, reproductive studies, etc. of endangered species as it applies to conservation medicine & conservation issues.
Facilities: 400-vol. library of conservation & wildlife books; 100-seat auditorium; restaurant; picnic areas; classroom; 1,300 sq. ft. exhibit space; nature center; field research station; wilderness camping. Gift & nature items for sale.
Activities: guided tours; lectures; organized education programs for children, adults & undergraduate or graduate students; special overnight conservation activities for schools and scout groups; internship programs in conservation medicine, education and animal care.
Publications: scientific research papers; monthly email newsletter to members.
Hours & Admission Prices: March 9-Sept. 30 Mon.-Fri. 11-4, Sat.-Sun. 10-4:30; Oct. Mon.-Fri. 11-4, Sat.-Sun. 10-5; Nov. 1-March 8 Mon., Tues., Thurs. & Fri. 11-4, Wed. 10-4, Sat.-Sun. 10-5. Off-Season Nov.-Feb.: adults $17.95, seniors 62 & over $14.95, children 3-11 $11.95. Peak Season March-Oct. weekdays: adults $20.95, seniors 62 & over $17.95, children 3-11 $14.95; March-Oct. weekends: adults $23.95, seniors 62 & over $20.95, children 3-11 $17.95. Discounts Tues. Wed. & Thurs. during the Peak Season and to groups; members and children 2 & under no charge. Closed Thanksgiving; Christmas Eve & Day. &
Attendance: 135,000 (accurate)

SOMERVELL COUNTY HISTORICAL MUSEUM, Elm & Vernon Sts., Glen Rose, TX 76043. Mailing Address: Box 669, Glen Rose, TX 76043-0669. Tel.: 254-898-0640.
Founded: 1966.
Congressional District: 17
Key Personnel: C.E.O., Mary Lee Lilly; Chm. (V) & Museum Shop Mgr., Barbara Sloan.
Personnel Profile: Full-Time Volunteers 2; Part-Time Volunteers 2.
Volunteer Hours: 1,040
Governing Authority: nonprofit organization. Parent Institution: Somervell Historical Society. Tax-exempt: 501(c)(3).
Institution Type/Description: Historical Society Museum.
Collections: agriculture; archaeology; archives; geology; Indian artifacts; dinosaur foot tracks; fossils displays from local River Paluxy Whiskey Still; early settlers artifacts.
Facilities: Museum-related items for sale.
Activities: permanent exhibitions.
Publications: city & county map.
Hours & Admission Prices: Wed.-Sat. 11-4. No charge; donations accepted. Closed Thanksgiving; Christmas. &
Attendance: 5,000 (estimated)
Membership: Individual $10; Contributing $15; Sustaining $50; Life $100.

Goliad

GOLIAD STATE PARK, 108 Park Rd. 6, Goliad, TX 77963-3206. Tel.: 361-645-3405. Fax: 361-645-8538.
E-mail: brenda.justice@tpwd.state.tx.us
Web Site: www.tpwd.state.tx.us
Founded: 1931.
Congressional District: 15
Key Personnel: Exec. Dir., Carter Smith; Supt., Brenda Justice; Chm. Commission, T. Dan Friedkin; Dir. Parks, Walt Dabney; Education Devel., Beth Ellis.
Personnel Profile: Full-Time Paid 11; Part-Time Paid 3; Part-Time Volunteers 4.
Governing Authority: state. Parent Institution: Texas Parks & Wildlife Dept. 4200 Smith School Rd., Austin, TX 78744. Tel. 512-479-4800. Tax-exempt.
Institution Type/Description: Historic Sites: 1749-1830 sites of Spanish mission Espiritu Santo de Zuniga.
Collections: archaeological artifacts; Spanish Colonial & early Texas items & memorabilia. Historic Sites: ruins of mid to late 18th-century Mission Nuestra Senora del Rosario; 1829 birthplace of Gen. Ignacio Zaragoza.
Research Fields: Spanish Colonial history.
Facilities: history & nature trails; picnic area; camping facilities.
Activities: guided tours; formally organized education programs for children; permanent exhibitions; demonstrations; crafts workshops; living history programs. Museum Sponsors: two spring concerts; Spanish Tracks & Trails; Christmas concert.

Publications: various natural & cultural resource topics.
Hours & Admission Prices: Daily 8-5. Adults $3. Closed Christmas. &
Attendance: 150,000 (estimated)
Membership: One Cardholder $60; Two Cardholders $75.

PRESIDIO LA BAHIA, Refugio Hwy., 1 mile south of Goliad on Hwy. 183, Goliad, TX 77963. Mailing Address: P.O. Box 57, Goliad, TX 77963-0057. Tel.: 512-645-3752. Fax: 512-645-1706.
E-mail: presidiolabahia@goliad.net
Web Site: www.presidiolabahia.org
Founded: 1966.
Key Personnel: Dir., Newton M. Warzecha.
Personnel Profile: Full-Time Paid 1; Part-Time Paid 2; Part-Time Volunteers 25.
Governing Authority: church. Parent Institution: Diocese of Victoria. Tax-exempt.
Institution Type/Description: Historic Site Museum: Spanish Fort on lower San Antonio River, part of the 1772 Presidio Line, scene of Goliad Massacre.
Collections: artifacts showing 9 levels of civilization; artifacts of habitation, including pre-colonial, Spanish, Mexican & Texan era.
Research Fields: pre-Columbian, Spanish, colonial & Mexican cultural materials; military specimens; Spanish, Colonial architecture, archaeology & documents.
Facilities: library; archives.
Activities: tours; living history programs.
Publications: book, Presidio La Bahia; video, Presidio La Bahia and It's Place in the History of Texas.
Hours & Admission Prices: Daily 9-5. Adults $4, senior citizens $3.50, children 5-11 $1; members no charge. Closed New Year's Day; Easter; Thanksgiving; Christmas. &
Attendance: 27,000 (accurate)
Membership: Individual $25; Family $50; Supporting $150; Caballero $250; Empresario $500; Alcalde $1,000.

Gonzales

GONZALES MEMORIAL MUSEUM, 414 Smith, Gonzales, TX 78629. Tel.: 830-672-6350. Fax: 830-672-2813.
E-mail: curator@cityofgonzales.org
Web Site: www.cityofgonzales.org
Founded: 1936.
Congressional District: 14
Key Personnel: Chm. (V), Kay Bakken; Dir. Economic Devel., Carolyn Gibson.
Personnel Profile: Full-Time Paid 1; Part-Time Volunteers 10.
Governing Authority: city. Tax-exempt.
Institution Type/Description: History Museum.
Collections: original documents & relics dating to 1660 dealing with Texas, Mexico & the Great Southwest; manuscripts; cannon used in the first battle in Texas' fight for independence from Mexico; George Washington Davis' journal.
Activities: guided tours; inter-museum loan & permanent exhibitions.
Hours & Admission Prices: Tues.-Sat. 10-5, Sun. 1-5. No charge; donations accepted. Closed Christmas. &
Attendance: 7,000 (accurate)

Graham

OLD POST OFFICE MUSEUM & ART CENTER, 510 Third St., Graham, TX 76450. Mailing Address: P.O. Box 1111, Graham, TX 76450. Tel.: 940-549-1470. Fax: 940-549-1478.
E-mail: info@opomac.com
Web Site: www.opomac.net
Key Personnel: Exec. Dir., Marlene Edwards
Institution Type/Description: Historic Building & Art Center: housed in a former U.S. Post Office; built in 1937. Listed on the National Register of Historic Places.
Collections: local history & artifacts; period furnishings; Native American artifacts; ranching history; Texas Cattle Raisers Association history; photographs; military artifacts; paintings; 1939 mural.
Facilities: Museum-related items for sale.
Activities: rental facilities; special events.
Hours & Admission Prices: Tues.-Fri. 10-4. No charge; donations accepted.

ROBERT E. RICHESON MEMORIAL MUSEUM, Graham Municipal Airport, CAF Cactus Squadron Hangar, U.S. 380, Graham, TX 76450. Tel.: 940-549-2790.
Institution Type/Description: History Museum.

Collections: local history; personal artifacts; uniforms; model aircraft; military artifacts.
Hours & Admission Prices: Thurs. 1-5.

Grand Saline

SALT PALACE MUSEUM, 100 W. Garland Ave., Grand Saline, TX 75140. Tel.: 903-962-5631.
Institution Type/Description: History Museum.
Collections: salt industry history; salt mining memorabilia & photographs; personal artifacts; period tools & equipment.
Facilities: Museum-related items for sale.
Activities: mining video.
Hours & Admission Prices: Tues.-Sat. 8:30-5. No charge; donations accepted.

Grapevine

GRAPEVINE HISTORICAL MUSEUM, (M), 705 S. Main St., Grapevine, TX 76051-5351. Mailing Address: Grapevine Historical Society, P.O. Box 995, Grapevine, TX 76099-0995. Tel.: 817-410-8145.
Founded: 1973.
Key Personnel: Pres. (V), Joann Standlee; Cur., Paul Ernst.
Personnel Profile: Part-Time Paid 1; Part-Time Volunteers 18.
Operating Expenses: 19,706
Operating Income: 24,138
Governing Authority: Parent Institution: Grapevine Historical Society. Tax-exempt.
Institution Type/Description: History Museum.
Collections: local history & culture; photographs; personal artifacts; period furnishings; archaeological artifacts.
Hours & Admission Prices: Mon.-Sat. 10-5, Sun. 1-5. No charge; donations accepted.
Attendance: 28,240 (estimated)
Membership: Individual $15; Family $25; Business/Organization $30; Historian $100; Preservation Leader $1,000.

Greenville

AUDIE MURPHY/AMERICAN COTTON MUSEUM, INC, (M), 600 I-30, E., Greenville, TX 75401. Mailing Address: P.O. Box 347, Greenville, TX 75403-0347. Tel.: 903-450-4502. Fax: 903-454-1990.
E-mail: amacm@att.net
Web Site: www.cottonmuseum.com
Founded: 1987.
Congressional District: 1
Key Personnel: Exec. Dir., Susan Lanning; Pres. (V), John Hanners; Asst. Dir., Linda Owens.
Personnel Profile: Full-Time Paid 2; Part-Time Paid 1; Part-Time Volunteers 32.
Operating Expenses: 181,916
Operating Income: 183,448
Governing Authority: board of trustees. Tax-exempt.
Institution Type/Description: Cotton & History Museum.
Collections: regional history; National cotton production; memorabilia from Audie Murphy's military service & movie career; military history; 10 ft. bronze statue of Audie Murphy.
Research Fields: Cotton industry & history; Greenville & Hunt county; military history.
Facilities: library; archives.
Activities: guided tours; monthly speaker; educational programs; family fun day programs; summer camp. Annual Events: Cotton History Conference; Audie Murphy Days; Easter After Dark; Military Gun Show.
Publications: brochure; monthly newsletter, The Compress; annual report.
Hours & Admission Prices: Tues.-Sat. 10-5. Adults $6, senior citizens, veterans, & college students $4, youth 6-18 $2; discounts to AAM & ICOM members; active military & members no charge. Closed major holidays. &
Attendance: 8,493 (accurate)
Membership: Senior Citizen $20; Individual $30; Family $70; Contributing $100; Sustaining $250; Sponsor $500; Benefactor $1,000; Lifetime $2,500.

Harlingen

HARLINGEN ARTS AND HERITAGE MUSEUM, 2425 Boxwood St., Harlingen, TX 78550. Tel.: 956-216-4901. Fax: 956-430-8502.
E-mail: rgvmuse@hiline.net
Web Site: hiline.net/rgvmuse

Founded: 1967.
Congressional District: 15
Key Personnel: Dir., Linn Keller.
Personnel Profile: Full-Time Paid 2; Part-Time Paid 5; Part-Time Volunteers 2.
Governing Authority: municipal; nonprofit. Tax-exempt.
Institution Type/Description: History Museum.
Collections: Valley history including Republic of Texas; natural history; period clothing, furniture & household objects; period archives of Civil War & other conflicts. Historic Buildings: c.1850 Paso Real Stage Coach Inn, c.1923 Harlingen Hospital, home of city founder; Historical Museum; Lon C. Hill Home.
Research Fields: historical events of Rio Grande Valley.
Facilities: auditorium & video facilities.
Activities: guided tours; lectures; permanent & temporary exhibitions; summer art programs; classes.
Hours & Admission Prices: Tues.-Sat. 10-4, Sun. 1-4. Adults $2, senior citizens & children $1. Closed Easter; Independence Day; Labor Day; Thanksgiving; Christmas. &
Attendance: 15,500 (estimated)
Membership: Personal $15; Family $25; Patron $50; Benefactor $100.

IWO JIMA MEMORIAL MUSEUM, 320 Iwo Jima Blvd., Harlingen, TX 78550. Tel.: 956-412-2207.
Institution Type/Description: Military History Museum.
Collections: World War II history; military artifacts & equipment; photographs; sculpture; personal artifacts.
Activities: video.
Hours & Admission Prices: Mon.-Sat. 10-4, Sun. 12-4.

Henderson

THE DEPOT MUSEUM COMPLEX, 514 N. High St., Henderson, TX 75652-5912. Tel.: 903-657-4303. Fax: 903-657-2679. Facebook: Depot Museum.
E-mail: depot@depotmuseum.com
Web Site: www.depotmuseum.com
Founded: 1979.
Congressional District: 1
Key Personnel: Dir., Vickie Armstrong; Asst. Dir., Jim White.
Personnel Profile: Full-Time Paid 2; Part-Time Paid 2; Part-Time Volunteers 35.
Volunteer Hours: 5,694
Governing Authority: county; nonprofit. Parent Institution: Rusk County. Subsidiary Institution: Rusk County Historical Commission. Tax-exempt: 501(c)(3).
Institution Type/Description: History Museum.
Collections: costumes; tools; photographs; rural southern; communication artifacts; local, state, pre-historic & modern artifacts. Historic Structures: 1901 Depot; 1841 log cabin; 1908 outhouse; c.1880 doctor's office; 1884 dog trot home; syrup mill; printing shop; saw mill; steam drilling rig; country store; oil derrick & pumping jack; c.1920 cotton gin.
Research Fields: 1830-1970 Texas life; culture, lifestyle, social institution & folk arts of Rusk County.
Facilities: archives; 7,900 sq. ft. exhibit space; 45-seat theater. Genealogical books & other museum-related items for sale.
Activities: guided tours; organized educational programs; participatory exhibits. Museum Sponsors: Heritage Folk Art Day for children; Grandparent's Day in July; Heritage Syrup Festival in November.
Publications: monthly newsletter, The Telegram.
Hours & Admission Prices: Mon.-Fri. 9-5, Sat. 9-1. Adults $3, seniors 55 & up $2, children $1; discounts to AAM, ICOM & Texas Museum Association members, bus tours & groups; members no charge. Closed state & federal holidays. &
Attendance: 26,413 (accurate)
Membership: Individual $10; Family $15; Club $25; Friend $50; Patron $100; Benefactor $300; Life $1,000.

HOWARD-DICKINSON HOUSE MUSEUM, 501 S. Main St., Henderson, TX 75654-3544. Mailing Address: P.O. Box 2434, Henderson, TX 75653-2434. Tel.: 903-657-7405 & 5256. Fax: 903-657-9283.
Founded: 1964.
Congressional District: 9
Key Personnel: Pres., Art Rousseau; Treas., Louise Slover.
Personnel Profile: Part-Time Volunteers 6.
Governing Authority: nonprofit. Rusk County Heritage Association. Tax-exempt: 501(c)(3).
Institution Type/Description: Historic House Museum: 1855 restored Howard-Dickinson House with 1905 frame-wing authentically restored & furnished.

Collections: furnishings; historical books; papers; paintings; manuscripts; clothing; linens; lace; dolls; china; glass; silver.
Research Fields: local history; genealogy.
Facilities: 240-vol. library of medical books, genealogy, history, cookbooks, fiction, poetry & biography available for use on premises; reading room.
Activities: guided group & school tours by appointment; facilities available for rent.
Publications: brochures; Howard-Dickinson House Cook Book.
Hours & Admission Prices: By appointment only. Adults $10, children $1. Closed major holidays. &
Attendance: 1,479 (estimated)

Henrietta

CLAY COUNTY 1890 JAIL MUSEUM, 116 N. Graham St., Henrietta, TX 76365. Mailing Address: P.O. Box 483, Henrietta, TX 76365. Tel.: 940-538-5655.
E-mail: 1890jailmuseum@gmail.com
Web Site: claycountyjailmuseum.org
Founded: 1997.
Congressional District: 13
Key Personnel: Pres. (V), Travis Childs; Museum Shop Mgr., Lucille Glasgow.
Personnel Profile: Full-Time Volunteers 12; Part-Time Volunteers 50.
Volunteer Hours: 4,855
Operating Expenses: 12,709
Operating Income: 17,224
Governing Authority: Parent Institution: Clay County Historical Society, Inc. Tax-exempt.
Institution Type/Description: History Museum.
Collections: local history; photographs; period furnishings; personal artifacts; early clothing & medical equipment; military memorabilia.
Research Fields: family & community history.
Facilities: archives.
Activities: temporary exhibitions; guided tours; fundraising events.
Publications: annual letter.
Hours & Admission Prices: Thurs.-Fri. 10-2, Sat. 1-4. No charge; donations accepted. Closed major holidays.
Attendance: 600 (estimated)
Membership: Student $5; Individual $10; Couple $20; Life $200; Couple Life $300.

Hereford

DEAF SMITH COUNTY MUSEUM, 400 Sampson, Hereford, TX 79045. Mailing Address: P.O. Box 1007, Hereford, TX 79045-1007. Tel.: 806-363-7070.
E-mail: deafsmithmuseum@wtrt.net
Web Site: www.deafsmithcountymuseum.org
Founded: 1966.
Congressional District: 19
Key Personnel: Exec. Dir., Paula Edwards.
Personnel Profile: Full-Time Paid 2; Part-Time Volunteers 25.
Governing Authority: county. Subsidiary Institution: Victorian Home, 508 W. Third, Hereford, TX. Tax-exempt.
Institution Type/Description: History Museum.
Collections: 1905 household items; ranching gear; 1898 dug-out; Santa Fe caboose; early farm machinery; jail cells from first county seat, La Plata; windmill; general store; fashions; textiles. Historic Structures: 1927 Catholic schoolhouse, former site of St. Anthony's Catholic school; 1909 Victorian Home, listed on the National Register of Historic Places; chapel.
Facilities: 100-vol. library of school books of late 1800s; novels; magazines; 1900s music study magazines & song books available for use by special arrangement.
Activities: guided tours; lectures; arts festivals; docent program or council; permanent & temporary exhibitions.
Publications: newsletter, The Deaf Smithsonian.
Hours & Admission Prices: Mon.-Fri. 10-12 & 1-5, Sat. 10-12 & 1-3; Sun. by appointment only. No charge; donations accepted. Closed major holidays.
Attendance: 3,750 (estimated)
Membership: Deaf Smith County Historical Society: Adult $15; Business $25; Life $250.

Hidalgo

OLD HIDALGO PUMPHOUSE, 902 S. Second St., Hidalgo, TX 78557-2703. Tel.: 956-843-8686. Fax: 956-843-6519.
E-mail: iduran08@sbcglobal.net
Web Site: www.theworldbirdingcenter.com/Hidalgo.html
Formerly: Hidalgo Pumphouse Heritage and Discovery Park
Founded: 1999.

Congressional District: 25
Key Personnel: Dir., Irma D. Duran; Administrative Asst., Viola Arismendez.
Governing Authority: municipal; nonprofit. Subsidiary Institution: World
　Birding Center. Tax-exempt.
Institution Type/Description: Agriculture Museum: housed in a restored
　pumphouse which operated from 1909 to 1983 pumping water from the Rio
　Grande for agricultural irrigation.
Collections: two 1911-1912 Worthington 60-inch pumps driven by two double
　action, double expansion Hamilton-Corliss steam engines; 1948 Worthing-
　ton diesel engine & pump; 1952 Ingersoll-Rand diesel engine & pump.
Research Fields: history of steam power in the industrial revolution & the
　machine age.
Facilities: 7,500 sq. ft. exhibit space; nature center. Museum-related items for
　sale.
Activities: guided group tours; participatory exhibits.
Hours & Admission Prices: Mon.-Fri. 10-5, Sun. 1-5. Adults $4, senior citizens
　$3, children & students $2. Closed national holidays. &

Attendance: 3,000 (estimated)

Hillsboro

HILL COUNTY CELL BLOCK MUSEUM, 200 N. Waco St.,
　Hillsboro, TX 76645-2140. Mailing Address: P.O. Box 555, Hills-
　boro, TX 76645-0555. Tel.: 254-582-8912.
Founded: 1998.
Key Personnel: Chm. (V), Joyce Ho Lungsworth; Pres. (V), Milton Peterson.
Personnel Profile: Part-Time Volunteers 35.
Volunteer Hours: 500
Governing Authority: nonprofit organization. Tax-exempt.
Institution Type/Description: History Museum: housed in the former county
　jail & sheriff's family home; built in 1893. Listed on the National Register
　of Historic Places.
Collections: local history & culture; period furnishings; personal artifacts;
　photographs.
Hours & Admission Prices: April-Oct. Sat. 10-4; other times by appointment.
　No charge; donations accepted.
Attendance: 300 (estimated)
Membership: General $15; Club $20; Business $30.

TEXAS HERITAGE MUSEUM, (M), 112 Lamar Dr., Hillsboro, TX
　76645-2711. Tel.: 254-659-7750. Fax: 254-580-9529.
E-mail: jversluis@hillcollege.edu
Web Site: www.hillcollege.edu
Founded: 1963.
Congressional District: 6
Key Personnel: Dir., John Versluis.
Personnel Profile: Full-Time Paid 2; Part-Time Paid 2.
Governing Authority: college. Parent Institution: Hill College. Subsidiary
　Institution: History Center. Tax-exempt.
Institution Type/Description: Military History Museum.
Collections: Hood's Texas Brigade; Civil War & WWI & WWII military guns;
　artifacts; manuscripts; Audie Murphy memorabilia.
Research Fields: Texas military history; weaponry history; World War II
　history; military history of Texas; War between the States.
Facilities: 7,000-vol. library of books on military history of the United States
　and of Texas.
Activities: self-guided tours of exhibit; research facilities for Civil War soldiers
　& units. Museum Sponsors: Annual History Symposium.
Publications: scholarly monographs.
Hours & Admission Prices: Memorial Day to Labor Day Mon.-Thurs. 8-4:30,
　Fri. 8-4, Sat. 10-5; Sept.-May Mon.-Thurs. 8-4:30, Fri. 8-4. No charge;
　donations accepted. Closed national holidays. &
Attendance: 6,800 (accurate)
Membership: Individual $25; Family $35; Business $50; Patron $100; Bene-
　factor $250.

Hondo

MEDINA COUNTY MUSEUM, 2202 18th St., Hondo, TX 78861.
　Mailing Address: P.O. Box 98, Hondo, TX 78861. Tel.: 830-741-
　2105.
Institution Type/Description: History Museum.
Collections: local history & culture; period furnishings; personal artifacts;
　photographs; school artifacts; steam engine.
Hours & Admission Prices: Thurs.-Sat. 10-4, Sun. 1-4.

Houston

ART CAR MUSEUM, 140 Heights Blvd., Houston, TX 77007. Tel.:
　713-861-5526.
E-mail: info@artcarmuseum.com
Web Site: www.artcarmuseum.com
Founded: 1998.
Personnel Profile: Full-Time Paid 4; Part-Time Paid 1.
Institution Type/Description: Art Museum.
Collections: works by contemporary artists; paintings; sculpture; art cars.
Hours & Admission Prices: Wed.-Sun. 11-6. No charge.
Attendance: 14,500 (estimated)

ART LEAGUE OF HOUSTON, 1953 Montrose Blvd., Houston, TX
　77006-1243. Tel.: 713-523-9530. Fax: 713-523-4053.
E-mail: alh@artleaguehouston.org
Web Site: www.artleaguehouston.org
Founded: 1948.
Congressional District: 18
Key Personnel: Pres., Cara Pauloski Rudelson.
Personnel Profile: Full-Time Paid 3; Part-Time Paid 1; Part-Time Volunteers
　100; Interns 2.
Governing Authority: nonprofit. Tax-exempt: 501(c)(3).
Institution Type/Description: Art Gallery.
Collections: various forms of art media.
Research Fields: Art Registry of Texas; Art League's Visual Arts.
Facilities: classrooms; studios.
Activities: lectures; workshops; formally organized educational programs;
　traveling exhibitions; temporary juried exhibits; selection of Texas artist of
　the year.
Publications: bimonthly newsletter; monthly E-Newsletter.
Hours & Admission Prices: Mon.-Fri. 9-5, Sat. 11-5. No charge, donations
　accepted. Closed national holidays. &
Attendance: 10,000 (estimated)
Membership: Artist $25; Individual $35; Family $60; Patron $100-$499;
　Benefactor $500-$999.

BAYOU BEND COLLECTION AND GARDENS, 6003 Memorial
　Dr., Houston, TX 77007. Mailing Address: P.O. Box 6826, Hous-
　ton, TX 77265-6826. Tel.: 713-639-7750. Fax: 713-639-7770.
E-mail: bayoubend@mfah.org
Web Site: www.mfah.org/bayoubend
Founded: 1957.
Congressional District: 7
Key Personnel: Dir., Bonnie Campbell; Chm. (V), Bobbie Nau; Cur. Collec-
　tions, Michael K. Brown; Dir. Education, Jennifer Hammond; Conservator,
　Steven Pine; Cur. Gardens, Bart Brechter; Museum Shop Mgr., Lisa Sugita.
Personnel Profile: Full-Time Paid 32; Part-Time Paid 8; Part-Time Volunteers
　291; Interns 1.
Governing Authority: nonprofit organization. Parent Institution: Museum of
　Fine Arts, Houston. Tax-exempt: 501(c)(3).
Institution Type/Description: Historic House & Site.
Collections: c.1620-1876 American decorative arts; American paintings;
　c.1700-1840 English ceramics; gardens; historic house.
Research Fields: decorative arts; paintings; American history & culture.
Facilities: library; 14 acres of gardens; visitor & education center. Museum-
　related items for sale.
Activities: guided & self-guided tours; audio tours of house & gardens;
　lectures; family & adult programs; formally organized education programs
　for children, undergraduate college students, schools & teachers; docent
　programs; permanent exhibitions; videos.
Publications: books, America's Treasures at Bayou Bend: Celebrating Fifty
　Years; Bayou Bend Gardens; Bayou Bend magazine, Inside Out; biennial
　symposium proceedings; exhibition catalogues; collection brochures.
Hours & Admission Prices: House: Tues.-Thurs. 10-11:30 & 1-2:45 (60 min.
　docent-guided tours), Fri.-Sat. 10-11:30am (60 min. docent-guided tours),
　Sun. 1-4 (60 min. self-guided audio tour). Self-Guided Garden Audio Tours:
　Tues.-Sat. 10-5, Sun. 1-5. Family Day: Sept.-May third Sun. of month 1-5.
　No charge. Gardens: adults $15, $12.50, $5; discounts to seniors, students,
　youth, and AAA & MFAH members; children 9 & under no charge. &
Attendance: 80,000 (estimated)
Membership: Friend $200; Patron $500; Founder $1,500; Fellow $2,500;
　Gallery $5,000; Director's Circle $10,000; Chairman's Circle $25,000.

**BLAFFER ART MUSEUM, THE ART MUSEUM OF THE
　UNIVERSITY OF HOUSTON, (M),** 120 Fine Arts Bldg., Hous-
　ton, TX 77204. Tel.: 713-743-9521. Fax: 713-743-9525.
Web Site: www.blafferartmuseum.org
Founded: 1973.

Congressional District: 18
Key Personnel: Dir., Claudia Schmuckli; Deputy Dir., James Rosengren; Chm., Ryan Gordon; Devel., Emily Church; Public Rels., Watt Johns; Registrar, Young Min Chung; Admin., Karen Zicterman; Cur. Education, Katherine Veneman; Cur. University Collections, Michael Guidry.
Personnel Profile: Full-Time Paid 10; Part-Time Paid 30; Interns 4.
Governing Authority: public university; nonprofit. Parent Institution: University of Houston. Tax-exempt: 501(c)(3).
Institution Type/Description: Art Gallery.
Collections: works by contemporary artists.
Facilities: 7,500 sq. ft. exhibit space.
Activities: docent program; formal educational programs; guided tours; lectures; traveling exhibitions; tours in foreign languages; young artist apprenticeship workshops & program; brown bag gallery tours; contemporary salon; summer arts program.
Publications: triannual newsletter, Blaffer; exhibition catalogs; education kits.
Hours & Admission Prices: Tues.-Sat. 10-5. No charge; donations accepted. Closed university holidays. &
Attendance: 45,000 (accurate)
Membership: Community Partner $35; Supporting Partner $100; Leading Partner $250; Founding Partner $500; Visionary Partner $1,000; Corporate or Director's Partners $2,500; Corporate or Director's Circle Partner $5,000.

BUFFALO SOLDIERS NATIONAL MUSEUM, 3816 Caroline St., Houston, TX 77004-5947. Tel.: 713-942-8920. Fax: 713-942-8912.
E-mail: matthews@buffalosoldiermuseum.com
Web Site: www.buffalosoldiermuseum.com
Institution Type/Description: Military Museum.
Collections: American military history from 1770 to 2000; photographs.
Activities: outreach programs; lectures; youth drill team; summer high school ROTC internship program.
Hours & Admission Prices: Mon.-Fri. 10-5, Sat. 10-4. Adults $10, students & seniors $5; children 5 & under no charge.

THE CHILDREN'S MUSEUM OF HOUSTON, (M), 1500 Binz, Houston, TX 77004-7112. Tel.: 713-535-7200. Fax: 713-522-5747.
E-mail: info@cmhouston.org
Web Site: www.cmhouston.org
Founded: 1980.
Congressional District: 18
Key Personnel: Exec. Dir., Tammie Kahn; Pres. Bd., Lance Gilliam; Controller, Mallory Rissmiller; Dir. Finance, Richard Daigneault; Assoc. Dir. Devel., Julie Oates; Dir. Education, Cheryl McCallum; Dir. Educational Technology & Devel., Keith Ostfeld; Receptionist, Yadira Juarez; Museum Admin., Merri Stinnett; Volunteer & Special Events Mgr., Krystal Harris; Gallery Mgr., Elaine Holthe; Dir. Facilities, Pete Lancaster; Dir. Public Rels., Henry Yau; Dir. Business Devel., Alexandra Vasquez; Box Office Supvr., Elizabeth Graven; Retail Operations Mgr., Connie Schnupp; Traveling Exhibit Coord., Sharon Smallwood; Public Rels. Assoc., Melissa Denman; Arts Educator, Bunmi Gaidi; Reservations Coord., Lydia Dungus.
Personnel Profile: Full-Time Paid 64; Part-Time Paid 120; Interns 3.
Governing Authority: nonprofit organization. Tax-exempt: 501(c)(3).
Institution Type/Description: Children's Museum.
Collections: participatory exhibits & programs in art, science, cultures, technology & environment.
Research Fields: evaluations.
Facilities: Museum-related items for sale.
Activities: guided tours; rotating exhibits; special events; kids committee; junior volunteer program; museum guild; birthday parties; overnights; performances; demonstrations; teacher & parent workshops and lectures; outreach programs; parent resource library.
Publications: newsletter, Calendar of Events; newsletter, In-Touch; exhibit catalogues; hands-on activity guides.
Hours & Admission Prices: Memorial Day to Labor Day Mon.-Wed. & Fri.-Sat. 10-6, Thurs. 10-8, Sun. 12-6; Sept.-May Tues.-Wed. & Fri.-Sat. 10-6, Thurs. 10-8, Sun. 12-6. Adults $9, senior citizens 65 & over and military $8; children under one, ATSC & museum members and Thurs. nights no charge. Closed New Year's Day; Easter; Thanksgiving; Christmas. &
Attendance: 800,000 (accurate)
Membership: Family I $75; Family Plus $95; Explorer $150; Discoverer $500; Inventor $1,000.

✱   **CONTEMPORARY ARTS MUSEUM HOUSTON, (M),** 5216 Montrose Blvd., Houston, TX 77006-6547. Tel.: 713-284-8250. Fax: 713-284-8275.
E-mail: info@camh.org

Web Site: www.camh.org
Founded: 1948.
Congressional District: 7
Key Personnel: Dir., Bill Arning; Chair, Sissy Kempner; Pres., William J. Goldberg; Dir. Community Engagement, Connie McAllister; Sr. Cur., Valerie Cassel Oliver; Cur., Dean Daderko; Deputy Dir. Devel. & Administration, Amber Winsor; Deputy Dir. Facilities & Risk Management, Michael Reed; Retail Operations Dir., Sue Pruden; Registrar, Tim Barkley.
Personnel Profile: Full-Time Paid 19; Part-Time Paid 47; Interns 3.
Governing Authority: nonprofit organization. Tax-exempt: 501(c)(3).
Institution Type/Description: Art Museum.
Collections: contemporary works of art.
Major Exhibits: Outside the Lines, 10/31/13-3/23/14.
Research Fields: contemporary art.
Facilities: education resource center. Museum-related items for sale.
Activities: lectures; gallery talks; school programs; audio tours; family activities; children's workshops; tours with information guides; temporary exhibitions.
Publications: exhibition catalogs; Gallery Notes; Young People's Guides.
Hours & Admission Prices: Tues.-Wed. & Fri. 10-7, Thurs. 10-9, Sat. 10-6, Sun. 12-6. No charge; donations accepted. Closed New Year's Day; Thanksgiving; Christmas. &
Attendance: 66,586 (accurate)
Membership: Artist, Student & Senior Citizen $35; Individual $50; Household $100; Partner $150; Sponsor $250; Advocate $500; Fellow $1,000.

CY TWOMBLY GALLERY, 1501 Branard St., Houston, TX 77006. Mailing Address: 1511 Branard, Houston, TX 77006-4721. Tel.: 713-525-9400. Fax: 713-525-9444.
E-mail: info@menil.org
Web Site: www.menil.org
Founded: 1995.
Key Personnel: Dir., Josef Helfenstein.
Governing Authority: Tax-exempt.
Institution Type/Description: Art Gallery.
Collections: Cy Twombly's contemporary abstract expressionism art work.
Hours & Admission Prices: Wed.-Sun. 11-7. No charge. &
Attendance: 30,000 (accurate)

CZECH CENTER MUSEUM HOUSTON, 4920 San Jacinto St., Houston, TX 77004. Tel.: 713-528-2060. Fax: 713-528-2017.
E-mail: czech@czechcenter.org
Web Site: www.czechcenter.org
Institution Type/Description: Cultural History Museum.
Collections: Czech culture & history; paintings; sculpture; personal artifacts; photographs; porcelain.
Facilities: library.
Activities: rental facilities.
Hours & Admission Prices: By appointment.

DIVERSEWORKS ART SPACE, 4102 Fannin St., Ste. 200, Houston, TX 77004. Tel.: 713-223-8346.
E-mail: info@diverseworks.org
Governing Authority: nonprofit organization.
Institution Type/Description: Art Gallery.
Collections: works by contemporary artists.
Activities: educational programs.
Hours & Admission Prices: Wed. 12-8, Thurs.-Sat. 12-6. No charge.

DOWNTOWN AQUARIUM, 410 Bagby St., Houston, TX 77002. Tel.: 713-223-3474.
Institution Type/Description: Aquarium.
Collections: over 200 species of aquatic life from around the world.
Facilities: restaurant. Aquarium-related items for sale.
Activities: hands-on exhibits.
Hours & Admission Prices: Sun.-Thurs. 10-9, Fri.-Sat. 10-11. Adults $9.25, seniors 65 & over $8.25, children 2-12 $6.25; children under 2 no charge.

THE HERITAGE SOCIETY, 1100 Bagby St., Houston, TX 77002-2504. Tel.: 713-655-1912, ext. 114. Fax: 713-655-9249.
E-mail: info@heritagesociety.org
Web Site: www.heritagesociety.org
Founded: 1954.
Congressional District: 18
Key Personnel: C.E.O. & Exec. Dir., Alice Collette; Pres., Evelyn Boatwright; Dir. Communications, Debbie Duty; Dir. Devel., Carol Watson; Dir.

Finance, Emison Lewis; Dir. Collections, Kimberly Wolfe; Cur. Collections, Wallace Saage; Registrar, Ginger Berni; Education/Program Coord., Elizabeth Martin; Devel. Asst., Membership, Jenifer Jackson.
Personnel Profile: Full-Time Paid 7; Part-Time Volunteers 805; Interns 8.
Governing Authority: nonprofit organization; bd. of directors. Tax-exempt: 501(c)(3).
Institution Type/Description: History Museum & Historic House Site.
Collections: Historic Structures: 1823 Old Place cabin; 1847 Kellum Noble house; 1850 Nichols-Rice Cherry house; 1868 Pillot house; 1868 San Felipe German Cottage; 1870 Jack Yates early African-American home; 1891 St. John rural church; 1905 Staiti house; 1860 Fourth Ward Cottage; 1893 Baker Playhouse.
Major Exhibits: Greatest Show on Earth, 11/13-2/14.
Research Fields: Houston history; decorative arts.
Facilities: library; archives; tea room. Museum-related items for sale.
Activities: guided tours; slide presentations; educational trips, lectures & programs; outreach programs; museum exhibits. Museum Sponsors: Annual Candlelight Tour.
Publications: quarterly, Panorama.
Hours & Admission Prices: Tues.-Sat. 10-4. Historical House Tour: Tues.-Sat. 10, 11:30, 1 & 2:30. Adults $15, senior citizens 65 & up $12, children 6-18 $6; discounts to members, Time Travelers, AAM, ICOM, AAA & TAM members; children 5 & under and members no charge. Closed major holidays. &
Attendance: 204,423 (accurate)
Membership: Individual $50; Family $60; Sustaining $100; Patron $250; Sponsor $500; Lifetime $750; Corporate $1,250.

✱ HOLOCAUST MUSEUM HOUSTON, (M), 5401 Caroline St., Houston, TX 77004-6804. Tel.: 713-942-8000, ext. 100. Fax: 713-942-7953.
E-mail: info@hmh.org
Web Site: www.hmh.org
Founded: 1996.
Congressional District: 18
Key Personnel: C.E.O., Kelly J. Zuniga; Chm. (V), Mark Mucasey; Cur. Permanent Exhibit, Carol Manley; Devel., Marci Dallas; Education, Mary Lee Webeck; Public Rels., Ira D. Perry; Security, Roger Henderson; Museum Shop Mgr., Pam Hamilton.
Personnel Profile: Full-Time Paid 21; Part-Time Volunteers 225; Interns 5.
Governing Authority: private; nonprofit organization. Tax-exempt.
Institution Type/Description: History Museum.
Collections: books; diaries; photographs; film reels; maps; documents; personal items; prison uniforms; bricks from Auschwitz & the Warsaw ghetto.
Major Exhibits: The Congressional Gold Medal, 12/21/13-1/26/14.
Facilities: 6,000-vol. library of history books; 11,768 sq. ft. exhibit space; 102-seat theater; 2 classrooms. Museum-related items for sale.
Activities: docent program; films; formal education programs for children; guided tours; lectures; participatory exhibits; theater. Annual Events: Guardian of the Human Spirit Award; LBJ Moral Courage Award.
Publications: quarterly newsletter; e-newsletter.
Hours & Admission Prices: Mon.-Fri. 9-5, Sat.-Sun. 12-5. No charge; donations accepted. Closed New Year's Day; 1st day of Rosh Hashana; Yom Kippur; Thanksgiving; Christmas. &
Attendance: 150,000 (accurate)
Membership: Student & Educator $25; Individual $50; Associate $100; Supporter $150; Sponsor $250; Contributor $500; Patron $1,000.

HOUSTON ARBORETUM & NATURE CENTER, 4501 Woodway Dr., Houston, TX 77024-7708. Tel.: 713-681-8433; 866-510-7219 (Toll Free). Fax: 713-681-1191.
E-mail: arbor@houstonarboretum.org
Web Site: houstonarboretum.org
Institution Type/Description: Arboretum & Nature Center.
Collections: hands-on exhibits; plants; trees; flowers; wildlife & their habitats.
Facilities: 155 acres; nature trails. Nature-related items for sale.
Activities: special events; summer camp; educational programs.
Hours & Admission Prices: Grounds: daily 7-7. Center: Tues.-Sun. 10-4.
Attendance: 200,000 (estimated)

HOUSTON CENTER FOR CONTEMPORARY CRAFT, 4848 Main St., Houston, TX 77002-9718. Tel.: 713-529-4848.
E-mail: mheadrick@crafthouston.org
Web Site: www.crafthouston.org
Founded: 2001.
Key Personnel: Exec. Dir., Julie Farr; Pres. Bd., Victoria Lightman; Dir. Communication, Mary Headrick; Asher Gallery Mgr., Suzanne Sippel.
Personnel Profile: Full-Time Paid 6; Part-Time Paid 7; Part-Time Volunteers 11.

Governing Authority: nonprofit organization. Tax exempt.
Institution Type/Description: Craft Museum.
Collections: works using craft media including fiber, metal, glass, clay & wood.
Facilities: Museum-related items for sale.
Activities: outreach to over 15,000 students; workshops.
Hours & Admission Prices: Tues.-Sat. 10-5, Sun. 12-5. No charge; donations accepted. Closed New Year's Day; Thanksgiving; Christmas.
Attendance: 20,000 (accurate)
Membership: Apprentice $45; Crafter $100; Artisan $150; Master $250.

HOUSTON CENTER FOR PHOTOGRAPHY, (M), 1441 W. Alabama, Houston, TX 77006-4103. Tel.: 713-529-4755. Fax: 713-529-9248.
E-mail: info@hcponline.org
Web Site: www.hcponline.org
Founded: 1981.
Congressional District: 7
Key Personnel: Exec. Dir., Bevin Bering Dubrowski; Chm. (V), Jereann Chaney.
Personnel Profile: Full-Time Paid 5; Part-Time Paid 6; Part-Time Volunteers 218; Interns 18.
Operating Expenses: 819,748
Operating Income: 838,815
Governing Authority: Tax-exempt.
Institution Type/Description: Art Gallery.
Collections: photographs.
Major Exhibits: 2014 Print Auction Exhibition, 1/17/14-2/12/14; Newsroom, 3/7/14-4/27/14; Nermine Hammom, 3/7/14-4/27/14; Maitha Bin Demithan: Ajyal, 3/7/14-4/27/14; 2014 Juried Fellowship Exhibition, 5/9/14-6/29/14; Collaborations XI & XII, 5/9/14-6/29/14; 32nd Annual Juried Membership Exhibition, 7/18/14-9/17/14.
Activities: temporary exhibits; special events; educational programs; workshops.
Publications: Spot Magazine.
Hours & Admission Prices: Wed. & Thurs.11-9, Fri. 11-5, Sat-Sun 11-7. No charge; donations accepted. Closed New Year's Day; Thanksgiving & day after; Christmas; banking holidays. &
Attendance: 16,000 (estimated)
Membership: Student, Military, & Senior $35; Individual $55; Household $80; Institutional & International $100; Artist/Photographer $250; Professional $500.

HOUSTON FIRE MUSEUM, 2403 Milam St., Houston, TX 77006-2359. Tel.: 713-524-2526. Fax: 713-520-7566.
E-mail: hfmi@houstonfiremuseum.org
Web Site: www.houstonfiremuseum.org
Founded: 1982.
Congressional District: 18
Key Personnel: Dir., Angela Rayne; Pres. (V), Tom McDonald.
Personnel Profile: Full-Time Paid 4; Part-Time Paid 2; Part-Time Volunteers 50.
Governing Authority: municipal; nonprofit. Tax-exempt: 501(c)(3).
Institution Type/Description: Fire-Fighting Museum: housed in 1899 Fire Station No. 7, active until 1969.
Collections: fire-fighting tools & equipment; fire trucks; badges; photographs; fire department artifacts.
Research Fields: Houston Fire Dept.; fire-fighting.
Facilities: 200-vol. library of fire service history material available to the public; classrooms. Gift items for sale.
Activities: self-guided tours; participatory & loan exhibitions. Museum Sponsors: annual parade & festival; special events.
Publications: quarterly newsletter, The Leather Bucket.
Hours & Admission Prices: Tues.-Sat. 10-4. Adults $5, senior citizens $4, children $3; discounts to AAM, ICOM & Texas Assoc. of Museums members; members & children under 2 no charge. Closed holidays.
Attendance: 18,379 (accurate)
Membership: Firefighter $15; Individual $30; Family $40; Patron $100; Life $500 & up.

HOUSTON MARITIME MUSEUM, 2204 Dorrington, Houston, TX 77030-3210. Tel.: 713-666-1910. Fax: 713-838-8557.
E-mail: contact@houstonmaritimemuseum.org
Web Site: www.houstonmaritimemuseum.org
Founded: 2000.
Congressional District: 7
Key Personnel: Exec. Dir., Diane Lipton; Dir. Operations, Heather Schiappa; Chm. (V), Niels Aalund; Master Modeler, Lorena Alvarez.

Personnel Profile: Full-Time Paid 3; Part-Time Paid 0; Part-Time Volunteers 33.
Governing Authority: nonprofit organization. Tax-exempt: 501(c)(3).
Institution Type/Description: Maritime Museum.
Collections: maritime history, artifacts, & navigation instruments; ship plans; ship models; Merchant Marine veterans; books.
Facilities: library. Museum-related items for sale.
Activities: lecture series; guided tours; Boy Scout merit badge workshops.
Publications: monthly e-newsletter.
Hours & Admission Prices: Tues.-Sat. 9-4:30. Adults $5, children 11 & under $3; members no charge. Closed New Year's Day; Thanksgiving; Christmas.
Attendance: 3,000 (estimated)
Membership: Single $50; Family $100; Contributing $200; Benefactor $500; Commander $1,500; Captain $2,500; Commodore $5,000; Admiral $10,000.

*** HOUSTON MUSEUM OF NATURAL SCIENCE, (M),** 5555 Hermann Park Dr., Houston, TX 77030-1749. Tel.: 713-639-4629. Fax: 713-523-4125. TDD: 713-639-4687.
Web Site: www.hmns.org
Founded: 1909.
Congressional District: 7
Key Personnel: C.E.O. & Pres., Joel A. Bartsch; Chm. (V), Peter Huddleston; C.F.O., Stephen Sachnik; Vice Pres. Devel. & Membership, Barbara Hawthorn; Vice Pres. Mktg. & Communications, Latha Thomas; Vice Pres. Astronomy & Physics, Dr. Carolyn Sumners; Vice Pres. Collections, Lisa Rebori; Vice Pres. Exhibits, Hayden Valdes; Dir. Horticulture, Dr. Nancy Greig; Vice Pres. IMAX Operations & Production, Charlotte Brohi; Cur. Anthropology, Dr. Dirk Van Tuerenhout; Cur. Vertebrate Zoology, Dr. Dan Brooks; Dir. Youth Education, Nicole Temple.
Personnel Profile: Full-Time Paid 159; Part-Time Paid 263; Part-Time Volunteers 850.
Governing Authority: nonprofit corporation. Subsidiary Institutions: George Observatory & Challenger Learning Center, Brazos Bend State Park, 21901 FM 762, Needville, TX 77461; HMNS at Sugar Land, 13016 University Blvd., Sugar Land, TX 77479. Tax-exempt: 501(c)(3).
Institution Type/Description: Natural Science Museum.
Collections: natural science; anthropology (pre-Columbian & Native American emphasis); astronomy; gems & minerals; entomology; malacology; paleontology; vertebrate zoology; Texas & African wildlife; space & petroleum technology.
Research Fields: malacology; mineralogy.
Facilities: 400-seat IMAX theatre; 232-seat planetarium; 400-seat auditorium; observatory; butterfly center; classrooms. Gift items for sale.
Activities: guided tours; lectures; films; formally organized educational programs; docent program or council; permanent & temporary exhibitions.
Publications: quarterly newsletter, Museum News; monthly calendar listings; annual report; brochures.
Hours & Admission Prices: Mon. & Wed.-Sun. 9-5, Tues. 9-8; call for additional & holiday hours. Exhibit Halls: adults $15, children $10; members no charge. IMAX: adults $11, children $9, members $6. Butterfly Center: adults $8, children $7, members $4. Planetarium: adults $8, children $7, members $4. Special Exhibits: call for pricing. Closed March 5.
Attendance: 2,700,000 (accurate)
Membership: Student & Senior $30; Individual $50; Dual $65; Family $85; Voyager $150; Discoverer $250; Benefactor $500.

HOUSTON POLICE MUSEUM, 1200 Travis, Houston, TX 77002. Tel.: 281-230-2353. Fax: 281-230-2314.
E-mail: hpdmuseum@cityofhouston.net
Web Site: www.houstontx.gov/police/museum/museum.htm
Founded: 1981.
Key Personnel: Dir., James Chapman.
Governing Authority: city. Houston Police Dept. Tax-exempt: 170(b)(1)(A).
Institution Type/Description: Police History Museum: located at the Houston Police Academy.
Collections: 1841-present day items pertaining to the history of the Police Dept.; helicopter; motorcycles; cars; uniforms; guns; photographs; material relating to the development of the department & famous cases.
Research Fields: Houston city history; Houston Police Dept.
Facilities: 600-vol. library pertaining to law enforcement available for research by appointment only; classrooms; 200-seat auditorium.
Activities: guided tours; lectures; films; gallery talks; formally organized education programs; permanent & temporary exhibitions.
Hours & Admission Prices: Mon.-Fri. 8-3; guided tours by appointment. No charge. Closed city holidays.

HOUSTON RAILROAD MUSEUM, 7390 Mesa Rd., Houston, TX 77028-3520. Mailing Address: Gulf Coast Chapter NRHS, P.O. Box 457, Houston, TX 77001-0457. Tel.: 713-631-6612.
E-mail: info@houstonrrmuseum.org
Institution Type/Description: Railroad Museum.
Collections: railroad locomotives, cars & artifacts; model railroad.
Activities: rental facilities. Annual Event: Open House in November.
Hours & Admission Prices: April to early Nov. Sat. 11-4. Adults $5, children 12 & under $2.50.

HOUSTON ZOO, INC., 6200 Hermann Park Dr., Houston, TX 77030-1603. Mailing Address: 1513 Cambridge St., Houston, TX 77030-1603. Tel.: 713-533-6500. Fax: 713-533-6755. Facebook: Houston Zoo.
E-mail: bhill@houstonzoo.org
Web Site: www.houstonzoo.org
Formerly: Houston Zoological Gardens
Founded: 1922.
Congressional District: 18
Key Personnel: Pres. & C.E.O., Deborah Cannon; Dir., Rick Barongi; Vice Pres. Animal Operations, Sharon Joseph; Vice Pres. Operations, Chris Lyons; Vice Pres. Education, Chance Sanford; Vice Pres. Conservation & Science and Cur. Natural Encounters, Peter Riger; Chief Mktg. Officer & Vice Pres. Advancement, David Brady; C.F.O., Leslie Forestier; Cur. Children's Zoo, Kevin Hodge; Cur. Birds & Natural Encounters, Hannah Bailey; Cur. Herpetology, Stan Mays; Cur. Aquarium, George Brandy; Cur. Primates & Carnivores, Beth Schaefer; Cur. Large Mammals, Daryl Hoffman; Public Rels., Brian Hill; Dir. Veterinary Svcs., Joe Flanagan; Museum Shop Mgr., Wanda Mitchell.
Personnel Profile: Full-Time Paid 380; Part-Time Paid 18; Part-Time Volunteers 340; Interns 2.
Governing Authority: municipal. Tax-exempt: 501(c)(3).
Institution Type/Description: Zoo.
Collections: 6,000 animals representing 800 species from around the world.
Research Fields: conservation programs; captive breeding & management of endangered species; field research & conservation; behavioral enrichment programs.
Facilities: children's zoo; education center; aquarium; restaurants. Gift items for sale.
Activities: formally organized educational programs; volunteer programs; lectures; tours; classes; camps; kits to loan; overnight programs; outreach van & programs; speakers' bureau; carousel.
Publications: bimonthly members' newsletter, WildLife; course catalogs.
Hours & Admission Prices: March 8-Nov. 1 daily 9-7; Nov. 2-March 7 daily 9-6. Adults 12-64 $14, children 2-11 $10, senior citizens 65 & over $7.50; discounts to school groups; Lone Star members; active military, children under 2, members & 1st Tues. Sept.-May 2-7 no charge. Closed Christmas.
Attendance: 2,100,000 (accurate)
Membership: Individual $65; Family $96; Family Plus & Grandparent $106; Supporting $159; Sustaining $250; Conservator $500.

JOHN C. FREEMAN WEATHER MUSEUM, 5104 Caroline St., Houston, TX 77004. Tel.: 713-529-3076.
Founded: 2006.
Institution Type/Description: Science Museum.
Collections: world climates; meteorological history; satellite & radar images of hurricanes, cyclones & typhoons; weather safety.
Facilities: restaurant. Museum-related items for sale.
Activities: simulated weather broadcasting studio; weather experiments; tornado chamber; 3-D weather sphere; weather talks, camps & labs; scout merit badge classes; teacher workshops.
Hours & Admission Prices: Mon.-Sat. 10-4. Adults $5, children & seniors $3; children under 3 no charge.

JOHN P. MCGOVERN MUSEUM OF HEALTH & MEDICAL SCIENCE, 1515 Hermann Dr., Houston, TX 77004-7126. Tel.: 713-942-7054. Fax: 713-526-1434.
E-mail: info@thehealthmuseum.org
Web Site: www.thehealthmuseum.org
Formerly: Museum of Health & Medical Science
Founded: 1969.
Congressional District: 25
Key Personnel: Pres. & C.E.O., Jon Iszard; Chm., Kenneth Mattox, M.D.; Vice Chm., Denton Cooley, M.D.; Vice Pres. Devel., Sue Walden; C.O.O., Anna Hawley; Controller, Sharon Campbell; Bd. Sec. & Admin. Mgr., Stacey Spears.
Personnel Profile: Full-Time Paid 15; Part-Time Paid 4; Part-Time Volunteers 500; Interns 10.

Governing Authority: Tax-exempt.
Institution Type/Description: Health Museum.
Collections: health & science of the human body; interactive exhibits.
Facilities: 4D theater; learning center.
Activities: programs & classes for school-aged children, but are accessible to adults; traveling exhibits; homeschool classes; scouts; spring & summer camps.
Publications: member newsletter, Muse News; annual report.
Hours & Admission Prices: Mon.-Sat. 9-5. Adults $8, senior citizens & children 3-12 $6; discounts to AAM & ICOM members; members & families Thurs. 2-5 no charge. ASTC reciprocal membership program. Closed Thanksgiving; Christmas. &
Attendance: 141,000 (accurate)
Membership: Professional Circle $35; Plus One $45; Young Professional Circle $50; Family $60; Family Plus $70; Associate $125; Contributing $250; Sustaining $500; Director's Circle $1,000; President's Circle $2,500; Chairman's Circle $5,000.

LAWNDALE ART CENTER, (M), 4912 Main St., Houston, TX 77002. Tel.: 713-528-5858. Fax: 713-528-4140. Facebook: Lawdale Houston.
E-mail: askus@lawndaleartcenter.org
Web Site: www.lawndaleartcenter.or
Founded: 1979.
Key Personnel: Exec. Dir., Christine Jelson West
Institution Type/Description: Contemporary Art Space.
Collections: works by contemporary artists.
Activities: special events; temporary exhibitions.
Hours & Admission Prices: Mon.-Fri. 10-5, Sat. 12-5. &
Attendance: 20,000 (estimated)

THE MENIL COLLECTION, 1533 Sul Ross, Houston, TX 77006. Mailing Address: 1511 Branard St., Houston, TX 77006-4721. Tel.: 713-525-9400. Fax: 713-525-9444. Facebook: Menil Collection.
E-mail: info@menil.org
Web Site: www.menil.org
Founded: 1987.
Key Personnel: Chm., Louisa Stude Sarofim; Pres., Janet Hobby; Dir., Josef Helfenstein; C.O.O., Sheryl Kolasinski; Cur. Modern & Contemporary Art, Toby Kamps; Chief Conservator, Brad Epley; Dir. Communications, Vance Muse; Coord. Special Events, Elsian Cozens; Coord. Membership, Amanda Shagrin; Registrar, Anne Adams; Dir. Programs, Karl Kilian; Security, Steve McConathy; Bookstore Mgr., Paul Forsythe.
Personnel Profile: Full-Time Paid 114; Part-Time Paid 4.
Governing Authority: nonprofit. Branch Museums: Dan Flavin Installation at Richmond Hall; Cy Twombly Gallery. Tax-exempt: 501(c)(3).
Institution Type/Description: Art Museum.
Collections: Paleolithic & antiquities; Byzantine & Medieval art; indigenous arts of Africa, Oceania & the Pacific Northwest; modern & contemporary art; Andy Warhol; Pablo Picasso; Henri Matisse; Fernand Leger; Robert Rauschenberg; Barnett Newman; Mark Rothko; Jasper Johns; Michael Heizer; Elsworth Kelly. Cy Twombly Gallery: paintings, sculpture & works on paper by American painter Cy Twombly.
Major Exhibits: Lee Bontecou: Drawn Worlds (T), 1/14-5/14; Magritte: The Mystery of the Ordinary, 1926-38 (T), 2/14-6/14; Memories of a Voyage: The Late Work of Rene Magritte (T), 2/14-7/14; A Thin Wall of Air: Charles James, 5/14-9/14; Dario Robleto: The Building of Life is Quietly Crossed (T), 8/14-11/14; Art and Truth: Gandhi and Images of Nonviolence (T), 10/14-1/15.
Research Fields: 20th-century; Rene Magritte Catalogue Raisonne; conservation; Byzantine; Image of the Black in Western Art; Catalogue Raisonne of the drawings of Jasper Johns.
Facilities: 35,000-vol. library.
Activities: loan, temporary & traveling exhibitions.
Publications: exhibition catalogues; collection monographs.
Hours & Admission Prices: Wed.-Sun. 11-7. No charge. Closed New Year's Day, Martin Luther King Jr. Day; Easter; Memorial Day; Independence Day; Labor Day, Thanksgiving; Christmas. &
Attendance: 180,500 (accurate)
Membership: Student $25; $100; $250; $500; $1,000; $5,000; $10,000.

MICHAEL E. DEBAKEY LIBRARY AND MUSEUM, Baylor College of Medicine, One Baylor Plaza, Ste. 177A, MS: BCM 506, Houston, TX 77030-3411. Tel.: 713-798-4710.
Key Personnel: Cur., JoAnn Pospisil
Institution Type/Description: Medical Museum.

Collections: Dr. DeBakey's life & career including instruments, inventions, awards, papers, photographs, & video displays; Baylor history; medical advances.
Facilities: library.
Hours & Admission Prices: During Academic Year: Mon.-Fri. 9-4. No charge. Closed holidays.

MUSEUM OF AMERICAN ARCHITECTURE AND DECORATIVE ARTS, Houston Baptist Univ., 7502 Fondren Rd., Houston, TX 77074-3298. Tel.: 281-649-3997. Fax: 281-649-3993.
E-mail: ssnoddy@hbu.edu
Web Site: www.hbu.edu
Founded: 1964.
Congressional District: 9
Key Personnel: Interim Dir., Sue Snoddy; Cur., Maggie Brown.
Personnel Profile: Full-Time Paid 1.
Governing Authority: college. Affiliated with Houston Baptist University. Tax-exempt: 501(c)(3).
Institution Type/Description: Social History & Doll Museum.
Collections: Schissler miniature furniture; household goods & decorative arts of the ethnic groups who established the Republic of Texas; furnished dog trot log cabin; the doll collection of Theo Redwood Blank; pre-Columbian art; African art.
Research Fields: dolls, history of decorative arts, tools & photographs of early Texas architecture.
Facilities: 150-vol. reference and research library available for use in reading room.
Activities: guided tours; lectures; films; permanent, temporary & traveling exhibitions. Museum Sponsors: special events in summer; Christmas exhibit.
Publications: books, Days of Colonial Texas; Native Houstonian.
Hours & Admission Prices: Museum of American Architecture and Decorative Arts: Mon.-Sat. 10-4; special tours by appointment. Adults $6, seniors $5, children under 12 $4, discounts to groups; members no charge. Closed Easter; Independence Day; Thanksgiving; Christmas vacation; university holidays. &
Attendance: 7,000 (estimated)
Membership: American Museum Society: Individual $50.

＊　**THE MUSEUM OF FINE ARTS, HOUSTON, (M),** 1001 Bissonnet, Houston, TX 77005-1896. Mailing Address: P.O. Box 6826, Houston, TX 77265-6826. Tel.: 713-639-7300. Fax: 713-639-7784. TDD: 713-639-7390.
E mail: visitorservices@mfah.org
Web Site: www.mfah.org
Founded: 1900.
Congressional District: 18
Key Personnel: Dir., Peter C. Marzio; Chm., Cornelia C. Long; Guild Pres., Kathy Shaw; Dir. Glassell School of Art, Joseph Havel; Assoc. Dir. Administration, Willard Holmes; Assoc. Dir. Investment & Finance, Gwendolyn H. Goffe; Assoc. Dir. Devel., Amy Purvis; Cur. Prints & Drawings and 20th-Century Art, Barry Walker; Cur. Renaissance, Baroque Painting & Sculpture, The Blaffer Collection, James Clifton; Audrey Jones Beck Cur. European Art, Edgar Peters Bowron; Cur. Modern & Contemporary Art, Alison de Lima Greene; Gus & Lyndall Wortham Cur. Photography, Anne Wilkes Tucker; Dir. Rienzi, Katherine S. Howe; Dir. Bayou Bend, Bonnie Campbell; Cur. Bayou Bend Collection, Michael K. Brown; Cur. Film & Video, Marian Luntz; The Wortham Cur. of Latin American Art & Dir. Intl. Center for the Arts of the Americas (ICAA), Mari Carmen Ramirez; Cur. American Painting & Sculpture, Emily Neff; Cur. The Glassell Collections, Frances Marzio; Cur. Modern & Contemporary Decorative Arts and Design, Cindi Strauss; Cur. Asian Art, Christine Starkman; Junior School Dean, Norma Dolcater; Dir. Communications & Mktg., Mary Haus; Mgr. Docent Program, Danielle Stephens; Mgr. Preparations, Michael Kennaugh; Mgr. Preparations, Richard Hinson; Librarian, Margaret Culbertson; Registrar, Julia Bakke; Chief Technology Officer, Shemor Bar-Tal; Controller, Marchell King; Curatorial Admin., Karen Vetter; Dir. Conservation, Wynne Phelan; Dir. Publications, Diane Lovejoy; Dir. Human Resources, Sheila Armsworth; Dir. Retail Operations, Patricia Smith; Photographic Svcs. Mgr., Marty Stein; Staff Photographer, Thomas DuBrock; Coord. Volunteer Svcs., Joi Maria Probus; Museum Shop Mgr., Suzanne Harrison.
Personnel Profile: Full-Time Paid 495; Part-Time Paid 97; Part-Time Volunteers 1,100; Interns 14.
Governing Authority: nonprofit organization. Branch Museums: Rienzi, 1406 Kirby Dr., Houston, TX 77019; The Glassell School of Art, 5101 W. Montrose Blvd., Houston, TX 77006-6534; Bayou Bend Collection & Gardens, 1 Westcott St., Houston, TX 77007. Tax-exempt: 501(c)(3).
Institution Type/Description: Art Museum.
Collections: European & American paintings; decorative arts; sculpture; graphics; pre-Columbian art & archaeology; photography; antiquities;

American Indian art; African & Oceanic art; Far Eastern art; textiles & costumes; prints & drawings.

Research Fields: European painting & sculpture; 20th-century art; decorative arts; photography; Far Eastern art; art of the Americas, Africa & Oceania; Renaissance art; textiles & costumes; prints & drawings; Latin American art.

Facilities: 90,000-vol. art library; archives; auditorium; cafe; sculpture garden. Museum-related items for sale.

Activities: guided tours; lectures; films; gallery talks; concerts; formally organized education programs; docent program or council; inter-museum, permanent, temporary & traveling exhibitions; outreach programs & exhibitions.

Publications: exhibition catalogues; bimonthly magazine, MFAH Calendar; quarterly magazine, MFAH Today; educational materials for adults & children.

Hours & Admission Prices: Mon. holidays & Tues.-Wed. 10-5, Thurs. 10-9, Fri.-Sat. 10-7, Sun. 12:15-7. Adults $7, students, seniors & children 6-18 $3.50; discounts to AAM & ICOM members; children with library cards Sat.-Sun., members & Thurs. no charge. Closed Thanksgiving; Christmas. &

Attendance: 2,643,901 (accurate)

Membership: Student $40; Individual $50; Dual $65; Family $85; Patron $150; Supporting $275; Sponsor $550; Benefactor $1,200.

THE MUSEUM OF SOUTHERN HISTORY, (M), Cultural Arts Center, Houston Baptist Univ., 7502 Fondren Rd., Houston, TX 77074-3298. Tel.: 281-649-3997. Fax: 281-649-3993.

E-mail: ssnoddy@hbu.edu
Web Site: www.hbu.edu
Founded: 1978.
Congressional District: 9
Key Personnel: Interim Dir., Suzie Snoddy; Cur., Maggie Brown; Mktg. Advancement, Charles Bacarisse.
Personnel Profile: Full-Time Paid 2; Part-Time Paid 1; Part-Time Volunteers 6.
Governing Authority: private; nonprofit. Parent Institution: Houston Baptist University. Tax-exempt: 501(c)(3).
Institution Type/Description: Regional History Museum.
Collections: US colonial era late 1700s southern history; Civil War artifacts, 1861-1865; weapons; uniforms; letters from soldiers; prints; paintings; furniture; clothing & medical supplies; The Terry's Texas Rangers; War of the Rebellion official records; southern historical society papers; early periodicals; books.
Research Fields: Texas & Confederates.
Facilities: 884-vol. library. Books, t-shirts, pictures, prints, posters, caps, historical DVDs for sale.
Activities: guided tours; lecture program; special exhibits. Museum Sponsors: Old South Ball; southern hospitality luncheon.
Publications: newsletter published thrice annually; brochures.
Hours & Admission Prices: Mon.-Sat. 10-4. Adults $6, senior citizens $5, children 6-12 $4; discounts to AAM members; members no charge. Closed major holidays. &
Attendance: 7,500 (accurate)
Membership: Individual $35; Family $50; Life $1,000; Corporate $2,500.

THE NATIONAL MUSEUM OF FUNERAL HISTORY, 415 Barren Springs Dr., Houston, TX 77090-5918. Tel.: 281-876-3063. Fax: 281-876-3907. Facebook: The National Museum of Funeral History.

E-mail: info@nmfh.org
Web Site: www.nmfh.org
Formerly: American Funeral Service Museum; Museum of Funeral History
Founded: 1992.
Congressional District: 18
Key Personnel: Pres. & C.O.O., Genevieve Keeney; C.E.O., Bob Boetticher; Museum Shop Mgr., Maria Gonzalez.
Personnel Profile: Full-Time Paid 2; Part-Time Paid 3; Part-Time Volunteers 5.
Governing Authority: nonprofit. Tax-exempt: 501(c)(3).
Institution Type/Description: History Museum.
Collections: concentration on funeral-related artifacts & memorabilia, from mid-19th century through 21st century America, embalming tools, documenting the history of American funeral customs.
Major Exhibits: Jazz Funerals, 1/14-2/14; Last Tip of My Hat (T), 2/14-3/14; A Honored Tribute, 5/14; Dracula, 10/14; Day of the Dead Altars, 11/14; Memorial Tree, 12/14.
Research Fields: thanatology; taphophilia; death studies; death care; hospice care; bereavement, funeral & mourning rituals.
Facilities: rental facilities; 140-seat auditorium.
Activities: docent program; rental facilities. Annual Events: Dracula's Grave-

yard & Family Friendly Haunted House in October; 6th Annual Halloween Classic Car Show in October.
Publications: monthly newsletter.
Hours & Admission Prices: Mon.-Fri. 10-4, Sat. 10-5, Sun. 12-5. Adults $10, veterans & senior citizens $9, members $8, children 3-12 $7; discount to groups & AAM members; children under 3 no charge. Closed New Year's Day; Easter; Thanksgiving; Christmas. &
Attendance: 7,000 (accurate)

1940 AIR TERMINAL MUSEUM, (M), William P. Hobby Airport, 8325 Travelair Rd., Houston, TX 77061-4716. Tel.: 713-454-1940. Fax: 713-454-1930.

E-mail: info@1940 airterminal.org
Web Site: www.1940airterminal.org
Key Personnel: C.E.O. & Pres. (V), Drew Coats; Dir., Amy Rogers.
Personnel Profile: Full-Time Paid 1; Part-Time Volunteers 40.
Governing Authority: Tax-exempt.
Institution Type/Description: Aviation History Museum.
Collections: aviation history & memorabilia; period aircraft.
Facilities: library; theater. Museum-related items for sale.
Activities: educational outreach program; rental facilities.
Publications: quarterly newsletter.
Hours & Admission Prices: Tues.-Sat. 10-5, Sun. 1-5. Adults $5, children $2; members, military, law enforcement, firefighters & their families no charge.
Attendance: 3,000 (estimated)
Membership: Observation Deck $75; Barnstormer $194; Travelair $500; Clipper $940; Starliner $1,940; Texanaire $5,000.

THE PRINTING MUSEUM, (M), 1324 W. Clay, Houston, TX 77019-4036. Tel.: 713-522-4652, ext. 202. Fax: 713-522-5694.

E-mail: akasman@printingmuseum.org
Web Site: www.printingmuseum.org
Formerly: The Museum of Printing History
Founded: 1983.
Congressional District: 2
Key Personnel: Exec. Dir., Ann Kasman; Chm. Bd., A. John Harper, III; Cur., Amanda Stevenson; Administrative Asst. & Museum Shop Mgr., Marilyn Davenport.
Personnel Profile: Full-Time Paid 3; Part-Time Paid 2; Part-Time Volunteers 25; Interns 3.
Governing Authority: private; nonprofit organization. Tax-exempt: 501(c)(3) and 509(a)(1).
Institution Type/Description: History & Art Museum.
Collections: history of communication beginning with writing on clay tablets to modern day newspapers; period workshops & bookbinding filled with period tools & machines; Bible & Bible leaves; great master printers & early printing; lithography from its beginnings through the Belle Epoch (Toulouse-Lautrec) & beyond; the Great Books of the World; history of graphic communication, arts & design; paper making facility.
Research Fields: history of books and printing.
Facilities: library; 20,000 sq. ft. exhibit space; 67-seat theater. Museum-related items for sale.
Activities: docent program; guided tours; hobby workshops; lectures; rental gallery; temporary & traveling exhibitions. Annual Events: Houston Book Fair; gala.
Publications: quarterly newsletter, The Printed Word; quarterly trifolds; exhibition brochures; brochure, Summer Book Arts Studio.
Hours & Admission Prices: Tues.-Sat. 10-5. Museum: no charge; donations accepted. Guided Tours: adults $7, seniors $5, students $3. Closed New Year's Day; Memorial Day; Independence Day; Labor Day; Thanksgiving; Christmas Eve & Day. &
Attendance: 25,000 (estimated)
Membership: Student & Senior Citizens $30; Individual $40; Family $65; Friend $100; Sponsor $250; Patron $500; Gutenberg Society $1,000; Master Printer $2,500; Imprimatur $5,000.

RICE UNIVERSITY ART GALLERY, 6100 Main St., Ground Fl., Sewall Hall, Houston, TX 77005-1892. Mailing Address: P.O. Box 1892, MS59, Houston, TX 77251-1892. Tel.: 713-348-6069. Fax: 713-348-5980.

E-mail: ruag@rice.edu
Web Site: www.ricegallery.org
Founded: 1971.
Congressional District: 22
Key Personnel: Dir., Kimberly Davenport.
Personnel Profile: Full-Time Paid 3; Part-Time Paid 22; Interns 1.
Governing Authority: Parent Institution: Rice University. Tax-exempt.
Institution Type/Description: Art Gallery.

Collections: works by contemporary artists.
Publications: exhibition catalog.
Hours & Admission Prices: Tues.-Wed. & Fri.-Sat. 11-5, Thurs. 11-7, Sun. 12-5. No charge; donations accepted. Closed university holidays; between exhibits. &

Membership: Catalyst $50; Friend $100; Advocate $250; Associate $500; Partner $1,000. Patrons: Site-Specific $3,000; New Genre $5,000; No Boundaries $10,000.

ROBERT A. VINES ENVIRONMENTAL SCIENCE CENTER,
955 Campbell Rd., Houston, TX 77024-2803. Tel.: 713-251-7975. Fax: 713-365-4178.
Web Site: www.springbranchisd.com
Founded: 1960.
Key Personnel: Dir., Dee Goldberg.
Governing Authority: public school district; nonprofit. Parent Institution: Spring Branch ISD. Tax-exempt.
Institution Type/Description: Natural Science Museum.
Collections: natural history of Houston, Texas region.
Hours & Admission Prices: June-Aug. Mon.-Thurs. 8:30-4:30; Sept.-May Mon.-Fri. 8:30-4:30. No charge; donations accepted. Closed Memorial Day; Independence Day; Labor Day; Thanksgiving; spring break; two weeks during Christmas break. &
Attendance: 5,000 (estimated)

ROTHKO CHAPEL,
1409 Sul Ross, Houston, TX 77006-4829. Tel.: 713-524-9839. Fax: 713-524-7461.
E-mail: info@rothkochapel.org
Web Site: www.rothkochapel.org
Founded: 1971.
Key Personnel: Dir., Emilee Whitehurst; Chm. (V), Gayle Ross DeGeurin.
Personnel Profile: Full-Time Paid 6; Part-Time Paid 5.
Governing Authority: Tax-exempt.
Institution Type/Description: Religious, Art & Architecture Museum.
Collections: 14 panels painted by Mark Rothko.
Hours & Admission Prices: Daily 10-6; groups by appointment. No charge; donations accepted. &
Attendance: 57,000 (accurate)

SPACE CENTER HOUSTON,
1601 NASA Pkwy., Houston, TX 77058-3199. Tel.: 281-244-2100. Fax: 281-283-7724. TDD: 713-283-7730.
E-mail: mjohnson@spacecenter.org
Web Site: www.spacecenter.org
Founded: 1992.
Congressional District: 25
Key Personnel: C.E.O. & Pres., Richard Allen; Chm. (V), Ron Kapche; Museum Shop Mgr., Sharon Glenn.
Personnel Profile: Full-Time Paid 148; Part-Time Paid 74; Part-Time Volunteers 25.
Governing Authority: nonprofit organization. Parent Institution: Manned Space Flight Education Foundation, Inc. Tax-exempt.
Institution Type/Description: Space Museum.
Collections: artifacts & exhibits pertaining to America's manned space program.
Facilities: mission simulation & training facility; rocket park; mission control center; IMAX theater; cafeteria. Museum-related items for sale.
Activities: guided tours; mission control center briefings; films; lectures; formally organized education programs for children.
Hours & Admission Prices: Winter: Mon.-Fri. 10-5, Sat.-Sun. & holidays 10-6. Summer: daily 9-7. Adults $22.95, seniors $21.95, children 4-11 $18.95. Education Groups: $10.95 2 weeks advance reservations, $12.95 late reservations. Children under 4 no charge. Closed Christmas. &
Attendance: 750,000 (accurate)
Membership: Annual Member $29.95; Annual Family $59.95.

Hubbard

PELHAM COMMUNITY HISTORY MUSEUM,
22535 FM 744 Pelham Rd., Hubbard, TX 76648. Tel.: 254-678-1850. Fax: 254-678-1850.
E-mail: info@pelham-museum.org
Institution Type/Description: History Museum: housed in the former Pelham School; built in 1890.
Collections: local African American heritage & history; personal artifacts; period furnishings; photographs.
Activities: lectures; educational programs; summer camps.
Hours & Admission Prices: Sun. & Wed. 3-5, Fri. 11-2. No charge; donations accepted.

Humble

HUMBLE BICENTENNIAL MUSEUM, INC.,
219 Main St., Humble, TX 77338. Tel.: 281-446-2130. Fax: 281-446-1964.
E-mail: humblemuseum@live.com
Web Site: www.humblemuseum.com
Founded: 1976.
Personnel Profile: Part-Time Paid 3.
Governing Authority: Subsidiary Institution: McKay Clinic Medical Museum. Tax-exempt.
Institution Type/Description: History Museum.
Collections: local history, heritage & culture; oil & lumber industries; early pioneer life; period furnishings; personal artifacts; photographs.
Hours & Admission Prices: Tues.-Fri. 10-4, Sat. 10-2. No charge; donations accepted. Closed New Year's Day; Good Friday; Independence Day; Thanksgiving; Christmas. &
Attendance: 1,800 (estimated)
Membership: Student $1; Individual $10; Business $25.

McKAY CLINIC MEDICAL MUSEUM,
110 N. Ave. C, Humble, TX 77338. Mailing Address: c/o Humble Museum, 219 Main St., Humble, TX 77338. Tel.: 281-446-2130.
Governing Authority: Parent Institution: Humble Bicentennial Museum, Inc. Tax-exempt.
Institution Type/Description: History Museum: housed in the former clinic of Dr. McKay, used from 1938-1996.
Collections: furniture & instruments from Dr. McKay, Sr; McKay family history; medical instruments & equipment; personal artifacts.
Hours & Admission Prices: By appointment. No charge; donations accepted.
Membership: Student $1; Individual $10; Business $25.

MERCER ARBORETUM & BOTANIC GARDENS,
22306 Aldine Westfield Rd., Humble, TX 77338-1071. Tel.: 281-443-8731. Fax: 281-443-6078. Facebook: The Mercer Society.
E-mail: mercerarboretum@hcp4.net
Web Site: www.hcp4.net/mercer
Founded: 1974.
Key Personnel: Dir., Darrin Duling; Pres. (V), Alan Raymond; Museum Shop Mgr., Kathy Reed.
Personnel Profile: Full-Time Paid 24; Part-Time Volunteers 328.
Governing Authority: Parent Institution: Harris County Precinct & Parks. Subsidiary Institution: The Mercer Society. Tax-exempt.
Institution Type/Description: Arboretum & Botanic Gardens.
Collections: plants; trees; flowers; herbarium; cryogenic seed.
Research Fields: botany; habitat restoration; endangered species conservation.
Facilities: library; visitor center; herbarium. Gift items for sale.
Activities: educational programs; weekend plant sales; autumn garden fair; annual garden party.
Publications: The Leaflet.
Hours & Admission Prices: Mar.-Oct. 8-dusk, Nov.-Feb. 8-5. No charge; donations accepted. Closed New Year's Day; Thanksgiving; Christmas Eve & Day.
Attendance: 200,000 (estimated)
Membership: The Mercer Society: Junior 18 & under & Senior 55 & over $25; Individual $35; Family $50; Supporting $100; Sustainer $500; Benefactor?$1,000.

Huntsville

H.E.A.R.T.S. VETERANS MUSEUM OF TEXAS,
463 State Hwy. 75, Huntsville, TX 77320-1119. Tel.: 936-295-5959. Fax: 936-295-0714.
E-mail: info@heartsmuseum.com
Web Site: www.heartsmuseum.com
Founded: 2000.
Key Personnel: Dir., C.E.O., & Chm., Richard Harris; Treas., Tom Oleinik; Sec., Reva Bishop; Dir. Events, Charlotte Oleinik; Museum Shop Mgr, Donna Barron.
Personnel Profile: Full-Time Paid 3; Full-Time Volunteers 1; Part-Time Paid 1; Part-Time Volunteers 65.
Governing Authority: nonprofit organization. Tax-exempt: 501(c)(3).
Institution Type/Description: Military History Museum.
Collections: military history & artifacts; photographs; women in the military; war timelines.
Facilities: library. Museum-related items for sale.
Activities: guided tours; temporary exhibitions.
Publications: monthly newsletter.
Hours & Admission Prices: Mon.-Sat. 10-5. Adults $7, seniors $5, veterans $4, students $2. Closed New Year's Day; Easter; Thanksgiving; Christmas. &

Attendance: 6,900 (accurate)

* **SAM HOUSTON MEMORIAL MUSEUM, (M),** 1836 Sam Houston Ave., Huntsville, TX 77341. Mailing Address: Box 2057, SHSU, Huntsville, TX 77341-2057. Tel.: 936-294-1832 & 1831. Fax: 936-294-3670.
E-mail: SMM_PBN@shsu.edu
Web Site: www.samhouston.memorial.museum
Founded: 1927.
Congressional District: 2
Key Personnel: Dir., Patrick B. Nolan; Cur. Collections, Mac Woodward; Cur. Education, Michael Sproat; Cur. Exhibits, Casey Roon; Interpreter, Helen Belcher; Interpreter, Elizabeth Barry; Interpreter, Rebecca Lewis; Registrar, Sandra Rogers; Coord. Mktg. & Museum Shop Mgr., Megan Buro; Administrative Asst., JoAnn Purvis.
Personnel Profile: Full-Time Paid 10; Part-Time Paid 5; Part-Time Volunteers 25; Interns 1.
Governing Authority: university. Parent Institution: Sam Houston State University; State of Texas. Tax-exempt.
Institution Type/Description: History Museum: 15-acre historical site.
Collections: personal possessions of General Houston, his family, Mexican President, Santa Ana; objects related to Texas history, two homes; law office; period kitchen; blacksmith shop; park & pond.
Research Fields: Texas history; genealogy.
Facilities: education center; memorial museum. Gifts & books for sale.
Activities: guided tours; lectures; formally organized educational programs.
Publications: books: Braving the Storm, The Houston's at Home, Woodland Home.
Hours & Admission Prices: Tues.-Sat. 9-4:30, Sun. 12-4:30. Adults $4, children 6-17 $2; discounts to AAM & ICOM members; members no charge. Closed New Year's Day; Thanksgiving; Christmas. &
Attendance: 46,267 (accurate)
Membership: Individual $35; Family $50; Contributing $100; Supporting $250; Sustaining $500.

TEXAS PRISON MUSEUM, 491 State Hwy. 75 N., Huntsville, TX 77320-1119. Tel.: 936-295-2155. Fax: 936-295-0205.
E-mail: jimwillett@sbcglobal.net
Web Site: www.txprisonmuseum.org
Founded: 1989.
Congressional District: 8
Key Personnel: Dir., Jim Willett; Pres. (V), Tommy Martin; Museum Shop Mgr., Riley Tilly.
Personnel Profile: Full-Time Paid 1; Part-Time Paid 8; Part-Time Volunteers 4; Interns 1.
Governing Authority: Parent Institution: Texas Prison Museum Board. Tax-exempt.
Institution Type/Description: Prison Museum.
Collections: history of the Texas penal system from 1848 to present; photographs; electric chair; contraband; prison hardware; prison uniform.
Activities: rental facility; special events.
Hours & Admission Prices: Mon.-Sat. 10-5, Sun. 12-5. Adults $4, seniors $3, children 6-17 $2; discounts to active military & AAA members; children under 6 no charge. &
Attendance: 30,000 (accurate)

Iraan

IRAAN MUSEUM, 9261 Alley Oop Lane, Iraan, TX 79744. Mailing Address: P.O. Box 95, Iraan, TX 79744-0095. Tel.: 432-639-2522. Fax: 432-639-2248.
E-mail: iraan.city@sbcglobal.net
Founded: 1965.
Congressional District: 21
Key Personnel: Chm. (V), Cur. & Museum Shop Mgr., Morine Collett; Asst. Cur., Francis McCormick; Pres. Bd., Jean Owens; Museum Shop Mgr., Edna Collett.
Personnel Profile: Part-Time Paid 2.
Governing Authority: Parent Institution: city of Iraan. Tax-exempt.
Institution Type/Description: Archaeological & Historic Museum: located on what was once the San Antonio-San Diego Stage Line.
Collections: stone tools & weapons; period Spanish artifacts; ranchlife & cattle trails displays; relics from military forts & stage coach stations; cretaceous fossils; minerals; cores from oil wells; steel oil derrick with shopmade wellhead; Lufkin wooden walking beam pumping unit, 1927 American pumping unit; Red Line pumping unit; print shop; linotype; Iraan newspaper; tack room; surgery table; 1926 fire extinguishers; hospital equipment; desks; books; photo record; year books; military photos, WWII; 112 ft.

windtower blade; flagtail & mule deer; redtail squirrels. Museum Park: over 60 species of native & migratory birds;
Research Fields: local Indian archaeology; fossils; oil; ranching; military.
Activities: Iraan Archaeology Society field trips to archaeological & historical sites; programs; recording & cataloging artifacts and sites. Museum Sponsors: Southwestern Federation Archaeological Symposium in April; Texas Archaeological Society Archaeological Awareness Month in October.
Publications: annual bulletin, Southwestern Federation of Archaeological Societies; Rock Art photos; Iraan Archeological Society monthly newsletter.
Hours & Admission Prices: Feb.-Dec. Thurs.-Sun. 1-5; groups & other times by appointment. No charge; donations accepted. Closed Easter; Thanksgiving; Christmas. &
Attendance: 800 (estimated)
Membership: Iraan Archaeological Society: Individual $10; Family $12.

Irving

IRVING ARTS CENTER, (M), 3333 N. MacArthur Blvd., Ste. 300, Irving, TX 75062-4497. Tel.: 972-252-7558. Fax: 972-570-4962. Facebook: Irving Arts Center.
E-mail: minman@cityofirving.org
Web Site: www.irvingartscenter.com
Founded: 1990.
Congressional District: 24
Key Personnel: Exec. Dir., Richard E. Huff; Chm. Bd. (V), Jo-Ann Bresowar; Asst. Dir., Rosemary Meng; Asst. Dir., Kass Prince; Dir. Exhibitions & Educational Programs, Marcie J. Inman.
Personnel Profile: Full-Time Paid 14; Part-Time Paid 16; Interns 3.
Governing Authority: nonprofit organization. Parent Institution: City of Irving. Tax-exempt.
Institution Type/Description: Art Museum & Center.
Collections: Sculpture Garden: works by Jesus Moroles, James Surls & Michael Manjarris.
Facilities: 8,000 sq. ft. exhibit space; classrooms; studios; 2 theaters; 2 rehearsal halls; 2 acre sculpture garden; visual & performing arts center.
Activities: concerts; dance recitals; docent program; formal education programs for children; guided tours; lectures; traveling & participatory exhibits; theater; musical theater
Publications: quarterly newsletter, Calendar of Events; exhibition brochures & catalogues.
Hours & Admission Prices: Mon.-Wed. & Fri. 9-5, Thurs. 9-8, Sat. 10-5, Sun. 1-5. No charge. Closed New Year's Day; Thanksgiving; Christmas. &
Attendance: 48,000 (estimated)
Membership: Student $10; Artist & Senior Citizen $30; Individual $40; Senior Citizen & Spouse $40; Smithsonian Affiliate & Family $75; Patron $125; Connoisseur $250; Benefactor $500; Director's Circle $1,000.

IRVING HERITAGE HOUSE, 303 S. O'Connor, Irving, TX 75060-2949. Mailing Address: P.O. Box 171572, Irving, TX 75017-1572. Tel.: 972-252-3838.
E-mail: irvingheritagesociety@yahoo.com
Web Site: www.irvingheritage.com
Founded: 1978.
Congressional District: 32
Key Personnel: Administrative Asst., Mary Higbie.
Governing Authority: Parent Institution: City of Irving. Operated by Irving Heritage Society. Tax-exempt.
Institution Type/Description: Historic House Museum: housed in the former home of C.P. Schulze, brother of co-founder of Irving; built in 1912. A Texas State Historical Landmark.
Collections: Schulze family & local history; life & works of author Washington Irving; period furnishings; photographs; personal artifacts; early 1900s artifacts.
Major Exhibits: Celebrating Irving: The City and the Man, 4/14; Vintage Bridal Gowns, 6/14.
Activities: Museum Sponsors: Celebrating the Life & Works of Washington Irving in April.
Publications: biannual newsletter, Whistlestop.
Hours & Admission Prices: March-Dec. 1st Sun. each month 3pm-5pm. No charge; donations accepted.
Attendance: 385 (estimated)
Membership: Student $10; Senior Couple $20; Couple $25; Corporate $250.

NATIONAL SCOUTING MUSEUM, (M), 1329 West Walnut Hill Lane, Irving, TX 75038-3027. Tel.: 800-303-3047; 972-580-2100. Fax: 972-580-2020.
E-mail: nsmuseum@scouting.org
Web Site: nationalscoutingmuseum.org

Founded: 1959.
Key Personnel: Exec. Dir., Janice Babineaux.
Personnel Profile: Full-Time Paid 8; Part-Time Paid 4; Part-Time Volunteers 30.
Governing Authority: nonprofit organization. Parent Institution: Boy Scouts of America. Tax-exempt: 501(c)(3).
Institution Type/Description: Scouting Museum.
Collections: over 500,000 items & artifacts; objects related to the history of the Boy Scouts of America & other youth & scouting organizations, including equipment & records of founders Baden-Powell, West, Beard & Seton; original Norman Rockwell paintings of scouting; Norman Rockwell scouting theme collection of original works of art.
Research Fields: history; scouting; youth organizations.
Facilities: 50,000 sq. ft. exhibit space; rental space. Museum-related items for sale.
Activities: virtual reality adventures; hands on learning experiences; school programs & tours; special public programs; interactive exhibits.
Publications: newsletter, Bridges & Trails; museum guide.
Hours & Admission Prices: Mon. 10-7, Tues.-Sat. 10-5, Sun. 1-5. Adults $8, seniors $7, children $6, Scout $5; discounts to AAM members; members, children under 4, Sun. & Mon. no charge. &
Attendance: 25,129 (accurate)
Membership: Scout $25; Individual $50; Family $100; Sustaining $250; Sponsor $500; Patron $1,000.

Jacksboro

FORT RICHARDSON STATE HISTORICAL PARK, 228 State Park Rd. 61, Jacksboro, TX 76458. Tel.: 940-567-3506. Fax: 940-567-5488.
Web Site: www.tpwd.state.tx.us
Founded: 1968.
Key Personnel: Park Supt., Robert Frie; Dir. Interpretation & Exhibits, Glenn Barnett.
Personnel Profile: Full-Time Paid 7; Part-Time Paid 3.
Governing Authority: state. Subsidiary Institution: Texas Parks and Wildlife Dept. Tax-exempt.
Institution Type/Description: Park Museum: located on the site of 1867 Old Cavalry Fort.
Collections: items pertaining to Fort Richardson, Jack County & Lost Battalion.
Facilities: Interpretive Visitor Center.
Activities: tours; Fort Richardson Days, Annual Living History event in April.
Hours & Admission Prices: Mon.-Thurs. & Sun. 8-5, Fri. 8-9, Sat. 8-8. Adults $3, children under 12 no charge. Closed Christmas. &
Attendance: 65,000 (accurate)

JACK COUNTY MUSEUM, 241 W. Belknap, Jacksboro, TX 76458. Mailing Address: P.O. Box 861, Jacksboro, TX 76458. Tel.: 940-567-5410.
Web Site: wwwjackcountymuseum.com
Founded: 1988.
Institution Type/Description: History Museum: housed in the former home of Mr. & Mrs. Stanley Cooper; built in 1882.
Collections: local history & culture; period furnishings; personal artifacts; photographs.
Publications: book, The Jack County History Book.
Hours & Admission Prices: Thurs.-Sun. 11-4; other times by appointment. No charge; donations accepted. &

Jefferson

EXCELSIOR HOUSE, 211 W. Austin St., Jefferson, TX 75657-2245. Tel.: 903-665-2513; 800-490-7270. Fax: 903-665-9389.
E-mail: jgoulds@aol.com
Web Site: www.theexcelsiorhouse.com
Founded: 1850.
Congressional District: 1
Key Personnel: Pres., Karl Frederickson.
Governing Authority: board of directors of Jessie Allen Wise Garden Club.
Institution Type/Description: Historic Hotel.
Collections: period furnishings.
Facilities: overnight guest ballroom & banquet facilities available.
Activities: guided tours.
Publications: Excelsior House Cookbook.
Hours & Admission Prices: Daily 7-9. Adults $4; children under 10 no charge. Closed Christmas Eve & Day.

JEFFERSON HISTORICAL SOCIETY AND MUSEUM, 223 W. Austin, Jefferson, TX 75657-2253. Tel.: 903-665-2775. Fax: 903-665-9017.
E-mail: jeffersonmuseum@yahoo.com
Web Site: www.jeffersonmuseum.com
Founded: 1948.
Congressional District: 1
Personnel Profile: Full-Time Paid 2; Part-Time Paid 2.
Governing Authority: society. Tax-exempt.
Institution Type/Description: Historical Society Museum: housed in 1888 old federal building.
Collections: Civil War artifacts; dolls; glass; medical items; paintings; farm implements; costumes; period furniture; bibles; Republic of Texas documents & money; manuscripts.
Facilities: 500-vol. library of genealogy & history books available for use on the premises; reading room.
Activities: guided tours for groups; permanent exhibitions.
Hours & Admission Prices: Daily 9:30-4:30. Adults 18-61 $7, senior over 62 $5, teens 13-17 $4, youth 6-12 $3; discounts to student groups; children under 5 no charge. Closed New Year's Eve; Easter; Thanksgiving; Christmas Eve & Day.
Attendance: 19,500 (estimated)
Membership: Individual $7.50; Family $15; Life $75; Sustaining $200; Patron $300; Benefactor $1,000.

Johnson City

THE EXOTIC RESORT ZOO, 235 Zoo Trail, Johnson City, TX 78636. Tel.: 830-868-4357. Fax: 830-868-7586.
E-mail: exotic@moment.net
Web Site: www.zooexotics.com
Institution Type/Description: Zoo.
Collections: over 500 animals including over 80 species.
Facilities: 137 acres.
Activities: petting zoo.
Hours & Admission Prices: Daily 9-6. Adults $11.95, seniors $10.95, children 2-12 $9.95.

LYNDON B. JOHNSON NATIONAL HISTORICAL PARK, 100 Ladybird Lane, Johnson City, TX 78636. Mailing Address: P.O. Box 329, Johnson City, TX 78636-0329. Tel.: 830-868-7128. Fax: 830-868-0810 & 7863.
E-mail: lyjo_superintendent@nps.gov
Web Site: www.nps.gov/lyjo
Founded: 1969.
Congressional District: 11 & 21
Key Personnel: Supt., Russ Whitlock; Volunteer Coord., Elizabeth Lindig.
Personnel Profile: Full-Time Paid 52; Part-Time Paid 18; Part-Time Volunteers 55.
Governing Authority: federal. Parent Institution: National Park Service, Dept. of Interior. Tax-exempt.
Institution Type/Description: Historic Site and Museum: 1901 Lyndon B. Johnson boyhood home, Johnson City, restored 1973-1974; 1867 Johnson Settlement, Johnson City, restored 1972-1974; 1888 Lyndon B. Johnson Birthplace, Stonewall, reconstructed 1964; Johnson Family Cemetery; 1894 Johnson Ranch, Stonewall.
Collections: furnishings; farm & ranch equipment; historic automobiles & vehicles; Lyndon B. Johnson & Lady Bird Johnson memorabilia & personal effects reflecting their lives in the hill country.
Research Fields: Texas settlement & ranching; life & presidency of Lyndon B. Johnson; collections management; care of living history site; pre-arranged graduate research projects related to contemporary material culture.
Facilities: materials relating to President Lyndon B. Johnson & Texas hill country life available for research in park library & archives; in-depth studies can be arranged with the Curator.
Activities: tours; self-guided vehicle tours of LBJ Ranch; movies; permanent exhibitions.
Publications: orientation brochures; newsletters, LBJ National Historical Park, LBJ State Park & LBJ Country.
Hours & Admission Prices: Visitor Center: daily 8:45-5. Johnson City Unit: daily 9-5. No charge. Ranch Unit: daily 9-4:30. Minimal charge. Closed New Year's Day; Thanksgiving; Christmas. &
Attendance: 98,200 (accurate)

Junction

KIMBLE COUNTY HISTORICAL MUSEUM, 101 N. 4th St., Junction, TX 76849-4705. Mailing Address: P.O. Box 271, Junction, TX 76849-0271. Tel.: 325-446-4219. Fax: 325-446-2871.
E-mail: fwyatt30@yahoo.com
Web Site: www.junctiontexas.net/museum.htm
Founded: 1966.
Congressional District: 21
Key Personnel: Cur., Frederica Wyatt.
Personnel Profile: Full-Time Volunteers 2; Part-Time Volunteers 3.
Governing Authority: county.
Institution Type/Description: History Museum.
Collections: farm & ranch tools; household items; guns; photographs; clothing.
Hours & Admission Prices: Mon.-Fri. 2-5; other times by appointment. No charge; donations accepted. Closed major holidays.
Attendance: 3,000 (estimated)

Katy

KATY HERITAGE MUSEUM, 6002 George Bush Dr., Katy, TX 77493. Tel.: 281-391-4884.
Institution Type/Description: History Museum.
Collections: local history & culture; early farming equipment & tools; photographs; period furnishings & artifacts.
Hours & Admission Prices: Tues.-Thurs. 10-2, 1st Sat. each month 10-3, 1st Sun. each month 12-3. No charge.

KATY VETERANS MEMORIAL MUSEUM, 6206 George Bush Dr., Katy, TX 77493-1806. Tel.: 281-391-8387.
Institution Type/Description: Military History & Memorial Museum.
Collections: war memorabilia; photographs; personal artifacts.
Hours & Admission Prices: Call for hours.

Kerrville

L.D. "BRINK" BRINKMAN FOUNDATION, 444 Sidney Baker St. S., Kerrville, TX 78028-5919. Tel.: 830-257-2000. Fax: 830-257-2030.
Founded: 1985.
Congressional District: 21
Key Personnel: C.E.O. & Dir., L.D. Brinkman; Dir., Pam Stone; Dir., Charles C. Thomas.
Personnel Profile: Part-Time Volunteers 5.
Governing Authority: nonprofit. Tax-exempt: 501(c)(3).
Institution Type/Description: Art Foundation: housed in LDB Corporation's headquarters.
Collections: Western art.
Activities: guided tours by appointment only.
Hours & Admission Prices: Mon.-Fri. 9-12. No charge. Closed major national holidays.
Attendance: 900 (estimated)

THE MUSEUM OF WESTERN ART, 1550 Bandera Hwy., Kerrville, TX 78028-9547. Mailing Address: P.O. Box 294300, Kerrville, TX 78029-4300. Tel.: 830-896-2553. Fax: 830-257-5206.
E-mail: gsimon@mowa.tx.com
Web Site: www.museumofwesternart.com
Formerly: National Center for American Western Art
Founded: 1983.
Congressional District: 21
Key Personnel: Chm. Bd. & Pres., Melissa Hoelscher; Librarian, Nan Stover; Office Coord., Gladys Simon.
Personnel Profile: Full-Time Paid 2; Part-Time Paid 1; Part-Time Volunteers 50.
Governing Authority: nonprofit organization. Tax-exempt: 501(c)(3).
Institution Type/Description: Art Museum.
Collections: paintings; drawings; sculpture.
Research Fields: written & photographic records.
Facilities: 2,500-vol. library pertaining to range cattle industry & Western American realistic art available for research by appointment on premises only; 80-seat auditorium. Prints, books, cards & other museum-related items for sale.
Activities: education programs for youth & adults; guided tours; lectures; gallery talks; loan & permanent exhibitions; art workshops for young professional artists; history workshop.
Publications: pamphlets; book, exhibit catalogs.
Hours & Admission Prices: Tues.-Sat. 10-4. Adults $7, senior citizens 65 & over $6, children 9-17 $5; discounts to groups; children 8 & under and members no charge. Closed New Year's Day; Easter; Memorial Day; Labor Day; Thanksgiving; Christmas. &
Attendance: 30,000 (estimated)
Membership: Cowboy $35; Settler $50; Wrangler $100; Rangerider $250; Wagonmaster $500; Ranger $1,000; Sheriff $2,500; Marshall $5,000.

RIVERSIDE NATURE CENTER, 150 Francisco Lemos, Kerrville, TX 78028-5211. Tel.: 830-257-4837. Fax: 830-257-4837.
E-mail: office@riversidenaturecenter.org
Web Site: www.riversidenaturecenter.org
Founded: 1989.
Congressional District: 21
Key Personnel: Exec. Dir., Cass Keen; Pres., Gloria Olsen; Museum Shop Mgr., Ann Laughlin.
Personnel Profile: Part-Time Paid 2; Part-Time Volunteers 50; Interns 1.
Governing Authority: private; nonprofit organization. Tax-exempt.
Institution Type/Description: Herbarium.
Collections: flora & fauna of the upper Guadalupe River watershed & Texas Hill Country.
Research Fields: coordinating information for a watershed map; collect data on flora & fauna species.
Facilities: library; botanical garden; nature center; educational facilities. Field guides, arts & crafts reflecting natural diversity of Hill Country & museum-related items for sale.
Activities: guided tours; lectures; formal educational programs; participatory exhibits. Museum Sponsors: Earth Day Celebration; Down By the Riverside Festival.
Publications: quarterly newsletter; trail guide.
Hours & Admission Prices: Tree Trail: daily dawn to dusk. Visitor's Center: Mon-Fri. 9-4, Sat.-Sun. 10-3. No charge; donations accepted. &
Attendance: 6,500 (estimated)
Membership: Individual $35; Family $50; Friend $100; Supporter $250; Patron $1,000.

SCHREINER MANSION - HISTORICAL SITE AND EDUCATION CENTER, 226 Earl Garrett St., Kerrville, TX 78028-5305. Tel.: 830-896-8633.
Formerly: The Hill Country Museum
Founded: 1983.
Congressional District: 21
Personnel Profile: Part-Time Paid 1; Part-Time Volunteers 50.
Governing Authority: Parent Institution: Schreiner University. Tax-exempt.
Institution Type/Description: Historic House: 1870s home of Capt. Charles Schreiner.
Collections: local history of Hill Country from 1850s to 1930s with special interest in Kerrville & Kerr County.
Hours & Admission Prices: Wed.-Sat. 11:30-3. No charge; donations accepted. Closed holidays.
Attendance: 10,600 (estimated)

Kilgore

EAST TEXAS OIL MUSEUM AT KILGORE COLLEGE, Hwy. 259 at Ross St., 1301 S. Henderson Blvd., Kilgore, TX 75662. Tel.: 903-983-8295. Fax: 903-983-8659.
E-mail: info@easttexasoilmuseum.com
Web Site: www.easttexasoilmuseum.com
Founded: 1980.
Congressional District: 1
Key Personnel: Dir., Joe L. White.
Personnel Profile: Full-Time Paid 3; Part-Time Paid 2; Part-Time Volunteers 40.
Governing Authority: college; not-for-profit. Parent Institution: Kilgore College. Tax-exempt: 501(c)(3).
Institution Type/Description: Oil Museum.
Collections: 1920s-1930s in East Texas when East Texas Oil Field was discovered & developed, largest oil field in the world at that time.
Hours & Admission Prices: April-Sept. Tues.-Sat. 9-5, Sun. 2-5; Oct.-March Tues.-Sat. 9-4, Sun. 2-5; call for special schedule Dec. 20-31. Adults $8, children 3-11 $5. Closed Easter; Thanksgiving; Christmas. &
Attendance: 50,000 (accurate)

Kingsbury

PIONEER FLIGHT MUSEUM, Old Kingsbury Aerodrome, Farm Rd. 1104, Kingsbury, TX 78638-2528. Tel.: 830-639-4162.
Web Site: pioneerflightmuseum.org
Institution Type/Description: Flight Museum.
Collections: early flight history; early aircraft & vehicles.

Hours & Admission Prices: By appointment.

Kingsville

JOHN E. CONNER MUSEUM, Texas A&M University-Kingsville, 905 W. Santa Gertrudis, Kingsville, TX 78363. Mailing Address: P.O. Box 2172, Station 1, Kingsville, TX 78363-8321. Tel.: 361-593-2810. Fax: 361-593-2112.
Founded: 1925.
Congressional District: 27
Key Personnel: Dir., Jonathan Plant; Educator, Brenda Canizalez; Administrative Asst., Cynthia F. Villalon.
Personnel Profile: Full-Time Paid 4; Part-Time Paid 3; Part-Time Volunteers 15.
Governing Authority: university. Parent Institution: Texas A&M University-Kingsville. Tax-exempt.
Institution Type/Description: General Museum.
Collections: south Texas & University history; Farm & Ranching tools & equipment; Native American & Mexican American cultural materials; south Texas & Meso American archaeology; south Texas natural history; geology; Graves Peeler mounted trophy specimens.
Research Fields: local, regional & natural history.
Facilities: Texana books for sale.
Activities: guided tours; field trips; lectures; films; videotapes; hobby workshops; formally organized education programs for children, graduate & undergraduate students affiliated with Texas A&M University - Kingsville; volunteer training programs; permanent, temporary & traveling exhibitions; school loan service; docent outreach programs on subjects related to the museum for local & area schools.
Publications: loose-leaf guide for docents, Las Manos; quarterly newsletter; handbook, South Texas Wildflowers: Collection I; color brochure; exhibit catalog, El Rancho in South Texas; natural history video, The Living Mosaic.
Hours & Admission Prices: Mon.-9-5, Sat. 9-4. No charge; donations accepted. Closed university holidays. &
Attendance: 20,000 (estimated)
Membership: Individual $15; Family $25; Participating $100; Sustaining $500; Patron $1,000; Life $2,000.

KING RANCH MUSEUM, 405 N. 6th St., Kingsville, TX 78363. Tel.: 361-595-1881.
Institution Type/Description: History Museum.
Collections: local history; photographs; saddles; guns; early carriages & cars; sculpture.
Hours & Admission Prices: Mon.-Sat. 10-4, Sun. 1-5. Closed Easter; Memorial Day; Independence Day; Thanksgiving; Christmas Eve & Day.

KINGSVILLE TRAIN DEPOT MUSEUM, 102 Kleberg Ave., Kingsville, TX 78363. Mailing Address: c/o KCVB, 1501 Hwy. 77, Kingsville, TX 78363. Tel.: 361-592-8516.
Institution Type/Description: Historic Building Museum: built in 1904.
Collections: local & depot history; photographs; period furnishings; telegraph; personal artifacts.
Facilities: Museum-related items for sale.
Hours & Admission Prices: Mon.-Fri. 10-4, Sat. 10-1; groups by appointment. No charge.

Kountze

BIG THICKET NATIONAL PRESERVE, 6044 FM 420, Kountze, TX 77625-7841. Tel.: 409-951-6700. Fax: 409-951-6717.
E-mail: leslie.dubey@nps.gov
Web Site: www.nps.gov/bith/
Founded: 1974.
Congressional District: 2
Key Personnel: C.E.O. & Supt., Todd Brindle; Chief Div. Interpretation, Leslie Dubey; Museum Shop Mgr., Cindy Reed.
Personnel Profile: Full-Time Paid 36; Full-Time Volunteers 34; Part-Time Paid 2; Part-Time Volunteers 10; Interns 2.
Governing Authority: federal. Parent Institution: National Park Service, Dept. of the Interior. Tax-exempt.
Institution Type/Description: Biological Preserve.
Collections: natural history; herbarium; skulls & skins; insects; cultural history; pioneer artifacts. Visitor Center: plant communities; interactive dioramas; ecology of the Big Thicket film.
Research Fields: biological.
Facilities: library available for use on premises; visitor center; nature trails. Museum-related items for sale.

Activities: self-guided trails; guided hikes; environmental education program; Jr. Ranger program.
Publications: pamphlets; checklists.
Hours & Admission Prices: Visitor Center: daily 9-5. Headquarters: Mon.-Fri. 8-4:30. No charge; donations accepted. &
Attendance: 115,000 (estimated)

Kyle

KATHERINE ANNE PORTER LITERARY CENTER, 508 Center St., Kyle, TX 78640. Tel.: 512-268-6637.
Institution Type/Description: History Museum.
Collections: local history & culture; period furnishings; personal artifacts; photographs.
Hours & Admission Prices: By appointment.

La Grange

FAYETTE HERITAGE MUSEUM & ARCHIVES, 855 S. Jefferson, La Grange, TX 78945-3230. Tel.: 979-968-6418. Fax: 979-968-5357.
E-mail: library@cityoflg.com
Web Site: www.cityoflg.com/library.html
Founded: 1978.
Congressional District: 14
Key Personnel: Dir. & Museum Shop Mgr., Sherie Knape.
Personnel Profile: Full-Time Paid 2; Part-Time Paid 1; Part-Time Volunteers 3.
Governing Authority: municipal; nonprofit. Parent Institution: City of LaGrange. Tax-exempt.
Institution Type/Description: History Museum.
Collections: local history items; photos; newspapers; maps; school, business, organization & personal papers.
Research Fields: local history.
Facilities: 500-vol. library available to the public; meeting room. Books for sale.
Hours & Admission Prices: Tues.-Thurs. 10-6, Fri. 10-5, Sat. 10-1, Sun. 1-5. No charge. Closed major holidays. &
Attendance: 2,500 (estimated)

NATHANIEL W. FAISON HOME AND MUSEUM, 822 S. Jefferson St., State Hwy. 77, La Grange, TX 78945. Mailing Address: P.O. Box 681, La Grange, TX 78945. Tel.: 979-968-9416 & 5756.
E-mail: mariewatts@cvtv.net
Web Site: www.faisonhouse.org
Founded: 1960.
Key Personnel: Arnold Romberg Suzy Romberg.
Personnel Profile: Part-Time Volunteers 12.
Governing Authority: nonprofit organization. Affiliated with the La Grange Garden Club. Tax-exempt.
Institution Type/Description: Historic House: 1840-1855 Nathaniel W. Faison Home.
Collections: paintings; history of Texas; original furniture; documents.
Facilities: senior citizen AARP community center; movie theatre; garden club; ballrooms; bowling alley.
Activities: guided tours; permanent exhibitions; educational programs & scholarships for children in the community. Annual Event: Christmas Open House in December.
Publications: brochure.
Hours & Admission Prices: 2nd Sat. each month 10-4; tours by appointment. Admission $3; children 11 & under no charge.
Attendance: 400 (estimated)
Membership: Active Members $10; Associate Members $25.

La Porte

BATTLESHIP TEXAS STATE HISTORIC SITE, 3523 Independence Pkwy. S., La Porte, TX 77571. Tel.: 281-479-2431, ext. 236 & 248. Fax: 281-479-5618.
E-mail: barbara.graf@tpwd.state.tx.us
Web Site: www.tpwd.state.tx.us/park/battlesh
Formerly: Battleship Texas State Historical Park
Founded: 1948.
Congressional District: 25
Key Personnel: Dir., Andy Smith; Pres. (V), Herb Powers; Cur., Winnie Trippet; Mgr. Collections, Janice Sniker; Museum Shop Mgr., Jerry Moak.
Personnel Profile: Full-Time Paid 20; Part-Time Volunteers 51; Interns 3.
Governing Authority: state. Parent Institution: Texas Parks & Wildlife Dept. Tax-exempt.

Institution Type/Description: Historic Warship.
Collections: photos, documents, furnishings, & artifacts relating to operation & history of the Battleship TEXAS, only surviving World War I era DREAD-NOUGHT; veteran of Vera Cruz Incident (1914), World War I (1917-18) & World War II (1941-45).
Research Fields: material pertaining to USS Texas (BB00) & (BB35) with some material relating to other U.S. Naval forces 1914-1945.
Facilities: battleship.
Activities: self-guided tour; guided hard hat tours. Museum Sponsors: Veterans Art & Essay Contest in November; Pearl Harbor Ceremony in December; Yuletide Texas - A Sailors Christmas in December.
Publications: The Dreadnought quarterly; bi-weekly e-newsletter, Battle Report.
Hours & Admission Prices: Daily 10-5; groups by appointment. Adults $12, seniors $6, school & youth groups $3; children 12 & under no charge. Closed Thanksgiving; Christmas Eve & Day.
Attendance: 111,000 (estimated)

SAN JACINTO MUSEUM OF HISTORY ASSOCIATION, (M), One Monument Circle, La Porte, TX 77571-9585. Tel.: 281-479-2421. Fax: 281-479-2428.
E-mail: sjm@sanjacinto-museum.org
Web Site: sanjacinto-museum.org
Founded: 1938.
Congressional District: 8
Key Personnel: C.O.O. & Pres., Larry Spasic; Chm. (V), Robert Hixon; Museum Shop Mgr., Josh Olivarez.
Personnel Profile: Full-Time Paid 15; Part-Time Paid 3.
Governing Authority: nonprofit organization. Tax-exempt: 501(c)(3).
Institution Type/Description: Historic Building: 1939 San Jacinto Monument.
Collections: history of region & Texas; artifacts; manuscripts; documents; visual arts.
Facilities: 25,000-vol. library of history books, newspapers & periodicals available for use by appointment only; 162-seat theater & audio-visual production. Texas history books, postcards, slides & museum-related items for sale.
Activities: permanent, temporary & traveling exhibitions.
Publications: occasional monographs; quarterly newsletter.
Hours & Admission Prices: Daily 9-6. No charge. Three Venues: adults $12; members no charge. Closed Thanksgiving; Christmas Eve & Day.
Attendance: 450,000 (estimated)
Membership: Student & Senior Citizen $25; Individual $50; Dual Senior Citizen $45; Dual Individuals $75; Family Freedom $125; Independence $275; Museum Circle $500; Monument $1,000. Corporate: Sam Houston $3,000; San Jacinto $5,000.

Lackland Air Force Base

USAF AIRMAN HERITAGE MUSEUM, 2051 George Ave., Bldg. 5206, Lackland Air Force Base, TX 78236-5218. Tel.: 210-671-3055. Fax: 210-671-0347.
Formerly: History and Traditions Museum
Founded: 1956.
Congressional District: 20
Personnel Profile: Full-Time Paid 1; Part-Time Volunteers 3.
Governing Authority: federal.
Institution Type/Description: Military Museum.
Collections: rare aeronautical equipment; 40 static aircraft painted to represent noted pilots & squadrons; engines, weapons & memorabilia related to enlisted training from WWI to present.
Research Fields: aviation & aerospace history.
Facilities: library of books, periodicals, photographs, news clippings.
Activities: guided tours; films; educational programs.
Publications: museum brochure.
Hours & Admission Prices: Mon.-Wed. & Fri. 8-4, Thurs. 9-5:30. No charge; donation accepted. Closed federal holidays.
Attendance: 30,000 (accurate)

USAF SECURITY FORCES MUSEUM, 1300 Femoyer St., Bldg. 10501, Lackland Air Force Base, TX 78236. Tel.: 210-671-2615.
E-mail: 37trw.sfmu@us.af.mil
Web Site: www.securityforcesmuseum.af.mil
Founded: 1977.
Key Personnel: Dir., Terri L. Bedore-Andersen.
Personnel Profile: Full-Time Paid 1; Part-Time Volunteers 1.
Governing Authority: federal government. Tax-exempt: 170(b)(1)(A).
Institution Type/Description: Military History Museum.
Collections: history & heritage of the Air Force Security Forces from Air Police in 1947 to present; 1,400 artifacts; over 3,000 archival materials including documents & photographs; Air Force Military Police; Security Police; Security Forces Career Field.
Research Fields: U.S. Air Force including Air Police, Military Police, Security Police, & Security Forces Career Field.
Facilities: research library. 60-seat auditorium; 7,500 sq. ft. exhibit space; 60-seat theater.
Activities: guided tours; participatory exhibits.
Hours & Admission Prices: Museum: Mon.-Wed. & Fri. 8-4, Thurs. 9-5:30; by appointment. Research Center: by appointment. No charge. Closed federal holidays.
Attendance: 16,885 (accurate)

Lago Vista

LAGO VISTA AIRPOWER MUSEUM, Rusty Allen Airport, Lago Vista, TX 78645. Mailing Address: 20624FM 1431, Ste. #8, Lago Vista, TX 78645. Tel.: 512-267-7952. Fax: 512-267-2338.
E-mail: info@lagovista.org
Web Site: www.lagovista.org
Institution Type/Description: History Museum.
Collections: military history & artifacts; WW I & II artifacts; photographs; oral histories; aircraft; engines; models; books.
Hours & Admission Prices: Sat. 1-5; other times by appointment. No charge; donations accepted.

Lake Jackson

LAKE JACKSON HISTORICAL MUSEUM, (M), 249 Circle Way, Lake Jackson, TX 77566-5232. Mailing Address: P.O. Box 242, Lake Jackson, TX 77566-0242. Tel.: 979-297-1570. Fax: 888-247-0046.
E-mail: director@lakejacksonmuseum.org
Web Site: lakejacksonmuseum.org
Founded: 1981.
Key Personnel: Dir., Jennifer Caulkins; Pres., Gayle Driskill; Treas., Harry Sargent; Museum Shop Mgr., Angela Villarreal.
Personnel Profile: Full-Time Paid 2; Part-Time Paid 1; Part-Time Volunteers 45.
Governing Authority: private; nonprofit organization. Tax-exempt: 501(c)(3).
Institution Type/Description: History Museum.
Collections: history of Lake Jackson; art; clothing; textiles; documents; photographs; audio & video; newspapers.
Facilities: 24-seat auditorium; 15,000 sq. ft. exhibit space. Museum-related items for sale.
Activities: docent program; formal education programs; guided tours; lectures; temporary & traveling exhibitions; theater; oral history programs.
Publications: quarterly newsletter, "The Lake Jackson Historian".
Hours & Admission Prices: Tues.-Sat. 10-4, Sun. 1-4. No charge.
Attendance: 6,000 (accurate)
Membership: Student & Individual Senior $20; Senior Family $35; Family $40; Nonprofits $50; Business $100; Life $500; Patron $1,000; Benefactor $5,000.

SEA CENTER TEXAS, 300 Medical Dr., Lake Jackson, TX 77566. Tel.: 979-292-0100.
Founded: 1996.
Key Personnel: Dir., David Abrego
Institution Type/Description: Aquarium.
Collections: sharks; eel; salt marsh aquarium; tropical fish; artificial reef; coral reef; over 150 species of birds; butterfly & hummingbird gardens.
Facilities: visitor center; fish hatchery. Aquarium-related items for sale.
Activities: educational programs; fishing pond; touch tank.
Hours & Admission Prices: Tues.-Sat. 9-4, Sun. 1-4. No charge; donations accepted.

Lamesa

DAL-PASO MUSEUM, 125 Main Ave., Lamesa, TX 79331. Mailing Address: P.O. Box 1445, Lamesa, TX 79331-1445. Tel.: 806-872-5007. Fax: 806-872-2181.
Founded: 1988.
Congressional District: 19
Key Personnel: C.E.O., Wayne C. Smith, Ed.D.; Pres., Walter Buckel; Museum Shop Mgr., Lorraine Johnson.
Personnel Profile: Part-Time Paid 1.
Governing Authority: municipal; county. Parent Institution: Lamesa-Dawson County Museum Association. Tax-exempt: 501(c)(3).
Institution Type/Description: General Museum: housed in 1925 restored Dal Paso Hotel.

Collections: general & historical interest to local viewers.
Research Fields: local history.
Activities: guided tours.
Hours & Admission Prices: Mon., Thurs. & Sat. 2-5. No charge. &
Attendance: 1,000 (estimated)

Lampasas

LAMPASAS COUNTY MUSEUM, 303 S. Western St., Lampasas, TX 76550. Mailing Address: P.O. Box 373, Lampasas, TX 76550. Tel.: 512-556-2224.
Formerly: Keystone Square Museum
Founded: 1979.
Governing Authority: Tax-exempt.
Institution Type/Description: History Museum.
Collections: county history & culture; period furnishings; personal artifacts; photographs
Activities: group tours; special events. Museum Sponsors: Needle Art and Quilt Show in March; Teddy Bear Tea Party in April; Honoring our Veterans in November; Christmas Tree and Gingerbread House Contests in December; Christmas at the Museum in December; Holiday Home Tour in December.
Hours & Admission Prices: Sat. 10-2; other times by appointment. &
Attendance: 1,100 (estimated)

Lancaster

COLD WAR AIR MUSEUM, 850 Ferris Rd., Lancaster, TX 75146. Tel.: 972-218-9700.
Governing Authority: nonprofit organization.
Institution Type/Description: Military History Museum.
Collections: military history, aircraft & artifacts.
Hours & Admission Prices: Mon.-Sat. 10-4; other times by appointment.

Langtry

JUDGE ROY BEAN VISITOR CENTER, Hwy. 90, W., Loop 25, Langtry, TX 78871. Mailing Address: P.O. Box 160, Langtry, TX 78871-0160. Tel.: 800-452-9292. Fax: 915-291-3366. Facebook: Langtry TIC.
E-mail: lytic@dot.state.tx.us
Web Site: www.dot.state.tx.us
Founded: 1939.
Congressional District: 21
Key Personnel: Supvr., Kenneth R. Fatheree.
Personnel Profile: Full-Time Paid 6.
Governing Authority: state. Tax-exempt.
Institution Type/Description: Historic Building Museum: c.1896 Judge Roy Bean Saloon.
Collections: relics pertaining to Judge Roy Bean, Law West of the Pecos.
Research Fields: history concerning Roy Bean.
Facilities: botanical garden.
Hours & Admission Prices: Daily 8-5. No charge. Closed New Year's Day; Easter; Thanksgiving; Christmas Eve & Day. &
Attendance: 70,000 (estimated)

Laredo

IMAGINARIUM OF SOUTH TEXAS, 5300 San Dario, Ste. 505, Mall del Norte, Laredo, TX 78041-3000. Tel.: 956-728-0404. Fax: 956-725-7776.
E-mail: info@imaginariumstx.org
Web Site: www.imaginariumstx.org
Formerly: Laredo Children's Museum
Founded: 1991.
Key Personnel: Dir., Sandra Cavazos.
Personnel Profile: Full-Time Paid 1; Full-Time Volunteers 0; Part-Time Paid 15; Part-Time Volunteers 100; Interns 3.
Governing Authority: nonprofit organization. Parent Institution: Laredo Children's Museum, Inc. Tax-exempt: 501(c)(3).
Institution Type/Description: Children's Museum: housed in c.1900 Fort McIntosh chapel & guardhouse.
Collections: grocery store; construction zone; toddler space; science; Imagination Station; Little Learner Library; Bubble Daze; NASA Station
Research Fields: science, technology, engineering and mathematics.
Facilities: 4,000 sq. ft. exhibit space; train car.
Activities: formal education programs for undergraduate or graduate students affiliated with Laredo Junior College & Texas A&M International Univ.; seasonal camps; community workshops; teacher-training workshops; outreaches.

Hours & Admission Prices: Wed.-Thurs. 10-7, Fri.-Sat. 10-8, Sun. 12-6. Children & adults $4; members no charge. Closed New Year's Day; Easter; Thanksgiving; Christmas Eve & Day. &
Attendance: 40,000 (accurate)
Membership: Student $50; Bright Choice $75; Texans $80; Passport $110; Lone Star $250; Rio Grande $500.

LAREDO CENTER FOR THE ARTS, 500 San Agustin Ave., Laredo, TX 78040-8103. Tel.: 956-725-1715. Fax: 956-725-1741.
E-mail: info@laredoartcenter.org
Web Site: www.laredoartcenter.org
Key Personnel: Exec. Dir., Gabriel Castillo
Institution Type/Description: Art Museum.
Collections: works by local & international artists.
Activities: workshops.
Publications: quarterly newsletters.
Hours & Admission Prices: Tues.-Sat. 11-4. No charge.
Membership: Warhol $35; Kahlo $40; Haring $50; Rothko $100; Rauschenberg $200; Picasso $500; O'Keefe $1,000; Matisse $2,000; Degas $3,000; Rembrandt $5,000.

League City

BUTLER LONGHORN MUSEUM AND HERITAGE PARK, (M), 1220 Coryell St., League City, TX 77573. Tel.: 281-332-1393.
E-mail: butler.longhornmuseuminfo@yahoo.com
Web Site: www.butlerlonghornmuseum.com
Founded: 2009.
Key Personnel: Dir., Monica Hughes.
Volunteer Hours: 2,000
Governing Authority: nonprofit organization.
Institution Type/Description: History Museum.
Collections: local history, ranching & farming; Butler family history; Longhorn artifacts; photographs; personal artifacts; period furnishings; western art & music.
Hours & Admission Prices: Tues.-Sat. 10-4.
Attendance: 100,000 (accurate)

WEST BAY COMMON SCHOOL CHILDREN'S MUSEUM, (M), 210 N. Kansas Ave., League City, TX 77573-2466. Tel.: 281-554-2994.
E-mail: catharin@orbitworld.net
Web Site: www.oneroomschoolhouse.org/
Key Personnel: Dir., Catharin Lewis.
Governing Authority: Parent: League City Historical Society.
Institution Type/Description: Children's Museum.
Collections: hands-on history learning environment.
Hours & Admission Prices: Mon.-Thurs. 9-4, Fri. 9-1; other times by appointment. History Program: $4 per person. Docent Talk: no charge.

Leakey

REAL COUNTY HISTORICAL MUSEUM, Evergreen St., Leakey, TX 78873. Mailing Address: P.O. Box 852, Leakey, TX 78873. Tel.: 830-232-5330.
E-mail: realmuseum@hctc.net
Web Site: www.realcountyhistoricalmuseum.com
Founded: 1986.
Institution Type/Description: History Museum.
Collections: local history & heritage; period furnishings; personal artifacts; photographs.
Hours & Admission Prices: March-Dec. Fri.-Sat. 10-2. Adults $1, children under 12 $.25.

Levelland

CHRISTINE DEVITT FINE ART CENTER, South Plains College, 1401 S. College Ave., Levelland, TX 79336. Tel.: 806-716-2261.
E-mail: lgreer@southplainscollege.edu
Founded: 2008.
Personnel Profile: Part-Time Paid 2; Interns 2.
Governing Authority: Parent Institution: South Plains College. Tax-exempt.
Institution Type/Description: Art Gallery.
Collections: works by Marjorie Merriweather Post; paintings.
Hours & Admission Prices: Call for hours. No charge. Closed school holidays. &
Attendance: 1,000 (estimated)

Liberty

SAM HOUSTON REGIONAL LIBRARY & RESEARCH CENTER, 650 FM 1011, Liberty, TX 77575-6841. Mailing Address: P.O. Box 310, Liberty, TX 77575-0310. Tel.: 936-336-8821. Fax: 936-336-7049.
E-mail: samhoustoncenter@tsl.state.tx.us
Web Site: www.tsl.state.tx.us
Founded: 1977.
Congressional District: 2
Key Personnel: Mgr., Alana Inman; Librarian, Darlene Mott; Archivist, Lisa Meisch; Archives Processor, Sandra Burrell.
Personnel Profile: Full-Time Paid 4; Part-Time Paid 1; Part-Time Volunteers 4.
Governing Authority: nonprofit; state. Parent Institution: Texas State Library & Archives Commission. Tax-exempt.
Institution Type/Description: Regional Museum & Archives.
Collections: Southeast Texas history; Indian artifacts; household goods; glass; Civil War items; photographs; local government records; blueprints & maps; newspapers; manuscripts. Historic Buildings: 1848 Gillard-Duncan House; Gov. Price Daniel Home & Archives; 1883 Norman House; 1898 St. Stephen's Episcopal Church; 1930 Hull Rotary.
Research Fields: pre-historic to modern times history of southeast Texas; Sam Houston; Price Daniel; Martin Dies; Jean Lafitte.
Facilities: research library of Southeast Texas history; educational facilities; exhibit space.
Activities: guided tours; lectures; workshops.
Hours & Admission Prices: Tues.-Fri. 8-5, Sat. 9-4. No charge; donations accepted. Closed major holidays. &
Attendance: 1,500 (accurate)

Lipscomb

WOLF CREEK HERITAGE MUSEUM, 13310 Hwy. 305, Lipscomb, TX 79056. Mailing Address: P.O. Box 5, Lipscomb, TX 79056. Tel.: 806-852-2123. Fax: 806-852-2172.
E-mail: wolfcrk@amaonline.com
Web Site: www.wolfcreekheritagemuseum.org
Founded: 1982.
Congressional District: 13
Key Personnel: Dir., Virginia Scott; Pres. (V), Anna Lee Barton; Museum Shop Mgr., Dorothy Schoenhals.
Personnel Profile: Part-Time Volunteers 7.
Governing Authority: Tax-exempt.
Institution Type/Description: History Museum.
Collections: local history & culture; photographs; personal artifacts; period furnishings; works by local artists.
Publications: quarterly newsletter; Lipscomb County History, Vol. I & II; Pioneers of Prairie; Panhandle Interstate 1887-1888.
Hours & Admission Prices: Mon.-Fri. 10-4; other times by appointment. No charge; donations accepted. &
Attendance: 1,686
Membership: Individual $25; Family $35; Friends $50-$399; Patrons $400-$999; Benefactors $1,000 & up.

Littlefield

DUGGAN HOUSE MUSEUM, 520 E. Waylon Jennings Blvd., Littlefield, TX 79339. Mailing Address: P.O. Box 734, Littlefield, TX 79339. Tel.: 806-385-9001.
Institution Type/Description: History Museum.
Collections: local history & culture; period furnishings; personal artifacts; photographs.
Activities: special events.
Hours & Admission Prices: Call for hours.

Livingston

POLK COUNTY MEMORIAL MUSEUM, 514 W. Mill St., Livingston, TX 77351. Tel.: 936-327-8192. Fax: 936-327-8192.
E-mail: museum@livingston.net
Web Site: www.livingston.net/museum
Founded: 1963.
Congressional District: 2
Key Personnel: Chm., Patricia Srook; Vice Chm., Joanne Westmoreland; Chmn., Wanda L. Bobinger.
Personnel Profile: Full-Time Paid 2; Part-Time Paid 1; Part-Time Volunteers 35.
Governing Authority: nonprofit organization. Parent Institution: Polk County Historical Commission. Tax-exempt.
Institution Type/Description: History Museum.

Collections: Polk County history; agriculture; archaeology; archives; folklore; glass; Indian artifacts; geological exhibits; textiles; military; numismatic; manuscript collections; Civil War artifacts.
Research Fields: local history.
Facilities: archives; tapes; microfilm; reference books available for use on premises; reading room.
Activities: guided tours; lectures; films; formally organized education programs for children; permanent & temporary exhibitions.
Publications: Civil War: There Never Were Such Men Before; Peebles; Polk County Pictorial History; Cemeteries in Polk County; The History of Polk County; Torched - Fire of 1902; Trains & Railroads of Polk County; Courthouses of Polk County.
Hours & Admission Prices: Mon.-Fri. 9-5. No charge; donations accepted. &
Attendance: 4,500 (accurate)

Llano

LLANO COUNTY HISTORICAL MUSEUM, 310 Bessemer Ave., Llano, TX 78643. Mailing Address: P.O. Box 434, Llano, TX 78643. Tel.: 325-247-3026.
E-mail: llanomuseum@verizon.net
Founded: 1965.
Congressional District: 24
Personnel Profile: Full-Time Paid 3.
Governing Authority: Tax-exempt.
Institution Type/Description: History Museum: housed in the former Bruhl Drugstore building.
Collections: local history & culture; period furnishings; polo artifacts; farm & ranch equipment; clothing; textiles; military artifacts; rocks; photographs. Historic Building: log cabin.
Hours & Admission Prices: Wed.-Sat. 11-5. No charge; donations accepted. &
Membership: Student $5; Individual $10; Family $25; Patron $50; Sustaining Patron $100.

Longview

GREGG COUNTY HISTORICAL MUSEUM, 214 N. Fredonia St., Longview, TX 75601-7222. Mailing Address: P.O. Box 3342, Longview, TX 75606-3342. Tel.: 903-753-5840. Fax: 903-753-5854.
E-mail: bill@gregghistorical.org
Web Site: gregghistorical.org
Founded: 1983.
Congressional District: 7
Key Personnel: Exec. Dir., Neina Kennedy; Pres. (V), Walter Northcutt.
Personnel Profile: Full-Time Paid 1; Part-Time Paid 1; Part-Time Volunteers 65; Interns 3.
Governing Authority: nonprofit organization; society. Parent Institutions: The Gregg County Historical Foundation & Gregg County Historical & Genealogical Society. Tax-exempt: 501(c)(3).
Institution Type/Description: County Museum: housed in 1910 brick Everett bank building.
Collections: artifacts, memorabilia & photographs illustrating the development of Gregg County from 1870-1930; exhibits progress chronologically from early settler's subsistence farming, to commercial agriculture to the growth of cities in early 1900s; theme rooms & areas: barn; railroad depot; dentist office; parlor; bedroom; general mercantile store; log cabin interior; newspaper presses; military collection; manuscript collections; Texas architecture; Caddo Indians.
Research Fields: genealogy; local history; county history; Texas architecture; archaeology.
Facilities: library of books & archives pertaining to local & county history available for research on premises only; reading room; meeting room; classroom. Museum-related items for sale.
Activities: guided tours; lectures; demonstrations; study clubs; organized education programs for children; volunteer program; docent organization; permanent exhibits; oral history program.
Publications: Gregg County Historical Museum Newsletter; activity books for children; books, Home Grown: How We Grew; Did You Know; Guide to Gregg County's Historical Markers; Traditions of the Land: The History of Gregg County, Texas.
Hours & Admission Prices: Tues.-Fri. 10-4, Sat. 12-4. Adults $2, children under 18 & senior citizens $1; discount to AAA members; members no charge; Texas Association of Museums reciprocal admissions program. Closed major holidays.
Attendance: 12,000 (estimated)
Membership: Senior $35; Individual $50; Family $75; Sustaining $100; Explorer $250; Pioneer $500; Builder $1,000; Patron $1,500; Director $2,500; Historian $5,000; Inventor $10,000.

LONGVIEW MUSEUM OF FINE ARTS, 215 E. Tyler St., Longview, TX 75601-7219. Mailing Address: P.O. Box 3484, Longview, TX 75606-3484. Tel.: 903-753-8103. Fax: 903-753-8217. Facebook: Longview Museum of Fine Arts.
E-mail: fineart@lmfa.org
Web Site: www.lmfa.org
Formerly: Longview Museum & Arts Center
Founded: 1958.
Congressional District: 4
Key Personnel: Dir., Renee Hawkins; Chm. (V), Glenn McCutchen; Pres. (V), Nancy Peralta; Dir. Education, Ann Werline; Dir. Education, Andy Don Emmons; Registrar, Michelle Newby; Museum Shop Mgr., Valencia Southerland.
Personnel Profile: Full-Time Paid 5; Part-Time Paid 1; Part-Time Volunteers 18; Interns 1.
Governing Authority: nonprofit organization. Tax-exempt: 501(c)(3).
Institution Type/Description: Art Museum.
Collections: paintings; sculpture; graphics; photographs.
Facilities: 2,500-vol. library of catalogs, art reference works, periodicals & photographs; educational facilities; sculpture garden.
Activities: guided tours; lectures; films; formally organized education programs; docent program; inter-museum loan, temporary & traveling exhibitions.
Publications: bimonthly newsletter; annual reports; gallery guides; brochures; checklists.
Hours & Admission Prices: Tues.-Fri. 10-4, Sat. 12-4. Adults $5; discounts to NARM members; members no charge. Closed national holidays. &
Attendance: 16,000 (estimated)
Membership: Student $10; Individual $25; General $50; Contributor $100; Advocate $250; Supporter $500; Patron $1,000; Sustainer $2,500; Benefactor $5,000; Promoter $10,000; Bestower $25,000; Angel $50,000.

Lubbock

AMERICAN WIND POWER CENTER, 1701 Canyon Lake Dr., Lubbock, TX 79403-4908. Tel.: 806-747-8734. Fax: 806-740-0668.
E-mail: sales@windmill.com
Web Site: windmill.com
Formerly: American Wind Power Center and Museum
Founded: 1993.
Key Personnel: Dir., Coy Harris; Chm. (V) & Museum Shop Mgr., Tanya Meadows.
Personnel Profile: Full-Time Paid 3; Part-Time Paid 2; Part-Time Volunteers 10.
Governing Authority: nonprofit organization. Tax-exempt: 501(c)(3).
Institution Type/Description: History Museum.
Collections: wind power history; wind mills; photographs; wind turbines; model trains.
Major Exhibits: Windy City, 11/14-12/14.
Research Fields: wind power.
Facilities: Museum-related items for sale.
Activities: educational programs.
Publications: quarterly newsletter.
Hours & Admission Prices: May-Sept. Tues.-Sat. 10-5, Sun. 2-5; Oct.-April Tues.-Sat. 10-5. Admission $5 per person; discounts to families; active military & their families no charge. &
Membership: Eclipse Windmiller Crew $40; Maud S. Windmiller Crew $50; Flowerdew Crew $125; Hummer Windmiller Crew $250; Flying Dutchman Crew $500; Twin Wheel Crew $1,000; Texas Pattern Crew $2,500.

BAYER MUSEUM OF AGRICULTURE, (M), 1121 Canyon Lake Dr., Lubbock, TX 79403-4911. Mailing Address: P.O. Box 505, Lubbock, TX 79408-0505. Tel.: 806-744-3786; 877-789-8335.
E-mail: amadirector@agriculturehistory.org
Web Site: www.agriculturehistory.org
Formerly: American Museum of Agriculture
Key Personnel: Dir., Lacee Hoelting.
Governing Authority: Tax-exempt.
Institution Type/Description: Agriculture Museum.
Collections: agricultural history; farm equipment; tractors; farming.
Hours & Admission Prices: Wed.-Sat. 10-5.
Membership: Individual Donor $35; Family Contributor $50; Top Hand $250; Crew Boss $500; Sod Buster $1,000; Homesteader $5,000; Pioneer $10,000; Master Farmer $25,000.

BUDDY HOLLY CENTER, (M), 1801 Crickets Ave., Lubbock, TX 79401-5128. Tel.: 806-775-3560. Fax: 806-767-0732. Facebook: Buddy Holly Center.
E-mail: info@buddyhollycenter.org
Web Site: www.buddyhollycenter.org
Founded: 1999.
Congressional District: 19
Key Personnel: Dir., Brooke Witcher.
Personnel Profile: Full-Time Paid 7; Part-Time Paid 4.
Governing Authority: Parent Institution: City of Lubbock. Tax-exempt.
Institution Type/Description: History Museum.
Collections: Buddy Holly Gallery: life & music of Buddy Holly; clothing; photographs; recording contracts; personal artifacts; artists & musicians of West Texas. Fine Arts Gallery: contemporary Art. Foyer Gallery: personal artifacts; photographs; audio; archives. Historic House: J.I. Allison House.
Facilities: Museum-related items for sale.
Activities: special events.
Hours & Admission Prices: Tues.-Sat. 10-5, Sun. 1-5; groups by appointment. Call for holiday hours. Adults $5, senior citizens 60 & over $3, children 7-17 & students $2; discounts to groups of 20 or more; active military, members and children 6 & under no charge. &

LOUISE HOPKINS UNDERWOOD CENTER FOR THE ARTS, (M), 511 Ave. K, Lubbock, TX 79401. Tel.: 806-762-8606.
Key Personnel: Exec. Dir., Karen Wiley
Institution Type/Description: Art Gallery.
Collections: works by contemporary artists; sculpture; paintings.
Facilities: 159-seat theatre.
Activities: performances; educational programs.
Hours & Admission Prices: Tues.-Sat. 9-5.

✱ MUSEUM OF TEXAS TECH UNIVERSITY, (M), 3301 4th St., Lubbock, TX 79403-4613. Mailing Address: P.O. Box 43191, Lubbock, TX 79409-3191. Tel.: 806-742-2442. Fax: 806-742-1136.
E-mail: museum.texastech@ttu.edu
Web Site: www.museum.ttu.edu
Founded: 1929.
Congressional District: 19
Key Personnel: Exec. Dir., Dr. Eileen Johnson; Pres., Dr. Guy Bailey; Chancellor, Kent Hance; Asst. Dir., Nicola Ladkin; Registrar, Matt Renick; Cur. Ethnology & Textiles, Mei Campbell; Cur. & Dir. Natural Science Research Lab., Dr. Robert Baker; Cur. Paleontology, Dr. Sankar Chatterjee; Cur. Education, Dr. Jill Hoffman; Cur. History, Henry B. Crawford; Cur. Art, Dr. Peter S. Briggs; Business Mgr., Jamie Looney; Museum Shop Mgr., Jouana Stravlo; Head Guard, Sara Armenta; Administrative Asst., Claudia Cory.
Personnel Profile: Full-Time Paid 62; Full-Time Volunteers 1; Part-Time Paid 32; Part-Time Volunteers 50; Interns 3.
Governing Authority: state; university. Parent Institution: Texas Tech University. Supporting Organization: Museum of TTU Association. Tax-exempt: 170(b)(l)(iii).
Institution Type/Description: General Museum.
Collections: art; clothing; textiles; historical furnishings & artifacts; ethnology; anthropology; mammalogy; paleontology; geology; ornithology; frozen tissues.
Research Fields: mammalogy; anthropology; paleontology; history; historical architecture; historical textiles; earth sciences; invertebrates; genetic resources.
Facilities: 10,000-vol. library; planetarium; archeological site; research labs. Museum-related items for sale.
Activities: Master of Arts Museum Science Program; Master of Museum Science with alternate track in Heritage Management; planetarium programs; school tours programs; docent guild; guided tours; gallery talks; films; symposia; permanent, temporary & traveling exhibits.
Publications: quarterlies by support organizations; museum journal; occasional papers; special publications; newsletter, MuseNews; electronic newsletter, eMuse; planetarium newsletter, What's Up.
Hours & Admission Prices: Tues.-Wed. & Fri.-Sat. 10-5, Thurs. 10-8:30, Sun. 1-5. Main building: no charge; donations accepted. Planetarium: general admission $5, children, seniors & university students $3; children 5 & under and active duty military & family no charge. Closed New Year's Eve & Day; Martin Luther King Jr. Day; Memorial Day; Independence Day; Thanksgiving; Christmas Break. &
Attendance: 172,356 (accurate)
Membership: Museum of TTU Association: Student $10; Educator $25; Individual $35; Family $40; Museum League $75; Patron $150; Benefactor $250; Director's Circle $500.

NATIONAL RANCHING HERITAGE CENTER, Texas Tech University, 3121 Fourth St., Lubbock, TX 79409. Tel.: 806-742-0498. Fax: 806-742-0616.
E-mail: ranchhc@hu.edu
Web Site: www.depts.ttu.edu/ranchhc
Founded: 1969.
Congressional District: 19
Key Personnel: Dir., Jim Pfluger.
Personnel Profile: Full-Time Paid 17; Part-Time Paid 2.
Governing Authority: Parent Institution: Texas Tech University. Tax-exempt.
Institution Type/Description: History Museum.
Collections: American ranching history & culture; period furnishings; photographs; 1923 Model T; guns; early toys; personal artifacts.
Research Fields: structure preservation; social-economic impact of wind farms on ranching.
Facilities: library.
Activities: workshops; guided tours.
Publications: quarterly, Ranch Record Magazine; occasional books related to ranching history & NRHC artifacts.
Hours & Admission Prices: Mon.-Sat. 10-5, Sun. 1-5. No charge; donations accepted. Closed New Year's Eve & Day; Martin Luther King Jr. Day; Easter; Memorial Day; Independence Day; Labor Day; Thanksgiving; Christmas Eve & Day. &
Attendance: 55,671
Membership: Junior Rough Riders $10 per child (10 & under); Individual $30; Family $50; Patron $250; Sustainer $500; Society of the Brand $1,000.

SCIENCE SPECTRUM, (M), 2579 S. Loop 289 #250, Lubbock, TX 79423-1400. Tel.: 806-745-2525. Fax: 806-745-1115.
E-mail: sandy@sciencespectrum.org
Web Site: www.sciencespectrum.org
Founded: 1986.
Congressional District: 19
Key Personnel: Dir., Cassandra L. Henry; Museum Shop Mgr., Kay Dudley.
Personnel Profile: Full-Time Paid 8; Part-Time Paid 25; Part-Time Volunteers 8; Interns 1.
Governing Authority: private; nonprofit organization. Tax-exempt: 501(c)(3).
Institution Type/Description: Science Museum.
Collections: hands-on displays of science, focusing on physical science, health, aquariums, economics & space flight; Kidspace, children's museum area.
Hours & Admission Prices: Mon.-Fri. 10-5, Sat. 10-6, Sun. 1-5. Museum: adults $7.50, senior citizens 60 & over and children 3-12 $6; discounts to groups; members no charge. Omni Theater: adults $8, senior citizens 60 & over and children 3-12 $6.50; members $6; discounts to groups. Combination tickets available. Closed Thanksgiving; Christmas. &
Attendance: 200,000 (accurate)
Membership: Discoverer (up to 2 people) $60; Explorer (up to 5 people) $100 ($15 each additional); Life $1,000.

SILENT WINGS MUSEUM, (M), 6202 N. Interstate 27, Ste. 2, Lubbock, TX 79403-7526. Tel.: 806-775-3049. Fax: 806-775-3337.
E-mail: bwitcher@mylubbock.us
Web Site: www.silentwingsmuseum.com
Formerly: Military Glider Pilots Association, Silent Wings Museum
Founded: 1979.
Key Personnel: Dir., Brooke Witcher; Asst. Dir., Eddy Grigsby; Cur., Donald Abbe; Museum Shop Mgr., David Seitz.
Personnel Profile: Full-Time Paid 6; Full-Time Volunteers 1; Part-Time Paid 5; Part-Time Volunteers 8; Interns 2.
Governing Authority: Parent Institution: City of Lubbock. Tax-exempt.
Institution Type/Description: Military Museum.
Collections: military history; historical aviation memorabilia; World War II gliders & airborne forces, 1940-1946.
Research Fields: military aviation history.
Facilities: 500-vol. library of fiction & non-fiction on action of World War II airborne forces; theater. Videos, books and unit & organization pins of airborne units for sale.
Activities: children's workshops; family activities.
Publications: quarterly newsletter, Museum News.
Hours & Admission Prices: Tues.-Sat. 10-5, Sun. 1-5. Adults $5; members, WWII glider pilots & spouses, and military personnel in uniform no charge. &
Attendance: 12,500 (accurate)
Membership: Bronze Wings $35; Silver Wings $50; Gold Wings $100 Corporate $1,000-$5,000.

ELLEN TROUT ZOO, 402 Zoo Circle, Lufkin, TX 75904-1345. Tel.: 936-633-0399. Fax: 936-633-0311.
E-mail: gordon@ellentroutzoo.com
Web Site: www.ellentroutzoo.com
Founded: 1967.
Congressional District: 2
Key Personnel: Zoo Dir., Gordon B. Henley, Jr.; Pres., Jamie Zayler; Gen. Cur., Celia K. Falzone; Dir. Educational Svcs., Charlotte Henley; Staff Veterinarian, Mike Nance, D.V.M
Personnel Profile: Full-Time Paid 23; Part-Time Paid 3; Part-Time Volunteers 15; Interns 1.
Governing Authority: municipal. Parent Institution: City of Lufkin, TX. Tax-exempt.
Institution Type/Description: Zoo.
Collections: variety of vertebrate animals, including amphibians, reptiles, birds & mammals.
Facilities: 2,000-vol. library of zoological & related books, available for research by special request; zoological park; classrooms.
Activities: guided tours; lectures; films; TV & radio programs; formally organized education programs for children; docent program and council; permanent exhibitions; mobile vans; in-school programs.
Publications: book, Wildlife on Wheels; teacher training manual, Zookeeper Training; employee manual, Docent Handbook.
Hours & Admission Prices: Daily 9-5. Adults $5; members, AZA, members & zoo reciprocity list no charge. &
Attendance: 121,854 (accurate)
Membership: Individual $25; Family & Grandparent $40; Tiger $100; Giraffe $200; Corporate $250; Hippopotamus $500; White Rhino $2,000.

THE MUSEUM OF EAST TEXAS, (M), 503 N. Second St., Lufkin, TX 75901-3013. Tel.: 936-639-4434. Fax: 936-639-4435.
E-mail: jmcdonald@metlufkin.org
Web Site: www.metlufkin.org
Founded: 1976.
Congressional District: 2
Key Personnel: Exec. Dir., J.P. McDonald; Pres., Misty Croley; Pres. Museum Guild, Susan Belasco; Cur. Education, Ann Reyes; Administrative Asst., Claudette Ratcliff; Museum Shop Mgr., Dixie Welch.
Personnel Profile: Full-Time Paid 4; Part-Time Paid 3; Part-Time Volunteers 42.
Governing Authority: nonprofit organization. Tax-exempt: 501(c)(3).
Institution Type/Description: Museum of Art & History.
Collections: regional history artifacts through the present; fine arts collection of American, European & regional artists; changing national, regional & international art & history exhibits.
Research Fields: local human history; regional art.
Facilities: photo archives; 14,000 sq. ft. exhibit space; 200-seat auditorium; public gardens; rental space. Museum-related items for sale.
Activities: lectures; art workshops; docent programs; guided tours; school tours; traveling exhibits; summer art camp for children; adult art classes.
Publications: exhibits catalog; collections catalogs.
Hours & Admission Prices: Tues.-Fri. 10-5, Sat.-Sun. 1-5. No charge; donations accepted. Closed major holidays. &
Attendance: 18,093 (accurate)
Membership: Individual $25; Family $50; Contributor $100; Sponsor $150; Sustainer $250; Patron $500; Guarantor $1,000; Benefactor $5,000; Life Member $10,000.

TEXAS FORESTRY MUSEUM, (M), 1905 Atkinson Dr., Lufkin, TX 75901-2505. Tel.: 936-632-9535. Fax: 936-632-9543.
E-mail: info@treetexas.com
Web Site: www.treetexas.com
Founded: 1972.
Congressional District: 2
Key Personnel: Dir., Rachel Collins; Pres. Bd., John Grigsby; Sec. & Treas., R.H. Hufford; Museum Shop Mgr., Laurie Vaughn.
Personnel Profile: Full-Time Paid 3; Part-Time Paid 1; Part-Time Volunteers 200.
Governing Authority: nonprofit organization. Tax-exempt.
Institution Type/Description: Forestry Museum.
Collections: logging equipment; tools; logging train; chain saws; fire lookout tower; photographs; paper industry. Historic Building: logging town railroad depot.
Research Fields: photographs.
Facilities: woodland trail. Gift items for sale.
Activities: guided tours; slide program for civic clubs & organizations.
Publications: brochure; membership quarterly, Crosscut.

Hours & Admission Prices: Mon.-Sat. 10-5. No charge; donations accepted. Closed New Year's Eve & Day; Easter; Thanksgiving; Christmas Eve & Day. &

Attendance: 10,000 (estimated)

Membership: Individual $35, $60, $100, $500; Life $1,000. Corporate $100, $250, $500, $1,000.

Luling

CENTRAL TEXAS OIL PATCH MUSEUM, 421 E. Davis St., Luling, TX 78648-2316. Mailing Address: P.O. Box 1002, Luling, TX 78648-1002. Tel.: 830-875-1922. Fax: 830-875-2082.

E-mail: oilmuseum@austin.rr.com

Web Site: www.oilmuseum.org

Key Personnel: Dir., Carol Voigt.

Personnel Profile: Part-Time Paid 1.

Institution Type/Description: History Museum.

Collections: local oil producing methods, equipment & workers; photographs; period artifacts; cultural history.

Activities: Museum Sponsors: Meet the Authors Book Signing in March; Roughneck Chili Cook Off and Oil City Car Show in April; Reflections of Texas Art Exhibit in September.

Hours & Admission Prices: Mon.-Sat. 10-4. No charge; donations requested. Closed most holidays.

Marble Falls

THE FALLS ON THE COLORADO MUSEUM, 2001 Broadway St., Marble Falls, TX 78654. Mailing Address: P.O. Box 1333, Marble Falls, TX 78654. Tel.: 830-798-2157. Facebook: Falls on the Colorado Museum.

E-mail: focmuseum@gmail.com

Web Site: www.fallsmuseum.org

Founded: 1998.

Key Personnel: Chmn. (V), William E. Becker.

Governing Authority: nonprofit organization. Tax-exempt: 501(c)(3).

Institution Type/Description: History Museum.

Collections: local history, heritage & culture; period furnishings; personal artifacts; photographs.

Activities: special events; group tours. Museum Sponsors: Black History Month; Children's Day Celebration; Scoutarama; Founders Day in July; Victorian Christmas in December.

Hours & Admission Prices: Thurs.-Sat. 10-5. No charge, donations accepted. &

Attendance: 1,240 (accurate)

Marfa

CHINATI FOUNDATION, One Cavalry Row, Marfa, TX 79843. Mailing Address: P.O. Box 1135, Marfa, TX 79843-1135. Tel.: 432-729-4362. Fax: 432-729-4597.

E-mail: information@chinati.org

Web Site: www.chinati.org

Founded: 1986.

Congressional District: 23

Key Personnel: Pres., Andrew Cogan; Museum Shop Mgr., Sandra Hinojos.

Personnel Profile: Full-Time Paid 16; Part-Time Paid 1; Part-Time Volunteers 50; Interns 4.

Governing Authority: public; nonprofit. Tax-exempt: 501(c)(3).

Institution Type/Description: Art Museum: housed in c.1919 & 1938 buildings of former Fort D.A. Russell.

Collections: installations by contemporary artists including Donald Judd, Carl Andre, Ingolfur Arnarsson, John Chamberlain, Dan Flavin, Roni Horn, Ilya Kabakov, Richard Long, Claes Oldenburg; Coosje Van Bruggen, David Rabinowitch & John Wesley.

Facilities: educational facilities; reading room; studio & exhibition space for artists in residence.

Activities: children's summer art classes; lectures & symposia. Annual Event: Open House Celebration; Community Day.

Publications: annual newsletter; symposia books.

Hours & Admission Prices: Guided Tours: Wed.-Sun. 10am & 2pm by appointment. Adults $25, students $10; discounts to AAM & ICOM members; members no charge. &

Attendance: 11,000 (estimated)

Membership: Student & Senior Citizen $50; Individual & Family $100; Supporting $250-$500; Sustaining $1,000-$2,500; Friend $5,000.

MARFA AND PRESIDIO COUNTY MUSEUM, 110 W. San Antonio St., Marfa, TX 79843. Mailing Address: P.O. Box 538, Marfa, TX 79843-0538. Tel.: 432-729-4140.

E-mail: marfapresidiocountymuseum@gmail.com

Web Site: www.marfamuseum.org

Founded: 1979.

Key Personnel: Pres., W.D. Edmonds

Institution Type/Description: History Museum.

Collections: fossils; Jumano & Native American tools; early ranching; rural life; military history; photographs.

Hours & Admission Prices: March-Dec. 14 Thurs.-Sat. 1-5. No charge; donations accepted. Closed major holidays.

Attendance: 1,200 (accurate)

Membership: Individual $10; Family $25.

Marshall

HARRISON COUNTY HISTORICAL MUSEUM, 1 Peter Whetstone Square, Marshall, TX 75670. Mailing Address: P.O. Box 1987, Marshall, TX 75671-1987. Tel.: 903-935-8417. Fax: 903-927-2534.

E-mail: info@harrisoncountymuseum.org

Web Site: www.harrisoncountymuseum.org

Formerly: Old Courthouse Museum/Harrison County Historical Museum

Founded: 1965.

Congressional District: 1

Key Personnel: C.E.O. & Dir., Janet Cook; Chm. (V), Alice Barron; Pres. (V), Jean Birmingham; Sec., Bea White; Museum Shop Mgr., Carla Nolan.

Personnel Profile: Full-Time Paid 1; Part-Time Paid 1; Part-Time Volunteers 4.

Governing Authority: society. Parent Institution: Harrison County Historical Society. Library: 117 E. Bowie, Marshall, TX. Tax-exempt: 170(b)(1)(A).

Institution Type/Description: Historical Museum: housed in the Harrison County Courthouse; built in 1901.

Collections: portraits; paintings; porcelains; jewelry; silverware; manuscripts; photographs; costumes; hand-painted china; cut & pressed glass; time pieces; Doctors' Room; Dentist's Office; photographs; Edison Room; 400 B.C.-1982 ceramics; 1890-1910 talking machines; c.1900 Edison recording machine; 1919-1935 radios; 1780s-1900s ethnic group heritage, including sport uniforms; string & wind instruments; folios; songbooks; needlecraft implements & specimens; pioneer implements; local industries; transportation & communications; mementoes of the Elks, Odd Fellows, Woodmen of the World & Masons; religious; politics; military; celebrities; school pictures; old text books; toys; flags; history; genealogy; Caddo Lake memorabilia; dolls; miniature soldiers & knights.

Research Fields: genealogy; local history; Civil War records.

Facilities: 2,000-vol. library of books on history, genealogy, art gallery & biography available for use on premises; children's room. Postcards & publications for sale.

Activities: tours; permanent exhibitions.

Publications: monthly newsletter.

Hours & Admission Prices: Call for hours. &

Attendance: 11,000 (estimated)

Membership: Individual $25; Family $35; Sustaining $50; Patron $100; Benefactor $250; Corporation $500; Individual Life $1,000.

MICHELSON MUSEUM OF ART, (M), 216 N. Bolivar, Marshall, TX 75670-3307. Mailing Address: P.O. Box 8290, Marshall, TX 75671-8290. Tel.: 903-935-9480. Fax: 903-935-1974.

E-mail: leomich@sbcglobal.net

Web Site: www.michelsonmuseum.org

Founded: 1985.

Congressional District: 1

Key Personnel: Dir., Susan Spears; C.E.O., Suzanne Planchard; Dir. Education, Bonnie Strauss; Pres. (V), Keith Feille.

Personnel Profile: Full-Time Paid 4; Part-Time Paid 1; Part-Time Volunteers 80.

Governing Authority: nonprofit organization. Tax-exempt: 501(c)(3).

Institution Type/Description: Art Museum.

Collections: 1887-1978 Russian-American works of Leo Michelson, including 100 oil paintings, 1,000 drawings, watercolors, lithographs, etchings with documentation; Kronenberg collection including works by Milton Avery, David Burliuk, Henri Matisse, Abraham Walkowitz, John Edward Costigan; Ramona & Jay Ward collection of African masks & artifacts; works by Milton Avery, Joseph Stella, Ralston Crawford, & Byron Browne.

Facilities: 5,000 sq. ft. exhibit space.

Activities: guided tours; docent program; loan, temporary & traveling exhibitions.

Publications: newsletters.

Hours & Admission Prices: Tues.-Fri. 10-4, Sat. 1-4. No charge; donations accepted. Closed Easter; Independence Day; Thanksgiving; Christmas. &

Attendance: 8,000 (estimated)

Membership: Student & Senior $15; Fellow $25; Supporter $50; Sustainer $100; Donor $200; Sponsor $500; Founder $1,000; Patron $2,500; Corporate $500-$2,500; Life $5,000; Benefactor $10,000.

Mason

MASON COUNTY MUSEUM, 321 Moody St., Mason, TX 76856. Mailing Address: Mason County Museum, P.O. Box 1473, Mason, TX 76856. Tel.: 325-347-6681.

Founded: 1966.

Congressional District: 21

Key Personnel: Pres. (V), Weldon Whittaker; Vice Pres., Mike Innis; Chm., Nancy Jordan; Co-Chm., Dolores Keller; Sec., Lou Fleming; Treas., Dane Phillips

Governing Authority: society. Parent Institution: Mason County Historical Society. Subsidiary Institution: Mason County Museum. Tax-exempt.

Institution Type/Description: Local History Museum: housed in 1887 two-storied sandstone school building.

Collections: local history & culture; photographs; period artifacts; Hall of Pioneers; first settlers; Fort Mason years; Robert E. Lee; Albert Sidney Johnston; 28 early officers famous in U.S. history; Native Americans; Mason County captives of Native Americans; schools; churches; artifacts; 20th century room.

Activities: guided tours with advanced request.

Hours & Admission Prices: Thurs.-Sat. 11-4. No charge; donations requested.

Attendance: 1,350 (estimated)

MASON SQUARE MUSEUM, 103 Fort McKavitt, Mason, TX 76856. Mailing Address: P.O. Box 203, Mason, TX 76856. Tel.: 325-347-0507.

E-mail: info@masonsquaremuseum.org

Web Site: www.masonsquaremuseum.org

Founded: 2006.

Congressional District: 11

Key Personnel: Pres. Bd., Dennis Evans.

Personnel Profile: Full-Time Volunteers 1; Part-Time Volunteers 150.

Governing Authority: Tax-exempt: 501(c)(3).

Institution Type/Description: History Museum.

Collections: local history & culture; ranching; banking; period furnishings; personal artifacts; photographs.

Facilities: Publications, local art & museum-related items for sale.

Activities: special events; reenactments.

Hours & Admission Prices: Thurs.-Sat. 10-4. No charge; donations accepted.

Attendance: 4,200 (accurate)

Membership: Individual $10.

Matador

MOTLEY COUNTY HISTORICAL MUSEUM, 828 Dundee St., Matador, TX 79244. Mailing Address: P.O. Box 523, Matador, TX 79244. Tel.: 806-347-2968.

Founded: 1988.

Congressional District: 13

Key Personnel: Chm. (V), Marisue Potts; Pres. (V), Bill Manney.

Personnel Profile: Part-Time Volunteers 2.

Governing Authority: Tax-exempt.

Institution Type/Description: History Museum: housed in the former Traweek Hospital; built in 1928.

Collections: local history & culture; hospital history & artifacts; veterans; ranching; Native American artifacts; saddle-making; personal artifacts; clothing; toys; cameras.

Research Fields: family & county history; Matador ranch; Native Americans; buffalo hunters; Confederate veterans settlements.

Activities: school children tours; archeological dig for middle school students.

Hours & Admission Prices: Mon., Wed. & Fri. 2-5; other times by appointment. No charge; donations accepted.

Attendance: 400 (estimated)

McAllen

*** INTERNATIONAL MUSEUM OF ART AND SCIENCE, (M),** 1900 Nolana, McAllen, TX 78504-4199. Tel.: 956-682-1564, ext. 104. Fax: 956-686-1813.

E-mail: jbravo@imasonline.org

Web Site: www.imasonline.org

Formerly: McAllen International Museum

Founded: 1967.

Congressional District: 15

Key Personnel: Exec. Dir., Joseph Bravo; Devel. Officer, Angela Maxwell; Pres., Tim Smith; Dir. Education, Susan Zwerling; Cur., Maria Elena Macias; Museum Shop Mgr., Jo Fisher; Dir Mktg., Michelle Rowe.

Personnel Profile: Full-Time Paid 24; Part-Time Paid 6.

Governing Authority: nonprofit organization. Tax-exempt.

Institution Type/Description: Arts & Sciences Museum.

Collections: science; geology; paleontology; fine & contemporary art; Mexican folk art.

Facilities: 2,000-vol. reference library available on the premises; classrooms; meeting rooms. Museum-related items for sale.

Activities: guided tours; lectures; films; gallery talks; docent program; permanent & traveling exhibitions; class program.

Publications: quarterly newsletter; annual report; brochures; postcards.

Hours & Admission Prices: Tues.-Wed. & Fri.-Sat. 9-5, Thurs. 9-8, Sun. 1-5. Adults $7, seniors & students $5, children 4-12 $3; children 3 & under and members no charge. Closed holidays. &

Attendance: 110,000 (estimated)

Membership: Winter Texan $35; Individual $42; Family $60; Smithsonian $25; Patron $500-$10,000; Corporate Sponsor $2,500; Corporate Benefactor $5,000; Corporate Director's Circle $10,000.

MCALLEN HERITAGE CENTER, INC., 301 S. Main St., McAllen, TX 78501-4806. Mailing Address: P.O. Box 1929, McAllen, TX 78505-1929. Tel.: 956-687-1906. Facebook: McAllen Heritage.

E-mail: mcheritage@att.net

Web Site: www.mcallenheritagecenter.com

Founded: 2006.

Key Personnel: Mng. Dir., Elva M. Cerda; Pres. (V), Nedra S. Kinerk; Museum Shop Mgr., Lily Hernandez.

Personnel Profile: Part-Time Paid 2; Part-Time Volunteers 4; Interns 2.

Volunteer Hours: 400

Operating Expenses: 6,600,000

Operating Income: 8,500,000

Governing Authority: Tax-exempt.

Institution Type/Description: History Museum.

Collections: antique wheelchair; 19th-century maps of the Rio Grande Valley; historic items; photographs; books; model rail display; sports; military.

Major Exhibits: Called to Higher Service, 11/13-7/30/14.

Activities: storytelling; children's activities & workshops; historical presentations; musical & arts exhibits.

Publications: publications, McAllen Centennial Magazine 2004; McAllen Leading the Way.

Hours & Admission Prices: Wed.-Fri. 1-5, Sat. 11-4. No charge; donations accepted. &

Attendance: 5,500 (estimated)

Membership: Student $10; Senior $25; Active $30; Family $60; Business $100.

SOUTH TEXAS COLLEGE - LIBRARY ART GALLERY, 3201 W. Pecan, McAllen, TX 78501-6661. Tel.: 956-872-3488.

Institution Type/Description: Art Gallery.

Collections: works by regional, national & international artists.

Hours & Admission Prices: Call for hours.

McCamey

MENDOZA TRAIL MUSEUM, Hwy. 67 E., McCamey, TX 79752. Mailing Address: P.O. Box 1409, McCamey, TX 79752-1409. Tel.: 432-652-3192.

Institution Type/Description: History Museum.

Collections: Native American artifacts; fossils; oil boom memorabilia; period furniture.

Hours & Admission Prices: By appointment only.

McKinney

CHESTNUT SQUARE HISTORIC VILLAGE - THE HERITAGE GUILD OF COLLIN COUNTY, 315 S. Chestnut St., McKinney, TX 75069-5607. Mailing Address: P.O. Box 583, McKinney, TX 75070-8139. Tel.: 972-562-8790. Fax: 972-562-8790.

E-mail: info@chestnutsquare.org

Web Site: www.chestnutsquare.org

Founded: 1973.

Congressional District: 3

Key Personnel: Dir., Cindy Johnson; Chm. (V), Pat Rodgers; Museum Shop Mgr., Kim Ducote.
Personnel Profile: Part-Time Paid 5; Part-Time Volunteers 100; Interns 1.
Governing Authority: Tax-exempt.
Institution Type/Description: History Museum.
Collections: area history & culture; five house; general store; chapel; stage coach inn; period furnishings.
Facilities: rental facilities.
Activities: special events; rental facilities. Museum Sponsors: Historic McKinney Farmers Market; Living History Days; Killis Melton Ice Cream Crank-Off; Spirit of the Cowboy; Legends of McKinney Ghost Walks; Holiday Tour of Homes.
Hours & Admission Prices: Tues., Thurs. & Sat. 11. Tours: adults $7, children under 12 $3. &
Attendance: 50,000 (estimated)
Membership: Senior $25; Single & Senior Family $35; Family $55; Heritage $150; Life $1,000.

COLLIN COUNTY FARM MUSEUM, 7117 County Rd. 166, McKinney, TX 75070. Tel.: 972-548-4792. Fax: 972-542-4594. Facebook: Collin County Farm Museum.
E-mail: ccfm@collincountytx.gov
Web Site: www.co.collin.tx.us/parks/myers/farm_museum.jsp
Founded: 1976.
Key Personnel: Park Mgr., Judy Florence; Museum Coord., Jennifer Rogers.
Personnel Profile: Full-Time Paid 1; Part-Time Paid 1; Part-Time Volunteers 50.
Governing Authority: county. Parent Institution: Collin County Government. Subsidiary Institution: Farm Museum Foundation of Collin County, Inc. Tax-exempt.
Institution Type/Description: Farm Museum.
Collections: agricultural tools & machinery; kitchen & household furnishings; 2 farmhouses.
Research Fields: agricultural heritage of Collin County from earliest settlement through 1940.
Facilities: 200-vol. library of farming & local history; granary; windmill.
Activities: docent program; formal educational programs; guided tours; hobby workshops; loan, participatory & traveling exhibitions; school loan service; children's day camps.
Publications: quarterly newsletter, Collin County Farm Museum News.
Hours & Admission Prices: March-June & Sept.-Dec. Fri.-Sat. 10-3; other times by appointment. Admission $1. Closed major holidays. &
Attendance: 3,000 (estimated)

HEARD-CRAIG CENTER FOR THE ARTS, 205 W. Hunt St., McKinney, TX 75069. Tel.: 972-569-6909. Fax: 972-542-5092.
E-mail: bjohnson@heardcraig.org
Web Site: heardcraig.org
Key Personnel: Exec. Dir., Barbara Johnson; Asst. Dir., Ashley Kennedy
Institution Type/Description: Historic House Museum: built in 1900.
Collections: family history, heirlooms, art, furnishings & personal artifacts.
Activities: educational programs; special events.
Hours & Admission Prices: Tues.-Sat. 8:30-4:30; other times by appointment.
Membership: Individual $30; Family $55; Patron $100; Lifetime $1,000.

HEARD NATURAL SCIENCE MUSEUM & WILDLIFE SANCTUARY, 1 Nature Place, McKinney, TX 75069-8840. Tel.: 972-562-5566. Fax: 972-548-9119.
E-mail: info@heardmuseum.org
Web Site: www.heardmuseum.org
Founded: 1967.
Congressional District: 4
Key Personnel: Dir., Sy Shahid.
Personnel Profile: Full-Time Paid 8; Part-Time Paid 12; Part-Time Volunteers 150.
Governing Authority: nonprofit organization. Tax-exempt: 501(c)(3).
Institution Type/Description: Natural Science Museum & Wildlife Sanctuary.
Collections: malachology; insects; herpetology; ornithology; mammalogy; rocks & minerals; fossils; anthropology; nature prints.
Research Fields: ornithology; prairie ecology; environmental assessments; wetlands.
Facilities: nature center; 289 acre wildlife sanctuary; classrooms; 2-acre native plant garden; science technology center. Museum-related items for sale.
Activities: lectures; formally organized educational programs; docent program; nature trails guided & self-guided; handicap trail; travel tours; summer camps; permanent & temporary exhibits; birding; museum ancillary groups: hobby beekeepers, Prairie & Timbers Audubon Club, Heard Nature Photographers Club; Collin County Archeological Society; Collin County Chapter of the Native Plant Society of TX; canoeing programs.

Publications: bimonthly email newsletter; quarterly print newsletter.
Hours & Admission Prices: Tues.-Sat. 9-5, Sun. 1-5. Adults $9, children 3-12 $6; discounts to AAM & ASTC members; members no charge. Closed New Year's Day; Thanksgiving; Christmas. &
Attendance: 100,000 (estimated)
Membership: Seniors, Student, Active Military & Educator $40; Individual $60; Family $80; Road Runner $150; Bobcat $500; Red-Tailed Hawk $1,500; Golden Eagle $2,500.

NORTH TEXAS HISTORY CENTER, 300 E. Virginia St., McKinney, TX 75069-4325. Tel.: 972-542-9457. Fax: 972-542-4594. Facebook: North Texas History Center.
E-mail: info@collincountyhistoricalsociety.org
Web Site: collincountyhistoricalsociety.org
Formerly: Collin County Historical Society, Inc. & History Museum
Founded: 1957.
Congressional District: 4
Key Personnel: Exec. Dir., Victoria Day.
Governing Authority: private; nonprofit organization. Subsidiary Museum: Collin County Farm Museum, 7117 County Rd. 166, McKinney, TX 75071. Tax-exempt: 501(c)(3).
Institution Type/Description: History Museum.
Collections: Collin County & North Texas history.
Research Fields: Collin County; North Texas; Civil War in Texas.
Facilities: 800-vol. library; educational facilities; 5,500 exhibit space. Museum-related items for sale.
Activities: docent program; lectures; traveling trunk K-1st grade; education programs for grades 2-5.
Publications: quarterly newsletter, Collin County History; books.
Hours & Admission Prices: Tues. & Thurs. 10-3.
Attendance: 10,000 (estimated)
Membership: Individual $25; Dual $40; Family $75; Corporate Contributor $100; Friend & Corporate Friend $150; Patron & Corporate Supporter $250; Corporate Partner $1,000; Corporate Benefactor $2,500.

McLean

DEVILS ROPE BARBED WIRE MUSEUM, 100 Kingsley St., McLean, TX 79057. Mailing Address: P.O. Box 290, McLean, TX 79057-0290. Tel.: 806-779-2225.
E-mail: barbwiremuseum@centramedia.net
Web Site: www.barbwiremuseum.com
Governing Authority: nonprofit organization. Tax-exempt: 501(a)(1).
Institution Type/Description: History Museum.
Collections: barbed wire history & artifacts; barbed wire specimens; tools; construction equipment; military artifacts; photographs; ranching history; cowboy memorabilia; illustrations; sculptures.
Activities: barbed wire-making demonstrations.
Hours & Admission Prices: Tues.-Sat. 10-4.

MCLEAN-ALANREED AREA MUSEUM, 116 Main St., McLean, TX 79057. Mailing Address: P.O. Box 354, McLean, TX 79057-0354. Tel.: 806-779-2731.
Founded: 1969.
Congressional District: 31
Key Personnel: Treas., Lynn D. Reeves; Pres. (V), Bob Glass; Mgr., Carol McClure; Vice Pres. (V), Nita Riemer; Secretary (V), Mickey Jackson.
Governing Authority: nonprofit. Tax-exempt: 501(c)(3).
Institution Type/Description: Historical Society Museum.
Collections: pioneer costumes; western exhibits including saddles, ranch tools, plows & Indian artifacts; photographs; vet supplies; documents; fire truck; Model T car; barber shop, doctor's office, POW camp (World War II) display; German POW camp display.
Facilities: 5,500 sq. ft. exhibit space.
Hours & Admission Prices: March-Nov. Tues.-Fri. 10-4. No charge; donation accepted. Closed major holidays. &
Attendance: 1,125 (estimated)
Membership: Individual $10; Family $25; Business $150; Silver $250; Gold $500; Platinum $1,000.

Mertzon

IRION COUNTY MUSEUM, 210 W. Sherwood Ave., Mertzon, TX 76941. Mailing Address: 598 Lindell Ave., Mertzon, TX 76941. Tel.: 325-835-7771. Fax: 325-835-2008.
Founded: 1980.
Congressional District: 51
Key Personnel: Museum Shop Mgr., Lily Flores.
Personnel Profile: Full-Time Paid 1; Full-Time Volunteers 2; Part-Time Paid 1.

Governing Authority: Parent Institution: Irion County. Tax-exempt.
Institution Type/Description: History Museum.
Collections: local history & culture; period furnishings; personal artifacts; photographs.
Hours & Admission Prices: Sat.-Mon. 1-5. No charge; donations accepted. &
Attendance: 20

Mesquite

FLORENCE RANCH HOMESTEAD, (M), 1424 Barnes Bridge Rd., Mesquite, TX 75150-4206. Mailing Address: P.O. Box 850137, Mesquite, TX 75185-0137. Tel.: 972-216-6468. Fax: 972-329-8340.
E-mail: corr@ci.mesquite.tx.us
Web Site: historicmesquite.org
Founded: 1987.
Congressional District: 3
Key Personnel: Exec. Dir., Charlene Orr; Chm. (V), Kelly Baird; Museum Shop Mgr., Becky Allen.
Personnel Profile: Full-Time Paid 1; Part-Time Paid 2; Part-Time Volunteers 5.
Governing Authority: Historic Mesquite, Inc. Tax-exempt.
Institution Type/Description: Historic Site.
Collections: artifacts from 1880-1920. Historic Buildings: Ranch House; Homestead.
Facilities: Museum-related items for sale.
Hours & Admission Prices: Tues. & Thurs.-Fri. 12-4, Wed. 1-5, 2nd Sat. of month 10-1; other times by appointment. No charge; donations accepted. &
Attendance: 1,100 (estimated)
Membership: Student $15; Pioneer $25; Settler $50; Homesteader $100; Frontiersman $250; Trailblazer $500.

Miami

ROBERTS COUNTY MUSEUM, 120 E. Commercial St., Miami, TX 79059. Mailing Address: P.O. Box 306, Miami, TX 79059-0306. Tel.: 806-868-3291. Fax: 806-868-3381.
E-mail: robertscomuseum@amaonline.net
Web Site: robertscountymuseum.org
Founded: 1979.
Congressional District: 31
Key Personnel: Exec. Dir., Emma Bowers.
Personnel Profile: Full-Time Paid 1; Part-Time Volunteers 15.
Governing Authority: county. Tax-exempt.
Institution Type/Description: Historic Building & Museum: housed in c.1888 Santa Fe Depot.
Collections: East Room: photographs; medical toys; school memorabilia; American War artifacts. West Room: cattle brands; barbed wire; period tools & supplies from covered wagons. Ferguson Room: animals native to Roberts County in their natural habitat; typical pioneer home with period clothing, furnishings & paintings; general store; railroad displays; quilts & handwork; wedding dresses. Payne Memorial Barn: blacksmith shop; tin shop; buggies; saddles; chuck wagon. Dugout & farm equipment. Paleontology & Native American Room: 1920 silk screen prints by Kiowa and other artists; 1200 pottery, rugs, baskets & blankets of the Southwestern Indian tribes; fossils & skeletal remains of the Clovis culture; Mead collection of rocks, minerals & arrowheads; Pioneer Miami 1900-1910.
Facilities: 208-vol. library pertaining to county history available for research on premises only. Books pertaining to Texas history-local & statewide available for check-out. Books & museum-related items for sale.
Activities: guided tours; permanent & temporary exhibitions.
Publications: brochures; books, Roberts County History Book I and II; Miami Mammoth Kill Site.
Hours & Admission Prices: Tues.-Fri. 10-5, Sat.-Sun. call for hours. No charge; donations accepted. Closed New Year's Day; Memorial Day; Independence Day; Labor Day; Veterans Day; Thanksgiving; Christmas. &
Attendance: 3,000 (estimated)

Midland

✳ CAF AIRPOWER MUSEUM, 9600 Wright Dr., Midland, TX 79711. Mailing Address: P.O. Box 62000, Midland, TX 79711-2000. Tel.: 432-563-1000. Fax: 432-567-3047.
E-mail: dsanders@aahm.org
Web Site: www.airpowermuseum.org
Formerly: American Airpower Heritage Museum, Inc. & Commemorative Air Force Headquarters
Founded: 1957.
Key Personnel: Dir., Autumn Hicks; Research Specialist & Librarian, Keegan Chetwynd; Registrar, Aaron Steele; Museum Shop Mgr., Kara Thurman; Gift Shop Asst., Pat Moore; Patron Svcs. Specialist, Darrell Sanders.

Personnel Profile: Full-Time Paid 4; Part-Time Volunteers 20.
Governing Authority: nonprofit organization. Affiliated with The Commemorative Air Force. Tax-exempt 501(c)(3).
Institution Type/Description: World War II, Military Aviation Museum.
Collections: U.S. & foreign combat aircraft of World War II; weapons; uniforms & equipment; photographs; memorabilia of the era c.1939-1945.
Research Fields: World War II combat aircraft; militaria.
Facilities: library; archives by appointment; 60,000 sq. ft. hangar with restored aircraft; 14,000 sq. ft. exhibit hall. Museum-related gift items for sale.
Activities: guided tours; lectures; docent program; films; permanent, traveling & loan exhibitions; outreach programs; actively collecting World War II oral histories; aviation cadet academy for 10-18 year olds.
Publications: The Co-Pilot Communique; Dispatch.
Hours & Admission Prices: Tues.-Sat. 9-5. Adults $10, senior citizens & teens 13-18 $9, children 6-12 $7, school groups $2; discounts to groups, AAA, AAM, CAA, TAM & National Historical Society members; children under 6 & members no charge. Closed New Year's Eve & Day; Thanksgiving; Christmas Eve & Day. &
Attendance: 35,000 (estimated)
Membership: Student $45; Active $55; Patron $100; Family $150; Sustaining $250; Preservation Colonel $300; Corporate $1,500; Lifetime $2,400.

THE GEORGE W. BUSH CHILDHOOD HOME, 1412 W. Ohio Ave., Midland, TX 79701. Mailing Address: P.O. Box 8586, Midland, TX 79708-8586. Tel.: 432-685-1112; 866-684-4380. Fax: 432-684-7012.
E-mail: gwbhome@bushchildhoodhome.org
Web Site: www.bushchildhoodhome.org
Founded: 2001.
Key Personnel: Exec. Dir., Paul St. Hilaire; Asst. Dir., Jaclyn Woolf.
Personnel Profile: Full-Time Paid 2.
Volunteer Hours: 3,500
Operating Expenses: 240,800
Operating Income: 247,700
Governing Authority: Tax-exempt.
Institution Type/Description: Historic Home: housed in the childhood home of former President George W. Bush.
Collections: furnishings from 1952-1956; photographs; Bush family artifacts & memorabilia.
Hours & Admission Prices: Tues.-Sat. 10-5, Sun. 2-5. Adults $5. Closed New Year's Day; Easter; Thanksgiving; Christmas. &
Attendance: 5,000 (estimated)

MCCORMICK GALLERY, Midland College, Allison Fine Arts Bldg., 3600 N. Garfield, Midland, TX 79705. Tel.: 432-685-4770. Fax: 432-685-4721.
E-mail: mccormickgallery@midland.edu
Web Site: www.midland.edu/mccormick
Founded: 1978.
Congressional District: 11
Key Personnel: Dir., J. Don Wallace.
Personnel Profile: Full-Time Paid 2; Part-Time Paid 1; Part-Time Volunteers 8.
Governing Authority: Parent Institution: Midland College. Tax-exempt.
Institution Type/Description: Art Gallery.
Collections: paintings; photography; ceramics; sculpture.
Hours & Admission Prices: Mon.-Thurs. 8am-10pm, Fri. 8-5, Sat. 10-5, Sun. 1-5. No charge; donations accepted. Closed holidays. &
Attendance: 2,500 (accurate)

MIDLAND COUNTY HISTORICAL MUSEUM, 301 W. Missouri, Midland, TX 79701-5108. Tel.: 915-682-2931 & 688-8947.
Founded: 1930.
Congressional District: 19
Key Personnel: Pres. Midland County Historical Society (V), Mrs. Nancy R. McKinley.
Personnel Profile: Part-Time Volunteers 2.
Governing Authority: county. Tax-exempt: 501(c)(3).
Institution Type/Description: General Museum.
Collections: archives; archaeology; manuscripts; glass; original pictures of Midland; early obituary files; geology; Indian artifacts; medicine; history. Historic House: 1899 Taylor Brown-Sarah Dorsey Home.
Research Fields: archives; archaeology; Indian artifacts; history.
Facilities: Midland County history & classified news clipping file on all phases of Midland County growth available for use on premises.
Activities: guided tours; lectures; permanent & temporary exhibitions.
Publications: weekly newspaper article, Historical Markers.

Hours & Admission Prices: Wed. & Fri. 2-5. No charge, donations accepted. Closed New Year's Day; Independence Day; Labor Day; Thanksgiving; Christmas. &

Attendance: 1,000 (estimated)

Membership: Historical Society: Regular $10; Patron $25; Life $100.

*** MUSEUM OF THE SOUTHWEST, (M),** 1705 W. Missouri Ave., Midland, TX 79701-6516. Tel.: 432-683-2882. Fax: 432-684-9151.

E-mail: info@museumsw.org

Web Site: www.museumsw.org

Founded: 1965.

Congressional District: 21

Key Personnel: Exec. Dir., Brian Whisenhunt; Pres. Bd. Trustees, Jaime Alexander; Dir. Devel., Audrie Palmer; Dir. Mktg., Chelsea Dey; Dir. Blakemore Planetarium, Andrew Kerr; Dir. Operations, Angela Galvan; Mgr. Collections & Exhibitions, Jenni Opalinski; Cur. Durham Children's Museum, Annelorre Robertson; Cur. Education, Kristen Wagstrom; Cur. Collections & Exhibitions, Wendy Earle; Campus Mgr., Robin Pruett; Special Projects Asst., Coleman Bales; Security Chief, Ellis Brown.

Personnel Profile: Full-Time Paid 12; Part-Time Paid 7; Part-Time Volunteers 100; Interns 2.

Volunteer Hours: 1,300

Operating Expenses: 1,500,000

Operating Income: 1,650,000

Governing Authority: nonprofit. Tax-exempt: 501(c)(3).

Institution Type/Description: Regional Art, Children's Museum & Planetarium.

Collections: southwestern art, anthropology; archaeology; Margaret & C.E. Bud Bissell archaeological collection of West Texas; Fred T. & Novadean Hogan Taos Founders Collection; contemporary & historic southwestern art with emphasis on Texas & New Mexico; the Barrett Collection of contemporary Texas art; Audubon's Quadrupeds; Karl Bodmer collection; Native American; sculpture garden.

Major Exhibits: Estamos Aqui (T), 1/14-3/14; Barnaby Fitzgerald, 1/14-3/14; Pulled, Pressed & Screened (T), 2/14-4/14; Kathy Sosa, 4/14-6/14; Our Land, Our People, Our Images (T), 6/14-8/14; Katie Maratta, 7/14-8/14; Winslow Homer & The American Pictorial Press (T), 9/14-11/14; Step Into Art (T), 10/14-2/15; Orna Feinstein, 11/14-12/14; Thomas Nast - Santa Claus, 12/14-1/15.

Research Fields: Art of the American & Southwest.

Facilities: planetarium; multi-purpose room.

Activities: lectures; art socials; art classes; meet-the-artists receptions; workshops; temporary & traveling exhibitions; docent program. Museum Events: Septemberfest (Arts Festival); Summer Lawn Concerts; Christmas at the Mansion.

Publications: triennual newsletter; gallery materials; podcast; monthly e-newsletter.

Hours & Admission Prices: Tues.-Sat. 10-5, Sun. 2-5. Adults $5, children $3; discounts to seniors, active-duty military personnel & AAM members; children 2 & under and members no charge. &

Attendance: 70,000 (estimated)

Membership: Individual $40; Family $60; Reciprocal $125; Supporter $250; Patron $500; President's Club $1,200.

NITA STEWART HALEY MEMORIAL LIBRARY & J. EVETTS HALEY HISTORY CENTER, 1805 W. Indiana, Midland, TX 79701-6949. Tel.: 432-682-5785. Fax: 432-685-3512.

E-mail: haley-mail@att.net

Web Site: www.haleylibrary.com

Founded: 1976.

Congressional District: 21

Key Personnel: Trustee Chm., Brian T. McLaughlin; Trustee Vice Chm., Jeff Haley; Dir., J.P. "Pat" McDaniel; Administrative Asst., Glenna Gifford; Archives, Jim Bradshaw; Librarian, Nancy Jordan.

Personnel Profile: Full-Time Paid 2; Part-Time Paid 3.

Governing Authority: nonprofit organization. Trust Indenture. Tax-exempt: 501(c)(3).

Institution Type/Description: Research Library & History Center.

Collections: Western art; photography collection; archives; tack; western gear; maps; books; interviews pertaining to early range cattle history & frontier settlement; bronzes by Ed Fraughton, Glenna Goodacre, Veryl Goodnight, Buck McCain, Joe Beeler & others; paintings by Robert Lockheed, Charlie Dye & other masters of the West.

Research Fields: cattle industry in southwestern & western U.S., Canada, Mexico & South America; Indians; Texas Rangers; county histories; education; religion; politics; agriculture; horses.

Facilities: 30,000-vol. research library pertaining to the Southwest. Books, art prints, postcards & other related items for sale.

Activities: guided tours; organized educational programs for adults, under-

graduate & graduate college students affiliated with the Univ. of Texas of the Permian Basin; docent program; participatory, temporary & traveling exhibitions. Museum Sponsors: Annual Art Show & Sale in March. Library: 3 art fundraiser shows a year.

Publications: quarterly newsletter, The Haley Library Newsletter.

Hours & Admission Prices: Mon.-Fri. 9-5. No charge; donations accepted. Closed New Year's Day; Memorial Day; Independence Day; Labor Day; Thanksgiving; Christmas Eve & Day. &

Attendance: 2,000 (estimated)

Membership: Old Maude $35; Abilene Couple $60; XIT Cowboy $100; Jeff Milton Ranger $250; Alamo Mission Bell $500; Charles Goodnight Traildriver $1,000; J. Evetts Haley History Center Benefactor $5,000.

THE PETROLEUM MUSEUM, (M), 1500 Interstate 20 West, Midland, TX 79701-2041. Tel.: 432-683-4403. Fax: 432-683-4509.

E-mail: info@petroleummuseum.org

Web Site: www.petroleummuseum.org

Founded: 1967.

Congressional District: 19

Key Personnel: Pres. (V), Roy Williamson; Exec. Dir., Kathy Shannon; Dir. Education, Brenda Rathjen; Dir. Archives, Leslie Meyer; Museum Shop Mgr., Fifi Sanchez.

Personnel Profile: Full-Time Paid 10; Part-Time Paid 5; Part-Time Volunteers 200.

Governing Authority: nonprofit corporation. Tax-exempt: 501(c)(3).

Institution Type/Description: History & Technology Museum.

Collections: original paintings showing historical & oil industry subjects; period oil industry machinery; geology; paleontology; technology relating to petroleum; photos & documents relating to West Texas oil fields; 14 original paintings by two-time Prix de West Gold Medal Awardee, Tom Lovell; 7 Chaparral race cars by designer & racer Jim Hall.

Research Fields: regional petroleum industry & social history.

Facilities: 10,000-vol. library of books; 230,000 image photo collection; 250-seat auditorium. Petroleum & museum-related items for sale.

Activities: guided tours; films; docent program; permanent & temporary exhibitions; field trips; classes.

Publications: books, Oil In West Texas And New Mexico; Permian: A Continuing Saga; Tom Lovell, Storyteller With a Brush; Chaparral, Can Am & Prototype Race Cars.

Hours & Admission Prices: Mon.-Sat. 10-5, Sun. 2-5. Adults $8, senior citizens & students 12-17 $6, children 6-11 $5; discounts to Texas Hwy. Travel Passports & AAA members; children under 6 & members no charge. Closed New Year's Day Thanksgiving; Christmas Eve & Day. &

Attendance: 35,000 (estimated)

Membership: Senior $20; Individual $35; Family $50; Friend $100-$149; Associate $250-$499; Supporter $500-$999; Patron $1,000-$2,499; Sustaining $2,500-$4,999; Benefactor $5,000-$9,999; Underwriter $10,000-$20,000; Director $20,000 & up.

Z. TAYLOR BROWN-SARAH DORSEY HOUSE, 213 N. Weatherford, Midland, TX 79701. Mailing Address: c/o Midland Genealogical Society, 301 W. Missouri, Midland, TX 79701. Tel.: 432-682-2931.

Founded: 1899.

Congressional District: 21

Key Personnel: Interim Pres., Pat McDaniel.

Personnel Profile: Part-Time Volunteers 6.

Governing Authority: society; nonprofit. Parent Institution: Midland County Historical Society. Tax-exempt: 501(c)(3).

Institution Type/Description: Historic House: housed in c.1899 Z. Taylor Brown House.

Collections: furniture & furnishings; historical artifacts.

Activities: guided tours; lectures; films; colored slide programs.

Hours & Admission Prices: By special appointment for groups. No charge; donations accepted.

Attendance: 800 (estimated)

Membership: Individual $10; Sponsor $25; Benefactor $100.

Mingus

W.K. GORDON CENTER FOR INDUSTRIAL HISTORY OF TEXAS, (M), 65258 I-20, Mingus, TX 76463. Mailing Address: P.O. Box 218, Mingus, TX 76463-0218. Tel.: 254-968-1886. Fax: 254-968-1903.

Web Site: www.tarleton.edu/gordoncenter

Founded: 2002.

Congressional District: 31

Key Personnel: Dir., Dr. T. Lindsay Baker; Cur., LeAnna Biles Schooley.

Personnel Profile: Full-Time Paid 2; Part-Time Paid 4.

Governing Authority: public university. Parent Institution: Tarleton State University, Stephenville, TX. Tax-exempt: 501(c)(3).
Institution Type/Description: Industrial Museum.
Collections: Texas industrial history with emphasis on Thurber and the Texas & Pacific Coal Company.
Research Fields: Texas & Pacific Coal Company; Thurber & it's residents; Thurber brick.
Facilities: library; 7,200 sq. ft. exhibit space; 40-seat theater. Museum-related items for sale.
Activities: concerts; formal education programs; lectures; rental gallery; temporary exhibitions; theater.
Hours & Admission Prices: Tues.-Sat. 10-4, Sun. 1-4. Adults $4, children $2; discounts to groups; AAM members no charge. Closed New Year's Day; Easter; Thanksgiving; Christmas; university holidays. &
Attendance: 3,000 (estimated)

Mission

MISSION HISTORICAL MUSEUM, 900 Doherty Ave., Mission, TX 78572-5812. Tel.: 956-580-8646.
E-mail: lcontreras@missiontexas.us
Founded: 2002.
Key Personnel: Dir., Luis D. Contreras, II
Personnel Profile: Full-Time Paid 4; Part-Time Paid 2; Part-Time Volunteers 21; Interns 2.
Governing Authority: nonprofit organization. Tax-exempt.
Institution Type/Description: History Museum.
Collections: local history & culture; period clothing & furniture; photographs; sports memorabilia; military artifacts.
Hours & Admission Prices: Mon.-Fri. 10-4. Adults $2, seniors $1.50, children 6-18 $1; children under 6 no charge. &

Monahans

MILLION BARREL MUSEUM, 400 Museum Blvd., Monahans, TX 79756. Mailing Address: P.O. Box 114, Monahans, TX 79756. Tel.: 432-943-8401.
E-mail: millionbarrel_museum@monahans.org
Web Site: www.monahans.org/new/chamber/museums.html
Institution Type/Description: History Museum.
Collections: oil storage tank; original Monahans jail; period caboose; eclipse windmill; early farm equipment. Historic House: Holman House.
Hours & Admission Prices: Call for hours.

Mont Belvieu

BARBERS HILL - MONT BELVIEU MUSEUM, 11607 Eagle Dr., Mont Belvieu, TX 77580. Mailing Address: P.O. Box 1048, Mont Belvieu, TX 77580. Tel.: 281-576-2213.
Institution Type/Description: History Museum.
Collections: local history; photographs; period barber chairs & shaving equipment; early furnishings; period drug store & post office artifacts; memorial quilt; school desks; yearbooks.
Hours & Admission Prices: Tues. & Thurs. 10-2, 2nd Sat. each month 10-2. No charge.

Mount Vernon

FRANKLIN COUNTY HISTORICAL MUSEUM, (M), 107 Scott St., Mount Vernon, TX 75457. Mailing Address: P.O. Box 289, Mount Vernon, TX 75457-0289. Tel.: 903-537-4760 & 7012. Facebook: Franklin County Historical Assoc.
E-mail: fchadirector@mt-vernon.com
Web Site: www.fcha-online.org
Key Personnel: Dir., Elaine McFeely; Pres. (V), B.F. Hicks.
Volunteer Hours: 2,000
Governing Authority: Parent Institution: Franklin County Historical Association. Subsidiary Institutions: Old Fire Station Museum; Old Depot; Majors-Parchman House Museum; Thruston House Museum. Tax-exempt.
Institution Type/Description: History Museum.
Collections: area history; butterflies; bird eggs; hand-carved violins; Indian artifacts; Don Meredith sports memorabilia; agricultural implements.
Facilities: nature trail; nature reserve.
Activities: Museum Sponsors: Spring Youth Living Wax Museum; Birding Expo in Spring.
Publications: bimonthly newsletter.
Hours & Admission Prices: Office: Tues.-Fri. 9-3. Museum: Thurs.-Sat. 10-2; other times by appointment. No charge; donations accepted. &
Attendance: 2,500 (estimated)

Membership: Individual $15; Family $25; Patron $50; Sponsor $100.

Nacogdoches

DURST-TAYLOR HISTORIC HOUSE AND GARDENS, 304 North St., Nacogdoches, TX 75961-5002. Mailing Address: City of Nacogdoches, Historic Sites Dept., P.O. Box 635030, Nacogdoches, TX 75963-5030. Tel.: 936-560-4443. Fax: 936-560-4448. Facebook: Nac Historic Sites.
E-mail: historicsites@ci.nacogdoches.tx.us
Web Site: www.ci.nacogdoches.tx.us
Founded: 2006.
Congressional District: 2
Key Personnel: Asst. Historic Sites Mgr., Jessica Sowell.
Personnel Profile: Full-Time Paid 1; Part-Time Paid 3; Part-Time Volunteers 2; Interns 2.
Governing Authority: city; nonprofit. Tax-exempt: 501(c)(3).
Institution Type/Description: Historic House Museum: former home of Bennet Blake, delegate to the 1875 Constitutional Convention and later to Thomas J. Rusk, a signer of the Texas Declaration of Independence; built c.1830s. Listed on the National Register of Historic Sites. Recorded Texas Historic Landmark & State Archaeological Site.
Collections: local history; period furnishings; photographs; personal artifacts; blacksmith shop; smokehouse; gardens; sugarcane mill; chicken coop.
Facilities: visitors center; gardens.
Activities: guided tours. Annual Event: Sugar Cane Syrup Making.
Hours & Admission Prices: Tues.-Sat. 10-4. No charge; donations accepted. Closed major holidays. &
Attendance: 2,023 (accurate)

NACOGDOCHES FIRE MUSEUM, 202 S. Fredonia, Nacogdoches, TX 75963. Mailing Address: P.O. Drawer 630648, Nacogdoches, TX 75963. Tel.: 936-559-2541.
Congressional District: 1
Institution Type/Description: Firefighting History Museum.
Collections: firefighting history, equipment & memorabilia; early fire engine; photographs.
Activities: group tours.
Hours & Admission Prices: By appointment. No charge.

STERNE-HOYA HOUSE MUSEUM AND LIBRARY, 211 S. Lanana St., Nacogdoches, TX 75961-5148. Mailing Address: City of Nacogdoches, Historic Sites Dept., P.O. Box 635030, Nacogdoches, TX 75963-5030. Tel.: 936-560-5426. Fax: 936-569-9813. Facebook: Nac Historic Sites.
E-mail: historicsites@ci.nacogdoches.tx.us
Web Site: ci.nacogdoches.tx.us
Founded: 1959.
Congressional District: 2
Key Personnel: Asst. Historic Sites Mgr., Jessica Sowell.
Personnel Profile: Full-Time Paid 1; Part-Time Paid 3; Part-Time Volunteers 2; Interns 2.
Governing Authority: municipal. Parent Institution: City of Nacogdoches. Tax-exempt.
Institution Type/Description: History Museum: housed in Adolphus Sterne Home.
Collections: Texas artifacts; furniture; household fixtures.
Facilities: 5,000-vol. library of Texas history books available for use on premises.
Activities: historical tours.
Hours & Admission Prices: Mon. group tours only, Tues.-Sat. 10-4. No charge; donations accepted. Closed national holidays. &
Attendance: 3,645 (accurate)

STONE FORT MUSEUM, (M), 1808 Alumni Dr., N., Nacogdoches, TX 75962. Mailing Address: P.O. Box 6075, SFASU, Nacogdoches, TX 75962. Tel.: 936-468-2408. Fax: 936-468-7084.
E-mail: cspears@sfasu.edu
Web Site: www.sfasu.edu/stonefort
Founded: 1936.
Congressional District: 2
Key Personnel: Dir., Carolyn Spears; Pres. University, Dr. Baker Patillo; Chm. University Bd., Melvin R. White.
Personnel Profile: Full-Time Paid 1; Part-Time Paid 9; Part-Time Volunteers 1; Interns 1.
Governing Authority: state. Parent Institution: Stephen F. Austin State University. Tax-exempt: 501(c)(3).

Institution Type/Description: Local History Museum: housed in 1936 reconstruction of 1780s structure.
Collections: East Texas artifacts dating prior to 20th-century.
Research Fields: 19th-century technology; Spanish & Mexican periods in East Texas.
Activities: guided tours; permanent & temporary exhibitions.
Hours & Admission Prices: Tues.-Sat. 9-5, Sun. 1-5. No charge; donations accepted. Closed national & university holidays. &
Attendance: 7,909 (accurate)
Membership: Student $5-$34; Individual $35; Contributor $50; Patron $100; Sponsor $250; Bronze $1,000; Gold $2,500; Diamond $5,000.

Nederland

DUTCH WINDMILL MUSEUM, 1515 Boston Ave., Nederland, TX 77627. Tel.: 409-722-0279.
Web Site: www.nederlandtx.com
Founded: 1969.
Institution Type/Description: History Museum: housed in a replica Dutch windmill.
Collections: Dutch history & artifacts; wooden shoes; W.F. (Buddy) Davis' 1952 Gold Medal; singer, Tex Ritter memorabilia; western artifacts; personal artifacts.
Facilities: Museum-related items for sale.
Hours & Admission Prices: Summer: Tues.-Sun. 1-5; Winter: Thurs.-Sun. 1-5. No charge.

New Braunfels

MCKENNA CHILDREN'S MUSEUM, 801 W. San Antonio St., New Braunfels, TX 78130-5503. Tel.: 830-606-9525. Fax: 830-606-9535.
E-mail: ajewell@mckenna.org
Web Site: www.mckennakids.org
Formerly: The Children's Museum in New Braunfels
Founded: 1986.
Congressional District: 28
Key Personnel: Dir., Alice Jewell.
Personnel Profile: Full-Time Paid 4; Part-Time Paid 5; Part-Time Volunteers 10.
Governing Authority: nonprofit. Parent Institution: McKenna. Tax-exempt.
Institution Type/Description: Children's Museum.
Collections: hands-on exhibits.
Facilities: 18,000 sq. ft. exhibit space.
Activities: art, science, cooking, garden & health programming.
Publications: quarterly newsletter.
Hours & Admission Prices: Mon.-Sat. 10-5. Adults: Memorial Day to Labor Day $7.50; Labor Day to Memorial Day $5.50; members no charge. Closed New Year's Day; Easter; Memorial Day; Independence Day; Labor Day; Thanksgiving; Christmas. &
Attendance: 60,000 (accurate)
Membership: Family of Two $100; Family of Four $125; Family of Six $165.

NEW BRAUNFELS CONSERVATION SOCIETY, 1300 Church Hill Dr., New Braunfels, TX 78130-3205. Tel.: 210-629-2943.
E-mail: lnbcs@att.net
Web Site: www.nbconservation.org
Founded: 1964.
Congressional District: 21
Key Personnel: Dir., Martha Rehler; Pres., JoBeth Oestreich; Treas., Delitha Guenzel; Security, Josef Campos.
Personnel Profile: Part-Time Paid 1; Part-Time Volunteers 50.
Governing Authority: private; nonprofit organization. Tax-exempt: 501(c)(3).
Institution Type/Description: Historical & Preservation Society: located on 3 sites-Conservation Plaza (19 buildings & Rose Conservatory over three and a half acres), Lindheimer Home, Buckhorn Barber Shop & Museum.
Collections: handmade furniture; household items, c.1845-1900s. Conservation Plaza: 19 restored structures.
Facilities: botanical garden. Museum-related items for sale.
Activities: guided tours; lectures; rental gallery; participatory & temporary exhibitions. Annual Events: two membership appreciation dinners; Gartenfest & Historic Home Tour in March; Kaffee Haus-Conservation Plaza in November; Elderhostels.
Publications: monthly (except summer) newsletter; brochure.
Hours & Admission Prices: Lindheimer Haus: by appointment. Adults $2.50, students $.50; discounts to AAM & AAA members. Buckhorn Barber Shop & Museum: by appointment. Antique Rose Conservatory & Conservation Plaza: Tues.-Fri. 10-2:30, Sat.-Sun. 2-4. Closed Thanksgiving; Christmas-New Years.
Attendance: 5,000 (estimated)

Membership: Active $15; Patron $25; Family $30.

NEW BRAUNFELS MUSEUM OF ART & MUSIC, 1259 Gruene Rd., New Braunfels, TX 78130-3003. Tel.: 830-625-5636.
Founded: 1991.
Key Personnel: C.E.O. & Dir., Charles R. Gallagher; Devel., Cassey Parkey; Education, Ruth Sullivan; Pres. (V), Charles Teeter; Public Rels., Janelle Berger; Treas., Debbie Voorhees; Registrar, Tony Lyle; Cur., Craig Hillis; Museum Shop Mgr., Nancy Webb.
Personnel Profile: Full-Time Paid 5; Full-Time Volunteers 2; Part-Time Paid 5; Part-Time Volunteers 10; Interns 2.
Governing Authority: private; nonprofit organization. Tax-exempt: 501(c)(3).
Institution Type/Description: Music History Museum.
Collections: photographs; oral histories; recordings; film documentaries; visual art; folk art; music & crafts in Texas.
Research Fields: Texas musicology; oral histories of living Texas artists.
Facilities: lab; 299-seat restaurant; 15,000 sq. ft. exhibit space. Museum-related items for sale.
Activities: concerts; dance recitals; docent program; formal education programs. Annual Events: 4th Annual Music & Arts Festival; Lone Star Arts Award.
Hours & Admission Prices: Summer: Mon. & Wed.-Sat. 10-9, Sun. 12-6; Winter: Wed.-Sat. 10-6, Sun. 12-6. Adults $3, children $2.

NEW BRAUNFELS RAILROAD MUSEUM, 302 W. San Antonio St., New Braunfels, TX 78131. Mailing Address: P.O. Box 310475, New Braunfels, TX 78131-0475. Tel.: 830-627-2447.
E-mail: info@newbraunfelsrailroadmuseum.org
Web Site: www.newbraunfelsrailroadmuseum.org
Governing Authority: nonprofit organization.
Institution Type/Description: Railroad Museum.
Collections: railroad history & artifacts; HO & N scale model railroads; photographs.
Activities: Museum Sponsors: Annual Fall Train Show in November.
Hours & Admission Prices: Thurs.-Mon. 12-4. No charge; donations accepted.
Membership: Annual $25; NRHS $39.

SOPHIENBURG MUSEUM & ARCHIVES INC., (M), 401 W. Coll St., New Braunfels, TX 78130-5618. Tel.: 830-629-1572. Fax: 830-629-3906.
E-mail: director@sophienburg.com
Web Site: www.sophienburg.com
Founded: 1926.
Congressional District: 21
Key Personnel: Dir., Linda Dietert.
Personnel Profile: Part-Time Paid 10; Part-Time Volunteers 160.
Governing Authority: society. Affiliated with Sophienburg Museum & Archives. Tax-exempt.
Institution Type/Description: History Museum: housed in fieldstone veneer building on site of the headquarters of original German colony founded in 1845 in Republic of Texas.
Collections: local history; German immigration & settlement in Texas 1840 to present; manuscript materials; genealogy records; photographs; German & local artifacts; tools & equipment; guns & armament.
Research Fields: TX history; genealogy; German Immigration; photographs.
Facilities: library; archives.
Activities: educational services; guided tours. Museum Sponsors: Fourth of July Celebration; Weihnachtsmarkt, German Christmas Market in November; traditional visit of St. Nicholas in December.
Publications: cookbook, Guten Appetit; New Braunfels Comal County, Texas: A Pictorial History; The New Braunfels Sesquicentennial Minutes; It's Fair Time - History of the Comal County Fair, Kindermaskenball - Past and Present; War Between the States Participants from Comal County, Texas; War Between the States; Comal County Texas in the Civil War.
Hours & Admission Prices: Tues.-Sat. 10-4. Museum: adults $5, students 13-18 $2, children 6-12 $1; discounts to AAM members; active military & members no charge. Archives: adults & students $10. Closed New Year's Eve & Day; Good Friday; Independence Day; Thanksgiving & day after; Christmas Eve & Day. &
Attendance: 5,600 (estimated)
Membership: Student, Military & Teacher $30; Individual $50; Family $75; Business $200.

Newcastle

FORT BELKNAP MUSEUM AND ARCHIVES, INC., Fort Belknap Cir., Newcastle, TX 76372. Mailing Address: P.O. Box 444, Graham, TX 76450. Tel.: 940-846-3222.
Founded: 1851.

Congressional District: 17
Key Personnel: C.E.O., Dr. Harry Hewitt.
Governing Authority: county. Affiliated with Texas Wesleyan College. Tax-exempt: 501(c)(3).
Institution Type/Description: History & Military Museum.
Collections: archives; Indian artifacts; preservation project; photographs. Historic Houses: 1853, Corn House; officer's quarters; magazine; army barracks #1, #2 and #4; c.1854 commissary; household items from 1850s-1890s.
Research Fields: Texas history & government; frontier & Indian history; 19th-century military; frontier biography; genealogy; archaeology.
Facilities: 2,000-vol. library of rare books. Publications, curios and postcards for sale.
Activities: guided tours; lectures; films; arts festivals; study clubs; inter-museum loan & permanent exhibitions; rental space available.
Publications: annual books, Fort Belknap Society Yearbook; Fort Belknap Genealogical Association Bulletin.
Hours & Admission Prices: Museum: Mon.-Tues. & Thurs.-Sat. 9-12 & 1:30-5, Sun. 1-5. No charge; donations accepted. Archives: Sat. 8:30-5:30. Adults $10. &
Attendance: 30,000 (estimated)
Membership: Individual $5; Sustaining $25; Life $100.

Nocona

NORTH TEXAS SOCIETY OF HISTORY AND CULTURE DBA TALES 'N' TRAILS MUSEUM, 1522 E. Hwy. 82, Nocona, TX 76255. Tel.: 940-825-5330.
E-mail: contact@talesntrails.org
Web Site: talesntrails.org
Institution Type/Description: History Museum.
Collections: local history, heritage & culture; photographs; personal artifacts; oil & gas industry; leather industry; agriculture; western heritage; Native American artifacts.
Facilities: Museum-related items for sale.
Hours & Admission Prices: Call for hours.

Odessa

*　**ELLEN NOEL ART MUSEUM, (M),** 4909 E. University, Odessa, TX 79762-7960. Tel.: 432-550-9696. Fax: 432-550-9226. Facebook: Ellen Noel Art Museum.
E-mail: info@noelartmuseum.org
Web Site: noelartmuseum.org
Founded: 1985.
Congressional District: 19
Key Personnel: Exec. Dir., George Jacob.
Personnel Profile: Full-Time Paid 7; Part-Time Paid 2; Part-Time Volunteers 125.
Governing Authority: nonprofit organization. Tax-exempt: 501(c)(3).
Institution Type/Description: Art Museum.
Collections: temporary exhibitions of art.
Major Exhibits: Inner Vision: The Sculpture of Michael Naranjo, 12/13/13-3/2/14; The Color of Oil: Paintings by Margarete Bagshaw, 11/23/13-3/2/14; Russian Orthodox Art of Iconography, 3/14/14-5/18/14; Healing Blade, 5/30/14-8/24/14; Odessa Art Association Juried Show, 5/8/14-6/1/14; Crocheted Metal: The Art of Tracy Krumm, 6/6/14-8/24/14.
Facilities: art classrooms; sensory & sculpture garden; auditorium.
Activities: guided tours; lectures; organized education programs for children & adults; hobby workshops; community art days; fundraisers; art events; outreach programs.
Publications: newsletter; exhibit catalogs.
Hours & Admission Prices: Tues.-Sat. 10-5, Sun. 2-5. No charge; donations accepted. Closed national holidays. &
Attendance: 22,000 (estimated)
Membership: Student & Educator $30; Individual $60; Dual & Household $80; Smithsonian $150; Supporter $250; Patron $500; Connoisseur $1,000; Benefactor $2,500. Corporate sponsorships available.

THE ODESSA METEOR CRATER AND MUSEUM, 620 N. Grant Ave., Ste. 1204, Odessa, TX 79761-4549. Tel.: 432-381-0946. Fax: 432-332-1667.
E-mail: tomrodman@cableone.net
Web Site: www.nwol.net/virtdomains/meteorcrater
Founded: 2002.
Congressional District: 11
Key Personnel: Pres. (V), Thomas E. Rodman; Museum Shop Mgr., Francis Hankins
Governing Authority: county. Tax-exempt.
Institution Type/Description: Meteorite Museum.

Collections: meteorites; tektites; meteorite impact products; videos.
Hours & Admission Prices: Museum: Tues.-Sat. 10-5, Sun. 1-5. No charge; donations accepted. &
Attendance: 15,000 (accurate)

THE PRESIDENTIAL ARCHIVES AND LEADERSHIP LIBRARY, 4919 E. University Blvd., Odessa, TX 79762-8144. Tel.: 432-363-7737. Fax: 432-550-2851.
E-mail: presiden.museum@att.net
Web Site: thepresidentialmuseum.org
Founded: 1964.
Congressional District: 23
Personnel Profile: Part-Time Volunteers 10.
Governing Authority: Parent Institution: University of Texas of the Permian Basin. Subsidiary Institution: John Ben Shepperd Public Leadership Institute. Tax-exempt: 501(c)(3).
Institution Type/Description: History Museum.
Collections: 6,000 books, journals & papers pertaining to the presidency; photographs; campaign memorabilia; posters; campaign buttons; letters; personal artifacts from presidents & first ladies; miniature doll collection.
Research Fields: presidential campaigns; political parties; U.S. history.
Facilities: 6,000-vol. library.
Activities: guided tours; lectures; docent program or council; inter-museum loan, permanent, temporary & traveling exhibitions; workshops; film festivals & demonstrations.
Publications: newsletter.
Hours & Admission Prices: Tues.-Sat. 10-5. No charge; donations accepted. &
Attendance: 5,210 (accurate)
Membership: Student $10; Individual $25; Family: $40; Ambassador $100; Cabinet $250; President's Circle $500. Business & Corporate: Supporter $500; Patron $1,000; Benefactor $2,500; Advocate $5,000 & up.

Olton

SAND CRAWL MUSEUM, Olton Library, 701 Main St., Olton, TX 79064. Tel.: 806-285-7772. Fax: 806-285-7770.
Institution Type/Description: History Museum.
Collections: local history & culture; photographs; books; period artifacts.
Hours & Admission Prices: Mon. & Wed.-Fri. 9-12 & 1-5:30, Tues. 1-8.

Orange

HERITAGE HOUSE OF ORANGE COUNTY ASSOCIATION INC., 905 W. Division, Orange, TX 77630-6959. Tel.: 409-886-5385. Fax: 409-886-0917.
E-mail: hhmuseum@exp.net
Web Site: www.heritagehouseoforangecounty.com
Founded: 1977.
Congressional District: 19
Key Personnel: Pres. & Dir., Joyce Atkins; Chm., Linda Garrett.
Personnel Profile: Full-Time Paid 1; Part-Time Volunteers 30.
Governing Authority: nonprofit organization. Branch Museums: Heritage House Museum, 905 W. Division, Orange, TX 77630; Heritage History Museum of Orange County, 110 Border St., Orange, TX 77630. Tax-exempt: 501(c)(3).
Institution Type/Description: Historic House & History Museum: housed in 1902 turn-of-the-century house.
Collections: Orange County History from early Indian settlements to present day; period rooms; medical instruments; archival & photographic images; costumes & textiles; decorative arts; historic furnishings.
Research Fields: Orange County history; artifacts.
Facilities: library.
Activities: guided tours; lectures; organized education programs for children; docent program; training programs for professional museum workers; participatory, loan & temporary exhibitions.
Publications: quarterly newsletter, Nostalgic News.
Hours & Admission Prices: Temporarily closed. &
Attendance: 1,800 (estimated)
Membership: Regular & Civic Clubs $25; Patron $35; Benefactor $50; Sustaining & Corporate $100.

SHANGRI LA BOTANICAL GARDENS, 2111 W. Park Ave., Orange, TX 77630. Mailing Address: P.O. Box 1044, Orange, TX 77631. Tel.: 409-670-9113. Fax: 409-670-9341.
E-mail: info@shangrilagardens.org
Web Site: www.shangrilagardens.org
Institution Type/Description: Botanical Gardens.
Collections: over 300 plants species; sculpture; nesting birds; hands-on exhibits.

Facilities: laboratory; classrooms; theater; cafe. Garden-related items for sale.
Activities: educational programs; demonstrations.
Hours & Admission Prices: March-Oct. Tues.-Fri. 9-5, Sat. 9-7, Sun. 12-5; Nov.-Feb. Tues.-Sat. 10-5, Sun. 12-5. Closed New Year's Day; Thanksgiving; Christmas.

STARK MUSEUM OF ART, 712 Green Ave., Orange, TX 77630-5721. Tel.: 409-886-2787. Fax: 409-883-6361. Facebook: Stark Museum of Art.
E-mail: info@starkmuseum.org
Web Site: www.starkmuseum.org
Founded: 1974.
Congressional District: 2
Key Personnel: C.E.O., Nelda C. and H. J. Lutcher Stark Foundation, Walter G. Riedel, III; Dir., Dr. Sarah E. Boehme; Registrar, Allison H. Evans; Mgr. Collections & Exhibitions, Terri Fox; Chief Security, Tom Parks; Administrative Asst., Alicia Benitez-Booker; Librarian, Jenniffer Hudson Connors; Chief Educator, Dr. Elena Ivanova; Educator Studio & Family Programs, Amelia Wiggins; Retail Operations Mgr., Trang Tran.
Personnel Profile: Full-Time Paid 6; Part-Time Paid 27; Interns 2.
Governing Authority: nonprofit organization. Parent Institution: Nelda C. and H.J. Lutcher Stark Foundation. Tax-exempt: 501(c)(3).
Institution Type/Description: Art Museum.
Collections: 19th century Western American art; American Indian art; porcelain & crystal; rare books & manuscripts.
Major Exhibits: Tales & Travel, 7/13-1/14.
Research Fields: relating to the collections.
Facilities: 2,500-vol. library related to areas of collections. Books, catalogues, posters & postcards for sale.
Activities: guided tours; gallery talks; family days; lectures; educational programs.
Publications: catalogues; The Western Collection; Taos Portfolio; The American & British Birds of Dorothy Doughty; The Steuben Glass Collection; A Mirror Unto Nature; The Art and Life of W. Herbert Dunton, 1878-1936; First Artistic Traditions; A Guide to the Galleries.
Hours & Admission Prices: Tues.-Sat. 9-5. Adults $6; active military & their families and upper-level members of Museums West Consortium no charge. Closed New Year's Day; Easter; Independence Day; Thanksgiving; Christmas Eve & Day. ♿
Attendance: 14,247 (accurate)
Membership: Individual $60; Couple & Dual $90; Family $120. Benefactor: Star $250; Crescent Moon $500.

THE W.H. STARK HOUSE, (M), 610 W. Main Ave., Orange, TX 77630-5704. Mailing Address: P.O. Drawer 909, Orange, TX 77631-0909. Tel.: 409-883-0871. Fax: 409-883-3530.
E-mail: info@whstarkhouse.org
Web Site: www.whstarkhouse.org
Founded: 1981.
Congressional District: 2
Key Personnel: C.E.O., Walter Riedel, III; Dir., Patricia L. Herrington.
Personnel Profile: Full-Time Paid 4; Part-Time Paid 22.
Governing Authority: owned & operated by the Nelda C. & H.J. Lutcher Stark Foundation. Tax-exempt: 501(c)(3).
Institution Type/Description: Historic House: 1894 Victorian Home of William H. Stark & Miriam Lutcher Stark.
Collections: original furniture; rugs; family portraits; lace curtains; silver; ceramics; glass; oriental rugs.
Activities: guided tours.
Publications: video tour.
Hours & Admission Prices: Tues.-Sat. 9-3:30. Adults $6, students & seniors 65 & over $5; discounts to AAM members; children under 10 not admitted.
Attendance: 3,350 (accurate)

Ozona

CROCKETT COUNTY MUSEUM, 408 11th St., Ozona, TX 76943. Mailing Address: P.O. Box 1444, Ozona, TX 76943-1444. Tel.: 325-392-2837. Fax: 325-392-5654.
E-mail: ccmuseum@wcc.net
Founded: 1939.
Congressional District: 67
Key Personnel: Pres. (V), Jan Van Schoubrouek; Vice Pres., Cathy Carson; Coord., Emily Guerra; Aide, Roberta Schoenhals.
Personnel Profile: Full-Time Paid 1; Full-Time Volunteers 1; Part-Time Paid 2; Part-Time Volunteers 20.
Governing Authority: society; nonprofit organization. Parent Institution: Crockett County Historical Society. Tax-exempt: 170(b)(1)(A).
Institution Type/Description: History Museum.

Collections: prehistoric & historic Indian tools, weapons, & artifacts; geological specimens of the county; county wildflowers & plants; memorabilia of early ranchers, ranching industry & brands; photographs of early pioneers; guns; furniture.
Facilities: library of county history available on premises; reading room.
Activities: Museum Sponsors: Crockett County Style Show.
Publications: museum & county brochures; book; DVD, A History of Crockett County; video, This Rugged Land Called Crockett County.
Hours & Admission Prices: Mon.-Fri. 9-5, Sat. 10-3. Adults $2. Closed New Year's Day; Presidents' Day; Good Friday; Memorial Day; Independence Day; Labor Day; Veterans Day; Thanksgiving & day after; Christmas Eve & Day. ♿
Attendance: 2,100 (estimated)
Membership: Individual $20; Family $25.

Pampa

FREEDOM MUSEUM USA, 600 N. Hobart, Pampa, TX 79065-5235. Tel.: 806-669-6066.
E-mail: fusa@att.net
Web Site: freedommuseumusa.org
Founded: 1987.
Governing Authority: Tax-exempt.
Institution Type/Description: Military History Museum.
Collections: military history, equipment & artifacts; B25-D Bomber; Vietnam-era Bell UH-1F helicopter; photographs; F4E Phantom jet; M60 Patton tank; M110 Howitzer.
Hours & Admission Prices: Tues.-Sat. 12-4. No charge; donations accepted.

WHITE DEER LAND MUSEUM, 112-116 S. Cuyler, Pampa, TX 79065. Mailing Address: P.O. Box 1556, Pampa, TX 79066-1556. Tel.: 806-669-8041. Fax: 806-669-8030. Facebook: White Deer Land Museum.
E-mail: wdlmuseum@graycch.com
Web Site: www.pampamuseum.org
Founded: 1970.
Congressional District: 31
Key Personnel: Dir. & Cur., Courtney Oxley; Asst., Kay Lard.
Personnel Profile: Full-Time Paid 2; Part-Time Volunteers 6.
Governing Authority: county. Tax-exempt.
Institution Type/Description: General Museum: housed in 1916 White Deer Land Co. Building.
Collections: Rolla J. Sailor arrowheads; David F. Barry limited edition photographs; furniture; clothing; documents; manuscripts; toys; dolls; musical instruments; carriages; farm machinery; military items; wood carvings; replica of first wood-derrick oil well on White Deer lands; replica of first Gray County courthouse; maps; chapel furnished with items from early day churches of Pampa; Red River War artifacts 1874-1875.
Research Fields: history.
Facilities: library of M.K. Brown Range Life Series of ranch life in the Southwest & complete company records of White Deer Land Company available for use on premises.
Activities: guided tours; special programs for civic clubs, study clubs, scout troops elementary & secondary school classes; permanent & temporary exhibits & demonstrations.
Publications: brochure; books, Personal Diary of GCT, History of Pampa Post Office, For the Reason We Climb Mountains, Red River Expedition, White Deer Land Museum History Wall, Gray County History Book, The Log House on White Deer Creek, History of M.K. Brown, History of C.P. Buckler, History of T.D. Hobart, White Deer Land Building.
Hours & Admission Prices: Winter & Summer: Tues.-Fri. 1-4; group tours by appointment. No charge; donations accepted. Closed national holidays. ♿
Attendance: 2,000 (accurate)

Panhandle

CARSON COUNTY SQUARE HOUSE MUSEUM, (M), TX Hwy. 207 @ Fifth St., Panhandle, TX 79068. Mailing Address: P.O. Box 276, Panhandle, TX 79068-0276. Tel.: 806-537-3524. Fax: 806-537-5628.
E-mail: shm@squarehousemuseum.org
Web Site: www.squarehousemuseum.org
Founded: 1965.
Congressional District: 18
Key Personnel: Bus. Mgr., Shirlyne Grantham; Bd. Chm., Curtis Downs; Educator, Sandy Poteet; Admin. Asst. & Museum Shop Mgr., Janie Plumlee.
Personnel Profile: Full-Time Paid 1; Part-Time Paid 3; Interns 7.
Governing Authority: board of trustees. Tax-exempt: 501(c)(3).

Institution Type/Description: General Museum: located at the former terminus of the Santa Fe railroad.
Collections: agriculture; archaeology; archives; paintings; costumes; manuscripts; entomology; ethnology; folklore; glass; Indian artifacts; military, music; paleontology; transportation; period artifacts; furniture; guns. Historic Structures: mid-1880 Square House; caboose; reconstructed pioneer dugout; windmill; relocated 1912 community church; full-sized activity-specific diorama.
Research Fields: local & regional (Texas Panhandle) prehistory, history, natural history & art.
Facilities: research library of Texana books; photographic archives. Museum-related books & items for sale.
Activities: guided tours; lectures; films; study clubs; hobby workshops; formally organized educational programs; docent program; permanent & temporary exhibitions; school loan service; bus tours to archaeological & historic sites; 29 historic videos may be viewed at the museum or loaned to organizations or individuals.
Publications: 4-vol. book, A Time To Purpose; coloring book, Land of Coronado; poems, Voices of the Square House; cookbook, The Square House Cook Book.
Hours & Admission Prices: Mon.-Sat. 9-5, Sun. 1-5. No charge; donations accepted. Closed New Year's Day; Easter; Thanksgiving; Christmas Day. &
Attendance: 21,275 (estimated)

Paris

HAYDEN MUSEUM OF AMERICAN ART, 930 Cardinal Lane, Paris, TX 75460-6522. Tel.: 903-785-1925. Fax: 903-784-7631.
Founded: 1992.
Key Personnel: Dir. & Pres. (V), William deG. Hayden, M.D.
Institution Type/Description: American Art Museum.
Collections: history of American art from folk art to modern & contemporary; paintings; decorative arts; prints; photographs; sculpture.
Facilities: library.
Activities: seminar; lectures.
Hours & Admission Prices: By appointment. No charge.
Attendance: 2,500 (estimated)

SAM BELL MAXEY HOUSE STATE HISTORIC SITE, 812 S. Church St., Paris, TX 75460-7112. Tel.: 903-785-5716. Fax: 903-785-6716.
E-mail: sam-bell-maxey@thc.state.tx.us
Web Site: www.visitsbmh.com
Congressional District: 4
Personnel Profile: Full-Time Paid 4; Part-Time Volunteers 15.
Governing Authority: state. Parent Institution: Texas Historical Commission, Austin, TX 78744. Tel. 512-479-4882. Tax-exempt.
Institution Type/Description: Historic House Museum: 1868 High Victorian Italianate style home belonging to Civil War Confederate Gen. & U.S. Senator, Sam Bell Maxey.
Collections: household furnishings; 1830-1950 clothing, textiles & memorabilia of the Sam Bell Maxey family; manuscript & music collections.
Research Fields: 19th-century decorative arts; Texas & Civil War history; career of U.S. Senator, Sam Bell Maxey.
Facilities: 3,000-vol. library of books, programs and documents pertaining to Texas history and 19th & early 20th-century material culture available for qualified research by appointment.
Activities: guided tours; lectures; formally organized education programs for children & undergraduate students; special events.
Publications: visitors guide.
Hours & Admission Prices: Guided Tours: Tues.-Sun. 9-4, tours on the hour; groups of 10 or more by appointment. Adults $4, students 6-18 $3; children 5 & under no charge. Closed New Year's Eve & Day; Thanksgiving; Christmas Eve & Day. &
Attendance: 2,000 (estimated)

Parker

SOUTHFORK RANCH & VISITOR CENTER, 3700 Hogge Rd., Parker, TX 75002. Tel.: 972-442-7800.
Institution Type/Description: Historic Mansion: housed on the site of the filming for the television series "Dallas."
Collections: ranch history; "Dallas" series memorabilia & movie props; photographs; personal artifacts.
Facilities: Gift items for sale.
Activities: rental facilities.
Hours & Admission Prices: Daily 9-5. Adults $10.75, seniors $9, children 5-12 $7; discounts to groups of 15 or more; children 4 & under no charge. Closed Thanksgiving; Christmas.

Pasadena

ARMAND BAYOU NATURE CENTER, 8500 Bay Area Blvd., Pasadena, TX 77507. Mailing Address: P.O. Box 58828, Houston, TX 77258-8828. Tel.: 281-474-2551. Fax: 281-474-2552.
E-mail: abnc@abnc.org
Web Site: www.abnc.org
Founded: 1974.
Congressional District: 25
Key Personnel: Pres. Bd. Trustees (V), Peter Zollers; Exec. Dir., Tom Kartrude; Pres. (V), Tom Scarscella; Museum Shop Mgr., Barbara Baxter.
Personnel Profile: Full-Time Paid 6; Part-Time Paid 13; Part-Time Volunteers 185; Interns 6.
Governing Authority: nonprofit organization. Tax-exempt: 501(c)(3).
Institution Type/Description: Nature Center & Preserve: located on 2,500 acres of tallgrass prairie, forest and bayou including a demonstration turn-of-century farm, also includes several prehistoric archeological sites.
Collections: prairie, bayou & woodland preserves; native animal (birds, mammals, reptiles, amphibians, fish) study skins and specimens; early 1900's farmhouse; farm implements; furnishings.
Research Fields: Texas Upper Coast flora & fauna; prairie restoration & management; marsh restoration & management; local water shed water quality.
Facilities: 2,800 sq. ft. exhibit space; 100-seat auditorium; indoor classrooms. Gift items for sale.
Activities: guided tours; lectures; organized educational programs; docent program; participatory exhibits. Annual Events: Fall Festival; Creepy Crawlers; Third Sundays in Nature.
Publications: quarterly membership newsletter, Along the Bayou; volunteer monthly, The Bayou Foliage.
Hours & Admission Prices: Wed.-Sat. 9-5, Sun. 12-5. Adults $4, children 5-17 & seniors over 61 $2; discounts to groups; children under 5 & members no charge. &
Attendance: 35,448 (accurate)
Membership: Seniors $40; Individual $45; Family $55.

POMEROY HOUSE AT PASADENA HERITAGE PARK, 204 Main, Pasadena, TX 77506. Tel.: 713-472-0565.
Web Site: www.pasadenahistoricalsociety.org
Institution Type/Description: History Museum: built in 1906.
Collections: local history & culture; period furnishings; personal artifacts; photographs. Historic Buildings: 1928 Anna's House; Strawberry House.
Activities: group tours; special events. Museum Sponsors: Strawberry Festival in April; Victorian Tea & Fashion Show in July.
Hours & Admission Prices: Wed.-Fri. 9:30-2:30, Sat. 9:30-5, Sun. 1-5.

Pecos

WEST OF THE PECOS MUSEUM, (M), 120 E. Dot Stafford St., Pecos, TX 79772. Mailing Address: Box 1784, Pecos, TX 79772-1784. Tel.: 432-445-5076. Fax: 432-445-3149.
Web Site: www.westofthepecosmuseum.com
Founded: 1962.
Congressional District: 23
Key Personnel: Pres. (V), Bill Oglesby; Cur., Dorinda Millan.
Personnel Profile: Full-Time Paid 3; Part-Time Paid 3; Part-Time Volunteers 50.
Governing Authority: board of trustees; nonprofit organization. Tax-exempt.
Institution Type/Description: History Museum: housed in 1896 two-story red sandstone saloon & c.1904 three-story concrete block Orient Hotel.
Collections: local history memorabilia; saloon; railroad & telegraph memorabilia; western ranch life; school room; bridal suite; barber & beauty room; local pictures & history; Indian artifacts; horse-drawn wagons; chuck wagons; water wagon.
Research Fields: local history.
Facilities: courtyard; park.
Activities: self-guided tours; educational & cultural activities & exhibits. Museum Sponsors: Friends of the Museum; Old Timer's Reunion, Student Art Festival; Kid's Programs; Presidents exhibit in February; Art Show & Sale in July & October; Fall Fair; History Trivia Contest; Living Christmas Tree exhibit.
Publications: brochure (English & Spanish); cookbooks; local history books & brochures; newsletters.
Hours & Admission Prices: June-Aug. Mon.-Sat. 9-5, Sun. 1-4; Sept.-May Tues.-Sat. 9-5. Adults $4, senior citizens & tours $3, children 6-18 $1; discounts to AAM members; members, children under 6, tour bus driver & guides no charge. Closed Christmas week. &
Attendance: 10,000 (accurate)
Membership: Friends of the Museum: Individual $25; Family $50; Associate $100; Sponsor $250; Supporter $500; Partner $1,000.

Perryton

MUSEUM OF THE PLAINS, 1200 N. Main, Perryton, TX 79070-2314. Tel.: 806-435-6400.
E-mail: motp@ptsi.net
Web Site: www.museumoftheplains.com
Institution Type/Description: History Museum.
Collections: lives of settlers in the 1800s; period bottles; natural history; religion.
Facilities: Museum-related items for sale.
Hours & Admission Prices: Mon.-Fri. 9-5, Sat. 10-5, Sun. 1-5. No charge; donations accepted. Closed New Year's Day; Thanksgiving; Christmas Eve & Day.

Pflugerville

HERITAGE HOUSE MUSEUM, 901 Old Austin Hutto Rd., Pflugerville, TX 78660. Mailing Address: P.O. Box 2451, Pfulgerville, TX 78691-2451. Tel.: 512-251-5082.
E-mail: heritagehouse@pfulgervilletx.gov
Institution Type/Description: History Museum.
Collections: local history & culture; period furnishings; personal artifacts; photographs.
Activities: Museum Sponsors: Quilt Show in April.
Hours & Admission Prices: 1st Sun. each month 1-4.

Pharr

OLD CLOCK MUSEUM, 929 E. Preston St., Pharr, TX 78577-5013. Tel.: 956-787-1923.
Founded: 1968.
Key Personnel: Mgr., Gene Shawn.
Governing Authority: individual operation; nonprofit.
Institution Type/Description: Horological Museum.
Collections: clocks.
Research Fields: horological.
Facilities: 60-vol. library of historical books and catalogs on old clocks & watches and their makers available for use on premises. Postcards for sale.
Activities: private tours.
Hours & Admission Prices: Mon.-Fri. 1-5 by appointment. Adults $1.

Pittsburg

NORTHEAST TEXAS RURAL HERITAGE MUSEUM, 204 W. Marshall, Pittsburg, TX 75686-1312. Mailing Address: P.O. Box 157, Pittsburg, TX 75686-0157. Tel.: 903-856-1200.
E-mail: campcountymuseum@aol.com
Web Site: www.pittsburgtexasmuseum.com
Founded: 1989.
Congressional District: 4
Key Personnel: Dir., Fanny Hively; Pres., Robert Peoples; Treas. (V), Gary Bicknell; Cur., Glenda Brogoitti.
Personnel Profile: Part-Time Paid 3; Part-Time Volunteers 18.
Governing Authority: private; nonprofit organization. Tax-exempt: 501(c)(3).
Institution Type/Description: History Museum.
Collections: history of Northeast Texas rural life & inhabitants from prehistoric times to 1950s; Ezekiel Airship. Historic Buildings: 1901 farmstead, barn, outhouse, smokehouse, blacksmith shop, general store; 1900 Cotton Belt Depot & annex with caboose.
Facilities: garden. Museum-related items for sale.
Activities: docent program; films; formal education programs for children; guided tours; lectures; loan, participatory, temporary & traveling exhibitions; rental gallery. Annual Events: Samuel Morse Day; Black History Celebration, Mardi Gras for the Museum.
Publications: quarterly newsletter, Museum News.
Hours & Admission Prices: Thurs.-Sat. 10-4. Adults $4, senior citizens $3, students $2; discount to student groups; members no charge. Closed New Year's Eve & Day; Thanksgiving; Christmas Eve & Day. &
Attendance: 3,164 (accurate)
Membership: Senior & Student $10; Individual $25; Family $50; Century Club $100; Supporter $250; Sponsor $500; Benefactor $1,000.

Plainview

MALOUF ABRAHAM FAMILY ARTS CENTER - WAYLAND BAPTIST UNIVERSITY AKA ABRAHAM ART GALLERY, Mabee Learning Resources, 1900 W. 7th St., CMB #1249, Plainview, TX 79072. Tel.: 806-291-1083 (office) & 3710 (gallery). Fax: 806-291-1980. Facebook; Abraham Art Gallery.
E-mail: kellerc@wbu.edu
Founded: 1997.
Congressional District: 19
Key Personnel: Art Cur., Candace Keller; Museum Shop Mgr., Jeanette Curry.
Personnel Profile: Full-Time Paid 2; Part-Time Paid 5; Part-Time Volunteers 7; Interns 2.
Governing Authority: Parent Institution: Wayland Baptist University. Tax-exempt.
Institution Type/Description: Art Gallery.
Collections: works by regional & national artists; paintings; sculpture; photography.
Major Exhibits: West Texas Regional 2014 Scholastic Art Competition, 1/27/14-2/14/14; Our People, Our Land, Our Images (T), 2/24/14-3/28/14; WBU art Exhibition & Senior Art Practicum, 4/7/14-5/2/14; Plains Art Association 53RO, 5/26/14-6/29/14; American Watercolor Society Awards Exhibition (T), 7/11/14-10/31/14; Rafael Canizares-Yonez, 11/14-12/14.
Facilities: library.
Activities: guided tours.
Publications: brochures; calendars.
Hours & Admission Prices: Call for hours. No charge; donations accepted. &
Attendance: 7,000 (estimated)

MUSEUM OF THE LLANO ESTACADO, Wayland University J.E. & L.E. Mabee Rgnl. Heritage Ctr., 1900 W. 8th St., Plainview, TX 79072. Mailing Address: 1900 W. 7th St., #1226, Plainview, TX 79072-6900. Tel.: 806-291-3660. Fax: 806-291-1982.
E-mail: watsonr@wbu.edu
Web Site: www.wbu.edu/museum
Founded: 1976.
Congressional District: 19
Key Personnel: Dir., Rodney Watson; Administrative Asst., Elva Hipolito.
Personnel Profile: Full-Time Paid 2; Part-Time Paid 2; Part-Time Volunteers 6.
Governing Authority: university. Parent Institution: Wayland University. Tax-exempt.
Institution Type/Description: General Museum.
Collections: geological, archaeological development of the Llano Estacado region; natural history & history.
Research Fields: the Llano Estacado Region.
Facilities: 110-seat auditorium; seminar room.
Activities: lectures; traveling exhibits; programs to schools & other museums; guided tours; films; special lectureships & exhibits; courses in museum-related arts.
Publications: audio-visual programs; occasional reports of activities.
Hours & Admission Prices: April-Nov. Sat.-Sun. 1-5; Dec.-March Mon.-Thurs. 9-5, Fri. 9-4. No charge; donations accepted. Closed college holidays. &
Attendance: 6,200 (accurate)

Plano

* **HERITAGE FARMSTEAD MUSEUM, (M),** 1900 W. 15th St., Plano, TX 75075-7329. Tel.: 972-881-0140. Fax: 972-422-6481.
E-mail: director@heritgefarmstead.org
Web Site: www.heritagefarmstead.org
Founded: 1986.
Congressional District: 3
Key Personnel: Cur., Hillary Kidd; Cur. Exhibits, Lolisa Laenger; Dir. Education, Kathy Strobel; Mktg. & Publicity, Angie Carroll; Facilities, Alex Pelt; Business Mgr., Michelle Prengle; Education Asst., Victoria James.
Personnel Profile: Full-Time Paid 4; Part-Time Paid 6; Part-Time Volunteers 200.
Governing Authority: nonprofit organization. Tax-exempt: 501(c)(3).
Institution Type/Description: Historic House: 1891 Victorian farmhouse located on a 4-acre historic site.
Collections: one-room 1897 schoolhouse; period furnishings from late 19th century; farm implements (1890-1940); textiles; photographs; toys; games; windmill, barns, cisterns, smokehouse, old-fashioned flower, herb & vegetable gardens & farm animals.
Research Fields: Texas; Collin County; Blackland Prairie; pioneer farm life.

Facilities: grounds available for rental; party barn. Crafts & other museum-related items for sale.
Activities: concerts; docent program; formally organized educational program; guided tours; lectures; participatory & temporary exhibits. Annual Events: Lantern Light Tours; Heritage Scout Day: Spring Festival.
Publications: quarterly newsletter, Heritage Today; books, One Room Schools of Collin County, Texas, Hunter's New Adventure.
Hours & Admission Prices: Grounds: daily 10-4:30. Tours: Tues.-Sun. 1:30. Self Guided Tours: adults $2; discounts to AAM members. Guided Tours: adults $5, senior citizens 66 & over and children 3-18 $3.50; discounts to AAM members. Closed holidays. &
Attendance: 32,612 (accurate)
Membership: Individual $35; Family $65; Patron $125; Partner $300; Preservation $500; Corporate $1,000.

INTERURBAN RAILWAY MUSEUM, 901 E. 15th St., Plano, TX 75074-5807. Mailing Address: P.O. Box 861810, Plano, TX 75086-1810. Tel.: 972-941-2117. Fax: 972-941-2656.
E-mail: planoconservancy@earthlink.net
Web Site: www.planoconservancy.org
Formerly: Interurban Railway Station Museum
Founded: 1991.
Key Personnel: Co-Dir., Russell C. Kissick; Co-Dir., Jeffrey Campbell.
Personnel Profile: Full-Time Paid 2; Part-Time Volunteers 10; Interns 1.
Governing Authority: city. Tax-exempt.
Institution Type/Description: Transportation Museum.
Collections: electric rail & interurban transportation; working model O gauge.
Major Exhibits: Moving Picture Across the Prairie, 11/13-4/14.
Research Fields: electric rail transportation North Texas area 1900-1948.
Activities: Friday storytime for preschool children at 10:30am.
Publications: Plano And The Interurban Railway.
Hours & Admission Prices: Mon.-Fri. 10-2, Sat. 1-5. No charge; donations accepted. &
Attendance: 20,000
Membership: Individual $5-$250; Business $75-$1,000.

JC PENNEY ARCHIVES AND HISTORICAL MUSEUM, 6501 Legacy Dr., Plano, TX 75024-3698. Mailing Address: P.O. Box 10001, Dallas, TX 75301. Tel.: 972-431-7926. Fax: 972-431-7896.
Institution Type/Description: Company History Museum.
Collections: company history; 20th-century American clothes & artifacts; period records; photographs; company newspapers & catalogues; recreated first JC Penney store.
Hours & Admission Prices: Mon.-Fri. 8-5. No charge.

Pleasanton

LONGHORN MUSEUM, 1959 Hwy. 97 E., Pleasanton, TX 78064-6500. Tel.: 830-569-6313.
Web Site: www.pleasantontx.org/museum.html
Founded: 1976.
Key Personnel: Dir., Donna Rice; Asst., Valerie Purgason.
Personnel Profile: Full-Time Paid 2; Full-Time Volunteers 1; Part-Time Paid 1.
Governing Authority: nonprofit. Tax-exempt.
Institution Type/Description: History Museum.
Collections: local history & culture; ranching; personal artifacts; farming & industry.
Hours & Admission Prices: Mon.-Sat. 8-5; groups by appointment. No charge; donations accepted. Closed New Year's Day; Easter; Memorial Day; Independence Day; Labor Day; Thanksgiving; Christmas; Saturdays before Monday holidays. &
Attendance: 2,500 (estimated)

Port Arthur

MUSEUM OF THE GULF COAST, (M), 700 Procter St., Port Arthur, TX 77640-6521. Tel.: 409-982-7000. Fax: 409-982-9614.
E-mail: Shannon.Harris@lamarpa.edu
Web Site: www.museumofthegulfcoast.org
Founded: 1964.
Congressional District: 9
Key Personnel: Pres. (V), Dr. Sam Monroe; Dir., Shannon Harris; Education Coord., Carol Boethcher; Treas., Betty Herlin; Museum Shop Mgr., Peggy Arrant.
Personnel Profile: Full-Time Paid 4; Part-Time Paid 7; Part-Time Volunteers 25.
Governing Authority: society. Parent Institution: Port Arthur Historical Society, 1953 Lakeshore, Port Arthur, TX 77640. Subsidiary Institution: Lamar State College, Port Arthur, TX. Tax-exempt: 501(c)(3).

Institution Type/Description: Regional History Museum.
Collections: flora & fauna of the Gulf Coast area; geology; Paleo-Indian artifacts from McFadden Beach; social & cultural materials from the Port Arthur region; decorative arts; pop culture memorabilia.
Research Fields: Sabine Pass Battle 1863.
Facilities: music hall.
Activities: guided tours; enviro-kids summer camp; arts express after school program; lectures; workshops; school loan service. Museum Sponsors: fund-raising musicals.
Publications: newsletter, Mosquito Bytes.
Hours & Admission Prices: Mon.-Sat. 9-5, Sun. 1-5. Adults $4, senior citizens over 62 & college students $3, students 4-18 $2; discounts to groups over 20; children 3 & under and members no charge. Closed New Year's Eve afternoon & Day; Easter; Independence Day; Thanksgiving; Christmas Eve & Day. &
Attendance: 15,000 (accurate)
Membership: General $20-$1,000.

TEXAS ARTISTS' MUSEUM, 3501 Cultural Center Dr., Port Arthur, TX 77642. Tel.: 409-983-4881.
Founded: 1987.
Institution Type/Description: Art Museum.
Collections: works by Texas artists.
Activities: special events; temporary & permanent exhibitions; workshops; summer programs.
Hours & Admission Prices: Tues.-Sat. 12-4. No charge.
Membership: Student $10; Individual $25; Donor $50; Patron $100; Business & Supporter $200; Benefactor $500; Lifetime $700 & up.

Port Isabel

PORT ISABEL HISTORICAL MUSEUM, 317 Railroad Ave., Port Isabel, TX 78578-4107. Tel.: 956-943-7602.
Institution Type/Description: History Museum: housed in a building built by Charles Champion which served as a post office, US Customs house, railroad depot, general store, & restaurant with a residence upstairs; built in 1899.
Collections: local history & culture; period furnishings; personal artifacts; photographs.
Hours & Admission Prices: Call for hours.

Port Lavaca

CALHOUN COUNTY MUSEUM, (M), 301 S. Ann St., Port Lavaca, TX 77979-4205. Tel.: 361-553-4689. Fax: 361-553-4689.
E-mail: director@calhouncountymuseum.org
Web Site: www.calhouncountymuseum.org
Founded: 1964.
Congressional District: 14
Key Personnel: Dir., George Ann Cormier.
Personnel Profile: Full-Time Paid 1; Part-Time Paid 1.
Governing Authority: county; nonprofit. Tax-exempt.
Institution Type/Description: History Museum.
Collections: lens from the Matagorda Island Lighthouse; Calhoun County history.
Research Fields: photographs of Indianola, Port Lavaca, Seadrift, Point Comfort, Port O'Connor, Olivia; French influence in Calhoun County & LaSalle; natural history of area.
Activities: guided tours; temporary exhibitions.
Hours & Admission Prices: Tues.-Wed. 10:30-4:30, Thurs.-Fri. 10:30-5, Sat. 10-3. No charge; donations accepted. Closed county holidays.
Attendance: 1,785 (accurate)

Post

GARZA COUNTY HISTORICAL MUSEUM, 119 North Ave. N., Post, TX 79356-3105. Tel.: 806-495-2207.
E-mail: lgpuckett@gmail.com
Web Site: garzacountymuseum.org
Founded: 1970.
Key Personnel: Cur., Linda G. Puckett.
Governing Authority: county. Tax-exempt.
Institution Type/Description: History Museum.
Collections: local history & culture; photographs; period furnishings; personal artifacts; C.W. Post collection including his Post Cereal factory office, 19th century art, sculptures, Italian carved wood chairs & suit of armour.
Publications: Garza County; Post.
Hours & Admission Prices: Tues.-Sat. 10-5. No charge; donations accepted. &
Attendance: 8,000 (accurate)
Membership: Friends $35; Pioneer $100; Founder $250; Silver $500; Gold $1,000; Platinum $2,500 & up.

OS MUSEUM, 201 E. Main St., Post, TX 79356-3351. Tel.: 806-495-3570. Fax: 806-495-2288.
E-mail: osmuseum_mneff@yahoo.com
Formerly: OS Ranch Museum
Founded: 1991.
Congressional District: 28
Key Personnel: Dir., Marie T. Neff.
Personnel Profile: Full-Time Paid 1; Part-Time Paid 1.
Institution Type/Description: Fine Art Museum.
Collections: ranch history; Native American drawings; fossil sites; works by Western artists & cowboys; sculptures; photographs; fine art; artifacts from around the world.
Activities: Annual Events: Spring Exhibit Feb.-May; Summer Exhibit July-Sept.; Christmas Exhibit Nov.-Jan.
Hours & Admission Prices: Call for hours. No charge; donations accepted. &

Presidio

FORT LEATON STATE HISTORICAL SITE, 16953 E. FM170, Presidio, TX 79845. Mailing Address: P.O. Box 2439, Presidio, TX 79845-2439. Tel.: 432-229-3613. Fax: 432-229-4814.
Web Site: www.tpwd.state.tx.us
Founded: 1977.
Congressional District: 21
Key Personnel: C.E.O., Carter Smith; Chm. Commission, T. Dan Friedkin; Dir. Parks, Brent Leisure.
Personnel Profile: Full-Time Paid 4; Part-Time Paid 1; Part-Time Volunteers 8; Interns 2.
Governing Authority: state. Parent Institution: Texas Parks & Wildlife Dept., 4200 Smith School Rd., Austin, TX 78744. Tel. 512-479-4882. Tax-exempt.
Institution Type/Description: Historic Building & Museum: 1848 private 40-room adobe fortress built by Indian trader Benjamin Leaton, located on old Indianola-San Antonio-Chihuahua Trail & overlooking the Rio Grande.
Collections: 16th to late 19th-century artifacts & interpretive materials covering the history of La Junta de los Rios now known as the Presidio-Ojinaga region.
Facilities: picnic sites.
Activities: guided tours; permanent exhibitions.
Publications: brochure.
Hours & Admission Prices: Daily 8-4:30. Adults $5; children under 12 no charge. Closed Christmas.
Attendance: 3,600 (accurate)
Membership: Friends Group $15-$250.

Quitman

LIGHT CRUST DOUGHBOYS HALL OF FAME & MUSEUM - GOVERNOR JIM HOGG CITY PARK, 518 S. Main St., Quitman, TX 75783. Tel.: 903-763-2701. Fax: 903-763-2764.
E-mail: info@quitmanheritage.org
Web Site: quitmanheritage.org
Formerly: Governor Hogg Shrine State Park
Founded: 2005.
Key Personnel: Exec. Dir. & Museum Mgr., Rebecca Barrett; Chm. (V) & Pres., Gordon Stone; Vice Pres., Jo An Coker; Treas., Barry Carlson.
Personnel Profile: Full-Time Paid 1; Part-Time Paid 1; Part-Time Volunteers 10.
Governing Authority: state. Parent Institution: Texas Parks & Wildlife Dept., 4200 Smith School Road, Austin, TX 78744, 512-389-4889. Subsidiary Institution: City of Quitman. Tax-exempt.
Institution Type/Description: Historic Houses: 1869 Stinson Home; honeymoon cottage of Governor Hogg.
Collections: history items of Wood County & northeast Texas; period furnishings; Light Crust Doughboys collection; Gov. W. Lee (Pappy) O'Daniel.
Facilities: playground; picnic areas; Pony Truss Bridge on Nature Trail.
Activities: self-guided & guided tours; half mile nature & hiking trail. Park Sponsors: Western Swing Festival in May; Old Settlers Reunion in August.
Publications: quarterly newsletter, Quitman Heritage Foundation.
Hours & Admission Prices: Mon.-Sat. 9-4. Adults $3. Closed New Year's Day; Thanksgiving; Christmas Eve & Day. &
Attendance: 5,000 (estimated)
Membership: Quitman Heritage Foundation: Friend $30; Family $100; Supporter $250; Patron $500; Lifetime $1,000.

Ralls

RALLS HISTORICAL MUSEUM, 801 Main St., Ralls, TX 79357. Mailing Address: P.O. Box 384, Ralls, TX 79357-0384. Tel.: 806-253-2425. Fax: 806-253-2425.
E-mail: rallshistoricalmuseum@windstream.net
Founded: 1970.
Congressional District: 84
Key Personnel: Pres. (V), Dale Sedgwick; Treas., Mary Helen Jamerson; Registrar, Jeanette Wilson.
Personnel Profile: Full-Time Paid 1.
Governing Authority: bd. of directors; nonprofit organization. Tax-exempt: 501(c)(3).
Institution Type/Description: General Museum: housed in 1918 First National Bank.
Collections: art objects; history; archaeology; science; natural history; ethnology; military.
Research Fields: history.
Activities: guided tours; permanent & temporary exhibitions. Museum Sponsors: Pioneer Family Days, quilt shows, doll shows.
Hours & Admission Prices: Tues.-Fri. 10-12 & 1-3. No charge; donations accepted.
Attendance: 474 (accurate)
Membership: Student $1; Family $5; Contributor $10-$25; Life $100; Sustaining $250; Benefactor $500.

Rankin

RANKIN MUSEUM, 100 W. Main St., Rankin, TX 79778. Mailing Address: P.O. Box 22, Rankin, TX 79778-0022. Tel.: 915-693-2758 & 2422. Fax: 915-693-2303.
Founded: 1974.
Congressional District: 21
Key Personnel: Pres. (V), Donna Bell.
Governing Authority: nonprofit organization. Parent Institution: Rankin Museum Association. Tax-exempt.
Institution Type/Description: Local History Museum.
Collections: costumes; geology; glass; machinery; brands; restored 1940 fire truck, fully operative; photographs.
Research Fields: geology; local history; genealogy.
Facilities: library of local period ledgers.
Activities: visits from school groups; exhibits by local artists; planned programs for local organizations & school; building used for special functions.
Publications: articles in local newspaper; occasional write-ups in area publications.
Hours & Admission Prices: Thurs.-Fri. 2-5, Sat. 1-5; other times by appointment. No charge; donations accepted. &
Attendance: 1,000 (estimated)

Refugio

REFUGIO COUNTY MUSEUM, 102 W. West St., Refugio, TX 78377-2433. Tel.: 361-526-5555.
E-mail: brefugiomuseum@aol.com
Founded: 1983.
Key Personnel: C.E.O. & Pres. (V), Bart Wales; Chm. (V), Maxine H. Reilly.
Personnel Profile: Full-Time Paid 1; Part-Time Paid 2.
Governing Authority: society; nonprofit. Parent Institution: Refugio County Historical Society. Tax-exempt: 501(c)(3).
Institution Type/Description: Local History Museum.
Collections: history & culture of Refugio County; period artifacts; medical artifacts; nostalgia items.
Research Fields: local history; genealogy.
Facilities: 520-vol. library, including 5,000 newspapers; 1,500 sq. ft. exhibit space. Local history books, cookbooks & other items of local interest for sale.
Activities: guided tours; lectures; films; organized educational programs; participatory, loan & temporary exhibitions.
Publications: newsletter.
Hours & Admission Prices: Tues.-Fri. 12-4, Sat. 1-5. No charge; donations accepted. Closed major holidays. &
Attendance: 2,500
Membership: Individual $10.

Richardson

THE NATIONAL MUSEUM OF COMMUNICATIONS, 2001 Plymouth Rock, Richardson, TX 75081-3946. Tel.: 972-690-3636; 214-616-6562. Fax: 972-889-2329.
E-mail: billbragg@mail.com
Web Site: www.yesterdayusa.com
Founded: 1979.
Congressional District: 4
Key Personnel: Founder & Exec. Cur., William J. Bragg; Chm. Bd. (V), Kim Bragg; C.E.O., Walden Hughes; Chief Engineer, Roger Wenzel; Tour Dir., Don Richards; Museum Shop Mgr., Mike Handy.
Personnel Profile: Full-Time Paid 5; Part-Time Paid 5; Part-Time Volunteers 3.
Governing Authority: nonprofit organization. Tax-exempt: 501(c)(3).
Institution Type/Description: Communications Museum.
Collections: vintage radio equipment; master control console from Voice of America in Washington, D.C.; vintage phonographs; radio & television sets; radio station transmitter, including audio control console, turntables & microphones; record-cutting lathe; replica of 1960s television studio; first type of color TV camera; Walter Cronkite's microphone; Thomas Edison's microscope; Edison mimeograph; wax cylinder dictating machines; records; news tapes; technical journals; linotype machine; rare books; photography & motion picture film equipment; working amateur radio station; period telegraph keys & telephones.
Facilities: 60,000-vol. library of books, films, video tapes, wire recordings, pictures, phonograph records for research; 65-seat auditorium; educational facilities.
Activities: guided tours; films; vintage TV & radio programs.
Publications: newsletter, The Transcription.
Hours & Admission Prices: Call for information on hours. Adults $9.95, senior citizens 65 & over $7.95; discounts to AAM & ICOM members. &

Attendance: 70,000 (estimated)

Richmond

✳ **FORT BEND COUNTY MUSEUM ASSOCIATION, (M),** 500 Houston, Richmond, TX 77469-3522. Mailing Address: P.O. Box 460, Richmond, TX 77406-0460. Tel.: 281-342-6478 & 1256. Fax: 281-342-3782. Facebook: Fort Bend Museum.
E-mail: info@fortbendmuseum.org
Web Site: www.fortbendmuseum.org
Founded: 1967.
Congressional District: 22
Key Personnel: Interim Dir., Claire Rogers.
Personnel Profile: Full-Time Paid 34; Part-Time Paid 45; Part-Time Volunteers 230.
Governing Authority: society. Subsidiary Institution: George Ranch Historical Park, 10215 FM 762, Richmond TX 77469, phone: 281-343-0218, fax: 281343-9316, Tax-exempt: 501(c)(3).
Institution Type/Description: Local History Museum.
Collections: settlement of Stephen F. Austin's colony; agriculture; crafts; industry; furniture; clothes; manuscripts; documents; photographs; livestock and artifacts of history of George Ranch, 1824-1950. Historic Buildings: 1883 John M. Moore Mansion; 1901 Railroad Station; 1856 Alexander McNabb House; 1840 Jane Long House; 1896 County Jail; 1890s Kochn-Reed Home.
Research Fields: 1821-1836 culture and material culture of Austin's colony; county history; biographies of local leaders.
Facilities: auditorium; 150-vol. library of books about area and local leaders available for research by appointment only. Books, original sketches and postcards for sale.
Activities: guided tours; lectures; films; docent program; living history demonstration; spinning & weaving demonstrations; permanent & temporary exhibitions.
Publications: book, Sowell's History of Fort Bend County; cookbook, Czech Cookbook; book, Wharton's History of Fort Bend County.
Hours & Admission Prices: Ranch: Tues.-Sat. 9-5. Adults $10, senior citizens (62+) $9, children (5-15) $5, 4 & under no charge. Museum: Tues.-Fri. 9-5, Sat. 10-5, Adults $5, senior citizens (62+) $4, children (5-15) $3, 4 & under & members no charge; discounts to AAM members. Closed New Year's Day; Easter; Independence Day; Thanksgiving & day after; Christmas. &
Attendance: 91,265 (accurate)
Membership: Colonist $50; Texian $100; Ranger $500; Empresario (lifetime) $2,500.

Rockport

AQUARIUM AT ROCKPORT HARBOR, 702 Navigation Cir., Rockport, TX 78382. Tel.: 361-729-2328.
Web Site: www.rockportaquarium.com

Founded: 2007.
Institution Type/Description: Aquarium.
Collections: fish; coral; seagrass; posters; marine artifacts; photographs.
Facilities: Gift items for sale.
Activities: rental facilities; special events.
Hours & Admission Prices: Thurs.-Mon. 1-4; other times by appointment. No charge.

FULTON MANSION STATE HISTORIC SITE, 317 Fulton Beach Rd., Rockport, TX 78382. Tel.: 361-729-0386. Fax: 361-729-6581.
E-mail: fulton-mansion@thc.state.tx.us
Web Site: www.visitfultonmansion.com
Founded: 1983.
Congressional District: 18
Key Personnel: Site Mgr., Marsha Hendrix.
Personnel Profile: Full-Time Paid 5; Part-Time Paid 3; Part-Time Volunteers 40.
Institution Type/Description: Historic House Museum: c.1877 French Second Empire style Mansion.
Collections: period furniture; middle to late 19th-century decorative arts.
Research Fields: 19th-century decorative arts; material culture; landscaping & mechanical systems; ranching.
Facilities: 2.3 acres of grounds; maintenance complex.
Activities: guided tours; lectures; study clubs; docent program; formally organized educational programs; permanent & temporary exhibitions.
Publications: brochures; cookbook; newsletters.
Hours & Admission Prices: Education & History Center: Tues.-Sat. 9:30-4:30, Sun. 12:30-4:30. Adults $6, children 6-18 $4; children 5 & under no charge. Closed New Year's Eve & Day; Thanksgiving; Christmas Eve & Day. &
Attendance: 20,100 (accurate)
Membership: Individual $25; Family $50; Supporting $100; Corporate $250; Sustaining $500.

✳ **TEXAS MARITIME MUSEUM, (M),** 1202 Navigation Cir., Rockport, TX 78382-2773. Tel.: 361-729-6644 & 1271; 866-729-2469. Fax: 361-729-9938. Facebook: Texas Maritime Museum.
E-mail: klrd@pelicancoast.net
Web Site: www.texasmaritimemuseum.org
Founded: 1980.
Congressional District: 14
Key Personnel: C.E.O., Kathy Roberts-Douglass; Pres., Keith Hamilton; Pres., Rick McKinney; Museum Shop Mgr., Sally Reynolds.
Personnel Profile: Full-Time Paid 4; Part-Time Paid 2; Part-Time Volunteers 30.
Volunteer Hours: 7,000
Operating Expenses: 500,000
Operating Income: 500,000
Governing Authority: nonprofit. Tax-exempt: 501(c)(3).
Institution Type/Description: Texas Maritime Museum.
Collections: nautical equipment; commercial fishing equipment; ship & small boat building tools; photographs; oil & gas energy; lighthouse paintings.
Major Exhibits: From the Banks of the Sabine to the Rio Grande: U.S. Naval Presence Since 1845, 9/13-6/14; Local Shrimping, 7/14-3/15.
Facilities: 1,201-vol. library of maritime & nautical material; education center.
Activities: organized activities for children; temporary exhibits; educational programs for children; tours for Elder Hostel groups; River Barge tours.
Publications: quarterly newsletter, The Log Line.
Hours & Admission Prices: Tues.-Sat. 10-4, Sun. 1-4. Adults $8, seniors $6, active military $5, children 6-12 $3; discounts to AAM & ICOM members w/ID and groups of 10 or more with 2 weeks notice; children 5 & under and members no charge. Closed New Year's Day; Easter; Memorial Day; Thanksgiving; Christmas. &
Attendance: 16,232 (accurate)
Membership: Individual $25; Family $40; Sponsor $100; Benefactor $500.

Rosenberg

ROSENBERG RAILROAD MUSEUM, 1921 Avenue F, Rosenberg, TX 77471. Mailing Address: P.O. Box 369, Rosenberg, TX 77471-0369. Tel.: 281-633-2846.
Founded: 1999.
Congressional District: 26
Key Personnel: Pres. (V), Bill Rickert
Institution Type/Description: Railroad Museum.
Collections: railroad history, artifacts & memorabilia; railroad equipment; trains.
Activities: birthday parties; special events.
Hours & Admission Prices: Tues.-Sat. 10-5, Sun. 1-5; tours by appointment. Adults $5, senior citizens 55 & over $4, children 1-14 $3.

Attendance: 10,000

Round Top

BRISCOE CENTER FOR AMERICAN HISTORY-WINEDALE HISTORICAL CENTER, UNIVERSITY OF TEXAS AT AUSTIN, (M), 3738 FM Road 2714, Round Top, TX 78954. Mailing Address: P.O. Box 11, Round Top, TX 78954-0011. Tel.: 979-278-3530. Fax: 979-278-3531.
E-mail: winedale@austin.utexas.edu
Web Site: www.cah.utexas.edu
Formerly: Winedale, Center for American History, University of Texas at Austin
Founded: 1966.
Congressional District: 25
Key Personnel: Dir., Dr. Don Carleton; Administrative Mgr., Barbara B. White; Office Mgr., Beth Stewart.
Personnel Profile: Full-Time Paid 6; Part-Time Paid 8; Part-Time Volunteers 4.
Governing Authority: state. Parent Institution: University of Texas at Austin. Tax-exempt.
Institution Type/Description: Historic House.
Collections: decorative arts, folk art, furniture, tools & agricultural implements pertaining to German settlement of Texas; historic houses complete with c.1850s furnishings. Collections amassed by Miss Ima Hogg and donated by German emigrant descendants. Historic Buildings: Lewis Wagner House (1848); McGregor-Grimm House.
Research Fields: Texas-German cultural history; agricultural history; American social history; cabinetmaking; textiles; slavery in central Texas.
Facilities: 24-person dorm & dining facility; classrooms; interpretive center; visitor's center; conference facility; outdoor pavilion.
Activities: guided tours; lectures; workshops; permanent & temporary exhibits; plays; museum & library seminars. Museum Sponsors: American History Symposium; Shakespeare Festival; Christmas at Winedale-19th Century Folklife Live Reenactment.
Publications: quarterly newsletter, Quid Nunc; occasional papers; catalogs.
Hours & Admission Prices: Mon.-Fri. 8-5. Tours: by appointment. Adults $6. Closed holidays. &
Attendance: 9,580 (accurate)
Membership: Student $10; Contributing $25; Associate $50; Sustaining $100; Patron $250; Life Member $1,000.

FESTIVAL-INSTITUTE, JAMES DICK FOUNDATION, 248 Jaster Rd., Round Top, TX 78954-5445. Mailing Address: P.O. Box 89, Round Top, TX 78954-0089. Tel.: 979-249-3129. Fax: 409-249-5078.
E-mail: lamarl@festivalhill.org
Web Site: www.festivalhill.org
Founded: 1971.
Congressional District: 15
Key Personnel: Founder, Dir. & Pres., James Dick; Dir. Museum & Library Collections, Lamar Lentz; Treas., Richard R. Royall; Information Officer, Alain Declert.
Personnel Profile: Full-Time Paid 12; Full-Time Volunteers 7; Part-Time Paid 7; Part-Time Volunteers 62; Interns 1.
Governing Authority: public; nonprofit. Tax-exempt: 501(c)(3) & 170(b)(1)(A).
Institution Type/Description: Architecture & Art Museum: historic restorations include 1883 Edythe Bates Old Chapel; 1884 William Lockhart Clayton House; 1902 C.A. Menke House.
Collections: art; architecture; furniture; paintings; ceramics; glass; textiles; photographs; prints; metalworks; musical instruments; recordings; music; Festival Concert Hall with period rooms: David W. Guion museum room; Anders Gustav Fredrik & Josephine Oxehufwud Swedish museum room; Winfrey Toscanini collection; James and June Painter art collection.
Research Fields: architecture and decorative arts of the 19th & early 20th century; 19th-century Texas history; American and European art; 16th-century Northern Renaissance art; 16th-20th century Swedish Art & Decorative Arts; garden history; British Country House Guides; Country House Art Catalogues, 1750-1950; music.
Facilities: 20,000-vol. library of music, art, architecture, decorative arts, garden and landscape history, Texas history; 20,000 historic recordings; 1,000-seat auditorium; herb garden; educational facilities; conference facility. Compact discs, T-shirts, books, prints & note cards for sale.
Activities: concerts; films; formal educational programs; guided tours; lectures; permanent & temporary exhibitions; forums; walking tours.
Publications: exhibition catalogues.
Hours & Admission Prices: Mon.-Sat. by appointment. Tours: $5 per person; self-guided tours of grounds during day; discounts to AAM members. Closed Thanksgiving; Christmas. &
Attendance: 34,000 (estimated)

Membership: Contributor $35; Friend $150; Patron $500; Sponsor $1,000; Guarantor $5,000.

Salado

CENTRAL TEXAS AREA MUSEUM, INC., 423 S. Main St., Salado, TX 76571. Mailing Address: Box 36, Salado, TX 76571-0036. Tel.: 254-947-5232. Fax: 254-947-5232.
E-mail: office@ctam-salado.org
Web Site: www.ctam-salado.org
Founded: 1958.
Key Personnel: Pres., Susan Slye; 1st Vice Pres., Robert Rangel; 2nd Vice Pres., Beth Rangel; Librarian, Joy Dunaway; Librarian, Bob Dunaway; Historian Coord., Nancy Kelsey; Museum Shop Mgr., Mary Mendez.
Personnel Profile: Part-Time Paid 2.
Governing Authority: nonprofit. Subsidiary Institution: Wee Scots Shop. Tax-exempt: 501(c)(3).
Institution Type/Description: General Museum: housed in an early Central Texas store building.
Collections: genealogy; history; archives; Scottish folklore; Central Texas artifacts; Tonkawa artifacts; period artifacts.
Research Fields: central Texas history; Scottish heritage & other ethnic cultures.
Facilities: library of material on history with emphasis on Central Texas, including fine arts, genealogy, Scottish history & lore; auditorium. Books, Scottish merchandise, fine bone china, pressed and cut glass.
Activities: monthly programs; readers & writers roundtable ethnic shows. Museum Sponsors: Salado Scottish Games & Competitions.
Hours & Admission Prices: Tues.-Sat. 10-5. No charge; donations accepted.
Membership: Individual $15; Husband/Wife & Family with children under 18 $25; Sustaining $40; Benefactor $100; Life $400; Corporations $1,000.

Salt Flat

GUADALUPE MOUNTAINS NATIONAL PARK, 400 Pine Canyon Dr., Salt Flat, TX 79847-4755. Tel.: 915-828-3251. Fax: 915-828-3269. TDD: 915-828-3251.
E-mail: gumo_superintendent@nps.gov
Web Site: www.nps.gov/gumo
Founded: 1972.
Congressional District: 16
Key Personnel: Superintendent, Dennis Vasquez.
Governing Authority: federal. Parent Institution: National Park Service, Dept. of the Interior. Tax-exempt.
Institution Type/Description: National Park Museum.
Collections: natural history; local history; archaeology; historic sites.
Research Fields: geology; biology; history; archaeology.
Facilities: visitor centers; campground; back-country trails.
Activities: campfire programs; special walks; hikes; self-guiding nature trails.
Publications: brochure, The Guadalupes; Wilderness: The Guadalupe Mountains.
Hours & Admission Prices: Park: open year-round; Visitor Center: daily 8-4:30. 7-day individual fee $5. &
Attendance: 76,000 (accurate)

San Angelo

ANGELO STATE UNIVERSITY PLANETARIUM, Vincent Nursing-Physical Science Bldg., 1st Fl., San Angelo, TX 76909-0904. Mailing Address: ASU Station #10904, San Angelo, TX 76909-0904. Tel.: 325-942-2188 & 2136. Fax: 325-942-2188.
E-mail: msonntag@angelo.edu
Web Site: www.angelo.edu/dept/physics/planetarium.html
Institution Type/Description: Planetarium.
Collections: space science; astronomy; Sci-Dome HD digital projector.
Facilities: theater.
Activities: educational programs.
Hours & Admission Prices: Call for hours. Adults $3, children, senior citizens & active military $2; ASU students, faculty & staff no charge.

FORT CONCHO NATIONAL HISTORIC LANDMARK, (M), 630 S. Oakes St., San Angelo, TX 76903-7013. Tel.: 325-481-2646 & 657-4444. Fax: 325-657-4540.
E-mail: admin@fortconcho.com
Web Site: fortconcho.com
Founded: 1928.
Congressional District: 11
Key Personnel: Dir., Robert Bluthardt; Chm. (V), Harry Thomas; Librarian & Archivist, Evelyn Lemons; Dir. Education, Chris Morgan; Special Events, Carol Cummings; Mng. & Visitor Svcs., Cory Robinson.

Personnel Profile: Full-Time Paid 12; Part-Time Paid 2; Part-Time Volunteers 250; Interns 2.
Volunteer Hours: 20,000
Governing Authority: municipal. Parent Institution: City of San Angelo. Cooperates with Angelo State University & San Angelo Independent School District. Tax-exempt.
Institution Type/Description: Historic Landmark.
Collections: costumes; furniture; military artifacts; historic photographs; historic buildings.
Research Fields: Southern Plains Indians; military history of Plains Indian Wars; frontier settlement after 1864; San Angelo and West Texas history, 1870-1930.
Facilities: 4,000-vol. library; meeting space; activities space. Postcards & museum-related items for sale.
Activities: guided tours; lectures; educational programs; permanent, temporary & traveling exhibitions; festivals; frontier school; craft demonstrations; living history programs; concerts; holiday celebrations; summer children's programs; teacher training workshops.
Publications: quarterly newsletter, The Guidon; Fort Concho Museum Press publications.
Hours & Admission Prices: Mon.-Sat. 9-5, Sun. 1-5. Adults $3; discounts for senior citizens, military, groups, Texas Forts Trail, AAM, NTHP, AAA & TAM members; members & children under 6 no charge. Closed New Year's Day; Thanksgiving; Christmas. &

Attendance: 60,000 (accurate)
Membership: Lieutenant $35; Captain $75; Major $100; Lt. Colonel $150; Colonel $250; Brigadier General $500; Major General $1,000; Lt. General $2,500; General $5,000.

MISS HATTIE'S BORDELLO MUSEUM, 18 1/2 E. Concho Ave., San Angelo, TX 76903-6412. Tel.: 325-653-0112.
E-mail: mrksalot@wtxcoxmail.com
Web Site: www.misshatties.com/tour.html
Institution Type/Description: History Museum.
Collections: brothel history; period furnishings & clothing; photographs.
Activities: rental facilities.
Hours & Admission Prices: Thurs.-Sat. 1-4. Adults $5; groups of 10 or more by appointment.

RAILWAY MUSEUM OF SAN ANGELO, 703 S. Chadbourne, San Angelo, TX 76903-6931. Tel.: 325-486-2140.
E-mail: railmuseum@gmail.com
Web Site: www.railwaymuseumsanangelo.homestead.com
Founded: 1994.
Congressional District: 21
Key Personnel: Pres. (V), David Wood.
Personnel Profile: Part-Time Volunteers 35.
Governing Authority: Parent Institution: Historic Orient/Santa Fe Depot, Inc. Tax-exempt.
Institution Type/Description: Railway Museum: housed in the 1909 Orient-Santa Fe Passenger Depot built by the KCM&O.
Collections: railroad history & artifacts; trains; train memorabilia; model train layouts; photographs.
Activities: Model Railroad Club. Annual Events: Haunted Depot in October; Santa's Santa Fe Christmas in December.
Hours & Admission Prices: Sat. 10-4. Adults 12 & over $4, children 5-12 $2; children under 5 no charge. &

Attendance: 9,575 (accurate)
Membership: Brakeman $10; Engineer $50-$99; Conductor $100-$999; Station Manager $500 & up.

* **SAN ANGELO MUSEUM OF FINE ARTS, (M),** One Love St., San Angelo, TX 76903-6911. Tel.: 325-653-3333. Fax: 325-658-6800.
E-mail: museum@samfa.org
Web Site: www.samfa.org
Founded: 1981.
Congressional District: 11
Key Personnel: Dir., Howard Taylor; Pres., Debbie Cross; Exec. Asst., Gracie Fernandez; Collections Mgr., Laura Huckaby; Preparator, John Mattson; Bookkeeper, Janet Bingham; Cur. Education, Megan DiRienzo; Asst. Museum Educator, Rebekah Coleman; Weekend Supvr., Sylvia Grimaldo; Museum Shop Mgr., Betty Connally.
Personnel Profile: Full-Time Paid 6; Part-Time Paid 20; Part-Time Volunteers 40; Interns 3.
Governing Authority: nonprofit. Tax-exempt: 501(c)(3).
Institution Type/Description: Art Museum.
Collections: the work of Texas artists from 1945; ceramic arts emphasizing work created since 1945; American painting and sculptures; Mexican & Mexican-American art; European, Oriental & African; American glass.
Major Exhibits: National Ceramic Competition, 4/11/14-6/29/14.
Research Fields: West Texas art & cultural history, late 20th-century crafts, design & architecture.
Facilities: sculpture garden; education center. Museum-related items for sale.
Activities: guided tours; lectures; concerts; films; organized educational programs; docent program; changing exhibitions.
Publications: newsletter; exhibit catalogues; gallery guides.
Hours & Admission Prices: Tues.-Sat. 10-4, Sun. 1-4. Adults $2, senior citizens $1; discounts to groups of 10 or more, Texas Assoc. of Museums, AAM & ICOM members; members, children, students & military no charge. Closed national holidays. &

Attendance: 60,000 (estimated)
Membership: Individuals $10; Families $20.

SAN ANGELO NATURE CENTER, 7409 Knickerbocker Rd., San Angelo, TX 76904-7885. Tel.: 325-942-0121.
Institution Type/Description: Nature Center.
Collections: flowers; trees; plants; wildlife.
Facilities: library; nature trails.
Activities: special events; educational programs.
Hours & Admission Prices: Tues.-Sat. 12-5.

San Antonio

THE ALAMEDA NATIONAL CENTER FOR LATINO ARTS AND CULTURE, 101 S. Santa Rosa Ave., San Antonio, TX 78207-4509. Tel.: 210-299-4300. Fax: 210-299-4340.
E-mail: info@thealameda.org
Web Site: www.thealameda.org
Institution Type/Description: Latino Arts & Culture.
Collections: Latino arts & cultural heritage
Activities: performance; educational programs.
Hours & Admission Prices: Tues. & Thurs.-Sat. 10-6, Wed. 10-8, Sun. 12-6. Adults $4, seniors 55 & over, military and educators $3, students & children 4-11 $2; discounts to groups; children under 3 no charge. Closed New Year's Day; Easter; Battle of Flowers; Memorial Day; Independence Day; Thanksgiving; Christmas.

THE ALAMO, 300 Alamo Plaza, San Antonio, TX 78205-2606. Mailing Address: c/o Texas General Land Office, P.O. Box 12873, Austin, TX 78711-2873. Tel.: 210-225-1391. Fax: 210-354-3602.
Web Site: www.thealamo.org
Founded: 1905.
Congressional District: 20
Key Personnel: Commissioner, Texas Gen. Land Office, Jerry Patterson; Historian & Cur., Dr. Richard Bruce Winders.
Governing Authority: nonprofit organization. Parent Institution: Texas General Land Office. Tax-exempt: 501(c)(3).
Institution Type/Description: Historic Site & Complex: 1836 Mission San Antonio De Valero, site of the Battle of the Alamo.
Collections: rifles; pistols; cannon; knives; spurs; jewelry; glass; ceramics; documents; paintings; dioramas; utilitarian objects; textiles; furniture; clothing from the Texas Revolution & the Republic of Texas periods. Historic Structure: 18th-century Long Barrack & Alamo chapel.
Research Fields: Republic of Texas; Texas Revolution; genealogy; archeology; illustration; textbooks; publications; conservation preservation.
Facilities: botanical garden; 9,700 sq. ft. exhibit space; video system. Museum-related items for sale.
Activities: guided tours; lectures; films; organized education programs for children; temporary exhibitions. Museum Sponsors: Texas Independence; Pilgrimage to the Alamo; Fall at the Alamo.
Publications: book, The Long Barrack; The Wall of History
Hours & Admission Prices: Daily 9-5:30; Summer: daily 9-7. No charge; donations accepted. Closed Christmas Eve & Day. &

Attendance: 2,479,329 (accurate)

ARTPACE SAN ANTONIO, 445 N. Main Ave., San Antonio, TX 78205-1441. Tel.: 210-212-4900. Fax: 210-212-4990. blog.artpace-.org.
E-mail: info@artpace.org
Web Site: www.artpace.org
Founded: 1995.
Congressional District: 20
Key Personnel: Exec. Dir., Amada Cruz; Deputy Dir., Mary Heathcott; Grants, Marjory Newman; Membership & Devel. Assoc., Mike Martinez; Cur. Education, Kaela Hoskings; C.F.O., Tricia Peebles; Studio Dir., Riley Robinson.

Personnel Profile: Full-Time Paid 16; Part-Time Paid 6; Part-Time Volunteers 161; Interns 18.
Governing Authority: nonprofit organization. Tax-exempt: 501(c)(3).
Institution Type/Description: Art Museum.
Collections: contemporary art.
Facilities: library; educational facilities; 7,000 sq. ft. exhibit space. Museum-related items for sale.
Activities: arts festival; films; formal education programs; guided tours; lectures; participatory exhibits; teacher workshops & professional artists; summer art camps; international artist-in-residence program. Annual Events: Chalk It Up.
Publications: magazine, Artpace; annual exhibition catalogue.
Hours & Admission Prices: Wed.-Sun. 12-5; other times by appointment. No charge; donations accepted. Closed New Year's Day; Fiesta Day; Easter; Independence Day; Thanksgiving; Christmas Eve & Day. &
Attendance: 72,000 (estimated)
Membership: Student, Educator, Artist & Senior $25; Individual $40; Dual & Household $75; Associate $150; Connector $250; Supporter $500; Patron $750. Corporate: Enthusiast $1,000-$2,499; Executive $2,500-$4,999; Advocate $5,000-$9,999; Catalyst $10,000-$14,999; Leader $15,000-$24,999; Partner $25,000 & up.

BLUE STAR CONTEMPORARY ART CENTER, (M), 116 Blue Star, San Antonio, TX 78204-1713. Tel.: 210-227-6960. Fax: 210-229-9412.

E-mail: bill@bluestarart.org
Web Site: www.bluestarart.org
Founded: 1986.
Congressional District: 20
Key Personnel: Exec. Dir. & Pres., Bill FitzGibbons; Chm. (V), Edward Valdespino; Program Dir., Rebecca Geibel; Dir. Devel., Therese McDevitt; Mgr. Mosaic Studio, Alex Rubio; Preparator, Pedro Luera; Membership & Community Outreach Coord., Emily R. Barker; Program Asst., Brittany Parker; Gallery Liaison, Krisanne Frost; Devel. Asst., Allison Salinas.
Personnel Profile: Full-Time Paid 7; Part-Time Paid 2; Part-Time Volunteers 10; Interns 12.
Governing Authority: private; nonprofit organization. Parent Institution: Contemporary Art for San Antonio. Tax-exempt: 501(c)(3).
Institution Type/Description: Contemporary Art Museum
Collections: contemporary art by Alex Rubio, James Surls, Vincent Valdez, Jesus Morales, Dan Borris, Ricardo Legorreta & Graciela Iturbide.
Facilities: 13,000 sq. ft. exhibit space. Museum-related items for sale.
Activities: arts festivals; concerts; formal education programs for all ages; lectures; loan, participatory & traveling exhibitions; rental gallery.
Publications: quarterly newsletter.
Hours & Admission Prices: Gallery: Wed.-Sun. 12-6. Office: Mon.-Fri. 9-6. No charge; donations accepted. Closed New Year's Day; Martin Luther King Jr. Day; Memorial Day; Independence Day; Labor Day; Thanksgiving & day after; Christmas Eve & Day. &
Attendance: 150,000 (estimated)
Membership: Artist $25; Artist Family $35; Bluestar $40; Bluestar Family $60; Contemporaries $250; Contemporaries Family $350; Society de Cien $1,000.

BRISCOE WESTERN ART MUSEUM, (M), 210 W. Market St., San Antonio, TX 78205. Tel.: 210-299-4499. Fax: 210-299-4118. Facebook: The Briscoe.

E-mail: info@briscoemuseum.org
Web Site: www.briscoemuseum.org
Founded: 2013.
Congressional District: 20
Key Personnel: Exec. Dir., Steven M. Karr; Chm. (V), Debbie Montford; Museum Shop Mgr., Ann Miller.
Governing Authority: Parent Institution: National Western Art Foundation. Tax-exempt.
Institution Type/Description: Art Museum.
Collections: art, history, & culture of the American West.
Facilities: sculpture garden.
Activities: educational programs; public events; permanent & temporary exhibitions.
Hours & Admission Prices: Tues.-Thurs. 10-4, 1st Tues. each month 10-9, Fri.-Sun. 10-5. Adults $5; members no charge. &
Membership: Ranch Hand $35; Wrangler $50; Header and Heeler $65; Homesteader $80; Charro $125; Trail Boss $250; Wildcatter $500; Collector's Circle $1,500; Director's Circle $5,000; Governor's Circle $10,000.

BUCKHORN SALOON & MUSEUM, 318 E. Houston St., San Antonio, TX 78205-1816. Tel.: 210-247-4000. Fax: 210-247-4020. Facebook: Buckhorn Saloon & Museum.

E-mail: sales@buckhornmuseum.com
Web Site: www.buckhornmuseum.com
Founded: 1881.
Congressional District: 20
Key Personnel: Dir., David Phillips.
Governing Authority: county. Parent Institution: Gunnison County Pioneer & Historical Society. Subsidiary Institution: Pioneer Museum.
Institution Type/Description: Natural History Museum; Texas - San Antonio historic house site.
Collections: wildlife; marine; birds; cowboy collectables & gear; cowboy art. Historic House: Texas History Hall.
Facilities: 38,000 sq. ft. exhibit space; banquet hall.
Publications: brochures & footnotes of the Buckhorn.
Hours & Admission Prices: Call for hours. Adults $18.99, children $14.99; children 3 & under no charge. &
Attendance: 56,000 (accurate)

CASA NAVARRO STATE HISTORICAL PARK, 228 S. Laredo St., San Antonio, TX 78207-4544. Tel.: 210-226-4801. Fax: 210-226-4801.

E-mail: casa-navarro@thc.state.tx.us
Web Site: www.visitcasanavarro.com
Founded: 1964.
Congressional District: 20
Key Personnel: Site Mgr., Georgia Ruiz Davis; Chief Cur., Laura DeNormandie-Bass.
Governing Authority: state. Parent Institution: Texas Historical Commission, P.O. Box 12276, Austin, TX 78711-2276. Tel. 512-463-6100.
Institution Type/Description: Historic Site & House Museum: c.1850 home site of Texas patriot, Jose Antonio Navarro.
Collections: period furnishings; early Texas decorative arts; items pertaining to the lifestyle of a Tejano family of San Antonio.
Research Fields: career of Jose Antonio Navarro; Southwestern, Spanish Colonial & Mexican history.
Facilities: Publications for sale.
Activities: guided tours.
Publications: pamphlet.
Hours & Admission Prices: Tues.-Sat. 10-5, Sun. 12-5. Adults $4, students 6-18 $3; discounts to school & adult tour groups; children 5 & under no charge. Closed New Year's Eve & Day, Thanksgiving, Christmas Eve & Day. &
Attendance: 4,277 (accurate)

GUINNESS WORLD RECORDS MUSEUM, 329 Alamo Plaza, San Antonio, TX 78205-2667. Tel.: 210-226-2828.

Institution Type/Description: World Records Museum.
Collections: world record facts, feats & record holders; photographs; personal artifacts; movie memorabilia.
Hours & Admission Prices: Memorial Day to Labor Day Sun.-Thurs. 10-11, Fri.-Sat. 10 am-12 am; Sept.-May Sun.-Thurs. 10-8, Fri.-Sat. 10-10. Adults 12 & over $20.99, children 3-11 $12.99.

INSTITUTE OF TEXAN CULTURES, (M), 801 E. Cesar E. Chavez Blvd., San Antonio, TX 78205-3209. Tel.: 210-458-2300. Fax: 210-458-2205.

E-mail: itcweb@utsa.edu
Web Site: www.texancultures.com
Founded: 1968.
Congressional District: 21
Key Personnel: Asst. Exec. Dir., Aaron Parks; Dir. Research, Exhibits & Collections, Dr. Bryan Howard; Dir. Special Events & Festivals, Jo Ann Andera; Dir. Education & Interpretation, Lupita Barrera.
Personnel Profile: Full-Time Paid 36; Part-Time Paid 8; Part-Time Volunteers 250.
Governing Authority: state; university. Parent Institution: University of Texas at San Antonio. Tax-exempt.
Institution Type/Description: Educational center for the history & diverse cultures of Texas.
Collections: 3.5 million historical photographs.
Research Fields: humanities; ethnic history; history; folklore; multicultural studies.
Facilities: 6,500-vol. non-circulating research library available to the public; 200-seat auditorium; conference center with AV capabilities.
Activities: guided tours; lectures; gallery talks; cultural festivals; formally

organized education programs for children, adults, undergraduate & graduate college students; docent program; inter-museum loan, permanent, temporary & traveling exhibitions; outreach program to the state. Museum Sponsors: Texas Folklife Festival; Asian Festival.

Publications: pamphlet series, The Texians and Texans; books, The Melting Pot; Texans One and All; The Texas Rangers: Images & Incidents; The Swedish Texans; With Domingo Leal In San Antonio; The Irish Texans; Journey To Pleasant Hill: The Civil War Letters of Capt. E.P. Petty; Reflections on Texas, A Teacher's Guide To The Institute of Texan Cultures; Vaquero: Genesis of the Texas Cowboy; The Hungarian Texans; The German Texans; The Polish Texans; Exploration in Texas Ancient and Otherwise, With Thoughts on the Nature of Evidence; The Japanese Texans, The English Texans; Echoes of the Past: The Cowboy Poetry of Melvin Whipple; Texans: A Story of Texan Cultures for Young People; The Irish Texans.

Hours & Admission Prices: Mon.-Sat. 9-5, Sun. 12-5. Adults $8, seniors over 65 $7, military & children 3-11 $6; discounts to tours & school groups; children under 2 & members no charge. Closed New Year's Day; Easter; Thanksgiving; Christmas. &

Attendance: 200,000 (estimated)

Membership: Individual $40; Family $75; Smithsonian $125; Mockingbird $250; Blue Bonnet $500; Lone Star $1,000. Corporate memberships also available.

THE MAGIC LANTERN CASTLE MUSEUM, 1419 Austin Hwy., San Antonio, TX 78209-4337. Tel.: 210-805-0011. Fax: 210-822-1226.

E-mail: castle@magiclanterns.org
Web Site: www.magiclanterns.org
Founded: 1991.
Key Personnel: C.E.O. & Cur., Jack Judson.
Personnel Profile: Full-Time Volunteers 2; Part-Time Paid 6; Part-Time Volunteers 2.
Governing Authority: private sole ownership; nonprofit.
Institution Type/Description: Audiovisual slide & Film Museum.
Collections: concentration on magic lanterns from 1700s into the 20th-century; glass slides, prints, books, accessories & related paraphernalia; worldwide scientific instruments for optical projection.
Research Fields: history of the magic lantern throughout the world.
Facilities: library; 50-seat auditorium; 4,000 sq. ft. exhibit space; 1,000 sq. ft. research area.
Activities: guided tours; lectures; demonstrations.
Hours & Admission Prices: By appointment, request or invitation. No charge. Closed state & national holidays.
Attendance: 2,000 (estimated)

***   MCNAY ART MUSEUM, (M),** 6000 N. New Braunfels Ave., San Antonio, TX 78209-4618. Mailing Address: P.O. Box 6069, San Antonio, TX 78209-0069. Tel.: 210-824-5368. Fax: 210-824-0218.

E-mail: info@mcnayart.org
Web Site: www.mcnayart.org
Formerly: Marion Koogler McNay Art Museum
Founded: 1950.
Congressional District: 21
Key Personnel: Dir., William J. Chiego; Pres., Sarah E. Harte; Chief Cur., Cur. Art after 1945, Rene Paul Barilleaux; Dir. Education, Kate Carey; Chief Devel. Officer, Colleen Kelly; Head Librarian, Ann Jones; C.O.O. & Controller, Bryan Dome; Cur. Prints & Drawings, Lyle Williams; Cur. Tobin Collection of Theatre Arts, Jody Blake; Collections Mgr. & Exhibitions Coord., Heather Lammers; Mgr. Bldg. & Grounds, Robert Sanderson; Museum Store Mgr., Janet Goddard; Security & Visitor Svcs. Mgr., James Jones.
Personnel Profile: Full-Time Paid 78; Part-Time Paid 25; Part-Time Volunteers 250; Interns 2.
Governing Authority: nonprofit organization. Tax-exempt: 501(c)(3).
Institution Type/Description: Fine Arts Museum.
Collections: modern art; 19th- to 20th-century European & American paintings; graphic arts & sculpture; arts and crafts of New Mexico; Medieval & Renaissance art; theatre arts collection of books, sketches, paintings & maquettes relating to the opera, ballet & musical stage.
Research Fields: painting; sculpture; graphics; theatre arts.
Facilities: 30,000-vol. library of art reference material; 226-seat lecture hall; auditorium; learning centers; sculpture garden; teacher resource center.
Activities: lectures; films; gallery talks; concerts; seminars; teacher workshops.
Publications: annual report; exhibition catalogs; members magazine, Impressions.
Hours & Admission Prices: Tues.-Wed. & Fri. 10-4, Thurs. 10-9, Sat. 10-5, Sun. 12-5. Adults $10-$17; students, seniors 65 & over, & active military

$5-$12; discount to AAM members & groups; children 12 & under and members no charge. H-E-B Thurs. 4-9, AT&T 1st Sun. each month; no charge. Closed New Year's Day; Independence Day; Thanksgiving; Christmas. &

Attendance: 125,500 (estimated)

Membership: Individual $55; Family/Dual $85; Supporting $150; Contributing $275; Sustaining $550; Patron $1,000; Associate $1,500; Sponsor $2,500; Benefactor $5,000; Philanthropist $10,000; Director's Circle $25,000.

NSSA-NSCA MUSEUM & HALL OF FAME, 5931 Roft Rd., San Antonio, TX 78253. Tel.: 210-688-3371 & 2574. Fax: 210-688-9269.

E-mail: museum@nssa-nsca.com
Web Site: www.nssa-nsca.org
Key Personnel: Cur., Mike Brazzell; Asst. Cur., Jim Harris
Institution Type/Description: National Shooting Museum
Collections: sporting clays & skeet shooting history & artifacts; photographs; Hall of Fame.
Hours & Admission Prices: Thurs. 9-9, Sat. 9-5.

SAN ANTONIO ART LEAGUE MUSEUM, 130 King William St., San Antonio, TX 78204-1311. Tel.: 210-223-1140.

E-mail: saalm@att.net
Web Site: www.saalm.org
Founded: 1912.
Key Personnel: Chm. (V), Helen Fey
Institution Type/Description: Art Museum.
Collections: paintings; graphic art; photography; ceramics; sculpture; drawings; silverware; furniture; fabric; wall hangings.
Activities: lectures.
Hours & Admission Prices: Tues.-Sat. 10-2. No charge; donations accepted. &
Attendance: 500
Membership: Individual $35; Family $50; Supporting $100 & above; Sustaining $250 & above; Patron $500 & above.

SAN ANTONIO BOTANICAL GARDEN, 555 Funston Place, San Antonio, TX 78209-6631. Tel.: 210-207-3250. Fax: 210-207-3274. TDD: 210-207-3255.

E-mail: john.brackman@sanantonio.gov
Web Site: www.sabot.org
Founded: 1980.
Key Personnel: Bd. Chm. (V), Claire Alexander; Dir., Bob Brackman; Museum Shop Mgr., Cynthia Reed.
Personnel Profile: Full-Time Paid 28; Part-Time Paid 2; Part-Time Volunteers 100; Interns 1.
Governing Authority: municipal. Parent Institution: City of San Antonio, TX. Subsidiary Institution: San Antonio Botanical Center Society. Tax-exempt: 501(c)(3).
Institution Type/Description: Botanical Garden.
Collections: woody & herbaceous native plant material from the regions of East Texas, South Texas, & Edwards Plateau; neo-tropical flora in the Lucile Halsell Conservatory. Historic Buildings: 1840 Schumacher House; 1860 East Texas log cabin; 1896 Sullivan carriage house & stables; 1880 Auld House.
Research Fields: endangered plant species of Texas.
Facilities: Horticultural books & other botanical items for sale.
Activities: guided tours; lectures; concerts; children's garden; summer children's courses; docent program; family days.
Hours & Admission Prices: Daily 9-5. Adults $8, senior citizens, students & military $6, children 3-13 $5; discount to groups; AHS members & members no charge. Closed New Year's Day; Thanksgiving; Christmas. &
Attendance: 100,000 (estimated)
Membership: Individual $45; Family & Dual $65; Friend $125; Contributor $250; Patron $500; Director's Circle $1,000.

SAN ANTONIO CHILDREN'S MUSEUM, 305 E. Houston St., San Antonio, TX 78205-1802. Tel.: 210-212-4453. Fax: 210-242-1313.

E-mail: lupita@sakids.org
Web Site: www.sakids.org
Founded: 1995.
Key Personnel: Exec. Dir., Chris Sinick; Pres., Joan Wyatt.
Personnel Profile: Full-Time Paid 12; Part-Time Paid 10; Part-Time Volunteers 700.
Governing Authority: private; nonprofit organization. Tax-exempt: 501(c)(3).
Institution Type/Description: Children's Museum.
Collections: hands-on exhibits.
Hours & Admission Prices: Mon.-Fri. 9-5, Sat. 9-6, Sun. 12-5. Adults $7;

children under 2 & members no charge. Closed New Year's Day; Thanksgiving; Christmas. &

Attendance: 165,000 (accurate)

Membership: Family (1 year) & Grandparent (1 year) $65; Extended Membership (1 year) $90; ACM Reciprocal (1 year) $115; Family (2 years) & Grandparent (2 years) $120; Extended Membership (2 years) $170; ACM Reciprocal (2 years) $220; Shooting Star Society $250; Constellation Society $500.

SAN ANTONIO CONSERVATION SOCIETY, 107 King William St., San Antonio, TX 78204-1312. Tel.: 210-224-6163. Fax: 210-224-6168.

E-mail: conserve@saconservation.org

Web Site: www.saconservation.org

Founded: 1924.

Congressional District: 20

Key Personnel: Exec. Dir., Bruce MacDougal; Pres., Marcie Ince; Administrative Asst., Glory Bohne.

Personnel Profile: Full-Time Paid 18; Part-Time Paid 3.

Governing Authority: society. Branch Museum: 1876 Steves Homestead, 509 King William St., San Antonio, TX. Tel: 210-225-5924. Tax-exempt: 501(c)(3).

Institution Type/Description: Historic House Museums: Wulff House (library only; no tours); 1840-1860 Yturri-Edmunds Historic Site; 1876 Steves Homestead House Museum.

Collections: Steves Homestead: decorative arts; Victorian period furniture; family memorabilia & treasures. Yturri-Edmunds: family memorabilia; photos; local history, preservation & restoration projects.

Research Fields: local history; preservation & restoration projects; architectural history.

Facilities: 3,000-vol. library of books pertaining to history & preservation; slide collection available for research.

Activities: guided tours; permanent & temporary exhibitions; food festivals; historic preservation seminars; speakers bureau.

Publications: monthly, San Antonio Conservation Society Newsletter; brochures, King William Walking Tour; Texas Star Trail; Steves Homestead.

Hours & Admission Prices: Yturri Edmunds: tours by appointment only. Steves Homestead: daily 10-4:15. Adults $6, seniors & groups $4, students & military $3; children under 12 no charge. Closed most major holidays. &

Attendance: 11,000 (accurate)

Membership: Associate & Active $25.

SAN ANTONIO MISSIONS NATIONAL HISTORICAL PARK, Visitor Center, 6701 San Jose Dr., San Antonio, TX 78210. Mailing Address: 2202 Roosevelt Ave., San Antonio, TX 78210. Tel.: 210-534-8833. Fax: 210-534-1106.

Web Site: www.nps.gov/saan

Founded: 1978.

Congressional District: 28

Key Personnel: Park Supt., John Lujan.

Personnel Profile: Full-Time Paid 45; Part-Time Volunteers 120.

Governing Authority: federal. National Park Service, Dept. of Interior. Washington, DC. Tax-exempt.

Institution Type/Description: National Historic Park: comprised of the following four Spanish missions: 1720 San Jose Y San Miguel de Aguayo; 1731 La Purisima Concepcion; 1731 San Juan Capistrano; 1731 San Francisco de la Espada.

Collections: anthropology; archaeology; archives; Indian artifacts; Spanish Colonial collections; technology.

Research Fields: Spanish Colonial history & architecture; building stabilization & preservation.

Facilities: visitor center.

Activities: self-guided tours; teacher's guide; temporary & permanent exhibitions; ranger programs; film.

Publications: San Antonio Missions National Historical Park Brochure (English & Spanish).

Hours & Admission Prices: Daily 9-5. No charge; donations accepted. Closed New Year's Day; Thanksgiving; Christmas. &

Attendance: 1,400,000 (estimated)

✻　**SAN ANTONIO MUSEUM OF ART, (M),** 200 W. Jones Ave., San Antonio, TX 78215-1402. Tel.: 210-978-8100. Fax: 210-978-8182. Facebook: San Antonio Museum of Art.

E-mail: info@samuseum.org

Web Site: www.samuseum.org

Founded: 1981.

Congressional District: 1

Key Personnel: Dir., Katie Luber, Ph.D.; Dir. of Administration & Finance, Polly Beth Vidaurri; C.O.O., Emily Jones; Cur. Latin American Art, Dr.

Marion Oettinger, Jr.; Chm. Bd., Karen Hixon; Dir. Operations, Dan Walton; Registrar, Karen Baker; AT&T Dir. Education, Kathryn Suzanne Erickson; Dir. Mktg., Cary Marriott; Museum Shop Mgr., Christine Barnett; Dir. Exhibits, Tim Foerster.

Personnel Profile: Full-Time Paid 65; Full-Time Volunteers 2; Part-Time Paid 23; Part-Time Volunteers 85; Interns 10.

Governing Authority: nonprofit organization. Tax-exempt: 501(c)(3).

Institution Type/Description: Art Museum: housed in restored turn-of-the-century Lone Star Brewing Company.

Collections: American & contemporary art; American painting, sculpture, glass from 18th-, 19th- & 20th-centuries; Ancient Egyptian, Greek & Roman art; period glass; Asian art; contemporary Latin American art; 17th- to 20th-century European paintings & works on paper; Irish silver; Islamic art; Latin American folk art; Oceanic art; 20th-century photography; pre-Columbian art; Spanish Colonial art; Texas furniture, paintings & decorative arts.

Major Exhibits: Lethal Beauty, 9/13-1/14.

Research Fields: Latin American art; Asian art; Ancient Egyptian, Greek & Roman antiquities; American & Contemporary art.

Facilities: 188-seat auditorium. Museum-related items for sale.

Activities: guided tours; lectures; films; gallery talks; concerts; formally organized education programs for children & adults; docent program & council; loan, temporary, permanent & traveling exhibitions; annual endowed symposia.

Publications: exhibition catalogs; quarterly calendar & newsletter; teacher resource guides; special collection catalogues; collection handbook.

Hours & Admission Prices: Sun. 10-6, Tues. & Fri.-Sat. 10-9, Wed.-Thurs. 10-5. Adults $10; discounts to students, military, AAM & ICOM members; Tues. 4-9, Sun. 10 to noon, children 12 & under and members no charge. Closed New Year's Day; Easter; Fiesta Friday; Thanksgiving; Christmas. &

Attendance: 93,000 (estimated)

Membership: Senior Citizen 65 & over $35; Individual $45; Friends Support Group $50; Senior Family 65 & over $60; Family $75; Sponsor $125; Connoisseur $150; Associate $250; Arte-preneur Support Group $300; Patron $500; Society $1,000; Benefactor $5,000; Leader $10,000; Philanthropist $25,000. Corporate Levels: Corporate Society $1,000; Corporate Patron $2,500; Corporate Benefactor $5,000; Corporate Leader $10,000; Corporate Circle $25,000 & up.

SAN ANTONIO ZOOLOGICAL SOCIETY, 3903 N. Saint Mary's St., San Antonio, TX 78212-7183. Tel.: 210-734-7184. Fax: 210-734-7291.

E-mail: information@sazoo.org

Web Site: www.sazoo.org

Founded: 1914.

Congressional District: 20

Key Personnel: Exec. Dir., J. Stephen McCusker; Pres. (V), David S. Herrmann; Visitor Svcs. Mgr., Ron Kipp.

Personnel Profile: Full-Time Paid 300; Part-Time Paid 15; Part-Time Volunteers 225.

Governing Authority: society. Parent Institution: San Antonio Zoological Society. Tax-exempt: 501(c)(3).

Institution Type/Description: Zoo.

Collections: birds; mammals; reptiles; aquarium.

Research Fields: animal reproduction; behavioral research.

Facilities: 547-vol. library of technical & general reference material relating to zoos & zoo operation available for use on premises. Publications & gifts for sale.

Activities: guided tours; summer programs; elephant behavioral training demonstration; keeper connections presentations.

Publications: quarterly news magazine; e-newsletter.

Hours & Admission Prices: Daily 9-5. Adults $12, senior citizens 62 & over and children 3-11 $9.50; discount to school groups & reciprocal zoo members; members no charge. &

Attendance: 1,000,000 (estimated)

Membership: Senior Citizen $35; Individual & Senior Couple $45; Family $70; Family Plus One $80; Supporting $110; Contributing $130; Benefactor $250; Patron $500; Zoo Master $1,000.

SOUTHWEST SCHOOL OF ART, 300 Augusta St., San Antonio, TX 78205-1216. Tel.: 210-224-1848. Fax: 210-224-9337.

E-mail: information@swschool.org

Web Site: www.swschool.org

Founded: 1965.

Key Personnel: Dir., Paula Owen; Assoc. Cur., Kathy Armstrong; Museum Shop Mgr., Clare Watters.

Personnel Profile: Full-Time Paid 32; Part-Time Paid 120; Part-Time Volunteers 300; Interns 2.

Governing Authority: nonprofit organization. Tax-exempt: 501(c)(3).

Institution Type/Description: Contemporary Art Center & History Museum.
Collections: changing exhibitions; various artistic media.
Research Fields: studio arts.
Facilities: educational & studio facilities; 100-seat restaurant; 3,500 sq. ft. exhibition space.
Activities: guided tours; lectures; arts festivals; organized educational programs; classes; rental facilities.
Publications: brochures; invitations; catalogues.
Hours & Admission Prices: Exhibitions: Mon.-Sat. 9-5. Museum Shop: Mon.-Sat. 10-5. No charge; donations accepted. Closed major holidays. &
Attendance: 250,000 (estimated)
Membership: Individual $45; Family $75; Patron $150.

SPANISH GOVERNOR'S PALACE, 105 Plaza de Armas, San Antonio, TX 78205-2412. Tel.: 210-224-0601. Fax: 210-223-5562.
E-mail: charlotte.boord@sanantonio.gov
Web Site: www.spanishgovernorspalace.org
Founded: 1749.
Key Personnel: Dir. Downtown Operations, Paula X. Stallcup; Museum Asst., Charlotte Boord.
Personnel Profile: Full-Time Paid 2.
Governing Authority: municipal. Parent Institution: City of San Antonio. Subsidiary Institution: Dept. of Downtown Operations. Tax-exempt.
Institution Type/Description: Historic Building: 1722 Spanish Governor's Palace.
Collections: Spanish culture; history; paintings; decorative arts; agriculture; graphics; sculpture; interpretation.
Facilities: botanical garden. Postcards & historical literature for sale.
Activities: self-guided tours.
Publications: brochure.
Hours & Admission Prices: Tues.-Sat. 9-5, Sun. 10-5. Adults $4, seniors 60 & over, military & groups of 10 or more $3, children 7-13 $2; discounts to AAM & ICOM members; children under 7 no charge. Closed New Year's Day; Easter; Battle of Flowers Parade Friday; Thanksgiving; Christmas. &
Attendance: 22,092 (accurate)

STEVES HOMESTEAD, 509 King William St., San Antonio, TX 78204-1411. Tel.: 210-225-5924 & 227-9160. Fax: 210-223-9014.
E-mail: dchenoweth@saconservation.org
Web Site: www.saconservation.org/tours/steves.htm
Founded: 1924.
Congressional District: 20
Key Personnel: Pres., Rollette Schreckenghost; Exec. Dir., Bruce MacDougal; Administrative Asst., Glory Bohne; House Museum Mgr., Diana Chenoweth.
Personnel Profile: Full-Time Paid 2; Part-Time Paid 4; Part-Time Volunteers 10.
Governing Authority: society. Parent Institution: San Antonio Conservation Society Foundation, 107 King William St., San Antonio, TX 78204. Branch Museum: 1840-60 Yturri-Edmunds Historic Site, 128 Mission Rd., 210-534-8237. Tax-exempt: 501(c)(3).
Institution Type/Description: History Museum.
Collections: local history; preservation & restoration projects; photos; Victorian period furniture; decorative arts. Historic Houses: 1876 Steves Homestead; Wulff House (library only; no tours); 1840-1860 Yturri-Edmunds Historic Site.
Research Fields: local history; preservation & restoration projects; architectural history.
Facilities: 3,000-vol. library of books pertaining to history & preservation; visitor center. San Antonio Conservation Society Office: slide collection available for research.
Activities: guided tours; permanent & temporary exhibitions; food festivals; historic preservation seminars; speakers bureau.
Publications: monthly, San Antonio Conservation Newsletter; brochures; King William Walking Tour; Texas Star Trail; Steve Homestead brochure.
Hours & Admission Prices: Steves Homestead: daily 10-4:15; last tour 3:30. Adults $6, seniors & groups $5, students & military $4; children under 12 & members no charge. Wulff House: no tours; research library only. Mon.-Thurs. 10-3. Call for information for other sites. Yturri-Edmunds by appointment only. Closed most major holidays.
Attendance: 10,154 (accurate)
Membership: Associate & Active (voting) $25.

TEXAS AIR MUSEUM, Stinson Field, 1234 99th St., San Antonio, TX 78214. Tel.: 210-977-9885. Fax: 210-927-4447.
E-mail: info@texasairmuseum.org
Web Site: www.texasairmuseum.org
Institution Type/Description: Military Aviation History Museum.
Collections: military aviation history; aircraft; personal artifacts.

Hours & Admission Prices: Tues.-Sat. 10-5. Adults $4, military and seniors 55 & over $3, youth 12-16 $2, children 11 & under $1; discounts to school groups. Closed New Year's Eve & Day; Thanksgiving; Christmas Eve & Day.

TEXAS TRANSPORTATION MUSEUM, 11731 Wetmore Rd., San Antonio, TX 78247-3606. Tel.: 210-490-3554. Facebook: Texas Transportation Museum.
E-mail: hugh@txtransportationmuseum.org
Web Site: www.txtransportationmuseum.org
Founded: 1964.
Key Personnel: Dir. & Museum Shop Mgr., Hugh Hemphill; Chm. (V), Pat Halpin.
Personnel Profile: Full-Time Paid 1; Part-Time Paid 1; Part-Time Volunteers 200.
Governing Authority: Tax-exempt: 501(c)(3).
Institution Type/Description: Transportation Museum.
Collections: area transportation history; model railroads; automobiles; trucks; horse carriages.
Activities: train rides; fire truck rides.
Publications: The Dust Collector.
Hours & Admission Prices: Fri. 9-3, Sat.-Sun. 10-5. Fri: family $16, adults $6, children 2-12 $4; Sat.-Sun. family $24, adult $8, children 2-12 $6; discounts to military & senior citizens. &
Attendance: 10,000 (accurate)
Membership: Annual $35.

TOILET SEAT ART MUSEUM, 239 Abiso Ave., San Antonio, TX 78209-5103. Tel.: 210-824-7791.
Key Personnel: Owner, Barney Smith
Institution Type/Description: Art Museum.
Collections: over 1,000 painted toilet seats; cosmetic dentistry tools & equipment.
Activities: video.
Hours & Admission Prices: By appointment. No charge.

THE UNIVERSITY OF TEXAS AT SAN ANTONIO, ART GALLERY, Dept. of Art & Art History, Main Campus - 1604, One UTSA Cir., San Antonio, TX 78249-1130. Tel.: 210-458-4391.
E-mail: laura.crist@utsa.edu
Web Site: art.utsa.edu/galleries.art-gallery/
Founded: 1982.
Congressional District: 21
Key Personnel: Gallery Dir., Scott Scherer, Ph.D.; Gallery Coord., Laura Crist.
Personnel Profile: Full-Time Paid 1.
Governing Authority: university. Tax-exempt.
Institution Type/Description: University Art Gallery.
Collections: changing exhibits; portfolios of Elliott Erwitt & Manuel Alvarez Bravo.
Research Fields: contemporary regional & national art.
Facilities: security storage; preparation & photography area; fabrication shop.
Activities: guided tours; lectures; films; practicum in Gallery work; participatory, loan, temporary & traveling exhibitions.
Publications: catalogues, Texans Past and Present; The Soul of Mexico; Catherine Lee Catalogue; Biennial Faculty Exhibition; Aberrations; Mapping.
Hours & Admission Prices: Mon.-Fri. 10-4, Sat.-Sun. 1-4 & by appointment. No charge; donations accepted. Closed university holidays. &
Attendance: 3,000 (estimated)

VILLA FINALE & VILLA FINALE VISITOR CENTER, (M), 401 King William St. & 122 Madison St., San Antonio, TX 78204. Mailing Address: 122 Madison St., San Antonio, TX 78204. Tel.: 210-223-9800. Fax: 210-223-9802.
E-mail: info@villafinale.org
Web Site: www.villafinale.org
Founded: 2005.
Congressional District: 20
Key Personnel: Exec. Dir., Jane Lewis; Chm. (V), Toby Tate; Museum Shop Mgr., Syeira Budd.
Personnel Profile: Full-Time Paid 8; Part-Time Paid 1.
Governing Authority: Parent Institution: National Trust For Historic Preservation. Tax-exempt.
Institution Type/Description: Historic House Museum: housed in the former home of preservationist & civic leader, Walter Mathis. House construction began in 1876.
Collections: Mathis family history; personal artifacts; European furniture; fine & decorative arts; artifacts relating to Napoleon Bonaparte's life & death; Texas artists including Mary Bonner, and Julian & Robert Onderdonk;

Texas decorative arts including Bell silver, Texas furniture, and Texian campaign ceramics.

Activities: educational programs; workshops; guided tours.

Hours & Admission Prices: Tues.-Sat. 9:30-4:30. Adults $10; discounts to National Trust members; members no charge. Admission tickets sold at Visitor Center. &

Attendance: 8,500 (estimated)

Membership: Student $35; Individual $50; Family $75; Supporting $250; Sustaining $500; Patron $1,000.

* **WITTE MUSEUM, (M),** 3801 Broadway, San Antonio, TX 78209-6396. Tel.: 210-357-1900. Fax: 210-357-1882.

E-mail: witte@wittemuseum.org
Web Site: www.wittemuseum.org
Founded: 1926.
Congressional District: 9
Key Personnel: Pres. & C.E.O., Marise McDermott; Vice Pres. Devel., Heather Welder Russo; Vice Pres. External Affairs, Kim Biffle; Vice Pres. Admin. Svcs., Bea Abercrombie; Cur. Collections, Amy Fulkerson; Vice Pres. Exhibits, Randall Webster; Vice Pres. Visitor Experience, Ralph Voight.
Governing Authority: nonprofit organization. Tax-exempt: 501(c)(3).
Institution Type/Description: History Museum: located in Brackenridge Park.
Collections: Texana items; Texas paintings; textiles & costumes; wildlife & ecology exhibits & dioramas; Southwest Indian artifacts; dinosaurs; participatory exhibits; Texas stone artifacts; Filipino artifacts; historical photographs; Lower Pecos archaeology; Central Plains Indian artifacts; cowboy & western artifacts; Hertzberg Circus collection; firearms. Historic Houses: 1753 Ruiz House; 1835 Navarro House; 1840 Twohig House; two reconstructed log cabins; H-E-B Science Treehouse.
Research Fields: anthropology; basketry; archaeology of the lower Pecos; history: Texas art & decorative arts, costumes & textiles.
Facilities: library of historical photographs & 2,200 slides all pertaining to Texas & natural history, archaeology, decorative, primitive & Texas art available for research on premise by appointment only; 300-seat auditorium; native plant garden. Gift items for sale.
Activities: lectures; theatre; permanent & temporary exhibitions; Museum-School programs for children, adults & families; concerts; live science demonstrations.
Publications: books, Mary Bonner: Impressions of a Printmaker; San Antonio Was; Touring Texas: Through the Eyes of an Artists; Ancient Texans: Rock Art & Lifeways Along the Lower Pecos; Snakes of Bexar County; Art for History's Sake: The Texas Collection of the Witte Museum; Prehistoric Basketry of the Lower Pecos, Texas; Patterns With Potential; Tischlermeister Jahn; Iwonski in Texas, Painter and Citizen.
Hours & Admission Prices: Mon. & Wed.-Sat. 10-5, Tues. 10-8, Sun. 12-5. Adults $10, active duty military & senior citizens 65 & over $9, children 4-11 $7; discounts to groups; children 3 & under & members no charge. Closed third Mon. in Oct.; Thanksgiving; Christmas. &
Attendance: 200,000 (accurate)
Membership: Individual $40; Senior Family $70; Family $75; Individual Witte Society $110; Family Witte Society $200; Explorer $250; Voyager $500; Quillin Society Sustainer $1,000; Quillin Society Benefactor $5,000; Diamond Circle of Giving $7,500.

WOODEN NICKEL HISTORICAL MUSEUM, 345 Old Austin Rd., San Antonio, TX 78209-6933. Tel.: 210-829-1291. Fax: 210-832-8965.

E-mail: museum@wooden-nickel.net
Web Site: www.wooden-nickel.net
Founded: 1998.
Key Personnel: C.E.O., Herb Hornung.
Personnel Profile: Part-Time Volunteers 2.
Governing Authority: private; not for profit.
Institution Type/Description: General Museum.
Collections: wooden nickels 1931-present; printing plates; printing equipment.
Facilities: 2,000 sq. ft. exhibit space. Museum-related items for sale.
Hours & Admission Prices: Mon.-Fri. 9-5, Sat. by appointment. No charge.
Attendance: 400 (estimated)

San Elizario

LOS PORTALES MUSEUM & INFORMATION CENTER, 1521 San Elizario Rd., San Elizario, TX 78949. Mailing Address: P.O. Box 1090, San Elizario, TX 79849. Tel.: 915-851-1682. Fax: 915-851-0045.

E-mail: sanelizario@sbcglobal.net
Web Site: www.epcounty.com/sanelizariomuseum
Founded: 1962.

Key Personnel: Dir., Eloisa Levario; Pres. (V), Al Borrego; Museum Shop Mgr., Transito Macias
Governing Authority: Parent Institution: San Elizario Genealogy & Historical Society. Tax-exempt.
Institution Type/Description: History Museum.
Collections: artifacts & memorabilia pertaining to San Elizario.
Publications: quarterly, Los Portales Newsletter.
Hours & Admission Prices: Tues.-Sat. 10-2, Sun. 12-4. No charge; donations accepted. Closed major holidays.
Attendance: 5,000 (accurate)

San Marcos

AQUARENA CENTER, 951 Aquarena Springs Dr., San Marcos, TX 78666. Tel.: 512-245-7570.

Key Personnel: Dir., Ron Coley
Institution Type/Description: Natural Science Museum.
Collections: environmental science; ecosystem; aquarium; hands-on exhibitions.
Facilities: Center-related items for sale.
Activities: educational programs; group tours; classes; scout activities; birthday parties.
Hours & Admission Prices: March-May & Aug. 23-Sept. 3 daily 9:30-5; June-Aug. 22 & Sept. 4-Sept. 7 daily 9:30-6; Sept. 7-Oct. daily 10-5; Nov.-Feb. daily 10-4.

CALABOOSE AFRICAN AMERICAN HISTORY MUSEUM, 200 W. Martin Luther King Jr. Dr., San Marcos, TX 78666-5522. Tel.: 512-393-8421. Facebook: Calaboose African American History Museum.

Web Site: www.sanmarcosarts.com/mcal.htm
Key Personnel: Dir., Mrs. Johnnie Armstead
Institution Type/Description: History Museum.
Collections: local black history; Buffalo soldiers; period artifacts.
Hours & Admission Prices: Sat. 10-2; other times by appointment. Suggested Donation: $3.

COMMEMORATIVE AIR FORCE CENTRAL TEXAS WING, San Marcos Municipal Airport, 1814 Airport Dr., Bldg. 2249, San Marcos, TX 78666. Tel.: 512-396-1943. Fax: 512-396-1992.

E-mail: contact@cafcentex.com
Web Site: www.cafcentex.com
Founded: 1974.
Governing Authority: Commemorative Air Force. Tax-exempt.
Institution Type/Description: Military History Museum.
Collections: military aircraft & artifacts; paintings; books.
Facilities: library.
Activities: active vintage aircraft flying.
Hours & Admission Prices: Mon., Wed. & Fri.-Sat. 9-4. Suggested Donation: $3 per person.
Attendance: 7,000 (estimated)
Membership: CENTEX $50; CAF $200.

DICK'S CLASSIC GARAGE, 120 Stagecoach Trail, San Marcos, TX 78666. Mailing Address: P.O. Box 1422, San Marcos, TX 78667-1422. Tel.: 512-878-2406. Fax: 512-757-8908.

E-mail: info@dicksclassicgarage.org
Web Site: www.dicksclassicgarage.org
Formerly: CTMAH
Founded: 2009.
Congressional District: 32
Key Personnel: Dir., Brian Pauli; Pres. (V), Richard L. Burdick; Treas., Kathleen Cheatham; Cur., Thom Fortney; Museum Shop Mgr., Ray Terry.
Personnel Profile: Full-Time Paid 4; Part-Time Paid 1; Part-Time Volunteers 10.
Governing Authority: private; nonprofit organization. Parent Institution: CTMAH, Rosanky, TX. Tax-exempt: 501(c)(3).
Institution Type/Description: Automotive Museum.
Collections: automobiles from 1929 to 1959.
Research Fields: automobile collection.
Facilities: 20-vol. library. Museum-related items for sale.
Activities: guided tours. Annual Event: Car Show.
Hours & Admission Prices: Mon.-Sat. 10-5, Sun. 12-5. Adults $10, senior citizens & students $8, children $5. Closed Easter; Thanksgiving; Christmas. &
Attendance: 12,000 (estimated)

LBJ MUSEUM OF SAN MARCOS, 131 N. Guadalupe, San Marcos, TX 78666-5606. Mailing Address: P.O. Box 3, San Marcos, TX 78667-0003. Tel.: 512-353-3300. Fax: 512-353-3305.
E-mail: director@lbjmuseum.com
Web Site: www.lbjmuseum.com/contactus.htm
Founded: 1997.
Congressional District: 14
Key Personnel: Pres. (V), Ed Michalkanin.
Personnel Profile: Full-Time Paid 2.
Governing Authority: Tax-exempt.
Institution Type/Description: History Museum.
Collections: LBJ-related campaign memorabilia; presidential buttons; personal artifacts; sample voting machine; hats; photographs; manuscripts; periodicals; audiovisual materials; sculpture.
Facilities: library; archives; meeting areas.
Activities: school programs; internships; research; rental facilities; special events; presentations.
Publications: LBJ Connection.
Hours & Admission Prices: Thurs.-Fri. & Sun. 1-5, Sat. 10-5. No charge; donations accepted. &
Attendance: 1,716 (accurate)
Membership: Sponsor $100; Educator $250; Senatorial $1,000; Presidential $10,000.

THE WITTLIFF COLLECTIONS - TEXAS STATE UNIVERSITY-SAN MARCOS, Alkek Library, 7th Fl., San Marcos, TX 78666-4604. Mailing Address: Alkek Library, 601 University Dr., San Marcos, TX 78666-4604. Tel.: 512-245-2313. Fax: 512-245-7431.
E-mail: thewittliffcollections@txstate.edu
Web Site: www.thewittliffcollections.txstate.edu
Institution Type/Description: History Museum.
Collections: personal artifacts; manuscripts; books; diaries; photographs; works of art; period artifacts; paintings; journals; personal correspondence; memorabilia.
Activities: special events.
Hours & Admission Prices: Call for hours. No charge. Closed university holidays & breaks.

San Saba

SAN SABA COUNTY HISTORICAL MUSEUM, Mill Pond Park, San Saba, TX 76877. Mailing Address: 500 E. Wallace St., San Saba, TX 76877.
E-mail: info@sansabamuseum.org
Web Site: www.sansabamuseum.org
Institution Type/Description: History Museum.
Collections: local history & culture; period furnishings; personal artifacts; photographs.
Hours & Admission Prices: April-Sept. Sat.-Sun. 1:30-4. &

Sanderson

TERRELL COUNTY MEMORIAL MUSEUM, 203 E. Mansfield St., Sanderson, TX 79848. Mailing Address: P.O. Box 7, Sanderson, TX 79848-0702. Tel.: 432-345-2936.
E-mail: katiedroberts@hotmail.com
Web Site: www.sandersontx.info/pages/thingstodo/museum.html
Congressional District: 19
Key Personnel: Caretaker, Bill Smith; Chm. (V), Katie Roberts; Sec., Lea Haun.
Governing Authority: Tax-exempt.
Institution Type/Description: History Museum: former home of the Lemmons family.
Collections: railroad memorabilia; period costumes; cowboy & ranching relics; tools; pioneer furnishings; Terrell County history books.
Hours & Admission Prices: Mon.-Fri. 10-12 & 1-3. No charge, donations accepted.

Sarita

KENEDY RANCH MUSEUM OF SOUTH TEXAS, 200 E. La Parra Ave., Sarita, TX 78385. Mailing Address: P.O. Box 70, Sarita, TX 78385. Tel.: 361-294-5751. Fax: 361-294-5228.
E-mail: hsv@kenedy.org
Web Site: kenedyranchmuseum.org
Founded: 2003.
Congressional District: 34
Key Personnel: Coord., Homero S. Vera

Institution Type/Description: History Museum.
Collections: Mifflin Kenedy's family & career; period furnishings; personal artifacts; photographs; paintings.
Hours & Admission Prices: Tues.-Sat. 10-4, Sun. 12-4. Adults $3, seniors & children 13-18 $2. Closed New Year's Day; Good Friday; Easter; Independence Day; Thanksgiving; Christmas. &

Schulenburg

STANZEL MODEL AIRCRAFT MUSEUM, 311 Baumgarten St., Schulenburg, TX 78956-2101. Mailing Address: P.O. Box 6, Schulenburg, TX 78956-0006. Tel.: 979-743-6559. Fax: 979-743-2525.
E-mail: museum@stanzelmuseum.org
Web Site: www.stanzelmuseum.org
Founded: 1999.
Key Personnel: Pres., Robert Stanzel; Vice Pres., Theodore Stanzel; Museum Mgr., Archivist & Museum Shop Mgr., Eugenia Reeves.
Personnel Profile: Part-Time Paid 7.
Governing Authority: private; nonprofit organization. Tax-exempt: 501(c)(3).
Institution Type/Description: Aircraft Museum.
Collections: history of flight; model aircraft; videos; photographs; furnishings; personal artifacts; farmhouse; videos.
Facilities: ancestral home & garden. Museum-related items for sale.
Activities: children & adult guided tours.
Hours & Admission Prices: Mon., Wed. & Fri.-Sat. 10:30-4:30. Adults $4, senior citizens $2; school groups & children under 12 no charge. Closed New Year's Day; Martin Luther King Jr. Day; Presidents' Day; Good Friday; Mother's Day; Memorial Day; Independence Day; Labor Day; Veterans Day; Thanksgiving; Christmas Eve & Day. &
Attendance: 4,099 (accurate)

Seabrook

BAY AREA MUSEUM, 5000 Nasa Rd. I, Seabrook, TX 77586. Mailing Address: P.O. Box 58348, Webster, TX 77598-8348. Tel.: 281-326-5950. Fax: 281-326-5950.
E-mail: bayarea_museum@yahoo.com
Web Site: www.museumbayarea.org/index.html
Founded: 1984.
Congressional District: 25
Key Personnel: Pres., Carole Murphy; Dir., Diana Dorrak; Dir., MaryAnn Baxter.
Governing Authority: nonprofit. Parent Institution: Lunar Rendezvous, Inc. Tax-exempt.
Institution Type/Description: General Museum: housed in the church which Buzz Aldrin celebrated communion from the surface of the moon in 1969.
Collections: history of the area from late 19th century to the development of NASA; from Arrowheads to astronauts.
Research Fields: local history.
Activities: lectures; loan & temporary exhibitions.
Publications: quarterly newsletter.
Hours & Admission Prices: Sun. & Wed.-Fri. 1-5, Sat. 10-4. No charge; donations accepted. Closed federal holidays. &
Membership: Family $45; Sustaining $60.

Seagraves

SEAGRAVES-LOOP MUSEUM AND ART CENTER INC., Seagraves-Loop Div., 201 Main, Seagraves, TX 79359. Mailing Address: Box 1387, Seagraves, TX 79359-1387. Tel.: 806-546-2810. Fax: 806-546-2810.
Founded: 1974.
Congressional District: 19
Key Personnel: C.E.O., Dan Calfee; Dir., Treas. & Museum Shop Mgr., Leslie McConal; Sec., Penny Hayes.
Personnel Profile: Full-Time Paid 1; Part-Time Paid 1; Part-Time Volunteers 3.
Governing Authority: county government. Tax-exempt: 501(c)(3).
Institution Type/Description: Historic Building: built in 1926, survived a town fire.
Collections: barbed wire collections; Indian artifacts; early settlers artifacts; Santa Fe depot with caboose; military exhibit; restored 1937 fire truck.
Research Fields: local history; art; collections.
Activities: Museum Sponsors: arts & craft show; spring harvest festival; school tours in May.
Hours & Admission Prices: Mon.-Fri. 9-12 & 1-5. No charge; donations accepted. &
Membership: Student $1; Individual $5; Family $10; Associate $25; Contributing $50; Sustaining $100; Life $500.

Seguin

FIEDLER MEMORIAL MUSEUM, Texas Lutheran University, Seguin, TX 78155. Tel.: 830-372-8038. Fax: 830-372-8188.
Web Site: www.txlutheran.edu
Founded: 1973.
Congressional District: 23
Key Personnel: Dir., Evelyn Fiedler Streng.
Governing Authority: college & individual; nonprofit. Texas Lutheran University, Seguin, TX. Tax-exempt.
Institution Type/Description: Geology Museum.
Collections: rocks, minerals & fossils; New Guinea artifacts; Native American artifacts; geological & geographic slides.
Research Fields: geology.
Facilities: outdoor rock display; learning center.
Activities: guided tours; lectures; informally organized education programs for children; permanent exhibitions; slide shows.
Publications: Study Guides to Displays; Rock Walk Guide.
Hours & Admission Prices: Mon.-Fri. 1-5; other times by appointment. No charge. Closed school holidays. &

LOS NOGALES MUSEUM, 415 S. River, Seguin, TX 78155. Mailing Address: P.O. Box 245, Seguin, TX 78156-0245. Tel.: 830-379-3257; 830-372-6168. Fax: 830-379-4685.
Web Site: www.seguinconservation.org
Founded: 1952.
Congressional District: 25
Key Personnel: Dir. & Museum Shop Mgr., Greg Ander; Pres. (V), Marty Keil.
Governing Authority: society. Affiliated with Seguin Conservation Society, P.O. Box 245, Seguin, TX 78155. Tax-exempt: 501(c)(3).
Institution Type/Description: Conservation Society Museum: housed in 1849 building.
Collections: historical items; books; court minutes; early photos; doll house; log cabin; adobe building; calaboose. Historic House: 1894 Queen Anne home; first church built in Seguin as a church.
Research Fields: genealogy.
Activities: guided tours; programs pertaining to history of Seguin.
Publications: City/County History.
Hours & Admission Prices: May-Sept. Sun. 2-5; other times by appointment. No charge; donations accepted.
Attendance: 2,000 (estimated)
Membership: Individual $15; Family $25; Business $50.

Seminole

GAINES COUNTY MUSEUM, SEMINOLE DIVISION, 700 Hobbs Hwy., Seminole, TX 79360. Tel.: 432-758-4016. Fax: 432-758-6638.
E-mail: seminolemuseum@co.gaines.tx.us
Founded: 1976.
Key Personnel: Dir., Roy Lynn Barnes; Pres. (V), Paul Elam.
Personnel Profile: Full-Time Paid 1; Part-Time Paid 1.
Governing Authority: Parent Institution: Gaines County. Tax-exempt.
Institution Type/Description: History Museum.
Collections: local history & culture; period furnishings & clothing; photographs; personal artifacts; sculpture; paintings; pre-Columbian & Mesoamerican artifacts; pottery; early typewriters; military artifacts; school records; pre-Columbian Teotihuacan artifacts representing the 5 cultures & languages.
Publications: brochures.
Hours & Admission Prices: Mon.-Fri. 1-5. No charge; donations accepted.
Attendance: 4,000 (estimated)
Membership: Student $10; Adult $20; Corporate $100; Lifetime $200.

Seymour

BAYLOR COUNTY MUSEUM, 116 N. Washington St., Scymour, TX 76380-2557. Mailing Address: P.O. Box 544, Seymour, TX 76380-0544. Tel.: 940-889-6780.
E-mail: baylorcomuseum@srcaccess.net
Institution Type/Description: History Museum.
Collections: local history & culture; period furnishings; military artifacts; medical equipment; household items; Western artifacts; religious & school memorabilia.
Activities: group tours; temporary & permanent exhibitions.
Hours & Admission Prices: Mon.-Fri. call for hours.

Shamrock

PIONEER WEST MUSEUM, 204 N. Madden, Shamrock, TX 79079-2340. Tel.: 806-256-3941.
Institution Type/Description: History Museum: housed in the former Reynolds Hotel; built in 1925.
Collections: local history & culture; period furnishings; personal artifacts; photographs.
Activities: group tours.
Hours & Admission Prices: Daily 10-12 & 1-3; other times by appointment.

Sherman

C.S. ROBERTS HOUSE MUSEUM, 915 S. Crockett, Sherman, TX 75090. Mailing Address: P.O. Box 159, Sherman, TX 75091-1059. Tel.: 903-893-4067.
Institution Type/Description: Historic House Museum: built in 1896.
Collections: local & family history; period furnishings; personal artifacts.
Activities: rental facilities; lectures; special events; traveling trunk program.
Hours & Admission Prices: Sun. 1-4; other times by appointment. Adults 17 & over $5, senior citizens 55 & over $3, children 11-16 $2; children under 10 no charge.

THE SHERMAN MUSEUM, (M), 301 S. Walnut, Sherman, TX 75090-7152. Tel.: 903-893-7623.
E-mail: theshermanmuseum@verizon.net
Web Site: www.theshermanmuseum.org
Formerly: Red River Historical Museum
Founded: 1976.
Congressional District: 4
Personnel Profile: Full-Time Paid 2; Part-Time Paid 2; Part-Time Volunteers 1; Interns 1.
Governing Authority: private; nonprofit organization. Tax-exempt: 501(c)(3).
Institution Type/Description: Museum: housed in 1914 Andrew Carnegie Library building.
Collections: local & area historic photographs & memorabilia.
Facilities: library. Museum-related items for sale.
Activities: lectures; films; guided tours; hobby workshops; temporary exhibitions. Annual Events: fundraiser in January & July; Arts Festival in September; History Saturday; Craft Saturday.
Publications: newsletter, The Sentinel.
Hours & Admission Prices: Tues.-Sat. 10-4. Adults $5, children $2; members no charge. Closed Thanksgiving; Christmas.
Attendance: 3,000 (accurate)
Membership: Individual $25; Family $50; Sustainer $100; Sponsor $250; Patron $500; Benefactor $1,000.

Shiner

EDWIN WOLTERS MEMORIAL MUSEUM, 306 S. Ave. I, Shiner, TX 77984. Mailing Address: P.O. Box 308, Shiner, TX 77984-0308. Tel.: 512-594-3774 & 3362. Fax: 512-594-3566.
Founded: 1963.
Key Personnel: Cur., Bernard Siegel, Jr.
Personnel Profile: Full-Time Paid 1; Part-Time Paid 1.
Governing Authority: municipal. Tax-exempt.
Institution Type/Description: General Museum: housed in 1900 home of Edwin Wolters, founder of museum.
Collections: local history; industry; Indian artifacts; archaeology; costumes; botany; dolls; period furniture.
Facilities: old country store.
Activities: guided tours; permanent & temporary exhibitions; displays for children.
Hours & Admission Prices: Mon.-Fri. 8-5; groups by appointment. No charge; donations accepted. &
Attendance: 741 (accurate)

Sinton

WELDER WILDLIFE FOUNDATION & REFUGE, 10620 Hwy. 77 N., Sinton, TX 78387. Mailing Address: P.O. Box 1400, Sinton, TX 78387-1400. Tel.: 361-364-2643. Fax: 361-364-2650.
E-mail: welderfoundation@welderwildlife.org
Web Site: www.welderwildlife.org
Founded: 1954.
Key Personnel: Dir., Dr. Terry L. Blankenship; Asst. Dir., Dr. Selma N. Glasscock.
Personnel Profile: Full-Time Paid 7; Part-Time Paid 2; Interns 1.
Institution Type/Description: Wildlife Refuge.

Collections: reptiles & amphibians; wildlife paintings; 2,000 bird skins; 500 mammal skins; 300 mounted birds.
Research Fields: wildlife management.
Activities: educational workshops & classes; college courses; training programs for professional teachers; research.
Publications: biennial reports; books.
Hours & Admission Prices: Thurs. 3pm. No charge; donations accepted.
Attendance: 2,500 (accurate)

Slaton

TEXAS AIR MUSEUM, 12102 FM 400, Slaton, TX 79364. Mailing Address: P.O. Box 36, Slaton, TX 79364-4192. Tel.: 806-796-7618.
Web Site: www.thetexasairmuseum.org
Founded: 1986.
Congressional District: 27
Key Personnel: C.E.O., Mike Delano; Chm. (V), Malcolm Laing; Bd. Member & Treas., Kristin Snow; Museum Shop Mgr., Steve Oldham.
Personnel Profile: Full-Time Volunteers 2; Part-Time Volunteers 25.
Governing Authority: private; nonprofit organization. Parent Institution: Texas Air Museum, Inc. Subsidiary Institution: TAM Stinson & South Plains Chapter. Tax-exempt.
Institution Type/Description: Aeronautics Museum.
Collections: concentration on aircraft used for warfare, commerce & agriculture, 1904-present; emphasis on aircraft used in WWI & II, Korea, WASPS-Women Military Pilots of WWII & Vietnam.
Research Fields: WWII Axis aircraft & support equipment; female pilots in U.S. aviation.
Facilities: library of books, videos & other printed materials available to public on premises; 22,000 sq. ft. exhibit space; air strip. Museum-related items for sale.
Activities: films; guided tours; lectures; loan & temporary exhibitions; school loan service.
Publications: quarterly newsletter, Airspeed.
Hours & Admission Prices: Sat. 9-4. Adults $5, senior citizens $4, youth 12-16 $3, children 11 & under $2; discounts for groups, AAM & ICOM members; members no charge. Closed New Year's Day; Thanksgiving; Christmas. ♿
Attendance: 6,324 (accurate)
Membership: Annual $40; Individual $250; Corporate $2,000.

Snyder

* **SCURRY COUNTY MUSEUM, (M),** Western Texas College, 6200 College Ave., Snyder, TX 79549-6105. Tel.: 325-573-6107. Facebook: Scurry County Museum.
E-mail: scm@snydertex.com
Web Site: www.scurrycountymuseum.org
Founded: 1970.
Congressional District: 17
Key Personnel: Dir., Daniel Schlegel, Jr.; Pres. (V), John Rogotzky.
Personnel Profile: Full-Time Paid 3; Full-Time Volunteers 10; Part-Time Paid 1; Part-Time Volunteers 10.
Governing Authority: Parent Institution: Scurry County Museum Association. Tax-exempt: 501(c)(3).
Institution Type/Description: History Museum.
Collections: archives; farming & ranching equipment; frontier furnishings; county records; contemporary prints; Victorian lamps; costumes; local history memorabilia; oil production tools & equipment.
Research Fields: local & regional history; genealogy.
Facilities: library of law & medical books & county records dating from 19th century to 1960; meeting facilities; classrooms. Museum-related items for sale.
Activities: guided tours; lectures; films; gallery talks; arts festivals; docent program; permanent & temporary exhibitions. Museum Sponsors: Christmas special.
Publications: quarterly newsletter.
Hours & Admission Prices: Mon.-Fri. 9-6, Sat. 10-2. No charge; donations accepted. ♿
Attendance: 9,000 (accurate)
Membership: Personal: Basic $20; Associate $50; Friend $100; Supporter $250; Patron $500; Benefactor $1,000. Business: Community $100; Sponsor $150; Leader $250; Trailblazer $500; Benefactor $1,000.

Sonora

SUTTON COUNTY HISTORICAL SOCIETY - MIERS HOME MUSEUM, CAUTHORN MEMORIAL DEPOT AND OLD ICE HOUSE RANCH MUSEUM, 307 Oak St., Sonora, TX 76950-2647. Mailing Address: P.O. Box 885, Sonora, TX 76950-0885. Tel.: 325-387-5084.
E-mail: schs@sonoratx.net
Web Site: old-ice-house-ranch-museum.com
Founded: 1968.
Key Personnel: Pres. (V), Rex Ann Friess; Vice Pres., Betty F. Smith; Sec., Joyce Chalk; Treas., Robin Street; Museum Shop Mgr., David L. Smith.
Personnel Profile: Part-Time Paid 1; Part-Time Volunteers 5.
Governing Authority: society. Tax exempt.
Institution Type/Description: History Museum.
Collections: local & area history; furniture 1890s-1950s; period photographs; 100 years of The Devils River News; sheriff's office photos & archives; branding irons of county; period ranching artifacts. Historic Buildings: 1890 Miers Home Period Museum; 1928 West Texas Utility Building; 1930s Old Santa Fe Depot; John & Mildred Cauthorn Memorial Bldg.
Research Fields: Sutton County History Book 1887-1977; family histories.
Facilities: archives.
Activities: guided tours. Annual Event: Sutton County Days in August.
Publications: books: The True Story of Will Carver, Sutton County 1889-1890; Bad Old Days In and Around Sutton County 1889-1939; Sutton County Marriage Records 1890-1940; John Eatons Tales of Wild Bill Taylor; Johnny Ward Cowboy.
Hours & Admission Prices: Wed.-Fri. 1-4, Sat. 10am-12pm. No charge; donations accepted. ♿
Attendance: 876 (accurate)
Membership: Individual $25.

VETERANS OF ALL WARS & PIONEER RANCH WOMEN MUSEUM, 105 Concho St., Sonora, TX 76950. Mailing Address: c/o Texas Pecos Trail Region, P.O. Box 212, Sonora, TX 76950.
Institution Type/Description: History Museum.
Collections: local history & culture; period furnishings; military equipment; photographs; letters; uniforms; personal artifacts; clothing.
Hours & Admission Prices: Mon.-Fri. 10-4; other times by appointment.

South Padre Island

COASTAL STUDIES LABORATORY - THE UNIVERSITY OF TEXAS-PAN AMERICAN, 100 Marine Lab Dr., South Padre Island, TX 78597. Tel.: 956-761-2644. Fax: 956-761-2913.
E-mail: coastal@utpa.edu
Web Site: www.utpa.edu/csl
Institution Type/Description: Marine Science Museum.
Collections: marine science; local fauna & flora; hands-on exhibits.
Activities: educational programs; classes; guided tours.
Hours & Admission Prices: Mon.-Fri. 1:30-4:30. Lab: no charge; donations accepted. Park: $4 per car, $10-$15 per bus.

Spearman

STATIONMASTER'S HOUSE MUSEUM, 30 S. Townsend, Spearman, TX 79081-2644. Tel.: 806-659-3008.
Founded: 1975.
Key Personnel: Founder, Clementine Renner
Institution Type/Description: Historic House Museum: housed in the former station master's house; built in 1920.
Collections: local history; railroad artifacts; photographs; windmills.
Hours & Admission Prices: Tues.-Sat. 1-5; other times by appointment.

Spring

PEARL FINCHER MUSEUM OF FINE ARTS, (M), 6815 Cypresswood Dr., Spring, TX 77379-7705. Tel.: 281-376-6322. Fax: 281-376-2944.
E-mail: tnovak@mfah.org
Web Site: www.pearlmfa.org
Founded: 2007.
Congressional District: 2
Institution Type/Description: Art Museum.
Collections: paintings; drawings; sculpture.

Facilities: rental facilities; conference room.
Activities: educational programs including art classes; lectures; concerts; summer camps; traveling programs.
Hours & Admission Prices: Tues.-Wed. & Sat. 10-5, Thurs. 10-8, Fri. 10-6, Sun. 12-5. Suggested Donation: $3 per person. &
Membership: Senior & Student $25; Family $55.

Stamford

COWBOY COUNTRY MUSEUM, 113 S. Wetherbee St., Stamford, TX 79553. Tel.: 325-773-2411.
Founded: 1977.
Institution Type/Description: History Museum.
Collections: local history & culture; period furnishings; photographs; early medical equipment; farm & ranch artifacts; newspapers; telephone books; personal artifacts.
Hours & Admission Prices: Mon.-Fri. 9-12 & 1-5; other times by appointment.

Stanton

MARTIN COUNTY HISTORICAL MUSEUM, 207 E. Broadway, Stanton, TX 79782. Mailing Address: P.O. Box 929, Stanton, TX 79782-0929. Tel.: 432-756-2722.
E-mail: martincountymuseum@gmail.com
Founded: 1978.
Congressional District: 17
Key Personnel: Dir., Elodia Bravo; Pres., Steve Stalling.
Personnel Profile: Full-Time Paid 1; Part-Time Volunteers 15.
Governing Authority: board of trustees; nonprofit organization. Tax-exempt.
Institution Type/Description: County History Museum: includes old jail & Connell House.
Collections: history of Martin County; prehistoric & Native American artifacts; early settlers; railroad; Catholic heritage; farming, ranching & oil production; schools, churches; wedding dresses & fashions; military display; barbed wire collection; cowboy regalia; blacksmith shop; archives; old jail; Connell house.
Facilities: library of Martin County history; Martin County citizens' genealogy; Texas history; Texas county histories; Permian Basin history; U.S. & world history; reading room.
Activities: permanent & temporary exhibitions; Jr. Historian Clubs; special Memorial Day services & reception. Museum Sponsors: Old Settlers Day Family Reception in July.
Publications: quarterly newsletter.
Hours & Admission Prices: Mon.-Fri. 12:30-5:30; other times by appointment. No charge; donations accepted. Closed major holidays. &
Attendance: 1,000 (estimated)
Membership: Student $5; Individual $25; Family $35; Business, Commercial & Patron $100; Sustainer $250; Benefactor $500 & up.

Star

STAR HISTORICAL MUSEUM, 44 S. FM 1047, Star, TX 76880. Mailing Address: P.O. Box 356, Star, TX 76880. Tel.: 325-948-3677.
Web Site: www.startexasmuseum.org
Institution Type/Description: History Museum.
Collections: county history & culture; period furnishings; personal artifacts; photographs.
Activities: special events. Museum Sponsors: Quilt Show in June.
Hours & Admission Prices: By appointment.

Stephenville

STEPHENVILLE MUSEUM, (M), 525 E. Washington, Stephenville, TX 76401-4439. Mailing Address: P.O. Box 899, Buffalo Gap, TX 79508-0899. Tel.: 254-965-5880.
E-mail: llohr@our-town.com
Formerly: Stephenville Historical House Museum
Founded: 1965.
Congressional District: 17
Key Personnel: Pres., Betty Heath; Treas., Ben Baty.
Personnel Profile: Part-Time Paid 2; Part-Time Volunteers 20.
Governing Authority: municipal. Parent Institution: City of Stephenville. Tax-exempt.
Institution Type/Description: General Museum: housed in 1869 Berry House; blacksmith shop.
Collections: pressed glass; rocks; early day tools; period furnishings for a house of the late 19th century; children's toys; barbed wire; photographs. Historic Sites, Buildings & homestead cabin: 1858 homestead log cabin;

1899 Gothic church, 1870 dogtrot log cabin; 19th-century church; carriage house; 1861 log corncrib; 1890 two-room schoolhouse; Tarleton house (important to local university).
Research Fields: local history.
Facilities: church; school; 1870 ranch house; log cabins; blacksmith shop.
Activities: guided tours by arrangement.
Hours & Admission Prices: Tues.-Sat. 10-5, Sun. 1-5. No charge; donations accepted.
Attendance: 5,000 (estimated)

Sulphur Springs

SOUTHWEST DAIRY MUSEUM, 1210 Houston, Sulphur Springs, TX 75482-2310. Mailing Address: Southwest Dairy Farmers, P.O. Box 936, Sulphur Springs, TX 75483. Tel.: 903-439-6455. Fax: 903-439-1125.
Web Site: www.southwestdairyfarmers.com
Founded: 1982.
Congressional District: 4
Key Personnel: C.E.O., James Hill; Dir., Carolyn McKinney; Pres. (V), David DeJong; Museum Shop Mgr., Paula Tidwell.
Personnel Profile: Full-Time Paid 29; Part-Time Paid 1.
Governing Authority: Tax-exempt.
Institution Type/Description: Dairy Museum.
Collections: dairy industry history; period artifacts; documents; hands-on exhibits.
Activities: hands-on exhibits.
Hours & Admission Prices: Museum: Mon.-Fri. 9-4. Office: Mon.-Fri. 8-5. No charge; donations accepted. &
Attendance: 11,150 (accurate)

Sweetwater

CITY COUNTY PIONEER MUSEUM, 610 E. Third, Sweetwater, TX 79556-4643. Tel.: 325-235-8547.
Founded: 1968.
Congressional District: 5
Key Personnel: Dir., Mrs. Franzas Cupp; Chm. (V), Kent Boatright; Mgr., Beverly Puckett.
Personnel Profile: Part-Time Paid 3; Part-Time Volunteers 8.
Governing Authority: municipal; county. Tax-exempt.
Institution Type/Description: Local History Museum.
Collections: furniture; documents; artifacts; china; glassware; period gowns; ranch & farm equipment; saddles; photos; historic house.
Research Fields: local history.
Activities: guided tours; lectures; special exhibits; monthly bake sales; public school trunk shows; group & individual tours. Annual Events: quilt show in August; Christmas exhibit.
Hours & Admission Prices: Tues.-Sat. 1-5; other times by appointment. No charge; donations accepted. Closed holidays. &
Attendance: 3,500 (estimated)

THE NATIONAL WASP WWII MUSEUM, 210 Avenger Field Rd., Sweetwater, TX 79556. Mailing Address: P.O. Box 456, Sweetwater, TX 79556-0456. Tel.: 325-235-0099.
E-mail: waspmuseum@yahoo.com
Founded: 2005.
Congressional District: 19
Key Personnel: Pres. (V), Tom Henderson; Museum Shop Mgr., Carol C. Cain.
Personnel Profile: Full-Time Paid 1; Part-Time Paid 4; Interns 1.
Volunteer Hours: 10,230
Governing Authority: Parent Institution: Board of Directors. Tax-exempt.
Institution Type/Description: Military History Museum.
Collections: Steamau pt-17women Air Force service pilot history; photographs; posters; PT-19; uniforms; personal artifacts; aircraft; DVD's; free standing exhibits; link trainer; Stearman PT-17
Research Fields: WASP files; photographs; papers.
Facilities: historic hanger.
Activities: educational programs; homecoming; 5K run.
Publications: newsletter.
Hours & Admission Prices: Wed.-Sat. 10-5, Sun. 1-5. No charge, donations accepted.
Attendance: 4,800 (estimated)
Membership: Cadet $10; Aviator Adult $38; Bronze Wings $100-$499; Silver Wings $500-$999; One Star General $1,000-$2,499; Two Star General $2,500-$4,999; Three Star General $5,000-$9,999; Four Star General $10,000 & up.

Taft

BLACKLAND MUSEUM, 301 Green Ave., Taft, TX 78390. Mailing Address: P.O. Box 731, Taft, TX 78390. Tel.: 361-528-2206. Facebook: The Blackland Museum.
E-mail: blacklandmuseum@aol.com
Founded: 1979.
Key Personnel: Exec. Dir., Patrick King; Pres. (V), David Smith.
Personnel Profile: Part-Time Paid 1.
Governing Authority: Tax-exempt: 501(c)(3)
Institution Type/Description: History Museum.
Collections: local history & culture; period furnishings; personal artifacts; photographs; early farm & ranch equipment; fossil collections; geological displays; historical archives; Native American artifacts; shell collections.
Hours & Admission Prices: Thurs.-Fri. 10-4, Sat. 10-5; No charge; donations accepted.
Membership: Basic $25; Docent $100.

Taylor

GOV. DAN MOODY MUSEUM, 114 W. Ninth St., Taylor, TX 76574. Mailing Address: P.O. Box 669, Taylor, TX 76574. Tel.: 512-352-5990 & 365-7396. Facebook: Moody Museum.
E-mail: moodymuseum@gmail.com
Web Site: moodymuseum.com
Founded: 1975.
Congressional District: 31
Key Personnel: Chm. (V), Susan Komandosky.
Personnel Profile: Part-Time Volunteers 15.
Governing Authority: Parent Institution: City of Taylor. Subsidiary Institution: Moody Museum Advisory Bd. Tax-exempt.
Institution Type/Description: History Museum: housed in the boyhood home of Texas Governor, Dan Moody; served from 1927-1931.
Collections: Dan Moody's family & career; personal artifacts; period furnishings; photographs; local history.
Hours & Admission Prices: Sun. & Fri. 2-5, Tues.-Thurs. 8-5; other times by appointment. No charge; donations accepted.
Attendance: 400 (estimated)
Membership: Individual $25; Family $50; Patron $75; Century $100; Benefactor $150.

Teague

BURLINGTON-ROCK ISLAND RAILROAD AND HISTORICAL MUSEUM, 208 S. 3rd Ave., Teague, TX 75860-1645. Mailing Address: P.O. Box 604, Teague, TX 75860-0604. Tel.: 254-739-2145.
Web Site: therailroadmuseum.com
Founded: 1969.
Congressional District: 2
Key Personnel: Pres. (V), Benny Walker; Financial Dir., Gaylon Hall.
Personnel Profile: Full-Time Volunteers 20; Part-Time Volunteers 20.
Governing Authority: nonprofit. Parent Institution: City of Teague, TX. Tax-exempt: 501(c)(3).
Institution Type/Description: History Museum: housed in 1906 former Trinity & Brazos Valley Railway Depot, listed on National Register of Historic Places.
Collections: railroad memorabilia including 1925 Baldwin locomotive; 1928 fire engine; grist mill; loom; farming implements; dishes & cooking utensils; clothing; Indian artifacts; printing press; military items; Boy Scout memorabilia; medical; local history; two-room log house built in early 1850s; family history records; model trains; veterans memorabilia.
Research Fields: local & county historical research; genealogical research for the Teague Family Research Center; transportation & railroad history.
Facilities: Cookbooks, pins, postcards, railroad spikes, lanterns & other related items for sale.
Activities: guided tours; study clubs; loan, permanent & temporary exhibitions.
Publications: History of Freestone County, Texas.
Hours & Admission Prices: Sat.-Sun. 1-5; group tours by appointment. Adults $2, children $1; discounts to groups; members & school group tours no charge. Closed New Year's Eve & Day; Easter; Christmas. &
Attendance: 2,000 (estimated)
Membership: Individual $5; Family $10.

Temple

CZECH HERITAGE MUSEUM & GENEALOGY CENTER, 119 W. French Ave., Corner of Third St. and French Ave., Temple, TX 76501. Tel.: 254-899-2935. Fax: 254-774-7447.
E-mail: czechmuseum@yahoo
Web Site: www.czechmuseum.org
Formerly: SPJST Library Archives & Museum
Founded: 1971.
Congressional District: 11
Key Personnel: Dir., Rebecca Vajdak; Chm. (V), Jerry B. Milan.
Personnel Profile: Full-Time Paid 1.
Governing Authority: society. Parent Institution: SPJST. Tax-exempt.
Institution Type/Description: Library & Museum of Czech History, Culture & Genealogy.
Collections: artifacts & articles of early pioneer Czech-Texas settlers; housewares; musical instruments; agricultural implements; medical displays; pioneer kitchen; blacksmith shop; military displays; log replica of the first Czech Home built in Texas.
Research Fields: habits & living modes of Czech-Texas pioneers; history of the SPJST dating back to 1897.
Facilities: more than 22,000-vol. library mostly in the Czech language available for interlibrary loan & on premises.
Activities: guided tours; lectures; film strips; study clubs; inter-museum loan.
Publications: brochure, Vestnik Society Paper.
Hours & Admission Prices: Daily 8-12 & 1-5. Adults $4, seniors $3. Closed holidays. &
Attendance: 3,500 (estimated)

RAILROAD AND HERITAGE MUSEUM INC., (M), 315 W Ave. B, Temple, TX 76501-4226. Tel.: 254-298-5172. Fax: 254-298-5171.
Web Site: www.rrhm.org
Formerly: Railroad Pioneer Museum
Founded: 1973.
Congressional District: 44
Key Personnel: Dir., Judith A. Covington; Chm. (V), Mark Erskine; Vice Chm., Roberta Amos.
Personnel Profile: Full-Time Paid 3; Part-Time Paid 1; Part-Time Volunteers 75.
Governing Authority: board of directors. Tax-exempt.
Institution Type/Description: Transportation & History Museum.
Collections: railroad & pioneer items; documents; photographs; furniture; 1910 restored train station; clothing; tools; archives; steam engine & three cabooses; World War II Troop Sleeper; ATSF 2301 diesel locomotive; 1917 Pullman Sleeper.
Research Fields: railroad; pioneer history; local history.
Facilities: library; archives.
Activities: tours; educational programs.
Publications: newsletter, RRHM.
Hours & Admission Prices: Tues.-Sat. 10-4. Adults $4, senior citizens & military $3, children $2; discounts to AAM members; military & their family and children under 5 no charge.
Attendance: 20,000 (accurate)
Membership: Individual $35; Family & Sustaining $100; Sponsor $250; Patron $500; Benefactor $1,000.

Terrell

NO. 1 BRITISH FLYING TRAINING SCHOOL MUSEUM, INC., 119 Silent Wings Blvd., Terrell, TX 75160. Mailing Address: Box 6, Silent Wings Blvd., Terrell, TX 75160. Tel.: 972-551-1122. Fax: 972-551-1122. Facebook: No 1 BFTS.
E-mail: info@bftsmuseum.org
Web Site: www.bftsmuseum.org
Founded: 1993.
Congressional District: 5
Key Personnel: Chm. (V), Clifford N. Taylor; Pres. (V), L. Don Thurman; Museum Shop Mgr., W. Michael Grout.
Personnel Profile: Part-Time Paid 1; Part-Time Volunteers 14.
Governing Authority: Tax-exempt.
Institution Type/Description: Military History Museum.
Collections: WWII history; British & American military aviation; log books; photographs; training materials; WWII memorabilia & uniforms.
Facilities: 8,450 sq. ft. exhibition space.
Activities: Museum Sponsors: Annual Fly-In & Dinner Dance in March.
Publications: quarterly newsletter.

Hours & Admission Prices: Wed. & Fri.-Sat. 10-4; other times by appointment. No charge; donations accepted. Closed New Year's Day; Christmas. &
Attendance: 4,000 (estimated)
Membership: Senior $25; Individual $35; Family $50; Corporate $100.

TERRELL HERITAGE SOCIETY MUSEUM, 207 N. Frances, Terrell, TX 75160. Tel.: 972-524-6082.
E-mail: terrellheritage@sbcglobal.net
Web Site: terrellheritagemuseum.org
Founded: 1972.
Congressional District: 5
Governing Authority: Parent Institution: Terrell Heritage Society.
Institution Type/Description: Historical Society Museum.
Collections: loacl history & culture; period furnishings; photographs.
Hours & Admission Prices: Sat.-Sun. 1-4. No charge; donations accepted.

Texarkana

ACE OF CLUBS HOUSE, 420 Pine St., Texarkana, TX 75501-5513. Mailing Address: P.O. Box 2343, Texarkana, TX 75504-2343. Tel.: 903-793-4831. Fax: 903-793-7108. Facebook: Ace of Clubs House.
E-mail: aceofclubs@texarkanamuseums.org
Web Site: www.texarkanamuseums.org
Founded: 1985.
Congressional District: 1
Key Personnel: Pres., Ed Black; Cur., Jamie Simmons; Cur. Draughon-Moore Collection, Melissa Nesbitt.
Personnel Profile: Full-Time Paid 1.
Governing Authority: society; nonprofit. Parent Institution: Texarkana Museums System. Tax-exempt: 501(c)(3).
Institution Type/Description: Historic House: 1885 Italianate house in form of rectangle with three octagonal bays, club shape & surrounded by dry moat.
Collections: furnishings; personal artifacts; recreational artifacts, 1885-1950.
Research Fields: late 19th-century architecture & decorative arts; regional social history 1885-1950.
Facilities: 2,500-vol. library of reference materials; media room with video, film & slide facilities. Gift items for sale.
Activities: guided tours; lectures; workshops; organized educational programs; docent program; temporary exhibitions; school loan service; summer programs.
Hours & Admission Prices: Tues.-Fri. tours by appointment, Sat. 10-4 (last tour begins at 3). Tours at 10:15, 11:45, 1:30 & 3. Adults $6, seniors $5, children 5 & up $4; discounts for groups, active duty military & AAA members; children 4 & under and members no charge. Closed national holidays. &
Attendance: 1,459 (accurate)
Membership: Student $20; Individual $50; Family $75.

* **TEXARKANA MUSEUMS SYSTEM, (M),** 219 N. State Line Ave., Texarkana, TX 75501-5606. Mailing Address: P.O. Box 2343, Texarkana, TX 75504-2343. Tel.: 903-793-4831. Fax: 903-793-7108. Facebook: Texarkana Museums System.
E-mail: curator@texarkanamuseums.org
Web Site: www.texarkanamuseums.org
Founded: 1971.
Congressional District: 1
Key Personnel: Cur. Draughon-Moore House, Melissa Nesbitt; Cur., Jamie Simmons.
Personnel Profile: Full-Time Paid 3; Part-Time Paid 2.
Governing Authority: nonprofit organization. Parent Institution: Texarkana Museums System. Branch Museum: Discovery Place; The Ace of Clubs House; The Texarkana Museum of Regional History. Tax-exempt: 501(c)(3).
Institution Type/Description: History & Science Museum.
Collections: historical development of area; Caddo people; regional history; 11,000 photographs.
Research Fields: pertaining to collections.
Facilities: 2,500-vol. library of medical, genealogical & history books; archives; 50-seat theater; classrooms.
Activities: guided tours; lectures; films; docent program or council; permanent, temporary & traveling exhibitions; school loan service; classroom lectures in schools; annual arts & crafts festival; summer program for students. Museum Sponsors: programs in community centers.
Publications: quarterly newsletter, Artifacts.
Hours & Admission Prices: Tues.-Sat. 10-4. Admission varies for each site; discounts to AAA members & active duty military. Closed legal holidays. &
Attendance: 1,442 (accurate)

Membership: Student $20; Individual $50; Family $75; Patron $100; Contributing $250; Sustaining $500; Landmark $5,000; Corporate $10,000.

Texas City

TEXAS CITY MUSEUM, (M), 409 Sixth St. N., Texas City, TX 77590-7854. Tel.: 409-229-1660. Fax: 409-229-1636.
E-mail: lturner@texas-city-tx.org
Web Site: www.texas-city-tx.org
Founded: 1991.
Key Personnel: Cur., Linda Turner.
Personnel Profile: Full-Time Paid 1; Part-Time Paid 1; Part-Time Volunteers 6.
Governing Authority: city; nonprofit organization. Parent Institution: City of Texas City. Tax-exempt: 501(c)(3).
Institution Type/Description: History Museum.
Collections: history & artifacts of Texas City, TX from the 1900s-present.
Activities: children's discovery area; model train layouts of the Galveston County Model Railroad Club.
Hours & Admission Prices: Tues.-Sat. 10-4. Adults $5, seniors $3, students $2; members & children under 6 no charge. Railroad Club: Sat. 10-4. Closed Easter; Thanksgiving; Christmas Eve & Day. &
Attendance: 7,000 (estimated)

The Woodlands

THE WOODLANDS CHILDREN'S MUSEUM, 4775 W. Panther Creek Dr. #280, The Woodlands, TX 77381. Mailing Address: P.O. Box 130864, The Woodlands, TX 77393-0864. Tel.: 281-465-0955.
E-mail: acolton@woodlandschildrensmuseum.org
Web Site: www.woodlandschildrenmuseum.org
Institution Type/Description: Children's Museum.
Collections: hands-on exhibitions.
Activities: summer programs; educational programs; birthday parties; special events.
Hours & Admission Prices: Memorial Day to Labor Day Mon.-Sat. 10-5, Sun. 12-5; Sept.-May Tues.-Sat. 9:30-5.

Tomball

TOMBALL COMMUNITY MUSEUM CENTER, 510 N. Pine St., Tomball, TX 77375-4400. Mailing Address: P.O. Box 457, Tomball, TX 77377-0457. Tel.: 281-444-2449.
Web Site: www.tomballmuseumcenter.org
Founded: 1961.
Congressional District: 4
Key Personnel: Dir., Jean Alexander; Pres., Charles Hall; Treas., Mary McCoy; Sec., Lessie Upchurch.
Personnel Profile: Part-Time Paid 1; Part-Time Volunteers 6.
Governing Authority: society. Owned by Spring Creek County Historical Association. Tax-exempt.
Institution Type/Description: Historic House: 1860 Griffin Memorial House.
Collections: 19th-century decorative art collection; pioneer country doctor's office; 1800s cotton gin; farming implements & domestic artifacts to depict the rural & cultural heritage of Northwest Harris county; 1905 Trinity Evangelical Lutheran Church; c.1800 log house & corn crib; c.1866 German Pioneer farmhouse; 1940 Oil Camp House; 1920 country schoolhouse; Esther Bubley photographs taken in Tomball in 1945.
Activities: guided educational tours for adults, school & scout groups; lectures on topics of historical interest; permanent exhibitions. Museum Sponsors: Candlelight Tour; Heritage Tea.
Publications: fall, winter & spring newsletters.
Hours & Admission Prices: Thurs.-Sat. 10-2, Sun. 2-4; groups by appointment. Donations: $4 per person.
Attendance: 9,000 (estimated)
Membership: Individual $25; Family $35.

Tulia

SWISHER COUNTY LIBRARY ARCHIVES AND MUSEUM ASSOCIATION, 127 S.W. 2nd St., Tulia, TX 79088-2700. Tel.: 806-995-3447.
E-mail: swishercolib@gmail.com
Founded: 1965.
Congressional District: 13
Key Personnel: Pres., Jo Venhaus; Dir., Alan Glasscock; Treas., Joe Cowan.
Personnel Profile: Full-Time Paid 1; Part-Time Volunteers 3.
Governing Authority: nonprofit organization. Tax-exempt.
Institution Type/Description: History Museum & Archives: housed in community building.

Collections: local memorabilia; furniture; clothing; photographs; farm equipment; stagecoach; needle art; Indian artifacts; blacksmith; J.O. Bass; Quanah Parker; woodworking tools; dolls of the world; seashells; Swisher County memorabilia.

Research Fields: quilts; blacksmith; early history of Swisher County.

Facilities: 200-vol. library of publications dealing with Swisher County, cemetery & genealogy records, available for use on premises.

Activities: guided tours; permanent exhibitions; educational.

Hours & Admission Prices: Tues.-Sat. 10-4; other times by appointment. No charge; donations accepted. &

Attendance: 6,228 (accurate)

Membership: Individual $10.

Tyler

CALDWELL ZOO, 2203 W. Martin Luther King, Tyler, TX 75702-2954. Mailing Address: P.O. Box 4785, Tyler, TX 75712-4785. Tel.: 903-593-0121. Fax: 903-595-5083. Facebook: Caldwell Zoo.

E-mail: info@caldwellzoo.org

Web Site: www.caldwellzoo.org

Founded: 1953.

Key Personnel: Exec. Dir., Hayes Caldwell.

Personnel Profile: Full-Time Paid 80; Part-Time Paid 4.

Governing Authority: Parent Institution: Caldwell Foundation. Tax-exempt.

Institution Type/Description: Zoo.

Collections: North American, South American & East African species; 2,000 animals, including 250 species, covering 85 acres.

Hours & Admission Prices: March to Labor Day daily 9-5; after Labor Day-Feb. daily 9-4. Adults $10.50, senior 55 & up $9.25, children 3-12 $7; discounts to groups; members and children 2 & under no charge. Closed New Year's Day; Thanksgiving; Christmas. &

Attendance: 600,000 (estimated)

Membership: Individual $39; Family $89.

DISCOVERY SCIENCE PLACE, (M), 308 N. Broadway Ave., Tyler, TX 75702-5711. Tel.: 903-533-8011. Fax: 903-593-0300.

E-mail: phil@discoveryscienceplace.org

Web Site: www.discoveryscienceplace.org

Founded: 1993.

Congressional District: 1

Key Personnel: C.E.O., Phil Lindsey; C.O.O., Vel Williamson; Mgr. Programs, Emily Keane; Mgr. Guest Svcs., Latresa Jackson.

Personnel Profile: Full-Time Paid 8; Part-Time Paid 5.

Governing Authority: nonprofit. Tax-exempt.

Institution Type/Description: Children's Museum.

Collections: hands-on exhibits.

Hours & Admission Prices: March-Aug. Mon.-Sat. 9-5, Sun. 1-5; Sept.-Feb. Tues.-Sat. 9-5, Sun. 1-5; call for holiday Mon. hours. Admission $6; discounts to groups of 14 or more with reservations; children 2 & under and members no charge. Closed New Year's Day; Easter; Memorial Day; Independence Day; Labor Day; Thanksgiving; Christmas. &

Attendance: 60,000 (estimated)

Membership: Annual $75-$125.

GOODMAN MUSEUM, 624 N. Broadway Ave., Tyler, TX 75702-5344. Tel.: 903-531-1286.

E-mail: gmuseum@tylertexas.com

Founded: 1962.

Congressional District: 4

Key Personnel: Cur., Patricia J. Heaton.

Personnel Profile: Full-Time Paid 1.

Governing Authority: municipal & nonprofit organization. Operated by City of Tyler, Parks & Recreation Dept., P.O. Box 2039, Tyler, TX 75710. Tax-exempt.

Institution Type/Description: Historic House: 1859 4-room cottage remodeled in 1880 into 2-story Texas Colonial by Dr. William J. Goodman, remodeled in 1924 into a Greek Revival Mansion by Dr. Goodman's daughter, Sallie Goodman LeGrand.

Collections: house furnishings of the Goodmans; cradles; china imported from England & France.

Activities: guided tours; lectures; loan, permanent & temporary exhibitions.

Publications: brochure.

Hours & Admission Prices: Tues.-Sat. 10-4. Suggested Donation: $2 per person. Closed New Year's Day; Memorial Day; Independence Day; Thanksgiving; Christmas. &

Attendance: 6,000 (estimated)

HISTORIC AVIATION MEMORIAL MUSEUM, Tyler Pounds Airport, 150 Airport Dr., Tyler, TX 75704-6642. Mailing Address: 150 Airport Dr., Box 2-7, Tyler, TX 75704-6600. Tel.: 903-526-1945. Fax: 903-526-0946.

E-mail: cjverver@aol.com

Web Site: www.tylerhamm.org

Founded: 1985.

Key Personnel: Pres., Carolyn Verver; Museum Shop Mgr., Dave Verver.

Personnel Profile: Part-Time Paid 3; Part-Time Volunteers 50.

Governing Authority: nonprofit organization. Tax-exempt.

Institution Type/Description: Aviation Museum.

Collections: aircraft; aviation artifacts.

Facilities: library. Museum-related items for sale.

Activities: Annual Events: Hangar Dance; Air Show; Host Fly-Ins.

Publications: monthly newsletter.

Hours & Admission Prices: Tues.-Sat. 10-5, Sun. 1-5. Adults $5, teens 13-18 $3; discounts to senior citizens & groups; active military & members no charge. Closed New Year's Day; Easter; Christmas Eve & Day. &

Attendance: 6,000 (accurate)

Membership: Individual $40; Family $75; Life $400.

SMITH COUNTY HISTORICAL SOCIETY, 125 S. College Ave., Tyler, TX 75702-7216. Tel.: 903-592-5993. Fax: 903-526-0924.

E-mail: info@smithcountyhistoricalsociety.org

Web Site: www.smithcountyhistoricalsociety.org

Founded: 1959.

Congressional District: 4

Key Personnel: Pres., Jerry Shamburger; Vice Pres., Randal B. Gilbert; Treas., Sam Kidd.

Personnel Profile: Full-Time Volunteers 1; Part-Time Volunteers 10; Interns 1.

Governing Authority: nonprofit organization. Tax-exempt: 501(c)(3).

Institution Type/Description: History Museum: housed in c.1904 Carnegie Library.

Collections: Smith County history; Native American items; Civil War items; Gilded Age; clothing; 1930s kitchenware; World War I & II uniforms; Camp Fannin; WWII training camp; Camp Ford Confederate Training Camp & Union Prisoner of War Camp; audiovisual & film; decorative arts; textiles; murals of Smith County History; late 19th-century furnishings; games & toys; wireless telegraphic equipment; typewriters; writing implements; period toys; business & institutional documents; newspapers; archives; photographs.

Research Fields: Smith County & east Texas history; Smith County history.

Facilities: 1,500-vol. library; archives; 75-seat auditorium.

Activities: guided tours; docent program; educational programs for school children; monthly society programs; outreach programs to schools & retirement homes; films.

Publications: monthly newsletter; biannual, Chronicles of Smith County, Texas; brochure; Camp Ford lithograph; books, The History of Smith County; Born In Dixie; Uncovering Camp Ford - Archaeological Interpretations of a Confederate Prisoner-of-War Camp in East Texas, a Chronological History of Smith County, Texas; Never in Doubt, a History of Delta Drilling Company.

Hours & Admission Prices: Tues.-Sat. 10-4; other times by appointment. No charge; donations accepted. &

Attendance: 3,600 (estimated)

Membership: Student $7.50; Individual $25; Family $35; Sustaining $50; Patron $100; Life $500; Life Patron $750; Benefactor $1,000.

* **TYLER MUSEUM OF ART, (M),** 1300 S. Mahon Ave., Tyler, TX 75701-3438. Tel.: 903-595-1001. Fax: 903-595-1055.

E-mail: info@tylermuseum.org

Web Site: www.tylermuseum.org

Founded: 1969.

Congressional District: 4

Key Personnel: Dir., Christopher M. Leahy; Facility Mgr., Robert Owen.

Personnel Profile: Full-Time Paid 12; Part-Time Paid 3; Part-Time Volunteers 27; Interns 2.

Governing Authority: nonprofit organization. Tax-exempt: 501(c)(3).

Institution Type/Description: Art Museum.

Collections: Williford Foundation Collection of 29th-century American art; Boeckman Collection of Contemporary Mexican Folk Art; early to contemporary Texas art.

Research Fields: general fine arts; early Texas art; Mexican folk art; 19th century American art.

Facilities: 4,000-vol. library of art surveys & periodicals available upon request on premises; classrooms; cafe. Museum-related items for sale.

Activities: guided tours; lectures; gallery talks; formally organized education programs for children, adults & college students; temporary & traveling exhibitions.

Publications: exhibition catalogs; gallery brochures; preview calendar.
Hours & Admission Prices: Adults $7; museum professionals, children 12 & under and AAM & museum members no charge. Closed national holidays. &
Attendance: 25,000 (estimated)
Membership: Student: $10; Educator $25; Individual $50; Individual Plus & Family $75; Supporter $100; Sustainer $300; Guarantor $500; Patron $1,000; Benefactor $1,500; Corporate $1,500, $2,500 & $5,000; Director's Circle $5,000; Collector's Circle $10,000 & up.

TYLER ROSE MUSEUM, 420 Rose Park Dr., Tyler, TX 75702-6859. Mailing Address: P.O. Box 8224, Tyler, TX 75711. Tel.: 903-597-3130. Fax: 903-597-3031.
E-mail: info@texasrosefestival.com
Web Site: tylermuseum.com
Key Personnel: Exec. Dir. & Cur., Julie Dawson
Institution Type/Description: Rose Museum.
Collections: over 40,000 rose bushes & 500 varieties of roses.
Facilities: Museum-related items for sale.
Hours & Admission Prices: March-Oct. Mon.-Fri. 9-4:30, Sat. 10-4:30, Sun. 1:30-4:30; Nov.-Feb. Mon.-Fri. 9-4:30, Sat. 10-4:30. Adults $3.50, children 3-11 $2.
Attendance: 100,000

Uvalde

AVIATION MUSEUM AT GARNER FIELD, Uvalde Municipal Airport, 201 Airport Blvd., Uvalde, TX 78802. Mailing Address: P.O. Box 453, Uvalde, TX 78802. Tel.: 830-278-2552.
E-mail: avmusgarner@yahoo.com
Governing Authority: nonprofit organization. Tax-exempt: 501(c)(3).
Institution Type/Description: Aviation History Museum.
Collections: local & regional aviation history; WWII aircraft & memorabilia; Liaison-4; a 1945 Piper L-4 Grasshopper; Burt Rutan Vari-Viggan; PT-19 primary trainer; 1959 German sailplane.
Hours & Admission Prices: Tues. & Fri. 9-4. No charge; donations accepted.
Attendance: 500 (estimated)

BRISCOE-GARNER MUSEUM, 333 N. Park St., Uvalde, TX 78801-4658. Tel.: 830-278-5018. Fax: 830-279-0512.
Web Site: www.cah.utexas.edu/museums/garner.php
Formerly: John Nance Garner Museum
Founded: 1952.
Congressional District: 23
Key Personnel: C.E.O., Dr. Don Carleton.
Personnel Profile: Full-Time Paid 2.
Governing Authority: Parent Institution: University of Texas at Austin. Subsidiary Institution: Dolph Briscoe Center for American History. Tax-exempt.
Institution Type/Description: Historic House: 1920 Home of Vice President John N. Garner.
Collections: photographs; political material; furniture, clothing & other items pertaining to the life & career of John Nance Garner.
Research Fields: life, career & political achievements of John Nance Garner.
Activities: permanent & temporary exhibitions.
Hours & Admission Prices: Tues.-Sat. 9-4.
Attendance: 3,500 (accurate)

Van Horn

CLARK HOTEL HISTORICAL MUSEUM, 112 W. Broadway, Van Horn, TX 79855-0231. Mailing Address: Box 231, Van Horn, TX 79855-0231. Tel.: 432-283-8028.
E-mail: wstxdsrt@yahoo.com
Web Site: http://clarkhotelmuseum.com
Formerly: Culberson County Historical Museum
Founded: 1975.
Congressional District: 16
Key Personnel: Pres. (V), Larry Simpson; Dir., Patricia Golden; Sec., Ellen Lipsey.
Personnel Profile: Full-Time Volunteers 1.
Governing Authority: Parent Institution: Culbertson County Historical Society. Tax-exempt: 501(c)(3).
Institution Type/Description: General Museum: housed in 1906 two-story adobe & cement block, Clark Hotel Building.
Collections: Indian artifacts; pioneer articles; costumes; local art; photographs; old furniture; antique bar; rocks & fossils; mineral & mine exhibit.
Research Fields: local history.

Facilities: 200-vol. library of history books available for research on premises; reading room.
Activities: guided tours; lectures; films; docent program or council; permanent exhibitions.
Publications: brochure.
Hours & Admission Prices: No charge; donations accepted. &
Attendance: 1,200 (estimated)
Membership: Individual $10; Family $25; Sustaining $50; Contributing $100; Memorial $500 & up.

Vanderpool

LONE START MOTORCYCLE MUSEUM, 36517 Hwy. 187 N., Vanderpool, TX 78885. Tel.: 830-966-6103.
E-mail: awjohncock@swtexas.net
Web Site: lonestarmotorcyclemuseum.com
Institution Type/Description: Motorcycle Museum.
Collections: motorcycles from around the world dating back to 1910; photographs.
Facilities: cafe.
Hours & Admission Prices: March-Nov. Fri.-Sun. 10-5. Adults $5, seniors 65 & over $4; discounts to groups of 10 or more; children under 15 no charge.

Vernon

RED RIVER VALLEY MUSEUM, (M), 4600 College Dr., Vernon, TX 76384-4052. Mailing Address: P.O. Box 2004, Vernon, TX 76385-2004. Tel.: 940-553-1848. Fax: 940-553-1849.
E-mail: rrvml@yahoo.com
Web Site: www.redrivervalleymuseum.org
Founded: 1963.
Congressional District: 13
Key Personnel: Exec. Dir., Sherry Yoakum; Pres. (V), Judy Heatly; Pres. Bd. Dir., Staley Heatly.
Personnel Profile: Part-Time Paid 4; Part-Time Volunteers 25.
Operating Expenses: 99,000
Operating Income: 99,000
Governing Authority: nonprofit organization. Tax-exempt: 501(c)(3).
Institution Type/Description: History, Science & Fine Arts Museum.
Collections: J. Henry Ray Indian artifacts; early Wilbarger Co. history; Electra Waggone Bigg studio exhibit; original sculptor's models of bronzes by Electra Waggoner Biggs; Wm. A. Bond Big Game collection; Adrian Martinez mural; Texas ranching history; Great Western Cattle Trail; Jack Teagarden; Gideon dollhouses.
Major Exhibits: Ron Gordon Phot, 1/14-3/14; International Juried Art Show, 5/14-6/14; Art of the American West, 7/14-8/14; Youth Art Show, 9/14-10/14; Gideon Doll Houses, 12/14.
Facilities: Commemorative coins, local history book, pamphlets & museum-related items for sale.
Activities: guided tours; loan, permanent & traveling exhibitions; videos; educational trunk shows.
Hours & Admission Prices: Tues.-Sun. 1-5; other times by appointment. General Admission: no charge; donations accepted. Large Tours: $2 per person. Closed some major holidays. &
Attendance: 8,000 (estimated)
Membership: Active $15; Associate $20.

Victoria

CHILDREN'S DISCOVERY MUSEUM OF THE GOLDEN CRESCENT, 204 N. Main St., Victoria, TX 77901. Tel.: 361-485-9140. Fax: 361-485-9141.
E-mail: cdm204@sbcglobal.net
Web Site: www.cdmgoldencrescent.com
Founded: 1999.
Key Personnel: Chm. (V), Betty Jo Elder.
Personnel Profile: Part-Time Paid 1; Part-Time Volunteers 28.
Governing Authority: Tax-exempt.
Institution Type/Description: Children's Museum.
Collections: hands-on exhibitions.
Activities: educational programs; special events; birthday parties; field trips; summer camps.
Hours & Admission Prices: Thurs.-Sat. 10-5. Admission $5; children under 2 & members no charge. &
Attendance: 23,754 (accurate)
Membership: Family of 2 $85; Family of 4 $100; Family of 6 or more $150.

MUSEUM OF THE COASTAL BEND, (M), The Victoria College, 2200 E. Red River, Victoria, TX 77901-4442. Tel.: 361-582-2511. Fax: 361-582-2437.
Web Site: www.museumofthecoastalbend.org
Founded: 2003.
Congressional District: 14
Key Personnel: Dir., Sue Prudhomme; Chm. (V), Amy Mundy; Museum Shop Mgr., Cheryl Beran.
Personnel Profile: Full-Time Paid 2; Part-Time Paid 4; Part-Time Volunteers 30.
Governing Authority: Parent Institution: The Victoria College. Tax-exempt.
Institution Type/Description: Heritage Museum.
Collections: local history & culture; maritime artifacts; ranching industry; archaeology.
Research Fields: archaeology.
Publications: quarterly members' newsletter, Museum News.
Hours & Admission Prices: Tues.-Sat. 10-4. Adults $3.50, senior citizens $2.50, students $2; members & children under 4 no charge. Closed major holidays. &

Attendance: 4,900 (accurate)

NAVE MUSEUM, 306 W. Commercial, Victoria, TX 77901-6602. Mailing Address: P.O. Box 1776, Victoria, TX 77902. Tel.: 361-575-8227. Facebook: Nave Museum.
E-mail: info@navemuseum.com
Web Site: www.navemuseum.com
Founded: 1976.
Congressional District: 14
Key Personnel: Pres., Julie McCan; Exec. Dir., Amy Leissner; Gallery Attendant, Aileen Cruz; Gallery Attendant, Stella Ramirez.
Personnel Profile: Full-Time Paid 1; Part-Time Paid 3; Part-Time Volunteers 30.
Governing Authority: nonprofit organization. Parent Institution: Victoria Regional Museum Assoc., P.O. Box 1776, Victoria, TX 77902. Tax-exempt: 501(c)(3).
Institution Type/Description: Art Museum.
Collections: early 20th century paintings by Royston Nave.
Major Exhibits: Eye to Eye Portrait Show, 1/9/14-2/23/14; Amber Eagle (T), 3/6/14-4/27/14; Art Car Victoria (T), 5/17/14; Irv Tepper (T), 5/17/14-6/29/14; Carlos Donju and Marilyn Jolly, 7/17/14-8/14; Sharon Koprilla, 9/11/14-10/19/14; Dia de los Muertos, 10/30/14-12/14/14.
Facilities: educational facilities.
Activities: participatory, loan & temporary exhibitions.
Publications: Nave News.
Hours & Admission Prices: Tues.-Wed. & Fri.-Sun. 12-4, Thurs. 12-7. Admission: Pay What you Want. Closed New Year's Eve & Day; Easter; Thanksgiving; Christmas Eve & Day. &
Attendance: 7,500 (estimated)
Membership: Friend $60; Supporter $125; Family Supporter $150; Collectors Society $250; Patron $500; Benefactor $1,000.

THE TEXAS ZOO, 110 Memorial Dr., Victoria, TX 77901-6334. Tel.: 361-573-7681. Fax: 361-576-1094.
E-mail: arocha@texaszoo.org
Web Site: www.texaszoo.org
Founded: 1976.
Congressional District: 14
Key Personnel: Exec. Dir., Andrea Blomberg; Dir. Programs, Amanda Rocha; Cur. Animals, Michael Magaw; Office Mgr., Dlorah Maddox.
Personnel Profile: Full-Time Paid 12; Part-Time Paid 4; Part-Time Volunteers 30.
Governing Authority: South Texas Zoological Society. Tax-exempt: 501(c)(3).
Institution Type/Description: Zoo.
Collections: wildlife indigenous to Texas; animals from around the world; natural history displays; dinosaur bones & teeth; modern day birds & eggs; endangered Texans Caboose; animal kingdom exhibits.
Facilities: educational station & classroom; concession stand. Gift items for sale.
Activities: programs; teacher in-service training; hobby workshops; formally organized educational programs; docent program; permanent exhibitions; family playroom; school parties; special events. Museum Sponsors: Annual Amateur Animal Photo Contest; summer camps; Enrichment Day; Wild About Wine; Texas Tunes; Annual Pow Wow; Annual Earth Day; Breakfast with Bunny; Haunted Zoo, ZooBoo; Breakfast With Santa.
Hours & Admission Prices: Daily 9-5. Adults $6, children 3-12 $5, seniors 55 & over $4.50; discounts to reciprocal members; members & children 2 & under no charge. Closed New Year's Day; Thanksgiving; Christmas Eve & Day. &
Attendance: 50,000 (estimated)

Membership: South Texas Zoological Society: Senior $25; Individual $30; Senior Couple $35; Individual Plus Guest $45; Family & Grandparents $55; Family Plus Guest $70. (Add one adult $20; Add a child $10).

Volente

ANDERSON MILL GARDENERS, 13974 FM 2769, Volente, TX 78641. Tel.: 512-258-2613.
E-mail: amgc@volente.org
Web Site: www.volente.org/amgc
Formerly: Anderson Mill and Robinson Museum
Founded: 1955.
Key Personnel: Pres. (V), Roger D. Shull.
Personnel Profile: Part-Time Volunteers 40.
Governing Authority: Parent Institution: Anderson Mill Gardeners, Inc.; Subsidiary Institution: Anderson Mill Memorial, Inc. Tax-exempt.
Institution Type/Description: History Museum.
Collections: local history & culture; period furnishings; personal artifacts; photographs.
Facilities: Anderson Grist Mill & Anderson Cotton Gin replicas.
Activities: corn grinding; water wheel demonstration.
Publications: book, A Legend Collection: Fact & Fantasy.
Hours & Admission Prices: March-Oct. 4th Sun. each month 2-5; groups by appointment. No charge; donations accepted. &
Attendance: 400 (estimated)

Waco

ARMSTRONG BROWNING LIBRARY, Baylor University, Eighth & Speight Sts., Waco, TX 76706. Mailing Address: One Bear Place, #97152, Waco, TX 76798-7152. Tel.: 254-710-3566. Fax: 254-710-3552.
E-mail: jessica_mcadoo@baylor.edu
Web Site: www.browninglibrary.org
Founded: 1918.
Congressional District: 11
Key Personnel: Dir. & Cur. Manuscripts, Rita Patteson; Public Rels. & Facilities Supvr., Jessica McAdoo; Cur. Books & Printed Materials, Cynthia Burgess.
Personnel Profile: Full-Time Paid 6; Part-Time Paid 4.
Governing Authority: university. Parent Institution: Baylor University, Waco, TX. Tax-exempt.
Institution Type/Description: Library of Browningiana.
Collections: items pertaining to Robert & Elizabeth Barrett Browning & the Victorian Age, including original letters, manuscripts, books, periodical articles, pamphlets & other publications; personal artifacts.
Research Fields: Robert & Elizabeth Barrett Browning; the Victorian Age.
Facilities: 26,000-vol. library available for use on premises only; reading rooms. Publications of the library, paperweights, slides, postcards & souvenir items for sale.
Activities: guided tours; lectures; films; permanent & temporary exhibitions. Library Sponsors: annual Pied Piper Tours for 5th grade students.
Publications: annual periodical, Studies in Browning & His Circle; semiannual newsletter, Armstrong Browning Library Newsletter; periodical, Baylor Browning Interests; books, Browning's Old Schoolfellow - The Artistic Relationship Between Two Robert Brownings; The Browning Collections: A Reconstruction with other memorabilia; The Letters of Elizabeth Barrett Browning to Mary Russell Mitford, 1836-1854; Armstrong Browning Library; More Than Friend: The Letters of Robert Browning to Katharine de Kay Bronson; Robert Browning's Flowers; Elizabeth Barrett Browning at The Mercy of Her Publishers; descriptive catalogs, Meeting the Brownings; Browning Music; Robert Browning-A Telescopic View, 1812-1889; Elizabeth Barrett Browning: Life in a New Rhythm.
Hours & Admission Prices: Mon.-Fri. 9-5, Sat. 10-2. No charge; donations accepted. Closed University holidays. &
Attendance: 22,446 (accurate)
Membership: Individual $50; Life $1,000.

THE ART CENTER OF WACO, 1300 College Dr., Waco, TX 76708-1401. Tel.: 254-752-4371. Fax: 254-752-3506.
E-mail: info@artcenterwaco.org
Web Site: www.artcenterwaco.org
Founded: 1972.
Congressional District: 11
Key Personnel: Exec. Dir., Mark Arnold; Business Office Mgr., Jennifer Warren.
Personnel Profile: Full-Time Paid 2; Full-Time Volunteers 2; Part-Time Paid 4; Part-Time Volunteers 40; Interns 1.
Governing Authority: nonprofit corporation. Tax-exempt: 501(c)(3) & 170(b)(1)(A).

Institution Type/Description: Art Center.
Collections: Historic House: 1924 William Cameron summer house.
Facilities: library; studio classrooms; courtyard. Museum-related items for sale.
Activities: guided tours; lectures; films; education programs; traveling exhibitions; weekend programs for children.
Publications: catalogs; newsletter.
Hours & Admission Prices: Tues.-Sat. 10-5, Sun. 1-5. Suggested Donation: adults $2, educator & children 5-12 $1. Closed New Year's Day; Martin Luther King Jr. Day; Memorial Day; Independence Day; Labor Day; Thanksgiving; Christmas. &
Attendance: 120,000 (estimated)
Membership: Student/Educator $20; Friend/Family $35; Master $100; Patron $250; Benefactor $500; Corporate $1,000.

CAMERON PARK ZOO, 1701 N. 4th St., Waco, TX 76707-2463. Tel.: 254-750-8400. Fax: 254-754-8430.
E-mail: jimf@ci.waco.tx.us
Web Site: www.cameronparkzoo.com
Founded: 1955.
Congressional District: 11
Key Personnel: Zoo Dir., Jim Fleshman; Pres., Ben Lacy; Gen. Cur., Johnny Binder; Cur. Exhibits & Programs, Terri Cox; Membership Svcs. & Museum Shop Mgr., Michael Davis; Membership Mgr., Kristi Hemrick; Cur. Education, Connie Kassner; Office Mgr., Laney Van Antwerp; Mktg. Mgr., Duane McGregor.
Personnel Profile: Full-Time Paid 65; Part-Time Paid 13; Part-Time Volunteers 35; Interns 4.
Governing Authority: municipal. Parent Institution: City of Waco. Subsidiary Institution: Cameron Park Zoological Society. Tax-exempt: 501(c)(3).
Institution Type/Description: General Museum.
Collections: mammals; birds; reptiles; fish; amphibians; plants; zoological species.
Research Fields: animal breeding; conservation projects in Caribbean, Africa & Texas.
Facilities: 200-vol. library of magazines, reference books & books on management of zoo animals available for research on premises only. Zoo-related items for sale.
Activities: guided tours; lectures; films; formally organized educational programs; docent program or council.
Publications: quarterly magazine, Zoo News; bimonthly newsletter.
Hours & Admission Prices: Mon.-Sat. 9-5, Sun. 11-5. Adults $9, senior citizens & active military $8, children 3-12 $6; discount to groups; members, children under 3 & AZA members no charge. Closed New Year's Day; Thanksgiving; Christmas. &
Attendance: 249,897 (accurate)
Membership: Individual $30; Family $75; Patron $150; Business $175 & up.

DR PEPPER MUSEUM AND FREE ENTERPRISE INSTITUTE, (M), 300 S. 5th St., Waco, TX 76701-2115. Tel.: 254-757-1025. Fax: 254-757-2221.
E-mail: dp-info@drpeppermuseum.com
Web Site: www.drpeppermuseum.com
Founded: 1989.
Key Personnel: Dir., Jack N. McKinney; Assoc. Dir., Joy Summar-Smith; Dir. Visitor Svcs. & Communications, Jennie Sheppard; Exhibits Mgr., Gabe Schooley.
Governing Authority: private; nonprofit. Subsidiary Institution: DP Museum Enterprise, Inc., a for-profit company. Tax-exempt: 501(c)(3).
Institution Type/Description: History Museum: housed in a 3-story, 18,000 sq. ft. 1906 building, which was the home of Dr Pepper. Structure is of architectural significance to Waco, reflecting the popularity of Richardsonian Romanesque architecture in Texas.
Collections: soft drink industry; machinery; advertising; from fountain to factory; Dr Pepper artifacts; historical & current soft drinks brands; operational historic soda fountain.
Research Fields: developing a new definition of soft drinks to meet current market developments; Waco tornado research; Big Red; international soft drink advertising & brands.
Facilities: 50-vol. bound magazine library for use on premises; printed advertising art; photos; personal & industry files; 6,000 sq. ft. exhibit space; classroom. Museum-related items for sale.
Activities: films; formal education programs for children; guided tours; lectures; temporary exhibitions; training programs for museum workers & junior high school & high school students.
Publications: quarterly newsletter, Bottlecaps; brochure, Soda Pop.
Hours & Admission Prices: Mon.-Sat. 10-4:15, Sun. 12-4:15; Adults $8, senior citizens $6, students & children $5; members no charge. Closed New Year's Day; Easter; Thanksgiving; Christmas Day. &
Attendance: 58,089 (accurate)

THE EARLE-HARRISON HOUSE & PAPE GARDENS, 1901 N. 5th St., Waco, TX 76708-3603. Tel.: 254-753-2032.
E-mail: earleharrisonpapegardens@gmail.com
Web Site: www.earleharrison.com
Formerly: The Earle-Harrison House and Gardens on 5th Street
Founded: 1956.
Congressional District: 11
Key Personnel: Property Mgr., Kathy Riggs.
Personnel Profile: Full-Time Paid 3; Part-Time Paid 5.
Governing Authority: private; nonprofit. Parent Institution: The G.H. Pape Foundation, Waco, TX. Tax-exempt: 501(c)(3).
Institution Type/Description: Historic House: 1858-1859 Greek Revival antebellum house.
Collections: Earle & Harrison family heirlooms; pieces from the Victorian period c.1850-1900.
Activities: docent program; guided tours. Annual Events: Gardening on 5th Street; Christmas Season at the Earle-Harrison House.
Publications: The Columns.
Hours & Admission Prices: By appointment. Admission $5; discounts to AAM members; children under 12 & members no charge. Closed New Year's Eve & Day; Good Friday; Easter weekend; Memorial Day; Independence Day; Labor Day; Thanksgiving; Christmas Eve & Day. &
Attendance: 19,100 (accurate)
Membership: Garden Sprout $5; Family Sprout, Student & Senior Citizens $10; Individual $30; Friends of the Earles and Harrisons $50; Dr. Earle $100; The Column Society $250.

* **HISTORIC WACO FOUNDATION,** 810 S. Fourth St., Waco, TX 76706-1036. Tel.: 254-753-5166. Fax: 254-714-1242.
E-mail: hwf@hot.rr.com
Web Site: www.historicwaco.org
Founded: 1967.
Congressional District: 11
Key Personnel: Exec. Dir., Donald B. Davis; Administrative Asst., Deborah Vardiman.
Personnel Profile: Full-Time Paid 2; Part-Time Paid 2; Part-Time Volunteers 65.
Governing Authority: society. Branch Museums: 1858 Earle-Napier-Kinnard House, 814 S. 4th St.; 1872, East Terrace, 100 Mill St.; 1868 Fort House, 503 S. Fourth St.; 1866 McCulloch House, 407 Columbus Ave. Tax-exempt.
Institution Type/Description: Historic Foundation & Historic Houses.
Collections: period furnishings & clothing.
Research Fields: Old Waco.
Facilities: 60-seat lecture hall.
Activities: guided tours; lectures; docent program; permanent exhibitions; two major fundraising events.
Publications: semiannual, Waco Heritage & History; bimonthly newsletter.
Hours & Admission Prices: Earle-Napier-Kinnard: Sat.-Sun. 2-5. East Terrace Tues.-Fri. 11-3, Sat.-Sun. 2-5. McCulloch House & Fort House Sat.-Sun. 2-5. Adults $3, seniors $2.50, students $2, children over 5 $1; discounts to military, AAM members & visiting more than one house; children under 5 & members no charge. &
Attendance: 20,292 (accurate)
Membership: Student $15; General $35; Contributor $50; Sponsor $100; Sustainer $250; Patron $500; Benefactor $1,000.

MARTIN MUSEUM - BAYLOR UNIVERSITY, (M), Hooper-Schaefer Fine Arts Center, Waco, TX 76798. Tel.: 254-710-6390.
E-mail: martin_museum@baylor.edu
Institution Type/Description: Art Gallery.
Collections: paintings; sculpture; photographs.
Hours & Admission Prices: Tues.-Fri. 10-5, Sat. 12-5; call to confirm. No charge.

MASONIC GRAND LODGE LIBRARY AND MUSEUM OF TEXAS, 715 Columbus, Waco, TX 76701-1349. Mailing Address: P.O. Box 446, Waco, TX 76703-0446. Tel.: 254-753-7395. Fax: 254-753-2944.
E-mail: gs@grandsecretaryoftx.org
Web Site: grandlodgeoftexas.org
Founded: 1936.
Congressional District: 11
Key Personnel: Cur. & Librarian, Barbara Mechell; Grandlibrarian, Tommy D. Guest.
Governing Authority: bd. dirs. Tax-exempt: 501(c)(3).
Institution Type/Description: History Museum: Educational Masonic Library.
Collections: history & biography of Masons & Texas; bibliographical & general reference tools; Masonic artifacts including jewels, symbolic tools, aprons, gavels, tilers, swords, Bibles, certificates.

Research Fields: Texas; Masonic.
Facilities: 33,491-vol. library of Masonic & Texas materials available for research on premises; reading room.
Activities: permanent exhibits.
Hours & Admission Prices: Jan. 3-Dec. 22 Mon.-Fri. 8:30-4. No charge; donations accepted. Closed national holidays. &
Attendance: 10,000 (estimated)
Membership: Supporting $25; Sam Houston Hall of Fame $1,000; Patron $5,000; Gold Seal Patron $10,000; Platinum Patron $25,000.

MAYBORN MUSEUM COMPLEX, 1300 S. University Parks Dr., Baylor University, Waco, TX 76706-1221. Mailing Address: One Bear Place #97154, Baylor Univ., Waco, TX 76798-7154. Tel.: 254-710-1110. Fax: 254-710-1173.
Web Site: www.baylor.edu/mayborn/
Formerly: Strecker Museum Complex
Founded: 1893.
Congressional District: 11
Key Personnel: Dir., Dr. Ellie Caston; Asst. Dir. Visitor Experience, Lesa Bush; Mgr. Museum Operations, Patricia Pack; Asst. Dir. Promotions & Events, Mark Smith; Asst. Dir. Facilities & Collections, Tom Haddad; Collections Mgr., Anita Benedict; Dir. Community Rels., Sarah Levine; Assoc. Prof., Interim Chm. Museum Studies & Dir. Academic Programs, Dr. Ken Hafertepe.
Personnel Profile: Full-Time Paid 28; Part-Time Paid 11; Part-Time Volunteers 200.
Governing Authority: Parent Institution: Baylor University. Tax-exempt: 501(c)(3).
Institution Type/Description: Natural Science & History Museum.
Collections: natural science & cultural history; science; zoology; herpetology; archaeology; anthropology; entomology; paleontology; botany; geology; mineralogy; interactive themed rooms; historic village 1880-1910.
Research Fields: archaeology; herpetology; natural history; zoology; 1880-1910 rural central Texas towns.
Facilities: 143,000 sq. ft. exhibit space; nature pathway.
Activities: guided tours; lectures; formally organized education programs for children; museum studies program leading to M.A. degree; volunteer program; inter-museum loan & permanent exhibitions.
Publications: monthly e-newsletter; Mammoths in Waco: Exploring the Mystery.
Hours & Admission Prices: Mon.-Wed. & Fri.-Sat. 10-5, Thurs. 10-8, Sun. 1-5. Adults $6, seniors 65 & over $5, children 18 months to 12 yrs. $4; discounts to AAM members. Closed New Year's Day; Easter weekend; Thanksgiving; Christmas Day. &
Attendance: 94,944 (accurate)
Membership: Individual $30; Family $75; Contributing $100; Sustaining $250; Patron $500; Benefactor $1,000.

TEXAS RANGER HALL OF FAME AND MUSEUM, 100 Texas Ranger Trail, Waco, TX 76706-1209. Mailing Address: P.O. Box 2570, Waco, TX 76702-2570. Tel.: 254-750-8631. Fax: 254-750-8629.
E-mail: info@texasranger.org
Web Site: www.texasranger.org
Founded: 1964.
Congressional District: 11
Key Personnel: Dir. & CEO, Byron A. Johnson; Museum Shop Mgr., Lisa Daniel.
Personnel Profile: Full-Time Paid 13; Part-Time Paid 4; Part-Time Volunteers 2.
Governing Authority: municipal. Parent Institution: City of Waco, TX. Tax-exempt.
Institution Type/Description: Western History Museum: specializing in Texas Ranger History.
Collections: Texas Ranger artifacts including arms, saddles, badges, Colt & Winchester firearms; Western cattle range; photographs; historical papers; oral history; Western history, paintings, sculpture & costumes.
Research Fields: Texas frontier history items; Texas Colonial items; Texas law enforcement; American west; Mexico.
Facilities: 3,500-vol. library; archives; multimedia show area; banquet facility; education facility. Museum-related items for sale.
Activities: guided tours; lectures; special events.
Publications: online journal, Dispatch; monthly newsletter.
Hours & Admission Prices: Daily 9-5; guided tours by appointment. Adults $7, seniors & military $6, children 6-12 $3; discounts to AAM & ICOM members, groups, & law enforcement personnel; children 5 & under no charge. Closed New Year's Day; Thanksgiving; Christmas. &
Attendance: 80,000 (accurate)

Membership: Silver Star $55; Gold Star $250 ($55 annual renewal); Corporate Club $500, $1,500, $2,500, & $5,000; Captain's Circle $5,000; 3rd Century Club $10,000.

TEXAS SPORTS HALL OF FAME, 1108 S. University Parks Dr., Waco, TX 76706-1223. Tel.: 254-756-1633; 800-567-9561. Fax: 254-756-2384.
E-mail: phyllis.trice@tshof.org
Web Site: www.tshof.org
Founded: 1949.
Congressional District: 11
Key Personnel: Exec. Dir., Steve Fallon; Pres. (V), Carroll Dawson; Cur., Jay Black; Operations Mgr., Sales & Mktg. Coord., Phyllis Trice; Collections Mgr., Paige Davis.
Personnel Profile: Full-Time Paid 4; Part-Time Paid 5; Part-Time Volunteers 1.
Governing Authority: private; nonprofit.
Institution Type/Description: Sports Museum.
Collections: personal artifacts & memorabilia related to Texas high school, college & professional sports figures.
Facilities: library of books, magazines, programs, media guides & photographs; 11,000 sq. ft. exhibit space; 50-seat theater. Sports-related items for sale.
Activities: films; guided tours; loan exhibitions; broadcast programs. Annual Events: induction ceremony; horse race gala; Bob Lilly Celebrity Golf Tournament.
Hours & Admission Prices: Mon.-Sat. 9-5, Sun. 12-5. Adults $7, seniors 60 & over $6, students & children over 5 $3; discounts to groups of 10 or more; active military & children under 6 no charge. Closed New Year's Day; Easter; Thanksgiving; Christmas. &
Attendance: 30,000 (accurate)
Membership: Student $20; Individuals $25 & $50; Family $50; Patron $100; Supporting $250; Corporate $250 & up; Sustaining $500.

Washington

BARRINGTON LIVING HISTORY FARM, Washington-on-the-Brazos State Historic Site, 23400 Park Rd. 12, Washington, TX 77880. Mailing Address: Washington-on-the-Brazos State Park Assn., P.O. Box 305, Washington, TX 77880-0305. Tel.: 936-878-2213 & 2214. Fax: 936-878-2810.
E-mail: info@birthplaceoftexas.com
Web Site: www.birthplaceoftexas.com
Founded: 1936.
Congressional District: 10
Key Personnel: Complex Mgr., Bill Irwin; Site Mgr., Cathy Nolte.
Governing Authority: state. Parent Institution: Texas Parks & Wildlife Dept., 4200 Smith School Rd., Austin, TX 78744. Tax-exempt.
Institution Type/Description: Historic House: 1844-57 Barrington, home of Anson Jones, fourth & last president of the Republic of Texas; house was moved to Washington State Park, site of the signing of the Texas Declaration of Independence.
Collections: Anson Jones family personal artifacts; home furnishings from 1840-1860.
Research Fields: Republic of Texas, 1836-1846; Texas agriculture 1840-1860.
Facilities: reconstructed Anson Jones Farm, Barrington.
Activities: site tours; living history demonstrations & special events.
Hours & Admission Prices: Farm: daily 10-4:30. Park: daily 8am to sundown. Family $15; adults $5, children 7 & up $3. &
Attendance: 265,000 (estimated)

* **STAR OF THE REPUBLIC MUSEUM, (M),** 23200 Park Rd. 12, Washington, TX 77880. Mailing Address: P.O. Box 317, Washington, TX 77880-0317. Tel.: 936-878-2461. Fax: 936-878-2462. Facebook: Star of the Republic.
E-mail: star@blinn.edu
Web Site: www.starmuseum.org
Founded: 1970.
Congressional District: 14
Key Personnel: Dir., Houston McGaugh; Cur. Collections, Dr. Shawn Carlson; Cur. Education, Anne McGaugh; Coord. Visitor Svcs., Elaine Platt; Office Mgr., Effie Wellmann.
Personnel Profile: Full-Time Paid 6; Part-Time Paid 6; Part-Time Volunteers 10; Interns 2.
Governing Authority: college; Blinn College, Brenham, TX. Tax-exempt.
Institution Type/Description: History Museum: located on the site of the signing of the Texas Declaration of Independence, twice the capital of the Republic of Texas.
Collections: cultural, social, economic & political history of pre-1850 Texas.
Research Fields: The Republic of Texas 1836-1846.

Facilities: library; theater.
Activities: temporary exhibitions; group presentations; school educational programs; audiovisual interpretive programs; Texas Independence Day celebration.
Publications: quarterly newsletter; exhibition catalogs.
Hours & Admission Prices: Daily 10-5. Family $15, adults $5, students $3; discounts to AAM & Texas Assoc. of Museums members. Closed New Year's Eve & Day; Thanksgiving; Christmas Eve, Day & week. &
Attendance: 35,000 (accurate)

Waxahachie

ELLIS COUNTY MUSEUM, INC., 201 S. College, Waxahachie, TX 75165-3711. Mailing Address: P.O. Box 706, Waxahachie, TX 75168-0706. Tel.: 972-937-0681.
E-mail: ecmuseum@sbcglobal.net
Web Site: elliscountymuseum.org
Founded: 1967.
Congressional District: 6
Key Personnel: Dir. & Cur., Shannon Simpson; Pres. (V), Glinda Felty.
Personnel Profile: Full-Time Paid 1; Part-Time Paid 1; Part-Time Volunteers 200.
Governing Authority: nonprofit. Tax-exempt: 501(c)(3).
Institution Type/Description: History Museum: housed in 1889 Masonic Lodge Hall Bldg.
Collections: relics; photographs; manuscripts; period furniture & furnishings.
Research Fields: local & county history.
Facilities: Arts & crafts, Gingerbread Trail items & other museum-related items for sale.
Activities: Annual Event: Gingerbread Trail Home Tour.
Publications: quarterly newsletter; book: Ellis County, A Photohistory, 1993; booklet: Ellis County Courthouse, 1995.
Hours & Admission Prices: Mon.-Sat. 10-5, Sun.1-5. No charge; donations accepted.
Attendance: 9,000 (accurate)
Membership: Individual $15; Family $25; Patron $100; Sponsor $200; Advocate $500; Benefactor $1,000; Organization & Business $50-$100.

Weatherford

CLARK GARDENS BOTANICAL PARK, 567 Maddux Rd., Weatherford, TX 76088. Mailing Address: P.O. Box 276, Mineral Wells, TX 76068. Tel.: 940-682-4856. Fax: 940-682-4078.
E-mail: info@clarkgardens.org
Web Site: www.clarkgardens.org
Governing Authority: nonprofit organization. Tax-exempt: 501(c)(3).
Institution Type/Description: Botanical Gardens.
Collections: local history; plants; flowers; trees; butterfly garden; photographs.
Facilities: 35-acre park; rental facilities. Gift items for sale.
Activities: special events; rental facilities; classes; group tours; train rides. Annual Events: Spring Festival in March; Run From the Ducks in September; Fall Festival in October; BOOtanical in October; Holiday Festival in December.
Hours & Admission Prices: Mon.-Sat. 7:30-6, Sun. 10-5. Adults $7, seniors 65 & over and children 5-12 $5; discounts to groups of 25 or more; children 4 & under no charge. &
Membership: Individual $50; Family $150.

DOSS HERITAGE AND CULTURE CENTER, 1400 Texas Dr., Weatherford, TX 76086. Mailing Address: P.O. Box 215, Weatherford, TX 76086-0215. Tel.: 817-599-6168. Fax: 817-599-6193.
E-mail: info@dosscenter.org
Web Site: www.dosscenter.org
Founded: 2006.
Key Personnel: Exec. Dir., Heather Castagna.
Governing Authority: Tax-exempt.
Institution Type/Description: History Museum.
Collections: local history & culture; period furnishings; personal artifacts; photographs; Native American artifacts; tools.
Activities: hands-on exhibits.
Publications: quarterly newsletter, The Windmill.
Hours & Admission Prices: Tues.-Sat. 10-5, Sun. 1-5. Adults $5, senior citizens & students $3; members & children under 6 no charge. &
Membership: Subscriber $25; Patron $50; Family $100; Director's Circle $250; Benefactor $500.

MUSEUM OF THE AMERICAS, 216 Fort Worth Hwy., Weatherford, TX 76086. Tel.: 817-341-8668.
E-mail: museumam@sbcglobal.net
Web Site: www.museumoftheamericas.com
Founded: 2001.
Congressional District: 12
Institution Type/Description: History Museum; Ethnographic Museum.
Collections: Native American history, culture & artifacts; period furnishings; clothing; paintings; photographs; carvings; costumes; folk & indigenous craft of Mexico, Central & South America.
Facilities: Museum-related items for sale.
Hours & Admission Prices: Feb.-July & Sept.-Dec. 23 Tues.-Fri. 10-5, Sat. 11-4; tours by appointment. Suggested Donation: $2 per person.

THE NATIONAL VIETNAM WAR MUSEUM, 12685 Mineral Wells Hwy., Weatherford, TX 76088. Mailing Address: P.O. Box 146, Mineral Wells, TX 76068. Tel.: 940-325-4003.
E-mail: info@nationalvnwarmuseum.org
Web Site: nationalvnwarmuseum.org
Founded: 1997.
Key Personnel: Dir., Ed Luttenberger; Devel., Charles Bogle; Treas., Jim Messinger.
Personnel Profile: Part-Time Volunteers 85.
Governing Authority: private; nonprofit organization. Tax-exempt.
Institution Type/Description: Military Museum.
Collections: Vietnam era artifacts; personal artifacts; photographs; military weapons & uniforms.
Activities: guided tours; loan exhibitions.
Publications: newsletter.
Hours & Admission Prices: Memorial Gardens: daily dawn to dusk. Visitor Center: call for hours. No charge; donations accepted. Closed New Year's Day; Thanksgiving; Christmas. &
Attendance: 20,000 (estimated)

Wellington

COLLINGSWORTH COUNTY MUSEUM, 824 East Ave., Wellington, TX 79095. Mailing Address: P.O. Box 495, Wellington, TX 79095-0495. Tel.: 806-447-5327.
E-mail: collingsworthmuseum@windstream.net
Web Site: www.collingsworthcountymuseum.org
Founded: 1971.
Key Personnel: Chm. (V), Bettye Baumgardner; Vice Pres. & Dir., W. Doris Stallings.
Personnel Profile: Part-Time Paid 2; Part-Time Volunteers 30.
Governing Authority: Tax-exempt.
Institution Type/Description: Historic Buildings.
Collections: Tyler House: agricultural & ranching artifacts. Pruden Building: works by local artists. Sullivan Buildings: historical artifacts; schoolroom; Templeton Law Office; soda fountain; Mothuskek square piano; military artifacts; quilts.
Hours & Admission Prices: Mon.-Fri. 9-5; other times by appointment. No charge; donations accepted. Closed holidays. &
Attendance: 549 (accurate)
Membership: Individual $10; Angel $100; Patron $1,000.

Weslaco

FRONTERA AUDUBON SOCIETY, 1101 S. Texas Blvd., Weslaco, TX 78596-7001. Tel.: 956-968-3275. Fax: 956-968-1388.
E-mail: fronteraaudubon@gmail.com
Web Site: www.fronteraaudubon.org
Founded: 1974.
Congressional District: 15
Key Personnel: Dir., Sarah Williams; Chm. (V), Jim Chapman; Museum Shop Mgr., Christine Warren.
Personnel Profile: Full-Time Paid 1; Part-Time Paid 3; Part-Time Volunteers 5.
Governing Authority: Tax-exempt
Institution Type/Description: Wildlife Refuge.
Collections: agricultural developments; human influences; organic orchard; historical homestead.
Hours & Admission Prices: Tues.-Sat. 8-4, Sun. 12-4. Adults $5, seniors $4; members & children no charge. &
Attendance: 2,351 (accurate)
Membership: Individual $20; Family $25; Group $35; Black-bellied

Whistling-Duck $50; Kiskadee $70; Buff-bellied Hummingbird $100; Chachalaca $250; Altamira Oriole $500; Life $1,000.

LOWER RIO GRANDE VALLEY NATURE CENTER, 301 S. Border Ave., Weslaco, TX 78596-5815. Mailing Address: P.O. Box 8125, Weslaco, TX 78599-8125. Tel.: 956-969-2475. Fax: 956-969-9915.
E-mail: info@valleynaturecenter.org
Web Site: www.valleynaturecenter.org
Founded: 1984.
Key Personnel: Exec. Dir., Lydia Cavazos Guerra; Pres., Mark Gibbs; Office Mgr., Cindy Flores; Dir. Education, Allie Zamora; Dir. Education Asst., Susan Hoehne; Park Technician, Marissa Latigo; Park Tech Asst., Paul Garza.
Personnel Profile: Full-Time Paid 7; Part-Time Paid 2.
Governing Authority: Subsidiary Institution: Rio Grande Valley Bird Observatory. Tax-exempt.
Institution Type/Description: Nature Center.
Collections: Nature Center: species of Rio Grande Valley of Texas including flora & fauna; shells of the Gulf; birds' nests; butterflies; reptiles; insects; mammals; native woods; hand-carved birds & small collection of wildlife photography and paintings. Exhibit Hall & Theater/meeting room: connected to 5 1/2 acre nature park with signs identifying trees & native plants; migrating & resident birds; native lizards & tortoises; butterfly garden.
Research Fields: cultivation of native plant species.
Facilities: library; 6 acres nature preserve; visitors center; meeting room. Museum-related items for sale.
Activities: nature trails; learning activities. Annual Events: Spring Fest in February; Dragonfly Days in May.
Publications: monthly newsletter; Lower Rio Grande Valley bird checklist; wildlife brochure series.
Hours & Admission Prices: Tues.-Fri. 9-5, Sat. 8-5, Sun. 1-5. Adults $3, seniors 55 & over $2.50, children 12 & under $1; members no charge.
Attendance: 10,000 (accurate)
Membership: Student $15; Individual $25; Family $35.

WESLACO MUSEUM, 500 S. Texas, Weslaco, TX 78596-6202. Mailing Address: P.O. Box 8062, Weslaco, TX 78599-8062. Tel.: 956-968-9142. Fax: 956-447-0955. Facebook: Weslaco Museum.
E-mail: info@weslacomuseum.org
Web Site: weslacomuseum.org
Formerly: Weslaco Bicultural Museum
Founded: 1971.
Congressional District: 15
Key Personnel: Dir., Yuridia Sanchez; Pres., Jesse Colin; Programs, Maria Arrieta.
Personnel Profile: Full-Time Paid 1; Full-Time Volunteers 60; Part-Time Paid 1; Part-Time Volunteers 15; Interns 2.
Governing Authority: private; nonprofit organization. Parent Institution: City of Weslaco. Tax-exempt: 501(c)(3).
Institution Type/Description: Local History & Cultural Art Museum.
Collections: early 20th century local & border history artifacts & archives; photographs.
Research Fields: local history.
Facilities: 150-vol. library; 2,500 sq. ft. exhibit space.
Activities: guided tours; docent program; lectures; formal education programs for University of Texas-Pan American students; loan, traveling & temporary exhibitions; study clubs; cultural & art programs; historical programs.
Publications: quarterly newsletter; Images of America Series (through Arcadia Publishing).
Hours & Admission Prices: Tues.-Sat. 10-4. Adults $4, seniors & students $3, children 5-17 $2; members & children under 5 no charge. Closed New Year's Day; Good Friday; Memorial Day; Independence Day; Labor Day; Thanksgiving; Christmas. &
Attendance: 5,000 (accurate)
Membership: Student $10; Individual $20; Family $50; Sponsor $250; Patron $500; Champion $1,000.

West Columbia

COLUMBIA HISTORICAL MUSEUM, 247 E. Brazos Ave., West Columbia, TX 77486. Mailing Address: P.O. Box 867, West Columbia, TX 77486-0867. Tel.: 979-345-6125.
Founded: 1990.
Governing Authority: Tax-exempt.
Institution Type/Description: History Museum.
Collections: local history & culture; period furnishings; personal artifacts; photographs.
Facilities: Museum-related items for sale.

Hours & Admission Prices: Thurs.-Sat. 10-2; other times by appointment. No charge; donations accepted. Groups of 10 or more $3. &

VARNER-HOGG PLANTATION STATE HISTORIC SITE, 1702 N. 13th St., West Columbia, TX 77486. Mailing Address: P.O. Box 696, West Columbia, TX 77486-0696. Tel.: 979-345-4656. Fax: 979-345-4412.
E-mail: susan.miller@thc.state.tx.us
Web Site: www.visitvhp.com
Founded: 1958.
Congressional District: 22
Key Personnel: Exec. Dir. THC, Mark Wolfe; Site Mgr., Sue Miller; Pres. (V), Janet Dahse.
Personnel Profile: Full-Time Paid 7; Part-Time Paid 1; Part-Time Volunteers 25.
Governing Authority: state. Parent Institution: Texas Historical Commission, P.O. Box 12276., Austin, TX 78711-2276. Tax-exempt.
Institution Type/Description: State Historic Site.
Collections: decorative arts; agriculture; archaeology; mementos relating to Gov. James Stephen Hogg & early events in Texas history; furniture from 1850's-1860's; Texian campaignware & other china; cemetery; sugar mill equipment; barn; slave quarters. Historic Building: c.1835 mansion.
Research Fields: history; furniture; decorative arts; Texas-made furniture.
Facilities: picnic area. Museum-related items for sale.
Activities: guided house tours; permanent exhibitions; grounds tour; folkway programs.
Publications: site leaflet, Interpretive Guide.
Hours & Admission Prices: Tues.-Sun. 8-5. Tours: Tues.-Sun. 10-3. Site: $1 entrance fee. Mansion: adults $6, students & children over 5 $2. Plantation House Tours: adults $6, children & students $4. Closed New Year's Eve & Day; Easter; Thanksgiving; Christmas Eve & Day.
Attendance: 8,000 (accurate)

Wharton

WHARTON COUNTY HISTORICAL MUSEUM, 3615 N. Richmond Rd., Wharton, TX 77488-2022. Mailing Address: P.O. Box 349, Wharton, TX 77488-0349. Tel.: 979-532-2600. Fax: 979-532-0871.
E-mail: wchm@awesomenet.net
Web Site: www.whartoncountymuseum.org
Founded: 1979.
Congressional District: 14
Key Personnel: C.E.O. & Museum Shop Mgr., Marvin Albrecht; Pres. (V), Linda Joy Stovall.
Personnel Profile: Full-Time Paid 3; Part-Time Paid 2; Part-Time Volunteers 32.
Governing Authority: nonprofit organization. Tax-exempt: 501(c)(3).
Institution Type/Description: History Museum: housed in former Marshall & Lillie A. Johnson residence.
Collections: over 6,000 historic photos of Wharton County; land grants & maps dating from 1824-1830; listing of doctors of Wharton County; medical equipment; silver, china, quilts, dressing apparel & furniture; sulphur mine exhibit; handmade dolls; military records, service decorations, clippings & photos; Shanghai Pierce Ranch exhibit; business journals; city & county record books; Indian artifacts; military Post West Bernard artifacts; barbed wire; telephone company switchboard; sports, including Hall of Famers; two Medal of Honor recipients exhibits: M/Sgt Roy P. Benavidez (Army) & SFC Johnnie D. Hutchins (Navy); Dan Rather shotgun house; Pulitzer Prize & Emmy award winning author & playwright Horton Foote; Marshall Johnson's big game trophy room.
Research Fields: local history.
Activities: guided tours; formally organized education programs for children; docent program or council; training programs; temporary exhibitions.
Publications: newsletter; newspapers; brochures.
Hours & Admission Prices: Mon.-Fri. 9:30-4:30, Sat.-Sun. 1-5; appointments available in evenings for special groups. No charge; donations accepted. &
Attendance: 6,000 (estimated)
Membership: Associate $25; Friend $50; Supporter $100; Patron $250; Sponsor $500; Benefactor $1,000.

White Settlement

WHITE SETTLEMENT HISTORICAL MUSEUM, INC., 8320 Hanon Dr., White Settlement, TX 76108-2317. Tel.: 817-246-9719.
E-mail: wshm@sbcglobal.net; hanontx@lycos.com
Web Site: www.wsmuseum.com
Founded: 1991.
Congressional District: 12
Key Personnel: Mgr., Carol L. Davis.

Personnel Profile: Part-Time Paid 1; Part-Time Volunteers 10.
Volunteer Hours: 180
Operating Expenses: 24,500
Operating Income: 24,500
Governing Authority: White Settlement Historical Museum board of directors. Tax-exempt: 501(c)(3).
Institution Type/Description: History Museum.
Collections: local history & culture; photographs; period artifacts & documents; 1940s windmill; farm milk shed; 1927 Farmall tractor; WWI horse & mule drinking trough; farm equipment; blacksmith tools; B-36 Peacemaker/7th Bomb Wing room; weapons; uniforms & equipment from Mexican War, Civil War, WWI, WWII & 1st Gulf War; local aviation industry history. Historic Buildings: 1864 Allen log cabin; 1940s playhouse; 1929 Model A Ford farm truck.
Research Fields: local history & genealogy; Liberator Village & Convair - Consolidated Bomber Plant #4; WWII.
Facilities: library.
Activities: special events. Annual Event: 15th Texas Cavalry/2nd MO US Civil War Reenactors; Early American & Celtic Musician Trio.
Publications: Memories of Liberator Village, 3rd edition.
Hours & Admission Prices: Tues.-Sat. 10-3. No charge; donations accepted. Closed New Year's Day; Independence Day; Thanksgiving; Christmas. &
Attendance: 1,200 (estimated)

Wichita Falls

THE JUANITA HARVEY ART GALLERY - MIDWESTERN STATE UNIVERSITY, 3410 Taft Blvd., Wichita Falls, TX 76308-2099. Tel.: 940-397-4264.
Institution Type/Description: Art Gallery.
Collections: paintings; sculpture; photographs.
Hours & Admission Prices: Mon.-Fri. 9-12 & 2-5.

KELL HOUSE MUSEUM, 900 Bluff St., Wichita Falls, TX 76301-3203. Tel.: 940-723-2712. Fax: 940-723-6592. Facebook: Kell House.
E-mail: kellhouse1909@yahoo.com
Web Site: www.wichita-heritage.org
Founded: 1981.
Congressional District: 13
Key Personnel: Exec. Dir., Delores Culley; Cur., Stacie Crosetto Flood.
Personnel Profile: Full-Time Paid 2; Part-Time Paid 1; Part-Time Volunteers 50; Interns 1.
Volunteer Hours: 2,919
Operating Expenses: 285,486
Operating Income: 172,593
Governing Authority: nonprofit organization. Parent Institution: Wichita County Heritage Society. Tax-exempt: 501(c)(3).
Institution Type/Description: Historic House: 1909 home of Frank Kell.
Collections: Wichita Falls history from 1900-1980; furniture; decorative arts; fabrics; costumes; documents; photographs.
Research Fields: local history.
Facilities: 4,800 sq. ft. exhibit space; gardens.
Activities: guided tours; organized education programs for children; docent program; temporary exhibitions. Annual Events: Mother/Daughter Garden Tea in May; July 4th celebration; Santa House.
Publications: quarterly newsletter.
Hours & Admission Prices: Mon.-Tues. & Thurs.-Fri. 10-3, Sat.-Sun. 2-4. No charge; donations accepted. Closed major holidays.
Attendance: 6,885 (estimated)
Membership: Heritage $50; Preservation $100; Conservation $250; Restoration $500; Revitalization $1,000.

KEMP CENTER FOR THE ARTS, 1300 Lamar, Wichita Falls, TX 76301-7031. Tel.: 940-767-2787. Fax: 940-767-3956.
E-mail: info@kempcenter.org
Web Site: www.kempcenter.org
Formerly: Arts Council Wichita Falls Area
Founded: 1995.
Key Personnel: Dir., Carol Sales; C.E.O., Carlana Fitch; Pres. (V), Vern Huffines; Mgr. Gallery, Gary R. Kingcade.
Personnel Profile: Full-Time Paid 4; Part-Time Paid 3.
Governing Authority: Tax-exempt.
Institution Type/Description: Community Art Center.
Collections: works by regional artists.
Facilities: 250-seat performance hall; sculpture garden. Museum-related items for sale.
Activities: classes; temporary exhibitions. Museum Sponsors: Arts Festival; 3 Performing Arts Series; Home & Garden Festival.
Publications: quarterly newsletter; annual sculpture catalogue.

Hours & Admission Prices: Mon.-Fri. 9-5. No charge; donations accepted. Closed New Year's Day; Good Friday; Memorial Day; Independence Day; Labor Day; Thanksgiving; Christmas. &
Attendance: 32,000 (estimated)
Membership: Individual $40; Family $75; Performer $100; Collector $250; Director $500; Benefactor $1,000.

MUSEUM OF NORTH TEXAS HISTORY, 720 Indiana St., Wichita Falls, TX 76301-6512. Mailing Address: P.O. Box 1619, Wichita Falls, TX 76307-1619. Tel.: 940-322-7628.
Web Site: www.month-ntx.org
Founded: 2000.
Congressional District: 13
Key Personnel: Dir., Lita Watson; Pres. (V) & Museum Shop Mgr., Jim Newson.
Personnel Profile: Full-Time Paid 1; Part-Time Paid 1; Part-Time Volunteers 10.
Governing Authority: Branch Museum: Call Field Army Air Base, 4515 Jacksboro Hwy. (Kickapoo Airport), Wichita Falls, TX. Tax-exempt.
Institution Type/Description: History Museum.
Collections: North Texas history; photographs; military; petroleum; over 500 western hats.
Research Fields: local & family history.
Facilities: archives.
Activities: school tours; temporary & permanent exhibits; temporary exhibits.
Publications: newsletter.
Hours & Admission Prices: Tues.-Fri. 10-12 & 1-4, Sat. 10-2; other times by appointment. No charge; donations accepted. Closed all major holidays. &
Attendance: 6,137 (accurate)
Membership: Individual $35; Family $50; Business & Contributor $100; Patron $150; Pioneer $250; Benefactor $500; Historian $1,000.

WICHITA FALLS MUSEUM OF ART, (M), Two Eureka Cir., Wichita Falls, TX 76308-2998. Tel.: 940-692-0923. Fax: 940-696-5358.
E-mail: mary.maskill@mwsu.edu
Web Site: www.mwsu.edu/wfma
Founded: 1964.
Congressional District: 13
Key Personnel: Chm. (V), Jane Spears; Dir. Operations & Museum Shop Mgr., Mary Maskill; Mgr. Facility and Interactive Exhibits, Jeff Desborough; Cur. Collections & Exhibits, Danny Bills.
Personnel Profile: Full-Time Paid 4; Part-Time Paid 6; Part-Time Volunteers 4.
Governing Authority: nonprofit. Parent Institution: Midwestern State University. Tax-exempt: 501(c)(3).
Institution Type/Description: General Museum.
Collections: contemporary American prints; art galleries.
Research Fields: local & regional history.
Facilities: 1,500-vol. library of art & exhibit reference books available by request and on the premises; classrooms.
Activities: guided tours; lectures; films; gallery talks; formally organized educational programs; docent program or council; inter-museum, temporary & traveling exhibitions; school loan service; Museum School-Humanities curriculum for ages 3-17.
Publications: weekly e-newsletter.
Hours & Admission Prices: Tues.-Fri. 10-5, Sat. 1-5. No charge; donations accepted. Closed university holidays. &
Attendance: 20,000 (estimated)
Membership: Student $15; Friend $55; Dual & Family $110; Associate $200; Collector's Circle $500; Benefactor $1,000.

Wimberley

OLD WEST MUSEUM, 333 Wayside Dr., Wimberley, TX 78676-5117. Tel.: 512-847-3338.
E-mail: cowboymuseum@earthlink.net
Web Site: cowboymuseum.net
Formerly: Cowboy Museum
Founded: 1956.
Key Personnel: Owner & Cur., Jack N. Glover; Co-Owner, Cherie Glover.
Governing Authority: individual operation.
Institution Type/Description: General Museum.
Collections: Indian artifacts; agriculture; history; early farming; anthropology; paintings; sculpture; graphics; archaeology; folklore; medical; military; natural history; oil field.
Research Fields: pertaining to collections.
Facilities: 500-vol. library of Indian & frontier books available for use on premises. Antiques, Indian artifacts, ranch items, original art & arrow heads for sale.

Activities: guided tours; lectures; formally organized education programs for children, adults, undergraduate & graduate college students; permanent & traveling exhibitions.
Publications: monthly magazine, International Barbwire Gazette; books, Barbwire Bible, VI; Sex Life of American Indian; Glovers Illustrated Letters.
Hours & Admission Prices: Memorial Day to Labor Day daily 10-6. No charge. &

Attendance: 2,500 (estimated)

PIONEER TOWN, 333 Wayside Dr., Wimberley, TX 78676-5117. Tel.: 512-847-3289. Fax: 512-847-6705.
Founded: 1956.
Congressional District: 10
Key Personnel: Sec. & Treas., John D. White.
Personnel Profile: Full-Time Volunteers 1; Part-Time Paid 4.
Governing Authority: individual operation. Parent Institution: Pioneer Museum of Western Art.
Institution Type/Description: Village Museum: authentic reproduction of c.1880 old West Town.
Collections: artifacts of the Old West; complete Frederic Remington sculpture museum of all 22 sculptures; Russell bronze sculptures; sculpture & metal art of the West. Historic House: 1861 Texas style log cabin.
Research Fields: culture of the Old West; bronze sculptures.
Facilities: 500-vol. library of Old West, history, material of people & places of the Old West available for use on premises.
Activities: tours by appointment.
Publications: newsletters; brochures; catalogs; books, Frederic Remington 1861-1909: He Knew the Horse, What's A Bronze.
Hours & Admission Prices: by appointment. No charge; donations accepted.
Attendance: 35,000 (estimated)

WIMBERLEY VALLEY ART LEAGUE AND GALLERY, 14068 Ranch Rd. 12, Wimberley, TX 78676. Mailing Address: P.O. Box 1652, Wimberley, TX 78676. Tel.: 512-826-4286.
Web Site: www.visitwimberley.com/artleague
Institution Type/Description: Art Gallery.
Collections: works by national & international artists.
Activities: special events; workshops; demonstrations.
Hours & Admission Prices: Mon.-Fri. 8:30-4:30, Sat.-Sun. 1-5.

Wink

ROY ORBISON MUSEUM, Texas 115, Wink, TX 79789. Mailing Address: P.O. Box 621, Wink, TX 79789-0621. Tel.: 432-527-3622.
Key Personnel: Cur., Dorothy Wolf
Institution Type/Description: History Museum.
Collections: artifacts & memorabilia from Roy Orbison's childhood & music career.
Activities: Museum Sponsors: Annual Festival.
Hours & Admission Prices: By appointment only. No charge; donations accepted.

Woodville

ALLAN SHIVERS MUSEUM, 302 N. Charlton, Woodville, TX 75979-4806. Tel.: 409-283-3709. Fax: 409-283-5258.
E-mail: rbunch75979@yahoo.com
Founded: 1963.
Congressional District: 2
Key Personnel: Pres. (V), R.E. Mosey-Bunch; Dir. Library & Museum, Rosemary Bunch.
Personnel Profile: Part-Time Paid 4; Part-Time Volunteers 1.
Governing Authority: county. Tax-exempt.
Institution Type/Description: Historic House Museum: 1881 restored building.
Collections: period furnishings; mementos of the family of Allan Shivers, governor of Texas, including documents, photographs, inaugural ball gowns.
Research Fields: state history.
Facilities: 1,500-vol. library of historical & political books available for use on premises.
Activities: guided tours; films; permanent exhibitions.
Hours & Admission Prices: Mon.-Fri. 9-4:30, Sat. 10-1:30. Adults $3, seniors $2, schoolchildren $1; preschool no charge. Closed major holidays.
Attendance: 2,500 (estimated)

HERITAGE VILLAGE MUSEUM, 157 PR 6000, Woodville, TX 75979. Mailing Address: Tyler County Heritage Society, P.O. Box 888, Woodville, TX 75979-0888. Tel.: 409-283-2272; 800-323-0389. Fax: 409-283-2194.
E-mail: hvillagemuseum@att.net
Web Site: www.heritage-village.org
Key Personnel: Dir. & Museum Shop Mgr., Ofeira Gazzaway; Pres., Elizabeth Toliver.
Personnel Profile: Part-Time Paid 5.
Governing Authority: private; nonprofit organization. Tax-exempt: 501(c)(3).
Institution Type/Description: Pioneer Village Museum.
Collections: 1840-1900 early East Texas Pioneer farming village including 36 buildings; costumes; tools.
Research Fields: Whitmeyer Research & Genealogy Library.
Activities: skills demonstrations.
Hours & Admission Prices: Mon.-Fri. 9-3, Sat.-Sun. 9-5. Adults $4, children under 12 $2; members no charge. Closed New Year's Eve & Day; Easter, Thanksgiving; Christmas. &
Attendance: 15,000 (estimated)
Membership: Individual $20; Family $30; Small Business $60.

Woodway

CARLEEN BRIGHT ARBORETUM, 9001 Bosque Blvd., Woodway, TX 76712-3486. Mailing Address: 1 Pavilion Way, Woodway, TX 76712. Tel.: 254-399-9204. Fax: 254-399-9216.
E-mail: arboretum@woodway-texas.com
Web Site: www.woodway-texas.com/lev1.cfm/9
Key Personnel: Dir., Janet Schaffer
Institution Type/Description: Arboretum.
Collections: trees; plants.
Facilities: nature trail.
Activities: rental facility.
Hours & Admission Prices: Daily 8am to dark. No charge; donations accepted.

Yorktown

YORKTOWN HISTORICAL MUSEUM, 144 W. Main St., Yorktown, TX 78164. Mailing Address: P.O. Box 1284, Yorktown, TX 78164-1284. Tel.: 361-564-9115. Facebook: Yorktown Historical Museum.
E-mail: bcbruns@wildblue.net
Founded: 1978.
Congressional District: 15
Key Personnel: Chm. (V) & Pres. (V), Beverly Bruns; Vice Pres., Shirley Mueller; Treas., Dorothy Mayfield; Sec., Marie Metting-Boles.
Personnel Profile: Part-Time Volunteers 45.
Governing Authority: Parent Institution: Yorktown Historical Society. Tax-exempt: 501(c)(3).
Institution Type/Description: Historical & Preservation Society: housed in 1876 C. Eckhardt & Sons Store.
Collections: period furnishings; clothing; musical instruments; documents; pictures; books; archeological dig findings; Indian arrowheads; early tools & farm implements; school memorabilia.
Research Fields: history & artifacts of the City of Yorktown.
Facilities: 50-seat auditorium. Museum-related items for sale.
Activities: guided tours; lectures; films; wildflower & historic cemetery tours; handwork & art displays.
Publications: book, Yorktown, TX, It's History, 1848-1989; postcards; revised book, 1848-1998 Yorktown Texas - 150 Year Anniversary.
Hours & Admission Prices: Thurs.-Sat. 10-2. No charge; donations accepted. &
Attendance: 700 (accurate)
Membership: Individual $5; Family $10; Business $25; Life $100.

UTAH

(125 listings)

Alpine

ALPINE ART CENTER, 450 S. Alpine Hwy., Alpine, UT 84004-1508. Tel.: 801-763-7173. Fax: 801-763-9799.
E-mail: steves@alpineartcenter.com
Web Site: alpineartcenter.com
Key Personnel: Event Dir., Steve Streadbeck
Institution Type/Description: Art Center.
Collections: paintings; sculpture.
Hours & Admission Prices: Art Center: Mon.-Fri. 10-5. Sculpture Park & Gardens: daily.

American Fork

TIMPANOGOS CAVE NATIONAL MONUMENT, Alpine Loop, Hwy. 92, American Fork, UT 84003-9803. Mailing Address: R.R. 3, Box 200, American Fork, UT 84003-9803. Tel.: 801-756-5239. Fax: 801-756-5661.
Web Site: www.nps.gov/tica/
Founded: 1922.
Congressional District: 3
Key Personnel: Supt., Jim Ireland; Administrative Officer, Shannon Stephens.
Governing Authority: federal. Parent Institution: U.S. Dept. of the Interior, National Park Service.
Institution Type/Description: Park Museum.
Collections: photographs; geological, botanical & paleontological items; historical artifacts.
Research Fields: speleogenesis; speleomorphology; cave biology, hydrology & hydrochemistry; Timpanogos Cave history.
Facilities: library with natural history books & articles related to the monument; archives.
Activities: temporary exhibitions & special interpretive programs dealing with the history of the monument.
Publications: trail guide; book, Timpanogos Cave: Window Into the Earth; Park brochure.
Hours & Admission Prices: May to Sept. daily. Cave Tours: adults $7; junior 6-15 $5; child 3-5 $3; discounts to seniors; infants 0-2 no charge. &
Attendance: 12 (accurate)

Blanding

THE DINOSAUR MUSEUM, 754 S. 200 W., Blanding, UT 84511-3909. Tel.: 435-678-3454.
E-mail: dinos@dinosaur-museum.org
Web Site: www.dinosaur-museum.org
Founded: 1992.
Governing Authority: Tax-exempt.
Institution Type/Description: Dinosaur History Museum.
Collections: paleontology; dinosaur history; skeletons; fossilized skin, eggs & footprints; sculptures; feathered dinosaurs.
Facilities: theater.
Activities: tours; movies.
Hours & Admission Prices: April 15-Oct. 15 Mon.-Sat. 9-5. Adults $3.50, senior citizens $2.50, children $2; discounts to AAA & AAM members & groups of 10 or more; members no charge. &
Membership: Individual $20; Family $25.

EDGE OF THE CEDARS STATE PARK MUSEUM, 660 West, 400 N., Blanding, UT 84511. Tel.: 435-678-2238. Fax: 435-678-3348.
E-mail: edgeofthecedars@utah.gov
Web Site: parks.state.ut.us/parks/www1/edge.htm
Founded: 1978.
Congressional District: 3
Key Personnel: Museum Dir., Teri Paul; Museum Shop Mgr., Kathrina Perkins; Cur. Collections, Deborah Westfall; Maintenance & Historic Replication, Andrew Goodwin; Cur. Education, Rebecca Stoneman.
Personnel Profile: Full-Time Paid 5; Part-Time Paid 3; Part-Time Volunteers 4; Interns 1.
Governing Authority: state. Parent Institution: State of Utah Div. of Parks & Recreation, P.O. Box 146001, Salt Lake City, UT 84114-6001. Tax-exempt.
Institution Type/Description: Native American Cultural Museum: site of ancestral Puebloan village occupied A.D. 700-1220.
Collections: prehistoric Paleoindian, Archaic, & Ancestral Puebloan artifacts; Navajo & Ute artifacts; Navajo oral histories; paintings; photographs; sculptures; textiles.
Research Fields: anthropology; ethnology.
Facilities: 2,100-vol. library of archaeology and Indian culture; 4,200 sq. ft. exhibit area; labs; 50-seat indoor auditorium; 100-seat outdoor amphitheater & Indian dance plaza; visitor information center. Museum-related items for sale.
Activities: archaeology & Native American craft workshops & programs; self-guided & guided tours; lectures; permanent & temporary exhibitions.
Publications: Spirit Windows: Native American Rock Art of Southeastern Utah.
Hours & Admission Prices: mid-April to mid-Sept. daily 9-6; mid-Sept. to mid-April daily 9-5. Adults $5; discounts to Fun Tag members 65 & over; children no charge. Closed New Year's Day; Thanksgiving; Christmas. &
Attendance: 19,777 (accurate)

HUCK'S MUSEUM AND TRADING POST, 1243 S. Main St., Blanding, UT 84511-3204. Tel.: 435-678-2329.
Founded: 1976.
Institution Type/Description: History Museum.
Collections: Native American culture & history; pottery; beads; arrowheads.
Hours & Admission Prices: Daily 8-5. Adults $3, children $2.

Boulder

ANASAZI STATE PARK MUSEUM, (M), 460 N. Hwy. 12, Boulder, UT 84716. Mailing Address: P.O. Box 1429, Boulder, UT 84716-1429. Tel.: 435-335-7308. Fax: 435-335-7352.
E-mail: nrdpr.ansp@state.ut.us
Web Site: www.stateparks.utah.gov/parks/anasazi
Founded: 1970.
Congressional District: 1
Key Personnel: Park Supt., Mike Nelson; Div. Dir., Mary Tullis; Cur., Bill Latady; Cur., Don Montoya; Museum Shop Mgr., Brenda Woolsey.
Personnel Profile: Full-Time Paid 3; Part-Time Paid 1; Part-Time Volunteers 1.
Governing Authority: state. Utah State Div. of Parks & Recreation, 1636 W. North Temple, Salt Lake City, UT. 84116.
Institution Type/Description: Historic Site: 1050-1200 A.D., excavated Anasazi Indian Village.
Collections: artifacts representative of the Kayenta Anasazi culture during the period 1050-1200 A.D.; replica of Coombs Village.
Research Fields: the Coombs site; primitive technology manufacture methods.
Facilities: Publications & museum-related items for sale.
Activities: guided tours; lectures; films.
Hours & Admission Prices: Memorial Day-Labor Day daily 8-6; Sept.-May daily 9-5. Admission: $3 per person; $5 per car; discount to groups; children under 6 & senior citizens from State of Utah no charge. Closed New Year's Day; Thanksgiving; Christmas. &
Attendance: 35,000 (accurate)

Bountiful

BOUNTIFUL HISTORICAL MUSEUM, 845 S. Main St., Ste. B5, Bountiful, UT 84010-6482. Tel.: 801-296-2060.
Web Site: www.bountifulutah.gov/HistoricalCommission/A_index01.html
Institution Type/Description: History Museum.
Collections: local history & culture; period artifacts.
Hours & Admission Prices: Wed. 2-4, Sat. 1-3.

Brigham City

BRIGHAM CITY MUSEUM-GALLERY, (M), 24 N. 300 W., Brigham City, UT 84302-2030. Mailing Address: P.O. Box 583, Brigham City, UT 84302-0583. Tel.: 435-226-1439. Facebook: Brigham City Museum.
Web Site: www.brighamcitymuseum.org
Founded: 1970.
Congressional District: 1
Key Personnel: Dir., Kaia Landon; Dir. Research, Mary Alice Hobbs.
Personnel Profile: Full-Time Paid 1; Part-Time Paid 2; Interns 4.
Governing Authority: municipal government. Parent Institution: Brigham City Corporation. Tax-exempt.
Institution Type/Description: Art Gallery & History Museum: collections span Box Elder County history from 1851-present.
Collections: local & state art; 1855-1900 furniture, furnishings, guns, tools, documents relating to the history of Brigham City; 1900-1960 photographs & documents.
Research Fields: local artists; economic, political & folk history of Brigham City and Box Elder county.
Activities: lectures; films. Annual Event: Two County High School Art Competition in April.
Publications: brochures; Historic Tour of Brigham City; Polygamy in Lorenzo Show's Brigham City: An Architectural Tour; Mayors of Brigham City.
Hours & Admission Prices: Tues.-Fri. 11-6, Sat. 1-5. No charge; donations accepted. &
Attendance: 10,000 (estimated)

Bryce

BRYCE WILDLIFE ADVENTURE - BRYCE CANYON MUSEUM, 1945 W. Utah State Hwy. 12, Bryce, UT 84764. Mailing Address: P.O. Box 640049, Bryce, UT 84764-0049. Tel.: 435-834-5555.
E-mail: terri@brycewildlifeadventure.com
Web Site: www.brycewildlifeadventure.com

Formerly: Paunsaugunt Wildlife Museum
Founded: 1995.
Key Personnel: Owner & Cur., Robert Driedonks; Owner & Cur., Terri Driedonks
Institution Type/Description: History Museum.
Collections: over 900 animals in a natural setting; Native American artifacts; butterflies; birds of prey; bugs; ocean fish; seashells; endangered Utah prairie dogs; early Western, Indian & African artifacts.
Facilities: Museum-related items for sale.
Activities: hand feed deer.
Hours & Admission Prices: April-Nov. 15 daily 9-9. Adults $8, children 6-12 $5; discounts to AAM members; children under 3 no charge. &

Bryce Canyon

BRYCE CANYON NATIONAL PARK VISITOR CENTER, Bryce Canyon National Park, Hwy. 63 Bryce #1, Bryce Canyon, UT 84717. Mailing Address: P.O. Box 170001, Bryce, UT 84764-0201. Tel.: 435-834-5322. Fax: 435-834-4102. TDD: 435-834-5322.
E-mail: brca_reception_area@nps.gov
Web Site: www.nps.gov/brca
Founded: 1959.
Congressional District: 1
Key Personnel: Supt., Eddie Lopez; Chm. (V), Dan Ng; Museum Shop Mgr., Gayle Pollock.
Personnel Profile: Full-Time Paid 52; Part-Time Volunteers 60; Interns 4.
Governing Authority: federal. Parent Institution: U.S. Dept. of Interior, National Park Service, Washington, DC. Subsidiary Institution: Bryce Canyon Natural History Association. Tax-exempt.
Institution Type/Description: Natural History Museum.
Collections: geology; herbarium; mammal skins; American Indian artifacts; insects; birds; fauna & flora.
Research Fields: Utah prairie dog; peregrine falcon; baseline inventories.
Facilities: library of natural history; 120-seat auditorium. Publications & post cards for sale.
Activities: guided tours; interpretive talks; films; permanent exhibitions.
Publications: A Kid's Guide to Bryce Canyon; A Natural History Guide to Bryce Canyon National Park; Queen's Garden at Sunrise Point Guide; Bryce Canyon National Park; Bryce Canyon Auto & Hiking Guide; Bryce Canyon Discovery; Shadows of Time: The Geology of Bryce Canyon National Park; Wildflowers of Southwestern Utah.
Hours & Admission Prices: Visitors Center: April & Oct. daily 8-6; May-Sept. daily 8-8; Nov.-March daily 8-4:30. Park: daily. Park: $12 per individual hiking or biking; $25 per car. Visitors Center: closed Thanksgiving; Christmas. &
Attendance: 1,600,000 (estimated)

Castle Dale

MUSEUM OF THE SAN RAFAEL, 70 N. 100 E., Castle Dale, UT 84513. Mailing Address: P.O. Box 1088, Castle Dale, UT 84513. Tel.: 435-381-5252. Fax: 435-381-2863. Facebook: Museum of the San Rafael.
E-mail: museum@co.emery.ut.us
Web Site: museumsanrafael.org
Formerly: Emery County Pioneer Museum
Founded: 1969.
Institution Type/Description: Pioneer Museum.
Collections: pioneer life & history; farm equipment; period clothing; personal artifacts; photographs; local outlaws; taxidermy; archaeological artifacts.
Hours & Admission Prices: Mon.-Fri. 10-4, Sat. 12-4.Donations Accepted

Cedar City

BRAITHWAITE FINE ARTS GALLERY, (M), 351 W. Center St., Cedar City, UT 84720-2470. Tel.: 435-586-5432. Fax: 435-865-8012. Facebook: Braithwaite Fine Arts Gallery.
E-mail: gallery@suu.edu
Web Site: www.suu.edu/pva/artgallery
Founded: 1976.
Congressional District: 1
Key Personnel: Dir., Reece Summers.
Personnel Profile: Full-Time Paid 1; Part-Time Paid 4.
Governing Authority: college. Parent Institution: Southern Utah University, Cedar City, UT 84720. Tax-exempt: 501(c)(3).
Institution Type/Description: College Art Gallery.
Collections: 19th- & 20th-century American art.
Activities: guided tours; illustrated lectures; art films; gallery talks; formally organized education programs; temporary & traveling exhibitions.

Hours & Admission Prices: June-Aug. Mon.-Sat. 10-8; Sept.-May Tues.-Sat. 12-7. No charge; donations accepted. Closed major holidays; academic breaks. &
Attendance: 10,000 (estimated)
Membership: Student $20; Individual $60; Sponsor $250; Patron $500; Corporate $1,000.

CEDAR BREAKS NATIONAL MONUMENT, 2390 W. Hwy. 56, Ste. 11, Cedar City, UT 84720-4151. Tel.: 435-586-9451 & 0787. Fax: 435-586-3813. Facebook: Cedar Breaks National Monument.
Web Site: www.nps.gov/cebr
Founded: 1933.
Congressional District: 1
Key Personnel: Supt., Paul Roelandt; Chief Park Ranger, Matthew Harrison.
Personnel Profile: Full-Time Paid 1; Part-Time Paid 6; Part-Time Volunteers 1; Interns 1.
Governing Authority: federal. Parent Institution: U.S. Dept. of Interior, National Park Service. Tax-exempt.
Institution Type/Description: Park Museum.
Collections: flora; fauna; geology specimens.
Research Fields: geology; flora.
Facilities: visitor center; viewing terrace.
Activities: guided tours; lectures.
Publications: books; guides; pamphlets; brochures; maps.
Hours & Admission Prices: Visitor Center: late May to mid-Oct. daily 9-6. 7-day individual fee $4; children 15 & under no charge. &
Attendance: 60,000 (estimated)

FRONTIER HOMESTEAD STATE PARK & MUSEUM, 635 N. Main, Cedar City, UT 84721-6179. Tel.: 435-586-9290.
E-mail: frontierhomestead@utah.gov
Web Site: www.stateparks.utah.gov
Formerly: Iron Mission State Park & Museum
Founded: 1973.
Congressional District: 8
Key Personnel: Park Manager, Todd Prince; Cur., Ryan Paul.
Personnel Profile: Full-Time Paid 2; Part-Time Paid 3; Interns 2.
Governing Authority: state. Utah State Div. of Parks & Recreation, 1594 W. North Temple, Ste. 116, Box 146001, Salt Lake City, UT 84114-6001. Tax-exempt.
Institution Type/Description: Pioneer History Museum.
Collections: Gronway Parry collection; horse drawn vehicle; early pioneer displays.
Research Fields: Iron County history; 1851-1924 agrarian history.
Facilities: library.
Activities: guided tours; lectures; living history.
Publications: Homestead News - quarterly.
Hours & Admission Prices: Mon.-Sat. 9-5. Adults $3; member Adults $1.50; children under 6 no charge. Closed New Year's Day; Thanksgiving; Christmas. &
Attendance: 20,000 (estimated)
Membership: Individual $12; Family $36.

Coalville

SUMMIT COUNTY HISTORICAL MUSEUM, 60 N. Main St., Coalville, UT 84017-9809. Mailing Address: P.O. Box 128, Coalville, UT 84017-0128. Tel.: 435-336-3200 & 3015.
E-mail: nvernon@summitcounty.org
Web Site: summitcounty.org
Founded: 2004.
Governing Authority: Tax-exempt.
Institution Type/Description: History Museum.
Collections: local history & culture; photographs; period furnishings; personal artifacts.
Activities: research; school & guided tours. Museum Sponsors: Echo Canyon Tour in June.
Publications: county self-guided tour brochures; history booklets; coffe table book of photographs.
Hours & Admission Prices: Daily 8-5; other times by appointment. No charge; donations accepted. &
Attendance: 1,000 (estimated)

Delta

GREAT BASIN HISTORICAL SOCIETY & MUSEUM, 45 W. Main St., Delta, UT 84624. Mailing Address: P.O. Box 550, Delta, UT 84624-0550. Tel.: 435-864-5013. Fax: 435-864-2446.
E-mail: gbm@frontiernet.net

Web Site: www.greatbasinmuseum.com
Founded: 1988.
Key Personnel: Dir. & Pres., Owen Neilsen; Treas. & Devel., Linda Neilsen; Cur. & Sec., Sindy McMichael.
Personnel Profile: Part-Time Paid 1; Part-Time Volunteers 17; Interns 1.
Governing Authority: private; nonprofit organization. Tax-exempt: 501(c)(3).
Institution Type/Description: Historical Society Museum.
Collections: geological specimens; photographs; personal artifacts; tools & implements for geology.
Hours & Admission Prices: Mon.-Sat. 10-5, call for additional hours. No charge; donations accepted. &
Attendance: 3,500 (accurate)
Membership: Individual & Family $10; Institutional $50; Contributing $100.

Ephraim

SNOW COLLEGE ART GALLERY, 150 E. College Ave., Ephraim, UT 84627-1550. Tel.: 435-283-7416.
E-mail: adam.larsen@snow.edu
Web Site: www.snow.edu/art/gallery/index.html
Key Personnel: Dir., Adam Larsen
Institution Type/Description: Art Gallery.
Collections: works by student & faculty.
Hours & Admission Prices: Mon.-Fri. 9-5; other times by appointment.

Eureka

TINTIC MINING MUSEUM, Main St., Eureka, UT 84628. Mailing Address: P.O. Box 218, Eureka, UT 84628-0218. Tel.: 435-433-6915 (Festival info). Fax: 435-433-6891.
Founded: 1974.
Congressional District: 2
Key Personnel: C.E.O., Dir. & Registrar, J. L. McNulty; Chm. (V), Deborah Treloar; Asst. Dir., E.C. McNulty; Museum Shop Mgr., Joan Morris.
Personnel Profile: Full-Time Volunteers 4; Part-Time Volunteers 5.
Governing Authority: nonprofit organization. Parent Institution: Utah State Historical Society, 300 Rio Grande, Salt Lake City, UT 84101. Tax-exempt: 501(c)(3).
Institution Type/Description: Mining Museum: housed in 1899 Eureka City Hall.
Collections: mining artifacts; complete mineral display; photos of the area. Historic Building: 1924 Union Pacific Railroad Depot.
Research Fields: mines.
Facilities: library of Eureka Reporter newspapers from 1902 to 1942, available for use on premises; history room.
Activities: self-guided tours. Museum Sponsors: Tintic Silver Festival.
Hours & Admission Prices: May-Sept. Sat.-Sun. 3-5; other times by appointment. No charge; donations accepted. &
Attendance: 1,500 (estimated)
Membership: Annual $5; Lifetime Individual $35; Lifetime Couple $50.

Fairfield

CAMP FLOYD/STAGECOACH INN STATE PARK, (M), 18035 W. 1540 N., Fairfield, UT 84013-9612. Tel.: 801-768-8932. Fax: 801-768-2794.
E-mail: marktrotter@utah.gov
Web Site: www.stateparks.utah.gov/parks/camp-floyd
Founded: 1964.
Key Personnel: Park Supt., Mark A. Trotter.
Personnel Profile: Full-Time Paid 2; Part-Time Paid 3; Part-Time Volunteers 3; Interns 1.
Governing Authority: state. Parent Institution: Utah State Parks, 1636 W. North Temple, Salt Lake City, UT 84116.
Institution Type/Description: Historic Site: 1858 site of Camp Floyd, former army camp of Utah.
Collections: local history & culture; period artifacts.
Activities: guided tours; lectures; youth camps; special events; school field trips & programs.
Hours & Admission Prices: Mon.-Sat. 9-5. Family $6, adults $2; children 5 & under no charge. Closed New Year's Day; Thanksgiving; Christmas. &
Attendance: 16,000 (accurate)

Fairview

FAIRVIEW MUSEUM OF HISTORY & ART, 85 N. 100 E., Fairview, UT 84629. Mailing Address: P.O. Box 157, Fairview, UT 84629-0157. Tel.: 435-427-9216.
E-mail: fvmuseum@cut.net

Web Site: www.sanpete.com
Founded: 1966.
Congressional District: 3
Key Personnel: Dir., Erma Lee Hansen; Pres. (V), Branch Cox.
Personnel Profile: Part-Time Paid 1; Part-Time Volunteers 17.
Governing Authority: nonprofit; board of trustees. Tax-exempt.
Institution Type/Description: General Museum & Pioneer Park.
Collections: pioneer relics & histories; Indian artifacts; miniature carvings; arts & crafts; Indian bead & leather art; engines; historic farm machines & vehicles; Columbian mammoth; statuary by Avard T. Fairbanks; geology & rocks; paintings & prints; local school children's artwork; photographs; history archives.
Research Fields: Indian & pioneer histories; geology.
Facilities: archives; garden.
Activities: guided tours.
Hours & Admission Prices: Summer: Mon.-Sat. 10-6; Winter: Mon.-Sat. 10-5; other times by appointment. No charge; donations accepted. &
Attendance: 20,000 (accurate)
Membership: Contributing $25; Supporting $50; Sustaining $100; Sponsor $250; Patron $500; Benefactor $1,000.

Farmington

BOUNTIFUL/DAVIS ART CENTER, 28 E. State St., Farmington, UT 84025. Mailing Address: P.O. Box 0221, Farmington, UT 84025. Tel.: 801-451-3660.
E-mail: info@bdac.org
Web Site: www.bdac.org
Founded: 1974.
Congressional District: 1
Key Personnel: Exec. Dir. & Museum Shop Mgr., Emma J. Dugal; Chm. (V), Aida Mattingley.
Personnel Profile: Full-Time Paid 2; Part-Time Paid 3; Part-Time Volunteers 25.
Governing Authority: Tax-exempt.
Institution Type/Description: Art Center.
Collections: paintings; drawings; sculpture; photography; printmaking.
Activities: education programs.
Hours & Admission Prices: Tues.-Fri. 10-6, Sat. 2-5. No charge; donations accepted. Closed holidays.
Attendance: 15,000 (estimated)
Membership: Bronze $35-$499; Silver $500-$4,999; Gold over $5,000 (annual), $10,000 (lifetime member)

PIONEER VILLAGE, 375 N. Lagoon Lane, Farmington, UT 84025-2502. Mailing Address: P.O. Box 696, Farmington, UT 84025-0696. Tel.: 801-451-8050. Fax: 801-451-8015.
Founded: 1954.
Congressional District: 1
Key Personnel: Dir., Peter Freed; Deputy Dir., Howard Freed.
Personnel Profile: Full-Time Paid 5; Part-Time Paid 30.
Governing Authority: nonprofit organization. Lagoon Corp., Box N, Farmington, UT 84025.
Institution Type/Description: History Museum.
Collections: carriages; guns; Indian artifacts; pioneer artifacts; toys; coins & silver. Historic Buildings: log & stone buildings; stores; homes.
Facilities: 700-vol. library; restaurant. Museum-related items for sale.
Activities: temporary & permanent exhibits.
Hours & Admission Prices: April-May & Sept.-Oct. Sat.-Sun. 10-7; June-Aug. daily 10-8. Admission $10; seniors no charge. &
Attendance: 500,000 (accurate)

S & S SHORTLINE TRAIN PARK & MUSEUM, 575 N. 1525 W., Farmington, UT 84025-2615. Mailing Address: P.O. Box 604, Farmington, UT 84025. Tel.: 801-451-0222.
Institution Type/Description: Railroad Museum.
Collections: railroad artifacts & memorabilia.
Hours & Admission Prices: May-Sept. 1st Sat. of month 10-4. No charge.

Fillmore

TERRITORIAL STATEHOUSE STATE PARK & MUSEUM, 50 W. Capitol Ave., Fillmore, UT 84631-5556. Tel.: 435-743-5316. Fax: 435-743-4723.
Founded: 1930.
Key Personnel: Curator, Carl Camp.
Personnel Profile: Full-Time Paid 1; Part-Time Paid 2; Part-Time Volunteers 15.

Governing Authority: state. Utah State Division of Parks & Recreation, 1636 W. North Temple, Salt Lake City, 84116. Tax-exempt.

Institution Type/Description: Regional History Museum: housed in Utah's first territorial Capitol 1855-1858.

Collections: pioneer furniture, tools, pictures, handicrafts; Indian artifacts; music; cotton; pioneer farm implements; paintings; sculpture; costumes; decorative arts. Historic Buildings: 2 pioneer log cabins; 1867 rock schoolhouse.

Facilities: 100-vol. library of books pertaining to history of people & localities, schoolbooks, personal histories, ancestor photographs, Bibles of several countries & songbooks available for research on premises. Books & gift items for sale.

Activities: guided tours; lectures; hands-on activities; summer youth camps; Citizenship in the Nation merit badge overnight camp for Boy Scouts. Museum Sponsors: pioneer dances; Arts & Living History Festival; Shadows of the Past Tour in October; Old Fashioned Christmas program.

Hours & Admission Prices: Mon.-Sat. 9-5; other times by appointment. Family $6, adults 12 & over $2, children 6-11 $1; members & children under 6 no charge. Closed New Year's Day; Thanksgiving; Christmas. &

Attendance: 45,000 (estimated)

Membership: Friends of the Territorial Statehouse: Individual $10; Family $25.

Fort Douglas

FORT DOUGLAS MILITARY MUSEUM, (M), 32 Potter St., Fort Douglas, UT 84113-5046. Tel.: 801-581-1251. Fax: 801-581-9846.

E-mail: admin@fortdouglas.org

Web Site: www.fortdouglas.org

Founded: 1974.

Congressional District: 2

Key Personnel: C.E.O. & Dir., Robert S. Voyles; Pres. (V), Brent Ashworth; Cur., Beau Burgess; Historian/Visitor Svcs. Mgr., Su Richards.

Governing Authority: state. Parent Institution: Utah National Guard, Box 1776, Draper, UT 84020-1776. Tax-exempt.

Institution Type/Description: Military Museum: housed in 1875 Quartermaster Victorian Infantry Barracks Building, located in Fort Douglas, founded in 1862 by California Volunteers to protect the Overland Mail & Telegraph lines.

Collections: military uniforms, accoutrements, memorabilia, documents; insignia of the Army 1857-present; military equipage of Navy, Marines, Air Corps and Coast Guard from 1860 to present; military vehicles; tanks; artillery pieces.

Research Fields: military history of Utah, the Utah National Guard & militia.

Facilities: 1,500-vol. library of military history of Utah and Fort Douglas and the history of military in Utah, available for use on premises; reading room. Books of military history for sale.

Activities: guided tours; lectures; films; gallery talks; study clubs; permanent & temporary exhibitions; 3rd California Infantry Fife, Drum and Bugle Corp. available for community events & museum activities.

Publications: monographs, Col. Patrick Edward Connor, Stephen Douglas, Utah's Navy Ships, and Utah & the Air Force Connection; Battle of Bear River; U.S. Army Pioneers; The Black Soldier in Utah; Opening of Uintah Indian Reservation; Prisoners at Fort Douglas; Patrick Edward Connor, A Closer Look; The Daily Union Vedette; Who Really Was Bonneville?

Hours & Admission Prices: Tues.-Sat. 12-5. No charge; donations accepted. Closed federal holidays weekends. &

Attendance: 6,000 (accurate)

Membership: Annual $35; Family $50; Family $60; Life $300.

Fort Duchesne

CULTURAL RIGHTS AND PROTECTION DEPARTMENT/UTE INDIAN TRIBE, 910 South 7500 East, Fort Duchesne, UT 84026. Mailing Address: P.O. Box 190, Fort Duchesne, UT 84026-0190. Tel.: 435-722-5141. Fax: 435-722-2083.

Web Site: www.utetribe.com

Founded: 1976.

Congressional District: 3

Key Personnel: Dir., Betsy Chapoose.

Governing Authority: tribal. Tax-exempt.

Institution Type/Description: Indian Museum: located on site related to the era of U.S. Cavalry and Old Fort Duchesne.

Collections: Indian-produced artwork in various media; old books from nearby military fort; Indian artifacts.

Research Fields: early Western American history; Ute history; archaeological; personal interviews with elderly to document verbal Indian history.

Facilities: 160-vol. library of books & research papers on Indian & early Western American history available for use on premises, or by request to research staff; reading room.

Activities: lectures.

Publications: book, A History of the Northern Ute People.

Hours & Admission Prices: Mon.-Thurs. 8-4:30.

Grantsville

DONNER-REED PIONEER MUSEUM, 90 N. Cooley, Grantsville, UT 84029. Tel.: 435-884-0824 & 3411.

Web Site: www.donner-reed-museum.org

Institution Type/Description: History Museum.

Collections: local history & culture; period furnishings; personal artifacts; photographs.

Hours & Admission Prices: By appointment.

UTAH FIREFIGHTERS MUSEUM, 37 N. Church, Grantsville, UT 84029. Mailing Address: P.O. Box 1128, Grantsville, UT 84029-1128. Tel.: 435-884-6680.

Institution Type/Description: Fire Museum.

Collections: firefighting history; period fire trucks & equipment; personal artifacts; photographs.

Hours & Admission Prices: Fri.-Sat. 11-3; other times by appointment.

Green River

JOHN WESLEY POWELL MUSEUM, 1765 E. Main St., Green River, UT 84525. Mailing Address: P.O. Box 620, Green River, UT 84525-0620. Tel.: 435-564-3427. Fax: 435-564-3526.

E-mail: director@johnwesleypowell.com

Web Site: www.jwprhm.com

Founded: 1990.

Key Personnel: Dir., Rey Lloyd Hatt; Chm. (V), Penney Riches; Museum Shop Mgr., JoAnn Wetherington.

Personnel Profile: Full-Time Paid 1; Part-Time Paid 9; Part-Time Volunteers 1.

Governing Authority: Tax-exempt.

Institution Type/Description: River History Museum.

Collections: works by local & regional artists; southern Utah history; riverboats; Native Americans; mountain men; explorers; River Runners Hall of Fame.

Facilities: 200-seat auditorium. Museum-related items for sale.

Hours & Admission Prices: Call for hours & admission prices. &

Attendance: 30,000

Heber City

COMMEMORATIVE AIR FORCE UTAH WING MUSEUM, CAF Hangar - Russ McDonald Field, 620 Airport Rd. D38, Heber City, UT 84068. Tel.: 435-657-1826.

Web Site: www.cafutahwing.org

Institution Type/Description: Military History Museum.

Collections: early military & commercial aircraft; women in aviation.

Hours & Admission Prices: May-Oct. Thurs.-Sun. 10-5. No charge; donations accepted.

Helper

WESTERN MINING + RAILROAD MUSEUM, 296 S. Main St., Helper, UT 84526. Mailing Address: P.O. Box 221, Helper, UT 84526-0221. Tel.: 435-472-3009.

E-mail: helpermuseum@helpercity.net

Web Site: www.wmrrm.org

Founded: 1963.

Key Personnel: Dir. & Museum Shop Mgr., Stephanie Fitzsimons; Chm. (V), Pat Kokal.

Personnel Profile: Part-Time Paid 2; Part-Time Volunteers 8.

Governing Authority: Parent Institution: Helper City. Tax-exempt.

Institution Type/Description: History Museum: housed in the Old Helper Hotel, built c.1913.

Collections: railroad, mining & cultural history; photographs; coal mining tools & equipment; railroad office & artifacts.

Facilities: Museum-related items for sale.

Activities: special events.

Hours & Admission Prices: May-Sept. Mon.-Sat. 10-5; Oct.-April Tues.-Sat. 11-4. No charge; donations accepted.

Attendance: 9,357 (accurate)

Hill Air Force Base

HILL AEROSPACE MUSEUM, 7961 Wardleigh Rd., Hill Air Force Base, UT 84056-5842. Tel.: 801-777-6868 & 6818. Fax: 801-777-6386.
Web Site: www.hill.af.mil/library/museum/index.asp
Founded: 1985.
Key Personnel: Dir., Scott Wirz; Chm. (V), Marc Reynolds; Museum Shop Mgr., Lorrie Slade.
Personnel Profile: Full-Time Paid 5; Part-Time Volunteers 100; Interns 1.
Governing Authority: federal; nonprofit. Parent Institution: Hill Air Force Base, USAF. Tax-exempt: 501(c)(3).
Institution Type/Description: Aerospace Museum.
Collections: 97 aerospace vehicles from pre-WWI to present; 3,500 artifacts; WWII chapel & barracks; archives.
Research Fields: Hill AFB history; USAF technology.
Facilities: library; archives.
Activities: education programs.
Publications: foundation newsletter, Volunteer Gazette.
Hours & Admission Prices: Daily 9-4:30. No charge; donations accepted. Closed New Year's Day; Thanksgiving; Christmas. &
Attendance: 180,000 (accurate)

Hurricane

HURRICANE VALLEY HERITAGE PARK MUSEUM, 35 W. State, Hurricane, UT 84737-1961. Mailing Address: P.O. Box 91, Hurricane, UT 84737-0091. Tel.: 435-635-3245. Fax: 435-635-4696.
E-mail: hurricanemuseum@hotmail.com
Web Site: www.hurricane-pioneer.org
Founded: 1989.
Key Personnel: C.E.O. & Pres. (V), Gregory Lawton; Treas., Verna Hinton; Cur., Phyllis Lawton; Education, Stella Shamo; Registrar, Lee Beaty.
Personnel Profile: Full-Time Paid 1; Full-Time Volunteers 2; Part-Time Paid 3; Part-Time Volunteers 6.
Governing Authority: private; nonprofit organization. Tax-exempt: 501(c)(3).
Institution Type/Description: General Museum.
Collections: local pioneer & Indian culture artifacts; pioneer homes; photographs; dolls; tools; family histories; history of town military men. Buildings: 2 homes; barn; blacksmith shop.
Research Fields: pioneer homes, activities, pictures & personal histories.
Facilities: library; 60-seat auditorium; 2,000 sq. ft. exhibit space. Museum-related items for sale.
Activities: arts festival; concerts; films; guided tours; lectures. Museum Sponsors: Outdoor Historical Pageant.
Publications: History of Homes of Hurricane.
Hours & Admission Prices: Mon.-Sat. 9-5. No charge; donations accepted. Closed New Year's Day; Thanksgiving; Christmas. &
Attendance: 10,000 (estimated)
Membership: Individual $10; Family $20; Business $50; Patron $500.

Hyrum

HYRUM CITY MUSEUM, (M), 83 W. Main St., Hyrum, UT 84319-1297. Tel.: 435-245-0208.
Key Personnel: Cur, Jeff McBride
Institution Type/Description: History Museum.
Collections: dinosaur bones; Egyptian artifacts; 19th century tools; minerals & rocks; pioneer artifacts; Mormon memorabilia; photographs.
Hours & Admission Prices: Tues., Thurs. & Sat. 3-5 by appointment. No charge.

Kanab

KANAB HERITAGE MUSEUM & JUNIPER FINE ARTS GALLERY, 13 S. 100 E., Kanab, UT 84741. Mailing Address: City Office, 76 N. Main, Kanab, UT 84741. Tel.: 435-644-3966.
Personnel Profile: Part-Time Paid 1.
Institution Type/Description: History Museum & Art Gallery.
Collections: local history & culture; period furnishings; photographs; personal artifacts; works by local artists.
Hours & Admission Prices: Summer: Mon.-Fri. 1-5. No charge; donations accepted.

Kaysville

KAYSVILLE LECONTE STEWART GALLERY OF ART, 44 N. Main, Kaysville, UT 84037-1949. Tel.: 801-544-2826. Fax: 801-544-5646.
E-mail: admin@kaysvillecity.com
Institution Type/Description: Art Gallery.
Collections: works by LeConte Stewart.
Hours & Admission Prices: Call for hours.

Layton

HERITAGE MUSEUM OF LAYTON, 403 N. Wasatch Dr., Layton, UT 84041-3238. Tel.: 801-336-3930.
Web Site: www.laytoncity.org/public/museum/default.aspx
Founded: 1980.
Personnel Profile: Full-Time Paid 1; Part-Time Volunteers 20.
Governing Authority: Parent Institution: Layton City Corporation. Tax-exempt.
Institution Type/Description: History Museum.
Collections: local history & culture; photographs; personal artifacts; documents; industry; Native American artifacts; newspapers.
Publications: Heritage Horizon, Spring & Fall.
Hours & Admission Prices: Tues.-Fri. 11-6, Sat. 1-5. No charge. Closed holidays. &
Attendance: 6,500 (accurate)

Lehi

JOHN HUTCHINGS MUSEUM OF NATURAL HISTORY, 55 N. Center St., Lehi, UT 84043-1826. Tel.: 801-768-7180. Fax: 801-768-9409. Facebook: www.facebook.com/pioneersgo.
E-mail: hmuseum@lehi-ut.gov
Web Site: www.lehi-ut.gov/discover/hutchings-museum
Founded: 1955.
Congressional District: 1
Key Personnel: Dir., Ben Woodruff.
Personnel Profile: Full-Time Paid 1; Part-Time Paid 6; Part-Time Volunteers 9; Interns 1.
Governing Authority: non-profit organization. Parent Institution: John Hutchings Museum of Natural History Board of Directors and Lehi City Corp. Tax-exempt: 501(c)(3).
Institution Type/Description: Natural History Museum.
Collections: rocks; minerals; fossils; shells; birds; eggs; pioneer tools & household items; Native American weapons; tools; baskets; guns; Porter Rockwell; Johnston's army; 19th century pharmaceuticals & inventions.
Research Fields: archaeology; geology; biology.
Facilities: library.
Activities: guided tours; lectures; workshops; summer camps; permanent & changing exhibitions; inter-museum loan; special events. Museum Sponsors: night@themuseum in February; Round Up Week in June; Museum Day & Pirate Night in September; Family Week in November; Christmas Workshop & Gifts of Nature in December.
Publications: e-mail bulletin.
Hours & Admission Prices: Tues.-Sat. 11-5. Adults $4, senior citizens, children & students $3; discount to groups; children under 2 no charge. Closed national holidays. &
Attendance: 10,000 (estimated)
Membership: Individual $25; Family $50.

NORTH AMERICAN MUSEUM OF ANCIENT LIFE, 2929 Thanksgiving Way, Lehi, UT 84043-3740. Tel.: 801-766-5000.
Web Site: www.thanksgivingpoint.com
Institution Type/Description: Paleontology Museum.
Collections: local history; mounted dinosaurs; fossils; skeletons; hands-on exhibits.
Facilities: cafe. Museum-related items for sale.
Activities: educational programs; field trips.
Hours & Admission Prices: Mon.-Sat. 10-8. Exhibits: adults $10, senior citizens 65 & over and children 3-12 $8. Exhibits & Movie: adults $15, senior citizens 65 & over and children 3-12 $12. Closed Thanksgiving; Christmas.

Logan

INTERMOUNTAIN HERBARIUM, UTAH STATE UNIVERSITY, Dept. of Biology, Utah State Univ., Logan, UT 84322. Mailing Address: 5305 Old Main Hill, Utah State University, Logan, UT 84322-5305. Tel.: 435-797-0061 & 1584. Fax: 435-797-1575.
E-mail: mary@biology.usu.edu
Web Site: herbarium.usu.edu/
Founded: 1931.
Key Personnel: Dir., Dr. Mary E. Barkworth.
Personnel Profile: Full-Time Paid 2; Part-Time Paid 4; Part-Time Volunteers 2.
Governing Authority: state; university. Parent Institution: Utah State University. Subsidiary Institution: Dept. of Biology. Tax-exempt.
Institution Type/Description: Herbarium.
Collections: 257,000 specimens: mostly vascular plants of the Intermountain Region, some mosses, fungi & lichens; seeds; photographic slides.
Research Fields: flora of the inter-mountain region; plant systematics particularly grasses; fungal systematics.
Facilities: 2,000-vol. library of plant taxonomy books available for use on premises; reading room.
Activities: research; inter-museum loan & exchanges; temporary exhibitions; teaching.
Hours & Admission Prices: Mon.-Fri. 8-5. No charge; donations accepted. Closed state & national holidays.
Attendance: 300 (accurate)

* **NORA ECCLES HARRISON MUSEUM OF ART, (M),** Utah State University, 650 N. 1100 E., Logan, UT 84322. Mailing Address: 4020 Old Main Hill, Logan, UT 84322-4020. Tel.: 435-797-0163. Fax: 435-797-3423.
E-mail: rachel.hamm@usu.edu
Web Site: www.artmuseum.usu.edu
Founded: 1982.
Congressional District: 1
Key Personnel: C.E.O. & Dir., Victoria Rowe Berry; Assoc. Cur. Education, Elizabeth Benson; Registrar, Casey Allen; Cur. Education, Nadra E. Haffar; Cur. Programs & Exhibitions, Deborah Banerjee.
Personnel Profile: Full-Time Paid 5; Part-Time Paid 13; Part-Time Volunteers 3.
Governing Authority: university. Parent Institution: Utah State University. Tax-exempt: 170(b)(1)(A).
Institution Type/Description: Art Museum.
Collections: 20th- & 21st-century West coast American paintings, sculpture & drawings; 20th- & 21st-century West coast ceramic vessels; Native American arts representing Pueblo, Hopi & Navajo tribes; Vogel Collection: 50 Works for 50 States.
Research Fields: related to collections.
Facilities: temporary & permanent exhibition galleries.
Activities: guided tours; lectures; participatory, permanent & temporary exhibitions; docent tours.
Publications: exhibition brochures; catalogues.
Hours & Admission Prices: Mon.-Fri. 10-5, Sat. 11-4. Suggested Donations: adults $3; members no charge. Closed holidays. &
Attendance: 29,000 (accurate)
Membership: Student & Senior Citizen $10; Individual $20; Family $35; Contributor $50-$99; Patron $100-$199; Associate $200-$249; Benefactor $250-$499; President's Circle $500 & up.

STOKES NATURE CENTER, 2696 E. Hwy. 89, Logan, UT 84323. Mailing Address: P.O. Box 4204, Logan, UT 84323-4204. Tel.: 435-755-3239. Fax: 435-755-6586.
E-mail: nature@logannature.org
Web Site: www.logannature.org
Founded: 1997.
Key Personnel: Exec. Dir., Bob Green; Dir. Education, Andrea Liberatore; Dir. Operations, Ru Mahoney
Institution Type/Description: Nature Center.
Collections: natural science; local history; hands-on exhibitions.
Activities: school & community programs for all ages, toddler-adult.
Publications: biweekly e-newsletter, Bird Call; quarterly members' newsletter, The Dipper; brochure, The History of Lore of Logan Canyon.
Hours & Admission Prices: Wed.-Fri. 10-4; call to confirm. No charge; donations accepted. &
Membership: Individual $25; Family $45; Student/Senior $15.

UTAH STATE UNIVERSITY'S MUSEUM OF ANTHROPOLOGY, 730 Old Main Hill, Logan, UT 84322-0730. Tel.: 435-797-7545. Fax: 435-797-1240.
E-mail: anthro.museum@usu.edu
Web Site: www.usu.edu/anthro/museum
Founded: 1963.
Key Personnel: Dir., Dr. Bonnie Pitblado; Cur., Mary Kay Gabriel.
Personnel Profile: Full-Time Paid 1; Part-Time Paid 11; Part-Time Volunteers 6; Interns 6.
Governing Authority: public university. Tax-exempt: 170(b)(1)(A).
Institution Type/Description: Anthropology Museum.
Collections: archaeological includes the Great Basin, Ancestral Puebloan people, Mayan civilization, and Petra Jordon; ethnographic includes Africa, India, the west coast from California to Alaska, Polynesia, New Zealand, & Peru; Native American artifacts.
Facilities: 1,600 sq. ft. exhibit space.
Activities: docent program; formal education programs; guided tours; school loan service; community programs. Museum Sponsors: biannual Archaeology Merit Badge sessions for BSA in Cache Valley.
Hours & Admission Prices: Mon.-Fri. 8-5, Sat. 10-4. No charge; donations accepted. Closed university holidays; federal holidays. &
Attendance: 7,000 (estimated)

WILLOW PARK ZOO, 419 W. 700 S., Logan, UT 84321-5599. Tel.: 435-716-9265. Fax: 435-716-9254.
E-mail: dharvey@loganutah.org
Web Site: www.loganutah.org/parks_and_rec/willow_park/index.cfm
Founded: 1970.
Key Personnel: Supt., Rod Wilhelm.
Personnel Profile: Full-Time Paid 3; Part-Time Paid 8; Part-Time Volunteers 4; Interns 1.
Governing Authority: Parent Institution: city of Logan. Tax-exempt.
Institution Type/Description: Zoo.
Collections: over 100 species of birds; 12 species of mammals; fish; turtles.
Hours & Admission Prices: Daily 9am to sunset. Adults $2, children 12 & under $1; AZA & museum members no charge. Closed New Year's Day; Thanksgiving; Christmas.
Attendance: 100,000 (estimated)
Membership: Individual $25; Individual Plus One $30; Family & Grandparents $40; Family & Friends Circle $50.

Magna

MAGNA ETHNIC AND MINING MUSEUM, 9056 W. Magna Main St., Magna, UT 84044-1149. Mailing Address: P.O. Box 742, Magna, UT 84044-0324. Tel.: 801-250-5656.
Institution Type/Description: Mining Museum.
Collections: local history & culture; mining industry; photographs.
Hours & Admission Prices: Tues. & Thurs.-Fri. 11-3, Wed. 11-5; other times by appointment.

Midvale

MIDVALE HISTORICAL SOCIETY MUSEUM, 7697 S. Main St., Midvale, UT 84047-7107. Tel.: 801-569-8040.
E-mail: mid_museum@xmission.com
Founded: 1979.
Governing Authority: Parent Institution: Midvale Historical Society. Tax-exempt.
Institution Type/Description: History Museum.
Collections: local history & culture; photographs; period artifacts; early pioneers; mining; transportation.
Hours & Admission Prices: Tues.-Wed. & Sat. 12-4. No charge; donations accepted. &
Attendance: 580 (accurate)

Moab

ARCHES NATIONAL PARK VISITOR CENTER, N. Hwy. 191, Moab, UT 84532. Mailing Address: P.O. Box 907, Moab, UT 84532-0907. Tel.: 435-719-2100 & 2299. Fax: 435-719-2305. TDD: 435-719-2319.
E-mail: archinfo@nps.gov
Web Site: www.nps.gov/arch
Congressional District: 3
Key Personnel: Park Supt., Kate Cannon.
Governing Authority: federal. Parent Institution: National Park Service. Tax-exempt.
Institution Type/Description: Natural History Museum.

Collections: herbarium; geology; botany; archaeology; entomology. Historic House: 1906 Wolfe Cabin.
Facilities: picnic area; campsite.
Activities: guided tours; organized education programs.
Publications: Visitor's Guide to Arches National Park.
Hours & Admission Prices: Park: daily 24 hours a day. Visitor Center: daily 8-4:30. 7-day vehicle $10, 7-day individual $5. Closed Christmas. &
Attendance: 733,000 (accurate)

DAN O'LAURIE MUSEUM OF MOAB, 118 E. Center St., Moab, UT 84532-2430. Tel.: 435-259-7985.
E-mail: moabmuseum@frontiernet.net
Web Site: www.moabmuseum.org
Formerly: Dan O'Laurie Canyon Country Museum
Founded: 1958.
Congressional District: 3
Key Personnel: Dir., Travis Schenck; Pres. (V), Lloyd Holyoak.
Governing Authority: county; nonprofit. Parent Institution: Southeast Utah Society of Arts & Sciences. Tax-exempt.
Institution Type/Description: General Museum.
Collections: archaeology; geology; early Moab history.
Research Fields: archaeology; oral history.
Activities: permanent, traveling & temporary exhibits.
Publications: Canyon Legacy.
Hours & Admission Prices: March-Oct. Mon.-Fri. 10-5, Sat.-Sun. 12-5; Nov.-Feb. Mon.-Sat. 12-5. Family $10, adult $5; children under 17 with adult & members no charge. Closed New Year's Day; Memorial Day; Independence Day; Labor Day; Thanksgiving; Christmas. &
Attendance: 7,000 (accurate)
Membership: Contributing $25; Sustaining $50; Donor $100; Associate $250; Patron $500.

DEAD HORSE POINT STATE PARK, Hwy. 313, Moab, UT 84532. Mailing Address: P.O. Box 609, Moab, UT 84532-0609. Tel.: 435-259-2614. Fax: 435-259-2615. Facebook: Dead Horse Point State Park.
E-mail: deadhorsepoint@utah.gov
Web Site: www.stateparks.utah.gov
Founded: 1959.
Key Personnel: Park Mgr., Megan Blackwelder; Asst. Mgr., Crystal Carpenter
Governing Authority: state. Parent Institution: State of Utah, Div. of Parks & Recreation, 1636 W. North Temple, Salt Lake City, UT. 84116. Tax-exempt.
Institution Type/Description: State Park Visitor Center.
Collections: anthropology; arboretum; archaeology; ethnology; folklore; geology; history; Indian artifacts; mineralogy; natural history; zoology.
Facilities: visitor center.
Activities: guided walks; campfire programs; junior ranger program.
Hours & Admission Prices: Visitor Center: mid-March to mid-Oct. 8-6; mid-Oct. to mid-March 9-5. Park: daily. Daytime: $10 per car; Camping: $25 per night; call for campground reservations. Visitor Center: closed New Year's Day; Thanksgiving; Christmas. &
Attendance: 182,419 (accurate)

HOLE N' THE ROCK, 11037 S. Hwy. 191, Moab, UT 84532-3969. Tel.: 435-686-2250. Fax: 435-686-9959.
E-mail: hnrock@citlink.net
Web Site: www.theholeintherock.com
Founded: 1945.
Key Personnel: Dir., Wyndee Hansen; Pres., Erik Hansen; Museum Shop Mgr., Malaine Wareham.
Personnel Profile: Full-Time Paid 8.
Institution Type/Description: Historic Home: housed in the home, carved out of rock by Albert and Gladys Christenson.
Collections: family history; personal artifacts; furnishings; paintings; sculpture.
Facilities: Museum-related items for sale.
Activities: petting zoo.
Hours & Admission Prices: Daily 9-5. Tours: adults $5, children 5-10 $3.50; children under 5 no charge. &
Attendance: 50,000 (estimated)

MOAB MUSEUM OF FILM & WESTERN HERITAGE, Red Cliffs Lodge, Mile Post 14., Hwy. 128, Moab, UT 84532-9618. Tel.: 435-259-2002; 866-812-2002. Fax: 435-259-5050.
Web Site: www.redcliffslodge.com/museum
Founded: 2002.
Institution Type/Description: History & Film Museum.
Collections: early cowboy ranching & local movie memorabilia.

Hours & Admission Prices: March 5-Dec. 11 daily 6am 10pm. No charge. &
Attendance: 15,000 (estimated)

Monticello

FRONTIER MUSEUM, 216 S. Main, Monticello, UT 84535. Mailing Address: P.O. Box 763, Monticello, UT 84535-0763. Tel.: 435-587-3401.
E-mail: ging0209@gmail.com
Web Site: www.utahscanyoncountry.com/en/entities/226/
Governing Authority: Tax-exempt.
Institution Type/Description: History Museum.
Collections: local history & culture; early pioneers; vintage clothing; household items; early telephones; period artifacts; remains from the Home of Truth.
Hours & Admission Prices: Winter: Oct.-March Fri.-Sun. 10-6. No charge, donations accepted. &

Mount Carmel

THUNDERBIRD FOUNDATION FOR THE ARTS, 2200 S. State St., Mount Carmel, UT 84755. Mailing Address: P.O. Box 5555, Mount Carmel, UT 84755-5555. Tel.: 435-648-2653.
Web Site: www.thunderbirdfoundation.com
Founded: 2001.
Congressional District: 2
Key Personnel: Dir., Susan Bingham; Chm. (V), Paul Bingham; Pres. (V), Daniel Shea; Treas., Emily Hollingshead; Devel., Bruce Bell; Security, Richard Anderson.
Personnel Profile: Full-Time Paid 1; Full-Time Volunteers 2; Part-Time Paid 4; Part-Time Volunteers 4.
Governing Authority: private; nonprofit organization. Tax-exempt: 501(c)(3).
Institution Type/Description: Art Museum: housed in the summer home of American painter, Maynard Dixon.
Collections: Maynard Dixon's works, furnishings & personal artifacts; photographs by Jack Hillers.
Activities: formal education programs for adults. Annual Event: Maynard Dixon Country Invitational August to September.
Publications: quarterly newsletter.
Hours & Admission Prices: March-Oct. daily 10-5; other times by appointment. Self-guided Tour: $10 per person. Guided Tour: $20 per person; discounts to large groups. &
Attendance: 3,000 (estimated)

Mount Pleasant

MT. PLEASANT PIONEER MUSEUM, 146 S. State St., Mount Pleasant, UT 84647. Tel.: 435-462-2456. Fax: 435-462-2581.
Institution Type/Description: Historic House Museum.
Collections: local history & culture; photographs; period furnishings.
Hours & Admission Prices: Mon.-Tues. 10-2, Wed.-Sat. 10-6. No charge; donations accepted.

Ogden

ECCLES COMMUNITY ARTS CENTER, (M), 2580 Jefferson, Ogden, UT 84401-2411. Tel.: 801-392-6935. Fax: 801-392-5295.
E-mail: eccles@ogden4arts.org
Web Site: www.ogden4arts.org
Founded: 1957.
Congressional District: 1
Key Personnel: C.E.O., Pat Poce; Chm., Colette Torghele; Vice Pres., Steve Poorman; Museum Shop Mgr., Arlene Muller.
Personnel Profile: Full-Time Paid 3; Part-Time Paid 2; Part-Time Volunteers 20.
Governing Authority: nonprofit organization. Tax-exempt: 501(c)(3).
Institution Type/Description: Art Center: housed in c.1893 David Eccles Home.
Collections: work of local artists.
Major Exhibits: Work by Palette Clobot Ogden Members, 1/14; 14th Black & White Competition, 2/14-3/14; Weber School District 2nd Annual Student Competition, 4/14; Paintings by Meri DeCaria, 5/14; Watercolors by Shaunna Alcord & Lola Kartchner, 6/14; Trace of the West, 7/14; 40th Annual Statewide Competition, 8/14; Art by David W. Jackson, 10/14; Paintings by Keith Dagley, 11/14; Paintings by Aaron Fritz, Pottery by Johnny Hughes, 12/14.
Facilities: library of art & art-related literature available for use by the Ogden community; classrooms. Gift-related items for sale.
Activities: guided tours; lectures; gallery talks; piano recitals; temporary exhibitions.

Publications: quarterly newsletter, class schedule.
Hours & Admission Prices: Mon.-Fri. 9-5, Sat. 9-3. No charge; donations accepted. Closed national holidays. &

Attendance: 30,000 (estimated)
Membership: Friend $25; Organization $30; Donor $50; Sponsor $100; Corporate $200; Special $500; Life $1,000.

FORT BUENAVENTURA, 2450 A Ave., Ogden, UT 84401. Mailing Address: 1181 N. Fairgrounds, Ogden, UT 84404-3100. Tel.: 801-399-8099.
Web Site: www.co.weber.ut.us/parks/fortb
Founded: 1980.
Key Personnel: Park Mgr., Jim Carter.
Personnel Profile: Full-Time Paid 1; Part-Time Paid 3.
Governing Authority: state. Parent Institution: Weber County Corp., 2380 Washington Blvd., Ogden, UT 84401. Tax-exempt: 170 (b)(1)(A).
Institution Type/Description: Historic Site: housed in an 1846 fort & 1874 Browning home.
Collections: stockade; cabins.
Hours & Admission Prices: Easter to Oct. daily 8-8. Admission $2 per person with education program, $1 per person without education program; discounts to family & school groups; children under 5 no charge. Season Pass: family $100, individual $30. &
Attendance: 14,700 (estimated)

THE MARY ELIZABETH DEE SHAW GALLERY, Kimball Visual Arts Ctr., 2001 University Cir., Ogden, UT 84408. Tel.: 801-626-6420. Fax: 801-626-6976.
E-mail: katherinelee@weber.edu
Web Site: www.weber.edu/shawgallery
Formerly: Weber State University Art Gallery
Founded: 1960.
Key Personnel: Gallery Dir., Katie Lee Koven.
Personnel Profile: Full-Time Paid 1; Part-Time Paid 7.
Operating Income: 100,000
Governing Authority: university. Parent Institution: Weber State University. Tax-exempt.
Institution Type/Description: Art Gallery.
Collections: contemporary prints, photos, drawings & ceramics.
Major Exhibits: School Days: Annual Juried Student Art Exhibition, 1/17/14-2/8/14; Return to the Sea: Saltworks by Motoi Yamamoto (T), 2/24/14-4/12/14; Spring BFA Thesis Exhibition, 4/18/14-5/3/14.
Activities: films; formal education programs for undergraduate & graduate college students; lectures.
Hours & Admission Prices: Mon.-Fri. 11-5, 1st Fri. of month 11-9, Sat. 12-5. No charge. &
Attendance: 4,000 (estimated)

MUSEUM OF NATURAL SCIENCE, Weber State University, 3848 Harrison Blvd., Ogden, UT 84408-2509. Tel.: 801-626-6160.
E-mail: ajohnston@weber.edu
Web Site: www.community.weber.edu/sciencemuseum
Founded: 1969.
Institution Type/Description: Natural Science Museum.
Collections: exploration; native cultures; geoscience; animal world; planet world; physical world.
Hours & Admission Prices: Mon.-Fri. 8-5. No changes.
Attendance: 18,000 (estimated)

OGDEN NATURE CENTER, 966 W. 12th St., Ogden, UT 84404-5410. Tel.: 801-621-7595. Fax: 801-621-1867. Facebook: Ogden Nature Center.
E-mail: info@ogdennaturecenter.org
Web Site: www.ogdennaturecenter.org
Founded: 1970.
Key Personnel: Dir., Mary McKinley; Chm. (V), Robert Lindquist; Museum Shop Mgr., Linda Page.
Personnel Profile: Full-Time Paid 8; Part-Time Paid 8; Part-Time Volunteers 600; Interns 1.
Governing Authority: Tax-exempt.
Institution Type/Description: Nature Center.
Collections: live native Utah wildlife; bird of prey; natural history; treehouses; over 100 birdhouses; spotting scopes.
Facilities: 152-acre nature preserve; walking trails; bird blinds; observation tower. Museum-related items for sale.
Activities: school programs; workshops; summer camps; classes; special events. Annual Events: Annual Birdhouse Competition; Earth Day; Fly

With The Flock 5K Fun Run/Walk; Garden Tour; summer concerts; Sunshine Breakfast; Creatures of the Night; Holiday Gift Shop Open House.
Publications: biannual newsletter, The Nature Log.
Hours & Admission Prices: Mon.-Fri. 9-5, Sat. 9-4. Adults 12-64 $4, seniors 65 & over $3, children 2-11 $2; members no charge. Closed major holidays.
Attendance: 38,000 (accurate)
Membership: Student $15 & up; Individual $30-$44; Family $45-$99; Grandparents $45-$99; Discovery Club $100-$249; Trails Alliance $250-$499; Preservation Grove $1,000-$4,999; Lifetime $5,000 & up.

OGDEN UNION STATION MUSEUMS, (M), 25th & Wall Ave., Union Station, Ogden, UT 84401. Mailing Address: 2501 Wall Ave., Ogden, UT 84401-1359. Tel.: 801-393-9886. Fax: 801-621-0230. Facebook: Ogden Union Station Museums.
E-mail: museums@theunionstation.org
Web Site: www.theunionstation.org
Founded: 1975.
Congressional District: 1
Key Personnel: Dir., Roberta Beverly; C.E.O., Leon Jones; Pres. (V), Julie Lewis; Museum Shop Mgr., Kathryn Johnson.
Personnel Profile: Full-Time Paid 6; Part-Time Paid 3; Part-Time Volunteers 100; Interns 2.
Governing Authority: Parent Institution: Union Station Foundation. Tax-exempt.
Institution Type/Description: History Museum: housed in 1924 Ogden Union Depot.
Collections: inventor's models & prototypes of Browning firearms, including milling machine, lathes, handtools, furniture & memorabilia; railroad artifacts including a steam-powered rotary snowplow, #6916 locomotive & wooden caboose, a derrick from Union Pacific, SP GP-9 locomotive, SP caboose; Moonglow observation car from 1947 Train of Tomorrow; photographs; Kimball-Browning vintage cars including 9 classic automobiles; 1,200 sq. ft. HO scale model railroad; VP gas turbine; art; replicas of Jupiter.
Research Fields: railroad; American automobiles; Browning guns.
Facilities: reading room; 600-seat auditorium; theater; classrooms;. Museum-related items for sale.
Activities: guided tours; films; gallery talks; concerts; dance recitals; drama; formally organized education programs; docent program; temporary & permanent exhibitions.
Publications: newsletter, All Aboard Annual
Hours & Admission Prices: Mon.-Sat. 10-5. Adults $5, seniors $4, children 2-12 $3; discounts to groups with reservations. Closed New Year's Day; Thanksgiving; Christmas Eve & Day. &
Attendance: 60,000 (estimated)
Membership: Individual & Student $35; Double $40; Single Family $45; Family & Grandparent $50; Family Plus $60.

OGDEN'S GEORGE S. ECCLES DINOSAUR PARK, 1544 E. Park Blvd., Ogden, UT 84401-0803. Tel.: 801-393-3466. Fax: 801-399-0895.
E-mail: info@dinosaurpark.org
Web Site: www.dinosaurpark.org
Founded: 1993.
Congressional District: 1
Key Personnel: Park Dir., Casey Allen; Bd. Chair (V), Cindy Purcell; Museum Shop Mgr., Bridgett Tasker.
Personnel Profile: Full-Time Paid 2; Part-Time Paid 30; Part-Time Volunteers 267.
Governing Authority: George S. Eccles Dinosaur Park and the Elizabeth Dee Shaw Stewart Dinosaur Museum.
Institution Type/Description: Dinosaur Park & Museum.
Collections: Museum: hands-on exhibits; paleontology; natural history; gems & rocks. Outdoor Park: full size dinosaur replicas in natural settings.
Facilities: lecture hall; paleontology lab. Museum-related items for sale.
Activities: special events; children's sand pit; lectures; birthday parties.
Publications: quarterly newsletter.
Hours & Admission Prices: Memorial Day to Labor Day Mon.-Sat. 10-8, Sun. 10-6; Sept.-May Mon.-Sat. 10-6. Adults $7, senior citizens 62 & over and students $6, children 2-12 $5; discounts to groups of 15 or more with reservation; children one & under no charge. Closed New Year's Eve & Day; Thanksgiving & day after; Christmas Eve & Day. &
Attendance: 125,000 (accurate)
Membership: Individual $25; Double $40; Family & Grandparent $60.

OTT PLANETARIUM, Weber State University, 3750 Harrison Blvd., Ogden, UT 84408. Mailing Address: 2508 University Circle, Ogden, UT 84408-2508. Tel.: 801-626-6871.
E-mail: planetarium@weber.edu
Web Site: community.weber.edu/planetarium
Key Personnel: Interim Dir., Dr. Stacy Palen
Institution Type/Description: Planetarium.
Collections: space science.
Hours & Admission Prices: By appointment. Call or see website for admission prices.

TREEHOUSE MUSEUM, 347 22nd St., Ogden, UT 84401. Tel.: 801-394-9663.
E-mail: treehouse@treehousemuseum.org
Web Site: www.treehousemuseum.org
Founded: 1992.
Key Personnel: exec. Dir., Lynne H. Goodwin.
Personnel Profile: Full-Time Paid 7; Part-Time Paid 18; Part-Time Volunteers 100; Interns 2.
Governing Authority: nonprofit organization. Tax-exempt.
Institution Type/Description: Children's Museum.
Collections: hands-on exhibits.
Facilities: theater.
Activities: special programs; birthday parties; facility rental.
Hours & Admission Prices: Mon. Sept.-May 10-3 & Mon. June-Aug. 10-5, Tues.-Thurs. & Sat.10-5, Fri. 10-8. Children 1-12 $6, 13 & up $5; discounts to scheduled groups; babies under 1 & members no charge. Closed New Year's Day; Independence Day; July 24; Thanksgiving; Christmas. &
Attendance: 169,621 (accurate)
Membership: One Child Family $50; 2 Child Family $80; 3 or more Child Family $100.

Orem

WOODBURY ART MUSEUM, (M), 575 E. University Pkwy., #250, Orem, UT 84097-7400. Tel.: 801-863-4200. Fax: 801-426-6218.
E-mail: uvmuseum@uvu.edu
Web Site: www.uvu.edu/museum
Founded: 2001.
Key Personnel: Interim Dir., Melissa Hempel; Registrar, Rebekah Monahan; Graphic Designer, Amanda Luker; Preparator, Chris Juber; Visitor Svcs., Katherine Hall; Museum Asst., Tia Mickelson.
Personnel Profile: Full-Time Paid 1; Part-Time Paid 6; Part-Time Volunteers 3.
Governing Authority: Parent Institution: Utah Valley University.
Institution Type/Description: Art Museum.
Collections: artwork.
Hours & Admission Prices: Tues. 11-8, Wed.-Sat. 11-5. No charge.
Attendance: 5,670 (accurate)

Park City

ALF ENGEN SKI MUSEUM, 3419 Olympic Pkwy., Park City, UT 84098. Mailing Address: P.O. Box 980187, Park City, UT 84098. Tel.: 435-658-4240. Fax: 435-658-4258.
Key Personnel: Exec. Dir., Connie Nelson
Institution Type/Description: History Museum.
Collections: ski history; history of ski & snow sport pioneers & athletes; photographs; personal artifacts.
Hours & Admission Prices: Daily 9-6. Adults $7, seniors & youth 3-17 $5.

JOE QUINNEY WINTER SPORTS CENTER, Olympic Legacy Plaza, Olympic Pkwy., Park City, UT 84098. Mailing Address: P.O. Box 980187, Park City, UT 84098. Tel.: 435-658-4233.
Institution Type/Description: Sports Museum.
Collections: Olympic sports history & competition sites.
Hours & Admission Prices: Daily 10 6. Closed New Year's Day; Easter; Thanksgiving; Christmas.

KIMBALL ART CENTER, 638 Park Ave., Park City, UT 84060-5106. Mailing Address: P.O. Box 1478, Park City, UT 84060 1478. Tel.: 435-649-8882. Fax: 435-649-8889. Facebook: Kimball Art Center.
E-mail: director@kimballartcenter.org
Web Site: www.kimballartcenter.org
Founded: 1976.
Key Personnel: Dir., Robin Marrouche; Chm. (V), Matt Mullin.

Personnel Profile: Full-Time Paid 6; Part-Time Paid 6; Part-Time Volunteers 450; Interns 10.
Governing Authority: Tax-exempt.
Institution Type/Description: Art Center.
Collections: works by regional & national artists.
Major Exhibits: The Art of the Timepiece, 2/8/14-4/6/14; Wasatch Back Student Art Show, 4/12/14-5/25/14; Elliott Erwitt: Dog Dogs (T), 5/31/14-8/10/14.
Activities: classes; A.R.T.S. school tour program; temporary exhibitions; cafe. Museum Sponsors: Park City Kimball Arts Festival.
Publications: monthly e-newsletter.
Hours & Admission Prices: Mon.-Thurs. 10-5, Fri. 10-7, Sat. 12-7, Sun. 12-5. No charge; donations accepted. Closed major holidays. &
Attendance: 118,000 (estimated)
Membership: Student, Teacher & Military $25; Individual $50; Family $95.

PARK CITY MUSEUM, (M), 528 Main St., Park City, UT 84060. Mailing Address: P.O. Box 555, Park City, UT 84060-0555. Tel.: 435-649-7457. Fax: 435-649-7384.
E-mail: museum@parkcityhistory.org
Web Site: www.parkcityhistory.org
Formerly: Park City Historical Society & Museum
Founded: 1984.
Congressional District: 2
Key Personnel: Exec. Dir., Sandra Morrison.
Personnel Profile: Full-Time Paid 4; Full-Time Volunteers 1; Part-Time Paid 6; Part-Time Volunteers 150.
Governing Authority: nonprofit organization. Tax-exempt.
Institution Type/Description: History Museum.
Collections: historic artifacts of Western Summit County from 1860s-present; photos; archives.
Research Fields: early mining history; fires; railroads; early settlers; early businesses; history of skiing in the area; immigration; 2002 Winter Olympic Games; Sundance Film Festival.
Facilities: research library.
Activities: guided walking tours; videos; temporary exhibitions of own collections; docent-led tours of museums; outreach school programs; family backpacks; lecture series; living history seasonal events.
Publications: quarterly newsletter; cemetery tour brochure; self-guided walking tour; newspaper articles, Way We Were.
Hours & Admission Prices: Mon.-Sat. 10-7, Sun. 12-6. Adults $10, children $5; discounts to AAM members; members no charge. Closed Thanksgiving; Christmas Day. &
Attendance: 73,505 (accurate)
Membership: Individual $55; Family $100; Museum Guild $300.

Parowan

PAROWAN OLD ROCK CHURCH MUSEUM, 90 S. Main, Parowan, UT 84761. Mailing Address: P.O. Box 576, Parowan, UT 84761-0576. Tel.: 435-477-3549.
Institution Type/Description: Historic Church: housed in a church built by hand by early pioneers in 1865.
Collections: local history & culture; photographs; music boxes; swords; period clothing & furnishings; books; dolls.
Hours & Admission Prices: Memorial Day to Labor Day Mon.-Sat. 1-5; other times by appointment. No charge.

Payson

HISTORIC PETEETNEET MUSEUM, CULTURAL ARTS AND SOCIAL CENTER, 10 N. 600 E., Payson, UT 84651-2359. Mailing Address: P.O. Box 603, Payson, UT 84651-0603. Tel.: 801-465-5265. Fax: 801-465-9427.
Institution Type/Description: Historic Building: housed in a Victorian school building built in 1901.
Collections: local history & culture; early pioneer artifacts; history of writing; period clothing; paintings; sculptures; blacksmith shop.
Hours & Admission Prices: Mon.-Fri. 10-4. No charge.

Price

COLLEGE OF EASTERN UTAH ART GALLERY, GALLERY EAST, 451 E. 400 North, Price, UT 84501-2699. Tel.: 435-613-5241. Fax: 435-613-4102.
E-mail: robert.degroff@ceu.edu
Web Site: www.ceu.edu
Founded: 1937.
Congressional District: 1

Key Personnel: Dir., Nole Carmack.
Governing Authority: college. Tax-exempt.
Institution Type/Description: Art Gallery.
Collections: paintings; sculpture; photographs.
Facilities: 25,000-vol. reference library available for use within the state.
Activities: guided tours; lectures; gallery talks; arts festivals; formally organized education programs; traveling & temporary exhibits.
Hours & Admission Prices: mid-Sept. to May Mon.-Fri. 9-5; call for special arrangements. No charge. Closed major holidays. &
Attendance: 10,000 (estimated)

∗ UTAH STATE UNIVERSITY - COLLEGE OF EASTERN UTAH PREHISTORIC MUSEUM, (M), 155 E. Main St., Price, UT 84501-3033. Mailing Address: 451 E. 400, N., Price, UT 84501-2699. Tel.: 435-613-5060; 800-817-9949. Fax: 435-637-2514.
E-mail: ken.carpenter@ceu.edu
Web Site: www.ceu.edu/museum
Founded: 1961.
Congressional District: 3
Key Personnel: Dir. & Cur. Paleontology, Dr. Kenneth Carpenter; Chm. (V), Michael Fleck; Museum Shop Mgr., Christine Trease.
Personnel Profile: Full-Time Paid 5; Part-Time Paid 2; Part-Time Volunteers 32.
Governing Authority: state. Parent Institution: Utah State University - College of Eastern Utah. Tax-exempt.
Institution Type/Description: Anthropology, Paleontology & Geology Museum.
Collections: dinosaurs; minerals; fossils of Utah; American Indian artifacts; paleontology; geology.
Research Fields: paleontology, Jurassic & Cretaceous dinosaurs; archaeology, Fremont Indians, proto-historic period.
Facilities: classroom; children's area. Museum-related items for sale.
Activities: guided tours; lectures; gallery talks; permanent & temporary exhibitions; school loan service.
Publications: newsletter, Raptor Review.
Hours & Admission Prices: April-Sept. daily 9-5; Oct.-March Mon.-Sat. 9-5. Family $15, adult $5, seniors $4, children 2-12 $2; members no charge. &
Attendance: 40,000 (accurate)
Membership: Individual $30; Saber Tooth $100; Utahraptor $500; Eolambia $1,000; Mammoth $5,000; Allosaur $10,000; Tyrannosaurus $25,000.

Promontory

GOLDEN SPIKE NATIONAL HISTORIC SITE, 6200 N. 22300 W., Promontory, UT 84307. Mailing Address: P.O. Box 897, Brigham City, UT 84302-0897. Tel.: 435-471-2209, ext. 29. Fax: 435-471-2341.
E-mail: gosp_interpretation@nps.gov
Web Site: www.nps.gov/gosp
Founded: 1965.
Congressional District: 1
Key Personnel: Supt., Leslie Crossland; Chief Operations, Tammy Benson; Museum Shop Mgr., Gary Willden.
Personnel Profile: Full-Time Paid 12; Part-Time Paid 5; Part-Time Volunteers 15.
Volunteer Hours: 3,000
Governing Authority: federal. U.S. Dept. of the Interior, National Park Service.
Institution Type/Description: Golden Spike National Historic Site: first transcontinental railroad was completed here on May 10, 1869.
Collections: materials relating to the railroad, railroad construction, & railroad workers; manuscripts; engine house.
Research Fields: first transcontinental railroad; the effects of the railroad on the development of the West & America; Promontory Summit, Utah.
Facilities: visitor center; Engine House; bookstore.
Activities: self-guided trail; audio tours; interpretive talks & walks; open caption videos; locomotive demonstrations; winter Engine House tours; Jr. ranger program; Boy & Girl Scout activities; school group activities; summer re-enactment. Special Events: May 10th Celebration; Railroaders Festival in August; Steam Festival in December.
Publications: guides, Promontory Trail; Jr. Ranger Program; Golden Spike Handbook; park brochure; handouts.
Hours & Admission Prices: Daily 9-5. Summer May 1-Columbus Day: $7 per vehicle. Winter Mid-Oct.-April 30 $5 per vehicle; Federal Interagency passes honored. Closed New Year's Day; Thanksgiving; Christmas. &
Attendance: 47,500 (accurate)
Membership: Western National Parks Association $25.

BRIGHAM YOUNG UNIVERSITY MUSEUM OF ART, (M), N. Campus Dr., Provo, UT 84602-1400. Mailing Address: 492 MOA, N. Campus Dr., Provo, UT 84602-1400. Tel.: 801-422-8287. Fax: 801-422-0527.
E-mail: moa@byu.edu
Web Site: moa.byu.edu
Founded: 1993.
Congressional District: 3
Key Personnel: Dir., Dr. Campbell B. Gray; Assoc. Dir., Ed Lind; Cur., Paul Anderson; Cur., Dawn Pheysey; Cur., Dr. Marian Wardle; Cur., Diana Turnbow; Exhibition Design, Jeff Barney; Educator, Campus, Rita Wright; Education Research, Dr. Herman du Toit; Head Cur., Dr. Cheryll May; Educator K-12, Lynda Palma; Cur., Jeff Lambson; Mgr. Exhibition Fabrication, John Adams; Head Security, Randy O'Hara; Museum Shop Mgr., Bethany Kramer; Cafe Mgr., Rebecca Armstrong; Exec. Asst., Hannah Diamond; Registrar, Emily Poulsen; Mktg. & Communications Mgr., Christopher Wilson; Designer, Brian Bird; Custodial Supvr., Suzanne Barney; Grounds Supvr., Steven Roylance; Business Mgr., Braden Burgon; Event Scheduler, Edie Zambrano.
Personnel Profile: Full-Time Paid 31; Full-Time Volunteers 1; Part-Time Paid 80; Part-Time Volunteers 210; Interns 5.
Governing Authority: college. Parent Institution: Brigham Young University. Tax-exempt: 501(c)(3).
Institution Type/Description: Art Museum.
Collections: 19th- & 20th-century American collection representing the Hudson River School, American Impressionism, California Regionalism, Ashcan School & Western including such artists as John Singer Sargent, Daniel Ridgway Knight, Francis David Millet, J. Alden Weir, Mahonri M. Young, Minerva Teichert & Maynard Dixon; American photography; 19th- & 20th-century Utah artists including C.C.A Christensen, John W. Clawson, Edwin Evans, J.B. Fairbanks. Rose Hartwell, J.T. Harwood, B.F. Larsen, Lee Greene Richards, John Hafen, Cyrus Dallin; 19th- & 20th-century religious work; European and American prints & drawings.
Research Fields: American art; photography; Baroque & Renaissance prints; religious art; musicology.
Facilities: family interactive center; print study room; auditorium; 2 lecture rooms.
Activities: lectures; gallery talks; formally organized education programs for undergraduate & graduate students; docent program; loan, inter-museum loan, permanent, temporary & traveling exhibitions; continuing education for adults; academic support for university classes; symposia; children's programs and activities.
Publications: catalogues, J. Alden Weir, An American Printmaker, The American Image 1830-1940: Selections from the Museum's Collection; New Directions in Mormon Art; Cipriano de Rore's Venus Motet; 150 Years of American Painting, 1794-1944; Rembrandt; Rembrandt?; Sacred Images: A Vision of Native American Rock Art; A Song of Joys: The Biography of Mahonri Mackintosh Young; Escape to Reality: The Western World of Maynard Dixon; American Women Modernists: The Legacy of Robert Henri, 1910-1945 (co-published with Rutgers University Press); Beholding Salvation: The Life of Christ in Words and Art; Minerva Teichert: Pageants in Paint.
Hours & Admission Prices: Mon.-Wed. & Fri. 10-6, Thurs. 10-9, Sat. 12-5. No charge; donations accepted. Closed Independence Day; Thanksgiving; Christmas Eve & Day. &
Attendance: 327,000 (accurate)
Membership: Students & Seniors $30; Museum Passport $50; Silver Passport $100; Gold Passport $300; Platinum Passport $1,000; Corporate according to employee number.

BRIGHAM YOUNG UNIVERSITY MUSEUM OF PALEONTOLOGY, 1683 N. Canyon Rd., Provo, UT 84602. Mailing Address: 140 ESM, BYU, P.O. Box 23300, Provo, UT 84602-3300. Tel.: 801-422-3939. Fax: 801-378-7919.
E-mail: rod_scheetz@byu.edu
Web Site: cpms.byu.edu/esm/
Formerly: Brigham Young University Earth Science Museum
Founded: 1987.
Congressional District: 3
Key Personnel: C.E.O. Brigham Young Univ., Gordon B. Hinckley; Pres. Brigham Young Univ., Cecil O. Samuelson; Cur., Rod Scheetz, Ph.D.; Vertebrate Paleontologist, Brooks Britt, Ph.D.
Personnel Profile: Full-Time Paid 1; Part-Time Paid 12; Part-Time Volunteers 4; Interns 2.
Governing Authority: private university; not-for-profit. Parent Institution: Latter Day Saints Church. Subsidiary Institution: Brigham Young Univ. Tax-exempt.
Institution Type/Description: Paleontology Museum.

Collections: fossils from most geologic periods; research collections of dinosaurs & other vertebrates, especially from Jurassic and early Cretaceous Periods.
Research Fields: Tertiary period mammals from the U.S. & Mexico; Mesozoic period dinosaurs of Utah, Colorado & surrounding states.
Facilities: 250-vol. library of books on vertebrate paleontology; 4,500 papers on vertebrate paleontology, available for inter-library loan; educational facilities; 3,600 sq. ft. exhibit space.
Activities: docent program; films; formal education programs for undergraduate or graduate students; guided tours; lectures; loan, temporary & traveling exhibitions; school loan service.
Publications: BYU Geology Studies.
Hours & Admission Prices: Mon.-Fri. 9-5. No charge; donations accepted. Closed state holidays. &
Attendance: 25,000 (estimated)

THE CRANDALL HISTORICAL PRINTING MUSEUM, 275 E. Center St., Provo, UT 84606-3133. Tel.: 801-377-7777. Fax: 801-375-5555.
E-mail: lou_crandall@yahoo.com
Web Site: crandallprintingmuseum.org
Founded: 1996.
Congressional District: 3
Key Personnel: Pres. (V), Louis E. Crandall; Chm. (V), Brent Ashford; Museum Shop Mgr., Wallace Saling.
Personnel Profile: Full-Time Paid 1; Part-Time Volunteers 3; Interns 1.
Governing Authority: Tax-exempt: 501(c)(3).
Institution Type/Description: Writing & Printing History Museum.
Collections: writing & printing history; type casting; printing; bookbinding; paper making; wood block prints; Gutenberg Press; documents & books from 1900 BC; cuneiform clay tablets; pages & letters from European & American historical figures; rare books & first print Bibles; early English common press; Acorn press; linotype machine; 19th century book binding equipment; monotype machine; 20th century presses.
Activities: monthly lecture series; mobile exhibit to schools. Museum Sponsors: Utah Printer's Hall of Fame Recognition Banquet in February; Colonial Days Celebration in July.
Publications: quarterly newsletter, The Printed Word.
Hours & Admission Prices: Mon.-Fri. 9-2. Adults $2; discount to Boy Scouts working on certificate or patch. Closed major holidays.
Attendance: 6,000 (estimated)

✤ **MONTE L. BEAN LIFE SCIENCE MUSEUM, (M),** 290 MLBM Bldg., Brigham Young University, Provo, UT 84602. Tel.: 801-422-5052. Fax: 801-422-0093.
E-mail: secretary@museum.byu.edu
Web Site: mlbean.byu.edu/
Founded: 1978.
Key Personnel: Dir., Larry L. St. Clair; Assoc. Dir., Dr. Jack Sites; Asst. Dir., Marta Adair; Museum Shop Mgr., Patty Jones.
Personnel Profile: Full-Time Paid 9; Full-Time Volunteers 3; Part-Time Paid 38; Part-Time Volunteers 6.
Governing Authority: university. Parent Institution: Brigham Young University. Subsidiary Institution: The Lytle Nature Preserve (an off-site preserve managed by the museum located in Utah's Mojave Desert in the southwest corner of the state. Tax-exempt.
Institution Type/Description: Life Science Museum.
Collections: herbarium; entomology; herpetology; shells; ichthyology; ornithology; mammalogy.
Research Fields: pertaining to collections.
Facilities: Museum-related items for sale.
Activities: guided tours; lectures; films; formally organized education programs for children and undergraduate college students; temporary & permanent exhibitions.
Publications: quarterly scientific journal, Western North American Naturalist.
Hours & Admission Prices: Mon.-Fri. 10-9, Sat. 10-5. No charge. Closed New Year's Day; Thanksgiving; Christmas. &
Attendance: 210,000 (estimated)
Membership: Student $10; Regular $15; Family $25; School & Society $50; Business & Corporation $100; Donor $500.

MUSEUM OF PEOPLES AND CULTURES, BYU, 700 N. 100 E., Allen Bldg., Provo, UT 84602. Mailing Address: 105 Allen-BYU, Provo, UT 84602. Tel.: 801-422-0020. Fax: 801-422-0026.
E-mail: mpc_programs@byu.edu
Web Site: mpc.byu.edu
Founded: 1946.
Congressional District: 3

Key Personnel: Dir., Paul Stavast; Cur. Education, Kari Nelson.
Personnel Profile: Full-Time Paid 1; Part-Time Paid 14; Interns 3.
Governing Authority: church affiliated. Parent Institution: Brigham Young University. Tax-exempt.
Institution Type/Description: Anthropology and Ethnology Museum.
Collections: archaeology & ethnology.
Major Exhibits: Concealing Faces Revealing Expressions, 4/12-4/14.
Research Fields: Mesoamerica; Southwest; Polynesia; Great Basin; Andean cultures.
Facilities: 2,500-vol. library; field research station; laboratory; classroom.
Activities: guided tours; temporary exhibits; instruction in museum practices through BYU Dept. of Anthropology; research; outreach education programs.
Publications: Publications in Archaeology; Museum of Peoples and Cultures Occasional Papers; Popular Series.
Hours & Admission Prices: Mon.-Fri. 9-5. No charge. Closed New Year's Day; Presidents' Day; Civil Rights Day; Memorial Day; Independence Day; Labor Day; Thanksgiving; Christmas. &
Attendance: 25,255 (accurate)

Salt Lake City

BEEHIVE HOUSE, 67 E. South Temple, Salt Lake City, UT 84150-9719. Tel.: 801-240-2681. Fax: 801-240-2695. TDD: 801-240-2672.
Founded: 1961.
Key Personnel: Dir., Mr. McLea.
Personnel Profile: Full-Time Volunteers 6.
Governing Authority: church. Parent Institution: Latter-Day Saints Church, 50 E. North Temple, Salt Lake City, UT 84150. Tax-exempt.
Institution Type/Description: Historic House: 1854-1877 Brigham Young's residence & office.
Collections: 1850-1870s household furnishings.
Activities: tours.
Publications: brochure.
Hours & Admission Prices: Daily 9-8:30. No charge. Closed New Year's Day; Thanksgiving; Christmas.
Attendance: 200,000 (estimated)

CHASE HOME MUSEUM OF UTAH FOLK ARTS, center of Liberty Park (approx. 600 E. 1100 S.), Salt Lake City, UT 84105. Mailing Address: c/o Utah Arts Council, 617 E. South Temple, Salt Lake City, UT 84102-1101. Tel.: 801-533-5760. Fax: 801-533-4202. TDD: 800-346-4128.
E-mail: cedison@utah.gov
Web Site: www.folkartsmuseum.net
Founded: 1986.
Congressional District: 1
Key Personnel: Dir., Carol Edison; Folk Arts Coord., Craig Miller.
Personnel Profile: Full-Time Paid 2; Part-Time Paid 3; Part-Time Volunteers 1; Interns 1.
Governing Authority: state government. Parent Institution: Utah Arts Council. Tax-exempt: state agency.
Institution Type/Description: Folk Arts Museum: 1853 two-story adobe structure built by Isaac Chase and Mormon leader Brigham Young, sold to city in 1880, renovated in 2000.
Collections: folk & ethnic art by Utah residents; quilts; saddles; Indian beadwork; needlework; rugs; woodcarving; horse tack; Indian basketry; cultural communities featured include Anglo pioneer, Native American (Goshute, Paiute, Shoshone, Ute & Navajo), Tongan, Hispanic & ranching. Ethnographical; textiles.
Research Fields: cultural communities of Utah including Anglo pioneer, Native American (Goshute, Paiute, Shoshone, Ute & Navajo), Tongan, Hispanic & ranching.
Activities: guided tours on request; archive of photographs & sound recordings; concerts. Annual Events: concerts in July & August.
Publications: booklet, Hecho en Utah (Made in Utah); recording, Listening In: Utah Storytelling; Willow Stories: Contemporary Navajo Baskets; Social Dance in the Mormon West; calendar, Annual Utah Traditions.
Hours & Admission Prices: Spring & Fall Sat.-Sun. 12-5; Memorial Day-Labor Day: Mon.-Thurs. 12-5, Fri.-Sun. 2-7. No charge. &
Attendance: 15,000 (estimated)

CHURCH HISTORY MUSEUM, 45 N. West Temple St., Salt Lake City, UT 84150-0902. Tel.: 801-240-4615. Fax: 801-240-5342.
E-mail: churchmuseum@ldschurch.org
Web Site: www.history.lds.org
Formerly: Museum of Church History and Art
Founded: 1869.

Congressional District: 2
Key Personnel: Dir., Kurt Graham; Mgr. Education, Ray K. Halls; Exhibits Mgr., Maryanne Andrus; Historic Sites Researcher, T. Michael Smith; Mgr. Collections, Carrie Snow; Conservator, Jennifer Hadley; Product Devel., Craig Rohde; Museum Store Mgr., Annette Burdette; Administrative Asst., Karen Westenskow.
Personnel Profile: Full-Time Paid 15; Part-Time Paid 3; Part-Time Volunteers 325; Interns 5.
Governing Authority: Parent Institution: The Church of Jesus Christ of Latter-day Saints. Tax-exempt: 501(c)(3).
Institution Type/Description: History & Art Museum.
Collections: art & artifacts emphasizing the history & culture of The Church of Jesus Christ of Latter-day Saints from the 1820s-present; fine art; folk art; native arts; furniture; costumes; decorative arts; tools; crafts; historical memorabilia.
Major Exhibits: Portraits of Childhood, 1/14-10/5/14; A Good Turn Daily, 1/14-10/5/14; A Book of Mormon Celebration, 1/14-10/5/14; No Greater Love, 1/14-10/5/14.
Research Fields: LDS Church history, historic sites; art; artists.
Facilities: 2,000-vol. library pertaining to the history, art & artifacts of The Church of Jesus Christ of Latter-day Saints; 180-seat auditorium. Books, art reproductions & other related items for sale.
Activities: guided tours; lectures; films; theater; organized education programs for children & adults; docent program; temporary exhibitions of your own collections; gallery talks; School Outreach Programs.
Publications: exhibit catalogues; brochures; fine art prints; posters; postcards; slides; notecards.
Hours & Admission Prices: Mon.-Fri. 9-9, Sat.-Sun. & holidays 10-5. No charge. Closed New Year's Day; Easter; Thanksgiving; Christmas Eve & Day. &
Attendance: 209,261 (accurate)

CLARK PLANETARIUM, 110 S. 400 W., Salt Lake City, UT 84101-1145. Tel.: 801-456-7827, ext. 0. Fax: 801-456-4928.
Web Site: www.clarkplanetarium.org
Formerly: Hansen Planetarium
Founded: 1965.
Congressional District: 2
Key Personnel: C.E.O. & Dir., Seth Jarvis; Museum Shop Mgr., Mike Sheehan.
Personnel Profile: Full-Time Paid 24; Part-Time Paid 65.
Governing Authority: Salt Lake County. Tax-exempt.
Institution Type/Description: Planetarium, Space Science Museum.
Collections: meteorites; space technology; Apollo 15 Moon rock; large scale model solar system; moonscape; marsscape; George Rhoades' kinetic rolling ball sculpture; Foucault pendulum; 75-inch earth globe.
Facilities: classroom; 217-seat theatre; 3-D IMAX theater. Books, executive toys & educational instruments for sale.
Activities: star programs; guided tours; lectures; science demonstrations; films; gallery talks; arts festivals; drama; study clubs; hobby workshops; TV & radio programs; formerly organized education programs for children & adults; permanent & temporary exhibitions; traveling science demonstration service.
Hours & Admission Prices: Mon.-Wed. 10:30-8, Thurs. 10:30-9, Fri.-Sat. 10:30-11, Sun. 10:30-6. Star Show: adults $8, children 12 & under $6; discounts to shows before 5pm; ASTC reciprocal admission; members no charge. Call 801-456-7827 for current showtimes & prices. &
Attendance: 350,000 (accurate)
Membership: Duo $49; Family $99; Super Family $149.

CLASSIC CARS INTERNATIONAL, 355 W. 700 South St., Salt Lake City, UT 84101-2609. Tel.: 801-322-5509; 582-6883 & 201-1683. Fax: 801-322-5509.
E-mail: classiccarsintl@hotmail.com
Web Site: www.classiccarsintl.net
Founded: 1975.
Key Personnel: C.E.O., Stacy Williams.
Personnel Profile: Full-Time Paid 1; Full-Time Volunteers 2; Part-Time Paid 1.
Governing Authority: private; nonprofit organization.
Institution Type/Description: Automobile Museum.
Collections: over 300 restored automobiles 1907-1970.
Hours & Admission Prices: Mon.-Sat. 9-4; Sun. by appointment. Adults $6, senior citizens & children $4; discounts to AAA, AAM & ICOM members. Closed Christmas to New Year's. &
Attendance: 1,000 (estimated)

DAUGHTERS OF UTAH PIONEERS PIONEER MEMORIAL MUSEUM & INTERNATIONAL SOCIETY DAUGHTERS OF UTAH PIONEERS, (M), 300 N. Main St., Salt Lake City, UT 84103-1699. Tel.: 801-532-6479, ext. 201. Fax: 801-532-4436.
E-mail: info@dupinternational.org
Web Site: www.dupinternational.org
Founded: 1901.
Key Personnel: C.E.O. & Pres. (V), Maurine P. Smith; 1st Vice Pres., Joleen Barker; Sec., Kathryn Brimhall; Custodian Museum Artifacts, Kari Main; Museum Shop Mgr., Gayle Sheffield.
Personnel Profile: Full-Time Paid 3; Full-Time Volunteers 3; Part-Time Paid 6; Part-Time Volunteers 125.
Volunteer Hours: 28,518
Governing Authority: society. Tax-exempt.
Institution Type/Description: Pioneer History Museum.
Collections: pioneer histories, vehicles & artifacts; crafts; art; relics of the Utah pioneer period, 1847-1900; manuscripts; photographs.
Research Fields: Utah pioneer history.
Facilities: auditorium. Museum-related items for sale.
Activities: permanent exhibitions; VCR program; outreach programs.
Publications: pamphlets; books, An Enduring Legacy; Our Pioneer Heritage; Chronicles of Courage; reproduction cards; Pioneer Pathways; Museum Memories.
Hours & Admission Prices: June-Aug. Mon.-Sat. 9-5, Sun. 1-5; Sept.-May Mon.-Sat. 9-5. No charge; donations accepted. Closed national holidays. &
Attendance: 36,263 (accurate)
Membership: Annual $10.

DISCOVERY GATEWAY CHILDREN'S MUSEUM, 444 W. 100 S., Salt Lake City, UT 84101-1195. Tel.: 801-456-KIDS (5437). Fax: 801-456-5440.
Web Site: www.discoverygateway.org
Formerly: The Children's Museum of Utah
Founded: 1979.
Congressional District: 2
Key Personnel: C.E.O., Maria Farrington; Chm. (V), Dorothy Pleshe; C.O.O. & C.F.O., Victoria Bernier; Dir. Devel. & External Rels., Kirsta Albert.
Personnel Profile: Full-Time Paid 25; Part-Time Paid 25; Part-Time Volunteers 50.
Governing Authority: private; nonprofit organization. Tax-exempt: 501(c)(3).
Institution Type/Description: Children's Museum & Discovery Center.
Collections: interactive, hands-on exhibits emphasizing science art & technology.
Facilities: 60,000 sq. ft. interactive exhibit space.
Activities: hands-on discovery center for children. Annual Events: Bumblebee Bash; Spooktacular; fall fundraiser; Breakfast with Santa.
Publications: monthly newsletter; monthly calendar of events.
Hours & Admission Prices: Mon.-Thurs. 10-6, Fri.-Sat. 10-8, Sun. 12-6. Admission $8.50; discounts to ASTC members; members no charge. Closed Easter; Independence Day; Thanksgiving; Christmas. &
Attendance: 285,000 (estimated)
Membership: Family Memberships: 3 Person $95; 4 Person $115; 5 Person $135; 6 Person $155 & $20 for each additional person.

FINCH LANE GALLERY, 54 Finch Lane, Salt Lake City, UT 84102-1809. Tel.: 801-596-5000.
Key Personnel: Dir., Nancy Bosskoff
Institution Type/Description: Art Gallery.
Collections: mixed media; photography; paintings.
Hours & Admission Prices: Call for hours. No charge. Closed holidays; between scheduled exhibits.

HELLENIC CULTURAL MUSEUM, 279 S. 300 W., Salt Lake City, UT 84101-1703. Tel.: 801-328-9681.
E-mail: tpmcgrath1@gmail.com
Founded: 1986.
Key Personnel: C.E.O., Jon Pezely; Trustee & Treas., T. McGrath
Institution Type/Description: Greek Heritage Museum.
Collections: Utah's Greek heritage & history; personal artifacts; clothing; costumes; photographs.
Facilities: Museum-related items for sale.
Activities: video; special events.
Hours & Admission Prices: Wed. 9-12, Sun. after church services; groups of 30 or more by appointment. No charge; donations accepted.
Attendance: 223 (accurate)

THE LEONARDO, 209 E. 500 S., Salt Lake City, UT 84111. Tel.: 801-531-9800. Fax: 801-531-9801.
E-mail: info@theleonardo.org
Web Site: www.theleonardo.org
Founded: 2002.
Key Personnel: Exec. Dir., Alexandra Hesse; Chm. (V), Nicholas Gibbs; Dir. Devel., Katie Smith; Dir. Visitor Svcs. & Museum Shop Mgr., Mike Aguilar; Community Rels., Angelina Kendzior
Governing Authority: Parent Institution: The Library Square Foundation for Art, Culture & Science. Tax-exempt.
Institution Type/Description: Art Museum.
Collections: art, science & technology.
Major Exhibits: Dead Sea Scrolls (T), 1/14-4/27/14.
Activities: workshops.
Hours & Admission Prices: Sun.-Thurs. & Sun. 10-5, Thurs.-Sat. 10-10. Adults $9, seniors, youth 13-17, students & military $8, children 3-12 $7; discounts to ASTC members; toddlers 2 & under no charge. &
Attendance: 96,000 (accurate)
Membership: Individual $45; Dual $65; Family $85; Curious $150; Insatiably Curious $250; Friend of the Leo $500; DaVinci Visionary $1,000.

THE MUSEUM OF UTAH ART & HISTORY, 125 S. Main St., Salt Lake City, UT 84111. Mailing Address: 825 N. 300 W., Ste. W109, Salt Lake City, UT 84103. Tel.: 801-355-5554. Fax: 801-355-5222.
Key Personnel: Exec. Dir., Kandace Steadman
Institution Type/Description: Art Museum.
Collections: Utah art & history; paintings; sculpture.
Hours & Admission Prices: Gallery: Tues.-Sat. 12-5, 3rd Fri. each month 6pm-9pm. Office: Mon.-Fri. 8-5.

✱ NATURAL HISTORY MUSEUM OF UTAH, (M), 301 Wakara Way, Salt Lake City, UT 84108. Tel.: 801-581-6927 & 4303. Fax: 801-585-3684.
E-mail: sgeorge@umnh.utah.edu
Web Site: nhmu.utah.edu
Founded: 1963.
Congressional District: 2
Key Personnel: Dir., Sarah B. George; Assoc. Dir. Community Rels., Ann Hanniball; Chief Cur., Duncan Metcalfe; Dir. School Programs, Madlyn Runburg; Dir. Public Programs, Becky Menlove; Dir. Mktg., Janet Frasier; Mgr. Public Rels., Patti Carpenter; Dir. Devel., Chris Eisenburg; Cur. Paleontology, Randall Irmis; Cur. Vertebrates, Eric Rickart; Cur. Herbarium, Mitchell Power; Operations Mgr. & Museum Shop Mgr., Tony Millet.
Personnel Profile: Full-Time Paid 45; Part-Time Paid 54; Part-Time Volunteers 200; Interns 10.
Governing Authority: university. Parent Institution: University of Utah. Tax-exempt: 501(c)(3) & 170(b)(A).
Institution Type/Description: Natural History Museum.
Collections: Utah geology; Jurassic & Cretaceous dinosaurs; Great Basin & Colorado Plateau ethnology & archaeology; Norton Hall of Minerals; mammal dioramas; fossil mammals; Garrett herbarium; research collections of biology, geology, paleontology & anthropology.
Research Fields: mammalian systematics; dinosaur systematics; plant & mammal biogeography; textile analysis; human foraging ecology; Fremont archaeology; plant paleoecology.
Facilities: classrooms; education wing. Books & native arts for sale.
Activities: guided tours; lectures; formally organized education programs for children & adults; docent program; teacher in-service workshops; permanent, temporary & traveling exhibitions; school outreach programs; Junior Science Academy; statewide exhibit & education outreach; at-risk youth program; Natural History Now.
Publications: quarterly newsletter; adults/children education course listing, Adventures.
Hours & Admission Prices: Wed. 10-9, Thurs.-Tues. 10-5. Adults $9, senior citizens 65 & over and youth 13-24 $7, children 3-12 $6; university students, faculty & staff, children 2 & under and members no charge. Closed Thanksgiving; Christmas. &
Attendance: 75,500 (accurate)
Membership: Individual $25; Duo $35; Family $55; Explorer $100; Adventurer $250; Patron $500.

PRICE FAMILY HOLOCAUST MEMORIAL, I.J. and Jeanne Wagner Jewish Community Center, 2 N. Medical Dr., Salt Lake City, UT 84113-1101. Tel.: 801-581-0098. Fax: 801-581-0718.
Web Site: www.slcjcc.org
Institution Type/Description: History Museum.

Collections: Holocaust history; Jewish life before the war; Nazi regime; Diaspora.
Hours & Admission Prices: Call for hours. No charge. &

RED BUTTE GARDEN & ARBORETUM, University of Utah, 300 Wakara Way, Salt Lake City, UT 84108. Tel.: 801-585-0556. Fax: 801-587-5887.
E-mail: information@redbutte.utah.edu
Web Site: www.redbuttegarden.org
Founded: 1961.
Congressional District: 1
Key Personnel: Exec. Dir., Gregory Lee; Chm. (V), David Gee; Dir. Devel., Kathryn Atwood; Volunteer Coord., Meghan Eames; Mktg. & Public Rels., Bryn Ramjoue; Visitor Svcs. Dir. & Museum Shop Mgr., Derrek Hanson; Horticulture Education Dir., Patrick Newman; Museum Shop Mgr., Dianne Crosby.
Personnel Profile: Full-Time Paid 27; Part-Time Paid 50; Part-Time Volunteers 200; Interns 2.
Governing Authority: university. Parent Institution: University of Utah. Tax-exempt: 501(c)(3).
Institution Type/Description: Botanical Garden & Arboretum.
Collections: Botanical Garden: 100 acres at mouth of Red Butte Canyon including 25 acres of formal gardens; Arboretum: 100 acres with over 9,000 specimens of trees & shrubs from around the world; hybrid oak grove; conifers; day lily garden; Dyke's Award iris collection; ornamental grasses; sage brush; penstemon; lilacs; crabapples; woody shrubs; aquatic plants & perennial display; medicinal, herbs & fragrant plant display; natural riparian area; 18 endangered species.
Research Fields: oak hybridizing; urban tree evaluation; cultivation of endangered plant species; Great Basin & Intermountain regional representative for center for plant conservation activity.
Facilities: visitor center; classroom; 60,000 sq. ft. children's garden; 3,000-seat public amphitheater; 25 acres of botanical gardens; four miles of mountain trails. Museum-related items for sale.
Activities: guided tours; lectures; temporary exhibitions; botany & horticulture classes; workshops; school field trips; concerts; plant sales; theatre; children's programs.
Publications: quarterly newsletter; quarterly catalog of courses & events; Self Guided Tree Tour; State Arboretum of Utah Pub. #1, Oak Hybridization at the University of Utah 1982; periodic papers; nature interpretation trail guides.
Hours & Admission Prices: April & Sept. daily 9-7:30; May-Aug. daily 9-9; Oct.-March daily 9-5; garden closed at 5pm on concert days. Adults $10, students & seniors $8; discounts to APGA & AHS reciprocal garden program; children under 3 & members no charge. Closed New Year's Eve & Day; Thanksgiving; Christmas Eve, Day & week. &
Attendance: 162,854 (accurate)
Membership: Basic $35; Individual $45; Duo $55; Family $65; Circle of Friends $75. Corporate & Garden Club memberships, call 801-585-5658 for information.

SALT LAKE ART CENTER, 20 S. West Temple, Salt Lake City, UT 84101-1406. Tel.: 801-328-4201. Fax: 801-322-4323.
E-mail: saras@slartcenter.org
Web Site: www.slartcenter.org
Founded: 1931.
Congressional District: 2
Key Personnel: Exec. Dir., Heather Ferrell; Pres. (V), Erik A. Christiansen; Controller, Erin M. Call; Dir. Devel., John B. Wither; Cur. Exhibits, Jay Heuman; Finance & Operations Admin., Sara South; Dir. Communications, Marlow Hoffman; Devel. Mgr., Patti Hanson; Visitor Svcs. Mgr., Kate Ithurralde; Co-Dir. Art Center School, Rodger Newbold; Co-Dir. Art Center School, Steve Fredrick.
Personnel Profile: Full-Time Paid 8; Part-Time Paid 4; Part-Time Volunteers 18; Interns 4.
Governing Authority: nonprofit. Tax-exempt: 501(c)(3).
Institution Type/Description: Art Museum & Center.
Collections: paintings; photographs.
Research Fields: contemporary art; contemporary Utah art.
Facilities: auditorium; classrooms. Books & periodicals for sale.
Activities: guided tours; changing exhibition programs; art workshops; performing arts; films; formally organized education programs & art classes for children & adults; lecture series.
Publications: quarterly newsletter; monthly calendar; exhibition catalogs.
Hours & Admission Prices: Tues.-Thurs. & Sat. 11-6, Fri. 11-9. No charge; donations accepted. Closed holidays. &
Attendance: 18,000 (estimated)
Membership: Student, Educator & Senior $20; Working Artist $30; Individual

$35; Companion $50; Enthusiast $100; Advocate $250; Friends of Contemporary Art $500; Director's Circle $1,000.

THIS IS THE PLACE HERITAGE PARK, 2601 E. Sunnyside Ave., Salt Lake City, UT 84108-1453. Tel.: 801-582-1847. Fax: 801-583-1869.
E-mail: eivory@thisistheplace.org
Web Site: www.thisistheplace.org
Founded: 1947.
Congressional District: 2
Key Personnel: Chm. (V), Ellis Ivory; Vice Pres. Finance, Lorin Cummings; Programming Coord., Cliff Harris.
Personnel Profile: Full-Time Paid 20; Part-Time Paid 60; Part-Time Volunteers 180.
Governing Authority: state. Parent Institution: Utah State Div. of Parks and Recreation, 1636 W. North Temple, Salt Lake City, UT 84116. Tax-exempt: 501(c)(3).
Institution Type/Description: Park Visitor Center, Monument & Living History Museum.
Collections: site of This Is The Place Monument; Brigham Young's Forest Farm House & Heritage Village, a pioneer village depicting the culture & heritage of Utah's pre-railroad (1869) era; Pony Express monument; Journey's End monument.
Research Fields: Utah social history; mid-19th century furnishing patterns.
Facilities: visitor center; picnic area; rental facilities.
Activities: self-guided tours; educational events; special holiday events; living history programs. Annual Event: Haunted Village; Pioneer Christmas Tour.
Hours & Admission Prices: Visitor Center & This Is The Place Monument: daily call for hours. Old Deseret Village: May-Sept. Mon.-Sat. 9-5. Adults $10, seniors & children 3-11 $7; members no charge. Closed New Year's Day; Thanksgiving; Christmas. &
Attendance: 300,000 (estimated)
Membership: Subject to change.

TRACY AVIARY, 589 E. 1300 S., Salt Lake City, UT 84105-1111. Tel.: 801-596-8500. Fax: 801-596-7325.
E-mail: info@tracyaviary.org
Web Site: www.tracyaviary.org
Founded: 1938.
Key Personnel: C.E.O. & Exec. Dir., Tim Brown; Chm. (V), Thomas Barton.
Personnel Profile: Full-Time Paid 25; Part-Time Paid 5; Part-Time Volunteers 20; Interns 5.
Governing Authority: municipal. Parent Institution: Friends of Tracy Aviary. Tax-exempt.
Institution Type/Description: Aviary.
Collections: birds from throughout the world; indoor rainforest; owl forest; wetlands.
Facilities: zoological park.
Activities: guided tours; bird shows; classes; camps; interactive Parrot & Pelican encounter.
Publications: monthly newsletter, Inside Your Aviary.
Hours & Admission Prices: Daily 9-5. Adults $7, students & seniors $6, children $5; discount to groups; members & children under 2 no charge. Closed Thanksgiving; Christmas. &
Attendance: 117,000 (estimated)
Membership: Individual $20; Family $60; (additional member $12.50).

* **UTAH MUSEUM OF FINE ARTS,** Marcia and John Price Museum Bldg., University of Utah, 410 Campus Center Dr., Salt Lake City, UT 84112-0360. Tel.: 801-585-5163. Fax: 801-585-5198.
E-mail: umfa.publicrelations@utah.edu
Web Site: www.umfa.utah.edu
Founded: 1951.
Congressional District: 2
Key Personnel: Exec. Dir., Gretchen Dietrich; Dep. Dir., George Lindsey; Dir. Collections & Exhibits, David Carroll; Dir. Education & Engagement, Kerry O'Grady; Dir. Special Projects, Sonja Lunde.
Personnel Profile: Full-Time Paid 25; Part-Time Paid 48; Part-Time Volunteers 163; Interns 6.
Governing Authority: state; university. Parent Institution: University of Utah, 204 Park Building. Tax-exempt.
Institution Type/Description: Art Museum.
Collections: features over 19,000 works; ethnographic materials; prints; photographs; works on paper; paintings; glass; ceramic; furniture; tapestries; sculpture; video; Pre-Columbian, Utah, Western, African, American, American Indian, Ancient, Classical, Asian, Dutch, Flemish, European, Modern & Oceanic art; Greek, Roman & Egyptian antiquities; The Val A. Browing Memorial Collection of 500 Years of European Masterworks; The Marriner S. Eccles Collection of Masterworks.
Major Exhibits: Exploring Sustainability, 10/13-7/14; Alfred Lambourne, 2/13-6/14; CLUI: Great Salt Lake, 1/14-5/14; Tacita Dean (T), 1/14-5/14; Salt 9, 1/14-6/14; Art of the Bingham Canyon Mine, 5/14-9/14; Fazal Sheikh (T), 7/14-11/14; Faculty Show, 10/14-1/15; Art Speaks, 12/14-6/15.
Facilities: 500-vol. library of reference books; 263-seat auditorium.
Activities: guided tours; lectures; films; gallery talks; concerts; formally organized education programs for public school children & undergraduate college students; inter-museum loan, permanent, temporary & traveling exhibitions; Art in a Box program; Teacher Resource Center; Museum Experiences for Senior Citizens; docent program or council.
Publications: Exhibition catalogues: Albert Tissandier: Drawings of Nature and Industry in the United States; American Women at Work: Prints by Women Artists of the Nineteen Thirties; The Big Print; The Bungalow Lifestyle and the Arts & Crafts Movement in the Intermountain West; Earl Jones; George Dibble Drawings; Images of the Great Salt Lake; In Support of the Arts in Utah: An Eccles Family Tradition; J. George Midgley 1882-1979: Utah Photographer; Lee Green Richards; Paging Through Medieval Lives. Exhibition brochures: Innovation and Tradition in Japanese Woodblock Prints; Photography by Jim Frankoski: The Restoration of the Cathedral of the Madeleine; Prints by the Nabis: Vuillard and His Contemporaries; Utah Art from the Permanent Collection: An Exhibition Honoring the Utah Statehood Centennial 1896-1996. American Art: Challenging Traditional Interpretations; Enlightened by Peace, Amsterdam Inspires the Arts: A Painted Allegory by Domenicus Van Wijnen (1661-ca. 1700); The Marriner S. Eccles Collection of Masterworks; The Utah Museum of Fine Arts: Selected Works; The Val A. Browning Memorial Collection of 500 Years of European Masterworks; The Val A. Browning Collection: A Selection of Old Master Paintings.
Hours & Admission Prices: Tues. & Thurs.-Fri. 10-5, Wed. 10-8, Sat.-Sun. 11-5. Museum: adults $7, youth 6-18 & seniors $5; members, Utah higher education students, staff & faculty, NARM members, military families no charge. Closed major holidays. &
Attendance: 131,190 (estimated)
Membership: Student $15; Senior $30; Individual $40; Duo $50; Family $65; Patron $100; Grand Patron & Young Benefactor $250; Benefactor $500; Connoisseur's Circle $1,000; Collector's Circle $2,500; Director's Circle $5,000.

UTAH STATE HISTORICAL SOCIETY, 300 Rio Grande St., (450 West), Salt Lake City, UT 84101-1182. Tel.: 801-245-7225. Fax: 801-533-3503. TDD: 801-533-3502.
Web Site: heritage.utah.gov
Founded: 1897.
Congressional District: 8
Key Personnel: Dir. & State Historic Preservation Officer (SHPO), Brad Westwood; Chm. (V), Michael Homer; Research Center Mgr., Greg Walz; Archaeology Record Mgr., Arie Leeflang; Contracts & Grants Mgr., Debbie Dahl; Research/Collections Coord., Doug Misner; Office Specialist, Janice Reed Campbell; Dep. SHPO & Antiquities Coord., Barbara Murphy; Historical Architect, Don Hartley; NPS Grant, Londi Rowley; Media Inquiries, Bd. of State History, Alicia Aldrich; Historical Collections Cur., Melissa Ferguson; SHPO Compliance, Preservation, Chris Hansen; Digitization Project Mgr., Kristen Jensen; National Register, Cory Jensen; Tax Credit Program, Nelson Knight.
Governing Authority: state. Tax-exempt.
Institution Type/Description: Historical Society Museum.
Collections: furnishings; historic prints; photographs; drawings; paintings; artifacts of 20th-century Utah; period artifacts; porcelain glass; object d'art collection; specimens; maps; books.
Research Fields: Utah, Western, Mormon & related history; archaeological & paleontological research.
Facilities: 23,000-vol. library of books, 20,000 pamphlets, 30,000 maps, 300,000 photographs & manuscripts available for use on premises. Publications for sale.
Activities: self-guided tours; docent tours; lectures; films; temporary & traveling exhibitions.
Publications: journal, Utah Historical Quarterly; books on Utah history; annual, Utah Preservation; annual, Beehive History; bimonthly, Utah State Historical Society Newsletter.
Hours & Admission Prices: Office: Mon.-Fri. 8-5, Research Center: Mon.-Fri. 9-4. No charge; donations accepted. Closed national & state holidays. &
Attendance: 75,000 (estimated)
Membership: Student & Senior Citizen $25; Individual $30; Institution/Business & Sustaining $40; Patron $60; Sponsor $100; Life $500.

UTAH'S HOGLE ZOO, 2600 E. Sunnyside Ave., (840 South), Salt Lake City, UT 84108-1454. Tel.: 801-582-1631. Fax: 801-584-1770. Facebook: Hogle Zoo.
Web Site: www.hoglezoo.org
Founded: 1931.
Congressional District: 2
Key Personnel: Dir., Craig Dinsmore; Asst. Dir. Programs, Kimberly Davidson.
Governing Authority: society. Parent Institution: Utah Zoological Society. Tax-exempt: 170(b)(1)(A).
Institution Type/Description: Zoo.
Collections: general animal collection.
Research Fields: snake venom research; captive breeding.
Facilities: 1,107-vol. library of animal life & natural history available for inter-zoo loan & for use on premises; picnic areas. Animal-related items for sale.
Activities: guided tours; lectures; films; broadcast programs; formally organized education programs for children, adults & undergraduate college students; docent program or council; animal walk and talks; exhibit interpretation by Eco-explorer staff; seasonal Bird Show and Discovery Theater Animal Show. Zoo Sponsors: in-school lectures with animals.
Publications: biannual newsletter, Safari.
Hours & Admission Prices: March-Oct. daily 9-5; Nov.-Feb. daily 9-4. May-Sept. adults $12.75, children 3-12 & senior citizens $9.75; Oct.-April adults $9.75, children 3-12 & senior citizens $7.75; discounts to groups of 20 or more; children under 2 no charge. Closed New Year's Day; Christmas. &
Attendance: 847,831 (accurate)
Membership: Duo $67; Basic $87; Plus $103; Zoo Booster $131; Booster Deluxe $197; Friend $350; Partner $500; Benefactor $750; Champion $1,000.

WHEELER HISTORIC FARM, 6351 South 900 E., Salt Lake City, UT 84121-2438. Tel.: 385-468-1755. Fax: 385-468-1754.
E-mail: wheeler1@slco.org
Web Site: www.wheelerfarm.com
Founded: 1976.
Congressional District: 2
Key Personnel: Facility Mgr., Kathleen Bailey; Communications & Pub. Rels. Mgr., Callie Birdsall; Program Coord., Raegan Scharman; Office Coord., Randi Morishita; Museum Cur., Sara Roach; Head Farmer, Richard Snow.
Governing Authority: county; nonprofit. Parent Institution: Salt Lake County. Tax-exempt: 170(b)(1)(A).
Institution Type/Description: Preservation Project: 1898 Wheeler Farm House, representing the initial statehood period & typical of Utah agriculture in 1898.
Collections: agricultural, dairy & ice industry collections maintained as living history; family life of the period; photographs. Historic Buildings: c.1870 granary, of adobe construction; barn; ice house.
Research Fields: community, agricultural & economic history relating to Utah dairy farms; ice industry; Settlement Salt Lake Valley.
Facilities: classrooms. Reproductions of historic clothing, household goods & farm items for sale.
Activities: guided tours; lectures; films; docent program; wagon rides; nature walks; demonstrations.
Hours & Admission Prices: Daily dawn-dusk. No charge to enter. Wagon Rides 2 & up: $2; Historic Farm House Tours: adults $4, children $2; Cow Milking: $1. Closed New Years Day; Christmas Day. &
Attendance: 395,735 (accurate)

Sandy

HILL GALLERY & SCULPTURE PARK, Canyon Ridge Center, 9045 S. 1300 E., Sandy, UT 84094-3134. Tel.: 801-562-9242.
Web Site: www.danhillsculpture.com/sculpturepark.phtml
Institution Type/Description: Gallery and Sculpture Park.
Collections: works by local & regional artists; life-size outdoor sculptures; carvings; paintings; giclee prints; photographs; kaleidoscopes; lamps; baskets; cards; books.
Hours & Admission Prices: Tues.-Fri. 12-5; other times by appointment.

LIVING PLANET AQUARIUM, 725 E. 10600 S., Sandy, UT 84094-4409. Tel.: 801-355-3474.
E-mail: info@thelivingplanet.com
Web Site: www.thelivingplanet.com
Institution Type/Description: Aquarium.
Collections: marine life including endangered & threatened species.
Facilities: cafe.
Activities: children's programs.

Hours & Admission Prices: Summer: Fri.-Sat. 10-8, Sun.-Thurs. 10-7; Winter: Fri.-Sat. 11-7, Sun.-Thurs. 11-6. Adults $8, seniors, military & students $7, children $6; children 2 & under no charge.

SANDY MUSEUM, 8744 S. 150 E., Sandy, UT 84070-1404. Tel.: 801-566-0878. Fax: 801-566-9608.
E-mail: sandymuseum@hotmail.com
Founded: 1987.
Congressional District: 2
Key Personnel: Dir., Sherry Worthen; C.E.O., Penny McLaughlin.
Personnel Profile: Full-Time Paid 2; Part-Time Volunteers 12.
Governing Authority: private; nonprofit organization. Tax-exempt: 501(c)(3).
Institution Type/Description: History Museum: built 1890.
Collections: local history & culture.
Facilities: library; classroom; turn of the century parlor, pantry & kitchen.
Activities: films; formal education programs for children; guided tours; lectures.
Publications: walking tour guide.
Hours & Admission Prices: Tues.-Thurs. & Sat. 1-5; other times by appointment. No charge; donations accepted. Closed Independence Day; Thanksgiving; Christmas Eve & Day. &
Attendance: 2,000 (accurate)
Membership: Individual $25.

Santa Clara

JACOB HAMBLIN HOME, 3325 Hamlin Dr., Santa Clara, UT 84770. Mailing Address: St. George Temple Visitor's Center & Historic Sites, 490 S. 300 E., Saint George, UT 84770-3665. Tel.: 435-673-5181. Fax: 435-652-9589.
E-mail: vcsgeorge@ldschurch.org
Founded: 1975.
Congressional District: 2
Key Personnel: Cur. Exhibits & Historic Sites, Don Enders; Cur. Art & Artifacts, Richard G. Oman; Registrar, T. Michael Smith.
Personnel Profile: Full-Time Paid 1; Part-Time Volunteers 10.
Governing Authority: church. LDS Church, 50 East North Temple, Salt Lake City, UT. 84150. Tax-exempt.
Institution Type/Description: Historic House: 1862-64 Jacob Hamblin Home, church leader & pioneer of the area.
Collections: household furnishings typical of a pioneer home in the 1860s, some belonging to Jacob Hamblin & his family.
Activities: guided tours.
Publications: brochure.
Hours & Admission Prices: Winter: daily 9-5; Summer: daily 9-7. No charge. &
Attendance: 22,592 (accurate)

Sevier

FREMONT INDIAN STATE PARK AND MUSEUM, 3820 W. Clear Creek Canyon Rd., Sevier, UT 84766-6058. Tel.: 435-527-4631.
E-mail: fremontindian@utah.gov
Web Site: stateparks.utah.gov/stateparks/parks/fremont/
Founded: 1987.
Institution Type/Description: Park Museum.
Collections: local history & culture; photographs; period artifacts.
Hours & Admission Prices: Visitor Center & Museum: Summer: daily 9-6; Winter: daily 9-5. Adults $3, $6 per vehicle up to 10 people; discounts to Utah resident seniors. Closed New Year's Day; Thanksgiving; Christmas. &
Attendance: 80,000 (estimated)

Springdale

ZION NATIONAL PARK, ZION HUMAN HISTORY MUSEUM, Zion National Park, Springdale, UT 84767-1099. Tel.: 435-772-3256. Fax: 435-772-3426.
E-mail: zion_park_information@nps.gov
Web Site: www.nps.gov/zion
Formerly: Zion National Park Museum
Founded: 1919.
Congressional District: 1
Key Personnel: Supt., Jock Whitworth; Chief Park Ranger, Cindy Purcell; Interpretation & Visitor Services Chief, Aly Baltrus.
Governing Authority: federal. Parent Institution: U.S. Dept. of the Interior National Park Service.
Institution Type/Description: Natural History and Human History Museum.
Collections: archaeological objects c.6500 B.C. - 1900 A.D. related to Archaic,

Ancestral Puebloan, Southern Paiute & Mormon cultures; historic objects of the Euro-American settlement of Zion Canyon; excavation equipment; tools; construction projects of 1920s & 1930s; archival materials of park's administrative history & resource management; herbarium; zoological; geological; paleontological; photographs c.1910 to present.

Research Fields: various aspects of park ecology.

Facilities: 2,700-vol. library of history; natural history; geology; Indian books available for inter-library loan & for use on premises. Books, slides, postcards & maps for sale.

Activities: Zion National Park: guided hikes; lectures; films; permanent exhibitions. Museum Sponsors: Junior Ranger Program; Zion Nature Center for children ages 6-12 Memorial Day to Labor Day.

Publications: maps; Geological Cross Section of the Cedar Breaks-Zion-Grand Canyon Region; books, Canyon Overlook Trail Guide; Zion National Park; The Sculpturing of Zion; Zion Album; The Outstanding Wonder-Zion Canyon's Cable Mountain Draw Works; Why the North Star Stands Still; Zion, The Story Behind The Scenery; Discover Zion; Posters, Temples & Towers of the Virgin; color slides.

Hours & Admission Prices: National Park: 24 hours a day. Human History Museum: March-Nov. Daily 10-5. 7-day vehicle entrance fee $25, 7-day motorcycle & individual entrance fee $12. Martin Luther King, Jr. Day, National Park Week, National Park Service Day, National Public Lands Day & Veterans Day Weekend no charge. &

Attendance: 2,500,000 (estimated)

Membership: Zion Natural History Assoc. $15.

Springville

SPRINGVILLE MUSEUM OF ART, (M), 126 E. 400 S., Springville, UT 84663-1953. Tel.: 801-489-2727. Fax: 801-489-2739.

E-mail: vswanson@smofa.org

Web Site: www.smofa.org

Founded: 1903.

Congressional District: 3

Key Personnel: Dir., Dr. Vern G. Swanson; Pres. (V), David Cook; Assoc. Dir., Natalie Petersen; Asst. Dir., Dr. Virgil Jacobsen; Pres. (V), Debbie Balzotti; Pres. (V), Ron Forbeck; Pres. (V), Sharee Forbeck.

Personnel Profile: Full-Time Paid 5; Full-Time Volunteers 105; Part-Time Paid 9; Part-Time Volunteers 5; Interns 5.

Governing Authority: Art Assoc. board of directors, nonprofit organization. Parent Institution: Springville City. Tax-exempt: 501(c)(3).

Institution Type/Description: State Museum of Utah Art.

Collections: 1,900-piece permanent collection of 19th- to 20th-century Utah art; collection of Cyrus E. Dallin, sculptor; John Hafen, painter, specializing in Utah art history; 20th-century American Realist paintings & sculpture; 20th century Soviet Socialist Realist paintings.

Research Fields: history of Utah art; 20th-century American Realist painting & Soviet Socialist Realism.

Facilities: library; classrooms; reception area; courtyard.

Activities: guided tours; lectures; gallery talks; concerts; arts festivals; intern program; docent program or council; permanent & traveling exhibitions; art workshops; films.

Publications: catalogs; exhibitions catalogs; books, biography of Cyrus E. Dallin; The History of Utah Art Through the Springville Museum's Collection.

Hours & Admission Prices: Tues. & Thurs.-Sat. 10-5, Wed. 10-9, Sun. 3-6. No charge; donations accepted. Closed holidays. &

Attendance: 110,000 (estimated)

Membership: Individual $20; Family $30; Sustaining $100; Patron $250; Friend $500; Benefactor $1,000.

St. George

BRIGHAM YOUNG'S WINTER HOME, 67 West 200 North, St. George, UT 84770. Mailing Address: St. George Temple Visitor's Center & Historic Sites, 490 S. 300 E., St. George, UT 84770-3665. Tel.: 435-673-5181.

E-mail: vcsgeorge@ldschurch.org

Web Site: www.stgeorgetemplevisitorcenter.org/byounghome.html

Founded: 1975.

Congressional District: 2

Key Personnel: Dir. Exhibits & Visitors' Centers, LDS Church, Tom Peterson; Cur. Exhibits & Historic Sites, Don Enders; Cur. Art & Artifacts, Richard G. Oman; Registrar, T. Michael Smith.

Personnel Profile: Full-Time Paid 1; Full-Time Volunteers 10.

Governing Authority: church. Parent Institution: LDS Church, 50 East North Temple, Salt Lake City, UT 84150. Tax-exempt.

Institution Type/Description: Historic House: 1869-70 Brigham Young Winter Home.

Collections: household furnishings & other items typical of the West in the 1870s, some belonging to Brigham Young.

Facilities: botanical garden.

Activities: guided tours.

Publications: brochure.

Hours & Admission Prices: Winter: daily 9-5. Summer: daily 9-7. No charge.

Attendance: 39,746 (accurate)

ROSENBRUCH WILDLIFE MUSEUM, 1835 Convention Center Dr., Ste. B, St. George, UT 84790-5843. Tel.: 435-656-0033 & 986-6697. Fax: 435-986-6694.

E-mail: angieh@rosenbruch.org

Web Site: www.rosenbruch.org

Founded: 2001.

Key Personnel: Exec. Dir. Museum & Foundation, Angie Rosenbruch-Hammer; Pres. (V), Jimmie C. Rosenbruch; Cur., Dustin Hammer; Dir. Mktg., Melissa Young.

Personnel Profile: Full-Time Paid 3; Part-Time Paid 6; Part-Time Volunteers 5.

Governing Authority: private; nonprofit organization. Tax-exempt: 501(c)(3).

Institution Type/Description: Wildlife Museum.

Collections: over 400 species of animals from Africa, Asia, the South Pacific, Europe & North America; insects; wildlife art.

Facilities: 180-seat auditorium; 30,000 sq. ft. exhibit space. Museum-related items for sale.

Activities: films; guest speakers; guided tours; lectures; temporary & traveling exhibitions; theater.

Publications: newsletter.

Hours & Admission Prices: Mon. 10-8, Tues.-Sat. 10-6. Adults $8, senior citizens $6, children 3-12 $4; discounts to Utah Museum Assoc. members. Annual Pass available. Closed New Year's Day; Thanksgiving; Christmas. &

Attendance: 40,500 (accurate)

ST. GEORGE ART MUSEUM, (M), 47 E. 200 N., St. George, UT 84770-2843. Tel.: 435-627-4525.

E-mail: museum@sgcity.org

Web Site: www.sgartmuseum.org

Founded: 1997.

Congressional District: 2

Key Personnel: Dir., Deborah Reeder; Chm. (V), Joe Viers; Museum Shop Mgr., Valerie Sullivan; Museum Asst., April Cummings.

Personnel Profile: Full-Time Paid 1; Part-Time Paid 4.

Governing Authority: Parent Institution: St. George City. Tax-exempt.

Institution Type/Description: Art Museum.

Collections: works by artists of the West primarily from Utah.

Facilities: Museum-related items for sale.

Activities: school & adult tours by appointment; adult study center; children's culture center; care for your art center; family discovery center. Museum Sponsors: art conversations 3rd Thursday.

Publications: catalog, A Century of Sanctuary: The Art of Zion National Park.

Hours & Admission Prices: Mon.-Sat. 10-5. Adults $3, children 3-11 $1; discounts to AAM & ICOM members; museum professionals, children under 3 & members no charge. &

Attendance: 8,908 (accurate)

Membership: Student $15; Individual $35; Family $50; Supporting $75-$1,000; Corporate $3,000 & up.

ST. GEORGE DINOSAUR DISCOVERY SITE AT JOHNSON FARM, 2180 E. Riverside, St. George, UT 84790-2483. Tel.: 435-574-3466. Fax: 435-627-0340.

E-mail: tracksofdinos@gmail.com

Web Site: www.utahdinosaurs.com

Founded: 2000.

Key Personnel: Dir., Rusty Salmon; Pres. (V), Gary Watts; Cur., Andrew R. C. Milner.

Personnel Profile: Part-Time Paid 4; Part-Time Volunteers 30; Interns 1.

Governing Authority: city. Subsidiary Institution: Dinosaur Ah! Torium Foundation. Tax-exempt.

Institution Type/Description: Paleontology Museum.

Collections: dinosaur fossil replicas; dinosaur tracks; fossil fish; plants.

Research Fields: Triassic-Jurassic transition; vertebrate ichnology; late Triassic/early Jurassic paleontology & geology.

Publications: quarterly member newsletter.

Hours & Admission Prices: Mon.-Sat. 10-6, Sun. 11-5. Adults $6, children 4-11 $3; members & children under 4 no charge. Closed New Year's Day; Thanksgiving; Christmas. &

Attendance: 37,500 (accurate)

Membership: Triassic $35; Jurassic $55; Cretaceous $75.

WESTERN SKY AVIATION WARBIRD MUSEUM, 4196 S. Airport Pkwy., St. George, UT 84790. Mailing Address: 2050 W. Canyon View Dr. #2, Saint George, UT 84770. Tel.: 435-669-0655.
Founded: 2006.
Governing Authority: nonprofit organization. Tax-exempt: 501(c)(3).
Institution Type/Description: Military Aviation Museum.
Collections: aviation Warbird history; aircraft.
Hours & Admission Prices: Fri.-Sat. 10-4.

Stansbury Park

BENSON GRIST MILL, 325 State Rd. 138, Stansbury Park, UT 84074. Mailing Address: 2930 W. Hwy. 112, Tooele, UT 84074. Tel.: 435-882-7678. Fax: 435-882-6003.
E-mail: bensonmill@trilobyte.net
Web Site: www.bensonmill.org
Founded: 1854.
Congressional District: 3
Key Personnel: Dir., Mark McKendrick; Museum Shop Mgr., Suzy Wall.
Personnel Profile: Full-Time Paid 1; Part-Time Paid 11.
Governing Authority: Parent Institution: Tooele County Parks & Recreation. Tax-exempt.
Institution Type/Description: History Museum: listed on the National Register of Historic Sites.
Collections: 150 year old mill; area history.
Hours & Admission Prices: Closed until further notice. &
Attendance: 12,000 (estimated)

Syracuse

SYRACUSE MUSEUM & CULTURAL CENTER, 1891 West 1700 South, Syracuse, UT 84075. Tel.: 801-825-3633. Fax: 801-825-3001.
Institution Type/Description: History Museum.
Collections: Syracuse history & culture; period furnishings; photographs.
Activities: special events.
Hours & Admission Prices: Tues.-Thurs. 2-5, Fri. 1-4.

Tooele

OQUIRRH MOUNTAIN MINING MUSEUM, 47 S. Maine, Tooele, UT 84074-2148. Tel.: 435-843-4000.
Institution Type/Description: Mining Museum.
Collections: local mining history & operations; mines; Gold Rush era.
Hours & Admission Prices: Call for hours. No charge.

TOOELE PIONEER MUSEUM, 47 E. Vine St., Tooele, UT 84074-2133. Tel.: 435-843-0771 & 882-1092.
E-mail: pioneer@wirelessbeehive.com
Web Site: tooelepioneermuseum.com
Founded: 2001.
Key Personnel: Dir. & Chm. (V), James Bevan; Museum Shop Mgr., Russell Hammond.
Personnel Profile: Part-Time Volunteers 3.
Governing Authority: Tax-exempt.
Institution Type/Description: History Museum.
Collections: local history & culture; photographs; period furnishings; personal artifacts; Native American artifacts.
Hours & Admission Prices: May-Sept. 25 Mon. 6pm-8pm, Tues., Fri.-Sat. & holidays 10-4; Sept. 26-April Mon. 6pm-8pm, Tues. & Sat. 10-4; other times by appointment. No charge. &
Attendance: 1,382 (accurate)

TOOELE VALLEY RAILROAD MUSEUM, 35 N. Broadway, Tooele, UT 84074. Mailing Address: 90 N. Main St., Tooele, UT 84074-2139. Tel.: 435-882-2836. Fax: 435-643-7888.
Web Site: www.tooelecity.org/parks&recreation/railroadmuseum.asp
Founded: 1980.
Key Personnel: Chm., Larry Deppe, Ph.D.; Dir. & Chm. (V), Jean Mogus; Pres., Bruce Grim.
Personnel Profile: Part-Time Paid 1; Part-Time Volunteers 7.
Governing Authority: Parent Institution: Tooele City Corp. Tax-exempt.
Institution Type/Description: History Museum.
Collections: industrial, railroad, mining, & smelting history; photographs; medical artifacts; rolling stock; simulated mine; school photographs; military artifacts; coach with HO scale layout.
Hours & Admission Prices: Memorial Day to Labor Day Tues.-Sat. 1-4. No charge; donations accepted. Closed Independence Day. &

Attendance: 3,000 (accurate)

UTAH STATE FIREFIGHTERS MUSEUM, Deseret Peak Complex, 2930 W. Hwy. 112, Tooele, UT 84074. Mailing Address: P.O. Box 1128, Grantsville, UT 84029. Tel.: 435-843-4040. Fax: 435-830-6556.
E-mail: curator@utahfiremuseum.com
Web Site: www.utahfiremuseum.com
Key Personnel: Cur., Dave Hammond
Institution Type/Description: Fire-Fighting Museum.
Collections: fire-fighting artifacts & memorabilia including over 50 fire engines from different eras & locations throughout Utah.
Hours & Admission Prices: Call for hours.

Torrey

CAPITOL REEF NATIONAL PARK VISITOR CENTER, 16 Scenic Dr., Torrey, UT 84775. Mailing Address: Capitol Reef National Park, HC 70 Box 15, Torrey, UT 84775. Tel.: 435-425-3791, ext. 4111. Fax: 435-425-3026.
Web Site: www.nps.gov/care
Founded: 1968.
Congressional District: 1 & 2
Key Personnel: Supt., Dave Worthington; Chief Interpretation, Lori Rome; Chief Visitor & Resource Protection, Scott Brown.
Personnel Profile: Part-Time Paid 1.
Governing Authority: federal. National Park Service. Tax-exempt.
Institution Type/Description: Natural History & Native American Ethnology Museum.
Collections: archaeology; geology; history; natural history; Fremont Indian culture; paleontology, Mormon history. Historic Building: 1890s, Old Fruita Schoolhouse.
Facilities: picnic area; camp grounds; nature center. Books for sale.
Activities: guided tours; amphitheater programs; junior ranger; junior geologist.
Publications: Geologic Cross Section; Geologic Map of Park; Geology of Capitol Reef; History of Fruita; Plant Ecology of Park; Dwellers of the Rainbow.
Hours & Admission Prices: Summer season daily 8-6; winter season daily 8-4:30. Park: 7-day vehicle pass $5, 7-day individual pass $3. Closed some holidays. &
Attendance: 750,000 (estimated)

Tremonton

TREMONTON FIREFIGHTERS MUSEUM, 95 S. 100 W., Tremonton, UT 84337. Mailing Address: 102 S. Tremont St., Tremonton, UT 84337. Tel.: 435-257-2625.
Institution Type/Description: Firefighting History Museum.
Collections: firefighting history & equipment; photographs; uniforms; helmets.
Hours & Admission Prices: By appointment.

Vernal

UTAH FIELD HOUSE OF NATURAL HISTORY STATE PARK, 496 E. Main St., Vernal, UT 84078-2610. Tel.: 435-789-3799. Fax: 435-789-4883.
Web Site: www.stateparks.utah.gov
Founded: 1948.
Key Personnel: Park Mgr., Steven D. Sroka; Cur. Education, Mary Beth Bennis-Smith; Cur. Collections, Heather Finlayson; Museum Maintenance, Craig Gerber; Retail Sales Mgr., Colleen Lawson.
Personnel Profile: Full-Time Paid 5; Part-Time Paid 3; Part-Time Volunteers 3.
Governing Authority: state. Parent Institution: Utah State Div. of Parks & Recreation, 1636 W. North Temple, Salt Lake City, UT 84116. Tax-exempt: 501(c)(3).
Institution Type/Description: Natural History Museum.
Collections: geology; fossils; archaeology; natural history of the Uinta Basin; 18 lifesize prehistoric replicas in garden setting.
Research Fields: geology; paleontology.
Facilities: Postcards, books, dinosaur models & souvenirs for sale.
Activities: lectures; tours; State Information Center.
Hours & Admission Prices: Mon.-Sat. 9-5. Adults $6, children 6-12 $3; children under 6 no charge. Closed New Years; Thanksgiving; Christmas. &
Attendance: 45,000 (accurate)

WESTERN HERITAGE MUSEUM, (M), 328 East 200 S., Vernal, UT 84078-3220. Tel.: 435-789-7399. Fax: 435-789-9798.
E-mail: whm@co.uintah.ut.us
Web Site: www.westernheritagemuseum-uc-ut.org
Founded: 1991.
Congressional District: 1
Personnel Profile: Full-Time Paid 1; Part-Time Paid 2.
Governing Authority: county. Parent Institution: Uintah County Government. Tax-exempt.
Institution Type/Description: History Museum.
Collections: early settlers, Fremont & Ute Indian artifacts; blacksmith display; barbershop; country store; 1890-1900 ladies fashions; one-room schoolhouse; Gilsonite exhibit; early rifles; saddles; tack & leather; First Ladies of the White House dolls.
Hours & Admission Prices: Memorial Day to Labor Day Mon.-Fri. 9-6, Sat. 10-4; Sept.-May Mon.-Fri. 9-5, Sat. 10-2. No charge; donations accepted. Closed New Year's; Human Rights Day; President's Day; Memorial Day; Independence Day; Pioneer Day; Labor Day; Columbus Day; Veteran's Day; Thanksgiving; Christmas. &
Attendance: 7,744 (accurate)

Washington

SOUTHERN UTAH AIR MUSEUM, 400 W. Telegraph Rd., Washington, UT 84780. Tel.: 435-669-7768.
Web Site: www.suam.org
Institution Type/Description: Aviation Museum.
Collections: 59-2579 Boeing B-52G Stratofortress cockpit; over 100 fine aviation art prints including aircraft from World War II, Korea and Viet Nam and X-Planes from the 50's, 60's and 70's.
Hours & Admission Prices: Sat.-Sun. 10-4.

Wellsville

AMERICAN WEST HERITAGE CENTER, 4025 S. Hwy., 89-91, Wellsville, UT 84339. Tel.: 435-245-6050; 800-225-3378. Fax: 435-245-6052.
E-mail: info@awhc.org
Web Site: www.awhc.org
Formerly: Ronald V. Jensen Living Historical Farm
Founded: 1979.
Congressional District: 1
Key Personnel: Exec. Dir., Steve Delong; Chm. (V), Gary Anderson; Cur., Reece Summers; Program Admin., Lorraine Bowen.
Personnel Profile: Full-Time Paid 3; Part-Time Paid 12; Part-Time Volunteers 80; Interns 6.
Governing Authority: university. Parent Institution: American West Heritage Foundation. Subsidiary Institution: Utah State University. Branch Museum: Jensen Historical Farm. Tax-exempt.
Institution Type/Description: Living Historical Farm: 1917 farm based on operations of Scandinavian & British emigrants influenced by extension programs of Utah Agricultural College.
Collections: 1917 farm with outbuildings; implements of the period; work horses; dairy cows; sheep; hogs; poultry; crops; pioneer setting of 1850; Native American Shoshone encampment 1835; Mountaineer 1830; 1895 print shop; 1880 woodworking shop.
Research Fields: rural life in Utah, World War I era; lifeways from 1820-1920, including Native cultures, Mountain Men, Pioneers.
Facilities: library; Farm: 40 acres of irrigated fields; classroom; picnic shelter; welcome center; livery stable.
Activities: demonstrations of early 20th-century farm practice & life; recreation of family farm life in Utah 1917; workshops. Museum Sponsors: over 15 special events during year.
Publications: quarterly newsletter, Hay Derrick.
Hours & Admission Prices: Jan.-May & Sept.-Oct. Mon.-Fri. 10-4; Memorial Day-Labor Day Tues.-Sat. 10-4. Adults $7, senior citizens & students $6, children $5; discounts to members. Closed New Year's Eve & Day; Christmas Eve, Day & week. &
Attendance: 60,000 (accurate)
Membership: Individual $25; Family $50.

Wendover

BONNEVILLE SPEEDWAY MUSEUM, 900 E. Wendover Blvd., Wendover, UT 84083. Tel.: 775-664-4400. Facebook: Bonneville Speedway Museum.
Web Site: bonnevillespeedmuseum.com
Institution Type/Description: Racing Museum.
Collections: speed racing memorabilia & films.
Hours & Admission Prices: Closed for renovation of new site.

HISTORIC WENDOVER AIRFIELD MUSEUM, 345 S. Airport Apron, Wendover, UT 84083. Tel.: 435-665-2308.
Web Site: www.wendoverairbase.com
Personnel Profile: Part-Time Volunteers 12.
Governing Authority: Tax-exempt.
Institution Type/Description: Military Museum.
Collections: WWII Army Air Force history; military artifacts; photographs.
Facilities: Museum-related items for sale.
Activities: video; special events. Annual Event: Wendover Air Show.
Hours & Admission Prices: Daily 8-6. No charge; donations accepted. &
Attendance: 12,000

VERMONT

(167 listings)

Addison

CHIMNEY POINT STATE HISTORIC SITE, 8149 VT Rte. 17W., Addison, VT 05491-8751. Tel.: 802-759-2412. Fax: 802-759-2547. Facebook: Vermont State Historic Sites.
E-mail: elsa.gilbertson@state.vt.us
Web Site: www.historicsites.vermont.gov/chimneypoint
Founded: 1968.
Congressional District: 1
Key Personnel: Historic Site Operations Chief, John P. Dumville; Site Admin., Elsa Gilbertson.
Personnel Profile: Full-Time Paid 1; Part-Time Paid 1.
Governing Authority: state. Parent Institution: State of Vermont, Div. for Historic Preservation, National Life Bldg., 6th Fl., Montpelier, VT 05620-0501. Tax-exempt.
Institution Type/Description: Historic Building: late 18th-century tavern.
Collections: furnishings; memorabilia; historic structures; Native American artifacts; archaeological; French colonial history.
Research Fields: Native Americans in Vermont, French settlement of Champlain Valley; early Champlain Valley history.
Facilities: visitors' center. Museum-related items for sale.
Activities: temporary exhibits; school program; Northeastern Open Atlatl Championship; special events.
Publications: guide.
Hours & Admission Prices: late May to mid-Oct. Wed.-Sun. & Mon. holidays 9:30-5. Adults $3; discounts to AAM members. &
Attendance: 2,300 (accurate)

Barnard

BARNARD HISTORICAL SOCIETY, Village School, VT Rte. 12, Barnard, VT 05031. Mailing Address: P.O. Box 234, Barnard, VT 05031-0234. Tel.: 802-457-3020. Fax: 802-234-9080.
E-mail: babrdh@gmail.com
Key Personnel: Pres., Caz Rozonewski
Institution Type/Description: Historical Society Museum: housed in the former Village School; c.1850.
Collections: local history & culture; photographs; period furnishings; personal artifacts; agricultural industry.
Hours & Admission Prices: Memorial Day to Oct. Sat. 12-3; other times by appointment.

Barnet

BARNET HISTORICAL SOCIETY, 24 Goodwillie Rd., Barnet, VT 05821-9555. Mailing Address: P.O. Box 34, Barnet, VT 05821. Tel.: 802-633-3831.
Founded: 1967.
Key Personnel: Pres., Dylan Ford; Treas., Ruth Anderson; Sec., Alan Boye.
Personnel Profile: Part-Time Volunteers 6.
Governing Authority: society. Tax-exempt.
Institution Type/Description: Historical Society Museum: housed in c.1790 Goodwillie House.
Collections: early pictures of the area; household & farm equipment; store ledgers; Roy Brothers Croquet factory.
Research Fields: local history.
Activities: guided tours by appointment; temporary exhibitions.
Publications: biannual newsletter.
Hours & Admission Prices: July-Sept. 10-4; tours by appointment. No charge; donations accepted. &
Attendance: 200 (estimated)
Membership: Student $1; Family $10; Sustaining $100; Life $200.

Barre

STUDIO PLACE ARTS, 201 N. Main St., Barre, VT 05641-4125. Tel.: 802-479-7069.
Key Personnel: Exec. Dir., Sue Higby
Institution Type/Description: Art Gallery.
Collections: paintings; photographs; sculpture.
Activities: art classes; temporary & permanent exhibits; workshops; lectures.
Hours & Admission Prices: Call for hours. No charge.

VERMONT GRANITE MUSEUM & STONE ARTS SCHOOL, 7 Jones Brothers Way, Barre, VT 05641. Mailing Address: P.O. Box 282, Barre, VT 05641-0282. Tel.: 802-476-4605. Fax: 802-476-6866.
E-mail: info@stoneartsschool.org
Web Site: www.granitemuseum.org
Founded: 1997.
Key Personnel: Chm. (V), Patricia L. Meriam; Treas., Paul Hutchins.
Governing Authority: private; nonprofit organization. Tax-exempt.
Institution Type/Description: Cultural Heritage Museum.
Collections: cultural heritage; granite manufacturing & technology; stone trades; plaster models; carved granite roses; photographs; oral histories.
Activities: films; guided tours; lectures; stone repair training program. Annual Event: Granite Festival.
Hours & Admission Prices: By appointment. No charge; donations accepted. ♿
Attendance: 350 (estimated)

Barton

CRYSTAL LAKE FALLS HISTORICAL ASSOCIATION, 97 Water St., Barton, VT 05822. Mailing Address: 536 Breezy Hill Rd., Barton, VT 05822-8641. Tel.: 802-525-6276.
E-mail: dhath@gaw.com
Formerly: Barton Museum
Founded: 1984.
Key Personnel: Pres., Earle Randall; Vice Pres., Patti Bondor; Treas., Bill May; Sec., Dorothy Hathaway.
Personnel Profile: Part-Time Volunteers 15.
Operating Expenses: 3,414
Operating Income: 3,626
Governing Authority: private; nonprofit organization. Tax-exempt: 501(c)(3).
Institution Type/Description: Historical and Industrial Museum: housed in a c.1820 structure.
Collections: artifacts from education in Barton; Barton's industries.
Major Exhibits: Old Barton Golf Course (T), 1/12-12/14; Barton's Contribution to Vermont's Effort in the Civil War (T), 10/13-12/15.
Research Fields: Barton's history, industries & education.
Facilities: 30-vol. library of books & memorabilia on Gov. Emerson; 600 sq. ft. exhibit space. Museum-related items for sale.
Activities: video; guided tours; lectures; talks & story telling to school children. Annual Event: Open House in June.
Publications: biannual newsletter.
Hours & Admission Prices: June-Aug. Sun. 1-4; other times by appointment. No charge; donations accepted. ♿
Attendance: 127 (accurate)
Membership: Student $.50; Adult $5; Lifetime $100.

Bellows Falls

ADAMS OLD STONE GRIST MILL, Mill St., Bellows Falls, VT 05101. Mailing Address: 47 Atkinson, Bellows Falls, VT 05101-1675. Tel.: 802-463-3734.
Founded: 1965.
Key Personnel: Pres. Historical Society, Dennis Ladd.
Governing Authority: nonprofit. Operated by the Bellows Falls Historical Society. Tax-exempt.
Institution Type/Description: Historical Society Museum.
Collections: 19th-century milling equipment; locally manufactured machinery; farming implements; railroad items; documents & records; household furnishings. Historic Structure: 1831 gristmill (160th year anniversary).
Activities: guided tours; permanent exhibitions.
Hours & Admission Prices: June-Oct. Sat.-Sun. 1-4; other times by appointment. No charge; donations accepted.
Attendance: 340

ROCKINGHAM FREE PUBLIC LIBRARY AND MUSEUM, 65 Westminster St., Bellows Falls, VT 05101-1555. Tel.: 802-463-4270.
E-mail: rockingham@dol.state.vt.us
Web Site: www.rockingham.lib.vt.us
Founded: 1909.
Key Personnel: Dir., Celina Houlne.
Personnel Profile: Full-Time Paid 4; Part-Time Paid 5.
Governing Authority: municipal. Tax-exempt.
Institution Type/Description: History Museum.
Collections: photographs; Civil War artifacts; local histories; local newspapers; machinery manufactured in town; articles owned by local citizens from 1700 to 1900; stereopticon views & viewers.
Facilities: 47,000-vol. library including children's books, available for inter-library loan; reading room.
Activities: temporary exhibitions; genealogy research; local history events.
Hours & Admission Prices: Mon.-Wed. 10-7, Thurs.-Fri. 10-5:30, Sat. 10-2. ♿
Attendance: 386 (accurate)

Belmont

MOUNT HOLLY COMMUNITY HISTORICAL MUSEUM, Tarbelville Rd., Belmont, VT 05730. Mailing Address: P.O. Box 17, Belmont, VT 05730-0017. Tel.: 802-259-2460.
Web Site: www.mounthollyvtmuseum.org
Formerly: Community Historical Museum of Mount Holly
Founded: 1968.
Congressional District: 4
Key Personnel: Chm. (V), Dennis Devereux; Vice Chm., Lory Doolittle; Sec., Maggie Blane; Treas., Linda Nexon; Cur., Robin Eatmon.
Personnel Profile: Part-Time Volunteers 18.
Governing Authority: nonprofit organization. Tax-exempt.
Institution Type/Description: General & Historical Society Museum: housed in 1834 blacksmith shop.
Collections: quilts; clothing; old farm tools; early records; apothecary bottles; blacksmith equipment; local photographs; early records; cemetery records; local history; post office.
Research Fields: local history.
Activities: permanent & temporary exhibitions; demonstrations of crafts; lectures; dinners with guest speakers.
Publications: books, Mount Holly, Its Early Days: History of Mt. Holly; annual newsletter.
Hours & Admission Prices: July-Aug. Sat.-Sun. 2-4. No charge; donations accepted. ♿
Attendance: 550 (estimated)
Membership: Individual $10; Family $25; Patron $50; Sustaining $100; Life $500.

Bennington

BENNINGTON CENTER FOR THE ARTS, 44 Gypsy Lane at VT Rte. 9, Bennington, VT 05201-9692. Mailing Address: P.O. Box 260, Bennington, VT 05201-0260. Tel.: 802-442-7158.
E-mail: shirley@thebennington.org
Web Site: www.thebennington.org
Formerly: Laumeister Center for the Arts
Founded: 1992.
Key Personnel: Dir., Shirley Hutchins; C.E.O., Bruce Laumeister; Pres. (V) & Museum Shop Mgr., Elizabeth Small.
Personnel Profile: Full-Time Paid 3; Full-Time Volunteers 2; Part-Time Paid 2; Part-Time Volunteers 4.
Governing Authority: Tax-exempt.
Institution Type/Description: Art & Culture Center.
Collections: Native American artifacts; Navajo weavings; Navajo, Hopi & Zuni jewelry; pottery; sculptures; paintings; wildlife art; bird carvings.
Major Exhibits: Art of the Animal Kingdom, 6/14-7/14; Oil Painters of America, 6/14-8/14; American Artists Abroad, 7/14-8/14; Impressions of New England, 7/14-9/14; Laumeister Fine Art Competition, 7/14-10/14; Plein Air Vermont, 9/14-12/14.
Hours & Admission Prices: Call for hours. Adults $9, seniors & students $8; discounts to Mass. teachers; children under 12 no charge. ♿
Attendance: 48,000 (estimated)
Membership: Annual $50.

✳ **THE BENNINGTON MUSEUM, (M),** 75 Main St., Bennington, VT 05201-2885. Tel.: 802-447-1571. Fax: 802-442-8305.
E-mail: info@benningtonmuseum.org
Web Site: www.benningtonmuseum.org
Founded: 1852.

Key Personnel: Exec. Dir. & Dir. Public Programs, Deana Mallory; Chm. Bd., Raymond Bolton; Cur., Jamie Franklin; Devel. Assoc., Joy Danila; Public Rels., Susan Strano; Mgr. Collections, Callie Stewart; Visitor Svcs., Karen Harrington; Museum Educator, Elizabeth Kane; Mgr. Devel., Denise Lariscy.
Personnel Profile: Full-Time Paid 7; Part-Time Paid 9; Part-Time Volunteers 128; Interns 3.
Governing Authority: nonprofit corporation. Tax-exempt: 501(c) (3).
Institution Type/Description: Art, Decorative Arts & History Museum.
Collections: regional painting & sculpture; New England early Vermont furniture; Bennington pottery-Rockingham, Flint Enamel & Parian pottery ware; early American glass; American silver & Vermont coins; regional historical materials; c.1920 Bennington made Martin-WASP touring car; flags; costumes & uniforms; dolls; toys; paintings & memorabilia of Grandma Moses. Historic Building: 1838 Grandma Moses Schoolhouse.
Research Fields: Bennington pottery; Bennington regional history; New England painting & furniture; Vermont silver; Grandma Moses.
Facilities: 5,002-vol. library of genealogy books, records & related materials available for use on premises; reading room. Museum-related reproductions & Vermont-made items for sale.
Activities: guided tours; lectures; permanent, temporary & traveling exhibitions; workshops; school out-reach.
Publications: annual report, The Bennington Museum Notes. exhibition catalogues; Highlights from the Bennington Museum; Norton Stoneware & American Redware, The Bennington Museum Collection; The Best The Country Affords: Vermont Furniture, 1765-1858.
Hours & Admission Prices: Feb.-June & Nov.-Dec. Thurs.-Tues. 10-5; July -Oct. daily 10-5. Adults $10, senior citizen & students $9, groups of 10 or more $8:50 per person; discounts to NARM & Consortium of New England Art Museums; members & children under 18 no charge. Closed New Year's Day; Thanksgiving Day; Christmas. &
Attendance: 34,752 (accurate)
Membership: Individual $50; Family $75; Contributing $100; Sustaining $200; Director's Circle $300 & up. Business Levels: Nonprofit $80; Sponsor $100; Patron $200; Director's Circle Corporate $300 & up.

THE BURGHDORF GALLERY AT SOUTHERN VERMONT COLLEGE, 982 Mansion Dr., Bennington, VT 05201-9269. Tel.: 802-447-6316. Fax: 802-447-4695.
E-mail: gwinter@svc.edu
Formerly: Southern Vermont College Art Gallery
Founded: 1979.
Key Personnel: Dir., Greg Winterhalter.
Governing Authority: college. Parent Institution: Southern Vermont College, Monument Ave., Bennington, VT 05201. Tax-exempt: 501(c)(3).
Institution Type/Description: College Art Gallery: housed in 1910 Everett Estate, built in the style of a 14th century English-Norman castle. Listed on the National Register of Historic Places.
Collections: photo collection of 1910 Everett Mansion construction.
Activities: temporary exhibits.
Hours & Admission Prices: Mon.-Fri. 9-3. No charge; donations accepted. Closed major holidays. &
Attendance: 5,000

SUZANNE LEMBERG USDAN GALLERY, Bennington College, 1 College Dr., Bennington, VT 05201-6003. Tel.: 802-442-5401.
Institution Type/Description: Art Gallery.
Collections: works by professional artists, college faculty, graduate students & graduating seniors.
Activities: lectures; workshops.
Hours & Admission Prices: March-June & Sept.-Dec. Tues.-Sat. 1-5. No charge.

VERMONT COVERED BRIDGE MUSEUM, 44 Gypsy Lane, Bennington, VT 05201-9692. Mailing Address: P.O. Box 260, Bennington, VT 05201-0260. Tel.: 802-442-7158.
E-mail: vtcolor@yahoo.com
Web Site: www.benningtoncenterforthearts.org/CBMHome.htm
Founded: 2003.
Key Personnel: Chm. (V), C.E.O., Bruce R. Laumeister; Pres. (V), Elizabeth Small; Museum Shop Mgr., Jana Lillie.
Personnel Profile: Full-Time Paid 3; Full-Time Volunteers 1; Part-Time Paid 1.
Governing Authority: Parent Institution: Bennington Center for the Arts. Tax-exempt.
Institution Type/Description: History Museum.
Collections: covered bridge history; truss designs; transportation history; portal styles; photographs; paintings; tools.

Hours & Admission Prices: Mon. & Wed.-Sun. 10-5. Families $20, adults $9, students & seniors $8; discounts to Mass. teachers; children under 12 no charge. &
Attendance: 12,255 (accurate)
Membership: Annual $50.

Bethel

BETHEL HISTORICAL SOCIETY, Main St., Bethel, VT 05032. Mailing Address: P.O. Box 25, Bethel, VT 05032. Tel.: 802-234-5064. Facebook: Bethel Historical Society.
E-mail: nick@nikolaidis.com
Web Site: bethelvt.com
Founded: 1972.
Institution Type/Description: Historical Society Museum: housed in the former Town Hall; built in 1892.
Collections: local history & culture; photographs; paintings; period artifacts.
Publications: quarterly newsletter.
Hours & Admission Prices: July-Aug. Sat. 10-2; other times by appointment. No charge; donations accepted. &
Attendance: 300 (estimated)
Membership: Annual $5.

Bradford

BRADFORD HISTORICAL SOCIETY INC., Bradford Academy Bldg., Main St., Bradford, VT 05033. Mailing Address: 67 Summer St., Bradford, VT 05033-9142. Tel.: 802-222-4011 & 4423.
E-mail: lccoffin@charter.net
Founded: 1959.
Key Personnel: Pres., Lawrence Coffin; Vice Pres., Wayne Kenyon; Sec., Jeanette Nordham; Treas., Diane Smarro; Cur., Karen DeRosa.
Personnel Profile: Part-Time Volunteers 8.
Governing Authority: board of directors; nonprofit. Tax-exempt.
Institution Type/Description: Historical Society Museum: housed in 1894 Woods School.
Collections: Civil War artifacts; items manufactured in Bradford; photographs; scrapbooks; store signs; china; 19th-century costumes; jewelry; furniture; cemetery survey; town & school reports; Admiral Charles Clark memorabilia; James Wilson, manufacturer of first world globe produced in USA, 1810.
Research Fields: genealogical.
Facilities: 300-vol. library of town reports, histories, Vermont history & folklore. Notepaper, prints, medals & commemorative items for sale.
Activities: guided tours; lectures; films; formally organized education programs; permanent, temporary & loan exhibitions.
Publications: annual newsletter.
Hours & Admission Prices: March-Dec. Fri. 10-12; other times by appointment. No charge; donations accepted. Closed holidays. &
Attendance: 300 (estimated)
Membership: Individual $5; Family $10; Contributing $25; Life $100.

Brattleboro

BRATTLEBORO HISTORICAL SOCIETY, Municipal Center, 230 Main St., Ste. 301, Brattleboro, VT 05301-2880. Tel.: 802-258-4957.
E-mail: histsoc@sover.net
Web Site: www.brattleborohistoricalsociety.org
Founded: 1982.
Institution Type/Description: Local History Museum.
Collections: local history & culture; photographs; period artifacts.
Facilities: history center.
Activities: research.
Hours & Admission Prices: Thurs. 2-4, Sat. 10am to noon, other times by appointment; Brattleboro History Center 196 Main St. Brattleboro, VT: Thurs & Fri. 2-4, Sat. 11-3.
Membership: Individual $20; Family $30; Contributing $50.

BRATTLEBORO MUSEUM & ART CENTER, 10 Vernon St., Brattleboro, VT 05301-3390. Tel.: 802-257-0124. Fax: 802-258-9182.
E-mail: info@brattleboromuseum.org
Web Site: www.brattleboromuseum.org
Founded: 1972.
Congressional District: 4
Key Personnel: Dir., Danny Lichtenfeld; Chief Cur., Mara Williams; Education Cur., Susan Calabria; Operations Mgr., Emily Wergin; Devel. Mgr., Melanie McDonald; Exhibits & Events Mgr., Margaret Shipman.

Governing Authority: private; nonprofit. Tax-exempt: 501(c)(3).
Institution Type/Description: Visual Art Center: housed in 1915 former Union Railroad Station.
Collections: changing visual arts exhibitions; art of our times.
Facilities: Museum-related items for sale.
Activities: school programs; lectures; performance workshops; temporary visual art exhibitions based on annual theme; public openings; family events.
Publications: exhibition brochures; catalogues; posters; newsletters.
Hours & Admission Prices: Sun.-Mon. & Wed.-Thurs. 11-5; Fri. 11-7; Sat. 10-5. Adults $8, senior citizens $6, students $4; children under 6 & members no charge. Closed New Year's Day; Independence Day; Thanksgiving; Christmas. &
Attendance: 23,421 (estimated)
Membership: Artists, Educators, Seniors & Students $40; Individual $45; Household $80; Friend $150; Sponsor $300; Patron $500; Director's Club $1,000.

ESTEY ORGAN MUSEUM, 108 Birge St., Brattleboro, VT 05301-6460. Tel.: 802-246-8366.
E-mail: info@esteyorganmuseum.org
Web Site: www.esteyorganmuseum.org
Founded: 2002.
Governing Authority: nonprofit organization. Tax-exempt: 501(c)(3).
Institution Type/Description: Organ Museum.
Collections: company history; reed, pipe & electronic organs; photographs; Estey tools, advertising, & catalogs; personal artifacts.
Activities: workshops; lectures; special events.
Hours & Admission Prices: Summer: Sat.-Sun. 2-4; other times by appointment. Admission $5; members no charge. &
Membership: Student & Senior $15; Individual $25; Family $35; Business $50.

Bristol

BRISTOL HISTORICAL SOCIETY MUSEUM, Howden Hall Community Center, 19 West St., Bristol, VT 05443-1227. Tel.: 802-453-3429 & 2888.
E-mail: lscoffin@guavt.net
Key Personnel: Pres., Sylvia Coffin; Vice Pres., Gerald Heffernan
Institution Type/Description: Historical Society Museum.
Collections: local history; photographs; maps; postcards; newspapers.
Hours & Admission Prices: June 15-Oct. 15 Mon.-Fri. 10-4 by appointment. No charge; donations accepted.

Brookfield Center

HISTORICAL SOCIETY OF BROOKFIELD-MARVIN NEWTON HOUSE, 1133 Ridge Rd., Brookfield Center, VT 05036. Mailing Address: P.O. Box 447, Brookfield, VT 05036-0447. Tel.: 802-276-3959 & 3497. Fax: 802-276-3023.
Web Site: brookfieldhistoricalsociety.wordpress.com
Founded: 1935.
Key Personnel: Pres., Michael Dempsey; Treas., Mary Waldo; Trustee, Linda Runnion; Trustee, Gary Lord; Trustee & News Editor, Greg Sauer; Trustee, Bonnie Fallon; Trustee, Joanna Boden Weber; Cur., Jacalin Wilder; Historian & Genealogist, Elinor Gray.
Personnel Profile: Part-Time Volunteers 15.
Governing Authority: society. Tax-exempt.
Institution Type/Description: Historic House: 1835, Marvin Newton house.
Collections: 19th-century furnishings & artifacts.
Activities: guided tours; inter-museum loan & permanent exhibitions.
Publications: newsletter; annual pictorial calendar.
Hours & Admission Prices: July-Aug. Sun. 2-5; other times by appointment. Adults $2.
Attendance: 135 (estimated)
Membership: Individual annual $15; Family annual $25; Individual life $150; Family life $250.

Brownington

ORLEANS COUNTY HISTORICAL SOCIETY & THE OLD STONE HOUSE MUSEUM, 109 Old Stone House Rd., Brownington, VT 05860-4420. Tel.: 802-754-2022. Fax: 802-754-9336. Facebook: Old Stone House Museum.
E-mail: information@oldstonehousemuseum.org
Web Site: oldstonehousemuseum.org
Founded: 1916.
Congressional District: 1

Key Personnel: Dir., Peggy Day Gibson; Chm. Bd. (V), John Eby; Museum Shop Mgr., Linda Child.
Personnel Profile: Full-Time Paid 1; Part-Time Paid 5; Part-Time Volunteers 25; Interns 1.
Governing Authority: society. Parent Institution: Orleans County Historical Society, Inc., Brownington, VT. Tax-exempt: 501(c)(3).
Institution Type/Description: Local History Museum: housed in 1836 Old Stone House built by Rev. Alexander Twilight, originally used as school dormitory.
Collections: period furniture; school texts & reference material on Orleans County history; paintings; early farm; household & military artifacts; Vermont imprints; antique farm equipment; textiles; folk art; tools, toys.
Research Fields: Orleans County history.
Facilities: 2,000-vol. library of local and state history; old textbooks; newspaper collection from 1860 available for research on premises by arrangement with librarian; reading room.
Activities: guided tours; permanent exhibitions; educational programs; classes in traditional skills.
Publications: quarterly bulletin; books.
Hours & Admission Prices: May 15-Oct. 15 Wed.-Sun. 11-5. Adults $8, Orleans County Residents & AAM members $7, students $5; discount to groups and NEMA, VMGA & AAA members; members no charge. &
Attendance: 3,000 (estimated)
Membership: Individual $15; Family $25; Contributing $35; Sustaining $50; Life $150.

Brownsville

WEST WINDSOR HISTORICAL SOCIETY, The Grange Hall, Rte. 44, Brownsville, VT 05037. Mailing Address: P.O. Box 12, Brownsville, VT 05037-0012. Tel.: 802-436-2262.
Web Site: www.westwindsorvt.govoffice2.com
Key Personnel: Pres., Genevieve Lemire
Institution Type/Description: Historical Society Museum.
Collections: local history & culture; photographs; personal artifacts.
Hours & Admission Prices: Call for hours. No charge.
Membership: Students & Seniors $5; Regular: Individual $10, Couple $15, Family $20; Sustaining: Individual $25, Family $50; Patron: Individual $100, Family $200; Life: Individual $250, Family $400.

Burlington

BCA CENTER, 135 Church St., Firehouse Center for the Visual Arts, Ground Fl., Burlington, VT 05401-8415. Tel.: 802-865-7166. Fax: 802-865-5839.
Web Site: www.burlingtoncityarts.org/BCAcenter
Formerly: The Firehouse Gallery
Founded: 1981.
Key Personnel: Exec. Dir., Doreen Kraft; Cur., Chris Thompson.
Governing Authority: Tax-exempt.
Institution Type/Description: Art Museum.
Collections: painting; sculpture; photography.
Hours & Admission Prices: Call for hours. No charge; donations accepted. &
Attendance: 60,000 (accurate)
Membership: Student $25; Basic $40; Family $75; Supporter $125; Director's Circle $1,000.

ECHO LAKE AQUARIUM AND SCIENCE CENTER/LEAHY CENTER FOR LAKE CHAMPLAIN, One College St., Leahy Center for Lake Champlain, Burlington, VT 05401-5215. Tel.: 802-864-1848; 877-ECHOFUN. Fax: 802-864-6832. Facebook: ECHO Vermont.
E-mail: info@echovermont.org
Web Site: www.echovermont.org
Founded: 1995.
Key Personnel: Chm. (V), Mark Saba; Exec. Dir., Phelan R. Fretz; Dir. Mktg. & Communications, Gerianne Smart; Dir. New Product Devel., Julie Silverman, M.A.T.; Dir. Finance, Chris Miller, Ph.D., CPA; Dir. Animal Care & Facilities Mgmt., Steve Smith; Treas., Tim Davis.
Personnel Profile: Full-Time Paid 24; Part-Time Paid 6; Part-Time Volunteers 200; Interns 30.
Governing Authority: Parent institution: Leahy Center for Lake Champlain, Inc., dba ECHO Lake Aquarium and Science Center; private; nonprofit organization. Tax-exempt: 501(c)(3).
Institution Type/Description: Lake Aquarium & Science Center
Collections: history, ecology & culture of the Lake Champlain basin; hands on interactive exhibits; live aquatic animals.
Major Exhibits: Alice's Wonderland (T), 1/18/14-5/11/14; Keva Planks (T), 5/24/14-9/1/14; Coffee (T), 9/13/14-1/4/15.

Facilities: library; educational facilities; aquarium; laboratory; resource room, movie theater; cafe. Museum-related items for sale.
Activities: formal education programs; guided tours; lectures; participatory exhibits; demonstrations; outreach programs; environmental monitoring classes; kit rental, adult after dark programming; films.
Publications: educational program flyer.
Hours & Admission Prices: Daily 10-5. Adult 18-59 $13.50, seniors 60 & over and college students with ID $11.50, children 3-17 $10.50; discount to military & ASTC members; children under 3 & members no charge. Closed Thanksgiving; Christmas Eve & Day. &
Attendance: 153,000 (accurate)
Membership: Single $35; Dual $60; Small Family $100; Large Family $120; Family $150.

ETHAN ALLEN HOMESTEAD MUSEUM & HISTORIC SITE, **(M),** 1 Ethan Allen Homestead, Burlington, VT 05408-1141. Tel.: 802-865-4556. Fax: 802-865-0661.
E-mail: info@ethanallenhomestead.org
Web Site: www.ethanallenhomestead.org
Founded: 1988.
Key Personnel: Exec. Dir., Daniel O'Neil; Pres. (V), Roger Marshall.
Personnel Profile: Full-Time Paid 1; Part-Time Volunteers 53; Interns 1.
Governing Authority: Parent Institution: Ethan Allen Homestead Foundation. Tax-exempt.
Institution Type/Description: Historic Site.
Collections: life of Ethan Allen; Native American artifacts; local history; gardens.
Facilities: interpretive center. Museum-related items for sale.
Activities: children's programs; educational programs; live demonstrations; reenactments; family events.
Hours & Admission Prices: May-Oct. Thurs.-Mon. 10-4; groups by appointment. Adults $7 nonresident, $5 VT resident, seniors $5, children 6-12 $3; children under 6 & members no charge.
Attendance: 4,000 (accurate)

FRANCIS COLBURN GALLERY, University of Vermont, Dept. of Art & Art History, Williams Hall, 72 University Pl., Burlington, VT 05405-0168. Tel.: 802-656-2014. Fax: 802-656-2064.
E-mail: artdept@uvm.edu
Web Site: www.uvm.edu/~artdept
Founded: 1975.
Key Personnel: Administrative Asst., Simone Blaise.
Governing Authority: university.
Institution Type/Description: University Art Gallery: housed in 1896 campus building.
Collections: various forms of art media exhibited by students, faculty & visiting artists.
Activities: lectures; films; gallery talks; formally organized education programs for undergraduate college students; loan & temporary exhibitions.
Publications: exhibition announcements.
Hours & Admission Prices: Sept.-May Mon.-Fri. 9-4:30. No charge. &

FROG HOLLOW GALLERIES, 85 Church St., Burlington, VT 05401-4420. Tel.: 802-863-6458. Fax: 802-860-6506.
Web Site: www.froghollow.org
Governing Authority: nonprofit organization. Tax-exempt: 501(c)(3).
Institution Type/Description: Art Gallery.
Collections: works by Vermont artists.
Activities: art & craft classes for students.
Hours & Admission Prices: Mon.-Wed. 10-6, Thurs.-Sat. 10-8, Sun. 11-6.

PERKINS GEOLOGY MUSEUM AT THE UNIVERSITY OF VERMONT, Delehanty Hall - Trinity Campus, 180 Colchester Ave., Burlington, VT 05405-1758. Tel.: 802-656-8694.
E-mail: geology@uvm.edu
Web Site: www.uvm.edu/perkins/
Key Personnel: Chm., Dept. of Geology, Andrea Lini
Institution Type/Description: Geology Museum.
Collections: geological artifacts; rocks; fossils; hands-on exhibits.
Activities: hands-on exhibits; educational programs.
Hours & Admission Prices: Academic Year: Mon.-Fri. 9-6, Sat.-Sun. 11-6; Call for summer hours. No charge. &

＊   **ROBERT HULL FLEMING MUSEUM, (M),** Univ. of Vermont, 61 Colchester Ave., Burlington, VT 05405. Tel.: 802-656-0750. Fax: 802-656-8059.
E-mail: fleming@uvm.edu
Web Site: www.flemingmuseum.org

Founded: 1931.
Key Personnel: Dir., Janie Cohen; Cur., Aimee Marcereau DeGalan; Education & Public Programs Cur., Christina Fearon; Public Rels. & Mktg. Mgr., Chris Dissinger; Mgr. Collections & Exhibitions, Margaret Tamulonis; Exhibition Designer & Preparator, Perry Price; Financial Mgr., Stephanie Glock; Museum Shop Mgr., Kristen Kilbashian.
Personnel Profile: Full-Time Paid 8; Part-Time Paid 2.
Governing Authority: university. Parent Institution: University of Vermont. Tax-exempt.
Institution Type/Description: Art & Anthropology Museum.
Collections: American & European paintings; sculpture, prints & drawings, medieval to modern; African, Oceanic, Asian, pre-Columbian, ancient; decorative arts; costumes; textiles; anthropological & archeological artifacts; Native American.
Research Fields: pertaining to collections & exhibitions.
Facilities: 2,000-vol. library on art history; on-premises research; reading room; classrooms; auditorium; seminar study room. Museum-related items for sale.
Activities: guided tours; lectures; gallery talks; concerts; arts festivals; permanent, temporary & traveling exhibitions; school services program; community outreach; public programs; films; symposia; workshops.
Publications: Seasonal Calendars; exhibition catalogs; brochures; posters.
Hours & Admission Prices: May to Labor Day. Tues.-Fri. 12-4, Sat.-Sun. 1-5; Sept.-April Tues. & Thurs.-Fri. 9-4, Wed. 9-8. Adults $5, seniors & students $3; discounts to AAM members; members no charge. Closed major holidays. &
Attendance: 25,000 (estimated)
Membership: Individual $30; Family & Dual $45; Contributing $100; Patron $250; Benefactor $500; Director's Circle $1,000; $5 reduced rate for artists, educators, and seniors.

Cabot

CABOT HISTORICAL SOCIETY, 193 McKinistry, Cabot, VT 05647-9755. Mailing Address: P.O. Box 275, Cabot, VT 05647-0275. Tel.: 802-563-2547.
E-mail: bonniesd@together.net
Web Site: cabothistory.org
Founded: 1966.
Key Personnel: Pres., Bonnie S. Dannenberg; Cur., Eric Ginette.
Governing Authority: nonprofit. Tax-exempt: 501(c)(3).
Institution Type/Description: Historical & Preservation Society: housed in 1845 schoolhouse in Cabot Village.
Collections: artifacts; pictures; documents; manuscripts.
Research Fields: local history.
Facilities: 25-vol. library of newspapers & account books available for research by permission of society officers. Local maps, church history & other museum-related items for sale.
Activities: guided tours; lectures; permanent & temporary exhibitions.
Hours & Admission Prices: Special local holidays & by appointment. No charge; donations accepted.
Attendance: 500 (estimated)
Membership: General $1.

Castleton

CASTLETON HISTORICAL SOCIETY, The Higley Homestead, 407 Main St., Castleton, VT 05735-0219. Mailing Address: P.O. Box 219, Castleton, VT 05735-0219. Tel.: 802-468-5105.
E-mail: blueshoehh@hotmail.com
Founded: 1947.
Key Personnel: Pres. (V) & Museum Shop Mgr., Holly Hitchcock.
Personnel Profile: Part-Time Volunteers 10.
Governing Authority: private; nonprofit organization. Tax-exempt: 501(c)(3).
Institution Type/Description: Village Museum: main office housed in 1811 Federal-style Georgian house, The Higley Homestead, listed on National Register of Historic Places.
Collections: decorative arts; furnishings; personal artifacts.
Activities: guided tours; temporary exhibits.
Publications: book, Castleton Looking Back; Castleton Scenes of Yesterday.
Hours & Admission Prices: May-Dec. by appointment. No charge; donations accepted. &
Attendance: 150 (estimated)
Membership: Adult $8.

CHRISTINE PRICE GALLERY, Castleton State College, Castleton, VT 05735. Tel.: 802-468-5611. Fax: 802-468-1440.
E-mail: william.ramage@castleton.edu
Institution Type/Description: Art Gallery.
Collections: paintings; photographs; sculpture; drawings.

Activities: temporary exhibitions.
Hours & Admission Prices: Academic Year: Mon.-Fri. 9-5.

Cavendish

CAVENDISH HISTORICAL SOCIETY MUSEUM, Main St., Rte. 131, Cavendish, VT 05142. Mailing Address: P.O. Box 472, Cavendish, VT 05142-9647. Tel.: 802-226-7807.
E-mail: margoc@tds.net
Web Site: cavendishhistory.org
Key Personnel: Coord., Margo Caulfield
Institution Type/Description: Historical Society Museum: housed in a former 19th-century town hall.
Collections: period history & culture; photographs; farm equipment & tools; household utensils; costumes; textiles;
Hours & Admission Prices: late June to mid-Oct. Sun. 2-4; other times by appointment. No charge.

Chester

CHESTER ART GUILD, The Green, Main St., Chester, VT 05143. Mailing Address: P.O. Box 154, Chester, VT 05143-0154. Tel.: 802-875-3767.
E-mail: dorisingram@mymailstation.com
Founded: 1960.
Key Personnel: Pres. & Chm. (V), Doris Ingram; Treas., Molly Ferris; Membership Coord., Dale O'Brien; Public Rels., Nancy Ball.
Personnel Profile: Full-Time Volunteers 70; Part-Time Volunteers 70.
Governing Authority: volunteer group. Tax-exempt.
Institution Type/Description: Art Association Gallery: restored former elementary school.
Collections: paintings by local artists.
Facilities: library relating to artists & art history; 800 sq. ft. exhibit space; classrooms & studio. All exhibited items for sale, except permanent collection.
Activities: lectures; workshops; educational programs; two high school scholarships for Governor's Institute on Art summer school at Castleton State College; demonstrations; outdoor shows; theme shows; college scholarship; classes. Annual Events: Holiday Sale.
Publications: quarterly members newsletter, Argus.
Hours & Admission Prices: mid-June to mid-Oct. Fri. & Sun. 1-4, Sat. 9 4. Art Gallery: June-Oct. Fri. & Sun. 1-4, Sat. 9-4. Studio Group: Tues.-Sat. 9-12. No charge; donations accepted. &
Attendance: 2,400 (estimated)
Membership: Student $2; Patron $6; Artist & Corporate or Sustaining $20; Family $35; Life $200.

Colchester

COLCHESTER HISTORICAL SOCIETY - LOG SCHOOLHOUSE MUSEUM, 119 Wintergreen Dr., Colchester, VT 05446. Mailing Address: 245 Bluebird Dr., Colchester, VT 05446. Tel.: 802-879-0042.
Founded: 2007.
Key Personnel: Dir. & Cur., Tom Mulcahy; Pres. (V), Dr. H. Clinton Reichard; Museum Shop Mgr., Carol Reichard.
Personnel Profile: Full-Time Volunteers 30.
Governing Authority: Parent Institution: Colchester Historical Society. Tax-exempt.
Institution Type/Description: Historical Society Museum: housed in a log schoolhouse; c.1815.
Collections: local history & culture; period furnishings; personal artifacts.
Activities: special events.
Hours & Admission Prices: Memorial Day to Labor Day Fri. Mon. 11-3. No charge; donations accepted.
Attendance: 1,750 (estimated)

MCCARTHY ARTS CENTER GALLERY, Saint Michael's College, One Winooski Park, Colchester, VT 05439. Tel.: 802-654-2246.
Web Site: www.smcvt.edu/academics/finearts
Institution Type/Description: Art Gallery.
Collections: paintings; contemporary graphics.
Hours & Admission Prices: Mon.-Fri. 9-5. No charge.

Concord

CONCORD HISTORICAL SOCIETY MUSEUM, Concord Town Hall, Concord, VT 05824. Mailing Address: P.O. Box 195, Concord, VT 08524-0195.
Key Personnel: Pres. (V), Kathleen Fisher
Institution Type/Description: Historical Society Museum.
Collections: replica of local doctor's office & smoking room; period post office; schoolroom; furnishings; tools; clothing; toys.
Publications: quarterly newsletter, Concord Historical Society; Concord "Then and Now."
Hours & Admission Prices: Sept. last Sat.-Sun.; other times by appointment. No charge; donations accepted.

Cuttingsville

SHREWSBURY HISTORICAL SOCIETY, 5419 Rte. 103, Cuttingsville, VT 05738. Tel.: 802-492-3324.
E-mail: ruthcon1@vermontel.net
Web Site: shrewsburyhistoricalsociety.com
Founded: 1970.
Key Personnel: Pres. (V), Conrad Winkler.
Governing Authority: Parent Institution: Shrewsbury Historical Society, Inc. Tax-exempt.
Institution Type/Description: Historical Society Museum.
Collections: local history & culture; photographs; period furnishings; costumes; books; toys; videos.
Publications: annual newsletter.
Hours & Admission Prices: July-Oct. Sun. 1-3. No charge; donations accepted.
Attendance: 75 (estimated)
Membership: Single $10; Family $15; Contributing $30; Life $125.

Dorset

DORSET HISTORICAL SOCIETY, Rte. 30 at Kent Hill Rd., Dorset, VT 05251. Mailing Address: P.O. Box 52, Dorset, VT 05251-0052. Tel.: 802-867-0331. Fax: 802-867-0412.
E-mail: info@dorsetvthistory.org
Web Site: www.dorsetvthistory.org
Founded: 1963.
Key Personnel: Chm. (V) & Pres. (V), Richard Hittle.
Personnel Profile: Part-Time Paid 2; Part-Time Volunteers 20.
Governing Authority: Tax-exempt.
Institution Type/Description: Historical Society Museum: housed in Bley House.
Collections: local history; genealogy; farm & household artifacts; photographs; pottery; paintings; decorative arts.
Research Fields: family histories.
Facilities: genealogical library.
Activities: lectures; programs.
Publications: quarterly newsletter; Dorset In The Shadow of the Marble Mountain; Quabbin to Dorset; Walking Tour - Driving Tour.
Hours & Admission Prices: Dec.-April 14 Wed.-Sat. 10-2; April 15-Nov. Wed. 10-12, Thurs.-Sat. 10-4; other times by appointment. No charge; donations accepted. Closed Christmas. &
Attendance: 1,000 (accurate)
Membership: Regular $35; Supporter $50; Sustaining $100; Patron $250; Benefactor $500.

East Calais

NDAKINNA CULTURAL CENTER & MUSEUM, 34 Moscow Woods Rd., East Calais, VT 05650. Mailing Address: P.O. Box 7, East Calais, VT 05650-0007. Tel.: 802-456-8884.
Institution Type/Description: Native American Museum.
Collections: Native American culture, heritage & history; photographs; personal artifacts; Abenaki baskets; clothing; ceremonial dress; beads; dream catchers.
Facilities: Museum-related items for sale.
Activities: classes; workshops.
Hours & Admission Prices: Jan.-March Sat. 9-4, Sun. 12-4; April-Dec. 24 Wed.-Fri. 12-5, Sat. 9-3. No charge; donations accepted.

East Montpelier

BRAGG FARM SUGAR HOUSE, 1005 VT Rte. 14 N., East Montpelier, VT 05651. Tel.: 802-223-5757; 800-376-5757.
E-mail: braggfarmmaple@aol.com
Web Site: www.braggfarm.com

Institution Type/Description: Farm History Museum.
Collections: farm history; maple sugaring; farm animals; period farm tools & equipment.
Facilities: nature trails. Gift items for sale.
Activities: maple syrup taste testing; educational programs; videos; walking trail.
Hours & Admission Prices: June-Aug. daily 8:30-8; Sept.-May daily 8:30-6.

East Poultney

POULTNEY HISTORICAL SOCIETY MUSEUM, On-the-Green, East Poultney, VT 05741. Mailing Address: P.O. Box 605, East Poultney, VT 05741-0605. Tel.: 802-287-5252.
E-mail: info@poultneyhistoricalsociety.org
Web Site: poultneyhistoricalsociety.org
Founded: 1935.
Key Personnel: Pres., Ina Smith.
Personnel Profile: Part-Time Volunteers 12.
Governing Authority: nonprofit. Affiliated with the Poultney Historical Society, Inc. Tax-exempt.
Institution Type/Description: Antiques Museum: housed in 1800 Old Blacksmith Shop & Melodeon Factory; 1791 brick schoolhouse; 1895 Victorian schoolhouse.
Collections: early Poultney history; costumes; farm & home implements; archives; cemetery records; town records; Horace Greeley items; melodeons; period artifacts; genealogical records; Native American artifacts; Civil War medical kit.
Major Exhibits: Poultney At Work: Agricultural and Forestry Tools, 7/14-9/14; The Valiant 13, 7/14-9/14.
Research Fields: local history; genealogy.
Activities: guided tours; formally organized education programs for children; permanent exhibitions.
Publications: History of Poultney; brochures & podcasts, Poultney historical audio walking & driving tours; DVD, A Tale of Two Villages: Poultney Village; quarterly member newsletter.
Hours & Admission Prices: Memorial Day to Labor Day Sun. 1-4; other times by appointment. No charge; donations accepted.
Attendance: 350 (estimated)
Membership: Individual $20; Family $35; Supporting $50; Sustaining $100.

Fairfield

PRESIDENT CHESTER A. ARTHUR HISTORIC SITE, 4588 Chester Arthur Road, Fairfield, VT 05455. Mailing Address: Historic Preservation, National Life Bldg., 6th Fl., Montpelier, VT 05633. Tel.: 802-828-3051. Fax: 802-828-3206.
E-mail: john.dumville@state.vt.us
Web Site: www.historicsites.vermont.gov/arthur
Founded: 1953.
Key Personnel: Historic Sites Operations Chief, John P. Dumville.
Personnel Profile: Full-Time Paid 1; Part-Time Paid 1.
Governing Authority: state. Parent Institution: State of Vermont. Subsidiary Institution: Division for Historic Preservation, One National Life Dr., 6th Fl., Montpelier 05602. Tax-exempt.
Institution Type/Description: Historic Houses: c.1830 Chester A. Arthur birthplace; 1820 Brick Church.
Collections: memorabilia. Historic Building: c.1820 Old Brick Church.
Facilities: picnic area.
Activities: permanent exhibitions.
Publications: brochures, Birthplace of Chester A. Arthur; Guide to Historic Sites In Vermont.
Hours & Admission Prices: June to mid-Oct. Wed.-Sun. 11-5. No charge; donations accepted. &

Attendance: 3,000 (estimated)

Ferrisburgh

ROKEBY MUSEUM, (M), 4334 Rte. 7, Ferrisburgh, VT 05456-9779. Tel.: 802-877-3406. Fax: 802-877-3406.
E-mail: rokeby@comcast.net
Web Site: www.rokeby.org
Formerly: Rokeby (Ancestral Estate of Rowland Evans Robinson)
Founded: 1962.
Key Personnel: Pres. (V), JoAnne C. LaBerge; Dir., Jane Williamson.
Personnel Profile: Part-Time Paid 2; Part-Time Volunteers 10.
Governing Authority: nonprofit organization. Tax-exempt.
Institution Type/Description: Historic House: c.1784 Rokeby.
Collections: 18th & 19th century Vermont furnishings; writings of Rowland Robinson; underground railroad station; local history; folklore; photographs; art collections; letters; Indian artifacts & spiritualism; manuscript collections; agricultural implements.

Research Fields: local & state history; social history; 18th- & 19th-century literature; 19th- & 20th-century agriculture; abolition Quakers.
Facilities: 3,000-vol. library of the family available for use by appointment; reading room.
Activities: guided tours; permanent exhibitions; special events.
Publications: newsletter, Rokeby Messenger.
Hours & Admission Prices: May-Oct. daily 10-5. Guided House Tours: Fri.-Mon. 11:30 & 2. Adults $10, seniors $9, students $8; discounts to AAM members, groups & Vermont museum employees; members & children under 5 no charge. &
Attendance: 2,400 (accurate)
Membership: Individual $25; Family $40; Life $500.

Glover

BREAD & PUPPET MUSEUM, 753 Heights Rd., Rte. 122, Glover, VT 05839-9637. Tel.: 802-525-6972 & 3031. Fax: 802-525-3618.
E-mail: breadpup@together.net
Web Site: www.breadandpuppet.org
Founded: 1975.
Key Personnel: Dir., Peter Schumann; Sec. & Museum Shop Mgr., Elka Schumann.
Personnel Profile: Part-Time Paid 1; Part-Time Volunteers 2.
Governing Authority: individual operation; nonprofit organization. Parent Institution: Bread & Puppet Theater. Tax-exempt: 170(b)(1)(A).
Institution Type/Description: Puppet Theater & Museum: housed in a 100 year old barn.
Collections: Peter Schumann works; puppets of all sizes; masks; banners; reliefs; paintings; graphics; stages; posters & literature on Bread & Puppet Theater; theater artifacts.
Research Fields: puppetry; theater; masks; street theater; pageants.
Facilities: library of printed matter concerning the works of the Bread & Puppet theater.
Activities: guided tours.
Publications: Bread and Puppet Museum. Please write to us for our annual mail order catalog of publications, posters, postcards & videos.
Hours & Admission Prices: June-Nov. 1 daily 10-6; other times by appointment. No charge; donations accepted.
Attendance: 20,000 (estimated)

Grafton

GRAFTON HISTORICAL SOCIETY, (M), 147 Main St., Grafton, VT 05146. Mailing Address: P.O. Box 202, Grafton, VT 05146-0202. Tel.: 802-843-2584.
E-mail: grafhist@vermontel.net
Web Site: www.graftonhistoricalsociety.com
Founded: 1962.
Key Personnel: Pres., Hardy Merrill; Sec., Harold Tincher.
Personnel Profile: Part-Time Paid 1; Part-Time Volunteers 45.
Governing Authority: nonprofit organization. Tax-exempt: 501(c)(3).
Institution Type/Description: Local History Museum.
Collections: Grafton history; photographs; soapstone; writing accessories; furniture; tools; textiles.
Research Fields: genealogy; local history.
Activities: permanent exhibitions; special exhibits of historical interest; lectures on Vermont & local history.
Publications: book, $5 and a Jug of Rum, a history of Grafton, Vermont 1754-2000; booklets, 125 Years with the Grafton Cornet Band; Barrett Store & Customers, 1816-1830; Innkeeping in Grafton 100 Years Ago; map, The First Map of Grafton 1764; Releasing Rebecca, An Exploration of Life, Death and Gravestone Art in Early Vermont; History of Grafton; Life of a Vermont Farmer; The Grafton Quilt Coloring Book; Grafton's Founding Century 1754-2004.
Hours & Admission Prices: Mon. & Thurs-Fri. 10-4; weekends Memorial Day-Columbus Day 10-4; other times by appointment. No charge; donations accepted.
Attendance: 1,700 (accurate)
Membership: Individual $10; Household $25; Sustaining $50; Household Sustaining $75; Patron $200; Household Patron $300.

THE NATURE MUSEUM, 186 Townshend Rd., Grafton, VT 05146. Mailing Address: P.O. Box 38, Grafton, VT 05146-0038. Tel.: 802-843-2111.
E-mail: info@nature-museum.org
Web Site: www.nature-museum.org
Founded: 1989.
Key Personnel: Pres. (V), Laurie Danforth; Dir. Operations, Noralee Hall; Treas., Steven Davis; Dir. Education, Beth Roy.
Personnel Profile: Part-Time Paid 5; Part-Time Volunteers 20; Interns 2.

Governing Authority: private; nonprofit organization. Tax-exempt: 501(c)(3).
Institution Type/Description: Natural History Museum.
Collections: emphasis on the mammals, birds & geology of New England.
Facilities: educational facilities; 2,000 sq. ft. exhibit space; nature trails; wildlife gardens. Museum-related items for sale.
Activities: educational programs; lectures; participatory exhibits; nature trails; wild life gardens; outreach school programs.
Publications: quarterly calendar & newsletter; annual school & library program brochure.
Hours & Admission Prices: Memorial Day to Columbus Day Thurs. & Sat.-Sun. 10-4; Oct.-May Thurs. 10-4. Call for additional hours. Adults $5, seniors $4, children 3-12 $3; discounts to VMGA members & families; members no charge. Closed major holidays.
Attendance: 4,550 (accurate)
Membership: Senior & Student $20; Individual $25; Family $50; Supporting $75; Sustaining $100; Patron $250.

Grand Isle

GRAND ISLE HISTORICAL SOCIETY, U.S. Rte. 2, Grand Isle, VT 05458. Mailing Address: P.O. Box 23, Grand Isle, VT 05458-0023. Tel.: 802-372-8339.
E-mail: dfchamb@aol.com
Formerly: Grand Isle County Historical Society
Founded: 2000.
Key Personnel: Pres. (V), Fay P. Chamberlin.
Governing Authority: Tax-exempt.
Institution Type/Description: Historical Society Museum.
Collections: local history & culture; photographs; furniture; clothing; period furnishings. Historic Buildings: c.1784 Hyde Log Cabin; Block Schoolhouse built in 1814.
Activities: antique appraisal. Museum Sponsors: Antique Show & Sale.
Hours & Admission Prices: Memorial Day to Columbus Day Fri.-Sun. 11-5. Adults $3; children under 14 no charge. &
Membership: Annual $10; Contributing $25; Business & Patron $50; Life $150; Heritage $1,000.

Graniteville

ROCK OF AGES VISITOR CENTER, 558 Graniteville Rd., Graniteville, VT 05654. Mailing Address: P.O. Box 482, Barre, VT 05641-0482. Tel.: 802-476-3119. Fax: 802 476 2110.
E-mail: visitor@rockofages.com
Web Site: tours.rockofages.com
Founded: 1885.
Key Personnel: Dir., Todd Paton
Institution Type/Description: Granite Industry Museum.
Collections: granite manufacturing history; photographs; granite artifacts & sculptures.
Facilities: theatre. Museum-related items for sale.
Activities: granite quarry demonstrations; video; outdoor granite bowling; make your own stone gift.
Hours & Admission Prices: Visitors Center: mid-May to mid-Sept. Mon.-Sat. 9-5; mid-Sept. to Oct. daily 9-5. Quarry Tours: Memorial Day to mid-Sept. Mon.-Sat. 9:15-3:35; mid-Sept. to mid-Oct. daily 9:15-3:35. Closed Independence Day. &

Guilford

GUILFORD HISTORICAL SOCIETY, Guilford Center Rd., Guilford, VT 05301. Mailing Address: 236 School Rd., Guilford, VT 05301. Tel.: 802-257-0147 & 254-5910.
E-mail: dabonvil@sover.net
Founded: 1973.
Key Personnel: Chm. (V) & Cur., Ann Bonneville; Pres. (V), Richard A. Austin; Asst. Cur., Michelle Frehsee.
Personnel Profile: Part Time Volunteers 6.
Governing Authority: Tax-exempt.
Institution Type/Description: Historical Society Museum: housed in the Town Hall building; built in 1822.
Collections: local history & culture; photographs; tools; household artifacts.
Facilities: Note cards & town history book for sale.
Activities: special events; educational programs.
Publications: newsletter, Guilford Slate.
Hours & Admission Prices: June-Sept. Tues.-Sat. 10-2. No charge; donations accepted. &
Attendance: 200 (accurate)

Hardwick

GRACE - GRASS ROOTS ART AND COMMUNITY EFFORT, 59 Mill St., Hardwick, VT 05843. Mailing Address: P.O. Box 960, Hardwick, VT 05843. Tel.: 802-472-6857. Fax: 802-472-9578.
E-mail: grace@vtlink.net
Web Site: www.graceart.org
Founded: 1975.
Key Personnel: Dir., Carol Putnam; Chm. (V), Stephen Farber.
Personnel Profile: Full-Time Paid 1; Part-Time Paid 4; Interns 2.
Institution Type/Description: Art Gallery: housed in the Old Firehouse; built in 1885. Listed on the National Register of Historic Sites.
Collections: paintings.
Activities: art workshops; outreach programs; special events; exhibition receptions. Museum Sponsors: Annual Open House.
Publications: annual newsletter.
Hours & Admission Prices: Tues.-Thurs. 10-4. No charge; donations accepted. &
Attendance: 1,200 (estimated)

Hartford

HARTFORD HISTORICAL SOCIETY - GARIPAY HOUSE MUSEUM, 1461 Maple St., Hartford, VT 05047. Mailing Address: P.O. Box 547, Hartford, VT 05047-0547. Tel.: 802-296-3132.
E-mail: info@hartfordhistory.org
Web Site: www.hartfordhistory.org
Founded: 1987.
Key Personnel: Chm. (V), Dorothy Yamashita; Pres. (V), Susanne Abetti.
Personnel Profile: Part-Time Volunteers 9.
Governing Authority: Tax-exempt.
Institution Type/Description: Historical Society Museum: housed in Dr. Garipay's home.
Collections: local history & culture; medical artifacts; documents; photographs; period furnishings; personal artifacts.
Activities: lectures.
Publications: newsletter.
Hours & Admission Prices: May-Oct. 1st Tues. each month 6 pm-8 pm, 2nd Sun. each month 1:30-4; other times by appointment. No charge; donations accepted.
Attendance: 30 (estimated)
Membership: Senior $10; Individual & Senior Family $15; Family $20; Commercial $25.

Hartland

HARTLAND HISTORICAL SOCIETY, Rte. 12, Hartland, VT 05048. Mailing Address: P.O. Box 297, Hartland, VT 05048-0297. Tel.: 802-436-1703.
E-mail: info@hartlandhistory.org
Web Site: www.hartlandhistory.org
Founded: 1916.
Key Personnel: Pres., Carol Mowry
Governing Authority: Tax-exempt.
Institution Type/Description: Historical Society Museum
Collections: local history & culture; photographs; letters; diaries; church records; uniforms; period furnishings.
Publications: newsletter 3 times a year.
Hours & Admission Prices: Mon. 1-4, Fri. 9-11am; other times by appointment. No charge; donations accepted. Closed holidays.
Attendance: 100 (estimated)
Membership: Individual $5; Family $10.

Holland

HOLLAND HISTORICAL SOCIETY MUSEUM, INC., Gore Rd., Holland, VT 05830. Mailing Address: 120 School Rd., Derby Line, VT 05830. Tel.: 802-895-4440.
Founded: 1972.
Congressional District: 17
Key Personnel: Dir., Harvey McDonald; Dir., Martha Judd; Dir., Albert Hauver; Dir., Laurel Mosher; Pres. (V), Diane Judd; Vice Pres., Melody Ricard; Sec., Bea Nelson.
Personnel Profile: Part-Time Volunteers 8.
Governing Authority: nonprofit organization. Tax-exempt: 170(b)(1)(A).
Institution Type/Description: Historical Society Museum: housed in c.1848 Congregational Church, birth town of Horace Tabor, the Silver king of Colorado.
Collections: period artifacts; pews; lecterns; local history artifacts; tools; 19th-century paintings & portraits; maps; organs; horse shed replica.

Research Fields: local family histories.
Facilities: archives.
Activities: permanent & temporary exhibitions; dinners; meetings. Museum Sponsors: Open House by appointment; Annual Old Home Day in August; Old Timer's Day in September.
Publications: annual newsletter; financial report.
Hours & Admission Prices: Aug. 1st Sun.; other times by appointment. No charge; donations accepted.
Attendance: 175 (estimated)
Membership: Annual $5; Life $100.

Hubbardton

HUBBARDTON BATTLEFIELD STATE HISTORIC SITE, 5696 Monument Hill Rd., Hubbardton, VT 05749. Mailing Address: Historic Preservation, National Life Bldg., 6th Fl., Montpelier, VT 05620. Tel.: 802-759-2412 & 828-3051. Fax: 802-828-3206.
E-mail: john.dumville@state.vt.us
Web Site: www.historicsites.vermont.gov/hubbardton
Founded: 1948.
Key Personnel: Historic Sites Operations Chief, John P. Dumville; Site Admin., Elsa Gilbertson.
Personnel Profile: Full-Time Paid 1; Part-Time Paid 2.
Governing Authority: state. Parent Institution: State of Vermont, Div. for Historic Preservation, National Life Bldg., 6th Fl., Montpelier 05620. Tax-exempt.
Institution Type/Description: State Historic Site: Hubbardton Battlefield.
Collections: military; Revolutionary War.
Research Fields: Revolutionary War in Vermont & environs.
Facilities: visitor's center.
Activities: reenactment program.
Publications: brochure; books, The Battle of Hubbardton; The American Rebels Stem the Tide.
Hours & Admission Prices: mid-May to mid-Oct. Wed.-Sun. 11-5. Adults $5; discounts to AAM members; registered school groups & children under 14 no charge. &

Attendance: 5,000 (estimated)

Huntington

BIRDS OF VERMONT MUSEUM, 900 Sherman Hollow Rd., Huntington, VT 05462-9420. Tel.: 802-434-2167.
E-mail: museum@birdsofvermont.org
Web Site: www.birdsofvermont.org
Founded: 1987.
Key Personnel: Dir., Erin Talmage; Chm. (V), Shirley Johnson
Governing Authority: private; nonprofit organization.
Institution Type/Description: Woodcarving Museum.
Collections: over 500 life-size carved birds representing 255 species; butterfly garden.
Research Fields: field ornithology.
Facilities: picnic area; walking trails.
Activities: school field trips; nature walks; carving classes; special programs & events.
Publications: quarterly newsletter, Chip Notes.
Hours & Admission Prices: May-Oct. daily 10-4; Nov.-April by appointment. Adults $6, seniors $5, children 3-17 $3; discounts to VT Public TV, AAA, VMGA members; members no charge. &
Attendance: 4,500 (accurate)
Membership: Individual $25; Family $40; Contributing $50; Supporting $100; Sponsor $250; Patron $500; Spear Society $1,000.

Isle La Motte

ISLE LA MOTTE HISTORICAL SOCIETY, 283 School St., Isle La Motte, VT 05463-9808. Mailing Address: P.O. Box 18, Isle La Motte, VT 05463-0018. Tel.: 802-928-3077.
E-mail: gloilm@yahoo.com
Founded: 1925.
Key Personnel: Pres., Robert McEwen; Vice Pres., Mary Jane Tiedgen; Treas., Lilian Masters; Cur., Gloria McEwen; Sec., Marty Dale.
Personnel Profile: Part-Time Volunteers 12.
Governing Authority: society. Tax-exempt: 501(c)(3).
Institution Type/Description: Local History Museum.
Collections: 18th to 19th-century artifacts; Indian culture. Historic Buildings: 19th-century blacksmith shop; 19-century stone schoolhouse; 19th century log/slab house.
Research Fields: genealogy.
Publications: annual newsletter; History of Isle Le Motte, VT; 75th Anniversary Booklet of Isle Le Motte Historical Society; History of Isle La Motte.

Hours & Admission Prices: July-Aug. Sat. 1-4, or by appointment. No charge, donations accepted.
Attendance: 500 (estimated)
Membership: Student $5; Adult $10; Couple $15; Family $20; Lifetime $100.

Jacksonville

WHITINGHAM HISTORICAL MUSEUM, 669 Reed Hill Rd., Jacksonville, VT 05342-9733. Mailing Address: P.O. Box 125, Jacksonville, VT 05342-0125. Tel.: 802-368-2448.
Founded: 1973.
Congressional District: 1
Key Personnel: Pres., Stella Stevens; Sec., Corrinne Boyd.
Personnel Profile: Part-Time Volunteers 20.
Governing Authority: nonprofit organization. Affiliated with Whitingham Historical Society, Jacksonville, VT 05342. Tax-exempt: 501(c)(3).
Institution Type/Description: History Museum.
Collections: local costumes; tools; historical papers; photographs; textiles.
Activities: guided tours; lectures; formally organized education programs for children; School Come to Museum; school group programs. Annual Events: One Room School House in fall.
Publications: annual newsletter, Sadawga Springs.
Hours & Admission Prices: June-Oct. Sun. 2-4; school groups & other times by appointment. No charge; donations accepted. &
Attendance: 375 (estimated)
Membership: Individual $7; Family $10; Life $75.

Jeffersonville

BRYAN MEMORIAL GALLERY, 180 Main St., Jeffersonville, VT 05464. Mailing Address: P.O. Box 340, Jeffersonville, VT 05464-0340. Tel.: 802-644-5100. Fax: 802-644-8342.
E-mail: info@bryangallery.org
Web Site: www.bryangallery.org
Key Personnel: Exec. Dir., Mickey Myers
Institution Type/Description: Art Gallery.
Collections: paintings; photographs.
Activities: workshops; educational programs.
Hours & Admission Prices: Spring & Fall: Thurs.-Sun. 11-4; Summer: daily 11-5; other times by appointment.
Membership: Artist & Individual $40; Family $50; Business & Associate $100; Patron $250; Contributor $500; Supporter $750; Director's Circle $1,000.

Jericho

EMILE A. GRUPPE GALLERY, 22 Barber Farm Rd., Jericho, VT 05465-9795. Tel.: 802-899-3211.
Institution Type/Description: Art Gallery: housed in an 1860s English Sheep barn at the home of Emile's daughter, Emilie Gruppe Alexander & her husband.
Collections: paintings; photographs.
Hours & Admission Prices: Thurs.-Sun. 10-3; other times by appointment.

JERICHO HISTORICAL SOCIETY, 4A Red Mill Dr., Jericho, VT 05465. Mailing Address: P.O. Box 35, Jericho, VT 05465-0035. Tel.: 802-899-3225.
Web Site: jerichohistoricalsociety.org
Founded: 1971.
Congressional District: 1
Key Personnel: Pres. (V), Ann Squires; Archives Chm., Wayne Howe; Sales Shop Mgr., Gail Prior.
Governing Authority: society; nonprofit. Tax-exempt: 501(c)(3).
Institution Type/Description: Historical Society Museum; housed in 1885 Chittenden Mills, five-story building once used as a grist mill.
Collections: 150 photographs & slides made from Wilson Bentley's original glass photographic plates along with many articles by & about Bentley; films; early photographs & memorabilia; Chitterden Mills historical material; roller milling process; period industrial artifacts. Historic Building: 1859 Mill House.
Research Fields: journals; accounts; Bentley archives; milling information.
Facilities: Locally made crafts, jewelry, china, glass & stationery for sale.
Activities: films; school loan service; temporary & loan exhibitions.
Publications: books, Old Mill Coloring Book; Old Mill Cook Book; Local Town History; History of Jericho, Vol. I (reprint) & Vol. II.
Hours & Admission Prices: Jan.-March Wed. & Sat. 10-5, Sun. 11:30-4; April-Dec. Mon.-Sat. 10-5, Sun. 11:30-4. Call to confirm summer hours. No charge. Closed Easter; Independence Day; Thanksgiving; Christmas. &
Attendance: 14,000 (estimated)
Membership: Active $10; Family & Business $15; Supporting $25; Benefactor $50; Life $100.

Johnson

JULIAN SCOTT MEMORIAL GALLERY, Johnson State College, Dibden Center, 337 College Hill, Johnson, VT 06565. Tel.: 800-635-2356.
Key Personnel: Dir., Leila Bandar
Institution Type/Description: Art Gallery.
Collections: paintings; drawings; sculptures.
Hours & Admission Prices: Summer: Tues.-Fri. 12-6, Sat. 12-4; Winter: call for hours.

RED MILL GALLERY - VERMONT STUDIO CENTER, 80 Pearl St., Johnson, VT 05656. Mailing Address: P.O. Box 613, Johnson, VT 05656-0613. Tel.: 802-635-2727.
Founded: 1984.
Key Personnel: Gallery Mgr., G. Todd Haun.
Governing Authority: Parent Institution: Vermont Studio Center. Tax-exempt.
Institution Type/Description: Art Gallery.
Collections: works by national & international artists.
Activities: lectures; special events.
Hours & Admission Prices: Call for hours. No charge.
Attendance: 1,000 (estimated)

Lincoln

LINCOLN HISTORICAL SOCIETY, 88 Quaker St., Lincoln, VT 05443-9253. Tel.: 802-453-7502.
E-mail: rhutster@gmail.net
Founded: 1986.
Key Personnel: Dir., Sandra Rhodes; Pres., Rhonda Hutchins; Vice Pres., Eleanor Menzer; Treas., Larry Masterson.
Personnel Profile: Part-Time Volunteers 6.
Governing Authority: Tax-exempt.
Institution Type/Description: Historical Society Museum.
Collections: local history. Historic Buildings: 18th-century farmhouse; 19th-century barn.
Activities: Museum Sponsors.
Publications: Updated Lincoln History.
Hours & Admission Prices: Memorial Day to mid-Oct. 2nd & 4th Sun. 1-5; other times by appointment. No charge; donations accepted. &
Attendance: 100 (estimated)

Ludlow

BLACK RIVER ACADEMY MUSEUM, High St., Ludlow, VT 05149. Mailing Address: P.O. Box 73, Ludlow, VT 05149-0073. Tel.: 802-228-5050. Fax: 802-228-7444.
E-mail: glbrehm@tds.net
Web Site: bramvt.org
Founded: 1972.
Congressional District: 1
Key Personnel: Pres. (V), Susan Pollender; Dir., Georgia Brehm.
Personnel Profile: Part-Time Paid 2; Part-Time Volunteers 8.
Governing Authority: society; nonprofit. Parent Institution: Black River Historical Society, Inc. Tax-exempt.
Institution Type/Description: Historical Society Museum: housed in Black River Academy, from which Pres. Calvin Coolidge, 30th president, graduated in 1890.
Collections: paintings; portraits; photographs; economic, cultural, political & domestic area history.
Research Fields: local history.
Facilities: library of educational, political & historical material, available for use on premises.
Activities: concerts; lectures; performances; permanent & changing exhibitions.
Publications: History of Ludlow; The History of Black River Academy; Ludlow Village Walking Tour.
Hours & Admission Prices: June to Labor Day Tues.-Sat. 12-4; Sept. to Columbus Day Sat.-Sun. 12-4. Adults $2; children under 2 no charge. &
Attendance: 1,100
Membership: Single $25; Dual $45; $100; $500; $1,000.

Lyndon Center

SHORES MEMORIAL MUSEUM, 202 Center St., Lyndon Center, VT 05850. Mailing Address: P.O. Box 85, Lyndon Center, VT 05850-0085.
Web Site: www.shoresmuseum.org
Key Personnel: Chm. (V), Pres. (V) & Museum Shop Mgr., Eric Paris; Cur., Chris Raymond.

Personnel Profile: Part-Time Volunteers 3.
Governing Authority: city. Parent Institution: Lyndon Historical Society.
Institution Type/Description: History Museum: housed in 1896 Queen Anne style home built by James Shores.
Collections: wooden plates & bowls; a mortar & pestle; wire basket for collecting eggs; long-handled bedwarmer; sadiron for pressing clothes; photographs; period organs; sheet music; musical instruments; trophies; period clothing; war memorabilia; dolls.
Activities: tours; living history classes; research.
Publications: quarterly, Lyndon Legacy.
Hours & Admission Prices: By appointment. Donations requested.

Lyndonville

QUIMBY GALLERY, 1001 College Rd., Lyndonville, VT 05851. Mailing Address: P.O. Box 919, Lyndonville, VT 05851. Tel.: 802-626-6487.
E-mail: barclay.tucker@lyndonstate.edu
Governing Authority: Parent Institution: Lyndon State College.
Institution Type/Description: Art Gallery.
Collections: works by contemporary artists.
Activities: art shows.
Hours & Admission Prices: Academic Year: Mon.-Fri. 8-4. No charge.

Manchester

✳ **THE AMERICAN MUSEUM OF FLY FISHING, (M),** 4070 Main St., Manchester, VT 05254. Mailing Address: P.O. Box 42, Manchester, VT 05254-0042. Tel.: 802-362-3300. Fax: 802-362-3308.
E-mail: ccomar@amff.com
Web Site: www.amff.com
Founded: 1968.
Congressional District: 1
Key Personnel: Chm., David Walsh; Pres. (V), Richard Tisch, M.D.; Exec. Dir., Cathi Comar; Deputy Dir., Yoshi Akiyama; Coord. Events, Christina Cole; Devel. Asst., Sarah Foster; Dir. Visual Communications, Sara Wilcox; Editor, Kathleen Achor; Coord. Membership, Laura Napolitano; Accounts Mgr., M. Patricia Russell.
Personnel Profile: Full-Time Paid 4; Part-Time Paid 2; Part-Time Volunteers 3.
Operating Expenses: 751,670
Operating Income: 797,494
Governing Authority: nonprofit organization. Tax-exempt: 501(c)(3).
Institution Type/Description: Sports Museum.
Collections: reels; fly rods; flies; fly boxes; paintings; memorabilia; rare books.
Major Exhibits: The Wonders of Fly Fishing, 1/14-12/14; .
Research Fields: sporting & conservation history.
Facilities: 7,000-vol. library. Gift items for sale.
Activities: inter-museum loan; permanent & temporary exhibitions; tours; seminars; demonstrations; stocked casting pond. Annual Events: Fly Fishing Festival; Ice Cream Social; Heritage Award Dinner; Angling & Art Benefit Art, Anglers Club of NY Dinner.
Publications: quarterly, The American Fly Fisher; books, American Fly Fishing: A History, A Treasury of Reels: The Fishing Reel Collection of the American Museum of Fly Fishing; A Graceful Rise: Women in Fly Fishing.
Hours & Admission Prices: June-Oct. Tues.-Sun. 10-4; Nov.-May Tues.-Sat. 10-4 Family $10, adults $5, children $3; discounts to AAM & ICOM members; members no charge. Closed major holidays. &
Attendance: 2,731 (accurate)
Membership: Associate $50; Benefactor $100; Business $250; Sponsor $500; Friends $1,000, $5,000 & $10,000.

HILDENE, THE LINCOLN FAMILY HOME, (M), 1005 Hildene Rd., Manchester, VT 05254. Mailing Address: P.O. Box 377, Manchester, VT 05254-0377. Tel.: 802-362-1788. Fax: 802-362-1564.
E-mail: info@hildene.org
Web Site: www.hildene.org
Founded: 1978.
Congressional District: 1
Key Personnel: Chm. (V), Kenneth Moriarty; Exec. Dir., Seth B. Bongartz; Dir. Education, Diane Newton; Deputy Dir., Laine Dunham; Volunteer Coord., Paula Maynard; Accountant, Ann Dailey; Farm Mgr., Peggy Galloup; Grounds Maintenance, Cary Lewis; Buildings Maintenance, T.J. Lillie; Dir. Private Functions, Sheila Burks; Museum Shop Mgr., Carol Korzelius.
Personnel Profile: Full-Time Paid 25; Part-Time Paid 9; Part-Time Volunteers 340; Interns 2.
Governing Authority: nonprofit organization. Parent Institution: Friends of Hildene, Inc. Tax-exempt: 501(c)(3).

Institution Type/Description: Historic House: 1905 Robert Todd Lincoln Home.
Collections: Lincoln family history; furniture; President Lincoln; restored 1903 Pullman car.
Research Fields: Lincoln family; President Lincoln; Civil War; Captains of Industry; Gilded Age.
Facilities: 2,500-vol. library of Lincoln's family, general reading material available for research by arrangement; nature center; welcome center. Lincoln-related items & Vermont items for sale.
Activities: tours; lectures; concerts; docent program; working farm; observatory; walking trails; picnicking; cross country skiing; permanent & temporary exhibitions. Museum Sponsors: Holiday Open Houses; Symposia; lecture series.
Publications: newsletter, News From Historic Hildene; calendar; book, No Braver Deeds; 4 Marys and A Jessie; Robert Todd Lincoln: A Man In His Own Right; Mary, Wife of Lincoln; Mr. Lincoln's Gift.
Hours & Admission Prices: Daily 9:30-4:30. Adults $16, children 6-14 $5; discounts to groups with reservation and AAM & AAA members; children under 6 & members no charge. Closed Easter; Thanksgiving; Christmas. &
Attendance: 35,000 (estimated)
Membership: Individual $50; Family $80; Sustaining $150; Associate $500; Preservation Society $1,000; Life $5,000.

MANCHESTER HISTORICAL SOCIETY, 48 West Rd., Manchester, VT 05254. Mailing Address: P.O. Box 363, Manchester, VT 05254-0363. Tel.: 802-362-3708.
Founded: 1898.
Governing Authority: Tax-exempt.
Institution Type/Description: Historical Society Museum.
Collections: local history & culture; photographs.
Hours & Admission Prices: Thurs. 1-3; other times by appointment.

SOUTHERN VERMONT ARTS CENTER, West Rd., Manchester, VT 05254. Mailing Address: P.O. Box 617, Manchester, VT 05254-0617. Tel.: 802-362-1405. Fax: 802-362-3279.
E-mail: info@svac.org
Web Site: www.svac.org
Founded: 1929.
Congressional District: 1
Key Personnel: Exec. Dir., Seline Skoug; Gallery Dir., Chester Kasnowski; Mgr. Devel. & Communications, Barbara Lundy; Chm. (V), Stan Stroup.
Personnel Profile: Full-Time Paid 8; Full-Time Volunteers 1; Part-Time Paid 3; Part-Time Volunteers 300; Interns 4.
Governing Authority: nonprofit. Tax-exempt.
Institution Type/Description: Art Center, museum & performing arts.
Collections: paintings; sculpture; graphics; photographs.
Research Fields: New England artists.
Facilities: 1,000-vol. library of art books; music; films available in members' room; botany trail; nature walks.
Activities: lectures; films; gallery talks; summer art classes; music festival; temporary & traveling exhibitions; school loan service; international art tours; docent tours; Artists in the Schools program.
Publications: annual catalog, Festival of the Arts; brochures, Calendar of Events; Summer Study Programs; SVA Music Festival.
Hours & Admission Prices: Tues.-Sat. 10-5, Sun. 12-5. Adults $8, students 3; children under 13 & members no charge. Special exhibits $10. &
Attendance: 35,000 (accurate)
Membership: Individual $55; Artist & Family $75; Donor $150-$499; Patron $500-$999; Benefactor $1,000-$2,999; Collectors Guild $3,000-$4,999; Lucioni Circle $5,000 & up.

Marlboro

MARLBORO HISTORICAL SOCIETY, 364 South Rd., Marlboro, VT 05344. Mailing Address: P.O. Box 242, Marlboro, VT 05344-0242. Tel.: 802-258-2568.
E-mail: forrest810@gmail.com
Web Site: www.marlboro.vt.us
Founded: 1968.
Key Personnel: Pres., Forrest Holzapfel; Vice Pres., Donald Sherefkin; Treas., Cathy Fuller; Clerk, Augusta Bartlett.
Personnel Profile: Part-Time Volunteers 10.
Governing Authority: society. Tax-exempt.
Institution Type/Description: Local History Museum, Historical & Preservation Society: housed in 1814 Rev. Ephraim Holland Newton House & 1895 Houghton Schoolhouse.
Collections: household items; farm implements; furniture; crockery; quilts. Historic House: 1895 Houghton Schoolhouse.
Research Fields: local genealogy.

Facilities: library of books available for research on premises; reading room.
Activities: guided tours; demonstrations of crafts; lectures; concerts; walks to historic sites & buildings.
Publications: local newsletter; annual calendar.
Hours & Admission Prices: June to Labor Day Sat. 12-3. Research: by appointment, call 802-254-2172. No charge; donations accepted.
Attendance: 100 (estimated)

Middlebury

HENRY SHELDON MUSEUM OF VERMONT HISTORY, One Park St., Middlebury, VT 05753-1101. Tel.: 802-388-2117. Fax: 802-388-2112.
E-mail: info@henrysheldonmuseum.org
Web Site: henrysheldonmuseum.org
Founded: 1882.
Congressional District: 1
Key Personnel: Exec. Dir., William F. Brooks, Jr.; Vice Pres., Marnie Wood; Vice Pres., Pat Mayo; Assoc. Dir., Mary Ward Manley; Education Coord., Susan Peden; Bookkeeper, Rachael Gosselin.
Personnel Profile: Part-Time Paid 5; Part-Time Volunteers 140; Interns 4.
Volunteer Hours: 4,987
Operating Expenses: 220,000
Operating Income: 220,000
Governing Authority: nonprofit organization. Tax-exempt: 501(c)(3).
Institution Type/Description: History Museum: housed in 1829 Judd-Harris House.
Collections: 30,000 letters; manuscript collections; 1,200 photographs; Middlebury newspapers 1801 to present; 19th-century Vermont furniture: pianos, clocks, portraits, china, pewter, kitchen utensils, tools & toys; musical instruments; Vermont portraits & landscapes; 1888 Barn; cabinet of curiosities.
Research Fields: state & local history.
Facilities: research center. Museum-related items for sale.
Activities: guided tours; permanent & temporary exhibitions; lectures; concerts; art exhibits; workshops; school program.
Publications: quarterly newsletter; annual report; A Walking History of Middlebury.
Hours & Admission Prices: Tues.-Sat. 10-5. Families $12, adults $5, senior citizens $4.50; discounts to AAM, NEMA, Vermont Museum & Gallery Alliance members; children 18 & under, students with ID and members no charge. Closed New Year's Day; Independence Day; Thanksgiving; Christmas. &
Attendance: 10,000 (estimated)
Membership: Individual $35; Family $50; Friend $100; Supporter $150; Patron $500; Benefactor $1,000 & up.

* **MIDDLEBURY COLLEGE MUSEUM OF ART, (M),** Mahaney Center for the Arts, Middlebury, VT 05753-6177. Tel.: 802-443-5235 & 5007. Fax: 802-443-2069.
Web Site: museum.middlebury.edu
Founded: 1968.
Key Personnel: Dir., Richard H. Saunders; Head Cur., Emmie Donadio; Chm. Friends of Art, Ray Hudson; Exhibit Designer, Kenneth Pohlman; Registrar, Margaret Wallace; Cur. Education, Sandra Olivo; Admin. Operations Mgr., Douglas Perkins; Museum Preparator, John Houskeeper; Museum Preparator, Chris Murray; Bookstore & Receptionist Coord., Mikki Lane.
Personnel Profile: Full-Time Paid 6; Part-Time Paid 3; Interns 5.
Governing Authority: college. Parent Institution: Middlebury College. Tax-exempt.
Institution Type/Description: Art Museum.
Collections: prints; drawings; paintings; sculpture; photographs.
Research Fields: 19th-century sculpture.
Facilities: 5,200 sq. ft. exhibit space.
Activities: public lectures & gallery talks; programs for students of Middlebury College & local school groups; permanent & traveling exhibitions.
Publications: newsletter; brochures; exhibition catalogues; annual report.
Hours & Admission Prices: Sept. to mid-Aug. Tues.-Fri. 10-5, Sat.-Sun. 12-5. No charge. Closed New Year's Eve & Day; Christmas Day & week. &
Attendance: 18,446 (accurate)
Membership: Student $15; Individual $30; Family & Couple $50; Contributor $100; Sponsor $250; Patron $500.

NATIONAL MUSEUM OF THE MORGAN HORSE, 34 Main St., Middlebury, VT 05753. Mailing Address: P.O. Box 101, Middlebury, VT 05753. Tel.: 802-388-1639.
E-mail: morgans@together.net
Web Site: www.morganmuseum.org
Founded: 1988.

Congressional District: 1
Personnel Profile: Part-Time Paid 1; Part-Time Volunteers 4.
Governing Authority: private; nonprofit organization. Parent Institution: American Morgan Horse Institute. Tax-exempt: 501(c)(3).
Institution Type/Description: Equine Museum.
Collections: concentration on the Morgan breed of horses & associated people from mid-18th century to present; photographs, paintings, ephemera & items related to equine world.
Research Fields: Morgan horse; equine; civilian military use of horses, including transportation, agricultural & warfare.
Facilities: 700-vol. library of various equine titles; 1,100 sq. ft. exhibit space. Museum-related items for sale.
Activities: self-guided tours; rotating exhibits featuring items from the archive; occasional solo exhibits by contemporary artists.
Publications: quarterly newsletter.
Hours & Admission Prices: Tues.-Sat. 10-5. Donation Requested. Closed major holidays. &
Attendance: 4,500 (estimated)
Membership: Contributor $25-$99; Supporter $100-$499; Sponsor $500-$999; Patron $1,000 & up.

VERMONT FOLKLIFE CENTER, 88 Main St., Middlebury, VT 05753-1425. Tel.: 802-388-4964. Fax: 802-388-1844.
E-mail: info@vermontfolklifecenter.org
Web Site: www.vermontfolklifecenter.org
Founded: 1983.
Key Personnel: Exec. Dir., Brent Bjorkman; Chm. (V), Bill Schubart; Dir. Education, Gregory Sharrow; Operations Mgr., Sarah Stahl.
Personnel Profile: Full-Time Paid 4; Part-Time Paid 3.
Governing Authority: private; nonprofit organization. Tax-exempt: 501(c)(3).
Institution Type/Description: Folk Life Center & Archive.
Collections: 5,000 taped interviews on the heritage & traditions of Vermont from settlement in the 18th-century to present; historic & contemporary photographs related to Vermont folklife; written documents, including family histories, diaries & letters; musical recordings.
Research Fields: agricultural, ethnic heritage; maple sugaring; logging; philanthropy renewable energy pioneers arts and crafts movement.
Facilities: media resource center; 600 sq. ft. exhibit space. Books, audio tapes, Vermont crafts & fair trade items from around the world for sale.
Activities: traveling exhibitions; special workshops & demonstrations; documentary gathering; traveling exhibits; traditional arts apprenticeship program; K-16 school outreach.
Publications: monthly e-newsletter.
Hours & Admission Prices: Gallery & Shop: Mon. Sat. 10 5, Sun. 11-4. Archive & Research Center: Mon.-Fri. 10-4. No charge; donations accepted. Closed Christmas through New Year's week. &
Attendance: 12,000 (accurate)
Membership: Individual $35; Family $50; Supporter $100; Patron $250; Benefactor $500; Founder $1,000.

VERMONT SOAPWORKS DISCOUNT FACTORY OUTLET AND SOAP MUSEUM, 616 Exchange St., Middlebury, VT 05753-1181. Tel.: 802-388-4302; 866-762-7482 (toll free). Fax: 802-388-7471.
E-mail: info@vtsoap.com
Web Site: www.vermontsoap.com
Founded: 1992.
Key Personnel: Gen. Mgr., Hilde Whalley.
Personnel Profile: Full-Time Paid 25; Part-Time Paid 2.
Governing Authority: Parent Institution: Vermont Country Soapworks.
Institution Type/Description: Soap Museum.
Collections: history of soap; soap products; period washing machine.
Facilities: Museum-related items for sale.
Activities: soap making demonstrations; children's programs; special events.
Hours & Admission Prices: Mon.-Fri. 9-5. No charge. Closed most holidays.
Attendance: 10,000 (estimated)

Milton

MILTON HISTORICAL MUSEUM, 13 School St., Milton, VT 05468-3632. Tel.: 802-893-1604.
E-mail: miltonhistorical@yahoo.com
Formerly: Milton Museum
Founded: 1978.
Key Personnel: Museum Dir. (V), Lorinda Henry; Pres. (V) Milton Historical Society, Bill Kaigle.
Personnel Profile: Part-Time Volunteers 10.
Governing Authority: society. Parent Institution: Milton Historical Society. Tax-exempt: 501(c)(3).
Institution Type/Description: Local History Museum.

Collections: wedding gowns of early Milton brides; doctor's chair & medical bag; maps; old gazetteers; pictures; tools; mantle clock; kitchen items; Indian artifacts; bicentennial quilt; furniture; books; Native American artifacts; Civil War Soldiers' Monument; Civil War artifacts.
Research Fields: Milton history.
Facilities: 300-vol. library of books including Walton's Register & history of early Milton & Vermont available for research on premises.
Activities: guided tours; loan, permanent & temporary exhibitions; bi-monthly historical program. Annual Event: Holiday Open House in December.
Publications: annual newsletter; Town Reports; Historic Timeline & Population Graph (1878-2000); 200 years pictorial exhibit & documentation.
Hours & Admission Prices: April-Oct. 1st & 3rd Sat.-Sun. 1-4. Other times by appointment, please call 802-893-1604. No charge; donations accepted. &
Attendance: 750 (estimated)
Membership: Honorary 80 & over no charge; Individual $10; Family $20.

Montpelier

MORSE FARM MAPLE SUGARWORKS & OUTDOOR FARM LIFE MUSEUM, 1168 County Rd., Montpelier, VT 05602-8135. Tel.: 800-242-2740.
E-mail: maple@morsefarm.com
Web Site: www.morsefarm.com
Institution Type/Description: History Museum.
Collections: Morse Farm history & maple process; period furnishings & farm equipment; historic buildings; carvings.
Facilities: nature trails; theater. Museum-related items for sale.
Activities: sugar house tours & tasting; video; special events. Annual Event: Sugarin' time in March.
Hours & Admission Prices: Call for hours.

T.W. WOOD GALLERY & ARTS CENTER, (M), 46 Barre St., Montpelier, VT 05602. Tel.: 802-262-6035.
E-mail: info@twwoodgallery.org
Web Site: twwoodgallery.org
Founded: 1891.
Key Personnel: C.E.O. & Acting Exec. Dir., William Pelton; Bd. Trustees Pres. (V), Elizabeth Reardon.
Personnel Profile: Full-Time Paid 1; Part-Time Paid 2; Part-Time Volunteers 60; Interns 2.
Governing Authority: nonprofit organization. Tax-exempt.
Institution Type/Description: Art Gallery.
Collections: 438 oils, drawings & watercolors by Thomas Waterman Wood, & his contemporaries including A.H. Wyant, A.B. Durard, J.G. Brown, DeHass; American artists of the 1920s & 30s; 60 works in WPA collection.
Research Fields: T.W. Wood; WPA art & artists.
Activities: films; loan, permanent & changing exhibitions; lectures; art classes; exhibition workshop.
Publications: booklet, Thomas Waterman Wood PNA.
Hours & Admission Prices: Tues.-Sun. 12-4. No charge; donations accepted. Closed major holidays. &
Attendance: 7,000 (estimated)
Membership: Individual $35; Family $50; Contributor $100 & up; Sustainer $250 & up; Benefactor $500 & up.

USS MONTPELIER MUSEUM, 39 Main St., 2nd Fl., Montpelier, VT 05602-3064. Mailing Address: 11 Greenfield Terr., Montpelier, VT 05602. Tel.: 802-223-9502. Fax: 802-223-9519.
Web Site: www.montpelier-vt.org/community/367.html
Founded: 1998.
Key Personnel: Dir., Chuck Karparis
Governing Authority: Parent Institution: CL-57 Association. Tax-exempt.
Institution Type/Description: Naval History Museum.
Collections: Naval history of ships named Montpelier & their crews; photographs; documents.
Hours & Admission Prices: Call for hours. No charge. Closed holidays. &
Attendance: 200 (estimated)

VERMONT HISTORICAL SOCIETY MUSEUM, (M), 109 State St., Montpelier, VT 05609-0002. Tel.: 802-828-2291. Fax: 802-828-1415.
E-mail: museum@state.vt.us
Web Site: www.vermonthistory.org
Founded: 1838.
Key Personnel: Exec. Dir., Mark Hudson; Pres. (V), Sarah Dopp; Registrar, Mary Labate Rogstad; Cur., Jacqueline Calder; Librarian, Paul Carnahan; Asst. Librarian, Marjorie Strong.
Personnel Profile: Full-Time Paid 13; Part-Time Paid 7.
Governing Authority: society. Parent Institution: Vermont Historical Society.

Library: Vermont History Center, 60 Washington St., Barre, VT 05641-4209. Tel.: 802-479-8500. Tax-exempt: 501(c)(3).
Institution Type/Description: History Museum.
Collections: state history from 1600 to present; pewter; tools; furniture; glass; costumes; paintings; textiles; film; fine & decorative arts; manuscripts.
Research Fields: Vermont history; New England genealogy.
Facilities: 40,000-vol. library of genealogy, Vermont & New England history books available for inter-library loan & use on premises; reading room; 5,000 sq. ft. exhibit space. Museum-related items for sale.
Activities: school tours; lectures; workshops; family programming.
Publications: biannual journal, Vermont History; books; quarterly newsletter, History Connections; exhibition catalogs; teacher guides & curricular aids.
Hours & Admission Prices: Museum: May-Oct. Tues.-Sat. 10-4, Sun. 12-4; Nov.-April Tues.-Sat. 10-4. Library: Tues.-Fri. 9-4:30, 2nd Sat. each month 9-4. Adults $5; discounts to AAM & ICOM members; members no charge. Closed major holidays. &
Attendance: 18,000 (estimated)
Membership: Senior & Institutional $35; Individual $40; Household $50.

THE VERMONT STATE HOUSE, 115 State St., Montpelier, VT 05633-0004. Tel.: 802-828-2228. Fax: 802-828-2424.
Web Site: www.leg.state.vt.us
Founded: 1808.
Key Personnel: Cur., David Schutz; Chief of Police, Leslie R. Dimick.
Personnel Profile: Full-Time Paid 2; Part-Time Paid 1; Part-Time Volunteers 115.
Governing Authority: state government; nonprofit. Tax-exempt: 501(c)(3).
Institution Type/Description: Historic Building.
Collections: period furnishings; paintings; sculpture; decorative arts; costumes; textiles.
Facilities: 200-seat cafeteria; 175 sq. ft. exhibit space. Vermont products & State House items for sale.
Activities: concerts; docent program; guided tours; lectures; temporary exhibitions of our own collection; broadcast programs.
Publications: biannual newsletter, Friends Proceedings; tour brochure; history guidebook.
Hours & Admission Prices: Mon.-Fri. 8-4. Guided Tours: July to mid-Oct. Mon.-Fri. 10-3:30, Sat. 11-2:30; Self-Guided Tours: mid-Oct. to June Mon.-Fri. 9-3; other times by appointment. No charge; donations accepted. Closed state & federal holidays. &
Attendance: 100,000 (estimated)
Membership: Individual & Family $30; Gold Dome $50; Cedar Creek $100; Supreme Court $300; Governor's Cabinet $500; Ammi Young $1,000. Corporate Support $100 & up.

Morrisville

NOYES HOUSE MUSEUM, 122 Lower Main St., Morrisville, VT 05661. Mailing Address: Morristown Historical Society, P.O. Box 1299, Morrisville, VT 05661-1299. Tel.: 802-888-7617.
Web Site: noyeshousemuseum.org
Founded: 1952.
Key Personnel: Dir., Scott A. McLaughlin; Pres., Jill Mudgett.
Personnel Profile: Part-Time Paid 2.
Governing Authority: Parent Institution: Morristown Historical Society. Tax-exempt.
Institution Type/Description: Historic House Museum: built in the early 19th-century by the Safford family.
Collections: local & regional history; photographs; furnishings; toys; household & farm tools; quilts; costumes; military artifacts. Historic Buildings: 1820s house; 1840s carriage barn.
Research Fields: local history; archives; collections.
Activities: guided tours; lectures; family events. Annual Event: Open House.
Hours & Admission Prices: June-Aug. Fri.-Sat. 10-4; Sept.-Oct. Sat. 10-4; groups by appointment. No charge; donations accepted. &
Attendance: 500 (estimated)
Membership: Individual $20; Family $30; Business $60; Life $500.

Moscow

LITTLE RIVER HOTGLASS STUDIO & GALLERY, 593 Moscow Rd., Moscow, VT 05662. Mailing Address: P.O. Box 1504, Stowe, VT 05672-1504. Tel.: 802-253-0889. Fax: 802-253-4128.
Institution Type/Description: Art Gallery.
Collections: blown glass sculptures.
Activities: glass blowing demonstrations.
Hours & Admission Prices: Wed.-Mon. 10-5.

Newfane

HISTORICAL SOCIETY OF WINDHAM COUNTY, Rte. 30, Newfane, VT 05345. Mailing Address: P.O. Box 246, Newfane, VT 05345-0246. Tel.: 802-365-4148.
E-mail: info@historicalsocietyofwindhamcounty.org
Web Site: www.historicalsocietyofwindhamcounty.org
Founded: 1927.
Personnel Profile: Part-Time Paid 1; Part-Time Volunteers 1.
Governing Authority: Tax-exempt.
Institution Type/Description: Historical Society Museum.
Collections: local history & culture; photographs; personal artifacts; period furnishings; portraits; folk art; Civil War artifacts; West River Railroad.
Activities: educational programs.
Publications: quarterly newsletter.
Hours & Admission Prices: late May to mid-Oct. Wed. & Sat.-Sun. 12-5. No charge; donations accepted.
Attendance: 500 (estimated)
Membership: Individual $10; Family $15; Contributing & Organization $25; Supporting $50; Life $300.

Newport

MEMPHREMAGOG HISTORICAL SOCIETY OF NEWPORT, Emory Hebard State Office Bldg., 2nd Fl., 100 Main St., Newport, VT 05855-5543. Mailing Address: 96 Stagecoach Dr., Newport, VT 05855-9153. Tel.: 802-334-6195.
Founded: 1991.
Institution Type/Description: Historical Society Museum.
Collections: local history & culture; photographs; Native American artifacts; period furnishings; 19th-20th century photographs; Lake Memphremagog's history, steamboats & genealogy; archives.
Activities: self-guided walking tours.
Publications: books, Crossroads Before & Beyond; Newport and the Northeast Kingdom; Around Lake Memphremagog.
Hours & Admission Prices: Mon.-Fri. 9-5; other times by appointment. No charge. &
Membership: Individual $5; Supporting $10; Business $20; Lifetime $50.

North Bennington

HISTORIC PARK-MCCULLOUGH, One Park St., North Bennington, VT 05257. Mailing Address: P.O. Box 388, North Bennington, VT 05257-0388. Tel.: 802-442-5441. Fax: 802-442-5442.
E-mail: info@parkmccullough.org
Web Site: www.parkmccullough.org
Founded: 1968.
Congressional District: 1
Key Personnel: Pres. (V), Katherine Traver; Bookkeeper, David Adams.
Personnel Profile: Full-Time Paid 3; Part-Time Paid 3; Part-Time Volunteers 35; Interns 2.
Governing Authority: nonprofit organization. Parent Institution: Park-McCullough House Assoc. Tax-exempt: 501(c)(3).
Institution Type/Description: Historic House Museum: housed in 1865 Second Empire mansion built for Trenor & Laura Hall Park which was also the home of two Vermont Governors.
Collections: books; furniture; carriages; art; papers; documents; Victorian artifacts & clothes; Chinese porcelain; photographic library including works of Carlton Watkins; Album prints of Yosemite.
Research Fields: genealogy of family; 19th-century technology; Vermont history; costumes; California history: 1850s & 1860s; carriage barn; New York City history; photography.
Facilities: library of family books available on premises for research by qualified people with references; children's playhouse; Victorian garden; carriage house; fountain; stable; fishpond.
Activities: guided tours; lectures; films; gallery talks; concerts; study clubs; formally organized educational programs; temporary exhibitions; Open Garden Days; tours & programming related to the museum's formal gardens; rental facilities.
Publications: booklets; monthly calendar; newsletter; The Park-McCullough House Historic House & Museum guide book; book, Within One's Memory.
Hours & Admission Prices: Hourly Tours: mid-May to Oct. daily 9-5. Adults $10, senior citizens $9, students $7; discounts to groups, AAM, VMGA, AAA & NEMA members; children under 12 & members no charge. &
Attendance: 6,500 (estimated)
Membership: Senior Citizen & Student $25; Individual $40; Family & Dual $60; Supporting $100; Sustaining $250; President's Circle $500; Corporate $1,000.

North Danville

DANVILLE HISTORICAL SOCIETY, North Danville School, North Danville, VT 05828. Mailing Address: P.O. Box 274, Danville, VT 05828-0274. Tel.: 802-684-3857.
Key Personnel: Pres., Mary Pryor
Institution Type/Description: Library & Archives.
Collections: books; local history.
Hours & Admission Prices: Mon., Wed. & Fri. 2-4; other times by appointment.

North Troy

MISSISQUOI VALLEY HISTORICAL SOCIETY, Main St., North Troy, VT 05859. Mailing Address: P.O. Box 237, North Troy, VT 05859-0237. Tel.: 802-988-4656.
E-mail: missisco@hotmail.com
Founded: 1976.
Key Personnel: Dir. & Pres. (V), Nancy L. Allen; Vice Pres., John Starr; Treas., Roy Barnett
Governing Authority: society; nonprofit. Tax-exempt.
Institution Type/Description: Historical Society Museum: housed in 1883 former St. Augustine Church.
Collections: general store memorabilia; photographs; household utensils; textiles; tools; cobblers tools; 19th century furniture; saw mill; medical instruments; local records & early school books.
Activities: rotating & loan exhibitions; research.
Publications: book, Memories of the Early Days in the Town of Troy.
Hours & Admission Prices: Memorial Day, Labor Day & Alumni Weekend; other times by appointment. Donations accepted. ఉ
Attendance: 20 (estimated)
Membership: Individual $2.50; Family $5; Life $25.

Northfield

NORTHFIELD FIRE DEPARTMENT MUSEUM, 128 Wall St., Northfield, VT 05663. Mailing Address: Northfield Municipal Bldg., 51 S. Main St., Northfield, VT 05663. Tel.: 802-279-7931. Fax: 802-485-8426.
E-mail: nfdchief@trans-video.net
Institution Type/Description: Firefighting History.
Collections: local firefighting history & equipment; personal artifacts; photographs; 1870s leather hose; 1836 hand pumper; model fire trucks.
Hours & Admission Prices: By appointment.

SULLIVAN MUSEUM AND HISTORY CENTER, (M), 158 Harmon Dr., Northfield, VT 05663-1000. Tel.: 802-485-2183. Fax: 802-485-2749.
E-mail: smhc@norwich.edu
Web Site: www.norwich.edu/museum
Formerly: Norwich University Museum
Founded: 1819.
Congressional District: 1
Key Personnel: Dir., Sarah E. Henrich; Museum Registrar, Candace Truso; University Historian, Gary T. Lord; Exhibitions Assoc., Phillip Whitman; Museum Asst., Suzanne Desch.
Personnel Profile: Full-Time Paid 2; Part-Time Paid 2; Part-Time Volunteers 10.
Governing Authority: university. Affiliated with Norwich University. Tax-exempt: 501(c)(3).
Institution Type/Description: University Museum.
Collections: history of Norwich exhibits; personal memorabilia of founder Alden Partridge; achievements of alumni such as: Adm. George Dewey, Gen. Alonzo Jackman, Gen. Grenville Dodge, builder of the Union-Pacific railroad; Norwich uniforms; flags; weapons; military accoutrements; 19th & 20th-century academic paraphernalia.
Research Fields: military history; Norwich history.
Facilities: archives.
Activities: permanent & temporary exhibitions; research.
Publications: biannual newsletter, The Rotunda Report.
Hours & Admission Prices: Mon.-Fri. 8-4, Sat. 11-4; other times by appointment. No charge. Closed holidays. ఉ
Attendance: 18,000 (accurate)
Membership: Student $5; Individual $25; Patron $100; Sponsor $500; Founder's $1,000.

Norwich

MONTSHIRE MUSEUM OF SCIENCE, INC., 1 Montshire Rd., Norwich, VT 05055-9334. Tel.: 802-649-2200. Fax: 802-649-3637.
E-mail: montshire@montshire.org
Web Site: www.montshire.org
Founded: 1975.
Key Personnel: Dir., David Goudy; Chm. (V), Margaret Rightmire; Dir. Education, Greg DeFrancis; Dir. Mktg. & Communications, Beth Krusi; Dir. Devel., Jennifer Rickards; Museum Shop Mgr., Barbara Mathewson.
Personnel Profile: Full-Time Paid 24; Part-Time Paid 10; Part-Time Volunteers 155; Interns 4.
Governing Authority: nonprofit organization. Tax-exempt: 501(c)(3).
Institution Type/Description: Science Museum.
Collections: interactive science exhibits; native plant landscape.
Facilities: classroom; nature area; garden; nature trails; outdoor science park.
Activities: lectures; workshops; organized educational programs; docent program; intern & work-study programs; public issues programs & forums; special events; permanent & temporary exhibits on physical & natural sciences.
Publications: quarterly newsletter; e-newsletter.
Hours & Admission Prices: Daily 10-5. Adults $14, children 2-17 $11; members & children under 2 no charge. Closed Thanksgiving; Christmas. ఉ
Attendance: 137,000 (accurate)
Membership: 2 Person $90; 4 Person $110; 6 Person $130.

NORWICH HISTORICAL SOCIETY, 277 Main St., Norwich, VT 05055. Mailing Address: P.O. Box 1680, Norwich, VT 05055-1680. Tel.: 802-649-0124.
E-mail: info@norwichhistory.org
Web Site: www.norwichhistory.org
Founded: 1951.
Key Personnel: Pres., Nancy Hoggson; Vice Pres., Melinda Stricker; Treas., Mike Woods; Museum Shop Mgr., Martha Howard.
Personnel Profile: Part-Time Paid 1; Part-Time Volunteers 14.
Governing Authority: private; not-for-profit organization. Tax-exempt.
Institution Type/Description: Historical Society Museum: located in Lewis House. Listed on the National Register of Historic Places.
Collections: artifacts dating from 1769 to present; textiles; documents; photographs; maps; tools; samplers; costumes.
Facilities: library; 500 sq. ft. exhibit space.
Activities: lectures; temporary exhibitions; broadcast programs.
Publications: newsletter, Norwich Historical Society.
Hours & Admission Prices: Memorial Day to Oct. Wed. 10-4, Sat. 10am to noon; other times by appointment. No charge; donations accepted. ఉ
Attendance: 500 (estimated)
Membership: Individual $1-$250.

Old Bennington

BENNINGTON BATTLE MONUMENT, 15 Monument Cir., Old Bennington, VT 05201-2134. Mailing Address: Historic Preservation, National Life Bldg., 6th Fl., Montpelier, VT 05620. Tel.: 802-828-3051. Fax: 802-447-0550 (call first).
E-mail: john.dumville@state.vt.us
Web Site: www.historicsite.vermont.gov/bennington
Founded: 1891.
Key Personnel: Historic Sites Operations Chief, John P. Dumville; Site Admin., Marylou Chicote.
Personnel Profile: Full-Time Paid 1; Part-Time Paid 7.
Governing Authority: state. Parent Institution: State of Vermont. Subsidiary Institution: Div. for Historic Preservation, One National Life Dr., 6th Fl., Montpelier 05602. Tax-exempt.
Institution Type/Description: State Historic Site: located near the site of the Bennington Battle of the Revolutionary War.
Collections: diorama depicting Bennington Battle of the Revolutionary War; General Burgoyne's camp kettle.
Facilities: observation room.
Activities: guided tours; permanent exhibitions.
Publications: brochure, A Guide to Historic Sites in Vermont; Bennington Battle Monument.
Hours & Admission Prices: April-Oct. daily 9-5. Adults $3.50, children $.50; registered school groups no charge. ఉ
Attendance: 50,000 (estimated)

Orwell

MOUNT INDEPENDENCE STATE HISTORIC SITE, Mount Independence Rd., Orwell, VT 05760. Mailing Address: Historic Preservation, National Life Bldg., 6th Fl., Montpelier, VT 05620. Tel.: 802-828-3051. Fax: 802-828-3206.
E-mail: john.dumville@state.vt.us
Web Site: www.historicsites.vermont.gov/independence
Founded: 1950.
Key Personnel: Operations Chief, John P. Dumville; Site Admin., Elsa Gilbertson.
Personnel Profile: Full-Time Paid 1; Part-Time Paid 1.
Governing Authority: state. Parent Institution: State of Vermont., Division for Historic Preservation, One National Life Dr., 6th Fl., Montpelier 05602. Tax-exempt.
Institution Type/Description: Historic Site: site of major Revolutionary War fort.
Collections: archaeological items from site.
Research Fields: Revolutionary War; Archaeological Field Schools.
Facilities: trails.
Activities: self guided tours.
Publications: brochure, A Guide To Historic Sites In Vermont & Mount Independence; booklet, Mount for Dependence; book, Mount Independence and the American Revolution, 1776-1777.
Hours & Admission Prices: mid-May to mid-Oct. daily 9:30-5:30. Adults & children 15 & over $5; discounts to AAM members; children 14 & under no charge. &

Attendance: 8,000 (accurate)

Peacham

PEACHAM HISTORICAL ASSOCIATION, 643 Bayley-Hazen Rd., Peacham, VT 05862. Mailing Address: P.O. Box 101, Peacham, VT 05862-0101. Tel.: 802-592-3571.
Web Site: peachamhistorical.org
Founded: 1916.
Key Personnel: Pres. (V), Jutta R. Scott; Cur., Lorna Quimby.
Personnel Profile: Part-Time Volunteers 12.
Governing Authority: bd. Tax-exempt.
Institution Type/Description: Historical Society Museum.
Collections: period artifacts; tools; paintings; photographs; maps; furniture; costumes; quilts; manuscripts. Historic Buildings: house; Ashbel Goodenough blacksmith shop; 1858 Peacham Hollow schoolhouse.
Research Fields: Vermont history; Civil War; Caledonia County grammar school, Peacham Academy.
Facilities: archives.
Activities: ghost walk.
Publications: Peacham Patriot; Peacham Anthology; Historic Homes of Peacham; A Vermont Hill Town in the Civil War: Peacham's Story.
Hours & Admission Prices: House: late June to Sept. Sun. 2-4; other times by appointment. Archives: Mon. 9 am-11:30 am, Thurs. 2-4. No charge; donations accepted. &

Attendance: 400 (estimated)
Membership: Senior $5; Individual $10; Family $15; Family Life $250.

Pittsford

NEW ENGLAND MAPLE MUSEUM, 4578 Rt. 7, Pittsford, VT 05763. Mailing Address: P.O. Box 131, Pittsford, VT 05763-0131. Tel.: 802-483-9414. Fax: 802-483-2101.
E-mail: newenglandmaplemuseum@yahoo.com
Web Site: www.maplemuseum.com
Founded: 1977.
Key Personnel: Pres., Michael Blanchard; Dir. & Museum Shop Mgr., Mary Blanchard.
Governing Authority: corporation.
Institution Type/Description: History Museum.
Collections: maple sugaring artifacts; wooden sap buckets; period sugar tubs; maple sugaring paintings by Vermont artists.
Facilities: theater. Maple products & Vermont crafts for sale.
Activities: lectures; permanent exhibitions.
Hours & Admission Prices: mid-March to Dec. daily 9:30-5:30. Adults $5, tours & children 12 and under $1; discounts to members. &

Attendance: 35,000

PITTSFORD HISTORICAL SOCIETY MUSEUM, U.S. Rte. 7 #3399, Pittsford, VT 05763. Mailing Address: P.O. Box 423, Pittsford, VT 05763-9774. Tel.: 802-483-2040.
E-mail: peggy.armitage@gmail.com
Web Site: www.pittsfordhistorical.com
Founded: 1960.
Congressional District: 1
Personnel Profile: Part-Time Volunteers 20.
Governing Authority: private; nonprofit organization. Parent Institution: Pittsford Historical Society, Inc. Tax-exempt.
Institution Type/Description: Historical Society Museum: located in Eaton Hall, former Masonic Hall.
Collections: concentration on the history of Pittsford from the mid-1700s to present.
Research Fields: vital statistics of local people; genealogies of Pittsford residents.
Facilities: Printed material for sale.
Activities: formal education program for adults & children; guided tours; permanent & temporary exhibitions.
Publications: Pittsford's Second Century, 1872-1997; postcards; pamphlet, Gleanings; Around Pittsford.
Hours & Admission Prices: April-June Tues. 9-4; July-Oct. Tues. 9-4, Sun. 1-4. No charge; donations accepted. &

Attendance: 275 (accurate)
Membership: Student $10; Individual $15; Family $20; Sponsor $50; Life $200.

Plymouth

THE CALVIN COOLIDGE MEMORIAL FOUNDATION, INC., 3780 Rte. 100A, Plymouth, VT 05056. Mailing Address: Box 97, Plymouth, VT 05056-0097. Tel.: 802-672-3389. Fax: 260-672-3289.
E-mail: info@calvin-coolidge.org
Web Site: www.calvin-coolidge.org
Founded: 1960.
Key Personnel: Exec. Dir., Stephen Woods; Chm., Frank Jay Barrett; Pres., Robert P. Kirby.
Personnel Profile: Full-Time Paid 2; Part-Time Paid 1; Part-Time Volunteers 18.
Governing Authority: nonprofit executive committee. Parent Institution: The Calvin Coolidge Memorial Foundation. Tax-exempt: 501(c)(3).
Institution Type/Description: Historic Foundation.
Collections: items relating to Coolidge Era & Plymouth history; books; newspapers; magazines; photographs; postcards; philatelic envelopes; Coolidge ephemera, memorabilia & materials. Historic Structure: 1840 Union Christian Church.
Research Fields: Coolidge Era; local history of Plymouth.
Facilities: 800-vol. reference library; reading room; exhibit space.
Activities: lectures; school visits & organized educational programs; speaker series; loan & temporary exhibitions.
Publications: periodic newsletter, books, The Autobiography of Calvin Coolidge; Meet Calvin Coolidge; Homestead Inaugural; Calvin Coolidge's Unique Vermont Inauguration; Coolidge & the Historians; Calvin Coolidge, Jr.; Growing Up in Plymouth, Notch, Vermont, 1872-1895; Return to These Hills: Calvin Coolidge in Vermont; From Plymouth Notch to President; A Plymouth Album; Calvin Coolidge Memorial Foundation.
Hours & Admission Prices: Mon.-Fri. 9-4:30. No charge; donations accepted. Closed federal holidays except Independence Day. &

Attendance: 70,000 (estimated)
Membership: Individual $35; Family $50; Contributing $100; Supporting $250; Sustaining $500; Benefactor $1,000 & up.

Plymouth Notch

CALVIN COOLIDGE STATE HISTORIC SITE, 3780 Rte. 100A, Plymouth Notch, VT 05056. Mailing Address: Historic Preservation, National Life Bldg., 6th Fl., Montpelier, VT 05620. Tel.: 802-828-3051& 672-3773. Fax: 802-828-3206.
E-mail: john.dumville@state.vt.us
Web Site: www.historicsites.vermont.gov/coolidge
Founded: 1947.
Key Personnel: Historic Sites Operations Chief, John P. Dumville.
Personnel Profile: Full-Time Paid 1; Part-Time Paid 15; Interns 2.
Governing Authority: Parent Institution: State of Vermont, Division for Historic Preservation, National Life Bldg., 6th Fl., Montpelier, VT 05620. Tax-exempt.
Institution Type/Description: Historic House Museum: Calvin Coolidge birthplace & homestead.
Collections: furnishings used on Aug. 23, 1923, the night Calvin Coolidge was sworn in by his father as President of the United States; Coolidge general store; photographs; Presidential memorabilia; Wilder barn with agricultural exhibits; Plymouth Notch Village.
Research Fields: President Calvin Coolidge era.

Facilities: visitors center; restaurant.
Activities: permanent exhibitions; self-guided tours.
Publications: brochure Plymouth Notch; booklet, Coolidge Homestead.
Hours & Admission Prices: mid-May to mid-Oct. daily 9:30-5:30. Adults over 13 $7.50; discounts to AAM members; registered school groups no charge. &

Attendance: 40,000 (estimated)

Proctor

VERMONT MARBLE MUSEUM, 52 Main St., Proctor, VT 05765-1177. Mailing Address: P.O. Box 607, 52 Main St., Proctor, VT 05765-0607. Tel.: 802-459-2300; 800-427-1396. Fax: 802-459-2948.
E-mail: info@vermont-marble.com
Web Site: www.vermont-marble.com
Founded: 1933.
Key Personnel: C.E.O., Marsha Hemm; Mgr., Robert Pye; Museum Shop Mgr., Cathy Miglorle.
Governing Authority: private; profit.
Institution Type/Description: Mining & Geology Museum.
Collections: sculpture; photographs; rocks; minerals.
Facilities: 70-seat theater; cafe. Marble-related items for sale.
Hours & Admission Prices: mid-May to Oct. daily 9-5:30. Adults $7, senior citizens $5, teens 15-18 $4, groups $3.50 per person; children 12 & under with parent no charge. &
Attendance: 45,000 (estimated)

WILSON CASTLE, 2708 West St., Proctor, VT 05765. Mailing Address: P.O. Box 290, Center Rutland, VT 05736-0290. Tel.: 802-773-3284. Fax: 802-773-3284.
E-mail: wilsoncastle@aol.com
Web Site: www.wilsoncastle.com
Founded: 1961.
Key Personnel: Dir., Denise Davine; Entertainment Dir., Rusty Trombley.
Governing Authority: nonprofit organization; Wilson Family Foundation.
Institution Type/Description: Historic Building: 1867 Wilson Castle, Victorian Building.
Collections: European & Oriental objects d'art; stained glass; furniture; period artifacts.
Facilities: Gift items for sale.
Activities: guided tours; lectures; concerts; permanent & temporary exhibitions.
Publications: brochure.
Hours & Admission Prices: Memorial Day Weekend to late Oct. daily 9-5. Adults $10, children 6-12 $6; discounts to AAA, AAM & ICOM members; children 5 & under no charge. &
Attendance: 75,000

Putney

PUTNEY HISTORICAL SOCIETY, Putney Town Hall, 127 Main St., Rte. 5, Putney, VT 05346. Mailing Address: P.O. Box 260, Putney, VT 05346-0260. Tel.: 802-387-5862, ext. 10.
E-mail: info@putneyhistory.us
Web Site: www.putneyhistory.us
Founded: 1959.
Congressional District: Vermont
Key Personnel: Cur., Laura Heller; Cur., Barbara A. Taylor.
Governing Authority: nonprofit organization. Tax-exempt.
Institution Type/Description: History Museum: housed in 1871 Town Hall.
Collections: artifacts relating to the town; Indian artifacts; photographs; costumes; memorabilia of John Humphrey Noyes.
Research Fields: genealogy; John Humphrey Noyes & Perfectionists; local Native American culture.
Facilities: 50-vol. library of Vermont and Putney history, available for use on premises; reading room.
Activities: lectures; changing exhibits.
Publications: newsletter; brochures.
Hours & Admission Prices: June-Aug. Tues. & Sat. 10-2; other times by appointment. No charge; donations accepted. &
Attendance: 250 (estimated)
Membership: Senior $5; Individual $15; Sustaining $25; Benefactor $50; Patron $100.

Quechee

VERMONT INSTITUTE OF NATURAL SCIENCE, 6565 Woodstock Rd., Rte. 4, Quechee, VT 05059. Mailing Address: P.O. Box 1281, Quechee, VT 05059. Tel.: 802-359-5000. Fax: 802-359-5001.
E-mail: info@vinsweb.org
Web Site: www.vinsweb.org
Founded: 1972.
Key Personnel: Dir., Mary Davidson Graham; C.E.O., Elizabeth Miller; Pres., John Dolan; Mgr. Nature Center Programs, Chris Collier; Nature Store Mgr., Joseph Morvan.
Personnel Profile: Full-Time Paid 22; Part-Time Paid 3; Part-Time Volunteers 60; Interns 3.
Governing Authority: Tax-exempt: 501(c)(3).
Institution Type/Description: Nature Center.
Collections: largest collection of birds of prey in the Northeast including bald eagles, hawks, owls & falcons.
Facilities: walking trails; classroom. Museum-related items for sale.
Activities: educational programs; walking trails; snowshoeing; nature camps; lectures; local artists exhibition; Museum Sponsors: Wildlife Rehab in Action.
Hours & Admission Prices: Jan. 6-April 15 two weekend programs 10-4; April 12-June 17 two programs daily 10-5; June 18-Oct. three programs daily 10-5:30; Nov.-Jan. 2 one program daily 10-4. Adults $12, students & seniors 65 & over $11, youth 3-18 $10; discount to AAA, AAM, ICOM & VAA members; children under 4 & members no charge. Closed Thanksgiving; Christmas.
Attendance: 35,000 (estimated)
Membership: Student $30; Individual $55; Friend $80; Partner $130; Supporter $250; Sustainer $500; Patron $1,000.

VERMONT TOY & TRAIN MUSEUM, 5573 Woodstock Rd., Quechee, VT 05059. Mailing Address: P.O. Box 730, Quechee, VT 05059-0730. Tel.: 802-295-1550, ext. 104. Facebook: Vermont Toy & Train Museum.
E-mail: quechee@quecheegorge.com
Web Site: www.quecheegorge.com
Founded: 2003.
Key Personnel: Dir., Gary Neil; Museum Shop Mgr., Robin Neil
Governing Authority: Parent Institution: Quechee Gorge Village.
Institution Type/Description: Toy Museum.
Collections: period toys; lunch boxes; model trains; hands-on exhibits.
Facilities: Museum-related items & vintage toys for sale.
Activities: train layouts; Seek & Find; train rides; carousel rides.
Hours & Admission Prices: Daily 10-5. No charge; donations accepted.
Attendance: 35,000 (estimated)

Randolph

RANDOLPH HISTORICAL SOCIETY, INC., Salisbury St., Randolph, VT 05060. Mailing Address: 9 Pleasant St., #304, Randolph, VT 05060. Tel.: 802-728-6677.
E-mail: hatchasse@earthlink.net
Founded: 1960.
Key Personnel: C.E.O., Laurence Leonard; Cur. & Museum Shop Mgr., Harriet Chase.
Personnel Profile: Part-Time Volunteers 15.
Governing Authority: society; nonprofit organization. Tax-exempt: 170(b)(1)(A).
Institution Type/Description: History Museum.
Collections: 1890 drug store & soda fountain; photo studio; period rooms; barber shop; early farm & home tools; costumes; pipe organ; early musical instruments; quilts; photographs; manuscripts; military artifacts; school books; Raleigh bottles; printing artifacts; Justin Morgan horse & man documents; 1943 B-17 crash site artifacts.
Research Fields: local history.
Facilities: research room; genealogical file.
Activities: permanent & temporary exhibitions; films; lectures. Annual Event: Independence Day Celebration.
Publications: books, Potash & Pine; Rustic Rhymes; Randolph Historical Sketches; Jonathan Carpenters Journal - the diary of a Revolutionary soldier and pioneer of VT. in conjunction with Greenhills books; Randolph's Beginnings; The 27 Flood; Randolph VT 1777-1927; The Golden Years 1850-1979.
Hours & Admission Prices: May-Sept. 3rd Sun. each month 2-4; July 4 call for hours; other times by appointment. No charge; donations accepted.
Attendance: 550 (estimated)
Membership: Individual $10; Family $15; Business $50; Patron $75.

Reading

READING HISTORICAL SOCIETY, Main St., Rte. 106, Reading, VT 05062. Mailing Address: P.O. Box 252, Reading, VT 05062-0252. Tel.: 802-674-2649.
Founded: 1953.
Key Personnel: Pres., Jonathan Springer; Trustee, Howard Sanderson, Jr.; Treas., Esther Allen.
Governing Authority: executive committee. Tax-exempt.
Institution Type/Description: General Museum.
Collections: history; military; medical; music; manuscripts; textiles. Historic House: Felchville Village.
Facilities: 100-vol. library of books, photographs, town reports, school registers, maps; reading room.
Activities: permanent & temporary exhibitions.
Hours & Admission Prices: By appointment only. No charge; donations accepted.
Membership: Senior $5; Regular $10; Family $15; Sustaining $100; Life $200.

Readsboro

READSBORO HISTORICAL SOCIETY, Main St., Readsboro, VT 05350. Mailing Address: 152 Glen Ave., Readsboro, VT 05350-9798. Tel.: 802-423-5432.
Founded: 1972.
Key Personnel: Pres. (V), Betty Bolognani; Vice Pres., Eunice Crowell; Treas., Priscilla Margola; Sec., Priscilla Thayer.
Personnel Profile: Part-Time Volunteers 12.
Governing Authority: society; nonprofit. Tax-exempt.
Institution Type/Description: Historical Society Museum: housed in 1840 frame structured church.
Collections: day books of the area dating back to the 1700s; 19th- & 20th-century photographs; period artifacts; spinning wheels; quilts; fireman's hose wheel; 19th-century costumes; c.1900 infant clothes.
Research Fields: historical sites.
Facilities: Museum-related items for sale.
Activities: lectures; films; hobby workshops; permanent & temporary exhibitions.
Publications: Down Thru the Years; computer CDs of photos of Early Readsboro; over 40 interview tapes of older residents.
Hours & Admission Prices: June-Oct. Sun. 1-3; other times by appointment. No charge; donations accepted.
Attendance: 100 (accurate)
Membership: Individual $3; Family $5; Lifetime $25.

Richmond

OLD ROUND CHURCH - RICHMOND HISTORICAL SOCIETY, 25 Round Church Rd., Richmond, VT 05477. Mailing Address: P.O. Box 453, Richmond, VT 05477-0453. Tel.: 802-434-3654. Facebook: Old Round Church.
E-mail: rhs@oldroundchurch.com
Web Site: www.oldroundchurch.com
Founded: 1973.
Key Personnel: Pres. (V), Frances Thomas.
Personnel Profile: Part-Time Volunteers 30.
Governing Authority: Parent Institution: Richmond Historical Society. Tax-exempt: 501(c)(3).
Institution Type/Description: Historic Building: housed in a 16-sided church built in 1812. A National Historic Landmark.
Collections: local history & culture; church history; religious artifacts; photographs.
Activities: summer concerts; weddings; community programs.
Publications: book, Richmond, Vermont - A History of More Than 200 Years; booklet, The Richmond Round Church, 1813-2013.
Hours & Admission Prices: Summer & Fall daily 10-4. No charge; donations accepted. &
Attendance: 2,000 (estimated)
Membership: Family $20; Life $200.

Rochester

ROCHESTER HISTORICAL SOCIETY, Rochester Library, 2nd Fl., Main St., Rochester, VT 05767. Mailing Address: P.O. Box 428, Rochester, VT 05767-0428. Tel.: 802-767-4453.
E-mail: admin@rochesterhistorical.org
Institution Type/Description: Historical Society Museum.
Collections: local history; farm tools; military artifacts; clothing; photographs.
Hours & Admission Prices: By appointment.

Royalton

ROYALTON HISTORICAL SOCIETY, 4184 Rte. 14, Royalton, VT 05068-5084. Tel.: 802-828-3051. Fax: 802-828-3206.
E-mail: john.dumville@state.vt.us
Founded: 1967.
Key Personnel: Pres., John P. Dumville; Asst. Dir., Ralph Eddy; Business Officer & Publications Dir., Richard L. McGovern.
Personnel Profile: Part-Time Volunteers 6.
Governing Authority: society.
Institution Type/Description: Historical Society Museum: housed in 1840 Royalton Town House located on Town Common.
Collections: photographs, manuscripts, printed matter, furniture, clothing & general artifacts. Historic Building: 1844 Royalton Center School; 1840 Town House; 1836 Episcopal Church.
Research Fields: local history.
Facilities: 200-vol. library of Royalton & Vermont history, available for use on premises; reading room; 300-seat auditorium.
Activities: guided tours; lectures; school loan service; permanent & temporary exhibits.
Publications: book, Royalton Vermont.
Hours & Admission Prices: Summer by appointment. No charge; donations accepted.
Attendance: 1,000 (estimated)
Membership: Individual $2; Life $100.

Rutland

NORMAN ROCKWELL MUSEUM OF VERMONT, 654 Rte. 4 E., Rutland, VT 05701. Tel.: 877-773-6095 (toll free). Fax: 802-775-2440.
E-mail: sales@normanrockwellvt.com
Web Site: www.normanrockwellvt.com
Key Personnel: Mgr., Rachel Lynes-Bells
Institution Type/Description: Art Museum.
Collections: Norman Rockwell's art, prints & posters; magazine covers; advertisements; calendars; books.
Facilities: Museum-related items for sale.
Hours & Admission Prices: Daily 9-4. Adults $5.50, seniors 62 & above $5, children $2.50; discount to AAA members & groups 10 and over.

RUTLAND AREA ART ASSOCIATION DBA CHAFFEE ART CENTER & CHAFFEE DOWNTOWN, 16 S. Main St. & 75 Merchants Row, Rutland, VT 05701-4136. Mailing Address: P.O. Box 1447, Rutland, VT 05701-1447. Tel.: 802-775-0356. Facebook: Chaffee Art Center.
E-mail: info@chaffeeartcenter.org
Web Site: www.chaffeeartcenter.org
Formerly: Chaffee Center for the Visual Arts
Founded: 1961.
Key Personnel: Exec. Dir., Margaret Barros; Gallery Coord., Richelle Franzoni; Asst. Gallery Coord., Beth Seck; Pres. (V), W. Tracy Carris, Esq.
Personnel Profile: Full-Time Paid 1; Part-Time Paid 2; Part-Time Volunteers 14; Interns 2.
Volunteer Hours: 1,250
Operating Expenses: 190,547
Operating Income: 171,457
Governing Authority: Tax-exempt: 501(c)(3).
Institution Type/Description: Art Center: housed in a preserved building from 1892-94.
Collections: works by Vermont artists.
Major Exhibits: Annual Full House Exhibit, 12/31/13-2/28/14; Annual Student Art Show, 3/14/14-5/2/14; Emerging Artists-New Juried Artists, 5/9/14-5/31/14; 7th Annual Photography Contest: Farm & Food, Collaboraton with RAFFL, 6/27/14-7/26/14; Fiber Artists of Vermont, 7/25/14-9/12/14; Director's Choice: Fran Bull, Collaboration with Castleton Downtown Gallery, 9/24/14-10/25/14; Oil & Water: Peter Huntoon & Mareva Millarc, 11/4/14-11/29/14; Call to New England Artists, 11/11/14-12/16/14.
Facilities: art center; downtown art gallery.
Activities: classes; special events. Museum Sponsors: Annual Fundraiser; Art in the Park in August & October; week-long Mascarade Party & Halloween Exhibit.
Hours & Admission Prices: Downtown Gallery: Tues.-Thurs. 11-6, Fri.-Sat. 11:30-7. Art Center: March-Nov. Thurs.-Sat. 12-6. No charge; donations accepted. &
Attendance: 23,000 (estimated)
Membership: Student $25; Basic $75; Family $100; Contributing $200-$449; Patron $450-$999; Sponsor $999-$2,499; Benefactor $2,500 & up.

RUTLAND HISTORICAL SOCIETY, 96 Center St., Rutland, VT 05701-4023. Tel.: 802-775-2006. Facebook: Rutland Historical Society.
E-mail: rutlandhistory@comcast.net
Web Site: rutlandhistory.com
Founded: 1969.
Key Personnel: Chm. (V), Pam Johnson; Pres. (V), Carolynn Ranftle.
Personnel Profile: Part-Time Volunteers 30.
Governing Authority: Tax-exempt.
Institution Type/Description: Historical Society: housed in the former Nickwackett Firehouse, built in 1860.
Collections: local history; manuscripts; documents; books; photographs; costumes.
Publications: quarterly magazine.
Hours & Admission Prices: Mon. 6-9pm, Sat. 1-4pm; other times by appointment. No charge.
Attendance: 200 (estimated)
Membership: Student & Senior $8; Regular $10; Contributing $20; Sponsor $50; Life $200.

Saint Albans

ST. ALBANS HISTORICAL MUSEUM, (M), 9 Church St., Saint Albans, VT 05478-1675. Mailing Address: P.O. Box 722, Saint Albans, VT 05478-0722. Tel.: 802-527-7933.
E-mail: stamuseum.history@myfairport.net
Web Site: www.stamuseum.com
Founded: 1971.
Key Personnel: Dir., A J McDonald; Pres. (V), Warren C. Hamm; Treas., Danielle Manahan; Museum Shop Mgr., Betty Anderson.
Personnel Profile: Full-Time Paid 1; Part-Time Paid 1; Part-Time Volunteers 50.
Governing Authority: nonprofit organization. Parent Institution: St. Albans Historical Society. Tax-exempt: 501(c)(3).
Institution Type/Description: Historical Society Museum: housed in 1861 three-story brick Franklin County Grammar School.
Collections: Saint Albans Raid Data; historical reference books; photographs; china & glass; maps; paintings; gowns; country doctor's office; furniture; jewelry; quilts; linens; laces; paintings; Central Vermont railway articles; military items; sports equipment; toys; dolls; craft items; made-to-scale historic diorama northwest corner of Vermont; St. Albans & Champlain Valley history.
Major Exhibits: Sol Levenson Civil War Murals, 12/11-1/15.
Research Fields: Vermont & local history.
Facilities: library & reference room.
Activities: guided tours; meetings; history classes.
Publications: newsletter; local history pamphlets.
Hours & Admission Prices: May 25-Oct. 8 Tues.-Fri. 1-4, Sat. 10-2; other times by appointment. Adults $5, children 6-14 $2; members & children under 6 no charge. &
Attendance: 2,000 (accurate)
Membership: St. Albans Historical Society: Individual $35; Family $40; Life $1,000.

Saint Johnsbury

*   **FAIRBANKS MUSEUM AND PLANETARIUM, (M),** 1302 Main St., Saint Johnsbury, VT 05819-2224. Tel.: 802-748-2372. Fax: 802-748-1893. Facebook: Fairbanks Museum.
E-mail: info@fairbanksmuseum.org
Web Site: www.fairbanksmuseum.org
Founded: 1889.
Key Personnel: Exec. Dir., Adam Kane; Chm. (V), Tracy Zschau; Museum Shop Mgr., Virginia Platt.
Personnel Profile: Full-Time Paid 10; Part-Time Paid 10; Part-Time Volunteers 140; Interns 16.
Governing Authority: private; nonprofit organization. Tax-exempt: 501(c)(3).
Institution Type/Description: General Museum & Planetarium.
Collections: 2,600 mounted birds & mammals; regional toys, tools, furniture, agriculture & industrial equipment; North American, Far Eastern, Middle Eastern, Polynesian & African ethnology; sculpture; herbarium; regional photographs; maps; newspapers; documents; weather records & natural history records.
Major Exhibits: Exploration Station, 1/1/14-12/31/14.
Research Fields: Vermont's natural environments & history; historic preservation; science & history education; weather forecasting.
Facilities: 2,500-vol. library of reference books; archives center; garden; planetarium; NOAA & FAA weather stations; meeting rooms; classrooms.
Activities: lectures; daily weather & information broadcasts; special exhibitions; formally organized education programs for children & adults; field trips.
Publications: e-newsletter; information sheets; brochures; Annual Report.
Hours & Admission Prices: Museum: May-Sept. Mon.-Sat. 9-5, Sun. 1-5; Oct.-April Tues.-Sat. 9-5. Adults $8, senior citizens & children 5-17 $6; discounts to groups & AAM members; members no charge. Planetarium: July-Aug. daily 11 & 1:30; Sept.-June Sat.-Sun. 1:30; other times by appointment. $3 per person; discounts to groups, AAM, VMGA, ASTC & NEMA members. Closed New Year's Day; Easter; Thanksgiving; Christmas. &
Attendance: 65,000 (accurate)
Membership: Individuals $50; Family $75; Sustaining $100; Friends of the Museum $100 & up.

MAPLE GROVE SUGARHOUSE MUSEUM AND GIFT SHOP, 1052 Portland St., Saint Johnsbury, VT 05819-2041. Tel.: 802-748-5141, ext. 5547. Fax: 802-748-0844.
E-mail: maple@maplegrove.com
Web Site: www.maplegrove.com
Formerly: Maple Grove Museum and Factory
Founded: 1915.
Institution Type/Description: Historic Building: housed in a former sugarhouse.
Collections: period & modern sugaring equipment.
Activities: video.
Hours & Admission Prices: mid-March to May Mon.-Fri. 8-5; June-Dec. Mon.-Fri. 8-5, Sat.-Sun. 9-5. No charge. Closed Thanksgiving; Christmas.

NORTHEAST KINGDOM ARTISANS GUILD, 430 Railroad St., #2, Saint Johnsbury, VT 05819-1727. Tel.: 802-748-0158.
Institution Type/Description: Art Gallery.
Collections: handmade crafts & fine art including baskets, clay, fiber, glass, metal, paper, & wood; prints; watercolors; oils; photographs.
Facilities: Museum-related items for sale.
Activities: temporary exhibits.
Hours & Admission Prices: Mon.-Sat. 10:30-5:30.

STEPHEN HUNECK GALLERY AT DOG MOUNTAIN, 143 Parks Rd., Saint Johnsbury, VT 05819-8907. Tel.: 800-449-2580; 802-748-2700. Fax: 802-748-3075.
E-mail: info@dogmt.com
Web Site: www.dogmt.com
Founded: 1998.
Key Personnel: Dir. & C.E.O., Gwendolyn Huneck; Chm. (V) & Pres. (V), Lisa Nelson; Museum Shop Mgr., Amanda McDermott.
Personnel Profile: Full-Time Paid 8; Part-Time Paid 1; Part-Time Volunteers 4.
Institution Type/Description: Art Gallery.
Collections: sculptures; paintings; photographs.
Facilities: outdoor sculpture garden; chapel; hiking trails.
Activities: dog parties.
Publications: newsletter, Dog Mountain.
Hours & Admission Prices: Mon.-Sat. 10-5, Sun. 11-4. No charge; donations accepted. &
Attendance: 10,000 (estimated)

Saxtons River

SAXTONS RIVER HISTORICAL SOCIETY, Main St., Saxtons River, VT 05154. Mailing Address: P.O. Box 18, Saxtons River, VT 05154-0018. Tel.: 802-869-2566.
Institution Type/Description: Historical Society Museum: housed in the former Congregational Church, built in 1836.
Collections: local history & culture; photographs; toys; farm implements & tools; period furnishings; genealogical records.
Hours & Admission Prices: Summer: Sun. 2-4:30; other times by appointment. No charge.

Shaftsbury

ROBERT FROST STONE HOUSE MUSEUM, 121 Historic Rte. 7A, Shaftsbury, VT 05262. Tel.: 802-447-6200.
E-mail: ffriends@sover.net
Web Site: www.frostfriends.org
Founded: 2002.
Institution Type/Description: Historic House Museum: housed in the former home of American poet, Robert Frost.
Collections: Frost's life & works; personal artifacts.
Hours & Admission Prices: Oct. daily 10-5; Nov. Tues.-Sun.11-4 by appointment. Adults $5, students under 18 $2.50; children under 6 no charge.

SHAFTSBURY HISTORICAL SOCIETY, 3542 VT Rte. 7A, Shaftsbury, VT 05262. Mailing Address: P.O. Box 401, Shaftsbury, VT 05262-0401. Tel.: 802-375-6376.
E-mail: gronning@sover.net
Founded: 1967.
Congressional District: 1
Key Personnel: Pres. (V), Norman D. Gronning; Vice Pres., Angie Abbatello; Vice Pres., Robert Millington; Treas., David Curtis; Sec., Ruth Levin.
Personnel Profile: Part-Time Paid 1; Part-Time Volunteers 30.
Governing Authority: nonprofit organization. Tax-exempt: 170(b)(1)(A).
Institution Type/Description: Historical Society Museum: housed in 1846 Meeting House, oldest Baptist Church in Vermont.
Collections: tools & furnishings; historical documents; decorative arts; crafts; costumes; industries; clothing; records & artifacts of early settlers; manuscripts; Bibles.
Research Fields: genealogy; local buildings; cemeteries & early gravestones; 19th-century economic & social history.
Facilities: 300-vol. library of old books, diaries & newspapers available for use by request; 3,000 historical & genealogical documents; 200-seat auditorium. Museum-related items for sale.
Activities: guided tours; lectures; temporary exhibitions.
Publications: pamphlets, Shaftsbury Historic House Tour; Introduction to the Shaftsbury Historical Society Museum; Shaftsbury Historical Map. Books, Ordinary Heroes-The Story of Shaftsbury; Shaftsbury Gravestone Records; History of the Baptist Church in Shaftsbury, Vermont.
Hours & Admission Prices: June 15-Oct. 15 Mon., Wed. & Fri. 1-4, Sat.-Sun. 2-4. No charge; donations accepted.
Attendance: 200 (estimated)
Membership: Student $2; Adult $5; Family $10; Associate $25; Patron $100.

Shelburne

FURCHGOTT SOURDIFFE GALLERY, 86 Falls Rd., Shelburne, VT 05482-6208. Tel.: 802-985-3848.
Institution Type/Description: Art Gallery.
Collections: paintings; pottery.
Hours & Admission Prices: Tues.-Fri. 9:30-5:30, Sat. 10-5.

SHELBURNE MUSEUM, INC., 5555 Shelburne Rd., Shelburne, VT 05482-7491. Mailing Address: P.O. Box 10, Shelburne, VT 05482-0010. Tel.: 802-985-3346. Fax: 802-985-2331.
E-mail: info@shelburnemuseum.org
Web Site: www.shelburnemuseum.org
Founded: 1947.
Congressional District: 1
Key Personnel: Chm., James Pizzagalli; Co Vice Chm., Peter Martin; Co Vice Chm., Caroline Almy Gerry; Dir., Thomas Denenberg; Dir. Devel., Sam Ankerson; Dir. Finance & Admin., Lois Nial; Dir. Preservation & Conservation, Richard Kerschner; Dir. Bldgs., Chip Stulen; Dir. Protection Svcs., Grounds & Gardens, Rick Peters; Merchandising Mgr., Lee Wheeler; Mktg. & Public Rels. Mgr., Leslie Wright; Sr. Cur., Jean Burks.
Personnel Profile: Full-Time Paid 51; Part-Time Paid 103; Part-Time Volunteers 160; Interns 10.
Governing Authority: nonprofit. Tax-exempt: 501(c)(3).
Institution Type/Description: Art & Design Museum.
Collections: 38 historic buildings; American folk art; objects of everyday life; tools; horse-drawn vehicles; toys; decorative arts; circus collection; dolls; flower, herb & heritage vegetable gardens; orchard; arboreta; American paintings; European impressionist paintings.
Research Fields: American folk art; decorative art; textiles; horse-drawn vehicles; Vermont history, early American life.
Facilities: 5,000-vol. library; archives; cafeteria; family activity center. Museum store with publications, photos, prints & reproductions for sale.
Activities: lectures; tours; gallery talks; children's programs; special events; annual symposium; workshops.
Publications: catalogs; newsletters; calendars; annual report; occasional publications.
Hours & Admission Prices: mid-May to Oct. daily 10-5; Nov.-Dec. Tues.-Sun. & Mon. holidays 10-5. &
Attendance: 103,913 (accurate)
Membership: Individual $50; Dual & Family $75; Family Plus $100; Sustaining $250; Patron $500; Benefactor $1,000.

Shoreham

SHOREHAM HISTORICAL SOCIETY, Rte. 22-A, Old Stone Schoolhouse, Shoreham, VT 05770. Mailing Address: P.O. Box 235, Shoreham, VT 05770-0235. Tel.: 802-897-2572.
Key Personnel: Pres., Dale Birdsall; Cur., Ginny Spadaccini

Institution Type/Description: Historical Society Museum.
Collections: Shoreham history & culture; photographs; documents; arrowheads; genealogy; sleighs.
Hours & Admission Prices: By appointment. No charge.

South Hero

SOUTH HERO BICENTENNIAL MUSEUM, Rte. 2, South Hero, VT 05486. Mailing Address: 19 Phelps Ln., South Hero, VT 05486-4802. Tel.: 802-372-5259.
E-mail: lmjshvt@comcast.net
Founded: 1974.
Key Personnel: Chm. (V), Hazel Quelch; Pres. (V), Lorraine Janick.
Personnel Profile: Part-Time Volunteers 5.
Governing Authority: town; nonprofit. Tax-exempt.
Institution Type/Description: Local History Museum.
Collections: bibles; local historic pieces.
Activities: loan, permanent & temporary exhibitions; hobby workshops; arts & crafts exhibits & demonstrations.
Hours & Admission Prices: June-Aug. Mon. & Thurs. 1:30-3:30. No charge.
Attendance: 75 (estimated)

South Royalton

JOSEPH SMITH BIRTHPLACE MEMORIAL & VISITORS CENTER, 357 LDS Lane, South Royalton, VT 05068. Tel.: 802-763-7742.
Founded: 1905.
Personnel Profile: Full-Time Paid 4; Full-Time Volunteers 10.
Institution Type/Description: History Museum: the birthplace of Joseph Smith, the first president & prophet of The Church of Jesus Christ of Latter-day Saints.
Collections: Joseph Smith's life & career; religious artifacts; sculpture; photographs.
Hours & Admission Prices: May-Oct. Mon.-Sat. 9-7, Sun. 1:30-7; Nov.-April Mon.-Sat. 9-5, Sun. 1:30-5. No charge.
Attendance: 6,000 (estimated)

South Woodstock

GREEN MOUNTAIN PERKINS ACADEMY & HISTORICAL ASSOCIATION, 32 Academy Circle, VT Rte. 106, South Woodstock, VT 05071. Mailing Address: P.O. Box 143, South Woodstock, VT 05071-0143. Tel.: 802-457-3779.
E-mail: marymaple13@gmail.com
Web Site: www.greenmountainperkinsacademy.org
Key Personnel: Pres., Mary McCuaig
Institution Type/Description: Historic House Museum: housed in a building that served as a private high school until 1898; built in 1848.
Collections: academy history; school furnishings & books; wood stoves; maps; a loom; weaving supplies; Windsor backed chairs; musical instruments; sheet music; photographs; school records.
Hours & Admission Prices: July-Aug. Sat. 2-5; other times by appointment.
Attendance: 300 (estimated)

Springfield

EUREKA SCHOOL HOUSE, 470 Charlestown Rd., Rte. 11, Springfield, VT 05156. Mailing Address: Historic Preservation, National Life Bldg., 6th Fl., Montpelier, VT 05620. Tel.: 802-828-3051. Fax: 802-828-3206.
E-mail: john.dumville@state.vt.us
Web Site: www.historicsites.vermont.gov/eureka
Founded: 1968.
Key Personnel: Historic Sites Operations Chief, John P. Dumville.
Personnel Profile: Part-Time Paid 1.
Governing Authority: state. Parent Institution: State of Vermont, Division for Historic Preservation, One National Life Dr., 6th Fl., Montpelier, VT 05602. Tax-exempt.
Institution Type/Description: Historic Buildings: 1785 Eureka School House; Lattice Truss Covered Bridge.
Collections: period furnishings; memorabilia; historic structures.
Facilities: schoolhouse; covered bridge.
Publications: A Guide to Historic Sites in Vermont; brochure, Eureka Schoolhouse.
Hours & Admission Prices: mid-May to mid-Oct. daily 9-5. No charge; donations accepted. &
Attendance: 3,000 (estimated)

JAMES HARTNESS-RUSSELL PORTER ASTRONOMY MUSEUM, 30 Orchard St., Springfield, VT 05156-2612. Tel.: 800-732-4789; 802-885-2115.
Institution Type/Description: Astronomy Museum.
Collections: astronomy history & instruments; paintings; drawings; Porter Garden Telescope; Cassegrain and Coude telescopes; photographs; Hartness sundial.
Activities: guided tours.
Hours & Admission Prices: Call for hours.

SPRINGFIELD ART & HISTORICAL SOCIETY - MILLER ART CENTER, 9 Elm Hill, Springfield, VT 05156-0313. Mailing Address: P.O. Box 313, Springfield, VT 05156-0313. Tel.: 802-885-2415.
E-mail: info@millerartcenter.com
Web Site: www.millerartcenter.com
Founded: 1956.
Key Personnel: Dir., Maureen Bolduc; Bd. Pres., Leonard Bolduc.
Personnel Profile: Part-Time Paid 1; Part-Time Volunteers 10.
Governing Authority: society. Parent Institution: Springfield Art & Historical Society. Tax-exempt: 501(c)(3).
Institution Type/Description: Art Museum: housed in 1865 house last occupied by Edward W. Miller & family.
Collections: Richard Lee pewter; machine tool industry photos; Vermont novelty works; dolls; carriages; toys; primitive portraits; Bennington pottery; local artist collections; textile (costumes); archives.
Facilities: historical records available for inter-library loan & on premises; classrooms. Handcrafts & paintings for sale.
Activities: guided tours; lectures; gallery talks; arts festivals; hobby workshops; formally organized education programs; permanent, temporary & traveling exhibitions.
Publications: annual calendar of events; brochure; quarterly newsletter; workshops announcements.
Hours & Admission Prices: Thurs.-Fri. 11-5, Sat. 11-4. Adults $3; members no charge. &
Attendance: 1,500 (estimated)
Membership: Individual $20; Family $35; Sponsor $50; Corporate $100.

St. Johnsbury

ST. JOHNSBURY ATHENAEUM, 1171 Main St., St. Johnsbury, VT 05819-2289. Tel.: 802-748-8291. Fax: 802-748-8086.
E-mail: inform@stjathenaeum.org
Web Site: www.stjathenaeum.org
Founded: 1871.
Congressional District: 3
Key Personnel: Exec. Dir., Matthew Powers; Head Librarian, Lisa Von Kann; Chm. (V), William Marshall.
Personnel Profile: Full-Time Paid 4; Part-Time Paid 9; Part-Time Volunteers 50; Interns 1.
Governing Authority: nonprofit organization. Tax-exempt: 501(c)(3).
Institution Type/Description: Art Museum and National Historic Landmark.
Collections: 100 works of art, primarily 19th-century American; paintings & sculpture reflect collectors' tastes 1865-1890; Hudson River school landscapes.
Facilities: 45,000-vol. public library collection; reading room.
Activities: poetry readings; school tours; interpretive arts & culture; lectures; arts & humanities programs; children's literacy program.
Publications: catalogue.
Hours & Admission Prices: Mon.-Fri. 10-5:30, Sat. 9:30-5. Art Gallery: adults $8; members no charge. Closed national holidays. &
Attendance: 70,000 (estimated)
Membership: Basic $15.20.

Stannard

STANNARD HISTORICAL SOCIETY, Old Methodist Church, Stannard Mountain Rd., Stannard, VT 05842. Mailing Address: 92 Old Pasture Rd., Greensboro Bend, VT 05842-2100. Tel.: 802-533-2561.
Key Personnel: Pres., Jan Lewandoski
Institution Type/Description: Historical Society Museum: housed in an 1888 church. Listed on the National Register of Historic Sites.
Collections: religious artifacts; period furnishings.
Activities: Museum Sponsors: Old Home Day in August.
Hours & Admission Prices: By appointment.

Stowe

GREEN MOUNTAIN FINE ART GALLERY, 64 S. Main St., Stowe, VT 05672. Mailing Address: P.O. Box 1384, Stowe, VT 05672-1384. Tel.: 802-253-1818. Fax: 802-253-6837.
Institution Type/Description: Art Gallery.
Collections: watercolors; oils; pastels; prints; mixed media; photography.
Hours & Admission Prices: Wed.-Mon. 10-6.

HELEN DAY ART CENTER, 90 Pond St., Stowe, VT 05672. Mailing Address: P.O. Box 411, Stowe, VT 05672-0411. Tel.: 802-253-8358. Fax: 802-253-2703. Facebook: Helen Dat Art Center.
E-mail: mail@helenday.com
Web Site: www.helenday.com
Founded: 1981.
Key Personnel: Exec. Dir., Nathan Suter; Chm. (V), Elizabeth Brown; Asst. Dir., Rachel Moore; Education & Facilities Coord., Sarah Beggs; Exec. Asst., Anahi Costa.
Personnel Profile: Full-Time Paid 1; Part-Time Paid 4; Part-Time Volunteers 40; Interns 4.
Governing Authority: nonprofit organization. Tax-exempt: 501(c)(3).
Institution Type/Description: Art Center: housed in 1863 Greek Revival Building used as a school for 100 years.
Collections: paintings; sculpture; itinerant exhibitions.
Major Exhibits: Andrea Lilienthal, 2/14; Student Art Show, 5/14; Exposed - Sculpture Exhibition, 7/14-10/14; Magic Lantern Show Fil Festival, 10/31/14; Members Art Show, 12/14.
Facilities: Exhibit-related items for sale; artwork.
Activities: guided tours; lectures; concerts; film festival; organized education programs; docent program; participatory & loan exhibitions; public programs.
Publications: seasonal newsletter; exhibit brochures; sculpture catalogue, Exposed.
Hours & Admission Prices: Wed.-Sun. 12-5 & by appointment. No charge; donations accepted. Closed New Year's Day; President's Day; Easter; Memorial Day; Independence Day; Thanksgiving; Christmas. &
Attendance: 20,000
Membership: Artist $30; Individual $35; Individual plus 1 $50; Family $60.

ROBERT PAUL GALLERY, 394 Mountain Rd., Stowe, VT 05672. Mailing Address: P.O. Box 1413, Stowe, VT 05672-1413. Tel.: 800-873-3791.
Institution Type/Description: Art Gallery.
Collections: paintings; sculpture; photography.
Hours & Admission Prices: Mon.-Sat. 10-6, Sun. 10-5.

STOWE HISTORICAL SOCIETY MUSEUM, 90 School St., Stowe, VT 05672. Mailing Address: P.O. Box 730, Stowe, VT 05672-0730. Tel.: 802-253-1518.
E-mail: info@stowehistoricalsociety.org
Web Site: www.stowehistoricalsociety.org
Founded: 1956.
Congressional District: 1
Key Personnel: Pres. (V), Barbara Baraw.
Volunteer Hours: 880
Operating Expenses: 10,672
Operating Income: 23,797
Governing Authority: Tax-exempt.
Institution Type/Description: Historical Society Museum.
Collections: local history & culture; photographs; Civil War artifacts; personal artifacts.
Activities: educational programs; research.
Hours & Admission Prices: Tues. & Thurs. 2-5, Sat. 12-3; other times by appointment. No charge; donations accepted. Closed New Year's Eve & Day; Christmas Eve, Day & week; state holidays. &
Attendance: 300 (estimated)
Membership: Individual $10; Family $20.

VERMONT SKI AND SNOWBOARD MUSEUM, The Perkins Bldg., One S. Main St., Stowe, VT 05672. Mailing Address: The Perkins Bldg., P.O. Box 1511, Stowe, VT 05672-1511. Tel.: 802-253-9911. Fax: 802-253-2616.
E-mail: info@vtssm.com
Web Site: www.vtssm.com
Founded: 1988.

Key Personnel: Dir. & Cur., Meredith Scott; Chm., Rick Hamlin; Museum Shop Mgr., Susi Clark.
Personnel Profile: Full-Time Paid 1; Part-Time Paid 3; Part-Time Volunteers 10; Interns 2.
Governing Authority: Tax-exempt.
Institution Type/Description: History Museum.
Collections: Vermont's ski history; period clothing; equipment; ski lifts 1934 to present; personal artifacts; Vermont racing history; snowboarding; films; Hall of Fame.
Major Exhibits: Kick & Glide: Vermont's Nordic Skiing History, 10/13-10/14; VTSD, 10/13-10/14.
Facilities: library & archives by appointment. Museum-related items for sale.
Activities: tours; films. Annual Events: Vermont Ski Museum Hall of Fame Induction; Vermont Antique Ski Race; Stowe Mountain Film Festival.
Publications: quarterly newsletter.
Hours & Admission Prices: June-Oct. 30 & Dec.-March Wed.-Mon. 12-5. Suggested Donations: family $5, individual $3; discounts to groups; members no charge. &
Attendance: 8,000 (estimated)
Membership: Partner $60; Historian $100; Collector $150; Educator $250; Patron $500; Curator $750; Benefactor's Circle $1,000.

WEST BRANCH GALLERY & SCULPTURE PARK, 17 Towne Farm Lane, Stowe, VT 05672-4138. Mailing Address: P.O. Box 250, Stowe, VT 05672-0250. Tel.: 802-253-8943.
Institution Type/Description: Art Gallery.
Collections: paintings; sculpture; photographs.
Hours & Admission Prices: Wed.-Sun. 11-6; other times by appointment.

Strafford

JUSTIN SMITH MORRILL STATE HISTORIC SITE, 214 Justin Morrill Memorial Hwy., Strafford, VT 05072. Mailing Address: Historic Preservation, National Life Bldg., 6th Fl., Montpelier, VT 05620. Tel.: 802-828-3051. Fax: 802-828-3206.
E-mail: john.dumville@state.vt.us
Web Site: www.historicsites.vermont.gov/morrill
Founded: 1969.
Key Personnel: Historic Site Operations Chief, John P. Dumville.
Personnel Profile: Full-Time Paid 1; Part-Time Paid 7.
Governing Authority: state. Parent Institution: State of Vermont, Division for Historic Preservation, National Life Bldg., 6th Fl., Montpelier 05620-0501. Tax-exempt.
Institution Type/Description: Historic House Museum: c.1849 Justin Smith Morrill Homestead, Gothic Revival Homestead with seven Agricultural Buildings.
Collections: original furnishings & memorabilia; historic structures.
Research Fields: period landscape, land grant colleges.
Publications: A Guide to Historic Sites in Vermont; brochure, The Justin Smith Morrill Homestead.
Hours & Admission Prices: mid-May to mid-Oct. Wed.-Sun. 9:30-5:30. Adults $5; discounts to AAM members; registered school groups no charge. &
Attendance: 2,000 (estimated)

Swanton

ABENAKI CULTURAL CENTER, 49 Church St., Swanton, VT 05488. Tel.: 802-868-2559.
Formerly: Abenaki Tribal Museum & Cultural Center
Institution Type/Description: Native American Museum.
Collections: Abenaki history, culture & life; personal artifacts; clothing; photographs.
Hours & Admission Prices: Call for hours.

Thetford

HUGHES BARN MUSEUM, 2274 Rte. 113, Thetford, VT 05074. Mailing Address: Thetford Historical Society, P.O. Box 33, Thetford, VT 05074. Tel.: 802-785-2068.
Institution Type/Description: History Museum.
Collections: local history & culture; period furnishings; personal artifacts; farm equipment & tools; household artifacts; local business & industry.
Hours & Admission Prices: Aug. to Labor Day Sun. 2-5. No charge.

THETFORD HISTORICAL SOCIETY LIBRARY AND MU-SEUM, Bicentennial Bldg., 16 Library Rd., Thetford, VT 05074. Mailing Address: P.O. Box 33, Thetford, VT 05074-0033. Tel.: 802-785-2068.
E-mail: info@thetfordhistoricalsociety.org
Founded: 1943.
Key Personnel: Pres. (V) & Librarian, Charles Latham; Asst. Dir., Martha Howard.
Personnel Profile: Part-Time Paid 1; Part-Time Volunteers 12.
Governing Authority: society. Tax-exempt.
Institution Type/Description: Library & Agriculture Museum.
Collections: Clara Siprell photographs; library; portraits; furniture; tools; agricultural implements.
Research Fields: town history; genealogy; agricultural history; town industry.
Facilities: library of books, pamphlets, pictures & advertisements relating to local history, genealogy, tools & agriculture available for research by permission of curator.
Activities: permanent & temporary exhibitions; program for local elementary school students & students at Thetford Academy; special exhibits at Thetford Hill on Fair Day.
Publications: books, History & Folklore of Post Mills; Short History of Thetford; Beloved Village; The Mills & Villages of Thetford, VT; The Life of Asa Burton; bicentennial map of Thetford; Green Mountain Copper.
Hours & Admission Prices: Library: Mon. & Thurs. 2-4, Tues. 10-12. Barn Museum: Aug. Sun. 2-5. No charge; donations accepted.
Attendance: 300 (estimated)
Membership: Individual $3; Family $5; Sustaining $10; Honorary $25.

Vergennes

BIXBY MEMORIAL LIBRARY, 258 Main St., Vergennes, VT 05491-1056. Tel.: 802-877-2211. Fax: 802-877-2411.
E-mail: bixby_verg@vals.state.vt.us
Web Site: www.bixbylibrary.org
Key Personnel: Dir., Jane Spencer; Bd. Chm., Kitty Oxholm; Youth Svcs. Librarian, Rachel Plant; Library Asst., Carolyn Tallen
Institution Type/Description: Library & History Museum.
Collections: books; Native American artifacts; paintings by Vermont artists; maps; documents; manuscripts; Vermont stamps & covers.
Facilities: library & archives.
Activities: Annual Event: Summer Art Show.
Hours & Admission Prices: Mon. 12:30-7, Tues. & Fri. 12:30-5, Wed. 10-5, Thurs. 10-7, Sat. 9-2.

LAKE CHAMPLAIN MARITIME MUSEUM, 4472 Basin Harbor Rd., Vergennes, VT 05491-9192. Tel.: 802-475-2022. Fax: 802-475-2953.
E-mail: info@lcmm.org
Web Site: lcmm.org
Founded: 1985.
Key Personnel: Exec. Dir, Erick Tichonuk; Chm. (V), Darcey Hale; Sr. Advisor, Arthur B. Cohn; C.F.O., Susan Jones; Dir. Archaeological Projects, Chris Sabick; Dir. Exhibits & Collections, Eloise Beil; Dir. Boat-building & Outdoor Education, Nick Patch; Museum Shop Mgr., Lisa Percival.
Personnel Profile: Full-Time Paid 14; Full-Time Volunteers 1; Part-Time Paid 14; Part-Time Volunteers 40; Interns 5.
Governing Authority: private; not-for-profit organization. Subsidiary Institution: Maritime Research Institute. Tax-exempt: 501(c)(3).
Institution Type/Description: Maritime & Nautical Archaeology Museum.
Collections: pre-17th century to mid-20th century artifacts, reflecting the lake's colonial through commercial periods; Native American artifacts; full-size working replica vessel from Revolutionary War, 54-ft. square-rigged Philadelphia II; replica 1862 class sailing canal boat. Historic Buildings: c.1818 stone schoolhouse; c.1920 Adirondack-style camp; Westport Winch House.
Major Exhibits: Star Spangled Nation: Celebrating the Bicentennial of the War of 1812 (T), 7/14-9/14.
Research Fields: nautical archaeology; historic shipwrecks of Lake Champlain; effects of non-native nuisance species, specifically the zebra mussel; history of Lake Champlain Basin.
Facilities: library; education center; conservation laboratory; nautical archaeology center; 12,000 sq. ft. exhibit space; visitor center. Gift items for sale.
Activities: guided & self-guided tours; maritime skills workshops; lectures; films; education programs for adults, children & Univ. of Vermont, Middlebury College and Texas A&M college students; docent program; participatory & traveling exhibits. Special Events: Blacksmith's "Hammer-In"; lake cruises; Native American Heritage Festival in June; Kids Pirate Festival in June; Small Boat Festival in July; Lake Champlain Challenge

Race in July; Gala Raffle in July; Rabble in Arms/Living History Weekend in August; Archaeology Month in September.

Publications: biannual newsletter, LCMM Compass; technical reports; books, Lake Studies: Meditations on Lake Champlain; Lake Champlain's Sailing Canal Boats.

Hours & Admission Prices: late May to mid-Oct. daily 10-5. Adults $10, senior citizens $9, students 5-17 $6; discounts to military veterans, AAM, AAA & VT museum & Gallery Alliance members; Council of American Maritime Museums, children under 5 & members no charge. &

Attendance: 14,800 (accurate)

Membership: Student $20; Single $30; Dual $40; Family $50; Family Plus 4 $60; Family Plus 8 $125; Family Plus $500; Friend of Museum $1,000.

Vernon

VERNON HISTORIANS, INC., Vernon Historical Museum, 567 Governor Hunt Rd., Vernon, VT 05354-9484. Tel.: 802-257-0292.

Founded: 1968.

Key Personnel: Pres., Dale Gassett.

Personnel Profile: Part-Time Volunteers 20.

Governing Authority: nonprofit organization. Branch Museum: Pond Road Chapel, 634 Pond Rd., Vernon, VT 05354. Tax-exempt: 501(c)(3).

Institution Type/Description: Historical Museum & Chapel.

Collections: books, bibles, postcards, photographs and diaries of early pioneers of the town; school desks; school-related material; early kitchen utensils; farm tools; costumes; manuscripts. Historic Buildings: 1860 Pond Road Chapel; 1848 Red Brick School House.

Research Fields: local history.

Facilities: library of Bibles and Vermont history books available by permission.

Activities: guided tours; lectures; films; drama; formally organized education programs; permanent exhibitions.

Publications: occasional newsletter.

Hours & Admission Prices: Museum: June-Sept. Sun. 2-4. Pond Road Chapel: by appointment. No charge; donations accepted.

Attendance: 200 (estimated)

Membership: Junior (under 18) $1; Adult $5; Sustaining $10; Institutional $25; Life $100.

Waitsfield

WAITSFIELD HISTORICAL SOCIETY AT GENERAL WAIT HOUSE, 4061 Main St., Waitsfield, VT 05673. Mailing Address: P.O. Box 816, Waitsfield, VT 05673-0816. Tel.: 802-496-2027.

Web Site: www.waitsfieldhistoricalsociety.com

Founded: 1970.

Key Personnel: Pres. & Treas., Lois De Heer; Vice Pres., Peter Laskowsky; Sec., Barbara Mansfield; Archivist, Judy Dodds.

Personnel Profile: Part-Time Volunteers 20.

Governing Authority: board. Parent Institution: Waitsfield Historical Society. Tax-exempt.

Institution Type/Description: Historic Site: original home of General Wait c.1793.

Collections: local history & culture; personal artifacts; period furnishings; photographs.

Research Fields: Vermont history.

Facilities: visitor's center; garden; restored parlor & hall.

Activities: barn tours; historic home walking tours; temporary exhibitions.

Publications: annual newsletter.

Hours & Admission Prices: Daily 9-5. No charge. &

Attendance: 1,500 (estimated)

Membership: Waitsfield Historical Society Membership $15, $25 & $50.

Waterbury

GREEN MOUNTAIN COFFEE VISITOR CENTER, 1 Rotarian Place, Waterbury, VT 05676-1582. Tel.: 877-879-2326.

Institution Type/Description: Visitor Center: housed in an 1867 Amtrak station.

Collections: company history; process of growing, roasting, & packaging coffee.

Facilities: cafe. Museum-related items for sale.

Activities: self-guided tours.

Hours & Admission Prices: Memorial Day to Labor Day daily 7-7; Winter: daily 7-6; groups by appointment. Suggested Donation: $1. Closed New Year's Day; Thanksgiving; Christmas.

Waterbury Center

GREEN MOUNTAIN CLUB, INC., 4711 Waterbury Stowe Rd., Waterbury Center, VT 05677-8325. Tel.: 802-244-7037. Fax: 802-244-5867.

E-mail: gmc@greenmountainclub.org

Web Site: greenmountainclub.org

Founded: 1910.

Key Personnel: Pres. (V), Richard Windish; Dir. Devel., Shawn Keeley, Jr.; Dir. Finance, Arthur Goldsweig.

Personnel Profile: Full-Time Paid 8; Part-Time Paid 3; Part-Time Volunteers 800; Interns 4.

Governing Authority: private; nonprofit organization. Tax-exempt: 501(c)(3).

Institution Type/Description: History Museum.

Collections: historical photographs, artifacts & documents relating to the Green Mountain Club (formed in 1910), Vermont's Long Trail hiking trail and hiking in Vermont & the northeast.

Facilities: 100-vol. library relating to hiking & outdoor activity information; 150-seat auditorium; hiking & nature trails. Gift items for sale.

Activities: lectures. Annual Events: meeting of members; James P. Taylor Series-educational & participation events; workshops; conferences.

Publications: quarterly newsletter, Long Trail News; guidebooks; maps; history books.

Hours & Admission Prices: Memorial Day-Oct. Mon.-Fri. 9-5, Sat.-Sun. 8-4; Nov.-May Mon.-Sat. 10-5. No charge; donations accepted. Closed federal holidays; Christmas Eve.

Attendance: 5,000 (estimated)

Membership: Limited Income $22; Adult $40; Family $50; Business & Corporation $150; Life $1,000.

Weathersfield

REVEREND DAN FOSTER HOUSE, MUSEUM OF THE WEATHERSFIELD HISTORICAL SOCIETY, 2656 Weathersfield Center Rd., Weathersfield, VT 05156. Mailing Address: P.O. Box 126, Perkinsville, VT 05151-0126. Tel.: 802-263-5230. Fax: 802-263-9263.

E-mail: ellen.clattenburg@dresden.us

Founded: 1951.

Key Personnel: Pres., Ginger Wimberg; Vice Pres., Karen McGee; Cur., Ellen F. Clattenburg.

Personnel Profile: Part-Time Volunteers 5.

Governing Authority: society. Parent Institution: Weathersfield Historical Society. Tax-exempt: 501(c)(3).

Institution Type/Description: Local History Museum.

Collections: 1787 barn replica; local artifacts: costumes; furniture; pictures; Civil War items; old cobbler's tools; early American farm tools; blacksmith shop; genealogic & photographic records; quilts; coverlets; musical instruments; post office equipment; photographs; ephemera.

Research Fields: local history; genealogy of Weathersfield inhabitants.

Facilities: 500-vol. library of historical books & Bibles with family records; catalogued collection of photographs.

Activities: guided tours; temporary exhibitions; various programs.

Publications: triannual newsletter; annual report; books, Weathersfield History Vols. I & II; The Weathersfield Burying Grounds; The Inhabitants of Weathersfield (reissued 1990); The Life of Consul William Jarvis; The Warren Family & Weathersfield, VT; three videos: A Lot of Water Over The Dam - the removal of lower Perkinsville in 1959 to make way for a flood control dam; The Weathersfield Meeting House Rebuilt - 1985-87 following a devastating fire in August of 1985; A Crash Course in the history of Weathersfield; Weathersfield Historical Society cookbook, Spider Bread, Cider Pie & Rhubarb Wine 2001.

Hours & Admission Prices: June-Oct. by appointment. No charge; donations accepted.

Attendance: 200 (estimated)

Membership: Individual $10; Family $15; Contributing $25; Sustaining $50; Life $150.

West Addison

DAR JOHN STRONG MANSION MUSEUM, 6656 VT Rte. 17 W., West Addison, VT 05491-8893. Tel.: 802-759-2309.

Web Site: www.northshirecomputer.com/VTDAR

Founded: 1934.

Key Personnel: Cur., Maureen Labenski.

Personnel Profile: Part-Time Volunteers 30.

Governing Authority: society. Parent Institution: Vermont Society, Daughters of the American Revolution. Tax-exempt.

Institution Type/Description: Historic House.

Collections: period furniture, 1790-1860; art; textiles; decorative artifacts; household implements.

Facilities: 30-vol. collection of genealogy books available for reference on premises.
Activities: educational events & speakers.
Publications: brochure.
Hours & Admission Prices: Memorial Day weekend to Labor Day weekend. Sat.-Sun. 10-5. Family $10, adults $5, seniors & students $3; discount to groups, AAM, ICOM & VMGA members; members no charge. &

Attendance: 500 (accurate)

West Halifax

HALIFAX HISTORICAL SOCIETY MUSEUM, 98 Branch Rd., West Halifax, VT 05358. Mailing Address: P.O. Box 94, West Halifax, VT 05358. Tel.: 802-368-7490.
Founded: 1978.
Key Personnel: Pres. (V), Arthur Copeland; Museum Shop Mgr., Doug Parkhurst.
Volunteer Hours: 100
Operating Expenses: 1,400
Operating Income: 2,500
Governing Authority: bd. of trustees.
Institution Type/Description: Historical Society Museum: housed in a former two-room schoolhouse.
Collections: local history & culture; period furnishings; clothing; quilts; photographs.
Activities: demonstrations.
Hours & Admission Prices: Summer: Sat. 2-4; other times by appointment. No charge.
Attendance: 40 (estimated)

West Marlboro

SOUTHERN VERMONT NATURAL HISTORY MUSEUM, Hogback Mt. Overlook, 7599 Vermont Rte. 9, West Marlboro, VT 05363. Tel.: 802-464-0048.
E-mail: museum@sover.net
Web Site: www.vermontmuseum.org
Founded: 1962.
Congressional District: 1
Key Personnel: Pres. (V) & Exec. Dir., Edward C. Metcalfe.
Personnel Profile: Full-Time Paid 2; Full-Time Volunteers 2; Part-Time Paid 2; Part-Time Volunteers 6; Interns 1.
Governing Authority: private; nonprofit organization. Tax-exempt: 501(c)(3).
Institution Type/Description: Natural History Museum.
Collections: natural history of the northeast U.S.
Facilities: aquarium; nature & conservation center. Museum-related items for sale.
Activities: hiking on 600 area preserve.
Publications: quarterly newsletter, Wildlife Notes.
Hours & Admission Prices: Memorial Day-Columbus Day daily 10-5; extended foliage season hours; Winter call for hours. Adults 13 & over $5, senior citizens $3, children 5-12 $2; children under 5, NEMA, UMGA, SPNHC, AAM & ICOM members no charge. Closed Thanksgiving; Christmas. &
Attendance: 15,000 (estimated)
Membership: Student $10; Individual $15; Family $25; Supporting $50; Sustaining $100; Benefactor $500.

West Rutland

CARVING STUDIO & SCULPTURE CENTER, 636 Marble St., West Rutland, VT 05777. Mailing Address: P.O. Box 495, West Rutland, VT 05777-0495. Tel.: 802-438-2097. Fax: 802-438-2020.
E-mail: info@carvingstudio.org
Web Site: www.carvingstudio.org
Institution Type/Description: Sculpture Center.
Collections: sculptures.
Facilities: sculpture garden.
Activities: workshops; temporary & permanent exhibitions; special events.
Hours & Admission Prices: Call for hours.

Westminster

WESTMINSTER HISTORICAL SOCIETY, Main St., Westminster, VT 05158. Mailing Address: P.O. Box 2, Westminster, VT 05158-0002.
Web Site: www.westminsterVThistory.org
Founded: 1966.
Key Personnel: C.E.O. & Pres. (V), Virginia Lisai; Vice Pres., Richard

Michelman; Treas., Linda Fawcett; Sec., Barbara Greenoe; Dir., Ruth Grandy; Dir., Bob Haas; Dir., Pat Haas; Dir., Karen Walters; Dir., Karen Larsen.
Personnel Profile: Part-Time Volunteers 20.
Governing Authority: nonprofit. Tax-exempt.
Institution Type/Description: Historical Society Museum: housed in c.1890 Old Town Hall.
Collections: history.
Research Fields: genealogy; history of the Abenaque Machine Works, 1893-1930.
Activities: permanent exhibitions.
Publications: book, Abenaque Machine Works; Vignettes of Westminster, Vermont; Around Bellows Falls; Westminster Vermont 1735-2000 Township Number One.
Hours & Admission Prices: July-Sept. Sun. 2-4. No charge; donations accepted.
Attendance: 100 (estimated)
Membership: Individual $10; Couple $20; Business $50; Life $200.

Weston

FARRAR-MANSUR HOUSE & OLD MILL MUSEUM, Main St., Weston, VT 05161. Mailing Address: P.O. Box 247, Weston, VT 05161-0247. Tel.: 802-824-5294. Fax: 802-824-5294, ext. 51.
E-mail: morlind@comcast.net
Founded: 1933.
Key Personnel: Pres. (V) Weston Historical Society, Robert Brandt; Dir., Jean Lindman.
Personnel Profile: Part-Time Paid 1; Part-Time Volunteers 7.
Volunteer Hours: 120
Governing Authority: society. Parent Institution: The Weston Community Association. Affiliated with The Weston Historical Society. Tax-exempt.
Institution Type/Description: Historic House Museum.
Collections: local period artifacts; murals; dolls; portraits; family bibles; school & church records; 19th-century house furnishings; period mill tools; 19th-century band wagon, uniforms & instruments; 18th-19th century New England Village life; guns; pianos; melodeons; loom; spinning wheels; niddy noddies; saltware pottery. Historic Buildings: 1785 grist mill; 1797 house.
Major Exhibits: Historic Costume, 7/13-10/14.
Facilities: library of local history & genealogy books available for use on premises. Museum-related items for sale.
Activities: guided tours; permanent & temporary exhibitions; school groups; programs; architectural & historical walking tours of town. Museum Sponsors: Open Houses; Hearth Cooking Candlelight Dinners.
Publications: newsletter.
Hours & Admission Prices: Call for hours. No charge; donations accepted.
Attendance: 2,000 (estimated)
Membership: Individual $10; Family $25; Supporting $50; Sustaining $50; Associate $75; Friends of Farrar-Mansur $100 & up.

White River Junction

THE MAIN STREET MUSEUM, 58 Bridge St., White River Junction, VT 05001-7040. Tel.: 802-356-2776.
Web Site: www.mainstreetmuseum.org/wiki
Founded: 1992.
Key Personnel: Dir., David Fairbanks Ford; Pres. (V), Bunny Harvey; Museum Shop Mgr., Christopher W. Comperry.
Personnel Profile: Full-Time Volunteers 1; Part-Time Volunteers 4; Interns 5.
Governing Authority: nonprofit organization. Tax-exempt: 501(c)(3).
Institution Type/Description: General Museum.
Collections: local history & culture; photographs; sculptures; paintings; natural history.
Research Fields: flora; fauna; rocks; generations.
Activities: music; movies; films; lectures.
Publications: The Electric Organ.
Hours & Admission Prices: Thurs.-Sun. 1-6. Suggested Donation: $3-$5; discounts to groups; members no charge. &
Attendance: 5,000 (accurate)
Membership: Individual $35.

NEW ENGLAND TRANSPORTATION INSTITUTE AND MUSEUM, 100 Railroad Row, White River Junction, VT 05001-7042. Mailing Address: P.O. Box 4211, White River Junction, VT 05001-4211. Tel.: 802-291-9838.
E-mail: netim@myfairpoint.net
Founded: 2001.
Key Personnel: Dir., Philip Rentz; Chm. (V), Eugene Vigneault
Institution Type/Description: Transportation Museum.

Collections: transportation history; railroad, river, & air transportation; equipment; period artifacts; photographs.
Activities: educational programs.
Hours & Admission Prices: By appointment.

Williamstown

WEATHERED BARN DOLL MUSEUM, 452 George Rd., Williamstown, VT 05679-9403. Tel.: 802-433-6077.
Institution Type/Description: Doll Museum.
Collections: over 5,000 dolls.
Hours & Admission Prices: May-Dec. call for hours.

Windsor

AMERICAN PRECISION MUSEUM, INC., (M), 196 Main St., Windsor, VT 05089-1312. Mailing Address: P.O. Box 679, Windsor, VT 05089-0679. Tel.: 802-674-5781. Fax: 802-674-2524.
E-mail: info@americanprecision.org
Web Site: www.americanprecision.org
Founded: 1966.
Congressional District: 1
Key Personnel: C.E.O., Ann Lawless; Chm. (V), Gilbert Whittemore.
Personnel Profile: Full-Time Paid 4; Part-Time Paid 3; Part-Time Volunteers 15; Interns 5.
Governing Authority: nonprofit organization. Tax-exempt: 501(c)(3).
Institution Type/Description: Industrial History Museum: housed in 1846 Robbins & Lawrence Armory. National Historic Landmark.
Collections: machine tools; gun-making machines; wood-working machines; metal-working machinery; hand tools; firearms; models; typewriters.
Activities: permanent & temporary exhibitions; special programs; traveling education kits for schools.
Publications: newsletter, Tools & Technology.
Hours & Admission Prices: Memorial Day-Oct. daily 10-5. Family $18, adults $6, students $4; members & Sun. no charge. New England Museum Association reciprocal admissions program. &
Attendance: 4,597 (accurate)
Membership: Individual $35; Dual & Family $55; Associate $100; Patron $250; Steward $500; Benefactor $1,000.

OLD CONSTITUTION HOUSE - STATE HISTORIC SITE, 16 N. Main St., Windsor, VT 05089-1307. Mailing Address: Historic Preservation, National Life Bldg., 6th Fl., Montpelier, VT 05620. Tel.: 802-672-3773. Fax: 802-828-3206.
E-mail: john.dumville@state.vt.us
Web Site: www.historicsites.vermont.gov/constitution
Founded: 1961.
Key Personnel: Historic Sites Operations Chief, John P. Dumville.
Personnel Profile: Full-Time Paid 1; Part-Time Paid 1.
Governing Authority: state. Parent Institution: State of Vermont, Division for Historic Preservation, One National Life Dr., 6th Fl., Montpelier, VT 05602. Tax-exempt.
Institution Type/Description: Historic House Museum: 1777 Old Constitution House.
Collections: Colonial & Civil War periods; pottery; furniture.
Research Fields: Republic of Vermont (1777-1791), Vermont constitution.
Activities: permanent exhibitions.
Publications: brochure; guide.
Hours & Admission Prices: mid-May to mid-Oct. daily 10-5. Adults $2; discounts to AAM members; registered school groups no charge. &
Attendance: 2,000 (estimated)

Winooski

HERITAGE WINOOSKI MILL MUSEUM, Champlain Mill on Winooski Falls Way, Winooski, VT 05404. Mailing Address: Heritage Winooski, Box 181, Saint Michael's College, Winooski Park, Colchester, VT 05439. Tel.: 802-985-2431.
E-mail: millmuseum@yahoo.com
Founded: 1998.
Congressional District: 1
Key Personnel: Dir. & C.E.O., Laura Krawitt; Pres., John Warshow
Governing Authority: Parent Institution: Saint Michael's College. Tax-exempt.
Institution Type/Description: History Museum: mill listed on the National Historic Register.
Collections: mill history & construction; machinery; tools; period artifacts; waterpower technology; photographs 1910-1940s.
Research Fields: Winooski Falls mill economy; textile manufacture; mill owners; Yankee & immigrant workforce; working & living conditions; waterpower technology; environmental impact.

Activities: educational programs; self-guided school tours; internship program; teachers' workshops; lectures.
Publications: semi-annual newsletter, Mill Run; The Mills at Winooski Falls, Illustrated Essays & Oral Histories; interdisciplinary curriculum guide, Clickity Clack: Wool & Waterpower.
Hours & Admission Prices: Mon.-Fri. 9-5, Sat. 10-1. Closed holidays. No charge. &
Attendance: 600 (estimated)

MCCARTHY GALLERY, McCarthy Arts Center, St. Michael's College, Winooski, VT 05404. Tel.: 802-654-2246.
Institution Type/Description: Art Gallery.
Collections: paintings; photographs; sculpture.
Hours & Admission Prices: Mon.-Fri. 3-5 & 7:30-9:30, Sat.-Sun. 1-5.

Woodstock

BILLINGS FARM & MUSEUM, (M), 53 Elm St., Woodstock, VT 05091. Mailing Address: P.O. Box 489, Woodstock, VT 05091-0489. Tel.: 802-457-2355. Fax: 802-457-4663.
E-mail: info@billingsfarm.org
Web Site: www.billingsfarm.org
Founded: 1976.
Congressional District: 1
Key Personnel: Pres., David A. Donath; Vice Pres., Darlyne S. Franzen; Farm Mgr., Jason Johnson; Admin. Officer, Marian E. Koetsier; Mgr. Collections, Boden Harris; Facilities Mgr., David V. Ferrero; Coord. Education & Interpretation, Megan Campbell; Public Rels. & Events Coord., Susan Plump; Sec., Marjorie Wakefield; Farm Worker, Michael Birkett; Farm Worker, Charlie Ferrero; Farm Worker, Ashley Koetsier; Farm Worker, Paul LeBlanc.
Personnel Profile: Full-Time Paid 12; Part-Time Paid 35; Part-Time Volunteers 25.
Governing Authority: nonprofit organization. Parent Institution: Woodstock Foundation, Inc. Tax-exempt: 501(c)(3).
Institution Type/Description: History Museum & 320 acre dairy farm: 1890 restored & furnished farm house, creamery & ice house.
Collections: 16,000 objects relating to agriculture & folklife in east-central Vermont during the late 19th century. Historic Buildings: 1890 farm house; 19th-century farm barns.
Major Exhibits: 28th Annual Quilt Exhibition, 8/14-9/21/14.
Research Fields: agriculture & rural life in east-central Vermont during the late 19th century; The roles of George Perkins Marsh, Frederick Billings, and Laurance S. Rockefeller as conservationists.
Facilities: 7,500-vol. library including archives, photos & microforms for research use; visitor center; classroom; 100-seat theatre; 230-acre dairy farm. Local crafts, books & products for sale.
Activities: special events; year-round educational program; volunteer program; special activities; temporary exhibitions; daily livestock & agricultural programs.
Publications: education brochures; visitor guides; book, The Vermont Farm Year; annual report.
Hours & Admission Prices: May-Oct. daily 10-5; Nov.-Feb. Sat.-Sun. 10-3:30, call for additional hours. Adults $12, seniors $11, children 5-15 $6, children 3-4 $3; discounts to AAM & ICOM members; members & children 2 & under no charge. &
Attendance: 53,045 (accurate)
Membership: Individual $30; Family $50; Friend $100; Supporter $175; Subscriber $250; Sustaining $500; Benefactor $1,000.

WOODSTOCK HISTORICAL SOCIETY, INC., 26 Elm St., Woodstock, VT 05091-1024. Tel.: 802-457-1822. Fax: 802-457-2811.
E-mail: info@woodstockhistorical.org
Web Site: www.woodstockhistorical.org
Founded: 1943.
Key Personnel: Dir., Jack Anderson; Chm., Chuck Wise; Guide Admin., Gina Moore; Coord. Education, Jennie Shurtleff.
Personnel Profile: Full-Time Paid 1; Part-Time Paid 10; Part-Time Volunteers 4.
Governing Authority: board of trustees; nonprofit organization. Tax-exempt: 501(c)(3).
Institution Type/Description: History Museum: housed in 1807 Federal style house with brick gables, designed by Nathaniel Smith.
Collections: portraits; John Taylor Arms etchings; 1740-1900 furniture; toys; dolls; doll houses; silver & glass; costumes; artisan & craft tools; winter sports equipment; Woodstock's Royal charter; manuscripts, photographs, & business records relating to Woodstock & environs. Historic Building: 1807 Dana House.

Research Fields: Woodstock history & genealogy.
Facilities: 800-vol. research library of local history, biography of local residents, local imprints, genealogical reference, books & papers of the Dana family; 50-seat auditorium.
Activities: lectures; temporary & permanent exhibitions; education program for children.
Publications: Walking Guide to the Village of Woodstock; The Long Light of Those Days: Recollections of a Vermont Village at Mid-century; The Hills Were Full of Dairy Farmers: Memories of Farming in the Region of Woodstock, VT; Woodstock's Heritage - A Brief History of the Shire Town.
Hours & Admission Prices: Office: Mon.-Fri. 9-5. Library: Wed.-Thurs. & Sat. 10-3. Dana House: call for hours. Adults $5; discounts to VMGA, AAA, AAM & NEMA members; members and youth 16 & under no charge. &
Attendance: 4,450 (accurate)
Membership: Student & Senior $15; Individual $25; Family $35; Contributing $75; Subscribing $150; Sustaining $250; Patron $500.

VIRGINIA

(393 listings)

Abingdon

HISTORICAL SOCIETY OF WASHINGTON COUNTY, VIR-GINIA, 306 Depot Square, Abingdon, VA 24210-3102. Mailing Address: P.O. Box 484, Abingdon, VA 24212-0484. Tel.: 276-623-8337.
E-mail: office@hswcv.org
Web Site: hswcv.org
Founded: 1935.
Congressional District: 9
Key Personnel: Pres. & Bulletin Editor, Eleanor Grasselli; Treas., Mike Shaffer; Recording Sec., Doris Wells; Corresponding Sec., Ina Stephenson-Marbury; Membership, Riley Clark; Local History, Joella Barbour; Newsletter Editor, Greg McMillian; Database Mgr., Jack Niemann; Photo Digitization, Jane Oakes; Library Mgr., Melissa Watson.
Personnel Profile: Full-Time Paid 1; Part-Time Volunteers 30; Interns 1.
Governing Authority: Tax-exempt.
Institution Type/Description: Historical Society Library.
Collections: local history pertaining to Washington County & Abingdon; genealogy files; photographs.
Research Fields: genealogy; local history.
Facilities: library; archives.
Activities: educational programs. Museum Sponsors: Virginia Highlands Festival activities; tour of local homes in December.
Publications: newsletters; HSWC Bulletin.
Hours & Admission Prices: Jan. 15-March & Dec. 1-Dec. 15 Mon.-Fri. 10-4; April-Nov. Mon.-Fri. 10-4 & 1st and 3rd Sat. 11-4. No charge; donations accepted.
Attendance: 2,500 (estimated)

✳ **WILLIAM KING MUSEUM CENTER FOR ART AND CULTURAL HERITAGE, (M),** 415 Academy Dr., Abingdon, VA 24210-2617. Mailing Address: P.O. Box 2256, Abingdon, VA 24212-2256. Tel.: 276-628-5005. Fax: 276-628-3922.
E-mail: mmiller@wkmuseum.org
Web Site: www.williamkingmuseum.org
Formerly: William King Regional Arts Center
Founded: 1979.
Congressional District: 9
Key Personnel: Exec. Dir., Marcy Miller; Pres. Bd. (V), Evelyn Goldston; Vice Pres., Joe Lyle; Treas., John Jeter; Treas, Doris Shuman; Sec., Pam Kramer; Cur., Leila Cartier.
Personnel Profile: Full-Time Paid 7; Part-Time Paid 15; Part-Time Volunteers 10; Interns 2.
Governing Authority: private; nonprofit organization. Tax-exempt: 501(c)(3).
Institution Type/Description: Art Museum: housed in 1913 school building.
Collections: decorative arts.
Major Exhibits: From These Hills, 10/18/13-2/17/14; Heroes and Villains: The Comic Art Collection of Shelton Drum, 1/17/14-6/29/14; There/Here: Architectural Projects, 2/7/14-6/29/14; Artist by Trade, 3/7/14-8/24/14; Civil War in Virginia (T), 8/9/14-2/1/15.
Research Fields: cultural heritage of Southwest Virginia.
Facilities: classrooms; 7,500 sq. ft. exhibit space. Museum-related items for sale.
Activities: guided tours; lectures; exhibition receptions; community family days; facility rental; adult & children's educational programs; resident studio artists.
Publications: gallery guides; exhibition catalogs & posters; reception announcement cards; annual reports; program calendars; newsletter.

Hours & Admission Prices: Tues.-Wed. & Fri. 10-5, Thurs. 10-9, Sat.-Sun. 1-5. Adults $5; members no charge. Closed New Year's Eve & Day; Easter; Memorial Day; Independence Day; Labor Day; Thanksgiving; Christmas Eve & Day. &
Attendance: 20,000 (accurate)
Membership: Individual $40; Family $65; Dali $250 & up; Matisse $750 & up; Picasso $1,500 & up; Monet $2,500 & up; Rembrandt $5,000 & up.

Alexandria

✳ **ALEXANDRIA ARCHAEOLOGY MUSEUM, (M),** 105 N. Union St., # 327, Alexandria, VA 22314-3217. Tel.: 709-746-4399. Fax: 703-838-6491.
E-mail: archaeology@alexandriava.gov
Web Site: www.alexandriaarchaeology.org
Founded: 1977.
Congressional District: 8
Key Personnel: Preservation Archaeologist, Francine Bromberg; Education Coord., Ruth Reeder; Admin. Support, Jennifer Barker.
Personnel Profile: Full-Time Paid 2; Part-Time Paid 2; Part-Time Volunteers 103.
Governing Authority: municipal. Parent Institution: City of Alexandria. Subsidiary Institution: Office of Historic Alexandria. Tax-exempt.
Institution Type/Description: Archaeology Museum.
Collections: 18th- & 19th-century artifacts from archaeological digs in Alexandria; field notes; photographs; research files associated with collections; prehistoric stone tools.
Research Fields: urban archaeological research; historic neighborhoods; Alexandria waterfront local history.
Facilities: 300-vol. library pertaining to local history, archaeology & antiques; research laboratory.
Activities: guided tours; lectures; interpreter program; organized education program for undergraduate or graduate college students affiliated with George Washington University; temporary exhibitions; volunteer program; museum & outreach programs for elementary school children; family programs in the museum & on site; Alexandria Archaeology Institute for adults.
Publications: brochure has 43 listings including, Geographical Methods in Urban Preservation Planning, Historical Methods in Urban Preservation Planning; The Volunteer in Alexandria Archaeology; A Field Manual for Alexandria Archaeology; A Laboratory Manual for Alexandria Archaeology; A Guide to the Alexandria Archaeological Research Museum; the Potter's Art: Salt-glazed Stoneware of 19th-Century Alexandria; papers, Approaches to Preserving a City's Past, The Alexandria Waterfront Forum: Birth & ReBirth 1730-1983, Alexandria Antiquity 1984, Across the Fence, But a World Apart; Archaeologists at Work: A Teacher's Guide to Classroom Archaeology; catalogue, Artifacts, Advertisements & Archaeology; monthly newsletter, Alexandria Archaeology Volunteer News; Alexandria, Virginia; Walking with Washington; Walk and Bike the Alexandria Heritage Trail.
Hours & Admission Prices: Tues.-Fri. 10-3, Sat. 10-5, Sun. 1-5. No charge; donations accepted. &
Attendance: 35,798 (accurate)
Membership: Friends of Alexandria Archaeology: Individual $20; Family & Group $25; Sponsor $50; Benefactor $100; Corporate $500.

✳ **ALEXANDRIA BLACK HISTORY MUSEUM, (M),** 902 Wythe St., Alexandria, VA 22314-1839. Tel.: 703-838-4356. Fax: 703-706-3999.
E-mail: blackhistory@alexandriava.gov
Web Site: www.alexblackhistory.org
Formerly: Alexandria Black History Resource Center
Founded: 1983.
Congressional District: 8
Key Personnel: Dir., Louis Hicks; Asst. Dir & Cur., Audrey P. Davis; Cur., Lillian Patterson; Sec., Jewel Plummer.
Personnel Profile: Full-Time Paid 2; Full-Time Volunteers 2; Part-Time Paid 2; Part-Time Volunteers 50; Interns 6.
Governing Authority: municipal. Parent Institution: Office of Historic Alexandria. Tax-exempt: 501(c)(3).
Institution Type/Description: History Museum: located in former public library.
Collections: photographs; documents.
Research Fields: local African-American history.
Activities: lectures; slide presentations; workshops; walking tours.
Publications: book, Black Word Find.
Hours & Admission Prices: Tues.-Sat. 10-4. Admission $2. Closed New Year's Day; Martin Luther King Jr. Day; Easter; Independence Day; Thanksgiving; Christmas. &
Attendance: 10,300 (estimated)

Membership: Alexandria Society for the Preservation of Black Heritage $20.

ALEXANDRIA LIBRARY - LOCAL HISTORY & SPECIAL COLLECTIONS, 717 Queen St., Alexandria, VA 22314-2420. Tel.: 703-838-4577. Fax: 703-706-3912.
E-mail: gkcombs@alexandria.lib.va.us
Web Site: www.alexandria.lib.va.us/branches/lhsc.html
Founded: 1976.
Congressional District: 8
Personnel Profile: Full-Time Paid 2; Part-Time Paid 1.
Governing Authority: Tax-exempt.
Institution Type/Description: Library.
Collections: local history & culture; photographs; genealogy; books; microfilm; manuscripts.
Research Fields: local history; genealogy.
Facilities: library.
Activities: research; special events.
Hours & Admission Prices: Mon. 1-9, Tues. 2-7, Wed. & Fri. 10-7, 1st Sat. of month 10-5. No charge.
Attendance: 14,865 (accurate)

***   CARLYLE HOUSE HISTORIC PARK, (M),** 121 N. Fairfax St., Alexandria, VA 22314-3229. Tel.: 703-549-2997. Fax: 703-549-5738.
E-mail: carlyle@NVRPA.org
Web Site: www.carlylehouse.org
Founded: 1976.
Congressional District: 8
Key Personnel: Dir., Jim Bartlinski; Cur. Education, Heather Dunn; Cur., Sarah Coster.
Personnel Profile: Full-Time Paid 3; Part-Time Paid 14; Part-Time Volunteers 95.
Governing Authority: nonprofit organization. Operated by the Northern Virginia Regional Park Authority, 5400 Ox Rd., Fairfax Station, VA 22039. Tax-exempt: 501(c)(3).
Institution Type/Description: Historic House: 1753 Carlyle House, Georgian Palladian style, used by General Braddock as headquarters for planning early campaigns of French & Indian War.
Collections: 18th-century decorative arts; archaeological collection.
Research Fields: local history; archaeology; furnishings.
Facilities: classroom.
Activities: guided tours; formally organized education programs; docent program; special events.
Publications: brochures; book, Who Built Alexandria Architects in Alexandria, 1750-1900; Colo. John Carlyle, Gent. 1720-1780.
Hours & Admission Prices: Tues.-Sat. 10-4, Sun. 12-4. Adults $5, children 11-17 $3; discount to AAM members; members & children 10 & under no charge. Closed New Year's Day; Thanksgiving; Christmas Eve & Day. &
Attendance: 22,000 (accurate)
Membership: Member $25; Contributor $50; Patron $100; Benefactor $500; Braddock Society $1,000.

COLLINGWOOD LIBRARY AND MUSEUM ON AMERICANISM, 8301 E. Boulevard Dr., Alexandria, VA 22308-1399. Tel.: 703-765-1652. Fax: 703-765-8213.
E-mail: pfrank@collingwoodlibrary.org
Web Site: collingwoodlibrary.org
Founded: 1977.
Congressional District: 8
Key Personnel: Pres. (V), Kent S. Webber; Exec. Dir., Paul A. Frank; Treas., Bill Williamson; Sec., John Mayers.
Personnel Profile: Full-Time Paid 1; Part-Time Paid 2.
Governing Authority: nonprofit organization. Tax-exempt: 501(c)(3).
Institution Type/Description: American History Museum: housed in 1785 structure used as overseer's house for George Washington's River Farm.
Collections: Indian artifacts; coins, china; state & colonial flags; Revolutionary War items; militaria; replicas of historic documents & artifacts.
Research Fields: local history; military; Masonic.
Facilities: 7,000-vol. library pertaining to American history for public use; 19,000 microfiche.
Activities: guided tours; lectures; loan exhibitions.
Publications: quarterly newsletter.
Hours & Admission Prices: Mon. & Wed.-Sat. 10-4, Sun. 1-4. No charge; donations accepted. &
Attendance: 9,000 (estimated)

***   FORT WARD MUSEUM AND HISTORIC SITE, (M),** 4301 W. Braddock Rd., Alexandria, VA 22304-1007. Tel.: 703-838-4848. Fax: 703-671-7350.
E-mail: fort.ward@alexandriava.gov
Web Site: oha.alexandriava.gov/fortward/
Founded: 1964.
Congressional District: 8
Key Personnel: Dir., Susan G. Cumbey; Cur., Walton H. Owen.
Personnel Profile: Full-Time Paid 2; Part-Time Paid 3; Part-Time Volunteers 15.
Governing Authority: municipal. Parent Institution: City of Alexandria. Subsidiary Institution: Office of Historic Alexandria. Tax-exempt: 501(c)(3).
Institution Type/Description: Military Museum: located on the site of 1861-65 Fort Ward, built to protect Washington DC during the Civil War.
Collections: Civil War artifacts; library collection; historic site: preserved Union Fort featuring restored Northwest Bastion.
Research Fields: The Civil War; mid-19th-century American military, local & social history.
Facilities: 2,000-vol. library on Civil War available for research on premises; reading room; park setting. Museum-related items for sale.
Activities: guided tours; lectures; permanent & temporary exhibitions; outreach program: Life During the Civil War; interpretive programs.
Publications: newsletter; site & exhibition brochures.
Hours & Admission Prices: Museum: Tues.-Sat. 10-5, Sun. 12-5. Park: daily 9-sunset. Office: Tues.-Sat. 9-5. No charge; donations accepted. Closed New Year's Day; Thanksgiving; Christmas. &
Attendance: 34,000 (estimated)
Membership: Individual $10; Organization $25; Supporting $35.

FRANK LLOYD WRIGHT'S POPE-LEIGHEY HOUSE, 9000 Richmond Hwy., Alexandria, VA 22309. Mailing Address: P.O. Box 15097, Alexandria, VA 22309-0097. Tel.: 703-780-4000. Fax: 703-780-8509.
E-mail: woodlawn@savingplaces.org
Web Site: www.woodlawnpopeleighey.org
Founded: 1964.
Congressional District: 10
Key Personnel: Coun. Chm., Peter Christensen; Exec. Dir., John Riley; Visitor Svcs. Mgr., Meredith Mitchell.
Personnel Profile: Full-Time Paid 3; Part-Time Paid 45; Part-Time Volunteers 50.
Governing Authority: nonprofit. Property of the National Trust for Historic Preservation, 1785 Massachusetts Ave., N.W., Washington, DC 20036. Tax-exempt: 501(c)(3).
Institution Type/Description: Historic House: 1940 Frank Lloyd Wright Usonian house, located on the grounds of Woodlawn.
Collections: Frank Lloyd Wright furniture & furnishings.
Research Fields: works of Frank Lloyd Wright.
Facilities: 1,250 sq. ft. exhibit space; nature trails; wildflower landscape. Books & museum related items for sale.
Activities: guided tours; lectures; formally organized educational programs; Tech Tours; book club tours.
Hours & Admission Prices: March-Nov. Fri.-Mon. 12-4. Adults $10, seniors & active-duty military $8, students K-12 $5.
Attendance: 42,000
Membership: Friend $30; Family $40; Contributing $50; Sustaining $100; Supporting $500; Pope-Leighey Forum $1,000 & up.

***   FRIENDSHIP FIREHOUSE, (M),** 107 S. Alfred St., Alexandria, VA 22314-3001. Tel.: 703-746-3891 & 4994. Fax: 703-838-4997.
E-mail: friendship@alexandriava.gov
Web Site: www.friendshipfirehouse.org
Founded: 1993.
Congressional District: 8
Personnel Profile: Part-Time Paid 1.
Governing Authority: municipal; nonprofit. Parent Institution: City of Alexandria, VA. Subsidiary Institution: Office of Historic Alexandria. Tax-exempt.
Institution Type/Description: Firehouse Museum: housed in c.1855 Italianate-style brick building which has a first-floor engine room for storing apparatus & a second-floor meeting room for social & ceremonial activities.
Collections: history of Friendship Fire Company & local firefighting history; firefighting apparatus, equipment, & uniforms; letters; photographs; period furnishings.
Research Fields: history of Friendship Fire Company & other local firefighting history.
Facilities: 500 sq. ft. exhibit space.
Activities: Museum Sponsors: Birthday Party/Public Safety Festival in August.

Hours & Admission Prices: Sat.-Sun. 1-4. Adults $2; discounts to AAM & ICOM members. Closed New Year's Day; Thanksgiving; Christmas. &

Attendance: 3,935 (accurate)

Membership: Friendship Veterans Fire Engine Association $30.

＊ GADSBY'S TAVERN MUSEUM, (M), 134 N. Royal St., Alexandria, VA 22314-3226. Tel.: 703-746-4242. Fax: 703-838-4270.

E-mail: gadsbys.tavern@alexandriava.gov

Web Site: gadsbystavern.org

Founded: 1976.

Congressional District: 8

Key Personnel: Dir., Gretchen M. Bulova; Asst. Dir., Lizabeth Williams; Cur. Collections, Callie Stapp; Cur. Education, Michele Longo; Museum Shop Mgr., Sue Walker.

Personnel Profile: Full-Time Paid 2; Part-Time Paid 14; Part-Time Volunteers 150; Interns 1.

Governing Authority: municipal. Parent Institution: City of Alexandria, VA. Subsidiary Institution: Office of Historic Alexandria. Tax-exempt: 501(c)(3).

Institution Type/Description: Historic Buildings: c.1785 tavern; 1792 Federal style City Hotel.

Collections: 18th- & early 19th-century furnishings; decorative arts.

Research Fields: late 18th- & early 19th-century furnishings & decorative arts; taverns & travel accommodations, 1770-1810.

Facilities: library; research room; restaurant.

Activities: guided tours; seminars; workshops; concerts; special exhibits & events; private rentals available.

Publications: brochure; history pamphlet; interpretive bulletins; Furnishings Plan Book.

Hours & Admission Prices: April-Oct. Sun.-Mon. 1-5, Tues.-Sat. 10-5; Nov.-March Wed.-Sat. 11-4, Sun. 1-4. Adults $5, children 5-12 $3; discounts to groups, and AAM, VAM, HHMC & ICOM members; children under 5 no charge. Closed major holidays.

Attendance: 25,000 (accurate)

Membership: Gadsby's Tavern Museum Society: Student $15; Individual $25; Family $35; Merchant $50; Patron $100; Lifetime $1,000.

GEORGE WASHINGTON MASONIC NATIONAL MEMORIAL, 101 Callahan Dr., Alexandria, VA 22301-2751. Tel.: 703-683-2007. Fax: 703-519-9270.

E-mail: gseghers@gwmemorial.org

Web Site: www.gwmemorial.org

Formerly: George Washington Masonic Memorial

Founded: 1910.

Congressional District: 10

Key Personnel: Exec. Dir., George D. Seghers; Dir. Collections, Mark A. Tabbert; Dir. Communications, Shawn E. Eyer; Special Events Administration, Radka Mavrova; Chm. (V), Donald G. Hicks, Jr.

Personnel Profile: Full-Time Paid 8; Part-Time Paid 26; Part-Time Volunteers 1; Interns 1.

Governing Authority: nonprofit organization. Tax-exempt: 501(c)(3).

Institution Type/Description: History Museum.

Collections: 18th- & early 19th-century George Washington related artifacts; furniture; paintings; books; ceramics; glass; prints & manuscripts; Masonic regalia.

Research Fields: George Washington & his family; Masonic affiliations of Washington & his associates.

Facilities: 20,000-vol library; archive collection; auditorium; observation deck; 36-acre site including Fort Ellsworth.

Activities: guided tours; permanent & temporary exhibitions.

Publications: newsletter; annual report.

Hours & Admission Prices: Daily 10-4. General Admission $5. Guided Tower Tours $8. Closed New Year's Day; Veterans Day; Memorial Day; Independence Day; Labor Day; Thanksgiving; Christmas. &

Attendance: 89,357 (accurate)

Membership: Silver Craftsman $100; Gold Master $250; Platinum Presidential $500; 21st Century $1,000; Millennium Architect $5,000; Millennium Master Architect $10,000; Millennium Builder $25,000; Millennium Master Builder $50,000; Millennium Grand Master Builder $100,000.

JEROME "BUDDIE" FORD NATURE CENTER, 5750 Sanger Ave., Alexandria, VA 22311-5602. Tel.: 703-838-4829; 746-5559.

E-mail: mark.kelly@alexandriava.gov

Web Site: alexandriava.gov/recreation/info/default.aspx?id=12362

Founded: 1979.

Key Personnel: Dir., Mark S. Kelly.

Personnel Profile: Full-Time Paid 2; Part-Time Paid 3; Part-Time Volunteers 2; Interns 1.

Governing Authority: municipal. Parent Institution: City of Alexandria, Dept. of Recreation & Parks. Tax-exempt.

Institution Type/Description: Nature Center.

Collections: nature displays; aquariums; live animals; prehistoric Native Americans; rocks; minerals; fossils.

Research Fields: natural history.

Facilities: 250-vol. natural history library available for use on premises; nature & conservation center.

Activities: formally organized environmental education programs; interpretive tours of the park; lectures; films; nature day camp.

Publications: monthly calendar of events, Nature News; self-guiding trail brochure; fliers of special events.

Hours & Admission Prices: April-Nov. Wed.-Sat. 10-5, Sun. 1-5; Oct.-March Wed.-Sat. 10-5. No charge; donations accepted. Closed holidays. &

Attendance: 23,500 (accurate)

LEE-FENDALL HOUSE MUSEUM AND GARDEN, (M), 614 Oronoco St., Alexandria, VA 22314-2308. Tel.: 703-548-1789. Fax: 703-229-6350. Facebook: Lee Fendall House Museum & Garden.

E-mail: contact@leefendallhouse.org

Web Site: www.leefendallhouse.org

Founded: 1974.

Congressional District: 8

Key Personnel: Exec. Dir., Erin Adams.

Personnel Profile: Full-Time Paid 1; Part-Time Volunteers 12; Interns 3.

Governing Authority: nonprofit organization. Parent Institution: Virginia Trust for Historic Preservation. Tax-exempt: 501(c)(3).

Institution Type/Description: Historic House: 1785 Lee-Fendall House, Lee family home & former home of labor leader John L. Lewis, located in Alexandria's Old Town Historic District.

Collections: Lee family heirlooms; 18th- & 19th-century decorative arts.

Research Fields: Lee family history; Alexandria history; John L. Lewis; social history of the Victorian period.

Facilities: garden.

Activities: guided tours; education program for children; docent program; intern program; available to rent for meetings or parties; temporary & permanent exhibitions.

Publications: brochure; newsletters.

Hours & Admission Prices: Feb. to mid-Dec. Wed.-Sat. 10-3, Sun. 1-3. Adults $5, children 5-17 $3; discounts to NTHP, AAM & ICOM members; children 4 & under no charge. Closed major holidays; private events.

Attendance: 9,500 (accurate)

Membership: Individual $35; Family $60; Contributor $100; Supporter $200; Patron $225 & up.

＊ THE LYCEUM, ALEXANDRIA'S HISTORY MUSEUM, (M), 201 S. Washington St., Alexandria, VA 22314-3697. Tel.: 703-746-4994. Fax: 703-838-4997.

E-mail: lyceum@alexandriava.gov

Web Site: www.alexandriahistory.org

Founded: 1974.

Congressional District: 8

Key Personnel: Dir., James C. Mackay; Asst. Dir., Kristin B. Lloyd; Facilities Coord., Bob Schurk; Visitor Svcs., Pamela Budde.

Personnel Profile: Full-Time Paid 3; Part-Time Paid 10; Part-Time Volunteers 40.

Governing Authority: municipal. Parent Institution: City of Alexandria, VA. Administered by the Office of Historic Alexandria. Tax-exempt: 501(c)(3).

Institution Type/Description: History Museum: housed in 1839 The Lyceum.

Collections: 18th-, 19th- & 20th-century artifacts relating to Alexandria & northern Virginia.

Research Fields: Alexandria history; Lyceum history; Northern Virginia history; Benjamin Hallowell.

Facilities: lecture hall; meeting room. Museum-related items for sale.

Activities: education programs for elementary students; older adult tour programs; special events; changing exhibits; docent programs; concerts; summer group programs; history camp.

Publications: Guide to Historic Alexandria; 3 Centuries of Alexandria Silver; Changing Perceptions: Charting Alexandria, 1590-1999.

Hours & Admission Prices: Mon.-Sat. 10-5, Sun. 1-5. Adults $2; discounts to AAM & ICOM members. Closed New Year's Day; Thanksgiving; Christmas Eve & Day. &

Attendance: 23,986 (accurate)

Membership: The Lyceum Company: Individual $20; Family & Couple $30; Patron $100; Sponsor $500; Benefactor $1,000.

NATIONAL INVENTORS HALL OF FAME, 600 Dulany St. Madison W., Alexandria, VA 22314. Mailing Address: 3701 Highland Park, N.W., North Canton, OH 44720. Tel.: 571-272-0095. Fax: 703-706-0484.
E-mail: museum@invent.org
Web Site: www.invent.org
Founded: 1973.
Key Personnel: Exec. Dir., Rini Paiva; Dir. Merchandising & Store Operations, Mitch Scott.
Personnel Profile: Full-Time Paid 60; Part-Time Paid 25; Interns 5.
Governing Authority: private; nonprofit. Parent Institution: Invent Now, Inc.
Institution Type/Description: Science & History Museum.
Collections: personal artifacts; hall of fame; science & history.
Research Fields: inventors.
Facilities: 5,000 sq. ft. exhibit space.
Activities: participatory & traveling exhibits. Annual Events: Induction Ceremony; Collegiate Inventors Competition; Camp Invention; Club Invention.
Hours & Admission Prices: Mon.-Fri. 9-5, Sat. 12-5. No charge. Closed federal holidays. &
Attendance: 70,000 (estimated)

THE NORTHERN VIRGINIA FINE ARTS ASSOCIATION AT THE ATHENAEUM, 201 Prince St., Alexandria, VA 22314-3313. Tel.: 703-548-0035. Fax: 703-548-0456.
E-mail: admin@nvfaa.org
Web Site: www.nvfaa.org
Founded: 1961.
Congressional District: 8
Key Personnel: Exec. Dir., Catherine Aselford; Pres., Twig Murray.
Personnel Profile: Full-Time Paid 1; Part-Time Paid 1; Part-Time Volunteers 68.
Governing Authority: nonprofit; a Virginia corp. Tax-exempt: 501(c)(3).
Institution Type/Description: Art Gallery: housed in 1851, restored Greek Revival Building.
Collections: paintings; photographs; prints; sculpture.
Research Fields: American artists and their impact on the times in which they work.
Facilities: lecture area; walled sculpture garden.
Activities: art exhibits; lectures; annual juried show; dance classes for children and adults; music, dance, & poetry performances.
Publications: newsletter; calendar; brochures pertaining to exhibits.
Hours & Admission Prices: Thurs.-Fri. & Sun. 12-4, Sat. 1-4. Lecture & performance fees discounted for members. Closed major holidays.
Attendance: 41,000 (estimated)
Membership: Student & Senior $25; Contributor $40; Sponsor $100; Patron $500; Benefactor $1,000.

✱ OFFICE OF HISTORIC ALEXANDRIA, (M), 220 N. Washington St., Alexandria, VA 22314-2521. Tel.: 703-746-4554. Fax: 703-838-6451.
E-mail: historicalalexandria@alexandriava.gov
Web Site: oha.alexandriava.gov/contactus
Founded: 1978.
Congressional District: 8
Key Personnel: Dir. Office of Historic Alexandria, J. Lance Mallamo; Dir. The Lyceum, Alexandria's History Museum & Friendship Firehouse, James Mackay, III; Dir. Gadsby's Tavern Museum & Stabler-Leadbeater Apothecary Museum, Gretchen M. Bulova; Dir. Alexandria Archaeology, Francine Bromberg; Dir. Alexandria Black History Museum, Audrey P. Davis; Dir. Fort Ward Museum & Historic Site, Susan G. Cumbey.
Personnel Profile: Full-Time Paid 16; Part-Time Paid 60; Part-Time Volunteers 316; Interns 2.
Governing Authority: municipal. Parent Institution: City of Alexandria. Subsidiary Institution: Office of Historic Alexandria. Branch Museums: Alexandria Archaeology; Archives & Records Center; Alexandria Black History Museum; Fort Ward Museum & Historic Site; Friendship Firehouse; Gadsby's Tavern Museum; The Lyceum, Alexandria's History Museum; Stabler-Leadbeater Apothecary Museum. Tax exempt: 501(c)(3).
Institution Type/Description: Historical Agency.
Collections: local history.
Research Fields: urban history.
Activities: guided tours; lectures; annual seminar on historic preservation; docent program; organized educational programs for children, adults & undergraduate or graduate college students affiliated with George Washington University; loan, temporary & traveling exhibitions; volunteer program; archaeological digs with community participation.
Hours & Admission Prices: For times & admissions see individual listings. &
Attendance: 162,700 (accurate)

RAMSAY HOUSE VISITORS CENTER, 221 King St., Alexandria, VA 22314-3209. Tel.: 703-746-3301. TDD: 703-838-6494.
Web Site: visitalexandriava.com/
Founded: 1962.
Congressional District: 8
Key Personnel: Pres. & C.E.O., Stephanie Brown; Mgr., Renee Cardone.
Personnel Profile: Full-Time Paid 9; Part-Time Paid 10; Part-Time Volunteers 3.
Governing Authority: municipal. Administered by the Alexandria Convention & Visitor's Center. Parent Institution: City of Alexandria. Tax-exempt: 501(c)(3).
Institution Type/Description: Historic House & Visitor Center: c.1724 home of William Ramsay, First Lord Mayor, first postmaster.
Collections: local history & culture; period artifacts; photographs.
Facilities: historical material on Alexandria & Scottish events available for research to travel writers & others promoting the city; visitor information; botanical garden. Gift items for sale.
Publications: book, Occupied City: Portrait of Civil War Alexandria, Virginia; calendar of events; restaurant, shops & hotel guides; group tour manual; meeting planner guide.
Hours & Admission Prices: Daily 10-8. No charge; donations accepted. Closed New Year's Day; Thanksgiving; Christmas. &
Attendance: 194,256

✱ STABLER-LEADBEATER APOTHECARY MUSEUM, (M), 105-107 S. Fairfax St., Alexandria, VA 22314. Tel.: 703-746-3852. Fax: 703-838-4270.
E-mail: apothecary.museum@alexandriava.gov
Web Site: www.apothecarymuseum.org
Founded: 1939.
Congressional District: 8
Key Personnel: Site Mgr., Lauren Gleason.
Personnel Profile: Full-Time Paid 1; Part-Time Paid 6; Part-Time Volunteers 12; Interns 1.
Governing Authority: Parent Institution: City of Alexandria. Tax-exempt.
Institution Type/Description: Pharmaceutical Museum: housed in 1796 drugstore building.
Collections: pharmaceutical equipment; apothecary bottles; pharmaceutical artifacts.
Facilities: Museum-related items for sale.
Activities: guided tours; permanent exhibitions.
Hours & Admission Prices: April-Oct. Sun.-Mon. 1-5, Tues.-Sat. 10-5; Nov.-March Wed.-Sat. 11-4, Sun. 1-4. Adults $5, children 5-12 $3; discounts to AAM, VAM & HHMC members; members no charge. Closed Thanksgiving; Christmas.
Attendance: 10,000 (estimated)

TORPEDO FACTORY ART CENTER, 105 N. Union St., Alexandria, VA 22314-3217. Tel.: 703-838-4565. Fax: 888-882-7695. Facebook: Torpedo Factory Art Center.
E-mail: marketing@torpedofactory.org
Web Site: www.torpedofactory.org
Founded: 1974.
Congressional District: 8
Key Personnel: Interim C.E.O., Harry Mahon; Pres. (V), Susan Corrigan; Mgr. Mktg., Tracy Baetz; Dir. Target Gallery, Allison Nance; Dir. Operations & Museum Shop Mgr., Richard Johnson; Mktg., Administrative Asst. & Volunteer Coord., Rebecca Lasky; Target Gallery Asst. & Graphic Designer, Kaitlyn Ward; Dir. Special Events, Tara Zimnick-Calico; Mgr. Resource Devel., Lori Ann Eshbaugh.
Personnel Profile: Full-Time Paid 5; Part-Time Paid 2; Part-Time Volunteers 55; Interns 2.
Governing Authority: Parent Institution: Torpedo Factory Art Center Board. Subsidiary Institution: Torpedo Factory Artists' Association. Tax-exempt.
Institution Type/Description: Art Center & Archaeology Museum: housed in World War II torpedo factory.
Collections: history of the torpedo factory; U.S. Navy loaned torpedoes; art collection; works by 82 studio artists.
Facilities: educational facilities. Museum related items for sale.
Activities: special exhibits; weekly open tours; special tours; educational programs; 85 artists studios open to public to watch & interact with artists at work; art school; rental facilities. Art League School, The Art League Gallery Annual Events: Patron's Show; Art on the Rocks; ArtFest; 2nd Thursday Art Nights.
Publications: Torpedo Factory Art Center Visitor's Guide and Directory; brochure, Torpedo Factory Art Center; The Art League Catalog; monthly event calendars; exhibition postcards & materials.
Hours & Admission Prices: Thurs.10-9, Fri.-Wed. 10-6. No charge. Closed New Year's Day; Easter; Independence Day; Thanksgiving; Christmas. &

Attendance: 500,000 (estimated)
Membership: Bronze $35; Silver $75; Gold $150; Collectors $250; Platinum $500; Artist's Circle $1,000.

WOODLAWN, 9000 Richmond Hwy., Alexandria, VA 22309. Mailing Address: P.O. Box 15097, Alexandria, VA 22309. Tel.: 703-780-4000. Fax: 703-780-8509.
E-mail: woodlawn@savingplaces.org
Web Site: www.woodlawnpopeleighey.org
Founded: 1951.
Congressional District: 10
Key Personnel: Chm. Council, Peter Christensen; Exec. Dir., John Riley; Visitor Svcs. Mgr., Meredith Mitchell.
Personnel Profile: Full-Time Paid 3; Part-Time Paid 75; Part-Time Volunteers 50.
Governing Authority: nonprofit organization. Parent Institution: National Trust for Historic Preservation, 1785 Massachusetts Ave., N.W. Washington, DC 20036. Tax-exempt: 501(c)(3).
Institution Type/Description: Historic House: 1800-1805 Woodlawn Home of Maj. Lawrence & Eleanor Parke Custis Lewis, nephew and granddaughter of George & Martha Washington.
Collections: 19th-century decorative arts; furniture; paintings; china; glassware; needlework; textiles; Washington family memorabilia.
Major Exhibits: 51st Annual Needlework Show, 3/1/14-3/31/14.
Research Fields: Civil War; Quaker community; slavery; archaeology.
Facilities: library; archives; reconstructed gardens; nature trails; select rooms and grounds available for rental. Books & museum-related items for sale.
Activities: guided tours; lectures; programs. Museum Sponsors: Annual Needlework Exhibit.
Hours & Admission Prices: March-Nov. Fri.-Mon. 12-4. Adults $10, seniors & active-duty military $8, students K-12 $5.
Attendance: 58,000 (accurate)
Membership: Individual $30; Family $40; Contributing $50; Sustaining $100; Supporting $500; Woodlawn Assembly $1,000.

Altavista

AVOCA MUSEUM, 1514 Main St., Altavista, VA 24517-1161. Tel.: 434-369-1076. Fax: 434-369-1077 (call first).
E-mail: avocamuseums@embarqmail.com
Web Site: www.avocamuseum.org
Founded: 1922.
Congressional District: 5
Key Personnel: Dir., Michael Hudson.
Governing Authority: Parent Institution: Avoca Historical Society.
Institution Type/Description: History Museum.
Collections: local history & culture; personal artifacts; historical artifacts; arrowheads; period pieces; photographs.
Hours & Admission Prices: mid-April to Oct. Thurs.-Sat. 11-3, Sun. 1:30-4:30; call to confirm hours. Adults $5, seniors $4, children under 18 $2; children under 6 & members no charge.

Amherst

AMHERST COUNTY MUSEUM & HISTORICAL SOCIETY, 154 S. Main St., Amherst, VA 24521. Mailing Address: P.O. Box 741, Amherst, VA 24521-0741. Tel.: 434-946-9068.
E-mail: staff@amherstcountymuseum.org
Web Site: www.amherstcountymuseum.org/
Founded: 1973.
Congressional District: 6
Key Personnel: Pres. (V), Bonnie Limbrick; Dir., Holly Mills.
Personnel Profile: Full-Time Paid 1; Part-Time Volunteers 10.
Governing Authority: nonprofit organization. Tax-exempt.
Institution Type/Description: County History Museum: housed in 1907 Georgian Revival house; reconstructed one-room schoolhouse.
Collections: aspects of former Amherst life-styles; maps & books.
Research Fields: Amherst County; central Virginia; family history & genealogy.
Facilities: 400-vol. library on family histories, Amherst County, Virginia, Civil War, & U.S. Census available for research; multipurpose room.
Activities: permanent & temporary exhibitions; educational programs.
Publications: bimonthly newsletter, Newsletter of ACHM; brochure.
Hours & Admission Prices: Tues.-Sat. 10-12 & 1-5. No charge; donations accepted. Closed major holidays.
Attendance: 1,300 (accurate)
Membership: Student $10; Senior $15; Regular & Senior Household $20; Regular Household $25; Business $50; Corporate $100.

Appomattox

APPOMATTOX COURT HOUSE NATIONAL HISTORICAL PARK, VA Rte. 24, Appomattox, VA 24522. Mailing Address: P.O. Box 218, Appomattox, VA 24522-0218. Tel.: 434-352-8987, ext. 26. Fax: 434-352-8330.
E-mail: joe_williams@nps.gov
Web Site: www.nps.gov/apco
Founded: 1940.
Congressional District: 5
Key Personnel: Supt., H. Reed Johnson; Historian, Patrick Schroeder; Chief Museum Svcs., Joe Williams.
Personnel Profile: Full-Time Paid 14; Part-Time Volunteers 3.
Governing Authority: federal. Affiliated with the National Park Service, Interior Building, Washington, DC. Tax-exempt.
Institution Type/Description: Historic Village Museum: located on the site where General Lee surrendered to General Grant to end the Civil War.
Collections: furnishings; Civil War objects. Historic Buildings: 1846 The Courthouse; 1848 The McLean House; Meeks General Store; The Woodson Law Office; 1819 Clover Hill Tavern; The County Jail; The Jones Law Office; The Peers House; The Mariah Wright House; The Isbell House.
Research Fields: Civil War.
Facilities: 2,000-vol. library on the Civil War available for use on the premises by appointment; 72-seat auditorium. Civil War publications for sale.
Activities: guided tours; films; formally organized education programs for children; permanent exhibitions. Museum Sponsors: living history program.
Publications: book, Appomattox Court House.
Hours & Admission Prices: Visitor Center: daily 8:30-5. Call for admission prices. Closed New Year's Day; Martin Luther King Jr. Day; Presidents' Day; Thanksgiving; Christmas.
Attendance: 200,000 (accurate)

Arlington

ARLINGTON ARTS CENTER, 3550 Wilson Blvd., Arlington, VA 22201-2348. Tel.: 703-248-6800. Fax: 703-248-6849.
E-mail: information@arlingtonartscenter.org
Web Site: www.arlingtonartscenter.org
Founded: 1974.
Congressional District: 10
Key Personnel: Exec. Dir., Stefanie Fedor; Pres., Jerrie Bethel; Dir. Education, Penelope Nunes; Exhibitions Coord., Catherine Satterlee; Administrative & Mktg. Coord., Samantha Marques-Moudkofsky.
Personnel Profile: Full-Time Paid 3; Part-Time Paid 1; Part-Time Volunteers 10; Interns 2.
Governing Authority: nonprofit organization. Tax-exempt: 501(c)(3).
Institution Type/Description: Art Center: housed in 1910 Matthew F. Maury School.
Collections: visual arts.
Facilities: 1,550 sq. ft. exhibit space; program & education room; one acre outdoor park.
Activities: lectures; panel discussions; art classes for children & adults; solo & group, juried & curated exhibitions; workshops; resident artist program.
Publications: newsletter, Programs; exhibition announcements & checklists; occasional catalogs; brochures.
Hours & Admission Prices: Wed.-Fri. 1-7, Sat.-Sun. 12-5. No charge; donations accepted.
Attendance: 30,000 (estimated)
Membership: Student, Senior Citizen, & Artist $35; Friend $50; Family $60; Studio Circle $100; Curators Circle $250; Directors Circle $500; Collectors Circle $1,000; Gallerists Circle $2,000; Patrons Circle $5,000; Benefactors Circle $10,000.

ARLINGTON HISTORICAL MUSEUM & SOCIETY, INC., 1805 S. Arlington Ridge Rd., Arlington, VA 22202-1628. Mailing Address: P.O. Box 100402, Arlington, VA 22210-3402. Tel.: 703-892-4204. Facebook: Arlington Historical Society.
E-mail: info@arlingtonhistoricalsociety.org
Web Site: www.arlingtonhistoricalsociety.org
Founded: 1956.
Congressional District: 8
Key Personnel: Dir., Dr. Mark Benbow; Pres. (V), John P. Richardson; Museum Shop Mgr., Eleanor Pourron
Governing Authority: Parent Institution: Arlington Historical Society, Inc. Subsidiary Institutions: Arlington Historical Museum; Ball-Sellers House Museum. Tax-exempt.
Institution Type/Description: Historical Society & Museum: housed in the former Hume School; built in 1891.
Collections: local history artifacts from English exploration period to present.
Major Exhibits: Civil War Sesquicentennial, 6/11-6/15.

Research Fields: local Arlington history.
Facilities: Publications for sale.
Activities: five Society meetings annually; guest lecturers; Arlington Reunion - 4 sessions per season: group oral history - neighborhood area discussions; permanent & temporary exhibitions; annual banquet; special events; receptions for local & regional authors; County Fair booth.
Publications: annual, The Arlington Historical Magazine; quarterly newsletter; Arlington County, Virginia: A History; Washington & Old Dominion Railroad (reprint); periodic leaflets for society & museum collections.
Hours & Admission Prices: Sat.-Sun. 1-4; other times by appointment. No charge; donations accepted.
Attendance: 1,000
Membership: Adult $25; Family $35; Sponsor $75; Donor $125; Corporate $125; Life $500.

ARLINGTON HOUSE, THE ROBERT E. LEE MEMORIAL,
Arlington National Cemetery, Arlington, VA 22211. Mailing Address: National Park Service-George Washington Memorial Pkwy., Turkey Run Park, McLean, VA 22101. Tel.: 703-235-1530. Fax: 703-235-1546.
E-mail: gwmp-arlingtonhouse@nps.gov
Web Site: www.nps.gov/arho
Founded: 1925.
Congressional District: 2
Key Personnel: Rgnl. Dir., Steve Whitesell; Cur., Maria Angela Capozzi; Field Park Supt., Dottie Marshall; Site Mgr., Brandon Bies; Volunteer Coord., Delphine Gross.
Personnel Profile: Full-Time Paid 11; Part-Time Paid 5; Part-Time Volunteers 15.
Governing Authority: federal. Parent Institution: National Park Service, 18th between D & C, Washington, DC 20240. Subsidiary Institution: Dept. of Interior. Tax-exempt.
Institution Type/Description: Historic House Museum: residence of Gen. Robert E. Lee, built by George Washington Parke Custis, foster son of George Washington, restored to 1861 appearance as a national memorial to General Lee.
Collections: decorative arts; archives; music; manuscripts; 18th & 19th-century furnishings; furnishings & memorabilia of the Robert E. Lee & G.W.P. Custis families; museum exhibit on R.E. Lee's life with artifacts. Historic Buildings: house & two slave quarters.
Research Fields: life & activities of R.E. Lee & family members; furnishings; gardens; Arlington House structural history & restoration; slave life.
Facilities: 600-vol. library of history related to R.E. Lee & George Washington Parke Custis available for use by appointment with curator. Books, slides & postcards for sale.
Activities: self-guided & guided tours; formally organized education programs for grades K-12 by appointment only; volunteer program; permanent & temporary exhibitions; applied history internship program; flower garden.
Publications: pamphlet, Arlington House; historic handbook, Arlington House.
Hours & Admission Prices: April-May daily 9-5; June to Labor Day daily 9-5:30; Sept.-March daily 9:30-4:30. No charge. &

Attendance: 500,000 (accurate)

BALL-SELLERS HOUSE,
5620 Third St. S., Arlington, VA 22204-1118. Mailing Address: P.O. Box 100402, Arlington, VA 22210-3402. Tel.: 703-892-4204. Facebook: Ball-Sellers House.
E-mail: info@arlingtonhistoricalsociety.org
Web Site: www.arlingtonhistoricalsociety.org
Founded: 1975.
Key Personnel: Chm. (V), Annette Benbow.
Personnel Profile: Part-Time Volunteers 20.
Governing Authority: Parent Institution: Arlington Historical Society, Inc. Tax-exempt.
Institution Type/Description: Historic House Museum: housed in the log cabin built by John Ball c.1750s; addition built in 1880.
Collections: 18th century artifacts; period furnishings; personal artifacts; photographs.
Facilities: Publications for sale.
Activities: permanent & temporary exhibitions; special events.
Publications: book, The House that John Built (1993).
Hours & Admission Prices: April-Oct. Sat. 1-4. No charge; donations accepted.
Attendance: 250 (estimated)

BLUEMONT HISTORICAL RAILROAD JUNCTION,
601 N. Manchester St., Arlington, VA 22205. Mailing Address: 2100 Clarendon Blvd., Ste. 414, Arlington, VA 22201. Tel.: 703-525-0294.
Web Site: geocities.com/yosemite/trails/9401/railroad.html#cabooses
Founded: 1992.

Congressional District: 8
Key Personnel: Ranger, Sedgewick Moss; Ranger Coord., Lynne Everly.
Personnel Profile: Part-Time Paid 1; Part-Time Volunteers 3.
Governing Authority: county government; nonprofit. Parent Institution: Arlington County Dept. of Parks & Recreation. Tax-exempt: 170(b)(1)(A).
Institution Type/Description: Transportation Museum: housed in former Southern Railway Caboose X-441, built in 1972, on the W&OD trail.
Collections: history of W&OD, RF&P and trolley line area communities; photographs & artifacts of local rail history.
Activities: guided tours.
Hours & Admission Prices: May-Sept. Sat. & holidays 10-6, Sun. 1-5. No charge; donations accepted.
Attendance: 5,000 (estimated)

GULF BRANCH NATURE CENTER,
3608 N. Military Rd., Arlington, VA 22207-4830. Tel.: 703-228-3403.
Web Site: www.arlingtonva.us
Key Personnel: Dir., Denise Chauvette
Institution Type/Description: Nature Center.
Collections: hands-on exhibits; restored log cabin; live animals.
Facilities: 20-seat classroom.
Activities: environmental education programs.
Hours & Admission Prices: Tues.-Sat. 10-5, Sun. 1-5. No charge.

LEE ARTS CENTER,
5722 Lee Hwy., Arlington, VA 22207. Tel.: 703-228-0560. Fax: 703-228-0559.
E-mail: leearts@arlingtonva.org
Web Site: www.arlingtonarts.org
Key Personnel: Dir., Steven Munoz
Institution Type/Description: Art Gallery.
Collections: works by local & national artists; pottery; sculpture; prints.
Activities: workshops.
Hours & Admission Prices: Mon. & Fri. 9:30-6, Tues.-Thurs. 9:30-9, Sat. 9:30-5.

LONG BRANCH NATURE CENTER,
625 S. Carlin Springs Rd., Arlington, VA 22204-1000. Tel.: 703-228-6535. Fax: 703-845-2654.
E-mail: aabugattas@arlingtonva.us
Key Personnel: Acting Dir., Alonso Abugattas
Institution Type/Description: Nature Center.
Collections: native plants; gardens; live animals.
Facilities: 17 acres; 40-seat classroom; gardens; nature center; trails; amphitheater.
Hours & Admission Prices: Tues.-Sat. 10-5, Sun. 1-5. No charge. &

THE NATURE CONSERVANCY,
4245 N. Fairfax Dr., Ste. 100, Arlington, VA 22203-1606. Tel.: 703-841-5300. Fax: 703-841-1283.
Web Site: www.nature.org
Founded: 1951.
Key Personnel: Pres. & C.E.O., Mark R. Tercek; Coord., Maria Fisher.
Governing Authority: nonprofit organization. Tax-exempt: 501(c)(3).
Institution Type/Description: Conservation Area: educational & passive recreational opportunities on largest private nature preserve system in country.
Collections: national data bank on rare & endangered species.
Research Fields: rare & endangered plant & animal species.
Facilities: nature conservation center.
Activities: annual events.
Publications: magazine, Nature Conservancy.
Hours & Admission Prices: Mon.-Fri. 9-5. No charge; donations accepted. Closed federal holidays.
Attendance: 200,500 (estimated)
Membership: Individual $25-$1,000; Corporate $1,000.

Ashland

FLIPPO GALLERY,
Randolph-Macon College, Pace-Armistead Hall, 211 N. Center St., Ashland, VA 23005. Tel.: 804-752-7200.
Key Personnel: Cur., Katie Shaw
Institution Type/Description: Art Gallery: housed in Pace-Armistead Hall c.1876, listed on the National Register of Historic Places.
Collections: student & faculty artwork; exhibits by national & state artists.
Hours & Admission Prices: Mon.-Fri. 10-4, Sat.-Sun. by appointment.

Bastian

WOLF CREEK INDIAN VILLAGE & MUSEUM, (M), 6394 N. Scenic Hwy., Bastian, VA 24314-5202. Tel.: 276-688-3438. Fax: 276-688-2496.
E-mail: info@indianvillage.org
Web Site: indianvillage.org
Founded: 1996.
Congressional District: 9
Key Personnel: General Mgr., Sam Wright; Museum Program Coord. WCIV, Denise A. Smith; Pres. EDA (V), David Dillow.
Personnel Profile: Full-Time Paid 4; Part-Time Paid 4; Part-Time Volunteers 12.
Governing Authority: Parent Institution: Economic Development Authority; County of Bland. Tax-exempt.
Institution Type/Description: American Indian History Museum.
Collections: Eastern Woodland Indian history; personal artifacts.
Research Fields: eastern Woodland Indians; Bland County VA history.
Facilities: archives; nature trails; picnic area. Museum-related items for sale.
Activities: tours; special events; demonstrations; local history programs.
Publications: Wolf Creek Times.
Hours & Admission Prices: Tues.-Sat. 10-5. Adults $10, children 5-16 $6; discounts to AAA, AAM, & ICOM members & groups of 10 or more; children under 5 & members no charge. Closed Thanksgiving; Christmas Eve & Day. ♿
Attendance: 30,000 (estimated)
Membership: Single $25; Family $60; Sustaining $100-$999; Benefactor $1,000 & up; Lifetime $5,000.

Beaverdam

PATRICK HENRY'S SCOTCHTOWN, 16120 Chiswell Lane, Beaverdam, VA 23015-1726. Mailing Address: 204 W Franklin, Richmond, VA 23220. Tel.: 804-227-3500. Fax: 804-227-3559.
E-mail: scotchtown@preservationvirginia.org
Web Site: www.preservationvirginia.org/scotchtown
Founded: 1719.
Congressional District: 1
Key Personnel: Site Mgr., Ann Reid; Exec. Dir., Elizabeth Kostelny; Dir. Properties, Louis J. Malon; Assoc. Dir. Museum Operations, Jennifer Hurst-Wender; Museum Operations Asst., Sienna Fennell.
Personnel Profile: Part-Time Paid 10; Part-Time Volunteers 2; Interns 1.
Operating Expenses: 74,330
Operating Income: 38,810
Governing Authority: nonprofit organization. Parent Institution: Preservation Virginia, 204 W. Franklin St., Richmond, VA 23223. Tax-exempt.
Institution Type/Description: Historic House Museum: housed in the home of Patrick Henry, 1771-1778, the first elected governor of Virginia and the residence from where he rode to St. John's Church to give his Give Me Liberty or Give Me Death speech.
Collections: 18th-century furnishings; Henry Map table.
Research Fields: Henry Patrick.
Facilities: Books, handmade & museum-related items for sale.
Activities: guided tours; wildlife & birding trail; Saturday hands-on educational programs.
Publications: Ventures, biannual.
Hours & Admission Prices: March-Nov. Fri.-Sat. 10-5, Sun. 1-5; other times by appointment. Adults $8, seniors $6, students $4; discount to groups & AAA members; Preservation Virginia members no charge. Closed Easter; Christmas Eve & Day.
Attendance: 4,500 (estimated)
Membership: Student & Teacher $35; Individual $50; Individual Plus One $60; Family $70; Organization $75.

Bedford

BEDFORD MUSEUM AND GENEALOGICAL LIBRARY, (M), 201 E. Main St., Bedford, VA 24523-2012. Tel.: 540-586-4520.
E-mail: bccm-info@bedfordvamuseum.org
Web Site: www.bedfordvamuseum.org
Formerly: Bedford City/County Museum
Founded: 1932.
Congressional District: 5
Key Personnel: Dir., Doug Cooper; Chm. (V) & Cur., Annie Polland; Aide, Jeremy Loftis; Museum Shop Mgr., Shirley Wheeler.
Personnel Profile: Part-Time Paid 4; Part-Time Volunteers 75.
Volunteer Hours: 4,500
Operating Expenses: 100,000
Operating Income: 100,000
Governing Authority: nonprofit organization. Tax-exempt: 501(c)(3).
Institution Type/Description: History Museum & Historic Building: c.1895 three-story Masonic building.
Collections: Indian relics; tools & implements; Revolutionary War artifacts; Civil War artifacts; clothing; military uniforms; household articles; furniture; linens; photographs; 19th-century general store.
Research Fields: Indian life; Revolutionary War; Civil War; genealogy records of local families; local history.
Facilities: 1,000-vol. library pertaining to genealogy & local history.
Activities: guided tours; lectures; films; organized education programs for adults; participatory, loan & temporary exhibitions; school loan service; genealogy classes.
Publications: newsletter, Museum News.
Hours & Admission Prices: June-Sept. Mon.-Sat. 10-5. No charge; donations accepted. ♿
Attendance: 8,725 (accurate)
Membership: Donor $10; Sustaining Patron $50; Annual Patron $100; Building Facility Patron $500; Benefactor $1,000.

PEAKS OF OTTER VISITOR CENTER, 85919 Blue Ridge Pkwy., Bedford, VA 24523-3795. Tel.: 540-586-4357. Fax: 540-586-9445.
Web Site: www.nps.gov/blri
Key Personnel: District Ranger, Paulette Mullinox; Interpretive Specialist, Randy Sutton; Park Ranger, Bobby Miller.
Governing Authority: federal. Affiliated with National Park Service, Blue Ridge Pkwy., 700 Northwestern Bank Bldg., Asheville, NC. 28807. Tax-exempt.
Institution Type/Description: Natural History & Cultural Museum.
Collections: Indian era rural farm life; 1920s cultural & natural history items & artifacts.
Facilities: visitors center.
Activities: guided walks & tours; evening programs; living history demonstrations.
Hours & Admission Prices: late April to May Fri.-Tues. 9-5; Memorial Day to Oct. daily 9-5. No charge.

Berryville

CLARKE COUNTY HISTORICAL ASSOCIATION, INC., 32 E. Main St., Berryville, VA 22611-1338. Mailing Address: P.O. Box 306, Berryville, VA 22611-0306. Tel.: 540-955-2600. Fax: 540-955-0285.
E-mail: admin@clarkehistory.org
Web Site: www.clarkehistory.org
Founded: 1939.
Congressional District: 7
Key Personnel: Dir., Laura Christiansen; Pres. (V), Howard Means; Treas., Lucia Henderson; Archivist, Mary T. Morris.
Personnel Profile: Full-Time Paid 1; Part-Time Paid 2; Part-Time Volunteers 60.
Governing Authority: society. Branch Museum: Burwell-Morgan Mill. Tax-exempt: 501(c)(3).
Institution Type/Description: Historical Society Museum: housed in 19th-century home.
Collections: Clarke Co. archives; photographs; portrait photos; early Fairfax land grants; newspapers; Civil War artifacts; costumes. Historic Building: 1785 Burwell-Morgan Mill.
Research Fields: lands & families, 1738-present.
Facilities: 1,000-vol. library of local history books, available for use by public; 1,000 sq. ft. exhibit space. Society publications for sale.
Activities: organized educational programs; temporary exhibitions.
Publications: biannual historical journal, Clarke Co. Proceedings; quarterly newsletter.
Hours & Admission Prices: Museum: call for hours. Archives: Mon.-Fri. 10-5. Museum: no charge; donations accepted. Archives: Adults $5; members & locals no charge. Closed Fri. before Christmas to Jan. 2. ♿
Attendance: 1,000 (estimated)
Membership: Individual $25; Family $50.

Big Stone Gap

HARRY W. MEADOR, JR. COAL MUSEUM, East Third and Shawnee Ave., Big Stone Gap, VA 24219. Mailing Address: 505 E. 5th St. S., Big Stone Gap, VA 24219-3050. Tel.: 276-523-9209.
E-mail: tfranklin@bigstonegap.org
Web Site: www.bigstonegap.org/attract/coal.htm
Key Personnel: Dir., Tammy Franklin.
Personnel Profile: Part-Time Volunteers 1.
Institution Type/Description: Coal Mining Museum.
Collections: exhibits & objects collected by the late Harry W. Meador.

Hours & Admission Prices: Wed.-Sat. 10-5, Sun. 1-5; other times by appointment. No charge.

SOUTHWEST VIRGINIA MUSEUM HISTORICAL STATE PARK, 10 W. 1st St. N., Big Stone Gap, VA 24219-2528. Mailing Address: P.O. Box 294, Big Stone Gap, VA 24219. Tel.: 276-523-1322. Fax: 276-523-6616.
E-mail: swvamuseum@dcr.state.va.us
Web Site: www.dcr.state.va.us/parks/swvamus.htm
Founded: 1948.
Congressional District: 9
Key Personnel: Dir., Sharon B. Ewing.
Personnel Profile: Full-Time Paid 3; Part-Time Paid 11; Part-Time Volunteers 7.
Governing Authority: state. Parent Institution: Commonwealth of Virginia. Subsidiary Institution: Div. of State Parks. Tax-exempt.
Institution Type/Description: State Park & Historic House: 1888-95 Ayers Mansion.
Collections: early history & pioneer period of Southwest Virginia through the industrial development, coal boom & preceding years; Victorian furnishings.
Facilities: rental facilities. Museum-related items for sale.
Activities: guided tours with prior reservations; permanent & temporary exhibitions; interpretive programs; children's & school programs; monthly programs; workshops.
Publications: General Museum; General State Park; brochures, Museum Self Guided Grounds, Victorian Parlor.
Hours & Admission Prices: March-May & Sept.-Dec. Tues.-Thurs. 10-4, Fri. 9-4, Sat. 10-5, Sun. 1-5; Memorial Day-Labor Day Mon.-Thurs. 10-4, Fri. 9-4, Sat. 10-5, Sun. 1-5. Adults $4, children 6-12 $2; discounts to groups, AAM & ICOM members; children under 6 no charge. Closed Thanksgiving; Christmas.
Attendance: 19,984 (accurate)
Membership: Child $3; Adult $5; Family $15.

Blacksburg

HISTORIC SMITHFIELD, 1000 Smithfield Plantation Rd., Blacksburg, VA 24060. Tel.: 540-231-3947. Fax: 540-231-3006.
E-mail: info@smithfieldplantation.org
Web Site: smithfieldplantation.org
Formerly: Smithfield Plantation
Founded: 1964.
Congressional District: 9
Key Personnel: Dir. Bd., Joann Sutphin; Administrative Dir., W. David McKissack; Museum Shop Mgr., Diane Hoover.
Personnel Profile: Full-Time Paid 1; Part-Time Paid 5; Part-Time Volunteers 120; Interns 3.
Governing Authority: society; nonprofit. Parent Institution: Association for the Preservation of Virginia Antiquities, 204 W. Franklin St., Richmond, VA 23220. Tax-exempt.
Institution Type/Description: Historic House: 1774 Smithfield Plantation built by Revolutionary War hero, William Preston & later the home of three Virginia Governors.
Collections: furnishings from the Colonial & Federal periods. Historic Building: 1774 house.
Research Fields: local history; slave life at Smithfield; plantation crops.
Facilities: historic gardens; orchard.
Activities: children's activities; junior interpreter program; adult group tours; school tours; guest speakers; children's summer history camps; monthly guild meetings. Museum Sponsors: Opening Day; Juneteenth - A Commemoration of Slaves Lives at Smithfield in June; Independence Day Celebration; Holidays at Smithfield in December.
Publications: newsletter; Smithfield Review.
Hours & Admission Prices: April to 1st weekend in Dec. Mon.-Tues. & Thurs.-Sat. 10-5, Sun. 1-5; special tours by appointment. Adults $7, students $4, children 5-11 $3; discounts to Time Traveler (VA state program), AAA members & groups; APVA members no charge.
Attendance: 5,500 (estimated)
Membership: Association for the Preservation of Virginia Antiquities: Individual $40; Mr. & Mrs. $50; Family $60.

MUSEUM OF GEOSCIENCES, Virginia Tech MC0420, 1405 Perry St., Blacksburg, VA 24061. Tel.: 540-231-6894. Fax: 540-231-3386.
E-mail: llyn@vt.edu
Web Site: www.outreach.geos.vt.edu
Formerly: Museum of the Geological Sciences
Founded: 1969.
Congressional District: 9

Key Personnel: Dir., Dr. Robert J. Tracy; Coord., Llyn Sharp.
Personnel Profile: Part-Time Paid 3; Part-Time Volunteers 3; Interns 1.
Governing Authority: university. Parent Institution: Dept. of Geosciences, Virginia Tech. Tax-exempt.
Institution Type/Description: Geology & Mineralogy Museum.
Collections: Dr. C.A. Michael gems & minerals; Dr. A.A. Kirk cut gemstones; paleontology; D. Murray minerals from Australia & Tsumeb Namibia; J. Hearn collection.
Research Fields: minerals & crystallography; geosciences.
Facilities: 30,000-vol. library of geology books, available for inter-library loan; laboratory; classrooms.
Activities: tours; lectures; permanent & temporary exhibits; teacher workshops; educational material loans.
Hours & Admission Prices: Mon.-Fri. 8-5; call to confirm. No charge; donations accepted. Closed holidays; university breaks. &
Attendance: 6,000 (accurate)

PERSPECTIVE GALLERY, Virginia Tech/Aquires Student Center, Blacksburg, VA 24061. Tel.: 540-231-4053. Fax: 540-231-5430.
E-mail: squiresgallery@ut.edu
Web Site: www.studentcenters.vt.edu/perspectivegallery/index.php
Key Personnel: Art Dir., Robin Scully Boucher
Institution Type/Description: Art Gallery.
Collections: works by local, regional, national & international artists.
Hours & Admission Prices: Academic Year: Tues.-Sat. 12-9, Sun. 1-5. No charge. Closed during academic breaks. &
Attendance: 6,785 (accurate)

Boyce

ORLAND E. WHITE ARBORETUM, Blandy Experimental Farm, 400 Blandy Farm Lane, Boyce, VA 22620-2117. Tel.: 540-837-1758. Fax: 540-837-1523.
E-mail: blandy@virginia.edu
Web Site: blandy.virginia.edu
Founded: 1927.
Congressional District: 7
Key Personnel: Dir., Dr. David E. Carr; Cur., Dr. T'ai H. Roulston; Dir., Foundation of State Arboretum, Martha Bjelland; Pres., Foundation of State Arboretum, Bruce Downing; Arborist, Robert D. Arnold; Public Rels. Coord., Tim Farmer; Dir. Education, Candace Lutzow-Felling; Dir. Pub. Programs, Dr. Steven B. Carroll; Landscape Architect, Nancy Takahashi; Bldg. Supt., Dennis Heflin.
Personnel Profile: Full-Time Paid 19; Part-Time Paid 8; Part-Time Volunteers 180; Interns 1.
Governing Authority: university. Parent Institution: University of Virginia, Charlottesville. Subsidiary Institution: Blandy Experimental Farm. Tax-exempt.
Institution Type/Description: Arboretum.
Collections: Buxus types; Quercus species & hybrids; Pinus species conifers of Northern hemisphere; 1,000 species of plants.
Research Fields: environmental sciences; biology; pollination; plant ecology & conservation; plant-herbivore interactions.
Facilities: library; meeting facilities for small groups; greenhouses; laboratories; herbarium.
Activities: adult education classes; accredited university classes; horticultural fair; tours; lectures; educational programs for school groups.
Publications: newsletter, Arbor-Vitae; Guide to the Natural Forms of Boxwood; calendar, Education Programs/Spring & Fall; brochure; trail guides
Hours & Admission Prices: Daily dawn to dusk; tours by appointment. Office: Mon.-Fri. 9-4. No charge; donations accepted.
Attendance: 150,000 (estimated)
Membership: Individual $35; Family, Nonprofit & Business $50; Sustainer $150; Benefactor $300; Life $600; Dual & Spousal Life $1,000; Corporate $1,500.

Bridgewater

REUEL B. PRITCHETT MUSEUM, (M), Bridgewater College, 402 E. College St., Bridgewater, VA 22812. Tel.: 540-828-5462. Fax: 540-828-5482.
E-mail: sgardner@bridgewater.edu
Web Site: www.bridgewater.edu
Founded: 1954.
Congressional District: 9
Key Personnel: Dir. Library & Museum, Andrew Pearson; Cur., Stephanie Gardner.
Personnel Profile: Full-Time Paid 1.
Governing Authority: board of trustees. Parent Institution: Bridgewater College, Bridgewater, VA 22812. Tax-exempt: 501(c)(3).

Institution Type/Description: College Museum.
Collections: rare books & Bibles; coin & currency; bottles & jugs; weaving looms, spinning wheels & accessories; tools; folk items from the Philippines, India, Africa & China; guns; glass, pottery & chinaware.
Activities: educational outreach programs; temporary exhibits.
Hours & Admission Prices: Temporary closed. &

Brookneal

RED HILL-PATRICK HENRY NATIONAL MEMORIAL, 1250 Red Hill Rd., Brookneal, VA 24528-3302. Tel.: 434-376-2044. Fax: 434-376-2647. Facebook: Patrick Henry's Red Hill.
E-mail: redhill@redhill.org
Web Site: www.redhill.org
Founded: 1944.
Congressional District: 5
Key Personnel: Pres., Mark Holman, Jr.; Admin., Hope Marstin.
Personnel Profile: Full-Time Paid 3; Part-Time Paid 5; Part-Time Volunteers 45.
Governing Authority: private. Patrick Henry Memorial Foundation, Inc. Tax-exempt.
Institution Type/Description: Historic Site: Last home & burial place of Patrick Henry.
Collections: Rothermel painting of Patrick Henry's Stamp Act speech; 18th-century furnishings; Patrick Henry artifacts & furniture; jewelry & musical instruments of Patrick Henry & his family; 18th- & 19th-century law books, paintings & engravings; land grants.
Research Fields: Red Hill plantation; career & works of Patrick Henry; genealogy of Patrick Henry & his descendants.
Facilities: library containing law books of Patrick Henry & William Wirt Henry and files of Dr. Robert D. Meade, the author of 2 volumes of Patrick Henry history, available for use on premises; reading room; picnic area. Museum-related items for sale.
Activities: guided tours; temporary & permanent exhibitions; videotape showings; slide lectures; educational tours; living history demonstrations. Museum Sponsors: July 4th celebration; naturalization ceremony; Bluegrass Barbecue & Brew Festival; tastings on the terrace; Christmas open house.
Publications: quarterly newsletter, Elvira Henry's Cooking Book; Proceedings of the Virginia Convention of 1775; Patrick Henry, The Last Years 1789-1799; Patrick Henry: Economic, Domestic and Political Life in Eighteenth Century Virginia; Patrick Henry: Prophet of the Revolution; Patrick Henry's Thoughts on Life, Liberty (or Death) and the Pursuit of Happiness; Patrick Henry Essays; Patrick Henry and Thomas Jefferson; Patrick Henry's Virginia; The Demosthenes of His Age; Five more Minutes with Mr. Henry; Give Me Poetry.
Hours & Admission Prices: April-Oct. Mon.-Sat. 9-5, Sun. 1-5; Nov.-March Tues.-Sat. 9-4, Sun. 1-4, Mon. by appointment. Adults $6, students $2; discounts to groups & VA Assoc. members. Closed New Year's Day; Thanksgiving; Christmas. &
Attendance: 9,289 (accurate)
Membership: Student & Teacher $10; Individual $25; Family $50; Burgess $100; Orator $250; Governor $500; Patriot $1,000; Founding Father $2,500.

Cape Charles

CAPE CHARLES MUSEUM AND WELCOME CENTER, 814 Randolph Ave., Cape Charles, VA 23310. Mailing Address: The Cape Charles Historical Society, P.O. Box 11, Cape Charles, VA 23310-0011. Tel.: 757-331-1008.
E-mail: ccmuseum@hughes.net
Web Site: www.smallmuseum.org/capechas.html
Founded: 1986.
Congressional District: 2
Key Personnel: Pres. (V), Marion Naar; Museum Shop Mgr., Linda Schulz.
Personnel Profile: Part-Time Paid 2; Part-Time Volunteers 10.
Volunteer Hours: 500
Operating Expenses: 24,639
Operating Income: 43,546
Governing Authority: Tax-exempt.
Institution Type/Description: History Museum.
Collections: Cape Charles history; Native American artifacts; photographs; railroad & steamer artifacts; 18th century Arlington mansion artifacts; Chesapeake Bay impact crater, bridge-tunnel.
Facilities: research archives. Eastern Shore history, travel books & postcards for sale.
Activities: fundraisers. Museum Sponsors: Spring Shrimp Boil; Fall Oyster Roast & Live Band.
Publications: newsletters 3 times a year

Hours & Admission Prices: April-Nov. Mon.-Fri. 10-2, Sat. 10-5, Sun. 1-5. No charge; donations accepted. &
Attendance: 2,936 (accurate)
Membership: Single $20; Household $25; Business $35.

Chantilly

*** STEVEN F. UDVAR-HAZY CENTER - NATIONAL AIR AND SPACE MUSEUM, (M),** 14390 Air & Space Museum Pkwy., Chantilly, VA 20151. Tel.: 703-572-4118.
E-mail: nasm-visitorservices@si.edu
Web Site: airandspace.si.edu/udvarhazy
Founded: 2003.
Key Personnel: Dir., John Dailey; Chm. (V), James Guyette; Vice Chm. (V), Randall A. Greene; Museum Shop Mgr., Phyllis Killeen.
Governing Authority: Parent Institution: Smithsonian Institution.
Institution Type/Description: Aviation & Space Museum.
Collections: aviation & space artifacts; aircraft; engines; helicopters; ultralights; experimental flying machines; space shuttle Discovery; spacecraft; rockets; satellites; photographs; Wall of Honor.
Facilities: IMAX theater. Museum-related items for sale.
Activities: flight simulators; guided tours; educational programs; school group tours.
Publications: email newsletter, NASM Now.
Hours & Admission Prices: Memorial Day to Labor Day daily 10-6:30; Sept.-May daily 10-5:30. (All visitors screened upon entry). Center: no charge. Parking: $15. Closed Christmas. &
Attendance: 1,400,000 (accurate)

WALNEY VISITOR CENTER-AT ELLANOR C. LAWRENCE PARK, 5040 Walney Rd., Chantilly, VA 20151-2306. Tel.: 703-631-0013. Fax: 703-631-8319.
Web Site: www.fairfaxcounty.gov/parks
Founded: 1971.
Congressional District: 10
Key Personnel: Park & Visitor Center Mgr. and Naturalist, John Shafer.
Personnel Profile: Full-Time Paid 4; Part-Time Paid 12; Part-Time Volunteers 55; Interns 1.
Governing Authority: county; nonprofit. Parent Institution: Fairfax County Park Authority. Tax-exempt.
Institution Type/Description: Visitor Center: housed in 1780 structure, located on 650-acre park.
Collections: 19th-century local history & natural history; stone barn & ice house ruins; reconstructed smoke house. Historic Structures: 19th century dairy; 18th century mill.
Research Fields: local history; natural & cultural resource inventory; cultural & natural history.
Facilities: classroom; 18th-century gristmill available for rental; demonstration agricultural field with gardens. Books, clay pipes & native-oriented objects for sale.
Activities: guided tours; lectures; films; concerts; organized interpretive programs for children & adults; docent program; temporary exhibitions of your own collections; live animal displays; hands-on exhibitions.
Publications: brochures; newsletters; quarterly calendar of events.
Hours & Admission Prices: Jan.-Feb. Wed.-Mon. 12-5; March-Dec. Mon. & Wed.-Fri. 9-5, Sat.-Sun. 12-5. No charge; donations accepted. Closed New Year's Day; Thanksgiving; Christmas. &
Attendance: 47,830 (accurate)
Membership: National Association for Interpretation memberships available.

Charles City

BERKELEY PLANTATION, 12602 Harrison Landing Rd., Charles City, VA 23030-3339. Tel.: 804-829-6018; 888-466-6018. Fax: 804-829-6757.
E-mail: info@berkeleyplantation.com
Web Site: www.berkeleyplantation.com
Founded: 1619.
Congressional District: 1
Key Personnel: Owner & Operator, Malcolm E. Jamieson.
Personnel Profile: Full-Time Paid 2; Part-Time Paid 10.
Governing Authority: individual operation.
Institution Type/Description: Historic House: 1726 Berkeley Plantation.
Collections: period artifacts; Indian relics; Civil War relics.
Facilities: Brass, glass, pottery, china, relics, wood articles for sale.
Activities: guided tours; slide presentation; gardens.
Publications: Berkeley Brochure, The Army of the Potomac at Berkeley Plantation.
Hours & Admission Prices: Tours daily 9-5. Adults $11, students 13-16 $7.50, children 6-12 $6; discounts to groups, senior citizens, military & AAA members. Closed Thanksgiving; Christmas.

NORTH BEND PLANTATION, 12200 Weyanoke Rd., Charles City, VA 23030. Tel.: 804-829-5176.
Founded: 1801.
Institution Type/Description: Historic House Museum: built in 1801. Listed on the National Register of Historic Places.
Collections: local history & culture; period furnishings; personal artifacts; photographs. Historic Buildings: 1819 smoke house; 1819 daily house; 1819 ice house.
Hours & Admission Prices: House Tours: by appointment only, Adults $10; Grounds: daily 9-5, Adults $3.

SHERWOOD FOREST PLANTATION, 14501 John Tyler Memorial Hwy., Charles City, VA 23030. Mailing Address: P.O. Box 8, Charles City, VA 23030-0008. Tel.: 804-829-5377. Fax: 804-829-2947.
E-mail: ktyler@sherwoodforest.org
Web Site: www.sherwoodforest.org
Founded: 1975.
Key Personnel: Dir., Harrison R. Tyler; Pres. (V), Frances P.B. Tyler
Institution Type/Description: Historic House: c.1730, home of Pres. John Tyler.
Collections: original furniture, silver, china of Pres. Tyler; George Moreland; G.P.A. Healy, Scarboro, Edouart & Sully paintings; original furniture used in the White House; Virginia & Southern furniture; 19th century Oriental rug collection; milk house; smoke house; garden house; slave house; 19th century Overseer's House; 1660-1848 kitchen, laundry & wine house; longest frame house in U.S. (300 ft.); 80 trees (36 varieties not indigenous to America), including many more than 250 years old; 25 acres of landscaped lawn & trees; 11 original plantation buildings (1660-1846).
Research Fields: history of John Tyler & authentication of his belongings; history of Charles City County, VA.
Facilities: meeting room; private rentals available. Museum-related items for sale.
Activities: Annual Events: birthday celebration for President John Tyler in March; Historic Garden Week in April; Christmas activities.
Publications: James River Plantation Cookbook; Virginia Presidential Homes Cookbook.
Hours & Admission Prices: Grounds: daily 9-5. Adults $10; children under 15 no charge. House Tours: by appointment. Adults $35. Closed Thanksgiving; Christmas. &
Attendance: 5,000 (estimated)

SHIRLEY PLANTATION FOUNDATION, 501 Shirley Plantation Rd., Charles City, VA 23030-2907. Tel.: 804-829-5121. Fax: 804-829-6322.
E-mail: info@shirleyplantation.com
Web Site: www.shirleyplantation.com
Founded: 2009.
Congressional District: 3
Key Personnel: Dir., Janet Appel.
Personnel Profile: Full-Time Paid 3; Part-Time Paid 14.
Governing Authority: public foundation.
Institution Type/Description: Historic Site: Virginia's first plantation, 1613.
Collections: 17th- to 19th-century furniture includes items by craftsmen Robert Walker & John Seldon; 17th- to 18th-century portraits by Sir Godfrey Kneller, Charles Bridges & John Wollaston; 18th-century crested English silver collection; Queen Anne style house c.1723 with carved woodwork & three story square flying staircase; kitchen; laundry; tool barn; icehouse; stable; smoke house; pump house; root cellar; corn crib; dovecote; Historic House: Great House; 6 generations of family portraits.
Research Fields: history of the Hill & Carter families; 18th-century Queen Anne architecture; southern agriculture (1613-present); collection of Carter papers including rare books, letters, journals, pictures, drawings housed in the Rockefeller Library in Colonial Williamsburg, Virginia; black history; Civil War; Robert E. Lee.
Facilities: Museum-related items for sale.
Activities: guided tours; hands-on activities; educational programs for students. Annual Events: Historic Garden Week in April; The Christmas activities.
Publications: Shirley Plantation - Home to A Family and A Business for 11 Generations.
Hours & Admission Prices: Daily 9:30-4:30. Adults $11, senior 60 & over $10, youth 6-18 $7.50; discounts to AAA & AAM members, groups of 20 or more, & military w/ID; children under 6 no charge. Closed Thanksgiving; Christmas.
Attendance: 50,000 (accurate)
Membership: Friends of Shirley $40; Patrons of Shirley $100; 1638 Business Founders $250; 1723 Carter Society $500; 1613 Crown Grant Society $1,000.

Charlottesville

ALBEMARLE CHARLOTTESVILLE HISTORICAL SOCIETY, McIntire Bldg., 200 Second St., NE, Charlottesville, VA 22902-5245. Tel.: 434-296-1492. Fax: 434-296-4576.
E-mail: info@albemarlehistory.org
Web Site: albemarlehistory.org
Formerly: The Albemarle County Historical Society
Founded: 1940.
Key Personnel: Pres., Steven G. Meeks; Librarian, Margaret M. O'Bryant; Communications & Collections Mgr., Keri Matthews.
Governing Authority: Tax-exempt: 501(c)(3).
Institution Type/Description: Historical Society Museum.
Collections: over 1,500 artifacts pertaining to Charlottesville & Albemarle County; photographs; manuscripts; books.
Research Fields: central Virginia.
Facilities: research library.
Activities: walking tours; spirit walk; quarterly meetings.
Publications: annual magazine, Albemarle County History; quarterly bulletin.
Hours & Admission Prices: Library: Mon.-Fri. 9-5, Sat. 10-1. No charge; donations accepted. Closed holidays. &
Membership: Single $35; Family $50.

ASH LAWN-HIGHLAND, 1000 James Monroe Pkwy., Charlottesville, VA 22902-7505. Mailing Address: 2050 James Monroe Pkwy., Charlottesville, VA 22902. Tel.: 434-293-8000. Fax: 434-979-9181.
E-mail: info@al-h.us
Web Site: ashlawnhighland.org
Founded: 1930.
Congressional District: 7
Key Personnel: Exec. Dir., Sara Bon-Harper; Consultant, David B. Voelkel; Summer Festival Gen. Mgr., Judy Walker; Museum Shop Mgr., Barbara Hensley; Accountant, Marie Edwards; Office Mgr., Baylor Reinhart.
Personnel Profile: Full-Time Paid 5; Part-Time Paid 55; Part-Time Volunteers 100; Interns 6.
Governing Authority: college. Parent Institution: College of William & Mary, Williamsburg, VA 23187. Tax-exempt.
Institution Type/Description: Historic House: 1799 Ash Lawn-Highland house built by James Monroe with 535-acre working plantation.
Collections: Monroe American & French furnishings & period artifacts 1799-1823; garden pavilion. Historic Buildings: Monroe House & outbuildings.
Research Fields: American history, American & European decorative arts.
Facilities: picnic area; mountain trails; ornamental & kitchen gardens. Gifts, handicrafts, herbs & reproductions for sale.
Activities: guided tours; lectures; concerts; summer festival of operas, plays, chamber music & family entertainment; craft demonstrations; hands-on educational workshops. Museum Sponsors: Monroe Farm Tour for children K-2; Spring Festival of Virginia Wines.
Publications: booklet, Ash Lawn-Highland, Home of James Monroe; coloring book, Ash Lawn; Elizabeth Kortright Monroe; Monroe & the Constitution; Monroe Family Recipes; Monroe USA; Monroe On...; Monroe Portraits; Monroe & Music; The Religion of the Founding Fathers; Life of James Monroe; The Presidency of James Monroe, 1817-1825.
Hours & Admission Prices: March-Oct. daily 9-6; Nov.-Feb. daily 10-5. Adults $10, children 6-11 $5; discounts to groups, AAM, ICOM & AAA members; members no charge. Closed New Year's Day; Thanksgiving; Christmas. &
Attendance: 68,000 (accurate)
Membership: Supporter $100; Sponsor $250; Patron $500; Sustainer $1,000; Investor $2,500; Benefactor $5,000; Advocate $10,000 & up.

✻ THE FRALIN MUSEUM OF ART AT THE UNIVERSITY OF VIRGINIA, (M), 155 Rugby Rd., Charlottesville, VA 22903-2427. Mailing Address: P.O. Box 400119, Charlottesville, VA 22904-4119. Tel.: 434-924-3592. Fax: 434-924-6321.
E-mail: bab8sa@virginia.edu
Web Site: www.virginia.edu/artmuseum
Formerly: Bayly Art Museum of the University of Virginia; University of Virginia Art Museum
Founded: 1935.
Congressional District: 5
Key Personnel: Dir., Bruce Boucher; Admin., David Chennault; Cur., Jennifer Farrell; Mgr. Collections, Jean Collier; Dir. Devel., Elizabeth Wright; Mgr. Exhibitions, Ana Marie Liddell.
Personnel Profile: Full-Time Paid 15; Part-Time Paid 9; Part-Time Volunteers 130; Interns 5.
Governing Authority: university. Parent Institution: University of Virginia. Tax-exempt.
Institution Type/Description: Art Museum.

Collections: 14th to 20th-century American & European paintings, sculpture & works on paper; ancient art; pre-Columbian art; Asian art; African sculpture; American Indian art; contemporary American art; photography.
Activities: inter-museum loan, permanent, temporary & traveling exhibitions; guided tours.
Publications: exhibition catalogs; annual reports; calendars.
Hours & Admission Prices: Tues.-Sun. 12-5. No charge; donations accepted. Closed Thanksgiving; Christmas. &
Attendance: 25,000 (estimated)
Membership: Senior 65 & over and UVA Faculty & Staff $40; Basic $75; Sponsor $200; Patron $500; Benefactor $1,000; Curator's Circle $2,500; Director's Circle $5,000.

KLUGE-RUHE ABORIGINAL ART COLLECTION, U. VA, (M), 400 Worrell Dr., Pantops, Peter Jefferson Pl., Charlottesville, VA 22911-8691. Tel.: 434-244-0234. Fax: 434-244-0235.
E-mail: kluge-ruhe@virginia.edu
Web Site: www.kluge-ruhe.org
Founded: 1998.
Key Personnel: Dir. & Cur., Margo Smith; Mgr. Collections & Registrar, Nicole Wade; Adjunct Cur., Howard Morphy; Education & Program Coord., Lauren Maupin.
Personnel Profile: Full-Time Paid 3; Part-Time Paid 2; Part-Time Volunteers 10; Interns 3.
Governing Authority: Parent Institution: University of Virginia.
Institution Type/Description: Art & Culture Museum.
Collections: Australian Aboriginal art; library & archival materials of Professor Edward L. Ruhe.
Facilities: reference library & archives. Handmade Aboriginal objects & gift items for sale.
Activities: temporary exhibits; lectures; education programs; artist residencies.
Publications: newsletter.
Hours & Admission Prices: Tues.-Sat. 10-4, Sun. 1-5. Guided Tour: Sat. 10:30. No charge; donations accepted. &
Membership: $25 & up.

LEANDER J. MCCORMICK OBSERVATORY, Dept. of Astronomy, 530 McCormick Rd., The University of Virginia, Charlottesville, VA 22904. Mailing Address: P.O. Box 400325, Charlottesville, VA 22904-4325. Tel.: 434-924-7494. Fax: 434-924-3104. TDD: 434-982-4327.
E-mail: dept@mail.astro.virginia.edu
Web Site: www.astro.virginia.edu
Founded: 1885.
Congressional District: 5
Key Personnel: Dir., Ed Murphy.
Personnel Profile: Full-Time Paid 1; Part-Time Paid 2; Part-Time Volunteers 5.
Governing Authority: university. Parent Institution: University of Virginia. Subsidiary Institution: Department of Astronomy. Tax-exempt.
Institution Type/Description: Observatory.
Collections: photographic archive; 26-inch Alvan Clark refracting telescope.
Research Fields: astrometry of nearby stars & planets.
Facilities: 45-seat auditorium; classrooms.
Activities: guided tours; organized education programs for undergraduate or graduate college students affiliated with the Univ. of Virginia.
Hours & Admission Prices: April-Oct. 1st & 3rd Fri. each month 9pm-11pm; Nov.-March 1st & 3rd Fri. each month 7pm-9pm. No charge.
Attendance: 4,000 (estimated)
Membership: Individual $35; Family $60; Stone Fellow $100; Mitchell Contributor $250; Alden Benefactor $500; Fredrick Assoc. $1,000; McCormick Trustee $2,500.

MCGUFFEY ART CENTER, 201 Second St., N.W., Charlottesville, VA 22902-5012. Tel.: 434-295-7973. Fax: 434-295-0322.
E-mail: mcguffey@mcguffeyartcenter.com
Web Site: www.mcguffeyartcenter.com
Institution Type/Description: Art Center.
Collections: photographs; sculpture; paintings.
Facilities: Museum-related items for sale.
Activities: art, dance & theater classes.
Hours & Admission Prices: Tues.-Sat. 10-6, Sun. 1-5. No charge. Closed New Year's Day; Independence Day; Thanksgiving; Christmas.

MICHIE TAVERN CA. 1784, 683 Thomas Jefferson Pkwy., Rt. 53, Charlottesville, VA 22902-7145. Tel.: 434-977-1234. Fax: 434-296-7203.
E-mail: info@michietavern.com
Web Site: www.michietavern.com

Founded: 1928.
Congressional District: 5
Key Personnel: Gen. Mgr., Gregory L. MacDonald; Asst. Mgr., Sam Morris; Cur., Cynthia Conte; Museum Shop Mgr., Wendy Pugh.
Governing Authority: individual operation.
Institution Type/Description: Historic Tavern: housed in c.1784 structure, relocated to present site in 1927 at the height of the Colonial Revival era; grist mill c.1797, Piney River Cabin c.1790, Sowell House c. 1820; 1784 tavern; 1822 rural Virginia house.
Collections: 18th- to 19th-century Southern furniture & artifacts.
Facilities: Period items for sale.
Activities: guided tours. Museum Sponsors: living history from April to October.
Publications: book, Cooking Treasures of the Past.
Hours & Admission Prices: Daily 9-5. Adults $9, senior citizens $8, children 6-11 $4.50; children under 6 & members no charge. Closed New Year's Day; Christmas. &

MONTICELLO, HOME OF THOMAS JEFFERSON, THOMAS JEFFERSON FOUNDATION, INC., 931 Thomas Jefferson Pkwy., Charlottesville, VA 22902-7148. Mailing Address: P.O. Box 316, Charlottesville, VA 22902-0316. Tel.: 434-984-9800. Fax: 434-977-7751.
Web Site: www.monticello.org
Formerly: Thomas Jefferson Memorial Foundation
Founded: 1923.
Congressional District: 5
Key Personnel: Pres., Leslie Greene Bowman; Exec. Vice Pres., Ann H. Taylor; Vice Pres. & C.F.O., Victoria W. Jones; Dir. Archaeology, Fraser D. Neiman; Richard Gilder Sr. Cur. & Vice Pres. Museum Programs, Susan R. Stein; Foundation Librarian, Jack Robertson; Robert H. Smith Dir. Restoration, Robert L. Self; Dir. Devel., Joshua Scott; Dir. Gardens & Grounds, Gabriele Rausse; Shannon Sr. Historian, Gaye Wilson.
Governing Authority: private. Parent Institution: Thomas Jefferson Foundation, Inc. Tax-exempt: 501(c)(3).
Institution Type/Description: Historic House Museum and Plantation: 1769-1826 Monticello, designed by Thomas Jefferson.
Collections: Jeffersonian furniture; memorabilia; art objects; manuscripts; books & personal items; slavery artifacts.
Research Fields: fine arts of the Colonial & early National Periods; Thomas Jefferson's personal life; slavery & plantation life.
Facilities: botanical garden; visitor center; research center. Books, reproduction items, curios, china and silver for sale.
Activities: guided tours; permanent & temporary exhibitions; educational programs.
Publications: brochures; guidebook; publications related to Thomas Jefferson.
Hours & Admission Prices: Daily 10-5; Nov.-Feb. daily 9-4:30. Adults $18-$25, children 5-11 $8; discounts to groups; children under 6 no charge. Closed Christmas. &
Attendance: 450,112 (accurate)

THE ROTUNDA, UNIVERSITY OF VIRGINIA, 1826 University Ave., Charlottesville, VA 22904-0305. Mailing Address: P.O. Box 400305, Charlottesville, VA 22904-4305. Tel.: 434-924-7969 & 1019. Fax: 434-924-3817.
E-mail: rotunda@virginia.edu
Web Site: www.virginia.edu/~urelat/Tours/rotunda/rotunda.html
Founded: 1819.
Key Personnel: C.E.O., Michael Strine; Admin., Leslie M. Comstock; Pres., Teresa Sullivan.
Personnel Profile: Full-Time Paid 1; Part-Time Paid 12; Part-Time Volunteers 9.
Governing Authority: university. Parent Institution: University of Virginia. Tax-exempt.
Institution Type/Description: Historic Buildings: site of Thomas Jefferson's original academical village, which includes the Rotunda, pavilions, student rooms & the lawn (1817-1826).
Collections: the founding of the University & its architectural history; prints & engravings of the grounds, early 19th-century American furniture.
Facilities: 50-vol. library pertaining to Thomas Jefferson & history of University of Virginia.
Activities: guided tours.
Hours & Admission Prices: mid-Jan. to mid-Dec. daily 9-4:45. Historical Tours: daily 10, 11, 2, 3, 4. No charge. &
Attendance: 135,000 (accurate)

SECOND STREET GALLERY, 115 Second St., S.E., Charlottesville, VA 22902-5270. Tel.: 434-977-7284. Fax: 434-979-9793.
E-mail: members@secondstreetgallery.org

Web Site: www.secondstreetgallery.org
Founded: 1973.
Congressional District: 7
Key Personnel: Exec. Dir., Steve Taylor; Chm., Charlotte Dammann; Pres. (V), Claire Holmann Thompson; Mgr. Operations & Outreach Coord., Andrew Greeley.
Personnel Profile: Full-Time Paid 2; Part-Time Paid 2; Part-Time Volunteers 30; Interns 3.
Governing Authority: nonprofit organization. Tax-exempt.
Institution Type/Description: Contemporary Art Gallery.
Collections: contemporary art in all media.
Facilities: 1,880 sq. ft. exhibit space.
Activities: guided tours; lectures; workshops.
Publications: exhibition catalogues; brochures.
Hours & Admission Prices: Tues.-Sat. 11-6. Suggested Donation: $3. &
Attendance: 15,000 (estimated)
Membership: Individual $35; Friend $100; Patron $250; Benefactor $500.

THE VIRGINIA DISCOVERY MUSEUM, 524 E. Main St., East End of the Downtown Mall, Charlottesville, VA 22902. Mailing Address: P.O. Box 1128, Charlottesville, VA 22902-1128. Tel.: 434-977-1025. Fax: 434-977-9681.
E-mail: operationsdirector@vadm.org
Web Site: www.vadm.org
Founded: 1981.
Congressional District: 5
Key Personnel: Exec. Dir., Amy Wicks-Horn; Bd. Chair, Ames Winter; Dir. Operations, Lindsay Jones; Mgr. Education, Kaitlin Clear; Gallery Mgr., Diana Goodwin; Asst. Gallery Mgr., Jamie Frieling; Mgr. Mktg. & Exhibits, Matt Berman.
Personnel Profile: Full-Time Paid 6; Part-Time Paid 2; Part-Time Volunteers 132; Interns 1.
Governing Authority: nonprofit organization. Tax-exempt: 501(c)(3).
Institution Type/Description: Children's Museum.
Collections: 400 brass animals; puppets; Nigerian collection; Pueblo artifacts; participatory exhibits including 200-year old log house, VA heritage.
Facilities: educational facilities.
Activities: educational programs; school tours; camps; special events; portable planetarium (Starlab); portable educational trucks with lesson plans for teachers.
Publications: program calendar; annual report.
Hours & Admission Prices: Mon.-Sat. 10-5. Admission $6; discounts to AAA, AARP, military, ACM & ASTC members and groups. Pay what you wish the 1st Wed. each month. Closed major holidays; New Year's Day; Memorial Day; Independence Day; Labor Day; Thanksgiving & day before; Christmas Eve & Day. &
Attendance: 59,395 (estimated)
Membership: Family $75; Explorer $125; Discovery $250.

Chase City

MACCALLUM MORE MUSEUM AND GARDENS, (M), 603 Hudgins St., Chase City, VA 23924-1237. Mailing Address: P.O. Box 104, Chase City, VA 23924-0104. Tel.: 434-372-0502. Fax: 434-372-3483.
E-mail: mmmg@verizon.net
Web Site: www.mmmg.org
Founded: 1991.
Congressional District: 5
Key Personnel: Exec. Dir., Amber Bradford; Pres. (V), Diana Ramsey; Treas., Dr. Earle Moore; Public Rels., Joe Epps.
Personnel Profile: Full-Time Paid 1; Part-Time Paid 2; Part-Time Volunteers 25.
Governing Authority: private; nonprofit organization. Tax-exempt: 501(c)(3).
Institution Type/Description: General Museum & Botanical Garden.
Collections: Arthur Robertson Indian artifact exhibit dating from 9500 BC to 1600 AD; artifacts in garden from around the world: 1st-century Roman bust, 13th-century Italian wellhead, Spanish cloister, eight fountains; Mecklenberg Hotel memorabilia, known for its curative waters; Thyne Institute, an African American boarding and day school established in Chase City in 1876; herb, rose, all white & all pink gardens; native plants; backyard wildlife habitat.
Facilities: arboretum; birding & wildlife trail; Botanical Gardens. Garden, environmental, museum & nature items for sale.
Activities: guided tours; concerts; seasonal workshops; monthly garden lectures. Museum Sponsors: Native American Day for school children; Archaeology Day; Spring Concert; Music in the Gardens; Fall Fundraiser (Chili Cook Off); Christmas Bazaar; Grand Illumination.
Publications: brochure with map, points of interest, and brief history.
Hours & Admission Prices: Museum, Office & Gift Shop: Mon.-Fri. 10-5, Sat.

10-1. Gardens: daily 10-5. Museum & Gardens: adults $5, children under 12 $2.50; discounts to military, AAA, AAM & ICOM members; members no charge. Gardens (after hours): $2 donation. &
Attendance: 6,500 (estimated)
Membership: Individual $25; Family $35; Friend $50-$99; Patron $100-$499; Sponsor $500-$999; Corporate $1,000 & up.

Chesapeake

CHESAPEAKE PLANETARIUM, 310 Shea Dr., Chesapeake, VA 23328. Mailing Address: 312 Cedar Rd., Chesapeake, VA 23328-6496. Tel.: 757-547-0153, ext. 208.
E-mail: hittrja@cps.k12.va.us
Founded: 1963.
Key Personnel: C.E.O., Dr. Robert J. Hitt.
Governing Authority: public school district. Tax-exempt.
Institution Type/Description: Planetarium & Space Science Museum.
Collections: astronomy; space science.
Research Fields: meteor showers; solar eclipses; sun spot activity.
Activities: lectures; films; formally organized education programs; permanent & traveling exhibitions; telescope available for observing on clear nights after programs.
Publications: monthly newsletter, Astronomy News.
Hours & Admission Prices: Winter: Mon.-Wed. & Fri. 10:30-4:30, Thurs. 10:30-4:30 & 8 p.m.; June & Aug. Thurs. 8pm; other times by appointment. Group lectures or demonstrations $45; Chesapeake school groups no charge. &
Attendance: 40,000 (estimated)

PORTLOCK GALLERIES AT SONO, 3815 Bainbridge Blvd., Chesapeake, VA 23324-1607. Tel.: 727-502-4901.
E-mail: nbenson@cityofchesapeake.net
Web Site: www.portlockgalleries.com
Founded: 2005.
Key Personnel: Gallery Dir., Nicole Benson.
Governing Authority: city.
Institution Type/Description: Art Gallery: housed in a 1908 four-room schoolhouse.
Collections: art exhibitions.
Hours & Admission Prices: Tues.-Fri. 10-5, Sat.-Sun. 12-4. No charge. Closed holidays. &

Chesterfield

CHESTERFIELD HISTORICAL SOCIETY OF VIRGINIA - CHESTERFIELD COUNTY MUSEUM COMPLEX, 6813 Mimms Loop, Chesterfield, VA 23832. Mailing Address: P.O. Box 40, Chesterfield, VA 23832-0040. Tel.: 804-768-7311. Fax: 804-777-9643.
E-mail: admin@chesterfieldhistory.com
Web Site: www.chesterfieldhistory.com
Founded: 1981.
Congressional District: 3
Key Personnel: Admin., Diane Dallmeyer; Pres. (V), Therese Wagenknecht; Historic Sites Specialist, Bryan Truzzie; Cur. Magnolia Grange House & Museum Shop Mgr., Tamara Evans; Cur. County Museum, Pat Roble.
Personnel Profile: Part-Time Paid 3; Part-Time Volunteers 65; Interns 3.
Governing Authority: Tax-exempt: 501(c)(3).
Institution Type/Description: Historical Society Museum.
Collections: replica 1750 Courthouse; photographs; stone implements used by the Appomattox & Monocan Indians; Sir Thomas Dale's portrait; WWII artifacts; 18th-century kitchen, household & farm tools; 1749 Commission of Peace; coal mine artifacts; Sidney King painting; fossils; Civil War weapons, household items, uniforms; period loom; archaeological artifacts recovered from the site of the 1732 Ware Bottom Church, built by Captain Thomas Jefferson, grandfather of Pres. Jefferson. Historic Structures: 1892 Jail; 1828 Court Green clerk's office; monuments to imprisoned Baptist ministers & Chesterfield's Confederate soldiers; 1822 Federal period Magnolia Grange house; 1817 federal period Castlewood plantation house containing local history & genealogy research library.
Research Fields: local history & genealogy.
Facilities: research library.
Activities: guided tours; organized education programs for children; docent program; fundraising events & programs; winter lecture series.
Publications: quarterly, The Messenger.
Hours & Admission Prices: Tues.-Fri. 10-4, Sat. 10-2. Magnolia Grange: adults $5, seniors $4, students $2; members no charge. County Museum: Suggested Donation: adults $2; discounts to AAA members; members no charge. Historic Jail: $1. Closed Chesterfield holidays.
Attendance: 10,517 (accurate)

Membership: Student $10; Senior $20; Senior Couple $25; Individual $25; Household $35; Life $300; Benefactor $500.

Chincoteague

MUSEUM OF CHINCOTEAGUE ISLAND, 7125 Maddox Blvd., Chincoteague, VA 23336. Mailing Address: P.O. Box 352, Chincoteague, VA 23336-0352. Tel.: 757-336-6117.
E-mail: chincoteaguemuseum@verizon.net
Formerly: The Oyster and Maritime Museum of Chincoteague
Founded: 1966.
Congressional District: 1
Key Personnel: Dir., William Borges; Pres., John Jester; Officer, Kelly Conklin; Officer, William Spann; Officer, Christian Young.
Personnel Profile: Full-Time Paid 1; Part-Time Paid 2; Part-Time Volunteers 16.
Governing Authority: nonprofit organization. Tax-exempt: 170(b)(1)(A).
Institution Type/Description: Local History Museum.
Collections: local history; photographs; local artifacts; shellfish farming implements; diorama of oyster industry; first order Fresnel Lens from Assateague Lighthouse; boat models.
Facilities: 100-vol. library of books, videos & materials related to island history & culture.
Activities: self-guided tour; films. Annual Event: Road Scholar Host.
Hours & Admission Prices: Tues.-Sun. 10-5. Adults $3; children 12 & under no charge. &
Attendance: 13,000 (accurate)
Membership: Individual $25; Family $35.

Christiansburg

CAMBRIA DEPOT MUSEUM, 630 Depot St., N.E., Christiansburg, VA 24073. Tel.: 540-382-6431.
Institution Type/Description: Historic Building: built in 1868. Listed on the National Register of Historic Places.
Collections: local & railroad history; photographs; period furnishings.
Facilities: Museum-related items for sale.
Hours & Admission Prices: Fri.-Sat. 10-5, Sun. 1-5; other times by appointment.

MONTGOMERY MUSEUM & LEWIS MILLER REGIONAL ART CENTER, 300 S. Pepper St., Christiansburg, VA 24073-3537. Tel.: 540-382-5644. Facebook: Montgomery Museum and Lewis Miller Regional Art Center.
E-mail: director@montgomerymuseum.org
Web Site: www.montgomerymuseum.org
Founded: 1983.
Congressional District: 9
Key Personnel: Exec. Dir., Sue Farrar; Chm. (V) & Pres. (V), Kim Harich; Mgr. Collections, Sherry Wyatt; Treas., Nancy Miller; Museum Shop Mgr., Susan Keith.
Personnel Profile: Part-Time Paid 2; Part-Time Volunteers 108; Interns 2.
Governing Authority: nonprofit organization. Tax-exempt: 501(c)(3).
Institution Type/Description: Historic House: 1850 Presbyterian Manse.
Collections: photographs; genealogical; New River Valley artifacts; Civil War items; prehistoric Indian artifacts; local history items; archaeological site artifacts; art work of regional historic artifacts, local artists; native plants garden; history books containing area histories.
Major Exhibits: Lewis Miller's New River Valley, 4/20/13-1/14.
Research Fields: genealogy & local history.
Facilities: library; community room; picnic area; garden. Books & museum-related items for sale.
Activities: guided tours; art talks; history chats; day trips; craft shows; arts festivals; docent program; participatory, loan & temporary exhibitions. Annual Events: Wilderness Trail Festival; Heritage Day.
Publications: booklet, self-guided tour of Christiansburg; quarterly newsletter; book, The Montgomery County Story 1776-1957; history book, Virginia's Montgomery County.
Hours & Admission Prices: Tues.-Sat. 10:30-4:30. Tours: adults $2, children under 12 $1; members no charge. Closed New Year's Eve & Day; Thanksgiving & two days after; Christmas Eve, Day & day after. &
Attendance: 875 (accurate)
Membership: Individual $30; Family $40; Business $100; Patron $250.

Clarksville

OCCONEECHEE STATE PARK, 1192 Occoneechee Park Rd., Clarksville, VA 23927-2946. Tel.: 434-374-2210.
E-mail: occoneechee@dcr.virginia.gov
Web Site: www.dcr.virginia.gov/state_parks/occ.shtml
Founded: 1968.
Congressional District: 5
Key Personnel: VA State Parks Dir., Joe Elton.
Personnel Profile: Full-Time Paid 7.
Governing Authority: Parent Institution: VA Dept. of Conservation/Recreation. Subsidiary Institution: State Parks. Tax-exempt.
Institution Type/Description: Native American History Museum.
Collections: Native American history; personal artifacts.
Facilities: nature trails. Museum-related items for sale.
Hours & Admission Prices: Parking: Mon.-Fri. $2, Sat.-Sun. $3. &
Attendance: 202,000 (accurate)

PRESTWOULD FOUNDATION, 429 Prestwould Dr., Clarksville, VA 23927. Mailing Address: P.O. Box 872, Clarksville, VA 23927-0872. Tel.: 434-374-8672. Fax: 434-374-3060.
Founded: 1963.
Congressional District: 5
Key Personnel: C.E.O., Dr. Julian D. Hudson.
Personnel Profile: Full-Time Volunteers 1.
Governing Authority: nonprofit organization. Tax-exempt: 501(c)(3).
Institution Type/Description: Local History Museum: house built in 1795 Prestwould House.
Collections: archives; costumes; original furniture; original French scenic wallpaper; manuscripts.
Facilities: 1,000-vol. library of Lady Jean Skipwith, housed at Prestwould; Colonial gardens.
Activities: guided tours; temporary exhibitions.
Publications: Life by the Roaring Roanoke.
Hours & Admission Prices: April 15-Oct. Thurs.-Sat. 12:30-3, Sun. 1:30-3. Adults $10, seniors over 65 $8, children 6-12 $4; discounts to groups of 15 or more. Grounds only $4.
Attendance: 7,000 (accurate)
Membership: Individual $40; Family $50; Summer House Society $100-$499; Garden Club $500-$999; Manor House Society $1,000 & up.

Clifton Forge

ALLEGHANY HIGHLANDS ARTS & CRAFTS CENTER, INC., 439 E. Ridgeway St., Clifton Forge, VA 24422-1326. Mailing Address: P.O. Box 273, Clifton Forge, VA 24422-0273. Tel.: 540-862-4447.
E-mail: info@HighlandsArtsandCrafts.com
Web Site: highlandsartsandcrafts.com
Founded: 1984.
Congressional District: 6
Key Personnel: Exec. Dir., Nancy Newhard-Farrar; Pres. (V), Carolyn O. Conner; Museum Shop Mgr., Madelyn Miller.
Personnel Profile: Full-Time Paid 1; Full-Time Volunteers 5; Part-Time Paid 1; Part-Time Volunteers 90.
Governing Authority: nonprofit organization. Tax-exempt: 501(c)(3).
Institution Type/Description: Arts & Crafts Center: housed in early 1900s building.
Collections: works produced by Highlands & other regional artists & crafts people.
Facilities: Art & crafts work for sale.
Activities: temporary exhibitions; classes; off site visual arts residency at area high schools; workshops.
Publications: bimonthly, Volunteer Voice; triannual newsletter, Center News.
Hours & Admission Prices: Jan.-April Tues.-Sat. 10-4:30; May-Dec. Mon.-Sat. 10-4:30; groups by appointment. No charge; donations accepted. Closed Thanksgiving; Christmas Eve & Day. &
Attendance: 13,928 (accurate)
Membership: Student $8; Individual $20; Family $30.

Clintwood

RALPH STANLEY MUSEUM, 249 Main St., Clintwood, VA 24228. Mailing Address: P.O. Box 456, Clintwood, VA 24228-0456. Tel.: 276-926-8550 & 5591. Fax: 276-926-8693.
E-mail: tammy@ralphstanleymuseum.com
Web Site: www.ralphstanleymuseum.com
Founded: 2004.
Key Personnel: Dir., Tammy Hill
Institution Type/Description: History Museum.
Collections: memorabilia from the life & career of Ralph Stanley; audio-visual displays; photographs.
Hours & Admission Prices: Mar.-Dec. 24 Tues.-Sat. 10-4, Sun. 1-5; Dec. 25-April 1 Wed.-Sat. 10-4, Sun. 1-5. Adults $5; children 12 & under no charge. Closed New Year's; Memorial Day; Thanksgiving; Christmas.

Colonial Beach

GEORGE WASHINGTON BIRTHPLACE NATIONAL MONU-MENT, 1732 Popes Creek Rd., Colonial Beach, VA 22443-5115. Tel.: 804-224-1732. Fax: 804-224-2142.
Web Site: www.nps.gov/gewa
Founded: 1930.
Congressional District: 1
Key Personnel: Supt., Tarona Armstrong; Administrative Officer, John Storke.
Governing Authority: federal. Parent Institution: U.S. National Park Service, Northeast Regional Office, U.S. Customs House, 200 Chestnut St., Fifth Floor, Philadelphia, PA 19106. Tax-exempt.
Institution Type/Description: Historic Site: birthplace of George Washington, 1730-1750.
Collections: Washington family colonial history, 1657-1730; colonial furnishings; living farm.
Research Fields: Washington family history; colonial tidewater plantation life; lands owned by John Washington.
Facilities: picnic area; visitor center; nature trail. Postcards & publications for sale.
Activities: guided tours; lectures; orientation film; colonial craft demonstrations; living colonial farm.
Publications: book, Popes Creek Plantation.
Hours & Admission Prices: Daily 9-5. No charge. Closed New Year's Day; Thanksgiving; Christmas. &
Attendance: 131,000 (accurate)

Colonial Heights

VIOLET BANK MUSEUM, 303 Virginia Ave., Colonial Heights, VA 23834. Tel.: 804-520-9395.
E-mail: woodburnr@colonial-heights.com
Web Site: www.colonial-heights.com
Founded: 1968.
Personnel Profile: Full-Time Paid 1; Part-Time Paid 1; Part-Time Volunteers 1.
Governing Authority: Tax-exempt.
Institution Type/Description: History Museum: housed in a manor house, built in 1815. Former headquarters of Gen. Robert E. Lee.
Collections: Civil War artifacts; period furniture from 1815-1873; textiles & ceramics.
Hours & Admission Prices: Tues.-Sat. 10-5, Sun. 1-6. No charge; donations accepted.

Courtland

RAWLS MUSEUM ARTS, (M), 22376 Linden St., Courtland, VA 23837-1143. Tel.: 757-653-0754. Fax: 757-653-0341.
E-mail: leighanne@rawlsart.com
Web Site: www.rawlsarts.com
Founded: 1958.
Congressional District: 4
Key Personnel: Pres. Bd., Pat Hartman; Exec. Dir., Leigh Anne Chambers.
Personnel Profile: Full-Time Paid 1; Part-Time Paid 2; Part-Time Volunteers 24; Interns 1.
Governing Authority: nonprofit organization. Tax-exempt: 501(c)(3); 170(b)(1)(A).
Institution Type/Description: Art Museum & Visual Arts Center.
Collections: contemporary regional painting; glass; Indian artifacts; seashells.
Facilities: Museum-related items for sale.
Activities: concerts; arts festivals; organized educational programs; loan & temporary exhibitions; lectures.
Publications: newsletter; flyers; RMA Happenings.
Hours & Admission Prices: Tues. & Sat.-Sun. 1-5, Wed.-Fri. 10-5. No charge; donations accepted. &
Attendance: 6,749 (estimated)
Membership: Student $12; Individual $30; Family $42; Patron $50; Corporate $250; Sustaining $500.

Critz

REYNOLDS HOMESTEAD, 463 Homestead Lane, Critz, VA 24082-3044. Tel.: 276-694-7181. Fax: 276-694-7183.
E-mail: jws@vt.edu
Web Site: www.reynoldshomestead.vt.edu
Founded: 1970.
Congressional District: 5
Key Personnel: Dir., Julie Walters Steele; Sr. Program Coord., Lisa Martin; Asst. Program Coord., Casey Hudgins; Historical Svcs. Asst., Beth Ford; Coord. Bldgs. & Grounds, Steve Isley.
Personnel Profile: Full-Time Paid 3; Part-Time Paid 4; Part-Time Volunteers 10.

Governing Authority: state. Parent Institution: Virginia Tech. Tax-exempt.
Institution Type/Description: Historic House: 1843 boyhood home of R.J. Reynolds, founder of Reynolds Tobacco.
Collections: furnishings of Reynolds' family; 19th-century furnishings; c.1840 rosewood grand piano made by Henry Gaehle; Victorian furniture; photograph album; paintings; silver.
Activities: guided tours; lectures; concerts; art exhibits; theater; study clubs; workshops; docent program; non-credit programs for children & adults throughout the region.
Publications: historic brochure; quarterly Calendar of Events; hiking trail brochure.
Hours & Admission Prices: April-Oct. Sat.-Sun. 1-4pm. Adults $5, students $3; RJR Tobacco Inc. employees & former employees no charge. Group tours scheduled. Closed New Year's Day; Thanksgiving; Christmas. &
Attendance: 1,500 (estimated)

Culpeper

THE MUSEUM OF CULPEPER HISTORY, 803 S. Main St., Culpeper, VA 22701-3213. Mailing Address: P.O. Box 951, Culpeper, VA 22701-0951. Tel.: 540-829-1749. Fax: 540-829-9698.
E-mail: director@culpepermuseum.com
Web Site: www.culpepermuseum.com
Formerly: Culpeper Cavalry Museum
Founded: 1975.
Congressional District: 30
Key Personnel: Bd. Trustees Pres. (V), Philip L. Thornton, IV; Exec. Dir., Lee Langston-Harrison; Museum Coord., Linda B. Montgomery.
Personnel Profile: Full-Time Paid 1; Part-Time Paid 2; Part-Time Volunteers 35; Interns 2.
Governing Authority: nonprofit organization. Tax-exempt: 501(c)(3).
Institution Type/Description: History Museum.
Collections: fossils; 215 million year old dinosaur tracks; Native American artifacts; American Revolution & Civil War artifacts; art; ethnography; photos; prints; textiles; commercial memorabilia. Historic Building: Burgandine House c.1800s.
Research Fields: local history.
Facilities: Museum-related items for sale.
Activities: guided tours; educational programs; monthly special events; permanent & temporary exhibits. Annual Events: Remembrance Days; Downtown Holiday Open House; History Alfresco; Bluegrass, Vittles & Brew Fest.
Publications: educational booklets; newsletter; web site.
Hours & Admission Prices: Feb.-Dec. Mon.-Sat. 10-5, Sun. 1-5; tours by appointment. Adults $5; discounts to museum professionals with ID, AAM, AARP, VAM & AAA members; children, members, and town & county residents no charge. &
Attendance: 15,000 (accurate)
Membership: Minutemen $50-$199; Colonials $200-$599; Explorers $600-$999; Heritage $1,000-$4,999; Council $5,000 & up.

Danville

DANVILLE MUSEUM OF FINE ARTS & HISTORY, (M), 975 Main St., Danville, VA 24541-1822. Tel.: 434-793-5644. Fax: 804-799-6145.
E-mail: artandhistory@danvillemuseum.org
Web Site: danvillemuseum.org
Founded: 1974.
Congressional District: 5
Key Personnel: Exec. Dir., Lynne Bjarnesen; Pres. (V), Jack Neal, Jr.; Education Coord., Patsi Compton; Office Mgr., Gerry Scearce; Visitor Svcs., Sarah Latham; Museum Shop Mgr., Tim Stowe.
Personnel Profile: Full-Time Paid 2; Part-Time Paid 10; Part-Time Volunteers 100.
Governing Authority: nonprofit organization. Tax-exempt.
Institution Type/Description: Art & History Museum: located in the Sutherlin Mansion where confederate President Jefferson Davis stayed when the confederacy fled Richmond in 1865 and issued the last proclamation of the confederacy.
Collections: paintings; prints; Victorian furnishing, textiles; American costumes; historic documents & artifacts pertaining to the history of Danville.
Research Fields: biographies of area citizens who have achieved national or international prominence; collection items; contemporary art; local architecture; local history.
Facilities: meeting rooms; auditorium; art studios; classroom.
Activities: art classes; camps; lecture series; changing exhibitions; annual juried art show; history workshops; Civil War history camp. Museum Sponsors: Art on the Lawn; Victorian Holiday; May Day.

Publications: quarterly newsletter; Last Capitol of the Confederacy.
Hours & Admission Prices: Tues.-Sat. 10-5, Sun. 2-5. Adults $8, senior citizens 55 & over $6, students $4; children 6 & under and members no charge. &
Attendance: 14,264 (accurate)
Membership: Individual $35; Family $65; Sponsor $100; Patron $250-$999; Benefactor $1,000; Corporate $1,000-$5,000.

✻ **DANVILLE SCIENCE CENTER, (M),** 677 Craghead St., Danville, VA 24541-1503. Tel.: 434-791-5160. Fax: 434-791-5168.
E-mail: dscstaff@smv.org
Web Site: www.dsc.smv.org
Founded: 1995.
Congressional District: 5
Key Personnel: Chm. Trustees, Robert O. Satterfield; Pres. DSC, Inc. Directors, Margie E. Wilkinson; Exec. Dir., Jeff Liverman; Asst. Dir., Sonya Wolen; Education Coord., Robin H. Bailey; Exec. Dir. DSC, Inc., Deborah L. Anderson.
Personnel Profile: Full-Time Paid 4; Part-Time Paid 5; Part-Time Volunteers 125.
Governing Authority: state. Parent Institution: Science Museum of Virginia. Tax-exempt.
Institution Type/Description: Science Museum.
Collections: astronomy; physics; chemistry; rocks & minerals; computers; zoology.
Research Fields: science education in a museum environment; environmental studies.
Facilities: early childhood learning center; seasonal butterfly greenhouse; classrooms; ZOOM zone; Outdoor River lab. Museum-related items for sale.
Activities: outreach & StarLab programs; education programs for pre-K through high school level; science lectures & events; exhibit hall demonstrations; hands-on exhibits.
Publications: annual report; quarterly newsletter for members; group visit planning guide; program brochures.
Hours & Admission Prices: Tues.-Sat. & holiday Mon. 9:30-5, Sun. 1-5. Adults $7, seniors 60 & over, college students and active military $6, youth 4-18 $5; discounts to groups & AAA; members, ASTC members and children 3 & under no charge. Closed Thanksgiving; Christmas. &
Attendance: 20,000 (estimated)
Membership: Individual $25; Family $40; Associate $100; Contributor $250; Benefactor $500; Sustainer $1,000. Volunteer -25 hours of volunteer service annually-free.

TANK MUSEUM, 3401 U.S. Hwy. 29B, Danville, VA 24540-1429. Tel.: 434-836-5323. Fax: 434-836-3532.
E-mail: aaftank@gamewood.net
Web Site: www.aaftankmuseum.com
Founded: 1981.
Personnel Profile: Full-Time Paid 2; Full-Time Volunteers 2; Part-Time Paid 3.
Governing Authority: Tax-exempt: 501(c)(3).
Institution Type/Description: Military Museum.
Collections: military history; tanks; cavalry artifacts.
Hours & Admission Prices: Jan.-March Sat. 10-4; April-Dec. Wed.-Sat. 10-4. Adults $10; members no charge. Closed Thanksgiving; Christmas.
Attendance: 22,000 (accurate)
Membership: Individual $25; Family $50; Associate $80.

Dayton

HARRISONBURG-ROCKINGHAM HISTORICAL SOCIETY, (M), 382 High St., Dayton, VA 22821. Mailing Address: P.O. Box 716, Dayton, VA 22821-0716. Tel.: 540-879-2616 & 2681. Fax: 540-879-2616.
E-mail: heritage@heritagecenter.com
Web Site: www.heritagecenter.com
Formerly: Shenandoah Valley Folk Art and Heritage Center
Founded: 1895.
Congressional District: 6
Key Personnel: Pres. (V) & C.E.O., Dale MacAllister; Admin., Mary Nelson.
Personnel Profile: Full-Time Paid 1; Part-Time Paid 3; Part-Time Volunteers 35.
Governing Authority: nonprofit organization. Parent Institution: Harrisonburg-Rockingham Historical Society. Tax-exempt.
Institution Type/Description: Historical Society Museum.
Collections: local folklore; folk art of the Society; electric map of Stonewall Jackson's Valley campaign; music exhibitions; archives; Civil War items; photographs; costumes; glass; ceramics; firearms; Shenandoah Valley history, 1700s-present.

Research Fields: genealogical; old homes; wills & deeds; history pertaining to Harrisonburg, Rockingham Co. & Shenandoah Valley.
Facilities: approx. 800-vol. library of local history books; reading room. Gift items for sale.
Activities: slide shows; speakers; permanent & temporary exhibitions; genealogical research; community activities.
Publications: quarterly newsletter, The Rockingham Recorder.
Hours & Admission Prices: April-Nov. Tues.-Sat. 10-5, Sun. 1-5; Dec.-March Tues.-Sat. 10-5. Adults $5; youth under 18 & members no charge. &
Attendance: 7,493 (accurate)
Membership: Individual & Family $25; Friends of Society $50-$99; Associate $100-$249; Patron $250-499; Sponsor $500 & up.

Deltaville

DELTAVILLE MARITIME MUSEUM & HOLLY POINT NATURE PARK, 287 Jackson Creek Rd., Deltaville, VA 23043. Mailing Address: P.O. Box 466, Deltaville, VA 23043-0466. Tel.: 804-776-7200.
E-mail: museumpark@oonl.com
Web Site: deltavilleva.com/museumpark
Founded: 2002.
Congressional District: 1
Key Personnel: Dir. & Museum Shop Mgr., Raynell Smith; Pres. (V), Bob Kates.
Personnel Profile: Full-Time Paid 3; Full-Time Volunteers 5; Part-Time Paid 2; Part-Time Volunteers 200; Interns 3.
Governing Authority: Parent Institution: Deltaville Community Assoc. Tax-exempt.
Institution Type/Description: Maritime Museum.
Collections: Deltaville's history & culture; photographs; restored period boats; ship models.
Facilities: library.
Activities: special events; eight farmers' markets; eight summer concerts; kayak tour. Museum Sponsors: Family Boat Building Week; Historical Reenactments; Art Show.
Publications: monthly newsletter.
Hours & Admission Prices: Call for hours. Adults $5; members no charge. &
Attendance: 20,000 (accurate)
Membership: Student $5; Individual $15; Family $25; Sustaining $100; Patron $250; Sponsor $500; Benefactor $1,000.

Duffield

NATURAL TUNNEL STATE PARK, 1420 Natural Tunnel Pkwy., Duffield, VA 24244-3672. Tel.: 276-940-1643. Fax: 276-940-2029.
E-mail: megan.france@dcr.virginia.gov
Web Site: www.virginiastateparks.gov
Governing Authority: Branch Museums: The Daniel Boone Wilderness Trail Association - The Blockhouse; Wilderness Road Blockhouse Interpretive Center.
Institution Type/Description: State Park.
Collections: local natural history & culture; folk culture; costumes & textiles; photographs; audiovisual & film; furnishings; personal artifacts; recreational artifacts; structures; tools & equipment.
Facilities: nature trails; visitor center. Museum-related items for sale.
Activities: guided tours; hiking; educational programs; demonstrations.
Hours & Admission Prices: Visitor Center: April-May & mid-Sept.-Oct. Sat.-Sun. 10-6; Memorial Day-Labor Day Thurs.-Mon. 10-6. The Blockhouse: May-Oct. Sat.-Sun. 2-4. No charge. &

Fairfax

✻ **FAIRFAX COUNTY PARK AUTHORITY, RESOURCE MANAGEMENT DIVISION, (M),** 12055 Government Center Pkwy., #927, Fairfax, VA 22035-1118. Tel.: 703-324-8702. Fax: 703-324-3996.
Web Site: www.fairfaxcounty.gov/parks
Founded: 1972.
Congressional District: 10
Key Personnel: Div. Dir., Cindy Walsh; Mgr. Cultural Resources Protection, Liz Crowell; Mgr. Colvin Run Mill, Mike Henry; Mgr. Sully, Carol McDonnell; Site Operations Branch Mgr., Todd Brown; Mgr. Green Spring Gardens, Mary Olien; Mgr. Ellanor C. Lawrence Park, Leon Nawojchik; Mgr. Huntley Meadows Park, Kevin Munroe; Mgr. Hidden Oaks Nature Center, Michael McDonnell; Mgr. Hidden Pond Nature Center, Jim Pomeroy; Mgr. Riverbend, Marty Smith.
Personnel Profile: Full-Time Paid 95; Part-Time Paid 93; Part-Time Volunteers 700; Interns 10.
Governing Authority: county board of supervisors. Parent Institution: Fairfax

County Government. Subsidiary Institution: Fairfax County Park Authority Board. Tax-exempt.

Institution Type/Description: Historic Sites & House Museums.

Collections: period furniture & domestic furnishings; decorative arts; agricultural machinery; tools; country store merchandise. Historic Buildings: 1760 Green Spring; 1794 Sully Plantation; 1780 Walney; 1811 Colvin Run Mill; 1815 Colvin Run Mill Millers House; 1750 Cabells Mill; 1820 Cabells Mill House; 1823 Dranesville Tavern. Historic Sites: 1902 Clark House; Ash Grove 1792; Lahey Lost Valley 1760; Green Spring Farm; Frying Pan Farm; Dranesville Tavern; Freedom Hill Fort; Walney; Middlegate; Hunter House; Wakefield Chapel; Sully; c.1820 Huntley House. Heritage Resource Sites: Lanes Mill; Summers Cemetery; Frying Pan Baptist Meeting House; Mt. Air; Union Mills; Ox Hill Battlefield.

Research Fields: pertaining to collections; historic preservation easements; historic sites; natural & cultural resource management.

Facilities: Books & reproductions for sale.

Activities: lectures; films; concerts; arts festivals; interpretive programs; preservation easement program; formally organized educational programs; docent program; Museum Sponsors: handicapped interpretation; training program in historic crafts; archaeological services; preservation consulting services.

Publications: books: Sully, Biography of a House; Sully, An Architectural Study; History of Fairfax County; Historic District Publications; Huntley, A Study for Preservation; Dranesville Tavern; Colvin Run Mill; A History Program for Fairfax County; Green Spring Farm; brochures; Michael Straight on Green Spring Farm; Stories from Floris; Resources.

Hours & Admission Prices: Call for hours. Grounds: no charge. Closed New Year's Day; Thanksgiving; Christmas. ♿

Attendance: 403,939 (accurate)

FAIRFAX MUSEUM & VISITOR CENTER, (M), 10209 Main St., Fairfax, VA 22030-2403. Tel.: 703-385-8414 & 8415. Fax: 703-385-8692.

E-mail: sgray@fairfaxva.gov
Web Site: www.fairfaxva.gov
Founded: 1992.
Congressional District: 11
Key Personnel: Cur. & Visitor Svcs. Mgr., Susan Inskeep Gray.
Personnel Profile: Full-Time Paid 2; Part-Time Paid 3; Part-Time Volunteers 40.
Governing Authority: nonprofit organization. Parent Institution: City of Fairfax. Tax-exempt.
Institution Type/Description: History Museum & Historic Site: housed in 1873 historic Fairfax elementary school, the first brick public school in Fairfax County.
Collections: personal & archaeological artifacts; photographs.
Research Fields: history & educational system of northern Virginia & City of Fairfax; Civil War history; Fairfax County Courthouse.
Facilities: 2,000 sq. ft. exhibit space. Educational books & museum-related items for sale.
Activities: guided tours. Annual Event: Civil War weekend.
Hours & Admission Prices: Daily 9-5. No charge; donations accepted. Closed New Year's Day; Easter; Thanksgiving; Christmas. ♿
Attendance: 10,200 (accurate)
Membership: Individual, Family & Nonprofit Organizations $25; Profit Organizations $50.

GALLERY 123 - GEORGE MASON UNIVERSITY FINE ARTS GALLERY, Rm. 123, Johnson Center, Fairfax, VA 22030. Tel.: 703-993-8888.

Key Personnel: Dir., Walter Kravitz
Institution Type/Description: Art Gallery.
Collections: paintings; photographs; sculpture.
Hours & Admission Prices: Mon.-Fri. 9-9; other times by appointment.

NATIONAL FIREARMS MUSEUM, (M), 11250 Waples Mill Rd., Fairfax, VA 22030-7400. Tel.: 703-267-1620. Fax: 703-267-3913.

E-mail: nfmstaff@nrahq.org
Web Site: nramuseum.com
Founded: 1871.
Congressional District: 11
Key Personnel: Pres., John Sigler; Exec. Vice Pres., Wayne LaPierre; Museum Dir., Jim Supica; Museum Cur., Doug Wicklund; Museum Cur., Phil Schreier; Museum Shop Mgr., Benjamin Van Scoyoc.
Personnel Profile: Full-Time Paid 7; Part-Time Paid 4; Part-Time Volunteers 15.
Governing Authority: Parent Institution: National Rifle Association. Associate Institution: The NRA Foundation. Tax-exempt: 501(c)(3).

Institution Type/Description: Firearms Museum focus on American society from 1350 to present.
Collections: over 4,500 firearms including period & modern; over 2,500 non-firearm artifacts.
Research Fields: history of firearms development.
Facilities: library; laboratory.
Activities: self-guided tours; permanent & traveling exhibitions.
Publications: National Firearms Museum Brochures; Five Hundred Years of North American Freedom & Liberty; Firearms, Freedom, and the American Experience: The National Firearms Museum Exhibit Guide; Illustrated History of Firearms; Treasures of the National Firearms Museum.
Hours & Admission Prices: Daily 9:30-5. No charge; donations accepted. Closed some holidays. ♿

Falls Church

CHERRY HILL FARMHOUSE & BARN, 312 Park Ave., Falls Church, VA 22046-3301. Tel.: 703-248-5171. Fax: 703-536-8150.

E-mail: recreation@fallschurchva.gov
Web Site: www.fallschurchva.gov
Institution Type/Description: Historic Site: listed on the National Register of Historic Places, built in 1845.
Collections: 19th century furniture, tools, & farming; period farm implements. Historic Building: 1856 timber barn.
Hours & Admission Prices: April-Oct. Sat. 10-1; Nov.-March Mon.-Thurs. 10-3. No charge; donations accepted. ♿
Membership: 2 Years: $30; Lifetime $250.

Farmville

∗ LONGWOOD CENTER FOR THE VISUAL ARTS, (M), 129 N. Main St., Farmville, VA 23901-1305. Tel.: 434-395-2206. Fax: 434-392-6441. TDD: 800-828-1120.

E-mail: robertsbm@longwood.edu
Web Site: www.longwood.edu/lcva/
Founded: 1971.
Congressional District: 5
Key Personnel: Dir., Scott Habes; Chm. (V), Julie K. Heyn.
Personnel Profile: Full-Time Paid 6; Part-Time Paid 5; Part-Time Volunteers 8; Interns 3.
Governing Authority: college. Parent Institution: Longwood University. Tax-exempt: 501(c)(3).
Institution Type/Description: College Art Museum.
Collections: 19th-century American art; African & Chinese art; contemporary Virginia artists collection; regional crafts.
Facilities: activity room; education classroom; volunteer center.
Activities: lectures; gallery talks; formally organized education programs; loan, permanent & traveling exhibitions; community art school.
Publications: exhibit catalogs; exhibition gallery handouts.
Hours & Admission Prices: Galleries: Mon.-Sat. 11-5. Administrative: Mon.-Fri. 8:30-5. No charge; donations accepted. Closed college holidays; Thanksgiving; Christmas. ♿
Attendance: 39,000 (accurate)
Membership: Friend $1-$99; Advocate $100-$249; Fellow $250-$499; Collector $500-$749; Connoisseur $750-$1,249; Benefactor $1,250-$2,499; Champion $2,500-$4,999; Patron $5,000 & up.

ROBERT RUSSA MOTON MUSEUM, (M), 900 Griffin Blvd., Farmville, VA 23901-2236. Mailing Address: P.O. Box 908, Farmville, VA 23901-0908. Tel.: 434-315-8775. Fax: 434-392-8568.

E-mail: info@motonmuseum.org
Institution Type/Description: History Museum.
Collections: Civil Rights & U.S. history; photographs; period furnishings.
Facilities: Museum-related items for sale.
Hours & Admission Prices: By appointment. No charge.

Ferrum

BLUE RIDGE INSTITUTE AND MUSEUM, 20 Museum Dr., Ferrum College, Ferrum, VA 24088. Mailing Address: P.O. Box 1000, Ferrum, VA 24088-9001. Tel.: 540-365-4412. Fax: 540-365-4419.

E-mail: brl@ferrum.edu
Web Site: www.blueridgeinstitute.org
Founded: 1971.
Congressional District: 5
Key Personnel: Dir., J. Roderick Moore; Office Mgr., Jenny Rorrer; Asst. Dir., Vaughan Webb; Head Interpreter, Rebecca Austin.
Personnel Profile: Full-Time Paid 4; Part-Time Paid 7.

Governing Authority: college. Parent Institution: Ferrum College, Ferrum, VA 24088. Tax-exempt: 501(c)(3).
Institution Type/Description: History Museum.
Collections: items from the Blue Ridge Mountains from the late 18th century to the present; farm tools & implements; textiles; furniture. Historic Building: 1800 German Farm.
Research Fields: folklore & folklife; music; crafts; regional history & folkways.
Facilities: 1,500-vol. library on folklore & folklife, available for research; field research station; classrooms.
Activities: guided tours; lectures; concerts; festivals; TV & radio programs; formally organized education programs for children; training program for professional museum workers; temporary exhibitions; research.
Publications: BRI records, Virginia Traditions Series BRI001-BRI010.
Hours & Admission Prices: BRI Museum & Blue Ridge Heritage Archive: daily 10-4. Closed holidays. Farm Museum: May-Aug. Sat. 10-5, Sun. 1-5. Farm tour $5, special tours, senior citizens & children 6-15 $4; discount to ICOM, AAM & VA Assoc. of Museums members; children under 6 no charge. &
Attendance: 15,000 (estimated)

Fincastle

BOTETOURT COUNTY HISTORICAL SOCIETY, 3 W. Main St., Fincastle, VA 24090. Mailing Address: P.O. Box 468, Fincastle, VA 24090-0468. Tel.: 540-473-8394.
E-mail: info@bothistsoc.org
Web Site: www.bothistsoc.org
Founded: 1966.
Key Personnel: Exec. Dir., Weldon L. Martin.
Personnel Profile: Part-Time Paid 5.
Governing Authority: Tax-exempt.
Institution Type/Description: Historical Society Museum.
Collections: personal artifacts; period furniture; photographs.
Facilities: Gift items for sale.
Publications: quarterly newsletter.
Hours & Admission Prices: Mon.-Sat. 10-2, Sun. 2-4. No charge; donations accepted.
Membership: Individual $10; Friend $25; Patron $100; Life $500.

Floyd

FLOYD COUNTY HISTORICAL SOCIETY, 217 N. Locust St., Floyd, VA 24091. Mailing Address: P.O. Box 292, Floyd, VA 24091-0292. Tel.: 540-745-3247.
E-mail: floydhistoricalsociety@gmail.com
Web Site: www.floydhistoricalsociety.org
Founded: 1976.
Key Personnel: Museum Shop Mgr., Rhonda F. Smith.
Personnel Profile: Part-Time Paid 1; Part-Time Volunteers 20.
Governing Authority: Tax-exempt.
Institution Type/Description: Historical Society Museum: housed in a former hospital built by Lather Hylton for Dr. Martin L. Dalton who practiced there from 1914-1923.
Collections: local history & culture; period furnishings; personal artifacts; photographs; medical equipment.
Activities: group tours.
Publications: quarterly newsletter.
Hours & Admission Prices: Thurs.-Sat. 12-5. No charge; donations accepted. &
Attendance: 500 (estimated)
Membership: Individual $15; Family $25.

Forest

THOMAS JEFFERSON'S POPLAR FOREST, 1542 Bateman Bridge Rd., Forest, VA 24551. Mailing Address: P.O. Box 419, Forest, VA 24551-0419. Tel.: 434-525-1806. Fax: 434-525-7252.
Web Site: www.poplarforest.org
Founded: 1983.
Congressional District: 6
Key Personnel: Pres. & C.E.O., Jeffrey L. Nichols; Chm. Bd. Directors, Madeline Miller; Dir. Communications, Kelcey Thurman; Dir. Interpretation & Education, Octavia Starbuck; Mgr. Visitor Svcs. & Volunteers, Dianne Kinney; Dir. Archaeology & Landscapes, Jack Gary; Assoc. Archaeologist, Eric Proebsting; Dir. Architectural Restoration, Travis C. McDonald; Museum Shop Mgr., Kyle Tello; Dir. Devel., Alyson Ramsey; Dir. Institutional Advancement, Wayne Gannaway.
Personnel Profile: Full-Time Paid 18; Full-Time Volunteers 1; Part-Time Paid 18; Part-Time Volunteers 137.

Governing Authority: nonprofit organization. Parent Institution: The Corporation for Jefferson's Poplar Forest. Tax-exempt: 501(c)(3).
Institution Type/Description: Historic House: 1806 octagon house Thomas Jefferson designed & used as his personal retreat; architectural restoration in progress.
Collections: Jefferson Letters.
Research Fields: history of Poplar Forest & Jefferson's life at his retreat; archaeological research focusing on Jefferson, slavery & landscape.
Facilities: Museum-related items for sale.
Activities: guided tours; docent program; archaeology & restoration field schools; school programs; concerts; family programming; hands on activities for children & adults; democracy broadcasts to schools; circulating curriculum kits. Museum Sponsors: Independence Day Celebration; Annual Conversations with Thomas Jefferson; Thomas Jefferson Wine Festival in November.
Publications: spring & fall newsletters; brochures; visitor's guide.
Hours & Admission Prices: March 15-Dec. 15 daily 10-4; groups by appointment. Adults $14, military & senior citizens $12; youth & students 12-18 $6, children 6-11 $2; discount to groups & AAA members; children 5 & under no charge. Closed Easter; Thanksgiving. &
Attendance: 22,473 (accurate)
Membership: Individual $40; Scholar's Society $40-$249; Burgess' Society $250-$499; Ambassador's Society $500-$999; Governor's Society $1,000-$2,499; Secretary of State's Society $2,500-$4,999; President's Society $5,000-$9,999; Regent's Society $10,000-$24,999; Thomas Jefferson Society $25,000 & up.

Fort Defiance

THE AUGUSTA MILITARY ACADEMY MUSEUM, (M), 1640 Lee Hwy., Fort Defiance, VA 24437. Mailing Address: P.O. Box 100, Fort Defiance, VA 24437-0100. Tel.: 540-248-3007. Fax: 540-248-4533.
E-mail: augustamilitaryacademy@verizon.net
Web Site: www.amaalumni.org
Founded: 2000.
Key Personnel: Exec. Dir., Crysta Stephenson; Chm. (V), Frank Williamson; Pres. (V), Jorge Rovirosa.
Personnel Profile: Full-Time Paid 1; Part-Time Volunteers 4.
Governing Authority: Parent Institution: AMA Alumni Foundation. Subsidiary Institution: AMA Alumni Association. Tax-exempt.
Institution Type/Description: Military History Museum: housed in Roller-Robinson House.
Collections: period furnishings; recreation of a cadet barracks room; uniforms; period military school artifacts.
Research Fields: genealogy.
Activities: Annual Event: Alumni Reunion.
Publications: quarterly, The Bayonet.
Hours & Admission Prices: Tues.-Sun. 10-4; other times by appointment. No charge; donations accepted. Closed major holidays. &
Attendance: 2,000 (estimated)

Fort Eustis

U.S. ARMY TRANSPORTATION MUSEUM, 300 Washington Blvd., Besson Hall, Fort Eustis, VA 23604-5260. Tel.: 757-878-1115 & 1182. Fax: 757-878-5656.
E-mail: david.hanselman@us.army.mil
Founded: 1959.
Congressional District: 1
Key Personnel: Dir., David S. Hanselman; Pres., Col. James Rockey, (Ret.); Cur., Marc W. Sammis; Asst. Cur., James E. Atwater; Museum Shop Mgr., Trish Wright.
Personnel Profile: Full-Time Paid 5; Part-Time Paid 1; Part-Time Volunteers 2.
Governing Authority: federal. Parent Institution: Dept. Army, Army Transportation Center. Tax-exempt.
Institution Type/Description: Military Transportation Museum.
Collections: transportation; military.
Research Fields: transportation; military.
Facilities: archives includes official army documents, studies & histories dealing with the history & development of army transportation available for research by written request; auditorium. Museum-related items for sale.
Activities: film series; permanent & temporary exhibitions.
Publications: Red Book, listing of retired Transportation Corps personnel; museum brochure; marine, rail, aircraft & cargo-handling display guide pamphlets; Transportation Corps-An Illustrated History; Memorializations on Fort Eustis & Fort Story, VA.
Hours & Admission Prices: Tues.-Sun. 9-4:40. No charge; donations accepted. Closed federal holidays; Easter. &
Attendance: 78,000 (accurate)

Membership: $100; $500; $1,000; $5,000; $10,000; $25,000.

Fort Lee

U.S. ARMY ORDNANCE TRAINING & HERITAGE CENTER, 2221 Adams Ave., Bldg. 5020, Fort Lee, VA 23801. Tel.: 804-734-4878.
E-mail: usarmy.lee.tradoc.mbx.ordnance-museum@mail.mil
Web Site: www.goordnance.apg.army.mil/museum
Founded: 1919.
Congressional District: 2
Personnel Profile: Full-Time Paid 5; Part-Time Volunteers 2; Interns 1.
Governing Authority: federal. Affiliated with U.S. Army Center of Military History. Parent Institution: U.S. Army Ordnance Center & School. Tax-exempt.
Institution Type/Description: History & Military Museum.
Collections: ordnance equipment from the principal powers with emphasis on WWI and WWII; foreign military equipment both captured and donated; small arms from 16th century to the present; ammunition; archives; outdoor collection of 250 pieces of artillery & tanks.
Research Fields: weaponry.
Facilities: 5,000-vol. library of books, pamphlets, field & technical manuals pertaining to ordnance material available for research on premises by appointment. Museum-related items for sale.
Activities: self-guided tours; permanent & temporary exhibitions.
Publications: brochure.
Hours & Admission Prices: Closed for relocation. &
Attendance: 75,000 (estimated)

＊ THE UNITED STATES ARMY QUARTERMASTER MU-SEUM, (M), 1201 22nd St., Fort Lee, VA 23801-1601. Tel.: 804-734-4203. Fax: 804-734-4359.
Web Site: www.qmmuseum.lee.army.mil
Founded: 1957.
Congressional District: 4
Key Personnel: Cur., Luther D. Hanson; Exhibits Technician, Patrick Fisher; Cur. Education, Laura Baghetti; Office Svcs. Asst., Susan Tatum; Museum Shop Mgr., Paulette Bordwell.
Personnel Profile: Full-Time Paid 4.
Governing Authority: federal. Parent Institution: USA Quartermaster School. Tax-exempt.
Institution Type/Description: Military & History Museum.
Collections: uniforms; flags; insignia; equestrian equipment; weapons; historical military dioramas; paintings; photographs; personal artifacts; clothing; furnishings.
Research Fields: U.S. Army uniforms; quartermaster equipment; unit insignia; Quartermaster history; military manuals, publications & photos.
Facilities: research library by appointment; 100-seat auditorium. Museum-related items for sale.
Activities: inter-museum loan, permanent & temporary exhibitions; guided tours; training programs.
Publications: pamphlet; brochure, QM Museum.
Hours & Admission Prices: Tues.-Fri. 10-5, Sat.-Sun. & holidays 11-5. No charge; donations accepted. Closed New Year's; Thanksgiving; Christmas. &
Attendance: 50,000 (estimated)

U.S. ARMY WOMEN'S MUSEUM, (M), 2100 A Ave., Fort Lee, VA 23801-2100. Tel.: 804-734-4327. Fax: 804-734-4337. Facebook: U.S. Army Women's Museum.
E-mail: usarmy.lee.tradoc.mbx.leee-awmweb@mail.mil
Web Site: www.awm.lee.army.mil
Founded: 1955.
Congressional District: 4
Key Personnel: Pres. Bd., Lt. Col. Pat Sigle; Museum Dir., Dr. Francoise Bonnell.
Personnel Profile: Full-Time Paid 3; Part-Time Volunteers 10.
Governing Authority: federal. Parent Institution: Center of Military History. Subsidiary Institution: Friends of the Army Women's Museum Association. Tax-exempt.
Institution Type/Description: Military History Museum.
Collections: history, tradition & development of the WAAC/ WAC, army women of today and beyond to include the Army Nurse Corps and the Women Air Service Pilots (WASPs); military uniforms; war art; costumes; music; archival material; flags; guidons; books; pamphlets; photographs pertaining to the WAAC/WAC & women in the Army; 140 oral histories; 1.5 million archival documents; recreational artifacts; tools & equipment for communication.
Research Fields: women in the Army; women's history all era's.
Facilities: 410-vol. library of books, videos & albums relating to U.S.

Women's Army Corps available for use on premises with consent of curator; research area; theater.
Activities: student intern program; guided tours; lectures; films; permanent, temporary & loan exhibitions; living history program. Annual Events: Open House in May; Flames of War Miniature War Gaming.
Publications: brochure, History of Women in the Army; museum guides; special event pamphlets.
Hours & Admission Prices: Gallery: Tues.-Fri. 10-5, Sat. 11-5; call to confirm; other times by appointment. No charge; donations accepted. Closed New Year's Day; Thanksgiving; Christmas; Federal holidays. Administrative Hours: Mon.-Sat. 8-5. &
Attendance: 40,000 (estimated)

Fort Monroe

CASEMATE MUSEUM, (M), 20 Bernard Rd., Fort Monroe, VA 23651-1004. Mailing Address: P.O. Box 51341, Fort Monroe, VA 23651-0341. Tel.: 757-788-3391. Fax: 757-788-3886.
E-mail: claire.samuelson@us.army.mil
Founded: 1951.
Congressional District: 1
Key Personnel: Pres. Foundation, Robert Wood; Coord., Earle Richards; Cur., Claire Samuelson; Museum Specialist, David J. Johnson; Museum Shop Mgr., Rosalinda Watson.
Personnel Profile: Full-Time Paid 3; Part-Time Paid 7; Part-Time Volunteers 25.
Governing Authority: U.S. Army. Parent Institution: U.S. Army Center of Military History. Subsidiary Institution: Fort Monroe Authority. Tax-exempt.
Institution Type/Description: Military Museum: built in 1826 casemates in Fort Monroe, VA.
Collections: restored cell where Jefferson Davis was imprisoned in 1865; models; dioramas; documents on the Monitor & Merrimack; documents, pictures & artifacts on Jefferson Davis, Grant, Poe, Black Hawk, Robert E. Lee; prints; relics of Fort Monroe history, Civil War; military; U.S. Army Coast Artillery Museum.
Research Fields: history of Fort Monroe, VA; Civil War in southeast Virginia; Coast Artillery.
Facilities: 4,000-vol. research library on the history of Fort Monroe & Civil War & Coast Artillery. Postcards, historical pamphlets, books, models & prints for sale.
Activities: guided tours; lectures; gallery talks; audiovisual program; permanent exhibitions; monthly military film series.
Publications: pamphlets, Tales of Old Fort Monroe, Nos. 1-15; Dr. Craven & the Captivity of Jefferson Davis; Harrison Phoebus: From Farm to Fortune; Annual History of Fort Monroe; Controversial Ben Butler; The Shackling of Jefferson Davis; Fort Wool; Highlights of Black History at Fort Monroe; museum guidebook.
Hours & Admission Prices: Daily 10:30-4:30. No charge; donations accepted. Closed New Year's Day; Thanksgiving; Christmas. &
Attendance: 37,747 (accurate)
Membership: Member $12-$34; Donor $35-$99; Patron $100-$499; Participating Patron $500-$999; Contributing Patron $1,000-$4,999; Sustaining Patron $5,000-$9,999; Distinguished Patron $10,000-$24,999; Gallery Patron $25,000 & up.

Fort Myer

THE OLD GUARD MUSEUM, ANOG-OGM (Museum), 201 Lee Ave., Fort Myer, VA 22211-1203. Tel.: 703-696-6670. Fax: 703-696-4256. Facebook: Old Guard Museum.
E-mail: kirk.heflin1@us.army.mil
Web Site: www.army.mil/oldguard/museum
Founded: 1962.
Key Personnel: Dir., Kirk Heflin.
Personnel Profile: Full-Time Paid 3.
Governing Authority: federal government. Parent Institution: U.S. Army. Subsidiary Institution: U.S. Army Museum System. Tax-exempt.
Institution Type/Description: Military Museum: housed in late 19th-century building originally used as barracks.
Collections: military uniforms & weapons; accoutrements of the 3rd INF Regiment 1784-present.
Research Fields: history of the 3rd U.S. Infantry; history of Military District of Washington; history, traditions & values of the U.S. Army.
Facilities: 600-vol. library pertaining to United States Army history; professional development of soldiers. Gift items for sale.
Activities: guided tours; films.
Hours & Admission Prices: Temporarily closed for renovations & relocation.
Attendance: 8,000 (estimated)

Fredericksburg

CENTRAL RAPPAHANNOCK HERITAGE CENTER, 900 Barton St., Unit 111, Fredericksburg, VA 22401-5784. Tel.: 540-373-3704.
E-mail: crhc@verizon.net
Web Site: www.crhcarchives.org
Founded: 1997.
Key Personnel: Dir., Barbara Barrett.
Personnel Profile: Part-Time Volunteers 50.
Governing Authority: Tax-exempt.
Institution Type/Description: History Museum.
Collections: area history; documents; archives.
Publications: quarterly newsletter.
Hours & Admission Prices: Tues.-Thurs. 10-4, 1st Sat. of each month 9-12; other times by appointment. No charge; donations accepted. &
Membership: Senior $20; Individual $25; Family $35; Benefactor $100; Corporate $250.

FREDERICKSBURG & SPOTSYLVANIA NATIONAL MILITARY PARK, 120 Chatham Lane, Fredericksburg, VA 22405-2508. Tel.: 540-371-0802; 373-6122 (visitor's center). Fax: 540-371-1907.
Web Site: www.nps.gov/ffrsp
Founded: 1927.
Congressional District: 1 & 3
Key Personnel: Supt., Russell P. Smith; Chief Historian, Robert K. Krick; Staff Historian, Donald C. Pfanz.
Governing Authority: federal. Affiliated with National Park Service, Washington, DC. Branch Museums: Chancellorsville Visitor Center, Chancellor; Fredericksburg Battlefield Visitor Center, Fredericksburg; Jackson Shrine, Guinea.
Institution Type/Description: Military Park Museum: located on the site of the Battlefields of Fredericksburg, Chancellorsville, Wilderness & Spotsylvania.
Collections: military artifacts; manuscripts. Historic Houses: pre-Civil War Innis House; 1771 Chatham; 1790 Ellwood; 1844 Salem Church; 1828 Chandler office.
Research Fields: Civil War military history.
Facilities: 5,000-vol. library of books on the Civil War available by permission; 100-seat auditorium. Civil War books for sale.
Activities: guided tours; lectures; films; permanent & temporary exhibitions.
Publications: brochures; pamphlets on Civil War events.
Hours & Admission Prices: Fredericksburg Battlefield Visitor Center, Chancellorsville Visitor Center: Mon.-Fri. 9-5, Sat.-Sun. 9-6. Chatham Manor: daily 9-4:30. Jackson Shrine: daily 9-5. Adults $3; children 16 & under no charge. Closed New Year's Day; Christmas. &
Attendance: 230,000 (accurate)

＊ FREDERICKSBURG AREA MUSEUM & CULTURAL CENTER, INC., (M), 1001 Princess Anne St., Fredericksburg, VA 22401. Mailing Address: P.O. Box 922, Fredericksburg, VA 22404-0922. Tel.: 540-371-3037. Fax: 540-371-1001.
E-mail: info@famcc.org
Web Site: www.famcc.org
Founded: 1985.
Congressional District: 7
Key Personnel: Dir., Pres. & C.E.O., Ellen Killough; Exec. Vice Pres. & C.O.O., Christa Stabler; Dir. Collections & Exhibitions, Christopher Uebelhor; Museum Store Mgr., Ellen Fortunato; Volunteer Coord., D. Janelle Kennedy; Dir. Education & Public Programs, Tramia Jackson; Membership & Events Dir., Heidi Krofft; Visitor Services Director, Mary Garrett; Membership & Events Asst., Melanie Johnson; Devel. Asst., Lisa King; Office Mgr., Darlene Davis.
Personnel Profile: Full-Time Paid 5; Part-Time Paid 13; Part-Time Volunteers 75.
Governing Authority: nonprofit organization. Tax-exempt: 501(c)3.
Institution Type/Description: History Museum.
Collections: Native American artifacts; Civil War items; decorative arts; glass; ceramic; photographs; furniture; silver. Historic Buildings: c.1816 town hall & market house; 1927 bank.
Research Fields: Fredericksburg history & culture.
Activities: guided tours; lectures; films; concerts; organized education programs for adults, children, undergraduate & graduate college students affiliated with Mary Washington College-Center for Historic Preservation; participatory & temporary exhibitions.
Publications: quarterly newsletter.
Hours & Admission Prices: Mon.-Sat. 10-5, Sun. 12-5; Adults $7, students $2; discounts to AAA & AARP members; members & children under 6 no charge. Closed New Year's Day; Thanksgiving; Christmas. &

Attendance: 22,900 (accurate)
Membership: Individual $35; Family $45; Benefactor $100; Town Hall Council $250; General Lafayette Circle $500; General Washington Circle $1,000.

＊ GARI MELCHERS HOME AND STUDIO, (M), 224 Washington St., Fredericksburg, VA 22405-2360. Tel.: 540-654-1015. Fax: 540-654-1785.
E-mail: belmont@umw.edu
Web Site: www.garimelchers.org
Formerly: Belmont, The Gari Melchers Estate and Memorial Gallery
Founded: 1975.
Congressional District: 7
Key Personnel: Dir., David S. Berreth; Cur., Joanna D. Catron; Mktg. & Museum Shop Mgr., Susan Taylor-Schran; Mgr. Education & Communications, Michelle Dolby; Mgr. Site Preservation, Beate Jensen; Mgr. Special Events, Betsy Labar.
Personnel Profile: Full-Time Paid 6; Part-Time Paid 26.
Governing Authority: state. Parent Institution: University of Mary Washington. Tax-exempt.
Institution Type/Description: Art Museum: housed in 18th-century Belmont, the home & studio of the American artist Gari Melchers, 1860-1932.
Collections: 1871-1932 paintings & sketches by Gari Melchers; European antiquities; china & furnishings; 17th to 20th- century European & American paintings; correspondence of Mr. & Mrs. Gari Melchers.
Research Fields: art of Gari Melchers.
Facilities: orientation theatre; 200-seat public event room; formal gardens; walking trails; visitor's center. Museum-related items for sale.
Activities: guided tours; lectures; gallery talks; docent program; school-age aesthetics program; education programs & internships for undergraduate college students affiliated with University of Mary Washington; permanent & temporary exhibitions; rentals for weddings & receptions.
Publications: quarterly newsletter; brochures; catalogues, True & Clear: The Story of Gari Melchers; True & Clear: The Gift of Belmont.
Hours & Admission Prices: Thurs.-Tues. 10-5. Adults $10; discounts to museum, ICOM, & AAM members; children 18 & under, other museum staff, volunteers, students, faculty & staff of University of Mary Washington no charge. Closed New Year's Eve & Day; Easter; Independence Day; Thanksgiving; Christmas Eve & Day. &
Attendance: 18,000 (accurate)
Membership: Friends of Belmont: Individual $45; Couple $65; Household $75; Artisan $150; Supporter $250; Patron $500; Advisory Council $1,000.

THE GEORGE WASHINGTON FOUNDATION, HISTORIC KENMORE & GEORGE WASHINGTON'S BOYHOOD HOME AT FERRY FARM, 1201 Washington Ave., Fredericksburg, VA 22401-3747. Tel.: 540-373-3381. Fax: 540-371-6066.
E-mail: mailroom@kenmore.org
Web Site: www.kenmore.org
Formerly: George Washington's Fredericksburg Foundation
Founded: 1922.
Congressional District: 1
Key Personnel: Dir. & C.E.O., William E. Garner; Chm. (V), Fielding L. Cocke; Vice Chm., Samuel C. Harding, Jr.; Museum Shop Mgr., Susan Bailey.
Personnel Profile: Full-Time Paid 23; Part-Time Paid 45; Part-Time Volunteers 43; Interns 2.
Volunteer Hours: 5,000
Operating Expenses: 2,423,794
Operating Income: 2,431,859
Governing Authority: board of trustees. Parent Institution: The George Washington Foundation. Tax-exempt: 501(c)(3).
Institution Type/Description: Historic Houses: Kenmore, the 18th century home of Revolutionary War patriot Fielding Lewis and his wife, Betty, sister of George Washington. Ferry Farm: George Washington's boyhood home from 6 to 20 years old.
Collections: 18th-century decorative & fine arts; ornamental plaster ceilings & overmantels; archaeological artifacts.
Major Exhibits: Patristo Lewis: What Would You Give?, 1/14-12/14.
Research Fields: Washington youth.
Facilities: gardens.
Activities: guided house tours at Kenmore; park tour at Ferry Farm; hands-on educational programs; special events.
Publications: monthly e-news.
Hours & Admission Prices: March-Oct. daily 10-5; Nov.-Dec. daily 10-4. Kenmore: adults $10, seniors 60 & over $9, children 6-17 $5; discounts to groups, trolley passengers, Time Travelers, AAA, DAR members & active military; children under 6 no charge. Ferry Farm: adults $8, seniors 60 & over $7, children 6-17 $4; discounts to groups, trolley passengers, Time

Travelers, AAA, DAR members & active military; children under 6 no charge. Closed New Year's Eve & Day; Easter; Thanksgiving; Christmas Eve & Day. &
Attendance: 28,791 (accurate)

HUGH MERCER APOTHECARY SHOP, 1020 Caroline St., Fredericksburg, VA 22401-3814. Mailing Address: 1200 Charles St., Fredericksburg, VA 22401. Tel.: 540-373-3362. Fax: 540-373-1569. Facebook: Hugh Mercer Apothecary Shop.
E-mail: hmas@washingtonheritagemuseums.org
Web Site: www.washingtonheritagemuseums.org
Founded: 1761.
Congressional District: 7
Key Personnel: Pres., Gail G. Braxton; Mgr., Genevieve Bugay.
Personnel Profile: Part-Time Paid 10.
Governing Authority: nonprofit corporation. Parent Institution: Washington Heritage Museums, 1200 Charles St., Fredericksburg, VA 22401. Tax-exempt.
Institution Type/Description: Historic Building: 1761 Hugh Mercer Apothecary Shop.
Collections: pharmaceutical implements; medical implements; historic papers.
Activities: guided tours.
Publications: brochure.
Hours & Admission Prices: Call for hours. Adults $5, children $2; discount to members, AAA members, military families & groups; active military & WHM members no charge. Closed New Year's Eve & Day; Thanksgiving; Christmas Eve & Day.
Attendance: 12,000 (accurate)
Membership: Student & K-12 Teacher $20; Individual $30; Dual $40; Family $50; Contributing $100.

JAMES MONROE MUSEUM AND MEMORIAL LIBRARY, (M), 908 Charles St., Fredericksburg, VA 22401-5801. Tel.: 540-654-1043. Fax: 540-654-1106. TTY: 800-828-1120.
E-mail: sharris4@umw.edu
Web Site: www.jamesmonroemuseum.org
Founded: 1927.
Congressional District: 1
Key Personnel: Dir., Scott H. Harris; Cur., Jarod Kearney; Membership & Events Coord., Adele Uphaus-Conner; Office Mgr., Lynda Allen.
Personnel Profile: Full-Time Paid 3; Part-Time Paid 17; Part-Time Volunteers 4; Interns 4.
Governing Authority: state; nonprofit organization, administered by University of Mary Washington. Tax-exempt.
Institution Type/Description: Presidential Historical Museum & Library.
Collections: personal furnishings; jewelry; possessions of President & Mrs. James Monroe; manuscripts; rare books; U.S. presidential collection; 27,000 historic documents, prints, maps, manuscripts & family papers.
Research Fields: life of James Monroe; Virginia families; U.S. presidents; foreign policy; 18th- & 19th-century American history; papers & books of James Monroe.
Facilities: 10,000-vol. library of Monroe's life & times available by appointment only. Museum-related items for sale.
Activities: guided tours; internship education programs for undergraduate college students; permanent exhibitions; changing annual & temporary exhibits; scholarly & special events. Museum Sponsors: annual James Monroe Lecture; Monroe Birthday Celebration; Christmas with the Monroes.
Publications: Quotations of James Monroe: On the Subjects of his Family, Friends, Private Affairs, and Public Policy; James Monroe: An Illustrated History; A Presidential Legacy; The Life of James Monroe; The Making of a Revolutionary; The Presidency of James Monroe; Elizabeth Kortright Monroe.
Hours & Admission Prices: March-Nov. Mon.-Sat. 10-5, Sun. 1-5; Dec.-Feb. daily 10-4. Adults $5, students & children $1; discounts to senior citizens, groups, AAM, AAA, ICOM, VAM & Timeless Ticket to Fredericksburg members; MW college students & museum members no charge. Closed New Year's Eve & Day; Thanksgiving; Christmas Eve & Day. &
Attendance: 10,000 (estimated)
Membership: The Friends of James Monroe Museum: Individual $25; Couples & Households $35; Representatives $50; Senators $100; Diplomats $500; Presidential $1,000 & up.

MARY WASHINGTON HOUSE, 1200 Charles St., Fredericksburg, VA 22401-3706. Tel.: 540-373-1569. Fax: 540-373-1569. Facebook: Mary Washington House.
E-mail: mwhouse@washingtonheritagemuseums.org

Web Site: www.washingtonheritagemuseums.org
Founded: 1772.
Congressional District: 7
Key Personnel: Pres., Gail G. Braxton; Administrator, Anne Darron; Museum Shop Mgr., Jan Swager.
Personnel Profile: Part-Time Paid 15; Part-Time Volunteers 30.
Governing Authority: nonprofit organization. Parent Institution: Washington Heritage Museums, 1200 Charles St., Fredericksburg, VA 22401. Tax-exempt.
Institution Type/Description: History Museum: 1772-1789 home of Mary Ball Washington.
Collections: 18th-century furnishings.
Research Fields: pertaining to collections.
Facilities: Museum-related items for sale.
Activities: guided tours of house & gardens.
Publications: brochure.
Hours & Admission Prices: Call for hours. Adults $5, children $2; discounts to AAA members, military families, and groups of 10 & over; active military & WHM members no charge. Closed New Year's Eve & Day; Thanksgiving; Christmas Eve & Day.
Attendance: 10,000 (accurate)
Membership: Student & K-12 Teacher $20; Individual $30; Dual $40; Family $50; Contributing $100.

THE NATIONAL CIVIL WAR LIFE MUSEUM, 829 Caroline St., Fredericksburg, VA 22401-5805. Mailing Address: The National Civil War Life Foundation, 4712 Southpoint Pky., Fredericksburg, VA 22407. Tel.: 540-834-1859. Fax: 540-834-1859.
E-mail: civilwarlife@yahoo.com
Web Site: www.civilwarlife.org
Formerly: Civil War Life - The Soldier's Museum
Founded: 2000.
Congressional District: 1
Key Personnel: Exec. Dir., Terry Thomann.
Personnel Profile: Full-Time Paid 2; Part-Time Paid 2.
Governing Authority: Parent Institution: National Civil War Foundation. Tax-exempt.
Institution Type/Description: Military Museum.
Collections: Civil War life, weapons, equipment & personal artifacts; photographs.
Research Fields: Civil War veterans; flag passions of the Confederate South.
Facilities: 300-vol. library; 20-seat auditorium; 2,000 sq. ft. exhibit space; 20-seat theater. Museum-related items for sale.
Activities: films; guided tours; lectures.
Hours & Admission Prices: Call for hours. Museum: adults $5, children 7-16 $2.50; discounts to active military, seniors & AAM members. 3-D Theater: $3. Combo: adult $7, child $4.50. Closed Thanksgiving; Christmas. &
Attendance: 10,000 (accurate)

RISING SUN TAVERN, 1304 Caroline St., Fredericksburg, VA 22401-3704. Mailing Address: 1200 Charles St., Fredericksburg, VA 22401. Tel.: 540-371-1494. Fax: 540-373-1569. Facebook: Rising Sun Tavern.
E-mail: rst@washingtonheritagemuseums.org
Web Site: www.washingtonheritagemuseums.org
Founded: 1760.
Congressional District: 7
Key Personnel: Pres., Gail G. Braxton; Mgr., Jo Atkins.
Personnel Profile: Part-Time Paid 12.
Governing Authority: nonprofit organization. Parent Institution: Washington Heritage Museums, 1200 Charles St., Fredericksburg, VA 22401. Tax-exempt.
Institution Type/Description: Historic House Museum: c.1760 built by Charles Washington as his home; later used as tavern.
Collections: American and English pewter; 18th-century furniture, instruments & music; Tap Room.
Facilities: Museum related items for sale.
Activities: guided tours.
Publications: brochure.
Hours & Admission Prices: Call for hours. Adults $5, children $2; discounts to AAA members, military families & groups of 10 or more; active military & WHM members no charge. Closed New Year's Eve & Day; Thanksgiving; Christmas Eve & Day.
Attendance: 10,000 (accurate)
Membership: Students & K-12 Teacher $20; Individual $30; Dual $40; Family $50; Contributing $100.

ST. JAMES' HOUSE, 1300 Charles St., Fredericksburg, VA 22401-3708. Mailing Address: 1200 Charles St., Fredericksburg, VA 22401. Tel.: 540-373-1569. Fax: 540-373-1569. Facebook: St. James' House.
E-mail: mwhouse@washingtonheritagemuseums.org
Web Site: www.washingtonheritagemuseums.org
Founded: 1760.
Congressional District: 7
Key Personnel: Dir., Gail G. Braxton; Resident Mgr., Elizabeth Butler.
Personnel Profile: Part-Time Volunteers 12.
Governing Authority: nonprofit corporation. Parent Institution: Washington Heritage Museums, 1200 Charles St., Fredericksburg, VA 22401. Tax-exempt.
Institution Type/Description: Historic House: c.1770 home of Hon. James Mercer.
Collections: private collection of the donors, William H. Tolerton and Daniel J. Breslin.
Activities: guided tours during Garden Week in April & October and by appointment.
Publications: brochure.
Hours & Admission Prices: Garden Week: April & 1st week Oct.; other times by appointment. Adults $3, children 6-18 $1; discount to groups; WHM members no charge.
Attendance: 175 (estimated)
Membership: Student & K-12 Teacher $20; Individual $30; Dual $40; Family $50; Contributing $100.

UNIVERSITY OF MARY WASHINGTON GALLERIES, 1301 College Ave. at Seacobeck St., Fredericksburg, VA 22401-5358. Tel.: 540-654-1013. Fax: 540-654-1171. TDD: 540-654-1104.
E-mail: gallery@umw.edu
Web Site: galleries.umw.edu
Founded: 1956.
Congressional District: 7
Key Personnel: Dir., Anne Timpano; Coord. Visitors Svcs., Allison Long Hardy; Asst. Cur. & Registrar, Ashley Anttila; Office Mgr., Angela Whitley.
Personnel Profile: Full-Time Paid 1; Part-Time Paid 3; Interns 2.
Governing Authority: nonprofit. Parent Institution: University of Mary Washington. Tax-exempt.
Institution Type/Description: College Art Galleries.
Collections: 19th & 20th-century European & American art; Asian art.
Research Fields: American Art; 20th century art; contemporary art.
Activities: guided tours; lectures by visiting speakers; internships.
Publications: exhibition catalogs; members newsletter.
Hours & Admission Prices: Museum: Mon., Wed. & Fri. 10-4, Sat.-Sun. 1-4 during college session only. Office: Mon.-Fri. 8-5. No charge. Closed New Year's Day; Thanksgiving; Christmas; university holidays & breaks.
Attendance: 7,000 (accurate)
Membership: Student $5; Individual $25; Family $50; Contributor $125; Patron $500; Ridderhof Martin Society $1,000.

Front Royal

WARREN RIFLES CONFEDERATE MUSEUM, 95 Chester St., Front Royal, VA 22630-3368. Mailing Address: P.O. Box 1304, Front Royal, VA 22630-0027. Tel.: 540-636-6982 & 660-0941.
E-mail: warrenriflescmm@gmail.com
Founded: 1959.
Congressional District: 7
Key Personnel: Dir. & Pres. (V), Suzanne W. Silek; Museum Shop Mgr., Frances Woodward.
Personnel Profile: Part-Time Paid 3; Part-Time Volunteers 8.
Governing Authority: society. Parent Institution: Chapter 934, United Daughters of the Confederacy. Tax-exempt.
Institution Type/Description: History & Military Museum: located on one of oldest streets in Front Royal.
Collections: relics of the War between the States including guns, uniforms, flags, swords, cutlasses, bayonets; letters & other signed documents, including a signed photograph of Belle Boyd; rare books; chromolithographs & engravings; furniture & furnishings of the period; military artifacts.
Research Fields: the War between the States.
Facilities: 500-vol. library of history books, available for use by appointment. Books & other items relating to U.S. history for sale.
Activities: guided tours; permanent collections.
Publications: brochures.
Hours & Admission Prices: mid-April to Nov. Mon.-Sat. 9-4, Sun. 12-4; other times by appointment. Groups: call for admission prices; discounts to members, AAA, AAM & AARP members; students no charge.

Attendance: 1,000 (estimated)

Galax

JEFF MATTHEWS MEMORIAL MUSEUM, 606 W. Stuart Dr., Galax, VA 24333-2718. Tel.: 276-236-7874.
E-mail: info@jeffmatthewsmuseum.org
Web Site: www.jeffmatthewsmuseum.org
Founded: 1974.
Congressional District: 27
Key Personnel: Chm. (V), Bobby Thomson, Jr.; Cur., Tony Burcham.
Personnel Profile: Part-Time Paid 2.
Governing Authority: municipal. Parent Institution: City of Galax. Tax-exempt. 170(b)(1)(A).
Institution Type/Description: History Museum.
Collections: over 10,000 Indian artifacts; 1,000 knives; tools; over 50 animal trophies; rugs of North America; receipts maps; land grants; two reconstructed 19th century log cabins with period furnishings; Miss America display; over 100 African artifacts; pictures of local Civil War veterans; 1860s covered wagon. Historic Building: 1834 Log Cabin.
Facilities: 8,000 sq. ft. exhibit space.
Activities: guided tours.
Publications: brochure.
Hours & Admission Prices: Wed.-Sat. 11-4; other times by appointment. No charge; donations accepted. Closed New Year's Day; Easter; Thanksgiving; Christmas.
Attendance: 4,253 (accurate)
Membership: Annual $10.

Glen Allen

THE CULTURAL ARTS CENTER AT GLEN ALLEN, 2880 Mountain Rd., Glen Allen, VA 23060-2121. Mailing Address: P.O. Box 1249, Glen Allen, VA 23060-1249. Tel.: 804-261-2787.
E-mail: info@artsglenallen.com
Web Site: www.artsglenallen.com
Key Personnel: Pres., K. Alferio; Performing Arts Mgr. & Technical Dir., Richard Koch; Visual Arts Mgr., Lauren Hall; Dir. Mktg. & Public Rels., Anita Waters
Institution Type/Description: Cultural Arts Center.
Collections: art exhibitions.
Major Exhibits: An Almost Tactile Memory, 1/9/14-3/9/14; Dell 'Oceano: From the Sea, 3/13/14-5/11/14; Seen in Virginia - Pastel Society of VA, 5/15/14-7/13/14.
Hours & Admission Prices: Call for hours.

THE MUSEUM IN MEMORY OF VIRGINIA E. RANDOLPH, 2200 Mountain Rd., Glen Allen, VA 33060-2232. Mailing Address: P.O. Box 90775, Henrico, VA 23273-0775. Tel.: 804-360-2071.
Institution Type/Description: Historic Building: housed in the former office of vocation school teacher, Virginia Randolph; built in 1937. A National Register Landmark.
Collections: Virginia Randolph's life & career; period furnishings; personal artifacts; photographs.
Hours & Admission Prices: Call for hours. No charge.

Gloucester

GLOUCESTER MUSEUM OF HISTORY, 6539 Main St., Gloucester, VA 23061. Mailing Address: P.O. Box 5, White Marsh, VA 23183-0005. Tel.: 804-693-1234. Fax: 804-693-1234.
E-mail: bdeal@gloucesterva.info
Web Site: www.gloucesterva.info/museum/historyhome.htm
Founded: 1990.
Congressional District: 1
Key Personnel: Dir., Betty Jean Deal.
Personnel Profile: Part-Time Paid 2; Part-Time Volunteers 30.
Governing Authority: Tax-exempt.
Institution Type/Description: History Museum: built ca 1770.
Collections: local history & culture; photographs; period furnishings.
Hours & Admission Prices: Mon.-Sat. 10-3; tours by appointment. No charge; donations accepted. Closed holidays.
Attendance: 3,000 (accurate)
Membership: Friends of Museum $10.

ROSEWELL RUINS, 5113 Old Rosewell Lane, Gloucester, VA 23061. Mailing Address: P.O. Box 1456, Gloucester, VA 23061. Tel.: 804-693-2585. Facebook: Rosewell Fouondatioin.
E-mail: rosewell@inna.net

Web Site: www.rosewell.co
Key Personnel: Pres. (V), Sandra Pait; Museum Shop Mgr., Beverly Egan; Museum Shop Mgr., Jennifer Griffin.
Personnel Profile: Part-Time Paid 2; Part-Time Volunteers 3.
Governing Authority: Parent Institution: Rosewell Foundation. Tax-exempt: 501(c)(3).
Institution Type/Description: Historic House Museum: the ruins of the Page family mansion; built in 1725.
Collections: Page family life & history; mansion ruins.
Facilities: visitor center. Museum-related items for sale.
Activities: educational programs.
Hours & Admission Prices: Summer: Mon.-Thurs. 10-4, Sat. 10-4, Sun. 1-4; Winter: call for hours; groups by appointment. Adults $4, students $3, children 6-12 $2; discount to Student Time Travelers; children 5 & under no charge. Closed Fridays.

WARNER HALL GRAVEYARD, 4750 Warner Hall Rd, Gloucester, VA 23061-4507. Tel.: 804-648-1889.
E-mail: apva@apva.org
Web Site: www.apva.org/warnergraveyard
Congressional District: 1
Key Personnel: Exec. Dir., Elizabeth Kostelny.
Governing Authority: nonprofit society. Parent Institution: Association for the Preservation of Virginia Antiquities, 2300 E. Grace St., Richmond, VA 23223. Tax-exempt.
Institution Type/Description: Historic Site: graveyard containing the tombs of the Warner & Lewis Families; including that of Augustine Warner, the first Warner to settle in Gloucester County & the forefather of George Washington.
Collections: tombstones of the Colonial period.
Research Fields: genealogy; history.
Hours & Admission Prices: Daily dawn to dusk. No charge. &
Attendance: 500 (estimated)

Gloucester County

WALTER REED BIRTHPLACE, At the corner of Hwy. 614 & 616, Gloucester County, VA 23061. Mailing Address: P.O. Box 160, Gloucester, VA 23061-0160. Tel.: 804-693-3663.
E-mail: ccbzanoni@gmail.com
Founded: 1927.
Congressional District: 1
Key Personnel: Chm. (V), Ceci Brown.
Personnel Profile: Full-Time Volunteers 1; Part-Time Paid 1; Part-Time Volunteers 5.
Volunteer Hours: 30
Governing Authority: nonprofit society. Parent Institution: Assoc. for Preservation of Virginia Antiquities (APVA), 2300 E. Grace St., Richmond, VA 23223. Tax-exempt: 501(c)(3).
Institution Type/Description: Historic House: three-room frame house, the birthplace of Walter Reed, September, 1851, a Major in the U.S. Army & the surgeon who is known as the conqueror of yellow fever.
Collections: period furniture; mid-19th century lifestyle.
Research Fields: history; architecture.
Activities: Annual Events: Historic Garden Week in April by appointment only.
Publications: brochure on facility.
Hours & Admission Prices: By appointment only. Suggested Donation: adults $5; APVA members, children under 17 & members no charge.
Attendance: 50 (accurate)
Membership: Student $15; Annual $25; Sustaining & Benefactor $500; Life $1,000.

Gloucester Point

VIRGINIA INSTITUTE OF MARINE SCIENCE, Rte. 1208, Greate Rd., Gloucester Point, VA 23062. Mailing Address: P.O. Box 1346, Gloucester Point, VA 23062-1346. Tel.: 804-684-7000 & 7285. Fax: 804-684-7097.
E-mail: jmusick@vims.edu
Web Site: www.vims.edu
Founded: 1969.
Congressional District: 1
Key Personnel: Dean & Dir., John Wells; Chief Administrative Officer, Jennifer LaTour; Dir. Communications, Dave Malmquist; Dir. Library, Carl Coughlin; Bibliographic Svcs. Librarian, Marilyn Lewis; Cur., Paul Gerdes.
Governing Authority: college. College of William & Mary, Williamsburg, VA 23185. Tax-exempt.
Institution Type/Description: Marine Research Institute: located on Colonial Village archaeological site.
Collections: 100,000 specimens of preserved fishes; marine fishes from Nova Scotia to Florida; deep sea fishes; deep-sea sharks; Appalachian freshwater fishes; scientific Ichthyology collection.
Research Fields: systematics & ecology of fishes & herps.
Facilities: 29,000-vol. library of bound journals & books on marine science, available for inter-library loan; aquarium; field research station; 100-seat auditorium; classrooms.
Activities: formally organized education programs for graduate students affiliated with William & Mary College; systematic fish collections for scientific study.
Publications: VIMS special scientific reports series.
Hours & Admission Prices: Mon.-Fri. 9-4:30. Open to qualified professional scientists. No charge. &

Goldvein

THE GOLD MINING CAMP MUSEUM, (M), 14421 Gold Dust Pkwy., Goldvein, VA 22720. Tel.: 540-752-5330. Fax: 540-752-5325.
E-mail: monroepark@fauquiercounty.gov
Web Site: www.goldvein.com
Founded: 1998.
Personnel Profile: Full-Time Paid 1; Part-Time Paid 1; Part-Time Volunteers 30.
Governing Authority: Parent Institution: Fauquier County Parks & Recreation Dept. Tax-exempt.
Institution Type/Description: History Museum.
Collections: gold mining artifacts; 3 reconstructed mine camp buildings.
Facilities: 14 acre park.
Activities: group programs; gold panning demonstrations.
Hours & Admission Prices: Wed.-Sat. 9:30-5, Sun. 12-4. No charge; donations accepted. Closed New Year's Day; Easter; Independence Day; Thanksgiving; Christmas. &

Goochland

GOOCHLAND COUNTY MUSEUM & HISTORICAL CENTER, 2875 River Rd. W., Rte. 6, Goochland, VA 23063. Mailing Address: P.O. Box 602, Goochland, VA 23063-0602. Tel.: 804-556-3966. Fax: 804-556-3966. TDD: 804-556-5300.
E-mail: goochlandhistory@verizon.net
Web Site: www.goochlandhistory.org
Founded: 1968.
Congressional District: 3
Key Personnel: Pres. (V), Peter Rippe; Exec. Dir., Phyllis Silber.
Personnel Profile: Part-Time Paid 2; Part-Time Volunteers 25; Interns 1.
Governing Authority: society. Goochland County Historical Society. Tax-exempt: 501(c)(3).
Institution Type/Description: Local History Center: housed in 1836 old jail.
Collections: archives; folklore; archaeology; artifacts; maps; documents; medical instruments.
Research Fields: pre-history-Indians; histories of Goochland medicine, schools & churches; genealogy.
Facilities: workroom; library.
Activities: lectures; oral history; scholarship program with schools. Annual Event: tour historic houses in October.
Publications: annual magazine; newsletter; books.
Hours & Admission Prices: Historical Center: Winter Wed.-Fri. 10-3; Spring & Summer Tues.-Fri. 10-3. Jail Museum: by appointment. No charge; donations accepted. Closed legal holidays. &
Attendance: 500 (accurate)
Membership: Library & Institution $15; Friends $25; Supporting & Business $50; Courthouse $50-99; Gold Mine $100-$249; Byrd Creek $250-$499; James River $500 & up.

Gordonsville

CIVIL WAR MUSEUM AT THE EXCHANGE HOTEL, (M), 400 S. Main St., Gordonsville, VA 22942. Mailing Address: P.O. Box 542, Gordonsville, VA 22942-0542. Tel.: 540-832-2944.
E-mail: hgiexchangehotel@gmail.com
Web Site: www.hgiexchange.org
Founded: 1971.
Congressional District: 7
Key Personnel: Dir. & Museum Shop Mgr., Angel May; Pres. (V), Christopher Stephens.
Personnel Profile: Full-Time Paid 1; Part-Time Volunteers 2.
Governing Authority: nonprofit organization. Parent Institution: Historic Gordonsville, Inc.
Institution Type/Description: Civil War Museum: built in 1860 railroad hotel used as a Confederate receiving hospital during the Civil War.

Collections: Civil War uniforms, weapons & other items; period artifacts from hotel; medical instruments & equipment; railroad memorabilia; research library.
Facilities: 30-vol. library of material relating to the town of Gordonsville & Civil War as it pertains to the hotel available to the public. Museum-related items for sale.
Activities: guided tours; organized educational programs; loan exhibitions. Museum Sponsors: Medical & Military Living History, spring & fall; Christmas Open House.
Publications: quarterly newsletter.
Hours & Admission Prices: Mon.-Thurs. & Sat. 10-4, Sun. 1-4. Adults $5, children 8-12 $3; children 7 & under no charge.
Attendance: 4,400 (accurate)
Membership: Senior Citizen $15; Individual $25; Family $40.

Gum Springs

GUM SPRINGS MUSEUM & CULTURAL CENTER, 8100 Fordson Rd., Gum Springs, VA 22306. Tel.: 703-375-9825.
Web Site: www.gshsfcva.org/gshs05.htm
Key Personnel: Pres., Ron Chase
Institution Type/Description: History Museum.
Collections: history of the Gum Springs community; photographs of Gum Springs' residents & founding families.
Hours & Admission Prices: Call for hours.

Gwynn's Island

GWYNN'S ISLAND MUSEUM, Old Ferry Rd., Gwynn's Island, VA 23066. Tel.: 804-725-7949.
Web Site: www.gwynnsislandmuseum.org
Institution Type/Description: History Museum.
Collections: local history & culture; period furnishings; early clothing; photographs; weapons; uniforms; school room.
Hours & Admission Prices: April-Oct. Fri.-Sun. 1-5; other times by appointment. No charge; donations accepted.

Hampden-Sydney

THE ESTHER THOMAS ATKINSON MUSEUM, College Rd., Hampden-Sydney, VA 23943. Mailing Address: P.O. Box 745, Hampden-Sydney, VA 23943-0745. Tel.: 434-223-6134. Fax: 434-223-6344.
E-mail: away@hsc.edu
Web Site: www.hsc.edu/Museum/
Founded: 1968.
Congressional District: 5
Key Personnel: Chm. Program Bd., Frank B. Atkinson; Dir. & Cur., Angela Way.
Personnel Profile: Full-Time Paid 1; Part-Time Volunteers 20.
Governing Authority: college; nonprofit. Parent Institution: Hampden-Sydney College. Tax-exempt: 501(c)(3).
Institution Type/Description: College Museum.
Collections: Hampden-Sydney college history from 18th century to present; Draper camera.
Research Fields: museum collections.
Activities: guided tours; lectures; temporary exhibitions; workshops; special events.
Publications: newsletter; museum brochure; personalized brick brochure.
Hours & Admission Prices: Tues.-Fri. 10-12 & 1-5; other times by appointment. No charge; donations accepted. Closed school holidays. &
Attendance: 3,100 (estimated)
Membership: Contributor up to $199; 1775 Club $200-$499; Preservation Circle $500-$999; Curator's Circle $1,000-$2,499; Spencer Patron Society $2,500-$4,999; Bowman Heritage Society $5,000-$9,999; Atkinson Leadership Society $10,000 & up; Sponsorship $500-$15,000.

Hampton

CHARLES H. TAYLOR ARTS CENTER, 4205 Victoria Blvd., Hampton, VA 23669-4243. Tel.: 757-727-1490. Fax: 757-727-1167.
E-mail: artscom@hampton.gov
Web Site: www.hamptonarts.net
Founded: 1989.
Congressional District: 2
Key Personnel: Dir., Michael P. Curry; Gallery Mgr., James Warwick Jones; Chm. (V), Ross A. Mugler.
Personnel Profile: Full-Time Paid 5; Part-Time Paid 7; Part-Time Volunteers 100.

Governing Authority: municipal government. Tax-exempt: 170(b)(1)(A).
Institution Type/Description: Visual Arts Center.
Collections: regional & national artists.
Activities: guided tours; lectures; films; performing arts productions & festivals; organized education programs; workshops.
Publications: bimonthly Diversions magazine, includes calendar; monthly Calls for Entries.
Hours & Admission Prices: Tues.-Fri. 10-6, Sat.-Sun. 1-5. No charge; donations accepted. Closed New Year's Day; Presidents' Day; Memorial Day; Independence Day; Labor Day; Thanksgiving; Christmas; city holidays. &
Attendance: 10,348 (accurate)
Membership: Artists Organization $30.

HAMPTON HISTORY MUSEUM, 120 Old Hampton Lane, Hampton, VA 23669-4096. Tel.: 757-727-1610. Fax: 757-727-6712.
E-mail: gdrummond@hampton.gov
Web Site: www.hampton1610.com
Founded: 2003.
Congressional District: 2
Key Personnel: Operations Mgr. & Grants Admin., Gaynell Drummond; Museum Assn. Pres., Tim Smith; Museum Assn. Treas., Robert Allsbrook; Museum Educator & Public Rels., Winette Jeffery; Registrar, Bethany Austin; Cur., Michael Cobb; Museum Shop Mgr., Vivian Tanzer; Administrative Asst., Gloria Jones.
Personnel Profile: Full-Time Paid 3; Part-Time Paid 5; Part-Time Volunteers 20; Interns 1.
Governing Authority: municipal government. Parent Institution: City of Hampton. Tax-exempt.
Institution Type/Description: History Museum.
Collections: area history from 1607 to present.
Research Fields: American history including settlement, colonial, revolution, antebellum, Civil War, modern, Native American, African American, women, military, & maritime.
Facilities: 7,000 sq. ft. exhibit space; rental hall. Museum-related items for sale.
Activities: docent program; formal education programs; guided tours; lectures; loan, participatory, traveling & temporary exhibitions; rental gallery; lecture series.
Publications: quarterly newsletter.
Hours & Admission Prices: Mon.-Sat. 10-5, Sun. 1-5. Adults $5, senior citizens, students & children $4; discount to groups & AAM members; members no charge. Closed New Year's Day; Thanksgiving; Christmas. &
Membership: Student $10; Individual $20; Couple $30; Family $50; Kecoughtan Society $100-$299; Fort Algernon Society $300-$499; Elizabeth City Society $500-$999; Port Hampton Society $1,000 & up.

HAMPTON UNIVERSITY MUSEUM, Hampton University, Hampton, VA 23668. Tel.: 757-727-5308. Fax: 757-727-5170.
E-mail: museumeducation@hamptonu.edu
Web Site: museum.hamptonu.edu
Founded: 1868.
Congressional District: 1
Key Personnel: C.E.O., William R. Harvey; Dir., Nashid Madyun; Cur. Collections, Vanessa Thaxton-Ward; Office Mgr., Brenda Carpenter; Visitor Svcs., Robert Jondreau; Editor International Review of African American Art, Juliette Harris; Asst. to the Archivist, Donzella Maupin; Archivist Staff, Cynthia Poston; Archivist Staff, Andreese Scott.
Personnel Profile: Full-Time Paid 8; Part-Time Paid 1; Part-Time Volunteers 100.
Governing Authority: university. Parent Institution: Hampton University. Tax-exempt: 501(c)(3).
Institution Type/Description: General Museum.
Collections: traditional African, Asian, Oceanic & American Indian art; 19th- & 20th-century African American art; contemporary African art; Hampton University history.
Research Fields: African, North American Indian & African American Art.
Facilities: 1,000-vol. library of books on traditional art & ethnology available for use upon request.
Activities: guided tours; lectures; gallery talks; formally organized education programs for children; permanent, temporary & traveling exhibitions.
Publications: books, To Lead & to Serve: American Indian Education at Hampton Institute 1878-1923; Five Decades: John Biggers & the Hampton Tradition in the Arts; The Frederick Douglass & Harriet Tubman Series of 1938-40; Magazine: The International Review of African American Art; book, A Taste for the Beautiful: Zairian Art from the Hampton University Museum; The Murals of John Thomas Biggers: American Muralist, African American Artist; Elizabeth Catlett: Works On Paper 1944-1992.

Hours & Admission Prices: Mon.-Fri. 8-5, Sat. 12-4. No charge; donations accepted. Closed national holidays; campus holidays. &

Attendance: 30,000 (estimated)

Membership: Student & Senior Citizen $15; Individual $25; Dual $30; Family $40; Organization $60; Contributor $100; Supporting $250; Sustaining $500; Benefactor $1,000.

ST. JOHN'S CHURCH AND PARISH MUSEUM, 100 W. Queens Way, Hampton, VA 23669-4014. Tel.: 757-722-2567. Fax: 757-722-0641.

E-mail: office@stjohnshampton.org
Web Site: www.stjohnshampton.org
Founded: 1976.
Congressional District: 1
Key Personnel: Parish Historian & Cur., Beverly F. Gundry.
Personnel Profile: Part-Time Volunteers 8.
Governing Authority: church. Parent Institution: Diocese of Southern VA. Tax-exempt.
Institution Type/Description: Historic Buildings: c.1728 church, fourth site of worship in Elizabeth City Parish, established in 1610, the oldest parish in continuous existence in the English-settled United States. Museum housed in 1889, Parish Hall. Two 17th century sites are owned by the church & contain exhibits of interest; gravestones & pictorial displays.
Collections: 16th to 18th-century prayer books & bible; artifacts from the 1623-4 church site; ceramics; photographs & illustrations; memorabilia.
Research Fields: parish history & archaeological research.
Activities: guided tours.
Publications: Cemetery Inscriptions.
Hours & Admission Prices: Mon.-Fri. 9-3:30, Sat. 9-12. No charge; donations accepted. Closed holidays. &
Attendance: 1,969 (estimated)

VIRGINIA AIR & SPACE CENTER, 600 Settlers Landing Rd., Hampton, VA 23669-4033. Tel.: 757-727-0900. Fax: 757-727-0898. Facebook: Virginia Air & Space Center.

E-mail: bdeprofio@vasc.org
Web Site: www.vasc.org
Formerly: Virginia Air and Space Center and Hampton Roads History Center
Founded: 1991.
Congressional District: 1
Key Personnel: Interim Exec. Dir., Brian DeProfio; Pres. (V), James Reade Chisman; Dir. Education, Richard Byles; Dir. Administrative Svcs., Jenny Kelly; Cur. & Dir. of Exhibits & Collections, Allen Hoilman; Museum Shop Mgr., Danielle Price.
Personnel Profile: Full-Time Paid 25; Part-Time Paid 60; Part-Time Volunteers 174; Interns 3.
Governing Authority: nonprofit organization. Tax-exempt: 501(c)(3).
Institution Type/Description: Air, space, science & technology museum.
Collections: aerospace accomplishments in the region; military & civilian aircraft; research aircraft; NASA spacecraft & other space artifacts.
Research Fields: aerospace history; math & science education techniques; visitor studies.
Facilities: 600-vol. library of aerospace material; 300-seat IMAX 3D theatre; educational facilities; 50,000 sq. ft. exhibit space; cafe. Museum-related items for sale.
Activities: docent program; films; formal education programs for children; guided tours; loan, temporary, traveling & participatory exhibits; rental gallery; IMAX 3D theatre.
Hours & Admission Prices: Jan. 3-March 14 & Sept. 12-Dec. Tues.-Sat. 10-5, Sun. 12-5, call for additional Mon. hours; March 15-May 25 Mon.-Sat. 10-5, Sun. 12-5; May 26-Sept. 5 Mon.-Wed. 10-5, Thurs.-Sun. 10-7. Exhibits: adults $11.50, senior citizens $10.50, children $9.50. IMAX: adults $9, senior citizens $8, children $7. IMAX feature films: adults $13, senior citizens $12, children $11; discounts for AAA, AAM & ASTC members, military & NASA; members no charge for exhibits. Combination tickets available. Closed Thanksgiving; Christmas. &
Attendance: 402,169 (accurate)
Membership: Cosmic Kids Club & Commuter $65; First Class $95; Gold Preferred $120; Executive $275; Million Miler $500; Corporate $1,000.

Hanover

HANOVER HISTORICAL SOCIETY MUSEUM - OLD STONE JAIL, Hwy. 301, Court Green, Hanover, VA 23069. Mailing Address: c/o Hanover County Historical Society, P.O. Box 91, Hanover, VA 23069-0091. Tel.: 804-537-6262.

Founded: 1967.
Congressional District: 7
Personnel Profile: Full-Time Volunteers 2; Part-Time Volunteers 2.

Governing Authority: society. Parent Institution: Hanover Historical Society. Tax-exempt.
Institution Type/Description: Historical Society Museum.
Collections: local historical material; 1860 school maps; portraits. Historic Buildings: 1835 county jail; 1735 courthouse; 1727 tavern.
Research Fields: pre-1865 homes & buildings.
Activities: lectures; permanent & temporary exhibitions; tours by appointment. Museum Sponsors: Old Hanover Day.
Publications: Hanover in Retrospect; Hanover Historical Society Bulletin; books, History of Hanover County; Old Homes of Hanover County Virginia; Portraits in Courthouse; Names on Confederate Monument; A Child's History of Hanover; Bulletin Vol. I.; Hanover County Graveyards Vol. I & II; biannual members bulletin.
Hours & Admission Prices: By appointment. No charge; donations accepted.
Attendance: 8,042 (estimated)
Membership: Annual $10; Life $100.

Hardy

BOOKER T. WASHINGTON NATIONAL MONUMENT, 12130 Booker T. Washington Hwy., Hardy, VA 24101-3968. Tel.: 540-721-2094. Fax: 540-721-8311 & 5128. Facebook: Booker T. Washington National Monument.

Web Site: www.nps.gov/bowa
Founded: 1956.
Congressional District: 5
Key Personnel: C.E.O., Carla Whitfield; Bookstore Mgr., L. Betsy G. Haynes; Volunteer Coord., Janet Blanchard.
Personnel Profile: Full-Time Paid 10; Part-Time Paid 2; Part-Time Volunteers 20; Interns 1.
Governing Authority: federal. Parent Institution: Department of the Interior, National Park Service, Washington, DC. Tax-exempt.
Institution Type/Description: National Monument: located on the site of Burroughs Plantation, birthplace & early home of Booker T. Washington.
Collections: plantation equipment & furniture; blacksmith tools; archaeological artifacts.
Research Fields: period crops & clothing; life & influence of Booker T. Washington.
Facilities: 1,000-vol. library of books on agriculture, biographical, slavery, Black history; living history farm; 48-seat auditorium; 90-seat multipurpose room & exhibition hall; 1.5 mile National Recreation Trail. Postcards, videos & books for sale.
Activities: guided tours; lectures; slide presentation; living history programs; special events.
Publications: brochures; Rack Cards.
Hours & Admission Prices: Daily 9-5. No charge; donations accepted. Closed New Year's Day; Thanksgiving; Christmas. &
Attendance: 20,000 (estimated)

Harrisonburg

D. RALPH HOSTETTER MUSEUM OF NATURAL HISTORY, Eastern Mennonite University, 1200 Park Rd., Harrisonburg, VA 22802-2462. Tel.: 540-432-4400 & 4000. Fax: 540-432-4488.

E-mail: dossc@emu.edu
Web Site: http://www.emu.edu/sciencecenter
Founded: 1968.
Congressional District: 7
Key Personnel: Museum Educator, Christine C. Hill; Educational Dir., Maureen Gallon; Cur., James Yoder.
Personnel Profile: Part-Time Paid 4.
Governing Authority: college. Parent Institution: Eastern Mennonite University. Tax-exempt.
Institution Type/Description: Natural History Museum.
Collections: geology; mineralogy; zoology; anthropology; paleontology; fluorescent specimens.
Activities: guided tours; permanent exhibitions; study programs for elementary, high school, & college groups.
Publications: yearly brochure.
Hours & Admission Prices: Academic Year: Sun. 2-4; groups by appointment. &
Attendance: 6,000 (accurate)

EXPLORE MORE DISCOVERY MUSEUM, 150 S. Main St., Harrisonburg, VA 22803. Mailing Address: P.O. Box 957, Harrisonburg, VA 22803. Tel.: 540-442-8900.

E-mail: info@iexploremore.com
Institution Type/Description: Children's Museum.
Collections: hands-on exhibitions.
Activities: educational programs; special events; birthday parties.

Hours & Admission Prices: Tues.-Sat. 9:30-5. Admission $5 per person; children under one no charge.

JOHN C. WELLS PLANETARIUM, James Madison University, c/o Physics Dept., Harrisonburg, VA 22807. Mailing Address: James Madison University, Miller Hall Rm 102, MSC-4502, Harrisonburg, VA 22807. Tel.: 540-568-2312. Fax: 540-568-2800. Facebook: JMU Planetarium.
Web Site: www.jmu.edu/planetarium
Founded: 1975.
Congressional District: 6
Key Personnel: Dir., Shanil N. Virani.
Personnel Profile: Full-Time Paid 1.
Governing Authority: state. Commonwealth of Virginia. Parent Institution: James Madison University. Tax-exempt.
Institution Type/Description: Planetarium.
Collections: meteorites; 6 Meade telescopes; Chronos Star projector; Digistar 5.
Facilities: planetarium theater.
Activities: public programs; special groups; public observations; school groups.
Hours & Admission Prices: Visit website for hours. No charge; donations accepted. ら
Attendance: 20,000 (accurate)

SAWHILL GALLERY, JAMES MADISON UNIVERSITY, Main & Grace Sts., Duke Hall, Rm. 101, Harrisonburg, VA 22807. Mailing Address: MSC 7101, Duke Hall, Rm. 101, Harrisonburg, VA 22807. Tel.: 540-568-6407. Fax: 540-568-5862.
E-mail: freebugl@jmu.edu
Founded: 1967.
Congressional District: 6
Key Personnel: Dir., Gary L. Freeburg.
Personnel Profile: Full-Time Paid 1; Part-Time Paid 3; Interns 12.
Governing Authority: university. Affiliated with Art Department, James Madison University. Tax-exempt.
Institution Type/Description: Art Gallery.
Collections: Ernest Staples Collection; Indonesian works; Sawhill collection; pre-classical & classical items; modern works of art.
Activities: temporary exhibitions.
Hours & Admission Prices: Academic Year Mon.-Fri. 10-5, Sat. 12-5; Summer: call for hours. No charge. Closed university holidays. ら
Attendance: 10,000 (estimated)

VIRGINIA QUILT MUSEUM, 301 S. Main St., Harrisonburg, VA 22801-2606. Tel.: 540-433-3818. Fax: 540-433-3818.
E-mail: info@vaquiltmuseum.org
Web Site: www.vaquiltmuseum.org
Founded: 1992.
Congressional District: 26
Personnel Profile: Full-Time Paid 1; Full-Time Volunteers 1; Part-Time Paid 4.
Governing Authority: bd of directors; nonprofit organization.
Institution Type/Description: Quilt Museum.
Collections: early & contemporary quilts; quilting; sewing machines.
Facilities: Museum-related items for sale.
Activities: classes; programs.
Publications: quarterly newsletter.
Hours & Admission Prices: Feb.-Dec. Tues.-Sat. 10-4. Adults $7, students 5-18 $5; children under 5 no charge. Closed major holidays; between exhibits. ら
Attendance: 5,962 (accurate)
Membership: Basic $25; Contributor $50; Patron & Guild $100; Sponsor $250.

Heathsville

NORTHERN NECK FARM MUSEUM, 12705 Northumberland Hwy., Heathsville, VA 22473. Mailing Address: P.O. Box 365, Heathsville, VA 22473-0365. Tel.: 804-443-1118.
Institution Type/Description: Farm Museum.
Collections: agricultural history; farming equipment & tools; photographs.
Activities: school groups; educational programs; special events.
Hours & Admission Prices: May-Oct. Sat.-Sun.

Henrico

ARMOUR HOUSE AND GARDENS AT MEADOWVIEW PARK, 4001 Clarendon Rd., Henrico, VA 23223. Mailing Address: Henrico County Historical Society, P.O. Box 90775, Henrico, VA 23273-0775. Tel.: 804-343-3506.
E-mail: ola@co.henrico.va.us
Institution Type/Description: Historic House Museum: built in 1915.
Collections: local history & culture; period furnishings; personal artifacts; photographs; gardens.
Hours & Admission Prices: Mon.-Fri. 9-4:30. ら

Herndon

KIDWELL FARM AT FRYING PAN PARK, (M), 2709 W. Ox Rd., Herndon, VA 20171-3807. Tel.: 703-437-9101.
Web Site: www.fairfaxcounty.gov/parks/fpp/kidwell.htm
Institution Type/Description: Living History Museum: depicting a family dairy farm from 1920-1950.
Collections: period farm equipment; dairy; smokehouse; corn cribs; equipment sheds; chicken house.
Facilities: Museum-related items for sale.
Activities: special events; educational programs. Museum Sponsors: hayrides March to November.
Hours & Admission Prices: Park: daily dawn to dusk. Farm: daily 9-5.

Hood

ROARING TWENTIES ANTIQUE CAR MUSEUM, Rte. 230, W., Hood, VA 22723. Mailing Address: 1445 Wolftown-Hood Rd., Hood, VA 22723-9802. Tel.: 540-948-6290. Fax: 540-948-6290. Facebook: Roaring Twenties Antique Car Museum.
E-mail: info@roaring-twenties.com
Web Site: www.roaring-twenties.com
Founded: 1967.
Congressional District: 5
Key Personnel: C.E.O. & Owner, Clarissa Dudley; Cur. & Museum Shop Mgr., Martha Dudley.
Governing Authority: private.
Institution Type/Description: Transportation Museum.
Collections: classic cars of 1920s & 1930s with emphasis on rare body styles & models.
Facilities: Museum-related items for sale.
Hours & Admission Prices: Wed.-Thurs. 9-5; other times by appointment. Adults & students $10, children 6-12 $3; discounts to AAM, ICOM, AACA, Car Club members & groups of 2 or more; children under 6 no charge.
Attendance: 225 (estimated)

Independence

GRAYSON CROSSROADS MUSEUM AND CULTURAL EXHIBITS, 107 E. Main St., Independence, VA 24348. Mailing Address: P.O. Box 336, Independence, VA 24348-0336. Tel.: 276-773-3711.
E-mail: 1908courthouse@gmail.com
Founded: 1994.
Congressional District: 9
Key Personnel: Bd. Pres., Laura Bryant.
Personnel Profile: Part-Time Paid 1; Part-Time Volunteers 4.
Governing Authority: Parent Institution: Historic 1908 Courthouse Foundation. Tax-exempt.
Institution Type/Description: Local History Museum: housed in the former county courthouse, 1908.
Collections: Grayson County history & culture; period artifacts; photographs; furnishings; personal artifacts.
Facilities: Museum-related items for sale.
Activities: music jams. Museum Sponsors: Mountain Foliage Festival in October; Christmas Parade.
Hours & Admission Prices: Mon.-Fri. 10-5, Sat. 10-4. No charge; donations accepted. Closed holidays. ら
Attendance: 3,000 (estimated)

Irvington

STEAMBOAT ERA MUSEUM, 156 King Carter Dr., Irvington, VA 22480. Mailing Address: P.O. Box 132, Irvington, VA 22480-0132. Tel.: 804-438-6888. Fax: 804-438-6598.
E-mail: director@steamboatmuseum.org
Web Site: www.steamboateramuseum.org
Congressional District: 1
Key Personnel: Exec. Dir., Terri Thaxton; Pres., Eric Nost.
Personnel Profile: Full-Time Paid 1; Part-Time Volunteers 50.
Governing Authority: Tax-exempt.
Institution Type/Description: Steamboat Era Museum.
Collections: steamboat history & artifacts; Chesapeake Bay; film; oral histories; photographs.
Hours & Admission Prices: Thurs.-Sat. 10-4, Sun. 1-4; other times by appointment. No charge; donations accepted. &
Attendance: 2,000 (accurate)
Membership: Upperdeck $30; Cabin $50; Stateroom $125; Captain's Table $250; Wheelhouse $500; Commodore $1,000.

Isle of Wight

BOYKIN'S TAVERN MUSEUM, 17130 Monument Cir., Isle of Wight, VA 23397. Tel.: 757-365-9771.
Institution Type/Description: Historic Building: built in 1762. Listed on the National Register of Historic Places.
Collections: local history & culture; photographs; period artifacts.
Activities: special events.
Hours & Admission Prices: Feb. 4 to late Dec. Thurs.-Sat. 11-4, Sun. 1-5. Closed Easter; Thanksgiving; Christmas Eve, Day & week.

Jamestown

HISTORIC JAMESTOWNE, 1365 Colonial Pkwy., Jamestown, VA 23081. Tel.: 757-229-4997. Fax: 757-564-3844.
E-mail: info@historicjamestowne.org
Web Site: www.historicjamestowne.org
Formerly: Jamestown National Historic Site
Founded: 1607.
Congressional District: 1
Key Personnel. Exec. Dir. Preservation Virginia, Elizabeth Kostelny; Dir. Research & Interpretation, Preservation Virginia, Dr. William Kelso; Dir. Public Programs & Operations, Sheryl Kingery Mays; Supt., CNHP, Dan Smith; Cur. Archaeology, Preservation Virginia, Beverly Straube; Information Officer, CNHP, James Perry; Retail Mgr., Historic Jamestowne, Carrie Wiggins.
Personnel Profile: Full-Time Paid 28; Part-Time Paid 18; Part-Time Volunteers 90.
Governing Authority: nonprofit society. Joint venture of Preservation Virginia and Colonial National Historic Park, National Park Service. Tax-exempt.
Institution Type/Description: Historic Site & Preservation Project: first permanent English settlement in America.
Collections: interpretive markers; monuments; memorials; 17th-century artifacts. Historic Buildings: 17th-century church tower; 1907 memorial church.
Research Fields: 17th-century Virginia history; the remains of 1607 fort.
Facilities: visitors center; restaurant.
Activities: archaeological excavations; tours.
Publications: Jamestown Rediscovery Monographs; Jamestown: The Buried Truth; The Jamestown Archaeological Assessment; The Archaearium: Rediscovering Jamestown 1607-1699; Historic Jamestowne Guidebook.
Hours & Admission Prices: Daily 8:30-4:30. Adults $10; children under 15, Preservation Virginia & National Park Service members no charge. Closed New Year's Day; Thanksgiving; Christmas. &
Attendance: 216,000 (accurate)

Kilmarnock

KILMARNOCK MUSEUM, 76 N. Main St., Kilmarnock, VA 22482. Mailing Address: P.O. Box 1371, Kilmarnock, VA 22482. Tel.: 804-436-9100.
Institution Type/Description: History Museum.
Collections: local history, business, commerce, & culture; photographs; period furnishings; personal artifacts.
Hours & Admission Prices: Thurs.-Sat. 11-3.

King George

KING GEORGE MUSEUM AND RESEARCH CENTER, 9483 Kings Hwy., King George County Courthouse, King George, VA 22485. Mailing Address: P.O. Box 424, King George, VA 22485-0424. Tel.: 540-775-9477.
Web Site: www.kghistory.org
Governing Authority: Parent Institution: King George County Historical Society.
Institution Type/Description: Historical Society Museum.
Collections: local history, heritage, & culture; photographs; genealogy; period artifacts.
Facilities: library.
Activities: educational programs; hands-on activities for children; research.
Hours & Admission Prices: March-Oct. Thurs. & Sat. 10-2; Nov.-Feb. Sat. 10-2; other times by appointment. Closed New Year's Eve & Day; Easter; Memorial Day; Independence Day; Labor Day; Thanksgiving weekend; Christmas Eve, Day & week.

King William

KING WILLIAM HISTORICAL MUSEUM, 227 Horse Landing Rd., King William, VA 23086. Mailing Address: P.O. Box 233, King William, VA 23086-0233. Tel.: 804-769-9619.
E-mail: kwits@kingwilliamhistory.org
Web Site: www.kingwilliamcounty.us
Founded: 2005.
Congressional District: 1
Key Personnel: Dir., Carl Fischer; Chm. (V), Lloyd Huckstep; Museum Shop Mgr., Pat Fitzgerald.
Personnel Profile: Full-Time Volunteers 4; Part-Time Volunteers 12; Interns 1.
Governing Authority: Tax-exempt.
Institution Type/Description: History Museum.
Collections: county history & culture; African American, Colonial & Native American artifacts; murals; photographs; period artifacts.
Publications: quarterly newsletters.
Hours & Admission Prices: Sat.-Sun. 1-5; other times by appointment. No charge; donations accepted.
Attendance: 645 (estimated)

PAMUNKEY INDIAN MUSEUM, (M), 175 Lay Landing Rd., King William, VA 23086-2126. Tel.: 804-843-4792.
Web Site: www.pamunkey.net/museum.html
Key Personnel: Mgr., Joyce Krigsvold
Institution Type/Description: American Indian Museum.
Collections: life & culture of Pamunkey Indians.
Facilities: Museum-related items for sale.
Hours & Admission Prices: Tues.-Sat. 10-4, Sun. 1-4. Adults $2.50, seniors $1.75, children 6-12 $1.25; children under 6 no charge.

Kinsale

KINSALE FOUNDATION AND MUSEUM, 449 Kinsale Rd., Kinsale, VA 22488. Mailing Address: P.O. Box 307, Kinsale, VA 22488-0307. Tel.: 804-472-3001.
Institution Type/Description: History Museum.
Collections: local history & culture; photographs; period artifacts.
Facilities: Museum-related items for sale.
Hours & Admission Prices: May-Sept. Fri.-Sat. 10-5, Sun. 2-5; Oct.-April Fri.-Sat. 10-5. No charge.

Lancaster

MARY BALL WASHINGTON MUSEUM & LIBRARY, INC., (M), 8346 Mary Ball Rd., Lancaster, VA 22503. Mailing Address: Box 97, Lancaster, VA 22503-0097. Tel.: 804-462-7280. Fax: 804-462-6107.
E-mail: history@mbwm.org
Web Site: www.MBWM.org
Founded: 1958.
Congressional District: 1
Key Personnel: Mgr., Valencia Keeve; C.E.O., Karen Hart; Pres. (V), Carolyn H. Jett.
Personnel Profile: Full-Time Paid 1; Part-Time Paid 1; Part-Time Volunteers 40.
Volunteer Hours: 4,800
Operating Expenses: 125,000
Operating Income: 125,000
Governing Authority: nonprofit organization. Tax-exempt: 501(c)(3).

Institution Type/Description: Local History Museum.
Collections: 18th to 20th century artifacts pertaining to Virginia's Northern Neck region including textiles; furniture; paintings; family & regional memorabilia; oral histories; folklife relics; Civil War artifacts; archives; genealogical & family research center; genealogy. Historic Buildings: 1821 jail; 1797 clerks office; 1830 Lancaster House.
Research Fields: local history; Virginia genealogy; regional folk culture; regional historic architecture.
Facilities: 8,000-vol. research library; museum exhibit galleries; historic buildings.
Activities: permanent & temporary exhibitions; historic site surveys; oral history; public & membership programs; Virginia standards of learning based school programs.
Publications: newsletter; 1850 Census of Lancaster County, Abstracts of Wills of Lancaster County WB-1796-1839, Millenbeck Report; Abstracts of Wills of Lancaster County WB 29 & 30, 1840-1925; Queenstown, Early Port Town of Lancaster County; The Thomas Carters of Lancaster, Virginia; James Gordon & His Family of Lancaster; Ball Family Outline; 1860 Census, Lancaster County, Virginia; Where the River Meets the Bay; Civil War Roster, Lancaster County, 1860-1864; Land Between Waters; occasional paper series, "Echoes of Yesteryear".
Hours & Admission Prices: Museum: Wed.-Fri. 10-4; other times by appointment. Adults $3; discounts to AAM members; members no charge. Library: Tues.-Fri. 10-4, Sat. 11-3; other times by appointment. Research: $5. Closed major holidays.
Attendance: 1,000 (accurate)
Membership: Individual $35; Family $50; MBW Friends $100; MBW Supporters $300; Lifetime $1,000.

Lawrenceville

BRUNSWICK COUNTY MUSEUM, 228 N. Main St., Lawrenceville, VA 23868-1823. Mailing Address: P.O. Box 837, Lawrenceville, VA 23868. Tel.: 434-848-6773; 866-783-9768 (Toll Free). Fax: 434-848-8553.
Institution Type/Description: History Museum.
Collections: county's history & culture; Native American artifacts; Gov. Albertis Harrison; period dolls & clothing; furnishings.
Activities: special events.
Hours & Admission Prices: Tues. & Thurs. 10:30-1, Sat. 1:30-4; other times by appointment.

Leesburg

THE GEORGE C. MARSHALL INTERNATIONAL CENTER AT DODONA MANOR, 217 Edwards Ferry Rd., Leesburg, VA 20176-2305. Tel.: 703-777-1880 & 1301. Fax: 703-777-1889.
E-mail: info@georgecmarshall.org
Web Site: www.georgecmarshall.org
Founded: 2005.
Congressional District: 10
Key Personnel: Exec. Dir., Patricia Daly; Pres. (V), Gorham S. Clark, Esq.; Asst. Dir., Janet Vandervaart; Docent Dir., Tom Bowers
Institution Type/Description: Historic House Museum: housed in the former residence of General & Mrs. George C. Marshall. A National Historic Landmark.
Collections: period furnishings; personal artifacts; photographs.
Hours & Admission Prices: Tours: March-Dec. Sat. 10-5, Sun. and Memorial Day & Labor Day 1-5; groups by appointment. Adults $10; seniors & groups $8; students with ID & children 9-17 $5.
Attendance: 4,000 (estimated)
Membership: Friend of George C. Marshall up tp $99; Friend of the Marshall House $100 & up; George C. Marshall Honor Society $1,000 & up; The Marshall House Honor Society $5,000 & up; Marshall Legacy Trust $25,000 & up.

LOUDOUN MUSEUM, INC., 16 Loudoun St., S.W., Leesburg, VA 20175-2907. Tel.: 703-777-7427. Fax: 703-777-8873.
E-mail: info@loudounmuseum.org
Web Site: www.loudounmuseum.org
Founded: 1967.
Congressional District: 10
Key Personnel: Pres. (V), Elizabeth Whiting; Cur., Alana Blumenthal.
Personnel Profile: Full-Time Paid 1; Part-Time Paid 1; Part-Time Volunteers 15.
Governing Authority: nonprofit organization. Tax-exempt.
Institution Type/Description: Local History Museum: housed in mid-19th century buildings.

Collections: life in the county from prehistory-present; historic documents; manuscripts; decorative arts; costumes; flat textiles; photographs; furniture; tools.
Research Fields: local history; Civil War history.
Facilities: library of reference books & materials on county history available for study on premises. Books for sale.
Activities: lectures; tours; permanent & temporary exhibitions; school programs; special events; audiovisual program on county history.
Publications: book, Legends of Loudoun, A Walk Around Leesburg; The Lure of Loudoun; A Self-Guided Civil War Walking Tour.
Hours & Admission Prices: Fri.-Sat. 10-5, Sun. 1-5. Adults & students $1; discounts to military; children under 4, military & members no charge. Closed New Year's Day; Thanksgiving; Christmas Eve & Day. &
Attendance: 35,500 (estimated)
Membership: Teacher, Student & Senior Citizen $25; Individual $30; Family $40; Sustainer $100; Patron $250; Benefactor $500; Corporate $1,000.

MORVEN PARK, 17263 Southern Planter Lane, Leesburg, VA 20176-7131. Mailing Address: P.O. Box 6228, Leesburg, VA 20178-7433. Tel.: 703-777-6034. Fax: 703-771-9211.
E-mail: jshafagoj@morvenpark.org
Web Site: www.morvenpark.org
Founded: 1955.
Congressional District: 10
Key Personnel: Exec. Dir., Frank Milligan.
Governing Authority: nonprofit organization. Parent Institution: Westmoreland Davis Memorial Foundation. Tax-exempt: 501(c)(3).
Institution Type/Description: Historic House Museum.
Collections: preservation project; carriage & fox hunting museum; history; agriculture; 16th-century tapestries from Flanders; early 20th-century lifestyle & political history. Historic House: Governor's Mansion.
Research Fields: agriculture; early 20th-century Virginia political history; Civil War.
Facilities: nature trails; equestrian center; garden.
Activities: guided tours; public programs; group & school programs; equestrian events; wedding rentals; corporate events.
Publications: thematic brochures of Morven Park's history & inhabitants.
Hours & Admission Prices: Daily. Closed New Year's Day; Thanksgiving; Christmas. &
Attendance: 20,000 (accurate)
Membership: Individual $50; Dual & Family $85; Supporter $125; Patron $250; Donor $500; Governor's Circle $1,000-$10,000.

OATLANDS, 20850 Oatlands Plantation Lane, Leesburg, VA 20175-6572. Tel.: 703-777-3174. Fax: 703-777-4427.
E-mail: oatlands@erols.com
Web Site: www.oatlands.org
Founded: 1965.
Congressional District: 10
Key Personnel: Exec. Dir., Andrea McGimsey; Chm., Mike O'Conner; Dir. Operations, Carolyn McCarthy; Museum Shop Mgr., Carolyn Barnett.
Personnel Profile: Full-Time Paid 11; Part-Time Paid 24; Part-Time Volunteers 80; Interns 1.
Governing Authority: Parent Institution: National Trust for Historic Preservation, 1785 Massachusetts Ave., N.W., Washington, DC 20036. Subsidiary Institution: Oatlands Plantation. Tax-exempt: 501(c)(3).
Institution Type/Description: Historic House Museum: 1804 Oatlands mansion with English walled garden, built by George Carter; a portico with Corinthian capitals, carved by Henry Farnham, was added in 1827.
Collections: 18th- to 20th-century decorative arts.
Research Fields: decorative arts.
Facilities: dependencies; formal gardens; carriage house. Museum-related items for sale.
Activities: exhibits.
Publications: quarterly newsletter.
Hours & Admission Prices: April-Dec. Mon.-Sat. 10-5, Sun. 1-5. Adults $12, senior citizens $10; discounts to groups, National Trust for Historic Preservation members; members no charge. &
Attendance: 40,000 (estimated)
Membership: Single $50; Contributor $100; Supporter $250; Donor $500; Carter-Eustis $1,000; Corporate $2,500.

Lexington

GEORGE C. MARSHALL MUSEUM, 1600 VMI Parade Ground, Lexington, VA 24450. Mailing Address: P.O. Drawer 1600, Lexington, VA 24450-1600. Tel.: 540-463-7103, ext. 125. Fax: 540-464-5229.
E-mail: marshallfoundation@marshallfoundation.org

Web Site: www.marshallfoundation.org
Formerly: George C. Marshall Research Foundation
Founded: 1953.
Congressional District: 6
Key Personnel: Pres., Brian D. Shaw; Assoc. Dir. Leadership Programs, Marti Bissell; Dir. Admin., Carol E. Wheeler; Dir. Library & Archives, Paul B. Barron.
Personnel Profile: Full-Time Paid 15; Part-Time Paid 8; Part-Time Volunteers 1; Interns 8.
Governing Authority: nonprofit organization. Parent Institution: George C. Marshall Foundation. Tax-exempt: 501(c)(3).
Institution Type/Description: Library and History Museum.
Collections: artifacts of the life & times of Army General George C. Marshall; Marshall papers & artifacts; military history artifacts from 1880-1960; Noble Peace Prize; archives; textiles.
Research Fields: 20th-century American military history & diplomatic history; World War I & II.
Facilities: 25,000-vol. library of 20th-century American military history & diplomatic history of World War I & II and the Korean Conflict.
Activities: educational programs for student & adult groups by appointment; temporary exhibitions; teacher workshops. Museum Sponsors: George C. Marshall Lecture Series.
Publications: bulletin, The Strategist; book, The China Mission; Marshall Biography; Papers of George Catlett Marshall-Vols. I, II, III, IV & V; book, The George C. Marshall Interviews and Reminiscences for Forrest C. Pogue; In Search of a Useable Past: The Marshall Plan and Post War Reconstruction Today; The Marshall Plan - Lessons Learned for the 21st Century.
Hours & Admission Prices: Tues.-Sat. 11-4, Sun. 1-5. Adults $5, senior citizens $3, students $2; discounts to groups of 10 or more; children & active military no charge. Closed New Year's Eve & Day; Christmas Eve, Day & week, Easter. &

Attendance: 18,000 (estimated)

LEE CHAPEL & MUSEUM, (M), Washington & Lee University, Lexington, VA 24450-2116. Mailing Address: 11 University Place, Lexington, VA 24450-2116. Tel.: 540-458-8768. Fax: 540-458-5804.

E-mail: lwilkins@wlu.edu
Web Site: leechapel.wlu.edu
Founded: 1928.
Congressional District: 6
Key Personnel: Mgr., Linda Donald; Administrative Asst., Pat Larew; Museum Shop Mgr., Gloria Gorlin; Museum Shop Mgr., Margaret Samdahl.
Personnel Profile: Full-Time Paid 2; Part-Time Paid 20.
Governing Authority: private university; nonprofit. Parent Institution: Washington & Lee University. Tax-exempt: 501(c)(3).
Institution Type/Description: History Museum: housed in 1868 historic building constructed under the direction of R.E. Lee.
Collections: university history with emphasis on its namesakes; Washington, Custis & Lee portraits; Lee family artifacts & memorabilia; statue of Lee by Edward Valentine; statue chamber; Lee family crypt; Lee's office.
Facilities: 525-seat auditorium. Museum-related items for sale.
Activities: school & guided tours; permanent exhibitions; lectures.
Hours & Admission Prices: April-Oct. Mon.-Sat. 9-5, Sun. 1-5; Nov.-March Mon.-Sat. 9-4, Sun.1-4; call to verify. Suggested Donation: adults $5, children under 12 $3. Closed New Year's Eve & Day; Easter; Independence Day; Thanksgiving & weekend after; Christmas Eve, Day & week; university holidays. &

Attendance: 43,994 (accurate)

THE REEVES CENTER, WASHINGTON AND LEE UNIVERSITY, 204 W. Washington St., Lexington, VA 24450. Tel.: 540-458-8034 & 8476. Fax: 540-458-8741.

E-mail: pgrover@wlu.edu
Founded: 1982.
Congressional District: 6
Key Personnel: Dir., Peter Dun Grover; Assoc. Dir. & Cur. Collections, Patricia Hobbs; Mgr. & Cur., Ronald W. Fuchs; Coord. Collections, Kyra Swanson.
Personnel Profile: Full-Time Paid 4; Interns 6.
Governing Authority: private; nonprofit university. Parent Institution: Washington and Lee University. Tax-exempt: 501(c)(3).
Institution Type/Description: Art Museum: housed in two buildings - 1840 Greek Revival house on the front campus & a Palladian-style pavilion with two galleries.
Collections: Chinese ceramics, 2nd to 19th-centuries; European ceramics, 17th-to 19th-centuries; paintings, 17th- to 19th-centuries; Japanese tearoom with tea utensils, scrolls & chabaria containers.

Research Fields: Chinese export porcelain; English & European ceramics; 19th-century American art; decorative arts.
Facilities: 900-vol. library on ceramics, art history, decorative arts & history available to the public on site; seminar room; lecture hall; 3,500 sq. ft. exhibit space.
Activities: formal education programs for university students; guided tours; lectures; loan, temporary & traveling exhibitions.
Publications: catalogs; books, Chinese Export Porcelain in the Reeves Center Collections at Washington & Lee University; A Fragile Union: The Story of Louise Herreshoff.
Hours & Admission Prices: Mon.-Fri. 9-4:30, Sat.-Sun. by appointment. No charge. Closed Memorial Day; Independence Day; Thanksgiving; Christmas to New Year's. &

Attendance: 5,000 (estimated)

ROCKBRIDGE HISTORICAL SOCIETY, 101 E. Washington St., Lexington, VA 24450. Mailing Address: P.O. Box 1409, Lexington, VA 24450-1409. Tel.: 540-464-1058.

E-mail: rochist@hotmail.com
Web Site: rockhist.org
Founded: 1939.
Congressional District: 6
Key Personnel: Exec. Dir., Eric Wilson; Pres. (V), Dr. Charles A. Bodie.
Personnel Profile: Full-Time Paid 1; Part-Time Volunteers 31.
Governing Authority: society. Tax-exempt: 501(c)(3).
Institution Type/Description: General Museum: housed in c.1844 Campbell House, a 3-story brick home.
Collections: books; documents; journals; photographs; works of art; tools; furnishings; furniture relating to history of Rockbridge County with associated artifacts; genealogy.
Research Fields: local architecture; local history; genealogy.
Facilities: 600-vol. library of family papers, clippings, local newspapers located at the library of Washington-Lee University.
Activities: interpretive presentations; bimonthly meetings. Museum Sponsors: Traditional Folklife Festival; Victorian Christmas Open House; Ice Cream Social.
Publications: bimonthly newsletter; Roads of Rockbridge; RHS Proceedings.
Hours & Admission Prices: mid-April to mid-Oct. Mon.-Sat. 10-4, Sun. 12-4; mid-Oct. to mid-April Mon.-Sat. 10-1, Sun. 1-4. No charge; donations accepted. Closed New Year's Eve & Day; Christmas Eve, Day & week. &

Attendance: 5,800 (accurate)
Membership: Student $10; Individual $20; Family $30; Corporate $50.

STANIAR GALLERY, Wilson Hall, 100 Glasgow St., Washington & Lee University, Lexington, VA 24450-2116. Tel.: 540-458-8861 & 8860. Fax: 540-458-8112.

E-mail: archerc@wlu.edu
Web Site: www.wlu.edu
Formerly: Dupont Gallery
Founded: 2006.
Congressional District: 6
Key Personnel: Dir., Clover Archer Lyle.
Personnel Profile: Full-Time Paid 1; Interns 1.
Governing Authority: college. Parent Institution: Washington and Lee University. Tax-exempt: 501(c)(3).
Institution Type/Description: University Art Gallery.
Collections: non-collecting contemporary art gallery.
Activities: lectures; panel discussions; cross-disciplinary symposia; visiting artists' workshops that link the gallery & the university's curriculum.
Publications: exhibition catalogs for selected exhibitions.
Hours & Admission Prices: Sept.-May Mon.-Fri. 9-5. No charge. &

STONEWALL JACKSON HOUSE, (M), 8 E. Washington St., Lexington, VA 24450-2529. Tel.: 540-463-2552. Fax: 540-463-4088.

E-mail: director@stonewalljackson.org
Web Site: www.stonewalljackson.org
Founded: 1954.
Congressional District: 6
Key Personnel: Exec. Dir., Michael Anne Lynn; Pres. (V), George H. Roberts, Jr.; Cur., Heidi Wing Sheldon.
Personnel Profile: Full-Time Paid 4; Part-Time Paid 15; Part-Time Volunteers 60; Interns 1.
Governing Authority: nonprofit organization. Parent Institution: Stonewall Jackson Foundation, 8 E. Washington St., Lexington, VA 24450. Tax-exempt: 501(c)(3).
Institution Type/Description: Historic House Museum: 1859-1861 Stonewall Jackson Home.

Collections: furniture & furnishings owned by T. J. Jackson & his family; period furniture appropriate to 1851-1861 small town life; manuscripts.
Research Fields: life & times of Gen. Thomas J. Jackson; Civil War; ante-bellum Virginia; social history.
Facilities: restored gardens. Museum-related items for sale.
Activities: guided tours; lectures; formally organized education programs for children; permanent & temporary exhibitions; biennial symposium on Stonewall Jackson; summer fellowships for graduate students.
Publications: books, Stonewall Jackson & the Virginia Military Institute: The Lexington Years; Stonewall Jackson in Lexington: The Christian Soldier.
Hours & Admission Prices: Summer: Mon.-Sat. 9-5, Sun. 1-5; Winter call for hours. Adults $6, children 18 & under $3; discounts to groups, and AAM, VAM & SEMC members. Closed New Year's Day; Easter; Thanksgiving; Christmas.
Attendance: 20,709 (accurate)

✳ VIRGINIA MILITARY INSTITUTE MUSEUM, (M), Virginia Military Institute, Jackson Memorial Hall, 415 Letcher Ave., Lexington, VA 24450-2194. Tel.: 540-464-7334. Fax: 540-464-7112. TDD: 540-464-7616.
E-mail: gibsonke@vmi.edu
Web Site: www.vmi.edu/museum
Founded: 1856.
Congressional District: 9
Key Personnel: Exec. Dir., Keith E. Gibson; Registrar, Barbara J. Blakey; Museum Shop Mgr., Betty E. Skillman.
Personnel Profile: Full-Time Paid 3; Part-Time Paid 5.
Governing Authority: university. Parent Institution: Virginia Military Institute. Branch Museum: ROTC Building, Kilbourne Hall. Tax-exempt.
Institution Type/Description: Military, National Historic District and General Museum: located on Virginia Military Institute campus.
Collections: The Henry Stewart Antique Firearms Collection; artifacts relating to the history of VMI & cadet life; alumni & faculty memorabilia; military weapons & uniforms.
Research Fields: history of VMI, alumni, faculty and vicinity.
Facilities: 500-vol. library of books available for use on premises; auditorium. Postcards, prints & books for sale.
Activities: guided tours; lectures; films; inter-museum loan, permanent & temporary exhibitions.
Hours & Admission Prices: Daily 9-5. No charge; donations accepted. Closed New Year's Eve, Day & day after; Thanksgiving; Christmas Eve, Day & week. &
Attendance: 40,000 (accurate)

Lorton

POHICK EPISCOPAL CHURCH, 9301 Richmond Hwy., Lorton, VA 22079-1519. Tel.: 703-339-6572. Fax: 703-339-9884.
E-mail: troknya@pohick.org
Web Site: www.pohick.org
Founded: 1774.
Key Personnel: Rector, Rev. Donald D. Binder.
Personnel Profile: Full-Time Paid 5; Part-Time Paid 3; Part-Time Volunteers 6; Interns 1.
Governing Authority: The Episcopal Church, Diocese of Virginia. Tax-exempt.
Institution Type/Description: Active Church: housed in 1774 parish church of George Washington & George Mason.
Collections: interior box pews; two baptismal fonts; c.1968 pipe organ made of 880 pipes, 13 stops & 17 ranks; Vestry book, containing records from 1732-1785; 1761 Big Prayer Book; 1796 2-vol. Bible; c.1737 Lee plate & cup of hammered silver; chalice; silver bread box & cruets; silver ewer; 2 silver communion chalices; one of the baptismal fonts was originally a mortar (late Saxon-early Norman).
Activities: guided tours; concerts; study clubs; organized educational programs; docent program; regular church programs.
Publications: book, Minutes of the Vestry, Truro Parish 1732-1785; brochure; book of genealogical records.
Hours & Admission Prices: Mon.-Sat. 9-4:30, Sun. 8-4:30. No charge; donations accepted. &

WORKHOUSE ARTS CENTER, 9601 Ox Rd., Lorton, VA 22079. Tel.: 703-584-2900. Fax: 703-690-1880.
E-mail: info@workhousearts.org
Key Personnel: C.E.O., John Mason; Chm. (V), Richard Hausler
Governing Authority: Parent Institution: Lorton Arts Foundation.
Institution Type/Description: Art Gallery.
Collections: works by regional artists; glass & glass blowing.
Facilities: ceramics & glass studios with state of the art infrastructure.
Activities: classes; workshops; special events; educational programs.

Hours & Admission Prices: Wed.-Sat. 11-7, Sun. 12-5. &
Attendance: 75,000 (estimated)
Membership: Friend $50-$149; Enthusiast $150-$249; Contributor $250-$499; Supporter $500-$999; Sustainer $1,000-$1,999; Patron $2,000-$4,999; Benefactor $5,000 & up.

Louisa

LOUISA COUNTY HISTORICAL SOCIETY, 214 Fredericksburg Ave., Louisa, VA 23093-6531. Mailing Address: P.O. Box 1172, Louisa, VA 23093-1172. Tel.: 540-967-5975. Facebook: Louisa County Historical Society.
E-mail: louisahistory@verizon.net
Web Site: www.louisahistory.org
Founded: 1966.
Congressional District: 7
Key Personnel: Pres. (V), Maren Smith.
Personnel Profile: Full-Time Paid 1.
Volunteer Hours: 806
Operating Expenses: 70,000
Operating Income: 70,000
Governing Authority: society. Tax-exempt.
Institution Type/Description: Historical Society Museum: housed in 1868 jail.
Collections: Louisa County history from prehistoric to modern times.
Facilities: 100-vol. library of local & Virginia history, available to the public.
Activities: educational & enrichment for all ages.
Publications: biannual, Louisa County Historical Magazine; quarterly newsletter.
Hours & Admission Prices: Jail Museum: April-Sept. Fri.-Sat. 10am-12pm. Sargeant Museum: Mon.-Sat. 10-4, call to confirm. No charge; donations accepted. &
Attendance: 950 (accurate)
Membership: Individual $35; Family $35; Supporting $50; Sustaining $200; Benefactor $1,000; Life 1,500.

Lovettsville

LOVETTSVILLE HISTORICAL SOCIETY INCORPORATED, 4 E. Pennsylvania Ave., Lovettsville, VA 20180. Mailing Address: P.O. Box 5, Lovettsville, VA 20180-0005. Tel.: 540-822-5499. Fax: 540-822-9797.
E-mail: info@lovettsvillehistoricalsociety.org
Web Site: lovettsvillehistoricalsociety.org
Founded: 1974.
Congressional District: 10
Key Personnel: Chm. (V), Thomas Bullock.
Volunteer Hours: 700
Operating Expenses: 5,200
Operating Income: 11,000
Governing Authority: Tax-exempt.
Institution Type/Description: Historical Society Museum.
Collections: Tax-exempt.
Research Fields: geneology; mapping Revolutionary & Civil wars.
Facilities: Books for sale; postcards.
Activities: lectures; archaeological digs; walking tours.
Publications: local history & heritage books.
Hours & Admission Prices: May-Dec. Sat. 1-4; other times by appointment. No charge; donations accepted.
Attendance: 890 (estimated)
Membership: Student & Senior Citizens (65 & over) $25; Individual $35; Family $50; Lifetime $500.

Luray

SHENANDOAH NATIONAL PARK, 3655 U.S. Hwy. 211 E., Luray, VA 22835-4702. Tel.: 540-999-3500. Fax: 540-999-3601.
E-mail: shen_superintendent@nps.gov
Web Site: www.nps.gov/shen
Founded: 1935.
Congressional District: 7
Key Personnel: Supt., Martha Bogle.
Personnel Profile: Full-Time Paid 180.
Governing Authority: federal. National Park Service, Interior Building, Washington, DC. Visitors Centers: Dickey Ridge, Front Royal, VA., 22630. Byrd Visitor Center. Tax-exempt.
Institution Type/Description: Park Museum & Visitor Center.
Collections: history; herbarium; geology; archives.
Facilities: Books, maps, guides, pictures, T-shirts for sale.
Activities: guided hikes; movies; evening programs; environmental education study areas.

Publications: booklets; books; maps; Everything Was Wonderful: A Pictorial History of the Civilian Conservation Corps. in Shenandoah National Park; In the Light of the Mountain Moon: An Illustrated History of Skyland; The Greatest Single Feature...A Sky-Line Drive: 75 Years of a Mountaintop Motorway.
Hours & Admission Prices: Park: daily 24 hours. $10-15 per car, camping $15-20 per night per site. Byrd Visitor Center: March 25-Nov. 27 daily 9-5. Dickey Ridge Visitor Center: March 31-Nov. 27 Thurs.-Mon. 9-5. Closed Thanksgiving. &
Attendance: 1,200,000 (estimated)
Membership: Shenandoah Natural History Association: Individual $25; Family $35; Supporting $75; Contributing $150; Patron $250; Old Rag Society $500; Stony Man Society $1,000; Hawksbill Society $1,500.

Lynchburg

AMAZEMENT SQUARE, THE RIGHTMIRE CHILDREN'S MUSEUM, (M), 27 Ninth St., Lynchburg, VA 24504-1422. Tel.: 434-845-1888. Fax: 434-845-5221. Facebook: Amazement Square.
E-mail: visitus@amazementsquare.org
Web Site: www.amazementsquare.org
Founded: 1993.
Congressional District: 6
Key Personnel: Pres. & C.E.O., Mort Sajadian, Ph.D.; Chm. (V), Hylan T. Hubbard, III; Treas., Thomas Pettyjohn, Jr.; Exhibit Fabricator, John Kastner; Museum Shop Mgr., Candace Farmer.
Personnel Profile: Full-Time Paid 14; Part-Time Paid 16; Part-Time Volunteers 64; Interns 3.
Governing Authority: private; nonprofit organization. Tax-exempt: 501(c)(3).
Institution Type/Description: Children's Museum.
Collections: hands-on exhibits; paintings; multicultural dolls.
Research Fields: Everyone is Special Partnership with the Laurel Regional School studying the impact & use of multisensory environments and adaptive technology on severely handicapped children.
Facilities: educational facilities; 22,000 sq. ft. exhibit space; education center; 600-seat outdoor theater. Museum-related items for sale.
Activities: arts festivals; docent program; films; formal education programs for children; guided tours; loan, participatory & traveling exhibitions; theater; mentoring programs; outreach programs to underserved communities. Annual Events: Ugly Bug Ball; Amazing Mile Children's Run.
Publications: quarterly newsletter; comic book series, Amazing Adventures of Scorpy Bug.
Hours & Admission Prices: Tues.-Sat. 10-5, Sun. 1-5; call for Mon. holiday hours. Offices. Mon.-Fri. 9-5. Admission 2 59 $9, senior citizens 60 & over $6; discounts to groups of 10 or more by appointment; members & children under 2 no charge. &
Attendance: 83,114 (accurate)
Membership: Flex Square $60; Mini Square $90; Fantastic Square $125; Amazing Square $235.

THE ANNE SPENCER MEMORIAL FOUNDATION, INC., 1313 Pierce St., Lynchburg, VA 24501-1935. Tel.: 434-845-1313.
Founded: 1977.
Congressional District: 6
Key Personnel: Chm., Hugh R. Jones; Chm. Tours, Liz Lovern.
Personnel Profile: Part-Time Volunteers 15.
Governing Authority: nonprofit organization. Tax-exempt.
Institution Type/Description: Historic House & Garden: housed in the former home of poet, Anne Spencer; built in 1903. Listed on the National Register of Historic Places.
Collections: photographs; Victorian furnishings; writings of Anne Spencer; book; stained glass window; African head; statue of Minerva; flowers & shrubs.
Research Fields: Victorian era.
Facilities: 300-vol. library. Museum-related items for sale.
Activities: guided tours; films.
Hours & Admission Prices: By appointment only. Adults $10, seniors & college students $5, children under 12 $3; discounts to groups.
Attendance: 1,000 (estimated)
Membership: Membership by donation.

CREATION HALL MUSEUM, 1971 University Blvd., Lynchburg, VA 24502. Tel.: 434-582-2209. Fax: 434-582-2488
Web Site: www.liberty.edu
Formerly: Museum of Earth and Life History
Founded: 1985.
Congressional District: 6
Key Personnel: Dir., David A. DeWitt, Ph.D.
Governing Authority: Parent Institution: Liberty University. Tax-exempt.
Institution Type/Description: Natural History Museum.

Collections: natural history.
Hours & Admission Prices: Mon.-Tues. & Thurs.-Fri. 8-8, Wed. 8-4, Sat. 9-6, Sun. 1-4. No charge. &
Attendance: 4,000 (estimated)

DAURA GALLERY, (M), Lynchburg College, 1501 Lakeside Dr., Lynchburg, VA 24501-3199. Tel.: 434-544-8343 & 8349. Fax: 804-544-8277.
E-mail: rothermel@lynchburg.edu
Web Site: www.lynchburg.edu/daura
Founded: 1974.
Congressional District: 6
Key Personnel: Dir., Barbara Rothermel; Asst. Dir., Steve Riffee.
Personnel Profile: Full-Time Paid 2; Part-Time Paid 2; Part-Time Volunteers 6; Interns 3.
Governing Authority: private college; nonprofit. Parent Institution: Lynchburg College. Tax-exempt.
Institution Type/Description: Art Gallery.
Collections: 2,000 pieces including 20th-century American and European artists; art of Catalan-American Modernist Pierre Daura.
Research Fields: museum studies.
Facilities: library.
Activities: guided tours by appointment; temporary exhibitions; rotating exhibitions of various artists.
Publications: newsletter; exhibition catalogs; brochures.
Hours & Admission Prices: Aug.-May Mon.-Fri. 9-4, Sun. call for hours; June-July by appointment. No charge; donations accepted. Closed college holidays; New Year's Day; Thanksgiving; Christmas. &
Attendance: 6,000 (estimated)
Membership: Individual $25; Family $50; Patron $100; Curator's Circle $250; Virginia Davis Society $500; Georgia Morgan Society $1,000; Pierre & Louis Daura Society $2,500 & up.

LEGACY MUSEUM OF AFRICAN-AMERICAN HISTORY, 403 Monroe St., Lynchburg, VA 24504-2808. Mailing Address: P.O. Box 308, Lynchburg, VA 24505-0308. Tel.: 434-845-3455. Fax: 434-845-9809.
E-mail: legacymuseum@ntelos.net
Web Site: www.legacymuseum.org
Key Personnel: Museum Admin., Cheryl Robinson
Institution Type/Description: History Museum.
Collections: local African American history & culture.
Activities: exhibit-related programs; special events.
Hours & Admission Prices: Wed.-Sat. 12-4, Sun. 2-4; other times by appointment. Adults $5, seniors $3, youth $2; children under 6 no charge. Closed major holidays. &

LYNCHBURG MUSEUM SYSTEM, (M), 901 Court St., Lynchburg, VA 24504-1603. Mailing Address: P.O. Box 529, Lynchburg, VA 24505-0529. Tel.: 434-455-6226. Fax: 434-528-0162.
E-mail: museum@lynchburgva.gov
Web Site: www.lynchburgmuseum.org
Founded: 1976.
Congressional District: 6
Key Personnel: Dir., Douglas K. Harvey; Pres. (V), Patsy Meyer; Chm. (V), Robert Craighill; Museum Shop Mgr., Laura Crumbley.
Personnel Profile: Full-Time Paid 4; Part-Time Paid 7; Part-Time Volunteers 55.
Volunteer Hours: 1,060
Operating Expenses: 438,440
Operating Income: 88,150
Governing Authority: municipal. Parent Institution: Lynchburg Museum Foundation. Tax-exempt.
Institution Type/Description: History Museum.
Collections: furniture; costumes; art exhibit; photographs; machinery; decorative arts. Historic Buildings: 1855 Old Court House; 1815 Point of Honor.
Major Exhibits: The James River, 6/14-12/15.
Research Fields: Lynchburg history.
Activities: volunteer programs; tours; permanent exhibits; living history program. Museum Sponsors: Fall Festival in October; Christmas Open House; Garden Day tour.
Publications: quarterly, MuseNews.
Hours & Admission Prices: Mon.-Sat. 10-4, Sun. 12-4. Historic House: adults $6. Museum: adults $6; discounts to members; museum professionals no charge. Closed New Year's Day; Thanksgiving; Christmas Eve & Day. &
Attendance: 16,700 (accurate)
Membership: Teacher & Student $25; Grandparents & Individual $50; Family $75.

MAIER MUSEUM OF ART AT RANDOLPH COLLEGE, (M),
One Quinlan St., Lynchburg, VA 24503-1519. Mailing Address:
2500 Rivermont Ave., Lynchburg, VA 24503-1526. Tel.: 434-947-
8136 & 8000. Fax: 434-947-8726. Facebook: Maier Museum.
E-mail: museum@randolphcollege.edu
Web Site: maiermuseum.org
Formerly: Maier Museum of Art, Randolph-Macon Woman's College
Founded: 1957.
Congressional District: 6
Key Personnel: Dir., Martha Kjeseth Johnson; Registrar, Deborah Spanich;
Office Mgr., Danni Schreffler; Acting Preparator & Security Guard, John
Spanich.
Personnel Profile: Full-Time Paid 2; Part-Time Paid 2; Part-Time Volunteers
60; Interns 4.
Governing Authority: college. Parent Institution: Randolph College. Tax-
exempt: 501(c)(3).
Institution Type/Description: Art.
Collections: American paintings; works on paper; photographs.
Research Fields: 19th, 20th & 21st century American art.
Activities: inter-museum loans, permanent & temporary exhibitions; docent
program; public lectures; gallery tours; museum education programs.
Publications: catalogue of the Collection; brochure; exhibition catalogues;
members newsletter; online collection catalog.
Hours & Admission Prices: late April to late Aug. Wed.-Sun. 1-4; late Aug. to
late April Tues.-Sun. 1-5. No charge; donations accepted. Closed academic
holidays, New Year's; Easter; Independence Day; Thanksgiving; Christmas.
Attendance: 7,200 (estimated)
Membership: Special $35; Individual $45; Family & Household $65; Friend
$100; Patron $250; Sponsor $500; Mary Frances Williams Society $1,000;
Benefactor $2,500; Maier Circle $5,000.

OLD CITY CEMETERY MUSEUMS & ARBORETUM, (M),
401 Taylor St., Lynchburg, VA 24501-1245. Tel.: 434-847-1465.
Fax: 434-856-2004.
E-mail: occ@gravegarden.org
Web Site: www.gravegarden.org
Founded: 1806.
Congressional District: 6
Key Personnel: Dir., D. Bruce Christian; Pres. (V), Betty Brown; Museum
Shop Mgr., Kathy Wise.
Personnel Profile: Full-Time Paid 3; Part-Time Paid 2; Part-Time Volunteers
100.
Governing Authority: Parent Institution: Southern Memorial Association.
Tax-exempt.
Institution Type/Description: Historic Landmark: listed on the National Reg-
ister of Historic Places.
Collections: local history & culture; mourning customs; hand tools; grave-
markers; gravestone carvers; railroad history.
Activities: programs; guided tours; events; museums & chapel equipped with
push-button audio.
Publications: books, Free Blacks of Lynchburg, Virginia 1805-1865; Food to
Die For, A Book of Funeral Food, Tips and Tales.
Hours & Admission Prices: Cemetery: daily dawn to dusk. Visitor Center &
Victorian Mourning Exhibit: May & Oct. daily 10-3; June-Sept. &
Nov.-April Mon.-Sat. 10-3. No charge; donations accepted.
Attendance: 30,844 (accurate)

POINT OF HONOR, 112 Cabell St., Lynchburg, VA 24504-1211.
Mailing Address: Lynchburg Museum System, P.O. Box 529,
Lynchburg, VA 24505-0529. Tel.: 434-455-6226. Fax: 434-528-
0162. Facebook: Lynchburg Museum System.
E-mail: museum@lynchburgva.gov
Web Site: www.pointofhonor.org
Founded: 1976.
Congressional District: 6
Key Personnel: Admin., Douglas K. Harvey; Museum Shop Mgr., Laura
Crumbley.
Personnel Profile: Full-Time Paid 4; Part-Time Paid 7; Part-Time Volunteers
55.
Volunteer Hours: 1,075
Governing Authority: municipal. Parent Institution: Point of Honor, Inc.;
Subsidiary Institution: Lynchburg Museum System. Tax-exempt: 501(c)(3).
Institution Type/Description: Historic House: home of Dr. George Cabell,
physician to Patrick Henry.
Collections: furnishings, accessories & decorative arts reflecting the lifestyle
of the 1800-1830 period.
Research Fields: Federal period Virginia furnishings & decorative arts;
Genealogical investigation of the former owners of Point of Honor.
Facilities: Museum-related items for sale.

Activities: guided tours; lectures; organized education programs for children;
docent program.
Publications: biannual newsletter, MUSE News.
Hours & Admission Prices: Mon.-Sat. 10-4, Sun. 12-4. Adults $6, seniors 60
& over $5, youth 6-17 $3; children under 6 no charge. Closed New Year's
Day; Thanksgiving; Christmas Eve & Day.
Attendance: 9,950 (accurate)
Membership: Teacher & Student $25; Individual & Grandparents $50; Family
$75.

SOUTH RIVER MEETING HOUSE, 5810 Fort Ave., Lynchburg,
VA 24502-1928. Tel.: 434-239-2548. Fax: 434-239-6071.
Web Site: qmpc.org/contents/meetinghousehistory.shtml
Founded: 1757.
Congressional District: 6
Key Personnel: Dir., Diane Baldwin.
Personnel Profile: Part-Time Volunteers 7.
Governing Authority: church. Parent Institution: Quaker Memorial Presbyte-
rian Church. Tax-exempt: 501(c)(3); 170(b)(1)(A); 509(a)(1).
Institution Type/Description: Historic Site: a restored Society of Friends
meeting house, c.1791. Listed on the National Register of Historic Places.
Collections: photographs; artifacts from 1864 Civil War battle; documents
relating to history of Quakers & Meeting House; Quaker cemetery.
Research Fields: local Quaker history.
Facilities: 25-vol. library of books on Quaker history, spirituality & genealogy
housed in adjacent Quaker Memorial Presbyterian Church.
Activities: guided tours; lectures; organized educational programs; interpreter;
training programs.
Publications: brochures; book, Lynchburg's Pioneer Quakers & Their Meeting
House.
Hours & Admission Prices: Mon.-Fri. 9-2; guided tours by appointment. No
charge; donations accepted. Closed major holidays. &
Attendance: 350

Machipongo

**EASTERN SHORE OF VIRGINIA BARRIER ISLANDS CEN-
TER,** 7295 Young St., Machipongo, VA 23405. Mailing Address:
P.O. Box 206, Machipongo, VA 23405-0206. Tel.: 757-678-5550.
Fax: 888-315-8780.
E-mail: barrierislandscenter@live.com
Web Site: www.barrierislandscenter.com
Institution Type/Description: History Museum.
Collections: Virginia's Barrier Island history & culture; maritime artifacts.
Facilities: Museum-related items for sale.
Activities: lectures; classes; special events.
Hours & Admission Prices: Tues.-Sat. 10-4. No charge; donations accepted. &

Manassas

MANASSAS MUSEUM SYSTEM, (M), 9101 Prince William St.,
Manassas, VA 20110-5615. Tel.: 703-368-1873. Fax: 703-257-
8406. TDD: 703-257-8255.
Web Site: www.manassasmuseum.org
Founded: 1973.
Congressional District: 8
Key Personnel: Dir., Elizabeth S. Via-Gossman; Chm. (V), Keith Mueller;
Museum Shop Mgr., Jane Riley.
Personnel Profile: Full-Time Paid 3; Part-Time Paid 6; Part-Time Volunteers
95.
Governing Authority: municipal government; nonprofit. Parent Institution:
City of Manassas, 9101 Prince William St., Manassas, VA 20110. Branch
Sites: The Manassas Museum; The Hopkins Candy Factory; Mayfield Fort;
Manassas Train Depot; Manassas Industrial School/Jennie Dean Memorial;
Liberia Plantation; Cannon Branch Fort; Speiden Carper House. Tax-
exempt: 170(b)(1)(A).
Institution Type/Description: History Museum.
Collections: photographs; local & regional industrial, commercial, personal &
household collections including buildings and their archaeological materi-
als.
Research Fields: Civil War; local history; African-American history; historical
architecture; modern life.
Activities: guided tours; living history programs; changing exhibitions; mu-
seum trunk outreach.
Publications: monthly newsletters, Word from the Junction; volunteer bulletin.
Hours & Admission Prices: Memorial Day to Labor Day daily dawn to dusk;
Sept.-May Tues.-Sun. dawn to dusk. Adults $5, seniors & students $4;
children under 6 no charge. Closed New Year's Day; Thanksgiving;
Christmas. &
Attendance: 17,571 (accurate)

Membership: Student & Senior Citizen $20; Individual $30; Family $40; Sustainer $50; Patron & Corporate $100 & up.

* **MANASSAS NATIONAL BATTLEFIELD PARK,** 6511 Sudley Rd., (Rt. 234), Manassas, VA 20109-2358. Mailing Address: 12521 Lee Hwy., Manassas, VA 20109. Tel.: 703-361-1339. Fax: 703-361-7106. TDD: 703-361-7075.
E-mail: mana_superintendent@nps.gov
Web Site: www.nps.gov/mana
Founded: 1940.
Congressional District: 10
Key Personnel: Supt., Edward W. Clark; Chm. (V), Henry Elliott; Chief Interpretation, Ray Brown; Collection Mgr., James Burgess; Museum Shop Mgr., Larry Swanson.
Personnel Profile: Full-Time Paid 28; Part-Time Paid 2; Part-Time Volunteers 605; Interns 3.
Governing Authority: federal. Parent Institution: U.S. Dept. of Interior. Subsidiary Institution: National Park Service. Tax-exempt.
Institution Type/Description: Park Museum & Visitor Center
Collections: historic artifacts & documentary materials relating to first & second Battles of Manassas; archaeological & architectural materials associated with the civilian & military occupation of the battlefield. Historic House: 1848 Stone house, L. Dogan house.
Research Fields: American Civil War.
Facilities: 2,000-vol. reference library of military history books & manuscripts open to researchers Mon.-Fri. by appointment only.
Activities: guided & self-guided tours; orientation film; battle map program; permanent & temporary exhibits; living history demonstrations.
Publications: folders; maps; site bulletins.
Hours & Admission Prices: Daily 8:30-5. Adults 16 & over $3; children & disabled no charge. Closed Thanksgiving; Christmas. &
Attendance: 586,476 (accurate)

Marion

SMYTH COUNTY MUSEUM, 105 E. Strother St., Marion, VA 24354-2707. Mailing Address: P.O. Box 710, Marion, VA 24354-0710. Tel.: 276-783-7286.
Founded: 1961.
Congressional District: 9
Key Personnel: Pres. (V), Brenda Gwyn.
Personnel Profile: Part-Time Volunteers 12.
Governing Authority: society. Parent Institution: Smyth County Historical & Museum Society, Inc. Headquarters for Smyth County Museum: Staley/Collins House, 109 W. Strother St., Marion, VA 24354, Tel.: 540-783-7286; Wed. & Thurs. 10-3. Tax-exempt: 170(b)(1)(A).
Institution Type/Description: General Museum: housed in 1838 schoolhouse.
Collections: artifacts pertaining to women's & men's culture; weaving; pictorial history of county; 1861-65 Civil War artifacts; period medical items from Southwestern State Hospital; country store; 1908 Marion High School.
Research Fields: local history; genealogy; costumes; archaeology.
Facilities: library & archives of over 3,000 newspaper, books & journals.
Activities: lectures; research.
Publications: Historical Gossip.
Hours & Admission Prices: Fri.-Sat. 10-4. No charge; donations accepted. &
Attendance: 3,500
Membership: Students $5; Individual $10; Family $20; Contributor $25; Sponsor $50; Patron $100; Benefactor $500; Sustaining $1,000.

Martinsville

* **PIEDMONT ARTS ASSOCIATION, (M),** 215 Starling Ave., Martinsville, VA 24112-3832. Tel.: 276-632-3221. Fax: 276-638-3963.
E-mail: kathyrogers@piedmontarts.org
Web Site: www.piedmontarts.org
Founded: 1961.
Congressional District: 5
Key Personnel: Interim Exec. Dir., Kathy Rogers; Mgr. Finance, Pam Allen; Pres., David Stone; Vice Pres., David Martin; Administrative Asst., Barbara Bradshaw; Dir. Exhibitions, Brandon Adams; Coord. Education, Heidi Pinkston; Dir. Programs, Barbara Parker; Maintenance, Nat Cooper.
Personnel Profile: Full-Time Paid 6; Part-Time Paid 9; Part-Time Volunteers 432.
Governing Authority: nonprofit organization. Tax-exempt.
Institution Type/Description: Arts Center: housed in c.1900 M.R. Schottland Estate.
Collections: works by national & regional artists and craftsmen.

Facilities: art studios; public meeting rooms; classroom/studio. Craft-items for sale.
Activities: guided tours; lectures; films; concerts; arts festivals; organized educational programs for children; docent program; loan & temporary exhibitions.
Publications: biannual newsletter; annual report; brochure; exhibition catalogs.
Hours & Admission Prices: Mon.-Fri. 10-5, Sat. 10-3. No charge; donations accepted. Closed holidays. &
Attendance: 36,000 (estimated)
Membership: Senior Citizen & Student $25; Single $35; Family $50; Patron $150; Major Patron $250; Sustaining Patron $500; Benefactor $1,000.

* **VIRGINIA MUSEUM OF NATURAL HISTORY, (M),** 21 Starling Ave., Martinsville, VA 24112-2921. Tel.: 276-634-4141. Fax: 276-634-4199. TDD: 276-634-4149.
E-mail: information@vmnh.virginia.gov
Web Site: www.vmnh.net
Founded: 1984.
Congressional District: 5
Key Personnel: Chm. Bd. Trustees, Missy Neff Gould; Pres. VMNH Foundation, Stacey Reed; Exec. Dir., Joe B. Keiper, Ph.D.; Dir. Education & Public Programs, Dennis A. Casey, Ph.D.; Dep. Dir., Ryan L. Barber; Dir. Research & Collections, Cur. Earth Sciences, Nancy Moncrief, Ph.D.; Cur. Archaeology, Elizabeth A. Moore, Ph.D.; Cur. Marine Biology, Judith E. Winston, Ph.D.; Visitor Svcs. Mgr., Diane Clark.
Personnel Profile: Full-Time Paid 37; Part-Time Paid 20; Part-Time Volunteers 200.
Operating Expenses: 3,048,115
Operating Income: 3,226,530
Governing Authority: Parent Institution: Commonwealth of Virginia Secretary of Natural Resources. Tax-exempt.
Institution Type/Description: Natural History Museum.
Collections: Mesozoic & Cenozoic vertebrate & plant fossils; Virginia archaeology; rocks & minerals; mounted Virginia vertebrates; insects; marine invertebrates; invertebrate fossils; bryozoans.
Major Exhibits: Living on the Water (T), 9/21/13-4/6/14; Dinosaur Discovery, 10/13-4/5/14; Farmers, Warriors, Builders, 4/26/14-7/5/14; Living on the Water (T), 7/26/14-4/4/15.
Research Fields: invertebrate biology; invertebrate paleontology; earth sciences; vertebrate paleontology; mammalogy; conservation of natural history specimens; bryology; archaeology.
Facilities: research library; lecture hall; educational facilities; laboratories; coffee shop. Museum-related items for sale.
Activities: guided tours; lectures; films; field trips; hobby workshops; organized education programs; docent program; temporary exhibitions.
Publications: scientific publications, including Memoirs of the Virginia Museum of Natural History; Jeffersonia; Myriapodologica.
Hours & Admission Prices: Mon.-Sat. 9-5; Adults $5, seniors 60 & over and college students $4, children 3-18 $3; children under 3 & members no charge. Closed Thanksgiving; Christmas Day; New Year's Day. &
Attendance: 31,827 (accurate)
Membership: Individual $40; Family & Grandparent $55; Smithsonian $100; Dogwood $250; Cardinal $500; Tiger Swallowtail $1,000.

Mason Neck

* **GUNSTON HALL PLANTATION, (M),** 10709 Gunston Rd., Mason Neck, VA 22079-3901. Tel.: 703-550-9220. Fax: 703-550-9480.
E-mail: Historic@GunstonHall.org
Web Site: www.gunstonhall.org
Founded: 1932.
Congressional District: 8
Key Personnel: First Regent (V), Mrs. Wylie G. Raab; Devel. Coord., Susan Blankenship; Administrative Asst., Lena McAllister; Education Asst., Frank Barker; Librarian & Archivist, Mark Whatford; Docent Chm., Mary Kay Ruwe; Archaeologist, David Shonyo; Museum Shop Mgr., Karen E. Bazzle.
Personnel Profile: Full-Time Paid 10; Part-Time Paid 35; Part-Time Volunteers 71; Interns 3
Governing Authority: state. Parent Institution: Commonwealth of Virginia. Subsidiary Institution: The National Society of The Colonial Dames of America. Tax-exempt.
Institution Type/Description: Historic House: 1755 Gunston Hall, plantation home of George Mason.
Collections: furniture & fine arts of the 18th century; rare books; decorative arts; archives; manuscripts; plantation tools.
Research Fields: 18th-century life styles, room use, George Mason.
Facilities: 5,000-vol. library of decorative arts & history books & 1,500-vol. rare book library available for use by appointment; 110-seat auditorium;

nature trail; Mason Family grave yard; meeting room; Boxwood gardens; picnic area. Museum-related items for sale.

Activities: guided tours; lectures; docent programs; special events; permanent & temporary exhibitions; active archaeological site.

Publications: special events programs flyers; brochures; books, The Five George Masons; George Mason & The Bill of Rights, DVD; The Recollections of John Mason.

Hours & Admission Prices: Daily 9:30-5. Adults $10, senior citizens 60 & over $8, children 6-18 $5; discount to AAM members; members and children under 6 no charge. Closed New Year's Day; Thanksgiving; Christmas. &

Attendance: 25,000 (accurate)

Membership: Individual $30; Family $45; Sponsor $75; Patron $100; Sustaining $250; Benefactor $500; Guardian $1,000 & up.

Mathews

MATHEWS COUNTY HISTORICAL SOCIETY, INC., Headquarters at Tompkins Cottage, 27 Brickbat Rd., Mathews, VA 23109. Mailing Address: P.O. Box 855, Mathews, VA 23109-0855.

Web Site: www.rootsweb.com/~vamchs/index.html

Founded: 1964.

Key Personnel: Pres. (V), Reed B. Lawson; Museum Shop Mgr., Martha Ellen Traband

Governing Authority: Parent Institution: Mathews County Historical Society, Inc. Tax-exempt.

Institution Type/Description: Historic House: housed in a former mercantile store owned by Christopher Tompkins, father of Captain Sally Tompkins, the first female commissioned officer in the U.S. military; built in 1815. Historic Store c. 1820 named Old James Store.

Collections: local history & culture; photographs; personal artifacts; early furnishings.

Facilities: Books for sale.

Activities: holiday openings; school & historical groups.

Publications: quarterly newsletter, History and Progress of Mathews County, VA; Tombstones of Mathews County; Mathews County Panorama; Historic Homes and Properties of Mathews County, Virginia.

Hours & Admission Prices: Spring & Summer Fri.-Sat. 10-1. No charge; donations accepted.

Membership: Individual $20; Couple $25.

McDowell

HIGHLAND COUNTY MUSEUM, 161 Mansion House Rd., McDowell, VA 24458. Mailing Address: P.O. Box 63, McDowell, VA 24458-0063. Tel.: 540-396-4478. Fax: 540-396-4478.

E-mail: highlandhist@mgwnet.com

Web Site: www.highlandcountyhistory.com

Founded: 2005.

Congressional District: 6

Key Personnel: Exec. Dir., Lorraine White; Chm., Sarah Samples; Vice Chm., Christopher Scott; Treas., James Blagg.

Personnel Profile: Part-Time Paid 1; Part-Time Volunteers 14.

Governing Authority: private; nonprofit organization. Parent Institution: Highland Historical Society, McDowell, VA 24458. Tax-exempt: 501(c)(3).

Institution Type/Description: History Museum: housed in an 1851 former hospital used during the Civil War which later became a hotel & stagecoach stop on the Staunton to Parkersburg Turnpike.

Collections: Battle of McDowell; local artifacts; period furnishings; Virginia history; Highland County.

Facilities: educational center; 1,200 sq. ft. exhibit space. Museum-related items for sale.

Activities: docent program; films; lectures; genealogy classes; related special events.

Publications: quarterly newsletter.

Hours & Admission Prices: Fri.-Sat. 11-4, Sun. 1-4. No charge; donations accepted.

Attendance: 1,935 (accurate)

Membership: Individual $15; Family & Business $25; Life $250.

McLean

NATIONAL PARK SERVICE-GREAT FALLS PARK, 9200 Old Dominion Dr., McLean, VA 22102. Mailing Address: 700 George Washington Memorial Pkwy., McLean, VA 22101. Tel.: 703-285-2966. Fax: 703-285-2223. Facebook: National Park Service-Great Falls Park; TDD: 703-285-2966.

Web Site: www.nps.gov/gwmp/grfa

Founded: 1966.

Congressional District: 10

Key Personnel: Park Supt., Alexcy Romero; Site Mgr. & Museum Shop Mgr., Brent O'Neill.

Personnel Profile: Full-Time Paid 9; Part-Time Volunteers 8.

Governing Authority: federal; nonprofit. Parent Institution: National Park Service. Tax-exempt.

Institution Type/Description: Park Museum: site of the 1785 Patowmack Canal developed by George Washington.

Collections: Patowmack Canal history & artifacts; wildlife; natural environment; American Indian artifacts; early amusement park era; recreation safety.

Major Exhibits: High Water, 1/14-12/14.

Facilities: 100-seat auditorium; nature & conservation center; snack shop; picnic area; 15 miles of trails. Books pertaining to canal history & natural sciences for sale.

Activities: guided tours; films; organized educational programs. Museum Sponsors: Patowmack Canal Festival.

Hours & Admission Prices: Mon.-Fri. 10-5, Sat.-Sun. 10-6; seasonal hours vary. $5 per vehicle, $3 per person; discounts to seniors 62 & over and disabled. Annual passes available. &

Attendance: 600,000 (estimated)

Middleburg

NATIONAL SPORTING LIBRARY & MUSEUM, (M), 102 The Plains Rd., Middleburg, VA 20117. Mailing Address: P.O. Box 1335, Middleburg, VA 20118-1335. Tel.: 540-687-6542. Fax: 540-687-8540.

E-mail: museum@nsl.org

Web Site: www.nsl.org

Founded: 1954.

Congressional District: 8

Key Personnel: Exec. Dir., Melanie Mathewes; Librarian, Lisa Campbell; Dir. Communications & Education, Maureen Gustafson; Cur., Claudia Pfeiffer; Asst. Cur., Nicole Stribling; Devel. Coord., Diana Kingsbury-Smith; Office Mgr., Judy Sheehan; Bookkeeper, Mary Deppa; Accountant, Jo Wolford.

Personnel Profile: Full-Time Paid 8; Part-Time Paid 4; Part-Time Volunteers 6; Interns 1.

Volunteer Hours: 800

Governing Authority: nonprofit. Tax-exempt: 501(c)(3).

Institution Type/Description: Library & Art Museum.

Collections: 25,000 books on horse & field sports from 16th-21st centuries; scrapbooks; manuscripts; photographs; audio visual materials; American, English & continental paintings & sculpture from 17th-21st centuries.

Research Fields: equestrian & field sports; sporting artists; sport history; animal studies; British & American studies; cultural & social history; environmental history.

Facilities: 25,000-vol. research library; archives; 10,000 sq. ft. museum exhibition space.

Activities: permanent & temporary exhibitions; lectures; special events; symposia; fellowships; book fair.

Publications: books; exhibition catalogs; quarterly newsletter.

Hours & Admission Prices: Museum: Wed.-Sat. 10-4, Sun. 12-4. Library: Tues.-Fri. 10-4. Sat. 1-4. No charge; donations accepted. Closed federal holidays. &

Attendance: 7,431 (accurate)

Middletown

BELLE GROVE PLANTATION, (M), 336 Belle Grove Rd., Middletown, VA 22645. Mailing Address: P.O. Box 537, Middletown, VA 22645-0537. Tel.: 540-869-2028. Fax: 540-869-9638.

E-mail: info@bellegrove.org

Web Site: www.bellegrove.org

Founded: 1964.

Congressional District: 10

Key Personnel: Exec. Dir., Kristen Laise; Coord. Programs & Events, Rich Coyle; Administrative Asst., Chelsea Kashani; Supt. Bldgs. & Grounds, Dennis Campbell; Bookkeeper, Renee Maines; Chm. (V) Belle Grove, Inc., John Adamson; Housekeeper, Dorothy Fletcher.

Personnel Profile: Full-Time Paid 3; Part-Time Paid 6; Part-Time Volunteers 80; Interns 3.

Governing Authority: nonprofit organization. Operated by Belle Grove, Inc. Property of National Trust for Historic Preservation, 1785 Massachusetts Ave., N.W., Washington, D.C. 20036. Parent Institution: National Trust for Historic Preservation. Subsidiary Institution: Belle Grove, Inc. Tax-exempt: 501(c)(3).

Institution Type/Description: Historic House: 1797 Belle Grove home of Revolutionary War officer Maj. Isaac Hite, Jr., brother-in-law of President James Madison, & 1864 Civil War headquarters of Gen. Philip Sheridan,

Civil War Battle of Cedar Creek was fought on Belle Grove's grounds. Thomas Jefferson assisted on design.

Collections: decorative arts; architecture; Shenandoah Valley furniture.

Research Fields: local history; rural folk cultures.

Facilities: library of local, architectural, crafts & agricultural history books; barn; demonstration garden; heritage apple orchard. Museum-related items for sale.

Activities: guided tours; school programs; special events; lectures.

Publications: calendar of events; newsletter; guidebook; Hite family letters; Hite and Bowman Cemetery Guides; cookbook, Belle Grove; book, The Women of Belle Grove.

Hours & Admission Prices: April-Oct. Mon.-Sat. 10-4 (last tour 3:15), Sun. 1-5 (last tour 4:15); Nov.-Dec. call for hours. Adults $12, senior citizens $10, students 6-12 $6; discounts to groups, AAA, National Park, Mobile Travel, National Trust, AAM & Belle Grove members; members no charge. Closed Easter; Memorial Day; Independence Day; Labor Day; Columbus Day; Thanksgiving; Christmas Eve & Day.

Attendance: 25,000 (estimated)

Membership: Belle Grove, Inc.: Individual $35; Family $40. National Trust for Historic Preservation: Individual $40; Family $50. Combination: Hite Society $100; Hunnewell Society $250; President's Society $500; Heritage Society $1,000; Madison Society $2,500.

Millwood

BURWELL-MORGAN MILL, 15 Tannery Ln., Millwood, VA 22646. Mailing Address: P.O. Box 306, Berryville, VA 22611-0306. Tel.: 540-837-1799. Fax: 540-955-0285.

E-mail: bmmill@clarkehistory.org

Web Site: www.clarkehistory.org/themill.htm

Founded: 1964.

Congressional District: 7

Key Personnel: Dir., Laura Christiansen; Pres. (V), Howard Means; Archivist, Mary Morris.

Personnel Profile: Full-Time Paid 1; Part-Time Paid 3; Part-Time Volunteers 30.

Governing Authority: society. Parent Institution: Clarke County Historical Assoc. Tel.: 540-955-2600. Tax-exempt: 501(c)(3).

Institution Type/Description: History Museum: 1782-85, operating, water-powered Merchant Grist & Flour Mill.

Collections: wooden water wheel and gear train; grinding stones; 1784 wooden mill equipment; iron gudgeon bearings; spindle shafts; colonial kitchen equipment; farm tools. Historic Building: 1840 Miller's House.

Research Fields: mill design and techniques; 18th century financial records.

Facilities: Mill products for sale.

Activities: guided tours; operation of mill; permanent exhibitions; art shows; antique shows; special tours for school children.

Publications: quarterly newsletter, CCHA.

Hours & Admission Prices: May-Nov. Sat. 10-5, Sun. 12-5. No charge.

Attendance: 10,000 (estimated)

Membership: Individual $25; Family $50; Corporate $100; Life $500.

Mineral

NORTH ANNA NUCLEAR INFORMATION CENTER, Rte. 700, 1022 Haley Dr., Mineral, VA 23117. Mailing Address: 1022 Haley Dr., Mineral, VA 23117-4527. Tel.: 804-771-3200 & 540-894-2029. Fax: 540-894-0379.

E-mail: mike.duffey@dom.com

Web Site: www.dom.com

Founded: 1973.

Congressional District: 7

Key Personnel: Information Representative, Debbie Seay; Coord., Michael Duffey.

Personnel Profile: Full-Time Paid 2.

Governing Authority: profit-making organization. Parent Institution: Dominion. Subsidiary Institution: Dominion Resources Services.

Institution Type/Description: Nuclear Energy Museum.

Collections: narrated & animated exhibits describing how electricity is generated by nuclear fuel.

Facilities: North Anna Power Station; two 100-seat auditoriums; 32-seat theater.

Activities: self-guided tours; lectures; films; organized education programs for children, adults & undergraduate college students.

Publications: booklet, Vepco & Nuclear Power, Now & For the Future; brochure, North Anna Nuclear Information Center.

Hours & Admission Prices: Mon.-Fri. 9-4. No charge. Closed major holidays. &

Attendance: 3,058 (accurate)

Monterey

HIGHLAND MAPLE MUSEUM, 61 Highland Center Rd., Monterey, VA 24465. Mailing Address: P.O. Box 223, Monterey, VA 24465-0223. Tel.: 540-468-2550. Fax: 540-468-2551.

E-mail: highcc@cfw.com

Web Site: www.highlandcounty.org

Founded: 1983.

Congressional District: 6

Key Personnel: C.E.O., Carolyn Pohowsky.

Personnel Profile: Part-Time Volunteers 2.

Governing Authority: nonprofit organization. Parent Institution: Highland County Chamber of Commerce. Tax-exempt.

Institution Type/Description: Agriculture Museum.

Collections: old & new tools; exhibits demonstrating techniques of maple syrup & sugar production from Indian times to present.

Facilities: 600 sq. ft. exhibit space.

Hours & Admission Prices: Daily. No charge; donations accepted. &

Attendance: 3,000 (estimated)

Montross

ARMSTEAD TASKER JOHNSON HIGH SCHOOL MUSEUM, 18849 King's Hwy., Montross, VA 22520. Mailing Address: P.O. Box 1149, Montross, VA 22520-1000. Tel.: 804-493-7070.

E-mail: atjohnsonmuseum1@verizon.net

Congressional District: 99

Key Personnel: Chm. (V), Marian Veney Ashton; Pres. (V), Dr. Lois Harrison-Jones.

Personnel Profile: Part-Time Volunteers 1; Interns 1.

Institution Type/Description: History Museum: housed in the first high school in the Northern Neck for African American students.

Collections: local history & culture; period furnishings; photographs; memorabilia.

Hours & Admission Prices: May-Sept. Wed.-Thurs. & Sat. 1-4; Sept.-Nov. Sat. 1-4; other times by appointment. No charge; donations accepted. &

Attendance: 300 (estimated)

WESTMORELAND COUNTY MUSEUM AND LIBRARY, INC., 43 Court Square, Montross, VA 22520. Mailing Address: P.O. Box 247, Montross, VA 22520-0247. Tel.: 804-493-8440. Fax: 804-493-1312.

E-mail: wcmuseum@verizon.net

Web Site: www.westmoreland-county.org

Founded: 1939.

Congressional District: 1

Personnel Profile: Part-Time Paid 3.

Institution Type/Description: History Museum.

Collections: local history & culture; paintings; photographs; period artifacts; furnishings; archaeological artifacts.

Hours & Admission Prices: Mon.-Sat. 10-4. No charge; donations accepted. Closed county holidays.

Membership: Individual $25; Family/Joint $45; Patriot $100.

Morattico

MORATTICO WATERFRONT MUSEUM, Morattico Rd., Morattico, VA 22523. Mailing Address: P.O. Box 80, Morattico, VA 22523-0080. Tel.: 804-462-0532.

E-mail: moratticowaterfrontmuseum@yahoo.com

Founded: 2003.

Governing Authority: Tax-exempt: 501(c)(3).

Institution Type/Description: History Museum: housed in the former Morattico General Store; built in 1901.

Collections: local history & culture; period furnishings; personal artifacts; photographs

Publications: members' newsletter.

Hours & Admission Prices: May-Oct. Sat. 12-4, Sun. 1-4. &

Membership: Individual $20; Family $40; Contributing $50; Sustaining $100.

Mount Vernon

GEORGE WASHINGTON'S MOUNT VERNON ESTATE, MUSEUM & GARDENS, (M), South End of George Washington Memorial Pkwy., Mount Vernon, VA 22309. Mailing Address: George Washington's Mount Vernon, P.O. Box 110, Mount Vernon, VA 22121-0110. Tel.: 703-780-2000. Fax: 703-799-8654. TDD: 703-799-8121.
E-mail: info@mountvernon.org
Web Site: www.mountvernon.org
Formerly: Mount Vernon: George Washington's Estate & Gardens
Founded: 1853.
Congressional District: 11
Key Personnel: Regent, Ann Bookout; Pres., Curtis Viebranz; C.F.O., Philip Manno; C.O.O., Barton Groh; Vice Pres. Advancement, Susan Magill; Vice Pres. Collections, Carol Borchert Cadon; Dir. Operations & Mgmt., Joe Sliger; Dir. Security, Curt Clevenger; Dir. Retail, Julia Mosley; Dir. Human Resources, Megan Dunn; Dir. Horticulture, Dean Norton; Mount Vernon Inn Mgr., William H. Robertson; Librarian, Joan Stahl.
Personnel Profile: Full-Time Paid 257; Part-Time Paid 264; Part-Time Volunteers 400.
Governing Authority: nonprofit organization. Subsidiary Institution Mount Vernon Inn, Inc.; Ford Orientation Center; Donald W. Reynolds Museum and Education Center. Tax-exempt: 170(b)(1)(a).
Institution Type/Description: Historic Plantation: 1735-1799 Mount Vernon, home of George Washington.
Collections: decorative arts; graphic arts; textiles; archaeology; fine arts; research & study collections; Washington memorabilia; furnishings; manuscripts; books; boxwood gardens; reconstructed greenhouse; kitchen building with larder; stable; coach house; wash house; storehouses; spinning house; slaves quarters; cobbler shops; smoke house; tombs of George & Martha Washington; pioneer farm site; reconstruction of 16-sided threshing barn & slave cabin; 18th century working farm; hands-on history children's area.
Research Fields: life & times of George Washington, his family & associates; history of Mount Vernon Estate from time of original patent to present, architectural development, horticulture, agriculture & household furnishings of the period of George Washington's ownership.
Facilities: 12,000-vol. general library available by appointment on premises; restaurant. Publications and museum-related items for sale.
Activities: in-school programs; garden tour; slave life tour; children's activities; films; permanent & temporary exhibitions; sightseeing cruises. Special Events: Colonial Days; public wreath-laying at tomb; Slave Memorial Commemoration; gardening days; Candlelight Tours; Wine Festivals; Spring Garden Party; George Washington's Birthday Celebration; Independence Day Celebration; 18th Century Craft Fair; Fall Harvest Family Days; Christmas at Mount Vernon.
Publications: annual periodical, Annual Report; semi-annual newsletter, Mount Vernon: Yesterday, Today & Tomorrow; handbook; specialized publications of the collections; photographic reproductions of paintings, prints & maps.
Hours & Admission Prices: March & Sept.-Oct. daily 9-5; April-Aug. daily 8-5; Nov.-Feb. 9-4. Adults $15, senior citizens $14, children 6-11, students & youth groups grades 1-12 $7; discount to groups of 20 or more; scouting groups in uniform no charge Nov. to 3rd Sun. in Feb. Annual Pass $25. &
Attendance: 1,136,793 (accurate)
Membership: Palladian Society $50-$99; Piazza Society $100-$249; Colonnade Society $250-$499; Cupola $500-$999; Regent's Circle $1,000-$4,999; Mount Vernon One Hundred $5,000-$9,999; Washington Council $10,000 & up.

Natural Bridge

NATURAL BRIDGE WAX MUSEUM, 70 W. Faulkner Hwy., Natural Bridge, VA 24578. Mailing Address: P.O. Box 85, Natural Bridge, VA 24578-0085. Tel.: 540-291-2426. Fax: 540-291-3785.
E-mail: johnwax@embarqmail.com
Web Site: www.naturalbridgeva.com
Founded: 1978.
Congressional District: 4
Key Personnel: Dir., John McFerren; Museum Shop Mgr., Sherry Crawford.
Personnel Profile: Full-Time Paid 4; Part-Time Paid 5.
Governing Authority: Parent Institution: Dorfman Museums, Inc. Tax-exempt.
Institution Type/Description: Wax Museum.
Collections: wax figures including Native Americans, explorers, & presidents; scenes of Virginia and Natural Bridge history.
Hours & Admission Prices: March-Nov. daily 10-6; Dec.-Feb. Sat.-Sun. 10-6. Adult $12, children $6. &
Attendance: 65,000 (estimated)

NATURAL BRIDGE ZOOLOGICAL PARK, Rte. 11, Natural Bridge, VA 24578. Mailing Address: P.O. Box 88, Natural Bridge, VA 24578-0088. Tel.: 540-291-2420. Fax: 540-291-1891.
E-mail: naturalbridgezoo@hotmail.com
Web Site: www.naturalbridgezoo.net
Founded: 1972.
Institution Type/Description: Zoo.
Collections: zoological.
Facilities: Museum-related items for sale.
Activities: elephant rides; petting zoo; photos with baby animals.
Hours & Admission Prices: March 15-May 22 daily 9-6; May 23-Sept. 7 Mon.-Fri. 10-5, Sat.-Sun. 9-7; Sept. 8-Oct. Mon.-Fri. 10-5, Sat.-Sun. 9-6; Nov. 1-Nov. 25 daily 10-5; Nov. 27-Nov. 29 daily 10-4; call to confirm. Adults $12, senior citizens $10, children 3-12 $8; children 2 & under no charge.

New Market

*** VIRGINIA MUSEUM OF THE CIVIL WAR, (M),** 8895 George Collins Dr., New Market, VA 22844. Mailing Address: P.O. Box 1864, New Market, VA 22844-1864. Tel.: 540-740-3101; 866-515-1864 (toll free). Fax: 540-740-3033. TDD: 703-464-7616.
E-mail: nmbshp@vmi.edu
Web Site: www.vmi.edu/newmarket
Formerly: New Market Battlefield State Historical Park
Founded: 1967.
Congressional District: 7
Key Personnel: Exec. Dir., Col. Keith E. Gibson; Dir., Major Troy D. Marshall; Visitor Svcs. Supvr., Judith Drury; Supvr. Historical Interpretation, Stacey R. Nadeau; Office Mgr., Brittney Phillips.
Personnel Profile: Full-Time Paid 7; Part-Time Paid 11; Part-Time Volunteers 28.
Governing Authority: college. Parent Institution: Virginia Military Institute, Lexington, VA 24450. Tax-exempt.
Institution Type/Description: Military History Museum: located on the site of the 1864 Battle of New Market.
Collections: Civil War artifacts; Historic House; (1830-1865) Bushong House & period farm collections.
Research Fields: Shenandoah Valley 1855-1870.
Facilities: 300-vol. library on secondary historical reference works available upon request for use on premises. Museum-related items for sale.
Activities: guided tours; Emmy Award-winning film; interpretive exhibits; self-guided battlefield tour.
Publications: magazine reprint, Battle of New Market.
Hours & Admission Prices: Daily 9-5. Adults $10, seniors $9, children 6-12 $6; discounts to groups, seniors, military, VAM & AAM members. Closed New Year's Day; Thanksgiving; Christmas Eve and Day. &
Attendance: 41,258 (accurate)

Newbern

WILDERNESS ROAD REGIONAL MUSEUM, State Rt. 611, 5240 Wilderness Rd., Newbern, VA 24126. Mailing Address: P.O. Box 373, Newbern, VA 24126-0373. Tel.: 540-674-4835.
Web Site: newriverhistoricalsociety.com
Founded: 1980.
Congressional District: 9
Key Personnel: C.E.O. & Pres. (V), Carolyn Mathews; Dir., Chm. (V), Dir. Public Rels. & Museum Shop Mgr., Mary Williams; Librarian, Elinor Farmer; Treas., Barbara Duncan; Administrative Asst., Rebecca Gunn.
Personnel Profile: Part-Time Paid 5; Part-Time Volunteers 25.
Governing Authority: nonprofit organizations. Parent Institution: New River Historical Society. Tax-exempt: 501(c)(3); 170(b)(1)(A).
Institution Type/Description: Historical Society Museum: housed in c.1810 2-story frame/log building located in Old Newbern National Historic District.
Collections: regional history of New River Valley from 1810-1875; dolls & quilts; archives of local history, Civil War items; photographs; farm & hand tools; buggy.
Research Fields: genealogy; local history.
Facilities: 700-vol. library of Virginiana & local historical publications available to the public; classrooms. Museum-related items for sale.
Activities: guided tours; arts festivals; organized educational programs; docent program; participatory & temporary exhibitions. Museum Sponsors: Civil War reenactments; 3 buffet dinners per year; Fall Festival of Arts & Crafts; genealogy workshop in May; annual used book sale; Open House; yard sale; Christmas Along The Wilderness Road.
Publications: annual Journal, N.R.H.S.; quarterly newsletter, Benchmarks.
Hours & Admission Prices: Mid-March to mid-Dec. Tues.-Sat. 10:30-4:30; mid-Dec. to mid-March by appointment only. No charge; donations

accepted. Closed New Year's Day; Easter; Mother's Day; Labor Day; Thanksgiving; Christmas. &
Attendance: 20,000 (estimated)

Newport News

✻ ENDVIEW PLANTATION, (M), 362 Yorktown Rd., Newport News, VA 23603-1017. Tel.: 757-887-1862. Fax: 757-888-3869.
E-mail: endview@nngov.com
Web Site: www.endview.org
Institution Type/Description: Historic House Museum: plantation built c.1769.
Collections: plantation history; Civil War artifacts; American culture & history; period furnishings; photographs.
Activities: educational programs.
Hours & Admission Prices: Jan.-March Thurs.-Sat. 10-4, Sun. 1-5; April-Dec. Mon. & Thurs.-Fri. 10-4, Sat. 10-5, Sun. 12-5. Adults $6, seniors 62 & over $5, children 7-18 $4; discounts to groups; children under 7 no charge.

GOLF MUSEUM, James River Country Club, 1500 Country Club Rd., Newport News, VA 23606-2840. Tel.: 757-595-3327. Fax: 757-596-4807.
Founded: 1932.
Congressional District: 1
Key Personnel: Dir. & C.E.O., William S. Hargette.
Personnel Profile: Full-Time Volunteers 1; Part-Time Volunteers 9.
Governing Authority: board of trustees; nonprofit. Tax-exempt.
Institution Type/Description: Sports Museum.
Collections: international golf history; golf clubs; balls; tees; books; engravings; paintings; photographs; silver; medals; tools; 16th-20th century golf artifacts; first printed reference to golf 1566; Bobby Jones' club (Brassie) used in achieving his 1930 Grand Slam.
Facilities: 1,000-vol. library available to researchers only.
Activities: guided tours; loan exhibitions.
Hours & Admission Prices: By appointment. No charge. Closed holidays. &
Attendance: 1,000 (estimated)

JAMES A. FIELDS HOUSE, 617 27th St., Newport News, VA 23607-4033. Tel.: 757-245-1991.
E-mail: jafieldshouse@hotmail.com
Founded: 2002.
Congressional District: 3
Key Personnel: Dir., Saundra N. Cherry.
Governing Authority: Tax-exempt.
Institution Type/Description: Historic House Museum: housed in the former home & law office of James A. Fields. Listed on the National Register of Historic Places.
Collections: local history & culture; period furnishings; personal artifacts; photographs.
Hours & Admission Prices: Tues.-Sat. 11-4. Admission $3.
Attendance: 250 (estimated)

✻ LEE HALL MANSION, (M), 163 Yorktown Rd., Newport News, VA 23603-1127. Tel.: 757-888-3371. Fax: 757-888-3373.
E-mail: bgutierr@ci.newport-news.va.us
Web Site: www.leehall.org
Key Personnel: Site Coord., Braxton Gutierrez.
Governing Authority: Parent Institution: the City of Newport News, Department of Parks and Recreation, Historical Services Division.
Institution Type/Description: Historic House Museum: housed in the former home of Richard Decauter Lee, built in 1859.
Collections: period artifacts & furnishings; weaponry.
Facilities: Museum-related items for sale.
Activities: special programs; lectures.
Hours & Admission Prices: Jan.-March Thurs.-Sat. 10-4, Sun. 1-5; April-Dec. Mon. & Thurs.-Fri. 10-4, Sat. 10-5, Sun. 12-5. Adults $8, senior citizens $5, children 7-18 $4; children under 7 no charge. Closed New Year's Day; Easter; Thanksgiving; Christmas. &

✻ THE MARINERS' MUSEUM, (M), 100 Museum Dr., Newport News, VA 23606-3759. Tel.: 757-596-2222; 800-581-7245. Fax: 757-591-7311.
E-mail: info@marinersmuseum.org
Web Site: www.marinersmuseum.org
Founded: 1930.
Congressional District: 1
Key Personnel: Pres. & C.E.O., Elliot H. Gruber; Chm. (V), John R. Lawson, II; Vice Pres. Museum Collections, Anna Holloway; Chief Cur., Lyles Forbes; Museum Shop Mgr. & Dir. Visitor Svcs., Cassi LeDuc.

Personnel Profile: Full-Time Paid 75; Part-Time Paid 20; Part-Time Volunteers 110; Interns 10.
Volunteer Hours: 13,241
Operating Expenses: 8,700,000
Operating Income: 11,500,000
Governing Authority: nonprofit organization. Tax-exempt: 501(c)(3).
Institution Type/Description: International & National Maritime History Museum.
Collections: Age of Exploration Gallery; Crabtree miniature ships; Chesapeake Bay Gallery; International Small Craft Center; navigational instruments; ship equipment; whaling & fishing equipment; naval armament; sailors' handiwork; figureheads; lighthouse & lifesaving equipment; maritime decorative arts; prints, paintings & drawings of maritime subject matter; archives; industrial history; folklore; Chris-Craft archives; rare books; journals; Haviland collection of maritime photographs, information, and news clippings; USS Monitor Center exhibition & conservation laboratory; Titanic exhibit and artifacts.
Major Exhibits: Oyster Wars!, 8/13-1/14; View from the Sea: Works of Art on Paper, 11/13-4/14; Between the States: Civil War Photography (T), 3/14-4/14; Abandon Ship: Stories of Survival, 5/14; Savage Ancient Seas (T), 5/14-12/14; Majestic Grace & Speed: The Marine Paintings of James Edward Buttersworth, 10/14.
Research Fields: regional, national & international maritime & naval history; history of technology; Civil War ironclads; small craft; maritime art.
Facilities: 1,750,000-vol. research library & archives; 80,000 sq. ft. exhibit space; 27-seat lecture room; 50-seat orientation theater; 2 classrooms; cafe; 550 acre park with lake & 5 mile walking trail; picnic area; scenic Lion's Bridge by the James River. Museum-related items for sale.
Activities: guided tours; lectures; gallery talks, costumed interpreters; formal education programs for children, adults & college students; permanent & temporary exhibitions; symposia; Civil War programming; paddleboats; concerts; internet video classes.
Publications: annual; books; monographs; brochures; quarterly newsletter.
Hours & Admission Prices: Mon. Mon.-Sat. 9-5, Sun. 11-5. Adults $12, students 13 & up $10, children 6-12 $7; discounts to senior citizens 65 & older, active duty military, AAA, AAM, ICOM & CAMM members; members and children 5 & under no charge. Closed Thanksgiving; Christmas. &
Attendance: 69,732 (accurate)
Membership: Individual $35; Individual+1 $50; Family $75; Premium $150.

THE NEWSOME HOUSE MUSEUM & CULTURAL CENTER, 2803 Oak Ave., Newport News, VA 23607-3713. Tel.: 757-247-2360. Fax: 757-928-6754.
E-mail: ddavis@nngov.com
Web Site: www.newsomehouse.org
Founded: 1991.
Congressional District: 3
Key Personnel: Historic Site Mgr., Mary Kayaselcuk.
Personnel Profile: Full-Time Paid 2; Part-Time Paid 1; Part-Time Volunteers 5.
Governing Authority: municipal government; nonprofit. City of Newport News. Parent Institution: Virginia War Museum. Tax-exempt: 501(c)(3).
Institution Type/Description: Historic House: built in 1899 The Newsome House is a modified Queen Anne structure, which was home to the Joseph Thomas Newsome family from 1906-1977.
Collections: personal papers of the Newsome family; local African-American history & culture; folkart of Anderson Johnson.
Research Fields: local African-American history.
Facilities: 500-vol. library on Black history & the books and papers of the Newsome family; meeting rooms; 1,500 sq. ft. exhibit space; banquet facilities; grounds available for picnics or musical programs. Lithographs for sale.
Activities: formal education programs for adults & children; internships; guided tours; workshops; lectures; temporary exhibits. Museum Sponsors: art exhibits & demonstrations; workshops on genealogy & Black history. Annual Events: Holiday Open House; A Newsome House Christmas.
Publications: book; A Life in Newport News: An Oral History of Inettie Banks Edwards; Huntington High School: Symbol of Community Hope and Unity.
Hours & Admission Prices: Thurs.-Sat. 10-5. Suggested Donation: adults $2. Closed New Year's Day; Easter; Thanksgiving, Christmas Eve & Day. &
Attendance: 5,000 (estimated)
Membership: Friends of the Newsome House: Student $5; Individual $10; Family $20; Patron $50; Lifetime $100; Organization $250; Sponsor $1,000.

✻ PENINSULA FINE ARTS CENTER, (M), 101 Museum Dr., Newport News, VA 23606-3758. Tel.: 757-596-8175. Fax: 757-596-0807. Facebook: Peninsula Fine Arts Center.
E-mail: info@pfac-va.org
Web Site: www.pfac-va.org

Founded: 1962.
Congressional District: 1
Key Personnel: Exec. Dir., Courtney Gardner; Pres. & Chm. (v), Chris Stewart; Dir. Programs, Michael Preble; Staff Accountant & Personnel Mgr., Debbie Hill; Dir. Mktg., Michael McGrann; Mgr. Exhibitions & Facilities, Fred Rich.
Personnel Profile: Full-Time Paid 5; Part-Time Paid 3; Part-Time Volunteers 214; Interns 2.
Governing Authority: nonprofit organization. Affiliate of Virginia Museum of Fine Arts. Tax-exempt: 501(c)(3).
Institution Type/Description: Arts Center.
Collections: temporary exhibitions during the year.
Research Fields: contemporary American art.
Facilities: sculpture courtyard. Local & regional art for sale.
Activities: guided tours; public programs; arts festivals; organized educational experiences; loan & traveling exhibitions; video gallery.
Publications: newsletter; class schedules; exhibition catalogs.
Hours & Admission Prices: Tues.-Sat. 10-5, Sun. 1-5. Adults $7.50, students, military and seniors 65 & over $6, children 6-12 $4; discounts to AAA members; members and children 5 & under no charge. Call for special exhibition charges. Closed New Year's Day; Thanksgiving; Christmas. &
Attendance: 40,000 (estimated)
Membership: Student, Military, Teacher & Senior Citizen $25; Individual or Senior Family, Military Family, Student Family, Teacher Family $40; Family $60.

* **VIRGINIA LIVING MUSEUM,** 524 J. Clyde Morris Blvd., Newport News, VA 23601-1999. Tel.: 757-595-1900. Fax: 757-599-4897. TDD: 757-595-1900.
E-mail: webmaster@thevlm.org
Web Site: www.thevlm.org
Founded: 1964.
Congressional District: 1
Key Personnel: Exec. Dir., Page Hayhurst; Deputy Dir., Fred Farris; Pres. (V), Dawn T. Hunt; Curatorial Dir., George K. Mathews, Jr.; Dir. Devel., Carolyn Cuthrell; Dir. Education, Chris Lewis; Dir. Finance & I.T., Dave Osman; Dir. Mktg., Virginia Gabriele; Dir. Volunteer Svcs., Shandran J. Thornburgh; Museum Shop Mgr., Sarah Wilcox.
Personnel Profile: Full-Time Paid 43; Part-Time Paid 56; Part-Time Volunteers 450.
Governing Authority: nonprofit organization. Tax-exempt: 501(c)(3).
Institution Type/Description: Specialized Natural Center.
Collections: live native animals, plants, birds, insects, mammals, reptiles, amphibians, marine & freshwater animals; rocks, minerals, fossils, botanical specimens.
Facilities: library; marine & freshwater aquariums; aviary; planetarium; observatory; discovery center; nature trail & boardwalk; classrooms.
Activities: lectures; films; hobby workshops; formally organized educational programs; docent program; permanent & temporary exhibitions; interpretive exhibits; classrooms; nature trail; behind the scenes tours.
Publications: quarterly newsletter, Paws & Reflect; annual report, Teachers Guide; program brochures; promotional brochure.
Hours & Admission Prices: Summer: daily 9-5; Winter: Mon.-Sat. 9-5, Sun. 12-5. Museum: adults $17, children 3-12 $13; discounts for senior citizens, groups & ASTC (restrictions apply) members; members & children under 3 no charge. Planetarium $4 additional fee; discounts to members. Closed New Year's Day; Thanksgiving; Christmas Eve & Day. &
Attendance: 197,894 (accurate)
Membership: Individual $40; Couple $65; Family & Grandparents with 3 Guests $85; Family with 3 Guests & Grandparents with 6 Guests $110.

THE VIRGINIA WAR MUSEUM, 9285 Warwick Blvd., Huntington Park, Newport News, VA 23607-1537. Tel.: 757-247-8523. Fax: 757-247-8627. Facebook: Virginia War Museum.
E-mail: virginiawarmuseum@nngov.com
Web Site: www.warmuseum.org
Founded: 1924.
Congressional District: 1
Key Personnel: Cur., G. Richard Hoffeditz; Registrar, Jerry Coggeshall; Public Rels. & Museum Shop Mgr., Colin Romanick.
Personnel Profile: Full-Time Paid 9; Full-Time Volunteers 2; Part-Time Paid 3; Part-Time Volunteers 12; Interns 3.
Governing Authority: municipal. Parent Institution: Newport News. Subsidiary Institution: Newsome House; Lee Hall Mansion, Newport News, VA; Endview Plantation, Newport News, VA. Tax-exempt: 501(c)(3).
Institution Type/Description: Military Museum.
Collections: weapons; uniforms; posters; prints; vehicles; tanks; insignia; artillery & accoutrements relating to United States military involvement from 1775 to present.

Research Fields: Military-United States & European.
Facilities: 20,000-vol. library on military, historical & photographic files available for research; 120-seat theater; educational facilities. Museum-related items for sale.
Activities: films; lectures; historical groups; permanent & temporary exhibitions; docent program; formal education programs; theater; broadcast programs. Annual Events: Military Vehicle Show; Toy Soldier Show; Tuskegee Airmen; Peal Harbor Day Ceremony.
Publications: quarterly newsletter, Dufflebag.
Hours & Admission Prices: Mon.-Sat. 9-5, Sun. 12-5. Adults $6, military & senior citizens $5, children 6-15 & students $4; discounts to AAA members; members no charge. Closed New Year's Day; Easter; Thanksgiving; Christmas Eve & Day. &
Attendance: 50,000 (estimated)
Membership: Student $10; Individual $20; Family $35; Supporter $100-$499; Patron $500-$999; Lifetime $1,000.

Norfolk

AFRICAN ART GALLERY, Norfolk State University, 700 Park Ave., Norfolk, VA 23504-8050. Tel.: 757-823-2002. Fax: 757-823-2005.
Web Site: www.nsu.edu/archives/gallery
Formerly: Lois E. Woods Museum
Key Personnel: Dir., Dr. Tommy L. Bogger; Asst. Dir., Annette Montgomery.
Personnel Profile: Full-Time Paid 2; Part-Time Volunteers 2.
Governing Authority: Parent Institution: Norfolk State University. Tax-exempt.
Institution Type/Description: African Art Gallery.
Collections: African art; African-American art & memorabilia.
Publications: gallery brochure.
Hours & Admission Prices: Mon.-Fri. 9-5. No charge; donations accepted. Closed holidays. &
Attendance: 1,300 (estimated)

BARON & ELLIN GORDON ART GALLERIES, OLD DOMINION UNIVERSITY, (M), 4509 Monarch Way, Norfolk, VA 23529. Mailing Address: 9000 Batten Arts and Letters, Norfolk, VA 23529. Tel.: 757-683-6271. Fax: 757-683-6776.
E-mail: fbayersd@odu.edu
Web Site: al.odu.edu/art/gallery/about.shtml
Founded: 1971.
Congressional District: 2
Key Personnel: Asst. Dean & Dir., Frederick S. Bayersdorfer; Cur., Ramona Austin.
Personnel Profile: Full-Time Paid 2; Part-Time Paid 3; Part-Time Volunteers 8; Interns 1.
Governing Authority: public university; nonprofit. Parent Institution: Old Dominion University. Subsidiary Institution: College of Arts & Letters. Tax-exempt.
Institution Type/Description: University Art Gallery
Collections: American self-taught art; contemporary art; paintings; prints; sculpture.
Research Fields: self-taught American art.
Activities: lectures; workshops.
Publications: newsletter; posters; invitations; catalogs.
Hours & Admission Prices: Jan. 7-Dec. 19 Tues.-Sat. 11-5, Sun. 1-5. No charge. &
Attendance: 7,743 (estimated)

* **CHRYSLER MUSEUM OF ART, (M),** One Memorial Place, Norfolk, VA 23510-1587. Tel.: 757-664-6200. Fax: 757-664-6201.
E-mail: museum@chrysler.org
Web Site: www.chrysler.org
Founded: 1939.
Congressional District: 2
Key Personnel: Pres. & Dir., William J. Hennessey; Dir. Finance & C.F.O., Dana Fuqua; Chm. Bd., Peter Meredith; Public Rels., Cindy Mackey; Dir. Education & Public Programs, Anne Corso; Chief Cur. & Acting Cur. American and Contemporary Art, Jefferson Harrison; Deputy Dir. Operations, Susan Leidy.
Personnel Profile: Full-Time Paid 63; Part-Time Paid 27; Part-Time Volunteers 250.
Governing Authority: private. Branch Museum: The Willoughby-Baylor House, 601 E. Freemason St., Norfolk, VA; Moses Myers House, 323 E. Freemason St., Norfolk, VA; Chrysler Museum Glass Studio 745 Duke St., Norfolk VA. Tax-exempt: 501(c)(3).
Institution Type/Description: Art Museum.
Collections: arts from Egypt, Greece, Rome, Near East, Middle East, Far East, Africa, Orient, pre-Columbian America; Italian, French, Dutch, Flemish, English, German, American paintings & sculptures of all periods; ancient to

modern glass from Asia, Africa, Europe, Sandwich, Tiffany, Galle, New England; American, English, French, Italian, German furniture; Ricau Collection; decorative arts; textiles; costumes; jewelry; photographs; silver; china; artifacts.

Major Exhibits: Libensky/Brychtova: Selections from the Anderson Collection, 4/13/14-8/14; In The Box: Plato's Alley (2008) by Charles Atlas, 6/19/14-8/14; 70 Years of Smokey Bear, 8/9/14-2/1/15; Peter Paul Rubens' The Departure of Lot and His Family from Sodom: A Baroque Masterpiece from the Ringling Museum, 9/16/14-3/1/15; In The Box: A Site-Specific Installation by Saya Woolfalk, 9/18/14-11/14; Worn to Be Wild: The Black Leather Jacket, 10/4/14-1/4/15; America's Eden: Thomas Cole and The Voyage of Life, 10/21/14-1/18/15; The Tulsa Portfolio by Larry Clark, 8/30/14-1/18/15; Picasso, Braque, and Cubism: Masterpieces from the National Gallery of Art, 10/14/14-2/15/15; In The Box: Black Righteous Space (2012) by Hank Willis Thomas, 12/9/14-2/1/15.

Facilities: 80,000-vol. library of art reference books; 375-seat theatre; restaurant. Museum-related items for sale.

Activities: guided tours; lectures; films; gallery talks; concerts; docent program; formally organized education programs for families & school groups; permanent & temporary exhibitions; art travel program; demonstrations; classes; visiting artist series. Museum Sponsors: Annual Juried Student Gallery Exhibition.

Publications: bimonthly magazine including calendar of events.

Hours & Admission Prices: Closed until April 2014. Museum: Tues.-Sat. 10-5, Sun. 12-5, 3rd Thurs. each month 10-10. Collection: no charge. Fee charged for some special exhibitions. Closed New Year's Day; Independence Day; Thanksgiving; Christmas. &

Attendance: 165,000 (accurate)

Membership: Student $25; Senior, Teacher & Military $45; Individual $55; Senior, Teacher & Military Household $65; Household $75; Associate $150; Friend $250; Patron $500; Director's Circle $1,000; Masterpiece Society $2,500 & up.

GENERAL DOUGLAS MACARTHUR MEMORIAL, (M), MacArthur Sq., Norfolk, VA 23510-2382. Tel.: 757-441-2965. Fax: 757-441-5389.

E-mail: macarthurmemorial@norfolk.gov
Web Site: www.macarthurmemorial.org
Founded: 1964.
Congressional District: 2
Key Personnel: Dir., William J. Davis; Administrative Asst., Janice S. Dudley; Archivist, James W. Zobel; Cur., Charles R. Knight; Museum Shop Mgr., Mitzi N. Van Horn.
Personnel Profile: Full-Time Paid 9; Part-Time Paid 6; Part-Time Volunteers 7.
Governing Authority: municipal. Parent Institution: City of Norfolk, VA. Tax-exempt: 501(c)(3).
Institution Type/Description: Military Museum: housed in 1850 courthouse building.
Collections: history; personal papers & mementoes.
Research Fields: life of Gen. MacArthur; military history of World War I, World War II, & Korean War; the occupation of Japan; 1898-1945 history of the Philippines; 1914-1960 American military & diplomatic history.
Facilities: 5,700-vol. personal library of Gen. MacArthur, including official & personal papers, photographs available for use on premises under supervision only; working archive of 2.5 million documents; 150-seat theater. Gen. MacArthur & MacArthur Memorial items for sale.
Activities: lectures; film; temporary, permanent & loan exhibitions; training programs for professional museum workers.
Publications: brochure, quarterly newsletter, General Douglas MacArthur Memorial; The MacArthur Report.
Hours & Admission Prices: Tues.-Sat. 10-5, Sun. 11-5. No charge; donations accepted. Closed New Year's Day; Thanksgiving; Christmas. &
Attendance: 35,378 (accurate)
Membership: National Member: $25, $50 & $75. Corporate Member: $250 & $500. Five Star Member: $100. MacArthur One Thousand Member: $1,000.

* HAMPTON ROADS NAVAL MUSEUM, (M), One Waterside Dr., Ste. 248, Norfolk, VA 23510-1607. Tel.: 757-322-2987 & 444-8971. Fax: 757-445-1867.

E-mail: elizabeth.poulliot@navy.mil
Web Site: www.hrnm.navy.mil
Founded: 1979.
Congressional District: 2
Key Personnel: Foundation Exec. Dir., Capt. Thomas H. Smith, USN (Ret.); Dir. Museum, Elizabeth A. Poulliot; Foundation Pres., John Griffing; Public Rels., Susanne Greene; Cur., Joseph M. Judge; Volunteer Coord., Thomas M. Dandes; Exhibits Specialist, Marta E. Joiner; Events Coord. & Educator, Laura Orr; Architectural Historian, Katherine Renfrew; Newsletter Editor,

Gordon B. Calhoun; Dir. Education, Lee Duckworth; Deputy Educator, Matthew Eng; Museum Shop Mgr., Andrea D. Condon-Auen.
Personnel Profile: Full-Time Paid 10; Part-Time Paid 3; Part-Time Volunteers 62.
Governing Authority: federal government. Parent Institution: Naval History and Heritage Command. Tax-exempt.
Institution Type/Description: Naval History Museum: housed on the second floor of NAUTICUS.
Collections: naval artifacts & artworks; ship & aircraft models, weaponry, electric maps; underwater artifacts; period photographs; naval artwork; ship memorabilia; naval uniforms. materials pertaining to historic battleship USS Wisconsin: WWII era, Iowa class battleship; artifacts; photographs; audiovisual history of ship; historic ship, maritime & military history.
Research Fields: U.S. Naval history of Hampton Roads region.
Facilities: 3,000-vol. library.
Activities: tours; lectures; slide presentations; videos; docent training; school & community outreach programs; off-site exhibits; interpretive docents in period reproduction uniforms; interpreter vignettes; traveling programs; speakers' bureau.
Publications: pamphlets; newsletter.
Hours & Admission Prices: Memorial Day to Labor Day daily 10-5; Sept.-May Tues.-Sat. 10-5, Sun. 12-5. No charge. Closed Thanksgiving; Christmas Eve & Day. &
Attendance: 446,312 (accurate)
Membership: Shipmate $35; Plankowner $125; Corporate $350.

HERMITAGE MUSEUM AND GARDENS, (M), 7637 North Shore Rd., Norfolk, VA 23505-1730. Tel.: 757-423-2052. Fax: 757-423-2410.

E-mail: info@thehermitagemuseum.org
Web Site: www.thehermitagemuseum.org
Founded: 1937.
Congressional District: 2
Key Personnel: Pres. Bd. Trustees, Robert E. Garris, Jr.; Exec. Dir., Jen Duncan; Education Programs Mgr., Melissa Ball; Coord. Weddings & Events, Lil Acosta; Cur. Gardens, Yolima Carr; Site Coord., Tom Allan; Mktg. Mgr., Jennifer Font; Mgr. Membership & Devel., Alanna Kibiloski; Cur. Collections, Colin Brady; Visual Arts Programs, Truly Matthews.
Personnel Profile: Full-Time Paid 8; Part-Time Paid 13; Part-Time Volunteers 30; Interns 2.
Governing Authority: nonprofit organization. Tax-exempt: 501(c)(3).
Institution Type/Description: Art Museum.
Collections: paintings, sculpture & decorative arts from Western & Oriental cultures; family papers & photographs.
Activities: guided tours; lectures; permanent exhibitions; facility rental. Museum Sponsors: art scholarships.
Publications: quarterly newsletter, Hermitage Press.
Hours & Admission Prices: Mon.-Tues. & Fri.-Sat. 10-5, Sun. 1-5. Adults $5.50, college students w/ID $3.30, children 6-18 $2.20; discounts to AAM, SERM & NARM members; children under 6, active military & members no charge. Closed New Year's Day; Memorial Day; Independence Day; Labor Day; Columbus Day; Veterans Day; Thanksgiving; Christmas. &
Attendance: 30,000 (estimated)
Membership: Basic: Individual $40; Household $60. Contributing: Caprice $100; Seaboard $250; Wainwright $500. Sloane Society: Volk $1,000; Poynter $2,500; Frishmuth $5,000; Turner $10,000.

HUNTER HOUSE VICTORIAN MUSEUM, 240 W. Freemason St., Norfolk, VA 23510-1221. Tel.: 757-623-9814.

E-mail: thequeen1894@hunterhousemuseum.org
Web Site: www.hunterhousemuseum.org
Key Personnel: Dir., Margaret H. Spencer
Institution Type/Description: Historic House Museum.
Collections: decorative arts; furnishings; paintings.
Facilities: Museum-related items for sale.
Activities: educational programs; tours; special events.
Hours & Admission Prices: April-Dec. Wed.-Sat. 10:30-3:30, Sun. 12:30-3:30; other times by appointment. Adults $5, senior citizens $4, children $1.

MOSES MYERS HOUSE, 323 E. Freemason St., Norfolk, VA 23510. Mailing Address: Chrysler Museum of Art, 245 W. Olney Rd., Norfolk, VA 23510-1587. Tel.: 757-441-1526. Fax: 757-333-1089.

E-mail: jchristiansen@chrysler.org
Web Site: www.chrysler.org
Founded: 1951.
Congressional District: 2

Key Personnel: Dir., Dr. William Hennessey; Historic House Mgr., John Christiansen.

Personnel Profile: Full-Time Paid 2; Part-Time Paid 10; Part-Time Volunteers 50; Interns 1.

Governing Authority: private nonprofit. Parent Institution: Chrysler Museum of Art, 245 W. Olney Rd., Norfolk. Tax-exempt: 501(c)(3).

Institution Type/Description: Historic House: c.1792 late Georgian, early Federal brick 2-story townhouse.

Collections: American & English furniture; glass; silver; china; textiles; costumes; copper implements; working kitchen; Jewish culture & history.

Research Fields: local history, Jewish history; American, French & English decorative arts; maritime commerce; architecture.

Activities: Jewish programs; guided tours; lectures; films; concerts; study clubs; hobby workshops; organized education programs; docent program; garden programs.

Publications: brochures; workbooks; educational materials.

Hours & Admission Prices: Wed.-Sat. 10-4, Sun. 12-4. No charge; donations accepted. Closed New Year's Day; Independence Day; Thanksgiving; Christmas.

Attendance: 7,500 (accurate)

Membership: Individual $15; Dual & Family $25.

NAUTICUS, One Waterside Dr., Ste. 100, Norfolk, VA 23510-1737. Tel.: 757-664-1000; 800-664-1080. Fax: 757-623-1287.

Web Site: www.nauticus.org

Formerly: Nauticus, The National Maritime Center

Founded: 1994.

Congressional District: 2

Key Personnel: Acting Dir., John Rhamstine; Chm. (V), Robert T. Taylor; Dir. Finance, Yvonne Eberflus; Dir. Education, Jennifer Petro; Deputy Dir. Education & Research, Rolf Johnson; Dir. Mktg. & Public Rels., Shelia Harrison; Enterprise Controller, Raymond McEvoy; Museum Shop Mgr., Liz Etheridge; Dir. Visitor Svcs., Christine Arrasate.

Personnel Profile: Full-Time Paid 35; Full-Time Volunteers 55; Part-Time Paid 30; Interns 8.

Governing Authority: municipal. Parent Institution: City of Norfolk (municipal). Subsidiary Institution: National Maritime Center Foundation. Tax-exempt: 501(c)(3).

Institution Type/Description: Maritime Museum.

Collections: computer & video interactives; commerce & military-related displays; USS Wisconsin & related exhibits; exotic aquaria; aquatic animals; exotic aquaria; Hampton Roads Naval Museum

Facilities: Hampton Roads Naval Museum; USS Wisconsin. Museum-related items for sale.

Activities: arts festivals; concerts; films; formal educational programs; guided tours; lectures; rental gallery; theater; traveling exhibitions; hands-on exhibits including computer & video interactives; touch pools; shark petting.

Publications: quarterly newsletter, Signals; annual report.

Hours & Admission Prices: Memorial Day to Labor Day daily 9-5; Sept.-May Tues.-Sat. 10-5, Sun. 12-5. Adults $10.95, senior citizens $9.95, children 4-12 $8.50; discount to AAA members, active military & groups; members and children 3 & under no charge. Closed New Year's Day; Thanksgiving; Christmas.

Attendance: 320,000 (accurate)

Membership: Senior 55 & over, Student & Teacher $20; Individual $30; Individual Plus One $40; Grandparents $50; Family $60; Family Plus $100; Schooner & Family Silver $250; Brigantine & Family Gold $500. Corporate Silver $500; Corporate Gold $1,000; Corporate Platinum $2,000.

*** NORFOLK BOTANICAL GARDEN,** 6700 Azalea Garden Rd., Norfolk, VA 23518-5337. Tel.: 757-441-5830, ext. 324. Fax: 757-853-8294.

E-mail: don.buma@nbgs.org

Web Site: www.norfolkbotanicalgarden.org

Founded: 1938.

Congressional District: 2

Key Personnel: Exec. Dir., Donald R. Buma; Pres., William C. Eisenbeiss; Dir. Visitor Svcs. & Special Events, Marcia Riley; Dir. Horticulture, Brian O'Neil; Dir. Science, Tim Motley; Dir. Education, Donna Krabill; Dir. Donor Rels., Cathy Fitzgerald; Dir. Finance, Douglas A. Ward; Museum Shop Mgr., Lynn Clark.

Personnel Profile: Full-Time Paid 45; Part-Time Paid 55; Part-Time Volunteers 865; Interns 2.

Governing Authority: society; nonprofit. Parent Institution: Norfolk Botanical Garden Society. Tax-exempt.

Institution Type/Description: Botanical Garden.

Collections: azaleas; rhododendrons; camellias; holly; roses; crapemyrtle; conifers; general; sculpture; VA native plants; hydrangea.

Research Fields: designated a National Camellia collection by the North American Association of Botanical Gardens & Arboreta.

Facilities: 1,910-vol. library of horticulture available for use on premises; reading room; 200-seat auditorium; classrooms; restaurant.

Activities: guided tours; narrated boat & train tours; lectures; hobby workshops; formally organized educational programs; temporary exhibitions. Museum Sponsors: Garden Lights in November & December.

Publications: quarterly, Norfolk Botanical Garden Society Newsletter; quarterly members publication, Bloom; quarterly educational programs publication, Grow.

Hours & Admission Prices: Gardens: April-Oct. 20 daily 9-7; Oct. 21-March daily 9-5. Children's Garden: April-Oct. 20 daily 9:30-6:30; Oct. 21-March daily 9:30-4:30. Butterfly House: June 14-Sept. 22 daily 10-5. Adults $11, seniors & military $10, children 3-18 $9; discounts to groups & AAA members; children 2 & under and members no charge. Closed New Year's Day; Thanksgiving; Christmas.

Attendance: 263,000 (accurate)

Membership: Individual $45; Family $75; Azalea Level $125; Camellia Level $300.

NORFOLK HISTORICAL SOCIETY, 810 Front St., Norfolk, VA 23510. Mailing Address: P.O. Box 6367, Norfolk, VA 23508-0367. Tel.: 757-640-1720.

E-mail: info@norfolkhistorical.org

Web Site: www.norfolkhistorical.org

Founded: 1965.

Congressional District: 2

Key Personnel: Pres. (V), Louis Guy.

Personnel Profile: Part-Time Volunteers 25.

Governing Authority: board of directors. Tax-exempt: 501(c)(3).

Institution Type/Description: Historic Site: 1810 fort.

Collections: history.

Research Fields: history of Norfolk & Hampton Roads area.

Facilities: 600-vol. library of books pertaining to the history of Norfolk & surrounding areas available for research on premises.

Activities: reenactment of War of 1812; Civil War; lecture series.

Publications: book, Norfolk Highlights, 1584-1881; quarterly newsletter, Norfolk Historical Society Courier.

Hours & Admission Prices: Self-Guided Tours: Mon.-Fri. call for hours. No charge; donations accepted.

Attendance: 500 (estimated)

Membership: Students $10; Individual $25; Family $35; Supporter $50; Contributor $100; Grand Old Sentinel $200.

NORFOLK HISTORY MUSEUM, 601 E. Freemason St., Norfolk, VA 23510-2404. Mailing Address: Chrysler Museum of Art, 245 W. Olney Rd., Norfolk, VA 23510-1587. Tel.: 757-441-1526. Fax: 757-333-1089.

E-mail: jchristiansen@chrysler.org

Web Site: www.chrysler.org

Formerly: Willoughby-Baylor House

Founded: 1962.

Congressional District: 2

Key Personnel: Dir., Dr. William Hennessey; Historic House Mgr., John Christiansen.

Personnel Profile: Full-Time Paid 2; Part-Time Paid 10; Part-Time Volunteers 5; Interns 1.

Governing Authority: municipal. Parent Institution: The Chrysler Museum, 245 W. Olney Rd., Norfolk. Tax-exempt: 501(c)(3).

Institution Type/Description: History Museum: housed in former home of Captain William Willoughby.

Collections: Norfolk history.

Research Fields: local history.

Activities: guided tours; lectures; films; concerts; study clubs; hobby workshops; organized education programs; docent program; gardening.

Publications: brochures.

Hours & Admission Prices: Wed.-Sat. 10-4, Sun. 12-4. No charge; donations accepted. Closed New Year's Day; Independence Day; Thanksgiving; Christmas.

Attendance: 4,000 (accurate)

Membership: Individual $15; Dual & Family $25.

OHEF SHOLOM TEMPLE ARCHIVES, 530 Raleigh Ave., Norfolk, VA 23507-2199. Tel.: 757-625-4295. Fax: 757-625-2775.

E-mail: archives@ohefsholom.org

Web Site: www.ohefsholom.org

Founded: 1992.

Congressional District: 2

Key Personnel: Chm. (V), Mark Friedman; Admin., Lynn Evans.
Personnel Profile: Part-Time Paid 1; Part-Time Volunteers 5.
Governing Authority: Parent Institution: Ohef Sholom Temple. Tax-exempt.
Institution Type/Description: Religious Museum.
Collections: history of congregation & local Jewry; photographs; documents; 24 4x5 framed panels.
Research Fields: Norfolk Jewry.
Hours & Admission Prices: Mon.-Fri. 9-5. No charge; donations accepted. &

Attendance: 2,000 (estimated)

PRETLOW PLANETARIUM, Old Dominion University, Dept. of Physics, OCNPS Bldg., Rm. 306, 4600 Elkhorn Ave., Norfolk, VA 23529. Tel.: 757-683-4619.
E-mail: ddepaor@odu.edu
Web Site: sci.odu.edu/physics/about/pretlow.shtml
Key Personnel: Dir., Dr. Declan DePaor
Institution Type/Description: Planetarium.
Collections: space science.
Activities: shows.
Hours & Admission Prices: Call for hours; groups by appointment.

VIRGINIA ZOOLOGICAL PARK, 3500 Granby St., Norfolk, VA 23504-1329. Tel.: 757-441-5227 & 2374. Fax: 757-441-5408.
E-mail: virginiazoo@norfolk.gov
Web Site: www.virginiazoo.org
Founded: 1902.
Congressional District: 2
Key Personnel: Pres. (V), Larry Brett; Zoo Supt., Gary D. Ochsenbein; Exec. Dir., Greg Bockheim; Supvr. Animal Svcs., Louise L. Hill; Asst. Supvr. Animal Svcs., Craig Pelke; Asst. Supvr. Animal Svcs., Joseph M. Roman; Horticulture Cur., Mark Schneider; Supvr. Maintenance, Charles E. Ryan; Administrative Asst., Stacey Connolly.
Personnel Profile: Full-Time Paid 55; Part-Time Paid 24; Part-Time Volunteers 30; Interns 2.
Governing Authority: municipal. Parent Institution: City of Norfolk, Executive Dept., 1101 City Hall Ave., 810 Union St., Norfolk, VA. 23510. Subsidiary Institution: Virginia Zoological Society. Tax-exempt.
Institution Type/Description: Zoology Museum.
Collections: zoological & botanical specimens.
Research Fields: reptile veterinary medicine & reproduction; photo biology of captive reptiles.
Facilities: 150-vol. library of books, primarily zoological, zoo biology, horticulture available for research by students of zoo biology on premises; zoological park. Gift items for sale.
Activities: guided tours; docent program; permanent exhibitions; school loan service.
Publications: quarterly newsletter, Virginia Zoo Review.
Hours & Admission Prices: Daily 10-5. Adults $7, senior citizens 62 & over $6, children 2-11 $5; children under 2 no charge. Closed New Year's Day; Thanksgiving; Christmas. &
Attendance: 285,000 (estimated)
Membership: Individual $35; Military Family $40; Zoo for Two $45; Family $55; Family Plus $65; Family & Guests $100; Zookeeper $250; Curator $500; Leadership Society: $1,000 & up.

Occoquan

OCCOQUAN HISTORICAL SOCIETY, 413 Mill St., Occoquan, VA 22125. Mailing Address: P.O. Box 65, Occoquan, VA 22125-0065. Tel.: 703-491-7525.
Web Site: www.occoquanhistorical society.org
Founded: 1969.
Congressional District: 8
Personnel Profile: Part-Time Paid 5.
Governing Authority: county; nonprofit. Tax-exempt: 501(c)(3).
Institution Type/Description: General Museum: housed in 1768 Miller's Cottage of Grist Mill.
Collections: Indian lore, Civil War history; old costumes & clothing; house hold items; farm items; toys; miscellaneous items.
Facilities: library; 250 sq. ft. exhibit space. Books, China cups, post cards & other items for sale.
Activities: guided tours; films; docent program; temporary exhibitions.
Publications: monthly newsletter.
Hours & Admission Prices: Daily 11-4. No charge; donations accepted. Closed New Year's Day; Thanksgiving; Christmas. &
Attendance: 9,300 (estimated)

Membership: Annual $20.

Onancock

KER PLACE, (M), 69 Market St., Onancock, VA 23417-4223. Mailing Address: P.O. Box 179, Onancock, VA 23417-0179. Tel.: 757-787-8012. Fax: 757-787-4271.
E-mail: kerplace@verizon.net
Web Site: www.shorehistory.org
Founded: 1957.
Congressional District: 1
Key Personnel: Exec. Dir., Jenny Barker; Pres., Susan Stinson; Treas., Ridgway Dunton.
Personnel Profile: Full-Time Paid 1; Part-Time Paid 4; Part-Time Volunteers 30.
Governing Authority: society. Parent Institution: Eastern Shore of Virginia Historical Society, Inc. Tax-exempt.
Institution Type/Description: Historical Society Museum: housed in c.1799 federal period 2-story building.
Collections: 1760-1875 furniture; glass & china; area reference books; c.1782 uniforms & furnishings of Gen. John Cropper; 17th-18th century portraits; local history; paintings.
Research Fields: Eastern shore history.
Facilities: 500-vol. library.
Activities: guided tours.
Publications: newsletter.
Hours & Admission Prices: March 15-Dec. 15 Tues.-Sat. 11-3. Adults $5, students $2; discounts to AAM, AARP & AAA members & groups; members no charge. Closed national holidays.
Attendance: 3,000 (estimated)
Membership: Student, Teacher, Researcher $5; Individual $30; Family $50; Friends of the Historical Society $100; Sustaining $250; Patron $500; Benefactor $1,000.

Orange

THE ARTS CENTER IN ORANGE, 129 E. Main St., Orange, VA 22960. Mailing Address: P.O. Box 13, Orange, VA 22960-0011. Tel.: 540-672-7311.
E-mail: theartsorange@aol.com
Web Site: www.artscenterorange.org
Key Personnel: Exec. Dir., Laura Thompson; Pres., Ed Harvey
Institution Type/Description: Art Museum.
Collections: works by emerging artists.
Facilities: Museum-related items for sale.
Activities: classes.
Hours & Admission Prices: Mon.-Sat. 10-5.

THE JAMES MADISON MUSEUM, 129 Caroline St., Orange, VA 22960-1532. Tel.: 540-672-1776. Fax: 540-672-0231.
E-mail: info@thejamesmadisonmuseum.org
Web Site: www.thejamesmadisonmuseum.org
Founded: 1976.
Congressional District: 7
Key Personnel: Pres. (V), Dr. Martha Neff Smith; Museum Shop Mgr., Sally O'Brien.
Personnel Profile: Full-Time Paid 1; Part-Time Paid 1; Part-Time Volunteers 12; Interns 1.
Governing Authority: nonprofit organization. Parent Institution: The James Madison Memorial Foundation. Tax-exempt: 501(c)(3).
Institution Type/Description: History Museum.
Collections: James & Dolley P. Madison artifacts; Orange County rural culture; 18th-20th century agriculture; cube house; transportation.
Major Exhibits: Bicentennial of War of 1812, 3/12-12/14.
Research Fields: James Madison; Orange County; agriculture; the Constitution.
Facilities: Books & items dealing with James Madison, agriculture & local history for sale.
Activities: loan, permanent & temporary exhibitions; workshops & programs related to exhibits; lectures; book signing events.
Publications: quarterly newsletter.
Hours & Admission Prices: Tues.-Sat. 10-4. Admission $5; local students & members no charge. &
Attendance: 4,500 (accurate)
Membership: Student & Teacher $15; Individual $25; Family $35; Friend $100; Sustainer $250; Contributor $500; Patron $1,000; Benefactor $5,000.

JAMES MADISON'S MONTPELIER - THE MONTPELIER FOUNDATION, 11395 Constitution Hwy., Orange, VA 22957. Mailing Address: P.O. Box 911, Orange, VA 22960-0551. Tel.: 540-672-2728. Fax: 540-301-2776.
E-mail: ewessel@montpelier.org
Web Site: www.montpelier.org
Founded: 1984.
Congressional District: 7
Key Personnel: Exec. Vice Pres. & C.O.O., Sean T. O'Brien; Montpelier Foundation Chm. (V), Greg May; Dir. Archaeology, Dr. Matthew Reeves; Dir. Finance & Administration, Sherida Hawthorne; Acting Cur., Meg Kennedy; Horticulturist, Sandy Mudrinich; Vice Pres. & C.O.O., Sean O'Brien; Dir. Restoration & Facilities, John Jeanes; Dir. Education & Visitor Engagement, Christian Cotz; Dir. Retail & Special Events, Rick Payne.
Personnel Profile: Full-Time Paid 60; Part-Time Paid 70; Part-Time Volunteers 80; Interns 5.
Governing Authority: nonprofit. Parent Institution: National Trust for Historic Preservation, 1785 Massachusetts Ave., N.W. Washington, DC 20036. Subsidiary Institution: The Montpelier Foundation. Tax-exempt: 501(c)(3).
Institution Type/Description: Historic House: lifelong home of President James Madison (1751-1836); 20th century home of William duPont family.
Collections: furnishings & memorabilia of the Madison & duPont families; photographs; extensive grounds including gardens, formal gardens, J. Madison Landmark Forest walking trails, race track, train station & post office.
Major Exhibits: The War of 1812, 6/12-1/14.
Research Fields: archaeology; architecture; Madison archival research; African American history; Civil War history.
Facilities: visitor center; theater; cafe; archaeology lab; Landmark Forest Trail; Civil War encampment site and trail; Gilmore Cabin; Annie du Pont formal garden. Gift items for sale.
Activities: audio tour; guided tours; orientation film. Museum Sponsors: Dolley's Kitchen Outdoor Cooking Demonstration; Hands on Restoration Tent April to October.
Publications: newsletter, Discovering Montpelier; monograph, Building A President's House; souvenir guidebook; commemorative copy of the Constitution.
Hours & Admission Prices: House: April-Oct. Tues.-Sun. & Mon. federal holidays 9:30-5:30, Nov.-March Wed.-Sun. & Mon.-Tues. federal holidays 10:30-4:30. Adults $18, National Trust members $9, children 6-14 $7; children under 6 & Friends of Montpelier no charge. Closed Thanksgiving; Christmas. &
Attendance: 125,000 (estimated)
Membership: Friends of Montpelier: Family $50; Mount Pleasant Friends $100-$249; Blue Ridge Friends $250-$499; Colonnade Circle $500-$999; Portico Circle $1,000-$2,499; Temple Circle $2,500-$4,999. Madison Cabinet: Delegates Council $5,000-$9,999; Congressional Council $10,000-$14,999; Secretary of State's Council $15,000-$24,999; President's Council $25,000 & up.

Palmyra

THE FLUVANNA COUNTY HISTORICAL SOCIETY OLD STONE JAIL MUSEUM, 14 Stone Jail St., Palmyra, VA 22963. Mailing Address: P.O. Box 8, Palmyra, VA 22963-0008. Tel.: 434-589-7910. Fax: 434-589-7910 (call first).
E-mail: info@fluvannahistory.org
Web Site: www.fluvannahistory.org
Founded: 1964.
Congressional District: 5
Key Personnel: Pres., Marvin Moss; Dir., Judith Mickelson.
Personnel Profile: Full-Time Paid 1; Part-Time Paid 1; Part-Time Volunteers 40; Interns 2.
Governing Authority: nonprofit society. Parent Institution: Fluvanna County Historical Society. Branch Museum: Holland Page Place Museum. Tax-exempt.
Institution Type/Description: Historical Society Museum.
Collections: firearms from Revolutionary War-WWI; uniforms from War of 1812-WW I; 19th- & early 20th-century textiles; 19th-century furniture, glassware & prints; books; photographs; household & farm items. Historic Buildings: 1828 stone jail; c.1865 log cabin, Holland Page Place.
Major Exhibits: Fluvanna's Enslaved Community, 10/13-10/14.
Research Fields: genealogy; local history.
Facilities: Museum-related items for sale.
Activities: guided tours; organized educational programs; docent program; temporary exhibitions; lecture series; book signings & talks.
Publications: Fluvanna County Historical Society History; semiannual newsletters; The Artist at Melrose.

Hours & Admission Prices: Archives: Tues.-Wed. 1-4. Museum: June-Oct. Wed. 1-4, Sun. 2-5; other times by appointment. No charge; donations accepted.
Attendance: 500 (estimated)
Membership: Family $25; Sustaining $50; Life $300.

Parksley

EASTERN SHORE RAILROAD MUSEUM, INC., 18468 Dunne Ave., Parksley, VA 23421. Mailing Address: P.O. Box 135, Parksley, VA 23421. Tel.: 757-665-7245.
E-mail: kparksley@aol.com
Web Site: easternshorerailwaymuseum.org
Founded: 1988.
Key Personnel: Pres. (V), Helena T. Killian.
Personnel Profile: Part-Time Paid 1; Part-Time Volunteers 30.
Governing Authority: Tax-exempt.
Institution Type/Description: Historic Buildings: housed in a former railroad station.
Collections: local history; early railroad cars & artifacts; tools.
Hours & Admission Prices: April-Nov. Thurs. & Sun. 1-4, Fri.-Sat. 11-4; other times by appointment. No charge; donations accepted.
Attendance: 1,500 (estimated)
Membership: Individual $25; Family $40; Corporate $50; Gold Spike $100.

Pearisburg

GILES COUNTY HISTORICAL SOCIETY, 208 N. Main St., Pearisburg, VA 24134-1626. Tel.: 540-921-1050.
E-mail: info@gilescountyhistorical.org
Web Site: www.gilescountyhistorical.org
Founded: 1978.
Congressional District: 9
Key Personnel: Exec. Dir. & Museum Shop Mgr., Terri Fisher; Pres. (V), J.R. Peck.
Personnel Profile: Part-Time Paid 3; Part-Time Volunteers 3.
Governing Authority: Tax-exempt.
Institution Type/Description: Historical Society Museum.
Collections: local history & culture; doctor's office; period furnishings; photographs. Historic House: Andrew Johnston House, 1829.
Major Exhibits: Civil War in Giles County, 10/12-12/14.
Facilities: genealogy research office. Museum-related items for sale.
Activities: research.
Hours & Admission Prices: March-Dec. Wed.-Fri. 12-5, Sat.-Sun. 2-5. No charge; donations accepted. &
Attendance: 1,500 (estimated)
Membership: Student $10; Active $20; Supporting $40; Century $100; Patron $200; Benefactor $500 & up.

Petersburg

BLANDFORD CHURCH & RECEPTION CTR. (THE PETERSBURG MUSEUMS), (M), 321 S. Crater Rd., Petersburg, VA 23803-3213. Mailing Address: Petersburg Museums, 15 W. Bank St., Petersburg, VA 23803-3213. Tel.: 804-733-2396. Fax: 804-863-0837.
E-mail: matkinson@petersburg-va.org
Web Site: www.petersburg-va.org
Founded: 1972.
Congressional District: 3
Key Personnel: Site Coord., Martha Atkinson; Acting Museum Mgr. & Cur. Collections, Laura Willoughby.
Personnel Profile: Full-Time Paid 3; Part-Time Paid 25.
Governing Authority: municipal. Parent Institution: City of Petersburg, VA. Subsidiary Institution: Ladies Memorial Association, Petersburg Museums Foundation, Historic Blandford Cemetery Foundation. Tax-exempt.
Institution Type/Description: Historic Building: 1735 Colonial Church of Bristol Parish, located in Blandford Cemetery, containing graves of 30,000 Confederate soldiers.
Collections: Tiffany stained-glass windows donated by the Confederate states, Maryland & Missouri, The Ladies Memorial Association of Petersburg & Tiffany; monuments; symbols related to death.
Research Fields: Church, Colonial, Tiffany & Confederate history; cemetery history.
Facilities: Museum-related items for sale.
Activities: guided tours; lectures. Annual Events: Confederate Memorial Day in June; Cemetery Halloween tour in October.
Publications: pamphlets, Walking Tour of Blandford Cemetery; Old Blandford Church; Blandford Cemetery, Death and Life at Petersburg, Virginia.
Hours & Admission Prices: Call for hours. Adults $5, senior citizens, active

duty military & children 7-12 $4. Closed New Year's Eve, Day & day after; Thanksgiving; Christmas Eve, Day & day after. &

Attendance: 10,000 (estimated)

CENTRE HILL MUSEUM (THE PETERSBURG MUSEUMS), (M), 1 Centre Hill Ave., Petersburg, VA 23803-3213. Mailing Address: Petersburg Museums, 15 W. Bank St., Petersburg, VA 23803-3213. Tel.: 804-733-2401. Fax: 804-863-0837.
E-mail: lwilloughby@petersburg.va.org
Web Site: www.petersburg-va.org
Founded: 1976.
Congressional District: 3
Key Personnel: Acting Museum Mgr. & Cur. Collections, Laura Willoughby.
Personnel Profile: Full-Time Paid 3; Part-Time Paid 25; Part-Time Volunteers 2.
Governing Authority: municipal government. Parent Institution: City of Petersburg. Part of Petersburg Museum System. Subsidiary Institution: Petersburg Museums Foundation. Tax-exempt.
Institution Type/Description: Historic House: c.1823 Centre Hill Mansion.
Collections: Federal, Greek revival & colonial revival architecture; period furnishings; architecture; decorative arts; archives & historic photographs relating to the history City of Petersburg.
Research Fields: 19th-century decorative arts & architecture; repository for archival & photographic material related to Petersburg's history.
Facilities: 500-vol. library emphasizing Petersburg history & decorative arts. Museum-related items for sale.
Activities: guided tours; lectures; temporary exhibitions; research by appointment only.
Hours & Admission Prices: Call for hours. Adults $5, Block Ticket to 3 Museums: Adults $11; discounts to seniors, children & military; Petersburg residents no charge. Closed New Year's Day; Thanksgiving; Christmas Eve & Day.
Attendance: 4,000 (estimated)

FARMERS BANK, 19 Bollingbrook St., Petersburg, VA 23803-4548. Mailing Address: 15 W. Bank St., Petersburg, VA 23803. Tel.: 804-733-2400. Fax: 804-863-0837 & 861-0883.
E-mail: dholmes@petersburg-va.org
Web Site: www.petersburg-va.org
Founded: 1974.
Congressional District: 3
Key Personnel: Sr. Interpreter, Dawn Holmes.
Personnel Profile: Part-Time Paid 42
Governing Authority: Parent Institution: Preservation Virginia. Operated by City of Petersburg Visitor Services. Tax-exempt.
Institution Type/Description: Bank Museum & City of Petersburg Visitor Center.
Collections: pre-Civil War banking; furnishings used in banks of the period; how bank notes were made; gold storage area.
Research Fields: banking history; general Petersburg.
Activities: tours.
Hours & Admission Prices: Sat. 10am-12pm. Adults $5, senior citizens, active duty military & children 7-12 $4; block tickets available; discounts to AAM & ICOM members. Closed New Year's Day; Thanksgiving; Christmas Eve & Day.

PAMPLIN HISTORICAL PARK AND THE NATIONAL MUSEUM OF THE CIVIL WAR SOLDIER, 6125 Boydton Plank Rd., Petersburg, VA 23803-7494. Tel.: 804-861-2408. Fax: 804-861-2820.
E-mail: generalmailbox@pamplinpark.org
Web Site: www.pamplinpark.org
Founded: 1994.
Congressional District: 4
Key Personnel: Exec. Dir., A. Wilson Greene; Pres. (V), Dr. Robert B. Pamplin, Jr.; Dir. Operations, Patrick A. Olienyk.
Personnel Profile: Full-Time Paid 14; Part-Time Paid 20; Part-Time Volunteers 100, Interns 2.
Volunteer Hours: 1,600
Governing Authority: private; nonprofit. Parent Institution: Pamplin Foundation, Portland, OR. Tax-exempt: 501(c)(3).
Institution Type/Description: Military History Museum: Park includes The National Museum of the Civil War Soldier; Tudor Hall Plantation & Field Quarter; The Banks House; re-created Military Encampment; Battlefield Center; The Breakthrough Battlefield, a National Historic Landmark; Civil War Adventure Camp; Hart Farm.
Collections: Civil War military artifacts with emphasis on enlisted soldiers; Antebellum furnishings, decorative arts, agricultural tools & equipment; kitchen/slave quarter; four historic houses; battlefield with earthworks.

Research Fields: Petersburg Campaign of 1864-65 & the Breakthrough of April 2, 1865; life of the Civil War soldier; life on antebellum Virginia plantations.
Facilities: banquet facilities; 100-seat education center. Museum-related items for sale.
Activities: formal educational programs; guided tours; living history demonstrations; children's day camps; annual historical symposium; overnight programs; special events; teacher institutes; interactive distance learning programs.
Publications: park guide, Pamplin Historical Park & the National Museum of the Civil War Soldier; Duty Called Me Here: The Soldier Comrades of the National Museum of the Civil War Soldier; Tudor Hall: The Boisseau Family Farm.
Hours & Admission Prices: See website for hours. Adults $12, children 6-12 $7; discounts to AAM members, schools & groups; members & children under 6 no charge. Closed New Year's Day; Thanksgiving; Christmas. &
Membership: Individual $42.65; Family $100; Pamplin Society $500.

PETERSBURG AREA ART LEAGUE, 7 E. Old St., Petersburg, VA 23803-4558. Tel.: 804-861-4611.
E-mail: galleryadmin@paalart.org
Web Site: www.paalart.org
Founded: 1932.
Key Personnel: Pres. (V), Ellen Ende; Membership & Corresponding Sec., Walt Smith.
Personnel Profile: Part-Time Paid 1; Part-Time Volunteers 1.
Governing Authority: nonprofit. Parent Institution: Virginia Museum of Fine Arts.
Institution Type/Description: Art Gallery: housed in c.1700 granary where Indians came to trade goods.
Collections: various forms of rotating art of local artists.
Activities: lectures; workshops; organized educational programs; participatory & temporary exhibitions. Annual Events: Poplar Lawn Art Festival; Trees of Christmas; fund-raisers.
Publications: newsletter.
Hours & Admission Prices: Tues.-Fri. 12-6, Sat. 10-4. No charge; donations accepted. Closed major holidays. &
Attendance: 4,300
Membership: Student $15; Single $30; Family & Joint $40; Patron $50-$99; Sponsor $100-$499; Corporate Sponsor $250 & up; Benefactor $500 & up.

PETERSBURG NATIONAL BATTLEFIELD, 1539 Hickory Hill Rd., Petersburg, VA 23803-4721. Tel.: 804-732-3531. Fax: 804-732-3615.
Web Site: www.nps.gov/pete
Founded: 1926.
Congressional District: 5
Key Personnel: Supt., Lewis Rogers.
Governing Authority: federal. Parent Institution: National Park Service, U.S. Dept. of Interior. Branch Locations: Grant's Headquarters at City Point, 1001 Pecan Ave., Hopewell, VA. Tel.: 804-458-9504; Five Forks Battlefield, 9840 Courthouse Rd., Dinwiddie, VA. Tel.: 804-469-4093; Eastern Front, 5001 Siege Rd., Petersburg, VA. Tel.: 804-732-3531; Poplar Grove National Cemetery, 8005 Vaughan Rd., Petersburg, VA. Tax-exempt.
Institution Type/Description: Military Museum: located on Petersburg Battlefield.
Collections: military history of campaign for Petersburg; artillery; maps.
Research Fields: Petersburg Campaign; Civil War.
Facilities: 2,000-vol. library by appointment only. Historical publications, postcards & other museum-related items for sale.
Activities: guided tours; lectures; films; permanent exhibits; 17-minute map presentation on Civil War Siege of Petersburg.
Publications: folder, Petersburg; 5 site bulletins: City Point; Poplar Grove National Cemetery; Five Forks Battlefield; African Americans at Petersburg; African-Americans on Lee's Retreat, April 1865.
Hours & Admission Prices: Battlefield: daily 8:30-dusk. Visitor Center: daily 9-5. Vehicle $5, individual $3. Closed New Year's Day; Thanksgiving; Christmas. &

PETERSBURG REGIONAL ART CENTER, 132 N. Sycamore St., Petersburg, VA 23803-3245. Tel.: 804-733-8200.
E-mail: angielong7@aol.com
Web Site: www.pracarts.com
Formerly: Shockoe Bottom Arts Center
Founded: 1993.
Key Personnel: Co Founder SBAC, Rusty Davis; Co Founder SBAC, Deanna Thomas; Dir., Donna Jacobs; Asst. Dir., Angela Long
Institution Type/Description: Art Center.
Collections: works by local artists.

Hours & Admission Prices: Wed.-Sat. 10-4. No charge. Closed holidays. &
Membership: Individual $50.

SIEGE MUSEUM (THE PETERSBURG MUSEUMS), (M), 15 W. Bank St., Petersburg, VA 23803-3213. Tel.: 804-733-2403. Fax: 804-863-0837.
E-mail: lwilloughby@petersburg.va.org
Web Site: www.petersburg.va.org
Founded: 1976.
Congressional District: 3
Key Personnel: Cur. Collections, Laura Willoughby.
Personnel Profile: Full-Time Paid 3; Part-Time Paid 25.
Governing Authority: municipal. Parent Institution: City of Petersburg. Subsidiary Institution: Petersburg Museum System. Tax-exempt.
Institution Type/Description: Historic Museum: housed in an 1839 Greek revival style agricultural exchange.
Collections: Nineteeth century artifacts; industrial objects; decorative arts; photographs.
Research Fields: Civil War in Petersburg; Petersburg history.
Facilities: theater.
Activities: children's scavenger hunt.
Hours & Admission Prices: Call or visit website for hours. Adults $5; discounts to seniors, children, military; Petersburg residents no charge. Closed New Year's Day; Thanksgiving; Christmas Eve & Day. &
Attendance: 6,000 (estimated)

Pocahontas

POCAHONTAS MINE & MUSEUM, 11 Centre St., Pocahontas, VA 24635. Mailing Address: P.O. Box 128, Pocahontas, VA 24635-0128. Tel.: 276-945-2134 & 9522. Fax: 276-945-9904.
E-mail: pocahontas@comcast.net
Institution Type/Description: History Museum: designated as a national historic landmark.
Collections: coal mining history.
Hours & Admission Prices: April to Oct. Mon.-Sat. 10-5, Sun. 1-5. Adults $7, children 6-12 $4.50; discounts to AAA members & groups; children under 6 no charge.

Portsmouth

CHILDREN'S MUSEUM OF VIRGINIA, (M), 221 High St., Portsmouth, VA 23704. Tel.: 757-393-5258.
E-mail: schweize@portsmouthva.gov
Web Site: childrensmuseumva.com
Founded: 1980.
Congressional District: 4
Key Personnel: Dir., Nancy S. Perry.
Personnel Profile: Full-Time Paid 25; Part-Time Paid 10; Part-Time Volunteers 75.
Governing Authority: city. Tax-exempt.
Institution Type/Description: Children's Museum.
Collections: over 10,000 model toys & trains.
Facilities: concession; rental facilities. Museum-related items for sale.
Activities: birthday parties; temporary exhibitions.
Publications: quarterly newsletter.
Hours & Admission Prices: Mon.-Sat. 9-5, Sun. 11-5. Adults 18 & over $11, children 2-17 $10; military & seniors $9; discounts to VAM & AAM members; members & children under 2 no charge. &
Attendance: 143,690 (accurate)
Membership: Friend $50; Enthusiast Plus 3 $80; Stargazer Plus 3 $125; Visionary Plus 5 $275.

THE HILL HOUSE, 221 North St., Portsmouth, VA 23704. Tel.: 757-393-0241.
Institution Type/Description: Historic House Museum.
Collections: Hill family history; period furnishings; personal artifacts; photographs.
Hours & Admission Prices: April-Dec. Wed. 12:30-4:30, Sat.-Sun. 1-5.

PORTSMOUTH ART & CULTURAL CENTER, 400 High St., Portsmouth, VA 23704. Mailing Address: 521 Middle St., Portsmouth, VA 23704. Tel.: 757-393-8543.
E-mail: paulg@portsmouth.gov
Web Site: www.portsmouthartcenter.com
Formerly: Courthouse Galleries Art Museum
Founded: 1984.
Congressional District: 3

Key Personnel: Dir., Nancy Perry; Cur., Gayle Paul; Pres. (V), Barbara Vincent; Museum Shop Mgr., Liona Bourgeault.
Personnel Profile: Full-Time Paid 4; Part-Time Paid 3; Part-Time Volunteers 20; Interns 2.
Governing Authority: city. Tax-exempt.
Institution Type/Description: Art Museum.
Collections: paintings; sculpture.
Facilities: Museum-related items for sale.
Activities: lectures; classes; workshops; performances; music; demonstrations.
Publications: quarterly newsletter.
Hours & Admission Prices: Memorial Day to Labor Day Mon.-Sat. 10-5, Sun. 1-5; Sept.-May Tues.-Sat. 10-5, Sun. 1-5. Adults $3; discounts to AAA & AAM members, senior citizens & military.
Attendance: 27,816 (accurate)
Membership: Friend $50; Enthusiast + 3 $80; Visionary + 5 $250.

PORTSMOUTH HISTORICAL ASSOCIATION, 221 North St., Portsmouth, VA 23704-2601. Tel.: 757-393-0241.
Founded: 1957.
Congressional District: 3
Key Personnel: C.E.O. & Pres. (V), Alice C. Hanes; 1st Vice Pres., Marshall W. Butt, Jr.; 2nd Vice Pres., Ms. Macon Williams.
Personnel Profile: Part-Time Volunteers 20.
Institution Type/Description: Historical House: c.1830 Hill House.
Collections: original period furnishings.
Activities: guided tours.
Hours & Admission Prices: April-Dec. Sat.-Sun. 1-5; tour groups by appointment at other times. Adults $3, children 6-12 $1; children under 6 with adult & members with dues card no charge.
Attendance: 601 (accurate)
Membership: Individual $5.

PORTSMOUTH MUSEUMS, (M), 521 Middle St., Portsmouth, VA 23704-3708. Tel.: 757-393-8983. Fax: 757-393-5228.
E-mail: perryn@portsmouthva.gov
Web Site: www.portsmouthva.gov
Founded: 1980.
Congressional District: 4
Key Personnel: Dir., Nancy Perry; Site Mgr. Children's Museum of Virginia, Al Schweizer; Cur. Portsmouth Art & Cultural Center, Gayle Paul; Naval Shipyards & Lightship Museum Cur., Corey Thornton; Education Coord., Christine Matyseck; Volunteer Coord., Barbara Pickett; Volunteer Coord., Stephen Grunett.
Personnel Profile: Full-Time Paid 26; Part-Time Paid 16; Part-Time Volunteers 60; Interns 1.
Governing Authority: municipal. City of Portsmouth, 801 Crawford St., Portsmouth, VA 23704-3266. Tel.: 757-393-8983. Branch Museums: Portsmouth Art & Cultural Center, High & Court St., Portsmouth, VA 23704-3266. Tel.: 757-393-8543; Children's Museum of Virginia, 221 High St., Portsmouth, VA 23704-3266. Tel.: 757-393-5258; Portsmouth Naval Shipyard Museum, 2 High St. Portsmouth, VA 23704-3266. Tel.: 757-393-8591; The Lightship Portsmouth Museum, Foot of London at Water St., Portsmouth, VA 23704-3266. Tel.: 757-393-8591. Tax-exempt.
Institution Type/Description: General Museum.
Collections: Portsmouth Art & Cultural Center: all exhibits are traveling or created. Children's Museum of Virginia: educational & participatory exhibits concentrating on science, art & the humanities. Lancaster Antique Toy & Train: includes over 10,000 objects & Planetarium. Naval Shipyard Museum: history of the Naval Shipyard & Portsmouth, models, uniforms, flags, arms, maps, military, research library. Lightship Portsmouth: restored & furnished c.1915 Coast Guard Lightship.
Research Fields: The Lightship Portsmouth: lightship service. Portsmouth Naval Shipyard Museum: naval & local history.
Facilities: Portsmouth Art & Cultural Center: gift items for sale. Portsmouth Naval Shipyard Museum: 5,000-vol. library of naval & local history available for research by appointment. Children's Museum of Virginia: gift items for sale.
Activities: Portsmouth Art & Cultural Center: traveling exhibitions; concerts; lectures; workshops; youth art programs. Children's Museum of Virginia: out-reach & education programs; workshops; teacher training; special events; boy scout & girl scout programs; camps. Naval Shipyard Museum: lectures; family programs.
Publications: quarterly newsletters.
Hours & Admission Prices: Portsmouth Art & Cultural Center: Tues.-Sat. 9-5, Sun. 1-5. Adults $11, children $10. Children's Museum of Virginia: Tues.-Sat. 9-5, Sun. 11-5. Portsmouth Naval Shipyard Museum: Tues.-Sat. 10-5, Sun. 1-5. Lightship Museum: Memorial Day to Labor Day Mon.-Sat.

10-5, Sun. 1-5; Sept.-May Sat. 10-5, Sun. 1-5. Admission $4; discounts to AAM & VAM members; members no charge. Combination tickets available.
Attendance: 230,836 (accurate)
Membership: Individual $50; Enthusiast $80; Stargazer $125; Visionary $250.

PORTSMOUTH NAVAL SHIPYARD MUSEUM, (M), 2 High St., Portsmouth, VA 23704. Tel.: 757-393-8591.
E-mail: contact@portsmouthnavalshipyardmuseum.com
Web Site: www.portsmouthnavalshipyardmuseum.com
Founded: 1949.
Governing Authority: Parent Institution: Portsmouth Museums.
Institution Type/Description: Naval History Museum.
Collections: U.S. Navy history; ship models; uniforms; military artifacts; photographs.
Hours & Admission Prices: Tues.-Sat. 10-5, Sun. 1-5. Tues.-Thurs. adults $2, seniors 62 & over and military $1.50, students 2-17 $1; children under 2 no charge. Fri.-Sun. (includes admission to the Lightship Portsmouth Museum): adults $4, seniors 62 & over and military $3, students 2-17 $2; children under 2 no charge.

VIRGINIA SPORTS HALL OF FAME & MUSEUM, 206 High St., Portsmouth, VA 23704-3720. Mailing Address: P.O. Box 370, Portsmouth, VA 23705-0370. Tel.: 757-393-8031. Fax: 757-393-8288. Facebook: VSHFM.
E-mail: info@vshfm.com
Web Site: www.vshfm.com
Founded: 1972.
Congressional District: 4
Key Personnel: Pres., Eddie Webb; Chm., Joel Rubin; Dir. Museum Operations, Elizabeth Goodwin; Mktg. & Sales Coord., Corry Gross; Education Coord., Elaina Trafny; Coord. Special Events, Jason Taylor.
Personnel Profile: Full-Time Paid 5; Part-Time Paid 13; Part-Time Volunteers 5; Interns 3.
Governing Authority: nonprofit organization. Tax-exempt: 501(c)(3).
Institution Type/Description: Sports Museum: located in Historic Olde Towne Portsmouth.
Collections: historical & contemporary Virginia sports figures.
Research Fields: Virginia sportsmen & sportswomen.
Facilities: 17,000 sq. ft. exhibit space.
Activities: guided tours; children's programs; video game tournaments; health fairs; facility rental. Museum Sponsors: annual Induction Banquet. Celebrity golf tournament.
Publications: annual banquet program; magazine, The Press Box.
Hours & Admission Prices: Summer: Mon.-Sat. 10-5, Sun. 1-5; Fall & Spring Tues.-Fri. 10-2, Sat. 10-5, Sun. 1-5. Adults $7; discounts to AAA members & military; members no charge. Closed New Year's Day; Thanksgiving; Christmas. &
Attendance: 70,000 (estimated)
Membership: Individual $35; Family, Colleges & Friends $100; Family Plus $150; University $200; Corporate Benefactor $1,000.

Pulaski

FINE ARTS CENTER FOR NEW RIVER VALLEY, 21 W. Main St., Pulaski, VA 24301-5015. Mailing Address: P.O. Box 309, Pulaski, VA 24301-0309. Tel.: 540-980-7363. Fax: 540-980-7363.
E-mail: info@facnrv.org
Web Site: www.facnrv.org
Founded: 1978.
Congressional District: 9
Key Personnel: Dir., Judy C. Ison; Pres. (V), Gary Hancock; Asst., Donna Rorrer.
Personnel Profile: Full-Time Paid 1; Part-Time Volunteers 20; Interns 2.
Governing Authority: nonprofit. Tax-exempt: 501(c)(3)
Institution Type/Description: Art Museum: housed in 1898 Victorian Commercial structure.
Collections: eclectic & contemporary works by local & regional artists.
Research Fields: area artists.
Facilities: 150-vol. library of visual & performing arts for public use; educational facilities; catering service available; meeting facilities. Paintings, pottery, weaving, basketry & other related items for sale.
Activities: guided tours; lectures; concerts; dance recitals; arts festivals; theater; hobby workshops; organized education programs; docent program; participatory, loan, temporary & traveling exhibitions; performance sponsorship.
Publications: monthly newsletter, Centerpiece.
Hours & Admission Prices: Mon.-Fri. 10-5, Sat. 11-3. No charge; donations accepted. Closed federal holidays. &

Attendance: 20,000 (estimated)
Membership: Student $10; Individual $15; Family $25; Patron $75; Sponsor $125; Sustaining $500. Business Memberships: Business Sponsor $150; Sustaining Corporation $550; Corporate Benefactor $1,000.

RAYMOND F. RATCLIFFE MEMORIAL MUSEUM, Pulaski Railroad Station, 124 S. Washington Ave., Pulaski, VA 24301. Tel.: 540-980-2055.
Institution Type/Description: History Museum: housed in the historic Pulaski Train Depot; built in 1886.
Collections: local history & culture; period furnishings; personal artifacts; photographs.
Hours & Admission Prices: Call for hours.

Radford

GLENCOE MUSEUM, 600 Unruh Dr., Radford, VA 24141-1501. Mailing Address: P.O. Box 3339, Radford, VA 24143-3339. Tel.: 540-731-5031.
E-mail: info@glencoemuseum.org
Web Site: www.glencoemuseum.org
Founded: 1998.
Congressional District: 9
Key Personnel: Dir., Scott L. Gardner; Chm. (V), Margaret Sproule.
Personnel Profile: Part-Time Paid 1; Part-Time Volunteers 30; Interns 1.
Governing Authority: Parent Institution: Radford Heritage Foundation. Tax-exempt.
Institution Type/Description: History Museum: housed in c.1870 home built by Gen. Gabriel Colvin Wharton.
Collections: period furnishings; decorative arts; photographs; historical documents.
Research Fields: local & regional history.
Activities: special events; educational programs; temporary exhibitions; summer history camp; Glencoe lecture series.
Hours & Admission Prices: Tues.-Sat. 10-4, Sun. 1-4. No charge; donations accepted. Closed national holidays. &
Attendance: 2,271 (accurate)
Membership: Student $10; Individual $25; Family $50; Sustaining $100; Patron $500; Leadership Giving $1,000.

RADFORD UNIVERSITY ART MUSEUM, (M), Corner of Jefferson & Downey Sts., Radford, VA 24142. Mailing Address: P.O. Box 6965, Radford, VA 24142-6965. Tel.: 540-831-5754. Fax: 540-831-6799.
E-mail: ruartmuseum@radford.edu
Web Site: www.radford.edu/rumuseum
Founded: 1985.
Congressional District: 9
Key Personnel: Dir., Steve Arbury; Registrar, Kim Cochran.
Personnel Profile: Full-Time Paid 1; Part-Time Paid 1; Interns 10.
Operating Expenses: 18,000
Operating Income: 18,000
Governing Authority: Parent Institution: Radford University. Tax-exempt.
Institution Type/Description: University Art Museum.
Collections: late 20th-century painting & sculpture; ancillary collections of Nigerian & Huichol art from Mexico.
Major Exhibits: James Thurman: Thermanite etc. New Creations, New Materials, 1/30-3/7/14; James & Umut Demirguc Thurman: Synergies, 1/30/14-3/7/14.
Research Fields: contemporary art.
Facilities: 2,300 sq. ft. exhibit space; 16,000 sq. ft. sculpture court; sculpture collection placed throughout campus.
Activities: national, regional & international artists; loan exhibitions; lectures & symposia in conjunction with exhibitions; public sculpture.
Publications: brochures; catalogs; Selections from the Permanent Collection; Dorothy Gillespie; Ibram Lassaw; Adolf Dehn.
Hours & Admission Prices: May-July Mon.-Fri. 10-4, Sat.-Sun. 12-4; Sept.-April Mon.-Fri. 10-5, Sat.-Sun. 12-4. No charge; donations accepted. Closed national holidays & school vacations. &
Attendance: 17,400 (accurate)
Membership: Associate $100-$249; Patron $250-$499; Sponsor $500-$999; Benefactor $1,000 & up.

Raphine

CYRUS H. MCCORMICK MEMORIAL MUSEUM, 128 McCormick Farm Circle, Raphine, VA 24472. Tel.: 540-377-2255. Fax: 540-377-5850.
E-mail: dafiske@vt.edu

Web Site: www.vaes.vt.edu/steeles/history.html
Founded: 1956.
Congressional District: 6
Key Personnel: Supt., David A. Fiske.
Governing Authority: university. Affiliated with Virginia Polytechnic Institute.
Institution Type/Description: Agricultural History Museum.
Collections: agriculture; McCormick reaper; c.1800 grist mill; replicas of first & subsequent reapers and mowing machines.
Research Fields: agriculture; livestock.
Facilities: picnic grounds.
Activities: permanent exhibitions. Museum Sponsors: Mill Day Festival in October.
Hours & Admission Prices: Daily 8-5. No charge; donations accepted. Closed during inclement weather; major holidays.
Attendance: 8,000 (estimated)

Reedville

REEDVILLE FISHERMEN'S MUSEUM, (M), 504 Main St., Reedville, VA 22539-4401. Mailing Address: P.O. Box 306, Reedville, VA 22539-0306. Tel.: 804-453-6529. Fax: 804-453-7159.
E-mail: office@rfmuseum.org
Web Site: www.rfmuseum.org
Founded: 1986.
Congressional District: 1
Key Personnel: Dir., Shawn Hall; Pres. (V), Ted Hower; Museum Shop Mgr., Jeri Brewer; Office Mgr., Karen Rogers.
Personnel Profile: Full-Time Paid 1; Part-Time Paid 2; Part-Time Volunteers 277.
Governing Authority: Parent Institute: Greater Reedville Association, Inc. Tax-exempt.
Institution Type/Description: Regional Maritime Museum.
Collections: tools & equipment related to commercial and recreational fishing, crabbing & oyster industries; photographs. Historic Buildings: The William Walker House built in the 1870's and furnished as a typical waterman's home of the early 1900's; The Covington Building: fishing industry, watermen's culture of Chesapeake Bay; regional history of the Northern Neck of Virginia; The Pendleton Building: boat & model making shops; The Butler House: museum offices, library & archives; in the water are displayed the Claud W. Somers, a 42-foot skipjack built in 1911 and the Elva C., a 55-foot traditional workboat built in 1922.
Research Fields: Menhaden fisheries; maritime history; lower Chesapeake Bay.
Facilities: library & archives documenting the Menhaden fishing industry 1870 century to the present day.
Activities: lecture series on topics related to life on the lower Chesapeake Bay; craft workshops for children & adults; boat building & model making classes; apprenticeship opportunities; monthly meetings of groups interested in wooden boats, educational cruises aboard the restored buy boat and skipjack; needlework. Annual Events: Antique & Classic Boat Show in September; Old fashioned Independence day celebration in July; Oyster Roast in November; Christmas on Cockrell's Creek in December.
Publications: quarterly newsletter, Starry Banner.
Hours & Admission Prices: March-April Sat.-Sun. 10:30-4:30; May-Oct. daily 10:30-4:30; Nov.-Jan. Fri.-Mon. 10-30-4:30. Adults $5, senior citizens over 60 $3; discounts to CAMM & AAM members; children under 12 & members no charge. ♿
Attendance: 14,500 (estimated)
Membership: Student $15; Individual $25; Family $35; Sustaining $100; Patron $250; Sponsor $500.

Reston

THE GREATER RESTON ARTS CENTER (GRACE), 12001 Market St., Ste. 103, Reston, VA 20190-6244. Tel.: 703-471-9242. Fax: 703-471-0952.
E-mail: info@restonarts.rog
Web Site: www.restonarts.org
Founded: 1974.
Congressional District: 8
Key Personnel: Dir., Damion Sinclair; Exhibitions Dir., Holly Koons McCullough.
Personnel Profile: Full-Time Paid 3; Part-Time Paid 4; Part-Time Volunteers 10.
Governing Authority: nonprofit organization. Tax-exempt: 501(c)(3).
Institution Type/Description: Civic Art & Cultural Center.
Collections: works of contemporary artists.
Research Fields: contemporary visual art.
Activities: lectures; gallery talks; lectures; children's workshops; docent program; Art-In-the-Schools program. Special Event: Northern Virginia Fine Arts Festival.

Publications: exhibition catalogues.
Hours & Admission Prices: Tues.-Sat. 11-5. No charge. Closed New Year's Day; Easter; Christmas. ♿
Attendance: 20,000
Membership: Educator & Artist $35; Individual $50; Family $100; Visionary Levels $250 & up.

RESTON MUSEUM, 1639 Washington Plaza, Reston, VA 20190-4305. Tel.: 703-709-7700. Fax: 703-709-6668.
E-mail: restonmuseum@gmail.com
Web Site: www.restonmuseum.org
Founded: 1997.
Congressional District: 8
Key Personnel: Chair, Shelley Mastran.
Personnel Profile: Part-Time Paid 2; Part-Time Volunteers 25; Interns 2.
Governing Authority: Tax-exempt.
Institution Type/Description: History Museum.
Collections: Reston history.
Facilities: Museum-related items for sale.
Activities: walking tours; author events; artist show and sale events; scout badge programs; community celebrations; home tours.
Hours & Admission Prices: Tues.-Sun. 12-5, Sat. 10-5. No charge; donations accepted. ♿
Attendance: 6,913 (accurate)
Membership: Senior & Student $20; Individual $30; Household $45; Patron $100; Benefactor $500.

Richmond

AGECROFT HALL, 4305 Sulgrave Rd., Richmond, VA 23221-3256. Tel.: 804-353-4241. Fax: 804-353-2151.
Web Site: www.agecrofthall.com
Founded: 1967.
Congressional District: 3
Key Personnel: Exec. Dir., Richard W. Moxley; Pres. (V), Richard B. Woodward; Business Officer & Museum Shop Mgr., Sieglinde F. Nix; Cur. Education, Jill Pesesky; Mgr. Tour Svcs., Katie Reynolds.
Personnel Profile: Full-Time Paid 8; Part-Time Paid 21.
Governing Authority: nonprofit organization. Tax-exempt.
Institution Type/Description: Historic House: 15th-century English Country Manor House disassembled in 1926, brought over & rebuilt; formerly located at Lancashire, England.
Collections: 16th- & 17th-century English furnishings & paintings; textiles; armor; musical instruments.
Research Fields: 16th- & early 17th-century English country life & related pursuits.
Facilities: herb garden; knot garden; formal gardens & Tradescant garden.
Activities: guided tours; children's programs; living history presentations; lectures, theatrical & musical presentations.
Publications: online newsletter.
Hours & Admission Prices: Tues.-Sat. 10-4, Sun. 12:30-5. Adults $8, senior citizens $7, students $5; discounts to AAA & active military. Closed legal holidays. ♿
Attendance: 16,957 (accurate)

THE AMERICAN CIVIL WAR CENTER AT HISTORIC TREDEGAR, (M), 490 Tredegar St., Richmond, VA 23219-4328. Tel.: 804-780-1865. Fax: 804-780-0264.
E-mail: info@tredegar.org
Web Site: www.tredegar.org
Formerly: Tredegar National Civil War Center Foundation
Founded: 2000.
Key Personnel: Pres., Christy S. Coleman; Dir. Finance, John Gould; Chief Devel., Lynn Meyer; Dir. Strategic Initiatives, Christie Ann Bieber; Coord. Education, Sean Kane; Coord. Public Rels., Penelope Wallace; Cur., Randy Klemm; Museum Shop Mgr., Gail Anderson
Institution Type/Description: History Museum.
Collections: Civil War history & artifacts; period artifacts; photographs; hands-on exhibits.
Activities: educational programs; teacher development workshops; guest historian lecturers; website training; Civil War history programs; monographs.
Hours & Admission Prices: Daily 9-5. Adults $8, seniors 62 & up $6, children 6-17 $4; groups and children 5 & under no charge. Closed New Year's Day; Thanksgiving; Christmas. ♿
Attendance: 116,244 (accurate)
Membership: Student & Senior $30; Senior Plus One $45; Individual $40; Individual Plus One $55; Family $65.

ANDERSON GALLERY, SCHOOL OF THE ARTS, VIRGINIA COMMONWEALTH UNIVERSITY, (M), 907 1/2 W. Franklin St., Richmond, VA 23284-2514. Mailing Address: Anderson Gallery/VCU, 907 1/2 W. Franklin St., P.O. Box 842514, Richmond, VA 23284-2514. Tel.: 804-828-1522. Fax: 804-828-8585.
Web Site: www.vcu.edu/arts/gallery
Founded: 1969.
Congressional District: 3
Key Personnel: Dir., Ashley Kister; Gallery Coord., Traci Garland; Mgr. Exhibitions, Michael Lease.
Personnel Profile: Full-Time Paid 3; Part-Time Paid 12; Part-Time Volunteers 4; Interns 4.
Governing Authority: Parent Institution: Virginia Commonwealth University. Subsidiary Institution: School of the Arts. Tax-exempt: 170(b)(1)(A).
Institution Type/Description: University Museum.
Collections: prints; photographs; contemporary painting & sculpture; folk art; Mayan contemporary textiles.
Research Fields: European & American graphic arts; photography; contemporary art.
Facilities: study collection of visual materials available by special arrangement. Books & art related items for sale.
Activities: lectures; films; gallery talks; training programs for professional museum workers; inter-museum loan, temporary & traveling exhibitions.
Publications: catalogs; posters; limited edition graphics.
Hours & Admission Prices: Tues.-Fri. 10-5, Sat.-Sun. 12-5; extended hours for student exhibitions. No charge. Closed state & university holidays.
Attendance: 30,000 (accurate)

ARTSPACE GALLERY, Zero E. 4th St., Richmond, VA 23224-4202. Tel.: 804-232-6464.
E-mail: artspaceorg@gmail.com
Web Site: artspacegallery.org
Founded: 1988.
Key Personnel: Pres. (V), Jessica Sims.
Personnel Profile: Part-Time Paid 1.
Governing Authority: nonprofit organization.
Institution Type/Description: Art Gallery.
Collections: works by contemporary artists; paintings; sculptures.
Activities: outreach programs; special events.
Hours & Admission Prices: Tues.-Sun. 12-4; other times by appointment. No charge; donations accepted. &
Attendance: 4,925 (accurate)

BETH AHABAH MUSEUM & ARCHIVES, (M), 1109 W. Franklin St., Richmond, VA 23220-3700. Tel.: 804-353-2668. Fax: 804-358-3451. Facebook: Beth Ahabah Museum.
E-mail: bama@bethahabah.org
Web Site: www.bethahabah.org
Founded: 1977.
Key Personnel: Exec. Dir., David Farris; Docent & Administrative Asst., Grace Zell; Admin., Bonnie Eisenman.
Personnel Profile: Part-Time Paid 3; Part-Time Volunteers 3; Interns 2.
Governing Authority: religious institution. Parent Institution: Cong. Beth Ahabah. Tax-exempt.
Institution Type/Description: Jewish History Museum.
Collections: records & items concerning KK Beth Shalome, Cong. Beth Ahabah & the Richmond Jewish Community.
Major Exhibits: That You'll Remember me: Jewish Voices of the Civil War, 1/13-12/15; Wheels of Justice - Sculptures by Linda Gissen, 11/13-10/14.
Research Fields: Jewish genealogy & history.
Facilities: library pertaining to Judaic genealogy & history. Museum-related items for sale.
Activities: guided tours; lectures; organized educational programs; docent program; participatory, loan & temporary exhibitions.
Publications: monthly inclusion, Temple Bulletin; News from the Archives; quarterly scholarly publication, Generations.
Hours & Admission Prices: Sun.-Thurs. 10-3; call to confirm. No charge; donations accepted. Closed Jewish & national holidays. &
Attendance: 1,500 (accurate)
Membership: Support $25-$79; Friend $80-$149; Donor $150-$499; Sponsor $500-$999; Patron $1,000 & up.

BLACK HISTORY MUSEUM & CULTURAL CENTER OF VIRGINIA, 00 Clay St., Richmond, VA 23219. Mailing Address: P.O. Box 61052, Richmond, VA 23261. Tel.: 804-780-9093. Fax: 804-780-9107.
E-mail: information.bhm@gmail.com
Web Site: www.blackhistorymuseum.org

Founded: 1981.
Key Personnel: Dir., Dr. Maureen Elgersman Lee.
Governing Authority: bd. of trustees. Tax-exempt.
Institution Type/Description: History Museum.
Collections: lives & accomplishments of Blacks in Virginia; documents; limited editions; prints; art & photographs; written records.
Facilities: Museum-related items for sale.
Hours & Admission Prices: Closed, reopening 2015.

CHASEN GALLERIES OF FINE ART, 3554 W. Cary St., Richmond, VA 23221-2729. Tel.: 800-524-2736 (US); 804-204-1048 (Intl.). Fax: 804-204-1049.
E-mail: art@chasengalleries.com
Web Site: www.chasengalleries.com
Founded: 1999.
Key Personnel: Dir., Lorelle Rau; Pres., Andrew Chasen
Institution Type/Description: Art Gallery.
Collections: fine art glass; paintings; sculpture.
Activities: special events.
Publications: newsletter.
Hours & Admission Prices: Mon.-Sat. 10-6. No charge. Closed holidays.
Attendance: 2,000 (accurate)

CHILDREN'S MUSEUM OF RICHMOND, (M), 2626 W. Broad St., Richmond, VA 23220-1904. Tel.: 804-474-7000 & CMOR. Fax: 804-474-7099.
E-mail: info@c-mor.org
Web Site: www.c-mor.org
Founded: 1977.
Congressional District: 9
Key Personnel: Dir., Karen Coltrane; Chm. (V), Iris Holliday; Treas., Mark Cross; Museum Shop Mgr., Jennifer Boyle.
Personnel Profile: Full-Time Paid 20; Part-Time Paid 40; Part-Time Volunteers 145; Interns 8.
Governing Authority: nonprofit organization; board directors. Subsidiary Institution: Children's Museum of Richmond - Short Pump. Tax-exempt: 501(c)(3).
Institution Type/Description: Children's Museum.
Collections: replica of Virginia limestone cave.
Facilities: celebration center; children's pavilion; performing arts area; art studio; resource center.
Activities: performances; organized education programs for children, participatory, hands-on exhibits focused on art, humanities, & special events; outreach field trips; backyard outdoor experience; birthday parties.
Publications: annual report; bimonthly email newsletter.
Hours & Admission Prices: Mon.-Sat. 9:30-5, Sun. 12-5. Adults & children $8, seniors $7; discounts to groups, AYM, VAM & AAM members and after 4pm; museum members & children under one no charge. Closed New Year's Day; Easter; Thanksgiving; Christmas. &
Attendance: 360,000 (accurate)
Membership: Family & Grandparent $100; Family Advantage $200; Patron $350.

CROSSROADS ART CENTER GALLERY, 2016 Staples Mill Rd., Richmond, VA 23230-3109. Tel.: 804-278-8950.
Key Personnel: Dir. Exhibitions, Jenni Kirby
Institution Type/Description: Art Gallery.
Collections: works by local, regional & national contemporary artists.
Activities: special events; permanent & temporary exhibitions.
Hours & Admission Prices: Call for hours.

DABBS HOUSE MUSEUM, 3812 Nine Mile Rd., Richmond, VA 23223-4848. Tel.: 804-652-3406.
Founded: 2008.
Governing Authority: Parent Institution: Henrico County.
Institution Type/Description: Historic House Museum: housed in the former field headquarters of General Lee during the summer of 1862.
Collections: local history; period furnishings; personal artifacts, photographs.
Facilities: Museum-related items for sale.
Activities: educational programs.
Hours & Admission Prices: Call for hours. No charge

EDGAR ALLAN POE MUSEUM, 1914-16 E. Main St., Richmond, VA 23223-6964. Tel.: 804-648-5523. Fax: 804-648-8729.
E-mail: info@poemuseum.org
Web Site: www.poemuseum.org
Founded: 1922.

Congressional District: 3
Key Personnel: Pres. (V), Harry Poe; Treas., Edward D. Campbell, Jr.; Exec. Dir., Katarina Spears.
Personnel Profile: Part-Time Paid 4; Part-Time Volunteers 8.
Governing Authority: literary foundation. Parent Institution: Poe Foundation. Tax-exempt: 170(b)(1)(A).
Institution Type/Description: Literary Museum: c.1737-1740 Old Stone House.
Collections: family ephemera; personal artifacts; Poe books & furnishings; manuscripts; memorabilia; illustrations; model of Old Richmond prior to 1849.
Research Fields: 19th-century American literature; Edgar Allan Poe; history of Richmond.
Facilities: 1,000-vol. library of Poe-related items & manuscripts available for research on premises by appointment. Gift items for sale.
Activities: guided tours; lectures; readings; performances; school programs. Museum Sponsors: Poe's Birthday Party in January.
Publications: magazine, The Poe Messenger; newsletter, Evermore; book, The Incredible Mr. Poe: Comic Book Adaptations of the Works of E.A. Poe, 1943-2007.
Hours & Admission Prices: Tues.-Sat. 10-5, Sun. 11-5; tours on the hour. Adults $6, senior citizens 60 & over and students 8 & over $5; discounts to AAA members; AAM members, children under 8 & members no charge. Group tours available, call 804-648-5523. Closed New Year's Day; Christmas Day.
Attendance: 17,000 (estimated)
Membership: Student & Teacher $15; Individual $25; Family $35; Contributing $100; Benefactor $250.

ELEGBA FOLKLORE SOCIETY'S CULTURAL CENTER, 101 E. Broad St., Richmond, VA 23219-1733. Tel.: 804-644-3900. Fax: 804-644-3919.
E-mail: gallery@efsinc.org
Web Site: www.efsinc.org
Founded: 1990.
Key Personnel: Found Pres. & Artistic Dir., Janine Bell.
Governing Authority: nonprofit organization. Tax-exempt.
Institution Type/Description: Art Museum.
Collections: paintings; sculpture; drawings; dolls; multicultural crafts.
Hours & Admission Prices: Mon.-Fri. 10-6, Sat. 12-4; other times by appointment. No charge; donations accepted.

FOLK ART SOCIETY OF AMERICA, 1904 Byrd Ave., #312, Richmond, VA 23230-3029. Mailing Address: P.O. Box 17041, Richmond, VA 23226-7041. Tel.: 804-285-4532. Facebook: Folk Art Society of America.
E-mail: fasa@folkart.org
Web Site: www.folkart.org
Founded: 1987.
Key Personnel: Pres. (V), Ann Oppenhimer; Financial Dir., William Oppenhimer; Administrative Asst., Barbara Hassett.
Personnel Profile: Full-Time Volunteers 1; Part-Time Paid 1; Part-Time Volunteers 6.
Governing Authority: private; nonprofit. Tax-exempt: 501(c)(3).
Institution Type/Description: Folk Art Museum.
Collections: field of folk, outsider & self-taught art; slide library; videos; photographs; archival materials.
Research Fields: folk & self-taught art with emphasis on late 20th-century contemporary American.
Facilities: library.
Activities: lectures. Annual Event: symposium conference.
Publications: three times annual journal, Folk Art Messenger.
Hours & Admission Prices: Sept.-July open by appointment only. No charge.
Attendance: 300 (estimated)
Membership: Individual $35; Family $50; Foreign & Patron $60; Contributor $100; Bronze Star $250; Silver Star $500; Gold Star $1,000.

GRAND LODGE AF & AM LIBRARY, MUSEUM AND HISTORICAL FOUNDATION - ALLEN E. ROBERTS MASONIC LIBRARY AND MUSEUM, 4115 Nine Mile Rd., Richmond, VA 23223-4926. Tel.: 804-222-3110.
E-mail: library@grandlodgeofvirginia.org
Web Site: www.grandlodgeofvirginia.org/library1.htm
Founded: 1778.
Personnel Profile: Part-Time Paid 1; Part-Time Volunteers 2.
Governing Authority: Parent Institution: Grand Lodge of Virginia, Library, Museum & Historical Foundation. Subsidiary Institution: Allen E. Roberts Masonic Library and Museum. Tax-exempt: 501(c)(3).
Institution Type/Description: Fraternal Museum & Freemasonry.

Collections: Virginia Masonic history; portraits; Grand Lodge commemorative & local Lodge artifacts; furniture; textiles; glassware; tokens; jewelry.
Major Exhibits: 150th Anniversary of Civil War and Virginia Freemasonry (1862), 11/12-10/15.
Research Fields: history of freemasonry in Virginia.
Facilities: library.
Publications: The Virginia Masonic Herald.
Hours & Admission Prices: Mon., Wed. & Fri. 9-12:30 & 1:30-4. No charge; donations accepted. Closed Presidents' Day; Easter Mon.; Memorial Day; Labor Day; Thanksgiving; Christmas.
Attendance: 300

HENRICO COUNTY HISTORIC PRESERVATION & MUSEUM SERVICES, 8600 Dixon Powers Dr., Richmond, VA 23228-2735. Mailing Address: P.O. Box 27032, Richmond, VA 23273. Tel.: 804-501-5736 & 7275. Fax: 804-501-5284.
E-mail: gre26@co.henrico.va.us
Web Site: www.co.henrico.va.us/rec
Founded: 1977.
Congressional District: 7
Key Personnel: History Supvr., Christopher M. Gregson; Cur., Kimberly Sicola; Asst. Cur., Alyson Rhodes-Murphy; Site Mgr., Anna Truong; Asst. Site Mgr., Linda Eikmeier.
Personnel Profile: Full-Time Paid 6; Part-Time Paid 3; Part-Time Volunteers 50; Interns 4.
Governing Authority: county. Parent Institution: Henrico County. Tax-exempt.
Institution Type/Description: History Museum: housed in c.1810 Meadow Farm, depicting mid-19th century rural life in southeastern Virginia.
Collections: c.1772-1960 furniture, textiles, decorative arts, kitchen equipment, porcelain, glassware, flat & hollow silverware; costumes; paintings; prints; photographs; military paraphernalia, including camping equipment, insignias, arms & cavalry equipment; agricultural tools & equipment; medical tools, books & manuscripts; family papers; archival material; contemporary folk art; local school photographs; school books, desks & supplies from 19th & early 20th century. Historic Buildings: Deep Run School, 1902 two room schoolhouse; Spring Park, 1890 Springhouse; Walkerton, 1820s Inn; c.1925 Courtney Road service station.
Research Fields: agriculture; material culture & socio-economics of southeastern Virginia in the 19th century; local history of people, places & events; local architectural history; 19th-20th century folk art.
Facilities: 1,000-vol. library of 19th-century books on medicine, history, health, housekeeping, schoolbooks & fiction available for scholarly research on premises; orientation center. Gift items for sale.
Activities: education programs for children, adults, undergraduate & graduate college students; docent program; loan, permanent & traveling exhibitions; military reenactments; special holiday events.
Publications: guidebook, Manuscript Collection of the Sheppard Family of Meadow Farm; calendar of events; semiannual newsletter; exhibit flyers; brochures; event posters.
Hours & Admission Prices: March to mid-Dec. Tues.-Sun. 12-4. No charge. &
Attendance: 60,000 (estimated)

THE JOHN MARSHALL HOUSE, 818 E. Marshall St., Richmond, VA 23219-1917. Mailing Address: P.O. Box 1098, Richmond, VA 23218-1098. Tel.: 804-648-7998. Fax: 804-648-5880.
E-mail: johnmarshallhouse@preservationvirginia.org
Web Site: www.preservationvirginia.org
Founded: 1911.
Congressional District: 3
Key Personnel: Exec. Dir. APVA, Elizabeth Kostelny; Cur., Catherine Dean; Coord. Education, Jennifer Hurst.
Personnel Profile: Part-Time Paid 12; Interns 1.
Governing Authority: nonprofit society. Parent Institution: Preservation Virginia, 204 W. Franklin St., Richmond, VA 23220. Tax-exempt.
Institution Type/Description: Historic Site & Historic House: 1790 home of U.S. Chief Justice John Marshall & only surviving 18th-century brick Federal house in Richmond.
Collections: furnishings associated with the Chief Justice; 18th to early 19th-century glass; textiles; porcelain; paintings; silver; musical instruments; writing implements; Chief Justice John Marshall's judicial robe.
Research Fields: John Marshall & 18th-century politics & architecture; Richmond city history, the law, social life in the 18th & early 19th century.
Facilities: Museum-related items for sale.
Activities: guided tours; special events.
Publications: newsletter, Preservation Virginia.
Hours & Admission Prices: Visit website for hours.
Attendance: 4,000 (accurate)
Membership: Student & Teacher $35; Individual $50; Individual Plus One $60; Family $70.

LEWIS GINTER BOTANICAL GARDEN, 1800 Lakeside Ave., Richmond, VA 23228-4700. Tel.: 804-262-9887. Fax: 804-262-9934.
E-mail: frankr@lewisginter.org
Web Site: www.lewisginter.org
Founded: 1984.
Congressional District: 3
Key Personnel: Exec. Dir., Frank L. Robinson; Chm. (V), Kitten Clarke; Pres. (V), William H. King, Jr.; Asst. Dir., Shane Tippet; Business Mgr., Freda M. Lushbaugh; Mgr. Horticulture, Neil Beasley; Dir. Devel., Jennifer Little; Mgr. Education, Randee Humphrey; Mgr. Public Rels., Beth Monroe; Gift Shop Mgr., Martha Anne Ellis.
Personnel Profile: Full-Time Paid 45; Full-Time Volunteers 130; Part-Time Paid 45; Part-Time Volunteers 220; Interns 3.
Governing Authority: nonprofit organization. Tax-exempt: 501(c)(3).
Institution Type/Description: Botanical Garden & Historic House: c.1888.
Collections: Lora & Claiborne Robins Tea House; Henry M. Flagler perennial garden; Grace Arents garden; Martha & Reed West Island garden; Lucy Payne Minor Memorial Garden; The Asian Valley; Cottage Garden; Children's garden; seasonal annuals; aquatic plants; rhododendron & azalea collections; Conservatory: exotic & unusual plants from around the world.
Research Fields: introduction of new & superior plants to zones 7 & 8.
Facilities: 3,500-vol. library on horticulture & gardening, available for use by the public; classroom; herbarium; 80 acres of grounds; Henry M. Flagler Perennial Garden, 3 acres; Lucy Payne Minor Memorial Garden; Children's Garden; Lora & Clairborne Robins Tea House; Grace Arents Garden; visitor center. Education & Library Complex: education center; classroom; laboratory; conference center; auditorium; meeting areas. Books & items related to the garden & gardening for sale.
Activities: guided tours; lectures; concerts; arts festivals; rental gallery; organized educational programs. Garden Sponsors: plant sales; major symposia; Daffodil Show; Charles F. Gillette Forum on landscape design & history; Mother's Day Concert; Gardenfest of Lights, a holiday illumination.
Publications: Garden Times; annual report.
Hours & Admission Prices: Daily 9-5; call for extended hours. Adults $11, seniors $10, children 3-12 $7; children under 3 & members no charge. Closed New Year's Day; Thanksgiving; Christmas Eve & Day. &
Attendance: 175,000 (accurate)
Membership: Teacher & Student $35; Senior Citizen $40; Individual $50; Dual $60; Family $70; Contributing $125; Supporting $250; Sponsoring $500.

THE LIBRARY OF VIRGINIA, 800 E. Broad St., Richmond, VA 23219-8000. Tel.: 804-692-3535. Fax: 804-692-3556. TTY: 804-692-3976.
E-mail: sandra.treadway@lva.virginia.gov
Web Site: www.lva.virginia.gov
Founded: 1823.
Congressional District: 3
Key Personnel: Dir. & State Librarian, Sandra G. Treadway; Chm. Library Bd. (V), Mark E. Emblidge; Deputy Administration, Connie Warne; Public Information & Policy Coord., Janice M. Hathcock; Dir. Public Svcs. & Outreach, Gregg Kimball; Dir. Archives, Records & Collection Svcs., John D. Metz; Museum Shop Mgr., Laura Curzi.
Personnel Profile: Full-Time Paid 138; Part-Time Paid 29; Part-Time Volunteers 31.
Governing Authority: state government agency. Parent Institution: Commonwealth of Virginia. Tax-exempt.
Institution Type/Description: State Library & Archives.
Collections: books; maps; personal papers; Virginiana; genealogy records; newspapers; local & state government records; deeds; governors' papers; photographs; Virginia portraiture; prints; broadsides; paintings; music sheets; federal records; broadsides; ephemera.
Major Exhibits: The Importance of Being Cute: Pet Photography in Virginia, 6/13-2/14; No Vacancy: Remnants of Virginia's Roadside Culture, 10/13-2/14.
Research Fields: Virginia history; colonial records; genealogy.
Facilities: library; archives; 2,000 sq. ft. exhibit space.
Activities: guided tours, lectures, participatory exhibits.
Publications: quarterly, Broadside; LVA e-newsletter.
Hours & Admission Prices: Mon.-Sat. 9-5. No charge; donations accepted. Closed most state holidays. &
Attendance: 246,000 (estimated)

MAGGIE L. WALKER NATIONAL HISTORIC SITE, 600 N. 2nd St., Richmond, VA 23219. Mailing Address: 3215 E. Broad St., Richmond, VA 23223-7517. Tel.: 804-771-2017. Fax: 804-771-2226.
E-mail: dave_ruth@nps.gov

Web Site: www.nps.gov/mawa
Founded: 1978.
Congressional District: 3
Key Personnel: Supt., David R. Ruth; Chief Interpreter, Beth Stern; Cur., Ethan P. Bullard; Supvr. Park Rangers, Ajena Rogers; Park Guide, Melissa Weissert; Park Guide, George Peeterse.
Personnel Profile: Full-Time Paid 5; Part-Time Paid 1; Part-Time Volunteers 10.
Governing Authority: federal. Parent Institution: National Park Service. Tax-exempt.
Institution Type/Description: Historic House: Victorian-Italianate home of Maggie Lena Walker, first woman founder & president of an African American bank, newspaper editor & African-American community leader; African-American & women's history.
Collections: period furniture, clothing; personal items of Maggie Lena Walker; family photos; furnishings & memorabilia; online virtual exhibit.
Research Fields: Maggie L. Walker.
Facilities: 50-vol. library pertaining to Black history for use on premises only.
Activities: guided tours.
Publications: brochure; curriculum guide for classroom study.
Hours & Admission Prices: March-Oct. Mon.-Sat. 9-5; Nov.-Feb. Mon.-Sat. 9-4:30. No charge; donations accepted. Closed New Year's Day; Thanksgiving; Christmas. &
Attendance: 12,000 (accurate)

MAYMONT, (M), 1700 Hampton St., Richmond, VA 23220-6899. Tel.: 804-358-7166, ext. 310. Fax: 804-358-9994.
E-mail: info@maymont.org
Web Site: www.maymont.org
Founded: 1925.
Congressional District: 3
Key Personnel: Exec. Dir., Norman O. Burns, II; Assoc. Exec. Dir., C. Fred Murray; Pres. (V), Amy McDaniel Williams; Dir. Historical Collections & Programs, Dale C. Wheary; Dir. Nature Center, Henry Bireline; Mgr. Zoology, Debbie Rea; Carriage Collections Mgr., Armistead Wellford; Mgr. Historical Programs, Carol Harris; Special Program Coord., Nancy Lowden; Rental Coord., Rebekah Davis; Dir. Finance, Ron Thompson; Administrative Dir., Ann Voss; Mgr. Environmental Education, Kate Jarrell; Dir. Devel., Carol Akin; Dir. Public Rels. & Mktg., Cathie Rosenberg; Asst. Dir. Public Rels., Carla Murray; Mgr. Historical Collections, Kathy Garrett Cox; Mgr. Horticulture, Peggy M. Singlemann; Dir. Special Events, Kim Pauley.
Personnel Profile: Full-Time Paid 54; Part-Time Paid 35; Part-Time Volunteers 470; Interns 4.
Governing Authority: nonprofit organization. Property & facilities owned by City of Richmond. Tax-exempt: 501(c)(3).
Institution Type/Description: Historic House & Gardens: housed in c.1890s estate of James H. Dooley.
Collections: turn-of-the-century furniture & decorative arts; 19th-century carriages; 150 species of native Virginia live animals; 200 species of trees & shrubs.
Research Fields: late Victorian & gilded age history; decorative arts & architecture landscape history & preservation.
Facilities: Japanese garden; Italian garden; zoological park; children's farm; arboretum; aquarium. Museum-related items for sale.
Activities: guided tours; lectures; films; concerts; workshops; formal historical & environmental educational programs for children; carriage rides; docent program; permanent exhibitions; annual special events.
Publications: photographic essay, Maymont; quarterly newsletter; events programs; guide to school programs.
Hours & Admission Prices: Maymont House: Tues.-Sun. 12-5. Suggested Donation: $5. Robins Nature Center & Grounds: daily 10-5. Exhibits: Tues.-Sun. 12-5. Suggested Donation $4. Children's Farm Barn: Tues.-Sun. 12-5. Tram: Tues.-Sun. 12-5. Adults $3, children $2; members no charge. Carriage Rides: Sun. 12-4. Closed New Year's Day; Thanksgiving; Christmas. &
Attendance: 500,000 (estimated)
Membership: Senior & Student $35; Individual $45; Family Senior $50; Family $60.

THE MUSEUM OF THE CONFEDERACY, (M), 1201 E. Clay St., Richmond, VA 23219-1615. Tel.: 804-649-1861. Fax: 804-644-7150.
E-mail: vyates@moc.org
Web Site: www.moc.org
Founded: 1896.
Congressional District: 3
Key Personnel: Pres. & C.E.O., S. Waite Rawls, III; Chm. (V), Matthew G. Thompson, Jr.; Vice Pres. Advancement, Otis C. Crowther, Jr.; Sr. Cur. &

Dir. Collections, Robert Hancock; Registrar, Shelly Berger; Cur., Cathy Wright; Dir. Museum Operations, Eric D. App; Historian, Dr. John M. Coski; Dir. Mktg. & Public Rels., Vickie Yates; Museum Retail Mgr., Patrick Schulz.

Personnel Profile: Full-Time Paid 20; Part-Time Paid 36; Part-Time Volunteers 69; Interns 12.

Governing Authority: nonprofit organization. Parent Institution: Confederate Memorial Literary Society. Branch Museum: Museum of the Confederacy - Appomattox. Tax-exempt: 501(c)(3).

Institution Type/Description: History Museum: historic site.

Collections: artifacts & documents pertaining to the Confederate States of America; southern history; the Confederate & Civil War years; the White House of the Confederacy; decorative arts; prints; photographs; Confederate militaria: flags, uniforms, weapons; Jefferson Davis Collection; Confederate bonds & currency; imprints.

Research Fields: Civil War; Confederate States of America; southern history.

Facilities: Brockenbrough Library. Gifts & books for sale.

Activities: tours; school tours; bus tours; permanent, temporary & traveling exhibitions; education program; lecture & film series; book awards; living history programs; outreach programs.

Publications: The Museum of the Confederacy Journal; occasional catalogues; checklists; books; magazine, The Museum of the Confederacy.

Hours & Admission Prices: Museum & White House: daily 10-5. White House or Museum: $10; discounts to AAM members. Combination Ticket: $15; discounts to AAM members. Closed New Year's Day; Thanksgiving; Christmas. ♿

Attendance: 65,009 (accurate)

Membership: Full-time Student $20; Senior Individual $30; Senior Family $35; Individual $40; Family $55; Sustaining $100; Benefactor $250; Patron $500; 1896 Society $1,896; White House Society $5,000.

PRESERVATION VIRGINIA, 204 W. Franklin St., Richmond, VA 23220-5012. Tel.: 804-648-1889. Fax: 804-775-0802.

Web Site: www.preservationvirginia.org
Formerly: Association for the Preservation of Virginia Antiquities
Founded: 1889.
Congressional District: 3
Key Personnel: Exec. Dir., Elizabeth Kostelny; Exec. Asst., Alexis Feria; Pres., Mr. Anne Geddy Cross; Dir. Preservation Svcs., Louis J. Malon; Controller, Cheryl Greenday; Assoc. Dir. Museum Operations & Education, Jennifer Hurst-Wender; Museum Operations Asst., Sienna Fennell.

Personnel Profile: Full-Time Paid 31; Part-Time Paid 150; Part-Time Volunteers 200; Interns 2.

Governing Authority: society; nonprofit organization. House Sites: early 18th century Scotchtown, home of Patrick Henry; c.1760 Smith's Fort Plantation, near Surry Courthouse; 1607 Historic Jamestowne; 1790 John Marshall House, Richmond; 1817 Farmers Bank, Petersburg; 1665 Bacon's Castle, Surry; 1800 Walter Reed Birthplace, Gloucester County; c.1760 Old Tobacco Warehouse, Urbanna; 1791 Cape Henry Lighthouse, Virginia Beach; Richmond; 1782 18th-century Holly Brook, Eastville. Tax-exempt: 501(c)(3).

Institution Type/Description: Historic Building & Site; Preservation Project.

Collections: 17th- to 19th-century furniture; decorative arts; textiles; metals; ceramics; glass; fine arts; photographs; archaeology.

Research Fields: architecture; period furnishings & decorative arts; 17th-century history; preservation technology; archaeology.

Activities: guided tours; lectures; participatory, permanent, temporary & loan exhibitions; education programs; concerts; docent program. Annual Events: Preservation Conference & Preservation Workshops; VA's Most Endangered Sites; VA Preservation Awards; Revolving Fund.

Publications: biannual, Journal; annual, Ventures; publications relating to sites.

Hours & Admission Prices: For hours & admission prices of House Museums see separate listings or web site.

Attendance: 400,000 (estimated)

Membership: Student & Teacher $35; Individual $50; Family $70; Organization $75.

RICHMOND NATIONAL BATTLEFIELD PARK, Chimborazo Medical Museum, 3215 E. Broad St., Richmond, VA 23223-7517. Tel.: 804-226-1981, ext. 3. Fax: 804-771-8522. Facebook: Richmond National Battlefield Park.

Web Site: www.nps.gov/rich
Founded: 1936.
Congressional District: 7
Key Personnel: Supt., Dave Ruth; Chief Interpretation, Elizabeth Stern; Chief Ranger, Timothy Mauch; Cur., Ethan P. Bullard.

Governing Authority: federal. Parent Institution: National Park Service, Dept. of Interior, Washington, DC 20240. Visitor Centers: Civil War Visitor Center at Tredegar Iron Works, 470 Tredegar St., Richmond, VA; Cold

Harbor Battlefield Visitor Center: 5515 Anderson-Wright Dr.; Fort Harrison Visitor Center, 8621 Battlefield Park Rd., Richmond, VA; Glendale/Malvern Hill Battlefield Visitor Center, 8301 Willis Church Rd. Tax-exempt: 170(b)(1)(A).

Institution Type/Description: Civil War Military Museum.

Collections: military artifacts; Civil War history; medical equipment & hospital life. Historic Houses & Sites: 1835 Watt House, Gaines' Mill battlefield; 1800 Garthright House, Cold Harbor & Malvern Hill battlefields; 1720s Shelton House, Totopotomoy Creek Battlefield.

Research Fields: African-American Civil War history; Civil War medicine; Richmond history; historic preservation.

Facilities: 600-vol. library of Civil War books available for use by special permission from the superintendent. Civil War manuscript material relative to Richmond; publications, postcards, slides, prints for sale.

Activities: self-guided tours & trails; lectures; films; formally organized education programs for children; living history programs.

Publications: pamphlets; folders & handbooks about the battles for Richmond.

Hours & Admission Prices: Tredgar Iron Works, Cold Harbor Visitor Center, & Chimborazo Medical Museum: daily 9-5. Glendale/Malvern Hill Visitor Center & Fort Harrison Visitor Center: seasonal. No charge; donations accepted. Closed New Year's Day; Thanksgiving; Christmas. ♿

Attendance: 245,504

RICHMOND RAILROAD MUSEUM, 102 Hull St., Richmond, VA 23224-4240. Mailing Address: P.O. Box 8583, Richmond, VA 23226-0583. Tel.: 804-233-6237.

Web Site: www.odcnrhs.org
Formerly: Old Dominion Railway Museum
Founded: 1994.
Congressional District: 3
Key Personnel: Pres. (V), Randy Ridgely; Chm. (V), Charles Curley; Chm. (V) & Archivist, Calvin Boles.

Personnel Profile: Full-Time Volunteers 40; Part-Time Volunteers 40.

Volunteer Hours: 3,512
Operating Expenses: 21,751
Operating Income: 16,509
Governing Authority: private; nonprofit organization. Parent Institution: National Railway Historical Society. Tax-exempt: 501(c)(3).

Institution Type/Description: Railroad History Museum: housed at site of 1915-1957 Southern Railway depot.

Collections: railroading history in Central Virginia; steam, caboose, passenger & freight equipment; locomotives & rolling stock from first half of 20th century; railroad company archives, especially the Seaboard Air Line, Richmond Fredericksburg & Potomac, and Richmond electric streetcar lines; photographs.

Research Fields: Hull Street Station (Richmond, VA) & environs.

Facilities: 500-vol. library including railroad company account books, law books, journals & manuals; 250 sq. ft. exhibit space. Museum-related items for sale.

Activities: formal education programs for children; guided tours; lectures; loan & temporary exhibitions; railroad excursions. Annual Events: Children's Day; annual picnic.

Publications: monthly newsletter, Highball.

Hours & Admission Prices: Sat. 11-4, Sun. 1-4. No charge; donations accepted. Closed Christmas Eve & Day. ♿

Attendance: 2,000 (accurate)

Membership: Adult $50; Family $56.

ST. JOHN'S EPISCOPAL CHURCH, (M), 2401 E. Broad St., Richmond, VA 23223-7128. Tel.: 804-648-5015. Fax: 804-649-0878.

E-mail: kpeninger@saintjohns.cc
Web Site: www.historicstjohnschurch.org
Founded: 1920.
Congressional District: 3
Key Personnel: Exec. Dir., Kay Peninger; Pres., Neill Goff; Treas., Everett Melton; Museum Shop Mgr., Trudy Russell.

Personnel Profile: Full-Time Paid 1; Part-Time Paid 21.

Governing Authority: church; nonprofit organization. Tax-exempt: 170(b)(1)(A).

Institution Type/Description: Historic Building & Site: 1741 church, location of the 2nd Virginia Convention where Patrick Henry addressed the Convention-1775 with his famous, Give me liberty or give me death speech; first cemetery in Richmond.

Collections: items pertaining to the church's 250 year history, including three-decker pulpit, rector's reading desk, mid-18th century chandeliers, Lotto rug, early 18th-century marble font, 1730-1773 manuscript of colonial vestry-book, 1700 Prayer-Book, a gift from King George VI.

Research Fields: church history.

Facilities: Items pertaining to the Revolution, Patrick Henry & Richmond history for sale.

Activities: guided tours. Museum Sponsors: reenactment of the 2nd Virginia Convention from May to September.

Hours & Admission Prices: Mon.-Sat. 10-4, Sun. 1-4. Adults $7, senior citizens 62 & over $6, students $5; discounts to groups of 10 or more; children under 7 no charge. Closed New Year's Eve & Day; Easter; Thanksgiving; Christmas Eve & Day. &

Attendance: 40,000 (accurate)

* **SCIENCE MUSEUM OF VIRGINIA, (M),** 2500 W. Broad St., Richmond, VA 23220-2057. Tel.: 804-864-1400; 800-659-1727. Fax: 804-864-1560.

E-mail: info@smv.org

Web Site: www.smv.org

Founded: 1970.

Congressional District: 3

Key Personnel: Chm., Roger L. Boeve; Pres. FDN Directors, Stanton Thalhimer; Dir. & C.E.O., Richard C. Conti; Dir. Science, Education, Eugene Maurakis, Ph.D.; Museum Shop Mgr., Jennifer Morehead.

Personnel Profile: Full-Time Paid 57; Part-Time Paid 55; Part-Time Volunteers 537; Interns 4.

Governing Authority: state. Parent Institution: Commonwealth of Virginia. Subsidiary Institutions: Danville Science Center; Virginia Aviation Museum. Tax-exempt.

Institution Type/Description: Science Museum.

Collections: rocks & minerals; astronomy; computers; aircraft; train cars; aviation; physics; chemistry; technology; zoology; submarine; life sciences.

Research Fields: science education in a museum environment; exhibit development through formative evaluation; environmental studies.

Facilities: large format films & planetarium shows in 250-seat Ethyl IMAX(R) DOME Theater; computer lab; stepped theater; 120-seat lecture hall; classrooms; railroad cars; birthday caboose; submarine; outdoor park; amphitheaters. Museum-related items for sale.

Activities: hands-on science exhibits; outreach van programs; education programs for Pre-K through high school levels; distance learning & university science courses & collaborations; science lectures & seminars; science events; exhibit hall demonstrations; live theatrical performances.

Publications: annual report; quarterly newsletter for members; group visit planning guide; program brochures; educator newsletters.

Hours & Admission Prices: Tues.-Sat. 9:30-5, Sun. 11:30-5; call for additional school holiday hours. Exhibits only: adults 13-59 $11, senior citizens 60 & over and youth 4-12 $10; ASTC & museum members no charge. IMAX(R)DOME Film only: $9. Exhibit & Film: adults 13-59 $16, senior citizens 60 & over and youth 4-12 $15, members $5; discounts to AAM, AAA & ASTC members; children 3 & under no charge. &

Attendance: 306,872 (accurate)

Membership: Star $55; Pulsar $65; Supernova $99; Galaxy $120.

1708 GALLERY, 319 W. Broad St., Richmond, VA 23220-4218. Mailing Address: P.O. Box 12520, Richmond, VA 23241-0520. Tel.: 804-643-1708. Fax: 804-643-7839.

E-mail: info@1708gallery.org

Web Site: 1708gallery.org

Key Personnel: Exec. Dir., Emily Smith; Gallery Admin., Jolene Giandomenico

Institution Type/Description: Art Gallery.

Collections: works by national & international artists.

Hours & Admission Prices: Tues.-Fri. 11-5, Sat. 1-5; other times by appointment. No charge. &

UNIVERSITY OF RICHMOND MUSEUMS, (M), 28 Westhampton Way, Richmond, VA 23173-. Tel.: 804-289-8276. Fax: 804-287-1894.

E-mail: museums@richmond.edu

Web Site: museums.richmond.edu

Formerly: Marsh Art Gallery, University of Richmond

Founded: 1830.

Congressional District: 3

Key Personnel: Exec. Dir., Richard Waller; Deputy Dir. & Cur. Exhibitions, Elizabeth Schlatter; Museum Preparator, Stephen Duggins; Museum Preparator, Henley Guild; Cur. Museum Programs, Heather Campbell; Coord. Museum Visitor & Tour Svcs., Denisse DeLeon; Mgr. Museum Operations, Katreena Clark; Asst. Collections Mgr., David Hershey.

Personnel Profile: Full-Time Paid 9; Part-Time Paid 28; Part-Time Volunteers 4; Interns 2.

Governing Authority: nonprofit. Parent Institution: University of Richmond. Branch Museums: Joel & Lila Harnett Museum of Art, 1968; Joel & Lila

Harnett Print Study Center, 2001; Lora Robins Gallery of Design from Nature, 1977. Tax-exempt: 501(c)(3).

Institution Type/Description: University Museums.

Collections: Joel & Lila Harnett Museum of Art: all media fine arts; Joel & Lila Harnett Print Study Center: Renaissance prints to present; I. Webb Surratt, Jr. print collection; Center Street Studio archives; Lora Robins Gallery of Design from Nature: minerals; gems; shells; fossils; ancient Greek & Roman through Byzantine gold, silver & bronze coins; Carver collection of Chinese ceramics; Rice collection of Oceanic Art; Inuit sculptures; cultural artifacts; decorative arts.

Major Exhibits: Chasing Bugs: Insects as Subject and Metaphor, 9/13-1/14; Virginia Rocks! Selections from the Collection, 9/13-6/14; American Odyssey: The Art of Julius J. Lankes, A Retrospective (T), 11/13-2/14; Threads of Silk and Gold: Chinese Textiles from the Qing Dynasty, 1/14-6/14; Virginia Rocks! Selections from the Collection, 1/14-6/15; No Eye Flowers: Paintings, Calligraphy, and Ceramics by Stephen Addiss, 2/14-5/14; 19th Century French and Russian Art: Works from the Virginia Museum of Fine Arts, 2/14-4/14; Gifts of Art: Recent Acquisitions in the University Museums, 8/14-10/14; The Temple of Flora: Prints by Jim Dine, 8/14-7/15; The 2014 Harnett Biennial of American Prints, 10/14-12/14; Wish You Were Here: Postcards from the Thomas and Donna Brumfield Collection, 10/14-4/15.

Research Fields: modern & contemporary art; historical fine arts; printmaking; photography; decorative arts; Asian art; geology; mineralogy; earth & natural sciences.

Facilities: sculpture courtyard; print study center; publications for sale.

Activities: guided tours; lectures; traveling & temporary exhibits; films; gallery talks; videos; workshops; symposia; concerts; formal education programs for undergraduate & graduate students; internships.

Publications: exhibition catalogues; brochures; posters.

Hours & Admission Prices: Joel and Lila Harnett Museum of Art: mid-Aug. to April Sun.-Fri. 1-5. Lora Robins Gallery of Design from Nature: mid-Aug. to April Sun.-Fri. 1-5; May-July Tues.-Fri. 1-5. Joel & Lila Harnett Print Study Center: mid-Sept. to mid-April Sun.-Fri. 1-3; other times by appointment. No charge. Closed spring break; fall break; Thanksgiving week; Easter weekend; semester breaks. &

Attendance: 15,223 (accurate)

* **VALENTINE RICHMOND HISTORY CENTER, (M),** 1015 E. Clay St., Richmond, VA 23219-1527. Tel.: 804-649-0711. Fax: 804-643-3510.

E-mail: info@richmondhistorycenter.com

Web Site: www.richmondhistorycenter.com

Formerly: Valentine Museum/Richmond History Center

Founded: 1892.

Congressional District: 3

Key Personnel: Chm., Pamela J. Royal, M.D.; Vice Chm., John Stanchima; Dir., William J. Martin; Dir. Finance, Donna Kolba; Dir. Educ., Pat Armbrust; Dir. Public Rels., Domenick Casuccio; Dir. Operations, Ken Myers; Dir. Devel., Ty Toepke; Dir. Archives, Collections & Interpretation, Meghan Glass Hughes; Registrar, Jackie Mullins; Dir. Tours, Linda Krinsky; Museum Shop Mgr., Jane Seaman.

Personnel Profile: Full-Time Paid 17; Part-Time Paid 30; Part-Time Volunteers 75; Interns 10.

Governing Authority: private; nonprofit organization. Tax-exempt: 501(c)(3).

Institution Type/Description: Urban History Museum.

Collections: neo-classical wall paintings; sculpture studio of E.V. Valentine; costumes, 1600-present; flat textiles; laces; quilts; coverlets; embroideries; decorative arts relating to Richmond history; portraits; landscapes; tools and industrial artifacts; items relating to minority history; 19th-century toys; photographs; manuscripts; prints on Richmond & Virginia history. Historic Building: Wickam house (1812).

Research Fields: American urban history; American social history; historic preservation; costumes; decorative arts; photographs.

Facilities: research library of books, pamphlets, manuscripts & photographs on Richmond available by appointment; reading room; conference & reception areas; kitchen; 19th-century formal garden; cafe. Books & museum-related items for sale.

Activities: self-guided tours of museum; guided tours of historic house; lectures; formally organized educational programs; inter museum loan; temporary & traveling exhibitions; museum intern program; minority intern program; guided walking & bus tours; special events.

Publications: newsletter; exhibition catalogs; monographs.

Hours & Admission Prices: Tues.-Sat. 10-5, Sun. 12-5. Adults $8, senior citizens 55 & over, students, children 4-18 and groups of 10 or more $7; discounts to AAM & ICOM members; teachers, children under 4 & members no charge. Court End Passport $10. Closed New Year's Day; Thanksgiving; Christmas Eve & Day. &

Attendance: 43,000 (accurate)

Membership: Senior, Teacher & Student $25; Individual $50; Dual & Family

$75; Clay Street Council $100; Richmond History Circle $150; Court End Circle $250; Centennial Circle $500; Director's Council $1,000; Trustee's Council $2,500; Founder's Circle $5,000; Wickham Society $10,000.

VIRGINIA ASSOCIATION OF MUSEUMS, 3126 W. Cary St. #447, Richmond, VA 23221-3504. Tel.: 804-358-3171. Fax: 804-358-3174.
E-mail: mcarlock@vamuseums.org
Web Site: vamuseums.org
Founded: 1968.
Congressional District: 3
Key Personnel: Exec. Dir., Margo Carlock; Pres., Tracy Jo Gillespie.
Personnel Profile: Full-Time Paid 3; Part-Time Paid 2; Part-Time Volunteers 2.
Governing Authority: private; nonprofit association. Tax-exempt.
Institution Type/Description: State Association: service organization to museum professionals throughout Virginia.
Collections: serving the museum community of Virginia & District of Columbia through education, technical assistance, & advocacy.
Facilities: reference library for members & job board.
Activities: training programs for professional museum workers. Annual Events: VAM Annual Conference (workshops, keynote addresses, concurrent sessions, exhibit hall; Workshop Series.
Publications: VAM quarterly newsletter; directory of Virginia museums; VAM membership directory.
Hours & Admission Prices: Mon.-Fri. 9-5. &
Membership: Student $25; Faculty & Staff of member institution $30; Individual $45; Patron & Business Individual $115; Company $170; Business Partner $575. Institutional Levels: Institution I $45; Institution 2 $115; Institution 3 $170; Institution 4 $230; Institution 5 $400; Institution 6 $515; Institution 7 $630; Institution 8 $745.

VIRGINIA CENTER FOR ARCHITECTURE, (M), 2501 Monument Ave., Richmond, VA 23220-2618. Tel.: 804-644-3041. Fax: 804-643-4607.
Web Site: www.virginiaarchitecture.org
Founded: 1954.
Key Personnel: Exec. Dir., Helene Combs Dreiling; Programs Coord., Lauren Stelzer; Museum Shop Mgr., Eric Knight.
Personnel Profile: Full-Time Paid 1; Part-Time Paid 3; Part-Time Volunteers 8; Interns 2.
Institution Type/Description: Architecture Museum.
Collections: architecture history & design.
Facilities: 28,000 sq. ft. mansion housing museum spaces; offices; collections facilities & conference rooms. Museum shop.
Activities: exhibitions; education programs; public programs; lectures; symposia & scholarships.
Hours & Admission Prices: Tues.-Fri. 10-5, Sat.-Sun. 1-5; groups by appointment. No charge; donations welcome. Closed New Year's Day; Easter; Thanksgiving; Christmas. &
Attendance: 8,600
Membership: $35.

VIRGINIA DEPARTMENT OF CONSERVATION AND RECREATION, 203 Governors' St., Richmond, VA 23219-2049. Tel.: 804-786-1712. Fax: 804-786-9294. TDD: 804-786-2121.
E-mail: pco@dcr.virginia.gov
Web Site: dcr.virginia.gov
Founded: 1926.
Key Personnel: Parks Dir., Joe Elton; Operations Dir., Nancy Healthman.
Governing Authority: state. Tax-exempt.
Institution Type/Description: State Parks & Historic Sites.
Collections: botany; entomology; geology; herbarium; herpetology. Historic Sites: 1700s Chippokes Plantation, Surry County; 1800s Sailor's Creek Battlefield Park & House, Amelia & Prince Edward Counties; 1800s Shot Tower, Wythe County; 1700s George Washington's Grist Mill, Alexandria; 1800s Southwest Virginia Museum, Big Stone Gap.
Research Fields: botany; entomology; geology; herbarium; herpetology.
Facilities: 150-vol. library of natural history books; visitor & nature center interpretive displays.
Activities: guided tours; lectures; films; interpretive & environmental education activities.
Publications: brochure; annual bulletin.
Hours & Admission Prices: Call for hours & admission fees, information differs for separate locations. &

VIRGINIA DEPARTMENT OF HISTORIC RESOURCES, 2801 Kensington Ave., Richmond, VA 23221-2470. Tel.: 804-482-6441. Fax: 804-367-2391. TDD: 804-367-2386.
E-mail: dee.deroche@dhr.virginia.gov
Web Site: www.dhr.state.va.us
Founded: 1966.
Congressional District: 7
Key Personnel: Dir., Kathleen Kilpatrick; Deputy Dir. Policy & Planning, Catherine Slusser; Deputy Dir. Community Svcs. Div., Robert A. Carter; State Archaeologist, Michael B. Barber; Chief Cur., Dee DeRoche.
Personnel Profile: Full-Time Paid 46; Part-Time Paid 2; Interns 5.
Governing Authority: state. Parent Institution: Commonwealth of Virginia, Secretariat of Natural Resources. Regional Preservation Offices: Capital Region, 2801 Kensington Ave., Richmond, VA 23221; Western Region, 962 Kime Lane, Salem, VA 24153; Tidewater Region, 14415 Old Courthouse Way, 2nd Fl., Newport News, VA 23608; Northern Region, 5357 Main St., P.O. Box 519, Stephens City, VA 22655. Tax-exempt.
Institution Type/Description: Preservation Agency.
Collections: archaeology; architecture; history.
Research Fields: history & architecture of Virginia; archaeology.
Facilities: 5,377-vol. library of books & publications pertaining to Virginia history, architecture & archaeology available upon written application; field research station; curation center for archaeology collections.
Activities: workshops providing technical assistance; preservation planning conferences.
Publications: annual journal, Notes on Virginia; books, Virginia Landmarks Register; Preserving a Legacy; brochures pertaining to agency & Open-Space Easements in Virginia; A Guide to Historic Highway Markers in Virginia; publications: Archaeological Bibliographies; Research Report Series; Handbook and Resource Guide for owners of Virginia's historic houses; Annual Virginia Archaeology Month poster.
Hours & Admission Prices: Mon.-Fri. 8:15-5. No charge. Closed state holidays. &
Attendance: 2,000 (estimated)

* **VIRGINIA HISTORICAL SOCIETY, (M),** 428 North Blvd., Richmond, VA 23220-3307. Mailing Address: P.O. Box 7311, Richmond, VA 23221-0311. Tel.: 804-358-4901. Fax: 804-342-9647. Facebook: Virginia Historical Society.
E-mail: jguild@vahistorical.org
Web Site: www.vahistorical.org
Founded: 1831.
Congressional District: 7
Key Personnel: Chm., Thomas G. Slater, Jr.; Vice Chm., E. Claiborne Robins, Jr.; Pres. & C.E.O., Dr. Paul A. Levengood; Vice Pres. Collections, E. Lee Shepard; Vice Pres. Programs, Nelson D. Lankford; Vice Pres. Operations & C.F.O., Richard S.V. Heiman; Vice Pres. Advancement, Pamela R. Seay; Sr. Grants Officer, Elaine Hagy; Sr. Officer Public Rels., Jennifer M. Guild; Retail Officer, Jessica Deruosi; Institutional Advancement Research Analyst, Dana Fariss.
Personnel Profile: Full-Time Paid 62; Part-Time Paid 13; Part-Time Volunteers 62.
Volunteer Hours: 4,947
Operating Expenses: 5,400,000
Operating Income: 5,400,000
Governing Authority: society, nonprofit organization. Tax-exempt: 501(c)(3) & 170(b)(1)(A).
Institution Type/Description: Virginia History Museum: Virginia House.
Collections: 10,000,000 manuscripts; 150,000 books (15,000 rare books); over 5,000 maps; 1,200 newspapers; 300 serials; over 4,000 pieces of sheet music; over 1,000 paintings; 200,000 photographs; textiles; silver; furniture; weaponry.
Research Fields: Virginia history; Colonial American history; Civil War history; English history; African American history, genealogy.
Facilities: 125,000-vol. library for genealogical & historical research; lecture hall; 25,500 sq. ft. exhibit space. Museum-related items for sale.
Activities: lectures; educational programs & outreach; temporary, permanent & traveling exhibits; membership function; films; behind the scenes tours.
Publications: Virginia Magazine of History and Biography; History Notes.
Hours & Admission Prices: Mon.-Sat. 10-5, Sun. 1-5. No charge; donations accepted. Closed New Year's Eve & Day; Easter; Independence Day; Thanksgiving; Christmas Eve & Day. &
Attendance: 60,000 (estimated)
Membership: Student, Teacher & Military $48; Senior over 65 $52; Individual $58; Senior Couple over 65, Individual Plus One & Military Family $68; Family $72.

VIRGINIA HOLOCAUST MUSEUM, (M), 2000 E. Cary St., Richmond, VA 23223-7032. Tel.: 804-257-5400. Fax: 804-257-4314. Facebook: Virginia Holocaust Museum.
Web Site: www.va-holocaust.com
Founded: 1997.
Congressional District: 3
Key Personnel: Dir. & Pres. (V), Dr. Charles W. Sydnor; Chm., Marcus M. Weinstein; Education, Megan Ferenczy; Research Librarian, Timothy Hensley; Chief Admin. Officer, Charles Coulomb.
Personnel Profile: Full-Time Paid 14; Full-Time Volunteers 5; Part-Time Paid 4; Part-Time Volunteers 30; Interns 10.
Governing Authority: private; nonprofit organization. Tax-exempt: 501(c)(3).
Institution Type/Description: History Museum.
Collections: European & Nazi wartime memorabilia; Holocaust history; personal artifacts; genocide history.
Major Exhibits: The Art of Morgot Blank, 1/14-2/14; The Trial of Adolf Eichmann (T), 3/14-5/14; The Art of Molly Robinson, 6/14-7/14; Perpetrators (T), 8/14-10/14; The Art of Morgot Blank II, 11/14-12/14.
Research Fields: survivors & liberators of the Holocaust; victims & survivors of genocide.
Facilities: research library; educational facilities; 10,000 sq. ft. exhibit space; 300-seat theater. Museum-related items for sale.
Activities: special events; tours; film series; lectures; docent program; formal education programs; guided tours; rental gallery; theater. Annual Events: Annual Gala.
Publications: quarterly newsletter, De Malyene.
Hours & Admission Prices: Mon.-Fri. 10-5, Sat.-Sun. 11-5; groups of 10 or more by appointment. No charge; donations accepted. Closed New Year's Day; Easter; First Day of Rosh Hashana; Yom Kippur; Thanksgiving; Christmas. &
Attendance: 61,000 (accurate)
Membership: Young Friend $25; Individual $36; Family $72; Supports $125; Educator $250; Witness $500; Memorial $1,000; Benefactor $5,000; Founder $10,000.

* **VIRGINIA MUSEUM OF FINE ARTS, (M),** 200 N. Boulevard, Richmond, VA 23220-4007. Tel.: 804-340-1400 & 1401; 800-943-8632. Fax: 804-340-1548. TDD: 804-340-1401.
E-mail: visitorservices@vmfa.museum
Web Site: www.vmfa.museum
Founded: 1934.
Congressional District: 3
Key Personnel: Dir. & C.E.O., Alexander Nyerges; Pres., Thurston R. Moore; Pres. Virginia Museum of Fine Arts Foundation, Dr. Monroe E. Harris, Jr.; C.F.O., Fern Spencer; Deputy Dir. Foundation, Administration & Government Rels., David B. Bradley; Mgr. Major Gifts & Planned Giving, Jillian Krupski; Donor Rels. Mgr., Chasity Miller; VMFA Fund Mgr., Katie Merritt; Dir. Membership, Jenna Mosman; Mgr. Corporate Rels., Amanda Tate; Deputy Dir. Art & Education, Robin Nicholson; Acting Mgr. Exhibitions, Aiesha Halstead; E. Rhodes and Leona B. Carpenter Cur. East Asian Art, Li Jian; Cur. Ancient Art, Dr. Peter Schertz; Assoc. Cur. American Art, Dr. Elizabeth O'Leary; Cur. African Art, Richard B. Woodward; Cochrane Cur. American Art & Curatorial Chair, Dr. Sylvia Yount; Assoc. Cur. South Asian & Islamic Art, Dr. John Henry Rice; Paul Mellon Cur. & Head Dept. of European Art, Dr. Mitchell Merling; Sydney and Frances Lewis Family Cur. Modern & Contemporary Art, John Ravenal; Asst. Cur. American Decorative Art, Dr. Susan J. Rawles; Asst. Cur. Ancient American Art, Dr. Lee Anne Hurt; Sydney and Frances Lewis Family Cur. Decorative Arts from 1890 to Present, Barry Shifman; Mgr. Photographic Resources, Howell Perkins; Deputy Dir. Collections & Facilities Management, Stephen Bonadies; Conservator Paintings, Carol Sawyer; Conservator Sculpture & Decorative Arts, Kathy Gillis; Asst. Librarian, Lee Viverette; Registrar, Elizabeth H. Hancock; The Thomas C. Gordon Jr. Dir. Studio School, Mary Holland; Chief Educator, Della Watkins; Mgr. Human Resources, Randy Webne; Coord. Volunteer Programs, Kim Frola; Dir. Community Affairs, Carmen F. Foster; Dir. Special Events & Food Svcs., Cathy Turner; Council Pres., Mary Peppiatt; Mgr. VMFA Shop, Barbara Lenhardt; Dir. Mktg., Bob Tarren; Chief Communications Officer, Suzanne D. Hall; Mgr. Budgeting, Anne Kenny Urban; Mgr. Security, Rick Pleasants; Deputy Dir. Sales & Mktg., Alexis Vaughn.
Personnel Profile: Full-Time Paid 197; Part-Time Paid 293; Part-Time Volunteers 853; Interns 20.
Governing Authority: state. State of Virginia. Tax-exempt: 501(c)(3).
Institution Type/Description: Art Museum.
Collections: ancient Egyptian, Greek, Etruscan, Roman, Byzantine, European medieval; ancient American; African; Chinese, Japanese, Korean, Indian, Tibetan & Nepalese art; American painting, sculpture & decorative arts;

European decorative arts, works on paper & old master paintings; the Lillian Thomas Pratt collection of Faberge; Mellon collection of British sporting art, French Impressionist & Post-Impressionist art; the Sydney & Frances Lewis collection of modern & contemporary art; art nouveau; arts & crafts; art deco; modern decorative arts; the Jerome & Rita Gans collection of English silver.
Facilities: 70,000-vol. art reference library; reading room; sculpture garden; restaurants; center for education and outreach; classrooms; banquet facilities; lecture hall; theater; studio school. Museum-related items for sale.
Activities: docent guided tours; teachers workshops; in-services; summer teachers' institute; performance art programs; lectures; films; gallery talks; concerts; dance, music recitals; arts festivals; education programs for children, adults, students; docent program; participatory, permanent, loan, temporary & traveling exhibitions; studio art & art history classes; school loan service; statewide outreach exhibitions & programs.
Publications: magazine, My VMFA; monthly e-newsletters; annual report; African Art: Virginia Museum of Fine Arts Collection, The Making of Virginia Architecture; Selections from the Virginia Museum of Fine Arts; Modern & Contemporary Art at the Virginia Museum of Fine Arts; The Jerome and Rita Gans Collection of English Silver; Art of Late Roman & Byzantium in the Virginia Museum of Fine Arts; British Sporting Paintings: The Paul Mellon Collection in the Virginia Museum of Fine Arts; Designed to Sell: Turn-of-the-Century American Posters; Video Series: Five African Art Facts; Tangible Spirits With Alison Saar; Women's Work: Urban Bush Women; An American's Dream: Jack Warner on Collecting American Art; Faberge: Shopping, Collecting, Remembering; Impressionist Paintings; What Color is Black...; The Fine Art of Life; Tigers and Sails and ABC Tales; Faberge: Virginia Museum of Fine Arts Collection; French Paintings: the Collection of Mr. & Mrs. Paul Mellon in the Virginia Museum of Fine Arts; Late 19th and Early 20th Century Decorative Arts: The Sydney & Frances Lewis Collection in the Virginia Museum of Fine Arts; German Expressionist Art: The Ludwig & Rosy Fischer Collection; Old-Russian Enamels; Three Masters of Landscape: Fragonard, Robert, Boucher; Champion Animals: Sculptures by Herbert Haseltine; American Dreams: Paintings & Decorative Arts from the Warner Collection; Ancient Art: Virginia Museum of Fine Arts; Selections: Virginia Museum of Fine Arts; Tigers and Sails and ABC Tales; Robert Lazzarini; James McNeil Whistler: Uneasy Pieces; William Blake: The Book of Job; The Arts of India; Outer and Inner Space: Pipolotti Rist, Shirin Neshat, Jane & Louise Wilson and the History of Video Art; Capturing Beauty: American Impressionist and Realist Paintings from the McGlothlin Collection; Rule Britannia! Art, Royalty & Power in the Age of Jamestown; Great British Watercolors from the Paul Mellon Collection; A Noble Feast: English Silver from the Jerome & Rita Gans Collection.
Hours & Admission Prices: Thurs.-Fri. 10-9, Sat.-Wed. 10-5; Holidays: 12-5. No charge. &
Attendance: 354,494 (accurate)
Membership: Student $10; Senior Individual, Out of Town Individual & Teacher $40; Individual $50; Dual $65; Family $75; Reciprocal $125; Supporter $250-$499; Curator's Circle $500-$999; Patron $1,000-$2,499; Fellows $2,500-$4,999; Faberge Society $5,000-$9,999; Founders $10,000-$24,999; Commonwealth Society $25,000 & up.

* **VIRGINIA WAR MEMORIAL, (M),** 621 S. Belvidere St., Richmond, VA 23220-6504. Tel.: 804-786-2060. Fax: 804-786-6652. Facebook: Virginia War Memorial.
E-mail: info@vawarmemorial.org
Web Site: www.vawarmemorial.org
Founded: 1956.
Congressional District: 7
Key Personnel: Exec. Dir., Jon C. Hatfield; Education Specialist, Candice L. Shelton; Museum Shop Mgr., Cebrina Sternberg.
Personnel Profile: Full-Time Paid 2; Part-Time Paid 3.
Governing Authority: Parent Institution: State of Virginia. Tax-exempt.
Institution Type/Description: Military Memorial.
Collections: Shrine of Memory includes names of nearly 12,000 Virginians killed in action from WWII to the Global War on Terror. Education Center features military artifacts & exhibits; films.
Facilities: 175-seat auditorium; 200-seat reception hall; 19-seat conference room; rental facilities; amphitheater.
Activities: educational programs; group tours; student seminars. Annual Events: Veterans Day Ceremony; Memorial Day Ceremony; Pearl Harbor Ceremony; Artifacts Roadshow.
Publications: quarterly newsletter, The Front Line; e-newsletters.
Hours & Admission Prices: Shrine of Memory: 5 am to midnight. Visitor Center: Mon.-Sat. 9-4, Sun. 12-4. No charge; donations accepted. &
Attendance: 60,000 (accurate)

THE VISUAL ARTS CENTER OF RICHMOND, 1812 W. Main St., Richmond, VA 23220. Tel.: 804-353-0094. Facebook: Visual Arts Center of Richmond.
E-mail: info@visarts.org
Web Site: visarts.org
Formerly: The Hand Workshop
Founded: 1963.
Key Personnel: Dir., Caroline Wright; C.E.O., Ava Spece
Institution Type/Description: Art Gallery.
Collections: works by contemporary artists.
Hours & Admission Prices: True F. Luck Gallery: Mon.-Fri. 9-9, Sat. 10-4, Sun. 1-4. No charge; donations accepted.
Attendance: 20,000 (accurate)

* **WILTON HOUSE MUSEUM, (M),** 215 S. Wilton Rd., Richmond, VA 23226-2212. Tel.: 804-282-5936. Fax: 804-288-9805.
E-mail: wiltonmuseum@comcast.net
Web Site: www.wiltonhousemuseum.org
Founded: 1934.
Congressional District: 7
Key Personnel: Dir., Keith D. Mac Kay; Pres. & Chm. (V), Laura R. Towers; Dir. Education & Public Rels., William Strollo; Dir. Devel., Elizabeth G. Johnson; Office Mgr., Elizabeth Gosack-Fleming; Mgr. Collections, Erica L. Borey.
Personnel Profile: Full-Time Paid 2; Part-Time Paid 12; Part-Time Volunteers 60; Interns 6.
Governing Authority: society. Parent Institution: National Society of the Colonial Dames of America in the Commonwealth of Virginia. Tax-exempt: 501(c)(3).
Institution Type/Description: Historic House: 1753 Georgian brick mansion, home of William Randolph, III & moved to present James River location in 1935.
Collections: fully paneled interior; period furniture; Randolph family portraits; archives; 18th & 19th-century decorative arts.
Major Exhibits: Randolph Family Reunion, 11/13-2/14.
Research Fields: Virginia genealogy; architecture; furniture; 18th century social & material culture history.
Facilities: meeting room; lecture hall.
Activities: guided tours; permanent & temporary exhibitions; meeting room; school tours with costumed docents; musical performance; lecture series; summer concerts; children's programs; family activities. Museum Sponsors: Garden Week.
Publications: quarterly, Dominion Dispatch.
Hours & Admission Prices: Tues.-Sat. 10-4:30. Adults $10, seniors $8, students $6; discounts to groups, National Trust for Historic Preservation, AASLH, AAA, VAM, ICOM & AAM members; Colonial Dames, members, active military & children under 6 no charge. Closed national holidays.
Attendance: 10,000 (accurate)
Membership: Friends of Wilton: Individual & Foot Soldier $25; Individual & Patriot $50; Family & VA Company $100; Signer $250; Historic Founder $500.

Richmond International Airport

* **VIRGINIA AVIATION MUSEUM, (M),** 5701 Huntsman Rd., Richmond International Airport, VA 23250-2416. Tel.: 804-236-3620 & 3622. Fax: 804-236-3623.
E-mail: mboehme@smv.org
Web Site: www.vam.smv.org
Founded: 1987.
Congressional District: 3
Key Personnel: C.E.O. & Dir. Science Museum of VA, Richard C. Conti; Exec. Dir., Michael P. Boehme; Chm. (V), Roger L. Boeve; Operations Supvr., Elaine Sutton; Museum Shop Mgr., Jennifer Morehead.
Personnel Profile: Full-Time Paid 1; Part-Time Paid 2; Part-Time Volunteers 45; Interns 1.
Governing Authority: state. Parent Institution: Science Museum of Virginia. Tax-exempt.
Institution Type/Description: Aviation Museum.
Collections: aircraft; aviation; WWII memorabilia.
Research Fields: aviation history relating to Virginia's rich aviation heritage; Civil War era ballooning.
Facilities: large video projection in 65-seat J.D. Benn Theater; aircraft renovation facility; Neilson J. November Observation deck; Kid's Ready Room. Museum-related items for sale.
Activities: hands-on exhibits; education program for preschoolers through 12th grade; wind tunnel & forces of flight demonstrations; flight simulators; SR-71 Forum; Wright Brothers Symposium. Annual Event: Tuskegee Airmen Presentation.

Publications: annual report (included within SMV); quarterly newsletter for members (included within SMV); group visit planning guide; program brochures.
Hours & Admission Prices: Tues.-Sat. 9:30-5, Sun. 12-5; call for additional hours. Adults $6.50, seniors 60 & over and youth 4-12 $5.50; discounts to groups, active military, VAHS members; children 3 & under and ASTC & SMV members no charge. Closed Thanksgiving; Christmas Eve & Day. ♿
Attendance: 20,400 (accurate)
Membership: Single $55; Family $99; Subscriber $100; Pilot $180; Senior Pilot $500; Command Pilot $750. Volunteer-100 hours of volunteer service annually-no charge.

Roanoke

CATHOLIC HISTORICAL MUSEUM OF THE ROANOKE VALLEY, 400 Campbell Ave., S.W., Roanoke, VA 24016-3627. Tel.: 540-982-0152. Fax: 540-982-0152.
E-mail: inuevik@msn.com
Web Site: chsrova.org
Founded: 1983.
Congressional District: 6
Key Personnel: Pres., Nick Centrone; Museum Shop Mgr., Marie Grewe.
Personnel Profile: Part-Time Volunteers 16.
Governing Authority: society; nonprofit organization. Tax-exempt: 501(c)(3).
Institution Type/Description: History Museum.
Collections: religious articles; photographs; Roanoke Valley religious historical data.
Research Fields: Local religious & education history; Diocese of Richmond.
Facilities: 300-vol. library of religious material available to the public. Museum-related items for sale.
Activities: research; guided tours; permanent & temporary exhibitions.
Publications: quarterly newsletter.
Hours & Admission Prices: Tues. 10-2; other times by appointment. No charge; donations accepted. ♿
Attendance: 500 (estimated)
Membership: Individual $15; Family $20; Patron $30; Sponsor $50; Benefactor $100.

ELEANOR D. WILSON MUSEUM AT HOLLINS UNIVERSITY, (M), 8009 Fishburn Dr., Roanoke, VA 24020-1679. Mailing Address: P.O. Box 9679, Roanoke, VA 24020-1679. Tel.: 540-362-6532. Fax: 540-362-6694. Facebook: Siddy Wilson.
E-mail: wilsonmuseum@hollins.edu
Web Site: www.hollins.edu/museum
Founded: 2004.
Key Personnel: Mgr. Museum Operations, Laura Jane Ramsburg; Exhibitions Coord., Janet Carty; Museum Coord., Karyn McAden.
Personnel Profile: Full-Time Paid 2; Part-Time Paid 1; Part-Time Volunteers 6; Interns 5.
Governing Authority: Parent Institution: Hollins University. Tax-exempt.
Institution Type/Description: Art Museum.
Collections: modern & contemporary paintings, photographs & prints.
Major Exhibits: Home Sweet Home, 1/14-2/14; Landscaped fromt he Collection, 1/14-2/14; Ben Grasso: Artist-in-Residence, 3/14-4/14; Kris Iden: Cadence, 3/14-4/14; Senior Art Majors, 5/14; Susan Cofer: Draw Near (T), 5/14-8/14.
Facilities: 2,800 sq. ft. exhibit space.
Activities: artists talks; lectures; workshops & demonstrations; exhibitions.
Publications: annual newsletter/events poster; exhibition catalogues.
Hours & Admission Prices: Tues.-Fri. 10-4, Sat. 1-5. No charge; donations accepted. Closed university breaks. ♿
Attendance: 10,000 (estimated)

HARRISON MUSEUM OF AFRICAN-AMERICAN CULTURE, 523 Harrison Ave., N.W., Roanoke, VA 24016-1740. Mailing Address: P.O. Box 12544, Roanoke, VA 24026-2544. Tel.: 540-345-4818.
E-mail: asbolden@harrisonmuseum.org
Web Site: www.harrisonmuseum.org
Key Personnel: Exec. Dir., Aletha Bolden; Exec. Asst., Donna Davis
Institution Type/Description: History Museum.
Collections: African-American history; photographs; African & contemporary art.
Facilities: Museum-related items for sale.
Activities: lectures. Museum Sponsors: Henry Street Heritage Festival in September.
Hours & Admission Prices: Tues.-Sat. 1-5. No charge; donations accepted.

HISTORY MUSEUM OF WESTERN VIRGINIA, (M), One Market Square, 3rd Fl., Roanoke, VA 24011-1429. Mailing Address: P.O. Box 1904, Roanoke, VA 24008-1904. Tel.: 540-342-5770. Fax: 540-224-1256.
E-mail: info@vahistorymuseum.org
Web Site: www.vahistorymuseum.org
Formerly: History Museum and Historical Society of Western Virginia
Founded: 1957.
Congressional District: 6
Key Personnel: Pres., Katherine Watts; Dir., Jeanne Bollendorf.
Personnel Profile: Full-Time Paid 2; Part-Time Paid 3; Part-Time Volunteers 90; Interns 2.
Governing Authority: society; nonprofit organization. Parent Institution: Historical Society of Western Virginia. Tax-exempt: 501(c)(3).
Institution Type/Description: History Museum.
Collections: archival & oral history repository; archeology; excavational railroad artifacts.
Research Fields: southwest Virginia history; oral history; genealogy.
Facilities: research library; 9,000 sq. ft. exhibit space.
Activities: guided tours; lectures; educational programs for children; historic bus tours.
Publications: monthly newsletter; books, Tour Games; Colonel William Fleming of Botetourt; Roanoke 1740-1982; The Journal of History Museum and Historical Society of Western Virginia; Iron Horses in the Valley; William Flemming, Patriot; The Visits of Lewis and Clark to Fincastle, Virginia; various historic maps; Notable Women West of the Blue Ridge; History of Roanoke; Charles Johnston's Frontier Adventure; Edward Beyer Travels in America; True Legacies of the Deyerle Builders.
Hours & Admission Prices: Tues.-Fri. 10-4, Sat. 10-5, Sun. 1-5. Adults $3, children & senior citizens over 60 $2; discounts to AAM & ICOM members and groups of 10 or more; members, children under 6, & Fri. 10-7 no charge. Closed New Year's Day; Martin Luther King Jr. Day; Memorial Day; Independence Day; Labor Day; Thanksgiving; Christmas. &
Attendance: 23,300 (accurate)
Membership: Senior $40; Individual $45; Family Senior $50; Family $55; Friend $51-$149; Associate $150-$249; Patron $250-$499; Sponsor $500-$999; Benefactor $1,000-$2,499; Angel $2,500 & up.

MILL MOUNTAIN ZOO, Pkwy. Spur Rd., Roanoke, VA 24034. Mailing Address: P.O. Box 13484, Roanoke, VA 24034-3484. Tel.: 540-343-3241. Fax: 540-343-8111.
E-mail: info@mmzoo.org
Web Site: www.mmzoo.org
Key Personnel: Exec. Dir., Sean Greene; Administration Mgr., Michaela Pace-Wilson
Institution Type/Description: Zoo.
Collections: 39 species; 135 animals.
Facilities: train. Museum-related items for sale.
Hours & Admission Prices: Winter: Thurs.-Sun. 10-4:30; Summer: daily 10-5. Adults $7.50, children 3-11 $5; children under 2 and members no charge. Closed Christmas. &

O. WINSTON LINK MUSEUM, (M), 101 Shenandoah Ave., N.E., Roanoke, VA 24016-2044. Tel.: 540-982-5465. Fax: 540-982-5683.
Web Site: www.linkmuseum.org
Founded: 2004.
Key Personnel: Dir., Kimberly Parker; Interim Chm. (V), Tucker Lemon; Pres. (V), David Helmer; Exec. Dir., Jeanne M. Bollendorf; Coord. Education, Shannon Lugar; Treas., Ron Sink; Devel. Officer, Monica Johnson; Security, Jack Stilton; Museum Shop Mgr., Jennifer Miller.
Personnel Profile: Full-Time Paid 4; Part-Time Paid 1; Part-Time Volunteers 90.
Governing Authority: private; nonprofit organization. Parent Institution: Historical Society of Western VA, Roanoke, VA. Tax-exempt: 501(c)(3).
Institution Type/Description: Photography Museum.
Collections: photographs; film; camera & lighting equipment; period artifacts.
Research Fields: Raymond Loewy.
Facilities: 75-seat auditorium; 17,000 sq. ft. exhibit space; 75-seat theater. Museum-related items for sale.
Activities: docent program; films; formal education programs; guided tours; hobby workshops; lectures; loan, temporary & traveling exhibitions. Annual Events: Celebration at the Station: Santa by Rail; Annual Norfolk Southern Calendar exhibit; Haunted Museum.
Publications: quarterly newsletter, Link News.
Hours & Admission Prices: Mon.-Sat. 10-5, Sun. 1-5. Adults $5, senior citizens $4.50, children $4; discounts to groups; members no charge. Closed New Year's Day; Easter; Thanksgiving; Christmas. &
Attendance: 18,679 (accurate)

Membership: Individual $30; Family $60; Photographer $90; Conductor $160; Engineer $250.

✲ **SCIENCE MUSEUM OF WESTERN VIRGINIA, (M),** 1 Market Sq., Roanoke, VA 24011. Tel.: 540-342-5710. Fax: 540-224-1240.
E-mail: frontdesk1@smwv.org
Web Site: www.smwv.org
Founded: 1970.
Congressional District: 6
Key Personnel: Exec. Dir., Jim Rollings; Business Mgr., Erma Williams; Devel. & Mktg. Dir., Michael Hemphill; Museum Shop Mgr., Megan Downing.
Personnel Profile: Full-Time Paid 14; Part-Time Paid 10; Part-Time Volunteers 160.
Volunteer Hours: 6,000
Operating Expenses: 670,000
Operating Income: 670,000
Governing Authority: nonprofit organization. Tax-exempt: 501(c)(3).
Institution Type/Description: Science Museum.
Collections: pertaining to earth science, health, energy, natural history & physical science.
Major Exhibits: Nature's Superpowers, 3/1/14-8/31/14; Roanoke/Blacksburg Science/Technology Festival, 9/14.
Facilities: 30,000 sq. ft. exhibit space; educational facilities. Science items for sale.
Activities: self-guided tours; panning for minerals; formally organized educational programs; permanent exhibitions.
Publications: curriculum guides; member newsletter.
Hours & Admission Prices: Call for hours. Museum: adults $10; discounts to AAM members; members no charge. Butterfly Garden: adults $4, member adults $2. &
Attendance: 50,000 (estimated)
Membership: Family & Grandparent $95.

✲ **TAUBMAN MUSEUM OF ART, (M),** 110 Salem Ave., S.E., Roanoke, VA 24011-1410. Tel.: 540-342-5760. Fax: 540-342-5798.
E-mail: info@taubmanmuseum.org
Web Site: taubmanmuseum.org
Formerly: Art Museum of Western Virginia
Founded: 1951.
Congressional District: 6
Key Personnel: Pres. & C.E.O., David Mickenberg; Chm. Bd. Trustees, Patricia Kermes; Vice Pres. Institutional Advancement, Kim Williamson; Vice Pres., Community & School Based Education, Cindy Petersen.
Personnel Profile: Full-Time Paid 20; Part-Time Paid 3; Part-Time Volunteers 200; Interns 5.
Governing Authority: nonprofit organization. Tax-exempt: 501(c)(3).
Institution Type/Description: Art Museum.
Collections: 19th-century to present American paintings; photography; sculpture; prints & drawings; contemporary folk art.
Research Fields: regional art, general fine arts, 19th- & 20th-century American art, outsider folk art; Thomas Eakins & His Circle.
Facilities: 2,600-vol. library; auditorium, theatre, cafe. Museum-related items for sale.
Activities: guided tours; lectures; gallery talks; concerts; arts festivals; education programs for children, adults, undergraduate & graduate college students; inter-museum loan, permanent, temporary & traveling exhibitions. Museum Sponsors: outreach education programs across region.
Publications: biannual newsletter; exhibition catalogs; monthly volunteer and membership newsletters.
Hours & Admission Prices: Tues.-Sat. 10-5, 1st Fri. each month 10-8, 1st Sun. each month call for hours. Adults $7, seniors $6, children 5-13 $3.75; members and children 4 & under no charge. Closed holidays. &
Attendance: 97,000 (accurate)
Membership: Student $10; Senior $25; Educator $30; Individual $35; Senior Dual $40; Family $65; 110 Society $110; Micheaux Society $250-$999; Sargent Society $1,000-$4,999; Eakins Society $5,000-$9,999; Founder's Society $10,000 & up.

VIRGINIA MUSEUM OF TRANSPORTATION, INC., 303 Norfolk Ave., S.W., Roanoke, VA 24016-3620. Tel.: 540-342-5670. Fax: 540-342-6898. Facebook: Virginia Museum of Transportation Inc.
E-mail: info@vmt.org
Web Site: www.vmt.org
Founded: 1962.
Congressional District: 6

Key Personnel: Dir., Beverly T. Fitzpatrick, Jr.; Museum Shop Mgr., Susan Loveman.
Personnel Profile: Full-Time Paid 7; Part-Time Paid 6; Part-Time Volunteers 150.
Governing Authority: private; nonprofit organization. Tax-exempt.
Institution Type/Description: Transportation Museum.
Collections: transportation equipment, vintage cars & carriages, locomotives & railcars, aviation pieces; interactive & historical exhibits; model layouts; historic railroad freight depot & railroad yard.
Research Fields: all models of transportation; technology.
Activities: guided tours; workshops; outreach programs; camps; permanent exhibitions; revolving displays; annual special events.
Publications: monthly email newsletter.
Hours & Admission Prices: Mon.-Sat. 10-5, Sun. 1-5. Adults $8, senior citizens $7, children 3-11 $6; discounts to AAA members; children under 3 & members no charge. Closed New Year's Eve & Day; Easter, Thanksgiving; Christmas Eve & Day. &
Attendance: 52,000 (accurate)
Membership: Individual $30; Dual $35; Grandparents $40; Family $50.

Salem

THE SALEM MUSEUM, (M), 801 E. Main St., Salem, VA 24153-4312. Tel.: 540-389-6760.
E-mail: info@salemmuseum.org
Web Site: www.salemmuseum.org
Founded: 1992.
Congressional District: 9
Key Personnel: Dir., John D. Long; Pres., William Robertson.
Personnel Profile: Full-Time Paid 1; Part-Time Paid 2; Part-Time Volunteers 50; Interns 2.
Governing Authority: private; nonprofit organization. Parent Institution: Salem Historical Society. Tax-exempt.
Institution Type/Description: History Museum: located in the c.1845 Williams-Brown House-Store.
Collections: concentration on history from prehistoric-modern in the Roanoke Valley of Virginia; emphasis on Salem; long term exhibits on the Civil War; Victorian Parlour; African American history; works by local artist, Walter Biggs.
Research Fields: local structures; pictorial history; Native American Settlements in the area; local history; Civil War; public history.
Facilities: 5,000 sq. ft. exhibit space; herb and kitchen garden. Museum-related items for sale.
Activities: internship program; guided tours; lectures; temporary & traveling exhibitions; weekend programs for children. Annual Events: Olde Salem Days Open House in September; Ghost Walk in October; Holiday Homes Tour in December.
Publications: monthly newsletter; seasonal historical newspaper featuring walking tour; occasional transcription & reproductions of local history documents; pictorial history.
Hours & Admission Prices: Tues.-Fri. 10-4, Sat. 10-3. No charge; donations accepted. Closed New Year's weekend; Independence Day; Thanksgiving; Christmas.
Attendance: 12,000 (estimated)
Membership: Individual $30; Family $45; Sustaining $120; Associate $275; Life $1,000.

Saltville

MUSEUM OF THE MIDDLE APPALACHIANS, (M), 123 Palmer Ave., Saltville, VA 24370. Mailing Address: P.O. Box 910, Saltville, VA 24370-0910. Tel.: 276-496-3633. Fax: 276-496-7033.
E-mail: museummoma@embarqmail.com
Web Site: www.museum-mid-app.org
Founded: 1998.
Congressional District: 9
Key Personnel: Pres., Jerry W. Catron; Pres., Ron Orr; Treas., Carl Rickman; Devel., Christine Helton; Museum Shop Mgr., Harry R. Haynes; Coord., Janice Orr.
Personnel Profile: Full-Time Paid 1; Full-Time Volunteers 1; Part-Time Paid 1; Part-Time Volunteers 20.
Governing Authority: private; nonprofit organization. Parent Institution: The Saltville Foundation, 123 Palmer Ave., P.O. Box 910, Saltville, VA 24370. Tax-exempt: 501(c)(3).
Institution Type/Description: Natural History Museum.
Collections: natural & cultural heritage; rocks; minerals; fossils; maps; Woodland Indians; Civil War artifacts; Ice Age fossils; photographs.
Research Fields: Ice Age paleontological digs; Civil War fortifications.
Facilities: library; educational facilities; field research station; 7,500 sq. ft. exhibit space; classroom; conference room. Museum-related items for sale.

Activities: lectures; temporary exhibitions; school & scout programs. Annual Events: Woolly Day; Kids Ice Age Dig.
Publications: quarterly newsletter.
Hours & Admission Prices: Mon.-Sat. 10-4, Sun. 1-4. Adults $3, senior citizens & children 6-12 $2, senior & student groups $1; members & children under 6 no charge. Closed New Year's Day; Easter; Thanksgiving; Christmas. &
Attendance: 12,000 (accurate)
Membership: Senior & Student $10 & up; Individual $15 & up; Family $35 & up; Sustaining $100 & up; Century $1,000 & up.

Scottsville

SCOTTSVILLE MUSEUM, 290 Main St., Scottsville, VA 24590. Mailing Address: 290 Main St., P.O. Box 101, Scottsville, VA 24590-0101. Tel.: 434-286-2247.
E-mail: smuseum@avenue.org
Founded: 1970.
Congressional District: 5
Key Personnel: Pres. (V), Evelyn Edson.
Personnel Profile: Part-Time Paid 1; Part-Time Volunteers 50; Interns 2.
Governing Authority: Tax-exempt.
Institution Type/Description: History Museum.
Collections: town history; James River transportation; the Civil War; Native American artifacts; school life; clothing; toys; furniture; photographs.
Facilities: theater.
Activities: group tours.
Publications: annual newsletter.
Hours & Admission Prices: April-Oct. Sat. 10-5, Sun. 1-5; other times by appointment. No charge; donations accepted.
Attendance: 2,000 (estimated)
Membership: Regular $10-$24.99; Supporting $25-$49.99; Sustaining $50-$99.99; Benefactor $100-$499.99; Life $500.

Smithfield

ISLE OF WIGHT COURTHOUSE, 130 Main St., Smithfield, VA 23430-1323. Mailing Address: 204 W. Franklin St., Richmond, VA 23220-5012. Tel.: 757-357-5182. Fax: 804-775-0802.
E-mail: info@preservationvirginia.org
Web Site: www.preservationvirginia.org/isleofwight
Congressional District: 4
Key Personnel: Exec. Dir., Elizabeth Kostelny; Pres. (V), Mr. Lacy Bennetward, Jr.; Chm. (V), Tom Mayes; Dir. Properties, Louis Malon.
Personnel Profile: Part-Time Paid 2; Part-Time Volunteers 2.
Governing Authority: nonprofit. Parent Institution: Preservation Virginia, 204 W. Franklin St., Richmond, VA 23220. Tax-exempt.
Institution Type/Description: Historic Building: restored Courthouse; built in 1750.
Collections: local history & culture; paintings; period furnishings.
Facilities: Books, handmade gifts & museum-related items for sale.
Activities: guided tours; workshops; symposiums; community events.
Publications: biannual, Ventures.
Hours & Admission Prices: Feb. Fri.-Sat. 10-4, Sun. 1-4; March-Dec. Tues.-Thurs. & Sun. 1-4, Fri.-Sat. 10-4. No charge; donations accepted. Closed Thanksgiving; Christmas.
Attendance: 5,000 (estimated)
Membership: Student & Teacher $35; Individual $50; Individual Plus One $60; Family $70; Organizational $75.

SCHOOLHOUSE MUSEUM, 516 Main St., Smithfield, VA 23430. Mailing Address: P.O. Box 1113, Smithfield, VA 23431. Tel.: 757-365-4789.
Institution Type/Description: Historic Building: housed in a former one-room schoolhouse built to educate county African American children.
Collections: local history & culture; period furnishings.
Hours & Admission Prices: Call for hours.

South Boston

SOUTH BOSTON-HALIFAX COUNTY MUSEUM OF FINE ARTS & HISTORY, (M), 1540 Wilborn Ave., South Boston, VA 24592-2400. Mailing Address: P.O. Box 383, South Boston, VA 24592-0383. Tel.: 434-572-9200. Fax: 434-572-8996. Facebook: South Boston Halifax County Museum of Fine Arts and History.
E-mail: sbhcm1@centurylink.net
Web Site: sbhcmuseum.org
Founded: 1981.
Congressional District: 5

Key Personnel: Pres., Paul Smith; Vice Pres., Linda Mercer; Dir., Beth Coates; Treas., Jane Jones; Sec., Louise Sheppard.
Personnel Profile: Full-Time Paid 1; Part-Time Paid 1; Part-Time Volunteers 60.
Governing Authority: nonprofit organization. Tax-exempt: 501(c)(3).
Institution Type/Description: History Museum.
Collections: artifacts from South Boston, Halifax County & Southside Virginia.
Facilities: library; 20,000 sq. ft. exhibit space; premises available for rental. Museum-related items for sale.
Activities: guided tours; films; concerts; temporary & traveling exhibitions; lectures.
Publications: annual, Museum Newsletter; quarterly newsletter; museum flyer; books, Civil War Letters; Black History in Halifax County; Black Schools in Halifax County 1940s; National Tobacco Festivals 1935-1941; Historic Delights Cookbook; Virginia Born Presidents.
Hours & Admission Prices: Wed.-Sat. 10-4. No charge; donations accepted. Closed New Year's Day; Thanksgiving; Christmas. ♿
Attendance: 10,000 (accurate)
Membership: Senior Citizen $15; Individual $20; Family $35; 100 Plus Club $100; Roundtable $250; Inner Circle $500.

Spotsylvania

SPOTSYLVANIA HISTORICAL ASSOCIATION AND MUSEUM, 9019 Old Battlefield Blvd., Spotsylvania, VA 22553. Mailing Address: P.O. Box 64, Spotsylvania, VA 22553-0064. Tel.: 540-507-7278.
E-mail: shainc@verizon.net
Founded: 1962.
Congressional District: 8
Key Personnel: Dir., Treas. & Cur., Jo Harding; Pres., John E. Pruitt, Jr.; Vice Pres., Stephen P. Lampert.
Personnel Profile: Part-Time Paid 4.
Governing Authority: state. nonprofit organization. Operated by the Spotsylvania Historical Association, Inc. Tax-exempt: 170(b)(1)(A).
Institution Type/Description: General Museum.
Collections: Indian relics; early Colonial artifacts; china; tools; pottery; dolls; maps; genealogies and manuscripts; diaries; Civil War artifacts; weapons; pharmaceutical collection; books. Historic Building: 1856 church; 1838 court house; 1855 jail.
Research Fields: local history; archaeology; military; Revolutionary and Civil War; genealogy.
Facilities: 2,500-vol. library of books, diaries and old maps of county available for use on premises. Commemorative coins & museum-related items for sale.
Activities: organized educational programs; permanent & temporary exhibitions.
Publications: manuscript, Court Houses of Spotsylvania County; books, Patriots 1775-1781; A History of Early Spotsylvania; booklets, Church Histories; Histories of Old Homes.
Hours & Admission Prices: Daily 9-5. No charge; donations accepted. Closed New Year's Day; Thanksgiving; Christmas Eve & Day. ♿
Attendance: 5,000 (estimated)
Membership: Individual & Family $15; Life $100.

Springfield

DEA MUSEUM & VISITORS CENTER, (M), 8701 Morrissette Dr., Springfield, VA 22152. Tel.: 202-307-3463. Fax: 202-307-8956.
Web Site: www.deamuseum.org
Founded: 1999.
Key Personnel: Dir., Sean T. Fearns; Pres. (V), William Alden; Museum Shop Mgr., Jim Lumsden.
Personnel Profile: Full-Time Paid 4; Part-Time Paid 2; Part-Time Volunteers 12; Interns 2.
Governing Authority: Parent Institution: Drug Enforcement Administration (DEA). Subsidiary Institution: DEA Educational Foundation. Tax-exempt.
Institution Type/Description: History Museum.
Collections: history of drug, drug addiction & drug enforcement in the U.S.
Activities: school & group programs; outreach programs.
Hours & Admission Prices: Tues.-Fri. 10-4; groups by appointment. No charge. ♿

Staunton

AUGUSTA COUNTY HISTORICAL SOCIETY, 20 S. New St., 3rd Fl., Staunton, VA 24401. Mailing Address: P.O. Box 686, Staunton, VA 24402-0686. Tel.: 540-248-4151.
E-mail: augustachs@ntelos.net
Web Site: www.augustacountyhs.org
Institution Type/Description: Historical Society Museum.
Collections: county history & culture; manuscripts; books.
Hours & Admission Prices: Tues. & Thurs.-Fri. 9-12; other times by appointment.

CAMERA HERITAGE MUSEUM, 1 W. Beverley St., Staunton, VA 24401. Tel.: 540-886-8535.
E-mail: campal7@verizon.net
Web Site: www.cameraheritagemuseum.com/index.html
Institution Type/Description: Camera History Museum.
Collections: camera & photographer history; early cameras & accessories; photographs.
Hours & Admission Prices: Mon.-Fri. 9-5, Sat. 9-2.

✻ **FRONTIER CULTURE MUSEUM OF VIRGINIA, (M),** 1290 Richmond Rd., Staunton, VA 24401-4976. Mailing Address: P.O. Box 810, Staunton, VA 24402-0810. Tel.: 540-332-7850. Fax: 540-332-9989. TDD: 540-332-7850.
E-mail: visitors.center@frontiermuseum.org
Web Site: www.frontiermuseum.org
Formerly: Museum of American Frontier Culture
Founded: 1986.
Congressional District: 6
Key Personnel: Chm. Public Bd. (V), Paul Vames; Chm. Private Bd., John Dod; Exec. Dir., G. John Avoli; Museum Shop Mgr., Kimi Wills.
Personnel Profile: Full-Time Paid 25; Part-Time Paid 15; Part-Time Volunteers 55; Interns 3.
Volunteer Hours: 11,000
Governing Authority: state. Parent Institution: Commonwealth of Virginia. Subsidiary Institution: Secretariat of Education. Tax-exempt: 501(c)(3).
Institution Type/Description: Outdoor Living History Museum.
Collections: German, Scotch-Irish, English, American, West African/Igbo folklife & culture; Native American; reconstructed 17th- to 19th-century farms & buildings; study collections; furniture & other related furnishings; textile-weaving, agricultural tools, domestic craft, trade materials & equipment; archives; ledgers; family papers.
Research Fields: European cultures from Ireland, England, Germany & America; immigration to America & Virginia; 18th & 19th century agricultural, social & economic life; West Africa/Igbo culture; Native American; woodworking.
Facilities: 4,000-vol. library pertaining to European & Virginian history, agriculture & architecture; research & educational facilities; 120 & 60 seat theater & lecture rooms. Museum-related items for sale.
Activities: living history demonstrations; outreach programs; guided tours; organized education programs; participatory exhibits; workshops; lecture series; special events. Museum Sponsors: traditional Oktoberfest; traditional holiday tours.
Publications: quarterly newsletter, News From The Frontier; annual calendar of events; museum guidebook.
Hours & Admission Prices: mid-March to Nov. daily 9-5; Dec. to mid-March daily 10-4; group & educational tours available. Adults $10, children 6-12 $6; discounts to AAM members; children under 6 & members no charge. Closed New Year's Day; Thanksgiving; Christmas. ♿
Attendance: 78,000 (accurate)
Membership: Student $20; Individual $30; Family or Grandparent $50; Contributor $100; Associate $250; Corporate Patron $500; Benefactor $1,000.

MARY BALDWIN COLLEGE/HUNT GALLERY, Market & Vine, Staunton, VA 24401. Mailing Address: Dept. of Art & Art History, Deming Hall, Mary Baldwin College, Staunton, VA 24401-3610. Tel.: 540-887-7196. Fax: 540-887-7139.
E-mail: pryan@mbc.edu
Web Site: www.mbc.edu/college/events/huntgallery.asp
Founded: 1842.
Key Personnel: Dir., Paul Ryan.
Personnel Profile: Interns 2.
Governing Authority: college; Mary Baldwin College. Tax-exempt.
Institution Type/Description: College Art Gallery & Museum.

Collections: 20th-century contemporary paintings, drawings, prints & photographs; contemporary works by regional artists & national artists connected to the college.
Activities: lectures; formal education programs for adults; participatory & loan exhibitions.
Publications: exhibition catalogs.
Hours & Admission Prices: Sept.-May Mon.-Fri. 9-5. No charge. &
Attendance: 1,200 (estimated)

MUSEUM OF BANK HISTORY AT SUNTRUST BANK, 2-14 W. Beverley St., Staunton, VA 24401. Tel.: 540-887-0174.
Institution Type/Description: History Museum.
Collections: Valley National Bank history & artifacts; banking industry; photographs.
Hours & Admission Prices: Mon.-Thurs. 9-5, Fri. 9-6. No charge.

STAUNTON AUGUSTA ART CENTER, 20 S. New St., Staunton, VA 24401-4308. Tel.: 540-885-2028. Fax: 540-885-6000.
E-mail: info@saartcenter.org
Web Site: www.saartcenter.org
Founded: 1961.
Congressional District: 6
Key Personnel: Exec. Dir., Beth Hodge; Pres. (V), Steve Grande; Cur., Hannah Scott.
Personnel Profile: Full-Time Paid 2; Part-Time Volunteers 100; Interns 3.
Governing Authority: nonprofit organization. Parent Institution: Virginia Museum of Fine Art. Tax-exempt: 501(c)(3).
Institution Type/Description: Art Association: housed in 19th-century pump house which once supplied water for the city of Staunton.
Collections: paintings; sculpture; photographs.
Activities: gallery talks; lectures; films; educational programs; workshops; annual outdoor art show. Museum Sponsors: summer studio art camp for children ages 4-15; Art for All Ages, an outreach program for area seniors; Outside the Line, program exploring issues of mental health & art.
Publications: biennial newsletter.
Hours & Admission Prices: Mon.-Fri. 10-5, Sat. 10-4. No charge; donations accepted. Closed major holidays; between exhibitions. &
Attendance: 14,000 (estimated)
Membership: Individual $35; Family $50; Pumphouse Club $51-$99; The Gallery $100-$249; Impressionist Society $250-$499; Renaissance Circle $500-$999; Ruth Owen Patron of the Arts $1,000 & up.

STAUNTON MILITARY ACADEMY MUSEUM & VIRGINIA WOMEN'S INSTITUTE FOR LEADERSHIP MUSEUM, Mary Baldwin College, 227 Kable St., Staunton, VA 24402. Mailing Address: SMA Alumni Assoc., P.O. Box 958, Staunton, VA 24402-0958. Tel.: 540-885-1309.
E-mail: smaoffice@sma-alumni.org
Web Site: www.sma-alumni.org/museum.htm
Founded: 2001.
Institution Type/Description: History Museum.
Collections: academy & institute history and artifacts; photographs; memorial.
Hours & Admission Prices: Wed. & Sat.-Sun. 1-4. No charge.

* **WOODROW WILSON PRESIDENTIAL LIBRARY, (M),** 20 N. Coalter St., Staunton, VA 24401-4332. Mailing Address: P.O. Box 24, Staunton, VA 24402-0024. Tel.: 540-885-0897. Fax: 540-886-9874.
E-mail: info@woodrowwilson.org
Web Site: www.woodrowwilson.org
Formerly: Woodrow Wilson Birthplace & Museum
Founded: 1938.
Congressional District: 6
Key Personnel: Pres., Don W. Wilson, Ph.D.; Chm. (V), Dr. Michael Dickens; Administrative Officer, Robin Von Seldeneck; Dir. Museum Operations, Marcene Molinaro; Dir. Library & Archives, Peggy L. Dillard; Museum Shop Mgr., Virginia M. Engleman.
Personnel Profile: Full-Time Paid 6; Part-Time Paid 13; Part-Time Volunteers 75; Interns 24.
Governing Authority: nonprofit organization. Tax-exempt: 501(c)(3).
Institution Type/Description: Historic House: 1846 former Presbyterian Manse & birthplace of Woodrow Wilson, 28th President of the U.S.
Collections: decorative arts; furnishings; costumes; uniforms; textiles; musical instruments; paintings, graphics; photographs; sculpture; manuscripts; rare books; his accomplishments as author, scholar, university president, governor & statesman. Historic Houses: 1880 Emily P. Smith Administration Bldg.; 1856 Dolores Lescure Center (Museum).

Research Fields: relating to collections & the life & times of President Woodrow Wilson, 1856-1924.
Facilities: 8,000-vol. research library; education center; carriage house; meeting rooms. Gift items for sale.
Activities: guided tours; films; lectures; gallery talks; on-site programs for school groups K-8; packet program for school groups 9-12; summer internships & informal educational programs for undergraduate college students; semester internships at graduate level in museology; permanent & temporary exhibitions; inter museum loans; digital archive.
Publications: newsletter; interpretive brochures; pamphlets.
Hours & Admission Prices: See website for hours. Adults $14, AAA members, seniors & active military $12, students 13 & over $7, children 6-12 $5; discounts to groups; members & children under 6 no charge. Closed New Year's Day; Easter; Thanksgiving; Christmas Eve & Day. &
Attendance: 21,500 (accurate)
Membership: Student $35; Individual $50-$99; Small Business $100-$249; Family $100-$499; Corporation $250 & up; Diplomat $500-$999; Ambassador $1,000-$2,499; Cabinet $2,500-$4,999; Emily Pancake Smith Circle $5,000 & up; Delores Lescure Circle $5,000 & up; Arthur S. Link Circle $5,000 & up; Woodrow Wilson Circle $5,000 & up.

Stephens City

NEWTOWN HISTORY CENTER, 5408 Main St., Stephens City, VA 22655-2829. Mailing Address: P.O. Box 143, Stephens City, VA 22655-0143. Tel.: 540-869-1700. Fax: 540-869-0400.
E-mail: info@newtownhistorycenter.org
Web Site: newtownhistorycenter.org
Formerly: Historic Stephensburg Museums
Founded: 1990.
Congressional District: 10
Key Personnel: Pres. (V), Linden A. Fravel; Dir. & Cur., Byron C. Smith; Treas., Mary S. Dyke; Mgr. Collections & Programs, Wayne A. Eldred.
Personnel Profile: Full-Time Paid 2; Part-Time Volunteers 11.
Governing Authority: private; nonprofit organization. Parent Institution: Stone House Foundation. Tax-exempt: 501(c)(3).
Institution Type/Description: History Museum.
Collections: lower Shenandoah Valley history for prehistoric to modern times; history of Stephens City, VA from early settlement to Civil War; folk culture; decorative arts; communication artifacts; recreatioinal artifacts.
Research Fields: freight wagons of the Shenandoah Valley; local African American history.
Facilities: 260-vol. library; classrooms; 3,416 sq. ft. exhibit space. Museum-related items for sale.
Activities: lectures; guided tours; study clubs. Annual Event: Newtown Heritage Festival.
Publications: quarterly newsletter, Museum Musings.
Hours & Admission Prices: June-Aug. Tues.-Sat. 10-4, Sun. 1-5; Sept.-Nov. Wed.-Sat. 10-4, Sun. 1-5; Dec.-May by appointment. Family $5, adults $2, children 6-18 $1; AASLH members no charge. Closed New Year's Day; Martin Luther King Jr. Day; Presidents' Day; Good Friday; Labor Day; Columbus Day; Thanksgiving & day after; Christmas Eve & Day.
Attendance: 371 (accurate)
Membership: Individual $10; Family $25; Contributing $55; Heirloom Society, Family History Patron.

Sterling

HERITAGE FARM MUSEUM OF LOUDOUN COUNTY, 21668 Heritage Farm Lane, Sterling, VA 20164-9207. Tel.: 571-258-3800. Fax: 571-258-3801.
E-mail: hfm.baileyboy@gmail.com
Web Site: www.heritagefarmmuseum.org
Founded: 1999.
Congressional District: 10
Key Personnel: Pres. (V), Su Webb; Co Dir., Katie Eichler Jones; Co Dir., Christie Love.
Personnel Profile: Full-Time Paid 2; Full-Time Volunteers 3; Part-Time Paid 2; Part-Time Volunteers 172.
Governing Authority: private; nonprofit organization. Tax-exempt: 501(c)(3).
Institution Type/Description: History Museum.
Collections: agricultural implements & machinery; livestock; gardening; horticulture; early 20th century general store items.
Research Fields: agriculture; history of Loudoun County.
Facilities: 200-vol. library; 7,500 sq. ft. exhibit space; classroom; 30-seat theater. Museum-related items for sale.
Activities: interactive exhibits for families; docent program; formal education programs for students & scouts; hobby workshops; livestock interactive animal programming; guided tours; loan, temporary & participatory exhibits; theater.

Hours & Admission Prices: Tues.-Sat. 9:30-4:30, Sun. 11:30-4:30. Adults $5, senior citizens $4, children $3. &

Attendance: 16,400 (accurate)

Membership: Quarterly $50; Weekday $75; Classic $100; Red Ribbon $125; Blue Ribbon $150.

Strasburg

CRYSTAL CAVERNS AT HUPP'S HILL HISTORIC PARK, 33231 Old Valley Pike, Strasburg, VA 22657-3715. Mailing Address: 3299 K St., N.W., Washington, DC 20007-4415. Tel.: 540-465-5884.

E-mail: wayside@shentel.net

Web Site: www.crystalcavernsofva.com

Founded: 1998.

Key Personnel: C.E.O., Babs B. Funkhouser; Pres. (V), Ami Aronson.

Personnel Profile: Full-Time Paid 1; Part-Time Paid 3.

Governing Authority: public; nonprofit foundation. Parent Institution: Wayside Foundation, Strasburg, VA. Tax-exempt: 501(c)(3).

Institution Type/Description: Geology Museum: housed in the first discovered cave in Virginia.

Collections: historical cavern artifacts; calcite crystals; cave drawings from Archaic Indian culture (on cave ceiling).

Research Fields: endangered species of amphipods & bats; historical use of caves.

Facilities: nature center; 200 sq. ft. exhibit space; walking trail. Museum-related items for sale.

Activities: formal education programs for children; guided tours; lantern tours. Museum Sponsors: living history events.

Publications: quarterly newsletter, By The Wayside.

Hours & Admission Prices: Daily 11-4. Adults $10, senior citizens, students, government, military, police, fire & rescue with ID; children $8; discounts to AAM & ICOM members. Closed New Year's Eve & Day; Easter; Thanksgiving; Christmas Eve & Day.

Attendance: 8,000 (estimated)

STRASBURG MUSEUM, 440 E. King St., Strasburg, VA 22657-2433. Mailing Address: P.O. Box 333, Strasburg, VA 22657-0333. Tel.: 540-465-3175 & 3728.

E-mail: gastick@shentel.net

Web Site: strasburgmuseum.org

Founded: 1970.

Congressional District: 6

Key Personnel: Pres. (V), Gloria Stickley; Vice Pres., John Adamson; Treas., Pat Clem; Sec., Tina Crabill; Museum Shop Mgr., Margo Hammock.

Personnel Profile: Part-Time Volunteers 75.

Governing Authority: nonprofit. Tax-exempt.

Institution Type/Description: History Museum: housed in 1891 pottery factory.

Collections: local history; Shenandoah Valley agriculture & crafts; Strasburg pottery 1830-1910; southern & B&O railroads

Facilities: Local craft items & history books for sale.

Activities: workshops; tours.

Publications: books, The Story of Strasburg, Virginia; A Strasburg Potter's Daughter; Strasburg Community Memories: A Pictorial Display in the Strasburg Museum.

Hours & Admission Prices: May-Oct. daily 10-4. Adults $3; members no charge. &

Attendance: 3,500 (estimated)

Membership: Individual $3; Family $7.50; Sustaining $20; Life $100.

Stratford

STRATFORD HALL, ROBERT E. LEE MEMORIAL ASSO-CIATION, INC., 483 Great House Rd., Stratford, VA 22558. Tel.: 804-493-8038. Fax: 804-493-0333.

E-mail: info@stratfordhall.org

Web Site: www.stratfordhall.org

Formerly: Stratford, Robert E. Lee Memorial Association, Inc.

Founded: 1929.

Congressional District: 1

Key Personnel: Exec. Dir., Paul C. Reber; Pres. (V), Custis S. Glover; Dir. Research & Library Collections, Judith S. Hynson; Dir. Education & Interpretation, Abby Newkirk; Dir. Mktg., Jim Schepmoes; Cur., Gretchen Goodell; Plantation Store Mgr., Janet Branson.

Personnel Profile: Full-Time Paid 28; Part-Time Paid 45; Part-Time Volunteers 20; Interns 6.

Governing Authority: nonprofit organization. Tax-exempt: 501(c)(3).

Institution Type/Description: Historic Site & General Museum: housed in 1738 Stratford Hall with operating plantation.

Collections: 18th- & 19th-century decorative arts; furniture, silver, ceramics,

textiles, portraits & prints; archaeology artifacts; rare books & manuscripts; archives; seven original & eight reconstructed plantation buildings; operating grist mill.

Research Fields: American history; 18th-century life.

Facilities: 8,000-vol. library of Virginiana available for research only; visitor center with museum; gardens; dining room. Museum-related items for sale.

Activities: guided tours; lectures; mill demonstrations; inter-museum loan, permanent & temporary exhibitions; elementary & secondary educational programs; seminars; conferences; archaeology field school; nature trails; introductory & interactive videos.

Publications: handbook; archaeology book; Paul Buchanan: Stratford Hall and other Architectural Studies; Robert E. Lee: Commemorative Essays on the Bicentennial of His Birth.

Hours & Admission Prices: Daily 9:30-4. Adults $10, senior citizens 60 & over, AAA members, groups of 20 or more, servicemen with valid ID $9, children 6-11 $5; discounts to the National Trust for Historical Preservation members & AAM members; children under 6, friends of Stratford members & Virginia Assoc. of Museum members no charge. Grounds pass: adults $5, children $3. Closed New Year's Eve & Day; Christmas Eve & Day. &

Attendance: 28,000 (estimated)

Membership: Friends of Stratford, various categories starting at $35.

Stuart

WOOD BROTHERS RACING MUSEUM, 21 Performance Dr., Stuart, VA 24171. Tel.: 276-694-2121.

Web Site: www.woodbrothersracing.com

Institution Type/Description: Racing Museum.

Collections: race cars; photographs; personal artifacts; racing memorabilia.

Activities: tours.

Hours & Admission Prices: Mon.-Fri. 9-12 & 1-5.

Suffolk

RIDDICK'S FOLLY HOUSE MUSEUM, 510 N. Main St., Suffolk, VA 23434. Tel.: 757-934-0822. Fax: 757-934-0822 (call first). Facebook: Riddick's Folly House Museum.

E-mail: rfcurator@verizon.net

Web Site: www.riddicksfolly.org

Formerly: Riddick's Folly

Founded: 1978.

Congressional District: 4

Key Personnel: Pres., James E. Butler, III; Dir., Cur. & Museum Shop Mgr., Edward L. King.

Personnel Profile: Part-Time Paid 2; Part-Time Volunteers 35.

Governing Authority: nonprofit. Tax-exempt: 501(c)(3).

Institution Type/Description: Historic House Museum: built in 1837 Greek revival home of the Riddick family which was commandeered by the Union army as a headquarters during the occupation of Suffolk.

Collections: Riddick family items; Suffolk history; Empire period furniture & decorative arts; Gov. Mills E. Godwin, Jr. memorabilia; Civil War artifacts; permanent Civil War exhibit.

Research Fields: Riddick family history; Suffolk history; 19th century American south.

Facilities: 1,165 sq. ft. exhibit space; multipurpose room. Books, artwork by local artists & craftsmen, brass reproductions & other items for sale.

Activities: guided tours; lectures; organized educational programs; temporary & traveling exhibitions; changing exhibits in art & history.

Publications: quarterly newsletter.

Hours & Admission Prices: Wed.-Fri. 10-5, Sat. 10-4, Sun. 1-5; groups by appointment. Adults $5, seniors 55 & over and active military $4, children 3-12 $3. Closed New Year's Day; Easter; Independence Day; Thanksgiving; Christmas. &

Attendance: 4,000 (estimated)

Membership: Student $10; Individual $15; Family $25; Associate $50; Sustainer $100; Patron $250; Fellow $500.

THE SUFFOLK MUSEUM, 118 Bosley Ave., Suffolk, VA 23434-5755. Tel.: 757-514-7284. Fax: 757-538-0833.

E-mail: nkinzinger@city.suffolk.va.us

Founded: 1986.

Congressional District: 4

Key Personnel: Dir., Nancy Kinzinger.

Personnel Profile: Full-Time Paid 1; Full-Time Volunteers 2; Part-Time Paid 2; Part-Time Volunteers 5; Interns 1.

Governing Authority: municipal; nonprofit. Parent Institution: City of Suffolk. Tax-exempt: 501(c)(3).

Institution Type/Description: Art Museum.

Collections: paintings; sculpture; photographs.

Research Fields: related to exhibitions.

Facilities: 3,000 sq. ft. exhibit space; educational studio room.

Activities: guided tours; lectures; concerts; theater; hobby workshops; organized education programs; participatory, loan & traveling exhibitions; demonstrations; classes.

Publications: quarterly newsletter; monthly exhibition program.

Hours & Admission Prices: Tues.-Sat. 10-5, Sun. 1-5. No charge. Closed New Year's Day; Lee-Jackson-King Day; Washington's Birthday; Easter; Memorial Day; Independence Day; Labor Day; Veterans Day; Thanksgiving; Christmas. &

Attendance: 18,000 (accurate)

SUFFOLK SEABOARD STATION RAILROAD MUSEUM, 326 N. Main St., Suffolk, VA 23434. Mailing Address: P.O. Box 1255, Suffolk, VA 23439-1255. Tel.: 757-923-4750. Fax: 757-923-4751.

Institution Type/Description: Historic Building: housed in a former railroad station; built in 1885.

Collections: local history; railroad memorabilia; HO-scale model; caboose.

Facilities: Museum-related items for sale.

Hours & Admission Prices: Thurs.-Sat. 10-4, Sun. 1-4.

Surry

CHIPPOKES FARM & FORESTRY MUSEUM, 868 Plantation Rd., Surry, VA 23883-2406. Tel.: 757-294-3439. Fax: 757-294-3550.

Founded: 1990.

Congressional District: 4

Key Personnel: Pres. (V), Sen. Frederick M. Quayle; Exec. Dir., Linda Guntharp; Museum Shop Mgr., Sarah Cosby.

Personnel Profile: Full-Time Paid 2; Part-Time Paid 6.

Governing Authority: nonprofit organization. Parent Institution: Chippokes Plantation Farm Foundation. Tax-exempt: 170(b)(1)(A).

Institution Type/Description: Farm & Forestry Museum: housed in series of buildings, two of which are historically significant, at Chippokes Plantation State Park.

Collections: period farm & forestry equipment; tools; housewares.

Activities: guided tours; steam & gas engine show.

Publications: annual newsletter, Plantation Quarterly; calendar of events.

Hours & Admission Prices: April-Oct. Mon. & Wed.-Fri. 10-3, Sat. 10-5, Sun. 12-5. No charge. Fee for special events. &

Attendance: 8,408 (accurate)

SMITH'S FORT PLANTATION, 217 Smith's Fort Lane, Rte. 31 (halfway between Jamestown Ferry & Surry C.H.), Surry, VA 23883. Mailing Address: Box 240, Surry, VA 23883-0240. Tel.: 757-294-3872. TDD: 757-294-3872.

Web Site: www.apva.org/smithsfort

Founded: 1925.

Congressional District: 4

Key Personnel: C.E.O., Elizabeth Kostelny; Dir., A. Kent Harrell; Mgr., Thomas Forehand.

Personnel Profile: Part-Time Paid 4.

Governing Authority: nonprofit organization. Parent Institution: Association for the Preservation of Virginia Antiquities, Cole Diggs House, 204 W. Franklin St., Richmond, VA 23220-5091. Tel. 804-648-1889. Tax-exempt.

Institution Type/Description: Historic House Museum: 17th-century fort site; earthworks remain of fort started in 1609 by Capt. John Smith; property a dower gift in 1614 from Powhatan to John Rolfe & Pocahontas; 18th-century brick dwelling a Faulcon family property 1754-1835.

Collections: Indian artifacts; 17th- to 18th-century furnishings; formal boxwood garden.

Research Fields: 17th- to 18th-century colonial Virginia history.

Facilities: picnic.

Activities: guided tours.

Hours & Admission Prices: March & Nov. Sat.-Sun. 12-5; April-Oct. Wed.-Sun. 12-5. Adults $8, AAA members $7, senior citizens $6, students $5, groups $4; children under 6 & APVA members no charge.

Attendance: 3,000 (estimated)

Membership: Individual $40; Family $60; Contributor $100; Sponsor $250; Benefactor $500.

SURRY NUCLEAR INFORMATION CENTER, 5570 Hog Island Rd., Surry, VA 23883. Tel.: 757-357-5410. Fax: 757-357-4711.

Web Site: www.dom.com

Institution Type/Description: Science Museum.

Collections: hands-on exhibits on energy, electricity, & nuclear power.

Hours & Admission Prices: Mon.-Fri. 9-4. Closed major holidays.

Surry County

BACON'S CASTLE, 465 Bacon's Castle Trail, Surry County, VA 23883-2213. Mailing Address: 204 W. Franklin St., Richmond, VA 23220. Tel.: 757-357-5976; 804-648-1889. Fax: 804-775-0802.

E-mail: baconscastle@preservationvirginia.org

Web Site: www.preservationvirginia.org/baconscastle

Founded: 1665.

Congressional District: 4

Key Personnel: Exec. Dir., Elizabeth Kostelny; Site Coord., Todd Ballance; Dir. Properties, Louis J. Malon; Assoc. Dir. Museum Operations, Jennifer Hurst-Wender; Controller, Cherly Greenday; Museum Operations Asst., Sienna Fennell.

Personnel Profile: Full-Time Paid 1; Part-Time Paid 7; Interns 3.

Operating Expenses: 85,868

Operating Income: 35,533

Governing Authority: nonprofit corporation. Parent Institution: Preservation Virginia, 204 W. Franklin St., Richmond, VA 23220. Tax-exempt 501(c)(3).

Institution Type/Description: Historic Site & Historic House: 1665 Jacobean manor house used by Nathaniel Bacon's troops as a fortress during the rebellion against Royal Governor William Berkeley in 1676.

Collections: 17th- to 18th-century American & English decorative arts; restored 17th-century garden.

Research Fields: 17th- to 19th-century political, social & economic history; historical archaeology; horticultural history.

Hours & Admission Prices: March & Nov. Sat.-Sun. 12-5. Adults $8, seniors $7, students $5; discounts to AAA & AAM members; children under 6 & members no charge. Closed Independence Day.

Attendance: 4,931 (accurate)

Membership: Teacher & Student $35; Individual $50; Individual Plus One $60; Family $70; Organizational $750.

Sweet Briar

SWEET BRIAR COLLEGE ART COLLECTION AND GALLERIES, (M), Sweet Briar College, Sweet Briar, VA 24595-1115. Mailing Address: Pannell 208, Sweet Briar, VA 24595. Tel.: 434-381-6248.

E-mail: klawson@sbc.edu

Web Site: sbc.edu/art-galleries

Founded: 1901.

Congressional District: 6

Key Personnel: Dir., Karol A. Lawson, Ph.D.; Chm. (V), Molly Sutherland Gwinn; Registrarial Asst., Nancy McDearmon.

Personnel Profile: Full-Time Paid 2; Part-Time Paid 15; Part-Time Volunteers 25; Interns 5.

Governing Authority: college. Parent Institution: Sweet Briar College. Tax-exempt: 501(c)(3).

Institution Type/Description: College Art Gallery: housed in 1906 historic landmark building.

Collections: 13th to 20th-century European & American paintings, drawings & prints; 18th & 19th-century Japanese prints; works by contemporary women.

Research Fields: 13th to 20th-century European & American art.

Activities: lectures; tours; special exhibitions.

Publications: catalogues; annual newsletter of the Friends of Art, Visions.

Hours & Admission Prices: Sept.-May Mon.-Thurs. 10-5, Fri. 10-2, Sun. 1-4; Summer by appointment. No charge; donations accepted. Closed college breaks; reading days; exams. &

Attendance: 2,500 (estimated)

Membership: Friends of Art: Student $10; Regular $25; Family $50; Contributor $100; Sponsor $250; Benefactor $500; Patron $1,000.

SWEET BRIAR MUSEUM, Sweet Briar College, Boxwood Alumnae House, Sweet Briar, VA 24595. Mailing Address: Sweet Briar College, Sweet Briar, VA 24595-1056. Tel.: 434-381-6246.

E-mail: museum@sbc.edu

Web Site: sbc.edu/museum

Founded: 1980.

Congressional District: 6

Key Personnel: Dir., Karol A. Lawson.

Personnel Profile: Full-Time Paid 2; Part-Time Paid 6; Interns 4.

Governing Authority: college; nonprofit organization. Parent Institution: Sweet Briar College. Tax-exempt.

Institution Type/Description: College History Museum: housed in renovated c.1940 college building.

Collections: alumnae memorabilia; founding family items, including 19th-century clothing, furniture, jewelry, silver, lace, decorative arts, letters, photos & farming implements; college memorabilia.

Research Fields: 19th-century decorative arts; history of the college.

Activities: permanent exhibitions; educational programs.
Publications: informational brochure; exhibition guides & checklists; booklets on college history.
Hours & Admission Prices: Academic Year: Tues.-Thurs. 1-4. No charge. Closed college breaks, reading days, exams. &
Attendance: 500 (estimated)

Tangier Island

TANGIER HISTORY MUSEUM & INTERPRETIVE CULTURAL CENTER, 16215 Main Ridge, Tangier Island, VA 23440. Mailing Address: P.O. Box 182, Tangier, VA 23440-0182. Tel.: 757-891-2374.
E-mail: thmicc@aol.com
Web Site: tangierhistorymuseum.org
Founded: 2008.
Congressional District: 2
Key Personnel: Dir., Debra Howard; Pres. (V), Edward V. Parks.
Personnel Profile: Full-Time Volunteers 2; Part-Time Volunteers 12.
Governing Authority: Tax-exempt.
Institution Type/Description: History Museum.
Collections: island history & culture; photographs; period furnishings; personal artifacts.
Facilities: library; nature trails.
Activities: historical slide shows; yoga; movie night for teens; kayak water trails.
Publications: e-newsletter.
Hours & Admission Prices: late April to late Oct. 11-4; other times by appointment. No charge; donations accepted. &
Attendance: 15,000 (accurate)

Tappahannock

ESSEX COUNTY MUSEUM & HISTORICAL SOCIETY, 218 Water Lane, Tappahannock, VA 22560. Mailing Address: P.O. Box 404, Tappahannock, VA 22560-0404. Tel.: 804-443-4690.
E-mail: info@ccmhs.org
Web Site: www.essexmuseum.org
Founded: 1996.
Key Personnel: Pres. (V), Suzanne Derieux; Museum Shop Mgr., Priscilla Vaughan
Institution Type/Description: Historical Society Museum.
Collections: County history & culture; personal artifacts; war memorabilia; photographs; Native American.
Hours & Admission Prices: Mon.-Tues. 10-3, Thurs.-Sat. 10-3. No charge; donations accepted. &
Membership: Individual $25; Family $45; 1692 Society $50; Tappahannock Society $100; Organization & Corporate Sponsorship $150; Essex Circle $500.

Tazewell

HISTORIC CRAB ORCHARD MUSEUM & PIONEER PARK, INC., Rts.19 & 460 at Crab Orchard Rd., 3663 Crab Orchard Rd., Tazewell, VA 24651-9200. Tel.: 276-988-6755. Fax: 276-988-9400. Facebook: Historic Crab Orchard Museum & Pioneer Park Inc.
E-mail: info@craborchardmuseum.com
Web Site: www.craborchardmuseum.com
Founded: 1978.
Congressional District: 9
Key Personnel: C.E.O. & Dir., Charlotte G. Whitted; Chm. (V), Martha Hurst; Dir. Museum Programs, I. Joan Yates; Museum Shop Mgr., Cindy Ringstaff; Cur., Elisabeth Hemsworth.
Personnel Profile: Full-Time Paid 5; Part-Time Paid 5; Part-Time Volunteers 180.
Governing Authority: nonprofit organization. Tax-exempt: 501(c)(3).
Institution Type/Description: Historic Houses & Site: located on Big Crab Orchard Archaeological & Historic Site.
Collections: paleo, archaic, & woodland Indian artifacts; manuscripts; 19th-century log structures; furnishings; farming implements; blacksmith, cobbler, cooper & weavers' tools; clothing; Revolutionary & Civil War military equipment; medical instruments; horse-powered equipment; barn; period hand-woven textiles.
Research Fields: history & genealogy.
Facilities: 800-vol. library pertaining to local & regional history, reference on archaeological & genealogical resources, available for research on premises. Books, crafts & other museum-related items for sale.
Activities: guided tours; lectures; films; gallery talks; holiday festivals; reading room; formally organized educational programs; permanent & temporary exhibitions; living history outreach to school classes. Museum Sponsors:

Skirmish at Jeffersonville Civil War reenactment; Tazewell County Old Time & Bluegrass Fiddler's Convention; Independence Day Celebration; Christmas on the Frontier.
Publications: quarterly newsletter, The Pisgah Pathfinder.
Hours & Admission Prices: Jan.-March Mon.-Fri. 9-5; April-May & Sept.-Dec. Mon.-Sat. 9-5; Memorial Day to Labor Day Mon.-Sat. 9-5, Sun. 1-5. Adults $4, senior citizens over 60 $3, children 6-12 $2; discounts to AAM, AAA, AARP & Time Travelers members; children under 6 & members no charge. Closed New Year's Day; Thanksgiving; Christmas. &
Attendance: 20,000 (accurate)
Membership: Students & Senior Citizens $25; Family $60; Sustaining $100; Patron $500; Benefactor $1,000.

The Plains

AFRO AMERICAN HISTORICAL ASSOCIATION OF FAUQUIER COUNTY, 4243 Loudoun Ave., The Plains, VA 20198-0340. Mailing Address: P.O. Box 340, The Plains, VA 20198-0340. Tel.: 540-253-7488. Fax: 540-253-5126.
E-mail: info@aahafauquier
Web Site: aahafauquier.org
Founded: 1992.
Congressional District: 10
Key Personnel: Pres., Karen Hughes White; Vice Pres., Karen King Lavore.
Personnel Profile: Part-Time Paid 3; Part-Time Volunteers 1; Interns 1.
Governing Authority: bd. of directors. Tax-exempt.
Institution Type/Description: History Museum.
Collections: African American history; personal artifacts; archives.
Publications: semiannual newsletters.
Hours & Admission Prices: Tues.-Wed. 10-3; other times by appointment. No charge; donations accepted. Closed holidays. &
Attendance: 2,188 (accurate)
Membership: Individual $25; Family $40; Nonprofit $50; Corporation $250; Lifetime $500.

Triangle

NATIONAL MUSEUM OF THE MARINE CORPS, 18900 Jefferson Davis Hwy., Triangle, VA 22172-1938. Tel.: 877-653-1775. Fax: 703-432-0029.
E-mail: info@usmcmuseum.org
Web Site: www.usmcmuseum.org
Founded: 2006.
Congressional District: 1
Key Personnel: Dir., Lin Ezell; Mgr. Visitor Svcs., Patrick Mooney; Museum Shop Mgr., Andy Pineau.
Personnel Profile: Full-Time Paid 49; Part-Time Paid 1; Part-Time Volunteers 170; Interns 17.
Governing Authority: Parent Institution: U.S. Marine Corps. Subsidiary Institution: Marine corps Univ. Tax-exempt.
Institution Type/Description: Military Museum.
Collections: military artifacts including 1775-19th century, WWI, WWII, Korean War, Vietnam; current conflicts.
Research Fields: military & US history; Marine Corps history.
Activities: hands-on activities; programs; classes. Museum Sponsors: Monthly Family Day.
Hours & Admission Prices: Daily 9-5. No charge; donations accepted. Closed Christmas. &
Attendance: 534,966 (accurate)

PRINCE WILLIAM FOREST PARK VISITOR CENTER, 18100 Park Headquarters Rd., Triangle, VA 22172-1644. Tel.: 703-221-7181 & 4706. Fax: 703-221-3258. TDD: 703-221-7181.
E-mail: prwi_info@nps.gov
Web Site: www.nps.gov/prwi
Founded: 1936.
Congressional District: 1, 10 & 11
Key Personnel: Asst. Supt., George Liffert; Chief Interpretation, Laura Cohen.
Personnel Profile: Full-Time Paid 1; Part-Time Paid 3; Part-Time Volunteers 3; Interns 1.
Governing Authority: federal. U.S. Dept. of Interior National Park Service. Tax-exempt.
Institution Type/Description: Park Museum.
Collections: zoological, botanical & geological specimens; cultural history artifacts; 2000 BC-1940 AD archeological artifacts; photographs; maps; archival papers.
Research Fields: Civilian Conservation Corps 1933-1942; World War II U.S. Army Office of Strategic Services, 1942-1945; Cabin Branch Pyrite Mine 1889-1920.
Facilities: auditorium. Books for sale.

Activities: organized education programs; interpretive programs & walks; docent program; participatory exhibits.
Publications: park map.
Hours & Admission Prices: March to mid-Dec. daily 9-5; mid-Dec. to March 1 Fri.-Mon. 9-5. Seven Day Pass: $5 per vehicle, $3 per person; children under 16 & permanently disabled or blind persons no charge. Closed New Year's Day; Thanksgiving; Christmas. &
Attendance: 30,000
Membership: Daily & Weekly Pass $5; Annual Pass $20; National Parks Pass $50.

University of Richmond

VIRGINIA BAPTIST HISTORICAL SOCIETY, Boatwright Library, University of Richmond, VA 23173. Mailing Address: P.O. Box 34, University of Richmond, VA 23173. Tel.: 804-289-8434. Fax: 804-289-8953.
Web Site: www.baptistheritage.org
Founded: 1876.
Congressional District: 3
Key Personnel: C.E.O., Fred Anderson; Pres. (V), Frank G. Schwall, Jr.
Personnel Profile: Full-Time Paid 2; Part-Time Paid 2; Part-Time Volunteers 8.
Governing Authority: society. Tax-exempt: 501(c)(3).
Institution Type/Description: Historical Society Museum.
Collections: Virginia Baptist history; University of Richmond archives.
Major Exhibits: Free Indeed! The Trials & Triumphs of Virginia's Enslaved People, 1/14-5/14.
Research Fields: Virginia Baptist history; University of Richmond archives.
Facilities: 20,000-vol. library pertaining to Virginia Baptist history available to the public.
Activities: guided tours; lectures; organized education programs for adults, children, undergraduate & graduate college students; temporary & traveling exhibitions.
Publications: annual journal, Virginia Baptist Register.
Hours & Admission Prices: Mon.-Fri. 9-12 & 1-4:30. No charge; donations accepted. Closed legal holidays.
Attendance: 5,000 (estimated)
Membership: Individual $25; Family $35; Institute $50.

Vienna

MEADOWLARK BOTANICAL GARDENS, 9750 Meadowlark Gardens Ct., Vienna, VA 22182. Tel.: 703-255-3631. Fax: 703-255-2392.
E-mail: ktomlinson@nvrpa.org
Web Site: www.nvrpa.org
Personnel Profile: Full-Time Paid 11; Part-Time Paid 12; Part-Time Volunteers 50; Interns 2.
Governing Authority: Parent Institution: NVRPA.
Institution Type/Description: Botanical Gardens.
Collections: native plants & wildflowers; cherry trees; irises; peonies; daylilies; conifers; salvias. Historic Building: c.1755 log cabin.
Research Fields: native plant conservation.
Facilities: 96 acres; nature trails; visitor center; rental facilities.
Activities: educational programs; workshops; concerts; tours; rental facilities.
Publications: newsletter, Dirca; quarterly e-newsletter.
Hours & Admission Prices: March & Oct. daily 10-6; April & Sept. daily 10-7; May daily 10-7:30; June-Aug. daily 10-8; Nov.-Feb. daily 10-5. Adults $5; discounts to APGA; members no charge. Closed New Year's Day; Thanksgiving; Christmas.
Attendance: 50,000 (accurate)
Membership: Family $35.

Virginia Beach

ADAM THOROUGHGOOD HOUSE, 1636 Parish Rd., Virginia Beach, VA 23455-4401. Mailing Address: Francis Land House, 3131 Virginia Beach Blvd., Virginia Beach, VA 23452-6923. Tel.: 757-460-7588.
E-mail: mreed@vbgov.com
Founded: 1961.
Congressional District: 2
Key Personnel: Admin., Mark A. Reed.
Personnel Profile: Part-Time Paid 8; Part-Time Volunteers 10.
Governing Authority: municipal. Parent Institution: City of Virginia Beach, VA-Dept. of Museums and Cultural Arts, 717 General Booth Blvd., Virginia Beach, VA 23451. Tax-exempt: 501(c)(3).
Institution Type/Description: Historic House: c.1680 Southern modified hall & parlor, one & one-half story brick structure located on the Grand Patent of 1636 which is part of the original land grant given to Adam Thoroughgood.

Collections: late 17th & early 18th century English artifacts including furniture, ceramics & pewter.
Research Fields: local history; American, French & English decorative arts.
Facilities: 17th-century garden. Museum-related items for sale.
Activities: guided tours; lectures; concerts; hobby workshops; organized education programs; docent program.
Publications: brochures.
Hours & Admission Prices: Call for hours.
Attendance: 6,899 (accurate)

ATLANTIC WILDFOWL HERITAGE MUSEUM, 1113 Atlantic Ave., Virginia Beach, VA 23451-3503. Tel.: 757-437-8432. Fax: 757-437-9055.
E-mail: atlanticwildfowl@verizon.net
Web Site: www.awhm.org
Founded: 1995.
Key Personnel: Dir., Thomas P. Beatty; Pres. (V), Ann Verhaagen; Museum Shop Mgr., Ann Smith.
Personnel Profile: Full-Time Paid 2; Part-Time Volunteers 15.
Governing Authority: private; nonprofit. Parent Institution: Back Bay Wildfowl Guild, Inc. Tax-exempt: 501(c)(3).
Institution Type/Description: Wildfowl Museum: located in an 1895 three-story beach house on the Atlantic Ocean; an example of Queen Anne architecture.
Collections: concentration on waterfowl of the Atlantic Flyway; special emphasis on turn-of-the-century decoys; hunting paraphernalia; carving; boatbuilding demonstrations.
Research Fields: ecohistory of Back Bay & Currituck Sound; gunning clubs of the region.
Facilities: library; educational facilities. Museum-related items for sale.
Activities: docent program; formal education programs for adults; guided tours; hobby workshops; lectures; participatory exhibits; school loan service. Annual Events: Back Bay Birding Club.
Hours & Admission Prices: Memorial Day-Oct. 1 Mon.-Sat. 10-5, Sun. 12-5; Oct. 2-May Tues.-Sat. 10-5, Sun. 12-5. No charge; donations accepted. Closed New Year's Day; Thanksgiving; Christmas. &
Attendance: 15,000 (accurate)
Membership: Individual $25; Family $35.

BACK BAY NATIONAL WILDLIFE REFUGE, 1324 Sandbridge Rd., Virginia Beach, VA 23456-4023. Tel.: 757-721-2412. Fax: 757-721-6141.
Institution Type/Description: Wildlife Refuge.
Collections: wildlife & their habitats including ducks, snow geese, loggerhead sea turtles, piping plovers, peregrine falcons, & bald eagles.
Facilities: 9,000 acre refuge; nature trails.
Activities: educational programs.
Hours & Admission Prices: Refuge: daily dawn to dusk. Visitor Center: April-Nov. Mon.-Fri. 8-4, Sat.-Sun. 9-4; Dec.-March Mon.-Fri. 8-4, Sun. 9-4.

CAPE HENRY LIGHTHOUSE, 583 Atlantic Ave., Fort Story, Virginia Beach, VA 23459-1048. Mailing Address: 204 W. Franklin St., Richmond, VA 23220-5012. Tel.: 757-422-9421. Facebook: Cape Henry Lighthouse.
E-mail: capehenry@preservationvirginia.org
Web Site: www.preservationvirginia.org/capehenry
Formerly: Old Cape Henry Lighthouse
Founded: 1791.
Congressional District: 2
Key Personnel: Exec. Dir. APVA, Elizabeth Kostelny; Pres. (V), Anne Geddy Cross, IV; Dir. Properties, Louis J. Malon; Site Coord., Stuart Nesbit; Operations Asst., Senna Fennell; Assoc. Dir. Operations & Education, Jennifer Hurst-Wender.
Personnel Profile: Full-Time Paid 2; Part-Time Paid 8; Part-Time Volunteers 5.
Operating Expenses: 151,939
Operating Income: 220,097
Governing Authority: nonprofit society. Parent Institution: Preservation Virginia, 204 W. Franklin St., Richmond, VA. 23220. Tax-exempt.
Institution Type/Description: Historic Building: 1791 first commissioned public works building in the United States, built near the monument marking the first landing of the Jamestown colonists.
Collections: local history; historic building.
Facilities: Museum-related items for sale.
Publications: Ventures, biannual.
Hours & Admission Prices: March 16-Oct. daily 10-5; Nov.-March 15 daily 10-4. Adults $5, seniors & military $4, children 3-12 $3; members &

children under 3 no charge. Walking Tour: $8. Closed New Year's Eve & Day; Thanksgiving; Christmas Eve, Day & week after.
Attendance: 63,073 (accurate)
Membership: Teacher & Student $35; Individual $50; Individual Plus One $60; Family $70; Organization $75.

FIRST LANDING STATE PARK, 2500 Shore Dr., Virginia Beach, VA 23451-1415. Tel.: 757-412-2300. Fax: 757-412-2315.
E-mail: firstlanding@dcr.virginia.gov
Web Site: www.virginiastateparks.gov
Founded: 1936.
Congressional District: 2
Key Personnel: Park Mgr., Bruce Widener; Asst. Park Mgr., Charlie Whalen; Volunteer Coord., Solans Kennedy.
Personnel Profile: Full-Time Paid 10; Part-Time Paid 40; Part-Time Volunteers 540.
Operating Expenses: 916,000
Operating Income: 1,600,000
Governing Authority: state. Parent Institution: Commonwealth of Virginia. Branch of First Landing State Park. Tel.: 757-412-2300. Tax-exempt.
Institution Type/Description: Park Museum.
Collections: local Indian artifacts; great blue heron; brown pelican; osprey; great horned owl; screech owl; cormorants; Bald Cypress knees exhibit; children's touch table; first landing of English colonists on their way to Jamestown; Chesapeake Bay marine environment.
Facilities: 150-vol. library of reference books available in interpretive services office under supervision of park interpreter; nature & conservation center. Interpretive book sales area.
Activities: guided tours; lectures; films.
Publications: pamphlet, Bald Cypress Nature Trail; Birds of First Landing State Park; Checklists of Species.
Hours & Admission Prices: Daily 8-4:30. No charge. Parking: Mon.-Fri. $4, Sat.-Sun. $5. Annual parking passes available. Closed New Year's Day; Thanksgiving; Christmas. ♿
Attendance: 1,500,000 (estimated)

FRANCIS LAND HOUSE HISTORIC SITE, 3131 Virginia Beach Blvd., Virginia Beach, VA 23452-6923. Tel.: 757-385-5100 & 5104.
E-mail: mreed@vbgov.com
Web Site: www.vbgov.com
Founded: 1986.
Congressional District: 2
Key Personnel: Chm. (V) Admin., Mark Reed; Museum Educator II, Nora Jean Corillo; Museum Educator II, Starr D. Donlon.
Personnel Profile: Full-Time Paid 4; Part-Time Paid 3; Part-Time Volunteers 85.
Governing Authority: municipal. City of Virginia Beach, 717 General Booth Blvd., Virginia Beach, VA 23451. Subsidiary Institution: Dept. of Museums and Cultural Arts. Tax-exempt.
Institution Type/Description: Historic Site & House: late 18th-early 19th century brick plantation home of gentry-class planters, built by later generation of Land family.
Collections: period furnishings.
Research Fields: Land family; 18th century history of Virginia Beach; material culture of gentry class plantation; agricultural methods; slavery & African-American history.
Facilities: meeting rooms; wedding & banquet facilities available; herb garden; 18th century vegetable garden; park & trail. Museum-related items for sale.
Activities: guided tours; lectures; concerts; organized educational programs; docent programs; training programs for professional museum workers; participatory exhibits; textile programs including flax growing & processing. Museum Sponsors: 12th Night interpretive program in January.
Publications: brochure; quarterly newsletter, The Francis Land House.
Hours & Admission Prices: Tues.-Sat. 10-4, Sun. 12-4. Adults $5, senior citizens $4, students 6 & over $3; discounts to groups, AAM & VAM members; members & children under 6 no charge. Closed most major holidays. ♿
Attendance: 13,807 (accurate)
Membership: Individual $15; Family $20; Contributing $25; Sustaining $50; Patron $100; Corporate $250; Life $500.

LYNNHAVEN HOUSE, 4409 Wishart Rd., Virginia Beach, VA 23455. Mailing Address: 4401 Wishart Rd., Virginia Beach, VA 23455-5524. Tel.: 757-460-7109.
E-mail: apva@apva.org
Web Site: www.apva.org
Founded: 1976.

Congressional District: 2
Key Personnel: C.E.O. (APVA), Elizabeth Kostelny; Chm. (V), William Chitty; APVA & Chm. Finance Committee (V), David Sparks; Chm. (V), Rosemary Wilson; Admin., Shirley S. Bueche; Museum Shop Mgr., Peggie Everett.
Personnel Profile: Part-Time Paid 3; Part-Time Volunteers 35.
Governing Authority: nonprofit society. Parent Institution: Association for the Preservation of Virginia Antiquities, 204 W. Franklin St., Richmond, VA 23220-5091. Subsidiary Institution: Southeastern Branch. Tax-exempt.
Institution Type/Description: Historic House: preserved c.1725 brick dwelling, that is an example of early 18th-century eastern Virginia vernacular architecture.
Collections: household furnishings.
Facilities: graveyard; herb garden; colonial garden; 3 flower beds indigenous to the time & area. Museum-related items for sale.
Activities: tours; school programs, on-site & at schools; special school tours; craft & skill demonstrations; music programs; Civil War encampments; 2 colonial experiences-hands on camps for kids; 2 drama camps. Museum Sponsors: Revolutionary encampment in May; medieval days in September; Craft Show in November; Christmas open house in December.
Publications: brochure, Lynnhaven House; docent handbook; schedule of annual special events.
Hours & Admission Prices: Tues.-Sat. 10-4, Sun. 12-4; extended hours for special events. Adults $5, seniors $4, children 6-18 $3; discount to groups & Time Travelers; children under 6 & APVA members no charge.
Attendance: 3,510 (accurate)
Membership: Teacher & Student $25; Assoc. Individual $30; Assoc. Individual Plus One $35; Individual & Assoc. Family $40; Individual Plus One $50; Family $60; Century $100; Sponsor $250; Benefactor $500.

MILITARY AVIATION MUSEUM, 1341 Princess Anne Rd., Virginia Beach, VA 23457-1542. Tel.: 757-721-7767. Fax: 757-497-8083.
Web Site: www.militaryaviationmuseum.org
Founded: 2006.
Congressional District: 2nd
Key Personnel: Dir., Gary Powers; C.E.O., Gerald Yegan; Museum Shop Mgr., Pattie Nelson.
Personnel Profile: Full-Time Paid 4; Part-Time Paid 2.
Governing Authority: Tax-exempt.
Institution Type/Description: Military Aviation Museum.
Collections: World War II warbirds; World War I aircraft replicas.
Research Fields: historical aviation.
Activities: monthly special events; guest speakers; flight demonstrations. Annual Event: Air Show in May & Sept.
Publications: Prop Noise quarterly members newsletter.
Hours & Admission Prices: Daily 9-5. Adults $10; discounts to seniors & active military & groups; WWII veterans & members & children 5 & under no charge. Closed Thanksgiving; Christmas. ♿
Attendance: 50,000 (estimated)
Membership: Individual $50; Family $80.

VIRGINIA AQUARIUM & MARINE SCIENCE CENTER, (M), 717 General Booth Blvd., Virginia Beach, VA 23451-4811. Tel.: 757-385-FISH (24 hour recording) & 385-7777 (office). Fax: 757-437-4976.
E-mail: fish@virginiaaquarium.com
Web Site: www.virginiaaquarium.com
Founded: 1986.
Congressional District: 2
Key Personnel: C.E.O., Lynn Clements; Pres. Virginia Aquarium Foundation, Dorcas Helfant-Browning; Deputy Dir., Stanley Burchfield; Controller, Donna Ellis; Dir. Devel., Russell Turner; Dir. Research & Conservation, Mark Swingle; Dir. Education, Chris Witherspoon; Dir. Retail Operations, Ruth Ann Steenburgh; Dir. Mktg., Linda Candler; Public Rels. Mgr., Joan Barns; Volunteer Resources Mgr., Kathleen Reed.
Personnel Profile: Full-Time Paid 76; Part-Time Paid 45; Part-Time Volunteers 916; Interns 15.
Governing Authority: city; nonprofit organization. Tax-exempt 501(c)(3).
Institution Type/Description: Marine Science Center & Aquarium.
Collections: 800,000 gallons of aquariums; 360 interactive exhibits; 12,000 live animals & their habitats representing over 700 species; two touch pools.
Research Fields: sea turtle conservation; marine animal strandings & rehabilitation.
Facilities: 1/3 mile nature trail; Virginia native gardens; aviary with 30 species; 300-seat 3D IMAX theater; classrooms; 800,000 gallons of aquariums; 88-seat interactive theater; restaurant; traveling ocean in motion truck. Aquarium-related items for sale.
Activities: K-12 curriculum-based school program; daily floor programs

including fish feedings, lectures, field trips & craft programs for adults & children; touch pool; interactive exhibits; ocean collection & coastal explorer pontoon boat trips; traveling ocean in motion truck; whale & dolphin watching; teacher in-services; graduate & undergraduate courses; volunteer program; marine animal stranding program; changing exhibits; harbor seal splash encounter program; harbor seal & sea turtles behind the scenes programs.
Publications: bimonthly members newsletter, SeaBrowser; quarterly volunteer newsletter, Owls Creek Gazette; quarterly newsletter from the Virginia Aquarium Stranding Response Program, Strandlines; marine science information pamphlets; K-12 curriculum packages; educational resource guide; program brochures.
Hours & Admission Prices: Memorial Day to Labor Day daily 9-6; Sept.-May daily 9-5. Aquarium: adults $17, senior citizens 62 & over $16, children 3-11 $12. IMAX Film: adults $8.50, senior citizens $8, children $7.50. Aquarium & IMAX Film: adults $23, senior citizens $22, children 3-11 $18; discounts to groups, active duty military & aquarium members; children 2 & under no charge. Closed Thanksgiving; Christmas. &
Attendance: 622,000 (estimated)
Membership: Otter $75; Crab $100; Hedgehog $150; Stingray $275; Seal $500.

VIRGINIA BEACH MARITIME MUSEUM, INC./THE OLD COAST GUARD STATION, (M), 24th St. & Atlantic Ave., Virginia Beach, VA 23451. Mailing Address: P.O. Box 1035, Virginia Beach, VA 23451-0035. Tel.: 757-422-1587. Fax: 757-491-8609.
E-mail: director@oldcoastguardstation.com
Web Site: www.oldcoastguardstation.com
Founded: 1981.
Congressional District: 2
Key Personnel: Exec. Dir., Kathryn A. Fisher; Pres. Bd. Directors, Jack Drescher; Admin. Dir. & Museum Store Mgr., Leslie Small; Dir. Education, Volunteers & Programs, Darcy J. Nelson.
Personnel Profile: Full-Time Paid 2; Part-Time Paid 4; Part-Time Volunteers 40.
Governing Authority: nonprofit organization. Tax-exempt: 501(c)(3).
Institution Type/Description: Maritime Museum: housed in 1903 former United States Life-Saving & Coast Guard Station.
Collections: ship models; photographs; maritime artifacts; audiovisual programs; uniforms.
Research Fields: shipwrecks off Virginia Coast; personal histories of the Surfmen; history of U.S. Life-saving Service & Coast Guard in Virginia.
Facilities: 500-vol. library pertaining to United States Life-Saving Service & Coast Guard history & related maritime topics available to researchers on premises. Nautical items, books, art, jewelry, children's corner, and nautical reproductions for sale.
Activities: guided tours; lectures; films; organized educational programs; docent program; participatory, loan, temporary & traveling exhibitions; special events; story time; ghost walks.
Publications: quarterly newsletter, The Keeper.
Hours & Admission Prices: Memorial Day to Sept. Mon.-Sat. 10-5, Sun. 12-5; Oct.-May Tues.-Sat. 10-5, Sun. 10-5. Adults $4, senior citizens & military $3, children 6-18 $2; discounts to AAM members; members & children under 6 no charge. Closed New Year's Eve & Day; Thanksgiving; Christmas Eve & Day.
Attendance: 15,000 (estimated)
Membership: Senior Citizen $15; Individual $25; Family $35; Associate $50; Friend $100; Bronze $250; Silver $500; Gold $1,000.

✳ **VIRGINIA MUSEUM OF CONTEMPORARY ART, (M),** 2200 Parks Ave., Virginia Beach, VA 23451-4062. Tel.: 757-425-0000, ext. 0. Fax: 757-425-8186. Facebook: Virginia MOCA.
E-mail: kate@virginiamoca.org
Web Site: www.virginiamoca.org
Formerly: Contemporary Art Center of Virginia
Founded: 1952.
Congressional District: 2
Key Personnel: Dir., Debra C. Gray; Chm. Bd., Dave Durham; Administrative Coord., Imee Wong; Dir. Exhibitions & Education, Alison Byrne; Dir. Facilities Mktg., Irene Tavenner; Dir. Security, Louis Cross; Accounting Assoc., Chris Hurst; Registrar & Preparator, Monee Bengtson; Dir. Operations, Kate Pittman; Dir. Devel., Amy Walton; Mgr. Gallery & Youth Programs, Rebecca Davidson; Assoc. Cur., Heather Hakimzadeh; Devel. Assoc., Kay Barbini; Dir. Art Show, Christie Kelly.
Personnel Profile: Full-Time Paid 17; Part-Time Paid 30; Part-Time Volunteers 445; Interns 6.
Governing Authority: private; nonprofit organization. Tax-exempt: 501(c)(3).
Institution Type/Description: Art Museum, Center & School.
Collections: rotating traveling exhibitions.

Major Exhibits: New Waves 2014, 1/31/14-4/27/14; Vik Muniz: Poetics of Perception (T), 1/31/14-4/27/14; Multiplicity (T), 5/19/14-8/17/14.
Research Fields: contemporary art.
Facilities: classrooms; studios; auditorium.
Activities: lectures; gallery talks; arts festivals; formally organized educational programs; loan, permanent & traveling exhibitions; film festivals; performing arts.
Publications: quarterly newsletter; catalogues.
Hours & Admission Prices: Tues. 10-9, Wed.-Fri. 10-5, Sat.-Sun. 10-4. Adults $7, students 5 & over, seniors, military and AAA members $5; discounts to AAM, ICOM & VAM members; members and children 4 & under no charge. Closed New Year's Day; Martin Luther King Jr. Day; Presidents' Day; Independence Day; Labor Day; Thanksgiving & day after; Christmas Eve & Day. &
Attendance: 517,000 (estimated)
Membership: Student, Senior, Military & Teacher $40; Individual $50; Student, Senior, Military & Teacher Household $55; Standard Household $65; Associate $125; Patron $250; Donor $500; Collector's Circle $1,250; Chairman's Circle $2,500.

Wallops Island

NASA WALLOPS FLIGHT FACILITY VISITOR CENTER, Bldg. J-17, Rte. 175, Wallops Island, VA 23337. Tel.: 757-824-2298.
Institution Type/Description: Science Museum.
Collections: space science; solar system; photographs.
Facilities: theater.
Hours & Admission Prices: March-June & Sept.-Nov. Thurs.-Mon. 10-4; July to Labor Day daily 10-4; Dec.-Feb. Mon.-Fri. 10-4. No charge. Closed New Year's Day; Martin Luther King Jr. Day; Presidents' Day; Columbus Day; Veterans Day; Thanksgiving; Christmas. &

Warm Springs

BATH COUNTY HISTORICAL SOCIETY, 99 Courthouse Hill Rd., Warm Springs, VA 24484. Mailing Address: P.O. Box 212, Warm Springs, VA 24484-0212. Tel.: 540-839-2543. Fax: 540-839-2566. Facebook: Bath County Historical Society.
E-mail: bathcountyhistory@tds.net
Web Site: www.bathcountyhistory.org
Founded: 1969.
Congressional District: 6
Key Personnel: Pres. (V), Richard Armstrong; Vice Pres., Paul Lancaster.
Personnel Profile: Part-Time Paid 3.
Governing Authority: Tax-exempt: 501(c)(3).
Institution Type/Description: Historical Society Museum.
Collections: Bath County history & culture from 1745 to present; photographs; clothing; period artifacts; historic buildings.
Research Fields: genealogy.
Publications: bimonthly newsletter.
Hours & Admission Prices: April-Dec. Wed.-Sat. 10-4; other times by appointment. No charge; donations accepted. &
Attendance: 300 (estimated)
Membership: Individual $50; Donor $100 & up.

Warrenton

MOSBY MUSEUM, (M), 173 Main St., Warrenton, VA 20186. Mailing Address: P.O. Box 3528, Warrenton, VA 20188. Tel.: 540-349-8606. Fax: 540-349-9299.
E-mail: jennifer@partnershipforwarrenton.org
Web Site: mosbymuseum.org
Formerly: Brentmoor: The Spilman-Mosby House
Institution Type/Description: Historic House Museum: housed in the former home of Judge Edward Spilman, Judge James Keith, Colonel John Singleton Mosby, & U.S. Senator Eppa Hunton; 1859-1902. Listed on the National Register of Historic Places.
Collections: local history & culture; personal artifacts; photographs; period furnishings.
Facilities: Museum-related items for sale.
Activities: special events.
Hours & Admission Prices: Mon.-Fri. 9-5.

THE OLD JAIL MUSEUM, 10 Ashby St. Courthouse Sq., Warrenton, VA 20186. Mailing Address: P.O. Box 675, Warrenton, VA 20188-0675. Tel.: 540-347-5525.
E-mail: theoldjailmuseum@yahoo.com
Web Site: www.fauquierhistory.com

Founded: 1964.
Congressional District: 10
Key Personnel: Dir., Frances Allshouse; Pres. (V), Yakir Lubowsky.
Personnel Profile: Full-Time Paid 1; Full-Time Volunteers 2; Part-Time Paid 2; Part-Time Volunteers 20; Interns 2.
Volunteer Hours: 1,000
Operating Expenses: 60,781
Operating Income: 44,305
Governing Authority: society; nonprofit. Parent Institution: Fauquier Historical Society. Tax-exempt.
Institution Type/Description: Local History Museum: housed in two buildings, 1808 jail and 1823 jail, the town's only existing jail until 1966.
Collections: Fauquier County history & Northern Virginia history; local artifacts; Civil War, World War I & II items; documents; photographs; kitchen items.
Major Exhibits: Fauquier County in the War of 1812, 5/14-10/14.
Research Fields: history of Fauquier County jails; Fauquier history.
Facilities: limited library of materials available to the public for use on premises.
Activities: self-guided tours; lectures; docent program; Time Travelers program participant; video tours & interactive virtual tours to visitors with disabilities.
Publications: biannual newsletter, News & Notes; monthly e-news.
Hours & Admission Prices: Wed.-Mon. 10-4; guided tours by appointment. Suggested Donation: $3 per person. Closed New Year's Day; Thanksgiving; Christmas.
Attendance: 12,408 (accurate)
Membership: Student $10; Basic $25; Family $35; Patron & Nonprofit $50; Sponsor $100; Corporate $100.

Warsaw

RICHMOND COUNTY MUSEUM, 5874 Richmond Rd., Warsaw, VA 22572. Mailing Address: P.O. Box 884, Warsaw, VA 22572. Tel.: 804-333-3607.
Institution Type/Description: History Museum: housed in a two-story brick county jail; built in 1872.
Collections: county history & culture; early artifacts; photographs; personal artifacts.
Activities: educational programs.
Hours & Admission Prices: Feb. to mid-Dec. Wed.-Sat. 11-3; other times by appointment. No charge. Closed holidays.

Waverly

MILES B. CARPENTER MUSEUM, 201 Hunter St., Waverly, VA 23890-2631. Mailing Address: 201 Hunter St., P.O. Box 1376, Waverly, VA 23890-1376. Tel.: 804-834-2151 & 3327. Fax: 804-834-3327.
Founded: 1986.
Congressional District: 4
Key Personnel: Financial Dir., Devel. & Mktg., Thelma Wyatt; Cur., Shirley S. Yancey; Sec., Deborah A. Davis
Governing Authority: nonprofit organization. Tax-exempt: 501(c)(3).
Institution Type/Description: Wood Products Museums.
Collections: Miles Carpenter art & memorabilia; history of peanuts; peanut farm machinery; history of area timber operations.
Major Exhibits: Yesteryear, 12/13-1/14.
Research Fields: folk art; peanuts.
Facilities: 185-vol. library; educational facilities; 2,689 sq. ft. exhibit space.
Activities: guided tours; lectures; films; concerts; arts festivals; theatre; formal education programs; docent program; temporary exhibitions of new artists. Annual Events: Cherry Blossom Time in April; Folk Art Festival in May; Peanut Harvest Time in November. ancestral toys December to January.
Publications: Cutting the Mustard; In and Around Waverly, Virginia.
Hours & Admission Prices: Thurs.-Mon. 2-5. No charge; donations accepted. Closed New Year's Day; Easter; Thanksgiving; Christmas.
Attendance: 6,000 (accurate)

Waynesboro

HUMPBACK ROCKS MOUNTAIN FARM & VISITOR CENTER, Blue Ridge Pkwy., Mile Post 5.9, Waynesboro, VA 24483. Mailing Address: 133 Whetstone Ridge Rd., Vesuvius, VA 24483-2113. Tel.: 540-377-2377 (Montebello Ranger Station). Fax: 540-377-6758.
Web Site: www.nps.gov/blri
Founded: 1939.
Congressional District: 6

Key Personnel: District Ranger, Bruce Bytnar; Interpretive Specialist, Randy Sutton; Supt. Blue Ridge Pkwy., Phil Francis; Admin. Tech., Susan Bryant.
Governing Authority: federal. Parent Institution: National Park Service, Dept. of Interior, Blue Ridge Pkwy., 200 Northwestern Bank Bldg., Asheville, NC 28807. Tax-exempt.
Institution Type/Description: Park Museum: 1880-1900 pioneer mountain farm.
Collections: mountain life.
Research Fields: oral history; Southern Appalachian rural life.
Facilities: library pertaining to natural & cultural history. Museum-related items for sale.
Activities: participatory exhibits; living history cultural demonstrations.
Publications: general, natural & cultural history of Southern Appalachians.
Hours & Admission Prices: late April to Nov. 1 daily 10-5. No charge.
Attendance: 150,000

P. BUCKLEY MOSS MUSEUM, 150 P. Buckley Moss Dr., Waynesboro, VA 22980. Tel.: 540-949-6476; 800-343-8643.
E-mail: mossmuseum@aol.com
Web Site: pbuckleymoss.com
Founded: 1989.
Key Personnel: Dir., Corrado Gabellieri; Pres., Jake Henderson; Museum Shop Mgr., Jo Cowherd.
Personnel Profile: Full-Time Paid 3; Part-Time Paid 20.
Governing Authority: Parent Institution: P. Buckley Moss Galleries Ltd.
Institution Type/Description: Art Museum.
Collections: works by P. Buckley Moss.
Hours & Admission Prices: Summer: Mon.-Sat. 10-5, Sun. 12:30-5. Winter: call for hours. No charge. Closed holidays.
Attendance: 19,056 (accurate)

Weems

FOUNDATION FOR HISTORIC CHRIST CHURCH, INC., (M), 420 Christ Church Rd., Weems, VA 22576. Mailing Address: P.O. Box 24, Irvington, VA 22480-0024. Tel.: 804-438-6855. Fax: 804-438-5186.
E-mail: info@christchurch1735.org
Web Site: www.christchurch1735.org
Founded: 1958.
Congressional District: 1
Key Personnel: Exec. Dir., Camille E. Bennett; Pres. (V), Rev. Hugh C. White, III; Office Mgr., Trish Geeson.
Personnel Profile: Full Time Paid 4; Part Time Volunteers 260.
Governing Authority: nonprofit organization. Tax-exempt: 501(c)(3).
Institution Type/Description: Historic Foundation.
Collections: original & transcriptions or copies of original documents, maps, photos & related materials concerning the church & social history of its people in Lancaster County; historic archaeological artifacts & support materials portraying colonial life.
Research Fields: social history of church & its builder Robert King Carter & archaeological research into the site surrounding the church; historic architecture; political & economic history of the colonial county, parish & population.
Facilities: 200-vol. library on Colonial history, genealogy & social life available for use on premises; 16-seat theater; community meeting facility.
Activities: guided tours; docent program; video; programs for school children; permanent exhibitions; continuing education; cultural events.
Publications: books, Historic Structure Report (1994); Christ Church Parish Vestry Book, 1739-1786 & 1832-1869; Christ Church, Lancaster County, Virginia (2001); Robert King Carter, Builder of Christ Church (2001); People in Profile, Christ Church Parish, 1720-1750 (2002); Landholders & Landholdings, Christ Church Parish (2004); activity book, Discover History at Historic Christ Church; Last Will & Testament of Robert Carter; Executors' Letters of Robert Carter.
Hours & Admission Prices: Historic Christ Church: April-Nov. Mon.-Sat. 10-4, Sun. 2-5; Dec.-March Mon.-Fri. 8:30-4:30. Carter Reception Center: April-Nov. Mon.-Sat. 10-4, Sun. 2-5; other times by appointment. Adults $5, senior citizens $4; active military & their families no charge. Closed New Year's Eve & Day; Thanksgiving; Christmas Eve, Day & week.
Attendance: 13,000 (estimated)

West Point

CHELSEA PLANTATION, 874 Chelsea Plantation Lane, West Point, VA 23181. Tel.: 804-843-2386. Fax: 804-843-2386 (call first).
E-mail: donmasterson2@gmail.com
Institution Type/Description: Historic House Museum: built in 1709.
Collections: local history; period furnishings; personal artifacts; photographs.

Hours & Admission Prices: Thurs.-Sun. 10-4:30; other times by appointment. Adults $15.

MATTAPONI INDIAN MUSEUM, 1271 Mattaponi Reservation Cir., West Point, VA 23181. Tel.: 804-769-2229 & 2194.
Key Personnel: Dir., Gertrude Custalow
Institution Type/Description: Native American Museum.
Collections: Mattaponi Indian life, culture & history; period artifacts.
Hours & Admission Prices: Sat.-Sun. 2-5. Admission $2.

Williamsburg

* **ABBY ALDRICH ROCKEFELLER FOLK ART MUSEUM, (M),** 326 W. Francis St., Williamsburg, VA 23185. Mailing Address: P.O. Box 1776, Williamsburg, VA 23187-1776. Tel.: 757-220-7554. Fax: 757-565-8804.
E-mail: museums@cwf.org
Web Site: www.colonialwilliamsburg.com/do/art-museums
Founded: 1957.
Congressional District: 1
Key Personnel: C.E.O. & Chm., Colin Campbell; Vice Pres. Conservation, Collections & Museums, Ronald Hurst; Dir., Richard Hadley; Cur., Barbara Luck; Museum Shop Mgr., Joanna Heitz; Education & Programs, Christina Westenberger; Exhibits Mgr., Jan Gilliam.
Personnel Profile: Full-Time Paid 7; Part-Time Paid 3; Part-Time Volunteers 50.
Governing Authority: nonprofit organization. Parent Institution: Colonial Williamsburg Foundation. Tax-exempt: 501(c)(3) & 170(b)(1)(A).
Institution Type/Description: Art Museum.
Collections: paintings; sculpture; drawings.
Research Fields: American folk art.
Facilities: 242-seat auditorium; cafe.
Activities: live, musical, interactive, family & educational programs; tours; lectures.
Publications: Colonial Williamsburg; The Journal of the Colonial Williamsburg Foundation.
Hours & Admission Prices: Jan.-March daily 10-5; April-Dec. daily 10-7. Museum: adults $9.95, youth 6-12 $4.95; children under age 6 no charge. Annual Museum Pass: adults $19.95; youth 6-12 $9.95. Admission also included in Colonial Williamsburg pass. &
Attendance: 200,000 (accurate)

* **BASSETT HALL, (M),** 522 E. Francis St., Williamsburg, VA 23185-4207. Mailing Address: c/o Colonial Williamsburg Foundation, P.O. Box 1776, Williamsburg, VA 23187-1776. Tel.: 757-220-7453. Fax: 757-220-7173.
E-mail: museums@cwf.org
Web Site: www.colonialwilliamsburg.com/do/art-museums/bassett-hall
Founded: 1979.
Congressional District: 1
Key Personnel: C.E.O. & Chm., Colin Campbell; Vice Pres. Museums & Collections, Ron Hurst; Site Mgr., Cynthya Nothstine.
Personnel Profile: Full-Time Paid 4; Part-Time Paid 3.
Governing Authority: nonprofit organization. Affiliated with the Colonial Williamsburg Foundation, Goodwin Bldg., Williamsburg 23187. Tel.: 804-229-1000. Tax-exempt: 501(c)(3).
Institution Type/Description: Historic House Museum & Grounds: built in c.1753, purchased in 1800 by Burwell Bassett, nephew of Martha Washington, and acquired by the Rockefellers in the 1920s.
Collections: furnishings from the 1930s belonging to Mr. & Mrs. John D. Rockefeller, Jr.; American folk art; ceramics; textiles; decorative arts; garden; nature trails. Historic Structures: 18th-century house, kitchen outbuilding, dairy & smokehouse.
Research Fields: American folk art; decorative arts; history of the property & the individuals associated with it.
Facilities: orientation video; 585-acre site; garden; two self-guided nature trails.
Activities: guided tours; interactive family art programs; character interpreters.
Publications: guidebook, Bassett Hall: The Williamsburg Home of Mr. & Mrs. John D. Rockefeller, Jr.
Hours & Admission Prices: :Wed., Thurs. & Sat. 9:30-4:30. Day ticket: adults $11.95, youth $5.95. Annual Pass: adults $21.95, youth $10.95. Admission is also included with Colonial Williamsburg tickets. &
Attendance: 19,738 (accurate)

* **COLONIAL WILLIAMSBURG, (M),** 134 N. Henry St., Williamsburg, VA 23185-4138. Mailing Address: P.O. Box 1776, Williamsburg, VA 23187-1776. Tel.: 757-229-1000 & 220-7286. Fax: 757-220-7702. TDD: 757-221-8939.
E-mail: rhurst@cwf.org
Web Site: www.colonialwilliamsburg.org
Founded: 1926.
Congressional District: 1
Key Personnel: C.E.O. & Pres., Colin G. Campbell; Pres. Colonial Williamsburg Hospitality Group, John Hallowell; Sr. Vice Pres., Robert Taylor; Vice Pres. Collections & Museums, Ronald L. Hurst; Vice Pres. Research, James Horn; Vice Pres. Devel., Michael Rierson; Vice Pres. Sec. & Gen., John S. Bacon; Vice Pres. Products, John Hallowell.
Personnel Profile: Full-Time Paid 1,893; Part-Time Volunteers 900; Interns 2.
Governing Authority: nonprofit organization. Administered by The Colonial Williamsburg Foundation. Tax-exempt: 501(c)(3).
Institution Type/Description: Art & History Museum District: an on-site preservation of the former capital of Virginia colony.
Collections: archaeology; archives; manuscripts; paintings; sculpture; graphics; decorative arts; costumes; folk art; history; military; music; textiles; transportation. Historic Buildings: over 400 buildings from 1693-1837.
Research Fields: 17th- & 18th-century social, cultural, architectural & decorative arts history of Tidewater, Virginia area.
Facilities: 55,000-vol. library of 17th- & 18th-century Virginia, Colonial & United States history; archival materials available for inter-library loan & research by special arrangement; 450-seat auditorium; theater; classrooms; restaurants. Reproduction & museum-related items for sale.
Activities: guided tours; lectures; films; gallery talks; concerts; drama; TV & radio programs; formally organized education programs for children, adults & undergraduate college students; permanent & temporary exhibitions. Foundation Sponsors: special weekend & monthly activities.
Publications: books; films; filmstrips; videotapes; recordings; CD's.
Hours & Admission Prices: Visitor Center: daily 8:45-5:45. Basic Ticket: adults $36, children 6-14 $18. Freedom Pass (annual): adults $49, children $24.50 (holders eligible for discount on evening programs). DeWitt Wallace Museum or Abby Aldrich Rockefeller Folk Art Museum: adults $9.95, children 6-17 $4.95. Annual Museums Pass to all three museums: adults $19.95, children $9.95; discounts to military, AAM & ICOM members & staff from other museums. &
Attendance: 780,000 (accurate)

* **DEWITT WALLACE DECORATIVE ARTS MUSEUM, (M),** 326 W. Francis St., Williamsburg, VA 23185. Mailing Address: P.O. Box 1776, Williamsburg, VA 23187-1776. Tel.: 757-220-7554. Fax: 757-565-8804.
E-mail: museums@cwf.org
Web Site: www.colonialwilliamsburg.com/do/art-museums
Founded: 1985.
Congressional District: 1
Key Personnel: C.E.O., Colin Campbell; Vice Pres. Conservation, Collections & Museum, Ronald Hurst; Dir., Richard Hadley; Education, Patricia Balderson; Exhibition Planning, Jan Gilliam; Exhibition Conservation, Patty Silence; Registrar, Enice Glosson; Security, Barbara Banks; Programs, Mary Cottrill; Museum Shop Mgr., Joanna Heitz.
Personnel Profile: Full-Time Paid 13; Part-Time Paid 13; Part-Time Volunteers 61; Interns 2.
Governing Authority: private; nonprofit organization. Subsidiary Museum: Colonial Williamsburg Foundation, Williamsburg, VA. Tax-exempt.
Institution Type/Description: Decorative Arts Museum.
Collections: British & American decorative arts from 1600-1830; furniture; metals; ceramics; glass; paintings; prints; maps; textiles.
Research Fields: decorative arts from 1600-1830.
Facilities: 240-seat auditorium; 62-seat cafe; 27,200 sq. ft. exhibit space. Museum-related items for sale.
Activities: musical concerts; docent program; formal education programs; guided tours; lectures; loan, temporary & traveling exhibitions; family programs.
Publications: Colonial Williamsburg; The Journal of the Colonial Williamsburg Foundation; exhibition catalogues, Painters & Paintings in the Early American South.
Hours & Admission Prices: Jan.-March daily 10-5; April-Dec. daily 10-7. Museum: adult $9.95, youth 6-12 $4.95; children under 6 no charge. Annual Museum Pass: adult $19.95, youth 6-12 $9.95. Admission also included in Colonial Williamsburg pass. &
Attendance: 200,000 (accurate)

* **JAMESTOWN SETTLEMENT, YORKTOWN VICTORY CENTER, THE JAMESTOWN-YORKTOWN FOUNDA-TION, (M),** Rte. 31 S., 2110 Jamestown Rd. (GPS), Williamsburg, VA 23185. Mailing Address: P.O. Box 1607, Williamsburg, VA 23187-1607. Tel.: 757-253-4838. Fax: 757-253-5299. TDD: 757-253-5110.
E-mail: laura.bailey@jyf.virginia.gov
Web Site: www.historyisfun.org
Founded: 1957.
Congressional District: 1
Key Personnel: Exec. Dir., Philip G. Emerson; Exec. Asst. to Bd., Laura W. Bailey; Deputy Exec. Dir. Administration, J. Jeffrey Lunsford; Acting Sr. Dir. Museum Operations & Education, James S. Holloway; Sr. Dir. Mktg. & Retail Operations, Susan K. Bak; Dir. Outreach & Special Svcs., Pamela J. Pettengell; Sr. Dir. Devel., Julie W. Basic; Curatorial Svcs. Mgr., Dr. Thomas E. Davidson; Mgr. Human Resources, Patrick O. Teague; Sr. Retail Operations Mgr., Gary T. Joyner.
Personnel Profile: Full-Time Paid 180; Part-Time Paid 250; Part-Time Volunteers 900; Interns 49.
Governing Authority: state. Parent Institution: Commonwealth of Virginia. Subsidiary Institution: Jamestown-Yorktown Foundation, Inc. Tax-exempt.
Institution Type/Description: History Museum.
Collections: 17th- to 18th-century Virginia archaeology; prehistoric Indian artifacts; 16th- to 18th-century English & American artifacts; military, medical objects, navigational instruments, domestic & decorative arts relating to the settling of Jamestown, the War of Independence & siege of Yorktown.
Major Exhibits: Jamestown's Legacy to the American Revolution, 3/12-1/14.
Research Fields: 17th-century Jamestown & Virginia Indian history; colonial history & culture; American Revolutionary War period; history of the town of York & Yorktown's sunken fleet.
Facilities: library; museum galleries & outdoor living history areas including re-creation of James Fort, Powhatan Indian Village, Riverfront Discovery area; 3 ships; Revolutionary War encampment & 1780s farm site; theater & film at each museum. Museum-related items for sale.
Activities: guided tours for school groups; living-history demonstrations; crafts program; special programs; films for schools; outreach programs for schools.
Publications: Jamestown-Yorktown Foundation Facts/Annual Highlights; Education Planner; Group Tour Planner; periodical newsletter, Dispatch; brochures: Special Programs; Jamestown Settlement & Yorktown Victory Center; Jamestown Settlement Ships; Jamestown Settlement & Yorktown Victory Center Museum Guide; pamphlet: 2007 Commemoration High-lights; America's 400th Anniversary-Jamestown 2007 Steering Committee Report; books, Anne of Jamestown; Mukambu of Ndongo; William-Ship's Boy; Winganusk of the Powhatans.
Hours & Admission Prices: Jamestown Settlement: June 15-Aug. 15 daily 9-6; Aug. 16-June 14 daily 9-5. Adults $15.50, children 6-12 $7.25. Yorktown Victory Center: June 15-Aug. 15 daily 9-6; Aug. 16-June 14 daily 9-5. Adults $16, children 6-12 $7.50. Combination: adults $20.50, children $10.25; discounts to groups & AAM members; residents of James City, York Counties, VA, and the city of Williamsburg including College of William and Mary no charge. American Heritage Annual Pass: adults $35, children 6-12 $17.50; children 5 & under no charge. Closed New Year's Day; Christmas. &
Attendance: 648,207 (accurate)

* **MUSCARELLE MUSEUM OF ART, (M),** Lamberson Hall, College of William and Mary, 603 Jamestown Rd., Rm. 1, Williamsburg, VA 23185. Mailing Address: Lamberson Hall, College of William and Mary, P.O. Box 8795, Williamsburg, VA 23187-8795. Tel.: 757-221-2710. Fax: 757-221-2711. Facebook: Muscarelle Museum of Art.
E-mail: museum@wm.edu
Web Site: www.wm.edu/muscarelle
Founded: 1982.
Congressional District: 1
Key Personnel: Chm. Bd. Directors, Janet Osborn; Dir. & C.E.O., Aaron H. De Groft, Ph.D.; Senior Assoc. Dir., Christina M. Carroll, Esq.; Asst. to Dir., Cindy Lucas; Head Collections & Exhibitions Mgmt., Melissa M. Parris; Facilities & Exhibitions Mgr., Kevin Gillian; Special Projects Admin., Ursula McLaughlin-Miller; Chief Cur. & Distinguished Scholar-in-Residence, John T. Spike, Ph.D.; Dir. Security, Larry Wright; Mgr. Institutional Advocate & Annual Giving, Patrick Slebonick.
Personnel Profile: Full-Time Paid 13; Part-Time Paid 3; Part-Time Volunteers 95; Interns 15.
Governing Authority: Parent Institution: College of William and Mary, Williamsburg, VA 23185. Subsidiary Institution: Muscarelle Museum of Art Foundation. Tax-exempt.

Institution Type/Description: Art Museum.
Collections: Colonial to post-modern paintings; expansive works on paper collection including major archives of J.J. Lankes, Hans Grohs & Japanese prints; non-western collection of sculpture including African, Asian & Islamic; native American pottery.
Major Exhibits: Dialogue of the Franciscans: Caravaggio's St. Francis in Meditatiion, 2/8/14-8/3/14; 21st Century Diplomacy: Ballet, Ballots & Bullets, 2/8/14-4/6/14; Curators at Work IV, 4/19/14-8/3/14; Toshi Yoshida and the Art of Japanese Hanga, 10/14-11/14.
Research Fields: 16th- to 20th-century European & American art.
Facilities: permanent & temporary exhibition galleries. Museum-related items for sale.
Activities: lectures; gallery talks; guided tours; docent program; school programs; loan, permanent & temporary exhibitions.
Publications: exhibition catalogs; bulletin; William D. Barnes: Three Decades of Still Life and Landscape; The Tsar's Cabinet: Two Hundred Years of Russian Decorative Arts Under the Romanovs from the Kathleen Durdin Collection of Russian Decorative Arts; Marlene Jack: A Journey in Clay, 1974-2011; Caravaggio's Still Life on a Stone Ledge, c.1603; Michelangelo: Anatomy as Architecture, Drawings by the Master, Sacred and Profane A Brush with Passion (1613-1699).
Hours & Admission Prices: Tues.-Fri. 10-5, Sat.-Sun. 12-4. Special Exhibitions: adults $15; members no charge. Closed all major holidays; New Year's Eve & Day; Christmas Eve, Day & week. &
Attendance: 88,000 (estimated)
Membership: University Membership $40; Affiliate $75; Subscriber $125; Contributor $250; Supporter $500; Patron $1,000; Sustainer $2,500; Benefactor $5,000; Lamberson Circle $10,000; Muscarelle Circle $25,000.

THIS CENTURY ART GALLERY, 219 N. Boundary St., Williamsburg, VA 23185-3610. Mailing Address: P.O. Box 388, Williamsburg, VA 23187-0388. Tel.: 757-229-4949. Fax: 757-258-5624.
E-mail: thiscenturyartgallery@verizon.net
Web Site: www.thiscenturyartgallery.org
Formerly: The Twentieth Century Gallery
Founded: 1959.
Congressional District: 1
Key Personnel: Exec. Dir., Kerry Mellette; Pres. (V), Michael Kirby; Vice Pres., Linda Caviness; Artistic Dir., Apryl Altman; Sec., Sharon Parker; Treas., Greg Spryn; Business Mgr., Charlene Zolad; Office Mgr., Mav Reyes.
Personnel Profile: Part-Time Paid 3; Part-Time Volunteers 100.
Governing Authority: nonprofit organization. Parent Institution: Williamsburg Art Center at paper Mill Creek, Ltd. Tax-exempt: 501(c)(3).
Institution Type/Description: Art Gallery.
Collections: arts & crafts of contemporary artists.
Facilities: Arts & crafts for sale.
Activities: lectures; workshops; organized educational programs; participatory exhibits; monthly exhibits; mobile vans.
Publications: annual brochure; quarterly newsletter.
Hours & Admission Prices: Tues.-Sun. 11-5. No charge; donations accepted. Closed Thanksgiving; Christmas. &
Attendance: 4,800 (estimated)
Membership: Student $10; Individual $40; Family $60; Friends $100-$249; Sustainer $250-$499; Patron $500-$999; Benefactor $1,000 & up.

Winchester

SHENANDOAH VALLEY DISCOVERY MUSEUM, (M), 54 S. Loudoun St., Winchester, VA 22601-4720. Tel.: 540-722-2020. Fax: 540-722-2189. Facebook: Shenandoah Valley Discovery Museum.
E-mail: business@discoverymuseum.net
Web Site: www.discoverymuseum.net
Founded: 1993.
Congressional District: 10
Key Personnel: Exec. Dir., Mary Braun; Chm., Phil Glaize; Dir. Programs, Jan Kirby; Business & Retail Mgr., Pamela Lam; Gallery Mgr., Mark Lawson; Paleontologist, Geb Bennett.
Personnel Profile: Full-Time Paid 5; Part-Time Paid 1; Part-Time Volunteers 10.
Governing Authority: Tax-exempt.
Institution Type/Description: Children's Discovery Museum.
Collections: fossils; dinosaurs; hands-on science, technology, culture, art & humanity exhibits.
Research Fields: paleontology in Hell Creek, Montana.
Activities: Museum Sponsors: Visiting Artist Series.
Publications: quarterly, Bright Ideas; biannual, Friends of Hell Creek.

Hours & Admission Prices: Mon.-Sat. 9-5, Sun. 1-5. Adults $6; discounts to ASTC, ACM, & AAA members; members no charge. Closed Federal holidays. ♿
Attendance: 41,000 (accurate)
Membership: Grandparent $65; Family $125.

WINCHESTER-FREDERICK COUNTY HISTORICAL SOCIETY, INC., 1360 S. Pleasant Valley Rd., Winchester, VA 22601-4447. Mailing Address: 1340 S. Pleasant Valley Rd., Winchester, VA 22601. Tel.: 540-662-6550. Fax: 540-662-6991.
E-mail: wfchs@verizon.net
Web Site: www.winchesterhistory.org
Founded: 1930.
Congressional District: 11
Key Personnel: Dir., Cissy Shull.
Personnel Profile: Full-Time Paid 1; Part-Time Paid 15; Part-Time Volunteers 5.
Volunteer Hours: 100
Governing Authority: society. Tax-exempt: 501(c)(3).
Institution Type/Description: History Museum Complex.
Collections: furniture; artifacts; Civil War relics; manuscripts; early surveying instruments. Historic Houses: 1755-1756 George Washington's Office; 1754 Abrams Delight; 1861-1862 Stonewall Jackson Headquarters.
Major Exhibits: Quilt Exhibit, 4/1/14-10/31/14.
Research Fields: Civil War; Washington 1748-1758.
Facilities: Museum-related items for sale.
Activities: guided tours; meetings. Museum Sponsors: Washington's Birthday Celebration; Candlelight Tour in December.
Publications: books, Men & Events of the Revolution; Civil War Battles in Winchester & Frederick County, VA. 1861-1865; Images of the Past; 1748-1758 George Washington & Winchester; What I Know About Winchester 1800-1891; Winchester-Frederick County Historical Society Journal, Vols. IV-XX; Journal, Vols. III-XXI.
Hours & Admission Prices: April-Oct. daily 10-4, Sun. 12-4. Abrams's Delight & Stonewall Jackson's Headquarters: adults $5, senior citizens $4.50, children $2.50. George Washington's Office: adults $5, children $1.75. Block Tickets: adults $12; members & children under 6 no charge.
Attendance: 15,000 (estimated)
Membership: Individual $30; Couple $40; Patron $100; Life Individual $500; Life Couple $700.

Wise

WISE COUNTY HISTORICAL SOCIETY, Wise County Courthouse, Rm. 250, Wise, VA 24293. Mailing Address: P.O. Box 368, Wise, VA 24293-0368. Tel.: 276-328-6451 & 6569.
E-mail: wchs_133@yahoo.com
Web Site: www.wisevahistoricalsoc.org
Founded: 1993.
Congressional District: 9
Key Personnel: Pres. (V), William C. Gobble; Vice Pres., Denver J. Osborne; Treas., Wanda Rose; Archivist, Fannie Steele; Museum Shop Mgr., Bill Porter.
Governing Authority: Tax-exempt.
Institution Type/Description: Historical Society.
Collections: over 5,000 items including books, manuscripts, old marriage records & photographs.
Publications: history books; photo books.
Hours & Admission Prices: Mon.-Thurs. 9-4, Fri. 9-12. No charge; donations accepted.
Attendance: 12 (estimated)
Membership: Individual $10.

Woodstock

WOODSTOCK MUSEUM OF SHENANDOAH COUNTY, INC., 104 S. Muhlenberg St., Woodstock, VA 22664. Mailing Address: P.O. Box 741, Woodstock, VA 22664-0741. Tel.: 540-459-5518.
E-mail: info@woodstockmuseumva.org
Web Site: woodstockmuseumva.org
Founded: 1969.
Congressional District: 7
Key Personnel: Pres., Jean Martin.
Personnel Profile: Part-Time Volunteers 20.
Governing Authority: nonprofit organization. Tax-exempt: 501(c)(3).
Institution Type/Description: Local History Museum.

Collections: local history; manuscripts.
Research Fields: genealogy.
Facilities: archives room.
Activities: guided tours; lectures; permanent exhibitions.
Publications: books, Yesterday in Woodstock According to Fred Painter; Glimpses of the Past in the Shenandoah Valley; Early Woodstock.
Hours & Admission Prices: May-Oct. Thurs.-Sat. 12-4. No charge; donations accepted. ♿
Attendance: 400 (estimated)
Membership: Single $25; Family $40; Business Supporter $50; Business Sponsor $100; Life $250.

Wytheville

THE EDITH BOLLING WILSON BIRTHPLACE, (M), 145 E. Main St., Wytheville, VA 24382-2319. Tel.: 276-223-3484 & 228-8474. Fax: 276-228-5987.
E-mail: info@edithbollingwilson.org
Web Site: www.edithbollingwilson.org
Founded: 2006.
Key Personnel: Dir., Leslie King; Chm. (V), William Smith.
Governing Authority: Parent Institution: The Edith Bolling Wilson Birthplace Foundation.
Institution Type/Description: Historic House Museum: housed in the birthplace of First Lady Edith Bolling Wilson, the second wife of the 28th President of the United States, Woodrow Wilson.
Collections: life & history of Edith Bolling Wilson, her family & President Wilson; furniture; books; letters; paintings; photographs; personal artifacts; presidential artifacts.
Activities: Annual Events: Mother's Day Tea in May; Edith Bolling Wilson Birthday Celebration in October.
Publications: newsletter; recipe booklet, Tea Time.
Hours & Admission Prices: Tues.-Sat. 10-5. Season House Tour: adults $5, seniors $4, students $2. Closed Thanksgiving; Christmas. ♿
Membership: Friend $25.

HALLER - GIBBONEY ROCK HOUSE MUSEUM, 205 Tazwell St., Wytheville, VA 24382-2313. Mailing Address: 115 W. Spiller St., Wytheville, VA 24382. Tel.: 276-223-3330. Fax: 276-223-3455. Facebook: Wytheville Museums.
E-mail: museum@wytheville.org
Web Site: museums.wytheville.org/museums.htm
Founded: 1970.
Congressional District: 9
Key Personnel: Museum Dir., Frances Emerson.
Personnel Profile: Full-Time Paid 3; Part-Time Paid 3; Part-Time Volunteers 30.
Governing Authority: nonprofit. Parent Institution: City of Wytheville Tax-exempt.
Institution Type/Description: Historic House: 1823 home of Dr. John Haller, the first resident physician of Wytheville.
Collections: furniture; artifacts; memorabilia of Wytheville.
Research Fields: genealogy.
Facilities: library.
Activities: guided tours. Museum Sponsors: Christmas Open House.
Publications: quarterly newsletter, Wythe County Historical Review; brochure.
Hours & Admission Prices: Tues.-Fri. 10-5 (last tour 3:30), 3rd Sat. each month 10-5 (last tour 3:30). Adults $4, children 6-12 $2.
Attendance: 3,000 (estimated)

THE THOMAS J. BOYD, 295 Tazwell St., Wytheville, VA 24382. Mailing Address: 115 W. Spiller St., Wytheville, VA 24382. Tel.: 276-223-3330. Fax: 276-223-3315. Facebook: Wytheville Museum.
E-mail: museum@wytheville.org
Web Site: museums.wytheville.org
Key Personnel: Museum Dir., Frances Emerson.
Personnel Profile: Full-Time Paid 3; Part-Time Paid 3; Part-Time Volunteers 30.
Governing Authority: municipal. Town of Wytheville.
Institution Type/Description: General History Museum.
Collections: children's artifacts; town's 1st fire truck; farming & mining equipment; Civil War artifacts; polio 1950.
Hours & Admission Prices: Tues.-Fri. 10-5, 3rd Sat. each month 10-5, last tour 3:30. Adults $4, children 6-12 $2.
Attendance: 3,000 (estimated)

Yorktown

COLONIAL NATIONAL HISTORICAL PARK: JAMESTOWN & YORKTOWN, Colonial Pkwy. & Rte. 238, Yorktown, VA 23690. Mailing Address: P.O. Box 210, Yorktown, VA 23690. Tel.: 757-898-2416. Fax: 757-898-6346.
E-mail: colo_interpretation@nps.gov
Web Site: www.nps.gov/colo/
Founded: 1930.
Congressional District: 1
Key Personnel: Supt., P. Daniel Smith; Chief Historian, Karen Rehm; Cur. Jamestown, Melanie Pereira; Cur. Yorktown, David Riggs; Museum Shop Mgr., Brenda Cummins.
Personnel Profile: Full-Time Paid 78; Part-Time Paid 3; Part-Time Volunteers 150; Interns 4.
Governing Authority: federal. Parent Institution: National Park Service; U.S. Dept. of Interior, Washington DC 20240. Tax-exempt.
Institution Type/Description: Historical Park: 1607-1781, first permanent English settlement at Jamestown Island & last major battle of American Revolution at Yorktown, VA.
Collections: 17th & 18th-century arms & artifacts on display at the Jamestown Museum & Yorktown Visitor Center; 18th-century Moore House & Nelson House.
Research Fields: 1607-1781 colonial American history.
Facilities: library of 17th & 18th century Virginia history; visitor centers; theaters. Books & museum-related items for sale.
Activities: guided tours; permanent exhibitions; informational & interpretive programs. Museum Sponsors: glass blowing demonstration at Glasshouse.
Publications: interpretive folders; handbooks, Jamestown & Yorktown.
Hours & Admission Prices: Yorktown: daily 9-5. Jamestown: daily 8:30-4:30. Yorktown & Jamestown: adult $10; children 16 & under no charge. Senior Interagency, Access Interagency & America The Beautiful passes honored. Closed New Year's Day; Thanksgiving; Christmas. &
Attendance: 3,100,000 (estimated)

GALLERY ON THE YORK - YORKTOWN ARTS FOUNDATION, 7907 George Washington Memorial Hwy., Yorktown, VA 23692-4857. Mailing Address: P.O. Box 657, Yorktown, VA 23690. Tel.: 757-898-3076.
Web Site: www.galleryontheyork.com
Formerly: On the Hill Cultural Arts Center
Founded: 1976.
Key Personnel: Chm. & Pres. (V), Gary Hess; Gallery Mgr., Brian Lobarr; Museum Shop Mgr., Helen C. Hughes.
Personnel Profile: Full-Time Volunteers 1; Part-Time Volunteers 25.
Governing Authority: nonprofit organization. Parent Institution: Yorktown Arts Foundation. Tax-exempt: 501(c)(3).
Institution Type/Description: Art Center & Gallery.
Collections: original art & crafts by Virginia & North Carolina artists; pottery; crafts.
Facilities: educational facilities.
Activities: guided tours; lectures; films; concerts; arts festivals; organized educational programs; participatory & temporary exhibitions. Museum Sponsors: four juried art shows; Small Works; Fall Family Festival; photographic show, Aperture.
Publications: quarterly newsletter; program brochure.
Hours & Admission Prices: Tues.-Sat. 10-5, Sun. 1-4. No charge; donations accepted. Closed New Year's Day; Mother's Day; Thanksgiving; Christmas. &
Attendance: 8,000 (estimated)
Membership: Patrons & Nonparticipating Artists $35; Participating Artist $50.

WATERMEN'S MUSEUM, 309 Water St., Yorktown, VA 23690. Mailing Address: P.O. Box 519, Yorktown, VA 23690-0519. Tel.: 757-887-2641. Fax: 757-888-2089.
E-mail: admin@watermens.hrcoxmail.com
Web Site: watermens.org
Founded: 1980.
Congressional District: 1
Key Personnel: Pres., John Hanna; Mng. Dir., David Niebuhr; Vice Pres., Dick Lane; Sec., Jim Smith; Treas., Fred Malvin; Founder, Marian Hornsby Bowditch; Coord. Education, Kathryn Hanna; Public Rels., Jim Baumgardner; Asst. to Dir., Linda Myers; Museum Shop Mgr., Joan Karafa; Gift Shop Mgr., Tina McManus.
Personnel Profile: Full-Time Paid 1; Part-Time Paid 5; Part-Time Volunteers 100.
Governing Authority: nonprofit organization. Tax-exempt: 501(c)(3).
Institution Type/Description: Seafood Industry Museum: housed in 1935 Colonial Revival house & outbuildings.
Collections: Watermen who fish the Chesapeake Bay, their work & history.
Research Fields: Virginia watermen's tools & experiences.
Facilities: children's museum, Minnows & Mates; outdoor wharf; pier; carriage house.
Activities: guided tours; lectures; arts festivals; organized educational programs; docent program; participatory & temporary exhibitions. Museum Sponsors: On-The-Water education program.
Publications: quarterly newsletter, The Ship's Log.
Hours & Admission Prices: April to late Nov. Tues.-Sat. 10-5, Sun. 1-5; late Nov. to March Sat. 10-5, Sun. 1-5. Adults $4, students $1; discounts for pre-arranged groups of 20 or more; members no charge. &
Attendance: 7,572 (accurate)
Membership: Deckhand $30; Mates $50; Patent Tonger $125; Captain $250; Marian J Crew $500; Marian Hornsby Bowditch Society $1,000.

YORK COUNTY HISTORICAL MUSEUM, 301 Main St., Yorktown, VA 23692-5431. Mailing Address: P.O. Box 2431, Yorktown, VA 23692-5431. Tel.: 757-890-3508.
E-mail: meredith@yorkcounty.gov
Founded: 2011.
Congressional District: 1
Key Personnel: Pres. (V), Jim Funk.
Personnel Profile: Part-Time Volunteers 10.
Governing Authority: Tax-exempt.
Institution Type/Description: History Museum.
Collections: local history & culture; Native American tools; Revolutionary & Civil War artifacts; USS Yorktown.
Publications: newsletter.
Hours & Admission Prices: Tues.-Sun. 1-3:30. No charge.
Attendance: 3,000 (accurate)
Membership: Friends $15.

WASHINGTON

(264 listings)

Aberdeen

ABERDEEN MUSEUM OF HISTORY, 111 E. Third St., Aberdeen, WA 98520-4002. Tel.: 360-533-1976.
E-mail: museum@aberdeen-museum.org
Web Site: www.aberdeen-museum.org
Key Personnel: Dir. & Cur., Dann Sears
Institution Type/Description: History Museum.
Collections: native cultures & artifacts; Pacific Northwest maritime documents; Grays Harbor history; WA state history; blacksmith shop; period fire engine; general store; photographs.
Facilities: Museum-related items for sale.
Hours & Admission Prices: Tues.-Sat. 10-5, Sun. 12-4. Suggested Donations: family $5, adults $2, students & seniors $1.
Membership: Student/Senior $10; Individuial $15; Family $30; Organization $50; Lifetime $150.

Anacortes

ANACORTES MUSEUM & MARITIME HERITAGE CENTER, (M), 1305 8th, Anacortes, WA 98221-1833. Tel.: 360-293-1915.
E-mail: coa.museum@cityofanacortes.org
Web Site: museum.cityofanacortes.org
Founded: 1957.
Key Personnel: Dir., Steve Oakley; Administrative Asst., Elaine Walker; Educator, Bret Lunsford; Cur. Collections, Judy Hakins; Museum Shop Mgr., Pam Bagnall.
Personnel Profile: Full-Time Paid 1; Part-Time Paid 8; Part-Time Volunteers 3.
Governing Authority: municipal. Parent Institution: City of Anacortes. Tax-exempt: 501(c)(3).
Institution Type/Description: History Museum: housed in 1910 Carnegie Library Building.
Collections: clothing of early Anacortes; pictures; photos; tools; ledgers; school books; military; furniture; glass; medical & dental instruments; musical instruments; artifacts, photos, documents related to history of Anacortes, Fidalgo & Guemes Islands; records of early Salmon Cannery. Historic Ship: W.T. Preston, 1939 sternwheel snagboat, National Historic Landmark.
Major Exhibits: We're Still Here: The Survival of Washington Indians (T), 1/14-4/14; All in the Same Boat: The Great Depression in Anacortes, 5/14-5/15.
Research Fields: history of Anacortes, & Fidalgo & Guemes Islands.
Facilities: research library. Museum-related items for sale.
Publications: books, Fidalgo Fishing; Exploration of Whidbey, Fidalgo and Guemes Islands; From Logs to Lumber; The Geology of Fidalgo Island.

Hours & Admission Prices: Gallery: Tues.-Sat. 10-4, Sun. 1-4. No charge; donations accepted. Vessel: April-May & Sept.-Oct. Sat. 10-4, Sun. 11-4; June-Aug. Tues.-Sat. 10-4, Sun. 11-4. Adults $3, seniors 65 & over $2, children 8-16 $1; children under 8 no charge. Closed New Year's Day; Easter; Thanksgiving; Christmas Eve & Day. &

Attendance: 5,278 (estimated)

Membership: Student & Senior $10; Individual $20; Family $35; Business & Organization $50; Sponsor $250.

Anderson Island

ANDERSON ISLAND HISTORICAL SOCIETY, 9306 Otso Point Rd., Anderson Island, WA 98303-9653.

Web Site: www.andersonislandhs.com

Founded: 1975.

Congressional District: 9

Key Personnel: Pres. (V), Ed Stephenson; Museum Shop Mgr., Jeanne Ditmore

Governing Authority: bd. of directors. Tax-exempt.

Institution Type/Description: Historical Society Museum.

Collections: local history & culture; personal artifacts; furnishings; quilts.

Publications: newsletter.

Hours & Admission Prices: Call for hours. No charge; donations accepted.

Attendance: 5,000 (estimated)

Membership: Household $15; Supporting $30; Sustaining $100.

Arlington

STILLAGUAMISH VALLEY PIONEER MUSEUM, 20722 67th Ave., N.E., Arlington, WA 98223. Tel.: 360-435-7289.

E-mail: stillypioneers@frontier.com

Web Site: www.stillymuseum.org

Founded: 1997.

Key Personnel: Pres. (V), Myrtle Rausch.

Governing Authority: Parent Institution: Stillaguamish Valley Pioneers. Tax-exempt.

Institution Type/Description: History Museum.

Collections: local history & culture; photographs; personal artifacts; period furnishings; logging; dairy industry; military; railroad; sports; medical; education; music.

Activities: special events.

Hours & Admission Prices: March-Oct. Wed. & Sat.-Sun. 1-4; other times by appointment. Adults $5, children 12 & under $2. Closed Easter; Mother's Day; Father's Day; Independence Day. &

Attendance: 1,000 (estimated)

Membership: Individual $10; Lifetime $100.

Ashford

MT. RAINIER NATIONAL PARK, 55210 238th Ave. E., Ashford, WA 98304. Tel.: 360-569-6784. Fax: 360-569-2170.

E-mail: brooke_childrey@nps.gov

Web Site: www.nps.gov/mora/

Congressional District: 8

Key Personnel: Supt., Dave Uberuaga.

Personnel Profile: Full-Time Paid 1; Part-Time Paid 6; Part-Time Volunteers 4.

Governing Authority: federal. Parent Institution: U.S. Dept. of Interior, National Park Service, Washington, DC 20240.

Institution Type/Description: Natural History Museum.

Collections: zoology; geology; botany; history; Indian artifacts. Historic Building: 1888 Longmire cabin.

Research Fields: geology; zoology; hydrology; glaciology; botany; history; park management; artifacts concerning Mount Rainier National Park.

Facilities: 3,000-vol. library of books on natural history, conservation, ecology, history of park; bibliography park archives.

Activities: guided tours; lectures; films; interpretive programs for children & adults; permanent exhibits.

Publications: park newspaper & guide to seasonal activities & facilities, Tahoma; park-related publications.

Hours & Admission Prices: Longmire Museum: June-Sept. daily 9-5; Oct.-May 9-4. Paradise Visitor Center: May-Oct. daily 9-6; Oct.-April Sat.-Sun. 10-5. Ohanapecosh Visitor Center: Memorial Day to mid-Oct. daily 9-6. Sunrise Visitor Center: July to early Sept. daily 9-6. Park: $20 per automobile, $5 per person, good for seven days. Annual Pass: $30. &

Attendance: 2,000,000 (accurate)

Auburn

NEELY MANSION, 12303 Auburn-Black Diamond Rd., Auburn, WA 98092. Mailing Address: P.O. Box 738, Auburn, WA 98071-0738. Tel.: 253-833-9404. Facebook: Neely Mansion Association.

Web Site: www.neelymansion.org

Founded: 1983.

Congressional District: 8

Governing Authority: Tax-exempt: 501(c)(3).

Institution Type/Description: Historic House Museum: built in 1894 by Aaron & David Neely.

Collections: period furnishings; personal artifacts; Japanese bathhouse.

Facilities: rental facilities.

Activities: public & school tours; Spring & Christmas Teas.

Publications: newsletter.

Hours & Admission Prices: Summer: Sat. 1-4; other times by appointment. No charge; donations requested. &

Membership: Student/Senior $15; Individual $20; Family $30; Nonprofit $40; Business $100.

WHITE RIVER VALLEY MUSEUM, 918 H St., S.E., Auburn Community Campus, Auburn, WA 98002-6112. Tel.: 253-288-7433. Fax: 253-931-3098.

Web Site: www.wrvmuseum.org

Founded: 1959.

Congressional District: 8

Key Personnel: C.E.O., Patricia Cosgrove; Chm. (V), Ronnie Beyersdorf.

Personnel Profile: Full-Time Paid 2; Part-Time Paid 6; Part-Time Volunteers 80.

Governing Authority: partnership between municipal government & private nonprofit organization. Tax-exempt: 501(c)(3).

Institution Type/Description: Historical Museum.

Collections: interactive exhibits; life of Native Americans & settlers from 1855-1920; history of Auburn, Washington with emphasis on railroading & farming. Historic Building: Mary Olson Farm, c.1879.

Research Fields: Japanese-American regional history; railroading social history.

Facilities: 1,300-vol. library of rare & local history books, scrapbooks & railroad manuals; 4,500 sq. ft. exhibit space; 50-seat auditorium; botanical garden. Museum-related items for sale.

Activities: docent program; guided tours; school loan service; loan, participatory, traveling & temporary exhibitions.

Publications: quarterly newsletter, White River Journal.

Hours & Admission Prices: Jan. 5-Dec. 20 Wed.-Sun. 12-4 by appointment for group tours & research. Adults $2, seniors & children $1; Wed. no charge. &

Attendance: 15,000 (accurate)

Membership: Steward $10; Brakeman $35; Switchman $50; Engineer $100; Conductor $250; Train Master $500.

Bainbridge Island

BAINBRIDGE ARTS AND CRAFTS GALLERY, 151 Winslow Way E., Bainbridge Island, WA 98110. Tel.: 206-842-3132. Fax: 206-780-8149.

E-mail: gallery@bacart.org

Web Site: www.bacart.org

Founded: 1948.

Key Personnel: Dir., Susan Jackson.

Personnel Profile: Full-Time Paid 2; Part-Time Paid 11; Part-Time Volunteers 37.

Governing Authority: trustees. Tax-exempt.

Institution Type/Description: Art Gallery.

Collections: works by local & regional artists.

Activities: special events; demonstrations; lectures; workshops.

Hours & Admission Prices: Mon.-Sat. 10-6, Sun. 11-5.

Attendance: 41,500 (accurate)

BAINBRIDGE ISLAND HISTORICAL MUSEUM, (M), 215 Ericksen Ave., N.E., Bainbridge Island, WA 98110-1855. Tel.: 206-842-2773. Fax: 206-842-0914.

E-mail: info@bainbridgehistory.org

Web Site: www.bainbridgehistory.org

Founded: 1948.

Congressional District: 6

Personnel Profile: Full-Time Paid 4; Part-Time Volunteers 100.

Governing Authority: nonprofit organization. Parent Institution: Bainbridge Island Historical Society. Tax-exempt: 501(c)(3).

Institution Type/Description: Historical Society Museum.

Collections: Bainbridge Island history & culture; photographs.
Research Fields: Japanese American internment.
Facilities: library. Museum-related items for sale.
Activities: school programs; research; special events.
Publications: Let It Go Louie; Picture Bainbridge; Port Blakely.
Hours & Admission Prices: Daily 10-4. Adults $4, seniors $3; members no charge. Closed New Year's Day; Easter; Thanksgiving; Christmas. &
Attendance: 11,000 (accurate)
Membership: Senior $25; Adult $40.

THE BLOEDEL RESERVE, 7571 N.E. Dolphin Dr., Bainbridge Island, WA 98110-1097. Tel.: 206-842-7631. Fax: 206-842-8970.
E-mail: email@bloedelreserve.org
Web Site: www.bloedelreserve.org
Key Personnel: Exec. Dir., Ed Moydell
Institution Type/Description: Wildlife Refuge.
Collections: native birds & plants.
Hours & Admission Prices: Tues.-Sun. 10-4. Adults $13, seniors 65 & over $9, children 13 & over and college students $5; discounts to groups; children 12 & under no charge. Closed New Year's Day; Thanksgiving; Christmas.

Bellevue

BELLEVUE ARTS MUSEUM, (M), 510 Bellevue Way, N.E., Bellevue, WA 98004-5014. Tel.: 425-519-0770. Fax: 425-637-1799. Facebook: Bellevue Arts Museum.
E-mail: info@bellevuearts.org
Web Site: www.bellevuearts.org
Founded: 1975.
Congressional District: 8
Key Personnel: Interim Dir., Linda Pawson; Dir. Strategic Mktg. & Engagement, Elizabeth Martin-Calder; Docent Pres., Lynne Allison; Controller, Jodi Scharlock; Dir., Curatorial Affairs & Artistic Dir., Stefano Catalani; Museum Shop Mgr., Nancy Whittaker.
Personnel Profile: Full-Time Paid 16; Part-Time Paid 10; Part-Time Volunteers 312; Interns 10.
Governing Authority: nonprofit organization. Tax-exempt: 501(c)(3).
Institution Type/Description: Art Museum.
Collections: contemporary art; craft; design.
Major Exhibits: A World of Paper - Isabelle de Borchgrave Meets Mariano Fortuny (T), 11/13-2/16/14; Crafting a Continuum: Rethinking the Contemporary Craft Field (T), 1/30/14-4/27/14; At Your Service (T), 2/14/14-9/21/14; Kathy Venter: Lfe (T), 3/7/14-6/15/14; Fragile Fortress: The Art of Dan Webb, 3/7/14-6/15/14; Folding Paper: The Infinite Possibilities of Origami (T), 5/16/14-9/21/14; The Art of Gaman: Arts & Crafts from the Japanese American Internment Camps 1942-1946 (T), 7/3/14-10/12/14; Nick Mount: The Fabric of Work (T), 10/10/14-1/25/15; Jason Walker: On the River, Down the Road, 10/10/14-3/1/15; BAM Biennial 2014: Knock on Wood, 10/31/14-3/29/15.
Research Fields: related to exhibitions.
Facilities: art studio.
Activities: guided tours; lectures; children's education program; docent program; volunteer program; temporary & traveling exhibitions. Museum Sponsors: arts fair; kids fair.
Publications: catalogues, Michael Peterson: Evolution/Revolution; The Book Borrowers; Tip Toland: Melt, The Figure in Clay; A Tapestry of Memories: The Art of Dinh Q. Le; Dim Sum at the On-On Tea Room: The Jewelry of Ron Ho.
Hours & Admission Prices: Tues.-Sun. 11-5. Family $25, adults $10, students & senior citizens $7; discounts to AAM members; first Fri. 11-8, children under 6 & members no charge. Closed New Year's Day; Martin Luther King Jr, Day; Easter; Independence Day, Labor Day; Thanksgiving; Christmas. &
Attendance: 50,000 (accurate)
Membership: Student, Teacher, Artist & Senior $40; Individual $55; Dual Senior $65; Dual $75; Family $90; Mixmedia $100; Patron $200; Sustainer $275; Benefactor $550; Bronze Circle $1,000; Silver Circle $2,500; Gold Circle $5,000; Platinum Circle $10,000.

THE BELLEVUE BOTANICAL GARDEN, 12001 Main St., Bellevue, WA 98005-3522. Tel.: 425-452-2750.
E-mail: nkartes@bellevuewa.gov
Web Site: www.bellevuebotanical.org
Founded: 1992.
Congressional District: 8
Governing Authority: city. Parent Institution: City of Bellevue. Subsidiary Institution: Bellevue Botanical Garden Society. Tax-exempt.
Institution Type/Description: Botanical Garden.
Collections: 53 acres of display gardens, woodlands, meadows & wetlands.

Publications: newsletter.
Hours & Admission Prices: Garden: daily dawn to dusk. Visitor Center: daily 9-4. No charge; donations accepted. &
Attendance: 300,000 (estimated)
Membership: Individual $30; Family $45.

EASTSIDE HERITAGE CENTER, (M), 2102 Bellevue Way, S.E., Bellevue, WA 98004. Mailing Address: P.O. Box 40535, Bellevue, WA 98015-4535. Tel.: 425-450-1049. Fax: 425-450-1050.
E-mail: director@eastsideheritagecenter.org
Web Site: eastsideheritagecenter.org
Founded: 1965.
Congressional District: 1
Key Personnel: Bd. Pres., Ross McIvor; Dir., Heather Trescases; Education Coord., Jane Morton; Archivist, Megan Carlisle.
Personnel Profile: Full-Time Paid 1; Part-Time Paid 3; Part-Time Volunteers 15; Interns 2.
Governing Authority: nonprofit organization. Tax-exempt: 501(c)(3).
Institution Type/Description: Historical Society Museum.
Collections: artifacts & archival material of the region dating from the 1850s; manuscripts; lumbering; costumes; room furnishings; fixtures; photographs.
Research Fields: local history.
Facilities: 200-vol. library of general reading books, school books, school records & photographs available for research. Museum Sponsors: annual strawberry festival. Books for sale.
Activities: guided tours; outreach loan program of historic photos & artifacts; permanent & temporary exhibitions; outreach loan program of slides & videos and hands-on loan boxes.
Publications: quarterly newsletter; books: Our Town Redmond, Willowmoor, The Story of Marymoor Park; Collected Memoirs of the Central School, Kirkland; Bellevue: Its First 100 Years; Eastside Historic Coloring Book; Bellevue Timeline; Lake Washington: The East Side; Culinary History of a Pacific Northwest Town: Bellevue, Washington; Generations: Kemper Freeman, Jr. and the Freeman Family.
Hours & Admission Prices: Winters House at 2102 Bellevue Way S.E. open Mon.-Fri. 10-4, Sat. 10-2. No charge; donations accepted. &
Attendance: 10,040 (accurate)
Membership: Student $15; Individual $25; Family $40; Contributor & Organization $100; Sponsor & Steward $250; Benefactor $1,000.

KIDSQUEST CHILDREN'S MUSEUM, 4091 Factoria Mall, S.E., Bellevue, WA 98006-6125. Tel.: 425-637-8100. Fax: 425-747-7178. Facebook: Kidsquest Children's Museum.
E-mail: info@kidsquestmuseum.org
Web Site: www.kidsquestmuseum.org
Formerly: iQuest Children's Museum
Founded: 1997.
Congressional District: 9
Key Personnel: Exec. Dir., Putter Bert; Co Chm. (V), Mike Hubbard; Co Chm. (V), Stacy Graven; Dir. Education & Exhibits, John Ito; Dir. Advancement, Shelley Saunders; Museum Shop Mgr., Alison Lewis.
Personnel Profile: Full-Time Paid 13; Part-Time Paid 16; Part-Time Volunteers 35.
Governing Authority: private; nonprofit organization. Tax-exempt: 501(c)(3).
Institution Type/Description: Children's Museum.
Collections: hands-on exploration of the arts, sciences, culture & life experiences.
Research Fields: Early Childhood Learning.
Facilities: 6,000 sq. ft. exhibit space.
Activities: education programs; workshops; day camps; science clubs.
Publications: newsletter, KidsQuest Quarterly.
Hours & Admission Prices: Call for hours. Adults $9; discounts to ACM, ASTC & NWAYM members; members no charge. &
Attendance: 151,000 (accurate)
Membership: Family Fun $95; Family Fun Deluxe $125; Family Fun Patron $250.

Bellingham

BELLINGHAM RAILWAY MUSEUM, 1320 Commercial St., Bellingham, WA 98225. Tel.: 360-393-7540.
Web Site: www.bellinghamrailwaymuseum.com
Founded: 2003.
Key Personnel: Dir., Fred Dodds; Pres. (V), John D. Stephens
Institution Type/Description: Railway Museum.
Collections: railway history; railroad artifacts; lanterns; period railroad equipment; tools; photographs; Lionel trains.
Hours & Admission Prices: Tues. & Thurs.-Sat. 12-5; other times by

appointment. Adults $4, children 2-16 $1; Fri. & children under 2 no charge. Closed Thanksgiving; Christmas. &
Attendance: 7,000

HERITAGE FLIGHT MUSEUM, 4165 Mitchell Way, Bellingham, WA 98226-. Tel.: 360-733-4422. Fax: 360-733-4423.
E-mail: admin@heritageflight.org
Web Site: www.heritageflight.org
Institution Type/Description: Military Aircraft Museum.
Collections: period military aircraft.
Activities: special events & programs.
Hours & Admission Prices: Call for hours.

SPARK MUSEUM OF ELECTRICAL INVENTION, 1312 Bay St., Bellingham, WA 98225-4322. Tel.: 360-738-3886. Fax: 360-733-2532.
E-mail: tana@sparkmuseum.org
Web Site: www.sparkmuseum.org
Formerly: American Museum of Radio and Electricity
Founded: 1996.
Congressional District: 2
Key Personnel: Dir., Tana Granack; C.E.O., John D. Jenkins.
Personnel Profile: Full-Time Paid 1; Full-Time Volunteers 3; Part-Time Paid 3; Part-Time Volunteers 20; Interns 1.
Governing Authority: private; nonprofit organization. Tax-exempt: 501(c)(3).
Institution Type/Description: Science Museum.
Collections: history of electricity from 1600s-1950s; radio, telegraph equipment; electrical apparatus; books; analog radio content; static electric machines; early television; over 10,000 vacuum tubes.
Facilities: 25,000 sq. ft. exhibit space. Museum-related items for sale.
Activities: hands-on exhibitions & presentations; lightning demonstrations; permanent & temporary exhibits; workshops; education programs; lectures; performances.
Publications: newsletter, Currents.
Hours & Admission Prices: Wed.-Sun. 11-5; other times by appointment. Adults $5, children 12 & under $2; members no charge. Closed national holidays. &
Attendance: 12,000 (estimated)
Membership: Individual $35; Family $80; Contributing $125; Supporting $250; Sustaining $500; Associate $1,000.

VIKING UNION GALLERY, 516 High St., Western Washington University, Bellingham, WA 98225-5946. Tel.: 360-650-6534. Fax: 360-650-7736.
E-mail: asp.vu.gallery@wwu.edu
Web Site: gallery.as.wwu.edu
Formerly: Viking Union Satellite Gallery
Founded: 1950.
Congressional District: 2
Key Personnel: Dir. & Coord., Allie Paul.
Personnel Profile: Part-Time Paid 1; Part-Time Volunteers 1.
Governing Authority: university. Western Washington University. Parent Institution: Associated Students of WWU. Tax-exempt.
Institution Type/Description: University Art Gallery.
Collections: student art collection of paintings, drawings & lithographs.
Activities: lectures; gallery talks; temporary exhibitions; visiting artists.
Hours & Admission Prices: Mon.-Fri. 11-5. No charge. Closed national holidays. &
Attendance: 10,000 (estimated)

WESTERN GALLERY, WESTERN WASHINGTON UNIVERSITY, (M), Fine Arts Complex, Bellingham, WA 98225-9068. Tel.: 360-650-3900 & 3963. Fax: 360-650-6878.
E-mail: sarah.clarklangager@wwu.edu
Web Site: www.westerngallery.wwu.edu
Founded: 1950.
Congressional District: 40
Key Personnel: Dir. & C.E.O., Sarah Clark-Langager, Ph.D.; Museum Preservation Specialist II, Paul Brower.
Personnel Profile: Full-Time Paid 1; Part-Time Paid 1; Interns 10.
Governing Authority: university. Parent Institution: Western Washington University. Tax-exempt: 501(c)(3).
Institution Type/Description: University Art Gallery.
Collections: 20th century prints & drawings; study collection of chairs by 20th century international designers; outdoor sculpture.
Research Fields: contemporary art & sculpture.
Facilities: 4,500 sq. ft. exhibit gallery.
Activities: films; formal education programs for undergraduate & graduate

students; guided tours; temporary, loan & traveling exhibitions; lectures; gallery talks; arts festivals.
Publications: exhibition catalogues if organized by Western Gallery; audiophone tour & catalogues of Outdoor Sculpture Collection; brochures.
Hours & Admission Prices: Academic Year: Mon.-Tues. & Thurs.-Fri. 10-4, Wed. 10-8, Sat. 12-4. No charge; donations accepted. &
Attendance: 55,000 (estimated)
Membership: Friends of Gallery: Individual $25; Family $50; Supporting $100; Sponsor $500; Patron $1,000 & up.

* **WHATCOM MUSEUM, (M),** 121 Prospect St., Bellingham, WA 98225-4497. Tel.: 360-778-8930. Fax: 360-778-8931.
E-mail: museuminfo@cob.org
Web Site: www.whatcommuseum.org
Formerly: Whatcom Museum of History and Art
Founded: 1940.
Congressional District: 2
Key Personnel: Dir., Patricia Leach; Cur., Barbara Matilsky; Educator, Mary Jo Maute; Photo Research, Jeff Jewell; Designer, Scott Wallin; Accountant, Judy Frost; Educator, Chris Brewer; Educator, Carrie Brooks; Collections Mgr., Rebecca Hutchins.
Personnel Profile: Full-Time Paid 20; Part-Time Paid 23; Part-Time Volunteers 300; Interns 10.
Governing Authority: municipal. Parent Institution: City of Bellingham. Subsidiary Institution: Whatcom Museum Society. Tax-exempt: 501(c)(3).
Institution Type/Description: Art, History & Children's Museum: housed in three buildings including the 1892 Old City Hall.
Collections: Northwest coast Indian ethnology; American art; Darius Kinsey photo collection; Wilbur Sandison photo collection; H.C. Hanson naval architecture; American Folk Art & Americana.
Major Exhibits: Treasures from the Trunk: The JJ Donavan Story (T), 10/13-5/14; Vanishing Ice: Alpine & Polar Landscapes in Art, 1775-2012, 11/13-3/14; Jordan Schnitzer Collection, 4/14-8/14; Reaching Beyond: NW Designer, 9/14-12/14.
Research Fields: historical on a regional basis; photography; contemporary art.
Facilities: photo archives with research & reproduction services. Museum-related items for sale.
Activities: permanent & temporary exhibitions; guided tours; lectures; national & international art & history tours; films; gallery talks; concerts; formally organized education programs for children; inter-museum loan; museum outreach program.
Publications: books, The Tree Project; John Cole Paintings; Dale Gottlieb; An Enduring Legacy: Women Painters of Washington, 1930-2005; Northwest Designer Craftsmen at 50; annual reports; The Thompson Language, Stylized Characters' Speech in Thompson Salish Narrative; A Grammar of Bella Coola; American Indian Linguistics and Ethnography in Honor of Laurence C. Thompson; Salish Etymological Dictionary; Cowlitz Dictionary and Grammatical Sketch.
Hours & Admission Prices: F16: Wed.-Sat. 10-5, Sun. 12-5; Lightcatcher: Wed., Fri. & Sun 12-5, Thurs. 12-8, Sat. 10-5; Old City Hall: Thurs.-Sun. 12-5. Admission $10, students, military & seniors $8, children under 5 $4.50; members no charge. Closed most holidays. &
Attendance: 90,000 (estimated)
Membership: Student, Senior, Educator & Military $30; Individual $50; Dual Senior, Dual Student & Dual Military $55; Grandfamily $65; Friends & Family $75; Smithsonian Affiliate $150; Contributor $250; Patron $500.

Beverly

WANAPUM HERITAGE CENTER, 15655 Wanapum Village Lane, S.W., Beverly, WA 99321-9705. Mailing Address: P.O. Box 878, Ephrata, WA 98823-0878. Tel.: 509-754-5088, ext. 2571. Fax: 509-766-2522 & 5020.
E-mail: wanapum@grantpud.org
Web Site: www.wanapum.org
Formerly: Wanapum Dam Heritage Center
Founded: 1966.
Key Personnel: Museum Dir., Angela Buck.
Personnel Profile: Full-Time Paid 3; Part-Time Paid 2.
Governing Authority: board of commissioners. Parent Institution: Public Utility District of Grant County. Tax-exempt.
Institution Type/Description: General Museum.
Collections: Native American artifacts; dioramas; archaeology; photos.
Research Fields: local history & archaeology.
Activities: guided tours.
Hours & Admission Prices: Mon.-Fri. 8:30-4:30, Sat.-Sun. 9-5. No charge. &
Attendance: 13,780 (accurate)

Bingen

WEST KLICKITAT COUNTY HISTORICAL SOCIETY - GORGE HERITAGE MUSEUM, 202 E. Humboldt, Bingen, WA 98605. Mailing Address: P.O. Box 394, Bingen, WA 98605-0394. Tel.: 509-493-3228.
E-mail: ghm@gorge.net
Web Site: community.gorge.net/ghmuseum/About.htm
Founded: 1984.
Congressional District: 4
Key Personnel: Pres. (V), Barbara Sexton.
Personnel Profile: Part-Time Paid 1; Part-Time Volunteers 20.
Governing Authority: Tax-exempt.
Institution Type/Description: Historic Building Museum: housed in the former Bingen Congregational Church; c.1912.
Collections: local history & culture; Native American artifacts; clothing; household items; tools; medical & surgical equipment; documents; newspapers.
Hours & Admission Prices: May-Sept. Fri.-Sun. 12-5; other times by appointment. Adults 16 & over $5; members no charge.
Attendance: 250 (estimated)
Membership: Individual $20; Family $35; Sustaining $100; Lifetime $1,000.

Black Diamond

BLACK DIAMOND MUSEUM & HISTORICAL SOCIETY, 32626 Railroad Ave., Black Diamond, WA 98010. Mailing Address: P.O. Box 232, Black Diamond, WA 98010-0232. Tel.: 360-886-2142.
E-mail: museum@blackdiamondmuseum.org
Web Site: www.blackdiamondmuseum.org
Founded: 1975.
Key Personnel: Pres., Keith Watson; Vice Pres., Ken Jensen.
Governing Authority: nonprofit organization. Tax Exempt: 501 (c)(3)
Institution Type/Description: Historic Building Museum: housed in the former train depot.
Collections: local history; period artifacts; photographs.
Hours & Admission Prices: Summer: Thurs. 9-4, Sat.-Sun. 12-4; Winter: Thurs. 9-4, Sat.-Sun. 12-3.

Bremerton

KITSAP COUNTY HISTORICAL SOCIETY MUSEUM, (M), 280 4th St., Bremerton, WA 98337-1813. Tel.: 360-479-6226. Fax: 360-415-9294.
E-mail: info@kitsaphistory.org
Web Site: www.kitsaphistory.org
Founded: 1948.
Congressional District: 6
Key Personnel: Dir., Patricia Drolet; Pres. (V), Scott Nelson; Vice. Pres., John Sledd; Exec. Asst., Juahela Leiter.
Personnel Profile: Full-Time Paid 1; Full-Time Volunteers 0; Part-Time Paid 3; Part-Time Volunteers 40.
Operating Expenses: 74,841
Operating Income: 145,500
Governing Authority: society. Parent Institution: Kitsap County Historical Society. Tax-exempt: 501(c)(3).
Institution Type/Description: History Museum: housed in a former bank building.
Collections: artifacts; archival papers; documents; photographs of Kitsap history.
Research Fields: Kitsap County; Native Americans; government housing projects; county records; county government & services; Edward & Asahel Curtis; Navy.
Facilities: library of Northwest history books & diaries available for use on premises; archives available for use by appointment. Books pertaining to local area for sale.
Activities: guided tours; lectures; permanent & temporary exhibitions; workshops; special heritage events.
Publications: local history books; newsletter.
Hours & Admission Prices: Tues.-Sat. 10-4, 1st Fri. Artwalk 5-8. Suggested Donations: adults $4, seniors 65 & up, military with ID & children 6-17 $3; discounts to AAM, AASLH & WMA members; children under 6 & members no charge. Closed New Year's Day; Independence Day; Thanksgiving; Christmas. &
Attendance: 6,000 (accurate)
Membership: Individual Senior & Student with ID $20; Individual $30; Senior Family $40; Family $50; Patron $100.

---

*   **PUGET SOUND NAVY MUSEUM, (M),** 251 First St., Bremerton, WA 98337-5612. Tel.: 360-627-2270. Fax: 360-627-2273.
E-mail: lindy.dosher@navy.mil
Web Site: www.pugetsoundnavymuseum.org
Formerly: Bremerton Naval Museum
Founded: 2008.
Congressional District: 23
Key Personnel: Dir., Lindy Dosher; Deputy Dir., Danelle Feddes; Cur., Andrea Lewis; Educator, Carolyn Lane; Mgr. Collections, Kathrine Young.
Personnel Profile: Full-Time Paid 4; Part-Time Volunteers 50; Interns 5.
Governing Authority: federal. Parent Institution: Naval History and Heritage Command.
Institution Type/Description: Naval History Museum.
Collections: Puget Sound naval history from 1840s to present; Puget Sound Naval Shipyard; paintings; photographs.
Research Fields: Puget Sound; US Navy; Puget Sound Naval Shipyard; Pacific Northwest Navy.
Facilities: library; theater. Gift items for sale.
Activities: self-guided tours; permanent & temporary exhibitions; public programming; family discovery room; dry dock theater.
Hours & Admission Prices: May-Sept. Mon.-Sat. 10-4, Sun. 1-4; Oct.-April Mon. & Wed.-Sat. 10-4, Sun. 1-4. No charge; donations accepted. Closed New Year's Day; Easter; Thanksgiving; Christmas. &
Attendance: 35,000 (accurate)
Membership: Student $10; Individual $25; Family & Club $35; Business $50; Life $500-$5,000.

USS TURNER JOY MUSEUM SHIP, 300 Washington Beach Ave., Bremerton, WA 98337-5668. Tel.: 360-792-2457. Fax: 360-377-1020.
E-mail: dd951@sinclair.net
Web Site: ussturnerjoy.org
Institution Type/Description: Military Ship Museum: housed on a Navy destroyer from the Vietnam War.
Collections: Navy & ship history; Navy men & women memorial; military artifacts; photographs; personal artifacts.
Facilities: Museum-related items for sale.
Activities: overnight program; guided tours.
Hours & Admission Prices: March to late Oct. daily 10-5; Winter: Wed.-Sun. 10-3:30 (weathering permitting). Adults $12, seniors 62 & over $10, children 5-12 $7; children under 5 no charge. Closed New Year's Day; Easter; Thanksgiving; Christmas.

Brewster

FORT OKANOGAN INTERPRETIVE CENTER, 14379 Hwy. 17, Brewster, WA 98812. Mailing Address: Alta Lake State Park, 1 B Otto Rd., Pateros, WA 98846-9618. Tel.: 509-689-6665. Fax: 509-923-2980.
Web Site: www.parks.wa.gov
Founded: 1962.
Key Personnel: Dir., Rex Derr; Park Mgr., Sharon Soelter.
Personnel Profile: Full-Time Volunteers 2; Part-Time Volunteers 2.
Governing Authority: state. Parent Institution: Alta Lake State Park. Affiliated with the Washington State Parks and Recreation Commission, 7150 Cleanwater Lane, Olympia, WA 98504. Tax-exempt.
Institution Type/Description: Historic Building & Site: history of fur trade & Indian Interpretive Center.
Collections: Indian & pioneer items: basketry; weapons.
Activities: permanent exhibitions.
Publications: books.
Hours & Admission Prices: May 14-Aug. Wed.-Sun. 9-5; other times by appointment only. No charge; donations accepted. &
Attendance: 9,000 (estimated)

Buckley

FOOTHILLS HISTORICAL SOCIETY & MUSEUM, 128 River Ave., Buckley, WA 98321. Mailing Address: P.O. Box 530, Buckley, WA 98321-0530. Tel.: 360-829-1291. Fax: 360-829-2107.
E-mail: foothillsmuseum@live.com
Founded: 1980.
Congressional District: 31
Key Personnel: Dir., Martha Olsen; Pres. (V), Nancy Stratton
Institution Type/Description: Historical Society Museum.
Collections: local history & culture; photographs; period artifacts.
Hours & Admission Prices: Tues.-Thurs. 12-4, Sun. 1-4. No charge.
Attendance: 130
Membership: Single $15; Family $25.

Burlington

CHILDREN'S MUSEUM OF SKAGIT COUNTY, 550 Cascade Mall Dr., Burlington, WA 98233. Tel.: 360-757-8888.
Founded: 2008.
Key Personnel: Exec. Dir., Cate Melcher
Institution Type/Description: Children's Museum.
Collections: hands-on exhibitions.
Activities: educational programs; special events.
Hours & Admission Prices: Mon.-Sat. 10-6, Sun. 12-6. Toddler Tuesdays: 8:30am-10am. Admission $5 per person; children under one no charge. Closed Easter; Memorial Day; Independence Day; Thanksgiving; Christmas.

Cashmere

CHELAN COUNTY HISTORICAL SOCIETY - CASHMERE MUSEUM & PIONEER VILLAGE, 600 Cotlets Way, Cashmere, WA 98815-1602. Mailing Address: P.O. Box 22, Cashmere, WA 98815-0022. Tel.: 509-782-3230. Fax: 509-782-3219.
E-mail: info@cashmeremuseum.org
Web Site: cashmeremuseum.org
Founded: 1956.
Congressional District: 8
Key Personnel: Mgr., Patricia Lynd; Mgr. Bldg. & Grounds, Fred Harvey.
Personnel Profile: Part-Time Paid 3; Part-Time Volunteers 54.
Governing Authority: nonprofit organization. Parent Institution: Chelan County Historical Society. Tax-exempt: 501(c)(3).
Institution Type/Description: History Museum & Village.
Collections: history of Central Washington from earliest inhabitants to present; Native American artifacts; natural history; Tonk paintings; memorial plaques; pioneer artifacts; Graham paintings; mineralogy; Great Northern wooden caboose; 1921 Toro Model B dumptruck; 1891 waterwheel. Historic Buildings: 1872 Horan cabin; 1889 log school; 1856-1863 log mission; 1879 Assay office; 1889 blacksmith shop; 1888 Richardson Cabin; 1872 post office; 1890 doctor & dentist office; 1886 Buckhorn Saloon; geology; 1891 Weythman Cabin; 1896 general store; 1895 barber shop; 1898 Mission Hotel; c.1885 millinery shop; c.1885 jail; 1895 depot; 1896 print shop.
Research Fields: archaeology.
Facilities: picnic pavilion; gardens. Books, crafts & other items for sale.
Activities: guided tours & lectures by appointment; permanent & temporary exhibits; treasure hunt for children. Museum Sponsors: Apple Days in October.
Publications: quarterly news bulletin, Cashmere Museum News.
Hours & Admission Prices: March-Nov. 1. daily 10:30-4:30. Adults $5.50, seniors 62 & over and students 13 & up $4.50; discounts to military & school tour groups; children under 6 & members no charge. Closed Easter. &
Attendance: 11,000 (accurate)
Membership: Junior & Senior $20; Individual $25; Family $40; Village Friend $75; Pioneer $125; Curator $250; Village Builder $500; Museum Builder $1,000.

Castle Rock

MOUNT ST. HELENS VISITOR CENTER AT SILVER LAKE, 3029 Spirit Lake Hwy., Castle Rock, WA 98611-8706. Tel.: 360-274-0962.
Institution Type/Description: Visitor Center.
Collections: Mount St. Helen's history.
Hours & Admission Prices: Summer: daily 9-5. Adults $5, youth 7-17 $2.50. Closed New Year's Day; Thanksgiving; Christmas.

Cathlamet

WAHKIAKUM COUNTY HISTORICAL SOCIETY MUSEUM, 65 River St. & Division, Cathlamet, WA 98612. Mailing Address: P.O. Box 541, Cathlamet, WA 98612-0541. Tel.: 360-846-1604. Wahkiakum History.
E-mail: wahkiakumhistory@gmail.com
Founded: 1958.
Congressional District: 19B, 10A
Key Personnel: Chm. (V) & Pres. (V), Judith Brawn; Cur., Kari Kandoll.
Personnel Profile: Part-Time Volunteers 10.
Governing Authority: society. Tax-exempt.
Institution Type/Description: Natural History Museum & Historic Site: housed in the former homestead of Judge William Strong, the first Territorial Judge in the Oregon Territory.

Collections: history of the Pacific Northwest; pioneer items of the lower Columbia River; Indian artifacts; early logging; early Americana; fishing.
Activities: guided tours; formally organized education programs for children.
Publications: biannual newsletter.
Hours & Admission Prices: May-Oct. Sat.-Sun. 1-4; other times by appointment. Adults $5, seniors $3; children & members no charge. &
Attendance: 1,000 (estimated)
Membership: Individual $20; Couple $30; Corporate $50.

Chehalis

LEWIS COUNTY HISTORICAL MUSEUM, 599 N.W. Front Way, Chehalis, WA 98532-2048. Tel.: 360-748-0831. Fax: 360-740-5646.
E-mail: director@lewiscountymuseum.org
Web Site: www.lewiscountymuseum.org
Founded: 1965.
Congressional District: 3
Key Personnel: Pres., Peter Lahmann; Dir. & Museum Shop Mgr., Steven (Andy) Skinner.
Personnel Profile: Full-Time Paid 1; Full-Time Volunteers 2; Part-Time Paid 1; Part-Time Volunteers 10.
Governing Authority: county; society. Parent Institution: Lewis Co. Historical Society. Tax-exempt: 170(b)(1)(A) & 501(c)(3).
Institution Type/Description: Local History Museum.
Collections: general county historical artifacts; Lewis County history archives; over 20,000 indexed photographs; Chehalis Indian artifacts; manuscripts; 400 oral history cassette tapes.
Research Fields: genealogy; cemeteries; Chehalis Indians; schools & churches; pioneer's histories; obituary file of county residents; Cowlitz Indians.
Facilities: 1,600-vol. research library on local & state history; Indian archive collection; census & cemetery records.
Activities: guided tours.
Publications: Postmarked Washington; Lewis & Cowlitz Counties; Territorial Marriages; 1871 Census; quarterly, The Historian; books, 1989 Centennial Cookbook; Lewis County Pictorial History.
Hours & Admission Prices: Tues.-Fri. 10-4, Sat. 10-2. Adults $5; members no charge. Closed major holidays. &
Attendance: 5,836 (accurate)
Membership: Student & Senior $18; Individual $25; Family $35; Sustaining $50; Supporting $150; Fellow $250; Patron $500.

VETERANS MEMORIAL MUSEUM, 100 S. W. Veterans Way, Chehalis, WA 98532-1100. Tel.: 360-740-8875.
E-mail: vmm@compprime.com
Web Site: www.veteransmuseum.org
Founded: 1995.
Congressional District: 20
Key Personnel: Dir., Lee T. Grimes; Pres. (V), Ernest Graichen.
Personnel Profile: Full-Time Paid 2; Part-Time Volunteers 35.
Institution Type/Description: Military Museum.
Collections: Armed Forces veterans; uniforms; personal artifacts; photographs; military equipment.
Facilities: banquet facilities.
Activities: special events; rental facilities.
Publications: quarterly, Veterans Museum News.
Hours & Admission Prices: Call for hours. Adults $5, children 6-18 $3; members no charge. &
Attendance: 20,000 (accurate)
Membership: Students $10; Veterans & Seniors $20; Adults $25; Business & Organization $100.

Chelan

CHELAN MUSEUM, 204 E. Woodin Ave., Chelan, WA 98816. Mailing Address: P.O. Box 1948, Chelan, WA 98816. Tel.: 509-682-5644.
E-mail: museum@chelanmuseum.com
Web Site: chelanmuseum.com
Founded: 1970.
Key Personnel: Dir., Samantha Lagge.
Governing Authority: Parent Institution: Lake Chelan Historical Society. Tax-exempt: 501(c)(3).
Institution Type/Description: Historical Society Museum: housed in the former Miners & Merchants Bank; built in 1907.
Collections: local history; period furnishings; personal artifacts; photographs.
Activities: rental facilities.
Publications: quarterly newsletter, Lake Chelan Historical Society History Notes.

Hours & Admission Prices: Winter: Mon.-Fri. 10-12 & 1-4; Summer: Mon.-Fri. 10-4, Sat. 10-3; other times by appointment. Adults $2; members no charge.
Attendance: 4,750 (accurate)
Membership: Family $20; Individual & Business Life $150.

Chinook

FORT COLUMBIA HOUSE MUSEUM, Fort Columbia State Park, Hwy. 101, Chinook, WA 98614. Mailing Address: P.O. Box 488, Ilwaco, WA 98624-0488. Tel.: 360-777-8221. Fax: 360-642-4216.
E-mail: lcic@parks.wa.gov
Founded: 1954.
Congressional District: 5
Governing Authority: Administered by Washington State Parks & Recreation Commission, 7150 Clean Water Lane, Olympia, WA 98504.
Institution Type/Description: Historic House: 1902 Commanding Officer's house at Fort Columbia State Park.
Collections: period furniture & appliances.
Hours & Admission Prices: Summer: daily 6:30 a.m.-9:30 p.m. Winter: daily 8-5. No charge; donations appreciated.

Clarkston

VALLEY ART CENTER, INC., 842-6th St., Clarkston, WA 99403-2013. Tel.: 509-758-8331. Facebook: Valley Art Center.
E-mail: artcenter@cableone.net
Founded: 1968.
Congressional District: 9
Key Personnel: Pres. Bd. Dirs., Paul Fuson; Museum Shop Mgr., H. Craig Whitcomb.
Personnel Profile: Full-Time Volunteers 1; Part-Time Volunteers 15.
Governing Authority: nonprofit organization. Tax-exempt: 501(c)(3).
Institution Type/Description: Art Center.
Collections: historical & Indian art; graphic art; sculpture.
Major Exhibits: The Art of Eve Rockwell, 1/14-2/14.
Facilities: library of art books & publications available for inter-library loan.
Activities: guided tours; lectures; gallery talks; arts festivals; hobby workshops; formally organized education programs for children & adults; temporary exhibitions; traveling community art show; affiliated with Walla Walla, Washington, Community College for accredited art classes; antique shows; mobile art workshop. Annual Events: Pacific Northwest Heritage Art Exhibit; summer art show; benefit gallery.
Publications: quarterly events calendar, Artist's Registryl; quarterly events calendar.
Hours & Admission Prices: Jan. 2 to Dec. 23 Tues.-Thurs. & Sat. 9-3, Fri. 12-6; other times for tours by appointment. No charge; donations accepted. ⅄
Attendance: 6,800 (estimated)
Membership: Associate $50; Patron $75; Supporting $100.

Cle Elum

CLE ELUM TELEPHONE MUSEUM, 221 E. 1st St., Cle Elum, WA 98922-1103. Mailing Address: 302 W. 3rd, Cle Elum, WA 98922. Tel.: 509-649-2880.
E-mail: info@highcountryartists.com
Web Site: www.nkcmuseums.org
Founded: 1967.
Congressional District: 13
Key Personnel: Pres., Bonnie Hawk.
Personnel Profile: Part-Time Volunteers 41.
Governing Authority: state; nonprofit organization. Parent Institution: Cle Elum Historical Society. Subsidiary: Telephone Museum. Tax-exempt.
Institution Type/Description: Communications Museum: housed in a former Bell Telephone building.
Collections: history of the phone; telephone parts; old & new phones; old manually used board; miner's equipment; history & pictures of area.
Facilities: Commemorative plates, tote bags & tiles for sale.
Activities: hands-on dial system for children.
Publications: brochure.
Hours & Admission Prices: Memorial Day-Labor Day Sat.-Sun. 12-4 by appointment. Suggested Donation: $1 per person.
Attendance: 1,301 (accurate)
Membership: Senior & Student $15; Individual $20; Family $35; Business & Corporate $50.

Coulee City

DRY FALLS INTERPRETIVE CENTER, 34875 Park Lake Rd., N.E., Coulee City, WA 99115-9607. Tel.: 509-632-5214. Fax: 509-632-5971.
E-mail: dry.falls@parks.wa.gov
Web Site: www.parks.wa.gov
Founded: 1965.
Key Personnel: Park Mgr., Denis Felton; Asst. Park Mgr., John Ashley; Chief Interpretive Svcs., Chris McCart.
Personnel Profile: Full-Time Paid 1; Part-Time Paid 2.
Governing Authority: state. Affiliated with the Washington State Parks and Recreation Commission, 9150 Cleanwater Lane, Olympia, WA 98504.
Institution Type/Description: Natural Science Museum.
Collections: local geological events; archaeological investigations; anthropology; ethnology; Indian artifacts.
Facilities: visitor center.
Activities: permanent exhibitions; geologic & archaeology guided tours.
Hours & Admission Prices: April-Oct. daily 9-5; Nov.-March Fri.-Wed. 9-4. Donation: $1. Closed holidays. ⅄
Attendance: 350,000 (estimated)

Coupeville

ADMIRALTY HEAD LIGHTHOUSE INTERPRETIVE CENTER, 1280 Engle Rd., Coupeville, WA 98239. Mailing Address: P.O. Box 5000, Coupeville, WA 98239-5000. Tel.: 360-240-5584. Fax: 360-678-4120.
E-mail: admiraltyheadlighthouse@gmail.com
Web Site: admiraltyhead.wsu.edu
Formerly: Fort Casey Interpretive Center
Founded: 1903.
Congressional District: 10
Key Personnel: Park Mgr., Jon Crimmins; Program Coord., Julie Pigott; Ranger II, Brett Bayne; Museum Shop Mgr., Cheryl Thomas
Governing Authority: state. Affiliated with the Washington State Parks & Recreation Commission & WSU Extension.
Institution Type/Description: Military/Lighthouse Museum: housed in late 1890s masonry lighthouse located within the former boundaries of Fort Casey, an Endicott period coastal fortification.
Collections: coast artillery artifacts & memorabilia including two 10 inch disappearing cannons; artifacts affiliated with natural history, sea life & park history.
Activities: permanent exhibitions.
Hours & Admission Prices: March Sat.-Sun. 11-5; April & Sept. Fri.-Mon. 11-5; May Thurs.-Mon. 11-5; June daily 11-5; Oct. to Thanksgiving weekend & Dec. Sat.-Sun. 11-4. No charge; donations accepted. Closed Christmas Eve & Day.
Attendance: 55,000 (accurate)
Membership: Individual $25; Family $50; Supporting $100; Lifetime $500; Corporate $1,000.

ISLAND COUNTY HISTORICAL SOCIETY MUSEUM, (M), 908 N.W. Alexander St., Coupeville, WA 98239. Mailing Address: P.O. Box 305, Coupeville, WA 98239-0305. Tel.: 360-678-3310. Fax: 360-678-1702.
E-mail: ed-ichs@whidbey.net
Web Site: wp.islandhistory.org
Founded: 1949.
Congressional District: 10
Key Personnel: Exec. Dir., Richard Castellano.
Governing Authority: nonprofit. Parent Institution: Island County Historical Society. Tax-exempt: 501(c)(3).
Institution Type/Description: Local History Museum: located in the historic town of Coupeville, founded in 1853.
Collections: Indian artifacts; sea captains; Pioneer Families; military history in Island County.
Research Fields: local and county history.
Facilities: 100-vol. library of general reference material available on special request; meeting room. Admiralty Head Lighthouse; Town of Coupeville; Island Country; Methodist Church; Ebey's National Reserve. Stationery and books for sale.
Activities: guided tours; formally organized education programs; permanent & temporary collections.
Publications: quarterly newsletter; books, A History of Whidbey's Island; A Particular Friend, Penn's Cove, Sails, Steamships & Sea Captains.
Hours & Admission Prices: May-Sept. Mon.-Sat. 10-5, Sun. 11-5; Oct.-April Mon.-Sat. 10-4. Family $6, adults $3, military, students & senior citizens $2.50; children under 5 & members no charge. ⅄
Attendance: 12,000 (estimated)

Membership: Student $15; Individual $30; Senior Couple $35; Couple $45; Family $50; Coupe Associate $90; NW Partner $120; Corporate $200.

PACIFIC NORTH WEST ART SCHOOL GALLERY, 15 N.W. Birch St., Coupeville, WA 98239. Tel.: 360-678-3396; 866-678-3396.
E-mail: info@pacificnorthwestartschool.org
Web Site: www.pacificnorthwestartschool.com
Institution Type/Description: Art Gallery.
Collections: paintings; fiber arts; photographs.
Activities: workshops.
Hours & Admission Prices: Call for hours.

Davenport

FORT SPOKANE VISITOR CENTER: NPS (NATIONAL PARK SERVICE), 44150 District Office Lane N., Davenport, WA 99122-9338. Tel.: 509-725-2715 & 633-3830. Fax: 509-633-3834. Facebook: Lake Roosevelt NRA.
Web Site: www.nps.gov/laro/index.htm
Founded: 1965.
Congressional District: 5
Key Personnel: Interpretation & Social Media, Clayton Hanson.
Governing Authority: federal. Parent Institution: National Park Service, U.S. Dept. of the Interior, Washington, DC.
Institution Type/Description: Military Museum: housed in 1892 guardhouse located on the site of Fort Spokane, former army reservation 1880-99 & Indian agency headquarters, school & hospital 1899-1929.
Collections: originals & machine copies of historic documents & photographs relating to Fort Spokane between 1880-1929; uniforms; anthropology; archaeology; Indian artifacts. Historic Buildings: 1883 quartermaster stable; 1888 powder magazine; 1889 reservoir house; 1892 guardhouse.
Facilities: library of manuscripts & documents from Fort Spokane during the army period 1880-1900 & Colville Indian Agency period 1900-1929, limited availability for inter-library loan; auditorium. Books & booklets on natural science & history for sale.
Activities: guided tours; lectures; films; permanent exhibitions.
Publications: book, Sentinel of Silence-History of Ft. Spokane; historical photos.
Hours & Admission Prices: Memorial Day to Labor Day call for hours. No charge; donations accepted. &
Attendance: 5,000 (accurate)

LINCOLN COUNTY HISTORICAL MUSEUM, 7th & Park, Davenport, WA 99122. Mailing Address: P.O. Box 585, Davenport, WA 99122-0585. Tel.: 509-725-6711.
Founded: 1972.
Congressional District: 5
Key Personnel: Pres. (V), John Coley; Treas., Nancy Ellis; Dir. Visitors Information Center & Sec., Tannis Jeschke.
Personnel Profile: Full-Time Paid 1; Part-Time Volunteers 3.
Governing Authority: nonprofit organization. Parent Institution: Lincoln County Historical Society. Tax-exempt: 600.437.715
Institution Type/Description: General Museum: located on the site of an early Indian campsite along Cottonwood Springs Crossroads for early pioneer trails.
Collections: early agricultural tools & equipment, combine harvester, 1899 Case steam engine; photographic history; cameras; printing tools & equipment; quilts; costumes; furniture; guns; Indian artifacts; dolls; archival history; permanent & rotating displays; fire engines; railroad material.
Facilities: 50-vol. limited desk library pertaining to the history of the area & genealogical studies. Museum-related items for sale.
Activities: Pioneer Day Living History, third weekend in July.
Publications: quarterly newsletter, Lincoln County Historical Society.
Hours & Admission Prices: May 1 to Sept. 30 Mon.-Sat. 9-5. No charge; donations accepted.
Attendance: 1,500 (estimated)
Membership: Individual $15; Family $20; Commercial $25; Life $200; Patron $500; Benefactor $1,000 & up.

Dayton

DAYTON HISTORICAL DEPOT SOCIETY, 222 E. Commercial St., Dayton, WA 99328-1313. Tel.: 509-382-2026. Fax: 509-382-2640.
Web Site: daytonhistoricdepot.org
Founded: 1974.
Congressional District: 5
Key Personnel: Pres. (V) & Museum Shop Mgr., Mary Laughery; Mgr., Mary Byrd; Treas., Eric Johnson.

Personnel Profile: Full-Time Paid 1; Part-Time Volunteers 30.
Governing Authority: society. Tax-exempt.
Institution Type/Description: Historical Society Museum: housed in 1881 railroad depot.
Collections: depot memorabilia; picture; clothes; furniture.
Research Fields: Columbia County pioneer history.
Activities: guided tours; arts festivals; temporary exhibitions. Society Sponsors: Victorian Christmas.
Publications: Columbia County & Dayton Washington Visitor Information Guide.
Hours & Admission Prices: Summer: Wed.-Sat. 10-12 & 1-5, Sun.1-4; Winter: Wed.-Sat. 11-12 & 1-4. Adults $5; members no charge. Closed holidays. &
Attendance: 3,650 (accurate)
Membership: Annual $15; Couples $25; Life Member $300.

DuPont

DUPONT HISTORICAL SOCIETY, 207 Barksdale Ave., DuPont, WA 98327-9001. Tel.: 253-964-2399 & 3492. Fax: 253-964-3554.
E-mail: info@dupontmuseum.com
Web Site: www.dupontmuseum.com
Formerly: Dupont Historical Museum
Founded: 1976.
Congressional District: 2
Key Personnel: Pres. (V), Lee McDonald; Treas., Joe Babb; Museum Mgr., Johanna Jones.
Personnel Profile: Full-Time Volunteers 1; Part-Time Volunteers 14.
Governing Authority: municipal; nonprofit. Tax-exempt.
Institution Type/Description: Historical Society Museum: housed in 1910 building constructed by DuPont Co. as part of a company town.
Collections: photographs, site of 1833 Ft. Nisqually; site of Nisqually Mission 1839-1842 (first US settlers in Western Washington); site of Wilkes Observatory 1841; site of 1843 Ft. Nisqually; maps & artifacts of Hudson Bay Co.; photos, artifacts of DuPont, the Company Town & DuPont Explosive Manufacturing Co., 1906-1976.
Research Fields: Hudson Bay in 1830 & DuPont in 1906; Nisqually Methodist Episcopal Mission 1839-41.
Activities: guided tours; lectures; films; celebration of first 4th of July west of Missouri River with personnel and Wilkes personnel participating.
Publications: DuPont-The Story of a Company Town.
Hours & Admission Prices: Wed.-Fri. & Sun. 1-4; other times by appointment. No charge; donations accepted. Closed holidays. &
Attendance: 1,000 (estimated)

Eatonville

NORTHWEST TREK WILDLIFE PARK, 11610 Trek Dr. E., Eatonville, WA 98328-9502. Mailing Address: Metro Parks Tacoma, 4702 S. 19th St., Tacoma, WA 98405. Tel.: 360-832-6117. Fax: 360-832-6118.
E-mail: whitney.dalbalcon@pdza.org
Web Site: www.nwtrek.org
Founded: 1975.
Congressional District: 6
Key Personnel: Dir., Gary Geddes; Mktg. & Public Rels. Mgr., Whitney DalBalcon.
Personnel Profile: Full-Time Paid 25; Part-Time Paid 50; Part-Time Volunteers 90; Interns 8.
Governing Authority: municipal. Parent Institution: Metropolitan Park Tacoma. Tax-exempt: 170(b)(1)(A).
Institution Type/Description: Wildlife Park.
Collections: native North American species.
Facilities: food service available. Gift items for sale.
Activities: guided tours; films; organized education programs for children; docent program; meeting room rentals.
Publications: brochure, Northwest Trek Wildlife Park; quarterly membership newsletter, Trek Tracks.
Hours & Admission Prices: Seasonal hours, please call or check website to confirm. Adults 13-64 $18.25, senior citizens 65 & over $16.75, youth 5-12 $12.95, tots 3 & 4 $9.25; discounts for Pierce County residents & military personnel; children 2 & under and members no charge. Closed Thanksgiving; Christmas. &
Attendance: 182,000 (accurate)
Membership: Individual $50; One + One $85; Household & Grandparent $155; Deluxe Household $145; Sponsor $265; Patron $350.

PIONEER FARM MUSEUM AND OHOP INDIAN VILLAGE,
7716 Ohop Valley Rd. E., Eatonville, WA 98328-9342. Tel.:
360-832-6300. Fax: 360-832-4533. Facebook: Pioneer Farm Museum.
E-mail: pioneer@mashell.com
Web Site: www.pioneerfarmmuseum.org
Founded: 1975.
Congressional District: 2
Key Personnel: Bd. Pres. (V), Merrilee McBride; Museum Mgr., Valerie Sivertson; Sec. & Treas., Lori Ramsey; Museum Shop Mgr., Nora Cady.
Personnel Profile: Part-Time Paid 13; Part-Time Volunteers 2.
Governing Authority: nonprofit organization. Tax-exempt: 501(c)(3).
Institution Type/Description: History Museum: hands-on living history c.1880 farm.
Collections: hands-on guided pioneer experience includes 6 buildings in a farm setting: 1887 trading post, 1888 two-story cabin, interpreters cabin, pole barn with animals & hay loft, working blacksmith shop, working wood work shop; 1892 replica of one room schoolhouse.
Activities: hands on living history guided tour, grind grain, churn butter, card wool, scrub clothing, curl hair, milking, pet farm animals, spud logs, jump in hay; hands on tour of the traditional Native American lifestyles of the Northwest; Pioneer Farm, pioneer folklore & crafts, program, schoolhouse lesson; nature trail walk; overnight Pioneer Living Experience; Native American Season tour, Indian lore & craft program; Co-Salish overnight program.
Publications: A Tale of Two Cabins.
Hours & Admission Prices: Public Tours: March-May & Sept. to mid-Nov. Sat.-Sun. 11-4; Father's Day-Labor Day daily 11-4. Group Tours: mid-March to Thanksgiving by reservation only. Adults $9, children $8; discounts to AAA members; members no charge. Native American Seasons Tour: Father's Day to Labor Day Fri.-Sun. 1 & 2:30. Adults $8.50, children $7.50; members no charge. ⓑ
Attendance: 26,000 (estimated)
Membership: Senior 61 & over $20; Individual $25; Family $75; Lifetime $1,000.

Edmonds

EDMONDS ART FESTIVAL MUSEUM, Frances Anderson Center, 700 Main St., Edmonds, WA 98020-3032. Mailing Address: Edmonds Arts Festival Foundation, P.O. Box 699, Edmonds, WA 98020-0699. Tel.: 425-771-1984.
E-mail: hardarmc@verizon.net
Web Site: www.eaffoundation.org
Founded: 1979.
Congressional District: 2
Key Personnel: Pres., Terry Vehrs; Cur., Darlene McLelland.
Personnel Profile: Part-Time Paid 1.
Governing Authority: nonprofit organization. Subsidiary Institution: Edmonds Arts Festival. Tax-exempt: 501(c)(3).
Institution Type/Description: Art Museum: located in Frances Anderson Center.
Collections: various forms of art media.
Activities: arts festival; organized education programs for undergraduate or graduate college students affiliated with Edmonds Community College; temporary & traveling exhibitions.
Hours & Admission Prices: Mon.-Fri. 9-9, Sat. 10-4. No charge. Closed federal holidays. ⓑ
Attendance: 250,000 (estimated)
Membership: Annual $2; Patron $60; Business $50; Sustaining $125; Life Patron $500.

EDMONDS SOUTH SNOHOMISH COUNTY HISTORICAL SOCIETY, INC., 118 Fifth Ave., N., Edmonds, WA 98020-3145. Mailing Address: P.O. Box 52, Edmonds, WA 98020-0052. Tel.: 425-774-0900. Fax: 425-774-6507.
E-mail: edmondsmuseum118@gmail.com
Web Site: www.historicedmonds.org
Founded: 1973.
Congressional District: 1
Key Personnel: Museum Dir., Tarin Erickson; Pres., Bill Lambert.
Personnel Profile: Part-Time Paid 2; Part-Time Volunteers 100.
Governing Authority: society; nonprofit organization. Tax-exempt: 501(c)(3).
Institution Type/Description: Historical Society Museum: housed in 1910 Carnegie Library.
Collections: business, industrial & domestic artifacts relating to the settlement & history of Edmonds & south Snohomish County; archival documents; photographs; permanent exhibit on history of Edmonds including working

shingle mill diorama & historic hotel room; Victorian parlor; 1910 town scene diorama; maritime gallery; model train layout & transportation artifacts.
Research Fields: local history.
Activities: guided tours; special events; permanent & temporary exhibitions.
Publications: quarterly newsletter.
Hours & Admission Prices: Wed.-Sun. 1-4. Suggested Donation: adults $5, children & students $2. Closed holidays. ⓑ
Attendance: 4,700 (estimated)
Membership: Bronze (Individual) $25; Silver (Family) $40; Business $75; Gold $125; Platinum $375.

Ellensburg

CLYMER MUSEUM OF ART, (M), 416 N. Pearl St., Ellensburg, WA 98926-3112. Tel.: 509-962-6416. Fax: 509-962-6424.
E-mail: clymermuseum@charter.net
Web Site: www.clymermuseum.org
Founded: 1988.
Congressional District: 13
Key Personnel: Dir., Mia Merendino.
Personnel Profile: Full-Time Paid 1; Part-Time Paid 3; Part-Time Volunteers 12; Interns 2.
Governing Authority: nonprofit organization. Tax-exempt.
Institution Type/Description: Art Museum: honoring John F. Clymer housed in 1901 building in historic downtown Ellensburg.
Collections: works of John Clymer.
Facilities: 7,100 sq. ft. exhibit space. Gift items for sale.
Activities: docent programs; guided tours; lectures; loan & temporary exhibitions. Annual Event: auction.
Hours & Admission Prices: Mon.-Fri. 10-5, Sat. 10-4. No charge; donations accepted. Closed major holidays. ⓑ
Attendance: 40,000 (accurate)
Membership: Individual $35; Family $50; friend $75; Associate $100; Patron $250; Premier $500; Benefactor $1,000 & up.

KITTITAS COUNTY HISTORICAL MUSEUM, (M), 114 E. Third Ave., Ellensburg, WA 98926-3346. Tel.: 509-925-3778. Facebook: Kittitas County Museum.
E-mail: kchm@kchm.org
Web Site: www.kchm.org
Founded: 1961.
Congressional District: 4
Key Personnel: Dir., Sadie Thayer.
Personnel Profile: Full-Time Paid 2; Part-Time Volunteers 6.
Volunteer Hours: 2,362
Operating Expenses: 157,589
Operating Income: 129,373
Governing Authority: Parent Institution: Kittitas County Historical Society. Tax-exempt.
Institution Type/Description: History Museum.
Collections: Kittitas County history & culture; household furnishings; decorative arts; clothing & textiles; Native American bags & baskets; archives; photographs; early businesses; ranching; farming; geological specimens; automobiles; dolls.
Facilities: over 8,000 sq. ft. exhibit space.
Activities: permanent & temporary exhibits; school tours; lecture series.
Hours & Admission Prices: Mon.-Sat. 10-4. No charge; donations accepted. ⓑ
Attendance: 5,880 (accurate)
Membership: Senior $15; Individual $25; Family $30; Supporter $50 & up; Contributor $100 & up; Patron $250 & up; Benefactor $500 & up.

MUSEUM OF CULTURE & ENVIRONMENT, 400 E. University Way, Ellensburg, WA 98926-7544. Tel.: 509-963-2313. Fax: 509-963-3215.
E-mail: museum@cwu.edu
Web Site: www.cwu.edu/~museum/
Formerly: Museum of Man, Anthropology Museum
Founded: 1972.
Congressional District: 4
Key Personnel: Dir., Kathleen Barlow.
Personnel Profile: Full-Time Paid 1.
Governing Authority: state. Parent Institution: Central Washington University. Tax-exempt.
Institution Type/Description: University Anthropology Museum.
Collections: ethnographic materials from New Guinea, Mexico, Western Plateau, San Blas Islands, Panama, Africa, Southwest & Northwest Coast U.S.
Research Fields: archaeology; ethnology.

Facilities: classrooms; research labs.
Activities: temporary exhibitions, some training opportunities.
Hours & Admission Prices: Mon.-Fri. 8-5 by appointment only. No charge; donations accepted. &
Attendance: 3,242 (accurate)

OLMSTEAD PLACE STATE PARK, 921 N. Ferguson Rd., Ellensburg, WA 98926-8109. Tel.: 509-925-1943. Fax: 509-925-1955.
E-mail: olmstead.place@parks.wa.gov
Web Site: www.parks.gov
Founded: 1968.
Congressional District: 13
Key Personnel: Park Ranger, Brandon Holkstra.
Personnel Profile: Full-Time Paid 1; Full-Time Volunteers 1; Part-Time Volunteers 5.
Governing Authority: state. Parent Institution: Washington State Parks & Recreation Commission. Tax-exempt.
Institution Type/Description: Historic Site: 160-acre 1875 homestead.
Collections: agricultural equipment.
Research Fields: agricultural history.
Facilities: 100-vol. library pertaining to agricultural; 1875 log cabin; 1908 farmhouse; 19th-century Seaton Schoolhouse; barns & outbuildings; hiking trail; picnic & play areas.
Activities: guided tours; participatory exhibits.
Publications: brochures: Olmstead Place, Altapes Creek Trail; coloring book: Olmstead Place; Seaton Cabin Schoolhouse.
Hours & Admission Prices: Summer: 6:30am to dusk; Winter: 8am to dusk. No charge; donations accepted.
Attendance: 43,000
Membership: Friend $10 or 4 hours of labor; Supporting Friend $100 or 40 hours of labor; Life Friend $1,000 or 400 hours of labor.

Enumclaw

ENUMCLAW PLATEAU HISTORICAL SOCIETY, 1837 Marion St., Enumclaw, WA 98022. Tel.: 360-825-3356.
Founded: 1995.
Key Personnel: Dir. & Pres. (V), Ronald Tyler; Treas., Robert Stygar.
Personnel Profile: Part-Time Volunteers 13.
Governing Authority: private; nonprofit organization. Tax-exempt: 501(c)(3).
Institution Type/Description: Historical Society Museum: housed in a former Masonic Hall, c.1909.
Collections: local history from mid-1800s to present; personal artifacts; furnishings; photographs; family records; obituaries, weddings, anniversary & business history catalogs from 1935 to present.
Research Fields: photographs of local homes & sites from 1950s to present.
Facilities: 1,500 sq. ft. exhibit space.
Activities: Annual Events: History Award Dinner; Holiday Bazaar; Antique Appraisal Event.
Publications: triannual newsletter, Links.
Hours & Admission Prices: Thurs. & Sun. 1-4. No charge; donations accepted. Closed most holidays. &
Attendance: 700 (estimated)
Membership: Senior & Student $5, Individual $15; Household $20; Business $30; Patron $100; Life $200.

Ephrata

GRANT COUNTY HISTORICAL MUSEUM, 742 Basin St., N.W., Ephrata, WA 98823-1635. Mailing Address: P.O. Box 1141, Ephrata, WA 98823-1141. Tel.: 509-754-3334. Fax: 509-754-2148.
E-mail: grantcomuseum@mail.com
Founded: 1951.
Congressional District: 13
Key Personnel: Pres. (V), Rita Mayrant; Dir. & Museum Shop Mgr., Pat Witham.
Personnel Profile: Full-Time Paid 5; Part-Time Paid 1; Part-Time Volunteers 16.
Governing Authority: bd. dirs. & county commissioners; nonprofit organization. Parent Institution: Grant County Historical Society. Tax-exempt.
Institution Type/Description: Local History Museum: 36 building village.
Collections: early history of county; furnishings; early cattleman's equipment; pioneer village & homestead with fixtures & furniture; replica of Grant County Fire Hall; clothing; archives; early dishes & utensils used by settlers; farm machinery; photographs; oral histories; documents; pictorial history of Grand Coulee Dam, 1933-completion; period car & farm truck. Historic Buildings: St. Rose of Lima Catholic church; land office; early gas station; replica of Wilson Creek State Bank; livery stable; Line Cabin; Marlin (Krupp) jail; 1902 homestead; one-room schoolhouse; saloon; dress

shop; barber shop; newspaper office; blacksmith shop; country store; line cabin; millinery shop; Justice of the Peace; country store; camera shop; old time doctor's & dentist's office; pharmacy; Grant County Journal; 1971 Burlington Northern Caboose.
Research Fields: history of Grant County; pioneer settlers; early county schools; Grand Coulee Dam.
Facilities: interpretive center; picnic area. Museum-related items for sale.
Activities: pre-arranged tours; permanent & temporary exhibitions. Museum Sponsors: living museum during Sage & Sun Festival in June; Pioneer Day in September; Old Time Political Rally in fall on election years.
Publications: quarterly newsletter; oral history compilation, Memories of Grant County, Washington.
Hours & Admission Prices: May-Sept. Mon.-Tues. & Thurs.-Sat. 10-5, Sun. 1-4. Adults $3.50, students 5-15 $2.50; children under 5 no charge. &
Attendance: 7,500 (estimated)
Membership: Student $5; Individual $10; Family $25; Caliche $100; Basalt $500; Granite $1,000.

Everett

FLYING HERITAGE COLLECTION, (M), 3407 109th St., S.W., Everett, WA 98204-1351. Tel.: 206-342-4242. Fax: 206-342-4235.
Web Site: www.flyingheritage.com
Key Personnel: Exec. Dir., Adrian Hunt; Museum Shop Mgr., Liz Davidson
Institution Type/Description: Military Aviation History.
Collections: aviation history; combat aircraft & artifacts.
Activities: special events.
Hours & Admission Prices: Memorial Day to Labor Day daily 10-5; Sept.-May Tues.-Sun. 10-5. Adults $12, military & seniors $10, youth 6-15 $8; discounts to groups of 15 or more; children 5 & under no charge. &

IMAGINE CHILDREN'S MUSEUM, 1502 Wall St., Everett, WA 98201-4008. Tel.: 425-258-1006. Fax: 425-258-5406.
E-mail: info@imaginecm.org
Web Site: www.imaginecm.org
Key Personnel: Exec. Dir., Nancy Johnson; Museum Shop Mgr., Lynndee Blair.
Personnel Profile: Full-Time Paid 9; Part-Time Paid 20.
Governing Authority: nonprofit organization. Tax-exempt: 501(c)(3).
Institution Type/Description: Children's Museum.
Collections: hands-on exhibits.
Activities: special events; parties; meetings.
Hours & Admission Prices: Tues.-Wed. 9-5, Thurs.-Sat. 10-5, Sun. 11-5. Admission $7.75; discounts Thurs. 3:30-5; children one & under no charge. Closed New Year's Day; Easter; Memorial Day; Labor Day; Thanksgiving; Christmas.

RUSSELL DAY GALLERY, Everett Community College, Parks Student Union Bldg., Rm. 219, 2000 Tower St., Everett, WA 98201-1352. Tel.: 425-388-9036.
E-mail: slepper@everettcc.edu
Web Site: www.everettcc.edu/russelldaygallery
Key Personnel: Dir., Sandra Lepper
Institution Type/Description: Art Gallery.
Collections: artwork.
Hours & Admission Prices: Mon. & Wed. 10-4, Tues. & Thurs. 12-4, Fri. 10-2.

SNOHOMISH COUNTY MUSEUM OF HISTORY, 3001 Oakes Ave., Everett, WA 98201-3657. Mailing Address: P.O. Box 5556, Everett, WA 98206. Tel.: 425-345-7349.
E-mail: bg.snocomuseum@comcast.net
Web Site: snocomuseum.org
Key Personnel: Exec. Dir., Barbara George
Institution Type/Description: History Museum.
Collections: local history & culture; period furnishings; personal artifacts; photographs.
Activities: educational programs.
Hours & Admission Prices: Call for hours.

Federal Way

PACIFIC RIM BONSAI COLLECTION, 33663 Weyerhaeuser Way S., Federal Way, WA 98001-9620. Mailing Address: P.O. Box 9777, Federal Way, WA 98063-9777. Tel.: 253-924-5206; 800-525-5400, ext. 5206. Fax: 253-924-3837.
E-mail: david.degroot@weyerhaeuser.com
Web Site: www.weyerhaeuser.com/bonsai
Founded: 1989.

Key Personnel: Cur., David De Groot

Personnel Profile: Full-Time Paid 2; Part-Time Paid 1; Part-Time Volunteers 20.

Governing Authority: Parent Institution: Weyerhaeuser Company.

Institution Type/Description: Conservatory.

Collections: over 60 bonsai trees from Canada, China, Japan, Korea, Taiwan & the U.S.

Facilities: outdoor & indoor display areas; office; lecture/event tent; restrooms.

Activities: special events.

Hours & Admission Prices: Tues.-Sun. 10-4. No charge. Closed New Year's Day; Thanksgiving; Christmas Eve & Day. ⅋

Attendance: 32,000 (accurate)

RHODODENDRON SPECIES BOTANICAL GARDEN, 2525 S. 336th St., Federal Way, WA 98003-7825. Mailing Address: P.O. Box 3798, Federal Way, WA 98063-3798. Tel.: 253-838-4646. Fax: 253-838-4686. Facebook: Rhododendron Species Botanical Garden.

E-mail: info@rhodygarden.org

Web Site: www.rhodygarden.org

Founded: 1964.

Congressional District: 30

Key Personnel: Exec. Dir & Cur., Steve Hootman; Program Mgr., Katie Swickard; Museum Shop Mgr., Heather Ford.

Personnel Profile: Full-Time Paid 6; Part-Time Paid 3; Part-Time Volunteers 80; Interns 3.

Governing Authority: nonprofit organization. Tax-exempt: 501(c)(3).

Institution Type/Description: Botanical Garden.

Collections: plants; species rhododendron.

Research Fields: species rhododendron.

Facilities: reference library of books, pamphlets, maps, vertical files on the genus rhododendron & other horticultural subjects available for research on site; slide library; botanical garden; reading room; 22-acre display garden. Museum-related items for sale.

Activities: guided tours; lectures; slide programs; docent program or council; permanent exhibitions; traveling display.

Publications: books, Rhododendrons of China; The Rhododendron Species, Volume I, Lepidotes; RSF newsletter; RSF Yearbook, Rhododendron Species.

Hours & Admission Prices: Tues.-Sun. 10-4. Adults $8, seniors & students $5; Weyerhaeuser employees, children under 12 & military no charge. ⅋

Attendance: 13,848 (accurate)

Membership: Student $25; Individual $35; Family $50; Supporting $100; Sustaining $250; Patron $500; Benefactor $1,000.

Fife

FIFE HISTORY MUSEUM, 2820 54th Ave. E., Fife, WA 98424-2140. Tel.: 253-896-4710.

E-mail: fifehistorymuseum1957@gmail.com

Web Site: https://sites.google.com

Founded: 2001.

Key Personnel: Pres. (V), Louise Hospenthal.

Personnel Profile: Part-Time Volunteers 11.

Governing Authority: nonprofit organization. Parent Institution: Fife Historical Society. Tax-exempt: 501(c)(3).

Institution Type/Description: History Museum: housed in the former home of Louis Dacca, a member of the original Fife City Council.

Collections: local history & culture; period furnishings; personal artifacts; photographs.

Hours & Admission Prices: Wed. 12-4:30, Fri. 9-4:30, Sat. 9-1; other times by appointment. No charge; donations accepted. ⅋

Attendance: 1,000 (estimated)

Membership: Student $10; Senior $15; Individual $20; Family $35; Patron & Business $100; Corporate 250 Plus Employees $250.

Forks

FORKS TIMBER MUSEUM, 1421 S. Forks Ave., Forks, WA 98331-9383. Mailing Address: P.O. Box 873, Forks, WA 98331-0873. Tel.: 360-374-9663.

Key Personnel: Mgr., Sherrill Fouts.

Personnel Profile: Full-Time Volunteers 5; Part-Time Paid 1.

Institution Type/Description: Logging History Museum.

Collections: local history; loggers; logging tools; photographs; personal artifacts; period furnishings.

Facilities: nature trails; garden.

Activities: hiking; tours.

Hours & Admission Prices: May-Oct. Tues.-Sat. 10-4; Winter: by appointment. Adults $3.

Attendance: 2,500 (estimated)

Membership: Single $10; Family $15; Service Organization $25; Business $50.

Fox Island

FOX ISLAND HISTORICAL SOCIETY, (M), 1017 Ninth Ave., Fox Island, WA 98333. Mailing Address: P.O. Box 242, Fox Island, WA 98333-0242. Tel.: 253-549-2835.

E-mail: foxislandmuseum@centurytel.net

Web Site: www.foxislandmuseum.org

Founded: 1895.

Congressional District: 6

Key Personnel: Pres. & Acting Dir., Marie Weis; Treas., Vera Hackett.

Personnel Profile: Part-Time Volunteers 30.

Governing Authority: nonprofit organization. Parent Institution: Washington State Historical Society. Tax-exempt: 501(c)(3).

Institution Type/Description: Historical Society Museum.

Collections: 3,000 artifacts pertaining to life on Fox Island & in Washington State; 500 pulley blocks; household goods; farm equipment; albums; photographs; textiles; Indian baskets; local history scrapbooks.

Research Fields: local history.

Facilities: 300-vol. library pertaining to Fox Island, State & natural history; auditorium. Books, local maps, postcards & other gift items for sale.

Activities: guided tours; organized education programs for children; docent program; participatory, loan & temporary exhibitions.

Publications: books, A History of Fox Island; Island in the Sound; Fox Island; Echoes of Yesterday; Fox Island in Pictures (Acadia Press).

Hours & Admission Prices: Wed. & Sat.-Sun. 1-4. No charge; donations accepted. Closed New Year's Day; Easter; Christmas. ⅋

Attendance: 1,000 (estimated)

Membership: Senior $5; Individual $10; Family $20; Sustaining $30.

Friday Harbor

SAN JUAN HISTORICAL SOCIETY, 405 Price St., Friday Harbor, WA 98250. Mailing Address: P.O. Box 441, Friday Harbor, WA 98250-0441. Tel.: 360-378-3949. Fax: 360-378-3949 (call first).

Web Site: www.sjmuseum.org

Founded: 1961.

Congressional District: 40

Key Personnel: Pres., Mary Jean Cahail; Exec. Dir., Kevin Loftus; Treas., Dave Hylton.

Personnel Profile: Part-Time Paid 1; Part-Time Volunteers 80; Interns 1.

Governing Authority: society. Tax-exempt: 501(c)(3).

Institution Type/Description: Historical Society Museum: housed in turn-of-the-century farmhouse.

Collections: San Juan Island history; Native American baskets. Historic Buildings: town jail; log cabin; farmhouse; barn; carriage house; milk house; root cellar.

Activities: guided tours; organized education programs for adults & children; public events; lectures.

Publications: quarterly newsletter.

Hours & Admission Prices: April & Oct. Sat. 1-4; May-Sept. Wed.-Sat. 10-4, Sun. 1-4; other times by appointment. Adults $5, seniors 60 & over $4, youth under 18 $3; members & children under 5 no charge.

Attendance: 6,500 (estimated)

Membership: Seniors 60 & over $10; Individual $15; Business & Family $25; Sponsor $50; Benefactor $100.

SAN JUAN ISLAND NATIONAL HISTORICAL PARK, 650 Mullis St., Ste. 100, Friday Harbor, WA 98250-7951. Mailing Address: P.O. Box 429, Friday Harbor, WA 98250-0429. Tel.: 360-378-2240, ext. 2233. Fax: 360-378-2615.

E-mail: sajh_interpretation@nps.gov

Web Site: www.nps.gov/sajh

Founded: 1966.

Congressional District: 2

Key Personnel: Supt., Steve Gibbons; Chief Ranger, Barry Lewis.

Personnel Profile: Full-Time Paid 6; Full-Time Volunteers 8; Part-Time Paid 6; Part-Time Volunteers 200.

Governing Authority: federal. Parent Institution: National Park Service. Tax-exempt.

Institution Type/Description: Park Museum: area includes American & English camps. Joint occupation of San Juan Island by British & Americans occurred until the final settlement of the water boundary dispute & the Pig War in 1872.

Collections: artifacts associated with the Pig War, 1859-72; U.S. Army &

British Royal Marines. Historic Buildings: Restored 1860 Block House, 1859 earthwork, Barracks & Commissary; Officers' Quarters at American Camp.

Research Fields: pertaining to collections.

Facilities: 200-vol. library of diplomatic, Pacific Northwest & military history 1846-1872 & environmental material available for use on premises.

Activities: guided & self guided tours; AV program; summer weekend interpretive programs; special events; Jackle's Lagoon Guided Walk.

Publications: brochures, A Historic Guided Walk: American Camp, A Historic Guided Walk: English Camp; Young Hill Guided Walk; South Beach Prairie Guided Walk; Bird Walk; Pig War Guided Walk; Belle Vue Sheep Farm Archaeology; Jakles Lagoon Nature Walk.

Hours & Admission Prices: June-Sept. 3 daily 8:30-5; Sept. 4-May Wed.-Sun. 8:30-4:30. No charge. Closed federal holidays in winter. &

Attendance: 274,000 (estimated)

SAN JUAN ISLANDS MUSEUM OF ART, (M), 540 Spring St., Friday Harbor, WA 98250. Mailing Address: IMA, P.O. Box 339, Friday Harbor, WA 98250-0339. Tel.: 360-370-5050. Fax: 360-370-5805. Facebook: San Juan Islands Museum of Art.

E-mail: info@sjima.org

Web Site: www.sjima.org

Formerly: Island Museum of Art, Westcott Bay Institute

Founded: 2000.

Key Personnel: Dir., Jennifer Elise.

Personnel Profile: Full-Time Paid 1; Part-Time Paid 1; Part-Time Volunteers 30; Interns 3.

Governing Authority: private; nonprofit organization. Tax-exempt: 501(c)(3).

Institution Type/Description: Art Museum.

Collections: works of art by Northwest & West Coast artists.

Facilities: 800 sq. ft. exhibit space; 19 acre sculpture park. Museum-related items for sale.

Activities: arts festivals; dance recitals; docent program; formal education programs; student internships; temporary exhibitions. Annual Events: 2 Festivals

Publications: annual newsletter.

Hours & Admission Prices: Thurs.-Sat. 11-5, Sun. 1-4. No charge; donations accepted. &

Attendance: 20,000 (estimated)

Membership: Individual $30; Family $50; Donor $100; Sponsor $250; Patron $500; Benefactor $1,000.

THE WHALE MUSEUM, (M), 62 1st St. N., Friday Harbor, WA 98250. Mailing Address: P.O. Box 945, Friday Harbor, WA 98250-0945. Tel.: 360-378-4710; 800-946-7227, ext. 30. Fax: 360-378-5790. Facebook: The Whale Museum.

E-mail: info@whalemuseum.org

Web Site: www.whalemuseum.org

Founded: 1976.

Congressional District: 2

Key Personnel: Dir., Jenny L. Atkinson; Bd. Pres., Richard Day; Finance Mgr., Elli Gull; Collections Cur., Jennifer Olson; Education Cur., Cindy Hansen; Museum Shop Mgr., Rena Eubanks.

Personnel Profile: Full-Time Paid 5; Part-Time Paid 5; Part-Time Volunteers 50; Interns 5.

Governing Authority: society; nonprofit. Tax-exempt: 501(c)(3).

Institution Type/Description: Natural History Museum.

Collections: interpretive graphics; life-size models; skeletons; biological specimens; ethnographical artifacts related to cetaceans; current research on resident orcas; wild whales.

Research Fields: natural history of the marine environment, with special emphasis on cetaceans, physiology & demographics of resident orcas, & human impacts on cetaceans.

Facilities: library of books & scientific reprints on natural history, ecology, communications, physiology, bioacoustics, general biology of marine mammals, museology available for research by request; exhibit hall; field research station; 50-seat auditorium. Books, original art, prints & other museum-related items for sale.

Activities: guided tours; lectures; films; gallery talks; concerts; formally organized education programs for children, adults & college students; permanent & temporary loans and traveling exhibits.

Publications: Cetus biannual newsletter; field guides; scientific papers; educational teaching packages.

Hours & Admission Prices: June-Sept. daily 9-6; Oct.-May daily 10-5. Adults $6, senior citizens $5, college student & children $3; children under 5 & members no charge. Closed New Year's Day; Thanksgiving; Christmas.

Attendance: 28,000 (accurate)

Membership: Individual $25; Individual Plus Adoption $35; Family $40;

Business $50 & up; Supporter $100; Contributor $250; Patron $500; Benefactor $1,000; Steward $2,500 & up.

Gig Harbor

HARBOR HISTORY MUSEUM, 4121 Harborview Dr., Gig Harbor, WA 98332. Mailing Address: P.O. Box 744, Gig Harbor, WA 98335-0744. Tel.: 253-858-6722. Fax: 253-853-4211.

E-mail: info@harborhistorymuseum.org

Web Site: www.harborhistorymuseum.org

Formerly: Gig Harbor Peninsula History Museum

Founded: 1963.

Congressional District: 6

Key Personnel: Exec. Dir., Sue Loiland; Pres. (V) Bd. Dirs., Frank Ruffo; Treas., Mark Caviness; Cur. Exhibitions & Collections, Victoria Gehl-Blackwell.

Personnel Profile: Full-Time Paid 3; Part-Time Paid 4; Part-Time Volunteers 60.

Governing Authority: articles of incorporation/board of trustees; nonprofit. Tax-exempt: 501(c)(3).

Institution Type/Description: Historical Society Museum.

Collections: Gig Harbor & local peninsulas social & occupational history; fishing; logging; farming; school records; personal archives; photographs; local biographies.

Research Fields: personal biographies; local history.

Facilities: permanent & temporary exhibitions; research room. Museum-related items for sale.

Activities: films; organized education programs for children & adults.

Publications: membership newsletter; Gig Harbor History Walk; An Excellent Little Bay - A History of Gig Harbor.

Hours & Admission Prices: Museum: Jan.-April 15 Wed.-Sun. 10-4; April 16-Dec. Tues.-Sun. 10-5. Research: by appointment. Adults $7, seniors 65 & over and military $6, youth 7-17 $5; members no charge. Closed Martin Luther King Jr. Day; Independence Day; Labor Day; Thanksgiving, Christmas. &

Attendance: 14,000 (accurate)

Membership: Individual $45; Dual $55; Family $60; Supporting $100; Sustaining $250; Patron $500; Benefactor $1,000; Collector $2,500; Visionary $5,000.

Goldendale

GOLDENDALE OBSERVATORY, 1602 Observatory Dr., Goldendale, WA 98620-3315. Tel.: 509-773-3141. Fax: 509-773-6929.

E-mail: goldendale.observatory@parks.wa.gov

Web Site: www.perr.com/gosp.html

Founded: 1973.

Congressional District: 4

Key Personnel: Area Mgr., Lem Pratt; Museum Mgr., Troy Carpenter.

Personnel Profile: Full-Time Paid 1; Part-Time Paid 1; Part-Time Volunteers 3.

Governing Authority: state. Affiliated with Washington State Parks & Recreation Commission, 1111 Israel Rd., Olympia, WA 98504-2650.

Institution Type/Description: Observatory.

Collections: astronomical displays; telescopes.

Facilities: telescopes available for use by public.

Activities: PowerPoint presentations; demonstrations; view astronomical objects through a telescope.

Hours & Admission Prices: April-Sept. Wed.-Sun. 10am-11:30pm; Oct.-March Mon.-Fri. 10-4. Access by Washington State Parks Discover Pass: One Day Pass $10 per vehicle; Annual Pass $30 per vehicle. &

Attendance: 25,000 (estimated)

KLICKITAT COUNTY HISTORICAL SOCIETY, 127 W. Broadway, Goldendale, WA 98620. Mailing Address: P.O. Box 86, Goldendale, WA 98620-0086. Tel.: 509-773-4303.

E-mail: presbymuseum@gorge.net

Web Site: www.presbymuseum.com

Founded: 1958.

Congressional District: 4

Key Personnel: Pres. (V), Bonnie Beeks; Sec., Mary Evans Childs; Treas., Dennis Birney; Museum Shop Mgr., Marilyn Enwards.

Personnel Profile: Part-Time Paid 1; Part-Time Volunteers 31.

Governing Authority: society; board of directors. Tax-exempt: 501(c)(3).

Institution Type/Description: History Museum: housed in 1903 W.B. Presby Mansion.

Collections: agriculture; clothing; glass; 1900 period house & furnishings; 126 coffee mills; turn-of-the-century school display; linens; photographs; newspaper & advertising copperplate page set up stereotype castings; toys; dolls; velocipede; printshop; cameras; period school documents; books; period dental chair & x-ray; late 18th- and early 19th-century woodworking tools;

record cylinders & records; pianola & player piano; military memorabilia including WWI & WWII. Historic Buildings: Presby House; carriage/school house.

Research Fields: county history; genealogy; cemeteries; early abstracts & surveys of county.

Activities: guided tours; permanent & temporary exhibitions. Museum Sponsors: Heritage Barn Tour in September; Quilt Show; Christmas Concert.

Publications: books, They Worked Hard; Citizens of the Century; Goldendale 1904; The White Salmon Valley; The Goldens of Goldendale; Klickitat County Death and Birth Records; 1860, 1870 and 1880 Census of Klickitat County, Washington; Blockhouse Bicentennial; Poems from a Saddle Bag; The Nichols - Sheppard Steam Engine; The Joys of Yesterday Can be Found In The Memories of Today; leaflets: Sam Hill, Stonehenge and Maryhill Town; The Story of Maryhill Museum; The Life of Samual Hill; Stonehenge; Who Were the Guests at the Central Hotel; The Old Red House; Andrew J. Bolon: U.S. Indian Agent; Winthrop B. Presby; annual magazine, Klickitat Heritage; The History of Klickitat County; Glimpses in the Past of Klickitat County - 1900-1940; Tragedies in Klickitat County 1900-1940; Klickitat County Deaths & Obituaries 1900-1935; Wagon Road; Boot Leggers; Still & The Law.

Hours & Admission Prices: May-Oct. daily 10-4; other times by appointment. Adults $5, students 6-12 & school tours $1; discount to groups; members no charge.

Attendance: 805 (accurate)

Membership: Active $25; Family $30; Patron $50; Sponsor $100.

* **MARYHILL MUSEUM OF ART, (M),** 35 Maryhill Museum Dr., Goldendale, WA 98620-4601. Tel.: 509-773-3733. Fax: 509-773-6138.

E-mail: maryhill@maryhillmuseum.org

Web Site: www.maryhillmuseum.org

Founded: 1923.

Congressional District: 4

Key Personnel: C.E.O. & Dir., Colleen Schafroth; Pres. (V), David Savinar; Cur. Art, Steven Grafe; Cur. Education, Carrie Clark-Peck; Collections Mgr., Anna Berg; Museum Shop Mgr., Jacque Francois; Operations & Finance Mgr., Leslie Wetherell.

Personnel Profile: Full-Time Paid 7; Part-Time Paid 5; Part-Time Volunteers 85.

Volunteer Hours: 2,742

Operating Expenses: 1,544,993

Operating Income: 2,146,951

Governing Authority: nonprofit organization. Tax-exempt: 501(c)(3).

Institution Type/Description: Art Museum: housed in a beaux-arts style concrete mansion; built in 1914. Listed on the National Register of Historic Places.

Collections: Rodin watercolors & sculptures; American Indian art & cultural materials; rare & modern chess sets; Queen Marie of Romania royal furnishings; European & American paintings, sculpture & decorative arts; 1945 French fashion mannequins & theatre sets; history of Sam Hill, founder; Russian icons; over 5,000 photographs; Stonehenge monument.

Major Exhibits: James Lee Hansen: Sculpture, 3/14-7/14; The Flip Side: Comic Art by New Yorker Cartoonists, 3/14-11/14; Angela Swedberg: Historicity, 3/14-11/14; African Art from the Mary Johnston Collection, 8/14-11/14.

Research Fields: Sam Hill documents, including Good Roads Movement; Pacific Northwest history; Loie Fuller papers; materials & papers of Queen Marie of Romania; Russian icons; ecclesiastical textiles; American Indian objects and photos; Theatre de la Mode.

Facilities: 26 acre park; cafe; 10,000 sq. ft. exhibition space; 5,300 acres of ranchlands; educational resource room; lecture hall. Museum-related gifts for sale.

Activities: self-guided tours; special tours by appointment; seminars; symposia; youth & adult art appreciation classes; performing arts events; teacher institutes; permanent, temporary exhibitions.

Publications: books & publications on the collections; exhibition catalogues.

Hours & Admission Prices: mid-March to mid-Nov. daily 10-5. Adults $9, senior citizens 65 & over $8, children 7-18 $3; discounts to AAM members; members no charge. &

Attendance: 46,712 (accurate)

Membership: Individual $50; Family $75; Sponsor $100; Patron $250; Sustaining $500; Benefactor $1,000.

Granite Falls

GRANITE FALLS HISTORICAL MUSEUM, 109 E. Union St., Granite Falls, WA 98252. Mailing Address: P.O. Box 1414, Granite Falls, WA 98252-1414. Tel.: 360-691-2603.

E-mail: info@gfhistory.org

Web Site: www.gfhistory.org

Personnel Profile: Part-Time Volunteers 20; Interns 3.

Governing Authority: Tax-exempt.

Institution Type/Description: History Museum.

Collections: local history & culture; logging; railroad; mining mills; transportation; schools; business; fashion; personal artifacts; period furnishings; photographs.

Activities: special events.

Hours & Admission Prices: Sun. 12-5; other times by appointment. No charge; donations accepted. &

Attendance: 2,000 (estimated)

Membership: Individual $15; Family $20; Life $100.

Greenbank

MEERKERK RHODODENDRON GARDENS, 3531 Meerkerk Lane, Greenbank, WA 98253. Mailing Address: P.O. Box 154, Greenbank, WA 98253-0154. Tel.: 360-678-1912.

E-mail: meerkerk@whidbey.net

Web Site: www.meerkerk gardens.org

Founded: 1979.

Congressional District: 2

Key Personnel: Dir., Joan Bell; Pres. (V), Don Lee.

Personnel Profile: Part-Time Paid 3; Part-Time Volunteers 30; Interns 1.

Volunteer Hours: 4,000

Operating Expenses: 122,000

Operating Income: 124,000

Governing Authority: bd. of directors. Tax-exempt.

Institution Type/Description: Garden.

Collections: plants; trees; flowers.

Facilities: nature trails. Museum-related items for sale.

Activities: educational programs; guided tours; youth programs.

Hours & Admission Prices: Daily 9-4. Adults $5; members & children under 16 no charge. &

Attendance: 6,000 (estimated)

Membership: Annual $50 & up.

ROB SCHOUTEN GALLERY, Greenbank Farm, C103, 765 Wonn Rd., Greenbank, WA 98253. Tel.: 360-222-3070.

Institution Type/Description: Art Gallery.

Collections: works by Rob Schouten including oil paintings, etchings, & Giclee prints; works by other artists include paintings, drawings, printmaking, & sculpture.

Hours & Admission Prices: Summer: daily 10-5; Winter: daily 11-4.

Greenwater

CATHERINE MONTGOMERY INTERPRETIVE CENTER, Federation Forest State Park, 49201 Hwy. 410, Greenwater, WA 98022-8015. Tel.: 360-663-2207. Fax: 360-663-0172.

Web Site: www.parks.wa.gov

Founded: 1949.

Congressional District: 8

Key Personnel: Park Mgr., Eric Lewis.

Personnel Profile: Full-Time Paid 1; Part-Time Paid 2; Part-Time Volunteers 2; Interns 1.

Governing Authority: state. Parent Institution: The Washington State Parks & Recreation Commission, 7150 Clean Water Lane, Olympia, WA 98504.

Institution Type/Description: Nature Center.

Collections: specimens located in natural settings surrounding the interpretive center.

Facilities: interpretive center; trails; picnic area; hiking trails.

Activities: guided tours; permanent exhibitions.

Hours & Admission Prices: April-Oct. 8 a.m to dusk. &

Attendance: 75,000 (estimated)

Hoquiam

POLSON MUSEUM, (M), 1611 Riverside Ave., Hoquiam, WA 98550-2739. Mailing Address: P.O. Box 432, Hoquiam, WA 98550-0432. Tel.: 360-533-5862.

E-mail: jbl@polsonmuseum.org

Web Site: www.polsonmuseum.org

Founded: 1976.

Congressional District: 6

Key Personnel: Dir., John Larson; Pres. (V), Rob Radford.

Personnel Profile: Full-Time Paid 1; Part-Time Paid 1; Part-Time Volunteers 25.

Governing Authority: nonprofit organization. Tax-exempt: 501(c)(3).

Institution Type/Description: Park & Museum: 1924 mansion located on 1884 homestead site of timber pioneer Alex Polson.

Collections: Grays Harbor history including photographs, maps & written materials; logging & sawmill artifacts; sports; houseware items; dolls; military items; fraternal organizations.
Research Fields: 1845-1915 genealogical study of the area.
Facilities: library pertaining local history & biographies; railroad camp. Books, art work, & gift items for sale.
Activities: guided tours.
Publications: quarterly bulletin, Polson Museum News.
Hours & Admission Prices: April-Dec. 23 Wed.-Sat. 11-4, Sun. 12-4; Dec. 27-March Sat.-Sun. 12-4. Families $10, adults $4, students $2, children $1; discounts to AASLH & AAM members; members no charge.
Attendance: 4,000 (estimated)
Membership: Senior Citizen & Student $15; Individual $20; Couple $30; Family $40; Sustaining $60 & up.

Ilwaco

COLUMBIA PACIFIC HERITAGE MUSEUM, (M), 115 S.E. Lake St., Ilwaco, WA 98624. Mailing Address: P.O. Box 153, Ilwaco, WA 98624-0153. Tel.: 360-642-3446. Fax: 360-642-4615.
E-mail: info@columbiapacificheritagemuseum.org
Web Site: columbiapacificheritagemuseum.org
Formerly: Ilwaco Heritage Museum
Founded: 1983.
Congressional District: 19
Key Personnel: Exec. Dir., Betsy Millard; Chm. Bd. & Pres. (V), Karla Nelson; Librarian, Carol Bell; Collections Mgr., Barbara Minard; Museum Shop Mgr., Rosemary Hickman.
Personnel Profile: Part-Time Paid 6; Part-Time Volunteers 30; Interns 1.
Governing Authority: nonprofit organization. Parent Institution: Ilwaco Heritage Foundation. Tax-exempt: 501(c)(3).
Institution Type/Description: History Museum.
Collections: southwest Washington history from prehistoric to modern times; Lewis & Clark's expedition stay in Nov. 1805; Chinook culture; pioneer household items & tools; railroad depot; train car.
Research Fields: history; regional Native American culture.
Facilities: Museum-related items for sale.
Activities: temporary exhibitions. Annual Events: Railroad Days Festival in July; Cranberrian Festival in October; Ocian in View Lecture Series in November.
Publications: quarterly newsletter; historical shipwreck map; books, Coast Country; They Remembered I, II, III, IV.
Hours & Admission Prices: Tues.-Sat. 10-4, Sun. 12-4. Adults $5, seniors $4, youths 12-17 $2.50; discounts to AAM & ICOM members; Thurs. & members no charge. Closed Thanksgiving; Christmas.
Attendance: 13,500 (accurate)
Membership: Individual $25; Family $40; Cranberry Club $100; Clamshell $250; Heritage Club $500; Columbia Club $1,000.

LEWIS & CLARK INTERPRETIVE CENTER, Cape Disappointment State Park, Ilwaco, WA 98624. Mailing Address: P.O. Box 488, Ilwaco, WA 98624-0488. Tel.: 360-642-3078. Fax: 360-642-4216.
E-mail: lcic@parks.wa.gov
Web Site: www.capedisappointment.org
Founded: 1976.
Key Personnel: Mgr., Aaron Webster.
Personnel Profile: Full-Time Paid 2; Part-Time Volunteers 24.
Governing Authority: state. Affiliated with Washington State Parks & Recreation Commission, 7150 Clean Water Lane, Olympia, WA 98504.
Institution Type/Description: Interpretive Center.
Collections: Lewis & Clark related items; coast artillery items; Coast Guard & lighthouse artifacts.
Research Fields: Lewis & Clark; coast artillery.
Activities: self-guided tour; permanent exhibits.
Hours & Admission Prices: Daily 10-5. Adults $5, children 7-17 $2.50; children 6 & under no charge. Closed Thanksgiving; Christmas.
Attendance: 48,000 (estimated)

Issaquah

ISSAQUAH HISTORY MUSEUMS, 165 S.E. Andrews & 150 First Ave., N.E., Issaquah, WA 98027. Mailing Address: P.O. Box 695, Issaquah, WA 98027-0026. Tel.: 425-392-3500. Fax: 425-392-4236.
E-mail: info@issaquahhistory.org
Web Site: www.issaquahhistory.org
Formerly: Gilman Town Hall Museum
Founded: 1972.
Congressional District: 8
Key Personnel: Pres., Ed Seil; Dir., Erica S. Maniez; Programs Coord., Lissa Kramer; Administrative Coord., Laura C. Hall; Mgr. Collections, Julie Hunter; Archival Specialist, Julia Belgrave.
Personnel Profile: Part-Time Paid 5; Part-Time Volunteers 25; Interns 3.
Governing Authority: municipal; society; nonprofit. Parent Institution: Issaquah Historical Society. Tax-exempt: 501(c)(3).
Institution Type/Description: Historical Society/Gilman Town Hall Museum & Train Depot.
Collections: artifacts; memorabilia. Historic Structures: 1889 Town Hall; 1889 Train Depot; 1914 Town Jail.
Research Fields: historical.
Facilities: 200-vol. library for public use; educational facilities. Books & other museum-related items for sale.
Activities: guided tours; lectures; slides; organized education programs for children; participatory, loan & temporary exhibitions; school loan service.
Publications: quarterly newsletter; Preserving the Stories of Issaquah; King County Lumber Index; Images of America: Issaquah, WA.
Hours & Admission Prices: Depot: Fri.-Sun. 11-3. Town Hall: Thurs.-Sat. 11-3. Adults $2, children $1; members no charge. Closed holidays & holiday weekends.
Attendance: 7,000 (accurate)
Membership: Student & Senior Citizen $10; Senior Family $15 Individual $20; Family $25; Corporate $50.

Joint Base Lewis-McChord

LEWIS ARMY MUSEUM, Constitution Ave. & Main St., Bldg. 4320, Joint Base Lewis-McChord, WA 98433-1001. Mailing Address: P.O. Box 331001, Joint Base Lewis-McChord, WA 98433-1001. Tel.: 253-967-7206. Fax: 253-966-3029.
E-mail: lewisdptmsmuseum@conus.army.mil
Web Site: www.lewis-mcchord.army.mil/dptms/museum.htm
Formerly: Fort Lewis Military Museum
Founded: 1972.
Key Personnel: Museum Shop Mgr., Col. Ian Larson.
Personnel Profile: Full-Time Paid 3; Full-Time Volunteers 2.
Governing Authority: Dept. of the Army. Tax-exempt.
Institution Type/Description: Military Museum.
Collections: uniforms, equipment, weapons, photographs related to the military history of the Northwest from Lewis & Clark to present day; military vehicles.
Research Fields: military organizational histories camp; Fort Lewis; Northwest army.
Facilities: artillery. Museum-related gifts for sale.
Activities: historical society.
Publications: brochures; quarterly journal, The Banner.
Hours & Admission Prices: Wed.-Sun. 12-4. No charge. Closed legal holidays.
Attendance: 15,000 (estimated)
Membership: Single $10; Family $15; Sustaining Single & Family $25; Life $100; Silver Life $500; Gold Life $1,000 & up.

Kelso

COWLITZ COUNTY HISTORICAL MUSEUM, (M), 405 Allen St., Kelso, WA 98626-4103. Tel.: 360-577-3119. Fax: 360-423-9987.
E-mail: freeced@co.cowlitz.wa.us
Web Site: www.co.cowlitz.wa.us/museum
Founded: 1953.
Congressional District: 3
Key Personnel: Pres. (V), Sam Wardle; Exec. Dir., David W. Freece; Cur., Bill Watson; Education Coord., Danielle Robbins; Administrative Asst., Jim Elliott.
Personnel Profile: Full-Time Paid 1; Part-Time Paid 4; Part-Time Volunteers 25.
Governing Authority: nonprofit. Parent Institution: Cowlitz County and Cowlitz County Historical Society. Tax-exempt: 501(c)(3).
Institution Type/Description: History Museum.
Collections: community histories; tools of pioneer logging industry; photographs; manuscripts; books; ceramics; glass; textiles; Indian artifacts; doll collection; costumes; World War I & World War II collections; city planning archives; toys; waterfowl decoys; Historic Building: 1884, Ben Beighle cabin.
Research Fields: local history; genealogy; biography; Northern Pacific Railroad; river transportation.
Facilities: 1,000-vol. library of reference material available for use on premises; manuscripts, photographs & archival materials. Museum-related items for sale.

Activities: guided tours, training programs for museum interns; permanent & temporary exhibitions; lectures; workshops.
Publications: magazine, Cowlitz Historical Quarterly; quarterly newsletter; books, They Came to Six Rivers: The Story of Cowlitz County; Cowlitz County: Then and Now.
Hours & Admission Prices: Tues.-Sat. 10-4. No charge; donations accepted. Closed holidays. &
Attendance: 12,000 (estimated)
Membership: Individual $35; Family $40; Contributing $75; Benefactor & Business $150-$499; Sponsor $500-$999; Life $1,500 & up.

Kennewick

EAST BENTON COUNTY HISTORICAL MUSEUM, 205 Keewaydin Dr., Kennewick, WA 99336-0602. Tel.: 509-582-7704.
E-mail: ebchs@frontier.com
Web Site: ebchs.org
Founded: 1982.
Key Personnel: Dir., Corene Hulse.
Personnel Profile: Part-Time Paid 1.
Governing Authority: Tax exempt.
Institution Type/Description: History Museum.
Collections: local history & culture; Native American petroglyphs; pioneer farm equipment & tools; photographs; personal artifacts; period furnishings; agriculture.
Activities: educational programs.
Hours & Admission Prices: Tues.-Sat. 12-4. Adults $4, seniors $3, children 5-17 $1; active military & members no charge. &
Attendance: 2,000 (estimated)
Membership: Single $25; Family $35.

Kent

GREATER KENT HISTORICAL SOCIETY MUSEUM, 855 E. Smith St., Kent, WA 98030-4623. Tel.: 253-854-4330.
E-mail: ctyofkent@msn.com
Web Site: www.kenthistoricalmuseum.org/index.html
Founded: 1992.
Congressional District: 7th
Key Personnel: Dir., Stephen Chandler; Pres. (V), Nancy Simpson
Institution Type/Description: Historical Society Museum: housed in Bereiter's home; built in 1907. A National Historic Landmark.
Collections: artifacts; historic books; historic public & personal documents; newspaper archive; photograph archive; map archive; e-media archive; reference books.
Major Exhibits: History of Fort Thomas, 1/14.
Publications: quarterly newsletter, The Repenten.
Hours & Admission Prices: Tues.-Sat. 12-4. No charge; donations accepted. &
Attendance: 1,750 (estimated)
Membership: Senior $15; Individual $20; Family $35; Organization $50; Business & Corporate $100.

HYDROPLANE & RACEBOAT MUSEUM, 5917 S. 196th St., Kent, WA 98032-2132. Tel.: 206-764-9453. Fax: 206-766-9620.
E-mail: ddw@thunderboats.org
Web Site: www.hydromuseum.org
Founded: 1982.
Key Personnel: Dir., David D. Williams; Museum Shop Mgr., Glenn Raymond; Pres. (V), Rick Lentz.
Personnel Profile: Full-Time Paid 3; Part-Time Paid 1; Part-Time Volunteers 20.
Volunteer Hours: 9,000
Operating Expenses: 395,000
Operating Income: 400,000
Governing Authority: Tax-exempt.
Institution Type/Description: Sport Museum.
Collections: hydroplanes including boats that have won 17 Gold Cups; hydroplane racing history; books, magazines, race programs, newspaper, photos, trophies, & memorabilia; hydroplane racing videos from 1940s to present.
Facilities: Museum-related items for sale.
Activities: Annual Event: H-1 Season Preview in April; Hydro Fever open house in May; Gala Dinner & Auction in August; Fall Wine Tasting in Nov.
Hours & Admission Prices: Tues. & Thurs. 10-8, Wed. & Fri.-Sat. 10-4. Adults $10, seniors & students $5; discounts to groups; members no charge. Closed holidays. &
Attendance: 3,000 (estimated)
Membership: Limited $50; General $100; Premium $200; Benefactor $500; Lifetime $2,500; Patron $5,000.

Kettle Falls

KETTLE FALLS HISTORICAL CENTER, 1188 Portage Rd., Kettle Falls, WA 99141. Mailing Address: P.O. Box 498, Kettle Falls, WA 99141. Tel.: 509-738-6964.
Institution Type/Description: History Museum.
Collections: local history; period artifacts; photographs; Native American artifacts; early furnishings.
Hours & Admission Prices: May-Sept. Wed.-Sat. 11-5.

Keyport

NAVAL UNDERSEA MUSEUM, 1 Garnett Way, Keyport, WA 98345-7600. Mailing Address: Navy Region Northwest, 1103 Hunley Rd., Silverdale, WA 98315-1103. Tel.: 360-396-4148.
Web Site: www.navalunderseamuseum.org
Founded: 1979.
Congressional District: 1
Key Personnel: Dir., Bill Galvani; Cur., Mary Ryan; Educator, John Buchinger; Exhibits, Ron Roehmholdt; Mgr. Collections, Jennifer Heinzelman; Mgr. Collections, Lorraine Scott; Mgr. Operations, Olivia Wilson; Museum Shop Mgr., Daina Birnbaums.
Personnel Profile: Full-Time Paid 5; Part-Time Paid 2; Part-Time Volunteers 80; Interns 1.
Governing Authority: Parent Institution: Naval History & Heritage Command, US Navy. Tax-exempt.
Institution Type/Description: Military Museum.
Collections: Navy history, science & operations; torpedoes; weapons; mine; hands-on exhibits; submarine technology; diving suits; submarine battle flags.
Major Exhibits: The War of 1812, 3/12-12/15.
Activities: educational programs; concerts; lectures; hands-on exhibits; videos.
Publications: The Undersea Quarterly.
Hours & Admission Prices: June-Sept. daily 10-4; Oct.-May Wed.-Mon. 10-4. No charge; donations accepted. Closed New Year's Day; Easter; Thanksgiving; Christmas. &
Attendance: 56,000 (accurate)

Kirkland

KIRKLAND ARTS CENTER, 620 Market St., Kirkland, WA 98033-5421. Tel.: 425-822-7161. Fax: 425-889-2963.
E-mail: info@kirklandartscenter.org
Web Site: www.kirklandartscenter.org
Founded: 1962.
Key Personnel: Exec. Dir., Christopher Shainin.
Personnel Profile: Full-Time Paid 4; Part-Time Paid 4; Interns 8.
Governing Authority: Tax-exempt.
Institution Type/Description: Art Museum.
Collections: paintings; sculpture; ceramics; prints.
Activities: classes.
Hours & Admission Prices: Mon.-Fri. 11-6, Sat. 11-5. No charge; donations accepted. Closed national holidays.
Membership: Individual $50; Family $75; Patron $150; Benefactor $250; Chuck Webster Forum $500; William Radcliffe Circle $1,000.

La Conner

LA CONNER QUILT & TEXTILE MUSEUM, (M), 703 S. 2nd St., La Conner, WA 98257. Mailing Address: P.O. Box 1270, La Conner, WA 98257-1270. Tel.: 360-466-4288. Fax: 360-466-1051.
E-mail: info@laconnerquilts.com
Web Site: laconnerquilts.com
Founded: 1996.
Key Personnel: Dir., Amy Green; Pres. (V), Susan Wells Hall.
Personnel Profile: Full-Time Paid 1; Part-Time Paid 1.
Governing Authority: nonprofit organization. Tax-exempt.
Institution Type/Description: Quilt & Textile Museum: housed in a Victorian mansion built in 1891.
Collections: national & international quilts with a focus on works from the Pacific Northwest.
Activities: monthly textile enrichment series; special events. Annual Events: Autumn Quilt Fest; Arts Alive In November.
Publications: quarterly newsletter.
Hours & Admission Prices: Wed.-Sun. 11-5; other times by appointment. Adults $7, students & military $5; members & children under 12 no charge. Closed Thanksgiving; Christmas. &
Attendance: 7,200 (estimated)
Membership: Individual $30; Family $40; Sponsor $100; Patron $500.

MUSEUM OF NORTHWEST ART, (M), 121 S. First St., La Conner, WA 98257. Mailing Address: P.O. Box 969, La Conner, WA 98257-0969. Tel.: 360-466-4446, ext. 109. Fax: 360-466-7431.
E-mail: timd@museumofnwart.org
Web Site: www.museumofnwart.org
Founded: 1981.
Congressional District: 40
Key Personnel: C.E.O. & Pres. (V), Jessica Pavish; Exec. Dir., Tim Detweiler; Cur., Kathleen Moles; Museum Shop Mgr., Jacque Chase.
Personnel Profile: Full-Time Paid 4; Part-Time Paid 7; Part-Time Volunteers 140.
Governing Authority: nonprofit organization. Tax-exempt: 501(c)(3).
Institution Type/Description: Art Museum.
Collections: Pacific Northwest regional art, early 20th century to present.
Research Fields: Pacific Northwest Art.
Facilities: library; 12,000 sq. ft. exhibit space. Gift items for sale.
Activities: permanent collection; guided tours; lectures; arts festivals; organized education programs for adults & children; loan & temporary exhibitions.
Publications: exhibition books.
Hours & Admission Prices: Sun.-Mon. 12-5, Tues.-Sat. 10-5. Adults $8, seniors $4, student $3; members & youth under 12 no charge. Closed Thanksgiving; Christmas. &
Attendance: 14,291 (accurate)
Membership: Student $15; Individual $35; Family & Small Business $50; Donor $100; Sponsor & Corporation $250; Patron $500; Benefactor $1,000.

SKAGIT COUNTY HISTORICAL MUSEUM, 501 S. 4th St., La Conner, WA 98257-0818. Mailing Address: P.O. Box 818, La Conner, WA 98257-0818. Tel.: 360-466-3365. Fax: 360-466-1611.
E-mail: museum@co.skagit.wa.us
Web Site: www.skagitcounty.net/museum
Founded: 1959.
Congressional District: 2
Key Personnel: Interim Dir. & Cur., Patricia L. Doran; Pres. (V), Jaci Turner; Tour Coord., Eileen Barnes; Librarian, Mari C. Anderson-Densmore; Fundraiser & Publicity Coord., Jo Wolfe; Maintenance & Security, Bob Skeele; Business Mgr., Kathy Pace.
Personnel Profile: Part-Time Paid 9; Part-Time Volunteers 60.
Governing Authority: Parent Institution: Skagit County Historical Society. Subsidiary Institution: Skagit County. Tax-exempt: 501(c)(3).
Institution Type/Description: History Museum.
Collections: library; archives; manuscripts; oral history; photographs; general artifact collections documenting local history topics: recreation; entertainment; art; domestic life; clothing; ethnicity; politics; communication; transportation; commerce; industry; agriculture. Historic Building: Rosario School.
Research Fields: Skagit county history.
Facilities: 1,500-vol. research library & 20,000 photographs available for use by appointment; video theater. Books & other museum-related items for sale.
Activities: guided tours; lectures; workshops; educational programs for children; long-term, temporary, traveling & hands-on exhibitions; video programs; outreach programs for children & adults. Museum Sponsors: Heritage Award; special events; Treaty Day Celebration; Rosario School Picnic.
Publications: books, Skagit County History (5 volumes); Chechacos All; Indians of Skagit County; Skagit Memories; Sternwheelers & the Skagit River; Skagit Settlers; Skagit County Grows Up, 1917-1941; Last Frontier in the North Cascades; Harvesting the Light: Images of Contemporary Skagit Farm Life, 2007; Harvesting the Light-Farming the Skagit Vallet; self-guided walking tour of La Conner; quarterly newsletter; brochures.
Hours & Admission Prices: Tues.-Sun. 11-5. Family $8, adults $4, seniors & children 6-12 $3; discounts to AAM & ICOM members; historical society members and children 5 & under no charge. &
Attendance: 10,000 (estimated)
Membership: Individual $25; Family $40; Business/Organization $50; Patron $80; Benefactor $250.

Lacey

LACEY MUSEUM, 829 Lacey St., S.E., Lacey, WA 98503. Mailing Address: 420 College St., S.E., Lacey, WA 98503. Tel.: 360-438-0209.
E-mail: equinnva@ci.lacey.wa.us
Web Site: www.ci.lacey.wa.us
Founded: 1980.
Key Personnel: Cur., Erin Quinn Valcho.
Personnel Profile: Full-Time Paid 1; Part-Time Volunteers 10.
Governing Authority: Parent Institution: City of Lacey.
Institution Type/Description: History Museum.
Collections: community history from the Oregon Trail days to the present.
Research Fields: history of Lacey area & WA state; local family genealogy.
Hours & Admission Prices: Thurs.-Fri. 11-3, Sat. 10-4 ; other times by appointment. No charge; donations accepted.
Attendance: 400 (accurate)

Lakewood

HISTORIC FORT STEILACOOM, 9601 Steilacoom Blvd. S.W. (on Western State Hospital grounds), Lakewood, WA 98498-7213. Mailing Address: P.O. Box 88447, Steilacoom, WA 98388-0447. Tel.: 253-582-5838.
E-mail: info@historicfortsteilacoom.org
Web Site: www.historicfortsteilacoom.org
Founded: 1983.
Congressional District: 10
Key Personnel: Pres., Lawrence Bateman; Treas., Michael McGuire; Dir. & Sec., Joseph W. Lewis.
Personnel Profile: Part-Time Volunteers 16.
Governing Authority: nonprofit organization. Tax-exempt: 501(c)(3).
Institution Type/Description: Historic Site: four original buildings remaining on the site of Fort Steilacoom.
Collections: artifacts; recreated furnishings.
Research Fields: Pierce County history; military history (1849-1868).
Facilities: meeting rooms; interpretive center. Gift items for sale.
Activities: guided tours; lecture series; docent program; events relating to local history; living history.
Publications: Historic Fort Steilacoom newsletter; brochure; tour guide.
Hours & Admission Prices: Jan.-May & Sept.-Dec. first Sun. each month 1-4; June-Aug. Sun. 1-4. No charge; donations accepted.
Membership: Senior Citizen & Student $15; Individual $20; Family $35; Patron $100; Life $300.

Langley

SOUTH WHIDBEY HISTORICAL SOCIETY, 312 Second St., Langley, WA 98260. Mailing Address: P.O. Box 612, Langley, WA 98260. Tel.: 360-221-2101.
Web Site: www.southwhidbeyhistory.org
Institution Type/Description: Historical Society Museum.
Collections: local history, heritage, & culture; period furnishings; personal artifacts; photographs.
Activities: Annual Event: Then and Now evening presentations.
Hours & Admission Prices: Feb.-May & Sept.-Nov. Sat.-Sun. 1-4; June-Aug. Fri.-Sun. 1-4; other times by appointment.

Leavenworth

LEAVENWORTH NUTCRACKER MUSEUM, 735 Front St., Leavenworth, WA 98826. Mailing Address: P.O. Box 129, Leavenworth, WA 98826. Tel.: 509-548-4573.
E-mail: curator@nutcrackermuseum.com
Web Site: www.nutcrackermuseum.com
Founded: 1995.
Congressional District: 4
Institution Type/Description: Nutcracker Museum.
Collections: over 6,000 nutcrackers dating back to Roman times.
Publications: monthly newsletter.
Hours & Admission Prices: May-Oct. daily 2-5; Nov.-April Sat.-Sun. 2-5; other times by appointment. Adults $2.50, students $1; discounts to groups; active military & their families and children 5 & under no charge. &
Attendance: 10,000 (estimated)

Lind

ADAMS COUNTY HISTORICAL SOCIETY MUSEUM, First St., Lind, WA 99341. Mailing Address: P.O. Box 526, Lind, WA 99341-0526. Tel.: 509-677-3393 & 3642.
E-mail: galeirma@ritzcom.net
Founded: 1963.
Congressional District: 5
Key Personnel: Dir. & Sec., Irma E. Gfeller; Pres., Allan Koch.
Personnel Profile: Part-Time Volunteers 20.
Governing Authority: county. Tax-exempt.
Institution Type/Description: Local History Museum.
Collections: local history & memorabilia; personal artifacts.
Research Fields: local history.

Activities: guided tours.
Publications: books, History of Adams Co. Washington Vol. I & II.
Hours & Admission Prices: By appointment only. No charge; donations accepted. ♿
Attendance: 400 (estimated)
Membership: Individual & Family $25; Organizations $100.

Long Beach

PACIFIC COAST CRANBERRY RESEARCH FOUNDATION MUSEUM & GIFT SHOP, 2907 Pioneer Rd., Long Beach, WA 98631-5011. Tel.: 360-642-5553.
E-mail: cranberries@willapabay.org
Web Site: www.cranberrymuseum.com
Governing Authority: nonprofit organization.
Institution Type/Description: Cranberry Farm Museum.
Collections: cranberry history, farming, varieties, & marketing.
Facilities: Museum-related items for sale.
Activities: tours; demonstrations; educational activities; crop harvesting in October.
Hours & Admission Prices: April-Dec. 15 daily 10-5; other times by appointment.

WORLD KITE MUSEUM & HALL OF FAME, 303 Sid Snyder Dr., Long Beach, WA 98631-3725. Mailing Address: P.O. Box 964, Long Beach, WA 98631-0964. Tel.: 360-642-4020. Fax: 360-642-4020.
E-mail: info@worldkitemuseum.com
Web Site: www.worldkitemuseum.com
Founded: 1988.
Congressional District: 19
Key Personnel: Dir., Chelsay Libby; Chm. (V), Blaine Walker.
Personnel Profile: Part-Time Paid 3; Part-Time Volunteers 21.
Governing Authority: private; nonprofit organization. Tax-exempt: 501(c)(3).
Institution Type/Description: Hobby Museum: promote fun, art & science of kiting by telling the history of kites, recording the present & honoring the people involved.
Collections: kites from 27 different countries; kite-flying accessories; stories of kite makers; archives for kites, kite makers & kiting events; history of kites & kite celebrations; significance of kites in cultures around the world.
Research Fields: kite as art; historical uses of kites & kite patterns; stories of kite makers; kite festivals; oral history program.
Facilities: 269-vol. library; 8,000-piece archives; 1,500 sq. ft. exhibit space. Museum-related items for sale.
Activities: lectures; hobby workshops; films; docent program; temporary exhibits; formal education programs for children & adults. Annual Events: Hall of Fame Announcement; Kite Auction; Windless Kites in January; Kite Making in February; Asian New Year in February; Spring Break, Family Fun in March & April; Washington State International Kite Festival in August; One Sky, One World Kite Festival in Oct.
Publications: quarterly newsletter, The Flyer; monthly e-letter.
Hours & Admission Prices: May-Sept. daily 11-5; Winter Fri.-Mon. 11-5. Adults $5, senior citizens $4, children $3; discounts to groups of 10 or more & AAM members. Closed New Year's Day; Thanksgiving; Christmas. ♿
Attendance: 11,000 (estimated)
Membership: Individual $35; Family & Donor $50; Benefactor $500.

Longview

THE ART GALLERY, LOWER COLUMBIA COLLEGE FINE ARTS GALLERY, 1600 Maple St., Longview, WA 98632-3907. Mailing Address: P.O. Box 3010, Longview, WA 98632-0310. Tel.: 360-442-2510. Fax: 360-577-6620.
E-mail: dbartlett@lowercolumbia.edu
Web Site: lowercolumbia.edu/gallery
Founded: 1978.
Congressional District: 18
Key Personnel: Dir., Diane Bartlett.
Personnel Profile: Part-Time Paid 1.
Governing Authority: college. Parent Institution: Lower Columbia College.
Institution Type/Description: College Art Gallery.
Collections: paintings; prints by Pacific Northwest artists; various types of art media exhibited by students, faculty & visiting artist.
Activities: permanent, temporary & traveling exhibitions; lecture workshop series.
Hours & Admission Prices: Mon.-Tues. & Fri. 10-4, Wed.-Thurs. 10-7. No charge. Closed legal holidays. ♿
Attendance: 5,437 (accurate)

Lopez Island

LOPEZ ISLAND HISTORICAL MUSEUM, (M), 28 Washburn Pl., Lopez Village, Lopez Island, WA 98261. Mailing Address: P.O. Box 163, Lopez Island, WA 98261-0163. Tel.: 360-468-2049.
E-mail: lopezmuseum@rockisland.com
Web Site: www.lopezmuseum.org
Founded: 1980.
Congressional District: 40
Key Personnel: Exec. Dir., Mark Thompson-Klein; Pres. (V), Ande Finley.
Personnel Profile: Part-Time Paid 2; Part-Time Volunteers 40; Interns 2.
Governing Authority: nonprofit. Parent Institution: Lopez Island Historical Society. Tax-exempt: 501(c)(3).
Institution Type/Description: Historical Society & Museum.
Collections: regional, native & pioneer history; agriculture; maritime; anthropology; quilts; photograph collection; archives.
Research Fields: genealogy.
Facilities: books for sale.
Activities: guided tours; lectures; films.
Publications: semiannual newsletter.
Hours & Admission Prices: May-Sept. Wed.-Sun. 12-4. Adults $2; members no charge. Closed Memorial Day Sun.; Independence Day. ♿
Attendance: 2,500 (accurate)
Membership: Individual $15; Family $25; Sponsor $100; Benefactor $1,000.

Lynden

LYNDEN PIONEER MUSEUM, 217 W. Front St., Lynden, WA 98264-1418. Tel.: 360-354-3675. Facebook: Lynden Pioneer Museum.
E-mail: lyndenpioneermuseum@gmail.com
Web Site: www.lyndenpioneermuseum.com
Founded: 1976.
Congressional District: 42
Key Personnel: Dir. & Cur., Troy Luginbill; Pres. (V), Clarence Zylstra; Treas., David Vos; Volunteer & Membership Mgr., Tammi Rylaarsdam.
Personnel Profile: Full-Time Paid 1; Part-Time Paid 3; Part-Time Volunteers 137; Interns 20.
Governing Authority: private; nonprofit organization. Parent Institution: Lynden Heritage Foundation. Tax-exempt: 501(c)(3).
Institution Type/Description: History Museum.
Collections: prehistory & settlement of historic Whatcom County; historic methods of travel & transportation from 1700s to present.
Research Fields: historical research of local communities, Whatcom Memories; a Photographic and Interview Project; oral histories & archival collections involving local community heritage.
Facilities: 50-vol. library of history & heritage of Whatcom County; 200-seat auditorium; 2,000 sq. ft. exhibit space. Museum-related items for sale.
Activities: docent program; formal education programs for adults, college students & children; guided tours; hobby workshops; lectures; participatory & temporary exhibits; rental gallery; school loan services; training programs for professional museum workers. Annual Event: Open House.
Publications: semiannual newsletter, Museum Musings; bimonthly historical journal, Whatcom Moments & Memoirs.
Hours & Admission Prices: Mon.-Sat. 10-4. Adults $7, senior citizens & students $4; discounts to groups, Washington Museum Association members, AAM & ICOM members; members, children 6 & under no charge.
Attendance: 28,000 (estimated)
Membership: Individual $25; Senior $35; Family & Donor $50; Sponsor $100; Patron $250; Benefactor $500; Visionary $1,000; Cornerstone $5,000.

Lynnwood

GENEALOGY RESEARCH LIBRARY, 19827 Poplar Way, Lynnwood, WA 98036-6940. Tel.: 425-775-6267.
Governing Authority: Tax-exempt.
Institution Type/Description: Library: housed in the Humble House.
Collections: books; periodicals; family heritage; local history; photographs.
Facilities: 1,500-vol. library.
Activities: research.
Hours & Admission Prices: Tues. 10-2, Thurs. 10-8, Sat. 10-3.

HERITAGE PARK MUSEUM AND INTERURBAN CAR 55, 19921 Poplar Way, Lynnwood, WA 98036-6940. Mailing Address: 19000 44th Ave. W., Lynnwood, WA 98036. Tel.: 425-744-6478.
Founded: 2004.
Congressional District: 1
Key Personnel: Dir., Parks Recreation & Cultural Arts, Lynn Sordel; Chm. (V), Laurie Cowan

Governing Authority: Parent Institution: City of Lynnwood. Subsidiary Institution: Parks Foundation. Tax-exempt.
Institution Type/Description: Transportation History Museum.
Collections: trolley; Lynnwood's transportation history & heritage; photographs.
Facilities: library of genealogy records; visitor information center; heritage resource center.
Activities: guided tours; research.
Publications: Tour of Historic Sites in Lynnwood.
Hours & Admission Prices: Park Museum: Mon.-Tues. & Thurs.-Fri. 9-5, Wed. & Sat.-Sun. 9-3. Trolley Tours: June-Sept. 1st Sat. each month 11-3. No charge. &
Attendance: 3,000 (estimated)

HERITAGE RESOURCE CENTER, 19903 Poplar Way, Lynnwood, WA 98036-6940. Mailing Address: Alderwood Manor Heritage Assoc., P.O. Box 2206, Lynnwood, WA 98036-2206. Tel.: 425-775-4694.
Institution Type/Description: Heritage Center: housed in the Alderwood Manor Heritage Cottage.
Collections: Alderwood Manor history, heritage & culture; photographs; newspapers; manuscripts; oral histories; books.
Activities: guided tours.
Hours & Admission Prices: Tues., Thurs. & Sat. 11-3.

SNOHOMISH COUNTY VISITOR INFORMATION CENTER, 19921 Poplar Way, Lynnwood, WA 98036-6940. Tel.: 425-776-3977.
Institution Type/Description: Visitor Center.
Collections: community, county, & state history; photographs.
Facilities: visitor center.
Activities: research.
Hours & Admission Prices: Sun., Tues. & Thurs. 9-3, Mon., Wed. & Fri.-Sat. 9-5.

TRANSPORTATION MUSEUM, 19921 Poplar Way, Lynnwood, WA 98036-6940. Tel.: 425-774-6478.
Institution Type/Description: Transportation Museum.
Collections: Lynnwood transportation history; photographs.
Hours & Admission Prices: Sun., Tues. & Thurs. 9-3, Mon., Wed., & Fri.-Sat. 9-5.

Maple Valley

MAPLE VALLEY HISTORICAL SOCIETY, 23015 S.E. 216th Way, Maple Valley, WA 98038-8412. Mailing Address: P.O. Box 123, Maple Valley, WA 98038-0123. Tel.: 425-432-3470.
E-mail: pilgrim.dskc@gmail.com
Web Site: www.maplevalleyhistorical.com
Founded: 1972.
Congressional District: 8
Key Personnel: Pres. & Museum Shop Mgr., Mona Pickering.
Personnel Profile: Part-Time Volunteers 15.
Governing Authority: society; nonprofit organization. Parent Institution: Maple Valley Historical Society, Inc. Tax-exempt: 501(c)(3).
Institution Type/Description: Historical Society Museum.
Collections: 1930-1964 school pictures; 1900s items & artifacts; 1920s fire engine. Historic Building: 1894 general store; c.1920 school building.
Facilities: Books for sale.
Activities: guided tours.
Publications: annual booklet.
Hours & Admission Prices: 1st Sat. of each month 10-2; other times by appointment. No charge; donations accepted.
Attendance: 1,250 (accurate)
Membership: Senior Citizen 62 & over and Individual $20; Family & Nonprofit Club $50; Business $100; Life $500.

Marysville

MARYSVILLE HISTORICAL SOCIETY, 1508-B Third St., Marysville, WA 98270-5002. Mailing Address: P.O. Box 41, Marysville, WA 98270. Tel.: 360-659-3090. Fax: 360-659-0725.
E-mail: info@marysvillehistory.org
Web Site: marysvillehistory.org
Founded: 1974.
Congressional District: 2
Key Personnel: Pres. (V), Ken Cage.
Personnel Profile: Part-Time Volunteers 20.

Governing Authority: Tax-exempt.
Institution Type/Description: Historical Society Museum.
Collections: local history & culture; photographs; period furnishings; personal artifacts; telephones.
Activities: special events.
Publications: quarterly newsletter.
Hours & Admission Prices: Mon.-Sat. 10-3. No charge; donations accepted.
Attendance: 2,000 (estimated)
Membership: Annual $25; Life $300.

McChord Field

MCCHORD AIR MUSEUM, Airforce Base, 100 Main St., McChord Field, WA 98438. Mailing Address: McChord Air Museum Foundation, P.O. Box 4205, Tacoma, WA 98438-0205. Tel.: 253-982-2419. Fax: 253-982-9560.
E-mail: raymond.jordan@mcchord.af.mil
Web Site: www.mcchordmuseum.org
Founded: 1983.
Key Personnel: Vice Pres., Randy Getz; Pres. (V), Tom Hansen; Museum Shop Mgr., Ernie White, II
Personnel Profile: Full-Time Volunteers 1; Part-Time Volunteers 60.
Governing Authority: private; nonprofit organization. Parent Institution: McChord Airforce Base. Tax-exempt.
Institution Type/Description: Military Museum.
Collections: history of the Air Force units associated with McChord.
Publications: quarterly, The Rip Chord.
Hours & Admission Prices: Wed.-Fri. 12-4; military ID required for entry. No charge; donations accepted.
Attendance: 5,000 (accurate)
Membership: Student & Airman $5; Regular $20; Silver Wing Contributor $50-$499; Silver Wing Patron $500-$999; Silver Wing Benefactor $1,000 & up.

Moclips

MUSEUM OF THE NORTH BEACH, 4658 State Rte. 109, Moclips, WA 98562. Mailing Address: P.O. Box 231, Moclips, WA 98562-0231. Tel.: 360-276-4441.
E-mail: kelly@moclips.org
Web Site: www.moclips.org
Key Personnel: Pres. Bd. Dirs., Kelly Calhoun.
Governing Authority: nonprofit organization. Parent Institution: Moclips-by-the-Sea Historical Society. Tax-exempt: 501(c)(3).
Institution Type/Description: History Museum.
Collections: local history & culture; photographs; period furnishings; personal artifacts.
Hours & Admission Prices: May-Oct. Thurs.-Mon. 11-4; Nov.-April Sat.-Sun. 11-4.

Monroe

MONROE HISTORICAL MUSEUM, 207 E. Main St., Monroe, WA 98272. Mailing Address: P.O. Box 1044, Monroe, WA 98272-4044. Tel.: 360-217-7223.
E-mail: info@monroehistoricalsociety.comcastbiz.net
Web Site: monroehistoricalsociety.org
Formerly: Monroe Historical Society Museum
Founded: 1976.
Congressional District: 39
Key Personnel: Pres. (V), Dexter Taylor; Dir. Museum Shop Mgr., Chris Bee.
Personnel Profile: Full-Time Volunteers 1; Part-Time Volunteers 20.
Governing Authority: Tax-exempt.
Institution Type/Description: Historical Society Museum.
Collections: local history & culture; photographs; personal artifacts; period furnishings; 1908 two-story cement block old city hall containing exhibit galleries.
Major Exhibits: Monroe Athletes, 1/2013-6/2013; Logging, 7/2013-12/2013.
Research Fields: Local & regional settlement; city development; genealogy; buildings; military history; racing; local schools; & athletes.
Facilities: Research library; archives; hands-on room; conference room; & two rental spaces.
Activities: Three to four membership meetings per year (programs on local history); quilt raffle, booth at local events & hosting of the Shannahan Cabin at the Evergreen State Fair every year.
Publications: quarterly newsletter, Heritage Herald; Monroe: The First Fifty Years; Monroe: The Next Thirty Years; Monroe Cemeteries; Monroe Area Pioneers, Old Timers & Other Noted Citizen; Local History Series Booklets: The First Twenty-five, The Next The next Twelve; booklets, The Frye Lettuce Farm, Snohomish County Fairs, Early Monroe, Monroe Fire Department, Early Monroe Arcadia.

Hours & Admission Prices. Mon. & Wed. 1-4, Sat. 11-3. No charge; donations accepted.
Attendance: 451 (accurate)
Membership: Individual $15; Family $25; Business $50; Lifetime $150.

Moses Lake

MOSES LAKE MUSEUM & ART CENTER, (M), 401 S. Balsam, Moses Lake, WA 98837-1933. Mailing Address: P.O. Drawer 1579, Moses Lake, WA 98837. Tel.: 509-764-3830. Fax: 509-764-3709.
E-mail: fhart@cityofml.com
Web Site: www.moseslakemuseum.com
Formerly: Adam East Museum
Founded: 1958.
Congressional District: 4
Key Personnel: Dir., Freya K. Liggett; Cur., Ann Schempp.
Personnel Profile: Full-Time Paid 2; Part-Time Paid 3; Part-Time Volunteers 75; Interns 1.
Governing Authority: municipal; nonprofit organization. Parent Institution: City of Moses Lake. Tax-exempt.
Institution Type/Description: History Museum & Art Center.
Collections: home of the Adam East Collection of Native American artifacts from the mid-Columbia.
Major Exhibits: Moses Lake Women's Club 1914-2014, 1/10/14-2/28/14; Subscapes: Russ Rummler, 1/10/14-2/28/14; Don Nutt, 3/7/14-4/25/14; Brent Blake & Kathy Keifer, 5/2/14-6/13/14; Feelings, Flora and Form: Martha Flores, 6/20/14-8/1/14; Libby Eastman Sullivan and Tania Gonzalez Ortega, 8/8/14-9/26/14; Frances Wood, 11/21/14-1/2/15.
Facilities: auditorium; classroom; council chambers.
Activities: history & art exhibits; cultural events; art workshops; school program; family activities.
Publications: quarterly activity schedule.
Hours & Admission Prices: Mon.-Sat. 11-5. No charge. Closed major holidays. &
Attendance: 12,000 (estimated)
Membership: Senior & Student $20; Individual $30; Family $40; Associate $55; Booster $100; Sponsor $250; Patron $500; Benefactor $1,000.

Mukilteo

FUTURE OF FLIGHT AVIATION CENTER & BOEING TOUR, 8415 Paine Field Blvd., Mukilteo, WA 98275-3239. Tel.: 425-438-8100, ext. 224. Fax: 425-265-9808.
E-mail: info@futureofflight.org
Web Site: www.futureofflightfoundation.org
Founded: 2005.
Congressional District: 1
Key Personnel: Dir. & C.E.O., Barry Smith; Pres. (V), Tom Sanger; Dir. Devel., Mary Brueggeman; Museum Shop Mgr., Peter Bro.
Personnel Profile: Full-Time Paid 12; Part-Time Paid 14; Part-Time Volunteers 240; Interns 19.
Volunteer Hours: 600
Operating Income: 1,075,000
Institution Type/Description: Aviation Center.
Collections: aviation history; airplanes; innovations in aviation; photographs; videos; biofuels.
Major Exhibits: Ball State University Recycling of Airplanes, 11/13-1/14.
Facilities: 240-seat theater; 125-seat cafe; strato deck. Gift items for sale.
Activities: design your own airplane; hands-on exhibits; assembly plant tour; rental facilities; educational program for middle school students; community events.
Hours & Admission Prices: Daily 8:30-5:30. Adults $20, children 5-15 $14, children under 5 no charge; discounts for online tickets. Closed Thanksgiving; Christmas. &
Attendance: 240,000 (estimated)
Membership: Friends $50; Explorers $100; Innovators $150; Researchers $250; Ambassadors $500; Leadership $1,000.

HISTORIC FLIGHT FOUNDATION, 10719 Bernie Webber Dr., Mukilteo, WA 98275. Tel.: 425-348-3200. Facebook: Historic Flight.
E-mail: airborne@historicflight.org
Web Site: historicflight.org
Founded: 2005.
Congressional District: 1
Personnel Profile: Part-Time Paid 4.
Institution Type/Description: Aviation History Museum.
Collections: aviation history; aircraft.
Activities: special events. Annual Events: Air Shows.
Hours & Admission Prices: Labor Day to Memorial Day Thurs.-Sun. 10-5;

May-Sept. Tues.-Sun. 10-5. Adults $12; discounts to seniors, military & AAA members; members no charge. Closed Thanksgiving; Christmas. &
Attendance: 25,000 (accurate)
Membership: Senior & Military $50; Adult $75; Family $100. Flight: Beaver $150; Staggerwing $300; Douglas DC-3 $350; B25D Mitchell $495; T-6 Texan $600; F7F Tigercat $3,500.

MUKILTEO LIGHT STATION & INTERPRETIVE CENTER, 608 Front St., Mukilteo, WA 98275. Mailing Address: Mukilteo Historical Society, 304 Lincoln Ave., Ste. 101, Mukilteo, WA 98275.
Institution Type/Description: Light Station Museum.
Collections: local history & culture; documents; photographs; papers; personal artifacts.
Facilities: Museum-related items for sale.
Activities: guided tours; special events; rental facilities.
Hours & Admission Prices: Grounds: daily. Lighthouse & Center: April-Sept. Sat.-Sun. & holidays 12-5; other times by appointment. No charge.

Neah Bay

MAKAH CULTURAL AND RESEARCH CENTER, 1880 Bayview Ave., Neah Bay, WA 98357. Mailing Address: P.O. Box 160, Neah Bay, WA 98357. Tel.: 360-645-2711. Fax: 360-645-2656.
Institution Type/Description: Native American History Museum.
Collections: Native American history, culture, & artifacts; photographs; dug-out canoes; whaling; sealing; fishing; basketry; tools.
Facilities: Museum-related items for sale.
Activities: educational programs.
Hours & Admission Prices: Daily 10-5. Adults $5, students, military & seniors $4; discounts to groups; children & under no charge.

Nine Mile Falls

SPOKANE HOUSE INTERPRETIVE CENTER, 9711 W. Charles, Nine Mile Falls, WA 99026-8648. Tel.: 509-465-5064 & 466-4747. Fax: 509-465-5571.
E-mail: riverside@parks.wa.gov
Web Site: www.riversidestatepark.org
Founded: 1950.
Congressional District: 6
Key Personnel: Chief Interpretive Svcs., Steve Wang; Agency Dir., Rex Derr; Park Mgr., Rene Wiley.
Personnel Profile: Part-Time Paid 1; Part-Time Volunteers 5.
Governing Authority: state. Parent Institution: Riverside State Park. Affiliated with the Washington State Parks & Recreation Commission, 7150 Clean Water Ln., Olympia, WA 98504.
Institution Type/Description: Historic Site Museum: 1810, trading post used as a fur trading post & operated at various times by British, Canadian & American interests.
Collections: archaeological investigations; American Indian culture & artifacts; Tribes of the Plateau.
Activities: permanent exhibitions.
Publications: brochure.
Hours & Admission Prices: Memorial Day to Labor Day Sat.-Sun. 10-4. No charge; donations accepted. &
Attendance: 7,000 (accurate)

North Bend

SNOQUALMIE VALLEY HISTORICAL MUSEUM, 320 Bendigo Ave., S., North Bend, WA 98045-8260. Mailing Address: P.O. Box 179, North Bend, WA 98045-0179. Tel.: 425-888-3200. Fax: 425-888-3200. Facebook: Snoqualmie Valley Historical Museum.
E-mail: info@snoqualmievalleymuseum.org
Web Site: www.snoqualmievalleymuseum.org
Founded: 1960.
Congressional District: 8
Key Personnel: Pres. (V), Kris Kirby; Treas., Vicki Bettes; Asst. Dir., Cristy Lake.
Governing Authority: Parent Institution: Snoqualmie Valley Historical Society. Tax-exempt.
Institution Type/Description: Local Historical Museum.
Collections: local pioneer artifacts and photographs; local Indian and northwest artifacts and history; photographic collection of Valley; pioneer diaries; manuscript collection. Historic Structures: 1912 vintage parlor; farm shed with vintage implements of agriculture, transportation & industry; early logging diorama.
Research Fields: local area history; valley forts; Snoqualmie Pass Wagon Road.

Facilities: history library available by appointment; early school texts; medical books available on premises or by appointment. Local historical books, booklets, map & tour guide brochure of Valley historic sites & other museum-related items for sale.

Activities: guided tours; lectures; docent program; informal education tours; mythology of Southern Puget Sound; slide programs.

Publications: books, A History of the Snoqualmie Valley; Fall City in the Valley of the Moon; booklet, Forts of the Snoqualmie Valley; brochure; historic tour guide; quarterly newsletter.

Hours & Admission Prices: April-Oct. Sat.-Tues. 1-5; Nov.-March Mon.-Tues. 12-4; other times by appointment. Suggested Donation $1. Closed national holidays. ♿

Attendance: 4,000

Membership: Individual Yearly $25; Individual Life $250.

Oak Harbor

PBY NAVAL HERITAGE CENTER, NAS Whidbey Island - Seaplane Base, Simard Hall Bldg. 12, 315 W. Pioneer Way, Oak Harbor, WA 98278. Mailing Address: P.O. Box 941, Oak Harbor, WA 98277-0941. Tel.: 360-240-9500.

Web Site: www.pbymf.org
Founded: 1998.
Congressional District: 10
Key Personnel: Dir., Will Stein; Chm. (V)., Richard Rezabek; Pres. (V), Win Stites; Museum Shop Mgr., George Love.
Personnel Profile: Part-Time Volunteers 12; Interns 2.
Governing Authority: Parent Institution: PBY Memorial Foundation; Subsidiary Institution: NAS Widbey Island & US Navy. Tax-exempt.
Institution Type/Description: Military History Museum; Heritage Center.
Collections: naval history; personal artifacts; military artifacts; photographs.
Hours & Admission Prices: Wed.-Sat. 11-5. (Photo ID, car registration & insurance papers required for entry). No charge; donations accepted. ♿
Attendance: 1,790 (accurate)
Membership: Individual $25; Family $35; Life $500.

Okanogan

OKANOGAN COUNTY HISTORICAL SOCIETY & FIRE HALL MUSEUM, 1410 2nd Ave. N., Okanogan, WA 98840-1129. Tel.: 509-422-4272.

E-mail: ochs@ncidata.com
Web Site: www.okanoganhistory.org
Institution Type/Description: History Museum.
Collections: local history; firefighting history, artifacts & equipment; photographs; personal artifacts; books; maps.
Publications: quarterly magazine, Okanogan County Heritage.
Hours & Admission Prices: Daily 10-4.

Olympia

BIGELOW HOUSE MUSEUM, 918 Glass Ave., N.E., Olympia, WA 98506-3976. Mailing Address: P.O. Box 1821, Olympia, WA 98507-1821. Tel.: 360-753-1215.

E-mail: bigelowhousemuseum@gmail.com
Web Site: www.bigelowhouse.org
Founded: 1992.
Congressional District: 3
Key Personnel: Pres. (V), Roger Easton.
Personnel Profile: Part-Time Volunteers 25.
Governing Authority: nonprofit organization. Administered by Bigelow House Preservation Association. Tax-exempt.
Institution Type/Description: Historic House Museum: ca. 1860 Bigelow House.
Collections: 19th-century furnishings; books; photographs; domestic arts.
Research Fields: Social, political & domestic history of Washington Territory.
Activities: guided tours; lectures.
Publications: book, Stories of the Oregon Trail; Workingman's Hill: History of an Olympia Neighborhood.
Hours & Admission Prices: May-Oct. Sat.-Sun. 12-4; other times by appointment. Adults $3, children 18 & under $1; discounts to Life Balance Program members; members no charge. ♿
Attendance: 930 (accurate)
Membership: Family $35; Pioneer $75; Territorial $150; Third Century $300; Lifetime $1,859.

EVERGREEN GALLERY, (M), The Evergreen State College, 2700 Evergreen Pkwy., N.W., Olympia, WA 98505. Tel.: 360-867-5125. Fax: 360-867-6794. Facebook: Evergreen Gallery Olympia WA.

E-mail: gallery@evergreen.edu
Web Site: www.evergreen.edu/gallery
Founded: 1970.
Congressional District: 10
Key Personnel: Dir., Ann Friedman.
Personnel Profile: Part-Time Paid 1.
Governing Authority: college. Affiliated with The Evergreen State College. Tax-exempt.
Institution Type/Description: College Art Gallery.
Collections: photography; painting; sculpture; prints; drawings; ceramics; Chicano posters.
Major Exhibits: Selected Druckworks: Books & Projects by Johanna Drucker, 1/14-3/14.
Facilities: gallery.
Activities: artist talks & workshops; loan, temporary & traveling exhibitions.
Hours & Admission Prices: Oct.-May Mon.-Thurs. 12-4. No charge. Closed national holidays. ♿

HANDS ON CHILDREN'S MUSEUM, 414 Jefferson St., N.E., Olympia, WA 98501. Tel.: 360-956-0818, ext. 0. Fax: 360-754-8626. Facebook: Hands On Children's Museum.

E-mail: hocm@hocm.org
Web Site: www.hocm.org
Founded: 1987.
Congressional District: 3
Key Personnel: C.E.O. & Dir., Patty Belmonte; Pres. Elect, Carrie Bell; Financial Dir., Jamin May; Dir. Exhibits & Facilities, Kathy Irwin.
Personnel Profile: Full-Time Paid 25; Part-Time Paid 20; Part-Time Volunteers 1,000; Interns 6.
Governing Authority: private; nonprofit organization. Tax-exempt: 501(c)(3).
Institution Type/Description: Children's Museum.
Collections: hands-on exhibits.
Major Exhibits: Raccoon Run by Patrick Dougherty, 2013-2015.
Facilities: 28,000 sq. ft.; art studio & makespace; half-acre outdoor discovery center; four field trip & party rooms; preschool; cafe; gift shop.
Activities: art & science workshops; guest artists, performers & scientists; field trips; guided tours; participatory exhibits; parents' night out; campouts; birthday parties; toddler time; preschool, summer & spring break camps; parenting workshops.
Publications: quarterly newsletter, Applause.
Hours & Admission Prices: Tues.-Sat. 10-5, Sun. & Mon. 11-5. Adults $9.95, seniors $7.95, toddlers $6.95; babies no charge. Closed New Year's Day; Easter; Independence Day; Thanksgiving; Christmas. ♿
Attendance: 260,000 (estimated)
Membership: One Plus One $75; Grandparent $95; Basic Family $130; Deluxe Family $165.

MONARCH CONTEMPORARY ART CENTER AND SCULPTURE PARK, 8431 Waldrick Rd., S.E., Olympia, WA 98501. Mailing Address: P.O. Box 1125, Tenino, WA 98589-1125. Tel.: 360-264-2408.

E-mail: sculpturepark@monarchartcenter.org
Web Site: monarchartcenter.org
Founded: 1994.
Congressional District: 9
Key Personnel: Founder, Dir. & Cur., Myrna Orsini.
Personnel Profile: Full-Time Volunteers 1.
Governing Authority: Tax-exempt.
Institution Type/Description: Art Center.
Collections: works by regional, national & international artists; over 100 contemporary sculptures; gardens.
Facilities: nature trails.
Activities: classes; workshops; special events; demonstrations.
Hours & Admission Prices: Outdoor Gallery: dawn to dusk. Indoor Gallery: June-Oct. 1 by appointment. No charge; donations accepted. ♿
Attendance: 5,000 (estimated)

OLYMPIC FLIGHT MUSEUM, 7637 A Old Hwy. 99, S.E., Olympia, WA 98501-5728. Tel.: 360-705-3925. Fax: 360-236-9839.

E-mail: info@olympicflightmuseum.com
Web Site: www.olympicflightmuseum.com/
Founded: 1998.
Key Personnel: Pres. & Founder, Brian Reynolds; Dir., Teri Thorning.
Personnel Profile: Full-Time Paid 1; Part-Time Paid 2; Part-Time Volunteers 50.

Governing Authority: nonprofit organization. Tax-exempt.
Institution Type/Description: Aviation History Museum.
Collections: aviation history; World War II artifacts & memorabilia; lithographs; paintings; inert weapon systems; aircraft models; WWII, Korean, & Vietnam era aircraft.
Facilities: rental facility. Gift items for sale.
Activities: special events; lectures; tours; rental facility; meetings. Annual Event: Olympic Air Show in June.
Publications: quarterly newsletter, Flight Line.
Hours & Admission Prices: Summer: daily 11-5; Winter: Tues.-Sun. 11-5. Adults $7, children 7-12 $5; discounts to AAA members; members and children 6 & under no charge. Closed Thanksgiving; Christmas. &
Attendance: 20,000 (accurate)
Membership: Family $60.

***   WASHINGTON STATE CAPITAL MUSEUM AND OUTREACH CENTER, (M),** 211 21st Ave., S.W., Olympia, WA 98501-2811. Tel.: 360-753-2580. Fax: 360-586-8322.
E-mail: srohrer@wshs.wa.gov
Web Site: www.washingtonhistory.org/scmoc/
Founded: 1941.
Congressional District: 3
Key Personnel: Cur. Education, Susan Rohrer; Administrative Asst., Chris Nicandri; Preservation & Museum Specialist, Mark Vessey; Coord. Women's History Consortium, Shanna Stevenson.
Personnel Profile: Full-Time Paid 5; Part-Time Volunteers 10; Interns 2.
Governing Authority: Parent Institution: Washington State Historical Society. Subsidiary Institution: Capital Museum Foundation. Tax-exempt: 501(c)(3).
Institution Type/Description: History Museum: housed in Lord Mansion, c.1923.
Collections: history & culture of Washington; Native American artifacts; living botanical collection; garden.
Activities: tours; lectures; formally organized education program for children & adults; school loan service; technical assistance to heritage organizations statewide; Women's History Consortium; Heritage Resource Center. Museum Sponsors: National History Day.
Hours & Admission Prices: Wed.-Sat. 11-3. Family $5, adults $2, senior citizens $1.75, children $1; discounts to AAM members; members no charge. Closed state holidays. &
Attendance: 12,000 (estimated)
Membership: refer to Washington State Historical Society information.

Orcas Island

ORCAS ISLAND HISTORICAL MUSEUM, (M), 181 N. Beach Rd., Orcas Island, WA 98245. Mailing Address: P.O. Box 134, Eastsound, WA 98245-0134. Tel.: 360-376-4849. Fax: 360-376-4869.
E-mail: orcasmuseum@rockisland.com
Web Site: orcasmuseum.org
Founded: 1950.
Congressional District: 42
Key Personnel: Administrative Asst., Heather Wallace.
Personnel Profile: Part-Time Paid 1; Part-Time Volunteers 25.
Governing Authority: society. Parent Institution: Orcas Island Historical Society. Tax-exempt: 501(c)(3).
Institution Type/Description: General Museum: housed in six original homestead cabins built between the 1880s & 1890s.
Collections: Native American artifacts; history of the islands first peoples, Lummi & Samish Nations; woodworking & farming implements; objects associated with early industries such as fruit farming; objects utilized by early homesteaders; homestead grants; local historical photos; various types of early records & documents.
Research Fields: family genealogies; aspects of Orcas Island History: fruit farming, homesteading, Native Americans from this area, lime kilns/production, water transportation, ferry history, communication, farming.
Activities: guided tours available by request; lectures; permanent exhibitions. Museum Sponsors: Historical Parade in July; Historical Day Fair in July.
Publications: quarterly newsletter, The Orcas Islander.
Hours & Admission Prices: Memorial Day to Sept. Tues.-Thurs. & Sat.-Sun. 10-3, Fri. 1-6; other times for groups by appointment. Family $10, adults $3, seniors & students $2, children 6-12 $.50; discounts for AAM & ICOM members; museum members & children under 6 no charge.
Attendance: 3,250 (accurate)
Membership: Senior 65 & up & Student $10; Individual $15; Family & Business $25; Sponsor $50; Benefactor $100.

Pasco

CHILDREN'S MUSEUM OF THE THREE RIVERS, Broadmoor Square Mall, 5220 Outlet Dr., Pasco, WA 99301-8969. Mailing Address: P.O. Box 5641, Pasco, WA 99302-5601. Tel.: 509-543-7866.
E-mail: trcmuseum@yahoo.com
Web Site: www.childrensmuseumtr.org
Governing Authority: nonprofit organization. Tax-exempt: 501(c)(3).
Institution Type/Description: Children's Museum.
Collections: hands-on exhibits.
Activities: birthday parties; classes; field trips.
Hours & Admission Prices: Wed.-Fri. 10-5, Sat. 12-5. Admission $3; discounts to groups of 18 or more. Closed major holidays. &
Attendance: 17,500 (estimated)
Membership: Basic $25; Family $50; Supporting $75; Patron $100.

FRANKLIN COUNTY HISTORICAL MUSEUM, 305 N. 4th Ave., Pasco, WA 99301-5324. Tel.: 509-547-3714. Fax: 509-545-2168.
Founded: 1982.
Congressional District: 8
Key Personnel: Pres., Anne Hayden; Treas., Hazel Hanson; Museum Shop Mgr., Gracie Cooper; Admin., Sherel Webb.
Personnel Profile: Part-Time Paid 5; Part-Time Volunteers 18.
Governing Authority: nonprofit organization. Parent Institution: Franklin County Historical Society. Tax-exempt: 501(c)(3).
Institution Type/Description: Historical Society Museum: housed in 1911 Andrew Carnegie library building.
Collections: Franklin County history; Indian culture; railroad history; homesteading; agriculture; river transportation; aviation.
Facilities: Gift items & historical publications available for sale.
Activities: organized educational programs; guided tours; lectures; films; docent program.
Publications: monthly newsletter, Franklin Express; quarterly historical publication, Franklin Flyer.
Hours & Admission Prices: Tues.-Fri. 12-4; other times by appointment. No charge; donations accepted. Society members 10% off gift shop purchases. Closed national holidays. &
Attendance: 6,280 (accurate)
Membership: Library & School $7.50; Individual $20; Couple $30; Business $50 & up; Lifetime $500.

SACAJAWEA INTERPRETIVE CENTER, Sacajawea State Park, 2503 Sacajawea Park Rd., Pasco, WA 99301-6413. Tel.: 509-545-2361.
Web Site: www.park.wa.gov/stewardship/sacajawea
Founded: 1940.
Key Personnel: Park Mgr., Reade Obern.
Governing Authority: state. Affiliated with the Washington State Parks & Recreation Commission, 7150 Cleanwater Lane, Ky-11, Olympia, WA 98504. Tax-exempt.
Institution Type/Description: Interpretive Center.
Collections: c.7000 B.C. to early 19th-century stone & bone tools of Columbia Plateau Indians; displays honoring Lewis & Clark Expedition; material culture of area Indians.
Research Fields: Indians of S.E. Washington; Lewis & Clark; NPRR town of Ainsworth, 1878-1886.
Facilities: picnic area; swimming area; boating facilities.
Activities: permanent exhibitions; interpretive & school programs; tours.
Publications: brochures; reading lists; educational packets.
Hours & Admission Prices: April-Nov. 1 daily 10-5. No charge; $1 donation suggested. &
Attendance: 650 (estimated)

WASHINGTON STATE RAILROADS HISTORICAL SOCIETY MUSEUM, 122 N. Tacoma Ave., Pasco, WA 99301. Mailing Address: P.O. Box 552, Pasco, WA 99301-0552. Tel.: 509-543-4159.
E-mail: email@wsrhs.org
Web Site: www.wsrhs.org
Founded: 1990.
Key Personnel: Pres. (V), Tom Gronewald; Museum Shop Mgr., James M Bowers.
Personnel Profile: Part-Time Volunteers 4.
Governing Authority: Tax-exempt.
Institution Type/Description: Railroad Museum.
Collections: railroad history & artifacts; photographs; steam locomotives.
Facilities: Books, pins & hats for sale.

Hours & Admission Prices: May to mid-Dec. Thurs.-Fri. 12-4, Sat. 9-3. Adults $2, teens & seniors $1; children & members no charge. &

Membership: Teen 13-18 and Senior 65 & over $12.50; Senior Family $15; Adult $25.

Port Angeles

CLALLAM COUNTY HISTORICAL SOCIETY, Museum at the Carnegie, 207 S. Lincoln St., Port Angeles, WA 98362. Mailing Address: P.O. Box 1327, Port Angeles, WA 98362-0244. Tel.: 360-452-2662. Fax: 360-452-2662. Facebook: Clallam County Historical Society.

E-mail: artifact@olypen.com

Web Site: clallamhistoricalsociety.com

Formerly: The Museum of Clallam Historical Society

Founded: 1948.

Congressional District: 3

Key Personnel: Pres. (V), John Hubbard; Exec. Dir., Kathryn M. Monds.

Personnel Profile: Full-Time Paid 1; Part-Time Volunteers 70.

Governing Authority: society; nonprofit. Affiliated with the Clallam County Historical Society, Port Angeles, WA 98362. Administrative Center, Research & Genealogy Libraries, 931-933 West 9th, Port Angeles, WA 98363. Tax-exempt: 501(c)(3).

Institution Type/Description: History Museum.

Collections: period furniture; photographs; artifacts; history. Historic Building: 1888, Beaumont Cabin.

Research Fields: local history; oral taping; genealogy.

Facilities: 6,590-vol. library.

Activities: workshops; organized education programs; lectures; guided tours; training programs for volunteer museum workers; docent program; temporary exhibitions.

Publications: quarterly bulletin, Strait News; brochures; pamphlets.

Hours & Admission Prices: Wed.-Sat. 1-4. Suggested donation: Family $5, adults $2. &

Attendance: 8,000 (estimated)

Membership: Senior Citizen $25; Individual $30; Family $35; Explorer $100; Pioneer $250; Homesteader $500; Visionary $1,000.

FEIRO MARINE LIFE CENTER, 315 N. Lincoln St., Port Angeles, WA 98362. Mailing Address: P.O. Box 625, Port Angeles, WA 98362-0112. Tel.: 360-417-6254. Facebook: Feiro Marine Life Center.

E-mail: deborahm@feiromarinelifecenter.org

Web Site: www.feiromarinelifecenter.org

Formerly: Arthur D. Feiro Marine Life Center c/o Peninsula College

Founded: 1981.

Congressional District: 6

Key Personnel: Pres., Betsy Wharton; Dir., Deborah Moriarty; Coord., Robert Campbell.

Personnel Profile: Full-Time Paid 2; Part-Time Paid 1; Part-Time Volunteers 35; Interns 2.

Volunteer Hours: 2,200

Operating Expenses: 243,427

Operating Income: 207,131

Governing Authority: nonprofit. Tax-exempt: 501(c)(3).

Institution Type/Description: Marine Life Center.

Collections: local Pacific Northwest marine life; plankton & jellyfish; giant pacific octopus; vertebrate; invertebrate; fossils.

Major Exhibits: Elwha River Restoration, 10/13-12/14.

Research Fields: marine baseline water conditions.

Facilities: aquarium; 1,800 sq. ft. exhibit space. Museum-related items for sale.

Activities: docent program; films; formal & informal education programs; guided tours; temporary exhibitions.

Hours & Admission Prices: Memorial Day to Labor Day daily 10-5; Sept.-May daily 12-4; other times by appointment. Adults $4, youth 4-17 $1; children 3 & under no charge. &

Attendance: 20,000 (estimated)

Membership: Individual $30; Family $50.

OLYMPIC NATIONAL PARK VISITOR CENTER, 3002 Mt. Angeles Rd., Port Angeles, WA 98362-6775. Mailing Address: 600 E. Park Ave., Port Angeles, WA 98362. Tel.: 360-565-3130 & 3000. Fax: 360-565-3147. Facebook: Olympic National Park.

E-mail: olym_interpretation@nps.gov

Web Site: www.nps.gov/olym

Founded: 1938.

Congressional District: 6

Key Personnel: Supt., Sarah Creachbaum; Chief Park Interpreter, Kathy Steichen; Museum Shop Mgr., Margaret Baker.

Personnel Profile: Full-Time Paid 1; Part-Time Paid 4; Part-Time Volunteers 8; Interns 4.

Governing Authority: federal. Parent Institution: National Park Service, Washington, DC. Tax-exempt: 501(c)(3).

Institution Type/Description: Visitor Center for National Park.

Collections: Historic House: 1880 Beaumont Cabin.

Facilities: 1,800-vol. library of books on natural history, history & anthropology available for use on premises. Books, maps, slides & postcards for sale.

Activities: lectures; films; formally organized education programs for children & adults; permanent exhibitions.

Publications: books; information handouts.

Hours & Admission Prices: Open daily. No charge. Closed Thanksgiving; Christmas. &

Attendance: 160,000 (accurate)

PORT ANGELES FINE ARTS CENTER & WEBSTER'S WOODS ART PARK, 1203 E. Lauridsen Blvd., Port Angeles, WA 98362-6630. Tel.: 360-457-3532 & 417-4590. Fax: 360-457-3532.

E-mail: pafac@olypen.com

Web Site: www.pafac.org

Founded: 1986.

Congressional District: 2

Key Personnel: C.E.O., Jake Seniuk; Chm. (V), Darlene Ryan; Pres., Jean Heessels-Petit; Cur. Education & Asst., Barbara Slavik.

Personnel Profile: Full-Time Paid 3; Part-Time Paid 1; Part-Time Volunteers 50; Interns 1.

Governing Authority: municipal government; nonprofit. Parent Institution: City of Port Angeles. Subsidiary Institution: PAFAC Foundation. Tax-exempt: 501(c)(3).

Institution Type/Description: Art Museum.

Collections: paintings & drawings from donor Esther Webster; Northwestern artists; sculpture garden.

Activities: lectures; readings; performances.

Publications: bimonthly newsletter, On Center.

Hours & Admission Prices: March-Oct. Thurs.-Sun. 11-5; Nov.-Feb. Thurs.-Sun. 10-4. No charge. Closed New Year's Eve & Day; Independence Day; Thanksgiving; Christmas Eve & Day. &

Attendance: 18,000 (estimated)

Membership: Friend $35-$49; Good Friend $50-$99; Close Friend $100-$499; Esteemed Friend $500-$999; Sustaining Friend $1,000 & up.

Port Gamble

PORT GAMBLE HISTORIC MUSEUM, 32400 Rainier Ave., N.E., Port Gamble, WA 98364. Mailing Address: P.O. Box 85, Port Gamble, WA 98364-0085. Tel.: 360-297-8078. Fax: 360-297-5616.

E-mail: ssmith@orminc.com

Web Site: www.portgamble.com

Founded: 1976.

Congressional District: 23

Key Personnel: C.E.O., Dave Nunes; Cur. & Museum Shop Mgr., Shana Smith.

Personnel Profile: Full-Time Paid 1; Part-Time Paid 3.

Governing Authority: Parent Institution: Pope Resources. Subsidiary Institution: Olympic Property Group.

Institution Type/Description: History Museum: housed in 1853 Port Gamble General Store.

Collections: founding family memorabilia; maritime; sales office; manuscripts; artifacts. Historic Buildings: 1853 U.S. Post Office; 1872 Masonic Temple; 1859 Thompson House; 1879 St. Paul's Episcopal Church; 1887 Walker-Ames House; 1870 M.S. Drew House.

Research Fields: corporate & family history.

Facilities: library of books on corporate archives available for research by request, reviewed on individual basis. Forest industry historical books for sale.

Activities: permanent & temporary exhibitions.

Publications: booklet, Port Gamble Historic Museum; brochure, Welcome to Historic Port Gamble; book, Pope Resources, Rooted in the Past, Growing in the Future; Walking Tour of Port Gamble.

Hours & Admission Prices: May-Oct. daily 9:30-5; Nov.-April Fri.-Sun. 10-4. Adults $4, students & senior citizens $3; children 5 & under no charge. &

Attendance: 7,800 (accurate)

Port Townsend

JEFFERSON MUSEUM OF ART AND HISTORY, (M), 540 Water St., Port Townsend, WA 98368-5725. Tel.: 360-385-1003.

E-mail: billtennent@jchswa.org

Web Site: www.jchsmuseum.org
Formerly: Jefferson County Historical Society Museum
Founded: 1951.
Congressional District: 3
Key Personnel: Pres. (V), Dorothy Cotton Banks; Dir., William Tennent; Museum Coord., Phyllis Snyder; Archivist, Marsha Moratti.
Personnel Profile: Full-Time Paid 1; Part-Time Paid 8; Part-Time Volunteers 200.
Governing Authority: (scope); nonprofit organization. Parent Institution: Jefferson County Historical Society. Subsidiary Institution: Research Center, 13694 Airport Cutoff Rd., Port Townsend, WA 98368. Tax-exempt: 501(c)(3).
Institution Type/Description: Art & History Museum: housed in 1892 City Hall focusing on Jefferson County history & prehistory (Native American & European).
Collections: local memorabilia; maritime heritage; Victorian Era & early Pioneer exhibits; Victorian furniture & clothing; Indian & Alaskan baskets & artifacts; 20,000 photographs; sailing ships; bound volumes of Port Townsend Leaders; family histories; preservation & restoration reference books; hearse from 1886-1912; ship rudder of sunken 1880 ship; old police court; jail; early fire hall; buttons; maps; deeds; genealogy. Historic Buildings: 1889 Fire Bell Tower; Rothschild House 1868.
Research Fields: census records; voting records; school records; family histories; property records; maritime records; military records; local newspapers from 1860.
Activities: permanent & temporary exhibits; lecture series; local preservation projects; docent program; awards program.
Publications: quarterly newsletter.
Hours & Admission Prices: Daily 11-4; special tours available. Adults $4, children under 12 $1; members no charge. Closed New Year's Day; Thanksgiving; Christmas. &
Attendance: 20,000 (estimated)
Membership: $50-$1,200.

KELLY ART DECO LIGHT MUSEUM, Vintage Hardware, 2000 Sims Way, Port Townsend, WA 98368-2229. Tel.: 360-379-9030. Fax: 360-379-9029.
E-mail: vhprs@earthlink.net
Web Site: www.thedecomuseum.com
Founded: 2004.
Key Personnel: Dir., Chm. & Founder, Ken Kelly
Institution Type/Description: Decorative Art Museum.
Collections: over 400 fixtures including hanging chandeliers, wall sconces & table lights from the Great Depression.
Research Fields: how the Great Depression changed lighting companies in the 30s; mktg. & product devel.; artistic companies that were lost during the depression.
Hours & Admission Prices: Daily 10-5. No charge; donations accepted.

NORTHWIND ARTS CENTER, 2409 Jefferson St., Port Townsend, WA 98368. Tel.: 360-379-1086.
Web Site: www.northwindarts.org
Founded: 2002.
Governing Authority: Tax-exempt: 501(c)(3).
Institution Type/Description: Art Center.
Collections: works by Northwest regional artists.
Activities: lectures; educational programs; workshops; studio tours; demonstrations; special events.
Publications: online newsletter.
Hours & Admission Prices: Thurs.-Mon. 12-5. No charge; donations accepted.
Attendance: 5,000
Membership: $35 & up.

PORT TOWNSEND AERO MUSEUM, 105 Airport Rd., Port Townsend, WA 98368. Mailing Address: P.O. Box 101, Chimacum, WA 98325-0101. Tel.: 360-379-5244.
Web Site: www.ptaeromuseum.com
Founded: 2001.
Key Personnel: Dir., G. F. Thuotte.
Personnel Profile: Full-Time Paid 1; Full-Time Volunteers 2; Part-Time Volunteers 12; Interns 15.
Governing Authority: tax-exempt.
Institution Type/Description: Aviation Museum.
Collections: over 30 aircraft; 100 pieces of aviation art; aviation photographs; 200 aircraft models.
Activities: special events.
Hours & Admission Prices: Wed.-Sun. 9-4; groups by appointment. Adults $10; members no charge. Closed Thanksgiving & Christmas &
Attendance: 5,000 (accurate)

Membership: Individual $35; Family $50; Supporter $250; Sponsor $500; Lifetime $1,000; Visionary $5,000.

PORT TOWNSEND MARINE SCIENCE CENTER, Fort Worden State Park, 532 Battery Way, Port Townsend, WA 98368-3431. Tel.: 360-385-5582. Fax: 360-385-7248.
E-mail: info@ptmsc.org
Web Site: www.ptmsc.org
Founded: 1982.
Congressional District: 24
Key Personnel: Dir., Anne Murphy.
Personnel Profile: Full-Time Paid 5; Part-Time Paid 4; Part-Time Volunteers 175; Interns 4.
Governing Authority: nonprofit organization. Parent Institution: Port Township Marine Science Society. Tax-exempt: 501(c)(3).
Institution Type/Description: Marine & Natural History Museum: housed at c.1900 Fort Worden.
Collections: live exhibits of local marine animals in touch tanks & glass aquariums; microscopes for viewing plankton; underwater video camera for viewing sealife; daily interpretive programs; natural history of Puget Sound; fossilized ancestors of local animals; sands from around the world; geological formation of Washington; interactive hydrophone station.
Research Fields: water quality; microplastics; killer whale vocalizations; toxics in the marine environment.
Facilities: 400-vol. library of books, field guides, reference books, scientific studies, periodicals & videotapes available to the public; aquarium.
Activities: guided tours; lectures; films; organized education programs for adults & children; docent program; participatory exhibits, touchtables; summer camps. Museum Sponsors: cruises to National Wildlife Refuge in spring, summer & fall.
Publications: newsletter, Octopress; monthly e-news.
Hours & Admission Prices: Summer, Spring & Fall: Two Exhibits. Adults $5; youth & members no charge. Winter: One Exhibit. Adults $3, youth $2; members no charge. &
Attendance: 20,000 (accurate)
Membership: Student $15; Individual $30; Family $45; Friend $75; Sustaining $100; Business $125; Octopress Sponsor $250; Benefactor $500; Sponsor $1,000.

PUGET SOUND COAST ARTILLERY MUSEUM AT FORT WORDEN, Bldg. 201, Fort Worden State Park, Port Townsend, WA 98368. Mailing Address: 200 Battery Way, Port Townsend, WA 98368-3621. Tel.: 360-385-0373.
E-mail: coastartillery@gmail.com
Web Site: pscoastartillerymuseum.org
Founded: 1976.
Congressional District: 6
Key Personnel: Pres., Alfred Chiswell; Vice Pres., Ron Novak; Treas. & Museum Shop Mgr., Mike Cornforth; Park Mgr., Allison Alderman.
Personnel Profile: Full-Time Volunteers 1; Part-Time Volunteers 12.
Governing Authority: nonprofit. Tax-exempt: 501(c)(3).
Institution Type/Description: Military Museum: housed in 1904, Enlisted Barracks, Bldg. 201 of the Harbor Defense of Puget Sound.
Collections: artifacts of seacoast artillery; 1890-1944 harbor defenses; personal artifacts; papers; photos; military rifle collection from 26 countries; Coast Artillery; Coast Defense Puget Sound.
Research Fields: coast artillery, harbor defense of Puget Sound & the United States.
Facilities: 700-vol. library of military related books, service manuals, official orders, private correspondence, related photographs available for research with permission from director-librarian on premises; theater.
Activities: guided tours; lectures; permanent & temporary exhibitions. Annual Event: Military Vehicle Show in September.
Publications: Fort Worden Guide; photo book, Then & Now (reprint).
Hours & Admission Prices: July-Aug. Sun.-Thurs. 11-4, Fri.-Sat. 10-5; Sept.-June daily 11-4. Adults $3, children $2; active military no charge. &
Attendance: 11,213 (accurate)
Membership: Individual $20; Family $30; Patron $100.

ROTHSCHILD HOUSE STATE PARK, Franklin St., Port Townsend, WA 98368. Mailing Address: Fort Worden State Park, 200 Battery Way, Port Townsend, WA 98368-3621. Tel.: 360-385-1003 & 344-4400. Fax: 360-385-7248.
E-mail: kate.burke@parks.wa.gov
Web Site: www.fortworden.net
Founded: 1959.
Congressional District: 6
Key Personnel: Park Mgr. Fort Worden, Kate Burke; Museum Mgr., Phyllis Snyder.

Personnel Profile: Full-Time Paid 1; Part-Time Paid 2; Part-Time Volunteers 2.
Governing Authority: state. Parent Institution: Fort Worden State Park. Subsidiary Institution: Jefferson County Historical Society. Tax-exempt.
Institution Type/Description: Historic House Museum: 1868 Rothschild House, built by the merchant D.C.H. Rothschild.
Collections: furnished 19th-century home; Rothschild family artifacts
Activities: guided tours; permanent exhibitions.
Publications: brochure.
Hours & Admission Prices: May-Sept. daily 11-4. Adults $4, children $1; discounts to JCHS members.
Attendance: 3,225 (accurate)

Poulsbo

POULSBO MARINE SCIENCE CENTER, 18743 Front St., N.E., Poulsbo, WA 98370. Mailing Address: P.O. Box 408, Keyport, WA 98345-0408. Tel.: 360-598-4460.
E-mail: info@poulsbomsc.org
Web Site: www.poulsbomsc.org
Founded: 2006.
Key Personnel: Dir. Aquarium, Patrick Mus; Dir. Education, Bruce Claiborne; Chm. (V), Bruce Harlow.
Personnel Profile: Full-Time Paid 1; Part-Time Paid 1; Interns 45.
Volunteer Hours: 4,200
Governing Authority: Parent Institution: Poulsbo Marine Science Foundation. Tax-exempt.
Institution Type/Description: Marine Science Center.
Collections: hands-on exhibits; aquariums.
Research Fields: marine science.
Facilities: library; aquariums.
Activities: educational programs; videos.
Hours & Admission Prices: Thurs.-Sun. 11-4. No charge; donations accepted. &
Attendance: 10,500 (estimated)

SUQUAMISH MUSEUM, (M), 15838 Sandy Hook Rd., Poulsbo, WA 98370-7867. Mailing Address: P.O. Box 498, Suquamish, WA 98392-0498. Tel.: 360-394-8496.
E-mail: jsmoak@suquamish.nsn.us
Web Site: www.suquamish.nsn.us
Founded: 1983.
Congressional District: 1
Key Personnel: Tribal Chm., Leonard Forsman; Tribal Council Sec., Nigel Lawrence; Dir., Marilyn Jones; Cur., Lydia Woods.
Personnel Profile: Full-Time Paid 4; Part-Time Paid 3; Part-Time Volunteers 10.
Governing Authority: nonprofit organization. Parent Institution: Suquamish Tribal Cultural Center. Tax-exempt: 501(c)(3).
Institution Type/Description: Native American Museum: located on the Port Madison Indian Reservation.
Collections: traditional culture of the Suquamish Tribe & other Puget Sound Tribes; 1792-present photographs & artifacts, including basketry, carvings, tools, mats, bowls, spoons & fishing gear.
Research Fields: history of federal Indian policy in respect to land & education; Indian Treaty Rights protection; traditional & contemporary religion; traditional uses of natural resources by Indians; basketry techniques.
Facilities: 120-vol. library pertaining to Puget Sound Indian history; 40-seat auditorium; 75-seat restaurant; educational facilities; 3,000 sq. ft. exhibit space.
Activities: guided tours; lectures; films; arts festivals; rental gallery; participatory, loan & traveling exhibitions; organized education programs for college students. Museum Sponsors: Native American Art Fair; Chief Seattle Days Celebration.
Publications: exhibit catalogue, The Eyes of Chief Seattle; quarterly newsletter.
Hours & Admission Prices: Daily 10-5. Call for pricing. Closed major holidays. &
Attendance: 10,000 (estimated)

Prosser

BENTON COUNTY HISTORICAL MUSEUM, 1000 Paterson Rd. (located in the city park), Prosser, WA 99350. Mailing Address: P.O. Box 1407, Prosser, WA 99350-0800. Tel.: 509-786-3842.
E-mail: prossermuseum@hotmail.com
Founded: 1968.
Congressional District: 4
Key Personnel: Pres. (V), Dick Sampson; Cur., Frankie Wallace.
Personnel Profile: Part-Time Paid 1; Part-Time Volunteers 10.

Governing Authority: society; nonprofit organization. Parent Institution: Benton County Museum and Historical Society, Inc. Tax-exempt.
Institution Type/Description: General Museum.
Collections: early Americana; Indian artifacts; natural history; guns; dolls; china; handwork; cavern crystals; corals & sea shells; oral histories; Homestead Shack; The Parlor; Farm Room, replica 1890 Holt Combine; hand-carved work horses in action; replica carousel with calliope music; doll house; general store; 1843-1920 gown collection; 1867 Chickering piano, playable; NASA's space items to moon & back.
Facilities: 5,000 sq. ft. exhibit space.
Activities: tours by appointment.
Publications: brochures.
Hours & Admission Prices: Tues.-Sat. 11-4, Sun. call for hours. Adults $3, children $1; students no charge. Closed New Year's Day; Easter; Thanksgiving; Christmas. &
Attendance: 2,400 (estimated)
Membership: Annual $10; Life Supporting $100; Silver Historian $250; Golden Pioneer $500.

Pullman

CHARLES R. CONNER MUSEUM, Washington State University, Pullman, WA 99164. Tel.: 509-335-3515. Fax: 509-335-3184.
E-mail: connermuseum@wsu.edu
Web Site: sbs.wsu.edu/connermuseum
Founded: 1894.
Congressional District: 5
Key Personnel: Dir., Larry Hufford; Cur., Dr. Kelly M. Cassidy.
Personnel Profile: Full-Time Paid 2; Part-Time Paid 1; Part-Time Volunteers 1.
Governing Authority: state. Parent Institution: Washington State University. Tax-exempt: 501(c)(3).
Institution Type/Description: Zoology Museum; Natural History Museum.
Collections: herpetology; ornithology; mammalogy.
Research Fields: systematics; ecology; zoogeography of vertebrate animals.
Facilities: 500-vol. library.
Activities: permanent exhibitions; research collections; school outreach.
Hours & Admission Prices: Daily 8-5. No charge; donations accepted. &
Attendance: 23,000 (estimated)

MUSEUM OF ANTHROPOLOGY, Department of Anthropology, Washington State University, College Hall, Pullman, WA 99164. Tel.: 509-335-3441. Fax: 509-335-3999.
E-mail: collinsm@wsu.edu
Founded: 1966.
Congressional District: 5
Key Personnel: Dir., Mary Collins.
Personnel Profile: Full-Time Paid 2; Part-Time Volunteers 10.
Governing Authority: university. Parent Institution: Washington State University.
Institution Type/Description: Anthropology Museum.
Collections: archaeological & ethnographic collections from Western North America, South America, Africa, Asia & Oceana; basketry from Western U.S.; material from China, S. America, W. Africa & New Guinea.
Activities: guided tours; permanent & temporary exhibitions.
Hours & Admission Prices: Academic Year: Mon.-Fri. 9-4; other times by appointment. No charge; donations accepted. Closed school holidays & vacations. &
Attendance: 4,000 (estimated)

MUSEUM OF ART, (M), 6077 Wilson Rd., Fine Arts Center, Pullman, WA 99164-7460. Mailing Address: P.O. Box 647460, Pullman, WA 99164-7460. Tel.: 509-335-1910. Fax: 509-335-1908.
E-mail: artmuse@wsu.edu
Web Site: museum.wsu.edu
Founded: 1973.
Congressional District: 5
Key Personnel: Dir., Chris Bruce; Assoc. Dir., Anna-Maria Shannon; Program Coord., Karri Dieken; Cur., Keith Wells; Asst. Cur., Zach Mazur; Public Rels. & Media Mgr., Debby Stinson; Dir. Devel., Jill Aesoph.
Personnel Profile: Full-Time Paid 7; Part-Time Paid 11; Part-Time Volunteers 13; Interns 5.
Governing Authority: university. Parent Institution: Washington State University.
Institution Type/Description: University Art Museum.
Collections: late 19th-century to contemporary American & European paintings & graphics; glass; photographs.
Research Fields: 19th- & 20th-century American & European art, with emphasis on contemporary & regional collections.
Facilities: 4,500 sq. ft. exhibition space.

Activities: temporary exhibitions; guided tours; lectures; films; gallery talks; education programs for undergraduate & graduate students; art workshops for children.

Publications: posters; catalogues include A Temporary Possession: The Human Image in 20th-Century Photography; Artist & Place: American Landscape Painting 1860-1914; Wendell Brazeau: 1910-1974; Diverse Directions: The Fiber Arts; Goya: Los Disparates; George Inness: Evening Landscape, 1863; Laisner; Norman Lundin; Recent Acquisitions - Ancient Art, the J. Paul Getty Museum; Rodin: The Maryhill Collection; Six From California; Margaret Tomkins; Works on Paper: American Art 1945-1975; Mel Katz Works 1971-1978; A Partial View: Young Photographers in the Northwest; Drawings 1900-1945: A Survey of American Works, 1979; Swords of the Samurai; Contemporary Metals: Focus on Idea; Noritake Art Deco Porcelains; Fabric Traditions of Indonesia; Gaylen Hansen: The Paintings of a Decade, 1975-1985; books, Extending the Artist's Hand: Contemporary Sculpture from the Walla Walla Foundry; Roy Lichtenstein Prints 1956-1997: From the Collections of Jordan D. Schnitzer and His Family Foundation; Art & Context: The 1950's and 60's; Gaylen Hansen: Three Decades of Paintings; Sherry Markovitz: Shimmer; Running the Numbers: An American Self-Portrait.

Hours & Admission Prices: July-Aug. Tues.-Sat. 12-4; Sept.-June Mon.-Wed. & Fri.-Sat. 10-4, Thurs.10-7. No charge. Closed during semester breaks & between exhibition installations. &

Attendance: 30,000 (estimated)

Membership: Student $15; Individual $35; Family $50; Associate $100; Patron $250; Sustaining $500; President's Associate $1,000; Silver PA $2,500; Crimson PA $5,000; Platinum PA $10,000.

PALOUSE DISCOVERY SCIENCE CENTER, 950 N.E. Nelson Ct., Pullman, WA 99163. Tel.: 509-332-6869.

Web Site: www.palousescience.org

Key Personnel: Exec. Dir., Victoria Scalise; Coord. Educ., Stephanie McNelis; Financial Mgr., Carter Shae; Education Specialist, Dona Abderhalden; Museum Shop Mgr., Meg Kelley; Program Coord., Heather Kelley

Institution Type/Description: Science Center.

Collections: hands-on science exhibitions.

Facilities: Museum-related items for sale.

Activities: summer camps; hands-on exhibitions; special events; birthday parties; nature walk. Museum Sponsors: Family Science Saturdays.

Hours & Admission Prices: Tues. 10-5, Wed.-Sat. 10-3. Adults 12-54 $6, seniors 55 & over $5, children 2-12 $4; children under 2 no charge.

Membership: Senior $30; Individual $35; Grandparent $50; Family $60.

Puyallup

THE FRED OLDFIELD WESTERN HERITAGE & ART CENTER, 110 9th S.W., Puyallup, WA 98371. Mailing Address: P.O. Box 1539, Puyallup, WA 98371-0216. Tel.: 253-752-9708; 866-445-9175 (Toll Free). Fax: 253-752-9708.

Founded: 2002.

Key Personnel: Dir., Joella Oldfield; Chm. (V), Glennis Golden.

Personnel Profile: Full-Time Paid 1; Part-Time Paid 1; Part-Time Volunteers 65.

Institution Type/Description: Heritage Center.

Collections: Western art; American West history & heritage; paintings; sketches; American Indian baskets & artifacts.

Activities: classes; tours; special events; educational programs; seminars.

Hours & Admission Prices: Sat. 12-4; other times by appointment. No charge; donations accepted. &

Attendance: 30,000 (estimated)

Membership: Seniors $20; Individual $30; Family $45; Associate $100.

PAUL H. KARSHNER MEMORIAL MUSEUM, 309 Fourth St., N.E., Puyallup, WA 98372-3062. Tel.: 253-841-8748. Fax: 253-840-8951.

E-mail: curator@karshnermuseum.org

Web Site: www.karshnermuseum.org

Founded: 1930.

Congressional District: 2

Key Personnel: Cur., Beth Bestrom.

Personnel Profile: Full-Time Paid 1; Part-Time Paid 2.

Governing Authority: public school district. Parent Institution: Puyallup School District. Tax-exempt.

Institution Type/Description: Children's History Museum: housed in c.1920 school building.

Collections: Indian artifacts; geology; paleontology; entomology; natural science; trade items, American pioneer clothing, utensils, tools & artifacts.

Facilities: 500-vol. library of historical, religious & text books available for use by arrangement.

Activities: guided tours; lectures; formally organized education programs for children; permanent & temporary exhibitions. Museum Sponsors: monthly Family Day.

Publications: annual booklet.

Hours & Admission Prices: Call for appointment. No charge; donations accepted. Closed national holidays. &

Attendance: 12,000 (estimated)

Membership: Individual $15; Family $25; Patron $40; Institution/Organization $100.

PUYALLUP HISTORICAL SOCIETY AT MEEKER MANSION, 312 Spring St., Puyallup, WA 98372. Mailing Address: P.O. Box 103, Puyallup, WA 98371-0011. Tel.: 253-848-1770.

E-mail: ezra@meekermansion.org

Web Site: www.meekermansion.org

Founded: 1970.

Congressional District: 9

Key Personnel: Historian, Andy Anderson; Dir. & Museum Shop Mgr., Sue Fass; Pres. (V), Robert Minnich.

Personnel Profile: Part-Time Paid 2.

Governing Authority: society; nonprofit organization. Parent Institution: Ezra Meeker Historical Society. Tax-exempt: 501(c)(3).

Institution Type/Description: Historic House Museum: 1890 Ezra Meeker Home, a Victorian mansion located at end of the Oregon Trail.

Collections: furnishings of the period; the books & writings of Ezra Meeker; textiles; clothing; art displays.

Research Fields: pioneers of the Northwest in print & photos.

Facilities: library of books written by Ezra Meeker available for research with the approval of the board for use on premises; antique rose garden; available for weddings.

Activities: guided tours; lectures. Museum Sponsors: Victorian Teas in May; Meeker Days in June; sourdough pancake breakfast; cider squeeze; Christmas at the Mansion in December.

Publications: book, Ezra Meeker; Bentley's Tale of the Oregon Trail; The Ox Team or the Old Oregon Trail; Uncle Ezra's Short Stories for Children; Ezra Meeker a Brief Resume of His Life & Adventures.

Hours & Admission Prices: March to mid-Dec. Wed.-Sun. 12-4; other times for special events. Adults $4, senior citizens & students $3, children $2; members no charge. Closed Easter; Thanksgiving. &

Attendance: 10,000 (accurate)

Membership: Individual $15; Family $25; Donor $40; Supporting $100; Business $150, Lifetime $1,000.

Quilcene

QUILCENE HISTORICAL MUSEUM, 151 E. Columbia St., Quilcene, WA 98376. Mailing Address: P.O. Box 574, Quilcene, WA 98376-0574. Tel.: 360-765-4848.

E-mail: quilcenemuseum@olypen.com

Web Site: quilcenemuseum.org & worthingtonparkquilcene.org

Founded: 1991.

Congressional District: 6

Key Personnel: Chm. (V), Mari Phillips; Museum Shop Mgr., Larry McKeehan.

Personnel Profile: Part-Time Volunteers 25.

Volunteer Hours: 625

Operating Expenses: 25,000

Operating Income: 25,000

Governing Authority: Tax-exempt.

Institution Type/Description: History Museum.

Collections: local history & culture; period artifacts, photographs; country store; early 1900s kitchen; millinery shop; early 1900s school; toys; business; logging; early pioneers; clubs & organizations.

Major Exhibits: Japanese Settlers of Jefferson County, 4/14-9/14; People in Uniform, 4/14-9/14; Worthington Family, 4/14-9/14.

Publications: Quilcene Cooks: Past and Present; Quilcene's Heritage: Looking Back; Dub of South Burlap; Timber Country Revisited; Brinnon: A Scrapbook of History; King of the Sea World; Home Cookin'...from Quilcene and other Northwest Kitchens; Discover Historic Washington State, Bunkerhouse Tales; Jim and Ana: Leland Munn Family; Conflict in Our National Forests.

Hours & Admission Prices: March-Sept. Fri.-Mon. 1-5. No charge; donations accepted. &

Attendance: 2,100 (estimated)

Membership: Individual $20; Family $30; Service Group $25; Business $60; Lifetime $200; Couple Lifetime $300.

Raymond

THE NORTHWEST CARRIAGE MUSEUM, 314 Alder St., Raymond, WA 98577-2434. Tel.: 360-942-4150. Facebook: Northwest Carriage Museum.
E-mail: lbowman@crescomm.net
Web Site: www.nwcarriagemuseum.org
Founded: 2002.
Key Personnel: Museum Shop Mgr., Amy Dennis.
Governing Authority: Tax-exempt.
Institution Type/Description: Transportation Museum.
Collections: over 28 horse drawn carriages, buggies & sleighs from the late 19th century.
Hours & Admission Prices: May-Sept. Sun.-Tues. 12-4, Wed.-Sat. 10-4; Oct.-April Wed.-Sat. 10-4, Sun. 12-4. Adults $4; discounts to AAA members; members no charge. &
Attendance: 3,500 (accurate)
Membership: Individual $25; Family $30; Business $50; Sponsor $100; Surrey $250-$499; Phaeton $500-$999; Brougham $1,000 & up.

WILLAPA SEAPORT MUSEUM, 310 Alder St., Raymond, WA 98577-2434. Tel.: 360-942-4149.
E-mail: jvwallace@centurytel.net
Web Site: willapaseaportmuseum.org
Formerly: Puget Sound Museum
Founded: 1966.
Congressional District: 3
Key Personnel: Dir., Pete Darrah; Pres. (V), Bill Coomer; Museum Shop Mgr., Virginia Wallace.
Personnel Profile: Part-Time Volunteers 20.
Governing Authority: Parent Institution: Frink Foundation. Tax-exempt.
Institution Type/Description: Maritime Museum.
Collections: Willapa Bay history; marine artifacts; logging industry; shipbuilding; life saving service; Spruce Division (WWI); Native Americans; lightships; lighthouses; 100 yr. old wheel house from Vansee fishing vessel.
Research Fields: Coast Artillery Corp.; lifesaving service; logging; sea service; lighthouse service.
Facilities: 5,000 sq. ft. museum.
Activities: tours; outboard motor meeting; survival camp.
Hours & Admission Prices: Open daily 12-4. No charge; donations accepted. &
Attendance: 1,500 (accurate)

Renton

RENTON HISTORICAL SOCIETY AND MUSEUM, (M), 235 Mill Ave. S., Renton, WA 98057-2133. Tel.: 425-255-2330. Fax: 425-255-1570. Facebook: Renton History Museum.
E-mail: estewart@rentonwa.gov
Founded: 1966.
Congressional District: 11
Key Personnel: Dir., Elizabeth P. Stewart; Pres., Theresa Clymer; Treas., Phyllis Hunt; Collection Mgr., Sarah Samson; Volunteer Coord., Dorota Rahn.
Personnel Profile: Full-Time Paid 2; Part-Time Paid 2; Part-Time Volunteers 60.
Governing Authority: nonprofit. Parent Institution: City of Renton. Tax-exempt: 501(c)(3).
Institution Type/Description: Historic Fire Station: 1942 art deco building.
Collections: more than 15,000 photographs; clothing; newspapers; cultural artifacts; coal mining artifacts; home furnishings; turn-of-the-century coal car & logging equipment; vintage neon movie theater sign.
Research Fields: local history.
Facilities: 500-vol. library for public use. Books & gift items for sale.
Activities: guided tours; lectures; organized education programs for children, adults, undergraduate & graduate college students; docent program; loan, temporary & traveling exhibitions.
Publications: newsletter, Renton Historical Quarterly.
Hours & Admission Prices: Tues.-Sat. 10-4. Adults: non-city residents $3, city residents $2, children ages 8-16 $1; discounts to AAM & ICOM members; members no charge. &
Attendance: 4,000 (accurate)
Membership: Student & Senior $12; Individual & Senior Couple $20; Family $30; Patron & Benefactor $100; Life $500.

Republic

STONEROSE INTERPRETIVE CENTER, 15-1 N. Kean St., Republic, WA 99166. Mailing Address: P.O. Box 987, Republic, WA 99166-0987. Tel.: 509-775-2295.
E-mail: srfossils@rcabletv.com
Web Site: www.stonerosefossil.org
Founded: 1989.
Key Personnel: Dir., Catherine Brown.
Governing Authority: Parent Institution: Friends of Stonerose Fossils.
Institution Type/Description: Paleontology Museum.
Collections: Pacific Northwest's geological & biological history; fossils.
Research Fields: paleobotany; paleoentomology.
Activities: fossil digging.
Publications: Stonerose Fossil: Guidebook Eocene Fossils of Republic, WA.
Hours & Admission Prices: May & Sept.-Oct. Wed.-Sun. 8-5; Memorial Day to Labor Day daily 8-5. Site Admission Sticker: adults $8, senior citizens & student $5; members no charge. &
Attendance: 8,000 (accurate)
Membership: Individual $20; Family $40; Family Plus $60; Business $75; Contributing $100; Sustaining $250; Corporate $500; Lifetime $1,000.

Richland

COLUMBIA RIVER EXHIBITION OF HISTORY, SCIENCE AND TECHNOLOGY, 95 Lee Blvd., Richland, WA 99352-4222. Mailing Address: P.O. Box 1890, Richland, WA 99352-6490. Tel.: 509-943-9000. Fax: 509-943-1770.
E-mail: crehstmuseum@crehst.org
Web Site: www.crehst.org
Founded: 1963.
Congressional District: 4
Key Personnel: Exec. Dir., Ellen Low; Pres. (V), Paul Schuler; Cur., Connie Estep.
Personnel Profile: Full-Time Paid 4; Part-Time Paid 5; Part-Time Volunteers 65.
Governing Authority: private foundation. Parent Institution: Environmental Science & Technology Foundation; under Contract to U.S. Dept. of Energy. Tax-exempt.
Institution Type/Description: History & Science Museum.
Collections: nuclear & alternate energy sources with hands-on exhibits, emphasizing Hanford programs; atomic marbles & nuclear waste storage & clean-up display; history of the mid-Columbia Basin & the Hanford Atomic site.
Research Fields: broad range of environmental sciences & historical exhibits.
Facilities: 50-seat auditorium. Museum-related items for sale.
Activities: guided tours; lectures; films & videos; formally organized education programs for children & adults.
Publications: video, Termination Winds; cookbook, history & photos, In the Shadow of Rattlesnake Mountain.
Hours & Admission Prices: Mon.-Sat. 10-5, Sun. 12-5. Adults $4, seniors & youth $3; ASTC, CREHST & museum members and children 6 & under no charge. Closed New Year's Day; Easter; Thanksgiving; Christmas. &
Attendance: 13,000 (estimated)
Membership: Educator, Senior & Student $25; Individual $30; Grandparent $45; Family $50; Museum & Business Donor $125-$249; Museum & Business Patron $250-$999; Gold Circle & Business Benefactor $1,000 & up.

HANFORD REACH INTERPRETIVE CENTER, 1766 Fowler St., Richland, WA 99352-4843. Mailing Address: P.O. Box 1160, Richland, WA 99352-1160. Tel.: 509-943-4100. Fax: 509-943-4133.
E-mail: kcamp@visitthereach.org
Web Site: visitthereach.org/contact.php
Formerly: Hanford Reach National Monument Heritage & Visitor Center (The Reach)
Founded: 2004.
Key Personnel: C.E.O., Kimberly Camp; Pres. (V), Eric Gerber.
Personnel Profile: Full-Time Paid 5.
Governing Authority: Parent Institution: Richland Public Facilities District. Tax-exempt.
Institution Type/Description: History Museum.
Collections: geological; ecological; cultural history.
Hours & Admission Prices: Opening summer 2014. Call for information. &

Seattle

* **BURKE MUSEUM OF NATURAL HISTORY AND CUL-TURE, (M),** University of Washington Campus, 17th Ave. N.E. & N.E. 45th St., Seattle, WA 98195. Mailing Address: Univ. of Washington, Box 353010, Seattle, WA 98195. Tel.: 206-543-5590. Fax: 206-685-3039.
E-mail: burke_museum@uw.edu
Web Site: www.burkemuseum.org
Founded: 1885.
Congressional District: 7
Key Personnel: Exec. Dir., Dr. Julie K. Stein; Dir. HR & Operations, Leslie Jones; Dir. Education, Diane Quinn; Assoc. Dir. Exhibits & Institutional Planning, Erin Younger; Dir. External Affairs, Alaina Smith; Registrar, Hollye Keister; Cur. Archaeology & Assoc. Dir. Research, Dr. Peter Lape; Cur. Herbarium, Dr. Richard Olmstead; Cur. Invertebrate Paleontology, Dr. Elizabeth Nesbitt; Cur. Native American Art, Dr. Robin K. Wright; Cur. Vertebrate Paleontology, Christian Sidor; Cur. Paleobotany, Caroline Stromberg; Cur. Fishes, Ted Pietsch; Cur. Birds, John Klicka; Cur. Genetic Resources & Herpetology, Adam Leache; Cur. Mammals, Sharlene Santana.
Personnel Profile: Full-Time Paid 60; Full-Time Volunteers 1; Part-Time Paid 40; Part-Time Volunteers 96.
Governing Authority: state; university. Parent Institution: University of Washington. Subsidiary Institution: Burke Museum Association. Tax-exempt: 501(c)(3).
Institution Type/Description: Anthropology & Natural History Museum.
Collections: Pacific Rim anthropology; Northwest Coast native art; paleontology; mineralogy; geology; zoology; entomology; mammalogy; ornithology; malacology; botany; ichthyology; genetic resources; herbarium; lepidoptera.
Research Fields: paleontology; geology; zoology; anthropology including archaeology; botany; genetics.
Facilities: classroom; events room; cafe. Museum-related items for sale.
Activities: temporary exhibits; docent guided tours; K-12 group tours & traveling study collections; lectures & gallery talks.
Publications: research reports; member's newsletter; exhibit-related publications.
Hours & Admission Prices: Call for hours. Adults $9.50, senior citizens $7.50, students and youth 5 & over $6; discounts to AAA, WMA, AAM & ICOM members; members & 1st Thurs. of month no charge. Closed New Year's Day; Independence Day; Thanksgiving; Christmas. &
Attendance: 109,000 (accurate)
Membership: Student $10; Senior Citizen $20; Individual $30; Dual Senior $35; UW Family $36; Family $55; Cascade Associate $100; Northwest Partner $250; Pacific Patron $500; Director's Circle $1,000.

CARL S. ENGLISH JR. BOTANICAL GARDENS, 3015 N.W. 54th St., Seattle, WA 98107-4213. Tel.: 206-789-2622. Fax: 206-782-3192.
Web Site: www.nws.usace.army.mil
Institution Type/Description: Botanical Gardens.
Collections: over 500 species & 1,500 varieties of plants from around the world.
Facilities: 7 acres.
Activities: garden; fish ladder; lock and dam.
Hours & Admission Prices: Grounds: daily 7am-9pm. Visitor Center: May-Sept. daily 10-6; Oct.-April Thurs.-Mon. 10-4. No charge. &

CENTER FOR WOODEN BOATS, (M), 1010 Valley St., Unit Main, Seattle, WA 98109-4444. Tel.: 206-382-2628. Fax: 206-382-2699.
E-mail: cwb@cwb.org
Web Site: www.cwb.org
Founded: 1978.
Congressional District: 1
Key Personnel: Pres., Alex Bennet; Founding Dir., Dick Wagner; Exec. Dir., Elizabeth Davis, Business Mgr., Katy Mathias; Business Asst., Laurie Leak; Events & Operations Mgr., Eldon Tam; Lead Boatwright, Heron Scott; Boat Sales Mgr. & Instructor, Patrick Gould; Waterfront Programs Mgr., Jake Beattie; Coord. Youth Field Trip, Tom Baltzell; Livery Mgr., Dock Master & Youth Sailing, Zach Carver; Boatwright & Workshop Coord., Edel O'Connor; Curriculum, Courtney Bartlett; Custodian, Bud Ricketts; Shipwright in Residence, Geoff Braden; Operations Asst., Sarah Salter; Sailing Instructor, Julia Makowski; Sail Now! Coord., Vern Velez.
Personnel Profile: Full-Time Paid 11; Full-Time Volunteers 1; Part-Time Paid 5; Part-Time Volunteers 690; Interns 2.
Governing Authority: nonprofit organization. Tax-exempt: 501(c)(3).
Institution Type/Description: Operational Maritime Museum.

Collections: traditional small craft, up to 35 ft. in length; historic photos; models of small craft; tools.
Research Fields: historic small craft of the world.
Facilities: 1,500-vol. library pertaining to maritime history; construction of voyages of small craft; technical information on wood & woodworking tools & techniques available to the public; working boatshop; botanical garden. Museum-related items for sale.
Activities: guided tours lectures; films; rowing, paddle a canoe; regattas; steam-bend a boat's rib; cast an oarlock; forge the hank for a foresail; splice a line; loft a hull; organized education programs for adults & children; various activities for members.
Publications: bimonthly newsletter, Shavings; monographs on historic small craft.
Hours & Admission Prices: March 21 to May & Sept.-Nov. 1 daily 10-6; Memorial Day to Labor Day daily 10-8; Nov. 2 to March 20 Tues.-Sun. 10-5. Fees for use of boats & maritime skills workshops. Closed Thanksgiving; Christmas. &
Attendance: 60,000 (estimated)
Membership: Senior Citizen & Student $15; Individual $35; Household $50; Contributing $100; Benefactor $250; Sustaining $500; Captain's Circle $1,000.

CENTER ON CONTEMPORARY ART, 6413 Seaview Ave., N.W., Seattle, WA 98107-2666. Tel.: 206-728-1980. Fax: 206-728-1980.
E-mail: info@cocaseattle.org
Web Site: www.cocaseattle.org
Founded: 1981.
Congressional District: 7
Key Personnel: Pres., Ray Freeman; Vice Pres. & Chm., Danilo Bonilla; Dir. Talent & Mktg., Lauren Collins.
Personnel Profile: Full-Time Volunteers 1; Part-Time Volunteers 65; Interns 2.
Governing Authority: private; nonprofit organization. Congressional District: WA 7th District. Tax-exempt: 501(c)(3).
Institution Type/Description: Art Museum.
Collections: contemporary art; performance art; multimedia & multidisciplinary programs.
Facilities: 4,000 sq. ft. exhibit space. Museum-related items for sale.
Activities: concerts; films; guided tours; lectures; participatory exhibits; members meetings. Annual Events: Northwest Annual: Juried Show; 24 House Painting Marathon & Auction; COCA Holiday Store; Seafair Ball & Art Invitational.
Publications: biannual newsletter, COCA Newsletter.
Hours & Admission Prices: Mon.-Fri. 10-5. No charge; donations accepted. Closed Independence Day; Thanksgiving; Christmas. &
Attendance: 15,000 (estimated)
Membership: Artist, Student & Senior Citizens $30; Individual $40; Couple $65; Family $100; Small Business $125; Donor $500; Millennium $1,000.

DAYBREAK STAR INDIAN CULTURAL CENTER - SACRED CIRCLE GALLERY, Discovery Park on Magnolia Hill, 5011 Bernie Whitebear Way, Seattle, WA 98199. Mailing Address: United Indians of All Tribes Foundation - Discovery Park, P.O. Box 99100, Seattle, WA 98139-0100. Tel.: 206-285-4425. Fax: 206-282-3640. Facebook: Daybreak Star Indian Cultural Center.
E-mail: info@unitedindians.org
Web Site: www.unitedindians.org
Founded: 1970.
Institution Type/Description: Art Gallery.
Collections: contemporary Native American art.
Hours & Admission Prices: Call for hours.

DES MOINES HISTORICAL SOCIETY MUSEUM, 730 S. 225th St., Seattle, WA 98198-6824. Mailing Address: P.O. Box 98055, Des Moines, WA 98198-0055. Tel.: 206-824-5226.
Governing Authority: nonprofit organization. Tax-exempt: 501(c)(3).
Institution Type/Description: History Museum.
Collections: local history & culture; photographs; personal artifacts; period furnishings.
Hours & Admission Prices: Memorial Day to Labor Day Sat. 1-4; other times by appointment. No charge; donations accepted.

EMP MUSEUM, (M), 325 5th Ave. N., Seattle, WA 98109-4630. Mailing Address: 300 8th Ave. N., Seattle, WA 98109. Tel.: 206-770-2700. Fax: 206-770-2727.
E-mail: experience@empmuseum.org
Web Site: www.empmuseum.org
Formerly: Experience Music Project

Founded: 1999.
Congressional District: 7
Key Personnel: Deputy Dir & Acting C.E.O., Patty Isacson Sabee; C.F.O., Traci Carman; Dir. Curatorial Affairs, EMP, Jasen Emmons; Dir. Mktg. & Audience Devel., Jessica Toon; Museum Shop Mgr., Chris Maresca.
Personnel Profile: Full-Time Paid 98; Part-Time Paid 68; Part-Time Volunteers 55; Interns 13.
Governing Authority: private; nonprofit organization. Parent Institution: Experience Learning Community. Tax-exempt: 501(c)(3).
Institution Type/Description: Pop Culture, Music & Science Fiction Museum.
Collections: Jimi Hendrix, instruments; Northwest music; roots of rock 'n' roll, hip-hop, punk, reggae; film; video; photography.
Research Fields: American pop music; blues; music videos; science fiction; fantasy; horror; film; audio oral histories.
Facilities: digital library; lounge; 2 learning labs; digital lab; 140,000 sq. ft. exhibit space; 191-seat theater; sound lab. Gift items for sale.
Activities: guided tours; lectures; concerts; films; formal education programs; family programs; permanent, temporary & traveling exhibitions; public programs. Annual Events: Founders Award: Science Fiction Hall of Fame Induction Ceremony; Science Fiction & Fantasy Short Film Festival; Pop Conference; Sound Off!
Publications: exhibition catalogues.
Hours & Admission Prices: Memorial Day to Labor Day daily 10-7; Sept.-May daily 10-5. Adults $20, adults online $18, senior citizens & students $17, military & youth 5-17 $14; discounts to AAM & ICOM members; members and children 4 & under no charge. Closed Thanksgiving; Christmas. &
Attendance: 525,800 (accurate)
Membership: Senior $45; Individual $50; Senior Dual $60; Dual $65; Family $95; Contributing $150; Supporting $250; Sustaining $500; Council Friend $1,200; Council Associate $2,500; Council Partner $5,000; Council Builder $10,000; Council Leader $15,000.

*   **FRYE ART MUSEUM, (M),** 704 Terry Ave., Seattle, WA 98104-2019. Tel.: 206-622-9250. Fax: 206-223-1707.
E-mail: info@fryemuseum.org
Web Site: www.fryemuseum.org
Founded: 1952.
Congressional District: 1
Key Personnel: Exec. Dir., Midge Bowman; Pres., David Buck; Deputy Dir., Jill Rullkoetter; Collections Mgr. & Registrar, Donna Kovalenko; Chief Cur., Robin Held; Museum Exhibitions Designer, Shane Montgomery; Museum Designer, Charla Reid; C.F.O., Donna DiFiore; Administration Coord., Roxanne Hadfield; Museum Shop Mgr., Karla Glanzman.
Personnel Profile: Full-Time Paid 22; Part-Time Paid 34; Part-Time Volunteers 40; Interns 3.
Governing Authority: nonprofit foundation. Tax-exempt: 501(c)(3).
Institution Type/Description: Art Museum.
Collections: 19th- & 20th-century European & American paintings; contemporary figurative & representational paintings.
Facilities: 3,000-vol. library; 12,000 sq. ft. exhibit space; art studio; 142-seat auditorium; restaurant. Museum-related items for sale.
Activities: guided tours; gallery talks; lectures; concerts; workshops; classes for adults & children; traveling exhibitions.
Publications: quarterly bulletin, FRYE; 4 exhibition catalogs annually.
Hours & Admission Prices: Tues.-Wed. & Fri.-Sat. 10-5, Thurs. 10-8, Sun. 12-5. No charge; donations accepted. Closed New Year's Day; Independence Day; Thanksgiving; Christmas. &
Attendance: 100,000 (estimated)
Membership: Student $25; Senior $30; Working Artist & Teacher $40; Dual Senior $45; Individual $50; Dual & Family $75; Supporter $150; Contributor $300; Patron $500; Frye Art Circle $1,000; Corporate $2,500; Frye Leadership Circle $5,000.

*   **HENRY ART GALLERY, (M),** 15th Ave. N.E. & N.E. 41st St., University of Washington, Seattle, WA 98195-1410. Mailing Address: U.W. Box 351410, Seattle, WA 98195. Tel.: 206-543-2280. Fax: 206-685-3123.
E-mail: info@henryart.org
Web Site: www.henryart.org
Founded: 1927.
Congressional District: 1
Key Personnel: Dir., Sylvia Wolf; Pres. (V), John Hoedemaker; Chm. (V), William True; Deputy Dir. External Rels., Nan Garrison; Deputy Dir. Finance & Operations, Anne Walsh; Deputy Dir. Art & Education, Luis Croquer; Assoc. Dir. Mktg., Communications & Public Rels., Dana Van Nest; Cur. Collections, Judy Sourakli; Exhibitions Mgr. & Registration, Susan Lewandowski; Head Preparator, Jim Rittimann; Exec. Asst. to Dir., Dustin Engstrom; Graphic Designer, Jayme Yen.

Personnel Profile: Full-Time Paid 30; Part-Time Paid 9; Part-Time Volunteers 134.
Governing Authority: Parent Institution: University of Washington. Subsidiary Institution: Henry Gallery Association. Tax-exempt.
Institution Type/Description: Art Museum.
Collections: 19th- to 20th-century American & European landscape paintings; paintings, prints, drawings & photographs; Japanese mingei ceramics; textiles & costumes, concentrated on hand-woven techniques; mid-19th century American & European dress.
Major Exhibits: Hands, 1/14-5/4/14; Katinka Bock: A and I, 2/14-5/4/14; Parallel Practices: Joan Jonas & Gina Pane, 3/14-6/8/14; The Brink: Anne Fenton (T), 3/14-6/15/14; UW School of Art MFA + MDes Thesis Exhibition, 5/24/6/22/14.
Research Fields: 19th- & 20th-century American art & photography history; contemporary art criticism; costume & textile history, design & techniques.
Facilities: print study room.
Activities: lectures & symposia; collection research; community art services.
Publications: David Hartt: Stray Light; Paul Laffoley: Premonitions of the Bauharoque; Out [o] Fashion Photography: Embracing Beauty; Like a Valentine: The Art of Jeffry Mitchell.
Hours & Admission Prices: Wed. & Sat.-Sun. 11-4, Thurs.-Fri. 11-9. Adults $10, senior citizens 62 & over $6; discounts to AAM, AAA & ICOM members; members, students & children no charge. Closed New Year's Day; Independence Day; Thanksgiving; Christmas Eve & Day. &
Attendance: 39,070 (accurate)
Membership: Individual $45 (20% off for UW alumni, seniors, artists & educators); Household $75; Henry + $150; Contemporaries $300; Patrons $1,500-$2,999; Director's Circle $3,000-$4,999; Chairman's Circle $5,000-$9,999; Artist's Circle $10, 000 & up.

HIRAM M. CHITTENDEN LOCKS VISITOR CENTER, 3015 N.W. 54th St., Seattle, WA 98107-4213. Tel.: 206-783-7059.
Institution Type/Description: History Museum.
Collections: history & operation of the ship canal & locks; gardens; fish ladders; Army Corps of Engineers.
Facilities: botanical gardens. Museum-related items for sale.
Activities: fish-viewing; locks raising & lowering boats.
Hours & Admission Prices: May-Sept. daily 10-6; Oct.-April Thurs.-Mon. 10-4.

KLONDIKE GOLD RUSH NATIONAL HISTORICAL PARK, 319 Second Ave. S., Seattle, WA 98104-2618. Tel.: 206-220-4240. Facebook: Klondike Gold Rush National Historical Park.
E-mail: klse_Ranger_Activities@nps.gov
Web Site: www.nps.gov/klse
Founded: 1979.
Congressional District: 7
Key Personnel: Supt., Jacqueline L. Ashwell; Cur. Record, Brooke Childrey.
Personnel Profile: Full-Time Paid 6; Part-Time Paid 6; Part-Time Volunteers 7; Interns 2.
Governing Authority: federal. Parent Institution: National Park Service, Washington, DC. Tax-exempt.
Institution Type/Description: Historic Building & Museum: housed in Seattle's Pioneer Square Historical District.
Collections: 1897-1898 artifacts from the Klondike stampede; 70,000 archival documents & photographs relating to Seattle history & the Klondike gold rush; equipment replicas & historic artifacts related to the gold rush.
Research Fields: historical Seattle as it related to Klondike Gold Rush; areas & people involved in Klondike gold rush.
Facilities: 70-seat auditorium.
Activities: guided tours; gold panning demonstrations; films; permanent & temporary exhibitions.
Publications: brochure.
Hours & Admission Prices: Memorial Day-Labor Day daily 9-5; Labor Day-Memorial Day daily 10-5. No charge; donations accepted. Closed New Year's Day; Thanksgiving; Christmas. &
Attendance: 70,000 (estimated)

LAST RESORT FIRE DEPARTMENT MUSEUM, 301 2nd Ave. S., Seattle, WA 98107. Mailing Address: 1433 N.W. 51st St., Seattle, WA 98107. Tel.: 206-783-4474. Fax: 206-784-1485.
E-mail: lastresortfd@hotmail.com
Web Site: www.lastresortfd.org/museum.htm
Institution Type/Description: Firefighting History Museum.
Collections: firefighting history, equipment & artifacts; fire trucks including 1834 hand pumper; 1899 horse-drawn steam pumper; 1940 Ford pumper; 1950 Kenworth pumper; 1958 Mack pumper; photographs; alarm system.
Hours & Admission Prices: Summer: Wed.-Thurs. 11-3; Winter: Wed. 11-3. No charge.

MUSEUM OF COMMUNICATIONS, 7000 E. Marginal Way S., Seattle, WA 98108-3411. Mailing Address: P.O. Box 81103, Seattle, WA 98108-1103. Tel.: 206-767-3012.
E-mail: qwest541@qwest.net
Web Site: www.museumofcommunications.org
Formerly: Vintage Telephone Museum
Founded: 1985.
Key Personnel: Dir., Don Ostrand.
Personnel Profile: Part-Time Volunteers 15.
Governing Authority: Parent Institution: Telecommunications History Group (THG). Tax-exempt.
Institution Type/Description: Communications Museum.
Collections: telephone history; Alexander Graham Bell's first communications device to modern technology; Seattle telephone directories from 1900 to present.
Research Fields: genealogy.
Hours & Admission Prices: Tues. 8:30-2, 1st Sun. each month 11:30-4; other times by appointment. No charge; donations accepted. &
Attendance: 875 (estimated)
Membership: THG: Individual $35.

＊　**THE MUSEUM OF FLIGHT, (M),** 9404 E. Marginal Way S., Seattle, WA 98108-4097. Tel.: 206-764-5700 (Admin.) & 5720 (Visitor Information). Fax: 206-764-5707.
E-mail: info@museumofflight.org
Web Site: www.museumofflight.org
Founded: 1964.
Congressional District: 7
Key Personnel: Pres. & C.E.O., Doug King; Vice Pres. & C.O.O., Laurie B. Haag; Vice Pres. & C.F.O., Matt Hayes; Chm. (V), Michael Hallman; Sr. Cur., Dan Hagedorn; Dir. Education, Seth Margolis; Dir. Devel., Dean McColgan; Dir. Exhibits, Chris Mailander; Dir. Mktg., Mike Bush; Dir. Sales, Rich Rime; Dir. Aircraft Collections, Tom Cathcart; Controller, Lynda King; Museum Shop Mgr., Mary Christensen; Dir. Facilities, Clark Miller.
Personnel Profile: Full-Time Paid 132; Part-Time Paid 44; Part-Time Volunteers 548.
Governing Authority: nonprofit organization. Tax-exempt: 501(c)(3).
Institution Type/Description: Aeronautics & Space Museum: c.1910 the first aircraft manufacturing facility in the region
Collections: over 125 historic aircraft; thousands of smaller artifacts including aircraft parts, garments, flight test instruments & model aircraft; space flight artifacts.
Research Fields: aviation & space industry history in the United States; history of fighter aviation in WWI & WWII.
Facilities: library of textbooks, technical, popular & historical books & publications; 268-seat auditorium; educational facilities; aircraft restoration center; 178,000 sq. ft. exhibit space. Books, model aircraft & other related items for sale.
Activities: guided tours; lectures; films; rental gallery; organized educational programs for children; docent program; participatory, loan & temporary exhibitions; outreach to schools taking mini-museum; special aerospace events; fly-ins of historic aircraft; Challenger Learning Center missions.
Publications: bimonthly newsletter, Aloft; books, monographs.
Hours & Admission Prices: Daily 10-5, first Thurs. each month 10-9. Adults $17, seniors $13, children 5-17 $8.50; discount to groups and Boeing employees, active military & AAM members; members no charge. Closed Thanksgiving; Christmas. &
Attendance: 458,000 (estimated)
Membership: Navigator Solo $50; Aviator Family $75; Captain Family $100; Flight Leader Family $250; Barnstormer Family $500; Barnstormer Gold Family $1,000.

＊　**MUSEUM OF HISTORY & INDUSTRY (MOHAI), (M),** 860 Terry Ave. N., Seattle, WA 98109. Tel.: 206-324-1126. Fax: 206-324-1346.
E-mail: information@seattlehistory.org
Web Site: www.seattlehistory.org
Founded: 1914.
Congressional District: 7
Key Personnel: Co Pres. Bd., Maggie Walker; Co Pres. Bd., Jerry Vandenberg; Dir., Leonard Garfield; Deputy Dir., Martha Aldridge; Librarian, Carolyn Marr; Historian, Dr. Lorraine McConaghy; Registrar, Kristin Halunen; Mgr. Collections, Betsy Bruemmer; Mgr. Public Programs, Helen Diujak; Mgr. Education, Martha Lindsey; Cur. Photography, Howard Giske; Cur. Textiles & Asst. Librarian, Mary Montgomery.
Personnel Profile: Full-Time Paid 32; Part-Time Paid 11; Part-Time Volunteers 130; Interns 5.

Governing Authority: Parent Institution: Historical Society of Seattle & King County. Tax-exempt: 501(c)(3).
Institution Type/Description: History & Industry Museum.
Collections: 800,000 items including mid-19th to 20th-century costumes; more than 2,000,000 photographs recording the history of the Pacific Northwest; textiles; decorative arts; furniture; silver; china; glassware; handmade items; folk art; paintings; tools & equipment related to communications & local industries: maritime, fishing, agriculture, logging, mining; recreation items: toys, musical instruments & sports; vintage vehicles: boats, airplanes & cable cars.
Research Fields: history of Seattle & Puget Sound Region from 1850 including pioneering, settlement, westward movement & activities & events thereof; preservation of photograph collection.
Facilities: library of Northwest Americana including books, photographs, films, videotapes; sound recordings, manuscripts, ephemeral materials, holdings of the Puget Sound Maritime Historical Society & Black Heritage Society of Washington State; 375-seat auditorium; meeting room. Books, note cards, toys, games, jewelry & other museum-related items for sale.
Activities: guided tours; lectures; workshops; rental gallery; formally organized education programs for children; docent programs; permanent & temporary exhibitions.
Publications: MOHAI newsletter, Old News.
Hours & Admission Prices: Thurs. 10-8, Fri. -Wed. 10-5. Adults $14, seniors 62 & over, students and military w/ID $12, youth 13-19 $5; discounts to AAM & AAA members; children under 13 & members no charge. Closed Thanksgiving; Christmas. &
Attendance: 62,857 (accurate)
Membership: Seniors 62 & over, Students and Teachers $10 discount for any membership level; Individual $45; Dual $55; Family $65; Friend $100-$249; Patron $250-$499; Benefactor $500-$999; Heritage Guild $1,000.

NORDIC HERITAGE MUSEUM, 3014 N.W. 67th St., Seattle, WA 98117-6215. Tel.: 206-789-5707. Fax: 206-789-3271. Facebook: Nordic Heritage Museum.
E-mail: nordic@nordicmuseum.org
Web Site: www.nordicmuseum.org
Founded: 1979.
Congressional District: 5
Key Personnel: Dir. & C.E.O., Eric Nelson; Cur. Collections, Lisa Hill-Festa; Devel. Dir., Jan Woldseth Colbrese; Children Education Coord., Alison Church; Mktg. & Communications Mgr., Erin M. Schadt.
Governing Authority: nonprofit organization. Tax-exempt: 501(c)(3).
Institution Type/Description: Heritage Museum. housed in 1907 Daniel Webster School.
Collections: folk costumes; textiles; woodcarvings; logging, fishing & mining equipment; furniture & household items; art by Nordic American artists.
Research Fields: Nordic immigration & settlement.
Facilities: 12,000-vol. library of Nordic language books; 150-seat auditorium; classrooms; 31,000 sq. ft. exhibit space; meeting rooms. Gift items for sale.
Activities: guided tours; language classes; folk art classes; lectures; films; concerts; arts festivals; study clubs; hobby workshops; organized education programs for children & adults; docent program; loan, temporary, & traveling exhibitions. Special Events: Christmas Crafts Fair; Summer Crafts/Music & Dance Festival; oral history project; annual summer camp for children.
Publications: bimonthly newsletter, Nordic News; Nordic Heritage Museum Historical Journal; Tasty Traditions; Voices of Ballard: Immigrant Stories from the Vanishing Generation.
Hours & Admission Prices: Tues.-Sat. 10-4, Sun. 12-4. Adults $6, senior citizens $5, children 5 & up $4; first Thurs. of the month; members & children under 5 no charge. Closed New Year's Day; Easter; Thanksgiving; Christmas Eve & Day. &
Attendance: 55,000 (estimated)
Membership: Student & Senior 62 & up $25; Individual $35; Senior Couple $40; Family $60; Organization $60/$120; Sustaining $150; Patron & Corporate $300; President's Club $1000 & up; Nordic Round Table $5,000 & up.

NORTH SEATTLE COMMUNITY COLLEGE ART GALLERY, 9600 College Way N., Instructional Bldg. Rm. 1322, Seattle, WA 98103-3599. Tel.: 206-528-4557.
E-mail: lynne.hull@seattlecolleges.edu
Key Personnel: Coord. & Cur., Lynne Hull
Institution Type/Description: Art Gallery.
Collections: works by local & regional artists.
Activities: workshops; special events; art group meetings.
Hours & Admission Prices: During Exhibitions: Mon.-Tues. & Fri. 11-3, Wed.-Thurs. 11-3 & 6-8.

NORTHWEST AFRICAN AMERICAN MUSEUM, (M), 2300 S. Massachusetts St., Seattle, WA 98144-3821. Tel.: 206-518-6000. Fax: 206-518-5665.
E-mail: info@naamnw.org
Web Site: www.naamnw.org
Founded: 2007.
Congressional District: 37
Key Personnel: Exec. Dir., Barbara Earl Thomas
Institution Type/Description: African American Museum.
Collections: African American history & culture; photographs; art; personal artifacts.
Activities: educational programs; school groups; facility rental.
Hours & Admission Prices: Wed. & Fri. 11-4:30, Thurs. 11-7, Sat. 11-4, Sun. 12-4. Adults $6, students & seniors $4; members, 1st & 2nd Thurs. each month, and children 5 & under no charge. Closed New Year's Eve, Day & day after; Independence Day; Thanksgiving & weekend after; Christmas Eve, Day & week. ♿
Membership: Student & Senior $20; Individual $35; Family $50; Leader $100; Patron $500; Platinum $1,000; Corporate $2,500; Benefactor $5,000.

NORTHWEST MUSEUM OF LEGENDS AND LORE, Inscape Bldg., 815 Airport Way S., Seattle, WA 98134. Mailing Address: P.O. Box 12213, Seattle, WA 98102-0213. Tel.: 206-328-6499.
E-mail: seattlemysterymuseum@gmail.com
Formerly: Seattle Museum of the Mysteries
Founded: 2003.
Key Personnel: Dir., Charlette Lefevre; Librarian, Philip Lipson
Institution Type/Description: Paranormal Science Museum.
Collections: UFO & ghost history; Bigfoot casts; photographs; movie memorabilia.
Facilities: library.
Activities: ghost tour; lectures; ghost poker; special events.
Hours & Admission Prices: Mon.-Fri. 12-8:30, Sat. 12-8, Sun. 1:30-5:30. Suggested Donation: adults $3, youth 9-17 $2; children 8 & under and members no charge. ♿
Attendance: 5,000
Membership: Individual $24.

ODYSSEY MARITIME DISCOVERY CENTER, 2205 Alaskan Way, Pier 66, Seattle, WA 98121-1604. Tel.: 206-374-4000. Fax: 206-374-4002.
E-mail: info@ody.org
Founded: 1998.
Congressional District: 7
Key Personnel: Pres. (V), Paul E. Stevens; Education Mgr., Cassandra Sandkam; Media Contact, Leigh Barer.
Personnel Profile: Full-Time Paid 5; Part-Time Paid 3; Part-Time Volunteers 22.
Governing Authority: private; nonprofit organization. Tax-exempt: 501(c)(3).
Institution Type/Description: Maritime Museum.
Collections: maritime history & heritage; hands-on exhibits; photographs.
Research Fields: visitor studies; museum education; museum governance.
Facilities: educational facilities; 20,000 sq. ft. exhibit space; waterfront sitting. Maritime-related items for sale.
Activities: docent program; participatory exhibits; hands-on exhibits.
Publications: quarterly newsletter, Voyages.
Hours & Admission Prices: Wed.-Thurs. 10-3, Fri. 10-4, Sat.-Sun. 11-5. Call 206-374-4000 for additional hours. Adults $7, senior citizens & students $5, children 2-4 $2; discounts for military, groups & AAM members; children under 2 & members no charge. Closed New Year's Day; Thanksgiving; Christmas. ♿
Attendance: 40,135 (accurate)
Membership: Individual $30; Family $60; Silver $100; Gold $200; Platinum $350; Lifetime $10,000. Admiral's Club: Petty Officer $250; Mater Chief $500; Lieutenant $1,000; Commander $2,500; Captain $5,000; Commodore $10,000.

∗ OLYMPIC SCULPTURE PARK, (M), 2901 Western Ave., Seattle, WA 98121-1025. Mailing Address: 1300 First Ave., Seattle, WA 98101-2003. Tel.: 206-654-3100 & 332-1377. TDD: 206-344-5267.
E-mail: webmaster@seattleartmuseum.org
Web Site: www.seattleartmuseum.org/visit/osp
Institution Type/Description: Sculpture Park Museum.
Collections: classic, modern & contemporary sculptures; native plants.
Facilities: 9 acre park; visitors pavilion.
Activities: temporary exhibits; educational programs.
Hours & Admission Prices: Sculpture Park: April-Oct. daily 6 am-10 pm;

Nov.-March daily 7-6. Pavilion: May to Labor Day Tues.-Sun. 10-5; Sept.-April Tues.-Sun. 10-4. No charge. Closed New Year's Eve & Day; Labor Day; Thanksgiving; Christmas Eve & Day.

PACIFIC SCIENCE CENTER, (M), 200 2nd Ave. N., Seattle, WA 98109-4895. Tel.: 206-443-2001. Fax: 206-443-3631. TDD: 206-443-2887.
Web Site: pacificsciencecenter.org
Founded: 1962.
Congressional District: 36
Key Personnel: C.E.O. & Pres., R. Bryce Seidl; C.F.O. & C.O.O., Michal Anderson; Guest Svcs. & Theater Operations, Diane Carlson; Vice Pres. Develop., Erik Pihl; Vice Pres. Science & Education, Ellen Lettvin; Museum Shop Mgr., Katherine Johnson; Vice Pres. Communications & Mktg., Crystal Clarity; Vice Pres. Exhibits, Diana Johns; Vice Pres. Strategic Programs, Meena Selvakumar.
Personnel Profile: Full-Time Paid 147; Part-Time Paid 280; Part-Time Volunteers 279; Interns 4.
Governing Authority: nonprofit organization. Tax-exempt: 501(c)(3).
Institution Type/Description: Science & Technology Museum.
Collections: astronomy; space sciences; anthropology; biology; geology; historical models; health sciences; life sciences.
Major Exhibits: The Photographs of Modernist Cuisine (T), 10/26/13-2/12/14; Spy: The Secret World of Espionage (T), 3/25/14-9/1/14.
Facilities: classrooms; 400-seat IMAX theater; 350-seat IMAX theater; 300-person laser theater; planetarium; 50,000 sq. ft. exhibit space; restaurant; rental facilities. Museum-related items for sale.
Activities: participatory exhibits; science demonstrations; film series; science enrichment classes for adults & children; lectures; field trips; in-school traveling education programs; intern & volunteer programs; summer camps; live animal area; science cafes; science forums; teacher education; research weekends; scientist spotlights; preschool.
Publications: bimonthly newsletter, Discover Pacific Science Center; weekly e-newsletter.
Hours & Admission Prices: Daily 10 a.m. Exhibits & IMAX: adults $22, seniors $20, youth 6-15 $17, children 3-5 $14. Exhibits: adults $18, senior citizens 65 & over $16, youth 6-15 $13, children 3-5 $10; members no charge. IMAX: adults $9, seniors $8, youth 6-15 $7; children 3-5 $6. Closed Thanksgiving; Christmas. ♿
Attendance: 1,000,000 (accurate)
Membership: Individual $60; Dual $80; Family $100; Platinum $250; Titanium $500.

PHOTO CENTER NW, 900 Twelfth Ave., Seattle, WA 98122. Tel.: 206-720-7222. Fax: 206-720-0306.
E-mail: pcnw@pcnw.org
Web Site: pcnw.org/gallery
Key Personnel: Dir., Ann Pallesen
Institution Type/Description: Art Gallery.
Collections: over 200 photographs by emerging & international artists.
Hours & Admission Prices: Mon.-Thurs. 11-10, Fri.-Sun. 12-8. No charge; donations accepted. Closed New Year's Day; Memorial Day; Independence Day; Labor Day; Thanksgiving; Christmas. ♿
Membership: Supporting $75; Sustaining $150; Ambassador $250; Benefactor $500; Patron $1,000.

PRATT FINE ARTS CENTER, 1902 S. Main St., Seattle, WA 98144-2206. Tel.: 206-328-2200. Fax: 206-328-1260.
E-mail: info@pratt.org
Web Site: www.pratt.org
Key Personnel: Exec. Dir., Michelle Bufano.
Governing Authority: Branch Institution: Tashire Kaplan Studios.
Institution Type/Description: Art Gallery.
Collections: works by instructors, scholarship recipients, studio renters, & students.
Activities: classes; lectures.
Hours & Admission Prices: Daily 9am-10pm. Closed New Year's Day; Memorial Day; Independence Day; Labor Day; Thanksgiving; Christmas.

SEATTLE AQUARIUM, 1483 Alaskan Way, Pier 59, Seattle, WA 98101-2015. Tel.: 206-386-4300. Fax: 206-386-4328.
E-mail: contactus@seattleaquarium.org
Web Site: www.seattleaquarium.org
Founded: 1977.
Congressional District: 7
Key Personnel: Pres. & C.E.O., Robert W. Davidson; Chm. (V), Jim Gurke; Dir. Public Affairs, Tim Kuniholm; Dir. Finance & Admin., Ryan Dean; Dir. Mktg., Membership & Guest Experience, Marsha Savery; Dir. Devel., Lori

Montoya, Dir. Human Resources, Veronica Smolen, Dir. Life Sciences, C.J. Casson; Dir. Facilities, Alan Maxey; Dir. Conservation & Education, Jim Wharton; Museum Shop Mgr., Teresa Reed; Food Svcs., Rose Van Ommen.

Personnel Profile: Full-Time Paid 153; Part-Time Paid 3; Part-Time Volunteers 1,146.

Governing Authority: nonprofit. Parent Institution: Seattle Aquarium Society. Tax-exempt: 501(c)(3).

Institution Type/Description: Aquarium and Marine Museum: located on downtown waterfront.

Collections: live collection of fishes, invertebrates, marine mammals & birds; from local saltwater environment & fresh, saltwater habitats over the world; environmental exhibits on Puget Sound.

Research Fields: biological fields relating to marine life; educational research on teacher use of field trip sites.

Facilities: library; classrooms; special event space.

Activities: self-guided tours for school groups; lectures; films; formally organized education programs for children & adults; outdoor beach interpretation; teacher workshops; special events.

Publications: quarterly newsletter, seasonal brochures, annual school program brochures; annual report; tide charts.

Hours & Admission Prices: Daily 9:30-5. Adults $21.95, youth 4-12 $14.95; children 3 & under and members no charge. Closed during fundraiser in June; Christmas. &

Attendance: 829,668 (accurate)

Membership: Individual $80; Family $95; Family Plus $120.

---

* **SEATTLE ART MUSEUM, (M),** 1300 First Ave., Seattle, WA 98101-2003. Tel.: 206-654-3100. Fax: 206-654-3135. TDD: 206-654-3137.

E-mail: pr@seattleartmuseum.org
Web Site: www.seattleartmuseum.org
Founded: 1933.
Congressional District: 1

Key Personnel: Illsley Ball Nordstrom Dir., Kimerly Rorschach; Chm. Bd., Charles Wright; Pres. Bd., Winnie Stratton; Cur. Art of Africa & Oceania, Pamela McClusky; Jon & Mary Shirley Cur. Modern & Contemporary Art, Catharina Manchanda; The Ruth J. Nutt, Cur. Decorative Arts, Julie Emerson; The Kayla Skinner Dir. Education & Public Programs, Sandra Jackson-Dumont; Dir. Mktg. & Public Rels., Cara Egan.

Governing Authority: nonprofit. Subsidiary Institution: Seattle Asian Art Museum; Olympic Sculpture Park. Tax-exempt: 501(c)(3).

Institution Type/Description: Art Museum.

Collections: Chinese, Japanese, Korean, Indian & South East Asian art; Chinese jades; Katherine White Collections of African Art; ancient American & Oceanic art; Northwest Coast; modern & contemporary European & American paintings, sculpture, prints & photography; European & American decorative arts; Near Eastern, Egyptian, Greek, Roman, Medieval, Renaissance & Baroque paintings; numismatics; textiles.

Research Fields:

Facilities: library; slide library; 70,000 sq. ft. exhibit space; auditorium & lecture hall (each with assistive listening devices); restaurant. Museum-related items for sale.

Activities: guided tours; lectures; films; gallery talks; demonstrations; tea ceremonies; workshops; concerts; formally organized education programs; outreach programs to children & adults; CD-ROM Audio Gallery Guide; interactive multimedia computer kiosk on permanent collection; curriculum packets & workshops for teachers; inter-museum loans; permanent, temporary & traveling exhibitions; docent tours; special preview events.

Publications: annual report; members' newsletter & program guides; permanent collection catalogs; exhibition catalogs; brochures; gallery guides (many in large type).

Hours & Admission Prices: Wed. & Fri.-Sun. 10-5, Thurs. 10-9. Suggested Admissions: adults $19.50, seniors 62 & over and military with ID $17.50; students with ID & teen 13-17 $12.50; discounts to AAM, ICOM, AAA & Entertainment members; members, children 12 & under and 1st Thurs. of month no charge. &

Attendance: 500,000 (estimated)

Membership: Individual $65; Dual $80; Family $95; Patron $200; Friend $300; Fellow $600.

---

* **SEATTLE ASIAN ART MUSEUM, (M),** 1400 E. Prospect, Volunteer Park, Seattle, WA 98112-3303. Tel.: 206-654-3100. Fax: 206-654-3191. TDD: 206-344-5267.

E-mail: webmaster@seattleartmuseum.org
Web Site: www.seattleartmuseum.org
Founded: 1933.
Congressional District: 1

Key Personnel: Dir. The llsley Ball Nordstrom, Derrick Cartwright; Chm. Bd.,

---

Jon Shirley; Pres. Bd., Susan Brotman; Librarian, Traci Timmons; Maintenance Supvr., Jim Haarsager; Foster Foundation Asst. Cur. Chinese Art, Josh Yiu.

Personnel Profile: Full-Time Paid 162; Full-Time Volunteers 512; Part-Time Paid 30; Interns 6.

Governing Authority: private; nonprofit organization. Parent Institution: Seattle Art Museum. Tax-exempt: 501(c)(3).

Institution Type/Description: Art Museum.

Collections: Japanese, Korean, Indian, Southeast Asian & Chinese art.

Research Fields:

Facilities: 3,500-vol. library of books on the arts & sciences of Asia; auditorium; activities room; art storage; teacher resource center. Asian-related merchandise including art books, cards & jewelry for sale.

Activities: guided tours; lectures; films; gallery talks; demonstrations; workshops; concerts; formally organized education programs for adults, families, senior citizens, children & undergraduate college students, teachers & members; outreach programs for children & adults; curriculum packets & workshops for teachers; inter-museum loan, permanent, temporary & traveling exhibitions; docent tours; special preview events.

Publications: annual report; member newsletter/program guides; permanent collection & exhibition catalogs; brochures; gallery guides (many in large type).

Hours & Admission Prices: Summer: Tues.-Wed. & Fri.-Sun. 10-5, Thurs. 10-9; Winter: Wed. & Fri.-Sun. 10-5, Thurs. 10-9. Suggested: adults $7 students, youth 13-17 & seniors $5; discounts to AAM & ICOM members; members no charge. Closed New Year's Eve & Day; Labor Day; Thanksgiving; Christmas Eve & Day. &

Attendance: 74,878 (accurate)

Membership: Student & Senior $30; Individual $55; Dual $70; Family $75; Patron $175; Friend $250; Fellow $550; Ambassadors Circle $1,000; Stewards Circle $2,500; Curators Circle $5,000; Benefactors Circle $10,000; Directors Circle $15,000; Presidents Circle $25,000; Chairman's Circle $50,000.

SEATTLE CHILDREN'S MUSEUM, 305 Harrison St., Seattle, WA 98109-4623. Tel.: 206-441-1768. Fax: 206-448-0910. Facebook: Seattle Children's Museum.

E-mail: infoplease@thechildrensmuseum.org
Web Site: www.thechildrensmuseum.org
Formerly: The Children's Museum, Seattle
Founded: 1981.
Congressional District: 7

Key Personnel: Exec. Dir., Donna Marie Bertrand; Chm., Joseph Hagar; Dir. Communication, Jennifer Ross; Museum Shop Mgr., Brittany Baurer

Governing Authority: nonprofit organization. Tax-exempt: 501(c)(3).

Institution Type/Description: Children's Museum.

Collections: hands-on exhibits, including a child-sized neighborhood; infant-toddler discovery area; multicultural exhibits; drop-in art center.

Facilities: Museum-related items for sale.

Activities: interactive programs for children from birth to 10 years old.

Hours & Admission Prices: Mon.-Fri. 10-5, Sat.-Sun. 10-6. Admission $7.50, seniors $6.50; discounts to groups of 10 or more; children under one no charge. Closed New Year's Day; Labor Day weekend; Thanksgiving; Christmas. &

Attendance: 232,000 (accurate)

SOUTHWEST SEATTLE HISTORICAL SOCIETY - LOG HOUSE MUSEUM, 3003 61st Ave., S.W., Seattle, WA 98116-2810. Tel.: 206-938-5293. Facebook: Log House Museum.

E-mail: loghousemuseum@comcast.net
Web Site: www.loghousemuseum.info
Founded: 1997.

Key Personnel: Museum Mgr., Sarah Baylinson; Pres., Marcy Johnsen.

Personnel Profile: Part-Time Paid 2; Part-Time Volunteers 32.

Governing Authority: Parent Institution: Southwest Seattle Historical Society. Tax-exempt.

Institution Type/Description: History Museum: log house; built in 1903.

Collections: history of the Duwamish Peninsula & west Seattle; photographs; archives.

Facilities: library; gardens. Museum-related items for sale.

Activities: heritage education trunk kits; school tour presentations; teen docent program; speaker series; heritage tours & walks.

Publications: newsletter, Footprints in the Sands of Time; All Aboard for the Luna Park; Memories of Southwest Seattle Businesses; Tell Me a Story: A Memory Book for Youth from the Log House Museum; West Seattle Memories - Alki.

Hours & Admission Prices: Thurs.-Sun. 12-4. Suggested Donation: adults $3, children $1. Closed holidays. &

Attendance: 1,813 (accurate)

Membership: Senior & Student $15; Individual & Nonprofit $20; Family $35.

SWEDISH FINN HISTORICAL SOCIETY ARCHIVES AND LIBRARY, 1920 Dexter Ave. N., Seattle, WA 98109-2718. Tel.: 206-706-0738. Fax: 206-782-5813.
Institution Type/Description: Historical Society Museum.
Collections: Swedish-Finn culture, tradition & history; photographs; books.
Facilities: library; archives.
Activities: research.
Hours & Admission Prices: Mon. & Thurs. 9:30-12:30, Wed. 2-5.

UNIVERSITY OF WASHINGTON BOTANIC GARDEN, 3501 N.E. 41st St., Seattle, WA 98105. Mailing Address: Box 354115, Seattle, WA 98195-4115. Tel.: 206-543-8616. Fax: 206-685-2692. Facebook: UW Botanic Gardens.
E-mail: uwbg@u.washington.edu
Web Site: depts.washington.edu/uwbg/index.php
Formerly: Washington Park Arboretum
Founded: 1934.
Congressional District: 1
Key Personnel: Dir. & Orin and Althea Soest Chair for Urban Horticulture, Sarah Reichard; Asst. Dir., Fred Hoyt; Cur. Living Collections, Ray Larson; Mgr. Horticulture & Plant Records, David Zuckerman; Arboretum Arborist/Gardener Lead, Christopher Watson.
Personnel Profile: Full-Time Paid 17; Part-Time Paid 22; Part-Time Volunteers 250; Interns 7.
Governing Authority: municipal; nonprofit organization. Parent Institution: University of Washington, City of Seattle. Tax-exempt: 170(b)(1)(A).
Institution Type/Description: Arboretum & Botanical Garden.
Collections: 230-acres containing approximately 4,400 taxa of woody plants. Historic Houses: 1936 Stone Cottage; 1936 Dawson Plan (Olmsted Brothers).
Research Fields: horticultural taxonomy; woody plant introduction & testing; plant conservation; urban ecology & forestry; horticultural education; habitat restoration.
Facilities: visitor center; lecture room; outdoor classrooms; greenhouse; reading room. Museum-related items for sale.
Activities: guided tours; lectures; films; formally organized education programs for adults; docent program; slide programs; primary school programs; primary school programs (saplings); family programs; tours. Arboretum Sponsors: Branching Out (inner city); summer day camp.
Publications: quarterly magazine & journal, Volunteer News Washington Arboretum Bulletin; Arboreta Foundation Newsletter; Center for Urban Horticulture: CUH Presents; Pro Hort.
Hours & Admission Prices: Arboretum: daily dawn-dusk. No charge. Visitors' Center: daily 9-5. Japanese Garden $5. Closed New Year's Day; Thanksgiving; Christmas.
Attendance: 400,000 (estimated)
Membership: Arboretum Foundation Annual $25.

WESTERN BRIDGE, 3412 Fourth Ave. S., Seattle, WA 98134-1905. Tel.: 206-838-7444.
E-mail: info@westernbridge.org
Web Site: www.westernbridge.org
Key Personnel: Dir., Eric Fredericksen
Institution Type/Description: Art Gallery.
Collections: works by contemporary artists.
Hours & Admission Prices: Call for hours.

WING LUKE MUSEUM OF THE ASIAN PACIFIC AMERICAN EXPERIENCE, 719 S. King St., Seattle, WA 98104-3035. Tel.: 206-623-5124. Fax: 206-623-4559. Facebook: Wing Luke Museum.
E-mail: folks@wingluke.org
Web Site: www.wingluke.org
Formerly: Wing Luke Asian Museum
Founded: 1967.
Congressional District: 7
Key Personnel: Exec. Dir., Beth Takekawa; Deputy Exec. Dir., Cassie Chinn; Deputy Dir. Operations, Ethelyn Abellanosa; Devel. & Mktg. Dir., Margaret Su; Finance Dir., Monica Day; Education Dir., Charlene Mano-Shen; Facilities Operations Dir., John Hom; Exec. & Trustee Assoc., Karen Kajiwara; Devel. Mgr., Josie Baltan; Grants Mgr., Elizabeth Bly; Community Programs Mgr., Vivian Chan; Collections Mgr., Robert Fisher; Exhibits Mgr., Michelle Kumata; Membership & Mktg. Mgr., Jennifer Maines; Education Mgr., Virgel Paule; Marketplace Mgr., Hanh Pham; Visitor Svcs. Mgr., Stacey Swamby; Exhibit Devel./YouthCAN Mgr., Mikala Woodward; Accountant, Troy Tsuchikawa; Education Coord., Roldy Ablao; Librarian/Community Programs Coord., Janet Aviado; Tour Coord., Andrea Kim Taylor; Exhibit Specialist, Jessica Rubenacker; Information Technol-

ogy Specialist, David Chattin-McNichols; Devel. Officer, Inmi Kim; Devel. Asst., Maria Martinez; YouthCAN Asst., Mario Pilapil.
Personnel Profile: Full-Time Paid 14; Part-Time Paid 7; Part-Time Volunteers 90; Interns 18.
Governing Authority: nonprofit organization. Tax-exempt: 501(c)(3).
Institution Type/Description: Asian Pacific American History, Art & Culture Museum.
Collections: Asian-American and Asian history, art & culture.
Major Exhibits: Alaskeros (Space 3) (George Tsutakawa Art Gallery), 10/12/12-9/7/14; Asian American Idols (New Dialogues Initiative), 9/6/13-4/13/14; Digging Deep: Archeology Heritage Sites (Special Exhib. Gallery), 12/13/13-10/19/14; Tribal Tattoos (Space 3) (George Tsutakawa Art Gallery), 10/3/14-9/13/15.
Research Fields: early Asians & Pacific islanders in the Pacific Northwest; northwest Asian-American history.
Facilities: 500-vol. library of material pertaining to the collection & research fields available on premises; community & social service use & sponsorship.
Activities: guided tours; lectures; formally organized education programs for adults & children; inter-museum loan & traveling exhibitions; video documentaries; film & discussion series; readings.
Publications: quarterly membership newsletter; books, Executive Order 9066: 50 Years Before & 50 Years After, Reflections of Seattle's Chinese-Americans: The First 100 Years, They Painted from Their Hearts: Pioneer Asian-American Artists, Beyond the Rock Garden: Craft Form for a New World; Divided Destiny: A History of Japanese Americans in Seattle; P.I. (Made in America): Filipino American Artists in the Pacific Northwest; CD: A Bridge Home.
Hours & Admission Prices: Tues.-Sun. 10-5, 1st Thurs & 3rd Sat. of month 10-8. Adults $12.50, seniors & students13-18 $9.95, children 5-12 $8.95; discounts to AAM & ICOM members; 1st Thurs. & 3rd Sat. of month, members & children under 5 no charge. Closed New Year's Day; Independence Day; Thanksgiving; Christmas Eve & Day. &
Attendance: 45,000 (estimated)
Membership: Individual $45; Friends $65; Family $75; Patron $125; Benefactor $250; Gallery Supporter $500; Leadership Circle $1,000.

WOODLAND PARK ZOO, (M), 5500 Phinney Ave. N., Seattle, WA 98103-5897. Mailing Address: 601 N. 59th St., Seattle, WA 98103-5858. Tel.: 206-548-2500. Fax: 206-548-1536. TTY: 206-548-2599.
E-mail: webkeeper@zoo.org
Web Site: www.zoo.org
Formerly: Woodland Park Zoological Gardens
Founded: 1900.
Congressional District: 32
Key Personnel: Pres. & C.E.O., Deborah B. Jensen, Ph.D.; Vice Pres. External Rels., David Wu; Vice Pres. Human Resources, Damian King; Dir. Mktg., Jim Bennett; C.O.O., Bruce Bohmke; Gen. Cur., Nancy Hawkes; Dir. Animal Health, Dr. Darin Collins; Vice Pres. Education, Jamie Creola; Vice Pres. Conservation, Fred Koontz; Zoo Store Mgr., Terry Blumer.
Personnel Profile: Full-Time Paid 300; Full-Time Volunteers 750; Part-Time Paid 50.
Governing Authority: Woodland Park Zoological Society, 601 N. 59th St., Seattle, WA 98103. Tax-exempt.
Institution Type/Description: Zoo.
Collections: birds; mammals; reptiles; amphibians; insects; exotic & native fauna & flora.
Research Fields: reproduction; psychology & behavior.
Facilities: zoological gardens.
Activities: guided tours; lectures; docent program; zoo education classes for general public; inter-museum loan, permanent & traveling exhibitions; educational self-guided tours.
Publications: self-guided tour brochures; books; zoo activity & teacher packets; member magazine.
Hours & Admission Prices: May-Sept. daily 9:30-6; Oct.-April daily 9:30-4. Summer: adults $17.75, children 3-12 $11.50; children 2 & under no charge. Winter: adults $11.75, children 3-12 $8.50; children 2 & under no charge. Closed Christmas. &
Attendance: 1,094,514 (accurate)
Membership: Children 3-10 $11; College Student $25; Adult $42; Flexible Guest Adult $55; Discovery Passport $125.

Sedro-Woolley

NORTH CASCADES NATIONAL PARK SERVICE COMPLEX, 810 State Route 20, Sedro-Woolley, WA 98284-1239. Tel.: 360-854-7200. Fax: 360-856-1934.
Web Site: www.nps.gov/noca
Founded: 1968.

Congressional District: 2
Key Personnel: Supt., Karen Taylor-Goodrich.
Governing Authority: federal. Parent Institution: U.S. Dept. of Interior. Subsidiary Institution: National Park Service. Tax-exempt.
Institution Type/Description: Park & Visitor Centers: Newhalem Visitor Center, Newhalem, WA; Golden West Visitor Center, Stehekin, WA.
Collections: archaeological; historical; herbarium; wet specimens; archives.
Research Fields: aquatic biology; archaeology; natural history; regional history.
Facilities: 1,000-vol. library pertaining to natural history, park management & environmental studies available to the public. Books for sale.
Activities: guided tours; lectures; organized education programs for adults.
Publications: trail guides.
Hours & Admission Prices: Visitor's Center: May-June & Sept.-Oct. daily 9-5, July-Sept. 9-6. &
Attendance: 10,000 (estimated)

SEDRO-WOOLLEY MUSEUM, 725 Murdock St., Sedro-Woolley, WA 98284-1450. Tel.: 360-855-2390.
Web Site: sedrowoolleymuseum.org
Founded: 1991.
Congressional District: 39
Key Personnel: Pres. (V) & Museum Shop Mgr., Carolyn Freeman; Vice Pres., Dale Robertson.
Personnel Profile: Full-Time Volunteers 12; Part-Time Volunteers 35.
Governing Authority: nonprofit organization. Tax-exempt.
Institution Type/Description: History Museum.
Collections: local history & culture; photographs; personal artifacts; logging; agricultural equipment.
Activities: Museum Sponsors: Founders' Day in September; Holiday Home Tour in December.
Publications: quarterly newsletter.
Hours & Admission Prices: Wed.-Thurs. 12-4, Sat. 9-4, Sun. 1-4; other times by appointment. Suggested Donation: adults $1.50, seniors & children $1; members no charge. &
Attendance: 3,750 (accurate)
Membership: Senior $10; Individual $15; Couple $20; Family $30; Business & Organization $45.

Sequim

DUNGENESS RIVER AUDUBON CENTER AT RAILROAD BRIDGE PARK, 2151 Hendrickson Rd., Sequim, WA 98382. Mailing Address: P.O. Box 2450, Sequim, WA 98382-2450. Tel.: 360-681-4076. Fax: 360-681-8060.
E-mail: rivercenter@olympus.net
Web Site: www.dungenessrivercenter.org
Formerly: Dungeness River Audubon Center
Key Personnel: Dir., Bob Boekelheide
Institution Type/Description: Audubon Center.
Collections: Olympic Peninsula; Dungeness River & watershed; local wildlife; native plants & trees.
Facilities: nature trail.
Activities: educational & interpretive programs; bird walks.
Hours & Admission Prices: April-Oct. Tues.-Sat. 10-4, Sun. 12-4; Nov.-March Tues.-Fri. 10-4, Sat. 12-4. No charge; donations accepted. &
Attendance: 21,500 (accurate)

MUSEUM AND ARTS CENTER IN THE SEQUIM DUNGE-NESS VALLEY, 175 W. Cedar St., Sequim, WA 98382-3318. Tel.: 360-683-8110. Fax: 360-681-2325. Facebook: Museum Arts Center in the Sequim Dungeness Valley.
Web Site: www.macsequim.org
Founded: 1977.
Congressional District: 6
Key Personnel: Exec. Dir., DJ Bassett; Communications Coord., Renee Mizar.
Personnel Profile: Full-Time Paid 1; Full-Time Volunteers 2; Part-Time Paid 3; Part-Time Volunteers 150; Interns 1.
Governing Authority: private; nonprofit organization. Parent Institution: Museum & Arts Center. Tax-exempt: 501(c)(3).
Institution Type/Description: General Museum.
Collections: Sequim area artifacts & photographs; archives; dairy farming tools & equipment; Manis Mastodon specimens.
Research Fields: pioneer genealogy; video histories of local historic sites & people; medicine; midwives; farming; local Native American-S'Klallam Tribal artifacts; morticians; mastodon & mammoth bones, tusks & teeth; pioneer artifacts, logging & dairy implements; regional photographic

archive; dairy industry artifacts & records; textiles with emphasis on clothing; timber industry; Manis Mastodon; John Cowan collection; local school district history.
Facilities: library for local genealogical and historical research; archives of historical photos.
Activities: North Olympic Peninsula history class; traveling exhibits to schools; Seniors Making Art classes; Art-iculate(R) art classes. Museum Sponsors: Community Independence Day Picnic; Elegant Flea Show & Fundraiser; Holiday Tea at Historic Dungeness School; First Friday Sequim Art Walk.
Publications: quarterly newsletter.
Hours & Admission Prices: Tues.-Sat. 10-3. No charge; donations accepted. Closed holidays. &
Attendance: 4,000 (estimated)
Membership: Senior (62 & up) $25; Individual $30; Family $45; Friend $100; Fellow $500; Benefactor $1,000; Life $3,000.

Shaw Island

SHAW ISLAND LIBRARY & HISTORICAL SOCIETY, Blind Bay Rd., Shaw Island, WA 98286. Mailing Address: P.O. Box 844, Shaw Island, WA 98286-0844. Tel.: 360-468-4068.
E-mail: rkg@rockisland.com
Web Site: www.shawislanders.org/others/library/library.htm
Founded: 1966.
Congressional District: 2
Key Personnel: Pres., Jennifer Swanson; Cur., Alex McCloud; Historian, Sheri Christiansen; Head Librarian, Jeb Nichols.
Governing Authority: society. Tax-exempt: 501(c)(3).
Institution Type/Description: Local History Museum: housed in c.1870 one room log cabin.
Collections: items of local historic interest.
Activities: permanent exhibitions.
Hours & Admission Prices: Tues. 2-4, Thurs. 11-1, Sat. 10-12 & 2-4; other times by appointment. No charge.
Membership: Individual $1; Life $25.

Shoreline

SHORELINE HISTORICAL MUSEUM, (M), 18501 Linden Ave., N., Shoreline, WA 98133-4801. Mailing Address: P.O. Box 55594, Shoreline, WA 98155-0594. Tel.: 206-542-7111.
E-mail: shm@shorelinehistoricalmuseum.org
Web Site: shorelinehistoricalmuseum.org
Founded: 1976.
Congressional District: 1
Key Personnel: Exec. Dir. & C.E.O., Victoria Stiles; Chm. & Pres. (V), Kevin Sill.
Personnel Profile: Full-Time Paid 1; Part-Time Paid 1; Part-Time Volunteers 70; Interns 3.
Volunteer Hours: 6,000
Operating Expenses: 100,000
Operating Income: 100,000
Governing Authority: board of trustees. Tax-exempt: 501(c)(3), 170(b)(1)(A).
Institution Type/Description: Historical Museum.
Collections: Shoreline, Lake Forest Park, north Seattle, northwest King County history; period home & business memorabilia; schoolroom; photographs; blacksmith shop; early Richmond Beach post office; country store & farmyard; Playland Amusement Park.
Major Exhibits: Duwamish Connection (T), 1/1/14-12/31/14; LinKing the Community Through Time, 1/1/14-12/31/14.
Research Fields: oral histories of area pioneers; written accounts; maps; books; tools; household items; pictures; early art & crafts; archival photography; church & school histories; businesses; historic locations; descriptions.
Facilities: classrooms; meeting rooms; multipurpose room.
Activities: guided tours; lectures; films; antique show; docent program or council; loan, permanent, temporary & traveling exhibitions; internships (volunteer).
Publications: quarterly newsletter; books, Shoreline Memories; Shoreline or Steamers, Stumps and Strawberries; Growing Up with Lake Forest Park, vols. 1; Growing Up with Lake Forest Park Vol. II; Cooking Up History; Country Store Nevermore; Broadview Memories; Once Upon a Time in Playland; Judge James T. Ronald Memoir.
Hours & Admission Prices: Tues.-Sat. 10-4. No charge; donations accepted. Closed New Year's Day; Independence Day; Thanksgiving; Christmas. &
Attendance: 8,000 (estimated)
Membership: Individual $15; Family $25; Business $50; Life $1,000.

Snohomish

BLACKMAN HOUSE MUSEUM, 118 Ave. B, Snohomish, WA 98290. Mailing Address: P.O. Box 174, Snohomish, WA 98291-0174. Tel.: 360-568-5235.
Web Site: www.snohomishhistoricalsociety.org
Founded: 1969.
Congressional District: 2
Key Personnel: Pres. (V), Warner Blake; Archivist, Middy Ruthruff.
Personnel Profile: Part-Time Paid 2; Part-Time Volunteers 14.
Governing Authority: society; nonprofit. Parent Institution: Snohomish Historical Society. Tax-exempt.
Institution Type/Description: Historical Society Museum: 1878 Blackman House, home of town's first mayor.
Collections: furniture & furnishings of the late 1800s; logging tools & artifacts; period clothing; photographs.
Research Fields: early town history.
Facilities: library of early community photography & printed matter available for research by private arrangement. Museum-related items for sale.
Activities: guided tours; temporary exhibitions. Museum Sponsors: annual Tour of Historic Homes; Holiday Parlor Tours of Historic Homes.
Publications: books, River Reflections-History of Snohomish from 1859 to 1910; River Reflections 1910-1959; Early Snohomish.
Hours & Admission Prices: April to mid-Dec. Sat.-Sun. 11-3. Donation requested. Closed Mother's Day.
Attendance: 1,000 (estimated)
Membership: Individual $10; Family $12; Business $25; Honorary over age 75 no charge.

Snoqualmie

NORTHWEST RAILWAY MUSEUM, (M), 38625 S.E. King St., Snoqualmie, WA 98065. Mailing Address: P.O. Box 459, Snoqualmie, WA 98065-0459. Tel.: 425-888-3030, ext. 7201. Fax: 425-888-9311.
E-mail: director@trainmuseum.org
Web Site: www.trainmuseum.org
Founded: 1957.
Congressional District: 8
Key Personnel: Exec. Dir., Richard R. Anderson; Pres., Susan Hankins; Chm. (V) & Pres. (V), Dennis Snook; Vice Pres., Cindy Walker; Sec., Sue Stewart; Treas., Jon Beveridge; Museum Shop Mgr., James Sackey.
Personnel Profile: Full-Time Paid 6; Part-Time Paid 3; Part-Time Volunteers 137.
Volunteer Hours: 14,710
Operating Expenses: 957,000
Operating Income: 1,200,000
Governing Authority: volunteer nonprofit organization. Tax-exempt: 501(c)(3).
Institution Type/Description: Railroad Museum & Historic Building: c.1890 railroad depot.
Collections: railway vehicles including steam & diesel locomotives, passenger coaches & freight cars, rotary snow plow, cranes, observation cars, maintenance-of-way equipment, kitchen car; logging railway artifacts.
Research Fields: maintenance & restoration of steam & diesel locomotives; role of railroads in development of the N.W.
Facilities: picnic area. Books & other railway-related items for sale.
Activities: permanent exhibitions; diesel passenger train ride; charter service available; school field trip program; self-guided walking tour.
Publications: bimonthly newsletter, Sounder; bimonthly children's newsletter, Jr. Sounder.
Hours & Admission Prices: Memorial Day to Labor Day daily 10-5. No charge. Railway Excursion: April-Oct. Sat.-Sun., occasional weekdays & holidays. Adults $18, seniors 62 & over $15, children 3-12 $10; discounts to groups & AAM members; children under 3 no charge. &
Attendance: 89,571 (accurate)
Membership: Cecil the Diesel Club (children 3-10) $35; Basic $40; Expanded $60; Double Track $75; Business $80; Patron $125; Benefactor $250.

South Bend

PACIFIC COUNTY HISTORICAL SOCIETY & MUSEUM (PCHS), 1008 W. Robert Bush Dr., South Bend, WA 98586-0039. Mailing Address: P.O. Box P, South Bend, WA 98586-0039. Tel.: 360-875-5224. Fax: 360-875-5224 (call first).
E-mail: museum@willapabay.org
Web Site: www.pacificcohistory.org
Founded: 1970.
Congressional District: 3
Key Personnel: Pres. (V), Steve Rogers; Mgr., Charlene Shute.
Personnel Profile: Full-Time Paid 1; Part-Time Volunteers 14.

Governing Authority: nonprofit organization. Parent Institution: Pacific County Historical Society. Tax-exempt: 501(c)(3).
Institution Type/Description: Local History Museum.
Collections: Indian artifacts & relics; records; documents; photographs; relics of pioneers; school records & photos; census records; manuscripts; archives; local history.
Research Fields: ethnic groups; industries; local history; maritime facts & lore.
Facilities: 600-vol. library of printed books; census records; school teaching records; business cash books; diaries; letters; scrapbooks; land office records available for research by appointment or special request; reading room. Books of Pacific Northwest history for sale.
Activities: guided tours; temporary exhibitions.
Publications: quarterly magazine, The Sou'wester.
Hours & Admission Prices: Daily 11-4. No charge; donations accepted. Closed Thanksgiving; Christmas. &
Attendance: 3,750 (estimated)
Membership: Single $25; Family $35; Contributing $50; Corporate $100; Benefactor $200; Grantor $300; Steward $500; Sponsor $1,000 & up.

South Cle Elum

DEPOT INTERPRETIVE CENTER/CASCADE RAIL FOUNDATION, 801 Milwaukee Rd., South Cle Elum, WA 98943. Mailing Address: P.O. Box 462, South Cle Elum, WA 98943-0462. Tel.: 509-674-5939. Fax: 509-674-1708.
E-mail: maryp@cleelum.com
Web Site: www.milwelectric.org
Formerly: Depot Museum/Cascade Rail Foundation
Founded: 2006.
Congressional District: 4
Key Personnel: Pres. (V), Bruce Reason; Vice Pres., David Newcomb; Treas., Mary Pittis; Education, Mark Borleske; Sec., Paul Krueger.
Personnel Profile: Full-Time Volunteers 9; Part-Time Volunteers 2.
Governing Authority: private; nonprofit organization. Parent Institution: Cascade Rail Foundation. Subsidiary Institution: Friends of the South Cle Elum Depot. Tax-exempt: 501(c)(3).
Institution Type/Description: Railroad Museum.
Collections: railroad artifacts; photographs; Milwaukee road's western expansion from 1906-1980.
Research Fields: 1946 Milwaukee caboose rehabilitation; original drawings & photographs.
Facilities: 2,000 sq. ft. exhibit space; 49-seat restaurant; interpretive trail. Museum-related items for sale.
Activities: Annual Events: Depot Days in June; Brewfest in July. Museum Sponsors: Milwaukee Modelers Meet in October.
Hours & Admission Prices: May-Oct. Sat.-Sun. 12-4. No charge; donations accepted. &
Attendance: 839 (accurate)
Membership: Gandy Dancer $35; Telegrapher $50; Engineer $75; Conductor $100; Dispatcher $250; Roadmaster $500; Trainmaster $750; Superintendent's Club $1,000; President's Club $2,500.

Spokane

CHASE GALLERY AT CITY HALL, Spokane City Hall, 808 W. Spokane Falls Blvd., Spokane, WA 99201-3301. Mailing Address: P.O. Box 8737, Spokane, WA 99203. Tel.: 509-321-9614.
E-mail: km.artsfund@visitspokane.city
Web Site: www.spokanearts.org/chase.aspx
Founded: 1984.
Congressional District: 3
Key Personnel: Dir., Shannon Halberstadt; Program Mgr., Karen Mobley.
Personnel Profile: Full-Time Paid 1; Part-Time Paid 1; Interns 2.
Governing Authority: Parent Institution: City of Spokane. Tax-exempt.
Institution Type/Description: Art Gallery.
Collections: works by local & regional artists.
Activities: evening artists receptions; exhibitions of contemporary regional art.
Hours & Admission Prices: Mon.-Fri. 8-5. No charge. &
Attendance: 24,000 (estimated)

CORBIN ART CENTER, 507 W. 7th Ave., Spokane, WA 99204-2709. Mailing Address: 808 W. Spokane Falls Blvd., 5th Fl. City Hall, Spokane, WA 99201-3301. Tel.: 509-625-6677.
E-mail: spokaneparks@spokanecity.org
Web Site: www.spokaneparks.org/recreation/CAC.htm
Key Personnel: Dir., Gary Lawton.
Governing Authority: Parent Institution: City of Spokane Parks and Recreation Department.
Institution Type/Description: Art Gallery: housed in D.C. Corbin house, built in 1898. Listed on the National Register of Historic Places.

Collections: paintings.
Activities: fine arts & crafts classes.
Hours & Admission Prices: Mon.-Thurs. 9-4. No charge.

JOHN A. FINCH ARBORETUM, 3404 W. Woodland Blvd., Spokane, WA 99224-2240. Mailing Address: Park Operations, 2304 E. Mallon, Spokane, WA 99202. Tel.: 509-363-5455. Fax: 509-363-5454. snittalo@spokanecity.org.
E-mail: ssullivan@spokanecity.org
Web Site: www.spokanecity.org/parks
Founded: 1947.
Key Personnel: Park Div. Mgr., Tony Madunich.
Personnel Profile: Full-Time Paid 1; Part-Time Paid 4.
Governing Authority: municipal. Affiliated with City of Spokane Park Board, 4th Fl., Municipal Bldg., W. 808 Spokane Falls Blvd., Spokane, WA. 99201.
Institution Type/Description: Arboretum.
Collections: 65-acres of trees & shrubs.
Facilities: 600-vol. library of plant books; reading room.
Activities: Arbor Day & Fall Leaf Festival.
Hours & Admission Prices: Mon.-Fri. 7:30-3:30 & 1-4. No charge; donations accepted.

JUNDT ART MUSEUM, (M), 502 E. Boone Ave., Spokane, WA 99258. Tel.: 509-313-6611. Fax: 509-313-5525. Facebook: Jundt Museum.
E-mail: manoguerra@gonzaga.edu
Web Site: www.gonzaga.edu
Founded: 1995.
Congressional District: 5
Key Personnel: Dir. & Cur., Dr. Paul A. Manoguerra; Cur. Education, Karen Kaiser; Program Coord., Anita Martello.
Personnel Profile: Full-Time Paid 3; Part-Time Volunteers 23; Interns 6.
Governing Authority: university. Parent Institution: Gonzaga University. Tax-exempt.
Institution Type/Description: Art Museum/Center.
Collections: student prints; Old Master's prints; photography prints; contemporary prints; Auguste Rodin sculptures; Chihuly glass installation.
Major Exhibits: Manzanar: The Wartime Photographs of Ansel Adams (T), 1/4/14-3/29/14; Legacy of the Kiln: The Works of Terry Gieber and His Former Students, 3/22/14-6/7/14; Andy Warhol: Photographs, 5/24/14-8/9/14; Views of Rome: Eighteenth Century Prints by Giovanni Battista, Piranes & His Contemporaries, 5/24/14-8/9/14; Close-In: Frank Werner, An Art of Deception, 6/21/14-9/13/14; David Hayes, 8/1/14-7/31/15; Amen, Amen: Religion & Southern Taught Artists in the Mullis Collection, 10/4/14-1/10/15.
Research Fields: printmaking survey.
Facilities: 118-seat auditorium; print study room.
Activities: lectures; gallery talks.
Hours & Admission Prices: Mon.-Sat. 10-4. No charge. Closed university holidays. &
Attendance: 25,000

MOBIUS KIDS CHILDREN'S MUSEUM, 808 W. Main, Lower Level, Spokane, WA 99201. Tel.: 509-624-5437. Fax: 509-624-6453.
E-mail: info@mobiusspokane.org
Web Site: www.mobiusspokane.org
Formerly: Children's Museum of Spokane
Founded: 1995.
Congressional District: 5
Key Personnel: Mgr., Karen Hudson; Pres., David Gruber; C.E.O., Chris Majer; Dir. Operations, Marty Gonzales.
Personnel Profile: Full-Time Paid 1; Part-Time Paid 7; Part-Time Volunteers 150; Interns 5.
Governing Authority: private; nonprofit organization. Parent Institution: Mobius Spokane. Tax-exempt: 501(c)(3).
Institution Type/Description: Children's Museum.
Collections: hands-on exhibits.
Publications: newsletter, Mobius Matters.
Hours & Admission Prices: Tues.-Sat. 10-5, Sun. 11-5. Adults $6, seniors & military $5; members & children under one no charge. Closed New Year's Day; Easter; Thanksgiving; Christmas. &
Attendance: 60,000 (estimated)
Membership: Family $65.

*** NORTHWEST MUSEUM OF ARTS & CULTURE (EASTERN WASHINGTON STATE HISTORICAL SOCIETY), (M),** W. 2316 First Ave., Spokane, WA 99201-5906. Tel.: 509-456-3931. Fax: 509-363-5303.
E-mail: betsy.godlewski@northwestmuseum.org
Web Site: www.northwestmuseum.org
Formerly: Eastern Washington State Historical Society, Cheney Cowles Museum
Founded: 1916.
Congressional District: 5
Key Personnel: Bd. Pres., Al Payne; Exec. Dir., Forrest B. Rodgers; C.F.O., John Drexel; Cur. History, Marsha Rooney; Cur. Collections, Val Wahl; Volunteer Coord., David Brum; Museum Support Operations, Lori Bertis; Museum Shop Mgr., Bobbi Cypher.
Personnel Profile: Full-Time Paid 22; Part-Time Paid 15; Part-Time Volunteers 300; Interns 3.
Governing Authority: state society. Parent Institution: Washington State. Subsidiary Institution: Art @ Work Gallery. Tax-exempt: 501(c)(3).
Institution Type/Description: General Museum.
Collections: American Indian cultural materials representing all Tribes of North America; regional history of Spokane, WA & the Inland NW region of WA, ID & MT; contemporary & historical art with concentration on the Inland N.W.; photographs. Historic House. Historic Building: 1898 Campbell house.
Major Exhibits: Patrick Siler: Meet Me at the Spot, 2/14-8/14; 100 Stories: A Centennial Exhibition, 2/14-2/16.
Research Fields: regional history; ethnology & art.
Facilities: 12,000-vol. reference library of inland Northwest history & the American Indian; manuscripts, archives, 200,000 historical photographs, oral history materials restricted to library use; 180-seat auditorium; audio-visual center. Books & other museum-related items for sale.
Activities: cultural guided tours; changing art, history & American Indian exhibitions; docent program; films; loan & temporary exhibitions; lectures; films; concerts; organized educational programs for elementary through graduate college levels; traveling art & history exhibits; internships by arrangement with colleges in area; rental gallery. Annual Events: ArtFest; Works From The Heart; American Indian Friendship Dance; Antique Appraisal Days; Mother's Day Historic Neighborhood Tour; Campbell House Holidays; Family MACFests October to March.
Publications: Campbell House: Age of Elegance; American Firearms; Changing Frontier; Cornhusk Bags of the Plateau Indians; Viewing the Past, an Interpretive Guide; Spokane's Historic Architecture; Guide to the Cutter Collection; A Guide to the Manuscript Collections in the Eastern Washington State Historical Society; Frank Palmer, Scenic Photographer, Lewis & Clark High School: The Hart Years; exhibition catalogues; Enchanted Visions: The Taos Society of Artists & Ancient Cultures.
Hours & Admission Prices: Wed.-Sat. 10-6. Adults $10, students & seniors $5; discount to National Trust for Historic Preservation members; reciprocal with NARM & Time Travelers; first Fri. of month 5-8pm by donation; members & children 5 & under no charge. Closed major holidays. &
Attendance: 140,000 (estimated)
Membership: Student & Senior $25; Individual $35; Dual Senior & Student $40; Dual $50; Family, Household & Grand Family $75; MAC Smithsonian Affiliate $150; Patron $500; Sustainer $1,000.

SPOKANE FALLS COMMUNITY COLLEGE FINE ART GALLERY, Fine Arts Bldg., Bldg. 6, 3410 W. Fort George Wright Dr., Spokane, WA 99224-5288. Tel.: 509-533-3710. Fax: 509-533-3484.
E-mail: tomo@spokanefalls.edu
Institution Type/Description: College Art Gallery.
Collections: works by regional, national & international artists.
Activities: workshop. Museum Sponsors: Visiting Artist Lecture Series.
Hours & Admission Prices: Sept.-May Mon.-Fri. 8-4, Sat. 11-2. No charge. Closed college holidays & breaks.

SPOKANE FIRE DEPARTMENT MUSEUM, 1618 N. Rebecca, Spokane, WA 99207. Tel.: 509-625-7062.
Institution Type/Description: Firefighting History Museum.
Collections: firefighting history, equipment & memorabilia; photographs.
Activities: public safety education.
Hours & Admission Prices: Call for hours.

WASHINGTON FIRE LOOKOUT MUSEUM, 123 W. Westview, Spokane, WA 99218-2226. Tel.: 509-466-9171.
E-mail: rkresek@webtv.net
Web Site: www.firelookouts.com/museum.html
Institution Type/Description: History Museum.
Collections: wildfire detection & firefighting history; Smokey Bear artifacts;

1953 pumper; firefinders; radios; telephones; fire tools; photographs; books; uniforms; badges; personal artifacts; videos. Historic Buildings: lookout towers; fire guard station; weather station.

Activities: group tours.

Hours & Admission Prices: March-Nov. by appointment. No charge.

Stanwood

STANWOOD AREA HISTORY MUSEUM & D. O. PEARSON HOUSE MUSEUM, 27108 102nd Ave., N.W., Stanwood, WA 98292. Mailing Address: P.O. Box 69, Stanwood, WA 98292-0069. Tel.: 360-629-6110.

Web Site: www.sahs-fncc.org

Institution Type/Description: History Museum.

Collections: local history & culture; period furnishings; personal artifacts; photographs. Historic Buildings: Pearson House c.1890; Tolin House 1880s.

Hours & Admission Prices: Wed., Fri. & Sun. 1-4. No charge; donations accepted. &

Steilacoom

STEILACOOM HISTORICAL MUSEUM ASSOCIATION, (M), Rainier & Main, Steilacoom, WA 98388. Mailing Address: P.O. Box 88016, Steilacoom, WA 98388-0016. Tel.: 253-584-4133.

E-mail: steilacoomhistorical@gmail.com

Web Site: www.steilacoomhistorical.org

Founded: 1970.

Congressional District: 6

Key Personnel: Pres. & Museum Shop Mgr., Marianne Bull; Dir., French Wetmore; Cur., Joan Curtis.

Personnel Profile: Part-Time Volunteers 100.

Operating Expenses: 100,000

Operating Income: 100,000

Governing Authority: society. Branch Museums: Nathaniel Orr Home & Pioneer Orchard, Steilacoom, WA; Bair Drug & Hardware Store, 1617 Lafayette St., Steilacoom, WA. Tax-exempt: 501(c)(3).

Institution Type/Description: Local History.

Collections: Steilacoom history & early pioneer memorabilia; historical orchard.

Research Fields: local history.

Facilities: 500-vol. library of books from first chartered library in the state; research library for local & regional history; restaurant; living museum. Notepaper, books, calendars & brochures of local interest for sale.

Activities: guided tours; programs of local interest; fund raising activities; school tours; the preservation of the Steilacoom Historical District.

Publications: quarterly newsletter.

Hours & Admission Prices: Spring-Fall Sat.-Sun. 12-4. Suggested Donation: $2. &

Attendance: 4,800 (estimated)

Membership: Individual Senior Citizen $25; Individual $30; Senior Family $35; Family $50; Friend of History $51 & up; Patron of History $100 & up; Benefactor of History $350 & up.

Stevenson

COLUMBIA GORGE INTERPRETIVE CENTER MUSEUM, 990 S.W. Rock Creek Dr., Stevenson, WA 98648. Mailing Address: P.O. Box 396, Stevenson, WA 98648-0396. Tel.: 509-427-8211. Fax: 509-427-7429.

E-mail: info@columbiagorge.org

Web Site: www.columbiagorge.org

Founded: 1959.

Congressional District: 17

Key Personnel: Dir., Sharon Tiffany; Chm. & Pres. (V), Jim Price; Museum Shop Mgr., KaCey Gunther.

Personnel Profile: Full-Time Paid 4; Full-Time Volunteers 1; Part-Time Paid 5; Part-Time Volunteers 25.

Governing Authority: society. Parent Institution: Skamania County Historical Society. Tax-exempt.

Institution Type/Description: Local History Museum.

Collections: Indian artifacts; pioneer artifacts; Rosary Collection; Corliss steam engine; logging artifacts; 1917 Curtiss JN-4 biplane.

Research Fields: local history via oral interviews.

Facilities: 11,000 sq. ft. exhibit space.

Activities: lectures; guided tours with advanced notice.

Publications: newsletter, Explorations.

Hours & Admission Prices: Daily 10-5. Family $30, adults $10, students & senior citizens $8, children 6-12 $6; discounts to AAM, ICOM, AAA members, media & tourism industry members; county residents on 1st Sat.

of month, members, children 5 & under, and during the anniversary celebration no charge. Closed New Year's Day; Thanksgiving; Christmas. &

Attendance: 19,542 (accurate)

Membership: Student & Senior 60 & over $15; Individual $25; Family $35; Sustaining $50; Supporting $100; Sponsor $500; Benefactor $1,000; Explorer $2,000; Guardian $2,500.

Sunnyside

SUNNYSIDE HISTORICAL MUSEUM, 704 S. 4th St., Sunnyside, WA 98944-2162. Mailing Address: Box 782, Sunnyside, WA 98944-0782. Tel.: 509-837-6010 & 2105.

E-mail: ssmuseum@bentonrea.com

Founded: 1972.

Congressional District: 4

Key Personnel: Pres., John Saras; Vice Pres. & Cur., Don Wade.

Personnel Profile: Part-Time Volunteers 10.

Governing Authority: nonprofit organization. Tax-exempt.

Institution Type/Description: Early Pioneer Museum.

Collections: pioneer kitchen; dining room; living room; quilts; costumes; period glassware & toys; paintings; painted mural; carvings; early irrigation; period photographs; mounted animals. Historic Structure: 1859 Snipes cabin.

Activities: guided tours.

Hours & Admission Prices: Thurs.-Sun. 1-4. No charge; donations accepted. &

Attendance: 1,000 (estimated)

Membership: Individual $5.

Tacoma

CHILDREN'S MUSEUM OF TACOMA, 1501 Pacific Ave., Ste. 202, Tacoma, WA 98402. Tel.: 253-627-6031. Fax: 253-627-2436.

E-mail: tandrews@playtacoma.org

Web Site: www.playtacoma.org

Founded: 1985.

Congressional District: 6

Key Personnel: Exec. Dir., Tanya Andrews; Pres. Bd., Dave Edwards; Dir. Communications & Operations, Brenda Morrison; Mgr. Experience, Deean Marsh; Mgr. Research & Assessment, Kimberly McKenney.

Personnel Profile: Full-Time Paid 6; Part-Time Paid 24; Part-Time Volunteers 100.

Governing Authority: nonprofit organization. Tax-exempt.

Institution Type/Description: Children's Museum.

Collections: hands-on exhibits.

Activities: education programs; parties; camps; workshops.

Publications: exhibit parent play guides.

Hours & Admission Prices: Wed.-Sun. 10-5. Pay As You Will. Closed New Year's Day; Easter; Independence Day; Labor Day; Thanksgiving; Christmas Eve & Day. &

Attendance: 120,000 (accurate)

Membership: Playful $100; Little Travelers $125.

FORT NISQUALLY LIVING HISTORY MUSEUM, (M), 5400 N. Pearl St., #11, Tacoma, WA 98407-3224. Tel.: 253-591-5339. Fax: 253-759-6184.

E-mail: fortnisqually@tacomaparks.com

Web Site: www.fortnisqually.org

Formerly: Fort Nisqually Historic Site

Founded: 1837.

Congressional District: 6

Key Personnel: Fort Nisqually Foundation Chm. (V), Glen Sutt; Metro Parks Tacoma Exec. Dir., Jack Wilson; Cur. Education, J. Michael McGuire; Cur., Bill Rhind; Museum Shop Mgr., Jill Stephenson; Education Specialist, Lane Sample; Special Projects, Peggy Barchi.

Personnel Profile: Full-Time Paid 3; Part-Time Paid 10; Part-Time Volunteers 100.

Governing Authority: municipal. Parent Institution: Metropolitan Park District. Subsidiary Institution: Fort Nisqually Foundation. Tax-exempt.

Institution Type/Description: Historic Site & History Museum Complex.

Collections: blueprints; maps; charts; photographs; Hudson's Bay Company trade goods; agricultural & farming implements; reference material; Puget Sound artifacts; 7 building replicas of Old Fort Nisqually. Historic Structures: c.1855 Gentlemen's Dwelling House; two original HBC buildings; c.1850 granary, c.1855 Factor's House; replica of c.1855 sale shop; blacksmith shop, large store, laborer's dwelling, kitchen, laundry & labor; two c.1850 defensive bastions.

Major Exhibits: Turning Drudgery into a Pastime, 12/7/13-4/15/14.

Research Fields: physical structures; ethnic cultures in fur trade era of Washington.

Facilities: reference books on the northwest fur trade & related maps, charts & documents. Museum-related items & replicas for sale.
Activities: special events; school & public group tours; docent training program.
Publications: quarterly journal, Occurrences; brochure, Self-Guided Tour of Fort Nisqually; informational pamphlet.
Hours & Admission Prices: April-May Wed.-Sun. 11-5; June-Aug. daily 11-5; Sept.-March Wed.-Sun. 11-4. Adults $6.50, children $4; discount to AAA members; members no charge. Closed New Year's Day; Thanksgiving; Christmas. &
Attendance: 30,000 (accurate)
Membership: Engage (Individual) $25; Master (Family) $35; Clerk $50; Trader $100; Factor $250; Governor $1,000.

FOSS WATERWAY SEAPORT, 705 Dock St., Tacoma, WA 98402-4625. Tel.: 253-272-2750. Fax: 253-272-3023.
E-mail: info@fosswaterwayseaport.org
Web Site: www.fosswaterwayseaport.org
Formerly: Working Waterfront Maritime Museum
Key Personnel: Exec. Dir., Tom Cashman
Institution Type/Description: Maritime Museum.
Collections: maritime crafts & skills.
Activities: demonstrations; educational programs.
Hours & Admission Prices: Mon.-Fri. 10-5, Sat.-Sun. 12-5. Adults $6, senior citizens 62 & over, students, and military $3; members no charge. Closed New Year's Day; Easter; Independence Day; Memorial Day; Thanksgiving; Christmas.
Membership: Family $50.

JAMES R. SLATER MUSEUM OF NATURAL HISTORY, University of Puget Sound, 1500 N. Warner St., #1088, Tacoma, WA 98416. Tel.: 253-879-2798.
E-mail: slatermuseum@pugetsound.edu
Web Site: www.ups.edu/slatermuseum.xml
Formerly: Puget Sound Museum
Founded: 1926.
Congressional District: 6
Key Personnel: Dir., Dr. Peter Wimberger; Mgr. Collections, Dr. Gary Shugart; Dir. Emeritus, Dr. Dennis R. Paulson.
Personnel Profile: Full-Time Paid 1; Part-Time Paid 1.
Governing Authority: University of Puget Sound. Tax-exempt.
Institution Type/Description: Natural History Museum.
Collections: Northwest flora & fauna, including approximately 30,000 mammal skins & skulls; 18,000 bird skins & skeletons; 4,400 bird wings; 4,600 sets of bird eggs; 8,500 reptiles & amphibians; 9,000 plants on herbarium sheets; 2,000 invertebrates; 1,200 bird nests.
Research Fields: mammalogy; ornithology; herpetology; marine invertebrate zoology; botany.
Facilities: 1,600-vol. library of books, monographs & journals, reprints & articles available for use on premises.
Activities: guided tours; acquisition, preparation & maintenance of research collections for visiting scientists & students; undergraduate teaching; exhibit programs.
Publications: Slater Museum of Natural History Occasional Papers.
Hours & Admission Prices: By appointment. No charge. Closed holidays. &

JOB CARR CABIN MUSEUM, 2350 N. 30th St., Tacoma, WA 98403-3323. Mailing Address: P.O. Box 7609, Tacoma, WA 98417-0609. Tel.: 253-627-5405.
E-mail: mbowlby@jobcarrmuseum.org
Web Site: www.jobcarrmuseum.org
Founded: 2000.
Congressional District: 6
Key Personnel: Exec. Dir., Mary Bowlby; Program Mgr., Holly Stewart.
Personnel Profile: Part-Time Paid 2.
Volunteer Hours: 650
Operating Expenses: 74,000
Operating Income: 78,900
Governing Authority: Tax-exempt.
Institution Type/Description: History Museum: housed in a replica of Job Carr's home built in 1865. He was Tacoma's first settler, Postmaster and Mayor.
Collections: period furnishings.
Research Fields: Tacoma history; Oregon Trail.
Activities: educational programs; special events; Traveling Trunk education program.
Publications: Eureka Times.

Hours & Admission Prices: Jan. by appointment, Feb.-May & Oct.-Dec. Wed.-Sat. 1-4; June-Sept. Wed.-Sat. 12-4. No charge; donations accepted. &
Attendance: 3,300 (accurate)
Membership: Student $12; Senior $15; Individual $20; Family & Nonprofit $35; Small Business $75; Corporate $125.

THE KARPELES MANUSCRIPT LIBRARY MUSEUM, 407 S. "G" St., Tacoma, WA 98405-4711. Tel.: 253-383-2575.
E-mail: kmuseumtaq@aol.com
Web Site: www.rain.org/~karpeles/taqfrm.html
Key Personnel: Dir., Thomas M. Jutilla
Institution Type/Description: History Museum.
Collections: original papers of historic importance.
Hours & Admission Prices: Tues.-Fri. 10-4. No charge.

KITTREDGE GALLERY, UNIVERSITY OF PUGET SOUND ART DEPT., 1500 N. Warner St., CMB 1072, Tacoma, WA 98416-0005. Tel.: 253-879-3701. Fax: 253-879-3500.
Web Site: www.pugetsound.edu/kittredge
Founded: 1950.
Congressional District: 6
Personnel Profile: Part-Time Paid 2.
Governing Authority: university; not-for-profit. Parent Institution: University of Puget Sound. Tax-exempt.
Institution Type/Description: Art Gallery.
Collections: Abby Williams Hill paintings & drawings; contemporary American ceramics; contemporary northwest art.
Facilities: 2,100 sq. ft. exhibit space.
Activities: lectures; temporary exhibitions.
Hours & Admission Prices: Sept. to mid-May Mon.-Fri. 10-5, Sat. 12-5. No charge. Closed holidays; semester breaks. &
Attendance: 4,000 (accurate)

LEMAY AMERICA'S CAR MUSEUM, 2702 E. D St., Tacoma, WA 98421. Mailing Address: P.O. Box 1117, Tacoma, WA 98401-1117. Tel.: 253-779-8490. Fax: 253-779-8499.
Web Site: www.lemaymuseum.org
Formerly: Harold E. LeMay Museum
Institution Type/Description: Transportation Museum.
Collections: automobiles; motorcycles; trucks.
Activities: Museum Sponsors: Annual Car Show in August.
Hours & Admission Prices: Guided Tours: Tues.-Sun. 10-5 by appointment. Admission to museum is by guided tour only. Members no charge. Closed major holidays.
Attendance: 400,000 (estimated)

* **MUSEUM OF GLASS, (M),** 1801 Dock St., Tacoma, WA 98402-3217. Tel.: 253-284-4750 & 4719; 866-468-7386. Fax: 253-396-1769. Facebook: www.facebook.com/museumofglass.
E-mail: info@museumofglass.org
Web Site: www.museumofglass.org
Founded: 1995.
Congressional District: 9
Key Personnel: Exec. Dir., Susan Warner; Chm. (V), Gail T. Weyerhaeuser; Dir. Operations & Finance, Jeff Ganung; Dir. External Affairs, Joanna Sikes; Dir. Visitor Svcs., Tom Findlay; Assoc. Dir. Communications, Hillary Ryan.
Personnel Profile: Full-Time Paid 24; Part-Time Paid 23; Part-Time Volunteers 35; Interns 6.
Governing Authority: private; nonprofit organization. Tax-exempt: 501(c)(3).
Institution Type/Description: Art Museum.
Collections: contemporary works of art in glass.
Major Exhibits: Irish Cylinders by Dale Chihuly & Seaver Leslie with Glass Drawings by Flora C. Mace from the George R. Stroemple Collection, 10/13-9/14; Caution! Fragile Irish Glass: Tradition in Translation, 11/13-9/14; Bohemian Boudior, 1/14-5/14; Kids Design Glass, 1/14-6/14; Echo Chamber, 2/14-9/14; Look! See?, 2/14-9/14; Hilltop Artist 10th Anniversary, 9/14-1/15; Links: Australian Glass and the Pacific Northwest, 10/14-1/15; Davide Salvadore, 10/14-4/15.
Facilities: cafe; 200-theater; 11,000 sq. ft. exhibit space. Museum-related items for sale.
Activities: docent led tours; dramatic performances; conversations with artists; live glassblowing demonstrations.
Publications: quarterly newsletter, Fuse.
Hours & Admission Prices: Adults $12; discounts to AAM members; members no charge. &
Attendance: 130,000 (estimated)
Membership: Artist, Teacher, Student $50 Individual $60; Senior $70; Dual

Senior $75; Dual & Family $85; Gather $175; Studio $250; Collector $500; Maestro $1,000. (10% discount to seniors.)

PACIFIC LUTHERAN UNIVERSITY GALLERY, Dept. of Art & Design, Ingram Hall, Tacoma, WA 98447. Tel.: 253-535-7150. Fax: 253-536-5063.
E-mail: soac@plu.edu
Web Site: www.plu.edu/~soac/
Key Personnel: Prof. Heather Mathews.
Governing Authority: Parent Institution: Pacific Lutheran University.
Institution Type/Description: University Art Gallery.
Collections: works by contemporary artists.
Hours & Admission Prices: Mon.-Fri. 9-4. No charge.

POINT DEFIANCE ZOO & AQUARIUM, 5400 N. Pearl St., Tacoma, WA 98407-3224. Tel.: 253-591-5337. Fax: 253-591-5448.
E-mail: pdzacomments@tacomaparks.com
Web Site: www.pdza.org
Founded: 1905.
Congressional District: 6
Key Personnel: Dir., Gary Geddes; Dep. Dir., John Houck; Pres., Zoo Society, David Panco; Exec. Dir., Zoo Society, Larry Norvell; Mktg. & Public Rels. Mgr., Whitney DalBalcon.
Personnel Profile: Full-Time Paid 68; Part-Time Paid 60; Part-Time Volunteers 160; Interns 5.
Governing Authority: municipal. Parent Institution: Metro Parks Tacoma.
Institution Type/Description: Zoo & Aquarium.
Collections: Pacific Rim zoo & aquarium.
Publications: member newsletter, Zoopoints.
Hours & Admission Prices: Hours vary throughout the year, call or visit Web site for detailed information. Adults $15, senior citizens 65 & up $14, youth 5-12 $13, tots 3-4 $8.75; discounts to military & Pierce County residents; children under 3 no charge. Closed Thanksgiving; Christmas.
Attendance: 500,000 (accurate)
Membership: Individual $50; One Plus One $75; Household & Grandparent $115; Deluxe Household & Grandparent $145; Sponsor $250; Patron $350.

SHANAMAN SPORTS MUSEUM, 2727 E. "D" St. (Tacoma Dome), Tacoma, WA 98421-1216. Mailing Address: 9908-63rd Ave. Ct. E., Puyallup, WA 98373-1170. Tel.: 253-627-5857.
E-mail: marc@tacomasportsmuseum.com
Web Site: www.tacomasportsmuseum.com
Founded: 1994.
Key Personnel: Pres., Marc Blau.
Personnel Profile: Part-Time Paid 1; Part-Time Volunteers 4.
Governing Authority: Tax-exempt.
Institution Type/Description: Sports Museum.
Collections: sports history; personal artifacts of recreational, amateur, & professional athletes, coaches & teams; photographs; sports officials, broadcasters & sportswriters.
Hours & Admission Prices: Open during sporting events & trade shows; other times by appointment. No charge when attending sporting event; donations accepted.
Attendance: 5,749

*   **TACOMA ART MUSEUM, (M),** 1701 Pacific Ave., Tacoma, WA 98402-3214. Tel.: 253-272-4258, ext. 3035. Fax: 253-627-1898.
E-mail: info@tacomaartmuseum.org
Web Site: www.tacomaartmuseum.org
Founded: 1935.
Congressional District: 6
Key Personnel: Pres. (V), Janine Terrano; Dir., Stephanie A. Stebich; Deputy Dir., Cameron Fellows; Dir. Devel., Kara Hefley; Dir. Education & Audience Devel., Paula McArdle; Dir. Curatorial Admin., Rock Hushka; Museum Shop Mgr., Kristie Worthey.
Personnel Profile: Full-Time Paid 30; Part-Time Paid 22; Part-Time Volunteers 129; Interns 4.
Governing Authority: nonprofit organization. Tax-exempt: 501(c)(3) & 170(b)(1)(A).
Institution Type/Description: Art Museum.
Collections: emphasis on northwest art of the modern period (1800-present): American Eight, Pacific Northwest; Dale Chihuly Retrospective Installation; 19th-century European paintings; Japanese Ukiyo-e prints.
Research Fields: exhibitions; Northwest art.
Facilities: 6,000-vol. library of art volumes available for use on premises; reading room; 25,000 sq. ft. exhibit space; 200-seat event space.
Activities: guided tours; lectures; films; gallery talks; concerts; ongoing art

making opportunities that complement formally organized education programs for children, adults, undergraduate & graduate college students; docent program; inter-museum loan, permanent & temporary exhibitions.
Publications: quarterly bulletins; exhibition catalogs; gallery guides.
Hours & Admission Prices: Summer: Tues.-Sat. 10-5, Sun. 12-5; Winter: Wed.-Sun. 10-5. Adults $9, student, military and seniors 65 & over $8; discounts to families, AAM & staff of other museums; third Thurs. of the month, members, and children 5 & under no charge. Closed New Year's Day; Martin Luther King Jr. Day; Thanksgiving; Christmas.
Attendance: 76,000 (accurate)
Membership: Educator, Senior, Student & Military $30; Individual $50; Family & Friends $75; Gallery $125; Studio $200; Collector $500; Patron's Circle $1,000 & up.

TACOMA HISTORICAL SOCIETY, 3712 S. Cedar St., Tacoma, WA 98409. Mailing Address: P.O. Box 1865, Tacoma, WA 98401-1865. Tel.: 253-472-3738. Facebook: Tacoma Historical Society.
E-mail: info@tacomahistory.org
Web Site: www.tacomahistory.org
Founded: 1990.
Congressional District: 6
Governing Authority: Tax-exempt: 501(c)(3).
Institution Type/Description: Historical Society.
Collections: local history & culture; photographs; personal artifacts; period furnishings.
Facilities: Museum-related items for sale.
Activities: special events; historic homes tour; lecture series.
Publications: newsletter; Herbert Hunt, History of Tacoma, 2003; quarterly newsletter, City of Destiny.
Hours & Admission Prices: Wed.-Sat. 12-5. No charge, donations accepted. discounts to WMA members.
Attendance: 150 (estimated)
Membership: Individual $25; Family $35; Corporate $150 & up.

TACOMA PUBLIC LIBRARY/THOMAS HANDFORTH GALLERY, 1102 Tacoma Ave., S., Tacoma, WA 98402-2098. Tel.: 253-292-2001 ext. 1111.
E-mail: ddomkoski@tacomapubliclibrary.org
Web Site: www.tpl.lib.wa.us
Founded: 1886.
Congressional District: 6
Key Personnel: Library Dir., Susan Odencrantz; Media Rels. Officer, David Domkoski.
Governing Authority: municipal government. Tax-exempt: 170(b)(1)(A).
Institution Type/Description: Public Library & Art Gallery: housed in 1903 Carnegie Building.
Collections: Thomas Handforth prints & sketches; photographic prints & negatives; World War I posters; Asahel Curtis, Edward S. Curtis, Turner Richards & Verna Haffer photo collections; rare book room; 1.2 million photographs; 36,000-vol. Pacific Northwest Americana; 30,000 maps; 15,000-vol. genealogy & local history books.
Facilities: 600,000-vol. library.
Activities: guided tours; lectures; films; study clubs; organized education programs for children & adults; participatory & temporary exhibitions; rotating exhibits of Pacific N.W. artists & craftspeople.
Publications: quarterly newsletters, Handforth Gallery Preview.
Hours & Admission Prices: Tues.-Wed. 11-8, Thurs.-Sat. 9-6. No charge.
Attendance: 1,600,000

W.W. SEYMOUR BOTANICAL CONSERVATORY, 316 South G St., Tacoma, WA 98405-4733. Tel.: 253-591-5330. Fax: 253-627-2192.
E-mail: wwseymour@tacomaparks.com
Web Site: www.metroparkstacoma.org/conservatory/
Founded: 1908.
Congressional District: 6
Key Personnel: Natural Resources Supvr., Mary Anderson; Horticulture Technician, Tyra Shenaurlt; Natural Resources Mgr., Joe Brady.
Governing Authority: municipal. Parent Institution: Metropolitan Park District of Tacoma.
Institution Type/Description: Botanical Garden: museum housed in Victorian styled conservatory with twelve-sided central dome containing two side wings & an entry wing.
Collections: over 200 species of exotic tropical plants including ornamental figs, tropical fruit trees, Bird of Paradise, bromeliads & orchids. Flower displays change monthly: spring bulbs, azaleas, Easter lilies, summer annuals, flowering houseplants, chrysanthemums, a Halloween pumpkin patch & Christmas poinsettias.
Facilities: botanical garden.

Activities: guided tours; organized education programs.
Publications: brochure, Seymour Botanical Conservatory; quarterly newsletter, Botanical Prints.
Hours & Admission Prices: Tues.-Sun. 10-4:30. Suggested donation $3. Closed New Year's Day; Thanksgiving; Christmas. &
Attendance: 78,000
Membership: Senior 65 & up $25; Individual $35; Senior 65 & up Friend/Couple $50; Friend/Couple $60; Organization $75; Azalea $100; Lily $250; Orchid $500; W.W. Seymour Circle $1,000.

* **WASHINGTON STATE HISTORICAL SOCIETY & HISTORY MUSEUM, (M),** 1911 Pacific Ave., Tacoma, WA 98402-3109. Tel.: 888-238-4373; 253-272-3500. Fax: 253-272-9518.
E-mail: ptobiason@wshs.wa.gov
Web Site: www.washingtonhistory.org
Founded: 1891.
Congressional District: 6
Key Personnel: Dir., Jennifer Kilmer; Pres., Larry Kopp; CAO & Supvr. Admissions, Misty Reese; Deputy Dir., Patricia Tobiason; Dir. Support Svcs., Mark Sylvester; Dir. Outreach Svcs. & Head State Capital Museum, Susan Rohrer; Head Collections, Lynette Miller; Head Special Collections, Edward V. Nolan; Asst. Librarian Manuscripts, Joy Werlink; Publications, Christina DuBois; C.F.O., Christopher Lee; Head Exhibits, Redmond J. Barnett; Dir. Devel. & Membership Svcs., Laura Berry; Head Education, Stephanie Lile; Natl. History Day Asst., Mark Vessey; Exhibits Designer, SueSan Chan; Cur., Maria Pascualy; Supvr. Admissions, Aaron Harris; Coord. School Programs, Gwen Perkins; Dir. Info. Tech., Tamara Georgick; Registrar & Digital Assets Mgr., Fred Poyner, IV; Museum Shop Mgr., Bonnie Gross.
Personnel Profile: Full-Time Paid 19; Part-Time Paid 23; Part-Time Volunteers 102.
Governing Authority: state. Subsidiary Institutions: State Capital Museum. Tax-exempt: 509(a) & 501(c)(3).
Institution Type/Description: History Museum.
Collections: Pacific Northwest history, paintings, furnishings, textiles & ethnology; rare books; 450,000 photographs, 300 manuscripts; maps, pamphlets & ephemera; over 1,000 microfilm.
Research Fields: Washington state & Pacific Northwest history.
Facilities: 12,000-vol. library of Pacific Northwest reference books; 106,000 sq. ft. Washington State history museum; outdoor amphitheatre; research center; reading room; cafe. Museum-related items for sale.
Activities: lectures; films & special events promoting exhibits; inter-museum loan, permanent & temporary exhibitions; formal education programs for elementary & high school students; internships for college students; group tours.
Publications: quarterly, Columbia, The Magazine of Northwest History; Columbia Kids.
Hours & Admission Prices: Wed.-Sun. 10-5, 3rd Thurs. each month 10-8. Adults $9.50; members & third Thurs. 2-8 no charge. Closed Memorial Day; Independence Day; Labor Day; Thanksgiving; Christmas. &
Attendance: 81,670 (accurate)
Membership: Subscription, Teachers & Student $35; Individual $45; Dual $50; Family $65; Sustaining $125; Business $300; Patron $500; Benefactor $1,000.

Tenino

TENINO DEPOT MUSEUM, 399 W. Park, Tenino, WA 98589. Mailing Address: P.O. Box 339, Tenino, WA 98589-0339. Tel.: 360-264-4321.
Web Site: www.teninodepotmuseum.org
Founded: 1974.
Congressional District: 3
Key Personnel: Pres., Bob Hill.
Personnel Profile: Part-Time Volunteers 14.
Volunteer Hours: 5,725
Operating Expenses: 3,597
Operating Income: 3,939
Governing Authority: nonprofit organization. Parent Institution: South Thurston County Historical Society. Tax-exempt: 501(c)(3).
Institution Type/Description: Local History Museum.
Collections: pioneer days-present; doctor's office; school room; artifacts; Tenino wooden money; wooden money printing press; sandstone display & tools; railroad room.
Facilities: depot train museum; Ticknor one-room school house; Rota farm building; Montgomery Mill display.
Activities: guided tours; temporary exhibitions. Museum Sponsors: Wooden Money Printed in July; Annual Dinner in October.
Publications: newsletter.

Hours & Admission Prices: mid-April to mid-Oct. Sat.-Sun. 12-4. No charge; donations accepted. &
Attendance: 1,800 (estimated)
Membership: Individual $10; Family $25; Life $100.

Toppenish

MARY L. GOODRICH LIBRARY & TOPPENISH HISTORICAL MUSEUM, One S. Elm, Toppenish, WA 98948-1574. Tel.: 509-865-3600.
Web Site: www.yvl.org
Formerly: Toppenish Museum
Founded: 1975.
Congressional District: 4
Key Personnel: Mng. Librarian, Krystal Corbray.
Personnel Profile: Full-Time Volunteers 1; Part-Time Paid 2; Part-Time Volunteers 12.
Governing Authority: municipal. Dept. of the City of Toppenish, WA. Tax-exempt.
Institution Type/Description: History Museum: housed in 1923 first Agency Building for the Yakima Indian Nation.
Collections: period artifacts; Indian baskets; regional contemporary arts & crafts.
Research Fields: local history.
Facilities: 18,500-vol. Toppenish Public Library located on lower floor.
Activities: gallery talks; permanent & traveling exhibitions.
Publications: quarterly newsletter; reprints of local history.
Hours & Admission Prices: Mon.-Thurs. 10-7, Fri.-Sat. 10-5.

NORTHERN PACIFIC RAILWAY MUSEUM, 10 Asotin Ave., Toppenish, WA 98948-1300. Mailing Address: P.O. Box 889, Toppenish, WA 98948-0889. Tel.: 509-930-7210 (Director).
Web Site: www.nprymuseum.org
Formerly: Yakima Valley Rail & Steam Museum
Founded: 1989.
Congressional District: 4
Key Personnel: Museum Dir., Cur. & Archivist, Larry Rice; Pres., Dennis Lee; Financial Dir. & Treas., Ken Severson; Gift Shop Mgr., Roger O'Dell; Gift Shop Mgr., Mary O'Dell.
Personnel Profile: Part-Time Volunteers 8.
Governing Authority: private; nonprofit organization. A division of The Yakima Valley Rail & Steam Museum Association. Tax-exempt: 501(c)(3).
Institution Type/Description: Railway Museum
Collections: railroad artifacts; 2 steam locomotives. Historic Building: c.1911 depot.
Facilities: 30-vol. library of railroad maintenance reference books; 51,560 sq. ft. exhibit space. Museum-related items for sale.
Activities: guided tours; caboose rides; engineer classes for diesel & steam. Annual Events: Railroad Days in June.
Publications: newsletter every 4 months, The Orderboard.
Hours & Admission Prices: May-Oct. Tues.-Fri. 10-4, Sun. 11-4. Adults $5, children 12 & under $3; discount to groups & Pacific Historical Association members; members no charge. &
Attendance: 4,700 (accurate)
Membership: Student & Senior $30; Annual Dues $40; Family (Husband & Wife) $50; Life $500.

YAKAMA NATION MUSEUM, 100 Spiel-yi Loop, Toppenish, WA 98948. Mailing Address: P.O. Box 151, Toppenish, WA 98948-0151. Tel.: 509-865-2800. Fax: 509-865-5749.
E-mail: pamela@yakama.com
Web Site: www.yakamamuseum.com
Founded: 1980.
Congressional District: 15
Key Personnel: Tribal Chm., Ralph Sampson; Mgr., Pamela K. Fabela; Photograph Collection, Liz Antelope; Registrar & Webmaster, Heather Hull.
Personnel Profile: Full Time Paid 3.
Governing Authority: tribe; nonprofit organization. Tax-exempt: 501(c)(3).
Institution Type/Description: Tribal Museum: located on the Yakima Indian Reservation.
Collections: Northwest Native American artifacts; items collected by Nipo Strongheart (1891-1966) during his career as a Hollywood movie consultant; Northwestern Plateau artifacts including buckskin & beaded shirts, pipes, spoons, baskets, fishing nets, mats, bags.
Research Fields: oral interviews.
Facilities: library; restaurant; theater. Handcrafted items for sale.
Activities: Klickitat basketry; Yakama beadwork, woodcarving & buckskin work by artists & craftspeople; lectures. Annual events: Treaty Day

Celebration commemorating the June 9 signing of the Yakama Nation & United States Treaty; Tiin Ma Gathering.
Publications: booklets, Mother Nature's Lesson; Time Ball.
Hours & Admission Prices: Mon.-Fri. 8-5, Sat.-Sun. 9-5. Family $15, adults $6, college students, senior citizens, active military, & students 11-18 $4, children 10 & under $2. Guided Tours: $25. Closed New Year's Day; Thanksgiving; Christmas. &
Attendance: 70,000 (estimated)

Union Gap

CENTRAL WASHINGTON AGRICULTURAL MUSEUM, 4508 Main St., Union Gap, WA 98903-2138. Tel.: 509-457-8735.
E-mail: info@centralwaagmuseum.org
Web Site: centralwaagmuseum.org
Founded: 1979.
Congressional District: 4
Key Personnel: Pres. (V), Nick Schultz; Vice Pres., Jim Warner; Treas., Dick Drew; Sec., Marty Humphrey.
Personnel Profile: Full-Time Volunteers 1; Part-Time Paid 1; Part-Time Volunteers 43.
Governing Authority: city; board of directors; nonprofit. Tax-exempt.
Institution Type/Description: Agricultural Museum.
Collections: large collection of farm equipment in the northwest includes plows; discs; cultivators; sprayers; hay equipment; dusters; steam engines; choppers; beet harvest equipment; planters; various period farm equipment; tractors; cold storage compressor; portable hop-picking machinery; pea picker; horse drawn machinery; hand tools; blacksmith shop; saw mill; apple picking line; log cabin; working windmill; railroad box car; trappers cabin.
Facilities: meeting room; picnic shelter.
Activities: drive-through display area. Annual Events: Truck Show in May; Tractor Pull in June; Farm Expo in August; Central WA Fair.
Publications: quarterly newsletter.
Hours & Admission Prices: Museum: April-Oct. Tues.-Fri. 9-3, Sat. 9-5, Sun. 1-4. Grounds: daily sunrise to sunset. No charge; donations accepted. &
Attendance: 8,500 (estimated)
Membership: Individual $15; Family $25; Sponsor $30-$40; Supporting $50-$99; Patron $100-$499; Benefactor $500 & up.

Vancouver

THE ARCHER GALLERY, Clark College, Penguin Student Union Building, 1933 Fort Vancouver Way, Vancouver, WA 98663-3501. Tel.: 360-992-2246. Fax: 360-992-2888.
E-mail: mhirsch@clark.edu
Web Site: www.clark.edu
Founded: 1978.
Key Personnel: Dir., Marjorie Hirsch.
Personnel Profile: Part-Time Paid 1.
Governing Authority: college; Parent Institution: Clark College. Tax-exempt.
Institution Type/Description: Art Gallery.
Collections: paintings; photographs; sculpture.
Activities: lectures; films.
Hours & Admission Prices: Tues.-Thurs. 10-7, Fri. 10-4, Sat. 1-5. No charge; donations accepted. &
Attendance: 5,000 (estimated)

CLARK COUNTY HISTORICAL SOCIETY & MUSEUM, (M), 1511 Main St., Vancouver, WA 98660-2945. Mailing Address: P.O. Box 61916, Vancouver, WA 98666. Tel.: 360-993-5679. Fax: 360-993-5683.
E-mail: cchm@pacifier.com
Web Site: www.cchmuseum.org
Founded: 1917.
Congressional District: 3
Key Personnel: Exec. Dir., Susan M.G. Tissot; Bd. Chm. & Pres., Joan Dengerink.
Personnel Profile: Full-Time Paid 1; Full-Time Volunteers 5; Part-Time Paid 3; Part-Time Volunteers 10.
Governing Authority: nonprofit organization. Parent Institution: Clark County Historical Society of Clark County, WA. Tax-exempt: 501(c)(3).
Institution Type/Description: General Museum.
Collections: Northwest history from pre-historic settlement to present including Native American exhibits; pre-native culture ceramics; railway equipment; Hudson Bay Company & textiles; household items; agriculture; oral history collection; archives; photographs; historical research library; Grant House collection.
Research Fields: Northwest & local history.
Facilities: 3,800-vol. library of letters, diaries, ledgers & books pertaining to local & Pacific Northwest history; reading room; classroom; archives.

Activities: guided tours; lectures; permanent & traveling exhibitions; seasonal walking tours; adult education classes; Mr. Carnegies Grand Tour of Washington. Annual Event: Harvest Fun Day.
Publications: annual book, Clark County History; four times a year newsletter, It's History; Naming Clark County; Darkness Next Door; Woven History; 12 Days in Clark County.
Hours & Admission Prices: Tues.-Sat. 11-4, 1st Thurs. of month call for additional hours. Adults $4, seniors & students $3, children 6-18 $2; discounts to WMA, AAM & ICOM members; members and children 5 & under no charge. Closed major holidays. &
Attendance: 15,000 (estimated)
Membership: Basic Membership: Student $25; Individual $40; Family $60; Friend of the Museum $100; Business $250; Historian $500; President's Circle $1,000.

FORT VANCOUVER NATIONAL HISTORIC SITE, 612 E. Reserve St., Vancouver, WA 98661-3897. Tel.: 360-816-6200. Fax: 360-816-6363.
E-mail: FOVA_superintendent@nps.gov
Web Site: www.nps.gov/fova
Founded: 1948.
Congressional District: 3
Key Personnel: Park Supt., Tracy Fortmann; Cur. Park, Theresa Langford; Chief Ranger, Greg Shine; Museum Shop Mgr., Mike True.
Governing Authority: federal. Parent Institution: National Park Service. Tax-exempt.
Institution Type/Description: Historic Site: site of Old Fort Vancouver, administrative headquarters & supply depot for the Hudson's Bay Company 1829-1860.
Collections: archaeology; military; fur trade; agriculture; manufacturing; ceramic spodeware.
Research Fields: Hudson's Bay Company in the Pacific Northwest; U.S. Army until 1945 in the Pacific Northwest.
Facilities: 1,400-vol. library on fur trade & early military history in the Pacific Northwest.
Activities: guided tours; lectures; films; formally organized education programs for children; permanent & temporary exhibitions.
Publications: Fort Vancouver Handbook.
Hours & Admission Prices: March-Oct. daily 9-5; Nov.-Feb. daily 9-4. Entrance fee: $3 per person, $5 per family. Closed Thanksgiving; Christmas Eve & Day. &
Attendance: 1,000,000 (estimated)

PEARSON AIR MUSEUM, 1115 E. 5th St., Vancouver, WA 98661-3802. Tel.: 360-694-7026. Fax: 360-694-0824.
E-mail: mike.trug@fortvan.org
Web Site: www.pearsonairmuseum.org
Founded: 1987.
Congressional District: 17
Key Personnel: Mgr. Operations, Michael Trug; Tourism Mgr., Brenna Beck; Museum Store Mgr., Deborah Bessette.
Personnel Profile: Full-Time Paid 3; Part-Time Paid 2; Part-Time Volunteers 35.
Governing Authority: private; nonprofit organization. Parent Institution: Vancouver National Historic Trust. Tax-exempt: 501(c)(3).
Institution Type/Description: Aeronautics Museum: located on the site of 1905 Pearson Field; a pre-WWII army air corps field.
Collections: period furnishings; historic flyable aircraft; photos; relics; art work; models; videos; 3 historic buildings c.1904.
Facilities: 20,000 sq. ft. exhibit space. Aviation-related items for sale.
Activities: Annual Events: annual summer dance; annual open cockpit day; Al Coupe summer camp.
Publications: newsletter, Fort Vancouver Times.
Hours & Admission Prices: Wed.-Sat. 10-5. Family $22, Adults $7, senior citizens & active military $5, students 6-17 $3; pre-schoolers & members no charge. Tour rates available. Closed New Year's Day; Christmas. &
Attendance: 60,000 (estimated)
Membership: Junior Flyer $5; Flyer $35; Co-Pilots $55; Squadron Family $75; Crew Chief $100; Barnstormer $250; Flying Ace $500; Aviator $1,000; Wing Commander $5,000.

Vantage

GINKGO PETRIFIED FOREST STATE PARK, Interstate 90, Exit 136, Vantage, WA 98950. Mailing Address: P.O. Box 1203, Vantage, WA 98950-1203. Tel.: 509-856-2700. Fax: 509-856-2294.
Web Site: parks.wa.gov
Founded: 1936.
Key Personnel: Area Mgr., Jim Mitchell.

Personnel Profile: Part-Time Paid 2.
Governing Authority: state: Affiliated with Washington State Parks & Recreation Commission, 7150 Clear Water Lane, Olympia, WA. 98504.
Institution Type/Description: State Park Museum.
Collections: Frank Bobo petrified wood collection; Professor George Beck's collections.
Facilities: 7,470-acre park; interpretive trails and center.
Activities: tours; permanent exhibitions.
Hours & Admission Prices: Summer: daily 6:30am to dusk. Winter: Sat.-Sun. 8am to dusk. Suggested Donation $1.
Attendance: 35,000 (accurate)

Vashon

VASHON MAURY ISLAND HERITAGE ASSOCIATION, (M), 10105 S.W. Bank Rd., Vashon, WA 98070-4645. Mailing Address: P.O. Box 723, Vashon, WA 98070-0723. Tel.: 206-463-7808.
E-mail: admin@vashonheritage.org
Web Site: www.vashonhistory.org
Founded: 1975.
Congressional District: 30
Key Personnel: Pres., Deb Damman; Vice Pres., Laurie Tucker; Treas., Steve Church; Sec., Jenna Everett; Corresponding Sec., Barbara Steen; Museum Shop Mgr., Yvonne Kuperberg.
Personnel Profile: Part-Time Volunteers 30.
Governing Authority: nonprofit organization. Tax-exempt: 501(c)(3).
Institution Type/Description: History Museum.
Collections: 1877 pre-settlement artifacts; industry; agriculture.
Research Fields: local history.
Facilities: 40-vol. library pertaining to Vashon & Maury Island history.
Activities: lectures; films; organized education programs for children. Museum Sponsors: slide shows; Vashon Festival.
Publications: quarterly newsletter.
Hours & Admission Prices: Wed.-Sun. 1-4. No charge; donations accepted. &
Attendance: 1,833 (accurate)
Membership: Senior Citizen & Student $15; Individual $20; Family $50; Supporting $75; Patron $100; Life $500; Benefactor $1,000 & up.

Walla Walla

FORT WALLA WALLA MUSEUM, 755 Myra Rd., Walla Walla, WA 99362-8035. Tel.: 509-525-7703. Fax: 509-525-7798.
E-mail: info@fortwallawallamuseum.org
Web Site: fortwallawallamuseum.org
Founded: 1968.
Congressional District: 5
Key Personnel: Exec. Dir., James Payne; Pres. (V), Steve Stevenson; Bookkeeper, Carolyn Burdine; Collection Mgr., Laura Schulz; Mgr. Operations, Don Locati; Mgr. Communications, Paul Franzmann; Tour Coord., Bill Lake; Programs, Anne Threlfall; Mgr. Bldgs. & Grounds, James Klees; Museum Shop Mgr., Cheryl Gibson; Exec. Asst., Nancy Pavvy.
Personnel Profile: Full-Time Paid 9; Part-Time Paid 2; Part-Time Volunteers 300; Interns 2.
Governing Authority: nonprofit organization. Parent Institution: Fort Walla Walla Museum, Walla Walla Valley Historical Society. Tax-exempt: 501(c)(3).
Institution Type/Description: Pioneer Settlement & Agricultural Museum: housed on the 1858 military reservation of Old Ft. Walla Walla. General history.
Collections: wooden combine hitched to 33 fiberglass mules surrounded by wheat harvest mural; 34,000 historical artifacts. Historic Buildings: 1859 Ransom Clark cabin; 1868 Union schoolhouse; c.1888 Babcock railway depot; block house; barber shop; 1903 Prescott Jail; horse era agriculture; pioneer & military artifacts; 40,000 historical artifacts.
Facilities: 22 buildings on 15 acres.
Activities: guided tours; lectures; permanent & rotating exhibitions.
Publications: quarterly newsletter, The Dispatch.
Hours & Admission Prices: Jan.-March Mon.-Fri. 10-4; April-Oct. daily 10-5; Nov.-Dec. daily 10-4; tours by appointment. Adults $7, senior citizens & students $6, children 6-12 $3; discounts to AAA members & Time Travelers; members and children 5 & under no charge. &
Attendance: 25,100 (accurate)
Membership: Student & Senior $27; Individual $35; Family $45; Fort Walla Walla $100; Pioneer $250; Explorer $500; Director's Circle $1,000.

KIRKMAN HOUSE MUSEUM, 214 N. Colville St., Walla Walla, WA 99362-1917. Tel.: 509-529-4373. Fax: 509-529-4373.
E-mail: khm@kirkmanhousemuseum.org
Web Site: www.kirkmanhousemuseum.org
Congressional District: 5

Key Personnel: Pres. (V), Donna Cook Gardner.
Personnel Profile: Full-Time Volunteers 2; Part-Time Paid 1; Part-Time Volunteers 12.
Governing Authority: Tax-exempt.
Institution Type/Description: Historic House Museum: built in 1880. Listed on the National Register of Historic Places.
Collections: period furnishings; personal artifacts.
Activities: Museum Sponsors: Sweet Home Walla Walla historic homes tour in April; Sheep to Shawl in September; Victorian Christmas Jubilee in December.
Publications: quarterly newsletter.
Hours & Admission Prices: Wed.-Sat. 10-4, Sun. 1-4. Adults $5; discounts to NARM members; members no charge.
Membership: Individual $25; Family $50; Supporter $100; Friend $250; Donor $500; Patron $1,000.

SHEEHAN GALLERY AT WHITMAN COLLEGE, Olin Hall, 814 Isaacs, Walla Walla, WA 99362. Mailing Address: 345 Boyer Ave., Walla Walla, WA 99362-2083. Tel.: 509-527-5249. Fax: 509-527-5039.
E-mail: forbesdm@whitman.edu
Web Site: www.whitman.edu/sheehan/sheehan_mission.html
Founded: 1972.
Congressional District: 16
Key Personnel: Interim Dir., Dawn Forbes; Pres. Whitman College, Thomas E. Cronin; Pres. Bd. Trustees, Charles E. Anderson; Dean Faculty, Patrick Keef, Ph.D.; Treas., Peter Harvey; Collections & Exhibitions Mgr., Kynde Kiefel.
Personnel Profile: Full-Time Paid 1; Part-Time Paid 1; Part-Time Volunteers 12; Interns 10.
Governing Authority: college; nonprofit. Parent Institution: Whitman College. Tax-exempt: 501(c)(3).
Institution Type/Description: Art Museum/Center.
Collections: Thomas Potter Davis/Seafirst Bank Asian art; Floyd Whittington ceramics; Pacific Northwest paintings & prints.
Research Fields: contemporary art; Asian art & art history; Buddhist art; 20th-century German art; Pacific Northwest modern & contemporary art.
Facilities: 2,400 sq. ft. exhibit space.
Activities: 6 exhibits per year; guided tours; lectures; films.
Publications: exhibition notices; posters; catalogues; Chinese Ceramics in the Thomas Davis/Seafirst Bank Collection; Paths to Enlightenment: Buddhist Art in the Davis/Seafirst Collection; The Doll Theater; Bunraku Puppets in the Davis/Seafirst Collection; Schwarzweiss: The Revival of the Print in Germany; Palm Leaves & Postcards; Material Culture & its Representation in Colonial Ceylon; Thomas O'Day: wake 1/wake 2; We the People: Satiric Prints; Contemporary English Crafts; Piranesi's Careri: Sources of Invention; Auto as Icon (With George Eastman house); The Japanese Artist's Book: Wood block Impressions from the 17th-20th Centuries; George Tsutakawa & Morris Graves; Wendall Brazeau, 1910-1974.
Hours & Admission Prices: Sept.-May Mon.-Fri. 12-5, Sat.-Sun. 12-4. No charge; donations accepted. Closed spring breaks; Christmas. &
Attendance: 12,000 (estimated)

WHITMAN MISSION NATIONAL HISTORIC SITE, 328 Whitman Mission Rd., Walla Walla, WA 99362-7299. Tel.: 509-522-6360. Fax: 509-522-6355. TDD: 509-522-6357.
Web Site: www.nps.gov/whmi
Founded: 1936.
Congressional District: 5
Personnel Profile: Full-Time Paid 8; Part-Time Paid 2; Part-Time Volunteers 5.
Governing Authority: federal. Affiliated with National Park Service, Interior Building, Washington, DC. Tax-exempt.
Institution Type/Description: Park Museum.
Collections: various exhibits dealing with history of Whitman & missionary era in Pacific Northwest.
Research Fields: Oregon territory mission stations.
Facilities: 1200-vol. library of books relating to the Whitman story available for inter-library loan & for use on premises; self-guiding trail; visitor center; auditorium. Printed matter, maps & cultural craft kits for sale.
Activities: lectures; films; formally organized education programs for children, adults & undergraduate college students; permanent & temporary exhibitions; cultural demonstrations.
Publications: handbook & minifolder, Whitman Mission.
Hours & Admission Prices: Oct. 1, 2013-May 25, 2014 Tues.-Sat. 9-4; May 26, 2014-Aug. 31,2014 Wed.-Sun. 9-4; Sept. 1, 2014-Nov. 30, 2014 Tues.-Sat. 9-4. No charge; donations accepted. Closed most holidays; Visitor Center closed December & January. &
Attendance: 70,000 (accurate)

Washougal

TWO RIVERS HERITAGE MUSEUM, 1 Durgan St., Washougal, WA 98671. Mailing Address: P.O. Box 204, Washougal, WA 98671-0204. Tel.: 360-835-8742.
E-mail: referenceroom@trhm.comcastbiz.net
Web Site: www.2rhm.org
Founded: 1978.
Congressional District: 3
Key Personnel: Pres., Curtis Hughey; Chm. (V), Lois Cobb.
Personnel Profile: Part-Time Volunteers 46.
Governing Authority: Parent Institution: Camas Washougal Historical Society. Tax-exempt.
Institution Type/Description: History Museum.
Collections: local history; over 8,000 photographs & 250 oral histories; land records & local families histories; Native American artifacts & period baskets; art; memorabilia; books, local craft items & keepsakes; early wagons; logging; physicians; mining; musical instruments; dolls; toys; player piano.
Research Fields: family histories; area town & communities; Eastern Clark county & Western Skamania county.
Facilities: Museum-related items for sale.
Activities: quarterly society members meetings; heritage day; ice cream social; plant & garden fair.
Publications: quarterly newsletter.
Hours & Admission Prices: Tues.-Sat. 11-3. Families $8, adults $3, seniors 60 & over $2, students $1; discounts to AAA members; members & children under 5 no charge. &
Attendance: 1,000 (estimated)
Membership: Individual $15; Family & Organization $25; Sustaining $50; Supporting $100; Contributing $250; Sponsoring $500.

Wenatchee

ROBERT GRAVES GALLERY, Wenatchee Valley College, 1300 5th St., Wenatchee, WA 98801-1741. Tel.: 509-682-6776.
E-mail: robertgravesgallery@wvc.edu
Web Site: www.wvc.edu
Formerly: Gallery '76
Founded: 1976.
Congressional District: 4
Key Personnel: Bd. Pres. (V) & Acting Coord., John Crew.
Personnel Profile: Full-Time Volunteers 1; Part-Time Volunteers 35.
Governing Authority: nonprofit organization. Tax-exempt.
Institution Type/Description: Art Gallery.
Collections: paintings; drawings; sculpture.
Activities: guided tours; lectures; docent program; organized education programs for undergraduate college students affiliated with Wenatchee Valley College; loan & traveling exhibitions.
Publications: Robert Graves Gallery News.
Hours & Admission Prices: Mon. 8-8, Tues.-Thurs. 9-1; other times by appointment. No charge; donations accepted. Closed holidays. &
Attendance: 4,500 (accurate)
Membership: Senior & Student $25; Supporting $50; Patron $150; Booster $250; Benefactor $500; Sponsor $1,000.

ROCKY REACH DAM, 5000 97A, Wenatchee, WA 98801-2011. Mailing Address: P.O. Box 1231, Wenatchee, WA 98807-1231. Tel.: 509-663-7522 & 661-4960, ext. 4960. Fax: 509-661-8149. Facebook: Rocky Reach.
E-mail: debbie.gallaher@chelanpud.org
Web Site: www.chelanpud.org
Founded: 1963.
Congressional District: 4
Key Personnel: C.E.O., Steve Wright; Museum Shop Mgr., Debbie Gallaher.
Personnel Profile: Full-Time Paid 5; Part-Time Paid 2; Part-Time Volunteers 1.
Governing Authority: county. Parent Institution: Public Utility District #1 of Chelan County.
Institution Type/Description: General Interpretive Museum.
Collections: paintings; graphics; Indian artifacts; Gallery of the Columbia; electrical antiques; Thomas Edison artifacts.
Research Fields: interpretive local geology; prehistory; history; electrical.
Facilities: fish viewing room; theater; picnic area; gardens; snack bar; rental facilities. Gifts for sale.
Activities: self-guided tours; guided tours during summer or when personnel available; films; permanent & temporary exhibitions; art, craft & hobby displays; geocaching.
Publications: brochures.
Hours & Admission Prices: March-Oct. daily 9-4. No charge. &
Attendance: 62,000 (accurate)

WENATCHEE VALLEY MUSEUM AND CULTURAL CENTER, (M), 127 S. Mission St., Wenatchee, WA 98801-3039. Tel.: 509-888-6240. Fax: 509-888-6256.
E-mail: info@wvmcc.org
Web Site: www.wvmcc.org
Formerly: North Central Washington Museum
Founded: 1939.
Key Personnel: Dir., Brenda Abney; Pres. (V), Darlene Spargo; Cur. Collections, Mark Behler; Cur. Exhibits & Programs, Bill Rietveldt; Education Coord., Selina Danko; Administrative Svcs. Mgr., Tammy Moad.
Governing Authority: municipal; nonprofit organization. Tax-exempt: 501(c)(3).
Institution Type/Description: General Museum.
Collections: north central Washington artifacts & memorabilia including pioneer & farm equipment; archaeological artifacts; printing paraphernalia; archives; manuscripts; fruit industry; fine arts; theater pipe organ; print shop; Great Northern Railroad model; 1920's apple packing warehouse; theater, AV multi-screen presentations.
Research Fields: regional Native American information; regional history; development of fruit industry; Great Northern Railroad; hydroelectric dams & irrigation; genealogy; heritage reference.
Facilities: genealogy library; archives & reference center; two theaters; auditorium; interpretive center; meeting rooms. Museum-related items for sale.
Activities: guided tours; traveling, permanent & temporary exhibits; school programs; demonstrations; lecture series; pipe organ programs; concert & silent movie; hands-on activities; bus tours; exhibit receptions; catering facilities.
Publications: quarterly newsletter; Walking Tour of Wenatchee; quarterly magazine, The Confluence.
Hours & Admission Prices: Tues.-Sat. 10-4. Adults $5, seniors & students $4, children 6-12 $2; members & children under 6 no charge. Closed major holidays. &
Attendance: 32,523 (accurate)
Membership: Student $20; Senior $30; Adult $40; Senior Couple $50; Family $55; Friend $100-$249; Patron $250-$499; Founder $500-$999; Director's Circle $1,000 & up.

West Seattle

SOUTH SEATTLE COMMUNITY COLLEGE ART GALLERY, South Seattle Community College, Jerry Brockey Student Center, West Seattle, WA 98106. Mailing Address: 6000 16th Ave., S.W., Seattle, WA 98106-1499. Tel.: 206-764-5337.
Web Site: www.southseattle.edu/student-life/art-gallery/
Key Personnel: Gallery Coord., Akiko Masker
Institution Type/Description: Art Gallery.
Collections: paintings.
Major Exhibits: How We Remember Our Veterans, 1/13/14-2/6/14.
Activities: Museum Sponsors: student & community art shows.
Hours & Admission Prices: Mon.-Fri. 10-4. No charge. Closed school holidays & breaks.

Westport

WESTPORT MARITIME MUSEUM, 2201 Westhaven Dr., Westport, WA 98595. Mailing Address: P.O. Box 1074, Westport, WA 98595-1074. Tel.: 360-268-0078. Facebook: Westport Maritime Museum.
E-mail: info@westportmaritimemuseum.com
Web Site: www.westportmaritimemuseum.com
Founded: 1985.
Congressional District: 3
Key Personnel: Pres. (V), Jeff Pence; Treas., Sue Thomas.
Personnel Profile: Full-Time Paid 2; Part-Time Paid 1; Part-Time Volunteers 95.
Governing Authority: private; nonprofit organization. Parent Institution: Westport-South Beach Historical Society, P.O. Box 1074, Westport, WA 98595. Tax-exempt: 501(c)(3).
Institution Type/Description: Maritime Museum.
Collections: Fresnel lens, 1st order & 3rd order; whale skeletons; local history; local Coast Guard history; local industries, fishing, logging, cranberry growing & ship building.
Research Fields: life in a maritime community; local logging camps; evolution of navigational aids.
Facilities: children's discovery room; 3,000 sq. ft. exhibit space; 90-seat auditorium; lighthouse. Museum-related items for sale.
Activities: docent program; guided tours; lectures; fireside chats; guest speakers. Annual Events: Run 4 The Light; Haunted Light Station Tour; Old Fashioned Independence Day; Santa by the Sea; Annual Arts Festival.

Publications: quarterly newsletter, Foghorn.
Hours & Admission Prices: Museum: April-Sept. Thurs.-Tues. 10-4. Museum: adults $5, children 5-15 $2. Lighthouse: Feb.-March Fri.-Mon.; April-Sept. Thurs.-Tues. Admission $5.
Attendance: 18,000 (accurate)
Membership: Individual $20; Couple $30; Family $40; Business $60, $120, $500, $1,000; Patron $75; Benefactor $100.

White Swan

FORT SIMCOE STATE PARK, 5150 Fort Simcoe Rd., White Swan, WA 98952-9745. Tel.: 509-874-2372. Fax: 509-874-2351.
Formerly: Fort Simcoe Interpretive Center
Key Personnel: Park Mgr., Jim Mitchell.
Governing Authority: state. Affiliated with Washington State Parks & Recreation Commission, 7150 Clean Water Lane, Olympia, WA 98504.
Institution Type/Description: Military Museum: housed in 1856-59 military outpost.
Collections: materials related to life at the fort; military & Indian exhibits & artifacts.
Activities: permanent exhibitions.
Hours & Admission Prices: Summer: 6:30am to dusk. No charge.

Winlock

JOHN R. JACKSON HOUSE, Lewis & Clark State Park, Winlock, WA 98596. Mailing Address: 4583 Jackson Hwy., Winlock, WA 98596-9646. Tel.: 360-864-2643.
Founded: 1915.
Key Personnel: Dir., Rex Derr; Chief Interpretive Svcs., Steve Wang.
Governing Authority: state. Affiliated with Washington State Parks & Recreation Commission, 7150 Clean Water Lane, Olympia, WA 98504. Tax-exempt.
Institution Type/Description: Historic House Museum: 1850 John R. Jackson Home.
Collections: owner's books and possessions.
Activities: guided tours; lectures.
Publications: brochure.
Hours & Admission Prices: House: April-Sept. 8am to dusk; tours by appointment. No charge; donations accepted. &

Winthrop

SHAFER HISTORICAL MUSEUM, 285 Castle Ave., Winthrop, WA 98862. Mailing Address: P.O. Box 46, Winthrop, WA 98862-0046. Tel.: 509-996-2712.
E-mail: staff@shafermuseum.com
Web Site: shafermuseum.com
Founded: 1974.
Congressional District: 5
Governing Authority: Parent Institution: Okanogan County Historical Society. Tax-exempt.
Institution Type/Description: Historic Buildings: listed in the National Register of Historical Places.
Collections: pioneer artifacts; mining equipment; period farm equipment & cars; photographs; historic buildings.
Facilities: Museum-related items for sale.
Publications: book, Bound for the Methow.
Hours & Admission Prices: Memorial Day to Labor Day Thurs.-Mon. 10-5. Adults $2; members no charge.
Attendance: 20,000 (estimated)
Membership: Individual $25.

Yakima

ALLIED ARTS OF YAKIMA, 5000 W. Lincoln Ave., Yakima, WA 98908-2657. Tel.: 509-966-0930. Fax: 509-966-0934.
E-mail: info@alliedartsyakima.org
Web Site: www.alliedartsyakima.org
Formerly: Peggy Lewis Gallery
Founded: 1962.
Congressional District: 14
Key Personnel: Exec. Dir., Michael Liddicoat; Pres. (V), John O'Rourke; Office Mgr., Helena Parrish.
Personnel Profile: Full-Time Paid 3; Part-Time Volunteers 19; Interns 2.
Governing Authority: nonprofit. Tax-exempt.
Institution Type/Description: Art Museum.
Collections: photographs; paintings.
Facilities: library; classroom; dance room; rental facilities.

Activities: summer camps; special events; classes; rental facilities; arts van rental.
Publications: quarterly newsletter, Art Scope.
Hours & Admission Prices: Mon.-Fri. 9-5. No charge; donations accepted.
Membership: Individual $50; Patron $100; Benefactor $250; Renaissance Club $500; De Medici Guild $1,000; Arts Leadership $2,500.

LARSON MUSEUM AND GALLERY, (M), S. 16th Ave. and Nob Hill Blvd., Yakima Valley Community College, Yakima, WA 98902. Mailing Address: P.O. Box 22520, Yakima, WA 98907-2520. Tel.: 509-574-4875. Fax: 509-574-6826. TDD: 509-574-4600.
E-mail: gallery@yvcc.edu
Web Site: www.larsongallery.org
Founded: 1949.
Congressional District: 4
Key Personnel: Chm. (V) & Asst. Dir., Denise Olsen; Dir., David Lynx; Pres. (V), Erwina Peterson.
Personnel Profile: Part-Time Paid 4; Part-Time Volunteers 20.
Governing Authority: nonprofit organization. Parent Institution: Yakima Valley Community College. Subsidiary Institution: Larson Gallery Guild.
Institution Type/Description: Art Gallery.
Collections: fine arts, ethnic arts & annual juried art.
Activities: guided tours; arts workshops; organized education programs for children; participatory exhibits. Museum Sponsors: summer workshops in June & July.
Publications: newsletter; tour of homes brochure; books, Central Washington Artists, 2007-2010 - Artist Archive Project (1st edition); Robert A. Fisher, Retrospective Portfolio 2011; Charles A. Smith, Retrospective Portfolio 2010; Joe Feddersen, Terrain - A Survey 2012.
Hours & Admission Prices: Sept.-July 14 Tues.-Sat. 10-5. No charge, donations accepted. &
Attendance: 10,000 (accurate)
Membership: Student & Senior $20; Supporter $40; Contributor $65; Family $70; Donor $120; Patron $275; Benefactor $500; Advocate $1,000.

MCALLISTER MUSEUM OF AVIATION, 2008 S. 16th Ave., Yakima, WA 98903. Tel.: 509-457-4933.
E-mail: mcallister@nwinfo.net
Web Site: mcallistermuseum.org
Institution Type/Description: Aviation Museum: housed in a former flight school owned by Charlie & Alister McAllister.
Collections: aviation history & artifacts; aircraft
Activities: video; flight simulator.
Hours & Admission Prices: Thurs.-Fri. 10-4, Sat. 9-4.

YAKIMA AREA ARBORETUM & BOTANICAL GARDEN, Yakima, WA 98901-8513. Tel.: 509-248-7337. Fax: 509-248-8197. Facebook: Yakima Area Arboretum.
E-mail: info@ahtrees.org
Web Site: www.ahtrees.org
Founded: 1967.
Key Personnel: Co.-Exec. Dir., Colleen Adams-Schuppe; Co.-Exec. Dir., Jheri Ketcham; Pres. (V), Leslie Wahl; Groundkeeper, Jeff Neal; Nature Science Dir., Jacob Belsher; Care Taker, Joy Howell; Care Taker, Bob Howell; Facility Mgr., Gaye McCarthy.
Personnel Profile: Full-Time Paid 5; Part-Time Paid 3; Part-Time Volunteers 1.
Volunteer Hours: 1,500
Operating Expenses: 301,000
Operating Income: 298,000
Governing Authority: Board of Directors. Tax-exempt.
Institution Type/Description: Arboretum & Botanical Garden.
Collections: native & exotic species of woody plants; trees; shrubs; gardens.
Facilities: display garden; banquet facilities. Gift items for sale.
Activities: educational classes. Annual Events: arbor festivals; garden tour; luminaria.
Publications: newsletter, Arborescent.
Hours & Admission Prices: Daily dawn to dusk. Visitor Center: Mon.-Sat. 9-5. No charge; donations accepted. &
Attendance: 10,000 (estimated)
Membership: Individual $30 & up.

*** YAKIMA VALLEY MUSEUM AND HISTORICAL ASSOCIATION, (M),** 2105 Tieton Dr., Yakima, WA 98902-3766. Tel.: 509-248-0747. Fax: 509-453-4890.
E-mail: info@yakimavalleymuseum.org
Web Site: www.yakimavalleymuseum.org
Founded: 1952.
Congressional District: 4

Key Personnel: C.E.O. & Dir., John A. Baule; Pres. (V), Sharon Harris; Cur. Collections, Michael Siebol; Cur. Exhibits & Programs, Andrew Granitto; Cur. Education, Jessica Carlton; Mktg. & Fund Devel., Kimberly Thompson; Museum Shop Mgr., Cathy Robinson.
Personnel Profile: Full-Time Paid 3; Part-Time Paid 4; Part-Time Volunteers 107; Interns 2.
Governing Authority: nonprofit organization. Subsidiary Institution: H. M. Gilbert Homeplace. Tax-exempt: 501(c)(3).
Institution Type/Description: Historical Museum.
Collections: archival repository; horse-drawn vehicles; Native American artifacts; textile & costume collections; household furnishings; agricultural artifacts; mineral collections; personal accessories; irrigation artifacts; The Children's Underground, a hands-on center focusing on natural & human history of the area; operating 1930's Art Deco ice cream soda fountain.
Research Fields: Yakima Valley & central Washington history.
Facilities: Museum-related gifts & books for sale.
Activities: guided tours; lectures; college classes & seminars; music & cultural programs.
Publications: quarterly newsletter; books; report to membership.
Hours & Admission Prices: March-Oct. Mon.-Sat. 10-5; Nov.-Feb. Tues.-Sat. 10-5. Families $12, adults $5, students & senior citizens $2.50; discounts to AASLH, AAA, ICOM & AAM members; members & children under 6 no charge. Children's Underground: March-Oct. Mon.-Sat. 10-5; Nov.-Feb. Tues.-Sat. 10-5. &
Attendance: 27,000 (estimated)
Membership: Individual $30; Family $40; Sponsor $75; Supporting $100; Patron $300; Benefactor $500.

WEST VIRGINIA

(134 listings)

Anstead

AFRICAN AMERICAN HERITAGE FAMILY TREE MUSEUM, Logtown Rd., Anstead, WV 25812. Mailing Address: P.O. Box 369, Amsted, WV 25812. Tel.: 304-658-5528.
E-mail: normanjordan@pocketmail.com
Institution Type/Description: History Museum.
Collections: African-American life, culture & history; Underground Railroad; artifacts of Thomas Friend, Booker T. Washington & Carter G. Woodson; photographs; personal artifacts; family histories; household artifacts; coal mining.
Hours & Admission Prices: Summer: by appointment.

Ansted

CONTENTMENT, Rte. 60, Ansted, WV 25812. Mailing Address: HC 66 Box 94B, Hico, WV 25854-7468.
Personnel Profile: Full-Time Volunteers 1.
Institution Type/Description: Historic House Museum: housed in the former home of Civil War Col. George Imboden.
Collections: period furnishings; personal artifacts; 3 historic buildings.
Activities: school tours; special events.
Hours & Admission Prices: June-Aug. Mon.-Sat. 10-4.

Arthurdale

ARTHURDALE HERITAGE, INC., Q & A Rds., Arthurdale, WV 26520. Mailing Address: P.O. Box 850, Arthurdale, WV 26520-0850. Tel.: 304-864-3959. Fax: 304-864-4602. Facebook: Arthurdale Heritage, Inc.
E-mail: ahi@arthurdaleheritage.org
Web Site: www.arthurdaleheritage.org
Founded: 1985.
Congressional District: 1
Key Personnel: Dir., Jeanne Goodman; Pres. (V), Randy Weaver; Museum Shop Mgr., Loretta Davis.
Personnel Profile: Full-Time Paid 1; Part-Time Volunteers 50.
Governing Authority: private; nonprofit. Tax-exempt: 501(c)(3).
Institution Type/Description: Historic Site: housed in several buildings constructed in 1930s when Arthurdale became the first New Deal homestead.
Collections: structures; tools & equipment; artisan crafts. Historic House: working 1930s homestead.
Research Fields: 1930s New Deal homestead communities; Arthurdale oral history, 1947-present.
Facilities: library; archives; 800 sq. ft. exhibit space.
Activities: guided tours; concerts; temporary exhibits; blacksmith & other craft demonstrations. Annual Events: New Deal Festival in July; Traditional Crafts in July; Old Fashioned Ice Cream Social; parade.

Publications: quarterly newsletter, Restoring Yesterday for Tomorrow.
Hours & Admission Prices: May-Oct. Tues.-Sat. 11-3, Sun. 12-4; Nov.-April Tues.-Fri. 11-3; other times by appointment. Adults $8, seniors $7, children $3; discounts to groups of 10 or more & AAA members; members and children 5 & under no charge. Closed holidays. &
Attendance: 5,000 (estimated)
Membership: Individual $15; Household $25; Business $50.

Athens

CONCORD COLLEGE ARTHUR BUTCHER ART GALLERY, Alexander Fine Arts Center, Athens, WV 24712. Mailing Address: P.O. Box 1000, Athens, WV 24712-1000. Tel.: 304-384-3115.
Institution Type/Description: Art Gallery.
Collections: paintings; sculpture.
Hours & Admission Prices: Mon.-Thurs. 9-4.

Aurora

AURORA AREA HISTORICAL SOCIETY, 23976 George Washington Hwy., Aurora, WV 26705. Mailing Address: P.O. Box 100, Aurora, WV 26705. Tel.: 304-288-6850.
Founded: 1999.
Institution Type/Description: Historical Society Museum: housed in a former general store; built c.1850.
Collections: local history & culture; personal artifacts; photographs; oral histories; documents; census & cemetery records.
Hours & Admission Prices: Call for hours.

Barboursville

DAUGHTERS OF THE AMERICAN REVOLUTION TOLL HOUSE MUSEUM, 731 Main St., Barboursville, WV 25705. Mailing Address: 214 Forestview Dr, Huntington, WV 25705. Tel.: 304-652-2922.
Institution Type/Description: Historic House Museum: housed in a former toll house originally located on the Guyan River bank to collect tolls from ferry riders; built in 1837.
Collections: local history & culture; period artifacts; photographs.
Hours & Admission Prices: By appointment.

Beckley

EXHIBITION COAL MINE & YOUTH MUSEUM, 513 Ewart Ave., Beckley, WV 25801. Tel.: 304-256-1747 & 252-3730.
Institution Type/Description: Mining Museum.
Collections: local coal mining history; mining artifacts & tools; coal mining camp; photographs; geological specimens.
Facilities: Museum-related items for sale.
Activities: underground mine tour.
Hours & Admission Prices: April-Nov. 1 daily 10-6.

RALEIGH COUNTY VETERANS MUSEUM, 1557 Harper Rd., Beckley, WV 25801-3307. Mailing Address: P.O. Box 3165, Beckley, WV 25801-1945. Tel.: 304-253-1775.
Web Site: rcvm.org
Institution Type/Description: Veterans Museum.
Collections: military history from Revolutionary War to present; photographs.
Hours & Admission Prices: April-Oct. Fri.-Sat. 1-7, Sun. 1-5; other times by appointment. Adults $2, children 12 & under $.50.

WILDWOOD HOUSE MUSEUM, 121 Laurel Ter., Beckley, WV 25801-4217. Mailing Address: P.O. Box 2514, Beckley, WV 25802-2514. Tel.: 304-252-3730. Fax: 304-252-3764.
E-mail: sparker@beckleymine.com
Web Site: beckley.org
Key Personnel: Dir., Sandi Parker; C.E.O., Leslie Baker.
Personnel Profile: Part-Time Volunteers 6.
Governing Authority: city. Tax-exempt.
Institution Type/Description: Historic House: housed in the former home of General Alfred Beckley; built in 1836. Listed on the National Register of Historic Places.
Collections: Beckley family history; period furnishings; personal artifacts.
Hours & Admission Prices: April-Nov. Sat.-Sun. 10-6 Adults $5.
Attendance: 1,000 (estimated)

YOUTH MUSEUM OF SOUTHERN WEST VIRGINIA, 509 Ewart Ave., Beckley, WV 25802. Mailing Address: P.O. Box 2514, Beckley, WV 25802-2514. Tel.: 304-252-3730. Fax: 304-252-3764.
E-mail: sparker@beckleymine.com
Web Site: beckleymine.com
Founded: 1977.
Congressional District: 3
Key Personnel: Exec. Dir. & C.E.O., Leslie Baker; Administrative Coord., Donna Clark Totten.
Personnel Profile: Full-Time Paid 2; Part-Time Paid 4; Part-Time Volunteers 2.
Governing Authority: The city of Beckley.
Institution Type/Description: Youth Museum.
Collections: used to interpret Appalachian heritage in the Mountain Homestead.
Research Fields: developing hands-on science curriculum.
Facilities: 120-vol. library; educational facilities; 1,500 sq. ft. exhibit space; planetarium; 200-seat amphitheater. Museum-related items for sale.
Activities: arts festivals; concerts; films; guided tours; lectures; participatory & traveling exhibitions. Annual Events: Tailgate Halloween; Appalachian Heritage Day; Christmas in the Homestead.
Publications: quarterly newsletter, Museum News.
Hours & Admission Prices: April-Nov. 1 daily 10-6; Nov. 2-March Tues.-Sat. 10-5. Call for rate schedule. Closed New Year's Day; Christmas. &
Attendance: 50,000 (estimated)
Membership: Individual $15; Couple $25; Family $35; Sponsor $50; Community $100; WV $250; Patron $500.

Benwood

CASTLE HALLOWEEN MUSEUM, 1595 Boggs Run Rd., Benwood, WV 26031-1050. Tel.: 304-233-1031.
E-mail: castlehalloween@comcast.net
Web Site: www.castlehalloween.com
Founded: 2005.
Key Personnel: Owner, Pamela Apkarian-Russell
Institution Type/Description: Halloween Social History Museum.
Collections: Halloween-related artifacts; arcade machines; costumes; paintings; toys; postcards; folk art; paintings; art; ephemera.
Activities: research.
Hours & Admission Prices: By appointment only. Tours: $8. &

Berkeley Springs

MUSEUM OF THE BERKELEY SPRINGS, Fairfax & Wilkes St., Berkeley Springs, WV 25411. Mailing Address: P.O. Box 99, Berkeley Springs, WV 25411-0099. Tel.: 800-447-8797.
E-mail: history@museumoftheberkeleysprings
Founded: 1984.
Congressional District: 2
Key Personnel: Pres., David Milburn; Vice Pres., Betty Lou Harmison; Treas., Susan Winkeler-Milburn; Exec. Dir. & Museum Shop Mgr., Tamme Marggraf.
Personnel Profile: Part-Time Paid 4; Part-Time Volunteers 1.
Governing Authority: private; nonprofit organization. Tax-exempt: 501(c)(3).
Institution Type/Description: History Museum: housed in the c.1820 Roman Bath building in Berkeley Springs State Park.
Collections: local history associated with the mineral springs at Berkeley Springs, the town & county.
Facilities: 1,250 sq. ft. exhibit space. Museum-related items for sale.
Activities: guided tours.
Publications: annual newsletter; pamphlets.
Hours & Admission Prices: Feb.-Dec. Sat.-Sun. 11-4; call for additional hours. No charge; donations accepted. &
Attendance: 8,476 (accurate)
Membership: Individual $15; Family $25; Trustee $100.

Bethany

HISTORIC BETHANY - ALEXANDER CAMPBELL MANSION, Bethany College, Rte. 67 E., Bethany, WV 26032. Mailing Address: P.O. Box 478, Bethany, WV 26032-0478. Tel.: 304-829-4258. Fax: 304-829-4258.
E-mail: historic@bethanywv.edu
Founded: 1840.
Key Personnel: C.E.O., Dr. G.T. Smith; Cur., Felicity Ruggiero.
Personnel Profile: Full-Time Paid 1; Full-Time Volunteers 6; Part-Time Paid 4; Part-Time Volunteers 10.
Governing Authority: Parent Institution: Bethany College; nonprofit. Tax-exempt.

Institution Type/Description: Historic Mansion: housed in the former home of Bethany College founder, Alexander Campbell; built c.1794. Listed on the National Register of Historic Places.
Collections: Alexander Campbell family house items; Campbell books & papers; Upper Ohio Valley collection; historic buildings.
Research Fields: Campbell & his relationship to early U.S. history; Thomas Barclay & family; early U.S. history.
Facilities: 210,000-vol. general resource materials & special collections library available for inter-library loan; educational facilities. Museum-related items for sale.
Activities: guided tours. Annual Event: Children's Heritage Day.
Publications: quarterly, newsletter, The Campbell Light.
Hours & Admission Prices: April-Oct. Tues.-Fri. 10-12 & 1-4; other times by appointment. Adults $4, youth 1-12 grade $2; discounts to AAM & AAA members. Closed holidays. &
Attendance: 3,500 (estimated)

Beverly

BEVERLY HERITAGE CENTER, One Court St., Beverly, WV 26253. Tel.: 304-637-7424.
Web Site: www.historicbeverly.org/bevhcent.htm
Institution Type/Description: Heritage Center.
Collections: local history & culture; period tools & furnishings; photographs. Historic Buildings: 1808 Randolph County Courthouse; 1854 Bushrod Crawford bldg.; 1900 Beverly Bank bldg.; 1907 Hill store Bldg.
Facilities: archives. Museum-related items for sale.
Hours & Admission Prices: Call for hours.

RANDOLPH COUNTY MUSEUM (AT BEVERLY), Main St., Beverly, WV 26253. Mailing Address: P.O. Box 1164, Elkins, WV 26241-1164. Tel.: 304-636-0841 & 1951.
E-mail: dlrice@suddenlink.net
Founded: 1924.
Congressional District: 2
Key Personnel: C.E.O. (V) & Pres. (V), Randy Allan; Acting Cur. & Archivist, Donald Rice.
Personnel Profile: Part-Time Volunteers 4.
Governing Authority: private; nonprofit organization. Parent Institution: Randolph County Historical Society. Tax-exempt: 501(c)(3).
Institution Type/Description: History Museum: housed in the Blackman-Bosworth Store building.
Collections: 200 years of Randolph County history.
Research Fields: Rich Mountain Civil War Battlefield; Beverly town & Randolph County history; genealogy.
Facilities: 1,600 sq. ft. exhibit space. Books for sale.
Activities: formal education programs for children; guided tours; lectures; temporary exhibitions; special events. Annual Event: Beverly Historic Days.
Publications: books; journals.
Hours & Admission Prices: Museum: mid-May to mid-Oct. Fri.-Sat. 12-4; other times by appointment. No charge; donations requested.
Attendance: 700 (estimated)
Membership: Society: Children $2; Seniors $10; Adult $20.

Bluefield

BUDDY'S COUNTRY STORE & MUSEUM, 295 Warden Ave., Bluefield, WV 24701. Tel.: 304-589-5659.
Institution Type/Description: History Museum.
Collections: local history; coal camp house replica; period artifacts & memorabilia; photographs.
Hours & Admission Prices: May-Sept. by appointment. No charge.

EASTERN REGIONAL COAL ARCHIVES & CRAFT MEMORIAL LIBRARY, 600 Commerce St., Bluefield, WV 24701. Tel.: 304-325-3943. Fax: 304-325-3702.
E-mail: cml@mail.mln.lib.wv.us
Web Site: craftmemorial.lib.wv.us
Institution Type/Description: History Museum.
Collections: local history & culture; coal industry; photographs.
Facilities: library; archives.
Hours & Admission Prices: Archives: by appointment. Library: Mon.-Thurs. 9:30-7, Fri.-Sat. 9:30-5. Closed holidays.

THE SCIENCE CENTER OF WEST VIRGINIA, 500 Bland St., Bluefield, WV 24701-4257. Tel.: 304-325-8855. Fax: 304-324-0513.
Web Site: www.sciencecenterwv.com

Founded: 1994.
Congressional District: 3
Key Personnel: Chm. (V) & Pres. (V), Patty Wilkinson; Dir., Thomas Willmiten; Museum Shop Mgr. & Administrative Asst., Pam R. Lester.
Personnel Profile: Full-Time Paid 2; Part-Time Paid 6; Part-Time Volunteers 4.
Governing Authority: nonprofit organization. Parent Institution: Alliance for the Arts, Ltd. Tax-exempt.
Institution Type/Description: Science Center/Museum.
Collections: interactive science exhibits.
Facilities: 10,000 sq. ft. exhibit space.
Hours & Admission Prices: June-Aug. Tues.-Sat. 10-5; Sept.-May Tues.-Thurs. 9-3, Sat. 10-4. Adults $5; discounts to groups & ASTC members; children under 2 & members no charge. Closed holidays. &
Attendance: 20,000 (estimated)
Membership: Individual $45; Family $65; Supporters $100; Benefactors $1,000.

Bramwell

COAL HERITAGE TRAIL INTERPRETIVE CENTER, 100 Station Sq., Bramwell, WV 24715. Mailing Address: P.O. Box 103, Bramwell, WV 24715. Tel.: 304-248-8595.
Web Site: www.coalheritage.org
Congressional District: 3
Key Personnel: Dir., Richard Bullins.
Personnel Profile: Full-Time Paid 1; Full-Time Volunteers 2.
Governing Authority: Parent Institution: Coal Heritage Highway Authority. Tax-exempt.
Institution Type/Description: Mining Museum: housed in a former train depot.
Collections: local history & coal heritage; mining artifacts & tool; miners & mine operators.
Facilities: Gift items for sale.
Activities: walking tour.
Publications: brochures; rack cards.
Hours & Admission Prices: Mon.-Sat. 10-4, Sun. 11-4. No charge; donations accepted. &

Capon Bridge

CAPON BRIDGE MUSEUM, Rte. 50, Capon Bridge, WV 26711. Tel.: 304-856-2661.
Institution Type/Description: History Museum: housed in the former dental office of Dr. Gardner.
Collections: local history & culture; photographs; period furnishings; personal artifacts; early clothing.
Hours & Admission Prices: Call for hours.

Ceredo

CEREDO MUSEUM, 501 Main St., Ceredo, WV 25507. Mailing Address: P.O. Box 691, Ceredo, WV 25507. Tel.: 304-453-3025.
Governing Authority: Parent Institution: Ceredo Historical Society.
Institution Type/Description: History Museum.
Collections: local history & culture; period furnishings; personal artifacts; Civil War artifacts; school memorabilia; N gauge railroad layout; books; photographs; genealogy; handblown glass.
Hours & Admission Prices: Tues. & Thurs. 9-4. &

RAMSDELL HOUSE, 1108 B St., Ceredo, WV 25507. Mailing Address: P.O. Box 446, Ceredo, WV 25507-0446. Tel.: 304-453-2482.
Institution Type/Description: Historic House: housed in the home of Union Capt. Z.D. Ramsdell; built in 1858.
Collections: Civil War records & memorabilia; period artifacts.
Activities: school tours; special events.
Hours & Admission Prices: By appointment. &

Charleston

* **CLAY CENTER FOR ARTS & SCIENCES WEST VIRGINIA, (M),** One Clay Sq., Charleston, WV 25301-2424. Tel.: 304-561-3552. Fax: 304-561-3598.
E-mail: info@avampatodiscoverymuseum.org
Web Site: www.theclaycenter.org
Formerly: Avampato Discovery Museum
Founded: 1961.
Congressional District: 3
Key Personnel: Chm., Melvin Jones; C.E.O. & Pres., Judith Wellington; C.F.O., Rebecca Gillespie; Dir. Exhibits & Art Cur., Barbara Racker; Dir.

Art & Science Education, Lewis Ferguson; Dir. Mktg. & Communications, Traci West-McCombs; Dir. Performing Arts, Lakin Cook; Museum Shop Mgr., Megan Douglas.
Personnel Profile: Full-Time Paid 51; Part-Time Paid 10; Part-Time Volunteers 175.
Governing Authority: private; nonprofit organization. Tax-exempt: 501(c)(3).
Institution Type/Description: Performing & Visual Arts & Science Museum.
Collections: 19th- & 20th-century American painting; sculpture; works on paper.
Research Fields: American art.
Facilities: GEMS teacher training center; large format theater; planetarium; cafe; classrooms; 2 performance halls.
Activities: guided tours; lectures; films; educational programs; workshops; temporary & permanent exhibitions; classes; special events; demonstrations; performances.
Publications: quarterly newsletter.
Hours & Admission Prices: Wed.-Sat. 10-5, Sun. 12-5; groups by appointment. Adults $7, senior citizens, teachers & children $5.50; discounts to groups & AAM members; museum members, children under 3 & ASTC members no charge. Closed national holidays. &
Attendance: 180,000 (accurate)
Membership: Individual Plus $55; Family $75; Film Lovers $100; Contributor $125; Art Enthusiasts $150; Supporter $300; Patron $600; Bronze $1,000; Silver $1,500; Gold $2,500; Platinum $5,000.

CRAIK-PATTON HOUSE, 2809 Kanawha Blvd. E., Charleston, WV 25311-1727. Mailing Address: P.O. Box 175, Charleston, WV 25321. Tel.: 304-925-5341. Facebook: Craik-Patton House.
E-mail: info@craik-patton.org
Web Site: www.craik-patton.org
Founded: 1986.
Congressional District: 2
Key Personnel: Dir., Brianne Jackson.
Personnel Profile: Full-Time Paid 1; Part-Time Paid 1; Part-Time Volunteers 10; Interns 1.
Governing Authority: Tax-exempt.
Institution Type/Description: Historic House Museum: built by James Craik in 1834; owned by George Smith Patton, a leader in the Confederate Army & grandfather of General George S. Patton.
Collections: southern West Virginia history & culture; period furnishings; gardens.
Facilities: picnic area.
Activities: rental facilities.
Hours & Admission Prices: Mon.-Fri. 10-4, Sat.-Sun. by appointment. No charge; donations accepted.
Attendance: 1,000 (accurate)
Membership: Individual $35.

WEST VIRGINIA STATE MUSEUM, (M), 1900 Kanawha Blvd., E., Charleston, WV 25305-0009. Tel.: 304-558-0220. Fax: 304-558-2779. TDD: 304-558-3562.
E-mail: charles.w.morris@wv.gov
Web Site: www.wvculture.org
Founded: 1894.
Congressional District: 3
Key Personnel: Commissioner, Randall Reid-Smith; Dir. West Virginia State Museum, Charles Morris; Dir. Archives & History, Joe Geiger; Coord. Exhibits, Betty Gay; Dir. Historic Preservation, Susan Pierce; Editor-Goldenseal, John Lilly; Dir. West Virginia Independence Hall, Travis Henline; Dir. Grave Creek Mound Historic Site, David E. Rotenizer; Dir. Museum in the Park, Elizabeth Williams; Dir. Camp Washington Carver, James Hess; Cur., James Mitchell.
Personnel Profile: Full-Time Paid 96; Part-Time Paid 3; Part-Time Volunteers 12; Interns 4.
Governing Authority: state. Parent Institution: State of West Virginia, WV Div. of Culture & History. Tax-exempt.
Institution Type/Description: Cultural Center; History Museum.
Collections: history & culture of West Virginia; archaeological artifacts; historic costumes & textiles; settlement & Civil War artifacts; firearms & edged weapons; early industrial tools & products; WV. Regimental flags; 19th century household goods & furnishings; fine arts; historic & contemporary WV folk arts & crafts; television newsfilm; WV travel & documentary films; numismatic & philatelic collections; archival collections of manuscripts, public documents, state & county records, photographs; Boyd Stutler Civil War collection; C.O. Ericksen coin & currency collection.
Major Exhibits: West Virginia 150, 1/13-1/14.
Research Fields: West Virginia history; state & regional genealogy; colonial America; Civil War; regional military history; Appalachian culture.

Facilities: historical research library; theater; video studio; meeting & conference rooms.
Activities: tours; arts festivals highlighting traditional music, arts & crafts, jazz, dance & theater; conferences on historic preservation, literary arts & arts management; marketing programs include WV. Juried Exhibition & Performing Arts showcase; concerts; films; recitals; theatrical productions; outreach programs & Arts & Humanities & Historic Preservation grants programs; oral interviews.
Publications: quarterly, checklist of State Publications; quarterly, Folklife Magazine; annual, Grants Brochure; annual, West Virginia History; periodic Elderberry books and records.
Hours & Admission Prices: Tues.-Sat. 9-5, Sun. 12-5. No charge. Closed Christmas; most holidays. &
Attendance: 120,000 (accurate)

Clifftop

WEST VIRGINIA CAMP WASHINGTON-CARVER, Rte. 41 S., Clifftop, WV 25831. Mailing Address: HC 35, Box 5, Clifftop, WV 25831-9601. Tel.: 304-438-3005. Fax: 304-438-3006.
E-mail: campwashingtoncarver@wvculture.org
Web Site: www.wvculture.org/sites/carver.html
Founded: 1979.
Key Personnel: Facility Mgr., George Sheaves.
Personnel Profile: Full-Time Paid 3.
Governing Authority: state. Parent Institution: West Virginia Dept. of Culture and History, Cultural Center, Capitol Complex, Charleston, WV 25305. Tax-exempt.
Institution Type/Description: Historic Site: Park Museum/Visitor Center; camp built as first 4-H Camp for blacks in U.S.
Collections: Historic Building: 1939-1940 chestnut building designed by J.R. Strock & built by the WPA.
Research Fields: history of Camp Washington-Carver.
Facilities: 5,500 sq. ft. open free space; 4 indoor exhibition spaces; indoor & outdoor performance areas.
Activities: changing exhibitions; performances; craft demonstrations.
Hours & Admission Prices: June-Oct. Mon.-Fri. call for appointment. No charge. &
Attendance: 10,000 (estimated)

Elizabeth

BEAUCHAMP NEWMAN MUSEUM, Court St., Elizabeth, WV 26143. Mailing Address: P.O. Box 621, Elizabeth, WV 26143-0621. Tel.: 304-275-3569.
Founded: 1950.
Key Personnel: Regent, Carole Menefee
Institution Type/Description: Historic House.
Collections: furniture; costumes; domestic artifacts.
Activities: guided tours.
Hours & Admission Prices: Holidays & by appointment. No charge; donations accepted.

Elkins

STIRRUP GALLERY, Davis and Elkins College, Myles Center for the Arts, Elkins, WV 26241-3971. Tel.: 304-637-1341. Fax: 304-637-1238.
E-mail: morganw@elkins.edu
Web Site: www.davisandelkins.edu
Formerly: Darby's Prehistoric and Early Pioneer's Art Museum
Key Personnel: Coord., Mark Lanham
Institution Type/Description: History Museum.
Collections: early pioneer history & culture; period clothing; personal artifacts.
Activities: school tours.
Hours & Admission Prices: Mon.-Fri. 9-5; other times by appointment. &

Fairmont

MARION COUNTY HISTORICAL SOCIETY MUSEUM, INC., 210 Adams St., Fairmont, WV 26554-2826. Mailing Address: P.O. Box 1636, Fairmont, WV 26555-1636. Tel.: 304-367-5398.
E-mail: marionhistorical@yahoo.org
Web Site: www.marionhistorical.org
Founded: 1986.
Key Personnel: Pres. (V), Dora Kay Grubb; Museum Shop Mgr., Betty Andrews.
Personnel Profile: Part-Time Paid 4; Part-Time Volunteers 2; Interns 1.
Governing Authority: Parent Institution: Marion County Historical Society, Inc. Tax-exempt.

Institution Type/Description: Historical Society Museum: housed in the Marion County Sheriffs' home from 1913-1985. Listed on the National Historic Register of Landmarks.
Collections: county history & culture; period artifacts; documents; historical papers.
Research Fields: Civil War; Marion County History.
Facilities: Museum-related items for sale.
Activities: school tours. Museum Sponsors: Founders' Month; WV's Birthday Historic Home Tour; Julie Pierpont Day in May; Holiday Historic Home Tour in November & December.
Publications: quarterly newsletter.
Hours & Admission Prices: Mon.-Sat. 10-2; tours by appointment. No charge; donations accepted.
Attendance: 2,000 (accurate)
Membership: Individual $10; Family $15.

PRICKETTS FORT, Rte. 3, Fairmont, WV 26554-9470. Mailing Address: Rte. 3, Box 407, Fairmont, WV 26554. Tel.: 304-363-3030. Fax: 304-363-3857.
E-mail: info@prickettsfort.org
Web Site: www.prickettsfort.org
Founded: 1976.
Congressional District: 1
Key Personnel: C.E.O., Greg Bray; Chm., Junie Mayle.
Personnel Profile: Full-Time Paid 3; Part-Time Paid 7; Part-Time Volunteers 50; Interns 2.
Governing Authority: not-for-profit organization. Parent Institution: Pricketts Fort Memorial Foundation. Tax-exempt.
Institution Type/Description: Historic Site & House: c.1859 Job Prickett House & 18th century civilian refuge fort.
Collections: artifacts of the late 18th & 19th centuries.
Facilities: 500-vol. library; 400-seat theater. Museum-related items for sale.
Activities: guided tours; concerts; education programs for children; participatory exhibits; theater; docent program; training program for professional museum workers; School of the Longhunter; reenactments.
Publications: quarterly newsletter, Gatepost; catalogues.
Hours & Admission Prices: mid-April to May & Sept.-Oct. Wed.-Sat. 10-5, Sun. 12-5; Memorial Day to Labor Day Mon.-Sat. 10-5, Sun. 12-5. Adults $8, senior citizens $6, children 6-12 $4; discounts to Frontiers to Mountaineers, tour groups, AAA & CAA members; members & children under 6 no charge. &
Attendance: 13,652 (accurate)
Membership: Individual $25; Family $35; Sustaining $75.

French Creek

WEST VIRGINIA STATE WILDLIFE CENTER, Rte. 20 S., French Creek, WV 26218. Mailing Address: P.O. Box 38, French Creek, WV 26218-0038. Tel.: 304-924-6211. Fax: 304-924-6781.
E-mail: eugene.r.thorn@wv.gov
Web Site: www.dnr.state.wv.us/wvwildlife/wildlifectr.htm
Founded: 1923.
Congressional District: 13
Key Personnel: Dir. WV Div. Natural Resources, Frank Jezioro; Biologist, Eugene Thorn; Museum Shop Mgr., Joann Kruk.
Personnel Profile: Full-Time Paid 5; Part-Time Paid 6.
Governing Authority: state. Parent Institution: WV Div. of Natural Resources, 324 4th Ave., South Charleston, WV 25303.
Institution Type/Description: Zoo.
Collections: animals & birds native to West Virginia.
Research Fields: wildlife.
Facilities: natural environmental enclosures.
Activities: guided tours; special events.
Publications: descriptive brochure.
Hours & Admission Prices: April & Sept.-Oct. daily 9-5; May-Aug. daily 9-6; Nov.-March daily 9-3. Adults $3, children 3-15 $1.50; discounts to groups & Golden Mountaineer members; children under 3 no charge. Annual passes also available. &
Attendance: 40,000 (estimated)
Membership: Individual $10; Family $25; Friends & Family $40.

Grafton

ANNA JARVIS BIRTHPLACE MUSEUM, Rte. 119/250 S., Grafton, WV 26354. Mailing Address: 3576 Webster Pike, Grafton, WV 26354-9643. Tel.: 304-265-5549.
E-mail: ajhouse26354@yahoo.com
Web Site: ww.annajarvismuseum.com
Founded: 1994.
Congressional District: 42

Key Personnel: Dir., Olive Dadisman.
Personnel Profile: Full-Time Volunteers 2; Part-Time Volunteers 8.
Volunteer Hours: 6,223
Operating Expenses: 31,311
Operating Income: 29,679
Governing Authority: Parent Institution: Thunder on the Tygart. Tax-exempt.
Institution Type/Description: Historic House Museum: housed in the birthplace of Anna Jarvis, the founder of Mother's Day.
Collections: family history; personal artifacts; period furnishings; photographs.
Hours & Admission Prices: April-Dec. Sun.-Mon. 10-3. Adults $5; discounts to AAA, AAM & ICOM members; children under 6 no charge.
Attendance: 500 (estimated)

INTERNATIONAL MOTHERS DAY SHRINE AND MUSEUM, 11 E. Main St., Grafton, WV 26354-1322. Mailing Address: P.O. Box 513, Grafton, WV 26354-0513. Tel.: 304-265-1589.
E-mail: info@mothersdayshrine.com
Web Site: www.mothersdayshrine.com/
Institution Type/Description: Historic Church: built in 1873. Listed on the National Register of Historic Places.
Collections: religious artifacts.
Facilities: Museum-related items for sale.
Activities: weddings; tour groups; rental facility.
Hours & Admission Prices: April 15-Oct. 15 Fri.-Sat. 10-4, Sun. 12-4; groups & other times by appointment. No charge; donations accepted. Closed holidays.

Green Bank

NATIONAL RADIO ASTRONOMY OBSERVATORY - GREEN BANK SCIENCE CENTER, Rte. 28/92, Green Bank, WV 24944-0002. Mailing Address: P.O. Box 2, Green Bank, WV 24944-0002. Tel.: 304-456-2011. Fax: 304-456-2229.
Institution Type/Description: Science Center.
Collections: science & technology hands-on exhibitions.
Facilities: cafe. Museum-related items for sale.
Activities: group tours; educational programs.
Hours & Admission Prices: Memorial Day to Labor Day daily 8:30-7; Sept.-Oct. Thurs.-Mon. 8:30-7. Closed New Year's Eve & Day; Easter; Thanksgiving; Christmas Eve & Day.

Harpers Ferry

THE APPALACHIAN TRAIL CONSERVANCY, 799 Washington St., Harpers Ferry, WV 25425-6587. Mailing Address: P.O. Box 807, Harpers Ferry, WV 25425-0807. Tel.: 304-535-6331. Fax: 304-535-2667.
E-mail: info@appalachiantrail.org
Web Site: www.appalachiantrail.org
Founded: 1925.
Congressional District: 2
Key Personnel: Exec. Dir., David N. Startzell; Chm. (V), Bob Almand; Treas., Kennard Honick; Dir. Devel., Royce Gibson; Foundation & Corp. Rels. Mgr., Amy McCormick; Gift Shop Mgr., Laurie Potteiger.
Personnel Profile: Full-Time Paid 40; Part-Time Paid 6; Part-Time Volunteers 15.
Governing Authority: nonprofit organization. Tax-exempt: 501(c)(3).
Institution Type/Description: Appalachian Trail Maintenance Association: housed in 1892 building.
Collections: natural resources & historical development of the volunteer-maintained Appalachian trail, including original maps & planning documents; library & memorabilia of regional planner E. Benton MacKay; archives of the Appalachian Trail Conference; photographs; publications.
Research Fields: state-by-state inventories of rare, threatened & endangered species within 500 ft. of the 2,160 mile Appalachian Trail; legislative history of the Appalachian National Scenic Trail & public & private cooperative management of it.
Facilities: 1,275-vol. library; 400 sq. ft. exhibit space; archival research area. Trail related items for sale.
Activities: loan & temporary exhibitions.
Publications: 6-times yearly "A.T. Journeys"; quarterly publication, The E-Register.
Hours & Admission Prices: Daily 9-5. No charge; donations accepted. Closed New Year's Day; Washington's Birthday; Columbus Day; Thanksgiving & day after; Christmas. &
Attendance: 29,000 (estimated)
Membership: Senior Citizen, Student or Affiliated Club (Individual) $30; Individual Adult $35; Supporting Organization $100; Life $1,000.

* **HARPERS FERRY NATIONAL HISTORICAL PARK, (M),** Fillmore St., Harpers Ferry, WV 25425. Mailing Address: P.O. Box 65, Harpers Ferry, WV 25425-0065. Tel.: 304-535-6224. Fax: 304-535-6244.
Web Site: www.nps.gov/hafe/home.htm
Founded: 1944.
Congressional District: 2
Key Personnel: Rgnl. Dir., Steve Whitesell; Supt., Rebecca L. Harriett; Chief Ranger, Jeffrie Woods; Cultural Resources Mgr., Mia Parsons; Museum Shop Mgr., Deborah Piscitelli; Lands Management Asst., Andrew Lee; Education Specialist, Catherine Bragaw; Visitor Svcs., Todd P.H. Bolton; Volunteers & Outreach, Jessica Liptak.
Personnel Profile: Full-Time Paid 99; Part-Time Paid 9; Part-Time Volunteers 769.
Governing Authority: federal. Parent Institution: National Park Service, U.S. Dept. of Interior, Washington, DC 20240.
Institution Type/Description: National Historic Park: approx. 56 restored buildings & Civil War fortifications.
Collections: artifacts specific to historical period; Harpers Ferry Arms; John Brown; Civil War; 19th-century African American education; 19th-century water-powered technology. Historic Buildings: 1775 Harper House; 1858 Master Armorer's House; 1826 Stagecoach Inn; exhibits pertaining to Storer College, Archeology, Restoration and Industry, Wetlands & Jefferson Rock.
Research Fields: John Brown raid; Civil War; industry; social; transportation; African American history; 19th-century America.
Facilities: 2,000-vol. research library specific to John Brown, Civil War, African American history and the industrial/transportation and social history of Harpers Ferry; visitor center. Gift items for sale.
Activities: Special tours offered daily Memorial Day-Labor Day on John Brown's Raid, Civil War, Stonewall's Brilliant Victory, C&O Canal Walk, Harpers Ferry Marsh Walk, Voices from Harpers Ferry and the Guns of Harper's Ferry. Seasonally scheduled Living History Exhibits & Interpretation. Education programs for schools throughout the year with reservations. Harpers Ferry Historical Association Sponsors: Elderhostel programs. Hiking & walking trails include Virginius Island Trail, Hamilton St., Jefferson Rock Trail, Maryland Heights Trail and intersecting at Harpers Ferry are the Appalachian Trail and the C&O Canal.
Hours & Admission Prices: Summer: daily 8-6; Winter: daily 8-5. Admission: $6 per car or $4 per person 17-62 (7 day pass); seniors and children 16 & under no charge. Annual Pass: $25. The Interagency Pass available and honored. &
Attendance: 350,000 (estimated)
Membership: Family $25.

JOHN BROWN WAX MUSEUM INC., 168 High St., Harpers Ferry, WV 25425. Mailing Address: P.O. Box 1417, Harpers Ferry, WV 25425. Tel.: 304-535-6342.
Web Site: www.johnbrownwaxmuseum.com
Formerly: National Historical John Brown Wax Museum
Founded: 1964.
Congressional District: 16
Governing Authority: private corporation.
Institution Type/Description: Wax Museum: housed in c.1820 historic town building.
Collections: wax figures & scenes; authentic costumes.
Research Fields: John Brown's life from boyhood to gallows.
Hours & Admission Prices: mid-March to mid-Dec. daily 9-5; call for additional hours. Adults $7, senior citizens 60 & up $6, child 6-12 $5; discounts to groups of 10 or more; children under 6 no charge.

Helvetia

HELVETIA MUSEUM, HC 76, Helvetia, WV 26224-9999. Mailing Address: HC 76 Box 200, Helvetia, WV 26224-0042. Tel.: 304-924-5455.
Founded: 1969.
Congressional District: 12
Key Personnel: Pres. & Chm. (V), Eleanor L. Betler; Mgr., Bruce Cressler.
Personnel Profile: Part-Time Volunteers 5.
Governing Authority: society. Parent Institution: Helvetia Restoration & Devel. Assoc. Tax-exempt.
Institution Type/Description: Historical & Preservation Society: housed in 1871 Betler Cabin.
Collections: Swiss immigration to West Virginia memorabilia; agricultural tools used in Helvetia; furniture of Swiss design.
Research Fields: Swiss immigration to West Virginia.
Activities: guided tours by appointment; gallery talks; permanent exhibitions.

Hours & Admission Prices: May-Sept. Sat.-Sun. 12-4; other times by special arrangement. No charge. &

Attendance: 500 (estimated)

Membership: Annual $2; Life $20.

Hillsboro

PEARL S. BUCK BIRTHPLACE FOUNDATION, Rte. 219 N., Hillsboro, WV 24946. Mailing Address: P.O. Box 126, Hillsboro, WV 24946-0126. Tel.: 304-653-4430.

E-mail: info@pearlsbuckbirthplace.com

Web Site: www.pearlsbuckbirthplace.com

Founded: 1974.

Congressional District: 2

Key Personnel: Pres., Gail Ratliff; Exec. Dir., Timothy VanReenen.

Personnel Profile: Full-Time Paid 2; Part-Time Paid 1.

Governing Authority: nonprofit. Tax-exempt: 501(c)(3).

Institution Type/Description: Historic House Museum: Sydenstricker house was Pearl S. Buck's fathers' childhood home; Stulting House, built by Pearl S. Buck's maternal ancestors & birthplace of Pearl S. Buck.

Collections: memorabilia of Pearl S. Buck; original family furniture in restored rooms; original manuscripts of Pearl S. Buck, housed at West Virginia Wesleyan College, available by appointment only. Historic Building: restored 19th century barn & machinery.

Research Fields: genealogical research into Pearl S. Buck ancestry.

Facilities: Pearl S. Buck books & West Virginia crafts for sale.

Activities: guided tours. Museum Sponsors: Pearl S. Buck Birthday celebration; Author's Day Ceremony.

Hours & Admission Prices: May-Oct. Mon.-Sat. 9-4:30. Adults $6, senior citizens $5, students $1; children under 6 no charge. Closed New Year's Day; Thanksgiving; Christmas. &

Attendance: 1,993 (accurate)

Membership: Individual $5; Sustaining $10; Organizational $15; Supporting $25; Contributing $50; Donor $100; Patron $500; Life $1,000.

Hinton

CAMPBELL FLANNAGAN MURRELL HOUSE, 422 Summer St., Hinton, WV 25951-2221. Mailing Address: P.O. Box 1504, Hinton, WV 25951-1504. Tel.: 304-466-1401.

Web Site: cfm-fmh.org

Founded: 1990.

Key Personnel: Pres., Dwight Emrich

Governing Authority: Tax-exempt.

Institution Type/Description: Historic House Museum: built in 1875. Listed on the National Register of Historic Places.

Collections: local history & culture; period furnishings; personal artifacts; period fashion doll by Pete Ballard.

Publications: biannual newsletter.

Hours & Admission Prices: Call for hours. No charge; donations accepted.

Attendance: 234 (accurate)

Membership: Single $15; Family $25; Life $1,000.

HINTON RAILROAD MUSEUM, 206 Temple St., Hinton, WV 25951-2331. Tel.: 304-466-5420. Fax: 304-466-5420.

Key Personnel: Dir., Dorothy Jean Boley

Institution Type/Description: Railroad Museum.

Collections: C&O Railway history; photographs; tools; books; uniforms.

Facilities: Museum-related items for sale.

Activities: school tours; special events.

Hours & Admission Prices: Summer: Mon.-Sat. 10-4. Winter: Mon.-Sat. 10-2. No charge; donations accepted. &

VETERANS MEMORIAL MUSEUM, 419 Ballengee St., Hinton, WV 25951. Mailing Address: P.O. Box 694, Hinton, WV 25951-0694. Tel.: 304-466-4443.

Institution Type/Description: Military History Museum.

Collections: military & war history; uniforms; military vehicles; General MacArthur's footlocker; photographs; personal artifacts.

Hours & Admission Prices: May-Nov. Fri.-Sat. 12-4; other times by appointment.

Huntington

BENJY'S HARLEY-DAVIDSON, 408 4th St., Huntington, WV 25701-1315. Tel.: 304-523-1340. Fax: 304-523-5474.

E-mail: info@benjyshd.com

Institution Type/Description: Motorcycle Museum.

Collections: motorcycles from 1916 to present.

Facilities: diner. Museum-related items for sale.

Activities: tours.

Hours & Admission Prices: Mon.-Thurs. 10-6, Fri. 10-9, Sat. 10-2:30. No charge.

BIRKE ART GALLERY - MARSHALL UNIVERSITY, Dept. of Art & Design, One John Marshall Dr., Huntington, WV 25755. Tel.: 304-696-2296.

Institution Type/Description: Art Gallery.

Collections: paintings; sculpture.

Hours & Admission Prices: Mon. 10-4 & 6-8, Tues.-Fri. 10-4.

COLLIS P. HUNTINGTON RAILROAD HISTORICAL SOCIETY, INC., 1323 8th Ave., Huntington, WV 25701-2919. Mailing Address: P.O. Box 393, Huntington, WV 25708-0393. Tel.: 866-639-7487; 304-523-0364. Fax: 304-523-0366. Facebook: New River Train.

E-mail: newrivertrain@aol.com

Web Site: www.newrivertrain.com

Formerly: Huntington Railroad Museum

Founded: 1959.

Key Personnel: Pres., Duane Legg; Vice Pres., Brian Cavender; Dir., Chris Lockwood; Dir., David Webb; Dir., Eugene Bosh; Dir., Skip Reinhard; Dir., Ernie Clay.

Personnel Profile: Full-Time Paid 2; Part-Time Paid 2; Part-Time Volunteers 40.

Governing Authority: private; nonprofit organization. Charter of National Railway Historical Society. Indoor: 1323 8th Ave., Huntington, WV 25701. Tax-exempt: 501(c)(3).

Institution Type/Description: Transportation Museum.

Collections: Chesapeake & Ohio 1949 steam locomotive #1308 (listed on the National Register of Historic Places) & 1950 caboose; 1960s CSX diesel locomotive cab; 1920s working pump handcar on track; lifesaver caboose; over 600 books on railroads; videos; railroad artifacts & photographs.

Facilities: 600-vol. library of railroad books, photographs, and 250-vol. video & DVDs.

Activities: guided tours; operating Garden G gauge railroad; railroad excursions. Museum Sponsors: Tri-state Railroad Show in March; Model Railroad Show in April; New River Train Excursions in October.

Publications: newsletter, Gondola Gazette.

Hours & Admission Prices: Open on call all year for groups & out of town visitors. No charge; donations accepted.

Attendance: 6,800 (estimated)

Membership: Students $26; Individual $60; Family $68.

GEOLOGY MUSEUM, Marshall University, One John Marshall Dr., Huntington, WV 25755. Mailing Address: Dept. of Geology, 176 Science Bldg., Huntington, WV 25755. Tel.: 304-696-6720. Fax: 304-696-3243.

Founded: 1837.

Key Personnel: Chm., Dr. William Niemann; Geology Dept., Dr. Ronald L. Martino; Geology Dept., Dr. Aley El-Shazly.

Governing Authority: university. Parent Institution: Marshall University.

Institution Type/Description: Geology Museum.

Collections: geology; minerals; rocks & fossils.

Activities: guided tours by appointment; permanent exhibitions.

Hours & Admission Prices: Mon.-Fri. 8-4:30. No charge. Closed university holidays. &

HERITAGE FARM MUSEUM & VILLAGE, 3300 Harvey Rd., Huntington, WV 25704-9112. Tel.: 304-522-1244. Fax: 304-523-6115.

E-mail: hfmv@comcast.net

Web Site: www.heritagefarmmuseum.com

Founded: 2002.

Congressional District: 3

Key Personnel: Chm. (V), A. Michael Perry; Pres. (V), Henrietta Perry

Institution Type/Description: History Museum.

Collections: farm history; personal artifacts; period furnishings; 15 restored buildings; period farm equipment; 1917 Model T; steam engines; 1908 electric truck; 1910 Sears Hi Wheeler.

Facilities: Museum-related items for sale.

Activities: tours; corporate retreats; educational outings; special events; petting zoo parties; weddings; bed & breakfast inns.

Hours & Admission Prices: March-Nov. Mon.-Sat. 10-3. Guided Tours: adults 13-64 $8, senior citizens 65 & over $7, children 3-12 $6; discounts to groups of 15 or more; children 2 & under no charge. Closed holidays. &

Attendance: 5,000 (estimated)

✱ **HUNTINGTON MUSEUM OF ART, INC., (M),** 2033 McCoy Rd., Huntington, WV 25701-4999. Tel.: 304-529-2701. Fax: 304-529-7447.
Web Site: www.hmoa.org
Founded: 1947.
Congressional District: 4
Key Personnel: Pres. Bd. Trustees, Brandon Roisman; Exec. Dir., Margaret Mary Layne; Dir. Finance, Billie Marie Karnes; Dir. Library, Chris Hatten; Sr. Cur., Jenine Culligan; Dir. Education, Katherine Cox; Dir. Public Rels., John Gillispie; Dir. Horticulture, Mike Beck, Ph.D.; Exec. Asst., Judy Clark; Dir. Devel., Carolyn Bagby; Dir. Facilities, Matt Matney; Appeals & Events Admin., Sandra Stone; Museum Shop Mgr., Patsy Lansaw.
Personnel Profile: Full-Time Paid 24; Full-Time Volunteers 3; Part-Time Paid 9; Part-Time Volunteers 600; Interns 3.
Governing Authority: nonprofit organization. Tax-exempt: 501(c)(3).
Institution Type/Description: Art Museum & Conservatory Gardens: includes 50 acres of woodlands & the C. Fred Edwards Conservatory.
Collections: 19th- & 20th-century American & European paintings, drawings, prints & sculpture; Appalachian folk art; American decorative arts & furniture; Ohio River Valley glass; American & European studio glass & art glass; Herman P. Dean firearms; English Georgian silver; Near Eastern decorative arts & furniture; Islamic prayer rugs; pre-Columbian pottery & artifacts; British Portraits; antique firearms; landscape sculpture; children's gallery.
Facilities: 20,000-vol. art research library; 3,000 sq. ft. plant conservatory; reading room; 1 1/2 miles of marked nature trails; 300-seat auditorium; open air stage.
Activities: guided & audio tours; lectures; films; gallery talks; concerts; arts festivals; drama; formally organized education programs for children & adults; docent program or council; inter-museum loan, permanent & temporary exhibitions.
Publications: member's magazine; exhibition catalogues; collection catalogues.
Hours & Admission Prices: Tues. 10-9, Wed.-Sat. 10-5, Sun. 12-5. Special Exhibitions: adults $5; discounts to SEMC & AAM members; members no charge. Closed New Year's Day; Independence Day; Thanksgiving; Christmas. &
Attendance: 45,448 (accurate)
Membership: Friend $25-$49; Donor $50-$99; Contributing $100-$249; Sponsor $250-$499; Benefactor $500-$999. Presidents Club: Bronze Door $1,000-$1,499; Silver Door $1,500-$2,499; Emerald Door $2,500-$3,749; Gold Door $3,750-$4,999; Platinum Door $5,000-$9,999; Diamond Door $10,000 & up.

MADIE CARROLL HOUSE PRESERVATION SOCIETY, INC., 234 Guyan St., Huntington, WV 25702-1526. Mailing Address: P.O. Box 3266, Huntington, WV 25702-0266. Tel.: 304-736-1655.
E-mail: knnance@comcast.net
Web Site: www.madiecarrollhouse.org
Founded: 1988.
Congressional District: 3
Key Personnel: Pres. (V), Johnny Nance; Vice Pres., Mary Jo Martin; Recording Sec., Karen N. Nance; Cur., Greg Miller; Corresponding Sec., Easter Miller; Treas., Robert Edmunds.
Personnel Profile: Part-Time Volunteers 25.
Governing Authority: nonprofit organization. Tax-exempt.
Institution Type/Description: Historic House Museum: built c.1810.
Collections: period furnishing; personal artifacts; archives.
Activities: bimonthly meetings; special events. Annual Events: Easter Event; Fall Heritage Event in October; Civil War Event in November; Christmas Event in December.
Publications: annual newsletter; books, Historic Madie Carroll House; Guyandotte Bicentennial; Civil War Diary of C. F. Ropes.
Hours & Admission Prices: By appointment. No charge; donations accepted.
Attendance: 2,000 (estimated)
Membership: Family $5; Supporting $25; Contributing $50; Benefactor & Corporate $100 & up.

MUSEUM OF RADIO & TECHNOLOGY, INC., 1640 Florence Ave., Huntington, WV 25701-4546. Tel.: 304-525-8890.
Web Site: www.mrtwv.org
Founded: 1992.
Key Personnel: Pres., Geoffrey Bourne; Vice Pres., Dave Bond; Treas., Judy Taylor; Sec., David Spears; Librarian, Bill Reich; Education, Fred Crews; Public Rels., Garry Ritchie; Public Rels., Jack Woodrum.
Personnel Profile: Full-Time Volunteers 2; Part-Time Volunteers 28.
Governing Authority: private; nonprofit organization. Tax-exempt: 501(c)(3).
Institution Type/Description: Technology Museum: housed in c.1931 school building.
Collections: working amateur radio station & communications display; radio & related artifacts; A.C. Gilbert toys; recreation of a 1920s radio store & shop; West Virginia Broadcasters Hall of Fame.
Facilities: library; 250-seat auditorium; classrooms & lab; 250-seat large screen theater; 10,000 sq. ft. exhibit space. Museum-related items for sale.
Activities: docent program; films; formal education programs for adults; guided tours; hobby workshops; lectures; participatory & temporary exhibitions; training programs for professional museum workers; community activities. Annual Events: auctions; banquets.
Publications: bimonthly newsletter, Radio Museum News.
Hours & Admission Prices: Jan.-April call for hours; May-Dec. Fri.-Sat. 10-4, Sun. 1-4. No charge; donations accepted. &
Attendance: 5,000 (estimated)
Membership: Institutional $25.

Lesage

JENKINS PLANTATION MUSEUM, 8814 Ohio River Rd., Lesage, WV 25537. Mailing Address: c/o West Virginia Division of Culture and History, 1900 Kanawha Blvd. E., Charleston, WV 25305-0300. Tel.: 304-762-1059.
Institution Type/Description: History Museum: housed in the former home of Captain William Jenkins and General Albert Gallatin Jenkins; built in 1863.
Collections: local history & culture; period furnishings; photographs; personal artifacts; military artifacts.
Hours & Admission Prices: Temporarily closed.

Lewisburg

CARNEGIE HALL MUSEUM, 105 Church St., Lewisburg, WV 24901-1303. Tel.: 304-645-7917. Fax: 304-645-5228.
E-mail: info@carnegiehallwv.com
Web Site: www.carnegiehallwv.com
Key Personnel: Exec. Dir., Susan Adkins; Artistic Dir., Lynn Creamer
Institution Type/Description: Art Museum.
Collections: works by local & regional artists.
Facilities: Museum-related items for sale.
Activities: temporary exhibits.
Hours & Admission Prices: Mon.-Fri. 9-4:30, Sat. 10-1. No charge. &

GREENBRIER HISTORICAL SOCIETY, INC. - NORTH HOUSE MUSEUM, (M), 301 W. Washington St., Lewisburg, WV 24901-1324. Tel.: 304-645-3398. Fax: 304-645-5201. Facebook: Greenbrier Historical Society.
E-mail: info@greenbrierhistorical.org
Web Site: www.greenbrierhistorical.org
Founded: 1963.
Congressional District: 3
Key Personnel: Dir., Elizabeth A. McMullen; Pres. (V), Margaret Hambrick; Treas., John B. Arbuckle, Jr.; Museum Coord., Toni Ogden; Archivist, James Talbert.
Personnel Profile: Full-Time Paid 1; Part-Time Paid 3; Part-Time Volunteers 30.
Volunteer Hours: 2,750
Governing Authority: nonprofit organization. Parent Institution: Greenbier Historical Society, Lewisburg, WV 24901. Tax-exempt: 501(c)(3).
Institution Type/Description: Historic House Museum: built c.1820.
Collections: Civil War & Indian artifacts; textiles; early tools; 18th- & 19th-century furnishings; period artifacts & vehicles, including Civil War; Greenbrier County artifacts; genealogy records.
Facilities: library of books & records; archives.
Activities: guided tours; educational programs & events; lectures; private group tours of house & downtown Lewisburg.
Publications: quarterly, Appalachian Springs; annual journal, Journal of the Greenbier Historical Society.
Hours & Admission Prices: Mon.-Sat. 10-4. No charge; donations accepted. Closed most major holidays. &
Attendance: 5,000 (estimated)
Membership: Junior $10; Adult $25; Family $35; Business $50; Life $400.

GREENBRIER MILITARY SCHOOL MEMORIAL MUSEUM, 400 N. Lee St., Lewisburg, WV 24901-1128. Mailing Address: P.O. Box 922, Lewisburg, WV 24901-0922. Tel.: 304-645-3247.
Web Site: www.gmsaa.org
Founded: 1982.
Congressional District: 3
Key Personnel: Cur., Mary Essig-Beatty.
Personnel Profile: Part-Time Paid 1; Part-Time Volunteers 1.
Governing Authority: Tax-exempt.

Institution Type/Description: Historical Museum.
Collections: school memorabilia & artifacts from 1812-1972.
Hours & Admission Prices: Mon.-Fri. No charge. Closed holidays. &
Attendance: 450 (estimated)

LOST WORLD CAVERNS VISITOR CENTER AND NATU-RAL HISTORY MUSEUM, 308 HC 34, Lewisburg, WV 24901. Tel.: 304-645-6677; 866-228-3778.
Web Site: www.lostworldcaverns.com
Institution Type/Description: Natural History Museum & Visitor's Center.
Collections: local natural history; dinosaur & fossil replicas; caverns.
Facilities: Museum-related items for sale.
Activities: cave tour; school groups; educational programs; gemstone mining.
Hours & Admission Prices: Jan.-March 1 Sat.-Sun. 10-4; March to May daily 10-5; Memorial Day to Labor Day daily 9-7; Sept.-Nov. 23 daily 9-5; Nov. 24-Dec. daily 10-4. Adults 13 & over $12, children 6-12 $6; discounts to groups; children under 6 no charge. Closed Thanksgiving; Christmas.

Logan

MUSEUM IN THE PARK, Chief Logan State Park, 376 Little Buffalo Creek Rd., Logan, WV 25601. Tel.: 304-792-7229.
Web Site: www.chiefloganstatepark.com/activities.html
Institution Type/Description: Park Museum.
Collections: local & regional history; photographs; early coal mining & railroad industries; period sewing machine; sewing implements.
Hours & Admission Prices: Wed.-Sat. 10-6, Sun. 1-6.

Lost Creek

WATTERS SMITH MEMORIAL STATE PARK & LIVING HISTORY MUSEUM, Duck Creek Rd., Lost Creek, WV 26385. Mailing Address: P.O. Box 296, Lost Creek, WV 26385-0296. Tel.: 304-745-3081. Fax: 304-745-3631.
Web Site: www.watterssmithstatepark.com
Founded: 1964.
Congressional District: 3
Key Personnel: Supt., Larry A. Jones.
Personnel Profile: Full-Time Paid 2; Interns 10.
Governing Authority: state. Owned by WV Dept. of Commerce, Div. of Parks & Recreation, 1800 Washington St., E., Charleston, WV 25305. Tel.: 304-558-2764. Tax-exempt.
Institution Type/Description: Historic House & State Park: located on site of 1876 Watters Smith Farm.
Collections: farm machinery; farm utensils; household tools & utensils; letters; bills; documents; buggies; manuscripts; historic house.
Facilities: visitor center; activity building; hiking trails; picnic area. Gift items for sale.
Activities: permanent exhibitions.
Hours & Admission Prices: Memorial Day-Labor Day daily 11-7. No charge; donations accepted. &

Lost River

LOST RIVER MUSEUM, Harper Barn, Rte. 259, Lost River, WV 26810. Mailing Address: P.O. Box 26, Lost River, WV 26810-0026. Tel.: 304-897-7242.
Congressional District: 2
Key Personnel: Dir., Dan W. Blumhagen, Ph.D.; C.E.O., Timothy Wheeler.
Personnel Profile: Part-Time Volunteers 20.
Governing Authority: Parent Institution: Lost River Educational Foundation. Tax-exempt.
Institution Type/Description: History Museum.
Collections: local history & culture; period furnishings; personal artifacts; early tools; photographs.
Hours & Admission Prices: Call for hours. No charge; donations accepted. &
Attendance: 400 (estimated)

Madison

BITUMINOUS COAL HERITAGE FOUNDATION MUSEUM, 347 Main St., Madison, WV 25130-1221. Tel.: 304-369-5180 & 9118. Fax: 304-369-9130.
E-mail: boonedevcorp@yahoo.com
Web Site: wvcoalmuseum.org
Key Personnel: Pres., Joy Underwood; Sec. & Treas., Larry V. Lodato
Institution Type/Description: History Museum.
Collections: southern WV coal fields; miner's tools; photographs; oral histories; company records; state mining history.

Hours & Admission Prices: Mon. Fri. 12 4.
Membership: Individual $15; Business & Professional $100; Corporate $250.

Mannington

WEST AUGUSTA HISTORICAL SOCIETY MUSEUM, 917 E. Main St., Mannington, WV 26582. Mailing Address: P.O. Box 414, Mannington, WV 26582. Tel.: 304-986-1252, 1298 & 2636.
Founded: 1971.
Key Personnel: Dir., Nany Ult; Chm. (V), Normz Wilcox; Pres. (V), Bevery Jones; Museum Shop Mgr., Esther Sturm.
Personnel Profile: Part-Time Paid 1; Part-Time Volunteers 25.
Volunteer Hours: 4,750
Governing Authority: Parent Institution: West Augusta Historical Society. Tax-exempt.
Institution Type/Description: Historical Society Museum: housed in the former Wilson School, built in 1912.
Collections: local history & culture; period furnishings; personal artifacts; photographs; musical instruments; early school artifacts. Historic Structures: 1912 caboose; 1870 log cabin; 1900s South Penn Oil Co. gas station; 1912 school house; 1912 round barn.
Activities: quilting. Museum Sponsors: Old Fashion Greenery Bazaar in December.
Hours & Admission Prices: May-Sept. Mon.-Fri. 9-2, Sun. 1-4; other times by appointment. No charges; donations accepted. &
Attendance: 230 (accurate)
Membership: Annual $5.

Marlinton

POCAHONTAS COUNTY MUSEUM, Seneca Trail, Marlinton, WV 24954. Mailing Address: 810 Second Ave., Marlinton, WV 24954-1091. Tel.: 304-799-4369. Fax: 304-799-6466.
E-mail: wpmcneel@gmail.com
Web Site: www.pocahontashistorical.org
Founded: 1962.
Congressional District: 3
Key Personnel: Pres. (V), Matt Tate; Librarian & Historian, William P. McNeel.
Personnel Profile: Part-Time Paid 1; Part-Time Volunteers 5.
Governing Authority: nonprofit organization. Affiliated with Pocahontas County Historical Society, Inc.
Institution Type/Description: Local History & Historical Society Museum: housed in the Frank & Anna Hunter Home.
Collections: history of Pocahontas County & surrounding area from the time of the Indians to present; historic photo collection; archives. Historic House: 1840s log cabin.
Research Fields: Pocahontas County History.
Facilities: 1,000-vol. local history library; reading rooms. West Virginia local crafts & books by West Virginian authors for sale.
Activities: school tours; meetings with programs on local history; art exhibits.
Publications: annual newsletter to members.
Hours & Admission Prices: Memorial Day-Labor Day Mon.-Sat. & holidays 11-5, Sun. 1-5. Families $10, adults $4, children 12-17 $2; discounts to groups; members & children under 12 no charge.
Attendance: 668 (accurate)
Membership: Regular $10.

Martinsburg

THE ARTS CENTRE, INC., 300 W. King St., Martinsburg, WV 25401-3202. Tel.: 304-263-0224. Fax: 304-263-0857.
Web Site: www.theartcentre.org
Key Personnel: Pres., Mary Lewis; Prog. Coord., Justis Saradji
Institution Type/Description: Art Museum.
Collections: works by local artists & craftsmen.
Activities: lectures; demonstrations; art classes.
Hours & Admission Prices: Call for hours.

BERKELEY COUNTY MUSEUM - BERKELEY COUNTY HISTORICAL SOCIETY, 136 E. Race St., Martinsburg, WV 25401-4310. Mailing Address: P.O. Box 1624, Martinsburg, WV 25401. Tel.: 304-267-4713.
E-mail: bchs@bchs.org
Web Site: bchs.org
Formerly: Belle Boyd House Museum - Berkeley County Historical Society
Founded: 1963.
Key Personnel: Pres. (V), Todd Funkhouser; Museum Shop Mgr., Nancy Crouse

Governing Authority: Tax-exempt.
Institution Type/Description: Historical Society Museum.
Collections: local history & culture; photographs; period furnishings; genealogies; commercial county history.
Research Fields: genealogy; architectural history; commercial history; significant events.
Facilities: library; archives.
Activities: research. Annual Event: Tour of Homes in September.
Publications: BCHS Newsletter; journals.
Hours & Admission Prices: Call for hours. Adults $5; children under 18 no charge.
Attendance: 1,546 (accurate)
Membership: Senior $15; Regular $20; Family $35; Patron $100 & up.

GENERAL ADAM STEPHEN HOUSE, 309 E. John St., Martinsburg, WV 25402. Mailing Address: General Adam Stephen Memorial Association, P.O. Box 1496, Martinsburg, WV 25402-1496. Tel.: 304-267-4434.
Web Site: www.orgsites.com/wv/adam-stephen
Founded: 1959.
Congressional District: 2
Key Personnel: Pres. (V), Martin Keesecker; Cur., Keith E. Hammersla.
Personnel Profile: Part-Time Paid 1; Part-Time Volunteers 12.
Governing Authority: municipal. Tax-exempt: 501(c)(3).
Institution Type/Description: Historic House: 1772-1789 Gen. Adam Stephen House.
Collections: period furniture.
Research Fields: Major Gen. Adam Stephen.
Facilities: library of letters, documents, pictures & maps available for use by appointment.
Activities: guided tours; lectures; temporary exhibitions.
Publications: booklets, General Adam Stephen, Founder; The Adam Stephen Memorial Project; single sheet, General Adam Stephen & His Home.
Hours & Admission Prices: May-Oct. Sat.-Sun. 2-5; other times by appointment. No charge; donations accepted. &

Attendance: 2,500 (estimated)
Membership: Individual $5; Family $10.

TRIPLE BRICK MUSEUM, 313 E. John St., Martinsburg, WV 25401. Mailing Address: c/o General Adam Stephen Memorial Association, P.O. Box 1496, Martinsburg, WV 25402. Tel.: 304-267-4434.
Web Site: www.orgsites.com/wv/adam-stephen
Institution Type/Description: History Museum: housed in a former apartment building; built c.1874.
Collections: local history & culture; period furnishings; personal artifacts; photographs; early surveying equipment; flax & wool spinning wheels; quilts; railroad artifacts.
Hours & Admission Prices: Call for hours.

Mathias

LOST RIVER STATE PARK, 321 Park Dr., Mathias, WV 26812-8088. Tel.: 304-897-5372. Fax: 304-897-5325.
E-mail: lostriversp@wv.gov
Web Site: www.lostriversp.com
Founded: 1968.
Congressional District: 2
Key Personnel: Park Supt., Mike Foster; Asst. Supt., Colby Caldwell.
Governing Authority: state. Affiliated with West Virginia Dept. of Commerce, Div. of Parks & Recreation, 324 4th Ave. S., Charleston, WV 25305. Tel.: 304-348-2764.
Institution Type/Description: Park Museum & Historic House: housed in Lee House Museum; built by Charles Carter Lee, as a summer retreat for son of Henry (Light Horse Harry) Lee; built in 1800.
Collections: old documents; clothing; tools.
Facilities: hiking trails; picnic area; restaurant. Gift items for sale.
Activities: guided tours. Park Sponsors: Heritage Weekend in September.
Hours & Admission Prices: Memorial Day-Labor Day Sat. 10-4, Sun.12-6. Office: Mon.-Fri. 8-4, Sat.-Sun. 10-3; other times by appointment. No charge.

Middlebourne

TYLER COUNTY HERITAGE & HISTORICAL SOCIETY, Dodd St., Middlebourne, WV 26149. Mailing Address: P.O. Box 317, Middlebourne, WV 26149-0317. Tel.: 304-758-2100 & 4288.
E-mail: tchandhs@verizon.net
Founded: 1994.

Key Personnel: Pres. (V), Ruth Moore; Chm. (V), Peggy Shields; Museum Shop Mgr., Lonnie Doak.
Personnel Profile: Part-Time Volunteers 5.
Governing Authority: Tax-exempt.
Institution Type/Description: Historic Building: housed in the former Tyler County High School; built in 1908.
Collections: Tyler County heritage; personal artifacts; documents; tools; military; genealogical books.
Facilities: Museum-related items for sale.
Publications: Heritage Windows; History of Tyler County Vol. 1 & 2; birth records; death records; marriage records; census records.
Hours & Admission Prices: May-Oct. Sun., Tues. & Thurs. 1-4; tours by appointment. No charge; donations accepted. &
Attendance: 1,200 (estimated)
Membership: Individual $10; Life $150.

Mineral Wells

NEW ERA ONE-ROOM SCHOOL - LIVING HERITAGE MUSEUM, 1838 Elizabeth Pike, Mineral Wells, WV 26150. Mailing Address: P.O. Box 340, Mineral Wells, WV 26150-0340. Tel.: 304-863-3583.
E-mail: lcarroll@access.k12.wv.us
Web Site: www.neweraoneroomschool.com
Governing Authority: Tax-exempt.
Institution Type/Description: History Museum: housed in one-room school built in 1884.
Collections: period furnishings; pot-bellied stove; desks; photographs.
Activities: school tours.
Hours & Admission Prices: May-Oct. Tues. 1-3; other times by appointment. No charge; donations accepted.

Morgantown

CHILDREN'S DISCOVERY MUSEUM OF WEST VIRGINIA, (M), 5000 Greenbag Rd., Morgantown, WV 26501. Tel.: 304-292-4646; 309-287-6271.
Governing Authority: Tax-exempt: 501(c)(3).
Institution Type/Description: Children's Museum.
Collections: hands-on exhibitions.
Facilities: Museum-related items for sale.
Activities: educational programs; special events; birthday parties.
Hours & Admission Prices: Tues.-Sat. 10-1. Admission $3.50; children under one no charge.

COOK-HAYMAN PHARMACY MUSEUM, 1132 Health Sciences North, 1st Fl., Morgantown, WV 26506. Mailing Address: P.O. Box 9500, Morgantown, WV 26506-9500. Tel.: 304-293-7806.
Institution Type/Description: Pharmacy Museum.
Collections: pharmacy history; pharmacists' period equipment; medicines; books.
Hours & Admission Prices: By appointment. &

CORE ARBORETUM, Monongahela Blvd., Rte. 7, WVU Evansdale Campus, Morgantown, WV 26506. Mailing Address: Dept. of Biology, P.O. Box 6057, Morgantown, WV 26506-6057. Tel.: 304-293-5201, ext. 31547. Fax: 304-293-6363.
E-mail: jweems@wvu.edu
Web Site: www.wvu.edu/biology/facility/arboretum.html
Founded: 1948.
Congressional District: 2
Key Personnel: Arboretum Specialist, Jonathan Weems.
Personnel Profile: Full-Time Paid 1; Part-Time Volunteers 3.
Governing Authority: state. Parent Institution: West Virginia University. Tax-exempt: 501(c)(3).
Institution Type/Description: Arboretum & Botanical Gardens.
Collections: botany.
Research Fields: botany; dendrology; plant ecology; population biology; native flora & ecosystems.
Facilities: 91-acres, 3.5 miles of woodland trails; small amphitheater.
Activities: lectures; guided tours by appointment; special wildflower tours in April; birdwalks in April & May.
Publications: brochures available on site; checklist of birds.
Hours & Admission Prices: Daily dawn-dusk. No charge.
Attendance: 30,000 (estimated)

EASTON ROLLER MILL, Easton Mill Rd., Morgantown, WV 26507-0127. Mailing Address: P.O. Box 127, Morgantown, WV 26507-0127. Tel.: 304-594-2290.
Governing Authority: Tax-exempt.
Institution Type/Description: History Museum: mill built in 1864. Listed on the National Register of Historic Places.
Collections: local history; early steam engine; roller mills used in grinding wheat; mill stones.
Activities: tours.
Publications: quarterly newsletter, annual proceedings.
Hours & Admission Prices: Call for hours. No charge; donations accepted.
Attendance: 250 (estimated)
Membership: Individual $15; Couple $25.

MONONGALIA ARTS CENTER, 107 High St., Morgantown, WV 26505-5412. Mailing Address: P.O. Box 239, Morgantown, WV 26507-0239. Tel.: 304-292-3325. Fax: 304-292-3326.
E-mail: info@monartscenter.com
Web Site: www.monartscenter.com
Founded: 1976.
Key Personnel: Exec. Dir., Ro Brooks; Gen. Admin. & Cur., Clint Fisher; Media & Adv. Coord., Lauren Riviello; Theatre Coord., Roger Banks; Museum Lobby Mgr., Danny Gibbons.
Governing Authority: Tax-exempt.
Institution Type/Description: Arts & Culture Center.
Collections: paintings.
Facilities: theater.
Activities: classes; community theater; art outreach.
Hours & Admission Prices: Mon.-Fri. 11-7, Sat. 11-4, Sun. by appointment. Closed major holidays. &
Membership: Individual & Artist $30; Family $60; MAC Friend $150; Sustaining $250; Patron $500; Benefactor $1,000.

MORGANTOWN HISTORY MUSEUM, 175 Kirk St., Morgantown, WV 26505. Tel.: 304-319-1800.
E-mail: morgantownmuseum@yahoo.com
Web Site: www.morgantownhistorymuseum.org
Founded: 2005.
Congressional District: 1
Key Personnel: Chm. (V), Pamela A. Ball.
Personnel Profile: Part-Time Paid 1; Part-Time Volunteers 12.
Governing Authority: municipal.
Institution Type/Description: History Museum.
Collections: local history, culture & industry; paintings; photographs; tools; glass.
Major Exhibits: Morgantown: 104 Years of Wild Blue Yondering, 1/21/14-9/30/14; Hometown Teams (T), 10/5/14-11/15/14.
Research Fields: history of Sterling Faucet Company & glass companies; development of Morgantown and Monongalia County.
Facilities: 3,600 sq. ft. exhibit space.
Activities: guided tours; films; formal education programs for adults & children; lectures.
Publications: biannual newsletter, Museum News.
Hours & Admission Prices: Tues.-Sat. 10-5. No charge; donations accepted. Closed New Year's Day; Memorial Day; Independence Day; Veterans Day; Thanksgiving; Christmas. &
Attendance: 3,450 (estimated)

THE ROYCE J. & CAROLINE B. WATTS MUSEUM, 334 Mineral Resources Bldg., Morgantown, WV 26506. Mailing Address: West Virginia University, Box 6070, Morgantown, WV 26506-6070. Tel.: 304-293-4609. Fax: 304-293-5708.
E-mail: wattsmuseum@mail.wvu.edu
Web Site: www.cemr.wvu.edu/wattsmuseum
Formerly: Comer Museum
Founded: 1990.
Congressional District: 2
Key Personnel: Dir., Danielle M. Petrak; C.E.O., Royce J. Watts.
Personnel Profile: Full-Time Paid 1; Part-Time Paid 1.
Governing Authority: public college; nonprofit. Parent Institution: West Virginia University. Tax-exempt: 501(c)(3).
Institution Type/Description: Industrial History Museum.
Collections: concentration on West Virginia with special emphasis on the coal, oil & natural gas industries & the social, technological & historical aspects of these industries.
Research Fields: social, cultural, & technological history of the coal, oil, & natural gas industries.

Facilities: classrooms, labs, amphitheater, 500 sq. ft. exhibit space; 300-seat auditorium.
Hours & Admission Prices: Mon., Wed. & Fri. 1-4; other times by appointment. No charge. &
Attendance: 2,000 (estimated)

WEST VIRGINIA UNIVERSITY-MESAROS GALLERIES, (M), Creative Arts Center, Evansdale Campus, West Virginia University, Douglas O. Blaney Lobby, Morgantown, WV 26506-6111. Mailing Address: P.O. Box 6111, Div. of Art, Morgantown, WV 26506-6111. Tel.: 304-293-4841, ext. 3210. Fax: 304-293-5731.
E-mail: bob.bridges@mail.wvu.edu
Web Site: artanddesign.wvu.edu/mesaros_galleries
Founded: 1968.
Congressional District: 1
Key Personnel: Dean & Dir., Bernard Schultz; Chm., Paul Krainik; Cur., Robert Bridges.
Personnel Profile: Full-Time Paid 1; Part-Time Paid 12.
Governing Authority: public university. Parent Institution: West Virginia University. Tax-exempt.
Institution Type/Description: College Art Gallery.
Collections: paintings; photographs; sculpture.
Facilities: classrooms; 200-seat auditorium; 2,200 sq. ft. exhibit space.
Activities: guided tours; lectures; formal education programs for undergraduate & graduate students in Art History & Studio Art. Annual Event: Visiting Artists Lecture Series.
Hours & Admission Prices: Mon.-Sat. 12-9:30. No charge. Closed university holidays. &
Attendance: 20,000 (accurate)

Moundsville

FOSTORIA GLASS MUSEUM, 511 Tomlinson Ave., Moundsville, WV 26041. Mailing Address: P.O. Box 826, Moundsville, WV 26041-0826. Tel.: 304-845-9188. Fax: 304-845-9188.
E-mail: fostoriaglassmuseum@frontier.com
Web Site: www.fostoriaglass.org
Founded: 1991.
Key Personnel: Dir. & Pres. (V), Jim Davis; Chm. & Museum Shop Mgr., Ralph C. Clark.
Personnel Profile: Part-Time Paid 1; Part-Time Volunteers 20.
Governing Authority: Parent Institution: Fostoria Glass Society of America. Tax-exempt.
Institution Type/Description: Glass Museum.
Collections: Fostoria glass; history of glass making.
Activities: Museum Sponsors: Annual Convention in June.
Publications: bimonthly member magazine, Facets of Fostoria.
Hours & Admission Prices: March-Nov. Wed.-Sat. 1-4. Suggested Donation: adults $6; members no charge. Closed holidays. &
Attendance: 1,000 (estimated)
Membership: Individual $25.

GRAVE CREEK MOUND ARCHAEOLOGICAL COMPLEX, 801 Jefferson Ave., Moundsville, WV 26041-2241. Mailing Address: P.O. Box 527, Moundsville, WV 26041-0527. Tel.: 304-843-4128. Fax: 304-843-4131.
E-mail: david.e.rotenizer@wv.gov
Web Site: www.wvculture.org/museum/GraveCreekmod.html
Formerly: Grave Creek Mound Historic Site
Founded: 1978.
Congressional District: 1
Key Personnel: Site Mgr., David E. Rotenizer.
Personnel Profile: Full-Time Paid 7; Part-Time Volunteers 5.
Governing Authority: Parent Institution: West Virginia Div. of Culture & History. Tax-exempt.
Institution Type/Description: Adena Culture Mound, Interpretive Museum, & State Cultural Facility.
Collections: West Virginia archaeology. Historic Site: 250-150 B.C. Grave Creek Mound.
Research Fields: West Virginia archaeology.
Facilities: 136-seat auditorium; 7 acre grounds; activity room.
Activities: permanent exhibitions; monthly art exhibits; cultural events; educational & other related programs. Annual Events: Archaeology Month Promotion; Student Art Show; West Virginia Day; Monthly Archaeological Lecture Series.
Hours & Admission Prices: Museum: Tues.-Sat. 9-5, Sun. 12-5. Mound & Gift Shop: Tues.-Sat. 9-4:30, Sun. 12-4:30. No charge; donations accepted. Closed holidays. &
Attendance: 30,000 (accurate)

MARX TOY MUSEUM, 915 2nd St., Moundsville, WV 26041-1422. Tel.: 304-845-6022.
Web Site: www.marxtoymuseum.com
Founded: 2001.
Key Personnel: Owner, Francis Turner.
Personnel Profile: Part-Time Paid 2.
Institution Type/Description: Toy Museum.
Collections: history of Marx toys; toys from 1920s to 1980s.
Facilities: Museum-related items for sale.
Activities: special events.
Hours & Admission Prices: April-Dec. Tues.-Sat. 11-5; other times by appointment. Adults $8.50, seniors $7.50, students $5; discounts to groups; children under 6 no charge. &

WEST VIRGINIA PENITENTIARY, 818 Jefferson Ave., Moundsville, WV 26041-2235. Tel.: 304-845-6200. Fax: 304-843-4146.
E-mail: cas@wvpentours.com
Web Site: www.wvpentours.com
Founded: 1995.
Key Personnel: Pres. (V), Sid Grisell; Dir., Paul Kirby; Museum Shop Mgr., Tom Stiles.
Personnel Profile: Full-Time Paid 2; Part-Time Paid 12; Part-Time Volunteers 14.
Governing Authority: Tax-exempt: 501(c)(3).
Institution Type/Description: Penitentiary Museum.
Collections: penitentiary history; Justice System; Deathhouse.
Facilities: Museum-related items for sale.
Activities: guided tours; night tours. Museum Sponsors: Elizabethtown Festival in May; Ghost Hunts; Dungeon of Horrors.
Hours & Admission Prices: April-Nov. Tues.-Sun. 11-4. Adults $10; discounts to senior citizens, school & church groups and US military. Closed holidays.
Attendance: 30,000 (accurate)

Mullens

TWIN FALLS STATE PARK & MUSEUM, RR 97, Mullens, WV 25882. Mailing Address: P.O. Box 667, Mullens, WV 25882-0667. Tel.: 304-294-4000. Fax: 304-294-5000.
E-mail: twinfallsinfo@wv.gov
Web Site: www.twinfallsresort.com
Founded: 1976.
Key Personnel: Supt., A. Scott Durham.
Governing Authority: society. Affiliated with the Wyoming County Women's Club, Box 706, Pineville, WV. 24874. Tax-exempt.
Institution Type/Description: State Park & Historic House: 1920 Old Severt Home.
Collections: furniture; household items; tools; books; clothing & artifacts of a local & general historical interest.
Activities: guided tours.
Publications: brochure.
Hours & Admission Prices: Memorial Day-Labor Day daily 10-6. No charge; donations accepted.

New Cumberland

HANCOCK COUNTY MUSEUM, 1008 Ridge Ave., New Cumberland, WV 26047-9501. Mailing Address: P.O. Box 672, New Cumberland, WV 26047-0672. Tel.: 304-564-4800 & 374-4884. Fax: 304-387-1427.
E-mail: kellerjfonf@hotmail.com
Founded: 2002.
Congressional District: 1
Key Personnel: Dir. & Pres. (V), Vivian Weigle; Pres. (V), Robert McNeil.
Volunteer Hours: 1,000
Operating Expenses: 12,550
Operating Income: 2,680
Governing Authority: bd. of directors. Tax-exempt: 501(c)(3).
Institution Type/Description: Historic House Museum: housed in the Victorian home of Oliver Sheridan Marshall, past president of the WV State Senate; built in 1887.
Collections: period furnishings; personal artifacts.
Activities: rental facilities; tours.
Hours & Admission Prices: Sun. 12:30-4; other times by appointment. No charge; donations accepted. &
Attendance: 500 (estimated)

Nitro

NITRO WORLD WAR I BOOMTOWN MUSEUM, 302 21st St., Nitro, WV 25143-1738. Mailing Address: P.O. Box 308, Nitro, WV 25143-0308. Tel.: 304-759-0200.
E-mail: nitrowarmuseum@yahoo.com
Web Site: nitrowarmuseum.com
Institution Type/Description: Military Museum.
Collections: military memorabilia; Nitro bungalow living room replica; historical artifacts.
Hours & Admission Prices: Fri. 10-4, Sat. 9-12.

Oceana

WYOMING COUNTY HISTORICAL MUSEUM, Cook Pkwy. & Logan St., Oceana, WV 24870. Mailing Address: P.O. Box 2041, Oceana, WV 24870. Tel.: 304-682-7448 & 5096.
E-mail: jimcook@jetbroadband.com
Web Site: wyomingcountymuseum.webs.com
Founded: 2008.
Congressional District: 3
Key Personnel: Dir., Jim Cook; C.E.O., Jesse Womack.
Governing Authority: Tax-exempt.
Institution Type/Description: Historical Society Museum.
Collections: local history & culture; period furnishings; personal artifacts; photographs.
Hours & Admission Prices: Sat. 12-4; other times by appointment. No charge; donations accepted. &
Attendance: 315 (accurate)

Parkersburg

THE BLENNERHASSETT MUSEUM OF REGIONAL HISTORY, 137 Juliana St., Parkersburg, WV 26101-5331. Tel.: 304-420-4800.
Web Site: www.blennerhassettislandstatepark.com/museum.html
Founded: 1983.
Institution Type/Description: History Museum.
Collections: period Indian tools; jewelry; weapons; household items; oil paintings; early clothing; guns; military artifacts; furniture; farm implements.
Hours & Admission Prices: Call for hours. Adults $3. &

HENRY COOPER CENTENNIAL CABIN MUSEUM, 2522 Grand Ave., Parkersburg, WV 26101. Mailing Address: P.O. Box 11527, Charleston, WV 25339.
Institution Type/Description: Historic Building Museum: housed in a two-story log house built by early settler, Henry Cooper; c.1805. Listed on the National Register of Historic Places.
Collections: local history & culture; period furnishings; personal artifacts; clothing; photographs; button collection.
Hours & Admission Prices: Memorial Day to Labor Day Sun. 1:30-4:30.

OIL & GAS MUSEUM, 119 Third St., Parkersburg, WV 26101-5310. Mailing Address: P.O. Box 1685, Parkersburg, WV 26102-1685. Tel.: 304-485-5446.
E-mail: dlmckain@yahoo.com
Web Site: oilandgasmuseum.com
Founded: 1989.
Key Personnel: Dir., David L. McKain
Governing Authority: Parent Institution: Oil, Gas & Industrial History Assoc. Tax-exempt.
Institution Type/Description: History Museum.
Collections: oil & gas industry; general industry; Civil War; West Virginia & Parkersburg history; gas engine history; photographs; furniture; military history & artifacts; industrial artifacts; videos.
Facilities: visitor's center.
Publications: newsletter.
Hours & Admission Prices: Mon.-Fri. 11-4, Sat. 11-5, Sun. 12-5. Adults $3, children $1.
Attendance: 4,000 (estimated)
Membership: Annual $15.

PARKERSBURG ART CENTER, 725 Market St., Parkersburg, WV 26101-4628. Tel.: 304-485-3859. Fax: 304-485-3850. Facebook: The Parkersburg Art Center.
E-mail: info@parkersburgartcenter.org
Web Site: parkersburgartcenter.org

Formerly: The Cultural Center of Fine Arts
Founded: 1938.
Congressional District: 1
Key Personnel: Exec. Dir., Abby Hayhurst; Education & Membership Dir., Jessie Siefert; Event Coord., M.J. Ayson; Facility Coord., Dwain Hartley.
Governing Authority: private; nonprofit organization. Tax-exempt: 501(c)(3).
Institution Type/Description: Arts Center.
Collections: works of art including all media; American Realism collection.
Facilities: library of books and magazines pertaining to art; reading room; classrooms; video library-artists, art history. Paintings & crafts by local artists for sale.
Activities: guided tours; lectures; films; gallery talks; art festivals; hobby workshops; formally organized education programs for children and adults; inter-museum & temporary exhibitions; school loan service; photography & print, painting & drawing competitions.
Publications: monthly newsletter; exhibition catalogs & brochures.
Hours & Admission Prices: Wed.-Sat. 10-5, Sun. 1-5; groups by appointment. Admission $2; discounts for AAM members; members & children under 12 no charge. Closed national holidays. &
Attendance: 18,000 (estimated)
Membership: Individual or Senior Citizen $55; Family $100; Sustaining $250; Patron $650; Contributor $1,500; Benefactor $3,000-$9,999; Life $10,000.

SUMNERITE AFRICAN AMERICAN HISTORY MUSEUM, 1016 Avery St., Parkersburg, WV 26101-4727. Mailing Address: P.O. Box 4426, Parkersburg, WV 26104-4426. Tel.: 304-485-1152 & 422-0985.
Institution Type/Description: African American History Museum.
Collections: African American history & culture; photographs; art; period artifacts.
Activities: special events.
Hours & Admission Prices: By appointment.

VETERANS MUSEUM OF MID-OHIO VALLEY, 1829 7th St., Parkersburg, WV 26101-4250. Tel.: 304-420-0332 & 0337. Fax: 304-420-0337.
E-mail: veteransmuseum@hotmail.com
Web Site: veteransmuseumofmidohiovalley.com
Founded: 2002.
Key Personnel: Dir., Gary Farris; Pres. (V), Ron Salter; Administrative Asst., Rejeana Jackson.
Personnel Profile: Full-Time Paid 1; Part-Time Paid 2; Part-Time Volunteers 3.
Governing Authority: board of directors. Tax-exempt
Institution Type/Description: Veterans Museum.
Collections: military history; veterans' memorabilia; personal artifacts.
Facilities: library. Museum -related items for sale.
Activities: school tours; docent program; reunion headquarters; model building classes; medal replacement program for veterans; mail care packages to troops; video documentaries; family support program; scholarship program for high school seniors. Annual Events: Welcome Home Vietnam Veterans Day in March; Veterans Memorial Day Cookout in March; Veterans Day Open House in November; Veterans Day Cook Out at Veterans Memorial Park in November.
Publications: monthly newsletter.
Hours & Admission Prices: Mon.-Sat. 9-5, Sun. by appointment. Suggested Donation: adults $5, children $3; school groups & members no charge. Closed New Year's Day; Christmas. &
Attendance: 1,025 (estimated)
Membership: Annual $29.95; Lifetime $150.

Pence Springs

GRAHAM HOUSE, Rte 3 & 12 at Lowell, Pence Springs, WV 24962. Mailing Address: HC 73 Box 158, Pence Springs, WV 24962-9700. Tel.: 304-466-3321 & 716-6430.
E-mail: jbowling44@yahoo.com
Web Site: www.grahamhouse.org
Key Personnel: Pres. (V), James Bowling.
Personnel Profile: Part-Time Paid 3; Part-Time Volunteers 12.
Governing Authority: Parent Institution: Graham House Preservation Society. Tax-exempt.
Institution Type/Description: Historic House Museum: housed in the 2 story log home of Colonel James Graham; built in 1770.
Collections: period furnishings; personal artifacts.
Facilities: Museum-related items for sale.
Activities: school tours; special events. Museum Sponsors: Heritage Chili Festival in September.
Hours & Admission Prices: Memorial Day to Labor Day Sat. 11-5, Sun. 1-5; Sept.-May by appointment only. Adults $2, children $.50; discounts to school children. &

Attendance: 1,642 (estimated)
Membership: Individual $15; Family $25.

Pennsboro

OLD STONE HOUSE MUSEUM, 310 Myles Ave., Pennsboro, WV 26415-1329. Tel.: 304-643-2738.
Key Personnel: Pres., David Scott.
Governing Authority: Parent Institution: Ritchie County Historical Society.
Institution Type/Description: Historic House Museum: built c.1810. Listed on the National Register of Historic Places.
Collections: local history & culture; photographs; personal artifacts; period furnishings.
Facilities: genealogy library.
Hours & Admission Prices: By appointment.

Petersburg

TOP KICK'S MILITARY MUSEUM, 149 Army Ln., Petersburg, WV 26847. Mailing Address: P.O. Box 152, Petersburg, WV 26847-0152. Tel.: 304-257-1392.
E-mail: topkicks@hardynet.com
Web Site: www.topkicksmilitarymuseum.com
Founded: 1995.
Congressional District: 48
Key Personnel: Owner & Cur., Gereald W. Bland.
Governing Authority: private.
Institution Type/Description: Military Museum.
Collections: military history, vehicles, equipment, & uniforms; photographs; personal artifacts; Iraqi, Civil War, WWII & Vietnam artifacts; arrowheads; Gold Rock; Foot Rint in Rock.
Publications: brochure.
Hours & Admission Prices: March-Nov. Mon.-Sat. 9 am to dusk, Sun. 12 to dusk; other times by appointment. Adults $5; discounts to groups; children under 12 no charge. &
Attendance: 1,000 (estimated)

Philippi

ADALAND MANSION AND HISTORIC BARN, Adaland Rd., Philippi, WV 26416. Mailing Address: P.O. Box 74, Philippi, WV 26416-0074. Tel.: 304-457-1587 & 2415. Fax: 304-457-2703.
E-mail: info@adaland.org
Web Site: www.adaland.org
Founded: 1999.
Congressional District: 1
Key Personnel: Chm. (V), Dr. Ann Serafin; Pres. (V), Okey F. Gallien, Jr.
Personnel Profile: Full-Time Paid 1; Part-Time Paid 8; Part-Time Volunteers 20.
Volunteer Hours: 2,020
Operating Expenses: 74,980
Operating Income: 70,810
Governing Authority: Tax-exempt.
Institution Type/Description: Historic House Museum: housed in the former home of an early county sheriff & bank president. Listed on the National Register of Historic Places.
Collections: period furnishings; personal artifacts; photographs; law office; tools; equipment. Historic Building: 1850 barn.
Activities: demonstrations; special events; school tours.
Publications: newsletters.
Hours & Admission Prices: Call for hours. Adults $10, children under 12 no charge; discounts to AAA & AAM members. &
Attendance: 5,000 (accurate)

BARBOUR COUNTY HISTORICAL SOCIETY MUSEUM, 200 N. Main St., Philippi, WV 26416-1100. Tel.: 304-457-4846.
Founded: 1985.
Key Personnel: Pres. (V), Edgar Brown; Sec., Treas. & Museum Shop Mgr., Doretta Brown.
Personnel Profile: Part-Time Volunteers 5.
Governing Authority: bd. of directors. Parent Institution: City of Philippi. Tax-exempt.
Institution Type/Description: Historical Society Museum: housed in a renovated railway station.
Collections: Barbour County history & culture; Civil War memorabilia; Philippi covered bridges; Philippi mummies; J.W. Myers dynasty; B&O railroad.
Research Fields: genealogy; Civil War history.
Hours & Admission Prices: May-Oct. Fri.-Sat. 11-4, Sun. 1-4; other times by appointment. No charge; donations accepted.

Attendance: 4,000 (accurate)
Membership: Annual $5; Lifetime $50.

ONE ROOM CAMPBELL SCHOOL, Alderson-Broaddus College, 101 College Hill Dr., Philippi, WV 26416. Tel.: 304-457-6322. Fax: 304-457-6239.
E-mail: fettyma@ab.edu
Web Site: www.ab.edu
Founded: 1871.
Congressional District: 1
Key Personnel: Dir., Annette Fetty
Governing Authority: Parent Institution: Alderson-Broaddus College. Tax-exempt.
Institution Type/Description: Historic Building: c.1865.
Collections: local history & culture; period furnishings; early 20th century learning materials.
Research Fields: early education; one room school houses.
Hours & Admission Prices: Call for hours. No charge; donations accepted.
Attendance: 150 (estimated)

Point Pleasant

POINT PLEASANT RIVER MUSEUM, 28 Main St., Point Pleasant, WV 25550-1026. Mailing Address: P.O. Box 412, Point Pleasant, WV 25550-0412. Tel.: 304-674-0144.
E-mail: museum@pprivermuseum.com
Web Site: www.pprivermuseum.com/
Key Personnel: Exec. Dir., Jack Fowler; Pres., Clifford "Butch" Leport
Institution Type/Description: History Museum.
Collections: local history & culture; period furnishings; personal artifacts; photographs.
Hours & Admission Prices: Tues.-Fri. 10-3, Sat. 11-4, Sun. 1-5. Adults $5, children $2; Life members no charge.

TU-ENDIE-WEI STATE PARK, 1 Main St., Point Pleasant, WV 25550-1025. Mailing Address: P.O. Box 486, Point Pleasant, WV 25550-0486. Tel.: 304-675-0869. Fax: 304-674-6162.
Web Site: www.tu-endie-weistatepark.com
Formerly: Point Pleasant Battle Monument State Park
Founded: 1901.
Congressional District: 3
Key Personnel: Park Supt., Doug Wiant.
Personnel Profile: Full-Time Paid 1; Part-Time Paid 5.
Governing Authority: state. Owned by the state of West Virginia; operated by Dept. of Commerce with collections belonging to the D.A.R. Chapter from Point Pleasant Area. Tax-exempt.
Institution Type/Description: Historic House & Site: 1796, The Mansion House, built by Walter Newman, first hand-hewn log house built in Kanawha Valley, where the first battle of the Revolution was fought.
Collections: heirlooms, relics; large square piano; authentic four poster beds; monument commemorating the first battle of the Revolutionary War; Chief Cornstalk.
Activities: guided tours.
Hours & Admission Prices: May-Oct. Mon.-Sat. 10-4:30, Sun. 1-4:30; tour groups by appointment. No charge; donations accepted. &
Attendance: 40,000 (estimated)

WEST VIRGINIA STATE FARM MUSEUM, 1458 Fairground Rd., Point Pleasant, WV 25550-3421. Tel.: 304-675-5737. Fax: 304-675-5430.
E-mail: wvsfm@wvfarmmuseum.org
Web Site: www.wvfarmmuseum.org
Founded: 1974.
Key Personnel: Acting Dir., Lloyd Akers; Financial Dir. & Treas., Dennis Brumfield.
Personnel Profile: Full-Time Paid 2; Part-Time Paid 3; Part-Time Volunteers 40.
Governing Authority: private; nonprofit organization. Tax-exempt.
Institution Type/Description: Agriculture Museum & Historic Village: over 31 buildings; largest historical farm museum east of the Mississippi River.
Collections: houses & buildings built in early 1800s; period artifacts; period tractor & steam engines.
Facilities: restaurant. Local arts & crafts and museum-related items for sale.
Activities: guided tours; docent program; formal education programs for children & adults; festivals. Annual Events: Steam & Gas Engine Show in May; Country Fall Festival in October; Christmas Light Show in December.
Hours & Admission Prices: April-Nov. 15 Tues.-Sat. 9-5, Sun. 1-5. No charge; donations accepted. Closed major holidays. &
Attendance: 100,000 (estimated)

Membership: Annual $2; Lifetime $25.

Princeton

DR. ROBERT B. MCNUTT HOUSE MUSEUM, 1522 N. Walker St., Princeton, WV 24740. Tel.: 304-487-1502. Fax: 304-425-0227.
E-mail: pmccc@frontiernet.net
Web Site: www.pmccc.com/mcnutt_history.htm
Governing Authority: Parent Institution: Princeton Mercer County Chamber of Commerce.
Institution Type/Description: History Museum: housed in the former home & office of Dr. McNutt; built in 1840. Later served as a hospital and headquarters for Lt. Col. Rutherford B. Hayes and Sgt. William McKinley in May 1862. Listed on the National Register of Historic Places.
Collections: local history & culture; photographs; period furnishings.
Hours & Admission Prices: Mon.-Fri. 8-5. No charge.

PRINCETON RAILROAD MUSEUM, 99 Mercer St., Princeton, WV 24740. Tel.: 304-487-5060.
Institution Type/Description: Railroad Museum.
Collections: railroading history; railroad art & artifacts; photographs; over 100 lanterns.
Hours & Admission Prices: May-Aug. Tues.-Sat. 12-5, Sun. 2-5; Sept.-April Fri.-Sat. 12-5, Sun. 2-5; call to confirm. Adults $5, seniors 62 & over $3, childlren under 10 $2.

THOSE WHO SERVED WAR MUSEUM, 1500 W. Main St., Princeton, WV 24740-2627. Tel.: 304-487-3670.
Key Personnel: Pres., Tony Whitlow; Sec. & Treas., Bill Blankenship
Institution Type/Description: Military History Museum.
Collections: military history & artifacts; photographs; uniforms; Civil War; Revolutionary War; WWI & WWII; Korean War; Vietnam; Desert Storm; Spanish-American War.
Hours & Admission Prices: March-Nov. Mon.-Fri. 10-4. No charge; donations accepted.

Ravenswood

WASHINGTON'S WESTERN LANDS MUSEUM & SAYRE LOG HOUSE, Old Lock Bldg. #22, 220 Riverfront Park, Ravenswood, WV 26164. Mailing Address: Jackson County Historical Society, 220 Riverfront Park, P.O. Box 324, Ravenswood, WV 26164. Tel.: 304-273-3316.
Founded: 1970.
Congressional District: 2
Key Personnel: Pres., Jackson County Historical Society, Bryan Thompson; Dir. & Sec., Jackson County Historical Society, Nancy Burford.
Personnel Profile: Part-Time Volunteers 10.
Volunteer Hours: 310
Operating Expenses: 672
Operating Income: 3,102
Governing Authority: society. Parent Institution: Jackson County Historical Society. Tax-exempt.
Institution Type/Description: Historical Society Museum: located on site of Ravenswood Land, once owned by George Washington.
Collections: local tools; clothing; old photos; Indian artifacts. Historic Building: 1870 house.
Activities: guided tours; permanent exhibitions.
Hours & Admission Prices: May-Oct. Sat.-Sun. 1-5. No charge; donations accepted.
Attendance: 1,800 (accurate)

Romney

TAGGART HALL CIVIL WAR MUSEUM & VISITOR'S CENTER, 91 S. High St., Romney, WV 26757. Tel.: 304-822-4320.
Institution Type/Description: History Museum.
Collections: Civil War history & memorabilia; photographs; period furnishings; personal artifacts.
Hours & Admission Prices: Call for hours.

Rowlesburg

ROWLESBURG AREA HISTORICAL SOCIETY, Buffalo St., Rowlesburg, WV 26425. Mailing Address: P.O. Box 605, Rowlesburg, WV 26425-0605. Tel.: 304-454-9303.
Institution Type/Description: Historical Society Museum.
Collections: local history & culture; photographs; personal artifacts; tools.

Hours & Admission Prices: By appointment. No charge.

Saint Albans

C & O DEPOT MUSEUM, 404 Fourth Ave., Saint Albans, WV 25177. Mailing Address: 69 Central Ave., Saint Albans, WV 25177-2411. Tel.: 304-727-4439.
Institution Type/Description: Railroad Depot Museum: housed in the Chesapeake & Ohio Railroad Depot.
Collections: railroad history & artifacts; photographs.
Facilities: Museum-related items for sale.
Hours & Admission Prices: By appointment only.

MORGAN'S KITCHEN PLANTATION MUSEUM, Rte. 60 MacCorkle Ave., Saint Albans, WV 25177. Mailing Address: 404 4th Ave., Saint Albans, WV 25177-2829. Tel.: 304-727-2654.
Web Site: stalbanshistory.com
Founded: 1972.
Key Personnel: Pres. (V), Bill Dean
Institution Type/Description: Plantation Museum: built in 1846.
Collections: kitchen artifacts; utensils; period artifacts.
Activities: school tours.
Publications: quarterly newsletter for society members.
Hours & Admission Prices: Memorial Day to Labor Day Sun. 2-4; other times by appointment. No charge.
Attendance: 300 (estimated)
Membership: Individual $7; Family $10.

Scarbro

WHIPPLE COMPANY STORE & APPALACHIAN HERITAGE EDUCATIONAL MUSEUM, 7485 Okey L. Patterson Rd., Scarbro, WV 25917. Mailing Address: P.O. Box 150, Scarbro, WV 25917-0150. Tel.: 304-465-0331.
E-mail: whipple@whipplecompanystore.com
Web Site: www.whipplecompanystore.com
Institution Type/Description: Company History Museum.
Collections: artifacts & memorabilia pertaining to the West Virginia Coal Mining family.
Hours & Admission Prices: May-Oct. Wed.-Mon. 11-6. Tours: adults $6; children 4 & under no charge.

Shepherdstown

HISTORIC SHEPHERDSTOWN MUSEUM, (M), 129 E. German St., Shepherdstown, WV 25443. Mailing Address: P.O. Box 1786, Shepherdstown, WV 25443-1786. Tel.: 304-876-0910. Fax: 304-876-0910.
E-mail: hsc1786@gmail.com
Web Site: www.historicshepherdstown.com
Founded: 1983.
Congressional District: 2
Key Personnel: Pres., Suni Johnson; HSC Admin., Cheryl Gregory.
Personnel Profile: Part-Time Paid 1; Part-Time Volunteers 18; Interns 4.
Governing Authority: private; nonprofit organization. Parent Institution: Historic Shepherdstown Commission. Tax-exempt 501(c)(3).
Institution Type/Description: Historic Building: housed in the 1786 Entler Hotel.
Collections: concentration on West Virginia-Jefferson County history with emphasis on town of Shepherdstown; model of Rumsey steam boat.
Research Fields: genealogy.
Activities: teacher training workshops.
Publications: Shepherdstown Vol. I, II, & III; Walking Tour of Shepherdstown; books, Historic Shepherdstown; History of Shepherdstown.
Hours & Admission Prices: April-Oct. Sat. 11-5, Sun. 1-4. Suggested donation $4; discounts to students; members no charge.
Attendance: 2,200 (accurate)
Membership: Individual & Family $25-$49; Sustaining $50-$99; Sponsor $100-$249; Patron $250-$499; Benefactor $500-$999; Preservation $1,000 & up.

Shinnston

BICE-FERGUSON MEMORIAL MUSEUM, 400 Pike St., Shinnston, WV 26431-1406. Mailing Address: 40 Main St., Shinnston, WV 26431-1199. Tel.: 304-203-8917. Fax: 304-592-1597.
E-mail: bfmuseum@gmail.com

Web Site: bice-fergusonmuseum.com
Founded: 2006.
Congressional District: 1
Key Personnel: Dir., Maxine Weser; Chm. (V), Woody Maley.
Personnel Profile: Part-Time Paid 1.
Institution Type/Description: History Museum.
Collections: Shinnston history; glass plants; potteries; businesses; churches; schools; oil wells; prominent local people; telephones; telephone memorabilia from 1890-1960; period furnishings.
Activities: history programming.
Hours & Admission Prices: May-Nov. Fri.-Sun. 2-6; additional hours for special programs. No charge; donations accepted. Closed holidays. &
Attendance: 1,200 (accurate)
Membership: Individual $15; Family $30.

LEVI SHINN LOG HOUSE MUSEUM, US Rte. 19, Shinnston, WV 26431. Mailing Address: 602 Highland Ave., Shinnston, WV 26431-1026. Tel.: 304-592-5631.
Governing Authority: Parent Institution: Shinnston Historical Association.
Institution Type/Description: Historic House: built in 1778.
Collections: local history & culture; period furnishings; personal artifacts.
Activities: guided tours.
Hours & Admission Prices: Wed. by appointment.

South Charleston

GORBY'S MUSEUM OF MUSIC, 214 Seventh Ave., South Charleston, WV 25303. Tel.: 304-744-9452; 800-642-3070.
E-mail: info@gorbysmusic.com
Web Site: www.gorbysmusic.com
Institution Type/Description: Musical Instrument Museum.
Collections: musical instrument history; string, woodwind, & brass instruments.
Activities: school music groups.
Hours & Admission Prices: By appointment.

SOUTH CHARLESTON MUSEUM FOUNDATION, 311 D St., South Charleston, WV 25303-3105. Mailing Address: P.O. Box 18226, South Charleston, WV 25303. Tel.: 304-744-9711. Fax: 304-720-3769.
E-mail: museum@cityofsouthcharleston.com
Founded: 1989.
Congressional District: 2
Key Personnel: Dir., C.E.O. & Chm. (V), Robert Pratt; Pres. (V), Peggy Thompson; Corresponding Sec. & Museum Shop Mgr., Judy Romano; Treas., Bill Breese; Asst. Treas., Lindell Griffith.
Personnel Profile: Full-Time Volunteers 20; Part-Time Paid 1; Part-Time Volunteers 10.
Governing Authority: nonprofit organization. Tax-exempt: 501(c)(3).
Institution Type/Description: History Museum.
Collections: history of Kanawha Valley, South Charleston Mound, historic Midland Trail; films on West Virginia; Belgian glass artifacts; American Indian artifacts; Native American artifacts; chemical industry artifacts.
Research Fields: Native American heritage in South Charleston & the surrounding area; Belgian heritage in South Charleston & West Virginia.
Facilities: research archives; 300-seat theater; 400 sq. ft. exhibit space.
Activities: docent program; guided tours; lectures; loan, temporary exhibitions; storytelling; demonstrations. Museum Sponsors: Film Festivals; Belgian Heritage Festival; monthly WV Film Series.
Publications: books, History of South Charleston; Pictorial History of South Charleston.
Hours & Admission Prices: Mon.-Fri. 9-4; groups by appointment. Archives by appointment. No charge. Special events: adults $5; discounts to members. Closed major holidays. &
Attendance: 5,000 (estimated)
Membership: Individual $10; Family $25; Friend $50. Corporate & Club: Sustaining $100; Patron $250; Benefactor $500 & up.

Summersville

CARNIFEX FERRY BATTLEFIELD STATE PARK & MUSEUM, 1194 Carnifex Ferry Rd., Summersville, WV 26651-4911. Tel.: 304-872-0825; 800-225-5982. Fax: 304-872-3820.
E-mail: carnifexferrysp@wv.gov
Web Site: www.carnifexferrybattlefieldstatepark.com
Congressional District: 3
Key Personnel: Supt., Samuel Cowell.
Governing Authority: state. West Virginia Dept. of Commerce, Labor & Environmental Resources, West Virginia Commerce Division Tourism & Parks, Charleston, WV 25305. Tel. 304-348-2764. Tax-exempt.

Institution Type/Description: State Park & History Museum: housed in c.1855 frame house, built by Henry Patterson.
Collections: Civil War relics, memorabilia, utensils & implements of that era.
Facilities: picnic area.
Activities: Museum Sponsors: Battle Reenactment in September.
Hours & Admission Prices: Memorial Day to Labor Day Sat.-Sun. 10-5. No charge.

Terra Alta

HISTORY HOUSE - PRESTON COUNTY HISTORICAL SOCIETY, 109 E. Washington Ave., Terra Alta, WV 26764-1203. Mailing Address: 100 Richfield Lane, Terra Alta, WV 26764. Tel.: 304-379-6612.
Founded: 1958.
Key Personnel: Pres. (V), Dave Thomas; Vice Pres. & Cur., Edna Britton.
Governing Authority: Tax-exempt.
Institution Type/Description: Historical Society Museum.
Collections: county history; period furnishings; photographs; Bucklew collection; historic house.
Activities: school tours.
Publications: semiannual, Now...and Long Ago; 1979 Preston County History.
Hours & Admission Prices: June-Sept. Sun.; other times by appointment. No charge; donations accepted.
Membership: Annual $7.

RECKART'S MILL, 17 Reckart Mill Rd., RR 2, Terra Alta, WV 26764. Tel.: 304-789-2225.
Institution Type/Description: Historic Building: grist mill powered by a 20 ft. Filz water wheel; built in 1865.
Collections: mill history; French millstones.
Facilities: Museum-related items for sale.
Activities: school tours; special events.
Hours & Admission Prices: April-Oct. Sun.-Thurs. by appointment; Fri.-Sat. 10-4. &

Union

REHOBOTH CHURCH AND MUSEUM, Rte. 3, Union, WV 24983. Mailing Address: H.C. 83, P.O. Box 154, Union, WV 24983. Tel.: 304-772-3518.
Institution Type/Description: Historic Church: built in 1786.
Collections: local history & culture; religious furnishings & artifacts.
Hours & Admission Prices: April-Oct. Thurs.-Sat. 11-5, Sun. 1-5.

Wellsburg

BROOKE COUNTY HISTORICAL MUSEUM & CULTURE CENTER, 704 Charles St., Wellsburg, WV 26070. Tel.: 304-737-4060.
Founded: 1975.
Congressional District: 1
Key Personnel: Pres. (V), Vickey Gallagher
Institution Type/Description: History Museum: housed in the former G.C. Murphy 5 & 10 from 1920 to 1991.
Collections: Brooke County history from 1797 to present; business & industry; early education; pioneer life.
Hours & Admission Prices: April-Oct. Fri. & Sun. 1-5; other times by appointment. No charge. &

BROOKE COUNTY PUBLIC LIBRARY MUSEUM BRANCH, 945 Main St., Wellsburg, WV 26070. Tel.: 304-737-1551. Fax: 304-737-1010.
E-mail: bcpl@lycos.com
Web Site: wellsburg.lib.wv.us
Founded: 2002.
Congressional District: 1
Key Personnel: Chm., David Hubbard; Devel. & Public Rels., George Wallace; Education, Kimberly Harless; Treas., Edward Jackfert; Registrar, Doris Tennant; Cur., Mary Kay Wallace; Security, Jeff Cionni; Archivist, Jane Kraina; Museum Shop Mgr., Dorothy Craig.
Personnel Profile: Part-Time Paid 2; Part-Time Volunteers 2; Interns 1.
Governing Authority: public; nonprofit. Parent Institution: Brooke County Public Library, Wellsburg, WV. Tax-exempt: 501(c)(3).
Institution Type/Description: History Museum.
Collections: WWII history including Defenders of the Philippines 1941-1945; military artifacts; photographs; personal artifacts; medals; ribbons; citations; books; audiotapes; prints; maps; videos.

Research Fields: Japanese atrocities; slave laborers; hell ships; Bataan Death March; Philippine scouts; residual effects of captivity as a POW; compensation by Japanese companies who benefited by slave labor.
Facilities: 32,305-vol. library; 1,000 sq. ft. exhibit space; field research station; 60-seat theater. Museum-related items for sale.
Activities: films; formal education programs for adults, children, and undergraduate & graduate college students of Franciscan University of Steubenville, OH; lectures; theater. Annual Events: POW Recognition Day; Memorial Day; Veterans Day; End of WWII in August.
Publications: quarterly magazine, The QUAN.
Hours & Admission Prices: Mon.-Thurs. 10-7, Fri. 10-5, Sat. 10-4. No charge; donations accepted. Closed New Year's Day; Easter; Thanksgiving; Christmas. &
Attendance: 635 (accurate)

Weston

THE MOUNTAINEER MILITARY MUSEUM, 345 Center Ave., Weston, WV 26452-2030. Tel.: 304-472-3943 & 516-0800.
E-mail: mountaineermilitarymuseum@yahoo.com
Web Site: mountaineermilitarymuseum.com
Founded: 2003.
Key Personnel: Dir., Ron McVaney; C.E.O., Barbara McVaney
Governing Authority: Tax-exempt.
Institution Type/Description: Military Museum: housed in the historic Colored School of Weston.
Collections: military history; personal artifacts; photographs.
Activities: Annual Event: Memorial Day Weekend Vet-Together.
Hours & Admission Prices: Memorial Day to Oct. Fri.-Sat. 10-5; Fall & Winter: Sat. 10-5; other times by appointment. No charge; donations accepted. &
Attendance: 1,500 (estimated)

MUSEUM OF AMERICAN GLASS IN WEST VIRGINIA, 230 Main Ave., Weston, WV 26452. Mailing Address: P.O. Box 574, Weston, WV 26452-0574. Tel.: 304-269-5006. Fax: 304-269-5006.
E-mail: wvmuseumofglass@aol.com
Web Site: wvmag.bglances.com
Founded: 1991.
Key Personnel: C.E.O., Dean Six; Pres. (V), Helen S. Jones.
Personnel Profile: Part-Time Paid 1.
Governing Authority: Tax-exempt.
Institution Type/Description: Glass Museum.
Collections: history of glass, tools, machines, glass factories & workers, 1900-1940.
Research Fields: glass; union.
Facilities: Museum-related items for sale.
Activities: special events. Annual Event: Glass Conference.
Publications: magazine, All About Glass.
Hours & Admission Prices: Mon.-Tues. & Thurs.-Sat. 12-4. No charge; donations accepted.
Attendance: 2,000 (estimated)
Membership: Annual $25; Supporting $35; Sustaining $50; Patron $100; Benefactor $500.

WVU JACKSON'S MILL FARMSTEAD, WVU Jackson's Mill State 4-H Conference Center, 160 WVU Jackson Mill Rd., Weston, WV 26452-8011. Tel.: 800-287-8206, ext. 7012. Fax: 304-269-3409.
E-mail: jacksons.mill@mail.wvu.edu
Web Site: www.jacksonsmill.wvu.edu
Formerly: Jackson's Mill Historic Area
Founded: 1968.
Congressional District: 3
Key Personnel: Heritage Program Specialist, Dean Hardman.
Personnel Profile: Full-Time Paid 3; Part-Time Paid 1; Part-Time Volunteers 10.
Governing Authority: university. Affiliated with West Virginia University. Tax-exempt.
Institution Type/Description: Historic Building: housed on the former boyhood home of Stonewall Jackson.
Collections: agricultural tools & artifacts; grist & saw milling equipment; blacksmith shop. Historic Buildings: 1796 Blaker's Mill (grinding wheat & corn); c.1800 Mary Conrad Cabin; c.1700 McWhorter Cabin; c.1800 Jackson's Mill.
Facilities: Museum-related items for sale.
Activities: guided tours; permanent exhibitions; hands-on activities.
Hours & Admission Prices: April-May & Sept.-Oct. Thurs.-Sun. 10-5; Memorial Day-Labor Day Tues.-Sun. 10-5. Call for admission prices. &

Attendance: 60,000 (estimated)
Membership: Jackson's Mill Heritage Foundation $10.

Wheeling

CHILDREN'S MUSEUM OF THE OHIO VALLEY, 1000 Main St., Wheeling, WV 26003. Tel.: 304-214-5437. Fax: 304-214-5437.
E-mail: cmovkids@gmail.com
Web Site: www.cmovkids.org
Founded: 2000.
Key Personnel: Dir., Patricia Croft; Pres., Amelia Parsons.
Personnel Profile: Full-Time Paid 1; Part-Time Paid 2; Part-Time Volunteers 6; Interns 2.
Governing Authority: Tax-exempt.
Institution Type/Description: Children's Museum.
Collections: hands-on interactive exhibits; science; art; literature; history.
Hours & Admission Prices: Tues.-Sat. 10-5, Sun. 12-5. Child $4, adult $2; discounts to ACM members & groups of 10 or more; members no charge.
Attendance: 7,500 (estimated)
Membership: ACCESS $55; Family $85; Reciprocal $125.

THE ECKHART HOUSE, 810 Main St., Wheeling, WV 26003-2530. Tel.: 888-700-0118; 304-232-5439.
E-mail: gfigaretti@gmail.com
Web Site: www.eckharthouse.com
Founded: 1990.
Institution Type/Description: Historic House Museum: housed in the former home of banker, George W. Eckhart, Jr.; built in 1892. Listed on the National Register of Historic Places.
Collections: local history; period furnishings.
Facilities: Museum-related items for sale.
Activities: guided tours, tea luncheons.
Hours & Admission Prices: Tours: May-Dec. Fri.-Sat. 1-3; other times by appointment. Tea Luncheons: May-Dec. Sat. by appointment. Tours: $3.50 per person. Luncheons: $15 per person.

KRUGER STREET TOY & TRAIN MUSEUM, 144 Kruger St., Wheeling, WV 26003-5158. Tel.: 304-242-8133; 877-242-8133 (Toll Free). Fax: 304-242-1925.
E-mail: museum@toyandtrain.com
Web Site: www.toyandtrain.com
Founded: 1998.
Congressional District: 1
Key Personnel: C.E.O. & Pres., Allan R. Miller; Cur., James M. Schulte; Museum Shop Mgr., Liz Hastings.
Personnel Profile: Full-Time Paid 2; Part-Time Paid 7.
Governing Authority: Parent Institution: The Eibel Corporation.
Institution Type/Description: Toy Museum: housed in a Victorian-era schoolhouse.
Collections: research & development; materials related to toy manufacturer; drawings; prototypes; mock-ups; display units; factory samples; toys; model trains (O, HO, G, lego scale & HO slot cars); restored railroad caboose.
Major Exhibits: Brick Works, 1/1/14-6/1/14; Hello Kitty, 6/1/14-9/1/14.
Facilities: rental facilities. Museum-related items for sale.
Activities: tours; interactive exhibits; student programs; scout badge programs; meetings; birthday parties. Annual Events: Annual West Virginia Mego Meet in June; Marx Toy & Train Collectors National Convention in June.
Publications: newsletter, to museum members.
Hours & Admission Prices: Jan.-May Fri.-Sun. 9-4; Memorial Day to New Year's Eve daily 9-4. Adults $10, senior citizens 65 & over $7.50, children 4-17 $5; discounts to AAA & active military; children 3 & under & members no charge. Closed New Year's Day, Easter, Thanksgiving, Christmas.
Attendance: 20,000 (estimated)
Membership: Individual $20; Family $50.

❋   THE MUSEUMS OF OGLEBAY INSTITUTE - MANSION MUSEUM & GLASS MUSEUM, (M), The Burton Center, Rte. 88, Oglebay Resort, Wheeling, WV 26003. Mailing Address: 1330 National Rd., Wheeling, WV 26003-5706. Tel.: 304-242-7272. Fax: 304-242-7287.
Web Site: www.oionline.com
Founded: 1930.
Congressional District: 1
Key Personnel: Dir., Christin Stein Byrum; Pres., Kathleen McDermott; Chm. (V), Beth Weaver; Asst. Dir., Mary Coffman; Cur., Lindsey Davis; Cur. Glass, Holly McCluskey.

Personnel Profile: Full-Time Paid 4; Part-Time Paid 10; Part-Time Volunteers 90; Interns 2.
Governing Authority: nonprofit organization. Parent Institution: Oglebay Institute. Tax-exempt: 501(c)(3).
Institution Type/Description: Decorative Arts & History Museum: housed in 1846 brick farm house renovated to be a mansion at the turn of the century.
Collections: Wheeling and Midwestern glass; Anglo-American china; Wheeling china, 1839-1912; period rooms, 1740-1850; decorative arts; history; manuscripts.
Research Fields: local history; glass; ceramics; decorative arts; pottery.
Facilities: 750-vol. library of local history & decorative arts books by appointment; 75-seat public programming area. Museum publications for sale.
Activities: guided tours; lectures; gallery talks; formally organized education programs for children & adults; permanent & temporary exhibitions.
Publications: Wheeling Glass (1829-1939) Collection of the Oglebay Institute Glass Museum.
Hours & Admission Prices: Feb.-March Sat.-Sun. 10-5; April-Oct. daily 10-5; Nov.-Dec. daily call for extended hours. One Museum: adults $6. Two Museums: adults $10; discounts to AAM & ICOM members; members and children 12 & under no charge. Closed New Year's Day; Thanksgiving; Christmas.
Attendance: 35,501 (accurate)
Membership: Individual $35; Dual $50; Family $75; Friend $100; Patron $250.

OGLEBAY'S GOOD ZOO, Oglebay Park, 465 Lodge Dr., Wheeling, WV 26003-9361. Tel.: 304-243-4027. Fax: 304-243-4110.
E-mail: pmiller@oglebay-resort.com
Web Site: www.oglebay-resort.com/goodzoo/
Founded: 1977.
Congressional District: 23
Key Personnel: C.E.O., Douglas Dalby; Dir., Penny Miller; Cur. Education & Dir. Theater, Vickie Markey-Tekely; Animal Mgr., Joe Greathouse; Gift Shop Mgr., Jill Neumann.
Personnel Profile: Full-Time Paid 20; Part-Time Volunteers 100; Interns 25.
Governing Authority: nonprofit organization. Parent Institution: The Wheeling Park Commission, Oglebay Park, Wheeling, WV 26003. Tax-exempt.
Institution Type/Description: Zoo.
Collections: goats, bears, river otters, waterfowl, snakes, fruit bats, monkeys, ocelot; red pandas; bald eagles; African Wild Dogs, lemurs, merkats, lorikeet landing, kangaroos, zebras, llamas, turtles, opposum.
Research Fields: Raptor rehabilitation-Isis.
Facilities: library of books on animals & planetariums, available for research on premises; zoological park; aquarium; planetarium; 133-seat auditorium; theater; refreshment stands. Zoo-related items for sale.
Activities: films; docent program; special events; camps; Master Naturalist Program; education programs for pre-school to elder hostel groups; scout programs.
Publications: newsletter, Good Zoo News.
Hours & Admission Prices: Daily 11-4. Adults $7.50, children 3-12 $5.50; children 2 & under and members no charge.
Attendance: 124,575 (accurate)
Membership: Individual $35; Family & Grandparents $55; Family Plus & Grandparents $75; Sustaining $90; Patron $105; Lifetime $1,000.

POINT OVERLOOK MUSEUM, 989 Grandview St., Wheeling, WV 26003-3048. Tel.: 304-232-3010.
Institution Type/Description: Tourist Center & Museum.
Collections: local history & culture; photographs; videos.
Hours & Admission Prices: Daily 10-3. Adults $3, senior $2; children 12 & under no charge.

SCHRADER ENVIRONMENTAL EDUCATION CENTER, The Burton Center, Wheeling, WV 26003. Tel.: 304-242-6855. Fax: 304-242-5197.
E-mail: ejanelsins@oionline.com
Web Site: www.oionline.com
Founded: 1930.
Key Personnel: Dir., Eriks Janelsins; Museum Shop Mgr., Jane Link.
Personnel Profile: Full-Time Paid 6; Part-Time Paid 20.
Governing Authority: Parent Institution: Oglebay Institute.
Institution Type/Description: Natural History Museum.
Collections: interactive & natural history exhibits; butterfly garden.
Research Fields: herpetology; astacology.
Facilities: nature trail; observatory.
Activities: school tours; special events.
Hours & Admission Prices: April-Oct. Mon.-Sat. 9:30-5, Sun. 12-5; Nov.-March daily 12-5.
Attendance: 45,000

WEST VIRGINIA INDEPENDENCE HALL, 1528 Market St., Wheeling, WV 26003-3532. Tel.: 304-238-1300. Fax: 304-238-1302.
E-mail: travis.l.henline@wv.gov
Web Site: www.wvculture.org
Founded: 1964.
Congressional District: 1
Key Personnel: Site Mgr., Travis Henline.
Personnel Profile: Full-Time Paid 3; Part-Time Paid 3.
Governing Authority: state. West Virginia Div. of Culture & History, Cultural Center, Capitol Complex, Charleston, WV 25305. Tax-exempt.
Institution Type/Description: History Museum & Historic Site.
Collections: period artifacts; documents & portraits from 1861-1863, Wheeling & Statehood Conventions; 1863, Statehood Movement; furnishings of Governor's office & anterooms; architectural artifacts & historic preservation records; Civil War battle flags; courtroom.
Facilities: theater; meeting rooms.
Activities: guided & self-guided tours; films; changing exhibitions; performances.
Publications: Goldenseal: West Virginia Traditional Life.
Hours & Admission Prices: Mon.-Sat. 10-4; groups of 10 or more by appointment. No charge; donations accepted. Closed major holidays. &
Attendance: 8,000 (estimated)

WEST VIRGINIA NORTHERN COMMUNITY COLLEGE ALUMNI ASSOCIATION MUSEUM, 1704 Market St., Wheeling, WV 26003-3643. Tel.: 304-233-5900, ext. 8817. Fax: 304-232-0965.
E-mail: alumni@northern.wvnet.edu
Web Site: www.wvnorthern.edu
Founded: 1984.
Key Personnel: Pres., Darryl Ruth; Sec. & Treas., Joan Weiskircher.
Governing Authority: private; nonprofit.
Institution Type/Description: History Museum.
Collections: artifacts as they relate to former inhabitants of the historic landmarks or to the history of the community; glass collection is located in the former headquarters of a glass company & on another campus where glass was produced; railroad collection is in a former passenger station; local history of Wheeling and Northern Panhandle, Panhandle of WV.
Research Fields: train engine photographs; train & other railroad photos, B&ORR history, engine history & other railroad history.
Facilities: 75-vol. library of railroad history; 200-seat auditorium; educational facilities.
Activities: guided tours; loan, temporary & traveling exhibitions; training programs for professional museum workers. Museum Sponsors: three openings per year.
Publications: semiannual newsletter, WVNCC Alumni Association Museum; annual brochures.
Hours & Admission Prices: Call for confirmation of hours. &
Attendance: 92,000 (accurate)

White Sulphur Springs

PRESIDENT'S COTTAGE MUSEUM-THE GREENBRIER, 300 W. Main St., White Sulphur Springs, WV 24986-2414. Tel.: 304-536-1110, ext. 7314 & 7198. Fax: 304-536-7854.
E-mail: the_greenbrier@greenbrier.com
Web Site: www.greenbrier.com
Founded: 1932.
Key Personnel: Cur., Dr. Robert S. Conte.
Personnel Profile: Full-Time Paid 1; Part-Time Paid 2.
Governing Authority: CSX Hotels Corporation, White Sulphur Springs.
Institution Type/Description: History Museum: housed in 1835-1858 summer cottage used as the vacation home & resort of Presidents Van Buren, Tyler, Pierce, Fillmore & Buchanan.
Collections: photographs; letters; official correspondence; furnishings typical of that era; memorabilia housed in adjacent archives; paintings; prints; material documenting the 200-year history of the Greenbrier resorts.
Activities: guided tours; lectures.
Publications: The History of the Greenbrier: America's Resort.
Hours & Admission Prices: April-Nov. Mon.-Sat. 10-5, Sun. 10-3; other times by appointment. Adults $30, children 10-17 $15.
Attendance: 15,000 (estimated)

Williamson

WILLIAMSON AREA RAILROAD MUSEUM, INC., 100 Prichard St., Williamson, WV 25661. Mailing Address: P.O. Box 466, Williamson, WV 25661-0466. Tel.: 304-235-0105. Fax: 304-235-4910.
Web Site: williamsonrailroadmuseum.com
Key Personnel: Pres. (V), R. Doyle Van Meter, II
Governing Authority: Tax-exempt.
Institution Type/Description: Railroad Museum.
Collections: N & W memorabilia; model train; railroad artifacts.
Facilities: library.
Hours & Admission Prices: Fri.-Sat. 10-4. No charge; donations accepted.
Membership: Student $15; Senior Citizen $25; Individual $35.

Williamstown

FENTON GLASS MUSEUM, 420 Caroline Ave., Williamstown, WV 26187-1121. Tel.: 304-375-7772; 800-319-7793. Fax: 304-375-6459.
E-mail: museum@fentongiftshop.com
Web Site: www.fentongiftshop.com/museum.asp
Founded: 1977.
Key Personnel: Pres., George Fenton; Museum Shop Mgr., Charles Mayer.
Governing Authority: private business. Parent Institution: Fenton Gift Shops Inc.
Institution Type/Description: Company Museum.
Collections: art glass made by glass companies in the upper Ohio Valley from Parkersburg, West Virginia to Steubenville, Ohio during the years from 1880 to 1980; history of Fenton Art Glass.
Research Fields: the glass industry; Fenton Art Glass Co.
Facilities: library; archives of records & files of National Association of Manufacturers of Pressed & Blown Glassware covering 1887-1965; 65-seat theater. Museum-related items for sale.
Activities: guided tours; lectures; films.
Hours & Admission Prices: June-Aug. Mon.-Fri. 8-7, Sat. 8-5, Sun. 12-5; Sept.-Dec. Mon.-Fri. 8-6, Sat. 8-5, Sun. 12-5. No charge. Closed New Year's Day; Easter; Thanksgiving; Christmas. &
Attendance: 50,000 (accurate)

WISCONSIN

(327 listings)

Albany

ALBANY HISTORICAL SOCIETY MUSEUM, 117-119 N. Water St., Albany, WI 53502. Tel.: 608-862-3423.
Institution Type/Description: History Museum.
Collections: local history & culture; period furnishings; personal artifacts; photographs; obituaries; family trees; local business history.
Hours & Admission Prices: June-Sept. Sat. 9-3; Oct.-May Sat. 9-12.

Alma

THE ALMA AREA MUSEUM, 505 S. 2nd St., Alma, WI 54610. Mailing Address: P.O. Box 473, Alma, WI 54610. Tel.: 608-685-3554.
E-mail: society@almahistory.org
Web Site: www.almahistory.org/almaareamuseum.html
Founded: 1978.
Governing Authority: Parent Institution: Alma Historical Society.
Institution Type/Description: History Museum: housed in the former Buffalo County Training School & Teachers College; built in 1902.
Collections: local history & culture; period furnishings & clothing; personal artifacts; photographs; logging industry.
Hours & Admission Prices: Memorial Day to early Oct. Thurs.-Sun. 1-4. No charge; donations accepted.

WINGS OVER ALMA NATURE & ART CENTER, 188 N. Main St., Alma, WI 54610. Mailing Address: P.O. Box 191, Alma, WI 54610. Tel.: 608-685-3303.
E-mail: center@wingsoveralma.org
Web Site: www.wingsoveralma.org
Institution Type/Description: Nature & Art Center.
Collections: wildlife & their habitats; plants; wild flowers; trees; art history; Gerhard Gesell photographs.

Activities: special events; educational programs.
Hours & Admission Prices: Daily 10-5. Closed Thanksgiving; Christmas.

Almond

ALMOND HISTORICAL SOCIETY - OLD BANK BUILDING MUSEUM, Main St., Almond, WI 54909. Mailing Address: 330 County Rd. A, Almond, WI 54909. Tel.: 715-366-8571. Fax: 715-366-4558.
E-mail: valmond@uniontel.net
Founded: 1979.
Congressional District: 72
Key Personnel: Pres. (V), Arthur Pagel; Dir., Brian Roehrborn; Dir., Amy Eppinger; Dir., Dianne Trebiatowski.
Personnel Profile: Part-Time Volunteers 20.
Volunteer Hours: 175
Governing Authority: Tax-exempt.
Institution Type/Description: Historical Society Museum.
Collections: local history & culture; period furnishings; photographs; personal artifacts; Civil War artifacts; early 20th century dentist chair & office equipment; period clothing, school & farm artifacts.
Facilities: old bank building.
Activities: Annual Event: Open House in July.
Hours & Admission Prices: June-Sept. Tues. 1:30-4; other times by appointment. No charge; donations accepted.
Attendance: 250 (accurate)
Membership: Youth $2; Voting Member $5; Individual Life $50; Couple Life $75.

Antigo

LANGLADE COUNTY HISTORICAL SOCIETY MUSEUM, 404 Superior St., Antigo, WI 54409-1855. Tel.: 715-627-4464.
E-mail: lchs@dwave.net
Web Site: langladehistory.com
Founded: 1929.
Congressional District: 8
Key Personnel: C.E.O., Dir. & Pres. (V), Joe Hermolin; Treas., Glenn Bugni; Museum Shop Mgr., Mary Kay Wolf.
Personnel Profile: Full-Time Paid 1; Part-Time Paid 1; Part-Time Volunteers 18.
Governing Authority: state. Parent Institution: Langlade County Historical Society, Inc. Tax-exempt.
Institution Type/Description: Historic Building & Museum.
Collections: farm and logging tools; domestic implements & furnishings; newspaper files; World War I memorabilia; medical & musical equipment; 440 locomotive & caboose. Historic House: c.1878 Deleglise cabin.
Research Fields: genealogy.
Activities: guided tours; permanent exhibitions; lectures.
Publications: semi-annual newsletter.
Hours & Admission Prices: Wed.-Sat. 9:30-3:30; other times by appointment. No charge; donations accepted.
Attendance: 16,000 (estimated)
Membership: Junior $3; Individual $25; Family $30; Business $100; Life $250.

Appleton

THE BUILDING FOR KIDS CHILDREN'S MUSEUM, 100 W. College Ave., Appleton, WI 54911-5749. Tel.: 920-734-3226. Fax: 920-734-0677.
E-mail: contact@buildingforkids.org
Web Site: www.thebuildingforkids.org
Formerly: Fox Cities Children's Museum
Founded: 1991.
Congressional District: 8
Key Personnel: Exec. Dir., Dana L. Thorpe; Mgr. Operations & Museum Shop Mgr., Julie Runnfeldt; Pres. (V), Sandra Began.
Personnel Profile: Full Time Paid 6; Part Time Paid 20; Part Time Volunteers 40; Interns 8.
Governing Authority: Tax-exempt.
Institution Type/Description: Children's Museum.
Collections: hands-on exhibits.
Research Fields: learning through play.
Activities: birthday parties; classes; overnights; special events.
Hours & Admission Prices: Tues.-Wed. & Fri.-Sat. 9-5, Thurs. 9-6, Sun. 12-5. Admission $7.25; discounts to ACM & ASTC members; children under one & members no charge. Closed New Year's Day; Easter; Memorial Day; Independence Day; Labor Day; Thanksgiving; Christmas. &
Attendance: 125,000 (accurate)

Membership: Caregiver add on $15; Child $70; Grandparent $80; Family $115; Preferred Family $135.

THE COTTAGE ANTIQUES MUSEUM, 1124 N. Mason St., Appleton, WI 54914. Tel.: 920-739-7642.
Founded: 2003.
Congressional District: 10
Institution Type/Description: History Museum.
Collections: local history & culture; miniature clocks; early toys & dolls; period furnishings; china; pottery; personal artifacts.
Hours & Admission Prices: Memorial Day to Labor Day Mon.-Fri. 10-4; other times by appointment. Adults $5. &

HEARTHSTONE HISTORIC HOUSE MUSEUM, 625 W. Prospect Ave., Appleton, WI 54911-6042. Tel.: 920-730-8204. Fax: 920-730-8266.
E-mail: hearthdirector@att.net
Web Site: www.hearthstonemuseum.org
Founded: 1986.
Congressional District: 8
Key Personnel: Pres. (V), Stephanie Malaney; Chm. (V), Caleb Rocke; Exec. Dir. & Dir. Programs, Tricia Adams; Financial Dir., Jennifer Thomas.
Personnel Profile: Full-Time Paid 1; Part-Time Paid 1; Part-Time Volunteers 80.
Governing Authority: nonprofit. Tax-exempt: 501(c)(3).
Institution Type/Description: Historic House Museum: first residence in the world to be lighted from a central hydroelectric power plant using the Edison system in 1882.
Collections: late 19th-century furnishings, decorative arts, fine arts & utilitarian items; original Edison light fixtures & switches; items related to early history of electrical lighting.
Research Fields: late 19th-century social & cultural history with an emphasis on the Fox River Valley and effects of residential electricity; decorative arts; art; architecture; restoration & conservation.
Facilities: research library.
Activities: guided tours; lectures; organized education programs for children, adults, undergraduate or graduate college students; volunteer program; participatory & temporary exhibitions. Annual Events: Haunted Hearthstone in October; Victorian Christmas in November & December.
Publications: quarterly newsletter, Upstairs & Downstairs; bimonthly volunteer newsletter.
Hours & Admission Prices: Tues.-Fri. 10-3:30, Sat. 11-3:30, Sun. 1-3:30; tours every half hour. Adults $6, children $3; discounts to groups of 15 or more. Haunted Hearthstone & Victorian Christmas: call for additional hours. Adults $7, children $4. Closed holidays.
Attendance: 5,500 (estimated)
Membership: Bricks & Mortar $30; Electric $50; Family $100; Hearthstone 1882 $250; Community Treasure $500; Heritage $1,000; Landmark $2,500; Legacy $5,000; Futures $10,000; Sustainers $25,000 & up.

THE HISTORY MUSEUM AT THE CASTLE, (M), 330 E. College Ave., Appleton, WI 54911-5715. Tel.: 920-735-9370. Fax: 920-733-8636.
E-mail: matt@myhistorymuseum.org
Web Site: www.myhistorymuseum.org
Formerly: Outagamie Museum
Founded: 1872.
Congressional District: 8
Key Personnel: Exec. Dir., Matthew J. Carpenter; Pres. (V), Monica Rico; Chief Cur., Nick Hoffman; Business Mgr., Sheila Ploeckelman; Community Engagement Mgr., Kathryn Voigt; Cur., Emily K. Rock.
Governing Authority: Outagamie County Historical Society, Inc. Branch Museums: Grignon Mansion, 1313 Augustine St., Kaukauna, WI 54130. Tax-exempt: 501(c)(3).
Institution Type/Description: Regional History Museum.
Collections: local history collections documenting the social, industrial, agricultural & political history of Wisconsin's Fox River Valley; famous & infamous locals including escape artist, Harry Houdini & Senator Joseph McCarthy and Pulitzer Prize winning author, Edna Ferber.
Research Fields: regional history; local history.
Facilities: library; archives; theater.
Activities: annual meetings; temporary exhibits; educational programs; school tours; slide shows; lecture series; papermaking demonstration; educational programs; genealogy classes; theater.
Publications: quarterly newsletter, History Today; quarterly newsletter.
Hours & Admission Prices: Tues.-Sun. 11-4. Family $20, adults $7.50, seniors 65 & up and students with ID $5.50, children 5-17 $3.50; discounts to AAA, AAM & ICOM members; children under 5 & members no charge. &
Attendance: 19,019 (accurate)

Membership: Senior Citizen & Student $15; Individual $25; Family $50; Professional $75; Benefactor $150; Life $1,000.

PAPER DISCOVERY CENTER, (M), 425 W. Water St., Appleton, WI 54911-6058. Tel.: 920-380-7491. Fax: 920-731-2704.
Web Site: www.paperdiscoverycenter.org
Founded: 2005.
Congressional District: 8
Key Personnel: Exec. Dir., Kathleen Lhost; Chm. (V), Harry Spiegelberg; Treas. (V), William H. Geenen; Education Coord., Amber Hamilton; Education Coord., Linda Werner.
Personnel Profile: Full-Time Paid 1; Part-Time Paid 5; Part-Time Volunteers 22.
Volunteer Hours: 2,100
Operating Expenses: 425,976
Operating Income: 440,143
Governing Authority: private; nonprofit organization. Parent Institution: Paper Industry International Hall of Fame, Inc., Appleton, WI. Tax-exempt: 501(c)(3).
Institution Type/Description: Paper Museum.
Collections: paper industry history; Atlas Mill history.
Facilities: 50-vol. library; educational facilities; 45-seat theater. Museum-related items for sale.
Activities: formal education programs for children; guided tours; hobby workshops; temporary exhibitions.
Publications: monthly newsletter, All Things Paper.
Hours & Admission Prices: Mon.-Sat. 10-4. Adults $5, seniors $4, students $3; discounts to families; members no charge. Closed holidays. &
Attendance: 10,000 (accurate)
Membership: Individual $45; Family $65; Friend $100; Partner $250; Benefactor $500; Patron $1,000.

THE TROUT MUSEUM OF ART, The Reigel Bldg., 111 W. College Ave., Appleton, WI 54911. Tel.: 920-733-4089. Fax: 920-733-4149.
E-mail: info@troutmuseum.org
Web Site: www.troutmuseum.org
Formerly: Appleton Art Center
Founded: 1960.
Congressional District: 8
Key Personnel: Exec. Dir., Pamela Williams-Lime; Gen. Mgr. Operations, Angie Bleck; Pres. (V), Michael Cisler.
Personnel Profile: Full-Time Paid 2; Part-Time Paid 4; Part-Time Volunteers 50; Interns 1.
Governing Authority: private; nonprofit. Tax-exempt: 501(c)(3).
Institution Type/Description: Art Museum.
Collections: visual arts exhibitions.
Research Fields: art history; contemporary artwork.
Facilities: educational facilities.
Activities: guided tours; education programs; arts festivals; participatory exhibits. Annual Event: Art in the Park.
Hours & Admission Prices: Tues.-Sat. 10-4, Sun. 12-4. Adults $6, students & seniors $4; discounts to AAM & ICOM members; members & children under 10 no charge. Closed major holidays. &
Attendance: 30,000 (estimated)
Membership: Student $25; Individual $45; Family & Artist $50; Apprentice $100-$249; Artisan $250-$499; Academy $500-$999; Guild $1,000-$2,499; Master $2,500-$4,999; Patron $5,000 & up.

WRISTON ART CENTER GALLERIES, (M), Lawrence University, 613 E. College Ave., Appleton, WI 54912-0599. Mailing Address: 711 E. Boldt Way, Appleton, WI 54911. Tel.: 920-832-6890. Fax: 920-832-7362.
E-mail: beth.a.zinsli@lawrence.edu
Web Site: www.lawrence.edu/dept/wriston
Founded: 1989.
Congressional District: 8
Key Personnel: Cur. & Galleries Dir., Beth A. Zinsli; Gallery & Collections Mgr., Leslie Walfish.
Personnel Profile: Full-Time Paid 2; Part-Time Paid 10; Interns 5.
Governing Authority: college; nonprofit organization. Parent Institution: Lawrence University. Tax-exempt.
Institution Type/Description: Art Gallery.
Collections: 19th & 20th-century paintings & prints; La Vera Pohl collection of German Expressionist art; Japanese prints & drawings; Greek & Roman coins.
Research Fields: catalogue of Ottilia Beurger's collection of Greek & Roman coins; catalogue of The La Vera Pohl Collection of German Expressionist Art.
Facilities: print study room.
Activities: films; formal education program for undergraduates & graduates; lectures; loan, temporary & traveling exhibitions.
Hours & Admission Prices: During academic year Tues.-Fri. 10-4, Sat.-Sun. 12-4. No charge. Closed months of Aug. & Dec. &
Attendance: 7,472 (accurate)

Ashippun

HONEY OF A MUSEUM, N. 1557 Hwy. 67, Ashippun, WI 53003. Mailing Address: P.O. Box 46, Ashippun, WI 53003. Tel.: 800-558-7745.
Institution Type/Description: Honey Bee History Museum.
Collections: honey bee history; beekeeping history; pollination, beeswax; bee activity.
Facilities: Museum-related items for sale.
Activities: video; close up view of honey bees; nature walk; honey tasting.
Hours & Admission Prices: May-Oct. Mon.-Fri. 9-3:30, Sat.-Sun. 12-4; Nov.-April Mon.-Fri. 9-3:30; groups by appointment. No charge.

Ashland

ASHLAND HISTORICAL SOCIETY MUSEUM, 509 W. Main St., Ashland, WI 54806-1513. Tel.: 715-682-4911. Facebook: Ashland Historical Society Museum.
E-mail: ashlandhistory@centurytel.net
Web Site: ashlandwihistory.com
Founded: 1954.
Congressional District: 7
Key Personnel: Pres. (V), Tory Stroshane; Cur., Amy Tromberg; Museum Shop Mgr., June Cleveland.
Personnel Profile: Part-Time Paid 2; Part-Time Volunteers 25.
Governing Authority: under management of Ashland Historical Society. Tax-exempt: 501(c)(3).
Institution Type/Description: General Museum.
Collections: costumes; history; military; transportation; Ashland history.
Research Fields: genealogy; local history.
Facilities: research library.
Activities: temporary exhibitions.
Publications: Garland City Gazette, newsletter of Ashland Historical Society.
Hours & Admission Prices: June to mid-Sept. Mon.-Fri. 10-4, Sat. 10-2; mid-Sept. to May Mon.-Fri. 10-4. No charge; donations accepted. Closed major holidays. &
Attendance: 1,140 (estimated)
Membership: Annual $10.

Ashwaubenon

ASHWAUBENON HISTORICAL SOCIETY, 737 Cormier Rd., Ashwaubenon, WI 54304-4825. Tel.: 920-429-2863.
Founded: 1971.
Governing Authority: Tax-exempt: 501(c)(3).
Institution Type/Description: Historical Society Museum.
Collections: local history & culture; photographs.
Publications: quarterly newsletter.
Hours & Admission Prices: April-Dec. Wed. & Sat. 1-4. Adults $1.
Attendance: 600 (estimated)
Membership: Single $10; Couple $18.

Augusta

1864 DELLS MILL HISTORICAL LANDMARK & MUSEUM, E18855 County Rd. V, Augusta, WI 54722. Tel.: 715-286-2714.
Institution Type/Description: Historic Building: mill built in 1864. Listed on the National Register of Historic Places.
Collections: local history & culture; period tools, furnishings & equipment; photographs.
Activities: guided tours.
Publications: book, The Civil War To America's Health Care War: 150 Years.
Hours & Admission Prices: May-Oct. daily 10-5. Adults $7, students $3.50.

Baileys Harbor

THE RIDGES SANCTUARY, INC., 8270 Hwy. 57, Baileys Harbor, WI 54202. Mailing Address: P.O. Box 152, Baileys Harbor, WI 54202-0152. Tel.: 920-839-2802 & 1101. Fax: 920-839-2234.
E-mail: info@ridgessanctuary.org
Web Site: www.ridgessanctuary.org
Founded: 1937.
Congressional District: 8
Key Personnel: Exec. Dir., Steve Leonard; Pres., Roy Thilly; Asst. Dir., Judy Drew; Mgr. Visitor Svcs., Kate LeRoy.
Governing Authority: nonprofit organization; board of directors. Tax-exempt: 501(c)(3).
Institution Type/Description: Nature Center.
Collections: 475 species of vascular plants in local area, including local mosses, lichens, & liverworts; 48 collection boxes housing moths, butterflies, wildflower sanctuary & other insects native to area; 1869 Range lights. Historic Structure: 1873 log building.
Research Fields: entomology; phenology of butterflies & moths, including information of life cycle; Ram's head lady slipper Orchid study; bird breeding.
Facilities: 550-vol. library pertaining to natural history & plant study; 1,440 acres; educational facilities; nature center. Bird houses & other museum-related items for sale.
Activities: guided tours; walking trail; lectures; organized educational programs for children & adults; participatory & temporary exhibitions.
Publications: quarterly newsletter, Ridges News; trail guide; History of the Ridges.
Hours & Admission Prices: Nature Center: May to Oct. Mon.-Sat. 9-4, Sun. 11-3. Trails: daily dawn to dusk. Adults $5; members & children under 18 no charge.
Attendance: 17,500 (estimated)
Membership: Individual $40; Family $65; Family Plus $95; Business $150.

Balsam Lake

POLK COUNTY MUSEUM, 120 Main St., Balsam Lake, WI 54810. Mailing Address: P.O. Box 41, Balsam Lake, WI 54810-0041. Tel.: 715-485-9269.
E-mail: polkcountymuseum@lakeland.ws
Web Site: www.polkcountymuseum.com
Founded: 1960.
Congressional District: 3
Key Personnel: Dir., Michelle Pedersen; Pres. (V), Muriel Pfeifer.
Personnel Profile: Full-Time Volunteers 20; Part-Time Paid 1; Part-Time Volunteers 37; Interns 1.
Governing Authority: county; nonprofit organization. Branch Museums: Polk County Red School House, St. Croix Falls-Fairgrounds; Rural Life Museum, Balsam Lake; Polk County Museum. Tax-exempt.
Institution Type/Description: General Museum.
Collections: area history; art & craft exhibits. Historic Building: 1899 courthouse; The Nye Bison Site Collection.
Research Fields: county history.
Facilities: 800-vol. historical library available for use on premises.
Activities: guided tours; lectures; films; formally organized education programs for adults & children.
Publications: books, Polk County Memories; Polk County's First Written History 1876; Building of the Dam at St. Croix Falls; Polk County Postal History & Directory.
Hours & Admission Prices: Memorial Day-Labor Day Thurs.-Mon. 12-4; tours by appointment. Adults $4, students 13-17 $2; children under 12 with adult no charge. &
Attendance: 2,500 (accurate)
Membership: Individual $15.

Baraboo

CIRCUS WORLD MUSEUM, 550 Water St., Baraboo, WI 53913-2578. Tel.: 608-356-8341. Fax: 608-356-1800. Facebook: Circus World Baraboo.
E-mail: ringmaster@circusworldmuseum.com
Web Site: www.circusworld.wisconsinhistory.org
Founded: 1959.
Congressional District: 2
Key Personnel: Chm. (V), Jonathan Lipp; Exec. Dir., Scott O'Donnell; Museum Shop Mgr., Ralph Pierce.
Personnel Profile: Full-Time Paid 7; Part-Time Paid 50; Part-Time Volunteers 50.
Governing Authority: nonprofit organization. Parent Institution: Wisconsin Historical Society, 816 State St., Madison, WI 53706. Tax-exempt: 501(c)(3).
Institution Type/Description: Circus Museum: site of 1884-1918 original winter quarters of Ringling Bros. Circus.
Collections: circus artifacts; circus wagons; railroad cars; archives; route books; programs; negatives; lithographs; couriers; heralds; business records.
Research Fields: circus & wild west.
Facilities: library & research center with books available for inter-museum loan.
Activities: guided tours; films; interactive programs; live circus performances including period circus music concerts, demonstrations; parades; circus dining department; calliope concerts; traveling displays; magic shows.
Hours & Admission Prices: Summer: daily 9-6; Fall: daily 10-4; Winter: by appointment; Spring Mon.-Fri. 10-4. Summer: adults $17.95, senior citizens $15.95, children 5-11 $7.95; discount to groups; children under 5 no charge. Fall, Winter & Spring: adults $9, seniors $8, children 5-11 $3.50; discounts for groups; children under 5 no charge. Library & Research Center: call for hours. Closed New Year's Day; Easter; Thanksgiving; Christmas Eve; Christmas; New Year's Eve. &
Attendance: 71,106 (accurate)
Membership: Membership available through the Wisconsin Historical Society and Friends group CWM, Inc.

THE INTERNATIONAL CLOWN HALL OF FAME & RE-SEARCH CENTER, INC., 102 4th Ave., Baraboo, WI 53913. Tel.: 608-355-0321.
E-mail: info@theclownmuseum.com
Web Site: www.theclownmuseum.com
Founded: 1986.
Congressional District: 1
Key Personnel: Dir., Greg DeSanto.
Personnel Profile: Full-Time Volunteers 1.
Governing Authority: nonprofit. Tax-exempt: 501(c)(3).
Institution Type/Description: Clown Museum.
Collections: history of clowns; 2,000 photographs; art.
Research Fields: history of clowns & healing power of humor.
Facilities: 1,000-vol. library of books; 2,500 sq. ft. exhibit area. Clown-related items for sale.
Activities: formal education programs for children & adults; guided tours; theatre; traveling exhibitions.
Publications: quarterly newsletter.
Hours & Admission Prices: Mon.-Fri. 10-3:30. Shows: $5. Museum: $2. Museum & Show: $8; members no charge. &
Attendance: 25,000 (estimated)
Membership: Clown/Supporter $35; Family $50; Joey/Patron $100; Jester/Advocate $500; Ringmaster/Founder $1,000.

INTERNATIONAL CRANE FOUNDATION, E. 11376 Shady Lane Rd., Baraboo, WI 53913-0447. Mailing Address: P.O. Box 447, Baraboo, WI 53913-0447. Tel.: 608-356-9462, ext. 118. Fax: 608-356-9465.
E-mail: cranes@savingcranes.org
Web Site: www.savingcranes.org
Founded: 1973.
Congressional District: 2
Key Personnel: Chair Emeritus (V), Mary E. Wickhem; Chm. Bd., Hall Healy; Vice Chm. Bd. & Co-Founder, George Archibald; C.E.O. & Pres., Richard Beilfuss; Treas., Charles Gibbons; Cur. Birds, Bryant Tarr; Field Ecologist, Jeb Barzen; Gift Shop Mgr., Darcy Love.
Personnel Profile: Full-Time Paid 41; Part-Time Paid 3; Part-Time Volunteers 35; Interns 15.
Governing Authority: nonprofit organization. Tax-exempt: 501(c)(3).
Institution Type/Description: Aviary & Ornithology Museum.
Collections: 140 cranes of 15 species; crane related artifacts from all over the world; research library, restored prairie.
Research Fields: cranes, their captive propagation, general maintenance, behavior; restoration of marsh, prairie & savanna.
Facilities: research library; 2,000 papers on cranes & related materials, available for research on premises with permission of the deputy director; nature center; field research station; 120-seat auditorium; zoo; trails. Museum-related items for sale.
Activities: guided tours; lectures; films; formally organized education programs for children, adults, undergraduate & graduate college students; docent program or council.
Publications: quarterly newsletter, The ICF Bugle; booklet, Cranes, Cranes, Cranes; books, Crane Research around the World; Reflections; Proceedings of the 1980 International Crane Workshop; Proceedings of the 1983 International Crane Workshop.

Hours & Admission Prices: April 15-Oct. daily 9-5. Guided Tours: April-May & Sept.-Oct. Sat.-Sun. 10, 1 & 3; Memorial Day to Labor Day daily 10, 1 & 3. Adults $9.50, senior citizens 62 & over and students $8, children 6-17 $5; children 5 & under and members no charge. ⅙

Attendance: 24,500 (accurate)

Membership: Student & Senior $25; Individual $35; Family & Foreign $50; Associate $100; Sustaining $250; Sponsor $500; Patron $1,000; Benefactor $2,000.

SAUK COUNTY HISTORICAL SOCIETY AND MUSEUM, 531 4th Ave., Baraboo, WI 53913-2034. Mailing Address: P.O. Box 651, Baraboo, WI 53913-0651. Tel.: 608-356-1001.

E-mail: history@saukcountyhistory.org

Web Site: www.saukcountyhistory.org

Founded: 1906.

Congressional District: 2

Key Personnel: Pres., Paul Wolter; Museum Keeper, Linda Levenhagen; Collections & Programs Mgr., Rebecca Dubey.

Personnel Profile: Part-Time Paid 3; Part-Time Volunteers 8.

Governing Authority: nonprofit organization. Branch Museums: Man Mound Park & Yellow Thunder Monument & Park. Tax-exempt: 501(c)(3).

Institution Type/Description: General Museum.

Collections: artifacts & photos from pioneer days to present; Indian artifacts; circus mementoes & displays; archaeological & geological materials; guns; quilts; textiles; toys; china.

Research Fields: county history; genealogy.

Facilities: 1,050-vol. library pertaining to local history available for use upon request.

Activities: guided tours; annual meeting. Museum Sponsors: annual banquet; annual social; tour of historic homes.

Publications: newsletter, Sauk Trails; Early Baraboo & Sauk County histories.

Hours & Admission Prices: Wed.-Sat. 12-4:30. No charge; donations accepted. Closed major holidays. ⅙

Attendance: 2,000 (estimated)

Membership: Individual $20; Family $35; Friend $50; Sponsor $100; Patron $250 and up; Benefactor $500 and up.

Bayfield

BAYFIELD HERITAGE CENTER, 30 N. Broad St., Bayfield, WI 54814. Mailing Address: P.O. Box 137, Bayfield, WI 54814. Tel.: 715-779-5958.

E-mail: bayfieldheritage@centurytel.net

Web Site: www.bayfieldheritage.org/index.html

Institution Type/Description: History Museum.

Collections: local history & culture; period furnishings; personal artifacts; photographs.

Activities: educational programs; special events.

Hours & Admission Prices: Tues.-Sat. 1-4.

Beaver Dam

BEAVER DAM AREA ARTS ASSOCIATION AT THE SEIPPEL HOMESTEAD AND CENTER FOR THE ARTS, 1605 N. Spring St., Beaver Dam, WI 53916-1103. Mailing Address: P.O. Box 442, Beaver Dam, WI 53916-0442. Tel.: 920-885-3635.

E-mail: bdaaa@seippelcenter.com

Web Site: www.bdaaa.org

Key Personnel: Exec. Dir., Karla R. Jensen; Pres. Bd. (V), Tom Helfert; Museum Shop Mgr., Betty Singer.

Governing Authority: Tax-exempt.

Institution Type/Description: Art Museum.

Collections: oil paintings.

Facilities: Museum-related items for sale.

Activities: art classes. Museum Sponsors: Youth Art monthly; Youth Art Camp in June; Annual Book Sale; Annual Secret Garden Tour.

Hours & Admission Prices: Wed.-Sun. 1-4. No charge.

Membership: Artist $20; Senior Citizen $25; Individual & Organization $30; Family $45; Mona Lisa $100; Picasso Patron $250; Michelangelo Angel $500; Van Gogh Benefactor $1,000.

DODGE COUNTY HISTORICAL SOCIETY MUSEUM, 105 Park Ave., Beaver Dam, WI 53916-2107. Tel.: 920-887-1266.

E-mail: dchs@powercom.net

Web Site: www2.powercom.net/~dchs

Formerly: Williams Free Library

Founded: 1938.

Congressional District: 9

Key Personnel: Cur., Mary Beth Jacobson; Pres. (V), Glen Link.

Personnel Profile: Part-Time Paid 2.

Governing Authority: nonprofit organization. Parent Institution: Dodge County Historical Society. Tax-exempt: 170(b)(A).

Institution Type/Description: General Museum.

Collections: arrowheads; replica of one-room schoolhouse; Celebrity Wall-Our Community Goes to War display from Civil War to present; Monarch ranges; 1903 Rambler.

Research Fields: Dodge County history; Native American; natural history; genealogy; Monarch Range Co.

Activities: guided tours; school group tours; slide presentation; permanent & temporary exhibits.

Publications: quarterly newsletter.

Hours & Admission Prices: Tues.-Sat. 1-4; tours by appointment. No charge; donations accepted. Closed holidays.

Attendance: 1,800 (estimated)

Membership: Student $1; Household $20; Affiliate $50-$99; Patron $100-$249; Sustaining $250-$499; Heritage $500 & up.

Belmont

FIRST CAPITOL - WISCONSIN HISTORICAL SOCIETY, 19101 Cty. Hwy. G, Belmont, WI 53510. Mailing Address: P.O. Box 270, Mineral Point, WI 53565. Tel.: 608-987-2122. Fax: 608-987-3738.

E-mail: firstcapitol@wisconsinhistory.org

Web Site: firstcapitol.wisconsinhistory.org

Key Personnel: Dir., Allen Schroeder; Cur., Tamara Funk.

Personnel Profile: Full-Time Paid 2; Part-Time Paid 21.

Governing Authority: Parent Institution: Wisconsin Historical Society. Tax-exempt.

Institution Type/Description: History Museum.

Collections: local history & culture; period furnishings; personal artifacts; photographs. Historic Buildings: Wisconsin Territorial Legislature Council House; lodging house.

Hours & Admission Prices: mid-June to Labor Day Wed.-Sun. 10-4. No charge; donations accepted. ⅙

Attendance: 1,000 (estimated)

Beloit

THE ANGEL MUSEUM, (M), 656 Pleasant St., Beloit, WI 53511-6242. Mailing Address: P.O. Box 816, Beloit, WI 53512-0816. Tel.: 608-362-9099.

E-mail: angelmuseum@gmail.com

Web Site: www.angelmuseum.org

Founded: 1998.

Congressional District: 2

Key Personnel: Dir. & Museum Shop Mgr., Ruth Carlson; Pres. (V), Shauna El-Amin; Museum Shop Mgr., Susan Sweetin.

Personnel Profile: Part-Time Volunteers 55.

Governing Authority: city; nonprofit organization. Tax-exempt.

Institution Type/Description: Historic Site: built in 1914; formerly St. Paul's Catholic Church. Listed on the Register for Historical Landmarks.

Collections: over 12,000 angel artifacts made from over 100 different materials from 50 different countries including 600 angels donated by Oprah Winfrey & the Berg Angel collection.

Facilities: gardens. Museum-related items for sale.

Activities: group tours; special events; facility rental.

Hours & Admission Prices: April-Dec. Tues.-Sat. 10-4. Adults $7; discounts to AAA & museum members. ⅙

Attendance: 17,000 (estimated)

Membership: Call for information.

BELOIT FINE ARTS INCUBATOR, 520 E. Grand Ave., Beloit, WI 53511-6314. Tel.: 608-313-9083.

E-mail: bfaiwi@yahoo.com

Web Site: beloitfineartsincubator.com

Key Personnel: Bd. Pres., Jerry Sveum; Bd. Vice Pres., Dean Folts; Volunteer, Ben Henthorn.

Personnel Profile: Part-Time Volunteers 1.

Governing Authority: Tax-exempt.

Institution Type/Description: Art Gallery.

Collections: works by resident & regional artists.

Major Exhibits: David Brickman, 1/14; Susan Swedlund, 2/14; Area High School Show, 4/14; Shelley Smith, 5/14; Wisconsin Regional Artists, 6/14; Dan Wuthrich, 7/14; Rock River Carvers, 8/14; Alice Blue & Susan Sweetin, 9/14; Mark Kosiba, 10/14; Paul Pinzarrone, 12/14.

Activities: traveling exhibits; photography, painting & ceramics classes.

Hours & Admission Prices: Mon.-Fri. 10-2. No charge; donations accepted

BELOIT HISTORICAL SOCIETY, 845 Hackett St., Beloit, WI 53511-5227. Tel.: 608-365-7835. Fax: 608-365-5999.
E-mail: pkerr@beloithistoricalsociety.com
Web Site: www.beloithistoricalsociety.com
Formerly: Hanchett Bartlett Homestead
Founded: 1910.
Congressional District: 1
Key Personnel: Dir., Paul K. Kerr; Pres. (V), William Yoss; Business Mgr., Scott Reichard; Volunteer Coord., Loretta Hatch.
Personnel Profile: Full-Time Paid 2; Part-Time Paid 3; Part-Time Volunteers 100; Interns 1.
Governing Authority: Beloit Historical Society. Affiliated with State Historical Society of Wisconsin. Branch Museum: Hanchett-Bartlett Homestead, 2149 St. Lawrence Ave., Beloit, WI 53511. Tax-exempt: 501(c)(3).
Institution Type/Description: Historic Site & History Museum: housed in 1857 restored homestead.
Collections: period furnishings; Stone barn & smokehouse; 1857 district 12 rural schoolhouse.
Research Fields: Beloit area individuals, architecture, businesses, industries, social & political history.
Activities: guided tours; lectures; formally organized education programs for children & adults; Hall of Fame for area leaders; Heritage Days; school group tours.
Publications: monthly newsletter, Lincoln in Beloit; Pioneer Beloit; Confluence.
Hours & Admission Prices: Hanchett-Bartlett Homestead: June-Oct. Sat.-Sun. 1-4. Office: Tues.-Fri. 12-4. Adults $3, seniors & children 12-18 $2; discounts to Heritage Rock County Consortium members; children under 12 & members no charge. &
Attendance: 8,500 (estimated)
Membership: Individual $20; Family $25.

* **LOGAN MUSEUM OF ANTHROPOLOGY, (M),** 700 College St., Beloit, WI 53511-5509. Tel.: 608-363-2677 & 2119. Fax: 608-363-7144. Facebook: Logan Museum.
Web Site: www.beloit.edu/logan
Founded: 1893.
Congressional District: 1
Key Personnel: Dir., William Green; Cur. Exhibits & Education, Dan Bartlett; Cur. Collections, Nicolette Meister; Museum Shop Mgr., Aaron Wilson.
Personnel Profile: Full-Time Paid 3; Part-Time Paid 10.
Governing Authority: Parent Institution: Beloit College. Tax-exempt: 170(b)(1)(A).
Institution Type/Description: Anthropology Museum.
Collections: American Indian, North, Meso & South American archaeology & ethnology; European & North African Paleolithic; Asian & New Guinea ethnology. Historic Building: 1869 Civil War Memorial Hall.
Research Fields: archeology & ethnology of Latin America, North America; Africa, New Guinea; European Paleolithic.
Facilities: laboratories; classrooms.
Activities: guided tours; lectures; films; gallery talks; cultural resource projects; education programs for adults & undergraduate college students; inter-museum loan, permanent & temporary exhibitions; school programs.
Publications: monographs, Occasional Contributions to Anthropology; guides; Logan Museum Bulletin.
Hours & Admission Prices: Tues.-Sun. 11-4. No charge; donations accepted. Closed college holidays. &
Attendance: 5,000 (accurate)

WRIGHT MUSEUM OF ART, BELOIT COLLEGE, (M), 700 College St., Beloit, WI 53511-5595. Tel.: 608-363-2095. Fax: 608-363-2248. Facebook: Wright Museum.
E-mail: museum@beloit.edu
Web Site: www.beloit.edu/wright/
Founded: 1892.
Congressional District: 1
Key Personnel: Dir., Joy Beckman; Cur., James F. Pearson; Office Coord., Aaron Wilson.
Personnel Profile: Full-Time Paid 2; Part-Time Paid 1; Part-Time Volunteers 5; Interns 1.
Governing Authority: Parent Institution: Beloit College. Tax-exempt: 170(b)(1)(A).
Institution Type/Description: Art Museum.
Collections: European & American paintings; works on paper; historic & contemporary photography; sculpture; Asian textiles, ceramics & other arts.
Major Exhibits: AfterWords / AfterWorlds, 12/3/13-2/5/14; Recent Acquisitions, 1/20/14-2/15/14; Hollensteiner Conservation Fund, 1/20/14-4/15/14;

Beloit College Art Faculty Show, 2/14/14-3/30/14; VSA - Creative Power (T), 2/14/14-3/30/14; Annual Senior Show, 4/18/14-5/18/14; Pan (T), 9/1/14-10/15/14.
Research Fields: American & European art; works on paper; Asian art; photography.
Facilities: auditorium; classrooms.
Activities: guided tours; lectures; films; gallery talks; concerts; arts festivals; drama; formally organized education programs for undergraduate college students; inter-museum loan, semi-permanent, temporary & traveling exhibitions; children & adult programs.
Publications: exhibitions catalogs; posters.
Hours & Admission Prices: Tues.-Sun. 11-4. No charge; donations accepted. Closed college holidays. &
Attendance: 5,000 (estimated)
Membership: Friends of the Beloit College Museums Memberships $10-$500.

Berlin

BERLIN AREA HISTORICAL SOCIETY MUSEUM OF LOCAL HISTORY, 111 S. Adams Ave., Berlin, WI 54923-2023. Mailing Address: P.O. Box 83, Berlin, WI 54923-0083. Tel.: 920-361-2460.
E-mail: lerdmann@centurytel.net
Web Site: www.berlinareahistoricalsociety.com
Founded: 1962.
Congressional District: 6
Key Personnel: Pres., Roberta Erdmann; Cur., Dan Freimark.
Personnel Profile: Part-Time Volunteers 30.
Governing Authority: nonprofit organization. Affiliated with Berlin Historical Society. Tax-exempt.
Institution Type/Description: Historic Building: 1868 Clark School, one room schoolhouse built in 1866; museum of local history built in 1888.
Collections: local school history; monument. Historic Buildings: one room schoolhouse; blacksmith shop; 1888 2-story brick building.
Activities: permanent exhibitions.
Publications: quarterly newsletter, Tales & Trails.
Hours & Admission Prices: Memorial Day to Labor Day 2nd & 4th Sun. 1-4; other times by appointment. No charge; donations accepted.
Attendance: 450 (estimated)
Membership: Individual $10; Family $15.

Birchwood

BIRCHWOOD AREA HISTORICAL SOCIETY LOG MUSEUM, Main St., Birchwood, WI 54817. Mailing Address: 121 N. Main St., Birchwood, WI 54817. Tel.: 715-354-3879.
Institution Type/Description: Historical Society Museum.
Collections: local history; logging industry; hand-carved miniature logging camp; photographs.
Hours & Admission Prices: Call for hours. No charge.

HOWARD MOREY HOUSE, Park Ave., Birchwood, WI 54817. Mailing Address: 121 N. Main St., Birchwood, WI 54817. Tel.: 715-354-3115.
Institution Type/Description: Historic House Museum: housed in the boyhood home of Howard Morey; built in 1901.
Collections: Howard Morey's life & career; photographs; personal artifacts; period furnishings.
Activities: special events.
Hours & Admission Prices: Memorial Day to Labor Day Wed. 11-3; other times by appointment. No charge.

Black Earth

BLACK EARTH DEPOT AND MUSEUM, 934 Mills St., Black Earth, WI 53515. Mailing Address: P.O. Box 214, Black Earth, WI 53515. Tel.: 608-767-2289.
Governing Authority: Parent Institution: Black Earth Historical Society. Tax-exempt.
Institution Type/Description: Historic Depot: c.1857.
Collections: local history & culture; period furnishings; personal artifacts; photographs.
Publications: quarterly members' newsletter.
Hours & Admission Prices: Memorial Day to Labor Day Sun. 1-4. No charge; donations accepted.
Attendance: 125 (estimated)
Membership: Individual $10; Family $20; Business $35.

Black River Falls

JACKSON COUNTY HISTORICAL SOCIETY, 13 S. 1st St. & 321 Main St., Black River Falls, WI 54615-0037. Mailing Address: P.O. Box 37, Black River Falls, WI 54615-0037. Tel.: 715-284-5314.
Web Site: www.blackriverfalls.com
Founded: 1916.
Congressional District: 7
Key Personnel: Pres. (V), Eugene Gutknecht; Vice Pres., Jerry Johnson; Museum Dir. & Treas., Gary Morris; Sec. & Cur. Ho Chunk Photos, Mildred Evenson; Cur. Ho Chunk Artifacts, Leona McKee; Cur. Clothing & Artifacts, Gloria Curran; Cur. Manuscripts, Clarice Phillips; Cur. Photos, Jo Anne Dougherty.
Personnel Profile: Full-Time Volunteers 12; Part-Time Volunteers 20.
Governing Authority: bd. of directors of society. Tax-exempt.
Institution Type/Description: Historical Society Museum: housed in the former 1885 Van Schaick Photograph Gallery and in the former 1915 Carnegie Library.
Collections: costumes; tools; school artifacts; collection of photographs 1880-1920; furniture; dairy & logging artifacts; kitchen & household items; manuscripts; dentist office; Victorian bedroom; Americana Music room; musicians of the area.
Research Fields: history of owner-operators of businesses in Black River Falls; identification of Winnebago Indian photos & artifacts; genealogies; flood of 1911.
Facilities: meeting room.
Activities: guided tours; lectures; films; gallery talks; permanent & temporary exhibitions.
Publications: Jackson County-A History.
Hours & Admission Prices: Fri.-Sat. 10-3; other times by appointment only. No charge; donations accepted.
Attendance: 790 (accurate)
Membership: Individual $6; Family $8; Life $50.

Blanchardville

BLANCHARDVILLE HISTORICAL SOCIETY & MUSEUM, 101 S. Main St., Blanchardville, WI 53516. Mailing Address: P.O. Box 62, Blanchardville, WI 53516. Tel.: 608-523-1220.
E-mail: blanchardvillehistorical@gmail.com
Web Site: www.blanchardville.com
Institution Type/Description: History Museum: housed in the former hydrostatic station used to help generate power for the city.
Collections: local history & culture; period furnishings; personal artifacts; photographs.
Hours & Admission Prices: March-Nov. Sat. 9-12.

Blue Mounds

CAVE OF THE MOUNDS -NATIONAL NATURAL LAND-MARK, 2975 Cave of the Mounds Rd., Blue Mounds, WI 53517-0148. Mailing Address: P.O. Box 148, Blue Mounds, WI 53517-0148. Tel.: 608-437-3038.
Founded: 1939.
Institution Type/Description: Natural History Museum.
Collections: local history; cave formations; fossils; rocks; geology.
Facilities: Museum-related items for sale.
Activities: guided tours.
Hours & Admission Prices: mid-March to May & Sept. to mid-Nov. Mon.-Fri. 10-4, Sat.-Sun. 9-5; Memorial Day to Labor Day daily 9-6; mid-Nov. to mid-March Mon.-Fri. 11 & 2 by appointment, Sat.-Sun. hourly 10-4. Adults $16, children 4-12 $8; children 3 & under no charge.

LITTLE NORWAY, INC., 3576 Hwy. JG N., Blue Mounds, WI 53517. Tel.: 608-437-8211. Fax: 608-437-7827.
E-mail: info@littlenorway.com
Web Site: www.littlenorway.com
Founded: 1926.
Congressional District: 2
Key Personnel: Pres., Scott Winner; Sec. & Treas., John D. Winner.
Governing Authority: individual operation.
Institution Type/Description: Historic Site Museum: housed in 1856 Norwegian pioneer homestead built by Austin Haugen.
Collections: original farmstead buildings; Norwegian & pioneer artifacts. Historic Building: 1893 replica of 12th-century Norwegian Christian Church.
Facilities: Museum-related items & items of Scandinavian origin for sale.
Activities: guided tours.

Hours & Admission Prices: May-June & Sept.-Oct. daily 9-5; July-Aug. daily 9-7. Adults $12, senior citizens $11, groups of 20 or more $10, children 5-12 $5; children under 5 no charge.
Attendance: 38,500

Boscobel

BOSCOBEL DEPOT MUSEUM, 800 Wisconsin Ave., Boscobel, WI 53805. Tel.: 608-375-2672.
Web Site: www.boscobelwisconsin.com/area-attractions.html
Institution Type/Description: Historic Building: housed in a former depot; built in 1857.
Collections: local history & culture; period furnishings; personal artifacts; photographs.
Hours & Admission Prices: Call for hours.

GRAND ARMY OF THE REPUBLIC MUSEUM, 102 Mary St., Boscobel, WI 53805. Mailing Address: 1006 Wisconsin Ave., Boscobel, WI 53805. Tel.: 608-375-5693 & 742-2589.
Institution Type/Description: Historic Building: listed on the National Register of Historic Places.
Collections: Civil War artifacts; personal artifacts; photographs; drums; flags; period furnishings.
Hours & Admission Prices: June-Aug. Sat. 12-3; other times by appointment.

Bowler

ARVID E. MILLER MEMORIAL LIBRARY/MUSEUM OF THE STOCKBRIDGE MUNSEE TRIBE, N8510 Moh-He-Con-Nuck Rd., Bowler, WI 54416. Mailing Address: P.O. Box 70, Bowler, WI 54416. Tel.: 715-793-4270. Fax: 715-793-4836.
E-mail: library.museum@mohicannsn.gov
Founded: 1976.
Personnel Profile: Full-Time Paid 2; Part-Time Volunteers 1.
Governing Authority: Tax-exempt.
Institution Type/Description: Research Library & Historic Museum.
Collections: Mohican history & culture; books; microfilm; maps; photographs; personal papers; portraits; missionary journals; government documents; tribal documents.
Facilities: library; archives.
Hours & Admission Prices: Mon.-Fri. 8-4:30. No charge; donations accepted.
Attendance: 700 (estimated)

Brandon

BRANDON HISTORICAL SOCIETY & MUSEUM, 102 E. Main St., Brandon, WI 53919. Mailing Address: P.O. Box 344, Brandon, WI 53919. Tel.: 920-346-2962.
Institution Type/Description: Historical Society Museum.
Collections: local history & culture; period furnishings; personal artifacts; photographs.
Hours & Admission Prices: By appointment.

Brillion

ARIENS(R) COMPANY MUSEUM, 109 Calumet St., Brillion, WI 54110-0157. Mailing Address: 655 W. Ryan St., Brillion, WI 54110. Tel.: 920-756-4273. Fax: 920-756-2407.
E-mail: lpahl@ariens.com
Web Site: www.ariens.com
Founded: 2003.
Institution Type/Description: Company Museum.
Collections: company history; power equipment; two-wheel tractors; snowblowers; lawnmowers; riding lawnmowers; tractors; commercial mowers & tillers; advertising; literature; prototypes.
Facilities: 7,000 sq. ft. exhibition space; auditorium; classroom.
Hours & Admission Prices: No charge. Call for hours. &

BRILLION HISTORY HOUSE AND MUSEUM, 110 N. Francis St., Brillion, WI 54110. Mailing Address: P.O. Box 35, Brillion, WI 54110. Tel.: 920-756-9294.
Founded: 1967.
Key Personnel: Pres. (V), Jane Fuhrmann.
Personnel Profile: Part-Time Volunteers 40.
Governing Authority: Parent Institution: Brillion Historical Society, Inc. Tax-exempt.
Institution Type/Description: History Museum: housed in the former Green Hotel; built in 1872.

Collections: local history & culture; period furnishings; personal artifacts; photographs; Brillion Victorian wallpaper; farm equipment; dairy industry.
Research Fields: local history & genealogy.
Activities: community celebrations.
Publications: newsletter.
Hours & Admission Prices: By appointment. No charge; donations accepted.
&

Attendance: 500 (accurate)

Brodhead

BRODHEAD HISTORICAL SOCIETY & DEPOT MUSEUM, 1108 1st Center Ave., Brodhead, WI 53520. Mailing Address: 707 9th St., Brodhead, WI 53520. Tel.: 608-897-4150.
E-mail: info@brodheadhistory.org
Web Site: www.brodheadhistory.org/depot.html
Institution Type/Description: Historical Society Museum: housed in the former Milwaukee Road Depot; built in 1881.
Collections: local history & culture; period furnishings; photographs; documents; Milwaukee Road locomotive & caboose; cheese-making equipment; musical instruments; local business artifacts.
Hours & Admission Prices: Memorial Day to Labor Day Wed. & Sat.-Sun. 1-4; other times by appointment.

Brookfield

DOUSMAN STAGECOACH INN MUSEUM, 1075 Pilgrim Pkwy., Brookfield, WI 53045. Mailing Address: Elmbrook Historical Society, P.O. Box 292, Brookfield, WI 53008. Tel.: 262-782-4057.
Institution Type/Description: History Museum: housed in a former stagecoach inn; built in 1857.
Collections: local history & culture; period furnishings; personal artifacts; photographs; blacksmith shop; smoke house; ice house; school bell tower.
Facilities: Museum-related items for sale.
Activities: special events.
Hours & Admission Prices: Tours: May-Oct. 1st & 3rd Sun. 1-4. Closed federal holidays.

PLOCH ART GALLERY AT THE WILSON CENTER, Brookfield's Mitchell Park, 19805 W. Capitol Dr., Brookfield, WI 53045. Tel.: 262-781-9470. Fax: 262-781-9798.
E-mail: rsvp@wilson-center.com
Web Site: www.wilson-center.com
Institution Type/Description: Art Gallery.
Collections: works by Wisconsin artists.
Activities: classes.
Hours & Admission Prices: Mon.-Sat. 8:30-5.

Browntown

BROWNTOWN HISTORICAL MUSEUM, 110 S. Mill St., Browntown, WI 53520. Tel.: 608-966-3514.
Institution Type/Description: History Museum.
Collections: local history & culture; community band & school memorabilia; Native American artifacts; agricultural equipment; personal artifacts; photographs.
Hours & Admission Prices: Memorial Day to Labor Day 1st & 3rd Sun. 1-3.

Burlington

BURLINGTON HISTORICAL SOCIETY, 232 N. Perkins Blvd., Burlington, WI 53105. Tel.: 262-767-2884.
Governing Authority: Subsidiary Institutions: Pioneer Log Cabin, Wehmhoff Sq., Burlington, WI; Whitman School, Beloit St., Burlington, WI.
Institution Type/Description: Historical Society Museum.
Collections: local history & culture; photographs; genealogy; Al-Vista panoramic cameras; Burlington Liars Club artifacts; Underground Railroad; household artifacts.
Hours & Admission Prices: Sun. 1-4; other times by appointment. No charge; donations accepted.

BURLINGTON HISTORICAL SOCIETY PIONEER - LOG CABIN MUSEUM, Wehmhoff Sq., 416 N. Perkins Blvd., Burlington, WI 53105. Mailing Address: 232 N. Perkins Blvd., Burlington, WI 53105. Tel.: 262-767-2884.
Institution Type/Description: Historical Society Museum: housed in an 1850 log cabin.
Collections: local history & culture; period furnishings; early tools; vintage garden; kitchen garden.

Activities: special events.
Hours & Admission Prices: mid-May to mid-Oct. Sat. 1-4; other times by appointment. No charge; donations accepted.

BURLINGTON HISTORICAL SOCIETY - WHITMAN SCHOOL, 401 W. Beloit St., Burlington, WI 53105. Mailing Address: 232 N. Perkins Blvd., Burlington, WI 53105. Tel.: 262-767-2884.
Institution Type/Description: Historic Building: built in 1840.
Collections: local history & culture; period furnishings; photographs.
Hours & Admission Prices: By appointment. No charge; donations accepted.

CHOCOLATE EXPERIENCE MUSEUM, 113 E. Chestnut St., Ste. B, Burlington, WI 53105. Tel.: 262-763-6044. Fax: 262-763-3631.
E-mail: info@burlingtonchamber.org
Web Site: www.burlingtonchamber.org
Key Personnel: Dir., Jan Ludtke.
Governing Authority: Tax-exempt: 501(c)(3).
Institution Type/Description: History Museum.
Collections: chocolate history; chocolate sculptures & artifacts.
Activities: special events. Annual Event: Chocolate Fest in May.
Publications: Discover the Treasures.
Hours & Admission Prices: Mon.-Fri. 9-5, Sat. 10-2. No charge; donations accepted. Closed federal holidays.

LOGIC PUZZLE MUSEUM, 533 Milwaukee Ave., Burlington, WI 53105. Tel.: 262-763-3946.
E-mail: logicpuzzlemuseum@hotmail.com
Web Site: www.logicpuzzlemuseum.org
Founded: 1995.
Personnel Profile: Part-Time Volunteers 3.
Institution Type/Description: Toy Museum; hands-on.
Collections: mechanical, brainteaser & logic puzzles from early days to present; Chinese checkerboards; Victorian parlor puzzles & toys; sliding block style puzzles.
Major Exhibits: Rebus Puzzles, 2/14; Hands-on Mystery Objects, 11/14.
Facilities: Museum-related items for sale.
Activities: 60 hands-on mechanical & logic puzzles; make a brainteaser puzzle; group tours; I Spy Hunt.
Hours & Admission Prices: Call for hours. Admission $8. &

SPINNING TOP & YO-YO MUSEUM, 533 Milwaukee Ave., Burlington, WI 53105. Tel.: 262-763-3946.
E-mail: thetopmuseum@hotmail.com
Web Site: www.topmuseum.org
Founded: 1987.
Institution Type/Description: Toy Museum.
Collections: spinning toy tops; yo-yos; gyroscopes; diabolos; trompos; trottoles; toupies; kreisels; komas; spinning top games; yo-yo awards & memorabilia; Cracker Jack tops; spinning top premiums; dreidels; peg tops; string spinners; MGM movie My Summer Story props; Fudge TV show memorabilia; ephemera; vintage Valentine's picturing spinning tops; magnetic tops; Carrom Boards; optical tops; whip tops; rattlebacks; postage stamps picturing toy tops; store display boxes; Chicago Siren King tops & ads; Whitman top spinning game; Duncan tops & yo-yos; pump tops, finger tops; advertising tops; youpay spinners; put n' take games; Gasing Malaysia; period & modern spinning tops & toys; hands-on exhibits.
Major Exhibits: Gyroscopes, 1/14-2/14; 1000 Yo-Yos & Memorabilia, 3/14.
Facilities: Museum-related items for sale.
Activities: presentations; classes; hands-on tops, top games & exhibits; yo-yo presentations; videos. Annual Event: Yo-Yo Convention in spring.
Hours & Admission Prices: By appointment; see website for hours & admission prices. &

Cable

CABLE NATURAL HISTORY MUSEUM, (M), 13470 County Hwy. M, Cable, WI 54821. Mailing Address: P.O. Box 416, Cable, WI 54821-0416. Tel.: 715-798-3890. Fax: 715-798-3828.
E-mail: info@cablemuseum.org
Web Site: cablemuseum.org
Founded: 1968.
Congressional District: 7
Key Personnel: Chm. (V), Ronald G. Anderson; Devel. Coord., Peter Mansfield; Devel. Asst. & Bookkeeper, Diane Cooper; Administrative Asst., Shari Cole; Administrative Dir., Deb Malesevich; Museum Shop Mgr., Penny Johnson.

Personnel Profile: Part-Time Paid 4; Part-Time Volunteers 100.
Governing Authority: nonprofit organization. Tax-exempt: 501(c)(3).
Institution Type/Description: Natural History Museum.
Collections: natural history items.
Facilities: nature trail; outdoor classroom.
Activities: lectures; organized education programs for children; workshops for children & adults; school outreach programs; field trips; annual scholarship; travel programs.
Publications: biannual, Museum Messenger; weekly newsletter.
Hours & Admission Prices: Tues.-Sat. 10-4. Adults $5; members & children no charge. &
Attendance: 25,000 (estimated)
Membership: Trillium $30-$99; Whitetail Deer $100-$249; Monarch Butterfly $250-$499; Painted Turtle $500-$999; Wood Duck $1,000-$2,499; Loon $2,500-$4,999; Dragonfly $5,000 & up.

Cambridge

CAMBRIDGE HISTORIC MUSEUM, 213 South St., Cambridge, WI 53523-9617. Mailing Address: P.O. Box 214, Cambridge, WI 53523-0214. Tel.: 608-423-3327.
E-mail: koplin1@msn.com
Web Site: www.cambridgehistoricmuseum.org
Key Personnel: Dir., Nancy Koplin; Pres. (V), Eileen M. Scott.
Personnel Profile: Full-Time Volunteers 1.
Governing Authority: Tax-exempt.
Institution Type/Description: Historic Building: housed in the former Cambridge school building built in 1906.
Collections: school history; Far East & African artifacts; farm equipment; photographs; personal artifacts.
Publications: newsletter.
Hours & Admission Prices: May-Oct. Tues.-Sat. 10-3; other times by appointment. No charge; donations accepted. Closed holidays.
Membership: Senior $15; Individual $20; Family $35; Organization $200.

Cameron

BARRON COUNTY HISTORICAL SOCIETY'S PIONEER VILLAGE MUSEUM, 1866 13 1/2 14th Ave., Cameron, WI 54822. Mailing Address: P.O. Box 242, Cameron, WI 54822-0242. Tel.: 715-458-2080.
E-mail: museum1@chibardun.net
Web Site: www.barroncountymuseum.com
Founded: 1960.
Congressional District: 75
Key Personnel: Pres., Jack Nedland; Dir., Caroline E. Olson.
Personnel Profile: Full-Time Paid 1; Part-Time Paid 1; Part-Time Volunteers 280.
Governing Authority: nonprofit organization. Affiliated with State Historical Society. Tax-exempt.
Institution Type/Description: state and local history.
Collections: agricultural, logging & household items; American Indian artifacts; books & manuscripts; dental, medical & veterinary tools; barbershop; depot; farmstead; dentists office; general store; post office; doctor's office; meeting house; filling station; town hall; Jerome Hall Display Bldg.; blacksmith shop; newspaper & printing office; law office; woodwork shop; leather shop-shoe & harness; cheese factory; saloon; period restored outboard motors; 1920 boat used by lumber baron, F.D. Stout; fishing equipment; 1958 Edsel; 1908 International Auto Wagon; 1934 Chevrolet; 1926 Pontiac; 1921 Ford. Historic Buildings: 1906 Joliet Schoolhouse; 1908 Ebenezer Lutheran Church; 1890 Hedin log house & contents; 1900 stone jail.
Activities: guided tours; formally organized education programs for children. Museum Sponsors: Side Kar Up Nort' Rally in June; Heritage Days in July; Bluegrass Festival in July; Quilt Show in July; Treadle-On in July; Treadle Sewing Machine Club in July; Voyageur Encampment in August; Classic Car Show in August; Judy Shaide Painting Demo & Art Show in August; Vintage Baseball in August.
Publications: quarterly, Pioneer Post.
Hours & Admission Prices: June-Labor Day Thurs.-Sun. 1-5. Adults $8; children 12 & under $4; discounts to groups of 35 or more; children under 5 no charge. &
Attendance: 5,000 (estimated)
Membership: Single $10; Couple $18; Supporting $50; Benefactor $100; Business Partnership $200.

Camp Douglas

WISCONSIN NATIONAL GUARD MUSEUM, 101 Independence Dr., Volk Field, Camp Douglas, WI 54618. Tel.: 608-427-1280. Fax: 608-427-1399.
E-mail: eric.lent@dva.state.wi.us
Web Site: www.wisvetsmuseum.com
Founded: 1984.
Congressional District: 6
Key Personnel: Dir., Michael Telzrow; Cur., Eric Lent.
Personnel Profile: Full-Time Paid 5; Full-Time Volunteers 2; Part-Time Paid 1; Part-Time Volunteers 3.
Governing Authority: state. Parent Institution: State of Wisconsin; Wisconsin Dept. of Veterans' Affairs; Wisconsin Veterans Museum. Tax-exempt.
Institution Type/Description: Military Museum: housed in 1896 rustic lodge made of white pine logs.
Collections: artifacts & archival materials relative to the history of the men & women of the Wisconsin National Guard from pre-Civil War times to present; general military items that help to describe the changes in warfare & weapons; aircraft & military equipment.
Research Fields: ongoing investigations to support exhibit work & programs.
Facilities: 2,500 sq. ft. exhibit space.
Activities: slide programs; guided tours; lectures; loan exhibitions; observe air operations on Volk Field. Museum Sponsors: Volk Field fly-in & open house in even number years. Annual Event: Veterans Day Winnebago Pow-Wow.
Publications: newsletter published 3 times annually, Volunteer; occasional information & guides, Panorama.
Hours & Admission Prices: Wed.-Sat. 9-4, Sun. 10-2; call to confirm hours. No charge. Closed New Year's Day; Easter; Thanksgiving; Christmas. &
Attendance: 18,500 (accurate)
Membership: Associate $10; Bronze Friends $25; Silver Friends $50; Gold Friends $75; Platinum Friends & Organizational $100; Corporate $1,000.

Campbellsport

HENRY S. REUSS ICE AGE VISITOR CENTER, DNR, Kettle Moraine State Forest - Northern Unit, N2875 Hwy. 67, Campbellsport, WI 53010. Mailing Address: N1765 County Hwy. G, Campbellsport, WI 53010. Tel.: 920-533-8322; 262-626-2116. Fax: 262-626-2117.
E-mail: jackie.scharfenberg@wisconsin.gov
Web Site: dnr.wi.gov/topic/parks/name/kmn/naturecenter.html
Founded: 1980.
Congressional District: 6
Key Personnel: Supt., Jerry Leiterman; Forest Naturalist, Jackie Scharfenberg.
Governing Authority: state. Dept. of Natural Resources, Box 7921, Madison, WI 53707. Affiliated area of National Park Service. Tax-exempt.
Institution Type/Description: Park & Geology Museum.
Collections: glacial geology; films.
Research Fields: glacial geology; forest & aquatic ecosystems.
Activities: guided tours; interpretive programs; films.
Publications: brochures; film, Night of the Sun.
Hours & Admission Prices: April-Oct. Mon.-Fri. 8:30-4, Sat.-Sun. 9:30-5; Nov.-March call for hours. No charge; donations accepted. Closed Thanksgiving; Christmas. &
Attendance: 28,000 (estimated)

Cassville

STONEFIELD HISTORIC SITE, 12195 County Rd. V V, Cassville, WI 53806. Mailing Address: P.O. Box 125, Cassville, WI 53806-0125. Tel.: 608-725-5210. Fax: 608-725-5919.
E-mail: stonefield@wisconsinhistory.org
Web Site: stonefield.wisconsinhistory.org
Founded: 1952.
Congressional District: 3
Key Personnel: Dir., Allen Schroeder.
Personnel Profile: Full-Time Paid 2; Part-Time Paid 10; Part-Time Volunteers 25.
Governing Authority: state. Parent Institution: State Historical Society of Wisconsin, 816 State St., Madison, WI 53706.
Institution Type/Description: State Agricultural Museum: housed in a re-created 1890 village located on the 1868 estate of Wisconsin's first state governor, Nelson Dewey. An Historic Site.
Collections: agricultural & trades tools; professional, commercial & domestic artifacts.
Facilities: 2,000 acre estate.
Activities: guided tours; special events.
Publications: Guide to Stonefield.

Hours & Admission Prices: Memorial Day to early Oct. daily 10-4. Family (up to two adults & two or more children) $24, adults $9, students & senior citizens $7.75, children 5-17 $4.50; discount to groups; children under 5 & History Lover members no charge.
Attendance: 7,200 (accurate)
Membership: Wisconsin Historical Society: Individual $45; Household $60; History Lover $100; History Guardian $250; History Ambassador $500; Heritage Circle $1,000.

Cecil

WISCONSIN BOWHUNTING HERITAGE FOUNDATION INC. MUSEUM, 5055 Co. Hwy. V, Cecil, WI 54111. Mailing Address: William Friede Complex, 17 E. Third St., P.O. Box 94, Clintonville, WI 54929-0094.
Founded: 2004.
Key Personnel: Pres. (V), Brian Tessmann; Treas., Chuck Matyska.
Volunteer Hours: 250
Governing Authority: private; nonprofit organization. Tax-exempt: 501(c)(3).
Institution Type/Description: Bowhunting History Museum.
Collections: bowhunting history, equipment & artifacts; archery; bows & cases; Wisconsin license tags; bow manufacturers; photographs.
Facilities: library of archery magazines; 1,750 sq. ft. exhibit space
Activities: educational programs; guided tours. Museum Sponsors: Open House in February.
Hours & Admission Prices: Mon.-Fri. 10-4, Sat. by appointment. No charge; donations accepted. Closed holidays. &
Attendance: 400 (estimated)

Cedarburg

GENERAL STORE MUSEUM, W61 N480 Washington Ave., Cedarburg, WI 53012-2426. Tel.: 262-375-3676.
E-mail: cccenter@ameritech.net
Web Site: www.cedarburgculturalcenter.org
Governing Authority: Parent Institution: Cedarburg Cultural Center.
Institution Type/Description: History Museum: housed in a restored 1860's era frame building.
Collections: The Roger C. Christensen Collection of antique packaging & advertising art; photographs.
Hours & Admission Prices: Call for hours.

KUHEFUSS HOUSE MUSEUM, W63 N627 Washington Ave., Cedarburg, WI 53012-1945. Mailing Address: W62 N546 Washington Ave., Cedarburg, WI 53012-0084. Tel.: 262-375-3676. Fax: 262-375-4120.
E-mail: cccmail@artmusichistory.org
Web Site: www.cedarburgculturalcenter.org
Key Personnel: Exec. Dir., Lauren Rose Hofland; Mktg. & Membership Coord., Laura Mandella; Facilities & Volunteer Mgr., Jean Lambo; Historic Programs & Graphics Coord., Sue Gyarmati; Education & Exhibits Mgr., Jeanette Gabrys.
Governing Authority: Parent Institution: Cedarburg Cultural Center.
Institution Type/Description: Historic House.
Collections: family photographs & memorabilia.
Hours & Admission Prices: Call for hours.

OZAUKEE ART CENTER, W 62 N 718 Riveredge Dr., Cedarburg, WI 53012-1337. Tel.: 262-377-8230; 262-377-7220.
E-mail: pjyank@pauljyank.com
Founded: 1971.
Congressional District: 9
Key Personnel: Dir., Paul Yank.
Governing Authority: nonprofit organization. Parent Institution: Wisconsin Fine Arts Association, Inc. Tax-exempt.
Institution Type/Description: Art Gallery: housed in 1843 Cedarburg brewery.
Collections: paintings; sculpture; prints; ceramics; jewelry; glass; photographs.
Facilities: studio. Art for sale.
Activities: guided tours; arts festivals; lectures; classes; formally organized education programs for adults & children.
Publications: monthly newsletter.
Hours & Admission Prices: Call for hours. &
Attendance: 10,000 (estimated)

WISCONSIN MUSEUM OF QUILTS & FIBER ART, (M), N50 W5050 Portland Rd., Cedarburg, WI 53012-2158. Mailing Address: P.O. Box 562, Cedarburg, WI 53012-0562. Tel.: 262-546-0300.
E-mail: info@wiquiltmuseum.com
Web Site: www.wiquiltmuseum.com
Founded: 2001.
Key Personnel: Exec. Dir., Melissa Wraalstad; Museum Shop Mgr., Betty Schmidt.
Personnel Profile: Full-Time Paid 1; Full-Time Volunteers 2; Part-Time Volunteers 35.
Governing Authority: Tax-exempt.
Institution Type/Description: History Museum.
Collections: artwork & quilts of mid-western artists.
Research Fields: quilt documentations; fiber arts.
Facilities: educational facilities; rental facilities. Museum-related items for sale.
Publications: biannual newsletter; weekly e-newsletter, Barn Blast.
Hours & Admission Prices: Wed.-Sat. 10-4, Sun. 12-4; other times by appointment. Adults $6, seniors & students $4; AAM & museum members no charge. Closed New Year's Day; Independence Day; Thanksgiving; Christmas. &
Attendance: 2,000 (accurate)
Membership: Student $15; Individual $30; Family $50; Donor $125; Patron $250; Benefactor $500; Founder $1,000.

Chilton

CALUMET COUNTY HISTORICAL SOCIETY, INC., 928 Wieting Ct., Chilton, WI 53014. Tel.: 920-849-4042.
Founded: 1963.
Congressional District: 6
Key Personnel: Pres., Terry Friederichs; Vice Pres., Chuck Schuknecht; Treas., Karen Gerhartz; Historian & Sec., Doris Zarling.
Personnel Profile: Part-Time Volunteers 4.
Governing Authority: nonprofit organization. Tax-exempt.
Institution Type/Description: Farm Museum.
Collections: agriculture; general.
Activities: guided tours; permanent & temporary exhibitions.
Hours & Admission Prices: June-Aug. Sun. 1-4; other times by appointment. No charge; donations accepted. &
Attendance: 200 (estimated)
Membership: Senior 65 & over $5; Individual $10; Family $15; Corporate $25; Lifetime $100.

Chippewa Falls

CHIPPEWA COUNTY HISTORICAL SOCIETY, 123 Allen St., Chippewa Falls, WI 54729-2898. Tel.: 715-723-4399.
Web Site: www.chippewacoountyhistoricalsociety.org
Founded: 1964.
Key Personnel: Pres. (V), Dave Gordon.
Personnel Profile: Part-Time Volunteers 18; Interns 1.
Volunteer Hours: 4,610
Operating Expenses: 28,000
Operating Income: 30,000
Governing Authority: Tax-exempt.
Institution Type/Description: Historical Society Museum: housed in the former Notre Dame Convent; built in 1883.
Collections: local history & culture; period furnishings; personal artifacts; photographs.
Activities: annual event: The Past Passed Here.
Publications: The Eagle Speaks quarterly newsletter.
Hours & Admission Prices: Tues. 9-4. No charge donations accepted.
Attendance: 510 (accurate)
Membership: Individual $20; Family $25; Contributing Member $50; Life Time Individual $125; Life Time Couple $175.

CHIPPEWA FALLS MUSEUM OF INDUSTRY AND TECHNOLOGY, 21 E. Grand Ave., Chippewa Falls, WI 54729-2560. Mailing Address: P.O. Box 711, Chippewa Falls, WI 54729-0711. Tel.: 715-720-9206. Fax: 715-720-9206.
E-mail: info.cfmit@gmail.com
Web Site: www.cfmit.org
Founded: 1998.
Congressional District: 3
Key Personnel: Dir., Sally Sweet; Pres. (V), Pamela Cernocky.
Personnel Profile: Part-Time Paid 1.
Governing Authority: Tax-exempt.

Institution Type/Description: Industry & Technology Museum.
Collections: local history of manufacturing & processing industries; photographs; computers.
Hours & Admission Prices: Thurs.-Sat. 10-3; other times by appointment. Adults 18 & over $5, children 13-17 $3, children under 12 $1; members no charge. Closed Independence Day; Thanksgiving; Christmas.
Membership: Homesteader $30; Explorer $100; Voyageur $250; Pathfinder $500; Heritage $1,000. Business Level: Bronze $100; Silver $250; Gold $500; Platinum $1,000; Patron $2,500.

COOK-RUTLEDGE MANSION, 505 W. Grand Ave., Chippewa Falls, WI 54729. Tel.: 715-723-7181.
E-mail: info@cookrutledgemansion.com
Web Site: www.cookrutledgemansion.com
Governing Authority: nonprofit organization.
Institution Type/Description: Historic House Museum: housed in the former home of Wisconsin Lt. Governor James Bingham; built in 1873.
Collections: local history & culture; period furnishings; personal artifacts; photographs.
Activities: guided tours; special events; rental facilities. Museum Sponsors: Christmas Open House in December.
Hours & Admission Prices: Tours: June-Aug. Thurs.-Sun. 2pm; Dec. call for hours. Adults $5.

Clear Lake

CLEAR LAKE AREA HISTORICAL MUSEUM, 450 Fifth Ave., Clear Lake, WI 54005. Mailing Address: P.O. Box 242, Clear Lake, WI 54005-0242. Tel.: 715-263-3050 & 2042.
Founded: 1977.
Congressional District: 10
Key Personnel: Pres. (V) & Cur., Charles T. Clark; Vice Pres., Tim Wyss; Treas. & Sec., Ardeth Clark.
Personnel Profile: Part-Time Volunteers 8.
Governing Authority: nonprofit organization. Tax-exempt: 501(c)(3).
Institution Type/Description: Historical Society Museum: housed in 1912 Old Brick Elementary School.
Collections: Clear Lake area from 1875 to present; Sen. Gaylord Nelson; Baseball Hall of Famer, Burleigh Grimes; films from Gaylord Nelson's career & Burleigh Grimes; military artifacts.
Research Fields: local history.
Facilities: library; reading room. Museum-related items for sale.
Activities: tours; permanent exhibitions; films.
Publications: brochure; monthly newsletter, Clear Lake Museum Chronicle.
Hours & Admission Prices: Memorial Day to Labor Day Tues. & Fri. 11-4, Sun. 1:30-4:30; other times by appointment. No charge; donations accepted.
Attendance: 1,200 (estimated)
Membership: Individual $25.

Clintonville

FOUR WHEEL DRIVE FOUNDATION, Foot of E. 11th St., Clintonville, WI 54929. Mailing Address: 105 E. 12th St., Clintonville, WI 54929-1518. Tel.: 715-823-2141, ext. 1209. Fax: 715-823-5768.
Founded: 1948.
Key Personnel: Pres., James M. Green.
Governing Authority: nonprofit organization. Parent Institution: FWD Corporation. Tax-exempt.
Institution Type/Description: Company Museum: housed in 1906, Zachow-Besserdick Machine Shop.
Collections: racing and passenger cars; trucks; fire engines; vintage vehicles; World War I ammunition carriers; 1911, passenger car; 1942, FWD/Eliason motor toboggan.
Activities: guided tours.
Hours & Admission Prices: By appointment only. No charge; donations accepted. &

Attendance: 500

Colfax

COLFAX RAILROAD MUSEUM, 500 E. Railroad Ave., Colfax, WI 54730. Mailing Address: P.O. Box 383, Colfax, WI 54730. Tel.: 715-962-2076.
E-mail: colfaxrr@wwt.net
Web Site: www.colfaxrrmuseum.org
Founded: 1999.
Congressional District: 3
Key Personnel: Chm. (V), Herbert Sakalaucks.

Personnel Profile: Full-Time Volunteers 2; Part-Time Volunteers 10.
Governing Authority: Tax-exempt.
Institution Type/Description: History Museum.
Collections: railroad equipment; Soo Line caboose #273; Barney & Smith heavyweight coach #991; Soo Line GP-30 & #703; lanterns; railroad china & paper-weights.
Hours & Admission Prices: May & Sept.-Oct. Sat.-Sun. 11-4; June-Aug. Thurs.-Sun. 11-4; other times by appointment. Adults $3.50; discounts to AAM members. &
Attendance: 1,800 (estimated)

Coon Valley

NORSKEDALEN NATURE & HERITAGE CENTER, INC., N455 O. Ophus Rd., Coon Valley, WI 54623. Mailing Address: P.O. Box 235, Coon Valley, WI 54623-0235. Tel.: 608-452-3424. Fax: 608-452-3424.
E-mail: info@norskedalen.org
Web Site: www.norskedalen.org
Founded: 1977.
Key Personnel: Business Mgr., Tammy Potaracke; Exec. Dir., Christine Hall; Pres. (V), Hans Schroeder; Mktg. & Education Coord., Tim Larson; Museum Shop Mgr., Kay Vance.
Personnel Profile: Full-Time Paid 3; Part-Time Paid 18; Part-Time Volunteers 80; Interns 2.
Governing Authority: Tax-exempt.
Institution Type/Description: Nature & Heritage Center.
Collections: hands-on nature exhibits; local natural & cultural heritage history. Historic Building: 1800s log homestead.
Facilities: 7 mile nature & hiking trails; nature center. Museum-related items for sale.
Activities: special events; classes; educational programs; guided tours; hiking.
Publications: quarterly members' newsletter, Crossings.
Hours & Admission Prices: Thrune Visitors' Center, Exhibits, Arboretum, Trails, Library: April & Nov.-Dec. Mon.-Fri. 8-4, Sun. 12-4; May-Oct. Mon.-Fri. 9-5, Sat. 10-5, Sun. 12-5. Farm: June-Aug. Fri. 9-5, Sat. 10-5, Sun. 12-5. Families $15, adults $6, children K-12 $3; discounts to AAA members & groups of 10 or more by appointment; members no charge. Closed New Year's Eve & Day; Good Friday; Easter; Thanksgiving & day After; Christmas Eve & Day. &
Attendance: 14,000
Membership: Student $20; Individual $30; Family $45; Trailblazer $100; Pathfinder $250; Pioneer $500; Life $1,000.

Crandon

FOREST COUNTY POTAWATOMI CULTURAL CENTER AND MUSEUM, 5460 Everybody's Rd., Crandon, WI 54520. Mailing Address: P.O. Box 340, Crandon, WI 54520-0340. Tel.: 800-960-5479.
Web Site: www.potawatomimuseum.com
Key Personnel: Dir., Mike Alloway, Sr.; Tribal Librarian, Kim Wensaut
Institution Type/Description: Cultural Center & Museum.
Collections: Potawatomi life & culture; area history; photographs.
Facilities: Museum-related items for sale.
Hours & Admission Prices: Mon.-Fri. 9-4, Sat. by appointment. Adults $3, children 5-12 & senior citizens over 55 $1; discounts to groups; children under 5 no charge.

Danbury

FORTS FOLLE AVOINE HISTORIC PARK, 8500 County Rd. U, Danbury, WI 54830-9351. Tel.: 715-866-8890. Fax: 715-866-8081.
E-mail: fahp@centurytel.net
Web Site: www.theforts.org
Founded: 1945.
Congressional District: 7
Key Personnel: Dir., Steve Wierschem; Pres., Dianne Gravesen; Museum Shop Mgr., Sue Long.
Personnel Profile: Full-Time Paid 2; Part-Time Paid 1; Part-Time Volunteers 317.
Governing Authority: county; society. Parent Institution: Burnett County Historical Society. Tax-exempt.
Institution Type/Description: Fur Trade Museum.
Collections: 18th & 19th-century material relating to North American fur trade; Native American culture of Woodlands, Ojibwe; blacksmith shop; fur trade post; log country school.
Research Fields: archaeology; fur trade; county history.
Facilities: library; visitors center.

Activities: historic site tours. Museum Sponsors: Fur Trade Rendezvous; BBQ Fest; Yellow River Echoes - Living History Event, 1802-1805; Christmas at the Fort.
Publications: The Power of Sand; Cecilia; Voices from Our Past; newsletter 3 times a year.
Hours & Admission Prices: Memorial Day to Labor Day Wed.-Sun. 10-4; Sept. Sat.-Sun. 10-4. Adults $7, children under 13 $5; children 5 & under and members no charge. &
Attendance: 12,000 (estimated)
Membership: Paddler $25; Trader $50; Explorer $100; Voyageur $500.

De Pere

ONEIDA NATION MUSEUM, W892 County Trunk EE, De Pere, WI 54115. Mailing Address: P.O. Box 365, Oneida, WI 54155-0365. Tel.: 920-869-2768. Fax: 920-869-2959.
Web Site: www.oneidanation.org/museum
Founded: 1979.
Congressional District: 8
Key Personnel: Dir., Rita Lara; Asst. Dir., Sara Summers-Luedtke; Administrative Asst., Susan Peterson.
Personnel Profile: Full-Time Paid 6.
Governing Authority: Indian Nation; Oneida Business Committee. Parent Institution: Oneida Tribe of Indians of Wisconsin. Tax-exempt.
Institution Type/Description: History & Cultural Museum: located on site of original Wisconsin reservation land.
Collections: Oneida Nation and Iroquois artifacts; Oneida & Iroquois history, culture & arts; Native American artifacts; tools; clothing; photographs; pre-Columbian to present history; formation of the League of the Iroquois; men's, women's & children's roles in society; Civil War artifacts; lace; dolls; quillwork; moose hair work; weapons; toys; medicines.
Research Fields: tribal history; genealogy; oral histories of Oneida elders; photographic collections; Oneida & Iroquois culture; regional research center: University of Wisconsin-Green Bay; State Historical Society.
Facilities: nature trail; medicinal garden; vegetable garden; picnic area. Gift items, local artwork, books & souvenirs for sale.
Activities: guided tours; workshops; lectures; gallery talks; hobby workshops; formally organized education programs; loan, permanent, temporary & participatory exhibitions; off-site hands-on program; annual cultural festival; children's activities in summer; special events.
Publications: newsletter; brochure.
Hours & Admission Prices: Feb.-May & Sept.-Dec. Tues.-Fri. 9-5; June-Aug. Tues.-Sat. 9-5; group tours by appointment. Adults $2, children $1; discounts to AAM members; Oneida Tribal members & employees no charge. Closed most major holidays. &
Attendance: 12,472 (accurate)

WHITE PILLARS MUSEUM, DE PERE HISTORICAL SOCIETY, 403 N. Broadway, De Pere, WI 54115-2511. Tel.: 920-336-3877.
E-mail: info@deperehistoricalsociety.org
Web Site: www.deperehistoricalsociety.org
Founded: 1970.
Congressional District: 8
Key Personnel: Pres. (V), Mary Ann Schumerth.
Personnel Profile: Part-Time Paid 1; Part-Time Volunteers 15.
Governing Authority: nonprofit organization. Tax-exempt: 501(c)(3).
Institution Type/Description: Local History Museum: housed in 1836 Greek Revival bank building.
Collections: archives; artifacts; local history; photographs dating back to 1870; funeral cards; city maps; yearbooks; tax records.
Research Fields: city & local history.
Facilities: library containing city & school records from 1857.
Activities: lectures; video presentations; picture shows.
Publications: newsletter.
Hours & Admission Prices: Mon.-Thurs. 2-6, Fri. 11-3; other times by appointment. No charge; donations accepted. Closed holidays. &
Attendance: 1,875 (accurate)
Membership: Sponsor $20; Benefactor $50; Patron $100; Founder $250; Millennium $500 & up.

Delafield

HAWKS INN HISTORICAL SOCIETY, INC., 426 Wells St., Delafield, WI 53018-1419. Mailing Address: P.O. Box 180104, Delafield, WI 53018-0104. Tel.: 262-646-4794.
Web Site: www.hawksinn.org
Founded: 1960.
Congressional District: 9
Key Personnel: Pres. (V), Mary Daniel; Museum Shop Mgr., Ruth Brehmer.

Personnel Profile: Part-Time Volunteers 25; Interns 1.
Governing Authority: Hawks Inn Historical Society, Inc. Affiliated with the Wisconsin State Historical Society. Tax-exempt.
Institution Type/Description: Historic Building Museum: 1846, restored stage coach inn.
Collections: 1846-1865; furnishings; historical artifacts from the early settlement period.
Research Fields: crafts & furniture of the 1800's; cooking & kitchen use; history of settlement period 1830-present; genealogical research of pioneer settlers of the Township Delafield & surrounding area.
Facilities: visitor's center; multi-purpose room. Museum-related items for sale.
Activities: guided tours; lectures; study clubs; docent program; program of developing craft programs based on daily activities of 1846-1863 pioneer Wisconsin; rental facilities.
Publications: quarterly newsletter, Hawks Inn Newsletter.
Hours & Admission Prices: May-Oct. Sat. 1-4; tours by appointment. No charge, donations accepted. &
Attendance: 1,100 (estimated)
Membership: Individual $15; Family $25; Business $50; Contributing $100; Patron $500.

ST. JOHN'S NORTHWESTERN MILITARY ACADEMY ARCHIVES & MUSEUM, 1101 Genesee St., Delafield, WI 53018-1411. Tel.: 262-646-7119. Fax: 262-646-7268.
E-mail: pkoller@sjnma.org
Web Site: www.sjnma.org
Founded: 1984.
Congressional District: 9th
Key Personnel: Supvr., Margaret H. Koller.
Personnel Profile: Part-Time Paid 1.
Governing Authority: private secondary school; nonprofit. Tax-exempt.
Institution Type/Description: History Museum: housed in 1884 military school.
Collections: academy artifacts, photographs, publications, uniforms, sabers; student items.
Facilities: 600-vol. library of yearbooks & catalogs; over 10,000 photographs; 845 books from Fred C. Best collection of Military Science & History; 1,400 sq. ft. exhibit space.
Activities: guided tours; lectures.
Publications: quarterly magazine, The Beacon.
Hours & Admission Prices: By appointment. No charge; donations accepted.

Dodgeville

IOWA COUNTY HISTORICAL SOCIETY MUSEUM, 1301 N. Bequette St., P.O. Box 44, Dodgeville, WI 53533-0044. Tel.: 608-935-7694.
E-mail: ichistory@mhtc.net
Web Site: iowacountyhistoricalsociety.org
Founded: 1976.
Congressional District: 2
Key Personnel: Dir., John Hess.
Personnel Profile: Full-Time Paid 12; Part-Time Paid 1; Part-Time Volunteers 12.
Governing Authority: Tax-exempt.
Institution Type/Description: Historical Society Museum.
Collections: local history & culture; photographs; period artifacts.
Research Fields: genealogy.
Facilities: Floyd school; Dodge mining camp cabin.
Publications: quarterly newsletter.
Hours & Admission Prices: Mon.-Fri. 1-4; other times by appointment. No charge; donations accepted.
Attendance: 500 (estimated)
Membership: Individual $10; Couple $15; Historian $25; Master Historian $50.

Eagle

OLD WORLD WISCONSIN, W372 S9727 Hwy. 67, Eagle, WI 53119-2004. Mailing Address: P.O. Box 69, Eagle, WI 53119-0069. Tel.: 262-594-6300. Fax: 262-594-6342.
E-mail: oww@wisconsinhistory.org
Web Site: www.oldworldwisconsin.org
Founded: 1976.
Congressional District: 9
Key Personnel: Dir., Dan Freas; Cur. Research, Martin Perkins; Cur. Interpretation, Jennifer Van Haaften; Volunteer Coord., Jeni Miller.
Personnel Profile: Full-Time Paid 15; Part-Time Paid 200; Part-Time Volunteers 200.
Governing Authority: state. Parent Institution: The State Historical Society of Wisconsin, 816 State St., Madison, WI 53706. Tax-exempt.

Institution Type/Description: Ethnic Museum & Historic Village.
Collections: over 65 original buildings built by 19th & early 20th-century immigrants; relocated, restored & arranged in 10 ethnic farmsteads & 1870s rural village.
Research Fields: 19th- & early 20th-century Wisconsin & American history with emphasis on immigration, architecture & social institutions; folkways; rural history.
Facilities: dining facilities in restored barn; picnic areas. Gifts items for sale.
Activities: orientation film; special events & exhibitions; living history interpretation; heirloom gardens; 19th-century farms; adult & children's workshops; shuttle tram transport; audio tour guides.
Publications: four booklets on Black, Danish, Finnish, German & Norwegian immigration & settlement in Wisconsin; book, Old World Wisconsin: America's Heartland, A Guide to Our Past.
Hours & Admission Prices: May-June 8 Mon.-Fri. 10-3, Sat. 10-5; June 9-June 30 Mon.-Fri. 10-4, Sat. 10-5, Sun. 12-5; July -Sept. 3 Mon.-Sat. 10-5, Sun. 12-5; Sept. 4-Oct. Mon.-Fri. 10-3, Sat. 10-5, Sun. 12-5. Adults $16, senior citizens 65 & over and students $14, children 5-17 $9; discounts to groups & members; children under 5 no charge. &
Attendance: 75,000 (estimated)
Membership: Individual $30; Individual & Guest or Family $50; Family & Guest $65; Sustainer $125; Sponsor $350; Patron $500; Benefactor $1,000.

Eagle River

NORTHWOODS CHILDREN'S MUSEUM, 346 W. Division St., Eagle River, WI 54521. Mailing Address: P.O. Box 216, Eagle River, WI 54521-0216. Tel.: 715-479-4623. Fax: 715-479-3289.
E-mail: ncm.er@frontier.com
Web Site: www.northwoodschildrensmuseum.com
Founded: 1998.
Key Personnel: Dir. & Museum Shop Mgr., Rouleen Gartner; Pres. (V), Stacy Richie.
Personnel Profile: Full-Time Paid 4; Part-Time Paid 8; Part-Time Volunteers 200.
Governing Authority: Tax-exempt.
Institution Type/Description: Children's Museum.
Collections: hands-on exhibits.
Facilities: Museum-related items for sale.
Activities: special events; birthday parties; programs; art & craft workshops.
Publications: newsletters.
Hours & Admission Prices: Memorial Day to Labor Day Mon.-Sat. 10-5, Sun. 12-5; Sept.-May Tues.-Sat. 10-5, Sun. 12-5. Admission $7 per person; ACM reciprocal membership; members no charge. Closed New Year's Day; Easter; Mother's Day; Independence Day; Labor Day; Thanksgiving; Christmas Eve & Day. &
Attendance: 25,000 (accurate)
Membership: Basic $75; Plus $110.

East Troy

EAST TROY ELECTRIC RAILROAD MUSEUM, 2002 Church St., East Troy, WI 53120-1302. Mailing Address: P.O. Box 943, East Troy, WI 53120-0943. Tel.: 262-642-3263. Fax: 262-642-3197.
E-mail: info@easttroyrr.org
Web Site: www.easttroyrr.org
Formerly: Wisconsin Trolley Museum, Inc.
Founded: 1975.
Key Personnel: Pres. (V), Ryan Jonas.
Personnel Profile: Part-Time Volunteers 60.
Governing Authority: nonprofit. Tax-exempt.
Institution Type/Description: Transportation Museum.
Collections: trolley cars; electric railway cars.
Facilities: Museum-related items for sale.
Activities: slide show; demonstration train rides; 10-mile train ride; dinner rides.
Publications: newsletter, Trolley Gazette.
Hours & Admission Prices: May-Aug. Fri. 10-4; June-Aug. Sat.-Sun. 11-4; Sept.-Oct. Sat.-Sun. 10-4. Adults $12.50, seniors $10.50, children $8; children under 3 & members no charge.
Attendance: 30,000 (estimated)
Membership: Individual $35; Family $70 (annual).

Eau Claire

CHILDREN'S MUSEUM OF EAU CLAIRE, 220 S. Barstow St., Eau Claire, WI 54701. Tel.: 715-832-5437. Fax: 715-832-5732. Facebook: Children's Museum of Eau Claire.
E-mail: info@cmec.cc

Web Site: www.cmec.cc
Founded: 2001.
Key Personnel: Dir., Darcy Way; Pres. (V), James Hanke.
Personnel Profile: Full-Time Paid 3; Part-Time Paid 14; Part-Time Volunteers 10.
Governing Authority: Tax-exempt.
Institution Type/Description: Children's Museum.
Collections: hands-on exhibitions.
Activities: birthday parties; educational programs; special events.
Hours & Admission Prices: Tues.-Wed. & Fri.-Sat. 9-5, Thurs. 9-7, Sun. 12-5. Admissions $5 per person; members & children under one no charge.
Attendance: 55,000 (estimated)
Membership: Family $80; Reciprocal & Grandparents $125; Day Care $130; (add a member $10).

* **CHIPPEWA VALLEY MUSEUM, INC., (M),** 1204 Half Moon Dr., Eau Claire, WI 54703. Mailing Address: P.O. Box 1204, Eau Claire, WI 54702-1204. Tel.: 715-834-7871. Fax: 715-834-6624.
E-mail: info@cvmuseum.com
Web Site: www.cvmuseum.com
Founded: 1966.
Congressional District: 3
Key Personnel: C.E.O. & Dir., Susan McLeod; Pres. (V), Dr. Larry Annett; Librarian, Eldbjorg Tobin; Education, Karen Jacobson; Cur., Carrie Ronnander; Volunteer Coord., Jill York; Researcher, Melissa Holmen; Facilities Mgr., Donalynn Hayden; Asst. Cur. & Office Mgr., Kathie Roy; Editor, Frank Smoot; Business Mgr., Dorie Boetcher; Mgr. Community Programs, Liz Reuter.
Personnel Profile: Full-Time Paid 8; Part-Time Paid 3; Part-Time Volunteers 346; Interns 1.
Governing Authority: private; nonprofit organization. Tax-exempt: 501(c)(3).
Institution Type/Description: Regional Historical Museum.
Collections: historical artifacts of the Chippewa Valley; manuscripts; oral histories; photographs. Historic Buildings: 1860 Lars Anderson log house; 1882 Sunnyview school; 1871-1906 Schlegelmilch House.
Research Fields: history of Chippewa Valley; Wisconsin Indian history; Hmong history; farm & rural life.
Facilities: Smoot library & archives. Museum-related items for sale.
Activities: guided tours; curriculum-related programs; mini-classes; children's activities; teacher institutes; traveling exhibits; traveling history kits. Museum Sponsors: Fourth of July celebration.
Publications: newsletter; book, Paths of the People: The Ojibwe in the Chippewa Valley; Settlement & Survival: Building Towns in the Chippewa Valley, 1850-1925; Hmong in America: Journey from a Secret War; Farm Life: A Century of Change for Farm Families & Their Neighbors; Farm Crossing: The Amazing Adventures of Addie & Zachery; Ralph Owen's Eau Claire: The Character of a City, 1884-1909; Ralph Owen's Eau Claire: The City Grows Up, 1920-1960.
Hours & Admission Prices: Memorial Day-Labor Day Mon.& Wed.-Sat. 10-5, Tues. 10-8, Sun. 1-5; Sept.-May Tues. 1-8, Wed.-Fri. & Sun. 1-5, Sat. 10-5. Adults $5, children 4-17 $2; discounts to AAM, ICOM & AASLH members; children under 4, members & Tues. 5-8 no charge. Closed New Year's Day; Thanksgiving; Christmas Eve & Day. &
Attendance: 22,012 (accurate)
Membership: Regular $30; Sustaining $60; Associate $80; Heritage Club $150; Pathfinder $250; Carson Club $500; Ingram Society $1,000.

FOSTER GALLERY, UNIVERSITY OF WISCONSIN-EAU CLAIRE, 121 Water St., Eau Claire, WI 54702-4004. Mailing Address: P.O. Box 4004, Eau Claire, WI 54702-4004. Tel.: 715-836-2328 & 3277. Fax: 715-836-4882.
E-mail: wagenetk@uwec.edu
Web Site: www.uwec.edu/art/foster
Founded: 1970.
Congressional District: 3
Key Personnel: Dir., Thomas K. Wagener; Financial Dir., Christos Theo.
Personnel Profile: Full-Time Paid 1; Part-Time Paid 9; Part-Time Volunteers 30; Interns 1.
Governing Authority: university; nonprofit organization. Parent Institution: University of Wisconsin-Eau Claire. Tax-exempt.
Institution Type/Description: University Gallery.
Collections: photographs; paintings; prints; sculpture.
Major Exhibits: Art & Tech, 1/14-2/14; Shading, 2/14-3/14; Hamilton Typeface Museum Works, 2/14-3/14; 57th Annual Student Art Show, 4/14; Hell is a Place We Made: Visions of Dante's Inferno, 10/14.
Activities: formal education programs for undergraduate & graduate college students affiliated with UW-Eau Claire; lectures; loan, participatory & traveling exhibitions.
Publications: posters & brochures for each exhibit.

Hours & Admission Prices: Mon.-Wed. & Fri. 10-4.30, Thurs. 10-4.30 & 6pm-8pm, Sat.-Sun. 1-4:30. No charge. Closed academic holidays. &
Attendance: 9,520 (accurate)

PAUL BUNYAN LOGGING CAMP MUSEUM, 1110 Carson Park Dr., Eau Claire, WI 54703. Mailing Address: P.O. Box 221, Eau Claire, WI 54702-0221. Tel.: 715-835-6200.
E-mail: blueox@clearwire.net
Web Site: www.paulbunyancamp.org
Founded: 1934.
Congressional District: 3
Key Personnel: Pres., Gordon Wall; Vice Pres., Larry Doyle; Exec. Dir. & Museum Shop Mgr., Diana Peterson; Sec., Edward Wells; Treas., John Burbank.
Personnel Profile: Part-Time Paid 7; Part-Time Volunteers 45.
Governing Authority: bd. of directors. Tax-exempt.
Institution Type/Description: Logging & Lumbering Museum.
Collections: 1890s logging camp including blacksmith shop, barn, cook shanty, bunkhouse, filer's shanty, foreman's office, wanigan, & heavy equipment shed.
Research Fields: logging & lumbering.
Facilities: interpretive center. Museum-related items for sale.
Activities: guided tours; permanent exhibitions; interactive children's room.
Publications: brochures for children & adults; booklet.
Hours & Admission Prices: May-Sept. daily 10-4:30. Adults $4, children 4-17 $2; discounts AAA members, WI teachers, & Big Brothers & Sisters. Closed Easter. &
Attendance: 12,000 (accurate)
Membership: Annual $30.

Edgerton

ALBION ACADEMY HISTORICAL MUSEUM, 605 Campus Ln., Edgerton, WI 53534. Mailing Address: 311 Park Ln., Edgerton, WI 53534. Tel.: 608-884-3896 & 6598.
Founded: 1959.
Congressional District: 2
Key Personnel: Chm. & Pres. (V), Robert Babcock; Treas., Rosemarie Burdick; Sec., Lucille Bartz.
Personnel Profile: Part-Time Volunteers 18.
Governing Authority: nonprofit society. Parent Institution: Wisconsin State Historical Society. Tax-exempt.
Institution Type/Description: Historical Society Museum: located on site of 1853, rebuilt Albion Academy, first coeducational institution of higher learning in the state of Wisconsin.
Collections: 1880s furniture; Academy books & records; school equipment; clothing; refurbished school bell; taxidermy artifacts. Historic Building: c.1860s one-room school.
Facilities: library; 100-seat auditorium; picnic grounds.
Activities: guided tours.
Publications: book, History of Albion Academy.
Hours & Admission Prices: June-Aug. Sun. 1-4. No charge; donations accepted.
Attendance: 267 (accurate)
Membership: Individual $10; Life $50.

Egg Harbor

CHIEF OSHKOSH NATIVE AMERICAN ARTS, 7631 State Hwy. 42, Egg Harbor, WI 54209-9548. Tel.: 920-868-3240. Facebook: Chief Oshkosh Native American Arts.
E-mail: chiefoshkosh97@yahoo.com
Founded: 1975.
Congressional District: 8
Key Personnel: Dir., Coleen Bins.
Governing Authority: Affiliated with the Door County Chamber of Commerce, P.O. Box 219, Sturgeon Bay, WI 54235. Tel.: 414-743-4456.
Institution Type/Description: Native American Art Gallery.
Collections: artwork produced by Native American artists.
Hours & Admission Prices: Daily 10-5. No charge; donations accepted. Closed most holidays.

CUPOLA HOUSE, 7836 Hwy. 42, Egg Harbor, WI 54209-9564. Tel.: 800-871-1871. Fax: 920-868-1710; 920-868-3922. Facebook: Cupola House.
E-mail: cupolahouse@gmail.com
Web Site: www.cupolahouse.com
Founded: 1871.
Key Personnel: C.E.O., Gloria Hansen.

Personnel Profile: Full-Time Paid 1; Part-Time Paid 4.
Governing Authority: Affiliated with the Door County Chamber of Commerce, Box 219, Sturgeon Bay, WI. 54235. Tel.: 414-743-4456.
Institution Type/Description: Historic House: 1871 house built by Levi Thorp, located on the Door County Peninsula.
Collections: period furnishings & furniture; pottery; Amish quilts; collectibles.
Facilities: restaurant. Gifts & collectibles for sale.
Activities: artist receptions; art shows. Annual Events: Amish Quilt Show; Pumpkin Patch Festival; Door County Director & Founder Christmas Tree Schooner Musical; History of Cupola House - Dr. Eames House.
Publications: brochure.
Hours & Admission Prices: Daily 10-6. No charge.
Attendance: 100,000 (estimated)

Elkhart Lake

ELKHART LAKE HISTORIC DEPOT MUSEUM, 104 S. Lake St., Elkhart Lake, WI 53020. Tel.: 920-876-2922.
Institution Type/Description: Historic Building Museum.
Collections: local history; period furnishings; photographs.
Hours & Admission Prices: Memorial Day to Labor Day call for hours.

HENSCHEL'S INDIAN MUSEUM, N8661 Holstein Rd., Elkhart Lake, WI 53020. Tel.: 920-876-3193.
E-mail: rosaliehenschel@hotmail.com
Web Site: www.henschelsindianmuseumandtroutfarm.com
Institution Type/Description: History Museum.
Collections: local history & culture; Native American artifacts; personal artifacts.
Hours & Admission Prices: Memorial Day to Labor Day Tues.-Sat. 1-5; other times by appointment. Adults $5, children $3. &

Elkhorn

WEBSTER HOUSE MUSEUM, 9 E. Rockwell, Elkhorn, WI 53121-1728. Mailing Address: P.O. Box 273, Elkhorn, WI 53121-0273. Tel.: 262-723-4248.
E-mail: walcohistory@tds.net
Web Site: walcohistory.org
Founded: 1955.
Congressional District: 1
Key Personnel: Pres. & C.E.O., Doris Reinke; Dir. Education, Barbara Shreves; Museum Shop Mgr., Linda Starks.
Personnel Profile: Full-Time Volunteers 6; Part-Time Volunteers 40.
Governing Authority: county; society. Parent Institution: Walworth County Historical Society. Tax-exempt.
Institution Type/Description: Historic House: 1830s Webster House.
Collections: children's museum; costumes; music of composer, Joseph P. Webster; history; Indian artifacts; mounted bird collection; general; agriculture. Historic Buildings: 1889 Blooming Prairie school; 1892 Sharon Town Hall; 1850s carriage barn of John W. Boyd.
Research Fields: Walworth county genealogy; history.
Facilities: 1,700-vol. library; resource center.
Activities: guided tours; lectures; permanent exhibitions; educational program for 4th & 5th graders, Morning in a One-Room School, by reservation only.
Publications: quarterly newsletter; annual pamphlets, Story of Sweet Bye and Bye; Walworth County History; Without Firing A Shot; school day essay collections, Treasured Moments; Golden Memories; North Walworth County History; audio tape, Sweet Bye and Bye; postcards; booklets; video, Webster House Museum; tape: Sweet By & By & Lorena.
Hours & Admission Prices: May to mid-Oct. Wed.-Sat. 1-5. Adults $5, student & children 6-12 $2; discounts to groups, AAM, AARP, AAA members, school tours, senior citizens & Walworth County residents; children under 6 & members no charge. &
Attendance: 6,000 (estimated)
Membership: Junior $5; Single $15; Family $20; Contributing $25; Patron $50.

Ellison Bay

DOOR COUNTY MARITIME MUSEUM (AT GILLS ROCK), 12724 Wisconsin Bay Rd., Ellison Bay, WI 54210-9796. Mailing Address: 120 N. Madison Ave., Sturgeon Bay, WI 54235-3416. Tel.: 920-743-5958. Fax: 920-743-9483. Facebook: Door County Maritime Museum.
E-mail: info@dcmm.org
Web Site: www.dcmm.org
Founded: 1969.
Congressional District: 8
Key Personnel: Exec. Dir., Bob Desh; Pres. (V), Dan Austad; Vice Pres. (V),

Jeff Weborg; Treas., Frank Forkert; Volunteer Coord., Jon Gast; Museum Shop Mgr., Jan Johnson.
Personnel Profile: Full-Time Paid 5; Part-Time Paid 25; Part-Time Volunteers 75.
Governing Authority: Parent Institution: Sturgeon Bay; registered with State Historical Society. Branch Museum: Sturgeon Bay Maritime Museum (see separate listing); Cana Island Lighthouse & Grounds. Tax-exempt.
Institution Type/Description: Maritime Museum.
Collections: artifacts of sailing, Coast Guard & commercial fishing.
Research Fields: shipbuilding & commercial fishing industries.
Facilities: Museum-related items for sale.
Activities: Annual Events: Classic Door County Fish Boil in September.
Publications: newsletter.
Hours & Admission Prices: May-Oct. daily 10-5. Adults $5, youth 5-17 $2; children 4 & under and active military no charge. &
Attendance: 5,000 (estimated)
Membership: Single $40; Two Adults $60; Family $70; Life $750.

NEWPORT STATE PARK, 475 Cty. Hwy. NP, Ellison Bay, WI 54210. Tel.: 920-854-2500. Fax: 920-854-1914.
E-mail: Michelle.Hefty@wisconsin.gov
Web Site: www.dnr.state.wi.us/Org/land/parks/specific/newport/
Key Personnel: Supt., Michelle M. Hefty.
Personnel Profile: Full-Time Paid 2; Part-Time Paid 6.
Governing Authority: state. Affiliated with the Door County Chamber of Commerce, Box 219, Sturgeon Bay, WI 54235. Tel. 414-743-4456. Tax-exempt.
Institution Type/Description: State Park: located on the Door County Peninsula.
Collections: rocks & minerals; photographs; birds & mammal (taxidermy).
Facilities: 28-mile hiking trails; 16 backpack campsites; picnic areas.
Activities: interpretive programs & hikes.
Hours & Admission Prices: Daily 6am-11pm. Call for entrance & camping fees. &

Ephraim

EPHRAIM HISTORICAL FOUNDATION, (M), 3060 Anderson Ln., Ephraim, WI 54211. Mailing Address: P.O. Box 165, Ephraim, WI 54211-0165. Tel.: 920-854-9688. Fax: 920-854-7232. Facebook: Ephraim Historical Foundation.
E-mail: info@ephraim.org
Web Site: ephraim.org
Founded: 1949.
Congressional District: 8
Key Personnel: Dir., Thea S. Thompson; Pres. (V), Marilyn Cushing; Dir. Program & Mktg., Kevin Free; Dir. Operations, Sally Jacobson.
Personnel Profile: Full-Time Paid 3; Part-Time Paid 5; Part-Time Volunteers 132.
Governing Authority: Parent Institution: The Ephraim Historical Foundation, Inc. Branch Museums: The Thomas Goodletson House; Pioneer School House; Anderson Store Museum; Anderson Barn Museum; Historic Iverson House. Tax-exempt.
Institution Type/Description: Historic House: c.1853, one of the first permanent buildings built on the Door County Peninsula.
Collections: 1853 restored home with period furnishings; original 1880 schoolhouse with period furnishings & costumes; 1850s log cabin with period furnishings; 1880s barn with Door County art exhibit & general local history & photo exhibits; 1857 general store with original fixtures.
Research Fields: Ephraim & Door county history.
Facilities: archives.
Activities: guided walking tours; Sunday night sing-alongs; child's play stories & crafts; MP3 tours; school tours; children's programming; Historic Tram Tours; Ephraim Moravian Church tours; A Cemetery Walk Through Ephraim's Past; Christmas holiday activities; children's encounter with history. Museum Sponsors: Historical Programs & Presentations in July.
Publications: quarterly newsletter; walking tour guide; cookbook, Amazing Grazing; books, Door County Letters; Did the Eagle Get You, Dr. Moss?; Images of America - Ephraim; Horseshoe Island - The Folda Years; 10 Women of Ephraim
Hours & Admission Prices: mid-June to Labor Day Tues.-Sat. 11-4; Sept. to mid-Oct. Fri.-Sat. 11-4. Adults $5, students 6-18 $3; children under 6 & members no charge. Tram Tours: adults $8, students $6. Closed Independence Day; Labor Day. &
Attendance: 5,100 (estimated)
Membership: Regular $75-$99; Supporting $100-$249; Sustaining $250-$499; Patron $500-$999; Founder's Circle $1,000 & up.

Fennimore

FENNIMORE DOLL & TOY MUSEUM, 1135 6th St., Fennimore, WI 53809. Mailing Address: P.O. Box 53, Fennimore, WI 53809. Tel.: 608-822-4100.
E-mail: dolltoy@fennimore.com
Web Site: www.dollandtoymuseum.com
Founded: 1991.
Key Personnel: Dir., Connie Neal
Institution Type/Description: Doll & Toy Museum.
Collections: collectible dolls & toys from around the world dating from 1800s to 2000; dollhouses; children's furniture.
Facilities: Museum-related items for sale.
Hours & Admission Prices: May-Oct. daily 10-4. Adults $3, students $1.50; children under no charge. &

FENNIMORE RAILROAD HISTORICAL SOCIETY MUSEUM, 610 Lincoln Ave., Fennimore, WI 53809-1559. Tel.: 608-822-6144.
Institution Type/Description: History Museum: housed in the former city power house and utility building.
Collections: local history & culture; period farm tools & equipment; military uniforms; war memorabilia; photographs; household artifacts; dinky locomotive.
Activities: miniature train rides.
Hours & Admission Prices: Memorial Day to Labor Day daily 10-4; Sept.-Oct. Sat.-Sun. 10-4; other times by appointment.

Fifield

OLD TOWN HALL MUSEUM, W7213 Pine St., Fifield, WI 54524-0156. Mailing Address: P.O. Box 156, Fifield, WI 54524-0156. Tel.: 715-339-2254.
Web Site: www.pricecountyhistoricalsociety.com
Founded: 1969.
Congressional District: 7
Key Personnel: Pres. (V), Therese Trojak; Chm. (V), Larry Wollner.
Personnel Profile: Part-Time Volunteers 20.
Governing Authority: society. Parent Institution: Price County Historical Society. Affiliated with State Historical Society, 816 State St., Madison, WI 83706. Tax-exempt.
Institution Type/Description: History Museum: housed in 1894 Old Town Hall.
Collections: artifacts of the logging era; Price County history; model living room & kitchen; one room school.
Facilities: auditorium. Crafts, books & museum-related items for sale.
Activities: lectures; demonstrations; school tours.
Publications: brochure; newsletter; books on local history.
Hours & Admission Prices: June to Labor Day Fri. & Sun. 1-5. No charge; donations accepted. &
Attendance: 714 (accurate)
Membership: Single $7; Family $9; Contributing $12; Sustaining $17; Life $150.

Fish Creek

EAGLE BLUFF LIGHTHOUSE MUSEUM, 9462 Shore Rd., Peninsula State Park, Fish Creek, WI 54212. Mailing Address: 9460 Maple Grove Rd., Fish Creek, WI 54212. Tel.: 920-421-3636. Fax: 920-839-9562.
E-mail: podge47@dcwis.com
Web Site: www.eagleblufflighthouse.org
Founded: 1963.
Congressional District: 8
Key Personnel: Pres. (V), George Evenson; Cur. & Museum Shop Mgr., Patti Podgers.
Personnel Profile: Full-Time Paid 1; Part-Time Paid 9; Part-Time Volunteers 5.
Governing Authority: society. Parent Institution: Door County Historical Society. Tax-exempt.
Institution Type/Description: Historic House: 1868 Eagle Bluff Lighthouse located on the Door County Peninsula.
Collections: period furnishings 1883-1926; original family furniture including log book.
Activities: lantern lit evening tours in early Oct.
Publications: monthly newsletter.
Hours & Admission Prices: Mid-May-late Oct. Sat.-Sun. 10-4; June-Oct. 21 daily 10-4. Adults $5, students 13-18 $2; youth 6-12 $1; children 5 & under no charge.
Attendance: 16,000 (accurate)
Membership: Individual $15; Family $25; Lifetime $150.

FRANCIS HARDY GALLERY, 3038 Anderson Ln., Fish Creek, WI 54212. Mailing Address: P.O. Box 394, Ephraim, WI 54211. Tel.: 920-854-5535 & 2210.
E-mail: info@thehardy.org
Web Site: thehardy.org
Key Personnel: Exec. Dir., Elizabeth Meissner-Gigstead; Mktg. & Community Rels. Dir., Melissa Ripp
Institution Type/Description: Art Gallery.
Collections: paintings; sculpture.
Hours & Admission Prices: mid-May to mid-Oct. daily 10-5; call for additional hours.

PENINSULA STATE PARK, 9462 Shore Rd., Fish Creek, WI 54212-9696. Tel.: 920-868-3258. Fax: 920-868-1931.
E-mail: kelli.bruns@wi.gov
Key Personnel: Supt., Kelli Bruns.
Governing Authority: state. Affiliated with the Door County Historical Society, Sturgeon Bay, WI. 54235.
Institution Type/Description: State Park: located on the Door County Peninsula.
Collections: geological, botanical & wildlife exhibits. Historic Structure: Eagle Lighthouse (see separate listing).
Facilities: picnic areas; concession stand; 467 campsites; amphitheater.
Activities: nature center; self guided nature & hiking trails; exercise trail; guided hikes; tennis; tours; lookout tower. Museum Sponsors: musical performances July-Aug.
Publications: brochure, Peninsula State Park.
Hours & Admission Prices: Daily 6am-11pm; call for reservations. Phone for entrance fees & camping fees, camping reservations in writing on state forms accepted Jan.-Sept. Nature Center: Memorial Day to Labor Day daily 10-2; Sept.-May call for hours.

Fond du Lac

CHILDREN'S MUSEUM OF FOND DU LAC, 75 W. Scott St., Fond du Lac, WI 54935. Tel.: 920-929-0707.
E-mail: info@cmfdl.org
Web Site: www.cmfdl.org
Founded: 2002.
Key Personnel: Dir., Andrea Walsh.
Personnel Profile: Full-Time Paid 4; Part-Time Paid 8; Part-Time Volunteers 40.
Governing Authority: Tax-exempt: 501(c)(3).
Institution Type/Description: Children's Museum.
Collections: hands-on exhibitions.
Facilities: Museum-related items for sale.
Activities: educational programs; special events; birthday parties; field trips.
Hours & Admission Prices: Tues.-Thurs. 9-5, Fri. 9-7, Sat.-Sun. 10-4. Admission $6; children under one & members no charge.
Attendance: 25,000 (accurate)

GALLOWAY HOUSE AND VILLAGE - BLAKELY MUSEUM, 336 Old Pioneer Rd., Fond du Lac, WI 54935-6126. Mailing Address: P.O. Box 1284, Fond du Lac, WI 54936-1284. Tel.: 920-922-1166. Facebook: Fond du Lac Historical Society.
E-mail: info@fdlhistory.com
Web Site: www.fdlhistory.com
Founded: 1948.
Congressional District: 6
Key Personnel: Pres., Mat Mueller; Vice Pres. & Sec., Tracy Qualmann; Treas., Tammy Thorton; Tour Chm., Carol Henke; Volunteer Coord., Polly May.
Personnel Profile: Full-Time Volunteers 50; Part-Time Paid 5; Part-Time Volunteers 225.
Governing Authority: nonprofit organization; society. Owned & operated by Fond du Lac County Historical Society. Branch Museum: Blakely Museum. Tax-exempt: 501(c)(3).
Institution Type/Description: Historic House & Village Museum: 30 historic buildings including Galloway house built in 1868.
Collections: Victorian household items; flower gardens; furniture; clothing; artifacts from log cabin times to late 1800s; Indian & Eskimo artifacts; pioneer tools; guns; items from American wars; early industrial tools & machines, paintings; prints; photographs; mounted animals; Susan Colman collection. Historic Buildings: mid & late 1800s village, one-room school; church; courthouse & law office; print shop; toy shop; general store; post office; blacksmith shop; photo shop; railroad depot; grist mill; carpenter shop; leather shop; carriage house; dress shop; quarry locomotive & stone car; railroad depot & caboose; fire house; gazebo; log cabin; drug store doctor's office; bank; town hall; barn; machine shop; farm machinery.

Research Fields: Fond du Lac County.
Facilities: research library on Fond du Lac County. Gifts, old fashioned candy & museum-related items for sale.
Activities: guided tours; permanent exhibitions. Museum Sponsors: Ice Cream Social; Wisconsin 3rd Infantry Regiment (Civil War); Halloween & Christmas at Galloway House; Christmas in the Postilion.
Publications: brochures & pamphlets; Fond du Lac County History Video & accompanying Book; newsletter;
Hours & Admission Prices: Memorial Day-Labor Day daily 10-4. Family $30, adults $9, seniors $7, children 5-17 $5; children under 4 & members no charge.
Attendance: 7,000 (estimated)
Membership: Student $10; Senior Citizen $20; Individual $25; Family $45; Life $500.

THELMA SADOFF CENTER FOR THE ARTS, 51 Sheboygan St., Fond du Lac, WI 54935-4219. Tel.: 920-921-5410. Facebook: Thelma Sadoff Center for the Arts.
E-mail: info@thelmaarts.org
Web Site: www.thelmaarts.org
Formerly: Windhover Center for the Arts
Founded: 2000.
Congressional District: 6
Key Personnel: Exec. Dir., Kevin Miller; Mktg. Dir., Jacqui Corsi; Operations Dir., Eric Hanrahan; Cur., Audra Gabrielson.
Personnel Profile: Full-Time Paid 6; Part-Time Paid 10; Part-Time Volunteers 20.
Operating Expenses: 375,000
Operating Income: 350,000
Governing Authority: nonprofit. Parent Institution: Thelma Center for the Arts, Inc. Tax-exempt.
Institution Type/Description: Art Gallery.
Collections: paintings; sculpture; light; drawings; photography.
Major Exhibits: Maia Flore, 1/6/14-3/30/14; Pamela Valfer, 11/13-2/2/14; Allen Brewer, 2/6/14-4/13/14.
Facilities: 3,000 sq. ft., two-level contemporary exhibition space attached to a transformed, three-level historic Masonic Temple; space for concerts, weddings & corporate gatherings.
Activities: concerts; film; temporary exhibitions; rental facilities.
Publications: weekly e-newsletter.
Hours & Admission Prices: Mon.-Wed. & Fri.-Sun. 10:30-5, Thurs. 10:30-8. Adults $5, first Thurs. of the month & members no charge.
Attendance: 50,000 (estimated)
Membership: Student $15; Senior Single $40; Single $60; Senior Couple $65; Couple/Family $100.

Fort Atkinson

HOARD HISTORICAL MUSEUM AND NATIONAL DAIRY SHRINE'S VISITORS CENTER, 401 Whitewater Ave., Fort Atkinson, WI 53538-2255. Tel.: 920-563-7769. Fax: 920-568-3203.
E-mail: oberle@hoardmuseum.org
Web Site: www.hoardmuseum.org
Founded: 1933.
Congressional District: 9
Key Personnel: Pres. (V), Tony Bolz; Dir., Kori Oberle; Cur., Karen O'Connor; Volunteer Coord., Tammy Doellstedt; Maintenance, Greg Misfeldt; Accountant, Linda Winn.
Personnel Profile: Full-Time Paid 2; Part-Time Paid 5; Part-Time Volunteers 150; Interns 1.
Governing Authority: municipal; society. Parent Institution: Fort Atkinson Historical Society. Tax-exempt: 170(b)(1)(A).
Institution Type/Description: Local History Museum.
Collections: history of Jefferson County, Wisconsin; Wisconsin archaeology; U.S. dairy history; early tools & crafts; quilts; dolls; local artwork; decorative arts; Civil War artifacts; books; archives. Historic Buildings: 1841 Dwight Foster House; 1864 Frank W. Hoard House.
Research Fields: local Native American history; Black Hawk War; local history & genealogy; Civil War.
Facilities: 3,500-vol. library; archives; reading room.
Activities: guided tours; bus tours; lectures; gallery talks; permanent & temporary exhibitions; Lincoln Era Library & Exhibit; audio walking tour of historic downtown Fort Atkinson; family program; fourth grade history essay exhibition. Museum Sponsors: local annual art shows; Ice Cream Social in July; Christmas Open House.
Publications: quarterly newsletter; maps; annual programs; book, The Mounds of Koshkonong; 3-vol. set on the Black Hawk War, Hunting A Shadow, The Battle of Wisconsin Heights, The Battle of Bad Axe.

Hours & Admission Prices: Tues.-Sat. 9:30-4:30. No charge; donations accepted. ⓐ
Attendance: 17,000 (estimated)
Membership: Individual $25; Family $50; Sponsor $125; Patron $250; Benefactor $500 & up.

Fountain City

ELMER'S AUTO & TOY MUSEUM, W903 Elmers Rd., Fountain City, WI 54629. Tel.: 608-687-7221.
Web Site: www.elmersautoandtoymuseum.com
Institution Type/Description: Auto & Toy Museum.
Collections: muscle & classic automobiles and truck from 1910 to present; Indian & Harley Davidson motorcycles; early bicycles & high wheel bikes; period racecars; pedal cars & tractors; pedal riding toys; early toys; farm, carpenter & mechanic tools; dolls including Shirley Temple, German & French, and paper mache.
Hours & Admission Prices: Call for hours. Adults $8, seniors 65 & over $7, students 6-17 $4; children under 6 no charge.

Fox Lake

FOX LAKE HISTORICAL MUSEUM, INC., 211 Cordelia St. & S. College Ave., Fox Lake, WI 53933. Mailing Address: P.O. Box 493, Fox Lake, WI 53933-0493. Tel.: 920-296-0254.
E-mail: scottfrankdesigns@hotmail.com
Founded: 1970.
Congressional District: 2
Key Personnel: C.E.O. & Pres. (V), Scott Frank; Museum Shop Mgr., Jim Clark.
Personnel Profile: Part-Time Volunteers 9.
Governing Authority: nonprofit. Parent Institution: City of Fox Lake. Tax-exempt: 170(b)(1)(A).
Institution Type/Description: Railroad Museum.
Collections: Fox Lake & railroad depot history; baggage wagons; time tables; railroad lanterns; miniature display depicting railroad loop; Elmwood Island Indians archaeological exhibit; period kitchen; working blacksmith shop; Fairbanks motorized hand car.
Major Exhibits: 175th Anniversary of Fox Lake, 6/15/13.
Research Fields: Main Street, Fox Lake.
Facilities: picnic area.
Activities: tours; permanent exhibitions.
Publications: brochures; newsletters.
Hours & Admission Prices: June-Oct. 1st & 3rd Sun. 1-4; other times by appointment. No charge; donations accepted. ⓐ
Attendance: 250 (estimated)
Membership: Junior $1; Individual & Senior Citizens $3; Family $5; Business $10; Industry $25.

Genesee Depot

TEN CHIMNEYS FOUNDATION, S43 W31575 Depot Rd., Genesee Depot, WI 53127. Mailing Address: P.O. Box 225, Genesee Depot, WI 53127-0225. Tel.: 262-968-4110. Fax: 262-968-4267. Facebook: Ten Chimneys Foundation.
E-mail: info@tenchimneys.org
Web Site: www.tenchimneys.org
Founded: 1996.
Key Personnel: Pres. & C.E.O., Randy Bryant; Chm. Bd., Judy Jorgenson; Programs, Kristine Weir-Martell; Mgr. Mktg. & Communications, Courtney Kihslinger; Mgr. Accounting, Rayanne Kobiske; Museum Store Mgr., Richard Quick.
Personnel Profile: Full-Time Paid 6; Part-Time Paid 4; Part-Time Volunteers 270.
Governing Authority: private; nonprofit organization. Tax-exempt: 501(c)(3).
Institution Type/Description: Historic House: housed in the estate of theatre legends, Alfred Lunt and Lynn Fontanne. Listed on the National Registry of Historic Places.
Collections: personal artifacts; cultural, historic, & artistic materials; historic buildings & furnishings.
Facilities: 2,400-vol. library. Museum-related items for sale.
Activities: docent program; guided tours; lectures; temporary exhibitions; play readings; lectures.
Hours & Admission Prices: Tours: May to mid-Nov. Tues.-Sat. reservations recommended. Office: Mon.-Fri. 9-5. Adults $28-$35; children under 12 not admitted. Closed New Year's Eve & Day; Easter; Memorial Day; Independence Day; Labor Day; Thanksgiving; Christmas. ⓐ
Attendance: 12,000 (estimated)
Membership: Grand Entrance in the Arrival Hall $50; Brunch w/Helen Hayes in the Garden Terrace $250; Game of Hearts w/Larry Olivier in the Library

$500; Chat w/Kate Hepburn in the Flirtation Room $1,500; Song w/Noel Coward in the Drawing Room $5,000; Dinner w/Lynn and Alfred in the Dining Room and Lunt-Fontanne Society Member $10,000.

Germantown

SILA LYDIA BAST BELL MUSEUM & FIRE HALL, W18780 Holy Hill Rd., Germantown, WI 53022. Mailing Address: P.O. Box 31, Germantown, WI 53022. Tel.: 262-628-3170.
E-mail: info@bastbellmuseum.com
Web Site: www.germantownhistoricalsociety.org
Institution Type/Description: History Museum.
Collections: local history; over 5,000 bells from around the world; Seagrave fire truck.
Facilities: Museum-related items for sale.
Activities: group tours.
Hours & Admission Prices: June-Nov. 1 Fri.-Sun. 1-4; other times by appointment. ⓐ

Gordon

GORDON-WASCOTT HISTORICAL MUSEUM, 9672 E. County Rd. Y, Gordon, WI 54838. Mailing Address: P.O. Box 222, Gordon, WI 54838. Tel.: 715-376-4249.
Key Personnel: Pres. (V), David D. Benson; Museum Shop Mgr., Pat Finstad
Institution Type/Description: Historical Society Museum.
Collections: local history & culture; photographs.
Hours & Admission Prices: Memorial Day-Labor Day Fri.-Mon. 10-4.

Green Bay

GREEN BAY BOTANICAL GARDEN, 2600 Larsen Rd., Green Bay, WI 54303-4841. Tel.: 920-490-9457. Fax: 920-490-9461.
E-mail: info@gbbg.org
Web Site: www.gbbg.org/index.htm
Founded: 1982.
Congressional District: 8
Key Personnel: Exec. Dir., Susan Garot; Dir. Horticulture, Mark A. Konlock.
Personnel Profile: Full-Time Paid 13; Part-Time Paid 13; Part-Time Volunteers 675; Interns 9.
Volunteer Hours: 675
Operating Expenses: 1,606,987
Operating Income: 1,428,840
Governing Authority: Tax-exempt.
Institution Type/Description: Botanical Garden.
Collections: plants; flowers; trees.
Facilities: library; 47-acres. Museum-related items for sale.
Activities: educational programs.
Publications: quarterly newsletter, What's Bloomin at the Garden.
Hours & Admission Prices: April-May & Sept.-Oct. daily 9-5; June to Aug. daily 9-8; Nov.-Dec. Mon.-Fri. 9-4; Jan.-Mar. Mon.-Sat. 9-4 Adults 13 & over $7, children 5-12 $2; discounts to seniors & AAA members; members and children 4 & under no charge. Closed New Year's Day; Thanksgiving; Christmas.
Attendance: 106,000 (estimated)
Membership: Single $40; Family $50. Varied Benefits $150-$2,500.

GREEN BAY PACKERS HALL OF FAME, 1265 Lombardi Ave., Green Bay, WI 54304-3997. Mailing Address: P.O. Box 10628, Green Bay, WI 54307-0628. Tel.: 920-569-7512. Fax: 920-569-7122.
Web Site: www.packers.com
Founded: 1969.
Congressional District: 8
Key Personnel: Hall of Fame & Stadium Tour Mgr., Krissy Zegers.
Personnel Profile: Full-Time Paid 3; Part-Time Paid 14.
Governing Authority: nonprofit organization. Parent Institution: Green Bay Packers. Tax-exempt: 501(c)(3).
Institution Type/Description: Sports Museum.
Collections: sports memorabilia of Green Bay Packers from beginning to present.
Facilities: 8 theaters.
Activities: self-guided tours; films; permanent & temporary exhibitions.
Publications: promotional brochure; exhibit guide.
Hours & Admission Prices: Mon.-Sat. 9-6, Sun. 9-5; hours vary on home game days. Adults $10, senior citizens, college students & youth $8, children 6-11 $5; discount to groups. Closed Easter; Thanksgiving; Christmas. ⓐ
Attendance: 100,000 (estimated)

HAZELWOOD HISTORIC HOME MUSEUM, 1008 S. Monroe Ave., Green Bay, WI 54301-3206. Mailing Address: P.O. Box 1411, Green Bay, WI 54305-1411. Tel.: 920-437-1840.
E-mail: bchs@netnet.net
Web Site: www.browncohistoricalsoc.org
Founded: 1995.
Key Personnel: Dir., Christine Dunbar; Pres. (V), Wendy Barsczc; Museum Shop Mgr., Joan Hogan.
Personnel Profile: Full-Time Paid 2; Part-Time Paid 2; Part-Time Volunteers 55.
Governing Authority: Parent Institution: Brown County Historical Society. Tax-exempt.
Institution Type/Description: Historic House Museum: built in 1837.
Collections: period furnishings; photographs.
Publications: Voyageur magazine; newsletter.
Hours & Admission Prices: May Sat.-Sun. 12-4; June-Aug. Thurs.-Sun. 12-4. Adults $4; discount to AAM members; members no charge.
Attendance: 1,500 (estimated)
Membership: Single $25; Family $35-$50; Partner $100-$199; Provider $200-$499; Benefactor $500-$999; Founder's Circle $1,000 & up.

HERITAGE HILL STATE HISTORICAL PARK, (M), 2640 S. Webster Ave., Green Bay, WI 54301-2997. Tel.: 920-448-5150, ext. 303. Fax: 920-448-5147.
E-mail: info@heritagehillgb.org
Web Site: www.heritagehillgb.org
Founded: 1977.
Congressional District: 8th
Key Personnel: Site Supvr., Rhonda Matzke; Program Coord., Kayla Filen; Supvr. Education, Patty Wagner; Preservation Specialist, Nick Backhaus; Museum Shop Mgr., Sue Storzer.
Personnel Profile: Full-Time Paid 12; Part-Time Paid 50.
Volunteer Hours: 9,800
Operating Expenses: 986,000
Operating Income: 985,000
Governing Authority: nonprofit organization. Parent Institution: Heritage Hill corporation under lease agreement with the Wisconsin Dept. of Natural Resources. Tax-exempt: 170(b)(1)(A), 509(a)(1) & 501(c)(3).
Institution Type/Description: Village & Park Museum: located on the site of 1820s Camp Smith, former army outpost.
Collections: Historic Buildings & Furnishings: Franklin Hose Co.; Y.M.C.A. library; 1897 Dewitt blacksmith shop; 1835 Baird law office; 1800-1820 cabin; 1912 Allouez town hall; 1817 cotton house; 1700 Tank College; 1851 Moravian church; 1905 Belgian farmstead; 1902 cheese factory; Franklin Hose Company;
Research Fields: area history relating to the museum's time periods & historic structures.
Facilities: Local history books & period reproductions for sale.
Activities: living history outdoor museum using first & third person interpretation; crafts programs; scenario reenactment of daily activities of the past; formally organized education programs for children.
Hours & Admission Prices: May 4-Sept. 1 Tues.-Sat. 10-4:30, Sun. 12-4:30; Sept. Sat. 10-4:30; Oct.-May Mon.-Fri. 10-4:30. Adults $7, children 5-17 $6; discount to AAA members; children 4 & under no charge.
Attendance: 56,000 (accurate)
Membership: Heritage Pass $65; Cotton Club $150.

LAWTON GALLERY, UNIVERSITY OF WISCONSIN-GREEN BAY, (M), 2420 Nicolet Dr. (TH 331), Green Bay, WI 54311-7003. Tel.: 920-465-2916. Fax: 920-465-2890.
E-mail: perkinss@uwgb.edu
Web Site: www.uwgb.edu/lawton
Key Personnel: Cur. Art, Stephen Perkins, Ph.D.; Asst. Cur., Erin Rose; Administrative Asst., Pamela Johnson.
Personnel Profile: Full-Time Paid 1; Part-Time Paid 1.
Governing Authority: Tax-exempt.
Institution Type/Description: Art Gallery.
Collections: local, regional, student & international artwork.
Publications: exhibition catalogues; occasional brochures.
Hours & Admission Prices: Sept. May Tues.-Sat. 10-3. No charge.
Attendance: 3,000 (accurate)

NATIONAL RAILROAD MUSEUM, 2285 S. Broadway, Green Bay, WI 54304-4832. Tel.: 920-437-7623. Fax: 920-437-1291. Facebook: Friends of the National Railroad Museum.
E-mail: staff@nationalrrmuseum.org
Web Site: www.nationalrrmuseum.org
Founded: 1956.
Congressional District: 8

Key Personnel: Pres. Bd. Dir., Gary Lockstein; Exec. Dir., Jacqueline D. Frank; Mgr. Operations & Cur., Daniel Liedtke; C.F.O., Robert Bloedorn; Dir. Education, Bob Lettenberger; Museum Shop Mgr., Allison Felchlin.
Personnel Profile: Full-Time Paid 9; Part-Time Paid 7; Part-Time Volunteers 300; Interns 2.
Operating Expenses: 1,400,000
Operating Income: 1,400,000
Governing Authority: nonprofit organization. Tax-exempt: 501(c)(3).
Institution Type/Description: Railroad Museum.
Collections: 20 steam & diesel-electric locomotives; 50 examples of rolling stock dating to 1890; Eisenhower equipment; railroad related papers, photographs & artifacts; 19th & 20th century maps.
Major Exhibits: Jay Christopher Railroad China, 11/13-1/14; Railroad Graffiti, 2/14-1/15.
Research Fields: 20th century development of railroads in the United States.
Facilities: library; archives; 33 acres including a 42,000 sq. ft. train pavilion; 36,000 sq. ft. exhibit center; 95-seat theater. Museum-related items for sale.
Activities: guided tours; lectures; films; permanent & temporary exhibitions; school programs.
Publications: quarterly newsletter, Rail Lines; collection catalog.
Hours & Admission Prices: Jan.-March Tues.-Sat. 9-5, Sun. 11-5; May-Dec. Mon.-Sat. 9-5, Sun. 11-5. Adults $9, senior citizens $8, children 3-12 $6.50; discounts to AAA members; children under & 3 no charge. Closed New Year's Day; Easter; Thanksgiving; Christmas Eve & Day.
Attendance: 75,000 (accurate)
Membership: Individual $35; Family $50; Deluxe $175; Lifetime $1,200.

＊ **NEVILLE PUBLIC MUSEUM OF BROWN COUNTY, (M),** 210 Museum Place, Green Bay, WI 54303-2780. Tel.: 920-448-7842. Fax: 920-448-4458.
E-mail: bc_museum@co.brown.wi.us
Web Site: www.nevillepublicmuseum.org
Founded: 1915.
Congressional District: 8
Key Personnel: Exec. Dir., Rolf E. Johnson; Chm. (V), Kramer Rock; Cur. Education, Matt Welter; Cur. Science, John Jacobs; Cur. Art, Marilyn Stasiak; Cur. Collections, Louise Pfotenhauer; Cur. History, Rebecca Looney; AV Technician, Larry LaMalfa; Exhibit Technician, Maggie Dernehl.
Personnel Profile: Full-Time Paid 12; Part-Time Volunteers 160; Interns 10.
Governing Authority: county. Tax-exempt.
Institution Type/Description: General Museum.
Collections: history items; natural history; anthropology; archaeology; glass; mineralogy; geology; costumes; paintings; sculpture; decorative arts; military; numismatic; historic negatives & photographs; manuscripts; anatomical mannequin; TV news film.
Facilities: 6,000-vol. library; meeting rooms; 132-seat theater. Gift items for sale.
Activities: guided tours; lectures; films; gallery talks; study clubs; formally organized education programs; observe artists at work; inter-museum loan, permanent, temporary & traveling exhibitions; teacher in-service workshops; docent program; TV & radio programs; field trips.
Publications: brochures; newsletters; e-newsletter, 'N Touch; biennial report; special events mailings.
Hours & Admission Prices: Sun. 12-5, Mon.-Tues. & Fri.-Sat. 9-5, Wed.-Thurs. 9-8. Adult $5, children 6-15 $3; discounts to AAM & ASTC members and school & youth groups; Thurs. 6-8 pm, members and children 5 & under no charge. Closed New Year's Day; Thanksgiving; Christmas.
Attendance: 64,508 (accurate)
Membership: Individual $35; Dual $45; Family $65; Pioneer $100; Explorer $250; Adventurer $500. Corporate: Bronze $600; Silver $1,000; Gold $2,500; Platinum $5,000.

NEW ZOO, 4378 Reforestation Rd., Green Bay, WI 54313. Tel.: 920-434-7841 & 448-6242.
Web Site: www.newzoo.org
Governing Authority: county.
Institution Type/Description: Zoo.
Collections: birds; fish; mammals; reptiles; amphibians; invertebrates.
Facilities: visitor center; restaurant.
Activities: rental facilities; special events; animal feedings; educational programs.
Hours & Admission Prices: Daily 9-6. Adults 16 & over $5, seniors 62 & over and children 3-15 $3; discounts to families; members and children 2 & under no charge.

Green Lake

DARTFORD DEPOT MUSEUM, 554 Mill St., Green Lake, WI 54941. Mailing Address: P.O. Box 638, Green Lake, WI 54941. Tel.: 920-294-6194.
E-mail: info@dartfordhistorical.org
Web Site: www.dartfordhistorical.org
Founded: 1956.
Congressional District: 6
Personnel Profile: Part-Time Volunteers 6.
Operating Expenses: 18,717
Operating Income: 15,276
Governing Authority: Tax-exempt.
Institution Type/Description: History Museum: housed in a former railroad depot; built in 1870s.
Collections: local history; railroad artifacts; photographs; period furnishings.
Hours & Admission Prices: Memorial Day to Labor Day Sat. 10-1. No charge; donations accepted.
Attendance: 400 (estimated)
Membership: Individual $15; Family $20.

DARTFORD HISTORICAL SOCIETY, 501 Mill St., Green Lake, WI 54941. Mailing Address: P.O. Box 638, Green Lake, WI 54941. Tel.: 920-294-6194.
E-mail: info@dartfordhistorical.org
Web Site: www.dartfordhistorical.org
Founded: 1956.
Congressional District: 6
Key Personnel: Pres. (V), Lawrence Behlen.
Personnel Profile: Part-Time Volunteers 15.
Volunteer Hours: 1,284
Operating Expenses: 18,717
Operating Income: 15,276
Governing Authority: Tax-exempt.
Institution Type/Description: Historical Society Museum: housed in the former public library.
Collections: local history & culture; school records; census records; town record; genealogy; newspapers 1860-1996.
Facilities: archives.
Activities: research.
Hours & Admission Prices: Fri. 10-4, Sat. 10-12; other times by appointment. No charge; donations accepted. &
Attendance: 300 (estimated)
Membership: Individual $15; Family $20; Supporting $50; Life $200.

Greenbush

WADE HOUSE HISTORIC SITE, W7824 Center St., Greenbush, WI 53026. Mailing Address: P.O. Box 34, Greenbush, WI 53026-0034. Tel.: 920-526-3271. Fax: 920-526-3626.
E-mail: wadehouse@wisconsinhistory.org
Web Site: www.wadehouse.org
Formerly: Wade House Stagecoach Inn & Wesley Jung Carriage Museum State Historic Site
Founded: 1953.
Congressional District: 6
Key Personnel: Dir., David Simmons; Cur. Education, Jeffrey Murray; Maintenance Supvr., Martin Keyport; Museum Shop Mgr., Jenni Laning.
Personnel Profile: Full-Time Paid 4; Part-Time Paid 29; Part-Time Volunteers 150.
Governing Authority: state. Parent Institution: Wisconsin Historical Society, 816 State St., Madison, 53706. Tax-exempt.
Institution Type/Description: Historic Site & Transportation Museum.
Collections: over 100 horse & hand-drawn vehicles of late 19th & early 20th century; 3-story Greek revival Stagecoach Inn; 2-story Greek revival home; blacksmith shop; Muley sawmill.
Research Fields: historic horse drawn transportation; carriage manufacturing industry; Yankee town builders; farming & agricultural practices; historic foodways; saw milling.
Facilities: picnic area; hiking trails; cafe. Museum-related items for sale.
Activities: guided and self-guided tours; summer day camp programs for children; historic foodway & dinner programs for adults; period games and amusements; candle-making. Special events throughout the season.
Publications: Education Plan; video, Site History.
Hours & Admission Prices: May 19-Oct. 14 daily 10-5, Sun. 12-5; additional hours for special events. Adults $11, students & seniors $9.25, children & SHSW member $5.50; discounts to groups, AAM & ICOM members, local museum professionals & docents. &
Attendance: 17,500 (estimated)
Membership: Individual $35; Individual & Guest or Family $60; Patron $100.

Greenfield

GREENFIELD HISTORICAL SOCIETY, 5601 W. Layton Ave., Greenfield, WI 53220. Mailing Address: 11745 W. Wooded Ct., Greenfield, WI 53228-1894. Tel.: 414-763-2675.
E-mail: bodamer@live.com
Founded: 1965.
Congressional District: 4
Key Personnel: Pres., Robert Roesler; Vice Pres., Anita Bodamer; Treas., Cindi Streicher.
Personnel Profile: Part-Time Volunteers 20.
Governing Authority: nonprofit organization. Parent Institution: Wisconsin Historical Society. Archives located in Greenfield Public Library. Tax-exempt: 501(c)(3).
Institution Type/Description: Historical Society Museum.
Collections: tools; housewares; clothes; furniture & other items pertaining to early history of City of Greenfield. Historic Structures: 1836 log cabin; 1856 Cream City Brick Home.
Facilities: archives by appointment.
Activities: guided tours.
Publications: quarterly newsletter; book, History of Greenfield 1841-1976; 27 Tales of Greenfield.
Hours & Admission Prices: Open during major events occurring at the civic facility. No charge; donations accepted.
Attendance: 200 (estimated)
Membership: Family $12; Business $25.

Hales Corners

FRIENDS OF BOERNER BOTANICAL GARDEN, INC., 9400 Boerner Dr., Hales Corners, WI 53130-2273. Tel.: 414-525-5653. Fax: 414-525-5668.
E-mail: jburaczewski@fbbg.org
Web Site: www.boernerbotanicalgardens.org
Founded: 1984.
Key Personnel: Pres. & C.E.O., Ellen Hayward; Bd. Chair, Lee A. Riordan.
Governing Authority: Parent Institution: Milwaukee County. Tax-exempt.
Institution Type/Description: Botanical Gardens.
Collections: over 500 varieties of roses; herbs; annuals; shrubs; daylilies; peonies; sculptures.
Facilities: 40 acres of formal gardens.
Activities: group tours; annual fundraisers; family events; craft fairs; guided evening garden walks; education classes.
Publications: newsletter.
Hours & Admission Prices: Garden: late April to early Oct. daily 8am to sunset; mid-April & Oct.-Nov. call for hours. Education & Visitor Center: late April to early Oct. daily 8-6; late Oct. to early April Mon.-Fri. 8-4, Sun. 9-3. Adults $5, seniors, disabled & students $4, youth 6-17 $3. &
Attendance: 100,000 (estimated)
Membership: Student $15; Educator $25; Individual $35; Individual Plus One $50; Family $65.

Hartford

WISCONSIN AUTOMOTIVE MUSEUM, 147 N. Rural St., Hartford, WI 53027-1407. Tel.: 262-673-7999. Facebook: Wisconsin Automotive Museum.
E-mail: info@wisconsinautomuseum.com
Web Site: www.wisconsinautomuseum.com
Formerly: Hartford Heritage Auto Museum
Founded: 1984.
Congressional District: 9
Key Personnel: Dir., Dawn Bondhus Mueller; Bd. Pres., Michael J. Mally.
Personnel Profile: Full-Time Paid 1; Part-Time Paid 6; Part-Time Volunteers 40.
Governing Authority: nonprofit organization. Tax-exempt: 501(c)(3).
Institution Type/Description: Transportation Museum.
Collections: 1906-1931 Kissel Motor Car Company automobiles; other vintage cars, trucks, fire engines, and auto-related memorabilia; outboard motors & period engines.
Research Fields: Kissel Motor Car Company.
Facilities: 165-vol. library of auto repair manuals; 78,000 sq. ft. exhibit space. Auto-related items for sale.
Publications: newsletter, Kissel Kar Klub.
Hours & Admission Prices: May-Sept. Mon.-Sat. 10-5, Sun. 12-5; Oct.-April Wed.-Sat. 10-5, Sun. 12-5. Adults $10, seniors over 62 $8, students 6-16 $6; discounts to groups; children 5 & under no charge. Closed New Year's Day; Easter; Thanksgiving; Christmas. &
Attendance: 10,000 (estimated)

Hayward

NATIONAL FRESH WATER FISHING HALL OF FAME, 10360 Hall of Fame Dr., Hayward, WI 54843. Mailing Address: P.O. Box 690, Hayward, WI 54843-0690. Tel.: 715-634-4440. Fax: 715-634-4440.
E-mail: fishhall@chegnet.net
Web Site: www.freshwater-fishing.org
Founded: 1960.
Congressional District: 7
Key Personnel: Exec. Dir., Emmett A. Brown, Jr.; Business Mgr., Kathy Polich.
Governing Authority: nonprofit organization. Tax-exempt: 501(c)(3).
Institution Type/Description: Freshwater & Marine Museum.
Collections: 450 classic outboard motors; 400 mounted fish; 1,000 reels; 5,000 lures; 300 rods; accessories.
Research Fields: fishing records; dated artifacts; world enshrinement for achievement.
Facilities: library of catalogues & books pertaining to fresh water fishing, conservation, history & development, available for research by visit to library by appointment; reading room; cafeteria. Insignia articles & other museum-related items for sale.
Activities: self-guided group tours; periodic seminars & clinics.
Publications: annual book, World Records on Fresh Water Fish; quarterly magazine, Splash.
Hours & Admission Prices: mid-April, May, Sept. & Oct. daily 9:30-4; June-Aug. daily 9:30-4:30. Nominal gate fee. Closed New Year's Eve & Day; Christmas Eve, Day & week. &
Attendance: 40,000 (estimated)
Membership: Bronze $30; Silver $35; Gold $45; Club or Business $50; Platinum $55; Memorial Wall Tile $150; Lifetime $300; Lifetime Corporate Club or Business $350.

Hilbert

MAIN STREET ART WORKS, 627 Main St., Hilbert, WI 54129. Mailing Address: P.O. Box 77, Hilbert, WI 54129-0077. Tel.: 920-853-7348.
E-mail: info.msaw@charter.net
Web Site: www.mainstreetartworks.com
Key Personnel: Dir., Bonnie de Arteaga
Institution Type/Description: Art Gallery.
Collections: works by local & regional artists including prints, photography, & paintings.
Activities: demonstrations
Hours & Admission Prices: Call for hours.

Hillsboro

HILLSBORO AREA HISTORICAL SOCIETY, Maple St. - City Park, Hillsboro, WI 54634. Mailing Address: P.O. Box 1, Hillsboro, WI 54634. Tel.: 608-489-3192.
Founded: 1958.
Congressional District: 3
Key Personnel: Pres., Tom Hotek; Sec., Arlene Schiefelbein; Treas., Nancy Hotek.
Personnel Profile: Part-Time Volunteers 30.
Governing Authority: society. Affiliated with Wisconsin State Historical Society, Madison. Tax-exempt: 501(c)(3).
Institution Type/Description: General Museum.
Collections: photographs; local history items; memorabilia; one-room school-house replica; machine shed. Historic Building: 1853 log cabin.
Research Fields: local history.
Activities: quilting scrapbook; meetings; guided tours.
Publications: Civil War Letters of Private Eno; The Lucky Kickapoo; Rockton; West Lima; Trippville; Greenwood; History of Hillsboro, Wisconsin; Hillsboro, the Friendly City, Vol. I & II.
Hours & Admission Prices: June-Labor Day Sun. 1-3; other times by appointment. No charge; donations accepted.
Attendance: 385 (accurate)
Membership: Individual $5; Family $10.

Horicon

SATTERLEE CLARK HOUSE, 322 Winter St., Horicon, WI 53032-1035. Mailing Address: Box 65, Horicon, WI 53032-0065. Tel.: 920-485-3200.
Founded: 1972.
Congressional District: 2

Key Personnel: Pres., Geri Pyrek; Vice Pres., Tom K. Jahnke; Sec., Margaret A. Drewa; Treas., Sandi Todd.
Personnel Profile: Part-Time Volunteers 6.
Governing Authority: society; nonprofit organization. Parent Institution: Wisconsin Historical Society. Tax-exempt: 501(c)(3).
Institution Type/Description: Historical Society Museum: housed in c.1863, Satterlee Clark Home.
Collections: period furnishings; Indian artifacts; blacksmith shop & forge; restored wooden working loom; old black iron cookware; red wing pottery; country schoolhouse.
Research Fields: genealogy.
Facilities: library of books, newspapers & microfilm pertaining to local history available for use by appointment; botanical garden.
Activities: guided tours; films; reading room; temporary exhibitions; map. Annual Events: Historical Encampment in May; Old Fashioned Christmas.
Publications: brochure, A Walking Tour of Historic Buildings.
Hours & Admission Prices: May-Oct. 4th Sun. each month 1-4; Dec. 1st Sun. each month 1-4; other times by appointment. No charge; donations accepted.
Attendance: 1,000 (estimated)
Membership: Active $5; Associate $10; Contributing $25; Sustaining $50; Patron $100.

Hudson

THE OCTAGON HOUSE, 1004 Third St., Hudson, WI 54016-1219. Tel.: 715-386-2654.
E-mail: octagonhousemuseum@juno.com
Founded: 1948.
Congressional District: 3
Key Personnel: Pres. (V), Dolores Taavola; Treas., Dorothy Wilson; Dir., Heidi Rushmann; Vice Pres., Helen Stoltz-Wood; Sec., LaVonne McCombie.
Personnel Profile: Part-Time Paid 1; Part-Time Volunteers 56; Interns 2.
Governing Authority: nonprofit society. Parent Institution: St. Croix County Historical Society. Tax-exempt: 170(b)(1)(A).
Institution Type/Description: Victorian Museum: housed in 1855 Octagonal House; carriage house & garden house.
Collections: furniture; books; dolls; tools; clothing; St. Croix household artifacts c.1800; carriage house. Historic Buildings: 19th century blacksmith shop, country store, summer kitchen.
Research Fields: local & county-wide history; genealogies of county residents.
Facilities: 600-vol. library of county history, novels, Bibles & other reference books available for study on premises.
Activities: lectures; training program for museum guides; tours of St. Croix Valley; tapings of old settlers of county; house tour for groups of 30 or more; city tours; tea tours; cemetery tour. Annual Events: Theatre in the garden July & August; Christmas Tour of Homes in November; Lite-up Night in December.
Publications: pamphlets, Westward to the St. Croix; Hudson in the Early Days; Historical Map of St. Croix County; The Life of a Boy in the Middle West; Hudson, 1900-1909, a Rail-River Town; History of the Octagon House Family; 50 Years on Petticoat Lane, Historic Hudson Revisited.
Hours & Admission Prices: May-Aug. Wed.-Sat. 12-4:30, Sun. 2-4:30; Sept.-Oct. & Dec. Sat. 12-4:30, Sun. 2-4:30. Adults $7, teens $3, children $2; discounts to AAA members. Closed national holidays.
Attendance: 3,000 (estimated)
Membership: Individual $10; Family $15.

THE PHIPPS CENTER FOR THE ARTS GALLERIES, 109 Locust St., Hudson, WI 54016. Tel.: 715-386-8409.
E-mail: info@thephipps.org
Web Site: www.thephipps.org
Institution Type/Description: Art Gallery.
Collections: paintings; sculpture.
Facilities: Gifts for sale.
Hours & Admission Prices: Mon.-Sat. 9-4:30, Sun. 12-4:30.

Hurley

IRON COUNTY HISTORICAL MUSEUM, 303 Iron St., Hurley, WI 54534-1356. Tel.: 715-561-2244.
E-mail: genec@chartermi.net
Formerly: Old Iron County Courthouse Museum
Founded: 1976.
Congressional District: 7
Key Personnel: C.E.O., Chm. & Museum Shop Mgr., Gene Cisewski; Pres. (V), Nick L. Zuvich; Treas., Minerva Stefani; Sec., Helen Zuvich.
Personnel Profile: Full-Time Volunteers 20; Part-Time Volunteers 20; Interns 2.

Governing Authority: society; nonprofit. Parent Institution: Wisconsin State Historical Society. Tax-exempt: 501(c)(3).
Institution Type/Description: General Museum: housed in 1893 Old Iron County Courthouse.
Collections: mining, farm, lumbering & household artifacts; textiles; period clothing & home furnishings; photographs; microfilm collection of county newspapers.
Research Fields: mining.
Facilities: library available for use on premises only; 200-seat auditorium. Hand-crafted & other museum-related items for sale.
Activities: guided tours; permanent exhibitions; ancient carpet weaving demonstrations.
Publications: brochures.
Hours & Admission Prices: Mon., Wed. & Fri.-Sat. 10-2; other times by appointment. No charge. Closed holidays. &
Attendance: 5,200 (estimated)
Membership: Individual $5; Family $10; Business & Clubs $25; Life $175.

Janesville

THE LINCOLN-TALLMAN RESTORATIONS, 440 N. Jackson St., Janesville, WI 53548. Mailing Address: P.O. Box 8096, Janesville, WI 53547-8096. Tel.: 608-752-4519 & 756-4509; 800-577-1859. Fax: 608-741-9596.
E-mail: jvanhaaften@rchs.us
Web Site: www.rchs.us
Founded: 1951.
Congressional District: 1
Key Personnel: C.E.O., Joel Van Haaften; Administrative Asst., Joyce Dodge; Collections Mgr., Laurel Fant.
Personnel Profile: Full-Time Paid 1; Part-Time Paid 7; Part-Time Volunteers 300; Interns 1.
Governing Authority: nonprofit organization. Parent Institution: The Rock County Historical Society. Tax-exempt: 501(c)(3).
Institution Type/Description: Historic House.
Collections: 19th-century decorative arts & furnishings; Tallman family furnishings & personal artifacts; Tallman ethnological collection. Historic Buildings: 1842 Stone House; 1855-57 The Tallman House; 1855-57 Tallman Horse Barn.
Research Fields: 19th-century life & customs.
Facilities: Museum-related items for sale.
Activities: guided tours; docent program; permanent & temporary exhibits; special events; children's educational programs; historic reenactments pertaining to Tallman family events.
Publications: information circulars; annual report; Preservation Ordinances in Action: Rock County; A Good & Caring Woman: The Life & Times of Nellie Tallman; There Stands Old Rock, County, Wisconsin & the War to Preserve the Union; The Recorder, quarterly newsletter.
Hours & Admission Prices: June-Sept. daily 9-4; Nov. 20-Dec. holiday tours daily. Adults $8, senior citizens $7.50, students 6-18 $4; discounts to groups, Time Travelers, AAA, AAM & ICOM members; members no charge. Closed all major holidays. &
Attendance: 4,480 (accurate)
Membership: Student $5; Senior Individual $15; Individual $25; Household $45; Patron $100; Benefactor $250; Corporate $500; Life $1,000.

ROCK COUNTY HISTORICAL SOCIETY, (M), 426 N. Jackson St., Janesville, WI 53548. Mailing Address: P.O. Box 8096, Janesville, WI 53547-8096. Tel.: 608-756-4509 & 800-577-1859. Fax: 608-741-9596.
E-mail: rchs@rchs.us
Founded: 1948.
Congressional District: 1
Key Personnel: C.E.O. & Exec. Dir., Joel Van Haaften; Pres. (V), Chuck Rydberg; Administrative Asst., Joyce Dodge; Collections Mgr., Laurel Fant.
Personnel Profile: Full-Time Paid 1; Part-Time Paid 8; Part-Time Volunteers 300; Interns 1.
Governing Authority: nonprofit organization. Branch Museum: 1853, Frances Willard Schoolhouse, 4-H Fairgrounds; The Lincoln-Tallman Restorations: 1855-1857, Tallman House & Horse Barn; Helen Jeffris Wood Museum Center 1910, 426 N. Jackson St.; 1842, Stone House, 440 N. Jackson, Janesville, WI. Tax-exempt: 501(c)(3).
Institution Type/Description: Historical Society Museum.
Collections: local history; 19th-century decorative arts & furnishings; social history & vehicles; manuscript & iconographic collections.
Research Fields: Rock County & Wisconsin history.
Facilities: 3,000-vol. library of material on Rock County & Wisconsin available for use on premises; meeting room. Historic publications & other museum-related items for sale.
Activities: guided tours; lectures; concerts; arts festival; formally organized

education programs for children; docent program; permanent & temporary exhibitions; school loan service.
Publications: irregular booklets & guides; information circulars; annual report; Preservation Ordinances in Action: Rock County; A Good & Caring Woman: The Life & Times of Nellie Tallman; There Stands Old Rock, Rock County, Wisconsin and the War to Preserve the Union; The Recorder, quarterly newsletter.
Hours & Admission Prices: Archives: Wed.-Thurs. 9-3, Fri. 12-3; other times by appointment. Museum: daily 9-4. Lincoln-Tallman House: adults $8, members & seniors $7.50, children $4; discounts to Beloit Historical Society, Milton Historical Society, AAA & AASLH members. &
Attendance: 16,000 (estimated)
Membership: Student $5; Senior Individual $15; Individual $25; Household $45; Patron $100; Benefactor $250; Corporate $500; Life $1,000.

Jefferson

AZTALAN MUSEUM, N. 6264 Hwy. Q, Jefferson, WI 53549. Mailing Address: P.O. Box 122, Lake Mills, WI 53551-0122. Tel.: 920-648-4362 & 8575 (off season).
Web Site: www.orgsites.com/wi/aztalan
Founded: 1941.
Congressional District: 9
Key Personnel: Bd. Member (V) & Sec., Dr. Cheryl D. Peterson; Pres. (V), Michael Ayers; Cur., Steve Steigerwald; Museum Shop Mgr., Bob Conlin; Museum Shop Mgr., Deb Conlin.
Personnel Profile: Full-Time Volunteers 2; Part-Time Volunteers 40.
Governing Authority: society. Parent Institution: Lake Mills - Aztalan Historical Society, Inc. Tax-exempt.
Institution Type/Description: General Museum.
Collections: American Indian artifacts; history; archaeology; Hansen Granary Tool Collection. Historic Buildings: 1852 Aztalan Baptist Church; 1843 Pettey cabin; 1849 Bornell Cabin; 1867 Zickert house; 1850s Mamre Church (log cabin construction, built by Moravians); 1918 Aztalan one room school.
Facilities: library of manuscripts, letters, maps, newspapers, books, & magazines pertaining to local history available to researchers at public library at Lake Mills. Society publications & other museum-related items for sale.
Activities: Museum Sponsors: Aztalan Day in July.
Publications: leaflets, The Aztalan Story; The Pioneer Aztalan Story; booklet, The Ancient Aztalan Story; annual program.
Hours & Admission Prices: mid-May to Sept. Thurs.-Sun. 12-4. Adults $3, children 6-17 $1; discounts to groups; members & children under 6 no charge.
Attendance: 6,000 (estimated)
Membership: Individual $10; Life $100.

Kaukauna

CHARLES A. GRIGNON MANSION, 1313 Augustine St., Kaukauna, WI 54130-1613. Mailing Address: 330 E. College Ave., Appleton, WI 54911-5715. Tel.: 920-735-9370. Fax: 920-733-8636.
E-mail: ochs@myhistorymuseum.org
Web Site: www.myhistorymuseum.org
Founded: 1837.
Congressional District: 8
Key Personnel: C.E.O., Terry Bergen; Pres. (V), Ed Bush.
Personnel Profile: Part-Time Volunteers 30.
Governing Authority: society. Parent Institution: Outagamie County Historical Society, Inc. Subsidiary Institutions: Outagamie Museum and Houdini Historical Center, 330 E. College Ave.; The History Museum, 330 E. College Ave., Appleton, WI 54911. Tax-exempt: 501(c)(3).
Institution Type/Description: Historic House: 1837 Charles A. Grignon Mansion.
Collections: Greek revival home furnished in pre-Civil War era photographs, furnishings, decorative & fine arts; books.
Research Fields: mid-19th century rural life; fur trade era.
Facilities: Museum-related items for sale.
Activities: guided thematic tours.
Publications: quarterly, History Today.
Hours & Admission Prices: By appointment only to groups of ten or more. Closed major holidays.
Attendance: 6,000 (accurate)

Kenosha

ANDERSON ARTS CENTER, 121 Sixty-Sixth St., Kenosha, WI 53143. Tel.: 262-653-0481. Fax: 262-657-2526. Facebook: Kemper Center.
E-mail: carolina@andersonartscenter.com
Web Site: www.andersonartscenter.com
Founded: 1992.
Key Personnel: Admin., Carolina Curi-Bado; Gen. Mgr., Nancy Weatherhead; Pres., Don Gillespie; Chm. (V), Marianne Ironside; Museum Shop Mgr., Candace Hoffmann.
Governing Authority: Parent Institution: Kemper Center, Inc. Tax-exempt.
Institution Type/Description: Arts Center.
Collections: works by local, regional & national artists; paintings; photographs; pottery.
Major Exhibits: Studio Art Quilts Assn. & Women's Journeys in Fiber, 1/26/14-3/23/14; Racine Art Guild, 4/13/14-6/1/14; Solo Show Winners, 4/13/14-6/1/14; League of Milwaukee Artists, 6/22/14-8/3/14; Chicago Society of Artists, 8/24/14-10/19/14; Annual Winter Juried Show, 11/9/14-1/5/15; Gallery of Trees, 11/30/14-12/8/14.
Facilities: Gift items for sale.
Activities: temporary exhibitions; summer art camps; art classes; educational programs.
Publications: Kemper Chronicle.
Hours & Admission Prices: Tues.-Sun. 1-4. No charge; donations accepted. Closed New Year's Eve & Day; Good Friday; Easter; Memorial Day; Independence Day; Christmas Eve & Day.
Attendance: 6,304 (accurate)

∗   **CIVIL WAR MUSEUM, (M),** 5400 First Ave., Kenosha, WI 53140-6508. Mailing Address: 5500 First Ave., Kenosha, WI 53140. Tel.: 262-653-4141. Fax: 262-653-4431. Facebook: The Civil War Museum.
Web Site: www.thecivilwarmuseum.org
Founded: 2008.
Congressional District: 1
Key Personnel: Dir., Dan Joyce; Chm. (V), Sally Heideman; Pres. (V), Shari Krewson; Vice Pres. (V), Lynda Bogdala; Deputy Dir., Peggy Gregorski; Civil War Cur., Doug Dammann; Sr. Cur. Education, Nancy Mathews; Cur. Education, Brett Lobello; Cur. Collections, Gina Radandt; Cur. Exhibits, Rachel Klees-Anderson; Coord. Operations, Ken Ade.
Personnel Profile: Full-Time Paid 13; Part-Time Paid 37.
Operating Expenses: 2,300,000
Operating Income: 2,300,000
Governing Authority: municipal; nonprofit organization. Parent Institution: Kenosha Public Museums. Tax-exempt.
Institution Type/Description: Military History Museum.
Collections: Civil War history & artifacts; hands-on exhibits; audio & videos; Veterans Memorial.
Research Fields: Civil War; genealogy.
Facilities: archives; 17,000 sq. ft. exhibition space; conference rooms; resource center. Gift items for sale.
Activities: audio & videos; special events; reenactments; lectures; classes; forums.
Publications: quarterly program guide; e-newsletter.
Hours & Admission Prices: Sun.-Mon. 12-5, Tues.-Sat. 9-5. Adults $7; discounts to Kenosha/Somers residents and ICOM & AAM members; members and children 15 & under no charge.
Attendance: 70,474 (accurate)
Membership: Individual $25; Family $40; Patron $75; Sponsor $150.

∗   **DINOSAUR DISCOVERY MUSEUM, (M),** 5608 Tenth Ave., Kenosha, WI 53140. Mailing Address: 5500 First Ave., Kenosha, WI 53140-3778. Tel.: 262-653-4450. Fax: 262-653-4437. Facebook: Dinosaur Discovery Museum.
E-mail: djoyce@kenosha.org
Web Site: www.dinosaurdiscoverymuseum.org
Founded: 2006.
Congressional District: 1
Key Personnel: Dir., Dan Joyce; Pres. (V), Shari Krewson; Vice Pres. (V), Lynda Bogdala; Deputy Dir., Peggy Gregorski; Sr. Cur. Education, Nancy Mathews; Cur. Education, Nick Wiersum; Cur. Exhibits, Rachel Klees Andersen; Cur. Collections, Gina Radandt.
Personnel Profile: Full-Time Paid 13; Part-Time Paid 37; Part-Time Volunteers 25; Interns 2.
Volunteer Hours: 10,058
Operating Expenses: 2,300,000
Operating Income: 2,300,000

Governing Authority: municipal; nonprofit. Parent Institution: Kenosha Public Museums, Kenosha, WI. Tax-exempt: 170(b)(1)(A).
Institution Type/Description: Dinosaur Museum.
Collections: life-sized carnivore dinosaur replicas; hands-on dinosaur & fossil exhibits; excavated specimens on exhibit.
Research Fields: evolutionary link between meat-eating theropods & modern birds.
Facilities: 1,000 sq. ft. classroom; lab; 2,500 sq. ft. exhibit space; field research station. Museum-related items for sale.
Activities: volunteer program; formal education program; temporary exhibitions; hands-on activities.
Publications: bimonthly newsletter, Friends of the Museum News.
Hours & Admission Prices: Tues.-Sun. 12-5. No charge; donations accepted. Closed New Year's Eve & Day; Martin Luther King Jr. Day; Good Friday; Memorial Day; Independence Day; Labor Day; Thanksgiving; Christmas Eve & Day.
Attendance: 37,922 (accurate)
Membership: Individual $15; Family $25; Family $40; Patron $75; Sponsor $100.

KENOSHA COUNTY HISTORICAL SOCIETY AND MUSEUM, INC., (M), 220 51st Place, Kenosha, WI 53140-2909. Tel.: 262-654-5770, ext. 102. Fax: 262-654-1730.
E-mail: kchs@kenoshahistorycenter.org
Web Site: www.kenoshahistorycenter.org
Founded: 1878.
Congressional District: 1
Key Personnel: Pres. (V), P. Mike Maki; Exec. Dir., Tom Schleif; Collections Mgr., Cynthia J. Nelson; Dir. Operations & Museum Shop Mgr., Don Shepard; Bookkeeper, Karen Gallion.
Personnel Profile: Full-Time Paid 2; Part-Time Paid 4; Part-Time Volunteers 65.
Governing Authority: private; nonprofit corporation. Tax-exempt: 501(c)(3).
Institution Type/Description: Local History Museum.
Collections: manuscripts; toys & dolls; American decorative arts; folk art & portraits; Kenosha automotive manufacturing; Kenosha County history.
Research Fields: city & county of Kenosha; southeastern Wisconsin; school curriculum development.
Facilities: 2,000-vol. library of local, state & U.S. history books & bound volumes of newspapers available for use on premises; 5,000 photographic images; reading room. Booklets for sale.
Activities: guided tours; permanent & temporary exhibitions; educational materials & kits; programs.
Publications: quarterly newsletter; occasional books & pamphlets; History in the Making.
Hours & Admission Prices: Tues.-Fri. 10-4:30, Sat. 10-4, Sun. 12-4. Archives: Wed.-Fri. 2-4:30. Suggested Donation: adults $1, children $.50. Closed holidays.
Attendance: 16,000 (accurate)
Membership: Senior $15; Individual $20; Family, Club & Corporate $30; Homesteader $100; Settler $250; Lighthouse Keeper $500; Rambler $1,000; Ambassador $5,000.

∗   **KENOSHA PUBLIC MUSEUM, (M),** 5500 First Ave., Kenosha, WI 53140-3778. Tel.: 262-653-4140. Fax: 262-653-4437. Facebook: Kenosha Public Library.
E-mail: djoyce@kenosha.org
Web Site: kenosha.org/museum
Founded: 1933.
Congressional District: 1
Key Personnel: Dir., Dan Joyce; Pres. (V), Shari Krewson; Vice Pres. (V), Lynda Bogdala; Sr. Cur. Education, Nancy Mathews; Cur. Exhibits, Rachel Klees Anderson; Cur. Collections, Gina Radandt; Deputy Dir., Peggy Gregorski; Museum Shop Mgr., Christine Schlater; Accounts Mgr., Steve Maraccini; Coord. Operations, Ken Ade.
Personnel Profile: Full-Time Paid 13; Part-Time Paid 37; Part-Time Volunteers 136; Interns 4.
Volunteer Hours: 10,058
Operating Expenses: 2,300,000
Operating Income: 2,300,000
Governing Authority: Parent Institution: City of Kenosha, Wisconsin. Subsidiary Institutions: Dinosaur Discovery Museum; Civil War Museum. Tax-exempt: 170(b)(1)(A).
Institution Type/Description: Natural Science & Fine Arts Museum.
Collections: works by local, national & international artists including Chagall, Picasso, & Dali; Wisconsin history; Kenosha County Woolly mammoth; Native American village; hands-on exhibits.
Research Fields: archaeology; anthropology; geology; megafauna research; mammoths; Paleo-Indian; museum education.

Facilities: 4,000-vol. library on art, cultural & natural history available on premises; 200-seat auditorium. Museum-related items for sale.

Activities: group tours; lectures; films; formally organized education programs for children & adults; outreach loan service; temporary & permanent exhibitions; children's hands-on field station; special events; rental facilities.

Publications: FOM newsletter; pamphlet, The Lorado Z. Taft Dioramas; Early Wisconsin Pottery; quarterly program schedule; Peter V. Bianchi: The Art of Science.

Hours & Admission Prices: March-Aug. Sun.-Mon. 12-5, Tues.-Sat. 9-5; Sept.-Feb. Tues.-Sat. 9-5, Sun. 12-5. No charge; donations accepted. New Year's Eve & Day; Martin Luther King Jr. Day; Good Friday; Memorial Day; Independence Day; Labor Day; Thanksgiving; Christmas Eve & Day. &

Attendance: 125,913 (accurate)

Membership: Annual $25; Family $40; Patron $75; Sponsor $150.

Kewaunee

KEWAUNEE COUNTY HISTORICAL JAIL MUSEUM, Court House Sq., Kewaunee, WI 54216. Mailing Address: 613 Dodge St., Kewaunee, WI 54216-1322. Tel.: 920-388-7176 & 3858.

Founded: 1970.
Congressional District: 8
Key Personnel: Pres. (V), Thomas Schuller; Vice Pres., Jerry Abitz; Treas., Arletta Bertrand; Sec., Julie Bloor.
Personnel Profile: Part-Time Volunteers 50.
Volunteer Hours: 680
Operating Expenses: 14,685
Operating Income: 15,248
Governing Authority: society. Affiliated with Wisconsin State Historical Society. Tax-exempt: 170(B)(1)(A).
Institution Type/Description: General Museum: housed in 1876 sheriff's residence & adjoining dungeon type jail cells.
Collections: furnished living room, bedroom, sheriff's office; tool shed; manuscripts; Indian artifacts; early settlers' artifacts; U.S.S. Pueblo display & replica; war mementos; silver; children's toys; writings by old settlers; wood carvings including one of Custer's Last Stand.
Research Fields: on our collections.
Activities: guided tours for schools & organizations on request; permanent & temporary exhibitions; annual fund drive.
Publications: quarterly newsletter to members; books on Kewaunee County History; audio & VHS tapes.
Hours & Admission Prices: Memorial Day-Labor Day daily 12-4. Suggested Donation: adults $2, students $1.
Attendance: 356 (accurate)

King

WISCONSIN VETERANS MUSEUM-KING, Wisconsin Veterans Home, Hwy. QQ, King, WI 54946. Mailing Address: 30 W. Mifflin St., Madison, WI 53703-2558. Tel.: 608-267-7207; 715-258-1486 & 5586. Fax: 608-264-7615; 715-258-5736.

Web Site: museum.dva.state.wi.us
Founded: 1935.
Congressional District: 6
Key Personnel: C.E.O., Kenneth Black; Dir., Michael Telzrew; Cur. Exhibitions & Collections Mgr., Jeff Kollatin; Museum Operations Mgr., Lynnette M. Wolfe; Registrar, Kristine Zickuhr; Cur., William Brewster; Museum Shop Mgr., Gregory Lawson.
Personnel Profile: Full-Time Paid 12; Full-Time Volunteers 6; Part-Time Paid 15; Part-Time Volunteers 6; Interns 2.
Governing Authority: state. Parent Institution: Wisconsin Dept. Veterans Affairs. Affiliated Museum: Wisconsin Veterans Museum-Madison. Tax-exempt.
Institution Type/Description: Military History Museum.
Collections: Civil War & World War II artifacts and displays including exhibits relating to Korean & Vietnam Wars; uniforms; arms; equipment.
Research Fields: Wars; military, veteran history.
Activities: permanent & temporary exhibitions.
Publications: brochures, Wisconsin in the World Wars; When America Fought Its Costliest War, Wisconsin Served in the Front Rank; Wisconsin at War.
Hours & Admission Prices: Mon.-Fri. 8-4, Sat.-Sun. 8-11 & 1-4. No charge. &
Attendance: 8,000 (estimated)

La Crosse

CHILDREN'S MUSEUM OF LA CROSSE, 207 Fifth Ave. S., La Crosse, WI 54601. Tel.: 608-784-2652. Fax: 608-784-6988. Facebook: Funmuseum.

E-mail: info@funmuseum.org
Web Site: funmuseum.org
Founded: 1999.
Key Personnel: Dir., Anne Snow; Museum Shop Mgr., Meg Steuer.
Personnel Profile: Full-Time Paid 3; Part-Time Paid 14; Part-Time Volunteers 60.
Governing Authority: Tax-exempt.
Institution Type/Description: Children's Museum.
Collections: hands-on exhibitions.
Facilities: Museum-related items for sale.
Activities: educational programs; special events; Boy & Girl Scout programs; summer day camps; preschool parent/child programs; holiday events.
Hours & Admission Prices: Tues.-Sat. 10-5, Sun. 12-5. Admission $6; discounts to ACM & ASTC members; children under one & members no charge. Closed major holidays. &
Attendance: 72,980 (accurate)
Membership: Grandparent $60; Explore $65; 6 Mix $90; Reciprocal $125.

HIXON HOUSE, 429 N. 7th St., La Crosse, WI 54601-3301. Mailing Address: La Crosse County Historical Society, P.O. Box 1272, La Crosse, WI 54602-1272. Tel.: 608-782-1980. Fax: 608-793-1359.

E-mail: lchs@centurytel.net
Web Site: lchsweb.org
Founded: 1898.
Congressional District: 3
Key Personnel: Exec. Dir., Daniel Moen; Pres. (V) & Chm. (V), Joe Kruse; Cur., Peggy Derrick.
Personnel Profile: Full-Time Paid 1; Full-Time Volunteers 1; Part-Time Paid 2; Part-Time Volunteers 32.
Governing Authority: society. Parent Institution: La Crosse Historical Society. Tax-exempt: 501(c)(3).
Institution Type/Description: Period Home: housed in c.1859 Hixon House.
Collections: artifacts that reflect life in La Crosse from 1860-1910; historical sketches; personal family collection reflecting upper class 19th century life style; furniture; paintings; fabrics & textiles; imported & domestic glassware & china.
Research Fields: local history.
Facilities: Museum-related items for sale.
Activities: guided tours; permanent & temporary exhibits; Christmas tours.
Publications: newsletter, Past, Present & Future; reprints of local histories.
Hours & Admission Prices: Memorial Day to Labor Day daily 11-5, group tours by appointment. Adults $8, seniors 55 & over $7, students $5, children 12 & under $4; discounts to groups; members no charge. &
Attendance: 8,000 (estimated)
Membership: Individual $35; Family $50; Century $125; Business $200; Patron $250; Benefactor $500; Heritage $1,000; Fellows Club $5,000.

PUMP HOUSE REGIONAL ARTS CENTER, 119 King St., La Crosse, WI 54601. Tel.: 608-785-1434. Fax: 608-785-0085.

E-mail: contact@thepumphouse.org
Web Site: www.thepumphouse.org
Founded: 1977.
Congressional District: 30
Key Personnel: Exec. Dir., Toni Asher; Chm. (V), Donald Smith.
Personnel Profile: Full-Time Paid 1; Part-Time Paid 2; Part-Time Volunteers 10; Interns 5.
Governing Authority: not-for-profit organization. Tax-exempt: 501(c)(3).
Institution Type/Description: Arts Center: housed in 1880 brick Romanesque revival water pumping building. Listed on National Register of Historic Buildings.
Collections: Midwestern art of regional artists & craftspersons Historical photographs.
Research Fields:
Facilities: 2,000 sq. ft. exhibit space.
Activities: workshops & classes for all ages; docent led tours; performing arts events.
Hours & Admission Prices: Tues.-Fri. 11-7, Sat. 12-4. No charge; donations accepted. Closed New Year's Day; Memorial Day; Independence Day; Labor Day; Thanksgiving; Christmas. &
Attendance: 12,000 (estimated)
Membership: Student $20; Individual $30; Household $40; Sustaining $75; Patron $125; Arts Angel $250; Director $500; Benefactor $1,000.

RIVERSIDE MUSEUM, (M), 410 E. Veterans Memorial Dr., La Crosse, WI 54601-4490. Mailing Address: P.O. Box 1272, La Crosse, WI 54602-1272. Tel.: 608-782-1980. Fax: 608-793-1359.
E-mail: lchs@centurytel.net
Web Site: lchsweb.org
Founded: 1990.
Congressional District: 3
Key Personnel: Dir., Jane M. Beseler; Pres. (V), Joe Kruse.
Personnel Profile: Full-Time Paid 1; Full-Time Volunteers 1; Part-Time Paid 2; Part-Time Volunteers 32.
Governing Authority: private; nonprofit. Parent Institution: La Crosse County Historical Society. Tax-exempt: 501(c)(3).
Institution Type/Description: History Museum: housed in an old fish hatchery building. Listed on the National Register of Historic Sites.
Collections: concentration on area history, from prehistoric to modern times with special emphasis on importance of river system.
Research Fields: riverboat travel.
Facilities: 2,000 sq. ft. exhibit space; 35-seat theater.
Activities: films; participatory exhibits. Annual Event: War Eagle Days-commemoration the sinking of the riverboat in 1870.
Publications: bimonthly newsletter, Past, Present and Future.
Hours & Admission Prices: Memorial Day-Labor Day Mon.-Sat. 10:30-4:30, Sun. 10:30-4; tours by appointment. Families $5, adults $2, children $1. &
Attendance: 12,000 (accurate)
Membership: Individual $35; Family $50; Century $125; Business $200; Patron $250; Sustaining $500; Heritage $1,000; Fellows Club $5,000.

SWARTHOUT MEMORIAL MUSEUM, 112 S. 9th St., La Crosse, WI 54602. Mailing Address: La Crosse County Historical Society, P.O. Box 1272, La Crosse, WI 54602-1272. Tel.: 608-782-1980.
E-mail: lchs@centurytel.net
Web Site: lchsweb.org
Founded: 1898.
Congressional District: 3
Key Personnel: C.E.O. & Dir., Dr. Carl Miller; Pres. (V), Dr. Erik Gundersen; Museum Coord., Lori Strom; Admin., Michelle Logan.
Personnel Profile: Full-Time Paid 2; Full-Time Volunteers 1; Part-Time Paid 9; Part-Time Volunteers 32.
Governing Authority: nonprofit organization. Parent Institution: La Crosse County Historical Society, Inc., P.O. Box 1272, La Crosse, WI. 54602. Tel.: 608-782-1980. Tax-exempt.
Institution Type/Description: Historical Society Museum.
Collections: Hixon collection representing 19th century lifestyle; Mississippi River community; lumbering; brewing; period costumes; photographs; textiles; decorative arts; local documents.
Research Fields: transportation; clothing; local history; brewing industry; lumbering; river lore.
Facilities: research library. Local histories & other museum related items for sale.
Activities: permanent, changing & traveling exhibits; traveling guided tours; heritage tour of city, La Crosse.
Publications: society newsletter, Past, Present and Future; reprints of local histories.
Hours & Admission Prices: Tues.-Fri. 10-5, Sat.-Sun. 1-5. No charge; donations accepted. &
Attendance: 15,000 (estimated)
Membership: Individual $35; Family $50; Century $125; Business $200; Patron $250; Sustaining $500; Heritage $1,000; Fellows Club $5,000.

La Pointe

MADELINE ISLAND MUSEUM, (M), Woods Ave. & Main St., La Pointe, WI 54850. Mailing Address: P.O. Box 9, La Pointe, WI 54850-0009. Tel.: 715-747-2415. Fax: 715-747-6985.
E-mail: madeline@wisconsinhistory.org
Web Site: www.madelineislandmuseum.org
Founded: 1958.
Congressional District: 7
Key Personnel: Dir. Div. Historic Sites, Alicia L. Goehring; Site Dir., Steven R. Cotherman; Cur., Sheree Peterson; Supt. Bldgs. & Grounds, Tim Eldred.
Personnel Profile: Full-Time Paid 2; Part-Time Paid 8.
Governing Authority: state. Parent Institution: Wisconsin Historical Society, 816 State St., Madison 53706. Tax-exempt.
Institution Type/Description: Historic Site & Museum.
Collections: area history; artifacts reflecting the exploration & settlement of the Apostle Islands, including Native American history; French & English exploration & fur trade; 19th-century industries including fishing, lumbering, shipping & mining; ephemera; photographs; cultural history of Madeline Island from 19th century to present.

Research Fields: 17th- & 18th-century exploration of Apostle Islands; Madeline Island Settlement; fur trade; lake & forest industries; Lake Superior history; Ojibwe cultural history.
Facilities: 5,200 sq. ft. exhibit space.
Activities: permanent exhibitions; formally organized educational programs for children; tours; island history video.
Hours & Admission Prices: Memorial Day weekend to 1st weekend in Oct. daily 10-5. Family $19, adults $7, senior citizens $6, children $3.50; discounts to groups. &
Attendance: 12,192 (accurate)
Membership: Wisconsin Historical Society: Individual $45; Household/Individual Plus $60; History Lover $100; History Guardian $250; History Ambassador $500; Heritage Circle $1,000.

Lac du Flambeau

OJIBWE MUSEUM AND CULTURAL CENTER, (M), 603 Peace Pipe Rd., Lac du Flambeau, WI 54538. Mailing Address: P.O. Box 804, Lac du Flambeau, WI 54538. Tel.: 715-588-3333. Fax: 715-588-2355.
E-mail: tmitchell@ldftribe.com
Web Site: www.ldfmuseum.com
Founded: 1989.
Key Personnel: Dir., Teresa Mitchell.
Personnel Profile: Full-Time Paid 1; Part-Time Volunteers 2.
Governing Authority: Parent Institution: Lac du Flambeau Tribe.
Institution Type/Description: Native American History Museum.
Collections: Ojibwe culture & History; canoes; traditional clothing; French fur trading post.
Facilities: Museum-related items for sale.
Activities: workshops; special events.
Hours & Admission Prices: Summer: Tues.-Thurs. 10-4, Sat. call for hours; Winter: Mon.-Fri. 10-4. Adults $4, seniors & children $3; discounts to AAA members. Closed holidays. &
Attendance: 2,500 (estimated)

Ladysmith

RUSK COUNTY HISTORICAL SOCIETY, Rusk County Fairgrounds, Hwy. 8, Ladysmith, WI 54848. Mailing Address: 408 Phillips Ave. E., Ladysmith, WI 54848-2333. Tel.: 715-532-5615.
E-mail: jplatteter@yahoo.com
Web Site: www.ruskcounty.org
Founded: 1961.
Congressional District: 7
Key Personnel: Pres., John Terrill; Vice Pres., Joe Baye; Treas., Melanie Meyer; Sec., Mary Lou Bisson; Cur., Janet Platteter; Chm. Genealogy Dept., Molla Clark.
Personnel Profile: Full-Time Volunteers 10; Part-Time Volunteers 5.
Governing Authority: society. Branch Museums: 1895 Apollonia Church, Bruce, WI; 1900's Little Red School House, Ladysmith. Tax-exempt.
Institution Type/Description: Historical Museum.
Collections: local historical artifacts & archives; Gates County courthouse artifacts; logging; machinery; horse-drawn machinery; 306 straight razors & related barbering items; log stamp hammers; rail display; Rusk County high school yearbooks; Rusk County's village & town histories; post cards from 1900 to present. Historic Buildings: c.1900 Little Red School House; information building from Flambeau Mine; early Glen Flora jail; log cabin; teacher's cabin.
Facilities: 300-vol. library. Cookbooks, booklets of recollections by area residents for sale.
Activities: children's classes; obituary research. Annual Events: Heritage Day; Log Cabin Day in June.
Publications: maps of Rusk County; books, History of Rusk County; History of Rural Schools of Rusk County; Glen Flora Pioneers; My Time in the Army; Glen Flora; Birth of Gates Co.; In Cub Chains; Volcanogenic Massive Sulfide Deposits in Northern Wisconsin; Travels with Sophie; book, Rusk County Centennial.
Hours & Admission Prices: Memorial Day-Labor Day, Sat.-Sun. 12:30-4:30. No charge; donations accepted. &
Attendance: 2,067 (estimated)
Membership: Annual $10; Life $75.

Lake Geneva

GENEVA LAKE MUSEUM, (M), 255 Mill St., Lake Geneva, WI 53147-1927. Tel.: 262-248-6060. Fax: 262-248-0661.
E-mail: staff@genevalakemuseum.org
Web Site: www.genevalakemuseum.org
Formerly: Geneva Lake History Buffs, Inc.

Founded: 1984.
Congressional District: 1
Key Personnel: Dir., Karen Jo Walsh; Pres. (V), James Gee; Museum Shop Mgr., Pat Gee.
Personnel Profile: Part-Time Paid 2; Part-Time Volunteers 50.
Governing Authority: Tax-exempt: 501(c)(3).
Institution Type/Description: History Museum: located in the 1929 Wisconsin Power and Light Building.
Collections: farm implements; architecturally-salvaged facades of historic Lake Geneva homes; log structure with Potawatomi Indian arrowheads & tools; furnishings; blacksmith shop with iron anvils & ice cutting tools; fire engine house with c.1890 hose wagon; Spring house for cooling milk; telephone switchboard; 1920's dental work station; general store; Chicago & North Western Railway memorabilia; Frank Lloyd Wright's Hotel Geneva artifacts; Ceylon Court exhibit; Yerkes observatory exhibit; Black Point exhibit; Mezges Sailing exhibit; Geneva law office; 1929 Model-A Ford; Potawatomi Woodland Indian; Jerseyhurst/Crane Estate stained glass windows.
Major Exhibits: Pickard China, 5/14-10/14; Teddy Bears, Summer 2014.
Publications: newsletter; books, Anals of Lake Geneva; History of Lake Geneva.
Hours & Admission Prices: Jan.-Feb. open Sat.; March-April & Nov.-Dec. Fri.-Sat. 10-4, Sun. 12-3; May-Oct. Mon. & Thurs.-Sat. 10-4, Sun. 12-3. Adults $7, seniors & students with ID $6; discounts to Blue Star Museum members; children 11 & under and members with ID no charge. &
Attendance: 10,000 (accurate)
Membership: Student $10; Senior $15; Regular $30; Commercial $50; Benefactor $100; Life $500.

Lake Tomahawk

LAKE TOMAHAWK HISTORICAL SOCIETY, 7247 Kelly Dr., Lake Tomahawk, WI 54539. Mailing Address: P.O. Box 325, Lake Tomahawk, WI 54539-0325.
E-mail: beverlyfagan@frontier.com
Formerly: Northland Historical Society, Inc.
Founded: 1957.
Congressional District: 7
Key Personnel: Pres. (V), Darlene Neuman; Treas., Beverly Fagan.
Governing Authority: nonprofit organization. Tax-exempt: 170(b)(1)(A).
Institution Type/Description: Historical Society Museum.
Collections: early costumes; pictures; maps; books; artifacts from settlers & Indians.
Research Fields: historical sites; buildings; travel routes of Indians, voyageurs, loggers & settlers.
Facilities: 500-vol. library of books on early logging, settlers, ethnic, sites & early maps available for research on premises only; 100-seat auditorium.
Activities: guided tours; lectures; films; formally organized education programs for adults; loan, permanent & temporary exhibitions; program meetings. Museum Sponsors: Northwoods Fall Conference in September.
Publications: The Deacon's Bench.
Hours & Admission Prices: Temporarily closed. &
Attendance: 50 (estimated)
Membership: Junior $2.50; Individual $10; Family $20; Contributing $25; Ebert Club $100 & up.

Lancaster

CUNNINGHAM MUSEUM, 129 E. Maple St., Lancaster, WI 53813. Tel.: 608-723-2239 & 4924.
E-mail: historicalsociety@tds.net
Web Site: grantcountyhistory.org
Institution Type/Description: History Museum.
Collections: local history & culture; period furnishings; personal artifacts; photographs; African American artifacts.
Hours & Admission Prices: Call for hours. No charge; donations accepted.

Laona

CAMP FIVE MUSEUM FOUNDATION, INC., 5480 Connor Farm Rd., Laona, WI 54541-9201. Mailing Address: P.O. Box 5, Laona, WI 54541-0005. Tel.: 715-674-3414. Fax: 715-674-7400.
E-mail: info@lumberjacksteamtrain.com
Web Site: www.lumberjacksteamtrain.com
Founded: 1969.
Congressional District: 7
Key Personnel: Pres. (V), Mrs. Edward J. Dellin; Dir. Education, Sara W. Connor.
Personnel Profile: Full-Time Paid 1; Part-Time Paid 40; Part-Time Volunteers 3.

Operating Expenses: 403,228
Operating Income: 373,403
Governing Authority: nonprofit. Subsidiary Institution: Laona & Northern Railway. Tax-exempt.
Institution Type/Description: Logging Museum & Ecology Complex: operates Laona & Northern Railway's Lumberjack Special steam train.
Collections: early farm tools; logging tools including pointer boats; early tractors; ice rutters; water wagons; railroad artifacts; Native American artifacts; surveying instruments; northern Wisconsin wildlife dioramas; videos; woodworking implements; lumber company tokens-money; tree display; harness shop; depot; active blacksmith shop; early logging history & artifacts. Historic Structures: 1916 Vulcan steam engine; 1920 coaches & cabooses; 1900 cracker barrel store.
Facilities: nature center; arboretum; audio-visual orientation center; picnic area; snack bar. Museum-related items for sale.
Activities: steam train video; audio-visual presentations; forest tour; hayrack & pontoon boat rides; petting corral.
Publications: coloring books; brochures; Forests for All: The Economics of Conservation, video & accompanying textbook for grades 3-12; DVD, Northwoods Logging Saga.
Hours & Admission Prices: mid-June to late Aug. Mon.-Sat. Trains at 11, 12, 1 & 2 to Camp Five Logging Museum Complex. Adults $21, children 4-12 $9; discounts to members, groups, AAA & AAM members, seniors & active military duty; children under 3 no charge. Hayrack & Pontoon Ride: adults $5, children $3. &
Attendance: 10,163 (accurate)
Membership: Brakeman $150; Conductor $300; Fireman $1,000; Engineer $2,000.

Madison

* **CHAZEN MUSEUM OF ART, (M),** 750 University Ave., Madison, WI 53706-1479. Tel.: 608-263-2246. Fax: 608-263-8188.
Web Site: chazen.wisc.edu
Formerly: Elvehjem Museum of Art
Founded: 1970.
Congressional District: 2
Key Personnel: Dir., Russell Panczenko; Registrar, Andrea Selbig; Registrar, Ann Sinfield; Preparator, Steve Johanowicz; Asst. Dir. Admin., Brian Thompson; Cur. Paintings, Sculpture & Decorative Arts, Maria Saffiotti Dale; Cur. Education, Anne Lambert; Cur., Prints & Drawings, Andrew Stevens; Exhibitions Coord., Mary Ann Fitzgerald; Exhibition Designer & Chief Preparator, Jerl Richmond.
Personnel Profile: Full-Time Paid 17; Part-Time Paid 25; Part-Time Volunteers 240; Interns 2.
Operating Expenses: 4,200,000
Operating Income: 4,200,000
Governing Authority: university. Parent Institution: University of Wisconsin at Madison. Tax-exempt.
Institution Type/Description: Art Museum.
Collections: paintings; prints; drawings; sculpture; decorative arts; archaeology; Watson collection of Indian miniatures; Van Vleck Japanese prints; Hall collection of European Medals; Davies collection of Russian icons, Russian & Soviet paintings; Lane collection of modern sculpture.
Major Exhibits: Karen LaMonte: Material and Light, 12/13-1/14; Ikeda Manabu and Tenmyouya, 12/13/13-2/16/14; St. John's Bible, 12/19/13-3/15/14; Changing Hands: Art Without Reservation, 3 Contemporary Native Art from the Northeast and Southeast (T), 2/7/14-4/27/14; Marginalia in Cartography, 3/1/14-5/18/14; Lichtenstein Prints, 5/16/14-6/30/14; Monotypes, 5/31/14-8/2/14; Hootkin Collection, 9/5/14-11/30/14.
Research Fields: pertaining to collections.
Facilities: 100,000-vol. library of art books available for inter-library loan & for loan by permission of librarian; reading room; 4 auditoriums; classrooms. Art objects, books, catalogs & other museum-related items for sale.
Activities: guided tours; lectures; gallery talks; concerts; docent program; formally organized education programs for undergraduate & graduate students affiliated with the University of Wisconsin & adults; permanent, temporary & traveling exhibitions; film program.
Publications: biannual bulletin; bimonthly calendar; catalogs of major exhibitions; semiannual newsletter; j-email newsletter.
Hours & Admission Prices: Tues.-Wed. & Fri. 9-5, Thurs. 9-9, Sat.-Sun. 11-5. No charge; donations accepted. Closed New Year's Day; Thanksgiving; Christmas Eve & Day. &
Attendance: 140,000 (accurate)
Membership: Student & Senior Citizen $25; Individual $35; Senior Couple $40; Family $50; Patron $75; Founder $125; Associate $250; Fellow $500; Director's Circle $1,000.

HELEN LOUISE ALLEN TEXTILE COLLECTION, (M), 1300
Linden Dr., Univ of Wisconsin, Madison, WI 53706-1524. Tel.:
608-262-1162. Fax: 608-265-5099.
E-mail: hlatc@mail.sohe.wisc.edu
Web Site: textilecollection.wisc.edu
Founded: 1968.
Congressional District: 2
Key Personnel: Cur., Maya Lea.
Personnel Profile: Full-Time Paid 1; Part-Time Paid 2.
Governing Authority: university. Parent Institution: University of Wisconsin-
Madison. Subsidiary Institution: Design Studios Dept. Tax-exempt:
170(b)(1)(A).
Institution Type/Description: Textile & Costume Collection.
Collections: 13,000 textiles including ethnographic textiles & costumes;
European & American home furnishings & apparel fabrics.
Research Fields: cultural anthropology; textile history, design, interior design
& textile science.
Facilities: 10,000-vol. research library.
Activities: lectures; educational programs; temporary exhibitions.
Publications: annual newsletter; exhibition catalogues.
Hours & Admission Prices: Closed for relocation. &
Membership: Friend $50; Sustainer $100; Patron $250; Benefactor $500.

HENRY VILAS PARK ZOO, 702 S. Randall Ave., Madison, WI
53715-1600. Tel.: 608-266-4733. Fax: 608-266-5923.
E-mail: zoo@countyofdane.com
Web Site: www.vilaszoo.org
Founded: 1911.
Key Personnel: Dir., Ronda Schwetz.
Personnel Profile: Full-Time Paid 20; Part-Time Paid 1; Interns 8.
Governing Authority: municipal; county. Dane County. Tax-exempt.
Institution Type/Description: Zoo.
Collections: general zoological park; children's zoo.
Facilities: concessions. Gift items for sale.
Activities: permanent exhibitions.
Hours & Admission Prices: Buildings: daily 10-4. Grounds: daily 9:30-5. Call
for holiday hours. No charge; donations accepted. &
Attendance: 700,000 (estimated)
Membership: Student & Senior Citizen $20; Single $25; Family $35; Sustain-
ing $75; Centennial $100; Patron $150; Steward $250; Protector $500;
Ambassador $1,000.

**JAMES WATROUS GALLERY OF THE WISCONSIN ACAD-
EMY OF SCIENCES, ARTS AND LETTERS,** Overture Center
for the Arts, 201 State St., 3rd Fl., Madison, WI 53703. Tel.:
608-265-2500. Fax: 608-265-3039.
Web Site: www.wisconsinacademy.org
Founded: 1994.
Key Personnel: Exec. Dir., Jane Elder; Dir., Martha Glowacki.
Personnel Profile: Part-Time Paid 3.
Governing Authority: Parent Institution: Wisconsin Academy of Sciences, Arts
& Letters. Tax-exempt.
Institution Type/Description: Art Gallery.
Collections: works by contemporary Wisconsin artists.
Major Exhibits: Patricia Callahan and Rhea Vedro, 1/14/14-3/2/14; Ida Wyman
and Kevin Miyazaki, 3/14/14-5/4/14; Don Friedlich and Dianne Soffa,
5/13/14-6/29/14; Tyler Robbins and Graham Yeager, 7/11/14-8/24/14; Paul
Vanderbilt, 9/14.
Facilities: 1,500 sq. ft. exhibition space.
Activities: temporary exhibitions.
Publications: quarterly newsletter, Wisconsin People and Ideas.
Hours & Admission Prices: June-Aug. Tues.-Thurs. 11-5, Fri.-Sat. 11-8, Sun.
1-5; Sept.-May Tues.-Thurs. 11-5, Fri.-Sat. 11-9, Sun. 1-5. No charge;
donations accepted. &
Attendance: 10,000 (estimated)
Membership: Intro $30; Continuing $40.

MADISON CHILDREN'S MUSEUM, INC., (M), 100 N. Hamilton
St., Madison, WI 53703-2116. Tel.: 608-256-6445. Fax: 608-268-
1398.
Web Site: www.madisonchildrensmuseum.org
Founded: 1980.
Congressional District: 1
Key Personnel: Exec. Dir., Ruth G. Shelly; Bd. Chair, Diane Ballweg; Exhibit
Dir., Brenda Baker; Guest Experience Mgr., Sarah Sosa-Acevedo.
Personnel Profile: Full-Time Paid 28; Part-Time Paid 31; Part-Time Volunteers
338.
Governing Authority: nonprofit organization. Tax-exempt: 501(c)(3).
Institution Type/Description: Children's Museum.

Collections: interactive exhibits & programs with arts, science & humanities
themes.
Research Fields: education.
Facilities: Museum-related items for sale.
Activities: hands-on exhibits & programs for children ages birth to 12 yrs.;
outreach programs to area schools & community groups; in-depth collabo-
ration with university schools & other groups.
Publications: quarterly newsletter; e-newsletter.
Hours & Admission Prices: Thurs. 9:30-5, Fri.-Wed. 9:30-5. Admission $7.95;
discounts to groups & those on public assistance; 1st Wed. each month 5-8,
members & children under one no charge. Closed some national holidays.
&
Attendance: 200,000 (estimated)
Membership: Dual $65; Family $95; Family Plus $135.

*** MADISON MUSEUM OF CONTEMPORARY ART, (M),**
227 State St., Madison, WI 53703-2214. Tel.: 608-257-0158. Fax:
608-257-5722.
E-mail: info@mmoca.org
Web Site: www.mmoca.org
Formerly: Madison Art Center
Founded: 1901.
Congressional District: 2
Key Personnel: Pres., Jim Yehle; Dir., Stephen Fleischman; Cur., Richard H.
Axsom; Cur. Education, Sheri Castelnuovo; Assoc. Cur. Exhibitions, Leah
Kolb; Business Mgr., Michael Paggie; Registrar, Marilyn Sohi; Dir. Devel.,
Elizabeth Tucker; Dir. Events & Volunteers, Annik Dupaty; Dir. Commu-
nications, Erika Monroe-Kane; Dir. Retail Operations, Leslie Genszler;
Devel. Assoc., Kaitlin Kropp; Supvr. Technical Svcs., Mark Verstegen.
Personnel Profile: Full-Time Paid 15; Full-Time Volunteers 1; Part-Time Paid
28; Part-Time Volunteers 450; Interns 8.
Governing Authority: nonprofit organization. Tax-exempt: 501(c)(3).
Institution Type/Description: Art Museum.
Collections: contemporary art; Rudolph E. Langer collection; American
paintings, sculpture & prints; photography; Mexican & European artworks;
works on paper; The Bill McClain Collection of Chicago Imagism.
Major Exhibits: Los Grandes del Arte Moderno Mexicano, 7/13-6/14; The
Mystery Beneath, 1/17/14-4/13/14; Real Surreal (from the Whitney Mu-
seum, NY), 1/24/14-4/27/14; Turn Turn Turn, 5/24/14-8/24/14; Story Back,
7/14-7/15; Jason Yi, 8/23/14-11/16/14; From Here to There: Alec Soth's
America (from the Walker Art Center MN), 9/14/14-1/24/15.
Research Fields: pertaining to permanent collections.
Activities: rotating exhibitions; guided tours; films; lectures; inter-museum
loans; docent programs; community & school outreach programs; family
resources & events; teacher resources; interactive website. Museum Spon-
sors: New Media & Performance Art; Art Fair on the Square; Arts Ball; bus
trips to neighboring museums.
Publications: member newsletters & brochures; exhibition catalogue, Chicago
Imagists.
Hours & Admission Prices: Tues.-Thurs. & Sun. 12-5, Fri. 12-8, Sat. 10-8. No
charge; donations accepted. Closed major holidays. &
Attendance: 175,000 (estimated)
Membership: Student $30; Individual $45; Family $65; Supporting $140;
Langer Society Donor $250-$499; Business Council $250 & up; Langer
Society Patron $500-$999; Director's Circle $1,000 & up.

OLBRICH BOTANICAL GARDENS, 3330 Atwood Ave., Madi-
son, WI 53704-5808. Tel.: 608-246-4550. Fax: 608-246-4719.
TDD: 608-267-4980; Facebook: Olbrich Botanical Gardens.
E-mail: rsladky@cityofmadison.com
Web Site: www.olbrich.org
Founded: 1952.
Congressional District: 2
Key Personnel: C.E.O., Roberta Sladky; Museum Shop Mgr., Cindy Sullivan.
Personnel Profile: Full-Time Paid 23; Part-Time Paid 16; Part-Time Volunteers
600; Interns 7.
Governing Authority: municipal. Parent Institution: City of Madison, WI,
Madison Parks Dept., County Bldg., 215 Martin Luther King Blvd. Tel.
608-266-4711. Tax-exempt: 501(c)(3).
Institution Type/Description: Botanical Garden.
Collections: flower, herb, rose & rock gardens; tropical; perennials; horticul-
tural library; tropical conservatory; Thai Garden; Thai Pavilion.
Facilities: 1,000 vol. library of horticultural books; botanical garden; class-
rooms; tropical conservatory. Gift items for sale.
Activities: flower shows; art exhibits; concerts; children's activities; classes;
horticulture & special interest groups meetings; tours; lectures.
Publications: newsletter, Olbrich Gardens Quarterly News.
Hours & Admission Prices: Garden: April-Sept. daily 8-8; Oct.-March daily
9-4. Conservatory: Mon.-Sat. 10-4, Sun. 10-5. Outdoor Gardens: no charge.

Conservatory: $2; members, Wed. & Sat. mornings no charge. Closed New Year's Day; Thanksgiving; Christmas. &

Attendance: 250,000 (accurate)

Membership: Olbrich Botanical Society: Garden Friend Plus One $50; Garden Family $55; Garden Family & Guests $65; Garden Contributor $100; Garden Patron $250.

STEENBOCK GALLERY, Wisconsin Academy of Sciences, Arts and Letters, 1922 University Ave., Madison, WI 53726. Tel.: 608-263-1692. Fax: 608-265-3039.

Institution Type/Description: Art Gallery.

Collections: paintings; sculpture.

Hours & Admission Prices: Mon.-Fri. 8:30-4:30.

UNIVERSITY OF WISCONSIN-MADISON ARBORETUM, 1207 Seminole Hwy., Madison, WI 53711-3726. Tel.: 608-263-7888. Fax: 608-262-5209.

E-mail: info@uwarboretum.org

Web Site: www.uwarboretum.org

Founded: 1934.

Congressional District: 2

Key Personnel: Mgr. Friends of the Arboretum, Sara Minkoff; Dir., Dr. Kevin McSweeney; Museum Shop Mgr., Molly Murray.

Personnel Profile: Full-Time Paid 12; Part-Time Paid 46; Part-Time Volunteers 610.

Governing Authority: university. Parent Institution: University of Wisconsin-Madison. Tax-exempt.

Institution Type/Description: Arboretum.

Collections: natural groupings of plants and animals in ecological communities; horticultural collections.

Research Fields: landscape architecture; ecology; wildlife & restoration ecology; botany; natural history; herpetology; paleontology; entomology; geology; phenology; zoology; horticulture; soil science; plant pathology.

Facilities: visitor center; outdoor classrooms for groups.

Activities: guided tours; restoration workshops & summer institutes for teachers & land managers.

Publications: maps; booklets; field guides; Newsleaf; prairie restoration materials for schools.

Hours & Admission Prices: Daily 7-10pm. No charge; donations accepted. &

Attendance: 650,000 (accurate)

Membership: Senior Citizen & Student $20; Single $25; Family $35; Business & Supporting $100; Patron $250 & up.

UNIVERSITY OF WISCONSIN ZOOLOGICAL MUSEUM, 250 N. Mills St., Lowell E. Noland Zoology Bldg., Madison, WI 53706-1708. Tel.: 608-262-3766. Fax: 608-262-5395.

Web Site: www.zoology.wisc.edu/uwzm/index.html

Founded: 1887.

Congressional District: 2

Key Personnel: Cur. Collections, Laura A. Halverson Monahan; Cur. Mammalogy & Ornithology, Paul M. Holahan; Registrar, M. Kathryn Jones; Distinguished Researcher & Cur. Emeritus Osteology, E. Elizabeth Pillaert; Cur. Emeritus, Exhibits & Vertebrate Paleontology, Dr. John E. Dallman; Adjunct Cur. Fish, Dr. John D. Lyons; Adjunct Cur. Herpetology, Dr. Gregory C. Mayer.

Personnel Profile: Full-Time Paid 2; Part-Time Paid 2; Part-Time Volunteers 2; Interns 15.

Governing Authority: university. Affiliated with the University of Wisconsin. Tax-exempt: 501(c)(3).

Institution Type/Description: Zoology Museum.

Collections: 25,000 ornithology (including osteology); 26,000 mammology (including osteology); 13,000 ichthyology; 8,000 herpetology; 17,500 osteology (all vertebrate classes); 2,000 paleontology; 250,000 malacology; 110,000 histological slide preparations & tissue fragments; H. W. Mossman collection of mammalian reproductive organs, 3,100 specimens; W.B. Quay collection of fluid mammals, 2,600 specimens; 800 period artifacts.

Research Fields: herpetology; osteology; paleontology; mammalogy; ornithology; ichthyology; archaeology.

Facilities: 5,900-vol. library & 45,300 reprints of zoological materials available on premises; reading room.

Activities: lectures; formally organized education programs for undergraduate and graduate college students affiliated with the University of Wisconsin; permanent & temporary exhibitions.

Hours & Admission Prices: Sept.-June Mon.-Fri. 8:30-12 & 1-4:30 by appointment. No charge; donations accepted. Closed national holidays. &

Attendance: 3,500 (estimated)

WISCONSIN HISTORICAL MUSEUM, (M), 30 N. Carroll St., Madison, WI 53703-2707. Tel.: 608-264-6555. Fax: 608-264-6575.

E-mail: museum@wisconsinhistory.org

Web Site: www.wisconsinhistory.org

Formerly: State Historical Museum of Wisconsin

Founded: 1846.

Congressional District: 2

Key Personnel: C.E.O., Dr. Ellsworth Brown; Dir., Jennifer Kolb; Exhibits Dir., Doug Griffin; Chief Cur., Paul Bourcier; Cur. Costumes & Textiles, Leslie Bellais; Cur. Business & Technology, David Driscoll; Museum Shop Mgr., Heather Groff; Cur. Domestic Life, Joe Kapler; Collections Mgr., Scott Roller; Special Events Coord., Katie Schumacher; Cur. Native American Collections, Jennifer Kolb.

Personnel Profile: Full-Time Paid 11; Part-Time Paid 26; Part-Time Volunteers 80; Interns 4.

Governing Authority: society. Parent Institution: Wisconsin Historical Society. Branch Museums & Historic Sites: First Capitol, Belmont; Villa Louis, Prairie du Chien; Old Wade House, Greenbush; Stonefield, Cassville; Madeline Island Historical Museum, LaPointe; Pendarvis, Mineral Point; Circus World Museum, Baraboo; Old World Wisconsin, Eagle; H.H. Bennett Studio & History Center, Wisconsin Dells; Reed School, Neillsville; Black Point, Lake Geneva. See separate listings for hours & admissions. Tax-exempt: 170(b)(1)(A).

Institution Type/Description: History Museum.

Collections: Wisconsin history from prehistoric to modern times; firearms; dolls; crafts; costumes; decorative arts; glass; ceramics; archives; photographs; toys.

Major Exhibits: Wisconsin Women of Style, 1/14-3/14; Seifert Landscapes, 4/14-8/14.

Research Fields: Wisconsin history, anthropology & archaeology.

Facilities: library of North American history; archives; reading rooms; auditorium.

Activities: lectures; films; gallery talks; workshops; formally organized education programs; dinner series; permanent & temporary exhibitions.

Publications: Wisconsin Magazine of History; Columns.

Hours & Admission Prices: Tues.-Sat. 9-4. Suggested Donation: family $10, adults $4, children $3; discounts to AAM members; members no charge. Closed national holidays. &

Attendance: 72,500 (accurate)

Membership: Senior 65 & over $30; Individual & Senior Citizen Family $40; Family $50; Institutional $65; Supporting $100; Sustaining $250; Patron $500; Life $1,000.

THE WISCONSIN UNION GALLERIES, UNIVERSITY OF WISCONSIN-MADISON, (M), 1308 W. Dayton St., Rm. 235, Madison, WI 53715. Tel.: 608-890-4432 & 262-7592. Fax: 608-890-4411. Facebook: Wudart.

E-mail: art@union.wisc.edu

Web Site: www.union.wisc.edu/wud/art-events.htm

Founded: 1928.

Congressional District: 2

Key Personnel: Gallery Dir., Cur. & Registrar, Robin Schmoldt; Union Dir., Mark Guthier.

Personnel Profile: Full-Time Paid 1; Part-Time Volunteers 20.

Governing Authority: state; college. Affiliated with the University of Wisconsin, Madison, WI 53706. Parent Institution: Wisconsin Union. Tax-exempt.

Institution Type/Description: Art Gallery.

Collections: 1,300 works of American, University of Wisconsin student or Wisconsin young professional, including photographs, prints, drawings, paintings, 3-D wood, fiber, metal & glass.

Major Exhibits: Patrick Earle Hammie, Emily Adams, Betsy Willliamson, 2/14-3/14; Alexandra Dooley, 2/14-4/14; Student Art Show (86th Annual), 2/14; MFA Exhibitions, 4/14-5/14.

Facilities: theater; indoor & outdoor music venues; meeting rooms.

Activities: guided tours; lectures; films; gallery talks; concerts; dance recitals; arts festivals; drama; hobby workshops; formally organized education programs for children, adults, undergraduate college students & graduate students affiliated with University of Wisconsin-Madison; loan, temporary & traveling exhibitions.

Publications: member newsletter, Terrace Views/Grapevine.

Hours & Admission Prices: Daily 10-8. No charge. Closed holidays & during school breaks. &

Attendance: 300,000 (estimated)

Membership: Union: Annual $50, Lifetime $250.

*** WISCONSIN VETERANS MUSEUM, (M),** 30 W. Mifflin St., Suite 200, Madison, WI 53703-2589. Tel.: 608-267-7207. Fax: 608-264-7615.

Web Site: wisvetsmuseum.com

Founded: 1901.
Congressional District: 2
Key Personnel: Dir., Michael Telzrow; Processing Archivist, Andrew Bara-
niak; Cur. Education, Jennifer Kollath; Reference Archivist, Russell Hor-
ton; Sr. Mktg. & Devel. Specialist, Jennifer Carlson; Asst. Dir., Kristine
Zickuhr; Collections Mgr., Andrea Hoffman; Registrar, Sarah Kapellusch;
Museum Shop Mgr., Greg Lawson; Cur. History, Gregory Krueger; Cur.
Public Programs & Research, Kevin Hampton; Exec. Staff Asst., Deb Ripp.
Personnel Profile: Full-Time Paid 12; Part-Time Paid 17; Part-Time Volunteers
102.
Governing Authority: state. Parent Institution: State of Wisconsin. Subsidiary
Institution: Wisconsin Dept. Veterans Affairs. Tax-exempt.
Institution Type/Description: State Military History Museum: located down-
town Madison.
Collections: state military & veterans' historical artifacts; aircraft; vehicles;
arms; battle flags; uniformed figures & military equipment; archives &
iconographic collections of Wisconsin's military & veterans organizations.
Major Exhibits: The Last Full Measure, 10/13-6/15.
Research Fields: Wisconsin military & veterans' history 1861-present.
Facilities: research library & archives; education center. Books & museum-
related items for sale.
Activities: permanent & temporary exhibitions; educational programs; lecture
series; public tours; living history; research facility. Documentary: Wiscon-
sin World War Two Stories; Wisconsin Korean War Series; Vietnam War
stories & series; Iraq & Afghanistan programs.
Publications: museum brochure, A Tribute to Freedom; book, Old Abe the War
Eagle; book, USS Wisconsin: The Story of Two Battleships; Flags of the
Iron Brigade; Wisconsin in the Civil War; Flags of the Iron Brigade;
Wisconsin At War; Wisconsin Grand Army of the Republic.
Hours & Admission Prices: April-Sept. Mon.-Sat. 9-4:30, Sun. 12-4; Oct.-
March Mon.-Sat. 9-4:30. Research Center: Mon.-Fri. 9-4. No charge.
Closed holidays. &
Attendance: 94,132 (accurate)
Membership: Individual $30; Family $45; Bronze Corp $350; Gold Corp
$1,000; Lifetime $1,200.

Manitowoc

LINCOLN PARK ZOO, 1215 N. 8th, Manitowoc, WI 54220.
Mailing Address: 330 Custer St., Manitowoc, WI 54220. Tel.:
920-683-4685 (Zoo) & 686-3580 (Office). Fax: 920-686-6525.
E-mail: dlarson@manitowoc.org
Web Site: www.manitowoc.org
Founded: 1935.
Key Personnel: Mgr. Parks & Zoo, Denise Larson.
Personnel Profile: Full-Time Paid 2; Part-Time Paid 2; Part-Time Volunteers
25; Interns 1.
Governing Authority: municipal. Parent Institution: City of Manitowoc Parks
Dept. Affiliated with Parks Dept., 2655 S. 35th St. Tax-exempt.
Institution Type/Description: Zoo.
Collections: animals & birds chiefly native to Wisconsin; exotic birds.
Facilities: picnic area.
Activities: permanent exhibitions; fish rearing pond containing Coho &
Chinook Salmon, Rainbow Trout; special events throughout the year.
Hours & Admission Prices: June-Aug. daily 7-7; Sept.-Oct. Mon.-Sat. 7-3,
Sun. 11-3; Nov.-May Mon.-Sat. 7-3. No charge; donations accepted. &
Attendance: 55,584 (estimated)

MANITOWOC COUNTY HISTORICAL SOCIETY, 1701 Michi-
gan Ave., Manitowoc, WI 54220-3137. Tel.: 920-684-4445. Fax:
920-684-0573.
E-mail: mchistsoc@lakefield.net
Web Site: www.mchistsoc.org
Founded: 1906.
Congressional District: 6
Key Personnel: Pres. Bd., Kathy Kowalski; Exec. Dir., Mike Maher.
Personnel Profile: Full-Time Paid 2; Part-Time Paid 1; Part-Time Volunteers
250; Interns 1.
Governing Authority: bd. of directors. Operates the Heritage Center and
Pinecrest Historical Village, 924 Pine Crest Lane, Manitowoc, WI. Tax-
exempt. 501(c)(3).
Institution Type/Description: History Museum.
Collections: 28 19th & early 20th-century buildings from Manitowoc County;
furnishings; household items; tools; agricultural machinery & equipment.
Research Fields: Manitowoc County.
Facilities: library.
Activities: guided school & motorcoach tours; special events & programs;
historical demonstrations.
Publications: newsletters; books; monographs; walking tour brochures.
Hours & Admission Prices: Heritage Center: Tues.-Fri. 9-4. Adults $5.

Pinecrest Historical Village: May-Oct. 24 daily 9-4. Families $18, adults $7,
seniors $6, children $5; discounts to MCHS members; members no charge.
Attendance: 7,000 (estimated)
Membership: Student $10; Senior $30; Individual $35; Senior Couple $40;
Family $50; Contributing $100.

* **RAHR WEST ART MUSEUM, (M),** 610 N. 8th St., Manitowoc,
WI 54220-3998. Tel.: 920-686-3090. Fax: 920-683-5047.
E-mail: rahrwest@manitowoc.org
Web Site: www.rahrwestartmuseum.org
Founded: 1950.
Congressional District: 6
Key Personnel: Exec. Dir., Greg Vadney; Chm., Amy Fricke-Weigel; Pres.
Foundation, William Pohlmann.
Personnel Profile: Full-Time Paid 2; Part-Time Paid 6.
Governing Authority: municipal. Parent Institution: the City of Manitowoc,
Manitowoc City Hall. Tax-exempt: 501(c)(3) & 170(b)(1)(A).
Institution Type/Description: Art Museum.
Collections: 19th, 20th & 21st century American paintings; period rooms; 19th
century American decorative arts & furnishings; dolls; Chinese ivory
carvings. Historic House: 1891, Joseph Vilas Home.
Research Fields: 20th century American art; historic buildings.
Facilities: 1,500-vol. library of history, archaeology, anthropology & art books
available for use on premises; reading room; classrooms.
Activities: guided tours; opening receptions; artist workshops; docent program;
traveling exhibitions; children's activity room.
Publications: special events program; catalogues; brochures, The Schwartz
Collection of Chinese Ivories; The Ruth & John D. West Art Collection.
Hours & Admission Prices: Mon.-Fri. 10-4, Sat.-Sun. 11-4. Suggested Dona-
tion: $5; members no charge. Closed holidays. &
Attendance: 25,000 (accurate)
Membership: Friend $50; Supporter $100; Advocate $250; Partner $500;
Patron $1,000; Leadership Circle $2,500; Legacy Circle $5,000.

* **WISCONSIN MARITIME MUSEUM, (M),** 75 Maritime Dr.,
Manitowoc, WI 54220-6823. Tel.: 920-684-0218. Fax: 920-684-
0219.
E-mail: museum@wisconsinmaritime.org
Web Site: www.wisconsinmaritime.org
Founded: 1969.
Congressional District: 6
Key Personnel: Exec. Dir., Norma Bishop; Pres. (V), Thomas Jagemann; Cur.,
David Beard; Education Coord., Wendy Lutzke, Financial Svcs. Mgr., Tom
Smith; Maintenance Supvr., Paul Rutherford; Education & Submarine
Program Coord., Karen Duvalle; Education Outreach, Kirsten Smith; Retail
Supvr., Marlys Schwartz; Devel. Asst. & Member Svcs., Bobbie Novak;
Receptionist, Toni Duvall; Visitor Svcs. Supvr., Cassi Adams.
Personnel Profile: Full-Time Paid 13; Part-Time Paid 40; Part-Time Volunteers
300.
Governing Authority: private; nonprofit. Tax-exempt: 501(c)(3).
Institution Type/Description: Maritime Museum.
Collections: submarine artifacts; photos; documents; ship tools; salvage
artifacts; models of Great Lakes ships; period artifacts; furniture; house
flags; manuscripts relating to history of Great Lakes; lake schooners;
steamships; car ferries; bulk carriers; submarine; life boat. Historic Submarine: 1943 U.S.S. Cobia; full-scale reproduction of 19th century schooner midship section. Historic Structures: Clipper City; former Coast Guard
utility boat, Icelander; Wisconsin built recreational boats.
Research Fields: Great Lakes maritime history; WWII submarines; Wisconsin
maritime history & culture.
Facilities: 8,000-vol. library & 40,000 photographs on Great Lakes maritime
and Manitowoc submarine history & submarines available for use on the
premises; reading room; classroom. Books, maritime gifts & other museum-
related items for sale.
Activities: guided tours; films; permanent, temporary & traveling exhibitions;
school loan service. Museum Sponsors: summer festival; picnic; Riverwalk
Festival; Annual submariners Memorial Service; family series of Maritime
related programs.
Publications: booklet, Manitowoc Submarines; quarterly newsletter, Anchor
News; books Fresh Water Submarines: The Manitowoc Story; The Nau Tug
Line.
Hours & Admission Prices: Museum & Submarine: May-Sept. daily 9-6;
Oct.-April daily 9-5. Museum: adult $12, children $10; children 5 & under
no charge. Museum & Submarine: adult $12, children $10; discounts to
family, veterans, senior citizens & AAA members; children 5 & under,
Council of American Maritime Museum & members no charge. Closed
New Year's Day; Easter; Thanksgiving; Christmas. &
Attendance: 46,525 (accurate)
Membership: Student & Senior Citizen $25; Senior Couple $35; Individual

$40; Family $60; Sustaining $100; Sponsor $250; Benefactor $500; Timothy J. Kelly Society $1,000; Edward Carus Society $2,500; Albert E. Goodrich Society $5,000; William W. Bates Society $10,000.

Marinette

MARINETTE COUNTY HISTORICAL MUSEUM, Stephenson Island, U.S. Hwy. 41, Marinette, WI 54143. Mailing Address: P.O. Box 262, Marinette, WI 54143-0262. Tel.: 715-732-0831.
Web Site: www.marinettecountyhistory.org
Founded: 1932.
Congressional District: 8
Key Personnel: Pres. (V), Frank Lauerman; Sec., Mary Falkenberg.
Personnel Profile: Part-Time Volunteers 30; Interns 1.
Governing Authority: society. Parent Institution: Marinette County Historical Society. Tax-exempt.
Institution Type/Description: History Museum: building & c.1895 log cabin.
Collections: logging & historical artifacts including miniature logging camp replica; copper culture; woodland Indian artifacts; manuscripts; fishing; lumbering; agriculture; early social history; ice road rutter rig exhibit; authentic 1889 sawmill. Historic Building: 1897 homesteader's cabin.
Research Fields: logging; copper culture.
Facilities: 200-vol. library of books on logging & county history available on premises only.
Activities: guided tours; formally organized education programs; permanent & temporary exhibitions.
Publications: annual, Historian.
Hours & Admission Prices: Memorial Day to Labor Day Mon.-Fri. 10-4. Adults $3, students grades 7-12 $1; discounts to groups; members & children under 12 accompanied by parent no charge. Closed Independence Day. &
Attendance: 2,000 (estimated)
Membership: Individual $15; Couple $20.

Marshfield

NEW VISIONS GALLERY, INC., (M), 1000 N. Oak Ave., Marshfield, WI 54449-5703. Tel.: 715-387-5562.
E-mail: newvisions.gallery@frontier.com
Web Site: www.newvisionsgallery.org
Founded: 1975.
Congressional District: 7
Key Personnel: C.E.O. & Dir., Betsy Tanenbaum; Pres., Debi McCullough; Treas., Rhonda Ackerman.
Personnel Profile: Full-Time Paid 1; Part-Time Paid 2; Part-Time Volunteers 70.
Volunteer Hours: 1,153
Operating Expenses: 164,006
Operating Income: 164,231
Governing Authority: nonprofit organization. Tax-exempt: 501(c)(3).
Institution Type/Description: Art Museum.
Collections: Japanese prints; West African masks & sculptures; Haitian paintings; Australian Aboriginal prints & paintings; contemporary prints; art glass; ceramics; contemporary sculpture.
Activities: guided tours; lectures; films; arts festivals; hobby workshops; docent program; participatory & loan exhibitions.
Publications: monthly brochures; occasional exhibition catalogues.
Hours & Admission Prices: Mon.-Fri. 9-5:30. No charge; donations accepted. Closed legal holidays. &
Attendance: 25,000 (estimated)
Membership: Art Partners: Basic $50; Friend $75; Sponsor $125; Donor $250; Patron $500; Benefactor $1,000; Sustaining $5,000 & up.

UPHAM MANSION - NORTH WOOD COUNTY HISTORICAL SOCIETY, 212 W. 3rd St., Marshfield, WI 54449-2706. Mailing Address: P.O. Box 142, Marshfield, WI 54449-0142. Tel.: 715-387-3322.
E-mail: uphammansion@frontier.com
Web Site: www.uphammansion.com
Founded: 1952.
Congressional District: 7
Key Personnel: Pres. (V), Shelby Weister; Coord., Kim Krueger.
Personnel Profile: Part-Time Paid 1; Part-Time Volunteers 20.
Governing Authority: nonprofit. Parent Institution: North Wood County Historical Society. Tax-exempt: 170(b)(1)(A).
Institution Type/Description: Historic Building/Site: Italianate-Victorian style home, built in 1880 by William Henry Upham, leading businessman of Marshfield till his death in 1924; governor of WI 1895-96.
Collections: pioneer artifacts & furnishings; clothing; household goods; utensils; toys; dolls; books; Pier mirror; Victorian furniture; curved glass curio cabinet; oval marble-topped table; Victorian cutter; horse-drawn fire hose cart; iron lung; scrap books; 32 varieties of roses in rose garden; photographs.
Research Fields: genealogy; early North Wood County history.
Facilities: 30-vol. library of books on early pioneers available on premises only; genealogical microfiche library; research center.
Activities: monthly meetings; four annual open houses; group tours; ice cream social. Museum Sponsors: Christmas Open House.
Publications: newsletter, Mansion News.
Hours & Admission Prices: Mansion: Wed. & Sat. 1:30-4; groups by appointment. Adults $2; members no charge. Heritage Rose Garden: June-Sept. daily. No charge. Closed holidays & holiday weekends.
Attendance: 1,800 (accurate)
Membership: Individual $35; Family $50; Business $100.

Mauston

THE BOORMAN HOUSE - THE JUNEAU COUNTY HISTORICAL SOCIETY, 211 N. Union St., Mauston, WI 53948-1418. Mailing Address: P.O. Box 321, Mauston, WI 53948-0321. Tel.: 608-847-4450 & 3294. Facebook: Boorman House.
E-mail: rclarkjco@gmail.com
Founded: 1963.
Congressional District: 6
Key Personnel: Pres. (V), Nancy McCullick; Historiographer & Museum Shop Mgr., Rose Clark.
Personnel Profile: Full-Time Volunteers 2; Part-Time Volunteers 70.
Volunteer Hours: 10,000
Operating Expenses: 49,000
Operating Income: 50,000
Governing Authority: society. Parent Institution: Wisconsin State Historical Society. Tax-exempt.
Institution Type/Description: Historical Site & Building. Listed on the National Register of Historic Places and on the Wisconsin Register of Historic Homes.
Collections: area history; photographs; Juneau County tax & country school records; Juneau County naturalization records; genealogy; historical documents c.1850; county officials from 1857; Native American artifacts.
Research Fields: Juneau County Native Americans; local genealogy & history.
Facilities: 1000-vol. library of books pertaining to state, local schools & war history and family genealogy available for use on premises.
Activities: guided tours; concerts; permanent & temporary exhibitions. Annual Event: Old Tyme Picnic in September. Museum Sponsors: ice cream social; spring event; fall tea & fashion show; Christmas Open House.
Publications: quarterly newsletter, Juneau County History Notes; book, Juneau County Wisconsin Reflections of 150 Years (2010).
Hours & Admission Prices: Memorial Day-Labor Day Sat.-Sun. 1-4 by appointment; Thanksgiving weekend & following weekend 1-8 by appointment; call historiographer Rose Clark. No charge; donations accepted. Research: donations accepted.
Attendance: 2,090 (estimated)
Membership: Youth $1; Regular $5; Life $100.

Mayville

MAYVILLE HISTORICAL SOCIETY, INC., 1 N. German St., Mayville, WI 53050. Mailing Address: P.O. Box 82, Mayville, WI 53050-0082. Tel.: 920-387-2420 & 5787.
Web Site: mayvillehistoricalsociety.org
Founded: 1968.
Congressional District: 9
Key Personnel: Pres. (V) & Membership Chm., Alyce Wurtz; Vice Pres., Barbara Larsen; Sec. & Museum Shop Mgr., Lois Gadow; Treas., Arlesse Groth.
Personnel Profile: Part-Time Volunteers 43.
Governing Authority: nonprofit organization. Affiliated with State Historical Society of Wisconsin, 816 State St., Madison, WI 53706. Tax-exempt: 170(b)(1)(A).
Institution Type/Description: Historical Society Museum.
Collections: local historical items; wagons; agricultural materials; glassware; clothing; religious articles; pictures; manuscripts. Historic Buildings: 1874 fire station; 1888 Wagon & Carriage Factory with attached 1873 residence; 1885 Cigar Factory.
Research Fields: iron industry; local history.
Facilities: 500-vol. library of early school books, German language books, law books. Mayville history material & other museum-related items for sale.
Activities: guided tours; temporary exhibitions. Society Sponsors: Membership Dinner in April; Ice Cream Social in June; Appreciation Luncheon in October.
Publications: quarterly newsletter.

Hours & Admission Prices: May-Oct. 2nd & 4th Sun. 1:30-4:30; tours by appointment only. No charge; donations accepted. Group Tours: $1 per person. &
Attendance: 385 (accurate)
Membership: Individual $10; Family, Business & Professional $20.

Mazomanie

MAZOMANIE HISTORICAL SOCIETY, 118 Brodhead St., Mazomanie, WI 53560. Mailing Address: Box 248, Mazomanie, WI 53560-0248. Tel.: 608-795-2992. Fax: 608-795-4576.
Web Site: www.mazomaniehistory.org
Founded: 1965.
Congressional District: 2
Key Personnel: Pres., Robert Dodsworth; Cur., Rita Frakes.
Personnel Profile: Part-Time Paid 1; Part-Time Volunteers 45.
Governing Authority: nonprofit organization. Affiliated with the State Historical Society, 816 State St., Madison, WI 53703. Tax-exempt: 501(3)(c).
Institution Type/Description: General Museum.
Collections: artifacts relating to the history of the area; woodworking; cobbler; printing; automotive; blacksmith tool collection. Historic Building: 1857 train depot, wooden depot.
Research Fields: local history.
Facilities: museum; research center. Gift items & stationery for sale.
Activities: guided tours; permanent & temporary exhibitions.
Publications: quarterly newsletter; local history pamphlets; books; annual research journal, Local Sheaves.
Hours & Admission Prices: May-Sept. Sun. 1-4, call to confirm. Suggested Donation: adults $3. &
Attendance: 334 (accurate)
Membership: Individual $6; Family $10; Business $25; Life $100; Life Couple $175.

McFarland

MCFARLAND HISTORICAL SOCIETY, 5814 Main St., McFarland, WI 53558. Mailing Address: P.O. Box 94, McFarland, WI 53558-0094. Tel.: 608-838-3992.
E-mail: bluebee@madtown.net
Web Site: www.mcfarlandhistorical.org
Founded: 1964.
Congressional District: 2
Key Personnel: Pres. (V), Dale Marsden; Treas., Jackie Utter.
Governing Authority: nonprofit organization. Tax-exempt.
Institution Type/Description: Historical Society Museum: housed in a log cabin.
Collections: local artifacts.
Research Fields: local buildings; early citizens; artifacts.
Activities: tours. Annual Events: Summer opening Bake Sale in May; Pioneer Day-Family Festival.
Publications: semi-annual newsletter.
Hours & Admission Prices: Memorial Day to late Sept. Sun. 1-4, tours by appointment. No charge; donations accepted.
Attendance: 2,000 (estimated)
Membership: Individual $10; Family $15; Life $150; Supporting & Business $250.

McNaughton

RIVERRUN CENTER FOR THE ARTS, 6938 Bridge Rd., McNaughton, WI 54543. Mailing Address: P.O. Box 95, McNaughton, MI 54543-0095. Tel.: 715-277-4224.
E-mail: joanslack@wildblue.net
Web Site: www.riverrunarts.com
Key Personnel: Dir. & Owner, Joan Molloy Slack
Institution Type/Description: Art Gallery.
Collections: works by regional artists including sculpture, paintings, prints, & woodworking.
Activities: workshops; classes; demonstrations; performances. Annual Events: Summer Art Tour in July; Christmas Festival of the Arts in December.
Hours & Admission Prices: Call for hours.

Menasha

WEIS EARTH SCIENCE MUSEUM, (M), University of Wisconsin-Fox Valley, 1478 Midway Rd., Menasha, WI 54952-1224. Tel.: 920-832-2925. Facebook: Weis Earth Science Museum.
Web Site: www.uwfox.uwc.edu/wesm

Key Personnel: Museum Dir., Joanne Kluessendorf
Institution Type/Description: Earth Science Museum.
Collections: interactive & hands-on exhibits; video displays; colorful graphics; fossils; minerals & rocks.
Hours & Admission Prices: Mon.-Fri.12-3, Sat. 10-5, Sun. 1-5. Adults $2, seniors 60 & over and teens 13-17 $1.50, children 3-12 $1; children under 3, UW Fox students, faculty & staff with ID no charge.

Menomonee Falls

OLD FALLS VILLAGE, N. 96 W. 15791 County Line Rd., Menomonee Falls, WI 53051-1537. Mailing Address: P.O. Box 91, Menomonee Falls, WI 53052-0091. Tel.: 262-250-5096. Fax: 262-250-5097.
E-mail: jsteliga@wi.rr.com
Web Site: www.menomonee-falls.org/index.aspx?NID=157
Founded: 1966.
Congressional District: 9
Key Personnel: Pres. & Museum Devel. Dir., Rev. Warren Granke; Museum Shop Mgr., Christine Schultz.
Personnel Profile: Part-Time Volunteers 30; Interns 2.
Governing Authority: municipal; society. Affiliated with Menomonee Falls Historical Society, Inc. Tax-exempt: 501(c)(3).
Institution Type/Description: Historic Houses.
Collections: primitive and country furniture 1800-1900; 18th & 19th-century decorative arts. Historic Houses: 1890 Railroad Depot; 1858 Miller Davidson House; 1845 Umhoefer log cabin; 1851 first schoolhouse; 1873 Farm House; 1842 log cabin; 1840 log cabin used as Catholic church.
Research Fields: local history.
Activities: guided tours. Museum Sponsors: Old Falls Village Days in June; Civil War Encampment in July; Quilt Show in August; Halloween Family Fun Night in October; Silver Tea in December.
Publications: newsletter.
Hours & Admission Prices: Call for hours. Adults $3, children 6-16 $1.
Attendance: 6,000 (estimated)
Membership: Students $6; Seniors 65 & up $7; Individuals $15; Family $25; Pioneer $50; Homesteader $100.

Menomonie

DUNN COUNTY HISTORICAL SOCIETY, (M), 1820 Wakanda St., Menomonie, WI 54751-1631. Mailing Address: P.O. Box 437, Menomonie, WI 54751-0437. Tel.: 715-232-8685. Fax: 715-232-8687.
E-mail: info@dunnhistory.org
Web Site: www.dunnhistory.org
Founded: 1950.
Congressional District: 3
Key Personnel: Pres. (V), Roy S. Ostenso; Vice Pres., Rose Mary Stoll; Treas., Steve Cole; Education, Rich Sterry; Public Rels., Don Steffen; Registrar & Museum Supvr., Carol Thibado.
Personnel Profile: Full-Time Volunteers 8; Part-Time Paid 6; Part-Time Volunteers 30.
Governing Authority: private; nonprofit organization. Subsidiary Institutions: Russell J. Rassbach Heritage Museum & Hilkrest Rural School, 1820 Wakanda St., Menomonie, WI; Empire in Pine Museum, County Rd. C, Downsville, WI; Fulton & Edna Holtby Science & Technology Museum. Tax-exempt: 501(c)(3).
Institution Type/Description: Historical Society Museum.
Collections: Wisconsin history from prehistoric to modern times with special emphasis on the history of the county of Dunn County & City of Menomonie. Historic Building: one-room school.
Facilities: library; 100-seat auditorium; 2,500 sq. ft. exhibit space; meeting room. Museum-related items for sale.
Activities: guided tours; lectures.
Publications: newsletter, The Dunn County Historian; pamphlet, Caddie Woodlawn: A Pioneer Girl on Wisconsin's Frontier; James Huff Stout - An Illustrated Timeline.
Hours & Admission Prices: Heritage Museum: Summer: Wed.-Sun. 10-5; Winter: Fri.-Sun. 12-4. Empire in Pine Museum: May-Sept. Sat.-Sun. 12-5. Adults $5, students $3; discount to AAM members; members no charge. Closed New Year's Day; Easter; Christmas. &
Attendance: 5,000 (accurate)
Membership: Senior $15; Individual & Senior Couple $25; Family $35; Supporting $100; Sustaining $250; Patron $500; Life $2,000.

JOHN FURLONG GALLERY, Micheels Hall, 415 13th Ave. E., Menomonie, WI 54751-3279. Mailing Address: Unniversity of Wisconsin-Stout, 712 S. Broadway St., 309C Applied Arts Bldg., Menomonie, WI 54751. Tel.: 715-232-2261 & 1097. Fax: 715-232-1669.
E-mail: furlong@uwstout.edu
Web Site: furlonggallery.uwstout.edu
Founded: 1965.
Congressional District: 3
Key Personnel: Dir., Susan Hunt; Asst. Dir., Debra D'souza.
Governing Authority: university. Parent Institution: University of Wisconsin. Tax-exempt.
Institution Type/Description: University Art Gallery.
Collections: 20th-century paintings, drawings, sculpture & prints.
Activities: guided tours; lectures; films; organized education programs for children; participatory, loan & temporary exhibitions.
Hours & Admission Prices: Mon.-Wed. 9-5:30, Thurs.-Fri. 9:30-6, Sat. 10-3. No charge; donations accepted. Closed national holidays. &
Attendance: 7,000 (estimated)

WILSON PLACE MANSION, 101 Wilson Circle, Menomonie, WI 54751-1860. Tel.: 715-235-2283. Facebook: Wilson Place Mansion.
Formerly: Wilson Place Museum
Key Personnel: Preservationist, Tim Dotseth; Educational Dir., Melissa Kneeland
Institution Type/Description: History Museum.
Collections: local history; period furnishings.
Hours & Admission Prices: mid-Nov. to mid.-Dec. daily 1-5. Adults $5, seniors $4.50, students $3.50, children 12 & under $2.

Mequon

CONCORDIA UNIVERSITY-WISCONSIN, ART GALLERY, 12800 N. Lake Shore Dr., Mequon, WI 53097-2418. Tel.: 262-243-4552. Fax: 262-243-4351.
Key Personnel: Dir., Jeff Shawhan.
Personnel Profile: Part-Time Paid 4.
Governing Authority: university; nonprofit organization. Parent Institution: Lutheran Church-Missouri Synod. Tax-exempt.
Institution Type/Description: Art Gallery.
Collections: religious art; Russian bronzes; Chinese porcelain.
Activities: formal education programs for undergraduate & graduate college students; lectures; temporary exhibitions.
Hours & Admission Prices: Sun.-Wed. & Fri. 12-4, Thurs. 12-4 & 6-8. No charge. Closed major holidays. &
Attendance: 600 (estimated)

Merrill

MERRILL HISTORICAL SOCIETY, 100 E. Third St., Merrill, WI 54452-2321. Tel.: 715-536-5652.
E-mail: merrillhs@frontier.com
Web Site: merrillhistory.org
Founded: 1978.
Congressional District: 7
Key Personnel: Pres., Beatrice Lebal; Treas., Patricia Burg; Cur., Beverly King; Security & Museum Shop Mgr., Erin McCarthy.
Personnel Profile: Part-Time Paid 1; Part-Time Volunteers 8.
Governing Authority: private; nonprofit organization. Parent Institution: Merrill Historical Society, Inc., 102 E. Third St., Merrill, WI 54452. Subsidiary Institutions: Brickyard School Museum, Merrill, WI; Hermien Livingston Collections Building. Tax-exempt: 501(c)(3).
Institution Type/Description: Historic Society Museum.
Collections: Wisconsin history from 1800 to present, special emphasis on Merrill Area, both rural & city; books; documents; early notebooks audio tapes, video tapes (transcribed from early movies).
Research Fields: materials to document village of Jenny 1847-1881, Merrill 1882-1998, Lincoln Co. Townships
Facilities: 975-vol. library; 360 sq. ft. exhibit space. Museum-related items for sale.
Activities: arts festivals; concerts; docent program; films; guided tours; lectures; participatory & traveling exhibitions.
Publications: bimonthly newsletter, Northwoods Historian.
Hours & Admission Prices: Tues.-Fri. 9-1; other times by appointment. No charge; donations accepted.
Attendance: 6,915 (accurate)
Membership: Senior $14; Individual $17; Senior Family $19; Family $22; Business $26; Life Individual $200; Life Family $300.

Middleton

NATIONAL MUSTARD MUSEUM, 7477 Hubbard Ave., Middleton, WI 53562-3117. Tel.: 800-438-6878.
E-mail: curator@mustardmuseum.com
Web Site: www.mustardmuseum.com
Formerly: Mount Horeb Mustard Museum
Founded: 1986.
Congressional District: 2
Key Personnel: Cur., Barry Levenson; Dir. Mktg. & Tours, Patti Levenson.
Personnel Profile: Part-Time Volunteers 3.
Governing Authority: Tax-exempt.
Institution Type/Description: Food Museum.
Collections: over 6,000 jars, bottles, & tubes from all 50 states and over 80 countries; period mustard pots & tins; vintage mustard advertisements & memorabilia.
Facilities: Museum-related items for sale.
Activities: mustard tastings & samples.
Hours & Admission Prices: Daily 10-5. No charge; donations accepted. Closed New Year's Day; Easter; Thanksgiving; Christmas. &
Attendance: 30,000 (estimated)

Milton

MILTON HOUSE MUSEUM HISTORIC SITE, (M), 18 S. Janesville St., Hwys. 26 & 59, Milton, WI 53563-1527. Mailing Address: P.O. Box 245, Milton, WI 53563-0245. Tel.: 608-868-7772. Fax: 608-868-1698.
E-mail: miltonhouse@miltonhouse.org
Web Site: www.miltonhouse.org
Founded: 1948.
Congressional District: 1
Key Personnel: Pres., Tom Sveum; Exec. Dir., Cori Olson.
Personnel Profile: Full-Time Paid 1; Part-Time Paid 1; Part-Time Volunteers 60.
Governing Authority: society. Parent Institution: Milton Historical Society. Tax-exempt: 501(c)(3).
Institution Type/Description: History Museum: housed in a stagecoach inn used as a stop on the Underground Railroad.
Collections: mid-1800 pioneer items; Civil War artifacts; costumes; clocks; tools. Historic Buildings: 1844 Stage Coach Inn, first poured grout (lime mortar) building in U.S.; 1837 Log Cabin; 1867 Goodrich House; Livery stable; 1876 buggy shed; smoke house; country store; blacksmith shop.
Research Fields: local history; genealogy; underground railroad.
Facilities: 2,000-vol. research library of books & manuscripts.
Activities: guided tours; education programs for children; permanent & special exhibits. Museum Sponsors: annual arts & crafts show; annual Pioneer Dinner.
Publications: quarterly newsletter; pamphlets.
Hours & Admission Prices: May Sat.-Sun. 10-4; Memorial Day-Labor Day daily 10-4; other times by appointment. Adults 13 & over $6, seniors 62 & over $5, children 5-12 $3; discounts to groups, AAA & AAM members; children 5 & under and members no charge. Closed New Year's Eve & Day; Easter; Thanksgiving & weekend after; Christmas Eve, Day & week. &
Attendance: 10,000 (estimated)
Membership: Senior Citizen $10; Individual $12; Senior Family $15; Family $20; Conductor $30; Business $50; Station Master $100-$249; Abolitionist $250-$499; North Star Friend $500 & up.

Milwaukee

BETTY BRINN CHILDREN'S MUSEUM, 929 E. Wisconsin Ave., Milwaukee, WI 53202-5406. Tel.: 414-390-KIDS & 291-0888, ext. 200. Fax: 414-291-0906.
E-mail: questions@bbcmkids.org
Web Site: www.bbcmkids.org
Founded: 1995.
Key Personnel: C.E.O., Fern Shupeck; Pres. Bd. (V), Melissa Nelsen; Treas., Jane Frank; Exhibits, Jim Toth; Devel., Carrie Wettstein; Education, Carolyn Rydlewicz; Education, Lisa Balster; Public Rels., Kristen Adams; Museum Shop Mgr., Joel Cencius.
Personnel Profile: Full-Time Paid 17; Part-Time Paid 18; Part-Time Volunteers 12; Interns 3.
Governing Authority: private; nonprofit organization. Tax-exempt: 501(c)(3).
Institution Type/Description: Children's Museum.
Collections: hands-on exhibits; Walt Disney cell art; Warner animation cells.
Research Fields: early childhood development.
Facilities: library; educational facilities; children's theater; 17,000 sq. ft. exhibit space. Museum-related items for sale.
Activities: guided tours; participatory & traveling exhibits; formal education

programs for adults & children; birthday parties. Annual Events: July 3rd Event; Teacher of the Year Awards; Valentine's Day event in February; Halloween event in October; gala fundraiser.

Publications: member newsletter 3 times a year; brochure 3 times a year; monthly e-newsletter.

Hours & Admission Prices: June-Aug. Mon.-Sat. 9-5, Sun. 12-5; Sept.-May Tues.-Sat. 9-5, Sun. 12-5. Adults $6; discounts to Association of Children's museum members & groups; children under one & members no charge. Closed Memorial Day; Independence Day; Labor Day; Thanksgiving; Christmas Eve & Day. &

Attendance: 160,000 (estimated)

Membership: A $65; B $85; C $120; D $250; E $500; F $1,000; G $2,500.

THE CAPTAIN FREDERICK PABST MANSION, 2000 W. Wisconsin Ave., Milwaukee, WI 53233-2004. Tel.: 414-931-0808. Fax: 414-931-1005.

E-mail: info@pabstmansion.com

Key Personnel: Exec. Dir., Dawn M. Day Hourigan; Dir. Devel., John C. Eastberg; Dir. Visitor Svcs., Rikki Thompson; Cur., Jodi Rich-Bartz.

Personnel Profile: Full-Time Paid 4; Part-Time Paid 3; Part-Time Volunteers 50.

Governing Authority: Tax-exempt.

Institution Type/Description: Historic House: former home of Captain Frederick Pabst, built in 1892.

Collections: local history & culture; period artifacts; furnishings; photographs.

Activities: tours; temporary & permanent exhibitions; special events; educational programs; rental facilities.

Publications: Heritage Newsletter.

Hours & Admission Prices: mid-Jan. to Feb. Tues.-Sat. 10-4, Sun. 1-4; March to mid-Jan. Mon.-Sat. 10-4, Sun. 1-4. Adults $8, seniors & students $7, children 6-17 $4; children under 6 no charge. Closed New Year's Day; Easter; Thanksgiving; Christmas. &

Attendance: 25,000 (estimated)

Membership: First Mate $40; Sea Captain $60; Brewmaster $125; Gambrinus $250; Beer Baron $500; Blue Ribbon $1,500.

CHARLES ALLIS ART MUSEUM, (M), 1801 N. Prospect Ave., Milwaukee, WI 53202-1933. Tel.: 414-278-8295. Fax: 414-278-0335.

E-mail: info@cavtmuseums.org

Web Site: www.cavtmuseums.org

Founded: 1945.

Congressional District: 5

Key Personnel: Exec. Dir., Maria Costello; Cur., Jenille Junco; Mgr. Mktg., John Sterr; Events Mgr., Ann Steinabach.

Governing Authority: Parent Institution: Milwaukee County War Memorial Corp., Inc., 750 N. Lincoln Memorial Dr., Milwaukee, WI 53202. Branch Museum: Villa Terrace Decorative Arts Museum, 2220 N. Terrace Ave., Milwaukee, WI 53202. Tax-exempt 501(c)(3).

Institution Type/Description: Art Museum: housed in 1909 Tudor style mansion designed by Alexander Eschweiler.

Collections: Chinese porcelains from Han thru Ching Dynasties; 19th-century French Barbizon paintings; 19th-century American landscape paintings; 16th-19th century bronzes; English garden; ceramics.

Research Fields: Oriental & Asian decorative arts; Barbizon School & 19th-century American landscapes paintings; 16th-19th century bronzes; Wisconsin art.

Facilities: meeting rooms; lecture hall; rental facilities.

Activities: guided tours; lectures; films; gallery talks; concerts; classes & workshops; teas; performances; permanent & temporary exhibitions; docent & educational programs.

Publications: exhibition catalogs; quarterly newsletter.

Hours & Admission Prices: Wed.-Sun. 1-5. Adults $7, seniors over 62, military & students $3; children under 12 & members no charge. Closed New Year's Eve & Day; Thanksgiving; Christmas Eve & Day. &

Attendance: 1,657 (accurate)

Membership: Student $25; Senior Citizen $35; Individual $45; Senior Couple $55; Family $60; Bronze $125; Silver $250; Gold $500; Platinum $1,000.

DISCOVERY WORLD, 500 N. Harbor Dr., Milwaukee, WI 53202-5601. Tel.: 414-765-9966. Fax: 414-765-0311.

E-mail: info@discoveryworld.org

Web Site: www.discoveryworld.org

Formerly: Discovery World - The James Lovell Museum of Science, Economics and Technology

Founded: 1984.

Key Personnel: Exec. Dir., Paul J. Krajniak; Pres., Joel Brennan; Dir. Mktg., Richard Cieslak; Dir. Exhibit Devel., Carl Schoettel; Museum Shop Mgr., Deena Thompson.

Personnel Profile: Full-Time Paid 75; Part-Time Paid 10; Part-Time Volunteers 100; Interns 3.

Governing Authority: nonprofit organization. Tax-exempt: 501(c)(3).

Institution Type/Description: Science, Economics & Technology Museum.

Collections: interactive science & technology exhibits.

Facilities: Gift items for sale.

Activities: 140 hands-on exhibits; live theater productions; educational workshops; labs; progressive academic partnerships; fresh & salt water aquariums.

Publications: monthly newsletter; electronic newsletter.

Hours & Admission Prices: Tues.-Fri. 9-4, Sat.-Sun. 10-5. Adults $16.95, seniors $14.95, children 3-17 $12.95; discounts to AAA & ASTC members, Entertainment cardholders & groups; members & children under 2 no charge. Closed Independence Day; Thanksgiving; Christmas. &

Attendance: 400,000 (estimated)

Membership: College Student $15; Young Adult $25; Educator $35; Senior Citizen $40.50; Individual $45; Family & Senior Deluxe Family $65; Deluxe Family $100; Researcher $200; Explorer $350; Innovator $500; Discoverer's Society $1,000.

DRS. JOSEPH AND JAMES ENGLANDER SCHOOL OF DENTISTRY MUSEUM, Milwaukee, WI 53201. Mailing Address: Marquette University School of Dentistry, P.O. Box 1881, Milwaukee, WI 53201-1881. Tel.: 262-242-0931.

Key Personnel: Cur., Dr. Peter Jacobsohn

Institution Type/Description: Dental Museum.

Collections: artifacts pertaining to the history of dental education & practice throughout the years.

Hours & Admission Prices: By appointment.

THE EISNER - AMERICAN MUSEUM OF ADVERTISING & DESIGN, Milwaukee, WI 53202. Mailing Address: 5307 S. 92nd St., Hales Corners, WI 53130. Tel.: 414-847-3290. Fax: 414-847-3299.

Institution Type/Description: Art Gallery.

Collections: paintings; sculpture.

Activities: rental facilities.

Hours & Admission Prices: Temporarily closed.

GREENE MEMORIAL MUSEUM, UNIVERSITY OF WISCONSIN-MILWAUKEE, 3209 N. Maryland Ave., Milwaukee, WI 53211-3102. Mailing Address: P.O. Box 413, Milwaukee, WI 53201. Tel.: 414-229-4561. Fax: 414-229-5452.

Web Site: www.geology.uwm.edu

Founded: 1913.

Congressional District: 5

Key Personnel: Chm., Tim Grundl; Sec., Lisa Alzalde; Cur., Stephen Dornbos.

Personnel Profile: Part-Time Paid 1.

Governing Authority: university. Parent Institution: University of Wisconsin, Milwaukee. Tax-exempt: 501(c)(3).

Institution Type/Description: Geology Museum.

Collections: Devonian & Silurian fossils from Illinois & Wisconsin region; paleontology; mineralogy; conchology.

Research Fields: paleontology; mineralogy.

Activities: guided tours; lectures; formally organized education programs for children & college students; permanent exhibitions.

Publications: monographs.

Hours & Admission Prices: Fall Mon. & Wed.-Thurs. 10-2, Tues. 10-2 & 4-5. No charge; donations accepted. Closed major holidays.

GROHMANN MUSEUM, 1000 N. Broadway, Milwaukee, WI 53202. Mailing Address: 1025 N. Broadway, Milwaukee, WI 53202-3109. Tel.: 414-277-2300. Facebook: Grohmann Museum.

E-mail: grohmannmuseum@msoe.edu

Web Site: www.msoe.edu/museum

Founded: 2007.

Key Personnel: Dir., James Kieselburg; Museum Shop Mgr., Ann Rice.

Personnel Profile: Full-Time Paid 3; Part-Time Paid 12; Part-Time Volunteers 8.

Governing Authority: Parent Institution: Milwaukee School of Engineering. Tax-exempt.

Institution Type/Description: Art Gallery.

Collections: paintings; sculptures.

Major Exhibits: Trains that Passed in the Night, 1/17/14-4/27/14; Art Shay: Working, 5/14-8/14; Ron Hoyle: Portraits of Work, 9/14-12/14.

Research Fields: the art of industry.

Facilities: Gallery-related items for sale.

Hours & Admission Prices: Mon.-Fri. 9-5, Sat. 12-6, Sun. 1-4. Adults $5, students & seniors $3; discounts to AAM members; MSOE students, alumni, faculty & staff, museum members & children under 12 no charge. ♿

Attendance: 16,000 (accurate)
Membership: Individual $25; Dual $50; Friend $75; Patron $150; Sponsor $250 & up.

HARLEY-DAVIDSON MUSEUM, 400 W. Canal St., Milwaukee, WI 53203-3208. Tel.: 414-343-4056; 877-HD MUSEUM.
Web Site: www.h-dmuseum.com
Key Personnel: Vice Pres., Museum & Factory Tours, Bill Davidson
Institution Type/Description: Motorcycle Museum.
Collections: Harley-Davidson history, products & culture; riders, employees, dealers, & suppliers stories; photographs.
Facilities: Museum-related items for sale.
Hours & Admission Prices: May-Sept. Mon.-Wed. & Fri.-Sun. 9-6, Thurs. 9-8; Oct.-April Mon.-Wed. & Fri.-Sun. 10-6, Thurs. 10-8. Adults $18, senior citizens 65 & over, military and students with ID $12, children 5-17 $10; members & children under 5 with adult no charge.
Membership: Online $40; Individual $60; Family $75; Couple and a Friend $90; Current Model Year Purchasers $200; Premier $500.

INSTITUTE OF VISUAL ARTS, UNIVERSITY OF WISCONSIN-MILWAUKEE, 2155 N. Prospect Ave., Milwaukee, WI 53202. Mailing Address: Box 413, Milwaukee, WI 53201. Tel.: 414-229-5070. Fax: 414-229-6785.
E-mail: inova@uwm.edu
Web Site: www.arts.uwm.edu/inova
Founded: 1982.
Congressional District: 5
Key Personnel: Dir., Sara Krajewski; Assoc. Dir., Bruce Knackert.
Personnel Profile: Full-Time Paid 2; Part-Time Paid 11; Interns 1.
Governing Authority: state. Parent Institution: University of Wisconsin-Milwaukee. Tax-exempt.
Institution Type/Description: Art Museum.
Collections: works by emerging & established artists.
Research Fields: contemporary art.
Activities: lectures; gallery talks; symposia; changing contemporary art exhibits.
Publications: catalogs.
Hours & Admission Prices: Wed. & Fri.-Sun. 12-5, Thurs. 12-8. No charge; donations accepted. Closed holidays. ♿
Attendance: 18,000 (accurate)

JEWISH MUSEUM MILWAUKEE, (M), 1360 N. Prospect Ave., Milwaukee, WI 53202. Tel.: 414-390-5730. Fax: 414-390-5755.
E-mail: info@jewishmuseummilwaukee.org
Web Site: www.jewishmuseummilwaukee.org
Founded: 2008.
Congressional District: 4
Key Personnel: Exec. Dir., Kathie Bernstein.
Personnel Profile: Full-Time Paid 3; Part-Time Paid 2; Part-Time Volunteers 230; Interns 5.
Governing Authority: Parent Institution: Milwaukee Jewish Federation. Tax-exempt.
Institution Type/Description: Jewish History Museum.
Collections: Jewish history, culture & heritage; photographs; personal artifacts; oral histories.
Facilities: archives.
Publications: annual newsletters.
Hours & Admission Prices: Mon.-Thurs. 10-4, Fri. 10-2, Sun. 12-4. Adults $6, seniors $5, students $3; discounts to AAM, ICOM & groups; children under 6 & members no charge. Closed Jewish holidays.
Attendance: 5,500 (estimated)
Membership: Student $15; Individual 35; Dual & Family $50; Friend $100; Associate $250; Donor $500; Supporter $1,000; Patron $2,500; Benefactor $5,000.

✳　**MILWAUKEE ART MUSEUM, (M),** 700 N. Art Museum Dr., Milwaukee, WI 53202-4098. Tel.: 414-224-3200 & 3220. Fax: 414-271-7588. TDD: 414-271-6678.
E-mail: mam@mam.org
Web Site: www.mam.org
Founded: 1888.
Congressional District: 5
Key Personnel: Dir., Daniel Keegan; Chief Cur., Brady Roberts; Sr. Dir. Devel., Mary Albrecht; Registrar, Dawn Gorman Frank; Dir. Programs & Education, Brigid Globensky; Dir. Mktg. & Communications, Vicki Scharfberg; C.F.O., Jane Wochos; Librarian & Archivist, Heather Winter.

Personnel Profile: Full-Time Paid 119; Part-Time Paid 83; Part-Time Volunteers 150.
Governing Authority: bd. of trustees, nonprofit organization. Tax-exempt: 501(c)(3).
Institution Type/Description: Art Museum.
Collections: over 30,000 works; 19th- to 20th-century American & European art, contemporary art, American decorative arts, Old Master works, folk & self-taught art.
Major Exhibits: Thomas Sully: Imagination & Innovation, 10/13-1/14.
Research Fields: art in permanent collection, exhibitions & prospective acquisitions.
Facilities: gardens; educational & public programming facilities.
Activities: guided tours; lectures; gallery talks; arts festival; education programs for children & adults; satellite specialization center for Milwaukee schools; docent program & council; inter-museum loan, permanent, temporary, & traveling exhibitions; school loan service; 10 special interest & support groups.
Publications: exhibition catalogs; calendar; collection handbook; annual report.
Hours & Admission Prices: Memorial Day to Labor Day Thurs. 10-8, Fri.-Wed. 10-5; Sept.-May Tues.-Wed. & Fri.-Sun. 10-5, Thurs. 10-8. Adults $15, senior citizens & students $12; discounts to AAM & ICOM members; members & children 12 and under no charge. Closed Thanksgiving; Christmas. ♿
Attendance: 400,000 (estimated)
Membership: Student $25; Individual & Teacher $60; Family & Dual $75; Friends of Art $150; Donor $350; Patron $500; Partner $1,000; Benefactor $2,500; Philanthropist $5,000; Sustaining Philanthropist $10,000; Director's Patron $25,000; Calatrava Society $50,000; Peg Bradley Society $100,000.

MILWAUKEE COUNTY HISTORICAL SOCIETY, 910 N. Old World Third St., Milwaukee, WI 53203-1501. Tel.: 414-273-8288. Fax: 414-273-3268.
E-mail: info@milwaukeehistory.net
Web Site: www.milwaukeehistory.net
Founded: 1935.
Congressional District: 5
Key Personnel: Pres., Randy Bryant; Cur. Research Collections, Steven L. Daily; Cur. Museum Collections, Michael Reuter.
Personnel Profile: Part-Time Paid 2; Part-Time Volunteers 30; Interns 6.
Governing Authority: nonprofit organization. Branch Museums: Kilbourntown House, Milwaukee; Lowell Damon House, Wauwatosa; Jeremiah Curtin House, Greendale; Trimborn Farm, Greendale. Tax-exempt: 501(c)(3).
Institution Type/Description: History Museum: housed in 1913 bank.
Collections: county history; brewery materials; manuscripts; transportation artifacts; panorama paintings; early settlers materials; aeronautics; archives; paintings; costumes; decorative arts; marine; military; technology; theater. Historic Buildings: 1846 Jeremiah Curtin house; 1847 Lowell Damon House; 1844 Kilbourntown House; 1847 Trimborn Farm.
Research Fields: history of Milwaukee County.
Facilities: 10,000-vol. library of historical material available for use on premises; county archives & court records, including naturalization papers; reading room.
Activities: guided tours; lectures; films; gallery talks; radio programs; formally organized education programs for undergraduate students; permanent & temporary exhibitions; school loan service; home tours; ethnic dinners; wine & beer tasting; tours of historic public & private buildings in Milwaukee County.
Publications: quarterly magazine, Milwaukee History; newsletter; books.
Hours & Admission Prices: Mon.-Fri. 9:30-5, Sat. 10-5, Sun. 1-5. Adults $3, seniors 62 & over and children 6 & over $2; county residents on Mon. & members no charge. Closed national holidays. ♿
Attendance: 65,000 (estimated)
Membership: Student & Senior Citizen $30; Individual $35; Family $40; Contributing $75; Sustaining $125; Associate $250; Benefactor $500.

MILWAUKEE COUNTY ZOOLOGICAL GARDENS, 10001 W. Bluemound Rd., Milwaukee, WI 53226-4384. Tel.: 414-771-3040. Fax: 414-256-5410. TDD: 414-771-1180.
E-mail: jdilibertishea@milwcnty.com
Web Site: www.milwaukeezoo.org
Founded: 1958.
Congressional District: 5
Key Personnel: Exec. Dir., Charles Wikenhauser; Deputy Dir. Animal, Bruce Beehler, D.V.M.; Public Affairs & Svcs. Dir., Laura Pedriani; Operations Mgr., Bldgs. & Grounds Mgr., Karl Hackbarth.
Personnel Profile: Full-Time Paid 125; Part-Time Paid 450; Part-Time Volunteers 680.

　　　　　　　　　　　　　　　

Governing Authority: Parent Institution: Milwaukee County. Tax-exempt.
Institution Type/Description: Zoo.
Collections: mammals; birds; fish; amphibians & reptiles of the world.
Research Fields: primate behavior; avian reptilian reproduction; elephant behavior; Humboldt penguin reproduction; snow leopard reproduction.
Facilities: approx. 500-vol. library of zoological reference books not available for outside use; education center; picnic area; restaurant. Museum-related items for sale.
Activities: formally organized education programs for children; permanent & temporary exhibitions; camel & pony rides; giraffe feeding; train; zoomobile; sea lion show.
Publications: quarterly newsletter, Alive (published by the Zoological Society of Milwaukee County).
Hours & Admission Prices: Jan.-Feb. & Nov.-Dec. Mon.-Fri. 9:30-2:30, Sat.-Sun. 9:30-4:30; March-May 24 & Sept. 3-Oct. daily 9-4:30; May 25-Sept. 2 daily 9-5. April-Oct. adults $14.25, seniors $13.25, children $11.25; discount to Milwaukee County residents; members & children 2 and under no charge. Nov.-March adults $11;75, seniors $10.25, children 3-12 $8.75; discount to Milwaukee County residents; members & children 2 and under no charge. Parking: buses $16, cars $12. &
Attendance: 1,300,000 (accurate)
Membership: See Zoological Society of Milwaukee County: www.zoosociety-.org.

MILWAUKEE FIRE HISTORICAL SOCIETY, LTD., 1615 W. Oklahoma Ave., Milwaukee, WI 53215. Mailing Address: 711 W. Wells St., Milwaukee, WI 53233. Tel.: 414-286-5272.

E-mail: jhley@sbcglobal.net & katpatward@aol.com
Founded: 1981.
Key Personnel: Chm. & Pres. (V), Warren Skonieczny.
Governing Authority: Parent Institution: Milwaukee Fire Department Historical Society. Tax-exempt: 501(c)(3).
Institution Type/Description: Fire Museum & Educational Center: housed in the former headquarters of Engine 23; built in 1927.
Collections: working fire alarm telegraph system; personal artifacts; photographs; fire equipment; 1943 Pirsch engine; two 1927 MFD engines; 1930 MFD city service truck; 1947 Cadillac ambulance; 1957 Mack/Magirus rear mount 100' aerial.
Hours & Admission Prices: 1st Sun. each month 1-4 (excluding holidays) & by appointment. No charge; donations accepted. &

✻ MILWAUKEE PUBLIC MUSEUM, (M), 800 W. Wells St., Milwaukee, WI 53233-1478. Tel.. 414-278-2700. Fax: 414-278-6100.

E-mail: webmaster@mpm.edu
Web Site: www.mpm.edu
Founded: 1882.
Congressional District: 5
Key Personnel: Chm., Mike Jones; Pres. & C.E.O., Jay Williams; Sr. Vice Pres. Finance & C.F.O., Michael Bernatz; Sr. Vice Pres. Devel., Karen Spahn; Academic Dean & Sr. Vice Pres. Museum Programs, Ellen Censky, Ph.D.; Sr. Vice Pres. Mktg. & Communications, Mary Bridges; Dir. Theater & Planetarium, Bob Bonadurer; Dir. Human Rels. & Labor Rels., Judith Atkinson; Dir. Security, Ralph Jones; Dir. Facility Operations, Larry Bannister; Dir. Information Svcs., Linda Gruber; Mgr. Education Programs, Art Montgomery; Dir. Admissions & Reservations, Laura Wake Wiesner; Asst. Dir. Exhibit Programs, Tom Shea; Gift Shops Operations Mgr., Catherine Wallberg.
Personnel Profile: Full-Time Paid 87; Part-Time Paid 53; Part-Time Volunteers 484; Interns 55.
Governing Authority: nonprofit. Tax-exempt: 501(c)(3).
Institution Type/Description: Natural & Human History Museum.
Collections: anthropology, including all major sub-disciplines; botany; geology-paleontology; invertebrate zoology; vertebrate zoology; history-worldwide including local history; fine, decorative & folk arts; business machines; firearms; photos, film; sound recordings; bibliographic; archival collections & sound recordings.
Major Exhibits. Body Worlds: The Cycle of Life (T), 1/14 6/14; Alien Worlds & Androids (T), 9/14 1/15.
Research Fields: metropolitan Milwaukee & Wisconsin flora; natural areas preservation; invertebrate paleontology; evolution & extinction of Paleozoic faunas; firearms; natural history, general ecology & population biology of butterflies; ecology & conservation; behavior & ecology of West Indies reptiles; Egyptian mummies; U.S. military; local & Wisconsin history.
Facilities: 125,000-vol. library of natural & human history; 148,178 sq. ft. exhibit space; 300-seat lecture hall; dome theater; classrooms; restaurant; butterfly garden; planetarium. Museum-related items for sale.
Activities: guided tours; lecture series; educational programs; special events; special exhibits; films & planetarium shows; docent program; volunteer program; public programming; summer camps; museum studies program for college students; private rental facilities.
Publications: membership newsletter & magazine; guides; pamphlets; annual report; teacher's guides; exhibition posters; postcards; note cards.
Hours & Admission Prices: Daily 9-5. Adults $15, senior citizens & students 13-17 $11, children 3-12 $10; discounts to handicapped visitors & AAM members; ASTC reciprocal; Milwaukee County residents on Mon. & members no charge. Closed Independence Day; Thanksgiving; Christmas. &
Attendance: 600,000 (estimated)
Membership: Basic: Individual $40; Dual $60; Pathfinder $75; Pathfinder Plus $140. Enrichment Club: Bronze $150-$249; Silver $250-$499; Gold $500-$749; Platinum $750-$999.

MITCHELL GALLERY OF FLIGHT, (M), General Mitchell International Airport, c/o Milwaukee County Airport Div., 5300 S. Howell Ave., Milwaukee, WI 53207-6156. Tel.: 414-747-4503. Fax: 414-747-4525. Facebook: Mitchell Gallery of Flight.

E-mail: flymitchell@mitchellgallery.org
Web Site: www.mitchellgallery.org
Founded: 1984.
Congressional District: 4
Key Personnel: Pres. (V) & Dir., Bill Streicher; Financial Dir., Anthony Snieg; Devel. & Membership, Jackie Stueck.
Personnel Profile: Part-Time Paid 1; Part-Time Volunteers 15.
Governing Authority: nonprofit organization. Parent Institution: Mitchell International Airport. Tax-exempt.
Institution Type/Description: Aviation Museum.
Collections: full size 1911 Curtiss Pusher aircraft; model airplanes; aviation art; miniature dioramas; photographs; aviation engines; medals & memorabilia on Gen. Billy Mitchell; historical archives; items relating to aviation in Wisconsin, Mitchell Airport, commercial & military aviation & lighter-than-air transportation.
Research Fields: history of local aviation; Gen. Mitchell.
Facilities: 1,400 sq. ft. exhibit space
Activities: self-guided tours. Museum Sponsors: Gen. Mitchell Open golf meet.
Publications: quarterly newsletter, Flightlines; General William Mitchell, Air Power Pioneer; General Mitchell International Airport - A Record of Progress.
Hours & Admission Prices: Gallery: daily 6am-10pm. Airport: daily 24 hrs. No charge; donations accepted. &
Attendance: 200,000 (estimated)
Membership: Individual $10; Family $15; Contributing $25; Sustaining $100; Bronze Corporate Sponsor $250; Silver Corporate Sponsor $500; Gold Corporate Sponsor $1,000.

MITCHELL PARK HORTICULTURAL CONSERVATORY, 524 S. Layton Blvd., Milwaukee, WI 53215-1236. Tel.: 414-257-5600. Fax: 414-649-8616.

Web Site: www.countyparks.com/horticulture
Founded: 1961.
Congressional District: 4
Key Personnel: Dir., Sandy Folaron; Horticulturist II, Amy Thurner; Horticulturist I (Tropical), Patrick Kehoe; Horticulturist I, Marian French; Horticulturist I (Display), Mary Braunreiter; Operating Engineer, David Loosemore.
Personnel Profile: Full-Time Paid 9; Part-Time Paid 8.
Governing Authority: county. Parent Institution: Milwaukee County Park System, 9480 W. Watertown Plank Rd., Wauwatosa, WI 53226. Tel.: 414-257-6100. Tax-exempt.
Institution Type/Description: Botanical Garden & Conservatory: housed on site of the Jacques Vieau settlement, one of the first permanent buildings built in the Milwaukee area.
Collections: tropical Bromeliads; 400 species/hybrid orchids; ferns; 550 species general tropicals; 75 species & forms of lithops; agaves; more than 1,000 cacti & succulents; Madagascar succulent collection; emphasis on tropical fruits; economic plants; birds; reptiles.
Research Fields: experiments on raising horticulture materials; conservation work; biological pest management
Facilities: 250-vol. library of informational material for staff available for research; botanical garden. Plants & other museum-related items for sale.
Activities: guided tours; occasional concerts; hobby workshops. Museum Sponsors: seasonal shows; Holiday Shows; Ethnic Festivals.
Publications: various cultural brochures.
Hours & Admission Prices: Mon.-Fri. 9-5, Sat.-Sun. 9-4. Adults $6.50, Milwaukee senior citizens with ID, juniors 6-18 & handicapped $5; discounts to groups of 20 or more; members & Milwaukee County students & residents Mon. 9-12 no charge &
Attendance: 200,000 (accurate)
Membership: Senior 62 & over or Student $25; Student & One Guest or Senior

62 & over and One Guest $30; Individual $40; Individual and Three Guests or Family $50; Contributing $125; Supporting $250; Sustaining $500; Benefactor or Corporate $1,000.

MOUNT MARY COLLEGE HISTORIC COSTUME COLLECTION, 2900 N. Menomonee River Pkwy., Milwaukee, WI 53222-4597. Tel.: 414-258-4810. Fax: 414-256-0172.
E-mail: gastone@mtmary.edu
Web Site: www.mtmary.edu
Founded: 1928.
Key Personnel: Cur., Elizabeth Gaston.
Personnel Profile: Part-Time Paid 2; Part-Time Volunteers 3; Interns 3.
Governing Authority: college; nonprofit organization. Parent Institution: Mount Mary College. Tax-exempt.
Institution Type/Description: Costume Museum.
Collections: costume: 1750s-present; special emphasis on American & European designers.
Research Fields: historic pattern reproduction; Calvin Klein archival project site; Bonnie Cashin; Valentina; Lynn Fontanne; Hildegarde.
Facilities: library featuring fashion periodicals; educational facilities.
Activities: formal education programs for undergraduate & graduate college students; loan & temporary exhibitions.
Hours & Admission Prices: Sept. to mid-May Mon.-Fri. 10-4; call for summer hours. No charge; donations accepted. Closed Easter; Thanksgiving; Christmas. &
Membership: Trendsetter $35; Private Label $50; Pret-a-Porter $100; Haute Couture $500.

THE PATRICK & BEATRICE HAGGERTY MUSEUM OF ART, (M), Marquette University, 13th & Clybourn, Milwaukee, WI 53233. Mailing Address: P.O. Box 1881, Milwaukee, WI 53201-1881. Tel.: 414-288-7290. Fax: 414-288-5415.
Web Site: www.marquette.edu/haggerty
Founded: 1984.
Congressional District: 5
Key Personnel: Dir., Wally Mason; Assoc. Dir., Lee Coppernoll; Assoc. Cur., Dr. Annemarie Sawkins; Cur. Education, Lynne Shumow; Registrar, John Loscuito; Administrative Asst., Mary Wagner; Curatorial Asst., Jerome Fortier; Preparator, Dan Herro; Asst. Preparator, Ric Stultz; Communications Asst. & Administrative Sec., Mary Dornfeld; Chief Security Officer, Clayton Ray.
Personnel Profile: Full-Time Paid 8; Part-Time Paid 23; Part-Time Volunteers 50; Interns 2.
Governing Authority: university. Parent Institution: Marquette Univ., Milwaukee. Tax-exempt: 501(c)(3).
Institution Type/Description: Art Museum.
Collections: European & American paintings from the 15th- to 20th-centuries; works by major international 20th-century artists; prints; photographs; sculpture; global arts.
Research Fields: European painting & prints; aesthetics & contemporary art; photography.
Facilities: library; 8,000 sq. ft. exhibit space. Museum-related gifts & books for sale.
Activities: participatory, loan, traveling & temporary exhibitions; guided tours; lectures; films; paintings & drawings; education programs for students & adults; performance art; concerts; docent training; Friends Assoc.; Old Master Collectors Group. Annual Events: spring gala in June; fall gala in October.
Publications: exhibition catalogs.
Hours & Admission Prices: Mon.-Wed. & Fri.-Sat. 10-4:30, Thurs. 10-8, Sun. 12-5; call to confirm. No charge; donations accepted. Closed some holidays. &
Attendance: 16,000
Membership: Student $15; Individual $40-$59; Family $60-$99; Sustaining $100-$249; Patron $250-$499; Benefactor $500-$999; Director's Circle $1,000-$2,499; Museum Partners $2,500-$4,999; Diamond Partners $5,000-$9,999; Platinum Partner $10,000 & up.

UWM UNION-ART GALLERY, 2200 E. Kenwood Blvd., Milwaukee, WI 53211-3361. Mailing Address: UWM Union P.O. Box 413, Milwaukee, WI 53201. Tel.: 414-229-6310. Fax: 414-229-6709.
E-mail: agallery@uwm.edu
Web Site: www.aux.uwm.edu/union/artgal.htm
Founded: 1972.
Congressional District: 5
Key Personnel: Mgr., Molly Evans.
Personnel Profile: Part-Time Paid 6.
Governing Authority: university. Affiliated with University of Wisconsin-Milwaukee. Tax-exempt.

Institution Type/Description: Art Gallery.
Collections: various art media by local, state, & national artists.
Activities: gallery lectures; concerts; poetry readings; multi-media exhibitions & performances; art & dance programs; temporary & traveling exhibitions; symposiums; special events.
Publications: flyers, posters & mailings.
Hours & Admission Prices: Sept.-May Mon.-Wed. & Fri.-Sat. 12-5, Thurs. 12-7. No charge. Closed holidays; university vacation periods. &
Attendance: 13,000

VILLA TERRACE DECORATIVE ARTS MUSEUM, 2220 N. Terrace Ave., Milwaukee, WI 53202-1216. Tel.: 414-271-3656. Fax: 414-271-3986.
E-mail: jsterr@cavtmuseums.org
Web Site: www.cavtmuseums.org
Founded: 1967.
Congressional District: 5
Key Personnel: Exec. Dir., Marie Costello; Events Mgr., Judith Hooks; Mgr. Mktg., Erica Haouchine; Cur., Martha Monroe.
Personnel Profile: Full-Time Paid 4; Part-Time Paid 10; Part-Time Volunteers 35.
Governing Authority: Parent Institution: Milwaukee County War Memorial Corp. Branch Museum: Charles Allis Art Museum, 1801 N. Prospect Ave., Milwaukee, WI 53202. Tax-exempt: 501(c)(3).
Institution Type/Description: Italian style villa designed by David Adler in 1923 for Lloyd & Agnes Smith of the A.D. Smith Co.; European & fine decorative arts.
Collections: European decorative arts from 15th-19th century; 16th-19th century European paintings and wrought iron master pieces by Cyril Colnik; Renaissance garden.
Major Exhibits: Species and Specimens, 3/14; Tell Me A Story, 11/14.
Research Fields: Italian Renaissance art & architecture; European fine & decorative arts; garden design & history; David Adler, wrought iron, Cyril Colnik.
Facilities: non-circulating library related to research fields; rental facilities.
Activities: guided tours; concerts; lectures; gallery talks; seminars; permanent & temporary exhibitions; docent program.
Publications: exhibition catalogues; quarterly newsletter.
Hours & Admission Prices: Wed.-Sun. 1-5. Adults $7, seniors & students $5; children under 13 & members no charge. Closed major federal holidays. &
Attendance: 4,465 (accurate)
Membership: Student $25; Senior Citizen $35; Individual $45; Senior Couple $55; Family $60; Bronze $125; Silver $250; Gold $500; Platinum $1,000.

WALKER'S POINT CENTER FOR THE ARTS, 839 S. 5th St., Milwaukee, WI 53204-1730. Tel.: 414-672-2787. Fax: 414-755-1960.
E-mail: staff@wpca-milwaukee.org
Web Site: www.wpca-milwaukee.org
Founded: 1987.
Key Personnel: Exec. Dir., Gary Tuma; Chm. (V), Lona Long Velasco.
Personnel Profile: Full-Time Paid 2; Part-Time Paid 4; Interns 6.
Institution Type/Description: Art Gallery.
Collections: paintings; sculpture.
Hours & Admission Prices: Tues.-Sat. 12-5. No charge; donations accepted. &
Attendance: 8,000 (accurate)
Membership: Individual $25.

WISCONSIN BLACK HISTORICAL SOCIETY/MUSEUM, 2620 W. Center St., Milwaukee, WI 53206-1155. Tel.: 414-372-7677. Facebook: Wisconsin Black Historical Society.
E-mail: clayborn@execpc.com
Web Site: www.wbhsm.org
Key Personnel: Exec. Dir., Clayborn Benson
Institution Type/Description: Historical Society Museum.
Collections: local history & culture pertaining to the historical heritage of African descent in Wisconsin.
Hours & Admission Prices: Mon.-Fri. 11-4, Sat. 10-12:30. Adults & college students with ID $5, seniors $4; children 12 & under $3; members no charge. Closed New Year's Day; Thanksgiving; Christmas.
Membership: Senior/Student $15; Individual $25; Family $50.

WISCONSIN MARINE HISTORICAL SOCIETY, 814 W. Wisconsin Ave., Milwaukee, WI 53233-2385. Tel.: 414-286-3074.
E-mail: info@wmhs.org
Web Site: www.wmhs.org
Founded: 1959.
Key Personnel: Exec. Dir., Suzette J. Lopez.
Personnel Profile: Part-Time Paid 1; Part-Time Volunteers 18.

Governing Authority: Tax exempt.
Institution Type/Description: Marine Historical Society.
Collections: artifacts relating to Great Lakes maritime history; photographs; period artifacts; historical marine information.
Research Fields: Great Lakes marine.
Publications: quarterly newsletter, SOUNDINGS; book, Maritime Milwaukee; book, Schooner Days In Door County by Walter & Mary Hirthe.
Hours & Admission Prices: Call for hours. No charge. &
Membership: Sailor $45; Crew $55; Engineer $75; Captain $100.

Mineral Point

PENDARVIS HISTORIC SITE, (M), 114 Shake Rag St., Mineral Point, WI 53565-1063. Mailing Address: P.O. Box 270, Mineral Point, WI 53565-0270. Tel.: 608-987-2122. Fax: 608-987-3738.
E-mail: pendarvis@wisconsinhistory.org
Web Site: pendarvis.wisconsinhistory.org
Founded: 1971.
Congressional District: 2
Key Personnel: Dir., Allen Schroeder; Cur., Tamara Funk.
Personnel Profile: Full-Time Paid 2; Part-Time Paid 21.
Governing Authority: society. Parent Institution: State Historical Society of Wisconsin, 816 State St., Madison WI 53706. Tax-exempt.
Institution Type/Description: Historic House: 1841-1852 miners' cottages built by immigrants from Cornwall, England.
Collections: 1840-1875 household furnishings; 19th-century mining tools & equipment; personal artifacts.
Research Fields: mining & smelting technology c.1825-1900; British immigration to mining frontiers c.1825-1900.
Facilities: Museum-related items for sale.
Activities: guided tours; special events.
Publications: site brochure; books, Mineral Point: A History, reprint 1986; On the Shake Rag: Mineral Point's Pendarvis House 1935-1970.
Hours & Admission Prices: mid May to Oct. daily 10-5. Adults $10, members & children 5-17 $5, senior members $4; discounts to senior citizens & groups; children under 5 no charge.
Attendance: 10,000 (accurate)

Mishicot

MISHICOT HISTORICAL MUSEUM, 411 Buchanan, Mishicot, WI 54228. Mailing Address: P.O. Box 237, Mishicot, WI 54228. Tel.: 920-755-2525 & 3317.
Web Site: www.mishicot.org
Institution Type/Description: History Museum: housed in a former two-room schoolhouse, built in 1874.
Collections: artifacts of the Mishicot area history from pioneer days to the present; collections on rural schools & Potawatomi Indians; photographs.
Hours & Admission Prices: June-Aug. Sat.-Sun. 12-4; other times by appointment.

Monroe

MONROE ARTS CENTER, 1315 11th St., Monroe, WI 53566. Mailing Address: P.O. Box 472, Monroe, WI 53566-0472. Tel.: 608-325-5700. Fax: 608-325-5701.
E-mail: info@monroeartscenter.com
Web Site: monroeartscenter.com
Institution Type/Description: Art Gallery.
Collections: paintings; sculpture.
Hours & Admission Prices: Tues.-Sat. 10-5.

Mount Horeb

MT. HOREB AREA MUSEUM, 100 S. Second St., Mount Horeb, WI 53572-2106. Mailing Address: 138 E. Main St., Mt. Horeb, WI 53572-2138. Tel.: 608-437-6486.
E-mail: mthorebmuseum@mhtc.net
Web Site: www.mthorebhistory.org
Founded: 1975.
Key Personnel: Pres. (V), Brian Bigler; Dir. & Museum Shop Mgr., Laurie Boyden; Archivist, Shan Thomas.
Personnel Profile: Part-Time Paid 1; Part-Time Volunteers 33.
Governing Authority: Tax-exempt.
Institution Type/Description: History Museum.
Collections: local cultural history; photographs; period artifacts; paintings.
Major Exhibits: Natural Wonders & Human Interaction, 5/14-12/16.
Facilities: archives.
Publications: quarterly, Mt. Horeb Area Past Times.

Hours & Admission Prices: May-Dec. Fri.-Sat. 10-4, Sun. 12.30-4. No charge; donations accepted. &
Attendance: 3,000 (accurate)
Membership: Individual $20; Household $35; Contributing $75; Sustaining $150; Patron $500.

Neenah

*** BERGSTROM-MAHLER MUSEUM OF GLASS, (M),** 165 N. Park Ave., Neenah, WI 54956-2956. Tel.: 920-751-4658. Fax: 920-751-4755. TDD: 920-751-4658.
E-mail: answers@bergstrom-mahlermuseum.com
Web Site: www.bergstrom-mahlermuseum.com
Formerly: Bergstrom-Mahler Museum
Founded: 1954.
Congressional District: 6
Key Personnel: Dir., Jan Smith; Education, Chelisa Behm; Mktg. Dir., Jen Stevenson; Business Dir., Ben Fauske; Museum Shop Mgr., Laureen Endter.
Personnel Profile: Full-Time Paid 3; Part-Time Paid 6; Part-Time Volunteers 70.
Volunteer Hours: 2,736
Operating Expenses: 612,020
Operating Income: 852,918
Governing Authority: nonprofit organization. Parent Institution: Bergstrom-Mahler Museum Inc. Tax-exempt: 501(c)(3).
Institution Type/Description: Art Museum.
Collections: glass paperweights; Germanic & contemporary glass; Victorian glass baskets.
Research Fields: European & American antique paperweights & related glass objects, contemporary paperweights, Germanic glass.
Facilities: Museum-related items for sale.
Activities: permanent & temporary art exhibits; guided tours; lectures; gallery talks; summer arts festival; films; art events; formally organized education programs for children & adults; glass studio classes.
Publications: newsletter; exhibition catalogues; collection catalogues of paperweights & Germanic glass; annual report.
Hours & Admission Prices: Tues.-Sat. 10-4:30, Sun. 1-4:30. No charge; donations accepted. Closed legal holidays. &
Attendance: 22,698 (accurate)
Membership: Senior Citizen $25; Individual $35; Family $50; Associate $125; Apprentice $250; Partner $500.

NEENAH HISTORICAL SOCIETY, HIRAM SMITH, NEENAH'S HERITAGE PARK, OCTAGON HOUSE & WARD HOUSE, 343 Smith St., Neenah, WI 54956-2434. Mailing Address: P.O. Box 343, Neenah, WI 54957-0343. Tel.: 920-729-0244.
E-mail: neenahhistoricalsociety@gmail.com
Web Site: www.focol.org/neenahhistorical
Founded: 1948.
Congressional District: 6
Key Personnel: Pres., JoEllen Wollangk; Treas., Peter Wick.
Personnel Profile: Full-Time Paid 1; Part-Time Volunteers 4.
Governing Authority: private; nonprofit organization. Tax-exempt: 501(c)(3).
Institution Type/Description: General Museum: housed in 1850s octagon house.
Collections: period clothing & furnishings; personal artifacts; photographs; newspapers.
Facilities: library; archives; 3,000 sq. ft. exhibit space. Museum-related items for sale.
Activities: guided tours; lectures; loan & temporary exhibitions. Museum Sponsors: 4th & 5th grade students historical bike tour stop; annual meetings; Memorial Day Parade; Community Fest Parade; cemetery walk; Old Fashioned Country Fair; Christmas Open House.
Publications: bimonthly newsletter, The Society Times.
Hours & Admission Prices: By appointment. No charge; donations accepted. &
Membership: Individual $15; Family $35; Contributing $50; Supporting $100; Benefactor $500; Sponsor $1,000.

New Berlin

THE NEW BERLIN HISTORICAL SOCIETY, 19885 W. National Ave., New Berlin, WI 53146. Tel.: 262-643-8855.
E-mail: djtotten@earthlink.net
Web Site: newberlinhistoricalsociety.org
Founded: 1965.
Congressional District: 5
Key Personnel: Pres. (V) & Publicity, Dave Totten.

Personnel Profile: Part-Time Volunteers 30.
Governing Authority: municipal. Parent Institution: Wisconsin State Historical Society, 816 State St., Madison, WI 53706. Tax-exempt.
Institution Type/Description: Historical Society Museum.
Collections: area historical artifacts; farm implements & veterinary medicines from local farm; 1840 immigrant trunk; mid-19th century oil painting of New Berlin farmstead; general household items; accoutrements; organs; fainting couch; rope bed; toys; Murphy bed; general store; windmill; country store. Historic Buildings: 1863 schoolhouse; 1900 carriage barn; 1870 Winton-Sprengel farm house; 1850 pioneer log house; 1847 Winton/Martin house.
Research Fields: genealogy & local history.
Activities: guided tours; slide show; permanent exhibitions; classes held in 1863 one-room schoolhouse. Museum Sponsors: Ice Cream Social in July; Historic Days in September; annual Apple Fest in October; open house & special entertainment.
Publications: annual magazine, The New Berlin Almanack; quarterly newsletter.
Hours & Admission Prices: May-Sept. call for hours; group tours by appointment. No charge; donations accepted.
Attendance: 1,200 (estimated)
Membership: Family $15.

New Glarus

CHALET OF THE GOLDEN FLEECE, 618 2nd Ave., New Glarus, WI 53574. Mailing Address: P.O. Box 564, New Glarus, WI 53574-0564. Tel.: 608-527-2095. Fax: 608-527-4991.
Founded: 1955.
Congressional District: 2
Key Personnel: Village Admin., Nicholas Owen; Mgr., Pete Etter.
Personnel Profile: Part-Time Paid 5.
Governing Authority: municipal. Parent Institution: Village of New Glarus, WI. Tax-exempt.
Institution Type/Description: Historic House Museum: Swiss chalet.
Collections: period Swiss & early American furniture, glass & china; weapons; jewelry; Swiss carvings; period parchments; prints; French prints; Swiss doll collection.
Activities: guided tours.
Publications: brochures.
Hours & Admission Prices: May-Oct. by appointment. Adults $7, students $2; children under 6 no charge.
Attendance: 4,000 (estimated)

SWISS HISTORICAL VILLAGE, 612 7th Ave., New Glarus, WI 53574. Mailing Address: P.O. Box 745, New Glarus, WI 53574-0745. Tel.: 608-527-2317. Fax: 608-527-2302. TDD: 608-527-2317.
Web Site: www.swisshistoricalvillage.org
Founded: 1938.
Key Personnel: Pres., John F. Marty; Museum Shop Mgr., Gail Beal.
Personnel Profile: Part-Time Volunteers 18.
Governing Authority: society. Parent Institution: New Glarus Historical Society. Tax-exempt.
Institution Type/Description: Historic Village.
Collections: tools, period artifacts & replicas of buildings used by the early settlers; Hall of History; printshop; fire station; log church; log cabin; blacksmith shop; cheese factory; country store; school; bee house.
Activities: guided tours; permanent & temporary exhibitions.
Publications: membership newsletter.
Hours & Admission Prices: May-Oct. 15 daily 10-4. Adults $9, children 6-13 $3; discounts to bus groups if paid in advance, AAM & AAA members. ♿
Attendance: 10,000 (estimated)
Membership: Single $20; Couple $30; Family $40.

New Holstein

TIMM HOUSE HISTORIC SITE, 1600 Wisconsin Ave., New Holstein, WI 53061-1340. Mailing Address: P.O. Box 144, New Holstein, WI 53061-0144. Tel.: 920-898-5900. Fax: 920-898-5879.
E-mail: tethiessen@frontier.com
Web Site: www.newholsteinhistory.info
Formerly: Pioneer Corner Museum
Founded: 1961.
Congressional District: 8
Key Personnel: Pres., Terry Thiessen; Vice Pres., Jerry Hallstrom; Sec., Kay Nett.
Personnel Profile: Part-Time Volunteers 30.
Volunteer Hours: 4,000
Operating Expenses: 36,600
Operating Income: 39,500
Governing Authority: nonprofit organization. Parent Institution: New Holstein Historical Society. Branch Museums: The Timm House, 1600 Wisconsin Ave., New Holstein, WI 53061; The Pioneer Corner Museum, 2103 Main St., New Holstein, WI 53061. Tax-exempt.
Institution Type/Description: Historical Society Museum: built in the 1870s with an addition added in 1892.
Collections: Victorian furnishings; farm machinery; furnished rooms; manuscripts; pictures; old tools; dolls; toys; clothing; guns; rocks; china; theme room displays & settings; general store; post office. Historic House: 1892 Timm house.
Major Exhibits: New Holstein and the Civil War, 5/1/14-10/1/14.
Activities: guided tours; permanent & temporary exhibitions.
Publications: book, When I Was a Boy in New Holstein; Centennial Book; The New Holstein Story; Pioneers' Corner; Land of Peace and Plenty; Die Familie Thiel; Memories of the First Years of the Settlement of New Holstein.
Hours & Admission Prices: Pioneer Corner Museum: Memorial Day-Labor Day Sat.-Sun. 1-4; other times by appointment. Timm House Historic Site: Memorial Day to Labor Day Thurs.-Fri. 1-3, Sat.-Sun. 1-4; Sept.-May Sat.-Sun. 1-4; other times by appointment. Pioneer Corner Museum: adults $7, children under 12 $3. Both Houses: adults $10; children under 12 $7. ♿
Attendance: 825 (estimated)
Membership: Individual $20; Family $30; Friend of the Society $65; Sustaining $250; Founder's Circle $500.

New London

NEW LONDON PUBLIC MUSEUM, (M), 406 S. Pearl St., New London, WI 54961-1441. Tel.: 920-982-8520. Fax: 920-982-8617.
E-mail: christinec@newlondonwi.org
Web Site: www.newlondonwi.org/museum.htm
Founded: 1917.
Congressional District: 8
Key Personnel: Pres. Bd. (V), Ron Steinhorst; Dir., Christine Cross.
Personnel Profile: Full-Time Paid 1; Part-Time Paid 2; Part-Time Volunteers 10.
Volunteer Hours: 130
Operating Expenses: 88,000
Operating Income: 5,000
Governing Authority: municipal. Parent Institution: City of New London, WI. Tax-exempt.
Institution Type/Description: General Museum.
Collections: historical items; taxidermied specimens with emphasis on birds; Native American artifacts; geological specimens; world culture artifacts.
Major Exhibits: History & Science of the Pinewood Derby, 3/17-6/23/14; Extinct: 100 Years Without the Passenger Pigeon, 9/6/14-2/15; Folk: Tradition & Art, 7/21-1/15.
Facilities: 1,000-vol. library of bird & historical reference books available for use on the premises.
Activities: in-house & outreach educational programs; temporary & permanent exhibitions; historic home & cemetery self-guided walking tours. Science days curiosity series.
Hours & Admission Prices: Memorial Day to Labor Day Mon.-Fri. 10-5; Sept.-May Mon.-Fri. 10-5, Sat. 10-1. No charge; donations accepted. ♿
Attendance: 5,000 (accurate)

New Richmond

NEW RICHMOND HERITAGE CENTER, 1100 Heritage Dr., New Richmond, WI 54017-1741. Tel.: 715-246-3276. Fax: 715-246-3115.
E-mail: info@nrheritagecenter.org
Web Site: nrheritagecenter.org
Formerly: New Richmond Preservation Society
Founded: 1982.
Congressional District: 3
Key Personnel: Dir., Irv Sather; Bd. Asst., Rachel Stanbuck; Pres., Cheryl Emerson; Treas., Gary Knutson; Cur., Mary Sather.
Personnel Profile: Full-Time Paid 1; Full-Time Volunteers 2; Part-Time Paid 1; Part-Time Volunteers 17.
Volunteer Hours: 9,655
Operating Expenses: 117,000
Operating Income: 111,000
Governing Authority: private; nonprofit organization. Tax-exempt: 501(c)(3).
Institution Type/Description: Heritage Center.
Collections: local history; agricultural artifacts. Historic Buildings: 1884 Marcus Sears Bell farmstead; 1916 barn & granary; 1887 Norwegian log cabin; 1890 Northside house; 1902 Camp Nine school & outhouse; 1933 Ubet store; 1875 Blacksmith log building; 1891 Heritage church.

Facilities: Heritage Church farmhouse & pavilion available for rental; nature pathways. New Richmond related items for sale.
Activities: formal education programs for children; historical & cultural activities; guided tours; hobby workshops; temporary & traveling exhibitions; spring & fall programs. Annual Events: Heritage Days; Antique Show & Tell; Flea Market May to October; Farmer's Market July to October; Christmas Open House.
Publications: quarterly newsletter, Heritage Centerpieces.
Hours & Admission Prices: May-Oct. Mon.-Fri. 10-4, Sat. 7:30-2, Sun. 12-4; Nov.-April Mon.-Fri. 10-4. Adults $5, children $1; discounts to AAM & AAA members; members no charge. Closed New Year's Day; Memorial Day; Independence Day; Labor Day; Thanksgiving; Christmas. &
Attendance: 10,000 (estimated)
Membership: Student $5; Senior $10; Individual $15; Family $25; Contributing $30; Supporting $50; Benefactor $100; Patron $101-$999; Lifetime $1,000.

North Freedom

MID-CONTINENT RAILWAY MUSEUM, (M), E8948 Diamond Hill Road, North Freedom, WI 53951-9699. Mailing Address: P.O. Box 358, North Freedom, WI 53951-0358. Tel.: 608-522-4261. Fax: 608-522-4490.
E-mail: jeff@midcontinent.org
Web Site: www.midcontinent.org
Founded: 1959.
Congressional District: 2
Key Personnel: C.E.O., Chm. (V) & Pres. (V), Jeffrey B. Bloohm; Museum Shop Mgr. (V), Jeffrey Haertlien.
Personnel Profile: Full-Time Paid 1; Full-Time Volunteers 1; Part-Time Paid 15; Part-Time Volunteers 120.
Governing Authority: society; nonprofit organization. Parent Institution: Mid-Continent Railway Historical Society. Tax-exempt.
Institution Type/Description: Railway Museum.
Collections: railway equipment: 1900 era steam locomotives, wood passenger and freight cars; artifacts; photos. Historic Building: c.1894 C & NW depot.
Research Fields: north central U.S. railroad history.
Facilities: library. Railway-related items for sale.
Activities: train rides. Museum Sponsors: Snow Train in February; Autumn Color in October; Pumpkin Special in October; Santa Express in November.
Publications: periodical magazine, Mid-Continent Railway Gazette; annual calendar, Whistle on the Wind; Rail Heritage Book series, guide, Tour of the Yards; Mid-Continent Compendium; Sauk County Mining; Louis W. Hill's Business Car, Great Northern A-22; Copper Range RR.
Hours & Admission Prices: mid-May to Labor Day daily 9:30-5; day after Labor Day to mid-Oct. Sat.-Sun. 9:30-5. Adults $11, senior citizens $10, children $6; discounts to groups & AAA members; members no charge. &
Attendance: 29,329 (accurate)
Membership: Associate $35; Regular $40; Life $1,000.

Oak Creek

OAK CREEK HISTORICAL SOCIETY-PIONEER VILLAGE, S. 15th Ave. & E. Forest Hill Ave., Oak Creek, WI 53154. Mailing Address: P.O. Box 243, Oak Creek, WI 53154-0243. Tel.: 414-529-0196.
Web Site: ochistorical.freeservers.com
Founded: 1964.
Congressional District: 1
Key Personnel: Pres., Elroy Honadel; Vice Pres., Henry Kohler; Sec., Joyce Willms; Treas., Dick Raatz; Cur., Marge Berres; Archivist & Genealogists, Judy Salchow.
Personnel Profile: Part-Time Volunteers 12.
Governing Authority: society; nonprofit organization. Parent Institution: State Historical Society of Wisconsin. Subsidiary Institution: Wisconsin Council for Local History, Oak Creek, WI. 53154. Tax-exempt.
Institution Type/Description: History Museum.
Collections: hand sketches of the Village of Oak Creek of 1860 from the Henry E. Rile collections; china service set; farm shed; horse drawn equipment; 1924 McCormick tractor; hand operated farm field tools; period artifacts; genealogical resources. Historic Buildings: 1840 Hughes log house; 1874 Oak Creek Town Hall; 1886 Wohlust Blacksmith Shop; 1902 print shop & cobbler shop; 1900 Franke farm summer kitchen.
Research Fields: genealogical information; archives.
Facilities: meeting rooms; classrooms.
Activities: guided tours; lectures; permanent exhibitions; demonstrations.
Publications: cookbook of old recipes; brochures; pamphlets; booklet, Water Color Sketchings; book, Images of America - Oak Creek, Wisconsin.
Hours & Admission Prices: Memorial Day-Labor Day Sun. 2-4. No charge; donations accepted.

Attendance: 750 (estimated)
Membership: Single $8; Family $15; Business $50; Life $250.

Oconomowoc

OCONOMOWOC AREA HISTORICAL SOCIETY & MUSEUM, 103 W. Jefferson St., Oconomowoc, WI 53066-3633. Mailing Address: P.O. Box 969, Oconomowoc, WI 53066-0969. Tel.: 262-569-0740.
E-mail: oahs-m@sbcglobal.net
Web Site: www.oconomowochistoricalsociety.com
Key Personnel: Admin., Nancy Lins
Institution Type/Description: Historical Society Museum.
Collections: local history & culture; period artifacts; photographs.
Hours & Admission Prices: Jan.-April by appointment; May-Dec. Fri.-Sun. 1-5.

Oconto

BEYER HOME, OCONTO COUNTY HISTORICAL SOCIETY MUSEUM, 917 Park Ave., Oconto, WI 54153-1641. Mailing Address: Box 272, Oconto, WI 54153-0272. Tel.: 920-835-5733.
E-mail: ocrl@bayland.net
Web Site: ocontocountyhistsoc.org
Founded: 1940.
Congressional District: 8
Key Personnel: Pres., Peter Stark.
Personnel Profile: Part-Time Paid 5.
Governing Authority: society. Subsidiary Institutions: Holt & Balcom Logging Camp, Lakewood, WI; Copper Culture Park. Tax-exempt: 501(c)(3).
Institution Type/Description: Historic House Museum: 1868 Beyer Home, Victorian brick mansion, carriage house & G.E. Hall Annex.
Collections: Native American artifacts; period furnishings; fur trade; fishing & lumbering; period vehicles; Oconto County area photographs; hand painted china. Historic Building: 1800s log cabin.
Research Fields: Oconto county history.
Facilities: 50-vol. library of historical books available on premises.
Activities: guided tours; socials; ice cream social. Museum Sponsors: Area Art Show in August; Nighttime Open House in October.
Publications: quarterly newsletter, News from the Northwoods; pamphlets, Historic Oconto; Lumber Era Oconto; Holt & Balcom Logging Camp; Copper Culture People; Christian Science Church (first in world); books; Would You Believe It; Recollections of Oconto County; A History of Oconto, Bicentennial Recollections of Oconto County; John & Almira Volk; A Walking Tour of the Historic West; Main Street of Oconto.
Hours & Admission Prices: Beyer Home & Museum Annex: June-Labor Day Mon.-Fri. & Sun. 12-4. Family $10, adults $4, students 6-18 $2; discount to groups & AAA members; members & children under 6 no charge.
Attendance: 1,000 (estimated)
Membership: Senior Citizen $5; Adult $10; Family $15; Business & Sustaining $25.

COPPER CULTURE MUSEUM, Mill St., Oconto, WI 54153. Mailing Address: Oconto Historical Society, 917 Park Ave., Oconto, WI 54153-1641. Tel.: 920-834-6206.
E-mail: ocm@bayland.net
Governing Authority: Parent Institution: Oconto Historical Society.
Institution Type/Description: Historic Site: listed on the National Registry of Historic Places.
Collections: local history & culture; Native American artifacts; burial ground of North America's earliest metal users.
Hours & Admission Prices: June to Labor Day Sat.-Sun. & holidays 10-3. No charge.

Oshkosh

*** EAA AIRVENTURE MUSEUM,** 3000 Poberezny Rd., Oshkosh, WI 54902-8900. Mailing Address: EAA Aviation Center, P.O. Box 3086, Oshkosh, WI 54903-3086. Tel.: 920-426-4818. Fax: 920-426-6765.
E-mail: museum@eaa.org
Web Site: www.airventuremuseum.org
Founded: 1963.
Congressional District: 4
Key Personnel: Librarian, Susan Lurvey; Cur. Collections, Ron Twellman.
Personnel Profile: Full-Time Paid 3; Part-Time Volunteers 175; Interns 4.
Governing Authority: Parent Institution: Experimental Aircraft Association. Tax-exempt: 501(c)(3).
Institution Type/Description: Aviation Museum.

Collections: over 250 aircrafts: Eagle hangar-all World War II aircraft & memorabilia; historical, military, civilian amateur built, rotor craft, gliders, engines, props; photographs; paintings; prints; video presentations; hands-on displays; restoration shop.

Research Fields: aeronautics.

Facilities: 15,000-vol. library including collection of periodicals, magazines & pictures on aeronautics; field research station; 120-seat auditorium; educational facilities; nature center; leadership center. Aviation publications, related material & models for sale.

Activities: seasonal flying program at Pioneer Airport on museum grounds; lectures; participatory, temporary & permanent exhibitions; docent program; formal education programs; guided tours; hobby workshops; forums; aircraft rides; theater. Museum Sponsors: EAA Air Academy for youth & adults; Young Eagles youth program. Annual Events: World's Largest Air Show - EAA Air Venture.

Publications: magazines; Sport Aviation; Vintage Aviation; Sport Aerobatics; Warbirds.

Hours & Admission Prices: Mon.-Sat. 8:30-5, Sun. 10-5. Family $31, adults $12.50, senior citizens 62 & over $10.50, students 6-17 $9.50; discounts to groups & AAM members; children under 6 & EAA members no charge. Closed New Year's Day; Easter; Thanksgiving; Christmas. &

Attendance: 126,342 (accurate)

Membership: Individual $40; Family $50; International Individual $56.

MILITARY VETERANS MUSEUM, INC., (M), 375 City Center, Ste. K, Oshkosh, WI 54901-4999. Mailing Address: P.O. Box 2194, Oshkosh, WI 54903-2194. Tel.: 920-426-8615. Fax: 920-426-1828.

E-mail: mvm@athenet.net

Web Site: www.mvmwisconsin.com

Founded: 1985.

Congressional District: 8

Key Personnel: Pres. & Special Projects, Ron Metz; Vice Pres., Mark Ropella; Sec., Lynn Beck; Treas., Displays & Special Events, Julie Nikolaus; Operations, Kate Robinson; Librarian, Luida Sanders; Shop Coord., Ralph Beck; Education, Donna Stammer; Education, John Pieper.

Personnel Profile: Full-Time Paid 1; Part-Time Volunteers 30.

Governing Authority: private; nonprofit. Tax-exempt: 501(c)(3).

Institution Type/Description: Military Museum.

Collections: military history & artifacts including the role of the citizen soldier in the U.S. military; patriotism; USS Wisconsin Recommissioning plaque; USS California's shiplog at Pearl Harbor on Dec. 7, 1941; a piece of the USS Arizona.

Publications: quarterly newsletter, News From Home.

Hours & Admission Prices: Temporarily closed.

Membership: Individual $25; Organization $100; Life $225-$325.

MORGAN HOUSE - WINNEBAGO COUNTY HISTORICAL AND ARCHAEOLOGICAL SOCIETY, 234 Church Ave., Oshkosh, WI 54901. Tel.: 920-232-0260.

Web Site: www.winnebagocountyhistoricalsociety.com

Key Personnel: Pres. (V), Julie Johnson.

Governing Authority: Tax-exempt.

Institution Type/Description: Historic House Museum: built in 1884. Listed on the National Register of Historic Places.

Collections: local history & culture; period furnishings; personal artifacts; paintings.

Hours & Admission Prices: Sat. 9-1; other times by appointment. Adults $5; children no charge.

✳   **OSHKOSH PUBLIC MUSEUM, (M),** 1331 Algoma Blvd., Oshkosh, WI 54901-2799. Tel.: 920-236-5799. Fax: 920-424-4738.

E-mail: museum@ci.oshkosh.wi.us

Web Site: www.oshkoshmuseum.org

Founded: 1924.

Congressional District: 6

Key Personnel: Dir., Bradley Larson; Pres. (V), Gary Huffeman; Cur., Debra Daubert; Registrar, Joan Lloyd; Mktg. & Membership, Karla Szekeres; Archivist, Scott Cross.

Personnel Profile: Full-Time Paid 10; Part-Time Paid 3; Part-Time Volunteers 55; Interns 4.

Volunteer Hours: 2,850

Operating Expenses: 1,100,000

Operating Income: 1,100,000

Governing Authority: municipal. Parent Institution: City of Oshkosh. Tax-exempt.

Institution Type/Description: General Museum.

Collections: decorative arts textiles; pressed glass; china; Wisconsin Indian archaeology & ethnology; local 19th-century social, military, cultural &

industrial history; fur trade; horse-drawn firefighting vehicles, Tiffany stained glass windows & interior; military; manuscripts; natural history specimens.

Major Exhibits: In Company With Angels (T), 2/14-4/14; Living With Tiffany, 2/14-4/14; Attack of the Bloodsuckers (T), 6/14-8/14; Deck the Halls, 11/14-1/15.

Research Fields: military history; local history; decorative arts.

Facilities: 8,000-vol. library of books & pamphlets for reference use on premises & inter-library loan; classroom. Books, DVDs & other museum-related items for sale.

Activities: lectures; formally organized education programs for children & adults; internship for undergraduate college students affiliated with the University of Wisconsin, Oshkosh; inter-museum loan, traveling, permanent & temporary exhibitions.

Publications: newsletter; information leaflets; books, Like a Deer Chased by the Dogs: The Life of Chief Oshkosh; A Self-Guided Walking Tour of Riverside Cemetery: From the Newspapers: Graphic Stories of Death.

Hours & Admission Prices: Tues.-Sat. 10-4:30, Sun. 1-4:30. Adults $7, seniors & students $5, children 6 & over $3.50; children under 6 & members no charge. Closed national holidays. &

Attendance: 28,233 (accurate)

Membership: Senior & Student $25; Individual $30; Dual $40; Family $55; Cornerstone $150.

✳   **PAINE ART CENTER AND GARDENS, (M),** 1410 Algoma Blvd., Oshkosh, WI 54901-7708. Tel.: 920-235-6903, ext. 21. Fax: 920-235-6303.

E-mail: info@thepaine.org

Web Site: www.thepaine.org

Founded: 1947.

Congressional District: 6

Key Personnel: Exec. Dir., Aaron Sherer; Public Rels., Connie Pirner.

Personnel Profile: Full-Time Paid 8; Part-Time Paid 11; Part-Time Volunteers 160; Interns 1.

Governing Authority: nonprofit organization. Tax-exempt: 101(6); 501(c)(3).

Institution Type/Description: Art Museum & Arboretum: housed in 19th-century Tudor Revival Manor house.

Collections: 19th- & 20th-century American paintings, sculpture & prints; 18th- to 20th-century European paintings, sculpture & prints; decorative arts; botanical art; Oriental rugs; gardens including annuals, native & exotic trees, shrubs & herbaceous plants which grow in Wisconsin.

Research Fields: 18th-20th century painting, sculpture & prints; American & European decorative arts & architecture; English gardens; historic landscapes.

Facilities: reference library on art, art history, decorative arts, architecture & horticulture available for on premises research. Museum-related items for sale.

Activities: guided tours; lectures; films; gallery talks; concerts; formally organized education programs for children & adults; bus tours; docent training program; permanent, temporary & traveling exhibitions. Museum Sponsors: Nutcracker in the Castle from November to January.

Publications: bimonthly newsletter; exhibition catalogues; annual report.

Hours & Admission Prices: Tues.-Sun. 11-4. Adults $7, senior citizens $6, students $5, children 5-12 $4; discount to AAM members; children under 5 & members no charge. Closed national holidays. &

Attendance: 55,915 (accurate)

Membership: Individual $40; Family $60; Sustaining $100; Supporting $250; Benefactor $500. Patron: Director's Circle $1,200; President's Circle $2,500; Founder's Circle $5,000.

Peshtigo

PESHTIGO FIRE MUSEUM, 400 Oconto Ave., Peshtigo, WI 54157-1299. Mailing Address: P.O. Box 26, Peshtigo, WI 54157. Tel.: 715-582-3244.

E-mail: contact@peshtigo.info

Web Site: www.peshtigofire.info/museum.htm

Founded: 1962.

Key Personnel: Pres. (V) & Cur., Jerry Devroy; Sec. & Cur., Margaret Wood; Treas. & Cur., Marion Devory; Cur., Sally Kahl; Cur., Lois Johnson; Cur., Sharon Schounard; Cur., Rosemary Leslie; Cur., Joan Berth; Cur., Pauline King.

Governing Authority: society. Parent Institution: State Historical. Tax-exempt.

Institution Type/Description: History Museum: Housed in 1878 Old Church located on the 1871 site of the worst forest fire in U.S. history.

Collections: period furnishings; glass; agriculture; blacksmith shop; school room; old country store.

Activities: permanent exhibitions.

Publications: books, The Great Peshtigo Fire; Embers of October.

Hours & Admission Prices: May 17-Oct. 8 daily 10-4:30. No charge; donations accepted.
Attendance: 7,500 (accurate)
Membership: Individual $5.

Pewaukee

CLARK HOUSE MUSEUM, 206 E. Wisconsin Ave., Pewaukee, WI 53072. Tel.: 262-691-0233.
Governing Authority: private; nonprofit organization. Parent Institution: Pewaukee Area Historical Society. Tax-exempt: 501(c)(3).
Institution Type/Description: History Museum.
Collections: local history & culture; period furnishings; Native American artifacts; photographs.
Activities: special events.
Hours & Admission Prices: Memorial Day to Oct. Sun. 1-4, Wed. 1-4 & 7-9; other times by appointment.

Phillips

JUMP RIVER VALLEY HISTORICAL SOCIETY & MUSEUM, N6882 The Loop, Phillips, WI 54555. Tel.: 715-339-2642 & 474-6775.
E-mail: koerner2@pctcnet.net
Institution Type/Description: Historical Society Museum.
Collections: local history & culture; period furnishings; personal artifacts; photographs; early 1800s homemaking, schooling, logging, farming & business; quilt.
Activities: special events.
Hours & Admission Prices: June-Sept. 2nd & 4th Sat. 10-3. No charge. &

Platteville

PLATTEVILLE MINING MUSEUM, (M), 385 E. Main St., Platteville, WI 53818-3204. Mailing Address: P.O. Box 780, Platteville, WI 53818-0780. Tel.: 608-348-3301. Fax: 608-348-4640.
E-mail: museums@platteville.org
Web Site: www.mining.jamison.museum
Founded: 1965.
Congressional District: 3
Key Personnel: Dir., Stephen J. Kleefisch; Pres. (V), Clay Shaffer; Cur., Stephanie Saager-Bourret; Education Coord., Mary Huck.
Personnel Profile: Full-Time Paid 3; Part-Time Paid 12; Part-Time Volunteers 6.
Volunteer Hours: 1,049
Operating Expenses: 257,600
Operating Income: 257,600
Governing Authority: municipal. Parent Institution: City of Platteville. Tax-exempt.
Institution Type/Description: Mining Museum: housed in 1863 schoolhouse & 1845 lead mine.
Collections: geology; mining; maps; archives; photographs; full-size replica head frame; 1840s lead mine.
Research Fields: early European/American mining settlements in the southwest Wisconsin area.
Facilities: library of mining related materials.
Activities: guided tours; lectures; train ride; formally organized education programs for children & undergraduate college students; permanent exhibitions.
Hours & Admission Prices: May-Oct. daily 9-5; Nov.- April Mon.-Fri. 9-4; group tours by appointment. May-Oct. family $24, adults $9, senior citizens $7.75, children $4.50; discount to groups, AAM & ICOM members; members no charge. Nov.-April adults $3, children $1. Closed New Year's Day; Veterans Day; Thanksgiving & day after; Christmas. &
Attendance: 9,552 (accurate)
Membership: Individual $10; Family $25.

ROLLO JAMISON MUSEUM, 405 E. Main St., Platteville, WI 53818-2834. Mailing Address: P.O. Box 780, Platteville, WI 53818-0780. Tel.: 608-348-3301. Fax: 608-348-4640.
E-mail: museums@platteville.org
Web Site: www.mining.jamison.museum
Founded: 1981.
Congressional District: 3
Key Personnel: Dir., Stephen J. Kleefisch; Pres. (V), Clay Shaffer; Cur., Stephanie Saager-Bourret; Education Coord., Mary C. Huck.
Personnel Profile: Full-Time Paid 3; Part-Time Paid 12; Part-Time Volunteers 8.

Volunteer Hours: 1,049
Operating Expenses: 257,600
Operating Income: 257,600
Governing Authority: municipal. Parent Institution: City of Platteville. Tax-exempt.
Institution Type/Description: History Museum.
Collections: historical artifacts of southwest Wisconsin; photographs; local history.
Research Fields: early European/American settlement in the southwest Wisconsin area; artifacts in the Rollo Jamison collection.
Facilities: meeting space. Gift items for sale.
Activities: guided tours; educational programs for children & adults.
Hours & Admission Prices: May-Oct. daily 9-5; Nov.-April Mon.-Fri. 9-4; group tours by appointment. May-Oct. family $25, adults $9, senior citizens $7.75, children $4.50; discount to groups, AAM & ICOM members; members no charge. Nov.-April adults $3, children $1. Closed New Year's Day; Thanksgiving & day after; Christmas. &
Attendance: 9,552 (accurate)
Membership: Individual $10; Family $25.

Plymouth

BRADLEY GALLERY OF ART, W. 3718 South Dr., Plymouth, WI 53073. Mailing Address: P.O. Box 359, Sheboygan, WI 53082-0359. Tel.: 920-565-2111 & 1280. Fax: 920-565-1206.
Web Site: www.lakeland.edu
Founded: 1988.
Key Personnel: Co-Dir., Denise Presnell-Weidner; Co-Dir., William Weidner.
Personnel Profile: Part-Time Paid 2; Interns 4.
Governing Authority: private college; nonprofit. Parent Institution: Lakeland College. Tax-exempt.
Institution Type/Description: Art Gallery.
Collections: paintings & prints from the Bick collection.
Facilities: 265 sq. ft. exhibit space.
Activities: lectures.
Hours & Admission Prices: Sept.-May Mon.-Fri. 9-5. No charge. Closed school holidays; semester breaks. &
Attendance: 3,000 (estimated)

JOHN G. VOIGT HOUSE, W 5639 Anokijig Lane, Plymouth, WI 53073. Tel.: 920-893-0782. Fax: 920-893-0873.
Founded: 1850.
Congressional District: 6
Key Personnel: Camp Dir., Jim Scherer.
Governing Authority: non-profit organization. Affiliated with Friends of Camp Anokitig Inc.
Institution Type/Description: Historic House: 1850 Log Cabin.
Collections: log cabin artifacts.
Activities: guided tours; wool-spinning demonstrations.
Hours & Admission Prices: By appointment only. No charge; donations accepted.

Port Edwards

ALEXANDER HOUSE, 1131 Wisconsin River Dr., Port Edwards, WI 54469-1039. Tel.: 715-887-3442. Facebook: Alexander House.
E-mail: clark@wctc.net
Web Site: www.alexanderhouseonline.org
Founded: 1990.
Key Personnel: Dir., Joe Clark; Dir., Dave Thiel; Dir., Karen Thiel; Dir., Joan Palen.
Personnel Profile: Part-Time Volunteers 18.
Governing Authority: Parent Institution: Alexander Charitable Foundation. Tax-exempt.
Institution Type/Description: Art & History Museum.
Collections: local history; Nekoosa-Edwards Paper Company artifacts; period furnishings; works by local & national artists; early lumbering; papermaking history.
Hours & Admission Prices: Tues., Thurs. & Sun. 1-4; other times by appointment. No charge. &
Attendance: 3,000 (estimated)

Port Washington

JUDGE EGHART HOUSE, 302 W. Grand Ave., Port Washington, WI 53074. Mailing Address: P.O. Box 87, Port Washington, WI 53074. Tel.: 262-284-2584.
E-mail: jjones5@esls.lib.wi.us
Institution Type/Description: Historic House Museum: built in 1872.

Collections: local & family history; period furnishings; personal artifacts; photographs.
Hours & Admission Prices: Memorial Day to Labor Day Sun. 1-4. Adults $2, children $1; discounts to families.

Portage

FORT WINNEBAGO SURGEONS QUARTERS, 1824 E. State Rd. 33, Portage, WI 53901-1466. Tel.: 608-239-1335.
Founded: 1938.
Congressional District: 2
Key Personnel: State Regent, Nancy Lesh
Governing Authority: restored & maintained by Wisconsin Society, National Society of the Daughters of the American Revolution. Tax-exempt.
Institution Type/Description: History Museum: housed in 1828 surgeons quarters.
Collections: relics of Fort Winnebago; furniture of fort period and old country school; surgical instruments; archives; military. Historic Building: 1850 Garrison School.
Research Fields: pioneer, territorial & military history of the area & its significance; national & world history; pioneer/Indian relations; significance of the Fox-Wisconsin Waterway.
Facilities: library of medical & school text books & general material.
Activities: guided tours; permanent exhibitions.
Publications: books, Fort Winnebago & The Surrender of Red Bird; Nicky's Bugle.
Hours & Admission Prices: May 15-Oct. 15 Mon.-Sat. 10-4, Sun. 11-4. Family $15, adults $6, senior citizens $5, children 7-17 $3; discounts to AAA members; FWSQ Life members no charge. Tours by appointment only.
Attendance: 3,000 (estimated)

HISTORIC INDIAN AGENCY HOUSE, 1490 Agency House Rd., Portage, WI 53901-0084. Mailing Address: P.O. Box 84, Portage, WI 53901-0084. Tel.: 608-742-6362.
E-mail: destineekae@hotmail.com
Web Site: www.agencyhouse.org
Formerly: Old Indian Agency House
Founded: 1932.
Congressional District: 2
Key Personnel: Chm., Barbara Meyer; Pres., Dr. Anne Vrarick; Dir., Destinee Udelhoven.
Personnel Profile: Full-Time Paid 1; Part-Time Paid 6.
Governing Authority: society. Parent Institution: The National Society of The Colonial Dames in the State of Wisconsin. Tax-exempt: 501(c)(3).
Institution Type/Description: Historic House: 1832 Historic Indian Agency House.
Collections: decorative arts of the American Empire period; period furnishings; Indian artifacts; Civil War memorabilia; early farm tools; objects from archaeological dig.
Research Fields: local history; archaeology of area.
Facilities: Indian crafts & other museum-related items for sale.
Activities: guided tours; nature trail; special events. Museum Sponsors: Hands-on History Days; summer speaker series.
Publications: book, Wau-Bun, The Early Day in the North-West; biannual newsletter, Waubun Express.
Hours & Admission Prices: May 15 to Oct. 15 Mon.-Sat. 10-4, Sun. 11-4; other times by appointment. Family $15, adults $6, senior citizens & AAA members $5, students 5-18 $3; discounts to groups, AAM, & National Trust for Historic Preservation members; members no charge. Nature Trails: $2; discounts with tour.
Attendance: 3,000 (estimated)
Membership: Individual $15; Family $30; Silver Patron $100; Gold Patron $250; Platinum Patron $500; Benefactor $1,000.

Potosi

POTOSI BREWING COMPANY TRANSPORTATION MUSEUM, 209 S. Main St., Potosi, WI 53820. Mailing Address: P.O. Box 177, Potosi, WI 53820. Tel.: 608-763-4002, ext. 106.
E-mail: info@potosibrewery.com
Institution Type/Description: Transportation Museum.
Collections: Potosi Brewery history; transportation & the brewing process; photographs; company memorabilia.
Hours & Admission Prices: Call for hours.

Poynette

MACKENZIE ENVIRONMENTAL EDUCATION CENTER, W7303 County Rd. CS & Q, Poynette, WI 53955-9690. Tel.: 608-635-8105. Fax: 608-635-2743.
Web Site: www.mackenziecenter.com
Founded: 1961.
Congressional District: 2
Key Personnel: Dir., Ruth Ann Lee; Pres. Friends of Mackenzie (V), Reggie Finn; Animal Keeper, Anna Lynn Hammond; Maintenance Foreman, Dan Lee.
Personnel Profile: Full-Time Paid 5; Part-Time Paid 5; Part-Time Volunteers 60.
Governing Authority: state. Parent Institution: State of Wisconsin, Dept. of Natural Resources. Tax-exempt: 170(b)(1)(A).
Institution Type/Description: Natural History Museum.
Collections: live native Wisconsin wildlife; Nelson cabin; logging history museum, pictures & tools from Wisconsin lumbering days; conservation. Historic Building: c.1880 log cabin.
Facilities: classrooms; picnic area; nature trails; trails to accommodate senior & disabled citizens; arboretum.
Activities: organized education programs for children & adults; workshops. Center Sponsors: Public Maple Fest in April.
Publications: brochures; trail guides.
Hours & Admission Prices: Exhibits & Zoo: May-Nov. 1 daily 8-4; Nov.-April Mon.-Fri. 8-4; guided tours by appointment. Grounds: daily dawn-dusk. No charge; donations accepted.
Attendance: 46,000 (estimated)
Membership: Friends of MacKenzie Environmental Center: Individual $20; Family $30; Organization $25; Life $150.

Prairie du Chien

FORT CRAWFORD MUSEUM, c/o Prairie du Chien Historical Society Inc., 717 S. Beaumont Rd., Prairie du Chien, WI 53821. Mailing Address: P.O. Box 298, Prairie du Chien, WI 53821-0298. Tel.: 608-326-6960.
E-mail: ftcrawmu@mhtc.net
Web Site: www.fortcrawfordmuseum.com
Formerly: Prairie du Chien Museum at Fort Crawford
Founded: 1996.
Congressional District: 3
Key Personnel: Pres. Historical Society (V), Mary Antoine; Sec. Historical Society, Janet Finn.
Personnel Profile: Part-Time Paid 4; Part-Time Volunteers 70.
Governing Authority: Parent Institution: Prairie du Chien Historical Society. Tax-exempt.
Institution Type/Description: History Museum: a National Historical Landmark.
Collections: military, medical & Native American artifacts; Indian treaties; Dr. William Beaumont's experiments; cultural, economics, religious, & political history; regional late 19th-century; photographs; early Crawford County Wisconsin manuscripts; dentist's office & pharmacy of the 1890s; Mississippi River's history of employment opportunities; hospital ward of Dr. Beaumont's era; bridges of Prairie du Chien; Native American artifacts; Fort Crawford. Prairie du Chien Museum: local artifacts.
Research Fields: Fort Crawford 1829-1870 & the US military on the frontier; Indian Agency for Winnebago; Civil War; Dr. William Beaumont & early medicine; Prairie du Chien & Crawford County, WI.
Facilities: theater. Museum-related items for sale.
Activities: guided tours; special events; programs; inter-museum loan.
Publications: brochures, Fort Crawford Museum & Local History Monographs; Prairie du Chien History; newsletter, The Union; book, Prairie du Chien French, British & American Monographs on Fort Crawford; special historic editions of newsletters; book, Old Fort Crawford & The Frontier.
Hours & Admission Prices: May-Oct. daily 9-4. Family $15, adults $5, seniors $4, children 12 & under $3; discounts to tour groups; members no charge.
Attendance: 6,000 (accurate)
Membership: Senior & Student $15; Individual Adult $20; Senior Couple $30; Family $35; Sustaining $100; Founder's Club $250, $500, $1,000.

VILLA LOUIS HISTORIC SITE, 521 Villa Louis Rd., Prairie du Chien, WI 53821-1333. Mailing Address: P.O. Box 65, Prairie du Chien, WI 53821-0065. Tel.: 608-326-2721. Fax: 608-326-5507.
E-mail: villalouis@wisconsinhistory.org
Web Site: villalouis.wisconsinhistory.org
Founded: 1936.
Congressional District: 3

Key Personnel: Site Dir., Susan Caya-Slusser; Facilities Mgr., Jacob Koresh; Program Asst., M. Susan Witters.

Governing Authority: state. Parent Institution: Wisconsin Historical Society, 816 State, Madison, WI 53706. Tax-exempt.

Institution Type/Description: History Museum: housed in 1870 Villa Louis, home of family of fur trader Hercules Dousman, on the site of 1814 Fort Shelby & 1816-1829 Fort Crawford.

Collections: history; fur trade manuscripts; Victorian decorative arts; furnishings. Historic Structure: 1851 general store.

Research Fields: fur trade in old Northwest; Victorian life in upper Midwest.

Facilities: 2,000-vol. library of art history books & Dousman family collection available on premises. Wisconsin history books & museum-related items for sale.

Activities: guided tours; special events.

Publications: brochure, Wisconsin Historic Sites.

Hours & Admission Prices: Tours: Winter by appointment only; Spring Wed.-Sun. 11, 1 & 3; Summer & Fall daily on the hour 10-4. Adults $10, seniors 65 and over & students $8.50, children 5-17 $5; discounts to groups; Wisconsin Historical Society members no charge. Closed New Year's Day; Easter; Christmas Eve & Day. &

Attendance: 16,000 (accurate)

Membership: Wisconsin Historical Society: Individual $45; Household $60; History Lover $100; History Guardian $250; History Ambassador $500; Heritage Circle $1,000.

Prairie du Sac

SAUK PRAIRIE AREA HISTORICAL SOCIETY, INC., 565 Water St., Prairie du Sac, WI 53578-1128. Tel.: 608-644-8444. Fax: 680-644-8444.

E-mail: spahs@frontier.com

Web Site: www.saukprairiehistory.org

Founded: 1961.

Congressional District: 2 & 3

Key Personnel: Pres., Jody Kapp; Mgr. & Correspondence Sec., Jack Berndt; Sec., Barb Wolfe; Treas., Marie Goddard.

Personnel Profile: Part-Time Volunteers 10.

Governing Authority: society. Subsidiary Institution: State Historical Society of Wisconsin, Office of Local History. Branch Museums: Firehouse Museum, 717 John Adams St.; Tripp Museum, 565 Water St., Prairie du Sac, WI 53578; Our Lady of Loretto Church Museum, Hwy. C, North Freedom, WI 53951; Salem Ragatz Church Museum, Hwy. PF, Prairie du Sac, WI 53578. Tax-exempt: 501(c)(3).

Institution Type/Description: Historical Society Museum.

Collections: household items of the period, 1850-2010.

Research Fields: local history.

Facilities: library of local history books.

Activities: meetings; Indian & Pioneer Day open to all 4th graders. Museum Sponsors: Annual Pie Contest in September; Christmas Open House.

Publications: 4X-newsletter; book, Pictorial History of Sauk City & Prairie du Sac; Fire Department History book; Historic Houses of Prairie du Sac; Book on History of Milk Hauling & Creameries of the Past; Historical Book on Prairie du Sac Businesses; Lives Lived Here: History of Sauk City.

Hours & Admission Prices: Fri.-Sat. 9-1; other times by appointment. No charge; donations accepted.

Attendance: 3,500 (estimated)

Membership: Individual $25; Family $50; Friend $75; Sponsor $100; Patron $250; Benefactor $500; Historian $1,000 & up.

Racine

FIREHOUSE NO. 3 MUSEUM, 700 Sixth St., Racine, WI 53403. Mailing Address: P.O. Box 081042, Racine, WI 53403. Tel.: 414-637-7395.

Key Personnel: Dir., Steve Hansen

Institution Type/Description: Firefighting History Museum: housed in the former fire station; built in 1882.

Collections: firefighting history & equipment; 1882 Stephen Freemen steamer; hydrants; nozzles; helmets; hand-drawn hose cart; trophies; awards.

Activities: group tours.

Hours & Admission Prices: By appointment.

∗ RACINE ART MUSEUM (RAM), (M), 441 Main St., Racine, WI 53403-1030. Mailing Address: P.O. Box 187, Racine, WI 53401-0187. Tel.: 262-638-8300. Fax: 262-898-1045.

E-mail: raminfo@ramart.org

Web Site: www.ramart.org

Formerly: Charles A Wustum Museum of Fine Arts

Founded: 1941.

Congressional District: 1

Key Personnel: C.E.O., Bruce W. Pepich; Pres. (V), Susan Boland; Facilities Mgr., Jim Sheppard; Controller, Barb Namowicz; Dir. Devel., Laura D'Amato; Cur. Education, Tricia Blasko; Principal Guest Experience & Retail Division, Lisa Englander; Devel. Coord., Susan K. Buhler-Maki; Bldg. Svcs. Asst., John Coley; Curatorial Asst., David Zaleski; Education Asst., Maureen Fritcher; Education Asst., Susan Silver; Registrar, Elizabeth Frozena; Mktg. & Publications Mgr., Jessica Z. Schafer; Mktg. Asst., Laura Gillespie; Librarian, Nancy Elsmo; Exhibition Preparator, Janelle Cairo; Exhibition Preparator II, Joseph Church; Vol. Coord., Angie Glenn.

Personnel Profile: Full-Time Paid 14; Part-Time Paid 44; Part-Time Volunteers 150; Interns 3.

Governing Authority: nonprofit organization. Parent Institution: The Racine Art Museum Assoc., Inc.; Branch Museum: Charles A. Wustum Museum of Fine Arts, 2519 Northwestern Ave., Racine, WI 53404-2299. Tel. 262-636-9177. Tax-exempt: 501(c)(3).

Institution Type/Description: Arts Center.

Collections: contemporary American works on paper; 20th-century American crafts.

Major Exhibits: Martha Glowacki in RAM's Windows on 5th Gallery, 8/13-7/14; Collection Focus: Ken Lieber, 9/13-1/14; Watercolor Wisconsin 2013, 12/13-4/14.

Research Fields: WPA art; regional, national & international artists; contemporary American crafts.

Facilities: 3,500-vol. library of art books and periodicals available for use upon request; classrooms; studios. Museum-related items for sale.

Activities: guided tours; lectures; gallery talks; art sales & rental gallery; study clubs; hobby workshops; formally organized educational programs; permanent, temporary & traveling exhibitions.

Publications: members newsletter; exhibition brochures & catalogues.

Hours & Admission Prices: RAM: Tues.-Sat. 10-5, Sun. 12-5. Wustum: Tues.-Sat. 10-5. Adults $5; discounts to NARM members; members no charge. NARM reciprocal membership over $100. Closed federal holidays; Easter. &

Attendance: 51,994 (accurate)

Membership: National Associate $30; Senior Citizen & Student $40; Individual $50; Family $70; Sustaining $125; Supporting $250; Benefactor $500; Master $1,000; Corporate Partner $1,500; RAM Society $1,000-$10,000.

RACINE HERITAGE MUSEUM, (M), 701 S. Main St., Racine, WI 53403-1211. Tel.: 262-636-3926. Fax: 262-636-3940.

E-mail: inquire@racineheritagemuseum.org

Web Site: www.racineheritagemuseum.org

Founded: 1960.

Congressional District: 1

Key Personnel: Exec. Dir., Christopher Paulson; Cur. & Asst. Dir., Karen Braun; Archivist, Mary Kay Nelson; Dir. Mktg. & Programs, Sally Orth; Cur. Education, Patty Wagner; Educator, Cheryl Maraccini.

Personnel Profile: Full-Time Paid 2; Part-Time Paid 9; Part-Time Volunteers 139.

Governing Authority: private; nonprofit organization. Tax-exempt: 501(c)(3).

Institution Type/Description: Southeast Wisconsin Industrial, Cultural, Invention & Product History: housed in 1904 Carnegie Library building.

Collections: 1840-present county, social & industrial history; archives; photographs. Historic Building: 1888 Bohemian schoolhouse.

Research Fields: county social & industrial history; genealogy.

Facilities: archives.

Activities: permanent, temporary & traveling exhibitions; school loan service; regularly scheduled classes for school groups in 1888 Bohemian schoolhouse.

Publications: quarterly journal, The Outlook; Boat Manufacturing In Racine; book, Invention City: The Sesquicentennial History of Racine, Wisconsin; historical brochures.

Hours & Admission Prices: Tues.-Fri. 9-5, Sat. 10-3, Sun. 12-4. No charge; donations accepted. Closed national holidays.

Attendance: 30,000 (accurate)

Membership: Student $30; Senior $30; Individual $40; Senior Couple $50; Family $60; Supporting $100; Contributing $250; Sustaining $500; Benefactor $1,000.

RACINE ZOOLOGICAL SOCIETY, 2131 N. Main St., Racine, WI 53402-4795. Mailing Address: 200 Goold St., Racine, WI 53402-4795. Tel.: 262-636-9189. Fax: 262-636-9307.

E-mail: info@racinezoo.org

Web Site: www.racinezoo.org

Founded: 1923.

Congressional District: 1

Key Personnel: Pres. & C.E.O., Jay R. Christie; Chm. Bd., Chris Eperjesy.

Personnel Profile: Full-Time Paid 19; Part-Time Paid 2; Part-Time Volunteers 185; Interns 11.
Governing Authority: municipal. Parent Institution: Racine Zoological Society. Tax-exempt.
Institution Type/Description: Zoo.
Collections: 403 specimens; 85 species.
Activities: education programs; tours.
Publications: quarterly newsletter, Racine Zoocine.
Hours & Admission Prices: Memorial Day-Labor Day daily 9-7; Sept.-May daily 9-4:30. Adults $6.50, seniors $5.25, children $4; discounts to Association of Zoo & Aquarium members; reciprocating AZA institutions; members no charge. &
Attendance: 100,000 (estimated)
Membership: One Plus One Zoo Pass & Zoologist $50; Family Zoo Pass $60; Family Plus Zoo Pass & Curator $100; Conservationist $500; Safari Club $1,000.

THE SOUTHEAST WISCONSIN AVIATION MUSEUM, EAA Chapter 838 Batten International Airport, 3333 N. Green Bay Rd., Racine, WI 53404. Tel.: 262-634-7575.
Web Site: www.eaa838.org/museum.asp
Institution Type/Description: Aviation Museum.
Collections: local aviation history; aircraft.
Hours & Admission Prices: May-Oct. 2nd Sat. each month 8:30am-12pm; other times by appointment.

Rhinelander

PIONEER PARK HISTORICAL COMPLEX, Pioneer Park-Martin Lynch Dr., Rhinelander, WI 54501. Mailing Address: P.O. Box 1304, Rhinelander, WI 54501. Tel.: 715-369-5004.
Formerly: Rhinelander Logging Museum
Founded: 1932.
Congressional District: 7
Governing Authority: municipal. Tax-exempt.
Institution Type/Description: Pioneer Logging Industry Museum.
Collections: logging history & artifacts; replica of 1870s lumber camp; printed materials; photographs; miniature saw mill; early fire equipment; narrow gauge railroad, engine & rail cars; HO model railroad; one-room schoolhouse; Civilian Conservation Corps (CCC) barracks. Historic Building: 1892 SOO Line restored depot.
Research Fields: restored 1892 Soo Line Depot with HO scale model railroad.
Facilities: blacksmith shop; schoolhouse; bunk house dining room; Civilian Conservation Corps Building (CCC); antiques fire engines in barn. Gift items for sale.
Activities: self-guided tours.
Publications: tour guides; brochures.
Hours & Admission Prices: Memorial Day to Labor Day Tues.-Sun. 10-5; Sept. Fri.-Sat. 10-5. No charge; donations accepted. &
Attendance: 14,378 (accurate)

Ripon

LITTLE WHITE SCHOOLHOUSE, 303 Blackburn St., Ripon, WI 54971-1524. Mailing Address: P.O. Box 305, Ripon, WI 54971-0305. Tel.: 920-748-6764. Fax: 920-748-6784.
E-mail: chamber@ripon-wi.com
Web Site: www.littlewhiteschoolhouse.org
Founded: 1951.
Congressional District: 6
Key Personnel: Exec. Dir., Paula T. Price; Pres. (V), Tom Moniz; Mktg. & Events Coord., Jason Mansmith.
Personnel Profile: Part-Time Paid 4.
Governing Authority: Parent Institution: Ripon Area Chamber of Commerce. Tax-exempt.
Institution Type/Description: Historic Building: c.1854 Little White Schoolhouse, birthplace of Republican Party.
Collections: Republican presidential memorabilia; political history; pioneer tools; furnished typical of 1800s schoolhouse. 1850 educational materials.
Facilities: Museum-related items for sale.
Activities: tours; docent presentations.
Publications: brochure, Little White Schoolhouse; booklet, A History of Ripon's Little White Schoolhouse, Birthplace of the Republican Party 1853-2005.
Hours & Admission Prices: May & Sept.-Oct. Sat.-Sun. 10-4; June to Labor Day daily 10-4; other times by appointment. Adults $2. &
Attendance: 3,500 (accurate)

River Falls

GALLERY 101, University of Wisconsin-River Falls, Fine Arts, 410 S. Third St., River Falls, WI 54022-5010. Tel.: 715-425-3266. Fax: 715-425-0657.
E-mail: susan.m.zimmeer@uwrf.edu
Web Site: www.uwrf.edu/art
Founded: 1973.
Key Personnel: Chm. Art Dept., Randy Johnson.
Governing Authority: Branch of University of Wisconsin. Tax-exempt.
Institution Type/Description: Art Gallery.
Collections: W.P.A. graphics; contemporary prints; regional artists.
Activities: gallery talks.
Hours & Admission Prices: Sept.-May Mon.-Fri. 9-5 & 7-9, Sun. 2-4. No charge. &

Saint Croix Falls

ST. CROIX NATIONAL SCENIC RIVERWAY, 401 N. Hamilton St., Saint Croix Falls, WI 54024-9214. Tel.: 715-483-2274. Fax: 715-483-3288.
E-mail: chris_stein@nps.gov
Web Site: www.nps.gov/sacn
Founded: 1968.
Congressional District: 3
Key Personnel: Supt., Chris Stein.
Governing Authority: federal. National Park Service, Dept. of the Interior, Washington DC. Tax-exempt.
Institution Type/Description: National Park & Museum.
Collections: aquarium; natural history concepts (riparian, floodplain forest & watershed).
Research Fields: history & natural history of the St. Croix & Namekagon Rivers.
Facilities: reference library. Books & museum-related items for sale.
Activities: group interpretive programs.
Hours & Admission Prices: mid-April to late Oct. daily 9-5. No charge. &
Attendance: 17,700 (accurate)

Saint Germain

SNOWMOBILE HALL OF FAME AND MUSEUM, 8481 W. Hwy. 70, Saint Germain, WI 54558. Mailing Address: P.O. Box 720, Saint Germain, WI 54558-0720. Tel.: 715-542-4463. Fax: 715-542-4260.
E-mail: info@snowmobilehalloffame.com
Web Site: www.snowmobilehalloffame.com
Institution Type/Description: History Museum.
Collections: race sleds; trophies; clothing; photographs; racing videos.
Facilities: theater. Museum-related items for sale.
Activities: special events.
Hours & Admission Prices: Off Season: Thurs.-Fri. 10-5, Sat. 10-3; Winter: call for hours.

Saukville

OZAUKEE COUNTY PIONEER VILLAGE, 4880 County Hwy. I, Saukville, WI 53080. Mailing Address: P.O. Box 206, Cedarburg, WI 53012-0206. Tel.: 262-377-4510. Fax: 262-377-4510.
E-mail: jean.steinke@gmail.com
Web Site: www.co.ozaukee.wi.us/ochs
Formerly: Ozaukee County Historical Society Pioneer Village
Founded: 1960.
Congressional District: 9
Key Personnel: Pres. (V), Jean Steinke; Volunteer Coord., Tom Oliver; 1st Vice Pres., Curt Gruenwald; 2nd Vice Pres., Allen Buchholz; Archivist, Dr. Nina Look; Sec., Trevor Weis; Treas., Tom Hogan.
Personnel Profile: Part-Time Paid 6; Part-Time Volunteers 100.
Governing Authority: society. Affiliated with State Historical Society of Wisconsin. Parent Institution: Ozaukee Co. Historical Society. Tax-exempt: 501(c)(3).
Institution Type/Description: Pioneer Village: over 20 buildings ranging from mid-1840 to 1907.
Collections: pioneer history; folklore; archives; agriculture; household equipment; photographs; railroad artifacts & memorabilia; 18th-century buildings & structures; newspapers; archives; tax records; school records; church histories; Ozaukee County history books. Historical Building: original one-room Stony Hill School, birthplace of National Flag Day.
Research Fields: genealogy; local history.
Facilities: library.

Activities: guided tours; demonstrations; school tour program; quarterly programs; special events; research. Annual Events: National Flag Day ceremony & celebration.
Publications: quarterly newsletter, TimeLines.
Hours & Admission Prices: Memorial Day to 2nd Sun. in Oct. Sat.-Sun. 12-5; other times groups by appointment. Family $16, adults $6, senior citizens & students 12-18 $4, children 6-12 $3; children 5 & under and members no charge. Additional charge for special events. &
Attendance: 4,500 (accurate)
Membership: Individual $15; Family $30; Business & Professional $50; Life: Individual $200, Couple $350.

SAUKVILLE AREA HISTORICAL SOCIETY - SAUKVILLE CROSSROADS MUSEUM, 200 N. Mill St., Saukville, WI 53080. Tel.: 262-692-9425.
Institution Type/Description: History Museum: housed in the former Saukville Firehouse Station; built in 1912.
Collections: local history & culture; period furnishings; early fire fighting equipment; photographs; personal artifacts.
Activities: special events.
Hours & Admission Prices: Call for hours.

Seymour

SEYMOUR COMMUNITY MUSEUM, 133 Depot St., Seymour, WI 54165. Mailing Address: P.O. Box 237, Seymour, WI 54165-1331. Tel.: 920-833-2868.
E-mail: pma@billcollar.com
Web Site: www.seymourhistory.org
Founded: 1976.
Congressional District: 8
Key Personnel: Pres. (V), Bill Collar; Vice Pres., Lois Dalke; Historian, Marge Coenen; Business Officer, Janice Eick; Asst., Mike Keyzers.
Personnel Profile: Part-Time Volunteers 12.
Governing Authority: municipal; nonprofit organization. Affiliated with the Wisconsin State Historical Society. Tax-exempt.
Institution Type/Description: Historic Building & Site: housed in 1879 Lumber Sales Building.
Collections: local history items; hamburger history & artifacts. Historic Building: 1914 Railroad Depot.
Research Fields: local history.
Facilities: 8,880 sq. ft. facility.
Activities: lectures; art shows; fashion show. Museum Sponsors: Music in the Park June-August.
Publications: quarterly, Museum Muse; quarterly bulletin, Seymour History.
Hours & Admission Prices: Summer: Tues.-Sat. No charge; donations accepted. &
Attendance: 3,512 (accurate)
Membership: Single $5; Family $10; Lifetime $50.

Shawano

SHAWANO COUNTY HISTORICAL SOCIETY, INC., 524 N. Franklin St., Shawano, WI 54166-1933. Tel.: 715-526-3323.
E-mail: schsociety@granitewave.com
Founded: 1940.
Congressional District: 8
Key Personnel: Pres., Ron Schumacher.
Personnel Profile: Part-Time Volunteers 25.
Governing Authority: society. Branch Museum: Railroad Station Museum, Gresham, WI 54128. Tax-exempt.
Institution Type/Description: General Museum: located in Heritage Park on the site where the first white man in Shawand County settled in 1848.
Collections: furniture; lamps; kitchen utensils; children's nursery; clothing; glassware and dishes; newspapers; books; 1895 one room country school; school-room artifacts; Indian artifacts; early judges stand; log cabin restoration; dairy exhibit; cheese-making display; replica of Zachow, WI depot.
Activities: guided tours; permanent exhibitions. Society Sponsors: meetings for members and friends in April & Oct.
Hours & Admission Prices: Mon.-Thurs 9-4, call to confirm. Tours: Thurs. 1:30-4, Sat. 9:30 am to noon; other times by appointment. Adults $3.
Attendance: 1,000 (estimated)
Membership: Active $20; Sustaining $30; Life $1,000.

Sheboygan

ABOVE & BEYOND CHILDREN'S MUSEUM, 902 N. 8th St., Sheboygan, WI 53081-4005. Tel.: 920-458-4263. Fax: 920-458-3402. Facebook: Above & Beyond Children's Museum.
E-mail: abcm@abkids.org
Web Site: www.abkids.org
Founded: 1992.
Key Personnel: Exec. Dir., Jeff Mehn.
Personnel Profile: Full-Time Paid 1; Part-Time Paid 8; Part-Time Volunteers 50; Interns 1.
Governing Authority: Tax-exempt.
Institution Type/Description: Children's Museum.
Collections: hands-on interactive exhibits including an historic Port of Sheboygan, a fire house, a tree house, a grocery store, & a doctor's office.
Facilities: 10,000 sq. ft. exhibit space.
Activities: birthday parties; school groups; educational programs; special events; fundraisers; Hands-On Family Fun.
Hours & Admission Prices: Tues.-Fri. & Sun. 10-6, Sat. 10-7. Admission $6; members & children one & under no charge.
Attendance: 30,000 (estimated)
Membership: Grandparent $65; Family $75; Family Plus $95; Premier $130.

GREAT LAKES AEROSPACE SCIENCE AND EDUCATION CENTER, 516 Broughton Dr., Sheboygan, WI 53081. Mailing Address: P.O. Box 904, Sheboygan, WI 53082-0904. Tel.: 920-889-7148.
E-mail: danielb@spaceportsheboygan.com
Web Site: www.spaceportsheboygan.com
Institution Type/Description: Science Center.
Collections: space science & history; hands-on exhibitions.
Facilities: planetarium theater; labs.
Activities: simulators; special events; educational programs; summer camp.
Hours & Admission Prices: June 9-Aug. 10 Tues.-Sat. 11-5. Planetarium Shows: daily 1 & 3. Adults 12 & over $5, children 5-12 $2; children under 5 no charge.

JOHN MICHAEL KOHLER ARTS CENTER, (M), 608 New York Ave., Sheboygan, WI 53081-4507. Tel.: 920-458-6144. Fax: 920-458-4473. Facebook: JMKAC.
E-mail: rkohler@jmkac.org
Web Site: www.jmkac.org
Founded: 1967.
Congressional District: 9
Key Personnel: Dir., Ruth DeYoung Kohler; Pres. Bd., Michael Cisler; Pres., Friends of Art, Mona Strean; Deputy Dir. Operations, Patti Sherman-Cisler; Cur., Alison Ferris; Chief Devel. Officer, Cynthia Echols; Registrar, Larry Donoval; Librarian & Archivist, Andrew Hunt; Human Resources Mgr., Anne Stauber Tritz; Arts Industry Coord., Kristin Plucar; Controller, Kelley Renzelmann; Museum Shop Mgr., Mary Kopp; Deputy Dir. Programming, Amy Horst; Community Affairs Coord., Andrea Avery; Senior Mgr. Public Programs, Ann Brusky; Education Program Mgr., Cate Bayles; Asst. Cur., Karen Patterson; Lead Registrar, Stephen Gorman.
Personnel Profile: Full-Time Paid 50; Part-Time Paid 20; Part-Time Volunteers 450; Interns 15.
Governing Authority: nonprofit organization. Affiliated with John Michael Kohler Arts Center, Inc. Tax-exempt: 501(c)(3).
Institution Type/Description: Visual & Performing Arts Center.
Collections: works by self-taught, folk, & vernacular artists; contemporary American art; vernacular art environments; crafts; arts industry; American furniture & artifacts from the J.M. Kohler home. Historic House: 1882 John Michael Kohler Home.
Major Exhibits: Joseph Yoakum: Unfolding Landscape, 9/13-2/2/14; Ray Yoshida's Museum of Extraordinary Values, 9/22/13-2/23/14; John Shimon & Julie Lindemann: We Go From Where We Know, 10/6/13-2/9/14; AAIEEE!, 11/13-2/14; Alter Egos, 11/13-3/14; Arts/Industry 40th Anniversary, 2/14-1/15.
Research Fields: vernacular environment builders; folk art; self-taught artists; contemporary American art; craft; photography.
Facilities: indoor & outdoor performance spaces; 5 studio-classrooms; a drop-in art making space; resource center of arts books & periodicals; meeting rooms; theater. Museum-related items for sale.
Activities: Up to 20 temporary and traveling exhibitions curated by Arts Center staff; collections exhibitions; Arts/Industry annual residency program for 8-16 artists from around the world with 2-6 month residencies at a nearby industrial pottery & iron foundry; Connecting Communities long-term residency program for visual, performing, literary, and media artists collaborating on major works with underserved constituencies and public; Ex Air 1-2 week residency program for exhibition artists working with

schools; Footlights performing arts residency series of 5 dance, music, and interdisciplinary ensembles annually at JMKAC and in schools; 3 other performing arts series; 3 annual festivals based on area cultures; Midsummer Festival of Arts; July 4 Art Armada boat making/racing inspired by a current exhibition; cinema series; docent tours; gallery talks; lectures; 2-year arts-infused licensed preschool; broad range of other education programs, camps, and classes with many based on 2-3 annual exhibition per year; immersive Teacher Training Institutes.

Publications: exhibitions catalogs; bimonthly newsletter; annual report;

Hours & Admission Prices: Mon., Wed. & Fri. 10-5, Tues. & Thurs. 10-8, Sat.-Sun. 10-4. No charge; donations accepted. &

Attendance: 200,000 (estimated)

Membership: Full-time Student $30; Senior Citizen Individual $40; Individual $48; Senior Citizen Dual $55; Family/Dual $60; Donor $100-$249; Supporter $250-$499; Sustainer $500-$999; Benefactor $1,000-$2,499; Silver Benefactor $2,500-$4,999; Gold Benefactor $5,000-$9,999; Platinum Benefactor $10,000 & up.

SHEBOYGAN COUNTY HISTORICAL MUSEUM, (M), 3110 Erie Ave., Sheboygan, WI 53081-3660. Tel.: 920-458-1103. Facebook: Sheboygan County Historical Museum.

E-mail: nancy.koeppen@sheboygancounty.com

Web Site: www.sheboygancounty.com/government/ departments-f-q/historical-museum

Founded: 1954.

Congressional District: 9

Key Personnel: Exec. Dir. & C.E.O., Travis Gross; Pres. (V), Mary Novak.

Personnel Profile: Full-Time Paid 2; Part-Time Paid 6; Interns 3.

Operating Expenses: 205,877

Operating Income: 235,865

Governing Authority: society. Parent Institutions: Sheboygan County Historical Society. Tax-exempt.

Institution Type/Description: History Museum.

Collections: artifacts of local history & local circus history; farm implements; household articles; medical aids. Historic House: Weinhold family homestead; 1864 log house; 1867 cheese factory; Schuchardt barn; Taylor House.

Major Exhibits: Sheboygan County Serves, 4/14-10/14; Holiday Memories 2014, 12/14.

Research Fields: local history.

Activities: guided tours; permanent & temporary exhibitions.

Publications: quarterly newsletters; magazine; annual report; books: When Then Was Now; The Branded Hand-Struggles of an Abolitionist; Historic Sheboygan County; And That's The Way It Was; The Promise of Prosperity-The Story of Greenbush Wisconsin & its Waterpowered Sawmill; pamphlet, Brick House On A Hill; pamphlet, Touring Historical Sheboygan County; pamphlet, Reflections of a Sheboygan County Farmwife; pamphlet, Sheboygan Indian Mound Park; pamphlet, Baseball in Sheboygan 1886-1986; 1868-69 Sheboygan City Directory; Cheese Factories of Sheboygan County.

Hours & Admission Prices: April-Oct. Mon.-Fri. 10-5; Nov. 28-Dec. 30 daily 12-5. Adults $4, children $2; members no charge. Closed Memorial Day; Independence Day; Labor Day; Christmas Eve & Day. &

Attendance: 12,000 (estimated)

Membership: Individual $25; Family $35; Supporting $75; Sustaining $150; Sponsor & Corporate $300; Patron $500; Benefactor $1,000.

Sheboygan Falls

AVIATION HERITAGE CENTER OF WISCONSIN, N6191 Resource Dr., Sheboygan Falls, WI 53085. Tel.: 920-467-2043.

E-mail: takeflight@ahcw.org

Web Site: www.ahcw.org

Founded: 2005.

Key Personnel: Dir., Jon Helminiak; Pres. (V), Greg Cayon.

Personnel Profile: Part-Time Paid 3; Part-Time Volunteers 3.

Governing Authority: Tax-exempt.

Institution Type/Description: Aviation History Museum.

Collections: aviation history; aircraft; aircraft hangar.

Facilities: library; laboratory.

Activities: special events; educational programs.

Publications: annual newsletter.

Hours & Admission Prices: April-Oct. Mon.-Fri. 10-5, Sat. 9-1; Nov.-March Thurs.-Fri. 10-5, Sat. 9-1; other times by appointment. No charge; donations accepted. &

Attendance: 6,000 (estimated)

Shell Lake

MUSEUM OF WOODCARVING, 539 Hwy. 63, Shell Lake, WI 54871-4438. Mailing Address: P.O. Box 371, Shell Lake, WI 54871-0371. Tel.: 715-468-7100.

Web Site: www.roadsideamerica.com/attract/WISHEwood.html

Founded: 1950.

Congressional District: 7

Key Personnel: Owner, Cur. & Museum Shop Mgr., Maria McKay.

Personnel Profile: Part-Time Paid 2.

Governing Authority: private.

Institution Type/Description: Woodcarving Museum.

Collections: over 100 life-sized carvings including one of The Last Supper; over 400 miniatures of animals, birds & historical events; Joseph Barta's tools.

Facilities: library; 10,000 sq. ft. exhibit space. Museum-related items for sale.

Activities: guided tours; lectures.

Hours & Admission Prices: May-Oct. daily 9-6. Adults $6.50, children under 12 $4.50; discounts to groups of 20 or more, AARP, AAA & AAM members. &

Attendance: 25,000 (estimated)

WASHBURN COUNTY HISTORICAL SOCIETY MUSEUM, 102 W. 2nd Ave., Shell Lake, WI 54871. Mailing Address: P.O. Box 366, Shell Lake, WI 54871-0366. Tel.: 715-468-2982.

Founded: 1954.

Congressional District: 7

Key Personnel: Pres. (V), Cathy Wahlstrom.

Personnel Profile: Part-Time Volunteers 25.

Governing Authority: county. Affiliated with the State Historical Society of Wisconsin, 816 State St., Madison, WI 53706. Additional Location: Hwy. 63, Springbrook. Tax-exempt.

Institution Type/Description: Historic Society Museum: housed in three buildings.

Collections: logging tools from county logging era; photographs; crafts; costumes; railroad & farming displays; household period artifacts; manuscripts; church period artifacts; old photographs & school books; furniture; items from the Civil War & World War I; bottles; 1888 clothing; music; button shoes; 70-year-old beaded Indian leather dress; teacher's desk with old schoolbooks; extensive Indian arrowhead display. Historic Buildings: 2 churches; one-room schoolhouse; sky watch building.

Research Fields: local memorabilia; vital statistics to 1940.

Facilities: 400-vol. library of mainly Wisconsin history material with specialties in Washburn County available for research on premises. History books & pamphlets for sale.

Activities: guided tours; permanent & temporary exhibitions.

Publications: books, Spooner 7 Vols.; Shell Lake Trego; Springbrook; Historical Collections of Washburn County; reprint booklet, 1915 Atlas of Washburn County; calendar, 1983 Historical Calendar; 4th Historical Collection out for sale; reprint, Stouffers' History of Shell Lake.

Hours & Admission Prices: Memorial Day-Labor Day Fri.-Sat. 11-4. No charge; donations accepted. &

Attendance: 500 (estimated)

Membership: Individual $7; Couple $12; Family $15.

Shullsburg

BADGER MINE AND MUSEUM, 279 W. Estey St., Shullsburg, WI 53586. Mailing Address: P.O. Box 580, Shullsburg, WI 53586. Tel.: 608-965-4860.

Founded: 1964.

Congressional District: 2

Personnel Profile: Full-Time Paid 1; Part-Time Paid 4.

Governing Authority: Parent Institution: Shullsburg Community Development Corporation.

Institution Type/Description: General Museum.

Collections: early mining tools; farm equipment; Indian artifacts; toys; country store items; photographs. Historic Structure: 1827 lead mine.

Activities: guided tours; 1/4-mile mining tour; lectures.

Publications: brochure, Hidden Valley.

Hours & Admission Prices: Memorial Day-Labor Day Wed.-Thurs. 12-4, Fri.-Sun. 11-4. Museum: adults $3, seniors over 65 $2, children under 10 $1.50. Mine & Museum: adults $5, seniors over 65 $4, children under 10 $3. &

Attendance: 2,300 (accurate)

Membership: Lifetime Single $25; Lifetime Couple $45.

South Milwaukee

SOUTH MILWAUKEE HISTORICAL SOCIETY MUSEUM, 717 Milwaukee Ave., South Milwaukee, WI 53172-2113. Tel.: 414-762-5214.
Web Site: www.southmilwaukee.org
Founded: 1972.
Congressional District: 4
Key Personnel: C.E.O. & Pres., Robert M. Pfeiffer; Vice Pres., Lois L. Schreiter; Sec., Sue Ziarek; Treas., Richard Raatz.
Personnel Profile: Part-Time Volunteers 10.
Governing Authority: State of Wisconsin. Parent Institution: State Historical Society. Tax-exempt.
Institution Type/Description: Historical Society Museum: housed in Victorian home.
Collections: pictures; genealogies; quilts; paintings; clothing; Lincoln Library; artifacts relating to South Milwaukee.
Research Fields: local history.
Facilities: library.
Activities: guided tours; slide program; talks to groups or school classes; temporary exhibitions.
Publications: quarterly, South Milwaukee Historical Society Newsletter; Images of America Series - South Milwaukee; South Milwaukee Then To Now.
Hours & Admission Prices: Memorial Day-Labor Day first Thurs. of the month 1-3; other times by appointment. No charge; donations accepted.
Attendance: 600 (estimated)
Membership: Senior Citizen $6; Individual $10; Family $15; Business & Sustaining $25; Life $100.

Sparta

THE DEKE SLAYTON MEMORIAL SPACE & BICYCLE MUSEUM, (M), 200 W. Main St., Sparta, WI 54656. Tel.: 608-269-0033.
E-mail: dekeslayton@centurytel.net
Web Site: www.dekeslaytonmuseum.com
Institution Type/Description: Transportation Museum.
Collections: transportation history bicycles to space flight; Wright Brothers' history & artifacts; aviation history; Donald "Deke" Slayton memorial & personal artifacts; photographs.
Activities: special events.
Hours & Admission Prices: Mon.-Sat. 10-4:30

MONROE COUNTY LOCAL HISTORY ROOM & LIBRARY, (M), 200 W. Main St., Sparta, WI 54656-2141. Tel.: 608-269-8680. Fax: 608-269-8921.
E-mail: mclhr@centurytel.net
Web Site: www.mclhr.org
Founded: 1976.
Congressional District: 3
Key Personnel: Chm. Bd. Trustees, Carolyn Habelman; Dir., Jarrod Roll.
Personnel Profile: Full-Time Paid 1; Part-Time Paid 1; Part-Time Volunteers 25.
Governing Authority: county government; nonprofit. Tax-exempt.
Institution Type/Description: Historical Society Museum: located in the former Masonic Temple, Monroe County, Sparta, WI.
Collections: emphasis on genealogy of Monroe County people, 1850-present.
Facilities: 400-vol. library.
Hours & Admission Prices: Mon.-Fri. 9-4:30, Sat. 10-4:30. No charge; donations accepted. Closed Memorial Day; Independence Day; Labor Day; Thanksgiving; Christmas. &
Attendance: 8,500 (estimated)

Spooner

WISCONSIN CANOE HERITAGE MUSEUM, (M), 312 N. Front St., Spooner, WI 54801. Mailing Address: P.O. Box 365, Spooner, WI 54801. Tel.: 715-635-5002.
E-mail: info@wisconsincanoeheritagemuseum.com
Web Site: www.wisconsincanoeheritagemuseum.com
Founded: 2010.
Key Personnel: Exec. Dir., Jed Malischke; Cur., Mike Johnson
Governing Authority: nonprofit organization. Tax-exempt.
Institution Type/Description: Heritage Museum: housed in the former Baker Grain Elevator company building; built c.1912.
Collections: cultural heritage of canoes & canoeing in North America; canoes; canoe-related artifacts; photographs.
Facilities: exhibit hall; canoe shop.
Activities: canoe construction & restoration instruction; special events.
Publications: quarterly newsletter, Canoe Current.
Hours & Admission Prices: Summer: Wed.-Sat. 10-4, Sun. 11-3; Sept. Sat.-Sun. 11-3; other times by appointment. Suggested Donations: adults $4, youth 13-18 $2; children 12 & under no charge. &
Attendance: 1,500 (estimated)
Membership: Individual $30; Family $40; Voyageur $100.

Spring Green

THE HOUSE ON THE ROCK, 5754 State Rd. 23, Spring Green, WI 53588-8912. Tel.: 608-935-3639. Fax: 608-935-9472.
E-mail: information@thehouseontherock.com
Web Site: www.thehouseontherock.com
Founded: 1961.
Key Personnel: Owner, Art Donaldson; Pres., Susan Donaldson; Asst. Mgr., Paula Widdish; Mktg. Mgr., Betty Smith.
Governing Authority: privately owned.
Institution Type/Description: Historical House Museum & Complex: original structure rests on a 60 foot chimney of rock jutting high above Wyoming Valley of Southwestern Wisconsin; multi-building complex on different levels & various outbuildings.
Collections: gardens including 285 types of plants; cannons; paintings; orchestrions & music machines; sculpture; stained glass; bronzes; ceramics; ivory & porcelain; guns; mechanical banks; steam engines; scrimshaw; taxidermy; bisque dolls; wooden dollhouses; suits of armor; carousel animals; theatre organ consoles; period artifacts; memorabilia; circus costumes; miniature circuses; entomology.
Facilities: concessions. Museum-related items for sale.
Activities: walking ramp over treetops.
Publications: brochures; guide book; pictorial book; Alex Jordan biography; video & audio tapes.
Hours & Admission Prices: May-Aug. daily 9-6; Sept.-April daily 9-5. One Tour: adults $12.50, children $7.50. Additional tour packages available. &
Attendance: 500,000

Stevens Point

CENTRAL WISCONSIN CHILDREN'S MUSEUM, 1100 Main St., Stevens Point, WI 54481. Tel.: 715-344-2003.
E-mail: cwcm@cwchildrensmuseum.org
Web Site: cwchildrensmuseum.org
Founded: 1994.
Key Personnel: Dir., Katy Matthai.
Personnel Profile: Full-Time Paid 1; Part-Time Paid 9; Part-Time Volunteers 20; Interns 2.
Institution Type/Description: Children's Museum.
Collections: hands-on exhibits.
Hours & Admission Prices: Tues.-Wed. & Fri. 9-4, Thurs. 9-8, Sat. 10-4, Sun. 12-4. Adults $5; children under one & members no charge. &
Attendance: 21,000 (accurate)
Membership: Grandparent $50; Family $75; Enhanced $125.

PORTAGE COUNTY HISTORICAL SOCIETY, 1475 Water St., Stevens Point, WI 54481-2920. Mailing Address: P.O. Box 672, Stevens Point, WI 54481-0672.
Founded: 1952.
Congressional District: 3
Key Personnel: Pres., Tim Siebert; Vice Pres., Mark Seiler; Treas., Jeanne Regnier; Sec., Karen J. Zinda; Volunteer Coord., Julie Richards.
Personnel Profile: Part-Time Volunteers 45.
Governing Authority: private; nonprofit organization. Village located at Washington Ave., Plover, WI 54467; Rising Star Mill, Hwy. 161, Nelsonville, WI 54458; Synagogue Museum, 1475 Water St., Stevens Point, WI; Firehouse #2, 1949 Strongs Ave., Stevens Point, WI. Tax-exempt: 501(c)(3).
Institution Type/Description: History Museum.
Collections: area domestic life; military; photographs; local Jewish heritage; fire fighting apparatus; horse drawn farm equipment.
Major Exhibits: The Local Circus, 6/14-9/14.
Facilities: 200-vol. library of local history books.
Activities: guided tours; concerts; art show; temporary exhibitions. Museum Sponsors: Heritage Days; Civil War Re-enactment.
Publications: quarterly newsletter, Portage County Historical Society Newsletter; local history books & booklets.
Hours & Admission Prices: Memorial Day Weekend-Labor Day Weekend Sat.-Sun. 1-4. Admission $2. No charge; donations accepted.
Attendance: 3,000 (estimated)
Membership: Student $15; Senior $20; Individual $25; Family $40; Business $50; Pioneer $100; Patron $200; Sustaining $300; Benefactor $400; Life $500.

THE UWSP MUSEUM OF NATURAL HISTORY, (M), 900 Reserve St., University of Wisconsin, Stevens Point, WI 54481-1962. Tel.: 715-346-2858 & 2821. Fax: 715-346-2367.
E-mail: rreser@uwsp.edu
Web Site: www.uwsp.edu/museum/
Founded: 1966.
Congressional District: 7
Key Personnel: Dir., Ray P. Reser; Museum Shop Mgr., Mary Bartkowiak.
Personnel Profile: Full-Time Paid 1; Part-Time Paid 1; Part-Time Volunteers 10.
Governing Authority: university. Parent Institution: University of Wisconsin-Stevens Point. Tax-exempt.
Institution Type/Description: Natural History Museum.
Collections: mammals; birds; Schoenebeck egg collection; reptiles; fossils; herbarium; Native American artifacts.
Facilities: Museum-related items for sale.
Activities: formally organized education programs for children, adults, undergraduate & graduate students.
Publications: reports of the UWSP Museum of Natural History.
Hours & Admission Prices: Academic year: Mon.-Thurs. 7:45am-12am, Fri. 7:45am-9pm, Sat. 9-9, Sun., 11-4. No charge; donations accepted. &
Attendance: 12,000 (estimated)
Membership: Basic $25; Associate $50; Advocate $100; Sponsor $250; Patron $500; Benefactor $1,000.

Stoughton

STOUGHTON HISTORICAL SOCIETY, 324 S. Page St., Stoughton, WI 53589-2166. Mailing Address: 901 Hwy. 51, Stoughton, WI 53589. Tel.: 608-873-8005.
E-mail: catherine_haynes@sbcglobal.net
Web Site: stoughtonhistoricalsociety.com
Founded: 1960.
Congressional District: 2
Key Personnel: Pres. (V), David Kalland.
Personnel Profile: Part-Time Volunteers 8.
Governing Authority: nonprofit. Tax-exempt.
Institution Type/Description: Historical Society Museum: housed in 1858 Universalist Church.
Collections: land grant papers; tools; surveying equipment; glass; costumes; musical instruments; Norwegian Rosemaling; military; portraits; wagon industry.
Activities: guided tours; lectures; formally organized education programs for children; permanent & traveling exhibitions; special events.
Publications: descriptive brochure.
Hours & Admission Prices: mid-May to Sept. Sun. 1-4. Requested Donation: $2.
Attendance: 1,500 (estimated)
Membership: Student $5; Senior 65 & over $10; Individual $15; Family $25; Supporting $50; Lifetime Century $100.

Sturgeon Bay

DOOR COUNTY HISTORICAL MUSEUM, 18 N. 4th Ave., Sturgeon Bay, WI 54235-2423. Tel.: 920-743-5809.
E-mail: dcmuseum@co.door.wi.us
Web Site: map.co.door.wi.us/museum/
Founded: 1939.
Congressional District: 9
Key Personnel: Cur., Margaret S. Weir; Asst. Cur., Ann Jinkins.
Personnel Profile: Part-Time Paid 4; Part-Time Volunteers 7.
Governing Authority: county. Tax-exempt: 501(c)(3).
Institution Type/Description: Historical Museum.
Collections: farm & dairy; orchards; geology; blacksmith; turn-of-century storefronts; Belgian & Scandinavian settlers; replica fire department containing restored local fire trucks; schools' Door County wildlife; local history videos.
Hours & Admission Prices: May-Oct. daily 10-4:30. No charge; donations accepted. &
Attendance: 10,000 (estimated)

DOOR COUNTY MARITIME MUSEUM (AT STURGEON BAY), (M), 120 N. Madison Ave., Sturgeon Bay, WI 54235-3416. Tel.: 920-743-5958. Fax: 920-743-9483. Facebook: Door County Maritime Museum.
E-mail: info@dcmm.org
Web Site: www.dcmm.org
Founded: 1969.
Congressional District: 8

Key Personnel: Exec. Dir., Bob Desh; Pres., Dan Austad; Vice Pres., Jeff Weborg; Dir. Devel., Trudy Herbst; Treas., Frank Forkert; Museum Shop Mgr., Jan Johnson; Cur., Bob Desh; Asst. Cur., June Larsen; Volunteer Coord., Jon Gast.
Personnel Profile: Full-Time Paid 5; Part-Time Paid 25; Part-Time Volunteers 75.
Governing Authority: nonprofit organization. Door County Maritime Museums Inc., affiliated with Wisconsin State Historical Society, Madison, WI. Branch Museum: Door County Marine Museum at Gills Rock (see separate listing). Tax-exempt.
Institution Type/Description: Maritime Museum.
Collections: actual boats displayed from 1900; earliest shipping container; marine books; ship operating records; shipbuilding pictures; records; restored Great Lakes Pilothouse; fishery artifacts; ship models; lighthouse artifacts.
Major Exhibits: Pirates Ship to Shore, 10/13-2/15.
Facilities: library of miscellaneous papers, pictures, books from the shipyards & steamship operation & marine-oriented charts available for use on premises. Marine-related books for sale.
Activities: Museum Sponsors: Door County Lighthouse Festival in June; Classic & Wooden Boat Festival in August.
Publications: brochures; postcards; quarterly newsletter.
Hours & Admission Prices: July to Labor Day daily 9-5; Sept.-June daily 10-5. Adults $12.50, youth 5-17 $9; children 4 & under and active military no charge. &
Attendance: 15,000 (estimated)
Membership: Single $40; Two Adults $60; Family $70; Life $750.

THE FARM, 4285 Hwy. 57, Sturgeon Bay, WI 54235. Mailing Address: P.O. Box 44, Sturgeon Bay, WI 54235-0044. Tel.: 920-743-6666. Fax: 920-743-2266.
E-mail: info@thefarmindoorcounty.com
Web Site: www.thefarmindoorcounty.com
Founded: 1965.
Congressional District: 8
Key Personnel: Owner, David Tanck; Vice Pres., Jeff Tanck; Museum Shop Mgr., Jenny Tanck; Museum Shop Mgr., Shirley Tanck.
Governing Authority: Affiliated with the Door County Chamber of Commerce, P.O. Box 219, Sturgeon Bay, WI 54235.
Institution Type/Description: Historical Farm Museum.
Collections: farm tools & equipment; gardens & crops; live animals. Historic Structures: c.1856 woodshed; granary; sugar shack.
Facilities: nature center; nature trails; picnic area. Country gift items for sale.
Activities: visit farm animals in natural surroundings; feeding smaller animals; milk a nanny goat; watch chicks hatch.
Publications: tabloid, Down on the Farm; brochure.
Hours & Admission Prices: Memorial Day to mid-Oct. daily 9-5. Adults $8, children 4-12 $4; discounts to AAA members, senior citizens & groups; children under 3 no charge. &
Attendance: 35,000 (estimated)

HERITAGE VILLAGE AT BIG CREEK, 2041 Michigan St., Sturgeon Bay, WI 54235. Mailing Address: P.O. Box 71, Sturgeon Bay, WI 54235. Tel.: 920-421-2332.
E-mail: doorcountyhistoricalsociety@yahoo.com
Web Site: doorcountyhistoricalsociety.com
Founded: 1993.
Key Personnel: Mgr., Dan G. Olson; Pres., George Everson.
Personnel Profile: Part-Time Paid 2; Part-Time Volunteers 15.
Governing Authority: private; nonprofit organization. Parent Institution: Door County Historical Society, Sturgeon Bay, WI 54235. Tax-exempt: 501(c)(3).
Institution Type/Description: Historical Society Museum: village consists of 9 restored buildings.
Collections: rural life from 1880-1910; period furnishings; personal artifacts; photographs.
Facilities: 75-seat auditorium; educational facilities. Museum-related items for sale.
Activities: concerts; docent program; guided tours; lectures; participatory exhibits.
Hours & Admission Prices: mid-June to mid-Oct. daily 1:30-3:30. No charge; donations accepted.
Attendance: 5,000 (estimated)
Membership: Individual $20.

MILLER ART MUSEUM, (M), 107 S. 4th Ave., Sturgeon Bay, WI 54235-2203. Tel.: 920-746-0707 (Offices).
E-mail: bmam@dcwis.com
Web Site: millerartmuseum.org

Founded: 1975.
Congressional District: 8
Key Personnel: Dir. & Museum Shop Mgr., Bonnie Hartmann; Chm. (V), Sharon Virlee; Pres. (V), Kristi Roenning; Cur. Exhibits & Permanent Collections, Deborah Rosenthal.
Personnel Profile: Full-Time Paid 1; Part-Time Paid 3; Part-Time Volunteers 145.
Volunteer Hours: 5,000
Operating Expenses: 185,000
Operating Income: 153,000
Governing Authority: county. Parent Institution: Miller Art Center Foundation, Inc. Tax-exempt 501(c)(3).
Institution Type/Description: Art Museum: housed in Door County Library.
Collections: 20th Century Wisconsin artists.
Major Exhibits: The Wizards of Pop-Up Books - Sabuda & Reinhart (T), 1/25/14-4/14/14; 40th Annual Salon of High School Art, 4/19/14-5/27/14; Drawings: Group Invitational, 5/31/14-7/22/14; Artists on the Road, 7/26/14-9/16/14; 39th Annual 5-County Juried, 9/20/14-11/4/14; Hands in Clay, 11/8/14-12/30/14.
Facilities: classroom. Museum-related items for sale.
Activities: guided tours; lectures; gallery talks; docent program; music & performing arts programs; temporary exhibitions; classes.
Publications: catalogues for changing exhibits.
Hours & Admission Prices: Mon. 10-8, Tues.-Sat. 10-5. No charge; donations accepted. Closed major holidays & 3 days preceding each exhibit. &
Attendance: 19,000 (accurate)
Membership: Volunteer $20; Sustaining $50 & up.

POTAWATOMI STATE PARK, 3740 County PD, Sturgeon Bay, WI 54235. Tel.: 920-746-2890. Fax: 920-746-2896.
Web Site: www.wiparks.net
Key Personnel: Supt., Don McKinnon.
Governing Authority: state. Affiliated with the Wisconsin Dept. of Resources, P.O. Box 7921, Madison, WI 53707.
Institution Type/Description: State Park: located on the Door County Peninsula.
Collections: wildlife & their habitats; natural science; geology.
Facilities: nature trails; 75 ft. observation tower; picnic areas; campsites.
Activities: hiking.
Hours & Admission Prices: Daily 6am-11pm. Entrance fee & camping fee. Reservations accepted.

WHITEFISH DUNES STATE PARK, 3725 Clark Lake Rd., Sturgeon Bay, WI 54235. Tel.: 920-823-2400. Fax: 920-823-2640.
E-mail: wiparks@dnr.state.wi.us
Web Site: www.dnr.state.wi.us
Founded: 1967.
Key Personnel: Park Ranger, Tony Knipfer; Naturalist, Carolyn Rock; Pres. Friends Group (V), Dick Weidman
Governing Authority: state. Parent Institution: Dept. of Natural Resources. Affiliated with Door County Chamber of Commerce, Sturgeon Bay, WI. Tax-exempt.
Institution Type/Description: Nature Center.
Collections: local history & culture; ecology; geology; archaeology; Native American artifacts.
Activities: interpretive programs.
Publications: pamphlet, People of the Dunes; species lists; interpretive trail guide, Brachiopod Trail.
Hours & Admission Prices: Daily 8-8. State Park admission sticker required, fees apply. Closed winter holidays. &
Attendance: 250,000 (estimated)

Sun Prairie

CITY OF SUN PRAIRIES HISTORICAL LIBRARY & MUSEUM, 115 E. Main St., Sun Prairie, WI 53590-2222. Mailing Address, 300 E. Main St., Sun Prairie, WI 53590-2222. Tel.: 608-837-2511. Fax: 608-825-6879.
E-mail: pklein@cityofsunprairie.com
Web Site: www.sunprairie.com
Founded: 1967.
Congressional District: 2
Key Personnel: Chm. Bd., Phyllis Buskager; Dir. & Cur., Peter Klein; Chm. (V), Paul Esser; Registrar, Travis Brimmer; Historian, Rev. Ardin Lapor; Asst., Shirley Thompson; Chm. Dept. Economic Growth & Devel., Neil Stechschulte; Recreation, Jana Stephens.
Personnel Profile: Full-Time Volunteers 1; Part-Time Paid 3; Part-Time Volunteers 118; Interns 1.

Governing Authority: government agency. Parent Institution: City of Sun Prairie. Department of Recreation Museum Division. Tax-exempt: 501(c)(3).
Institution Type/Description: museums development of city government, townships & daily life of citizens of Sun Prairie area.
Collections: artifacts; photographic documents relating to government services & the life of its citizens.
Research Fields: genealogy; local history; city government.
Facilities: archives of local history; documents, city records, school records, organizational records relating to Sun Prairie available for research on premises under supervision.
Activities: guided tours; docent program; permanent, temporary & traveling exhibitions; speakers bureau; local history classes for members & tour guides; outreach programs; research; traveling children programs.
Publications: quarterly newsletter, Sun Prairie Historical Newsletter; pamphlets, Walking Tours; History Pamphlet of O'Keeffe Family; History of Sun Prairie 1830s-1940s; Death List: a compilation of 16,000 burials in area cemeteries available on website.
Hours & Admission Prices: May-Nov. Wed. & Fri.-Sat. 2-4, Sun.-Mon. 6:30-8:30; research by appointment. No charge. &
Attendance: 9,526 (accurate)

Superior

DOUGLAS COUNTY HISTORICAL SOCIETY, 1101 John Ave., Superior, WI 54880-1640. Tel.: 715-392-8449. Facebook: Douglas County Historical Society.
E-mail: dchs@douglashistory.org
Web Site: www.douglashistory.org
Founded: 1854.
Congressional District: 7
Key Personnel: Interim Pres. (V), Nancy Minahan; Dir., Kathy Laakso.
Personnel Profile: Full-Time Paid 1; Part-Time Paid 1; Part-Time Volunteers 19.
Governing Authority: Parent Institution: Wisconsin State Historical Society. Tax-exempt.
Institution Type/Description: County Historical Society Museum.
Collections: c.1890 furniture & furnishings; Ojibwa Chippewa Indian crafts; David F. Barry collection of Sioux Indian portraits; archives & manuscripts; photographs & clippings of local history; oral histories on audiotape; President Calvin Coolidge artifacts including a desk, scrapbook, pictures of Coolidge & a plaque from the Class of 1929 donated by the former Central High School.
Research Fields: local history of Douglas County.
Facilities: 600-vol. library of area history.
Activities: lectures; arts festivals; loan, permanent & temporary exhibitions; school loan service by request; history theatre.
Publications: newsletter, History News; Central A to Z: History of a Superior School.
Hours & Admission Prices: Tues.-Thurs. 11-5, Fri. 10-2. Adults $3; members no charge. &
Attendance: 790 (estimated)
Membership: Students & Seniors $10; Individual $20; Family $30; Conserving $50; Preserving $100; Supporting $250; Maintaining $500; Sustaining $1,000.

FAIRLAWN MANSION & MUSEUM, 906 E. 2nd St., Superior, WI 54880-3245. Tel.: 715-394-5712. Fax: 715-394-2043.
E-mail: info@superiorpublicmuseums.org
Web Site: www.superiorpublicmuseums.org
Founded: 1999.
Key Personnel: Dir., Sara Blanck; Museums Coord., Stacie Buchanan.
Personnel Profile: Full-Time Paid 3; Part-Time Paid 20; Part-Time Volunteers 30.
Governing Authority: Tax-exempt.
Institution Type/Description: Historic House Museum: housed in 42 room Queen Anne Victorian style mansion, c.1891; home of Martin & Grace Pattison until 1918 when it was donated to be used as a home & refuge for children and young women.
Collections: period furnishings; photographs, children's home exhibit.
Research Fields: orphanage; children's home; foster care.
Facilities: Victorian-inspired gardens. Museum-related items for sale.
Activities: special events; guided tours.
Publications: quarterly newsletter; Fairlawn, Restoring the Splendor.
Hours & Admission Prices: mid-May to mid-Oct. Mon.-Sat. 9-4, Sun. 11-4; Winter: Sun.-Fri. 12-3, Sat. 10-3. Adults $9, seniors & students 6-18 $7.50; children under 6 no charge. &
Attendance: 8,000 (accurate)
Membership: Member $20; Patron $50; Sustaining $100; Donor $250.

OLD FIREHOUSE AND POLICE MUSEUM, 402 23rd Ave. E.,
Superior, WI 54880. Mailing Address: 906 E. 2nd St., Superior, WI
54880-3245. Tel.: 715-394-5712.
Web Site: superiorpupblicmuseums.org
Institution Type/Description: History Museum: housed in Superior's former
firehouse, Station No. 4; built in 1898.
Collections: fire & police department history; firefighting equipment; 1906
horse drawn Ahrens steam pumper; 1944 Mack fire engine; 1919 American
LaFrance ladder truck; 1965 Pirsch pumper truck; Wisconsin Fire & Police
Hall of Fame; photographs; personal artifacts.
Hours & Admission Prices: mid-May to Aug. Thurs.-Sat. 10-5, Sun. 11-5;
Sept. to mid-Oct. Sat. 10-5, Sun. 12-5. No charge.

RICHARD I. BONG VETERANS HISTORICAL CENTER, 305
Harbor View Pkwy., Superior, WI 54880-6845. Tel.: 715-392-7151.
Fax: 715-395-5526.
E-mail: fuhrman@bvhcenter.org
Web Site: www.brhcenter.org
Formerly: Richard I. Bong WWII Heritage Center
Founded: 2002.
Key Personnel: Exec. Dir., Robert B. Fuhrman; Chm. (V), Terry Lunberg;
Museum Shop Mgr., Donna O'Kash.
Personnel Profile: Full-Time Paid 3; Part-Time Paid 3; Part-Time Volunteers
80; Interns 2.
Governing Authority: private; nonprofit organization. Parent Institution: Bong
P-38 Fund, Inc. Tax-exempt: 501(c)(3).
Institution Type/Description: History Museum.
Collections: military artifacts from WWII to present.
Research Fields: World War II to present conflicts.
Facilities: library; classrooms; 14,000 sq. ft. exhibit space; 62-seat theater.
Museum-related items for sale.
Activities: formal education programs; internships; guided tours; lectures;
study clubs; temporary & traveling exhibitions; book study club; public
programs & lectures.
Hours & Admission Prices: May-Oct. Mon.-Sat. 9-5, Sun. 12-5; Winter:
Tues.-Sat. 9-5. Adults $9, senior citizens & students $8, children $7;
discounts to groups; veterans on designated days, active military &
members no charge. Closed New Year's Day; Easter; Thanksgiving;
Christmas. &
Attendance: 10,000 (estimated)
Membership: Aviator $40; Aviator Plus $55; Squadron $75; Squadron Plus
$100; Pilot $250; Command Pilot $500; Ace $1,000.

S.S. METEOR MARITIME MUSEUM, 300 Marina Dr., Barker's
Island, Superior, WI 54880-3287. Mailing Address: 906 E. 2nd St.,
Superior, WI 54880-3245. Tel.: 715-394-5712. Fax: 715-394-2043.
E-mail: info@superiorpublicmuseums.org
Web Site: www.superiorpublicmuseums.org
Founded: 1999.
Congressional District: 7
Key Personnel: Dir., Sara Blanck; Museums Coord., Stacie Buchanan.
Personnel Profile: Full-Time Paid 3; Part-Time Paid 12; Part-Time Volunteers
40.
Governing Authority: nonprofit organization. Parent Institution: Superior
Public Museums, Inc. Tax-exempt: 501(c)(3).
Institution Type/Description: Historic Ship Museum: housed in the hull &
quarters of the 1896 S.S. Meteor, last of the whalebacks.
Collections: whaleback artifacts; boat models; ship equipment; ship building
history; Seamen's Memorial Statue, in part dedicated to 29 hands aboard the
Edmund Fitzgerald; ship models.
Facilities: Gift items for sale.
Activities: guided tours; formally organized education programs for children;
loan, permanent & temporary exhibitions.
Publications: brochures; quarterly newsletter, Superior Public Museums; book,
Pigboat: The Story of the Whalebacks; annual, The Whaleback Log.
Hours & Admission Prices: Mid-May-Sept. Mon.-Sat. 10-4, Sun. 11-4. Adults
$7, students 6-18 & senior citizens 62 & over $6; discount to groups;
children under 6 no charge. &
Attendance: 6,000 (estimated)
Membership: Member $20; Patron $50.

Two Rivers

HAMILTON WOOD TYPE & PRINTING MUSEUM, 1816 10th
St., Two Rivers, WI 54241-3066. Tel.: 920-794-6272. Facebook:
Hamilton Wood Type Printing Museum.
E-mail: info@woodtype.org
Web Site: www.woodtype.org
Key Personnel: Museum Dir., Jim Moran, Sr.

Governing Authority: Parent Institution: Two Rivers Historical Society.
Institution Type/Description: Printing History Museum.
Collections: printing history; 1.5 million pieces of wood type including over
1,000 styles & sizes of patterns.
Activities: demonstrations; field trips; workshops.
Hours & Admission Prices: May-Oct. Tues.-Sat. 10-5, Sun. 1-5; Nov.-April
Tues.-Sat. 10-5. Adults $5, senior citizens, military veterans & children 12
& under $3.

Viroqua

VERNON COUNTY MUSEUM, 410 S. Center Ave., Viroqua, WI
54665-2001. Mailing Address: P.O. Box 444, Viroqua, WI 54665-
0444. Tel.: 608-637-7396. Fax: 608-637-8736. Facebook: Vernon
County Museum.
E-mail: vcmuseum@frontiernet.net
Web Site: www.vernoncountyhistoricalsociety.org
Founded: 1942.
Congressional District: 3
Key Personnel: Pres. (V), Julie Malone; Cur., Kristen Parrott; Asst. Cur., Carol
Krogan.
Personnel Profile: Part-Time Paid 2; Part-Time Volunteers 110.
Volunteer Hours: 2,000
Governing Authority: Parent Institution: Vernon County Historical Society.
Affiliated with Wisconsin State Historical Society. Tax-exempt.
Institution Type/Description: Local History Museum: housed in Vernon
County Normal School.
Collections: general store; tobacco farming; Vernon County schools; Indian
artifacts; Governor Rusk; genealogy & research room. Historic Buildings:
c.1900 church; 1889 rural school; 1870 Sherry-Butt House; 1919-1972
Teacher's college/museum.
Major Exhibits: Civil War, 1/11-1/15.
Research Fields: genealogy & Vernon County history.
Facilities: library of local history books & census records; local newspapers on
microfilm; research room.
Activities: guided tours; Round Barn Driving tour; Black Hawk Trail.
Publications: books, Viroqua's Main St. History, 1846-1996; Vernon County
Cemetery Locations & Histories; Vernon County, WI, Tombstone Inscrip-
tions; Index to 1878 Vernon County, WI, Plat Map.
Hours & Admission Prices: Museum: April-May & Sept.-Oct. Mon.-Fri. 12-4;
June-Aug. Mon.-Sat. 12-4; Nov.-March Tues.-Thurs. 12-4; other times by
appointment. No charge; donations accepted. Sherry-Butt House: Memorial
Day to Labor Day Sat.-Sun. 1-5. Adults $5; members no charge. Closed
holidays. &
Attendance: 3,600 (accurate)
Membership: Individual $15; Family $25; Supporting & Business $50;
Dawson Heritage Club $100; Life $250 & up.

Warrens

WISCONSIN CRANBERRY DISCOVERY CENTER, 204 Main
St., Warrens, WI 54666. Tel.: 608-378-4878.
E-mail: director@discovercranberries.com
Web Site: www.discovercranberries.com
Key Personnel: Dir., Barbara Hendricks
Institution Type/Description: History Center.
Collections: cranberry industry; Wisconsin cranberry history; woodworking
equipment; Native American artifacts including a dugout canoe.
Activities: special events.
Hours & Admission Prices: May-July & Nov.-Dec. Mon.-Sat. 10-4; Aug.-Oct.
daily 9-5; other times by appointment. Adults $4, seniors 65 & over $3.50,
students K-12 $3; discounts to families; children 5 & under no charge.

Washburn

**WASHBURN HISTORICAL MUSEUM & CULTURAL CEN-
TER, INC. AKA WASHBURN CULTURAL CENTER,** 1 E.
Bayfield St., Washburn, WI 54891-4401. Mailing Address: P.O.
Box 725, Washburn, WI 54891-0725. Tel.: 715-373-5591.
E-mail: washburnw@centurytel.net
Web Site: washburnculturalcenter.art.officelive.com
Founded: 1991.
Congressional District: 7
Key Personnel: C.E.O. & Chm. (V), Richard Olson; Pres. (V), A.H. (Tony)
Woiak; Treas., Lois Tetzner; Public Rels., Dora Kling; Museum Shop Mgr.,
Joanne Weister.
Personnel Profile: Part-Time Paid 3; Part-Time Volunteers 15.
Governing Authority: private; nonprofit organization. Tax-exempt: 501(c)(3).
Institution Type/Description: Historical Museum & Cultural Arts Center.
Collections: Washburn area history, industries & people; paintings.

Facilities: 15-vol. library of historical books; 100 bound newspapers 1886-1989; 2,800 sq. ft. exhibit space.
Activities: concerts; films; formal education programs; guided tours; hobby workshops; lectures; loan & temporary exhibitions; rental gallery; classes; meetings; receptions.
Publications: annual, Brownstone Notes.
Hours & Admission Prices: Jan.-April Mon.-Sat. 11-3; May-June & Sept.-Dec. Mon.-Sat. 10-4, July-Aug. Mon.-Sat. 10-4, Sun. 11-3. No charge; donations accepted. Closed New Year's Day; Thanksgiving; Christmas. &
Attendance: 3,608 (accurate)
Membership: Individual $5; Family $10; Business $15.

Washington Island

ROCK ISLAND STATE PARK, 1924 Indian Point Rd., Washington Island, WI 54246-9078. Tel.: 920-847-2235.
Web Site: www.dnr.state.wi.us/parks/
Founded: 1965.
Congressional District: 8
Key Personnel: Supt., Kirby Foss.
Personnel Profile: Part-Time Paid 1.
Governing Authority: state. Parent Institution: State of WI, Dept. of Natural Resources. Tax-exempt.
Institution Type/Description: State Park: located on the Door County Peninsula.
Collections: local history & culture; period furnishings; historic buildings.
Facilities: 900-acres state park; trails; 40 campsites; picnic area; motorized vehicles & horses not allowed. Accessible by boat only; public boat ramp at Jackson Harbor & ferry service from Washington Island.
Activities: nature programs & nature hikes.
Hours & Admission Prices: Memorial Day-Columbus Day daily. Park: No charge; donations accepted. Camping fee, camping reservations recommended, to reserve a campsite, call Reserve America at 888-947-2747.
Attendance: 30,000
Membership: Friends of Rock Island: Individual $10; Family $25; Lifetime $300.

Watertown

OCTAGON HOUSE-FIRST KINDERGARTEN IN AMERICA, 919 Charles St., Watertown, WI 53094-5001. Tel.: 920-261-2796.
Web Site: www.watertownhistory.org
Founded: 1933.
Congressional District: 9
Key Personnel: C.E.O. & Museum Shop Mgr., Linda Werth; Pres., Melissa Lampe.
Personnel Profile: Full-Time Paid 1; Part-Time Paid 4; Part-Time Volunteers 20; Interns 1.
Governing Authority: society; nonprofit organization. Parent Institution: Watertown Historical Society. Tax-exempt: 501(c)(3).
Institution Type/Description: History Museum: housed in 1854 Victorian Octagon House belonging to John & Eliza Richards and family; 1856 first kindergarten building in America.
Collections: memorabilia of early Watertown; archives; folk art; Civil War materials; children's museum; decorative arts; folklore; manuscripts. Historic Structure: 1853 pioneer barn.
Research Fields: genealogies; kindergarten; octagon houses.
Facilities: 400-vol. library of historical, biographical and kindergarten materials available for inter-library loan and for use on premises. Museum-related items for sale.
Activities: guided tours; gallery talks; concerts; permanent & temporary exhibitions. Museum Sponsors: annual Ice Cream Social in August; Christmas Play in November & December; Christmas Open House.
Publications: books, Heritage of Homes; John Richards: The Hill and the Mill; Reprint of Margaret Schurz Biography; Handi & Pussy Go to Kindergarten; monograph on the Octagon House; monograph on first kindergarten; monograph, Froebel's gifts.
Hours & Admission Prices: May & Sept.-Oct. daily 11-3; Memorial Day to Labor Day 10-4. Adults $8, senior citizens & AAA members $7, students 6-17 $4; discounts to AAM & AAA members.
Attendance: 6,000 (accurate)
Membership: Individual $12; Couple$18; Family $25; Patron $50 & up; Benefactor $100 & up.

Waukesha

WAUKESHA COUNTY MUSEUM, 101 W. Main St., Waukesha, WI 53186-4811. Tel.: 262-521-2859. Fax: 262-521-2865.
Web Site: www.waukeshacountymuseum.org
Formerly: Waukesha County Historical Society & Museum

Founded: 1914.
Congressional District: 4
Key Personnel: Exec. Dir., Kirsten Lee Villegas; Pres. (V), Dave Frazer; Dir. Devel. & Mktg., Jim Hahn, CFRE; Dir. Education, Kristen Matlick; Cur., Elisabeth Engel; Archivist, Eric Vanden Heuvel; Dir. Visitor Svcs., Marjorie Mahler.
Personnel Profile: Full-Time Paid 6; Part-Time Paid 2; Part-Time Volunteers 100; Interns 4.
Governing Authority: Tax-exempt: 501(c)(3).
Institution Type/Description: History Museum: housed in 1893 Waukesha County Courthouse.
Collections: county history including manuscripts & visual images; textiles; house & farm ware; Springs Era artifacts; military; Native American; stonework; Waukesha County history from Ice Age to the present.
Research Fields: local history; genealogy; historic preservation.
Facilities: research center.
Activities: permanent & temporary exhibits; spring & summer camps; scout days; guided tours; programs; events; school loan of Discovery Boxes.
Publications: magazines; quarterly publication, "Landmark"; books, From Farmland to Freeways: A History of Waukesha County, Discovering Waukesha County.
Hours & Admission Prices: Exhibits: Tues.-Sat. 10-4:30. Adults $3, seniors 62 & over $2, students $1; Waukesha County residents on Sat. & members no charge. Research Center: Tues. & Fri.-Sat. 10-12 & 12:30-4:30, Thurs. 12:30-4:30. Adults $3; members no charge. Closed major holidays. &
Attendance: 16,683 (accurate)
Membership: Institution & Preservationist $30; Builder $50; Educator $100; Visionary $250.

Waupaca

HOLLY HISTORY CENTER & HUTCHINSON HOUSE MUSEUM, 321 S. Main St., Waupaca, WI 54981-1745. Tel.: 715-258-5958 (museum) & 256-9980 (history center).
E-mail: wauphistsoc@waupacaonline.net
Web Site: www.waupacahistory.org
Founded: 1953.
Congressional District: 6
Key Personnel: Pres., Dennis Lear; Vice Pres., Mike Kirk; Treas., Robert Kessler; Sec., Betty Stewart; Dir., Julie Hintz.
Personnel Profile: Part-Time Paid 2; Part-Time Volunteers 16.
Governing Authority: society. Parent Institution: Waupaca Historical Society, Waupaca 54981. Subsidiary Institution: Waupaca Train Depot. Branch Museum: Holly History & Genealogy Center. Tax-exempt: 501(c)(3).
Institution Type/Description: Historic House & Preservation Project: 1854 Hutchinson House.
Collections: Victorian home furnishings & clothing; vintage china & glassware; early quilts, lace, linens, & samplers; magazines Victorian-1930s; photographs; manuscripts. Holly History & Genealogy Center: genealogy & history research section; Royal Doulton character jugs; historical photographs; history of Waupaca; Plat books; maps; family histories; Bibles; historical postcards; train depot.
Major Exhibits: Victorian Secrets, 6/14-9/14.
Research Fields: genealogy; photography; Victorian Americana; Waupaca history; Wisconsin history.
Facilities: library.
Activities: guided tours; arts festivals; permanent & temporary exhibitions; genealogy assistance; history programs.
Publications: quarterly member newsletter.
Hours & Admission Prices: History Center: Summer: Wed. & Fri. 12-4, Sat. 9-12; Winter: Wed. & Fri. 12-3. No charge; donations accepted. Museum: Memorial Day-Labor Day Sat.-Sun. & holidays 1-4; special group tours by appointment. No charge.
Attendance: 1,200 (estimated)
Membership: Waupaca Historical Society: Single $15; Family $30; Lifetime Single $150; Lifetime Family $220, Supporting $300.

Wausau

CENTER FOR THE VISUAL ARTS GALLERY, 427 N. 4th St., Wausau, WI 54403. Tel.: 715-842-4545.
E-mail: cvawausau@gmail.com
Web Site: www.cvawausau.org
Founded: 1982.
Congressional District: 7
Institution Type/Description: Art Gallery.
Collections: paintings; sculpture.
Facilities: Museum-related items for sale.
Activities: special events; classes; educational programs.
Hours & Admission Prices: Tues.-Fri. 10-5, Sat. 10-4. No charge; donations accepted. Closed holidays. &

* **LEIGH YAWKEY WOODSON ART MUSEUM, (M),** 700 N. 12th St., Wausau, WI 54403-5007. Tel.: 715-845-7010. Fax: 715-845-7103.
E-mail: museum@lywam.org
Web Site: www.lywam.org
Founded: 1973.
Congressional District: 7
Key Personnel: Dir., Kathy Kelsey Foley; Cur. Exhibitions, Andrew J. McGivern; Cur. Collections, Jane Weinke; Cur. Education, Jayna Hintz; Cur. Education, Catie Anderson; Administrative Svcs. Mgr., Shari Schroeder; Facilities Mgr., Joe Ruelle; Business Mgr., Diane Wendt; Mgr. Mktg. & Communications, Amy Beck.
Personnel Profile: Full-Time Paid 10; Part-Time Paid 18; Part-Time Volunteers 115.
Governing Authority: nonprofit organization. Tax-exempt: 501(c)(3).
Institution Type/Description: Art Museum.
Collections: paintings, sculpture, drawings & graphics depicting the natural world, with an emphasis on avian life; porcelain; glass.
Research Fields: artistic depictions of birds.
Facilities: 2,000-vol. library of art books available to staff & docents at all times & to the general public by appointment only.
Activities: sculpture garden; guided tours; docent & volunteer program; lectures; garden concerts; demonstrations; family programs; Art Park (family interactive area); videos; inter-museum loan, permanent, temporary & traveling exhibitions; audio tours.
Publications: exhibition catalogs; quarterly newsletter.
Hours & Admission Prices: Tues.-Fri. 9-4, 1st Thurs. each month 9-7:30, Sat.-Sun. 12-5. No charge; donations accepted. Closed national holidays. &
Attendance: 58,000 (estimated)
Membership: Household $50; Partner $75; Associate $150; Patron $250; Friend $350; Connoisseur $500; Collector $1,000; Benefactor $2,500.

MARATHON COUNTY HISTORICAL SOCIETY, (M), 410 McIndoe St., Wausau, WI 54403-4745. Tel.: 715-842-5750. Fax: 715-848-0576.
E-mail: director@marathoncountyhistory.org
Web Site: www.marathoncountyhistory.org
Founded: 1952.
Congressional District: 7
Key Personnel: Dir., Mary Forer; Pres., John Hattenhauer; Cur. Events & Public Rels., Sara Goetsch; Librarian, Gary Gisselman; Cur. Education, Anna Straub; Yawkey House Attendant, Gary Walters.
Personnel Profile: Part-Time Paid 6; Part-Time Volunteers 60.
Governing Authority: bd. of directors. Branch Museum: Yawkey House Museum, 403 McIndoe St., Wausau, WI.
Institution Type/Description: History Museum.
Collections: Wisconsin history; lumbering artifacts; period furnishings; pioneer artifacts; memorabilia; writings & photographs; period documents. Historic Houses: Yawkey House & Carriage House, built in 1901; A.P. Woodson House, built in 1914.
Major Exhibits: Marking Time: Voyage to Vietnam, 4/14-11/14.
Research Fields: logging; Native American; genealogy; photographs; history of Marathon County; German immigrants.
Facilities: research library of books on the history of Wausau, Marathon County, Wisconsin; meeting rooms; conference room; rental facilities.
Activities: tours; lectures; films; slides; programs for children & adults; permanent, temporary & traveling exhibitions; school programs; research; rental facilities; cemetery tours. Annual Events: Victorian Valentine's Tea in February; Living History Festival.
Publications: quarterly, Wanigan.
Hours & Admission Prices: Tues.-Fri. 9-4:30, Sat.-Sun. 1-4:30. Woodson History Center & Library: no charge. Yawkey House Museum: adults $7, seniors $6, students $5; members no charge. Closed national holidays. &
Attendance: 20,000 (estimated)
Membership: Student $10; Individual $35-$49; Family $50-$99; Pinery $100-$249; Pioneer $250-$499; Lumberjack $500-$999; Lumber Baron $1,000 & up.

Wauwatosa

LOWELL DAMON HOUSE, 2107 Wauwatosa Ave., Wauwatosa, WI 53213-1730. Mailing Address: 910 N. Old World 3rd St., Milwaukee, WI 53203-1501. Tel.: 414-273-8288. Fax: 414-273-3268.
E-mail: info@milwaukeehistory.net
Web Site: www.milwaukeehistory.net
Founded: 1941.
Congressional District: 4

Key Personnel: Exec. Dir., Robert T. Teske; Pres. (V), Randy Bryant; Cur. Collections, Michael Reuter.
Personnel Profile: Full-Time Paid 9; Part-Time Paid 1.
Governing Authority: Milwaukee County Historical Society. Tax-exempt.
Institution Type/Description: Historic House.
Collections: 1840-1880 furnishings.
Activities: guided tours.
Publications: brochure.
Hours & Admission Prices: Wed. 3-5, Sun. 1-5. No charge; donations accepted. Closed major holidays.
Membership: Senior Citizen & Student $20; Individual $30; Family $35.

West Allis

WEST ALLIS HISTORICAL SOCIETY MUSEUM, 8405 West National Ave., West Allis, WI 53227-1733. Tel.: 414-541-6970.
E-mail: wahs8405@gmail.com
Web Site: www.westallishistory.org
Founded: 1966.
Congressional District: 4
Key Personnel: Pres. (V), Devan Gracyalny, Jr.; 1st Vice Pres., Betty M. Hartwig; Treas., Helen Lundquist; Sec., Ed Wilkommen.
Personnel Profile: Part-Time Volunteers 20.
Governing Authority: nonprofit. Affiliated with State Historical Society, 816 State St., Madison 53706. Tax-exempt.
Institution Type/Description: General Museum: housed in 1887 cream city brick Romanesque style school building.
Collections: toys; model steam engine; motors; 1900 dental office; 1835 pioneer room & workshop with pioneer implements; turn-of-the-century grocery & post office; clothing from Civil War to present; manuscripts.
Research Fields: cemetery; government; pioneer families; artifacts; churches.
Facilities: 1,000-vol. library on local history available on premises; reading room; classrooms.
Activities: guided tours; lectures. Museum Sponsors: annual picnic on museum grounds in June; annual banquet in October; Holiday Open House on Sundays in December.
Publications: quarterly bulletin, Historic Buzz.
Hours & Admission Prices: Tues. 7pm-9pm, Sun. 2-4; groups by appointment. No charge; donations accepted. Closed New Year's Eve & Day; Memorial Day; Independence Day; Thanksgiving; Christmas. &
Attendance: 831 (accurate)
Membership: Student $5; Individual $8; Patron $50 & up; Business, Civic, Professional, Organization $50; Life $100.

West Bend

MUSEUM OF WISCONSIN ART, (M), 205 Veterans Ave., West Bend, WI 53095-3413. Tel.: 262-334-9638. Fax: 262-334-8080.
E-mail: greid@wisconsinart.org
Web Site: www.wisconsinart.org
Formerly: West Bend Art Museum
Founded: 1961.
Congressional District: 9
Key Personnel: Exec. Dir., Thomas D. Lidtke; Asst. Dir., Graeme Reid; Registrar, Andrea Waala; Dir. Devel., Joan Rudnitzki; Cur. Education, Courtney Spousta.
Personnel Profile: Full-Time Paid 6; Part-Time Paid 11; Part-Time Volunteers 100.
Governing Authority: nonprofit organization. Tax-exempt.
Institution Type/Description: Art Museum.
Collections: works of late 19th- & early 20th-century American expatriate artist Carl Von Marr consisting of more than 400 works of art; early Wisconsin art (from Euro-American settlement to 1950); photographs; books; video tapes; audio recordings.
Research Fields: Carl von Marr; Wisconsin art.
Facilities: archives of Wisconsin art history.
Activities: guided tours; lectures; films; gallery talks; permanent & temporary exhibitions; tours; adult & children humanities classes; educational outreach programs for grade schools.
Publications: monthly newsletter; catalog, Carl von Marr: German American Artist; early Wisconsin art exhibition catalogues.
Hours & Admission Prices: Wed.-Sat. 10-4:30, Sun. 1-4:30. Adults $5, seniors & students $3; discounts to AAM & ICOM members; members & children under 12 no charge. Closed holidays. &
Attendance: 14,000 (accurate)
Membership: Student $15; Seniors & Teacher $25; Individual $30; Family $50; Friends of Art $125; Sustaining $300; Exceptional Friend $500; Carl von Marr Society $1,000.

WASHINGTON COUNTY HISTORICAL SOCIETY, 320 S. 5th Ave., West Bend, WI 53095-3333. Tel.: 262-335-4678. Fax: 262-335-4612.
E-mail: wchs@historyisfun.com
Web Site: www.historyisfun.com
Founded: 1937.
Congressional District: 9
Key Personnel: Exec. Dir., Patricia Lutz; Pres. (V), John Best; Cur., Janean Mollet Van Beckum; Cur. Education, Jessica Sawinski; Research Supvr., Heather Przybylski.
Personnel Profile: Full-Time Paid 5; Part-Time Paid 1; Part-Time Volunteers 130; Interns 3.
Governing Authority: nonprofit organization. Subsidiary Institution: St. Agnes Historic Site. Tax-exempt: 501(c)(3).
Institution Type/Description: History Museum: housed in 1889 County Courthouse; 1886 Old County Jailhouse Museum.
Collections: period artifacts; photographs; maps; genealogies; manuscripts; death & burial records; Veterans records; Walter A. Zinn Dollhouse.
Research Fields: genealogy & history.
Facilities: 7,000-vol. archives; reading room. Gift items for sale.
Activities: guided tours; concerts; docent program; formal education programs for adults & children; hobby workshops; lectures; permanent & temporary exhibitions; bimonthly meetings; school tours; living history encampment. Annual Events: Old Settlers' Club Banquet; Pioneer Kids Day; Haunted Fun in October; Christmas Open House; Vintage Baseball Game.
Publications: quarterly newsletter, The Court Reporter; Annual Calendar of Events; catalogs; brochure
Hours & Admission Prices: Wed.-Fri. 11-5, Sat. 9-5, Sun. 1-4:30. Jailhouse Tour: adults 17-61 $5, children 6-16 $4; children under 6 & members no charge. Closed New Year's Eve & Day; Easter; Independence Day; Labor Day; Christmas Eve & Day. &
Attendance: 9,300 (estimated)
Membership: Student $12; Individual $15; Family $25; Patron $75; Sustaining $150; Benefactor $300; Sponsor $550; Corporate $1,000.

West Salem

HAMLIN GARLAND HOMESTEAD, 357 W. Garland St., West Salem, WI 54669-1146. Mailing Address: P.O. Box 884, West Salem, WI 54669-0884. Tel.: 608-786-1399.
Web Site: www.westsalemhistoricalsociety.org
Founded: 1973.
Congressional District: 3
Key Personnel: C.E.O. & Pres., Errol Kindschy.
Personnel Profile: Part-Time Volunteers 55.
Governing Authority: nonprofit organization. Owned and operated by West Salem Historical Society, Inc. Tax-exempt: 501(c)(3).
Institution Type/Description: Historic House: 1857-60 Hamlin Garland Homestead.
Collections: local historical materials; musical items; fans. Historic House: 1856 Gullickson Octagonal Home; Garland items of all kinds.
Research Fields: Garland materials; history of West Salem area.
Facilities: 70-vol. library of books by Garland and about Garland; letters; magazines available by request. Museum-related items for sale.
Activities: guided tours; lectures; films; school loan service.
Publications: monthly newsletter.
Hours & Admission Prices: Memorial Day to Labor Day Mon.-Sat. 10-4, Sun. 1-4; other times by appointment. Family $2.50, adults $1, students $.50; members no charge. Discounts to AAA members.
Attendance: 771 (accurate)
Membership: Individual $3; Family $5; Individual Life $75; Couple Life $125.

PALMER/GULLICKSON OCTAGON HOME, 360 N. Leonard, West Salem, WI 54669-1238. Mailing Address: P.O. Box 884, West Salem, WI 54669-0884. Tel.: 608-786-1399.
Founded: 1973.
Congressional District: 3
Key Personnel: C.E.O. & Pres., Errol Kindschy.
Personnel Profile: Part-Time Volunteers 50.
Governing Authority: nonprofit organization. Owned & operated by West Salem Historical Society, Inc. Tax-exempt.
Institution Type/Description: Historical Society Museum.
Collections: local historical items. Historic Building: Palmer/Gullickson Octagon Home.
Publications: monthly newsletter.
Hours & Admission Prices: Memorial Day-Labor Day Mon.-Sat. 10-4, Sun. 1-4; other times by appointment. Family $2.50, adults $1, students $.50; discounts to AAA members; members no charge.
Attendance: 419 (accurate)

Membership: Individual $3; Family $5; Individual Life $75; Couple Life $125.

Westfield

MARQUETTE COUNTY HISTORICAL SOCIETY, 125 Lawrence St., Westfield, WI 53964-9030. Mailing Address: P.O. Box 172, Westfield, WI 53964-0172. Tel.: 608-296-4700.
E-mail: mcgwin@frontier.net
Web Site: www.marqcohistorical.org
Formerly: Cochrane-Nelson House
Founded: 1962.
Congressional District: 6
Key Personnel: Pres., LeRoy Stublaski; Vice Pres., Ed Thalacker; Sec., Karen Watkins; Treas. & Cur., Joannie Ingraham.
Governing Authority: county. Tax-exempt: 170(b)(1)(A).
Institution Type/Description: Historical Society Museum: housed in c.1903 Cochrane-Nelson House.
Collections: photographs; manuscripts; textiles; farm & shop tools; home, schoolroom, war, business & church artifacts pertaining to Marquette County; granite quarrying; replica of first Ferris wheel; dental equipment c.1950; 19th-century dolls; railroad artifacts.
Research Fields: Fox River in Wisconsin; cemeteries; ethnic groups; county personages, businesses, schools, post offices & creameries; obituaries.
Facilities: library available for research at museum.
Activities: guided tours; temporary exhibitions.
Publications: brochure, Imprints On The Sands of Marquette County.
Hours & Admission Prices: Wed. 1-4; other times by appointment. Adults $1. &
Attendance: 1,000 (estimated)
Membership: Individual $10; Family & Sustaining $25; Life $200.

Weyauwega

LITTLE RED SCHOOL HOUSE MUSEUM, Weyauwega Community Park, 411 W. High St., Weyauwega, WI 54983. Mailing Address: P.O. Box 294, Weyauwega, WI 54983-0294. Tel.: 920-867-2500 & 2630.
E-mail: weyauwegaareahistoricalsociety@yahoo.com
Web Site: www.cityofweyauwega-wi.gov
Founded: 1970.
Congressional District: 7
Key Personnel: Caretaker, Suzanne Dyer.
Governing Authority: county. Tax-exempt.
Institution Type/Description: Historic Building: housed in 1861 Old Wood School House.
Collections: old school desks; books; hand made toys; old wood heater; area maps from 1871.
Activities: tours; student demonstrations; senior citizen discussions.
Hours & Admission Prices: Memorial Day-Labor Day Sun. 1-4; other times by appointment. No charge; donations accepted.
Membership: Annual $5.

Whitewater

CROSSMAN GALLERY, UW-WHITEWATER, 950 W. Main St., Whitewater, WI 53190. Mailing Address: 800 W. Main Street, Whitewater, WI 53190-1705. Tel.: 262-472-5708. Fax: 262-472-2808.
E-mail: flanagam@uww.edu
Web Site: www.uww.edu
Founded: 1970.
Key Personnel: Dir., Michael Flanagan.
Personnel Profile: Part-Time Paid 1; Interns 2.
Governing Authority: university; nonprofit organization. Parent Institution: University of Wisconsin-Whitewater. Tax-exempt.
Institution Type/Description: University Art Gallery.
Collections: student teaching collection; contemporary American & European; American outsider and folk art.
Research Fields: contemporary American art.
Facilities: 1,300-seat auditorium; educational facilities; 2,400 sq. ft. exhibit space; 400-seat theater.
Activities: temporary & traveling exhibitions. Museum Sponsors: biannual faculty show; 1-4 person & group invitationals. Museum Hosts: annual ceramics invitational; annual fiber show.
Publications: exhibit brochures & catalogs.
Hours & Admission Prices: Academic Year Mon.-Fri. 10-5, Sat. 1-4; Summer: call for hours. No charge. Closed Easter; Christmas. &
Attendance: 9,500 (accurate)

WHITEWATER HISTORICAL MUSEUM, 301 W. Whitewater
St., Whitewater, WI 53190. Mailing Address: W7646 Hackett Rd.,
Whitewater, WI 53190-4354. Tel.: 262-473-6820.
E-mail: ccart@idcnet.com
Founded: 1974.
Congressional District: 1
Key Personnel: Pres., Ellen Penwell.
Personnel Profile: Part-Time Volunteers 1.
Governing Authority: society. Affiliated with the Whitewater Historical Soci-
ety. Tax-exempt.
Institution Type/Description: History Museum: housed in 1890 Chicago,
Milwaukee & St. Paul Railroad depot.
Collections: china; glass; furniture; clothing & tools; books; genealogies; local
artifacts.
Research Fields: local history; settlement of village; early industries; early
homes; presidents.
Facilities: Museum-related items for sale.
Activities: guided tours; permanent exhibitions.
Publications: book, The Rile Collection; Sketches and Autobiography of
Henry E. Rile's Whitewater Years 1856-1862.
Hours & Admission Prices: Memorial Day-Labor Day Thurs. 5:30pm-7:30pm,
Sun. 1-4; other times by appointment. No charge; donations accepted. ♿
Attendance: 400 (estimated)
Membership: Annual $1; Life $25.

Wild Rose

PIONEER MUSEUM & WILD ROSE HISTORICAL SOCIETY,
477 & 479 Main St., Wild Rose, WI 54984. Mailing Address: P.O.
Box 63, Wild Rose, WI 54984-0063.
Founded: 1964.
Congressional District: 6
Key Personnel: Pres. & Cur., Pam Anderson; Vice Pres., Rodney Radloff;
Treas., Helen Cox; Museum Shop Mgr., Mary Ann Erdman.
Personnel Profile: Part-Time Volunteers 30.
Governing Authority: Wild Rose Historical Society. Parent Institution: State
Historical Society of Wisconsin. Tax-exempt.
Institution Type/Description: Pioneer Museum.
Collections: agriculture; costumes; medical; children's museum; textiles;
manuscripts; Pioneer Hall & Carriage House. Historic Buildings: 1880
Elisha Stewart house; 1880 tool & buggy shed; 1860 smokehouse &
blacksmith shop; 1860 cobblers shop & weaving room; 1860 drugstore-
country store; one-room schoolhouse.
Research Fields: local history.
Facilities: Museum-related items for sale.
Activities: guided tours; lectures; permanent & temporary exhibitions; demon-
strations.
Publications: two centennial booklets; postcards.
Hours & Admission Prices: mid-June to Labor Day Wed. & Sat. 1-3. No
charge; donations accepted. Gift Shop: Wed. & Sat. 1-3. Closed Indepen-
dence Day.
Attendance: 350 (estimated)
Membership: Annual $5.

Wisconsin Dells

BEAVER SPRINGS PUBLIC AQUARIUM, 600 Trout Rd., Wis-
consin Dells, WI 53965. Mailing Address: P.O. Box 1, Wisconsin
Dells, WI 53965. Tel.: 608-254-2735. Fax: 608-253-9446.
Institution Type/Description: Aquarium.
Collections: over 1,000 fish & marine life.
Activities: hands-on exhibits; videos.
Hours & Admission Prices: April-May & Sept.-Oct. daily 10-5; June-Aug.
daily 9-7; Nov.-March daily 10-4. Adults $7.99, children $5.99.

H.H. BENNETT STUDIO, 215 Broadway, Wisconsin Dells, WI
53965. Mailing Address: P.O. Box 147, Wisconsin Dells, WI
53965. Tel.: 608-253-3523. Fax: 608-253-4635.
E-mail: hhbennett@wisconsinhistory.org
Web Site: www.hhbennettstudio.org
Founded: 2000.
Key Personnel: Site Dir., Alan Hanson.
Personnel Profile: Part-Time Paid 4; Part-Time Volunteers 4.
Governing Authority: Parent Institution: Wisconsin Historical Society. Tax-
exempt.
Institution Type/Description: Photography Museum.
Collections: Bennett's life & family history; photographs; glass plate nega-
tives; photography inventions; cameras & equipment.
Major Exhibits: H.H. Bennett Then & Now, 1/14-12/14.

Facilities: Museum-related items for sale.
Activities: self guided & guided tours; photography workshops.
Hours & Admission Prices: Adults $7, students and seniors 65 & over $6,
children 5-17 $3.50; children under 5 no charge. ♿

PARSON'S INDIAN TRADING POST & MUSEUM, 370 Wiscon-
sin Dells Pkwy., Wisconsin Dells, WI 53965. Tel.: 608-254-8533;
866-281-8704. Fax: 608-253-6766.
E-mail: parsonitp@hotmail.com
Founded: 1918.
Institution Type/Description: Native American History Museum.
Collections: Native American crafts; clothing weapons; household utensils;
religious artifacts; jewelry.
Facilities: Museum-related items for sale.
Hours & Admission Prices: Summer: daily 9-9; Winter: daily 9-5. ♿

RIVERSIDE & GREAT NORTHERN RAILWAY, N115 County
Rd. N., Wisconsin Dells, WI 53965-9124. Tel.: 608-254-6367. Fax:
608-254-5628.
Institution Type/Description: Railway History Museum.
Collections: railway history, equipment, & artifacts; photographs; R&GN's
locomotives & cars.
Facilities: Museum-related items for sale.
Activities: train rides.
Hours & Admission Prices: Call for hours.

TIMBAVATI WILDLIFE PARK AT STORYBOOK GARDENS,
2220 Wisconsin Dells Pkwy., Wisconsin Dells, WI 53965. Mailing
Address: P.O. Box 68, Wisconsin Dells, WI 53965-0068. Tel.:
608-253-2391.
Institution Type/Description: Wildlife Park.
Collections: animals from around the world including giraffe, zebra, kangaroo,
birds, & white lions.
Facilities: Museum-related items for sale.
Activities: wildlife show; camel rides; pet & feed.
Hours & Admission Prices: Memorial Day to Labor Day daily 9-8. Adults
$18.95; children 2-12 $12.95; military & children under 2 no charge.

**TOMMY BARTLETT EXPLORATORY - INTERACTIVE SCI-
ENCE CENTER,** 560 Wisconsin Dells Pkwy., Wisconsin Dells,
WI 53965. Tel.: 608-254-2525.
E-mail: bartlett@tommybartlett.com
Web Site: www.tommybartlett.com
Formerly: Exploratory Interactive Science Center
Institution Type/Description: Science Center.
Collections: science; technology; Mercury space capsule; Russian Space
Station MIR; static electricity; hands-on exhibits; puzzles; holograms; lift a
full-size vehicle off the ground with a giant lever.
Activities: hands-on exhibits.
Hours & Admission Prices: Summer: daily 9-9; Spring, Fall, & Winter: daily
10-4. Adults $15, seniors 65 & over $12, children 5-11 $12; discounts to
groups of 20 or more; children 4 & under no charge. ♿

WISCONSIN DEER PARK, 583 Wisconsin Dells Pkwy., Wisconsin
Dells, WI 53965. Tel.: 608-253-2041.
Institution Type/Description: Nature Center.
Collections: wildlife including White-tail deer, American elk & bison, Euro-
pean Fallow deer, Japanese Silka deer, goats, birds, llamas, horses, emus, &
pigs.
Facilities: nature trails.
Activities: hiking.
Hours & Admission Prices: May & Sept.-Oct. daily 10-4; Memorial Day to
Labor Day daily 9-7. Adults 12 & over $12, children 3-11 $8; children 2 &
under no charge. ♿

Wisconsin Rapids

SOUTH WOOD COUNTY HISTORICAL CORP., 540 Third St.
S., Wisconsin Rapids, WI 54494-4352. Tel.: 715-423-1580.
E-mail: lori@swch-museum.com
Web Site: www.swch-museum.com
Founded: 1955.
Congressional District: 7
Key Personnel: Pres., Philip M. Brown; Dir., Dave Engel; Admin., Lori Brost.
Personnel Profile: Part-Time Paid 5; Part-Time Volunteers 2.
Governing Authority: nonprofit organization. Tax-exempt: 501(c)(3).
Institution Type/Description: Historical Society Museum: housed in 1907
mansion.

Collections: railroad depot; lumbering & blacksmith items; toys; country kitchen; country school; tools, machinery, equipment & archives relating to the cranberry industry from 1870-present day; local business; Gwim Natwick.
Research Fields: local genealogy; cranberry industry; paper making.
Activities: guided tours; permanent & temporary exhibitions.
Publications: quarterly newsletter.
Hours & Admission Prices: Memorial Day to Labor Day Sun. & Tues.-Thurs. 1-4. No charge; donations accepted.
Attendance: 1,300 (estimated)
Membership: Individual $20.

WYOMING

(127 listings)

Afton

CALLAIR MUSEUM, 150 S. Washington St., Afton, WY 83110. Mailing Address: Call Air Foundation, P.O. Box 1029, Afton, WY 83110-1029. Tel.: 307-885-2759.
E-mail: info@starvalleychamber.com
Web Site: www.starvalleychamber.com
Founded: 2008.
Institution Type/Description: Company History Museum.
Collections: CallAir company history; passenger & agricultural cropduster spray planes.
Hours & Admission Prices: Mon.-Fri. 9-5; other times by appointment. No charge.

DAUGHTERS OF UTAH PIONEERS, LINCOLN COMPANY - STAR VALLEY PIONEER AND HISTORICAL MUSEUM, 138 S. Washington, Afton, WY 83110. Mailing Address: P.O. Box 301, Thayne, WY 83127-0301.
Founded: 2001.
Key Personnel: Pres. (V), Carlie C. Jensen; Dir. & Chm. (V), Ruth H. Petersen
Governing Authority: Parent Institution: Daughters of Utah Pioneer International, Salt Lake City, UT. Tax-exempt.
Institution Type/Description: History Museum.
Collections: local history; Mormon history; period furnishings; photographs; early pioneer artifacts; Native American artifacts; clothing; dolls; buggies; beds; quilts; kitchen & homemaking items.
Activities: Annual Event: Lincoln County Fair Week.
Hours & Admission Prices: June-Aug. Fri.-Sat. 1-5, May & Sept. by appointment. No charge; donations accepted.
Attendance: 1,100 (accurate)

Banner

FORT PHIL KEARNY, 528 Wagon Box Rd., Banner, WY 82832-9604. Tel.: 307-684-7629. Fax: 307-684-7967.
Web Site: www.philkearny.vcn.com
Founded: 1913.
Congressional District: 1
Key Personnel: Site Supt., Robert C. Wilson; Cur., Sonny Reisch; Dept. Dir., Milward Simpson; State Parks & Historic Sites Div. Head, Dominic Bravo.
Personnel Profile: Full-Time Paid 3; Part-Time Paid 3.
Governing Authority: state. Parent Institution: Wyoming State Parks & Cultural Resources Department. Subsidiary Institution: State Parks & Historic Sites Division. Tax-exempt.
Institution Type/Description: Historic Site & Visitor Center: Fort Phil Kearny, the Wagon Box Fight site & the Fetterman Fight site.
Collections: Native American & military history.
Research Fields: Indian wars; military history.
Facilities: visitor center.
Activities: living history programs; tours.
Publications: information guide.
Hours & Admission Prices: April to mid-May & Oct.-Nov. Wed.-Sun. 12-4, weather permitting; mid-May to Sept. daily 8-6. Adults $4 (residents), $2 (non-residents).
Attendance: 24,000 (accurate)

Big Horn

BOZEMAN TRAIL MUSEUM, 335 Johnson St., Big Horn, WY 82833. Mailing Address: P.O. Box 566, Big Horn, WY 82833. Tel.: 307-674-6363.
Founded: 1981.
Key Personnel: Dir. & Museum Shop Mgr., Kevin Knapp; Chm. (V), Judy Slack; Pres. (V), Mike Kuzara.

Personnel Profile: Part-Time Paid 2; Part-Time Volunteers 2.
Volunteer Hours: 100
Operating Expenses: 3,000
Operating Income: 3,000
Institution Type/Description: Historic Building Museum: housed in a former blacksmith shop; built in 1879.
Collections: local history & culture; period furnishings; personal artifacts; photographs; Native American artifacts; pioneer clothing; books; blacksmith tools.
Publications: monthly newsletter; local history books.
Hours & Admission Prices: Memorial Day to Labor Day Sat.-Sun. 11-4; other times by appointment. No charge; donations accepted.
Attendance: 400 (estimated)
Membership: Individual $10.

BRADFORD BRINTON MEMORIAL & MUSEUM, (M), 239 Brinton Rd., Big Horn, WY 82833. Mailing Address: P.O. Box 460, Big Horn, WY 82833-0460. Tel.: 307-672-3173. Fax: 307-672-3258.
E-mail: kschuster@bbmandm.org
Web Site: www.bbmandm.org
Founded: 1961.
Congressional District: 1
Key Personnel: Dir. & Chief Cur., Kenneth L. Schuster; Facility Mgr., Jon Teigland; Assoc. Cur., Barbara Schuster; Registrar & Librarian, Margaret Brenneman; Annual Giving Coord., Cynthia Clark.
Personnel Profile: Full-Time Paid 4; Part-Time Paid 10; Part-Time Volunteers 6; Interns 1.
Governing Authority: privately endowed. Parent Institution: The Helen Brinton Trust. Tax-exempt.
Institution Type/Description: Historic House and Art Museum: on a gentleman's working ranch established in 1892.
Collections: Western paintings, prints, drawings & sculpture by Frederic Remington, Charles M. Russell, Edward Borein, Frank Tenney Johnson, Winold Reiss, E. W. Gollings, Frank W. Benson & Hans Kleiber; Indian arts & crafts including Navajo rugs & blankets; American Indian ethnology; American West history. Historic House: 1892 Quarter Circle A ranch & ranchhouse; ranch hand quarters; Little Goose Creek Lodge.
Research Fields: American West artists; early ranching in Wyoming.
Facilities: 3,600-vol. library of art & history books available for use on-site by appointment. Books, reproductions & original art for sale.
Activities: temporary & permanent exhibitions; gallery talks; lectures.
Publications: annual catalogs & monographs; newsletter.
Hours & Admission Prices: May 15-Sept. 3 Mon.-Sat. 10-5, Sun. 12-5; Sept. 8-Oct. Wed.-Sat. 10-5, Sun. 12-5, other times by appointment. Adults $4, seniors & students over 13 $3; discounts to AAM, ICOM, NARM & AAA members; children 13 & under, AAM & museum members no charge.
Attendance: 10,000 (estimated)
Membership: Senior & Student $15; Individual $25; Family $50; Contributor $75; Sponsor $100; Collector $500; Patron $750; Benefactor $1,000; Little Goose Creek Lodge Club $3,000.

Big Piney

GREEN RIVER VALLEY MUSEUM, 206 N. Front St., Big Piney, WY 83113. Mailing Address: P.O. Box 12, Big Piney, WY 83113-0012. Tel.: 307-276-5343.
Web Site: www.grvm.com
Founded: 1991.
Key Personnel: Dir., Jeannie Lockwood; Pres. (V), Joe Kozeal; Bookkeeper, Karen Taylor.
Personnel Profile: Full-Time Paid 2; Interns 2.
Institution Type/Description: History Museum.
Collections: history & culture of Green River Valley; prehistoric Indian artifacts; early ranching & brands; ranch equipment; area oil & gas history; period oil field tools; oral histories; homesteader cabin.
Activities: tours.
Hours & Admission Prices: June 15-Oct. 15 Tues.-Sat. 10-4. No charge; donations accepted.
Attendance: 600 (estimated)
Membership: Member $25; Sponsor $50; Patron $100; Benefactor $500.

Buffalo

*** JIM GATCHELL MEMORIAL MUSEUM, (M),** 100 Fort St., Buffalo, WY 82834. Mailing Address: P.O. Box 596, Buffalo, WY 82834-0596. Tel.: 307-684-9331.
E-mail: director@jimgatchell.com
Web Site: www.jimgatchell.com
Founded: 1957.

Key Personnel: C.E.O., John Gavin; Pres. (V), David Osmundsen; Museum Educator, Jennifer Romanoski; Registrar, Sylvia Bruner.

Personnel Profile: Full-Time Paid 4; Part-Time Paid 12; Part-Time Volunteers 30.

Governing Authority: county. Tax-exempt.

Institution Type/Description: History Museum.

Collections: military, pioneer, & Indian history in Wyoming & the West; Indian artifacts; archaeology; archives; mineralogy; military; paintings; period vehicles including chuck wagon, bed roll & supply wagon; sheep wagon; road wagon; buggies.

Major Exhibits: Out of the Frying Pan...Into the Fire, 5/13-3/14.

Research Fields: Bozeman Trail; Indian wars; Johnson County Cattle War.

Activities: guided tours; lectures; Bozeman trail map.

Publications: quarterly newsletter, The Sentry; book, Jim Gatchell the Man and the Museum.

Hours & Admission Prices: Memorial Day to Labor Day Mon.-Sat. 9-6, Sun. 12-6; Sept.-May Mon.-Sat. 9-4. Family $12, adults $5, children 6-16 $3; discounts for active military, AAM, ICOM, AASLH, MPMA & CWAM members; members & children under 6 no charge. &

Attendance: 7,500 (accurate)

Membership: Level 1 $35; Level 2 $60; Level 3 $150; Patron $2,500.

MUSEUM OF THE OCCIDENTAL, 10 N. Main St., Buffalo, WY 82834-1815. Mailing Address: P.O. Box 383, Buffalo, WY 82834-0383. Tel.: 307-684-0451, Fax: 307-684-5980.

E-mail: info@occidentalwyoming.com

Web Site: www.occidentalwyoming.com

Key Personnel: Dir., Dawn Wexo

Institution Type/Description: History Museum.

Collections: photographs; documents; period artifacts.

Hours & Admission Prices: Daily 10-6. No charge.

Casper

CASPER PLANETARIUM, 904 N. Poplar St., Casper, WY 82601-1348. Tel.: 307-577-0310. Fax: 307-577-6750.

E-mail: michele_wistien@ncsd.k12.wy.us

Web Site: www.natronaschools.org/planetarium

Key Personnel: Supvr., Michelle Wistisen

Institution Type/Description: Planetarium.

Collections: astronomy-related exhibits.

Activities: educational astronomy-related programs.

Hours & Admission Prices: Summer: Tues.-Sat. 8:30 am-9:30 pm; Winter: Sat. 7pm & 8 pm. Closed major holidays.

CRIMSON DAWN MUSEUM & PARK, 1620 E. Crimson Dawn Rd., Casper, WY 82601-9740. Mailing Address: P.O. Box 2578, Mills, WY 82644. Tel.: 307-235-1303 & 9311.

Web Site: www.crimsondawnpark.com

Founded: 1978.

Personnel Profile: Part-Time Paid 1.

Institution Type/Description: History Museum.

Collections: local history & culture; period furnishings; personal artifacts; ranching; photographs.

Facilities: nature trails.

Activities: walking trails.

Hours & Admission Prices: Mid-June to mid-Sept. Sat.-Sun. 10-7. No charge; donations accepted.

Attendance: 2,000 (estimated)

FORT CASPAR MUSEUM, (M), 4001 Fort Caspar Rd., Casper, WY 82604-2923. Tel.: 307-235-8462. Fax: 307-235-8464.

E-mail: ryoung@cityofcasperwy.com

Web Site: www.fortcasparwyoming.com

Founded: 1936.

Congressional District: 1

Key Personnel: C.E.O., Richard L. Young; Cur., Michelle Bahe; Cur. Education, Clifon Corkern, III; Administrative Support Tech, Anne Holman.

Personnel Profile: Full-Time Paid 4; Part-Time Paid 3; Part-Time Volunteers 15; Interns 1.

Governing Authority: municipal. Parent Institution: City of Casper. Subsidiary Institution: Fort Caspar Museum Association. Affiliated with the Advisory Guidance of Leisure Services Board of Directors. Tax-exempt.

Institution Type/Description: Social History Museum: Listed on the National Register of Historic Places.

Collections: Civil War, Indian wars, Indian & pioneer artifacts; central Wyoming materials; reconstructed c.1865 fort buildings.

Research Fields: Indian war periods; local & state history.

Facilities: interpretative center.

Activities: lectures; TV & radio programs; formally organized education programs for children; permanent & traveling exhibitions.

Publications: brochures; books, Fort Caspar; Bison Hunters to Black Gold; Frontier Crossroads: A History of Fort Caspar and the Upper Platte Crossing; The Life & Letters of Caspar W. Collins; Fort Caspar Activity Book; Natrona County; People, Place & Time.

Hours & Admission Prices: May-June & Aug.-Sept. daily 8-6; July daily 8-7; Oct.-April Tues.-Sat. 8-5. Adults $3, youth 13-18 $2; children 12 & under and members no charge. &

Attendance: 35,000 (estimated)

Membership: Student & Senior Citizen $10; Individual $15; Family $30; Business $250; Patron $500; Benefactor $1,000.

IDA GOODSTEIN VISUAL ARTS CENTER, (M), Casper College, 125 College Dr., Casper, WY 82601-4612. Tel.: 307-268-2060. Fax: 307-268-3337.

E-mail: veggemeyer@caspercollege.edu

Web Site: www.caspercollege.edu

Key Personnel: Dir., Valerie Innella

Institution Type/Description: Art Museum.

Collections: drawings; paintings; photography; ceramics; printmaking; graphic design; sculpture; jewelry.

Hours & Admission Prices: Mon.-Thurs. 9-4. No charge. Closed holidays.

NATIONAL HISTORIC TRAILS INTERPRETIVE CENTER, (M), 1501 N. Poplar St., Casper, WY 82601-1375. Mailing Address: P.O. Box 397, Casper, WY 82602. Tel.: 307-265-8030. Fax: 307-265-0986.

E-mail: nhtcf@hotmail.com

Web Site: www.blm.gov/wy/st/en/nhtic.html

Founded: 2002.

Key Personnel: Exec. Dir., Holly Turner; BLM Dir., Mike Abel; Pres. (V), Bill Mortimer.

Personnel Profile: Part-Time Paid 1.

Governing Authority: Parent Institution: BLM Foundation. Tax-exempt: 501(c)(3).

Institution Type/Description: History Museum.

Collections: local history & culture; period furnishings; personal artifacts; photographs; hands-on exhibits; Oregon Trail; Mormon Trail; California Trail; Pony Express Trail.

Activities: group tours; hands-on exhibits.

Hours & Admission Prices: May to Labor Day daily 8-5; Sept.-April Tues.-Sat. 9-4:30. Adults $6, seniors 62 & over $5, students 16 & over $4; discounts to AAM members; children under 16 no charge. Closed New Year's Day; Easter; Thanksgiving; Christmas. &

Attendance: 25,000 (accurate)

*** NICOLAYSEN ART MUSEUM AND DISCOVERY CENTER, (M),** 400 East Collins Dr., Casper, WY 82601-2815. Tel.: 307-235-5247. Fax: 307-235-0923.

E-mail: info@thenic.org

Web Site: www.thenic.org

Founded: 1967.

Congressional District: 1

Key Personnel: Exec. Dir., Brooks Joyner; Registrar, Lisa Fujita; Cur., Eric Wimmer; Deputy Dir., Jan DeBeer; Master Teacher, Jim Kopp; Special Events & Volunteer Coord., Lori Klatt; Master Teacher, Lisa Vlastos.

Personnel Profile: Full-Time Paid 11; Part-Time Paid 5; Part-Time Volunteers 50; Interns 3.

Operating Expenses: 1,089,000

Operating Income: 1,136,000

Governing Authority: private; nonprofit organization. Tax-exempt: 501(c)(3).

Institution Type/Description: Art Museum.

Collections: regional contemporary art.

Research Fields: contemporary art.

Facilities: discovery center.

Activities: traveling exhibitions; workshops; lectures; hands-on; self guided; art making; formally organized education programs for children & adults.

Publications: newsletters; catalogs.

Hours & Admission Prices: Tues.-Sat. 10-5, Sun. 12-4. Adults $5; members no charge. Closed holidays. &

Attendance: 50,000 (accurate)

Membership: Individual $35; Family $60; Business $250.

THE SCIENCE ZONE, INC., 111 W. Midwest Ave., Casper, WY 82601. Mailing Address: P.O. Box 2701, Mills, WY 82644-2701. Tel.: 307-473-9663.

E-mail: carrie.tsz@hotmail.com

Web Site: www.thesciencezone.org

Founded: 1998.
Key Personnel: Exec. Dir., Carrie O. Schroeder; Chm. (V), Christine Stack.
Personnel Profile: Full-Time Paid 4; Part-Time Paid 3; Part-Time Volunteers 12.
Governing Authority: Tax-exempt.
Institution Type/Description: Science Center.
Collections: hands-on science exhibits.
Facilities: classroom; laboratory.
Activities: educational programs; after-school clubs; summer camps.
Hours & Admission Prices: Tues.-Sat. 10-5. Adults $4, children $3; discounts to groups of 10 or more; ASTC members & children under 2 no charge. &
Attendance: 25,000 (accurate)
Membership: Family $75.

TATE GEOLOGICAL MUSEUM, (M), Casper College, 125 College Dr., Casper, WY 82601-4612. Tel.: 307-268-2447. Fax: 307-268-3308.
E-mail: dschaff@caspercollege.edu
Web Site: www.caspercollege.edu/tate
Founded: 1980.
Key Personnel: Dir., Deanna K. Schaff; Public Rels., Lisa Icenogle; Museum Shop Mgr., Kristin Taylor.
Personnel Profile: Full-Time Paid 4; Part-Time Paid 1; Part-Time Volunteers 36.
Governing Authority: public college. Parent Institution: Casper College. Tax-exempt.
Institution Type/Description: Geology Museum.
Collections: paleontological & mineralogical heritage of Wyoming.
Publications: bimonthly newsletter, Tate Museum Geological Times.
Hours & Admission Prices: Mon.-Fri. 9-5, Sat. 10-4. No charge; donations accepted.
Attendance: 18,000 (accurate)
Membership: Individual $18; Family $36; Sponsor $1,000.

WERNER WILDLIFE MUSEUM, (M), 405 E. 15th St, Casper, WY 82601-4612. Mailing Address: Casper College, 125 College Dr., Casper, WY 82601. Tel.: 307-235-2108.
Founded: 1970.
Key Personnel: Dir., Deanna Schaff; Pres. Caper College, Dr. Walter Nolte; Museum Assoc., Sue Easton.
Personnel Profile: Full-Time Paid 2; Part-Time Volunteers 2.
Governing Authority: Parent Institution: Casper College. Tax-exempt.
Institution Type/Description: Wildlife Museum.
Collections: wildlife from Wyoming & around the world.
Activities: tours.
Hours & Admission Prices: Mon.- Fri. 8:30-5. No charge; donations accepted. &
Attendance: 6,000 (accurate)

WYOMING VETERANS' MEMORIAL MUSEUM, 3740 Jourgensen Ave., Natrona County Intl. Airport, Casper, WY 82604. Mailing Address: 4001 Ft. Caspar Rd., Casper, WY 82604. Tel.: 307-472-1857.
E-mail: john.goss@wyo.gov
Web Site: www.wyo.gov
Founded: 2001.
Key Personnel: Dir., John G. Goss; Cur., Eric Wimmer.
Personnel Profile: Full-Time Paid 2; Part-Time Volunteers 7.
Governing Authority: Parent Institution: Wyoming Military Department. Subsidiary Institution: Wyoming Veterans Commission. Tax-exempt.
Institution Type/Description: Military History Museum.
Collections: Casper Army Air Base heritage; oral military history & equipment; photographs; veterans memorial; archives; photographs.
Research Fields: military history & personnel; Casper Army Air Base heritage.
Facilities: library; archives.
Activities: lectures; workshops; guided tours.
Hours & Admission Prices: Summer: Tues.-Sat. 9-4. No charge; donations accepted. &
Attendance: 1,600 (accurate)

Centennial

NICI SELF HISTORICAL MUSEUM, (M), 2734 Hwy. 130, Centennial, WY 82055. Mailing Address: P.O. Box 201, Centennial, WY 82055-0201. Tel.: 307-742-7763. Facebook: Nici Self Historical Museum.
E-mail: cvha@live.com
Web Site: www.cvha.qwestoffice.net

Founded: 1974.
Congressional District: 1
Key Personnel: Chm. (V) & Pres. (V), Jim Chase; Vice Pres., Joe Witt; Sec., Nancy Taft; Treas., Cecily Goldie.
Personnel Profile: Part-Time Paid 1; Part-Time Volunteers 30.
Governing Authority: nonprofit organization. Parent Institution: Centennial Valley Historical Association. Tax-exempt: 501(c)(3).
Institution Type/Description: History Museum: housed in 1907 Railroad Depot located at the base of the Medicine Bow Mountains.
Collections: railroad equipment; 1944 Union Pacific caboose; gold & platinum mining; lumbering & ranching; Native American artifacts.
Research Fields: newspapers & books of the area.
Activities: educational & research activities; special tours. Annual Event: art show.
Publications: book, Centennial, Wyoming 1876-1976: The Real Centennial; annual letter to members.
Hours & Admission Prices: Memorial Day to Labor Day Thurs.-Mon. 12-4. No charge; donations accepted.
Attendance: 1,450 (estimated)

Cheyenne

CHEYENNE BOTANIC GARDENS, (M), 710 S. Lions Park Dr., Cheyenne, WY 82001-7503. Tel.: 307-637-6458. Fax: 307-637-6453.
E-mail: info@botanic.org
Web Site: www.botanic.org
Founded: 1976.
Key Personnel: Dir., Shane Smith; Chm. (V), Cindy Pomeroy; Asst. Dir., Claus Johnson; Dir. Devel., Darcee Snider; Head Horticulture, Steve Scott; Asst. Education Dir., Tyler Mason; Exec. Sec., Trudy Fox.
Personnel Profile: Full-Time Paid 6.
Governing Authority: municipal government. Tax-exempt.
Institution Type/Description: Botanical Garden.
Collections: plants; trees; flowers.
Facilities: library; educational facilities.
Activities: guided tours; lectures; broadcast programs; passive solar energy demonstration; Wetland Discovery area; horticultural therapy for seniors & youth at risk. Annual Events: Stain Glass Show; Perennial Plant Exchange.
Publications: quarterly newsletter, The Cheyenne Garden Gazette.
Hours & Admission Prices: Grounds: daily dawn-dusk. Greenhouse: Mon.-Fri. 8-4:30, Sat.-Sun. 11-3:30. Children's Village: Tues.-Sat. 9-5:30, Sun 10-4. No charge; donations accepted. Closed New Year's Day; Easter; Thanksgiving; Christmas. &
Attendance: 42,525 (accurate)
Membership: Individual $35; Couple/Dual $45; Family $75; Business Associate $125; Sustaining $150; Business Circle $250; Ponderosa $500.

CHEYENNE DEPOT MUSEUM, Number One Depot Sq., 115 W. 15th, Cheyenne, WY 82001. Mailing Address: P.O. Box 2160, Cheyenne, WY 82003-2160. Tel.: 307-632-3905.
E-mail: james@cheyennedepotmuseum.org
Web Site: www.cheyennedepotmuseum.org
Formerly: Wyoming Transportation Museum
Founded: 2002.
Key Personnel: Dir., James Bowers.
Personnel Profile: Full-Time Paid 3; Full-Time Volunteers 1; Part-Time Paid 9; Part-Time Volunteers 11.
Governing Authority: Tax-exempt.
Institution Type/Description: History Museum.
Collections: period artifacts.
Hours & Admission Prices: Mon.-Fri. 9-7, Sat. 9-5, Sun. 11-5. &
Attendance: 120,000 (estimated)

CHEYENNE FRONTIER DAYS OLD WEST MUSEUM, (M), 4610 N. Carey Ave., Cheyenne, WY 82001-7505. Mailing Address: P.O. Box 2720, Cheyenne, WY 82003-2720. Tel.: 307-778-7290. Fax: 307-778-7288.
E-mail: info@oldwestmuscum.org
Web Site: oldwestmuseum.org
Founded: 1978.
Key Personnel: Dir., Amiee S. Reese; Exhibits, David Phelps; Cur. Collections, Mike Kassel; Facilities Mgr., Ron Duckworth; Dir. Arts Devel, Tiffany Smith; Volunteer Coord., Janet Wampler; Weekend Security, Wil Madrid.
Personnel Profile: Full-Time Paid 6; Part-Time Paid 2; Part-Time Volunteers 250.
Governing Authority: nonprofit organization. Tax-exempt: 501(c)(3).
Institution Type/Description: History Museum.
Collections: Cheyenne frontier days rodeo artifacts, horse-drawn vehicles, Western art, Cheyenne & Wyoming history; carriages.

Major Exhibits: C.B. Irwin Family Exhibit, 11/2/12-10/15/14.
Research Fields: Cheyenne Frontier Days; Cheyenne; Laramie County.
Facilities: 1,000-vol. library of books on Cheyenne Frontier Days, western history & Cheyenne-Wyoming history; educational facilities. Educational & souvenir items pertaining to western history, rodeo & Cheyenne Frontier Days for sale.
Activities: guided tours; school loan service; formally organized education programs. Museum Sponsors: Western Spirit Art Show & Sale; Western Invitational Art Show & Sale.
Publications: quarterly newsletter, The Stageline.
Hours & Admission Prices: Mon.-Fri. 9-5, Sat.-Sun. 10-5. Adults $10, seniors & military $8, students $5; discounts to AAM members & groups; children under 12 & members no charge. Closed New Year's Day; Presidents' Day; Easter; Thanksgiving; Christmas. &
Attendance: 45,000 (estimated)
Membership: Volunteer $25; Friend $35; Family $55; Business $75; HOF $125; Partner $250.

COWGIRLS OF THE WEST, 203-205 W. 17th St., Cheyenne, WY 82001-4411. Mailing Address: P.O. Box 525, Cheyenne, WY 82003. Tel.: 307-638-4994.
Founded: 1994.
Key Personnel: Chm. (V), Stacey Gierisch; Pres. (V), Sharon Russell; Museum Shop Mgr., Pam Cooper.
Personnel Profile: Part-Time Paid 1; Part-Time Volunteers 50.
Governing Authority: nonprofit organization. Tax-exempt.
Institution Type/Description: History Museum.
Collections: history of the western cowgirl & pioneering women; western culture; photographs; personal artifacts.
Activities: special events. Museum Sponsors: Frontier Days in July.
Publications: monthly, Cowgirl Update.
Hours & Admission Prices: Jan. 16-Dec. Tues.-Fri. 11-4, Sat. 11-3. No charge; donations accepted.
Attendance: 4,000 (accurate)

THE ESTHER AND JOHN CLAY FINE ARTS GALLERY, Laramie County Community College, Fine Arts Bldg., 1400 E. College Dr., Cheyenne, WY 82007-3204. Tel.: 307-777-1158.
Web Site: www.lccc.wy.edu
Institution Type/Description: Art Gallery.
Collections: paintings; sculpture; ceramics; drawings.
Hours & Admission Prices: Mon.-Fri. 8-5. No charge.

HISTORIC GOVERNORS' MANSION, 300 E. 21st. St., Cheyenne, WY 82001-3712. Mailing Address: Dept. of State Parks & Cultural Resources, Barrett Bldg., Cheyenne, WY 82002. Tel.: 307-777-7878. Fax: 307-635-7077.
E-mail: sphs@state.wy.us
Web Site: www.artsparkhistory.com
Founded: 1904.
Key Personnel: Dir. Div. Parks, Domenic Bravo; Supt., Deborah Amend.
Personnel Profile: Full-Time Paid 1; Part-Time Volunteers 2; Interns 1.
Governing Authority: state. Parent Institution: Dept. of State Parks & Cultural Resources, Barrett Bldg., Cheyenne, WY 82002. Tax-exempt.
Institution Type/Description: Historic Building: 1904 Colonial Revival Mansion & Carriage House.
Collections: turn-of-the-century art, photographs, quilts; 1937 Chicago Furniture Mart furnishings; furnishings belonging to the first families of Wyoming including Nellie Taylor Ross, the first elected woman governor of the US in 1925.
Research Fields: political history of the state of Wyoming.
Facilities: limited on-site research materials.
Activities: guided tours; public programs; historic hospitality house; video tour; touch screen kiosks; audio tours; cell phone tours.
Publications: information guide; self-guided brochure.
Hours & Admission Prices: June-Aug. Mon.-Sat. 9-5, Sun. 1-5; Sept.-May Tues.-Sat. 9-5; groups by appointment. No charge; donations accepted. Closed most holidays. &
Attendance: 10,189 (accurate)

NELSON MUSEUM OF THE WEST, (M), 1714 Carey Ave., Cheyenne, WY 82001-4420. Tel.: 307-635-7670. Fax: 307-778-3926.
E-mail: director@nelsonmuseum.com
Web Site: www.nelsonmuseum.com
Founded: 1998.
Key Personnel: Dir., Robert Nelson; Pres. (V), Beth Nelson.
Personnel Profile: Part-Time Volunteers 6.

Governing Authority: Tax-exempt.
Institution Type/Description: History Museum.
Collections: cowboy & American Indian artifacts; Western fine art; rodeo; paintings; sculpture; beadwork; pottery; weavings; spurs; baskets; chaps; saddles; period firearms; military; outlaws & lawmen; wildlife mounts from around the world; U.S. Cavalry artifacts including uniforms, saddles, & weapons; U.S. Army & Air Force officers uniforms.
Research Fields: military history.
Hours & Admission Prices: May-Aug. Mon.-Sat. 9-4:30; Sept.-Nov. 1 Mon.-Fri. 9-4:30; other times by appointment. Adults $4, seniors 65 & over $2; discounts to groups and AAM members & school groups; children under 12 no charge. &
Attendance: 5,000 (accurate)

WYOMING ARTS COUNCIL GALLERY, 2320 Capitol Ave., Cheyenne, WY 82002. Tel.: 307-777-7742. Fax: 307-777-5499.
E-mail: rita.basom@wyo.gov
Web Site: www.wyomingartscouncil.org
Founded: 1976.
Congressional District: 7
Key Personnel: Program Mgr., Rita Basom.
Personnel Profile: Full-Time Paid 8; Part-Time Paid 1.
Governing Authority: state. Subsidiary Institution: Department of State Parks and Cultural Resources, Division of Cultural Resources, Cheyenne, WY. Tax-exempt: 501(c)(3).
Institution Type/Description: Art Gallery.
Collections: works by Wyoming artists from the Governor's Capitol Art collection.
Facilities: 700 sq. ft. exhibit space; meeting area.
Activities: lectures; temporary exhibitions.
Hours & Admission Prices: Mon.-Fri. 8-5. No charge; donations accepted. Closed major holidays. &
Attendance: 600 (estimated)

* **WYOMING STATE MUSEUM, (M),** 2301 Central Ave., Barrett Bldg., Cheyenne, WY 82001-3110. Tel.: 307-777-7022. Fax: 307-777-5375.
E-mail: mvigil@state.wy.us
Web Site: wyomuseum.state.wy.us
Founded: 1895.
Key Personnel: Museum Supvr., Manny Vigil; Pres. (V), Carolyn Turbiville; Supvr. Interpretation & Education, Heyward Schrock; Supvr. Collections, Jennifer Alexander; Supvr. Collections, Jim Allison; Education Cur., Sarah Ligocki; Cur. Collections, Mandy Langfald; Cur. Collections, Mariah Emmons; Cur. Education, Sarah Wigocki; Registrar, Dominique Schultes; Exhibits Designer, Larry Lujan; Museum Store Mgr., Beth Miller.
Personnel Profile: Full-Time Paid 10; Part-Time Volunteers 60; Interns 1.
Governing Authority: state. Parent Institution: State of Wyoming. Subsidiary Institution: Dept. of State Parks & Cultural Resources. Tax-exempt.
Institution Type/Description: State History Museum.
Collections: over 100,000 artifacts related to Wyoming's heritage; textiles; firearms; household artifacts; Native American artifacts.
Research Fields: History Museum.
Facilities: meeting room. Museum-related items for sale.
Activities: guided tours; programs; monthly lecture series; monthly children's programs; temporary & traveling exhibitions; discovery trunks for schools statewide; demonstrations; workshops.
Publications: exhibition brochures & handouts.
Hours & Admission Prices: May-Oct. Mon.-Sat. 9-4:30; Nov.-April Mon.-Fri. 9-4:30, Sat. 10-2. No charge; donations accepted. Closed state & federal holidays. &
Attendance: 25,000 (estimated)
Membership: Active $12; Associate $25; Benefactor $50-$100, $100-$200, $200 & up.

Chugwater

CHUGWATER MUSEUM, Main St., Chugwater, WY 82210. Mailing Address: P.O. Box 33, Chugwater, WY 82210-0033. Tel.: 307-422-3509.
E-mail: ruthvaughn@hotmail.com
Founded: 1986.
Key Personnel: Dir., Ruth Vaughn.
Personnel Profile: Part-Time Volunteers 24.
Governing Authority: Parent Institution: Chugwater Historical Unity Group.
Institution Type/Description: History Museum.
Collections: local history; period furnishings; tools; horsedrawn farm equipment; sheep wagon; caboose; early office & business equipment; maps.
Activities: research.

Hours & Admission Prices: Memorial Day to Labor Day Sat.-Sun. & holidays 1-4; other times by appointment. No charge; donations accepted.
Attendance: 400 (estimated)

Clearmont

UCROSS FOUNDATION ART GALLERY, 30 Big Red Lane, Clearmont, WY 82835-9723. Tel.: 307-737-2291. Fax: 307-737-2322.
E-mail: info@ucross.org
Institution Type/Description: Art Gallery.
Collections: works by contemporary artists; local history. Historic House: 1880s ranch house.
Hours & Admission Prices: Mon.-Fri. 8:30-4, Sat. by appointment. No charge; donations accepted. Closed major holidays.

Cody

BIG HORN GALLERIES, 1167 Sheridan Ave., Cody, WY 82414-3627. Tel.: 307-527-7587. Fax: 307-527-7586.
E-mail: bhgcody@aol.com
Institution Type/Description: Art Gallery.
Collections: western & wildlife art and landscapes.
Hours & Admission Prices: Call for hours.

BUFFALO BILL DAM VISITOR CENTER, 47 Lakeside Dr., Cody, WY 82414-8501. Tel.: 307-527-6076.
E-mail: manager@bbdvc.com
Web Site: www.bbdvc.com
Governing Authority: nonprofit organization.
Institution Type/Description: History Museum.
Collections: dam & Big Horn Basin history; photographs.
Facilities: Museum-related items for sale.
Hours & Admission Prices: May & Sept. Mon.-Fri. 8-6, Sat.-Sun. 9-5; June-Aug. Mon.-Fri. 8-7, Sat.-Sun. 9-5. No charge; donations accepted.

* **BUFFALO BILL HISTORICAL CENTER, (M),** 720 Sheridan Ave., Cody, WY 82414-3428. Tel.: 307-587-4771, ext. 0. Fax: 307-587-5714.
E-mail: rustyh@bbhc.org
Web Site: www.bbhc.org
Founded: 1917.
Congressional District: 1
Key Personnel: Exec. Dir., Bruce Eldredge; Chm. (V), Barron G. Collier, II; Dir. Devel., Tom Roberson; Cur. Whitney Gallery of Western Art, Mindy Besaw; Cur. Plains Indian Museum, Emma Hansen; Cur. Draper Museum of Natural History, Dr. Charles Preston; Cur. Research Library, Mary Robinson; Cur. Buffalo Bill Museum, John Rumm; Accounting Mgr., Meg Kath; Facilities Mgr., Paul Brock; Museum Shop Mgr., Kelly Webber.
Personnel Profile: Full-Time Paid 85; Part-Time Paid 55; Part-Time Volunteers 150; Interns 5.
Governing Authority: nonprofit organization. Owned and operated by Buffalo Bill Memorial Association. Tax-exempt.
Institution Type/Description: Art, History, & Natural Science Museum.
Collections: Western memorabilia including personal possessions of Buffalo Bill Cody & his Wild West show. Whitney Gallery of Western Art: paintings & sculpture. Plains Indian Museum: ethnological & archaeological materials. Cody Firearms Museum: 16th-century to present American-European firearms; study gallery of firearms; two sculpture gardens; Powwow garden. Draper Museum of Natural History: mammalogy; ornithology, geology & paleontology of the Yellowstone region. Historic Buildings: 1842 Buffalo Bill's boyhood home; 1905 Joseph Henry Sharp cabin.
Research Fields: American firearms; Western art & history; Buffalo Bill; Native American studies; vertebrate ecology & conservation biology in Yellowstone and Rocky Mountain West.
Facilities: 20,000-vol. library of books, manuscripts & 500,000 photographs pertaining to Western Americana available for research on premises; archives; 190,000 sq. ft. exhibit space; restaurant. Books & museum-related items for sale.
Activities: inter-museum loan exhibitions; permanent & loan exhibitions of Western art and Americana; public programs; seminars; symposiums; receptions & fund-raising events; natural science field trips; lecture series; sleepovers; mentoring; live raptor education. Museum Sponsors: Larom Institute of Western American Studies; Plains Indian Seminar; Powwow.
Publications: newsletter, Points West; catalogues; annual report; exhibition & gallery brochures.
Hours & Admission Prices: March & Nov. daily 10-5; April & Sept. 16-Oct. daily 8-5; May-Sept. 15 daily 8-6; Dec.-Feb. Thurs.-Sun. 10-5. Adults $18, seniors $16, students 18 & over with ID $13, youth 6-17 $10; discounts to

groups and AAA & AAM members; children under 6 & patron members no charge. Closed New Year's Day; Thanksgiving; Christmas. &
Attendance: 200,000 (accurate)
Membership: Individual $45; Family $65; Centennial $100; Cody Firearms $150-$1,000; Sponsor $250; Sustaining $500; Benefactor $1,000; Pahaska League $2,000; Corporate membership available.

THE CODY DUG UP GUN MUSEUM, 1020 12th St., Cody, WY 82414. Tel.: 307-587-3344.
E-mail: codydugupgunmuseum@hotmail.com
Web Site: www.codydugupgunmuseum.com
Founded: 2008.
Key Personnel: Dir., Hans Kurth; C.E.O., Eva Kurth.
Personnel Profile: Full-Time Paid 2.
Institution Type/Description: History Museum.
Collections: local history; period guns & weapons; period artifacts.
Hours & Admission Prices: May-Oct. daily 9-9; call for additional hours. No charge; donations accepted.
Attendance: 17,500 (estimated)

MUSEUM OF THE OLD WEST & OLD TRAIL TOWN, 1831 DeMaris Dr., Cody, WY 82414. Mailing Address: P.O. Box 546, Cody, WY 82414-0546. Tel.: 307-587-5302.
E-mail: oldtrailtown@tctwest.net
Web Site: www.oldtrailtown.org
Institution Type/Description: Historic Buildings: 26 buildings built 1879-1901.
Collections: local history & culture; guns; period furnishings & clothing; over 100 horse drawn vehicles; Native American artifacts. Historic Buildings: saloon; livery barn; blacksmith shop; general stores; post office; schools; Butch Cassidy and the Sundance Kid's cabin.
Hours & Admission Prices: mid-May to Sept.

SIMPSON GALLAGHER GALLERY, 1161 Sheridan Ave., Cody, WY 82414-3627. Tel.: 307-587-4022.
E-mail: sue@simpsongallaghergallery.com
Web Site: www.simpsongallaghergallery.com
Founded: 1994.
Institution Type/Description: Art Gallery.
Collections: landscape paintings; wildlife sculpture; western intaglios.
Hours & Admission Prices: Call for hours. &

TECUMSAH'S OLD WEST MINIATURE VILLAGE & MUSEUM, 140 W. Yellowstone Ave., Cody, WY 82414. Tel.: 307-587-5362.
Web Site: www.tecumsehs.com
Key Personnel: Owner, Jerry Fick
Institution Type/Description: History Museum.
Collections: miniature village depicting 19th century frontier history in Wyoming; Geronimo's bow, arrows, & quiver; guns; Native American & cowboy artifacts.
Activities: school tours.
Hours & Admission Prices: mid-May to mid-Sept. daily 8-8; Winter: call for hours. Adults $5, seniors & students 13-18 $3, children 7-12 $2; children 6 & under no charge.

Colter Bay Village

GRAND TETON NATIONAL PARK, COLTER BAY VISITOR CENTER, Colter Bay Visitor Center, Grand Teton National Park, Colter Bay Village, WY 83012. Mailing Address: P.O. Drawer 170, Moose, WY 83012-0170. Tel.: 307-739-3594. Fax: 307-739-3504.
Web Site: www.nps.gov/grte/
Formerly: Grand Teton National Park, Colter Bay Indian Arts Museum
Founded: 1972.
Key Personnel: Supt., Mary Gibson.
Governing Authority: federal. Parent Institution: National Park Service. Subsidiary Institution: Grand Teton National Park. Tax-exempt: 501(c)(3).
Institution Type/Description: National Park.
Collections: Interpretive collection of Native American artifacts.
Facilities: theater. Descriptive & interpretive items on natural & local history for sale.
Activities: guided walks; illustrated talks; films; Indian cultural demonstrations.
Publications: booklets; leaflets; paperbacks.
Hours & Admission Prices: May-June & Sept. daily 8-5; June-Sept. daily 8-7; first two weeks of Oct. 9-5. &

Devils Tower

DEVILS TOWER VISITOR CENTER, Devils Tower National Monument, State Hwy. 110, Bldg. 170, Devils Tower, WY 82714. Mailing Address: P.O. Box 10, Devils Tower, WY 82714-0010. Tel.: 307-467-5283. Fax: 307-467-5350.
E-mail: detointerpretation@nps.gov
Web Site: www.nps.gov/deto
Founded: 1906.
Congressional District: 1
Key Personnel: Chief Resources Mgmt. & Cur., Angela Wetz; Supt., Dorothy Firecloud.
Personnel Profile: Full-Time Paid 9; Part-Time Paid 15; Part-Time Volunteers 5.
Governing Authority: federal. U.S. Dept. of the Interior, National Park Service. Tax-exempt.
Institution Type/Description: National Monument Museum; 1930s CCC/WPA log building.
Collections: geology; history; botany; Indian artifacts.
Research Fields: history; geology; botany; local Indian artifacts.
Facilities: library by appointment.
Activities: guided walks; talks in summer; cultural program series.
Publications: brochures on local rock climbing history, natural history & geology.
Hours & Admission Prices: April-Oct. call for hours. Monument: $10 per vehicle. Visitor Center: no charge; donations accepted. &
Attendance: 358,000
Membership: Devils Tower Natural History Association: Single $10; Family & Associate $25.

Douglas

DOUGLAS RAILROAD INTERPRETIVE MUSEUM, 121 Brownfield Rd., Douglas, WY 82633-2558. Tel.: 307-358-2950. Fax: 307-358-2972.
Web Site: www.douglaschamber.com
Governing Authority: Parent Institution: Douglas Chamber of Commerce.
Institution Type/Description: Historic Building: housed in the former Fremont, Elkhorn and Missouri Valley Railroad depot; built in 1886.
Collections: railroad history & artifacts; photographs; period furnishings.
Hours & Admission Prices: Summer: Mon.-Fri. 9-8, Sat.-Sun. 10-5; Winter: Mon.-Fri. 9-5, Sat.-Sun. 11-4. No charge; donations accepted.
Attendance: 8,200 (estimated)

FORT FETTERMAN STATE HISTORIC SITE, 752 Hwy. 93, Douglas, WY 82633-9267. Mailing Address: P.O. Box 911, Douglas, WY 82633-0911. Tel.: 307-358-2864 & 9288. Fax: 307-358-2864 & 9293.
E-mail: aeklan@state.wy.us
Formerly: Fort Fetterman State Museum
Founded: 1963.
Key Personnel: C.E.O., Dominic Bravo; Site Supvr., Arlene Ekland-Earnst; Museum Shop Mgr., Peg Fetterman.
Personnel Profile: Full-Time Paid 1; Part-Time Paid 2; Part-Time Volunteers 1.
Governing Authority: state. Parent Institution: Wyoming Dept. of State Parks & Cultural Resources. Subsidiary Institution: Wyoming State Parks & Historic Sites. Tax-exempt: 170(b)(1)(A).
Institution Type/Description: Military Museum: housed in 1867-1882 historic Fort Fetterman Officer's Quarters & ordnance building on original site.
Collections: military artifacts & equipment; archaeology; early wagons. Historic Building: Ordnance Building.
Research Fields: history of Fort Fetterman.
Facilities: Publications & museum-related items for sale.
Activities: living history demonstrations & crafts. Museum Sponsors: Ft. Fetterman Days.
Publications: information guide; site brochure.
Hours & Admission Prices: Memorial Day to Labor Day daily 9-5. Non-residents $4, residents $2; children under 17 no charge. &
Attendance: 10,000 (accurate)

WYOMING PIONEER MEMORIAL MUSEUM, Wyoming State Fairgrounds, 400 W. Center St., Douglas, WY 82633. Mailing Address: P.O. Box 911, Douglas, WY 82633-0911. Tel.: 307-358-9288. Fax: 307-358-9293.
E-mail: aeklan@state.wy.us
Web Site: www.spacr.state.wy.us
Founded: 1956.
Key Personnel: Dir. & Cur., Arlene E. Earnst.
Personnel Profile: Full-Time Paid 2; Part-Time Paid 1; Part-Time Volunteers 2.

Governing Authority: state. Parent Institution: State of Wyoming. Tax-exempt.
Institution Type/Description: General Museum.
Collections: Indian artifacts; pioneer relics; textiles; agriculture; pioneer costumes. Historic Houses: 1886 schoolhouse; 1896 first log cabin museum; original 1925 log cabin museum.
Activities: permanent & temporary exhibitions.
Hours & Admission Prices: Summer: Mon.-Fri. 8-5, Sat. 1-5; Winter: Mon.-Fri. 8-5. No charge. Closed national holidays. &
Attendance: 20,000

Dubois

DUBOIS MUSEUM, 909 West Ramshorn, Dubois, WY 82513. Mailing Address: P.O. Box 896, Dubois, WY 82513-0896. Tel.: 307-455-2284. Fax: 307-455-2912.
E-mail: duboismuseum@gmail.com
Web Site: duboismuseum.org
Formerly: Wind River Historical Center Dubois Museum
Founded: 1976.
Key Personnel: Dir., Katrina Krupicka.
Personnel Profile: Full-Time Paid 2; Part-Time Paid 6.
Governing Authority: county; nonprofit. Subsidiary Institution: Fremont County Museum, 450 N. Second St., Rm. 320, Lander, WY 82520. Tax-exempt.
Institution Type/Description: Local History & Natural History Museum and Interpretive Center.
Collections: history of the Wind River Valley in northwest Wyoming; Scandinavian tie-hack history; Mountain Shoshone archaeology; Indian artifacts; timbering; wildlife mounts; ranch equipment; rocks; historical documents; photographs; saddles; historic buildings; oral history file; manuscripts; CM Guest Ranch collection.
Research Fields: local & natural history.
Facilities: library of local history, historical photographs, letters, newspapers, wild game available to writers, students & researchers upon request; reading room.
Activities: guided tours; formally organized education programs.
Hours & Admission Prices: Winter: Tues.-Sat. 10-4. Summer: daily 9-6. No charge; donations accepted.
Attendance: 5,000 (accurate)
Membership: Single $10; Family $15; Contributing $25; Century Club $100.

NATIONAL BIGHORN SHEEP INTERPRETIVE CENTER, (M), 907 W. Ramshorn St., Dubois, WY 82513. Mailing Address: P.O. Box 1435, Dubois, WY 82513-1435. Tel.: 307-455-3429. Facebook: National Bighorn Sheep Interpretive Center.
E-mail: info@bighorn.org
Web Site: www.bighorn.org
Founded: 1993.
Congressional District: 1
Key Personnel: Exec. Dir., Suzan Moulton; Pres. (V), Mark Hinschberger; Museum Shop Mgr., Ramona Finley.
Personnel Profile: Full-Time Paid 1; Full-Time Volunteers 2; Part-Time Paid 1.
Volunteer Hours: 480
Operating Expenses: 130,000
Operating Income: 140,000
Governing Authority: private; nonprofit organization. Tax-exempt: 501(c)(3).
Institution Type/Description: Nature Center.
Collections: life-size dioramas; interactive exhibits; mounted specimens of wild sheep of the world.
Facilities: 4,000 sq. ft. exhibit space; nature center; 30-seat large screen theater. Museum-related items for sale.
Activities: guided wildlife tours; lectures; school loan service. Annual Events: birthday in July; fundraiser banquet in November.
Publications: quarterly newsletter, Sheep Tracks.
Hours & Admission Prices: Memorial Day to Labor Day daily 9-6; Sept.-May Mon.-Sat. 9-5. Adults $2.50, children $.75; members no charge. Closed New Year's Day; Easter; Thanksgiving; Christmas. &
Attendance: 10,000 (estimated)
Membership: Individual $30; Family $60; Business $125; Partner $250; Patron's Circle $500; Summit Club $1,000.

Encampment

GRAND ENCAMPMENT MUSEUM, INC., 807 Barnett Ave., Encampment, WY 82325. Mailing Address: P.O. Box 43, Encampment, WY 82325-0043. Tel.: 307-327-5308.
E-mail: gemdirector@gemuseum.com
Web Site: www.gemuseum.com
Founded: 1965.
Key Personnel: Pres. (V), John Varner; Exec. Dir., Judy Stepp.

Personnel Profile: Full-Time Paid 1; Part-Time Paid 10; Part-Time Volunteers 8.
Governing Authority: nonprofit organization. Tax-exempt: 501(c)(3).
Institution Type/Description: Regional Museum.
Collections: ranching; homesteading; transportation; local history; costumes; uniforms; glass; Indian artifacts; mining; manuscripts; U.S. Forest Service; Andrikopolis hat cleaning shop; Civil War, World War I, World War II & Korean War artifacts; Civil Conservation Corps; schoolhouse; lookout tower; air force; Tromway Towers.
Major Exhibits: Journey Stories (T), 10/14-12/14.
Research Fields: local history; U.S. Forest Service; World War I & World War II; Civil Conservation Corps.
Facilities: 250-vol. library of history books, manuscripts, letters available for use on premises. Maps, brochures, books, bottles for sale.
Activities: guided tours; permanent exhibitions; special events.
Publications: annual newsletter; brochure, Encampment Museum; maps: Map of Grand Encampment & Riverside; The Grand Encampment; Relief Map of the Grand Encampment Mining District 1903.
Hours & Admission Prices: Memorial Day to Christmas daily 9-5:30. No charge; donations accepted. &
Attendance: 6,000 (accurate)

Evanston

CHINESE JOSS HOUSE MUSEUM, 920 Front St., Evanston, WY 82930-3464. Mailing Address: 1200 Main St., Evanston, WY 82930-3316. Tel.: 307-783-6320. Fax: 307-783-6390.
E-mail: museum@nglconnection.net
Web Site: www.uintacounty.com/index.asp?NID=195
Founded: 1990.
Key Personnel: Dir., Donald Westfall.
Personnel Profile: Full-Time Paid 1; Part-Time Paid 1.
Governing Authority: city. Subsidiary Institution: Urban Renewal Agency. Tax-exempt.
Institution Type/Description: Chinese History Museum: housed in a replica of the 19th century Chinese temple that stood in Evanston's Chinatown.
Collections: Chinese history; photographs; ceramic & metal artifacts; medicinal materials.
Activities: Annual Event: Chinese New Year in February.
Hours & Admission Prices: Mon.-Fri. 9-5, Sat. 10-4. No charge; donations accepted. &
Attendance: 1,500 (estimated)

UINTA COUNTY MUSEUM, (M), 1020 Front St., Evanston, WY 82930-3437. Tel.: 307-789-8248.
Founded: 1987.
Key Personnel: Dir., Kay Rossiter; Cur., Mary Walberg.
Personnel Profile: Full-Time Paid 1; Part-Time Paid 5.
Governing Authority: Uinta County. Tax-exempt.
Institution Type/Description: History Museum.
Collections: Uinta County history & culture; photographs.
Hours & Admission Prices: Mon.-Fri. 9-5, Sat. 10-4. No charge; donations accepted. &
Attendance: 8,000 (accurate)

Evansville

RESHAW EXHIBIT, 71 Curtis St., Evansville Community Center, Evansville, WY 82636. Mailing Address: P.O. Drawer 158, Evansville, WY 82636-0158. Tel.: 307-234-6530. Fax: 307-266-5109.
E-mail: townclerk@evansvillewy.com
Web Site: www.townofevansville.gov
Founded: 1963.
Governing Authority: municipal. Tax-exempt.
Institution Type/Description: History Museum.
Collections: Indian & pioneer artifacts.
Activities: permanent exhibitions.
Hours & Admission Prices: Mon.-Fri. 11-1. No charge. Closed holidays. &
Attendance: 225 (estimated)

Fort Bridger

FORT BRIDGER STATE MUSEUM, 37,000 Business Loop I-80, Fort Bridger, WY 82933. Mailing Address: P.O. Box 35, Fort Bridger, WY 82933-0035. Tel.: 307-782-3842. Fax: 307-782-7181.
E-mail: linda.newman-byers@wyo.gov
Web Site: www.artsparkshistory.com
Founded: 1843.
Key Personnel: Dir., Milward Simpson; Dept. Dir., Mr. Domenic Bravo; Cur.,

Cecil Sanderson; Pres. (V), Martin Lammers; Historic Site Supt., Linda N. Byers; Museum Shop Mgr., Martha Powers.
Personnel Profile: Full-Time Paid 4; Full-Time Volunteers 2; Part-Time Paid 15; Part-Time Volunteers 10; Interns 1.
Governing Authority: state. Parent Institution: State Parks and Cultural Resources. Subsidiary Institution: Fort Bridger Historical Association. Tax-exempt: 170(b)(1)(A).
Institution Type/Description: General Museum: housed in 1888-1890 enlisted men's barracks located at historic Fort Bridger.
Collections: living history; military life; officers quarters; commanding officer's life style & quarters; relating to historical Fort Bridger; westward expansion exhibits on Indian mountain men; emigrants; turn-of-the-century ranching; Mormon pioneers.
Research Fields: historic; pioneer lifestyle; military.
Facilities: Books, pamphlets, videos for viewing & for sale.
Activities: self-guided tours & guided tours at nominal cost; special summer events. Museum Sponsors: Mt. Man Rendezvous in September.
Hours & Admission Prices: April & Oct. Sat.-Sun. 9-4:30; May-Sept. daily 9-5:30. Non-residents $4, residents $2. &
Attendance: 97,000 (accurate)
Membership: Individual $9; Family $12.

Fort Laramie

FORT LARAMIE NATIONAL HISTORIC SITE, 965 Gray Rocks Rd., Fort Laramie, WY 82212-7625. Tel.: 307-837-2221. Fax: 307-837-2120.
Web Site: www.nps.gov/fola
Founded: 1938.
Key Personnel: Supt., Mitzi Frank.
Governing Authority: federal. Parent Institution: National Park Service.
Institution Type/Description: Historic Site.
Collections: 19th-century military objects & accoutrements; plains Indian artifacts & cultural material; Victorian period furnishings; small arms, & ordnance vehicles & equipment; archeological materials. Historic Buildings: Bedlam 1855-1864; Post Trader's Store 1876; Surgeon's quarters 1880; Captain's quarters 1872; guardhouse 1866; enlisted men's bar, officer's club & post office 1883; Lieutenant Colonel's quarters 1888.
Research Fields: Trans-Mississippi Westward Expansion.
Facilities: 5,000-vol. library of books on the history of the Westward movement.
Activities: permanent & temporary exhibitions; tours; history talks; living history programs during the summer including military & civilian activities of a 19th-century frontier Army post.
Hours & Admission Prices: Labor Day to mid-June daily 8-4:30; mid-June to Labor Day daily 8-8. Adults $3; children no charge. Closed New Year's Day; Thanksgiving; Christmas. &
Attendance: 50,820 (accurate)
Membership: National Park Service: Single $3; Golden Age $10; Annual $15; NPS Pass $80.

Fort Washakie

SHOSHONE TRIBAL CULTURAL CENTER, 90 Ethete Rd., Fort Washakie, WY 82514-1008. Tel.: 307-332-9106. Fax: 307-332-3055.
E-mail: glendatrosper@washakie.net
Founded: 1988.
Key Personnel: C.E.O., Glenda Trosper; Museum Shop Mgr., Marsha Allen.
Personnel Profile: Full-Time Paid 1; Full-Time Volunteers 1.
Governing Authority: tribal government. Tax-exempt.
Institution Type/Description: Cultural Center.
Collections: Shoshone tribal history from the reservation treaty establishment, emphasizing tribal art, historic tribal individuals, historic research for Shoshone prehistory, and contemporary times.
Activities: tours; cultural arts & crafts classes; Shoshone language classes; multimedia development.
Publications: brochures.
Hours & Admission Prices: Mon.-Fri. 9-4. Tours $15-$300. Closed all legal holidays; American Indian Day; holidays for annual tribal ceremonies in August.
Attendance: 1,500 (accurate)

Frances E. Warren Air Force Base

WARREN ICBM & HERITAGE MUSEUM, 7405 Marne Loop, 90th SW/MU, Bldg. 210, Frances E. Warren Air Force Base, WY 82005-2865. Mailing Address: 90 SW/MU, Francis E. Warren Air Force Base, WY 82005. Tel.: 307-773-2980. Fax: 307-773-2791.
E-mail: paula.taylor@warren.af.mil

Web Site: www.warrenmuseum.com
Founded: 1967.
Key Personnel: Dir., Paula Bauman Taylor; Cur., Larry Sprague.
Personnel Profile: Full-Time Paid 3.
Governing Authority: U.S. Air Force Museum, Wright Patterson Air Force Base, OH. Tax-exempt.
Institution Type/Description: Military Museum: housed in c.1900 Military Post Headquarters Building located on the site of 1867 Fort D. A. Russell.
Collections: military uniforms; equipment dating from 1840-present; ICBM missile; period rooms of old army life; memorabilia & artifacts of the 90th Bomb Group-The Jolly Rogers of WWII.
Research Fields: Wyoming territory military history.
Activities: Modern ICBM exhibits.
Hours & Admission Prices: Mon.-Fri. 8-4; tours by appointment. No charge; donations accepted. Closed holidays. ♿
Attendance: 31,000 (estimated)

Gillette

CAMPBELL COUNTY ROCKPILE MUSEUM, (M), 900 W. 2nd St., Gillette, WY 82716-3405. Tel.: 307-682-5723 & 686-8551. Fax: 307-686-8528.
E-mail: rockpile@vcn.com
Web Site: www.rockpilemuseum.com
Founded: 1974.
Congressional District: 6
Key Personnel: Dir., Terry Girouard; Chm., Denise Tugman; Education Coord. & Museum Shop Mgr., Penny Schroder; Registrar, Robert Henning.
Personnel Profile: Full-Time Paid 4; Part-Time Paid 2; Part-Time Volunteers 12.
Governing Authority: county. Parent Institution: Campbell County; Wyoming government. Tax-exempt.
Institution Type/Description: History Museum.
Collections: Native American weapons & tools; agricultural equipment; farming & ranching materials; saddles; rifles & guns; textiles & clothing; transportation materials including chuckwagon, sheepwagon, early automobiles; Burlington northern caboose; technological materials including phonographs; radios; printing press; linotype; blacksmith tools; local document & photographic archives; windmill exhibit & one-room rural schoolhouse all located on museum grounds; kitchen furnishings; homestead cabin.
Research Fields: local history; ranching, homesteading & American Indian artifacts.
Activities: lectures; educational programs.
Publications: brochures; informational sheets; newsletter.
Hours & Admission Prices: Mon.-Sat. 9-5. No charge; donations accepted. Closed holidays. ♿
Attendance: 11,000 (accurate)
Membership: Students & Seniors $5; Individual $15; Family $20; Supporting $50-$99; Contributing $100-$199; Sustaining $200 & up; Corporate $500 & up.

Glendo

GLENDO HISTORICAL MUSEUM, Town Hall on Yellowstone Ave., Glendo, WY 82213. Mailing Address: P.O. Box 396, Glendo, WY 82213-0396. Tel.: 307-735-4242.
Founded: 1967.
Personnel Profile: Part-Time Volunteers 1.
Governing Authority: Tax-exempt.
Institution Type/Description: History Museum.
Collections: paleontology; Native American artifacts; area history.
Activities: traveling exhibits.
Hours & Admission Prices: Mon.-Fri. 8-12 & 1-4. No charge; donations accepted. Closed holidays. ♿

Glenrock

GLENROCK DEER CREEK MUSEUM, 935 W. Birch St., Glenrock, WY 82637. Mailing Address: P.O. Box 417, Glenrock, WY 82637-0417. Tel.: 307-436-2810.
Founded: 1998.
Key Personnel: Chm. (V), Rosalie R. Goff.
Personnel Profile: Part-Time Volunteers 12.
Institution Type/Description: History Museum.
Collections: local history & culture; period voting records & building permits; early maps; Native American artifacts; photographs; clothing; hand tools; jewelry.
Hours & Admission Prices: Memorial Day to Labor Day Tues.-Sat. 10-4; other times by appointment. No charge; donations accepted. ♿
Attendance: 400 (estimated)

GLENROCK PALEONTOLOGICAL MUSEUM, 506 W. Birch St., Glenrock, WY 82637. Mailing Address: P.O. Box 1362, Glenrock, WY 82637-1362. Tel.: 307-436-2667. Fax: 307-436-5477.
Key Personnel: Dir., Stuart I. McCrary
Institution Type/Description: Paleontology Museum.
Collections: prehistoric life; dinosaurs; period sea animals; Oligocene mammals; modern animals; scientific papers about prehistoric animals.
Research Fields: paleontology & geology of local area.
Facilities: library; children's education center. Museum-related items for sale.
Activities: movies; games.
Hours & Admission Prices: May-Aug. Mon.-Sat. 11-4; Sept.-April Tues., Thurs. & Sat. 11-4. No charge; donations accepted.

Green River

SEEDSKADEE NATIONAL WILDLIFE REFUGE, Hwy. 372, Green River, WY 82935. Mailing Address: P.O. Box 700, Green River, WY 82935. Tel.: 307-875-2187. Fax: 307-875-4425.
Institution Type/Description: Wildlife Refuge.
Collections: wildlife & their habitat; ecology; photographs.
Facilities: education center.
Activities: educational programs.
Hours & Admission Prices: Call for hours. No charge. ♿
Attendance: 8,000 (estimated)

SWEETWATER COUNTY HISTORICAL MUSEUM, (M), 3 E. Flaming Gorge Way, Green River, WY 82935-4239. Tel.: 307-872-6435. Fax: 307-872-3234.
E-mail: swchm@sweetwater.net
Web Site: sweetwatermuseum.org
Founded: 1967.
Key Personnel: Dir., Ruth Lauritzen; Pres. (V), Donna Mundschenk; Cur., Mark Nelson; Exhibits Coord., Gary Perkins; Museum Shop Mgr., Cyndi McCullers.
Personnel Profile: Full-Time Paid 4; Part-Time Paid 2; Part-Time Volunteers 1; Interns 1.
Governing Authority: county. Parent Institution: Sweetwater County. Subsidiary Institution: Sweetwater County Museum Foundation. Tax-exempt: 170(b)(1)(A).
Institution Type/Description: History Museum.
Collections: Indian & pioneer artifacts; industry, ranching & Chinese mementos; guns; photographs; items of historical significance to Sweetwater County & Southwestern Wyoming.
Research Fields: local history.
Facilities: 400-vol. library of local history books available for use on premises; 400 vertical files; 100-seat auditorium. Historic books, commemorative coins, gifts, & post cards for sale.
Activities: guided tours; lectures; formally organized education programs for children; permanent & temporary exhibitions.
Publications: quarterly newsletter, Overland & Underground.
Hours & Admission Prices: Mon.-Sat. 10-6. No charge; donations accepted. Closed holidays. ♿
Attendance: 4,550 (accurate)

Greybull

GREYBULL MUSEUM, 325 Greybull Ave., Greybull, WY 82426-2049. Mailing Address: Box 348, Greybull, WY 82426-0348. Tel.: 307-765-2444.
Founded: 1968.
Key Personnel: Dir., Wanda L. Bond.
Personnel Profile: Part-Time Paid 3; Part-Time Volunteers 2.
Governing Authority: municipal. Tax-exempt: 501(c)(3).
Institution Type/Description: General Museum.
Collections: geology; history; Indian artifacts; fossils & minerals.
Facilities: auditorium. Books, minerals, rocks, fossil material, raw gem stone material & jewelry for sale.
Activities: guided tours; field trips; lectures; films; formally organized education programs for children; permanent & temporary exhibitions.
Publications: brochure.
Hours & Admission Prices: April-May & Sept.-Oct. Mon.-Fri. 1-5; June-Aug. Mon.-Fri. 10-8, Sat. 10-6; Nov.-March Mon., Wed. & Fri. 1-4. No charge; donations accepted. ♿
Attendance: 7,000 (accurate)

THE MUSEUM OF FLIGHT AND AERIAL FIREFIGHTING, South Big Horn County Airport, Greybull, WY 82426. Mailing Address: South Big Horn County Airport, P.O. Box 412, Greybull, WY 82426-0412. Tel.: 307-765-4322.
E-mail: flight@tctwest.net
Key Personnel: Dir., Bob Hawkins; Tour Dir. & Museum Shop Mgr., Lorraine Reiner.
Personnel Profile: Full-Time Volunteers 1.
Governing Authority: Tax-exempt.
Institution Type/Description: Aviation Museum.
Collections: aviation history; aerial firefighting; aircraft; retardant systems.
Activities: walking tours; school group tours.
Hours & Admission Prices: mid-May to Sept. Wed. & Fri.-Sun. 10-4. No charge; donations accepted.
Attendance: 2,000 (estimated)

Guernsey

LAKE GUERNSEY MUSEUM-GUERNSEY STATE PARK, Interstate 25, exit 92 to US 26 to State Rte. 270, Guernsey, WY 82214. Mailing Address: P.O. Box 429, Guernsey, WY 82214-0429. Tel.: 307-836-2334 (office) & 2900 (museum). Fax: 307-836-3088.
Web Site: wyoparks.state.wy.us/parks/guernsey/index.asp
Founded: 1936.
Key Personnel: Dir. Wyoming Dept. of State Parks & Cultural Resources, Milward Simpson; Parks Supt., Todd Stevenson.
Personnel Profile: Part-Time Paid 1.
Governing Authority: Wyoming State Parks & Historic Sites, Barrett Bldg., 2301 Central Ave., Cheyenne, WY 82002. Tax-exempt.
Institution Type/Description: Historical Museum: housed in 1930s building constructed by the CCC.
Collections: agriculture; anthropology; ethnology; industry; transportation; military; historical museum building & 1930s exhibits; botany; archaeology.
Research Fields: pertaining to collections & history of facility.
Facilities: picnic area with pavilion.
Activities: permanent exhibits by Dr. John C. Ewers; temporary exhibits from local area; art, photo display.
Hours & Admission Prices: May-Oct. daily 10-6. Nonresident $4, resident $2. &
Attendance: 5,000 (estimated)
Membership: Friends of Guernsey Park.

Hanna

HANNA BASIN MUSEUM, Old Community Hall, Front St., Hanna, WY 82327. Mailing Address: P.O. Box 252, Hanna, WY 82327-0252. Tel.: 307-324-3915.
Key Personnel: Cur., Nancy Anderson
Institution Type/Description: History Museum.
Collections: Hanna history; coal camps; railroad; homesteading; ranching.
Activities: special events.
Hours & Admission Prices: Summer: Fri.-Sun. 1-5; Winter: Fri. 1-5; other times by appointment. No charge; donations accepted.

Jackson

THE CLUBHOUSE - THE JACKSON HOLE CHILDREN'S MUSEUM, 174 N. King St., Jackson, WY 83001. Mailing Address: P.O. Box 995, Jackson, WY 83001. Tel.: 307-733-3996.
E-mail: info@jhchildrensmuseum.org
Web Site: www.jhchildrensmuseum.org
Key Personnel: Founder & C.F.O., Craig Morris; Founder & Dir., K.J. Morris; Co Education Dir., Yvette Werner; Co Education Dir., Julie D'Amours
Institution Type/Description: Children's Museum.
Collections: hands-on exhibitions.
Activities: special events; birthday parties.
Hours & Admission Prices: Tues.-Sat. 10-6. Admission $7.50, children under 2 no charge.

JACKSON HOLE HISTORICAL SOCIETY AND MUSEUM, (M), 225 N. Cache St., Jackson, WY 83001. Mailing Address: P.O. Box 1005, Jackson, WY 83001-1005. Tel.: 307-733-2414. Fax: 307-734-8171.
E-mail: jhhsm@wyom.net
Web Site: www.jacksonholehistory.org
Founded: 1958.
Congressional District: 1
Key Personnel: C.E.O. & Dir., Lokey Lytjen; Pres., Jackie Montgomery; Pres. (V), Steve Ashley; Devel. Assoc., Liz Jacobson; Cur. Collections, Shannon Sullivan; Education & Outreach, Karen Reinhart; Museum Shop Mgr., Jean Hansen; Asst. to Exec. Dir., Brenda Roberts.
Personnel Profile: Full-Time Paid 6; Part-Time Paid 2; Part-Time Volunteers 100.
Governing Authority: Jackson Hole Historical Society & Museum. Tax-exempt.
Institution Type/Description: History Museum & Research Center.
Collections: archaeology; regional prehistory; Plains Indians; historical photographs; fur trade memorabilia; big game heads; local pioneer history; maps; regional & local history; conservation history; tourism; oral history; western Americana.
Research Fields: archaeology; history; folklore; traditions.
Facilities: library; archives; photograph archives.
Activities: lectures; permanent & rotating exhibits; summer walking tours; summer field trips; school & youth programs; volunteer program; publications program; oral histories.
Publications: newsletter; books, Jackson Hole: Crossroads of the Western Fur Trade 1807-1840, David E. Jackson, Field Captain of the Rocky Mountain Fur Trade, Landmarks of the Rocky Mountain Fur Trade: Two One-day Self-guided Tours from Jackson, Wyoming; And That's The Way It Was in Jackson's Hole; Historic Downtown Jackson: Self-Guided Walking Tour; Windows to the Past: Early Settlers in Jackson Hole.
Hours & Admission Prices: Mon.-Sat. 10-6, Sun. 12-5; call for off-season hours. Walking Tours: call for information. Family $18, adults $6, senior citizens $5, children & students 5-18 $4; children under 5 & members no charge. &
Attendance: 16,000 (accurate)
Membership: Senior Citizen $25; Individual $35; Senior Family $50; Family $60; Friend $100; Contributing $250; Sustaining $500; Benefactor $1,000.

* **NATIONAL MUSEUM OF WILDLIFE ART, (M),** 2820 Rungius Rd., Jackson, WY 83001. Mailing Address: P.O. Box 6825, Jackson, WY 83002-6825. Tel.: 307-733-5771. Fax: 307-733-5787.
E-mail: info@wildlifeart.org
Web Site: www.wildlifeart.org
Formerly: Wildlife of the American West
Founded: 1987.
Key Personnel: Pres. & C.E.O., James C. McNutt, Ph.D.; Chm. (V), Bill Mingst; C.F.O., Lisa Holmes; Sugden Family Cur. Education, Jane Lavino; Cur. Art, Adam Duncan Harris, Ph.D.; Facilities & Security, Joe Bishop; Dir. Operations, Steve Seamons; Museum Shop Mgr., Debra Ross Vassar.
Personnel Profile: Full-Time Paid 30; Part-Time Paid 8; Part-Time Volunteers 67; Interns 2.
Volunteer Hours: 3,353
Operating Expenses: 6,643,835
Operating Income: 6,295,156
Governing Authority: nonprofit organization. Tax-exempt: 501(c)(3).
Institution Type/Description: Art Museum.
Collections: 2500 B.C.-present fine art that depicts humanity's relationship to nature, focused on images of wildlife.
Major Exhibits: Harmless Hunter: Charlie Russell, Summer 2014; Hirschfield Plains Indian Collection, Summer 2014; Stone to Glass: 4000 Years of Animal Sculpture, Summer 2014; Audubon and the AA of Birds, Fall 2014 (T).
Research Fields: Fine wildlife art & artists; museum education research & program development.
Facilities: 3,000-vol. library of art history & artists' biography books available to the public by appointment; 15,000 sq. ft. exhibit space; 200-seat auditorium; 2 classrooms; 45-seat cafe. Museum-related items for sale.
Activities: guided tours; lectures; films; theatre; organized education programs for children, adults & college students; docent program; training programs for museum professionals; participatory, loan, temporary & traveling exhibitions; school loan service. Annual Event: Western Visions.
Publications: magazine, annual; calendar 2 times per year.
Hours & Admission Prices: Daily 9-5. Adults $12, senior citizens $10, children 5-18 $6; discounts to AAM & Museum's West members; children under 5 & members no charge. Closed Veterans Day; Thanksgiving; Christmas. &
Attendance: 58,989 (accurate)
Membership: Pika $35; Otter $65; Pronghorn $100; Caribou $250; Elk $500; Paintbox Society $1,000; Rungius Society $3,000.

Kaycee

HOOFPRINTS OF THE PAST MUSEUM, 344 Nolan Ave., Kaycee, WY 82639. Mailing Address: P.O. Box 114, Kaycee, WY 82639-0042. Tel.: 307-738-2381. Fax: 307-738-2381.
E-mail: curator@hoofprintsofthepast.org
Web Site: www.hoofprintsofthepast.org
Founded: 1990.
Key Personnel: Dir., Laurel Foster; Pres. (V), Wade Curuchet.
Personnel Profile: Part-Time Paid 3; Part-Time Volunteers 8.
Volunteer Hours: 500
Operating Expenses: 30,000
Operating Income: 40,000
Governing Authority: Tax-exempt.
Institution Type/Description: History Museum.
Collections: Johnson County history; Dull Knife Battlefield; Fort Reno; Bozeman Trail; Johnson County Invasion; Southern Johnson County history.
Research Fields: Johnson County war; Southern Johnson County history.
Facilities: Gift items for sale.
Activities: tours; talks; plays; special events; archival research.
Publications: newsletter; photo book.
Hours & Admission Prices: May-Oct. Mon.-Sat. 9-5, Sun. 1-5. No charge; donations accepted. &
Attendance: 3,500 (accurate)
Membership: Senior $7; Individual $10; Business $30.

Kelly

THE MURIE MUSEUM, 1 Ditch Creek Rd., Kelly, WY 83011. Mailing Address: P.O. Box 68, Kelly, WY 83011-0068. Tel.: 307-734-5657, ext. 3106. Fax: 307-739-9388.
E-mail: info@tetonscience.org
Web Site: www.tetonscience.org
Founded: 1973.
Key Personnel: C.E.O., John Shea; Chm. (V), Dick Jones; Dir., April Landale; Museum Mgr. & Dir. Research, Dr. Dale Gentry.
Personnel Profile: Part-Time Paid 1; Interns 2.
Governing Authority: private; nonprofit organization. Parent Institution: Teton Science Schools. Tax-exempt.
Institution Type/Description: Natural History Museum.
Collections: 600 bird study skins; over 1,000 mammal study skins; North American mammal skulls; plant specimens.
Research Fields: ecology & conservation biology.
Facilities: 4,000-vol. library of journals & books on natural history; classrooms; laboratories; field research station; nature & conservation center.
Activities: formal education programs for adults & children; wildlife, ecology, ornithology & animal behavior programs; lectures.
Hours & Admission Prices: Mon.-Fri. 8-4:30 by appointment, Sat.-Sun. by appointment. No charge; donations accepted. Closed Christmas.
Attendance: 2,500 (estimated)

Kemmerer

FOSSIL BUTTE NATIONAL MONUMENT VISITOR CENTER, 864 Chicken Creek Rd., Kemmerer, WY 83101. Mailing Address: P.O. Box 592, Kemmerer, WY 83101. Tel.: 307-877-4455. Fax: 307-877-4457.
E-mail: arvid_aase@nps.gov
Web Site: www.nps.gov/fobu
Founded: 1972.
Key Personnel: Dir., Nancy Skinner.
Governing Authority: Parent Institution: National Park Service. Tax-exempt.
Institution Type/Description: Paleontology Museum.
Collections: over 300 fossils including fishes, crocodile, turtles, bats, plants & birds; geology; habitats; photographs.
Research Fields: paleontology.
Activities: educational programs; fossil preparation demonstrations; video programs; ranger programs. Annual Event: active fossil excavation from mid-June to mid-August.
Hours & Admission Prices: Monument: daily sunrise to sunset. Visitor Center: May-Sept. daily 9-5:30; Oct.-April daily 8-4:30. No charge; donations accepted. Closed winter holidays. &
Attendance: 16,000 (accurate)

FOSSIL COUNTRY MUSEUM, (M), 400 Pine, Kemmerer, WY 83101. Mailing Address: P.O. Box 854, Kemmerer, WY 83101-0854. Tel.: 307-877-6551. Fax: 307-877-6552.
E-mail: museum@hamsfork.net
Web Site: www.hamsfork.net/~museum
Founded: 1989.
Congressional District: 3
Key Personnel: Dir., Judy Julian; Pres., Sue Giorgis; Vice Pres., Parry Baldwin.
Personnel Profile: Full-Time Paid 1; Part-Time Volunteers 3.
Governing Authority: private; nonprofit organization. Parent Institution: Fossil County Futures, Inc. Tax-exempt: 501(c)(3).
Institution Type/Description: History Museum.
Collections: history of Kemmerer/Diamondville, Wyoming from late 1890s-present; replica of Kemmerer coal mine; genuine whiskey stills.
Research Fields: southwest Wyoming history; history of coal mining in southwest Wyoming.
Facilities: 50-vol. library of books on Wyoming history; Kemmerer newspapers; rental facilities; 175-seat auditorium. Museum-related items for sale.
Activities: arts festivals; concerts; guided tours; hobby workshops; lectures; temporary & traveling exhibitions. Annual Event: Wyoming Heritage Auction.
Publications: Kemmerer History.
Hours & Admission Prices: Mon.-Sat. 10-4. No charge; donations accepted. Closed New Year's Day; Independence Day; Thanksgiving; Christmas. &
Attendance: 3,000 (accurate)
Membership: Individual $30.

J.C. PENNEY HOMESTEAD & HISTORICAL FOUNDATION, 109 J.C. Penney Dr., Kemmerer, WY 83101-2941. Tel.: 307-877-3164.
E-mail: swchm@sweetwater.net
Institution Type/Description: History Museum: housed in the Penney's first home & the site of the first J.C. Penney store which is still in operation today. Cottage is a National Historic Landmark.
Collections: period furnishings; photographs; personal artifacts.
Hours & Admission Prices: Memorial Day to Labor Day Mon.-Sat. 9-6, Sun. 1-6. No charge; donations accepted.

ULRICH'S FOSSIL GALLERY, Fossil Station #308, Kemmerer, WY 83101. Mailing Address: P.O. Box 308, Kemmerer, WY 83101-0308. Tel.: 307-877-6466. Fax: 307-877-3289.
Web Site: www.ulrichsfossilgallery.com
Founded: 1950.
Institution Type/Description: Fossil Gallery.
Collections: fish & plant fossils.
Activities: fossil digs; fossil preparation kits.
Hours & Admission Prices: Gallery: daily 8-5. Fossil Digs: by appointment. No charge. &

Lander

EVANS/DAHL MEMORIAL MUSEUM, 545 Main St., Lander, WY 82520-3075. Tel.: 307-332-8190; 800-768-7743.
Institution Type/Description: History Museum.
Collections: One Shot Antelope Hunt history; wildlife preservation; antelope habitats; photographs.
Facilities: Museum-related items for sale.
Hours & Admission Prices: Mon.-Fri. 10-5.

FREMONT COUNTY PIONEER MUSEUM, 1443 Main St., Lander, WY 82520-2649. Tel.: 307-332-4137. Fax: 307-332-6498.
E-mail: a.teamfcpm@gmail.com
Founded: 1908.
Congressional District: 1
Key Personnel: Cur., Connie Shannon.
Personnel Profile: Full-Time Paid 1; Part-Time Paid 2; Part-Time Volunteers 4.
Governing Authority: county. Tax-exempt.
Institution Type/Description: History Museum.
Collections: pioneer era artifacts from late 1840-1920; Indian artifacts; agricultural & transportation artifacts & implements; pioneer era artifacts.
Research Fields: local, regional & state history; Oregon Trail; Indians; women's suffrage.
Facilities: research library; picnic area. Books for sale.
Activities: permanent & temporary exhibitions; lectures; programs.
Hours & Admission Prices: Tues.-Sat. 10-6. No charge; donations accepted. &
Attendance: 10,000 (estimated)

LANDER ART CENTER, 224 Main St., Lander, WY 82520. Tel.: 307-332-5772.
E-mail: lisa@landerartcenter.com
Web Site: www.landerartcenter.com
Founded: 2002.

Personnel Profile: Full-Time Paid 2; Part-Time Paid 15; Part-Time Volunteers 100.
Volunteer Hours: 800
Governing Authority: nonprofit organization.
Institution Type/Description: Art Gallery.
Collections: works by local, regional & national artists; paintings; photographs; sculpture.
Major Exhibits: Through the Looking Glass (T), 1/7/14-2/22/14; Lander Valley High School & Middle Select Art Show, 2/28/14-3/29/14; Members Show, 4/14/14-5/10/14; Audubon Show, 5/16/14-6/21/14; Cheyenne Artist Group, 6/27/14-8/2/14.
Activities: temporary exhibits; classes; workshops; special events. Museum Sponsors: Summer of Arts Program.
Hours & Admission Prices: Call for hours. No charge; donations accepted. &
Attendance: 4,000 (estimated)
Membership: Friend $30-$74; Supporter $75-$124; Patron $125-$299; Benefactor $300-$499; Sustainer $500-$999; Connoisseur $1,000 & up.

LANDER CHILDREN'S MUSEUM, 465 Lincoln Ave., Lander, WY 82520-2831. Tel.: 307-332-1341.
E-mail: info@landerchildrensmuseum.org
Web Site: www.landerchildrensmuseum.org
Founded: 1999.
Key Personnel: Mgr., Jennifer O'Connor.
Personnel Profile: Full-Time Paid 1; Part-Time Paid 1; Part-Time Volunteers 15.
Institution Type/Description: Children's Museum.
Collections: hands-on exhibits.
Activities: special events; educational programs.
Hours & Admission Prices: Summer: Tues.-Fri. 10-3; Sept.-May Tues.-Fri. 10-1, Sat. 10-3. Adults $3; children under 2 & members no charge.
Attendance: 3,500
Membership: Family $50; Corporate $100.

MUSEUM OF THE AMERICAN WEST, 1445 W. Main St., Lander, WY 82520. Tel.: 307-335-8778.
E-mail: info@amwest.org
Web Site: museumoftheamericanwest.com
Institution Type/Description: History Museum.
Collections: local, natural & cultural history; photographs; personal artifacts.
Activities: educational programs; special events.
Hours & Admission Prices: Village: mid-May to Oct. Mon.-Sat. 9-4. Office: Mon.-Sat. 9-4.

Laramie

THE GEOLOGICAL MUSEUM, THE UNIVERSITY OF WYOMING, (M), 1000 E. University Ave., Laramie, WY 82071-3006. Tel.: 307-766-2646.
E-mail: geolmus@uwyo.edu
Web Site: www.uwyo.edu/geomuseum/index.html
Founded: 1887.
Key Personnel: Dir. & Cur., Brent H. Breithaupt.
Personnel Profile: Full-Time Paid 1; Part-Time Paid 3; Part-Time Volunteers 2; Interns 2.
Governing Authority: university. Parent Institution: University of Wyoming. Tax-exempt: 501(c)(3).
Institution Type/Description: Geology Museum.
Collections: vertebrate & invertebrate paleontology; rocks; minerals.
Research Fields: vertebrate & invertebrate paleontology.
Activities: guided tours; field trips: formally organized education programs for undergraduate & graduate college students; permanent & temporary exhibitions.
Publications: newsletter, Bronto.
Hours & Admission Prices: Mon.-Sat. 10-4. No charge; donations accepted. Closed university holidays. &
Attendance: 30,000 (estimated)
Membership: Individual $25; Family $30; Supporting $50; Contributing $100; Sponsor $200, Patron $500; Advanced Patron $1,000.

LARAMIE PLAINS MUSEUM, (M), 603 E. Ivinson Ave., Laramie, WY 82070-3243. Tel.: 307-742-4448.
E-mail: lpmdirector@bresnan.net
Web Site: www.laramiemuseum.org
Founded: 1966.
Congressional District: 1
Key Personnel: C.E.O., Dir. & Museum Shop Mgr., Mary Mountain; Pres. (V), Jacob Anfinson; Cur., Susan McGraw; Admin. Asst., Seth Lyle.

Personnel Profile: Full-Time Paid 1; Part-Time Paid 4; Part-Time Volunteers 100; Interns 2.
Governing Authority: nonprofit organization. Parent Institution: Board of Laramie Museum Association, Inc. Tax-exempt: 501(c)(3).
Institution Type/Description: Historical House Museum: housed in 1892 Victorian Mansion built by early pioneer banker, Edward Ivinson.
Collections: dishes, cookware, buggies, harness, tools & furniture of early pioneer families; Indian artifacts; photographs & maps of area; Victorian furnishings & fixtures; business records of Laramie City.
Research Fields: early history of Albany County; pioneer family histories; cattle & sheep industries; Union Pacific Railroad history.
Facilities: 200-vol. library of Western books, pamphlets & newspaper articles available for use on premises; adjacent community center.
Activities: guided tours. Museum Sponsors: Victorian Teas; special summer & holiday programs.
Publications: brochure; newsletter, Ghost Towns of Albany County.
Hours & Admission Prices: Feb.-May & Sept.-Dec. 14 Tues.-Sat. 1-4; June-Aug. Tues.-Sat. 9-5, Sun. 1-4. Adults $10, senior citizens $7, students $5; discounts to AAM members; children under 6, members & volunteers no charge. Closed major holidays. &
Attendance: 28,000 (estimated)
Membership: Seniors $40; Individual & Senior Family $60; Family $90.

ROCKY MOUNTAIN HERBARIUM, Aven Nelson Building-3rd Fl., 9th St., Laramie, WY 82071-3165. Mailing Address: University of Wyoming-Department of Botany, Dept. 3165, 1000 E. University Ave., Laramie, WY 82071-2000. Tel.: 307-766-2236 & 4393. Fax: 307-766-2851.
E-mail: rhartman@uwyo.edu
Web Site: www.rmh.uwyo.edu
Founded: 1893.
Key Personnel: Cur. Prof. Botany, Ronald L. Hartman; Herbarium Mgr., Burrell E. Nelson.
Personnel Profile: Full-Time Paid 2; Part-Time Paid 6.
Governing Authority: university. Parent Institution: University of Wyoming, Dept. of Botany. Tax-exempt.
Institution Type/Description: Herbarium.
Collections: botany; 780,000 plant specimens, including United States Forest Service National Herbarium.
Research Fields: plant systematics.
Facilities: library of research journals & monographs related to plant systematics.
Activities: temporary exhibitions.
Hours & Admission Prices: Academic year Mon.-Fri. 8-5; summer Mon.-Fri. 7:30-4:30. No charge; donations accepted. &
Attendance: 200

UNIVERSITY OF WYOMING, AMERICAN HERITAGE CENTER, 2111 Willett Dr., Centennial Complex, Laramie, WY 82071. Mailing Address: 1000 E. University Ave., Dept. 3924, Laramie, WY 82071-2000.
Web Site: ahc.uwyo.edu
Key Personnel: Dir., Mark Greene
Institution Type/Description: History Museum.
Collections: archives; books; manuscripts; Wyoming & American west history; mining & petroleum industries; U.S. politics; world affairs.
Activities: research.
Hours & Admission Prices: Center: Mon. 8 am-9 pm, Tues.-Fri. 8-5. Reference Services: Mon. 10-9, Tues.-Fri. 8-5.

UNIVERSITY OF WYOMING ANTHROPOLOGY MUSEUM, 12th & Lewis, Laramie, WY 82071. Mailing Address: 1000 E. University Ave., Dept. 3431 - Anthropology, Laramie, WY 82071-2000. Tel.: 307-766-2208. Fax: 307-766-2473.
E-mail: arrow@uwyo.edu
Web Site: uwadmnweb.uwyo.edu/anthropology/museum
Founded: 1966.
Key Personnel: Dir., Dr. Charles A. Reher.
Personnel Profile: Part-Time Paid 2; Part-Time Volunteers 3.
Governing Authority: university. Affiliated with the University of Wyoming. Tax-exempt.
Institution Type/Description: Anthropology Museum
Collections: ethnology; American Indian emphasis; archaeology; faunal, human osteology, & archaeological collections.
Research Fields: pertaining to collections.
Facilities: human osteology & archaeological laboratories.
Activities: formally organized education programs for undergraduate & graduate college students; permanent & temporary exhibitions; guided tours; display internships.

Hours & Admission Prices: mid-May to Aug. Mon.-Fri. 9-4; Sept. to mid-May Mon.-Fri. 9-5. No charge. Closed school vacations & holidays. &

Attendance: 10,000 (estimated)

✳ UNIVERSITY OF WYOMING ART MUSEUM, (M), 2111 Willett Dr., Laramie, WY 82071. Mailing Address: 1000 E. University Ave., Dept. 3807, Laramie, WY 82071-2000. Tel.: 307-766-6622. Fax: 307-766-3520.

E-mail: uwartmus@uwyo.edu
Web Site: www.uwyo.edu/artmuseum
Founded: 1968.
Congressional District: 1
Key Personnel: Dir., Susan Moldenhauer; Pres. Bd., Gary Negich; Cur. Education, Wendy Bredehoft; Cur. Collections, Nicole Crawford; Chief Preparator, Sterling Smith.
Personnel Profile: Full-Time Paid 9; Part-Time Paid 15; Part-Time Volunteers 55; Interns 3.
Governing Authority: university. Parent Institution: University of Wyoming. Tax-exempt.
Institution Type/Description: Art Museum.
Collections: American art; European art; art of Asia, Africa and the Americas; contemporary art; photography
Research Fields: 19th- & 20th-century American art.
Facilities: 20,000 sq. ft. outdoor sculpture terrace; 12,000 sq. ft. exhibition space; education studio. Museum-related items for sale.
Activities: loan, temporary & traveling exhibitions; lectures; gallery talks; workshops; symposia; films; formally organized education programs for adults, K-12 & college students; statewide outreach programs.
Publications: exhibition brochures & catalogs; quarterly newsletter.
Hours & Admission Prices: Feb.-April & Sept.-Nov. Mon. 10-9, Tues.-Sat. 10-5; May-Aug. & Dec. Mon.-Sat. 10-5. No charge; donations accepted. &
Attendance: 118,755 (accurate)
Membership: Student $25; Individual $40; Family $60; Sustaining $250; Donor $500; Patron $1,000 & up.

UNIVERSITY OF WYOMING GEOLOGICAL MUSEUM, S.H. Knight Geology Bldg., 1000 E. University Ave., Laramie, WY 82071-2000. Tel.: 307-766-2646.

Institution Type/Description: Geology Museum.
Collections: fossils; rocks; minerals.
Hours & Admission Prices: Mon.-Sat. 10-4. No charge. Closed university holidays.

UNIVERSITY OF WYOMING INSECT MUSEUM, Dept. of Renewable Resources, 100 E. University, Laramie, WY 82071-3354. Mailing Address: Department of Renewable Resources, P.O. Box 3354, Laramie, WY 82071-3354. Tel.: 307-766-1121.

Key Personnel: Dir., Scott Shaw
Institution Type/Description: Insect Museum.
Collections: over 250,000 specimens; hymenoptera; diptera; lepidoptera; coleoptera.
Hours & Admission Prices: By appointment. No charge.

WYOMING TERRITORIAL PRISON STATE HISTORIC SITE, 975 Snowy Range Rd., Laramie, WY 82070-6719. Tel.: 307-745-6161. Fax: 307-745-8620.

E-mail: tbeyer@state.wy.us
Web Site: wyoparks.state.wy.us
Formerly: Wyoming Territorial Park
Founded: 1986.
Congressional District: 1
Key Personnel: C.E.O. & Exec. Dir., Tom Lindmier; Pres. (V), Connie Kercher; Cur., Teresa Beyer; Museum Shop Mgr., Lynette Nelson.
Personnel Profile: Full-Time Paid 4; Full-Time Volunteers 5; Part-Time Paid 2; Part-Time Volunteers 20; Interns 1.
Governing Authority: private; not-for-profit organization. Parent Institution: Wyoming State Parks & Cultural Resources. Tax-exempt: 501(c)(3).
Institution Type/Description: Historic Site: 1872-1903 Wyoming Territorial Prison; only prison in North America to hold Butch Cassidy.
Collections: Historic Buildings: 1872-1903 Wyoming Territorial Prison; Wyoming Frontier Town; warden's house; Ranchland exhibit; horse barn dinner theatre; boxcar house; broom factory; country church.
Research Fields: archaeology; period folk arts & trades; historical personalities; historic ranches.
Facilities: 150-seat dinner theatre; nature trail. Museum-related items & Wyoming products for sale.
Activities: self-guided tours; concerts; theater; temporary exhibitions; educational programs for children. Annual Event: Ghost Tours.

Hours & Admission Prices: May-Oct. daily 8-7. Adults $5, youth 12-17 $2.50; children 11 & under no charge. &
Attendance: 47,000 (accurate)

Lingle

WESTERN HISTORY CENTER, 2308 U.S. Hwy. 26, Lingle, WY 82223-8527. Tel.: 307-837-3052.

E-mail: ggzzk@embarqmail.com
Founded: 1980.
Governing Authority: Tax-exempt.
Institution Type/Description: History Museum.
Collections: historic, prehistoric, & paleontological artifacts; archaeology; mining; the Texas Trail; paleontology; oral histories.
Research Fields: High Plains prehistory; Oregon Trail.
Facilities: laboratory.
Activities: research.
Publications: newsletter.
Hours & Admission Prices: Summer: Tues.-Sat. 9-4; Winter: Thurs.-Sat. 9-4 by appointment. No charge; donations accepted. &
Attendance: 10,000 (estimated)
Membership: Student $10; Regular $20; Family $30; Sustaining $100; Lifetime $500.

Lovell

BIGHORN CANYON NRA VISITOR CENTER, 20 Hwy. 14A E., Lovell, WY 82431. Tel.: 307-548-5406. Fax: 406-666-2415. Facebook: Bighorn Canyon NRA.

Web Site: www.nps.gov/bica
Founded: 1966.
Key Personnel: Supt., Cassity Bromley; Bookstore Mgr., Rhonda Wipf.
Governing Authority: Parent Institution: National Park Service. Tax-exempt.
Institution Type/Description: Visitor Center.
Collections: local history & culture; photographs; geology; life science; personal artifacts.
Research Fields: natural & cultural resources.
Facilities: 2 visitor centers.
Activities: touring; camping; boating; nature trails.
Publications: newspaper, Canyon Echoes.
Hours & Admission Prices: Memorial Day to Labor Day daily 8-6; Sept.-May daily 8:30-4:30. Park: $5 per person per day. National Federal Lands Pass accepted. Closed New Year's Day; Thanksgiving; Christmas. &
Attendance: 185,000 (estimated)

Lusk

NIOBRARA HISTORICAL SOCIETY, 322 S. Main, Lusk, WY 82225. Mailing Address: P.O. Box 367, Lusk, WY 82225-0367. Tel.: 307-334-3444.

E-mail: stagecoachmuseumlusk@gmail.com
Web Site: niobraracountylibrary.org/museum
Formerly: Cheyenne-Black Hills Stagecoach
Key Personnel: Dir., Wallace Petrie; Museum Shop Mgr., Rose Kremers
Institution Type/Description: History Museum: stagecoach used on the Cheyenne-Black Hills Stage and Express Line; built in 1860s by Abbott & Downing at Concord, NH.
Collections: stagecoach; pioneer & trail history.
Hours & Admission Prices: Mon.-Sat. 10-4:30. Adults $2; members no charge. &

Lyman

BRIDGER VALLEY HERITAGE MUSEUM, 100 E. Sage-Lyman Town Hall 2 Fl., Lyman, WY 82937. Mailing Address: P.O. Box 184, Lyman, WY 82937-0184. Tel.: 307-787-3525.

E-mail: bvhmuseum@union-tel.com
Formerly: Trona Mining Museum of Bridger Valley
Founded: 2005.
Key Personnel: Exec. Dir., Kay Rossiter.
Personnel Profile: Part-Time Paid 1.
Governing Authority: Parent Institution: Uinta County Museum. Tax-exempt.
Institution Type/Description: History Museum.
Collections: Trona mining; personal artifacts; western trails history; Lincoln Highway; pioneer trails.
Hours & Admission Prices: Mon.-Fri. 10-5; other times by appointment. No charge; donations accepted. Closed holidays & weekends. &
Attendance: 800 (accurate)

Medicine Bow

MEDICINE BOW MUSEUM, 405 Lincoln Hwy., Medicine Bow, WY 82329. Mailing Address: P.O. Box 187, Medicine Bow, WY 82329-0187. Tel.: 307-379-2383.
Web Site: www.medicinebow.org
Founded: 1983.
Institution Type/Description: History Museum: housed in the former old railroad depot, built in 1913. Listed on the National Register of Historical Places.
Collections: local history; cattle & sheep ranch brands; caboose; railroad history.
Hours & Admission Prices: Memorial Day to Labor Day Mon.-Fri. 10-5, Sun. 1-5. No charge; donations accepted.

Meeteetse

MEETEETSE BANK MUSEUM, 1033 Park Ave., Meeteetse, WY 82433. Mailing Address: P.O. Box 248, Meeteetse, WY 82433. Tel.: 307-868-2423.
Key Personnel: Dir., David Cunningham.
Personnel Profile: Part-Time Paid 1.
Governing Authority: Parent Institution: Meeteetse Museum District. Tax-exempt.
Institution Type/Description: Historic Building Museum: housed in a former bank building; built in 1900. Listed on the National Register of Historic Places.
Collections: local history; bank vault & teller cage; period furnishings; photographs.
Hours & Admission Prices: Call for hours. No charge; donations accepted.
Attendance: 2,270 (accurate)

MEETEETSE MUSEUMS, (M), 1947 State St., Meeteetse, WY 82433. Mailing Address: P.O. Box 248, Meeteetse, WY 82433. Tel.: 307-868-2423. Fax: 307-868-2423.
E-mail: info@meeteetsemuseums.org
Web Site: www.meeteetsemuseums.org
Formerly: Meeteetse Museum Inc. & Charles J. Belden Museum of Western Photography
Founded: 1974.
Congressional District: 1
Key Personnel: Pres. (V), Jim Allen; Vice Pres., Sharon Fech; Treas., Lili Turnell; Sec., Yvonne Renner; Cur. & Museum Shop Mgr., Paige Paisley.
Personnel Profile: Full-Time Paid 1; Part-Time Paid 3; Part-Time Volunteers 6.
Governing Authority: county; nonprofit organization. Tax-exempt: 501(c)(3).
Institution Type/Description: History Museum.
Collections: local history; mounted wildlife; ranching; photographs by Charles J. Belden; Harry Jackson sculptures.
Research Fields: local early families.
Facilities: Museum-related items for sale.
Activities: temporary & traveling exhibitions; guided tours. Annual Events: Kirwin Excursion in August; Arland Townsite & Old Meeteetse Cemetery tour in September; Legend Rock State Archaeological Site Tour; Medicine Lodge Creek State Archaeological Site Tour; Historic Double Dee Guest Ranch Tour.
Publications: quarterly newsletter.
Hours & Admission Prices: Feb.-April & Oct.-Dec. 22 Tues.-Sat. 10-3; May-Sept. Mon.-Sat. 9:30-4, Sun. 12-4. No charge.
Attendance: 7,000 (accurate)

Midwest

SALT CREEK MUSEUM, 531 Peake St., Midwest, WY 82643. Mailing Address: P.O. Box 190, Midwest, WY 82643-0190. Tel.: 307-437-6513. Fax: 307-437-6514.
E-mail: museum@rtconnect.net
Founded: 1980.
Key Personnel: Cur., Sandra Schutte; Museum Shop Mgr., William Howard.
Personnel Profile: Full-Time Volunteers 1; Part-Time Paid 1.
Governing Authority: town.
Institution Type/Description: History Museum.
Collections: Salt Creek oilfields from 1889 to present; local history; oilfield workers & their families; doctor's office; school room; kitchen; dining room; barber shop; household artifacts; oilfield tools.
Activities: research.
Hours & Admission Prices: By appointment. No charge; donations accepted.
Attendance: 500 (estimated)

Moorcroft

WEST TEXAS TRAIL MUSEUM, 100 E. Weston St., Moorcroft, WY 82721. Mailing Address: P.O. Box 497, Moorcroft, WY 82721. Tel.: 307-756-9300.
E-mail: wttmdirector@rtconnect.net

Founded: 1989.
Key Personnel: Dir., Justin Gaskin.
Personnel Profile: Full-Time Paid 1; Part-Time Paid 1; Part-Time Volunteers 12.
Governing Authority: county; nonprofit. Parent Institution: Crook County Museum District, Sundance, WY 82729. Tax-exempt: 501(c)(3).
Institution Type/Description: History Museum.
Collections: local history from 1860s to present; ranching; agricultural tools.
Facilities: library. Museum-related items for sale.
Activities: guided tours; lectures; temporary & traveling exhibitions.
Hours & Admission Prices: Mon.-Fri. 9-5. No charge; donations accepted.
Attendance: 750 (estimated)
Membership: Single $20; Family $30; Business $100.

Pine Bluffs

HIGH PLAINS ARCHAEOLOGY MUSEUM, 211 Elm St., Pine Bluffs, WY 82082. Mailing Address: P.O. Box 429, Pine Bluffs, WY 82082-0429. Tel.: 307-245-9372.
Institution Type/Description: History Museum.
Collections: local history & culture; photographs; archaeological artifacts.
Hours & Admission Prices: June-Aug. by appointment. No charge; donations accepted.

TEXAS TRAIL MUSEUM, 3rd & Market Sts., Pine Bluffs, WY 82082. Mailing Address: P.O. Box 545, Pine Bluffs, WY 82082-0545. Tel.: 307-245-3713.
Web Site: www.texastrailmuseumoflaramiecounty.org
Founded: 1986.
Key Personnel: Pres., Anthony J. Sacco, Sr.
Personnel Profile: Part-Time Paid 2; Part-Time Volunteers 2.
Governing Authority: Tax-exempt.
Institution Type/Description: History Museum.
Collections: Texas Trail history; early pioneer history & culture; 2 diesel fired generators; quilting; school house; U.P. Caboose; St. Mary's Catholic Church artifacts; homestead cabin.
Facilities: picnic area.
Activities: lectures. Museum Sponsors: Ice Cream Socials; Golf Tournament.
Publications: quarterly newsletter.
Hours & Admission Prices: Memorial Day to Labor Day Mon.-Sat. 11-4. No charge; donations accepted.
Attendance: 600 (accurate)
Membership: Individual $10; Family $15; Silver Shield $50; Gold Shield $100; Life $500.

Pinedale

MUSEUM OF THE MOUNTAIN MAN, 700 E. Hennick, Pinedale, WY 82941. Mailing Address: P.O. Box 909, Pinedale, WY 82941-0909. Tel.: 307-367-4101. Fax: 307-367-6768.
E-mail: director@mmmuseum.com
Web Site: www.museumofthemountainman.com
Founded: 1990.
Key Personnel: C.E.O., Laurie Hartwig; Pres., Jay Fear; Museum Shop Mgr., Mildred Pape.
Personnel Profile: Full-Time Paid 2; Part-Time Paid 6; Part-Time Volunteers 200; Interns 1.
Governing Authority: society. Parent Institution: Sublette County Historical Society. Tax-exempt: 501(c)(3).
Institution Type/Description: History Museum.
Collections: fur trade; Western exploration; Plains Indians; early settlement history of western Wyoming.
Research Fields: Rocky Mountain rendezvous; Alfred Jacob Miller art; fur trade bibliography; historic western sites.
Facilities: research library; 15,000 sq. ft. exhibit space; outdoor amphitheatre; picnic area. Museum-related items for sale.
Activities: education programs for children; guided tours; lectures; loan, traveling & temporary exhibitions; living history events; educational programs for adults & children; lectures; guided tours. Museum Sponsors: Green River Rendezvous weekend in July.
Publications: newsletter.
Hours & Admission Prices: May-Oct. 1 daily 9-5; Oct. 2-April by appointment. Adults $5, senior citizens $4, children $3; discounts to groups; members no charge.
Attendance: 12,000 (accurate)
Membership: Astorian (Individual, Couple & Family) $35; Trapper (Patron) $100; Rocky Mountain Fur Trade Co. (Business) $150; The Rendezvous (Benefactor) $500; Jim Bridger League $1,000.

Powell

HOMESTEADER MUSEUM, (M), 324 E. 1st St., Powell, WY 82435. Mailing Address: P.O. Box 54, Powell, WY 82435-0054. Tel.: 307-754-9481.
E-mail: homesteader@bresnan.net
Web Site: www.homesteadermuseum.com
Founded: 1968.
Key Personnel: Dir., Rowene Weems; Chm. (V), Steve Bailey; Registrar Collections, Brandi Wright.
Personnel Profile: Full-Time Paid 2; Part-Time Paid 3; Part-Time Volunteers 12.
Governing Authority: Parent Institution: Park County Museum Board. Tax-exempt.
Institution Type/Description: History Museum.
Collections: history of Powell and Shoshone Reclamation Project; homesteading; early 20th century Big Horn Basin settlers' memorabilia; homesteader cabins; CB&Q caboose; sheepherders wagon; outhouse.
Hours & Admission Prices: March-April & Oct.-Dec. Tues.-Fri. 10-4; May-Sept. Tues.-Fri. 10-5, Sat. 10-2. No charge; donations accepted. &
Attendance: 4,000 (estimated)
Membership: Senior $10; Single $15; Family $25; Business $50.

Ranchester

T-REX NATURAL HISTORY MUSEUM, 1116 Big Horn Dr., Ranchester, WY 82839. Mailing Address: P.O. Box 612, Ranchester, WY 82839-0612. Tel.: 307-655-3359.
Institution Type/Description: History Museum.
Collections: local history; fossils; dinosaurs; minerals; crystals.
Facilities: Museum-related items for sale.
Activities: lectures.
Hours & Admission Prices: Call for hours.

Rawlins

CARBON COUNTY MUSEUM, (M), 904 W. Walnut St., Rawlins, WY 82301-6556. Tel.: 307-328-2740. Facebook: Carbon County Museum.
E-mail: info@carboncountymuseum.org
Web Site: www.carboncountymuseum.org
Founded: 1940.
Key Personnel: Dir., Kelly Morris; Office Mgr., Gina Cabrera.
Personnel Profile: Full-Time Paid 5; Part-Time Paid 2.
Governing Authority: county; nonprofit. Parent Institution: Carbon County Commission. Tax-exempt.
Institution Type/Description: History Museum.
Collections: local history.
Research Fields: local history; genealogy.
Facilities: 25,000 sq. ft. exhibit space.
Activities: educational programs for adults & children.
Publications: semiannual newsletter.
Hours & Admission Prices: May-Sept. Tues.-Sat. 10-6; Oct.-April Tues.-Sat. 1-5; No charge; donations accepted. Closed New Years Eve, Day & day after, Martin Luther King Jr. Day, Presidents' Day, Memorial Day, Independence Day, Labor Day, Thanksgiving, Christmas Eve & Day. &
Attendance: 3,353 (accurate)

WYOMING FRONTIER PRISON MUSEUM, 500 W. Walnut St., Rawlins, WY 82301-4768. Tel.: 307-324-4422. Fax: 307-328-4004.
Key Personnel: Dir., Tina Hill
Institution Type/Description: Prison Museum: housed in the Wyoming Frontier Prison which operated from 1901 to 1981. Site for the filming of the 1987 movie, Prison.
Collections: prison history; gas chamber.
Activities: Museum Sponsors: Special Halloween Night Tours in October; Christmas in the Big House Craft Bazaar in December.
Publications: cookbook, Savory Recipes by Unsavory Characters; The Sweet Smell of Sagebrush.
Hours & Admission Prices: Guided Tours: Memorial Day to Labor Day daily 8:30-4:30; Sept.-May call for hours. Guided Tours: family (parents & minor children) $30, adults $7, seniors 60 & over and children 12 & under $6.
Attendance: 15,000
Membership: Friends of the Old Pen: Student $5; Single $15; Family $20; Corporate & Club and Inmate $75; Trusty $100; Guard $250; Warden $500-$1,000.

Riverton

RIVERTON MUSEUM, 700 E. Park Ave., Riverton, WY 82501-3657. Tel.: 307-856-2665.
E-mail: lrnjost@yahoo.com
Founded: 1956.
Congressional District: 1
Key Personnel: Dir., Loren Jost.
Personnel Profile: Full-Time Paid 3; Part-Time Volunteers 12.
Governing Authority: county; nonprofit organization. Tax-exempt.
Institution Type/Description: Local History Museum.
Collections: rotating exhibits; clothing; Indian artifacts; dentist material; saloon; mining; logging; books; trappers material; oil industry material; typewriters; musical instruments; beauty shop; nursery; library section; church; general store; shoes shop; flag display; sewing machines.
Research Fields: local & Wyoming history.
Facilities: 800-vol. library.
Activities: guided tours; school demonstrations; films; study clubs; work shops; tape interviews.
Publications: Wind River Mountaineer magazine.
Hours & Admission Prices: Tues.-Sat. 10-4. No charge; donations accepted. Closed holidays. &
Attendance: 7,500 (accurate)

ROBERT A. PECK ART CENTER, 2660 Peck Ave., Riverton, WY 82501-2215. Tel.: 307-855-2222; 800-735-8418, ext. 2222. Fax: 307-855-2090.
E-mail: gallery@cwc.edu
Institution Type/Description: Art Center.
Collections: works by local, regional & national artists including paintings & contemporary sculpture.
Facilities: 6,000 sq. ft. exhibit space; 940-seat theater.
Activities: performances; theater & music productions; permanent & temporary exhibitions.
Hours & Admission Prices: Mon.-Fri. 8 am-10 pm, Sat. call for hours. No charge. Closed holidays.
Attendance: 5,000 (estimated)

WIND RIVER HERITAGE CENTER, 1075 S. Federal Blvd., Riverton, WY 82501-4407. Mailing Address: P.O. Box 206, Riverton, WY 82501-0039. Tel.: 307-856-0706.
Founded: 1995.
Key Personnel: Dir., C.E.O., Chm. (V) & Pres. (V), Lewis B. Diehl.
Personnel Profile: Full-Time Volunteers 2; Part-Time Paid 2; Part-Time Volunteers 3.
Operating Expenses: 110,000
Operating Income: 100,000
Governing Authority: nonprofit organization. Tax-exempt.
Institution Type/Description: Heritage Center.
Collections: Wyoming wildlife full body mounts (60); Native American art; 250 period & rare animal traps; Indian wax figures. Historic Building: 1908 homestead cabin; wax figure diaramas in new 50' x 100' building.
Major Exhibits: Traditional Indian Dolls, 9/25/13-12/14.
Activities: Center Sponsors: Native American powwow dance program June to August; Mountain Man Rendezvous Reenactment.
Publications: annual newsletter.
Hours & Admission Prices: May-Dec. Mon.-Sat. 10-4. Donations: family $5, adults $3; school groups & seniors no charge. &
Attendance: 10,000 (estimated)
Membership: Pilgrim $25; Tenderfoot $26-$50; Explorer $51-$100; Guide $101-$250; Pathfinder $251-$500; Trailblazer $500-$2,000; Mountain Man over $2,000.

Rock Springs

COMMUNITY FINE ARTS CENTER, 400 C St., Rock Springs, WY 82901-6225. Tel.: 307-362-6212. Fax: 307-352-6657.
E-mail: cfac@sweetwaterlibraries.com
Web Site: www.cfac4art.com
Founded: 1966.
Congressional District: 1
Key Personnel: Dir., Debora Thaxton Soule; Chm. (V), Kim Loppicolo; Asst. to Dir., Jennifer Messer.
Personnel Profile: Full-Time Paid 2; Part-Time Paid 1; Part-Time Volunteers 10.
Governing Authority: county; Fine Arts Center Board. Affiliated with School District #1. Parent Institution: Sweetwater County Library System. Tax-exempt.
Institution Type/Description: Arts Center.
Collections: paintings; sculpture; drawing; graphics; photographs.
Facilities: meeting & workshop rooms.
Activities: guided tours; lectures; films; gallery talks; concerts; dance recitals; art

festivals; formally organized education programs for adults & children; inter-museum loan & permanent exhibitions.
Publications: quarterly newsletter, CFAC; circulars on exhibits & demonstrations by recognized artists.
Hours & Admission Prices: Mon.-Thurs. 10-6, Fri.-Sat. 12-5. No charge; donations accepted. &
Attendance: 9,978 (accurate)

NATURAL HISTORY MUSEUM OF WESTERN WYOMING COLLEGE, 2500 College Dr., Rock Springs, WY 82901-5802. Mailing Address: P.O. Box 428, Rock Springs, WY 82902. Tel.: 307-382-1600.
Web Site: wyshs.org/mus-wwcnathist.htm
Key Personnel: Dir., Kevin Thompson
Institution Type/Description: History Museum.
Collections: archaeology; ethnography; natural history.
Hours & Admission Prices: Daily 9am-10pm. No charge.

ROCK SPRINGS HISTORICAL MUSEUM, (M), 201 ”B“ St., Rock Springs, WY 82901-6250. Tel.: 307-362-3138. Fax: 307-352-1516.
Founded: 1988.
Key Personnel: Pres. (V), Marilynn Noble; Museum Coord., Bob Nelson; Exhibits Coord., Christina Shepard; Museum Tech, Janice Brown.
Personnel Profile: Full-Time Paid 1; Part-Time Paid 2; Part-Time Volunteers 20; Interns 1.
Governing Authority: municipal government. Parent Institution: City of Rock Springs, Div. of Finance & Administration. Tax-exempt.
Institution Type/Description: History Museum: 1894 Rock Springs City Hall, listed on the National Register of Historic Places, houses 1894 Rock Springs fire station & jail, stable & jail additions. A two-story sandstone structure, the building features two turreted bays, council chambers & mayor's balcony.
Collections: structure; furnishings; personal artifacts; tools & equipment.
Facilities: 2,500 sq. ft. exhibit space. Museum-related items for sale.
Activities: guided group tours on request; docent program; temporary exhibits. Annual Event: International Day.
Hours & Admission Prices: Mon.-Sat. 10-5. No charge; donations accepted. Closed Independence Day; major holidays. &
Attendance: 11,000 (accurate)

WEIDNER WILDLIFE MUSEUM, Western Wyoming Community College, 2500 College Dr., Rock Springs, WY 82901-5802. Tel.: 301-382-1600.
E-mail: emerrell@wwcc.wy.edu
Institution Type/Description: Wildlife Museum.
Collections: over 125 species of wildlife from around the world.
Hours & Admission Prices: Mon. & Wed. 10-1, Tues. & Thurs. 1-4. No charge; donations accepted. Closed major holidays; New Year's Eve & Day; Christmas Eve, Day & week.

WESTERN WYOMING COMMUNITY COLLEGE ART GALLERY, 2500 College Dr., Rock Springs, WY 82901-5802. Mailing Address: P.O. Box 428, Rock Springs, WY 82902-0428. Tel.: 307-382-1723. Fax: 307-382-7665.
E-mail: fmcewin@wwcc.wt.edu
Web Site: www.wwcc.wy.edu
Founded: 1989.
Key Personnel: Dir., Florence Alfano McEwin, Ph.D.
Governing Authority: Parent Institution: Western Wyoming College. Tax-exempt.
Institution Type/Description: Art Gallery.
Collections: contemporary paintings; sculpture; photographs.
Hours & Admission Prices: Daily 8-10. No charge. &
Attendance: 2,000 (estimated)

Saratoga

SARATOGA MUSEUM, 104 Constitution Ave., Saratoga, WY 82331. Mailing Address: P.O. Box 1131, Saratoga, WY 82331-1131. Tel.: 307-326-5511. Facebook: Saratoga Historical & Cultural Association.
E-mail: saratogamuseum@gmail.com
Web Site: www.saratoga-museum.com
Founded: 1978.
Key Personnel: Dir., Kimberly Givens; Pres. (V), Elizabeth Wood.
Personnel Profile: Full-Time Paid 1; Part-Time Volunteers 30.
Volunteer Hours: 130
Governing Authority: nonprofit organization. Parent Institution: Saratoga Historical & Cultural Association. Tax-exempt: 170(b)(1)(A).
Institution Type/Description: Historical Society Museum: housed in c.1890 Union Pacific Depot.

Collections: Union Pacific memorabilia; archaeology of Wyoming; items & artifacts of Carbon County. Historic Structures: renovated caboose; sheep wagon; blacksmith shop; geology exhibit.
Research Fields: history; archaeology; Union Pacific R.R.
Facilities: pavilion.
Activities: lectures; films; summer concerts; formally organized education programs for children; docent program; loan & permanent exhibitions. Museum Sponsors: historical trips; architectural awards presentation; Annual Dinner; Quilt Show; Wyoming Historical Trivias.
Publications: Lewis Shutterly Diary; Saratoga & Encampment, an Album of Family Histories; Window in Time; Winchester Williams.
Hours & Admission Prices: Memorial Day to Oct. 15 Tues.-Sat. 1-4. Tours after season available by request. No charge; donations accepted. &
Attendance: 1,800 (estimated)
Membership: Senior 65 & over $10; Individual $20; Family $30; Business $75.

Savery

LITTLE SNAKE RIVER MUSEUM, Rte. 70, Savery, WY 82332. Mailing Address: P.O. Box 13, Savery, WY 82332-0013. Tel.: 307-383-7262.
Web Site: www.littlesnakerivermuseum.com
Institution Type/Description: History Museum.
Collections: community history; personal artifacts; photographs.
Hours & Admission Prices: Memorial Day to Oct. daily 11-5. No charge; donations accepted. &

Sheridan

CUSTOM GROUP PROPERTIES LLC, 856 Broadway St., Sheridan, WY 82801-3623. Mailing Address: 901 S. 9th St., Broken Arrow, OK 74012. Tel.: 307-674-2178.
Web Site: www.sheridaninn.com
Formerly: Historic Sheridan Inn/Sheridan Heritage Center
Governing Authority: nonprofit organization. Tax-exempt: 501(c)(3).
Institution Type/Description: Historic Building: Sheridan Inn built c.1892.
Collections: Inn: period furnishings; photographs. Center: local history & culture.
Activities: rental facilities.
Hours & Admission Prices: Daily 10-8. Tours: daily 10-2. &

KING SADDLERY MUSEUM, 184 N. Main, Sheridan, WY 82801-3906. Tel.: 307-672-2702; 800-443-8919. Fax: 307-672-5235.
Key Personnel: Cur., Jean King
Institution Type/Description: Western History Museum.
Collections: Western leather work; cowboy & Western history; saddles; Indian artifacts; guns; photographs; braidwork.
Activities: research.
Publications: catalogue, A King Saddlery.
Hours & Admission Prices: Mon.-Sat. 8-5. No charge; donations accepted.

SHERIDAN COUNTY MUSEUM, 850 Sibley Cir., Sheridan, WY 82801-9626. Tel.: 307-675-1150. Fax: 307-675-1151.
E-mail: info@sheridancountyhistory.org
Web Site: www.sheridancountyhistory.org
Founded: 2006.
Key Personnel: Dir. & Educator, Nathan Doerr; Pres. (V), Judy Musgrave; Museum Shop Mgr., Rebecca Lincoln.
Personnel Profile: Full-Time Paid 1; Part-Time Paid 4; Part-Time Volunteers 45.
Governing Authority: Parent Institution: Sheridan County Historical Society. Tax-exempt.
Institution Type/Description: History Museum.
Collections: Sheridan County history; photographs; personal artifacts.
Activities: special events; rental facilities. Museum Sponsors: monthly programs on the Porch; Tidbit Tuesdays for Kids.
Publications: quarterly newsletter, The Log; exhibit series booklets.
Hours & Admission Prices: Museum: May & Sept.-Dec. 24 daily 1-5; June to Labor Day daily 10-6. Museum Store: Jan.-April Tues.-Sat. 1-5. Adults $4, seniors 60 & over $3, students $2; veteran, active military, Blue Star families and children 12 & under no charge. Closed New Year's Day; Thanksgiving. &
Attendance: 4,500 (accurate)
Membership: Senior $25; Single $30; Couple $50; Business $100; Corporate $500.

TRAIL END STATE HISTORIC SITE, 400 Clarendon Ave., Sheridan, WY 82801-4053. Tel.: 307-674-4589. Fax: 307-672-1720.
E-mail: trailend@state.wy.us
Web Site: www.trailend.org
Founded: 1982.
Key Personnel: Historic Program Mgr., Cynde Georgen; Cur., Sharie L. Mooney.
Personnel Profile: Full-Time Paid 2; Part-Time Paid 3; Part-Time Volunteers 15.

Governing Authority: state; nonprofit. Parent Institution: Wyoming Dept. of Parks & Cultural Resources, Division of State Parks & Historic Sites, Cheyenne, WY. Tax-exempt: 501(c)(3).
Institution Type/Description: Historic House.
Collections: early 20th-century social history with emphasis on the Kendrick family; local history from 1913-1933.
Research Fields: Kendrick family; Northern Plains agriculture; 1913-1933 Social history; history of entertainment in the 20th century.
Facilities: 1,300-vol. library; 80-seat auditorium. Site-related items for sale.
Activities: docent program; self-guided tours; guided group tours by appointment; temporary exhibitions; theater. Annual Events: fundraiser in July; Holiday Open House in December.
Publications: quarterly newsletter, End Notes; One Cowboy's Dream: John B. Kendrick, His Family, Home and Ranching Empire.
Hours & Admission Prices: April-May & Sept.-Dec. 14 daily 1-4; June-Aug. daily 9-6. Adults $2; discounts to Wyoming residents; children 17 & under no charge. Closed Veterans Day; Thanksgiving. &
Attendance: 14,000 (accurate)
Membership: Individual $5; Supporting $25; Sponsor $50; Patron $100; Benefactor $250; Cornerstone $500; Foundation $1,000.

Sinclair

PARCO SINCLAIR MUSEUM, 300 E. Lincoln Ave., Sinclair, WY 82334. Mailing Address: P.O. Box 247, Sinclair, WY 82334-0247. Tel.: 307-324-3058.
Institution Type/Description: History Museum: housed in the former First National Bank; built in 1924.
Collections: local history & culture; photographs; period furnishings.
Hours & Admission Prices: Mon.-Fri. 9-12 & 1-4:30. No charge.

South Pass City

SOUTH PASS CITY STATE HISTORIC SITE, 125 South Pass Main, South Pass City, WY 82520-8703. Tel.: 307-777-6323. Fax: 307-332-3688.
E-mail: jellis@state.wy.us
Web Site: www.southpasscity.com
Founded: 1967.
Key Personnel: Cur. Public Programs, Jon Lane; Supt., Joe Ellis.
Personnel Profile: Full-Time Paid 4; Part-Time Paid 2.
Governing Authority: state. Parent Institution: Wyoming State Parks & Historic Sites, 2301 Central Ave., Cheyenne, WY 82002. Tax-exempt.
Institution Type/Description: Historic Building & Site: c.1867-1910 gold-mining town consisting of 25 furnished buildings.
Collections: general store; hotel; school; bank; saloon; jail; dance hall; livery stable; gold mining memorabilia & artifacts; clothing; furnishings; bottles; photographs; guns; documents. Historic Buildings: 9 historic structures of the Carissa Mine including Mill house, hoist house, cook house, dorm, office, & residence.
Research Fields: history of mining; Oregon Trail history; community development; Woman Suffrage; 19th-century textiles; clothing; economic cycles.
Facilities: picnic grounds; concession desk.
Activities: guided tours; living history demonstrations; interpretive talks. Museum Sponsors: Gold Rush Days in July, Wyoming State Mining Championships & 1900 Vintage Baseball Tournament in July.
Publications: brochures, nature & historical.
Hours & Admission Prices: mid-May to Sept. daily 9-6. Adults: nonresident $4, resident $2; children 18 & under no charge. &
Attendance: 20,000 (accurate)
Membership: Individual $5.

Sundance

CROOK COUNTY MUSEUM & ART GALLERY, 309 Cleveland St., Sundance, WY 82729. Mailing Address: P.O. Box 63, Sundance, WY 82729-0063. Tel.: 307-283-3666. Fax: 307-283-1192.
E-mail: ccmuseum@rangeweb.net
Web Site: www.crookcountymuseum.com
Founded: 1971.
Key Personnel: Dir., Trudy Wadley; Chm., Rocky Courchaine.
Personnel Profile: Full-Time Paid 2.
Governing Authority: county. Parent Institution: Crook County. Tax-exempt.
Institution Type/Description: History Museum.
Collections: furniture; pictures; western historical items; land records.
Facilities: library.
Activities: special tours for individual groups; special activities for school groups & senior citizens.
Publications: brochure.
Hours & Admission Prices: June-Aug. Mon.-Sat. 8-4; Sept.-May Mon.-Fri. 8-4. No charge; donations accepted. Closed holidays. &
Attendance: 7,000 (estimated)

Membership: Single $30; Couple $45; Family $80; Corporate $500.

Ten Sleep

TEN SLEEP PIONEER MUSEUM, 436 S. Second St., Ten Sleep, WY 82442. Mailing Address: P.O. Box 93, Ten Sleep, WY 82442-0093. Tel.: 307-366-2759.
Key Personnel: Dir., Gloria Cutt
Institution Type/Description: History Museum.
Collections: pioneer life; tools; clothing; local family histories; books.
Activities: research.
Hours & Admission Prices: Daily 9-4; groups by appointment. No charge; donations accepted.

Thermopolis

HOT SPRINGS COUNTY MUSEUM AND CULTURAL CENTER, 700 Broadway, Thermopolis, WY 82443-2722. Tel.: 307-864-5183. Fax: 307-864-2974.
E-mail: hschistory@rtconnect.net
Web Site: hschistory.org
Founded: 1941.
Key Personnel: Dir. & Museum Shop Mgr., Ross R. Rhodes; Chm. (V), Kelly Andreen.
Personnel Profile: Full-Time Paid 1; Part-Time Paid 4; Part-Time Volunteers 4.
Volunteer Hours: 985
Operating Expenses: 110,000
Operating Income: 110,000
Governing Authority: Parent institution: Hot Springs County; county; nonprofit. Tax-exempt.
Institution Type/Description: History Museum.
Collections: county history from prehistoric to present times; 1880-1920s pioneering; arrowheads & related artifacts; ephemera & documentary archives; photographs.
Facilities: library; 80-seat auditorium; 18,000 sq. ft. exhibit space. Museum-related items for sale.
Activities: guided tours; lectures; temporary exhibitions.
Publications: brochure, Hot Springs County Historical Museum.
Hours & Admission Prices: Memorial Day-Labor Day Mon.-Sat. 8-5; Sept.-May Tues.-Sat. 9-4. Adults $4, senior citizens & children 6-12 $2; discounts to locals; current & former military & children 5 & under no charge. &
Attendance: 4,630 (accurate)

THE WYOMING DINOSAUR CENTER, 110 Carter Ranch Rd., Thermopolis, WY 82443-2457. Mailing Address: P.O. Box 868, Thermopolis, WY 82443-0868. Tel.: 307-864-2997; 800-455-3466. Fax: 307-864-5762.
E-mail: wdinoc@wyodino.org
Web Site: www.wyodino.org
Founded: 1995.
Congressional District: 20
Key Personnel: Gen. Mgr., Angie Guyon.
Governing Authority: privately owned. Parent Institution: Big Horn Prospecting. Branch Museum: Old West Wax Museum.
Institution Type/Description: Paleontology (Science) Museum.
Collections: 22 full size dinosaur mounts including Tyrannosaurus Rex, Triceratops & Camarasaurus; geology & prehistoric life on earth.
Research Fields: Paleontology-Jurassic, identification of primary dinosaur fossils, bone preparation and mounting.
Facilities: library; field research station; 18,000 sq. ft. exhibit area; bone preparation laboratory (viewable by visitors); bone storage; major dinosaur digsites comprising five acres on a 7,000-plus acre property (off-premises). Museum-related items for sale.
Activities: guided tours; Dig-for-a-day program for the public; kids' dinosaur dig; Elderhostel service programs.
Hours & Admission Prices: Mid-May to mid-Sept. daily 8-6; mid-Sept. to mid-May daily 10-5. Adults $10, senior citizens 60 and over, children 4-13 & veterans $5.50; discounts to groups & families; children 3 & under no charge. Closed New Year's Day; Thanksgiving; Christmas. &
Attendance: 32,000 (estimated)

Torrington

HOMESTEADERS MUSEUM, 495 Main St., Torrington, WY 82240. Mailing Address: P.O. Box 250, Torrington, WY 82240-0250. Tel.: 307-532-5612.
E-mail: museum@city-of-torrington.org
Founded: 1975.
Key Personnel: Dir., Dan Ringle; Pres. (V), Jean Dalton.
Personnel Profile: Full-Time Paid 1; Part-Time Paid 4; Interns 1.
Governing Authority: Parent Institution: City of Torrington. Tax-exempt.

Institution Type/Description: Historic House: built in 1910 by Ben Trout.
Collections: period artifacts & records; photographs; land claims.
Hours & Admission Prices: Call for hours. No charge; donations accepted. Closed New Year's Day; Presidents' Day; Memorial Day; Independence Day; Thanksgiving & day after; Christmas.
Attendance: 2,800 (accurate)

Upton

UPTON RED ONION MUSEUM, 203 Pine St., Upton, WY 82730. Mailing Address: P.O. Box 543, Upton, WY 82730-0543. Tel.: 307-468-2672. Fax: 307-468-2441.
Institution Type/Description: History Museum.
Collections: local history & culture; photographs; period furnishings; personal artifacts; paintings.
Activities: research; school group tours. Museum Sponsors: The Old Fashioned Christmas Celebration in December.
Hours & Admission Prices: Mon.-Fri. 9-5. No charge.

Wheatland

LARAMIE PEAK MUSEUM, 1601 16th St., Wheatland, WY 82201. Mailing Address: P.O. Box 451, Wheatland, WY 82201-0451. Tel.: 307-331-2961, 322-2309 & 3765.
Founded: 1982.
Key Personnel: Chm. (V), Marlin C. Marshall.
Personnel Profile: Part-Time Volunteers 22.
Volunteer Hours: 1,593
Institution Type/Description: History Museum.
Collections: area history; photographs; personal artifacts.
Hours & Admission Prices: Third Mon. of May to Sept. 13 Mon.-Fri. 10-5, Sat. 10-3. No charge; donations accepted. Closed holidays.
Attendance: 535 (accurate)
Membership: Individual $15.

WYOMING TRAILS GALLERY, 1004 16th St., Wheatland, WY 82201-2530. Tel.: 307-322-3300.
E-mail: barbara@wyomingtrailsgallery.com
Institution Type/Description: Art Gallery.
Collections: paintings; bronze; pottery; jewelry; folk arts.
Hours & Admission Prices: Wed.-Sat. 9-4; other times by appointment.

Worland

WASHAKIE MUSEUM, (M), 2200 Big Horn Ave., Worland, WY 82401-2932. Tel.: 307-347-4102. Fax: 307-347-4865.
E-mail: creichelt@washakiemuseum.org
Web Site: www.washakiemuseum.org
Founded: 1986.
Congressional District: 5
Key Personnel: Dir., Education & Museum Shop Mgr., Cheryl L. Reichelt; Pres., Christiane Geb; Vice Pres., Helen Koch; Treas., Dan Frederick; Sec., Christine Gee.
Personnel Profile: Full-Time Paid 3; Part-Time Paid 5; Part-Time Volunteers 38.
Governing Authority: private; nonprofit organization. Tax-exempt: 501(c)(3).
Institution Type/Description: General Museum.
Collections: history of Big Horn Basin; Paleo-Indian; paleontology; archaeology; geology; art; culture.
Major Exhibits: Washakie County Through the Decades Centennial Exhibit, 11/13-1/25/14; Portraits in Pastoralism (T), 4/14; Art of David Martin, 4/14; Shades of Greatness - Negro League Baseball Museum Collection (T), 6/14-7/14; Human Habitation in the Greater Yellowstone Area (T), 7/14-10/14; Artifact: A Cultural Heritage (T), 9/14-11/14; Fine Art Exhibit - Wyoming Artists, 12/14.
Facilities: 25,000 sq. ft. museum; two permanent galleries; one temporary gallery; event center.
Activities: art & history education for adult & children; performing arts; local artists exhibits; traveling, permanent & temporary exhibits.
Hours & Admission Prices: May 15 to Sept. 15 Mon.-Fri. 9-6, Sat. 9-5, Sun. 1-5; Sept. 16 to May 14 Tues.-Sat. 9-4. Adults $8; discount to AAA members; members no charge. &
Attendance: 14,088 (accurate)
Membership: Individual $35; Family $55; Mammoth Individual $100; Pronghorn Society $500; Meadowlark Society $1,000; Sundance Society $1,500.

Wright

WRIGHT CENTENNIAL MUSEUM, 104 Ranch Court, Wright, WY 82732. Mailing Address: P.O. Box 354, Wright, WY 82732-0354. Tel.: 307-464-1222.
Founded: 1990.
Key Personnel: Dir., Nolene Wright.
Volunteer Hours: 1,000
Operating Expenses: 30,000
Operating Income: 400
Institution Type/Description: History Museum.
Collections: homestead; period dishes; World War I artifacts; mining tools; woodworking artifacts; present-day oil drilling & coal mining; windmills.
Publications: brochure.
Hours & Admission Prices: May 20 to Oct. 15 Mon.-Fri. 10-5, Sat. 10-3. No charge; donations accepted. &
Attendance: 700 (estimated)

(U.S. TERRITORIES)
AMERICAN SAMOA

(1 listings)

Pago Pago

JEAN P. HAYDON MUSEUM, Fagatogo, Pago Pago, AS 96799. Mailing Address: Fagatogo, P.O. Box 1540, Pago Pago, AS 96799-1540. Tel.: 684-633-4347. Fax: 684-633-2059.
E-mail: ascach07@gmail.com
Founded: 1969.
Key Personnel: Dir., Leala E. Pili; Chm. (V), Pagofie A. Fiaigoa; Program Coord., Rexx Yandall; Museum Shop Mgr., Johnston Yardall.
Personnel Profile: Full-Time Paid 3.
Governing Authority: territorial government; nonprofit organization. Affiliated with American Samoa Arts Council. Tax-exempt.
Institution Type/Description: Historic Building & Site, General Museum & Art Museum: housed in 1900s old Post Office Building.
Collections: cultural, historical & environmental collections describing Samoan art & culture & its development to present times.
Research Fields: Samoan culture.
Facilities: classrooms; art gallery; herbarium. Books, tapes & other museum-related items for sale.
Activities: guided tours; lectures; films; formally organized education programs for children & adults.
Publications: Faasamoa PEA; Samoan Way.
Hours & Admission Prices: Mon.-Fri. 7:30-4. Special Programs: Sat. by appointment. No charge; donations accepted. &
Attendance: 30,000 (estimated)

COMMONWEALTH OF THE NORTH

(1 listings)

Saipan

THE NORTHERN MARIANA ISLANDS MUSEUM OF HISTORY AND CULTURE, Chalan Pale Arnold Rd., Garapan, Saipan, MP 96950. Mailing Address: P.O. Box 504570, Saipan, MP 96950-4305. Tel.: 670-664-2164.
E-mail: cnmimuseum@gmail.com
Web Site: cnmimuseum.wix.com/welcome#!
Founded: 1996.
Key Personnel: Exec. Dir., Robert Hunter.
Personnel Profile: Full-Time Paid 6; Full-Time Volunteers 1; Part-Time Volunteers 4.
Governing Authority: Parent Institute: CNMI Government. Tax-exempt.
Institution Type/Description: History & Archaeology Museum: housed in historic Japanese hospital.
Collections: cultural & personal artifacts; folk culture; archaeology; decorative arts.
Activities: temporary exhibits; workshops; cultural events; educational outreach.
Hours & Admission Prices: Mon.-Fri. 9-4. Off-island visitors 6 & up $2, CNMI residents with I.D. $1; discounts to groups; children under 6 no charge. Closed holidays & designated austerity Fridays. &

Attendance: 8,600 (estimated)

GUAM

(3 listings)

Hagatna

FANINADAHEN KOSAS GUAHAN-GUAM MUSEUM, Rm. 408, PNB Bldg., Hagatna, GU 96910. Mailing Address: P.O. Box 2950, Hagatna, GU 96932-2950. Tel.: 671-475-4229 & 4230. Fax: 671-475-4227.
E-mail: am_palomo@yahoo.com
Web Site: www.guam.nex/gov/museum/
Founded: 1932.
Key Personnel: Dir. & Cur., Anthony Ramirez; Pres., J. Lawrence Cruz; Chm. (V), Doring Duenas.
Personnel Profile: Full-Time Paid 5; Part-Time Paid 1; Part-Time Volunteers 25.
Governing Authority: nonprofit; U.S. unincorporated territory. Parent Institution: Department of Chamorro Affairs. Subsidiary: Guam Museum. Tax-exempt.
Institution Type/Description: General Museum: housed in 1776 Garden House.
Collections: artifacts from 1500 B.C.E.; Spanish Colonial historical artifacts from 17th-century to1899; historical artifacts from 1900-1945; Sgt. Shoichi Yokoi handmade survival gear.
Research Fields: pre-World War II Guam; prehistoric ceramics.
Facilities: 2,000-vol. library of Guam reference materials, available for use by public; 1,500 sq. ft. exhibit space. Postcard, books and posters for sale.
Activities: concerts; films; guided tours; lectures; participatory & temporary exhibits. Annual Events: Chamorro Week Feb.-March; Museum Week in May.
Publications: newsletter, Pappet.
Hours & Admission Prices: Daily 11-6. No charge; donations accepted. Closed holidays &
Attendance: 60,000 (accurate)

Mangilao

ISLA CENTER FOR THE ARTS AT THE UNIVERSITY OF GUAM, #15 Dean's Circle, Mangilao, GU 96923. Mailing Address: UOG Station, Mangilao, GU 96923. Tel.: 671-735-2965 & 2966. Fax: 671-735-2967.
E-mail: islacenter@gmail.com
Web Site: www.uog.edu/dynamicdata/classislacenterarts.aspx?siteid+l&p=191
Founded: 1980.
Key Personnel: Dir., Velma Yamashita; Extension Assoc., Gi Young Hwang; Extension Assoc., Liann Marie Castro.
Personnel Profile: Part-Time Paid 3.
Governing Authority: nonprofit. Parent Institution: Univ. of Guam.
Institution Type/Description: Art Museum.
Collections: Pacific artifacts; prints; paintings; sculpture.
Facilities: 1,000 sq. ft. exhibit space. Books, postcards & exhibition catalogs for sale.
Activities: films; hobby workshops; lectures; temporary & traveling exhibitions.
Publications: quarterly exhibition catalogs.
Hours & Admission Prices: Mon.-Fri. 10-5, Sat. 10-2. No charge; donations accepted. Closed federal & government of Guam holidays. &
Attendance: 5,000 (estimated)
Membership: Student $15; Senior Citizen $25; Contributor $50; Patron $100; Benefactor $250; Key Member $500; Director's Club $1,000; President's Circle $2,500; Regent's Circle $5,000.

Piti

WAR IN THE PACIFIC NATIONAL HISTORICAL PARK, 460 N. Marine Dr., Piti, GU 96922. Mailing Address: 135 Murray Blvd., Ste. 100, Hagatna, GU 96910-5104. Tel.: 671-477-7278 ext. 1001. Fax: 671-477-7281.
E-mail: wapa_administration@nps.gov
Web Site: www.nps.gov/wapa
Founded: 1982.
Key Personnel: Supt., Jim Richardson.
Governing Authority: federal. Dept. of the Interior. Parent Institution: National Park Service, Washington, DC. Tax-exempt.
Institution Type/Description: World War II Military Museum: located on the site of the American recapture of Guam.
Collections: items & artifacts related to the War in the Pacific from 1939-1945.
Research Fields: military history of the War in the Pacific; 1939-1945 Pacific Island history; cause & effects of World War II 1939-1945; underwater archaeology of WW II sites in Micronesia.
Facilities: 150-vol. library of collection and 600-vols. of documentary material relating to the War in the Pacific from 1937-1945, in English & Japanese, available for research on premises only.

Activities: guided tours upon special request; formally organized educational programs for children; permanent exhibits; audiovisual program in Japanese & English.
Publications: park brochures; site bulletins.
Hours & Admission Prices: Daily 9-4:30. No charge. Closed New Year's Day; Thanksgiving; Christmas. &
Attendance: 45,000 (accurate)

PUERTO RICO

(21 listings)

Barranquitas

LUIS MUNOZ RIVERA MUSEUM, 10 Munoz Rivera St., Barranquitas, PR 00794-1607. Tel.: 787-857-0230. Fax: 787-857-0230.
E-mail: npietri@icp.goblerno.pr
Web Site: www.icp.goblerno.pr
Founded: 1916.
Key Personnel: Exec. Dir., Mercedes Gomez Marrero; Dir., Nicole Pietri.
Personnel Profile: Full-Time Paid 1.
Governing Authority: state. Parent Institution: Puerto Rican Culture Institute. Tax-exempt.
Institution Type/Description: History Museum: housed in the birthplace of patriot, Luis Munoz Rivera.
Collections: mural representing the civic & political life of the patriot; articles, documents & photographs related to the life & death of Don Luis Munoz Rivera.
Research Fields: Puerto Rican literature, history & folklore.
Activities: guided tours; permanent exhibitions; expositions.
Hours & Admission Prices: Wed.-Sun. 8-4. No charge. &
Attendance: 5,000 (estimated)

Bayamon

DR. JOSE CELSO BARBOSA HOUSE MUSEUM, Calle Barbosa No. 16, Bayamon, PR 00961-6346. Mailing Address: Instituto de Cultura Puertorriquena, Apartado 9024184, San Juan, PR 00902-4184.
Web Site: www.icp.gobierno.pr
Founded: 1970.
Key Personnel: C.E.O., Prof. Mercedes Gomez; Dir., Nicole Pietri.
Personnel Profile: Full-Time Paid 1.
Governing Authority: state. Affiliated with Instituto de Cultura Puertorriquena; Tel. 787-724-0700. Tax-exempt.
Institution Type/Description: Historic House.
Collections: furniture; personal objects; documents.
Hours & Admission Prices: Tues.-Sat. 8:30-12 & 1-4:30. No charge.
Attendance: 1,534 (accurate)

Cayey

DR. PIO LOPEZ MARTINEZ ART MUSEUM, (M), University of Puerto Rico, Cayey Campus, 205 Antonio R. Barcelo Ave., Cayey, PR 00736-4127. Tel.: 787-738-2161, ext. 2209 & 2191. Fax: 787-738-0650.
E-mail: museo.cayey@upr.edu
Web Site: www.cayey.upr.edu
Founded: 1979.
Key Personnel: Dir., Prof. Humberto Figueroa.
Personnel Profile: Full-Time Paid 5; Part-Time Paid 3.
Governing Authority: Parent Institution: University of Puerto Rico. Tax-exempt.
Institution Type/Description: Art & History Museum.
Collections: paintings; drawings; prints; period artifacts.
Major Exhibits: Artist/Collector, 2/14-12/14; Letrismo, 2/14.
Publications: exhibition catalogs & brochures, Legado Frade; Homar Homo Humoris; Frade Arquitecto.
Hours & Admission Prices: Mon.-Fri. 8-4:30, Sat.-Sun. 11-5. No charge. &
Attendance: 4,360 (accurate)

Fajardo

LAS CABEZAS DE SAN JUAN NATURE RESERVE (EL FARO), Rte. 987, Km 5.9, Fajardo, PR 00738. Mailing Address: The Conservation Trust of Puerto Rico, P.O. Box 9023554, San Juan, PR 00902-3554. Tel.: 787-860-2560. Fax: 787-722-5872 & 860-1451.
E-mail: fideicomiso@fideicomiso.org
Web Site: www.fideicomiso.org
Founded: 1991.

Key Personnel: Exec. Dir., Fernando Lloveras San Miguel, Esq.; Chm., Jorge San Miguel, Esq.; Archivist, Rafael Lebron; Supt., Elizabeth Padilla.
Personnel Profile: Full-Time Paid 11; Part-Time Paid 3.
Governing Authority: private; nonprofit organization. Parent Institution: The Conservation Trust of Puerto Rico. Tax-exempt: 501(c)(3).
Institution Type/Description: Nature Reserve: 316-acre reserve on northeastern tip of Puerto Rico includes 1880 lighthouse (El Faro), second oldest of Puerto Rico's 14 lighthouses.
Collections: aquarium collections (nonliving): crustaceans, echinoderms, sponges, shells, coral, insects, rocks; aquarium collections (living): iguanas, hermit crabs, sea marine invertebrates, fish. Environmental collections: mangroves, lagoons, rocky beaches, dry forests, offshore cays, reefs; endangered species; archaeological artifacts; herbarium.
Research Fields: archaeological excavations; migration of South American Indians; marine science; neurobiology; ornithology; environmental studies.
Facilities: 100-vol. library on natural science available to scholars and researchers; aquarium; nature center; field research station; theater; educational facilities. Museum-related items for sale.
Activities: guided tours; lectures; educational programs for adults & children; assistance with school projects & science fairs; training programs for professional museum workers.
Publications: quarterly newsletter, Boletin El Faro.
Hours & Admission Prices: Oct. to mid-Aug. Wed.-Sun. by reservation only. Tours: 9:30, 10, 10:30 & 2. Adults $10, senior citizens & students $7; discounts to groups of 20 or more; children under 4 no charge. Closed New Year's Day; Epiphany; Good Friday; Mother's Day; Father's Day; Independence Day; Thanksgiving; Christmas.
Attendance: 43,830 (accurate)
Membership: Student K-12 & University Student 21 & under $10; Seniors 65 & over $30; Individual $40; Family $75.

Guaynabo

CAPARRA MUSEUM AND HISTORIC PARK, Villa Caparra, 212 Carretera No. 2, Guaynabo, PR 00966-1718. Mailing Address: Instituto de Cultura Puertorriquena, Apartado 9024184, San Juan, PR 00902-4184. Tel.: 787-781-4795. Fax: 787-723-7837.
Web Site: www.icp.gobierno.pr
Founded: 1970.
Key Personnel: Exec. Dir., Mercedes Gomez Morrero.
Personnel Profile: Full-Time Paid 2.
Governing Authority: state. Affiliated with Instituto de Cultura Puertorriquena; Tel. 787-724-0700. Tax-exempt.
Institution Type/Description: Historic Site: ruins of Caparra was the first spot of colonization in Puerto Rico founded by Ponce de Leon in 1508.
Collections: memorial plaques; period artifacts.
Hours & Admission Prices: Mon.-Fri. 8-12 & 1-4:30, Sat.-Sun. by appointment. No charge. &

Attendance: 1,876 (accurate)

Gurabo

TURABO UNIVERSITY, Carr. 189, Km. 3.1, Gurabo, PR 00778. Mailing Address: P.O. Box 3030, Gurabo, PR 00778-3030. Tel.: 787-743-7979, ext. 4135. Fax: 787-743-7979, ext. 4149.
Founded: 1980.
Personnel Profile: Part-Time Paid 1.
Governing Authority: private university; nonprofit.
Institution Type/Description: Archaeology Museum.
Collections: concentration on the archaeology of the eastern part of Puerto Rico; folkloric arts & crafts artifacts; ethnological representation of the region in the 19th-century.
Research Fields: archaeological studies of the region; urban development in 1940s; archaeological history in the 19th-century.
Facilities: 600-vol. library on art, history & archaeology available to the public; classroom; field research station.
Activities: temporary exhibitions. Annual Events: photography & art workshops.
Publications: annual bulletin, Center of Humanistic Studies.
Hours & Admission Prices: Aug.-May daily 8-12 & 1-5. No charge; donations accepted.
Attendance: 3,500

Mayaguez

UNIVERSITY OF PUERTO RICO DEPARTMENT OF MARINE SCIENCES MUSEUM, Dept. of Marine Sciences, Univ. of Puerto Rico, Mayaguez, PR 00681. Mailing Address: Call Box 9000, Mayaguez, PR 00681. Tel.: 787-832-4040, exts. 3443, 3447 & 3838. Fax: 787-899-5500 & 265-5408 & 832-3432.
Web Site: uprm.edu/cima
Founded: 1954.

Key Personnel: Dir., Dr. John M. Kubaryk; Cur. Marine Invertebrates, Dr. Nikolaos Schizas; Cur. Fish, Dr. Richard Appeldoorn.
Governing Authority: state. Affiliated with University of Puerto Rico, Mayaguez Campus. Tax-exempt.
Institution Type/Description: Marine Museum.
Collections: invertebrates; fishes; shells; tropical algae.
Research Fields: algae; fishes; invertebrates.
Facilities: research library of marine subjects; field research station; seawater systems; classrooms.
Publications: contributions of the Dept. of Marine Sciences.
Hours & Admission Prices: Call 787-899-2048 for appointment. No charge. &

Old San Juan

LA CASA DEL LIBRO, Callejon de la Capilla #199, Old San Juan, PR 00902. Mailing Address: P.O. Box 9023544, San Juan, PR 00902-3544. Tel.: 787-723-0354. Fax: 787-723-0354.
E-mail: lcdl@prw.net
Web Site: www.lacasadellibro.org
Founded: 1955.
Key Personnel: Pres. Bd. Dir., Jorge Rigau, Arq.; Admin., Marian Toledo; Coord., Victor Blay.
Personnel Profile: Part-Time Paid 2.
Governing Authority: Amigos Calle Del Cristo 255, Inc. Parent Institution: state. Subsidiary Institution: Instituto de Cultura Puertorriquena. Tax-exempt: 501(c)(3).
Institution Type/Description: Rare Book/Book Arts Museum.
Collections: typography & other arts related to bookmaking, printing, calligraphy, lettering, design, illustration, papermaking, binding; manuscripts; rare books: manuscripts, incunabula, 20th-century masterpieces, 14th- to 15th-century Spanish books. Historic Building: 18th-century colonial townhouse.
Research Fields: history of writing; history of printing; incunabula; maps; book-related research: bookbinding, calligraphy, typography, graphic arts, design, illustration, paper-making.
Facilities: 7,000-vol. library.
Activities: lectures; temporary exhibitions; storytelling for children; book arts workshops.
Publications: Libros Espanoles Siglos XV-XVI; Coleccion La Casa del Libro, La Temprana Imprenta Sevillana; Repetitiones I, II & III; Libros Venecianos S.XV-XVI; Coleccion Quijote.
Hours & Admission Prices: (Temporary location) call for hours. No charge; donations accepted.
Attendance: 30,000 (estimated)
Membership: Bernardo de Balbuena $35; Alonso Manso $100; Isabel la Catolica $500; El Quijote $1,000.

Ponce

HACIENDA BUENA VISTA, Rte. 123, Km. 16.8, Ponce, PR 00731. Mailing Address: The Conservation Trust of Puerto Rico, P.O. Box 9023554, San Juan, PR 00902-3554. Tel.: 787-722-5882. Fax: 787-841-5997.
E-mail: fideicomiso@fideicomiso.org
Web Site: www.fideicomiso.org
Founded: 1987.
Key Personnel: Exec. Dir., Fernando Lloveras San Miguel, Esq.; Chm., Jorge San Miguel, Esq.; Visitor Svcs. Mgr., Sandra Franqui.
Personnel Profile: Full-Time Paid 14; Part-Time Paid 5.
Governing Authority: private; nonprofit organization. Parent Institution: The Conservation Trust of Puerto Rico. Tax-exempt: 501(c)(3).
Institution Type/Description: Historic Site: reconstructed 1833 coffee plantation & corn mill in southern Puerto Rico.
Collections: Vives family records, documents, business papers & photographs; gardens; trails; reconstructed coffee processing machinery. Restored Buildings: manor house with period furniture, slave quarters; corn mill.
Research Fields: history of slavery in relation to Hacienda Buena Vista.
Facilities: 587-vol. library of Spanish & English catalogues, medical books, musical scores, accounting records, European & New World magazines, manuscripts & other documents; 350 sq. ft. exhibit space; gardens; trails. Museum-related items for sale.
Activities: guided tours; lectures; workshops; concerts; temporary & traveling exhibits; educational programs; docent program; training program for professional museum workers.
Publications: LaBuena Vista 1833-1904.
Hours & Admission Prices: Wed.-Sun. by appointment. Adults $8, senior citizens & students $5; discounts to groups of 20 or more; children 4 & under no charge. Closed New Year's Day; Epiphany; Good Friday; Independence Day; Thanksgiving; Christmas.
Attendance: 25,130 (accurate)
Membership: Student $7; University Student $15; Senior Citizen $25; Adult $30; Family $60.

✱ **MUSEO DE ARTE DE PONCE,** 2325 Ave. Las Americas, Ponce, PR 00918. Mailing Address: P.O. Box 9027, Ponce, PR 00732-9027. Tel.: 787-200-7090 & 7091; 787-848-0505 (Temporary location). Fax: 787-200-7094; 787-841-7309 (Temporary location).
E-mail: map@museoarteponce.org
Web Site: www.museoarteponce.org
Founded: 1959.
Key Personnel: C.E.O. & Dir., Dr. Agustin Arteaga; Volunteer Coord., Ms. Coral Cosals; Deputy Dir. Communications, Denise Berlingeri; Financial & Administration Dir., Miriam B. Quintero-Casanovas; Dir. Education, Ana Margarita Hernandez; Chief Cur., Cheryl Hartup; Registrar, Zorali De Feria; Dir. Conservation Laboratory, Lidia Aravena; Mgr. Human Resources, Nancy Colon; Facilities, Security & IT Mgr., Emilio Ruiz; Museum Shop Mgr., Mrs. Lourdes Vargas.
Personnel Profile: Full-Time Paid 36; Part-Time Paid 6; Part-Time Volunteers 100; Interns 1.
Governing Authority: nonprofit. Parent Institution: Luis A. Ferre Foundation, Inc. Tax-exempt: 501(c)(3).
Institution Type/Description: Art Museum.
Collections: European art from 14th-19th centuries with emphasis on Victorian Italian & Spanish painting; Latin American art from 19th century to present; Puerto Rican art from 18th century to present; paintings; sculptures; archaeology; Oriental & pre-Columbian ceramics; glassware.
Research Fields: Puerto Rican & Latin American art; European painting & sculpture.
Facilities: 5,000-vol. library of art books for research available for use on premises; conservation laboratory; reading room. Pamphlets, slides, postcards, reproductions & photographs of paintings for sale.
Activities: guided tours; lectures; films; gallery talks; concerts; workshops for students of all ages; formally organized education programs for children & undergraduate college students; inter-museum loan, permanent, temporary & traveling exhibitions; summer camps; collecting program; open houses.
Publications: book, Catalogues of Paintings & Sculptures; exhibition catalogues.
Hours & Admission Prices: Temporary location: 2325 Ave. Las Americas, Ponce, PR 00717-0076. Mon.-Fri. 11-7, Sat. 11-8, Sun. 11-5. Adults $6, children under 12 $3; discounts to groups, ICOM & AAM members; members no charge. Closed New Year's Day; Good Friday; Thanksgiving; Christmas. &
Attendance: 87,028 (accurate)
Membership: Students, Clergy, Senior Citizens & Handicapped $20; Individual $35; Family $50; Miguel Pou Category $100; Ramon Frade $250. Companies: Francisco Oller Category $500; Jesus Maria Sanroma Category $1,000; Jose Campeche Category $2,500; Mecenas Club $5,000; Gran Mecenas $10,000; Medici $20,000.

San Juan

CASA BLANCA MUSEUM, Calle San Sebastian No. 1, San Juan, PR 00901-1156. Mailing Address: Instituto de Cultura Puertorriquena, Apartado 9024184, San Juan, PR 00902-4184. Tel.: 787-725-1454. Fax: 787-723-7837.
Web Site: www.icp.gobierno.pr
Founded: 1974.
Key Personnel: Exec. Dir., Mercedes Gomez Marrero; Museum Guide, Angrette Merced.
Personnel Profile: Full-Time Paid 2; Part-Time Volunteers 1.
Governing Authority: state. Affiliated With Instituto de Cultura Puertorriquena; Tel.: 787-724-0700. Tax-exempt.
Institution Type/Description: Historic House: c.1521 building constructed for the sons of Juan Ponce de Leon & inhabited by their descendants until mid-18th century.
Collections: domestic life in San Juan during the first three centuries of Spanish colonization; furniture; household decorations.
Publications: Casa Blanca, residence of the descendants of Juan Ponce de Leon colonizer of Puerto Rico.
Hours & Admission Prices: Wed.-Sun. 8:30-12 & 1-4:30. Adults $2, children $1; seniors no charge. &
Attendance: 12,300 (accurate)

LUIS TORRES DIAZ PHARMACY MUSEUM, Pharmacy and Deanship of Students Bldg., Medical Sciences Campus, University of Puerto Rico, San Juan, PR 00936. Mailing Address: P.O. Box 365067, San Juan, PR 00936-5067. Tel.: 787-758-2525. Fax: 787-751-5680.
E-mail: farma@rcm.upr.edu
Web Site: farmacia.rcm.upr.edu
Institution Type/Description: Pharmacy Museum.
Collections: 210 porcelain apothecary jars; mortars; pestles; pharmaceutical artifacts from Spain, France & other countries.
Hours & Admission Prices: Tues.-Fri. 1-4.

MUSEO DE LAS AMERICAS, Cuartel de Ballaja, at the entrance of El Morro, 2nd Fl., San Juan, PR 00901. Mailing Address: P.O. Box 9023634, San Juan, PR 00902-3634. Tel.: 787-724-5052. Fax: 787-722-2848.
E-mail: museolasamericas@gmail.com
Web Site: www.museolasamericas.org
Founded: 1992.
Key Personnel: Dir., Maria Angela Lopez-Vilella; Museum Shop Mgr., Walleska Rivera.
Personnel Profile: Full-Time Paid 8; Part-Time Paid 10; Part-Time Volunteers 2.
Governing Authority: private; nonprofit organization.
Institution Type/Description: Folk Art Museum: housed in the 19th-century Ballaja Barracks.
Collections: history, economy, culture, religion & military of Puerto Rico dating back to 1511; folk art from the Americas; Native American artifacts from North, Central, & South America & the Caribbean; African anthropology.
Facilities: Museum-related items for sale.
Activities: children's workshops; arts & crafts; demonstrations; documentaries; lectures.
Publications: exhibition catalogues; biannual newsletter.
Hours & Admission Prices: Tues.-Sat. 9-12 & 1-4, Sun. 11-4. Adults $3, students $2. Closed New Year's Eve & Day; Good Friday; Mother's Day; Father's Day; Thanksgiving; Christmas Eve & Day. &
Attendance: 30,000 (estimated)

MUSEO DEL NINO, Calle Cristo 150, San Juan, PR 00901-1509. Mailing Address: P.O. Box 9022467, San Juan, PR 00902-2467. Tel.: 787-722-3791. Fax: 787-723-2058.
E-mail: info@museodelninopr.org
Founded: 1987.
Key Personnel: C.E.O., Ms. Carmen L. Vega; Bd. Member, Tere Bolivar.
Personnel Profile: Full-Time Paid 4; Part-Time Paid 1; Part-Time Volunteers 35; Interns 5.
Governing Authority: private; nonprofit organization. Tax-exempt: local exemption.
Institution Type/Description: Children's Museum.
Collections: hands-on exhibits.
Research Fields: AIDS, with the purpose of developing an exhibit aimed at kids.
Facilities: 25-seat theater; 8,000 sq. ft. exhibit space. T-shirts, hats, bags, books & museum-related items for sale.
Activities: lectures; participatory exhibits; theater; broadcast programs. Annual Event: fundraiser.
Publications: biannual newsletter, Boletin del Museo del Nino.
Hours & Admission Prices: Tues.-Thurs. 9-3:30, Fri. 9-5, Sat.-Sun. 12:30-5. Adults $5, children 14 & under $4; discount to members. Closed major holidays. &
Attendance: 45,000 (accurate)
Membership: Kids $12; Family of six $100.

✱ **MUSEUM OF HISTORY, ANTHROPOLOGY AND ART, (M),** University of Puerto Rico, San Juan, PR 00931. Mailing Address: P.O. Box 21908 UPR, San Juan, PR 00931-1908. Tel.: 787-764-0000, ext. 2452 & 763-3939. Fax: 787-763-4799.
E-mail: museo.universidad@upr.edu
Founded: 1940.
Key Personnel: Dir. & Cur. Art Collections, Flavia Marichal Lugo; Cur. Archaeological Collections, Ivan Mendez; Designer, Lionel Ortiz-Melendez; Educator, Lisa Ortega; Art Photographer, Jesus E. Marrero; Registrar, Chakira Santiago; Administrative Sec., Yolanda Vasquez; Admin., Maritza Rodriguez.
Personnel Profile: Full-Time Paid 13; Part-Time Paid 2.
Governing Authority: university. Parent Institution: University of Puerto Rico. Tax-exempt.
Institution Type/Description: Art, History & Archaeology Museum.
Collections: 18th- to 21st-century Puerto Rican paintings, sculpture, prints & drawings; archaeology; ethnography; international graphics; Puerto Rican painters of the past & present; archaeology; history; popular arts; numismatics; documents & memorabilia.
Research Fields: Puerto Rican art, history & archaeology.
Facilities: library of Puerto Rican exhibition catalogs from 1950-present, artists archives & slides from art collection available for research on premises; 4,582 sq. ft. exhibit space.
Activities: guided tours; concerts; inter-museum loan, permanent, temporary & traveling exhibitions.
Publications: exhibition catalogues.
Hours & Admission Prices: Sun. 11-5, Mon.-Tues. & Fri. 9-4:30, Wed. 9-8:30. No charge. Closed holidays; university recesses. &
Attendance: 20,000

MUSEUM OF THE PUERTO RICAN FAMILY OF THE 19TH CENTURY, 319 Fortaleza St., San Juan, PR 00901-1715. Mailing Address: Instituto de Cultura Puertorriquena, P.O. Box 9024184, San Juan, PR 00902-4184. Tel.: 787-723-1762. Fax: 787-723-7837.
Web Site: www.icp.gobierno.pr
Founded: 1964.
Key Personnel: Dir., Nicole Pietri; C.E.O., Prof. Mercedes Gomez.
Personnel Profile: Full-Time Paid 2.
Governing Authority: state. Affiliated with Instituto de Cultura Puertorriquena; Tel. 787-724-0700. Tax-exempt.
Institution Type/Description: Furniture Museum.
Collections: Puerto Rican 19th-century furniture.
Hours & Admission Prices: Wed.-Sun. 1-4:30. No charge.
Attendance: 2,024 (estimated)

PABLO CASALS MUSEUM, 101 San Sebastian St., San Juan, PR 00901. Mailing Address: P.O. Box 41227, San Juan, PR 00940-1227. Tel.: 787-723-9185. Fax: 787-722-3338 & 723-5843.
Founded: 1977.
Key Personnel: C.E.O. & Gen. Mgr., Evangelina Colon; Admin., Anibal Ramirez.
Personnel Profile: Full-Time Paid 1; Part-Time Paid 4.
Governing Authority: Theatrical Musical Arts Corp. Parent Institution: Corp. Artes Escenicas Musicales. Institution subsidized by Puerto Rican government.
Institution Type/Description: Classical Music Museum.
Collections: photographs & memorabilia of Pablo Casals; videos of past Casals concerts.
Research Fields: classical music history related to Pablo Casals.
Hours & Admission Prices: Tues.-Sat. 9:30-5. Adults $1, senior citizens & children $.50; discounts to AAM members.
Attendance: 15,000 (estimated)

PUERTO RICO MUSEUM OF CONTEMPORARY ART, (M), Rafael M. Labra Bldg., Ponce de Leon Ave., Corner of Roberto H. Todd (Stop 18), San Juan, PR 00909. Mailing Address: P.O. Box 362377, San Juan, PR 00936-2377. Tel.: 787-977-4030. Fax: 787-977-4036.
E-mail: adm1@museocontemporaneopr.org
Web Site: www.museocontemporaneopr.org
Founded: 1984.
Key Personnel: Exec. Dir., Marianne Ramirez Aponte; Pres. Bd., Frankie Vazquez; Pres. (V), Luis Fernando Rodriguez; Accountant, Walmy Rivera; Registrator, Rene Sandin; Admin., Michelle Dilan; Education & Public Programming Officer, Evita Busa; Museum Shop Mgr., Jorge L. Pardo.
Personnel Profile: Full-Time Paid 12; Part-Time Paid 11; Part-Time Volunteers 2; Interns 2.
Governing Authority: private; nonprofit organization. Tax-exempt.
Institution Type/Description: Art Museum.
Collections: works by contemporary artists from Latin America, Puerto Rico & the Caribbean from 1940 to the present; paintings; drawings; sculpture; prints; installation art; photographs; mixed media; video art.
Major Exhibits: Exhibicion de Bill Viola, 9/13-2/14; Raul Recio: Who is la Salsa?, 10/13-1/14; Taller Vivo: 440 dibujos de Abdiel Segarra/Orfeon, 10/13-2/14; Orfeon San Juan Bautista: Subject and Object, 11/13-2/14; Art Competition Sa Juan: Capital City of Latin America, 11/13-3/14; Wilderness Science & Art Collaboration, 2/14-10/14; Off the Beaten Path: Violence, Women and Art, 6/14-10/14; Osvaldo Budet and Shonah Trescott: Drawn into the Light, 11/14-3/15.
Research Fields: Puerto Rican, Caribbean & Latin American Art produced since 1940.
Facilities: 15,000-vol. library; 7,180 sq. ft. exhibit space; documentation center; conservation & restoration laboratory; digital workshop; education studio; administrative offices. Museum-related items for sale.
Activities: concerts; dance recitals; docent program; films; formal education programs for adults, school students & college students; guided tours; lectures; loan, participatory & traveling exhibitions; training programs for professional museum workers & Title I teachers; theater; summer camp.
Publications: members bulletin; exhibition catalogues; Puerto Rican art history; book series; art education brochures for teachers & the general public; multimedia productions (cd-roms), Contemporary Puerto Rican Art (1940-1999).
Hours & Admission Prices: Tues.-Fri. 10-4, Sat. 11-5, Sun. 1-5. Suggested Donations: adults $5, students with ID, children 5 & up, seniors 60-75 & persons with disabilities $3; discount to AAM & ICOM members; MAC members & seniors 75 & up no charge. Closed New Year's Eve & Day; Three Kings Day; Good Friday; Independence Day; Constitution Day; Discovery of America Day; Thanksgiving; Christmas. &
Attendance: 26,000 (estimated)
Membership: Student & Senior $20; Teachers $30; Individual $45; Familiar $100; Associate $250; Collaborator $500; Benefactor $1,000; Corporate $2,000; Design Own Membership $2,000 & up.

SAN JUAN NATIONAL HISTORIC SITE, 501 Norzagaray St., San Juan, PR 00901-1213. Tel.: 787-729-6777. Fax: 787-289-7165.
E-mail: felix_j_lopez@nps.gov
Web Site: www.nps.gov/saju/
Founded: 1949.
Key Personnel: Chief Cultural Resources, Felix J. Lopez; Supt., Walter Chavez; Eastern National Store Mgr., Angie Alicea; Park Historian, Eric Lopez; Archives Tech., Javier Martinez
Governing Authority: federal. Parent Institution: National Park Service. Tax-exempt.
Institution Type/Description: Historic Buildings: 16th to 19th-century Spanish fort located in Old San Juan, 3 miles of city walls.
Collections: Old Spanish uniforms & artifacts; weapons; coins; ceramics; construction plans; rare books; period furnishings; troop quarters; archives on the history of the forts of Old San Juan including: 5,000 maps & plans; 10,000 photos; 50,000 manuscripts; microfilms & documents.
Research Fields: Spanish-colonial military history; preservation of masonry structures.
Facilities: 2,000-vol. library pertaining to the history of Europe & Latin America, available for use by researchers or institutions under special circumstances; Fort San Cristobal.
Activities: guided tours; 30-min. video history of the forts of San Juan.
Hours & Admission Prices: Daily 9-6. Adults $3, children 15 & under and seniors 65 and over with Golden Age Pass no charge. Closed New Year's Day; Thanksgiving; Christmas. &
Attendance: 1,300,000 (accurate)

Santurce

* **MUSEO DE ARTE DE PUERTO RICO, (M),** 299 De Diego Ave., Stop 22, Santurce, PR 00909-1766. Mailing Address: P.O. Box 41209, San Juan, PR 00940-1209. Tel.: 787-977-6277. Fax: 787-977-4444.
E-mail: info@mapr.org
Web Site: www.mapr.org
Founded: 2000.
Key Personnel: Exec. Dir., Ms. Lourdes Ramos Rivas, Ph.D.; Attorney, Arturo Garcia Sola; Devel., Myrna Z. Perez; Cur., Juan Carlos Lopez; Education, Doreen Colon; Treas., Sonia Dominguez, CPA; Registrar, Sandra Cintron; Mgr. Exhibitions, Jaquelina Rodriguez; Mgr. Public Rels., Yetzenia Alvarez.
Personnel Profile: Full-Time Paid 42; Part-Time Paid 7; Part-Time Volunteers 85; Interns 26.
Governing Authority: private; nonprofit organization. Tax-exempt: 501(c)(3).
Institution Type/Description: Art Museum.
Collections: 18th to 21st century Puerto Rican paintings, sculpture, prints & drawings; installations.
Research Fields: Puerto Rican art, painting & graphics from colonial art to present.
Facilities: library; botanical garden; 92-seat restaurant; 32-seat cafeteria; 41,962 sq. ft. exhibit space; 400-seat theater; classrooms; workshops; computer lab; seminar room. Museum-related items for sale.
Activities: arts festivals; concerts; docent program; films; formal education programs; guided tours; hobby workshops; lectures; loan, traveling & temporary exhibitions; theater; training programs for professional museum workers. Annual Events: MAPR Annual Gala; MAPR Annual Auction.
Publications: annual report; exhibition catalogues; brochures.
Hours & Admission Prices: Tues. & Thurs.-Sat. 10-5, Wed. 10-8, Sun. 11-6. Adults $6, senior citizens, children & students $3; tax not included; Wed. 2-8 & members no charge. Closed New Year's Day; Good Friday; Election Day; Thanksgiving; Christmas. &
Attendance: 125,923 (accurate)
Membership: Students, Teachers, Seniors & Handicapped $25; Individual $50; Family $100; Friend $250; Collaborated $500; Executive $1,000.

Vieques

MUSEO FUERTE CONDE DE MIRASOL DE VIEQUES, Apartado 71, Vieques, PR 00765-0071. Mailing Address: P.O. Box 71, Vieques, PR 00765-0071. Tel.: 787-741-1717.
E-mail: bieke@prdigital.com
Web Site: www.icp.gobierno.pr/
Founded: 1991.
Key Personnel: Dir., Roberto L. Rubin Siegal.
Personnel Profile: Full-Time Paid 7; Part-Time Volunteers 10; Interns 8.
Governing Authority: state government; nonprofit. Parent Institution: Instituto del Cultura Puertorriquena. Tax-exempt.
Institution Type/Description: History Museum: housed in a two-story 19th-century Spanish fortress constructed of brick & local wood.
Collections: archeological material from pre-Hispanic groups indigenous to Vieques & main island of Puerto Rico; artifacts, photos & maps documenting Vieques' history, 1514-present; artifacts, photos & tools relative to Vieques' historic architecture, 1844-1940; itinerant plastic arts exhibits: photos, paintings; artisanry, sculpture, textiles.

Research Fields: Vieques' archaeology; African slavery in Vieques, 19th century; historic relations between Vieques & St. Croix, U.S. Virgin Islands; environmental & other impacts of the U.S. Navy presence on Vieques, 1940-present; Vieques' historic architecture.

Facilities: library of historical texts; local newspapers; microfilm archives of varied documents; programs/bulletins from local festivities & community organizations; 100-seat auditorium; classrooms. Museum-related items for sale.

Activities: cultural festival; concerts; docent program; films; formal education programs for undergraduate & graduate college students; guided tours; lectures; school loan service; study clubs; temporary & traveling exhibitions; theater.

Hours & Admission Prices: Museum: Wed.-Sun. 9-5:30. Historic Archives: Mon.-Fri. by appointment. Adults $3, seniors over 60 & children under 12 $1; discounts to AAM & ICOM members; children under 12 no charge. Closed Good Friday; Mother's Day; Father's Day.

Attendance: 15,000 (estimated)

VIRGIN ISLANDS

(4 listings)

Saint John

VIRGIN ISLANDS NATIONAL PARK, 1300 Cruz Bay Creek, Saint John, VI 00830-6108. Tel.: 340-776-6201, ext. 238.

E-mail: susanna-pershern@nps.gov
Web Site: www.nps.gov/viis
Founded: 1956.
Key Personnel: Supt., Mark Hardgrove; Chief Interpretive Ranger, Paul Thomas.
Personnel Profile: Full-Time Paid 4; Part-Time Volunteers 10.
Governing Authority: federal. National Park Service, Southeast Field area, 75 Spring St., S.W. Atlanta, GA. 30303. Tel. 404-331-5187. Headquarters: Virgin Islands National Park No. 10 Estate Nazareth, St. Thomas, VI 00802. Tax-exempt.
Institution Type/Description: National Park Museum & Visitor Center.
Collections: archaeological; history; marine. Historic Structures: c.1717-late 1800s, sugar mills & estate houses.
Research Fields: tropical, marine & terrestrial West Indian & Danish colonial history.
Facilities: 1,000-vol. library pertaining to Caribbean & natural history available for research on premises. Books & other museum-related items for sale.
Activities: guided tours; lectures; films; formally organized education programs for children; permanent & temporary exhibits.
Publications: Annaberg Plantation self-guided tour brochures; safe boating brochure; trail guide.
Hours & Admission Prices: Visitor Center: daily 8-4:30. &
Attendance: 250,000 (estimated)

St. Croix

CHRISTIANSTED NATIONAL HISTORIC SITE, 2nd Fl., Danish Custom House, Kings Wharf Christiansted, St. Croix, VI 00820. Mailing Address: 2100 Church St. #100, Danish Custom House, Christiansted, St. Croix, VI 00820-5402. Tel.: 340-773-1460. Fax: 340-773-5995.

E-mail: CHRI_superintendent@nps.gov
Web Site: www.nps.gov/chri
Founded: 1952.
Key Personnel: Chief Resource Management, Zandy Hillis-Starr; Park Supt., Joel A. Tutein; Chief Interpretation, David J. Goldstein.
Personnel Profile: Full-Time Paid 18.
Governing Authority: federal. National Park Service, Southeast Region, 100 Alabama St., S.W., 1924 Bldg., Atlanta, GA. 30303. Tel. 404-562-3327. Tax-exempt.
Institution Type/Description: Historic Site & Historic Houses: depicts Danish Colonial development in the West Indies.
Collections: Folmer Andersen collection of over 16,000 prehistoric artifacts; architectural specimens; Danish military uniforms; Colonial Bottle Collection. Historic Houses: 1753 Steeple Building; 1738 Fort Christiansvaern; 1855-56 Scale House; 1748-49 Danish West India & Guinea Company Ware House; 1751 Old Custom House.
Research Fields: pre-Columbian cultures in the Virgin Islands; Danish West Indian military history; Colonial history; African American heritage.
Activities: guided tours.
Publications: brochures, Christiansted National Historic Site, Fort Christiansvaern; Folmer Andersen; Steeple Building; Government House; Buck Island Reef National Monument; Salt River Bay National Historic Park & Ecological Preserve.
Hours & Admission Prices: Mon.-Fri. 8-5, Sat.-Sun. 9-5. Adult $3; senior citizens with Golden Age Pass & youth under 16 no charge. Closed Thanksgiving; Christmas.
Attendance: 100,000 (accurate)
Membership: Golden Age Pass $10.

WHIM PLANTATION MUSEUM, 52 Estate Whim, Frederiksted, St. Croix, VI 00840-3744. Tel.: 340-772-0598. Fax: 340-772-9446.

E-mail: info@stcroixlandmarks.org
Web Site: www.stcroixlandmarks.com
Institution Type/Description: History Museum: housed on an early 18th century sugar plantation.
Collections: sugar plantation life & history; restored windmill; sugar factory; gardens.
Facilities: library; 12 acre plantation.
Activities: special events; rental facilities.
Hours & Admission Prices: Mon.-Sat. 10-4. Adults $10, seniors $5, children 6-12 $4; VI residents on Sat. & children under 6 no charge. &

St. Thomas

SEVEN ARCHES MUSEUM, Off Kongens Gade, Government Hill, St. Thomas, VI 00802. Mailing Address: P.O. Box 6456, St. Thomas, VI 00804-6456. Tel.: 340-774-9295.

E-mail: sevenarchesmuseum@yahoo.com
Web Site: www.sevenarchesmuseum.com
Founded: 1993.
Key Personnel: Dir. & Cur., Barbara Demaras
Institution Type/Description: History Museum.
Collections: local history & culture; "Welcoming Arms" staircase, gun turret slots in the walls, letters, photographs & postcards from the 1800s; Danish porcelain; 18th-19th century West Indian furnishings. Historic Buildings: Danish West Indian home, c.1857; cook house, c.1795.
Publications: books.
Hours & Admission Prices: By appointment. Tours: $7 donation; school groups no charge.
Attendance: 1,000 (estimated)

Index to
Institutions

A

B–*Continued*

Now Available Online at: www.officialmuseumdirectory.com

C

C–Continued

Now Available Online at: www.officialmuseumdirectory.com

C–Continued

Now Available Online at: www.officialmuseumdirectory.com

C–Continued

D

G—Continued

H

H–*Continued*

I–*Continued*

K

L

L–Continued

N–*Continued*

N–Continued

P–Continued

R–Continued

S

S–Continued

S–*Continued*

X

Y

Index to Institutions by Category

List of Categories

ART

Art Associations, Councils and Commissions, Foundations and Institutes

Art Museums and Galleries

Art Museums and Galleries–Continued

Art Museums and Galleries–Continued

Art Museums and Galleries–Continued

Arts and Crafts Museums

Civic Art and Cultural Centers–Continued

Decorative Arts Museums

Folk Art Museums

CHILDREN'S MUSEUMS

CHILDREN'S MUSEUMS–Continued

CHILDREN'S MUSEUMS–Continued

COLLEGE AND UNIVERSITY MUSEUMS

COLLEGE AND UNIVERSITY MUSEUMS–Continued

COLLEGE AND UNIVERSITY MUSEUMS–Continued

EXHIBIT AREAS–Continued

Now Available Online at: www.officialmuseumdirectory.com

GENERAL MUSEUMS

GENERAL MUSEUMS–Continued

GENERAL MUSEUMS–Continued

GENERAL MUSEUMS–Continued

GENERAL MUSEUMS–Continued

HISTORY

Historic Agencies, Councils, Commissions, Foundations and Research Institutes

Historic Houses and Historic Buildings

Historic Houses and Historic Buildings–Continued

Historic Houses and Historic Buildings–Continued

Historic Houses and Historic Buildings–Continued

Historic Houses and Historic Buildings–Continued

Historic Houses and Historic Buildings–Continued

Historic Houses and Historic Buildings–Continued

Historic Sites–Continued

Historic Sites–Continued

Historic Sites—Continued

Historical and Preservation Societies

Historical Society Museums–Continued

Historical Society Museums–Continued

Historical Society Museums—Continued

Historical Society Museums–Continued

History Museums

History Museums–Continued

History Museums–Continued

History Museums–Continued

History Museums–Continued

History Museums–Continued

History Museums–Continued

History Museums–Continued

History Museums—Continued

History Museums–Continued

History Museums–Continued

History Museums–Continued

History Museums–Continued

History Museums—Continued

Maritime, Naval Museums and Historic Ships

Military Museums

Military Museums–Continued

Preservation Projects

LIBRARIES HAVING COLLECTIONS OF BOOKS

LIBRARIES HAVING COLLECTIONS OTHER THAN BOOKS–Continued

NATIONAL AND STATE AGENCIES, COUNCILS AND COMMISSIONS

NATURE CENTERS

NATURE CENTERS–Continued

NATURE CENTERS–Continued

PARK MUSEUMS AND VISITOR CENTERS

PARK MUSEUMS AND VISITOR CENTERS–Continued

SCIENCE

Academies, Associations, Institutes and Foundations

Aeronautics and Space Museums

Anthropology and Ethnology Museums

Aquariums, Marine Museums and Oceanariums

Entomology Museums

Geology, Mineralogy and Paleontology Museums

Now Available Online at: www.officialmuseumdirectory.com

SPECIALIZED

Agriculture Museums

Culturally Specific–Continued

Culturally Specific–Continued

Forestry Museums

Hobby Museums

Horological Museums

Industrial Museums

Money and Numismatics Museums

Musical Instruments Museums

Now Available Online at: www.officialmuseumdirectory.com

Technology Museums

Typography Museums

Village Museums

Products & Services Suppliers

Index to Advertisers

AUDIO & VIDEO-GENERAL

RBH MULTIMEDIA, INC.
12 Hatch Ter.
Dobbs Ferry, NY 10522
Tel: 914-693-8755; FAX: 914-693-3539
E-mail: rbh@rbhmedia.com
Web Site: www.rbhmedia.com

Type of Business:
  RBH Multimedia, Inc. is a media production company dedicated to developing and producing audiovisual experiences and exhibits. We bring imagination, skill, flair and years of experience to designing and producing audio, video and multiple media exhibits for museums, historic sites, visitor and interpretive centers, and cultural institutions. We also design and program websites and computer interactive exhibits. RBH is particularly known for producing signature multi-sensory, multimedia "experience" theater shows. Other services include hardware system design, engineering and installation, with expertise in videoconference technology. For more information about RBH Multimedia, Inc., please visit us online at www.rbhmedia.com.

Personnel:
  Steve Brosnahan (Partner)
  Nancy Haffner (Partner)
  Edgardo J. Resto (Associate)

Audio Visual Presentations

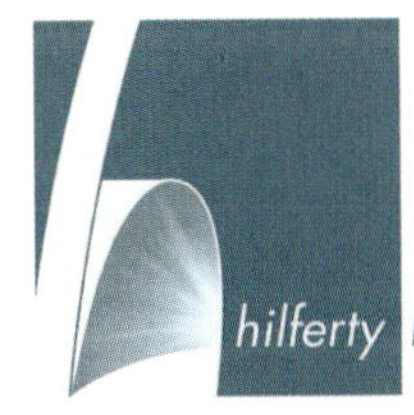

HILFERTY
  14240 State Rte. 550
  Athens, OH 45701
  Tel: 740-448-3821; FAX: 740-448-2331
  E-mail: gha@hilferty.com
  Web Site: www.hilferty.com

Type of Business:
  For more than thirty years, Hilferty has provided a full range of interpretive services to museums, visitor centers, nature centers, botanical gardens, historic sites, zoos/aquariums, and halls of fame. We excel at interpretive and facility master planning; concept development; content research; label writing; exhibit and graphic design; and providing donor recognition and funding support materials. From social histories to natural sciences, football to physics, presidents to children's gardens, every Hilferty exhibit vividly captures the stories that engage visitors' minds and hearts. Each project is approached with a unique mix of techniques chosen to best bring the subject to life. Our innovative portfolio spans the spectrum of intimate displays of precious objects, captivating immersive environments, enthralling interactive components, and breathtaking theatrical presentations. In short, Hilferty creates memorable museum experiences.

Personnel:
  Gerard Hilferty (President & Creative Director)
  Dean Clouse (CEO)

MONADNOCK media

MONADNOCK MEDIA, INC.
  112 Amherst Rd.
  Sunderland, MA 01375
  Tel: 413-665-1390; FAX: 413-665-1394
  E-mail: steve@monadnock.org
  Web Site: www.monadnock.org

Type of Business:
  REINVENTING Multimedia Design and Production for the Museum Environment. When exhibit designers and museums hire Monadnock Media for the design and production of multimedia experiences, they are hiring a group of passionate and talented individuals who work in close collaboration with the project team. We bring to the table a genuine interest in content, a real desire to find the most powerful stories and the clearest ways to express complex ideas. We also have the experience and know-how to bring the material to life no matter the multimedia format - high-tech interactives, immersive environments, theatrical experiences, and social media. We use a wide range of technologies and strategies to help present fresh and innovative approaches to storytelling - with Monadnock, you will never see your exhibit at another museum. Above all, working with us will be fun.

Personnel:
  Steve Bressler (Director)

RBH MULTIMEDIA, INC.
  12 Hatch Ter.
  Dobbs Ferry, NY 10522
  Tel: 914-693-8755; FAX: 914-693-3539
  E-mail: rbh@rbhmedia.com
  Web Site: www.rbhmedia.com

Type of Business:
  RBH Multimedia, Inc. is a media production company dedicated to developing and producing audiovisual experiences and exhibits. We bring imagination, skill, flair and years of experience to designing and producing audio, video and multiple media exhibits for museums, historic sites, visitor and interpretive centers, and cultural institutions. We also design and program websites and computer interactive exhibits. RBH is particularly known for producing signature multi-sensory, multimedia "experience" theater shows. Other services include hardware system design, engineering and installation, with expertise in videoconference technology. For more information about RBH Multimedia, Inc., please visit us online at www.rbhmedia.com.

Personnel:
  Steve Brosnahan (Partner)
  Nancy Haffner (Partner)
  Edgardo J. Resto (Associate)

Fiber Optics

NOUVIR LIGHTING
20915 Sussex Hwy.
Seaford, DE 19973
Tel: 302-628-9933; FAX: 302-628-9932
Web Site: www.nouvir.com

Type of Business:
  NoUVIR invented fiber optic, museum, preservation lighting. Fifty different miniature floodlights, spotlights, across-the-room pinspots, eyeballs, wall washers, pendants…all pure-white, stone-cold light with no UV and no IR…NoUVIR®. Our 24 U.S. patents guarantee superior performance. A 10-year warranty on optical hardware and fiber guarantees quality. Save energy, save money and save our art and heritage with NoUVIR. Call for conservation lighting textbooks, seminar information or for free product specifications, photometry and lighting design information.

Personnel:
  Miss Ruth Ellen Miller (President)
  Mr. Matthew S. Miller (Vice President Marketing)

Producers

HILLMANN & CARR INCORPORATED
2233 Wisconsin Ave., N.W., Ste. 425
Washington, DC 20007
Tel: 202-342-0001; FAX: 202-342-0117
E-mail: michalcarr@hillmanncarr.com
Web Site: www.hillmanncarr.com

Type of Business:
  Full-service creative storytellers. Producers of award winning media: video, 3D, film, computer interactive, audio environments, television broadcast, new media, video tours, immersive experiences and object theaters, graphic and multimedia presentations. Extensive 35-year expertise in science, technology, history and art museums, visitor centers, corporate exhibit media, expositions and multi-language international projects created in unique formats. Producers and project managers experienced in AV systems design, integration, installation, distribution. Wide-ranging experience in audiovisual media master planning and management, creative concept development, content research, script to screen production. WOSB.

Personnel:
  Alfred Hillmann (President)
  Michal Brand Carr (Vice President)

RBH MULTIMEDIA, INC.
12 Hatch Ter.
Dobbs Ferry, NY 10522
Tel: 914-693-8755; FAX: 914-693-3539
E-mail: rbh@rbhmedia.com
Web Site: www.rbhmedia.com

Type of Business:
  RBH Multimedia, Inc. is a media production company dedicated to developing and producing audiovisual experiences and exhibits. We bring imagination, skill, flair and years of experience to designing and producing audio, video and multiple media exhibits for museums, historic sites, visitor and interpretive centers, and cultural institutions. We also design and program websites and computer interactive exhibits. RBH is particularly known for producing signature multi-sensory, multimedia "experience" theater shows. Other services include hardware system design, engineering and installation, with expertise in videoconference technology. For more information about RBH Multimedia, Inc., please visit us online at www.rbhmedia.com.

Personnel:
  Steve Brosnahan (Partner)
  Nancy Haffner (Partner)
  Edgardo J. Resto (Associate)

Signage, Electronic Programmable

SUNRISE SYSTEM, INC.
720 Washington St.
Pembroke, MA 02359
Tel: 781-826-9706; FAX: 781-826-0061
E-mail: sales@sunrisesystems.com
Web Site: www.sunrisesystems.com

Type of Business:
  Sunrise Systems, Inc. is a manufacturer of custom LED signage. Founded in 1976, Sunrise has been a lifelong producer of fascinating public art, first-in-the-industry technological advances and uniquely tailored information systems. Our LED displays are installed in countries all over the world and in industries across the map. Transit system integrators, stock tickers, retail applications and academic exhibits are just a few well-established areas of Sunrise's expertise. Past installments are prominently featured in many permanent exhibits at prestigious museums around the globe. Our ability to meet even the most complex architectural, electrical and mechanical requirements keeps us at the forefront of the industry. A dedication to innovative technology and customer support make us a good place to start planning your next LED display project.

Personnel:
  Henry C. Appleton (President, Director Sales)
  Eric Harrington (Director, Engineering & General Manager)
  Leah Sylvia (Marketing Director & Project Manager)

Videos

RBH MULTIMEDIA, INC.
12 Hatch Ter.
Dobbs Ferry, NY 10522
Tel: 914-693-8755; FAX: 914-693-3539
E-mail: rbh@rbhmedia.com
Web Site: www.rbhmedia.com

Type of Business:

RBH Multimedia, Inc. is a media production company dedicated to developing and producing audiovisual experiences and exhibits. We bring imagination, skill, flair and years of experience to designing and producing audio, video and multiple media exhibits for museums, historic sites, visitor and interpretive centers, and cultural institutions. We also design and program websites and computer interactive exhibits. RBH is particularly known for producing signature multi-sensory, multimedia "experience" theater shows. Other services include hardware system design, engineering and installation, with expertise in videoconference technology. For more information about RBH Multimedia, Inc., please visit us online at www.rbhmedia.com.

Personnel:

Steve Brosnahan (Partner)
Nancy Haffner (Partner)
Edgardo J. Resto (Associate)

BUILDING & FACILITIES – GENERAL

Architectural Design Firms

Bergmeyer
Architecture and Interiors

BERGMEYER ASSOCIATES, INC.
51 Sleeper St.
Boston, MA 02210
Tel: 617-542-1025; FAX: 617-542-1026
E-mail: marketing@bergmeyer.com
Web Site: www.bergmeyer.com

Type of Business:
Bergmeyer is an award winning architecture and interior design firm. We
have specialized expertise in the development of public spaces and
visitor amenities including museum stores, exhibit shops, restaurants,
and admissions and information centers. Our services include
architecture and interior design, identity development, wayfinding and
graphic design, merchandising, strategic planning, business plan
development, and consumer/visitor research and analysis. We are
committed to design excellence and we create architecture that
supports our clients' brand and business. With forty years of
experience in a culture that challenges us to think in new ways, clients
know that Bergmeyer's design solutions will be unique and specific to
their needs.

Personnel:
David Tubridy, AIA, LEED AP (President)
Joseph P. Nevin, Jr. (Senior Principal)
Michael R. Davis, FAIA, LEED AP (Vice President)
Lewis Muhlfelder, AIA, LEED AP (Principal)
Matthew Hyatt, AIA, IIDA, LEED AP (Principal)
Anne Johnson (Director of Marketing)

CAMBRIDGE SEVEN ASSOCIATES, INC.
1050 Massachusetts Ave.
Cambridge, MA 02138
Tel: 617-492-7000; FAX: 617-492-7007
E-mail: joltman@c7a.com
Web Site: www.c7a.com

Type of Business:
At Cambridge Seven Associates, Inc. we design distinctive museums and
experiential exhibits that make learning stimulating, accessible, and
engaging. We seek to create immersive environments which spark
informal and multimodal learning in a captivating way. Cambridge
Seven endows museums, aquariums, zoos, and visitor centers with a
sense of place, wonder, and discovery, and these characteristics have
made our projects internationally- and nationally-renowned
destinations. Our services include architectural design, exhibit and
habitat design, feasibility assessments, site analysis, master planning,
space programming, institutional and operational planning, exhibit
installation coordination, graphic design and signage programs,
production of fund raising materials, and construction administration.

Personnel:
Peter Kuttner, FAIA (President)
Steven Imrich, AIA, LEED AP (Principal)
Patricia E. Intrieri, AIA (Principal)
Timothy D. Mansfield, AIA (Principal)
Peter Sollogub (Associate Principal)
Penny Sander (Associate Principal/Exhibit Planner)
Doug Simpson, LEED AP (Associate/Exhibit Designer)
Douglas Flandro, LEED AP (Exhibit Designer)
Jo Oltman (Marketing Manager)

EWING COLE

EWINGCOLE
100 N. 6th St.
Philadelphia, PA 19106
Tel: 215-625-4880; FAX: 215-574-9163
E-mail: jhirsch@ewingcole.com
Web Site: www.ewingcole.com/cultural

Type of Business:
EwingCole is a multi-discipline architecture and engineering firm
specializing in design for cultural institutions. We work with national
and regional museums for the fine arts, history, natural sciences and
science and technology. Our Cultural Practice experts apply deep
industry knowledge to enhance the visitor experience and to develop a
design closely aligned with the museum's strategic plan. Our
process engages the board and staff at all levels to create solutions that
are intellectually stimulating, fiscally responsible and environmentally
sustainable.

Personnel:
Jeffrey Hirsch, AIA, LEED AP (Principal, Director of Cultural Practice)

GWWO, INC./ ARCHITECTS
800 Wyman Park Dr., Ste. 300
Baltimore, MD 21211
Tel: 410-332-1009; FAX: 410-332-0038
E-mail: lwerther@gwwoinc.com
Web Site: www.gwwoinc.com

Type of Business:
GWWO specializes in the planning and design of museum, visitor center
and related facilities, with emphasis on quality design that is both
inspirational and evocative. We work in partnership with our clients to
develop individualized design solutions that respond to the unique
mission, character and collections of each institution. Our services
span all project stages, from site selection, master planning and
programming through final design and construction administration.
Clients include George Washington's Mount Vernon Estate & Gardens,
the Museum of the Rockies, the Children's Museum & Theatre of
Maine, the Brandywine River Museum, Morven Museum & Garden
and the Cade Museum of Creativity + Invention.

Personnel:
Alan E Reed, FAIA, LEED AP (Principal)
David G. Wright, FAIA, LEED AP (Principal)
Paul L. Hume, AIA, LEED AP (Principal)
Mark A. Lapointe, AIA (Principal)
Terry Squyres, AIA, LEED AP (Principal)
Laura M. Werther (Principal)

THE PORTICO GROUP

THE PORTICO GROUP
1500 4th Ave., 3rd Fl.
Seattle, WA 98101-1670
Tel: 206-621-2196; FAX: 206-621-2199
E-mail: portico@porticogroup.com
Web Site: www.porticogroup.com

Type of Business:
Celebrating 30 years of design, The Portico Group creates informal
learning opportunities in memorable settings-including natural and
cultural museums, zoos and aquaria, and public gardens. We are a
talented team of interpretive exhibit designers, architects, and
landscape architects capable of guiding a project from concept to

(Continued on next page)

(Continued from previous page)

conclusion. We seek opportunities to touch lives in meaningful ways - to connect people with nature and culture. Recent projects include: Hands On Children's Muscum (Olympia, WA), Assiniboine Park Zoo International Polar Bear Conservation Centre (Winnipeg, Manitoba), Los Angeles Zoo Living Amphibians, Invertebrates, and Reptiles (Los Angeles, CA), Pearl Harbor Visitor Center (Honolulu, HI), Smithsonian National Zoological Park American Trail (Washington, DC).

Personnel:
Michael Hamm, FASLA (President & CEO)
Charles G. Mayes, AIA (Principal)
Keith McClintock, ASLA (Principal)
Dennis Meyer, ASLA, LEED AP (Principal)
Alissa Rupp, AIA, LEED AP BD+C (Principal)
Richard Larson (Associate Principal)
Allison CraigSundine (Senior Associate)
Justin Lyon (Senior Associate)

LEE H. SKOLNICK ARCHITECTURE + DESIGN PARTNERSHIP
75 Broad St., Ste. 2700
New York, NY 10004
Tel: 212-989-2624; FAX: 212-727-1702
E-mail: mail@skolnick.com
Web Site: www.skolnick.com

Type of Business:
Established in 1980, Lee H. Skolnick Architecture + Design Partnership is an award-winning integrated architectural, museum exhibit and communication design firm. Museum facilities design services include: interpretive master planning, program development, conceptual and complete architectural services for new institutions, site planning, adaptive re-use of existing structures, historic preservation, adaptive reuse, feasibility studies. Exhibition programming and museum education services include:conceptual development, script and copy development, interpretive exhibition design, exhibition project management, evaluation of exhibits, educational programming. Communication design services include: exhibition and print graphics, identity and logo design, architectural signage and wayfinding and web design. LHSA+DP is also often called upon to create visionary renderings and fund-raising materials for prospective projects in the early concept phase.

Personnel:
Jo Ann Secor (Principal & Director Museum Services)
Scott Briggs (Senior Associate Museum Services)

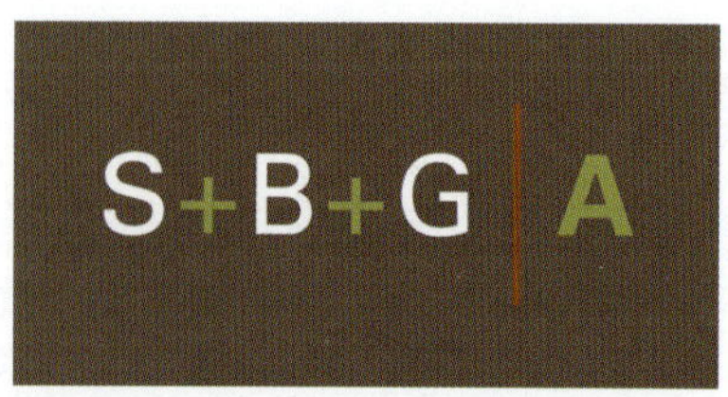

SOLOMON + BAUER + GIAMBASTIANI ARCHITECTS INC.
63 Pleasant St.
Watertown, MA 02472
Tel: 617-924-8200; FAX: 617-924-6685
E-mail: lbauer@sbgarch.com
Web Site: www.sbgarch.com

Type of Business:
Award-winning Architecture and Interior Design Firm with over forty projects for museums. Master-planning; programming; feasibility studies; design of new museums; renovation and adaptive reuse of existing and historic structures; and exhibit design support. Special expertise in collections storage analysis, planning, and design. Museums for which recent collection storage projects have been undertaken include Autry National Center, Blanton Museum of Art,

Dallas Museum of Art, Harvard Art Museums, Museum of Art Rhode Island School of Design, Museum of Fine Arts Houston, Princeton Art Museum, Saint Louis Art Museum, and The Broad.

Personnel:
Lawrence C. Bauer, AIA (Principal)
Karin Aslaksen (Associate)

STUDIOMUSARX LLC
2601 Pennsylvania Avenue, Ste. 8
Philadelphia, PA 19130
Tel: 215-232-3489
E-mail: info@studiomusarx.com
Web Site: www.studiomusarx.com

Type of Business:
studioMUSarx is a forward-thinking studio with special expertise in planning, architecture and exhibit design for new and existing museums and visitor centers. We are dedicated to a process of discovery acknowledging best practices while lending a friendly ear to client preferences in keeping with the shared values and aspirations of the studio. Each project undertaken is directed by the studio's principal, who collaborates closely with design team members to create strong working partnerships with clients. For over 40, years Joe Nicholson has assisted museums and interpretive centers with feasibility studies, master plans, space programming, signage programs, architectural and exhibit development, and design for sites, buildings, and interpretive exhibits providing full services through construction administration. We are open to the ongoing challenge of touching upon the extraordinary, embracing the artful, and going beyond constraints to deliver meaningful outcomes free from preconceptions of what the final design product should be.

Personnel:
Joseph A. Nicholson, AIA, NCARB (Principal)
Susan Hutton DeAngelus (Exhibit and Graphic Designer)
Juliet Geldi, RA (Planner, Architect and Exhibit Designer)
Sherman Lai, Assoc. AIA (Architectural and Exhibit Designer)

Engineering Design Firms

EWINGCOLE
100 N. 6th St.
Philadelphia, PA 19106
Tel: 215-625-4880; FAX: 215-574-9163
E-mail: jhirsch@ewingcole.com
Web Site: www.ewingcole.com/cultural

Type of Business:
EwingCole is a multi-discipline architecture and engineering firm specializing in design for cultural institutions. We work with national and regional museums for the fine arts, history, natural sciences and science and technology. Our Cultural Practice experts apply deep industry knowledge to enhance the visitor experience and to develop a design closely aligned with the museum's strategic plan. Our

(Continued on next page)

(Continued from previous page)

process engages the board and staff at all levels to create solutions that are intellectually stimulating, fiscally responsible and environmentally sustainable.

Personnel:
Jeffrey Hirsch, AIA, LEED AP (Principal, Director of Cultural Practice)

LANDMARK FACILITIES GROUP, INC.
252 East Ave.
Norwalk, CT 06855
Tel: 203-866-4626; FAX: 203-866-8019
E-mail: info@lfginc.com

Type of Business:
Landmark is a professional engineering firm offering HVAC, mechanical, electrical, plumbing, and fire protection engineering design, focusing its services on the special needs of museums, libraries, archives, historic structures, and their collections. The staff has a special capability in environmental controls and is well acquainted with facilities where the needs of preservation and sensitivity to architectural features are governing priorities. Services include monitoring, testing, long-range planning, systems design, constructions support, operations training, and commissioning.

Personnel:
Thomas E. Newbold, P.E., CEM, CPMP, LEED AP (Principal)
Gerard J. Rauth, P.E., LEED AP (Principal)

Maintenance & Equipment

SWIZZ-STYLE INC.
Professional humidification for your valuable assets
165 West Broadway
Dover, OH 44622
Tel: 1-877-663-7895
E mail: info@swizz-style.com
Web Site: www.swizz-style.com/brune

Type of Business:
Provider of high output capacity mobile humidifier, made in Germany. Brune is the market-leader in Europe and purveyor of humidification for most European museums. The B 500 humidifies areas up to 4200 square feet is ready to use, and requires only normal tap water. It features a 5-step whisper quiet automatic fan, an environmentally friendly organic filter and UV sterilization built-in. It is the only mobile humidifier with direct water supply possibility and self-diagnosing system display for maintenance and carries an 11 gallon water reserve basin.

Theater & Auditorium Design

EWINGCOLE
100 N. 6th St.
Philadelphia, PA 19106
Tel: 215-625-4880; FAX: 215-574-9163
E-mail: jhirsch@ewingcole.com
Web Site: www.ewingcole.com/cultural

Type of Business:
EwingCole is a multi-discipline architecture and engineering firm specializing in design for cultural institutions. We work with national and regional museums for the fine arts, history, natural sciences and science and technology. Our Cultural Practice experts apply deep industry knowledge to enhance the visitor experience and to develop a design closely aligned with the museum's strategic plan Our process engages the board and staff at all levels to create solutions that are intellectually stimulating, fiscally responsible and environmentally sustainable.

Personnel:
Jeffrey Hirsch, AIA, LEED AP (Principal, Director of Cultural Practice)

Theater Planning

CHICAGO SCENIC STUDIOS, INC.
1315 N. North Branch St.
Chicago, IL 60642
Tel: 312-274-9900; FAX: 312-274-9901
E-mail: dLanghorst@chicagoscenic.com
Web Site: www.chicagoscenic.com

Type of Business:
Chicago Scenic Studios provides museums, science & technology centers, zoos, aquariums and visitor centers more than 35 years of multi-disciplinary experience in themed environment design, fabrication, engineering and installation. Our highly skilled staff of project managers, designers, and skilled craftspeople partner with your team to provide safe, educational, immersive environments and exciting experiences through custom exhibits, retail, restaurant, and admission spaces. Chicago Scenic can serve as your general contractor, fabricator or project manager in support of your design or ours. Our background in exhibits, retail, theater, special events and corporate spaces provides us with the ability to cater to differing needs in unique spaces. We have enjoyed partnering with world-class institutions such as The Museum of Science and Industry, Chicago; The Great Lakes Science Center, the Chicago Children's Museum, The John G. Shedd Aquarium, the Brookfield Zoo, the Field Museum of Natural History and many more to bring exciting, educational and successful stories to life. We'd like to help tell your story as well.

Personnel:
Robert Doepel (President)
Diane Langhorst (Business Development)

BUSINESS/RETAIL SERVICES-GENERAL

HUB INTERNATIONAL NORTHEAST
1065 Avenue of Americas
New York, NY 10018
Tel: 212-338-2534 (Marilyn Brown) & 212-338-2524 (Norman Newman); FAX: 212-338-2520
E-mail: Marilyn.Brown@hubinternational.com; Norman.Newman@hubinternational.com
Web Site: http://northeast.hubinternational.com/industries/fine-arts/

Type of Business:
HUB International Northeast is a leading insurance brokerage that has a highly experienced team of seasoned professionals specializing in arranging coverage on all forms of valuable collectibles for museums and other non-profits, private collectors, exhibitions and galleries and dealers. With over 45 years of experience in serving this industry, HUB builds customized insurance and risk management needs to fit your unique requirements.

Personnel:
Marilyn Brown (Assistant Vice President, Fine Art & Special Risk Department)
Norman Newman (First Vice President, Fine Art & Special Risk Department)

TAM RETAIL, DIV. OF LODE DATA SYSTEMS, INC.
10609 West 159th St.
Orland Park, IL 60467
Tel: 888-THE-14POS (888-843-1476) & 708-460-0999; FAX: 708-460-1253
E-mail: sales@tamretail.com
Web Site: www.nonprofitpos.com

Type of Business:
TAM Retail is the sole provider of The Assistant Manager™ (TAM), a turn-key retail management solution that meets the growing needs of non-profits throughout the U.S. and internationally. TAM provides the knowledge of the things you cannot easily see for yourself. It provides you with a solution for knowing the most critical aspects of your organization's operations including exactly what products and services have sold, attendance levels, details on each transaction, associate and volunteer performance, perpetual value of your inventory and how it is performing, what is hot and what is not, helps you know your members better, what members like and don't like, and promotes increased donations. All the information needed to make the many important decisions that affect your organization is provided by TAM. TAM provides this at an affordable cost of ownership from a proven and trusted provider to help your organization get the most return on your investment.

Personnel:
Bruce H. Lode (Executive Vice President of Marketing and Sales)

Food Service

TAM RETAIL, DIV. OF LODE DATA SYSTEMS, INC.
10609 West 159th St.
Orland Park, IL 60467
Tel: 888-THE-14POS (888-843-1476) & 708-460-0999; FAX: 708-460-1253
E-mail: sales@tamretail.com
Web Site: www.nonprofitpos.com

Type of Business:
TAM Retail is the sole provider of The Assistant Manager™ (TAM), a turn-key retail management solution that meets the growing needs of non-profits throughout the U.S. and internationally. TAM provides the knowledge of the things you cannot easily see for yourself. It provides you with a solution for knowing the most critical aspects of your organization's operations including exactly what products and services have sold, attendance levels, details on each transaction, associate and volunteer performance, perpetual value of your inventory and how it is performing, what is hot and what is not, helps you know your members better, what members like and don't like, and promotes increased donations. All the information needed to make the many important decisions that affect your organization is provided by TAM. TAM provides this at an affordable cost of ownership from a proven and trusted provider to help your organization get the most return on your investment.

Personnel:
Bruce H. Lode (Executive Vice President of Marketing and Sales)

Fundraising/Membership Services

SCHULTZ & WILLIAMS
325 Chestnut St., Ste. 700
Philadelphia, PA 19106
Tel: 215-625-9955; FAX: 215-625-2701
E-mail: mail@schultzwilliams.com
Web Site: www.schultzwilliams.com

Type of Business:
Schultz & Williams provides comprehensive consulting services to the arts, museums, cultural institutions and other nonprofits of all types and sizes across the nation. Since 1987, the success and integrity of our work has been driven by one simple philosophy: development, management and marketing strategies must be fully integrated to achieve financial stability, operational excellence and greater mission impact. Our senior professionals have extensive hands-on experience from both sides of the desk. The strength of our consulting services is based on the skills we have gained in managing, operating and supporting major nonprofit organizations, including: Development: Capital campaigns, major gift and planned giving strategies; Direct Response: Full service digital and offline communications to raise funds through membership, annual giving, mid-level and monthly giving programs; Planning: Financial planning, facility and master plan management, leadership development; Marketing: Branding, planning, collateral; S&W StaffSolutions: Executive and management level interim, project, or start-up staffing.

(Continued on next page)

(Continued from previous page)

Personnel:
L. Scott Schultz (President)
M. Jane Williams (Principal)
Jessica Harrington (Vice President, S & W Direct)
Dell Fascione (Vice President)

TAM RETAIL, DIV. OF LODE DATA SYSTEMS, INC.
10609 West 159th St.
Orland Park, IL 60467
Tel: 888-THE-14POS (888-843-1476) & 708-460-0999; FAX: 708-460-1253
E-mail: sales@tamretail.com
Web Site: www.nonprofitpos.com

Type of Business:
TAM Retail is the sole provider of The Assistant Manager™ (TAM), a turn-key retail management solution that meets the growing needs of non-profits throughout the U.S. and internationally. TAM provides the knowledge of the things you cannot easily see for yourself. It provides you with a solution for knowing the most critical aspects of your organization's operations including exactly what products and services have sold, attendance levels, details on each transaction, associate and volunteer performance, perpetual value of your inventory and how it is performing, what is hot and what is not, helps you know your members better, what members like and don't like, and promotes increased donations. All the information needed to make the many important decisions that affect your organization is provided by TAM.

TAM provides this at an affordable cost of ownership from a proven and trusted provider to help your organization get the most return on your investment.
Personnel:
Bruce H. Lode (Executive Vice President of Marketing and Sales)

Insurance

ARTHUR J. GALLAGHER & CO.
250 Park Ave., 3rd Fl.
New York, NY 10177
Tel: 212-994-7100; FAX: 212-994-7047
E-mail: Ellen.Ross@ajg.com
Web Site: www.ajgrms.com

Type of Business:
The Arthur J. Gallagher & Co. Fine Arts Team with over 100 years of combined experience has been referred to as the "Masters in the Art of Insuring Art." We specialize in providing significant guidance and innovative solutions to meet the changing needs of the museum world. Our professional staff provides comprehensive insurance programs in all lines of business, including Fine Arts, Property & Casualty, Workers Compensation and Benefits.

HUB INTERNATIONAL NORTHEAST
1065 Avenue of Americas
New York, NY 10018
Tel: 212-338-2534 (Marilyn Brown) & 212-338-2524 (Norman Newman); FAX: 212-338-2520
E-mail: Marilyn.Brown@hubinternational.com; Norman.Newman@hubinternational.com
Web Site: http://northeast.hubinternational.com/industries/fine-arts/

Type of Business:
HUB International Northeast is a leading insurance brokerage that has a highly experienced team of seasoned professionals specializing in arranging coverage on all forms of valuable collectibles for museums and other non-profits, private collectors, exhibitions and galleries and dealers. With over 45 years of experience in serving this industry, HUB builds customized insurance and risk management needs to fit your unique requirements.

Personnel:
Marilyn Brown (Assistant Vice President, Fine Art & Special Risk Department)
Norman Newman (First Vice President, Fine Art & Special Risk Department)

HUNTINGTON T. BLOCK
INSURANCE AGENCY, INC.

HUNTINGTON T. BLOCK INSURANCE AGENCY, INC. AN AON GROUP COMPANY
1120 20th St., N.W., Suite 600
Washington, DC 20036-3406
Tel: 800-424-8830 & 202-223-9853; FAX: 202-331-8409
E-mail: Jeff.Minett@aon.com
Web Site: www.huntingtontblock.com

Type of Business:
HTB is a leading provider of insurance services for museums & exhibitions, universities, galleries and collectors. As a full service brokerage firm, we are the only agency with an in-house claims department, proprietary museum collection and temporary loans policy forms and have the largest underwriting authority available. Office locations include Washington DC, New York City, San Francisco, Houston and representation in London.

Personnel:
Jeff Minett (New York, Fine Art Insurance)
Lynn Marcin (Baltimore MD, Fine Art Insurance)
Adrienne Reid (Houston, Fine Art Insurance)
Sarah Barr (San Francisco, Fine Art Insurance)
Richard Mercado (Washington DC, Commercial Lines Insurance)

WILLIS OF MARYLAND INC.
Washington DC Office, 6700 Rockledge Dr., 5th Fl.
Bethesda, MD 20817-1824
Tel: 301-581-4247 & 800-456-3162; FAX: 301-897-7302
E-mail: Robert.Salmon@willis.com
Web Site: www.willis.com

Type of Business:
The Willis Fine Art, Jewelry & Specie Division (FAJS) is one of the leading specialist art and collections insurance brokers in the world. With a team of very experienced professionals in Chicago, New York, Washington DC, London, and Zurich, clients include many prominent museums, institutions, universities, exhibitions and private collections. Willis FAJS provides a flexible, broad insurance program tailored specifically to suit the special needs of your museum at a competitive cost. We are an acknowledged authority to the extent that many other brokers in the USA come to Willis to use our services in placing coverage for this specialized class of business.

Personnel:
Robert F. Salmon, ACII (Managing Director Willis Fine Art, Jewelry & Specie)

Management Consultants

SCHULTZ & WILLIAMS
325 Chestnut St., Ste. 700
Philadelphia, PA 19106
Tel: 215-625-9955; FAX: 215-625-2701
E-mail: mail@schultzwilliams.com
Web Site: www.schultzwilliams.com

Type of Business:
Schultz & Williams provides comprehensive consulting services to the arts, museums, cultural institutions and other nonprofits of all types and sizes across the nation. Since 1987, the success and integrity of our work has been driven by one simple philosophy: development, management and marketing strategies must be fully integrated to achieve financial stability, operational excellence and greater mission impact. Our senior professionals have extensive hands-on experience from both sides of the desk. The strength of our consulting services is based on the skills we have gained in managing, operating and supporting major nonprofit organizations, including: Development: Capital campaigns, major gift and planned giving strategies; Direct Response: Full-service digital and offline communications to raise funds through membership, annual giving, mid-level and monthly giving programs; Planning: Financial planning, facility and master plan management, leadership development; Marketing: Branding, planning, collateral; S&W StaffSolutions: Executive and management level interim, project, or start-up staffing.

Personnel:
L. Scott Schultz (President)
M. Jane Williams (Principal)
Jessica Harrington (Vice President, S & W Direct)
Dell Fascione (Vice President)

Museum Shop Merchandise

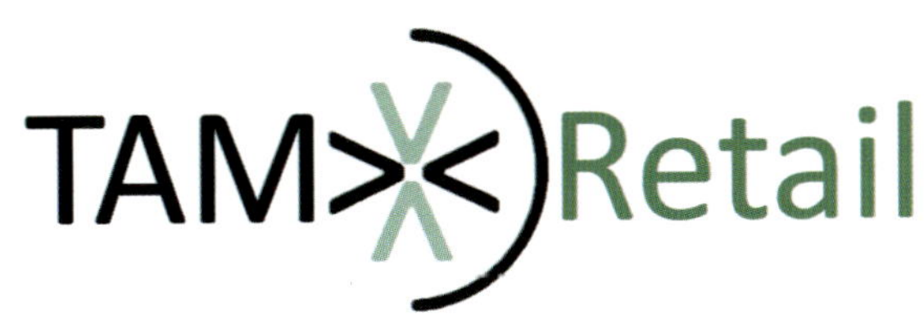

TAM RETAIL, DIV. OF LODE DATA SYSTEMS, INC.
10609 West 159th St.
Orland Park, IL 60467
Tel: 888-THE-14POS (888-843-1476) & 708-460-0999; FAX: 708-460-1253
E-mail: sales@tamretail.com
Web Site: www.nonprofitpos.com

(Continued on next page)

(Continued from previous page)

Type of Business:
TAM Retail is the sole provider of The Assistant Manager™ (TAM), a turn-key retail management solution that meets the growing needs of non-profits throughout the U.S. and internationally. TAM provides the knowledge of the things you cannot easily see for yourself. It provides you with a solution for knowing the most critical aspects of your organization's operations including exactly what products and services have sold, attendance levels, details on each transaction, associate and volunteer performance, perpetual value of your inventory and how it is performing, what is hot and what is not, helps you know your members better, what members like and don't like, and promotes increased donations. All the information needed to make the many important decisions that affect your organization is provided by TAM. TAM provides this at an affordable cost of ownership from a proven and trusted provider to help your organization get the most return on your investment.

Personnel:
Bruce H. Lode (Executive Vice President of Marketing and Sales)

Professional Services

CONSULTECON, INC.
545 Concord Ave., Ste. 210
Cambridge, MA 02138
Tel: 617-547-0100; FAX: 617-547-0102
E-mail: info@consultecon.com
Web Site: www.consultecon.com

Type of Business:
ConsultEcon, Inc. provides services to clients in the areas of project and plan concept development, market and financial evaluation, visitor surveys, economic impact and project implementation. We are dedicated to serving museums and visitor attractions of all sizes and types, and have worked over many years with clients responding to a broad spectrum of issues, ranging from the economics of operations to strategic planning.

Personnel:
Thomas J. Martin (President)
Robert E. Brais (Vice President)
Elena Kazlas (Principal)
Jason Drebitko (Senior Associate)
James Stevens (Senior Associate)

HUB INTERNATIONAL NORTHEAST
1065 Avenue of Americas
New York, NY 10018
Tel: 212-338-2534 (Marilyn Brown) & 212-338-2524 (Norman Newman); FAX: 212-338-2520
E-mail: Marilyn.Brown@hubinternational.com; Norman.Newman@hubinternational.com
Web Site: http://northeast.hubinternational.com/industries/fine-arts/

Type of Business:
HUB International Northeast is a leading insurance brokerage that has a highly experienced team of seasoned professionals specializing in arranging coverage on all forms of valuable collectibles for museums and other non-profits, private collectors, exhibitions and galleries and dealers. With over 45 years of experience in serving this industry, HUB builds customized insurance and risk management needs to fit your unique requirements.

Personnel:
Marilyn Brown (Assistant Vice President, Fine Art & Special Risk Department)
Norman Newman (First Vice President, Fine Art & Special Risk Department)

WHITE OAK ASSOCIATES, INC.
17 Essex St.
Marblehead, MA 01945
Tel: 781-639-0722; FAX: 781-639-2491
E-mail: info@whiteoakassoc.com
Web Site: www.whiteoakassoc.com

Type of Business:
White Oak Associates is dedicated to high quality analysis and strategic planning services for museums. Our mission is to collaborate with museums to create sustainable institutions providing essential community services. With over 40 years of very satisfied museum clients and hundreds of significant commissions, White Oak has earned respect for its collaborative, well-informed and thoughtful planning approach, including: community needs assessments, performance assessments, feasibility studies, concept development plans, strategic master plans, business plans, market studies, architectural program planning, and staff and operating plans. Using analysis of your budgets, attendance, facility size and other data, we integrate your conceptual vision with sustainable economic models. White Oak also offers implementation management from concept through post-opening for museum reinventions, expansions and new museums. We are flexible to fit your exact needs in team make-up, scope, schedule and budget. Visit our website and "Talk To Us" about your museum analysis and planning needs.

Personnel:
John W. Jacobsen (President)
Jeanie Stahl (Vice President)
Victor A. Becker (Director, Program Development)
Karen Hefler (Production & Logistics)
Rebecca Robison (Production & Logistics)

Ticketing Services

TAM RETAIL, DIV. OF LODE DATA SYSTEMS, INC.
10609 West 159th St.
Orland Park, IL 60467
Tel: 888-THE-14POS (888-843-1476) & 708-460-0999; FAX: 708-460-1253
E-mail: sales@tamretail.com
Web Site: www.nonprofitpos.com

Type of Business:
TAM Retail is the sole provider of The Assistant Manager™ (TAM), a turn-key retail management solution that meets the growing needs of non-profits throughout the U.S. and internationally. TAM provides the knowledge of the things you cannot easily see for yourself. It provides

(Continued on next page)

(Continued from previous page)

you with a solution for knowing the most critical aspects of your organization's operations including exactly what products and services have sold, attendance levels, details on each transaction, associate and volunteer performance, perpetual value of your inventory and how it is performing, what is hot and what is not, helps you know your members better, what members like and don't like, and promotes increased donations. All the information needed to make the many important decisions that affect your organization is provided by TAM. TAM provides this at an affordable cost of ownership from a proven and trusted provider to help your organization get the most return on your investment.

Personnel:
Bruce H. Lode (Executive Vice President of Marketing and Sales)

CONSERVATION/EDUCATION – GENERAL

Conservation

DORFMAN
MUSEUM FIGURES, INC.

DORFMAN MUSEUM FIGURES, INC.
6224 Holabird Ave.
Baltimore, MD 21224
Tel: 800-634-4873 & 410-284-3248; FAX: 410-284-3249
E-mail: info@museumfigures.com
Web Site: www.museumfigures.com

Type of Business:
We've been standing still for 50 years! Dorfman is the leader in creating lifelike, life-size human figures, and ETHAFOAM™ Conservation Forms. Choose non-ident, generic realistic figures with rigid or flexible foam bodies, or recognizable historic characters for your exhibit needs. Or choose from our line of Conservation Forms, Dress or Suit Forms, or our new Economy Ethafoam mannequin for your archival display and storage needs. Dorfman Museum Figures, Inc. has been in business since 1957. Our Museum Figures and Conservation Forms are used in museums internationally.

Personnel:
Robert Dorfman (President)
Penny Clifton (Project Manager)

EXHIBITS-GENERAL

MCCUNE DESIGN
6836 Valjean Ave.
Van Nuys, CA 91406
Tel: 818-779-1920; FAX: 818-781-9108
E-mail: Marketing@McCune-Design.com
Web Site: www.mccune-design.com

Type of Business:
McCune Design is an Academy Award winning design, fabrication and effects business serving the display, entertainment and advertising industries for over 30 years. Seen in museums, theme parks, hotels, and nature centers our exhibit work includes interactives, scenery and sets, specialty props, animatronics, robotics, working prototypes, practical effects, habitats, enclosures, scenics, dioramas, murals, display cases and precision replicas of all kinds. We create realistic, detailed miniatures and models, as well as full size/oversize props, set pieces and environments. Our eclectic work often includes complex electro-mechanical engineering, processor driven motor control systems, sculpture, complex castings, real and faux finishes. Our 30 years in motion picture effects include iconic historical films such as Star Wars, Star Trek, Caddyshack, and Spaceballs, and modern action blockbusters such as X-Men, Iron Man and Spiderman. We produce quality, custom objet d'art in unique, eclectic limited editions or exclusive works of art.

Personnel:
Katherine McCune (President)
Cole McCune (Vice President)
Debora Galloway (Producer/Marketing Director)
Monty Shook (Project Manager)

SPLIT ROCK STUDIOS
2071 Gateway Blvd.
St. Paul, MN 55112
Tel: 651-631-2211 & 800-433-9599; FAX: 651-631-0707
E-mail: info@splitrockstudios.com
Web Site: www.splitrockstudios.com

Type of Business:
Split Rock Studios is a full-service Exhibit firm specializing in designing and developing interpretive exhibits for museums and related institutions. We have the in-house capabilities to produce all facets of a major exhibition: Exhibit Design and Development; Custom Exhibit Casework and Furniture; Themed Environments and Dioramas; Hand-Painted Interpretive Murals; Realistic Animal and Human Sculpture; Graphic Design and Production.

Personnel:
Craig Sommerville
Anna Kling
Ann Pappas
John Gunning

TAYLOR STUDIOS INC.
1320 Harmon Dr.
Rantoul, IL 61866
Tel: 217-893-4874 & 800-707-2047; FAX: 217-893-1998
E-mail: sales@taylorstudios.com
Web Site: www.taylorstudios.com/omd

Type of Business:
Museums, nature centers, zoos, and similar institutions all have engaging stories to tell. They hire Taylor Studios to reveal these stories. Through detailed interpretive planning, design, and fabrication, these provocative stories become exhibits that capture and spark visitors' imaginations. For 23 years, our clients have elected to work with our team because we possess a rare balance: creativity and time-tested processes and procedures that maintain project schedules and budgets. This balance at first may sound over-rated, but contracting a firm with proven project management and financial processes is essential to making any exhibit project as stress-free as possible. Our exhibits are also backed by an unmatched-in-the-industry five-year warranty.

Personnel:
Betty Brennan (President)
Drew Levan (Account Executive)
Kara Vanskike (Marketing Manager)

Animation/Animatronics

MCCUNE DESIGN
6836 Valjean Ave.
Van Nuys, CA 91406
Tel: 818-779-1920; FAX: 818-781-9108
E-mail: Marketing@McCune-Design.com
Web Site: www.mccune-design.com

Type of Business:
McCune Design is an Academy Award winning design, fabrication and effects business serving the display, entertainment and advertising industries for over 30 years. Seen in museums, theme parks, hotels, and nature centers our exhibit work includes interactives, scenery and sets, specialty props, animatronics, robotics, working prototypes, practical effects, habitats, enclosures, scenics, dioramas, murals, display cases and precision replicas of all kinds. We create realistic, detailed miniatures and models, as well as full size/oversize props, set pieces and environments. Our eclectic work often includes complex electro-mechanical engineering, processor driven motor control systems, sculpture, complex castings, real and faux finishes. Our 30 years in motion picture effects include iconic historical films such as Star Wars, Star Trek, Caddyshack, and Spaceballs, and modern action blockbusters such as X-Men, Iron Man and Spiderman. We produce quality, custom objet d'art in unique, eclectic limited editions or exclusive works of art.

Personnel:
Katherine McCune (President)
Cole McCune (Vice President)
Debora Galloway (Producer/Marketing Director)
Monty Shook (Project Manager)

RBH MULTIMEDIA, INC.
12 Hatch Ter.
Dobbs Ferry, NY 10522
Tel: 914-693-8755; FAX: 914-693-3539
E-mail: rbh@rbhmedia.com
Web Site: www.rbhmedia.com

Type of Business:

RBH Multimedia, Inc. is a media production company dedicated to developing and producing audiovisual experiences and exhibits. We bring imagination, skill, flair and years of experience to designing and producing audio, video and multiple media exhibits for museums, historic sites, visitor and interpretive centers, and cultural institutions. We also design and program websites and computer interactive exhibits. RBH is particularly known for producing signature multi-sensory, multimedia "experience" theater shows. Other services include hardware system design, engineering and installation, with expertise in videoconference technology. For more information about RBH Multimedia, Inc., please visit us online at www.rbhmedia.com.

Personnel:

Steve Brosnahan (Partner)
Nancy Haffner (Partner)
Edgardo J. Resto (Associate)

Control Equip. for Temperature and Humidity

CASE[WERKS], LLC
1501 Saint Paul St., Ste. 116
Baltimore, MD 21202
Tel: 410-332-4160 & 800-810-2852 (toll free); FAX: 410-332-4106
E-mail: info@casewerks.com
Web Site: www.casewerks.com

Type of Business:

Case[werks] is a sales and service firm offering a range of products uniquely designed to meet exhibit requirements for original art, artifacts and special collections in all types of interior environments. Providing consultative sales support for institutions large & small, Case[werks] is THE museum professional's source for archival display cases & exhibit furnishings, library furnishings, gallery accessories, art hanging hardware, signage & graphic display products as well as conservation equipment including miniClima Humidity Control Systems. Many of our best-selling gallery accessories are in stock, ready to ship; major credit cards are accepted. An extensive catalog is available and custom inquiries are always welcome. Case[werks] is the North American representative for Vitrinen- und Glasbau REIER (Germany).

Personnel:

Matt Malaquias (Principal)
Bill Beitel (Principal)

SWIZZ-STYLE INC.
Professional humidification for your valuable assets
165 West Broadway
Dover, OH 44622
Tel: 1-877-663-7895
E-mail: info@swizz-style.com
Web Site: www.swizz-style.com/brune

Type of Business:

Provider of high output capacity mobile humidifier, made in Germany. Brune is the market-leader in Europe and purveyor of humidification for most European museums. The B 500 humidifies areas up to 4200 square feet is ready to use, and requires only normal tap water. It features a 5-step whisper quiet automatic fan, an environmentally friendly organic filter and UV sterilization built-in. It is the only mobile humidifier with direct water supply possibility and self-diagnosing system display for maintenance and carries an 11 gallon water reserve basin.

Dioramas

CHASE STUDIO, INC.
205 Wolf Creek Rd.
Cedar Creek, MO 65627
Tel: 417-794-3303; FAX: 417-794-3741
E-mail: chasestudio@chasestudio.com
Web Site: www.chasestudio.com

Type of Business:
Designers and builders of natural history and environmental science exhibits; paleontological reconstructions, zoological models, botanical reproductions, dioramas and habitat groups, illustration and mural painting, graphic design, photography, lighting design, fine cabinetwork, taxidermy, research, script writing, exhibit planning and consultation. Exhibits for over 200 museums worldwide since 1973.

Personnel:
Dr. Terry L. Chase (Director)
David F. Darby (Assistant Director Exhibits)
William L. Talbot (Assistant Director Business)

MCCUNE DESIGN
6836 Valjean Ave.
Van Nuys, CA 91406
Tel: 818-779-1920; FAX: 818-781-9108
E-mail: Marketing@McCune-Design.com
Web Site: www.mccune-design.com

Type of Business:
McCune Design is an Academy Award winning design, fabrication and effects business serving the display, entertainment and advertising industries for over 30 years. Seen in museums, theme parks, hotels, and nature centers our exhibit work includes interactives, scenery and sets, specialty props, animatronics, robotics, working prototypes, practical effects, habitats, enclosures, scenics, dioramas, murals, display cases and precision replicas of all kinds. We create realistic, detailed miniatures and models, as well as full size/oversize props, set pieces and environments. Our eclectic work often includes complex electro-mechanical engineering, processor driven motor control systems, sculpture, complex castings, real and faux finishes. Our 30 years in motion picture effects include iconic historical films such as Star Wars, Star Trek, Caddyshack, and Spaceballs, and modern action blockbusters such as X-Men, Iron Man and Spiderman. We produce quality, custom objet d'art in unique, eclectic limited editions or exclusive works of art.

Personnel:
Katherine McCune (President)
Cole McCune (Vice President)
Debora Galloway (Producer/Marketing Director)
Monty Shook (Project Manager)

SPLIT ROCK STUDIOS
2071 Gateway Blvd.
St. Paul, MN 55112
Tel: 651-631-2211 & 800-433-9599; FAX: 651-631-0707
E-mail: info@splitrockstudios.com
Web Site: www.splitrockstudios.com

Type of Business:
Split Rock Studios is a full-service Exhibit firm specializing in designing and developing interpretive exhibits for museums and related institutions. We have the in-house capabilities to produce all facets of a major exhibition: Exhibit Design and Development; Custom Exhibit Casework and Furniture; Themed Environments and Dioramas; Hand-Painted Interpretive Murals; Realistic Animal and Human Sculpture; Graphic Design and Production.

Personnel:
Craig Sommerville
Anna Kling
Ann Pappas
John Gunning

TAYLOR STUDIOS INC.
1320 Harmon Dr.
Rantoul, IL 61866
Tel: 217-893-4874 & 800-707-2047; FAX: 217-893-1998
E-mail: sales@taylorstudios.com
Web Site: www.taylorstudios.com/omd

Type of Business:
Museums, nature centers, zoos, and similar institutions all have engaging stories to tell. They hire Taylor Studios to reveal these stories. Through detailed interpretive planning, design, and fabrication, these provocative stories become exhibits that capture and spark visitors' imaginations. For 23 years, our clients have elected to work with our team because we possess a rare balance: creativity and time-tested processes and procedures that maintain project schedules and budgets. This balance at first may sound over-rated, but contracting a firm with proven project management and financial processes is essential to making any exhibit project as stress-free as possible. Our exhibits are also backed by an unmatched-in-the-industry five-year warranty.

Personnel:
Betty Brennan (President)
Drew Levan (Account Executive)
Kara Vanskike (Marketing Manager)

Display Fixtures

CASE[WERKS], LLC
  1501 Saint Paul St., Ste. 116
  Baltimore, MD 21202
  Tel: 410-332-4160 & 800-810-2852 (toll free); FAX: 410-332-4106
  E-mail: info@casewerks.com
  Web Site: www.casewerks.com

Type of Business:
  Case[werks] is a sales and service firm offering a range of products
  uniquely designed to meet exhibit requirements for original art,
  artifacts and special collections in all types of interior environments.
  Providing consultative sales support for institutions large & small,
  Case[werks] is THE museum professional's source for archival display
  cases & exhibit furnishings, library furnishings, gallery accessories, art
  hanging hardware, signage & graphic display products as well as
  conservation equipment including miniClima Humidity Control
  Systems. Many of our best-selling gallery accessories are in stock,
  ready to ship; major credit cards are accepted. An extensive catalog is
  available and custom inquiries are always welcome. Case[werks] is the
  North American representative for Vitrinen- und Glasbau REIER
  (Germany).

Personnel:
  Matt Malaquias (Principal)
  Bill Beitel (Principal)

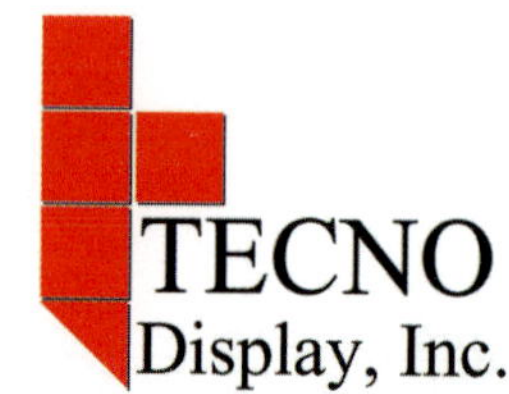

TECNO DISPLAY, INC.
  2277 National Ave.
  Hayward, CA 94545
  Tel: 800-255-3536; FAX: 510-782-5003
  E-mail: tecno@tecnodisplay.com
  Web Site: www.tecnodisplay.com

Type of Business:
  Manufacturers of pre-assembled tempered glass display showcases, wall
  units, free standing towers, exhibit display stands, pedestals, museum
  glass showcases, counter cases. Standard & custom showcases
  available. Small runs of custom showcases can be manufactured. We
  work with all types of laminates and solid wood veneers, stains and
  moldings. Halogen, fluorescent, LED lighting available. We have built
  our reputation on quality products, reliable delivery and excellent
  customer service.

Personnel:
  Patrick Lowe (Sales Manager)
  Cathie Harvey (Customer Service)
  John Wyatt (Customer Service)

Display Stands

PANELOCK SYSTEMS LIMITED
  P.O. Box 729
  Woodbury, CT 06798
  Tel: 877-315-1998; FAX: 203-643-2060
  E-mail: sales@panelock.com
  Web Site: www.panelock.com

Type of Business:
  Designers and manufacturers of truly movable, environmentally friendly
  exhibition display walls. Panelock's museum quality units offer
  designers the opportunity to re-create display footprints to complement
  exhibition themes. They transform a gallery to a multi-functional space
  which can house exhibitions and be transformed to use for special
  functions or events. The internationally patented retractable wheel
  system in our units is the secret to Panelock's unique relocatable
  functionality. See the dynamic flexibility of Panelock Systems 100,
  200, 600 and System 400 (preferred by architects worldwide) for
  yourself by viewing "The Panelock Movie" on our website. Panelock!
  Where sensitivity to the environment, functionality and beauty meet.

Personnel:
  Maureen Moreau

PANELOCK SYSTEMS LIMITED
  P.O. Box 729
  Woodbury, CT 06798
  Tel: 877-315-1998; FAX: 203-643-2060
  E-mail: sales@panelock.com
  Web Site: www.panelock.com

Type of Business:
  Designers and manufacturers of truly movable, environmentally friendly
  exhibition display walls. Panelock's museum quality units offer
  designers the opportunity to re-create display footprints to complement
  exhibition themes. They transform a gallery to a multi-functional space
  which can house exhibitions and be transformed to use for special
  functions or events. The internationally patented retractable wheel
  system in our units is the secret to Panelock's unique relocatable
  functionality. See the dynamic flexibility of Panelock Systems 100,
  200, 600 and System 400 (preferred by architects worldwide) for
  yourself by viewing "The Panelock Movie" on our website. Panelock!
  Where sensitivity to the environment, functionality and beauty meet.

Personnel:
  Maureen Moreau

Environmental Equipment

CASE[WERKS], LLC
 1501 Saint Paul St., Ste. 116
 Baltimore, MD 21202
 Tel: 410-332-4160 & 800-810-2852 (toll free); FAX: 410-332-4106
 E-mail: info@casewerks.com
 Web Site: www.casewerks.com

Type of Business:
 Case[werks] is a sales and service firm offering a range of products
 uniquely designed to meet exhibit requirements for original art,
 artifacts and special collections in all types of interior environments.
 Providing consultative sales support for institutions large & small,
 Case[werks] is THE museum professional's source for archival display
 cases & exhibit furnishings, library furnishings, gallery accessories, art
 hanging hardware, signage & graphic display products as well as
 conservation equipment including miniClima Humidity Control
 Systems. Many of our best-selling gallery accessories are in stock,
 ready to ship; major credit cards are accepted. An extensive catalog is
 available and custom inquiries are always welcome. Case[werks] is the
 North American representative for Vitrinen- und Glasbau REIER
 (Germany).

Personnel:
 Matt Malaquias (Principal)
 Bill Beitel (Principal)

LIGHTING SERVICES INC
 2 Holt Dr.
 Stony Point, NY 10980
 Tel: 800-999-9574 & 845-942-2800; FAX: 845-942-2177
 E-mail: Sales@mailLSI.com
 Web Site: www.LightingServicesInc.com

Type of Business:
 Lighting Services Inc (LSI) is the premier manufacturer of track, accent,
 display and LED lighting systems for museum environments. Since
 1958, LSI has been dedicated to designing, engineering and
 manufacturing lighting fixtures of the highest quality. Our reputation
 for creativity and innovative design, coupled with specification grade
 products and intelligent personalized service, has made us the
 manufacturer of choice amongst the most discriminating specifiers of
 lighting for museums and galleries.

Personnel:
 Daniel Gelman (President)

SWIZZ-STYLE INC.
 Professional humidification for your valuable assets
 165 West Broadway
 Dover, OH 44622
 Tel: 1-877-663-7895
 E-mail: info@swizz-style.com
 Web Site: www.swizz-style.com/brune

Type of Business:
 Provider of high output capacity mobile humidifier, made in Germany.
 Brune is the market-leader in Europe and purveyor of humidification
 for most European museums. The B 500 humidifies areas up to 4200
 square feet is ready to use, and requires only normal tap water. It
 features a 5-step whisper quiet automatic fan, an environmentally
 friendly organic filter and UV sterilization built-in. It is the only
 mobile humidifier with direct water supply possibility and
 self-diagnosing system display for maintenance and carries an 11
 gallon water reserve basin.

Exhibit Cases

CASE[WERKS], LLC
 1501 Saint Paul St., Ste. 116
 Baltimore, MD 21202
 Tel: 410-332-4160 & 800-810-2852 (toll free); FAX: 410-332-4106
 E-mail: info@casewerks.com
 Web Site: www.casewerks.com

Type of Business:
 Case[werks] is a sales and service firm offering a range of products
 uniquely designed to meet exhibit requirements for original art,
 artifacts and special collections in all types of interior environments.
 Providing consultative sales support for institutions large & small,
 Case[werks] is THE museum professional's source for archival display
 cases & exhibit furnishings, library furnishings, gallery accessories, art
 hanging hardware, signage & graphic display products as well as
 conservation equipment including miniClima Humidity Control
 Systems. Many of our best-selling gallery accessories are in stock,
 ready to ship; major credit cards are accepted. An extensive catalog is
 available and custom inquiries are always welcome. Case[werks] is the
 North American representative for Vitrinen- und Glasbau REIER
 (Germany).

Personnel:
 Matt Malaquias (Principal)
 Bill Beitel (Principal)

CHICAGO SCENIC STUDIOS, INC.
1315 N. North Branch St.
Chicago, IL 60642
Tel: 312-274-9900; FAX: 312-274-9901
E-mail: dLanghorst@chicagoscenic.com
Web Site: www.chicagoscenic.com

Type of Business:

Chicago Scenic Studios provides museums, science & technology centers, zoos, aquariums and visitor centers more than 35 years of multi-disciplinary experience in themed environment design, fabrication, engineering and installation. Our highly skilled staff of project managers, designers, and skilled craftspeople partner with your team to provide safe, educational, immersive environments and exciting experiences through custom exhibits, retail, restaurant, and admission spaces. Chicago Scenic can serve as your general contractor, fabricator or project manager in support of your design or ours. Our background in exhibits, retail, theater, special events and corporate spaces provides us with the ability to cater to differing needs in unique spaces. We have enjoyed partnering with world-class institutions such as The Museum of Science and Industry, Chicago; The Great Lakes Science Center, the Chicago Children's Museum, The John G. Shedd Aquarium, the Brookfield Zoo, the Field Museum of Natural History and many more to bring exciting, educational and successful stories to life. We'd like to help tell your story as well.

Personnel:
Robert Doepel (President)
Diane Langhorst (Business Development)

DACOBE ENTERPRISES, LLC
325 Lafayette St.
Utica, NY 13502
Tel: 315-368-0093; FAX: 315-368-0096
E-mail: gthorp@dacobe.com
Web Site: www.dacobe.com

Type of Business:

Manufacturers of pre-assembled acrylic and wood display and exhibit cases. Free standing, wall hanging, countertop, corner, and wall anchored units. Economically priced to fit most project budgets. We have been in the fabrication business since 1992 and our dedicated workforce includes individuals that have been in fabrication for over 20 years. Large and small runs can be made in our 20,000 square foot facility. Custom sizes made to your specifications. Quick delivery to meet even the most stringent time constraints. Our "No Headache" philosophy and our dependability keep our customers coming back year after year. Done right and done on time, that is the only way - The DACOBE way! And most importantly to us, all of our products are made 100% in the USA. For examples of our work and other capabilities, visit our website.

Personnel:
Dan Beal (President)
Geoff Thorp (Vice President of Sales)

GLASBAU HAHN AMERICA, LLC
15 Little Brook Lane
Newburgh, NY 12550
Tel: 845-566-3331 & 877-452-7228 (GLASBAU); FAX: 845-566-3176
E-mail: info@glasbau-hahn.com
Web Site: www.glasbau-hahn.com

Type of Business:

Renowned glass specialist and display case manufacturer GLASBAU HAHN was founded in Frankfurt/Main in 1829, and ranks among the global market leaders in museum exhibition and design construction. Having set standards for security (including seismic technology), micro-environmental control, innovative light systems, conservation, accessibility and design, HAHN quality stands for both tradition and innovation. With our certified product "HAHN PURE" one more step towards perfection has been achieved: HAHN PURE display cases are completely built from emission-tested materials (www.hahnpure.de/en). GLASBAU HAHN award winning cases have been installed in more than 1,600 museums, galleries, libraries, and cultural institutions around the world, including over 150 institutions in the U.S. Technical advances include award-winning innovations like the ALLGLASS Display Case, 3-Way Sliding Doors, Hinged Openings, the HAHN Swing-Door and the PROTECTOR Case (the ultimate safeguard for the display and transportation of hanging art). GLASBAU HAHN offers a full range of unique solutions to large and small museum projects, and is known for incorporating the vision of architects and designers with the expediency of functionality, low maintenance and ideal art conservation. A network of offices in Europe, Middle East, USA, Japan, China, and additional representatives worldwide, bring GLASBAU HAHN's standards and innovations to each exhibition. GLASBAU HAHN: Trusted with the World's Treasures for over 180 years! Home Office: GLASBAU HAHN GmbH, Hanauer Landstrasse 211, 60314 Frankfurt am Main, Germany. Tel: 011-49-69-944-1753, Fax: 011-49-69-944-1761.

Personnel:
Isabel Hahn (President GH GmbH)
Till Hahn (President GHA)
Jamie J. Ponton (Vice President GHA)
Norbert Leonhardt (Construction GH GmbH)
Catherine Lima (Office Manager GHA)

HELMUT GUENSCHEL, INC.
10 Emala Ave.
Baltimore, MD 21220
Tel: 410-686-5900 & 800-852-2525; FAX: 410-687-9342
E-mail: info@guenschel.com
Web Site: www.guenschel.com

Type of Business:

A purely American company for over 45 years and recognized internationally for creating display cases of the highest quality and for the innovative engineering behind the most advanced display case system available, HGI products are specified by architects, designers and used by many of the most celebrated museums. Our exhibit products address the full range of technical requirements and collections issues including all facets of active and passive conservation, the use of proven environmentally safe materials,

(Continued on next page)

(Continued from previous page)

state-of-the-art security and effective seismic anchoring. Engineered and manufactured to the highest standards, Helmut Guenschel cases are designed to provide decades of continuous use.

Personnel:
Helmut Guenschel (President)
Cynthia Shaffer (Vice President)

MALONE DESIGN/FABRICATION
5403 Dividend Dr.
Decatur, GA 30035
Tel: 770-987-2538; FAX: 770-987-0326
E-mail: twright@maloneinc.com
Web Site: www.maloneinc.com

Type of Business:
From inspiration to installation, Malone Design/Fabrication provides complete exhibition development and production services to the museum community. These services include interpretive planning, design, project management, fabrication and installation. Our fabrication capabilities include graphics, display cases, interactives, multimedia, scenic and more. Malone has nearly 50 years of experience designing and fabricating exhibits and store fixtures for all types of museums and visitor centers.

Personnel:
Tom Wright (CEO)
Brad Parker (Account Executive)

MCCUNE DESIGN
6836 Valjean Ave.
Van Nuys, CA 91406
Tel: 818-779-1920; FAX: 818-781-9108
E-mail: Marketing@McCune-Design.com
Web Site: www.mccune-design.com

Type of Business:
McCune Design is an Academy Award winning design, fabrication and effects business serving the display, entertainment and advertising industries for over 30 years. Seen in museums, theme parks, hotels, and nature centers our exhibit work includes interactives, scenery and sets, specialty props, animatronics, robotics, working prototypes, practical effects, habitats, enclosures, scenics, dioramas, murals, display cases and precision replicas of all kinds. We create realistic, detailed miniatures and models, as well as full size/oversize props, set pieces and environments. Our eclectic work often includes complex electro-mechanical engineering, processor driven motor control systems, sculpture, complex castings, real and faux finishes. Our 30 years in motion picture effects include iconic historical films such as Star Wars, Star Trek, Caddyshack, and Spaceballs, and modern action blockbusters such as X-Men, Iron Man and Spiderman. We produce quality, custom objet d'art in unique, eclectic limited editions or exclusive works of art.

Personnel:
Katherine McCune (President)
Cole McCune (Vice President)
Debora Galloway (Producer/Marketing Director)
Monty Shook (Project Manager)

NOUVIR LIGHTING
20915 Sussex Hwy.
Seaford, DE 19973
Tel: 302-628-9933; FAX: 302-628-9932
Web Site: www.nouvir.com

Type of Business:
Ten years of research gives you practical, effective, case RH and pollution control for under $500.00. The perfect compliment to NoUVIR's perfect zero UV, zero IR fiber optic museum lighting. NoUVIR invented museum preservation lighting. NoUVIR developed Reflected Energy Matching, extending exhibit life 70 to 100 times. Now NoUVIR gives you safe, affordable RH control plus protection from all chemical and particulate pollutants. Technical support, seminars, textbooks and research materials are available. Call for details.

Personnel:
Miss Ruth Ellen Miller (President)
Mr. Matthew S. Miller (Vice President Marketing)

STANDEX STRUCTURAL ALUMINUM DESIGNS
12935 Arroyo St.
Sylmar, CA 91342
Tel: 818-365-6464; FAX: 818-365-6465
E-mail: eclange@msn.com
Web Site: www.system-standex.dk; www.standexdesign.com

Type of Business:
Distribution and fabrication of museum exhibits and displays using System Standex, a square aluminum tubing system. Standex Designs has been meeting the exhibition needs of museums for over 20 years. Our lightweight, durable, and attractive material is perfect for showcases, exhibits, signage, and much more. Services provided include design, engineering, and fabrication of all projects. Tube sizes are 13/16 in., 1 in., & 1-3/16 in. square anodized aluminum with channels and flanges and are available in a variety of finishes. We offer three types of connectors: Permanent (glue-in or mechanical assembly), Semi-permanent (rap-in assembly), and Flexible (twist-lock for quick assembly/disassembly). Call or write for free catalog. eclange@msn.com www.system-standex.dk

Personnel:
Eric C. Lange (President)
Thelma Lyden (Office Manager)

TECNO DISPLAY, INC.
2277 National Ave.
Hayward, CA 94545
Tel: 800-255-3536; FAX: 510-782-5003
E-mail: tecno@tecnodisplay.com
Web Site: www.tecnodisplay.com

Type of Business:
Manufacturers of pre-assembled tempered glass display showcases, wall units, free standing towers, exhibit display stands, pedestals, museum glass showcases, counter cases. Standard & custom showcases available. Small runs of custom showcases can be manufactured. We work with all types of laminates and solid wood veneers, stains and

(Continued on next page)

(Continued from previous page)

moldings. Halogen, fluorescent, LED lighting available. We have built our reputation on quality products, reliable delivery and excellent customer service.

Personnel:
Patrick Lowe (Sales Manager)
Cathie Harvey (Customer Service)
John Wyatt (Customer Service)

Exhibit Design Firms

ANDREW MERRIELL & ASSOCIATES, LLC
7198 Old Santa Fe Trail
Santa Fe, NM 87505
Tel: 505-982-3950; FAX: 505-820-6674
E-mail: andy@merriell.com
Web Site: www.merriell.com

Type of Business:
Andrew Merriell & Associates plans and designs experiences for visitors to museums, visitor centers, parks, zoos, and public gardens. We identify opportunities to advance beyond the usual and predictable interpretive offerings, finding innovative ways to immerse visitors in exhibit stories. Each project provides visitors with memorable, meaningful experiences they can find nowhere else. Services include feasibility studies, interpretive master plans, concept studies, exhibit content development, and exhibit design.

Personnel:
Andrew Merriell (Principal)
Rebecca Shreckengast (Senior Designer)
Hella Rader (Project Manager)

RALPH APPELBAUM ASSOCIATES INCORPORATED
88 Pine St., 29th Fl.
New York, NY 10005
Tel: 212-334-8200; FAX: 212-334-6214
E-mail: frontdesk@raany.com
Web Site: www.raany.com

Type of Business:
Exhibition planning and design; masterplanning; media, interactive, website and publication design and production; graphic design and scriptwriting; photo research and acquisition; architectural integration. Major projects include United States Holocaust Memorial Museum, Fossil Halls and Rose Center for Earth and Space at the American Museum of Natural History, Newseum, Indiana State Museum, Country Music Hall of Fame, National Constitution Center, Clinton Presidential Library, National World War I Museum, US Capitol Visitor Center, Chemical Heritage Foundation, Singapore Discovery Centre, Richmond Canal Walk, Museum of the Portuguese Language, Craig Thomas Discovery Center at Grand Tetons National Park, Thomas Edison National Historic Site, London Transport Museum, Rutgers University Visitor Center, Culloden Battlefield Memorial, NASCAR Hall of Fame, China Shipbuilding and United Arab Emirates Pavilions at the Shanghai Expo, Smithsonian Arctic Studies Center, Bishop Museum Halls, National Museum of Scotland, Royal Albert Memorial Museum, Maritime Experiential Museum & Aquarium, Nationwide Children's Hospital, IBM THINK Centennial Exhibition, Jewish Museum and Tolerance Center, Canadian Museum for Human Rights, and National Museum of African American History and Culture.

Personnel:
Ralph Appelbaum (President)
Deborah Wolff (Chief of Staff)
Casey Lynn (Marketing Coordinator)

CAMBRIDGE SEVEN ASSOCIATES, INC.
1050 Massachusetts Ave.
Cambridge, MA 02138
Tel: 617-492-7000; FAX: 617-492-7007
E-mail: joltman@c7a.com
Web Site: www.c7a.com

Type of Business:
At Cambridge Seven Associates, Inc. we design distinctive museums and experiential exhibits that make learning stimulating, accessible, and engaging. We seek to create immersive environments which spark informal and multimodal learning in a captivating way. Cambridge Seven endows museums, aquariums, zoos, and visitor centers with a sense of place, wonder, and discovery, and these characteristics have made our projects internationally- and nationally-renowned destinations. Our services include architectural design, exhibit and habitat design, feasibility assessments, site analysis, master planning, space programming, institutional and operational planning, exhibit installation coordination, graphic design and signage programs, production of fund raising materials, and construction administration.

Personnel:
Peter Kuttner, FAIA (President)
Steven Imrich, AIA, LEED AP (Principal)
Patricia E. Intrieri, AIA (Principal)
Timothy D. Mansfield, AIA (Principal)
Peter Sollogub (Associate Principal)
Penny Sander (Associate Principal/Exhibit Planner)
Doug Simpson, LEED AP (Associate/Exhibit Designer)
Douglas Flandro, LEED AP (Exhibit Designer)
Jo Oltman (Marketing Manager)

CHASE STUDIO, INC.
205 Wolf Creek Rd.
Cedar Creek, MO 65627
Tel: 417-794-3303; FAX: 417-794-3741
E-mail: chasestudio@chasestudio.com
Web Site: www.chasestudio.com

Type of Business:
Designers and builders of natural history and environmental science exhibits; paleontological reconstructions, zoological models, botanical reproductions, dioramas and habitat groups, illustration and mural painting, graphic design, photography, lighting design, fine cabinetwork, taxidermy, research, script writing, exhibit planning and consultation. Exhibits for over 200 museums worldwide since 1973.

Personnel:
Dr. Terry L. Chase (Director)
David F. Darby (Assistant Director Exhibits)
William L. Talbot (Assistant Director Business)

CHICAGO SCENIC STUDIOS, INC.
1315 N. North Branch St.
Chicago, IL 60642
Tel: 312-274-9900; FAX: 312-274-9901
E-mail: dLanghorst@chicagoscenic.com
Web Site: www.chicagoscenic.com

Type of Business:

Chicago Scenic Studios provides museums, science & technology centers, zoos, aquariums and visitor centers more than 35 years of multi-disciplinary experience in themed environment design, fabrication, engineering and installation. Our highly skilled staff of project managers, designers, and skilled craftspeople partner with your team to provide safe, educational, immersive environments and exciting experiences through custom exhibits, retail, restaurant, and admission spaces. Chicago Scenic can serve as your general contractor, fabricator or project manager in support of your design or ours. Our background in exhibits, retail, theater, special events and corporate spaces provides us with the ability to cater to differing needs in unique spaces. We have enjoyed partnering with world-class institutions such as The Museum of Science and Industry, Chicago; The Great Lakes Science Center, the Chicago Children's Museum, The John G. Shedd Aquarium, the Brookfield Zoo, the Field Museum of Natural History and many more to bring exciting, educational and successful stories to life. We'd like to help tell your story as well.

Personnel:
Robert Doepel (President)
Diane Langhorst (Business Development)

EISTERHOLD ASSOCIATES, INC.
19310 N.W. Farley Hampton Rd.
Kansas City, MO 64153
Tel: 816-330-3276; FAX: 816-330-3278
E-mail: eai@eisterhold.com
Web Site: www.eisterhold.com

Type of Business:

Designers of philosophy-driven interpretive experiences that combine narrative, evidence, media, and space. We create memorable experiences, unique to their purpose and place. Our approach embodies a fundamental, protean examination of the content, allowing exploration to go beyond founding notions. We embrace and portray difficult, relevant, and current issues by creating interdisciplinary experiences that employ various media. Consequently, we connect with audiences of all kinds. Our in-depth exploration and creativity ensures that visitors are intrigued, excited, entertained, and encouraged to discover and learn. Projects include: Rosa Parks Museum, Montgomery; National Civil Rights Museum, Memphis; National Hurricane Museum and Science Center, Lake Charles; Living Tributes to America's Warriors at Walter Reed; North Carolina Civil War Museum; President's House, Independence Mall; African American Museum in Philadelphia; The Sixth Floor Museum, Dallas; International Civil Rights Center & Museum, Greensboro; currently completing National Museum of the United States Army, Fort Belvoir; NAACP, Baltimore; the National Museum of the Marine Corps, Quantico.

Gallagher & Associates

GALLAGHER & ASSOCIATES
8665 Georgia Ave.
Silver Spring, MD 20910
Tel: 301-656-7575; FAX: 301-656-5455
E-mail: gcoss@gallagherdesign.com
Web Site: www.gallagherdesign.com

Type of Business:

Gallagher & Associates is an internationally recognized museum planning and design firm with offices in Washington, D.C., San Francisco, and Asia. The firm specializes in museum master planning and exhibition design, media design and programming, environmental graphics, and brand development. Expertise includes a wide spectrum of visitor experiences such as public- and private-sector museums, visitor centers, hall of fame exhibits, science and learning centers, traveling exhibitions and corporate experiences.

Personnel:
Patrick Gallagher (President)
Cybelle Jones (Principal)
Gretchen Coss (Director of Business Development)

HADLEY EXHIBITS, INC.
1700 Elmwood Ave.
Buffalo, NY 14207
Tel: 716-874-3666, ext. 3018; FAX: 716-874-9994
E-mail: pwarner@hadleyexhibits.com
Web Site: www.hadleyexhibits.com

Type of Business:

Award-winning exhibit design, fabrication, and installation services for museums, visitor centers, zoos, historic sites, special events, and expositions. Additional skills and expertise in graphic design, graphics production, custom display cases, model making, and artifact mounting, to name a few. Significant experience with interactive environments and ADA compliance. In business over 65 years, Hadley Exhibits is housed in a well-equipped 180,000 square foot facility.

Personnel:
Ted Johnson (President)
Paul Warner (National Account Executive)

HILFERTY
14240 State Rte. 550
Athens, OH 45701
Tel: 740-448-3821; FAX: 740-448-2331
E-mail: gha@hilferty.com
Web Site: www.hilferty.com

Type of Business:

For more than thirty years, Hilferty has provided a full range of interpretive services to museums, visitor centers, nature centers, botanical gardens, historic sites, zoos/aquariums, and halls of fame. We excel at interpretive and facility master planning; concept development; content research; label writing; exhibit and graphic design; and providing donor recognition and funding support materials. From social histories to natural sciences, football to physics, presidents to children's gardens, every Hilferty exhibit vividly captures the stories that engage visitors' minds and hearts. Each project is approached with a unique mix of techniques chosen to best bring the subject to life. Our innovative portfolio spans the spectrum of intimate displays of precious objects, captivating immersive environments, enthralling interactive components, and breathtaking theatrical presentations. In short, Hilferty creates memorable museum experiences.

Personnel:
Gerard Hilferty (President & Creative Director)
Dean Clouse (CEO)

JRA (JACK ROUSE ASSOCIATES)
600 Vine St., Ste. 1700
Cincinnati, OH 45202-1100
Tel: 513-381-0055; FAX: 513-381-2691
E-mail: smccoy@jackrouse.com
Web Site: www.jackrouse.com/museum/index.cfm

Type of Business:

Named by The Wall Street Journal as "one of the world's more prominent design firms," JRA (Jack Rouse Associates) is a multi-disciplinary firm that has been structured to conceive, visualize and realize unique cultural experiences around the globe. Specific services include master planning, exhibit design, graphic design, media production, art direction and project management. From dramatic history museums to cutting-edge science centers, playful children's museums to innovative zoo exhibits, JRA is known for creating interpretive exhibits that are engaging as they are educational. Whether it's master planning a new museum or designing a breakthrough exhibit for an existing facility, JRA has a proven record of creating environments and experiences that truly connect with people.

Personnel:
Shawn McCoy (Vice President Marketing & Business Development)

MALONE DESIGN/FABRICATION
5403 Dividend Dr.
Decatur, GA 30035
Tel: 770-987-2538; FAX: 770-987-0326
E-mail: twright@maloneinc.com
Web Site: www.maloneinc.com

Type of Business:

From inspiration to installation, Malone Design/Fabrication provides complete exhibition development and production services to the museum community. These services include interpretive planning, design, project management, fabrication and installation. Our fabrication capabilities include graphics, display cases, interactives, multimedia, scenic and more. Malone has nearly 50 years of experience designing and fabricating exhibits and store fixtures for all types of museums and visitor centers.

Personnel:
Tom Wright (CEO)
Brad Parker (Account Executive)

MCCUNE DESIGN

6836 Valjean Ave.
Van Nuys, CA 91406
Tel: 818-779-1920; FAX: 818-781-9108
E-mail: Marketing@McCune-Design.com
Web Site: www.mccune-design.com

Type of Business:

McCune Design is an Academy Award winning design, fabrication and effects business serving the display, entertainment and advertising industries for over 30 years. Seen in museums, theme parks, hotels, and nature centers our exhibit work includes interactives, scenery and sets, specialty props, animatronics, robotics, working prototypes, practical effects, habitats, enclosures, scenics, dioramas, murals, display cases and precision replicas of all kinds. We create realistic, detailed miniatures and models, as well as full size/oversize props, set pieces and environments. Our eclectic work often includes complex electro-mechanical engineering, processor driven motor control systems, sculpture, complex castings, real and faux finishes. Our 30 years in motion picture effects include iconic historical films such as Star Wars, Star Trek, Caddyshack, and Spaceballs, and modern action blockbusters such as X-Men, Iron Man and Spiderman. We produce quality, custom objet d'art in unique, eclectic limited editions or exclusive works of art.

Personnel:

Katherine McCune (President)
Cole McCune (Vice President)
Debora Galloway (Producer/Marketing Director)
Monty Shook (Project Manager)

THE PORTICO GROUP

THE PORTICO GROUP

1500 4th Ave., 3rd Fl.
Seattle, WA 98101-1670
Tel: 206-621-2196; FAX: 206-621-2199
E-mail: portico@porticogroup.com
Web Site: www.porticogroup.com

Type of Business:

Celebrating 30 years of design, The Portico Group creates informal learning opportunities in memorable settings-including natural and cultural museums, zoos and aquaria, and public gardens. We are a talented team of interpretive exhibit designers, architects, and landscape architects capable of guiding a project from concept to conclusion. We seek opportunities to touch lives in meaningful ways - to connect people with nature and culture. Recent projects include: Hands On Children's Museum (Olympia, WA), Assiniboine Park Zoo International Polar Bear Conservation Centre (Winnipeg, Manitoba), Los Angeles Zoo Living Amphibians, Invertebrates, and Reptiles (Los Angeles, CA), Pearl Harbor Visitor Center (Honolulu, HI), Smithsonian National Zoological Park American Trail (Washington, DC).

Personnel:

Michael Hamm, FASLA (President & CEO)
Charles G. Mayes, AIA (Principal)
Keith McClintock, ASLA (Principal)
Dennis Meyer, ASLA, LEED AP (Principal)
Alissa Rupp, AIA, LEED AP BD+C (Principal)
Richard Larson (Associate Principal)
Allison CraigSundine (Senior Associate)
Justin Lyon (Senior Associate)

STEPHEN SAITAS DESIGNS

123 Fourth Ave., 3rd Fl.
New York, NY 10003
Tel: 212-388-0997; FAX: 212-388-0816
E-mail: ssaitas@aol.com

Type of Business:

Specializing in museum exhibition design, Stephen Saitas Designs provides services from schematic design through installation supervision for projects ranging from the design of a single pedestal through complete reinstallations of permanent collections. Clients have included museums, galleries, libraries, and historic houses. Since being established in 1982, SSD has planned, designed, and supervised over 200 permanent installations and temporary exhibitions, and provided related graphic design services to over forty institutions across the country.

Personnel:

Stephen Saitas (Principal)

LEE H. SKOLNICK ARCHITECTURE + DESIGN PARTNERSHIP

75 Broad St., Ste. 2700
New York, NY 10004
Tel: 212-989-2624; FAX: 212-727-1702
E-mail: mail@skolnick.com
Web Site: www.skolnick.com

Type of Business:

Established in 1980, Lee H. Skolnick Architecture + Design Partnership is an award-winning integrated architectural, museum exhibit and communication design firm. Museum facilities design services include: interpretive master planning, program development, conceptual and complete architectural services for new institutions, site planning, adaptive re-use of existing structures, historic preservation, adaptive reuse, feasibility studies. Exhibition programming and museum education services include:conceptual development, script and copy development, interpretive exhibition design, exhibition project management, evaluation of exhibits, educational programming. Communication design services include: exhibition and print graphics, identity and logo design, architectural signage and wayfinding and web design. LHSA+DP is also often called upon to create visionary renderings and fund-raising materials for prospective projects in the early concept phase.

Personnel:

Jo Ann Secor (Principal & Director Museum Services)
Scott Briggs (Senior Associate Museum Services)

SPLIT ROCK STUDIOS

2071 Gateway Blvd.
St. Paul, MN 55112
Tel: 651-631-2211 & 800-433-9599; FAX: 651-631-0707
E-mail: info@splitrockstudios.com
Web Site: www.splitrockstudios.com

Type of Business:

Split Rock Studios is a full-service Exhibit firm specializing in designing and developing interpretive exhibits for museums and related institutions. We have the in-house capabilities to produce all facets of a major exhibition: Exhibit Design and Development; Custom Exhibit Casework and Furniture; Themed Environments and Dioramas; Hand-Painted Interpretive Murals; Realistic Animal and Human Sculpture; Graphic Design and Production.

(Continued on next page)

(Continued from previous page)

Personnel:
  Craig Sommerville
  Anna Kling
  Ann Pappas
  John Gunning

studio**MUS**arx LLC

museum planning + **arch**itecture + e**xh**ibit design

STUDIOMUSARX LLC
  2601 Pennsylvania Avenue, Ste. 8
  Philadelphia, PA 19130
  Tel: 215-232-3489
  E-mail: info@studiomusarx.com
  Web Site: www.studiomusarx.com

Type of Business:
  studioMUSarx is a forward-thinking studio with special expertise in planning, architecture and exhibit design for new and existing museums and visitor centers. We are dedicated to a process of discovery acknowledging best practices while lending a friendly ear to client preferences in keeping with the shared values and aspirations of the studio. Each project undertaken is directed by the studio's principal, who collaborates closely with design team members to create strong working partnerships with clients. For over 40, years Joe Nicholson has assisted museums and interpretive centers with feasibility studies, master plans, space programming, signage programs, architectural and exhibit development, and design for sites, buildings, and interpretive exhibits providing full services through construction administration. We are open to the ongoing challenge of touching upon the extraordinary, embracing the artful, and going beyond constraints to deliver meaningful outcomes free from preconceptions of what the final design product should be.

Personnel:
  Joseph A. Nicholson, AIA, NCARB (Principal)
  Susan Hutton DeAngelus (Exhibit and Graphic Designer)
  Juliet Geldi, RA (Planner, Architect and Exhibit Designer)
  Sherman Lai, Assoc. AIA (Architectural and Exhibit Designer)

TAYLOR STUDIOS INC.
  1320 Harmon Dr.
  Rantoul, IL 61866
  Tel: 217-893-4874 & 800-707-2047; FAX: 217-893-1998
  E-mail: sales@taylorstudios.com
  Web Site: www.taylorstudios.com/omd

Type of Business:
  Museums, nature centers, zoos, and similar institutions all have engaging stories to tell. They hire Taylor Studios to reveal these stories. Through detailed interpretive planning, design, and fabrication, these provocative stories become exhibits that capture and spark visitors' imaginations. For 23 years, our clients have elected to work with our team because we possess a rare balance: creativity and time-tested processes and procedures that maintain project schedules and budgets. This balance at first may sound over-rated, but contracting a firm with proven project management and financial processes is essential to making any exhibit project as stress-free as possible. Our exhibits are also backed by an unmatched-in-the-industry five-year warranty.

Personnel:
  Betty Brennan (President)
  Drew Levan (Account Executive)
  Kara Vanskike (Marketing Manager)

Thinc™

THINC DESIGN
  435 Hudson St., 8th Fl.
  New York, NY 10014
  Tel: 212-741-3844; FAX: 212-741-9413
  E-mail: mail@thincdesign.com
  Web Site: www.thincdesign.com

Type of Business:
  Thinc is a leading exhibition design firm serving clients nationally and around the globe, including China, South Africa, and Europe. Inspired by each project's unique historical, social, cultural, and physical contexts, we craft designs that enable people to construct highly personalized meanings for themselves. We are particularly interested in creating transformative environments, installations, and exhibits—ones that fundamentally alter the way we look at and think about the world, which also build and sustain active communities. Founded by Tom Hennes in 1995, Thinc has successfully completed projects for aquaria, zoos, science centers, natural history and other types of museums, botanical gardens, heritage sites and museums, and theme parks. Our services include strategic and conceptual planning, exhibition and graphic design, media development, and project implementation. Located in New York City, Thinc has a multidisciplinary staff that includes 3-D designers, graphic designers, writers, content developers, researchers, project managers, administrative personnel, and a LEED-certified licensed architect.

Personnel:
  Tom Hennes (Principal)

xibitz

Experiential Spaces for Work and Life

XIBITZ
  7604 Harwood Ave., Ste. 202
  Milwaukee, WI 53213
  Tel: 414-727-4699 & 616-247-3500 (Home Office: Grand Rapids, MI);
  FAX: 414-727-4883
  E-mail: ezuern@xibitz.com
  Web Site: www.xibitz.com

Type of Business:
  Xibitz is dedicated to producing exhibits and environments that inform, educate and inspire for Museums, Visitor Center and other Cultural Institutions. Xibitz is a proven partner, delivering innovative and creative solutions; providing value in Planning, Design, Preproduction, Production and Installation. We offer a highly integrated turnkey approach, leveraging more than 20 years of experience. Friendly, Flexible and Committed, our passion is working hand-in-hand with our clients to discover, innovate and implement creative solutions.

Personnel:
  Erich Zuern (Producer - EZuern@Xibitz.com)
  Sarah Doty (Assistant Producer - SDoty@Xibitz.com)
  James Hungerford (CEO - JHungerford@Xibitz.com)

Exhibit Fabricators

ART GUILD, INC.
300 Wolf Dr.
West Deptford, NJ 08086
Tel: 856-853-7500, ext. 123; FAX: 856-853-0916
E-mail: kpaonessa@artguildinc.com
Web Site: www.artguildinc.com

Type of Business:
Art Guild's team of museum professionals works with design firms and museums nationwide building exceptional exhibit environments, which include casework, graphics, interactives, AV, and specialty fabrications. Working in two state of the art facilities, totaling over 400,000 square feet, we employ an expert project management team, experienced craftsmen, and utilize the latest technologies to transform your exhibit concepts into built reality. Art Guild, because the experience always matters. Contact Kim Paonessa to speak with one of our museum professionals.

Personnel:
David M. Egner (Director of Museum Services, Museums and Environments)
Kim M. Paonessa (Division Manager, Museums and Environments)

CHASE STUDIO, INC.

CHASE STUDIO, INC.
205 Wolf Creek Rd.
Cedar Creek, MO 65627
Tel: 417-794-3303; FAX: 417-794-3741
E-mail: chasestudio@chasestudio.com
Web Site: www.chasestudio.com

Type of Business:
Designers and builders of natural history and environmental science exhibits; paleontological reconstructions, zoological models, botanical reproductions, dioramas and habitat groups, illustration and mural painting, graphic design, photography, lighting design, fine cabinetwork, taxidermy, research, script writing, exhibit planning and consultation. Exhibits for over 200 museums worldwide since 1973.

Personnel:
Dr. Terry L. Chase (Director)
David F. Darby (Assistant Director Exhibits)
William L. Talbot (Assistant Director Business)

CHICAGO SCENIC STUDIOS, INC.
1315 N. North Branch St.
Chicago, IL 60642
Tel: 312-274-9900; FAX: 312-274-9901
E-mail: dLanghorst@chicagoscenic.com
Web Site: www.chicagoscenic.com

Type of Business:
Chicago Scenic Studios provides museums, science & technology centers, zoos, aquariums and visitor centers more than 35 years of multi-disciplinary experience in themed environment design, fabrication, engineering and installation. Our highly skilled staff of project managers, designers, and skilled craftspeople partner with your team to provide safe, educational, immersive environments and exciting experiences through custom exhibits, retail, restaurant, and admission spaces. Chicago Scenic can serve as your general contractor, fabricator or project manager in support of your design or ours. Our background in exhibits, retail, theater, special events and corporate spaces provides us with the ability to cater to differing needs in unique spaces. We have enjoyed partnering with world-class institutions such as The Museum of Science and Industry, Chicago; The Great Lakes Science Center, the Chicago Children's Museum, The John G. Shedd Aquarium, the Brookfield Zoo, the Field Museum of Natural History and many more to bring exciting, educational and successful stories to life. We'd like to help tell your story as well.

Personnel:
Robert Doepel (President)
Diane Langhorst (Business Development)

DIMENSIONAL COMMUNICATIONS INC.
1595 MacArthur Blvd.
Mahwah, NJ 06430-3601
Tel: 201-767-1500; FAX: 201-767-9696
E-mail: info@dimcom.com
Web Site: www.dimcom.com

Type of Business:
Dimensional Communications Inc. creates visitor experiences-trade shows, events, retail displays, museum exhibits and corporate interiors-that help our clients connect more productively with their customers. These are high-impact moments that can only happen in immersive environments. Increasingly, the impact is powered by Multimedia Technologies. Multimedia Systems design and implementation has been one of our foundation stones since our first project in 1964. Today it means custom kiosk design and production, indoor and outdoor LED displays, Augmented Reality, Multitouch, 3-D Technologies and beyond. With our full-service process, we make memorable experiences while making the client experience unexpectedly satisfying and cost-effective.

Personnel:
Douglas Fixell (President)
Robert Sneed (Vice President Multimedia Technologies)

HADLEY EXHIBITS, INC.
1700 Elmwood Ave.
Buffalo, NY 14207
Tel: 716-874-3666, ext. 3018; FAX: 716-874-9994
E-mail: pwarner@hadleyexhibits.com
Web Site: www.hadleyexhibits.com

Type of Business:
Award-winning exhibit design, fabrication, and installation services for museums, visitor centers, zoos, historic sites, special events, and expositions. Additional skills and expertise in graphic design, graphics production, custom display cases, model making, and artifact

(Continued on next page)

(Continued from previous page)

mounting, to name a few. Significant experience with interactive environments and ADA compliance. In business over 65 years, Hadley Exhibits is housed in a well-equipped 180,000 square foot facility.

Personnel:
Ted Johnson (President)
Paul Warner (National Account Executive)

MALONE DESIGN/FABRICATION
5403 Dividend Dr.
Decatur, GA 30035
Tel: 770-987-2538; FAX: 770-987-0326
E-mail: twright@maloneinc.com
Web Site: www.maloneinc.com

Type of Business:
From inspiration to installation, Malone Design/Fabrication provides complete exhibition development and production services to the museum community. These services include interpretive planning, design, project management, fabrication and installation. Our fabrication capabilities include graphics, display cases, interactives, multimedia, scenic and more. Malone has nearly 50 years of experience designing and fabricating exhibits and store fixtures for all types of museums and visitor centers.

Personnel:
Tom Wright (CEO)
Brad Parker (Account Executive)

MCCUNE DESIGN
6836 Valjean Ave.
Van Nuys, CA 91406
Tel: 818-779-1920; FAX: 818-781-9108
E-mail: Marketing@McCune-Design.com
Web Site: www.mccune-design.com

Type of Business:
McCune Design is an Academy Award winning design, fabrication and effects business serving the display, entertainment and advertising industries for over 30 years. Seen in museums, theme parks, hotels, and nature centers our exhibit work includes interactives, scenery and sets, specialty props, animatronics, robotics, working prototypes, practical effects, habitats, enclosures, scenics, dioramas, murals, display cases and precision replicas of all kinds. We create realistic, detailed miniatures and models, as well as full size/oversize props, set pieces and environments. Our eclectic work often includes complex electro-mechanical engineering, processor driven motor control systems, sculpture, complex castings, real and faux finishes. Our 30 years in motion picture effects include iconic historical films such as Star Wars, Star Trek, Caddyshack, and Spaceballs, and modern action blockbusters such as X-Men, Iron Man and Spiderman. We produce quality, custom objet d'art in unique, eclectic limited editions or exclusive works of art.

Personnel:
Katherine McCune (President)
Cole McCune (Vice President)
Debora Galloway (Producer/Marketing Director)
Monty Shook (Project Manager)

PACIFIC STUDIO
5311 Shilshole Ave., N.W.
Seattle, WA 98107
Tel: 206-783-5226; FAX: 206-783-5409
E-mail: mburns@pacific-studio.com
Web Site: www.pacific-studio.com

Type of Business:
Pacific Studio creates highly engaging and interactive experiences for museums, visitor centers and public spaces throughout the United States. Our artisans and craftspeople specialize in exhibit-grade cabinetry, custom metal fabrication, prototyping, interactive displays, hand-painted murals, and sculpting and casting.

Personnel:
Al Salm (General Manager)
Marc Burns (Business Development Director)

SPLIT ROCK STUDIOS
2071 Gateway Blvd.
St. Paul, MN 55112
Tel: 651-631-2211 & 800-433-9599; FAX: 651-631-0707
E-mail: info@splitrockstudios.com
Web Site: www.splitrockstudios.com

Type of Business:
Split Rock Studios is a full-service Exhibit firm specializing in designing and developing interpretive exhibits for museums and related institutions. We have the in-house capabilities to produce all facets of a major exhibition: Exhibit Design and Development; Custom Exhibit Casework and Furniture; Themed Environments and Dioramas; Hand-Painted Interpretive Murals; Realistic Animal and Human Sculpture; Graphic Design and Production.

Personnel:
Craig Sommerville
Anna Kling
Ann Pappas
John Gunning

STUDIO DISPLAYS, INC.
10600 Southern Loop Blvd.
Pineville, NC 28134
Tel: 704-588-6590; FAX: 704-588-6391
E-mail: mmatthews@studiodisplays.com
Web Site: www.studiodisplays.com

Type of Business:
Studio Displays is a full-service museum exhibit production company located in Charlotte, NC with over 33 years of experience. Studio Displays has worked with a diverse range of museums, visitor centers

(Continued on next page)

Exhibition Equipment & Services

(Continued from previous page)

and cultural attractions including the production of permanent, temporary, and traveling exhibitions for cultural history, natural history, science, corporate, sports, and children's-projects throughout the country. Our comprehensive services include project management, exhibit design collaborations, detailing, engineering, budgeting, value engineering, graphic production, fabrication, interactive development, audio-visual integration, scenic production, sculptural fabrication, artifact mounting, and installation.

Personnel:
Mark Matthews (Director, Museum Services)
Lori Pope (Director, Sales & Marketing)

TAYLOR STUDIOS INC.
1320 Harmon Dr.
Rantoul, IL 61866
Tel: 217-893-4874 & 800-707-2047; FAX: 217-893-1998
E-mail: sales@taylorstudios.com
Web Site: www.taylorstudios.com/omd

Type of Business:
Museums, nature centers, zoos, and similar institutions all have engaging stories to tell. They hire Taylor Studios to reveal these stories. Through detailed interpretive planning, design, and fabrication, these provocative stories become exhibits that capture and spark visitors' imaginations. For 23 years, our clients have elected to work with our team because we possess a rare balance: creativity and time-tested processes and procedures that maintain project schedules and budgets. This balance at first may sound over-rated, but contracting a firm with proven project management and financial processes is essential to making any exhibit project as stress-free as possible. Our exhibits are also backed by an unmatched-in-the-industry five-year warranty.

Personnel:
Betty Brennan (President)
Drew Levan (Account Executive)
Kara Vanskike (Marketing Manager)

XIBITZ
7604 Harwood Ave., Ste. 202
Milwaukee, WI 53213
Tel: 414-727-4699 & 616-247-3500 (Home Office: Grand Rapids, MI);
FAX: 414-727-4883
E-mail: ezuern@xibitz.com
Web Site: www.xibitz.com

Type of Business:
Xibitz is dedicated to producing exhibits and environments that inform, educate and inspire for Museums, Visitor Center and other Cultural Institutions. Xibitz is a proven partner, delivering innovative and creative solutions; providing value in Planning, Design, Preproduction, Production and Installation. We offer a highly integrated turnkey approach, leveraging more than 20 years of experience. Friendly, Flexible and Committed, our passion is working hand-in-hand with our clients to discover, innovate and implement creative solutions.

Personnel:
Erich Zuern (Producer - EZuern@Xibitz.com)
Sarah Doty (Assistant Producer - SDoty@Xibitz.com)
James Hungerford (CEO - JHungerford@Xibitz.com)

CASE[WERKS], LLC
1501 Saint Paul St., Ste. 116
Baltimore, MD 21202
Tel: 410-332-4160 & 800-810-2852 (toll free); FAX: 410-332-4106
E-mail: info@casewerks.com
Web Site: www.casewerks.com

Type of Business:
Case[werks] is a sales and service firm offering a range of products uniquely designed to meet exhibit requirements for original art, artifacts and special collections in all types of interior environments. Providing consultative sales support for institutions large & small, Case[werks] is THE museum professional's source for archival display cases & exhibit furnishings, library furnishings, gallery accessories, art hanging hardware, signage & graphic display products as well as conservation equipment including miniClima Humidity Control Systems. Many of our best-selling gallery accessories are in stock, ready to ship; major credit cards are accepted. An extensive catalog is available and custom inquiries are always welcome. Case[werks] is the North American representative for Vitrinen- und Glasbau REIER (Germany).

Personnel:
Matt Malaquias (Principal)
Bill Beitel (Principal)

GLASBAU HAHN AMERICA, LLC
15 Little Brook Lane
Newburgh, NY 12550
Tel: 845-566-3331 & 877-452-7228 (GLASBAU); FAX: 845-566-3176
E-mail: info@glasbau-hahn.com
Web Site: www.glasbau-hahn.com

Type of Business:
Renowned glass specialist and display case manufacturer GLASBAU HAHN was founded in Frankfurt/Main in 1829, and ranks among the global market leaders in museum exhibition and design construction. Having set standards for security (including seismic technology), micro-environmental control, innovative light systems, conservation, accessibility and design, HAHN quality stands for both tradition and innovation. With our certified product "HAHN PURE" one more step towards perfection has been achieved: HAHN PURE display cases are completely built from emission-tested materials (www.hahnpure.de/en). GLASBAU HAHN award winning cases have been installed in more than 1,600 museums, galleries, libraries, and cultural institutions around the world, including over 150 institutions in the U.S. Technical advances include award-winning innovations like the ALLGLASS Display Case, 3-Way Sliding Doors, Hinged Openings, the HAHN Swing-Door and the PROTECTOR Case (the ultimate safeguard for the display and transportation of hanging art). GLASBAU HAHN offers a full range of unique solutions to large and small museum projects, and is known for incorporating the vision of architects and designers with the expediency of functionality, low maintenance and ideal art conservation. A network of offices in Europe, Middle East, USA, Japan, China, and additional representatives worldwide, bring GLASBAU HAHN's standards and innovations to each exhibition. GLASBAU HAHN: Trusted with the World's Treasures for over 180 years! Home Office: GLASBAU HAHN GmbH, Hanauer Landstrasse 211, 60314 Frankfurt am Main, Germany. Tel: 011-49-69-944-1753, Fax: 011-49-69-944-1761.

(Continued on next page)

(Continued from previous page)

Personnel:
Isabel Hahn (President GH GmbH)
Till Hahn (President GHA)
Jamie J. Ponton (Vice President GHA)
Norbert Leonhardt (Construction GH GmbH)
Catherine Lima (Office Manager GHA)

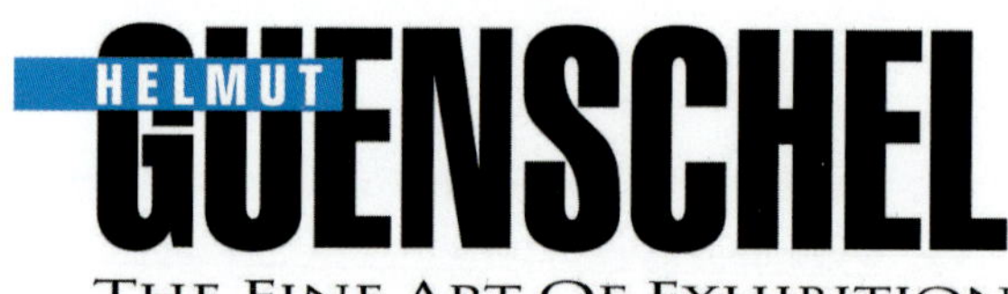

HELMUT GUENSCHEL, INC.
10 Emala Ave.
Baltimore, MD 21220
Tel: 410-686-5900 & 800-852-2525; FAX: 410-687-9342
E-mail: info@guenschel.com
Web Site: www.guenschel.com

Type of Business:
A purely American company for over 45 years and recognized
internationally for creating display cases of the highest quality and for
the innovative engineering behind the most advanced display case
system available, HGI products are specified by architects, designers
and used by many of the most celebrated museums. Our exhibit
products address the full range of technical requirements and
collections issues including all facets of active and passive
conservation, the use of proven environmentally safe materials,
state-of-the-art security and effective seismic anchoring. Engineered
and manufactured to the highest standards, Helmut Guenschel cases
are designed to provide decades of continuous use.

Personnel:
Helmut Guenschel (President)
Cynthia Shaffer (Vice President)

LIGHTING SERVICES INC
2 Holt Dr.
Stony Point, NY 10980
Tel: 800-999-9574 & 845-942-2800; FAX: 845-942-2177
E-mail: Sales@mailLSI.com
Web Site: www.LightingServicesInc.com

Type of Business:
Lighting Services Inc (LSI) is the premier manufacturer of track, accent,
display and LED lighting systems for museum environments. Since
1958, LSI has been dedicated to designing, engineering and
manufacturing lighting fixtures of the highest quality. Our reputation
for creativity and innovative design, coupled with specification grade
products and intelligent personalized service, has made us the
manufacturer of choice amongst the most discriminating specifiers of
lighting for museums and galleries.

Personnel:
Daniel Gelman (President)

Interactive Multimedia Programs & Exhibits

DIMENSIONAL COMMUNICATIONS INC.
1595 MacArthur Blvd.
Mahwah, NJ 06430-3601
Tel: 201-767-1500; FAX: 201-767-9696
E-mail: info@dimcom.com
Web Site: www.dimcom.com

Type of Business:
Dimensional Communications Inc. creates visitor experiences-trade
shows, events, retail displays, museum exhibits and corporate
interiors-that help our clients connect more productively with their
customers. These are high-impact moments that can only happen in
immersive environments. Increasingly, the impact is powered by
Multimedia Technologies. Multimedia Systems design and
implementation has been one of our foundation stones since our first
project in 1964. Today it means custom kiosk design and production,
indoor and outdoor LED displays, Augmented Reality, Multitouch, 3-D
Technologies and beyond. With our full-service process, we make
memorable experiences while making the client experience
unexpectedly satisfying and cost-effective.

Personnel:
Douglas Fixell (President)
Robert Sneed (Vice President Multimedia Technologies)

HILLMANN & CARR INCORPORATED
2233 Wisconsin Ave., N.W., Ste. 425
Washington, DC 20007
Tel: 202-342-0001; FAX: 202-342-0117
E-mail: michalcarr@hillmanncarr.com
Web Site: www.hillmanncarr.com

Type of Business:
Full-service creative storytellers. Producers of award winning media:
video, 3D, film, computer interactive, audio environments, television
broadcast, new media, video tours, immersive experiences and object
theaters, graphic and multimedia presentations. Extensive 35-year
expertise in science, technology, history and art museums, visitor
centers, corporate exhibit media, expositions and multi-language
international projects created in unique formats. Producers and project
managers experienced in AV systems design, integration, installation,
distribution. Wide-ranging experience in audiovisual media master
planning and management, creative concept development, content
research, script to screen production. WOSB.

Personnel:
Alfred Hillmann (President)
Michal Brand Carr (Vice President)

MCCUNE DESIGN
6836 Valjean Ave.
Van Nuys, CA 91406
Tel: 818-779-1920; FAX: 818-781-9108
E-mail: Marketing@McCune-Design.com
Web Site: www.mccune-design.com

Type of Business:

McCune Design is an Academy Award winning design, fabrication and effects business serving the display, entertainment and advertising industries for over 30 years. Seen in museums, theme parks, hotels, and nature centers our exhibit work includes interactives, scenery and sets, specialty props, animatronics, robotics, working prototypes, practical effects, habitats, enclosures, scenics, dioramas, murals, display cases and precision replicas of all kinds. We create realistic, detailed miniatures and models, as well as full size/oversize props, set pieces and environments. Our eclectic work often includes complex electro-mechanical engineering, processor driven motor control systems, sculpture, complex castings, real and faux finishes. Our 30 years in motion picture effects include iconic historical films such as Star Wars, Star Trek, Caddyshack, and Spaceballs, and modern action blockbusters such as X-Men, Iron Man and Spiderman. We produce quality, custom objet d'art in unique, eclectic limited editions or exclusive works of art.

Personnel:
Katherine McCune (President)
Cole McCune (Vice President)
Debora Galloway (Producer/Marketing Director)
Monty Shook (Project Manager)

RBH MULTIMEDIA, INC.
12 Hatch Ter.
Dobbs Ferry, NY 10522
Tel: 914-693-8755; FAX: 914-693-3539
E-mail: rbh@rbhmedia.com
Web Site: www.rbhmedia.com

Type of Business:

RBH Multimedia, Inc. is a media production company dedicated to developing and producing audiovisual experiences and exhibits. We bring imagination, skill, flair and years of experience to designing and producing audio, video and multiple media exhibits for museums, historic sites, visitor and interpretive centers, and cultural institutions. We also design and program websites and computer interactive exhibits. RBH is particularly known for producing signature multi-sensory, multimedia "experience" theater shows. Other services include hardware system design, engineering and installation, with expertise in videoconference technology. For more information about RBH Multimedia, Inc., please visit us online at www.rbhmedia.com.

Personnel:
Steve Brosnahan (Partner)
Nancy Haffner (Partner)
Edgardo J. Resto (Associate)

Lighting

MONADNOCK media

MONADNOCK MEDIA, INC.
112 Amherst Rd.
Sunderland, MA 01375
Tel: 413-665-1390; FAX: 413-665-1394
E-mail: steve@monadnock.org
Web Site: www.monadnock.org

Type of Business:

REINVENTING Multimedia Design and Production for the Museum Environment. When exhibit designers and museums hire Monadnock Media for the design and production of multimedia experiences, they are hiring a group of passionate and talented individuals who work in close collaboration with the project team. We bring to the table a genuine interest in content, a real desire to find the most powerful stories and the clearest ways to express complex ideas. We also have the experience and know-how to bring the material to life no matter the multimedia format - high-tech interactives, immersive environments, theatrical experiences, and social media. We use a wide range of technologies and strategies to help present fresh and innovative approaches to storytelling - with Monadnock, you will never see your exhibit at another museum. Above all, working with us will be fun.

Personnel:
Steve Bressler (Director)

LIGHTING SERVICES INC
2 Holt Dr.
Stony Point, NY 10980
Tel: 800-999-9574 & 845-942-2800; FAX: 845-942-2177
E-mail: Sales@mailLSI.com
Web Site: www.LightingServicesInc.com

Type of Business:

Lighting Services Inc (LSI) is the premier manufacturer of track, accent, display and LED lighting systems for museum environments. Since 1958, LSI has been dedicated to designing, engineering and manufacturing lighting fixtures of the highest quality. Our reputation for creativity and innovative design, coupled with specification grade products and intelligent personalized service, has made us the manufacturer of choice amongst the most discriminating specifiers of lighting for museums and galleries.

Personnel:
Daniel Gelman (President)

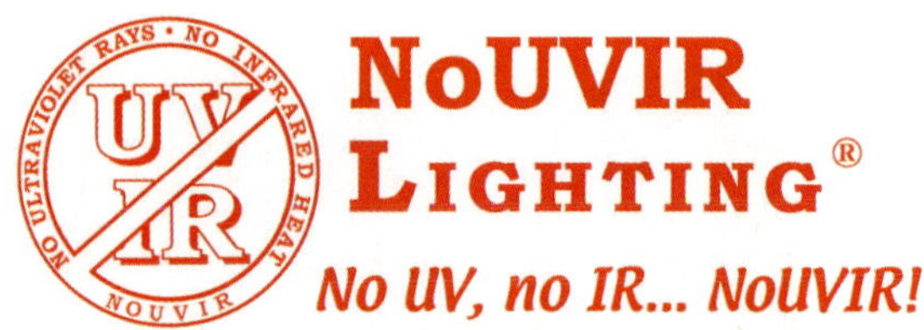

NOUVIR LIGHTING
20915 Sussex Hwy.
Seaford, DE 19973
Tel: 302-628-9933; FAX: 302-628-9932
Web Site: www.nouvir.com
Type of Business:
Pure-white, stone-cold, perfectly controlled lighting with no UV and no
IR from NoUVIR Research radically reduces photochemical and
photomechanical damage to exhibits and can give you documented
70% gallery energy savings. Fifty different miniature floodlights,
spotlights, across-the-room pinspots, eyeballs, wall washers, pendants
and more. Our 24 U.S. patents guarantee superior performance. A
10-year warranty on hardware and fiber guarantees quality. Call for
conservation textbooks, seminar information or free product
specifications, photometry and lighting design information.
Personnel:
Miss Ruth Ellen Miller (President)
Mr. Matthew S. Miller (Vice President Marketing)

RAMBUSCH
SINCE 1898

LIGHTING | CUSTOM LIGHTING | ARCHITECTURAL CRAFT

**RAMBUSCH LIGHTING, INC. - A FAMILY BUSINESS SINCE
1898**
160 Cornelison Ave.
Jersey City, NJ 07304
Tel: 201-333-2525
E-mail: edwinr@rambusch.com
Web Site: www.rambusch.com
Type of Business:
Rambusch since 1898. For over one hundred years, Rambusch has been
at the forefront of innovation in the manufacture and restoration of
standard and custom engineered and decorative lighting fixtures for
museums across America. A fourth generation family-owned business,
Rambusch's success is the result of giving individualized care and
attention to each project in our own workshops. No matter the size or
scope, Rambusch makes use of the newest, most effective lamp
technology. Recent projects include the Frick Collection, Freer Gallery
of Art, J. Paul Getty Decorative Arts Museum, Metropolitan Museum
of Art, Smithsonian National Portrait Gallery and the YALE
UNIVERSITY ART GALLERY. We welcome your inquiry.
Personnel:
Edwin P. Rambusch (President)
Martin V. Rambusch (Chairman)

MCCUNE DESIGN
6836 Valjean Ave.
Van Nuys, CA 91406
Tel: 818-779-1920; FAX: 818-781-9108
E-mail: Marketing@McCune-Design.com
Web Site: www.mccune-design.com

Type of Business:

McCune Design is an Academy Award winning design, fabrication and
effects business serving the display, entertainment and advertising
industries for over 30 years. Seen in museums, theme parks, hotels,
and nature centers our exhibit work includes interactives, scenery and
sets, specialty props, animatronics, robotics, working prototypes,
practical effects, habitats, enclosures, scenics, dioramas, murals,
display cases and precision replicas of all kinds. We create realistic,
detailed miniatures and models, as well as full size/oversize props, set
pieces and environments. Our eclectic work often includes complex
electro-mechanical engineering, processor driven motor control
systems, sculpture, complex castings, real and faux finishes. Our 30
years in motion picture effects include iconic historical films such as
Star Wars, Star Trek, Caddyshack, and Spaceballs, and modern action
blockbusters such as X-Men, Iron Man and Spiderman. We produce
quality, custom objet d'art in unique, eclectic limited editions or
exclusive works of art.
Personnel:
Monty Shook (Project Manager)
Katherine McCune (President)
Cole McCune (Vice President)
Debora Galloway (Producer/Marketing Director)

TAYLOR STUDIOS INC.
1320 Harmon Dr.
Rantoul, IL 61866
Tel: 217-893-4874 & 800-707-2047; FAX: 217-893-1998
E-mail: sales@taylorstudios.com
Web Site: www.taylorstudios.com/omd

Type of Business:

Museums, nature centers, zoos, and similar institutions all have engaging
stories to tell. They hire Taylor Studios to reveal these stories.
Through detailed interpretive planning, design, and fabrication, these
provocative stories become exhibits that capture and spark visitors'
imaginations. For 23 years, our clients have elected to work with our
team because we possess a rare balance: creativity and time-tested
processes and procedures that maintain project schedules and budgets.
This balance at first may sound over-rated, but contracting a firm with
proven project management and financial processes is essential to
making any exhibit project as stress-free as possible. Our exhibits are
also backed by an unmatched-in-the-industry five-year warranty.

Personnel:
Betty Brennan (President)
Drew Levan (Account Executive)
Kara Vanskike (Marketing Manager)

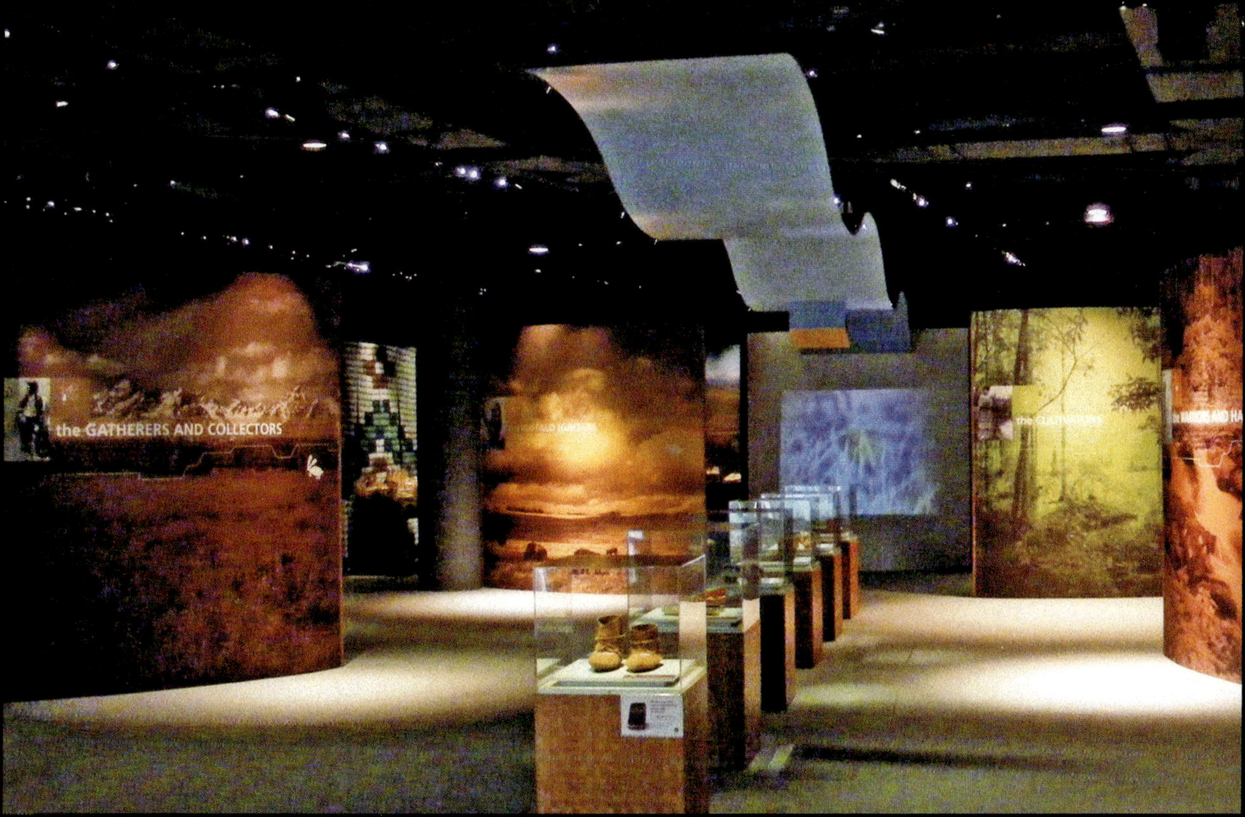

Models & Mannequins

CHASE STUDIO, INC.

CHASE STUDIO, INC.
205 Wolf Creek Rd.
Cedar Creek, MO 65627
Tel: 417-794-3303; FAX: 417-794-3741
E-mail: chasestudio@chasestudio.com
Web Site: www.chasestudio.com

Type of Business:

Designers and builders of natural history and environmental science exhibits; paleontological reconstructions, zoological models, botanical reproductions, dioramas and habitat groups, illustration and mural painting, graphic design, photography, lighting design, fine cabinetwork, taxidermy, research, script writing, exhibit planning and consultation. Exhibits for over 200 museums worldwide since 1973.

Personnel:

Dr. Terry L. Chase (Director)
David F. Darby (Assistant Director Exhibits)
William L. Talbot (Assistant Director Business)

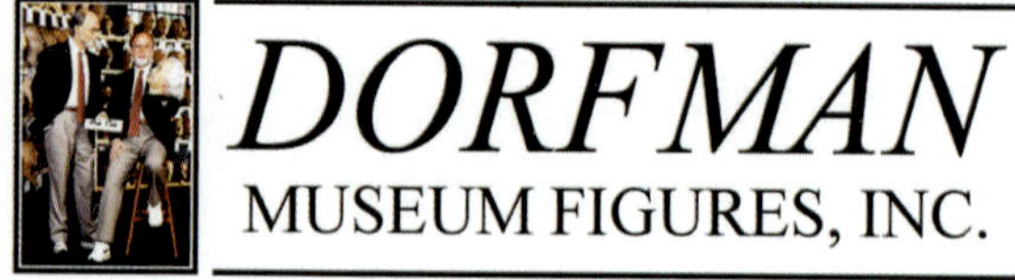

DORFMAN MUSEUM FIGURES, INC.
6224 Holabird Ave.
Baltimore, MD 21224
Tel: 800-634-4873 & 410-284-3248; FAX: 410-284-3249
E-mail: info@museumfigures.com
Web Site: www.museumfigures.com

Type of Business:

We've been standing still for 50 years! Dorfman is the leader in creating lifelike, life-size human figures, and ETHAFOAM™ Conservation Forms. Choose non-ident, generic realistic figures with rigid or flexible foam bodies, or recognizable historic characters for your exhibit needs. Or choose from our line of Conservation Forms, Dress or Suit Forms, or our new Economy Ethafoam mannequin for your archival display and storage needs. Dorfman Museum Figures, Inc. has been in business since 1957. Our Museum Figures and Conservation Forms are used in museums internationally.

Personnel:

Robert Dorfman (President)
Penny Clifton (Project Manager)

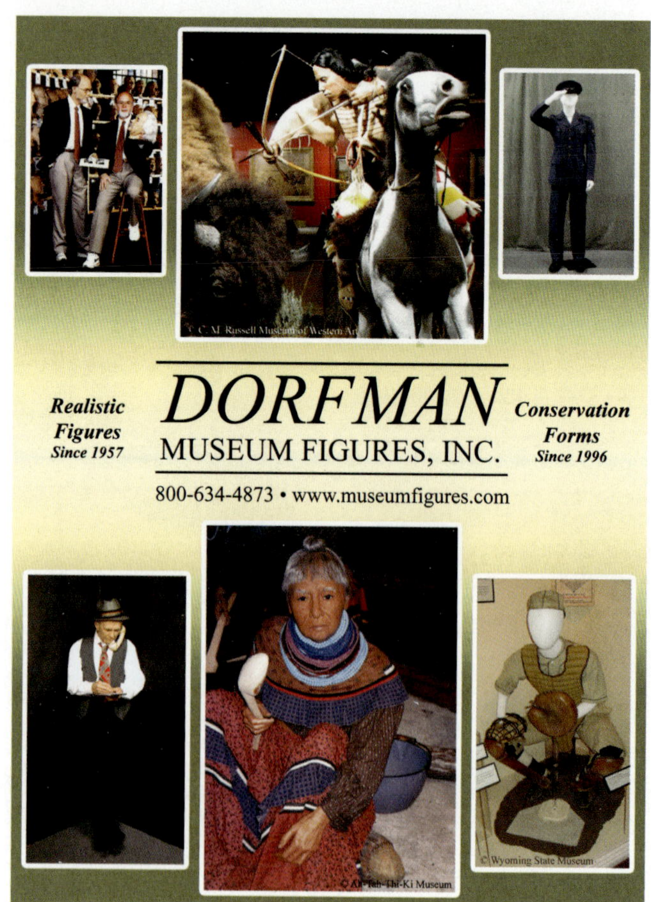

MCCUNE DESIGN
6836 Valjean Ave.
Van Nuys, CA 91406
Tel: 818-779-1920; FAX: 818-781-9108
E-mail: Marketing@McCune-Design.com
Web Site: www.mccune-design.com

Type of Business:

McCune Design is an Academy Award winning design, fabrication and effects business serving the display, entertainment and advertising industries for over 30 years. Seen in museums, theme parks, hotels, and nature centers our exhibit work includes interactives, scenery and sets, specialty props, animatronics, robotics, working prototypes, practical effects, habitats, enclosures, scenics, dioramas, murals, display cases and precision replicas of all kinds. We create realistic, detailed miniatures and models, as well as full size/oversize props, set pieces and environments. Our eclectic work often includes complex

(Continued on next page)

(Continued from previous page)

electro-mechanical engineering, processor driven motor control systems, sculpture, complex castings, real and faux finishes. Our 30 years in motion picture effects include iconic historical films such as Star Wars, Star Trek, Caddyshack, and Spaceballs, and modern action blockbusters such as X-Men, Iron Man and Spiderman. We produce quality, custom objet d'art in unique, eclectic limited editions or exclusive works of art.

Personnel:
Katherine McCune (President)
Cole McCune (Vice President)
Debora Galloway (Producer/Marketing Director)
Monty Shook (Project Manager)

TAYLOR STUDIOS INC.
1320 Harmon Dr.
Rantoul, IL 61866
Tel: 217-893-4874 & 800-707-2047; FAX: 217-893-1998
E-mail: sales@taylorstudios.com
Web Site: www.taylorstudios.com/omd

Type of Business:
Museums, nature centers, zoos, and similar institutions all have engaging stories to tell. They hire Taylor Studios to reveal these stories. Through detailed interpretive planning, design, and fabrication, these provocative stories become exhibits that capture and spark visitors' imaginations. For 23 years, our clients have elected to work with our team because we possess a rare balance: creativity and time-tested processes and procedures that maintain project schedules and budgets. This balance at first may sound over-rated, but contracting a firm with proven project management and financial processes is essential to making any exhibit project as stress-free as possible. Our exhibits are also backed by an unmatched-in-the-industry five-year warranty.

Personnel:
Betty Brennan (President)
Drew Levan (Account Executive)
Kara Vanskike (Marketing Manager)

Murals

CHASE STUDIO, INC.

CHASE STUDIO, INC.
205 Wolf Creek Rd.
Cedar Creek, MO 65627
Tel: 417-794-3303; FAX: 417-794-3741
E-mail: chasestudio@chasestudio.com
Web Site: www.chasestudio.com

Type of Business:
Designers and builders of natural history and environmental science exhibits; paleontological reconstructions, zoological models, botanical reproductions, dioramas and habitat groups, illustration and mural

painting, graphic design, photography, lighting design, fine cabinetwork, taxidermy, research, script writing, exhibit planning and consultation. Exhibits for over 200 museums worldwide since 1973.

Personnel:
Dr. Terry L. Chase (Director)
David F. Darby (Assistant Director Exhibits)
William L. Talbot (Assistant Director Business)

MCCUNE DESIGN
6836 Valjean Ave.
Van Nuys, CA 91406
Tel: 818-779-1920; FAX: 818-781-9108
E-mail: Marketing@McCune-Design.com
Web Site: www.mccune-design.com

Type of Business:
McCune Design is an Academy Award winning design, fabrication and effects business serving the display, entertainment and advertising industries for over 30 years. Seen in museums, theme parks, hotels, and nature centers our exhibit work includes interactives, scenery and sets, specialty props, animatronics, robotics, working prototypes, practical effects, habitats, enclosures, scenics, dioramas, murals, display cases and precision replicas of all kinds. We create realistic, detailed miniatures and models, as well as full size/oversize props, set pieces and environments. Our eclectic work often includes complex electro-mechanical engineering, processor driven motor control systems, sculpture, complex castings, real and faux finishes. Our 30 years in motion picture effects include iconic historical films such as Star Wars, Star Trek, Caddyshack, and Spaceballs, and modern action blockbusters such as X-Men, Iron Man and Spiderman. We produce quality, custom objet d'art in unique, eclectic limited editions or exclusive works of art.

Personnel:
Katherine McCune (President)
Cole McCune (Vice President)
Debora Galloway (Producer/Marketing Director)
Monty Shook (Project Manager)

Structural Panels

HIGHMARK TECHSYSTEMS
8343 Clinton Park Dr.
Fort Wayne, IN 46825
Tel: 260-483-0012; FAX: 260-484-9481
Web Site: www.highmarktech.com

Type of Business:
Versatile in design and application, Highmark Structural panels are lighter and setup faster than other systems. MAX offers the ultimate in versatility and exteriors, ExZact's simple in-fill system allows you the option of changing looks on a whim, and ExTTreme's seamless laminate finish is both elegant and cost effective when compared to traditional temporary wall construction. Features include: Light weight and strong aluminum construction, Sintra and other traditional substrate in-fills via velcro or slide in application, Acrylic in-fills via extrusion retainer or slide in application, Seamless laminate exteriors, Fabric in-fills via SEG extrusion retainer per panel or in an array,

(Continued on next page)

(Continued from previous page)

Panels can be stacked for taller design applications, Straight or curved panels of multiple sizes and radii, Capable of housing internal lighting and wiring Let us configure a MAX, ExZact®, or ExTTreme™ solution for your next exhibit design.

PANELOCK SYSTEMS LIMITED
P.O. Box 729
Woodbury, CT 06798
Tel: 877-315-1998; FAX: 203-643-2060
E-mail: sales@panelock.com
Web Site: www.panelock.com

Type of Business:

Designers and manufacturers of truly movable, environmentally friendly exhibition display walls. Panelock's museum quality units offer designers the opportunity to re-create display footprints to complement exhibition themes. They transform a gallery to a multi-functional space which can house exhibitions and be transformed to use for special functions or events. The internationally patented retractable wheel system in our units is the secret to Panelock's unique relocatable functionality. See the dynamic flexibility of Panelock Systems 100, 200, 600 and System 400 (preferred by architects worldwide) for yourself by viewing "The Panelock Movie" on our website. Panelock! Where sensitivity to the environment, functionality and beauty meet.

Personnel:
Maureen Moreau

Theming

MCCUNE DESIGN
6836 Valjean Ave.
Van Nuys, CA 91406
Tel: 818-779-1920; FAX: 818-781-9108
E-mail: Marketing@McCune-Design.com
Web Site: www.mccune-design.com

Type of Business:

McCune Design is an Academy Award winning design, fabrication and effects business serving the display, entertainment and advertising industries for over 30 years. Seen in museums, theme parks, hotels, and nature centers our exhibit work includes interactives, scenery and sets, specialty props, animatronics, robotics, working prototypes, practical effects, habitats, enclosures, scenics, dioramas, murals, display cases and precision replicas of all kinds. We create realistic, detailed miniatures and models, as well as full size/oversize props, set pieces and environments. Our eclectic work often includes complex electro-mechanical engineering, processor driven motor control systems, sculpture, complex castings, real and faux finishes. Our 30 years in motion picture effects include iconic historical films such as Star Wars, Star Trek, Caddyshack, and Spaceballs, and modern action blockbusters such as X-Men, Iron Man and Spiderman. We produce quality, custom objet d'art in unique, eclectic limited editions or exclusive works of art.

Personnel:
Katherine McCune (President)
Cole McCune (Vice President)
Debora Galloway (Producer/Marketing Director)
Monty Shook (Project Manager)

TAYLOR STUDIOS INC.
1320 Harmon Dr.
Rantoul, IL 61866
Tel: 217-893-4874 & 800-707-2047; FAX: 217-893-1998
E-mail: sales@taylorstudios.com
Web Site: www.taylorstudios.com/omd

Type of Business:

Museums, nature centers, zoos, and similar institutions all have engaging stories to tell. They hire Taylor Studios to reveal these stories. Through detailed interpretive planning, design, and fabrication, these provocative stories become exhibits that capture and spark visitors' imaginations. For 23 years, our clients have elected to work with our team because we possess a rare balance: creativity and time-tested processes and procedures that maintain project schedules and budgets. This balance at first may sound over-rated, but contracting a firm with proven project management and financial processes is essential to making any exhibit project as stress-free as possible. Our exhibits are also backed by an unmatched-in-the-industry five-year warranty.

Personnel:
Betty Brennan (President)
Drew Levan (Account Executive)
Kara Vanskike (Marketing Manager)

Traveling Exhibitions

CHICAGO SCENIC STUDIOS, INC.
1315 N. North Branch St.
Chicago, IL 60642
Tel: 312-274-9900; FAX: 312-274-9901
E-mail: dLanghorst@chicagoscenic.com
Web Site: www.chicagoscenic.com

Type of Business:

Chicago Scenic Studios provides museums, science & technology centers, zoos, aquariums and visitor centers more than 35 years of multi-disciplinary experience in themed environment design, fabrication, engineering and installation. Our highly skilled staff of project managers, designers, and skilled craftspeople partner with your team to provide safe, educational, immersive environments and exciting experiences through custom exhibits, retail, restaurant, and admission spaces. Chicago Scenic can serve as your general contractor, fabricator or project manager in support of your design or ours. Our background in exhibits, retail, theater, special events and corporate spaces provides us with the ability to cater to differing needs in unique spaces. We have enjoyed partnering with world-class institutions such as The Museum of Science and Industry, Chicago; The Great Lakes Science Center, the Chicago Children's Museum, The

(Continued on next page)

(Continued from previous page)

John G. Shedd Aquarium, the Brookfield Zoo, the Field Museum of Natural History and many more to bring exciting, educational and successful stories to life. We'd like to help tell your story as well.

Personnel:
Robert Doepel (President)
Diane Langhorst (Business Development)

EVERGREEN EXHIBITIONS
7979 Broadway, Ste. 107
San Antonio, TX 78209
Tel: 210-582-0015; FAX: 210-590-1071
E-mail: christi@evergreenexhibitions.com
Web Site: www.evergreenexhibitions.com

Type of Business:
EVERGREEN EXHIBITIONS is a premier provider of interactive educational exhibitions with over 21 years of experience touring science, natural history, art and object exhibitions. Highly respected in the museum community for its commitment and dedication to quality and education, Evergreen Exhibitions delivers immersive experiences to over 200 natural history museums, science centers and art museums worldwide. TRAVELING EXHIBITIONS include Vatican Splendors, MathAlive, Gridiron Glory, Leonardo da Vinci: Machines in Motion, Space: A Journey to Our Future, Brain: The World Inside Your Head, Genome: The Secret of How Life Works, Extreme Deep: Mission to the Abyss, Microbes: Invisible Invaders…Amazing Allies, Masters of the Night: The True Story of Bats, The Robot Zoo, and Spies, Traitors, Saboteurs: Fear and Freedom in America. More to come soon.

Personnel:
Mark Greenberg (President)
Anne Kinsey (Vice President Exhibitions)
Christi Klingelhefer (Venue Sales Manager)

GRAPHIC DESIGN/PUBLISHING – GENERAL

Exhibit Graphics

studio**MUS**arx LLC

museum planning + **ar**chitecture + **ex**hibit design

STUDIOMUSARX LLC
2601 Pennsylvania Avenue, Ste. 8
Philadelphia, PA 19130
Tel: 215-232-3489
E-mail: info@studiomusarx.com
Web Site: www.studiomusarx.com

Type of Business:
 studioMUSarx is a forward-thinking studio with special expertise in
 planning, architecture and exhibit design for new and existing
 museums and visitor centers. We are dedicated to a process of
 discovery acknowledging best practices while lending a friendly ear to
 client preferences in keeping with the shared values and aspirations of
 the studio. Each project undertaken is directed by the studio's
 principal, who collaborates closely with design team members to
 create strong working partnerships with clients. For over 40, years Joe
 Nicholson has assisted museums and interpretive centers with
 feasibility studies, master plans, space programming, signage
 programs, architectural and exhibit development, and design for sites,
 buildings, and interpretive exhibits providing full services through
 construction administration. We are open to the ongoing challenge of
 touching upon the extraordinary, embracing the artful, and going
 beyond constraints to deliver meaningful outcomes free from
 preconceptions of what the final design product should be.

Personnel:
 Joseph A. Nicholson, AIA, NCARB (Principal)
 Susan Hutton DeAngelus (Exhibit and Graphic Designer)
 Juliet Geldi, RA (Planner, Architect and Exhibit Designer)
 Sherman Lai, Assoc. AIA (Architectural and Exhibit Designer)

Graphic Design Firms

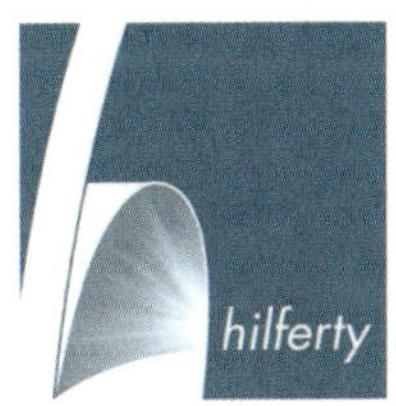

HILFERTY
14240 State Rte. 550
Athens, OH 45701
Tel: 740-448-3821; FAX: 740-448-2331
E-mail: gha@hilferty.com
Web Site: www.hilferty.com

Type of Business:
 For more than thirty years, Hilferty has provided a full range of
 interpretive services to museums, visitor centers, nature centers,
 botanical gardens, historic sites, zoos/aquariums, and halls of fame.
 We excel at interpretive and facility master planning; concept
 development; content research; label writing; exhibit and graphic
 design; and providing donor recognition and funding support materials.
 From social histories to natural sciences, football to physics, presidents
 to children's gardens, every Hilferty exhibit vividly captures the
 stories that engage visitors' minds and hearts. Each project is
 approached with a unique mix of techniques chosen to best bring the
 subject to life. Our innovative portfolio spans the spectrum of intimate
 displays of precious objects, captivating immersive environments,
 enthralling interactive components, and breathtaking theatrical
 presentations. In short, Hilferty creates memorable museum
 experiences.

Personnel:
 Gerard Hilferty (President & Creative Director)
 Dean Clouse (CEO)

Publishing & Distribution Services

AMERICAN ALLIANCE OF MUSEUMS
1575 Eye St., N.W., Suite 400
Washington, DC 20005-1105
Tel: 202-289-1818; FAX: 202-289-6578
Web Site: www.aam-us.org

Type of Business:
 The American Alliance of Museums (AAM) is dedicated to promoting
 excellence within the museum community. The only organization
 representing the entire museum field, AAM's mission is to enhance the
 value of museums to their communities through leadership, advocacy
 and service. Since its founding in 1906, AAM has grown to nearly
 21,000 members, including more than 17,000 individual members, 300
 corporate members, and more than 3,100 museums. For more about the
 Alliance, please visit www.aam-us.org.

Personnel:
 Dewey Blanton (Director, Strategic Communications)

We are the national champion for all museums.

At the American Alliance of Museums, we are committed to championing museums, and believe advocacy is a year-round job. To succeed, we need every voice to join the chorus in support of America's museums. Here's how you can participate:

- Sign up for Advocacy Alerts from the Alliance, keeping you apprised of museum-related developments in Congress and letting you know when to weigh-in.

- Find out who represents you in Congress, with staff and contact information, and identify your state legislators, too.

- Take action when the Alliance calls for you to contact Congress, and use our templates to easily send letters on issues that matter to museums.

- "Write One Letter" – Encourage students and visitors to you museum to let their elected officials know how special the museum is to them. Use our system and it's just a few clicks.

- Learn about the issues affecting museums, and Alliance positions on them with Alliance Issue Briefs, testimony and more.

- Get all the information about Museums Advocacy Day, the annual field-wide effort to take the value of museums to Capitol Hill, and how you can participate in DC or wherever you are.

You can take all of these steps and more at the Alliance website. Visit www.aam-us.org/advocacy for the tools you'll need and join our national advocacy efforts today.

Your **FREE** Alliance membership is waiting . . .

If your museum is a Tier 3 member and opts for the All-Staff package, here's what's waiting for you:

- **free** registration for online learning opportunities
- customized research assistance through the Information Center
- discounts at the Bookstore
- ability to join any or all of the Alliance's 22 Professional Networks
- free subscriptions to a range of Alliance publications and e-newsletters, including *Museum* magazine and *Dispatches* from the Center for the Future of Museums
- **free** registration for Museums Advocacy Day, held each February in Washington, D.C.
- your own membership card that many museums honor with free/reduced admissions

Activate your individual membership to gain access to your benefits today!

Visit www.aam-us.org.

MARKETING/PUBLIC RELATIONS – GENERAL

Marketing

ConsultEcon, Inc.

Economic and Management Consultants

CONSULTECON, INC.
545 Concord Ave., Ste. 210
Cambridge, MA 02138
Tel: 617-547-0100; FAX: 617-547-0102
E-mail: info@consultecon.com
Web Site: www.consultecon.com

Type of Business:
 ConsultEcon, Inc. provides services to clients in the areas of project and plan concept development, market and financial evaluation, visitor surveys, economic impact and project implementation. We are dedicated to serving museums and visitor attractions of all sizes and types, and have worked over many years with clients responding to a broad spectrum of issues, ranging from the economics of operations to strategic planning.

Personnel:
 Thomas J. Martin (President)
 Robert E. Brais (Vice President)
 Elena Kazlas (Principal)
 Jason Drebitko (Senior Associate)
 James Stevens (Senior Associate)

S&W Schultz & Williams

development, management, marketing

SCHULTZ & WILLIAMS
325 Chestnut St., Ste. 700
Philadelphia, PA 19106
Tel: 215-625-9955; FAX: 215-625-2701
E-mail: mail@schultzwilliams.com
Web Site: www.schultzwilliams.com

Type of Business:
 Schultz & Williams provides comprehensive consulting services to the arts, museums, cultural institutions and other nonprofits of all types and sizes across the nation. Since 1987, the success and integrity of our work has been driven by one simple philosophy: development, management and marketing strategies must be fully integrated to achieve financial stability, operational excellence and greater mission impact. Our senior professionals have extensive hands-on experience from both sides of the desk. The strength of our consulting services is based on the skills we have gained in managing, operating and supporting major nonprofit organizations, including: Development: Capital campaigns, major gift and planned giving strategies; Direct Response: Full-service digital and offline communications to raise funds through membership, annual giving, mid-level and monthly giving programs; Planning: Financial planning, facility and master plan management, leadership development; Marketing: Branding, planning, collateral; S&W StaffSolutions: Executive and management level interim, project, or start-up staffing.

Personnel:
 L. Scott Schultz (President)
 M. Jane Williams (Principal)
 Jessica Harrington (Vice President, S & W Direct)
 Dell Fascione (Vice President)

MUSEUM RESOURCES/TECHNICAL INFORMATION-GENERAL

AMERICAN ALLIANCE OF MUSEUMS
1575 Eye St., N.W., Suite 400
Washington, DC 20005-1105
Tel: 202-289-1818; FAX: 202-289-6578
Web Site: www.aam-us.org

Type of Business:

The American Alliance of Museums (AAM) is dedicated to promoting excellence within the museum community. The only organization representing the entire museum field, AAM's mission is to enhance the value of museums to their communities through leadership, advocacy and service. Since its founding in 1906, AAM has grown to nearly 21,000 members, including more than 17,000 individual members, 300 corporate members, and more than 3,100 museums. For more about the Alliance, please visit www.aam-us.org.

Personnel:
Dewey Blanton (Director, Strategic Communications)

LIGHTING SERVICES INC
2 Holt Dr.
Stony Point, NY 10980
Tel: 800-999-9574 & 845-942-2800; FAX: 845-942-2177
E-mail: Sales@mailLSI.com
Web Site: www.LightingServicesInc.com

Type of Business:

Lighting Services Inc (LSI) is the premier manufacturer of track, accent, display and LED lighting systems for museum environments. Since 1958, LSI has been dedicated to designing, engineering and manufacturing lighting fixtures of the highest quality. Our reputation for creativity and innovative design, coupled with specification grade products and intelligent personalized service, has made us the manufacturer of choice amongst the most discriminating specifiers of lighting for museums and galleries.

Personnel:
Daniel Gelman (President)

Associations

AMERICAN ALLIANCE OF MUSEUMS
1575 Eye St., N.W., Suite 400
Washington, DC 20005-1105
Tel: 202-289-1818; FAX: 202-289-6578
Web Site: www.aam-us.org

Type of Business:

The American Alliance of Museums (AAM) is dedicated to promoting excellence within the museum community. The only organization representing the entire museum field, AAM's mission is to enhance the value of museums to their communities through leadership, advocacy and service. Since its founding in 1906, AAM has grown to nearly 21,000 members, including more than 17,000 individual members, 300 corporate members, and more than 3,100 museums. For more about the Alliance, please visit www.aam-us.org.

Personnel:
Dewey Blanton (Director, Strategic Communications)

Auctioneers & Liquidators

CHRISTIE'S
20 Rockefeller Plaza
New York, NY 10020
Tel: 212-636-2620; FAX: 212-636-2370
E-mail: MuseumServicesNY@christies.com
Web Site: www.christies.com

Type of Business:
Christie's is available to assist museums in a variety of areas including: appraising collections, objects and bequests and establishing values for loans; managing sales of major objects and minor collections; assistance with indemnification appraisals; and consulting on buying and selling at auction. We welcome collector circle visits, and our specialists are available for participation in lectures, symposia and scholarly talks.

Personnel:
Allison Whiting (Senior Vice President & Director Museum Services)
Vanessa Fusco (Associate Vice President, Account Manager)

DOYLE NEW YORK
Auctioneers & Appraisers
175 E. 87th St.
New York, NY 10128
Tel: 212-427-2730; FAX: 212-369-0892
E-mail: info@DoyleNewYork.com
Web Site: www.DoyleNewYork.com

(Continued on next page)

(Continued from previous page)

Type of Business:

Doyle New York is one of the world's foremost auctioneers and appraisers of fine art, jewelry, furniture, decorations, books, manuscripts, prints and a variety of other categories. Headquartered in New York City, the global capital of the auction market, Doyle offers approximately forty sales each year that attract a broad base of buyers and consignors from around the world. Doyle New York regularly works with prominent museums and distinguished institutions, providing individualized appraisal and auction services for collections of all sizes. Museums, universities, libraries and other institutions rely on Doyle New York's experience and expertise in handling all aspects of the auction process with discretion, from the initial call through the final settlement.

Personnel:

Kathleen M. Doyle (Chairman & CEO)
Joanne Porrino Mournet (Executive Vice President, Appraisals & Consignments)
David A. Gallager (Senior Vice President, Museum Services)
Anne Cohen DePietro (Vice President, Museum Services)

Collection Management & Development

ADLIB INFORMATION SYSTEMS
99 Fifth Avenue, Ste. 214
Ottawa, ON, Canada K1S 5P5
Tel: +1 (312) 2390597
E-mail: sales@selagodesign.com
Web Site: www.selagodesign.com/adlib

Type of Business:

Adlib Information Systems is Europe's leading developer and distributer of collections management software. Our product is specially designed for Museums, Libraries and Archives. We provide solutions tailored to your institution or a fully integrated package combining any of our products. All our solutions are based on international standards such as Spectrum and EAD. The standard functionality of the core application can be easily expanded using our designer tool, our API and several functional modules. Adlib has over 25 years of experience in the collections management field which has provided a solid foundation for our products. Over the years, Adlib has become the most widely used collections management tool which now benefits over 1600 different customers.

Personnel:

Andrea Boyes (Director CEO Selago Design)
Bert Degenhart Drenth (Director CEO Adlib Information Systems)

Computer Software

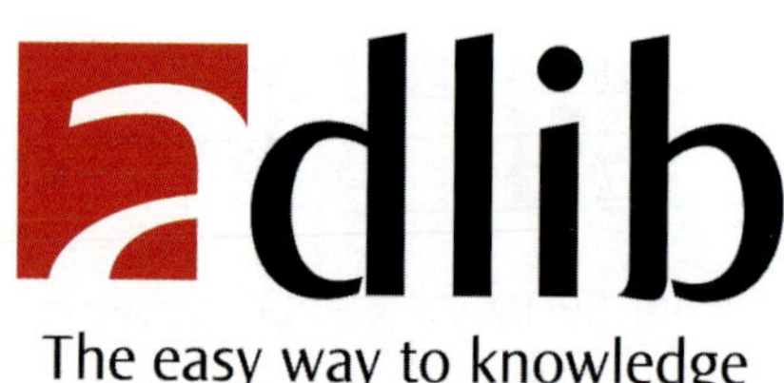

ADLIB INFORMATION SYSTEMS
99 Fifth Avenue, Ste. 214
Ottawa, ON, Canada K1S 5P5
Tel: +1 (312) 2390597
E-mail: sales@selagodesign.com
Web Site: www.selagodesign.com/adlib

Type of Business:

Adlib Information Systems is Europe's leading developer and distributer of collections management software. Our product is specially designed for Museums, Libraries and Archives. We provide solutions tailored to your institution or a fully integrated package combining any of our products. All our solutions are based on international standards such as Spectrum and EAD. The standard functionality of the core application can be easily expanded using our designer tool, our API and several functional modules. Adlib has over 25 years of experience in the collections management field which has provided a solid foundation for our products. Over the years, Adlib has become the most widely used collections management tool which now benefits over 1600 different customers.

Personnel:

Andrea Boyes (Director CEO Selago Design)
Bert Degenhart Drenth (Director CEO Adlib Information Systems)

TAM RETAIL, DIV. OF LODE DATA SYSTEMS, INC.
10609 West 159th St.
Orland Park, IL 60467
Tel: 888-THE-14POS (888-843-1476) & 708-460-0999; FAX: 708-460-1253
E-mail: sales@tamretail.com
Web Site: www.nonprofitpos.com

Type of Business:

TAM Retail is the sole provider of The Assistant Manager™ (TAM), a turn-key retail management solution that meets the growing needs of non-profits throughout the U.S. and internationally. TAM provides the knowledge of the things you cannot easily see for yourself. It provides you with a solution for knowing the most critical aspects of your organization's operations including exactly what products and services have sold, attendance levels, details on each transaction, associate and volunteer performance, perpetual value of your inventory and how it is performing, what is hot and what is not, helps you know your members better, what members like and don't like, and promotes increased donations. All the information needed to make the many important decisions that affect your organization is provided by TAM. TAM provides this at an affordable cost of ownership from a proven and trusted provider to help your organization get the most return on your investment.

Personnel:

Bruce H. Lode (Executive Vice President of Marketing and Sales)

Executive Search Organizations

MUSEUM MANAGEMENT
CONSULTANTS, INC.

MUSEUM MANAGEMENT CONSULTANTS, INC.
120 Green St., Ste. 200
San Francisco, CA 94111
Tel: 415-982-2288
E-mail: mmc@museum-management.com
Web Site: www.museum-management.com

Type of Business:
Museum Management Consultants, Inc. (MMC) specializes in
organizational assessment, institutional planning, business models,
board development, executive search, audience research, and
professional coaching. Founded in 1987 and based in San Francisco,
MMC has provided consulting services to hundreds of museums and
cultural organizations throughout the United States and abroad.
MMC's mission is to help our clients thrive in a competitive and
changing environment. We help museums accentuate their strengths,
address critical issues, and move strategically into the future. MMC
supports the highest ethical standards in our relationships with our
clients and their communities. We believe every organization is unique
in its culture, circumstance, and constituencies, and thus our process is
tailored to meet each client's individual needs and objectives.

Personnel:
Adrienne Horn (President)
Stephen Horn (Senior Vice President)
Katie Sevier (Vice President)
Georgianna de la Torre (Vice President)

Museum Planners

ANDREW MERRIELL & ASSOCIATES, LLC
7198 Old Santa Fe Trail
Santa Fe, NM 87505
Tel: 505-982-3950; FAX: 505-820-6674
E-mail: andy@merriell.com
Web Site: www.merriell.com

Type of Business:
Andrew Merriell & Associates plans and designs experiences for visitors
to museums, visitor centers, parks, zoos, and public gardens. We
identify opportunities to advance beyond the usual and predictable
interpretive offerings, finding innovative ways to immerse visitors in
exhibit stories. Each project provides visitors with memorable,
meaningful experiences they can find nowhere else. Services include
feasibility studies, interpretive master plans, concept studies, exhibit
content development, and exhibit design.

Personnel:
Andrew Merriell (Principal)
Rebecca Shreckengast (Senior Designer)
Hella Rader (Project Manager)

CONSULTECON, INC.
545 Concord Ave., Ste. 210
Cambridge, MA 02138
Tel: 617-547-0100; FAX: 617-547-0102
E-mail: info@consultecon.com
Web Site: www.consultecon.com

Type of Business:
ConsultEcon, Inc. provides services to clients in the areas of project and
plan concept development, market and financial evaluation, visitor
surveys, economic impact and project implementation. We are
dedicated to serving museums and visitor attractions of all sizes and
types, and have worked over many years with clients responding to a
broad spectrum of issues, ranging from the economics of operations to
strategic planning.

Personnel:
Thomas J. Martin (President)
Robert E. Brais (Vice President)
Elena Kazlas (Principal)
Jason Drebitko (Senior Associate)
James Stevens (Senior Associate)

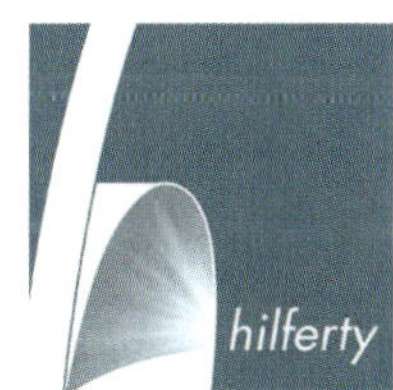

HILFERTY
14240 State Rte. 550
Athens, OH 45701
Tel: 740-448-3821; FAX: 740-448-2331
E-mail: gha@hilferty.com
Web Site: www.hilferty.com

Type of Business:
For more than thirty years, Hilferty has provided a full range of
interpretive services to museums, visitor centers, nature centers,
botanical gardens, historic sites, zoos/aquariums, and halls of fame.
We excel at interpretive and facility master planning; concept
development; content research; label writing; exhibit and graphic
design; and providing donor recognition and funding support materials.
From social histories to natural sciences, football to physics, presidents
to children's gardens, every Hilferty exhibit vividly captures the
stories that engage visitors' minds and hearts. Each project is
approached with a unique mix of techniques chosen to best bring the
subject to life. Our innovative portfolio spans the spectrum of intimate
displays of precious objects, captivating immersive environments,
enthralling interactive components, and breathtaking theatrical
presentations. In short, Hilferty creates memorable museum
experiences.

Personnel:
Gerard Hilferty (President & Creative Director)
Dean Clouse (CEO)

MUSEUM MANAGEMENT CONSULTANTS, INC.

120 Green St., Ste. 200
San Francisco, CA 94111
Tel: 415-982-2288
E-mail: mmc@museum-management.com
Web Site: www.museum-management.com

Type of Business:

Museum Management Consultants, Inc. (MMC) specializes in organizational assessment, institutional planning, business models, board development, executive search, audience research, and professional coaching. Founded in 1987 and based in San Francisco, MMC has provided consulting services to hundreds of museums and cultural organizations throughout the United States and abroad. MMC's mission is to help our clients thrive in a competitive and changing environment. We help museums accentuate their strengths, address critical issues, and move strategically into the future. MMC supports the highest ethical standards in our relationships with our clients and their communities. We believe every organization is unique in its culture, circumstance, and constituencies, and thus our process is tailored to meet each client's individual needs and objectives.

Personnel:

Adrienne Horn (President)
Stephen Horn (Senior Vice President)
Katie Sevier (Vice President)
Georgianna de la Torre (Vice President)

studioMUSarx LLC

museum planning + architecture + exhibit design

STUDIOMUSARX LLC

2601 Pennsylvania Avenue, Ste. 8
Philadelphia, PA 19130
Tel: 215-232-3489
E-mail: info@studiomusarx.com
Web Site: www.studiomusarx.com

Type of Business:

studioMUSarx is a forward-thinking studio with special expertise in planning, architecture and exhibit design for new and existing museums and visitor centers. We are dedicated to a process of discovery acknowledging best practices while lending a friendly ear to client preferences in keeping with the shared values and aspirations of the studio. Each project undertaken is directed by the studio's principal, who collaborates closely with design team members to create strong working partnerships with clients. For over 40, years Joe Nicholson has assisted museums and interpretive centers with feasibility studies, master plans, space programming, signage programs, architectural and exhibit development, and design for sites, buildings, and interpretive exhibits providing full services through construction administration. We are open to the ongoing challenge of touching upon the extraordinary, embracing the artful, and going beyond constraints to deliver meaningful outcomes free from preconceptions of what the final design product should be.

Personnel:

Joseph A. Nicholson, AIA, NCARB (Principal)
Susan Hutton DeAngelus (Exhibit and Graphic Designer)
Juliet Geldi, RA (Planner, Architect and Exhibit Designer)
Sherman Lai, Assoc. AIA (Architectural and Exhibit Designer)

TAYLOR STUDIOS INC.

1320 Harmon Dr.
Rantoul, IL 61866
Tel: 217-893-4874 & 800-707-2047; FAX: 217-893-1998
E-mail: sales@taylorstudios.com
Web Site: www.taylorstudios.com/omd

Type of Business:

Museums, nature centers, zoos, and similar institutions all have engaging stories to tell. They hire Taylor Studios to reveal these stories. Through detailed interpretive planning, design, and fabrication, these provocative stories become exhibits that capture and spark visitors' imaginations. For 23 years, our clients have elected to work with our team because we possess a rare balance: creativity and time-tested processes and procedures that maintain project schedules and budgets. This balance at first may sound over-rated, but contracting a firm with proven project management and financial processes is essential to making any exhibit project as stress-free as possible. Our exhibits are also backed by an unmatched-in-the-industry five-year warranty.

Personnel:

Betty Brennan (President)
Drew Levan (Account Executive)
Kara Vanskike (Marketing Manager)

Museum Planners and Producers

WHITE OAK ASSOCIATES, INC.

17 Essex St.
Marblehead, MA 01945
Tel: 781-639-0722; FAX: 781-639-2491
E-mail: info@whiteoakassoc.com
Web Site: www.whiteoakassoc.com

Type of Business:

White Oak Associates is dedicated to high quality analysis and strategic planning services for museums. Our mission is to collaborate with museums to create sustainable institutions providing essential community services. With over 40 years of very satisfied museum clients and hundreds of significant commissions, White Oak has earned respect for its collaborative, well-informed and thoughtful planning approach, including: community needs assessments, performance assessments, feasibility studies, concept development plans, strategic master plans, business plans, market studies, architectural program planning, and staff and operating plans. Using analysis of your budgets, attendance, facility size and other data, we integrate your conceptual vision with sustainable economic models. White Oak also offers implementation management from concept through post-opening for museum reinventions, expansions and new museums. We are flexible to fit your exact needs in team make-up, scope, schedule and budget. Visit our website and "Talk To Us" about your museum analysis and planning needs.

Personnel:

John W. Jacobsen (President)
Jeanie Stahl (Vice President)
Victor A. Becker (Director, Program Development)
Karen Hefler (Production & Logistics)
Rebecca Robison (Production & Logistics)

Props

Sculptor/Sculptures

MCCUNE DESIGN
  6836 Valjean Ave.
  Van Nuys, CA 91406
  Tel: 818-779-1920; FAX: 818-781-9108
  E-mail: Marketing@McCune-Design.com
  Web Site: www.mccune-design.com

Type of Business:
  McCune Design is an Academy Award winning design, fabrication and effects business serving the display, entertainment and advertising industries for over 30 years. Seen in museums, theme parks, hotels, and nature centers our exhibit work includes interactives, scenery and sets, specialty props, animatronics, robotics, working prototypes, practical effects, habitats, enclosures, scenics, dioramas, murals, display cases and precision replicas of all kinds. We create realistic, detailed miniatures and models, as well as full size/oversize props, set pieces and environments. Our eclectic work often includes complex electro-mechanical engineering, processor driven motor control systems, sculpture, complex castings, real and faux finishes. Our 30 years in motion picture effects include iconic historical films such as Star Wars, Star Trek, Caddyshack, and Spaceballs, and modern action blockbusters such as X-Men, Iron Man and Spiderman. We produce quality, custom objet d'art in unique, eclectic limited editions or exclusive works of art.

Personnel:
  Katherine McCune (President)
  Cole McCune (Vice President)
  Debora Galloway (Producer/Marketing Director)
  Monty Shook (Project Manager)

MCCUNE DESIGN
  6836 Valjean Ave.
  Van Nuys, CA 91406
  Tel: 818-779-1920; FAX: 818-781-9108
  E-mail: Marketing@McCune-Design.com
  Web Site: www.mccune-design.com

Type of Business:
  McCune Design is an Academy Award winning design, fabrication and effects business serving the display, entertainment and advertising industries for over 30 years. Seen in museums, theme parks, hotels, and nature centers our exhibit work includes interactives, scenery and sets, specialty props, animatronics, robotics, working prototypes, practical effects, habitats, enclosures, scenics, dioramas, murals, display cases and precision replicas of all kinds. We create realistic, detailed miniatures and models, as well as full size/oversize props, set pieces and environments. Our eclectic work often includes complex electro-mechanical engineering, processor driven motor control systems, sculpture, complex castings, real and faux finishes. Our 30 years in motion picture effects include iconic historical films such as Star Wars, Star Trek, Caddyshack, and Spaceballs, and modern action blockbusters such as X-Men, Iron Man and Spiderman. We produce quality, custom objet d'art in unique, eclectic limited editions or exclusive works of art.

Personnel:
  Katherine McCune (President)
  Cole McCune (Vice President)
  Debora Galloway (Producer/Marketing Director)
  Monty Shook (Project Manager)

SHIPPING/STORAGE-GENERAL

HAHN BROS. FIREPROOF WAREHOUSES, INC.
622 Communipaw Ave.
Jersey City, NJ 07304
Tel: 212-926-1505; FAX: 201-432-9547
E-mail: info@hahnbros.com
Web Site: www.hahnbros.com

Type of Business:
Over 100 years of fine art services to the museum and art community. Full service specialists offering customized crating and traveling cases and freight forwarding for collections and shows. Packing, moving, relocation and installation. Registrar, photography and gallery services available. Secure, climate-controlled storage with the ability to set up to client specifications.

Personnel:
Karen O. Dowling (President)
Marianne Mikulka (Fine Art Director)

Archival Storage Equipment

DELTA DESIGNS LTD.
P.O. Box 1733
Topeka, KS 66601
Tel: 785-234-2244 & 800-656-7426; FAX: 785-233-1021
E-mail: bdanielson@deltadesignsltd.com
Web Site: www.deltadesignsltd.com

Type of Business:
Custom and standard museum storage cabinets, free standing or mounted on high density mobile storage systems. Cabinets available for your collection: Natural History, Works on Paper, Textiles, Art Objects, Historical Artifacts. Design and Installation services provided. Quality through Incremental Change.

Personnel:
Bruce Danielson (President)

MONTEL INC.
225 4th Ave.
Montmagny, QC, Canada G5V 3S5
Tel: 877-935-0236; FAX: 418-248-7266
E-mail: system@montel.com
Web Site: www.montel.com

Type of Business:
The Art of Storage by Montel for Museums & Archival Collections: The Montel high density fixed and mobile storage systems allows each system to be custom-designed to the particular safety storage requirements for each type of collection. Montel has developed a line of efficient and safe products which respect the integrity and value of all types of art work regardless of their size & shape. Specializing in storage systems since 1924, Montel offers its expertise to insure planning solutions that maximize capacity, cost savings, security and accessibility. We offer our complete line for your particular collection requirement, including art racks, full line of cabinetry, textile racks, shelving, drawer cabinets and mobile storage systems.

Personnel:
Joey P. Boudreau (Marketing Director)

VIKING METAL CABINET COMPANY
a division of Austin-Westran
24047 West Lockport St., Ste. #209
Plainfield, IL 60544
Tel: 800-776-7767 & 815-782-8108; FAX: 815-267-6914
E-mail: sales@vikingmetal.com
Web Site: www.vikingmetal.com

Type of Business:
Viking manufactures a full line of quality museum cabinets that meet the storage requirements for all types of conservation, preservation, and archival collections. Major institutions and universities throughout the world rely on Viking cabinets to protect their most precious specimens, artifacts and documents. Versatility is the hallmark of Viking's product line. Our counter and full height conservation cabinets are offered in a wide range of sizes along with unique and optional accessories. We also offer a standard line of specialty cabinets for botanical, geological, and entomological collections along with archival flat files for artistic and historical works on paper. If your storage application is unique, Viking is ready to modify standard products or design custom cabinets that meet your particular needs with great attention to detail. Trust Viking…we've been designing and manufacturing museum storage products of unparalleled quality for over half a century.

Personnel:
Jim Dolan (Vice President Sales)

Art Storage Equipment

DELTA DESIGNS LTD.
P.O. Box 1733
Topeka, KS 66601
Tel: 785-234-2244 & 800-656-7426; FAX: 785-233-1021
E-mail: bdanielson@deltadesignsltd.com
Web Site: www.deltadesignsltd.com

(Continued on next page)

(Continued from previous page)

Type of Business:
Custom and standard museum storage cabinets, free standing or mounted on high density mobile storage systems. Cabinets available for your collection: Natural History, Works on Paper, Textiles, Art Objects, Historical Artifacts. Design and Installation services provided. Quality through Incremental Change.

Personnel:
Bruce Danielson (President)

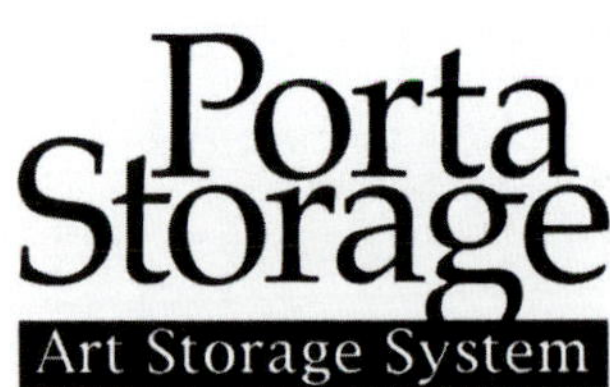

CHARLES J. DICKGIESSER & CO. INC.
257 Roosevelt Dr.
Derby, CT 06418-0475
Tel: 203-734-2553; FAX: 203-734-9221
E-mail: nancy.dickgiesser@sbcglobal.net
Web Site: Portastoragesystems.com

Type of Business:
Designers and manufacturers of museum storage systems. Porta Storage™ is a versatile, cost-effective system for displaying and storing textiles, paintings, photos and other art objects. Our pull-out art storage panels are floor supported for greater stability, maintenance free and custom designed to your specific needs. Call us today for more information.

Personnel:
Nancy L. Dickgiesser (President)

MONTEL INC.
225 4th Ave.
Montmagny, QC, Canada G5V 3S5
Tel: 877-935-0236; FAX: 418-248-7266
E-mail: system@montel.com
Web Site: www.montel.com

Type of Business:
The Art of Storage by Montel for Museums & Archival Collections: The Montel high density fixed and mobile storage systems allows each system to be custom-designed to the particular safety storage requirements for each type of collection. Montel has developed a line of efficient and safe products which respect the integrity and value of all types of art work regardless of their size & shape. Specializing in storage systems since 1924, Montel offers its expertise to insure planning solutions that maximize capacity, cost savings, security and accessibility. We offer our complete line for your particular collection requirement, including art racks, full line of cabinetry, textile racks, shelving, drawer cabinets and mobile storage systems.

Personnel:
Joey P. Boudreau (Marketing Director)

VIKING METAL CABINET COMPANY
a division of Austin-Westran
24047 West Lockport St., Ste. #209
Plainfield, IL 60544
Tel: 800-776-7767 & 815-782-8108; FAX: 815-267-6914
E-mail: sales@vikingmetal.com
Web Site: www.vikingmetal.com

Type of Business:
Viking manufactures a full line of quality museum cabinets that meet the storage requirements for all types of conservation, preservation, and archival collections. Major institutions and universities throughout the world rely on Viking cabinets to protect their most precious specimens, artifacts and documents. Versatility is the hallmark of Viking's product line. Our counter and full height conservation cabinets are offered in a wide range of sizes along with unique and optional accessories. We also offer a standard line of specialty cabinets for botanical, geological, and entomological collections along with archival flat files for artistic and historical works on paper. If your storage application is unique, Viking is ready to modify standard products or design custom cabinets that meet your particular needs with great attention to detail. Trust Viking…we've been designing and manufacturing museum storage products of unparalleled quality for over half a century.

Personnel:
Jim Dolan (Vice President Sales)

Crating (See also Shipping and Packing Services)

HAHN BROS. FIREPROOF WAREHOUSES, INC.
622 Communipaw Ave.
Jersey City, NJ 07304
Tel: 212-926-1505; FAX: 201-432-9547
E-mail: info@hahnbros.com
Web Site: www.hahnbros.com

Type of Business:
Over 100 years of fine art services to the museum and art community. Full service specialists offering customized crating and traveling cases and freight forwarding for collections and shows. Packing, moving, relocation and installation. Registrar, photography and gallery services available. Secure, climate-controlled storage with the ability to set up to client specifications.

Personnel:
Karen O. Dowling (President)
Marianne Mikulka (Fine Art Director)

Packing Services/Materials

ARTPACK SERVICES
24650 Crestview Ct.
Farmington Hills, MI 48335
Tel: 248-478-8946; FAX: 248-478-9588
E-mail: info@artpack.com
Web Site: www.artpack.com

Type of Business:
Artpack was founded in 1981 to provide professional services to museums, auction houses, galleries, corporate and private collectors for fine arts and antiques. We specialize in fine arts storage, custom crating and packing, installation, local and long distance shipping including a monthly New York shuttle. Our operations are climate controlled and staff trained to museum standards. Artpack also offers collection/project management and courier services, mount and pedestal fabrication, rigging and conservation services.

Personnel:
Ted Lee Hadfield (President)
Wendy MacGaw (Vice President)

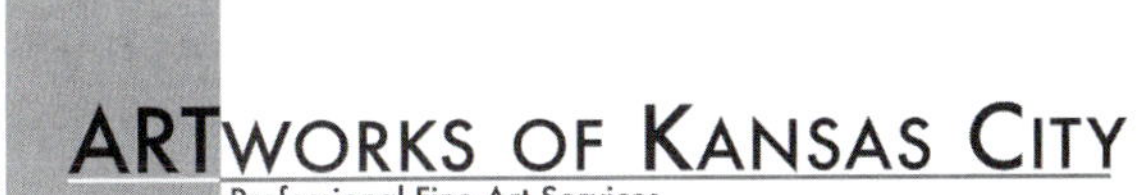

ARTWORKS OF KANSAS CITY FINE ART SERVICES
3017 Gillham Rd.
Kansas City, MO 64108
Tel: 816-753-4005 & 800-481-9856; FAX: 816-753-4007
E-mail: mike@artworkskc.com
Web Site: www.artworkskc.com

Type of Business:
Professional fine art packer and shipper providing museum-quality services nationwide. ARTworks' Midwest Regional Art Shuttle serves St. Louis, Chicago, Milwaukee, Minneapolis, Des Moines, Omaha, Lincoln, Kansas City - and all points in between. Air-ride, climate service with double drivers. AWKC offers custom EU-compliant crating, worldwide shipping, climate-controlled storage and installation. State-of-the-art facilities with professional staff.

Personnel:
Michael G. Otto (President)
Kris L. Luke (General Manager)

FINE ART SERVICES AND TRANSPORTATION
100 W. Forest
Detroit, MI 48201
Tel: 313-832-3278; FAX: 313-832-3293
E-mail: paul@fineartsolutions.com; chris@fineartsolutions.com
Web Site: www.fineartsolutions.com

Type of Business:
F.A.S.T is a full service, museum-quality art handling company which specializes in custom fine art crating, packing and installation. F.A.S.T also offers regional pick up and delivery service with climate-control. Freight forwarding, rigging services and much more are provided by our museum-trained staff.

Personnel:
Paul Smith (President)
Chris McInnis (Office Coordinator)

HAHN BROS. FIREPROOF WAREHOUSES, INC.
622 Communipaw Ave.
Jersey City, NJ 07304
Tel: 212-926-1505; FAX: 201-432-9547
E-mail: info@hahnbros.com
Web Site: www.hahnbros.com

Type of Business:
Over 100 years of fine art services to the museum and art community. Full service specialists offering customized crating and traveling cases and freight forwarding for collections and shows. Packing, moving, relocation and installation. Registrar, photography and gallery services available. Secure, climate-controlled storage with the ability to set up to client specifications.

Personnel:
Karen O. Dowling (President)
Marianne Mikulka (Fine Art Director)

Shelving Storage Equipment

MONTEL INC.
225 4th Ave.
Montmagny, QC, Canada G5V 3S5
Tel: 877-935-0236; FAX: 418-248-7266
E-mail: system@montel.com
Web Site: www.montel.com

Type of Business:
The Art of Storage by Montel for Museums & Archival Collections: The Montel high density fixed and mobile storage systems allows each system to be custom-designed to the particular safety storage requirements for each type of collection. Montel has developed a line of efficient and safe products which respect the integrity and value of all types of art work regardless of their size & shape. Specializing in storage systems since 1924, Montel offers its expertise to insure planning solutions that maximize capacity, cost savings, security and accessibility. We offer our complete line for your particular collection requirement, including art racks, full line of cabinetry, textile racks, shelving, drawer cabinets and mobile storage systems.

Personnel:
Joey P. Boudreau (Marketing Director)

Shipping & Moving Services

ARTPACK SERVICES
24650 Crestview Ct.
Farmington Hills, MI 48335
Tel: 248-478-8946; FAX: 248-478-9588
E-mail: info@artpack.com
Web Site: www.artpack.com

Type of Business:
Artpack was founded in 1981 to provide professional services to museums, auction houses, galleries, corporate and private collectors for fine arts and antiques. We specialize in fine arts storage, custom crating and packing, installation, local and long distance shipping including a monthly New York shuttle. Our operations are climate controlled and staff trained to museum standards. Artpack also offers collection/project management and courier services, mount and pedestal fabrication, rigging and conservation services.

Personnel:
Ted Lee Hadfield (President)
Wendy MacGaw (Vice President)

ARTWORKS OF KANSAS CITY FINE ART SERVICES
3017 Gillham Rd.
Kansas City, MO 64108
Tel: 816-753-4005 & 800-481-9856; FAX: 816-753-4007
E-mail: mike@artworkskc.com
Web Site: www.artworkskc.com

Type of Business:
Professional fine art packer and shipper providing museum-quality services nationwide. ARTworks' Midwest Regional Art Shuttle serves St. Louis, Chicago, Milwaukee, Minneapolis, Des Moines, Omaha, Lincoln, Kansas City - and all points in between. Air-ride, climate service with double drivers. AWKC offers custom EU-compliant crating, worldwide shipping, climate-controlled storage and installation. State-of-the-art facilities with professional staff.

Personnel:
Michael G. Otto (President)
Kris L. Luke (General Manager)

FINE ART SERVICES AND TRANSPORTATION
100 W. Forest
Detroit, MI 48201
Tel: 313-832-3278; FAX: 313-832-3293
E-mail: paul@fineartsolutions.com; chris@fineartsolutions.com
Web Site: www.fineartsolutions.com

Type of Business:
F.A.S.T is a full service, museum-quality art handling company which specializes in custom fine art crating, packing and installation. F.A.S.T also offers regional pick up and delivery service with climate-control. Freight forwarding, rigging services and much more are provided by our museum-trained staff.

Personnel:
Paul Smith (President)
Chris McInnis (Office Coordinator)

HAHN BROS. FIREPROOF WAREHOUSES, INC.
622 Communipaw Ave.
Jersey City, NJ 07304
Tel: 212-926-1505; FAX: 201-432-9547
E-mail: info@hahnbros.com
Web Site: www.hahnbros.com

Type of Business:
Over 100 years of fine art services to the museum and art community. Full service specialists offering customized crating and traveling cases and freight forwarding for collections and shows. Packing, moving, relocation and installation. Registrar, photography and gallery services available. Secure, climate-controlled storage with the ability to set up to client specifications.

Personnel:
Karen O. Dowling (President)
Marianne Mikulka (Fine Art Director)

THE ICON GROUP, INC.
2747 W. Taylor St.
Chicago, IL 60612-4047
Tel: 773-533-1800; FAX: 733-533-1900
E-mail: info@icongroup.us
Web Site: www.icongroup.us

Type of Business:
Since 1980 The Icon Group has provided museum quality fine art services to museums, collectors, galleries, artists and auction houses. ICON provides air-ride, climate-control transportation serving the Chicago, Midwest and Northeast regions; we offer a semi-monthly shuttle Service to New York and points-in-between, as well as exclusive use transport to any destination. ICON's 'Stand Alone', 95,000 square foot Storage Facility offers high security, climate-controlled private and mixed-client storage areas. Other services include custom crating, packing, as well as installation, rigging and national/international freight forwarding.

Personnel:
Bruce MacGilpin (President)
Ingrid Fassbender (Director)
Walt Solomon (Director, Long Distance Services)
Mike Gamis (Storage Services)
Colby Starck (Long Distance Services)
Kevin Brosnan (Local Services)
Eric Dimas (Crating & Packing Services)
Andrew Conaway (Facilities Manager)

NAGLEE FINE ARTS
1525 Grand Central Ave.
Elmira, NY 14901
Tel: 800-950-4533; FAX: 607-733-4850
E-mail: nfa@nagleegroup.com
Web Site: www.nagleefinearts.com

Type of Business:
Conservation quality storage and transportation services for artistic and historic objects. High security, precision climate-controlled facility: 68 degrees Fahrenheit at 45% relative humidity. Professional, museum-trained staff offering quality art handling, installation and crating services for individual artists and institutions alike. Certified Cargo Screening Facility (CCSF). Naglee Fine Arts is a division of Naglee Moving & Storage Inc., a full-service United Van Lines agent specializing in commercial moves for libraries and school systems.

Personnel:
Matthias H. Smith (Director, Naglee Fine Arts)
Scott M. Hoose (Vice President, Naglee Moving & Storage)
Allen C. Smith (Consultant)

Specimen Storage Equipment

DELTA DESIGNS LTD.
P.O. Box 1733
Topeka, KS 66601
Tel: 785-234-2244 & 800-656-7426; FAX: 785-233-1021
E-mail: bdanielson@deltadesignsltd.com
Web Site: www.deltadesignsltd.com

Type of Business:
Custom and standard museum storage cabinets, free standing or mounted on high density mobile storage systems. Cabinets available for your collection: Natural History, Works on Paper, Textiles, Art Objects, Historical Artifacts. Design and Installation services provided. Quality through Incremental Change.

Personnel:
Bruce Danielson (President)

MONTEL INC.
225 4th Ave.
Montmagny, QC, Canada G5V 3S5
Tel: 877-935-0236; FAX: 418-248-7266
E-mail: system@montel.com
Web Site: www.montel.com

Type of Business:
The Art of Storage by Montel for Museums & Archival Collections: The Montel high density fixed and mobile storage systems allows each system to be custom-designed to the particular safety storage requirements for each type of collection. Montel has developed a line of efficient and safe products which respect the integrity and value of all types of art work regardless of their size & shape. Specializing in storage systems since 1924, Montel offers its expertise to insure planning solutions that maximize capacity, cost savings, security and accessibility. We offer our complete line for your particular collection requirement, including art racks, full line of cabinetry, textile racks, shelving, drawer cabinets and mobile storage systems.

Personnel:
Joey P. Boudreau (Marketing Director)

VIKING METAL CABINET COMPANY
a division of Austin-Westran
24047 West Lockport St., Ste. #209
Plainfield, IL 60544
Tel: 800-776-7767 & 815-782-8108; FAX: 815-267-6914
E-mail: sales@vikingmetal.com
Web Site: www.vikingmetal.com

Type of Business:
Viking manufactures a full line of quality museum cabinets that meet the storage requirements for all types of conservation, preservation, and archival collections. Major institutions and universities throughout the world rely on Viking cabinets to protect their most precious specimens, artifacts and documents. Versatility is the hallmark of Viking's product line. Our counter and full height conservation cabinets are offered in a wide range of sizes along with unique and optional accessories. We also offer a standard line of specialty cabinets for botanical, geological, and entomological collections along with archival flat files for artistic and historical works on paper. If your storage application is unique, Viking is ready to modify standard products or design custom cabinets that meet your particular needs with great attention to detail. Trust Viking…we've been designing and manufacturing museum storage products of unparalleled quality for over half a century.

Personnel:
Jim Dolan (Vice President Sales)

Storage Design Consultants

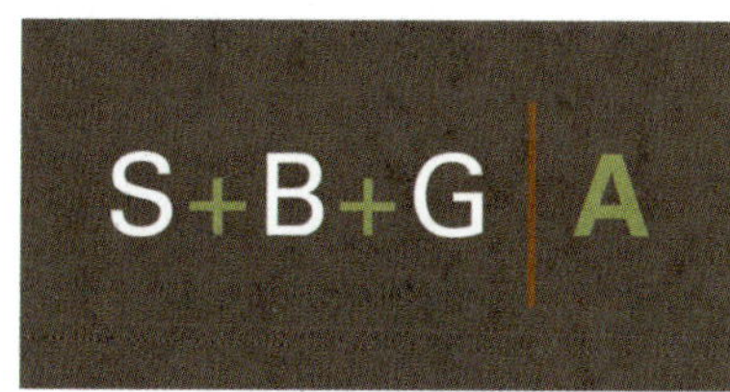

SOLOMON + BAUER + GIAMBASTIANI ARCHITECTS INC.
63 Pleasant St.
Watertown, MA 02472
Tel: 617-924-8200; FAX: 617-924-6685
E-mail: lbauer@sbgarch.com
Web Site: www.sbgarch.com

Type of Business:
Award-winning Architecture and Interior Design Firm with over forty projects for museums. Master-planning; programming; feasibility studies; design of new museums; renovation and adaptive reuse of existing and historic structures; and exhibit design support. Special expertise in collections storage analysis, planning, and design.

(Continued on next page)

(Continued from previous page)

Museums for which recent collection storage projects have been undertaken include Autry National Center, Blanton Museum of Art, Dallas Museum of Art, Harvard Art Museums, Museum of Art Rhode Island School of Design, Museum of Fine Arts Houston, Princeton Art Museum, Saint Louis Art Museum, and The Broad.

Personnel:
Lawrence C. Bauer, AIA (Principal)
Karin Aslaksen (Associate)

Storage Equipment

DELTA DESIGNS LTD.
P.O. Box 1733
Topeka, KS 66601
Tel: 785-234-2244 & 800-656-7426; FAX: 785-233-1021
E-mail: bdanielson@deltadesignsltd.com
Web Site: www.deltadesignsltd.com

Type of Business:
Custom and standard museum storage cabinets, free standing or mounted on high density mobile storage systems. Cabinets available for your collection: Natural History, Works on Paper, Textiles, Art Objects, Historical Artifacts. Design and Installation services provided. Quality through Incremental Change.

Personnel:
Bruce Danielson (President)

CHARLES J. DICKGIESSER & CO. INC.
257 Roosevelt Dr.
Derby, CT 06418-0475
Tel: 203-734-2553; FAX: 203-734-9221
E-mail: nancy.dickgiesser@sbcglobal.net
Web Site: Portastoragesystems.com

Type of Business:
Designers and manufacturers of museum storage systems. Porta Storage™ is a versatile, cost-effective system for displaying and storing textiles, paintings, photos and other art objects. Our pull-out art storage panels are floor supported for greater stability, maintenance free and custom designed to your specific needs. Call us today for more information.

Personnel:
Nancy L. Dickgiesser (President)

MONTEL INC.
225 4th Ave.
Montmagny, QC, Canada G5V 3S5
Tel: 877-935-0236; FAX: 418-248-7266
E-mail: system@montel.com
Web Site: www.montel.com

Type of Business:
The Art of Storage by Montel for Museums & Archival Collections: The Montel high density fixed and mobile storage systems allows each system to be custom-designed to the particular safety storage requirements for each type of collection. Montel has developed a line of efficient and safe products which respect the integrity and value of all types of art work regardless of their size & shape. Specializing in storage systems since 1924, Montel offers its expertise to insure planning solutions that maximize capacity, cost savings, security and accessibility. We offer our complete line for your particular collection requirement, including art racks, full line of cabinetry, textile racks, shelving, drawer cabinets and mobile storage systems.

Personnel:
Joey P. Boudreau (Marketing Director)

VIKING METAL CABINET COMPANY
a division of Austin-Westran
24047 West Lockport St., Ste. #209
Plainfield, IL 60544
Tel: 800-776-7767 & 815-782-8108; FAX: 815-267-6914
E-mail: sales@vikingmetal.com
Web Site: www.vikingmetal.com

Type of Business:
Viking manufactures a full line of quality museum cabinets that meet the storage requirements for all types of conservation, preservation, and archival collections. Major institutions and universities throughout the world rely on Viking cabinets to protect their most precious specimens, artifacts and documents. Versatility is the hallmark of Viking's product line. Our counter and full height conservation cabinets are offered in a wide range of sizes along with unique and optional accessories. We also offer a standard line of specialty cabinets for botanical, geological, and entomological collections along with archival flat files for artistic and historical works on paper. If your storage application is unique, Viking is ready to modify standard products or design custom cabinets that meet your particular needs with great attention to detail. Trust Viking…we've been designing and manufacturing museum storage products of unparalleled quality for over half a century.

Personnel:
Jim Dolan (Vice President Sales)

Precious little survived the
Impact of 2058

But once a great city thrived here. And a museum. Shattered columns, fractured statues and fragmentary artifacts tell us so. Not much to go on.

But what's this?

"We've discovered another!" we cheer. Eagerly we unseal it, confident that marvelous treasures await. Securely nestled inside. Butterflies. Brilliantly beautiful. Fearfully fragile. And wholly intact. Imagine!

We know little of the culture that produced this safe-guarding wonder. But we know the maker's name: **Delta Designs.** If only everyone had used these ...

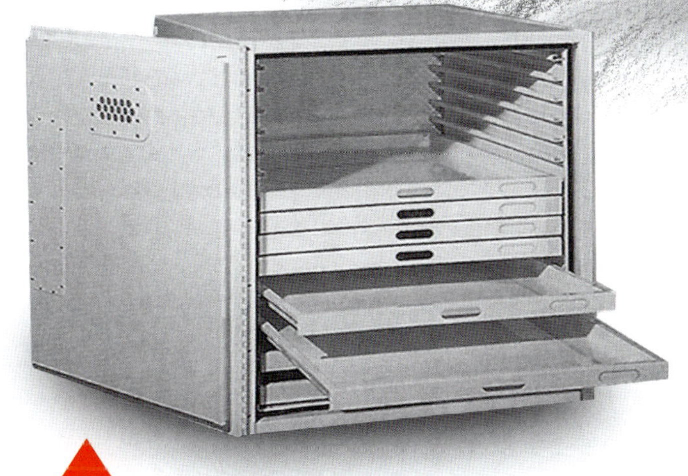

The Standard for Collection Storage

Storage Services

ARTPACK SERVICES INC.

ARTPACK SERVICES
24650 Crestview Ct.
Farmington Hills, MI 48335
Tel: 248-478-8946; FAX: 248-478-9588
E-mail: info@artpack.com
Web Site: www.artpack.com

Type of Business:
Artpack was founded in 1981 to provide professional services to museums, auction houses, galleries, corporate and private collectors for fine arts and antiques. We specialize in fine arts storage, custom crating and packing, installation, local and long distance shipping including a monthly New York shuttle. Our operations are climate controlled and staff trained to museum standards. Artpack also offers collection/project management and courier services, mount and pedestal fabrication, rigging and conservation services.

Personnel:
Ted Lee Hadfield (President)
Wendy MacGaw (Vice President)

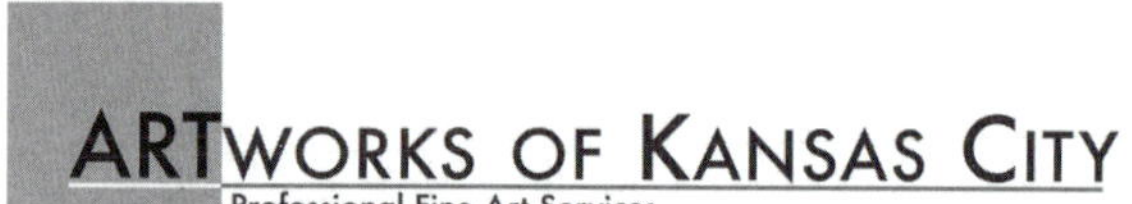

ARTWORKS OF KANSAS CITY FINE ART SERVICES
3017 Gillham Rd.
Kansas City, MO 64108
Tel: 816-753-4005 & 800-481-9856; FAX: 816-753-4007
E-mail: mike@artworkskc.com
Web Site: www.artworkskc.com

Type of Business:
Professional fine art packer and shipper providing museum-quality services nationwide. ARTworks' Midwest Regional Art Shuttle serves St. Louis, Chicago, Milwaukee, Minneapolis, Des Moines, Omaha, Lincoln, Kansas City - and all points in between. Air-ride, climate service with double drivers. AWKC offers custom EU-compliant crating, worldwide shipping, climate-controlled storage and installation. State-of-the-art facilities with professional staff.

Personnel:
Michael G. Otto (President)
Kris L. Luke (General Manager)

HAHN BROS. FIREPROOF WAREHOUSES, INC.
622 Communipaw Ave.
Jersey City, NJ 07304
Tel: 212-926-1505; FAX: 201-432-9547
E-mail: info@hahnbros.com
Web Site: www.hahnbros.com

Type of Business:
Over 100 years of fine art services to the museum and art community. Full service specialists offering customized crating and traveling cases and freight forwarding for collections and shows. Packing, moving, relocation and installation. Registrar, photography and gallery services available. Secure, climate-controlled storage with the ability to set up to client specifications.

Personnel:
Karen O. Dowling (President)
Marianne Mikulka (Fine Art Director)

THE ICON GROUP, INC.
2747 W. Taylor St.
Chicago, IL 60612-4047
Tel: 773-533-1800; FAX: 733-533-1900
E-mail: info@icongroup.us
Web Site: www.icongroup.us

Type of Business:
Since 1980 The Icon Group has provided museum quality fine art services to museums, collectors, galleries, artists and auction houses. ICON provides air-ride, climate-control transportation serving the Chicago, Midwest and Northeast regions; we offer a semi-monthly shuttle Service to New York and points-in-between, as well as exclusive use transport to any destination. ICON's 'Stand Alone', 95,000 square foot Storage Facility offers high security,

(Continued on next page)

(Continued from previous page)

climate-controlled private and mixed-client storage areas. Other services include custom crating, packing, as well as installation, rigging and national/international freight forwarding.

Personnel:
Bruce MacGilpin (President)
Ingrid Fassbender (Director)
Walt Solomon (Director, Long Distance Services)
Mike Gamis (Storage Services)
Colby Starck (Long Distance Services)
Kevin Brosnan (Local Services)
Eric Dimas (Crating & Packing Services)
Andrew Conaway (Facilities Manager)

NAGLEE FINE ARTS
1525 Grand Central Ave.
Elmira, NY 14901
Tel: 800-950-4533; FAX: 607-733-4850
E-mail: nfa@nagleegroup.com
Web Site: www.nagleefinearts.com

Type of Business:
Conservation quality storage and transportation services for artistic and historic objects. High security, precision climate-controlled facility: 68 degrees Fahrenheit at 45% relative humidity. Professional, museum-trained staff offering quality art handling, installation and crating services for individual artists and institutions alike. Certified Cargo Screening Facility (CCSF). Naglee Fine Arts is a division of Naglee Moving & Storage Inc., a full-service United Van Lines agent specializing in commercial moves for libraries and school systems.

Personnel:
Matthias H. Smith (Director, Naglee Fine Arts)
Scott M. Hoose (Vice President, Naglee Moving & Storage)
Allen C. Smith (Consultant)

Textiles Storage Equipment

DELTA DESIGNS LTD.
P.O. Box 1733
Topeka, KS 66601
Tel: 785-234-2244 & 800-656-7426; FAX: 785-233-1021
E-mail: bdanielson@deltadesignsltd.com
Web Site: www.deltadesignsltd.com

Type of Business:
Custom and standard museum storage cabinets, free standing or mounted on high density mobile storage systems. Cabinets available for your collection: Natural History, Works on Paper, Textiles, Art Objects, Historical Artifacts. Design and Installation services provided. Quality through Incremental Change.

Personnel:
Bruce Danielson (President)

CHARLES J. DICKGIESSER & CO. INC.
257 Roosevelt Dr.
Derby, CT 06418-0475
Tel: 203-734-2553; FAX: 203-734-9221
E-mail: nancy.dickgiesser@sbcglobal.net
Web Site: Portastoragesystems.com

Type of Business:
Designers and manufacturers of museum storage systems. Porta Storage™ is a versatile, cost-effective system for displaying and storing textiles, paintings, photos and other art objects. Our pull-out art storage panels are floor supported for greater stability, maintenance free and custom designed to your specific needs. Call us today for more information.

Personnel:
Nancy L. Dickgiesser (President)

MONTEL INC.
225 4th Ave.
Montmagny, QC, Canada G5V 3S5
Tel: 877-935-0236; FAX: 418-248-7266
E-mail: system@montel.com
Web Site: www.montel.com

Type of Business:
The Art of Storage by Montel for Museums & Archival Collections: The Montel high density fixed and mobile storage systems allows each system to be custom-designed to the particular safety storage requirements for each type of collection. Montel has developed a line of efficient and safe products which respect the integrity and value of all types of art work regardless of their size & shape. Specializing in storage systems since 1924, Montel offers its expertise to insure planning solutions that maximize capacity, cost savings, security and accessibility. We offer our complete line for your particular collection requirement, including art racks, full line of cabinetry, textile racks, shelving, drawer cabinets and mobile storage systems.

Personnel:
Joey P. Boudreau (Marketing Director)

VIKING METAL CABINET COMPANY
a division of Austin-Westran
24047 West Lockport St., Ste. #209
Plainfield, IL 60544
Tel: 800-776-7767 & 815-782-8108; FAX: 815-267-6914
E-mail: sales@vikingmetal.com
Web Site: www.vikingmetal.com

(Continued on next page)

(Continued from previous page)

Type of Business:

Viking manufactures a full line of quality museum cabinets that meet the storage requirements for all types of conservation, preservation, and archival collections. Major institutions and universities throughout the world rely on Viking cabinets to protect their most precious specimens, artifacts and documents. Versatility is the hallmark of Viking's product line. Our counter and full height conservation cabinets are offered in a wide range of sizes along with unique and optional accessories. We also offer a standard line of specialty cabinets for botanical, geological, and entomological collections along with archival flat files for artistic and historical works on paper. If your storage application is unique, Viking is ready to modify standard products or design custom cabinets that meet your particular needs with great attention to detail. Trust Viking…we've been designing and manufacturing museum storage products of unparalleled quality for over half a century.

Personnel:

Jim Dolan (Vice President Sales)

The Official Museum Directory (OMD), the most comprehensive and current single-reference source covering the museum community for over 40 years is also available online, offering:

POWERFUL, flexible searches

Official Museum Directory Online is a robust search engine that provides true search results in the blink of an eye. 19 unique search fields afford the option of broad based or more refined searches…all quick and easy to execute.

Data CURRENCY

Official Museum Directory Online is refreshed daily from our editorial system, immediately offering the most current and accurate data available.

QUICK look ups

Our data structure and consistency of format make quick look ups and discovery convenient and easy.

BREADTH of collection

Over 15,000 on the nations' museums, searchable among 89 different institution types and 31 collection specialties (from Arboretums to Zoos and everything in between).

Here are just **5** ways **The Official Museum Directory Online** can help your organization:

- Benchmark your institution against similar local, regional, or national institutions (e.g. staff size, attendance, admission prices, hours of operations, facilities, and more…)
- Identify new programming and event ideas to help grow attendance
- Locate museum staff by specialty for networking (e.g. identify curators for specific types of collections)
- Locate companies that provide specific products and services to the museum community
- **SAVE time and money!** Eliminate inefficient research using multiple, unreliable and unverified sources in favor of the most thorough, single-reference source covering the museum field

Special Discounted Pricing Available for OMD Book Buyers

For more information: Call: 1-800-473-7020 • Email: info@officialmuseumdirectory.com

www.officialmuseumdirectory.com

omdonline114

Save the Date! May 18–21, 2014!

The 108th Alliance Annual Meeting & MuseumExpo is coming back to Seattle in 2014. It's been 20 years since we brought the largest annual gathering of museum professionals in the world to the Emerald City. Seattle's vibrant museum and business communities are eager to welcome us.

Call **202.289.1818** for more information.